Lloyd's Register of British and Foreign Shipping [1890]: Universal Register [Volume 1 of 2]

Committee Of Register Of British And Foreign Shipping

Annotated by Nimble Books AI

Publishing Information

ISBN: 978-0-9799205-7-8

In the Year 1890 No. 3

Table of Contents

LIST OF CONTENTS

OF THE

UNIVERSAL REGISTER.

(To facilitate reference, subordinate lists of Contents are placed at the commencement of several sections, such as those referring to War Ships, Signal Letters, &c.)

The Register of Steamers in this Volume includes ships with names *A.C. Cheney – Mytilene.* The Register continues in Volume 2. with ships names beginning N-Z.

Publisher's Note: In The Year 1890

The inspiration for this series is *Winter's Guide to the Warships of the World 1890* by Dimitry Malkov, whose first volume was published in June 2023 by Nimble Books LLC. The two volumes of *Winter's Guide* together provide detailed illustrations and specifications for every single warship in service in every nation of the world as of January 1, 1890. Lloyd's *Universal Register* of 1890, republished here in two volumes of 900 and 980 pages as books 3 and 4 of *In The Year 1890*, extends this comprehensive wealth of detail to nearly the entire the maritime world of the late 19th century. The *Register* provides details on every self-propelled ship of over 100 tons in operation between April 1, 1889 and March 31, 1990.

For maritime historians and shipping enthusiasts, the *Register* offers factual, detailed information about the vessels that were once integral to global trade and exploration. Each entry includes specifications about construction, machinery, tonnage, and ownership, thus serving as a reliable reference point for research and historical study. This contemporaneous primary source material was verified by Lloyd's. The Lloyd's Register Foundation Heritage & Education Centre[1] has an excellent set of online guides on its website under Archive & Library > Interpreting the Register Books[2].

For engineers and industrial designers, this book provides a glimpse into the state of technology and mechanical design at the end of the 19th century. The wide range of vessels describes encompasses every aspect of the naval engineering of the era, offering a chance to understand the technological advances of the period.

Writers for readers of historical fiction may find inspiration within these pages. These records can serve as a factual foundation for constructing narratives set in this time period. There is no substitute for getting the details right.

Researchers into genealogy may also find the Lloyd's *Register* useful. The extensive records of ship ownership can provide leads to family histories linked to maritime endeavors, potentially offering new avenues for genealogical research. For those whose family histories include transoceanic voyages, there is a thrill in visualizing exactly which ship an ancestor might have taken en route to a new home or travel destination.

Those interested in databases, information design and software will take note that more than 130 years ago, the editors of the *Register* were able to convey an enormous amount of tabular information in a readily accessible format without the benefit of SQL, React or R!

For the book collector or those with a fondness for historical artifacts, this facsimile edition offers a tangible connection to the past. As a meticulous record of a specific year in maritime

---

[1] lrfoundation.org.uk

[2] https://hec.lrfoundation.org.uk/archive-library/interpreting-the-register-books

history, it's an interesting piece of our collective heritage. As a very large and very thick book full of proper nouns and unexpected details, it amply repays casual browsing.

Lloyd's *Register* not just a catalogue of facts and figures. It's a window into a world that has shaped our own, making it a noteworthy addition to the libraries of anyone keen to explore the maritime past.[3]

Fred Zimmerman
Nimble Books LLC

[3] Finally, it goes without saying that time travelers will find this a crucial tool for planning and coordinating trips into the Victorian Era.

Join An AI-Powered Chat About the 1890s

A shareable link to an OpenAI ChatGPT will get a good conversation going with the AI (or, as we called them in the 1890s, "clockwork intelligence") and with anyone else who has joined the conversation.

You:

Let's discuss the maritime world of 1890.

ChatGPT:

Certainly! The maritime world of 1890 was an exciting and transformative period in history, marked by significant advancements in shipbuilding, navigation, trade, and exploration. Here are some key aspects to consider:

1. Shipbuilding: The late 19th century witnessed the transition from sail-powered vessels to steamships. Steam engines became more efficient and reliable, leading to the construction of larger, faster, and more powerful ships. Steel hulls began replacing wooden ones, increasing durability and enabling ships to withstand harsh ocean conditions.

2. Transatlantic Travel: The advent of steamships revolutionized transatlantic travel. Companies like Cunard Line and White Star Line operated luxurious ocean liners that offered regular passenger services between Europe and North America. Prominent examples include the RMS Titanic, launched in 1911, which tragically sank in 1912 during its maiden voyage.

3. Exploration and Scientific Expeditions: The late 19th century saw numerous maritime expeditions aimed at exploration and scientific research. Notable examples include expeditions by Robert Falcon Scott and Ernest Shackleton to Antarctica, where they mapped previously uncharted territories and conducted scientific studies.

4. Colonial Trade and Empire: European colonial powers, such as Britain, France, and Germany, had extensive maritime networks for trade and transport. These powers established colonies and trading posts around the world, facilitating the exchange of goods, resources, and ideas. Steamships played a crucial role in connecting these colonies to their respective imperial powers.

5. Naval Power and Competition: The late 19th century witnessed intense naval competition among major global powers. The British Royal Navy was the dominant force at the time, with its fleet serving as a symbol of the British Empire's strength and influence. Other powers, including the United States, Germany, and Japan, also sought to build powerful naval forces.

6. Fishing and Whaling: Fishing and whaling were prominent maritime industries in 1890. Coastal communities relied on fishing for sustenance and trade, while whaling ships ventured into the open seas in search of valuable whale oil and other products.

7. Lighthouses and Navigational Aids: Lighthouses played a vital role in guiding ships safely through treacherous waters. In 1890, many lighthouses were manned by keepers who maintained the lights and sounded fog signals to warn approaching vessels of potential dangers.

8. Communication and Telegraphy: The late 19th century witnessed significant advancements in telegraphy, enabling faster and more reliable communication between ships at sea and the shore. The introduction of undersea telegraph cables connected continents and facilitated instant messaging across vast distances.

9. Maritime Accidents and Safety: Despite technological advancements, maritime accidents were still a concern. Shipwrecks, collisions, and storms posed significant risks to ships and sailors. The International Convention for the Safety of Life at Sea (SOLAS), established in 1914, aimed to improve safety standards and minimize maritime disasters.

Figure 1. OpenAI chatting about the world of 1890. Join the chat by entering this URL and find kindred spirits! https://chat.openai.com/share/81e27d74-399b-4043-b4ed-b2567a4b21dd

LLOYD'S REGISTER OF BRITISH AND FOREIGN SHIPPING.

UNIVERSAL REGISTER.

PUBLICATIONS ISSUED BY THE COMMITTEE OF LLOYD'S REGISTER.

OFFICE: 2, WHITE LION COURT, CORNHILL, LONDON.

LLOYD'S REGISTER OF SHIPPING, issued annually on the 1st of July. This Book contains the Names, CLASSES, and detailed information concerning the Vessels classed by Lloyd's Register and the late Underwriters' Registry for Iron Vessels; and in addition the Names, Dimensions, and other particulars of all Vessels of 100 tons and upwards registered in the United Kingdom, and of many ships of large tonnage owned abroad.

In the Appendix will be found a List of Owners of Ships recorded in the Register Book; details of the Dry and Wet Docks, Floating Docks, Pontoons, Patent Slipways, Tidal Harbours, &c., at all ports in the World; a List of Telegraphic Addresses of all Firms, &c., connected with Shipping (with the Telegraphic Codes severally employed), so far as ascertained, &c.

TERMS OF SUBSCRIPTION:—For Individuals or Firms, £3 3s. per annum; for Public Companies or Public Establishments (not engaged in Marine Insurance), £10 10s. per annum. The Subscription of Marine Insurance Companies is regulated by the Committee on special application in each case, but is not less than £10 10s. per annum.

The Register Books of London Subscribers are periodically posted by type, with additions and corrections throughout the year; but in the case of Books thus posted, supplied on a Subscription of £3 3s. per annum, an additional charge of £1 1s. per annum is made.

For the convenience of Subscribers not resident in London, or not having their Books posted, a Supplement, containing the additions to, and corrections made in the Register Book, is printed fortnightly and forwarded gratis to Subscribers.

The Rules for the Building and Classification of Ships are supplied with the above work.

THE RULES FOR THE BUILDING AND CLASSIFICATION OF SHIPS are published also separately.

PRICE:—Rules for Wood and Composite Ships, 5s.; Rules for Iron and Steel Ships, 5s.; Complete Rules, 10s.

UNIVERSAL REGISTER, issued annually on the 1st of April. This work contains, as far as possible, particulars of EVERY SEA-GOING VESSEL (including Yachts) in the world of 100 tons and upwards; an alphabetical LIST OF ALL SHIPOWNERS, with their addresses and the names and tonnages of their vessels; a LIST OF TELEGRAPHIC ADDRESSES OF ALL FIRMS, &c., connected with Shipping (with the Telegraphic Codes severally employed), so far as ascertained; particulars of the WAR VESSELS BELONGING TO ALL NATIONS; details of the DRY AND WET DOCKS, FLOATING DOCKS, PONTOONS, PATENT SLIPWAYS, TIDAL HARBOURS, QUAYS, &c., at all ports in the World; together with various statistical and other useful information.

TERMS OF SUBSCRIPTION:—For Individuals or Firms, or Public Companies, &c., other than those noted below, £3 3s. per annum. For Public Companies engaged in Marine Insurance, Public Libraries, Exchanges, Chambers of Commerce, Underwriters' Associations, &c., &c., £5 5s. for first copy, and £3 3s. for each additional copy.

YACHT REGISTER, issued annually on the 1st of May. The book contains, in addition to the Names, CLASSES, and other useful information relating to Yachts classed by the Society, the Names, Dimensions, &c., of all British and Foreign Yachts, the particulars of which are known: also, in an Appendix, a List of the various Yacht Clubs and Illustrations of their respective Flags, an Illustrated List of the RACING FLAGS of Yachts, an Index of Signal Letters, a List of Yachts the names of which have been changed, a List of Yachts arranged according to Tonnage, an Alphabetical List of the Names and Addresses of Yacht Owners so far as they can be ascertained, and a List of Yacht Builders and Designers showing the Yachts which they have respectively built and designed, besides Statistical and other interesting information.

The Society's Rules for the Building and Classification of Iron, Steel, Wood, and Composite Yachts are supplied with the above work.

TERMS OF SUBSCRIPTION:—15s. per copy for Individuals or Firms; and for Yacht Clubs, Marine Assurance Companies, and other public bodies, not less than £2 2s. per annum. For the latter Subscription two copies of the work will be supplied, if required.

THE RULES FOR THE BUILDING AND CLASSIFICATION OF YACHTS are published also separately.

PRICE:—2s. 6d. per copy.

LIST OF DRY AND WET DOCKS, PONTOONS, PATENT SLIPWAYS, TIDAL HARBOURS, ETC., AT ALL PORTS IN THE WORLD, giving dimensions, height of tides, &c., and including a Table showing the weight in ounces per cubic foot of samples of sea and river water taken from ports in the United Kingdom.

PRICE:—2s. 6d. per copy.

PARTICULARS OF THE WAR VESSELS OF THE WORLD, with which are included Statistics of Merchant Shipping and a List of Dry and Wet Docks, &c., extracted from the "Universal Register."

PRICE:—5s. per copy.

REPORT ON MASTING, being a Report made to the Committee of Lloyd's Register of British and Foreign Shipping by the Society's Chief Surveyor and his Assistants concerning the DISMASTING OF LARGE IRON SAILING SHIPS.

PRICE:—15s. per copy.

LLOYD'S REGISTER OF BRITISH AND FOREIGN SHIPPING.

UNIVERSAL REGISTER.

From the 1st *of APRIL*, 1889, *to the* 31st *of MARCH*, 1890.

This work—which has been compiled by the COMMITTEE OF LLOYD'S REGISTER OF BRITISH AND FOREIGN SHIPPING solely for the information of subscribers—contains as far as possible particulars of EVERY SEA-GOING VESSEL (including Yachts) in the world of 100 tons and upwards; an alphabetical LIST OF ALL SHIPOWNERS, with their addresses and the names and tonnages of their vessels; particulars of the WAR VESSELS BELONGING TO ALL NATIONS; details of the DOCKS, SLIPWAYS, ETC., IN ALL COUNTRIES, together with various statistical and other useful information. A Table of Contents will be found immediately following the Preface.

A KEY to the several contractions used in the Book, will be found to precede the Register of Sailing Vessels.

LONDON:
PRINTED BY WILLIAM CLOWES & SONS, LIMITED,
STAMFORD STREET AND CHARING CROSS
1889.

LONDON
PRINTED BY WILLIAM CLOWES AND SONS, LIMITED.
STAMFORD STREET AND CHARING CROSS.

PREFACE.

The Committee of Lloyd's Register of British and Foreign Shipping have the pleasure to place in the hands of Subscribers the Fourth Annual Edition of the Universal Register.

The principal additions to be noted in the present edition are the following:—

List of Builders of Marine Engines at home and abroad,

List of Shipowners' Trade Defence Associations,

List of Fast Merchant Steamers.

In effecting these additions, as well as in keeping the Universal Register abreast of the changes which have occurred in connection with the Merchant Shipping of the world, the Committee have received from many quarters most courteous and valuable aid, for which they take the present opportunity of returning their best thanks. Especially do the Committee feel bound to express their sense of the liberal action of the Marine Departments of the British, and of many Foreign, Governments in continuing to supply them, for the use of the public, with information drawn from the official records of the merchant shipping of their respective countries.

Without recapitulating the list of contents, attention may be invited to the following sections of the book which are of a somewhat special character:—

I. Statistical Tables.—The statistics which have been compiled for the present work will, it is thought, prove not only valuable to all who are connected with shipping, but also of considerable interest to the general public. They contain much information which had not, so far as the Committee know, been published prior to the issue of the Universal Register; and it will be seen that their practical usefulness has been much increased by the elaborate system of analysis which has been adopted. Tables are introduced showing the additions and deductions which have been made during the past year to the principal merchant navies of the world, as recorded in the Universal Register.

II. List of War Ships of the World.—The list has been compiled from the best available sources of information, and the Committee believe it will be found to be the most complete that has yet been published. The supplementary Tables, described in detail in the memorandum at the commencement of the list, afford means of comparison of the naval power of different nations. For the assistance which has kindly been rendered them in the preparation of this section of the work, the Committee beg to offer their thanks.

Although no effort has been spared to render the UNIVERSAL REGISTER both accurate and complete, the Committee are conscious that the compilation of so vast a mass of information can scarcely be unattended by errors and omissions. They will therefore most highly appreciate any aid with which, in their endeavour to meet the business requirements of the present day, they may be favoured, either by the rectification of errors or by the supply of further information.

It only remains to be stated that the UNIVERSAL REGISTER will in no way interfere with the publication of LLOYD'S REGISTER BOOK and the YACHT REGISTER, both of which will continue to be issued annually as heretofore.

2, White Lion Court,
Cornhill,
March 28th, 1889.

PRÉFACE.

Le Comité du Lloyd's Register of British and Foreign Shipping a le plaisir de livrer aux souscripteurs la quatrième édition annuelle de l'Universal Register.

Les additions les plus importantes à noter cette année, sont les suivante:

Liste des Constructeurs de Machines du Royaume-Uni et de l'étranger.

Liste des Associations d'Armateurs ayant pour but la défense de leurs intérêts.

Liste des Navires à Vapeur de Commerce de Grande Vitesse.

Afin de pouvoir faire, non seulement ces additions, mais encore les corrections nécessaires par suite des changements survenus dans la marine marchande, durant l'exercice écoulé, le Comité a été honoré du concours précieux et bienveillant de nombreuses personnes, auxquelles il est de ce chef infiniment reconnaissant. Il profite de cette occasion pour leur offrir ses meilleurs remerciments.

Il s'empresse de mentionner tout spécialement, à ce sujet, l'appui généreux que les départements de la Marine des Gouvernements anglais et étrangers ont continué à lui accorder; et de leur exprimer toute sa gratitude pour les renseignements officiels et précieux, qui lui ont été fournis sur leurs marines marchandes respectives, en vue d'être utile au public.

Sans faire la récapitulation de la table des matières, le Comité désire attirer l'attention sur les parties suivantes d'un caractère plus particulier.

I. Tables de Statistiques.—Ces Tables spécialement composées pour le présent ouvrage ne seront pas seulement, le Comité en a l'espoir, d'une utilité réelle pour le commerce maritime, mais présenteront certainement un grand intérêt pour le public en général. Elles contiennent maintes informations qui, pour autant que le Comité a pu s'en rendre compte, n'ont jamais été publiées avant que l'Universal Register eût paru. Leur utilité pratique est encore rehaussée par le système d'analyse adopté. Des tables ont été introduites, ayant pour but de montrer les additions et les diminutions faites, durant le courant de l'année dernière, aux principales marines marchandes du monde, ainsi que les renseigne l'Universal Register.

II. Liste des Navires de Guerre du Monde.—Cette section a été formée par la compilation de renseignements obtenus aux meilleures sources possibles, et ce travail sera certainement trouvé, le Comité en a l'espoir, l'ouvrage le plus complet de ce genre qui ait jamais été publié. Comme il est dit, d'une façon détaillée, dans la note qui précède cette section, les tables supplémentaires permettent de comparer les forces navales des différentes nations. Le Comité présente ses meilleurs remercîments pour l'assistance dont il a été favorisé lors de la formation de cette partie du livre.

Inutile d'assurer finalement que rien n'a été négligé pour rendre l'UNIVERSAL REGISTER aussi exac et aussi complet que possible ; le travail embrassant toutefois une quantité si considérable de renseigne ments, le Comité fait appel à l'indulgence du public pour les erreurs qui auraient pu s'y glisser malgr l'attention apportée à l'ouvrage.

Il serait heureux si, en lui signalant ces erreurs ou en lui fournissant de nouvelles information on voulait l'aider à atteindre le but qu'il s'est proposé, d'élever son livre au rang d'un ouvrage répondai sous tous les rapports aux exigences de nos jours.

Il reste à mentionner que l'UNIVERSAL REGISTER n'apportera aucun changement dans la publicati du LLOYD'S REGISTER BOOK et du YACHT REGISTER, qui continueront à être publiés séparément comme p le passé.

2, White Lion Court,
Cornhill, Londres,
28 Mars, 1889.

LIST OF CONTENTS

OF THE

UNIVERSAL REGISTER.

(To facilitate reference, subordinate lists of Contents are placed at the commencement of several sections, such as those referring to War Ships, Signal Letters, &c.)

TABLE DES MATIÈRES
DE
L'UNIVERSAL REGISTER.

NOTA.—Pour la facilité des Abonnés, des sous-tables des matières figurent au commencement de plusieurs Sections, telles que celles des Navires de Guerre, des Signaux Distinctifs, etc.

COMMITTEE OF MANAGEMENT.

1888–1889.

WILLIAM HENRY TINDALL, Esq., *Chairman.*
LEONARD C. WAKEFIELD, Esq., *Deputy-Chairman.*
THOMAS B. WALKER, Esq., *Chairman of the Sub-Committees of Classification.*

MEMBERS ELECTED IN LONDON.

H. J. BRISTOW, Esq.
MARMADUKE H. BROOKING, Esq.
JOHN CORRY, Esq.
JAMES DIXON, Esq.
GEORGE DUNCAN, Esq.
JOHN GLOVER, Esq.
C. A. HARDMAN, Esq.
CHARLES W. HARRISON, Esq.
GEORGE LIDGETT, Esq.
H. E. MONTGOMERIE, Esq.
GERARD FREDERICK MILLER, Esq.
FREDERIC B. B. NATUSCH, Esq.
HENRY NIXON, Esq.
JAMES PARK, Esq.
JAMES CARR SAUNDERS, Esq.
WM. FREDERICK SAUNDERS, Esq.
T. SCRUTTON, Esq.
CHARLES R. TATHAM, Esq.
JOHN HENRY TOD, Esq.
J. WILLIS, Esq.
WILLIAM YOUNG, Esq.

Right Hon. EDWARD CHARLES BARING, LORD REVELSTOKE, *Chairman of the Committee of Lloyd's.*
J. G. S. ANDERSON, Esq., *Chairman of the General Shipowners' Society.*

MEMBERS ELECTED AT THE PRINCIPAL OUTPORTS.

FOR LIVERPOOL.

SAMUEL CROSS, Esq., *Chairman of the Liverpool Committee.*
JOHN RANKIN, Esq., *Deputy Chairman* *ditto.*

JOHN S. ALLEN, Esq.
ROBERT N. DALE, Esq.
HENRY H. GRAYSON, Esq.
DONALD KENNEDY, Esq.
THOMAS R. SHALLCROSS, Esq.
H. T. WALLACE, Esq.
JOHN WILLIAMSON, Esq.
J. HALL WORTHINGTON, Esq.

GLASGOW — WILLIAM ADAMSON. Esq.; WALTER EASTON, Esq.; GEORGE M. KERR, Esq.; ALEXANDER MITCHELL, Esq.

SUNDERLAND — RALPH M. HUDSON, Esq.; JAMES LAING, Esq.

HARTLEPOOL DISTRICT — WILLIAM GRAY, Esq.; JOHN HALL, Esq.

BRISTOL—MARK WHITWILL, Esq.
LEITH DISTRICT—CHARLES BARRIE, Esq. *(Residing at Dundee.)*
GREENOCK—DUGALD MACDOUGALL, Esq.
TYNE DISTRICT — R. S. DONKIN, Esq., M.P.; J. D. MILBURN, Esq.; E. H. WATTS, Esq.
CARDIFF DISTRICT—COL. E. S. HILL, C.B., M.P.
HULL—HENRY J. ATKINSON, Esq., M.P.

TRUSTEES.

GEORGE DUNCAN, Esq.
HENRY NIXON, Esq.
WILLIAM HENRY TINDALL, Esq.
JOHN HENRY TOD, Esq.
LEONARD C. WAKEFIELD, Esq.
THOMAS B. WALKER, Esq.

BANKERS.

THE NATIONAL PROVINCIAL BANK OF ENGLAND (*Ld.*)

SECRETARY.

BERNARD WAYMOUTH, Esq.

ASSISTANT SECRETARY.

A. G. DRYHURST, Esq.

2, White Lion Court, Cornhill, London,
27th February, 1889.

LIVERPOOL BRANCH.

1888-1889.

COMMITTEE.

Samuel Cross, Esq., *Chairman.*
John Rankin, Esq., *Deputy-Chairman.*

John S. Allen, Esq.
James T. Binning, Esq.
Robert N. Dale, Esq.
David Fernie, Esq.
W. S. Graves, Esq.
Henry H. Grayson, Esq.
David Jardine, Esq.
Donald Kennedy, Esq.
Sir James Poole.
Thomas R. Shallcross, Esq.
Thomas James Storey, Esq.
H. T. Wallace, Esq.
John Williamson, Esq.
J. Hall Worthington, Esq.

MEMBER OF THE COMMITTEE EX-OFFICIO.

Thomas B. Royden, Esq., M.P., *Chairman of the Shipbuilders' Association.*

BANKERS.

The National Provincial Bank of England (*Ld.*).

SECRETARY.

John Frederick Light, Esq.

Office, 12, *Oriel Chambers, Liverpool.*
27*th February*, 1889.

LLOYD'S REGISTER OF BRITISH AND FOREIGN SHIPPING.

SURVEYORS—UNITED KINGDOM.

THE SURVEYORS AT THE FOLLOWING PORTS ARE EXCLUSIVELY THE OFFICERS OF THE SOCIETY, AND ARE NOT PERMITTED TO ENGAGE IN ANY OTHER BUSINESS OR EMPLOYMENT WHATSOEVER.

INSPECTEURS—ROYAUME-UNI.

LES INSPECTEURS ÉTABLIS AUX PORTS SUIVANTS SONT EXCLUSIVEMENT OFFICIERS DE LA SOCIÉTÉ, ET IL NE LEUR EST PAS PERMIS DE S'ENGAGER DANS D'AUTRES AFFAIRES OU EMPLOI QUELCONQUE.

Port		Surveyors
LONDON		William C. Davey, *Principal Surveyor for London District.*
	Benjamin Martell, Esq., *Chief Surveyor of Lloyd's Register*	Cooper, George P.
		Jordan, Chas. H.
	Harry J. Cornish, Thomas Edwards, *Assistants to Chief Surveyor*	Mares, George R.
		Read, Thomas C.
		Tierney, E. J.
	William Parker, Esq., *Chief Engineer Surveyor of Lloyd's Register*	Truscott, James H.
		Warren, Thomas S.
	William Allison, *Assistant to Chief Engineer Surveyor*	Wilkinson, Henri
	Engineer Surveyors	Stromeyer, Charles E.; Wilkinson, George E.
	Ship and Engineer Surveyors	Gray, T. L.; Jones, A. L.
	Draughtsman	Hunter, David S.
ABERDEEN	*with Banff, Peterhead and Fraserburgh ... (Office, 29, Regent Quay, Aberdeen) Ship and Engineer Surveyor ...*	Hindmarsh, G. L.
BANGOR	*with Carnarvon and Holyhead ... ... (Address, 5, Glan Dwr Terrace)*	Devonald, Thomas
BARROW	*and Ulverstone, Ship and Engineer Surveyors; also Engineer Surveyors for Whitehaven (Office, 8, Strand, Barrow)*	Ritchie, Duncan; Dykes, J.
BELFAST	*... ... ... (Office, Victoria Chambers, Waring Street)*	Turpin, James
	Ship and Engineer Surveyor ...	Maxton, James
BIDEFORD	*Barnstaple, and Appledore ... ... (residing at Bideford)*	Fittock, Charles
BRISTOL	*Gloucester, and Bridgwater ... ... (Office, 53, Queen Square, Bristol) Ship and Engineer Surveyor ...*	Coomber, R. W.
CARDIFF	*... ... ... (Offices, Merchants' Exchange)*	Andrew K. Hamilton, *Principal Surveyor.*; Rule, J. G. G.
	Ship and Engineer Surveyors for Cardiff; also Engineer Surveyors for the Swansea District	Ashton, H. H.; Fargie, W. A.; Mancor, J. H.; Taylor, J. E.

Port	Particulars	Surveyors
CHANNEL ISLANDS	... ... (*residing at Jersey*)	Chinneck, T. J.
DUBLIN	... ... (*Address, 23, Ballybough Road*)	Franklin, J.
DUNDEE	*with Arbroath, Montrose, and Perth* (*Office, 54, Commercial Street, Dundee*) *Ship and Engineer Surveyor*	Keydell, A. E.
FALMOUTH	*with all Ports on the coast between Looe and Padstow, including Fowey, Porthleven, Penzance, St. Ives, Hayle, Portreath, and New Quay; also the Scilly Islands* (*Address, 2, Park Terrace, Falmouth*)	Sandry, T. H.
	Ship and Engineer Surveyor for Falmouth District (*residing at Falmouth*) ...	Dyer, C. M. B.
GLASGOW	*with Dumbarton, Fairlie, Ardrossan, Irvine, Troon, Ayr, and Girvan* ... ... (*Office, 36, Oswald Street, Glasgow*)	William T. Mumford, *Principal Surveyor.* Burney, Charles E. Courtier-Dutton, W. T Dawkins, John Dodd, Thomas J. Dove, Herbert W. Edwards, Charles Hand, Henry Johnstone, William Roberts, J. T. Thearle, S. J. P. Thomson, Joseph
	Engineer Surveyors for Glasgow District	Mollison, James Robson, Walter E.
	Ship and Engineer Surveyors	Beveridge, R. J. Cooper, C. Harvey, J. Kidd, A. Ryder, C. L. Salmon, E. M. Sanderson, John Sibun, William Smith, W. J. Stewart, A.
	Inspector of Forgings for the Clyde District	Newcomb, George
GREENOCK	*and Port Glasgow, with Rothesay* (*Office, 13, Hamilton Street, Greenock*)	George Stanbury, *Principal Surveyor* Andrews, W. Fowling, Charles
	Ship and Engineer Surveyor for Greenock District	Heron, Andrew C.
HARTLEPOOL,	*Stockton, and Middlesbrough, also Whitby and Scarborough* ... ... (*Office, Dock Office Buildings, Victoria Terrace, West Hartlepool*)	Charles Davidson, *Principal Surveyor* Phillips, Thomas Williams, H. M. Wilson, Allison B.
	Engineer Surveyor for the Tees District	Stoddart, J. E.
	Ship and Engineer Surveyors	Austin, W. E. Blackie, T. R. Buchanan, J. H. Morrison, W. Pollock, J. Ritson, M. Shoosmith, H.
HULL	*Gainsborough, Goole, Selby, Grimsby, Burton Stather, and Knottingley* ... ... (*Office, Bank Chambers, Land of Green Ginger, Hull*)	Williamson, Robert
	Ship and Engineer Surveyor for the Hull District	Innes, James

Port	District	Surveyors
IPSWICH	... ... (*Address, 35, Lower Orwell Street*)	Robertson, Ebenezer J.
LEITH	*and Ports in the Frith of Forth, with Berwick-upon-Tweed* ... ... (*Office, 56, Constitution Street, Leith*)	Paulson, William
	Ship and Engineer Surveyor	Darling, William J.
LIVERPOOL	*with Lancaster and all intermediate places, the River Mersey, Chester, and River Dee; also the Isle of Man* (*Office, 12, Oriel Chambers, Liverpool*)	John F. Light, *Secretary and Principal Surveyor.* Champness, Edward C. Lawrence, John Moverly, William Rutherford, John Wheeler, Edward C.
	Engineer Surveyors for the Liverpool District	McGregor, Peter Milner, G. A.
MILFORD HAVEN,	*Pembroke, and Tenby, with district extending as far as Newquay inclusive* ... (*Address, Marina, Neyland, R. S. O., South Wales*)	*Harris, W.

* Mr. Harris is not an exclusive Officer of the Society.

Port	District	Surveyors
NEWCASTLE-ON-TYNE	*with North and South Shields; also Blyth and Hartley* ... ... (*Office, 3, St. Nicholas' Buildings, Newcastle-on-Tyne*)	Henry J. Boolds, *Principal Surveyor.* Cooke, Thomas H. McNeil, James Scullard, J. W. Sharpe, W. L. Shilston, Thomas Sibun, James Skentelbery, C.
	Engineer Surveyors for the Tyne District	Hirst, Richard Walliker, J. F.
	Ship and Engineer Surveyors	Barclay, J. Hake, G. A. Napier, Robert J. Rogers, H. M.
NEWPORT, MON.	*with Chepstow, Ship and Engineer Surveyor for the Newport District* ... ... (*Office, 30, Dock Street, Newport, Mon.*)	Heck, John H.
PLYMOUTH	*Ship and Engineer Surveyor* ... ... (*Office, 13, Exchange*)	Davey, W. M.
PORTMADOC	*and Barmouth, with a district extending as far as, but not including, Newquay* ... (*residing at Portmadoc*)	John, William
QUEENSTOWN	*with Cork, Kinsale, and Limerick* .. (*Address, 1, King's Terrace, Queenstown*)	Weight, George
SOUTHAMPTON	*and South Coast from Shoreham to Bridport inclusive, also the Isle of Wight* ... (*Office, 9, Oriental Place, Southampton*)	Sinnette, James L.
	Ship and Engineer Surveyor	Stevens, John B.
SUNDERLAND	*and Seaham* ... ... (*Office, 56, John Street, Sunderland*)	Richard J. Reed, *Principal Surveyor.* Bath, William Buchanan, Charles Harrison, George Keen, Joseph Williams, Jesse
	Engineer Surveyor for the Wear District	Salmon, Patrick
	Ship and Engineer Surveyor	Findlay, J. T.
	Inspector of Forgings for the North-East Coast	Cameron, Hugh
SWANSEA	*with Neath and Llanelly* (*Office, Adelaide Chambers, Adelaide Street, Swansea*)	Ashton, Thomas

(*For Engineer Surveyors for Swansea District see* CARDIFF.)

Port	District	Surveyors
WHITEHAVEN,	*Workington, Harrington, and Maryport, with Dumfries and Annan* ... ... (*residing at Whitehaven*)	House, Thomas J.

(*For Engineer Surveyors, see* BARROW.)

THE SURVEYORS AT THE FOLLOWING PORTS DO NOT HOLD APPOINTMENTS AS THE EXCLUSIVE SERVANTS OF THE SOCIETY.

GUERNSEY	... ...		
LYNN	*Boston, Wells, and intermediate Ports*		
ORKNEYS	... ...	*(residing at Stromness)*	Baillie, Geo. Gunn
RAMSGATE	*and Margate, with Deal and Dover*	*(residing at Ramsgate)*	Jones, Edward
SLIGO	... ...		Pollexfen, William
WATERFORD	*Ship and Engineer Surveyor*		Horn, Andrew
WEXFORD	... ...		Sparrow, R. J.

LIST OF FOREIGN AND COLONIAL SURVEYORSHIPS OF LLOYD'S REGISTER

(GEOGRAPHICALLY ARRANGED).

LISTE DES INSPECTEURS DU LLOYD'S REGISTER À L'ÉTRANGER ET AUX COLONIES

(PAR ORDRE GÉOGRAPHIQUE DES PORTS).

Port	Office	Surveyor
	RUSSIA.	
CRONSTADT AND ST. PETERSBURG	*Ship and Engineer Surveyor* (*Address, Officerskaia, No.* 57, *St. Petersburg*)	Wessblad, A.
HELSINGFORS	*Ship and Engineer Surveyor* ... ... (*Address,* 23, *Brunnspark*)	Kolster, R.
ODESSA	... ...	
RIGA	... ...	Lindemann, C. A.
	Engineer Surveyor ... ... (*Address, Pychau Dampfsagemühle*)	Hillbring, F. W.
SEBASTOPOL	*Ship and Engineer Surveyor*	Corry, John E.
TAGANROG	*Ship and Engineer Surveyor*	Bell, Charles
	NORWAY.	
ARENDAL AND GRIMSTAD	... ...	
BERGEN	*Ship and Engineer Surveyor*	Hougland, E.
	SWEDEN.	
CARLSCRONA	... ...	
GEFLE	... ...	
GOTHENBURG	*Ship and Engineer Surveyor* ... (*Address, Lazarettsgatan, No.* 101 *b*)	Möller, Carl Axel.
	Assistant Surveyor to Mr. Möller:—	
STOCKHOLM	*Ship and Engineer Surveyor* (*With a district extending from Gefle to Norrköping inclusive*)	Isakson, Albert
	DENMARK.	
COPENHAGEN	*Ship and Engineer Surveyor* ... ... (*Office,* 18, *Fredericiagade, K.*)	Kindler, P. Fred.
	Assistant Surveyors to Mr. Kindler:—	
AALBORG	*Ship Surveyor*	Jepmond, A. H.
	Engineer Surveyor	Lange, L. S.
COPENHAGEN	*Ship and Engineer Surveyor*	Herrmann, W. B.
NORDBY, FANÖ	*Ship Surveyor*	Hansen, L. J.
ODENSE	*Engineer Surveyor*	Allerup, E. S. V.
RANDERS	*Ship Surveyor*	Petersen, Poul F.
	Engineer Surveyor	Nøddelund, J. C.
SVENDBORG	*Ship Surveyor*	Kaas, L.
	GERMANY.	
BREMERHAVEN	*Ship and Engineer Surveyor for Weser District* ... ... (*Office, Lloyd Strasse, No.* 22)	Thomsen, F. H. T.
DANZIG	... ... (*Address,* 9, *Brabank*)	Domke, Robt.
HAMBURG	*with the river Elbe, and Ports in Holstein* (*Office, Steinhöft No.* 3, *Hamburg*)	Padderatz, Emil.
	Engineer Surveyor for the Hamburg District ... (*Office, Stubbenhuk No.* 20)	Berendt, M.
ROSTOCK	*with a district extending from Lubeck to Stettin inclusive*	Cordes, W.

HOLLAND.

AMSTERDAM	*with Nieuwdiep and neighbouring Ports; Ship and Engineer Surveyor for Amsterdam; also Engineer Surveyor for Ports in Holland ... ... (Address, Hoogte Kadijk 35)*	*Van Ollefen, W.F.D
ROTTERDAM	*with Dordrecht, Schiedam, and surrounding places, also Zeeland ... ... (Address, Westerhaven 6, Rotterdam)*	Loos, Jan C. W.
VEENDAM	*with Groningen, Zwolle, Harlingen, Embden, and surrounding places, including Hanover... ... ... (Address, Kerklaan H 785, Veendam)*	Hazewinkel, H. P.

* Mr. Van Ollefen is exclusively an officer of Lloyd's Register.

BELGIUM.

BELGIUM	*also the Ports in the River Scheldt, including Terneuzen and neighbouring Ports (Office, Rue d'Amsterdam, Antwerp)*	Paasch, Heinrich.
	Engineer Surveyor for the Belgium District ... (residing at Antwerp)	Demblon, Francis.

FRANCE.

BORDEAUX	... ... *(Address, 16, Rue Blanc Dutrouilh)*	Vandercruyce, Jules.
	Engineer Surveyor ... ... (Address, 108, Rue du Palais Gallien)	Douzelle, A.
DUNKIRK	... ...	De Wulf, H.
HAVRE	*with a District, including Boulogne and Barfleur... ... ... Ship and Engineer Surveyor (Address, 25, Place de l'Hôtel de Ville, Havre)*	Le Laidier, A.
LA ROCHELLE	... ...*(Address, 6, Place de la Préfecture)*	Garnier, André Louis.
MARSEILLES	*Ship and Engineer Surveyor ... ... (Address, 5, Rue Suffren)*	Westerman, Francis.
NANTES	*Ship and Engineer Surveyor ... (Address, 21, Quai de L'ile Gloriette)*	L'Heureux, J.
ST. MALO	*with a District extending from Brest to Cherbourg inclusive ... ... Ship and Engineer Surveyor ... ... (Address, 12, Rue d'Asfeld)*	Gallais, Jean.

SPAIN AND PORTUGAL.

BARCELONA	*Ship and Engineer Surveyor ... ... (Address, Pórtico Xifré, 10, 1°)*	Browne, J. J.
BILBAO	... ... *(Address, Muelle de Ripa, 3)*	De Bareno, German.
CADIZ	*Ship and Engineer Surveyor ... ... (Address, 25, Aduana)*	Cochrane, James.
CARTHAGENA	*Ship and Engineer Surveyor*	Sanchez, José Hernand
HUELVA	*Ship and Engineer Surveyor*	Langdon, William.
LISBON	*Ship and Engineer Surveyor (Address, Rua do Jardin do Tabaco 33, 22°)*	Westwood, J.
MALAGA	*Ship and Engineer Surveyor*	Lappe, Enrique.
OPORTO	... ... *(Office, Rua do Rebobeira, 13, 2°)*	Dos Santos, José Mari
	Engineer Surveyor	Ennor, C. J.

GIBRALTAR.

GIBRALTAR	... ...	

ITALY AND AUSTRO-HUNGARY.

ANCONA	... ...	Emett, F. G.
GENOA	... ... *(Office, Via S. Luca, No. 2, Prima Scala)*	Schiaffino, Francesco.
	Engineer Surveyor for Genoa and Leghorn (Address, 5, Rue Suffren, Marseilles)	Westerman, Francis.
LEGHORN	... ...	Gori, Costantino.
NAPLES	... ...	
TRIESTE	*with District of the Austro-Hungarian Coast ... ... (Office, Via Carradori, No. 1, Trieste)*	Florio, Elias.
	Engineer Surveyor for Trieste District	Schnabl, Frederick.
	Assistant Surveyor at Fiume	Bonetich, Ignazio.
	Ditto at Lussino	Tarrabocchia, Antonio
	Ditto at Venice	Fabro, Matteo.

MALTA.

MALTA	*Ship and Engineer Surveyor*	Wright, C. H.

GREECE, TURKEY, TURKEY IN ASIA AND ROUMANIA.

CONSTANTINOPLE	... ...	Admiral Woods Pasha.
	Engineer Surveyor	Warren, G. R.
GALATZ AND BRAILA	... ... *(residing at Braila)*	Archbold, T. H.
PIRÆUS	*Ship and Engineer Surveyor*	Graham, D.
SYRA	... ...	

BRITISH NORTH AMERICA.

HALIFAX, N.S. ...	... ...	Hunter, David.
	Engineer Surveyor	MacDonald, A.
MONTREAL ...	... ... *(Address, Port Warden's Office)*	Shaw, J. G.
PRINCE EDWARD ISLAND	... ... *(residing at Charlotte Town)*	Welsh, H. P.
QUEBEC	*and the River St. Lawrence*	Dick, John.
ST. JOHN, N.B.	*with Miramichi and Northern District of New Brunswick*	Coker, Charles R.
TORONTO	... ...	Harbottle, Thomas.
VANCOUVER, BRITISH COLUMBIA	... ...	Clarke, W. R.

UNITED STATES.

NEW YORK	*Principal Surveyor for the United States, and Surveyor for the Port and District of New York* ... ... *(Office, Kemble Buildings, Whitehall Street, New York)*	*Congdon, Thomas.

* Mr. Congdon is exclusively an Officer of Lloyd's Register.

BALTIMORE	... ... *(Office, 11, South Gay Street)*	Sanford, Edward H.
	Engineer Surveyor for Baltimore	Wells, Richard.
BOSTON	*Ship and Engineer Surveyor* *(Office, 24, White Street, East Boston)*	Shaw, Oliver L.
NEW ORLEANS	... *(Address, c/o Messrs. Marshall, J. Smith & Co., 64, Baronne Street)*	Turley, J. K.
PHILADELPHIA	*Ship and Engineer Surveyor* ... ... *(Office, 206, Walnut Place)*	Haug, John.
PORTLAND, OREGON	... ...	Pope, George.
SAN FRANCISCO	... ... *(Office, 219, Sansome Street)*	Freeman, J. H.
WILMINGTON, N.C.	... ...	Beery, B. W.

WEST INDIES.

KINGSTON, JAMAICA	... ...	

SOUTH AMERICA.

BAHIA	*Ship and Engineer Surveyor*	Woods, A. R. T.
BUENOS AYRES	*Ship and Engineer Surveyor*	Hunter, W. E.
DEMERARA	... ...	Duncan, Alexander.
MONTE VIDEO	... ... *(Address, c/o Messrs. Carlisle, Smith & Co., 29, Calle Rincon)*	Crocker, F.
RIO DE JANEIRO	... ...	
VALPARAISO	*Ship and Engineer Surveyor* ... ... *(Address, Casilla No. 4)*	Ivoi, C. C.

NORTH AFRICA.

ALGIERS, ORAN AND CONSTANTINE	... ... *(residing at Algiers)*	Burke, F. M.
SUEZ	... ...	Weston, A. R.
	Engineer Surveyor	Campbell, J.

South Africa.

Cape Town	...	Stephen, W.
Port Natal	...	Airth, Alexander.

Mauritius.

Mauritius	... (*Residing at Port Louis*)	Cowin, John.

India, Ceylon, Burmah, &c.

Akyab	...	Brown, G. B.
	Engineer Surveyor ...	McLaggan, C.
Bombay	... (*Office, 10, Meadow Street*)	Clarke, A. C.
	Engineer Surveyor ... (*Address, Bombay Steam Navigation Co., Frere Road, Bombay*)	Moir, James.
Calcutta	...	McKellar, D.
Colombo	...	Donnan, J.
Kurrachee	*Ship and Engineer Surveyor* ... (*Address, Persian Gulf Telegraphs, Manora, Kurrachee*)	Hughes, J.
Moulmein	*Ship and Engineer Surveyor* ...	Campbell, P.
Rangoon	...	Winter, R. R.
Singapore	...	Fittock, Charles.
	Engineer Surveyor ... (*Address, c/o Mr. Charles Fittock*)	Park, Robert.

East Indian Archipelago.

Batavia	*Acting Surveyor* ...	't Hoen, J. A.
Cheribon	...	Campbell, C.
Manila	*and Ports in the Philippine Islands* ... *Ship and Engineer Surveyor* ...	Sawyer, Frederick H.
Samarang	...	Vogel, W.
Sourabaya	...	Vader, P.

China and Japan.

Foochow	...	
Hong Kong	...	Burnie, Edward.
	Engineer Surveyor ... (*Address, Bank Buildings*)	Johnston, Andrew.
Kobé, Hiogo	*Ship and Engineer Surveyor* ...	Ellerton, James.
Nagasaki (*See* Kobé).		
Shanghai	*Ship and Engineer Surveyor* ...	Sonne, H.
Yokohama	...	Efford, John J.

Australia, Tasmania, and New Zealand.

Adelaide, S.A.	...	Begg, William.
Auckland, N.Z.	...	Clayton, M. T.
	Engineer Surveyor ...	Lodder, W.
Brisbane, Queensland	...	Brown, William B.
Christchurch, N.Z.	...	Ticehurst, Robert.
Dunedin, N.Z.	...	Thomson, W.
Hobart, Tasmania	...	Macmillan, Donald.
Melbourne, Victoria	...	Watson, William.
Napier, N.Z.	...	Davidson, A.
Newcastle, N.S.W.	...	Brooks, Thomas.
	Engineer Surveyor ...	Rorison, James.
Sydney, N.S.W.	... (*Office, 22, Bridge Street, Sydney*)	Pockley, Robert F.
	Engineer Surveyor ...	Pollock, R.
Townsville	...	Grainger, T. L.
Wellington, N.Z.	...	Bendall, William.

LIST OF SURVEYORSHIPS OF LLOYD'S REGISTER

(ALPHABETICALLY ARRANGED).

THE SURVEYORS AT THE PORTS MARKED * ARE EXCLUSIVELY THE OFFICERS OF THE SOCIETY, AND ARE NOT PERMITTED TO ENGAGE IN ANY OTHER BUSINESS OR EMPLOYMENT WHATEVER.

LISTE DES PORTS OÙ RÉSIDENT LES INSPECTEURS DU LLOYD'S REGISTER

(PAR ORDRE ALPHABÉTIQUE).

LES INSPECTEURS ÉTABLIS AUX PORTS MARQUÉS D'UN * SONT EXCLUSIVEMENT OFFICIERS DE LA SOCIÉTÉ, ET IL NE LEUR EST PAS PERMIS DE S'ENGAGER DANS D'AUTRES AFFAIRES OU EMPLOI QUELCONQUE.

Port	Office	Surveyor
*LONDON	Benjamin Martell, Esq., *Chief Surveyor of Lloyd's Register*	William C. Davey, *Principal Surveyor for London District.*
	Harry J. Cornish, Thomas Edwards, *Assistants to Chief Surveyor*	Cooper, George P. Jordan, Chas. H. Mares, George R. Read, Thomas C. Tierney, E. J.
	William Parker, Esq., *Chief Engineer Surveyor of Lloyd's Register*	Truscott, James H. Warren, Thomas S.
	William Allison, *Assistant to Chief Engineer Surveyor*	Wilkinson, Henri
	Engineer Surveyors	Stromeyer, Charles E. Wilkinson, George E.
	Ship and Engineer Surveyors	Gray, T. L. Jones, A. L.
	Draughtsman	Hunter, David S.
AALBORG	*Ship Surveyor* / *Engineer Surveyor* ... ... (*Assistants to Mr. Kindler at Copenhagen*)	Jepmond, A. H. Lange, L. S.
*ABERDEEN	*Ship and Engineer Surveyor* ... ... (*Office, 29, Regent Quay*)	Hindmarsh, G. L.
ADELAIDE, S.A.	... ...	Begg, William.
AKYAB	... ...	Brown, G. B.
	Engineer Surveyor	McLaggan, C.
ALGIERS, ORAN AND CONSTANTINE	... ... (*residing at Algiers*)	Burke, F. M.
*AMSTERDAM	*Ship and Engineer Surveyor for Amsterdam; also Engineer Surveyor for ports in Holland* ... ... (*Address, Hoogte Kadijk 35*)	Van Ollefen, W. F. D.
ANCONA	... ...	Emett, F. G.
ANTWERP	... ... (*Office, Rue d'Amsterdam*)	Paasch, Heinrich.
	Engineer Surveyor for Belgium ... ... (*residing at Antwerp*)	Demblon, Francis.
ARENDAL	... ...	
AUCKLAND, N.Z.	... ...	Clayton, M. T.
	Engineer Surveyor	Lodder, W.
BAHIA	*Ship and Engineer Surveyor*	Woods, A. R. T.
BALTIMORE	... ... (*Office, 11, South Gay Street*)	Sanford, Edward H.
	Engineer Surveyor	Wells, Richard.

Port	Appointment	Address	Surveyor
*BANGOR		(*Address*, 5, *Glan Dwr Terrace*)	Devonald, Thomas.
BARCELONA	*Ship and Engineer Surveyor*	(*Address*, *Portico Xifré*, 10, 1°)	Browne, J. J.
*BARROW	*Ship and Engineer Surveyors*	(*Office*, 8, *Strand*)	Ritchie, Duncan. Dykes, J.
BATAVIA	*Acting Surveyor*		'tHoen, J. A.
*BELFAST		(*Office*, *Victoria Chambers*, *Waring Street*)	Turpin, James.
	Ship and Engineer Surveyor		Maxton, James.
BERGEN	*Ship and Engineer Surveyor*		Hougland, E.
*BIDEFORD			Fittock, Charles.
BILBAO		(*Address*, *Muelle de Ripa*, 3)	De Bareno, German.
BOMBAY		(*Office*, 10, *Meadow Street*)	Clarke, A. C.
	Engineer Surveyor	(*Address*, *Bombay Steam Navigation Co.*, *Frere Road*, *Bombay*)	Moir, James.
BORDEAUX		(*Address*, 16, *Rue Blanc Dutrouilh*)	Vandercruyce, Jules.
	Engineer Surveyor	(*Address*, 108, *Rue du Palais Gallien*)	Donzelle, A.
BOSTON	*Ship and Engineer Surveyor*	(*Office*, 24, *White Street*, *East Boston*)	Shaw, Oliver L.
BRAILA (*See* GALATZ).			
BREMERHAVEN	*Ship and Engineer Surveyor*	(*Office*, *Lloyd Strasse*, *No.* 22)	Thomsen, Ferd. H. T.
BRISBANE, QUEENSLAND			Brown, William B.
*BRISTOL	*Ship and Engineer Surveyor*	(*Office*, 53, *Queen Square*)	Coomber, R. W.
BUENOS AYRES	*Ship and Engineer Surveyor*		Hunter, W. E.
CADIZ	*Ship and Engineer Surveyor*	(*Address*, 25, *Aduana*)	Cochrane, James.
CALCUTTA			McKellar, D.
CAPE TOWN			Stephen, W.
*CARDIFF		(*Offices*, *Merchants' Exchange*)	Andrew K. Hamilton, *Principal Surveyor.* Rule, J. G. G.
	Ship and Engineer Surveyors		Ashton, H. H. Fargie, W. A. Mancor, J. H. Taylor, J. E.
CARLSCRONA			
CARTHAGENA	*Ship and Engineer Surveyor*		Sanchez, José Hernandez
*CHANNEL ISLANDS		(*residing at Jersey*)	Chinneck, T. J.
CHERIBON			Campbell, C.
CHRISTCHURCH, N.Z.			Ticehurst, Robert.
COLOMBO			Donnan, J.
CONSTANTINE (*See* ALGIERS).			
CONSTANTINOPLE			Admiral Woods Pasha.
	Engineer Surveyor		Warren, G. R.
COPENHAGEN	*Ship and Engineer Surveyor*	(*Office*, 18, *Fredericiagade*, *K.*)	Kindler P. Fred.
	Assistant Ship and Engineer Surveyor		Herrmann, W. B.
CRONSTADT AND ST. PETERSBURG	*Ship and Engineer Surveyor*	(*Address*, *Officerskaia*, *No.* 57, *St. Petersburg*)	Wessblad, A.
DANZIG		(*Address*, 9, *Brabank*)	Domke, Robert
DEMERARA			Duncan, Alexander.
*DUBLIN		(*Address*, 23, *Ballybough Road*)	Franklin, J.
*DUNDEE	*Ship and Engineer Surveyor*	(*Office*, 54, *Commercial Street*)	Keydell, A. E.
DUNEDIN, N.Z.			Thomson, W.
DUNKIRK			De Wulf, H.
*FALMOUTH		(*Address*, 2, *Park Terrace*)	Sandry, T. H.
	Ship and Engineer Surveyor		Dyer, C. M. B.

Port	Position	Address	Surveyor
FIUME (TRIESTE DISTRICT)	*Assistant Surveyor*		Bonetich, Ignazio
FOOCHOW			
GALATZ		*(residing at Braila)*	Archbold, T. H.
GEFLE			
GENOA		*(Office, Via S. Luca No. 2, Prima Scala)*	Schiaffino, Francesco.
	Engineer Surveyor	*(residing at Marseilles)*	Westerman, Francis.
GIBRALTAR			
*GLASGOW		*(Office, 36, Oswald Street)*	William T. Mumford, *Principal Surveyor.* Burney, Charles E. Courtier-Dutton, W. T. Dawkins, John. Dodd, Thomas J. Dove, Herbert W. Edwards, Charles. Hand, Henry. Johnstone, William. Roberts, J. T. Thearle, S. J. P. Thomson, Joseph.
	Engineer Surveyors		Mollison, James. Robson, Walter E.
	Ship and Engineer Surveyors		Beveridge, R. J. Cooper, C. Harvey, J. Kidd, A. Ryder, C. L. Salmon, E. M. Sanderson, John. Sibun, William Smith, W. J. Stewart, A.
	Inspector of Forgings for the Clyde District		Newcomb, George.
GOTHENBURG	*Ship and Engineer Surveyor*	*(Address, Lazarettsgatan No. 101 b)*	Möller, Carl Axel.
*GREENOCK		*(Office, 13, Hamilton Street)*	George Stanbury, *Principal Surveyor.* Andrews, W. Fowling, Charles.
	Ship and Engineer Surveyor		Heron, Andrew C.
GUERNSEY			
HALIFAX, N.S.			Hunter, David.
	Engineer Surveyor		MacDonald, A.
HAMBURG		*(Office, Steinhöft No. 3)*	Padderatz, Emil.
	Engineer Surveyor	*(Office, Stubbenhuk, No. 20)*	Berendt, M.
*HARTLEPOOL		*(Office, Dock Office Buildings, Victoria Terrace, West Hartlepool)*	Charles Davidson, *Principal Surveyor.* Phillips, Thomas. Williams, H. M. Wilson, Allison B.
	Engineer Surveyor		Stoddart, J. E.
	Ship and Engineer Surveyors		Austin, W. R. Blackie, T. R. Buchanan, J. H. Morrison, W. Pollock, J. Ritson, M. Shoosmith, H.
HAVRE	*Ship and Engineer Surveyor*	*(Address, 25, Place de l'Hôtel de Ville)*	Le Laidier, A.

Port	Office / Address	Surveyor
HELSINGFORS	*Ship and Engineer Surveyor* ... ... (*Address*, 23, *Brunnspark*)	Kolster, R.
HOBART, TASMANIA	... ...	Macmillan, Donald.
HONG KONG	... ...	Burnie, Edward.
	Engineer Surveyor ... ... (*Address*, *Bank Buildings*)	Johnston, Andrew.
HUELVA	*Ship and Engineer Surveyor*	Langdon, William.
*HULL	... ... (*Office*, *Bank Chambers*, *Land of Green Ginger*, *Hull*)	Williamson, Robert.
	Ship and Engineer Surveyor	Innes, James.
*IPSWICH	... ... (*Address*, 35, *Lower Orwell Street*)	Robertson, Ebenezer J.
KINGSTON, JAMAICA	... ...	
KOBÉ, HIOGO	*Ship and Engineer Surveyor*	Ellerton, James.
KURRACHEE	*Ship and Engineer Surveyor* ... ... (*Address*, *Persian Gulf Telegraphs*, *Manora*, *Kurrachee*)	Hughes, J.
LA ROCHELLE	... ... (6, *Place de la Préfecture*)	Garnier, Andre Louis
LEGHORN	... ...	Gori, Costantino.
*LEITH	... ... (*Office*, 56, *Constitution Street*)	Paulsen, William.
	Ship and Engineer Surveyor	Darling, William J.
LISBON	*Ship and Engineer Surveyor* ... ... (*Address*, *Rua do Jardin do Tabaco* 33, 22)	Westwood, J.
*LIVERPOOL	... ... (*Office*, 12, *Oriel Chambers*)	John F. Light, *Secretary and Principal Surveyor.* Champness, Edward C. Lawrence, John. Moverly, William. Rutherford, John. Wheeler, Edward C.
	Engineer Surveyors	McGregor, Peter. Milner, G. A.
LUSSINO (TRIESTE DISTRICT)	... *Assistant Surveyor*	Tarrabocchia, Antonio E.
LYNN	... ...	
MALAGA	*Ship and Engineer Surveyor*	Lappe, Enrique.
MALTA	*Ship and Engineer Surveyor*	Wright, C. H.
MANILA	*Ship and Engineer Surveyor*	Sawyer, Frederick H.
MARSEILLES	*Ship and Engineer Surveyor* ... ...(*Address*, *Rue Suffren* 5) ...	Westerman, Francis.
MAURITIUS	... ... (*residing at Port Louis*)	Cowin, John.
MELBOURNE, VICTORIA	... ...	Watson, William.
MILFORD HAVEN, *Pembroke and Tenby*, *with district extending as far as Newquay inclusive*	... (*Address*, *Marina*, *Neyland*, *R. S. O.*, *South Wales*)	*Harris, W.
	* Mr. Harris is not an exclusive officer of the Society.	
MONTE VIDEO	... ... (*Address*, *c/o Messrs. Carlisle*, *Smith & Co.*, 29, *Calle Rincon*)	Crocker, F.
MONTREAL	... ... (*Address*, *Port Warden's office*)	Shaw, J. G.
MOULMEIN	*Ship and Engineer Surveyor*	Campbell, P.
NAGASAKI (*See* KOBÉ)		
NANTES	*Ship and Engineer Surveyor* ... (*Address*, 21, *Quai de L'ile Gloriette*)	L'Heureux, J.
NAPIER, N.Z.	... ...	Davidson, A.
NAPLES	... ...	
NEW ORLEANS	... (*Address*, *c/o Messrs. Marshall*, *J. Smith & Co.*, 64, *Baronne Street*)	Turley, J. K.
*NEW YORK	*Printing Surveyor for the United States* ... (*Office*, *Kemble Buildings*, *Whitehall Street*)	Congdon, Thomas.

Port	Description	Surveyors
*NEWCASTLE-ON-TYNE	... ... (*Office*, 3, *St. Nicholas' Buildings*)	Henry J. Boolds, *Principal Surveyor.* Cooke, Thomas H. McNeil, James. Scullard, J. W. Sharpe, W. L. Shilston, Thomas. Sibun, James. Skentelbery, C.
	Engineer Surveyors	Hirst, Richard. Walliker, J. F.
	Ship and Engineer Surveyors	Barclay, J. Hake, G. A. Napier, Robert J. Rogers, H. M.
NEWCASTLE, N.S.W.	... ...	Brooks, Thomas.
	Engineer Surveyor	Rorison, Jas.
*NEWPORT, MON.	*Ship and Engineer Surveyor* ... ... (*Office*, 30, *Dock Street*)	Heck, John H.
NORDBY, FANÖ	(*Assistant to Mr. Kindler at Copenhagen*)	Hansen, L. J.
ODENSE	*Engineer Surveyor* (*assistant to Mr. Kindler at Copenhagen*)	Allerup, E. S. V.
ODESSA	... ...	
OPORTO	... ... (*Office*, *Rua do Rebobeira*, 13, 2°)	Dos Santos, José Maria.
	Engineer Surveyor	Ennor, C. J.
ORAN (*See* ALGIERS).		
ORKNEYS	... ... (*residing at Stromness*)	Baillie, Geo. Gunn.
PHILADELPHIA	*Ship and Engineer Surveyor* ... ... (*Office*, 206, *Walnut Place*)	Haug, John.
PIRÆUS	*Ship and Engineer Surveyor*	Graham, D.
*PLYMOUTH	*Ship and Engineer Surveyor* ... ... (*Office*, 13, *Exchange*)	Davey, W. M.
PORT NATAL	... ...	Airth, Alexander.
PORTLAND, OREGON	... ...	Pope, George.
*PORTMADOC	... ... (*residing at Portmadoc*)	John, William.
PRINCE EDWARD ISLAND	... ... (*residing at Charlotte Town*)	Welsh, H. P.
QUEBEC	... ...	Dick, John.
*QUEENSTOWN	... ... (1, *King's Terrace*)	Weight, George.
RAMSGATE	... ...	Jones, Edward.
RANDERS	*Ship Surveyor* *Engineer Surveyor* (*assistants to Mr. Kindler at Copenhagen*)	Peterson, Poul F. Nøddelund, J. C.
RANGOON	... ...	Winter, R. R.
RIGA	... ...	Lindemann, C. A.
	Engineer Surveyor ... ... (*Address*, *Pychau Dampfsagemühle*)	Hillbring, F. W.
RIO DE JANEIRO	... ...	
ROSTOCK	... ...	Cordes, W.
ROTTERDAM	... ... (*Address*, *Westerhaven* 6)	Loos, Jan C. W.
	Engineer Surveyor for Rotterdam ... ... (*residing at Amsterdam*)	Van Ollefen, W.F.D.
ST. JOHN, N.B.	... ...	Coker, Charles R.
ST. PETERSBURG	(*See* CRONSTADT).	
ST. MALO	*Ship and Engineer Surveyor* ... ... (*Address*, 12, *Rue d'Asfeld*)	Gallais, Jean.
SAMARANG	... ...	Vogel, W.
SAN FRANCISCO	... ... (*Office*, 219, *Sansome Street*)	Freeman, J. H.
SEBASTOPOL	*Ship and Engineer Surveyor*	Corry, John E.
SHANGHAI	*Ship and Engineer Surveyor*	Sonne, H.

Place	Description	Surveyor
SINGAPORE	... ...	Fittock, Charles.
	Engineer Surveyor ... ... (*Address, c/o Mr. Charles Fittock*)	Park, Robert.
SLIGO	... ...	Pollexfen, William.
SOURABAYA	... ...	Vader, P.
*SOUTHAMPTON	... ... (*Office*, 9, *Oriental Place*)	Sinnette, James L.
	Ship and Engineer Surveyor	Stevens, John B.
STOCKHOLM (GOTHENBURG DISTRICT)	*Assistant Ship and Engineer Surveyor*	Isakson, Albert
SUEZ	... ...	Weston, A. R.
	Engineer Surveyor	Campbell, J.
*SUNDERLAND	... ... (*Office*, 56, *John Street*)	Richard J. Reed, *Principal Surveyor* Bath, William. Buchanan, Charles. Harrison, George. Keen, Joseph. Williams, Jesse.
	Engineer Surveyor	Salmon, Patrick.
	Ship and Engineer Surveyor	Findlay, J. T.
	Inspector of Forgings for the North-East Coast	Cameron, Hugh.
SVENDBORG	(*Assistant to Mr. Kindler at Copenhagen*)	Kaas, L.
*SWANSEA	... ... (*Office, Adelaide Chambers, Adelaide Street*)	Ashton, Thomas.
	(*For Engineer Surveyors for Swansea District see* CARDIFF.)	
SYDNEY, N.S.W.	... ... (*Office*, 22, *Bridge Street, Sydney*)	Pockley, Robert F.
	Engineer Surveyor	Pollock, R.
SYRA	... ...	
TAGANROG	*Ship and Engineer Surveyor*	Bell, Charles.
TORONTO	... ...	Harbottle, Thos.
TOWNSVILLE	... ...	Grainger, T. L.
TRIESTE	... ... (*Office, Via Carradori, No.* 1)	Florio, Elias.
	Engineer Surveyor for Trieste District	Schnabl, Frederick.
VALPARAISO	*Ship and Engineer Surveyor* ... ... (*Address, Casilla, No.* 4)	Ivol, C. C.
VANCOUVER, BRITISH COLUMBIA	... ...	Clarke, W. R.
VEENDAM	... ... (*Address, Kerklaan H.* 785, *Veendam*)	Hazewinkel, H. P.
VENICE (TRIESTE DISTRICT)	... *Assistant Surveyor*	Fabro, Matteo
WATERFORD	*Ship and Engineer Surveyor*	Horn, Andrew.
WELLINGTON, N.Z.	... ...	Bendall, William.
WEXFORD	... ...	Sparrow, R. J.
*WHITEHAVEN	... ...	House, Thomas J.
	(*For Engineer Surveyors, see* BARROW).	
WILMINGTON, N.C.	... ...	Beery, B. W.
YOKOHAMA	... ...	Efford, John J.

LIST OF SUBSCRIBERS TO THE UNIVERSAL REGISTER.

1889–1890.

LISTE DES ABONNÉS AU UNIVERSAL REGISTER.

1889–1890.

Alliance Marine Assurance Company (*Ld.*)	Capel Court, London; K 11 & 12, Exchange Buildings E, Liverpool; and at Sydney.
Anglo-Nevada Assurance Corporation	410, Pine Street, San Francisco.
Association of Underwriters and Insurance Brokers (R. Cooper Rundell, *Secretary*)	Glasgow.
Association of Underwriters	Hull.
Association of Underwriters	Liverpool.
Atlantic Mutual Insurance Company	51, Wall Street, New York.
Australian General Assurance Company (Watt, Gilchrist & Co., *Marine Agents*)	Princes Street Chambers, Princes Street, London.
Badische Rück und Mitversicherungs Gesellschaft	Mannheim.
Bengal Chamber of Commerce	102, Clive Street, Calcutta.
Bergens Börs	Bergen.
Biblioteca Direction de Hidrografia	Madrid.
Bristol Commercial Rooms	Corn Street, Bristol.
British and Foreign Marine Insurance Company (*Ld.*)	26, Cornhill, London.
British and Foreign Marine Insurance Company (*Ld.*)	Liverpool.
British and Foreign Marine Insurance Company (*Ld.*)	New York.
Chamber of Commerce	Port Elizabeth.
Club du Constantinople	Constantinople.
Colonial Maritime Insurance Company of Mauritius...	Port Louis, Mauritius.
Commercial Union Assurance Company (*Ld.*)	19, Cornhill, London.
Firemans' Fund Insurance Company	401, California Street, San Francisco.
Folger, G. H. (*Secretary Board of Underwriters*)...	50, State Street, Boston, U.S.A.
Globe Marine Insurance Company (*Ld.*)	79, Cornhill, London.
Home and Colonial Marine Insurance Company (*Ld.*)	8, Royal Exchange, London.
House of Commons Library	London.
Hull Exchange Company	Hull.
Indemnity Mutual Marine Assurance Company (*Ld.*)...	1, Threadneedle Street, London.
Insurance Company of North America (Fire and Marine)	232, Walnut Street, Philadelphia.
International Marine Insurance Company (*Ld.*) ... (*2 copies*)	Liverpool.
"La Española" Compañia de Seguros Maritimos	Bilbao.
Libau Exchange Committee	Libau.
Liverpool Exchange Company	News Room, Liverpool.

Liverpool Free Public Library ... ... Liverpool.
Lloyd's, Committee of ... ... (3 copies) London.
London Assurance Corporation ... ... 7, Royal Exchange, London.
London and Provincial Marine Insurance Company (*Ld.*) ... ... 2, Royal Exchange Buildings, London.
Mackenzie, Evan (*General Insurance Agent*) ... ... Piazza S. Lorenzo 15, Genoa.
Marine Insurance Company (*Ld.*) ... ... 20, Old Broad Street, London.
Marine Insurance Company ... ... Liverpool.
Mingotti, Giuseppe (*Underwriter and Director of the Compagnie "Italia"*) Piazza Scuole Pie 10, Genoa.
National Fire and Marine Insurance Company of New Zealand ... Dunedin.
Netherlands-India Sea and Fire Insurance Company ... ... Batavia.
Newcastle Commercial Exchange Company (*Ld.*) (William Miller, *Manager*; B. Plummer, jun., *Secretary*) ... ... Newcastle-on-Tyne.
Ocean Marine Insurance Company (*Ld.*) ... ... 2, Old Broad Street, London.
On Tai Insurance Company (*Ld.*) ... ... Hong Kong.
Otago Marine Underwriters' Association ... ... Dunedin, N.Z.
Reliance Marine Insurance Company ... ... Liverpool.
Riga Exchange Committee ... ... Riga.
Royal Agricultural and Commercial Society (Library and Exchange Rooms) ... ... Georgetown, Demerara.
Royal Exchange ... ... Glasgow.
Royal Exchange Assurance Corporation ... ... Royal Exchange, London.
Schött & Clarke (*Agents for United Swiss Marine Insurance Companies*) 3, Royal Exchange Buildings, London.
Sea Insurance Company (*Ld.*) ... ... (2 copies) Liverpool.
Shanghai Club ... ... Shanghai.
Sinclair & Boyd (Ulster Marine Insurance Company, *Ld.*) ... ... Belfast.
South British Fire and Marine Insurance Company ... ... London.
Southern Insurance Company (*Ld.*) ... ... Melbourne.
Straits Insurance Company (*Ld.*) ... ... 25, Cornhill, London.
Straits Insurance Company (*Ld.*)... ... ... Singapore.
Sydney Exchange (Alexander Campbell, *Chairman*) ... ... Sydney.
Thames and Mersey Marine Insurance Company (*Ld.*) ... ... 1, Royal Exchange Buildings, London.
Thames and Mersey Marine Insurance Company (*Ld.*) ... ... Liverpool.
Thames and Mersey Marine Insurance Company (*Ld.*) ... ... Manchester.
Underwriting and Agency Association (*Ld.*) ... ... 20, Bucklersbury, London.
Union Insurance Company... ... ... 416, California Street, San Francisco.
Union Insurance Society of Canton (*Ld.*) ... ... 9, Royal Exchange, London.
Union Insurance Society of Canton (*Ld.*) ... ... Paris.
Union Marine Insurance Company (*Ld.*) ... ... Liverpool.
Verein Hamburger Assecuradeure ... ... Hamburg.
Western Assurance Company (J. J. Kenny, *Managing Director*) ... Toronto.
Yangtsze Insurance Association (*Ld.*) ... ... 68 and 69, Cornhill, London.

Aberdeen Glen Line Steamship Co., *Shipowners* ... ... 48, Market Street, Aberdeen.
Aberdeen Steam Navigation Co. ... ... 87, Waterloo Quay, Aberdeen.
Adair & Co., *Engineers, Boiler Makers, Pump Makers, Ship Repairers, &c.* ... ... Liverpool.
Adam Brothers ... ... 17, Gracechurch Street, London.
Adamson, Gilfillan & Co. ... ... 2, Billiter Avenue, London.
Adamson & Maracan, *Merchants* ... ... Akyab.
Administration des Phares Impériales Ottomanes ... ... Constantinople.
Administration Sanitaire de l'Empire Ottoman ... ... Constantinople.
Admiralty ... ... London.
Admiralty Library ... ... London.
Admiralty (Naval Intelligence Department) ... ... London.
Admiralty (Transport Department) ... ... London.
"Adria" Hungarian Sea Navigation Co., *Steamship Owners* ... ... Fiume.
Agius & Co., *Foreign Coaling Exporters* ... ... 1, Cail's Buildings, Newcastle-on-Tyne.
Agius, E. T., & Co., *Steamship Agents* ... ... Malta.
Aicardi, Heynes & Co., *Steamship Agents* ... ... 30, Great St. Helens, London.
Akers Mekaniske Werkstadt, *Iron Shipbuilders* ... ... Christiania.
Alexander & Carr ... ... Rue Beauvau 4, Marseilles.
Allan, Rae & Co., *Steamship Agents* ... ... Quebec.
Ameln, Theodor ... ... Bergen.
Amos, J. H., *Secretary to the River Tees Conservancy Commissioners* ... ... Stockton-on-Tees.
Amos & Smith, *Engineers* ... ... Hull.
Amsterdamsche Droogdok Maatschappij (Dry Dock Company) ... ... Amsterdam.
Andersen, J., *Shipbroker* ... ... 19, Ladybellgate Street, Gloucester.
Anderson & Murison ... ... Cape Town.
Anderson, Wright & Co., *Merchants* ... ... 22, Strand, Calcutta.
André, Auguste, *Shipbroker* ... ... Quai St. Aldegonde 20, Antwerp.
Angus, George, & Co., *Leather and India Rubber Merchants* ... ... Grainger Street, Newcastle-on-Tyne.
Ansaldo, G., & Co. ... ... Sestri Ponente.
Apcar & Co., *Merchants and Shipowners* ... ... 19, Radha Bazar, Calcutta.
Appelberg, Axel ... ... Gothenburg.
Arauco, R. C. ... ... Valparaiso.
Ardrossan Shipbuilding Co., *Iron and Steel Yacht and Ship Builders; Slip and Graving Docks; Engineers, Boiler Makers, and General Repairers* ... ... Ardrossan.
Argus Printing and Publishing Co. of Cape Town ... ... 25, Cornhill, London.
Armstrong, Sir W. G., Mitchell & Co. (*Ld.*), *Shipbuilders* ... ... Walker Shipyard, Newcastle-on-Tyne.
Asfar & Co., *Merchants* ... ... Bussorah.
Atkinson, J. J. ... ... 42, Albert Road, Kingstown, Ireland.
Austin, S. P., & Son, *Iron and Steel Shipbuilders and Repairers* ... ... Wear Dock Yard, Sunderland.
Avis Steamships (*Ld.*), *Shipowners* ... ... 101, Leadenhall Street, London.
Aznar & Astigarraga, *Shipbrokers* ... ... Bilbao.
Azzopardi, F., & Co. ... ... Port Said.
Bachke & Co., *Shipbrokers* ... ... Drontheim.
Baessler, John, *Shipowner* ... ... Shanghai.
Bailey & Leetham, *Steamship Owners* ... ... Hull and Newcastle.

Balfour, Guthrie & Co., *Merchants* ... 306, California Street, S Francisco.

Balfour, Williamson & Co., *Merchants* ... Liverpool.

Baring Brothers & Co., *Merchants* ... 8, Bishopsgate Street Withi London.

Barnes, Guthrie & Co., *Shipowners and Brokers* ... Cardiff.

Barnett, J., & Co., *Steamship Agents and Marine Insurance Brokers* ... Liverpool.

Barrow Hematite Steel Co., Ld., *Iron and Steel Works* ... Barrow-in-Furness.

Barry Graving Dock & Engineering Co. ... Exchange Buildings, Cardiff.

Baxter, S., & Co. ... 18 & 19, Great St. Helen London.

Bee, H., & Son, *Coal Merchants, Stevedores and Contractors* ... Docks, Southampton.

Bell, Robert, & Co., *Steamship Owners and Brokers, Coal Exporters, &c.* ... Quayside, Newcastle-on-Tyn

Bellamy & Co., *Ship Agents and Owners* ... Plymouth.

Bellamy, J. H., & Co., *Merchants* ... Rio de Janeiro.

Bell's Asbestos Co. (*Ld.*), *Asbestos and India Rubber Manufacturers* ... Liverpool.

Bennett, C. T., *Shipowner* ... 66, Queen's Square, Bristol.

Bensaude & Co., *Merchants* ... Lisbon.

Bensaude & Co., *Merchants* ... St. Michael's, Azores.

Bergeret, F., *Shipbroker* ... Algiers.

Bergsunds Mekaniska Werkstads Aktiebolag, *Shipbuilders and Engineers* ... Finnboda Slip, Stockholm.

Bernard & Wood ... Yokohama.

Best, John P., & Co., *Shipbrokers, Shipowners, and General Agents* ... Place Verte 36 and 37, A werp.

Billard, Emile, *Shipbroker* ... Grand Quai 67, Havre.

Black, John, & Co., *Merchants and Shipowners* ... 204, Elliot Street, Glasgow.

Blackwood, Blackwood & Co. ... Calcutta.

Blackwood, Bryson & Co. ... 11 & 12, Great Tower Street London.

Blair & Co. (*Ld.*), *Marine Engineers* ... Stockton-on-Tees.

Bland, M. H., & Co., *Coal Merchants, Steamship Owners and Agents* ... Gibraltar.

Blohm & Voss, *Shipbuilders and Engineers* ... Steinwärder, Hamburg.

Blum, Baldwin & Girvin, *Merchants* ... 307, California Street, San Francisco.

Blyth Dry Dock Co. (*Ld.*), *Ship Repairers, Engineers, &c.* ... Blyth.

Board of Trade ... London.

Bolstorff, C. A. ... Elsasser Strasse 25, Berlin.

Bombay-Burmah Trading Corporation (*Ld.*) ... Rangoon.

Bombay & Persia Steam Navigation Co. ... Bombay.

Bond, *Marine and General Engineer and Boiler Maker* ... Constantinople.

Booth & O'Halloran, *Engineers and Boiler Makers* ... Rue du Brésil 13, Antwerp.

Bornholdt, P., & Co., *Steamship Agents* ... Libau.

Bornholdt, P., & Co., *Steamship Agents* ... Riga.

Bornholdt, P., & Co., *Steamship Agents* ... St. Petersburg.

Borniche, Edouard, *Insurance Agent* ... Place de Meir 10, Antwerp.

Borradaile, Thomas S., *Merchant* ... 16, St. Helen's Place, Londo

Borries, Craig & Co., *Shipbrokers* ... Lombard Street, Newcastle- Tyne.

Bossut-Plichon, Monsieur ... Roubaix.

Boston Tow Boat Co. *Towing, Wrecking, Transporting Merchandise, &c., &c.* ... Boston, U.S.A.

Bouhoulle, Alphonse, *Engineer and Boiler Maker* ... Quai Godefroid 66, Antwerp

Bowley & Bristow ...	68 & 69, Cornhill, London.
Boxwell, J. H., *Merchant* ...	Pernambuco.
Boyd & Co., *Engineers and Shipbuilders* ...	Shanghai.
Brandt, Gustav, & Co., *Timber, Wine and General Merchants* ...	Huelva.
Brassey, Lord ...	London.
Bremer, A. A. W. ...	Batavia.
Bridges, T. H., *Engineer* ...	Rochester Buildings, 138, Leadenhall Street, London.
British Association ...	Oporto.
British India and Queensland Agency Co. (*Ld.*), *Steamship Agents* ...	Brisbane.
British India Steam Navigation Co. ...	203, West George Street, Glasgow.
Broad, William, & Sons, *Ship Agents and Merchants, Lloyd's Agents* ...	Falmouth.
Brown, A., & Co. ...	Newcastle, N.S.W.
Brown, Emil, *Chartering Agent* ...	St. Petersburg.
Brown, E. T., & Co., *Shipbrokers* ...	22, Great St. Helen's, London.
Brown, J. & A., *Shipowners & Brokers, Insurance Brokers, Colliery Proprietors, Merchants, &c.*	1, Fenchurch Avenue, London.
Brown & McCabe, *Stevedores* ...	25, B Street, Portland, Oregon.
Brown, P., *Shipowner and Shipbroker* ...	Newquay, North Shields.
Bruna Brothers, *Steamship Brokers* ...	Genoa.
Bucknall, Henry, & Sons, *Merchants and Shipowners* ...	22, Crutched Friars, London.
Bulloch, J., *Shipowner* ...	13, Fenchurch Avenue, London.
Bulman & Tiffen, *Coal Exporters and Merchants* ...	Cardiff.
Bureau Veritas, Administration of ...	Paris.
Burke, John & Co., *Shipbrokers* ...	Belfast.
Burlage & Schumacher, *Insurance Brokers* ...	Binnenkant 8, Amsterdam.
Burmeister & Wain's (Maskin & Skibsbyggeri), *Engineers and Shipbuilders* ...	Copenhagen.
Burnard & Alger, *Warehousemen, Wharfingers, Chemical Manure Manufacturers, &c.* ...	Plymouth.
Burns, Philp & Co. ...	Townsville.
Butcher, Matthew, *Merchant, Shipping Agent and Insurance Broker* ...	Great Yarmouth.
Bute Docks Co. (*Ld.*), (Collectors' Department) ...	Cardiff.
Caird & Co. (*Ld.*), *Engineers and Shipbuilders* ...	Dalrymple Street, Greenock.
Cairns, R. & D., *Steamship Owners and Brokers* ...	Leith.
Cameron & Co., *Lloyd's Agents, &c.* ...	St. Thomas, W.I.
Cammell, Charles, & Co., *Steel Manufacturers* ...	Sheffield.
Cammell, Woolf & Haigh, *Shipowners, &c.* ...	Hull.
Canadian Pacific Railway Co. ...	88, Cannon Street, London.
Capitania del Puerto de Manila y Cavite ...	Manila.
Carara, John, & Sons, *Steam Coal Merchants* ...	Gibraltar.
Carara, John, & Sons, *Steam Coal Merchants* ...	Port Said.
Carassarini, A., & Co. ...	Sebastopol.
Cardiff Junction Dry Dock and Engineering Co. (*Ld.*) ...	Junction Dry Dock, Cardiff.
Carrel, F. P., *Shipbroker, &c.* ...	25, Mount Stuart Square, Cardiff.
Caspar, C., *Shipping and Custom House Agent, Agent for General S.N. Co., British India S.N. Co., Bailey & Leetham, Furness Line, &c.* ...	Rue Dauphine 2, Havre.
Cattley, C. R., *Lloyd's Agent* ...	Reval.
Catton, Bell & Co., *Merchants and Lloyd's Agents* ...	406, California Street, San Francisco.
Cay, Hall & Co., *Shipbrokers, &c.* ...	Mount Stuart Square, Cardiff.
Cellier, Eugène, *Shipbroker* ...	Hamburg.
Chambers, James, & Co., *Shipowners and Ship Store Dealers* ...	Liverpool.

Channel Coaling Co. ... Dartmouth.
China and Japan Trading Co. ... Kobé.
Christensen, Alfred, *Steamship Agent and Coal Merchant* ... Osterbrogade 1 & Holbergsgade 10, Copenhagen; and Elsinore.
Christophersen, P. ... Buenos Ayres.
Chubb & Son, *Marine Insurance Agents* ... 77, Beaver Street, New York.
Clark, George, *Engineer* ... Southwick Engine Works, Sunderland.
Clarke, Chapman, Parsons & Co., *Engineers* ... Victoria Works, Gateshead-on-Tyne.
Clarke (Stephenson) & Co. ... 4, St. Dunstan's Alley, London.
Cleland, William, & Co. (*Ld.*), *Engineers and Ship Repairers* ... Willington Quay, Newcastle-on-Tyne.
Clyde Shipping Co., *Steam Tug Owners* ... Pier Head, Cardiff.
Clyde Shipping Co. ... Queenstown.
Cobb, F. E., *Lloyd's Agent; Colonial Manager of the Falkland Islands Company* ... Stanley, Falkland Islands.
Cochran & Co., *Steam Yacht and Launch Builders* ... Birkenhead.
Cook, Robert, *Shipbuilder* ... Appledore, North Devon.
Coia, C., & Co., *Stevedores, Purveyors, &c.* ... 18, Sparks Street, Rangoon.
Collier Brothers, *Lloyd's Agents; and Ship and Steamship Agents* ... 53, South Side Street, Plymouth.
Collin, Emile ... 10, Rue Auber, Paris.
Commercial Agency ... Travancore.
Commercial Cable Co., *Atlantic Telegraph* ... 23, Royal Exchange, Liverpool.
Commercial Dock and Engine Works ... Howrah.
Compañia de Diques Flotantes de Valparaiso ... Valparaiso.
Compañia Esplotadora de Lota y Coronel, *Coal Miners, Copper Smelters, Shipowners, &c.* Valparaiso.
Compañia Nacional de Remolcadors (P. A. McKellar, *Manager*) ... Valparaiso.
Compañia Trasatlantica, *Shipowners* ... Cadiz.
Compton, F. E. ... Auckland.
Cooper, Edward Ernest, & Co., ... Lloyd's; and 4, Bishopsgate Street Within, London.
Corbett, John L. ... Londonderry.
Cory Bros. & Co. (*Ld.*), *Colliery Proprietors* ... Bute Docks, Cardiff.
Courtes, I., *Insurance Broker* ... Rue St. Ferreol, Marseilles.
Couves, A. J., *Master Mariner* ... 38, The Grove, Gravesend.
Coverley, Charles, & Co. ... Oporto and Lisbon.
Coverley & Westray, *Shipowners* ... 37, Leadenhall Street, London.
Coyte, G., *Average Adjuster* ... 23, Birchin Lane, Cornhill, London.
Craig, W. I., *Docks Manager* ... Passage West, Cork.
Cramp, The William, & Sons' Ship and Engine Building Co. ... Beach & Ball Streets, Philadelphia.
Cran, John, & Co., *Shipbuilders and Engineers* ... Leith.
Crawford & Co., *Steamship Owners and Brokers* ... Grangemouth; and 135, Buchanan Street, Glasgow.
Cruickshank, W. D., *Chief Engineer Surveyor to New South Wales Government* ... Marine Board Office, Sydney.
Cummins, N. & I., & Bros., *Lloyd's Agents* ... Cork and Queenstown.
Cunard Steamship Co. (*Ld.*), *Shipowners* ... 21, Quai d'Orléans, Havre.
Currie, Donald, & Co., *Merchants* ... Cape Town.
Currie, James, & Co., *Shipping Agents* ... 16, Bernard Street, Leith.
Daendels, J., & Co., *Shipbrokers* ... Sourabaya.
Darlington Forge Co. (*Ld.*), *Makers of Iron and Steel Forgings, Castings, &c.* ... Darlington.

Davie, George T., *Shipbuilder* ...	Lovis, Quebec.
Davison, Bisset & Co., *Ship Store Dealers and Provision Merchants* ...	Liverpool.
Dawson, J., & Co., *Shipping Agents* ...	Queenstown.
De Baerdemaecker, G. C., *Agent Maritime, Commissionaire-Expéditeur* ...	Rue Fiévé 17, Ghent.
De Beer, S., & Co. ...	Melbourne.
De Burgos Cañizares, José ...	Almeria.
De Caen, P. D. ...	Adelaide.
De Hart, J., *Merchant and Shipowner* ...	Maritime Buildings, Newcastle-on-Tyne.
De Jong, B. C., *Lloyd's Agent* ...	Batavia.
De Keyser, R., & Co. (late H. J. A. Telghuys), *Shipbrokers* ...	Canal des Brasseurs 36, Antwerp.
De Wit Schipper & Co. *Merchants* ...	Sourabaya.
Delprat, G. D., *Manager of Bede Metal Co. in Spain* ...	Huelva.
Denny, William, & Brothers, *Iron and Steel Shipbuilders* ...	Leven Ship Yard, Dumbarton.
Deppe, Adolf, *Shipowner* ...	Rue de Bordeaux 16, Antwerp.
Deulofeu, hijo, & Co., *Commission Merchants and Steamship Agents* ...	Havana.
Dewrance & Co. ...	158, Great Dover St., London.
Dick, W. B., & Co. (*Ld.*), *Oil Merchants, &c.* ...	20, Bucklersbury, London.
Dickinson, John, *Marine Engineer* ...	Palmer's Hill Engine Works, Sunderland.
Dixon & Corbitt & R. S. Newall & Co. (*Ld.*), *Hemp and Wire Rope Manufacturers* ...	Teams Rope Works, Gateshead-on-Tyne.
Dixon, John, *Harbour Superintendent* ...	Swansea.
Dixon (Raylton) & Co., *Iron and Steel Shipbuilders and Engineers* ...	Cleveland Dockyard, Middlesbro'-on-Tees.
Dobell, R. R., & Co., *Lumber Merchants* ...	Quebec.
Dobrillovich, E. V., *Steamship Agent* ...	Trieste.
Dobson, William, & Co., *Iron and Steel Shipbuilders* ...	Low Walker-on-Tyne.
Dodero, Gio. Batta fu D. ...	Piazza delle Pigne 3, Genoa.
Dodson & Craig, *Steamship Owners and Brokers, &c.* ...	3, Bernard Street, Leith.
Donaldson, William, *Shipping Agent and Steamship Broker, and Pitwood Importer* ...	Bo'ness.
Dorman, Philip, *Lloyd's Agent* ...	Casilla Correo 281, Buenos Ayres.
Douglas, R. R., *Insurance Agent* ...	Liverpool.
Drost, J., Dzn. ...	Amsterdam.
Drysdale, Thomas, & Co., *Merchants* ...	Buenos Ayres.
Duclos, C., *Merchant, Shipbroker, and Custom House Agent* ...	Huelva.
Duinker & Goedkoop, *Shipbrokers* ...	Amsterdam.
Duminy, E. ...	Rue Dicquemare 39, Havre.
Duncan, George, & Co. ...	2, East India Avenue, London.
Duncan, Robert, & Co., *Shipbuilders* ...	Port Glasgow.
Duncker, Arthur ...	Brodschrangen 26, Hamburg.
Dunn & Co. ...	East London.
Dunn, W., & Co. ...	95, Bishopsgate Street Within, London.
Dunn, W., & Co. ...	Durban.
Duprat, Joseph ...	Paris.
Durand, Prosper, *Coal Merchant* ...	Algiers.
Durham, Churchill & Co., *Engineers and Makers of Patent Marine Engine Governors and Pistons* ...	23, Leadenhall Street, London.
East & West India Dock Co....	London.
Ebbw Vale Steel, Iron & Coal Co. (*Ld.*) ...	Newport, Mon.

Edelfelt, E. G. ... ... Gothenburg.
Edinburgh Roperie & Sailcloth Co. (*Ld.*) ... ... Leith.
Edwards, Robertson & Co., *Steamship Owners* ... ... Bute Docks, Cardiff.
Elias, F. X., *Shipbroker* ... ... Rua da Remolars, Lisbon.
Ellertsen, F. J., & Co., *Shipbrokers and Steamship Agents* ... ... Bergen.
Ellis, Edward, *Surveyor to Bureau Veritas* ... ... Liverpool.
Ellis & Morrison, *Shipbrokers* ... ... Canal des Brasseurs 64, Antwerp.
Emson, C. W. ... ... 1, St. Michael's House, Cornhill, London.
Engelbert & Co., *Lubricant Manufacturers, &c.* ... ... 70 & 71, Bishopsgate Street, London.
Engels, Arnold Frederick, *Lloyd's Agent* ... ... Rue au Sucre 2, Antwerp.
Errington, Wm. ... ... Auckland, N.Z.
Essayian, O. M., *Bankers, Merchants, and Shipowners* ... ... 9, Mehemed Ali Pasha Khan, Galata, Constantinople.
Evans, W., & Son, *Engineers* ... ... Old Dock, Newport, Mon.
Exham, R., *Lloyd's Agent* ... ... Rarotonga.
Fairfield Shipbuilding & Engineering Co. (*Ld.*) ... ... Govan, Glasgow.
Farnham, S. C., & Co., *Engineers and Shipbuilders* ... ... Shanghai.
Felixstowe Dock and Railway Co. ... ... Felixstowe.
Fenning, G. & J., *Underwriters* ... ... Lloyd's; & 74 Cornhill, London.
Ferlin, Oscar ... ... 6, Vasagatan, Stockholm.
Fisher, James, & Sons, *Shipowners and Shipbrokers; Lloyd's Agents* ... ... Barrow-in-Furness.
Fisher, Renwick & Co., *Steamship Brokers* ... ... Cardiff.
Fishwick, Alexander, *Shipping Agent* ... ... Cronstadt.
Fleming, R. T., *Lloyd's* ... ... London.
Flensburger Schiffsbau-Gesellschaft ... ... Flensburg.
Forrester, George & Co. (*Ld.*), *Engineers* ... ... Liverpool.
Forster, J. J. & C. M., *Steamship Owners and Brokers; Coke and Coal Exporters* ... Quayside, Newcastle-on-Tyne.
Foscolo, Mango & Co., *Steamship Owners and Steamship Agents* ... ... Constantinople.
Fox, G. C., & Co., *Merchants and Ship Agents* ... ... Falmouth.
Fraser, Eaton & Co., *Merchants* ... ... Sourabaya.
Fraser, Farley & Co., *Merchants* ... ... 143, Yokohama, Japan.
Frisk, Aug., *Shipbroker* ... ... Riga.
Fulton Iron Works ... ... 213, Fremont St., San Francisco.
Furness, Christopher, *Steamship Owner and Loading Broker* ... ... Queen Street, Newcastle-on-Tyne.
Furness Railway Co. ... ... Barrow-in-Furness.
Galbraith and Moorhead, *Shipbrokers* ... ... Billiter House, London.
Galbraith, Pembroke & Co., *Shipbrokers* ... ... 8, Austin Friars, London.
Galliano, G. fu L. (*late* A. Bonanati & Co.), *Ship and Steamship Broker and Agent* ... Genoa.
Gambetta, Domenico ... ... Piazza Scuole Pie, Genoa.
Ganapathy, Pillay & Co., *Merchants* ... ... Negapatam.
Garland & Flexman, *Shipbrokers, Commission and Forwarding Agents* ... ... King's Lynn.
Garland, Laidley & Co., *Commission Agents* ... ... Lisbon.
Garnett (Alexander) & Co., *Merchants* ... ... Liverpool.
Garrad & Co., *Brokers for the Sale and Purchase of Steamers and Sailing Vessels* ... 15, Fenchurch Street, London.
Gaskell, G. ... ... London.
Gastaldi, G. B., & Co., *Steamship Brokers and Agents* ... ... Genoa.
Gellatly, Hankey, Sewell & Co., *Shipbrokers* ... ... 109, Leadenhall St., London.

Gellatly, Hankey, Sewell & Co., *Shipbrokers and Agents* ... Canal St. Pierre 1, Antwerp.
George, E., *Steamship Agent* ... Lisbon.
"Germania," Schiff-und Maschinenbau Actien Gesellschaft ... Kiel.
Gibb, Livingston & Co., *Lloyd's Agents; Merchants and Shipping Agents* ... Shanghai.
Gibson, George, & Co. ... 64, Commercial Street, Leith.
Gibson & Greenop, *Ship and Insurance Brokers* ... Whitehaven, Workington, and Maryport.
Gibson, John, & Sons, *Shipbrokers and Chandlers* ... Fleetwood.
Gillchrest, R. L., & Co., *Shipbrokers* ... Liverpool.
Gismondi, A., & Co., *Steamship Agents and Shipbrokers* ... Genoa.
Glafcke & Hennings, *Shipbrokers* ... Admiralitätstrasse 68, Hamburg.
Glamorgan Coal Co. (*Ld.*), *Colliery Proprietors* ... Cardiff.
Glamorgan Coal Co. (*Ld.*), *Colliery Proprietors* ... Constantinople.
Glamorgan Coal Co. (*Ld.*), *Colliery Proprietors* ... Piræus.
Goedkoop, D., jun., *Shipbuilder* ... Amsterdam.
Goldsmith, John E., *Merchant* ... 3, Rue Nadjenskia, Odessa.
Goldstück, Hainze & Co., *Superintendents of Grain and Produce Cargoes* ... 88, Bishopsgate Street Within, London.
Gollcher, O. F., & Sons, *Merchants and Shipowners; Lloyd's Agents; Consuls for Sweden and Norway, and for the Netherlands* ... Valetta, Malta.
Gondrand Bros., *Ship and Steamship Agents* ... Dunkirk.
Grace & Co., *Merchants* ... Valparaiso.
Graham, W. A., & Co., *Merchants* ... Bombay.
Graham, W. K., & Son ... Blomfield House, 85, London Wall, London.
Grant, Charles, *Merchant* ... Queen's Road, Hong Kong.
Gray, Dawes & Co. ... 13, Austin Friars, London.
Grayson, H. & C., *Shipbuilders and Engineers* ... Liverpool.
Green, F., & Co., *Shipbrokers* ... 13, Fenchurch Avenue, London.
Green & Johnson, *Shipbrokers and Agents* ... Buenos Ayres.
Green, W. A., *Manager of the Manchester, Sheffield and Lincolnshire Railway at Grimsby* Grimsby.
Greenock Steamship Co. (*Ld.*), *Steamship Owners* ... 1, Cross-shore Street, Greenock.
Grieg, Joachim, *Shipbroker & Steamship Agent* ... Bergen.
Grierson, C. ... Sebastopol.
Grisar & Marsily, *Shipbrokers* ... Quai St. Aldegonde 6, Antwerp.
Gunn & Cameron, *Newspaper Proprietors* ... 106, Union Street, Glasgow.
Haggie (R. Hood) & Son, *Patent Hemp and Wire Rope Manufacturers* ... Newcastle-on-Tyne, London, Cardiff, and Hull.
Hainsworth, Watson & Co., *Merchants* ... Liverpool.
Hall, John, jun., & Co., *Shipowners and Shipbrokers* ... 1, New London Street, London.
Hamilton, W., & Co., *Shipbuilders and Shipowners*... Port Glasgow.
Handels Vereeniging (Batavia Exchange) ... Batavia.
Handren & Robins, *Engineers and Dry Dock Owners* ... 126, Washington Street, New York.
Hanham, Sir John A., Bart. ... 116, Buckingham Palace Road, London.
"Hansa," Deutsche Dampfschiffahrts Gesellschaft ... Bremerhaven.
Hansen, C. K., *Shipowner, Agent, and Coal Merchant* ... Toldbodvej, Copenhagen, and Elsinore.
Hansen, Otto F., *Shipbroker* ... Bergen.

Hansen & Schwartz, *Shipbrokers* ... Quebec.
Harbour Board ... Table Bay.
Harbour Master (C. G. I. Barnaart) ... Batavia.
Harbour Master (Capt. Robert White) ... 16, Robertson Street, Glasgow
Harbour Master (H. M. Liardet) ... Liverpool.
Harbour Master (Capt. Usborne) ... Queenstown.
Harden & Co., *Agents* ... Gopalpore.
Harland & Wolff (*Ld.*), *Shipbuilders and Engineers* ... Belfast.
Harms, R. J., *Marine Surveyor* ... Rue Basse 29, Antwerp.
Harper, G. T., Agent to Powell-Duffryn Coal Co., *Shipowner and Shipbroker* ... Oriental Place, Southampton.
Harris, E., & Co., *Steamship Brokers and Lloyd's Agents* ... Middlesbro'-on-Tees.
Harrison, H. C., & Co. ... Port Adelaide.
Harrison, J. & C. ... 66, Mark Lane, London; also at Cardiff and Gravesend.
Hawthorn, R. & W., Leslie & Co. (*Ld.*), *Engineers and Shipbuilders* ... Hebburn Shipyard, Newcastle-on-Tyne.
Hawthorn, R. & W., Leslie & Co. (*Ld.*), *Engineers and Shipbuilders* ... St. Peter's Works, Newcastle-on-Tyne.
Hay & Co., *Ship and Insurance Brokers* ... 11, St. Mary Axe, London.
Hay, J. & J., *Shipbrokers* ... Grangemouth.
Haycraft & Gilfillan, *Marine Insurance Brokers* ... Lloyd's; and 3, Great Winchester Street, London.
Haynes, Thomas, *Merchant, Steamship Owner, Agent and Broker* ... Gibraltar.
Hazen & Compton, *Merchants and Marine Insurance Agents* ... Valparaiso.
Head, Henry, & Co. ... (*2 copies*) 85, Gracechurch Street, London
Hecksher & Son, *Lloyd's Agents and Brokers* ... Copenhagen.
Hedger, Philip, *Secretary, Southern Dock Co.* ... Southampton.
Hedger, Philip, T. F., *Solicitor* ... 1, Hastings Street, Calcutta.
Heidmann, H. W. ... Hafenstr. 93, Hamburg.
Helme, J., *Shipowner* ... 43, Church Street, Lancaster
Helmsing & Grimm, *Steamship Agents and Owners* ... Riga.
Helmsing & Grimm, *Steamship Agents and Owners* ... Libau.
Helsingørs Jernskibs & Maskinbyggeri (*Shipbuilders and Engineers*) ... Elsinore.
Henderson Brothers, *Steamship Owners and Merchants* ... Mersey Chambers, Liverpool.
Henderson, David & William, & Co., *Shipbuilders and Engineers* ... Partick, Glasgow.
Henderson & Glass, *Iron Merchants* ... Liverpool.
Henderson & Macfarlane ... Auckland.
Henderson & McIntosh, *Steamship Owners and Brokers* ... 44, Bernard Street, Leith.
Henderson & McKean, *Stevedores, Contractors, and Steam Tug Owners* ... 2, Tower Street, Leith.
Hepple & Co., *Engineers and Iron Shipbuilders* ... Slipway, North Shields.
Heredia, Hijos de M. A., *Merchants* ... Malaga.
Herinex, H., *Tugboat Director* ... Canal St. Pierre 2, Antwerp.
Heyn, G., & Sons, *Shipbrokers* ... Belfast.
Hill, Arthur, *Lloyd's Agent* ... Athens.
Hill, J. E. Gray, *Solicitor* ... Liverpool.
Hill's Dry Dock & Engineering Co., (*Ld.*) ... Cardiff.
Hine Brothers, *Shipowners, Ship and Insurance Agents* ... Maryport and Workington.
Hoffman & Bielby, *Steamship Chartering Agents and Coal Importers* ... Libau.
Hoffman, C., & Co., *Shipbrokers and Steamship Agents* ... 11, Hart Street, Mark Lane, London.
Hogg & Henderson ... Middlesbro'-on-Tees.

Holland, Arthur, & Co. ... 2, East India Avenue, London.
Holter, A., *Sworn Shipbroker* ... La Rochelle.
Hong Kong and Kowloon Wharf and Godown Co. (*Ld.*) ... Hong Kong.
Hong Kong and Whampoa Dock Co. (*Ld.*), *Shipbuilders* ... Hong Kong.
Hood, John, & Co., *Stevedores and Naval Contractors* ... 8, Barr Street, Rangoon.
Hooper, Campbell & Co., *Steamship Owners* ... Cardiff.
Hopkins (Manley), Son, & Cookes, *Adjusters of Averages* ... 32, Cornhill, London.
Houghton & Co., *Lloyd's Agents, Merchants, &c.* ... Sourabaya.
Houlder Brothers & Co., *Ship and Insurance Brokers* ... 146, Leadenhall Street, London, and at Liverpool and Glasgow.
Howard, C., & Son, *Shipowners and Brokers* ... 17, Philpot Lane, London.
Howell & Co., *Commission Merchants* ... Hakodate, Japan.
Howie, H., *Shipowner and Commission Agent* ... Whitehaven, and Harrington.
Hoyland, Wooldridge & Co., *Successors of Hoyland Brothers*, 1868 ... Taganrog.
Hubbuck, Thomas, & Son, *Paint Manufacturers* ... 24, Lime Street, London.
Hudig & Blokhuijzen, *Shipbrokers* ... Amsterdam.
Hudson, T. S., *Merchant and Shipowner* ... West Hartlepool.
Hugo, C., *Shipbroker and Steamship Agent* ... Holländische Brook 23, Hamburg.
Hunt, Samuel, *Shipbroker* ... 7, Mark Lane, London.
Hunting & Pattison, *Steamship Owners and Ship and Insurance Brokers* ... 32, Great St. Helens, London.
Huth, Frederick, & Co. ... 12, Tokenhouse Yard, London
Hutter, J. J. ... Rue Frey 55, Havre.
Huysmans & Bulcke, *Shipbrokers* ... Rue Pré 11/1, Antwerp.
Imperial Ottoman Bank ... Smyrna.
Inglis, A. & J., *Engineers and Shipbuilders* ... Pointhouse, Glasgow.
Irrawaddy Flotilla Co. (*Ld*), *Inland Carriers* ... Phayre Street, Rangoon.
Isle of Man Steam Packet Co., *Steamship Owners* ... Douglas, Isle of Man.
Itsuki, Captain ... Japan.
James, C. F., & Co., *Ship, Timber and Insurance Brokers* ... Fort Street, Auckland.
James, Inglis & Co., *Merchants* ... Iquique.
James, Inglis & Co., *Merchants* ... Valparaiso.
James, W. T. ... Hamilton, Bermuda.
Janson, J. W. & D. ... Lloyd's, London.
Japp & Kirby, *Shipbrokers* ... Liverpool.
Jardine, Matheson & Co., *Merchants* ... 1, Yokohama, Japan.
Jenkins, S. D. & Son, *Shipbrokers* ... Cardiff.
Jones, Edmund, & Co., *Merchants* ... 8, Strand, Rangoon.
Joosting & Van Eijk, *Insurance Brokers* ... Frederik's Plein, West Einde Amsterdam.
Kaiserliche Admiralität ... Berlin.
Kellock, C. W., & Co., *Brokers for the Sale of Ships and Steamers; Ship Valuers and Shipowners* ... Liverpool.
Kelso (S. Wright) & Co., *Ship and Insurance Brokers* ... Liverpool.
Kelway, H., & Sons, *Ship and Insurance Agents, and Steam Coal Merchants, Foreign Consuls, &c.* ... Milford Haven.
Kennedy, Hunter & Co., *Shipbrokers* ... Canal St. Pierre 6, Antwerp.
Ker, Bolton & Co. ... 7, East India Avenue, London.
Ker & Co. ... Manila.
Kilby & Churley, *Stevedores* ... 2, Queen's Square Avenue, Bristol.

Killick, Nixon & Co., *Merchants*	Bombay.
King & Sons, *Shipping Agents, &c.*	Durban, Natal.
King, C. J., & Sons, *Stevedores*	12, Princes Street, Bristol.
Kirkaldy, J., *Marine Engineer and Maker of Distilling Plant and Feed Water Heaters*	40, West India Dock Road, London.
Kirke, R. & J. J., *Merchants and Shipbrokers*	16, Mark Lane, London.
Kirkman, T., *Marine Insurance Agent*	4, Oldhall Street, Liverpool.
Kjolner, Charles, *Ship Chandler*	Lisbon.
Klett & Weber., *Insurance Brokers*	6, Gracechurch Street, Londo
Kley, W.	M. Gladbach.
Klingenberg, Carl. Joh. & Co., *Ship and Steamship Brokers, Bankers, &c.*	Bremen.
Knowles, Rawes & Co.	Lisbon.
Koninklijke Nederlandsche Stoomboot Maatschappij	Prins Hendrik Kade 16 Amsterdam.
La Platense Flotilla Co. (*Ld.*), *Shipowners*	Buenos Ayres.
Laeisz, F., *Shipowner*	Neueburg 14, Hamburg.
Lagos, *The Officer Administering the Government of* ... (*2 copies*)	
Lamplough, W., & Co., *Insurance Brokers*	72, Cornhill, London.
Lancaster, Speir & Co., *Colliery Proprietors*	Russell's Wharf, Newport, Monmouth.
Lane & Macandrew, *Shipbrokers*	26, Great St. Helen's, London.
Lang & Williamson, *Ship Repairers and Engineers*	Brithdir Wharf, Newport, Mo
Langridge, Henry, & Co.	16, Great St. Helen's, London.
Laurie, L. G., & Co., *Engineers and Ship Repairers*	Mill Parade, Newport, Mon.
Lawrence, T. C., & Co., *Shipbrokers*	Eastcheap Buildings, London.
Learmouth, F. A.	Cape Town.
Leary, C., & Co., *Timber Brokers*	19, Old Broad Street, London.
Leeds Forge Company	Leeds.
Leeuwen, J. F. van	Batavia.
Leeuwon, J. F. van, & Co., *Merchants and Shipbrokers*	Sourabaya.
Leffler, A., & Son, *Shipbrokers and Steamship Agents*	Gothenburg.
Leitch, A., & Co., *Shipbrokers*	Commercial Street, Dundee.
Lemoine, A.	Rouen.
Leroy, Alfred, *Ship and Steamship Broker*	Dunkirk.
Lidgett, John, & Sons, *Shipowners and Brokers*	6, Lime Street Square, Londo
Liebig's Extract of Meat Co. (*Ld.*)	9, Fenchurch Avenue, Londo
Lindbergs Werkstad, *Shipyard and Engineering Works*	Stockholm.
Lindholmens Werkstad, *Engineering Works, Shipyard, Slip and Dry Dock*	Gothenburg.
Little, J., & Co., *Shipbrokers and Steamship Owners*	17, Water Street, Liverpool.
Lloyd Austro-Ungarico (Arsenal)	Trieste.
Lloyd Austro-Ungarico (L. Ptack, *Naval Director*)	Trieste.
London and Edinburgh Shipping Co. (Thomas Aitken, *Manager*)	8 & 9, Commercial Street, Leit
London and Glasgow Engineering and Iron Shipbuilding Co. (*Ld.*)	172, Lancefield Street, Glasgo
Long, Captain, R.N.	H.M. Docky'd, Pembroke Doc
Lorentzen, J. H., & Co.	Hamburg.
Lowndes, R., Son & Wallace, *Average Adjusters*	Liverpool.
Ludwig, C., & Co., *Shipbrokers*	54, Regent Quay, Aberdeen.
Lundberg, A., *Shipbroker*	Oscarshamn.
Lund, I. Theod., *Consul and Lloyd's Agent*	Elsinore.
Lynch, Beech & Co., *Insurance and Exchange Brokers*	10, George Yard, Lombar Street, London.

Lyon, Charles, *Shipbroker* ...	Valparaiso.
Lyon, R., & Co., *Marine Salvage Surveyors and Brokers* ...	2, St. Peter's Alley, Cornhill, London.
McArthur, D., *Contractor and Stevedore* ...	104, Lower Commercial Street, Middlesbro'.
McCombie, James, & Co., *Shipowners* ...	Peterhead.
Macconnachie, W., & Co., *Shipbrokers and Shipowners* ...	Fraserburgh.
McCracken, Fenwick & Co., *Shipbrokers* ...	38, Fenchurch Street, London.
MacDonald & Morrison, *Shipbrokers* ...	35, Grant Street, Inverness.
McDowall & Barbour, *Shipbuilders and Engineers* ...	Vulcan Engine Works, Piræus.
MacGill, R., *Engineer and Boiler Maker, General Contractor* ...	Constantinople.
McGillivray, J., *Manager of Downie Boiler Incrustation Preventive Co.* ...	Liverpool.
McGregor, D., & Co., *Nautical Instrument Makers* ...	37, Clyde Place, Glasgow.
McGregor, Gow & Co., *Shipowners* ...	1, East India Avenue, London.
MacIlwaine & MacColl (*Ld.*), *Shipbuilders* ...	Belfast.
Mackay, A. & A. Y., *Shipowners and Shipbrokers* ...	Grangemouth.
McKenzie, D. H. ...	Auckland, N.Z.
Mackie, Dunn & Co. ...	Port Elizabeth.
Mackie, Koth, & Co., *Steamship Brokers and Coal Exporters* ...	54, Bernard Street, Leith.
Mackinnon, Wm. ...	Balinakill, Clachan, Argyllshire.
Maclaine, Watson & Co., *Merchants* ...	Batavia.
McMillan, Archibald & Son, *Shipbuilders* ...	Dockyard, Dumbarton.
MacNab, Rougier & Co., *Ship Agents* ...	Constantinople.
MacPherson, H. ...	Cadiz.
Macri, John ...	Kustendjie.
Made, J. D. van der, *Engineer* ...	Amsterdam.
Madoorny Brothers, *Stevedores, Agents, &c.* ...	2, Phayre Street, Rangoon.
Maibücher, Chs., *Insurance Broker* ...	Rue Grammey 6, Antwerp.
Manila Slipway Co. (El Varadero de Manila) ...	Manila.
Marin-Dubuard, Baron ...	Rue de Seine 37, Paris.
Marshall, David, *Steamship Owner and Broker* ...	Leith.
Marshall & Gerling, *Shipbrokers* ...	Canal des Brasseurs 64, Antwerp.
Marten, F. W., *Underwriter* ...	Lloyd's, and 23, Rood Lane, London.
Massey & Sawyer, *Shipowners and Brokers* ...	7, Quay Street, Hull.
Master Attendant ...	Madras.
Masuda, K. ...	Tokio.
Mateos, A., & Sons, *Steamship Agents and Coal Merchants* ...	Gibraltar.
Mathers, Isaac H., *Ship Agent and Commission Merchant, Vice-Consul for Sweden and Norway*	Halifax, N.S.
Mathews, Maclay & Manson, *Oil Merchants, Paint, Colour and Varnish Manufacturers* ...	104, Hyde Park Street, Glasgow.
Meinich, Jens, & Co., *Shipbrokers and Steamship Agents* ...	Christiania.
Melchers & Co., *Merchants* ...	Hong Kong.
Mercantile Dry Dock Co. (*Ld.*), *Ship Repairers and Graving Dock Owners* ...	Tyne Dock, South Shields.
Messageries Maritimes de France ...	97, Cannon Street, London.
Messageries Maritimes de France ...	Quai de la Joliette 2, Marseilles.
Meyer & Co., *Shipbrokers* ...	Prins Hendrik Kade 145, Amsterdam.
Middle Dock Company, *Ship Repairers, Graving Dock Owners, &c.* ...	South Shields.

Middleton & Co., *Ship Store Merchants* ... ... 213, Jackson Street, San Francisco.
Milford Docks Co., *Dock Proprietors* ... ... Milford Haven.
Miller, C., *Steamship Broker* ... ... Grangemouth and Bo'ness.
Miller, W., Robley & Co., *Steamship Agents* ... ... (2 *copies*) Leghorn and Genoa.
Millwall Docks Company ... ... 1, Railway Place, Fenchurch Street, London.
Minchin, F. J. V., *Lloyd's Agent* ... ... Gopalpore.
Miners Oil Supply Co. (*Ld.*), *Ship Store Merchants* ... ... 189, Pearl Street, New York.
Minister of Marine ... ... Rome.
Ministère de la Marine ... ... Paris.
Ministerie van Koloniën ... ... The Hague.
Minton, R. R., & Co., *Paint Manufacturers and Oil Merchants* ... ... Liverpool.
Mitchell & Co., *Merchants and Lloyd's Agents* ... ... Riga.
Mitsui Bussan Kaishia, *Commission Agents* (Y. Wooyeda, *Manager*) ... ... Shanghai.
Mitsui Bussan Kaishia, *Commission Agents* ... ... Nagasaki.
Montgomerie & Workman ... ... 36, Gracechurch Street, London.
Moore, H. C., *Managing Engineer of La Platense Co.* ... ... Buenos Ayres.
Moralee Brothers, *Iron and Wood Ship Repairers; and Dry Docks at Thrift Street, Holborn, and Tyne Dock* ... ... South Shields.
Mörch, P., *Steamship Owner and Broker* ... ... St. Petersburg.
Mordey, Jones, & Co., *Shipbrokers and Coal Factors* ... ... Dock Street, Newport, Mon.
Morehead, B. D., & Co. ... ... Brisbane.
Morris & Co. ... ... Deichstrasse 7, Hamburg.
Morris & Wright, *Shipbrokers* ... ... Manila.
Morrison, Thomas, & Co., *Merchants and Bankers* ... ... Huelva.
Moses, John, & Co., *Merchants and Shipbrokers* ... ... Rupena Street, Newport, Mon.
Mosses & Mitchell, *Merchants* ... ... 68–71, Chiswell St., London.
Mourilyan, Heimann & Co. ... ... Kobé.
Moutier, C. ... ... Paris.
Muir & Caldwell, *Engineers and Boiler Makers* ... ... Elliot Street, Glasgow.
Mullock & Sons, *Shipbrokers, Commission Agents, and Coal Merchants* ... ... Limerick.
Murray, James, *Ship Chandler* ... ... 8, Brown Street, Glasgow.
Nagasaki Dockyard and Engine Works ... ... Nagasaki.
Napier, Robert, & Sons, *Engineers and Shipbuilders* ... ... Lancefield House, Govan, Glasgow.
Narvaez, Manuel, *Sworn Broker and Ship Agent* ... ... Huelva.
Natal Harbour Board (*London Agent*, W. Peace, 21, *Finsbury Circus*) ... ... Natal.
National Mortgage and Agency Company of New Zealand (*Ld.*) ... ... Christchurch, N.Z.
National Mortgage and Agency Company of New Zealand (*Ld.*) ... ... Dunedin, N.Z.
Nautilus Steam Shipping Co. (*Ld.*), *Steamship Owners* ... ... 54, Fawcett Street, Sunderland.
Naval Construction and Armaments Co. (*Ld.*), *Shipbuilders* ... ... Barrow-in-Furness.
Navigazione Generale Italiana ... ... (2 *copies*) Via del Corso 385, Rome; and at Via Balbi, Genoa.
Nederlandsche Amerikaansche Stoomvaart, Maatschappij ... ... Prins Hendrikskade, Rotterdam.
Neill, Robert, *Commission Agent and Merchant* ... ... 10, South Hanover Street, Glasgow.
"Neptun" Bergnings och Dykeribolaget, *Salvage and Diving Company* ... ... Stockholm.
Nicol, A., & Co., *Shipowners* ... ... Guild Street Bldgs., Aberdeen.
Nicoll, Arch. ... ... 157, St. James Street, Montreal.
Nieuw Praauwenveer, *Boat Office* ... ... Sourabaya.

Nippon Yusen Kaisha, *Japan Mail Steamship Co.* (A. R. Browne, *Managing Director*) ... Tokio, Japan.
Nixon, Henry, *Underwriter and Insurance Broker* ... Lloyd's, London.
Ocean Coal Co., The (*Ld.*), *Colliery Proprietors* ... Cardiff.
Olsen, N., Chr., *Attorney* ... Rönne.
Olson & Wright, *Shipbrokers and Steamship Agents* ... Stockholm.
Ommeren, Ph. van., *Shipbroker* ... Rotterdam.
Onetti, J., & Sons, *Merchants and Steamship Agents* ... Gibraltar.
Orlando Brothers, *Shipbuilders* ... Leghorn.
Oswald, Mordaunt & Co., *Engineers and Shipbuilders* ... Woolston, Southampton.
Otto, George, *Steamship Owner and Ship Store Merchant* ... Borough Road, North Shields.
Page & Co., *Merchants* ... Cardiff.
Palmers' Shipbuilding & Iron Co. (*Ld.*), *Engineers and Shipbuilders* ... Jarrow-on-Tyne.
Park Brothers ... 52, Gracechurch St., London.
Parrott & Co., *Merchants* ... 306, California Street, San Francisco.
Paterson, A. S., & Co., *Merchants* ... Box 368, Dunedin, N.Z.
Paterson & Simons ... 39, Lime Street, London.
Peet, J., & Co., *East India Merchants* ... 26, Bush Lane, London.
Peirce, Becker & Bardi ... Messina.
Penarth Shipbuilding and Ship Repairing Co. (*Ld.*) ... Penarth Dock, Cardiff.
Philippucci & Condili, *Ship Chandlers* ... Constantinople.
Phillips, J., *Lloyd's Agent, and Ship and Insurance Agent* ... Milford Haven.
Pickford & Black, *General Commission and Steamship Agents* ... Halifax, N.S.
Pile & Co., *Steamship Brokers and Contractors for the Building of Ships and Steamers* ... 34, Great St. Helens, London.
Pinillos, Saenz & Co. ... Cadiz.
Plaideau, E., *Shipbroker* ... Dunkirk.
Poole, John, & Sons ... 27, Clement's Lane, London.
Pope & Talbot, *Lumber Merchants* ... 204, California Street, San Francisco.
Port Commissioners (Capt. H. S. Ransom, *Assistant Conservator*) ... Rangoon.
Port Officer ... Bombay.
Port Officer (Captain Heath, R.N.) ... Brisbane.
Port Officer ... Manora, Kurrachee.
Port Warden's Office ... Montreal.
Pott, H. W., & Körner, *Shipbrokers* ... Hamburg.
Pott & Smaal, *Shipbrokers* ... Delfzÿl.
Potter, W. H., & Sons, *Shipbuilders* ... Liverpool.
Preston, John, *Broker for Purchase and Sale of Steamers of all descriptions* ... 59 & 60, Cornhill, London.
Price Brothers & Co., *Lumber Merchants* ... Corner of Peter and Paul Streets, Quebec.
Puget Sound Stevedoring Co., *Stevedores* ... Box P, Tacoma, W.T.
Pyman, Watson & Co., *Merchants and Shipbrokers* ... Dock Street, Newport, Mon.
Quebec Steamship Co. ... Quebec.
Rabien & Stadtländer, *Steamship Owners and Brokers* ... Bremerhaven.
Ramage & Ferguson, *Shipbuilders and Engineers* ... Leith.
Ramos, José ... Alicante.
Raspe, M., & Co., *Merchants* ... Yokohama, Japan.
Rayment & Sharp, *Ship Chandlers* ... Hull.
Read, J., junr., *Ship and Boat Builder* ... Portsmouth.
Redaction der Mittheilungen aus dem Gebiete des Seewesens, *Austrian Navy Technical Journal* ... Pola, Austria.

Rees, D., ... Llanelly.
Reid & Co., *Proprietors of the Hooghly Dock* ... Hooghly Dock, Howrah.
Reid, Francis F., *Shipowner ; and Ship and Insurance Broker* ... 2, Baltic Street, Leith.
Reid, George, *Belgian Consul* ... Queen Street, Newcastle-on-Tyne.
Reid, J., & Co., *Shipbuilders* ... Port Glasgow.
Reinhold, F. G., *Steamship Owner and Shipbroker* ... Danzig.
Renck & Hessenmüller, *Shipowners and Shipbrokers* ... Harburg a.d. Elbe.
Reppen, C. J., *Ship Agent* ... Constantinople.
Ressich, McLaren & Co., *Shipbrokers and Coal Exporters* ... 53, Bernard Street, Leith.
Rettig, Rudolph, *Shipowner* ... Neuerwall 7, Hamburg.
Rhederei Independent ... Mannheim.
Richards, William, & Son, *Average Adjusters* ... New City Chambers, London.
Richardson, James, & Co., *Merchants and Shipowners* ... 2, Church Place, Greenock.
Rickards, G. H., & Son, *Steamship Agents and Shipbrokers, &c.* ... Messina.
Rickmers, R. C., *Rice Importer and Exporter, and Mill Owner* ... Bremen.
Ridley, John, Son, & Tully, *Merchants, Shipowners, and Brokers* ... Exchange Buildings, Newcastle-on-Tyne.
Rimmer, A., & Co., *Ship and Insurance Brokers* ... 2, Exchange Street East, Liverpool.
Roberts, John P., *Marine Surveyor* ... Shanghai.
Roberts, W., & Co., *Shipbrokers* ... Weymouth and Portland.
Roche, W., *Steamship and Coaling Agent* ... Roche's Wharf, Halifax, N.S.
Roeis, P., *Shipwright* ... Rue du Brésil 9, Antwerp.
Rogers & Bright, *Merchants* ... Liverpool.
Rollo, David, & Sons, *Engineers, &c.* ... Liverpool.
Rose, Thomson, Young & Co., *Insurance Brokers* ... 19, Change Alley, London.
Rosenfeld's Sons, J., *Merchants* ... 202, Sansome Street, San Francisco.
Rostocker Actien-Ges. für Schiff und Maschinenbau ... Rostock.
"Rotterdam" Internationale Crediet & Handelsvereeniging, *Merchants* ... Sourabaya.
Royal Marine Department ... Stockholm.
Rugeroni, Joseph, & Son, *Steamship Agents and Coal Merchants* ... Gibraltar.
Rugg, C. H., & Co. ... Mildmay Chambers, 82, Bishopsgate Street, London.
Ruhaak & Co., *Ship Chandlers* ... Sourabaya.
Russell & Co., *Shipbuilders* ... Port Glasgow.
Russell & Co., *Merchants* ... Hong Kong.
Russell & Co., *Merchants* ... Shanghai.
Russell, James... ... Newcastle, N.S.W.
Rustomjee, H. J. ... Kurrachee.
Rutherford, A., & Co., *Yacht Builders, &c.* ... Birkenhead.
Ruys & Co., *Shipbrokers* ... Quai aux Charbons 33, Antwerp.
Salvage Association ... 17 & 18, Cornhill, London.
Salvage Association ... Liverpool.
Salvesen, Chr., & Co., *Shipbrokers, Steamship Owners and Coal Exporters*... 29, Bernard Street, Leith.
Salvesen, J. T., & Co., *Steamship Owners and Brokers and Timber Merchants* ... Grangemouth.
Sanderson, G. R., & Co., *Shipowners and Brokers* ... Hull.
Satterthwaite & Platt, *Marine Insurance Agents* ... 16, Exchange Place, New York.

Sattler, L., *Shipbroker and Agent*	Casilla Correo 185, Buenos-Ayres.
Saunders, W. F.	Lloyd's, London.
Scaife, A. H., *Lloyd's Agent*	Constantinople.
Schack, Steenberg & Co., *Agents and Coal Merchants*	Copenhagen.
Scheur Brothers, *Shipbrokers*	Amsterdam.
Schiaffino, A., *Shipowner*	Boulevard République, Algiers.
Schiaffino & Solari, *Steamship Owners and Brokers*	Genoa.
Schlesinger, Davis & Co., *Iron and Steel Shipbuilders*	Wallsend-on-Tyne.
Schuldt, H., *Shipbroker*	Flensburg.
Schultz, John F., *Shipchandler*	Manila.
Schwalbe, Franz, *Merchant and Shipowner*	Baumwall 10, Hamburg.
Scott, Ernest, & Co., *Engineers*	Close Works, Newcastle-on-Tyne.
Scott, James, *Shipowner, Ship and Insurance Broker*	4, Royal Exchange Place, Dundee.
Scott, James, & Co., *Merchants and Ship Agents*	Queenstown and Cork.
Scott, John, & Co., *Engineers and Shipbuilders*	Kirkcaldy.
Searight, J., & Co., *Merchants*	Cape Town.
Searight, James, & Co.	7, East India Avenue, London.
Sedgwick, H. B., *Underwriter*	Lloyd's, London.
Seeamt	Admiralitätstrasse, Hamburg.
Serck, P., *Shipbroker*	St. Petersburg.
Seropian, C. H., *Insurance Agent*	Constantinople.
Sevestre, W. A.	New City Chambers, 121, Bishopsgate Street, London.
Sharp & Co., *Steamship Owners and Shipbrokers*	King St., Newcastle-on-Tyne.
Shaw Brothers, *Merchants*	Buenos Ayres.
Shepherd, W. & P., *Steamship Agents*	Civita Vecchia.
Sheryn, John, *Engineer*	Alexandra Dock, Newport, Mon.
Shipowners and Merchants Tug Co.	302, California Street, San Francisco.
Siemssen & Co., *General Merchants*	Shanghai.
Simpson, Howden & Co., *Ship and Insurance Brokers*	45, Cross Street, Manchester and at Liverpool.
Simpson, Spence & Young	12, Great St. Helens, London.
Simpson, Spence & Young, *Steamship Agents and Ship Brokers*	Quayside, Newcastle on-Tyne.
Sinclair, John, *Steamship Owner and Broker*	4, Church Street, West Hartlepool.
loman, M. I., *Shipping, Insurance, and General Broker*	14, Market Buildings, Flinders Lane West, Melbourne.
Sloman, R. M., junr., *Shipbroker*	Baumwall, Hamburg.
Smith, Bell & Co., *East India Merchants*	Manila.
Smith, Imossi & Co., *Coal Merchants, Steamship Owners and Agents*	Gibraltar.
Smith, John, *Broker*	105, Leadenhall Street, London.
Smith, Sundius & Co., *Ship and Insurance Brokers*	158, Leadenhall Street, London.
Smith, Sundius & Co., *Ship and Insurance Brokers*	11, Millbay Road, Plymouth.
Smith, Thomas & William, *Shipbuilders, &c.*	Newcastle and North Shields.
Smith, Wade & Co., *Lumber Merchants*	Quebec.
Société Anonyme Anversoise pour le Déchargement des Céréales	Rue de l'Etalage 23, Antwerp.
Société Commerciale d'Affretements et de Commission	Rue Lafayette 7, Paris.

Société John Cockerill, *Shipbuilders* ... Hoboken and Antwerp.
Someren, W. van, Greve & Co., *Ship Chandlers* ... Sourabaya.
Sönne, H., *Engineer Surveyor to Lloyd's Register, &c.* ... Shanghai.
Soutar & Co. ... Kingston, Jamaica.
South Derwent Colliery Office (Douglas Cooper, *Manager*), *Colliery Owners and Exporters* 8, Lombard Street, Newcas on-Tyne.
Sparke, George, *Iron and Coal Agent* ... Tredegar Wharf, Newport, M
Speirs, G. M., *Commercial Agent, Rio Tinto Co. (Ld.)* ... Huelva.
Spencer, John, & Sons (*Ld.*), *Steel Manufacturers* ... Newburn Steel Works, N castle-on-Tyne.
Spencer, John, & Sons (*Ld.*), *Iron and Steel Crank Shaft Manufacturers* ... Ouseburn Forge, Newcastle- Tyne.
Spiers, D., *Shipbroker and Steamship Agent* ... Dunkirk.
Spiller & Co., *Millers* ... Cardiff.
Spinck, J., & Co. ... Riga.
Sproston Dock & Foundry Co., *Contractors* ... Georgetown, Demerara.
Stappers, L., *Engineer and Boiler Maker* ... Rue Nassau 10, Antwerp.
State Steamship Co. (*Ld.*) ... 65, Great Clyde Street, Glasg
Stavanger Stöberi & dok, *Shipbuilders* ... Stavanger.
Stephen, Alexander, & Sons, *Shipbuilders and Engineers* ... Linthouse, Govan, Glasgow.
Stephenson, Robert, & Co. (*Ld.*), *Engineers and Shipbuilders* ... Newcastle-on-Tyne.
Stevens, Geo. R., & Co., *Merchants* ... Hong Kong.
Stevenson, D. M., & Co., *Shipowners and Coal Exporters* ... 12, Waterloo Street, Glasgow
Stevenson & Drynan, *Shipowners and Coal Exporters* ... Leith.
Stevenson, J., & Co., *Marine Insurance Brokers* ... (2 *copies*) Liverpool.
Stockton, H. J. ... 16, Philpot Lane, London.
Stone, C. B. ... Auckland, N.Z.
Storer, James, & Co., *Oil and Colour Manufacturers* ... Bridgeton, Glasgow.
Storm, William, *Ship Chandler* ... Durban, Port Natal.
Stothert, G. K., & Co., *Engineers and Shipbuilders* ... Hotwell Road, Bristol.
Strick, Frank C., & Co. (*Ld.*), *Shipbrokers and Chartering Agents* ... Woolpack Buildings, Gra church Street, London.
Strybos, Paul, & Co., *Coal Merchants* ... Rue du Fagot 17, Antwerp.
Stülcken, H. C., Wwe., *Shipbroker* ... Hamburg.
Sundheim & Doetsch, *Bankers and General Merchants* ... Huelva.
Surrey Commercial Dock Co. ... 106, Fenchurch Street, London.
Swainston, George, & Co., *Brokers for the Sale and Charter of Ships and Steamers Engineer Surveyors, Arbitrators and Valuers* ... West Sunniside, Sunderland.
Swan, Allan, *Steamship Agent, &c.* ... Custom House Buildin Greenock.
Swan, C. S., & Hunter, *Shipbuilders and Repairers* ... Wallsend-on-Tyne.
Swansea Dry Docks and Engineering Co. (*Ld.*) ... Swansea.
Swanzy, F. A., & Co. ... 147, Cannon Street, London.
Taff Vale Railway Co. (Collector's Department) ... Cardiff.
Tait, W. C., & Co., *Royal Mail Steam Packet Co., Compie. Bordelaise de Nav. à Vap., Adria Hungarian Sea Nav. Co.* ... Oporto.
Tasmanian Steam Navigation Co. ... Hobart, Tasmania.
Tattarachic, N., *Merchant* ... Mersyna.
Taylor, Tregent & Co., *Paint Manufacturers, Oil Merchants, &c.* ... Liverpool.

Temperley, J., & Co. ...	9, Gracechurch St., London.
Temperleys, Carter & Darke ...	London.
Tennent, J. & R., *Brewers* ...	Wellpark Brewery, Glasgow.
Thames Iron Works and Shipbuilding Co. (*Ld.*) ...	Orchard Yard, Blackwall, London.
Tharsis Sulphur and Copper Co. (*Ld.*) ...	53, Renfield Street, Glasgow.
Theobald, James H., *Ship and Insurance Broker* ...	86, Leadenhall Street, London.
Theoderidi, C. A., & Co., *Ship Agents* ...	Constantinople.
Thomas, E. R....	Dock Street, Newport, Mon.
Thompson, G. W., *Ship Repairer* ...	Mill Parade, Newport, Mon.
Thompson, Joseph L., & Sons, *Iron and Steel Shipbuilders and Repairers* ...	North Sands Ship Yard, Sunderland.
Thomson, James & George, *Engineers and Shipbuilders* ...	Clydebank, Glasgow.
Thomson, Watson & Co. ...	Cape Town.
Thorburn, John, *Coal Fitter*...	King St., Newcastle-on-Tyne.
Thorl, John, & Co., *Insurance Brokers* ...	Lloyd's, & 48, Cornhill, London.
Tindall, W. H., & Co. ...	77, Gracechurch Street, London.
Tolstoy, Comte Constantin ...	Odessa.
Tonnelier, G., *Shipbroker* ...	Rue St. Paul 25, Antwerp.
Tozer, Hewitt & Co., *Insurance Brokers* ...	27, Leadenhall Street, London.
Trapani, Alexis, *General Commission Agent* ...	Odessa.
Tresselt, C. ...	Bergen.
Trumpy, H. ...	Genoa.
Turner & Co., *Merchants* ...	Liverpool.
Turner, Morrison & Co., *Merchants and Shipbrokers* ...	Calcutta.
Tweedie Bros., *Insurance Brokers* ...	78, Cornhill, London.
Tylor, A., & Co., *Colliery Proprietors and Shippers* ...	Cardiff.
Tyne Boiler Works Co. (*Ld.*), *Boiler Makers* ...	Newcastle-on-Tyne.
Tyne Pontoons & Dry Docks Co. (*Ld.*), *Steamship Repairers, Engineers and Boiler Makers*	Wallsend-on-Tyne.
Union Iron Works, *Marine Engineers and Shipbuilders* ...	Corner of First and Mission Streets, San Francisco.
Union Steamship Co. (*Ld.*) ...	11, Leadenhall Street, London.
United States Navy Department (F. E. Chadwick) ...	103, Victoria Street, London.
Urraza, A., *Shipbroker* ...	Bilbao.
Vaills, J., *Agent of the Société Générale de Transports Maritimes* ...	Algiers.
Vickers, Sons & Co. (*Ld.*), *Steel Makers* ...	Sheffield; and 50, Old Broad Street, London.
Vinoelst & Peterson, *Shipwrights and Blacksmiths* ...	Vinoelst Place St. Jean 40, Antwerp.
Vivian & Sons, *Metal Merchants* ...	Liverpool.
Walcot & Co., *Timber Merchants* ...	17, Gracechurch Street, London.
Walford & Co., *Shipbrokers and Shipowners* ...	Rue de la Chapelle des Batalieres 2, Antwerp.
Walker, Thomas B., *Merchant and Shipowner* ...	39, Fenchurch Street, London.
Wallace Brothers ...	8, Austin Friars, London.
Wallace Brothers, *Coal Merchants* ...	Liverpool.
Wallace & Co. ...	Bombay.
Wallace, R. B. ...	Newcastle, N.S.W.
Wallsend Pontoon Co. (*Ld.*) ...	Cardiff.
Wallsend Slipway and Engineering Co. (*Ld.*), *Engineers, Boiler Makers, Ship Repairers, &c.*	Wallsend, Newcastle-on-Tyne.
Ward, John N., & Son, *Shipowners* ...	Fleetwood.
Watkins, William, *Tug Owner* ...	75, Mark Lane, London.

Watson, James, & Co., *Pig Iron Merchants*	Grangemouth.
Watson & Youell, *Lloyd's Agents, Steamship Brokers*	Braila.
Watson & Youell, *Lloyd's Agents, Steamship Brokers*	Galatz.
Watson & Youell, *Steamship Brokers*	Sulina.
Watts, Ward & Co., *Steamship Owners*	2, Whittington Avenue, Leaden-hall Street, London.
Watts, Ward, & Co., *Steamship Owners*	Exchange Buildings, Cardiff.
Webb, W. R., *Underwriter*	Liverpool.
Webster, F. I.	Cape Town.
Webster, G. D., & Son, *Paint and Composition Manufacturers*	5, 6, & 7, Great Union St., Hul
Weekes, Phillips & Co.	Plymouth.
Welch & Co., *Shipping and Commission Merchants*	220, California St., San Francis
Wellenstein, Krause & Co., *Merchants*	21, Mincing Lane, London.
Wellenstein, Krause & Co., *Merchants*	Sourabaya.
Wendt & Co.	4 & 6, Throgmorton Avenu London.
Westacott, John, *Shipbuilder*	Appledore, North Devon.
Westcott & Laurance, *Shipbrokers, &c.*	Thorner's Chambers, Ingra Court, Fenchurch St., Londo
Westoll, James, *Shipowner*	12, John Street, Sunderland.
Westray, J. B., & Co., *Shipbrokers*	112, Fenchurch Street, Lond
White, James, *Optician*	209, Sauchiehall St., Glasgow
Whitehaven Shipbuilding Co. (*Ld.*)	Whitehaven.
Wichhorst, I. H. N., *Shipbuilder*	Arningstr. 11, Hamburg.
Wiener, L., *Merchant*	Cape Town.
Wightman & Co., *Steamship Brokers*	19, Gracechurch Street, Londo
Wilks & Huskisson, *Engineers*	Alexandra Dock, Newport, M
Williams, John, *Ship and Insurance Broker*	9, Billiter Square, London.
Williams, W., & Sons, *Shipwrights*	Old Dock, Newport, Mon.
Williams, W. E., *Coal Exporter and Shipbroker*	Old Dock, Newport, Mon.
Wilkie, P. & J., *Timber Merchants, Shipbrokers, and Shipping Agents*	Grangemouth.
Williamson, Milligan & Co., *Steamship Owners*	Liverpool.
Willis, Henry, & Co., *Insurance Brokers*	South Sea House, Threadnee Street, London.
Wilson, Son & Co., *Steamship and Insurance Agents*	Liverpool.
Wilson, Thomas, Sons & Co., *Steamship Owners*	Hull.
Wintringham & Son, *Timber Merchants*	Grimsby.
Withy, E., & Co., *Iron and Steel Shipbuilders and Ship Repairers*	Middleton Ship Yard, We Hartlepool.
Workman, Clark & Co. (*Ld.*), *Shipbuilders*	Belfast.
Worlee, Fr.	Hamburg.
Worms, Josse & Co.	15, Allée de Chartres, Bordeau
Wreaks & Chubb, *Average Adjusters and Insurance Brokers*	16, Exchange Place, New Yo
Wright, Heaton & Co.	Townsville.
Wright, William, *Ship Painter*	Rue des Brasseurs 10, Antwer
Wyllie, J., & Co., *East India Merchants*	93, Bishopsgate Street With London.
Yorkshire Coal and Steamship Co. (*Ld.*), *Steamship Owners and Coal and Coke Exporters*	Goole.
Young, J. Lyle	Papeete, Tahiti.
Young, J. & R., & Co., *Shipbrokers*	111, Union Street, Glasgow.
Ziesmer, Robert, & Co., *Coal Merchants*	Canal St. Pierre 14, Antwerp

LIST OF SUBSCRIBERS TO THE UNIVERSAL REGISTER

(GEOGRAPHICALLY ARRANGED).

LISTE DES ABONNÉS AU UNIVERSAL REGISTER

(GÉOGRAPHIQUEMENT ÉTABLIE).

LONDON.

Alliance Marine Assurance Company (*Ld.*) … … … .. … … … … Capel Court

Australian General Assurance Company *London Agents*: Watt, Gilchrist & Co. … Prince's Street Chambers, Prince's Street.

British & Foreign Marine Insurance Company (*Ld.*) … … … … … … 26, Cornhill

Commercial Union Assurance Company (*Ld.*) … … … … … … … … 19, Cornhill

Globe Marine Insurance Company (*Ld.*) 79, Cornhill

Home & Colonial Marine Insurance Company … … … … … … … 8, Royal Exchange

House of Commons Library

Indemnity Mutual Marine Assurance Company (*Ld.*) … … … … … … 1, Threadneedle Street

Lloyd's, Committee of … … (*3 copies*)

London Assurance Corporation … … 7, Royal Exchange

London & Provincial Marine Insurance Company (*Ld.*) 2, Royal Exchange Buildings

Marine Insurance Company (*Ld.*) … … 20, Old Broad Street

Ocean Marine Insurance Company (*Ld.*) 2, Old Broad Street

Royal Exchange Assurance Corporation Royal Exchange

Schott & Clarke, *Agents for* United Swiss Marine Insurance Companies 3, Royal Exchange Buildings

South British Fire & Marine Insurance Company

Straits Insurance Company (*Ld.*) of Singapore … … … … … … … 1, Whittington Avenue

Thames & Mersey Marine Insurance Company (*Ld.*) 1, Royal Exchange Buildings

Underwriting & Agency Association (*Ld.*) … … … … … … … … 20, Bucklersbury

Union Insurance Society of Canton (*Ld.*) … … … … … … … … 9, Royal Exchange

Yangtsze Insurance Association (*Ld.*) … 68 & 69, Cornhill

Adam Brothers … … … … … … 17, Gracechurch Street

Adamson, Gilfillan & Co. … … … … 2, Billiter Avenue

Admiralty

Admiralty Library

Admiralty (Naval Intelligence Department)

Admiralty (Transport Department)

Aicardi, Hevnes & Co. *Steamship Agents*… 30, Great St. Helens

Argus Printing and Publishing Company of Cape Town … … … … … … 25, Cornhill

Axis Steamships (*Ld.*) *Shipowners*… … 101, Leadenhall Street

Baring Brothers & Co. *Merchants* … … 8, Bishopsgate Street Within

Baxter, S., & Co. … … … … … … 18 & 19, Great St. Helens

Blackwood, Bryson & Co. … … … … 11 & 12, Great Tower Street

Board of Trade.

Borradaile, Thomas S. *Merchant* … … 16, St. Helen's Place

Bowley & Bristow… … … … … … 68 & 69, Cornhill

Brassey, Lord

Bridges, T. H. *Engineer* … … … … Rochester Buildings, 138, Leadenhall Street

Brown, E. T., & Co. *Shipbrokers* … … 22, Great St. Helens

Brown, J. & A., *Shipowners and Shipbrokers, Colliery Proprietors, Merchants, &c.* … 1, Fenchurch Avenue

Bucknall, H., & Sons *Merchants and Shipowners* … … … … … … … … 22, Crutched Friars

Bulloch, J., *Shipowner*… … … … … 13, Fenchurch Avenue

Canadian Pacific Railway Co. … … … 88, Cannon Street

Clarke (Stephenson) & Co. … … … … 4, St. Dunstan's Alley

Cooper, Edward Ernest, & Co. … … … Lloyd's; and 4, Bishopsgate Street Within

Coverley & Westray *Shipowners* … … 37, Leadenhall Street

Coyle, G. *Average Adjuster* … … … 23, Birchin Lane, Cornhill

Dewrance & Co. … … … … … … 158, Great Dover Street

Dick, W. B., & Co. (*Ld.*) *Oil Merchants, &c.* 20, Bucklersbury

Duncan, George, & Co. … … … … 2, East India Avenue

Dunn, W., & Co. … … … … … … 95, Bishopsgate St. Within

Durham, Churchill & Co. *Engineers and Makers of Patent Marine Engine Governors and Pistons* … … … … … … 23, Leadenhall Street

East & West India Dock Co.

Emson, C. W.,

Engelbert & Co. *Lubricant Manufacturers, &c.* 70 & 71, Bishopsgate Street

Fenning, G. & J. *Underwriters* … … … Lloyd's; and 74, Cornhill

Fleming, R. T. … … … … … … Lloyd's

Galbraith & Moorhead *Shipbrokers*… … Billiter House

Galbraith, Pembroke & Co. *Shipbrokers*… 8, Austin Friars

Garrad & Co. *Brokers for the Sale and Purchase of Steam and Sailing Vessels*… … 15, Fenchurch Street

Gaskell, G.

Gellatly, Hankey, Sewell & Co. *Shipbrokers*… … … … … … … … 109, Leadenhall Street

Goldstück, Hainzé & Co. *Superintendents of Grain and Produce Cargoes* … … 88, Bishopsgate St. Within

Graham, W. K., & Son… … … … … Blomfield House, 85, London Wall

Gray, Dawes & Co.	13, Austin Friars
Green, F., & Co. *Shipbrokers*	13, Fenchurch Avenue
Hall, John, jun., & Co. *Shipowners and Shipbrokers*	1, New London Street
Hanham, Sir John A., Bart.	116, Buckingham Palace Road
Harrison, J. & C.	66, Mark Lane; also at Cardiff and Gravesend
Hay & Co. *Ship and Insurance Brokers*	11, St. Mary Axe
Haycraft & Gilfillan *Marine Insurance Brokers*	Lloyd's, and 3, Great Winchester Street
Head, Henry, & Co. (2 copies)	85, Gracechurch Street
Hoffman, C., & Co. *Shipbrokers and Steamship Agents*	11, Hart Street, Mark Lane
Holland, Arthur, & Co.	2, East India Avenue
Hopkins (Manley), Son, & Cookes. *Adjusters of Averages*	32, Cornhill
Howard, C., & Son *Shipowners and Brokers*	17, Philpot Lane
Hubbuck, Thos., & Son *Paint Manufacturers*	24, Lime Street
Hunt, Samuel. *Shipbroker*	7, Mark Lane
Hunting & Pattison *Steamship Owners and Ship and Insurance Brokers*	32, Great St. Helens
Huth, Frederick, & Co.	12, Tokenhouse Yard
Janson, J. W. & D.	Lloyd's
Ker, Bolton & Co.	7, East India Avenue
Kirkaldy, J. *Marine Engineer and Maker of Distilling Plant and Feed Water Heaters*	10, West India Dock Road
Kirke, R. & J. J. *Merchants and Shipbrokers*	16, Mark Lane
Klett & Weber *Insurance Brokers*	6, Gracechurch Street
Lamplough, W., & Co. *Insurance Brokers*	72, Cornhill
Lane & Macandrew *Shipbrokers*	26, Great St. Helens
Langridge, Henry, & Co.	16, Great St. Helens
Lawrence, T. C., & Co. *Shipbrokers*	Eastcheap Buildings
Leary, C., & Co. *Timber Brokers*	19, Old Broad Street
Lidgett, John, & Sons *Shipowners and Brokers*	6, Lime Street Square
Liebig's Extract of Meat Co. (*Ld.*)	9, Fenchurch Avenue
Lynch, Beech & Co. *Insurance and Exchange Brokers*	10, George Yard, Lombard Street
Lyon, R., & Co. *Marine Salvage Surveyors and Brokers*	2, St. Peter's Alley, Cornhill
McCracken, Fenwick & Co. *Shipbrokers*	38, Fenchurch Street
McGregor, Gow & Co. *Shipowners*	1, East India Avenue
Marten, F. W. *Underwriter*	Lloyd's; & 23, Rood Lane
Messageries Maritimes de France	97, Cannon Street
Millwall Docks Co.	1, Railway Place, Fenchurch Street
Montgomerie & Workman	36, Gracechurch Street
Mosses & Mitchell *Merchants*	68 & 71, Chiswell Street
Nixon, Henry *Underwriter and Insurance Broker*	Lloyd's
Park Brothers	52, Gracechurch Street
Paterson & Simons	39, Lime Street
Peet, J., & Co. *East India Merchants*	26, Bush Lane
Pile & Co. *Steamship Brokers, and Contractors for the Building of Ships and Steamers*	31, Great St. Helens
Poole, John, & Sons	27, Clement's Lane
Preston, John *Broker for purchase and sale of Steamers of all descriptions*	59 & 60, Cornhill
Richards, William, & Son *Average Adjusters*	New City Chambers
Rose, Thomson, Young & Co. *Insurance Brokers*	19, Change Alley
Rugg, C. H., & Co.	Mildmay Chambers, 82, Bishopsgate Street
Salvage Association	17 & 18, Cornhill
Saunders, W. F.	Lloyd's
Searight, James, & Co.	7, East India Avenue
Sedgwick, H. B. *Underwriter*	Lloyd's
Sevestre, W. A.	New City Chambers, B Bishopsgate Street
Simpson, Spence & Young	12, Great St. Helen's
Smith John *Broker*	105, Leadenhall Street
Smith, Sandius & Co. *Ship and Insurance Brokers*	158, Leadenhall Street
Stockton, H. J.	16, Philpot Lane
Strick, Frank C., & Co. (*Ld.*) *Shipbrokers and Chartering Agents*	Woolpack Buildings, Gracechurch Street
Surrey Commercial Dock Co.	106, Fenchurch Street
Swanzy, F. A., & Co.	147, Cannon Street
Temperley, J., & Co.	9, Gracechurch Street
Temperleys, Carter and Darke	
Thames Iron Works & Shipbuilding Co. (*Ld.*)	Orchard Yard, Blackwa
Theobald, James H. *Ship and Insurance Broker*	86, Leadenhall Street
Thorl, John, & Co. *Insurance Brokers*	Lloyd's; and 18, Corn
Tindall, W. H., & Co.	77, Gracechurch Street
Tozer, Hewitt & Co. *Insurance Brokers*	27, Leadenhall Street
Tweedie Brothers *Insurance Brokers*	78, Cornhill
Union Steamship Co. (*Ld.*)	11, Leadenhall Street
United States Navy Department (F. E. Chadwick)	103, Victoria Street
Walcot & Co.	17, Gracechurch Street
Walker, Thomas B. *Merchant and Shipowner*	39, Fenchurch Street
Wallace Brothers	8, Austin Friars
Watkins, William *Tug Owner*	75, Mark Lane
Watts, Ward & Co. *Steamship Owners*	2, Whittington Aven Leadenhall Street
Wollenstein, Krause & Co. *Merchants*	21, Mincing Lane
Wendt & Co.	4 & 6, Throgmorton Avenue
Westcott & Laurance *Shipbrokers, &c.*	Thorner's Chambers, gram Court, Fenchu Street
Westray, J. B., & Co. *Shipbrokers*	112, Fenchurch Street
Wightman & Co. *Steamship Brokers*	19, Gracechurch Street
Williams, John *Ship and Insurance Broker*	9, Billiter Square
Willis, Henry, & Co. *Insurance Brokers*	South Sea House, Thre needle Street
Wyllie, J., & Co. *East India Merchants*	93, Bishopsgate Stree Within

ABERDEEN.

Aberdeen Glen Line Steamship Co., *Shipowners*	48, Market Street
Aberdeen Steam Navigation Co.	87, Waterloo Quay
Ludwig, C., & Co. *Shipbrokers*	54, Regent Quay
Nicol, A., & Co. *Shipowners*	Guild Street Buildings

ADELAIDE.

De Caen, P. D.

AKYAB.

Adamson & Maracan *Merchants*

ALGIERS.

Bergeret, F. *Shipbroker*	
Durand, Prosper *Coal Merchant*	
Schiaffino, A. *Shipowner*	Boulevard République
Vaills, J. *Agent of the Société Générale de Transports Maritimes*	

ALICANTE

Ramos, José

ALMERIA.

D. Burgos Cañizares, José

AMSTERDAM.

Amsterdamsche Droogdok Maatschappij (Dry Dock Co.)
Burlage & Schumacher *Insurance Brokers* Binnenkant 8
Drost, J., Dzn
Dunker & Goedkoop *Shipbrokers*
Goedkoop, D., jun. *Shipbuilder*
Hudig & Blokhuijzen *Shipbrokers*
Joosting & Van Eijk *Insurance Brokers* Frederiks Plein, West Einde
Koninklijke Nederlandsche Stoomboot Maatschappij ... ... Prins Hendrik Kade 161
Made, J. D. van der *Engineer*
Meyer & Co. *Shipbrokers* ... ... Prins Hendrik Kade 145
Scheur Brothers *Shipbrokers*

ANTWERP.

André, Auguste *Shipbroker* ... ... Quai St. Aldegonde 20
Best, John P., & Co. *Shipbrokers, Shipowners, and General Agents* ... ... Place Verte 36 & 37
Booth & O'Halloran *Engineers and Boilermakers*... ... ... Rue du Brésil 13
Berniche, Edouard *Insurance Agent* ... Place de Meir, 10
Bonhoulle, Alphonse *Engineer and Boilermaker* ... ... Quai Godefroid 66
De Keyser, R., & Co. (late H. J. A. Telghuys) *Shipbrokers*... ... ... Canal des Brasseurs 36
Deppe, Adolf *Shipowner* ... ... Rue de Bordeaux
Ellis and Morrison *Shipbrokers* .. ... Canal des Brasseurs, 61
Engels, Arnold Frederick. *Lloyd's Agent* Rue au Sucre 2
Gellatly, Hankey, Sewell & Co. *Shipbrokers and Agents* ... ... Quai St. Pierre 5
Grisar & Marsily *Shipbrokers* ... ... Quai St. Aldegonde 6
Harms, R. J. *Marine Surveyor* ... ... Rue Basse 29
Henriex, H. *Tugboat Director* ... ... Canal St. Pierre 2
Huysmans & Buleke *Shipbrokers* ... ... Rue Pré 11/1
Kennedy, Hunter & Co. *Shipbrokers* ... Canal St. Pierre 6
Maihacher, Chs. *Insurance Broker* ... ... Rue Grammey 6
Marshall & Gerling *Shipbrokers* ... ... Canal des Brasseurs 64
Roels, P. *Shipwright* ... ... Rue du Brésil 9
Ruys & Co. *Shipbrokers* ... ... Quai aux Charbons 33
Société Anonyme Anversoise pour le Déchargement des Céréales ... ... Rue de l'Etalage 23
Société John Cockerill *Shipbuilders* ... Hoboken and Antwerp
Stappers, L. *Engineer and Boilermaker* Rue Nassau 10
Strybos, Paul, & Co. *Coal Merchants* ... Rue du Fagot 17
Tonnelier, G. *Shipbroker* ... ... Rue St. Paul 25
Vinoelst & Petersen *Shipwrights and Blacksmiths*... ... ... Vinoelst Place, St. Jean 10
Walford & Co. *Shipbrokers and Shipowners* Rue de la Chapelle des Bateliers 2
Wright, William *Ship Painter* ... ... Rue des Brasseurs 40
Ziesmer, Robert, & Co. *Coal Merchants* ... Canal St. Pierre 14

APPLEDORE (NORTH DEVON).

Cock, Robert *Shipbuilder*
Westacott, John *Shipbuilder*

ARDROSSAN.

Ardrossan Shipbuilding Co. *Iron and Steel Yacht and Ship Builders; Slip and Graving Docks; Engineers, Boiler Makers, and General Repairers.*

ATHENS.

Hill, Arthur *Lloyd's Agent*

AUCKLAND, N.Z.

Compton, F. E.
Errington, Wm.
Henderson & Macfarlane
James, C. F., & Co. *Ship, Timber and Insurance Brokers*... ... ... Fort Street
McKenzie, D. H.
Stone, C. B.

BARROW-IN-FURNESS.

Barrow Hematite Steel Co. (*Ld.*) *Iron and Steel Works*
Fisher, James, & Sons *Shipowners and Shipbrokers; Lloyd's Agents*
Furness Railway Company
Naval Construction and Armaments Co. (*Ld.*) *Shipbuilders*

BATAVIA.

NETHERLAND-INDIA SEA & FIRE INSURANCE COMPANY
Bremer, A. A. W.
De Jong, B. C. *Lloyd's Agent*
Handels Vereeniging (Batavia Exchange)
Leeuwen, J. F. van
Maclaine, Watson & Co. *Merchants*

BELFAST.

SINCLAIR & BOYD (Ulster Marine Insurance Co., *Ld.*)
Burke, John, & Co *Shipbrokers*
Harland & Wolff (*Ld.*) *Shipbuilders and Engineers*
Heyn, G., & Sons *Shipbrokers*
MacIlwaine & MacColl (*Ld.*). *Shipbuilders*
Workman, Clark & Co. (*Ld.*) *Shipbuilders*,

BERGEN.

BERGENS BÖRS.
Ameln, Theodor.
Ellertsen, F. J., & Co, *Shipbrokers and Steamship Agents*
Grieg, J., *Shipbroker and Steamship Agent*
Hansen, Otto F., *Shipbroker*
Tresselt, C.

BERLIN.

Bolstorff, C. A. ... ... Elsasser Strasse 25
Kaiserliche Admiralität

BILBAO.

"La Española" Compañia de Seguros Maritimos
Azuar & Astigarraga *Shipbrokers*
Urraza, A. *Shipbroker*

BLYTH.

Blyth Dry Dock Company (*Ld.*) *Ship Repairers, Engineers, &c.*

BOMBAY.

Bombay & Persia S. N. Co.
Graham, W. & A., & Co., *Merchants*
Killick, Nixon & Co., *Merchants*
Port Officer
Wallace & Co.

BO'NESS.

Donaldson, William *Shipping Agent and Steamship Broker, and Pitwood Importer*

BORDEAUX.

Worms, Josse & Co. ... ... 15, Allée de Chartrès

BOSTON, U.S.A.

Folger, G. H., Secretary Board of Underwriters ... ... 50, State Street
Boston Towboat Co. *Towing, Wrecking, Transporting Merchandise, &c., &c.*

BRAILA.

Watson & Youell *Steamship Brokers*

BREMERHAVEN.

"Hansa" Deutsche Dampfschiffahrts Gesellschaft
Klingenberg, Carl, Job & Co. *Ship and Steamship Brokers, Bunkers, &c.*
Rabien & Stadtländer *Steamship Owners and Brokers*
Rickmers, R. C. *Rice Importer and Exporter, and Mill Owner*

BRISBANE.

British India and Queensland Agency Co. (*Ld.*) *Steamship Agents*
Morehead, B. D., & Co.
Parbury, Lamb & Co. *Merchants*
Port Officer (Capt. Heath, R.N.)

BRISTOL.

Bristol Commercial Rooms ... ... Corn Street
Bennett, C. T. *Shipowner* ... ... 66, Queen's Square
Kilby & Churley *Stevedores* ... ... 2, Queen's Square Avenue
King, C. J., & Sons *Stevedores* ... ... 12, Princes Street
Stothert, G. K., & Co. *Engineers and Shipbuilders* ... ... Hotwell Road

BUENOS AYRES.

Christophersen, P.
Dorman, Philip *Lloyd's Agent* ... ... Casilla Correo 281
Drysdale, Thomas & Co. *Merchants*
Green & Johnson *Shipbrokers and Agents*
La Platense Flotilla Co. (*Ld.*) *Shipowners*
Moore, H. C. *Managing Engineer of La Platense Co.*
Sattler, L. *Shipbroker and Agent* ... ... Casilla Correo 185
Shaw Brothers *Merchants.*

BUSSORAH.

Asfar & Co. *Merchants*

CADIZ.

Compañia Trasatlantica *Shipowners*
Macpherson, H.
Pinillos, Saenz & Co.

CALCUTTA.

Bengal Chamber of Commerce ... ... 102, Clive Street
Anderson, Wright & Co. *Merchants* ... ... 22, Strand
Apcar & Co. *Merchants and Shipowners* ... 19, Radha Bazar
Blackwood, Blackwood & Co.
Hedger, Philip T. F. *Solicitor* ... ... 1, Hastings Street
Turner, Morrison & Co. *Merchants and Shipbrokers*

CAPE TOWN.

Anderson & Murison
Currie, Donald & Co. *Merchants*
Learmouth, F. A.
Searight, J., & Co. *Merchants*
Thomson, Watson & Co.
Webster, F. I.
Wiener, L. *Merchant*

CARDIFF.

Barnes, Guthrie & Co. *Shipowners and Brokers*
Barry Graving Dock & Engineering Co. ... Exchange Buildings
Bulman & Tiffen, *Coal Exporters and Merchants*
Bute Docks Company (*Ld.*) (Collector's Department)
Cardiff Junction Dry Dock & Engineering Co. (*Ld.*) ... ... Junction Dry Dock
Carrel, F. P. *Shipbroker, &c.* ... ... 23, Mount Stuart Squa
Cay, Hall & Co. *Shipbrokers, &c.* ... ... Mount Stuart Square
Clyde Shipping Co. (*Ld.*) *Steam Tug Owners* The Pier Head
Cory Brothers & Co. (*Ld.*). *Colliery Proprietors* Bute Docks
Edwards, Robertson & Co. *Steamship Owners* ... ... Bute Docks
Fisher, Renwick & Co. *Steamship Brokers*
Glamorgan Coal Co. (*Ld.*) *Colliery Proprietors*
Hill's Dry Dock and Engineering Co. (*Ld.*)
Hooper, Campbell & Co. *Steamship Owners*
Jenkins, S. D., & Son. *Shipbrokers*
Ocean Coal Co. (*Ld.*) *Colliery Proprietors*
Page & Co. *Merchants*
Penarth Ship Building & Ship Repairing Co. (*Ld.*) ... ... Penarth Dock
Spiller & Co. *Millers.*
Taff Vale Railway Co. (Collector's Department)
Tylor, A., & Co. *Colliery Proprietors and Shippers*
Wallsend Pontoon Co. (*Ld.*)
Watts, Ward & Co. *Steamship Owners* ... Exchange Buildings

CHRISTCHURCH, N.Z.

National Mortgage and Agency Co. of New Zealand (*Ld.*)

CHRISTIANIA.

Akers Mekaniske Werkstadt. *Iron Shipbuilders*
Meinich, Jens, & Co., *Shipbrokers and Steamship Agents*

CIVITA VECCHIA.

Shepherd, W. & P. *Steamship Agents*

CLACHAN, ARGYLLSHIRE.

MacKinnon, W. ... ... Balinakill

CONSTANTINOPLE.

Club de Constantinople
Administration des Phares Impériales Ottomanes
Administration Sanitaire de l'Empire Ottoman
Band *Marine and General Engineer and Boiler Maker*
Foscolo, Mango & Co. *Steamship Owners and Steamship Agents*
Glamorgan Coal Co. (*Ld.*) *Colliery Proprietors*
MacGill, R. *Engineer and Boiler Maker, General Contractor*
MacNab, Rougier & Co. *Ship Agents*
Philippucci & Condili *Ship Chandlers*
Reppen, C. J. *Ship Agent*
Scaife, A. H. *Lloyd's Agent*
Seropian, C. H. *Insurance Agent*
Theodoridi, C. A., & Co. *Ship Agents*

COPENHAGEN.

Burmeister & Wain's Maskin & Skibsbyggeri *Engineers and Shipbuilders*
Christensen, Alfred *Steamship Agent and Coal Merchant* Osterbrogade I, and Holbergsgade 10
Hansen, C. K. *Shipowner, Agent and Coal Merchant* Toldbodvej
Hecksher & Son *Lloyd's Agents and Brokers*
Schack, Steenberg & Co. *Agents and Coal Merchants*

CORK.

Craig, W. I. *Docks Manager* ... ... Passage West

CRONSTADT.

Fishwick, Alexander *Shipping Agent*

DANZIG.

Reinhold, F. G. *Steamship Owner and Shipbroker*

DARLINGTON.

Darlington Forge Co. (*Ld.*) *Makers of Iron and Steel Forgings, Castings, &c.*

DARTMOUTH.

Channel Coaling Co.

DELFZYL.

Pott & Smeel *Shipbrokers*

DRONTHEIM.

Baehke & Co. *Shipbrokers*

DUMBARTON.

Denny, Wm., & Bros. *Shipbuilders* ... ... Leven Shipyard
McMillan, Archibald, & Son *Shipbuilders* Dockyard

DUNDEE.

Leitch, A., & Co. *Shipbrokers* ... ... Commercial Street
Scott, James *Shipowner, Ship and Insurance Broker* 4, Royal Exchange Place

DUNEDIN.

National Fire and Marine Insurance Company of New Zealand
Otago Marine Underwriters' Association
National Mortgage & Agency Company of New Zealand (*Ld.*)
Paterson, A. S., & Co. *Merchants* ... ... Box 368

DUNKIRK.

Goudrand Brothers *Ship and Steamship Agents*
Leroy, Alfred *Ship and Steamship Broker*
Plaideau, E., *Shipbroker*
Spiers, D. *Shipbroker and Steamship Agent*

DURBAN.

Dunn, W., & Co.
King & Sons *Shipping Agents, &c.*
Storm, William *Ship Chandler*

EAST LONDON.

Dunn & Co.

ELSINORE.

Helsingors Jernskibs & Maskinbyggeri *Shipbuilders and Engineers*
Lund, I. Theod. *Consul and Lloyd's Agent*

FALMOUTH.

Broad, William, & Sons *Ship Agents and Merchants, Lloyd's Agents*
Fox, G. C., & Co. *Merchants and Ship Agents*

FELIXSTOWE.

Felixstowe Dock and Railway Co.

FIUME.

"Adria" Hungarian Sea Navigation Co. *Steamship Owners*

FLEETWOOD.

Gibson, John, & Sons *Shipbrokers and Chandlers*
Ward, John N., & Son *Shipowners and Ship and Insurance Brokers*

FLENSBURG.

Flensburger Schiffsbau Gesellschaft
Schuldt, H. *Shipbroker*

FRASERBURGH.

Macconnachie, W., & Co. *Shipbrokers and Shipowners*

GALATZ.

Watson & Youell *Steamship Brokers, &c.*

GENOA.

MACKENZIE, EVAN *General Insurance Agent* ... Piazza S. Lorenzo 15
MINGOTTI, GIUSEPPE ... ... Piazza Scuole Pie 10
Brunn Brothers *Steamship Brokers*
Dalero, Gio Batta fu D. ... ... Piazza della Vigne 3
Galliano, G. fu L. (*late* A. Bonanati & Co.) *Ship and Steamship Broker and Agent*
Gambetta, Domenico ... ... Piazza Scuole Pie
Gastaldi, G. B., & Co. *Steamship Brokers and Agents*
Gismondi, A., & Co. *Steamship Agents and Shipbrokers*
Miller, W. Robley & Co. *Steamship Agents*
Navigazione Generale Italiana ... ... Via Balbi
Schiaffino & Solari *Steamship Owners and Brokers*
Trumpy, R.

GEORGETOWN, DEMERARA.

ROYAL AGRICULTURAL & COMMERCIAL SOCIETY (LIBRARY & EXCHANGE ROOMS)
Sproston Dock & Foundry Co. *Contractors.*

GHENT.

De Baerdemaecker, G. C. *Agent Maritime Commissionnaire Expéditeur* ... ... Rue Fiévé 17

GIBRALTAR.

Bland, M. H., & Co. *Coal Merchants, Steamship Owners and Agents*
Carara, John, & Sons *Steam Coal Merchants*
Haynes, Thomas *Merchant, Steamship Owner, Agent and Broker*
Mateos, A., & Sons *Steamship Agents and Coal Merchants*
Onetti, J., & Sons, *Merchants and Steamship Agents*
Rugeroni, Joseph & Son *Steamship Agents and Coal Merchants*
Smith, Imossi & Co. *Coal Merchants, Steamship Owners and Agents*

GLASGOW.

ASSOCIATION OF UNDERWRITERS AND INSURANCE BROKERS (R. Cooper Rundell, *Secretary*)
ROYAL EXCHANGE
Black, J., & Co. *Merchants and Shipowners* 201, Elliot Street
British India Steam Navigation Co. ... ... 203, West George Street
Fairfield Shipbuilding and Engineering Co. (*Ld.*) ... ... Govan
Gunn & Cameron *Newspaper Proprietors* 106, Union Street
Harbour Master (Capt. Robert White) ... 16, Robertson Street
Henderson, David & William, & Co. *Shipbuilders and Engineers* ... ... Partick
Inglis, A. & J. *Engineers and Shipbuilders* Pointhouse
London and Glasgow Engineering and Iron Shipbuilding Co. (*Ld.*) ... ... 172, Lancefield Street
McGregor, D., & Co. *Nautical Instrument Makers* ... ... 37, Clyde Place
Mathews, Maclay & Manson *Oil Merchants, Paint, Colour and Varnish Manufacturers* ... ... 104, Hyde Park Street
Muir & Caldwell *Engineers and Boilermakers* ... ... Elliot Street
Murray, James, *Ship Chandler* ... ... 8, Brown Street
Napier, Robert, & Sons *Engineers and Shipbuilders* ... ... Lancefield House, Gov[illegible]
Neill, Robert *Commission Agent and Merchant* ... ... 10, South Hanover Str[illegible]
State Steamship Co. (*Ld.*) ... ... 65, Great Clyde Street
Stephen, Alex., & Sons *Shipbuilders and Engineers* ... ... Linthouse, Govan
Stevenson, D. M., & Co. *Shipowners and Coal Exporters* ... ... 12, Waterloo Street
Storer, James, & Co. *Oil and Colour Manufacturers* ... ... Bridgeton
Tennent, J. & R. *Brewers* ... ... Wellpark Brewery
Tharsis Sulphur and Copper Co. (*Ld.*) ... 53, Renfield Street
Thomson, James & George *Engineers and Shipbuilders* ... ... Clydebank
White, James *Optician* ... ... 209, Sauchiehall Street
Young, J. & R., & Co. *Shipbrokers* ... ... 111, Union Street

GLOUCESTER.

Andersen, J. *Shipbroker* ... ... 19, Ladybellgate Street

GOOLE.

Yorkshire Coal and Steamship Co. (*Ld.*) *Steamship Owners and Coal and Coke Exporters*

GOPALPORE.

Harden & Co. *Agents*
Minchin, F. J. V. *Lloyd's Agent*

GOTHENBURG.

Appelberg, Axel
Edelfelt, E. G.
Leffler, A., & Son *Shipbrokers and Steamship Agents*
Lindholmens Werkstad *Engineering Works, Shipyard, Slip and Dry Dock*

GRANGEMOUTH.

Crawford & Co. *Steamship Owners and Brokers*
Hoy, J. & J. *Shipbrokers*
Mackay, A. & A. Y. *Shipowners and Brokers*
Miller, C. *Steamship Broker*
Salvesen, J. T., & Co. *Steamship Owners and Brokers and Timber Merchants*
Watson, James, & Co. ... ... Pig Iron Merchants
Wilkie, P. & J. *Timber Merchants, Shipbrokers and Shipping Agents*

GRAVESEND.

Coxes, A. J. *Master Mariner* ... ... 38, The Grove

GREENOCK.

Caird & Co. (*Ld.*) *Engineers and Shipbuilders* ... ... Dalrymple Street
Greenock Steamship Co. (*Ld.*). *Steamship Owners* ... ... 1, Cross Shore Street
Richardson, James, & Co. *Merchants & Shipowners* ... ... 2, Church Place
Swan, Allan *Steamship Agent, &c.* ... ... Custom House Buildings

GRIMSBY.

Green, W. A. *Manager for the Manchester, Sheffield and Lincolnshire Railway at Grimsby*
Wintringham & Son *Timber Merchants*

HAGUE (THE).

Ministerie van Koloniën

HAKODATE, JAPAN.

Howell & Co. *Merchants*

HALIFAX, N.S.

Mathers, Isaac H. *Ship Agent and Commission Merchant; Vice-Consul for Sweden and Norway*
Pickford & Black. *General Commission and Steamship Agents*
Roche, W. *Steamship and Coaling Agent* Roche's Wharf

HAMBURG.

Verein Hamburger Assecuradeure
Blohm & Voss *Shipbuilders and Engineers* Steinwärder
Cellier, Eugène *Shipbroker*
Duncker, Arthur ... ... Brodschrangen 26
Gläfcke & Hennings *Shipbrokers* ... ... Admiralitätstrasse 68
Heidmann, H. W. ... ... Hafenstr. 93
Hugo, C. *Shipbroker and Steamship Agent* Holländische Brook 23
Laeisz, F. *Shipowner* ... ... Neueberg 14
Lorentzen, J. H., & Co.
Morris & Co. .. .. Deichstrasse 7
Pott, H. W., & Körner *Shipbrokers*
Rettig, Rudolph *Shipowner* ... ... Neuerwall 7
Schwalbe, Franz *Merchant and Shipowner* Baumwall 10
Scaunt ... ... Admiralitätstrasse
Sloman, R. M., jun. *Shipbroker* ... ... Baumwall
Stülcken, H. C., Wwe. *Shipbroker*
Wichhorst, I. H. N. *Shipbuilder* ... ... Arningstr. 11
Worlée, Fr.

HAMILTON (BERMUDA).

James, W. T.

HARBURG a.d. ELBE.

Renck & Hessenmüller *Shipowners and Shipbrokers*

HAVANA.

Deulofeu, hijo & Co. *Commission Merchants and Steamship Agents*

HAVRE.

Billard, Emile *Shipbroker* ... ... Grand Quai 67
Caspar, C. *Shipping and Custom House Agent; Agent for General S.N. Co., British India S.N. Co., Bailey & Leetham, Furness Line, &c.* ... ... Rue Dauphine 2
Cunard Steamship Co. (*Ld.*) *Shipowners* Quay d'Orléans 21
Duminy, E. ... ... Rue Dicquesmare 39
Hutter, J. J. ... ... Rue Prouy 55

HOBART, TASMANIA.

Tasmanian Steam Navigation Co.

HONG KONG.

On Tai Insurance Company (*Ld.*)
Grant, Chas. *Merchant* ... ... Queen's Road
Hong Kong & Kowloon Wharf, & Godown Co. (*Ld.*)
Hong Kong & Whampoa Dock Co. (*Ld.*) *Shipbuilders*
Melchers & Co. *Merchants*
Russell & Co. *Merchants*
Stevens, Geo. R., & Co. *Merchants.*

HOWRAH.

Commercial Dock and Engine Works
Reid & Co. *Dock Proprietors* ... ... Hooghly Dock

HUELVA.

Brandt, Gustav, & Co. *Timber, Wine and General Merchants*
Delprat, G. D. *Manager of Bede Metal Co. in Spain*
Duclos, C. *Merchant, Shipbroker and Custom House Agent*
Morrison, Thomas, & Co. *Merchants and Bankers*
Narvaez, Manuel *Sworn Broker and Ship Agent*
Speirs, G. M. *Commercial Agent* ... ... Rio Tinto Co. (*Ld.*)
Sundheim & Doetsch *Bankers and General Merchants*

HULL.

Hull Exchange Co.
Hull Underwriters' Association.
Amos & Smith *Engineers*
Bailey & Leetham *Steamship Owners*
Cammell, Woolf & Haigh *Shipowners, &c.*
Massey & Sawyer *Shipowners and Brokers* 7, Quay Street
Rayment & Sharp *Ship Chandlers*
Sanderson, G. R., & Co. *Shipowners and Brokers*
Webster, G. D., & Son *Paint and Composition Manufacturers* ... ... 5, 6 & 7, Great Union St.
Wilson, Thomas, Sons, & Co. *Steamship Owners*

INVERNESS.

MacDonald & Morrison *Shipbrokers* ... 35, Grant Street

IQUIQUE.

James, Inglis & Co. *Merchants*

ISLE OF MAN.

Isle of Man Steam Packet Co. *Steamship Owners* ... ... Douglas

KIEL.

Germania," Schiff- und Maschinenbau Actien Gesellschaft

KING'S LYNN.

Garland & Flexman *Shipbrokers, Commission and Forwarding Agents*

KINGSTON (JAMAICA).

Soutar & Co.

KINGSTOWN (IRELAND).

Atkinson, J. J. ... ... 42, Albert Road.

KIRKCALDY.

Scott, John, & Co. *Engineers and Shipbuilders*

KOBÉ.

China and Japan Trading Co.
Mourilyan, Heimann & Co.

KURRACHEE.

Port Officer ... ... Manora.
Rustomjee, H. J.

KUSTENDJIE.

Macri, John

LA ROCHELLE.

Holter, A. *Sworn Shipbroker*

LAGOS.

The Officer Administering the Government (*2 copies*)

LANCASTER.

Helme, J. *Shipowner* ... ... 43, Church Street

LEEDS.

Leeds Forge Co.

LEGHORN.

Miller, W., Robley, & Co. *Steamship Agents*
Orlando Brothers. *Shipbuilders*

LEITH.

Cairns, R. & D. *Steamship Owners and Brokers*
Cran, John, & Co. *Shipbuilders and Engineers*
Currie, James, & Co. *Shipping Agents* ... 16, Bernard Street
Dodson & Craig *Steamship Owners and Brokers, &c.* ... ... 3, Bernard Street
Edinburgh Roperie and Sailcloth Co. (*Ld.*)
Gibson, George, & Co. ... ... 64, Commercial Street
Henderson & McIntosh *Steamship Owners and Brokers* ... ... 44, Bernard Street
Henderson & McKean *Stevedores, Contractors and Steam Tug Owners* ... ... 2, Tower Street
London and Edinburgh Shipping Co. (Thos. Aitken, *Manager*) ... ... 8 & 9, Commercial St
Mackie, Koth & Co *Steamship Brokers and Coal Exporters* ... ... 54, Bernard Street
Marshall, David *Steamship Owner and Broker*
Ramage & Ferguson *Shipbuilders and Engineers*
Reid, Francis F. *Shipowner and Ship and Insurance Broker* ... ... 2, Baltic Street
Ressich, McLaren & Co. *Shipbrokers and Coal Exporters* ... ... 53, Bernard Street
Salvesen, Chr., & Co. *Shipbrokers, Steamship Owners, and Coal Exporters*... .. 29, Bernard Street
Stevenson & Drynan *Shipowners and Coal Exporters*

LIBAU.

LIBAU EXCHANGE COMMITTEE
Bornholdt, P., & Co. *Steamship Agents*
Helmsing & Grimm *Steamship Agents and Owners*
Hoffman & Bielby *Steamship Chartering Agents and Coal Importers*

LIMERICK.

Mullock & Sons *Shipbrokers, Commission Agents and Coal Merchants*

LISBON.

Bensaude & Co. *Merchants*
Elias, F. X. *Shipbroker* ... ... Rua da Remolars
Garland, Laidley & Co. *Commission Agents*
George, E. *Steamship Agent*
Kjolner, C. *Ship Chandler*
Knowles, Rawes & Co.

LIVERPOOL.

BRITISH AND FOREIGN MARINE INSURANCE COMPANY (*Ld.*)
INTERNATIONAL MARINE INSURANCE COMPANY (*Ld.*) (*2 Copies*)
LIVERPOOL EXCHANGE CO. News Room
LIVERPOOL FREE PUBLIC LIBRARY
LIVERPOOL UNDERWRITERS' ASSOCIATION
MARINE INSURANCE COMPANY
RELIANCE MARINE INSURANCE COMPANY
SEA INSURANCE COMPANY (*2 Copies*)
THAMES AND MERSEY MARINE INSURANCE COMPANY (*Ld.*)
UNION MARINE INSURANCE COMPANY (*Ld.*)
Adair & Co. *Engineers, Boiler Makers, Pump Makers, Ship Repairers, &c.*
Balfour, Williamson & Co. *Merchants*
Barnett, J., & Co. *Steamship Agents and Marine Insurance Brokers*
Bell's Asbestos Co. (*Ld.*) *Asbestos and India Rubber Manufacturers*
Chambers, James, & Co. *Ship Owners and Ship Store Dealers*

Cochran & Co. *Steam Yacht and Launch Builders* ... ... Birkenhead
Commercial Cable Co. *Atlantic Telegraph* 23, Royal Exchange
Davison, Bisset & Co. *Ship Store Dealers and Provision Merchants*
Douglas, R. R. *Insurance Agent*
Ellis, Edward *Surveyor to "Bureau Veritas"*
Forrester, George, & Co. (*Ld.*) *Engineers*
Garnett (Alexander) & Co. *Merchants*
Gillchrest, R. L., & Co. *Shipbrokers*
Grayson, H. & C. *Shipbuilders and Engineers*
Hainsworth, Watson & Co. *Merchants*
Harbour Master (H. M. Lunclet)
Henderson Bros. *Steamship Owners and Merchants* ... ... Mersey Chambers
Henderson & Glass *Iron Merchants*
Hill, J. E. Gray *Solicitor*
Houlder Bros. & Co. *Ship and Insurance Brokers*
Japp & Kirby *Shipbrokers*
Kellock, C. W., & Co. *Brokers for the Sale of Ships and Steamers, Ship Valuers and Shipowners*
Kelso (S. Wright) & Co. *Ship and Insurance Brokers*
Kirkman, T. *Marine Insurance Agent* ... 4, Old Hall Street
Little, J., & Co. *Shipbrokers and Steamship Owners* ... ... 17, Water Street
Lowndes, R., Son & Wallace *Average Adjusters*
McGillivray, J. *Manager of Downie Boiler Incrustation Preventive Co.*
Minton, R. R., & Co. *Paint Manufacturers and Oil Merchants*
Potter, W. H., & Sons *Shipbuilders*
Rimmer, A., & Co. *Ship and Insurance Brokers* ... ... 2, Exchange Street East
Rogers & Bright *Merchants*
Rollo, David, & Sons *Engineers, &c.*
Rutherford, A., & Co. *Yacht Builders, &c.* Birkenhead.
Salvage Association
Stevenson, J., & Co. *Marine Insurance Brokers (2 Copies)*
Taylor, Tregent & Co. *Paint Manufacturers Oil Merchants, &c.*
Turner & Co. *Merchants*
Vivian & Sons *Metal Merchants*
Wallace Brothers *Coal Merchants*
Webb, W. R. *Underwriter*
Williamson, Milligan & Co. *Steamship Owners*
Wilson, Son & Co. *Steamship and Insurance Agents*

LLANELLY.

Rees, D.

LONDONDERRY.

Corbett, John L.

MADRAS.

Master Attendant

M. GLADBACH.

Kley, W.

MADRID.

Biblioteca Direction de Hidrografia

MALAGA.

Heredia, Hijos de M. A. *Merchants*

MALTA.

Agius, E. T., & Co. *Steamship Agents*
Gollcher, O. F., & Sons *Merchants and Shipowners; Lloyd's Agents; Consuls for Sweden and Norway, and for the Netherlands* ... ... Valetta

MANCHESTER.

Thames and Mersey Marine Insurance Company (*Ld.*)
Simpson, Howden & Co. *Ship and Insurance Brokers* ... ... 15, Cross Street

MANILA.

Capitania del Puerto de Manila y Cavité
Ker & Co.
Manila Slipway Co. (El Varadero de Manila)
Morris & Wright *Shipbrokers*
Schultz, John F. *Shipchandler*
Smith, Bell & Co. *East India Merchants*

MANNHEIM.

Badische Rück und Mitversicherungs Gesellschaft.
Rhederei Independent

MARSEILLES.

Alexander & Carr ... ... Rue Beauvau 4
Courtes, J. *Insurance Broker* ... ... Rue St. Ferreol
Messageries Maritimes de France ... ... Quai de la Joliette 2

MARYPORT.

Hine Bros. *Shipowners, Ship and Insurance Agents*

MELBOURNE.

Southern Insurance Company (*Ld.*)
De Beer, S., & Co.
Sloman, M. I. *Shipping, Insurance and General Broker* 14, Market Buildings Flinders Lane West

MERSYNA.

Tattarachie, N. *Merchant*

MESSINA.

Peirce, Becker & Ilardi
Rickards, G. H., & Son *Steamship Agents and Shipbrokers, &c.*

MIDDLESBROUGH.

Dixon (Raylton) & Co. *Iron and Steel Shipbuilders and Engineers* ... ... Cleveland Dockyard

Harris, E., & Co. *Steamship Brokers and Lloyd's Agents*
Hogg & Henderson
McArthur, D. *Contractor and Stevedore* ... 104, Lower Commercial Street

MILFORD HAVEN.

Kelway, H., & Sons *Ship and Insurance Agents, Steam Coal Merchants, Foreign Consuls, &c.*
Milford Docks Company ... ... Dock Proprietors
Phillips, J. *Lloyd's Agent and Ship and Insurance Agent*

MONTREAL.

Nicoll, Arch. ... ... 157, St. James' Street
Port Warden's Office

NAGASAKI.

Mitsui Bussan Kaishia *Commission Agents*
Nagasaki Dockyard and Engine Works

NATAL.

Natal Harbour Board (*London Agent*, W. Peace, 21, Finsbury Circus)

NEGAPATAM.

Ganapathy, Pillay & Co. *Merchants*

NEW YORK.

Atlantic Mutual Insurance Company 51, Wall Street
British and Foreign Marine Insurance Company (*Ld.*)
Chubb & Son *Marine Insurance Agents* ... 77, Beaver Street
Handren & Robins *Engineers and Dry Dock Owners* ... ... 126, Washington Street
Miners Oil Supply Co. (*Ld.*) *Ship Store Merchants* ... ... 189, Pearl Street
Satterthwaite & Platt *Marine Insurance Agents* ... ... 16, Exchange Place
Wreaks & Chubb *Average Adjusters and Insurance Brokers* ... ... 16, Exchange Place

NEWCASTLE, N.S.W.

Brown, A., & Co.
Russell, James
Wallace, R. B.

NEWCASTLE-ON-TYNE.

Newcastle Commercial Exchange Co. (*Ld.*) (*Manager*, William Miller; *Secretary*, B. Plummer, junr.)
Agius & Co. *Foreign Coal Exporters* ... 1, Cail's Buildings
Angus, George, & Co. *Leather and India-Rubber Merchants* ... ... Grainger Street
Armstrong, Sir W. G., Mitchell & Co. (*Ld.*) *Shipbuilders* ... ... Walker Shipyard
Bell, Robert, & Co. *Steamship Owners and Brokers, Coal Exporters, &c.* ... ... Quayside
Borries, Craig & Co. *Shipbrokers* ... ... Lombard Street
Clark, Chapman, Parsons & Co. *Engineers* Victoria Works, Gateshead
Cleland, William, & Co. (*Ld.*) *Engineers and Ship Repairers* ... ... Willington Quay
De Hart, J. *Merchant and Shipowner* ... Maritime Buildings
Dixon & Corbitt, & R. S. Newall & Co. (*Ld.*) *Hemp and Wire Rope Manufacturers* Teams Rope Works, Gateshead
Dobson, William, & Co. *Iron and Steel Shipbuilders* ... ... Low Walker
Forster, J. J. & C. M. *Steamship Owners and Brokers, Coke and Coal Exporters* ... Quayside
Furness, Christopher *Steamship Owner and Loading Broker*
Haggie (R. Hood & Son) *Patent Hemp and Wire Rope Manufacturers*
Hawthorn, R. & W., Leslie & Co. (*Ld.*) *Engineers and Shipbuilders* ... ... Hebburn Shipyard
Hawthorn, R. & W., Leslie & Co. (*Ld.*) *Engineers and Shipbuilders* ... ... St. Peter's Works
Palmers' Shipbuilding & Iron Co. (*Ld.*) *Engineers and Shipbuilders* ... ... Jarrow
Reid, George *Belgian Consul* ... ... Queen Street
Ridley, John, Son & Tully *Merchants Ship Owners and Brokers* ... ... Exchange Buildings
Schlesinger, Davis & Co. *Iron and Steel Shipbuilders* ... ... Wallsend
Scott, Ernest, & Co. *Engineers* ... ... Close Works
Sharp & Co. *Steamship Owners and Shipbrokers* ... ... King Street
Simpson, Spence & Young *Steamship Agents and Brokers* ... ... Quayside
Smith, Thomas & William *Shipbuilders, &c.*
South Derwent Colliery Office (Douglas Cooper) *Colliery Owners and Exporters* 8, Lombard Street
Spencer, John, & Sons (*Ld.*) *Steel Manufacturers* ... ... Newburn Steel Works
Spencer, John, & Sons (*Ld.*) *Iron and Steel Crankshaft Manufacturers* ... ... Ouseburn Forge
Stephenson, Robert, & Co. (*Ld.*) *Engineers and Shipbuilders*
Swan, C. S., & Hunter *Shipbuilders and Engineers* ... ... Wallsend
Thorburn, John *Coal Fitter* ... ... King Street
Tyne Boiler Works Co. (*Ld.*) *Boilermakers*
Tyne Pontoons & Dry Docks Co. (*Ld.*), *Steamship Repairers, Engineers and Boilermakers* ... ... Wallsend
Wallsend Slipway & Engineering Co. (*Ld.*) *Engineers, Boiler Makers, Ship Repairers, etc.* Wallsend

NEWPORT, MON.

Ebbw Vale Steel Iron and Coal Co. (*Ld.*)
Evans, W., & Son *Engineers* ... ... Old Dock
Lancaster, Speir & Co. *Colliery Proprietors* Russell's Wharf
Lang & Williamson *Ship Repairers and Engineers* ... ... Brithdir Wharf
Laurie, L. G., & Co. *Engineers and Ship Repairers* ... ... Mill Parade
Mordey, Jones & Co. *Shipbrokers and Coal Factors* ... ... Dock Street
Moses, John, & Co. *Merchants and Shipbrokers* ... ... Ruperra Street
Pyman, Watson & Co. *Merchants and Shipbrokers* ... ... Dock Street
Sherwyn, J. *Engineer* ... ... Alexandra Dock
Sparke, G. *Iron and Coal Agent* ... Tredegar Wharf
Thomas, E. R. ... ... Dock Street
Thompson, G. W. *Ship Repairer* ... ... Mill Parade

Wilks & Huskisson *Engineers* ... ... Alexandra Dock
Williams, W. E. *Coal Exporter and Shipbroker* ... ... Old Dock
Williams, W., & Sons *Shipwrights*

NORTH SHIELDS.

Brown, P. *Shipowner and Shipbroker* ... Newquay
Hepple & Co. *Engineers and Iron Shipbuilders* ... ... Slipway
Otto, George *Steamship Owner and Ship Store Merchant* ... ... Borough Road

ODESSA.

Goldsmith, John E. *Merchant* ... ... 3, Rue Nadjenskia
Tolstoy, Comte Constantin
Trapani, Alexis *General Commission Agent*

OPORTO.

British Association
Coverley, Charles, & Co.
Tait, W. C., & Co. *Royal Mail Steam Packet Co., Compie. Bordelaise de Nav. à Vap. Adria Hungarian Sea Nav. Co.*

OSCARSHAMN.

Lundberg, A. *Shipbroker*

PARIS.

Union Insurance Society of Canton (*Ld.*)
Bureau Veritas, Administration of,
Collin, Emile... ... ... Rue Auber 10
Duprat, Joseph
Marin-Dubuard, Baron ... ... Rue de Seine 37
Ministère de la Marine
Moutier, C.
Société Commerciale d'Affretements et de Commission ... ... Rue Lafayette 7

PEMBROKE DOCK.

Long, Captain, R.N. ... ... H.M. Dockyard

PERNAMBUCO.

Boxwell, J. H. *Merchant*

PETERHEAD.

McCombie, James & Co. *Shipowners*

PHILADELPHIA.

Insurance Company of North America *Fire and Marine Insurance* ... ... 232, Walnut Street
Cramp, The William, & Sons, Ship and Engine Building Co. ... ... Beach and Ball Streets

PIRÆUS.

Glamorgan Coal Company (*Ld.*)
McDowall & Barbour *Shipbuilders and Engineers* ... ... Vulcan Engine Works

PLYMOUTH.

Bellamy & Co. *Ship Agents and Owners*
Burnard & Alger *Warehousemen, Wharfingers, Chemical Manure Manufacturers, &c.*
Collier Brothers *Lloyd's Agents, Ship and Steamship Agents* ... ... 53, Southside Street
Smith, Sundius & Co. *Ship and Insurance Brokers* ... ... 11, Millbay Road
Weekes, Phillips & Co.

POLA, AUSTRIA.

Redaction der Mittheilungen aus dem Gebiete des Seewesens *Austrian Navy Technical Journal*

PORT ADELAIDE.

Harrison, H. C., & Co.

PORT ELIZABETH.

Chamber of Commerce
Mackie, Dunn & Co.

PORT GLASGOW.

Duncan, Robert, & Co. *Shipbuilders*
Hamilton, W., & Co. *Shipbuilders and Shipowners*
Reid, J., & Co. *Shipbuilders*
Russell & Co. *Shipbuilders*

PORT LOUIS, MAURITIUS.

Colonial Maritime Insurance Company of Mauritius.

PORT SAID.

Azzopardi, F., & Co.
Carara, John, & Sons *Steam Coal Merchants*

PORTLAND, OREGON.

Brown & McCabe *Stevedores* ... ... 25, B Street

PORTSMOUTH.

Read, J., junr. *Ship and Boat Builder*

QUEBEC.

Allan, Rae & Co. *Steamship Agents*
Davie, George T. *Shipbuilder* ... ... Levis
Dobell, R. R., & Co. *Lumber Merchants*
Hansen & Schwartz *Shipbrokers*
Price Bros. & Co. *Lumber Merchants* ... Corner of Peter and Paul Streets
Quebec Steamship Co.
Smith, Wade & Co. *Lumber Merchants*

QUEENSTOWN.

Clyde Shipping Co.
Cummins, N. & I., & Bros. *Lloyd's Agents*
Dawson, J., & Co. *Shipping Agents*
Harbour Master (Captain Usborne)
Scott, James, & Co. *Merchants and Ship Agents*

RANGOON.

Bombay-Burmah Trading Corporation (*Ld.*)
Coin, C., & Co. *Stevedores, Purveyors, &c.* ... 18, Sparks Street
Hood, John, & Co. *Stevedores and Naval Contractors* ... ... 8, Barr Street
Irrawaddy Flotilia Co. (*Ld.*) *Inland Carriers* ... ... Phayre Street
Jones, Edmund, & Co. *Merchants* ... ... 8, Strand
Mahomay Brothers *Stevedores, Agents, &c.* .. 2, Phayre Street
Port Commissioners (Captain H. S. Ransome, *Assistant Conservator*)

RAROTONGA.

Exham, R. *Lloyd's Agent*

REVAL.

Cattley, C. R. *Lloyd's Agent*

RIGA.

RIGA EXCHANGE COMMITTEE
Bornholdt, P., & Co. *Steamship Agents*
Frisk, Aug. *Shipbroker*
Helmsing & Grimm *Steamship Agents and Owners*
Mitchell & Co. *Merchants and Lloyd's Agents*
Spinck, J., & Co.

RIO DE JANEIRO.

Bellamy, J. H., & Co. *Merchants*

ROME.

Minister of Marine
Navigazione Generale Italiana ... ... Via del Corso 385

RONNE.

Olsen, N. Chr. *Attorney*

ROSTOCK.

Rostocker Actien-Ges. für Schiff und Maschinenbau

ROTTERDAM.

Nederlandsche Amerikaansche Stoomvaart Maatschappij ... ... Prins Hendrikskade
Ommeren, P. H. van *Shipbroker*

ROUBAIX.

Bossut-Plichon, Monsieur

ROUEN.

Lemoine, A.

ST. MICHAEL'S (AZORES).

Bensaude & Co. *Merchants*

ST. PETERSBURG.

Bornholdt, P., & Co. *Steamship Agents*
Brown, Emil *Chartering Agent*
Mörch, P. *Steamship Owner and Broker*
Serck, P. *Shipbroker*

ST. THOMAS, W.I.

Cameron & Co. *Lloyd's Agents, &c.*

SAN FRANCISCO.

ANGLO-NEVADA ASSURANCE CORPORATION 410, Pine Street.
FIREMAN'S FUND INSURANCE CO. ... ... 401, California Street
UNION INSURANCE CO. ... ... 416, California Street
Balfour, Guthrie & Co. *Merchants* ... ... 306, California Street
Blum, Baldwin & Girvin *Merchants* ... 307, California Street
Catton, Bell & Co. *Merchants and Lloyd's Agents* ... ... 106, California Street
Fulton Iron Works ... ... 213, Fremont Street
Middleton & Co. *Steel Merchants* ... ... 213, Jackson Street
Parrott & Co. *Merchants* ... ... 306, California Street
Pope & Talbot *Lumber Merchants* ... ... 204, California Street
Rosenfeld's Sons, J. *Merchants*... ... ... 202, Sansome Street
Shipowners' and Merchants' Tug Co. ... 302, California Street
Union Iron Works *Marine Engineers and Shipbuilders* Corner of First & Mis[sion] Streets

Welch & Co. *Shipping and Commission Merchants* ... ... 220, California Street

SAVONA.

Bandini & Bozano *Steamship Agents; Agents for Navigazione Generale Italiana* Via Guarda Inferiore

SEBASTOPOL.

Carassarini, A., & Co.
Grierson, C.

SESTRI PONENTE.

Ansaldo, G., & Co.

SHANGHAI.

SHANGHAI CLUB
Baessler, John *Shipowner*
Boyd & Co. *Engineers and Shipbuilders*
Farnham, S. C., & Co. *Engineers and Shipbuilders*
Gibb, Livingston & Co. *Lloyd's Agents, Merchants and Shipping Agents*
Mitsui Bussan Kaishia, *Commission Agents* (Y. Wooyeda, *Manager*)
Roberts, John P. *Marine Surveyor*
Russell & Co. *Merchants*
Siemssen & Co. *General Merchants*
Sönne, H. *Engineer Surveyor to Lloyd's Register, &c.*

SHEFFIELD.

Cammell, Chas., & Co. *Steel Manufacturers*
Vickers, Sons & Co. (*Ld.*), *Steel Makers* And 50, Old Broad St[reet], London

SINGAPORE.

STRAITS INSURANCE CO. (*Ld.*)

SMYRNA.

Imperial Ottoman Bank

SOURABAYA.

Daendels, J., & Co. *Shipbrokers*
De Wit Schipper & Co. *Merchants*
Fraser, Eaton & Co. *Merchants*
Houghton & Co. *Lloyd's Agents, Merchants, &c.*
Leeuwen, J. F. van, & Co. *Merchants and Shipbrokers*

Nieuw Praamwerver ... ... Boat Office.
"Rotterdam" Internationale Crediet & Handels Vereeniging *Merchants*
Rubaak & Co. *Ship Chandlers*
Someren, W. van, Greve & Co. *Ship Chandlers*
Wellenstein, Krause & Co. *Merchants*

SOUTH SHIELDS.

Mercantile Dry Dock Co. (*Ld.*) *Ship Repairers and Graving Dock Owners* ... Tyne Dock
Middle Dock Co. *Ship Repairers, Graving Dock Owners, &c.*
Mordue Brothers *Iron and Wood Ship Repairers, and Dry Docks at Thrift Street, Holborn, and Tyne Dock*

SOUTHAMPTON.

Bee, H., & Son. *Coal Merchants, Stevedores and Contractors* ... ... Docks
Harper, G. T. *Agent to Powell-Duffryn Coal Co., Shipowner & Shipbroker* ... ... Oriental Place
Hedger, P. *Secretary, Southern Dock Co.*
Oswald, Mordaunt & Co. *Engineers and Shipbuilders* ... ... Woolston

STANLEY (FALKLAND ISLANDS).

Cobb, F. E. *Lloyd's Agent; Colonial Manager of the Falkland Islands Co.*

STAVANGER.

Stavanger Stöberi & dok. *Shipbuilders*

STOCKHOLM.

Agir Marine Insurance Co. (*Ld.*)
Bergsunds Mekaniska Werkstadsaktiebolag *Shipbuilders and Engineers* ... ... Finnboda Slip
Ferlin, Oscar ... ... 6, Vasagatan
Lindberg's Werkstad *Shipyard and Engineering Works*
"Neptun" Bergnings och Dykeribolaget *Salvage and Diving Company*
Olson & Wright *Shipbrokers and Steamship Agents*
Royal Marine Department

STOCKTON-ON-TEES.

Amos, J. H. *Secretary to the River Tees Conservancy Commissioners*
Blair & Co. (*Ld.*) *Marine Engineers*

SULINA.

Watson & Youell *Steamship Brokers*

SUNDERLAND.

Austin, S. P., & Son *Iron and Steel Shipbuilders and Repairers* ... ... Wear Dockyard
Clark, George, *Engineer* ... ... Southwick Engine Works
Dickinson, John *Marine Engineer* ... ... Palmer's Hill Engine Works
Nautilus Steam Shipping Co. (*Ld.*) *Steamship Owners* ... ... 54, Fawcett Street
Swainston, George, & Co. *Brokers for the Sale and Charter of Ships & Steamers, Engineer Surveyors, Arbitrators & Valuers* West Sunniside
Thompson, Joseph L., & Sons *Iron and Steel Shipbuilders and Repairers* ... ... North Sands Shipyard
Westoll, James *Shipowner* ... ... 12, John Street

SWANSEA.

Dixon, John *Harbour Superintendent*
Swansea Dry Docks and Engineering Co. (*Ld.*)

SYDNEY.

Sydney Exchange (Chairman, Alexander Campbell)
Cruickshank, W. D. *Chief Engineer Surveyor to the New South Wales Government* ... ... Marine Board Office

TABLE BAY.

Harbour Board

TACOMA, W. T.

Puget Sound Stevedoring Co. *Stevedores* ... Box F

TAGANROG.

Hoyland, Wooldridge & Co. (Successors of Hoyland Brothers, 1868)

TAHITI.

Young, J. Lyle ... ... Papeete

TOKIO.

Masuda, K.
Nippon Yusen Kaisha (Japan Mail Steamship Company; *Managing Director,* A. R. Browne)

TORONTO.

Western Assurance Company (J. J. Kenny, *Managing Director*)

TOWNSVILLE.

Burns, Philp & Co.
Wright Heaton & Co.

TRAVANCORE.

Commercial Agency.

TRIESTE.

Arsenal of Lloyd Austro-Ungarico
Dobrillovich, E. V. *Steamship Agent*
Ptack, L. *Naval Director of Lloyd Austro-Ungarico*

VALPARAISO.

Arauco, R. C.
Compañia de Diques Flotantes de Valparaiso Box 356
Compañia Esplotadora de Lota y Coronel *Coal Miners, Copper Smelters, Shipowners, &c.*
Compañia Nacional de Remoleadors (Manager, P. A. McKellar)
Grace & Co. *Merchants*
Hazen & Compton *Merchants and Marine Insurance Agents*
James, Inglis & Co. *Merchants*
Lyon, C. *Shipbroker*

WEST HARTLEPOOL.

Hudson, T. S. *Merchant and Shipowner*
Sinclair, John *Steamship Owner and Broker* 4, Church Street
Withy, E., & Co. *Iron and Steel Shipbuilders and Ship Repairers* ... ... Middleton Ship Yard

WEYMOUTH.

Roberts, W., & Co. *Shipbrokers*

WHITEHAVEN.

Gibson & Greenop *Ship and Insurance Brokers*
Howie, H. *Shipowner and Commission Agent*
Whitehaven Shipbuilding Co. (*Ld.*)

YARMOUTH.

Butcher, Matthew *Merchant, Shipping Agent and Insurance Broker*

YOKOHAMA, JAPAN.

Bernard & Wood
Fraser, Farley & Co. *Merchants*
Jardine, Matheson & Co. *Merchants*
Raspe, M., & Co. *Merchants*

STATISTICAL TABLES.

TABLES STATISTIQUES.

LIST OF STATISTICAL TABLES.

Note.—As regards Tonnage, these Tables are throughout based on the Net Tonnage of Sailing Vessels and the Gross Tonnage of Steamers.

LISTE DES TABLES STATISTIQUES.

NOTE.—Pour le tonnage, ces tables sont entièrement basées sur le tonnage net quand il s'agit de Navires à Voiles, et sur le tonnage total pour les Navires à Vapeur.

Table No. 1.—Nombre, tonnage et description des **navires à vapeur** de 100 tonneaux bruts et au-dessus, **appartenant aux divers pays du monde,** ainsi que les renseigne l' "Universal Register."

Table No. 2.—Nombre, tonnage et description des **navires à voiles** de 100 tonneaux nets et au-dessus, **appartenant aux divers pays du monde,** ainsi que les renseigne l' "Universal Register."

Sommaire des Tables Nos. 1 et 2.

Table No. 3.—Nombre, tonnage et description des navires ajoutés dans l'édition actuelle, par suite d'achat ou de construction, aux marines marchandes anglaise, française, allemande, italienne, norvégienne, suédoise, ainsi que le renseigne l' "Universal Register," avec indication des lieux de construction.

Table No. 3a.—Nombre, tonnage et description des navires déduits dans l'édition actuelle, par suite de naufrage, dépècement, vente à l'étranger, etc., des marines marchandes anglaise, française, allemande, italienne, norvégienne, et suédoise, ainsi que les renseigne l' "Universal Register."

Table No. 4.—Nombre, tonnage et description de tous les navires de 100 tonneaux et au-dessus, **construits dans chacun des divers pays du monde,** et encore existant, ainsi que les renseigne l' "Universal Register."

Table No. 5.—Nombre, tonnage et description de tous les navires de 100 tonneaux et au-dessus, **construits dans les principaux pays de construction maritime,** et encore existant, ainsi que les renseigne l' "Universal Register;" avec désignation respective (1) de ceux qui appartiennent au pays où ils ont été construits, (2) de ceux qui appartiennent au Royaume-Uni, et (3) de ceux qui appartiennent à d'autres pays.

Table No. 6.—Indiquant les **pays auxquels appartiennent actuellement les navires** de 100 tonneaux et au-dessus, **construits dans le Royaume-Uni,** et encore existant; et indiquant aussi le nombre, le tonnage et la description de ces navires, ainsi que les renseigne l' "Universal Register."

Table No. 7.—Nombre, tonnage et description de tous les navires de 100 tonneaux et au-dessus, **construits par chacun des divers pays du monde pendant l'année 1888,** ainsi que les renseigne l' "Universal Register."

Analyse de la Table No. 6, indiquant les pays pour lesquels les navires renseignés ont été construits.

Table No. 8.—Nombre, tonnage et description de tous les navires **classés par chacune des principales sociétés de classification du monde.**

Table No. 9.—Nombre, tonnage et description des navires de 100 tonneaux et au-dessus **construits en 1888, et classés par le Lloyd's Register,** comparés à ceux construits dans le monde et dans le Royaume-Uni pendant la même année.

[T]ABLE No. 3.—Showing Number, Description and Tonnage of Vessels added, by new construction [illegible] in the present Edition of the "Universal Register."

Country	Where Built.	Whether Steam or Sail.	New Vessels. Wood and Composite. No.	Tons.	Iron. No.	Tons.	Steel. No.	Tons.	Total. No.	Tons.	Bought from Foreigners, &c. Wood and Composite. No.	Tons.	Iron. No.	Tons.	Steel. No.	Tons.	Total. No.	Tons.	Grand T[otal]. No.	Tons.
UNITED KINGDOM.	Built at Home	Sail	10	1,177	10	14,782	42	64,204	62	80,163	2	705	2	3,677	..	..	4	4,382	66	84,[illegible]
		Steam	3	1,663	63	34,050	304	625,916	370	661,629	..	..	25	38,817	6	12,771	31	51,588	401	713,2[illegible]
		Total	13	2,840	73	48,832	346	690,120	432	741,792	2	705	27	42,494	6	12,771	35	55,970	467	797,7[illegible]
	Foreign Built	Sail	1	314	1	547	..	..	2	861	7	2,037	2	724	..	..	9	2,761	11	[illegible]
		Steam	..	..	1	108	..	..	1	108	..	..	..	..	..	..	..	..	1	[illegible]
		Total	1	314	2	655	..	..	3	969	7	2,037	2	724	..	..	9	2,761	12	3,7[illegible]
	Total	Sail	11	1,491	11	15,329	42	64,204	64	81,024	9	2,742	4	4,401	..	..	13	7,143	77	88,[illegible]
		Steam	3	1,663	64	34,158	304	625,916	371	661,737	..	..	25	38,847	6	12,771	31	51,588	402	713,[illegible]
		Grand Total	14	3,154	75	49,487	346	690,120	435	742,761	9	2,742	29	43,218	6	12,771	44	58,731	479	801,4[illegible]
FRANCE.	Built at Home	Sail	5	1,098	..	..	..	..	5	1,098	1	106	..	..	..	..	1	106	6	1,[illegible]
		Steam	..	..	3	486	10	16,995	13	17,481	..	..	..	..	..	..	..	..	13	17,4[illegible]
		Total	5	1,098	3	486	10	16,995	18	18,579	1	106	..	..	..	..	1	106	19	18,6[illegible]
	Built in the United Kingdom	Sail	..	..	1	1,054	3	6,936	4	7,990	7	2,089	2	1,496	..	..	9	3,585	13	11,[illegible]
		Steam	..	..	..	..	5	5,140	5	5,140	..	..	8	12,211	2	706	10	12,917	15	18,0[illegible]
		Total	..	..	1	1,054	8	12,076	9	13,130	7	2,089	10	13,707	2	706	19	16,502	28	29,6[illegible]
	Built in other Foreign Countries	Sail	..	..	..	..	..	..	..	..	19	6,183	..	..	..	..	19	6,183	19	6,1[illegible]
		Steam	..	..	..	..	..	..	..	..	..	..	2	558	..	..	2	558	2	[illegible]
		Total	..	..	..	..	..	..	..	..	19	6,183	2	558	..	..	21	6,741	21	6,74[illegible]
	Total	Sail	5	1,098	1	1,054	3	6,936	9	9,088	27	8,378	2	1,496	..	..	29	9,874	38	18,9[illegible]
		Steam	..	..	3	486	15	22,135	18	22,621	..	..	10	12,769	2	706	12	13,475	30	36,0[illegible]
		Grand Total	5	1,098	4	1,540	18	29,071	27	31,709	27	8,378	12	14,265	2	706	41	23,349	68	55,0[illegible]
GERMANY.	Built at Home	Sail	1	1,663	1	1,462	3	4,505	5	7,630	..	..	..	..	..	..	..	..	5	7,6[illegible]
		Steam	..	..	5	2,042	22	27,495	27	29,537	..	..	..	..	..	..	..	..	27	29,[illegible]
		Total	1	1,663	6	3,504	25	32,000	32	37,167	..	..	..	..	..	..	..	..	32	37,16[illegible]
	Built in the United Kingdom	Sail	..	..	2	3,311	8	13,584	10	16,895	2	997	24	23,257	..	..	26	24,254	36	41,1[illegible]
		Steam	..	..	..	..	12	33,135	12	33,135	..	..	12	10,742	1	2,454	13	13,196	25	46,[illegible]
		Total	..	..	2	3,311	20	46,719	22	50,030	2	997	36	33,999	1	2,454	39	37,450	61	87,4[illegible]
	Built in other Foreign Countries	Sail	..	..	..	..	..	..	..	..	10	4,323	..	..	..	..	10	4,323	10	4,3[illegible]
		Steam	..	..	1	611	..	..	1	611	..	..	1	628	1	895	2	1,523	3	2,1[illegible]
		Total	..	..	1	611	..	..	1	611	10	4,323	1	628	1	895	12	5,846	13	6,45[illegible]
	Total	Sail	1	1,663	3	4,773	11	18,089	15	24,525	12	5,320	24	23,257	..	..	36	28,577	51	53,10[illegible]
		Steam	..	..	6	2,653	34	60,630	40	63,283	..	..	13	11,370	2	3,349	15	14,719	55	78,0[illegible]
		Grand Total	1	1,663	9	7,426	45	78,719	55	87,808	12	5,320	37	34,627	2	3,349	51	43,296	106	131,104

[T]ABLE No. 3a.—Showing Number, Description, and Tonnage of Vessels, deducted owing to loss, breakin[g] [illegible] Norway, and Sweden, in the presen[t] [illegible]

Country.	Whether Steam or Sail.	Lost, Broken Up, &c. Wood and Composite. No.	Tons.	Iron. No.	Tons.	Steel. No.	Tons.	Total. No.	Tons.	Sold Foreign. Wood and Composite. No.	Tons.	Iron. No.	Tons.	Steel. No.	Tons.	Total. No.	Tons.	Grand Total. No.	Tons.
UNITED KINGDOM	Sail	150	62,330	59	60,534	3	4,110	212	126,974	97	60,136	28	22,756	..	..	125	82,892	337	209,86[illegible]
	Steam	7	1,150	144	156,625	6	6,827	157	161,602	1	239	69	74,267	21	22,304	91	96,810	248	264,41[illegible]
	Total	157	63,480	203	217,159	9	10,937	369	291,576	98	60,375	97	97,023	21	22,304	216	179,702	585	471,27[illegible]
FRANCE	Sail	88	24,134	3	2,030	..	..	91	26,164	20	9,626	4	1,931	..	..	24	11,557	115	37,72[illegible]
	Steam	..	..	11	13,107	4	4,051	15	17,158	..	..	10	14,779	1	519	11	15,298	26	32,4[illegible]
	Total	88	24,134	14	15,137	4	4,051	106	43,322	20	9,626	14	16,710	1	519	35	26,855	141	70,17[illegible]
GERMANY	Sail	137	50,217	5	3,343	..	..	142	53,560	93	41,150	7	3,083	..	..	[illegible]	[illegible]	242	[illegible]
	Steam	..	..	9	7,204	1	121	10	7,325	..	..	6	7,381	..	..	6	7,381	16	[illegible]
	Total	137	50,217	1[illegible]	10,547	1	121	152	60,885	93	41,150	13	10,464	..	..	109	51,614	258	112,4[illegible]

…hase, to the Merchant Navies of the United Kingdom, France, Germany, Italy, Norway and Sweden, …stinguishing also the Place of Build.

Where Built	Whether Steam or Sail	New Vessels: Wood and Composite No.	Tons	Iron No.	Tons	Steel No.	Tons	Total No.	Tons	Bought from Foreigners, &c.: Wood and Composite No.	Tons	Iron No.	Tons	Steel No.	Tons	Total No.	Tons	Grand Total No.	Tons
Built at Home	Sail	6	768	1	157	..	..	7	925	1	127	..	..	..	..	1	127	8	1,052
	Steam			2	432	1	133	3	565	..	..	..	..	..	..	..	..	3	565
	Total	6	768	3	589	1	133	10	1,490	1	127	..	..	..	..	1	127	11	1,617
Built in the United Kingdom	Sail									2	627	..	..	..	..	2	627	2	627
	Steam	..	..	..	..	5	9,202	5	9,202	..	..	9	17,797	..	..	9	17,797	14	26,999
	Total	..	..	..	..	5	9,202	5	9,202	2	627	9	17,797	..	..	11	18,424	16	27,626
Built in other Foreign Countries	Sail	..	..	..	..	..	..	..	..	1	150	..	..	..	..	1	150	1	150
	Steam	..	..	..	..	..	..	..	..	..	..	3	3,674	..	..	3	3,674	3	3,674
	Total	..	..	..	..	..	..	..	..	1	150	3	3,674	..	..	4	3,824	4	3,824
Total	Sail	6	768	1	157	..	..	7	925	4	901	..	..	..	..	4	901	11	1,829
	Steam	..	..	2	432	6	9,335	8	9,767	..	..	12	21,471	..	..	12	21,471	20	31,238
	Grand Total	6	768	3	589	6	9,335	15	10,692	4	904	12	21,471	..	..	16	22,375	31	33,067
Built at Home	Sail	16	9,107	..	..	..	..	16	9,107	2	503	..	..	..	..	2	503	18	9,610
	Steam	..	..	3	2,630	1	112	4	2,742	..	..	..	..	..	..	..	..	4	2,742
	Total	16	9,107	3	2,630	1	112	20	11,849	2	503	..	..	..	..	2	503	22	12,352
Built in the United Kingdom	Sail	..	..	..	..	..	..	..	..	30	12,384	1	890	..	..	31	13,274	31	13,274
	Steam	..	..	2	1,350	7	12,153	9	13,503	..	..	12	7,730	..	..	12	7,730	21	21,233
	Total	..	..	2	1,350	7	12,153	9	13,503	30	12,384	13	8,620	..	..	43	21,004	52	34,507
Built in other Foreign Countries	Sail	..	..	..	..	..	..	..	..	114	83,514	..	..	..	..	114	83,514	114	83,514
	Steam	..	..	..	..	..	..	..	..	4	1,917	2	975	1	401	7	3,293	7	3,293
	Total	..	..	..	..	..	..	..	..	118	85,431	2	975	1	401	121	86,807	121	86,807
Total	Sail	16	9,107	..	..	..	..	16	9,107	148	96,401	1	890	..	..	147	97,291	163	106,398
	Steam	..	..	5	3,980	8	12,265	13	16,245	4	1,917	14	8,705	1	401	19	11,023	32	27,268
	Grand Total	16	9,107	5	3,980	8	12,265	29	25,352	150	98,318	15	9,595	1	401	166	108,314	195	133,666
Built at Home	Sail	4	1,115	..	..	..	..	4	1,115	1	139	..	..	..	..	1	139	5	1,254
	Steam	1	184	6	1,184	2	330	9	1,698	..	..	..	..	..	..	..	..	9	1,698
	Total	5	1,299	6	1,184	2	330	13	2,813	1	139	..	..	..	..	1	139	14	2,952
Built in the United Kingdom	Sail	..	..	..	..	..	..	..	..	5	2,546	..	..	..	..	5	2,546	5	2,546
	Steam	..	..	..	..	1	765	1	765	..	..	5	7,304	..	..	5	7,304	6	8,069
	Total	..	..	..	..	1	765	1	765	5	2,546	5	7,304	..	..	10	9,850	11	10,615
Built in other Foreign Countries	Sail	..	..	..	..	..	..	..	..	35	11,563	..	..	..	..	35	11,563	35	11,563
	Steam	..	..	..	..	..	..	..	..	..	..	1	169	..	..	1	169	1	169
	Total	..	..	..	..	..	..	..	..	35	11,563	1	169			36	11,732	36	11,732
Total	Sail	4	1,115	..	..	..	..	4	1,115	41	14,248	..	..	..	..	41	14,248	45	15,363
	Steam	1	184	6	1,184	3	1,095	10	2,463	..	..	6	7,473	..	..	6	7,473	16	9,936
	Grand Total	5	1,299	6	1,184	3	1,095	14	3,578	41	14,248	6	7,473	..	..	47	21,721	61	25,299

…sale to Foreigners, &c., from the Merchant Navies of the United Kingdom, France, Germany, Italy, …ion of the "Universal Register."

…Country	Whether Steam or Sail	Lost, Broken Up, &c.: Wood and Composite No.	Tons	Iron No.	Tons	Steel No.	Tons	Total No.	Tons	Sold Foreign: Wood and Composite No.	Tons	Iron No.	Tons	Steel No.	Tons	Total No.	Tons	Grand Total No.	Tons
…aly	Sail	115	46,011	..	..	..	..	115	46,011	34	12,075	..	..	..	..	34	12,075	149	58,086
	Steam	..	..	6	6,665	..	..	6	6,665	..	..	5	6,697	..	..	5	6,697	11	13,362
	Total	115	46,011	6	6,665	..	..	121	52,676	34	12,075	5	6,697	..	..	39	18,772	160	71,448
…orway	Sail	248	99,077	..	..	..	..	248	99,077	12	5,130	..	..	..	..	12	5,130	260	104,207
	Steam	..	..	5	2,712	..	..	5	2,712	..	..	2	739	..	..	2	739	7	3,451
	Total	248	99,077	5	2,712	..	..	253	101,789	12	5,130	2	739	..	..	14	5,869	267	107,658
…eden	Sail	45	12,739	..	..	..	..	45	12,739	16	4,428	..	..	..	..	16	4,428	61	17,267
	Steam	1	368	1	186	..	..	2	554	6	2,500	5	2,048	3	1,662	14	6,270	16	6,724
	Total	46	13,107	1	186			47	13,293	22	6,928	5	2,048	3	1,662	30	10,698	77	23,991

TABLE No. 4.—Showing Number, Tonnage, and Description of all Vessels of 100
record

Country in which built	Whether Steam or Sail	Wood: Under 200 tons	200 to 399	400 to 599	600 to 799	800 to 999	1000 to 1499	1500 to 1999	2000 to 2999	3000 to 3999	4000 to 4999	5000 tons & above	Total	Composite: Under 200	200 to 399	400 to 599	600 to 799	800 to 999	1000 to 1499	1500 to 1999	2000 to 2999	3000 to
United Kingdom	Sail	1186	816	238	65	47	27	..	..	..	..	..	2379	9	10	26	31	26	17	..	1	..
	Steam	139	39	18	2	1	..	..	..	..	..	..	199	3	11	8	4	1	1	..	..	..
	Total	1325	855	256	67	48	27	..	..	..	..	..	2578	14	21	34	35	27	18	..	1	..
British Colonies	Sail	1112	634	261	228	174	344	67	5	..	..	..	2825	..	1	1	..	..	..	..	..	..
	Steam	94	68	18	12	5	4	..	..	..	..	..	201	3	3	1	..	..	3	..	..	..
	Total	1206	702	279	240	179	348	67	5	..	..	..	3026	3	4	2	..	..	3	..	..	..
America (United States)	Sail	1068	1188	625	280	101	337	104	29	..	..	..	3732	..	..	..	..	..	..	..	..	..
	Steam	42	64	55	26	26	42	25	9	..	..	..	289	..	..	1	1	..	..	..	..	..
	Total	1110	1252	680	306	127	379	129	38	..	..	..	4021	..	..	1	1	..	..	..	..	..
Austro-Hungary	Sail	69	113	94	37	13	3	..	..	..	..	..	329	..	..	..	..	..	..	..	..	..
	Steam	4	1	2	..	..	..	..	..	..	..	..	7	2	..	..	..	..	..	..	..	..
	Total	73	114	96	37	13	3	..	..	..	..	..	336	2	..	..	..	..	..	..	..	..
Denmark	Sail	390	87	5	2	1	1	..	..	..	..	..	486	..	..	..	..	..	..	..	..	..
	Steam	1	1	1	..	..	..	..	..	..	..	..	3	..	..	..	..	..	..	..	..	..
	Total	391	88	6	2	1	1	..	..	..	..	..	489	..	..	..	..	..	..	..	..	..
France	Sail	484	258	75	31	5	3	1	..	..	..	..	857	..	1	3	1	..	..	..	..	..
	Steam	2	..	..	..	..	..	..	..	..	..	..	2	2	..	..	..	..	..	..	..	..
	Total	486	258	75	31	5	3	1	..	..	..	..	859	2	1	3	1	..	..	..	..	..
Germany	Sail	397	644	260	83	38	31	3	..	..	..	..	1456	1	..	4	..	..	..	..	..	..
	Steam	..	5	1	..	..	..	..	..	..	..	..	6	1	1	..	..	..	..	..	..	..
	Total	397	649	261	83	38	31	3	..	..	..	..	1462	2	1	4	..	..	..	..	..	..
Greece	Sail	434	385	87	1	..	..	..	..	..	..	..	907	..	..	..	..	..	..	..	..	..
	Steam	..	..	..	..	..	..	..	..	..	..	..	..	..	..	..	..	..	..	..	..	..
	Total	434	385	87	1	..	..	..	..	..	..	..	907	..	..	..	..	..	..	..	..	..
Holland and Belgium	Sail	152	69	22	42	18	18	..	..	..	..	..	321	..	3	..	1	1	9	..	1	..
	Steam	..	..	1	..	..	..	..	..	..	..	..	1	..	..	..	..	..	..	..	..	..
	Total	152	69	23	42	18	18	..	..	..	..	..	322	..	3	..	1	1	9	..	1	..
Italy	Sail	335	362	443	177	92	26	..	..	..	..	..	1435	..	..	..	..	..	..	..	..	..
	Steam	3	..	..	..	..	..	..	..	..	..	..	3	..	..	..	..	..	..	..	..	..
	Total	338	362	443	177	92	26	..	..	..	..	..	1438	..	..	..	..	..	..	..	..	..
Norway	Sail	324	748	441	104	34	17	..	..	..	..	..	1668	..	..	..	..	..	..	..	..	..
	Steam	7	22	15	..	..	..	..	..	..	..	..	44	..	..	..	..	..	..	..	..	..
	Total	331	770	456	104	34	17	..	..	..	..	..	1712	..	..	..	..	..	..	..	..	..
Russia	Sail	260	342	130	41	12	6	..	..	..	..	..	791	1	..	..	..	..	..	..	..	..
	Steam	1	3	1	..	..	..	..	..	..	..	..	5	2	..	..	..	..	..	..	..	..
	Total	261	345	131	41	12	6	..	..	..	..	..	796	3	..	..	..	..	..	..	..	..
Spain	Sail	295	137	21	5	3	..	..	..	..	..	..	461	..	..	..	..	..	..	..	..	..
	Steam	4	2	..	..	..	..	..	..	..	..	..	6	..	..	..	..	..	..	..	..	..
	Total	299	139	21	5	3	..	..	..	..	..	..	467	..	..	..	..	..	..	..	..	..
Sweden	Sail	321	303	105	58	10	3	..	..	..	..	..	802	1	1	..	..	..	..	..	..	..
	Steam	28	15	9	7	1	..	..	..	..	..	..	58	44	40	9	1	..	..	..	..	..
	Total	349	318	114	65	11	3	..	..	..	..	..	860	45	41	9	1	..	..	..	..	..

Note.—The figures for Sailing Vessels

...els, **BUILT** by each of the several Countries of the World, and still in existence, as ...iversal Register."

IRON.											STEEL.												GRAND TOTAL.	Whether Steam or Sail.	COUNTRY IN WHICH BUILT.
[illegible] to 399 tons	[illegible] to 599 tons	600 to 799 tons	800 to 999 tons	1000 to 1499 tons	1500 to 1999 tons	2000 to 2999 tons	3000 to 3999 tons	4000 to 4999 tons	5000 and above	Total	Under 200 tons	200 to 399 tons	400 to 599 tons	600 to 799 tons	800 to 999 tons	1000 to 1499 tons	1500 to 1999 tons	2000 to 2999 tons	3000 to 3999 tons	4000 to 4999 tons	5000 & above	Total			
100	120	209	202	547	292	96	1	..	..	1683	8	16	21	2	12	33	44	25	4	..	..	165	4347	Sail ..	
777	[illegible]	533	[illegible]	1093	911	740	223	65	18	6222	79	119	112	72	65	164	174	321	97	52	30	1285	7736	Steam.	UNITED KINGDOM.
877	748	742	[illegible]	1640	1203	836	224	65	18	7905	87	135	133	74	77	197	218	346	101	52	30	1450	12083	Total.	
[illegible]	[illegible]	[illegible]	[illegible]	[illegible]	..	..	..	..	..	37	..	..	..	..	..	..	..	..	..	..	..	..	2864	Sail ..	
[illegible]	[illegible]	[illegible]	[illegible]	[illegible]	..	..	..	..	..	70	..	4	1	..	1	..	..	1	..	..	..	7	288	Steam.	BRITISH COLONIES.
22	18	25	5	6	..	..	..	..	..	107	..	4	1	..	1	..	..	1	..	..	..	7	3152	Total.	
[illegible]	1	..	..	..	[illegible]	..	..	..	..	5	..	..	..	..	..	..	..	..	..	..	..	..	3737	Sail ..	
[illegible]	[illegible]	9	25	27	21	63	20	..	1	218	..	2	..	1	4	9	3	1	..	..	..	20	529	Steam.	AMERICA (UNITED STATES).
21	19	9	25	27	24	63	20	..	1	223	..	2	..	1	4	9	3	1	..	..	..	20	4266	Total.	
[illegible]	[illegible]	..	..	..	..	..	..	..	..	1	..	..	1	..	..	..	..	..	..	..	..	1	334	Sail ..	
[illegible]	[illegible]	4	1	10	9	3	1	..	..	66	2	..	1	1	..	1	..	2	3	..	..	10	82	Steam.	AUSTRO-HUNGARY.
21	4	4	1	10	9	3	1	..	..	67	2	..	2	1	..	1	..	2	3	..	..	11	416	Total.	
..	..	..	..	..	..	..	..	..	..	..	..	..	..	..	..	..	..	..	..	..	..	..	486	Sail ..	
17	10	3	2	16	8	2	..	..	..	67	..	2	3	..	1	6	3	..	..	..	..	15	85	Steam.	DENMARK.
17	10	3	2	16	8	2	..	..	..	67	..	2	3	..	1	6	3	..	..	..	..	15	571	Total.	
[illegible]	7	6	2	2	1	1	..	..	..	23	..	..	..	..	..	..	..	..	..	..	..	..	885	Sail ..	
21	11	10	10	19	17	22	35	11	..	176	11	6	1	1	..	3	4	14	8	..	5	53	233	Steam.	FRANCE.
25	18	16	12	21	18	23	35	11	..	199	11	6	1	1	..	3	4	14	8	..	5	53	1118	Total.	
[illegible]	[illegible]	13	17	32	5	2	..	..	..	93	..	1	..	..	..	4	2	..	..	..	..	7	1561	Sail ..	
[illegible]	[illegible]	[illegible]	16	55	41	14	1	..	..	425	7	12	5	13	3	9	10	10	1	3	..	73	506	Steam.	GERMANY.
78	82	76	63	87	46	16	1	..	..	518	7	13	5	13	3	13	12	10	1	3	..	80	2067	Total.	
..	..	..	..	..	..	..	..	..	..	..	..	..	..	..	..	..	..	..	..	..	..	..	907	Sail ..	
..	..	1	..	..	..	..	..	..	..	1	..	..	..	..	..	..	..	..	..	..	..	..	1	Steam.	GREECE.
..	..	1	..	..	..	..	..	..	..	1	..	..	..	..	..	..	..	..	..	..	..	..	908	Total.	
[illegible]	1	3	4	9	3	..	..	..	..	25	..	..	..	..	..	1	..	..	..	..	..	1	362	Sail ..	
16	8	22	8	8	4	5	1	..	..	93	4	5	..	4	4	2	2	1	1	..	..	23	117	Steam.	HOLLAND AND BELGIUM.
20	9	25	12	17	7	5	1	..	..	118	4	5	..	4	4	3	2	1	1	..	..	24	479	Total.	
2	..	..	..	..	..	..	..	..	..	4	..	..	..	..	..	..	..	..	..	..	..	..	1439	Sail ..	
4	1	2	1	..	4	3	..	..	..	17	1	..	..	..	..	..	..	..	..	..	..	1	21	Steam.	ITALY.
6	1	2	1	..	4	3	..	..	..	21	1	..	..	..	..	..	..	..	..	..	..	1	1460	Total.	
..	1	..	1	2	..	..	..	..	..	4	..	..	..	..	..	..	..	..	..	..	..	..	1672	Sail ..	
17	24	10	3	3	2	1	..	..	..	85	3	2	3	..	2	1	..	..	..	..	..	11	140	Steam.	NORWAY.
17	25	10	4	5	2	1	..	..	..	89	3	2	3	..	2	1	..	..	..	..	..	11	1812	Total.	
..	..	..	..	..	..	..	..	..	..	..	..	..	..	..	..	..	..	..	..	..	..	..	792	Sail ..	
9	4	2	..	..	..	..	..	..	..	17	1	2	..	..	1	..	..	..	..	..	..	4	28	Steam.	RUSSIA.
9	4	2	..	..	..	..	..	..	..	17	1	2	..	..	1	..	..	..	..	..	..	4	820	Total.	
..	..	..	..	..	..	..	..	..	..	..	..	..	..	..	..	..	..	..	..	..	..	..	461	Sail ..	
1	..	..	..	..	..	..	..	..	..	6	..	..	..	..	..	..	..	..	..	..	..	..	12	Steam.	SPAIN.
1	..	..	..	..	..	..	..	..	..	6	..	..	..	..	..	..	..	..	..	..	..	..	473	Total.	
2	..	..	..	..	..	..	..	..	..	2	..	..	..	..	..	..	..	..	..	..	..	..	806	Sail ..	
101	61	29	15	10	..	..	..	..	..	333	12	13	18	5	6	7	2	..	..	..	..	63	548	Steam.	SWEDEN.
103	61	29	15	10	..	..	..	..	..	335	12	13	18	5	6	7	2	..	..	..	..	63	1354	Total.	

...Net Tonnage, and for Steamers on the Gross.

TABLE No. 5.—Showing Number, Tonnage, and Description of all Vessels of 100 Tons and upwards, . . .
and respectively distinguishing those now owned in the Country in . . .

COUNTRY IN WHICH BUILT.	TONNAGE.	BUILT IN COUNTRY, BUT OWNED IN THE UNITED KINGDOM.											BUILT AND OWNED IN COUNTRY. . . .								
		SAIL.					STEAM.					Total.	SAIL.					STEAM.			
		Wood.	Comp.	Iron.	Steel.	Total.	Wood.	Comp.	Iron.	Steel.	Total.		Wood.	Comp.	Iron.	Steel.	Total.	Wood.	Comp.	Iron.	Steel.
UNITED KINGDOM .. (For details, see Table 6.)	Total	..	..	..	..	..	..	..	..	..	..		1504	81	1445	146	3176	128	15	4167	949
BRITISH COLONIES ..	Under 200	241	..	..	..	241	..	..	2	..	2	243	782	..	1	..	783	90	3	36	..
	200-399	85	..	4	..	89	..	..	1	2	3	92	346	1	1	..	348	68	3	15	2
	400-599	30	..	5	..	35	..	..	4	..	4	39	105	1	1	..	107	17	1	..	1
	600-799	25	..	21	..	46	..	..	1	..	1	47	101	..	..	..	101	12	..	1	..
	800-999	22	..	2	..	24	1	..	2	..	3	27	101	..	..	..	101	4	..	1	1
	1000-1499	68	..	2	..	70	..	..	1	..	1	71	199	..	..	..	199	4	1	2	..
	1500-1999	8	..	..	..	8	..	..	..	..	..	8	54	..	..	..	54	..	..	..	..
	2000-2999	..	..	..	..	..	..	..	..	1	1	1	3	..	..	..	3	..	..	..	..
	3000-3999	..	..	..	..	..	..	..	..	..	..	..	..	..	..	..	..	..	..	..	..
	4000-4999	..	..	..	..	..	..	..	..	..	..	..	..	..	..	..	..	..	..	..	..
	5000 & above	..	..	..	..	..	..	..	..	..	..	..	..	..	..	..	..	..	..	..	..
	Total	479	..	34	..	513	1	..	11	3	15	528	1691	2	3	..	1696	195	8	55	4
AMERICA (UNITED STATES)	Under 200	10	..	..	..	10	1	..	..	..	1	11	966	..	..	..	966	22	..	9	..
	200-399	9	..	..	..	9	..	..	..	..	..	9	965	..	1	..	966	48	..	11	2
	400-599	1	..	..	..	1	..	..	..	..	..	1	542	..	1	..	543	46	..	12	..
	600-799	2	..	..	..	2	1	..	..	..	1	3	226	..	..	..	226	18	..	7	1
	800-999	1	..	..	..	1	..	..	..	..	..	1	72	..	..	..	72	23	..	23	1
	1000-1499	10	..	..	..	10	..	..	..	..	..	10	204	..	..	..	204	40	..	26	4
	1500-1999	..	..	..	..	..	..	..	..	..	..	..	93	..	3	..	96	20	..	19	1
	2000-2999	..	..	..	..	..	..	..	..	..	..	..	27	..	..	..	27	6	..	61	1
	3000-3999	..	..	..	..	..	..	..	..	..	..	..	..	..	..	..	..	..	..	20	..
	4000-4999	..	..	..	..	..	..	..	..	..	..	..	..	..	..	..	..	..	..	..	..
	5000 & above	..	..	..	..	..	..	..	..	..	..	..	..	..	..	..	..	..	..	1	..
	Total	33	..	..	..	33	2	..	..	..	2	35	3095	..	5	..	3100	223	..	189	10
FRANCE..........	Under 200	16	..	..	..	16	..	..	..	..	..	16	432	..	..	..	432	2	2	15	8
	200-399	..	..	..	..	..	..	..	3	..	3	3	196	1	4	..	201	..	..	13	3
	400-599	..	..	1	..	1	..	..	1	..	1	2	42	3	4	..	49	..	..	7	..
	600-799	..	..	1	..	1	..	..	1	..	1	2	16	1	4	..	21	..	..	7	1
	800-999	..	..	..	..	..	..	..	..	..	..	..	..	..	2	..	2	..	..	10	..
	1000-1499	..	..	..	..	..	..	..	..	..	..	..	2	..	2	..	4	..	..	19	2
	1500-1999	..	..	1	..	1	..	..	..	..	..	1	..	..	..	..	..	..	..	16	1
	2000-2999	..	..	1	..	1	..	..	..	..	..	1	..	..	..	..	..	..	..	21	12
	3000-3999	..	..	..	..	..	..	..	..	..	..	..	..	..	..	..	..	..	..	35	7
	4000-4999	..	..	..	..	..	..	..	..	..	..	..	..	..	..	..	..	..	..	11	..
	5000 & above	..	..	..	..	..	..	..	..	..	..	..	..	..	..	..	..	..	..	..	5
	Total	16	..	4	..	20	..	..	5	..	5	25	688	5	16	..	709	2	2	154	39
GERMANY..........	Under 200	6	..	..	..	6	..	..	1	..	1	7	222	1	1	..	224	..	1	55	3
	200-399	3	..	1	..	4	1	..	..	..	1	5	383	..	4	1	388	..	..	40	11
	400-599	1	..	..	..	1	..	..	2	..	2	3	256	..	10	..	266	..	..	48	3
	600-799	..	..	..	..	..	..	..	..	..	..	..	58	..	11	..	69	..	..	56	10
	800-999	..	..	..	..	..	..	..	..	..	..	..	26	..	17	..	43	..	..	38	5
	1000-1499	..	..	..	..	..	..	..	..	..	..	..	23	..	32	4	59	..	..	42	3
	1500-1999	1	..	..	..	1	..	..	..	..	..	1	2	..	5	1	8	..	..	18	10
	2000-2999	..	..	..	..	..	..	..	..	..	..	..	..	..	2	..	2	..	..	13	10
	3000-3999	..	..	..	..	..	..	..	..	..	..	..	..	..	..	..	..	..	..	1	1
	4000-4999	..	..	..	..	..	..	..	..	..	..	..	..	..	..	..	..	..	..	..	3
	5000 & above	..	..	..	..	..	..	..	..	..	..	..	..	..	..	..	..	..	..	..	..
	Total	11	..	1	..	12	1	..	3	..	4	16	970	1	82	6	1059	..	1	311	59
SWEDEN..........	Under 200	3	..	..	..	3	..	..	..	..	..	3	268	1	..	..	269	20	41	94	9
	200-399	..	..	..	..	..	..	..	..	..	..	..	221	1	2	..	224	2	24	74	11
	400-599	..	..	..	..	..	..	..	..	..	..	..	70	..	..	..	70	2	9	39	3
	600-799	..	..	..	..	..	..	..	..	..	..	..	25	..	..	..	25	2	1	20	[illegible]
	800-999	..	..	..	..	..	..	..	..	..	..	..	5	..	..	..	5	..	..	12	3
	1000-1499	..	..	..	..	..	..	..	..	..	..	..	5	..	..	..	5	..	..	4	2
	1500-1999	..	..	..	..	..	..	..	..	..	..	..	..	..	..	..	..	..	..	..	1
	2000-2999	..	..	..	..	..	..	..	..	..	..	..	..	..	..	..	..	..	..	..	..
	3000-3999	..	..	..	..	..	..	..	..	..	..	..	..	..	..	..	..	..	..	..	..
	4000-4999	..	..	..	..	..	..	..	..	..	..	..	..	..	..	..	..	..	..	..	..
	5000 & above	..	..	..	..	..	..	..	..	..	..	..	..	..	..	..	..	..	..	..	..
	Total	3	..	..	..	3	..	..	..	..	..	3	594	2	2	..	598	26	75	243	36

NOTE.—The figures for Sailing Vessels . . .

[…] of the principal Shipbuilding Countries, and still in existence, as recorded in the "Universal Register,

[…] built, those owned in the United Kingdom, and those owned in other Countries.

Built in Country, but owned in other Countries exclusive of the United Kingdom — Sail					Steam					Total	Total Number of Vessels Built — Sail					Steam					Grand Total	Tonnage	Country in which built
	Comp.	Iron	Steel	Total	Wood	Comp.	Iron	Steel	Total		Wood	Comp.	Iron	Steel	Total	Wood	Comp.	Iron	Steel	Total			
5	39	238	19	1171	71	15	2055	336	2477	3648	2379	120	1683	165	4347	199	30	6222	1285	7736	12083	Total.	United Kingdom. (For details, see Table 6.)
				89	4	..	..	..	4	93	1112	..	1	..	1113	94	3	38	..	135	1248	Under 200	British Colonies.
				203	..	..	1	..	1	204	631	1	5	..	640	68	3	17	1	92	732	200-399	
				126	1	..	..	..	1	127	261	1	6	..	268	18	1	4	1	24	292	400-599	
				102	..	..	2	..	2	104	228	..	21	..	249	12	..	4	..	16	265	600-799	
				48	..	..	..	..	..	48	174	..	2	..	176	5	..	3	1	9	185	800-999	
				77	..	2	1	..	3	80	344	..	2	..	346	4	3	4	..	11	357	1000-1499	
				8	..	..	..	..	..	8	67	..	..	..	67	..	..	..	..	..	67	1500-1999	
				2	..	..	..	..	..	2	5	..	..	..	5	..	..	..	1	1	6	2000-2999	
	..	..	..	..	..	..	..	..	..	..	..	..	..	..	..	..	..	..	..	..	..	3000-3999	
	..	..	..	..	..	..	..	..	..	..	..	..	..	..	..	..	..	..	..	..	..	4000-4999	
	..	..	..	..	..	..	..	..	..	..	..	..	..	..	..	..	..	..	..	..	..	5000 & above	
45				655	5	2	4	..	11	666	2825	2	37	..	2864	201	10	70	7	288	3152	Total.	
92				92	19	..	5	..	24	116	1068	..	..	..	1068	42	..	14	..	56	1124	Under 200	America (United States).
11				211	16	..	9	..	25	236	1188	..	1	..	1189	61	..	20	2	86	1275	200-399	
82				82	9	1	6	..	16	98	625	..	1	..	626	55	1	18	..	74	700	400-599	
52				52	7	1	2	..	10	62	280	..	..	..	280	26	1	9	1	37	317	600-799	
28				28	3	..	2	3	8	36	101	..	..	..	101	26	..	25	4	55	156	800-999	
23				123	2	..	1	5	8	131	337	..	..	..	337	42	..	27	9	78	415	1000-1499	
11				11	5	..	2	2	9	20	104	..	3	..	107	25	..	21	3	49	156	1500-1999	
2				2	3	..	2	..	5	7	29	..	..	..	29	9	..	63	1	73	102	2000-2999	
				..	..	..	..	..	..	..	..	..	..	..	..	..	..	20	..	20	20	3000-3999	
				..	..	..	..	..	..	..	..	..	..	..	..	..	..	..	..	..	..	4000-4999	
				..	..	..	..	..	..	..	..	..	..	..	..	..	..	1	..	1	1	5000 & above	
04				604	64	2	29	10	105	709	3732	..	5	..	3737	289	2	218	20	529	4266	Total.	
36	..	..	..	36	..	..	5	3	8	44	484	..	..	..	484	2	2	20	11	35	519	Under 200	France.
22	..	..	..	62	..	..	5	3	8	70	258	1	4	..	263	..	..	21	6	27	290	200-399	
33	..	2	..	35	..	..	3	1	4	39	75	3	7	..	85	..	..	11	1	12	97	400-599	
5	..	1	..	16	..	..	2	..	2	18	31	1	6	..	38	..	..	10	1	11	49	600-799	
5			..	5	..	..	..	..	..	5	5	..	2	..	7	..	..	10	..	10	17	800-999	
1			..	1	..	..	..	1	1	2	3	..	2	..	5	..	..	19	3	22	27	1000-1499	
1			..	1	..	..	1	3	4	5	1	..	1	..	2	..	..	17	4	21	23	1500-1999	
				..	..	..	1	2	3	3	..	..	1	..	1	..	..	22	14	36	37	2000-2999	
			..	..	..	..	..	1	1	1	..	..	..	..	..	..	..	35	8	43	43	3000-3999	
				..	..	..	..	..	..	..	..	..	..	..	..	..	..	11	..	11	11	4000-4999	
				..	..	..	..	..	..	..	..	..	..	..	..	..	..	..	5	5	5	5000 & above	
3		3		156	..	..	17	14	31	187	857	5	23	..	885	2	2	176	53	233	1118	Total.	
5			..	169	..	..	12	4	16	185	397	1	1	..	399	..	1	68	7	76	475	Under 200	Germany.
58	..	5	..	263	4	1	28	1	34	297	644	..	10	1	655	5	1	68	12	86	741	200-399	
3	4	3	..	10	1	..	19	2	22	32	260	4	13	..	277	1	..	69	5	75	352	400-599	
25		2	..	27	..	..	7	3	10	37	83	..	13	..	96	..	..	63	13	76	172	600-799	
2			..	12	..	..	8	..	8	20	38	..	17	..	55	..	..	46	3	49	104	800-999	
8				8	..	..	13	4	17	25	31	..	32	4	67	..	..	55	9	64	131	1000-1499	
		..	1	1	..	..	23	..	23	24	3	..	5	2	10	..	..	41	10	51	61	1500-1999	
				..	..	..	1	..	1	1	..	..	2	..	2	..	..	14	10	24	26	2000-2999	
				..	..	..	..	..	..	..	..	..	..	..	..	..	..	1	1	2	2	3000-3999	
				..	..	..	..	..	..	..	..	..	..	..	..	..	..	..	3	3	3	4000-4999	
				..	..	..	..	..	..	..	..	..	..	..	..	..	..	..	..	..	..	5000 & above	
5	4	10	1	490	5	1	111	14	131	621	1456	5	93	7	1561	6	2	425	73	506	2067	Total.	
50	..	..	..	50	8	3	23	3	37	87	321	1	..	..	322	28	44	117	12	201	523	Under 200	Sweden.
82	..	..	..	82	11	16	27	2	56	138	305	1	2	..	308	13	40	101	13	167	475	200-399	
35	..	..	..	35	7	..	22	15	44	79	105	..	..	..	105	9	9	61	18	97	202	400-599	
33	..	..	..	33	5	..	9	2	16	49	58	..	..	..	58	7	1	29	5	42	100	600-799	
5	..	..	..	5	1	..	3	3	7	12	10	..	..	..	10	1	..	15	6	22	32	800-999	
	..	..	..	..	..	..	6	1	7	7	3	..	..	..	3	..	..	10	7	17	20	1000-1499	
	..	..	..	..	..	..	..	1	1	1	..	..	..	..	..	..	..	..	2	2	2	1500-1999	
	..	..	..	..	..	..	..	..	..	..	..	..	..	..	..	..	..	..	..	..	..	2000-2999	
	..	..	..	..	..	..	..	..	..	..	..	..	..	..	..	..	..	..	..	..	..	3000-3999	
	..	..	..	..	..	..	..	..	..	..	..	..	..	..	..	..	..	..	..	..	..	4000-4999	
	..	..	..	..	..	..	..	..	..	..	..	..	..	..	..	..	..	..	..	..	..	5000 & above	
05	..	..	..	205	32	19	90	27	168	373	802	2	2	..	806	58	94	333	63	548	1354	Total.	

[…] N. Tonnage, and for steamers on the Gross.

TABLE No. 6.—Showing the Countries in which all Vessels, of 100 Tons and upwards, BUILT IN THE U… such Vessels, as r…

Country in which Owned	Tonnage	Sail: Wood	Sail: Composite	Sail: Iron	Sail: Steel	Sail: Total	Steam: Wood	Steam: Composite	Steam: Iron	Steam: Steel	Steam: Total	Total Sail & Steam
United Kingdom	Under 200	1007	9	32	8	1056	165	5	535	57	762	1818
	200–399	389	5	70	14	478	16	7	442	86	551	1029
	400–599	78	17	127	21	243	7	..	278	60	345	588
	600–799	12	21	176	2	211	..	2	308	48	358	572
	800–999	11	13	172	12	208	..	1	310	49	360	568
	1000–1499	7	12	497	27	543	..	..	766	109	875	1418
	1500–1999	..	..	284	36	320	..	..	701	132	833	1153
	2000–2999	..	1	86	23	110	..	..	536	261	797	907
	3000–3999	..	..	1	3	4	..	..	165	79	244	248
	4000–4999	..	..	..	..	..	..	..	54	46	100	100
	5000 & above	..	..	..	..	..	..	..	12	22	34	34
Total		1504	81	1445	146	3176	128	15	4167	949	5259	8435
British Colonies	Under 200	73	..	2	..	75	4	..	70	10	84	159
	200–399	104	2	17	..	123	10	1	99	13	123	246
	400–599	24	1	11	..	36	7	3	47	23	80	116
	600–799	4	3	5	..	12	1	..	32	11	44	56
	800–999	1	2	5	..	8	..	..	25	6	31	39
	1000–1499	4	1	18	1	24	..	..	36	10	46	70
	1500–1999	..	..	3	..	3	..	..	20	14	34	37
	2000–2999	..	..	..	..	..	..	..	8	12	20	20
	3000–3999	..	..	..	..	..	..	..	3	..	3	3
	4000–4999	..	..	..	..	..	..	..	..	..	..	..
	5000 & above	..	..	..	..	..	..	..	..	..	..	..
Total		210	9	61	1	281	22	4	340	99	465	746
America (United States)	Under 200	1	..	..	..	1	..	..	1	..	1	2
	200–399	3	..	..	..	3	..	..	..	..	..	3
	400–599	1	..	..	..	1	..	..	1	..	1	2
	600–799	..	..	..	..	..	..	..	2	..	2	2
	800–999	..	..	1	..	1	..	..	2	..	2	3
	1000–1499	..	..	..	..	..	..	..	4	1	5	5
	1500–1999	..	..	..	..	..	..	..	2	1	3	3
	2000–2999	..	..	..	..	..	..	..	..	..	..	..
	3000–3999	..	..	..	..	..	..	..	..	..	..	..
	4000–4999	..	..	..	..	..	..	..	..	..	..	..
	5000 & above	..	..	..	..	..	..	..	..	..	..	..
Total		5	..	1	..	6	..	..	12	2	14	20
Austro-Hungary	Under 200	1	..	..	..	1	..	..	2	1	3	4
	200–399	6	..	..	..	6	..	..	8	1	9	15
	400–599	7	..	..	..	7	..	..	3	..	3	10
	600–799	3	2	..	..	5	..	..	7	..	7	12
	800–999	1	2	..	..	3	..	..	3	..	3	6
	1000–1499	..	1	..	..	1	..	..	10	1	11	12
	1500–1999	..	..	..	..	..	..	..	18	1	19	19
	2000–2999	..	..	..	..	..	..	..	3	2	5	5
	3000–3999	..	..	..	..	..	..	..	..	..	..	..
	4000–4999	..	..	..	..	..	..	..	..	..	..	..
	5000 & above	..	..	..	..	..	..	..	..	..	..	..
Total		18	5		..	23			54	6	60	83

Country in which Owned	Tonnage	Sail: Wood	Sail: Composite	Sail: Iron	Sail: Steel	Sail: Total	Steam: Wood	Steam: Composite	Steam: Iron	Steam: …
Denmark	Under 200	5	..	..	..	5	1	..	6	
	200–399	24	..	1	..	25	..	..	7	1
	400–599	5	1	5	..	11	..	..	11	
	600–799	4	1	1	..	6	..	..	13	
	800–999	..	..	1	..	1	..	..	8	
	1000–1499	1	..	..	..	1	..	..	7	
	1500–1999	..	..	..	..	..	..	..	4	1
	2000–2999	..	..	..	..	..	..	..	..	
	3000–3999	..	..	..	..	..	..	..	2	
	4000–4999	..	..	..	..	..	..	..	..	
	5000 & above	..	..	..	..	..	..	..	..	
Total		39	2	8	..	49	1	..	58	2
France	Under 200	16	..	..	..	16	1	..	35	
	200–399	22	..	..	1	23	..	..	28	
	400–599	7	2	7	..	16	..	..	29	
	600–799	4	..	13	..	17	..	..	19	
	800–999	1	..	1	..	2	..	..	11	
	1000–1499	..	..	5	1	6	..	..	57	1
	1500–1999	..	..	1	..	1	..	..	13	
	2000–2999	..	..	4	1	5	..	..	29	
	3000–3999	..	..	..	1	1	..	..	10	1
	4000–4999	..	..	..	..	..	..	..	2	
	5000 & above	..	..	..	..	..	..	..	2	
Total		50	2	31	4	87	1	..	266	2…
Germany	Under 200	3	..	3	..	6	3	..	10	..
	200–399	26	2	2	..	30	..	..	22	..
	400–599	13	1	31	..	45	..	..	8	..
	600–799	2	1	10	..	13	..	..	13	1
	800–999	6	6	19	..	31	..	..	11	..
	1000–1499	2	1	21	2	26	..	..	33	
	1500–1999	..	..	3	6	9	..	..	13	
	2000–2999	..	..	5	1	6	..	..	31	2…
	3000–3999	..	..	..	..	..	..	..	14	
	4000–4999	..	..	..	..	..	..	..	2	1
	5000 & above	..	..	..	..	..	..	..	3	
Total		52	11	94	9	166	3	..	166	3…
Greece	Under 200	9	..	..	..	9	4	..	4	..
	200–399	24	1	..	..	25	2	..	7	..
	400–599	9	1	..	..	10	..	..	7	..
	600–799	1	..	..	..	1	..	..	10	..
	800–999	..	..	..	..	..	..	..	8	
	1000–1499	..	..	..	..	..	..	..	14	
	1500–1999	..	..	..	..	..	..	..	14	
	2000–2999	..	..	..	..	..	..	..	..	
	3000–3999	..								
	4000–4999	..								
	5000 & above	..					..	..		
Total		43	2			45	6	..	64	

NOTE.—The figures for Sailing Vessels are based on the Net Tonnage, and for Steamers on the Gross.

…INGDOM, and still in existence, are now owned; and showing also the Number, Tonnage, and Description of … the "Universal Register."

Country in which Owned	Tonnage	Sail: Wood	Sail: Comp.	Sail: Iron	Sail: Steel	Sail: Total	Steam: Wood	Steam: Comp.	Steam: Iron	Steam: Steel	Steam: Total	Grand Total of all Vessels
…land and …Bel…	Under 200	3	..	..	..	3	2	..	12	..	14	17
	200-399	22	..	..	..	22	..	..	10	1	11	33
	400-599	5	..	2	..	7	..	1	12	1	14	21
	600-799	1	..	2	..	3	..	1	7	..	8	11
	800-999	1	..	..	..	1	..	..	8	1	9	10
	1000-1499	1	..	2	..	3	..	1	19	5	25	28
	1500-1999	..	..	..	1	1	..	..	13	5	18	19
	2000-2999	..	..	..	..	..	..	..	33	3	36	36
	3000-3999	..	..	..	..	..	..	..	9	1	10	10
	4000-4999	..	..	..	..	..	..	..	1	..	1	1
	5000 & above	..	..	..	..	..	..	..	..	2	2	2
Total		33	..	6	1	40	2	3	124	19	148	188
…any	Under 200	3	..	..	..	3	1	..	12	..	13	16
	200-399	6	..	..	..	6	..	..	14	..	14	20
	400-599	1	..	..	..	1	..	1	27	1	29	30
	600-799	2	..	..	..	2	..	..	11	1	12	14
	800-999	1	1	..	..	2	..	..	9	..	9	11
	1000-1499	..	..	1	..	1	..	..	33	3	36	37
	1500-1999	..	..	1	1	2	..	..	15	3	18	20
	2000-2999	..	..	..	..	..	..	..	29	2	31	31
	3000-3999	..	..	..	..	..	..	..	9	1	10	10
	4000-4999	..	..	..	..	..	..	..	3	3	6	6
	5000 & above	..	..	..	..	..	..	..	1	..	1	1
Total		16	1	2	1	20	1	1	160	14	176	196
…orway	Under 200	21	..	..	..	21	..	..	10	..	10	31
	200-399	68	..	..	1	69	1	..	11	..	12	81
	400-599	45	..	1	..	46	..	..	21	..	21	67
	600-799	15	..	..	..	15	..	..	21	..	21	36
	800-999	21	1	1	..	23	..	..	6	2	8	31
	1000-1499	10	..	2	1	13	..	..	3	2	5	18
	1500-1999	..	..	..	..	..	..	..	13	..	13	13
	2000-2999	..	..	1	..	1	..	..	2	4	6	7
	3000-3999	..	..	..	..	..	..	..	..	..	..	..
	4000-4999	..	..	..	..	..	..	..	..	..	..	..
	5000 & above	..	..	..	..	..	..	..	..	..	..	..
Total ..		180	1	5	2	188	1	..	87	8	96	284
…ugal	Under 200	2	..	..	..	2	1	..	3	..	4	6
	200-399	10	..	..	..	10	..	..	6	..	6	16
	400-599	1	1	1	..	3	..	..	4	..	4	7
	600-799	..	..	1	..	1	..	..	2	..	2	3
	800-999	..	..	1	..	1	..	..	1	..	1	2
	1000-1499	..	..	..	..	..	..	..	3	1	4	4
	1500-1999	..	..	..	..	..	..	..	2	..	2	2
	2000-2999	..	..	..	..	..	..	..	2	..	2	2
	3000-3999	..	..	..	..	..	..	..	..	1	1	1
	4000-4999	..	..	..	..	..	..	..	..	..	..	..
	5000 & above	..	..	..	..	..	..	..	..	..	..	..
Total		13	1	3	..	17	1	..	23	2	26	43

Country in which Owned	Tonnage	Sail: Wood	Sail: Comp.	Sail: Iron	Sail: Steel	Sail: Total	Steam: Wood	Steam: Comp.	Steam: Iron	Steam: Steel	Steam: Total	Grand Total of all Vessels
Russia ..	Under 200	4	..	..	..	4	1	..	5	..	6	10
	200-399	10	..	1	..	11	..	..	25	4	29	40
	400-599	3	..	..	..	3	..	..	23	4	27	30
	600-799	..	..	..	..	..	..	..	6	2	8	8
	800-999	..	..	..	..	..	..	..	5	1	6	6
	1000-1499	..	..	..	..	..	..	..	12	6	18	18
	1500-1999	..	..	..	..	..	..	..	4	..	4	4
	2000-2999	..	..	..	..	..	..	..	7	1	8	8
	3000-3999	..	..	..	..	..	..	..	2	1	3	3
	4000-4999	..	..	..	..	..	..	..	..	..	..	..
	5000 & above	..	..	..	..	..	..	..	..	..	..	..
Total		17	..	1	..	18	1	..	89	19	109	127
Spain ..	Under 200	6	..	..	..	6	5	1	46	2	54	60
	200-399	13	..	4	..	17	2	1	35	1	39	56
	400-599	8	..	2	..	10	2	..	31	2	35	45
	600-799	..	..	..	..	..	..	..	37	1	38	38
	800-999	..	..	..	..	..	..	..	27	..	27	27
	1000-1499	1	..	..	..	1	..	..	51	2	53	54
	1500-1999	..	..	..	..	..	..	..	20	4	24	24
	2000-2999	..	..	..	..	..	..	..	10	5	15	15
	3000-3999	..	..	..	..	..	..	..	9	5	14	14
	4000-4999	..	..	..	..	..	..	..	2	2	4	4
	5000 & above	..	..	..	..	..	..	..	..	2	2	2
Total ..		28	..	6	..	34	9	2	298	26	335	369
Sweden ..	Under 200	5	..	..	..	5	..	..	1	..	1	6
	200-399	49	..	..	..	49	1	..	1	..	2	51
	400-599	8	..	..	..	8	..	..	1	..	1	9
	600-799	9	..	..	..	9	..	..	6	2	8	17
	800-999	3	..	..	..	3	..	..	8	..	8	11
	1000-1499	..	1	..	..	1	..	..	9	2	11	12
	1500-1999	..	..	..	..	..	..	..	2	..	2	2
	2000-2999	..	..	..	..	..	..	..	1	..	1	1
	3000-3999	..	..	..	..	..	..	..	..	..	..	..
	4000-4999	..	..	..	..	..	..	..	..	..	..	..
	5000 & above	..	..	..	..	..	..	..	..	..	..	..
Total ..		74	1	..	..	75	1	..	29	4	34	109
Other Countries ([illegible] Vessels owned in Central and South America.)	Under 200	27	..	8	..	35	11	1	23	6	41	76
	200-399	40	..	5	..	45	7	2	64	12	85	130
	400-599	20	2	4	..	26	2	1	57	15	75	101
	600-799	8	..	1	..	9	1	1	39	3	44	53
	800-999	1	1	1	..	3	1	..	19	2	22	25
	1000-1199	1	1	1	1	4	..	..	36	13	49	53
	1500-1999	..	..	..	..	..	..	..	30	7	37	37
	2000-2999	..	..	..	..	..	..	..	16	8	24	24
	3000-3999	..	..	..	..	..	..	..	..	2	2	2
	4000-4999	..	..	..	..	..	..	..	1	..	1	1
	5000 & above	..	..	..	..	..	..	..	..	..	..	..
Total ..		97	4	20	1	122	22	7	285	68	380	502

The Total Number of Vessels built in the United Kingdom will be found in Table 5.

TABLE No. 7.—Showing Number, Tonnage, and Description of all Vessels of 100 tons and upwards

Country in which Built	Whether Steam or Sail	Wood: Under 200 tons	200 to 399 tons	400 to 599 tons	600 to 799 tons	800 to 999 tons	1000 to 1499 tons	1500 to 1999 tons	2000 to 2499 tons	2500 to 2999 tons	3000 tons & upwards	Total No.	Total Tonnage	Composite: Under 200 tons	200 to 399 tons	400 to 599 tons	600 to 799 tons	800 to 999 tons	1000 to 1499 tons	1500 to 1999 tons	2000 to 2499 tons	2500 to 2999 tons	3000 tons & upwards
United Kingdom	Sail	8	1	..		..	..					9	1,282										
	Steam											..					2						
	Total	8	1									9	1,282				2						
British Colonies	Sail	34	7	2	3	..	1	..	..	..	..	47	10,992	..									
	Steam	7	9									16	3,518		1								
	Total	41	16	2	3	..	1					63	14,510	..	1								
America (United States)	Sail	10	9	11	7	5	3	..	..	..	..	45	22,702	..	..	..	..	..	..	..	..	..	..
	Steam	2	17	5	..	1		..	..	..	..	25	8,421										
	Total	12	26	16	7	6	3					70	31,123										
Austro-Hungary	Sail	..	1	..	..	..	..	..	..	..	..	1	268	..	..	..	..	..	..	..	..	..	..
	Steam	..																					
	Total		1									1	268										
Denmark	Sail	..	1	..	..	..	..	..	..	..	..	1	219	..	..	..	..	..	..	..	..	..	..
	Steam		..																				
	Total	..	1									1	219										
France	Sail	2	1	1	..	..	..	..	..	..	..	4	1,098	..	..	..	..	..	..	..	..	..	..
	Steam	..	..	..	..	..	..	..	..	..	..	..	..	..	..								
	Total	2	1	1	..	..	..	..	..	..	..	4	1,098	..	..	..	..						
Germany	Sail	..	..	..	..	..	..	1	..	..	..	1	1,663	..	..	..	..	..	..	..	..	..	..
	Steam	..	..	..	..	..	..	..	..	..	..	..	..	..	..								
	Total	..	..	..	..	..	..	1	..	..	..	1	1,663	..	..	..	..						
Greece	Sail	7	8	..	..	..	..	..	..	..	..	15	3,086	..	..	..	..	..	..	..	..	..	..
	Steam	..	..	..	..	..	..	..	..	..	..	..	..	..	..								
	Total	7	8	..	..	..	..	..	..	..	..	15	3,086	..	..	..	..	..	..				
Holland	Sail	..	..	..	..	..	..	..	..	..	..	..	..	..	..	..	..	..	..	..	..	..	..
	Steam	..	..	..	..	..	..	..	..	..	..	..	..	..	..	..	..	..	..	..	..	..	..
	Total	..	..	..	..	..	..	..	..	..	..			..	..	..	..	..	..	..	..	..	..
Italy	Sail	5	..	..	..	..	..	..	..	..	..	5	703	..	..	..	..	..	..	..	..	..	..
	Steam	..	..	..	..	..	..	..	..	..	..												
	Total	5	..	..	..	..	..	..	..	..	..	5	703	..	..	..	..	..	..	..	..	..	..
Norway	Sail	2	3	4	1	4	1	..	..	..	..	15	8,691	..	..	..	..	..	..	..	..	..	..
	Steam	..	..	..	..	..	..	..	..	..	..	..	..	..	..	..							
	Total	2	3	4	1	4	1	..	..	..	..	15	8,691	..	..	..	..	..	..	..	..	..	..
Russia	Sail	3	2	..	..	..	..	..	..	..	..	5	1,083	..	..	..	..	..	..	..	..	..	..
	Steam	..	..	..	..	..	..	..	..	..	..	..	..	..	..								
	Total	3	2	..	..	..	..	..	..	..	..	5	1,083	..	..	..	..	..	..	..	..	..	..
Sweden	Sail	..	3	..	..	..	..	..	..	..	..	3	860	..	1	..	..	..	..	..	..	..	..
	Steam	..	..	..	..	..	..	..	..	..	..	..	..	1	..								
	Total	..	3	..	..	..	..	..	..	..	..	3	860	1	1	..	..	..	..	..	..	..	..
Other Countries	Sail	3	..	..	..	..	1	..	..	..	..	4	1,663	..	..	..	..	..	..	..	..	..	..
	Steam	..	7	..	..	..	..	..	..	..	..	7	2,051	..	..	..	..	..	..	..	..	..	..
	Total	3	7	..	..	..	1	..	..	..	..	11	3,714	..	..								
Totals	Sail	74	36	18	11	9	6	1	..	..	..	155	54,310	..	1	..	..	..	..	..	..	..	..
	Steam	9	33	5	..	1	..	..	..	..	..	48	13,990	1	1	..	2	..	..	..	..	..	..
	Total	83	69	23	11	10	6	1	..	..	..	203	68,300	1	2	..	2	..	..	..	..	..	..

NOTE.—The figures for Sailing Vessels

ANALYSIS of above Table, showing Vessels built by other Countries than the United Kingdom, in 1888, **EXCLUDING THOSE BUILT FOR THEMS** recorded in the "Universal Register."

Whether Steam or Sail	Wood: Country in which Built	No.	Tons.	Country for which Built	Composite: Country in which Built	No.	Tons	Country for which Built	Iron: Country in which Built	No.	Tons	Country for which Built	Steel: Country in which Built	No.	Tons	Country for which Built
Sail	Austria	1	268	Greece	..	..	..	..	..	..	..	..	..	..	..	..
	Japan	1	130	British Colonies	..	..	..	..	..	..	..	..	..	..	..	..
	Total	2	398	..	..	..	..	..	..	..	..	..	..	..	..	..
Steam	America (U. States)	1	441	Hawaii	..	..	..	..	Denmark	1	[illegible]	Germany	[illegible]	1	[illegible]	[illegible]
														1	341	Denmark
													Germany	1	[illegible]	[illegible]
														1	[illegible]	[illegible]
													Sweden	2	[illegible]	[illegible]
	Total	1	[illegible]	..	..					1	[illegible]			[illegible]	[illegible]	
Total Steam and Sail		[illegible]	[illegible]	..	..	..	..		..	1	[illegible]			[illegible]	[illegible]	

each of the several Countries of the World DURING 1888, as recorded in the "Universal Register."

IRON 200 to 399 tons	IRON 400 to 599 tons	IRON 600 to 799 tons	IRON 800 to 999 tons	IRON 1000 to 1499 tons	IRON 1500 to 1999 tons	IRON 2000 to 2999 tons	IRON 3000 to 3999 tons	IRON 4000 tons and above	IRON Total No.	IRON Total Tonnage	STEEL Under 200 tons	STEEL 200 to 399 tons	STEEL 400 to 599 tons	STEEL 600 to 799 tons	STEEL 800 to 999 tons	STEEL 1000 to 1499 tons	STEEL 1500 to 1999 tons	STEEL 2000 to 2999 tons	STEEL 3000 to 3999 tons	STEEL 4000 tons and above	STEEL Total No.	STEEL Total Tonnage	GRAND TOTAL No.	GRAND TOTAL Tonnage	COUNTRY IN WHICH BUILT
		..	3	1	2	5	..		14	22,103	..	4	5	2	..	12	7	8	1	..	39	53,995	62	77,380	United Kingdom.
8	4	3	4	7	4	4	1	..	64	38,383	20	22	19	16	9	37	63	131	30	9	356	659,710	422	699,613	
8	4	3	7	8	6	9	1	..	78	60,486	20	26	24	18	9	49	70	139	31	9	395	713,705	484	776,993	
1	2								3	1,249	..	..	..	..	..			..	..	..		..	50	12,241	British Colonies.
				..					..						1						1	961	18	4,865	
1	2		..						3	1,249	..	..	..		1						1	961	68	17,106	
																					..		45	22,702	America (United States).
			..			1	..		1	2,867			..	..		1	..	1	..	..	2	4,208	28	15,496	
			..			1			1	2,867		..		..	..	1		1		..	2	4,208	73	38,198	
											..	..	1	..	..	..				..	1	152	2	720	Austro-Hungary.
							..	..	2	248										1	1	1,070	3	4,318	
								..	2	248		..	1						..	1	2	4,522	5	5,038	
											..	..	..	..	..	..	..	..	..	..	..	..	1	219	Denmark.
		1						..	1	634	..	..	..				3	..	..		3	4,868	4	5,502	
		1						..	1	634		..	..	..			3				3	4,868	5	5,721	
		..	..	..	..				..		..	..	..	..	..	..	..	..	..	..	..		4	1,098	France.
1	..			..		..		..	2	328	4	..	..	1	..	..	..	2	1	..	8	9,295	10	9,623	
1						..		..	2	328	4			1	..		..	2	1	..	8	9,295	14	10,721	
			..	..					..		..	..			..	2	1	..	..	..	3	4,497	4	6,160	Germany.
1	1	..	1		..				5	2,042	..	4	1	10	1	1	1	7	..	..	28	31,792	33	33,834	
1	1	..	1		..			..	5	2,042	..	4	1	10	1	3	2	7	..	..	31	36,289	37	39,994	
		..	..	..	..	..	..	..	..	..	..	..	..	..	..	..	..	..	..	..	..	..	15	3,086	Greece.
..		..	..	..	..	..	..	..	..	..	..	..	..	..	..	..	..	..	..	..	..	..	..	..	
		..	..	..	..	..	..	..	..	..	..	..	..	..	..	..	..	..	..	..	..	..	15	3,086	
				1	..	..	..	..	1	1,007	..	..	..	..	..	..	..	..	..	..	..	..	1	1,007	Holland.
			..	..	1	..	..	..	1	1,642	..	..	..	..	..	..	..	1	..	..	1	2,507	2	4,149	
		..	..	1	1		..	..	2	2,649		..	..	..	..	..	..	1	..	..	1	2,507	3	5,156	
..	..	..	..	..	..	..	..	..	..	..	..	..	..	..	..	..	..	..	..	..	..	..	5	703	Italy.
2	..	1	..				..	..	3	1,095	..	..	..	..	..	..	..	..	..	..	..	..	3	1,095	
2		1	..	..	..	..	..	..	3	1,095	..	..	..	..	..	..	..	..	..	..	..	..	8	1,798	
	..	..	..	..	..	..	..	..	..	..	..	..	..	..	..	..	..	..	..	..	..	..	15	8,691	Norway.
..	1	..		..	..	1	..	..	3	2,630	1	..	..	..	..	..	..	..	..	..	1	112	4	2,742	
	1		..	..	..	1	..	..	3	2,630	1	..	..	..	..	..	..	..	..	..	1	112	19	11,433	
		..	..	..	..	..	..	..	..	..	..	..	..	..	..	..	..	..	..	..	..	..	5	1,083	Russia.
		..	..	..	..	..	..	..	..	..	..	2	..	..	..	..	..	..	..	..	2	630	2	630	
		..	..	..	..	..	..	..	..	..	..	2	..	..	..	..	..	..	..	..	2	630	7	1,713	
	..	..	..	..	..	..	..	..	..	..	..	..	..	..	..	..	..	..	..	..	..	..	4	1,115	Sweden.
1		..	..	..	..	..	..	..	6	1,154	2	1	..	..	1	..	..	..	..	..	4	1,635	11	2,973	
1		..		..			..		6	1,154	2	1	..	..	1	..	..	..	..	..	4	1,635	15	4,088	
..	..	..	..	..	..	..	..	..	..	..	..	..	..	..	..	..	..	..	..	..	..	..	4	1,663	Other Countries.
..	..	..	..	..	..	..	..	..	..	..	..	..	..	..	..	..	1	..	..	..	1	1,764	8	3,815	
	..	..	..	..	..	..	..	..	..	..	..	..	..	..	..	..	1	..	..	..	1	1,764	12	5,478	
1	2	..	3	5	2	5	..	..	18	24,359	..	4	6	2	..	14	8	8	1	..	43	58,944	217	137,868	Totals.
13	6	5	5	4	5	6	1	..	88	51,023	27	29	20	27	15	39	68	142	31	10	408	721,552	548	788,655	
14	8	5	8	9	7	11	1	..	106	75,382	27	33	26	29	15	53	76	150	32	10	451	780,496	765	926,523	

Net Tonnage, and for Steamers on the Gross.

ANALYSIS of above Table, showing Vessels built in the United Kingdom in 1888, FOR OTHER COUNTRIES, as recorded in the "Universal Register."

	WOOD No.	WOOD Tons.	WOOD Country for which built.	IRON No.	IRON Tons.	IRON Country for which Built.	STEEL No.	STEEL Tons.	STEEL Country for which Built.	STEEL—continued. No.	STEEL—continued. Tons.	STEEL—continued. Country for which Built.	Grand Total No.	Grand Total Tons.
....	..	..	..	1	863	Hawaii.	2	[illegible]	France.	..	..	..	..	..
	..	..	..	1	1,034	France.	6	[illegible]	Germany.	(See	previous	column.)	..	..
	..	..	..	1	2,231	Germany.	1	[illegible]	Hawaii.	..	..	..	..	..
Total...	..	..	..	3	4,128	..	9	14,237	..	..	..	..	12	18,365
	..	..	..	2	1,438	Belgium.	2	541	Austro-Hungary.	3	5,010	Russia.	..	..
	..	..	..	3	5,346	Greece.	3	4,144	Belgium.	12	20,838	Spain.	..	..
	..	..	..	2	1,350	Norway.	5	3,140	France.	1	737	Sweden.	..	..
	..	..	..	1	156	Portugal.	7	16,904	Germany.	7	[illegible]	British Colonies.	..	..
...eam..........	..	..	..	1	1,372	Spain.	1	1,721	Holland.	9	11,045	China & Japan.	..	..
	..	..	..	..	..	..	4	7,345	Italy.	1	[illegible]	Hawaii.	..	..
	..	..	..	..	..	..	5	10,138	Norway.	7	[illegible]	South America.	..	..
	..	..	..	..	..	..				..	..	..	..	..
		..	..	..	..	..			(see continuation.)	..	..	..	..	..
Total...	..	..	..	9	9,662	..				67	110,968		76	120,570
Total ...		..	..	12	13,790	..	..	..	..	76	125,145	..	88	138,935

TABLE No. 8.—Showing Number, Tonnage, and Description of all Vessels CLASSED by each of the principal Classification Societies of the World.

NAME OF CLASSIFICATION SOCIETY.	TONNAGE	Sailing Vessels.					Steamers.				
		Wood.	Comp.	Iron.	Steel.	Total.	Wood.	Comp.	Iron.	Steel.	Total.
LLOYD'S REGISTER ...... The late Liverpool Underwriters' Registry is now amalgamated with Lloyd's Register. These figures do not include vessels classed in the Society's Yacht Register.	Under 200	1037	3	34	5	1079	18	2	310	51	411
	200-599	476	29	232	6	743	3	3	536	119	661
	600-999	123	46	402	11	585	1	1	632	97	731
	1000-1999	34	11	889	78	1015	..	..	1578	288	1866
	2000-2999	..	1	96	23	120			543	283	826
	3000 & above	..	..	2	2	4	..		144	132	276
	Total .	**1670**	**93**	**1656**	**128**	**3546**	**52**	**6**	**3743**	**970**	**4771**
BUREAU VERITAS	Under 200	1821	..	14	..	1835	23	..	67	19	109
	200-599	1751	1	46	1	1802	23	3	196	36	258
	600-999	672	4	62	..	738	5	..	166	29	200
	1000-1999	708	3	65	7	783	3	..	224	43	270
	2000-2999	31	..	12	1	44		..	109	33	142
	3000 & above	..	..	..	..	..	..	..	65	23	88
	Total .	**4986**	**8**	**199**	**9**	**5202**	**54**	**3**	**827**	**183**	**1067**
GERMANISCHER LLOYD ..	Under 200	738	..	8	..	746	13	2	45	8	68
	200-599	452	1	2	..	455	3	26	58	9	96
	600-999	84	2	13	..	99	..	..	44	4	48
	1000-1999	64	2	16	..	82	..	1	23	7	31
	2000-2999	2	..	1	..	3	..	..	16	2	18
	3000 & above	..	..	..	..	..	..	..	12	4	16
	Total .	**1340**	**5**	**40**	..	**1385**	**16**	**29**	**198**	**34**	**277**
NEDERLANDSCHE VEREENIGING VAN ASSURADEUREN ..	Under 200	338	..	6	..	344	..	..	8	..	8
	200-599	431	2	25	1	459	2	..	95	2	99
	600-999	218	3	56	..	277	..	..	201	2	203
	1000-1999	156	6	109	7	278	..	..	343	20	363
	2000-2999	..	..	2	..	2	..	..	102	13	115
	3000 & above	..	..	1	..	1	.	..	10	1	11
	Total .	**1143**	**11**	**199**	**8**	**1361**	**2**	..	**759**	**38**	**799**
NORSKE VERITAS......	Under 200	683	..	..	..	683	16	3	58	7	84
	200-599	1693	..	..	..	1693	45	2	69	14	130
	600-999	345	..	..	..	345	2	..	33	1	36
	1000-1999	95	..	2	1	98	..	..	26	2	28
	2000 & above	1	..	..	..	1	..	..	4	3	7
	Total .	**2817**	..	**2**	**1**	**2820**	**63**	**5**	**190**	**27**	**285**
RECORD OF AMERICAN AND FOREIGN SHIPPING	Under 200	373	..	..	..	373	6	..	5	..	11
	200-599	996	1	2	..	999	19	..	10	..	29
	600-999	309	1	2	..	312	7	1	16	..	24
	1000-1999	255	..	5	..	260	13	..	36	1	50
	2000-2999	21	..	..	..	21	..	..	72	2	74
	3000 & above	..	..	..	..	..	..	..	..	..	..
	Total .	**1954**	**2**	**9**	..	**1965**	**45**	**1**	**139**	**3**	**188**
REGISTRO ITALIANO	Under 200	249	..	3	2	254	1	..	5	3	9
	200-599	391	..	..	..	391	..	..	21	1	22
	600-999	206	..	..	..	206	..	..	10	..	10
	1000-1999	22	..	2	1	25	..	..	39	2	41
	2000-2999	..	..	..	..	..	..	..	28	1	29
	3000 & above	..	..	..	..	..	..	..	11	3	14
	Total .	**868**	..	**5**	**3**	**876**	**1**		**114**	**10**	**125**
VERITAS AUSTRO-UNGARICO....	Under 200	759	..	..	..	759	..	..	4	..	4
	200-599	398	2	..	..	400	..	..	3	1	4
	600-999	51	..	..	..	51	..	..	3	1	4
	1000-1999	3	1	1	..	5	..	..	11	2	13
	2000-2999	..	..	..	..	..	..	..	..	..	..
	3000 & above	..	..	..	..	..	..	..	..	..	..
	Total .	**1211**	**3**	**1**	..	**1215**		..	**21**	**4**	**25**
VERITAS ELLENICO	Under 200	142	..	..	..	142	4	..	1	..	5
	200-599	134	..	..	..	134	..	..	1	..	1
	600 & above	1	..	..	..	1	..				..
	Total .	**277**	..	..	..	**277**	**4**	..	**2**	..	**6**

NOTE.—The figures for Sailing Vessels are based on the Net Tonnage, and for Steamers on the Gross.

ABLE No. 9.—Showing the Number, Tonnage, and Description of Vessels of 100 tons and upwards **BUILT IN 1888, AND CLASSED BY LLOYD'S REGISTER**, in comparison with those built in the World and in the United Kingdom in the same year.

Material of Construction.	Whether Steam or Sail.	Built in the World in 1888.		Built in the United Kingdom in 1888.		Vessels built in 1888, and CLASSED BY LLOYD'S REGISTER.	
		No.	Tonnage.	No.	Tonnage.	No.	Tonnage.
Steel	Sail . . .	43	58,944	39	53,995	30	52,045
	Steam . . .	408	721,552	356	659,710	323	629,160
	Total . .	451	780,496	395	713,705	353	681,205
Iron	Sail . . .	18	24,359	14	22,103	15	23,110
	Steam . . .	88	51,023	64	38,383	62	38,363
	Total . .	106	75,382	78	60,486	77	61,473
Wood and Composite	Sail . . .	156	54,565	9	1,282	13	2,714
	Steam . . .	52	16,080	2	1,520	3	1,546
	Total . .	208	70,645	11	2,802	16	4,260
Total	Sail . . .	217	137,868	62	77,380	58	77,869
	Steam . . .	548	788,655	422	699,613	388	669,069
	Grand Total .	765	926,523	484	776.993	446	746,938

KEY TO THE UNIVERSAL REGISTER

N ENGLISH, DUTCH, FRENCH, GERMAN, ITALIAN, NORWEGIAN, AND SPANISH).

LEF DES MATIÈRES DE L'UNIVERSAL REGISTER

(EN ANGLAIS, HOLLANDAIS, FRANÇAIS, ALLEMAND, ITALIEN, NORVÉGIEN, ET ESPAGNOL).

EXPLANATION AND KEY TO CONTENTS OF "UNIVERSAL REGISTER."

REGISTER OF SAILING VESSELS.

COLUMN I.

Official Registry Numbers of Vessels.

COLUMN II.

Consecutive Numbers of Vessels in the "Universal Register."

COLUMN III.

Symbols denoting the Classification Societies (if any) with which Vessels are classed.

COLUMN IV.

Ships' Names, Materials of Construction, Rigs, &c. Masters and Dates of Appointment to Present Owners' Service and to Ships. Ships' Late Names.

COLUMN V.

Flags under which Vessels respectively sail.

COLUMN VI.

Registered Net Tonnages. The tonnages given are the Registered Tonnages of vessels according to the measurements of the countries to which they respectively belong, except in the cases of vessels which have been measured by the British Tonnage Surveyors, when the British Tonnages have been recorded.

COLUMN VII.

Dimensions. The Dimensions given are of English feet, and are as far as possible the Registered Dimensions of Vessels according to the Measurement of the countries to which they respectively belong, except in the case of Vessels which have been measured in Britain, when the British Measurements have (where practicable) been given.

A Record of Depths, thus, $\frac{18 \cdot 6}{25 \cdot 6}$ signifies that the depth to Main Deck is 18·6 and to Upper Deck 25·6.

COLUMN VIII.

Places of Build.

COLUMN IX.

Dates of Build. So far as known, the year in which the Vessel was launched is given as the date of Build.

An entry, thus, 18^{57}_{84}, signifies that the Vessel was built in 1857 and re-built in 1884; an entry thus, 18_{84}, signifies that the date of the Vessel's original Build is unknown, but that she was re-built in 1884.

COLUMN X.

Builders' Names.

COLUMN XI.

Dates of Metal Sheathing. The dates of Metal Sheathing are based on the records of the Books in which vessels are respectively classed, or on the reports of the Surveyors of "Lloyd's Register."

COLUMN XII.

Names of Owners or Managers. In the case of Companies which own a single Vessel *only*, and are named after that Vessel, and in the case of "Actieselskabs," &c., the name of the Manager is inserted thus: "Actieselskab" (Anton Bruun), Ship "Falls of Bruar" Co. (Ld.)—(Wright & Breakenridge). In these cases the Vessels will be found in the "List of Owners" under the names of the Managers.

COLUMN XIII.

Ports of Registry.

REGISTER OF STEAMERS.

COLUMN I.

Consecutive Numbers of Vessels in "Universal Register," a[...] Registry Numbers.

COLUMN II.

Symbols denoting the Classification Societies (if any) with [...] Vessels are classed; also, in the case of Vessels holding a Passeng[...] from the Board of Trade, the last date of bottom-sighting by the [...] Trade Surveyors.

COLUMN III.

Ships' Names, Materials of Construction, Modes of Propulsio[...] Masters and Dates of Appointment to Present Owners' Service a[...] Ships' Late Names, Number of Decks, &c.

COLUMN IV.

Flags under which Vessels respectively sail; also their Ports of [...]

COLUMN V.

Net and Gross Tonnages. See explanation of Column VI. of [...] Sailing Vessels.

COLUMN VI.

Dimensions. See explanation of Column VII. of Register [...] Vessels.

In cases, however, where the Registered Dimensions have bee[...] convey an inaccurate idea of the vessel's size, the entries in the [...] been made from the best available sources of information.

Erections on Deck.

COLUMN VII.

Places of Build. Builders' Names.

COLUMN VIII.

Dates of Build. See explanation of Column IX. of Sailing Vess[...]

COLUMN IX.

Numbers of Bulkheads. The Bulkheads recorded are complete [...] Bulkheads, extending to the upper deck, and are only given in [...] Vessels classed, now or formerly, with Lloyd's Register or the [...] Underwriters' Registry.

Particulars of Water Ballast.

The Dates of the Metal Sheathing of Wood and Composite St[...] also inserted in this column.

COLUMN X.

Particulars of Engines. The Nominal Horse-power from [...] Register is given as far as possible; but this is, of course, no [...] the real power.

Names and Addresses of Makers of Engines.

The figures before "Cy." indicate the Number of Cylinders; [...] following, the Diameters of Cylinders and length of stroke in Engl[...] See explanation of "Cy."

COLUMN XI.

Names of Owners or Managers. See explanation of Colu[...] Register of Sailing Vessels.

KEY TO THE ABBREVIATIONS IN "UNIVERSAL REGISTER."

[illegible]	[illegible]	MEANING.
		A
[illegible]	2	Lloyd's Register of British and Foreign Shipping.
[illegible]	2	Veritas Austro-Ungarico.
..	9	Aft. *See* explanation of "DB."
[illegible]	9	After Peak Tank.
[illegible]	3	Auxiliary.
[illegible]	3	Awning Deck.
		B
..	3	Tiers of Beams; thus—2 *Dks*3 *B* = Three Tiers of Beams of which two are laid with Decks.
..	6	Bridge Deck; thus—*B*66 *f.* = Bridge Deck 66 feet long.
..	9	Boiler space. *See* explanation of "DB."
4	3	Brig.
[illegible]	9	Bulkheads; thus—5 Blhds. = Five Bulkheads.
4	3	Barque.
4	3	Barquentine.
4	3	Bombarde.
4	3	Brigantine.
..	2	Board of Trade; thus—BT8.85 = Date of report of bottom sighting by Board of Trade Surveyors.
		C
11	9	Sheathed with Copper; thus—C5.85 = Sheathed with copper in May, 1885.
4	3	Composite (iron frame, wood planked).
[illegible]	9	Double Bottom of Cellular Construction; thus—CDB518t = Double Bottom of Cellular Construction, having a water capacity of 518 tons.
[illegible]	10	Engines Compounded; thus—Comp. '85, C. D. Holmes & Co., Hull = Engines Compounded in 1885, by Messrs. C. D. Holmes & Co., of Hull.
4	3	Cutter.
..	10	Cylinders; thus—2Cy.18"&36" = Two Cylinders with diameters of 18 & 36 English inches respectively; 4Cy.18"&36" = Four Cylinders of which two have diameters of 18 English inches each, and two of 36 English inches each; 3Cy.18"&(2)36" = Three Cylinders of which one has a diameter of 18 inches, and the other two have diameters of 36 inches.
		D
3	2	Nederlandsche Vereeniging van Assuradeuren.
..	9	Double Bottom; thus—DB*a&f*54t = Double Bottom aft and forward, with total water capacity of 54 tons; DB*a&f&eb*500t = Double Bottom aft and forward and under engines and boilers, with total water capacity of 500 tons; DB*a*26*ft f*25*ft* = Double Bottom aft 26 feet long, and forward 25 feet long.
4	3	Dhoy.
[illegible]	3	Deck or Decks; thus—2*Dks* = Two Decks; 1*Dk*& *Awn Dk*3*B* = Three Tiers of Beams of which one is laid with an Awning Deck, and the second with the Main Deck.
[illegible]	9	Deep Tank.
4	3	Dogger (rig).
4	3	Dandy.
		E
..	9	Engine space. *See* explanation of "DB."
		F
..	6	Forecastle; thus—*F*26 *t.* = Forecastle 26 tons; *F*26 *f.* = Forecastle 26 feet long.
..	9	Forward. *See* explanation of "DB."
..	6 & 9	Feet (English).
..	9	Fore Peak Tank.
		G
3	2	Germanischer Lloyd.
4	3	Galliot.
		H
3	2	Veritas Ellenico.
..	6	Deck House or Houses; thus—*H* 27 *t.* = Deck House or Houses 27 tons.

Contractions.	Register of Sailing Vessels in which used.	Register of Steamers in which used.	MEANING.
			H—*continued.*
HP	..	10	Horse Power. The Nominal Horse Power from the Vessels' Register is given as far as possible, but this is, of course, no indication of the real power.
H. Dk	..	3	Hurricane Deck.
			I
I	3	2	Registro Italiano.
I	4	3	Iron.
			K
K	4	3	Ketch.
Kf	4	3	Keff.
			L
Lng.	..	6	Lengthened.
Lr	4	3	Lugger.
			M
m	4	3	Masts; thus—3m = Three Masts.
MT	..	9	Midship Deep Tank; thus—MT200t = Midship Deep Tank, with water capacity of 200 tons.
			N
N	3	2	Norske Veritas.
			P
P	..	6	Poop; thus—*P* 28 *f.* = Poop 28 feet long.
Pad	..	3	Propelled by Paddles.
Pol	4	3	Polacre.
pt	11	3 & 9	Part.
			Q
Q	..	6	Raised Quarter Deck or Break Deck; thus—*Q* 18 *t.* = Raised Quarter Deck or Break Deck 18 tons.
			R
R	3	2	American Record.
			S
S	4	3	Ship.
Sc	..	3	Propelled by Screw.
Shd Dk	..	3	Shade Deck.
Sk	4	3	Smack.
Sp	4	3	Sloop.
Spr Dk	..	3	Spar Deck.
Sq	4	3	Square-rigged.
Sr	4	3	Schooner.
Stl	4	3	Steel.
Sw	4	3	Snow.
			T
t	..	6 & 9	Tons.
Trb	4	3	Trabacolo.
Tre	4	..	Trecantira.
Twn	..	3	Twin.
			U
U	3	2	Liverpool Underwriters' Registry for Iron Vessels (now amalgamated with Lloyd's Register of British and Foreign Shipping).
			V
V	3	2	Bureau Veritas.
			W
W	4	3	Wood.
WB	..	9	Water Ballast.
			Y
Y	11	9	Sheathed with Yellow Metal; thus—Y8.85 = Sheathed with Yellow Metal in August, 1885.
Yt	4	3	Yacht.
			Z
Z	11	9	Sheathed with Zinc.
2m3m4m	4	3	Two, three, or four-masted.

VERKLARING VAN EN SLEUTEL VOOR DE INHOUD VAN HET "UNIVERSAL R

ZEILSCHEPEN.—(Sailing Vessels.)

Kolom I.

Officieel Getal van de Schepen.

Kolom II.

Volgnummer van de Schepen in het "Universal Register."

Kolom III.

Opgave door Symbolen van de Registers, waarin de Schepen eventueel geclasseerd zijn.

Kolom IV.

Scheepsnamen, Grondstof waarvan gebouwd, Tuig enz. Kapiteins, en datums waarop zij als zoodanig bij hunnen tegenwoordige Reeders in dienst kwamen en op hunnen schepen aangesteld werden. Vroegere namen der schepen.

Kolom V.

Vlaggen waaronder de Schepen varen.

Kolom VI.

Geregistreerde Netto Tonnenmaat. De hier gegevene tonnenmaat is de geregistreerde tonnenmaat van de schepen volgens de meting van de landen waartoe zij behooren, uitgenomen in geval de schepen door Britsche scheepsmeters gemeten zijn en derhalve de Britsche tonnenmaat gegeven is.

Kolom VII.

Afmetingen. De hier gegevene afmetingen zijn in Engelsche voeten, en zoover mogelijk de geregistreerde afmetingen volgens de meting van de landen waartoe de schepen behooren, uitgenomen in geval zij in Brittanje gemeten zijn en alsdan de Britsche meting (waar doenlijk) gegeven is.

De vermelding van holten, aldus, $\frac{18·6}{25·6}$ beteekend dat de holte tot het middendek 18·6 voet is en tot het bovendek 25·6 voet.

Kolom VIII.

Plaatsen van Aanbouw.

Kolom IX.

Datum van Aanbouw. Voor zoover bekend, is het jaar waarin het schip van stapel liep als de datum van aanbouw opgegeven.

De vermelding, aldus, 18^{57}_{84}, beteekend dat het schip in 1857 gebouwd is en in 1884 herbouwd; de vermelding, aldus, 18_{84}, beteekend dat de datum van aanbouw van het schip onbekend is, doch dat hetzelve in 1884 herbouwd werd.

Kolom X.

Namen van de Bouwmeesters.

Kolom XI.

Datum van het koperen. De datums van koperen of met zink bekleeden der schepen zijn gegrond op de vermeldingen daarvan in de boeken waarin dezelve zijn geclasseerd, of op de rapporten van de experts van Lloyd's Register.

Kolom XII.

Namen van Eigenaars of Boekhouders. In het geval van maatschappijen welke slechts een enkel schip bezitten en diezelfde naam voeren, en in het geval van maatschappijen op aandeelen zijn de namen van Directeuren of Boekhouders opgegeven, aldus: "Actieselskab" (Anton Bruun), Ship "Falls of Bruar" Co. (Ld.)—(Wright & Breakenridge). In dergelijke gevallen worden de schepen in de "Lijst van Eigenaars," gevonden onder de namen van de Directeuren of Boekhouders.

Kolom XIII.

Haven waartoe de Schepen behooren.

STOOMSCHEPEN.—(Steamers.)

Kolom I.

Volgnummers van de Schepen in het "Universal Regist[er] Getallen.

Kolom II.

Opgave door Symbolen van de Registers waarin de Sch[epen] geclasseerd zijn; tevens voor het geval dat de schepen [...] Certificaat hebben, van de laatste datum van onderzoek van [...] Opzichters van de "Board of Trade."

Kolom III.

Scheepsnamen, Grondstoffen van den Bouw, Wijze va[n] Tuig van de Schepen. Kapiteins, en datums waarop zij [...] hunnen tegenwoordige Reeders in dienst kwamen en op h[...] aangesteld werden. Vroegere namen der schepen. Aantal D[...]

Kolom IV.

Vlaggen waaronder de Schepen varen, en Havens waarto[e ...]

Kolom V.

Netto en Gross Tonnenmaat. *Zie* verklaring van kolom [...] schepen.

Kolom VI.

Afmetingen. *Zie* verklaring van kolom VII. voor Zeilsche[pen].

In die gevallen echter waar bevonden werd dat d[e] afmetingen een onnauwkeurig denkbeeld van de grootte [...] zoude geven—zijn de inschrijvingen geschied naar inlicht[ingen] beste ten dienste staande bronnen.

Bovenbouw op Dek.

Kolom VII.

Plaatsen van Aanbouw. Namen van de Bouwmeesters.

Kolom VIII.

Datum van Aanbouw. *Zie* verklaring van kolom IX. voor[...]

Kolom IX.

Aantal Schotten. De vermelde schotten zijn geh[eel] schotten reikende tot het opperdek, en alleen van die sch[epen] geclasseerd zijn of vroeger geclasseerd waren in "Lloyd's Re[gister]" "Liverpool Underwriters' Registry."

Bijzonderheden van den Waterballast.

De Datums van het Koperen van Houten en Composite Ste[...] eveneens in deze kolom vermeld.

Kolom X.

Bijzonderheden der Machines. De Nominale Paa[rdekracht] zooveel doenlijk, gegeven volgens het scheepsregister; [...] maatstaf voor de werkelijke kracht der machines.

Namen en Adressen van de Machinefabriekanten. De cij[fers] geven het aantal cilinders aan, de volgende cijfers de [...] cilinders en de slaglengte in Engelsche duimen.

Zie verklaring van "Cy."

Kolom XI.

Namen van Eigenaars of Directeuren.

Zie verklaring van kolom XII. voor Zeilschepen.

Zie in het Reg. voor Zeilschepen kolom	Zie in het Reg. voor Stoomschepen kolom	VERKLARING.
		A
	3	Lloyd's Register of British and Foreign Shipping.
	2	Veritas Austro-Ungarico.
	9	Achter. *Zie* verklaring van "DB."
	9	Achter Piek Tank.
	6	Hulp.
	3	Awning Dek.
		B
	3	Rijën Balken; aldus—2*Dk* 3*B* = Drie Rijën Balken, waarvan twee met Dekken.
..	6	Brugdek; aldus—*B* 66 *f*. = Brugdek 66 voet lang.
	9	Ketelruim. *Zie* verklaring van "DB."
4	3	Brik.
	9	Schotten; aldus—5 Bhds. = Vijf Schotten.
4	3	Bark.
4	3	Barkentijn.
4	3	Bombarde.
4	3	Brigantijn.
..	2	Board of Trade; aldus—BT 8.85 = Datum van onderzoek van den bodem door Opzichters van de "Board of Trade."
		C
11	6	Gekoperd; aldus—C5.85 = Gekoperd in Mei 1885.
4	3	Composite (ijzeren spanten, houten huid).
	9	Dubbele Bodem volgens het Cellen Systeem; aldus—CDB518t = Dubbele Bodem volgens het Cellen Systeem met 518 ton waterruimte.
..	14	Machines in Compound veranderd; aldus—Comp.'85, C. D. Holmes & Co., Hull = Machines in Compound veranderd in 1885 door de firma C. D. Holmes & Co., te Hull.
4	3	Kotter.
..	14	Cilinders; aldus—2Cy. 18" & 36" = Twee Cilinders met middellijnen van 18 & 36 Engelsche duimen respectievelijk; 4 Cy. 18" & 36" = Vier Cilinders, waarvan twee met eene middellijn van 18 Engelsche duimen ieder en 2 van 36 Engelsche duimen ieder; 3 Cy. 18" & (2) 36" = Drie Cilinders, waarvan eene 18 Engelsche duimen middellijn heeft en de beide anderen 36 Engelsche duimen.
		D
3	2	Nederlandsche Vereeniging van Assuradeuren.
..	9	Dubbele Bodem; aldus—DB*a*&*f*54t = Dubbele bodem achter en voor, met waterruimte voor 54 ton; DB*af*&*eb*500t = Dubbele bodem achter, voor en onder machines en ketels met totale waterruimte voor 500 ton; DB*a*26*ft* *f*25*ft* = Dubbele bodem achter 26 voet lang en voor 25 voet lang.
4	3	Dhoy.
..	3	Dek of Dekken; aldus—2*Dks* = Twee Dekken; 1*Dk*& *AwnDk* 3*B* = Twee dekken waarvan een awning dek met balken in het ruim.
	9	Diepe Tank.
4	3	Dogger (tuig).
4	3	Dandy.
		E
..	9	Machineruim. *Zie* verklaring van "DB."
		F
..	6	Bak; aldus—*F* 26 *t*. = Bak 26 ton.
..	9	Voor. *Zie* verklaring van "DB."
..	6 & 9	Voet (Engelsch).
..	9	Voor Piek Tank.
		G
3	2	Germanischer Lloyd.
4	3	Galjoot.
		H
3	2	Veritas Ellenico.
..	6	Dekhuizen; aldus—*H* 27 *t*. = Dekhuizen 27 ton.

Verkortingen.	Zie in het Reg. voor Zeilschepen kolom	Zie in het Reg. voor Stoomschepen kolom	VERKLARING.
			H—*Voortzetting.*
HP	..	10	Paardenkracht. De Nominale Paardenkracht volgens het Scheeps Register wordt zoo veel doenlijk gegeven, doch is dit natuurlijk geen maatstaf voor de werkelijke stoomkracht van het schip.
HurDk	..	3	Hurricane Dek.
			I
I	3	2	Registro Italiano.
I	4	3	Yzer.
			K
K	4	3	Ketch.
Kf	4	3	Kof.
			L
Len	..	6	Verlengd.
Lr	4	3	Logger.
			M
m	4	3	Masten; aldus—3m = Drie Masten.
MT	..	9	Midscheepsche Diepe Tank; aldus—MT 200 ton = Midscheepsche Diepe Tank met waterruimte voor 200 ton.
			N
N	3	2	Norske Veritas.
			P
P	..	6	Kampanje; aldus—*P* 28 *f*. = Kampanje 28 voet lang.
Pad	..	3	Bewogen door Raderen.
Pol	4	3	Polak.
pt	11	8 & 9	Gedeeltelijk.
			Q
Q	..	6	Verhoogd Halfdek; aldus—*Q* 18 *t*. = Verhoogd Halfdek 18 ton.
			R
R	3	2	American Record.
			S
S	4	3	Volschip.
Sc	..	3	Bewogen door de Schroef.
ShdDk	..	3	Shade Dek.
Sk	4	3	Smak.
Sp	4	3	Sloep.
SprDk	..	3	Spar Dek.
Sq	4	3	Vierkante Zeilen.
Sr	4	3	Schooner.
Stl	4	3	Staal.
Sw	4	3	Snow.
			T
t	..	6 & 9	ton.
Trb	4	3	Trabacolo.
Trc	4	..	Trecantira.
Twn	..	3	Tweeling.
			U
U	3	2	"Liverpool Underwriters' Registry for Iron Vessels" (nu vereenigd met "Lloyd's Register of British and Foreign Shipping").
			V
V	3	2	Bureau Veritas.
			W
W	4	3	Hout.
WB	..	9	Water Ballast.
			Y
Y	11	6	Bekleed met "Yellow Metal"; aldus—Y 8.85 = Bekleed met "Yellow Metal" in Augustus 1885.
Yt	4	3	Yacht.
			Z
Z	11	9	Bekleed met Zinc.
2m3m4m	4	3	Twee, drie of vier Masten.

EXPLICATION ET CLEF DES MATIÈRES DE L'"UNIVERSAL REGIST

REGISTRE DES NAVIRES À VOILES.—(Sailing Vessels.)

COLONNE I.

Numéros matricule d'Enregistrement des Navires.

COLONNE II.

Numéros consécutifs des Navires dans l' "Universal Register."

COLONNE III.

Signes indiquant les Sociétés de Classification (s'il y a lieu) chez lesquelles le navire est coté.

COLONNE IV.

Noms, Matériaux de Construction, et Types du Gréement des Navires. Capitaines et dates d'entrée au service des Armateurs actuels et à bord des Navires. Anciens Noms des Navires.

COLONNE V.

Pavillons respectifs des Navires.

COLONNE VI.

Tonnages nets. Les tonnages donnés sont les tonnages officiels calculés d'après les mesures des pays auxquels les navires appartiennent; cependant, si les navires ont été jangés par les contrôleurs de la jauge anglaise, le tonnage enregistré est anglais.

COLONNE VII.

Dimensions. Les dimensions des navires sont données en pieds anglais, et autant que possible ce sont celles des pays auxquels les navires appartiennent; cependant, si les navires ont été mesurés en Angleterre, dans ce cas, celles qui sont données sont (si la chose est possible) les mesures anglaises.

L'annotation suivante pour les creux, $\frac{18·6}{25·6}$ signifie que le creux jusqu'au pont principal est de 18·6 pieds et que le creux jusqu'au pont supérieur est de 25·6 pieds.

COLONNE VIII.

Lieux des Constructions.

COLONNE IX.

Dates des Constructions. Autant que la date de la mise à l'eau a pu être vérifiée, cette date est donnée, et l'année du lancement est considérée comme année de la construction.

L'annotation $18\frac{57}{84}$ signifie que le navire a été construit en 1857 et reconstruit en 1884; 18_{84} signifie que la date de construction est inconnue, mais que le navire a été reconstruit en 1884.

COLONNE X.

Noms des Constructeurs.

COLONNE XI.

Date du Doublage en Métal. Les dates du doublage en métal ressortent des livres dans lesquels les navires sont respectivement classés ou des rapports des Inspecteurs du Lloyd's Register.

COLONNE XII.

Noms des Armateurs ou des Directeurs. Pour les compagnies qui ne possèdent qu'un navire et qui se nomment d'après le nom du navire, le nom du directeur est inscrit à côté de celui de la compagnie. Soit pour le "Actieselskab," etc., le nom du directeur est inscrit ainsi: "Actieselskab" (Anton Braun), Navire "Falls of Bruar" Co. (Ld.)—(Wright & Breakenridge). Dans ces conditions, les noms des navires seront trouvés dans la "Liste des Armateurs" au-dessous des noms des directeurs.

COLONNE XIII.

Ports d'Attache.

REGISTRE DES NAVIRES À VAPEUR.—

COLONNE I.

Numéros consécutifs des Navires dans l' "Universal Re[illegible] matricule d'Enregistrement.

COLONNE II.

Signes indiquant les Sociétés de Classification (s'il y a l[illegible] le navire est coté; et si le navire a un Certificat du "Board [illegible] le Transport des Passagers, il est fait mention de la date [illegible] inspection qui a été faite des fonds du navire par les Exper[illegible] Trade."

COLONNE III.

Noms, Materiaux de Construction, Modes de Propulsi[illegible] Gréement des Navires. Capitaines et dates d'entrée au servic[illegible] actuels et à bord des Navires. Anciens Noms des Navir[illegible] etc.

COLONNE IV.

Pavillons respectifs des Navires et leur Port d'Attache.

COLONNE V.

Tonnages nets et bruts. *Voir* les explications de la [illegible] Registre des Navires à Voiles.

COLONNE VI.

Dimensions. *Voir* les explications de la Colonne VII [illegible] Navires à Voiles.

Cependant, dans le cas où les dimensions enregistrées ne [illegible] une idée exacte de la capacité des navires, les indication[illegible] viennent néanmoins des meilleures sources d'informations.

Érections sur le Pont.

COLONNE VII.

Lieu de la Construction. Noms des Constructeurs.

COLONNE VIII.

Dates des Constructions. *Voir* les explications de la C[illegible] Navires à Voiles.

COLONNE IX.

Nombre des Cloisons. Les cloisons mentionnées sont [illegible] plètement étanches, régnant jusqu'au pont supérieur; elles [illegible] pour les navires actuellement ou antérieurement classés da[illegible] Register" ou dans le "Liverpool Underwriters' Registry."

Renseignements sur le lest d'eau.

Les Dates du Doublage en Métal des navires à vapeur [illegible] posites sont aussi insérées dans cette colonne.

COLONNE X.

Particularites relatives aux Machines à Vapeur. La [illegible] chevaux est donnée, autant que possible, d'après le Registre [illegible] elle n'est pas une indication de la force réelle ou effective.

Noms et Adresses des Constructeurs de Machines.

Les chiffres devant "Cy." indiquent le Nombre des [illegible] qui suivent indiquent le Diamètre des Cylindres et la longu[illegible] du piston en pouces anglais. *Voir* l'explication de "Cy."

COLONNE XI.

Nom des Armateurs ou des Directeurs. *Voir* les exp[illegible] Colonne XII du Registre des Navires à Voiles.

CLEF DES ABRÉVIATIONS DANS L'"UNIVERSAL REGISTER."

[illegible]	[illegible]	[illegible]	EXPLICATION.
			A
[illegible]	[illegible]	[illegible]	Lloyd's Register of British and Foreign Shipping.
A	[illegible]	[illegible]	Veritas Austro-Ungarico.
[illegible]		[illegible]	Arrière. *Voir* l'explication de "DB."
AFT		[illegible]	Cale à eau arrière.
[illegible]		[illegible]	Auxiliaire.
[illegible]		[illegible]	Pont Abri.
			B
[illegible]		3	Rangées de Baux; ainsi—*2Dks3B* = Trois Rangées de Baux et deux Ponts bordés.
[illegible]		6	Pont Passerelle; ainsi—*B 66 f.* = Un Pont Passerelle de 66 pieds de long.
[illegible]		9	Chambre des Chaudières. *Voir* l'explication de "DB."
[illegible]	4	3	Brick.
[illegible]	..	9	Cloisons.
[illegible]	4	3	Trois Mats Barque.
[illegible]	4	3	Barquentin.
[illegible]	4	3	Bombarde.
[illegible]	4	3	Brigantin.
BT		2	Board of Trade; ainsi—BT8.85 = Que le fond a été visité, à cette date, par les Experts du "Board of Trade," et que leur rapport a été déposé.
			C
C	11	9	Doublé en Metal Rouge; ainsi—C5.85 = Que le navire a été doublé en cuivre rouge en Mai 1885.
C	4	3	Composite (membrure en fer, bordé en bois).
CDB	.	9	Double-Fonds de Construction Cellulaire; ainsi—CDB518t = Double-Fonds de Construction Cellulaire, pouvant contenir un volume d'eau égal à 518 tonneaux.
[illegible]	..	10	Machines Compound; ainsi—Comp. '85, C. D. Holmes & Co., Hull = Machines Compound faites en 1885 par Messrs. C. D. Holmes & Co., de Hull.
[illegible]	4	3	Côtre.
[illegible]	..	10	Cylindres; ainsi—2Cy.18"&36" = Deux Cylindres ayant respectivement 18 & 36 pouces anglais de diamètre; 4Cy.18"&36" = Quatre Cylindres dont deux ont un diamètre de 18 pouces anglais, et deux de 36 pouces anglais chacun; 3Cy.18"&(2)36" = Trois Cylindres dont un a un diamètre de 18 pouces, et les deux autres un diamètre de 36 pouces chacun.
			D
D	3	2	Nederlandsche Vereeniging van Assuradeuren.
DB	..	9	Double-Fond; ainsi—DB*a&f*54t = Un Double-Fond à l'arrière et un Double-Fond à l'avant, pouvant contenir ensemble un volume d'eau de 54 tonneaux; DB*af&eb*500t = Double-Fond arrière, avant et sous les machines et les chaudières, pouvant contenir ensemble un volume d'eau de 500 tonneaux; DB*a26ft f25ft* = Double-Fond arrière ayant une longueur de 26 pieds et à l'avant un Double-Fond de 25 pieds de long.
[illegible]	4	3	Dhoy.
[illegible]		3	Pont ou Ponts; ainsi—*2Dks* = Deux Ponts; *1Dk & AwnDk3B* = Trois Étages de Baux avec un Pont Principal bordé, et un Pont Awning (ou abri) bordé et calfaté.
DT	..	9	Cale à eau.
Dr	4	3	Dogre (Greement).
Dy	4	3	Dandy.
			E
e	..	9	Chambre des Machines. *Voir* l'explication des "DB."
			F
F	..	6	Gaillard d'avant; ainsi *F 26 t.* = Gaillard d'avant de 26 tonneaux.
f		9	En avant. *Voir* l'explication de "DB."
[illegible]	..	6 & 9	Pieds (anglais).
FFT	..	9	Cale à eau avant.
			G
G	3	2	Germanischer Lloyd.
[illegible]	4	3	Galiotte.
			H
H	3	2	Veritas Hellénique.
H		6	Rouffle ou Rouffles; ainsi—*H 27 t.* = Un Rouffle ou des Rouffles cubant 27 tonneaux.
			H—*suite.*
HP		10	Chevaux de Force. La force nominale de chevaux est donnée aussi exacte que possible d'après le Registre du Navire, mais cette force n'est point la force réelle effective de la machine.
HurDk	..	3	Hurricane-Deck.
			I
I	3	2	Registro Italiano.
I	4	3	Fer.
			K
K	4	3	Ketch.
Kf	4	3	Koff.
			L
Lr	4	3	Lougre.
Len	..	6	Allongé.
			M
m	4	3	Mâts; ainsi—3m = 3 Mâts.
MT	..	9	Cale à eau au Milieu; ainsi—MT200t = Qu'à la partie centrale il y a une Cale à eau pouvant contenir un volume d'eau égal à 200 tonneaux.
			N
N	3	2	Norske Veritas.
			P
P	..	6	Dunette; ainsi—*P 28 f.* = Dunette ayant 28 pieds de long.
Pad	..	3	Mû par des Aubes (ou Roues).
Pol	4	3	Polacre.
pt	11	3 & 9	Partie ou Part.
			Q
Q	..	6	Demi-Dunette ou Coupée; ainsi—*Q 18 t.* = Demi-Dunette ou Coupée de 18 tonneaux.
			R
R	3	2	American Record.
			S
S	4	3	Trois Mats carré.
Sc	..	3	L'Hélice comme Propulseur.
ShdDk	..	3	Pont-Taud.
Sk	4	3	Smack.
Sp	4	3	Sloop.
SprDk	..	3	Spar Deck.
Sq	4	3	Mâté en quarré.
Sr	4	3	Goëlette.
Stl	4	3	Acier.
Sw	4	3	Senau.
			T
t	..	6 & 9	Tons.
Trb	4	3	Trabaccoli (Navires de la Mer Adriatique).
Tre	4	..	Trécantira.
Twn	..	3	Jumelles.
			U
U	3	2	"Liverpool Underwriters' Registry for Iron Vessels" qui a été fondu dans le "Lloyd's Register of British and Foreign Shipping."
			V
V	3	2	Bureau Veritas.
			W
W	4	3	Bois.
WB	..	9	Lest d'eau.
			Y
Y	11	9	Doublé en Métal Jaune; ainsi—Y8.85 = Doublé en Métal Jaune (ou Cuivre Jaune) dans le mois d'août 1885.
Yt	4	3	Yacht.
			Z
Z	11	9	Doublé en Zinc.
2m3m4m	4	3	A deux, trois ou quatre mâts.

ERKLÄRUNG UND SCHLUSSEL ZUM INHALT DES "UNIVERSAL REGIS[TER]

REGISTER DER SEGELSCHIFFE.—(Sailing Vessels.)

RUBRIK I.

Officielle Register-Nummern der Schiffe.

RUBRIK II.

Laufende Nummer der Schiffe im "Universal Register."

RUBRIK III.

Bezeichnungen für die Classifications-Gesellschaften, bei welchen die Schiffe eventuell classificirt sind.

RUBRIK IV.

Namen der Schiffe, Constructions-Material. Takelung u.s.w.; Capitäne, und Datum ihres Dienst-Antritts bei gegenwärtiger Rhederei und auf den betreffenden Schiffen. Frühere Namen der Schiffe.

RUBRIK V.

Flagge, unter welcher die Schiffe segeln.

RUBRIK VI.

Register-Netto-Tonnengehalt. Der Tonnengehalt ist den Vermessungen der Heimathsländer, der Schiffe gemäss gegeben. In den Fällen jedoch, wo Schiffe von Britischen Zollbeamten vermessen sind, ist der Britische Tonnengehalt verzeichnet.

RUBRIK VII.

Dimensionen. Die Dimensionen sind in Englischen Fuss und, soweit als möglich, den registrirten Dimensionen der Schiffe nach den Vermessungen der betr. Heimathländer gemäss gegeben. Ausgenommen hiervon sind diejenigen Fälle, in welchen Schiffe in Grossbritannien vermessen wurden, und sind dann die Englischen Vermessungs-Dimensionen (soweit wie thunlich) angegeben.

Ein Vermerk über Tiefe, als $\frac{18\cdot6}{25\cdot6}$, bedeutet, dass dieselbe bis zum Hauptdeck 18·6 und bis zum Oberdeck 25·6 Fuss beträgt.

RUBRIK VIII.

Ort der Erbauung.

RUBRIK IX.

Datum der Erbauung. Soweit es bekannt, ist das Jahr, in welchem das Schiff vom Stapel gelassen wurde, als Erbauungs-Datum angegeben.

Eine Eintragung, wie $18\frac{57}{84}$, bedeutet, dass das Schiff 1857 gebaut und 1884 umgebaut wurde; eine solche, wie 18_{84}, bedeutet, dass das Erbauungs-Datum des Schiffes unbekannt, dasselbe jedoch 1884 umgebaut worden ist.

RUBRIK X.

Name des Erbauers.

RUBRIK XI.

Datum des Metall-Beschlages. Die Daten des Metall-Beschlags basiren auf den Angaben der Register-Bücher, in welchen die Schiffe respective classificirt sind oder auf den Rapporten der Besichtiger des Lloyd's Register.

RUBRIK XII.

Namen der Rheder oder Correspondenz-Rheder. In Fällen wo Gesellschaften, Eigner nur eines Schiffes sind und dessen Namen führen, oder wenn es Actien-Gesellschaften u.s.w. sind, ist der Name des Correspondenz-Rheder angegeben: z. B. "Actieselskab" (Anton Bruun), oder, Ship "Falls of Bruar" Co. (Ld.)—(Wright & Breakenridge). In solchen Fällen findet man die Schiffe in der Rhederliste unter dem Namen der betr. Correspondenz-Rheder, u.s.w.

RUBRIK XIII.

Heimathshäfen.

REGISTER DER DAMPFSCHIFFE.—(Stea[m]

RUBRIK I.

Laufende Nummer im "Universal Register" und Offic[ielle] Nummern.

RUBRIK II.

Bezeichnungen für die Classifications-Gesellschaften, bei [welchen die] Schiffe eventuell classificirt sind; und ist, im Fall ein Schiff ei[n] Certificat des "Board of Trade" führt, auch das Datum [der] besichtigung durch einen "Board of Trade Surveyor," angegeben.

RUBRIK III.

Namen der Schiffe, Constructions-Material, Takelung u.s.w.; Ca[pitäne] Datum ihres Dienst-Antritts bei gegenwärtiger Rhederei u[nd] betreffenden Schiffen. Frühere Namen der Schiffe, Anzahl der [...]

RUBRIK IV.

Flagge, unter welcher die Schiffe segeln, sowie die Namen ihre[r Heimath]häfen.

RUBRIK V.

Netto- und Brutto-Tonnengehalt. *Siehe* Erklärung der Rub[rik ...] Registers für Segelschiffe.

RUBRIK VI.

Dimensionen. *Siehe* Erklärung der Rubrik VII. des R[egisters für] Segelschiffe.

In Fällen, wo die registrirten Dimensionen eine ungenaue [...] von der Grösse des Schiffes gaben, sind die Eintragungen au[f ...] best-möglichsten Quellen gemacht.

Aufbauten auf Deck.

RUBRIK VII.

Ort der Erbauung. Name des Erbauers.

RUBRIK VIII.

Datum der Erbauung. *Siehe* Erklärung der Rubrik IX. de[s Registers] der Segelschiffe.

RUBRIK IX.

Anzahl der Schotten. Die vermerkten Schotten sind völlig w[...] erstrecken sich bis zum Oberdeck, und sind nur in solchen Fälle[n ...] wo Schiffe jetzt oder früher im "Lloyd's Register," oder " Underwriters' Registry," classificirt sind oder waren.

Angaben über Wasser-Ballast.

Die Daten des Metall-Beschlages von Hölzernen und Comp[osite-]schiffen sind auch in dieser Rubrik angeführt.

RUBRIK X.

Näheres über Maschinen. Die nominale Pferdestärke ist [...] möglich, den officiellen Schiffs-Registern entnommen; jedoch [...] Maassstab für die wirkliche Kraft.

Namen und Adressen der Maschinen-Fabrikanten.

Die Zahlen vor "Cy." geben die Anzahl der Cylinder, [...] dahinter die Durchmesser der Cylinder und den Kolbenhub in [...] Zoll. *Siehe* Erklärung von "Cy."

RUBRIK XI.

Namen der Rheder oder Correspondenz-Rheder. *Siehe* E[rklärung der] Rubrik XII. des Registers für Segelschiffe.

SCHLÜSSEL ZU DEN ABKÜRZUNGEN IM "UNIVERSAL REGISTER."

Abkürzungen	Rubrik der Register der Segelschiffe	Rubrik der Register der Dampfschiffe	Bedeutung
			A
[illegible]	3	2	Lloyd's Register of British and Foreign Shipping.
[illegible]	3	2	Veritas Austro-Ungarico.
[illegible]		9	Hinten. *Siehe* Erklärung von "DB."
[illegible]		9	Hinter-Pick-Tank.
[illegible]		3	Hülfs-.
[illegible]		3	Sturm-Deck.
			B
[illegible]		3	Balkenreihen; z. B.—*2Dks3B* = Drei Reihen Balken von welchen zwei mit Decks belegt sind.
[illegible]	..	6	Brücken-Deck; z. B.—*B66 f.* = Brücken-Deck 66 Fuss lang.
[illegible]	..	9	Kesselraum. *Siehe* Erklärung von "DB."
[illegible]	4	3	Brigg.
[illegible]	..	9	Schotten.
[illegible]	4	3	Bark.
[illegible]	4	3	Schooner-Bark.
[illegible]	4	..	Bombardes.
[illegible]	4	3	Schooner-Brigg.
[illegible]	..	2	Board of Trade; z. B.—BTS.85 = Datum des Berichtes über Bodenbesichtigung von Seiten eines "Board of Trade Surveyors."
			C
[illegible]	11	9	Mit Kupfer beschlagen; z. B.—C5.85 = Mit Kupfer beschlagen im Mai 1885.
[illegible]	4	3	Composit (eiserne Spanten, Holz-Planken).
[illegible]		9	Doppel-Boden nach dem Zellen-System; z. B.—CDB518t = Doppel-Boden nach dem Zellen-System von 518 Tons Wasserinhalt.
[illegible]		10	Maschinen zu Compound-Maschinen umgebaut; z. B.—Comp.'85, C. D. Holmes & Co., Hull = Maschinen im Jahre 1885 von den Herren C. D. Holmes & Co. in Hull zu Compound-Maschinen umgebaut.
[illegible]	4	3	Kutter.
[illegible]	..	10	Cylinder; z. B.—2Cy.18"&36" = Zwei Cylinder mit resp. 18 & 36 Zoll engl. Durchmesser; 4Cy. 18"&36" = Vier Cylinder, von welchen zwei 18 und zwei 36 Zoll engl. Durchmesser haben; und 3Cy.18"&(2)36" = Drei Cylinder, von welchen einer 18 und zwei je 36 Zoll engl. Durchmesser haben.
			D
D	3	2	Nederlandsche Vereeniging van Assuradeuren.
[illegible]		9	Doppel-Boden; z. B.—DB*a*&*f*54t = Doppel-Boden hinten und vorne, mit Gesammt-Wasserinhalt von 54 Tons; DB*af*&*eb*500t = Doppel-Boden hinten, vorne und unter Maschinen- und Kesselraum, mit Gesammt-Wasserinhalt von 500 Tons; DB*a*26*ft* *f*25*ft* = Doppel-Boden hinten 26 Fuss lang und vorne 25 Fuss lang.
[illegible]	4	3	Dhoy.
[illegible]		3	Deck oder Decks; z. B.—2*Dks* = Zwei Decks; 1*Dk* &*AwnDk*3*B* = Drei Reihen Balken, von welchen die eine mit einem Sturm-Deck, die zweite mit dem Hauptdeck belegt ist.
[illegible]	..	9	Hoher Tank.
[illegible]	4	3	Dogger (Takelung).
[illegible]	4	3	Dandy (Takelung).
			E
[illegible]	..	9	Maschinenraum. *Siehe* Erklärung von "DB."
			F
[illegible]	..	6	Back; z. B.—*F* 26 *t.* = Back von 26 Tons Raumgehalt.
[illegible]	..	9	Vorder. *Siehe* Erklärung von "DB."
[illegible]	..	6 & 9	Fuss (Englisch Maass).
[illegible]	..	9	Vor-Pick-Tank.
			G
[illegible]	3	2	Germanischer Lloyd.
[illegible]	4	3	Galliot.
			H
H	3	2	Veritas Ellenico.
H	..	6	Deckhaus oder Häuser; z. B.—*H* 27 *t.* = Deckhaus oder Häuser von 27 Tons Raumgehalt.
			H—*Fortsetzung.*
[illegible]	..	10	Pferdestärke. Die nominellen Pferdestärken sind, soweit möglich, den officiellen Registern entnommen; diese Angabe ist jedoch kein Maassstab für die wirkliche Kraft.
HurDk	..	3	Hurricane Deck.
			I
I	3	2	Registro Italiano.
I	4	3	Eisen.
			K
K	4	3	Ketch.
Kf	4	3	Kuff.
			L
Len			Verlängert.
Lr	4	3	Lugger.
			M
m	4	3	Masten; z. B.—3m = Drei Masten.
MT	..	9	Hoher Mittschiffs-Tank; z. B.—MT200t = Hoher Mittschiffs-Tank, mit 200 Tons Wassergehalt.
			N
N	3	2	Norske Veritas.
			P
P	..	6	Hütte (Poop; z. B.—*P* 28 *f.* = Hütte 28 Fuss lang.
Pad	..	3	Fortbewegt durch Räder.
Pol	4	3	Polacker Brigg.
pt	11	3 & 9	Zum Theil.
			Q
Q	..	6	Erhöhtes Quarter-Deck oder Break-Deck; z. B.—*Q* 18 *t.* = Erhöhtes Quarter-Deck oder Break-Deck von 18 Tons Raumgehalt.
			R
R	3	2	American Record.
			S
S	4	3	Schiff.
Sc	..	3	Durch Schraube fortbewegt.
ShdDk	..	3	Schatten-Deck.
Sk	4	3	Smack.
Sp	4	3	Slup.
SprDk	..	3	Spar-Deck.
Sq	4	3	Mit Raasegeln betakelt.
Sr	4	3	Schooner.
Stl	4	3	Stahl.
Sw	4	3	Schnau.
			T
t	..	6 & 9	Tons.
Trb	4	3	Trabaccolo.
Trc	4	..	Trecantira.
Twn	..	3	Zwillings-.
			U
U	3	2	"Liverpool Underwriters' Registry for Iron Vessels" (jetzt vereinigt mit "Lloyd's Register of British and Foreign Shipping").
			V
V	3	2	Bureau Veritas.
			W
W	4	3	Holz.
WB	..	9	Wasser-Ballast.
			Y
Y	11	9	Mit "Yellow Metal" beschlagen; z. B.—Y8.85 = Mit "Yellow Metal" beschlagen im August 1885.
Yt	4	3	Yacht.
			Z
Z	11	9	Mit Zink beschlagen.
4m3m2m	4	3	Zwei-, drei- oder vier-mastiges Schiff.

SCHIARIMENTO E SPIEGAZIONE SUL CONTENUTO DEL "UNIVERSAL REGI

REGISTRO DEI VELIERI.—(Sailing Vessels.)

COLONNA I.

Numeri di matricola dei bastimenti.

COLONNA II.

Numeri progressivi dei bastimenti nel "Universal Register."

COLONNA III.

Simboli dinotanti le società di classificazione (se ve ne sono) nelle quali i bastimenti sono classificati.

COLONNA IV.

Nomi dei Bastimenti, Materiali di Costruzione, Attrezzature ecc.; Capitani e date di nomina al Servizio degli attuali Armatori e Bastimenti, Nomi antecedenti dei Bastimenti, Numero delle Coperte ecc.

COLONNA V.

Bandiera sotto cui i bastimenti rispettivamente navigano.

COLONNA VI.

Tonnellaggi netti di Registro. I tonnellaggi dati sono i tonnellaggi registrati dei bastimenti secondo le misure dei paesi a cui essi rispettivamente appartengono, eccetuato nei casi di bastimenti i quali vennero misurati dagli Ispettori del tonnellaggio Inglese, alloraquando i tonnellaggi Inglesi sono stati dati.

COLONNA VII.

Dimensioni. Le dimensioni date sono in piedi Inglesi, e sono al più giusto possibile le dimensioni registrate dei bastimenti secondo le misure dei paesi a cui essi rispettivamente appartengono, eccetuato nel caso di bastimenti che vennero misurati in Inghilterra, allorquando le misure Inglesi furono (se praticabile) date.

Le altezze inscritte, così $\frac{18{\cdot}6}{25{\cdot}6}$ significa che l'altezza al ponte ossia coperta principale è 18·6 piedi e al ponte ossia coperta superiore 25·6 piedi.

COLONNA VIII.

Luoghi di costruzione.

COLONNA IX.

Date della costruzione. Per quanto si conosca, l'anno in cui i bastimenti furono varati è dato quale quello della costruzione. L'inscrizione in questo modo, $18\frac{57}{84}$, significa che il bastimento fu construtto nel 1857 e ricostrutto nel 1884; l'inscrizione in questo modo, 18_{84}, significa che la data della costruzione del bastimento è sconosciuta, ma che esso venne ricostrutto nel 1884.

COLONNA X.

Nomi dei costruttori.

COLONNA XI.

Date del foderatura di metallo. Le date della foderatura di metallo si basano sulle inscrizioni dei Libri nei quali i Bastimenti sono rispettivamente classificati, oppure sui rapporti degli Ispettori del "Lloyd's Register."

COLONNA XII.

Nomi degli armatori od amministratori. Nel caso di compagnie le quali possedono un solo bastimento e prendono il nome da lui, e nel caso di "Actieselskab," ecc., il nome dell' amministratore viene inscritto, così "Actieselskab" (Anton Bruun), Bastimento "Falls of Bruar" Co. (Ld.)—(Wright & Breakenridge).

In questi casi il bastimento troverassi nella lista degli armatori sotto il nome dell' amministratore.

COLONNA XIII.

Porti a cui i bastimenti appartiengono.

REGISTRO DEI PIROSCAFI.—(Steamer

COLONNA I.

Numeri progressivi dei bastimenti nel "Universal Regist matricola.

COLONNA II.

Simboli dinotanti le società di classificazione (se ve ne so essi bastimenti sono classificati; nonchè—nel caso di bastim un certificato per passeggieri dal "Board of Trade"—l' ulti ispezione del fondo fatta dagli Ispettori del "Board of Trade."

COLONNA III.

Nomi, materiali di costruzione, sistema di propulsione, e dei bastimenti; Capitani e date di nomina al servizio degli att e bastimenti; nomi antecedenti dei bastimenti; nonchè il num ossia coperte ecc.

COLONNA IV.

Bandiera sotto la quale i bastimenti rispettivamente navi i porti a cui appartengono.

COLONNA V.

Tonnellaggi netti e lordi. *Vedi* spiegazioni alla Colonna VI dei Velieri.

COLONNA VI.

Dimensioni. Vedi spiegazioni alla Colonna VII del Regist

Nei casi però, dove trovossi che le dimensioni registrate da non troppo giusta della grandezza dei bastimenti, le inscriz vennero fatte colle migliori sorgenti d' informazioni che si poss

Costruzioni sopra coperta.

COLONNA VII.

Luoghi di costruzione. Nomi dei costruttori.

COLONNA VIII.

Date della costruzione. Vedi spiegazioni nella Colonna IX d

COLONNA IX.

Numero delle paratie stagne. Le paratie stagne inscritte completamente stagne all' acqua estendentesi al ponte superiore e sono esclusivamente inscritte per quei bastim attualmente o erano classificati nel "Lloyd's Register" oppure Underwriters' Registry."

Particolari sulla zavorra d' acqua.

Le date del fasciamento in metallo dei piroscafi in legno e anche inscrite in questa colonna.

COLONNA X.

Particolari delle macchine. La forza in cavalli nominali è che sia possibile dal Registro dei Bastimenti; però essa non la forza reale. Nomi e indirizzi dei costruttori delle macchine.

I numeri dinanzi "Cy." indicano il numero dei Cili seguenti il diametro dei Cilindri e la corsa del Pistone in p *Vedi* spiegazioni dei Cilindri.

COLONNA XI.

Nomi degli Armatori od Amministratori. *Vedi* spiegazioni all dei Velieri.

SPIEGAZIONE DELLE ABBREVIAZIONE DEL "UNIVERSAL REGISTER."

[illegible]	[illegible]	[illegible]	SIGNIFICA.
			A
[illegible]	[illegible]	2	Lloyd's Register of British and Foreign Shipping.
A	3	2	Veritas Austro-Ungarico.
[illegible]	..	9	Puppa. *Vedi* spiegazioni del "DB."
APT	..	9	Cisterna Poppiera.
[illegible]	..	3	Ausiliario.
[illegible]		3	Coperta di Manovra.
			B
B	..	3	Ordini di Bagli; così—*2Dks3B*=Tre Ordini di Bagli dei quali due sono coperti da ponte ossia coperta.
B	..	6	Ponte di Comando; così—*B 66 f.*=Ponte di Comando 66 piedi lungo.
[illegible]	..	9	Spazio delle Caldaie. *Vedi* spiegazioni del "DB."
[illegible]	4	3	Brigantino.
[illegible]	..	9	Paratio Stagno.
[illegible]	4	3	Brigantino a Pale.
[illegible]	4	3	Nave Goletta.
[illegible]	4	..	Bombarda.
[illegible]	4	..	Brigantino Goletta.
BT	..	2	Board of Trade; così—BT8.85=Data del rapporto di visita del fondo fatta dagli Ispettori del "Board of Trade."
			C
C	4	9	Fasciato con Rame; così—C5.85=Fasciato con Rame nel Maggio nel 1885.
C	4	3	Composito (ordinate in ferro, fasciame in legno).
CDB	..	9	Doppio Fondo di Costruzione Cellulare; così—CDB518t = Doppio Fondo di Costruzione Cellulare capace di contenere 518 tonnellate d'acqua.
Comp.	..	10	Macchine Trasformate; così — Comp. '85, C. D. Holmes & Co., Hull = Macchine Trasformate nel 1885 dai Signori C. D. Holmes e Co., di Hull.
Cr	4	3	Cutter.
Cy.	..	10	Cilindri; così—2Cy.18"&36"=Due Cilindri con diametro di 18 e 36 pollici Inglesi rispettivamente; 4Cy.18"&36"=Quattro Cilindri di cui due hanno il diametro di 18 pollici Inglesi e due di 36 per ciascheduno; 3Cy.18"&(2)36"=Tre Cilindri di cui uno ha il diametro di 18 pollici, e gli altri due hanno il diametro di 36 pollici per ciascheduno.
			D
D	3	2	Nederlandsche Vereeniging van Assuradeuren.
DB	..	3	Doppio Fondo; così—DB*a&f*54*t*=Doppio Fondo di poppa e di prua la di cui capacità è di 54 tonnellate d' acqua; DB*af&eb*500t=Doppio Fondo a poppa, a prua e sotto le macchine e le caldaie con una totale capacità di tonnellate 500 d' acqua; DB*a*26*ft* *f*25*ft*=Doppio Fondo a poppa lungo 26 piedi e a prua lungo 25 piedi.
Dhy	4	3	Dhoy.
Dk o *Dks*	..	3	Coperta o Coperte; così—*2Dks*=Due Coperte o Due Ponti; *1Dk&AwnDk3B*=Tre Ordini di Bagli dei quali uno è fasciato a coperta o ponte di manovra ed il secondo a coperta o ponte principale.
DT	..	9	Cisterna profonda.
Dr	4	3	Dogger (attrezzatura).
Dy	4	3	Dandy.
			E
e	..	9	Spazio delle Macchine. *Vedi* spiegazioni sul "DB."
			F
F	..	6	Castello di Prua; così—*F 26 t.*=Castello di Prua 26 tonnellate.
f	..	9	Pruavia. *Vedi* spiegazioni del "DB."
f. o *ft*	..	6 & 9	Piedi (Inglesi).
FPT	..	9	Cisterna Prodiera.
			G
G	3	2	Germanischer Lloyd (Lloyd Germanico).
Glt	4	3	Galleota.
			H
H	3	2	Veritas Ellenico.
H	..	6	Casetta o Casette in Coperta; così—*H 27 t.*=Casetta o Casette in Coperta 27 tonnellate.
			H—*continuate.*
HP	..	10	Cavalli Vapore. La Forza Nominale in Cavalli Vapore è data per quanto si può dal Registro dei Bastimenti, ma ciò naturalmente non indica la vera forza.
HurDk	..	3	Coperta d' Uragano.
			I
I	3	2	Registro Italiano.
I	4	3	Ferro.
			K
K	4	3	Ketch.
Kf	4	3	Koff.
			L
Len.	..	6	Allungato.
Lr	4	3	Louggre.
			M
m	4	3	Alberi; così—3m=Tre Alberi.
MT	..	9	Cisterna Profonda a Mezzania; così—MT200t=Cisterna Profonda a Mezzania con una capacità di tonnellate 200 d' acqua.
			N
N	3	2	Norske Veritas (Veritas Norvegino).
			P
P	..	6	Cassaretto; così—*P 28 f.*=Casseretto lungo 28 piedi.
Pad	..	3	Propulsione a Ruote.
Pol	4	3	Polacca.
pt	11	3 & 9	Parte.
			Q
Q	..	6	Cassero Elevato; così—Q 18 *t.*=Cassero Elevato 18 tonnellate.
			R
R	3	2	American Record.
			S
S	4	3	Nave.
Sc	..	3	Propulsione ad Elica.
ShdDk	..	3	Coperta Ombra.
Sk	4	3	Smake.
Sp	4	3	Sloop.
SprDk	..	3	Contro Coperta ossia Spar Deck.
Sq	4	3	Vele Quadre ossia attrezzatura quadra.
Sr	4	3	Schooner.
Stl	4	3	Acciajo.
Sw	4	3	Snow.
			T
t	..	6 & 9	Tonnellate.
Trb	4	3	Trabaccolo.
Tre	4	..	Trecantera.
Twn	..	3	Doppia.
			U
U	3	2	"Liverpool Underwriters' Registry for Iron Vessels" (adesso fuso col "Lloyd's Register of British and Foreign Shipping").
			V
V	3	2	Bureau Veritas.
			W
W	4	3	Legno.
WB	..	9	Zavorra d' Acqua.
			Y
Y	11	9	Fodrato con Metallo Giallo, così—Y8.85=Fodrato con Metallo Giallo nell' Agosto 1885.
Yt	4	3	Yacht.
			Z
Z	11	9	Fodrato con Zinco.
2m3m4m	4	3	Due, tre o quattro Alberi.

FORKLARING OG NOGLE TIL INDHOLDET I "UNIVERSAL REGISTER."

REGISTER FOR SEILSKIBE.—(Sailing Vessels.

Spalte I.

Skibenes officielle Register-Nummer.

Spalte II.

Löbenummeret i "Universal Register."

Spalte III.

Betegnelse for Klassifikations-Selskabe, hos hvilke Skibene har Klasse.

Spalte IV.

Skibenes Navne, Bygningsmateriale, Rig, etc. Capteinernes Navne og Datoen da de (1) förat kom i sin nuværende Rheders Tjeneste som Capteiner, og (2) blev Capteiner paa sine nuværende Skibe. Skibenes gamle Navne.

Spalte V.

Flag, under hvilke Skibene seiler.

Spalte VI.

Rumfanget i Register-Tons. Den anförte Tonnage er den Registrerede i Fölge Maalingen i de Lande til hvilke Skibene hörer, undtagen de er blevne maalte af Britiske Maale-Funktionærer, i hvilket Tilfælde den Britiske Tonnage er angivet.

Spalte VII.

Dimensioner. Disse er anförte i Engelske Fod og er saa vidt muligt de registrerede Dimensioner overeensstemmende med Maalingen i de forskjellige Lande, hvor Skibene hörer hjemme; dog i Tilfælde de er blevne maalte i Storbrittannien, er de Engelske Dimensioner angivne, naar de kunde opdrives.

Ved en Angivelse af Dybden i Rummet, som $\frac{18·6}{25·6}$, forstaaes, at Dybden fra Hoved-Dækket er 18·6, og fra Over-Dækket 25·6 Fod.

Spalte VIII.

Bygningssted.

Spalte IX.

Bygningsaar. Som Regel er Aaret, i hvilket Skibet löb af Stabelen anfört som Bygningsaaret.

En Angivelse, som 18^{37}_{81}, betyder, at Skibet blev bygget i 1837, og ombygget i 1881. En Angivelse, som 18_{81}, betyder, at Bygningsaaret er ukjendt, men at Skibet blev ombygget i 1881.

Spalte X.

Bygmesterens Navn.

Spalte XI.

Datoen for Metal-Forhudning er enten baseret paa Opgaver fra de Selskabers Registre, hvor de respektive Skibe er classeficerede, eller paa Indberetninger fra Lloyd's Besigtningsmænd.

Spalte XII.

Rederens eller Agentens Navn. Ved Selskabe, der kun er Eiere af et Skib, og er opkaldt efter dette, ligeledes ved "Actieselskabe" etc., er de bestyrende Rederes Navne angivne; f. E.: "Actieselskab" (Anton Bruun), eller, Ship "Falls of Bruar" Co. (Ld.)—(Wright & Breakenridge). I saadanne Tilfælde findes Skibene under de bestyrende Rederes Navne i Fortegnelsen over Eierne.

Spalte XIII.

Hjemsted.

REGISTER FOR DAMPSKIBE.—(Steamer

Spalte I.

Löbenummeret i "Universal Register" og officielle Register-N

Spalte II.

Betegnelse for Klassifikations-Selskabe, hos hvilke Skibene ogsaa Datoen for Bundens Besigtelse ved en af "Board of Besigtningsmænd i Tilfælde Skibet har Passager-Certificat fr of Trade."

Spalte III.

Skibenes Navne, Bygningsmateriale, Drivkraft, Rig, etc. Ca Navne og Datoen da de (1) först kom i sin nuværende Rheder som Capteiner, og (2) blev Capteiner paa sine nuhavende Skibe gamle Navne, Antal af Dæk, etc.

Spalte IV.

Flag, under hvilke Skibene seiler; Hjemsted.

Spalte V.

Netto og Brutto Tonnage. *See* Forklaringen over Spalte VI. i for Seilskibe.

Spalte VI.

Dimensioner. *See* Forklaringen over Spalte VII. i Reg Seilskibe.

De anförte Dimensioner er tagne fra de bedst mulige K de Registrerede fandtes at give en unöiagtig Forestilling Störrelse.

Overbygninger paa Dækket.

Spalte VII.

Bygningssted. Bygmesterens Navn.

Spalte VIII.

Bygningsaar. *See* Forklaring over Spalte IX. i Registeret for

Spalte IX.

Antal Skodder. De angivne Skodder er fuldstændig va rækker op til Overdækket, og er kun anförte ved de Skibe, de havde Klasse i "Lloyd's Register," eller i "Liverpool U Registry."

Oplysninger om Vand-Ballast.

Datoen for Træ- og Composit-Dampskibes Forbudning er ang i denne Spalte.

Spalte X.

Oplysninger om Maskinen. De nominelle Hestekræfter saavidt muligt fra Skibets Register; denne Angivelse er dog lestok for den virkelige Dampkraft.

Maskinbyggerens Navn og Adresse.

Tallet foran "Cy." betyder Antallet af Cylindre; Tallene Cylindernes Diametre og Slagets Længde i Engelske Tommer. Forklaringen over "Cy."

Spalte XI.

Rederens eller Agentens Navn. *See* Forklaringen over Sp Registeret for Seilskibe.

NÖGLE TIL FORKORTELSER I "UNIVERSAL REGISTER."

[illegible]	[illegible]	[illegible]	BETYDNING
[illegible]	3	2	Lloyd's Register of British and Foreign Shipping.
A	3	2	Veritas Austro-Ungarico.
[illegible]	..	9	Agter. See Forklaring over "DB."
APT	..	9	Tank i Skarpen agter.
A[illegible]	..	3	Hjelpe (Maskin).
[illegible]	..	3	Storm Dæk (Awning Dæk).
			B
B		3	Lag Bjælker; f. Ex.—*2Dks3B* = Tre Lag Bjælker, af hvilke to er belagt med Dæk.
B	..	6	Overbygning midtskibs; f. Ex.—*B66f.* = Overbygning midtskibs 66 Fod lang.
b	..	9	Kjedel-Rum. See Forklaring over "DB."
Bg	4	3	Brig.
Blds.	..	9	Vandtætte Skodder; f. Ex.—5 Blds = Fem vandtætte Skodder.
Bk	4	3	Bark.
Bkn	4	3	Skonnertskib.
Bmb	4	3	Bombarde.
Bn	4	3	Skonnertbrig.
BT		2	Board of Trade; f. Ex.—BT8.85 = Datoen for Bundens Besigtelse ved en af "Board of Trade's" Besigtningsmænd.
			C
C	11	9	Forhudet med Kobber; f. Ex.—C5.85 = Forhudet med Kobber i Mai 1885.
C	4	3	Composit (Træplanker paa Jernspant).
CDB	..	9	Dobbelt Celle-Bund; f. Ex.—CDB518t = Dobbelt Celle-Bund med et Vandindhold af 518 Tons.
Comp.	..	10	Maskine ombygget til Compound Maskine; f. Ex.—Comp. '85, C. D. Holmes & Co., Hull = Maskine ombygget til Compound Maskine i 1885 af d'Herrer C. D. Holmes & Co. i Hull.
Ct	4	3	Kutter.
Cy.	..	10	Cylindre; f. Ex.—2Cy.18"&36" = To Cylindre af resp. 18 og 36 engl. Tommers Diameter; 4Cy. 18"&36" = Fire Cylindre, af hvilke to har 18 og to 36 engl. Tommers Diameter; 3Cy.18"&(2)36" = Tre Cylindre af hvilke en har 18, og de to andre 36 engl. Tommers Diameter hver.
			D
D	3	2	Nederlandsche Vereeniging van Assuradeuren.
DB	..	9	Dobbelt Bund; f. Ex.—DB*a&f*54t = Dobbelt Bund agter og forud med et Vandindhold af 54 Tons; DB*af&eb*500t = Dobbelt Bund agter, forud og under Maskin- og Kjedel-Rum med et Vandindhold af tilsammen 500 Tons; DB*a26ft f25ft* = Dobbelt Bund agter 26 Fod lang og forud 25 Fod lang.
Dhy	4	3	Dhoy.
Dk eller *Dks*	..	3	Dæk; f. Ex.—*2Dks* = To Dæk; *1Dk&AwnDk3B* = Tre Lag Bjælker, af hvilke det överste er belagt med et Storm Dæk og det andet med Hoved Dækket.
DT	..	9	Dyb Tank.
Dr	4	3	Dogger (Rig).
Dy	4	3	Dandy (Rig).
			E
e	..	9	Maskin-Rum. *See* Forklaring over "DB."
			F
F	..	6	Bak; f. Ex.—*F26t.* = Bak 26 Tons; *F26f.* = Bak 26 Fod lang.
f	..	9	Forud. *See* Forklaring over "DB."
f eller *ft*	..	6 & 9	Fod (Engelske).
FPT	..	9	Tank i Skarpen forud.
			G
G	3	2	Germanischer Lloyd.
Glt	4	3	Galiot.
			H
H	3	2	Veritas Ellenico.
H	..	6	Dæk-Hus eller Huse; f. Ex.—*H 27 t.* = Dæk-Huse af 27 Tons Rumfang.

Forkortelse.	Rubrik i Registeret over Seilskibe.	Rubrik i Registeret over Dampskibe.	BETYDNING.
			H—*Fortsættelse.*
HP	..	10	Hestekræfter. De nominelle Hestekræfter er taget saavidt muligt fra de officielle Registre; denne Angivelse er dog ingen Maalestok for den virkelige Dampkraft.
HurDk	..	3	Orkan Dæk.
			I
I	3	2	Registro Italiano.
I	4	3	Jern.
			K
K	4	3	Ketch.
Kf	4	3	Koff.
			L
Len.	..	..	Forlænget.
Lr	4	3	Lugger.
			M
m	4	3	Maste; f. Ex.—3m = Tre Maste.
MT	..	9	Dyb Tank midtskibs; f. Ex.—MT200t = Dyb Tank midtskibs med et vandindhold af 200 tons.
			N
N	3	2	Norske Veritas.
			P
P	..	6	Hytte; f. Ex.—*P 28 f.* = Hytte 28 Fod lang.
Pad	..	3	Drevet ved Hjul.
Pol	4	3	Polacca.
pt	11	3 & 9	Delvis.
			Q
Q	..	6	Halvdæk eller Bræk; f. Ex.—*Q 18 t.* = Halvdæk eller Bræk af 18 Tons Rumfang.
			R
R	3	2	American Record.
			S
S	4	3	Fregatskib.
Sc	..	3	Drevet ved Skrue.
ShdDk	..	3	Shade Dæk.
Sk	4	3	Smakke.
Sp	4	3	Slup.
SprDk	..	3	Spardæk.
Sq	4	3	Takklet med Raaseil.
Sr	4	3	Skonnert.
Std	4	3	Snau.
Sw	4	3	Snov.
			T
t	..	6 & 9	Tons.
Trb	4	3	Trabaccolo.
Tre	4	..	Trecantira.
Twn	..	3	Tvilling (Skrue).
			U
U	3	2	"Liverpool Underwriters' Registry for Iron Vessels" (nu forenet med "Lloyd's Register of British and Foreign Shipping").
			V
V	3	2	Bureau Veritas.
			W
W	4	3	Wood.
WB	..	9	Vand-Ballast.
			Y
Y	11	9	Forhudet med Yellow Metal; f. Ex.—Y8.85 = Forhudet med Yellow Metal i August 1885.
Yt	4	3	Yacht.
			Z
Z	11	9	Forhudet med Zink.
2m3m4m	4	3	To, Tre og Fire-mastede Skibe.

EXPLICACION Y CLAVE DEL CONTENIDO DEL "UNIVERSAL REGISTER

REGISTRO DE BUQUES DE VELA.—(Sailing Vessels.)

COLUNA I.

Números de los Buques segun el Registro Oficial.

COLUNA II.

Números consecutivos de los Buques en el "Universal Register."

COLUNA III.

Signos que indican las Sociedades de clasificacion (si las hubiere) bajo las cuales se clasifican los Buques.

COLUNA IV.

Nombres de los buques, materiales de construccion, Aparejos, etc. Capitanes y fechas en que entráron al servicio de los dueños actuales y tomáron el mando de este buque. Nombres anteriores de los buques.

COLUNA V.

Banderas bajo las cuales navegan respectivamente los Buques.

COLUNA VI.

Portes netos registrados. Los portes que se dan son los registrados de buques segun las medidas de los países á que respectivamente pertenecen, salvo cuando los buques se hayan medido por los Inspectores Ingleses de Porte, en cuyo caso se anotan los Portes Ingleses.

COLUNA VII.

Dimensiones. Las dimensiones que se dan están en piés ingleses; y, en cuanto sea posible, son las registradas de los buques, segun las medidas de los países á que respectivamente pertenecen, salvo cuando los buques se han medido en Inglaterra, en cuyo caso van indicadas (cuando es posible) las medidas inglesas.

Un registro de profundidades, así $\frac{18·6}{25·6}$, significa que la profundidad hasta la cubierta principal es de 18·6 piés, y hasta la cubierta superior es de 25·6 piés.

COLUNA VIII.

Puntos de Construccion.

COLUNA IX.

Fechas de la Construccion. En cuanto sea conocido, el año en que fué botado al agua el buque, se dá como el año de su construccion.

La indicacion de fechas en esta forma $18\frac{57}{84}$, significa que el buque fué construido en 1857, y reconstruido en 1884; una anotacion, así 18_{84}, indica que la fecha de la construccion original de los buques se desconoce, pero que fué reconstruido en 1884.

COLUNA X.

Nombres de los Constructores.

COLUNA XI.

Fechas del Aforramiento de Metal. Las fechas del aforramiento de metal se toman únicamente de los registros en que los buques se hallan clasificados ó de las informes de los Inspectores de "Lloyd's Register."

COLUNA XII.

Nombres de los Dueños ó Directores. Hay buques que solo pertenecen á una compañía, la cual lleva entónces, el nombre del buque respectivo. Hay asimismo compañías escandinavias que no llevan otro nombre que el de "Actieselskab" que quiere decir compañía. En ambos el nombre del Director se indica así:—Ship "Falls of Bruar" Co. (Ld.)—(Wright & Breakenridge), "Actieselskab" (Anton Bruun). En dichos casos se encontrarán los buques en la "Lista de Dueños" bajo los nombres de los directores.

COLUNA XIII.

Puertos de Registro ó Matriculacion.

REGISTRO DE BUQUES DE VAPOR.—(Stea

COLUNA I.

Números consecutivos de los Buques en el "Universal R
Números del Registro Oficial.

COLUNA II.

Signos que indican las Sociedades de Clasificacion (si las h
las cuales se clasifican los Buques; y última fecha de la inspecc
por los Inspectores de la Junta de Comercio (Board of Trade
de que los buques tuvieren un Certificado de Pasajeros de l
Comercio.

COLUNA III.

Nombres, materiales de construccion, medios de propulsion, y
los buques. Capitanes y fechas en que entráron al servicio de
actuales y tomáron el mando de este buque. Nombres anterior
de cubiertas de los buques.

COLUNA IV.

Banderas bajo las cuales navegan respectivamente los Bu
Puertos de Matriculacion ó Registro.

COLUNA V.

Portes netos y brutos. *Véase* la explicacion en la Columna VI
de Buques de Vela.

COLUNA VI.

Dimensiones. *Véase* la explicacion en la Columna VII del R
Buques de Vela.

En los casos en que las dimensiones registradas dan una ide
del tamaño de un buque, los asientos en el libro han sido hechos p
que se han podido reunir.

Erecciones sobre Cubierta.

COLUNA VII.

Puntos de Construccion. Nombres de los Constructores.

COLUNA VIII.

Fechas de la Construccion. *Véase* la explicacion en la Colu
Buques de Vela.

COLUNA IX.

Números de Mamparas. Las mamparas registradas son
completas é impermeables, llegando hasta la cubierta superior
únicamente en el caso de buques clasificados, actual ó an
con el "Lloyd's Register" y con el "Liverpool Underwriters' Regi

Detalles del Lastre de Agua.

En esta Columna se dan igualmente las Fechas del Aforram
Buques de Vapor construidos de Madera ó Compuestos.

COLUNA X.

Detalles de las Máquinas. Se da, en cuanto sea posible, la fue
de caballos sacada del Registro de Buques, sin que esto repres
verdadera fuerza.

Nombres y Domicilios de los Fabricantes de las Máquinas.

Las cifras que preceden "Cy." indican el número de cilíndros
que siguen indican los diámetros de los cilíndros y la longitud d
pulgadas ingleses. *Véase* esplicacion de "Cy."

COLUNA XI.

Nombres de los Dueños ó Directores. *Véase* la explicacion de la C
del Registro de Buques de Vela.

CLAVE DE LAS ABREVIACIONES DEL "UNIVERSAL REGISTER."

Abreviaturas.	Columna del Registro de Buques de Vela en la cual se halla.	Columna del Registro de Vapores en la cual se halla.	SIGNIFICACIONES.
			A
	3	2	Lloyd's Register of British and Foreign Shipping.
	3	2	Veritas Austro-Húngaro.
	..	9	Á Popa. *Véase* explicacion de "DB."
	..	9	Tanque á Popa (After Peak Tank).
	..	3	Auxiliario.
	..	3	Toldo (Awning Deck).
			B
		3	Andanas de Baos; así—*2Dks3B* = Tres Andanas de Baos de los cuales dos llevan cubiertas.
	..	6	Cubierta de Puente; así—*B 66 f.* = Cubierta de Puente 66 piés de largo.
		9	Espacio de Caldera. *Véase* explicacion de "DB."
	4	3	Bergantin.
		9	Mamparas; así—5 Blhds. = Cinco Mamparas.
	4	3	Barca.
	4	3	Barquentina.
	4	3	Bombarda.
	4	3	Bergantin-goleta.
		2	Junta de Comercio; así—BT8.85 = Fecha de Informe de los Inspectores de la Junta de Comercio acerca de su exámen del fondo.
			C
	11	9	Aforrado de Cobre; así—C5.85 = Aforrado de Cobre en Mayo 1885.
	4	3	Compuesto (armazon de hierro, flanqueada con madera).
	..	9	Fondo Doble de Construccion Celular; así—CDB518t = Fondo Doble de Construccion Celular, teniendo una cabida de agua de 518 toneladas.
	..	10	Máquinas Compuestas; así—Comp. '85, C. D. Holmes & Co., Hull = Máquinas Compuestas en 1885 por los Sres Holmes y Cia., de Hull.
	4	3	Balandra.
		10	Cilíndros; así—2Cy.18"&36" = Dos Cilíndros con diámetros de 18 y 36 pulgadas inglesas, respectivamente; 4Cy.18"&36" = Cuatro Cilíndros, de los cuales dos tienen un diámetro de 18 pulgadas inglesas cada uno, y dos de 36 pulgadas inglesas respectivamente; 3Cy.18"&(2)36"—Tres Cilíndros de los cuales uno tiene un diámetro de 18 pulgadas y los dos otros tienen diámetros de 36 pulgadas cada uno.
			D
	3	2	Nederlandsche Vereeniging van Assuradeuren (Registro Holandés).
	..	9	Fondo Doble; así—DB*a&f*54t = Fondo Doble á popa y proa con una cabida de agua de 54 toneladas; DB*a,f&eb*500t = Fondo Doble á popa, proa y debajo de las máquinas y calderas con cabida total de 500 toneladas; DB*a26ft f25ft* = Fondo Doble á popa 26 piés de largo y á proa 25 piés de largo.
	4	3	Dhoy.
	..	3	Cubierta ó Cubiertas; así—*2Dks* = Dos Cubiertas; *1Dk&AwnDk3B* = Tres Andanas de Baos, de las cuales una está provista de una cubierta de toldo, y la segunda con la cubierta primera.
	..	9	Tanque Hondo.
	4	3	Dogger.
	4	3	Dandy.
			E
	..	9	Espacio de la Máquina. *Véase* explicacion de "DB."
			F
		6	Castillo de Proa; así—*F 26 t.* = Castillo de Proa 26 toneladas.
	..	9	Á Proa. *Véase* explicacion de "DB."
		6 & 9	Piés (Ingleses).
		9	Tanque á Proa (Fore Peak Tank).
			G
	3	2	Germanischer Lloyd (Registro Aleman).
	4	3	Galeota.
			H
	3	2	Veritas Helénico (Registro Helénico).
	..	6	Cámara ó Cámaras de Cubierta; así—*H 27 t.* = Cámara ó Cámaras de Cubierta 27 toneladas.

Abreviaturas.	Columna del Registro de Buques de Vela en la cual se halla.	Columna del Registro de Vapores en la cual se halla.	SIGNIFICACIONES.
			H—*continuando.*
HP	..	10	Fuerza de Caballo. Se da, en cuanto sea posible, la fuerza de caballo nominal del Registro del Buque mas esto por supuesto no da ninguna indicacion de la verdadera fuerza de vapor del buque.
Hur Dk	..	3	Cubierta de Huracan.
			I
I	3	2	Registro Italiano.
I	4	3	Hierro.
			K
K	4	3	Queche.
Kf	4	3	Koff.
			L
Len.	4	6	Alargado.
Lr		3	Lugger.
			M
m	4	3	Palos; así—3m = Tres Palos.
MT	..	9	Tanque Hondo de Mediania; así—MT200t = Tanque Hondo de Mediania con cabida de 200 toneladas de agua.
			N
N	3	2	Norske Veritas (Registro Noruego).
			P
P	..	6	Popa; así—*P 28 f.* = Popa 28 piés de largo.
Pad	..	3	Impelido por Ruedas.
Pol	4	3	Polacra.
pt	11	3 & 9	Purte.
			Q
Q	..	6	Alcázar Levantado; así—*Q 18 t.* = Alcázar Levantado 18 toneladas.
			R
R	3	2	American Record (Registro Americano)
			S
S	4	3	Buque ó Navío.
Sc	..	3	Impelido por Helice.
Shd Dk	..	3	Cubierta Entoldada (Shade Deck).
Sk	4	3	Esmaque.
Sp	4	3	Balandra.
Spr Dk	..	3	Spar Deck.
Sq	4	3	Aparejo Cruzado.
Sr	4	3	Goleta.
Stl	4	3	Acero.
Sw	..	..	Snow.
			T
t	..	6 & 9	Toneladas.
Trb	4	3	Trabacalo.
Tre	4	..	Trecantira.
Twn	..	3	Gemelo (dos hélices, una á cada lado del buque).
			U
U	3	2	"Liverpool Underwriters' Registry for Iron Vessels" (actualmente amalgamado con el "Lloyd's Register of British and Foreign Shipping").
			V
V	3	2	Bureau Veritas.
			W
W	4	3	Madera.
WB	..	9	Lastre de Agua.
			Y
Y	11	9	Aforrado de Metal Amarillo; así—Y8.85 = Aforrado de Metal Amarillo en Agosto 1885.
Yt	4	3	Yate.
			Z
Z	11	9	Aforrado de Zinc.
2m3m4m	4	3	Con dos, tres ó cuatro Palos

SAILING VESSELS.

NAVIRES À VOILES.

No. in Book	Society in which classed	Ship's Name	Master, and Dates of Appointment to present Owner's Service and to Ship; Ship's Late Name	Material and Rig	Flag	Net Tonnage	Length	Breadth	Depth	Built: Where	Built: When	Built: By whom	Date of Metalling	Owners or Managers	Port of Registry
1	NV	A. B. Bull	*Larsen*	W3mSr	Norweg'n	299	126·1	29·1	12·2	Kragerö	1885	J. Nielsen	V 10.85	J. A. Larsen & Co.	Kragerö
2	V	A. B. C.	*Ostrien*	WSr	French	136	121·0	17·3	15·1	CapeBreton	1869			Ostrien	St. Pierre, Miquelon
3	R	A. B. Crosby	*Smith* -87	W3mSr	British	175	89·2	27·9	9·9	Parrsboro, NS	1884	J. W. Dureme	V 1.88	Yule & Munro	Turks' Islands
4	V	A. B. D.	*Attiz*	WBg	French	149	85·7	22·6	11·2	Nantes	1867	Bojud Chaigeau		Cormerais frères	Nantes
5		A. B. Goodman		WSr	American	116	93·0	23·8	7·8	Laurel, Del.	1875			D. J. Hearn	Seaford, Del.
6		A. B. Perry	*Lock*	W3mSr	American	210	111·2	28·8	8·7	Addison, Me.	1873			J. Nash & Co.	Machias, Me.
7	R	A. B. Sherman	*Higgins*	W3mSr	American	581	167·7	34·9	15·3	Boston, Mass.	1883	R. Crosbie & Son		S. H. Walker	Fall River, Mass.
8	R	A. C. Bean	*Foster* -88	WBk	American	553	143·1	30·5	15·9	Deering, Me.	1872		V 2.87	J. S. Winslow & Co.	Portland, Me.
9	V	A. C. L.	*Savary* -88	WBg	French	242	96·1	23·9	12·4	Nantes	1875	J. Sevestre		Cie. de Transports Mar. [illegible]	Nantes
10		A. C. Lyons	*Bunnell*	W3mSr	American	254	122·0	30·0	8·8	Staten Is., N. Y.	1865			B. Jefferson	New York
11		A. C. Watson		WSr	British	112	76·5	21·0	8·0	Cambridge, N. B.	1872			D. H. Spragg	St. John, N. B.
12	G	A. C. de Wall	*Assing* [illegible] -83	WBk	German	276	114·5	21·1	14·9	Dundee	1857	Tay S. B. Co.		H. J. Assing	Grossefehn
13	✳	A. D. Bordes	*Kishpore* 12-87	I 4mS	French	2230	294·1	43·2	25·0	Glasgow	1884	W. B. Thompson		Ant. Dom. Bordes & fils	Bordeaux
14	R	A. D. Lamson	*Smith*	W3mSr	American	426	133·0	33·4	14·5	Leesburg, N. J.	1879			J. Russell	Somer's Point, N. J.
15		A. D. Whidden	*Ravaghan*	WBn	British	127	85·6	25·3	11·0	Nova Scotia	1869			S. N. Robinson	Dublin
16		A. Denike	*Townsend*	W3mSr	American	407	131·0	30·5	16·6	Greensb'ro, Md.	1866			J. A. Van Brunt	New York
17	D	A. E. Killam	*Rhode* -85	WBk	British	1032	179·6	37·0	21·5	N. Brunswick	1882	H. V. Crandall	V 10.87	J. T. Cowley	Liverpool
18		A. E. Sherwood		WBk	British	450	130·5	28·0	17·8	Maine, U. S.	1863			R. Shinn	Calcutta
19	V	A. F. Cosulich	*Martinolich* 1-88	IBk	Austrian	287	126·9	24·7	13·4	Lussinpiccolo	1886	N. Martinolich		C. Cosulich di A. & Co.	Lussinpiccolo
20	R	A. F. Crockett		W3mSr	American	412	141·2	33·1	12·3	Bath, Me.	1881	Goss & Sawyer		A. F. Crockett & Co.	Rockland, Me.
21		A. F. Kindberg	*Mount*	WSr	American	215	112·3	29·3	10·2	East Haven, Cn.	1865			J. W. Keough	Perth Amboy, N. J.
22	RV	A. G. Ropes	*Rivers*	WS	American	2342	258·2	44·7	28·5	Bath, Me.	1884	I. F. Chapman & C.	V 11.86	I. F. Chapman & Co.	New York
23		A. G. Ryan		WSr	British	112	92·6	22·5	7·8	Kingston, Ont.	1877	W. Power & Co.		Charles Dean	Kingston, Ont.
24		A. H. Howe	*Brown* -87	W3mSr	American	453	132·2	32·5	16·9	Smyrna, Del.	1871			A. J. Newbury	New York
25		A. H. Hurlburt	*Furness*	W3mSr	American	250	126·2	30·5	10·7	Middletwn, Cn.	1871	S. & E. S. Belden		G. J. Tarr	Gloucester, Mass.
26	✳D	A. H. van Bergen	*Efting*	WSr	Dutch	144	91·5	19·0	11·1	Hoogezand	1866	J. A. Hooites	Z 5.85	F. L. Drenth	Oude Pekela
27		A. Hammond	*Kilpatrick*	WSr	American	146	103·6	27·5	8·2	P. Jefferson, N. Y.	1854			R. F. C. Hartley	Saco, Me.
28		A. Hayford	*Jones*	WSr	American	145	102·0	27·5	7·5	Belfast, Me.	1872			D. Lane	Belfast, Me.
29		A. Heaton	*Latteriet*	W3mSr	American	448	130·5	31·3	16·5	New Haven, Cn.	1868			S. H. Mills	New York
30		A. Heaton	*Phinney*	WSr	American	138	97·0	27·2	8·7	E. Haven, Con.	1851			A. C. Guy	Rockland
31		A. J. Bentley	*Rankin*	W3mSr	American	276	124·0	30·0	10·0	Hartford, Con.	1865	S. & E. S. Belden		E. M. Williams	New York
32	RV	A. J. Fuller	*Colcord*	WS	American	1782	229·3	41·5	26·0	Bath, Me.	1881	McDonald	V 10.88	B. Flint	New York
33		A. J. Miller	*Jarvis*	WSr	American	105	91·0	27·2	6·1	Northport, N. Y.	1878			H. B. Leonard	Fall River, Mass.
34		A. L. Mitchell		WSr	American	139	96·0	28·0	7·8	Millbridge, Me.	1879	E. Dyer		L. H. Leighton	Machias, Me.
35		A. M. Brundrit		W3mSr	British	113	91·5	22·2	10·1	Runcorn	1882	Brundrit & Co.		W. Dawe	St. John's, NFl.
36	V	A. M. E. (LOST)	*Gravel*	WBg	French	159	81·7	21·6	12·0	Nantes	1867	Sevestre	Z 10.87	B. Ravilly	Nantes
37	G	A. M. Lotinga	*Hoppe* -88	WBg	German	292	107·6	26·7	14·0	Barth	1869	A. Witte		J. J. Wallis & Sohn	Barth
38	V	A. McCallum	*Tattersall* -87	WS	American	1878	215·0	42·0	29·4	Thomaston, Me.	1870	Edward O'Brien	V 2.87	Edward E. O'Brien	Thomaston, Me.
39	V	A. McLeod	*Florimon* -85	WBn	French	198	100·3	21·4	11·2	P. E. Island	1879	C. McDonald		Bordes & ses fils	S. Pierre, M'quc.
40		A. McNichol	*Libby*	WSr	American	116	93·0	24·7	7·9	E. Machias, Me.	1874			W. W. Libby	Calais, Me.
41		A. Muir		WSr	British	330	136·6	26·2	11·6	P. Dalhousie	1874	Muir Bros.		A. Muir & Bros.	St. Catharines
42	N	A. O. Vinje	*Jensen*	WBk	Norweg'n	324	119·0	29·0	15·0	Tvedestrand	1871	O. Gundersen	V 1.85	Ole Gundersen	Tvedestrand
43		A. P. Emerson	*Reynard*	W3mSr	American	232	124·7	29·1	9·5	Belfast, Me.	1873	Hen. McGilvery	V 11.83	A. P. Emerson & Co.	Bucksport
44	R	A. P. Nowell	*Conwell* 1-87	W3mSr	American	229	116·0	30·8	9·0	Milton, Del.	1874			T. Winsmore	Philadelphia
45		A. Patterson	*Haskie* 74-74	WBg	British	195	99·3	23·6	12·7	P. E. Island	1866	J. Douse, Sr.		J. A. Beeching	Dover
46		A. R. Falck	*Huig* -87	WBk	Dutch	599	127·6	33·8	21·6	Alblasserdam	1846			N. Brantjes	Purmerend
47		A. R. Tucker	*Baker*	WBk	American	138	92·0	23·0	11·0	Dartm'th, Mass.	1851	Matthews, Mashoe & Co.		J. & W. R. Wing	New Bedfrd, Mass.
48	R	A. R. Weeks	*Hexley*	W3mSr	American	423	131·2	31·9	14·6	Bath, Me.	1873	Robinson		J. E. Farr	Portland, Me.
49	R	A. T.	*Manning* -85	WSr	British	123	83·2	25·0	8·9	Granville, N. S.	1884	A. Turpie		Alex. Turpel	Annapolis, N. S.
50	D	A. T. Stallknecht (STRANDED)	*Sye?*	WBk	German	540	140·0	32·5	18·0	Apenrade	1875	N. Jacobsen	V 11.86	C. P. Schütt	Apenrade

[illegible] Lloyd's Register; D = Nederlandsche Vereeniging; G = Germanischer Lloyd; I = Registro Italiano; R = American Record; [illegible] Veritas Austro-Ungarico; H = Veritas Ellenico; N = Norske Veritas; V = Bureau Veritas.

In the 4th Column, C = Composite (iron frame, wood planked); I = Iron; Stl = Steel; W = Wood.

[illegible] Dates of Metalling (Col. 11) are based on the records of the Books in which Vessels are respectively classed, or on the Reports of the Surveyors of Lloyd's Register.

Further particulars of Vessels classed by Lloyd's Register will be found in Lloyd's Register Book and Lloyd's Yacht Register.

Official No.	No. in Book	Registry in which classed	Ship's Name. Material and Rig. Master, and Dates of Appointment to present Owner's Service and to Ship. Ship's Late Name.	Flag	Net Tonnage	Dimensions. Length	Breadth	Depth	Built. Where	When	By whom	Date of Metalling	Owners or Managers
	51		A. V. *Cabrera* WBk	Spanish	184	96·0	25·5	15·5	Ybiza	1855			F. E. Cardenas
	52		A. V. C. *Le Gal* -81 WBg	French	146	83·3	21·8	11·2	Nantes	1877	Alleau & Aubert		J. F. Legoff
103252	53		A. W. Ellis *Ferguson* 65-81 WSr	American	168	104·0	29·0	8·2	Belfast, Me.	1873			J. W. Frederick
	54	N	A. W. Singleton *Thiis* WBk	Norweg'n	579	145·8	30·4	18·6	Tusket, N.S.	1862			Chr. J. Thiis
	55	N	Aabine *Bliz* WBg	Norweg'n	258	122·8	26·9	14·1	Aalesund	1866	Carl Haasted	Y 5.86	L. L. Sunde
	56		Aagot *Jacobsen* WBk (ex Mase Louise) (WRECKED)	Norweg'n	689	159·8	31·9	19·8	Quebec	1870	P. V. Valin	Y 1.86	J. Simonsen
	57	N	Aalesund *Schage* -74 WBk	Norweg'n	278	121·6	28·1	13·2	Aalesund	1872	C. J. Haasted	Y 7.85	Brödrene Kraasbye
	58		Aallotar *Rotia* WSr	Russian	168	87·1	22·9	11·6	Björkö	1874	A. Karlin		K. Rotia
	59		Aallotar *Westergren* WSr	Russian	160	..	..	..	Björkö	1874			L. Kraki
	60		Aallotar *Gröndal* -86 WSr	Russian	159	..	..	..	Luvia	18 53/73			F. G. Gröndal
	61	NV	Aaloug *Rasmussen* 76-76 WSr	Norweg'n	127	84·0	22·1	10·5	Christiansand	1876	A. Martensen	V 10.85	C. Rasmussen
	62	GR	Aalto (ex Steamer) *Ahlsten* -83 CBk	Russian	376	158·0	26·0	13·0	Nystad	1872	H. Kgahlström		E. J. Savon
	63		Aalto *Arkman* -88 WBg	Russian	174	..	..	..	Uleaborg	18 53/87			G. & C. Bergbom
	64		Aalto *Klamo* -81 WBn	Russian	163	97·3	23·4	10·8	Joensuu	1876			G. Cederberg & Co.
106098	65	R	Aaron Reppard *Norton* -84 W3mSr	American	436	119·5	34·0	13·6	Philadelphia	1881	Doughty & Kappella		J. C. Cottingham
	66	N	Aarvak *Legaard* 80-80 WBk (ex Flexford)	Norweg'n	1061	176·3	36·5	23·2	Boston	1864	P. Curtis	V 3.87	I. Hesselberg
	67	❋	Aastroom *Siegel* 76-76 W3mSr	Dutch	202	108·6	25·8	12·0	Pekela	1876	J. J. Koerts	Y 9.86	P. P. Kolk
	68		Abahnah WBg	Turkish	209	..	..	..	Kinolce	1870			Abahoudlee Urtchlu-oglou Hadji Hussein
	69		Abahnah *Spahi Ahmed* WBg	Turkish	185	..	..	..	Greece	1868			Berber Mehemed Agha & Co.
	70		Abahnah WBg	Turkish	121	..	..	..	Alabnah	1868			Ackjee-Shehirlee Hadj Rashid oglou-Hadji Mehemed
66973	71	V	Abana *James* 78-87 WBk	British	1269	191·5	37·7	24·1	Portland, N.B.	1874	Hilyard Bros.	V 9.86	D. V. Roberts
	72	V	Abba *Harrebye* WSr	Danish	125	86·9	21·2	10·2	SonderboFano	1867	S. J. Sønnichsen		Akt. Selsk. (P. P. Harreby)
63204	73	❋	Abbey Holme *Rich* 82-82 I Bk	British	516	157·7	28·2	17·0	Sunderland	1869	J. Blumer & Co.		Hine Bros.
70976	74	❋	Abbey Town *Tullis* 82-86 I Bk	British	702	191·1	30·1	19·0	Port Glasgow	1875	W. Hamilton & C		Doward, Dickson & Co.
105812	75		Abbie *Jorgensen* WSr	American	139	98·5	29·5	8·3	Eureka, Cal.	1876	H. D. Bendixon		J. G. Jackson
609	76		Abbie Bradford (CONDEMNED) WSr	American	109	87·0	23·5	9·7	Scituate, Mass.	1860			J. Bourne
1436	77		Abbie Bursley *Crabtree* WSr	American	314	123·9	29·4	9·7	Camden, Me.	1864			C. A. Lovell & Son
106090	78	R	Abbie C. Stubbs *C. B. Pendleton* -85 W3mSr	American	328	155·0	33·0	11·5	Newhaven Con	1882	H. H. Hanscom		E. H. Weaver
214	79	R	Abbie Clifford *Storer* WBn	American	429	123·2	29·1	16·6	Stockton, Me.	1867	Mudgett & Co.	V	N. W. Bridge
6	80		Abbie E. Willard *Lewis* -87 WSr	American	129	89·4	25·8	8·2	Brewer, Me.	1860	S. Moulton		W. A. Swett & Co.
105905	81		Abbie H. Gheen *Gheen* (79-80) W3mSr	American	230	128·0	30·8	9·0	Frederica, Del.	1880	Nathaniel Lank		J. W. Hall
1611	82		Abbie H. Hodgman *Frye* WSr	American	145	89·7	26·7	8·7	Harrington Me	1868			J. Pendleton
1612	83		Abbie Ingalls *Ingalls* WSr	American	175	106·0	26·6	10·0	Machias, Me.	1868			G. F. Kelley
53	84		Abbie P. Cranmer *Izard* W3mSr	American	290	119·8	30·6	9·6	Baltimore, Md.	1867	J. S. Beacham & Bros.		T. C. Cranmer
83011	85	V	Abbie S. Hart *Goudley* WS	British	1450	209·0	38·8	23·4	Tusket, N.S.	1880	J. A. Hatfield	V 4.88	B. B. Law
106218	86	R	Abbie S. Walker *Dobbin* -83 W3mSr	American	181	103·7	28·4	8·5	Jonesboro' Me.	1883	Jonesboro' Lumber Co.		E. H. Lawrey
105400	87		Abbie & Eva Hooper *Willetts* -87 W3mSr	American	305	123·0	32·6	10·0	Tuckahoe, N.J	1875			J. Hooper
	88	R	Abbotsford *Olsgren* 88 WS (ex Prince of Wales)	Argentine	1178	176·3	38·2	24·7	Bath, U.S.	1860		Y 6.86	Mrs. M. Jones (Burgess & Co., Managers)
106075	89	R	Abbott W. Lewis *Peak* W3mSr	American	567	142·3	34·2	15·4	Boston, Mass.	1882	J. M. Brooks		C. Lovell & Son
105188	90		Abby K. Bentley *Baker* W3mSr	American	272	133·5	32·0	9·0	Hartford, Con.	1872			Bentley, Gildersleeve & Co.
1943	91		Abby L. Dow *Townshend* W3mSr	American	330	120·0	30·5	13·6	Tuckahoe, N.J	1870			J. B. Vandusen
82	92		Abby Thaxter *Hutchinson* WSr	American	183	108·8	24·9	9·8	Brewer, Me.	1853			A. E. Fernald
1887	93	R	Abby Wasson *Lord* WSr	American	134	93·4	25·7	8·2	Brooksville Me	1870			W. Wasson
	94		Abdallah *Redor* -87 WSr	French	118	76·4	20·3	10·2	Nantes	1860	T. Dubigeon		Le Bail
	95		Abdulsalam Bg	Turkish	122	..	..	..					Karakiri
	96	V	Abeille *Philippe* -88 WBk	French	280	105·9	24·3	13·9	Nantes	1872	Chaigneau	V 11.86	P. Ch. Grenet
	97		Abeille *Van Costenobel* WBk	French	212	90·9	23·9	13·4	St. Malo	1861	Lhotellier		L. Demalvilain
81982	98	❋	Abeja *Johnson* 84-84 WSr	British	190	102·8	23·1	12·2	Littlehampt'n	1881	J. & W. B. Harvey	V 5.86	J. Bartlett
	99	N	Abel *Danielsen* 72-83 WBk	Norweg'n	463	139·1	30·7	17·0	Tvedestrand	1879	E. P. Pedersen	V 5.86	J. M. A. Marcussen
	100	N	Abel *Danielsen* -88 WBk	Norweg'n	352	128·1	28·2	15·0	Porsgrund	1871	S. Johnsen	Y 4.87	A. Jansen

❋ = Lloyd's Register; D = Nederlandsche Vereeniging; G = Germanischer Lloyd; I = Registro Italiano; R = Ame[rican]; A = Veritas Austro-Ungarico; H = Veritas Ellenico; N = Norske Veritas; V = Bu[reau]

In the 4th Column, C = Composite (iron frame, wood planked); I = Iron; Stl = Steel; W = Wood.

The Dates of Metalling (Col. 11) are based on the records of the Books in which Vessels are respectively classed, or on the Reports of the Surveyors of Llo[yd's]

Further particulars of Vessels classed by Lloyd's Register will be found in Lloyd's Register Book and Lloyd's Yacht Register.

Nos dans le livre	Registres	Nom, Matériaux et Gréement. Capitaine. Ancien Nom.	Pavillon	Tonnage net	Longueur	Largeur	Creux	Construit: Où	Quand	Par qui	Date du doublage	Armateurs ou Directeurs-Gérants	Port d'Armement
101		Abel C. Buckley Crossman WSr	American	222	105·0	29·0	10·6	Greenport, NY	1869			A.C.Buckley	Greenport, N.Y
102		Abel W. Parker Dean W3mSr	American	193	116·0	32·2	7·9	Boothbay, Me	1873			W.H.H.Dean	FallRiver, Mass
103		Abelone Pedersen -89 WBn (ex Magdalene Sophie)	Danish	118	79·8	20·3	10·3	Aalborg	1853	J.W.Riis		Y.N.Pedersen	Sonderhö Fänö
104	✻	Abeona Leslie -88 IS	British	979	210·3	33·4	21·0	Glasgow	1867	A.Stephen&S'ns		A.Weir & Co.	Glasgow
105	✻	Abercarne Brown 81-81 IS	British	1088	208·5	35·1	21·8	Stockton	1875	Richardson, Duck&Co		J.McKiver	Greenock
106	✻	Abercarne Lewis IBk	British	711	186·1	31·1	18·4	Dumbarton	1875	McKellar, McMillan&Co.		J.D.Jones	Glasgow
107	✻	Abercorn Irvin -87 IBk (Jan T)	British	1292	238·5	36·1	21·0	Glasgow	1885	A.Stephen&S'ns		P.H.Dixon & Co.	Glasgow
108		Aberdonian Jeffrey WSr (Brooks C'y)	British	133	98·0	20·2	11·6	Aberdeen	1840	A. Hall & Co.			Aberdeen
109	✻	Aberfoyle Robertson Stl.S	British	1597	248·8	38·1	22·7	Dumbarton	1885	A.McMillan&Sn		G.Cowper	Glasgow
110	✻	Abergeldie Murray WS	British	1152	218·2	37·2	21·8	Aberdeen	1869	J.Duthie Sons &Co	Y 9.88	Duthie, Sons & Co.	Aberdeen
111	✻D	Aberlemno Daniel 75-88 IBk	British	735	184·4	31·0	18·5	Dumbarton	1876	Birrell, Stenhouse&Co		M. Tutton	Swansea
112	✻	Abernyte Campbell 85-85 IBk	British	700	186·2	31·0	18·5	Dumbarton	1875	McKellar, McMillan&Co.		N.British Shipg.Co.	Glasgow
113	✻	Aberystwith Castle Reaves IBk	British	1056	213·6	34·9	21·0	Liverpool	1875	R.&J.Evans&Co		Richards, Mills &Co.	Liverpool
114	DRV	Abiel Abbot Nickerson [82-88] WBkn	American	560	136·0	32·5	17·0	E.Boston	1874	A.&S.Thomps'n	Y 12.85	Abiel Abbot	New York
115	N	Able Isaksen -88 WBk	Norweg'n	576	142·5	31·2	18·1	Kragerö	1866	Brandis	Y 9.85	Thor Olsen	Kragerö
116	RV	Abner Coburn Nichols WS	American	1879	223·0	43·0	26·9	Bath, Me.	1882	W.Rogers	Y [illegible]	W.Rogers	Bath, Me.
117		Abner Taylor Dodge WSr	American	132	89·3	24·7	9·1	Brewer, Me.	18 51/68			Y.W.Dodge	Bangor, Me.
118	✻	Aboukir Bay Jack 83-83 IBk	British	1117	216·9	35·1	21·0	Greenock	1883	Russell& Co.		Hatfield, Cameron & Co.	Glasgow
119	A	Abraham Jelich WBk	Austrian	671	157·8	32·1	18·6	Trieste	1870	Tonello	Y 7.88	G.Jelich	Ragusa
120	D	Abraham Ridderikhoff [82-7] WBg (ex Betsey)	Dutch	347	129·7	27·9	14·5	Nieuwendam	1863	W.&A.H.Meursing		J.&J.Vinke	Amsterdam
121		Abraham Barker Smith WBk	American	389	117·7	28·9	18·9	Fairhaven, Mass	1845	ReubenFish		J.& W.R.Wing	NewBedfrd, Mass
122		Abraham Richardson Pollard 76-82 WSr	American	147	99·0	27·0	8·3	Tremont, Me.	1874	HenryNewman		George T.Osborn	Belfast, Me.
123		Abraham Rydberg Sellergren 75-79 WS	Swedish	144	121·3	22·3	11·5	Carlshamn	1879	D.R.Andersson		Rydbergska Stiftelsen (N. Paton)	Stockholm
124	N	Abraham Skalle (STRAND) Christensen -88 WBk	Norweg'n	343	119·4	28·1	15·9	Lillesand	1868	Ths. Asvaksen		A.Tönnesen	Lillesand
125	RV	Abram Young McLaughlin -88 WBk	British	756	151·8	33·9	20·0	Granville, N.S.	1870		Y 8.88	Troop & Son	St.John, N.B.
126	DV	Abyssinia Hatten 83-84 WBk	British	1127	186·5	36·7	22·4	St.John, N.B.	1885	O.Pittfield	Y 2.88	Taylor Bros.	St.John, N.B.
127	✻	Abyssinian Hughes -84 WS	British	1266	196·0	39·0	23·9	Quebec	1868	P.Baldwin	Y 9.84	R.Lines	Liverpool
128		Acacia Herbert WBk	British	233	118·0	24·0	12·0	Hobart	1871	Ross		Belbin & Co.	Hobart
129	R	Acacia Kennedy -88 WBn	British	186	98·4	25·6	10·1	PtMedway NS	1884	JohnHutt	Y 7.86	T.Forhan	PortMedway N.S.
130		Acacia Lohnes -87 WSr	British	117	80·6	24·7	9·5	Conquerall NS	1887	J.N.Rafuse		E.Lohnes	Lunenburg, N.S.
131	✻	Acacia Pearn 85-85 WSk	British	45	60·3	18·6	7·4	Plymouth	1880	Banks		D.Banks & Co.	Plymouth
132		Acadia W	British	375	150·0	26·5	11·6	Quebec	1875	F.Cantin		Montreal Transprtn.Co.	Montreal
133	R	Acadia Morrell -88 WBn	British	241	108·8	27·5	12·7	Picton, N.S.	1875	J.T.Paulin	Y 10.86	B.Davis	Picton, N.S.
134		Acadian Ellis 86-86 WSr	British	110	83·0	24·2	9·4	P.E.Island	1860	Bell		W.Barniston	Penzance
135	✻DV	Acadie Oldenburger 75-85 WBn (ex Alice)	Dutch	189	97·2	23·8	11·4	Geestemünde	1864	J.C.Tecklenborg	Y 4.87	J.H.Henkes	Rotterdam
136	V	Acancia Millet WPol	Spanish	169	88·9	26·0	12·3	Barcelona	1849		Y 2.87	Font & Rimdor	Barcelona
137	✻	Acapulco Simonsen -88 IBk (ex Callao)	German	549	168·0	29·2	17·6	Sunderland	1868	W.Pile& Co.		Oetling Gebrüder	Hamburg
138		Acara Cummings WSr	American	135	91·5	27·3	8·2	HarringtonMe	1873			R.P.Doyle	Machias, Me.
139	V	Acazo Godinho -87 WBk (ex Godin)	Portug'se	498	141·4	30·9	19·7	Savona	1873	G.Tixi	Y 3.87	J.T.de Silva Godinhon	Lisbon
140	✻	Accrington Fryke 85-85 IS (ex Strathmore, ex François Arago)	British	1831	243·4	38·8	30·9	Nantes	18 55/89	GuibertFils		Gracie, Beazley & Co.	Liverpool
141	✻	Achievement Thomas 88-88 IBk	British	499	154·0	27·9	17·3	Glasgow	1868	Dobie& Co.		Doward, Dickson & Co.	Liverpool
142	V	Achille Duward WBg	French	238	115·8	21·3	11·1	Dunkirk	1870	A.Lefebvre	Y [illegible]	A.Delrue	Dunkirk
143	I	Achille Faggioni WBn	Italian	112	88·6	23·4	11·4	Spezia	1873	G.Faggioni	Y 10.85	G.B.Salomone fu N.	Savona
144	V	Achille Célestine Delande 86-79 WBn	French	184	89·2	24·3	12·5	Nantes	1876	J& A.Lefrançois	Y 1.86	A.Lefrançois	Nantes
145	A	Achille F. Gelletich -87 WBk	Austrian	589	133·5	29·8	19·2	Portore	1877	L.Archianda	Y 9.85	Ved[a] Maria Franci	Fiume
146	✻	Achilles Robinson IS	British	1521	232·6	37·7	24·6	Liverpool	1866	T.Vernon& Son		A.Gibson & Co.	Liverpool
147	NV	Achilles Dahl -88 WBn (ex Morning Star)	Norweg'n	467	129·1	30·0	17·7	TwoRiversNS	1866		Y 11.85	Bladestad, Holta & Co.	Skien
148	N	Achilles Knudsen 87-87 W3mSr (ex Trinidad)	Norweg'n	311	123·4	24·4	16·0	Bristol	1854	C.Hill&Sons	Y 6.86	Act Selsk. "Achilles" (Gundersen)	Kragerö
149	N	Achilles Eriksen WBk	Norweg'n	291	100·1	26·9	15·6	Grimstad	1846	J.ErlandsenVig		Rederiselskabet "Achilles" (F. W. Wessel)	Drammen
150	V	Achilles Ekwall -87 WBn	Swedish	194	111·2	22·6	10·9	Oscarshamn	1877	P.Pettersson	Y 2.87	C.G.E.Ekvall	Oscarshamn

✻ = Lloyd's Register; D = Nederlandsche Vereeniging; G = Germanischer Lloyd; I = Registro Italiano; R = American Record;
A = Veritas Austro-Ungarico; H = Veritas Ellenico; N = Norske Veritas; V = Bureau Veritas.

Dans la 4me Colonne, C = Composite (membrure en fer, bordé en bois); I = Fer; Stl = Acier; W = Bois.

Les Dates du doublage en métal (Col. 11) ressortent des livres dans lesquels les navires sont respectivement classés ou des rapports des Inspecteurs du Lloyd's Register.

Pour plus amples renseignements sur les navires classés par le Lloyd's Register, voir le Lloyd's Register Book et le Lloyd's Yacht Register.

Official No.	No. in Book	Society in which classed	Ship's Name. Material and Rig. Master, and Dates of Appointment to present Owner's Service and to Ship. Ship's Late Name.	Flag	Net Tonnage	Dimensions: Length	Breadth	Depth	Built: Where	When	By whom	Date of Metalling	Owners or Managers
42	151		Achilleus WBk	Greek	368	119·1	28·0	..	Syra	1878	S.Cafulachi	[illegible]	J.Carissis
77768	152	R	Achsah *Massells* -88 WBn	British	139	90·0	23·8	10·6	Shelburne,N.S	1880	W.S.Kelly	[illegible]	Roop Brothers
27840	153	✱	Aconcagua *Kerr* 75-86 IBk	British	496	150·0	26·5	17·5	Liverpool	1859	J.Jones,Jr.		J.B.Walmsley & Co.
	154	✱F	Aconcagua *Irribaren* 76- IBk	French	691	186·5	30·2	18·3	Birkenhead	1870	Laird Bros.		Ant.Dom.Bordes & fils
28191	155		Acorn *Sarchet* WBg	British	167	90·5	23·0	13·9	Whitby	1854			P.Sarchet
	156		Acorn *Andersen* -85 W3mSr	Norweg'n	299	125·7	24·3	15·6	Dundee	1855	TayS.B.Co.		H.Andersen
52566	157		Actæa *Caudy* WBk	British	300	134·0	25·3	14·2	Aberdeen	1865	Burns&Smith		E.Edouard
	158	N	Actæon *Steen* 87-87 WBk	Norweg'n	555	137·5	30·5	19·1	Miramichi	1838	A.Gilmour&Co.	[illegible]	Steen & Ström
	159	N	Actie *Olsen* WBk (ex S. W. Fike, ex G. R. Mosell)	Norweg'n	541	141·5	29·2	18·0	Bath,Me.	1860		[illegible]	W.Boë
	160	V	Actif *Boursier* WBk	French	396	121·7	25·6	17·0	Nantes	1872	PJollet&LBabin	[illegible]	Aristide Briand
	161	V	Activ *Rifbjerg* 73-82 WSr	Danish	105	76·6	20·5	10·1	Nakskov	1865	Ridersborg& Trochmann		F.P.Rifbjerg
	162	G	Activ *Haase* -86 WBk	German	317	109·3	25·5	16·3	[illegible]	1861	H.Dierling		J.Haase
	163		Activ *Drost* -88 WBn	German	266	107·2	23·0	12·4	Hammelwarden	1873	A.F.Bosse	[illegible]	A.H.Arnold
	164	V	Activ *Orfel* 88-88 WBn (ex Blitz, ex Johann Gottfried)	German	206	108·5	23·7	11·6	Burg a.d Weser	1851 78	H.Bosse	[illegible]	H.Bauer
	165	GN	Activ *Andersen* -83 WBk (ex Home)	Norweg'n	706	151·5	34·1	20·1	St.John,N.B.	1866	Milledge	Y 12.87	J.Christophersen
	166		Activ *Christensen* -88 WBk (ex Vespucci)	Norweg'n	516	143·7	30·7	19·0	Sestri Ponente	1867	A.BriascofuG.B	Y4.8[illegible]	H.S.Jacobsen
	167	NV	Activ *Johannesen* W3mSr	Norweg'n	286	127·1	27·2	12·6	Stavanger	1876	K.Nielsen	Y [illegible]	H.J.Olsen
	168	N	Activ (FOUNDERED) *Jensen* WBg	Norweg'n	173	94·2	24·1	11·9	Risör	1861			S. Olsen
	169		Activ *Jacobsen* WSr	Norweg'n	118	82·4	22·1	11·0	Stavanger	1879			T.Amundsen
91771	170	✱	Active *Cheyne* WK	British	54	67·7	19·1	8·1	Fraserburg	1886	J.Webster&Co.		Fraserburg & Leith Trading Co.,Ld.
936	171		Active *Frisbie* WSr	American	251	104·0	28·0	9·0	Portsmth.N.H	1862	DanielMarcy		J. Peirce
	172	V	Active (WRECKED) *Joonekindt* WSr	French	136	90·9	21·6	11·5	Dunkirk	1877	A.Pyotte		G.Dinoir
	173	V	Active *Pédron* -88 WSr	French	128	88·3	20·0	11·0	St.Malo	1888	L.Tranchemer		L.Verry
	174		Active WBk	Russian	540	..	..	..	Gamla-Karlby	1863			F.Z.Durchmann
	175		Activo *Nercherin* WBk (ex William Mollhuish)	Chilian	654	179·8	30·3	19·1	Jersey	1859	F.C.Clarke		P.Nercherin
	176		Activo (ex Stena) WSr	Ven'zuel'n	179	94·2	25·1	10·3	P.E.Island	1876	Egan		F.Scandella
10578	177		Ada *Scott* 81-86 WBk	British	413	133·0	26·7	18·0	Yarmouth	1854		[illegible]	J.W.Waugh
32314	178		Ada *Beedle* 76-76 WBg	British	180	85·0	23·0	14·1	Newcastle	1845	Gaddy&Lamb		R.Harrowing
66314	179		Ada *Lewis* WBn	British	172	102·3	24·9	12·8	P.E.Island	1872	D.Egan		D.T.Sharper
52196	180		Ada *Magee* WSr	British	134	86·5	25·6	9·6	N.Brunswick	1866			J.Stewart
42472	181		Ada (ex Virginia Antonietta) WSr	British	123	89·1	24·5	7·5	Baltimore,U.S	1858			Chas.Bagnall
70484	182		Ada *Armstead* 69-77 WSr	British	115	88·6	22·3	11·5	Ulverston	1876	Post		J.Geldart
	183		Ada *Andersson* WSr	Russian	220	..	..	..	Kimito	1871			A.V.Andersson
105963	184	R	Ada A. Kennedy *Kennedy* W3mSr	American	633	155·3	35·4	15·4	Bath,Me.	1881	Goss&Sawyer		Cobb,Wight & Co.
1153	185		Ada Ames *Adams* WSr	American	190	100·6	26·6	8·6	Rockland,Me.	1860			J.L.Snow & Co.
106320	186	R	Ada Bailey *White* 83-86 W3mSr	American	496	160·0	33·2	12·3	Bath,Me.	1884	A.Sewall&Co.	[illegible]	A.Sewall & Co.
105135	187		Ada Barker *Rogers* W3mSr	American	220	113·0	30·0	9·3	Jonesport,Me.	1872	Barker		D.J.Sawyer
72631	188	DV	Ada Browne *Buick* 85-85 WBk	British	999	187·0	36·4	22·6	NovaScotia	1879	M'Dougall	[illegible]	Browne & Watson
75140	189	G	Ada C. Owen *Hicksen* 83-84 WBn	British	183	98·9	25·1	10·8	MahurangiNZ	1877	J.Darrach&Sons	Y 11.87	A.B.Donald
1309	190		Ada Carter (CONDEMNED) *Heustis* WBk	American	413	124·5	28·2	17·6	Harpswell,Me	1863			C.P.Heustis
105467	191	R	Ada G. Shortland *Hardine* W3mSr	American	216	117·4	30·5	9·3	Portl'nd,Conn	1874			Miller & Woodman
84033	192		Ada Gane *Waters* 86-88 WKBarge	British	110	96·0	22·4	8·4	Harwich	1882	J.H.Vaux		W.S.Gane & Co.
105877	193	R	Ada P. Gould *Henrahan* WBkn	American	495	131·6	31·2	17·5	Addison,Me.	1874		Y 11.88	W.B.Henrahan
73258	194	✱D	Ada Peard *Hocken* 72- W3mBn	British	241	119·3	25·6	13·8	Par	1875	S.Moss	[illegible]	N.Hocken
1580	195		Ada S. Allen *Dudley* WSr	American	142	94·0	27·0	18·6	Dennysvll,Me.	1867	Wyatt		G.W.Dudley
	196		Ada S. Babson WSr	American	147	96·5	26·4	10·6	E.Boothby,Me	1888	W.Adams&Son		D.C. & H.Babson
	197	V	Ada Stott (WRECKED) *Bruss* -82 WBk	German	505	139·6	30·3	18·4	Rostock	1875	O.Ludewig	[illegible]	Helm. Mentz,sen.
53637	198		Adah *Foster* WSr	British	125	85·4	24·5	10·0	Clement'sPort N.S.	1865			HowardB.Allen
2348	199		Adalgisa (ex Serita) *Durante* WBk	Italian	218	99·4	21·3	13·8	Varazze	1862			E.Ginocchio fu P.
	200												

✱ = Lloyd's Register; D = Nederlandsche Vereeniging; G = Germanischer Lloyd; I = Registro Italiano; R = Americ[an]
A = Veritas Austro-Ungarico; H = Veritas Ellenico; N = Norske Veritas; V = Bureau [Veritas]

In the 4th Column, C = Composite (iron frame, wood planked); I = Iron; Stl = Steel; W = Wood.

The Dates of Metalling (Col. 11) are based on the records of the Books in which Vessels are respectively classed, or on the Reports of the Surveyors of Lloyd['s]

Further particulars of Vessels classed by Lloyd's Register will be found in Lloyd's Register Book and Lloyd's Yacht Register.

N° dans le livre	Registre	Nom. Matériaux et Gréement. Capitaine et dates d'entrée au service des Armateurs actuels et à bord du navire. Ancien Nom.	Pavillon	Tonnage	Longueur	Largeur	Creux	Construit. Où.	Quand.	Par qui.	Date du doublage en métal	Armateurs ou Directeurs-Gérants.	Port d'Armement.
201		Adam Bowlby *Keith* W Sr	American	108	86·1	26·9	7·7	Ellsworth, Me.	1872			M. & B. W. Hall	Ellsworth, Me.
202		Adam Lodge *Abrahamsen* W S (ex SEMMERIS [illegible])	Norweg'n	607	130·8	32·5	23·0	Shields	1833			H. S. Jacobsen	Flekkefjord
203	V	Adam W. Spies *Pettit* W Bk	American	1171	185·0	38·4	22·8	Newburyport, [Mass.	1884	Atkinson & Son	Y 1.87	B. F. Atkinson	New York
204	✱ V	Adamant *Kirkland 81-* I Bk	German	815	174·9	30·0	19·2	Hull	1858	M. Samuels'n & C		C. M. D. Jörgensen	Hamburg
205	✱	Adamantine *Clark 76-82* I Bkn	British	222	121·5	24·0	11·4	Sunderland	1877	J. Blumer & Co.		C. T. Bowring & Co.	Liverpool
206		Adamastor W Bn	Brazilian	102	..	..	..					V. J. de Barros	Caraguatatuba
207		Adamastor *Ramos* W Bk (ex St. André)	Portug'se	263	104·9	23·3	13·8	Nantes	1853	Guibert		A. M. Dos Santos	Lisbon
208		Adami *Bernsen -71* W Bk	Norweg'n	250	97·9	27·2	15·2	Bremen	1839			Chr. Swane	Christiania
209	A	Adamo *Karstegl* W Bk	Austrian	442	119·1	29·6	17·4	Trieste	1865	Stabilimento [Tecnico	Y 7.86	G. Karstegl di B.	Sabioncello
210	R	Adda J. Bonner *Myrick* W Bkn	American	464	133·5	31·1	16·8	Baltimore	1872	Wellener & Buck	Y 4.87	W. B. Willson	Baltimore
211	✱	Adderley *Hains 88-88* Stl Bk	British	1198	225·6	36·3	22·0	P. Glasgow	1888	J. Reid & Co.		C. W. Kellock & Co.	Liverpool
212	R	Addie B. Bacon *Bacon* [82-] W 3m Sr	American	371	124·0	32·5	14·0	Philadelphia	1874	Birely, Hillman [& Streaker		Chas. A. Pettit	Philadelphia
213	R	Addie Benson *Carey -88* W Bn	British	328	122·5	29·0	12·3	Bear River, NS	1873		Y 8.88	Alex. M'Donald	Sydney, C.B.
214		Addie C. Hesseltine *Snow* W Sr	American	129	93·7	24·2	10·5	S. Francisco	1885	W. C. Woods		A. Crawford	S. Francisco
215	R	Addie E. Snow *Norton* W Sr	American	155	95·8	24·4	9·0	Rockland, Me.	1876	Snow & Farwell	Y 10.86	J. L. Snow & Co.	Rockland, Me.
216	✱	Addie Fuller *Hart* W Sr	American	206	107·0	26·0	10·6	Brewer, Me.	1867			M. Mehan	Boston, Mass.
217	R	Addie G. Bryant *Kendall* [82-88] W 3m Sr	American	225	122·6	28·0	9·8	Brewer, Me.	1871		Y 11.87	H. Lord	Bangor, Me.
218	R	Addie H. Cann *Litson -88* W Bk	British	650	152·0	31·0	19·0	Yarmouth, NS	1875	W. A. Cann		A. C. Haines	Yarmouth, N.S.
219	R	Addie Hale *Howes -88* W Bg (WRECKED)	American	334	120·3	27·7	12·0	Westbrook Me	1867			E. Nichols	Portland, Me.
220	R	Addie Jordan *Harriman* W 3m Sr	American	358	132·3	31·0	13·8	Saco, Me.	1873			R. F. C. Hartley	Saco, Me.
221	R	Addie Morrill *Andrews* W Bkn	American	631	160·0	35·3	12·0	Rockport, Me.	1882	Carleton, Nor- [wood & Co.	Y 1.89	Carleton, Norwood & Co.	Camden, Me.
222		Addie R. Warner W Sr	American	134	99·0	24·0	8·0	Duxb'ry, Mass.	1875			A. Merritt	Philadelphia
223		Addie Sawyer *Cook* W Sr	American	125	93·0	26·8	8·1	Calais, Me.	1873	Potter		Rideout & Lord	Calais, Me.
224		Addie Schlaeffer *Hoey -88* W Sr	American	169	105·6	28·6	8·5	P. Jefferson, LI	1874			Saml. E. Birdsall	New York
225		Adél *Eriksson -78* W 3m Sr	Russian	242	105·1	26·8	11·9	Sideby	1878	Hagström		N. Sittkoff	Mariehamn
226	✱	Adela *Evans -87* W Bg	British	168	102·7	22·7	12·5	Ipswich	1862	J. Cobbold	Y 3.88	G. Shepherd	Ipswich
227		Adéla *Fauxe -86* W Bg (CONDEMNED)	French	216	106·9	22·3	12·7	Grangemouth	1865	T. Adamson	C 11.86	J. L. Vincent	St. Malo
228		Adela *Quesada* W Bk (ex Annie Elise, ex Emma)	Spanish	327	125·9	28·9	18·3	Maitland, N.S.	1872	Douglas & Curry		Doña C. Echevarria	Havana
229		Adela (LOST) W Bn		137	94·5	23·6	10·9	Cavite, Luzon	1862				
230		Adela W Sr	Spanish	136	96·8	25·6	8·2	Blanes	1860			Mariano Julia	Barcelona
231		Adela *Cruañes* W Sr	Spanish	120	87·9	22·1	9.7	Vinaroz	1851			Jaime Guinaras	Vinaroz
232		Adela S. Hills *Jenkins* W Bk	British	463	140·7	34·7	14·3	United States	1874	Storrett	Y 7.86	E. Thomas	Aberystwith
233	G	Adelaar *De Boer -88* W 3m Sr (ex Alb[illegible])	Dutch	301	118·4	28·4	14·6	Stettin	1859	E. Zieske	C 75	M. van Neck	Nieuwediep
234	DG	Adelaar *De Boer* W Bn (ex Oprav)	Dutch	199	106·3	25·1	12·1	Quebec	1866	Paterson, Shaw & [Rousseau		H. Jans & Co.	Harlingen
235		Adelaida W Bn	Spanish	136	85·0	23·7	10·9	Blanes	1854			B. Florit	Barcelona
236		Adelaide (ex Maria & Elisabeth) W Bk	British	271	115·0	24·0	15·0	Rotterdam	1858			Geo. C. Ross	Singapore
237		Adelaide (BROKEN UP) W Bg	British	221	103·0	23·8	14·4	Guernsey	1830	J. Nandin		Mrs. P. Prévôt	Guernsey
238	G	Adelaide *Dunn -82* W Sr	British	217	144·8	23·3	8·9	Pyrm'nt NSW	1879	P. Allen	Y 4.80	C. B. Bond	Sydney, N.S.W.
239		Adelaide *Toms* W Bn	British	165	106·0	24·3	12·7	Padstow	1869	Tredwen		J. Merrifield	Fowey
240		Adelaide W Sr	British	136	86·8	24·0	9·5	Portneuf	1856			F. Thibaudeau	Quebec
241		Adelaide *Rumbold 84-84* W Sr	British	115	89·0	22·0	10·9	Middlefort	1866	H. Illum	Y 4.81	G. Kilner	Goole
242	V	Adelaide *Bailey* W Bk	American	371	129·0	28·5	14·4	Baltimore	1859	J. R. Robb	Y 10.88	T. M. Baudiel & Son	Baltimore
243		Adelaide *Good* W Sr	American	124	95·6	28·2	7·4	San Francisco	1883	Boole & Eaton		A. P. Lountzen	San Francisco
244		Adelaide W Sr	Brazilian	125	..	..	..					M. Garcia & Co.	Pelotas
245	V	Adélaide *Codet* W Bg	French	248	103·3	23·0	13·0	Nantes	1874	Sevestre frères	Y 1.86	Guegnier	Nantes
246	G	Adelaide *Braun 81-87* I S	German	1281	230·9	34·6	20·8	Vegesack	1881	H. F. Ulrichs		D. H. Wätjen & Co.	Bremen
247	I	Adelaide (ex Luigia Madre) *Basso 86-88* W Bk	Italian	597	131·1	30·1	18·7	Chiavari	1871	L. Gotuzzo	Y 1.81	F. Basso	Genoa
248		Adelaide *Lopez* W Sr (ex Matilda M.)	Portug'se	165	83·0	24·9	10·2	Parsboro', N.S	1871	W. Rogers	Y 1.80	J. J. Vera Cruz	Ilha do Sal, C.V.
249	V	Adelaide *Rodrigues -88* W Sr	Portug'se	142	83·0	21·3	9·1	Villa do Conde	1888	Lapa	Y 8.88	Bettencourt & Irmaõ	St. Michael's
250	DV	Adelaide *Josefsson 81-84* W Bk	Swedish	519	144·0	25·7	17·7	Uddevalla	1881	J. N. Sanne	Y 12.86	J. N. Sanne	Uddevalla

[illegible]= Lloyd's Register; D = Nederlandsche Vereeniging; G = Germanischer Lloyd; I = Registro Italiano; R = American Record; [illegible] = Veritas Austro-Ungarico; H = Veritas Ellenico; N = Norske Veritas; V = Bureau Veritas.

Dans la 4me Colonne, C = Composite (membrure en fer, bordé en bois); I = Fer; Stl = Acier; W = Bois.

Les Dates du doublage en métal (**Col. 11**) ressortent des livres dans lesquels les navires sont respectivement classés ou des rapports des Inspecteurs du Lloyd's Register.

Pour plus amples renseignements sur les navires classés par le Lloyd's Register, voir le Lloyd's Register Book et le Lloyd's Yacht Register.

1 Official No.	2 No. in Book	3 Society	4 Ship's Name. Material and Rig. Master, and Dates of Appointment to present Owner's Service and to Ship. Ship's Late Name.	5 Flag	6 Net Tonnage	7 Dimensions: Length	Breadth	Depth	8 Built: Where	9 When	10 By whom	11 Date of Metalling	12 Owners or Managers
[illegible]	251		Adelaide Baker (ex B. F. Carver) Mcgregor 85-88 W Bk	British	818	152·8	34·6	20·9	Brewer, Me.	1863		Y [illegible]	C. Barrie
[illegible]	252		Adelaide J. Alcott Johnson 82-83 W3mSr	American	363	146·6	35·6	10·4	City Isld., N.Y.	1876	John P. Hawkins		C. W. Alcott & Co.
[illegible]	253	✻D	Adelaide Mary Peppercill 81-87 I Bk	British	572	169·7	28·4	18·4	Liverpool	1866	R. & J. Evans & Co		J. B. Walmsley & Co.
[illegible]	254	RV	Adele Babin -88 W Bk	British	604	150·7	33·4	18·0	Maitland, N.S.	1880	J. Montieth	Y [illegible]	T. E. Kenny
[illegible]	255		Adele (ex Wissahicken) W Bk	American	387	166·0	28·5	10·5	Philadelphia	1861			J. J. Friend
	256		Adele (ex Gonyzo, ex Leao) Pinto -88 W Bg	Brazilian	174	..	..	..	England	1863		Y [illegible]	N. Mallburg
	257	✻	Adele (ex Agnes Marie) Pundt 81-86 I Bk	German	851	198·5	32·2	19·9	Pt. Glasgow	1869	R. Duncan & Co.		E. F. C. Tobias
	258	GV	Adele Balsen W Sr	German	139	87·1	20·7	10·8	Edewecht	1876	H. Kramer	Y [illegible]	F. G. Orth
[illegible]	259	IV	Adele Astarita W Bk	Italian	446	132·8	30·2	19·2	Meta	1878	G. Mauro	Y [illegible]	S. Cafiero & Co.
[illegible]	260	V	Adele Ballanir W Bk	Italian	395	118·1	26·6	18·3	Venice	1867	G. Zanon	Y [illegible]	Flli. Vianello-Moro
	261	NV	Adele (ex Josephine) Isaksen 81-82 W Bk	Norweg'n	693	150·9	31·8	19·6	Bremerhaven	1870	H. Ulrich	Y [illegible]	S. C. Larsen
[illegible]	262	I	Adele Accame Vignola W Bk	Italian	1180	192·5	33·1	23·4	Sestri Ponente	1875	G. Casanova	Y [illegible]	E. Accame & figli
[illegible]	263	DIV	Adele C. Isnardi 89-89 W Bk	Italian	798	164·0	33·2	22·5	Varazze	1877	A. Cerruti	Y [illegible]	Ant. Chiesa fu G. B.
[illegible]	264	I	Adele C. (ex Credo in Dio, ex Palestina G.) Gianuoni W Bg	Italian	178	106·3	24·7	13·5	Sestri Ponente	1878	Pozzani	Y [illegible]	Flli. Cignoni
	265	V	Adèle Catherine Halais W Bn	French	142	74·8	21·6	11·8	Nantes	1864	Boju & Clairgeau		Peillac
	266	GV	Adele Lubker Dade -83 W Bk	German	534	136·0	31·0	17·0	Rostock	1883	E. Burchard & Co.	Y 6.82	H. Bauer
[illegible]	267	R	Adele Thackara Kinsey -88 W3mSr	American	525	143·0	33·7	15·2	Milton, Del.	1881	Davidson		W. M. Thackara
[illegible]	268		Adele Trudell Fidler W Sr	American	149	104·4	24·7	8·3	Camden, N.J.	1865			Jeremiah Smith & Bro.
	269	R	Adèle & Louise Nilsson W Bk	Swedish	390	122·6	24·9	16·2	Nantes	1864			C. Pettersson
	270	N	Adele & Sabina Jorgensen 80-86 W Bk	Norweg'n	698	147·8	33·0	20·3	Genoa	1874		Y 6.86	Nils P. Höyer
[illegible]	271	R	Adelene Starken W3mSr	British	193	102·0	28·6	9·9	Rothesay, N.B.	1885	Henry Titus		G. F. Baird
	272		Adelfotis Nicoletto W Bk	Russian	316	103·3	28·5	13·6	Ile-de-Casos	1869	A. Emanoil		P. Nicoletto
	273	N	Adelgunde (ex Moira) Pedersen 81-83 W Bk	Norweg'n	834	162·2	33·2	22·6	Quebec	1854		Y 12.88	H. B. Johannesen
	274	N	Adelgunde Blichfeldt 82- W Bg	Norweg'n	222	91·4	25·3	14·6	Bremen	1827			O. Blichfeldt
	275	V	Adelheid Stahl -72 W Bg	German	213	94·8	23·6	13·0	Ueckermünde	1859	L. Wittenberg		Carl Wichards
	276	G	Adelheid Bittem -88 W Bn	German	167	101·0	23·0	12·0	Papenburg	1867	H. A. Klasen	Y 5.85	Herm. A. Klasen
	277		Adelheid A. Sulg W Sr	Russian	154	80·0	24·0	12·0	Arensburg	1880			P. Sulg
[illegible]	278		Adelheid & Bertha (CONDEMNED) Boos -87 W Bk	British	424	121·1	29·6	17·0	Elbing	1863		Y 8.84	Bevan, Evans & Co.
	279	N	Adelheim Andersen -82 W Bk	Norweg'n	393	121·8	29·4	17·2	Porsgrund	1866		Y 12.80	Chr. Th. Boe
[illegible]	280		Adelia Berry W Sr	British	163	96·2	25·7	10·5	Clements, N.S.	1869			J. E. Gilliatt
[illegible]	281		Adelia W Sr	British	138	77·1	22·3	9·1	Lussinpiccolo	1868			H. J. Whitlock
	282	A	Adelia I^a (ex Italia ex Rocha Cortelano) Pinto -87 W Sr	Portug'se	167	99·4	23·6	9·4	Villa do Conde	1869	J. M. d'Araujo	Y [illegible]	J. J. Barbosa
[illegible]	283		Adelia Felicia Blizzard W Sr	American	131	103·0	22·0	7·0	Bristol, Pa.	1853			B. Blizzard
	284	V	Adelina (ex Pacheco I) Dos Santos W Bg	Portug'se	239	93·8	25·3	13·1	Figu'ira da Foz	1856	Mendes & Irmaõs	Y [illegible]	J. N. Pinto
	285	GV	Adeline Biel W Bk	German	530	128·1	30·2	15·4	Elsfleth	1875	Fr. Detjeen	Y 3.86	Joh. G. Lübken
[illegible]	286		Adeline Burke (ex Liberty) W Bk	British	281	126·3	26·2	12·5	Maine, U.S.	1857			Henry Moss
[illegible]	287		Adeline Gibbs Reed W Bk	American	327	105·1	27·2	18·3	Fair Haven, [Mass.	1841			J. McCullough
[illegible]	288		Adeline Townsend Hart W Sr	American	220	100·0	30·2	9·6	Northport, NY	1854			J. S. Lounsberry
[illegible]	289		Adeliza M'Gregor W Sr	American	143	95·8	28·3	8·4	Tuckerton, NJ	1865			S. W. Haines
[illegible]	290		Adella Corson Corson W3mSr	American	379	120·5	31·4	9·7	Weatherley's [Landing, NY	1871			D. T. Corson
[illegible]	291	A	Adelphi Mitropulo -88 W Sr	Greek	239	104·9	27·2	14·6	Galaxidi	1886	J. Kanatas	Y [illegible]	J. D. Mitropoulos
[illegible]	292		Adelphi W Sr	Greek	238	104·6	27·2	14·4	Galaxidi	1886	J. Kanatas		J. D. Mitropoulos
	293	N	Adelphi Sivertsen W3mSr	Norweg'n	228	112·5	28·2	11·4	Arendal	1866	O. P. Olsen		Gunder Olsen
[illegible]	294	A	Adelphi Diamandidi Moschakis W Bk	Greek	190	90·5	24·9	..	Syra	1885	A. Livadara	Y [illegible]	M. Moschakis
[illegible]	295	AG	Adelphi Efstathiou (ex Evangeline) Siskos -79 W Bk	Greek	396	136·1	27·0	..	Jersey	1866	Le Visconte	Y [illegible]	D. Siskos
[illegible]	296	A	Adelphi Miliadi Evangeli W Sr	Greek	171	87·6	23·8	..	Syra	1878			P. Agapilos
	297		Adelphi Vagliano W Bg	Greek	312	107·9	27·6	..	Syra	1881			L. Rossolimos
	298	A	Adelphi Valieri (ex Delfino) W Bk	Greek	521	137·8	31·8	..	Sestri	1872	G. Traverso	Y 7.84	Valieri Bros.
	299	N	Adelphia (ex [illegible]) Marcussen W Bk	Norweg'n	465	120·1	31·8	18·5	Delfshaven	18[illegible]	H. D. Hoog	Y [illegible]	P. Smith
	300		Adelphoi Ballianoi W Bk	Russian	282	..	..	..		1852			Kampitsis

✻=Lloyd's Register; D=Nederlandsche Vereeniging; G=Germanischer Lloyd; I=Registro Italiano; R=Americ[an]; A=Veritas Austro-Ungarico; H=Veritas Ellenico; N=Norske Veritas; V=Bureau [Veritas]

In the 4th Column, C=Composite (iron frame, wood planked); I=Iron; Stl=Steel; W=Wood.

The Dates of Metalling (Col. 11) are based on the records of the Books in which Vessels are respectively classed, or on the Reports of the Surveyors of Lloyd's

Further particulars of Vessels classed by Lloyd's Register will be found in Lloyd's Register Book and Lloyd's Yacht Register.

1 No.	2 Cote	4 Nom. Matériaux et Gréement. Capitaine. Ancien Nom.	5 Pavillon	6 Tonnage	7 Dimensions. Longueur	Largeur	Creux	8 Construction. Où.	9 Quand.	10 Par qui.	11 Date du doublage en métal	12 Armateurs ou Directeurs-Gérants.	13 Port d'Armement.
301	H	Adelphotis *Chiras* WBn	Greek	218	89·9	24·6	..	[illegible]	1875	[illegible]		C.I.Chiras	Scopelos
302	A	Adelphotis *Niadho* WBn	Greek	13[illegible]	..	..	..	Cassos	1880			G.Niadho	Cassos
303	H	Adelphotis *Triketafilou* WSr	Greek	121	74·5	22·8	13·0	Scopelos	1876	M.Sarimitiasis		N.A.ChTriketafilou	Scopelos
304	A	Adelphotis *Ajoti* -88 WSr	Greek	122	73·8	22·3	..	Skiathos	1884	C.G.Mytilinaios		N.P.Agiotou	Skiathos
305	A	Adelphotis *Frango* WBn	Turkish	104	..	..	..	Syra	1868			S.Frango	Chios
306	❋	Adeona WSrYt	British	109	93·8	20·2	11·4	Cowes	1878	J.S.White	[illegible]	E.R.Bickersteth	Cowes
307	N	Adeona *Olsen* 88-88 WBk	Norweg'n	638	153·4	34·3	17·5	Arendal	1883	A.Henriksen	Y 7.88	J.E.Hanssen	Arendal
308		Adieu WBk	Norweg'n	278	112·0	24·5	13·3	Nova Scotia	1855			H.E.Röd	Tönsberg
309	N	Adiel *Jacobsen* W[illegible] (ex [illegible])	Norweg'n	272	118·1	27·8	12·7	Menudie,N.S.	1864	Gibben		E.H.Wahl	Fredriksstad
310		Adina *Danielson* 88 WSr	Swedish	16[illegible]	77·8	19·7	10·9	Glasgow	18 64/77	Gillies & Reid		P.Danielson	Kivik
311	N	Adjutor *Nielsen* WBk	Norweg'n	280	114·3	28·0	14·3	Arendal	1870	G.Thorbjörnsen	Y 7.8[illegible]	G.Thorbjörnsen	Arendal
312		Adjutor WSr	Norweg'n	113	83·1	22·5	10·8	Stavanger	1859	J.Johnsen		Brödrene Berner	Stavanger
313	GV	Adler *Brenken* -87 WBn	German	251	107·2	25·4	12·5	Hamm'lw'rdn	1873	C.Lühring	Y [illegible]	H.C.Meiners	Brake
314	V	Adler *Rüdigers* -88 WdmSr	German	2[illegible]8	105·2	23·2	12·2	Neustadt	1866	J.G.T[illegible]		W.Höveler	Papenburg
315	❋D	Admiraal de Ruyter WBn (*Boekhout* 86-88)	Dutch	203	107·2	22·4	12·5	Hoogezand	1874	J.A.Hooites	Y [illegible]	J.A.Hooites	Hoogezand
316	❋D	Admiraal Tromp WBg (*De Groot* [illegible])	Dutch	208	110·8	22·5	11·9	Hoogezand	1877	J.A.Hooites	Y [illegible]	J.A.Hooites	Hoogezand
317		Admiral WSr	British	269	161·0	23·0	10·2	Niagara	1842			E.S.Vinden	Quebec
318	GV	Admiral *Mollenhauer* [illegible] WBk	Norweg'n	732	143·1	33·3	20·9	Vegesack	1848	C.Lange	Y 6.88	Westergaard & Co.	Christiania
319	DG	Admiral Tegetthoff WBk (*Bohse* 71-7[illegible])	German	833	164·9	34·3	22·8	Bremerhaven	1871	J.C.Tecklenborg	Y 9.87	D.H.Wätjen & Co.	Bremen
320		Admiral Tromp WBk (*Christensen*)	Norweg'n	691	127·7	36·6	22·4	Amsterdam	1837			Chr.Swane	Christiania
321	NV	Admiralen *Olsen* 81-81 WBk (ex Admiral [illegible])	Norweg'n	380	136·0	29·1	15·5	Apenrade	1862		Y 3.87	A.Westerhuus	Christiansand
322		Adolf *Hansen* 84-84 WSr (ex [illegible])	Danish	122	80·9	19·5	11·0	Vegesack	1846		Y 12.82	P.Simonsen	Odense
323	GV	Adolf *Gram* -88 WBk	Norweg'n	1090	172·8	38·2	22·7	Bremerhaven	1865	F.W.Wencke	Y 10.85	J.C.Hansen	Tönsberg
324		Adolf (ex Edward) *Plath* -87 WBk	Russian	240	97·5	24·5	16·0	Scarbro	18[illegible]			Capt.Plath	St. Petersburg
325		Adolf *Carlsson* -81 WSr	Swedish	102	84·7	21·1	8·8	Misterhult	1872			S.Carlsson	Oscarshamn
326		Adolf Fredholm WBk (*[illegible]*)	Swedish	328	126·6	25·6	13·8	Stockholm	1865	S.G.Jensen		L.Påhlsson	Lerberget
327	V	Adolph *Westergaard* 76-85 WBk	German	867	176·8	31·5	20·8	Hamburg	1877	H.C.Stülcken, [Wwe.	Y 3.87	H.C.Stülcken,Wwe.	Hamburg
328	V	Adolph *Westendorff* 69-87 WBk	German	532	145·6	25·5	17·2	Hammelw'rdn	1868	C.Siwert	Y 6.86	A.Burchard	Rostock
329	G	Adolph *Stoeckius* -87 WBg	German	256	110·2	24·9	12·6	Brake	1870	J.D.Behrens	Y 8.56	Deutsche Westafrikanische Compagnie	Hamburg
330		Adolph *Hansen* 64-83 WBg	German	215	103·6	25·3	12·9	Stralsund	1864	C.W.Mohr		Seitz & Kindt	Stralsund
331		Adolph *Hamberg* 82-83 WBg	Russian	282	109·6	24·3	13·6	Hamburg	1857	Reiherstieg,S.B. [Co.		G.W.Hamberg	Hangö
332	V	Adolph *Sjöholm* 87-87 WBk	Swedish	458	139·7	27·4	16·3	Sundswall	1870		Y 3.87	Rederibolaget Union (G. O. Wallenberg)	Stockholm
333	V	Adolph Friedrich WBg (*[illegible]*)	German	197	100·0	25·6	12·3	Stralsund	1863	C.W.Mohr		C.A.Beug	Stralsund
334	G	Adolph Michels WBg (*Niemann* 80-88)	German	290	108·6	26·4	15·8	Rostock	1854	W.Zeltz		PaulKüchenmeister	Rostock
335	RV	Adolph Obrig *Carleton* WBk	American	1448	208·2	38·6	23·0	Rockport,Me.	1881	Carleton,Norwood&Co.	H 11.88	Carleton,Norwood&Co	Camden,Me.
336		Adolphe *Lemarchand* WBk	French	284	111·5	24·9	15·2	Fecamp	1853			J.Bial	Saint Malo
337	❋	AdolpheThiers *Fron* -86 IBk	French	782	170·0	31·5	20·0	Middlesboro'	1862	Richardson, [Duck&Co.		A.Postel & ses fils	Havre
338		Adolphe & Laura WBn (ex Charles Emmet) *Foster* 83-83	British	144	87·0	21·3	11·8	Nantes	1867	Ordronneau		SheernessCo-op.Soc.Ld.	Rochester
339		Adolphina (ex Mimosa) WSr	Dutch	116	82·0	19·6	8·0	Singapore	1875	HoAchong		C.W.R.Van Renesse Van Duivenbode	Ternate
340	V	Adolphine *Mosberg* 81-81 WSr (ex Apollo)	Danish	124	90·2	18·9	11·1	Svendborg	1864	CC&HP Hansen		C.Christensen	Aalborg
341		Adolpho W	Brazilian	124	..	..	..					L.M.de Souza	Aracaty
342		Adolpho WSr	Portug'se	100	71·4	20·3	8·4	Villa do Conde	1878	C.M.de Aranjo		M. dos Santos	Va. Nova de Portimão
343		Adonis *Hurley* WBn	British	108	92·4	20·7	8·3	JervisBay, [N.S.W.	1874	G.Dent,jun.		P.Hogan	Sydney,N.S.W.
344	V	Adonis *Jensen* 77-77 WSr	Danish	157	96·2	21·6	11·0	Thuro	1869	P.Bom		P.Bom	Svendborg
345	GV	Adonis *Braue* 86-86 WBk	German	411	134·7	27·9	15·3	Hamm'lw'rd'n	1878	W.Hagens	Y [illegible]	D.Haye	Brake
346	DGV	Adorna *Hashagen* 86-86 WS	Norweg'n	1410	192·9	39·0	22·7	Setauket	1870	D.B.Bayles	Y 5.87	Chr.Möller	Christiania
347		Adria *Wheldon* WBn	British	194	102·7	28·0	9·6	Portland, N.B	1871			J.F.Watson	St. John, N.B.
348		Adriana (ex Zephyr) WBk	Chilian	403	145·0	30·2	18·4	Sunderland	1874	J.Gill		Cia Esplotadora de Lota y Coronel	Valparaiso
349		Adriano *Pellegrini* -87 WBk (ex [illegible])	Italian	515	136·8	29·5	..	Fiume	1868			L.Ferraro fù A	Genoa
350		Adrianus *Burghout* 73- WBk (ex [illegible] (BROKEN UP))	Dutch	1075	..	..	..	Brunswick,Me	1854			H.J.Plaut & Co.	Schiedam

=Lloyd's Register; D=Nederlandsche Vereeniging; G=Germanischer Lloyd; I=Registro Italiano; R=American Record;
=Veritas Austro-Ungarico; H=Veritas Ellenico; N=Norske Veritas; V=Bureau Veritas.

Dans la 4me Colonne, C=Composite (membrure en fer, bordé en bois); I=Fer; Stl=Acier; W=Bois.

Les Dates du doublage en métal (Col. 11) ressortent des livres dans lesquels les navires sont respectivement classés ou des rapports des Inspecteurs du Lloyd's Register.

Pour plus amples renseignements sur les navires classés par le Lloyd's Register, voir le Lloyd's Register Book et le Lloyd's Yacht Register.

Official No.	No. in Book	Registry in which classed	Ship's Name. Material and Rig. Master, and Dates of Appointment to present Owner's Service and to Ship. Ship's Late Name.	Flag	Net Tonnage	Dimensions: Length	Breadth	Depth	Built: Where	When	By whom	Date of Metalling	Owners or Managers
	351	✻D	**Adriatic** *Schuth* 79-88 WSr	Dutch	158	95·5	22·7	12·3	Hoogezand	1870	J.A.Hooites	Y 3.88	J.A.Hooites
	352		**Adrien** WLr	French	112	72·2	21·3	10·1	Nantes	1858	Lemerle		A.Noblanc
90403	353	V	**Advance** *Bonnell* WBn	British	234	113·0	28·3	10·4	East No.4, N.S.	1885	E.Desamore	Y 2.87	W.R.Hutchings
45643	354		**Advance** *Stone* -79 WBn (ex L'Orient)	British	140	82·0	24·1	11·0	Antwerp	1855			E.G.Cralke
75247	355		**Advance** *Leally* -86 WSr	British	100	89·2	21·5	10·4	Garmouth	1876	J.Geddie,Jr.	Y 12.81	W.H.Steer
58966	356	✻	**Advancement** *James* 87-87 IBk	British	500	154·0	28·0	17·4	Glasgow	1868	Dobie&Co.		Doward,Dickson&Co.
	357	D	**Advena** *Visser* -87 WBn	Dutch	277	108·2	25·9	15·5	Sunderland	1867	Rawson		G.G.Albers
	358	N	**Advena** *Andersen* 86-88 WSmSr	Norweg'n	458	139·7	31·5	14·3	Arendal	1888	A.H.Friise	Y 8.88	J.M.A.Marcüssen
79378	359		**Adventure** I Lump	British	133	72·3	25·3	8·9	Liverpool	1875	W.H.Potter&Co		Mersey Dcks.&Harb.Bd.
70698	360	✻D	**Adventurer** *Jones* 85-85 IBk	British	698	186·4	31·1	18·5	Glasgow	1875	Dobie&Co.		Doward,Dickson&Co.
	361		**Adzuma Maru** WBkn	Japanese	489	143·6	29·1	15·2	Hiogo	1883	Ountsu Jingoro		Tatsunma Kiyo
	362	N	**Æolus** *Johnsen* -88 WBk	Norweg'n	584	149·3	32·6	17·9	Krageró	1881		Y 8.88	H.Biörn,jun.
	363		**Æolus** WBg	Norweg'n	283	95·9	26·8	15·6	Tvedestrand	1844	A.Smith		O.E.Ellingsen
	364	NV	**Æolus** *Krogh* 85-85 WSr	Norweg'n	220	105·9	24·0	12·1	Grimstad	1885	JacobBoe	Z 9.85	Jacob Boe
	365	N	**Æolus** *Jacobsen* -79 WSr	Norweg'n	141	94·3	25·0	11·2	Sogn	1869			K.H.Rönnevig
	366		**Æquivalent** WBk	Norweg'n	312	108·5	28·5	16·0	Tvedestrand	1853			O.Jörgensen
	367	N	**Æren** *Edvardsen* -88 WBk	Norweg'n	379	123·2	30·1	15·7	NyCarleby	18⁴⁸⁄₇₆			Actieselskabet "Æren" (J.Th.Bernhardsen)
	368	V	**Ærø** *Christensen* 84-84 W3mSr	Danish	224	113·9	23·7	12·0	Svendborg	1884	P.Troensegaard		E.E.Petersen
74549	369	✻	**Æthelberht** *Garriock* 88-88 IBk	British	768	201·0	31·0	19·1	Harrington	1876	R.Williamson&S.		Gracie,Beazley & Co.
	370		**Aemilius** *Anderson* 82 WSr (ex Minnehaha)	Swedish	204	103·7	21·8	12·7	NovaScotia	18⁶³⁄₇₅			C.Smith
89937	371		**Aeng** StlLrBarge	British	480	220·2	30·1	8·3	Dumbarton	1884	W.Denny&Bros		Irrawaddy FlotillaCo.
93184	372	✻	**Aeolus** *Brown* 79-86 IS	British	1610	259·7	38·2	23·0	Glasgow	1886	Barclay,Curle & Co.		A.& J.H.Carmichael&Co.
	373	DV	**Aeolus** *Leuning* 73-80 WSr	Dutch	135	89·9	21·8	10·8	Sappemeer	1866	C.Bakker		D.B.Dekker
	374	✻V	**Aeolus** *Putcker* 86-87 IBk (ex Bertram Rigby)	German	1245	209·0	34·2	23·5	Hull	1864	Earle & Co.		H.Bischoff & Co.
	375	GV	**Aeolus** *Reiners* 78-78 WBk	German	417	116·3	28·0	14·0	Hammelw'rdn	1872	C.Lühring	Y 8.86	Gebrüder Hustede
	376	GV	**Aequator** *Ewert* 88-88 WBk	German	547	132·5	30·7	17·8	Barth	1873	A.Witte	Y 8.87	C.A.Beug
	377	✻	**Aerial** *Simonsen* 80-86 WBn	Danish	156	93·0	23·9	12·0	Limekilns	1879	J.Whithead		J.C.Simonsen
9830	378	✻	**Aeron Lass** *Jones* 83-84 WSr	British	80	62·6	17·2	10·2	Llanddewy	1845	E.Jones		J.Jones
90876	379	V	**Aeronaut** *Helms* 87-87 WBn	British	446	139·5	31·5	13·0	Belliveau's Cove, N.S.	1886	W.D.Lovitt	Y 1.86	W.D.Lovitt
	380		**Aesah** *Mohamet Nasir* WBk	Dutch	223	93·8	25·6	..	Grissee	1859	Siemann	Y 7.81	Abdul Karim en Mohamed Taha
	381	DV	**Aethiopia** *Van Duys* 81-84 WBn (ex Italia)	Dutch	271	126·9	28·5	12·4	Bergen	1874	J.Trumpy&Sön	Y 8.86	Hk.Müller & Co.
	382	H	**Aetos** (ex Tessares Adelphi) WBg	Samos	140	..	..	..	Syra	1876			M.Isakiroglidis
	383	N	**Afær** *Nilsen* WSr	Norweg'n	109	89·9	21·2	10·4	Stavanger	1876	F.M.Haagenvik		E.Berentsen
	384		**Affehinia** *Gahlones* WBg	Turkish	135	..	..	..	Greece	1880			BulbulichleeVaughell Chatalidi & Co.
	385		**Affehti Devran** WBg	Turkish	187	..	..	..	Barten	1876			Lunuslee-MoussaKaiber
	386		**Affehti Bahri** WBg	Turkish	200	..	..	..	Kara Agatch	1881			Rizehlee Sas-Oglou Ibrahim & Co.
	387		**Affehti Bahri** *Ibrahim* WSr	Turkish	100	..	..	..	Aidos	1884			Haitab-tijar Hadji Shefki & Co.
965	388	DI	**Affezione** *Giulla* 88-88 WBk	Italian	1019	184·3	35·4	24·5	Chiavari	1884	L.Gotuzzo	Y 8.86	M.& A.Dall' Orso fu G.
	389		**Affonso Primeiro** WSr	Brazilian	288	..	..	..					M.Evora
79064	390	✻	**AfghanChief** *Evans* 84-84 WBkn	British	275	124·4	26·2	12·8	Kingston	1885	W.Kinloch	Y 1.85	W.Kinloch
93768	391	✻	**Afghanistan** *Cornell* 78-88 IfmS	British	2221	291·2	42·1	24·3	Stockton	1888	RichardsonDuck&Co.		British and Eastern Shipping Co., Ltd.
	392	DV	**Afiena** *Burhal* -88 WBn	Dutch	171	96·5	24·1	11·6	Pekela	1873	J.W.Kuiper	Y 1.86	J.J.Koerts & Zoon
	393	V	**Afma** *Ankersen* 85-85 WBn	Danish	190	130·5	22·3	10·1	Svendborg	1885	J.R.Andersen	Y 9.85	N.H.Svarrer
79062	394	DV	**Africa** *Buchanan* -87 WBk	British	1131	183·5	36·4	22·3	St.John,N.B.	1879	O.Pittfield	Y 6.87	Taylor Bros.
	395	V	**Africa** *Ellingsen* WBk (ex Seaward)	Norweg'n	668	144·6	33·2	18·8	MossGlen,N.B.	1871	Merritt & Wetmore	Y 11.87	Skibsaktieselskabet "Africa"
	396	V	**Africa** *Cardia* WBk	Portug'se	552	141·0	29·8	16·7	Oporto	1872	daSilvaSantos	Y 3.85	Azevodo & Rios
	397		**Africano** *Pereira* WSr	Brazilian	141	..	..	..					I. de Sant'Anna
	398		**Afrika** WBg	Turkish	200	..	..	..	Italy	1884			KerestejehMehemed Effendi
	399		**Afrika** WSr	Turkish	125	..	..	..	Barten	1871			Sarten-o-Tijar Hadji Ali Zade Hadji Hussein Agha
	400	N	**Aftenstjernen** *Hansen* WBk	Norweg'n	506	119·8	31·8	17·6	Drammen	1868	J.Jörgensen		Brödr.Jacobsen

✻ = **Lloyd's Register**; D = Nederlandsche Vereeniging; G = Germanischer Lloyd; I = Registro Italiano; R = Americ[an]; A = Veritas Austro-Ungarico; H = Veritas Ellenico; N = Norske Veritas; V = Bureau [Veritas]

In the 4th Column, C = Composite (iron frame, wood planked); I = Iron; Stl = Steel; W = Wood.

The Dates of Metalling (**Col. 11**) are based on the records of the Books in which Vessels are respectively classed, or on the Reports of the Surveyors of Lloyd['s]

Further particulars of Vessels classed by Lloyd's Register will be found in Lloyd's Register Book and Lloyd's Yacht Register.

1 Nº dans le livre	2	3 Registre dans lequel le navire est classé	4 Nom. Matériaux et Gréement. Capitaine et dates d'entrée au service des Armateurs actuels à bord du navire. Ancien Nom.	5 Pavillon.	6 Tonnage	7 Dimensions. Longueur	Largeur	Creux	8 Construit. Où	9 Quand	10 Par qui	11 Date du doublage	12 Armateurs ou Directeurs-Gérants.	13 Port d'Armement.
	401		Agauee Bettie W Bg	Turkish	248	..	..	..	Turkey	1865			Erphirlee-tijari-Hairie Mikhail Sakkha & Co.	Erphir
	402		Agastyr W Bk	Norweg'n	673	154·5	35·7	17·8	LaHave N.S.	1867		Y9.85	Et Interessentskab O. A. Leschbrandt	Mandal
	403		Agapithy W Bg	Turkish	110	..	..	..	Greece	1865			Costantin Coori-oglou	Constantinople
	404	V	Agapito (ex Angelina) Pagés W Bg	Spanish	237	96·8	26·9	14·6	Palamos	1855			C.Pagésy ManauAgapito	Barcelona
	405	N	Agat (ex Robert Cooper) Olsen ·77 W Bk	Norweg'n	594	143·5	31·1	18·9	Alma, Me.	1865	F.L.Carney	Y 10.88	R.M.Peterson	Moss
	406	N	Agat Bjerk ·75 W Bn	Norweg'n	166	95·8	23·1	11·8	Stavanger	1875	T.Amundsen	Y9.85	O.W.Bjerk	Stavanger
145	407	R	Agate Foster W Bk	American	595	146·0	31·1	18·5	Newburyport, [Mass.	1868	C.H.Currier&Co	Y6.87	Swan & Son	New York
[illegible]	408	*	Agatha (AUGUST) Blair ·87 W Bk	British	429	137·2	27·7	17·2	Sunderland	1866	Austin&Mills	Y6.85	E.Arnott	London
[illegible]	409		Agatha W Sw	British	187	92·9	23·6	13·5	Lynn	1859			R.Witten	Lynn
	410	V	Agatha Laty W Bg	French	191	110·1	22·7	12·1	Hoogezand	1873	J.A.Hooites		Ch.Surreau	Granville
	411	N	Agatha Röhr ·67 W Bk	Norweg'n	360	130·6	29·2	15·1	Holmestrand	1867	R.Roed	Y 10.88	J.Petersen	Holmestrand
	412	N	Agathe Hansen 56-81 W Bk	Norweg'n	854	164·1	35·1	21·8	Rotterdam	1856		Y 11.88	L.Hofgaard & Co.	Tönsberg
	413	N	Agathe (ex Excelsior) Ostenstad W Bk	Norweg'n	521	138·3	29·0	18·2	Richibucto	1860	Jardine		H.M.Ostenstad	Christiania
	414	A	Agathi Tichi Karidematti W Bn	Turkish	248	..	..	..	Galaxidi	1869			HaggiL.Onesti&Co	Cassos
	415	G V	Agda Strömberg W Bk	Swedish	357	130·2	26·1	14·1	Vätö	1874	L.Wänlund		J.A.Johanson	Vätö, Roslagen
	416	N	Agder (ex Tinker) Tobiassen 88-90 W Bk	Norweg'n	547	138·4	30·2	18·6	Newburyport, [Mass.	1869		Y8.88	Julius Olsen	Farsund
	417	N	Agder Flagestad 66-83 W Bk	Norweg'n	455	124·9	30·4	16·7	Grimstad	1867	JensTellefsen	Y1.89	S.Henriksen	Grimstad
	418	N	Agenda Simonsen ·87 W Bg	Norweg'n	220	98·5	27·0	12·9	Grimstad	1865	A.Omundsen		M.Simonsen	Fredriksstad
	419	N	Agenese Nielsen ·81 W Bg	Norweg'n	158	81·8	24·5	12·4	Sandefjord	18 62/73			T.Jacobsen Næss	Sandefjord
182	420	V	Agenor Frost ·87 W S	American	1414	202·3	39·8	24·2	E.Boston	1870	C.Smith&Co.	Y 10.86	B.B.Williams	Boston
50914	421		Agenora Bartley W Bg	British	215	110·0	25·2	13·8	Guernsey	1865	dePutron		W.Potter & Co.	Guernsey
8764	422		Agenoria Jenkins W Bn	British	176	98·2	27·7	11·8	Finland	1870			J.Williams	Liverpool
7123	423	*	Agenoria Masson 77-78 W Sr	British	168	99·3	23·3	12·0	Kingston	1879	Duncan	Y5.83	A.Marshall	Inverness
	424		Agent Lindell 71-81 W Bg	Russian	276	108·7	29·1	13·1	Nagu	1871	C.A.Lindblom		M.V.Michelson	Nagu
	425	V	Agent Lagoni Sørensen [85-85 W Bn	Danish	114	89·9	20·2	10·4	Faaborg	1872	R.Dyreborg		J.J.Larsen	Faaborg
	426	N	Ageroen Hansen ·88 W Bk	Norweg'n	382	117·2	29·8	16·0	Lillesand	1866	H.Jacobsen	Y 7.87	H.Hansen	Lillesand
40666	427		Aggie Davison W Sr	British	108	76·5	23·1	9·4	RichibuctoNB	1864			JohnR.Calhoun	Miramichi, N.B.
	428		Aghatomia W Bg	Turkish	162	..	..	..	Austria	1874			Vassilaki Lefter & Co.	Constantinople
266	429	V	Agia Markella Pissias W3mSr	Greek	265	101·0	26·2	17·7	Syra	1880	D.Chrisakis	Y 12.85	M.Pissias & Co.	Syra
	430	A	Agia Markella (ex Despina) Scoldi ·88 W Bg		236	92·2	26·2	16·6	Syra	1866	P.Bambis			
239	431	A	Agia Markella W Bg	Greek	203	89·6	25·6	..	Syra	1878		Y 2.83	S.Galounis	Syra
116	432		Agia Markella W Sr	Greek	138	72·2	21·3	11·5	Chios	1872			D.G.Andreadis	Zante
13	433		Agia Sophia W Bg	Greek	217	88·6	25·9	15·9	Syra	1869			A.C.Francescos	Hydra
21	434		Agia Trias W Bk	Greek	320	111·5	27·4	17·0	Scopelos	1875			C.G.Garoufalou	Scopelos
96	435	A	Agia Trias W Bn	Greek	249	109·5	24·6	17·1	Galaxidi	1867	E.Canata		G.Charopoulos	Piræus
	436		Agia Trias W Bg	Greek	229	92·2	24·4	14·0		1876			Z.Pithous	Syra
279	437		Agia Trias W Bg	Greek	216	89·9	25·4	..	Chios	1876			N.Stefanakis	Syra
3	438	A	Agia Trias W Bn	Greek	182	84·3	23·9	15·5	Piræus	1866		Y 9.81	B.C.Patrona & Co.	Syra
	439		Agia Trias W Sr	Greek	140	..	..	..	Skiathos	1882			G.Matarona	Skiathos
	440		Agia Trias W Bg	Greek	122	74·5	20·2	..	Syra					Spezzia
66	441		Agia Trias W Bg	Greek	121	74·1	22·0	12·8	Spezzia					Spezzia
	442		Agias Pelagias Momoridis W Bk	Russian	112	70·9	21·0	13·6	Cassos	1875			N.Stambulis	Odessa
70	443	A	Agii Apostoli W Bk	Greek	298	110·5	26·9	..	Galaxidi	1868	J.Canatas	Y 8.86	Heirs of Economou	Galaxidi
52	444		Agii Apostoli Pavlinos W Bn	Greek	184	82·7	23·8	15·2	Spezzia	1867		Y 4.80	A.Pavlinos	Spezzia
	445	V	Agile (ex May Flower) Boucher W Bn	French	138	88·9	24·9	10·0	P.MedwayNS	1881	E.Dolliver	Y 12.83	L.Coste & Co.	St.Malo
	446	V	Agile Deconninck W Sr	French	106	86·6	22·0	10·9	Dunkirk	1879	VanderieleFils		Govard	Dunkirk
21069	447		Agilis Noel W Bg	British	266	128·9	25·8	13·8	Guernsey	1860	P.Ogier		T.G.Robins	Jersey
	448		Agio Fagerlund ·88 W Bk	Russian	331	..	..	..	Nagu	1868			P.E.Grandell	Nagu
	449	A	Agios Anargyros W Bn	Greek	154	79·4	23·4	10·5	Cassos	1882			G.Pappadachis	Syra
10	450		Agios Andreas (ex Niala) W Bn	Greek	200	87·6	25·3	..	Savona	1860			G.Macheras	Corfu

*=Lloyd's Register; A=Veritas Austro-Ungarico; D=Nederlandsche Vereeniging; G=Germanischer Lloyd; H=Veritas Ellenico; I=Registro Italiano; N=Norske Veritas; R=American Record; V=Bureau Veritas.

Dans la 4me Colonne, C=Composite (membrure en fer, bordé en bois); I=Fer; Stl=Acier; W=Bois.

Les Dates du doublage en métal (Col. 11) ressortent des livres dans lesquels les navires sont respectivement classés ou des rapports des Inspecteurs du Lloyd's Register.

Pour plus amples renseignements sur les navires classés par le Lloyd's Register, voir le Lloyd's Register Book et le Lloyd's Yacht Register.

Official No.	No. in Book.	Registry in which classed.	Ship's Name. Material and Rig. Master, and Dates of Appointment to present Owner's Service and to Ship. Ship's Late Name.		Flag.	Net Tonnage.	Dimensions. Length.	Breadth.	Depth.	Built. Where.	When	By whom.	Date of Metalling.	Owners or Managers.
61	451		Agios Artemios *Mavromatis*	W Bk	Greek	326	110·8	28·5	17·2	Syra	1872	A.Livadaros	Y 8.85	G.Mavromatis
	452		Agios Cavassovich	W	Greek	414	128·9	30·1	..	Fiume	1872			P.Lascaridis
9	453	H	Agios Charalambos *Piangou*	W Bg	Greek	178	82·7	25·1	..	Syra	1868	S.Koufondakis		Piangou Bros.
	454		Agios Charalambos	W Bg	Greek	198	..	..	..	Syra	1865		C 7.75	Z.Goulandris
62	455		Agios Dimitrios	W Bg	Greek	197	88·9	22·0	..	Cephalonia	1862			Dem.Kouris
59	456	H	Agios Dimitrios	W Bg	Greek	194	87·3	25·6	15·1	Spezzia	1866	G.Raditi		D.A.Roucas
	457		Agios Dimitrios *Denaxas*	W Bg		185	89·9	24·1	..	Syra	1868			
4	458		Agios Dionyssios	W Bg	Greek	230	97·8	25·6	..	Galaxidi	1872			Spyridon Dadousis
	459		Agios Dionyssios *Dandouras*	W Bg	Greek	220	93·5	25·9	..	Galaxidi	1875			Banque Alkion
52	460	A H	Agios Dionyssios *Trigalas*	W Bg	Greek	127	73·5	24·9	..	Galaxidi	1877	J.Kanatas		Trigalas Bros.
35	461	A	Agios Georgios	W Bn	Greek	312	106·3	27·6	..	Cassos	1876		Y 1.86	I.M.Makris
32	462		Agios Georgios	W Bg	Greek	292	119·1	28·9	..	Syra	1874			G.A.Maplessos
	463		Agios Georgios *Catzoulis*	W Bg	Greek	275	..	..	..	Galaxidi	1871			HeirsofG.Stromboni
306	464		Agios Georgios	W Bg	Greek	263	87·9	26·1	..					G.Garifalou
41	465		Agios Georgios	W Bg	Greek	250	97·1	27·9	..	Syra	1877			S.I.Saris
	466		Agios Georgios *Manolessos*	W Bn	Greek	250	..	..	..	Syra	1866		C 9.80	G.A.Manolessos
69	467	H	Agios Georgios	W Bn	Greek	215	..	..	..	Galaxidi	1871	J.Kanatas		P.Gourgouris
	468		Agios Georgios *Selia*	W Bn	Greek	198	49·9	26·0	17·7	Syra	1882			B.Selia
38	469		Agios Georgios	W Bn	Greek	197	86·0	25·1	..	Syra	1871	A.Endimon		T.Nouicos
33	470	A	Agios Georgios	W Bg	Greek	191	86·6	24·3	..	Syra	1862			T.D.Saris
392	471		Agios Georgios *Alexandro*	W Bg	Greek	187	84·3	24·8	12·2					G.A.Alexandro
	472		Agios Georgios *Potomianos*	W Bn	Greek	186	..	..	..	Syra	1870		Z 6.82	N.G.Mavromatis
8	473		Agios Georgios	W Bg	Greek	173	86·9	24·9	..	Syra	1874			N.Straftis
	474		Agios Georgios	W Sr		156	89·6	22·3	..	Capo d'Istria	1876			
452	475		Agios Georgios *Macri*	W Sr	Greek	124	76·4	21·8	..	Cassos	1885	T.Tsicalas		H.Gerakarakis
18	476		Agios Georgios *Gourgouris*	W Bn	Greek	122	82·7	19·3	11·1	Galaxidi	1864			Heirs of Douroutis
	477		Agios Georgios	W Bg	Greek	120	..	..	..	Cassos	1868			D.Georgios
43	478	A	Agios Georgios	W Sr	Greek	114	75·1	22·1	11·0	Syra	1880			J.A.Dargentis
37	479	A	Agios Georgios	W Sr	Greek	107	74·1	22·1	10·8	Syra	1882		Y 6.86	J.Capiastas
29	480	A	Agios Georgios	W Sr	Greek	103	72·8	21·5	11·9	Skiathos	1880	C.G.Mytilinais		Damaskou Bros.
35	481		Agios Georgios	W Bg	Greek	102	62·3	20·7	11·1	Syra	1874			A.M.Alafouzos
	482	A	Agios Georgios	W Bn	Russian	187	..	..	..	Cassos	1874	H.Manoli		M.Zizinia
	483	A	Agios Giacomo (ex Vadjia) *Soppio -88*	W Bk	Greek	427	..	..	..	Lussinpiccolo	1870		Y 2.87	I.Carussis
19	484		Agios Ioanis Theologos *Iconomides 81-87*	W Bk	Greek	400	125·6	29·8	..	Syra	1874	Cufudache		G.Iconomides
	485	H	Agios I. Theologos *Simidopoulos*	W Bg		324	111·8	25·3	17·8	Spezzia	1872		C 12.83	
135	486		Agios I. Theologos (COLLISION)	W Bg	Greek	227	89·6	26·1	..	Hydra	1869			I.Faraclas
15	487		Agios Ioannis *Poulissos*	W Bk	Greek	272	115·8	22·0	..	Sunderland	1864			Callinikou Bros.
74	488		Agios Ioannis *Stavro*	W Bg	Greek	192	91·2	24·1	14·1	Galaxidi	1868	C.Papapetros		
132	489		Agios Ioannis	W Bg	Greek	169	86·9	23·9	13·4	Syra	1873			A.S.Potamianos
538	490		Agios Joannis	W Bg	Greek	169	84·3	23·9	13·4	Syra	1882			E.P.Kyriakos
	491	A	Agios Markos	W Bg	Greek	250	..	..	..	Chios	1879			N.Glyca
435	492	A	Agios Markos	W Bg	Greek	248	95·5	26·8	..	Syra	1869			Margaroti Bros.
	493		Agios Markos	W Bg	Greek	219	86·3	..	14·6	Syra	1860			D.Couphopakolis
332	494		Agios Mercurios	W Bn	Greek	186	93·8	24·6	..	Syra	1876	C.Cristali		C.N.Comninos & Co.
106	495		Agios Nicolaos (LOST) *A. Mitropulo*	W Bg	Greek	309	106·9	27·9	18·0	Galaxidi	1875		Y 3.87	A.Mitropulo
45	496	H	Agios Nicolaos	W Bn	Greek	276	103·6	24·0	16·1	Spezzia	1871			A.Lazarou
481	497	A	Agios Nicolaos	W Bg	Greek	266	107·9	26·2	..	Syra	1886	A.Livadaras		G.Papageorgiou
192	498	A	Agios Nicolaos *Daravello -88*	W Bn	Greek	254	102·3	27·6	16·1	Genoa	1869			E.Bozabardi
166	499		Agios Nicolaos (STRANDED)	W Bn	Greek	253	97·1	27·1	..	Syra	1875			P.M.Passaris
	500		Agios Nicolaos *D. Catsoulis*	W Bg	Greek	248	..	..	..	Galaxidi	1869			E.Catsoulis

✱ = Lloyd's Register; D = Nederlandsche Vereeniging; G = Germanischer Lloyd; I = Registro Italiano; R = Ameri[can]
A = Veritas Austro-Ungarico; H = Veritas Ellenico; N = Norske Veritas; V = Bur[eau]

In the 4th Column, C = Composite (iron frame, wood planked); I = Iron; Stl = Steel; W = Wood.

The Dates of Metalling (Col. 11) are based on the records of the Books in which Vessels are respectively classed, or on the Reports of the Surveyors of Lloy[d's]

Further particulars of Vessels classed by Lloyd's Register will be found in Lloyd's Register Book and Lloyd's Yacht Register.

1	No. dans le livre	Registre	Nom. Matériaux et Gréement. Capitaines et dates d'entrée au service des Armateurs actuels et à bord du navire. Ancien Nom.		Pavillon.	Tonnage net.	Dimensions. Longueur	Largeur	Creux	Construit. Où.	Quand.	Par qui.	Date du doublage en métal	Armateurs ou Directeurs-Gérants.	Port d'Armement.
141	501		Agios Nicolaos	W Bg	Greek	242	90·9	25·1	15·7	Galaxidi	1857			G.Paramythistis	Zante
45	502	A	Agios Nicolaos	W Sr	Greek	241	101·3	26·0	..	Cassos	1885	E.Sofos	Y 8.86	E.Giannagas	Syra
	503	H	Agios Nicolaos	W Bg		236	88·6	27·6	16·7	Galaxidi	1868	E.Canata			
	504		Agios Nicolaos	W Bg	Greek	230	90·9	28·9	..	Syra	1864			N.Anapliotis	Zante
23	505	A	Agios Nicolaos *Scourtis*	W Bn	Greek	221	94·8	25·6	15·2	Galaxidi	1873		Y 6.87	G.N.Scourtis	Spezzia
17	506		Agios Nicolaos	W Bk	Greek	219	88·6	25·9	..	Syra	1866			J.M.Maplessos	Santorin
	507		Agios Nicolaos *Nicea*	W Bn	Greek	216	..	..	..	Galaxidi	1873			E.Nicea	Syra
	508	V	Agios Nicolaos *Dourandzinos*	W Bk	Greek	207	89·9	23·6	15·7	Galaxidi	1872	G.Canata		E.Dourandzinos	Patras
3	509	A	Agios Nicolaos *Zoupas*	W Bg	Greek	203	90·5	24·0	15·1	Spezzia	1866	A.Petrussi		N.P.Zoupas	Spezzia
1	510	H	Agios Nicolaos *Vrichoropoulos*	W Bg	Greek	201	87·3	22·5	15·2	Spezzia	1866	G.Reditos		P.J.Vrichoropoulos	Spezzia
21	511		Agios Nicolaos	W Bg	Greek	196	86·6	23·8	..		1870			M.G.Loss	Zante
6	512	H	Agios Nicolaos	W Bg	Greek	196	88·9	24·6	15·0	Syra	1868			A.Petroutzis	Spezzia
40	513		Agios Nicolaos	W Bg	Greek	194	86·6	24·8	..	Syra	1864			M.Pissias	Syra
	514		Agios Nicolaos	W Bg		187	..	..	..	Galaxidi	1865				
19	515	H	Agios Nicolaos	W Bg	Greek	178	91·2	23·4	15·4	Galaxidi	1869	K.Papapetro		N.D.Sparos	Galaxidi
341	516	A	Agios Nicolaos	W Bn	Greek	169	81·0	24·4	..	Cassos	1881			G.Papadopoulos	Syra
60	517		Agios Nicolaos	W Bn	Greek	167	..	..	..	Galaxidi	1861			G.Drossopoulos	Galaxidi
	518	A	Agios Nicolaos *Moschakis*	W Sr		139	87·9	23·3	12·5	Syra	1881	A.Livadara	Y 4.83		
	519		Agios Nicolaos *Michopoulos*	W Bg	Greek	119	..	..	..	Galaxidi	1864			Heirs of Raftopoulou	Galaxidi
42	520		Agios Nicolaos	W Bn	Greek	112	70·2	21·2	..	Cassos	1874			T.G.Theodoron	Syra
30	521		Agios Nicolaos	W Trc	Greek	104	67·9	20·8	12·5	Spezzia	1863			A.I.Petrasos	Spezzia
	522		Agios Nicolaos *Janagas*	W Bk	Russian	228	93·1	25·3	14·1	Cassos	1869	M.Kaggis		N.Magripi	Odessa
	523		Agios Nicolaos *Nicolaos* (ex Agios Giorgios)	W Bk	Russian	226	89·6	23·3	16·2	Cassos	1871	N.Catramados		G.C.Papademitri	Eupatoria
	524		Agios Nicolaos *Calimeris*	W Bk	Russian	205	79·7	25·3	14·1	Syra	1864	N.Metelincas		G.Rosetti	Taganrog
	525	A	Agios Nicolaos *Niades*	W Bn	Russian	140	..	..	..	Cassos	1877			N.Niades	Cassos
	526	A	Agios Nicolaos	W Sr	Russian	102	..	..	..	PavlosckasDon	1880	N.Odinzow	Z 8.83	A.Durante	Berdiansk
74	527		Agios Panteleimon	W Bn		280	104·9	23·9	..	England	1865				
	528		Agios Spyridon *Rombos*	W Bk		477	..	..	..	Cherson	1872		Y 5.82		
	529		Agios Spyridon (ex Vennia, Nazoni) *Candarigi*	W Bk	Greek	397	..	..	..	Fiume	1856	Vranich		T.G.Candarigi	Piræus
	530	H	Agios Spyridon	W Bk	Greek	377	120·4	29·2	..	C'nstantin'ple	1856			N.Vassilachi	Piræus
84	531	A	Agios Spyridon *Nikas*	W Bg	Greek	319	115·4	28·0	..	Galaxidi	1884	J.Kanatis		D. & E.Nikas	Galaxidi
30	532		Agios Spyridon *J. Angelis*	W Bk	Greek	266	105·6	27·9	..	Galaxidi	1868			P.Angelis	Piræus
40	533	H	Agios Spyridon	W Bn	Greek	244	96·8	25·1	..	Galaxidi	1868	E.Canata		A. Mbakigiorgio	Zante
61	534		Agios Spyridon *Cosoglojanni*	W Bn	Greek	235	90·5	24·6	..	Galaxidi	1875	J.Kanatis	3 10.83	G.Mitropoulos	Galaxidi
53	535	A H	Agios Spyridon *Contogeorgopoulos*	W Bn	Greek	236	125·9	25·7	17·2	Galaxidi	1869	E.Canata	Y 8.82	P.Contogeorgopoulos	Galaxidi
69	536	A H	Agios Spyridon	W Sr	Greek	235	97·4	26·1	14·8	Galaxidi	1885	C.Papapetros	Y 3.85	G.Zagoras	Galaxidi
6	537		Agios Spyridon	W Bg	Greek	231	53·1	28·5	..	Skiathos	1868	G.Mytilinaios		Dem.Mitaronis	Zante
	538		Agios Spyridon *G. Marulas*	W Bn	Greek	219	95·1	23·6	14·0	Galaxidi	1867	P.Pedro	C 7.78	P.Marulas	Galaxidi
	539		Agios Spyridon	W Bg	Greek	215	..	..	..	Cassos	1863			J.Mantzarakis	Piræus
112	540		Agios Spyridon	W Bg	Greek	205	88·6	23·0	14·0	Galaxidi	1866			Evangelos Phokas	Zante
17	541		Agios Spyridon *Tsalanghiris*	W Bg	Greek	204	92·5	25·3	..	Galaxidi	1864			C.G.Tsalanghiris	Piræus
	542		Agios Spyridon *Manolas* (ex Costantino L.)	W Bn	Greek	194	..	..	..	Spezzia	1868		8.83	J.D.Mylonas	Spezzia
51	543		Agios Spyridon	W Bg	Greek	194	49·2	25·3	..	Piræus	1863	A.Kalpadis		Spyr.Kogellis	Zante
60	544	H	Agios Spyridon *Poriotis*	W Bg	Greek	190	85·6	24·1	15·0	Skiathos	1873			N.Poriotis	Skiathos
	545	H	Agios Spyridon *Pappadojani*	W Bn	Greek	188	..	..	..	Galaxidi	1876			A.Pappadojani	Galaxidi
5	546		Agios Spyridon	W Bg	Greek	177	85·6	24·6	13·4	Galaxidi	1875	Kanatas		Duruttis Bros.	Galaxidi
136	547		Agios Spyridon	W Bk	Greek	173	103·6	24·3	..	Fiume	1872			Jean S.Lazaro	Piræus
26	548		Agios Spyridon	W Bn	Greek	142	87·6	21·8	12·1	Syra	1862			G.Lichiardopoulos	Skiathos
27	549		Agios Spyridon	W Sr	Greek	133	82·3	20·9	..	Scopelos	1872	P.Bitunas		C.Tacharopoulos	Cumi
100	550		Agios Spyridon	W Bn	Greek	127	..	19·4	12·4	Galaxidi	1853			D.Rombas	Zante

✱ = Lloyd's Register; D = Nederlandsche Vereeniging; G = Germanischer Lloyd; I = Registro Italiano; R = American Record;
A = Veritas Austro-Ungarico; H = Veritas Ellenico; N = Norske Veritas; V = Bureau Veritas.

Dans la 4me Colonne, C = Composite (membrure en fer, bordé en bois); I = Fer; Stl = Acier; W = Bois.

Les Dates du doublage en métal (Col. 11) ressortent des livres dans lesquels les navires sont respectivement classés ou des rapports des Inspecteurs du Lloyd's Register.

Pour plus amples renseignements sur les navires classés par le Lloyd's Register, voir le Lloyd's Register Book et le Lloyd's Yacht Register.

Official No.	No. in Book.	Registry in which classed.	Ship's Name. Material and Rig. Master, and Dates of Appointment to present Owner's Service and to Ship. Ship's Late Name.	Flag.	Net Tonnage.	Length.	Breadth.	Depth.	Built. Where.	When.	By whom.	Date of Metalling.	Owners or Managers.
322	551	V	Agios Spyridon *Pappa* W Bg	Greek	119	69·9	22·1	12·8	Cassos	1880	E.Sofo		M.Papas
66	552	H	Agios Theodoros *Galetas* W Bg	Greek	249	94·8	25·2	15·6	Spezzia	1867		[illegible]	J.N.Galetas
	553	A	Agios Trifon *Anguras* W Bn	Greek	179	..	..	..	Falmouth	1872		[illegible]	C.Parlopulo
	554		Ägir (ex Washburn) *Andersson* W Bg	Swedish	196	98·7	21·6	12·9	NewBrunswk.	1850			A.Peilitz
	555	V	Aglaé & Charles *Autin* W Bg	French	163	84·0	21·3	11·1	Nantes	1866	Boju& Chaigeau	[illegible]	Autin
63920	556	✻	Aglaia *Shaw* I Bk	British	786	187·8	32·0	19·7	Belfast	1875	Harland& Wolff		Woodside & Workman
4	557	H	Aglaia *Sachtouris* W Sr	Greek	265	99·1	26·6	16·1	Syra	1873		[illegible]	Sachtouris & Kyriasis
	558	N	Aglaia *Larsen* -89 W Bg	Norweg'n	162	107·2	25·5	11·8	Bergen	1857	A.Dekke		C.Gjertsen
	559		Aglaia *Lindblom* W3mSr	Russian	227	..	..	..		1878			M.Eriksson
507	560	A	Aglaia Vagliano *Tavarbojulo* -88 W Bg	Greek	328	112·8	27·4	..	Syra	1888			A.M.Vagliano
79351	561	✻	Agnes *Joll* W Sr	British	59	72·3	20·7	8·0	Bude	1878	Stapledon		N.H.Tregaskes
	562	V	Agnes *Andreasen* 86-86 W Sr	Danish	128	91·5	20·8	10·2	Thuro	1874	P.Troensegaard		J.Jorgensen
	563	DG	Agnes *Hirdes* -67 W Bk	German	841	153·4	31·9	21·2	Vegesack	1862	J.Lange	[illegible]	R.M.Sloman & Co.
	564	✻D	Agnes (ex Dunholme) *Schepler* 84-84 W Bk	German	341	121·0	26·9	17·0	Sunderland	1865	J.Crown	[illegible]	J.J.Wallis & Sohn
	565		Agnes *Meinhahn* 65-77 W Bg	German	275	104·6	27·1	14·3	Ueckermünde	1865	L.Wittenberg		J.Rahlff & Sohn
	566	D	Agnes *Segebarth* 81-81 W Bg	German	249	104·0	25·2	15·0	Elsfleth	1865	G.Wempe	Y P 11.82 P 2.84	C.A.Beng
	567		Agnes (ex Cecoabo) W Bk	Norweg'n	666	163·7	32·6	17·9	P.E.Island	1882	A.McMillan	[illegible]	Eger & Sörensen
	568	N	Agnes *Pettersen* 74-81 W Sr	Norweg'n	124	93·7	22·7	11·3					T.O.Larsens Enke
	569	✻	Agnes (ex Tulmann) *Cunha* 58-86 I Bk	Portug'se	635	180·4	28·6	17·6	Dundee	1876	Thompson		J.H.Andresen
	570		Agnes *Seppa* 79-87 W Bg	Russian	310	111·3	26·8	16·6	Sunderland	1868			D.Brunila
	571	R	Agnes (ex Jane Tatham, ex Jane) *Jacobsson* 83-83 W Bg	Swedish	257	116·5	25·4	14·4	P.E.Island	1856			G.J.Sundberg
	572		Agnes (exTrieste) *Johansson* 78 W Bg	Swedish	197	94·2	22·3	13·1	Amsterdam	18 54/75			J.Berntsson
1998	573	R	Agnes Barton *Knight* W Bn	American	380	130·9	30·0	14·9	Baltimore	1871	Willend'r& B'ck	[illegible]	G.R.Heffner
38100	574		Agnes C. James *Barron* W Bkn	British	197	116·2	26·1	12·4	NovaScotia	1859	Gardner		J.Kelly
65787	575	✻	Agnes Cairns *Davies* W Bn	British	136	92·5	22·6	12·0	Limekilns	1873	J.Whitehead	[illegible]	J.Jones
	576	N	Agnes Campbell *Nielsen* W Bk	Norweg'n	675	158·1	32·2	19·1	Weym'th,NS	1870	R.Hankinson	[illegible]	J.C.&G.Knudson
88705	577	✻	Agnes Craig *Baird* [84-84] W3mSr	British	99	94·7	23·0	10·4	ConnahsQuay	1884	Fergus'n& Baird		R.Ferguson & Co.
70504	578	✻	Agnes Ellen *Libby* 83-83 W Sr	British	140	99·2	23·6	11·9	Porthcawl	1875	Martin		C.Allport & Co.
29033	579	✻	Agnes Fraser *Doughton* W Sr	British	98	80·4	20·9	10·9	Inverkeithing	1862	Ross		E.Doughton
88711	580	✻	Agnes Glover *Higham* W Sr	British	86	87·5	21·4	9·8	Tarleton	1883	Lund		R.Wright
105217	581	R	Agnes I. Grace *Seavy* W3mSr	American	358	136·0	31·5	11·9	WaldoboroMe	1873		[illegible]	Jas.Bliss & Co.
60446	582	✻	Agnes Lilian (ex City of Madrid) *Bevan* 88-88 I S	British	1191	231·6	35·3	22·2	Glasgow	1869	C.Connell & Co.		Goldberg & Co.
	583		Agnes Linck *Haack* 88-78 W Bk	German	672	130·7	33·5	22·5	Danzig	1865	J.W.Klawitter		Schoenberg& Domansky
106400	584	R	Agnes Manning *Birdsall* -88 W4mSr	American	833	185·6	38·0	17·4	Bath,Me.	1886	New England [S.B.Co.		Amos Birdsall
73829	585	✻D	Agnes Oswald *Henderson* [88-88 I S	British	1380	243·5	37·2	22·0	Port Glasgow	1876	H.Murray&Co.		T.Law & Co.
105289	586		Agnes R. Bacon (STRANDED) *Haley* W3mSr	American	376	128·7	32·0	9·7	Camden,N.J.	1873	Taylor& M'thias		A.Haley
1529	587		Agnes Repplier *Miller* W Sr	American	110	99·8	17·6	7·8	Hamburg,Pa.	1867			D.McDevitt
	588	✻	Agnes Stevenson (ex Emma) *Huertas* W Bk	C'staRic'n	318	118·7	26·0	14·9	P.E.Island	1876	J.Duncan&Co.	[illegible]	P.Terres
	589	V	Agnete *Møller* 85-77 W Bn	Danish	142	97·5	21·3	10·3	Svendborg	1877	J.R.Andersen		J.Anderskouv
2691	590	I	Agnesina A. *Lombardo* [-83 W Sr	Italian	115	90·2	21·3	10·3	TorredelGreco	1886	G.Guida		F.Ascione
	591	H	Agon *Mitropoulos* W Bg	Greek	140	..	..	..	Galaxidi	1873			A.Mitropoulos
	592	N	Agon *Syvertsen* 72-83 W Bk	Norweg'n	299	120·0	26·7	13·9	Stavanger	1872		[illegible]	H.J.HammersEnke
96	593	V	Agostino C. (ex Washington) *Cosulich* W Bk	Austrian	839	148·6	31·5	21·0	Lussinpiccolo	1876	N.Martinolich	a 4/.84	L.Cosulich di C.
1899	594	V	Agostino Felugo *Oarle* -88 W Bk	Italian	460	131·0	30·9	20·0	Voltri	1868	A.Canepa	[illegible]	A.Felugo
2836	595	I	Agostino Giuseppe *Casareggio* -86 W Bn	Italian	541	146·6	31·3	20·3	Savona	1877	B.Calcagno	[illegible]	A.Penco
3338	596	DIV	Agostino Rombo *Bosso* -88 W Bk	Italian	807	164·3	33·8	22·2	Sestri	1882	F.Cadenaccio	[illegible]	G.B.Vassallo
	597	V	Agra *Hansen* -88 W S	Norweg'n	924	172·5	34·8	22.6	Medford	1863	J.T.Foster	[illegible]	S.Hansen
	598	G	Agra *Abelt* -88 W Bk	Norweg'n	818	179·1	31·3	21·4	Liverpool	1858	J.Steel	[illegible]	Chr.Hannevig
	599	N	Agra *Hansen* W Bk	Norweg'n	618	142·9	32·7	20·9	Sunderland	1851	John Watson	[illegible]	Act. Selsk. "Agra" (A. Wilhelmsen)
43874	600		Agricola *Klene* 82-8- W Bg	British	279	118·7	26·0	14·9	Shoreham	1862	Bailey		J.&N.C.Bull

✻ = Lloyd's Register; D = Nederlandsche Vereeniging; G = Germanischer Lloyd; I = Registro Italiano; R = Americ… A = Veritas Austro-Ungarico; H = Veritas Ellenico; N = Norske Veritas; V = Bureau…

In the 4th Column, C = Composite (iron frame, wood planked); I = Iron; Stl = Steel; W = Wood.

The Dates of Metalling (**Col. 11**) are based on the records of the Books in which Vessels are respectively classed, or on the Reports of the Surveyors of Lloyd…

Further particulars of Vessels classed by Lloyd's Register will be found in Lloyd's Register Book and Lloyd's Yacht Register.

N°	Registre	Nom. Capitaine et dates d'entrée en service des Armateurs actuels et à bord du navire. Ancien Nom.	Matériaux et Gréement	Pavillon.	Tonnage	Longueur	Largeur	Creux	Construit. Où.	Quand.	Par qui.	Date du doublage	Armateurs ou Directeurs-Gérants.	Port d'Armement.
601		Agricola *Lees*	W Bg	British	151	95·8	22·3	12·4	Aberdeen	1857	A.Hall&Co.		A.Lees	Aberdeen
602	❊	Agricola *Kent*	W K	British	60	71·7	18·8	8·8	Jersey	1871	Bellot		J.Cantell & Co.	Jersey
603	V	Agricola *Kochet* (ASCENDEEN)	W Bk	French	311	116·1	24·9	15·9	Nantes	1870	Pastec	2.88	Béguinot & Co.	Bordeaux
604		Aguas I *Castello*	W Sr	Portuguese	185	..	..	..		1873		12.83	J.A.F.Aguas	Figueira
605		Aguia *Gaspar*	W Sr	Portuguese	133	85·6	20·9	10·0	Figueira	1863	F.G.Mendes		M.J. de Souza & Co.	Loanda
606		Aguila *Arrondi*	W Bk	Chilian	402	52·8	17·1	6·1	Apenrade	1865			P.E.Almeida	Valparaiso
607	N	Agur *Olsen 74-74*	W Bk	Norwegian	428	127·0	29·9	16·5	Grimstad	1874	OleRibe	V 8.88	O.Isaksen	Grimstad
608	V	Agur *Mathiesen -86*	W 3m Sr	Danish	300	121·4	24·7	14·9	Nordby	1886	S.Abrahamson	V 6.88	Actieselskab (S.B. Hansen)	Nordby Fano
609		Agustin Edwards *Le Troadec*	W Bk	French	1010	183·6	35·9	21·7	Bordeaux	1873	Chaigneau frères		Ant. Dom. Bordes & fils	Bordeaux
610	V	Agustina *Schumacher 86-88* (ex Wm. Reid)	W Bk	German	896	165·0	34·1	22·5	Portsmouth	1869	D.Marcy	V 1.88	C.Paulsen	Elsfleth
611		Agustina	W Pol	Spanish	168	78·7	21·9	10·0	Gerona	1837			A.C.Tellama	Barcelona
612	R V	Ahkera *Bellon -87* (WEE KEE)	W Bk	French	564	144·0	33·1	19·0	Ulealeaborg	1866	J.Bergelin	V 10.86	F.Guilhou	Marseilles
613		Ahkera *Bjorklund*	W Sr	Russian	204	..	..	..	Vöhsala	1874			K.A.Andersson	Lemland
614		Ahkera	W Glt	Russian	134	..	..	..	Leppävirta	1876			P.Kausanen	Leppävirta
615		Ahkera *Ivanoff*	W	Russian	114	..	..	..		1874			W.Hosainoff	Salmis
616		Ahlone	I Lr Barge	British	353	197·3	25·2	8·4	PortGlasgow	1883	R.Duncan&Co.		IrrawaddyFlotillaC°,Ld	Glasgow
617		Ahmassrah *Hussein*	W Bal	Turkish	162	..	..	..	Guetchdjur	1869			AhmassrahleeHadji Mehmet Ali & Co	Ahmassrah
618		Ahmassrah *Mehmet Kadri*	W Bg	Turkish	142	..	..	..	Chakkir [Aghiis	1868			AhmassrahleeHadji Mehmet Kadri	Ahmassrah
619		Ahmassrah	W Sr	Turkish	112	..	..	..	Ahmassrah	1872			Ahmassrahlee Ahmdar-Oglou Hassan & Bros.	Ahmassrah
620		Ahmassrah *Ali Reiss*	W Bg	Turkish	107	..	..	..	Chikraz	1869			Ahmassrahlee Tchi-Oglou Ali Reiss	Ahmassrah
621		Ahmassrah *Ahmed*	W Bg	Turkish	100	..	..	..	Ahmassrah	1869			Mehemed Effendi &Co.	Ahmassrah
622		Ahnatab Skellessi	W Bg	Turkish	162	..	..	..	Ahnatah [Skellessi	1866			AbahnableeIsmael&Co.	Abahnah
623	G	Ahti *Roos -86*	W Bk	Russian	338	122·0	29·0	15·0	Jacobstadt	18 $\frac{..}{76}$	C.J.Wukman		G.Duwaldt	Wasa
624		Ahti	W 3m Sr	Russian	317	117·1	26·9	14·9	Peeksala	1873	M.Strömberg		J.Bergroth	Peeksala
625		Ahti *Mattson*	W 3m Sr	Russian	312	..	..	..		1879			M.Mickelsson	Suna
626		Ahti *Niskanen*	W Glt	Russian	229	..	..	..	Impilaks	1883			J.Hämäläinen	Impilaks
627		Ahti *Andersson -73*	W Sr	Russian	158	95·2	25·0	10·4	Nystad	1873	A.Lindström		G.Grönholm	Nystad
628		Ahti	W Sr	Russian	132	83·2	24·5	10·3	Sastmola	1874	J.Elg		F.V.Gustafsson	Sastmola
629		Ahto *Danielsson 83-87*	W Bg	Russian	632	150·2	35·1	19·1	Brahestad	1867	J.Lang		J.Lang	Brahestad
630		Åhus *Bengtsson -87*	W Bg	Swedish	238	100·7	25·6	13·2	Finland	1864			N.Nilsson	Åhus
631		Ahyandon	W Bg	Turkish	100	..	..	..	Ahyandon	1866			Ali Oustah	Ahyandon
632		Ahyandon	W Bg	Turkish	190	..	..	..	Ahyandon	1869			Abainahlee Kuchuklee Oglou-Hadji Hussein & Co.	Abaina Kuchuk
633		Ai Giorgi	W Bg	Turkish	198	..	..	..					Sava Manolli	Alexandria
634		Ai Nicola	W Bg	Turkish	300	..	..	..					Sava Manolli	Alexandria
635		Ai Sava	W Bg	Turkish	207	..	..	..					Sava Manolli	Alexandria
636		Aid *Sjöström -88*	W Bg	Russian	291	116·4	29·1	13·1	Thoraswarf	1873	T.Lindqvist		K.L.Lindberg & Co.	Nagu
637	❊	Aigburth *Dowling 85-85*	I S	British	1798	266·7	39·0	23·6	Workington	1882	R.Williamson [& Sons		Sailing Ship Aigburth Co. Ld. (R. W. Leyland & Co.)	Liverpool
638		Aigle (BROKEN UP)	W Lr	French	103	..	..	..	Bayonne	1855	Lauga & Co.		Bailac	Bayonne
639		Aigle de Mer	W Sr	British	130	97·6	22·0	8·6	Batiscan	1864			Dan.Stirling	Quebec
640		Aigrette	W Bg	French	165	84·0	21·0	11·0	St.Malo	1865	L.Lhotelier		Société Générale des Cirages Français	Lorient
641		Aikoku Maru	W Sr	Japanese	356	120·9	26·7	13·9	Hiogo	1882	Omatsu Jugoro		Seiko Kwaisha	Hiogo
642		Aikoku Maru	W Sr	Japanese	167	89·3	21·7	11·7	Tokio	1882	OsakaShip-[buildingYard		Awoki Seikichi	Tokio
643	❊	Aikshaw *Humphreys 87-87*	I Bk	British	573	171·0	28·9	18·0	Sunderland	1875	W Doxford&Sns		Hine Bros.	Maryport
644		Ailo *Herlin*	W Bk	Russian	286	113·8	27·9	12·5	Nystad	1868			W.Akerblom	Nystad
645	❊	Ailsa *Allen 75-76*	C S	British	1061	213·1	35·2	19·8	Glasgow	1870	C.Connell & Co.	V 8.86	Sandbach, Tinne & Co.	Liverpool
646		Ailsa *Mc Farlane*	W Bk	British	455	136·2	29·0	15·9	NewBrunswk	1862	A.Anderson		J.Paton & Sons	Ayr
647		Aimable Hyacinthe	W Sr	French	107	68·2	20·7	10·9	Bayonne	1855			Tabary	Le Croisic
648	V	Aimable Lucie *Loquellec*	W Lr	French	115	74·1	20·0	10·4	Rosais	1865	Ollivaud		H.Joyau	Nantes
649	V	Aimée *Barel -88*	W Bg	French	213	97·8	25·3	12·5	Nantes	1875	Tillé		G.Dupeyrat	Bordeaux
650		Aimée *Tore*	W	French	145	..	..	..					Guibert & fils	Saint Malo

Lloyd's Register; D = Nederlandsche Vereeniging; G = Germanischer Lloyd; I = Registro Italiano; R = American Record; = Veritas Austro-Ungarico; H = Veritas Ellenico; N = Norske Veritas; V = Bureau Veritas.

Dans la 4me Colonne, C = Composite (membrure en fer, bordé en bois); I = Fer; Stl = Acier; W = Bois.

Les Dates du doublage en métal (Col. 11) ressortent des livres dans lesquels les navires sont respectivement classés ou des rapports des Inspecteurs du Lloyd's Register.

Pour plus amples renseignements sur les navires classés par le Lloyd's Register, voir le Lloyd's Register Book et le Lloyd's Yacht Register.

Official No.	No. in Book	Registry in which classed	Ship's Name. Material and Rig. Master, and Dates of Appointment to present Owner's Service and to Ship. Ship's Late Name.	Flag.	Net Tonnage	Dimensions. Length.	Breadth.	Depth.	Built. Where.	When	By whom.	Date of Metalling	Owners or Managers
	651		Aimée (ex Olso) WBn	French	119	..	..	..	Wallace,N.S.	1862		Y1.83	Boismenu fréres
	652	V	Aimée Emilie *Lechapelain* WSr	French	103	79·1	20·0	11·2	Dunkirk	1868	Vandoriele & [Mal		Vve.Dauphin
	653	V	Aimée Louise *Raoult* WBg	French	148	84·0	20·0	11·7	Cherbourg	1864	Hamel		Auguste Laronch
	654	✻	Aimée Marie (ex Harry Keslak) *E.Lelia* [86-86 WBn	French	246	116·0	25·2	12·5	London	1874	Turner	Y4.87	Amband fréres
	655	V	Aimée Victoire *Lamard* WSr	French	109	73·5	20·3	10·9	St.Vaast	1852			Briand
	656	R	Aimo *Smitsland* WBk	Russian	780	164·0	32·1	19·5	Abo	1866	C.J.F.Jörgensen	Y 12.85	G.A. Lindblom
	657		Aimo *Lindqvist 86-86* WSr	Russian	225	..	..	..	Nagu	1875			E.V.Lindqvist
	658		Aimo *Magnusson* W3mSr	Swedish	342	116·8	26·2	14·9	Christinestad	1865			J.C.Hulthén
	659		Aina *Sjövall* WBk	Russian	396	..	..	..	Lokalaks	1875			M.Östermau-Rand
	660	V	Aina (ex Alice Starrett) *Ursin -87* WBn	Russian	368	121·2	28·8	11·8	Rockland,Me.	1866			H.D.Ursin
	661		Aina *Eriksson* W3mSr	Russian	272	..	..	..	Hiitis	1874			G.Söderström
	662		Aina *Holmen* W3mSr	Russian	245	108·9	29·7	12·4	Laulaksvarf	1876	J. Roström		K.Junnila
	663		Aina *Sjödahl* WSr	Russian	115	79·0	23·5	10·0	Tenala	1873	Engblom		S.Sjödahl
	664		Aina *Petterson 88-88* WBn	Swedish	183	89·7	22·8	12·8	Kimito	1874			C.Hedin
	665	N	Aino *Olsen -88* WBk	Norweg'n	574	153·0	31·7	18·7	Christiansand	1882		Y3.86	Thr.B.Olsen
	666		Aino *Karlsson 88-88* WBg	Russian	522	140·6	34·3	18·4	Brahestad	1862	E.Danielsson		R.Mattsson
	667		Aino *Palander* WBg	Russian	521	..	..	..	Brahestad	1861			F.Sovelius
	668		Aino *Gardberg 72-72* Bkn	Russian	481	173·8	29·5	14·8	Sago	1872	J.Wigelius		H.N.Fager
	669		Aino *Groundstrom* WSr	Russian	165	..	..	..	Hvittisbofjard	1874			R.Groundstrom
	670		Aino *Pakkanen* WSr	Russian	139	..	..	..	Impilaks	1885			J.A.R.Lindberg
58748	671	✻	Aire *Daniels* WSr	British	86	77·6	18·2	10·2	Yarmouth	1871	Palmer		T.Ramsay
72546	672	✻	Airlie *Mackay 85-88* IS	British	1500	246·3	38·3	22·8	Glasgow	1875	A.Stephen&Sons		D.Bruce & Co.
76189	673		Airlie *Knight -89* W3mSr	British	237	114·9	23·7	12·4	Dundee	1876	TayShipBg.Co.	Y2.89	J.M.Ferguson
	674	N	Aise *Abrahamsen* WSr	Norweg'n	112	78·7	22·5	10·4	Sandefjord	1877			M.Abrahamsen
	675	R	Ajan *Tallgren -80* WBk	Russian	612	153·8	34·1	18·8	Avaisberg	1868	H.Kjäldström		V.A.Reins
	676	NV	Ajax *Pedersen* WS	Norweg'n	730	149·2	32·5	21·9	SouthShields	1862	MiddleDockCo.	Y7.84	T.Thommesen&Sö
	677		Ak. Kanad WBmb	Turkish	100	..	..	..	Barten	1871			Bartenlee Mehmed Reschid
82424	678	✻	Akaroa *Culbert 73-88* IBk	British	1298	221·0	36·5	22·7	Sunderland	1881	Osbourne,Gra-[ham&Co.		Shaw, Savill & Albion Co., Ld.
115352	679	R	Akbar *Huntley -88* WBn	American	415	127·0	28·4	16·5	Machias,Me.	1874		Y2.90	W.C.Holway
	680	N	Aker (ex Blanche Marie) *Jansen* WBk	Norweg'n	466	139·8	28·8	16·7	Motzen,Ger-[many	1864	D.Oltman Wwe.	Y 12.89	H.Oppegaard
	681	DV	Åkerhjelm *Johansen* WBk	Swedish	344	132·9	27·6	13·0	Landscrona	1880	P.Avenstrup	Y7.88	Wilh. Ohlsson för Varfsbolaget
	682	N	Akershus *Möller* WBk	Norweg'n	392	125·3	29·6	15·9	Christiania	1866	H.C.Hansen	Y6.87	T.Mydland
	683		Akjee Shehir WBg	Turkish	249	..	..	..	Alaplee	1868			Akjee-Shehirlee-Hadji-Mollah-Ogloo Osman & Co.
	684		Akjee Shehir WBg	Turkish	170	..	..	..	AkjeeShehir	1868			Akjee-Shehirlee Elias Zadeh Hadji Hussein & Son
	685		Akjee Shehir WBg	Turkish	150	..	..	..	AkjeeShehir	1870			Akjee Shehirlee Mollan-sadeh Osman Effendi & Co.
	686		Akjee Shehir WSr	Turkish	142	..	..	..	Barten	1868			Akjee-Shehirlee Yahran-Ogloo Mustapha & Co.
	687		Akjee Shehir WBg	Turkish	125	..	..	..	AkjeeShehir	1870			Akjee-Shehirlee Hadji Mustapha Zevjeed Emina Hanoum & Bros.
	688		Akjee Shehir WSr	Turkish	105	..	..	..	Barten	1869			Akjee Shehirlee Hadji Tahir & Sons
	689		Akjee Shehir W	Turkish	105	..	..	..	AkjeeShehir	1869			Akjee-Shehirlee Hadji Ali & Co.
	690		Akjee Shehir *Osman* W	Turkish	100	..	..	..	AkjeeShehir	1866			Katirlelee Angeli & Co.
	691		Akmé *Andersson 80-* W3mSr	Russian	241	102·9	28·3	12·3	Lamperland	1880	J.E.Rosenberg		J.E.Rosenberg
	692		Aktar Bahri *Ismael* WBg	Turkish	135	..	..	..	Abahnah	1884			Mettolee-OglouAli&Co.
	693		Aktarpolee *Dimitri* WBg	Turkish	224	..	..	..	Aktarpolee	1868			Aktarpoleiee-Zakhari Yorghaki & Sons
	694		Aktarpolee WSr	Turkish	181	..	..	..	Barten	1869			Aktarpoleiee-kiriaca Yanni & Co.
	695		Aktarpolee WBg	Turkish	177	..	..	..	Unia	1868			Aktarpoleiee-Yamandi Ogloo Nicola
	696		Aktarpolee *Costi* WBg	Turkish	100	..	..	..	Aktarpolee	1874			Vassilikoslee Nicola Costi
	697	V	Aktiv *Hassler 76-76* WBk	Swedish	314	134·8	25·1	12·9	Oscarshamn	1876	P.Pettersson	Y9.8	F.Lindström
	698		Aktiv *Hansson 82* WSr	Swedish	159	91·2	20·9	11·0	Kullen	1857			S.Cronberg
95060	699		Akyab StlLr	British	480	220·0	30·0	8·4	Dumbarton	1888	W.Denny & Bros		Irrawaddy Flotilla Co. Limited
1570	700	I	Akyab (ex Lidia) *Schiaffino* W3mSr	Italian	521	149·2	30·4	19·6	SestriPonente	1867	A.Briasco int.B.		G.Schiaffino

✻=Lloyd's Register; D=Nederlandsche Vereeniging; G=Germanischer Lloyd; I=Registro Italiano; R=Ame
A=Veritas Austro-Ungarico; H=Veritas Ellenico; N=Norske Veritas; V=Bure

In the 4th Column, C=Composite (iron frame, wood planked); I=Iron; Stl=Steel; W=Wood.

The Dates of Metalling (**Col. 11**) are based on the records of the Books in which Vessels are respectively classed, or on the Reports of the Surveyors of

Further particulars of Vessels classed by Lloyd's Register will be found in Lloyd's Register Book and Lloyd's Yacht Register.

No.	Cote	Nom. Capitaine et dates d'entrée en service des Armateurs actuels et à bord du navire. Ancien Nom	Matériaux et Gréement	Pavillon	Tonnage	Longueur	Largeur	Creux	Construit: Où	Quand	Par qui	Date du doublage en métal	Armateurs ou Directeurs-Gérants	Port d'Armement
701	N	Ala *Jensen -83*	W Bk	Norweg'n	459	127·8	30·8	18·0	Tvedestrand	1871		Y [illegible]	Chr.Th.Boe	Arendal
702	N	Ala *Areberg*	W Bk	Norweg'n	340	126·2	30·1	14·6	Abo	1867			J.Gunnerud	Drammen
703	N	Ala *Christiansen -85* (ex Anliem)	W Sr	Norweg'n	189	97·0	26·2	11·1	Herkmeri	1875	C.Sjuls		C.Möller	Christiania
704		Alabama	W	British	151	100·2	21·6	8·5	Saguenay	1871			J.Torrance	Quebec
705	✻	Alabama *Whitley 82-88*	W Sr	British	60	75·8	19·8	9·3	Jersey	1873	Vautier	[illegible]	T.De Faye & Co.	Jersey
706	R	Alabama *Dyer*	W Sr	American	105	85·0	24·0	7·0	Waldoboro' [Me.	1884			J.Murchie & Sons	Calais.Me.
707		Alabama *(ex Sarah Dowe)*	W Bk	Italian	425	127·0	28·7	12·1	Camden.Me.	1863			M.Rosaco	Genoa
708	N	Alabama *Olsen -87* (Wilhelm Ed.)	W Bk	Norweg'n	286	100·7	26·5	16·0	NovaScotia	1821		Y [illegible]	C.C.Sörensen	Tönsberg
709	I	Alabama P *Pellerano* (ex Nanno)	W Bk	Italian	636	148·4	29·9	19·7	Voltri	1872	C.Patrone	Y [illegible]	E.Piaggio fu G.B.	Genoa
710	✻	Aladdin *Shand*	I S	British	1640	244·8	38·3	24·2	Southampton	1865	Oswald. [Mordaunt&Co.		J.R.Haws & Co.	Liverpool
711		Aladdin	W Bk	British	264	98·0	23·4	18·6	Plymouth	1825			Geo.Salier	Hobart Town
712	V	Aladdin *Pedersen*	W Bk	Norweg'n	524	137·4	32·0	18·5	Königsberg	1864	L. Eggert	Y [illegible]	Brödrene Berner	Stavanger
713	N	Aladdin *Knudsen 76-76*	W Bk	Norweg'n	382	120·4	29·8	16·7	Tvedestrand	1867	M.Eldrup		Brödrene Thorvildsen	Tvedestrand
714		Alagna *(ex City of Halifax)*	W Bk	Norweg'n	859	169·0	35·8	20·5	Pugwash, N.S.	1869	J.Oulton		Chr.Stephansen,jr.	Arendal
715		Alagoana	W Sr	Brazilian	105	..	..	..					C.V.Vianna	Rio de Janeiro
716	RV	Alameda *Chapman -87*	W S	American	1400	211·3	40·0	24·0	Bath.Me.	1876	Goss & Sawyer	Y [illegible]	Chas. Davenport	Bath.Me.
717		Aland	W Sr	Russian	311	..	..	..		1873			E.P.Eriksson	Lemland
718		Alaplee *Mehmed*	W Sr	Turkish	262	..	..	..	Alaplee	1870			Jazidjee-zadeh Ahmed Bey Bros. & Co.	Alaplee
719		Alaplee *Meschid*	W Bg	Turkish	244	..	..	..	Alaplee	1868			AlapleeleeJazidjeeOglou Saleh & Co.	Alaplee
720		Alaplee	W Bg	Turkish	107	..	..	..	Alaplee	1870			Kaleh-Sultaniyeelee Vasell Pandari	Kaleh Sultaniyee
721		Alaplee	W Sr	Turkish	100	..	..	..	Alaplee	1878			AlapleeleeJakter-Oglou Hadji Mehemet & Son	Alaplee
722		Alarm	W Bg	British	195	94·8	21·3	14·2	P.E.Island	1854			UnionS.S.Co.ofNZ.	Dunedin.N.Z.
723		Alarm *Winterflood*	W Sr	British	117	..	16·9	10·6	Guernsey	1847		Y 11.84	C.S.Winterflood	Guernsey
724		Alarm	W Sr Yt	British	110	102·0	23·8	10·0	Lymington	1830	G.Inman		Executors of the late G.Batty	Southampton
725	✻	Alarm *Vince 73-81*	W Sr	British	99	96·3	16·7	9·9	Ipswich	1855	Bayley&Sons		E.J.Robertson	Ipswich
726		Alarm *Kingsland*	W Sr Yt	American	120	107·5	23·0	9·0	NewYork	1864	S.VanDuzen		G.L.Kingsland	NewYork
727	N	Alart *(ex Boyrain?) Olsen -87*	W Bk	Norweg'n	593	146·4	31·2	18·5	Granville,N.S.	1869	R.Mills		E.Olsen	Sandefjord
728	V	Alaska *Buck -84*	W Bn	British	529	132·0	30·3	17·6	Dorchester.N.B.	1874	R.A.Chapman	Y [illegible]	Taylor Bros.	Dorchester.N.B.
729	✻	Alaska *Lachance*	W Bn	British	149	92·6	25·6	10·3	Montague [Bridge.P.E.I.	1881	Geo.Wightman	Y 11.87	E.J.Price	Quebec
730		Alaska	W Sr	British	120	87·6	26·2	7·1	Maine,U.S.	1867			J.Stevenson	Moncton.N.B.
731	R	Alaska *McHeffey 84-86*	W Sr	British	118	94·0	29·9	8·5	Sackville	1884	T.Egan		Thomas Egan	Sackville.N.B.
732		Alaska	W Sk	British	87	76·3	20·1	11·0	Rye	1882	J.C.Hoad		J. & J.Emptage	Grimsby
733		Alaska *Parker 86-87*	W S	American	1250	188·2	39·0	23·8	KennebunkMe.	1867		Y [illegible]	S.Blair	San Francisco
734		Alaska *Fisher*	W Bk	American	340	122·0	30·0	18·0	Mattapoisett, [Mass.	1867			J.Bourne	N.Bedford,Mass.
735		Alaska *Clark*	W Sr	American	165	102·0	26·3	9·4	E.Machias,Me	1867			J.Talbot	Machias.Me.
736		Alaska *Wheeler*	W Sr	American	132	101·0	24·0	10·0	Pt.Townsend, [Wash.	1867			J.Lowrie	San Francisco
737	✻	Alastor *Glasebrook 71-75*	I Bk	British	824	196·6	31·7	18·6	Sunderland	1875	Mounsey& [Foster		R.H.Penney	Shoreham
738		Alba	W Sr	British	158	87·4	26·3	9·4	St.John,N.B.	1871			John W.Hayes	Arichat.C.B.
739	I	Alba *Sanguinetti -88*	W Bk	Italian	503	145·0	31·2	20·0	SestriPonente	1870	F.Tixi	Y 6.88	Maria Gardella	Genoa
740	DI	Alba *Ventura 85-85*	W 3m Sr	Italian	417	117·3	29·5	18·4	Sestri	1880	A.Carlevaro	Y 10.88	Em. Bambino	Genoa
741		Albacore	W Sr	British	328	136·5	26·3	11·0	P.Dalhousie	1872	Alex.Muir		John Ewart	S.Cath'rines.Ont
742		Albani	W Sp	British	159	107·5	22·6	7·4	Yamaska,P.Q	1883	P.Letendre		Paul Narcisse	Montreal
743		Albani	W Sr	British	110	77·5	23·9	9·8	LesEboule- [ments	1883	J.Tremblay		J.Dechene	Quebec
744	V	Albania *Brownell 73-85*	W S	British	1438	201·7	39·5	24·2	St.John,N.B.	1884	O.Pitfield	Y [illegible]	Taylor Bros.	St.John,N.B.
745	✻	Albany *Lindsay 87-87*	I Bk	British	499	155·5	26·7	17·2	Dundee	1867	Brown&Simpsn		Dundee Shipowners' Co.	Dundee
746		Albany	W Sr	British	132	97·5	23·5	8·6	Champlain	1874			D.E.Price	Quebec
747	DGV	Albatros *Soudenburg 187-* (ex Mikado)	W 3m Sr	Dutch	330	128·4	27·2	16·2	Oevelgönne	1866	J.P.Dircks	Y 10.85	Fontein & Tjallingii	Harlingen
748	DV	Albatros *Johannsen -88* (ex Vanessa)	W Bg	Dutch	222	102·6	25·4	12·1	Brake	1865	E.Nicolai	Y [illegible]	DeGroof&Co.	Flushing
749	✻	Albatros *Boon 70-87*	W Bg	Dutch	172	102·0	22·4	11·2	Hoogezand	1868	J.A.Hooites	Y 11.87	J.A.Hooites	Hoogezand
750		Albatros *Lallier* (CONDEMNED)	W Bg	French	235	97·8	23·9	12·8	Nantes	1872	Tillé		Pignot	Nantes

Lloyd's Register; D = Nederlandsche Vereeniging; G = Germanischer Lloyd; I = Registro Italiano; R = American Record;
Veritas Austro-Ungarico; H = Veritas Ellenico; N = Norske Veritas; V = Bureau Veritas.

Dans la 4me Colonne, C = Composite (membrure en fer, bordé en bois); I = Fer; Stl = Acier; W = Bois.

Les Dates du doublage en métal (Col. 11) ressortent des livres dans lesquels les navires sont respectivement classés ou des rapports des Inspecteurs du Lloyd's Register.

Pour plus amples renseignements sur les navires classés par le Lloyd's Register, voir le Lloyd's Register Book et le Lloyd's Yacht Register.

Official No.	No. in Book	Register / Character	Ship's Name. Material and Rig. Master, and Dates of Appointment to present Owner's Service and to Ship. Ship's Late Name.	Flag.	Net Tonnage.	Length	Breadth	Depth	Where.	When	By whom.	Date of Metalling.	Owners or Managers.
	751		Albatros *Renault* -88 WBn	French	124	84·0	20·3	10·2	Hammelwrdn.	1864	Gebr. Gross	[illegible]	L. Demalvilain
	752	V	Albatros *Siehr* WBk	German	318	121·0	25·5	15·3	Stettin	1847	F. Noske		Ferdinand Brumm
	753	N	Albatros *Olsen* -87 WBk	Norweg'n	509	137·2	31·2	17·7	Arendal	1880		[illegible]	N. C. Olsen
	754	N	Albatros *Clausen* -83 WBk	Norweg'n	344	125·1	28·6	14·8	Kragerö	1872	P. Thorsen	[illegible]	W. Hopp
	755	NV	Albatros *Christensen* [-88 W3mSr	Norweg'n	323	137·8	28·5	11·6	Christiansand	1885	M. Knudsen	[illegible]	H. Natvig
	756	N	Albatros *Sterndal* 87-88 WBg	Norweg'n	294	105·7	27·5	15·9	Lillesand	1851	G. Landvig	[illegible]	Krog & Henschien
	757	N	Albatros *Tallaksen* WBg	Norweg'n	184	102·5	24·9	12·2	Risör	1871	O. Jörgensen		T. Tallaksen
	758	V	Albatros *Ryberg* 82-82 WBk (ex Baron Nevale F.)	Swedish	937	173·5	31·8	22·4	Genoa	1874		[illegible]	Rederiaktiebolaget Ornen (J. F. Strömberg)
85501	759	V	Albatross *Chalmers* -88 WBkn	British	400	139·2	30·3	13·0	St. John, N.B.	1882	T. Mallery	[illegible]	Scholfield & Co., Ld.
75688	760		Albatross WSr	British	317	136·6	26·3	11·9	P. Dalhousie	1871	Alex. Muir		Bryce Muir
91431	761	*	Albatross *Ellis* WK	British	90	80·3	21·0	10·6	Brixham	1885	J. W. & A. Upham		R. Hellyer
	762	V	Albatross (ex Alida) WBg	Ecuador'n	386	123·3	26·9	15·1	Elsfleth	1871	J. Jürgens	[illegible]	Ecuadorian Lloyd's
	763		Albatross (CONDEMNED) WBk	German	348	117·8	26·6	17·0	Sunderland	1865	B. Hodgson		L. Krahmert
	764	GV	Albatross *Dierks* 81-87 WBk	German	316	126·9	26·5	12·5	Brake	1887	Oltmann Wwe	[illegible]	Seetzen Gebr.
	765	DV	Albatross *Hannes* 74-73 W3mSr	German	315	117·4	26·4	14·5	Barth	1873	Holzerland	[illegible]	C. A. Beug
	766	V	Albatross (LOST) *Pehrsson* WBk	Swedish	483	148·6	27·7	15·6	Hernösand	1875	Fr. Hägglund	[illegible]	H. Nordlander
810	767		Albemarle WSr	American	751	172·1	33·5	19·3	Brooklyn	1865			F. H. Chappell
105787	768	RV	Albemarle *Forbes* WBk	American	413	135·0	30·9	14·9	Baltimore	1878	Skinner & Sons	[illegible]	Whitby & Dickinson
8441	769	DI	Alberigo *Bruzzone* -88 WBk (ex Alberico Gentili)	Italian	823	166·6	33·5	23·6	Recco	1876	P. Rollandi [Saccomanno	[illegible]	G. Pabna
51553	770		Albert WBk	British	319	112·3	28·3	16·6	Salem, U.S.	1862			W. B. Forrest
44165	771		Albert *Walsh* WBkn (ex W. H. Gilliland)	British	242	121·2	29·3	10·9	Brooklyn, U.S.	1855			P. Gorman
34801	772		Albert *Chaney* 78-78 WBg	British	197	91·5	24·2	13·8	Stromness	1842	P. & J. Tait		G. Smee
33262	773		Albert WBn	British	166	98·7	25·0	11·4	Grondines	1856			Mrs. M. Hodgson
43281	774		Albert (ex Irene) WSr	British	110	78·9	22·7	7·0	New Jersey US				Thos. McWilliam
	775	DGV	Albert WSr (ex Albert Homan, ex Amphitrite)		133	86·3	22·2	10·3	Sappemeer	1864	J. G. Bodewes	[illegible]	
	776		Albert *Conrad* 84- WBg	German	221	96·1	27·7	13·6	Stettin	1850	E. Zieske		W. Lüdke & Co.
	777		Albert *Ahrens* 82- WSr	German	152	79·1	23·9	12·8	Barth	1858	G. Holzerland		E. W. Ohrloff Sohn
	778		Albert (CONDEMNED) (ex Barbaroux) WBk	Norweg'n	666	..	..	..	Freeport, Me.	1856			N. Selmer
	779		Albert *Widlund* WBk	Russian	340	..	..	..	Föglö	1876			N. Sittkoff
	780		Albert *Lundqvist* -86 WSr	Russian	235	..	..	..	Westanfjerd	1872			E. J. Hollsten
	781	G	Albert *Ulfberg* -87 WSr (ex Riecken ex Elise)	Swedish	114	80·3	18·8	10·3	Anclam	1866	G. Buchholtz		A. Johansson
[illegible]	782	*	Albert Baltzer *Morris* 63-67 WBn	British	162	91·1	23·0	12·7	Portmadoc	1867	Roberts	[illegible]	O. Morris & Co.
	783	V	Albert Ehrensvärd *Gadda* WBk	Swedish	590	156·1	28·2	19·3	Gothenburg	1874	F. O. Bergström	[illegible]	O. N. Liljegren
[illegible]	784	R	Albert H. Cross *Henderson* -85 W3mSr	American	340	133·6	32·7	12·0	Newburyport, [Mass.	1883	Atkinson & [Fillmore		Wm. Henderson
[illegible]	785		Albert H. Locke (WRECKED) *Grenatt* WBg	British	206	104·2	25·6	12·6	P. E. Island	1864	Locke		G. Judson
1828	786	R	Albert L. Butler *Foster* 85-85 W3mSr	American	327	119·0	23·8	10·1	Thomaston, Me.	1869	Walker & Dunn	[illegible]	J. S. Emery & Co.
	787	G	Albert M. *Vernon* -83 WBk (ex John F. Robertson)	Brazilian	398	136·9	25·8	14·5	P. E. Island	1874	A. McMillan	[illegible]	C. S. Sorensen
712	788		Albert Mason *Abbott* WSr	American	271	114·6	30·0	9·1	Patchogue, N.Y.	18[illegible]			Edwd. P. Kirk
	789	GV	Albert Neumann Berlin *Eichmann* 85-85 WBk	German	515	129·6	29·5	16·8	Rostock	1869	[illegible]	[illegible]	W. Maack
323	790		Albert Pharo *Cole* WSr	American	123	85·8	26·2	8·2	Tuckerton, N.J.	1853			E. B. Darling
	791	DV	Albert Reimann *V. d. Heyden* 76- WBg	German	225	95·3	24·2	13·4	Barth	1876	C. Holzerland	[illegible]	E. W. Ohrloff Sohn
105560	792	DRV	Albert Russell *Carver* [-88 WBk	American	724	151·2	33·2	20·1	Newburyport, [Mass.	1875	Atkinson & [Fillmore	[illegible]	D. F. Atkinson
105550	793		Albert Schultz *Thurlow* 81-81 WBkn	American	463	136·0	33·3	17·0	Calais, Me.	1875			S. R. Crowell
106158	794	R	Albert T. Stearns *Kelley* 16-81 WSr	American	483	152·2	35·1	13·4	Newburyport, [Mass.	1883	Geo. E. Currier		C. R. Kelley
71466	795		Albert T. Young *Gosden* 85-85 WBk	British	278	124·9	26·9	14·8	P. E. Island	1874	Young		Whitstable Shipping Co.
106286	796	R	Albert W. Smith *Berry* -80 W3mSr	American	572	145·5	35·5	16·9	Kennebunk Me	1873	N. Thompson	[illegible]	Albert W. Smith
47486	797	*	Albert William *Gurwell* 85-88 IBk	British	493	162·5	26·0	17·3	Sunderland	1863	J. Laing		Nicholson & McGill
45676	798		Albert the Good WBg	British	315	110·4	27·1	16·5	Sunderland	1863	W. Barkley & Co.		Melbourne Coal Ship. & Eng. Co. Ld.
	799	GV	Albert & Anna *Grabow* [-83 WSr	German	172	92·3	23·4	12·5	Wolgast	1863	E. Schelle		F. Neumann
	800												

*=Lloyd's Register; D=Nederlandsche Vereeniging; G=Germanischer Lloyd; I=Registro Italiano; R=Ameri[can]
A=Veritas Austro-Ungarico; H=Veritas Ellenico; N=Norske Veritas; V=Burea[u]

In the 4th Column, C=Composite (iron frame, wood planked); I=Iron; Stl=Steel; W=Wood.

The Dates of Metalling (**Col. 11**) are based on the records of the Books in which Vessels are respectively classed, or on the Reports of the Surveyors of Llo[yd's]

Further particulars of Vessels classed by Lloyd's Register will be found in Lloyd's Register Book and Lloyd's Yacht Register.

No.	Cote	Nom. Matériaux et Gréement. Capitaine et dates d'entrée au service des Armateurs actuels et à bord du navire. Ancien Nom.	Pavillon	Tonnage	Longueur	Largeur	Creux	Construit Où	Quand	Par qui	Doublage	Armateurs ou Directeurs-Gérants.	Port d'Armement.
801	V	Albert & Renée *Rogen* W Bn	French	199	92·2	22·1	12·2	Nantes	1876	E.Clergeau	25.8	Legal Chevreuil & frère	Nantes
802	R	Alberta *Dexter* W 3mBk	British	610	157·1	33·1	16·7	[illegible] N.S.	1888	T.A.Mosher	11.88	M.Curry	Windsor,N.S.
803		Alberta *Smith* W Bn	British	197	106·5	23·3	12·1	[illegible]	1860			J.Gunn,sen.	Faversham
804	❋ D	Albertha Romelingh W Sr	Dutch	121	92·0	20·8	9·8	Hoogezand	1857	J.Berg		J.J.Sijdzes	Farmsum
805	V	Albertina W Bn	French	163	110·8	20·7	11·8	Jersey	1868	Esnouf		Vve.Livois	Granville
806		Albertina W Sr	Spanish	161	91·9	25·6	11·8	Barcelona	1853			Goday y Gual	Barcelona
807		Albertina W Sr	Swedish	102	97·1	19·2	9·1	Sjötorp	1867	Andersson		G.Johansson	[illegible]
808	DG	Albertina Amelia W Bg	Dutch	262	99·7	24·0	13·5	Harlingen	1850	D.&L.Alta	Z 5.89	Fontein & Tjallingii	Harlingen
809	R	Albertine W Bk	British	1504	205·9	39·8	23·9	Brunswick	1864	McDonald	Y 5.86	Bk. Albertine Co. Lim. (J. S. Brown & Co.)	Leith
810	D	Albertine W Sr	Dutch	127	87·3	20·8	7·3	Pekela	1833			J.T.Leeuwe	Delfzijl
811		Albertine W Bk	French	277	124·2	26·2	12·3	Sestri Ponente	1875	[illegible]	Y 1.86	Herbay	Marseilles
812	D	Albertine Beatrice W Bk	Dutch	472	160·1	30·3	19·0	Sunderland	1868	Briggs & Sons	Y 1.85	P.Landberg & Zoon	Batavia
813	I	Alberto W 3mSr	Italian	136	95·5	20·8	10·8	Alimuri	1887	C.deSimone	Y 1.87	A.Cioffi	Castellamare
814		Alberto *Todd* W Bn	Venezul'n	143	86·5	20·3	10·5	Veendam	1860			G.A.Lutwosky	Maracaibo
815	V	Albin W 3mSr	Swedish	220	105·3	24·1	11·9	Timm'ruabb'n	1877	C.Nilsson	Y 4.86	N.Olsson	Höganäs
816		Albina W Sr	British	100	80·2	21·3	10·1	NwEdnbrgNS	1852			I.Jackson	Goole
817		Albina Emilia W Br	Italian	202	114·8	26·2	13·1	Sestri	1870	G.Casanova	Y 10.82	Roccatagliata & Bonsignore	Genoa
818	❋	Albion I Bk	British	334	165·5	24·2	11·5	Glasgow	1860	Tod&McGregor		Carr & Ashcroft	Liverpool
819		Albion W Bg	British	248	113·5	21·3	14·3	P.E.Island	1866		Y 7.87	Union S.S.Co.of N.Z.	Wellingtn.N.Z.
820		Albion *Harrison* W Bn	British	169	97·9	24·4	12·9	Portsmouth	1875	J.T.Crampton		J.T.Crampton	Portsmouth
821	N	Albion *Hansen* W Bk	Norweg'n	604	144·1	33·7	19·0	Drammen	1867	Ths.Lorentzen	Y 5.87	O.Hansen&Th.Thoresen	Drammen
822		Albion W Bk	Norweg'n	545	125·9	33·4	20·2	Rotterdam	1839	J.SmithFzn		P.C.Pedersen	Arendal
823	N	Albion W Bk	Norweg'n	478	117·7	30·4	19·6	Newcastle	1845	T.Metcalfe&Son	Y 4.84	L.Dahl	Tönsberg
824	N	Albion *Christensen* W Bk	Norweg'n	433	142·0	31·5	16·4	Fredrikshald	1872	J.Brevig	Y 9.85	J.Aalborg	Kragerö
825		Albion W Bk	Norweg'n	426	120·5	30·2	17·0	Brahestad	1858			A. K. Örvig	Brevig
826		Albion *Gustafsson* W Bk	Russian	435	130·8	31·4	17·7	Helsingfors	1871			C.W.J.Sundman	Helsingfors
827	V	Albis W Bn	German	240	105·9	22·6	12·7	Brake	1874	D.Rogge	Y 7.84	Frau B.W.P.Seele	Hamburg
828	N	Alborga W Bk	Norweg'n	647	153·6	32·1	18·3	Welford,N.B.	1878	Brown	Y 12.87	B.Jacobsen	Arendal
829	V	Albrecht W Sr	German	151	94·2	21·3	10·2	Grossefehn	1877	Schapp&Meyer	Y 6.88	Albertus Schapp	Grossefehn
830	❋	Albuera *Steven* Stl S	British	1502	236·6	39·2	22·8	Glasgow	1885	J.&G.Thomson		J.Hardie	Glasgow
831	❋	Albyn *Fisher* I 4mS	British	2095	284·0	40·3	24·4	Southampton	1883	Oswald Mordaunt & Co.		J.Houston	Liverpool
832	R	Alcalde *Crack* W 3mSr	American	305	127·0	33·0	10·0	Pt.Blakely	1882	W.C.Hall		Peter Crack	San Francisco
833		Alcedes *Helsberg* W Bg	Russian	276	..	..	..	Nagu	1866			J.Danielson	Korpo
834		Alcedo W Bn	Brazilian	231	104·5	23·3	13·0	P.E.Island	1874	J.E.McDonald	Y 1.83		
835	❋ V	Alcester *Bennet* I S	British	1597	257·0	38·2	23·2	Greenock	1883	Russell&Co.		J.R.Haws & Co.	Liverpool
836	❋	Alcestis *Norris* I Bk	British	398	149·6	27·2	16·1	Sunderland	1868	Hill Mounsey & Co.		C.B.Stone	Auckland.N.Z.
837		Alcibiades W Bn	Greek	503	141·0	28·5	..	Syra				L.Empirikos	Syra
838		Alcibiades W Bg	Greek	133	84·3	22·5	..	Syra	1874	S. Koufoudakis		C.Empiricos	Andros
839	V	Alcide W Sr	French	171	80·7	21·6	13·1	Méans	1868	Olliveaud		Pitre Legal	Nantes
840		Alcide W	French	115	..	..	..					F.Foci	Ajaccio
841	I	Alcide *Molinari* W Bk	Italian	473	133·2	31·4	20·0	SestriPonente	1871	G.Traverso	Y 2.87	RosaEnrigo-Negrotto	Genoa
842		Alcides W Sr	Swedish	118	78·3	26·7	9·4	Gefle	1844			H.Tillberg	Pukeberg Bruk, Kalmar län.
843	❋	Alcinous *Brown* I S	British	1576	259·0	38·2	23·3	Glasgow	1882	BarclayCurle&C		A. & J. H. Carmichael & Co.	Greenock
844	N	Alcyon *Christensen* W Bk	Norweg'n	309	120·3	26·1	14·1	Sundsvall	18 41/55			O.A.Föyen	Tönsberg
845	❋	Alcyone *Dale* W K Barge	British	108	95·0	22·5	8·2	Harwich	1884	J.H.Vaux		Hempson & Dale	Harwich
846		Alcyone W Sr	Swedish	106	89·2	26·2	9·5	Värnanäs	1859	Hahn		J.R.Vesterman	Blidö.Rslgn.
847	N	Alda W Sr	Norweg'n	121	94·4	25·5	10·0	Stavanger	1879	JohnNedrebo		Ths.S.Falck	Stavanger
848	❋	Aldborough *Franks* I Bk	British	1425	232·0	38·3	23·6	Stockton	1875	Richardson, Duck&Co.		British & Eastern Shipping Co., Ld.	Liverpool
849		Aldbro' *Pickering* W Sr	British	100	79·5	19·7	10·7	Yarmouth	1835			T. G. Pickering	Ipswich
850	❋	Aldebaran *Bruns* Stl S	German	1897	264·0	39·0	23·6	Workington	1888	R.Williamson & Son		Boyes & Ruyter	Bremen

❋=Lloyd's Register; D=Nederlandsche Vereeniging; G=Germanischer Lloyd; I=Registro Italiano; R=American Record; Z=Veritas Austro-Ungarico; H=Veritas Ellenico; N=Norske Veritas; V=Bureau Veritas.

Dans la 4me Colonne. C=Composite (membrure en fer, bordé en bois); I=Fer; Stl=Acier; W=Bois.

Les Dates du doublage en métal (Col. 11) ressortent des livres dans lesquels les navires sont respectivement classés ou des rapports des Inspecteurs du Lloyd's Register.

Pour plus amples renseignements sur les navires classés par le Lloyd's Register, voir le Lloyd's Register Book et le Lloyd's Yacht Register.

Official No.	No. in Book	Registry in which classed	Ship's Name. Material and Rig. Master, and Dates of Appointment to present Owner's Service and to Ship. Ship's Late Name.	Flag.	Net Tonnage	Dimensions. Length	Breadth	Depth	Built Where.	When	By whom.	[illegible]	Owners or Managers
	851	V	Aldebaran Anderson -88 W Bk (ex Wellamo)	Swedish	593	167·3	31·1	16·9	[illegible]	1876		[illegible]	J. M. Moller
	852		Aldebaran Olsson 81-84 W Bkn	Swedish	275	122·3	27·3	11·1	[illegible]	1884	S. [illegible]	[illegible]	C. A. Banck & Co.
1560	853	RV	Alden Besse Coulson 81-88 W Bk	American	813	165·0	35·0	20·0	Bath, Me.	1871	Goss & Sawyer	[illegible]	J. D. Spreckles & Bros.
70310	854	✻	Aldergrove Mull 85-85 I Bk	British	1271	234·7	37·0	21·9	P. Glasgow	1875	R. Duncan & Co.		T. C. Guthrie
33888	855		Aldersons Philliskirk W Bg	British	232	106·4	25·3	15·0	Sunderland	1865	W. Chilton		T. W. King
88525	856	RV	Aldine Curly 84-84 W Bn	British	344	131·0	29·5	12·8	Belliveau's Cove, N.S.	1884	W. D. Lovitt	Y 5.88	W. D. Lovitt
105320	857	R	Aldine Dennison W Sr	American	240	105·5	26·7	9·6	E. Machias M.	1873		[illegible]	Wm. Pope & Sons
	858		Aldjailanie W Bg	Dutch	110	..	..	..					Said Hoessin bin [illegible]
	859	V	Aldora (STRANDED) (ex Frigga) Gallus W Bk	Norweg'n	1134	173·0	35·7	23·4	[illegible]	1866	[illegible]	[illegible]	Chr. Gallus
88651	860	✻R	Aldwyth Wetmore W Bn	British	218	106·0	24·3	10·2	Alberton, PEI.	1884	E. Rogers	[illegible]	B. Rogers
48232	861		Alecto Allard W Bk	British	413	133·0	29·4	16·0	Kennebecassis, N.B.	1864	Milledge		Flli. M. Pulis
	862	DG	Alegonda Jacoba W 3m Sr (ex Maria Elisabeth ex Milka) Elas 72-77	Dutch	283	133·2	26·1	11·5	Canada	1856			Repko & Co.
	863		Alejandro (WRECKED) (ex Alejandro) Lorido W Bg	Spanish	107	76·4	21·0	10·4	[illegible]	1871		[illegible]	A. Portals
	864		Alejandro W Bn	Spanish	102	73·2	22·6	10·3	Corunna	1870			Manuel Romero
	865	V	Alejandro Bosch Argila-guet W Bk (ex Dos de Junio, ex Esmeralda)	Spanish	606	150·9	30·3	19·1	Quebec	1863		Y 5.88	Rómulo Bosch
	866		Alejo Vells 86- W Bn (ex Antonio Pomar)	Spanish	117	96·8	23·6	8·9	Palma	1870			J. Alejo Bonnati
	867	V	Alejo I. (ex Enrique) Pedrol W Sr	Spanish	164	86·0	23·7	11·9	Barcelona	1855		[illegible]	M. Carreras
105454	868	R	Alena Covert Minnis W Sr (ex E. W. B.)	American	176	94·2	27·0	10·9	Antigonish NS	1867	Z. [illegible]		Ellen F. Davidson
	869		Alencastro d'Oliveira W Bn (ex Ernesto)	Brazilian	173	..	..	..	Europe	1874			S. A. Braga
38326	870		Alert (ex Den Rebecdine) W Bn	British	170	108·5	24·0	10·1	Sweden	1854			A. Campbell
34213	871		Alert W Bg	British	147	92·9	23·3	12·6	Brigus, N.F.L.	1859			A. Munden
83511	872	✻	Alert Hughes W 3m S	British	147	103·8	23·6	11·1	Runcorn	1885	Brundrit & Co	Y 5.88	J. Foulkes, jr.
19345	873		Alert Walker W Bg	British	116	82·0	21·6	11·3	Peterhead	1853	J. Robertson		J. Duthie
	874	VN	Alert Olsen -88 W Bk	Norweg'n	904	193·8	35·9	20·6	Bergen	1877	S. K. Flage	[illegible]	Rasmus F. Olsen
	875		Alert Jensen 70-81 W Bk	Norweg'n	543	146·3	31·8	17·3	Arendal	1881	H. Kallevig	Y 6.88	Chr. Th. Boe
	876	V	Alert Gundersen -88 W 3m Sr	Norweg'n	267	119·9	27·7	11·8	Gefle	1879	O. A. Brodin	Y 1.88	[illegible]
	877	DN	Alert Berg W 3m Sr	Norweg'n	242	127·0	25·2	11·4	Altona	1862			[illegible]
	878	V	Alerte Turbot W Sr	French	131	91·2	21·3	11·1	Dunkirk	1879	Vancauwenberghe		A. Dinoir & fils
691	879		Alessandria Gazzolo W Sr	Italian	116	88·3	23·3	9·5	Sestri	1873	A. Briasco fu G.B.		Eredi Costa
2473	880		Alessandrina Ranci W Bg (ex Ettore)	Italian	143	86·6	22·8	12·8	Malta	1856			S. Ambrosano
633	881	I	Alessandro Ferraro -88 W Bk	Italian	654	153·5	32·7	21·6	Varazze	1873	D. Craviotto	[illegible]	G. Dall' Orso
1047	882	V	Alessandro Tonci W Bg	Italian	147	95·8	23·4	10·0	Viareggio	1879		Z 5.88	A. Raffaelli
	883	A	Alessandro [illegible] W Bg (ex [illegible])	Turkish	217	..	..	..	Fiume	1865			A. E. Kiosseoglu
139	884	V	Alessandro T. Marassi W Bk (ex Franciska T.)	Austrian	384	117·1	27·1	17·9	Buccari	1867	M. Mutterglian	Y 5.88	M. Dobrilovich fu G.
	885	N	Alette [illegible] -87 W Bk (ex Warjakka)	Norweg'n	681	161·5	34·4	18·6	Uleåborg	18 75/76	F. Pettersson	[illegible]	C. J. Gullichsen
	886	V	Alette Nordin W Bk	Swedish	492	133·2	28·7	16·2	Hernösand	1869	J. O. Strandberg		O. R. O. Nordin
	887	H	Alevras Elia W Bg	Greek	108	..	..	..	Syra	1874			L. Atanasovich Damianovich
79096	888	✻	Alexa Frances -88 I Bk	British	425	144·3	26·8	15·6	Sunderland	1878	Osbourne, Graham & Co.		C. A. Harris, Junr.
36435	889		Alexander W Sq	British	302	134·0	26·5	11·0	P. Dalhousie	1857			M. P. Hayes
38660	890		Alexander W Bn	British	238	118·0	27·5	9·7	United States				W. Thompson
105116	891	R	Alexander Blossom 80-86 W Bn	American	136	87·0	25·0	10·6	Cambridge, Md	1886	J. W. Crowell		W. Lewis
	892		Alexander Block -73 W Sr	German	132	88·9	21·4	10·4	Elsfleth	1853			F. W. Ogilvie
	893		Alexander Nass -88 W Bg	Norweg'n	190	100·5	24·4	11·9	Libau	1857	A. A. Bachhoff		H. C. M. Nass
	894	G	Alexander W Bk	Norweg'n	331	122·0	27·0	14·5	[illegible]	1864	D. Dargewitz	[illegible]	Rhederei Ges. "Linda"
427	895		Alexander Leelkaln -83 W 3m Sr	Russian	335	131·2	30·0	12·9	Riga	1883	A. Spirra		M. Leelkaln
	896		Alexander Waanaanen W Gt	Russian	247	..	..	..	Impilaks	1874			A. Waitlinen
247	897		Alexander Bertling W Sr	Russian	226	102·6	27·9	12·5	Nogallen	1877	K. Oschmutzneck		Bertling & Stahl
	898		Alexander Afanasjeff W Sc	Russian	222	..	..	..	Impilaks	1878			A. Nissin
	899		Alexander Sjoberg W Bg	Russian	214	113·4	26·3	12·8	[illegible]	18 64/75	J. Nyman		G. L. Söderström
417	900	V	Alexander Jacobsohn -86 W Sr	Russian	183	91·5	27·0	11·8	[illegible]	1882	Kirstein		J. Jansohn

✻ = Lloyd's Register; D = Nederlandsche Vereeniging; G = Germanischer Lloyd; I = Registro Italiano; R = Ame[rican]
A = Veritas Austro-Ungarico; H = Veritas Ellenico; N = Norske Veritas; V = Bu[reau]

In the 4th Column, C = Composite (iron frame, wood planked); I = Iron; Stl = Steel; W = Wood.

The Dates of Metalling (Col. 11) are based on the records of the Books in which Vessels are respectively classed, or on the Reports of the Surveyors of Llo[yd's]

Further particulars of Vessels classed by Lloyd's Register will be found in Lloyd's Register Book and Lloyd's Yacht Register.

No.	Registre	Nom. Matériaux et Gréement. Capitaine et dates d'entrée au service des Armateurs actuels et à bord du navire. Ancien Nom.	Pavillon.	Tonnage net.	Longueur	Largeur	Creux	Où.	Quand.	Par qui.	Date du doublage	Armateurs ou Directeurs-Gérants.	Port d'Armement.
901		Alexander Martinson WSr (ex [illegible])	[illegible]	179	[illegible]	26·7	11·[illegible]	[illegible]	1883			G.Martinson	Haynasch
902		Alexander WSr	Russian	1[illegible]	..	..	..		1860			W.Weldejeff	Sordavala
903		Alexander WSr	Russian	1[illegible]	..	..	..					E.Hanuihin	Kexholm
904	G	Alexander Andersson WBk (ex [illegible])	Swedish	717	147·1	[illegible]	20·[illegible]	[illegible]	1856		[illegible]	Rederiaktiebol Örnen (J. P. Strömberg)	Gothenburg
905		Alexander (BROKEN UP) WBg	[illegible]	1[illegible]	77·1	19·8	11·[illegible]	[illegible]	1838			P.Leffler	Gothenburg
906	R	AlexanderCampbell Fowler WBkn	American	451	140·0	31·[illegible]	14·[illegible]	[illegible] Me	1874		[illegible]	G.W.Bunker	Machias.Me.
907	V	Alexander Gibson [illegible] WS	American	[illegible]	247·3	42·[illegible]	[illegible]	Thomaston, Me	1877	EdwardO'Brien	[illegible]	E.E.O'Brien	Thomaston, Me.
908	DV	Alexander Keith McKenzie WS	British	628	139·2	32·7	17·7	St.John,N.B.	1874	D.Lynch	[illegible]	H.McLeod	St.John.N.B.
909	✻	Alexander Lawrence Crawley StlBk	British	12[illegible]	[illegible]	[illegible]	21·[illegible]	[illegible]	1886	A.M'Millan& [Son		A.M.Lawrence&Co.	London
910	RV	Alexander McNeil Fiske WS	American	10[illegible]	177·1	[illegible]	[illegible]	W[illegible] Me	1869	A.R.Reed	Y.[illegible]	J.D.Spreckels & Bros.	San Francisco
911		Alexander Marshall WS	American	1[illegible]	179·[illegible]	[illegible]	2[illegible]·8	New York	1860			C.H.Marshall	New York
912	G	Alexander Newton [illegible] WBk	[illegible]	309	151·8	25·[illegible]	12·[illegible]	ManningRiver [N.S.W.	1876	A.Newton	[illegible]	A.Newton	Sydney.N.S.W.
913	✻	Alexander Nicol [illegible] CmSr	[illegible]	266	126·5	25·2	13·9	Aberdeen	1876	J.Duthie Ss.&Co		Aberdeen Lime Co.	Aberdeen
914	V	Alexander Yeats [illegible] WS	[illegible]	15[illegible]	218·2	40·2	24·0	Portland,N.B.	1876	D.Lynch	Y.[illegible]	Alex. Yeats & Sons	St.John.N.B.
915		Alexander I. Holmsted (ex Anton) WBk	Russian	467	..	..	..	Wasa	18^{63}_{83}			A.Terichoff	Lovisa
916	G	Alexander II Lagerstam WBk	Russian	489	122·2	31·4	19·2	Nykarleby	18^{55}_{66}	Sherman	V 10.[illegible]	A.Terichoff	Lovisa
917	✻	Alexandra Prizee 84-85 IS	British	2462	304·7	41·3	25·2	Southampton	1884	Oswald,Mordaunt&Co.		VictoriaSteamboatAssociation [illegible]	London
918	✻	Alexandra Hunter 77-88 IS	British	1352	231·5	36·5	23·9	Belfast	1863	Harland&Wolff		Trinder Anderson&Co.	London
919	✻	Alexandra Nicholson WBk	British	42[illegible]	127·2	29·1	17·6	S.Shields	1863	Wilson&Crosthwaite	[illegible]	G.C.Pecket	Sunderland
920		Alexandra (ex America) W	British	255	102·0	25·0	11·4	Rotterdam	1851	VisserBros.		W.R.Cave	Adelaide.S.A.
921		Alexandra McFee 64 WBg	British	199	109·2	25·3	12·9	P.E.Island	1863	J.M'Dougall		J.Welsted	Yarmouth
922		Alexandra Tyrer 85-85 WSw	British	162	96·9	23·4	12·8	Sunderland	1865	J.Gardner		T.Jefferson	Aberdeen
923		Alexandra WS	British	107	85·3	23·6	9·7	NovaScotia				J.Niles	Georgetwn.Dem.
924		Alexandra Bass 85-85 WDy	British	102	77·2	21·1	10·4	Hamburg	1879	Dreyer		H.Dass	Grimsby
925		Alexandra (WRECKED) (ex [illegible]) Stee 83-84 WBk	Danish	451	122·8	24·4	19·0	London	18^{36}_{77}	M.Wigram		M.Christiansen	Ronne
926	AH	Alexandra Arvaniti WBg	Greek	266	..	..	..	Galaxidi	1875			S.Arvaniti	Galaxidi
927	✻N	Alexandra Larsen (ex Ann [illegible]) WS	Norweg'n	984	203·2	34·9	20·1	Aberdeen	1868	J.DuthieSns&C.	Y [illegible]	Morten Kallevig	Arendal
928	NV	Alexandra Gillensen -88 WBk	Norweg'n	583	147·5	31·7	18·5	Risör	1879	T.Jensen	[illegible]	J.W.Prebensen	Risör
929	N	Alexandra Christoffersen (ex Hansen) WBk	Norweg'n	439	132·8	27·8	17·8	Boston	1852			B.I.Larsen	Sandefjord
930	N	Alexandra Knudsen 88-88 (ex Anna [illegible]) WBk	Norweg'n	415	111·7	30·8	18·7	Voloska,Aust.	1854			Ths.W.Schwartz	Fredriksstad
931	N	Alexandra Thorsen -58 (ex [illegible]) WBg	Norweg'n	299	103·8	25·8	16·6	Sunderland	1865			L.Thoresen	Christiania
932	N	Alexandra Falkenberg [75-8[illegible]] WBk	Norweg'n	297	130·6	28·6	12·5	Drammen	1875		[illegible]	H.U.Singdahlsen	Laurvig
933	✻	Alexandre Lépé 87- IBk	French	686	181·5	30·0	18·9	Glasgow	1869	Aitken& Mansel		Ant.Dom.Bordes & fils	Bordeaux
934	V	Alexandre Boutreis -88 WBk	French	417	128·6	28·8	17·9	Nantes	1888	E.Clergeau	Y [illegible]	A.Bossard & fils	Nantes
935	V	Alexandre Corfmat -88 WBk	French	235	115·4	21·3	13·6	Dunkirk	1888	A.Lefebvre		J.Allemés & Co.	Dunkirk
936		Alexandre (WRECKED) WBk	French	213	90·2	23·3	13·9	St. Servan	1840			F. Fichet	St.Malo
937		Alexandre WBg	Turkish	147	..	..	..	[illegible]	1874			FiljonjeeKirher velideh Budres	Constantinople
938		Alexandre Clémence WBg	French	148	81·4	21·0	11·6	Nantes	1857	Chauvelon		Chauvelon & Co.	Nantes
939	V	Alexandre Herculano de Carvalho WBk	Portug'se	489	121·0	26·6	19·0	Oporto	18^{59}_{79}		Y [illegible]	F.A.de Lima & Co.	Oporto
940	V	AlexandreVagliano (ex [illegible] ex Fratelli Marasli) [illegible]	Greek	457	135·1	25·6	18·2	Varazze	1880	N.Cerruti	Y [illegible]	M.A.Vagliano	Syra
941		Alexandre & Marie WBg	French	281	..	..	..	Agde	1869			Pepin Veriés	Agde
942		Alexandre II. Joanne (BROKEN UP) WBg	French	111	67·9	19·4	10·9	St.Denis	1858			E.Corvee	Auray
943		Alexandria WBg	Turkish	150	..	..	..	Barten	1873			Tajar Costanti Polidi	Barten
944	V	Alexandrine Rickeles -61 WBk	German	425	126·9	29·8	18·2	Memel	1864	B.Pieper		J.E.G.Sternberg	Memel
945	H	Alexandros WBg	Greek	312	107·6	27·2	..	Austria	1870			D.N.Calopoulos	Syra
946		Alexandros WBg	Greek	249	..	..	..	Trieste	1866			C.Anatsitsos	Galaxidi
947		Alexandros WBg	Greek	223	98·8	18·4	..	Galaxidi	1875			NicholasPapapetros	Zante
948	H	Alexandros [illegible] WBg	Greek	185	..	..	..	[illegible]	1865		[illegible]	F.Boulounbasis	Hydra
949	A	Alexandros WSc	Greek	169	82·0	23·3	..	Chios	1873			G.Lichnos	Syra
950	H	Alexandros WBn	Greek	148	78·1	22·0	..	Piraeus	1863		[illegible]	P.Levidis	Zante

✻ = Lloyd's Register; D = Nederlandsche Vereeniging; G = Germanischer Lloyd; I = Registro Italiano; R = American Record;
A = Veritas Austro-Ungarico; H = Veritas Ellenico; N = Norske Veritas; V = Bureau Veritas.

Dans la 4me Colonne, C = Composite (membrure en fer, bordé en bois); I = Fer; Stl = Acier; W = Bois.

Les Dates de doublage en métal (Col. 11) ressortent des livres dans lesquels les navires sont respectivement classés ou des rapports des Inspecteurs du Lloyd's Register.

Pour plus amples renseignements sur les navires classés par le Lloyd's Register, voir le Lloyd's Register Book et le Lloyd's Yacht Register.

Official No.	No. in Book	Registry in which classed	Ship's Name. Master, and Dates of Appointment to present Owner's Service and to Ship. Ship's Late Name.	Material and Rig	Flag	Net Tonnage	Length	Breadth	Depth	Built: Where	When	By whom	Date of Metalling	Owners or Managers
	951		Alexandros	W Sr	Turkish	157	..	..	..	[illegible]	1874			Sakyzlee Yorghi Glass
134	952		Alexandros Vaglianos	W Bn	Greek	405	136·8	26·9	19·0	[illegible]	1881			M.A. Vagliano
	953	H	Alexandros III. *Anatsitos*	W Bg	Greek	522	..	..	..	[illegible]	1866		[illegible]	Anatsitos & Son
	954		Alexej	W Sr	Russian	249	..	..	..		1865			I. Judin
60050	955		Alexina	W Sr	British	114	92·1	22·5	8·5	[illegible]	1872			Cyrille Kironne
	956	DV	Alf *Rasmussen 80-84*	W Bn	Danish	158	100·5	21·4	10·9	[illegible]	1880	Ridersborg & Trochman		N. Petersen
	957	V	Alf (ex Thomas) *Rasmussen 84-84*	W Sr	Danish	133	93·5	20·2	10·4	[illegible]	1876	C.J. Jensen		L. Hansen
	958	N	Alf [illegible] *Bache 77-86*	W Bk	Norweg'n	637	158·5	32·8	18·6	[illegible]	1877	J. Olsen	[illegible]	O.L. Röed
	959	N	Alf *Jensen*	W Bk	Norweg'n	528	142·8	32·3	17·6	Tvedestrand	1878	O. Gundersen	[illegible]	M.H. Smith
	960	N	Alf (ex Chatillon) *Arvesen -88*	W Bk	Norweg'n	377	119·5	29·6	16·0	Bordeaux	18 51/74	Moulinié		A. Arvesen
	961	N	Alf *Lange*	W Bn	Norweg'n	299	124·8	27·7	12·3	Mandal	1877	C. Weyergang	[illegible]	Et Aktieselskab (M. Weyergang)
	962	N	Alf (ex Bernhart) *Thygesen 83-86*	W3mSr	Norweg'n	263	118·5	26·5	14·9	[illegible]	1864	Bailey		S.C. Larsen
2126	963	I	Alfa *Malatesta*	W Bk	Italian	540	138·4	31·7	21·0	Santa Margherita	1871	F. Rolla	[illegible]	Sigra F.M. Ansaldo
106812	964	R	Alfaretta Campbell *Campbell*	W3mSr	American	637	165·0	35·0	16·0	Rockland, Me.	1884	A.F. Ames		E.K. Glover
105907	965	R	Alfaretta S. Snare *Smith*	W3mSr	American	241	121·8	31·1	8·5	Bath, Me.	1880	Rogers	[illegible]	Fred. T. Hall
	966	N	Alfarin *Nielsen*	W Bk	Norweg'n	446	136·2	31·1	16·5	[illegible]	1873	Th. Tommesen	[illegible]	Fru A. L. Sörensen
	967	N	Alfen *Davidsen 63-83*	W Bk	Norweg'n	374	111·9	29·2	16·9	Grimstad	18 47/74		[illegible]	T. Gallus
	968	N	Alfen *Hiis 84-84*	W Bn	Norweg'n	116	92·0	22·7	10·1	Bergen	1856			O.N. Hiis
	969	NV	Alfhild (ex Charlotte Gladstone) *Salvesen 85-85*	W Bk	Norweg'n	1254	195·6	38·7	23·6	St. John, N.B.	1865	Fisher	[illegible]	Bull Kjöstered
	970	N	Alfhild (ex Ruuf, ex Alfred) *Olsen -6-87*	W Sr	Norweg'n	157	79·8	23·9	12·5	Sunderland	1847	H. Carr		C.E. Holter
	971		Alfhild *Berntsson*	W Bk	Swedish	412	133·5	26·1	16·5	Lubeck	1862			A. Lundberg
	972		Alfhild *Svensson*	W3mSr	Swedish	270	120·0	25·1	12·7	[illegible]	1874	I. Bergforss & Sn.	[illegible]	F. Dybäck
	973	V	Alfhild *Bengtsson 73-75*	W Sr	Swedish	185	97·1	22·8	12·2	[illegible]	1875	A. Ahlberg		J.H. Gren
	974	RV	Alfonso (ex Alfredo, ex L. & C., ex Flora) *Nörregaard*	W Bn	Mexican	246	111·8	25·3	12·8	[illegible]	1870	H. Deetjen	[illegible]	
	975		Alfonso (ex Francesco Gilberta)	W Bk	Spanish	301	105·6	29·0	16·2	Fiume	1856	A. Zanon		F. Torrontegui
79357	976	*	Alford *Vine*	W K	British	75	77·5	19·7	7·8	[illegible]	1885	H.M. Restarick		H.M. Restarick
26378	977		Alfred *Gray*	W S	British	1279	175·0	34·8	22·5	N. Brunswick	1852			Farnworth & Jardine
66277	978		Alfred *White*	W Bn	British	252	111·0	24·7	13·5	P.E. Island	1871			S.S. Ingham, sen.
	979	V	Alfred *Dyg -87*	W Sr	Danish	133	97·2	20·7	11·1	Nyborg	1876	Emil Bech		Oram & Wulff
	980	V	Alfred *Bertrand*	W Bg	French	190	94·2	21·6	12·0	[illegible]	1873	L. Leroy	[illegible]	Al Le Gualès de Mézaubran
	981		Alfred *Ruchelt*	W3mSr	French	132	81·0	20·7	10·2	Bordeaux	1859		[illegible]	Comolet frères & les fils de Paino
	982	V	Alfred (ex Georg) *Ybert*	W Sr	French	115	80·7	19·7	9·6	Fuhlendorf	1875	C. Schröder	[illegible]	A. Postel & ses fils
	983	GV	Alfred (Wilhelm) *Schade 84-84*	W3mSr	German	227	111·2	25·1	12·4	Reiherstieg	1869	Ernst. Dreyer	[illegible]	E.B. Crasemann
	984	N	Alfred *Mikkelsen -87*	W3mSr	Norweg'n	362	133·7	28·0	11·8	Stavanger	1875	R.O. Herrem	[illegible]	E. Schmidt, jun.
	985		Alfred *Petersen*	W3mSr	Norweg'n	243	103·0	24·7	11·5	Riga	1872	A. Ferle	[illegible]	J. Nyqvist
313	986	G	Alfred *Juschewsky -86*	W Sr	Russian	166	92·0	26·2	12·1	[illegible]	1878	P. Andersohn		J. Leeping
	987		Alfred	W Sr	Russian	150	..	..	..		1855			A. Karenius
105284	988		Alfred Brabook *Ferguson*	W3mSr	American	534	151·2	34·7	15·9	Bath, Me.	1873	Goss & Sawyer		J.M. Phillips
77977	989		Alfred Demers	W Sp	British	131	105·0	22·5	7·6	Quebec	1878	O. Arel		Harbour Commissioners Montreal
	990	N	Alfred Gibbs *Reimers -88*	W Bk	Norweg'n	374	126·9	27·4	16·3	Bath, Me.	1851	D. Word & Sons	[illegible]	Joh. P. Johnsen
29015	991		Alfred Hawley	W Bk	British	412	135·5	28·9	17·1	Sunderland	1860	W. Pile jr.		W. Llewellin
1227	992	R	Alfred Keen *Greely*	W Sr	American	230	116·0	28·0	10·8	Rockland, Me.	1865	[illegible]		A. F. Crockett & Co.
17436	993		Alfred Ray *Wilkinson 82-82*	W Bg	British	192	101·5	24·5	13·3	[illegible]	1857	[illegible]		J.T. Crampton
68350	994	*	Alfred Rooker *Matthews*	W K	British	70	71·0	19·5	9·2	[illegible]	1876	Darton		J. T. Mathews
90398	995	*	Alfred Trower *Tay*	I Sk	British	109	81·0	20·1	13·8	[illegible]	1884	Str'nd Slipw'y C.		J.W. Haylock
105114	996		Alfred W. Fiske *Waite*	W3mSr	American	376	136·3	32·1	14·3	Bath, Me.	1872			E. Kelly
	997	V	Alfrede (ex Astrea) *Jensen -88*	W Bn	Danish	177	105·8	22·2	11·9	Troense	1868	C.J. Jensen		N. Jensen
	998		Alfredo *Roig*	W Bn	Brazilian	295	..	..	..					M. Moreira
	999		Alfredo	W Sr	Brazilian	100	..	..	..					Maria J.C. Novaes
2381	1000	IV	Alfredo *[illegible] 80-82*	W Bk	Italian	910	176·5	34·4	22·7	Sestri Ponente	1875	A. Briasco fu L.	[illegible]	A. Nattini

* = Lloyd's Register; D = Nederlandsche Vereeniging; G = Germanischer Lloyd; I = Registro Italiano; R = Americ…
A = Veritas Austro-Ungarico; H = Veritas Ellenico; N = Norske Veritas; V = Burea…

In the 4th Column, C = Composite (iron frame, wood planked); I = Iron; Stl = Steel; W = Wood.

The Dates of Metalling (**Col. 11**) are based on the records of the Books in which Vessels are respectively classed, or on the Reports of the Surveyors of Lloyd…

Further particulars of Vessels classed by Lloyd's Register will be found in Lloyd's Register Book and Lloyd's Yacht Register.

No.	Classe	Nom. Capitaines et dates d'entrée. Ancien Nom.	Matériaux et Gréement	Pavillon	Tonnage net	Longueur	Largeur	Creux	Construit: Où	Quand	Par qui	Date du doublage	Armateurs ou Directeurs-Gérants	Port d'Armement
1001	V	Alfredo *Addi* (ex Amalf...)	W Bkn	Italian	412	126·6	30·0	16·4	Palermo	1869	A. LaRocca	Y	G. Massa	Genoa
1002	I	Alfredo *Poggi*	W Bn	Italian	256	109·5	25·8	15·0	Lerici	1877	A. Bardellini	Y	P. Poggi di G.B.	Spezia
1003	A	Alfredo (ex Felipe Colombo)	W Bg	Italian	256	99·7	26·8	16·4	Savona	1856	F. Berti	Y	F. Berti fu F.	Leghorn
1004		Alfredo (ex Cavalier)	W Bk	Peruvian	355	124·9	26·0	16·7	Southwick	1864	Hardie		Teso & Sanchez	Callao
1005	V	Alfredo *Roig*	W Bn	Spanish	310	117·7	28·0	15·7	SestriPonente	1872			L. Mir & Co.	Barcelona
1006		Alfredo	W Bn		117	86·6	20·2	9·0	Tayabas [Luzon	1876				
1007	V	Algeria *Tufts*	W Bk	British	620	118·8	31·7	18·8	Rockland, N.B.	1871		Y	F. Tufts	St. John, N.B.
1008		Algeria *Lohnes* ·84	W Sr	British	119	82·2	23·9	9·3	LaHave, N.S.	1884	S. Leary		E. Lohnes	Lunenburg, N.S.
1009	✻	Algoa Bay *Logie* 81·84	I Bk	British	1111	217·0	35·1	21·0	Greenock	1881	Russell & Co.		Hatfield, Cameron & Co.	Glasgow
1010	V	Algol *Binvel*	W Sr	French	142	79·4	23·0	11·2	Mesnes	1866	Malé		J.M. Binvel	Auray
1011	V	Algoma *Vero*	W S	British	1184	199·0	38·7	22·8	Maitland, N.S.	1872	A. McCallum	12.84	F. Curry	Maitland, N.S.
1012		Algorta *Costa*	W Bk	Peruvian	146	88·6	24·3	9·8					C. Mackehenie	Callao
1013	DN	Alhambra *Simoré*	W Bk	Norweg'n	468	140·4	31·1	16·8	Laurvig	1877	Treschow	Y	C. Berg	Fredrikshald
1014	N	Alia *Hansen*	W Bg	Norweg'n	266	103·9	27·5	14·3	Kragerö	1865			H. Pedersen	Kragerö
1015		Aliah el Rahman	W Bg	Turkish		..	..	..					Hajee Ali el Hakwani	Alexandria
1016	R	Alice *Stuart* ·87	W Bn	British	231	101·4	25·0	12·3	Mandal	1876		Y	W.R. Hatchings	Castries, S. Lucia
1017		Alice	W	British	163	108·0	22·1	7·1	Ottawa, Ont.	1878	A. Clarke		Margrt. W. Burll	Ottawa, Ont.
1018	R	Alice *Baxter*	W Bn	British	137	86·6	24·0	9·8	Cornwallis, NS	1877	S.H. Newcomb	Y	Albert Chase	Windsor, N.S.
1019		Alice	W Sr	British	116	82·0	24·3	9·5	LaHave, N.S.	1882	G. Leary		James Ross	Lunenburg, N.S.
1020		Alice	W Sr	British	107	83·0	20·5	8·6	Chicoutime	1864			E. Tremblay	Quebec
1021	V	Alice *Smith*	W Bk	American	819	161·2	34·8	20·5	Weym'th, Mas	1881	N.P. Kean	Y	G. Ropes	Boston, Mass.
1022	R	Alice *Kair* 83-87	W Bk	American	479	130·1	31·6	15·9	Deering, Me.	1874	G.W. Russell	Y 9.85	J.S. Winslow & Co.	Portland, Me.
1023		Alice *Clements* ·88	W Bn	American	296	116·8	27·1	15·0	Baltimore	1867		Y 6.86	T. Rollins	Baltimore
1024		Alice *Lewis*	W Sr	American	272	109·0	27·0	13·6	Lynchb'rg Tex	1880	H. Gibbs		Woodward, Wight & Co.	New Orleans
1025		Alice *Thornstrom* 75·	W Sr	American	221	115·0	31·3	10·4	Bean's Point, [Wash.	1874			J.A. Hooper	San Francisco
1026		Alice (ex Lepanto)	W Bg	American	176	90·7	27·2	13·2	Blanes, Spain	1848			Jacinto Miro & Miro	New Orleans, La
1027	A	Alice *Bumtilich* ·88 (ex ...)	W Bk	Austrian	747	153·2	34·8	20·1	Trieste	1870	Stabilimento Tecnico [Triestino	Y 8.86	P. Franassovich	Orebich
1028		Alice	W Bk	Brazilian	389	..	..	..					M.J.C. Junior	Bahia
1029		Alice *Agerlin* ·81 (ex Carthovan, ex H. H. ...mann)	W Bg	Danish	252	94·8	23·3	15·8	Hamburg	1850	J.F. Krohn		P.P. Schmidt	Drager
1030	DV	Alice *Kruysman* (ex A. H. van ...)	W S	Dutch	1333	205·6	39·7	25·0	Alblasserdam	1877	F H Von Lindern	11.81	Von Lindern & Van Vessem	Rotterdam
1031		Alice (ex Constance)	W Bn	Dutch	130	88·0	23·7	10·8	Halifax	1865			A.A.v. Romondt & Co.	St. Martin, W.I.
1032	V	Alice *Malmat*	W Sr	French	106	82·7	19·3	10·9	Paimpol	1878	L. Laboureur		Veuve S. Duval	Paimpol
1033	DG	Alice *Schmugel* 76-87 (ex F. B. Weldon)	W S	German	1463	208·1	40·4	23·8	Rockland, N.B.	1878	R.A. Chapman	Y 8.87	D.H. Wätjen & Co.	Bremen
1034	G	Alice *Rich* ·88	W Bg	German	168	93·8	23·6	12·8	Sunderland	1864			F.W. Ogilvie	Memel
1035	V	Alice *Gundersen* 68-88 (ex Woodville)	W Bk	Norweg'n	968	171·8	35·5	23·1	Quebec	1875	Gingras	Y 3.86	N.M. Gram	Drammen
1036	DN	Alice *Jacobsen* ·88 (ex Union)	W Bk	Norweg'n	935	169·9	35·0	21·7	Damariscotta, [Me.	1865	E. Norris & Co.	11.85	M. Jacobsen	Arendal
1037	N	Alice (ex BANDON)	W Bk	Norweg'n	758	153·2	36·8	20·9	Trieste	1856		Y 7.83	N.M. Gram	Drammen
1038	N	Alice *Knudsen*	W Bg	Norweg'n	274	100·6	24·4	15·7	Southampton	1847	H. Blaker		H.C. Torjusen	Kragerö
1039		Alice	W Bk	Portug'se	887	..	..	..	Halifax, N.S.	1869		Y 1.83	João da Silva Bôa	Lisbon
1040	G	Alice *Röhrich* ·81	W 3m Sr	Russian	273	139·5	26·4	11·6	Detroit, U.S.	1863			J. Röhrich	Riga
1041		Alice	W Bg	Turkish	110	..	..	..	Vassilikos	1883			Varnalee Challik-Oglou Vassil	Varna
1042	✻	Alice Ada *Casey*	W Bn	British	293	121·0	28·8	12·1	Pt. Elgin, N.B.	1873	E.W. Ogden	Y 9.88	E.W. Ogden	St. John, N.B.
1043	R	Alice Archer *Fletcher*	W 3m Sr	American	448	149·3	34·1	12·7	Bath, Me.	1882	A. Sewall	Y	A. Sewall & Co.	Bath, Me.
1044		Alice B. *Adams*	W Sr	American	216	124·1	25·1	9·1	Cleveland, Oho	1863			L.B. Simmons	Barnstable, Mass.
1045	R	Alice B. Phillips	W 3m Sr	American	591	160·6	35·8	16·5	Portsm'th, NH	1883	W.H. Fernald		H.O. Benton	Fall River, Mass.
1046	✻	Alice Bannister *Pritchard*	W Sr	British	99	88·0	22·2	10·1	Bridgwater	1877	Gough		J. Fisher & Sons	Barrow
1047	R	Alice Borda *Dukes*	W 3m Sr	American	398	124·4	32·2	11·6	Tuckahoe, N.J.	1872	J.C. Lee		G.L. Dilkes	Camden, N.J.
1048	R	Alice Bradshaw *Dahm*	W Bn	British	293	116·4	28·8	12·5	Kempt, N.S.	1875	B.D. Bradshaw	Y 1.87	E.E. Hutchings	Windsor, N.S.
1049	R	Alice C. Dickerman *Cook*	W Bkn	American	476	138·1	32·9	16·1	Boston, Mass.	1875	A. Gore	Y 10.85	W.C. Haskins & Sons	Boston, Mass.
1050		Alice Campbell *Ray*	W Bk	American	408	123·5	28·3	15·9	Boston, Mass.	1870	Taylor, Campbll [& Brooks	Y	D.A. Delina	New York

✻ = Lloyd's Register; D = Nederlandsche Vereeniging; G = Germanischer Lloyd; I = Registro Italiano; R = American Record;
A = Veritas Austro-Ungarico; H = Veritas Ellenico; N = Norske Veritas; V = Bureau Veritas.

Dans la 4me Colonne, C = Composite (membrure en fer, bordé en bois); I = Fer; Stl = Acier; W = Bois.

Les Dates du doublage en métal (**Col. 11**) ressortent des livres dans lesquels les navires sont respectivement classés ou des rapports des Inspecteurs du Lloyd's Register.

Pour plus amples renseignements sur les navires classés par le Lloyd's Register, voir le Lloyd's Register Book et le Lloyd's Yacht Register.

Official No.	No. in Book.	Registry in which classed.	Ship's Name. Material and Rig. Master, and Dates of Appointment to present Owner's Service and to Ship. Ship's Late Name.	Flag.	Net Tonnage.	Length.	Breadth.	Depth.	Built. Where.	When.	By whom.	Metalled.	Owners or Managers.
66485	1051	V	Alice Cooper *Williams 85-84* W Bk	British	863	168·5	35·0	20·6	Windsor, N.S.	1872		Y 6.86	S. Dimock
63932	1052		Alice Crookall *Rolleft* W Sr	British	129	97·0	22·8	11·0	Irvine	1871	Irvine Ship Co.		R. Rawlinson
105550	1053	D V	Alice D. Cooper *Saunders* W S	American	1345	200·6	38·3	24·0	Yarmouth, Me	1874	C. F. Sargent	Y 8.86	R. S. Powell
87783	1054	✳	Alice Gertrude *Foster* W K	British	80	76·0	20·3	10·9	Brixham	1884	S. Dewdney & Ss		J. Randall
56533	1055	✳	Alice Graham *Martin -88* C Bk	British	399	137·7	27·9	16·6	Middlesbro'	1867	Harkess	Y 7.86	R. J. Swyny
71447	1056		Alice H. *Hopkins* W Bn	British	199	103·2	24·0	12·9	P. E. Island	1874	McLaine		W. Kemp, jun.
1038	1057		Alice Jane *Williams 85-88* W 3m Bn (STEANFELD)	British	200	101·6	22·5	14·4	Liverpool	1836	Humble & Milchrist		Le Marquand, Stokes & Carver
105738	1058		Alice Kimball *Jensen* W Sr	American	102	89·0	27·0	7·2	L'tle River, Cal	1874			Allen & Co.
105830	1059		Alice Knowles *Foster* W Bk	American	288	116·0	32·0	17·0	N. Weymouth, Mass.	1878	R. P. Kean		J. P. Knowles
80805	1060	D V	Alice M. Claridge *Lockhart* W Bk	British	1129	192·2	39·0	23·4	Newport, N.S.	1880	John A. Harvie	Y 11.88	John A. Harvie
99182	1061	✳	Alice M. Craig *Johnston* I Bk	British	369	156·5	27·7	12·8	Belfast	1884	Workman, Clark & Co.		Bk "Alice M. Craig" Co. Ld. (Workman & Workman)
174	1062	R	Alice M. Minott *Colley -88* W S	American	1038	175·0	36·0	23·0	Phipsbrg, Me	1867	C. Minott	Y 8.86	C. V. Minott
106520	1063	R	Alice McDonald *Dalzell* W 3m Sr	American	624	175·0	35·2	14·1	Bath, Me	1888	J. McDonald		B. Flint
73639	1064		Alice Mary *Law -88* W Bk	British	361	147·0	28·7	12·9	Sunderland	1876	N. Gibbon & Co	Y 6.84	James Evans
	1065		Alice Maud W Bn	British	234	93·3	26·9	9·8	Maine, U.S.	1861			W. H. Chaffey
82979	1066	✳	Alice May *Daves 86-86* W 3m Sr	British	154	98·0	23·3	12·1	Neyland	1886	J. Mills	Y 1.86	J. Mills
109050	1067	R	Alice Montgomery W 3m Sr (ABANDONED) *Lavender Bros*	American	696	163·6	35·4	16·5	Bath, Me.	1882	Deering & Donnell		E. H. Atwood
65328	1068	✳	Alice Moor W Sr	British	74	75·6	19·8	9·4	Falmouth	1873	H. Trethowan		J. Rendle
61533	1069		Alice Muir *Fassell* W Bk	British	480	137·0	30·0	17·0	Shelburne, N.S	1875	J. & R. McGill		David Blair
78798	1070	✳	Alice Platt *Jones 83-85* I Bk	British	1090	216·5	34·5	20·5	Glasgow	1878	A. Stephen & Sons		J. Lloyd, jun.
105303	1071		Alice Reed *Stahl* W Bk	American	829	163·0	34·0	20·5	Waldebro' Me.	1873	Reed & Welt	Y 2.87	Yates & Porterfield
43762	1072		Alice Richardson *Kissell 79-88* W Bn	British	171	89·7	23·9	13·0	Sunderland	1861	Gray		J. T. Crampton
	1073	V	AliceRickmers *Lohmann 88-89* W Bk	German	1208	187·0	39·0	25·5	Geestemünde	1875	R. C. Rickmers	Y 8.87	R. C. Rickmers
19817	1074		Alice Smith (ex Lizzie J. Frost) W Bg	British	172	95·6	26·0	10·9	Connectic't U S				James N. Duck
105330	1075		Alice T. Boardman *Rice* W Sr	American	118	96·0	27·0	7·0	Calais, Me.	1873	W. Robinson		G. H. Boardman
38149	1076		Alice V. Goodhue *Bottingdale 81-83* W Bn	British	134	90·0	23·3	11·9	Salmon River, N.S.	1861			T. P. Cattley
67264	1077		Alice Watts *Meachen 79-77* W K	British	119	100·2	24·2	8·0	Ipswich	1875	Robertson		J. Watts
22789	1078		Alice Williams *Williams* W Sr	British	100	80·8	20·5	11·8	Llanelly	1854	Bevans		G. Sanders
	1079		Alicia *Brodhurst* W Bk (ex Anna Siebert)	Hawaiian	604	149·1	33·6	18·7	Apenrade	1877	J. Paulsen	Y 11.85	T. R. Foster
	1080	V	Alida *Nielsen 80-82* W Bk	Danish	234	118·8	23·8	12·4	Rudkjøbing	1875	Joh. Boas	Y 2.88	E. E. Nielsen
	1081	V	Alida *H. E. Nagel 69-69* W 3m Sr	Dutch	207	112·5	22·2	12·3	Muntendam	1869	Drenth & Co.	Y 7.87	C. Ten Horn
	1082	N	Alida *Bödtker 73-72* W Bk	Norweg'n	556	149·0	32·1	17·7	Drammen	1868	Svendsen	Y 5.85	J. Bödtker (M. Berner)
	1083		Alida *Steenfeldt -85* W Sr	Russian	136	95·5	25·1	11·8	Narva	1874	J. Kabel		J. Kabel
423	1084		Alida W Cr	Russian	102	76·3	27·5	9·8	Kesterzeem	1883	Andretz		G. Münder
	1085		Alida *Lycke* W Bg	Swedish	157	91·2	22·3	11·0	Neder-Kalix	1864			C. J. Hansson
	1086		Alida *Jonasson* W Sr	Swedish	146	88·6	20·0	10·2	Pekela	1848	Smith & Co.		J. Bengtsson
	1087	N	Alida Catharina *Wold* W Sr	Norweg'n	107	78·2	19·6	9·3	Pekela	1856			A. Schönings Sönner
4837	1088		Alide W Sr	Russian	173	93·5	23·5	12·7	Livland	1887			G. Lorenz
	1089		Alieh el Bari Bg	Turkish	182	..	..	..					Mohamed Effendi Mass
	1090		Aliki W 3m Sr	French	219	86·0	23·9	14·3	La Nouvelle	1858			Jean Musso
4957	1091	V	Alimuri *Cafiero -88* W Bk	Italian	331	124·6	26·9	17·7	Alimuri	1872	F. S. Mauro	Y 8.87	G. Starita
71184	1092		Alina (LOST) *Hansen* W Bn	British	190	103·7	24·5	12·9	P. E. Island	1874	J. Dunn and Co.		W. Ellis
	1093		Alina *Nielsen* W Bk	Norweg'n	328	120·0	28·5	11·8	Aland, Finland	1865			A. E. Nielsen
	1094		Alina W Bk	Russian	367	116·0	28·5	13·5	Christinestad	1866	E. Silwerberg		E. P. Erikson
	1095		Alina *Fauros* W 3m Sr	Russian	200	..	..	..	Sideby	1872			K. Ehrman
	1096		Alina *Vakeva* W Sr	Russian	109	..	..	..		1873			E. Vakeva
	1097	V	Alina *Riera* W Bn	Spanish	466	147·9	27·0	18·3	England	1857		10.84	Balsells & Subirau
72466	1098	✳	Aline *Henrichs 81-87* I Bk	British	718	185·9	30·2	18·8	Greenock	1875	R. Steele & Co.		J. Grieve, jun.
38473	1099		Aline W Sr Yt	British	114	107·4	21·8	11·3	Gosport	1860	Camper & Nicholson		H R H Prince of Wales
	1100	V	Aline *Tougeer* W Sr	French	119	88·6	19·3	10·5	St Malo	1883	L. Tranchemer		F. Le Goaster

✳ = Lloyd's Register; D = Nederlandsche Vereeniging; G = Germanischer Lloyd; I = Registro Italiano; R = Ameri[can]
A = Veritas Austro-Ungarico; H = Veritas Ellenico; N = Norske Veritas; V = Burea[u]

In the 4th Column, C = Composite (iron frame, wood planked); I = Iron; Stl = Steel; W = Wood.

The Dates of Metalling (**Col. 11**) are based on the records of the Books in which Vessels are respectively classed, or on the Reports of the Surveyors of Ll[oyd's]

Further particulars of Vessels classed by Lloyd's Register will be found in Lloyd's Register Book and Lloyd's Yacht Register.

N° dans le Livre	Registre dans lequel le navire est coté	Nom, Matériaux et Gréement. Capitaine et dates d'entrée au service des Armateurs actuels et à bord du navire. Ancien Nom.		Pavillon.	Tonnage net	Longueur	Largeur	Creux	Où.	Quand.	Par qui	Date de doublage	Armateurs ou Directeurs Gérants.	Port d'Armement.
1101	V	Aline Grabe... -84 (CONDEMNED)	W Bk	German	584	157·2	31·8	17·9	Hamburg	1864	E.Dreyer		Albr.Kossel	Rostock
1102		Alithis Adelphotis	W Bn	Greek	196	91·2	24·6	14·1	Sorrento	1867			J.Panazis	Calamne
1103		Alix	W Sr	French	142	79·4	21·0	12·0	Dunkirk	1860	Vande Zande		Le Borgne	Paimpol
1104	V	Alix Veiller -87	W Sr	French	100	82·0	18·7	10·8	Paimpol	1879	L.Labourear		A.Weiller	Trégnier
1105	N	Aljuca Andersen 87-87	W Bk	Norweg'n	1063	181·5	35·7	21·2	Tusket,N.S.	1874			J.M.Smith	Arendal
1106	G	Alkor Jorgensen	W Bg	Norweg'n	257	114·9	26·8	13·2	Arendal	1872	A.Johansen		T. Thommesen & Smith	Arendal
1107	V	Alku Wallin -87	W Bk	Russian	833	166·6	32·4	20·4	Nystad	1873	H.Kjuldström		Nystads Rederi Aktiebolag	Nystad
1108		Alku	W Bn	Russian	204	101·8	26·4	11·2	Nystad	1873	J.Balteros		J.H.Reinvall	Nystad
1109		Allah Kerim	W Bg	Turkish	166	..	..	..	Greece	1876			... Agar Hussein	Rizeh
1110		Allah Kerim	W Bg	Turkish	160	..	..	..	Greece	1870			...tahier Mollis	Mureftah
1111		Allah Kerim	W Bg	Turkish	163	..	..	..	Barten	1875			... Trebizondou ... Mehmed	Barten
1112		Allah Kerim	W Sr	Turkish	150	..	..	..	Barten	1872			... Massird... & Co.	Barten
1113		Allah Verdi	W Sr	Turkish	127	..	..	..	Russia	1876			...ezn-Mehmed	Constantinople
1114		Allan	W Bn	Russian	202	96·9	23·5	13·0	Kideskoken	1865	J.Loukkanen		H.G.Henrikson & Co.	Kumlingo
1115		Allan	W Sr	Russian	180	92·5	23·0	14·5	Pechorsag	1865	J.Loukkanen		A.J.Mustonen	Joensuu
1116		Allan Söderlund	W Sr	Russian	178	..	..	..		1875			J.D.Söderlund	Sund
1117		Allan Hämäläin	W Sr	Russian	166	..	..	..	Joensuu	1875			E.Hubbard & Co.	Joensuu
1118		Allanmyo	I Flat	British	777	197·0	25·7	8·3	PortGlasgow	1876	R.Duncan & Co.		Irrawaddy Flotilla Co. Ld.	Glasgow
1119	✻	Allanshaw	I S	British	1589	262·8	40·2	23·0	Renfrew	1874	W.Simons & Co.		J.Nourse	London
1120	R	Allanwilde	W Bkn	American	607	155·9	34·0	17·6	Bucksprt,Me.	1884	W.Beasley	Y 8.89	R.P.Buck & Co.	New York
1121		Allaplee	W Bg	Turkish	225	..	..	..	Allaplee	1865			Akjee-Shehirlee-Hadji-Mollah-Oglou Osman & Co.	Akjee-Shehir
1122	✻	Allegiance Roberts -81	I S	British	1180	217·5	36·7	22·0	Liverpool	1876	Potter & Co.		W.H.Owen	Liverpool
1123	N	Allegra Johansen -88	W3mSr	Norweg'n	214	114·8	25·8	11·6	Risör	1868	C.Stenersen		J.W.Prebensen,sen.	Risör
1124	N	Allegro Christiansen (ex Maria)	W Sr	Norweg'n	129	89·6	21·1	9·9	Elmshorn	1849	D.Kremer		M.Christiansen	Laurvig
1125		Allen A. Schage -88	W3mSr	American	350	160·0	35·0	9·4	Eureka	1888	Bendixon		A.A.Curtis	Eureka
1126	R	Allen Green Nickerson	W3mSr	American	488	139·5	32·5	16·0	Greenprt,N.Y.	1874	O.H.Bishop		C.&J.F.Baker	Providence,R.I.
1127	✻	Allerby Kakarukas	W Bk	Greek	383	130·6	26·5	17·1	Maryport	1878	Ritson		Z.Kakarukas	Syra
1128	✻V	Allerton Gyllenerentz 85-85	I S	British	2028	272·7	40·1	24·6	Southampton	1884	Oswald, Mordaunt&Co.		Sailing Ship "Allerton" Co. Ld. (R.W.Leyland & Co.)	Liverpool
1129	✻	Alletha Barber 19-88	W Bn	British	215	115·3	25·3	12·5	P.E.Island	1872	Glover		R.A.Peter	Plymouth
1130		Aliança D'Azevedo (ex Mathilde)	W Sr	Brazilian	204	..	..	..	Elmshorn	1869			Viuva de J. G. Carlos	Rio Grande
1131	V	Aliança Coelho (ex ...)	W Bk	Portug'ze	489	121·0	26·6	19·0	Oporto	1868	J.F.Lapa		Cia Alliança Maritima Portuense	Oporto
1132	✻	Alliance Williams -87	I Bk	British	977	202·0	34·4	20·6	Liverpool	1885	R.&J.Evans&Co		J.Evans	Liverpool
1133		Alliance Thomas	W Bk	British	338	119·2	27·5	17·0	Sunderland	1865	Hodgson		Great Grimsby Ice Co. Ld.	Grimsby
1134		Alliance Marquand 86-88	W Bn	British	115	90·0	19·5	11·4	Shippegn,N.B.	1877	J.Henry		A.Hacquoil	Chatham,N.B.
1135	✻	Alliance Summers -87	W K	British	45	63·3	17·6	9·0	Glmptn	1887	W.A.Gibbs		H.Summers	Ramsgate
1136	GV	Alliance Petersen 75-81	W3mSr	Danish	299	117·8	24·8	13·6	Elsfleth	1877	F.G.Deetjen		J.J.Petersen	Dragör
1137	V	Alliance Beckel -87	W Bk	French	263	105·0	25·3	14·9	Fécamp	1854			Comolet frères & les fils de l'aîné	Cette
1138	N	Alliance Berentsen	W Bk	Norweg'n	433	132·0	30·8	16·2	Drammen	1868	Jensen		S.Berentzens Enke	Fredrikstad
1139	NV	Alliance Salvesen (ex ...)	W Bk	Norweg'n	424	139·9	31·6	16·2	Apenrade	1865	J.Andersen		D.Davidsen	Farsund
1140		Alliance	W Bg	Swedish	300	112·4	25·6	16·3	Dumbarton	1865	A.McMillan&Ss.		N.E.Mellin	Ringarum
1141	V	Allianz Thussen 96-83	W Bg	Danish	232	109·1	23·4	14·4	Haderslev	1866	J.Nielsen		P.Berg	Nexo
1142		Allie B. Cathrall	W Sr	American	104	97·3	26·2	6·3	Milford,Del.	1885	J.W.Abbott&C.		C.R.Smith	Wilmingtn,Del.
1143		Allie H. Belden Kelsey	W Sr	American	161	101·0	28·5	7·0	RckyHl,Conn.	1870	Belden		E.H.Williams	Hartford,Conn.
1144		Allie Oakes	W Sr	American	121	79·7	23·3	7·0	Kennebunk,Me.	1867			C.B.Merriam	Camden,Me.
1145	R	Allie R. Chester	W3mSr	American	427	144·0	35·5	13·0	Northp'rt,N.Y	1883	J.Carll		J.Carll	New York
1146		Allies McCartney	W Bn	British	137	98·6	23·8	11·1	Nova Scotia	1855			H.McCartney	Maryport
1147		Alligator (STRANDED)	W Sr	American	117	92·0	26·0	7·0	Calais,Me.	1864	Hinds		J.G.Murchie	Calais, Me.
1148	V	Allina Dirks	W Sr	German	124	87·9	20·0	9·9	Jemgum	1879	G.&H.Aden		G.& H.Aden	Jemgum
1149		Alloa	W Bg	Norweg'n	199	86·5	24·3	14·7	St.Georg'sIsld				Johansen,Ritter & Co.	Christiania
1150	✻	Allonby Steele 80-88	I Bk	British	1400	240·5	38·0	21·7	Sunderland	1882	R.Thompsn&Ss.		P.Iredale & Son	Liverpool

✻ Lloyd's Register; D=Nederlandsche Vereeniging; G=Germanischer Lloyd; I=Registro Italiano; R=American Record; ... Veritas Austro-Ungarico; H=Veritas Ellenico; N=Norske Veritas; V=Bureau Veritas.

Dans la 4me Colonne, C=Composite (membrure en fer, bordé en bois); I=Fer; Stl=Acier; W=Bois.

Les dates du doublage en métal (Col. 11) ressortent des livres dans lesquels les navires sont respectivement classés ou des rapports des Inspecteurs du Lloyd's Register.

Pour plus amples renseignements sur les navires classés par le Lloyd's Register, voir le Lloyd's Register Book et le Lloyd's Yacht Register.

Official No.	No. in Book	Classed in	Ship's Name, Master, Material and Rig; Ship's Late Name	Flag	Net Tonnage	Length	Breadth	Depth	Built: Where	When	By whom	Date of Metalling	Owners or Managers
95	1151		Allston *Barbor 84-84* WSr	American	134	86·9	23·8	8·4	Bangor, Me.	1846			J.S.Emery & Co.
53516	1152		Allum Ghier WS	British	909	145·5	37·6	22·8	Cochin	1861	C.Gueirelar	V 12.87	Haddee Tharpe (?) Hussein Khaja
31876	1153		Alma WBk	British	250	121·6	29·0	11·9	Weymth., N.S.	1858			J.W.Bigelow
85802	1154		Alma (LOST) *Hegan* WSr	British	194	98·0	27·4	10·5	Alma, N.B.	1882	W.Stephenson		V.S.White
35888	1155		Alma (FOUNDERED) WSr	British	185	107·0	22·8	10·3	Quebec	1855			J.S.Sterry
13427	1156	*	Alma *St.Ledger* WSr	British	106	91·1	19·4	11·3	Bideford	1855	Johnson		W.F.Carvasso
105150	1157		Alma *Johnson* WSr	American	171	104·0	27·0	8·9	E.Machias, Me.	1872			J.R.Talbot
	1158	DGV	Alma *Misch 88-88* WBk	German	938	169·4	36·4	22·6	Bremerhaven	1871	F.W.Wencke	V 8.88	F.Misch
	1159	*	Alma (ex Ethel) *Tiemann 88-88* IBk	German	495	154·7	28·0	16·5	Sunderland	1875	R.Thompson, jr.		W.Stehr
	1160	V	Alma *Aden 73-73* WBg	German	140	83·1	19·7	10·3	Papenburg	1873	B.Tholen		Duis J.Aden
184	1161	V	Alma *Scarpa* WPol	Italian	349	117·7	29·5	18·6	Venice	1858	G.Zanon	V 1.86	Eredi P.Scarpa
	1162	N	Alma (ex Lucy) *Jacobsen* WBk	Norweg'n	593	146·8	31·6	19·5	Yarm'th, N.S.	1874			R.M.Peterson
	1163	N	Alma (ex Paoma) *Barker* WBk	Norweg'n	560	151·7	32·2	16·5	RverJohn, N.S.	1856	C.McLennan	V 1.85	J.Ihlen
	1164	N	Alma *Woxholt* WBk	Norweg'n	479	132·9	29·3	17·1	Grimstad	1867	O.Kittelsen	C 3.86	N.Jacobsen
	1165	N	Alma *Berg 72-82* WBk	Norweg'n	444	137·9	31·0	16·2	Tönsberg	1875	T.Hansen		P.S.Bergs Enke
	1166	N	Alma *Olsen 66-83* WBk	Norweg'n	284	99·6	27·5	15·3	Kragerö	1865			H.Poulsen & Co.
	1167	N	Alma *Gevertsen -88* WBg	Norweg'n	234	108·3	27·2	13·4	Ranmo	1849	J.S.Hafverman		J. C. Svendsen & Co.
	1168		Alma (ex Cedar) *Reed -88* WBg	Norweg'n	227	98·4	25·1	11·2	Sunderland	1857	R.Thompson		A. ... "Alma"
	1169	V	Alma *Rönnberg -83* WBk	Russian	750	162·4	31·8	18·6	Christinestad	1875	M.Hanniss	V 7.86	C.E.Carlström
	1170		Alma *Jansson* WBk	Russian	437	..	..	..	Geta	1867			E.Malmström
	1171		Alma *Isaksson* WBk	Russian	301	116·8	29·5	13·6	Korpo	1874	A.Tengström		J.Isaksson
	1172		Alma *Mattsson -84* WSr	Russian	243	104·9	28·0	12·9	Christinestad	1869	A.Granskog		E.Bengts
	1173		Alma *Isaksson* WSr	Russian	219	..	..	..	Westanfjerd	1872			G.Karlsson
	1174		Alma *Jansson* WSr	Russian	191	..	..	..	Nerpes	1869			J.Jansson
	1175		Alma *Björklund* W3mSr	Russian	185	101·3	25·7	12·0	Sagu	1867			G.Lindholm
	1176		Alma WSr	Russian	115	77·5	24·4	9·6	Pyttissoka	1873	H.Quarnström		H.Reinvall
	1177		Alma *Winberg* WSr	Russian	113	..	..	..	Bromarf	1883			J.F.Strandberg
	1178	V	Alma *Larsson 87-82* WBkn	Swedish	334	136·4	26·7	13·1	Bangsbo	1874	J.Bang	V 5.85	J.E.Olsson
	1179	V	Alma *Horndahl 85-85* WBk	Swedish	332	129·9	26·1	12·5	Carlshamn	1877	Andersson	V 1.85	P.N.Horndahl
	1180		Alma (ex Percy) *Bruce 78-81* WBk	Swedish	331	117·7	27·1	16·7	Sunderland	1858			M.Backelin
	1181	*	Alma (ex Annie Maria) *Olsson 80-85* WBn	Swedish	252	115·1	24·4	12·9	P.E.Island	1874	Peake Brothers	V 6.86	A.Nilsson
	1182		Alma *Österberg -87* WBn	Swedish	186	95·5	25·7	11·6	Finland	1865			O.Pettersson
	1183		Alma *Andersson -87* WSr	Swedish	136	85·0	21·8	10·9	Roslagen	1874			P.T.Olsson
47	1184		Alma Agnes WSr	Russian	140	87·0	23·0	8·0	NeuSalis	1875			A.von Vegesack
	1185		Almachlarie *Said Salem Madehie* WBg	Dutch	131	84·0	20·9	..	Rembang			V 1.88	Said Atoei bin Abdulla Madehie
	1186	NV	Almaria (ex Helen Marion) *Nilsen -88* WBk	Norweg'n	693	148·6	33·5	19·7	St.John, N.B.	1869	Lovitt & Son	V 5.86	B.Jacobsen
	1187		Almastor *Wa hama* WBg	Dutch	124	80·0	20·5	..					Daeng Mohamet Tahir bin Abdul ...
	1188		Almatra W3mSr	Russian	218	100·5	29·3	11·8	Torby	1876	J.Nordström		J.Nordström & Co.
105137	1189		Almeda WSr	American	154	99·0	28·1	8·8	Pembroke, Me.	1872			S.W.Theall
105363	1190	R	Almeda Willey *Copeland 74-83* W3mSr	American	520	147·9	33·5	12·1	Thomastn, Me.	1874	Dunn & Elliott	V 9.84	Dunn & Elliot
	1191	N	Almedia (ex Emma Ives) *Christensen 87-87* WBk	Norweg'n	600	152·0	31·9	17·3	RverJohn, N.S.	1870	Kitchen	V 11.84	Skibsselskabet "Almedia"
	1192		Almedina *Dias* W3mSr	Portug'se	355	131·8	29·5	12·5	St.Martinho	1862	M.Candido		F.Ricca
	1193	*	Almendral *Gouyet 85-85* IS	French	678	186·0	30·2	18·3	Birkenhead	1870	Laird Brothers		Ant.Dom. Bordes & ...
	1194	RV	Almina (ex Cato Paulie) *Thore* WSr	Swedish	168	88·5	20·3	12·3	Veendam	1868	K. & J.Wilkens	V 12.86	A.F.Andersson
1535	1195		Almira T. Rowland WSr	American	186	98·4	27·1	9·0	Port Jefferson, N.Y.	1863	T.Rowland		T.W.Rowland
1012	1196		Almira Wooley WSr	American	214	117·0	28·0	9·2	Greenprt, N.Y.	1869			E.E.Rackett
	1197		Almirante *Korfitzen 87-87* WBg	Spanish	179	95·8	23·7	11·9	Barcelona	1854			J.Romera
	1198	V	Almogavar (ex Dobo de Abril, ex E.P. Pout) WBg	Spanish	242	103·0	25·4	14·3	Corunna	1869			Compania Brugm...
159	1199		Almon Bacon WSr	American	113	86·0	27·2	7·0	East Haddam, Conn.	1853			J.Meacham
1307	1200		Almon Rowell WBg	American	287	105·8	27·2	11·0	Lubec, Me.	1859			S.S.Satterthewaite

*=Lloyd's Register; D=Nederlandsche Vereeniging; G=Germanischer Lloyd; I=Registro Italiano; R=Am...; A=Veritas Austro-Ungarico; H=Veritas Ellenico; N=Norske Veritas; V=Bure...

In the 4th Column, C=Composite (iron frame, wood planked); I=Iron; Stl=Steel; W=Wood.

The Dates of Metalling (**Col. 11**) are based on the records of the Books in which Vessels are respectively classed, or on the Reports of the Surveyors of Ll...

Further particulars of Vessels classed by Lloyd's Register will be found in Lloyd's Register Book and Lloyd's Yacht Register.

		Navire / Ancien Nom	Capitaines et dates d'entrée en service des Armateurs actuels et à bord du navire	Matériaux et Gréement	Pavillon	Tonnage	Longueur	Largeur	Creux	Où	Année	Par qui	Date du doublage	Armateurs ou Directeurs-Gérants	Port d'Armement
1201	G	Almuth Catharina		W Sr	German	106	91·3	17·2	8·8	Papenburg	1863	F.W.Meyer		F.W. Meyer Ww.	Papenburg
1202	DV	Almuth Elisabeth (ex ...)	Eekhoff -84	W Bk	German	151	88·6	21·9	11·3	Veendam	1864	K.&J.Wilken		R.G.Eekhoff	Iheringsfehn
1203	✻	Alney (ex "...")	Guiboux	W Bkn	French	213	114·7	25·6	12·8	Gloucester	1879	F. C. Hipwood	Y 1.89	Guiboux	St.Malo
1204		Alnoer		W Bk	Dutch	108	72·8	19·7	10·8					Successors of Sech Achmad bin Abee Baker Amate Ba Fadal	Samarang
1205		Alnoor	M. Assaed ...	W Sr	Dutch	128	96·5	21·0	34·8				Y 1.87	Saïd Alie bin Osman Isnahin	Palembang
1206	V	Alona		W 3m Sr	French	379	127·3	28·5	13·0	Hamburg	1881	J H N Wichhorst	Y 5.87	A.&E.Baux	Marseilles
1207	✻	Aloungpyah	Dyer	W K	British	87	79·4	20·3	10·9	Gt. Yarmouth	1884	Beeching Bros.		T.Kelsall & G.Beeching	Hull
1208	R	Alpha	Horn -88	W Bn	British	221	103·8	26·5	11·7	Lunenbrg.N.S.	1868			Isaac Rodenhiser	Lunenburg,N.S.
1209		Alpha		W Sr	British	137	111·3	22·1	8·4	Gravesend	1865			E.Goldfinch	Faversham
1210		Alpha	Fisher	W Bn	British	133	90·0	20·4	13·0	Workington	1851	C.Lamport		W.R.Kelly	Whitehaven
1211		Alpha	Wheatley	W Sr	British	100	79·3	21·3	9·9	Perth	1868	Perth Sh.Co.		T.Ramsey	London
1212	✻	Alpha		I Sk	British	90	79·4	20·3	10·1	Beverley	1884	Cochrane, Hamilton & Cooper		F.Hellyer	Hull
1213	✻	Alpha (ex Lady Robarts)	Norman 85-85	W Sr	British	87	77·0	20·5	10·5	Salcombe	1865	Vivian		H.G.Norman	Penzance
1214		Alpha		W Bg	German	139	85·3	22·5	12·9	Nexo	1854	J.H.Brandt		C.Petersen	Eckensund
1215	G	Alpha	Schulte -79	W Sr	German	112	83·6	19.9	9·7	Papenburg	1855	S.Beckmann		J.Dieckhaus	Papenburg
1216	N	Alpha	Eliassen -88	W Bk	Norweg'n	477	135·6	31·1	17·3	Laurvig	1866	Dekke	Y 3.88	L.C.Bjelke	Laurvig
1217	N	Alpha	Bidden	W Bk	Norweg'n	377	129·6	29·4	16·0	Namsos	1857			J.Capjon	Drammen
1218		Alpha	Olsen -88	W Bg	Norweg'n	280	114·8	27·9	14·6	Björneborg	1857			A.O.Olsen	Stavanger
1219		Alpha	Torgersen	W Bg	Norweg'n	269	110·1	25·5	13·9	Brevig	1865	G.J.Möe	Y 10.85	T.Torkildsen	Holmestrand
1220	DN	Alpha	Svendsen 86-86	W Bn	Norweg'n	239	112·6	27·8	12·0	Port Medway, N.S.	1865	E.Dolliver	Y 11.84	J.F.Juell	Risör
1221	N	Alpha	Nielsen	W Sr	Norweg'n	177	94·3	23·3	12·1	Tönsberg	18 55/76			O.Nielsen & Co.	Drammen
1222	N	Alpha (ex ...)	Jakobsen 82-82	W Sr	Norweg'n	163	96·5	23·2	9·9	Wiborg	1866			A.Jakobsen	Arendal
1223	N	Alpha	Larsen 84-84	W Sr	Norweg'n	125	93·6	23·4	9·9	Stordöen, Norway	1874	H.K.Haaland		H.K.Haaland	Haugesund
1224	V	Alpha	Olsson 83-83	W Bk	Swedish	336	136·8	26·1	12·1	Hernösand	1873	Fr.Hägglund	Y 8.87	Hernösands Lastings-Aktb (H.Nordlander)	Hernösand
1225		Alpha	Kihlman -86	W Sr	Swedish	140	82·0	22·1	11·7	GamlaLödöse	1870	G.A.Lundgren		C.J.Olsson	Gåsö
1226	R	Alpheta	Holley	W Bn	British	299	121·0	26·8	14·9	Mount Stewart, P.E.I.	1877	E.Coffin	Y 8.86	Geo.H.Taylor	Pt Hawkesbury,N.S.
1227	V	Alpheus Marshall	Kenney -88	W Bk	British	1096	191·3	36·8	22·2	BearRivr,N.S.	1881	R.Marshall	Y 7.88	Alpheus Marshall	Digby,N.S.
1228	V	Alphonse Elisa	Lancelot	W Bk	French	285	102·3	24·6	16·5	Nantes	1864	Bojn & Bertand	Y 8.88	F.Lancelot	Nantes
1229	V	Alphonsine	Pavie -88	W Bn	French	268	98·4	22·6	12·6	Marseilles	1874	Pillard	Y 8.88	A.Dor	Marseilles
1230	DV	Alphonsine Zélie	Bachelier	W Bk	French	326	117·7	24·6	14·8	Nantes	1874	J.Sevestre	Y 1.83	J.Bachelier	Nantes
1231		Alpine	Gray	W Sr	American	104	88·0	23·5	8·8	Bangor,Me.	1839			T.B.Pickering	Deer Island,Me.
1232	NV	Alrana	Olsen 85-85	W 3m Sr	Norweg'n	396	126·3	31·0	14·2	Arendal	1885	Olsen	Y 5.86	O.A.Olsen	Arendal
1233		Alroy	Redmond	W Bn	British	237	116·0	25·8	12·9	P.E.Island	1875	PeakeBros.		W.O.McCormick	London
1234		Alruna		W Bn	British	165	100·8	27·9	9·6	N.Brunswick	1871			W.Wallace	Dublin
1235		Alsace Lorraine (ex Annie ...)	Fèvre	W 3m Sr	French	354	128·2	29·2	16·8	Lussinpiccolo	1872	S.Cattarinich	Y 1.88	C.A.H.Chedru	Fécamp
1236	V	Alsace & Lorraine	...	W Bk	French	610	141·4	29·2	19·4	Bordeaux	1871	Chaigneaufrères	Y 8.86	P.Ch.Grenet	Nantes
1237	V	Alsen	Holz 79-84	W Bk	German	588	148·2	32·0	19·8	Danzig	1865	F.Devrient		A.Gibsone	Danzig
1238		Alster	Pettersson	W 3m Sr	Swedish	127	96·5	21·1	9·4	Alstersvarf	18 57/78			J.P.Pettersson	Gry.Roslagsbro
1239	N	Alsvid (ex John Elias)	Berthelsen	W Bk	Norweg'n	247	110·5	25·3	14·5	Aberystwith	1864	J.Evans		N.E.Berthelsen	Arendal
1240	NV	Alsylva (ex Caroline)	Rull	W Bk	Norweg'n	876	139·0	33·0	22·3	Vegesack	1861	J.Lange	Y 1.87	O.A.Olsen	Arendal
1241	R	Alta	Prints -87	W Sr	British	168	100·0	28·0	10·3	Antigonish, N.S.	1880	Kennedy & McDonald	Y 12.86	Andrew Muir	Halifax,N.S.
1242		Alta V. Cole	Cole	W Sr	American	193	106·0	27·0	9·3	Harrington, Me.	1874			J.E.Brett	New York
1243		Altagracia	Roig	W Bk	Spanish	555	138·1	30·3	17·5	Sackville,N.B.	1864			Jané & Co.	Barcelona
1244		Altai	Lindblad -86	W Bk	Russian	1021	139·8	36·1	20·5	Raumo	1876	J.Urnberg	Y 1.89	J.W.Söderlund	Raumo
1245		Altai	Johansson	W Bk	Russian	397	..	..	..	Kimito	1880			G.A.Wennström	Wardö
1246	✻	Altair	Munro 79-79	I Bk	British	400	140·7	27·2	16·5	Sunderland	1867	Iliff, Mounsey & Co.		T.C.Guthrie	Glasgow
1247		Altamaha (ex ...)	Saunders -88	W Bkn	American	326	139·0	24·6	14·0	Glasgow	18 45/70			E.H.Pray	Savannah,Ga.
1248	NH	Altana	Sideropulo -87	W Bg	Greek	234	100·4	26·7	..	Galaxidi	1868	C.Papapetros	Y 5.84	N.M.Pappapetros	Galaxidi
1249		Altcar	Magrath -87	Stl Bk	British	1283	209·5	36·0	23·6	Liverpool	1864	Jones, Quiggin & Co.		A.Burns	Sydney,N.S.W.
1250		Altha (ex Kaunia)		W Bg	British	259	103·0	25·0	14·0	Bangkok	1864	D.McLean&Co.		Crane Bros.	Singapore

=Lloyd's Register; D=Nederlandsche Vereeniging; G=Germanischer Lloyd; I=Registro Italiano; R=American Record;
=Veritas Austro-Ungarico; H=Veritas Ellenico; N=Norske Veritas; V=Bureau Veritas.

Dans la 4me Colonne, C=Composite (membrure en fer, bordé en bois); I=Fer; Stl=Acier; W=Bois.

Dates du doublage en métal (**Col. 11**) ressortent des livres dans lesquels les navires sont respectivement classés ou des rapports des Inspecteurs du Lloyd's Register.

Pour plus amples renseignements sur les navires classés par le Lloyd's Register, voir le Lloyd's Register Book et le Lloyd's Yacht Register.

Official No.	No. in Book	Registry in which classed	Ship's Name, Material and Rig. Master, and Date of Appointment to present Owner's Service and to Ship. Ship's Late Name.	Flag	Net Tonnage	Length	Breadth	Depth	Built: Where	When	By whom	Date of Metalling	Owners or Managers
	1251	V	Althæa *Rasmussen 87-87* WSr	Danish	111	86·3	20·1	9·8	Svendborg	1878	P.Troensegaard		C.Rasmussen
77796	1252	R	Althea *Melançon* (CONFERENCE) WSr	British	143	89·4	25·0	9·6	Pt.Gilbert,NS	1878	C.E.Everitt		A.Melançon
	1253	N	Althea *Wexholtt* WBk	Norweg'n	492	123·4	28·5	19·3	Medford, Mass	1847		Y 11.85	J.Börresen
93385	1254	✻	Altmore *Weeks 79-87* Stl S	British	1716	265·2	39·0	23·2	Pt.Glasgow	1887	R. Duncan & Co		Thom&Cameron
47858	1255	✻R	Alumbagh *Bagwell 88-88* WS	British	1138	193·0	36·2	23·8	Sunderland	1863	J.Laing	Y 3.86	J.Herron & Co.
50297	1256	✻	Alumina *Jones 83-83* I Bk	British	715	178·3	29·3	18·7	Pallion	1864	W.Doxford&Co.		C.H.Rees
	1257		Alvaro (ex Jose Pinto) *Freitas* WSr	Brazilian	158	..	..	..	Villa do Conde	1877		[illegible]	D.G. de Freitas
	1258	V	Alves *Gonçalves* W3mSr	Portug'se	299	126·6	24·9	11·8	Villa de Conde	1873	M.G.Rodrigues	Y [illegible]	L.Alves Rodrigues&Co
61577	1259	V	Alvin Kelly *Kelly* WBn	British	327	121·3	29·1	12·7	Shelburne,NS	1874	J.Kitchen	[illegible]	Thos.Johnson
	1260	G	Alwine *Fischer 85-* WBg	German	262	113·8	24·6	13·4	Stralsund	1863	J.P.Juhl		H.L.Juhl
	1261	GV	Alwine (LOST) *Ulrich 86-* WBn	German	215	101·0	25·6	12·5	Harburg	1866	G.Renck Sen.	Y 12.86	J.J.Rahmann
	1262	G	Alwine *Klaver -88* WKf	German	118	85·5	19·7	10·3	Papenburg	1853	H.W.Meyer		H.W.Meyer
	1263	G	Alwine *Andersson -84* WBk (ex Sarah F. / [illegible])	Russian	334	112·0	26·5	17·0	Sunderland	1855			F.W.Andersson
86526	1264	✻	Alwyn *Holten 85-85* WK	British	67	73·2	19·3	9·1	Plymouth	1885	Watson&Fox		W.H.Sarah
	1265		Alzira (ex Alegria) WBn	Brazilian	165	..	..	..	Laguna [illegible]	18[illegible]		Y [illegible]	J.J.P.de Ulyssea
	1266	V	Amable Antonia (ex Adriana) *Mas -88* WBg	Spanish	320	101·0	28·7	17·2	Naples	1856	E.Mauro	[illegible]	J.Ballet
	1267		Amable Teresa WBg	Spanish	193	95·1	27·8	13·3	Barcelona	1853			Goday y Gual
	1268	N	Amal *Tellefsen -88* WBk	Norweg'n	467	133·2	30·2	17·2	Arendal	1874	S.Johnson	[illegible]	NicolayOlsen
	1269		Amalia W3mSr	Brazilian	173	..	..	..					R.P.Marques & Co.
	1270		Amalia (ex Isaura) *Ribeiro* W3mSr	Brazilian	132	..	..	..		1869		Y [illegible]	R.P.de Magalhaes
	1271		Amalia *Petersen* WBk	Chilian	215	111·8	25·2	..	Boston,Mass.	1857			Sociedad Chilena de Fundiciones
3308	1272	I	Amalia (ex Angelo Ferrari) *Cela -88* WBn	Italian	420	138·7	26·4	16·8	SestriPonente	1878	A.Carlevaro	Y [illegible]	A.Chiappara
	1273		Amalia (ex Florida) *Andersen* WBk	Norweg'n	353	123·0	27·0	15·4	Salisbury N.E.	1851			H.H.Pettersen
	1274		Amalia (STRANDBO) *Gjertsen* WSr	Norweg'n	148	93·0	23·5	10·4	Wiborg	1863			A.R.Bull
	1275		Amalia *Silva 86-86* WBk	Peruvian	123	92·8	23·3	10·3	UnitedStates				J.Silva
	1276		Amalia *Hovi* WSr	Russian	110	76·3	21·0	10·5	Sikkjarvi	1877	S.Havie		J.Waln
	1277	V	Amalia (ex L. R. & H.) *Torres 77-77* WBk	Spanish	419	123·0	28·2	17·7	Sunderland	1867	Reay&Naisby	Y [illegible]	J.A.Gorocica
	1278		Amalia (ex Analyst) *Jimenez* WBk	Spanish	347	111·2	27·2	16·9	Sunderland	1863	Thompson	Y [illegible]	F.Arsuegind
	1279		Amalia WBg	Spanish	145	91·5	24·7	11·9	Barcelona	1855			Pablo Bertran
	1280	R	Amalia *Soderblom -88* WBk	Swedish	287	97·1	22·6	17·0	Shields	18 42/69			J.E.Olsson
	1281		Amalia *Hansson* W3mSr	Swedish	216	108·2	23·3	12·1	Timmernabben	1873	C.Bring		M.Hansson
	1282		Amalia *Nylander 75* WSr	Swedish	102	86·3	18·8	8·4	Nystad	1858			F.G.Nylander
	1283		Amalia WBg	Turkish	214	..	..	..	Russia	1865			Col.Mehemed Bey
	1284		Amalia G. I	Argentine	158	120·0	24·9	8·0	S.Nicolas de los Arroyos	1885	C.Geddes		Green & Johnson
978	1285	I	Amalia Gargiulo *Jaccarino* WBk	Italian	374	123·0	27·0	18·0	Alimuri	1873	- Mauro	Y [illegible]	A.Jaccarino & Co.
	1286		Amalie (ex Waverley) *Langlie* WBk	Norweg'n	786	150·2	31·8	22·5	Richmond,US	1857			Otto Jacobsen
463	1287	V	Amalie *Kukken 85-85* W3mSr	Russian	327	134·0	30·3	12·8	Koltzen	1885	Andersohn		G.Schmohr
	1288	G	Amalie *Lind -88* WBg	Swedish	206	87·4	23·4	14·1	Wismar	1842	I.C.Maren		C.M.Veklund
	1289		Amalie Charlotte *Olsen* WBk	Norweg'n	273	96·7	27·6	14·9	Kalmar	1825			W.Foss
	1290	V	Amalina *Jonassen -88* WBg	Norweg'n	181	102·8	23·7	12·5	Memel	1859	B.Pieper		Th.Kleiberg
	1291		Amalthea *Deutschbein -77* WBk	German	356	115·6	29·2	17·4	Memel	1863	B.Pieper		O.Deutschbein
	1292		Amalthea *Jansson* WSr	Russian	279	..	..	..	Nerpes	1870			G.R.Lindman
71750	1293	✻	Amana *Herbert 75-75* I S	British	1299	239·2	36·1	21·5	Glasgow	1875	A.Stephen&Sns		Smith & Service
30763	1294		Amanathoola WBg	British	155	90·1	20·9	12·5	Cochin	1865			Hadjee J.H.Moosa
	1295		Amanatool Rachman *Said Alie Albar* WBk	Dutch	235	107·2	21·3	1·9					Said Achmad bin Albar [illegible]
	1296		Amanatoor Rachman (ex [illegible]) *Said Alie Jusom* WBg	Dutch	156	86·3	20·5	..	Grisee,Java	1856		Y [illegible]	Said Hussein bin Achmad Alaydroos
83278	1297	V	Amanda *Scott 77-83* WBk	British	1073	184·0	37·7	21·3	Noel,N.S.	1883	O.O'Brien	Y [illegible]	O.O'Brien
64915	1298	R	Amanda *Clear -88* W3mS	British	350	123·0	28·3	12·9	Prince port,NS	1873	M.C.Norris	Y [illegible]	W.J.Menendez
53692	1299		Amanda *Pearson -82* WBn	British	177	99·8	25·4	12·5	P.E.Island	1865	Ross		W.B.Cooper
	1300												

✻ = Lloyd's Register; D = Nederlandsche Vereeniging; G = Germanischer Lloyd; I = Registro Italiano; R = Ame[rican]
A = Veritas Austro-Ungarico; H = Veritas Ellenico; N = Norske Veritas; V = Bu[reau Veritas]

In the 4th Column, C = Composite (iron frame, wood planked); I = Iron; Stl = Steel; W = Wood.

The Dates of Metalling (Col. 11) are based on the records of the Books in which Vessels are respectively classed, or on the Reports of the Surveyors of L[loyd's]

Further particulars of Vessels classed by Lloyd's Register will be found in Lloyd's Register Book and Lloyd's Yacht Register.

No.	Reg.	Nom. Capitaine (Ancien Nom)	Matériaux et Gréement	Pavillon	Tonnage net	Longueur	Largeur	Creux	Construit Où	Année	Par qui	Date de doublage en métal	Armateurs ou Directeurs-Gérants	Port d'Armement
1301		Amanda *Pritchard*	WBg	British	154	96·7	23·5	12·2	Portmadoc	1864	Owens		J.W.Taylor	Carnarvon
1302		Amanda (ex [illegible])	WBk	Brazilian	461	138·0	26·2	17·7	Maryport	1858		[illegible]		
1303	D	Amanda *Geertsema*	WBg	Dutch	258	103·3	26·7	15·0	Rostock	1854			B.H.Geertsema	Delfzijl
1304	GV	Amanda *Harms 88-*	WBn	German	225	106·1	25·2	12·9	Harburg	1871	G.Renck	Y 2.85	H.H.Witt	Hamburg
1305		Amanda *Eriksson*	WBk	Russian	382	176·0	30·0	16·0	Harstrom	1865	Beck		M.Toverlund	Mariehamn
1306		Amanda *Sundlöf*	W3mSr	Russian	282	..	..	..		1880			M.Eriksson	Finström
1307		Amanda *Söderlund*	WBn	Russian	243	109·3	28·2	12·3	Lökö	1879	J.Palmros		J.H.Malen	Lökö
1308		Amanda *Henricksson 87-87*	WSr	Russian	222	102·0	29·0	..	Kimito	1873	Kronström		C.G.Andersson	Abo
1309		Amanda *Sjölund*	WSr	Russian	154	98·1	27·9	11·2	Lavia	1870	J.Petillo		M.Eriksson	Fögiö
1310		Amanda *Pettersson*	WSr	Russian	121	..	..	..		1872			K.E.Pettersson	Brändö
1311	V	Amanda *Johansson* (ex J. A. Pierce)	WBg	Swedish	281	120·2	25·8	12·7	Annapolis,N.S	1863			N.Olsson	Glimsås,Oroust
1312	V	Amanda *Andersson*	WBg	Swedish	220	101·3	23·7	12·5	Oroust	1872	A.Samuelsson	Y7.84	N.M.Pahlsson	Helsingborg
1313	R	Amanda C. Parker *Lee*	W3mSr	American	449	138·6	35·1	14·2	May'sLdingNJ	1884	S.Gaskill		J.S.Lee	Somer'sPnt.NJ
1314	V	Amanda & Elisabeth *Ahler Sickt*	WBk	German	364	135·9	29·0	14·6	Kiel	1866	A.Bock	Y 3.87	Hansing & Co	Hamburg
1315		Amandina (ex [illegible]) *Bartin*	WBn	Italian	219	97·1	26·1	15·8	Loano	1861	B.Amico		O.Del Buono	P.Ferraio
1316	N	Amandus *Pettersen*	WBk	Norweg'n	331	116·3	28·7	14·5	Tvedestrand	1865	J.Larsen	Z 1.80	P.A.Pettersen	Risör
1317	V	Amarant *Thorbjørnsen* [79-79] (ex Ansleigh)	WBn	Norweg'n	280	116·9	25·4	12·8	Georgetown, P.E.I.	1868	Macdonald	Y 5.85	Holdt & Isachsen	Stavanger
1318		Amaranth *Crangle 88-88*	WBk	British	1198	182·5	36·7	23·7	Quebec	1865	Gilmour	Y 11.84	Amaranth Ship Co. Ld. (J. S. Wright & Co.)	Belfast
1319	✻	Amaranth *Burton 74-87*	WSr	British	120	94·3	23·4	11·7	Padstow	1886	Cowl&Sons		W.C.Phillips	Padstow
1320	V	Amaranth *Bertine 83-83*	WBk	German	934	178·6	35·5	21·5	Hammelwarden	1883	W.Hagens	Y 5.84	D.Haye	Brake
1321	DG	Amaranth *Maass -87*	WBk	German	898	161·7	33·7	21·6	Vegesack	1864	H.F.Ulrichs	Y 6.84	H.Maass	Rostock
1322	G	Amaranth *Schumann* [73-73]	WBg	German	243	95·7	27·1	13·7	Ueckermünde	18$\frac{52}{74}$	L.Wittenberg		Edwin Schültze	Swinemünde
1323	G	Amaranth *Kramer -87*	WBn	German	198	93·9	25·9	12·8	Rostock	1866	G.W.Tesdorpf		S.Stein	Hamburg
1324	N	Amaranth *Berntsen -88*	WBk	Norweg'n	356	128·0	28·2	15·2	Geestemünde	1861	Schau & Oltmanns	Y7.86	G.Thorbjörnsen	Arendal
1325	V	Amaranth *Hoppstrom 87-87*	WBk	Swedish	279	126·9	24·3	12·0	Timmernabbn	1874	C.Bring	Y 2.83	H.Andersson	Nyhamn
1326		Amarapoora	I Sk Barge	British	538	171·5	25·7	5·9	Rangoon	1871	IrrawaddyCo.		J.Heap & Sons	Glasgow
1327		Amasia	WSr	Russian	130	73·0	19·0	11·0	NeuSalis	1879	J.Jehkabson		G.Obsoling	Haynasch
1328		Amateur	WBk	British	239	118·0	21·3	13·8	NovaScotia	1856			Otago Harbour Board	Dunedin.N.Z.
1329		Amaury (WUNK) *Le Bret*	WBg	French	161	79·7	23·9	12·9	Redon	1857			C.Le Bret	Vannes
1330	✻	Amazon *McLaren 86-86*	I Bk	British	1998	286·7	42·5	24·1	Glasgow	1886	Barclay, Curle & Co.		R.Hill	Greenock
1331		Amazon	WBn	British	198	99·3	24·5	11·7	Parrsboro,NS	1867			J.H.Beatty	Sydney,C.B.
1332		Amazon *Baker*	WBk	American	218	116·2	25·8	10·6	Baltimore	1853	A.Flannigain		J.C M.Johnston	Baltimore
1333	N	Amazon *Bristendahl*	WBk	Norweg'n	351	132·3	30·5	12·7	Windsor,N.S.	1866	Lugatt		Chr.Steen	Holmestrand
1334	N	Amazon *Christiansen*	WBk	Norweg'n	339	118·1	27·8	15·0	Grimstad	1872	A.Bentsen	Y1.88	J.E.Einersen	Grimstad
1335	V	Amazon *Nordbäck*	WBk	Swedish	418	145·6	26·1	14·1	Gefle	1882	O.A.Brodin	Y8.87	P.C.Rettig & Co.	Gefle
1336	GV	Amazon *Andersson -82*	WBk	Swedish	364	123·0	26·6	16·5	Gothenburg	1855	A.Ahlberg	Y 3.84	A.B.Elfversson	Gothenburg
1337		Amazona *Roldes -87* (ex Tenedral)	WBkn	Spanish	449	128·6	29·1	17·6	Sestri Ponente	1866		Y 12.86	Pablo,E.Sur & Mir	Barcelona
1338	V	Amazone *Meyer* (ex Skiddaw)	WBk	German	356	130·0	26·5	16·3	Workington	1852	C.Lamport	Y5.86	F.Böttcher	Stralsund
1339		Amazone *Konow 80-*	WBk	German	324	118·0	28·1	16·2	Rostock	1857	W.Zeltz		F.Schultze	Rostock
1340	V	Amazone *Meyer 85-85*	WBg	German	316	124·5	28·7	13·5	Brake	1876	F.Nicolai	Y 12.88	J.H.Nicolai	Brake
1341	DN	Amazone *Larsen*	WBg	Norweg'n	176	106·0	26·3	9·7	Drammen	1873	H.F.Gjessing		H.F.Gjessing & Co.	Drammen
1342		Ambassadeur	WBn	French	133	..	..	..	Bastia	1867			Mattei frères	Bastia
1343	✻R	Ambassador *Guthrie 67-87*	CBk	British	692	176·0	31·3	19·1	London	1869	Walker	Y5.88	G.Milne & Co.	Aberdeen
1344		Ambassadress	WSrYt	American	221	131·6	28·4	12·3	ChyIsld.N.Y.	1877	D.Carll		N.Thayer	New York
1345	H	Ambelos *Papadaria -84*	WBg	Greek	276	100·0	24·2	14·3	Syra	1876		Y 1.88	T.Vavanatis	Galaxidi
1346	H	Ambelos *Logothetis*	W3mSr	Greek	185	..	..	..	Galaxidi	1871	A.Papapetros	Y 7.85	E.Logothetis	Galaxidi
1347		AmberNymph *Clarke 85-86*	WBg	British	254	129·5	25·6	12·8	Jersey	1859	Clarke	Y 4.82	Rider,Son&Andrew	London
1348		Ambrose Light *Jones*	WBn	American	215	110·8	27·1	10·2	Waldoboro'Me	1857		Y 9.86	G.W.Brown	New York
1349		Ambrosina	WBn	Brazilian	131	..	..	..	Fanö	1870			A.O.Guimaraes	Aracaya
1350	V	Ambrosio Sevastopoulos *Vergottis*	WBk	Greek	340	149·9	24·8	12·7	Bastholmen	1874	J.Johansson	Y 2.88	Sevastopoulos & Sons	Syra

✻=Lloyd's Register; D=Nederlandsche Vereeniging; G=Germanischer Lloyd; I=Registro Italiano; R=American Record;
✻=Veritas Austro-Ungarico; H=Veritas Ellenico; N=Norske Veritas; V=Bureau Veritas.

Dans la 4me Colonne, C=Composite (membrure en fer, bordé en bois); I=Fer; Stl=Acier; W=Bois.

Les Dates du doublage en métal (Col. 11) ressortent des livres dans lesquels les navires sont respectivement classés ou des rapports des Inspecteurs du Lloyd's Register.

Pour plus amples renseignements sur les navires classés par le Lloyd's Register, voir le Lloyd's Register Book et le Lloyd's Yacht Register.

1 Official No.	2 No. in Book	3 Registry in which classed	4 Ship's Name. Material and Rig. Master, and Dates of Appointment to present Owner's Service and to Ship. Ship's Late Name.	5 Flag	6 Net Tonnage	7 Dimensions: Length	Breadth	Depth	8 Built: Where	9 When	10 By whom	11 Date of Metalling	12 Owners or Managers
51631	1351		Ambulant *Tillman* WBg	British	191	102·5	20·3	11·9	Hoogezand	1870			Hossen Joomah
	1352	V	Amédée *Lehoerff -81* WSr	French	123	89·6	20·0	10·5	Granville	1878	L.Juhenne		Lehoerff
31398	1353	V	Amedeo *Capurro -86* WBk (ex Rubino, ex ...)	Italian	582	119·2	31·7	20·0	Recco	1869	P.Rolla	Y 1.86	N.Pescio
73608	1354	RV	Ameer *McKenzie -84* WBk	British	770	165·5	35·2	18·8	RiverJohn,NS	1879	J.Kitchin	[illegible]	Alex.C.Macdonald
74510	1355	✻	Amelia *Burrell 85-85* IBkn	British	294	145·3	25·4	12·4	Greenock	1876	J.E.Scott		Goodyear & Co.
11855	1356		Amelia *Burrell* WBg	British	236	109·2	25·8	12·7	RiverJohn,NS	1859			C.McLennan
35864	1357		Amelia WSr	British	118	85·0	23·0	10·4	St.Andre	1867			D.Desjardins
1074	1358	R	Amelia *Newhall* WBkn	American	378	153·0	37·0	10·0	CoosBay,Oreg	1870	R.Murray		S.C.Holmes
248	1359	A	Amelia *Cosulich* WS	Austrian	634	151·2	29·8	20·4	Fiume	1863	G.Brazzoduro	Y 8.86	Cas.Cav.Cosulich
	1360		Amelia WBk	Brazilian	338	..	..	..					J.J.Martins
	1361		Amelia WBn	Brazilian	190	..	..	..					A.C.dos Santos
	1362		Amelia *dos Reis* W3mSr (ex Arabella Rilda)	Brazilian	182	..	..	..	Bonces	1866		Y 86	R.P. de Magalhaes
	1363		Amelia WSr	Brazilian	140	..	..	..					Catharino M.de Souza
	1364		Amelia *da Silva* WBg	Brazilian	117	..	..	..	Westerwik			Y 9.84	A.R.Guimaraes
	1365	V	Amélia *Lepord 72-81* WSr	French	122	78·1	21·6	11·2	Nantes	1871	Sevestre		P.Ch.Grenet
	1366	V	Amelia *Warneke -3-83* WBk (ex Villeluna)	German	939	158·2	35·2	23·0	Bowdingham,[Me.	1866	R.Purrington	Y 1.84	J.D.Bischoff
	1367		Amelia WBk		474	129·6	31·0	19·3	Spotorno	1866	F.Cadenaccio	Y 1.84	
	1368	N	Amelia *Salvesen -85* WBk	Norweg'n	393	120·9	30·8	16·4	Helsingfors	1856			K.Finsdal
	1369		Amelia A. *Tejera* WBk (ex Susana, ex Concilias)	Spanish	350	117·1	25·2	15·0	Bilbao	1871	S.deArena		S.Aquiar
733	1370		Amelia G. Ireland *McDaid* W3mSr	American	270	124·0	30·0	12·0	Wicomico,Md.	1866			G.H.Vilas
161	1371		Amelia H. Leaming *Lee* WSr	American	165	97·0	27·4	9·0	Dennisville,NJ	1864			M.A.Ketchum
105313	1372		Amelia Hearn *Williams* WSr	American	108	94·2	23·9	9·0	Laurel,Del	1873			W.T.Moore
106237	1373		Amelia P. Schmidt *Pashley* W3mSr	American	266	128·9	32·2	8·9	DennisvilleNJ	1883	FrankWentzell		Frank Pashley
	1374		Amélie *Belin* WBg	French	145	79·4	23·4	12·1	Granville	1857	Gombert		A.Lemoine
	1375	V	Amélie *Moisson* WLr	French	104	74·4	21·3	13·4	Méans	1865	Mahé		C.Legloahec
	1376	N	Amelie *Henriksen 71-74* WBg	Norweg'n	307	114·2	27·2	13·0	Geffe	1866	J.A.Bang	Y 11.86	T.M.Wiel
	1377	N	Amelie *Rasmussen 85-85* WSr	Norweg'n	104	79·8	20·9	10·6	Kalmar	1854			R.Bugge
	1378	V	Amelie *Törnblom* WBk	Russian	495	136·4	27·6	16·9	Hammelwrdn.	1869	Menke& Botter	Y 86	V.Mannelin
76492	1379	✻D	America *Adams* IS	British	1236	229·9	35·8	20·9	Dumbarton	1877	A.McMillan&S.		Finn. Main & Montgomery
105430	1380	V	America *Gibson -87* WS	American	1952	232·8	43·1	28·0	QuincyPt.Mas	1874	Geo.Thomas	Y 86	C.Goodall
105732	1381		America *Armstrong* WBk	American	338	137·0	31·0	12·8	Madison,Conn	1877	Miner & Son	Y 11.82	H.Trowbridge's Sons
	1382		America WBk	Brazilian	134	..	..	..					J.J.A. e Silva
1149	1383	IV	America *Ferrari* WBk	Italian	695	152·8	30·5	20·5	Castellammare	1877	Trambarulo	Y 86	F.Starace
	1384	V	America WBk		109	87·3	24·8	8·5	Cassano	1865	A.Castellano	[illegible]	
	1385	V	America *Soares* WS	Portug'se	903	181·7	33·5	19·0	Oporto	1867	C.M.da Silva Santos	Y 1.86	[illegible] Portuense
	1386		America (ex Cabot) *Yzaguirre* WBk	Spanish	603	137·8	29·9	19·0	NwGlasgowNS	1861	Carmichael	Y 12.86	F.Sola
106077	1387		American Boy *Nelson* WSr	American	174	105·6	31·5	8·5	Seabeck,Wash.	1882			S.H.Harmon
830	1388		American Eagle (BARGE) WBk	American	290	118·1	30·0	11·8	Jacksonville,Fl	1858			W.B.Smith
1368	1389		American Eagle (WRECKED) WSr	American	140	99·0	20·0	8·0	WilmingtnDel	1864			
105399	1390		American Girl WSr	American	214	119·9	31·4	9·3	PLudlow,Wsh	1875	Hall Bros.		H.Moore
105600	1391		AmericanTeam *Handy* W3mSr	American	358	121·2	31·3	15·3	Lyme,Conn.	1875			S.N.Handy
90486	1392		American Union *Tinsford* WBg	British	263	110·5	26·1	10·7	BrewerMeU.S.	1860			Robert H.Cann
	1393		Americano WBn (ex Myrtle, ex Speedwell)	Argentine	199	103·3	24·3	13·0	P.E.Island	1875	McLaine	[illegible]	Eduardo Oliver & Em-Dupresso
850	1394	R	Americus *Patterson* WBg	American	384	121·3	29·4	16·4	Brooklyn,N.Y	1865		[illegible]	Swan & Son
707	1395		Americus WSr	American	284	109·4	27·0	9·0	Hoboken,N.J.	1854			M.J.Dailey
	1396	DNV	Amerika *Meling -78* WBk	Norweg'n	820	185·1	36·2	19·5	Silvea,Sweden	1874	C.P.Aberg	[illegible]	T.Meling
	1397	N	Amerika *Paulsen* WBk (ex Admiral Brommy)	Norweg'n	436	130·5	29·0	15·8	Bremen	1860	Gebrüder Gross	[illegible]	W.Wilhelmsen
	1398	N	Amerika *Tafle -87* WBk	Norweg'n	397	113·5	28·8	18·8	Boston				L.H.Christiansen
	1399	DV	Amete *Winther 81-84* W3mSr	Danish	292	122·0	25·2	13·5	Nordby,Fanø	1882	S.Abrahamsen		Act. Selskab. (P.N. Winther)
77740	1400	✻	Amethyst *Thomas 80-87* IBkn	British	194	128·2	21·6	11·6	Whitehaven	1876	WhthvnS.B.Co.		G.Nelson & Sons

✻=Lloyd's Register; D=Nederlandsche Vereeniging; G=Germanischer Lloyd; I=Registro Italiano; R=Ame[rican] A=Veritas Austro-Ungarico; H=Veritas Ellenico; N=Norske Veritas; V=Bur[eau]

In the 4th Column, C=Composite (iron frame, wood planked); I=Iron; Stl=Steel; W=Wood.

The Dates of Metalling (Col. 11) are based on the records of the Books in which Vessels are respectively classed, or on the Reports of the Surveyors of Ll[oyd's]

Further particulars of Vessels classed by Lloyd's Register will be found in Lloyd's Register Book and Lloyd's Yacht Register.

No.	Classe	Nom	Capitaine	Gréement	Pavillon	Tonnage	Longueur	Largeur	Creux	Construit: où	Année	Constructeur	Col. 11	Armateurs ou Directeurs-Gérants	Port d'Armement
1401	DV	Amicitia	v. d. Heulen -81	I Bk	Dutch	[illegible]	[illegible]	[illegible]	[illegible]	[illegible]	1885	Rijkee & Co.		P. van der Hoog	Krimpen a/d Lek
1402		Amicitia		W Bg	German	233	103·2	25·7	13·0	[illegible]	1848	O. Preizer		P. Grampp	Rostock
1403	V	Amicitia	Thomsen -87	W Bg	German	[illegible]	[illegible]	[illegible]	11·[illegible]	[illegible]	1860	M. Smidt		G. & H. Aden	Jemgum
1404	DN	Amicitia	Angell -82	W Bk	Norweg'n	585	152·0	32·2	18·5	SestriPonente	1875	Debarbieri & [Celesia]	[illegible]	Bergh & Helland	Bergen
1405	IV	Amicizia	[illegible]	W Bk	Italian	556	137·8	29·6	20·4	Lussinpiccolo	1875		[illegible]	Vedova Trapani	Castellammare
1406	I	Amicizia	Maragli -87	W Bk	Italian	545	141·0	31·2	20·5	[illegible]	1870	N. Cerruti	19.86	Marengo & Laviosa Eredi fu F.C.	Savona
1407		Amicizia (ex Fortuna & Concilia)	Marino	W Bk	Italian	418	126·9	30·0	19·2	Varazze	1867	N. Inosci	[illegible]	F. Marini & M. Raudeck	Naples
1408	I	Amicizia	Matarazzo	W Bn	Italian	153	87·6	24·0	12·[illegible]	[illegible]	1873	A. De Russo	[illegible]	G. Matarazzo	Gaeta
1409		Amicizia (ex Agostino & Federico)		W Bn		115	80·7	22·8	11·[illegible]	[illegible]	1864				
1410	R	Amicus	Harper -88	W Bk	British	516	153·7	28·9	17·7	[illegible]	1856		[illegible]	N. Rosa	Quebec
1411		Amie	Hermansson 79-	W Sr	Russian	201	95·6	29·7	11·9	[illegible]	1870	T. Lindqvist		K. H. Mattsson	Nagu
1412		Amie	Eriksson -80	W Sr	Swedish	199	92·5	25·9	12·0	[illegible]	1838			G. Holmberg	Vätö, Rslgn.
1413	V	Amilcare	Mazzocchi -87	W Bg	Italian	253	98·1	25·9	15·9	[illegible]	1857	A. Castellano	[illegible]	S. Mazzochi	Rimini
1414	V	Amilhujo	Riewerts 79-86	W Bn	German	146	91·7	23·9	11·0	[illegible]	18^{72}_{75}			L. Heymann	Wyk. Föhr
1415		Amiral de Mackau	Tréhet	W Bk	French	223	117·1	22·3	13·1	St. Vaast	1863	Costey & fils	[illegible]	Truscat	Havre
1416		Amisade		W Bn	Brazilian	225	..	..	..					J. A. S. Pereira	Desterro
1417		Amisade		W Sr	Brazilian	142	..	..	..					C. V. Vianna	Rio de Janeiro
1418		Amistad (ex Restaurador)	Roses	W Bg	Spanish	172	81·6	25·7	12·6	Barcelona	1850		V 12.85	Carlos Rosés	Barcelona
1419		Amistad	Villalonga	W Bn	Spanish	158	85·3	23·0	12·1	[illegible]	1865			B. C. Ferrer	Majorca
1420	V	Amitié	Le Rouzic	W Bg	French	158	81·1	23·6	12·0	[illegible]	1868	Rio fils		J. M. Le Rouzic	Auray
1421	V	Amitié	Germain -87	W Bg	French	114	82·3	22·9	10·2	Fécamp	1878			Beust père & fils	Granville
1422		Amizade (ex Lucie)		W Bk	Portug'se	349	111·1	23·3	16·7	Belgium				A. M. Dos Santos	Lisbon
1423		Amizade	Rocha	W Bg	Portug'se	269	110·8	25·6	12·1	Sundsvall	1867			C. J. Rodrigues & Co.	Oporto
1424		Ammonia		W Bg	Turkish	142	..	..	..	Russia	1876			Kerestah-tuzeri Hadji Seid Ahmed Effendi	Constantinople
1425	AV	Amor	Gaulich -87	W Bk	Austrian	504	135·5	29·5	17·0	Lussinpiccolo	1869	N. Martinolich	[illegible]	G. G. Premuda	Lussinpiccolo
1426	V	Amor	Hansen 83-89	W Bn	Danish	187	103·6	23·2	11·8	Marstal	1879	J. J. Bager		E. E. Petersen	Marstal
1427	V	Amor	Jensen	W Bn	Danish	160	94·2	22·5	10·8	Troense	1869	C. R. Møller	[illegible]	C. R. Møller	Svendborg
1428	GV	Amor	Kramer -75	W hm Sr	German	227	104·5	24·7	12·1	Papenburg	1875	R. Beckmann	[illegible]	A. Beckmann	Papenburg
1429	I	Amor	Rotkin -88	W Bk	Italian	685	160·7	30·8	21·8	Venice	1876	G. Zanon	[illegible]	F[lli] Vianello	Venice
1430	N	Amor	Møller 87-87	W Bk	Norweg'n	418	128·0	30·1	16·9	Risör	1876	Nils Olsen	[illegible]	Tellef Jensen	Kragerö
1431	V	Amor	Svendsen 87-87	W Bn	Norweg'n	245	100·0	23·2	12·1	Brake	1869	J. Oltmann's [Wwe.	[illegible]	Holdt & Isachsen	Stavanger
1432	N	Amor	Hansen -85	W Bg	Norweg'n	227	105·1	26·7	12·1	Arendal	1863	J. Amonsen		L. Tobiasen	Farsund
1433	V	Amor	Brockman	W Sr	Swedish	158	95·1	21·5	11·6	Oscarshamn	1880	P. Petersson	[illegible]	P. H. Hoffmann	Oscarshamn
1434		Amor	Kindberg 77-85	W Sr	Swedish	132	83·0	20·3	10·3	Oldenburg	1847			F. Nyström	Slite, Gotland
1435		Amor Divino		W Sr	Brazilian	171	..	..	..					T. A. N. de Brito	Rio de Janeiro
1436	R	Amorette	Charles -88	W Bg	British	265	122·0	26·2	14·5	P. E. Island	1872	M·Dougall	V 10.85	J. J. Jacobs & Co.	Swansea
1437	V	Amorosa	Arvidsson	W Bn	Swedish	199	97·1	23·0	12·6	Torskog	1877	P. Larsson		B. Samuelsson	Bäcken, Oronst
1438		Amorsios		W Bg	Turkish	275	..	..	..	Scio	1871			Leonaki Manol	Scio
1439		Amos Falkenburg	Green -86	W Sr	American	172	105·0	28·4	8·6	Fork'dRv. N.J.	1853			C. D. Luther	Fall River Mass.
1440	✻	Amoy (ex City of Amoy)	Crouch 87-87	I S	British	994	209·7	32·4	21·4	Glasgow	1863	Barclay, Curle [& Co.		Bramwell & Gardiner	Glasgow
1441		Amparo	Cordeiro	W Sr	Brazilian	143	..	..	..	Campos	1866		V 11.79	T. F. Vianna	Laguna
1442		Amparo		W Bg	Spanish	171	88·6	22·9	10·5	Mahon, Minca	1845			F. Oliveros	Havana
1443		Amphion		W Bg	British	125	..	..	..	Workington	1804			Mrs. J. Richardson	Whitehaven
1444	N	Amphion	Nyegaard -87	W Bk	Norwegian	295	120·3	29·2	14·1	[illegible]	1842			Rm. Nyegaard	Fredriksstad
1445		Amphion	Ström	W Sr	Russian	130	..	..	..	Lökö	1857			F. W. Ström	Raumo
1446	A	Amphissa		W Sr	Greek	104	75·1	21·0	..	Syra	1880	A. Livadara	[illegible]	G. G. Tselentis	Argostoli
1447	✻	Amphitrite	Anderson -87	I S	British	1708	264·9	39·5	23·4	Port Glasgow	1882	R. Duncan & Co.		C. S. Caird	Greenock
1448		Amphitrite		W Sr	British	182	96·0	24·0	10·8					W. Summerbell	Sydney, N.S.W.
1449		Amphitrite	Wright -81	W Bk	British	129	116·0	22·0	7·7	Jervis B., NSW	18^{63}_{75}			W. T. Yeager	Sydney, N.S.W.
1450															

[A = Veritas Austro-Ungarico; ✻ =]Lloyd's Register; D = Nederlandsche Vereeniging; G = Germanischer Lloyd; H = Veritas Ellenico; I = Registro Italiano; N = Norske Veritas; R = American Record; V = Bureau Veritas.

Dans la 4me Colonne, C = Composite (membrure en fer, bordé en bois); I = Fer; Stl = Acier; W = Bois.

Les Dates du doublage en métal (Col. 11) ressortent des livres dans lesquels les navires sont respectivement classés ou des rapports des Inspecteurs du Lloyd's Register.

Pour plus amples renseignements sur les navires classés par le Lloyd's Register, voir le Lloyd's Register Book et le Lloyd's Yacht Register.

1 Official No.	2 No. in Book	3 Registry in which classed	4 Ship's Name. Material and Rig. Master, and Dates of Appointment to present Owner's Service and to Ship. Ship's Late Name.	5 Flag.	6 Net Tonnage.	7 Dimensions. Length	Breadth	Depth	8 Built. Where.	9 When	10 By whom.	11 Date of Metalling	12 Owners or Managers.
[illegible]	1451	✳	Amphitrite WSr Yt	British	101	100·2	19·1	12·3	Gosport	1887	Camper & Nicholson	C [illegible]	Lt.Col.A.D.McGregor
	1452		Amphitrite WBk	Brazilian	517	..	..	..	Lökö				J.R. da Cunha
	1453	V	Amphitrite *Rowehl 81-84* WS (ex Defiance)	German	1814	251·3	46·3	26·2	Kennebunk	1875	N.Thompson	Y 7.84	Gildemeister & Ri[...]
9	1454	A	Amphitrite *Beristianos* WBg	Greek	195	98·4	25·5	..	Galaxidi	1875	Papapetros	Y 8.88	Fratelli Liaglio
41	1455	A	Amphitrite *Emiris* WBg	Greek	186	86·9	21·6	..	Cassos	1882			I.Minaculi
	1456	R	Amphitrite *Sahlstrom -87* WBk	Russian	679	151·0	36·0	20·0	Brahestad	18 69/80	J.G.Rein		J.Aspegren
	1457	V	Amphitrite *Påhlsson* WBk	Swedish	436	135·9	27·0	15·7	Carlshamn	1873	D.R.Andersson	Y 8.87	N.M.Pahlsson
	1458		Amphitrite *Hansen 78-85* WSr	Swedish	228	106·4	23·4	12·7	Bastholmen	1877			J. Johansson's Sterbhus
	1459		Amphitrite *Mikkal* WBmb	Turkish	112	..	..	..	[illegible]	1886			Zecche & Oyanitch
	1460		Ampthill WBk	Norweg'n	286	107·0	26·0	15·0	Sunderland	1856	S.Austin		C.Christoffersen
	1461		Ampurdanesa WSr	Spanish	101	83·7	21·7	9·4	Palamos	1851			Vicente Linares
	1462	D	Amstelstroom *Tensman* W3mSr (ex Acadia)	Dutch	531	153·2	33·2	19·2	St.John	1857		Y 8.89	P.H.Kaars Sypestein
	1463		Amsterdam *Olsen 77-83* WBk (ex Vlaak, ex Snaterand van Ewyck) (CONDEMNED)	Norweg'n	661	118·7	31·5	21·4	Amsterdam	1856	F.Haverkamp	Y 8.84	Otto Thoresen
	1464	N	Amund Tveten *Abrahamsen 81-86* WSr	Norweg'n	207	108·3	26·5	12·1	Porsgrund	1880	Ole Johnson		H.Thorsen
44506	1465		Amur (ex Agnes Holt) CBk	British	236	112·9	24·3	11·9	Sunderland	1862	Moore & Co.		W.E.Marmion
	1466		Amur *Eriksson* W3mSr	Russian	195	..	..	..	Laitakari	1859			J.Eriksson
64403	1467		Amy WBg	British	220	107·4	25·0	11·7	Terrigal, NSW	1872	T.Davis		C.G.Warburton
3929	1468		Amy *Ranaghan* W3mSr	British	180	91·1	23·4	15·4	Topsham	1839			S.N.Robinson
54293	1469		Amy *Young* WBn	British	178	105·6	24·8	12·0	P.E.Island	1866	Clow		J.Graham
62130	1470		Amy *Sanders* WBn	British	167	102·5	23·7	12·6	Teignmouth	1868	Mansfield	Y 1.90	W.C.Phillips
62143	1471		Amy *Laverton 80-86* WSr	British	126	86·0	20·7	11·4	Banff	1870	Watson		P.Willcock
106190	1472	RV	Amy *Klages -87* WBk	American	665	152·0	35·0	16·5	Bath, Me.	1883	Goss & Sawyer	Y 8.87	W.B.Wilson
	1473		Amy (ex Agios Andreas) (LIGHTER) *Bolt* WBg	Norweg'n	223	96·6	25·3	14·8	Galaxidi	1868			Johanson, Ritter & Co.
	1474		Amy (ex Exhibition) *Weidemann -88* WSr	Norweg'n	110	75·8	22·1	10·9	Sunderland	1851			Jens Gabrielsen
	1475		Amy *Ekelund* W3mSr	Swedish	214	111·5	25·2	10·7	America				G.Sjöberg
1717	1476	R	Amy A. Lane *Heath -88* WBn	American	388	119·0	27·1	17·0	Searsport, Me.	1867	Lane	Y 8.88	H.S.Lampher
85544	1477	R	Amy D. WSr	British	111	89·2	21·4	9·0	Meteghan, N.S.	1883	L.Dugan		Alex.Melancon
88147	1478	✳	Amy Isabel *Owers* WK	British	86	79·2	20·5	10·5	Brixham	1883	S.Dewdney & Son		J.Marr
106408	1479	R	Amy M. Sacker *Loving* WSr	American	114	85·0	22·5	9·5	Cambridge, Md	1886	J.W.Crowell	Y 1.88	W.N.Church
105674	1480	RV	Amy Turner *Johnson 67-88* WBk	American	961	174·0	35·4	21·6	East Boston	1877	Smith & T'wnsnd	Y 1.88	C.Brewer & Co.
69297	1481	✳	Amyone (now "Ocean") IS	British	1300	231·1	37·1	23·4	Port Glasgow	1873	J.Reid & Co.		W.Battersby
	1482		Ana WPol	Spanish	235	93·2	27·4	12·8	Barcelona	1850		[illegible]	F.Oliver y Alsina
	1483	V	Ana *Maristany* WBg	Spanish	223	98·4	28·4	14·0	Gerona	1853			M.P.Maristany
	1484		Ana Aurora (ex Copiapo, ex Hugo) WBk	Chilian	353	127·8	27·7	16·5	Bideford	1862	Johnson		D. Fuentes & B.Squella
40585	1485		Ana Luisa WSr	British	112	85·0	23·0	7·6	Maryland, U.S.	1849			E.Burnett
	1486	N	Anaconda *Johannessen -79* WSr	Norweg'n	173	98·0	24·2	10·9	St.John, N.B.	1856			H.C.Jahnsen
	1487		Anadolu WBg	Turkish	250	..	..	..	[illegible]	1885			Kharpootlee Zackiari Ismat
	1488		Anadolu WBg	Turkish	244	..	..	..	[illegible]	1886			Keresteh-Ijian Firman Oglou Elia
1	1489		Anagnostis (BURNT) *Bolonbassis* WBn	Greek	240	94·8	25·9	..	Syra	1878			D.I.Boulonbassis
1454	1490	V	Anahuac *Whiddon 88-88* WS	American	1221	192·3	36·3	23·9	East Boston	1867	R.E.Jackson	J 12.88	Estate of R.Baker, jr
	1491	V	Anais (FOUNDERED) *Crous 83-87* WSr	French	112	85·3	19·7	10·9	St Malo	1879	F.Gautier		Le Rochais
89091	1492	✳	Anamba *Stedman 83-88* IBk	British	1110	214·6	35·1	21·2	Greenock	1885	Russell & Co.		P.Denniston & Co.
50328	1493		Anapira *Speed* WSr	British	100	81·8	21·6	10·3	[illegible]	1867	Pottinger		H.Speed
50086	1494		Anastasia (WRECKED) WBg	British	177	106·2	24·5	12·5	P.E.Island	1868			R.Hennebury
570	1495		Anastasia *Ferrini* WBk	Italian	317	120·7	26·8	17·7	Sorrento	1854		Y 8.87	G.Tonietti fu Raff
	1496		Anastasia WBg	Turkish	130	..	..	..	Roumania	1877			Vassilikosloe Dimitri Yorghi
337	1497	A	Anastassia *Cuturandi* WBk	Greek	462	132·8	30·2	17·4	Syra	1881	S.Cofinaki	Y 8.88	N.Stathopoulo
1[illegible]	1498		Anastassia WBk	Greek	247	98·1	24·3	..	Sunderland	1864			E.D.Vlassopoulos
11[illegible]	1499		Anastassios WBg	Greek	245	97·1	25·8	11·4	Galaxidi	1868			E.P.Nika
[illegible]	1500		Anastassis WBn		225	94·9	24·6	..	Syra	1876			

✳ = **Lloyd's Register**; D = Nederlandsche Vereeniging; G = Germanischer Lloyd; I = Registro Italiano; R = A[...]
A = Veritas Austro-Ungarico; H = Veritas Ellenico; N = Norske Veritas; V = Bu[...]

In the 4th Column, C = Composite (iron frame, wood planked); I = Iron; Stl = Steel; W = Wood.

The Dates of Metalling (Col. 11) are based on the records of the Books in which Vessels are respectively classed, or on the Reports of the Surveyors of [...]

Further particulars of Vessels classed by Lloyd's Register will be found in Lloyd's Register Book and Lloyd's Yacht Register.

N°	Registre	Nom. Capitaine. Ancien Nom.	Matériel et Gréement	Pavillon	Tonnage	Longueur	Largeur	Creux	Où	Quand	Par qui	Doublage	Armateurs ou Directeurs Gérants	Port d'Armement
1501	H	Anastassis *[illegible]*	W Bn	[illegible]	[illegible]	..	..	..	Syra	1870	P. Butois		E. N. Alafuso	Taganrog
1502		Anatole *Leroy*	W Bg	French	272	87·8	22·9	12·1	St. Malo	1857	Clement		A. Lemoine	Saint Malo
1503	HV	Anatoli *Squiriano* (ex ...)	W Bg	Greek	217	102·0	25·9	..	Galaxidi	1872		[illegible]	G. Squiriano	Syra
1504		Anatoli *Condori Stan*	W	Turkish	128	..	..	..	[illegible]	1873			Sarafin, Simon & Co.	Constantinople
1505	A	Anatolikos Astir *[illegible]*	W Bg	Greek	218	117·4	29·3	..	Syra	1874		[illegible]	M. Domostinis	Syra
1506	✻	Anaurus *[illegible]*	I S	British	[illegible]	251·6	[illegible]	[illegible]	Glasgow	1882	Barclay, Curle & Co.		A. & J. H. Carmichael & Co.	Greenock
1507		Ancud *[illegible]* (ex ...)	W Bk	Chilian	[illegible]	[illegible]	[illegible]	[illegible]	[illegible]	1862	Smirthwait		F. Buckiman	Valparaiso
1508		Andalusa	W Bn	Brazilian	178	99·8	24·4	11·1	[illegible]	1872	J. M·Donald		M. C. dos Santos	RioGrande do Sul
1509	✻	Andaman *[illegible]*	I Bk	British	[illegible]	198·5	33·5	21·7	Port Glasgow	1865	J. Reid & Co.		Goldberg & Co.	Swansea
1510		Andelibi Bahri	W	Turkish	112	..	..	..	[illegible]	1869			Stambolee Kurchice Hadji Mehemet Bashi	Stambolee
1511		Andelibi Bahri	W Sr	Turkish	199	..	..	..	[illegible]	1876			Bartenlee Hussein & Co.	Barten
1512	✻	Andes *Lawrence 76-88*	I Bk	British	831	207·0	25·5	19·5	[illegible]	1874	Osbourne, Graham & Co.		W. Moran	Hull
1513		Andes	W Bk	British	328	125·0	25·8	17·0	Harrington	1859	Williamson		Ocean Steamship Co.	Liverpool
1514	✻	Andes *Cook -87*	W Sk	British	95	88·5	21·0	10·6	Rostock	1887	E. Burchard & Co		H. Cook	Hull
1515	✻V	Andola *Passmore 76-88*	I S	British	2045	275·6	40·7	24·4	Liverpool	1886	T. Royden & Sons		E. F. & W. Roberts	Liverpool
1516	✻V	Andora *Davies 85-88*	I S	British	1676	251·0	39·0	24·0	Stockton	1881	Richards'n Duck & Co.		E. F. & W. Roberts	Liverpool
1517		Andorinha	W Bn	Brazilian	120	..	..	..					V. J. de Barros	Caraguatatuba
1518	N	Andover *Otterstad*	W Bk	Norweg'n	272	101·7	26·8	14·9	St. Johns, NFL.	18 41/75			G. Ulrichsen	Holmestrand
1519		Andrade Neves	W Bk	Brazilian	196	..	..	..	Honfleur	1864		Y [illegible]	P. M. da Cunha	Pelotas
1520		Andrea *Bovrup -87* (ex Vigilant)	W Bk	Danish	398	122·8	26·4	15·4	Shoreham	1869	May		S. A. Bonnet	St. Thomas, W. J.
1521	I	Andrea (ex ...) *Ravetti 84-88*	W Bk	Italian	1027	185·0	32·4	23·4	Sestri	1876	Debarbieri & Celesta	Y 6.88	G. B. Beverino	Spezia
1522		Andrea *Scarpa*	W Bn	Italian	297	101·3	28·1	18·5	Sestri, Ponente	1852	L. Bianchi	Z 1.84	A. Scarpa	Venice
1523	I	Andrea A. Vagliano (ex ...) *Minnco -88*	W Bk	Italian	472	135·1	30·1	19·6	Savona	1867	G. Gustavino	Y 8.88	G. di Natale	Messina
1524	I	Andrea Antonio *Moscelli -86*	W Bk	Italian	599	149·5	32·4	19·9	Savona	1872	F. Sirello	Y 9.85	Viglianzone & Zanelli	Savona
1525	I	Andrea Domenico *Moresco*	W Bn	Italian	207	104·0	24·9	13·4	Sestri Ponente	1877	A. Carlevaro	Y 8.88	Flli Veroggio	Genoa
1526	I	Andrea Lo Vico *Bertarone*	W Bk	Italian	568	150·9	32·2	20·1	Sestri Ponente	1872	Canepa	Y 2.88	A. Lo Vico	Palermo
1527		Andrea P. *Faggioni*	W Bn	Italian	309	105·6	28·2	18·0	Loano	1857	G. Amico		G. Pertuso	Savona
1528	I	Andrea Padre *Lubrano* (ex Carmine) *[illegible] -86*	W Bg	Italian	296	115·4	28·4	16·9	Gaeta	1868	F. Contreras	Y 10.84	F. Accetta	Naples
1529	V	Andrea Papa *Vanari [illegible] -88*	W Bk	Italian	616	146·9	32·4	20·9	Sestri Ponente	1871	B. Canepa	Y 3.86	A. Vaccari	Spezia
1530	V	Andrea Wilhelmina *Pergsdelt 81-81*	W S	Swedish	855	177·8	33·4	18·6	Skellefteå	1873	E. A. Löfgren	Y 9.87	A. Markstedt & Söner	Skellefteå
1531	V	Andreas *Ericksen 84-84*	W Sr	Danish	135	95·5	20·4	10·4	Troense	1883	C. J. Jensen		P. E. Jensen	Æroeskjöbing
1532	GV	Andreas *Hoyer -77*	W Bk	German	424	138·5	28·5	16·7	Apenrade	1868	G. Raben	Y 1.87	C. Matthiessen	Hamburg
1533	A	Andreas *Palios*	W Bk	Greek	420	125·3	28·5	..	Syra	1877		Y 3.86	P. Margaronis	Syra
1534	N	Andreas *Backer*	W Bk	Norweg'n	388	115·4	29·5	16·9	Christiana	1861	L. Mathiesena	Y 11.84	Enkefru S. Gamborg	Christiania
1535	N	Andreas *Ellingsen 80-89* (ex Artistic)	W Sr	Norweg'n	178	105·1	21·4	12·5	Dundee	1862	Dundee S. B. Co.		A. Ellingsen	Holmestrand
1536		Andreas *Lassen 63-72*	W Bk	Russian	388	128·0	27·5	16·7	Dago	1863	P. Pargo		E. von Ungern Sternberg	Dago
1537		Andreas	W Cr	Russian	106	72·9	25·5	10·3	Augern	1881	J. Andretz		J. Thomel	Riga
1538		Andreas *Ljungström*	W Sr	Swedish	123	87·6	22·5	9·9	Väddö	1876	Bjork		J. L. Wennerberg	Uddevalla
1539	V	Andreas Linnemann *Nielsen [illegible]*	W Sr	Danish	121	92·5	20·5	10·1	Veile	1868	P. H. Blum	Y 11.83	AL's Rhederi (J P Møller)	Veile
1540	V	Andreas Riis *Jensen* (WREGKER...)	W Bk	Norweg'n	557	148·2	32·7	17·7	Grimstad	1875	J. Bang	Y 1.87	J. Bang	Grimstad
1541		Andreas Vaglianos	W Bg	Greek	260	99·4	26·6	..	Syra	1871			C. A. Levidi	Hydra
1542	V	Andrée Rickmers *Hess -88*	W Bk	German	1419	208·3	41·0	25·9	Geestemünde	1886	R. C. Rickmers	Y 7.88	R. C. Rickmers	Bremerhaven
1543		Andrei	W Sr	Russian	204	..	..	..		1861			I. Sidoroff	Salmis
1544		Andrei *Leontieff*	W Sr	Russian	198	..	..	..	Impilaks	1882			M. Korniloff	Sordavala
1545	✻V	Andreta *Hoben 83-86*	I S	British	1708	253·8	39·0	24·0	Stockton	1883	Richards'n Duck & Co.		Mrs. C. M. Roberts (E. F. & W. Roberts, Managers)	St. John, N. B.
1546	R	Andrew Adams *Adams -87*	W 3mSr	American	771	176·0	36·3	18·5	Wicasset, Me.	1887	J. M. Haynes		J. Manchester Haynes	Bath, Me.
1547		Andrew H. Edwards (ex FRANKLIN) *Dobbin*	W Sr	American	223	113·0	30·0	8·4	Camden, N. J.	1886			C. E. Gibson	Boston, Mass
1548		Andrew Hicks *Adams 176-81*	W Bk	American	288	111·3	27·1	15·7	Fair Haven, Mass.	1867	J. Delano		Wm. Lewis	New Bedford, Mass
1549		Andrew J. York *Littlejohn*	W 3mSr	American	217	114·9	30·0	9·4	Deering, Me.	1881			H. M. Sargent	Portland, Me.
1550	✻	Andrew Longmore *Mair 84-*	W Sr	British	135	93·4	23·0	11·4	Banff	1874	Watson		R. Morrison	Banff

✻ = Lloyd's Register; D = Nederlandsche Vereeniging; G = Germanischer Lloyd; I = Registro Italiano; R = American Record; A = Veritas Austro-Ungarico; H = Veritas Ellenico; N = Norske Veritas; V = Bureau Veritas.

Dans la 4me Colonne, C = Composite (membrure en fer, bordé en bois); I = Fer; Stl = Acier; W = Bois.

Les Dates du doublage en métal (Col. 11) ressortent des livres dans lesquels les navires sont respectivement classés ou des rapports des Inspecteurs du Lloyd's Register.

Pour plus amples renseignements sur les navires classés par le Lloyd's Register, voir le Lloyd's Register Book et le Lloyd's Yacht Register.

1 Official No.	2 No. in Book	3 Registry in which classed	4 Ship's Name. Material and Rig. Master, and Dates of Appointment to present Owner's Service and to Ship. Ship's Late Name.		5 Flag.	6 Net Tonnage	7 Dimensions. Length.	Breadth	Depth.	8 Built. Where.	9 When	10 By whom.	11	12 Owners or Managers
87410	1451	✻	Amphitrite	WSr Yt	British	101	100·2	19·1	12·3	Gosport	1887	Camper & Nicholson	C 11.88	Lt.Col.A.D.McGregor
	1452		Amphitrite	WBk	Brazilian	517	..	..	..	Lokö				J.R. da Cunha
	1453	V	Amphitrite *Rowehl* 81-84 (ex DeBand)	WS	German	1814	231·3	40·3	26·2	Kennebunk	1875	N.Thompson	V 1.86	Gildemeister & Ri[illegible]
9	1454	A	Amphitrite *Beristianos*	WBg	Greek	195	98·1	23·5	..	Galaxidi	1875	Papapetros	V 6.84	Fratelli Liaglio
41	1455	A	Amphitrite *Emiris*	WBg	Greek	186	86·9	24·6	..	Cassos	1882			I. Minaculi
	1456	R	Amphitrite *Sahlstrom* -87	WBk	Russian	679	154·0	36·0	20·0	Brahestad	18[illegible]	J.G.Rein		J.Aspegren
	1457	V	Amphitrite *Pählsson*	WBk	Swedish	436	133·9	27·0	15·7	Carlshamn	1873	D.R.Andersson	V 9.86	N.M.Pahlsson
	1458		Amphitrite *Hansen* 1885	WSr	Swedish	228	100·4	23·4	12·7	Bastholmen	1877			J. Johansson's Sterbhus
	1459		Amphitrite *Mikkal*	WBab	Turkish	112	..	..	..	Mareftah	1886			Zoeche & Oyanitch
	1460		Ampthill	WBk	Norweg'n	286	107·0	26·0	15·0	Sunderland	1856	S.Austin		C.Christoffersen
	1461		Ampurdanesa	WSr	Spanish	101	83·7	21·7	9·4	Palamos	1851			Vicente Linares
	1462	D	Amstelstroom (ex Acadia) *Tessman*	WBmSr	Dutch	531	153·2	33·2	19·2	St.John	1857		V 8.88	P.H.Knaars Sypestein
	1463		Amsterdam *Olsen* 77-83 (ex Ybosk, ex Stad mad van Ewyck) (CONDEMNED)	WBk	Norweg'n	661	118·7	34·5	21·1	Amsterdam	1856	F.Haverkamp	V 2.8[illegible]	Otto Thoresen
	1464	N	Amund Tveten *[illegible]* 84-88	WSr	Norweg'n	207	108·3	26·5	12·1	Porsgrund	1880	OleJohnsen		H.Thorsen
44500	1465		Amur (ex Agnes Holt)	CBk	British	236	112·9	24·3	11·9	Sunderland	1862	Moore&Co.		W.E.Marmion
	1466		Amur *Eriksson*	W3mSr	Russian	195	..	..	..	Leitakari	1859			J.Eriksson
64403	1467		Amy	WBg	British	220	107·4	25·0	11·7	Terrigal,NSW	1872	T.Davis		C.G.Warburton
39220	1468		Amy *Kanaghan*	W3mSr	British	180	91·1	23·4	15·4	Topsham	1839			S.N.Robinson
54269	1469		Amy *Young*	WBn	British	178	105·0	24·8	12·0	P.E.Island	1866	Clew		J.Graham
62136	1470		Amy *Sanders*	WBn	British	167	102·5	23·7	12·6	Teignmouth	1868	Mansfield	V 1.81	W.C.Phillips
63443	1471		Amy *Laverton* 89-89	WSr	British	126	86·0	20·7	11·4	Banff	1870	Watson		P.Willcock
106189	1472	RV	Amy *Klages* -87	WBk	American	665	152·0	35·0	16·5	Bath,Me.	1883	Goss&Sawyer	V 6.83	W.B.Wilson
	1473		Amy (ex Agios Andreas) (LIGHTER) *Bolt*	WBg	Norweg'n	223	96·6	25·3	14·8	Galaxidi	1868			Johansen,Rötter & Co.
	1474		Amy (ex Eskimo) *Weideman* -88	WSr	Norweg'n	110	75·8	22·1	10·9	Sunderland	1851			Jens Gabrielsen
	1475		Amy *Ekelund*	W3mSr	Swedish	214	111·5	25·2	10·7	America				G.Sjöberg
1715	1476	R	Amy A. Lane *Heath* -88	WBn	American	388	119·0	27·1	17·0	Searsport,Me.	1867	Lane	V 8.85	H.S.Lampher
85544	1477	R	Amy D.	WSr	British	111	89·2	21·4	9·0	Meteghan,N.S.	1883	L.Dugan		Alex.Melancon
88147	1478	✻	Amy Isabel *Owers*	WK	British	86	79·2	20·5	10·5	Brixham	1883	S.Dewdney&Son		J.Marr
106406	1479	R	Amy M. Sacker *Loring*	WSr	American	114	85·0	22·5	9·5	Cambridge,Md	1886	J.W.Crowell	V 5.86	W.N.Church
105074	1480	RV	Amy Turner *Johnson* 61-88	WBk	American	961	174·0	35·4	21·6	EastBoston	1877	Smith&T'wnsend	V 1.8[illegible]	C.Brewer & Co.
69295	1481	✻	Amyone (now "Ocean")	IS	British	1300	231·1	37·1	23·4	PortGlasgow	1873	J.Reid&Co.		W.Battersby
	1482		Ana	WPol	Spanish	235	93·2	27·4	12·8	Barcelona	1850		[illegible]	F.Oliver y Alsina
	1483	V	Ana *Maristany*	WBg	Spanish	223	98·4	28·4	14·0	Gerona	1853			M.P.Maristany
	1484		Ana Aurora (ex Copiapo, ex Hugo)	WBk	Chilian	353	127·8	27·7	16·5	Bideford	1862	Johnson		D. Fuentes & B.Squella
40585	1485		Ana Luisa	WSr	British	112	85·0	23·0	7·6	Maryland,U.S.	1849			E.Burnett
	1486	N	Anaconda *Johannessen* -79	WSr	Norweg'n	173	98·0	24·2	10·9	St.John,N.B.	1856			H.C.Jahnsen
	1487		Anadolu	WBg	Turkish	250	..	..	..	Greece	1885			Kharpootlee Zackhari Ialiont
	1488		Anadolu	WBg	Turkish	244	..	..	..	Greece	1886			Kovsteh-(t)ari Firman Oglou Elia
1	1489		Anagnostis (BURNT) *Bolonbassis*	WBn	Greek	240	94·8	23·9	..	Syra	1878			D.I.Boulombassis
1459	1490	V	Anahuac *Wheldon* 88-88	WS	American	1221	192·3	36·3	23·9	EastBoston	1867	R.E.Jackson	V 12.86	Estate of R.Baker,j[illegible]
	1491	V	Anaïs (FOUNDERED) *Conus* 83-87	WSr	French	112	85·3	19·7	10·9	St.Malo	1879	F.Gautier		Le Rochais
90091	1492	✻	Anamba *Siedman* 88-88	IBk	British	1110	214·6	35·1	21·2	Greenock	1885	Russell&Co.		P.Denniston & Co.
70028	1493		Anapira *Speed*	WSr	British	100	81·8	21·0	10·3	Anstruther	1867	Pottinger		H.Speed
50680	1494		Anastasia (WRECKED)	WBg	British	177	106·2	24·5	12·5	P.E.Island	1868			R.Hennebury
570	1495		Anastasia *Ferrini*	WBk	Italian	317	120·7	26·8	17·7	Sorrento	1854		V 8.[illegible]	G.Tonietti fu Raff
	1496		Anastasia	WBg	Turkish	130	..	..	..	Roumania	1877			Vassilikostas Dimitri Terzhi
33[illegible]	1497	A	Anastassia *Calacondi*	WBk	Greek	462	132·8	30·2	17·4	Syra	1881	S.Cefaláki	V 8.82	N.Stathopoulo
1[illegible]	1498		Anastassia	WBk	Greek	247	98·4	23·3	..	Sunderland	1864			E.D.Vlassopoulos
145	1499		Anastassios	WBg	Greek	245	97·1	23·8	14·4	Galaxidi	1868			E.P.Nika
5	1500		Anastassis	WBn		225	91·9	21·6	..	Syra	1876			

✻ = Lloyd's Register; D = Nederlandsche Vereeniging; G = Germanischer Lloyd; I = Registro Italiano; R = Am[illegible]
A = Veritas Austro-Ungarico; H = Veritas Ellenico; N = Norske Veritas; V = Bu[illegible]

In the 4th Column, C = Composite (iron frame, wood planked); I = Iron; Stl = Steel; W = Wood.

The Dates of Metalling (Col. 11) are based on the records of the Books in which Vessels are respectively classed, or on the Reports of the Surveyors of [illegible]

Further particulars of Vessels classed by Lloyd's Register will be found in Lloyd's Register Book and Lloyd's Yacht Register.

No.	Cote	Nom. Capitaine et dates d'entrée en service des Armateurs actuels et à bord du navire. Ancien Nom.	Matériaux et Gréement	Pavillon	Tonnage	Longueur	Largeur	Creux	Construit Où.	Quand	Par qui.	Date du doublage en métal	Armateurs ou Directeurs-Gérants.	Ports d'Armement.
1501	*	Anastassis *[illegible]*	W Bn	[illegible]	239	..	..	..	Syra	1870	P. Bunis		E.N. Alafuso	Taganrog
1502		Anatole *Leroy -87*	W Bg	French	172	87·8	22·9	12·1	St. Malo	1857	Clément		A. Lemoine	Saint Malo
1503	HV	Anatoli *Squiriano* (ex [illegible])	W Bg	[illegible]	247	102·6	25·9	..	Galaxidi	1872		Y [illegible]	G. Squiriano	Syra
1504		Anatoli *Condori-Effen*	W	Turkish	[illegible]	..	..	..	[illegible]	1873			Sarafin, Simon & Co.	Constantinople
1505	A	Anatolikos Astir *Zannis*	W Bg	[illegible]	[illegible]	117·4	28·5	..	Syra	1874		Y 6.82	M. Domestinis	Syra
1506	*	Anaurus *[illegible]*	I S	[illegible]	[illegible]	251·6	38·3	23·0	Glasgow	1882	Barclay, Curle & Co.		A. & J. H. Carmichael & Co.	Greenock
1507		Anend *Grandy* (ex George Canning)	W Bk	[illegible]	676	153·0	31·8	20·5	Sunderland	1862	Smurthwait		F. Bnckiman	Valparaiso
1508		Andalusa	W Bn	Brazilian	179	99·8	24·0	11·1	P.E.Island	1872	J. McDonald		M.C. des Santos	RioGrandedoSul
1509	*	Andaman *[illegible]*	I Bk	British	955	198·5	33·5	21·7	Port Glasgow	1865	J. Reid & Co.		Goldberg & Co.	Swansea
1510		Andelibi Bahri	W	Turkish	11[illegible]	..	..	..	[illegible]	1869			Stambolee Kurchise Hadji Mehemed Bashia	Stambolee
1511		Andelibi Bahri	W Sr	Turkish	[illegible]	..	..	..	[illegible]	1876			Bartenlee Hussein & Co.	Barten
1512	*	Andes *Lawrence 86-86*	I Bk	British	831	207·0	33·5	19·5	Sunderland	1874	Osbourne, Graham & Co.		W. Moran	Hull
1513		Andes	W Bk	British	[illegible]	123·0	25·6	17·0	Harrington	1859	Williamson		Ocean Steamship Co.	Liverpool
1514	*	Andes *Cook -87*	W Sk	British	95	88·5	21·0	10·6	[illegible]	1887	E. Burchard & Co		H. Cook	Hull
1515	*V	Andola *Fassmore 76-88*	I S	British	2045	275·6	40·7	24·4	Liverpool	1886	T. Royden & Sons		E.F. & W. Roberts	Liverpool
1516	*V	Andora *Davies 85-86*	I S	British	1676	251·0	39·0	24·0	Stockton	1881	Richardson Duck & Co.		E.F. & W. Roberts	Liverpool
1517		Andorinha	W Bn	Brazilian	120	..	..	..					V.J. de Barros	Caraguatatuba
1518	N	Andover *Otterstad*	W Bg	Norweg'n	272	104·7	26·3	14·9	St. Johns, NFL	18 71/76			G. Ulrichsen	Holmestrand
1519		Andrade Neves	W Bk	Brazilian	196	..	..	..	Honfleur	1864		Y 10.79	P.M. da Cunha	Pelotas
1520		Andrea *Precup -87* (ex Vigilant)	W Bk	[illegible]	398	122·8	26·4	15·4	Shoreham	1869	May		S.A. Bonnet	St. Thomas. W.I.
1521	I	Andrea (ex [illegible]) *Marsetti 84-84*	W Bk	Italian	1027	185·0	32·1	23·4	Sestri	1876	Debarbieri & Celesia	Y 6.88	G.B. Beverino	Spezia
1522		Andrea *Scarpa*	W Bn	Italian	297	104·3	28·1	18·5	Sestri, Ponente	1852	L. Bianchi	Z 1.84	A. Scarpa	Venice
1523	I	Andrea A. Vagliano (ex Annunziata) *Russo -88*	W Bk	Italian	472	135·1	30·1	19·6	Savona	1867	G. Gustavino	Y 8.88	G. di Natale	Messina
1524	I	Andrea Antonio *Monaldi -88*	W Bk	Italian	599	149·9	32·1	19·9	Savona	1872	F. Sirello	Y [illegible]	Viglianzone & Zanelli	Savona
1525	I	Andrea Domenico *Morlasi*	W Bn	Italian	297	104·0	24·9	13·1	Sestri Ponente	1877	A. Carlevaro	Y 8.83	F[lli] Veroggio	Genoa
1526	I	Andrea Lo Vico *Mortarano*	W Bk	Italian	565	150·9	32·2	20·1	Sestri Ponente	1872	Canepa	Y 2.88	A. Lo Vico	Palermo
1527		Andrea P. *Fuggioni*	W Bn	Italian	309	105·6	28·2	18·0	Loano	1857	G. Amico		G. Pertuso	Savona
1528	I	Andrea Padre *Lubrano -88* (ex Germinal)	W Bg	Italian	296	115·4	28·4	16·9	Gaeta	1868	F. Contreras	Y 10.84	F. Accetta	Naples
1529	V	Andrea Papa *Iannri -88*	W Bk	Italian	616	146·9	32·4	20·9	Sestri Ponente	1871	B. Canepa	Y 3.86	A. Vaccari	Spezia
1530	V	Andrea Wilhelmina *Fingstedt 84-84*	W S	Swedish	855	177·8	33·4	18·6	Skellefteå	1873	E.A. Löfgren	Y 9.87	A. Markstedt & Söner	Skellefteå
1531	V	Andreas *Erichsen 81-84*	W Sr	Danish	135	95·5	20·4	10·4	Troense	1883	C.J. Jensen		P.E. Jensen	Ærøskjøbing
1532	GV	Andreas *Hoyer -77*	W Bk	German	424	138·5	28·5	16·7	Apenrade	1868	G. Raben	Y 1.87	C. Matthiessen	Hamburg
1533	A	Andreas *Palios*	W Bk	Greek	420	125·3	28·5	..	Syra	1877		Y 3.88	P. Margaronis	Syra
1534	N	Andreas *Backer*	W Bk	Norweg'n	388	115·4	29·5	16·9	Christiana	1861	L. Mathiesena	Y 11.82	Enkefru S. Gamborg	Christiania
1535	N	Andreas *Ellingsen 85-89* (ex Arctic)	W Sr	Norweg'n	178	105·1	21·4	12·5	Dundee	1862	Dundee S.B. Co.		A. Ellingsen	Holmestrand
1536		Andreas *Lassen 63-72*	W Bk	Russian	388	128·0	27·5	16·7	Dago	1863	P. Pargo		E. von Ungern Sternberg	Dago
1537		Andreas	W Cr	Russian	106	72·9	25·5	10·3	Angern	1881	J. Andretz		J. Thomel	Riga
1538		Andreas *Ljungström*	W Sr	Swedish	123	87·6	22·5	9·9	[illegible]	1876	Bjork		J. L. Wennerberg	Uddevalla
1539	V	Andreas Linnemann *Nielsen 8-87*	W Sr	Danish	121	92·5	20·5	10·1	Veile	1868	P.H. Illum	Y 11.83	A L's Rhederi (J P Møller)	Veile
1540	V	Andreas Riis *Jensen* (WRECKED)	W Bk	Norweg'n	557	148·2	32·7	17·7	Grimstad	1875	J. Bang	Y 7.87	J. Bang	Grimstad
1541		Andreas Vaglianos	W Bg	Greek	26[illegible]	99·4	26·6	..	Syra	1871			C.A. Levidi	Hydra
1542	V	Andrée Rickmers *Hess 75-85*	W Bk	German	1419	208·3	41·0	25·9	Geestemünde	1886	R.C. Rickmers	Y 7.85	R.C. Rickmers	Bremerhaven
1543		Andrei	W Sr	Russian	204	..	..	..		1861			I. Sidoroff	Salmis
1544		Andrei *Leontieff*	W Sr	Russian	198	..	..	..	[illegible]	1882			M. Korniloff	Sordavala
1545	*V	Andreta *Hoben 83-86*	I S	British	1708	253·8	39·0	24·0	Stockton	1883	Richardson Duck & Co.		Mrs. C.M. Roberts (E.F. & W. Roberts, Managers)	St. John. N.B.
1546	R	Andrew Adams *[illegible]*	W Tr Sr	American	771	176·0	36·3	18·5	Wicasset, Me.	1887	J.M. Haynes		J. Manchester Haynes	Bath. Me.
1547		Andrew H. Edwards *Delrudge* (STRANDED)	W Sr	American	223	113·0	30·0	8·4	Camden, N.J.	1866			C.E. Gibson	Boston, Mass
1548		Andrew Hicks *Adams [76-83]*	W Bk	American	288	111·3	27·1	15·7	Fair Haven. Mass.	1867	J. Delano		Wm. Lewis	New Bedford. Mass
1549		Andrew J. York *Littlejohn*	W [illegible]	American	217	114·9	30·0	9·4	Deering, Me.	1881			H.M. Sargent	Portland. Me.
1550	*	Andrew Longmore *Muir 81-*	W Sr	British	135	93·4	23·0	11·4	Banff	1874	Watson		R. Morrison	Banff

* = Lloyd's Register; D = Nederlandsche Vereeniging; G = Germanischer Lloyd; I = Registro Italiano; R = American Record;
A = Veritas Austro-Ungarico; H = Veritas Ellenico; N = Norske Veritas; V = Bureau Veritas.

Dans la 4me Colonne, C = Composite (membrure en fer, bordé en bois); I = Fer; Stl = Acier; W = Bois.

Les Dates du doublage en métal (Col. 11) ressortent des livres dans lesquels les navires sont respectivement classés ou des rapports des Inspecteurs du Lloyd's Register.

Pour plus amples renseignements sur les navires classés par le Lloyd's Register, voir le Lloyd's Register Book et le Lloyd's Yacht Register.

Official No.	No. in Book.	Registry in which classed.	Ship's Name. Material and Rig. Master, and Dates of Appointment to present Owner's service and to Ship. Ship's Late Name.	Flag.	Net Tonnage.	Dimensions. Length.	Breadth.	Depth.	Built. Where.	When.	By whom.	Date of Metalling.	Owners or Managers.
[illegible]	1551	R	Andrew Nevinger *Smith* W3mSr	American	279	127·0	32·8	9·7	[illegible]	1873	[illegible]		P. Fitzpatrick
[illegible]	1552		Andrew Peters *Wa[illegible]* W Sr	American	112	91·1	23·2	8·1	[illegible]	1854			A.H.Sawyer
	1553	✻	Andrew Welch *Morrison 78-88* I Bk	Hawaiian	863	185·6	36·1	18·5	[illegible]	1888	[illegible] & Co.		C.Brewer & Co., Ld.
[illegible]	1554	✻V	Andrina *Campbell 72-86* I 4mS	British	2636	320·6	42·9	24·9	Southampton	1886	Oswald, Mordaunt & Co		Andrina Sailing Ship Co., Ld. (E.F. & W. Roberts)
	1555		Androklos (NOT SEAGOING)		400	..	..	..	Apenrade	1866	J. Paulsen	Y [illegible]	
[illegible]	1556	A	Andromachi *Christodoulopoulo -87* W Bn	Greek	194	..	..	..	Galaxidi	1865		[illegible]	A.Cazzulieris
	1557		Andromeda *Fugler 86-86* W Bk (ex Afrika)	Danish	326	102·7	26·5	17·1	Helsingfors	18 72/74			K.F.Marstrand's Efterfølgere
	1558	V	Andromeda *Bünger 86-86* CS (WRECKED)	German	1871	270·6	37·1	25·0	Liverpool	1864	Jones, Quiggin & Co.	[illegible]	W.A.Riedemann
	1559	N	Andromeda *Gregersen -71* W Bk (ex Blair Athol)	Norweg'n	441	130·5	29·0	16·2	Granville N.S.	1864			A.M.Gregersen
[illegible]	1560	✻V	Androsa *Lockhart 88-88* I S	British	1944	273·1	39·9	24·0	Southampton	1881	Oswald, Mordaunt & Co.		E.F. & W.Roberts
	1561	N	Andvaka *Jensen* W Bg	Norweg'n	167	94·4	21·4	11·4	Tvedestrand	1857	Chr. Olsen		A.Biering Olsen
	1562	V	Ane *Storm 72-72* W Bn	Danish	162	106·9	21·7	10·5	Troense	1866	C.R.Møller		N.R.Storm
	1563	V	Ane Cathrine *Andersen [illegible]* W Bg	Danish	193	104·0	22·7	11·8	Nordby, Fanø	1878	S.Arthausen	Y 3.87	Akt.Selsk.(N.H.Svarrer)
	1564	V	Ane Cathrine *Warrer [78-76]* W Bn	Danish	179	102·8	22·2	11·5	Nordby, Fanø	1877	N.Nielsen	Y 7.88	Akt.Selsk.(H.Nielsen)
	1565	V	Ane Cathrine *Jepsen 62-69* W Sr	Danish	120	85·6	20·5	10·4	Nordby, Fanø	1869	N.Nielsen		S.Mathisen
	1566	V	Ane Johanne *Sørensen [72-72]* W Bn	Danish	133	90·1	21·1	10·6	Nordby, Fanø	1872	J.N.Christensen	V 11.87	Akt.Selsk.(J.M.Sørensen)
	1567	G	Ane Kjerstine *Gregersen [77-77]* W Sr	Danish	124	87·5	21·6	10·3	Nordby, Fanø	1869	J.N.Christensen		Akt.Selsk.(N.H.Svarrer)
	1568	G	Ane Sophie *Hager -88* W Sr	Danish	117	79·3	20·2	10·7	Aarhus	1861	H.Thomsen		J.Hintze
	1569	V	Anémone *Fenoy* W Sr	French	126	95·8	22·3	11·2	St.Malo	1884	Gautier	Y [illegible]	C.Séheult
[illegible]	1570	R	Aneroid *Crosby -87* W S	British	983	173·0	34·8	21·2	Digby, N.S.	1873		Y 7.87	R.R.Crosby
[illegible]	1571	N	Aneroid *Coughlan 82-85* W Bn	British	212	105·5	25·9	13·7	P.E.Island	1874	Ramsay	Y [illegible]	T.H.Franks
	1572		Anette *Hansen* W Sr	Norweg'n	176	99·2	24·1	11·4	Sandefjord	1874			H.Hansen
	1573		Angantyr *Torsberg 88-88* W Bk	Swedish	675	186·6	26·8	16·3	Hernösand	1873	Fr.Hägglund		Hernösands Lastäng Aktb. (H.N. Bader)
[illegible]	1574	RV	Angara *Acker* W Bkn	British	630	139·2	33·0	21·2	Bridgwater NS	1884	R.A.Logan	Y [illegible]	J.Eisenhauer
	1575		Angel *Ramos* W Bn	Spanish	117	85·0	22·0	7·9	Philippines	1883			C.Aramburo
	1576	V	Angela *Barmann 79-79* W Sr	German	118	90·6	20·4	9·8	Elewecht	1879	Kramer		Joh.Pekeler
[illegible]	1577	DI	Angela *Mazzini* W Bk	Italian	872	165·0	33·8	23·0	Chiavari	1876	L.Gotuzzo	Y [illegible]	Flli. Sanguineti fu Giuseppe
[illegible]	1578	I	Angela Accame *Bellarino -86* W Bk	Italian	678	..	..	..	Loano	1876	B.Amico	Y [illegible]	N.Accame & figli
	1579		Angela Madre *Costa* W Bk (ex Giuseppina)	Argentine	384	..	..	..	P.E.Island	1874		[illegible]	N.Mihanovich & Co.
[illegible]	1580	I	Angela Maria *Mazzella -81* W3mSr (ex Salvatore L., ex Anacreto)	Italian	447	134·1	29·6	17·6	[illegible]	1880	A.Castellano	Y [illegible]	A.Mazzella
	1581		Angela Parodi *Daregno* W Bk	Austrian	314	120·7	30·0	14·1	Varazze	1873	S.Craviotto		Pio de Persic
[illegible]	1582	I	Angela Schiaffino *Schiaffino* W Bk (ex Angelie)	Italian	525	139·8	32·1	20·0	S.Margherita	1871	F.Rolla	Y [illegible]	E.Schiaffino
[illegible]	1583	V	Angela di Amore *Fravega* W Bk (ex Rivena)	Italian	782	164·0	32·8	24·2	Sestri Ponente	1865	Casanova	Y [illegible]	B.Fravega
	1584		Angèle (CONDEMNED) *Dolo* W Bk	French	393	126·6	28·9	16·4	Bordeaux	1878		[illegible]	D.Auger
	1585	V	Angèle *Créquer* W Bn	French	124	77·4	21·0	11·2	Méans	1875	E.Ollivaud		J.Créquer
	1586		Angeles *Garcia* W Bk	Spanish	377	130·5	27·4	15·7	Bilbao	1871	A. de Arana	Y [illegible]	José Goday
[illegible]	1587	IV	Angeli *Semidei 86-86* W Bk	Italian	1121	193·2	31·4	24·3	Sestri Ponente	1886	Flli. Bozzano	Y [illegible]	L.Peneo
[illegible]	1588	✻R	Angelia *Cleveland -88* W Bg	American	270	105·6	27·7	13·2	Cherryfield Me	1866			W.Cleveland
[illegible]	1589	I	Angelica Accame *Viggi* W Bk (ex Davide S.)	Italian	622	142·7	29·1	21·1	Varazze	1876	G.B.Baghetto	Y [illegible]	E.Accame & figli
	1590		Angelidah W Bg	Turkish	135	..	..	..	[illegible]	1882			Sakyzlee Yorghi Frangopulo
	1591		Angelika *Matisson* W Sr (ex Angelique, ex Rudolph Romelling, ex Thomas)	Swedish	133	72·2	19·5	12·8	[illegible]	1827	Calman		E.Jonsson
[illegible]	1592	H	Angeliki *Tsichilis* W Bg	Greek	225	95·8	24·6	..	Syra	1866			S.Lambrou
	1593	H	Angeliki *Vetropoulis* W Bn		223	93·2	26·2	16·0	Syra	1874	C.Cristali		
[illegible]	1594	A	Angeliki W Bg	Greek	190	91·2	25·7	..	Syra	1878			T.Archangelos
	1595	A	Angeliki *Emiri* W Sr	Greek	146	..	..	..	Syra	1887	C.Cristali		M.Emiri
	1596	H	Angeliki *Laimos* W Sr	Samos	110	..	..	..	Samos	1881			N.Laimos
	1597		Angeliki *Poxos* W Bn	Turkish	190	..	..	..	[illegible]	1878		[illegible]	C.G.Pass & Co.
	1598		Angeliki W Bg	Turkish	187	..	..	..	[illegible]	1871			[illegible] Hadji Ma[illegible]
[illegible]	1599	H	Angeliki Vagliano *Tsitsilianos* W Bn	Greek	256	100·1	26·7	..	Syra	1875		[illegible]	P.Tsitsilianos
	1600												

✻ = Lloyd's Register; D = Nederlandsche Vereeniging; G = Germanischer Lloyd; I = Registro Italiano; R = Ameri[can]; A = Veritas Austro-Ungarico; H = Veritas Ellenico; N = Norske Veritas; V = Bure[au Veritas].

In the 4th Column, C = Composite (iron frame, wood planked); I = Iron; Stl = Steel; W = Wood.

The Dates of Metalling (Col. 11) are based on the records of the Books in which Vessels are respectively classed, or on the Reports of the Surveyors of Lloyd['s Register].

Further particulars of Vessels classed by Lloyd's Register will be found in Lloyd's Register Book and Lloyd's Yacht Register.

N° dans le livre	Société dans laquelle le navire est classé	Nom. Matériaux et Gréement. Capitaine et date d'entrée au service des Armateurs actuels et à bord du navire. Ancien Nom.		Pavillon.	Tonnage net	Dimensions. Longueur.	Largeur.	Creux.	Construit. Où.	Quand.	Par qui.	Date du doublage en métal	Armateurs ou Directeurs-Gérants.	Port d'Armement.
1601	V	Angeliko *Calavasi*	W Bk	Greek	350	116·1	27·9	..	Syra	1883	A.Livadara	Y	A. Monarchidi	Syra
1602	H	Angeliko *Valas*	W Bg	Greek	133	73·5	23·0	..	Syra	1873			Valas & Mavromatis	Syra
1603	I	Angelina *Sampinato* -88	W Bn	Italian	218	96·8	25·8	16·1	Castellam. di [Stab.	1856		Y	C.Sampinato	Catania
1604	I	Angelina *Semeraglio*	W Sr	Italian	100	91·5	22·7	10·3	Torre del Greco	1887	S.Paolillo fu G.		R. di Majo di G.	Naples
1605		Angélique *Sauteraux*	W Bk	French	900	167·6	29·6	21·8	Quebec	1864	P.Valin	Y	E.Sauteraut	Bordeaux
1606		Angélique	W Bg	French	180	96·5	21·3	11·8	St.Malo	18 49/52			F.Fichet	Saint Malo
1607	I	Angolita *Moglia* (ex La Parcola)	W Bk	Italian	369	127·9	29·5	17·1	Lavagna	1871	G.Brigando	Y	F[lli] Moglia	Genoa
1608		Angelita	W Bg	Spanish	196	93·2	25·6	12·8	Lloret	1832			José Morugas	Barcelona
1609		Angelita	W Sr	Spanish	101	94·8	25·2	7·3	Barcelona	1859			Empresa de Fomento y Correspondencia del Sur.	Havana
1610		Angelita C. *Lubrano*	W Bk	Italian	372	108·9	27·2	17·6	Sestri	1866	G.Peragallo	Y	L.S.Carletti	Porto Ferraio
1611	DI	Angelo *Mortola*	W Bk	Italian	689	165·0	33·0	21·6	Savona	1875	E.Calamaro	Y	F[lli] Mortola fu A.	Genoa
1612	V	Angelo *Mucio*	W 3m St	Italian	265	110·5	23·4	15·2	Castellamare	1877	Frataburitio	Y	E. Mucio	Castellammare
1613		Angelo (ex La Lessa)	W Bn	Italian	118	82·3	22·8	11·6	Vietri	1864			Antonio Solari	Genoa
1614	IV	Angelo Castellano	W Bk	Italian	489	136·1	30·3	17·7	Equa	1878	Liberto & Serra	Y	T.Castellano Albert	Naples
1615	✻	Angerona *Barrow*	I S	British	1215	226·0	36·1	22·1	Whitehaven	1874	Whitehav'n Ship [Building Co.		Lowden, Edgar & Co.	Liverpool
1616	R	Angie L. Green	W 3m Sr	American	412	139·0	34·0	11·5	Cooper's Point. [N.J.	1882	Morris & Mathais		J. McElwee	Philadelphia
1617	I	Angiolette R. *Gaggino* (ex Angioletta Basso)	W Bn	Italian	687	157·5	29·8	21·5	Savona	1874	E.Calamaro	Y	G.Roncallo	Genoa
1618	I	Angiolina *Cassola* -88	W Bk	Italian	612	157·0	30·8	19·3	Sestri Ponente	1868	Briasco fu G.B.	Y 12.80	E.Piaggio fu G.B.	Genoa
1619	I	Angiolina B. *Pizzi* -88	W Bn	Italian	121	89·9	24·0	11·4	Cornigliano	1885	A.Carlevaro	Y	F.Pizzi	Spezia
1620	I	Angiolina C. *Scapillo*	W Bn	Italian	151	96·5	23·4	12·1	Castellammare [di Stab.	1882	D'Acampora	Y	Ditta Walser	Messina
1621		Angiolino *Bertisse*	W Bn	Italian	185	93·5	25·4	15·5	Lussinpiccolo	1849			A.Bertorella	Genoa
1622		Angiolino B. *Ghetto* (ex Girolamo)	W Bn	Italian	243	112·2	25·7	15·1	Fiume	1867		Y	A.Ballerin & Co.	Venice
1623	V	Anglesea *Bain*	W Bk	British	978	180·0	36·7	21·8	Newport N.S.	1879	J.A.Harvie	Y	A.Gibson	Windsor, N.S.
1624	✻	Anglesey *Rawden* 78-84	I S	British	1260	219·0	38·3	23·0	Liverpool	1875	T.Royden & Sons		Hughes & Co.	Liverpool
1625	R	Anglo *Mattson* -88	W Bn	British	298	126·0	28·5	12·5	Lunenbrg.NS.	1872		Y 12.87	J.Eisenhauer	Lunenburg, N.S
1626	V	Anglo America *Smith* 85-88	W S	British	1533	218·9	39·2	24·0	Portland, N.B.	1876	Hilyard Bros.	Y 10.86	D.V.Roberts	St.John, N.B.
1627	V	Anglo India	W S	British	1549	219·8	39·6	24·1	Portland, N.B.	1877	Hilyard Bros.	Y	D.V.Roberts	St.John, N.B.
1628		Anglo Indian	W Bk	British	444	133·4	28·5	18·2	Sunderland	1858	J.M.Reed	Y	C.H.C.Moller	Shanghai
1629	✻	Anglo Norman *Hopper* 81-81	I Bk	British	822	192·4	32·2	18·9	Port Glasgow	1876	Russell & Co.		Maxwell, Eastham & Co.	London
163[illegible]		Anglo Saxon	W Sr	British	253	133·0	26·0	11·0	Port Dalhousie	1864	Anderson & [Donaldson		D.D.Calvin	Sarnia, Ont.
1631		Ango *Grenkel* -86	W Bk	French	475	166·3	25·6	17·5	London	1862	Langley		Ulysse Bernard	Havre
1632	V	Angostura *Quast* 87-87	W Bk	German	418	142·3	27·6	17·2	Lübeck	1870	P.Heitmann	Y	A.H.Wappäus	Hamburg
1633		Anibal *Masot* -88	W Bk	Spanish	332	117·7	29·0	13·8	Palma Maj'rca	1867			M.Balaguer y Canovas	Majorca
1634	DIV	Aniello Galatola *Di Cottone* -88	W Bk	Italian	747	166·0	36·1	22·1	Ischia	1878	G.Bonifacio	Y	G.Galatola	Naples
1635	I	Animoso *Canossa*	W Sr	Italian	157	81·2	23·4	14·0	Varazze	1859	A.Corruti	Y	Nicolo Pescetto	Genoa
1636		Anita *Melanson* -88	W Sr	British	122	87·0	26·1	7·8	Clifton, N.B.	1882	G.H.Flewelling		J.M.Taylor	St.John, N.B.
1637	R	Anita *Small* -83	W 3m Sr	American	384	125·0	29·0	10·2	E.Machias, Me.	1872		Y 11.86	F. A. Small	Machias, Me.
1638		Anita	W Bn	Chilian	142	82·7	21·7	9·6					F.Court	Constitucion
1639	I	Anita *Carbonini* -87	W 3m Sr	Italian	315	136·1	26·9	12·2	Sestri Ponente	1878	A.Briasco fu G.B.	Y 2.87	Marré, Piombino & Co.	Genoa
1640	N	Anita *Hoggland*	W Bk	Norweg'n	371	134·9	25·2	15·7	Lübeck	1856		Y 12.80	A.Asbjörnsen	Arendal
1641	✻	Anita *Larrinaga* -88	I Bn	Spanish	333	130·1	26·1	13·3	Port Glasgow	1875	W Hamilton & Co		F. de Abaitera	Bilbao
1642		Anita *Fernandez* -88 (ex Ana Faura, ex Carmen)	W Bk	Spanish	205	100·4	24·8	13·1	Corunna	1871			Pedro Ferreiros	Villagarcia
1643		Anita	W Bn	Spanish	174	92·2	22·8	11·6	Albay, Luzon	1870			Inchausti & Co.	Manilla
1644		Anita	W Bn	Spanish	134	82·3	23·5	11·6	Barcelona	1849			Nicolas Company	Majorca
1645		Anita	W Sr	Spanish	111	74·8	21·5	9·6	Majorca	1852	O.Puig		M.Estela y Calafat	Mahon
1646	R	Anita Berwind *McBride* -71	W Bkn	American	633	159·5	36·0	17·5	Camden, N.J.	1877	R.Bisley & Co.	Y	R.W.McBride	Philadelphia
1647		Anita Owen	W Bg	American	465	121·3	30·0	16·9	Harpswell, Me	1869	Geo. Russell		J.S.Winslow & Co.	Portland Me.
1648	IV	Anita S. *De Rosa*	W Bk	Italian	373	116·1	26·6	18·5	Cassano	1876		Y	A.Castellano & Co.	Naples
1649		Anita T.	I	Argentine	158	120·0	24·9	8·0	S. Nicolas de [los Arroyos	1885	C.Geddes		Green & Johnson	Buenos Ayres
1650		Anita de Cárdenas	W Sr	Spanish	118	94·8	28·6	7·9	Palamos	1864			Ramón S.Garcia	Havana

✻ = Lloyd's Register; D = Nederlandsche Vereeniging; G = Germanischer Lloyd; I = Registro Italiano; R = American Record;
[illegible] = Veritas Austro-Ungarico; H = Veritas Ellenico; N = Norske Veritas; V = Bureau Veritas.

Dans la 4me Colonne, C = Composite (membrure en fer, bordé en bois); I = Fer; Stl = Acier; W = Bois.

Les Dates du doublage en métal (Col. 11) ressortent des livres dans lesquels les navires sont respectivement classés ou des rapports des Inspecteurs du Lloyd's Register.

Pour plus amples renseignements sur les navires classés par le Lloyd's Register, voir le Lloyd's Register Book et le Lloyd's Yacht Register.

Official No.	No. in Book.	Registry in which classed.	Ship's Name. Material and Rig. Master, and Dates of Appointment to present Owner's Service and to Ship. Ship's Late Name.	Flag.	Net Tonnage.	Length.	Breadth.	Depth.	Built. Where.	Built. When	Built. By whom.	Date of Metalling.	Owners or Managers.
	1651		Ankara WSr	Russian	218	95·8	24·9	13·[illegible]	[illegible]	1876			K.A.Bergroth
	1652		Ankara *Helin* WSr	Russian	211	..	..	..	[illegible]	1874			J.Helin
	1653		Anksartotos *Mo. & Chamandi* WBg	Turkish	158	..	..	..	[illegible]	1885			Keresteh-HjarChristophor Christophoulis
846	1654		Ann WBg	American	126	82·5	23·5	8·8	Guilford,Conn	1858			J.W.Aims
237	1655		Ann *Salisbury 70-86* WSr	American	103	80·0	23·6	7·1	Pittston,Me.	18 35/71	G.D.Clark		J.S.Emery & Co.
73092	1656	✻	Ann Clark *Wilberidge84-87* WSr (ex Eliza Ann)	British	127	88·6	22·6	11·2	Banff	1876	Geddie	[illegible]	R.Clark
68228	1657	✻	Ann Crewdson *Shaw 85-85* WSr	British	134	96·3	23·1	11·7	Ulverston	1873	Charnley		J.Gordon
1299	1658		Ann E. Valentine *Chambers* WSr	American	300	121·0	32·0	8·2	PortJefferson, [N.Y.	1866			J.A.Van Brunt
13334	1659	✻	Ann Elizabeth *Pearce [85-85* WK	British	34	58·1	16·0	6·4	Hastings	1856			J.E.Pearce
53132	1660		Ann Humphreys *Spurell 83-83* WBn	British	160	97·0	22·4	12·8	Whitehaven	1865	Shepherd		H.Kearon
106081	1661	R	Ann J. Trainer *Derrickson* W3mSr	American	406	141·8	35·2	11·7	Camden,N.J	1881	Tilton		H.D.May
21768	1662	✻	Ann Jane *Lewis -85* WSr	British	93	76·7	21·5	11·5	Nevin	1858	O.Griffiths		R.Lewis
1812	1663		Ann L. Lockwood *Evans -88* W3mSr	American	321	125·2	29·1	15·7	Norwalk,Conn	1869	G.S.Bill		W.E.Truslee
19694	1664		Ann Lucey *Noble* WBkn	British	248	124·7	23·5	12·9	Ross	1857	Hessell&Co.		W.Lucey
51489	1665	✻	Ann Millicent *Doherty (L.OST)* IBk	British	921	195·0	32·1	21·3	Hull	1865	HumberIronWks		R.W.Doherty
12498	1666		Ann Peat *Peacock* WBg	British	183	89·2	23·3	14·1	Arbroath	1850	W.Anderson		J.Hood
130	1667		Ann S. Brown WSr	American	167	105·0	28·0	8·0	Tuckerton,NJ	1860			J.Clark
6892	1668		Ann Sanderson WBg	British	266	94·0	25·2	16·1	Hartlepool	1852			P.Wheelan
53170	1669		Ann Walters *Walters* WSr	British	159	99·4	23·8	12·0	Padstow	1867	Tredwen		J.Walters,jun.
18796	1670	✻	Ann Warren *Davies* WSr	British	95	75·1	20·2	10·5	Newquay	1857	O.Owens		M.Davies
21764	1671		Ann & Ellen *Jones* WSr	British	110	79·0	21·4	11·1	Wales	1858	W.Jones		L.Williams,jun.
31777	1672		Ann & Jane WBk	British	252	97·2	24·0	16·0	Whitehaven	1836			Union S.S. Co. of New Zealand
27436	1673	✻	Ann & Jane Prichard *Roberts 78-* WSr	British	140	86·0	22·3	12·6	Pwllheli	1859	Prichard		R.A.Prichard
62105	1674		Ann & Mary *Paterson* WSr	British	220	106·5	24·1	13·7	Montrose	1869	Strachan		A.McLean
42578	1675		Ann & Mary *Stirling* W3mBk	British	185	109·4	23·1	12·0	Quaybach	1861	E.Daniels		R.Muirhead
74344	1676	RV	Anna *Beattie* WBn	British	629	156·0	34·8	16·0	RiverJohn,NS	1876	J.Kitchin	[illegible]	Mrs.E.Beattie
53583	1677		Anna WBn	British	242	106·0	26·0	11·9	Blackrock,NS	1865			Isaac Dart
105304	1678	R	Anna *Gillette -88* W3mSr	American	465	141·0	33·0	12·1	StonyBrook, [N.Y.	1874		[illegible]	H.Gillette
105887	1679	V	Anna *Williams 86-86* W3mSr	American	228	117·0	29·0	10·0	S.Francisco	1881	M.Turner		Oceanic S.S.Co.
	1680	GV	Anna *Jepsen 88-88* WBk	Danish	496	134·5	30·3	17·2	Greifswald	1876	A.Spruth	[illegible]	P.Christensen
	1681	V	Anna *Brening -87* W3mSr	Danish	313	123·4	27·1	13·1	Elmshorn	1876	J.Kremer	[illegible]	A.N.Brening
	1682	RV	Anna WBkn		278	128·7	25·0	11·0	Sundsvall	1873	V.Hägglund	[illegible]	
	1683		Anna (ex Lina) WBn	Swedish	223	98·0	26·1	12·4	Kinaito	1867			A.P.Anderson
	1684		Anna *Thurøe 81-81* WBn	Danish	176	94·4	22·1	13·2	Troense	1858	T.P.Petersen		A.C.Thurøe
	1685		Anna (ex C. A. Berg) WBn	Danish	158	83·1	20·4	11·6	Stralsund	1868	J.S.Kraeft		C.Nielsen
	1686	V	Anna *Jörgensen -88* WSr	Danish	135	95·8	20·3	10·5	Troense	1881	C.J.Jensen	[illegible]	C.J.Hœpfner
	1687	V	Anna *Petersen -87* WSr	Danish	123	90·4	20·2	10·7	Holbæk	1863	H.P.Mortensen		H.L.Smiths Enke
	1688		Anna *Petersen -87* WBn (ex Thorsenga-Bog)	Danish	118	81·6	20·7	11·8	Troense	1849	H.L.Schmidt		N.Petersen
	1689	V	Anna *Veltis -87* WBn	Dutch	170	93·9	24·0	11·7	OudePekela	1870	O.B.Drenth	[illegible]	J.J.Koerts & Zoon
	1690		Anna (WRECKED) *Hesseling* WSr (ex Clasina Aaltina)	Dutch	155	99·1	21·4	11·1	NieuweSchans	1850			K.D.Hesseling
	1691		Anna *Marcaut* WBk	French	340	117·1	25·3	15·4	Bayonne	1860	Langa&Dscamde		Ch.Cazalis
	1692	V	Anna WBk	French	281	120·0	27·2	14·0	Bayonne	1859			Jules Pannier
	1693	V	Anna *Jean* WBn	French	180	87·9	22·0	10·9	Nantes	1877	[illegible]		Jean
	1694		Anna *Noblanc -87* WBn	French	118	74·5	20·7	11·6	Bayonne	1866	PerezFrères		Petit
	1695	V	Anna *Bugault* WBn	French	110	78·4	21·6	9·9	St.Malo	1873	P.Hervot	[illegible]	P.Hervot
	1696	V	Anna *Brockhus 83-88* WS (ex Julia M. Clark)	German	1287	185·0	38·0	24·0	WaldoboroMe	1874	J.Clark & Sons	[illegible]	F.Roters
	1697	GV	Anna *Meinkens -78* WS (ex Elvo,HenryHandley,Quosnell)	German	1159	182·0	37·4	24·3	Medford	1861		[illegible]	Siedenburg,Wendt & Co
	1698	GV	Anna *Pieper 85-85* IBk	German	1099	203·0	34·4	19·1	Kiel	1885	GermaniaWerft		C.M.Matzen
	1699	G	Anna *Lohmeyer 71-87* IBk	German	942	176·9	32·1	19·8	Glasgow	1872	A.Stephen&Sns		D.H.Wätjen & Co.
	1700	DV	Anna *Zelek* WBk	German	448	131·5	30·1	16·8	Kiel	1871	A.Conradi&Sons	[illegible]	F.W.Fischer

✻ = Lloyd's Register; D = Nederlandsche Vereeniging; G = Germanischer Lloyd; I = Registro Italiano; R = Ame[rican]...
A = Veritas Austro-Ungarico; H = Veritas Ellenico; N = Norske Veritas; V = Bureau...

In the 4th Column, C = Composite (iron frame, wood planked); I = Iron; Stl = Steel; W = Wood.

The Dates of Metalling (Col. 11) are based on the records of the Books in which Vessels are respectively classed, or on the Reports of the Surveyors of ...

Further particulars of Vessels classed by Lloyd's Register will be found in Lloyd's Register Book and Lloyd's Yacht Register

No.	Cote	Nom. Matériaux et Gréement. Capitaines et dates d'entrée au service des Armateurs actuels et à bord du navire. Ancien Nom.	Pavillon.	Tonnage	Dimensions. Longueur	Largeur	Creux	Construit. Où.	Quand	Par qui.	Date du doublage en métal	Armateurs ou Directeurs-Gérants.	Port d'Armement.
1701	V	Anna *Heinecken* -86 W Bm Sr	German	347	128·2	24·9	12·9	Brake	1860	J.Oltmann Wwe.		H.P.Breckwoldt.jr.	Blankenese
1702	DV	Anna *Drügger* 82- W Bm Sr	German	273	114·8	22·7	13·0	Papenburg	1882	A.Beckmann		A.Beckmann	Papenburg
1703		Anna *Stark* 88-85 W Bg	German	269	98·0	26·7	14·6	Ueckermünde	18 13/64	L.Wittenberg		E.Patzig	Ueckermünde
1704	GV	Anna *Wöhle* -76 W Bn	German	224	103·9	25·3	12·4	Elsfleth	1871	J.Ahlers		H.G.Lübken	Elsfleth
1705		Anna W Bg	German	214	96·2	25·1	13·6	Wolgast	1855	W.Lubeke		E.Wendt.jun.	Danzig
1706		Anna (BROKEN UP) *Wasow* W Bg	German	190	88·0	25·3	13·1	Stralsund	1858	J.Kasten		H.Bartels	Stralsund
1707		Anna *Strmann* 77-77 W Bn	German	178	96·1	23·6	11·6	Barslenfleth	1862	D.Focke		C.A.Beng	Stralsund
1708	GV	Anna *Laarmann* 69-89 W Bn	German	177	98·7	24·9	11·0	Emden	1869	H.Klattenhoff		H.Kappelhoff Wwe.& Sohn	Emden
1709	G	Anna *Breymann* -82 W Sr	German	139	89·3	21·2	11·0	Papenburg	1853	F.W.Meyer		J.Dieckhaus	Papenburg
1710	V	Anna *Pallman* 86-86 W Bn (ex Ebbo Bodkey)	German	111	86·3	19·9	10·6	Sappemeer	1874	J.Berg.jun.		J.G.Sissingh	Weener
1711	G	Anna (WRECKED) *Feliru* 79-79 W Sr	German	107	81·0	19·5	9·5	Petkum	1866	J.J.Lucht		J.J.Lücht	Petkum
1712		Anna W Sr	Greek	234	112·5	24·3	..	Genoa	1880			P.Svoronos	Argostoli
1713	R	Anna *Hansen* 76-86 W Bk (ex Lady ...)	Norweg'n	967	177·0	37·7	22·0	Sunderland	1859	Haswell		J.M.Bryde	Sandefjord
1714	N	Anna *Gudmundsen* -84 W Bk (ex Elizabeth Yeo)	Norweg'n	828	167·5	35·2	21·3	Quebec	1856	Dinning		Kt Actieselskab (J.Ihlen)	Christiania
1715	N	Anna W Bk	Norweg'n	630	132·5	32·1	20·3	St.John.N.B.	18			Chr.Swane	Christiania
1716	N	Anna *Salvesen* 81-84 W Bk	Norweg'n	595	144·7	31·7	19·2	Frankfort.Me.	1859			S.Salvesen	Tvedestrand
1717	N	Anna *Unger* 86-86 W Bk	Norweg'n	547	113·8	30·1	18·3	Richibucto	1857	L.Des Brisay		C.Unger	Fredrikstad
1718	R	Anna (ex Sonnet) *Müller* W Bk	Norweg'n	522	130·7	32·5	19·6	Brahestad	1874			C.Moller	Arendal
1719	N	Anna *Nielsen* -88 W Bk	Norweg'n	458	130·7	30·3	16·8	Lillesand	1865	A.Stiansen		I.A.Henschien	Lillesand
1720	N	Anna *Johannesen* W Bk	Norweg'n	379	121·9	29·4	15·5	Lillesand	1863	J.Stiansen		M.Sundt	Moss
1721	N	Anna (WRECKED) *Ris* W Bk	Norweg'n	324	112·6	27·7	15·2	Grimstad	1866	Jorgen Tellefsen		G.Huseland	Grimstad
1722		Anna (BURNT) *Sorensen* W 3m Sr	Norweg'n	308	117·1	30·7	12·7	Christenstad	1882			H.U.Singdahlsen	Laurvig
1723	DN	Anna *Grön* -86 W Bk	Norweg'n	290	126·4	28·3	11·8	N.Brunswick	1858			P.A.Grön	Sandefjord
1724	N	Anna *Johnsen* -83 W Bg	Norweg'n	278	112·6	27·6	14·2	Tvedestrand	1874	T.Gundersen		J.P.Pedersen	Tvedestrand
1725	N	Anna *Wingaard* W Bk	Norweg'n	277	126·0	27·4	15·0	Christiansund	1869	J.Wessell		C.E.Rönneberg & Sönner	Aalesund
1726	N	Anna *Aas* W Bg	Norweg'n	219	116·8	26·0	11·1	Sarpsborg	1870	I.Olsen		P.Anker & Brödrene Amundsen	Sarpsborg
1727		Anna (ex Amalia) W Sr	Norweg'n	217	104·3	24·0	12·7	P.E.Island	1866			J.F.Johansen	Fredrikshald
1728	N	Anna *Arnesen* W Bg	Norweg'n	214	100·7	27·4	11·7	Riga	1866			O.Arnesen	Tönsberg
1729		Anna (ex Horder) *Solsberg* -74 W Bg	Norweg'n	198	102·5	23·5	13·1	Scotland	18 71			H.U.Singdahlsen	Laurvig
1730	N	Anna *Isaksen* 84-84 W Bg	Norweg'n	172	89·1	24·7	11·8	Bayonne	1852			S.Eeg	Brevig
1731	N	Anna *Andersen* 81-81 W Bg	Norweg'n	162	85·6	26·5	12·1	Sandefjord	1854			G.Gjertsen	Sandefjord
1732	N	Anna *Halvorsen* 84-84 W 3m Sr	Norweg'n	147	97·4	23·4	10·9	Gothenburg	1867			Ole Halvorsen	Brevig
1733		Anna (ex Ann) *Osmundsen* 86-86 W Sr	Norweg'n	122	77·9	22·7	10·7	Sunderland	1859	J.Robinson		Emil Syvertsen	Christianssand
1734		Anna *Sandström* -82 W Bk (ex Geflon)	Russian	518	145·0	33·1	17·1	Wasa	1867			M.V.Michelson	Nagu
1735		Anna *Wallenius* W Bk	Russian	420	127·6	29·2	14·6	Kimito	1874	O.Ylander		C.H.Lindroos	Kimito
1736		Anna *Westerlund* -86 W Bk	Russian	354	137·0	31·7	15·0	Kimito	1873	O.F.Olander		C.G.Holmberg	Åbo
1737		Anna *Sjölund* W 3m Sr	Russian	320	..	..	..		1879			E.Sjölund	Lemland
1738		Anna *Backman* W Bg	Russian	235	..	..	..		1869			K.Aladin	Fredrikshamn
1739	V	Anna *Stegge* -75 W 3m Sr	Russian	280	120·0	27·5	13·9	Riga	1867	Harimatschewsky		T.Renny & Co.	Riga
1740		Anna *Kuffschinoff* -83 W Sr	Russian	235	99·5	28·3	13·0	Åbo	1875	Kelman		J.Kuffschinoff	Åbo
1741		Anna *Janson* W Sr	Russian	132	83·3	25·0	10·9	Tolby	1875	K.Stuns		E.P.Akerlund	Lemland
1742		Anna W	Russian	119	..	..	..		1877			I.Judin	Impilaks
1743		Anna W Sr	Russian	110	76·0	22·0	8·0	Kärbis	1877			Wilhem von Aderkas	Haynasch
1744	DV	Anna *Uddenberg* 83 W Bk	Swedish	516	142·3	29·2	17·1	Umea	1883	A.Löfgren		Rederi-Bol.Union	Stockholm
1745		Anna *Johansson* 82-82 W Bk	Swedish	354	133·5	25·2	14·4	Hamburg	1860			P.A.Johansson	Florön Blhn.
1746	DR	Anna (ex ...) *Larsson* 83-84 W Bk	Swedish	315	108·6	24·5	16·1	Sunderland	1866			A.E.Österberg	Sölvesborg
1747		Anna (ex ...) *Mårtensson* W Bg	Swedish	295	104·0	24·6	16·5	Sunderland	18	O.Broden		A.Akesson	Cimbrishamn
1748		Anna *Ingvardson* 81-81 W Bk	Swedish	256	108·2	24·2	13·7	Rostock	1861	Joachim Möller		J.D.Björkegren	Cimbrishamn
1749		Anna *Jansson* -72 W Bg	Swedish	253	108·9	24·5	13·0	America	1865			O.Torbernisson	Kärehogen, Orust
1750	V	Anna *Pettersson* -87 W Bg	Swedish	243	115·8	25·5	12·9	Vestervik	1857			M.Ohlssohn	Oscarshamn

*=Lloyd's Register; D=Nederlandsche Vereeniging; G=Germanischer Lloyd; I=Registro Italiano; R=American Record; A=Veritas Austro-Ungarico; H=Veritas Ellenico; N=Norske Veritas; V=Bureau Veritas.

Dans la 4me Colonne, C=Composite (membrure en fer, bordé en bois); I=Fer; Stl=Acier; W=Bois.

Les Dates du doublage en métal (Col. 11) ressortent des livres dans lesquels les navires sont respectivement classés ou des rapports des Inspecteurs du Lloyd's Register.

Pour plus amples renseignements sur les navires classés par le Lloyd's Register, voir le Lloyd's Register Book et le Lloyd's Yacht Register.

Official No.	No. in Book	Registry in which classed	Ship's Name. Material and Rig. Master, and Dates of Appointment to present Owner's Service and to Ship. Ship's Late Name.	Flag.	Net Tonnage	Length	Breadth	Depth	Where	When	By whom	Date of Metalling	Owners or Managers
	1751		Anna WBg		230	94·2	24·3	13·7	Wismar	1866	Barmann		
	1752	G	Anna (ex Anna Dorn) Petersson -82 WSr	Swedish	227	103·0	26·2	12·1	Russia	1874			H.Pettersson
	1753		Anna (ex Maack) Moberg 77 WBg	Swedish	170	87·2	22·4	12·0	Königsberg	1865			B.Bengtsson
	1754		Anna Scanberg WSr	Swedish	167	90·9	23·0	11·3	Finland	1867			A.Lundberg
	1755	G	Anna* Johansson WSr	Swedish	164	100·0	21·4	11·4	Karlshamn	1862	D.R.Andersson		P.N.Horndahl
	1756		Anna Johanson -83 WSr	Swedish	163	90·6	22·7	11·6	Skedbo	1867	Isaksson		J.A.Samuelsson
	1757		Anna (STRANDED) (ex Nicola Hetton) Jönsson -88 WSr	Swedish	124	86·0	20·0	10·0	Hoogezand	1861	W.VanderWerf		H.P.Jönsson
	1758		Anna Andersson 82 WSr	Swedish	116	73·2	21·1	11·3	Strömstad	1852			C.Andersson
849	1759		Anna W3mSr	Swedish	105	98·2	20·5	7·2	Sjötorp (reblt.)	18 84			V.Vallin
	1760		Anna A. Holton Gilchrist -81 WSr	American	137	89·7	23·5	7·5	Wilmngtn,Del	1866			A.Levensaler
	1761	V	Anna Agathe Chenot -87 WSr	French	112	70·5	19·3	11·1	Nantes	1858	Bonnissant		A.Chapron
	1762	DV	Anna Aleida Blecker 87 I Bk	Dutch	1102	218·8	35·1	20·7	Krimpen	1887	J.&K.Smit		P.Van der Hoog
106172	1763	R	Anna B. Hutcheson Zeluff W3mSr	American	239	119·0	29·5	10·4	CityIsl.,N.Y.	1883	D.Carll		Miller,Bull & Co.
	1764	GV	Anna Bertha Nielsen -87 WBk	German	469	133·5	26·9	16·7	Kiel	1875	C.Thino	Y1.86	August Piehl
	1765	G	Anna Bertha Horndahl -81 W3mSr	Swedish	213	105·3	23·7	11·9	Mönsterås	1868	O.L.Nordstrand		P.N.Horndahl
74862	1766	*	Anna Braunschweig Jones WBkn	British	246	116·3	25·5	14·5	Portmadoc	1876	Roberts	Y 2.86	D.Jones & Co.
900	1767		Anna C. (ex Solo) W3mSr	Italian	464	139·4	28·7	19·9	Sestri Ponente	1863	D.Bianchi		V.Firpo
3362	1768	I	Anna C. (ex Maria Josepha) Villa WBn	Italian	141	108·2	22·0	11·6	Sestri Ponente	1884	A.Briasco fu L.	Y 9.86	A.&C.Liberti
	1769	DNV	Anna Camp Samuelsen -88 WS	Norweg'n	1202	182·4	36·0	21·1	Bath,Me.	1864	W.&J.Drummond	Y 6.86	C.Larsen
	1770		Anna Casper (ex Village Belle) Jans 18-78 WBk	Danish	612	142·9	31·7	20·3	Waldoboro,U.S.A.	1859			J.P.Schmidt
	1771		Anna Catharina Fisker WSr	Danish	134	92·4	20·6	10·5	Veile	1869	P.H.Illum		C.Nissen
220	1772		Anna Charlotte Lossmann WSr	Russian	223	108·5	24·4	12·7	Kürbis	1876	J.Grass		Schnore & Sauss
59958	1773		Anna Craig WSr	British	260	129·0	26·2	11·0	Quebec	1870			McArthur & Co.
54477	1774		Anna Currier WSr	British	100	84·0	24·6	7·5	Up.Gagetown,[N.B.	1867			R.A.Welsh
46735	1775		Anna D. Jordan WBn	British	334	110·9	27·9	11·2	Wildsboro,US	1858			G.H.Tobias
277	1776		Anna D. Torrey WBg	American	220	108·8	27·4	9·5	Stockton,Me.	1854			J.P.Ellicott
	1777	N	Anna Delius (ABANDONED) Iversen -81 WBk	Norweg'n	701	147·6	32·7	21·4	Bremen	1854			NOYoung & NTIversen
277	1778		Anna Dora WSr	Russian	104	69·2	20·1	9·5	Wandsen	1876			J.Mikkelsohn
	1779		Anna Dorothea Wenzel [88-87 WBk	German	523	129·5	31·2	19·5	Danzig	1869	J.W.Klawitter		Schoenberg&Domansky
	1780	V	Anna Dorothea Saathoff [-83 WSr	German	103	80·4	20·5	9·6	Veendam	1865	E.vanLinge		S.C.Campen
	1781		Anna Dorothea WBk	Japanese	343	129·0	29·5	11·5	Apenrade	1871	N.Jacobsen	Y 10.86	
105118	1782		Anna E. Babcock Lee W3mSr	American	379	121·0	31·0	14·8	Absecom,N.J.	1872			J.G.Babcock
522	1783		Anna E.Cranmer Cranmer WSr	American	203	110·1	29·8	9·6	Allowaytn,N.J	1867			A.R.Pharo
	1784	R	Anna E. J. Morse Lowell W3mSr	American	696	162·8	35·1	13·0	Bath,Me.	1886	B.W.&H.F.Morse		B.W. & H.F.Morse
105404	1785	R	Anna E. Ketchum Ketchum W3mSr	American	371	126·0	32·0	14·5	Chester,Pa.	1874			A.Ketchum
105225	1786		Anna E. Kranz Fevere (73-73 W3mSr	American	648	158·7	34·1	17·2	Bath,Me	1873	A.Hawthorne		R.R.Freeman
809	1787		Anna E. Safford Camp [-86 WSr	American	187	112·0	22·0	9·4	Philadelphia	1865	Philadel& Reading Railrd.Co.		Watson Malone & Sons
	1788	V	Anna Elisabeth Larsen [80-80 WSre	Danish	113	89·9	20·2	9·9	Svendborg	1879	P.Troensegaard		P.Troensegaard
	1789	N	Anna Elisabeth Olsen [80-82 WBn	Norweg'n	161	93·2	24·9	12·2	Stavanger	1879		Y3.87	T.Kleiberg
	1790		Anna Elisabeth Stahl WSr	Russian	124	82·2	25·2	9·6	Angern	1876	A.Kirstein		J.Stahl
	1791		Anna Elise Olsson 83-85 WSr	Swedish	167	91·9	22·6	11·8	Eckerna	1872	A.Ahlberg		E.Olsson
94	1792		Anna Elizabeth WSr	American	144	91·8	26·6	8·5	Hampden,Me.	1853			B.F.Smith
	1793		Anna Emilia Grant WSr	Russian	117	81·7	26·6	10·7	Orrenhof	1871	M.Grant		O.Grant
	1794		Anna Fanny Hult WSr	French	115	77·8	19·0	10·4	Granville	1861	Gombert		G.Thomas
1888	1795		Anna Frye WSr	American	122	93·0	26·1	8·0	Pembroke,Me	1869	Wilder		D.Gallagher
	1796	V	Anna Hauswedell (CONDEMNED) Schröder 85-86 W3mSr	German	362	120·0	28·0	14·5	Geestemünde	1875	S.Oltmann	Y 3.86	H.Hoh
86965	1797		Anna Helena WBn	British	160	102·1	11·1	12·1	Pekela	1866			J.Danby
97	1798	G	Anna Helena Lilienthal (80-86) WSr	Russian	203	99·1	26·8	12·4	Gutmannsbch	1880	P.Jöggi		J.Martinsohn
	1799		Anna Helena Nyström WSr	Swedish	137	80·0	19·1	12·3					J.E.Nystrom
	1800	V	Anna Herbertha (ex Kalliope) Wolter -86 WBk	Dutch	1090	181·7	36·7	23·8	Hamburg	1876	B.Wencke Söhne	Y 10.87	Merrein & LaPorte

*=Lloyd's Register; D=Nederlandsche Vereeniging; G=Germanischer Lloyd; I=Registro Italiano; R=[illegible]
A=Veritas Austro-Ungarico; H=Veritas Ellenico; N=Norske Veritas; V=[illegible]

In the 4th Column, C=Composite (iron frame, wood planked); I=Iron; Stl=Steel; W=Wood.

The Dates of Metalling (**Col. 11**) are based on the records of the Books in which Vessels are respectively classed, or on the Reports of the Surveyors of Lloyd's

Further particulars of Vessels classed by Lloyd's Register will be found in Lloyd's Register Book and Lloyd's Yacht Register.

No.	Classe	Nom et Capitaine	Gréement	Pavillon	Tonnage	Long.	Larg.	Creux	Construit: Lieu	Date	Par qui	Doublage	Armateurs ou Directeurs-Gérants.	Port d'Armement.
1801	G	Anna Jacoba *Mulder*	W Bg	Dutch	220	100·9	25·5	12·3	Arendal	1856	C.H.Christensen		A.W.Hutjes	Nieuwediep
1802		Anna Katharina	W Sr	[illegible]	282	106·8	25·0	13·1	Livland	1876	J.Strasding		J.Mikkelsohn	Haynasch
1803		Anna Katharina *Kalning*	W Sr	[illegible]	160	78·0	21·0	11·0	Haynasch	1875	J.Kalning		J.Kalning	Haynasch
1804		Anna Kirstine	W Sr	[illegible]	[illegible]	92·1	20·8	10·7	[illegible]	1868	I.C.Sørensen		N.Jensen	Svendborg
1805		Anna Kristina	W Sr	[illegible]	[illegible]	77·4	19·1	9·2	Mence				L.Hansson	Strömstad
1806	R	Anna L. Mulford *Tilton*	W3mSr	American	518	142·9	34·8	15·9	Millville,N.J.	1879	J.D.Bell		F.Smith	Philadelphia
1807	R	Anna L. Taylor *Proctor*	W Bk	British	314	129·0	26·6	18·0	Danzig	1865		Y 4.88	Mark Shaw	Turks'Island
1808		Anna Lina *Nitzmann*	W Sr	Russian	244	104·5	28·8	13·8	Poderage	1883	P.Krauss		J.Nitzmann	Riga
1809		Anna Lindsley *Kavanagh*	W Bn	British	193	104·0	27·0	11·5	Hopewell,N.J.	1868			W.W.Robinson	Dublin
1810	A	Anna M. *Percich*	W Bk	Austrian	390	135·5	25·7	17·0	Fiume	1867	Fili Schiavoni	Y	G.Minach, Jr.	Fiume
1811		Anna M. Dickinson *Hart*	W Sr	American	[illegible]	96·5	28·4	8·9	P.Jefferson,NY	1871			A.Hart	Edgartown
1812		Anna Madre *Arpe*	W Bn	Italian	199	73·5	21·2	12·2	Sestri Ponente	1860			L.Pendibene fu E.	Genoa
1813		Anna Margaretha *Oorlogg*	W Bk	Dutch	687	136·1	34·1	21·8	Zierikzee	1855	Maka		W.H.de Wolf	Amsterdam
1814	V	Anna Margaretha	W Sr	German	104	77·6	17·6	8·8	Papenburg	1886	F.W.Meyer Wwe		F.W.Meyer Wwe	Papenburg
1815	V	Anna Margrethe	W3mSr	Danish	229	109·0	22·7	12·2	Bremen	1862	G.Knickmann		H.J.Hansen	Kastrup
1816	*	Anna Maria *Fitzwalter*	W Bg	British	244	111·7	24·6	14·1	Kingsbridge	1868	Date		H.Grant & Co.	Salcombe
1817	R	Anna Maria	W3mSr	Danish	434	131·0	25·0	16·2	Brooklyn	18 71		Y 4.88	H.D.Prast	Dragör
1818	V	Anna Maria *Lage*	W Sr	French	118	77·8	20·0	10·3	Nantes	1876	T.Dubigeon & fils		T.Dubigeon & fils	Nantes
1819		Anna Maria	W Sr	Russian	192	93·0	24·0	11·0	Uzzeneem	1873	M.Leep		B. & M.Leep	Riga
1820	V	Anna Maria *Dietrichsohn*	W Sr	Russian	185	89·4	25·1	12·3	Upesgriwi	1887	Ado Sepp		J.Stahl	Riga
1821		Anna Maria *Leimann*	W Sr	Russian	183	97·7	26·7	11·8	Abegrief	1874	M.Leppe		K.Leimann	Riga
1822		Anna Maria *Kapkaln*	W Sr	Russian	124	89·1	22·8	11·1	Roja	1884	Sivingsohn		Kapkaln & Resewsky	Riga
1823		Anna Maria	W Sr	Russian	118	85·0	26·5	9·8	Uggenzeem	1881	O.Kirschstein		M.Leepa	Riga
1824		Anna Maria	W Sr	Russian	108	..	..	..	Säkkjärvi	1875			V.A.Sjölund	Helsingfors
1825	R	Anna Maria *Malmstrom*	W S	Swedish	506	138·1	25·9	18·7	N.Brunswick	1867		Y 12.85	J.D.Björkegren	Cimbrishamn
1826	I	Anna Maria D'Abundo	W Bk	Italian	558	141·0	29·2	19·7	Ischia	1876	L.Lubrano	Y 4.85	E.D'Abundo	Naples
1827		Anna Marie *Grewe*	W Sr	Russian	168	96·9	24·9	11·0	PetersKapelle	1867	M.Leep		J.Grewe	Riga
1828		Anna Mary	W Bg	British	184	89·1	22·9	14·4	Fraserburgh	1841			W.Wallace	Maryport
1829	D	Anna Olga	W3mSr	Russian	268	104·0	25·0	13·0	Kurbis	1879	G.Grass		Grass & Trautmann	Riga
1830		Anna Olga *Grewe*	W3mSr	Russian	258	111·0	27·7	13·1	Adiamunde	1874	J.Jacobsohn		J.Grewe	Riga
1831		Anna Olga *Behrsing*	W Sr	Russian	198	93·1	27·4	12·3	Wandsen	1877	P.Krause		H.&J.Damkaln	Riga
1832		Anna Olga *Wirsu*	W Sr	Russian	102	..	..	..		1877			A.Wirsu	Säkkjärvi
1833		Anna Ottilie *Salming*	W3mSr	Russian	296	121·8	28·8	13·7	Muhlgraben	1872	A.Ferle		H.Lyra	Riga
1834	DG	Anna Ottilie *Fischer*	W Sr	Russian	232	106·5	28·5	13·4	Kesterzeem	1883	J.Schall		K.Lohnfeld	Angern
1835	V	Anna Precht *Foss*	W Bk	German	429	127·7	29·5	18·1	Damgarten	1875	H.Dierling	Y 4.88	W.Maack	Rostock
1836	R	Anna R. Bishop *Ralon*	W3mSr	American	426	146·0	34·2	12·0	Wilmngtn,Del	1880	Jackson&SharpC	Y	J.E.Ridgway	Philadelphia
1837		Anna Rivarola	W Bn	Greek	273	111·8	25·3	15·2	Chiavari	1880	L.Gotuzzo	Y	IsvaronatoFocione	
1838		Anna Rosa *Giannelli*	W Bn	Italian	172	93·8	24·0	12·0	Voltri	1872	C.Patrone	Y 11.87	G.B.Giannelli	P.Ferraio
1839		Anna S. Murch *Adams*	W Sr	American	108	91·5	27·0	7·1	Ellsworth,Me.	1873			J.Davis	Ellsworth,Me.
1840		AnnaSarah *Pickering*	W Sr	British	141	86·5	21·5	11·0	Sunderland	1860	Metcalf		F.A.Christie	Ipswich
1841		Anna Shepard *Piper*	W Sr	American	166	94·4	28·0	7·6	Brooklyn	18 63/69			C.L.Gifford	Camden,Me.
1842	R	Anna T. Ebener *Springer*	W3mSr	American	474	138·2	34·2	15·4	Millville,N.J.	1883	J.D.Bell		H.Matthews	Philadelphia
1843	G	Anna Thormann *Witt*	W Bk	German	405	126·6	28·7	17·3	Rostock	1874	J.Möller	Y	J.C.Thormann	Wismar
1844	R	Anna V. Lamson	W3mSr	American	321	137·5	32·0	9·6	Milford,Del.	1888	J.W.Abbott & Co		E.A.Gaskill	Philadelphia
1845	DGV	Anna Victoria *Weide*	W Sr	Russian	234	98·0	23·0	12·0	Haynasch	1878	M.Magnus		G.Weide	Haynasch
1846	R	Anna W. Barker *Sargent*	W3mSr	American	284	105·5	27·4	12·4	Castine,Me.	1873		Y	J.W.Dresser	Castine,Me.
1847		Anna W. Collins	W Sr	American	142	95·0	27·0	7·0	P.Jefferson,NY	1864			R.Hardcastle	P.Jefferson,NY.
1848	V	Anna Wichhorst	W3mSr	German	385	130·0	30·0	13·0	Altona	1879	J. H. Wichhorst & Co.	Y	H.Holi	Blankenese
1849		Anna & Bertha	W Bk	German	336	108·0	28·2	16·6	Danzig	1856	J.M.Petermann		G.F.Focking	Danzig
1850		Anna & Ella Benton *Ireland*	W Sr	American	131	88·0	28·3	7·8	Tuckahoe,N.J.	1872			Benton,Sons & Co.	SomersPoint,NJ

* = **Lloyd's Register**; D = Nederlandsche Vereeniging; G = Germanischer Lloyd; I = Registro Italiano; R = American Record; A = Veritas Austro-Ungarico; H = Veritas Ellenico; N = Norske Veritas; V = Bureau Veritas.

Dans la 4me Colonne, C = Composite (membrure en fer, bordé en bois); I = Fer; Stl = Acier; W = Bois.

Les Dates du doublage en métal (Col. 11) ressortent des livres dans lesquels les navires sont respectivement classés ou des rapports des Inspecteurs du Lloyd's Register.

Pour plus amples renseignements sur les navires classés par le Lloyd's Register, voir le Lloyd's Register Book et le Lloyd's Yacht Register.

Official No.	No. in Index	Registry by which Classed	Ship's Name, Material and Rig. Master, and Dates of Appointment to present Owner's Service and to Ship. Ship's Late Name.	Flag.	Net Tonnage.	Length	Breadth	Depth	Where.	When	By whom.	Date of Metalling.	Owners or Managers.
	1851	V	Anna & Marie *Schurmberg 87-87* WBk	German	442	130·8	26·2	15·9	Wolgast	1864	W.Lütcke	[illegible]	J.J.Wallis & Sohn
	1852	R	Anna & Mary Carlisle *Keville* W3mSr	American	363	135·4	22·6	10·8	Milford, Del.	1884	Simpler & Keville		J.May & Son
	1853	V	Anna & Meta *Pahl 73* WBkn	German	248	112·2	26·5	13·5	Damgarten	1863	H.Dierling		W.Lange
55213	1854		Annabella Clark *Sharp [75-76* WBg	British	141	98·3	28·3	11·5	Ardrossan	1866	Barclay & [Shearer		A.Guthrie & Co.
[illegible]	1855		Annandale WSr	British	18[illegible]	111·8	24·9	9·1	Kingston,Ont	18 65/73	J.Carruthers		AlexanderPurser
[illegible]	1856		Annapolis *Bartaby* (WRECKED) WBk	British	904	175·1	35·4	20·8	Annapolis,N.S	1873		[illegible]	J.Woodworth
	1857		Annchen *Meeryanssen -87* (WRECKED) WBn	German	21[illegible]	103·4	25·0	12·7	Elsfleth	1865	J.Jürgens		G.J.Schulte's Wwe.
	1858	G	Annchen *Felt -77* WSr	German	167	92·0	21·3	11·6	Papenburg	1855	H.W.Meyer		W.Höveler
	1859	G	Annchen Lorenz *Moritz [74-77* WBg	German	19[illegible]	97·8	23·6	13·0	Barth	1864	Brakenwagen		W.A.Sarnow
53973	1860		Anne WSr	British	104	81·1	25·0	8·8	Geelong	1863			N.McLeod
75158	1861	❊	Anne (COLLISION) *Stribley 85* WK	British	59	70·5	19·8	8·8	[illegible]	1877	Burt		C.B.Kelway
	1862	V	Anne (WRECKED) *Pedersen -88* WBg	Danish	229	111·2	26·2	11·8	[illegible]	1872	C.A.Hahn	[illegible]	P.Brinch
	1863	V	Anne *Jensen* WBn	Danish	149	100·0	22·4	10·3	Marstal	1872	J.J.Bager		H.C.Christensens Enke
	1864	V	Anne *Jörgensen 86-86* WSr	Danish	124	87·9	21·8	10·[illegible]	Marstal	1876	L.J.&H.J.Bager		H.J.Bager
	1865		Anne WSr	Russian	131	84·9	26·7	10·4	[illegible]	1876	J.H.Sotka		A.Uusitupa
	1866	V	Anne Charlotte *Pedersen [76-76* WBn	Danish	153	96·5	22·3	11·1	[illegible]	1875	T.Schønan	[illegible]	Actieselskabet (J.N. Jensen)
	1867	R	Anne Cheshyre *Roullet 87-87* WBk	French	389	129·7	24·0	18·0	[illegible]	1853	Brundrit & [Whiteway	[illegible]	J.Lefauve
27905	1868	❊	Anne Davies *Parry 78-87* WSr	British	92	80·0	20·3	10·6	Aberdovey	1859	Jones		H.Parry
3187	1869	❊	Anne Duncan *Bryant [-80* WBkn	British	278	127·3	24·4	14·0	Ardrossan	1855	Barr & Shearer	[illegible]	Goodyear & Co.
	1870	N	AnneElisabeth *Johnsen-87* WBk	Norweg'n	435	115·7	30·1	16·7	[illegible]	1854	C.Klausen	[illegible]	L.Björnstad
[illegible]	1871	❊	Anne Holland *Griffith [83-83* WSr	British	128	84·7	23·1	12·0	Portmadoc	1868	Griffiths	[illegible]	O.Griffith
6875	1872		Anne Isobella *Allen* WBg	British	168	83·0	21·1	14·0	Montrose	1854	J.& D.Birnie		J.Boyes
	1873	V	Anne Jørgiane II. *Jessen -87* WBg	Danish	223	109·5	24·3	12·2	Nordby, Fanø	1875	N.Nielsen	[illegible]	AktSelsk(M.M.Andersen)
106208	1874	R	Anne Lord *Kendall 83-83* W3mSr	American	273	127·0	32·5	10·4	Brewer, Me.	1883	J.Nealy	[illegible]	H.Lord
[illegible]	1875	❊D	Anne Main *Black 81-81* IBk	British	499	156·6	28·0	16·8	Glasgow	1867	A.Stephen&Sns		A.Weir & Co.
	1876		Anne Margrethe *Bull* WBg	Norweg'n	258	91·2	26·0	15·8	Tvedestrand	1848			John Holst
	1877	V	Anne Marie *Clausen* (LOST) WBg	Danish	176	103·0	22·4	11·4	Nordby, Fanø	1872	J.N.Christensen	[illegible]	Act.S.(P.N.Harbye)
	1878	V	Anne Marie *Codet 86* (ex Marie) WBk	French	288	104·6	23·0	15·8	Bilbao	1870	A.deArena	[illegible]	K.Morvant
	1879		Anne Marie *Galesne -87* WBg	French	204	93·5	22·3	13·1	St. Malo	1867	L.Mallard		Guibert & fils
	1880	N	Anne Marie *Johnsen -86* WBg	Norweg'n	277	92·8	27·5	15·7	Risör	1838			C.Wroldsen
1810	1881		Anne Melhuish WBk	British	344	112·0	26·2	17·6	[illegible]	1849			D.Williams
89317	1882		Anne Powell WSr	British	123	109·0	24·4	6·9	Macleay River	1884	W.Piper		Ann.Powell
26278	1883		Anne & Sophia *Clothier 83-83* W3mBn	British	195	106·1	23·0	13·0	Cardiff	1845	Tredwen		F.W.Fairbrass
78360	1884	❊	Annesley *Ewart* (ex British Enterprise) IS	British	1635	246·0	40·1	23·9	Stockton	1876	Richardson [Duck & Co.		Johnston,Sproule & Co.
30	1885		Annetta WBg	British	251	99·8	27·0	15·0	[illegible]	1869	J.Canatus		P. DeCandia
1054	1886	A	Annetta (ex Esso) *Tulbo -88* WBk	Italian	420	127·3	30·8	17·9	Trieste	1867	E.Strudthoff	[illegible]	A.DeGioja
3052	1887	I	Annetta *Dodero 75-80* I3mSr	Italian	185	109·5	20·5	13·4	Sestri Ponente	1880	N.Odero e Ferro		F[lli] Bozzano
662	1888	I	Annetta *Sturlese* WBn	Italian	172	92·8	22·8	12·4	[illegible]	1874	A.Bardellini	[illegible]	S.Sturlese
	1889	A	Annetta (ex Salomone J.) WSr	Russian	102	..	..	..		1859		[illegible]	G.Damulachi
	1890		Annetta Covacevich WBk	Russian	510	..	..	..	[illegible]	1862			Spartali
	1891	V	Annette Jantiene (ex Elise) *van Dijk -87* WBk	Dutch	513	145·3	31·1	17·8	[illegible]	1878	G.Raben	[illegible]	F.T.Nijhoff
	1892	V	Anni (ex Garibaldi) *Portmann 77-88* WS	German	1336	192·2	39·4	28·2	[illegible]	1860	Mason & Fish	[illegible]	Boyes & Ruyter
	1893		Annibal WBg	Portug'se	279	102·3	25·0	13·9	St. Martinho	1853			A.D.S.Leal
50565	1894		Annie WBk	British	428	115·0	29·0	18·2	[illegible] N.B.	1864	Wright		John Panton
37741	1895		Annie (NOT NOW SEAGOING) WBk	British	248	109·0	25·5	13·1	[illegible] N.S.	1853			W.Elliott
62116	1896		Annie *Thomson* WBn	British	194	94·3	26·3	11·6	[illegible]	1865	McLeod		A.Thomson
71437	1897		Annie *A.McConaig* WBn	British	173	104·5	25·1	12·5	P.E.Island	1874	Egan		A.&J.McConaig
59753	1898		Annie *Wilcox* WBn	British	149	97·6	24·5	11·8	P.E.Island	1869	Carvell		T.Grenfell, jun.
58873	1899	❊	Annie *Evans 75-87* WSr	British	115	88·0	21·7	10·0	Peterhead	1867	Robins		H.Evans
91754	1900		Annie WK	British	106	81·6	22·3	7·7	Moulmein				C.C.Wyse

❊ = Lloyd's Register; D = Nederlandsche Vereeniging; G = Germanischer Lloyd; I = Registro Italiano; R = [illegible]
A = Veritas Austro-Ungarico; H = Veritas Ellenico; N = Norske Veritas; V = [illegible]

In the 4th Column, C = Composite (iron frame, wood planked); I = Iron; Stl = Steel; W = Wood.

The Dates of Metalling (Col. 11) are based on the records of the Books in which Vessels are respectively classed, or on the Reports of the Surveyors of Lloyd[illegible]

Further particulars of Vessels classed by Lloyd's Register will be found in Lloyd's Register Book and Lloyd's Yacht Register.

No.	Classé	Nom. Matériaux et Gréement. Capitaine et dates d'entrée au service des Armateurs actuels et à bord du navire. Ancien Nom.		Pavillon.	Tonnage net.	Longueur	Largeur	Creux	Où.	Quand	Par qui.	Dates du doublage	Armateurs ou Directeurs, Gérants.	Port d'Armement.
[illegible]	✱	Annie *Elder*	W Sr	British	[illegible]	88·6	21·2	10·5	Garmouth	1883	J. Geddie		T. Henderson	Inverness
[illegible]	✱	Annie *Cossale 11-85*	W Sr	British	86	75·0	21·8	9·4	P.E.Island	1861	D. Grundy		J.W.Walsh & P.Lambert	Wexford
[illegible]	✱	Annie *Finimore -82*	W Sr	British	65	75·0	19·5	8·8	Barnstaple	1872	Westacott		H. Finimore	Plymouth
1904	G V	Annie *Frage 72-87*	W Bk	German	505	126·1	30·1	15·1	Hammelw'rde	1875	[illegible] Lilbring	[illegible]	Gebrüder Hustede	Elsfleth
1905	✱	Annie *Steinbrugge -87*	I Bk	German	424	116·0	27·1	16·5	Sunderland	1869	W. Pile & Co.		E.C.H. Heuser	Hamburg
1906	N	Annie *Kernsen 71-8*	W Bk	Norweg'n	551	140·3	30·5	18·8	St.John,N.B.	1862	Andersen	[illegible]	Carl Wleugel	Christiania
1907	N	Annie *Bastiansen 83-84* ex Evening Star	W Bg	Norweg'n	303	107·7	26·9	16·0	Sunderland	1865	J. J. Gibbon		J.E. Bastiansen	Drammen
1908		Annie	W Sr	Norweg'n	371	116·7	28·0	12·7	NovaScotia	1867			C.S.& F.Pfützenreuter	Christiania
1909	✱	Annie *Jonsson 85*	W Bkn	Swedish	307	123·4	27·6	14·9	P.E.Island	1878	Henry	[illegible]	E.T. Norrman	Malmö
1910		Annie *Nilsson -86*	W Bg	Swedish	299	119·1	25·7	12·7	Yarmouth, NS	1856			G. Eklund	Trelleborg
1911	V	Annie *Nilsson*	W Sr	Swedish	1[illegible]	92·3	27·7	11·9	Hjellö	1871	J. Ohlsson		A. Olsson	Fiskebäckskil
1912	R	Annie A. Booth *Hansen*	W 3m Sr	American	192	113·8	29·0	9·0	P Jefferson NY	1874	Bayles	Y [illegible]		Pt Jefferson, N.Y.
1913		Annie A. Grady ex Maria White	W Bn	British	187	94·3	25·6	10·7	Maine,U.S.	1854			W. Killam	Nassau, N.P.
1914		Annie Ada *Good 84-84*	W Bk	British	238	116·5	26·9	19·9	NovaScotia	1865	Sutherland		J.J. Carroll	Dublin
1915	R	Annie Ainslie *McFadden*	W 3m Sr	American	288	128·2	31·8	10·1	Camden,N.J.	1883	J.& F. McDonnell		J. McDonnell	Philadelphia
1916	R	Annie Batcholder *Sanders -82*	W Bg	American	442	123·5	31·2	15·8	Philadelphia	1866	Fisher	[illegible]	H.V. Steelman	Philadelphia
1917	G V	Annie Berner *Schultz 83-76*	W Bk	German	398	129·2	28·4	16·8	Rostock	1876	J. Möller	[illegible]	A. Burchard	Rostock
1918		Annie Bingay *Pinkney 82-85*	W S	British	1049	176·7	36·0	21·7	Gilbert's Cove [N.S.	1871		Y 2.[illegible]	Jacob Bingay	Yarmouth, N.S.
1919		Annie Bliss *O'Donnell*	W 3m Sr	American	317	135·9	30·1	11·3	Thomaston Me	1871	Samuel Watts		J.J. Dale & Co.	Savannah, Ga.
1920		Annie Bow	W Sr	British	250	112·6	21·8	14·5	Garmouth	1869	A. Spence		Alex. White	Lyttelton, N.Z.
1921		Annie Brown *Maloney -83*	W Bn	British	16[illegible]	90·5	23·6	11·9	Perth	1867	Perth S.B.Co.	Y 11.82	John Bickers	Adelaide, S.A.
1922	V	Annie Burrill *Bain 85-86*	W Bk	British	897	166·8	34·8	21·4	Digby, N.S.	1878		Y 12.86	W. Burrill, jun.	Yarmouth, N.S.
1923	R	Annie C. Grace *Grace*	W 3m Sr	American	491	140·7	31·4	15·1	Cooper's Point	1880	Morris & Mathes		S.T. Grace	Philadelphia
1924	R	Annie C. Moore *Hewlett -81*	W Sr	British	113	81·0	21·9	9·5	BonneBay	1883	G.R. Moffat		C.P. Moffat	St.John, N.F.
1925	✱	Annie Christian *Allen*	W Sr	British	69	76·5	20·5	9·0	Barnstaple	1881	W W'stacott & Ss		J. Allen	Liverpool
1926	✱	Annie Crossfield *Edmondson 10-83*	W 3m Sr	British	100	91·3	22·5	9·5	Carrickfergus	1883	P. Rodgers		J. Fisher & Sons	Barrow
1927		Annie D. Merritt	W Sr	American	137	92·6	21·8	9·2	BoothBay, Me.	1874			Warner & Merritt	Philadelphia
1928	R	Annie E. Blackman *Blackman -87*	W Sr	American	264	118·6	32·0	9·4	Goshen, N.J.	1883	Garrison		J.J. Blackman	SomersPoint, N.J.
1929		Annie E. Hall (MISSISSY) *Queene 16-[illegible]*	W Sr	American	129	85·0	27·0	16·6	Frederica, Del	1881	N. Lank		J.W. Hall	Wilmington, Del.
1930	R	Annie E. Rickerson *Post 83-*	W 3m Sr	American	238	113·5	29·7	10·6	Franklin, Me.	1882	John P. Gordon	Y 11.86	C.L. Rickerson	NewYork
1931	R	Annie E. Rudolph (WRRL&Sth) *Lewis*	W 3m Sr	American	186	116·7	23·8	8·0	Camden, N.J.	1884	J & F McDonnell		J. & F. McDonnell	Philadelphia
1932		Annie E. Stevens *Davis*	W 3m Sr	American	228	116·0	28·6	8·7	Chester, Pa.	1869			Elwood Doran & Co.	Philadelphia
1933	V	Annie E. Wright *Wright*	W S	British	1847	237·9	43·0	21·2	Harvey, N.B.	1885	G.S. Turner	Y 12.86	J.N. Smith	Dorchester, N.B.
1934	R	Annie F. Conlon *Bennett 16-83*	W 3m Sr	American	561	121·8	31·0	17·0	Portsm'th, N.H.	1882	Fernald		C.E. Walker	Portsmouth, N.H.
1935		Annie Fairfax *Gordon -86*	W Sr	American	107	83·1	23·3	8·5	Essex, Mass.	1875			F.C. Butman	Boston, Mass.
1936		Annie Falconer	W Sr	British	175	108·0	24·3	9·0	Kingston	1867	Foster, Jenkins [& Co.		Thomas F. Taylor	Kingston, Ont.
1937	✱	Annie Fletcher *Simpson 85-86*	I Bk	British	491	158·0	26·2	16·9	Lancaster	1868	LaneSh.B.Co.		A. Rae	Liverpool
1938		Annie Florence *Ney 83-83*	W Sr	British	115	96·5	23·1	8·7	Emsworth	1878	Foster		S. Payn	Faversham
1939		Annie G. (STELLA) *Cameron -86*	W Sr	British	116	87·0	27·3	7·7	Dorchester NB	1886	T.B. Wilbur		T.B. Wilbur	Dorchester, N.B.
1940	R	Annie G. O'Leary *Boudrot*	W Sr	British	143	92·8	24·9	9·8	W.Quoddy, NS	1882	W. Rutledge		R. Boudrot	Halifax, N.S.
1941		Annie Gee *Miller*	W Sr	American	147	112·0	30·8	8·6	P.Ludlow Wsh	1874			W.S. Browne	SanFrancisco
1942	V	Annie Goudey *Sanders 18-83*	W S	British	1135	190·0	36·6	22·4	Meteghan, NS	1873	Germain & [Jenkins	Y 6.82	L.E. Baker	Yarmouth, N.S.
1943	R V	Annie H. Smith *Cook -88*	W S	American	1452	200·0	40·0	24·5	Calais, Me.	1876	Nickerson & [Rideout	Y 2.82	F.H. Smith & Co.	NewYork
1944	✱	Annie Harris *Mayers 83-85*	W Bn	British	148	103·0	22·2	11·2	Bideford	1870	Johnson	Y 8.82	W.S. Allport & Co.	Bristol
1945		Annie Hiapwan (ex Annie)	W Sr	Dutch	299	133·8	26·9	..	Siam				Lim Liang Tjioe	Grissee
1946		Annie Hill	W Sr	British	121	89·6	22·5	10·0	HuonRiver	1876			W.S. Waterston	Invercargill, N.Z.
1947	V	Annie J. Marshall *Foster -88*	W Bk	British	1100	181·0	36·5	22·4	BearRiver, NS	1877	A. Marshall	[illegible]	A. Marshall	Digby, N.S.
1948		Annie J. Pardee	W 3m Sr	American	682	158·2	35·4	15·6	Bath, Me.	1882	Goss, Sawyer & [Packard		F.W. Benedict	Newhaven, Con.
1949	V	Annie Johnston ex Ada Iredale *Emerson -88*	I Bk	American	947	212·1	34·1	21·6	Harrington, [England	1872	Williamson		A. Crawford & Co.	SanFrancisco
1950	V	Annie L. Henderson *Peters -88*	W 3m Sr	American	407	141·4	36·1	12·6	Camden, Me.	1880	H.M. Bean	Y 9.86	A. Henderson	Boston, Mass.

✱ = Lloyd's Register; D = Nederlandsche Vereeniging; G = Germanischer Lloyd; I = Registro Italiano; R = American Record;
A = Veritas Austro-Ungarico; H = Veritas Ellenico; N = Norske Veritas; V = Bureau Veritas.

Dans la 4me Colonne, C = Composite (membrure en fer, bordé en bois); I = Fer; Stl = Acier; W = Bois.

Les Dates du doublage en métal (Col. 11) ressortent des livres dans lesquels les navires sont respectivement classés ou des rapports des Inspecteurs du Lloyd's Register.

Pour plus amples renseignements sur les navires classés par le Lloyd's Register, voir le Lloyd's Register Book et le Lloyd's Yacht Register.

Official No.	No. in Book	Registry in which classed	Ship's Name. Material and Rig. Master, and Dates of Appointment to present Owner's Service and to Ship. Ship's Late Name.	Flag	Net Tonnage	Dimensions: Length	Breadth	Depth	Built: Where	When	By whom	Date of Metalling	Owners or Managers
10512[illegible]	1951		Annie L. McKeen *Mahoney 79-80* W3mSr	American	232	122·3	30·3	8·7	Belfast, Me.	1872			J. S. Preble & Co.
10542[illegible]	1952		Annie L. Palmer *Rogers-87* WSr	American	113	87·6	25·2	8·5	Bath, Me.	1874			W. H. Mitchill
[illegible]1099	1953		Annie Largie *[illegible]* WBg	British	229	101·1	25·2	14·4	Montrose	1866	Birnie		R. L. Manners
10600[illegible]	1954		Annie Larsen *Larsen* WSr	American	358	144·8	34·0	11·0	P. Blakely, Wsh	1881	Hall Bros.		James Tuft
105256	1955		Annie Lee *Cole 81-* WSr	American	114	89·0	26·2	7·4	Jonesboro' Me.	1873	Joseph Crandon		I. Sargent
105408	1956	R	Annie Lewis *Lewis 78-78* WBk	American	638	156·9	32·4	17·9	Deering, Me.	1874	R. Lewis & Co.	[illegible]	R. Lewis & Co.
70308	1957	✳	Annie Lloyd *Evans -88* WBn	British	149	95·3	23·5	12·1	Wales	1876	Hughes	[illegible]	J. Roberts & Co.
10524[illegible]	1958		Annie M. Allen *Brown* W3mSr	American	428	139·2	32·9	14·2	Kennebunk Me	1873			H. F. Witherel
71025	1959	V	Annie M. Law *Ryder -88* WS	British	1179	185·0	36·6	23·6	Argyle, N.S.	1875	M. W. & E. Forbes	[illegible]	William Law
1644	1960	V	Annie M. Smull *Bailey 87-87* WS	American	1010	181·9	37·2	21·9	Mystic, Conn.	1868	C. Mallory	[illegible]	E. Lawrence
85112	1961	✳	Annie McDonald *Sangster* I Bk	British	459	152·7	29·1	15·5	London	1882	W. Walker & Co.		McDonald, Scales & Co
	1962		Annie McNairn *Tayl* WBk	Danish	479	130·8	28·5	18·4	N. Brunswick	1868			S. J. Marval
49652	1963	R	Annie Maude (WRECKED) *Owen* WBn	British	136	88·4	22·6	11·5	Sunderland	1864	Lister	[illegible]	J. Edwards
83113	1964		Annie Minnes WSr	British	155	102·0	25·0	8·5	Portsm'th, Ont	18 67/81	Fraser & George		R. C. Smith, jr.
93111	1965	✳	Annie Park *Williams 58-86* Stl3mSr	British	192	134·2	24·1	10·2	Carrickfergus	1886	P. Rodgers		J. Fisher & Sons
105273	1966	R	Annie P. Chase *Poole* W3mSr	American	244	121·6	32·4	10·0	Sheepscott, Me	1873			G. J. Poole
106233	1967		Annie R. Kemp *Kemp* WSr	American	156	108·0	25·5	9·5	Essex, Mass.	1883	J. James & Co.		I. W. Quintard, trust[illegible]
105238	1968		Annie R. Lewis *Lewis 68-73* W3mSr	American	206	117·0	28·7	9·0	Bucksport, Me	1873	W. Beazley		S. P. Hall
1845	1969	R	Annie R. Storer *Adams -87* WBn	American	471	129·0	31·2	15·0	Waldoboro Me	1869	W. F. Storer	[illegible]	W. F. Adams
105635	1970	RV	Annie Reed *Warren -86* WBk	American	719	162·0	33·2	18·2	Waldoboro Me	1876	Reed & Welt	[illegible]	Nates & Porterfield
105347	1971	R	Annie S. Conant *Biddle 86-88* W3mSr	American	435	139·0	33·6	15·0	Duxbury, Mass.	1874	Standish, Merrit [& Bros.		P. H. Crowell
1815	1972		Annie S. Gaskill *Gaskill* WSr	American	174	102·0	29·0	8·1	May's Landing (N.J.)	1869			J. Gaskill
80012	1973	R	Annie Simpson *Petipas 71-86* WSr	British	161	92·7	27·3	10·2	Alma, N.B.	1880	N. H. Foster		A. Langlois
80094	1974	V	Annie Stafford *Veal -83* WBk	British	1297	199·0	38·4	23·5	Black Rvr, NB	1881	J. & R. McLeod	[illegible]	Joseph W. Penry
53062	1975	✳	Annie Story *Westray 74-76* I Bk	British	590	174·1	28·3	18·1	Glasgow	1867	A. Stephen & Sns.		R. G. Sharp & Co.
76770	1976	✳	Annie Stuart *Osborne 85-86* WSr	British	98	84·6	21·6	10·2	Kingston	1878	Duncan	[illegible]	J. McWilliams (R. Reid, Manager)
106007	1977	R	Annie T. Bailey *Marston -88* W3mSr	American	426	134·0	33·4	13·2	Bath, Me.	1882	Goss & Sawyer		J. A. Marson
	1978	NV	Annie Torrey *Jacobsen -88* WBk	Norweg'n	770	156·6	32·0	20·9	Richmond, Me	1869		[illegible]	S. Petersen
1473	1979		Annie V. Bergen *Odell -87* WSr	American	175	106·2	28·9	8·2	P. Jefferson NY	1865			J. E. Bayles
80040	1980		Annie W. Akers *McIntyre 80-83* WSr	British	124	94·7	25·4	8·0	Gagetown, NB	1880	E. & F. Williams		A. W. Lovett
286	1981		Annie W. Weston *Weston -88* WBk	American	703	149·7	31·6	19·6	Boston, Mass.	1867	Curtis, Smith & [Cushman	[illegible]	G. M. Winslow & Co.
78382	1982		Annie Wilson WSr	British	150	95·0	23·8	10·1	Aratapu, N.Z.	1879	J. Barbour	[illegible]	E. Mitchelson
105358	1983		Annie & Lily WBn	American	277	114·4	30·5	11·9	Cherryfield Me	1874	Hinckley		J. H. Flitner
105745	1984	R	Annie & Millard *Steelman* W3mSr	American	546	144·0	33·0	15·0	Milford, Del.	1877	T. Carlisle	[illegible]	J. D. Steelman
2611	1985	A	Annina *D'Ottone* WBn	Italian	142	78·2	23·3	9·2	Amalfi	1854			G. Leboffe
3096	1986		Annina *Seghezza -88* (ex Bella Maria) WBn	Italian	100	83·3	20·0	10·5	Dunkirk	1854			G. B. Capurro
3073	1987	I	Annina *Majo* WSr	Italian	100	86·9	20·4	10·2	Torre del Greco	1886	L. Paslillo		A. Mazza
	1988		Annina WBk	Spanish	1000	194·0	35·0	23·7	Sestri Ponente	1877	A. Briasco fu L.	[illegible]	J. Balcells & Co.
2647	1989	A	Annina B. *Accetta* (ex Alessandro, ex Augusto) WBg	Italian	148	88·2	22·0	13·3	Savona	1857		[illegible]	G. Boriello
	1990	I	Annina M. *Del Gatto* (ex Maddalena) (UNK) WBn	Italian	219	96·1	23·3	14·9	Sorrento	1858			G. Del Gatto & Co.
	1991		Anninha (ex Unione) WSr	Brazilian	125	..	..	..	S. João da [Barra	1852			G. P. Guimarães
5603	1992	VI	Annita Menotti *Fasce -85* WBk	Italian	913	177·1	34·1	23·3	Spertono	1875	G. B. Guastavino	[illegible]	G. Razzeto & S. C[illegible]
	1993		Annonciation WBg	French	104	67·6	21·3	9·8	Venice	1847			D. Martinetti
42666	1994	DV	Annot Lyle *Duncan 83-86* WS	British	1246	191·6	38·5	23·8	N. Brunswick	1861	McMorrn	[illegible]	W. T. Dixon & Sons
	1995	N	Anny Föyen *Danielsen 87-87* WBn	Norweg'n	155	107·9	25·2	11·1	Halsens, Norway	1868	Peder Roven	[illegible]	Brødrene Knudsen
	1996		Anoear Salim *Roeland* WBk	Dutch	192	87·9	23·6	..	Grisee	1852		[illegible]	Said Abdulrachman Alkaff
37028	1997		Ansdell (ex P. S. Heinsd) WBk	British	280	120·9	27·3	12·6	Franklin, U.S.				J. C. Ellis
	1998		Ansgar *Callesen 85-85* (ex Samb[illegible] Ellen) WBk	Danish	431	126·0	26·2	16·5	Vegesack	1849	J. Lange		J. A. Olsen
	1999		Ansgar *Pettersen 75-75* WBk	Norweg'n	448	145·5	30·1	16·0	Hudiksvall	1862			F. Bredt
	2000												

✳ = Lloyd's Register; D = Nederlandsche Vereeniging; G = Germanischer Lloyd; I = Registro Italiano; R = Ame[rican]
A = Veritas Austro-Ungarico; H = Veritas Ellenico; N = Norske Veritas; V = Bure[au Veritas]

In the 4th Column, C = Composite (iron frame, wood planked); I = Iron; Stl = Steel; W = Wood.

The Dates of Metalling (Col. 11) are based on the records of the Books in which Vessels are respectively classed, or on the Reports of the Surveyors of Lloyd[']s

Further particulars of Vessels classed by Lloyd's Register will be found in Lloyd's Register Book and Lloyd's Yacht Register.

No.	Reg.	Nom. Capitaine et date d'entrée au service des Armateurs actuels et à bord du navire. Ancien Nom.	Matériaux et Gréement	Pavillon	Tonnage net	Dimensions: Longueur	Largeur	Creux	Construit: Où	Quand	Par qui	Date du doublage	Armateurs ou Directeurs-Gérants	Port d'Armement
2001	N	Ansgar *Nielsen 85-88*	W Bk	Norweg'n	363	127·8	29·0	16·4	Risör	1867	Th. Gundersen		B.A.Olsen	Risör
2002	DN	Ansgar *Jacobsen*	W Bg	Norweg'n	216	99·0	24·9	12·8	Pomerania	1867			J.H.Nielsen	Christiania
2003		Ansgarius	W Bn	Norweg'n	129	101·3	22·7	10·9	Stavanger	1862	P.Haver		E.Berentsen	Stavanger
2004	G	Ansio *Grönbahl*	W Bk	Russian	456	139·3	30·2	16·1	Lokalaks	1876	J.Sandberg		E.Lindqvist	Lokolaks
2005	G	Ansio *Holmstrom -88*	W Bk	Russian	435	124·7	29·5	18·9	Raumo	1865			J.W.LindegrensEnka	Raumo
2006		Ansod *Jaani Pistanis*	W Bg	Turkish	253	..	..	..	Heraclea	1878			Bigadoslee, Teophilo, Jaani & Co.	Bigados
2007	✻	Antagonist *Peaven 86-88*	W Bk	British	575	170·3	29·4	17·9	P.E.Island	1878	J.Ramsay	Y.88	H.Hoskin	Swansea
2008	✻	Antarctic *Rowling 76-78*	I Bk	British	615	176·4	30·2	17·3	Glasgow	1875	D&W Henders'n		B.M.Hennessy	Swansea
2009		Antarctic *Bell*	W Sr	American	101	81·4	21·8	8·9	Scituate,Mass	18$^{51}_{77}$			W.A.Atkins	Provincetown,Mass.
2010	✻	Antares *Hutchins 85-88*	I Bk	British	821	196·6	31·7	18·1	Sunderland	1875	Mounsey& [Foster		R.H.Penney	Shoreham
2011	G	Antares *Stærke -86*	W Bn	Danish	177	87·7	21·6	12·4	Greifswald	1862	L.Wittenberg		H.R.Stærke	Ærøskjøbing
2012	V	Antarès *Guillo*	W Bn	French	138	74·1	21·0	11·9	Mesnas	1862	Ollivaud		Guillo	Nantes
2013	V	Antares *Hoffmann 85-88*	I S	German	1117	208·5	34·5	20·2	Vegesack	1878	H.F.Ulrichs		W.A.Fritze & Co.	Bremen
2014	G	Antares *Rohden*	W Bk	German	466	130·3	30·2	17·8	Danzig	1870	E.H.Gumm		W.Manck	Rostock
2015		Antelope	W Sr	British	334	138·6	26·3	11·4	P.Dalhousie	1873	A.Muir		B.Muir	S.CatharnesOnt
2016		Antelope	W Bg	British	180	109·2	24·0	12·0	Moulmein				Moothoo Caroppen Chitty	Penang
2017	✻	Antelope *Roche 78-88*	W Sr	British	119	91·5	24·7	10·2	Wexford	1880	DockYardCo.		Dock Yard Co.	Wexford
2018		Antelope	W K	British	116	100·0	23·2	8·0	Brightlingsea	1869			Mrs.Richmond	Harwich
2019		Antelope *Brigman -87*	W Sr	American	118	89·5	28·5	7·0	Benicia,Cal.	1887	M.Turner		G.W.Hume	SanFrancisco
2020		Antelope (ex H. W. M.) *Hut*	W Bk	Dutch	1030	187·0	34·8	21·2	Medford	1852		C.1.78	N.Brantjes	Purmerend
2021	GV	Antelope *Peters -76*	W Bn	German	170	97·1	23·9	9·1	Hamburg	18$^{64}_{81}$	C.Wriede	Y.88	J.Hubert	Hamburg
2022	V	Antelope *Abrahamsen*	W Bk	Swedish	372	123·0	26·9	16·3	Wasa	18$^{52}_{76}$		Y.81	A.B.Elfversson	Gothenburg
2023		Antelope *Ofverberg*	W3mSr	Swedish	178	101·3	21·8	11·4	Norrköping	1876	Heggblad		A.Ofverberg	Brunnby,Malmö
2024		Anthipi *Riggird*	W Bn	Greek	393	122·0	28·7	..	Syra	1875			M.Ch.Calimasias	Syra
2025		Anthon *Thomasen -87* (ex Barbara Hage ex Charlotte)	W Bg	Danish	183	92·2	23·4	13·6	Rønne	1847	S.P.Bech		N.W.Thomsen	Nykjøbing, Sjælland
2026		Anthon *Olsen 83-85* (ex Nora) (CONDEMNED)	W Bk	Norweg'n	728	177·2	28·8	19·9	Lübeck	1872	J.Meyer	Y.80	E.M.Olsen	Tönsberg
2027		Anthons	W Bn	British	133	103·6	25·0	9·1	Torquay,Tas.	1875	J.Drake		Alex.Robertson	Auckland,N.Z.
2028		Anthony Burton [illegible]	W Sr	American	151	111·2	22·2	7·5	Bristol,Pa.	1869			S.B.Segar	Stonington,Conn.
2029	N	Antilla *Petersen -78*	W Bg	Norweg'n	207	98·8	26·0	12·0	Bremen	1852			O.N.Hiis	Arendal
2030	✻	Antilles *Murdoch 84-88*	I Bk	British	498	158·5	27·8	16·5	Dumbarton	1869	A.McMillan&S.		A.Rae	Liverpool
2031	N	Antilope *Nielsen* (ex Soren Neilsen)	W Sr	Norweg'n	167	102·4	23·4	11·1	Aland	1874			T.Fr Linna	Langesund
2032	RV	Antioch *Hemingway 74-88*	W Bk	American	953	183·3	35·2	21·0	Kenneb'nk,Me	1876	D.Clark	Y.88	J.S.Emery & Co.	Boston,Mass.
2033	✻	Antiope *Massey 84-88*	I S	British	1443	242·3	38·4	23·7	PortGlasgow	1866	J.Reid&Co.		Australasian Shipping Co.Ltd.	Liverpool
2034		Antje (ex Pinjsden) *Diecks 83-*	W Sr	German	144	85·0	21·7	11·4	Wildervank	1865	Meihuisen&Zn.		J.Dirks	Emden
2035	V	Antje *Dykhoff -88*	W Sr	German	142	91·5	21·5	12·5	Leer	1875	C.Schwoon		J.G.Ukena	Leer
2036		Antje Visser *Visser 87-88* (ex Marianne) (TO BE BROKEN UP)	W Bn	Dutch	249	103·6	26·1	12·8	Halifax	1861	H.Paulin		F.L.Drenth	Oude Pekela
2037	✻	Antofagasta *Griffith -88*	I Bk	British	679	180·1	29·8	19·1	Sunderland	1875	W.Doxford&Ss.		Tomlinson,Hodgetts & Co.	Liverpool
2038		Antoine	W	French	103	..	..	..					J.D.Martinetti	Ajaccio
2039	V	Antoine Dor *Andersen -88* (ABANDONED)	W Bk	Norweg'n	460	128·2	26·3	18·2	Sables [d'Olonne	1864	S.Chaigneau	Y.1.84	N.Selmer	Arendal
2040	V	Antoinette *Ferguson*	W S	British	1118	189·0	37·0	22·7	Yarmouth,NS	1874	J.M.Gardiner	Y 11.86	J.Bingay	Yarmouth,N.S.
2041	V	Antoinette *Runje*	W Bk	British	884	183·0	32·8	19·5	Sweden	1878	A.Lafgren	Y.1.86	A.G.Morris	Hong Kong
2042	G	Antoinette *Oberg -88* (ex Queen of the Sea)	W Bk	Swedish	705	157·8	29·7	19·5	NovaScotia	1864	C.Lockhart	Y 10.81	P.A.Lindberg	Gothenburg
2043		Antolina *Lindqvist*	W Sr	Russian	198	..	..	..		1872			M.Eriksson	Lemland
2044	A	Anton *Stock -87* [illegible]	W3mSr	Austrian	246	108·6	23·0	12·5	Capo d'Istria	1881	L.Polli	Y.2.88	A.Covacevich & Co.	Orebich
2045	V	Anton *Heinemann -88*	W Bn	German	179	101·9	23·0	11·7	Haren	1877	A.&H.Klasen	Y 10.87	Rud.Klasen	Papenburg
2046		Anton *Pankow 74-79*	W3mSr	German	167	98·5	23·0	11·0	Ribnitz	1826	J.C.Peters		J.C.Holtz	Wolgast
2047	RV	Anton *Lange -88*	W Sr	German	154	91·7	22·7	10·6	Grossefehn	1880	T.J.Cassens	Y.88	Tonges J.Cassens & Sohn	Grossefehn
2048	G	Anton *Abrahamsen -87* (ex Antoine Vincent) (FOUNDERED)	W Bk	Norweg'n	443	128·4	28·2	18·6	Sunderland	1860	Douglas	Y.1.86	C.Carlsen	Lillesand
2049		Anton *Lindström*	W Bk	Russian	338	..	..	..	Sideby	1875			K.A.Matisson	Jomala
2050		Anton *Popov 83-*	W3mSr	Russian	283	113·6	24·3	11·0	Onega	1877	J.S.Markikoff		M.Kondakow	Archangel

✻ = Lloyd's Register; D = Nederlandsche Vereeniging; G = Germanischer Lloyd; I = Registro Italiano; R = American Record;
A = Veritas Austro-Ungarico; H = Veritas Ellenico; N = Norske Veritas; V = Bureau Veritas.

Dans la 4me Colonne, C = Composite (membrure en fer, bordé en bois); I = Fer; Stl = Acier; W = Bois.

Dates du doublage en métal (Col. 11) ressortent des livres dans lesquels les navires sont respectivement classés ou des rapports des Inspecteurs du Lloyd's Register.
Pour plus amples renseignements sur les navires classés par le Lloyd's Register, voir le Lloyd's Register Book et le Lloyd's Yacht Register.

1 Official No.	2 No. in Book.	3 Registry in which classed.	4 Ship's Name. Material and Rig. Master, and Dates of Appointment to present Owner's Service and to Ship. Ship's Late Name.	5 Flag.	6 Net Tonnage.	7 Dimensions. Length.	Breadth.	Depth.	8 Built. Where.	9 When	10 By whom.	11 Date of Metalling.	12 Owners or Managers.
	2051		Anton *Runström* W Sr	Russian	277	112·0	30·0	12·0	Finland	1872			J.Ahlström
428	2052		Anton *Rosenfeld -88* W S	Russian	261	96·3	26·3	12·1	Filsand	1883	M. [illegible]		B.Rosenfeld
	2053		Anton Günther (HULK) *[illegible]* W Bk	German	441	138·7	29·0	16·2	Brake	1857	J. D. [illegible]		O.Thyen
660	2054		Antonetta *Anastasio* (WRECKED) (ex [illegible] Filippo) W Bn	Italian	12?	86·9	23·9	11·4	Gaeta	1856	[illegible]		R.Minco
	2055	DV	Antonia *de Boer 86-88* W Bk	Dutch	967	181·4	35·4	21·2	Amsterdam	1876	W.&A.H.Meursing	Y 4.87	J.Veth
272	2056	A	Antonia *Barolini* W Bn	Italian	397	122·6	27·6	25·7	Fiume	1868		Y 4.87	A.Scarpa
	2057	R	Antonia *Castro* W Bk	Spanish	354	126·9	28·1	13·7	Bremen	1867	H.K[illegible]	Y 81	Carlos Casas
2867	2058	I	Antonia Madre (ex Madre Carruti) *Moffino -87* W Bk	Italian	567	151·5	31·9	20·3	Varazze	1871	[illegible]	Y 8.87	F.Mortola
105759	2059	RV	Antonia Sala *Pierce* W Bk	American	508	140·9	33·3	15·3	Bath, Me.	1878	William Rogers	Y 8.87	F.H.Mitchell
	2060	V	Antonia Sanchez *Vasquez* W Bk	Spanish	615	142·0	29·5	20·3	Lavagna	1875		Y 3.88	G.Sanchez
	2061		Antonie Emilio *Freimann [18-78* W Sr	Russian	244	104·1	27·4	13·6	Dondangen	1877	R.T[illegible]		C.Freimann
	2062		Antonieta W Pol	Spanish	119	87·6	24·7	6·6	Palamos	1851			Pedro Forgas
	2963		Antonieta W Sk	Spanish	104	65·9	20·5	7·9	Blanes, Spain	1876			S.Rocher
	2064		Antonietta (ex Lady Vere de Vere) *McLean* W S	Chilian	995	187·7	36·3	22·1	Maitland, N.S.	1873	Smith	Y 2.87	Gervasoni Hermanos
1288	2065		Antonietta *Ciano -87* W Bk	Italian	399	115·4	31·4	16·9	Arens, Spain	1843			P. de Logu
267	2066	A	Antonietta *Sinibaldi* W Bn	Italian	100	79·7	23·0	10·4	Chioggia	1883	[illegible]	6.88	A.Purisio
2040	2067	I	Antonietta Accame *Marchese -88* W Bk	Italian	653	150·5	29·5	20·0	Sestri Ponente	1871	[illegible]	Y 3.87	E.Accame fu L.
1967	2068	I	Antonietta B. *Traverso -88* W Bk	Italian	438	127·9	30·6	19·4	Voltri	1863	[illegible]	Y 8.88	F.Bertolotto di G.
1804	2069	I	Antonietta C. (ex Giacinto Caprile) *Barbaro* W Bk	Italian	566	146·0	32·0	18·0	Sestri Ponente	1868	A.Briasco fu L.	Y 8.87	V.Picasso
230	2070		Antonietta D. *Cecolini* W Bg	Italian	174	86·9	25·6	15·1	Sestri	1850			P.Delfino
2694	2071	I	Antonietta D. *Sorrentino* W Sr	Italian	100	83·0	20·0	9·8	Torre del Greco	1886	[illegible]		S.De Rosa Luise
3235	2072	IV	Antonietta O. *Schiaffino* W Bk	Italian	970	187·0	34·8	23·4	Savona	1882		Y 1.88	F.Ottone
1020	2073	I	Antonietta Q. (ex Luca Figlio [illegible]) *Queirolo* W Bk	Italian	525	145·6	32·5	19·9	Sestri	1878	F.Cadenaccio	Y 2.88	G.Queirolo
	2074	❋	Antonin *Michel 83-85* I Bk	French	700	186·3	30·5	18·9	Glasgow	1870	Dobie & Co.		Ant.Dom.Bordes & fils
	2075		Antoninha W S	Brazilian	130	..	..	..					J.P.S.Moreira
501	2076	A	Antonino *Regini* W Bn	Italian	216	92·5	26·0	16·3	Piano di Sorrento	1856	F. de Rosa	8 10.87	V.Regini fu L.
736	2077		Antonino (BROKEN UP) *Lola* W Bn	Italian	132	81·4	24·7	13·6	Varazze	1846	B.C[illegible]		Benfante f.lli
	2078		Antonino W Bn	Spanish	110	81·0	20·9	8·1	Tayabas-Luzon	1865			Feliciano Magsino
778	2079		Antonino Strino (ex Assur) *Arena* W Bn	Italian	122	81·0	22·1	12·3	Lussino	1873		11.81	O.Strino
91495	2080	❋	Antonio I Sk	British	90	79·2	20·3	10·1	Beverly	1885	Cochrane & Co.		C.Hellyer
	2081		Antonio (ex Elizabeth Hands) *Grima -87* W Bk	French	249	128·0	24·3	12·7	Jersey	1855		10.81	G.Queirolo
98	2082		Antonio (ex Chechino, ex Cecchino Danovaro) (BROKEN UP) *Chiesa -87* W Bk	Italian	438	135·1	30·1	19·0	Pra	1864	M.Tixi		S.Barrera fu S.
3582	2083	I	Antonio (ex Papa Felice) *Capurro -88* W Bn	Italian	289	121·0	26·8	15·6	Sestri Ponente	1876	A.Cadenaccio	Y 8.88	A.Degregori fu A.
478	2084	I	Antonio *D'Amico -88* W Bn	Italian	169	98·1	22·6	12·7	Gaeta	1862		Y 8.82	A.Lania & figli
1390	2085	I	Antonio *Martinelli -87* W Bn	Italian	115	90·9	22·5	11·1	Viareggio	1887	A.Raffaelli		Paolo Martinelli
3521	2086	I	Antonio A. (ex Vittoria Madre) *Alciatore -87* W Bg	Italian	358	118·1	29·5	18·2	Pra	1869	D.Tixi	Y 3.88	L.Alciatore
491	2087		Antonio C. (HULK) W Bg		188	..	..	..	C.d'Istria	1861		Y 1.86	
432	2088	I	Antonio C. (ex Juan de la Vega) *Castellano* W Sr	Italian	140	100·0	21·0	12·3	Aberdeen	1871	[illegible]	Y 10.88	A.Castellano
1843	2089	DV	Antonio Casabona *Schiaffino -88* W Bk	Italian	431	125·0	30·4	19·2	Sestri	1863	G.B.Cadenaccio	Y 8.86	B.Casabona
2205	2090	IDV	Antonio D'Abundo *Cammarota* W Bk	Italian	785	167·3	33·1	21·4	Castellammare	1878	[illegible]	[illegible]	E.D'Abundo
381	2091	❋	Antonio G. *Calabretta* W Bk	Italian	433	136·8	30·2	18·0	Trieste	1872	[illegible]	Y 2.88	G.Granata
	2092		Antonio Jané W Bk	Spanish	860	170·1	34·8	21·3	Patenniosen	1872			Jané & Co.
	2093		Antonio M. *Olivari* W Bk	Greek	519	136·1	31·7	20·3	Sestri	1868	S.[illegible]		D.Calimicho
3230	2094	A	Antonio M. (ex Vaglianos) *Marciani -88* W Bk	Italian	543	138·7	31·8	20·7	Varazze	1870	A.R[illegible]	Y 1.88	F.lli Mortola
399	2095	V	Antonio Padre (ex Raffaele) *Capodanno* W Bg	Italian	163	94·2	24·7	15·4	Gaeta	1870	[illegible]		L.Capodanno
	2096		Antonio Padre (ex San Spiridione) *Carrio [-87* W Bn	Italian	108	66·3	22·7	12·9	Molfetta	1863	[illegible]		S.Di Donna
	2097	GRV	Antonio Pallas (ex Bertha) *Martinez -87* W Bn	Mexican	299	115·2	26·9	12·9	Moltzen	1875	D.F[illegible]	Y 1.88	G.Pallas
	2998		Antonio Perez *Fernandez* W Bn	Spanish	191	101·9	25·1	11·6	Norway				Romualdo Perez
	2099		Antonio Ventura (ex [illegible]) W [illegible]	Spanish	231	113·1	24·3	13·4	St.Feliu	1850			J.Ventura
37	2100		Antonios W Sr	Greek	169	82·7	21·0	..	Syra	1869	S.Koufoudakis		A.Embiricos

❋ = **Lloyd's Register**; D = Nederlandsche Vereeniging; G = Germanischer Lloyd; I = Registro Italiano; R = Aus[…]
A = Veritas Austro-Ungarico; H = Veritas Ellenico; N = Norske Veritas; V = Bu[…]

In the 4th Column, C = Composite (iron frame, wood planked); I = Iron; Stl = Steel; W = Wood.

The Dates of Metalling (**Col. 11**) are based on the records of the Books in which Vessels are respectively classed, or on the Reports of the Surveyors of L[…]

Further particulars of Vessels classed by Lloyd's Register will be found in Lloyd's Register Book and Lloyd's Yacht Register.

Nos dans la Série	Registre dans lequel le navire est classé	Nom. Matériaux et Gréement. Capitaines et dates d'entrée au service des Armateurs actuels et à bord du navire. Ancien Nom.	Pavillon	Tonnage net	Longueur	Largeur	Creux	Construit: Où	Quand	Par qui	Date de doublage en métal	Armateurs ou Directeurs-Gérants	Port d'Armement
2101	GV	Antonita Gelpi ·· WPol	Spanish	148	82·3	25·1	11·5	Mataro	1854		Y 1.88	M.Gelpi	Barcelona
2102		Antoñito ·· WBk	Spanish	299	112·5	24·9	15·7	Bilbao	1858			J.P.Badiola	Bilbao
2103	✻	Antrim Martin ·84-85 I Bk	British	939	190·0	33·3	21·4	Sunderland	1864	W.Doxford&Ss.		R.Martin & Co.	Liverpool
2104	DV	Antwerp Lund ·84-85 WBk	British	1031	183·0	36·5	22·7	Newport,N.S	1874	J.A.Harvie	Y 1.87	Nelson Chambers	Windsor,N.S.
2105	G	AnunciacionFomento Sbert ·88 W Bn	Spanish	207	93·8	26·3	12·3	Blanes,Spain	1865			J.Amengual y Begavit	Barcelona
2106		Ap Min I LaBarge	British	392	199·5	34·1	5·9	PortGlasgow	1883	R.Duncan&Co.		IrrawaddyFlotillaCo.Ld	Glasgow
2107	A	Ape Premuda WSr	Austrian	109	76·1	20·3	9·1	Lussinpiccolo	1887	N.Martinolich	Y 4.87	G.F.Premuda	Lussinpiccolo
2108	✻	Aphrodita Tate 84-86 I S	British	1627	227·0	36·3	25·6	Liverpool	1858	J.Jones,jun.		W.Price&Co.	Liverpool
2109	✻	Aphrodite WSrYt	British	134	110·1	21·9	12·9	Cowes	1875	J.S.White	Cpt.87 11.87	Lord Porchester	Cowes
2110	NV	Aphrodite Thorsen WBk	Norweg'n	728	151·0	31·1	20·0	Cocaigne,N.B.	1869	G.R.Ray	Y 7.88	H.Mogensen	Tvedestrand
2111	N	Aphrodite Knudsen WSr	Norweg'n	117	74·9	21·1	10·9	Lillesand	1832			J.S.Jörgensen	Brevig
2112	A	Aphroessa Manolondis WBg	Greek	276	100·0	27·2	..	Syra	1870		Y 8	K.I.Manolondis	Syra
2113		Apipe I Sr	Argentine	247	134·0	30·3	7·8	PortGlasgow	1884	Murdoch & [Murray		La Platense Flotilla Co.	Buenos Ayres
2114		Apiros WSr	Greek	107	74·5	21·6	..	Syra	1879	S.Konfoudakis		N.Gialouris	Andros
2115		Apiros Bassan WBg	Turkish	149	..	..	..	Greece	1885			TijarHadji Ali Effendi	Constantinople
2116	V	Apoll Hohs ·74 WBn	German	306	120·0	25·5	13·5	Cappeln	1874	B.C.Steen	Y 3.88	G.Schade	Blankenese
2117	N	Apolline Nilsen WBk	Norweg'n	308	110·2	26·6	11·9	Bordeaux	1846			N.Engebretsen	Fredrikstad
2118	✻V	ApollineEmilie Evan 86-88 IBk	French	710	164·0	30·0	18·8	LaSeyne	1868	Forges&C.de la [Médit.		Ant.Dom.Bordes & fils.	Bordeaux
2119		Apollo WBk	British	696	156·0	30·7	19·6	Rangoon	1877		Y 3.86	Edwd.Hopper	Rangoon
2120	V	Apollo Christensen 82-82 WSr	Danish	192	111·2	25·3	11·8	Marstal	1876	J.J.Bager		H.C.Christensens Enke	Svendborg
2121	✻D	Apollo Hustede 76-84 I Bk	German	1156	215·4	35·0	20·7	WestHartlep'l	1884	W.Gray&Co.		A.Schiff	Elsfleth
2122	V	Apollo Rohn ·89 WBk	German	697	133·2	28·9	19·7	Rangoon	1877		Y 2.88	Peter Lindemann	Hamburg
2123	N	Apollo Nielsen ·88 WBk	Norweg'n	432	133·5	29·3	15·5	Vegesack	1854		Y 12.88	Ole Nielsen	Sandefjord
2124	N	Apollo (ex Teviot) Hytt WBk	Norweg'n	337	131·5	27·4	16·8	Scarboro'	1859	Tindall	Y 11.88	Act.Selsk."Achilles (Jens Olsen)	Kragerö
2125	N	Apollo Andreassen WBg	Norweg'n	258	101·8	27·1	13·8	Grimstad	1867	A.Omundsen		D.Ditlefsen	Christiansand
2126		Apollon WBg	French	110	..	..	..	Ancona	1837			E.Paoli	Marseilles
2127		Apollon WBn	Greek	386	109·5	26·2	17·7	Syra	1871			Cosmas Cypriotis	Zante
2128		Apollon Kerbajia WBg	Greek	229	95·5	24·9	..	Spezzia	1860			A.Afindachi	Piraeus
2129		Apollonia Janson ·86 WBk	Russian	377	..	..	..	Finström	1878			H.Eriksson	Finström
2130		Apollonia Olsson WSr	Swedish	209	94·8	25·3	12·7	Vätö	1858			A.Olsson	Fogdö,Roslagen
2131	A	Apostolis Cokino ·88 WBg	Greek	349	111·8	27·7	17·0	Syra	1876	P.Bonmi	Y 1.84	S.Cokino	Zante
2132		Apostolos Lambo Kibaria WSr	Turkish	137	..	..	..	Abalmah	1884			Marmorahlee Ligor Andon & Co.	Marmorah
2133		Apotheker Diesing (WRECKED) Stekt 85- WBk	German	388	115·8	28·8	17·0	Barth	1865	L.Freundt	Z 3.86	Carl Siebe	Stralsund
2134	R	Apphia & Amelia Wilford W3mSr	American	238	128·3	31·0	9·3	Yarmouth,Me.	1884	Loring&Cobb		H.M.Sargent	Portland,Me.
2135		Appia Oeregaard WBk	Norweg'n	475	113·0	29·6	17·3	Stavanger	1873	M.Magnus	Y 6.87	E.Berentsen	Stavanger
2136		Appolpa WBg	Turkish	176	..	..	..	Russia	1881			Yanni Mavrodi	Constantinople
2137	DV	Aquatic Halcrow 71-85 WBn	British	362	132·2	29·9	13·1	St.MartinsNB	1885	John Marr	Y 11.85	F.Tufts	St.John,N.B.
2138	R	Aquidneck Maristany ·87 (CONDEMNED) WBk	American	326	138·5	29·0	13·0	Mystic,Conn.	1865		Y 2.81	J.Slocum	Baltimore
2139	E	Aquila WBn	British	150	92·8	24·6	10·7	Georgetown,P.E.I	1888	W.Seneabaugh		M.McDonald	Charlottetown,P.E.I.
2140	A	Aquila Tibaldi WBk	Austrian	430	134·5	26·1	17·8	Fiume	1873	V.Brazzoduro	Y 3.86	Cas.Cav.Cosulich	Fiume
2141	V	Aquila (ex San Martino, ex Aquila) Chiesa WBk	Italian	321	115·4	27·4	17·8	SestriPonente	1860		Y 4.86	C.Renli	Leghorn
2142	I	Aquila (ex Archimede) Simeone WBg	Italian	170	81·0	25·1	14·2	Castellam.di [Stab.	1858			G.Cicconardi	Gaeta
2143	DN	Aquila (ex Silver Eagle) Siesten 70-83 WBk	Norweg'n	877	182·1	34·6	20·8	Troon	1861	PortlandS.B.Co.	Y 12.86	P.A.Christensen	Langesund
2144	NV	Aquila Johnsen 86-86 WBk	Norweg'n	485	134·1	30·8	17·5	Christiansand	1873	M.Knudsen	Y 9.86	M.F.Stray	Christiansand
2145		Aquila WSrYt	Portug'se	128	86·9	16·9	10·6	PontaDelgada	1877	E.L.Dos Santos		Baron de Fonte Bella	PontaDelgada
2146	NV	Ara Andersen ·87 WBg	Norweg'n	267	109·5	28·6	12·8	Fredrikshald	1869	E.Wiig	Y 7.83	H.Isachsen	Grimstad
2147		Arab W	British	294	122·9	26·8	9·4	Montreal	1876			William W.Tate	Montreal
2148	V	Arab Moss 80-80 WBk	German	541	147·5	32·7	18·1	Apenrade	1860	N.Jacobsen	Y 2.87	J.Matthiessen	Apenrade
2149	V	Arab Steed Bergstrom ·88 WBk	Swedish	664	158·5	32·0	19·5	Sunderland	1863	Pile Hay & Co.	Y 9.86	A.Liljeqvist	Gothenburg
2150		Arabell WBn	British	178	97·8	26·6	10·2	Kempt,N.S.	1871	R.Rose		T.Furniss	Waterford

✻ = Lloyd's Register; D = Nederlandsche Vereeniging; G = Germanischer Lloyd; I = Registro Italiano; R = American Record;
A = Veritas Austro-Ungarico; H = Veritas Ellenico; N = Norske Veritas; V = Bureau Veritas.

Dans la 4me Colonne, C = Composite (membrure en fer, bordé en bois); I = Fer; Stl = Acier; W = Bois.

✻ Dates du doublage en métal (**Col. 11**) ressortent des livres dans lesquels les navires sont respectivement classés ou des rapports des Inspecteurs du Lloyd's Register.

Pour plus amples renseignements sur les navires classés par le Lloyd's Register, voir le Lloyd's Register Book et le Lloyd's Yacht Register.

Official No.	No. in Book	Registry	Ship's Name (Ship's Late Name)	Master, and Dates of Appointment	Material and Rig	Flag	Net Tonnage	Length	Breadth	Depth	Built: Where	When	By whom	Date of Metalling	Owners or Managers
62860	2151	*	Arabella	Park 57-81	I Bk	British	665	177·6	31·0	17·4	Sunderland	1876	Bartram, Haswell & Co.		T.E.Hick
	2152	R	Arabella (MISSING)	Buck -84	W Sr	British	205	100·0	28·3	10·7	Harvey, N.B.	1887	G.S.Turner		A.Black
71048	2153	DV	Arabia	Hatfield -88	W Bk	British	957	173·8	34·5	21·0	Rockland, N.B.	1875	R.A.Chapman	Y 8.87	Taylor Bros.
52844	2154		Arabia		W Bk	British	309	131·6	25·8	11·7	Kingston, Ca.	18 62/63	G.N.Ault		C.A.Watson
106020	2155	RV	Arabia	McLean	W S	American	2024	238·9	43·2	27·6	Bath, Me.	1881	Houghton Bros.	Y 8.88	Houghton Bros.
105535	2156		Arabia		W S	American	1293	..	..	..	Kennebunk, Me	1863			T.Fuller
	2157	V	Arabia (ex Perda)	Tidemansen	W Bk	Norweg'n	973	178·4	35·6	20·8	Pictou, N.S.	1873	J.Kitchen	Y 11.87	Acktieselskabet "Arabia" (Chr. Rolsen & Co.)
53101	2158		Arabian	Coyde -83	W Bkn	British	271	108·5	27·1	16·4	Sunderland	1865	Barker		P.Taylor
62272	2159	*	Araby Maid	Stenhouse 86-88	I Bk	British	837	194·6	32·0	19·4	Greenock	1868	R.Steele		W.Thomson & Co.
	2160		Aracaju		W Bk	Brazilian	314	..	..	..					F.M.A. e Silva
	2161	V	Aracan	Hedberg	W Bk	Swedish	692	162·8	31·2	18·3	Sundsvall	1876	J.Sjölén	Y 1.88	Sundsvalls Skepps-Rederi-Aktiebolag (O. L. Wiklund)
	2162	G	Aradus	Teising	W Bg	German	250	108·6	24·3	12·5	Elsfleth	1866	C.F.Stindt	Y 1.87	Hinrich Wessels
48766	2163	*	Arafura (ex Bertha Marian)	Gale 86-88	C Bk	British	540	143·5	28·7	18·5	Sunderland	1864	Moore	Y 1.88	Trinder, Anderson & Co
1165	2164		Arago	Falkman 81	W Sr	American	176	103·0	28·0	9·0	Coos Bay, Oreg	1860			W.C.Lynde
	2165	V	Arago	Nicole -83	W Bn	French	131	78·1	21·6	11·5	Nantes	1874	W.Cassard	Z 3.86	H.Guillebot
	2166	V	Arakan	Loiseau 81	I Bk	French	392	137·4	24·9	15·4	Nantes	1868	Jollet & Babin		A.Viot
61472	2167	V	Araminta (WRECKED)	Svendson -88	W Bk	British	747	158·5	34·0	20·5	Nova Scotia	1870	Harvie	Y 6.88	Burgess & Co.
	2168	V	Aramis	Charruau	W Bg	French	227	98·8	22·6	12·8	Nantes	1874	J.Sevestre	Z 12.88	C.Simon
	2169		Arão (ex America)		W 3m Sr	Brazilian	159	..	..	..	Laguna	1856		Y 1.81	A.L.P.Valente
882	2170	A	Ararat	Ghersaglia	W Bk	Austrian	500	133·5	28·2	18·5	Fiume	1877	V.Brazzoduro	Y 5.88	Anna Calich
	2171	N	Ararat	Axelsen -77	W Bk	Norweg'n	445	132·6	30·3	16·9	Lillesand	1877	A.Reinertsen	Y 2.88	O.C.Heldahl
78354	2172		Aratapu	Bower 78-80	W Bn	British	122	95·0	23·7	9·0	Aratapu, N.Z.	1878	J.Barbour	Y 1.88	Union Steam Saw, Moulding, Sash & Door Co.
	2173		Arathusa	Dunn	W Bk	British	320	125·0	30·2	12·8	N.Brunswick	1865	Ogden		J.Dunn
45703	2174		Aratus	Adams	W Bn	British	173	109·6	23·0	12·1	P.E.Island	1862	Cox		J.Temple
44646	2175	*	Arauco	Blatchford 87-87	I Bk	British	572	176·0	28·0	18·8	Dundee	1862	Gourlay Bros.		Duncan, Fox & Co.
	2176		Arauco (ex Robert Burton)	Riera	W Bk	Spanish	422	118·7	27·6	18·3	Philadelphia	1846			J.M.Serra y hijo
	2177	V	Arauco (ex Ecuador)	Riera	W Bk	Spanish	268	115·1	26·8	14·0	San Sebastian	1879	A.Motiozabal		Basté Peira & Co.
	2178	*	Arauna	Lundgren	W Bk	Swedish	461	115·9	29·9	18·0	Sunderland	1866	Gardner	Y 5.88	C.Henckel
93074	2179	*	Arawa		W Sk	British	87	80·0	20·6	10·7	Hull	1886	G.Brown		W.L.Robins
83211	2180	D	Arbela	Smith	W S	British	1117	183·0	38·0	22·7	Cornwallis, NS	1881	G.Bigelow	Y 2.86	J.E.Bigelow
60373	2181		Arbitrator	Marker	W Bn	British	184	105·0	24·7	12·8	P.E.Island	1873	McWilliams		J.J.Shee
56114	2182		Arbutus	Bligh	W Bk	British	399	124·5	28·1	17·6	Sunderland	1867	Davison	Y 8.88	A.Eskdale
88436	2183	V	Arbutus	Fritz -88	W Bn	British	396	134·0	30·2	13·0	St.John, N.B.	1884	E.M.Gaggin	Y 11.87	H.J.Olive
	2184	DG	Arbutus	Mohrmann -8	W Bk	German	312	125·0	24·9	15·6	Dundee	1865	Dundee Shipbuilding Com.	Y 11.88	E.C.F.Mohrmann
66993	2185	V	Arcadia	McGonagle -88	W Bk	British	788	155·6	32·8	20·6	Rockland, N.B.	1873		Y 8.88	R.W.Doherty
18564	2186		Arcadia	Ewan 82-82	W Bk	British	417	136·4	27·3	17·1	Bristol	1863	Hill	Y 11.84	D.S.Ewan (per W. Findley)
106387	2187	R	Arcadia	Woodward	W Bn	American	283	120·0	29·0	12·0	Madison, Conn	1878	Miner & Son	Y 6.88	H. Trowbridge's Sons
36	2188		Arcadion	Vachlaris	W Bg	Greek	247	96·8	26·0	..	Galaxidi	1867	Papapetros		G.Vachlaris
2849	2189	I	Arcangelo (ex Serafino)	Consiglieri	W Bk	Italian	635	159·7	32·8	20·1	Sestri	1870	C.Tubino	Y 3.88	G.Consigliere
	2190	A	Arcangelo	Romeo	W Bg	Italian	276	115·1	24·6	20·0	Castellam.	1856	G.Bonifacio	Y [illegible]	G.Costagliola
911	2191		Arcangelo	Giordano -87	W Bg	Italian	166	89·5	24·1	14·0	Naples	1857			F^lli Scannapieco
1195	2192		Arcangelo Padre (ex August Marlet)	[illegible]	W Bk	Italian	549	135·8	27·8	19·9	Venice	1865		Y 6.88	A.Lubrano
	2193	H	Archangelos	Lazopulos -87	W Bk	Greek	370	115·1	26·9	..	Piræus	1872		Y 8.79	S.Lazapoulos
504	2194	A	Archangelos	Moraiti	W Bg	Greek	290	105·9	27·2	..	Syra	1888	M.Koufoundakis		D.G.Moraiti & Co.
340	2195	V	Archangelos	Costis	W Bg	Greek	283	106·9	26·9	16·7	Syra	1882	Koufoundakis	Y 7.88	M.I.Costis
	2196	H	Archangelos	D.Moraitis	W Bg		253	99·7	24·6	..	Syra	1866	S.Koufoundakis	Y 7.82	
226	2197	A	Archangelos	Cirojanni	W Bg	Greek	239	96·1	26·2	..	Syra	1865		Y [illegible]	Maria G.Cirojanni
74535	2198	*	Archer	Dawson 87-87	I Bk	British	765	189·1	32·0	18·7	Sunderland	1876	R.Thompson, jr.		P.Iredale & Son
1642	2199	R	Archer	Mitchell 84-86	W Bk	American	457	136·6	31·2	14·5	Westbro'k, Me	1868	R.Lewis & Co.	Y [illegible]	R.Lewis & Co.
954	2200		Archer & Reeves	Braymans	W Sr	American	240	112·3	31·0	9·7	Wilmington, Del.	1866			H.W.Hallowell

*=Lloyd's Register; D=Nederlandsche Vereeniging; G=Germanischer Lloyd; I=Registro Italiano; R=Amer
A=Veritas Austro-Ungarico; H=Veritas Ellenico; N=Norske Veritas; V=Bure

In the 4th Column, C=Composite (iron frame, wood planked); I=Iron; Stl=Steel; W=Wood.

The Dates of Metalling (Col. 11) are based on the records of the Books in which Vessels are respectively classed, or on the Reports of the Surveyors of Llo

Further particulars of Vessels classed by Lloyd's Register will be found in Lloyd's Register Book and Lloyd's Yacht Register.

No. dans le livre	Registre dans lequel le navire est coté	Nom	Capitaine et dates d'entrée au service des Armateurs actuels et à bord du navire. Ancien Nom.	Matériaux et Gréement	Pavillon	Tonnage net	Longueur	Largeur	Creux	Construit: Où	Quand	Par qui		Armateurs ou Directeurs-Gérants	Port d'Armement
2201		Archibald	Quinn	W Bn	British	221	101·8	25·0	12·9	P.E.Island	1877	T.P.Lewis		S.N.Robinson	Dublin
2202	*	Archie	Stribley -81	W K	British	63	74·8	19·3	8·1	Plymouth	1887	D.Banks&Co.		J.Hawken	Padstow
2203		Archimede	Filippini	W Bk	Italian	399	..	..	..	Venice	1868	G.Zanon	[illegible]	G.Sarpa	Venice
2204		Archimedes	Jones	W Sr	British	164	101·1	22·8	11·0	Cowes,I.of Wt.	1882	J.White	[illegible]	T.O'Dwyer	Maryborough
2205		Archimedes	Hansen 86-86	W Bg	Danish	243	102·8	25·2	13·5	Ribnitz	1859			H.P.Christiansen	Copenhagen
2206	H	Archimedes	Goulandris	W Bg	Greek	298	102·6	27·0	..	Syra	1878	S.Koufoudakis		D.&G.Goulandris	Andros
2207		Archipel	Pot -82	W Sr	Dutch	140	101·0	24·2	10·5	Vlaardingen	1863	L.VanDam&Co.		A.Hoogendijk,jzn.	Vlaardingen
2208	A	Archistratigh Michail	Arnirà	W Bg	Russian	110	..	..	..	Tireboli	1884	M.Tomadjoglu	[illegible]	Janni Arnirà	Odessa
2209	R	Architect	Liewen -88	W Bk	Swedish	370	121·0	25·6	16·5	P.E.Island	1865			J.C.Hulthén	Helsingborg
2210		Archos	Hansen -87	W Bk	British	449	135·8	28·9	17·7	Sunderland	1865	W.Adamson		R.H.Bruce	Shanghai
2211	A	Arciduca Rodolfo	Martinolich	W Bk	Austrian	863	164·0	32·2	21·5	Lussinpiccolo	1876	G.S.Cattarinich	[illegible]	B.R. Cattarinich di A.S.	Lussinpiccolo
2212	G V	Arcona	Lübken 86-86	W Bk	German	912	175·8	35·1	21·0	Elsfleth	1881	J.Ahlers	[illegible]	J.D.Ahlers	Elsfleth
2213	V	Arcona		W Bg	Norweg'n	186	91·2	25·6	13·1	Wollin	18 57/72	Schönemann		A.Eriksen	Haugesund
2214	R	Arcot	Cates	W Bn	American	381	127·0	29·6	15·7	E.Machias,Me.	1875		[illegible]	J.W.Pope & Co.	Machias,Me.
2215		Arctic	Allen 85-85	W Bn	British	125	90·8	23·9	11·0	New Harbour (N.F.)	1875			John Munn & Co.	St.Johns,N.F.
2216	N	Arctic	Henrichsen	W Bk	Norweg'n	537	150·6	28·3	19·6	Richmond,Me.	1852		[illegible]	L.Andreasen	Larsund
2217	N	Arctic	Hansen	W Bk	Norweg'n	262	120·1	27·5	12·5	Sandefjord	1876	SörenOlsen	[illegible]	C.Castberg	Sandefjord
2218		Arctic	Johansson	W Bk	Swedish	338	122·0	28·1	14·1	Gefle	1866	O.A.Brodin		S.N.Horndahl	Höganäs
2219	*	Arctic Stream	Bowen 86-86	I S	British	1482	250·5	38·1	23·0	PortGlasgow	1885	Russell&Co.		A.L.Polson	Glasgow
2220	R	Arcturus	Smith	W Bk	British	714	153·1	33·9	19·0	Granville,N.S.	1874			R.Mills	Annapolis,N.S.
2221	V	Arcturus	Hallmann 78-83	I S	German	1116	208·4	31·5	20·5	Vegesack	1878	H.F.Ulrichs		W.A.Fritze & Co.	Bremen
2222		Arcturus	Gibson -8	W Bk	American	1007	168·1	36·7	23·2	Kenneb'nk,Me	1866		[illegible]	R.Kendrick	Port Townsend
2223	*	Ardencaple	Guthrie 88-88	Stl S	British	1737	261·0	39·6	23·4	Glasgow	1885	A.Stephen&Sns		Edmiston & Mitchells	Glasgow
2224	*	Ardenclutha	Hamilton 86-86	I S	British	1222	229·8	35·5	21·2	Glasgow	1876	A.Stephen&Sns		Edmiston & Mitchells	Glasgow
2225	*	Ardencraig	Hampshire 86-86	I S	British	2072	277·7	40·0	24·9	Greenock	1886	Russell&Co.		Edmiston & Mitchells	Glasgow
2226	*	Ardendee	Thorburn 80-88	I Bk	British	610	176·0	30·2	17·3	Whitehaven	1876	Whthvn.SBCo.		A.Rae	Liverpool
2227	*	Ardgowan	Wilson 75-86	I Bk	British	1283	218·0	35·3	22·8	Greenock	1867	R.Steele&Co.		Adam, Hamilton & Co.	Greenock
2228		Ardita	Simare -83	W Bg	Italian	164	88·2	23·4	13·3	Fiume	1856			P.Scouzo	Palermo
2229	D V	Ardjoeno (ex Amsterdam)	Bays 87-87	C Bk	Dutch	1369	183·4	42·0	25·2	Amsterdam	1873	Meursing & Huijgens	[illegible]	A.Hendrichs & Co.	Amsterdam
2230	* D	Ardmillan	Mitchell	I S	British	1655	262·1	40·2	22·4	Glasgow	1878	Dobie&Co.		J.&A.Allan	Glasgow
2231	*	Ardmore	McVicar	I S	British	1092	219·6	35·1	21·2	Glasgow	1875	Barclay,Curle &Co.		W.Kidston & Sons	Glasgow
2232		Ardross	Minter -88	W Sr	British	184	100·0	22·9	12·9	Inverness	1864	Stewart		H.&E.Crundall	Dover
2233	*	Ardvar	Rae 79-81	I Bk	British	858	197·1	32·5	19·4	PortGlasgow	1875	R.Duncan&Co.		A.Rae	Liverpool
2234	N	Arendal	Josephsen	W Bk	Norweg'n	451	135·6	30·8	16·6	Arendal	1874	A.Aanonsen	[illegible]	P.Josephsen	Arendal
2235	N	Arendal	Gjermundsen 81-81	W Bg	Norweg'n	267	103·0	26·5	14·9	Arendal	1855			A.Arvesen	Arendal
2236	V	Areola	Peregrine 83-83	W Bk	British	947	171·1	35·0	21·0	Portland,N.B.	1877	HilyardBros.	[illegible]	D.V.Roberts	St.John,N.B.
2237	*	Arequipa	Lucas 88-88	I Bk	British	495	169·6	25·3	16·9	Liverpool	1862	W.C.Miller		S.Wakeham & Sons	Liverpool
2238	*	Ares	Cox 86-88	I Bk	British	514	162·8	28·2	17·2	Glasgow	1876	C.Connell&Co.		J.B.Walmsley & Co.	Liverpool
2239		Ares	(ex Agios Dionysios)	W Bg	Greek	230	..	27·7	..	Galaxidi	1872	C.Papapetros		Georgeos Skourlinis	Galaxidi
2240	*	Aretas	Hall -88	W Bk	British	241	127·7	25·3	12·9	Bideford	1871	Johnson	[illegible]	The DenneryCo.Ld.	London
2241	V	Aretas	Bidewe 82-	W Bn	German	391	131·8	26·8	13·0	Windsor,N.S.	1875	B.Smith	[illegible]	W.Höveler	Papenburg
2242	*	Arete	Tasker 68-88	I Bk	British	1286	237·2	35·7	21·2	Sunderland	1887	W.Pickersgill & Sons		T.Beynon & Co.	Newport
2243	*	Arethusa	Nixon 71-75	I S	British	1272	227·5	35·9	22·3	Greenock	1869	R.Steele&Co.		Hamilton Bros.	Liverpool
2244		Arethusa	Surkow 64-61	W Bk	German	323	104·0	28·6	17·1	Memel	1844	B.Pieper		F.Surkow	Memel
2245	I	Argentina		W Sr	Argentine	125	89·5	22·1	9·0	Cassano	1886	A.Castellano	[illegible]	N.Mihanovich fu P.	BuenosAyres
2246		Argestes	Kendall -88	W Bg	British	131	77·9	22·6	13·2	Douglas	1839	Williamson		J.Kelly	Whitehaven
2247	*	Argo	Thompson 80-81	I S	British	1490	218·0	37·9	23·0	Glasgow	1880	Barclay,Curle &Co.		A.&J.H.Carmichael&Co.	Greenock
2248		Argo	Harnett 87-87	W Bg	British	225	114·5	24·5	13·7	Jersey	1862	Grandin		R.Allix	Guernsey
2249		Argo	Knudsen -88	W Bn	Danish	142	96·3	20·7	10·8	Troense	1878	C.R.Møller		E.C.Jörgensen	Marstal
2250															

* = Lloyd's Register; D = Nederlandsche Vereeniging; G = Germanischer Lloyd; I = Registro Italiano; R = American Record;
A = Veritas Austro-Ungarico; H = Veritas Ellenico; N = Norske Veritas; V = Bureau Veritas.

Dans la 4me Colonne, C = Composite (membrure en fer, bordé en bois); I = Fer; Stl = Acier; W = Bois.

Dates du doublage en métal (Col. 11) ressortent des livres dans lesquels les navires sont respectivement classés ou des rapports des Inspecteurs du Lloyd's Register.

Pour plus amples renseignements sur les navires classés par le Lloyd's Register, voir le Lloyd's Register Book et le Lloyd's Yacht Register.

Official No.	No. in Reg.	Registry in which classed	Ship's Name. Material and Rig. Master, and Dates of Appointment to present Owner's Service and to Ship. Ship's Late Name.			Flag.	Net Tonnage	Dimensions. Length	Breadth	Depth	Built. Where	When	By whom	Date of Metalling	Owners or Managers
	2251	V	Argo	Steen 75-75	WSr	Danish	121	85·0	21·1	10·1	Nyborg	1875	Emil Beck		P.B.Goldmann
	2252	DV	Argo	Lenning -88	WSr	Dutch	146	93·5	21·9	10·8	Sappemeer	1864	K.Bakker		H.W.Homveld
	2253	✻V	Argo	Stack 73-82	I Bk	German	985	190·0	33·0	22·8	Hamburg	1875	ReiherstiegShip [Works		M.G.Amsinck
348	2254	HV	Argo	Vatis	WBk	Greek	597	147·9	31·9	..	Syra	1873	P.Beaum	[illegible]	I.B.Vatis
80	2255		Argo		WBg	Greek	269	98·8	26·9	..	Spezzia	1866			Heirs of Hadji Anargh
	2256	N	Argo	Wilhelmsen -88	WBk	Norweg'n	758	150·7	36·1	21·0	Björneborg	1859			O.Olsen
	2257	N	Argo	Arentsen 85-85	WBk	Norweg'n	616	154·3	33·1	18·1	Bergen	1868	J.Gran	[illegible]	Jörgensen & Jorkjaer
	2258	N	Argo (WRECKED)	Hoelstad	WBk	Norweg'n	430	118·5	30·9	18·3	WifstaVarf, [Sweden	1848	E.A.Hagglund		Chr.Swane
	2259	D	Argo	Rasmussen	WBk	Norweg'n	373	118·6	29·2	16·6	Kragerö	1861	T.Pedersen	[illegible]	B.Knoph
	2260	N	Argo	Nilsen -88	WBk	Norweg'n	307	114·2	29·0	14·6	Kalmar	1850			Actieselskabet "Argo"
	2261		Argo		WSr	Norweg'n	136	94·0	21·9	10·8	Brevig	1859	BrødrJörgensen		J. Fredriksen & Th. S[illegible]
	2262	R	Argo	Roos -88	WBk	Russian	582	136·1	34·1	19·8	Brahestad	18⁶⁵⁄₇₂	A.Westerlund		J.Lundberg & Co.
	2263	V	Argo	Bredelius 75-82	WBk	Swedish	661	174·8	32·2	17·7	Björneborg	1873	J.Elg	[illegible]	J.M.Möller
	2264		Argo	Petterson 45-49	WSr	Swedish	110	79·1	19·1	9·4	Groningen	18[illegible]			G.Pettersson
76415	2265	✻	Argomene	McAusland 85-85	I S	British	1662	241·0	38·7	24·0	Southampton	1877	T.R.Oswald		H.Fernie & Sons
72412	2266	✻	Argonaut	Bonner 86-86	I S	British	1488	254·4	38·6	23·2	Glasgow	1876	Barclay,Curle [&Co.		A.&J.H.Carmichael&Co
50727	2267	RV	Argonaut (CONDEMNED)	George 87-87	C Bk	British	1041	206·1	33·2	20·6	London	1866		[illegible]	J.I.Jacobs & Co.
105910	2268		Argonaut	Ducker 78-	WSr	American	185	105·0	30·8	9·4	SanFrancisco	1880			A.M.Simpson
	2269		Argonaute		WBk		477	131·2	26·2	18·5	Nantes	1866	Jollet&Babin	[illegible]	
64612	2270	DV	Argosy	Craigen 86-86	WBk	British	1061	174·8	37·6	22·7	Digby,N.S.	1872		[illegible]	James Milne
76170	2271	✻	Argus	Ramsay 88-88	I S	British	1543	255·5	38·7	23·1	Glasgow	1878	Barclay,Curle [&Co.		A.&J.H.Carmichael&Co
	2272	V	Argus	Svane 74-74	WBn	Danish	190	106·2	23·4	11·8	Marstal	1880	J.J.Bager		H.Hansen
	2273	G	Argus	Jörck -74	W3mSr	German	315	107·7	27·1	16·6	Memel	1856	B.Pieper		J.C.Janentzky
	2274	N	Argus	Larsen	WBg	Norweg'n	272	119·8	28·6	12·2	Arendal	1867	Dekke		J.C.Larsen
	2275	V	Argus	Seeberg	WSr	Norweg'n	159	97·8	23·7	11·1	Nordby,Fanø	1871	N.Nielsen	[illegible]	P.Wiborg
	2276		Argus	Bagas -88	W3mSr	Portug'se	268	119·7	22·6	12·7	Dundee	1873	DundeeS.B.Co.		Bensaude & Co.
63752	2277	✻	Argyleshire	Purdie 84-84	I Bk	British	708	185·8	31·2	18·4	Glasgow	1870	Dobie&Co.		T.Law & Co.
80030	2278	V	Argyll	Allen 86-86	WBk	British	1223	197·6	38·4	23·0	St.John,N.B.	1880	J.K.Dunlop	[illegible]	Wm.Thomson&Co
85975	2279		Argyll	Virgie -88	WBn	British	299	119·2	29·9	12·1	Harvey,N.B.	1882	G.S.Turner		G.S.Turner
89679	2280	✻	Ariadne	Crool 81-85	I Bk	British	1167	224·9	35·0	21·0	Dumbarton	1885	A.McMillan&Sn		A.M.Lawrence&Co
88653	2281	✻R	Ariadne (CONDEMNED)	Trainor	WBn	British	327	132·5	29·5	11·9	P.E.Island	1884	G.D.Longworth	[illegible]	G.D.Longworth
68571	2282		Ariadne (ex Boadicea)		WSr Yt	British	230	138·4	25·1	14·7	Gosport	1874	Camper & [Nicholson		Hon.Mrs.E.C.Meynell Ingram
105434	2283	R	Ariadne	Colby 80-82	W3mSr	American	378	126·1	30·5	11·5	Harpswell,Me.	1874	P.C.Merriman	[illegible]	W.S.Jordan & Co.
	2284	V	Ariadne	Frandsen -88	WBk	Danish	306	146·7	26·1	12·0	Lübeck	18⁶²⁄₆₈		[illegible]	Ths.Sonne & Co.
	2285	GV	Ariadne	Segebade 79-80	WBk	German	563	143·0	31·1	18·3	Flinthausen	1880	J.H.Strenge & [Sohn	[illegible]	J.G.Lubken
	2286		Ariadne (WRECKED)	Vogt 83-	WBk	German	321	104·8	26·1	16·9	Memel	1844	Pieper		W.A.Vogt
	2287	N	Ariadne	Johannsen	WBk	Norweg'n	668	152·0	31·9	19·7	Quebec	1861	Dinning	[illegible]	F.J.Hansen
	2288	N	Ariadne	Hansen -75	WBk	Norweg'n	480	134·6	31·0	17·0	Tvedestrand	1875	O.Gundersen	[illegible]	A.B.Olsen
	2289		Ariadne (BROKEN UP)	Thorsen	WBk	Norweg'n	317	112·1	27·4	16·6	Sunderland	1860	J.C.Robinson		R.Thorsen
	2290		Ariadne (ex Queen)	Nilsson 82-82	WBg	Swedish	244	89·2	23·5	15·5	Sunderland	18⁵⁵⁄₇₅			H.Svensson
207	2291		Ariadni (WRECKED)		WSr	Greek	118	76·4	22·1	..	Syra	1879			C.A.Antonelles
1	2292	H	Ariadni Coupa	Golmos	WBk	Greek	306	106·6	28·0	..	Trieste	1869		[illegible]	Gombou frères
	2293		Arianti		WSr	Greek	129	..	..	..	Syra	1879			A.Giorgio
44686	2294	✻	Arica	Netherton	I Bk	British	480	153·0	25·6	17·5	Liverpool	1862	W.C.Miller		S.Wakeham & Sons
	2295	R	Arica	Thomsen	WBk	Danish	383	128·8	27·2	16·4	Dundee	1865	Tay Ship Co.	[illegible]	H.Petersen
	2296	NV	Arica (ex Cathalina)	Svendsen	WBk	Norweg'n	621	152·9	33·2	18·3	Quebec	1871	L.Methot	[illegible]	C.Eyde & Hein
72736	2297		Ariel		WBk	British	543	136·3	30·4	20·2					Pestonjee Cowasjee
48961	2298		Ariel	Burt	WBn	British	178	102·7	23·2	12·8	Padstow	1864	Tredwen		W.S.Allport & Co.
57710	2299		Ariel		WBn	British	170	101·7	24·0	10·9	Barringt'n, NS	1868			P.Foley
69770	2300		Ariel	Turner -86	WBn	British	134	89·4	22·4	10·4	George'sRiver, [N.S.W.	1874	Geo.Thompson		James A.Grigg

✻=**Lloyd's Register**; D=Nederlandsche Vereeniging; G=Germanischer Lloyd; I=Registro Italiano; R=Ame[illegible]
A=Veritas Austro-Ungarico; H=Veritas Ellenico; N=Norske Veritas; V=Bur[illegible]

In the 4th Column, C=Composite (iron frame, wood planked); I=Iron; Stl=Steel; W=Wood.

The Dates of Metalling (Col. 11) are based on the records of the Books in which Vessels are respectively classed, or on the Reports of the Surveyors of [illegible]

Further particulars of Vessels classed by Lloyd's Register will be found in Lloyd's Register Book and Lloyd's Yacht Register.

No.	Signaux	Cote	Nom. Capitaines et Changements. Ancien Nom.	Matériau / Gréement	Pavillon	Tonnage reg.	Longueur	Largeur	Creux	Construit Où	Année	Par qui	Doublage	Armateurs ou Directeurs-Gérants.	Port d'Armement.
2301			Ariel *Evans*	W Bn	British	[illegible]	[illegible]	[illegible]	[illegible]	[illegible]	1862	Brocklebank		W.Tyrrell	Whitehaven
[illegible]			Ariel	W Bn	British	[illegible]	[illegible]	[illegible]	[illegible]	[illegible]	1846			P.Gorman	Youghal
[illegible]		✻	Ariel *Tucker 77-88*	W Bn	British	[illegible]	[illegible]	[illegible]	[illegible]	[illegible]	1873	Cox	[illegible]	C.T.Bowring & Co.	Liverpool
2304	DG		Ariel *Wiedenhoff -77*	W Bg	German	[illegible]	[illegible]	[illegible]	[illegible]	[illegible]	1852	A.Schönemann		E.C.Winter	Rostock
2305			Ariel *Witt 79-89*	W Sr	German	189	[illegible]	23·2	11·8	Elsfleth	1856			J.J.Wallis & Sohn	[illegible]
2306	DNV		Ariel *Schminke*	W S	Norweg'n	[illegible]	[illegible]	37·[illegible]	21·6	Bergen	1876	Jens Gran	[illegible]	M.Beyer	Bergen
2307	N		Ariel *Pedersen 77-89*	W Bk	Norweg'n	[illegible]	[illegible]	29·1	16·8	Holmestrand	1858			T.Holm	Holmestrand
2308	N		Ariel *Johnsen 87-*	W Bg	Norweg'n	[illegible]	[illegible]	25·8	12·4	Arendal	1873	O.Thorsen		John Johnsen	[illegible]
2309	N		Aries *Eide -86*	W Sr	Norweg'n	114	[illegible]	22·4	10·5	Laurvig	1875	C.Archer		Aktieselskabet "Oceana" (Chr. Christensen)	Sandefjord
2310	A		Ariete *Stipanovich*	W Bk	Austrian	578	[illegible]	[illegible]	19·7	Fiume	1876	V.Brazzoduro	V 11.84	Cas.Cav.Cosulich	Fiume
2311	H		Arietta *Vretta*	W Bn	Greek	240	..	..	..	Syra	1869		V 3.89	A.Vretta	Syra
2312			Arietta	W Sr	Greek	132	[illegible]	22·[illegible]	..	Syra	1877			S.Tsatsaronis	Syra
2313	✻		Arina *Williams*	I S	British	214	[illegible]	[illegible]	[illegible]	[illegible]	1869	W.Doxford & Sns.		T.S.Whitehead & Chambers	London
2314			Arion *Dringenberg -88*	W Bk	German	214	[illegible]	[illegible]	[illegible]	Vegesack	1863	H.F.Ulrichs		C.C.H.Bösch	Bremerhaven
2315	G		Ariosto *Rizzaeta*	W Bk	Italian	289	[illegible]	[illegible]	17·[illegible]	Meta	1857	Frat.Mauro	V 4.88	F.Collareta & Co.	Naples
2316	H		Aris *Stephanides -88* (ex Hedwig)	W Bg	Greek	[illegible]	..	..	..	[illegible]	1872		8.89	G.Gourletos	Galaxidi
2317			Aristide	W Bg	Turkish	137	..	..	..	Vassilikos	1869			Vassil Nicolaidi	Vassilikos
2318			Aristea	W Bn	Greek	289	[illegible]	26·0	..	Genoa	1858			N.Papalas	Syra
2319	V		Aristide *Coriton*	W Bk	French	329	125·6	28·3	16·5	Nantes	1873	J.Dubigeon & [fils]	V 6.88	Coriton	Nantes
2320	✻		Aristides *Spalding 82-87*	I S	British	1661	250·0	39·5	24·5	Aberdeen	1876	W.Hood & Co.		G.Thompson & Co.	Aberdeen
2321			Aristides *Papas*	W Bg	Greek	327	112·8	27·6	..	Syra	1875		[illegible]	J.D.Papas	Syra
2322			Aristides	W Bn	Greek	287	103·6	27·9	15·5	Syra	1867			P.Agapetos	Santorin
2323			Aristides *Stroumboulis*	W Bk	Greek	281	..	..	..	Syra	1867		C 76	J.M.Frarezozas	Syra
2324	F		Aristides *Th. Bachas*	W Bg	Greek	170	85·6	24·8	..	Syra	1877			Ant.Bachas	Syra
2325	N		Aristides *Björneboe*	W Bk	Norweg'n	458	135·6	31·0	16·6	Arendal	1860	S.Johansen	V 4.88	O.A.Jensen Björneboe	Arendal
2326	N		Aristides *Johansen 82-88*	W Bk	Norweg'n	305	122·3	26·6	14·5	Kragerö	1865	Chr.Jensen	V 5.88	Harald Larsen	Kragerö
2327	G		Aristides *Hansen*	W BmSr	Norweg'n	261	101·4	26·4	14·8	Rostock	1851	A.Schönemann		H.O.Hansen	Dröbak
2328	H		Aristipos *Baziotis*	W Bg	Greek	314	100·7	28·0	..	Syra	1875		C 4.86	A.Baziotis	Zante
2329			Aristodemo	W Bg	Italian	127	..	..	..	Leghorn	1847			E.Muti	PortoFerraio
2330	✻		Aristomene *Attridge 75-81*	I S	British	1759	263·0	39·1	24·1	Liverpool	1881	W.H.Potter & S.		H.Fernie & Sons	Liverpool
2331	N		Aristos *Axelsen -84*	W Bk	Norweg'n	528	145·3	32·4	17·9	Laurvig	1868	M.Treschow	V 8.86	O.C.Reinhardt	Christiansand
2332	AH		Aristotelis *Empiricos*	W Bk	Greek	499	145·3	30·5	..	Syra	1875		V 3.88	E.Empiricos	Andros
2333	A		Aristotelis Stamation *Zatzefilos*	W Sr	Greek	163	75·1	25·1	..	Cassos	1885	E.Sophos		M.Zatzefilos	Santorin
2334	V		Arizona *Robins -87*	W Bk	British	1085	183·3	36·5	22·4	St.John,N.B.	1882	O.Pittfield	V 3.88	Taylor Bros.	St.John,N.B.
2335			Arizona *Lohnes*	W Sr	British	115	84·0	25·0	9·8	P.Medway,N.S.	1883	Dalliver		A.Corkum	Lunenburg,N.S.
2336	NV		Arizona *Arnesen 75-88*	W Bk	Norweg'n	1302	189·5	37·7	24·1	St.John,N.B.	1869	T.E.Millidge & Co.	V 5.88	H.Kjösterud & Sönner	Drammen
2337	DN		Arizona *Rönneberg -87*	W Bk	Norweg'n	542	116·9	28·9	18·1	Stockton,Me.	1864		V 89	A.Brodersen	Mandal
2338			Ark	C Sr	British	521	176·5	35·6	10·0	P.Dalhousie	1875	A.Muir & Bro.		Bryce Muir	St.Cthrine's.Ont
2339	V		Arken *Höye 82-86*	W BmSr	Danish	184	110·7	25·7	9·4	Marstal	1886	J.O.Christensen	V 1.86	H.C.Christensens Enke	Svendborg
[illegible]			Arken (ex Iris?) *Halvorsen* (BUNK)	W Bg	Norweg'n	349	107·6	29·4	16·3	Grimstad	1863	O.Kittelsen		J.Olsen	Grimstad
2341	N		Arken *Amundsen*	W Sr	Norweg'n	189	96·1	25·2	11·6	Stavanger	1875			F.Amundsen	Stavanger
[illegible]	V		Arken *Hallström 84-4*	W Sr	Swedish	198	108·2	23·4	11·2	Oscarshamn	1884	C.Thorén	V 2.88	J.E.Hagström	Oscarshamn
[illegible]	V		Arklow *Farmer -88*	W S	British	1474	214·0	40·0	24·0	Newport,N.S.	1879	N.Mosher	V [illegible]	F.Curry	Windsor,N.S.
2344	V		Arklow *Pye*	W Bk	British	748	158·8	33·0	19·2	St.Martin's NB	1877	W.Osborne	V 3.88	O.Emery	St.John,N.B.
[illegible]			Arkwright *Jones 87-87*	W Bk	American	1216	197·0	35·5	24·4	Prtsmouth,NH	1855			Puget Sound Commercial Co.	Pt.Gamble,Wash
[illegible]	R		Arlington *Davis -87*	W S	British	[illegible]	165·0	34·2	19·5	P.Gilbert,N.S.	1868	J.Milbury	V 8.88	St. Clair Jones	Yarmouth,N.S.
2347	R		Arlington *Leach 82-85*	W Bkn	American	478	111·5	34·4	12·4	Bucksport,Me	1883	W.H.Genn	V 7.87	J.S.Emery & Co.	Boston,Mass.
[illegible]			Arlington *Hansen -88*	W Bk	Norweg'n	589	147·8	31·7	18·8	KennebunkMe	1863	TitcombPerkins		T.Thommesen & Sön	Arendal
[illegible]	✻		Armadale *Duncan 73-87*	I m Bk	British	[illegible]	285·2	40·5	23·8	Glasgow	1887	A.Stephen & Sns.		J.&A.Roxburgh	Glasgow
[illegible]	V		Armando *(Marcos?) Soares*	W Bg	Brazilian	472	116·4	26·8	17·1	Varase	1865		V 1.89	A. da C. Vianna & F. M. D. Moreira	Rio de Janeiro

✻ = Lloyd's Register; D = Nederlandsche Vereeniging; G = Germanischer Lloyd; I = Registro Italiano; R = American Record;
A = Veritas Austro-Ungarico; H = Veritas Ellenico; N = Norske Veritas; V = Bureau Veritas.

Dans la 4me Colonne, C = Composite (membrure en fer, bordé en bois); I = Fer; Stl = Acier; W = Bois.

Les Dates du doublage en métal (Col. 11) ressortent des livres dans lesquels les navires sont respectivement classés ou des rapports des Inspecteurs du Lloyd's Register.

Pour plus amples renseignements sur les navires classés par le Lloyd's Register, voir le Lloyd's Register Book et le Lloyd's Yacht Register.

Official No.	No. in Book	Registry in which Classed	Ship's Name. Master, and Dates of Appointment to present Owner's Service and to Ship. Ship's Late Name.	Material and Rig	Flag	Net Tonnage	Length	Breadth	Depth	Built: Where	When	By whom	Date of Metalling	Owners or Managers
	2351		Armando	W Sr	Brazilian	160	..	..	..					J.F.Praça
80064	2352	V	Armenia *Morse* -86	W Bk	British	1097	183·5	36·5	22·4	St.John,N.B.	1880	O.Pitfield		Taylor Bros.
92407	2353	R	Armenia *Bowen* -87	W Bk	British	680	152·3	32·5	21·2	Savona	1875	A.Cadeagro		Mrs.E.J.Hutchison
85339	2354		Armenia	W	British	142	110·3	22·2	7·6	Ottawa	1881	J.Neville		Thomas H.Kirby
105727	2355	RV	Armenia *Carter* -88	W S	American	1643	223·3	40·4	26·0	Bath,Me	1877	Houghton Bros.		Houghton Bros.
87423	2356	✻	Armida *Johnston* 79-85	Stl S	British	1624	261·9	38·3	23·1	Glasgow	1885	Barclay,Curle & Co.		W.Letham
126	2357	AV	Armida *Lar* -88	W S	Austrian	1098	191·6	34·5	24·0	Lussinpiccolo	1881	N.Martinolich		G.A.Scopinich
	2358	IV	Armida (ex Marsigliese) *Bozzo* -88	W Bk	Italian	552	134·4	30·8	19·6	Savona	1866	F.Calamaro		T.Bozzo
	2359	V	Armida *Casset* -88	W 3m Sr	Swedish	264	125·6	24·8	10·7	Sundsvall	1873	O.Högglund		C.G.Wickberg
	2360		Armida (STRANDED) *Pehrsson*	W Sr	Swedish	166	88·2	22·5	11·9	Söderhamn	1852			O.Pehrsson
	2361	F	Armide *Besrest*	W Sr	French	126	83·7	20·3	11·6	Dunkirk	1869	B.Derycke		Sebert aîné
	2362	G	Armin *Burhorn* 85-87	I Bk	German	842	174·3	31·9	21·5	Glasgow	1869	A.Stephen&Sns.		D.H.Watjen & Co.
24344	2363		Armistice *Hughes*	W Bk	British	331	118·5	26·5	16·6	Sunderland	1855			Charles G.Millar
	2364		Armokradi *Alexandiros*	W Bg	Turkish	210	..	..	..	Greece	1885			Kerestch-tljari Christo-phe Alexandiros & Co.
2017	2365	V	Armonia *Sablich*	W Bk	Austrian	410	125·3	27·2	17·9	Fiume	1868	F.lli Schiavoni		I.Bonetich
688	2366	I	Armonia (ex Rusaglo) *Ogilia* 86-88	W Bk	Italian	955	170·9	34·1	23·5	Chiavari	1875	L.Gotuzzo		M.&A.Dall'Orso fu G.
1087	2367	IV	Armonia *Cafiero*	W Bk	Italian	449	135·5	27·4	17·9	Alimuri	1875	F.S.Mauro		S.&A.Cafiero
207	2368	I	Armonia *Cormo*	W Bk	Italian	447	141·4	32·6	18·6	Gaeta	1866	F.Contreras		F.lli Matarazzo
27	2369		Armonia *Agretti*	W Bk	Italian	280	125·0	27·7	15·9	Leghorn	1855	G.Micheli		G.Tonietti fu Raff
	2370	N	Arndt (ex Havelock) *Gundersen* 86-86	W Bk	Norweg'n	464	136·8	27·3	17·4	Sunderland	1859	W.Briggs&Son		C.A.Leschbrandt
	2371	N	Arne (ex Alethea) *Eriksen* -86	W Bk	Norweg'n	418	129·4	27·0	17·2	Sunderland	1855			L.Hammer
58	2372	A	Arnea *Camenarovich*	W Bn	Austrian	134	81·6	20·3	11·2	Lussinpiccolo	1884	G.Giuricich		Pietro Lister fu M
	2373	V	Arnette Mathilde *Lauritzen*	W Sr	Danish	101	84·7	20·2	9·6	Nakskov	1864	Ridersborg & Trochman		P.C.Knudtzon&Son
	2374	N	Arnevig *Ellingsen* 65-65	W Bk	Norweg'n	291	110·6	28·3	14·7	Arendal	1865			S.E.Ellingsen
83183	2375	V	Arnguda *Green* 70-81	W Bk	British	978	188·0	35·6	21·9	NwGlsg'w,NS	1881	J.W.Carmichael		J.W.Carmichael
11862	2376	✻	Arno *King* 64-79	W Bn	British	107	80·0	20·9	11·1	S.Shields	1818			W.Bayley
	2377	V	Arnold (ex Eugenie) *Haverkamp* 78-86	I Bk	German	838	116·2	25·9	15·2	La Seyne	1868	Forges & C.de la Medit.		Adolph Schiff
	2378	V	Arnold von Bippen *Müller* 86-87	W Bk	German	446	126·1	30·5	17·7	Rostock	1877	W.Zeltz		W.Maack
627	2379		Arnolda *Marrine* 84-86	W Bk	American	340	104·9	28·1	18·0	Rochester,Mas	1844			L.Snow & Co.
	2380	NV	Aron *Christensen* 79-84	W Bk	Norweg'n	633	154·2	33·4	18·8	Arendal	1883	H.Stiansen		Carl Bech
	2381	DGN	Arracan (ex Darmstaedter Bank) *Rolfsen* -87	W Bk	Norweg'n	758	160·5	32·6	21·1	Bremerhaven	1856	J C Tecklenborg		D.M.Z.Bodin (J.Meyer)
	2382		Arrhowah	W Bg	Turkish	260	..	..	..	Greece	1868			Arrhowahlee-Batoom-oglou &c.
	2383		Arrow *Nestved*	W Sr	Russian	107	79·0	23·3	9·8	Dagö	1870	E.Pargo		N.Nestved
	2384		Arsène *Berthevin*	W Bg	French	164	85·6	22·6	12·6	St.Malo	1840	Picard		Leroux-Berthevin
	2385	FV	Arsène *Cocheril* -87	W Bn	French	163	90·5	22·0	12·1	St.Malo	1872	J.Parnet		Daugnet
	2386		Artemis *Rasmussen* 88-88	W Sr	Danish	115	84·1	20·2	10·8	Svendborg	1857	C.C.&H.P.Hansen,jun.		C.C.Hansen
	2387		Artemis	W Bg		258	..	..	..					
17	2388	A	Artemis	W Sr	Greek	146	83·7	22·6	..	Skiathos	1884	C.G.Mytilinaios		A.Giakoumis
	2389	A	Artemis	W Bg	Turkish	175	..	..	..	Scio	1870			Aghlikiro Mileadi
	2390	N	Artemis (ex Oriental) *Moe* -88	W Bk	Norweg'n	564	145·5	30·4	18·6	BlackRiverNB	1865	McLeod		Johs Röds Sönner
95088	2391		Artemisa *Gordon* -88	Stl Sr	British	219	115·2	28·0	9·2	Dumbarton	1888	W.Denny & Bros.		La Platense Flotilla Co. Ld
	2392	F	Artémise	W Sr	French	103	71·2	21·0	10·6	St.Vaast	1858			Le Cordier
1509	2393		Artemisia *D'Esposito* -88	I Bn	Italian	157	105·9	26·4	8·10	Leghorn	1888	F.lli Orlando		F.Capprillo
	2394		Artemisia *Tannis*	W Sr	Turkish	124	..	..	..	Vasilikos	1876			Vassilikos Yanni Yorghi
	2395	A	Artemisia (ex Evangelistria) *Paterra*	W Sr	Turkish	118	..	..	..	Sira	1879	E.Flumis		C.Paterra
136	2396		Artemus Tirrell *Morgan*	W Sr	American	196	107·0	27·7	9·3	Milford,Del.	1852			J.Davis & Son
88547	2397		Arthur (STRANDED)	W Sr	British	142	94·0	25·3	10·3	Shelburne,NS	1884	A.Hood		Hugh B.Cann
83170	2398		Arthur	W Sr	British	122	85·4	24·4	10·0	LaHave,N.S.	1881	J.Weagh		F.Desmond
91945	2399	✻	Arthur *Hargraves*	W Sr	British	92	83·5	21·5	10·1	Goole	1886	Cottingham Brs.		G.Kilner & Co.
	2400		Arthur (ex Palma, ex Johann)	W Sr	Brazilian	159	..	..	..	Germany	1866			A.F.G.Motta

✻=Lloyd's Register; D=Nederlandsche Vereeniging; G=Germanischer Lloyd; I=Registro Italiano; R=Americ
A=Veritas Austro-Ungarico; H=Veritas Ellenico; N=Norske Veritas; V=Bureau

In the 4th Column, C=Composite (iron frame, wood planked); I=Iron; Stl=Steel; W=Wood.

The Dates of Metalling (Col. 11) are based on the records of the Books in which Vessels are respectively classed, or on the Reports of the Surveyors of Lloy

Further particulars of Vessels classed by Lloyd's Register will be found in Lloyd's Register Book and Lloyd's Yacht Register.

No.		Noms, Matériaux et Gréement. Capitaines et dates d'entrée au service des Armateurs actuels et à bord du navire. Ancien Nom.	Pavillon	Tonnage	Longueur	Largeur	Creux	Construit Où	Quand	Par qui	Date du doublage en métal	Armateurs ou Directeurs-Gérants.	Port d'Armement.
2401	G	Arthur *Berger -88* W Bk	German	[illegible]	[illegible]	[illegible]	17·[illegible]	[illegible]	1862	C.F.Devrient		A. Haeger	Stolpe
2402	V	Arthur *[illegible]* W Sr	German	167	[illegible]	[illegible]	[illegible]	[illegible]	1857	J.Holzerland		J.C.Berg	Barth
2403	N	Arthur *Christensen* W 3mSr	Norweg'n	[illegible]	[illegible]	[illegible]	[illegible]	[illegible]	1884	A.Henriksen	[illegible]	H.H.Pettersen	Arendal
2404	V	Arthur *[illegible]* (ex [illegible]) W Bn	Portuguese	172	[illegible]	[illegible]	[illegible]	[illegible]	1876	J.G.M.Branco	[illegible]	Soares & Co.	Oporto
2405		Arthur *[illegible]* W Sr	Swedish	17[illegible]	87·7	22·[illegible]	12·6	Montrose	1867	J.Petrie		M.Mattsson	Wardö
2406		Arthur *[illegible] Andersson* W Bn	Swedish	1[illegible]	[illegible]	24·7	9·[illegible]	[illegible]	$18\frac{21}{80}$			A.Pehrsson	Skeppsmyra
2407	R	Arthur C. Wade (*[illegible]*) W Bkn	American	[illegible]	[illegible]	[illegible]	12·5	Bath Me	1884	W.Rogers	[illegible]	J.S.Winslow & Co.	Portland,Me.
2408	✱	Arthur Pelham *Douglas* W Sk	British	7[illegible]	71·[illegible]	[illegible]	[illegible]	[illegible]	1882	J.C.Hoad		G.Finch	Grimsby
2409	R	ArthurV.S.Woodruff *[illegible]* W 3mSr	American	184	[illegible]	27·[illegible]	[illegible]	[illegible]	1888			T.M.Nicholson	Bucksport
2410	✱	Arthurstone *[illegible]* I Bk	British	116[illegible]	[illegible]	[illegible]	21·[illegible]	Dundee	1876	Gourlay Bros.		D.Bruce & Co.	Dundee
2411	A	Artiero Giovanni *[illegible]* W Bk	Austrian	[illegible]	1[illegible]	[illegible]	1[illegible]	[illegible]	1874	V.Brazzodaro	[illegible]	A.Marussich fu G.	Fiume
2412		Artin W Bg	Turkish	111	..	..	..	Russia	1865			Kapamadji-oglou Artin	Constantinople
2413		Artis W Bg	Turkish	[illegible]	..	..	..	KaraAgatch	1869			OunjiZadehApikEffendi	Constantinople
2414	V	Artisan *Berger 12-81* W Bk	British	1109	189·6	37·7	22·4	St.John,N.B.	1881	Stew'rt&Ritchie	[illegible]	W.Thomson & Co.	St.John,N.B.
2415	✱	Artizan *Stephens [illegible]* W 3mBn	British	511	160·8	29·3	17·1	P.E.Island	1873	Richards	[illegible]	T.P.Richards & Co.	Swansea
2416	V	Artos *Forrest -88* W Bn	British	314	129·4	29·4	13·0	Moss Glen, King'sCo.NB	1882	G.Merritt	[illegible]	R.C.Elkin	St.John,N.B.
2417		Arturo *Henning* (ex [illegible]) W Bk	German	548	117·8	27·2	19·3	Voltri	1868	A.Canepa		H.Mentz,jun.	Rostock
2418	R	Arturo *Hansen -87* W 3mSr	Norweg'n	551	133·5	31·2	20·0	SestriPonente	1867	A.Briasco fu L.	[illegible]	Jörgen Hansen	Tönsberg
2419		ArturoPrat *[illegible]* W Bk	Chilian	231	108·2	24·9	9·7	Baltimore	1859			F.Caparro	Valparaiso
2420	✱	Arundel Castle *[illegible]* W K Barge	British	135	100·5	23·5	8·5	Littlehampt'n.	1888	J.&W.B.Harvey		Walker,Howard & Co.	London
2421		Arutdar I Lr	British	700	240·0	35·5	10·3	Pt.Glasgow	1884	Russell & Co.		IrrawaddyFlotillaCo. Ld.	Glasgow
2422	V	Arvid *Christiansson* W Bn	Swedish	200	103·0	23·2	11·4	Oscarshamn	1872	C.Pettersson	[illegible]	J.Christiansson	Brunsby, Malmöhus
2423	N	Arvilla (ex [illegible]) *Gundersen -87* W Bk	Norweg'n	1027	196·5	35·2	22·9	Whitehaven	1861	Brocklebank	[illegible]	FruA.L.Sörensen	Kragerö
2424	V	Arvio *Mannerström -88* W Bk	Russian	772	178·0	37·0	20·0	Nystad	1871	H.Kjeldstrom	[illegible]	V.A.Reims	Abo
2425		Arvio *Holmberg -88* W Sr	Russian	246	..	..	..					I.Esko	Björneborg
2426	✱	Arvon *Owens* W Sr	British	86	73·5	20·1	10·6	PortDinorwic	1860	Jones		O.Owens	Carnarvon
2427		Asahi Maru W Sr	Japanese	395	111·1	27·4	18·5	Hiogo	1882	OmatsuJugoro		SeikoKwaisha	Hiogo
2428		Asahi Maru W Sr	Japanese	334	116·2	25·9	15·2	Yechizen	1879	K.Tomojiro		K.Gohei	Osaka
2429	DV	Asante *Behrens 76-83* I Bg	German	29[illegible]	114·8	26·6	13·2	Vegesack	1876	H.F.Ulrichs		J.H.Hustede	Elsfleth
2430	✱	Ascalon *Langford 88-88* W Bk	British	938	201·0	34·0	20·5	Aberdeen	1868	Hood	[illegible]	Trinder,Anderson & Co.	Aberdeen
2431	I	Ascolta *Fone* W Bn	Italian	302	112·5	30·7	16·9	Sestri	1865	F.Cadenaccio	[illegible]	F[lli]. Marciani	Genoa
2432	R	Asenath A. Shaw *Martin* W 3mSr	American	529	155·0	35·6	15·6	Wilmingt'n,De.	1878	Jackson,Sharpe &Co		J.Holmes	NewYork
2433	G	AegerRyg *Hoeberg -86* (ex Juliet) W Bg	Danish	278	108·4	25·6	13·6	Griefswald	1862	J.W.Sadelkow		Chr.O.Ipsen	Copenhagen
2434	V	Ashantee *O'Brien -86* W Bk	British	700	158·0	34·4	18·5	Tatamago'che, N.S.	1874	A.Campbell	[illegible]	J.D.McGregor	Picton,N.S.
2435		Ashburton W Bk	British	506	134·0	28·0	18·6	Newcastle-on-Tyne	1853			W.K.Thomson	Melbourne
2436	V	Ashlow *Churchill* W Bk	British	639	147·7	32·[illegible]	18·6	St.MartinsNB	1880	J.Marr		O.Emery	St.John,N.B.
2437	✱	Ashmore *Whitmore 86-89* I S	British	1098	213·3	35·3	20·6	PortGlasgow	1877	J.Reid&Co.		J.Stewart & Co.	London
2438	V	Asia *Hilton -88* W S	British	1398	201·3	38·8	24·0	St.John,N.B.	1883	O.Pittfield	[illegible]	Taylor Bros.	St.John,N.B.
2439		Asia W S	British	1301	195·0	38·0	25·7	Bath,U.S.	1854			T.Lidbetter	Bombay
2440		Asia *Jones* W Bk	British	308	123·0	25·7	14·8	Montrose	1859	Cochar		A.McGregor	Hobart Town
2441	GV	Asia *Haack 76-81* W 3mSr	German	380	134·2	27·8	13·7	Lübeck	1881	J.Steffen	[illegible]	C.Brandt	Blankenese
2442		Asia *[illegible]* (ex Mehemet Ali) W Bk	German	317	99·5	26·0	17·5	[illegible]	1845			A.F.Juchter	Memel
2443	AH	Asia *[illegible] Logotheti* W Sr	Greek	117	..	..	..	Galaxidi	1880		[illegible]	G.Logotheti	Galaxidi
2444	I	Asia *Stingo -87* W Bo	Italian	131	95·2	23·6	8·5	Cassano	1887	A.Castellano	[illegible]	D.Ciampa	Castellamare
2445	✱	Asia *Christensen 85-88* (ex Aimwell) W Bk	Norweg'n	1163	189·2	37·9	23·0	St.John	1861	Gass	[illegible]	J.H.R.Christensen	Christiania
2446		Asia W Bk	Spanish	494	133·8	29·8	..	Stralsund	1857			Diego Ramos & Co.	Havana
2447	V	Asiana *Gray 84-84* W Bk	British	1185	189·7	38·8	23·7	St.John,N.B.	1864	D.V.Roberts	[illegible]	D.V.Roberts	St.John,N.B.
2448	A	Asimina *Kidoni* W Sr	Greek	152	86·0	24·5	..	Syra	1888	A.Livadaro		E.Kydonieus	Syra
2449		Asimoula W Bn	Greek	307	115·8	26·9	15·4	Galaxidi	1880	J.Canatas		Euth.Catzoulis	Galaxidi
2450	N	Askoy *Halvorsen -88* W Bk	Norweg'n	480	111·7	31·3	17·9	Risör	1872	J.Jörgensen	[illegible]	J.M.A.Marcussen	Risör

✱=Lloyd's Register; D=Nederlandsche Vereeniging; G=Germanischer Lloyd; I=Registro Italiano; R=American Record;
A=Veritas Austro-Ungarico; H=Veritas Ellenico; N=Norske Veritas; V=Bureau Veritas.

Dans la 4me Colonne, C=Composite (membrure en fer, bordé en bois); I=Fer; Stl=Acier; W=Bois.

Les Dates du doublage en métal (Col. 11) ressortent des livres dans lesquels les navires sont respectivement classés ou des rapports des Inspecteurs du Lloyd's Register.

Pour plus amples renseignements sur les navires classés par le Lloyd's Register, voir le Lloyd's Register Book et le Lloyd's Yacht Register.

Official No.	No. in Book.	Registry in which classed.	Ship's Name. Material and Rig. Master, and Dates of Appointment to present Owner's Service and to Ship. Ship's Late Name.	Flag	Net Tonnage	Dimensions: Length	Breadth	Depth	Built: Where.	When	By whom.	Date of Metalling	Owners or Managers.
	2451	N	Askur *Hansen* W Bk	Norweg'n	382	120·1	30·5	17·0	Tvedestrand	1859			B. Broch
	2452	N	Askur *Arnesen -87* W Bk	Norweg'n	305	104·6	28·4	16·0	Grimstad	1866	A. Omundsen	[illegible]	A. J. Arnesen
	2453	A	Aslani Bachri *Hussein* W Bn	Turkish	100	..	..	..	Dulcigno	1863	O. Hussein		M. Hussein
63647	2454		Aspasia *Patten 62-82* W Bkn	British	179	115·1	22·6	12·6	Bideford	1870	Rolle Canal Co.	[illegible]	J. Taylor & Co.
	2455		Aspasia *Scheranis* W Bk	Greek	336	..	..	..	Syra	1867		[illegible]	A. Cefalai
	2456		Aspasia *Vrazzano* W Bn	Greek	307	..	..	..	Syra	1882	L. Josipa		G. Vrazzano
	2457		Aspasia W Bn		305	104·6	28·5	..	Syra	1866			
268	2458		Aspasia W Bg	Greek	225	96·8	25·6	..	Syra	1869			A. Balas
	2459		Aspasia *Tsatsaronis* W Bg	Greek	225	..	..	..	Chios	1869		[illegible]	J. Tsatsaronis
118	2460		Aspasia *Beridis* W Bn	Greek	203	90·5	24·9	..	Castellamare	1868			Calliope Chritsa
73	2461		Aspasia W Bg	Greek	184	90·2	24·5	..	Syra	1864			C. Moschos
941	2462		Aspasia *Mercenaro -88* W Bk	Italian	424	130·6	30·2	18·4	SestriPonente	1864	G. B. Briasco fu [G. B.	[illegible]	G. B. Chiozzadi And.
	2463	A	Aspasia *Marco* W Bg	Russian	249	..	..	..	Syra	1870	P. Bani	[illegible]	D. Mussuri
	2464		Aspasia W Bg	Turkish	187	..	..	..	[illegible]	1885			Marmoralee Dimitri Raiti
	2465		Aspassia Ambanopoulou W Bn	Greek	397	124·0	..	..	Syra	1878			G. Ambanopoulos
75717	2466	V	Aspatogon *Scott* W Bk	British	861	161·5	35·6	20·0	[illegible], N.S.	1877	W. S. Kitchen	[illegible]	A. R. Durkee
	2467	N	Aspirant *Larsen* W Sr	Norweg'n	127	88·5	22·5	11·6	Stavanger	1864	B. Hoerem		T. Aarstad
74850	2468	V	Aspotogon *Taylor 87-87* W Bk	British	574	150·0	32·0	16·5	RiverJohn, NS	1876	McKenzie & [Sander	[illegible]	T. G. Taylor
62900	2469	✻	Assaye *McRitchie 69-89* I Bk	British	1281	227·1	35·9	22·4	Greenock	1868	R. Steele & Co.		D. McGillivray
73283	2470	✻	Assel *Gilmour 87-88* I Bk	British	795	190·4	33·0	19·2	Glasgow	1876	London & Glasgow [Co.		D. Hunter
	2471		Assistent *Hellissen 87-87* (CONDEMNED) W Bn	Norweg'n	187	104·5	24·8	12·1	Stavanger	1867	Nilsen	[illegible]	A. Meling
	2472	V	Asslan Bachri *Hassan* W Bn	Turkish	349	..	..	..					AhmetCapitan
	2473		Asslan Bahri W Bg	Turkish	375	..	..	..	Haraja Simon	1877			Heracleelee-Balukjee Zadeh Ahmed-eglou
	2474		Asslan i Bahri *Mehemed* W Bg	Turkish	375	..	..	..	KarajehSimon	1884			Heracleelee Balukjee Ahmed-eglou Ahmed
	2475	✻ F V	Assomption *Mercomble 84-86* C Bk	French	646	144·4	30·2	18·2	Bordeaux	1877	Bichon frères	[illegible]	H. Berguin & Co.
1316	2476	A	Assunta D. *Tomei -88* (ex Unle) W Bg	Italian	150	91·5	21·9	13·9	LeLegue	1850	L'hotelier		A. & G. B. Ducci
1001	2477	A	Assunta F *Ferrigni -88* (ex Paolina) W Bk	Italian	445	133·2	28·9	18·9	Leghorn	1868	L. Patrone	[illegible]	G. Ferrigni & Fˡˡⁱ
3291	2478	I	Assunta Marchese *Marchese -88* (ex Lucedia) W Bk	Italian	651	156·5	32·9	21·1	Varazze	1876	S. Craviotto	[illegible]	A. Marchese
327	2479	I	Assuntina *Turani* (ex Sophia Pate) W Bk	Italian	387	131·8	31·6	16·5	Leghorn	1854	G. Micheli	[illegible]	[illegible]
	2480	N	Assur *Kaisen* W 3m Sr	Norweg'n	189	108·9	24·7	12·0	Stavanger	1875		[illegible]	A. Bryne
92303	2481	D V	Assyria *Easton -87* W Bk	British	1148	189·1	36·7	22·5	St. John, N.B.	1887	O. Pittfield		J. F. Taylor
	2482	N	Asta *Gundersen -88* W Bk	Norweg'n	511	143·8	32·3	17·4	Drammen	1869	J. Jörgensen		P. Wiborg
	2483		Asta *(ex Emily) Augustinussen* W Bk	Norweg'n	351	127·9	29·1	12·6	Maitland, N.S.	1863	M'Dougall		A. S. Ohlsen
	2484	D N	Asta *Eeg -82* W Sr	Norweg'n	176	89·0	26·9	11·2	Drammen	1880	Antzen		T. Eeg
	2485	N	Aster *Olsen* W Bk	Norweg'n	402	133·7	30·6	15·9	Drammen	1869	J. Jörgensen	[illegible]	D. S. Olsen
62500	2486	✻	Asterion *Collingwood* I Bk	British	509	163·0	28·0	17·4	Sunderland	1869	Iliff		Shaw, Savill & Co.
	2487	H	Astir *Catsoulis* W Sr	Greek	195	..	..	..	Galaxidi	1882		[illegible]	E. Catsoulis
91158	2488	✻ D	Astoria *Moignard 72-86* I Bk	British	1429	240·7	38·0	24·7	Sunderland	1885	R. Thompson & Ss		P. Iredale & Son
91287	2489	✻ V	Astoria *Jackson 87-87* W S	British	1335	204·5	40·2	24·0	Bath, Me.	1875	Goss & Sawyer	[illegible]	"Astoria" Ship Co. Ld (J. Jackson)
54400	2490	V	Astracana *Richards 82-83* (WRECKED) W S	British	1193	190·1	37·8	23·3	St. John, N.B.	1866	D. V. Roberts	[illegible]	D. V. Roberts
40807	2491		Astræa *Rickman* W Bn	British	200	106·9	23·7	12·9	P. E. Island	1864	Jenkins		J. G. Hill
	2492	D V	Astræa *Rasmussen 81-81* W Sr	Danish	129	89·9	21·4	10·6	Thurø	1881	C. Bom		P. Bom
	2493	G V	Astraea *Osnabrück 88-88* W Bn	German	298	105·4	25·4	11·4	Bardenfleth	1876	D. Focke	[illegible]	D. Haye
	2494	N	Astræa *Jensen 78-82* W Bk	Norweg'n	641	157·5	34·5	18·5	Christinestad	1857			H. A. Lohre
	2495		Astrakan *Wolls* W Bk	Norweg'n	588	133·4	28·8	21·3	Boston				Winge & Co.
1??	2496		Astrapi *Coltsias* W Bg	Greek	263	101·7	26·6	..	Syra	1868		[illegible]	K. Zerves
	2497		Astre des Mers *Jouanjean* W Sr	French	103	76·8	16·4	8·5	St. Malo	1858	J. Clément fils		A. Lepésant
26166	2498		Astrea *Passmore* W Sr	British	122	85·0	20·5	11·5	Kingsbridge	1860	Date		H. Organ
	2499	D V	Astrea *Held -82* I Bg	German	291	128·2	28·5	14·1	Amsterdam	1864	W. & A. H. Meursing		E. F. C. Tobias
	2500	V	Astrea *Militzer -88* W Bn	German	222	110·0	26·0	12·0	Harburg	1872	C. Renck	[illegible]	J. Breckwoldt

✻ = **Lloyd's Register**; D = Nederlandsche Vereeniging; G = Germanischer Lloyd; I = Registro Italiano; R = Ameri
A = Veritas Austro-Ungarico; H = Veritas Ellenico; N = Norske Veritas; V = Burea

In the 4th Column, C = Composite (iron frame, wood planked); I = Iron; Stl = Steel; W = Wood.

The Dates of Metalling (**Col. 11**) are based on the records of the Books in which Vessels are respectively classed, or on the Reports of the Surveyors of Ll

Further particulars of Vessels classed by Lloyd's Register will be found in Lloyd's Register Book and Lloyd's Yacht Register.

N° du Lloyd	Cote	Nom. Matériaux et Gréement. Capitaine et dates d'entrée au service des Armateurs actuels et à bord du navire. Ancien Nom.	Pavillon	Tonnage	Longueur	Largeur	Creux	Construit: Où.	Quand	Par qui	Date du doublage en métal	Armateurs ou Directeurs-Gérants.	Port d'Armement.
2501	IV	**Astrea** *Gazzolo* WI&Sr	[illegible]	[illegible]	[illegible]	[illegible]	[illegible]	[illegible]	1866	D.Craviotto	V 7.88	T.Penco	Genoa
2502	A	**Astrea** *[illegible]* *Fiore* WBn	[illegible]	18?	[illegible]	21·?	[illegible]	[illegible]	1869	G.Poli	[illegible]	F.Viancllo Moro	Venice
2503	I	**Astrea** *Manca* -88 WBn	[illegible]	1??	..	..	..	[illegible]	1858	M.Tramparulo	V 7.88	V.Sardo	Castellamare
2504		**Astrea** WBk	[illegible]	4??	..	..	..	Kotjo	18??			A.Terichoff	Lovisa
2505		**Astrea** *Karlsson* WBg	[illegible]	[illegible]	..	..	..		1860			M.V.Jansson	Lumparland
2506	V	**Astrea** *Jonasson* 83-83 WBg	[illegible]	306	116·8	27·1	14·1	Wasa	1877		V 6.87	J.C.Holthén	Helsingborg
2507	DN	**Astrid** *Svensen* WBk (ex Mary Bond ex Fanny Crammond)	[illegible]	255	131·5	28·1	11·?	Baltimore	1861			J.M.Pauss	Drammen
2508	V	**Astronom** *Seeba* 88-88 WBk	[illegible]	879	154·0	36·0	21·0	Brag	1863	Gebr.Bosse	V 4.86	Carl Bölken	Bremen
[illegible]		**Asuncion de Nuestra Señora** I Bn (ex Homes)	[illegible]	193	105·2	22·6	12·6	S.Shields	1866	Readhead			Mexico
2510		**Asynjo** *Cappelen* 84-89 WBk (ex Ignes Kaufhart)	Norweg'n	395	142·6	30·3	15·2	Umeå	1860			A.H.Kiær	Fredrikstad
2511	✻	**Atahualpa** *Richards* 86-86 I Bk (MIS-ASO)	British	493	160·6	25·8	17·4	Liverpool	1865	W.C.Miller		S.Wakeham & Sons	Liverpool
2512	✻D	**Atalanta** *McBride* 85-85 I S	British	1696	265·1	39·7	23·4	Port Glasgow	1885	R.Duncan&Co.		N.Hill	Greenock
2513	✻	**Atalanta** *Jones* WBg	British	223	100·0	24·5	13·8	Pt.Dinorwic	1864	Jones	V 5.86	R.E.James	Carnarvon
2514		**Atalanta** *Anderson* 87-87 WBk	American	824	150·0	33·5	22·?	Amesbury, [Mass.	1851			Puget Sound Commercial Co.	P.Gamble, Wash.
2515	R	**Atalanta** *Aldrich* WBn	American	352	123·0	30·0	14·5	PtJeffers'nNY	1881		V 1.88	S.H.Mills	P.Jefferson, N.Y.
2516	V	**Atalanta** *Jepsen* 88-88 WBk (ex Bacchus ex Lorenzo)	Danish	1128	183·0	36·4	23·7	Portsm'th NH	1863	T.Littlefield	V 7.84	Actieselskabet (…)	Kolding
2517	GV	**Atalanta** *Schütte* -88 I Bk	German	1057	205·4	34·5	20·0	Kiel	1886	WerftGermania		M.G.Amsinck	Hamburg
2518	GV	**Atalanta** *Becker* 84-84 WBk	German	410	132·4	28·3	15·3	Elsfleth	1880	F.Deetjen	V 6.88	D.Haye	Brake
2519	DNV	**Atalanta** *Tønnesen* WBk	Norweg'n	569	146·5	34·0	17·5	Sikeå	1868	O.P.Aberg	V 10.86	T.Andr.Sleydal	Farsund
2520	N	**Atalanta** *Tønnesen* -88 WBg	Norweg'n	287	114·4	29·5	15·4	Norrköping	1852	Hammarsten	V 11.84	Johs.Röds Sönner	Tönsberg
2521		**Atalanta** *Brinkmann* -86 WBk (WIEBKE ED)	Russian	578	149·2	30·3	18·3	Vegesack	1857	H.F.Ulrichs		NavigationCo."Austra"	Riga
2522		**Atalante** *Backmann* 88 WS	Swedish	406	125·0	29·0	16·0	P.E.Island	1852	Froyland		J.Backman	Lysekil
2523		**Atalaya** *Brown* -87 WBg (WHEY E&D)	American	418	122·0	29·0	11·5	ColumbiaFal's [Me	1871			Miller & Houghton	Philadelphia
2524		**Ataulfo** *Gonzalez* -88 WBg	Spanish	258	102·3	27·9	12·8	Barcelona	1858			L.Mir & Co.	Barcelona
2525		**Atene** *Tarabochia* WBn	Austrian	126	78·4	22·1	10·4	Lussinpiccolo	1885	G.M.Peranovich	V 11.85	G.A.Tarabochia&Co	Lussinpiccolo
2526	N	**Athalia** *Stub* 87-87 WBk	Norweg'n	322	105·8	27·8	16·4	Tönsberg	1848			Actieselskabet "Athalia" (S. U. Larsen)	Brevig
2527	F	**Athalie** WBg		118	70·9	20·0	11·2	Granville	1853	Clément			
2528		**Athanassia Vaglianos** WBk	Greek	566	..	..	..	Chios	1874		V 5.88	Tsatsaronis & Tzolas	Syra
2529		**Atheleehmy** WBg	British	130	103·4	21·0	15·2	Valuertyturie	1876			Pana Lana Palaneappa-chitty	Jaffna
2530	V	**Athena** *Roberts* -88 WBkn	British	688	164·4	36·5	16·1	Newport, N.S.	1888	T.A.Mosher	V 10.88	M.Curry	Windsor,N.S.
2531		**Athena** *Gadd* 13-82 WBk	British	412	131·0	26·5	16·7	South Shields	1868	Middle Dock Co.	V 5.84	J.L.&W.A.Simpson	Adelaide,S.A.
2532	A	**Athena** *Couroussiotis* WBk (ex Erin)		436	140·5	29·5	17·2	P.E.Island	1877	Yeo	V 11.81		
2533	AH	**Athena** *Charopulo* WBk (ex Awalm …)	Greek	381	117·4	29·0	18·6	Sestri Ponente	1863	T.Tixi	V 2.88	G.Charopulo	Syra
2534	A	**Athena** *Petratza* WSr	Greek	223	92·8	26·7	..	Syra	1888	A.Livadara		E.D.Petratza	Syra
2535	A	**Athena** *Goulandris* WSr	Greek	149	87·9	22·3	..	Syra	1879	S.Cufudachi	V 3.82	L.Goulandris	Andros
2536	V	**Athene** *Hansen* 76-76 WSr	Danish	130	91·3	21·0	10·5	Thurø	1876	P.Bom		P. Bom	Svendborg
2537	GV	**Athene** *Bohne* -82 WBk	German	629	136·9	32·0	16·5	Elsfleth	1881	J.Jürgens	V 2.82	J.G.Lübken	Elsfleth
2538		**Athenian** *Abrahamsen* 88-88 WBk		318	117·9	27·3	16·9	Sunderland	1862	Haswell		J.S.Jorgensen	Brevig
2539	NV	**Athlet** *Natvig* 56-82 WBk	Norweg'n	682	164·9	34·1	18·9	Christiansand	1875	P.von der Lippe	V 6.88	S.Otto & Co.	Christiansand
2540	✻	**Athletæ** *Frear* 86-88 WSr	British	99	85·1	21·6	10·7	Goole	1867	Wake	V 2.88	C.Rowbotham	Goole
2541		**Athlitis** *Kypariss* WSr	Greek	245	102·0	21·6	..	Syra	1876		V 6.84	J.K.Valentas	Syra
2542	DV	**Athlon** *Dexter* WS	British	1406	200·1	40·8	23·7	Parrsboro',NS	1879	A.Soomer	V 11.87	S.Sheffield	Windsor,N.S.
2543	A	**Ati** *Ferrari* WBk	Austrian	420	127·3	27·4	18·1	Fiume	1868	Flli. Schiavoni	V 6.86	E.Ferrari	Lovrana
2544		**Atiat Rachman** WBk	Dutch	355	106·6	26·4	..	Java	1826			Said Salim bin Abdulla Alkaf	Palembang
2545		**Atiat Rachman** WBk (Said Omara Bassel)	Dutch	124	77·4	22·6	..	Java	1857		V 1.81	Said Hoesin bin Achmat Almashahir	Sourabaya
2546		**Atiatoor Rachman** WSr (ex Charlotte) Said Ipatag Barkla	Dutch	123	87·9	20·0	..	RembangJava	1878	Mering Begel & [Dunlop		Said Sech bin Abdul Rachman Barkla	Palembang
2547		**Atiet Rohoman** WBg	British	150	76·0	20·0	11·7	Surat	1846			Shaik Omer B.S.Saldon	Bombay
2548	R	**Atila** *(ex …) Bonnessen* -88 WBk	Norweg'n	1146	183·1	37·2	22·9	St.John	1860	Hilyard	V 5.87	J.L.Tönnesen	Lillesand
2549		**Atka** *Rancklius* WI&Sr	Russian	300	..	..	..	Kimito	1874			G.D.Hagman	Kimito
2550													

✻ = Lloyd's Register; D = Nederlandsche Vereeniging; G = Germanischer Lloyd; I = Registro Italiano; R = American Record;
A = Veritas Austro-Ungarico; H = Veritas Ellenico; N = Norske Veritas; V = Bureau Veritas.

Dans la 4me Colonne, C = Composite (membrure en fer, bordé en bois); I = Fer; Stl = Acier; W = Bois.

Les Dates du doublage en métal Col. 11 ressortent des livres dans lesquels les navires sont respectivement classés ou des rapports des Inspecteurs du Lloyd's Register.

Pour plus ample renseignements sur les navires classés par le Lloyd's Register, voir le Lloyd's Register Book et le Lloyd's Yacht Register.

Official No.	No. in Book	Registry in which classed	Ship's Name. Material and Rig. Master, and Dates of Appointment to present Owner's Service and to Ship. Ship's Late Name.	Flag.	Net Tonnage	Length	Breadth	Depth	Built. Where.	When	By whom.	Date of Metalling	Owners or Managers
[illegible]	2551		**Atlanta** W Bn	British	143	98·2	22·9	11·4	P.E.Island	1864	N.Smith		A.Munden
	2552	G	**Atlanta** *Nyberg* -75 W Bk	Russian	366	133·3	28·3	14·5	Karis	1875	E.Engblom		A.L.Boxberg
	2553		**Atlanta** *Jönsson* 74- W Bk	Swedish	434	136·8	27·1	15·3	Korpo,Finl'nd	1866			P.Jönsson
[illegible]	2554	✳	**Atlantic** *Belding* 81-85 I Bk	British	474	157·4	27·6	16·8	Glasgow	1869	A.Stephen&Sons		B.M.Hennessy
[illegible]	2555		**Atlantic** W Bk	British	284	99·3	25·9	17·3	Haddam,U.S.	1819			T.S.Pope
[illegible]	2556	✳	**Atlantic** *Coleman* -87 W Sk	British	86	78·4	20·5	11·1	Rye	1886	G.&T.Smith		J.Coleman
[illegible]	2557		**Atlantic** (ex Pacific) *Lanfare* W Bk	American	292	126·0	28·0	12·0	Picton,N.S.	1865			H.Trowbridge's Sons
	2558	DGV	**Atlantic** *Bavenborg* -88 (ex Joseph Fish) W Bk	German	1263	191·6	39·0	24·7	ThomastonNS	1866	S.Watts	[illegible]	D.Heinrichs
	2559	DGV	**Atlantic** *Stege* 76-76 W Bk	German	633	140·0	33·0	17·6	Elsfleth	1876	F.Deetjen	[illegible]	D.W.Stege
	2560	V	**Atlantic** *Liedke* 70-70 W Bk	German	512	145·0	30·5	18·9	Memel	1867	B.Pieper		A.Kahlan
	2561	G	**Atlantic** *Jaeger* 87-79 W Bk	German	465	127·3	30·5	17·6	Danzig	1866	W.R.Grott		Ernst Wendt
	2562	V	**Atlantic** *Schlottmann* 87-87 W Bg	German	307	110·7	27·9	16·3	Rostock	1867	C.Schönemann		W.Maack
	2563	GV	**Atlantic** *Janssen* 85-85 W Bg	German	291	117·1	25·8	14·0	[illegible]	1870	J.F.Strenge	Y 12.87	G.Eilers
	2564	V	**Atlantic** *Fuchs* -86 W Bn	German	173	95·1	20·0	11·5	Stralsund	1872	A.Juhl		H.L.Juhl
	2565	NV	**Atlantic** *Nielsen* 88-86 W Bk	Norweg'n	494	128·4	31·2	18·8	Vegesack	1856	J.Lange	Y 10.87	M.F.Stray
	2566	N	**Atlantic** *Olsen* 87-87 W Bk	Norweg'n	459	131·8	29·8	17·5	Bremen	1854		Y 5.86	N.J.F.Ugland
	2567	N	**Atlantic** (ex Hannibal) *Thomassen* W 3mSr	Norweg'n	388	120·4	28·6	16·8	Rostock	1855	W.Zeltz		M.Nielsen
	2568		**Atlantic** *Eriksson* W Bk	Russian	697	..	..	..	Sverige	18 57/73			N.Sittkoff
	2569		**Atlantic** *Törnroth* W Bk	Russian	549	133·3	33·8	19·4	Lantakari	1868			J.Gronfeldt
	2570	V	**Atlantic** *Söderström* W Bk	Swedish	996	187·0	33·5	21·6	Gefle	1876	O.A.Brodin	[illegible]	Engström & Luth
	2571		**Atlantico** *De Mello* -87 W Bn	Brazilian	178	..	..	..	S.JoãodaBarra	1880		[illegible]	Cia.Cabotagem de S.João da Barra
[illegible]	2572		**Atlantis** (ex Governor Langdon) (WRECKED) *Forbes* -87 W Bk	British	1127	186·2	37·7	23·3	Portsmth.U.S.	1854			H.A.Muirhead
	2573	GV	**Atlas** *Dade* 84-84 W Bg	German	282	113·4	27·2	14·6	Rostock	1877	W.Zeltz	Y 11.86	W.Maack
	2574		**Atlas** W Bn	Brazilian	182	..	..	..					Luiza A.Leoni
	2575	N	**Atlas** *Andersen* 87-87 W Bk	Norweg'n	602	146·2	33·7	17·3	Tönsberg	1863	P.Andersen	[illegible]	H.C.Seeberg
	2576	R	**Atlas** (ex Marion) *Pettersen* 86-86 W Bk	Norweg'n	599	136·7	30·3	20·0	Gardiner,Me	1847			E.Rönneberg
	2577	N	**Atlas** *Andersen* -88 W Bk	Norweg'n	487	152·5	31·7	18·3	Memel	1855			T.Mulvad
	2578	N	**Atlas** *Nielsen* W Bg	Norweg'n	214	94·1	25·0	13·4	Stettin	1846			H.Nielsen
	2579	N	**Atlas** *Rode* 88-88 W Bg	Norweg'n	168	88·8	25·2	11·6	Kragerö	1866	Brandis		S.C.Larsen
	2580	V	**Atle** *Akermark* 87-87 W 3mSr	Swedish	280	125·9	25·4	11·5	Hernösand	1873	Fr.Hägglund	[illegible]	Hernösands Lastång Aktieb (H.Ne. Ibsader)
43671	2581	✳	**Atossa** *Robinson* 82 W Bk	British	467	138·4	28·1	18·9	Sunderland	1863	R.Thompson&S.	[illegible]	G.& J.Robinson
47231	2582		**Atrato** *Cumben* W Bk	British	276	121·4	26·6	14·8	Shoreham	1863	May		R.H.Penney
[illegible]	2583		**Atromitos** *Michalakis* W Bn	Greek	157	78·7	24·7	13·4	Syra	1865		[illegible]	L.Michalakis
[illegible]	2584		**Atten hundrede og otte og firte (1848)** (HULK) W Bg		269	104·3	27·9	15·7	Grimstad	1853			
[illegible]	2585	A	**Attila** *Giadrossich* W Bn	Austrian	129	79·0	20·6	11·2	[illegible]	1885	G.Gerolimich	[illegible]	Gio de Antonio Giadrossich
[illegible]	2586	V	**Attilio** *Pietrolussi* -88 W Bk	Italian	1018	193·2	36·0	24·6	Sestri	1876	Casella	[illegible]	N.Parodi
	2587	A	**Attina** W Bg	Turkish	222	..	..	..	[illegible]	1882			Christo Affonosti & Co.
[illegible]	2588	V	**Attivo** (ABANDONED) *Rasi* -87 W Bk	Austrian	557	138·4	29·2	19·4	Capod'Istria	1878	L.Polli	[illegible]	G.Bielich
[illegible]	2589	A	**Attivo** *Comisso* -88 W Bg	Italian	284	104·9	25·9	16·6	Fiume	1863		[illegible]	S.Munzone fu V.
69095	2590	R	**Atwood** *McCrae* W Sr	British	118	84·6	24·0	8·5	Digby,N.S.	1876	J.A.Hughes		A.Holmes
	2591	R	**Au Sable** *Preston* -88 W Bkn	American	568	151·4	33·2	13·1	E.Deering,Me.	1866	Geo.Russell		J.S.Winslow & Co.
	2592	V	**Aubépine** *Menon* W Bk	French	677	157·4	28·2	18·0	La Seyne	1875		[illegible]	J.Martin
85730	2593		**Aubry A.** *Holland* -88 W Sr	British	108	81·5	24·0	9·3	MhoneBayNS	1883	A.Langill		B.Anderson
	2594		**Auburn** (LOST) *Schraeder* -87 W Bk	Dutch	357	119·1	27·0	16·4	Sunderland	1867	G.Bartram	[illegible]	G.G.Afters
105035	2595	RV	**Auburndale** *Wooster* 81-87 W Bk	American	574	149·3	32·6	17·3	Millbridge,Me	1880	J.W.Sawyer	[illegible]	J.S.Emery & Co.
71658	2596	✳	**Auckland** *James* 82-87 I S	British	1245	239·3	36·0	20·7	PortGlasgow	1874	R.Duncan & Co		[illegible]
48012	2597		**Auctioneer** (ex [illegible]) W Bk	British	172	99·0	24·0	10·7	Cohasset,U.S.				R.M.Hughes
	2598	N	**Audacia** *Johnsen* 88-88 W Bk	Norweg'n	367	114·4	29·3	16·2	[illegible]	1839			Krag & [illegible]
	2599	V	**Audacia** *Soares* W Bk	Portug'se	621	134·5	28·2	21·0	[illegible]	1872	[illegible]	[illegible]	F.P.Barbosa [illegible]
	2600	V	**Audacieuse** *Bolo* W Sr	French	121	91·5	19·7	11·1	Dunkirk	1878	A.Lefebvre		A.[illegible]

✳ = **Lloyd's Register**; D = Nederlandsche Vereeniging; G = Germanischer Lloyd; I = Registro Italiano; R = Ameri[can]
A = Veritas Austro-Ungarico; H = Veritas Ellenico; N = Norske Veritas; V = Bureau [Veritas]

In the 4th Column, C = Composite (iron frame, wood planked); I = Iron; Stl = Steel; W = Wood.

The Dates of Metalling (Col. 11) are based on the records of the Books in which Vessels are respectively classed, or on the Reports of the Surveyors of Ll[oyd's]

Further particulars of Vessels classed by Lloyd's Register will be found in Lloyd's Register Book and Lloyd's Yacht Register

No.	Registre	Nom. — Capitaine — Ancien Nom	Matériaux et Gréement	Pavillon	Tonnage	Longueur	Largeur	Creux	Construit: Où	Quand	Par qui	Doublage	Armateurs ou Directeurs-Gérants	Port d'Armement
2601		Audaz (ex Maria Augusta) Moreira	W Sr	Brazilian	221	..	..	..	P.E.Island	1874		[illegible]	A.Oliveira Maia	Pernambuco
2602		Augia, Williams	W Bkn	British	[illegible]	104·5	22·8	14·4	Jersey	1829	T.Silk		A.& F.Manuelle	Guernsey
2603	GV	August (ex Ross McNeill) Joburg	W S	German	1431	208·3	[illegible]	25·6	Thomaston Me	1871	S.Watts & Co.	[illegible]	Steenberg, Wendt & Co.	Bremen
2604	GV	August (ex [illegible]) -85	W Bk	German	342	115·4	29·3	15·1	Swinemünde	1858	T.Miether		A.Radmann	Swinemünde
2605	DV	August, Stockholder 66-72	W Bn	German	245	113·8	24·9	13·0	Finkenwärder	1867	C.Wrede	[illegible]	J.Nibbe	Cranz
2606		August (ex [illegible]) Hansen 87-	W Bk	Norweg'n	410	108·7	28·8	18·5	Newcastle	1846	J.Medcalf		J.A.Torpp	Fredrikshald
2607	N	August, Bach 86-86	W Bk	Norweg'n	396	115·6	30·3	19·9	Grimstad	1865	A.Omundsen		S.Eriksen	Arendal
2608		August, Graver 85-	W Bk	Russian	381	119·0	30·2	17·0	Riga	1852	G.Herbatschew [(ok)		Graver, Bientz & Noak	Riga
2609		August, Kohn	W Sr	Russian	244	104·0	23·0	12·0	Oerenhof	1874	M.& A.Koff		M.Koff	Haynasch
2610		August, Kaartinen	W Bk	Russian	109	..	..	..		1870			G.Kaartinen	Säkkjärvi
2611	✻	August, Hultman	I Bk	Swedish	670	175·2	29·1	16·0	Sjöberg	1881	Buchan		G.F.Cawallin	Sundsvall
2612	V	August, Johansson	W Bk	Swedish	309	122·[illegible]	24·6	15·1	Framnäs	1869	J.J.Strandberg	[illegible]	N.Olsson	Höganäs
2613	V	August, Nordlander 81-	W Bk	Swedish	362	131·2	26·1	12·5	Hernösand	1872	Fr.Hägglund	[illegible]	H.Nordlander	Hernösand
2614	V	August, Söderman 84-88	W Bkn	Swedish	109	95·5	21·5	12·4	Wolgast	1858	W.Löbke		P.Söderman	Rålmansö, Rslgn
2615		August, Nilsson	W Sr	Swedish	115	81·7	19·4	9·6	Martenshoeck	1876	W.G.Bodewes		G.Bachér	Slite, Gotland
2616	DGV	August Burchard, Prehnert	W Bk	German	474	136·4	29·7	17·9	Rostock	1875	O.Ludewig	[illegible]	A.Burchard	Rostock
2617		August Friedrich, Trost	W Bg	German	236	92·5	24·7	12·8	Anclam	1863	Heine		Heinrich Lauer	Anclam
2618	N	August Leffler, Olsen -89	W S	Norweg'n	542	137·4	30·1	17·5	Tvedestrand	1875		[illegible]	J.P.Pedersen	Tvedestrand
2619	N	August Smith, Christensen	W Bk	Norweg'n	621	155·0	33·4	17·7	Tvedestrand	1881	O.Gundersen	[illegible]	C.A.Smiths, Arvinger	Tvedestrand
2620	NV	August Tellefsen, Halvorsen	W Bk	Norweg'n	738	160·3	35·3	18·5	Tvedestrand	1883	F.A.Mörland	[illegible]	C.Wroldsen	Tvedestrand
2621	G	August Zancker (ex [illegible]) Rossil -84	W Bn	German	137	86·6	21·6	11·1	Aalborg	1855			Retzlaff & Schober	Stettin
2622		August & Eduard, Schwenn	W Bg	German	248	99·2	26·3	14·8	Rostock	1848	A.Warnke		A.Burchard	Rostock
2623	G	August & Marie, Grant	W Bk	German	256	108·0	26·0	13·0	Rostock	1863	W.Zeltz		A.Burchard	Rostock
2624	✻	Augusta, Holten	W Bk	British	528	150·3	30·2	18·9	Bristol	1868	Hill	[illegible]	G.H.Bridges	Bristol
2625		Augusta	W Sr	British	321	135·3	25·8	11·6	St. Catharines, Ont.	1872	L.Shickluna		S.Neelon	St. Catharines, Ont.
2626		Augusta, Dawe	W Bn	British	138	87·9	21·4	11·7	Salcombe	1877	H.Harnden		W.Dawe	S.Johns, N.F.L.
2627		Augusta (ex Adele), Schamvedel	W Bk	Chilian	845	166·0	34·6	20·8	Quebec	1868	Dunn		R.Moser	Valparaiso
2628		Augusta (ex Julie & Auguste)	W Bg	Danish	189	95·6	23·4	12·3	Wolgast	18 51/72	W.Löbke		N.R.Storm	Faaborg
2629		Augusta, Jensen 81-84	W Sr	Danish	130	86·0	23·2	10·8	Holbæk	1866	H.P.Mortensen		T.N.Brendstrup	Frederiksværk
2630	FV	Augusta, Loisne -88	W Bn	French	102	77·1	21·1	12·8	La Nouvelle	1869	A.Rouquette		J.L.Vincent	St.Malo
2631		Augusta (ex [illegible]) Jessen -86	W Bk	German	473	138·2	28·9	17·9	Sunderland	1866	W.Adamson	[illegible]	Aug.Piehl	Lübeck
2632	I	Augusta (ex Costantino) Ajello -87	W Bn	Italian	270	113·1	27·5	15·6	Castellamare	1879		[illegible]	A.Muscatello & Co.	Catania
2633	N	Augusta, Hansen -81	W Bk	Norweg'n	827	177·7	35·8	20·5	Sandefjord	1881	C.Christensen	[illegible]	A.F.Klaveness	Sandefjord
2634	DN	Augusta, Johnsen -79	W Bg	Norweg'n	208	97·2	24·5	13·5	Sandefjord	1853			Stang & Brecke	Fredriksstad
2635		Augusta, Sager -83	W Bg	Norweg'n	198	95·7	24·5	11·1	Piteå	1839			C.Johnsen	Christiansund
2636		Augusta, Eriksson 79-	W Bk	Russian	383	121·8	30·6	15·2	Vorrbäsby	1874	M.Jonas		K.Lundberg	Lemland
2637	F	Augusta, Irbiti	I Bk	Spanish	374	115·8	25·6	15·6	Hamburg	1868	Reiherstieg [Schiffswerfte		U.Ondarza	Bilbao
2638	V	Augusta, Westerberg	W Bk	Swedish	426	133·5	29·1	16·1	Gefle	1859		[illegible]	J.E.Waenerlund	Gothenburg
2639	R	Augusta (ex Sailor Prince) Ingeldsson 79-83	W Bk	Swedish	434	128·1	26·1	17·4	St.John, N.B.	1862		[illegible]	J.D.Björkegren	Cimbrishamn
2640	FV	Augusta (ex [illegible]) Olsson	W Bk	Swedish	418	126·6	23·8	17·4	Shelburne	1866			G.Olsson	Mollösund
2641		Augusta, Öhman	W Bg	Swedish	339	120·0	26·8	14·4	Sundsvall	1868	A.Lidberg		C.M.Åberg	Sundsvall
2642		Augusta (ex Pallas), Akerman 75	W Bk	Swedish	286	113·5	27·3	11·5	New York	1859			A.Åkerman	Landscrona
2643	G	Augusta, Elfversson -88	W Sr	Swedish	198	109·5	22·7	11·6	Carlshamn	1864	D.R.Andersson	[illegible]	A.Elfversson	[illegible]
2644		Augusta, Andersson 83-84	W Sr	Swedish	189	89·9	25·1	13·2	Öster Edsvik	1853	Pehr.Östenberg		A.F.Andersson	Vätö, Roslagen
2645		Augusta, Carlsson -87	W 3m Sr	Swedish	186	95·1	24·8	11·6	Grisslehamn	1867	A.Mattsson		M.Jansson	Veda, Roslagen
2646		Augusta (ex [illegible], ex Svalan) Osterlund 87-87	W Bkn	Swedish	152	96·8	23·3	9·9	Vätö, Rslgn.	1870	E.Söderlund		E.Ekstrom	Stockholm
2647		Augusta	W 3m Sr	Swedish	137	91·2	21·2	10·5	Wenersborg	1855	J.Olsson		L.O.Asberg	Strömstad
2648		Augusta, Levander	W Sr	Swedish	120	78·4	22·1	10·6	Wismar	1849			A.F.Levander	Klintehamn
2649	G	Augusta, Rydell 75-	W Sr	Swedish	119	80·0	21·8	9·7	Timmernabben	1875	C.Nilsson		V.Wahlberg & Co.	Calmar
2650	A	Augusta M. (ex Mio Padre) Mignone	W Bg	Italian	306	101·7	28·9	17·7	Sestri Ponente	1864	D.Fixi	[illegible]	G.Mignone	Savona

✻ = Lloyd's Register; D = Nederlandsche Vereeniging; G = Germanischer Lloyd; I = Registro Italiano; R = American Record; A = Veritas Austro-Ungarico; H = Veritas Ellenico; N = Norske Veritas; V = Bureau Veritas.

Dans la 4me Colonne, C = Composite (membrure en fer, bordé en bois); I = Fer; Stl = Acier; W = Bois.

Les Dates du doublage en métal (Col. 11) ressortent des livres dans lesquels les navires sont respectivement classés ou des rapports des Inspecteurs du Lloyd's Register.

Pour plus amples renseignements sur les navires classés par le Lloyd's Register, voir le Lloyd's Register Book et le Lloyd's Yacht Register.

Official No.	No. in Book	Registry in which classed	Ship's Name. Master, and Dates of Appointment to present Owner's Service and to Ship. Ship's Late Name.	Material and Rig	Flag	Net Tonnage	Length	Breadth	Depth	Built: Where	When	By whom	Date of Metalling	Owners or Managers
1088	2651	I *	Augusta Quinta *Starpoli 86-84*	W Bk	Italian	399	122·0	30·2	18·4	Castellamare	1875	G.Bonifacio	Y 6.87	G.B.Maresca
	2652	G	Auguste *Arkema 85-85*	W Bk	Dutch	762	115·0	34·1	21·1	Vegesack	1850	J.Lange	Y 8.88	S.Van der Hei
	2653	V	Auguste	W Sr	French	106	85·3	19·3	10·9	Paimpol	1878	L.Laboureur		F.Savin
	2654	G V	Auguste *Schumacher 77-85*	W Bk	German	1396	200·0	38·5	25·6	Bremerhaven	1878	C.Lange	Y 10.87	H.Addicks
	2655	*	Auguste *Fernsfeld 88-88* (ex Victoria Cross)	I Bk	German	1262	217·8	36·8	22·7	W.Hartlepool	1863	Pile,Spence&Co		J.H.Nicolai
	2656		Auguste *Schmit* (ex Thomas Whitney, ex Vicente)	W Bk	German	832	110·0	26·0	13·0	Yarmouth,NS	1857			M.G.Schmit
	2657	G	Auguste *Stubbe -88*	W Bg	German	273	103·2	26·9	16·0	Greifswald	1853	J.W.Sadelkow		Chr.Stabbe
	2658		Auguste *Wulf*	W Bg	German	188	86·6	25·3	13·6	Wismar	1840	J.Chr.Maren		C.Barmann
	2659		Auguste *Martens 72-87*	W Bn	German	135	84·4	24·0	13·3	Rostock	1836	E.Dethloff		Joh.Marteus
	2660	D V	Auguste *Frederiksen 86-86* (ex General Berry)	W S	Norweg'n	1316	187·9	38·0	25·7	Thomaston Me.	1863	S.Watts&Co.	Y 3.88	J.M.Bryde
	2661	V	Auguste Jeanette *Jurgens 81-*	W Bk	German	366	119·7	27·6	16·9	Colberg	1857	R.Lincke		H.Behrent
	2662		Auguste Louise *Bastard*	W Bn	French	137	83·3	21·3	11·4	Nantes	1874	T.Dubigeon		Paturel
	2663		Auguste Marie *Croisier*	W Bg	French	122	75·5	20·3	11·2	Cherbourg	1855	Nicolle		Laronche
	2664	V	Auguste Marie *Bareau*	W Bn	French	106	73·8	21·6	10·4	Means	1867	Olliveaud		E.Clergeau
	2665		Auguste Mathilde *Delahaye*	W Bk	French	274	117·1	25·3	12·5	Fünfhausen	1868			A.Demarle-Fetel
	2666		Auguste Mathilde *Seidel* (BROKEN UP)	W Bg	German	226	103·3	25·2	13·0	Greifswald	1830	J.F.Jahl		Herm.Bartels
	2667	V	Auguste Sophie *Dethloff 79-79*	W Bg	German	275	106·7	27·0	14·3	Rostock	1866	H.Rickmann		W.Maack
	2668	G	Auguste Teitge *Koch 182-*	W Bk	German	358	120·7	28·4	15·5	Stettin	1865	E.Zieske		J.H.Niemann
	2669	V	Augustine *Lebegue*	W Sr	French	128	90·5	20·3	11·8	Gravelines	1879	Legoulon fils		Legoulon Denève
217	2670	R	Augustine Kobbe *Chadbourne -88*	W Bk	American	506	136·0	30·0	16·9	Searsport, Me.	1866	G.A.Laur&Co.	Y 6.88	E.M.Conover
	2671	N	Augustinus *Halvorsen* (ex Il Romito de Caprera)	W Bk	Norweg'n	498	127·8	31·2	18·1	Sestri Ponente	1865		Y 2.86	Blackstad,Holta&Co
	2672		Augusto	W Bn	Brazilian	315	..	..	..					M.A. de Carvalho
1346	2673	I	Augusto *Cilento -88*	W 3m Sr	Italian	382	129·2	26·4	19·5	Cassono	1873	A.Castellano	Y 3.88	F.Cacace
100	2674	V	Augusto (ex Isaura) *Pacolini*	W Bg	Italian	157	107·2	22·5	12·2	Marciana	1859	Brasch	Y 8.84	S.Gargana
36000	2675		Augustus	W Bk	British	138	78·4	18·8	12·4	[illegible]	1814			Henry Moss
106130	2676	R	Augustus Hunt *Hall -88*	W 4m Sr	American	1141	208·0	40·7	20·6	Bath, Me.	1882	B W & H F Morse		B.W. & H.F.Morse
1448	2677	R	Augustus J.Fabens *Peck*	W Sr	American	191	109·5	27·5	9·2	Newburyport, Mass.	1868	C.Burnham		Farrand & Spear
[illegible]	2678		Augustus O. Gross (ex Oliver)	W Sr	American	148	97·3	25·8	9·0	Machias, Me.	$18\frac{59}{60}$			S.G.Haskell
	2679	D N V	Aukathor *Thorsen -87*	W Bk	Norweg'n	745	163·9	35·0	19·8	Grimstad	1869	S.Johnsen	Y 1.86	C.H.Johnsen
	2680	N	Aukathor *Evensen 84-85*	W Bk	Norweg'n	390	114·1	30·2	16·5	Grimstad	1867	S.Johnsen		A.& F.Smith
	2681		Aukathor	W 3m Sr	Norweg'n	139	92·2	23·6	11·4	Stavanger	1850	K.Kaisen		Th.Thorsen
	2682		Aunt Lizzie	W Bk	Russian	338	130·1	29·0	15·2	Altona	1855	O.Dolck		J.O.Pettersson
	2683		Aura (ex Achtma) *Dahlström*	W Bk	Russian	362	129·9	29·8	13·9	Achtma	1868	E.J.Nylund		J.G.Wikeström
	2684		Aura *Jansson -87*	W Sr	Russian	223	98·8	28·2	12·9	Sydeby	1868	A.Eckblom		F.A.Pettersson & Co
	2685		Aura	W Sr	Russian	211	..	..	..	Kimito	1871			J.Isaksson
90895	2686	*	Aurania	W K	British	90	80·9	21·0	10·5	Brixham	1886	S.Dewdney&Sns		J.Guzzwell
53130	2687	R	Aurea *Tornbohm -87*	W Bk	British	371	125·7	27·6	17·5	Sunderland	1866	R.Th'pson& Sns	Y 5.86	J.Tully
	2688		Aurea	W Sr	Spanish	109	82·0	22·4	8·3	Tayabas Luzon	1877			D.Inocencio
3587	2689		Aurelia *Bonifacio*	W Bg	Italian	276	104·9	31·5	16·7	[illegible]	1862	B.Delarbieri	Y 5.84	F.Olivari
	2690	V	Aurelia *Johannson*	W Bk	Swedish	473	145·6	27·1	15·3	Sundsvall	1876	J.N.Hägglund	Y 1.88	A.G.Johansen
582	2691	V	Aurelia R. *Molinari -88* (ex Timoshenko, ex Amelia R.)	W Bk	Italian	580	139·4	32·1	20·1	Chiavari	1869	M.Tappani	Y 4.87	F.Roncagliolo fu D.
	2692	G	Aurelius *Steinorth -75*	W Bk	German	282	106·6	27·3	14·2	Ziegenort	1868	F.Brüsewitz		J.N.Rodbertus
67632	2693	*	Aureola *Brien 86-86*	W Bkn	British	251	126·3	25·5	13·2	Bideford	1873	J.Johnson	Y 2.86	N.& S.R.March
1295	2694		Aureola *Saunders*	W Bk	American	785	150·0	33·0	20·0	Warren, R.I.	1863		Y 11.85	A.M.Simpson
88045	2695	* R	Auriga *Davis 78-84*	W Bk	British	887	193·0	35·2	19·9	Bideford, PEI	1884	Wm.Richards	Y 11.86	J.W.Richards (per T. P. Reynolds & Co.)
62537	2696	*	Auriga *Stone*	I Bk	British	518	164·5	28·1	17·3	Sunderland	1869	H.B.Mounsey &Co.		J.Hardie
	2697		Aurinko *Heed*	W 3m Sr	Russian	163	..	..	..		1874			Fr. Richardt
54837	2698	G	Aurora *Payne 85-88*	W Bk	British	341	118·7	27·6	17·1	Sunderland	1866	R.Th'pson& Sns	Y 1.86	T.D.Woodhead & Co
61273	2699		Aurora	W Bk	British	295	110·0	26·8	12·0	Bangkok	1865			Tan Hock Poh
66820	2700		Aurora *Taggart*	W 3m Bn	British	248	115·6	23·3	13·9	Montrose	1869	Petrie		J.H.Guillart & Co.

*=Lloyd's Register; D=Nederlandsche Vereeniging; G=Germanischer Lloyd; I=Registro Italiano; R=American [Record]
A=Veritas Austro-Ungarico; H=Veritas Ellenico; N=Norske Veritas; V=Bureau Ve[ritas]

In the 4th Column, C=Composite (iron frame, wood planked); I=Iron; Stl=Steel; W=Wood.

The Dates of Metalling (Col. 11) are based on the records of the Books in which Vessels are respectively classed, or on the Reports of the Surveyors of L[loyd's Register].

Further particulars of Vessels classed by Lloyd's Register will be found in Lloyd's Register Book and Lloyd's Yacht Register.

No.	Reg.	Nom	Capitaine	Mat. et Gréement	Pavillon	Tonnage net	Longueur	Largeur	Creux	Construit où	Quand	Par qui	Doublage	Armateurs ou Directeurs-Gérants	Port d'Armement
2701		Aurora	Kerr	WSr		234	125·0	26·2	10·5	Quebec	1867	McKay & Ware's		R.C.Smith, jun.	Port Hope, Ont.
2702		Aurora	Tyrell 80-82	WBn		232	111·8	27·0	12·5	NovaScotia	1862	M'Donald		R.Wyllie	Troon
2703		Aurora	Mahlkea	WBn		211		27·3	12·0	Bayfield,N.S.	1869			W.Vint & Sons	Belfast
2704		Aurora		WBn				21·3	12·2	Arenzano	1876			G.F.Mason	Sydney,N.S.W.
2705		Aurora	Little	WBg		191	..	..	..	Dundee	1833			J.F.Little	Workington
2706	V	Aurora		WBk							1874		V	A.De Freitas	Bahia
2707		Aurora		WBk			..	..	..					Machado & Povoas	Rio Grande
2708	D	Aurora	De Ween	WBg			125·0	28·6	12·8	Granville,N.S.	1863			R.J.de Ween	Pekela
2709	✱	Aurora		IBk		1377			22·2	Pt.Glasgow	1865	L.Hill & Co.		C.Paulsen	Elsfleth
2710	G	Aurora	Pöcker 75	WGt					9·5	Edewecht	1875	Gen.Deye		A.L.Pöcker	Ost-Rhauderfehn
2711	V	Aurora		WBk				30·8	20·0	Sorrento	1873	Paturzo	V 10.88	G.Jaccarino	Castellamare
2712	N	Aurora	Gundersen	WBk	Norwegian			32·3	19·1	Grimstad	1882	N.Osmundsen	Y 5.88	Joh.D.Havstad	Grimstad
2713	N	Aurora (ex John Peacock, ex Wasati)		WBk	Norwegian			32·8	17·9	Quebec	1868	Samson	Y	N.Bugge	Tönsberg
2714	N	Aurora	Pettersen	WBg	Norwegian	178		24·0	13·5	Baltimore				J.P.Pettersen	Fredrikshald
2715		Aurora		WBg	Norwegian	146		21·5	11·9	Copenhagen	1855			J.M.Halmsen	Christiania
2716		Aurora (ex Adela)		WSr	Peruvian	161			6·0	UnitedStates	1858			M.Cavassa	Callao
2717		Aurora	M.Grunberg 86	WSr	Russian	169		26·7	11·6	Angern	1885	J.Sarring		J.Jansohn	Riga
2718		Aurora	Mandelberg 87	WBkn	Russian	341	121·5		11·6	Åland	1875			Navigation Co. "Austra"	Riga
2719		Aurora	Conte	WBk	Spanish		141·7	30·9	20·9	Richmond	1852			R.Bosch y Alsina	Barcelona
2720		Aurora	Sevilla	WBk	Spanish	427	111·7	29·2	15·9	Vegesack	1864	H.F.Ulrichs	V 11.81		Canary Islands
2721	V	Aurora	Nordin 85-88	WBk	Swedish		146·9	28·7	17·0	Nordmaling	1877	Fr.Hägglund	Y 5.88	Nordmalings Skepps-Red-Bol. (C.M. Boden)	Nordmaling
2722	V	Aurora	Svensson	W3mSr	Swedish		128·6	25·0	11·9	Sundsvall	1875	J.N.Hägglund	Y 10.86	Essvikens Skeppsr.bol.	Sundsvall
2723		Aurore	Core	WBn	British	174	90·7	23·2	12·4	Dunkirk	1857		Y 2.8?	H.McCubbin	Swansea
2724		Aurore	Core	WSr	British	106	71·5	22·5	9·6	Rimouski	1879	E.Tremblay		E.Tremblay	Quebec
2725		Aurore	Stephany	WBn	French	139	76·4	20·7	11·8	Vannes	1864	Ridel		Vve Stephany	Vannes
2726	DV	Aurorita	Sørensen 85-76	WBk	Danish	570	148·0	27·9	18·0	Bilbao	1871	S.DeArana	Y 8.88	AktSelsk(M.Michaelsen)	Randers
2727		Auseklis (ex A. Klokmann)		WBk	Russian		136·1	27·9	17·6	Swinemünde	1864	O.Hartwig	Y 9.84	NavigationCo. "Austra"	Riga
2728		Ausonia (LOST)	Kennell 88	WBk	Chilian	616	144·3	29·3	19·1	SestriPonente	1873	Boasi		A.Sutherland	Valparaiso
2729		AustinBaldwin		WBk	Italian	674	170·9	31·4	23·5	Chiavari	1868	Tappani	Y 3.88	F.Dall'Orso di Davide	Spezia
2730	R	Austin D. Knight		W3mSr	American	372		32·8	12·0	Camden,Me.	1880	Coombs & Co.		J.&B.C.Adams	Camden,Me.
2731	V	Austra	Prinz 87	WBkn	Russian	342	138·6	30·7	12·9	Mühlgraben	1884	A.Ferle		Navigation Co. "Austra"	Riga
2732	✱	Australia	Will 86-86	IBk	British	2175	278·4	41·9	21·5	PortGlasgow	1886	Russell & Co.		P.Denniston & Co.	Glasgow
2733	V	Australia	Harris 87	WBk	British		180·2	35·0	21·1	Rockland,NB	1878	R.A.Chapman	Y	Taylor Bros.	Dorchester,N.B.
2734		Australia		WSr	British	118	76·0	20·0	8·5	Caraçao	1854			A.Cameron	Tortola
2735	✱N	Australia	Hergh 74-85	IBk	Norweg'n		221·1	35·6	20·7	Bergen	1886	Martens,Olsen&		Rasmus F. Olsen	Bergen
2736		Australia	Jacobsen 88	WBk	Norweg'n		175·3	31·6	21·4	Grimstad	1888			Joh.D.Havstad	Grimstad
2737	N	Australia	Christophersen	WBk	Norweg'n	696	158·3	36·1	18·5	Jacobstad,Fin.	1865	J.Lof	Y 4.87	C.Evensen	Hvidsten
2738	G	Australia	Sandin	WBk	Swedish	573	149·2	30·8	17·5	Gefle	1862	P.Elfstrand	Y	P.J.Hagerstrand	Gefle
2739	V	Austria	Foster	WS	British	1868	215·5	41·1	24·1	,NS	1884	G.W.Churchill	Y	G.W.Churchill	Windsor,N.S.
2740	V	Austria	Bakin 77-81	WBk	British	1106	183·6	36·5	22·5	St.John,N.B.	1881	D.Pittfield	V 10.87	Taylor Bros.	St.John,N.B.
2741	V	Austriana	McIntosh 81-81	WS	British	1519	215·2	39·4	24·0	Portland,N.B.	1875	HilyardBros.	V 10.88	D.V.Roberts	St.John,N.B.
2742	V	Autocrat	Knudsen 87	WBk	Norweg'n	665	152·0	32·6	20·1	,NS	1870	P.Young		A.S.Knudsen	Arendal
2743	V	Automne	87	WBk	French	315	113·2	23·3	13·2	St. Malo	1882	L.Mallard	Y	L.Mallard	St.Malo
2744		Avalon	88	WSr	British	121	94·9	28·2	7·9	,N.S.	1888	S.Smith		J.W.Smith	Windsor,N.S.
2745	V	Avance	Hansen 82-82	WSr	Danish	125	86·3	21·9	10·8	Marstal	1872	H.J.Bager		H.Hansen	Marstal
2746	N	Avance	Bergersen 76-76	WBk	Norweg'n	486	124·2	32·9	18·5		1853		V 10.82	Ths.&H.Lorentzen	Drammen
2747	N	Avance	Christiansen 76	WBk	Norweg'n	343	125·3	27·4	12·2	,Me.	1855		V 10.85	J.Beyleguard	Tönsberg
2748	N	Avance	Olsen	WBg	Norweg'n	211	97·0	24·2	13·7		1857	Bernt Tellefsen		B.Tellefsen	Risör
2749	GN	Avance	Andreassen	WS	Norweg'n		97·2	25·0	11·5	Arnis,Germany	1867	F.Johansen	V 11.87	Ths.Arbo Höeg	Laurvig
2750	N	Avance	Larsen 70-82	WSr	Norweg'n		87·1	23·6	11·8	Kragerö	1874	B.Knoph	Y 6.82	B.Knoph	Kragerö

✱=Lloyd's Register; D=Nederlandsche Vereeniging; G=Germanischer Lloyd; I=Registro Italiano; R=American Record. =Veritas Austro-Ungarico; H=Veritas Ellenico; N=Norske Veritas; V=Bureau Veritas.

Dans la 4me Colonne, C=Composite (membrure en fer, bordé en bois); I=Fer; Stl=Acier; W=Bois.

Dates du doublage en métal (Col. 11) ressortent des livres dans lesquels les navires sont respectivement classés ou des rapports des Inspecteurs du Lloyd's Register.

Pour plus amples renseignements sur les navires classés par le Lloyd's Register, voir le Lloyd's Register Book et le Lloyd's Yacht Register.

Official No.	No. in Book	Registry in which classed	Ship's Name. Material and Rig. Master, and Dates of Appointment to present Owner's Service and to Ship. Ship's Late Name.	Flag	Net Tonnage	Length	Breadth	Depth	Where	When	By whom	Date of Metalling	Owners or Managers
	2751	N	Avance Petersen -87 WSr (ex Dorothea)	Norweg'n	112	85·1	20·8	9·9	Blankenese	18 $\frac{62}{71}$			M.H.Astrup
	2752	DNV	Avanti Petersen 86-88 WS (ex Belle Morse)	Norweg'n	1274	192·8	36·7	23·7	Bath, Me.	1867	J.P.Morse	Y [illegible]	C.Zernichow
	2753	DNV	Avanti Bøge WBk	Norweg'n	589	141·7	32·6	18·9	Arendal	1877	Henriksen	Y [illegible]	Jens Lassen
365	2754	I	Avanti Savoia Parodi WBk (ex Corin. ex Oldbridge. [illegible])	Italian	952	180·5	34·8	23·5	Genoa	1876	G.Merello	Y [illegible]	G.Coxe
100	2755	V	Ave Drobillovich WSr	Austrian	169	87·9	23·2	11·5	Lussinpiccolo	1882	N.Martinolich	Y [illegible]	Olivetti & Commi
	2756		Avelinda W B	Spanish	489	..	..	..	[illegible]				B.Rucabado
	2757	✻	Avenger Ferguson 80 I S	British	1360	240·1	36·9	22·1	Whitehaven	1879	Whitehaven S.B. [Co.,Ld		W.Grice & Co.
	2758	V	Avenir Bon -87 WSr	Belgian	110	91·8	22·6	11·2	Antwerp	1883	H.Wappers		Vve de Mauroy
	2759	N	Avenir England 83-86 WBk (ex Sovereign)	Norweg'n	309	116·1	28·7	14·6	Åbo, Finland	1864	E.Blomgirst		A.Grön
	2760	N	Avenir Thorsen WBg (CONDEMNED)	Norweg'n	186	89·8	23·8	13·0	St Malo	1856	Nerve		C.Hvistendahl
	2761		Avenir Karlsson WBk	Russian	401	..	..	..	Eckerö	1870			E.Bengts
	2762		Aventuriere WSr (ex Alfred Victory)	French	122	92·0	20·7	11·0	Brixham	1860	Dewdney		E.Ollivaud
3335	2763	✻D	Aviemore Sorensen 86-86 WS	Norweg'n	1090	211·9	36·8	22·2	Aberdeen	1870	Hood	Y [illegible]	L.G.S.Larsen
	2764		Avio Johansson WSr	Russian	321	..	..	..		1874			G.F.Fagerström
85978	2765		Avis Farnsworth WSr	British	125	86·0	26·0	8·0	Rothesay, N.B.	1882	R.Titus		R.C.Elkin
3187	2766	I	AvoG. (ex Giovanni D.) Hari [-88 WBk	Italian	545	138·4	31·8	21·5	Lavagna	1870	G.Brignole	Y [illegible]	Franc.Razeto
91868	2767	✻	Avoca Hatch 81-85 I S	British	1625	257·5	38·2	23·2	Greenock	1885	Russell & Co.		J.Nourse
88302	2768	V	Avoca Mitchener 85-85 WBk	British	1430	209·0	41·0	24·0	Newport, N.S.	1885	T.A.Mosher	Y 12.85	T.Aylward
22917	2769		Avoca WSr	British	258	124·4	24·6	14·9	Waterford	1855	A.White		J.Allan
92403	2770	V	Avon Brady -87 WS	British	1590	222·8	41·5	24·0	Newport, N.S.	1887	T.A.Mosher	Y 12.87	A.Gibson & Co.
69302	2771		Avon WSr	British	205	97·3	27·0	11·2	Cornwallis, NS	1873			J.E.Newcomb
197	2772		Avon Park WSr	American	127	80·8	23·9	8·0	Searsport, Me.	18 $\frac{45}{65}$			W.Ross
92483	2773	V	Avonia Porter 74-86 WBk	British	1664	229·0	42·0	24·0	Horton, N.S.	1886	J.B.North		J.B.North
83209	2774	✻D	Avonmore Smith 81-86 WBk	British	1389	201·5	40·0	23·5	Horton, N.S.	1881	J.B.North	Y [illegible]	J.B.North
75451	2775	DV	Avonport Mitchner -86 WBk	British	987	174·9	36·0	22·6	Horton, N.S.	1876	J.B.North	Y [illegible]	J.B.North
62734	2776		Avonside Lord 67-87 WBk	British	254	114·8	25·4	15·1	Plymouth	1868	Hill		R. & W. King
135	2777	AH	Avra Filinis WBk	Greek	347	116·1	28·2	..	Syra	1876		Y [illegible]	G.Filinis
229	2778		Avra Limos WSr	Greek	129	77·4	24·6	..	Syra	1879			C.Limos
1548	2779	I	Avvenire Simonetti -87 WBk	Italian	399	121·0	30·5	19·0	Savona	1867	F.Sirello	Y [illegible]	G.Lavarello
3343	2780	IV	Avvenire C. WBk (ex Avvenir) Schiaffino -88	Italian	772	167·6	33·5	20·9	Gefle	1872	Bergforss & Son	Y [illegible]	G.B.Cavassa
	2781		Awajishima I Sr	Japanese	386	176·4	25·7	12·0	Glasgow	1861	Thos. Wingate [& Sons		Nippon Yusen Kaisha
	2782		Awunho Hovi W3mSr	Russian	180	104·1	23·9	11·2	Wederlax	1874	Hiiterlawarf		E.Hovi
	2783	V	Axel Olsen WSr	Danish	116	88·9	20·7	10·2	Holbæk	1870	Mortensen		H.L.Smiths Enke
	2784	GV	Axel Lundstedt 81-85 W3mSr	German	265	106·9	23·9	12·6	Rönnebeck	1875	H.Oltmann	Y [illegible]	H.B.Groth
	2785	✻N	Axel (ex St. Vincent) Nielsen CBk	Norweg'n	892	190·0	35·0	18·9	Sunderland	1865	Pile Hay & Co.	Y [illegible]	J.Boysen
	2786	N	Axel (ex Sea Gem) Nielsen -87 WBk	Norweg'n	568	149·8	31·2	17·8	St Andrews NB	1864			J.Gunnerud
	2787	DN	Axel Saanum 75-75 WBk	Norweg'n	535	133·6	33·7	18·3	Brahestad	1857			E.Kraft
	2788	N	Axel Andersen WBg	Norweg'n	253	102·9	25·9	14·7	Sunderland	1865			A.Andersen
	2789	N	Axel Feght WSr	Norweg'n	148	91·9	26·1	10·2	Svartvik	1845			H.Skabo
	2790		Axel Sandell WSr	Russian	241	106·8	28·2	13·0	Raumo	1872	J.Wigelius		H.V.Söderlund
	2791		Axel Andersson WSr	Russian	165	..	..	..	[illegible]	1873			A.Teir
	2792	V	Axel Carlsson 86-87 WBk	Swedish	369	127·6	26·8	14·4	[illegible]	1867	A.Liedberg	Y [illegible]	G.L.Widholm
	2793	DGV	Axel Nilsson -88 WBk (ex [illegible])	Swedish	358	117·7	28·0	16·7	[illegible]	1874	W.Zeltz	Y [illegible]	S.Hedberg
	2794		Axel Carlsson -87 WBk	Swedish	304	118·6	25·3	14·3	[illegible]	1864		Y [illegible]	Axel Pyk
	2795		Axel (ex Skelleftea) Sporsén WBg	Swedish	196	89·5	22·6	13·2	[illegible]	1838	[illegible] Thoren		C.A.Ideström
	2796		Axel Jansson WSr	Swedish	164	86·3	25·3	10·9	Finland	1872			C.E.Ersson
	2797		Axel Mauritz Åman WBg	Russian	283	..	..	..	Helsingfors	1857			O.Simolin
	2798	V	Axel Wästfelt Levengren 85-87 WBk	Swedish	549	142·0	29·3	18·2	Nordmaling	1881	Fr.Hägglund	Y [illegible]	Nordmalings Skepps [C.M.Bodén]
	2799		Aya Jonsson WSr	Swedish	112	77·4	20·3	10·3	Kiel	1852			J.Jonsson
	2800												

✻ = Lloyd's Register; D = Nederlandsche Vereeniging; G = Germanischer Lloyd; I = Registro Italiano; R = ...
A = Veritas Austro-Ungarico; H = Veritas Ellenico; N = Norske Veritas. V = Bure...

In the 4th Column, C = Composite (iron frame, wood planked); I = Iron; Stl = Steel; W = Wood.

The Dates of Metalling (Col. 11) are based on the records of the Books in which Vessels are respectively classed, or on the Reports of the Surveyors of ...

Further particulars of Vessels classed by Lloyd's Register will be found in Lloyd's Register Book and Lloyd's Yacht Register.

No. dans le livre	Caractère	Nom. Capitaine	Matériaux et Gréement	Pavillon	Tonnage	Longueur	Largeur	Creux	Construit Où	Quand	Par qui	Doublage	Armateurs ou Directeurs-Gérants	Port d'Armement
2801		Ayah Adelphi	WBg	Turkish	312	..	..	..	Greece	1882			Oonji Zade Apik Effendi	Constantinople
2802		Ayah Costanti	WSr	Turkish	200	..	..	..	England	1883			Tijar Anastasiadi	Constantinople
2803		Ayah Costanti	WSr	Turkish	200	..	..	..	England	1886			Sophia, daughter of Hadji Petro	Constantinople
2804	A	Ayah Giorgio *Lignos*	WSr	Turkish	112	..	..	..	Chios	1879			E. Lignos	Shios
2805		Ayah Haralambas	WBg	Turkish	126	..	..	..	Greece	1873			Hadji Dimitri Photi & Co	Constantinople
2806		Ayah Marco	WBg	Turkish	325	..	..	..	Heraclea	1865			Stephan Corec	Scio
2807		Ayah Nicola *Mavridi*	WBg	Turkish	325	..	..	..	Austria	1878			Limnilee Yorghi Mavridi	Lemnos
2808		Ayah Nicola	WBg	Turkish	300	..	..	..	Maiz Ada	1872			Costanti Hadji Nikolaki	Maiz Ada
2809		Ayah Nicola	WBg	Turkish	300	..	..	..	Maiz Ada	1872			Costanti Hadji Combalto	Maiz Ada
2810		Ayah Nicola	WBg	Turkish	300	..	..	..	Maiz Ada	1872			Costanti Coltamies	Maiz Ada
2811		Ayah Nicola	WBg	Turkish	300	..	..	..	Maiz Ada	1872			Koyoudelmos Hadji Dimitri	Maiz Ada
2812		Ayah Nicola	WBg	Turkish	280	..	..	..	Italy	1870			Simirelee Karokali Yorghi	Simire
2813		Ayah Nicola	WBg	Turkish	250	..	..	..	Aktarpoles	1883			Hatrieh-tijari Said Hadji Ahmed Effendi & Sons	Aktarpolee
2814		Ayah Nicola	WBg	Turkish	250	..	..	..	Greece	1884			Selanikleo Yorghi Hadji Adem & Co.	Selanik
2815		Ayah Nicola *Chalkis Mehmed*	WSr	Turkish	250	..	..	..	Aktarpolee	1884			Oonialee Kotch-Oglou Mustapha	Oonia
2816		Ayah Nicola	WBg	Turkish	235	..	..	..	Danube	1867			Limnilee Nicola Yorghi & Co.	Lemnos
2817		Ayah Nicola *Yani Nicola*	WBg	Turkish	228	..	..	..	Greece	1873			Manol Ninga & Co.	Constantinople
2818		Ayah Nicola	WSr	Turkish	225	..	..	..	Sakarieh	1881			Hattab-tujari Vassil Semphon & Co.	Sakarieh
2819		Ayah Nicola	WBg	Turkish	200	..	..	..	Cashot	1872			Yanni Nicola	Cashot
2820		Ayah Nicola	WBg	Turkish	200	..	..	..	Greece	1879			Mureftahlee Stellianos Dikhran & Co.	Mureftah
2821		Ayah Nicola	WBg	Turkish	198	..	..	..	Russia	1879			Mureftahlee Costanti Bindanos & Co.	Mureftah
2822		Ayah Nicola	WBg	Turkish	190	..	..	..	Greece	1866			Kerrasonlleo Panayotti	Kerassunda
2823		Ayah Nicola	WBg	Turkish	182	..	..	..	Sakarieh	1883			Kara Agop-Zadeh Issik Effendi	Sakarieh
2824		Ayah Nicola	WBg	Turkish	176	..	..	..	Greece	1874			Ipsaralee Melletta	Ipsara
2825		Ayah Nicola	WBg	Turkish	175	..	..	..	Maiz Ada	1874			Maizlee Vassilaki Saivah	Maiz Ada
2826		Ayah Nicola	WBg	Turkish	175	..	..	..	Greece	1885			Misslee Andon Nicola	Miss
2827		Ayah Nicola	WBg	Turkish	170	..	..	..	Greece	1881			Szusabollee Tchelebi-Oglou Yanghi & Co.	Szusaboli
2828		Ayah Nicola	WBg	Turkish	168	..	..	..	Greece	1874			Ipsarahlee Emmanoel Dimitri Porecaos	Ipsara
2829		Ayah Nicola	WBg	Turkish	153	..	..	..	Russia	1873			Pappa Christoppayuan	Constantinople
2830		Ayah Nicola *Chrissafes*	WBg	Turkish	153	..	..	..		1884			Kumur-tijari Christo Dimitri & Co.	Constantinople
2831		Ayah Nicola *Khristo*	WBg	Turkish	137	..	..	..	France	1868			Yanni Costantin & Co.	Lemnos
2832		Ayah Nicola	WBg	Turkish	136	..	..	..	England	1874				
2833		Ayah Nicola	WBg	Turkish	131	..	..	..	Greece	1877			Sakuzlee Yanni Scorroli & Bros.	Scio
2834		Ayah Nicola	WBg	Turkish	129	..	..	..					Mureftahlee Hadji Stelli-Simon	Mureftah
2835		Ayah Nicola	WSr	Turkish	125	..	..	..	Heraclea	1871			Bakal Christo	Heraclea
2836		Ayah Nicola	WSr	Turkish	125	..	..	..	Dardanelles	1871			Enoslee Yorghi Altanas	Enos
2837		Ayah Nicola	WBg	Turkish	125	..	..	..	Kooroojeh Shi	1884			Daoud Pashadar Beghir-[illegible]	Kooroojeh Shill
2838		Ayah Nicola	WBg	Turkish	124	..	..	..	Greece	1878			Bulbuljeelee Dimorani Nassal & Co.	Bulbuljeh
2839		Ayah Nicola	WBg	Turkish	122	..	..	..	Russia	1880			Tommak Mikkhal & Co.	Constantinople
2840		Ayah Nicola	WBg	Turkish	121	..	..	..	England	1883			Yanni Nicola Tekfoori	Constantinople
2841		Ayah Nicola *Vassil*	WBg	Turkish	120	..	..	..	Russia	1878			Ooncapanindah-Tijari-Kara-Alon-Oglou Haffi & Co.	Constantinople
2842		Ayah Nicola	WBmb	Turkish	120	..	..	..	Gallipoli	1883			Tashji-Zadeh Ahmed Kyamil Effendi	Gallipoli
2843		Ayah Nicola *Dimitri Voulous*	WBmb	Turkish	120	..	..	..	Gallipoli	1885			Daoud-Pashalee Ali Kaya & Co.	Pasha
2844		Ayah Nicola	WBg	Turkish	112	..	..	..	Cashot	1872			Andon Yorghi	Cashot
2845		Ayah Nicola	WBmb	Turkish	112	..	..	..	Muretta	1870			Mureftahlee-Hadji Simo Oglou Istrati	Mureftah
2846		Ayah Nicola	WBml	Turkish	103	..	..	..	Balletia Skel-[lesi	1885			Tulaji Seraphim Simon	BalletiaSkellesi
2847	A	Ayah Nicola	WSr	Turkish	102	..	..	..	Syra	1883			C. Ligno	Syra
2848		Ayah Nicola	WBmb	Turkish	100	..	..	..	Enos	1868			Enoslee Chaplan-Oglou Hadji Angeli & Co.	Enos
2849		Ayah Nicola	WBmb	Turkish	100	..	..	..	Enos	1869			Enoslee Khristophol-Oglou Andreas & Co.	Enos
2850		Ayah Pandalhemo	WBg	Turkish	236	..	..	..	Greece	1885			Nikarialee Costanti Karaiali	Nikaira

Æ = Lloyd's Register; D = Nederlandsche Vereeniging; G = Germanischer Lloyd; I = Registro Italiano; R = American Record; A = Veritas Austro-Ungarico; H = Veritas Ellenico; N = Norske Veritas; V = Bureau Veritas.

Dans la 4me Colonne, C = Composite (membrure en fer, bordé en bois); I = Fer; Stl = Acier; W = Bois.

Les Dates du doublage en métal (Col. 11) ressortent des livres dans lesquels les navires sont respectivement classés ou des rapports des Inspecteurs du Lloyd's Register.

Pour plus amples renseignements sur les navires classés par le Lloyd's Register, voir le Lloyd's Register Book et le Lloyd's Yacht Register.

Official No.	No. in Book.	Registry in which classed.	Ship's Name. Master, and Dates of Appointment to present Owner's Service and to Ship. Ship's Late Name.	Material and Rig.	Flag.	Net Tonnage.	Length.	Breadth.	Depth.	Built. Where.	When.	By whom.	Date of Metalling.	Owners or Managers.
	2851		Ayah Sigon	W Bnb	Turkish	150	..	..	..	Greece	1874			Alutjee Constanti
	2852		Ayah Spiridon *[illegible]*	W Bg	Turkish	198	..	..	..	Greece	1885			Nikarialee Petro Andas Mavrikoula
	2853		Ayah Spiridon	W Bg	Turkish	106	..	..	..	Greece	1882			[illegible]
	2854		Ayah Spiridon	W Sr	Turkish	100	..	..	..	Greece	1884			[illegible]
	2855		Ayah Spiro	W	Turkish	350	..	..	..	Cashot	1872			Costanti
	2856		Ayah Spiro	W Bg	Turkish	163	..	..	..	Italy	1873			Bulhedjedic Yanni Costanti
	2857		Ayah Spiro	W Bg	Turkish	110	..	..	..	Cashot	1872			Yorghi Nicola
	2858		Ayah Strati	W Bg	Turkish	212	..	..	..	Greece	1880			Aktarpoloee Joachim Pappa Strati
	2859		Ayah Yandalimo	W Bg	Turkish	163	..	..	..	Greece	1873			Simirelee Yannaki Mikali
	2860		Ayah Yorghi	W Bg	Turkish	350	..	..	..	Cashot	1872			Hadji Nicola Leon
	2861		Ayah Yorghi	W Bg	Turkish	350	..	..	..	Cashot	1872			Tuneyia Yaudi
	2862		Ayah Yorghi *Lucca Yorghi*	W Bg	Turkish	296	..	..	..	Greece	1884			Maizlee Prashko Y[illegible]
	2863		Ayah Yorghi	W Bg	Turkish	200	..	..	..	Cashot	1872			Manol Pappa Elia
	2864		Ayah Yorghi	W Sr	Turkish	182	..	..	..	Mukla[illegible]ah Ske	1878			Kerestabji Tujari Hassad Agop
	2865		Ayah Yorghi	W Bg	Turkish	177	..	..	..	Russia	1882			Tujar Christo Yorghi
	2866		Ayah Yorghi *Alexion*	W Sr	Turkish	177	..	..	..	Russia	1884			Serrerlee Tendori & Co.
	2867		Ayah Yorghi	W Bg	Turkish	175	..	..	..	Cashot	1872			Hadji Yorghi Kakah-Manol
	2868		Ayah Yorghi	W Bg	Turkish	154	..	..	..	Russia	1874			Touromah-Dimitri
	2869		Ayah Yorghi *Vassil*	W Bg	Turkish	125	..	..	..	Greece	1869			Simirelee Gharocala Vassil
	2870		Ayah Yorghi	W Sr	Turkish	120	..	..	..	Barten	1885			Kiyat Tijari Nessim Bameton
	2871		Ayas Tirias	W Sr	Turkish	346	..	..	..	Greece	1880			Nicola Sawn
	2872		Ayos Aftanios *Costanti Dimigenos*	W Sr	Turkish	150	..	..	..	Greece	1880			Sharaliji Yanni Kiriaki
	2873		Ayos Anastasios	W Bg	Turkish	207	..	..	..	Sinope	1869			Manol Pappa Strati & C[illegible]
	2874	A	Ayos Athanasias *Kandaulo*	W Bg	Turkish	147	..	..	..	Sinope	1875			Aslan Oglus
	2875		Ayos Barbios	W Bg	Turkish	225	..	..	..	Kerassunda	1871			Kynssakochish-Oglus Karaiannis & Bros.
	2876		Ayos Ephtanios *Chilikir Mehemed*	W Sr	Turkish	153	..	..	..	Greece	1884			Mureitahlee Costanti Dimitrios & Co.
	2877		Ayos Haralambos *Costanti Dimigenos*	W Bg	Turkish	120	..	..	..	Greece	1875			Dimitri Photi & Co.
	2878		Ayos Lijer	W	Turkish	155	..	..	..	Greece	1876			Limnilee Nicola Yorgo
	2879		Ayos Lijer	W Bg	Turkish	100	..	..	..	Russia	1876			Fuchcopo Mikhal Harinesa
	2880		Ayos Nicola	W Bg	Turkish	245	..	..	..	Greece	1876			Sakyzlee Stavri Yoann[illegible] kamzola
	2881		Ayos Nicolakis	W Bg	Turkish	250	..	..	..	Peljaı	1870			Costanti Calviotti
	2882		Ayos Pandaleemos *Nicola Yorghi*	W Bg	Turkish	193	..	..	..	Greece	1885			Limnilee Helleni & Co.
	2883		Ayos Yorghios	W Sr	Turkish	125	..	..	..	Iekkehumin	1874			Kerestejee Orkham vekeli Mikail
75643	2884		Ayr	W Sr	British	299	133·0	23·4	11·3	P.Dalhousie	1855	A.Muir & Bro.		A.McArthur
	2885		Ayr *Fowler -87*	W Sr	British	124	89·4	27·5	7·9	Bl'ckRiverN.B.	1886	J.&R.McLeod		J.McLeod
	2886		Ayrão	W Sr	Brazilian	302	..	..	..					Maria B. Veranny
93400	2887	✻	Ayrshire *Miller 82-87*	I S	British	1736	268·4	40·0	23·4	P.Glasgow	1887	Russell & Co.		T.Law & Co.
	2888	G	Azela *Knudsen -83*	W Bg	Danish	169	91·6	21·7	13·3	Newcastle	1858			J.A.Larsen
1462	2889		Azelda & Laura	W Sr	American	110	86·2	25·8	8·4	Essex.Mass.	1865			C.Jorgensen
	2890	V	Azha *Speilberg*	W Bg	Norweg'n	304	123·6	27·4	15·9	Arendal	1875		[illegible]	T.Thommesen & Smith
510	2891		Azoff *Olivari*	W Bg	Italian	210	90·9	25·8	16·2	Savona	1857		[illegible]	P.Olivari
[illegible]	2892	A	Azzardo *Ballarin*	W Bn	Italian	108	91·2	22·3	11·9	Chioggia	1886	I.Menetto	[illegible]	L.Ballarin

ADDITIONAL SHIPS ARE INSERTED IN THE SUPPLEMENT AT THE END OF THE REGISTER.

✻ = **Lloyd's Register**; D = Nederlandsche Vereeniging; G = Germanischer Lloyd; I = Registro Italiano; R = Ame[illegible]
A = Veritas Austro-Ungarico; H = Veritas Ellenico; N = Norske Veritas; V = Bur[illegible]

In the 4th Column, C = Composite (iron frame, wood planked); I = Iron; Stl = Steel; W = Wood.

The Dates of Metalling (Col. 11) are based on the records of the Books in which Vessels are respectively classed, or on the Reports of the Surveyors of [illegible]

Further particulars of Vessels classed by Lloyd's Register will be found in Lloyd's Register Book and Lloyd's Yacht Register.

No.	Registre dans lequel le navire est classé	Nom. Capitaine et dates d'entrée au service des Armateurs actuels et à bord du navire. Ancien Nom	Matériaux et Gréement	Pavillon	Tonnage	Dimensions Longueur	Largeur	Creux	Construit Où	Quand	Par qui	Date du doublage en métal	Armateurs ou Directeurs-Gérants	Port d'Armement
1		B. Fr. Peters [illegible] -84	W Bk	German	276	114·9	26·9	14·4	Rostock	1866	G. W. Tesdorff		W. Lange	Rostock
2	N	B.D. Metcalf Larsen	W S	Norweg'n	1243	185·4	36·6	24·1	Damariscotta, Me	1856	Metcalf & Norris	[illegible]	T. Thommesen & Sön	Arendal
3	R	B. Frank Neally [illegible]	W 3mSr	American	261	127·6	30·2	9·8	Bath, Me	1880	Goss, Sawyer & Pack.		B.F. Nealley	Portsmouth, N.H.
4		B. Havens Jones [illegible]	W Sr	American	205	108·0	29·2	9·8	P. Jefferson NY	1865	Mather		W. M. Conky	Perth Amboy N.J.
5	✻	B. L. Greenville ex [illegible]	W Sr	British	100	87·1	21·4	10·2	Shoreham	1874	May	[illegible]	Sailing Ship B.L. Co., Ld. (B. F. Hickman)	London
6		B. I. Hazard Smith	W 3mSr	American	373	130·8	31·1	14·6	P. Jefferson NY	1872			W. B. Dayton	P. Jefferson N.Y.
7		B. L. Burt Johnson 84-84	W 3mSr	American	720	163·2	35·6	15·6	Bath, Me.	1881	A. Sewell		J. B. Phillips	Fall River, Mass.
8		B. L. Eaton Grearson 74-75	W Sr	American	163	102·0	27·5	9·10	Calais, Me.	1872	Rideout & Lord		B. L. Eaton	Calais, Me.
9		B. L. Page	W Bn	French	159	94·0	24·6	10·9	Lockeport, NS	1883	S. Harlow		Folquet & Fils	St. Malo
10		B. L. Sherman Hallett	W 3mSr	American	170	150·6	26·3	8·5	E. Haven, Conn	1865			C. Lovell & Son	Barnstable, Mass.
11		B. N. Fullerton	W Sr	British	106	80·9	24·3	9·3	Parrsboro', N.S.	1880	W. Willeger		Charles S. Howard	Parrsboro', N.S.
12	R V	B. P. Cheney Humphreys 17-89	W S	American	1259	195·9	38·6	24·2	Bath, Me.	1874	Goss & Sawyer	Y [illegible]	J. H. Kimball	Bath, Me.
13	R	B. R. Woodside Weir -88	W 3mSr	American	535	158·0	33·5	12·0	Bath, Me	1883	Adams & Hitchcock		Adams & Hitchcock	Bath, Me.
14		B. S. Mimbelli	W Pol	British	323	111·5	29·3	12·8	Leghorn	1860		[illegible]	S. [illegible]	Malta
15		B. Vicuna Mackenna Smith		Chilian	100	75·5	19·4	8·7	Juan Fernandz	1883			Alfredo de Rodt	Valparaiso
16		B. W. Folger	W Sr	British	117	82·8	23·5	8·8	Kingston, Ont.	1871	W. Power & Co.		W. Dandy	Montreal
17	V	B. W. Morse Wylie	W 3mSr	American	531	156·0	35·0	14·0	Bath, Me.	1880	B. W. & H. F. Morse		B. W. & H. F. Morse	Bath, Me.
18	R V	B. Webster Kenney 78-89	W Bk	American	563	140·4	32·0	16·8	Bath, Me.	1873	Hager & Thurlow	Y [illegible]	B. Webster	Portland, Me.
19	D V	Baarn Brouwer 85-86	C Bk	Dutch	1029	192·2	37·1	21·7	Amsterdam	1880	W. H. Meursing	Y [illegible]	W. H. Meursing	Baarn
20		Babakar Amboina ex [illegible]	C Bk	Dutch	377	147·9	27·5	..	Amsterdam	1873			Sech Assan bin [illegible] bin Abdah Alkateri	Sourabaya
21		Babineau & Gaudry	W Sr	British	156	95·5	28·5	9·8	Grenadines	1864			Hiram Ives	Quebec
22	✻	Baccante P. di Giorgio	I Sr	Italian	65	83·5	16·1	8·3	Liverpool	1871	Bowdler, Chaffer & Co.		F. W. Woodhouse	Marsala
23	✻	Bacchante Nicholson 85-85	W Sk	British	85	77·7	21·3	10·7	Burton Stath'r	1885	J. Garside		Mrs. E. Jipson	Hull
24	✻	Bacchus Evans 83-86	I S	British	1250	216·0	36·2	23·7	Birkenhead	1867	G. R. Clover & Co.		Hughes & Co.	Liverpool
25	V	Bacchus Bjorkqvist (WEBSTER)	W Bk	Russian	781	171·5	31·2	20·5	Brahestad	1875	J. Wicklund	Y [illegible]	J. Lang	Brahestad
26	✻ I	Baciccino Revello Anselmo -87	W Bk	Italian	646	158·3	32·8	21·0	Sturla	1873	F. Parodi	Y 10.81	A. Revello	Genoa
27		Bacht Mastora Said Alwi Albaf	W Bk	Dutch	289	..	..	..	Grissee, Java	1837			Said Abdul Rachman bin Mohamed Albaf	Palembang
28	H	Bachtiar Tramouris	W Bk	Turkish	159	..	..	..	Syra	1868			Ali Bahtzet Bey	Constantinople
29	✻	Backworth Van der Werff	W Bg	Dutch	306	113·3	27·3	16·7	Blyth	1866	Blyth Fl. Dk. Co.	Y [illegible]	A. Sissingh	Farmsum
30	✻	Bactria Peterk n 82-87	I S	British	2112	279·0	40·2	24·3	Southampton	1885	Oswald, Mord'nt & Co.		T. & J. Brocklebank	Liverpool
31		Bahia	W Sr	Brazilian	301	..	..	..	Germany	1872			R. G. da Cunha	Rio Janeiro
32	✻ A	Bakran Lenusch -86	W Bk	Austrian	639	146·9	32·5	20·8	Buccari	1879	M. Matterglian	Y [illegible]	S. Stiglich	Buccari
33		Baktia Jorghi Sparo	W Bg	Turkish	144	..	..	..	Russia	1884			Pamboukli Emin Effendi & Sons	Constantinople
34	✻	Balaklava Palmer 83-83	I Bk	British	1310	223·0	36·7	22·7	Stockton [illegible]	1881	Osbourne Graham & Co.		Davison Bisset & Co.	Liverpool
35		Balance Fisher 85-85	W Bg	German	301	105·7	28·0	16·2	Rostock	1863	O. Ludewig		A. Burchard	Rostock
36		Balbina C. Villablanca ex Poland	W Bk	Chilian	564	136·4	34·1	21·3	Brahestad	1867	A. Westerlund		A. Bafico	Valparaiso
37	V	Balcarry Rumpff 81-86	I Bk	German	491	153·7	27·7	17·2	Glasgow	1867	Dobie & Co.		P. Breckwoldt, sen.	Blankenese
38	✻	Balclutha Constable 86-86	Stl S	British	1614	256·5	38·6	22·7	Glasgow	1886	C. Connell & Co.		R. M'Millan	Glasgow
39		Balclutha Constable -87	W Bn	British	133	93·6	23·7	11·6	Bridgeport N.F.	1871			W. A. Willcox	St. Johns, N.F.
40		Balder Nielsen	W Bk	Norweg'n	355	122·6	27·8	15·5	Risör	1860	J. Brundie		A. E. Olsen	Risör
41	N	Balder Thomassen -85	W Bk	Norweg'n	215	111·2	25·4	11·8	Kragerö	1867	P. Olsen	Y [illegible]	J. Thomesen	Kragerö
42	N	Balder Mathisen	W Bn	Norweg'n	162	98·9	25·5	11·6	Bergen	1871	S. Flago	[illegible]	Joachim Hansen	Bergen
43	N	Balder Agrée	W Sr	Norweg'n	130	89·1	23·5	10·9	Fredriksstad	1853			Gullik Jensen	Tönsberg
44		Balder Lie -84	W Sr	Norweg'n	114	86·5	21·4	10·4	Flekkefjord	18 51/71		Y [illegible]	D. H. Berner	Stavanger
45	V	Balder Ohlsson 88-88	W Bk	Swedish	420	157·4	21·9	16·1	Hernösand	1874	Fr. Hägglund	Y [illegible]	Hernösands Lastfing. Aktiebol. (H. Nordlander)	Hernösand
46		Balder Wahl -81	W Bk	Swedish	366	120·4	26·6	17·0	Söderhamn	1852			J. A. Kjellberg & Söner	Gothenburg
47	G V	Baldur Botes 83-83	W Bk	German	693	164·7	33·5	19·4	Hemmelwarde	1883	J. T. Strenge & Sohn	Y [illegible]	Gerh. Eilers	Brake
48	I	Balestrino Parodi	W Bk	Italian	484	138·1	31·2	19·5	Varazze	1871	F. Tassara	Y [illegible]	S. Balestrino	Genoa
49	✻ D	Balgay Parsons 85-86	W Bkn	British	221	114·0	24·2	12·2	Dundee	1872	Tay Sh. Bdg. Co.	Y [illegible]	J. R. Daniels	Dundee
50		Balgis Said Mohamed bin Alie Aibaf	W Bk	Dutch	349	108·2	29·1	..					Said Hossain bin Mohamed Shahab	Palembang

✻ Lloyd's Register; D = Nederlandsche Vereeniging; G = Germanischer Lloyd; I = Registro Italiano; R = American Record; A = Veritas Austro-Ungarico; H = Veritas Ellenico; N = Norske Veritas; V = Bureau Veritas.

Dans la 4me Colonne, C = Composite (membrure en fer, bordé en bois); I = Fer; Stl = Acier; W = Bois.

Les Dates du doublage en métal (Col. 11) ressortent des livres dans lesquels les navires sont respectivement classés ou des rapports des Inspecteurs du Lloyd's Register.

Pour plus amples renseignements sur les navires classés par le Lloyd's Register, voir le Lloyd's Register Book et le Lloyd's Yacht Register.

Official No.	No. in Book.	Registry in which classed.	Ship's Name. Material and Rig. Master, and Dates of Appointment to present Owner's Service and to Ship. Ship's Late Name.	Flag.	Net Tonnage.	Dimensions. Length.	Breadth.	Depth.	Built. Where.	When.	By whom.	Dates of Metalling.	Owners or Managers.
1	2	3	4	5	6	7			8	9	10	11	12
	51		Balgownie *Jepsen -83* W Bk	Dutch	305	123·8	24·2	16·5	Aberdeen	18 48/67	W. Hood & Co		J. P. Amsinga
	52		Balhaer W Bk	Dutch	130	94·9	23·3	..	Grissee, Java	1840		C1.79	Hadjee Mohamet Amp
54047	53	✳	Balkamah *Glover* I S	British	1318	220·4	38·2	22·8	Hull	1866	Humb. Irn Wks		E. Bates & Sons
4888	54	✳ A	Balkan *Franguel -86* W Bk	Austrian	564	133·8	31·8	19·5	Fiume	1879	Fili Schiavoni	Y 8.86	A. Sterk fu A.
	55	V	Ballester *Andreu* (ex Dolores, ex Ibera, ex Rima) W Bn	Spanish	148	85·6	22·0	11·1	Blanes	1878		Y 2.87	M. Ballester
83070	56	✳	Ballinbreich Castle *Fursien 86-86* W Sr	British	99	84·0	21·7	10·6	Perth	1879	Wallace		R. Clark
67930	57	✳	Ballochmyle *Tait 88-88* I S	British	1438	215·0	38·5	22·7	Sunderland	1873	W. Watson		D. Bruce & Co.
89671	58	✳	Ballumbie *Wood* I Bk	British	1131	215·4	35·5	20·7	W. Hartlepool	1885	W. Gray & Co.		F. H. Vaughan
30890	59		Bally Packet W Sr	British	119	71·7	22·8	10·6	Wexford	1830			Mahomed Meda Marican
1536	60		Balmarino W Bn	British	126	89·9	22·7	10·6	Canada	1853			J. Kelly
93727	61	✳	Balmoral *Gardiner 87-87* I S	British	2045	276·2	40·6	24·2	Liverpool	1887	T. Royden & Sons		IronSailingShip "Balmoral" Co. Ld. (Macvicar, Marshall & Co.)
75182	62	✳	Balmore *Mitchell -88* I Bk	British	725	193·3	31·5	17·9	Dundee	1876	Brown & Simpson		Albany Shipping Co. Ld
	63		Baltasar W Bg	Spanish	196	92·8	24·9	13·0	Italy	1861			B. Barrera
	64	V	Balthasar *Jantzen -85* W 3m Sr	German	275	115·2	26·0	13·0	Ribnitz	1877	J. H. Wilken	Y 4.88	B. Wahlen
75416	65	V	Baltic *Bradshaw* W Bn	British	438	129·6	30·7	16·9	P. Williams-NS	1876	C. Brinton	Y 11.85	J. W. Peury
65353	66	✳	Baltic *Dinnes 77-77* W Sr	British	100	83·5	21·7	10·6	Peterhead	1877	Carnegie		R. J. Kidd
10571	67	✳	Baltic *Williams 86-86* W Sr	British	79	78·0	20·7	9·0	Rhyl	1857	R. Jones		R. Hughes & Son
	68		Baltic *Busch 88-88* W Bk	German	310	116·2	27·3	15·3	Rügenwalde	1866	C. F. Zielke		J. J. Wallis & Sohn
	69	N	Baltic *Herlofsen -88* W Bk	Norweg'n	562	145·3	31·0	18·2	Arendal	1858	H. Herlofson	Y 5.88	H. Herlofsons Enke
	70	N	Baltic *Halvorsen 80-80* W Sr	Norweg'n	202	102·7	26·6	12·1	Drammen	1869	Ths. Lorentzen		B. O. Halden
	71	N	Baltic *Storhaug* W Sr	Norweg'n	118	89·9	22·5	10·9	Stavanger	1880	T. Amundsen		A. Birkeland
	72		Baltic (FOUNDERED) *Kihlberg 80-80* W Bg	Swedish	270	112·5	26·6	12·8	Gefle	1865	Bergfors		G. J. Sundberg
	73		Baltic (ABANDON'D) *Jonasson* W Bk	Swedish	265	99·1	23·8	16·1	Jersey	18 38/72			J. C. Hulthén
	74		Balticola *Kroger* W Bk	Norweg'n	558	133·6	28·9	18·5	Regwash, N.S.	1851			H. G. Kirsebom
3303	75	V	Baltimore *North -88* W Bk	American	696	163·0	35·0	17·2	Baltimore	1888	W. Skinner & Ss	Y 1.88	T. Rollins & Co.
	76	V	Baltimore (ex Volunteer) *Freese 76-82* W S	German	1081	184·5	36·5	22·8	Boston, Mass.	1863	W. F. Weld & Co.	Y 8.87	Joh. F. Arens
	77		Baluklie W Bg	Turkish	157	..	..	..	Greece	1869			Aktarpoiolee Christo Yelali Vasal
	78	N V	Bams *Andreasen* W Sr	Norweg'n	304	122·8	27·9	13·8	Christiansand	1872	O. P. Moe & Sön	Y 1.88	Ths. Isefjær
3511	79	✳ A	Ban Mazuranic *Ant -88* W Bk	Austrian	565	140·7	29·7	19·7	Buccari	1875	M. Matterglian	Y 1.88	G. Mikulicich
79364	80		Ban Righ *Smith* W Sr	British	100	87·4	21·7	10·4	Banff	1879	Geddie		J. Stephen
60235	81		Ban Seng Heng W Bn	British	172	95·0	22·0	9·0					Tan Yong Seeack
	82		Banas Sooed W Bg	Dutch	183	..	..	..					Sech A. Cdubmebim bin Abu Bansir
87647	83	✳	Banca *Haeberow 83-83* I Bk	British	999	207·2	33·6	20·0	P. Glasgow	1883	Russell & Co.		P. Denniston & Co.
	84	N	Banco *Wallentinsen* W 3m Sr	Norweg'n	99	87·0	20·8	10·3	Stavanger	1878		Z 3.88	E. Endersen
	85	V	Banco Mobiliario *Warmuth* I Bk	German	1061	208·4	34·1	19·9	Hamburg	1882	Blohm & Voss	Y 6.88	H. Fölsch & Co.
	86	✳	Banda (ex Marie) *Vierk -88* W 3m Sr	Dutch	280	113·1	24·8	13·2	Bremen	1865	F. Janssen	2.88	A. Köpcke
90018	87	✳	Bandaneira *Ahrens 81-85* I 4m Bk	British	1862	267·6	40·2	23·7	Greenock	1885	Russell & Co.		P. Denniston & Co.
84194	88	✳ D	Bandeeth *Young 85-87* I Bk	British	700	189·1	30·6	17·9	P. Glasgow	1882	Russell & Co.		W. & J. Lockett
124	89	V	Bandiera Moro *Boltini* W Bg	Italian	387	127·6	28·3	19·8	Venice	1869	G. Zanon	Y 1.88	Fili Vianello-Moro
58951	90		Banff *Crisp -86* W Sr	British	200	109·1	24·4	13·9	Inverness	1867	Stewart		W. Milburn
82319	91	✳	Banffshire (ex Southfield) *Thomson 88-88* I Bk	British	899	200·1	32·8	19·6	P. Glasgow	1880	Russell & Co.		T. Law & Co.
386	92	D	Banga *Wiktol -81* W Sr	Russian	225	118·7	27·4	12·5	Dondangen	1881	J. Katzmann		C. Wiktol
91035	93	✳	Bangalore *Congdon* I S	British	1699	260·2	39·9	23·0	Stockton	1886	Richardson, Duck & Co.		G. Croshaw & Co.
72500	94		Bangalore W Sr	British	296	136·0	26·2	12·0	Kingston, Ont.	1877	W. Power		Wm. Power
	95	V	Bangor *Sebastiani -87* W Bk	French	464	138·7	26·6	17·1	Nantes	1876	Jollet & Babin	Y 5.88	Knecht
	96		Banka (ex Grefvinnan Taube) *Tuwadin* W 3m Sr	Dutch	344	127·6	26·6	..	Sundsvall	1873		Y 8.88	Lie Boen Ping
87873	97	✳	Bankdale (BURNT) *Williams 79-85* I Bk	British	1308	222·0	37·2	22·6	Sunderland	1884	Osbourne, Graham & Co.		W. Just & Co.
74533	98	✳	Bankfields (ex James Beazley) *Wakeham 85-85* I Bk	British	835	195·4	32·4	19·7	Sunderland	1876	Osbourne, Graham & Co.		S. Wakeham & S…
87989	99	✳ D	Bankhall *Quinn 84-84* I Bk	British	1312	222·0	37·2	22·6	Sunderland	1884	Osbourne, Graham & Co.		W. Just & Co.
93770	100	✳	Banklands *Davenport 88-88* Stl Bk	British	1197	225·7	36·3	22·0	Pt. Glasgow	1888	J. Reid & Co.		Bank Shipping Co. Ld. (W. Just & Co.)

✳ = Lloyd's Register; D = Nederlandsche Vereeniging; G = Germanischer Lloyd; I = Registro Italiano; R = Ame…
A = Veritas Austro-Ungarico; H = Veritas Ellenico; N = Norske Veritas; V = Bure…

In the 4th Column, C = Composite (iron frame, wood planked); I = Iron; Stl = Steel; W = Wood.

The Dates of Metalling (Col. 11) are based on the records of the Books in which Vessels are respectively classed, or on the Reports of the Surveyors of L…

Further particulars of Vessels classed by Lloyd's Register will be found in Lloyd's Register Book and Lloyd's Yacht Register.

No.	Classe	Nom. Capitaine et dates d'entrée. Ancien Nom.	Matériaux et Gréement	Pavillon	Tonnage	Longueur	Largeur	Creux	Où	Quand	Par qui	Doublage	Armateurs ou Directeurs-Gérants.	Port d'Armement.
[illegible]	✻	Bankville [illegible] (ex [illegible])	I Bk	[illegible]	1314	211·4	[illegible]	[illegible]	Liverpool	1878	T.Royden&Sons		W.Just & Co.	Liverpool
102	✻	Bann Cutting [illegible]	I S	[illegible]	1467	250·1	38·9	24·2	Sunderland	1875	T.R.Oswald		J.Nourse	London
103		Banner	W Bn	[illegible]	132	85·3	25·5	9·0	P.Gilbert,N.S.	1872			St.Clair Jones	Digby,N.S.
104	✻	Bannockburn [illegible]	Stl 4mS	[illegible]	2006	287·0	42·5	24·1	Glasgow	1886	Barclay, Curle & Co.		R.Shankland & Co.	Greenock
105		Bannockburn	W Bn	[illegible]	111	91·[illegible]	22·2	8·2	C.Hawke,NSW	1880	H.T.White		R.S.Johnson	Sydney,N.S.W.
106		Bannockburn	W Bn	[illegible]	111	85·8	21·8	9·9	Auckland	1880	McQuarrie&McCallum		Cornelius Norris	Rockhampton
107		Banshee Mayo 15-80	W Bn	[illegible]	111	[illegible]	20·7	12·6	Barnstaple	1850	J.Westacott		H.E.Gape	Cowes
108		Banvard Leak	W Sr	American	1[illegible]	101·5	27·0	8·0	Haddam,Conn.	1848			A.H.Howes	New York
109		Bapandi	Bg	[illegible]	261	..	..	..					M.Dos	Alexandria
110		Baquio Zarragua 86-88	W Bk	Spanish	303	11[illegible]	[illegible]	15·0	Bilbao	1867	S.de Arana	V [illegible]	R.Goyeneches	Bilbao
111		Barba Azul Riera (ex [illegible])	W Sr	Spanish	243	1[illegible]·7	27·4	15·2	Rivadeo	1852			Romulo Bosch	Barcelona
112	✻D	Barbadian McLean 82-83	C S	[illegible]	799	17[illegible]	31·3	18·4	Dumbarton	1867	A.M'Millan&S.	V 10-85	J.Kerr & Co.	Greenock
113	✻	Barbara Bain 13-84	W Sr	British	98	85·2	21·8	10·4	[illegible]	1883	W.Geddie		A.Marr	Inverness
114		Barbara Petersen	W Sr	American	157	83·6	28·2	7·1	[illegible]	1877			L.G.Peterson	San Francisco
115	I	Barbara Gallino Pesenelli	W Bk	Italian	302	112·2	31·8	20·[illegible]	[illegible]	1872	N.Cerruti	V [illegible]	A.Gallino	Genoa
116		Barbara Hendrika Lehmann	W Bn	German	159	92·1	20·8	10·6	[illegible]	1866	OudeZijl	[illegible]	C.C.E.Lehmann	Hamburg
117		Barbara Hernster Rosebeck 87	W Sr	American	141	112·0	28·0	8·[illegible]	[illegible],Cal.	1887	Benedectsen		R.Sudden	San Francisco
118	I	Barbara Preve Campodonico	W Bk	Italian	719	159·[illegible]	3[illegible]·1	21·[illegible]	[illegible]	1875	A.Carlevaro	V [illegible]	F.Salvini	Savona
119	DG	Barbarossa Warnken [illegible] (ex Liberia)	W S	German	1313	194·9	38·7	24·0	Bath,Me.	1867	Lamont&Robinson	V [illegible]	Von Riegen & Stindt	Bremerhaven
120		Barbosa II.	W Bn	Portuguese	158	..	..	..	Vianna do Castello	1875			J.M.Barbosa	Vianna do Castello
121		Barcelona Stoles (ex Peter Plandeen)	W S	Spanish	878	21[illegible]·1	31·8	17·1	Mahon	1868	Padury freres		Jané & Co.	Barcelona
122	✻	Barcore Thomas -84	I S	British	2382	27[illegible]·6	40·8	24·5	[illegible]	1884	Richardson, Duck&Co		Eyre Evans & Co.	Liverpool
123	N	Barden Jensen 84-86	W Bg	Norweg'n	265	118·4	27·6	12·8	Porsgrund	1874		V [illegible]	J.Jeremiassen	Porsgrund
124		Barden Landberg	W Bk	Russian	527	17[illegible]·6	28·7	16·4	Nikolaistad	1867	H.Ahlbeck		R.Mattsson	Mariehamn
125	D	Barend Christiaan Valckenaer de Greeve (ex Biel Huling Haast ex Iscandria)	C Bk	Dutch	680	183·4	28·4	19·3	London	1857	T.Bilbe&Co.	V [illegible]	P.Landberg & Zoon & H.C. de Jong	Batavia
126		Barendine Osiris (ex Amstel) Tiddens [illegible]	C Bk	Dutch	1428	189·9	41·7	25·0	Amsterdam	1874	Meursing & Huijgens	V [illegible]	P.Landberg & Zoon	Batavia
127	✻	Bargany Milne 80-80	I S	British	1232	226·3	37·4	21·9	Dumbarton	1874	A.M'Millan&S.		J.Kerr & Co.	Greenock
128		Bariloche (ex Sarmiento) Ferro	W Bk	Argentine	3[illegible]0	123·3	25·7	16·7	Workington	1866	Fell	V [illegible]	S.Gianello & A.Fernandes	Buenos Ayres
129	V	Baring Brothers Watters 78-86	W S	American	2091	213·7	42·3	29·5	Thomaston,Me	1877	E.O'Brien	V 7.85	E.E.O'Brien	Thomaston,Me
130	✻D	Baron Blantyre Taylor [illegible]	I S	British	1585	257·0	39·9	23·5	Sunderland	1873	W.Watson		J.Neill	Greenock
131	✻D	Baron Colonsay Docherty [illegible]	I S	British	1632	269·0	40·1	23·5	Greenock	1875	J.E.Scott		J.Neill	Greenock
132	✻	Baron Hill Hughes 76-[illegible]	W 3mSr	British	209	119·0	25·4	13·0	Amlwich	1876	Thomas	V [illegible]	W.Postlethwaite	Liverpool
133	DN	Baron Holberg Olsen 69-89	W Bk	Norweg'n	383	126·5	29·3	15·7	Skien	1857			Niels Olsen	Laurvig
134	✻DV	Baron van Pallandt van Rosendael Schneider 72-84	C S	Dutch	1244	171·1	30·7	21·9	Amsterdam	1868	Meursing& Huijgens	V [illegible]	Huijgens & Hardenberg	The Hague
135	I	Barone Podesta Imperiale 86-87	W Bk	Italian	725	163·6	33·3	22·4	[illegible]	1873	G.Scala	V [illegible]	G.B.Degregori	Genoa
136		Baronet's Family (ex Washington)	W Bk	British	434	123·0	29·2	17·6	Nantes	1849			Pestonjee Johangee	Bombay
137	✻	Barossa Walker 81-73	I S	British	968	210·7	35·3	20·0	Sunderland	1873	W.Pile&Co.		T.B.Walker	London
138	✻	Barracouta Barber	I Bk	British	619	171·1	27·5	17·9	Liverpool	1864	R&J Evans&Co.		Staveley Taylor & Co.	Liverpool
139	✻	Barranca Morgan 86-89	C S	British	653	177·3	31·3	17·9	Aberdeen	1869	A.Hall&Co.	V [illegible]	W. Ross & Co.	Guernsey
140		Barsingerhorn Van der Woude (ex Maria Louisa Antoinette)	W Bk	Dutch	529	127·4	32·6	20·5	Kapelle	1847			S.H.van der Woude	Nieuwediep
141	V	Barso Sprensen	W 3mSr	Danish	141	97·8	20·9	9·7	Assens	1877	H.Christoffersen	V [illegible]	Actie Selskab (P.Poulsen)	Nordby,Fanø
142		Barten	W Sr	Turkish	228	..	..	..	Barten	1866			Dimitri Constantin & Co	Barten
143		Barten Hassan	W Bg	Turkish	216	..	..	..	Barten	1868			Hairich Tijaric Mollah Ahmed & Co.	Barten
144		Barten	W Sr	Turkish	192	..	..	..	Barten	1870			Mustapha Shakir's Daughters	Barten
145		Barten	W Bg	Turkish	190	..	..	..	Barten	1871			Alatjee Suleiman Agha	Barten
146		Barten	W Bg	Turkish	175	..	..	..	Barten	1871			Terebolielee Hadji Osman & Co.	Terebolie
147		Barten	W Bg	Turkish	175	..	..	..	Barten	1871			Terebolielee Shaban	Terebolie
148		Barten	W Sr	Turkish	162	..	..	..	Barten	1869			Bartenlee Dervisch Zadeh Ahmed & Co.	Barten
149		Barten Halil	W Sr	Turkish	147	..	..	..	Barten	1869			Bartenlee Kassap-oglou Hadji Halil	Barten
150														

Lloyd's Register; D = Nederlandsche Vereeniging; G = Germanischer Lloyd; I = Registro Italiano; R = American Record; Veritas Austro-Ungarico; H = Veritas Ellenico; N = Norske Veritas; V = Bureau Veritas.

Dans la 4me Colonne, C = Composite (membrure en fer, bordé en bois); I = Fer; Stl = Acier; W = Bois.

Dates du doublage en métal (Col. 11) ressortent des livres dans lesquels les navires sont respectivement classés ou des rapports des Inspecteurs du Lloyd's Register.

Pour plus amples renseignements sur les navires classés par le Lloyd's Register, voir le Lloyd's Register Book et le Lloyd's Yacht Register.

Official No.	No. in Book.	Registry in which classed.	Ship's Name. Material and Rig. Master, and Dates of Appointment to present Owners' Service and to Ship. Ship's Late Name.	Flag.	Net Tonnage.	Dimensions. Length.	Breadth.	Depth.	Built. Where.	When.	By whom.	Date of Metalling.	Owners or Managers.
1	2	3	4	5	6	7			8	9	10	11	12
	151		Barten W Sr	Turkish	137	..	..	..	Barten	1870			Erserourmlee [illegible] oglou Abdullah
	152		Barten W Sr	Turkish	139	..	..	..	Barten	1868			Katirleleo Sesiah oglou Yanandi & Co.
	153		Barten W	Turkish	127	..	..	..	Barten	1868			Bartenlee Hadji Mehemed Zadeh Mustapha
	154		Barten W Bg	Turkish	125	..	..	..	Barten	1868			Galata Kumurjee Hadji Mustapha
	155		Barten *Anastas* W Bg	Turkish	125	..	..	..	Barten	1868			Katirleiee Yamandi & Co.
	156		Barten W Sr	Turkish	123	..	..	..	Barten	1867			Selina dellalee Hadji Yanni & Co.
	157		Barten W Sr	Turkish	112	..	..	..	Barten	1866			Bartenlee Yenifu-oglou Hussein
	158		Barten W Sr	Turkish	112	..	..	..	Barten	1870			Chimitek-Yezohlee Hadji Osman Agha
	159		Barten W	Turkish	112	..	..	..	Barten	1870			Kerestiji Damad Aga
	160		Barten W Bmb	Turkish	106	..	..	..	Barten	1870			Bartenlee Veliyan [illegible] Mustapha & Co.
	161		Barten *Ahmed Reiss* W	Turkish	105	..	..	..	Barten	1870			Dervisch Zadeh Ahmed & Co.
	162		Barten W Bg	Turkish	100	..	..	..	Barten	1865			Tosson oglou Hadji Ali & Co.
	163		Barten W Bg	Turkish	100	..	..	..	Barten	1868			Vassili Kostiniache[illegible]
2262	164		Bartholomew Gosnold W Bk	American	365	106·6	28·0	18·4	Falmouth Mas.	1832			J.F.Tucker & Co.
	165	V	Bartimeus *Åkerman* 83 (ex Hermes) W Bn	Swedish	160	104·6	22·1	11·1	Rönne	1860	Bech		C.Smith
1076	166		Bartolomeo *Boriello* 36-88 W Bn	Italian	103	83·3	21·9	10·1	TorredelGreco	1874	G.Guido		R.Aurilia
1627	167		BartolomeoDanovaro *Bonanni* W Bk	Italian	572	144·3	32·6	19·9	SestriPonente	1867	G.Casanova		A.Danovaro
830	168	I	Bartolomeo Gagliardo *Solari* (ex Raffaele) W Bk	Italian	456	131·2	31·0	19·6	Sestri	1868	A.Briasco fu G.B.	Y 10.8[illegible]	F[lli] Gagliardo di F.
1092	169	I	Bartolomeo Marciani *Brignati* W Bk	Italian	626	155·8	32·5	20·8	Voltri	1875	G.Fava	Y 5.9[illegible]	Chichizzola Fr[lli]
47442	170	✻	Barunga *Thurston* 88-88 (ex Aprilia) I Bk	British	1030	216·7	32·0	20·6	London	1863	J.&W.Dudgeon		Trinder, Anderson & [illegible]
47539	171	✻	Bass Rock *Macfarlane* 87-87 (ex Rutlandshire) I Bk	British	1043	197·4	33·5	22·5	Birkenhead	1863	G.R.Clover & Co.		J.Cornfoot & Co.
	172	V	Basse Indre *Guillaume* W Bn	French	130	77·1	22·3	11·0	Nantes	1865	Bojn & Clairgeau		Guillaume
65706	173	✻ D	Basuto *Matthews* W Bk	British	348	132·7	28·2	14·9	Shoreham	1872	May		R.H.Bruce
67332	174	✻	Batavia *Ashton* W Sr	British	80	75·7	20·0	9·2	[illegible]	1880	Day		G.Kilner & Co.
	175		Batavia *Said Moessin Alkaf* W Bk	Dutch	208	81·3	24·5	..					Said Moessin Al[illegible]
	176	V	Batavia *Kronberg* (ex Johanna Antonia) W Bk	Swedish	359	137·0	25·9	13·8	Holland	1866	J.Pot	Y 2-[illegible]	C.Pettersson
	177	✻ V	Batavier *Prul* 88-88 I S	Dutch	1616	227·0	43·8	26·3	Kinderdyk	1876	L.Smit & Zoon		J.Koning & Van Dele[illegible]
	178		Batiuschka (WRECKED) *Blank* W Sr	Russian	112	70·5	21·9	11·1	Riga				Blank
	179	DGV	Bato *Brouwer* 86-86 (ex Albano) W3mSr	Dutch	257	118·9	25·7	13·0	Memel	1871	W.Rickeles	Z.2.87	De Voogd & Hordijk
2478	180	IV	Battaglia *Campodonico* W Bk	Italian	579	147·5	32·4	20·5	Varazze	1874	Gustavino	Y [illegible]	Eredi B.Schiaffino
1685	181	I	Battista *Villa* W Bn	Italian	365	122·7	29·2	18·3	Savona	1867	G.Tixi	Y 8.88	G.B.Bottino
2494	182	I	Battistina Madre *Olivari* 87 W Bk	Italian	782	165·5	30·7	22·2	Recco	1874	P.Rolla	Y 5.87	Lor.& And.Schiaffino
70941	183	✻	Battle Abbey *McKay* 88-88 I S	British	559	212·1	40·3	22·8	Liverpool	1875	T.Royden & Son		J.Poole & Co.
	184	DG	Baumeister Wilken *Stuben* 62-62 W Bg	German	247	102·3	26·6	13·8	Ribnitz	1862	J.H.Wilken		Rud.Josephy
	185	DN	Bauta *Larsen* 71-88 W3mSr	Norweg'n	264	121·5	26·9	12·8	Stavanger	1873	R.I.Hovem	Y 9.87	K.Knudsen
72595	186		Bavaria W Sr	British	361	115·0	26·1	12·5	GardenId.Ont.	1878	H.Roney		The Calvin Co.,Ld.
	187	V	Bavoletto *Belliard* 87-87 (ex Valkyrien) W Bn	French	164	104·6	25·2	11·8	Altona	1887		Y 11.87	H.Guillebot
2349	188		Bay State *Baker* W Sr	American	185	92·5	27·5	10·6	EastHaddam, [Conn.	1849			F.Bulkley
92804	189	✻	Bay Wreath W K Trawler	British	88	79·5	20·7	10·7	Brixham	1885	R.Jackman		J.Randall
73562	190	✻	Bay of Bengal *Latta* 86-86 I S	British	1527	260·3	39·3	23·1	Glasgow	1875	J.Elder & Co.		J.Bulloch
79878	191	✻	Bay of Cadiz *Davidson* 85-85 Stl S	British	1626	260·4	40·3	22·7	Glasgow	1878	J.&G.Thomson		J.Bulloch
83433	192	V	Bay of Fundy *Porter* 88-82 W Bk	British	1221	200·0	38·5	21·7	CornwallisNS	1882	J.E.Steele	Y 7.87	S.Sheffield
73583	193	✻	Bay of Naples *Buche* 84-84 I S	British	1621	266·7	39·3	23·8	Aberdeen	1875	A.Hall & Co.		J.Bulloch
89507	194	✻	Bay of Panama *Wright* 84-84 Stl 4m S	British	2282	294·0	42·3	24·3	Belfast	1883	Harland & Wolff		J.Bulloch
49891	195	✻	Bayard *Niles* 87-87 I Bk	British	1319	219·7	36·5	22·9	Liverpool	1864	T.Vernon & Son		Foley & Co.
	196		Bayard *Malleville* W Bk	French	255	104·0	24·9	14·4	Fecamp	1866	Brument & [Capot		H.Monnier & fils
	197		Bayard (WRECKED) W Bk		234	101·6	23·9	13·8	St.Vaast	1866	T.Lévêque		
	198	N	Bayard (ex Kelso) *Kittilsen* W Bk	Norweg'n	581	152·1	32·0	18·5	Sunderland	1861	W.Pile, Jun.	Y 4.88	A.D.Westergaard
	199	N	Bayard *Andersen* 81-81 W Bk	Norweg'n	420	122·7	30·6	17·1	Arendal	1855		Y 10.87	A.K.Orvig & P.Stians[illegible]
	200		Bayonnaise (BROKEN UP) *Guillon* 88 W Bk	French	227	99·4	25·3	13·9	Nantes	1869	Sevestre frères	Y 5.88	F.Boissière

✻ = Lloyd's Register; D = Nederlandsche Vereeniging; G = Germanischer Lloyd; I = Registro Italiano; R = Aus[illegible]
A = Veritas Austro-Ungarico; H = Veritas Ellenico; N = Norske Veritas; V = [illegible]

In the 4th Column, C = Composite (iron frame, wood planked); I = Iron; Stl = Steel; W = Wood.

The Dates of Metalling (Col. 11) are based on the records of the Books in which Vessels are respectively classed, or on the Reports of the Surveyors of [illegible]

Further particulars of Vessels classed by Lloyd's Register will be found in Lloyd's Register Book and Lloyd's Yacht Register.

No.	Classe	Nom	Capitaine / Ancien Nom	Matériaux et Gréement	Pavillon	Tonnage	Longueur	Largeur	Creux	Construit Où	Quand	Par qui	Doublage	Armateurs ou Directeurs-Gérants	Port d'Armement
201		Bazar	[illegible]	WBg	Swedish	[illegible]	[illegible]	23·1	[illegible]	Rostock				P.E.L.Wikholm	Oscarshamn
202	DV	Beaconsfield		WS	Norweg'n	1451	207·6	39·7	24·0	NovaScotia	1877	Jones		Troop & Son	London
203		Beagle		WBn	British	[illegible]	[illegible]	[illegible]	[illegible]	[illegible]	1874	McDonald		The Hull Ice Co.,Ld.	Hull
204		Beagle	[illegible]	WBg	British	131	[illegible]	[illegible]	[illegible]	Wivenhoe	1857			Newman,Hunt&Co	London
205		Beata	[illegible]	WSr	[illegible]	[illegible]	117·4	23·9	[illegible]	Timmernabbe	1870	Nordström		L.Olsson	Lerhamn
206	*	Beatrice	[illegible]	WBk	British	[illegible]	[illegible]	[illegible]	[illegible]	[illegible]	1868	Chaigneau		Maguire & Co.	Quebec
207		Beatrice	[illegible]	WSr	British	[illegible]	87·[illegible]	[illegible]	9·2	Yarmouth N.S.	1883	H.Jenkins		A.F.Stoneman	Yarmouth,N.S.
208	A	Beatrice	[illegible]	WBk	Austrian	[illegible]	133·8	31·2	18·2	Fiume	1866	G.tti Schiavoni		R.Barich	Fiume
209	I	Beatrice	[illegible]	WBk	Italian	[illegible]	[illegible]	[illegible]	[illegible]	[illegible]	1876	Bozzano		A.Benvenuto	Genoa
210	R	Beatrice Bernard	(ex [illegible])	WBk	Columbi'n	314	[illegible]	[illegible]	[illegible]	[illegible]	1864	TayS.B.Co.		Columbian Government	Bogota
211	*	Beatrice Forester		WK	British	84	76·2	20·7	10·7	[illegible]	1886	R.Jackman		C.Forester	Grimsby
212	HRV	Beatrice Havener	Havener	WBk	American	525	140·5	31·7	[illegible]	Searsport Me.	1874			J.W.Havener,jun.	Searsport,Me.
213	*	Beatrice Lines	Nielsen (ex Christiana [illegible])	WS	Norweg'n	1066	205·5	34·5	21·[illegible]	[illegible]	1866	W.Hood&Co.		Actieselskabet "Ellen Lines"	Fredrikstad
214	*	Beatrice Mabel	Edwards -87	WSk	British	87	[illegible]	21·5	10·6	Burton-Stather	1887	J.Garside		J.Robinson	Grimsby
215	G	Beatrice Suppicich	[illegible] 79	WBg	German	[illegible]	112·3	28·2	16·9	Rostock	1873	O.Ludewig		H.Bauer	Rostock
216	*	Beatrix	Davies 83-86	IBk	British	[illegible]	[illegible]	35·2	21·4	Sunderland	1886	N.of EnglandS.B.Co.		T.Beynon & Co.	Newport
217		Beatrix		WSrYt	British	109	92·9	18·8	9·9	Lymington	1860	G.Inman		H.Gerard Leigh	Southampton
218		Beatriz	[illegible]	WBn	Spanish	144	[illegible]	23·0	10·1	PalmaMaj'rca	1845			B.Cortes y Valeriola	Majorca
219		Beatriz		WFol	Spanish	128	79·4	21·9	10·5	Arens,Spain	1846			R.Lanuza	Valencia
220		Beauport	Knudt 77- (ex [illegible])	WBg	German	234	[illegible]	25·0	14·0	Shields	1860	Young		W.Höveler	Papenburg
221		Beaupre		WSp	British	163	107·[illegible]	23·7	9·8	Yamaska	1872			J.Dallaire	Quebec
222	*	Beauséant	Chalmers	WSrYt	British	113	88·0	19·0	11·5	Gosport	1876	B.Nicholson	C [illegible]	Edward Aiken	Portsmouth
223	*	Bebington	Officer -88	IBk	British	884	183·4	33·1	20·4	Belfast	1859	R.Hickson&Co.		C.Neill	Belfast
224	D	Bedford	[illegible] 86-86	WBk	British	1179	188·7	38·8	24·0	[illegible],N.S.	1877	J.Crow	Y [illegible]	M.Dickie	Halifax,N.S.
225	*	Bedfordshire	Hall -88	IBk	British	1155	201·9	34·1	23·4	Liverpool	1863	W.H.Potter&Co.		J.Tweedie	Liverpool
226		Bedmar	Da Cunha -84 (ex Tres de Dezembro)	WBk	Portug'se	212	108·2	24·6	10·2	Villa doConde	1858	J.M.d'Araujo		E.da Costa Corrêa Leite	Oporto
227	*	Beech Holm	Doughty -88	IBk	British	763	193·3	30·8	18·7	Sunderland	1876	J.Laing		R.H.Gayner	Sunderland
228	*	Beechdale	Prideaux	IBk	British	792	192·2	30·7	19·2	Liverpool	1877	W.H.Potter&Co		E.E.Broomhall	Liverpool
229	*	Beechwood	Thornton 85-86	IS	British	794	186·4	31·6	19·8	Liverpool	1875	T.Royden&Son		J.B.Walmsley & Co.	Liverpool
230	*	Beecroft	Christie	IS	British	1626	251·3	38·8	23·6	Sunderland	1875	T.R.Oswald		Hargrove & Hellon	Liverpool
231		Beeling		StlLr		480	220·0	30·0	8·4	Dumbarton	1888	W.Denny&Bros.		IrrawaddyFlotillaCo.,(Ld.)	Glasgow
232	*	Beemah	Mercier	IBk	British	964	209·7	33·2	19·9	PortGlasgow	1876	W.Hamilton&Co		Hargrove & Hellon	Liverpool
233	R	Beethoven	Smith	WS	British	969	172·5	36·2	22·5	Windsor,N.S.	1870		[illegible]	S.Dimock	Windsor,N.S.
234	V	Beethoven	Hansen 75-77	W3mSr	German	241	109·5	25·4	13·9	Lübeck	1865	J.A.Meyer	[illegible]	Gaedertz sen. & Co.	Lübeck
235		Bel Brandon	Simmons -88	WSr	British	100	79·0	23·0	9·0	SableRiverNS	1877	H.Harlow		L.P.Churchill	Shelburne,N.S.
236		Bela Scuttarina		WSr	Montengrn	148	93·5	20·2	12·2	Martinschizzo	1868			Hussein Malo Beg & Co.	Dulcigno
237	R	Belaetta	Flynn 74-86	WSr	British	150	90·0	25·8	9·6	St.MartinsNB	1880	J.Carson	Y [illegible]	J.M.Taylor	St.John,N.B.
238		Belem (WRECKED)	Galine	W	French	148	..	..	..	Belem				Guibert & fils	St.Malo
239	*	Belfast	Russell 79-88	IS	British	1865	260·5	40·2	24·2	Belfast	1874	Harland&Wolff		T. & J. Brocklebank	Liverpool
240	V	Belgium	Harvey -81	WBk	British	670	162·9	34·5	18·8	Kempe,N.S.	1870	G.Armstrong	Y [illegible]	G.Armstrong	Windsor,N.S.
241		Belisario	Serra	WBn	Spanish	179	88·2	25·7	13·1	Masnou	1845			F.Ginares	Barcelona
242		Belisario		WSr	Spanish	118	85·0	23·2	9·0	Palma	1867			Doña C.Bosch	Majorca
243		Belize	Johansson 75	WBkn	Swedish	207	96·8	21·5	12·1	NorthAmerica	1853			E.Andersson	Edeby,Väddö
244	*	Belkez	HusseinHadjiAli	WBg	Turkish	324	94·0	28·0	16·8	Heraclea	1873			SadykBey	Constantinople
245	I	Bell'Avvenire	DeAndreis	WBk	Italian	582	[illegible]	32·1	20·7	Prà	1872	L.Tixi	Y [illegible]	Tixi & De Andreis	Genoa
246	N	Bella	Christophersen 77-77 (ex [illegible])	WBk	Norweg'n	522	137·0	32·9	18·6	Alblasserdam	1856	C.Smit	Y [illegible]	F.Breda	Christainia
247	V	Bella Dolores	Pagés -88	WBn	Spanish	226	102·0	25·5	16·1	Lloret	18 56/76		[illegible]	J.Maristany	Barcelona
248	*	Bella Donna	Briault -86	WSr	British	63	73·2	19·2	8·9	Jersey	1877	Aubin	Y [illegible]	F.Le Sueur & Son	Guernsey
249	A	BellaPalmiraGallipoli	(ex [illegible])	WBn	Italian	143	93·2	21·2	12·1	Bostock	1877		[illegible]	F.& M.Caridi	Taranto
250	*	Bella Rosa	Bowden 79-84	WSr	British	157	98·1	22·4	12·7	Brixham	1874	Upham & Sons	[illegible]	A.Goodridge &Sons	St.Johns,N.F.

* = Lloyd's Register; D = Nederlandsche Vereeniging; G = Germanischer Lloyd; I = Registro Italiano; R = American Record; A = Veritas Austro-Ungarico; H = Veritas Ellenico; N = Norske Veritas; V = Bureau Veritas.

Dans la 4me Colonne, C = Composite (membrure en fer, bordé en bois); I = Fer; Stl = Acier; W = Bois.

Les Dates du doublage en métal (Col. 11) ressortent des livres dans lesquels les navires sont respectivement classés ou des rapports des Inspecteurs du Lloyd's Register.

Pour plus amples renseignements sur les navires classés par le Lloyd's Register, voir le Lloyd's Register Book et le Lloyd's Yacht Register

Official No.	No. in Book	Registry in which classed	Ship's Name. Material and Rig. Master, and Dates of Appointment to present Owner's Service and to Ship. Ship's Late Name.	Flag.	Net Tonnage	Length	Breadth	Depth	Where	When	By whom	Date of Metalling	Owners or Managers
1	2	3	4	5	6	7			8	9	10	11	12
[illegible]	251		**Bella Russell** *Steelman* W3mSr	American	211	118·0	29·5	8·6	Leesburg,N.J.	1872			J.Russell
	252		**Bella Vicenta** WBg	Spanish	200	87·6	22·8	14·1	Villajoyosa	1856			Torrientes herⁿ
	253	V	**Bellalie** *Poirier* 82 WSr	French	144	78·7	22·3	11·1	Nantes	1866	Sevestre frères		Sevestre frères
78121	254	✳	**Bellaport** *Stephens* 85-85 I Bk	British	1039	215·6	34·0	21·0	Harrington	1878	R.Williamson&Sn		T.Williamson
	255	NV	**Belle** *Fishwick* WBk	British	469	130·1	30·2	16·8	St.John,N.B.	1867	McLeod	[illegible]	W.Brooks
21075	256		**Belle** *Walker* WJury	British	198	86·0	23·5	14·5	Whitby	1858	T.Turnbull&Sns		E.Fisher
65048	257	✳	**Belle** *Parry* WSr	British	130	101·0	23·3	13·7	Preston	1871	W.Allsup		J.Walton & Co.
42406	258		**Belle** WBn	British	146	106·5	24·8	12·5	Nova Scotia	1861			J.Reid
2811	259	R	**Belle Brown** *Sawyer* -86 WSr	American	141	92·0	26·1	8·6	Bucksport,Me	1872	Snow&Farwell	[illegible]	Snow & Pearsons
	260		**Belle Brune** *Bourges* -86 WBg	French	171	89·2	23·0	12·5	Bayonne	1858			A.Lemoine
	261	V	**Belle Dijonnaise** *Lemaire* -87 WSr	French	139	93·8	21·3	12·0	Dunkirk	1870	Malo&Vande-[riele		A.Navet
54862	262	✳	**Belle Flower** *Brown* 85-86 WBk	British	375	125·2	28·2	17·1	Sunderland	1866	Robinson	[illegible]	Belle Flower Sailing Ship Co., Ld. (J.Winship)
2681	263	R	**Belle Halliday** *Nickerson* W3mSr	American	337	128·3	30·0	9·6	Camden,N.J.	1869			M.Nickerson
2782	264		**Belle Hardy** *Baker* W3mSr	American	343	125·0	31·4	14·0	E.Boston	1871	AbielGove		G.W.Rogers
	265	V	**Belle Hélène** *Jonneskindt* WSr	French	131	92·5	21·3	11·8	Dunkirk	1869	Malo& Vande-[riele		Jooneskindt & Deck
3127	266		**Belle Higgins** *Skolfield* [82-87] W3mSr	American	392	147·1	32·9	11·9	Bath,Me.	1880	A.Sewall&Co.		A.Sewall&Co.
2898	267	R	**Belle Hooper** *Gilkey* W3mSr	American	451	140·0	32·0	11·6	E.Boston	1874	AbielGove	[illegible]	Conant & Hatch
64702	268		**Belle Martiniquaise** WSr	British	108	89·5	23·3	9·5	Georgetown	1880			A.G.Clarke
2987	269	V	**Belle O'Brien** *Hodgman* -88 WS	American	1808	237·5	42·0	26·2	ThomastonMe	1875	E.O'Brien	[illegible]	E.E.O'Brien
3178	270	R	**Belle O'Neill** *Butler* [83-85] W3mSr	American	445	142·0	35·0	12·2	Bath,Me.	1881	W.Rogers	[illegible]	J.B.Drake
2358	271		**Belle R. Hull** *Brush* WSr	American	185	103·4	27·7	9·6	TomsRiverNJ	1858			C.E.Brush
2391	272		**Belle Seaman** *Steelman* WSr	American	105	101·0	17·7	7·8	Hamburg,Pa.	1863			M.W.Steelman
64040	273	R	**Belle Star** *Kearon* 85-85 WBkn	British	281	124·8	29·9	13·1	NovaScotia	1873	T.H.Rice	[illegible]	R.Kearon
59278	274	D	**Belle Vue** *Jansen* -88 WBk	Norweg'n	679	145·9	32·9	19·5	St.John,N.B.	1869	JohnMahony	[illegible]	J.Bugge
3080	275	V	**Belle Wooster** *Higgins* WBkn	American	456	129·2	32·8	15·4	Franklin,Me.	1878	J.Blaisdell	[illegible]	Swan & Son
74541	276	✳	**Belle of Arvon** *Evans* -87 I Bk	British	923	193·1	33·2	20·6	Sunderland	1876	Osbourne,Gra-[ham&Co		Tomlinson Hodgetts&Co.
3037	277	RV	**Belle of Bath** *Curtis* -88 WS	American	1347	203·9	39·0	24·3	Bath,Me.	1877	Goss&Sawyer	[illegible]	D.Nickels
70714	278	✳D	**Belle of Benin** *Payn* 80-88 IBk	British	285	144·7	25·0	11·5	Sunderland	1875	G.S.Gulston		T.Harrison & Co.
60092	279	✳	**Belle of Lagos** *Woolcock* [-88 I Bk	British	228	130·7	24·0	11·3	Glasgow	1868	A.Stephen&Sns.		D.W.Bain&Co.
3010	280	RV	**Belle of Oregon** *Matthews* WBk	American	1110	185·6	38·0	22·5	Bath,Me.	1876	Goss&Sawyer	[illegible]	W.H.Besse
52786	281		**Belle of Southesk** WBk	British	540	159·6	29·8	17·8	Montrose	1865	Birnie		Eastern & Australian S.S. Co.
2336	282	R	**Belle of the Bay** *Behrmann* WBn	American	342	120·0	31·0	14·8	Stonington,[Conn.	1860		[illegible]	J.D.Fish & Co.
69576	283	✳	**Belle of the Exe** *Clinton* -88 W3mBn	British	233	120·0	25·5	12·9	Exmouth,Dev.	1878	Redway	[illegible]	P.G.&L.Tessier
65698	284	✳D	**Belle of the Niger** *Osborne* 82-82 I Bk	British	283	141·2	25·0	11·6	Glasgow	1872	A.Stephen&Sns		T.Harrison & Co.
18583	285		**Belleisle** (ex Peter Symons) *Brownell* WBk	British	388	133·7	25·2	16·4	Bristol	1857	Symons		James C.Ellis
75480	286		**Bellewood** WBn	British	112	88·0	23·1	10·2	P.E.Island	1877	White		C.Neill
93518	287		**Bellinger** *Grenman* 86 WK	British	100	98·3	23·0	6·9	Bellinger [River, N.S.W	1886	F.Doepel		John Anderson
80808	288	V	**Bellona** *Warren* WBk	British	1123	188·8	38·4	22·7	Newport,N.S.	1880	Jas.Mosher	[illegible]	Geo.Mounce
70753	289	✳	**Bellona** (BURNT) *Logan* 81-81 I Bk	British	513	163·3	28·3	17·1	Glasgow	1877	C.Connell&Co.		T.C.Guthrie
	290		**Bellona** (ex G. A. Coonan) *Iversen* WBn	Danish	285	115·7	27·1	11·8	Roshea,N.B.	1865	S.V.Covnann		F.Iversen
	291	N	**Bellona** *Jørgensen* -81 WBk	Norweg'n	540	142·5	31·6	17·2	Arendal	1880	Guttormsen	[illegible]	Chr.Th.Boe
56487	292		**Bells** WBk	British	341	119·5	27·2	17·0	Sunderland	1867	W.Naizby	[illegible]	C.W.Turner
[illegible]	293		**Belmont** (ex P. F. Gibbon) WBk	British	371	117·5	27·2	18·4	[illegible]				S.H.Holbrook
[illegible]	294	R	**Belmont** (ex Betting Telek, ex Regina) *Hagan* -88 WBkn	American	513	135·5	30·5	17·5	Millbridge,Me	1873	E.Dyer	[illegible]	J.S.Emery & Co.
	295	V	**Belona** *Petersen* 78-78 WBn	Danish	170	102·5	22·6	11·0	Troense	1874	C.R.Møller	[illegible]	F.W.Valentin
[illegible]	296	✳	**Belpore** *Gadsden* 87- I Bk	British	1200	216·5	35·7	22·2	Glasgow	1866	BarclayCurle&[Co.		Eyre,Evans & Co.
[illegible]	297	D	**Belt** *Munroe* 85-85 WBk	British	1246	189·0	39·0	23·5	Cornwallis,Ns	1877	J.E.Woodworth	[illegible]	C.R.Burgess
[illegible]	298	✳	**Beltana** *Bright* 85-85 CBk	British	741	172·5	33·6	19·2	Sunderland	1869	J.Laing	[illegible]	A.L.Elder & Co.
[illegible]	299	✳	**Belted Will** *Godfray* 79-79 WSr	British	79	79·2	19·4	9·8	Jersey	1878	LeSueur		F.Godfray
	300	✳	**Belted Will** *Hammargren* WBk	Swedish	800	186·1	32·1	20·8	Workington	1863	Fell	[illegible]	J.C.Hultken

✳=Lloyd's Register; D=Nederlandsche Vereeniging; G=Germanischer Lloyd; I=Registro Italiano; R=Am[illegible]
A=Veritas Austro-Ungarico; H=Veritas Ellenico; N=Norske Veritas; V=Bu[illegible]

In the 4th Column, C=Composite (iron frame, wood planked); I=Iron; Stl=Steel; W=Wood.

The Dates of Metalling (Col. 11) are based on the records of the Books in which Vessels are respectively classed, or on the Reports of the Surveyors of Lloyd[illegible]

Further particulars of Vessels classed by Lloyd's Register will be found in Lloyd's Register Book and Lloyd's Yacht Register.

N° dans le Livre	Registre dans lequel le navire est coté	Nom, Matériaux et Gréement. Capitaine et dates d'entrée au service des Armateurs actuels et à bord du navire. Ancien Nom.	Pavillon	Tonnage net	Longueur	Largeur	Creux	Où	Quand	Par qui	Date de doublage en métal	Armateurs ou Directeurs-Gérants	Port d'Armement
301	❋	Beltrees W Bk	British	373	137·8	28·0	15·1	Ayr	1873	Weir	[illegible]	Thorburn & Tessier	Greenock
302		Belvedere W Bn	Italian	155	96·1	23·1	11·5	Sestri	1877	A.Brisasco fu L.		Flli Olivari	Genoa
303	I	Belvedere L. W 3m Sr	Italian	258	108·6	2?·6	15·3	Varazze	1875	G.Aschieri	[illegible]	P.Lavarello	Genoa
304		Belvidere W Sr	British	100	83·4	24·?	9·3	Parrsboro,N.S.	1882	W.R.Huntley		J.H.Newcombe	Parrsboro,N.S.
305	V	Belvidere W Bk	British	702	157·0	34·5	20·0	Windsor,N.S.	1876	S.Dimock	[illegible]	S.Dimock	Windsor,N.S.
306		Belvoir Castle W Sr	British	106	65·5	19·6	11·5	Wales	1801			J.Cochrane	Maryport
307	❋	Ben Aigen W Sr	British	98	84·9	21·4	10·0	Garmouth	1862	J.Geddie		J.Ross	Inverness
308	❋D	Ben Avon Stl Bk	British	[illegible]	[illegible]	[illegible]	[illegible]	Dumbarton	1885	Birrell, Stenhouse&Co.		Watson Bros.	Glasgow
309		Ben Bolt W Sr	British	[illegible]	81	[illegible]	8·0	St.Georges,N.B.	1863			S.Drake	S.Andrews,N.B.
310	❋	Ben Cruachan I S	British	[illegible]	[illegible]	[illegible]	21·7	Glasgow	1873	BarclayCurle&C		Watson Bros.	Glasgow
311	❋	Ben Douran I S	British	[illegible]	[illegible]	[illegible]	[illegible]	PortGlasgow	1881	H.Murray&Co.		Watson Bros.	Glasgow
312	❋	Ben Lawers I Bk	British	[illegible]	[illegible]	[illegible]	18·0	PortGlasgow	1878	Russell&Co.		H.Gourlay & Co.	Glasgow
313	❋D	Ben Macdhui I Bk	British	[illegible]	[illegible]	[illegible]	21·[illegible]	Glasgow	1877	BarclayCurle&C		Watson Bros.	Glasgow
314	❋	Ben More I Bk	British	536	166·5	28·1	17·[illegible]	Sunderland	1868	Iliff&Mounsey		H.Gourlay & Co.	Glasgow
315	❋	Ben More W Sr	British	61	71·5	19·5	8·8	Irvine	1876	Irvine S.B.Co.		G.Burns	London
316	❋	Ben Nevis I S	British	[illegible]	[illegible]	[illegible]	21·0	Glasgow	1868	BarclayCurle&C		Watson Bros.	Glasgow
317	❋	Ben Voirlich I S	British	1474	[illegible]	37·1	21·8	Glasgow	1873	BarclayCurle&C		Watson Bros.	Glasgow
318	❋D	Benan I S	British	1338	212·6	36·8	22·2	Glasgow	1875	BarclayCurle&C		W.Thomson & Co.	Leith
319	❋	Benares I 3m Bk	British	1646	[illegible]	[illegible]	[illegible]	PortGlasgow	1877	R.Murray&Co.		Watson Bros.	Glasgow
320	❋	Bencleuch I S	British	1274	216·9	37·2	22·0	Greenock	1875	R.Steele&Co.		W.Thomson & Co.	Alloa
321		Bender-Ereklee W Bg	Turkish	[illegible]	..	..	..	Heraclea	1870			Bender-Ereklelee Abdul Kadir & Bros.	Heraclea
322	I	Benedetto Cairoli W Bk	Italian	517	139·7	31·8	21·6	Lavagna	1869	L.Carbone	[illegible]	Luigi Bianchi fu G.	Genoa
323	V	Benedetto W Bk	Italian	376	118·1	27·8	17·8	Alimuri	1865	M.Mauro	[illegible]	Cafiero & Salinas	Castell. di Stab.
324		Benefactor W 3m Sr	American	544	151·2	34·0	18·4	NewYork	1857			F.H.Chappell	NewYork
325	A	Benefattore W Bk	Austrian	651	146·0	31·7	21·0	Fiume	1875	G.Brazzoduro	[illegible]	A.E.Descovich	Fiume
326	❋	Bengairn I Bk	British	590	172·2	27·9	17·4	Liverpool	1868	T.Royden&Son		A.Rae	Liverpool
327	❋	Bengal I S	British	1296	212·8	37·2	22·5	Liverpool	1868	G.R.Clover&Co.		E.F.Powell & Co.	Liverpool
328	V	Bengal *Westman* W Bk	Swedish	722	158·8	31·1	18·6	Gefle	1874	O.A.Brodin	[illegible]	J.Rettig	Gefle
329	V	Bengale *Scott* W Bk	French	544	111·0	30·0	16·6	St.Malo	1873	Gautier	[illegible]	A.Postel & Fils	Cherbourg
330	❋D	Bengalen W Bg	Dutch	274	106·9	26·1	14·6	Rotterdam	1854	C.Vlierboom&C.		A.W.Smit (J.P.Kremer)	Loppersum
331	❋	Bengollyun I S	British	1139	201·7	36·2	22·7	W.Hartlepool	1862	PileSpence&Co.		Foley&Co.	London
332	R	Bengt *Johansson* (ex Alert) W Sr	Swedish	156	88·6	21·3	11·0	Elmshorn	1864	Thormahlen	[illegible]	N.A.Johansson	Stockholm
333	NV	Benguela *Fuglesang* W Bk	Norweg'n	692	163·3	32·7	18·6	Quebec	1876	Charland,jr.	[illegible]	R.Ingvardsen	Flekkefjord
334	❋	Benicia *Taylor 79-83* I S	British	1810	261·0	39·3	24·1	Whitehaven	1883	Whitehaven, S.B.Co.Ld.		Lowden,Edgar&Co.	Liverpool
335		Benita W Bn	British	121	82·3	20·2	12·0	Harrington	1862	Williamson		T.MacIver	Whitehaven
336		Benjamin W Sr	Russian	216	110·2	25·8	12·4	Haynasch	1882	J.Grass		G.Ohsoling	Haynasch
337	GR	Benjamin W Bg	Swedish	298	108·6	25·2	16·2	St.John,N.B.	18 63/81	Anderson	[illegible]	J.E.Olsson	Gothenburg
338		Benjamin Adams W S	American	1479	184·0	36·8	33·2	Bath,Me.	1852			F.J.Chase	NewYork
339	R	Benjamin B. Church W 3m Sr	American	488	142·8	35·2	15·1	Bath,Me.	1873	Goss&Sawyer		Hart&Aiken	N.Bedford,Mass
340	N	Benjamin Bangs W S	Norweg'n	1118	182·0	36·4	23·8	Chelsea,Mass.	1860	J.Taylor	[illegible]	Akt.Selsk. "Minerva"	Christiania
341	R	Benjamin C. Cromwell W 3m Sr	American	585	161·4	35·4	16·4	Deering,Me.	1883	G.Russell		H.M.Sargent	Portland,Me.
342	R	Benjamin C. Terry W 3m Sr	American	247	125·5	31·0	9·9	Keyport,N.J.	1865			F.Davey	NewYork
343		Benjamin Carver W Bg	American	148	92·3	25·3	9·3	Searsport,Me.	1846			H.D.Gray	Bucksport,Me.
344	R	Benjamin Courtney W 3m Sr	American	476	140·7	32·3	16·0	E.Boston	1874	A.Gove		J.H.Conant & Co.	Boston,Mass.
345		Benjamin Dickerman W Bk	American	412	134·0	35·2	12·2	Duxb'ry,Mass.	1875	Keene	[illegible]	B.Dickerman	Boston,Mass.
346	RV	Benjamin F. Hunt, Jr. W Bk	American	1131	187·0	38·2	25·[illegible]	Newburyport, Mass	1881	G.E.Currier	[illegible]	W.A.Johnson	Newbryprt,Mass.
347		Benjamin F. Lee W 3m Sr	American	375	127·0	32·4	15·2	Leesburg,N.J.	1878	J.Russell		G.S.Marts	Bridgeton,N.J.
348	RV	Benjamin F. Packard W S	American	2026	244·2	43·3	26·7	Bath,Me.	1883	Goss,Sawyer& Packard	[illegible]	A.Sewall&Co.	Bath,Me.
349	R	Benjamin F. Poole W 3m Sr	American	1089	202·4	40·2	19·9	Bath,Me.	1886	W.Rogers		J.F. & C.A.Davis	Providence
350													

= Lloyd's Register; D = Nederlandsche Vereeniging; G = Germanischer Lloyd; I = Registro Italiano; R = American Record; = Veritas Austro-Ungarico; H = Veritas Ellenico; N = Norske Veritas; V = Bureau Veritas.

Dans la 4me Colonne, C = Composite (membrure en fer, bordé en bois); I = Fer; Stl = Acier; W = Bois.

Les Dates du doublage en métal (Col. 11) ressortent des livres dans lesquels les navires sont respectivement classés ou des rapports des Inspecteurs du Lloyd's Register.

Pour plus amples renseignements sur les navires classés par le Lloyd's Register, voir le Lloyd's Register Book et le Lloyd's Yacht Register.

Official No.	No. in Book	Registry in which classed	Ship's Name. Material and Rig. Master, and Dates of Appointment to present Owner's Service and to Ship. Ship's Late Name.	Flag	Net Tonnage	Length	Breadth	Depth	Built: Where	Built: When	Built: By whom	Date of Metalling	Owners or Managers
[illegible]	351	R	Benjamin Fabens *Gordon 84-86* W3mSr	American	687	153·5	35·0	18·4	East Boston	1882	J.M.Brooks	[illegible]	C.E.&B.H.Fabens
[illegible]	352		Benjamin Gartside *Sharp* W3mSr	American	326	130·9	29·2	13·7	Chester, Pa	1869			F.van Gilder & Co.
[illegible]	353	R	Benjamin Hale *Hale* W3mSr	American	567	162·0	35·7	16·2	Newburyport, [Mass.	1882	Atkinson & [Fillmore		J.Hall
	354	R	Benjamin Mollén *Månsson [67-78* W Bg	Swedish	213	102·3	23·1	12·4	Oroust	1878	A.Samuelsson		C.J.Odman
[illegible]	355	RV	Benjamin Sewall *Oliver* W S	American	1369	202·0	38·9	24·1	Brunswick, Me	1874	Pennell	[illegible]	S.J.Sewall
[illegible]	356		Benjamin T. Biggs *Webb* W3mSr	American	156	92·0	28·0	8·0	Milton, Del.	1870	W.F.Lamb		Scammell Bros.
[illegible]	357	✻	Benlarig *Flage 88-88* I S	British	1692	260·2	38·2	23·3	Dumbarton	1887	Birrell, Sten-[house & Co.		Watson Bros.
[illegible]	358	✻	Benmore *Jenkins 75-80* I S	British	1460	242·0	39·2	23·6	Port Glasgow	1870	J.Reid & Co.		Nicholson & M'Gill
	359	V	Bento de Freitas *Da Silva Pereira* W3mSr	Portug'se	253	121·3	24·3	10·8	Villa do Conde	1880	J.F.da Lapa	[illegible]	A.L.Duarte
[illegible]	360	✻	Benvenue *Mobbrel 86-86* I S	British	2033	289·2	42·4	24·1	Dumbarton	1884	Birrell, Sten-[house & Co.		Watson Bros.
[illegible]	361	✻I	Beppe *Bertolotto -87* (ex Maria Repetto Figlia) W Bk	Italian	861	173·1	33·6	23·5	Lerici	1876	P.Rollo	[illegible]	F.Schiaffino fu G.
[illegible]	362	I	Beppino R. *Parodi* W Bk	Italian	574	139·0	32·1	20·2	Chiavari	1871	M.Tappeni	[illegible]	G.Repetto fu G.
[illegible]	363	✻	Berar *Newcombe 87-87* I S	British	954	188·5	32·1	20·3	Sunderland	1863	W.Pile & Co.		Foley & Co.
[illegible]	364	✻D	Berbice (WRECKED) *Ross 81-85* C S	British	717	174·0	31·5	18·3	Dumbarton	1868	A.M'Millan & Sns	[illegible]	J.Kerr & Co.
[illegible]	365	✻	Berean *Wyrill 65-69* C Bk	British	526	160·5	30·2	17·2	Sunderland	1869	W.Pile & Co.	[illegible]	T.B.Walker
	366		Bereketti Bahri *Mehemed Reşad* W Bg	Turkish	167	..	..	..	Greece	1885			Anastas Oglou Andrea
	367		Bereketti Dadjah *Halil* W Bg	Turkish	236	..	..	..	Russia	1874			Arslan Zadeh Ismael [illegible]
	368		Bereketti Hhair W Bg	Turkish	263	..	..	..	Russia	1878			Inebolilee Kamahendi [illegible] Hadji Annetas, Sons, & Co.
	369		Bereketti Hhair W Sr	Turkish	225	..	..	..	Heraclea	1876			Pekmezjilee Kiryani Zade Ahmed Effendi
	370		Bereketti Hhair W Bg	Turkish	120	..	..	..	Greece	1869			Sakyzlee Kokoni Costanti Yanni & Co.
[illegible]	371	G	Berent *Blankenburg 83-88* W Bkn	Russian	301	122·0	26·5	14·4	Ueckermünde	1864	L.Wittenberg		J.Grell
	372	N	Bergenseren *Rolfson* W Bk	Norweg'n	528	148·2	30·9	17·6	Bergen	1865	Brunchorst & [Dekke	[illegible]	M.Pedersen
	373		Berguzar *Ali* W Bg	Turkish	162	..	..	..	Jeddah	1884			Woan Zadeh Mehemed Effendi
[illegible]	374		Berkeley Castle (ex Prince Consort) *Snyder* W Bk	British	1197	195·6	36·8	22·4	Quebec	1859	H.Dinning & Co.		T.Emerson
[illegible]	375	✻	Berkshire *Marshall 58-86* I Bk	British	1472	237·2	37·4	23·0	Glasgow	1867	Barclay, Curle & [Co.		G.Marshall & Sons
[illegible]	376	RV	Berlin *Whitemore* W S	American	1553	222·5	40·0	24·6	Phippsb'rg Me	1882	G.V.Minott	[illegible]	C.V.Minott
	377	G	Berlin *Hübner 85-* (ex Friedrich Wilhelm) W Bk	German	316	101·3	27·9	16·2	Elbing	1840	M.Mitzlaff		D.W.Hübner
[illegible]	378	IV	Berna *Chiozza* W Bk	Italian	975	181·0	33·7	23·1	Sestri Ponente	1876	F.Pezzani	[illegible]	U.Geisser
	379	V	Berna *Collberg 82* W Bk	Swedish	404	128·9	27·7	15·6	Gefle	1876	Bomhus Warf. [Aktie Bolag	[illegible]	V.T.Engvall
	380	DNV	Bernadotte *Torstensen -83* (ex Breadalbane) W S	Norweg'n	1427	207·2	39·1	24·0	St. John, N.B.	1872	Nevins Fraser & C	[illegible]	W.Wilhelmsen
	381	N	Bernadotte *Pettersen* W Bk	Norweg'n	371	110·0	30·6	17·7	Grimstad	1856	J.Vig		M.Smith Petersen
	382	V	Bernard *Korphage 67-* W Bn	German	163	88·9	21·3	11·8	Papenburg	1867	F.W.Meyer		F.W.Meyer Ww.
	383	G	Bernardina *Schoon -86* W Bg	German	196	103·4	23·8	12·0	Papenburg	1857	F.Freericks		B.Schoon
	384	V	Bernardino Bravo *Burach 83-87* W S	French	1022	191·2	36·1	21·5	Bordeaux	1874	Chaigneau frères	[illegible]	Ant.Dom. Bordes & fils
[illegible]	385	D	Bernardo (WRECKED) *Dapelo* W Bk	Italian	701	152·8	30·5	21·0	Varazze	1876	A.Bollo	[illegible]	G. B. Dezregori fu Bernardo
[illegible]	386	I	Bernardo Chiozza *Bellagalo 83-86* W Bk	Italian	941	165·4	34·0	23·4	Loano	1875	B.Anuco	[illegible]	F[lli] Chiozza
	387	D	Bernardus (ex Marinus Mas) (BROKEN UP) *Douwes* W Sr	Dutch	138	87·9	20·5	10·9	Hoogezand	1855	J.A.Hooites		J.Douwes
	388	DV	Berndina (ex L.Olivier) *Brussaart [-81* W Bn	Dutch	185	86·6	25·0	12·8	Lorient	1866		[illegible]	J.H.Houtman
	389		Bernhard (ex Johannes Anthonius) *Diekow 82-86* W Bk	German	658	137·8	34·2	21·0	Capelle	1856	Barkhuizen		F.Diekow
	390	G	Bernhard *Brahm 70-* W Bg	German	274	91·1	26·9	15·5	Barth	1857	J.Holzerland		E.W.Ohrloff Sohn
	391	V	Bernhard *Laning 64-66* W Bg	German	211	104·3	25·1	12·5	Papenburg	1867	F.W.Meyer	[illegible]	B.Breymann
224	392		Bernhard *Wihtol* W Sr	Russian	186	95·7	27·0	11·8	Ratern	1875	P.Andersohn		Andreitson, Ohsol & Sehmore
	393		Bernhard Carl [illegible] *Neumann 75* W Bk	German	431	136·6	30·2	16·6	Brake	1864	J.Oltman's Wwe	[illegible]	E.Burchard
	394	V	Bernhard Johann *Schaaf 82-* W Sr	German	130	85·6	19·4	11·9	Papenburg	1882	B.Tholen	[illegible]	Bernh.Tholen Ww.
297	395	D	Bernhard & Auguste *Stern -80* W3mSr	Russian	316	128·0	29·0	14·0	Riga	1878	A.Ferle		J.Dombrowsky
	396		Bernhardina *Simonsson* W Sr	Swedish	114	70·9	21·4	11·2	Oroust	1869			M.Jonsson
	397	G	Bernhardine *Bruarow -87* W Sr	German	123	83·7	21·7	11·0	Stralsund	1867	J.P.Juhl		H.L.Juhl
[illegible]	398		Bernicia *Curry 83-84* W Bn	British	111	75·0	22·0	11·1	Sunderland	1848	R.Thompson		R.Curry
	399		Bernigo (WRECKED) *Morlon 84-83* W Bg	French	155	82·0	21·3	14·1	Mesnes	1857			Prado frères
[illegible]	400	I	Bersagliere *Barbagelata* W Bk	Italian	1049	197·9	33·9	24·1	Foce	1877	S.Debarbieri	[illegible]	Frat. Pastorino

✻=Lloyd's Register; D=Nederlandsche Vereeniging; G=Germanischer Lloyd; I=Registro Italiano; R=Ame[illegible]
A=Veritas Austro-Ungarico; H=Veritas Ellenico; N=Norske Veritas; V=Bure[illegible]

In the 4th Column, C=Composite (iron frame, wood planked); I=Iron; Stl=Steel; W=Wood.

The Dates of Metalling (**Col. 11**) are based on the records of the Books in which Vessels are respectively classed, or on the Reports of the Surveyors of [illegible]

Further particulars of Vessels classed by Lloyd's Register will be found in Lloyd's Register Book and Lloyd's Yacht Register.

No. du livre	Cote	Nom. Matériaux et Gréement. Capitaine et dates d'entrée au service des Armateurs actuels, à bord du navire. Ancien Nom.	Pavillon.	Tonnage	Longueur	Largeur	Creux	Où.	Quand.	Par qui.	Doublage	Armateurs ou Directeurs-Gérants.	Port d'Armement.
401		Berta *Andersson 82-83* WSr	[illegible]	108	96·5	21·4	11·6	Bjorhöfda	1877	C. Andersson		J.E.Olsson	Sjorhöfda, [illegible]
402		Berta *Claussen* WSr	[illegible]	118	93·2	21·7	9·5	Calmar	1851			O.Johansson	QvilleBohuslän
403		Berta Maria *Olsson* WSr	[illegible]	304	114·8	25·0	12·4	Liverpool,N.S	1864			G.Olsson	Mollösund
404		Bertha *Crickett* WBk	[illegible]	918	[illegible]	[illegible]	[illegible]	[illegible]	1865	Hy.Hutchings		Benj.Webster	Portland.Me.
405		Bertha *Anderson*	[illegible]	[illegible]	..	..	..	[illegible]	1888	M.Turner			
406	R	Bertha *Cleveland* WBk	American	188	[illegible]	25·7	11·[illegible]	[illegible] Mass.	1878	S.Dammon	[illegible]	J.F.Tucker & Co.	NewBedford, Mass.
407	V	Bertha (ex [illegible]) *Jensen 85-86* WBk	[illegible]	597	134·3	35·9	18·5	Elsinore	1872	J.H.Love&Son	[illegible]	H.S.Hansen	Copenhagen
408		Bertha *Hansen* WBg	[illegible]	290	[illegible]	25·6	12·7	Ueckermünde	1857	L.Wittenberg		H.C.Hansen	Svendborg
409	DV	Bertha (ex [illegible]) *Pahnke* WBk	[illegible]	[illegible]	154·2	34·8	20·4	Schiedam	1867	Gips	[illegible]	Aug.Köpcke	Rotterdam
410	D	Bertha (ex Nina [illegible]) *Brouwer 59-68* WBk	[illegible]	651	155·7	34·1	21·3	Elshout	1857	PotBros.	[illegible]	Canne & Balwe	Amsterdam
411	D	Bertha (ex Jan [illegible]) *Siefken* WBk	[illegible]	[illegible]	139·8	31·4	18·3	Slikkerveer	1853	J.Smit		J.Meijjes Jer Zn	Amsterdam
412		Bertha (ex [illegible]) *Mousorie* WBk	[illegible]	458	125·9	30·1	19·7	NewYork	1855			VanEngers, Roelofs & Cº	Schiedam
413	V	Bertha *Ahrens* WBk	[illegible]	887	161·9	33·1	20·4	Brake	1883	F.F.Nicolai	[illegible]	U.C.Koch	Brake
414	GV	Bertha *Pritzwerd* WBk	[illegible]	583	155·7	31·5	18·7	Apenrade	1874	N.Jacobsen	Y 10.88	A.Burchard	Rostock
415	V	Bertha *Kruse* WBk	[illegible]	435	123·1	29·4	18·0	Barth	1866	A.Witte		J.N.Rodbertus	Barth
416	V	Bertha *Schult 71-72* WBk	[illegible]	402	118·3	28·6	17·4	Greifswald	1862	J.W.Sadelkow		J.J.Wallis & Sohn	Barth
417		Bertha *Kühne 85-88* WBn	German	221	99·7	23·6	12·0	Brake	1867	L.F.Paulsen	[illegible]	A.Tobias	Brake
418	V	Bertha *Nagel* WSr	German	153	85·9	20·7	11·3	Stralsund	1864	J.P.Jahl		Franz Böttcher	Stralsund
419	N	Bertha (ex Bertha [illegible]) *[illegible] 83-83* WBk	Norweg'n	597	156·7	32·1	17·8	Tynemth.N.B.	1867	J.Lovitt	[illegible]	C.Langaard	Christiania
420	N	Bertha *Hanevold* WBk	Norweg'n	439	137·2	27·2	17·0	Kingst'wn.N.B.	1864			O.Svendsen	Christiania
421		Bertha ([illegible]) *Alm* WBk	Norweg'n	396	138·6	28·1	17·2	Sunderland	1867	Watson	[illegible]	A.Alm	Tönsberg
422	N	Bertha (ex [illegible]) *Jacobsen* WBg	Norweg'n	345	122·6	29·5	16·5	Venice	1873		Y 11.85	N.Olsen	Arendal
423	NV	Bertha (ex [illegible]) *Steen 86-88* W3mSr	Norweg'n	32[illegible]	131·5	24·9	11·7	Alloa	1868	Adamson	[illegible]	Act.Selsk.Achilles (J.Olsen)	Kragerö
424	GN	Bertha *Fraatz 84* WBg	Norweg'n	299	111·7	25·2	11·5	Altona	1865	E.Dreyer	[illegible]	H.C.Hansen	Sandefjord
425	N	Bertha *Olsen 81-* W3mSr	Norweg'n	17[illegible]	91·8	23·6	12·0	Svartvik	18 53/70			P.Nielsen Föyen	Tönsberg
426		Bertha (ex Lorenzo) *de Costa* ([illegible]) WBg	Portug'se	267	107·6	23·6	13·4	Lübeck	1862			J.H.Andresen	Oporto
427		Bertha *Eriksson* WBk	Russian	317	131·1	30·0	14·5	Dragsfjerd	1875	Fredriksson		K.J.Blomberg	Dragsfjerd
428	G	Bertha Alwina *Kalning 79-83* W3mSr	Russian	249	119·0	27·4	13·7	Orrenleff	1883	J.J.Mangus		G.Kalning	Haynasch
429		Bertha Anderson ([illegible]) *Thomas* WBk	British	544	139·2	30·9	17·3	DorchesterNB	1877	W.Hickman	Y 10.86	Wm.Hickman	Dorchester.N.B.
430	R	Bertha D. Nickerson *Nickerson 79-83* WSr	American	136	90·6	24·6	9·7	BoothBay.Me.	1883	Hodgdon		S.Nickerson & Sons	BoothBay.Me.
431		Bertha Dean *Trippe 86* W3mSr	American	71[illegible]	160·1	35·0	15·0	Bath,Me.	1882	Packard & Haggett		W.B.Church	Fall River, Mass.
432		Bertha Dolbeer *Fairchild 81* W3mSr	American	231	125·0	31·5	9·6	Fairhaven	1881	H.D.Bendixen		G.D.Gray	San Francisco
433	R	Bertha E. Glover *Dyer 88* WSr	American	112	77·9	24·3	7·6	Rockland,Me.	1882			Chase Talbot & Co.	Rockland.Me.
434		Bertha F. Walker *Wengate* W3mSr	American	653	154·5	35·8	17·4	Bath,Me.	1879	Goss&Sawyer		S.H.Walker	FallRiver.Mass.
435	R	Bertha Warner *Lathwaite 83* W3mSr	American	42[illegible]	140·2	30·1	16·1	E.Deering,Me.	1883	G.Russell		J.S.Winslow & Co.	Portland.Me.
436	V	Berthe (ex Stord) *Rebel* WBn	French	174	99·7	22·2	11·7	Nyborg	1874	P.Peterson	I 11.87	A.Postel & ses fils	Havre
437	V	BertheCollet (ex Kingswell) *Pasquellie 87* WBk	French	351	124·3	27·2	14·9	P.E.Island	1878	Chiverie	[illegible]	J.L. & P. Collet-Taverne fils	Dunkirk
438	V	Berthe Emile *[illegible]* WSr	French	105	77·8	19·0	9·9	St.Malo	1870	Hervot	Z 11.87	E.Preuveille	St.Malo
439	N	Berthe Rod *Hansen* WBk	Norweg'n	293	135·8	27·3	12·6	Tönsberg	1869	J.Rød'sSønner	I 10.8[illegible]	O.Rød	Tönsberg
440		Berthe & Jeanne *Letrelaudic 77-8* WBn	French	222	98·1	22·3	13·6	P.E.Island	1868			Letrehudic	Nantes
441		Berthe & Lucy *Hutson* WBn	French	155	104·1	22·6	9·7	Calmar	1867	Roos		A.Cabissol & Co.	Marseilles
442	G	Berthel *Hansen 87* WBn	Danish	186	106·3	21·9	12·4	Brake	1859	J.F.Nicolai		H.C.Christensens Enke	Marstal
443	V	Berthold *Koert 85* WBg	Russian	27[illegible]	113·5	26·6	12·9	Stettin	1864	E.Zieske		Laiwing	Riga
444	✻	Bertie *Daniel 75-79* WBk	British	45[illegible]	141·6	27·8	18·3	Whitehaven	1863	Kennedy	[illegible]	D.Daniel	Aberystwith
445		Bertie IFt	British	107	81·6	19·3	8·5	Northwich	1869	J.Thompson&S.		J.Bradburn	Liverpool
446	✻	Bertie WK	British	99	84·0	21·5	10·8	Brixham	1886	S.Dewdney&Sns		T.White	Hull
447	V	Bertie Biglow *[illegible]* WBk	British	1142	192·2	37·6	22·5	BeaverRiverNS	1876	G.H.Jenkins	[illegible]	Wm.D.Lovitt	Yarmouth.N.S.
448		Bertie Minor *Barens* W3mSr	American	253	130·3	32·5	9·8	Fair Hav'n,Cal	1884	H.Bendixen		F.B.Ravens	Eureka.Cal.
449		Bertie&Maud *[illegible]* W3mSr	American	182	138·5	23·7	7·5	Camden.N.J.	1887	S.J.Tilton		W.H.Phillips	Bridgeton
450													

[illegible]=Lloyd's Register; D=Nederlandsche Vereeniging; G=Germanischer Lloyd; I=Registro Italiano; R=American Record;
[illegible]=Veritas Austro-Ungarico; H=Veritas Ellenico; N=Norske Veritas; V=Bureau Veritas.

Dans la 4me Colonne, C=Composite (membrure en fer, bordé en bois); I=Fer; Stl=Acier; W=Bois.

[illegible] Dates du doublage en métal (Col. 11) ressortent des livres dans lesquels les navires sont respectivement classés ou des rapports des Inspecteurs du Lloyd's Register.
Pour plus amples renseignements sur les navires classés par le Lloyd's Register, voir le Lloyd's Register Book et le Lloyd's Yacht Register.

1 Official No.	2 No. in Book	3 Registry in which classed	4 Ship's Name. Material and Rig. Master, and Dates of Appointment to present Owner's Service and to Ship. Ship's Late Name.	5 Flag.	6 Net Tonnage	7 Dimensions. Length	Breadth	Depth	8 Built. Where.	9 When	10 By whom.	11 Date of Metalling.	12 Owners or Managers
[illegible]	451	R	Bertram L. Townsend *Powell* W3mSr	American	609	153·5	35·1	16·0	Milford	1878	W.A.Scribner		S.L.Tunnell
	452	N	Bertrand *Bache -88* WBk	Norweg'n	557	148·5	32·1	17·4	Porsgrund	1880	LarsPedersen	[illegible]	L.H.Bache
84278	453	✳	Berwickshire *Furneaux 18-86* IBk	British	901	200·5	32·8	19·6	Greenock	1880	Russell&Co.		T.Law & Co.
80822	454	R	Bess *Smith -87* WSr	British	125	93·4	24·8	9·3	Meteghan,N.S.	1881	J.Cosman	[illegible]	James Cosman
[illegible]	455	✳	Bess Mitchell *Bennett 87-87* I Sr	British	99	82·8	21·4	10·0	Llanelly	1871	C.W.Nevill&Co.		R.Mitchell & Co.
	456	✳V	Bessel *Visser 86-86* (ex Ville d'Anvers) I Bk	German	460	153·0	27·6	16·0	South Shields	1869	Readhead, [Softley&Co.		C.H.H.Winters
93157	457	✳	Bessfield *Napton 86-86* I Bk	British	1292	239·7	36·2	21·1	Belfast	1886	Workman,Clark [& Co.		W.Porter & Sons
72613	458	G	Bessie (WRECKED) *Carsters -87* WBk	British	556	149·7	30·8	17·7	Moosebrook,N.S.	1876	A.T.Dalrymple	[illegible]	W.E.Meyer
[illegible]	459		Bessie *Rickets 82-86* WBn	British	228	116·4	25·7	12·8	P.E.Island	1873	Egan		John Bateman
67296	460	✳	Bessie *Clemens 83-83* W3mSr	British	177	102·0	23·5	12·9	Kingsbridge	1871	Date		J.Ennor
14893	461		Bessie (WRECKED) *Gendall 81-87* WSr	British	124	99·5	22·0	12·0	Polruan	1863	Butson		B.Spray
72553	462	✳	Bessie *Banfield 70-75* WSr	British	116	89·6	22·2	11·4	Falmouth	1875	Stephens		W.Lidgey & Co.
78868	463	✳	Bessie *Richards* WSr	British	54	64·3	18·4	8·7	Malpas	1879	Scoble		Scoble & Davies
	464		Bessie WBk	Brazilian	547	..	..	..					J.D.G.Junior
58186	435		Bessie Arnold WSr	British	114	86·8	22·2	11·4	Connah'sQu'y	1872	Ferguson		W.Morgan
83146	466		Bessie Barwick WBk	British	274	138·9	26·2	10·8	St.Catharines	1866	L.Shickluna		R.O.McKay
69706	467	✳	Bessie Bell (ABANDONED) WSr	British	99	93·0	22·6	10·8	Barnstaple	1874	Westacott		E.Hughes
56714	468		Bessie Belle WSr	British	128	92·2	23·1	11·1	Portreath	1867	Massey		J.L.Tregenza
3298	469	R	Bessie Brown *[illegible] -88* W3mSr	American	247	123·8	32·3	9·5	Bethel,Del.	1884	R.McIlvane		J.T.Vaughan
3134	470		Bessie C. Beach *Crosby* W3mSr	American	334	128·0	30·4	14·0	N. Haven,Con.	1880	H. H. Hanscomb		F.W.Benedict
69707	471	✳	Bessie Dodd *Jago 84-84* W3mSr	British	160	107·0	23·9	12·6	Barnstaple	1874	Westacott	[illegible]	W.Jago
90503	472		Bessie E. Crane WSr	British	184	105·6	27·6	10·7	Parrboro',N.S.	1885	N.B.Morris		W.Brown
75022	473	R	Bessie Gracie *Lovegrove* WBn	British	190	100·3	26·2	11·0	Shelburne,N.S.	1877	W.T.Kelly& Son	[illegible]	D.Cronan
3256	474	R	Bessie H. Rose *Adams* W3mSr	American	626	155·0	34·0	17·7	Wiscasset,Me.	1883	Haynes		J.M.Haynes
	475	R	Bessie Hamilton (ex Johanna) *Olsen -87* WBk	Norweg'n	941	173·2	32·8	22·9	Trieste	1881	G.Spadon	[illegible]	Edvard Olsen
73156	476	✳D	Bessie Jose *May* I Bk	British	595	170·6	29·0	17·2	Dumbarton	1876	Birrell,Sten- [house&Co.		J.Jose
77839	477	R	Bessie Louisa *Smith* WSr	British	138	90·0	24·0	10·0	BarringtonNS	1879		[illegible]	WilliamT.Smith
75894	478	V	Bessie Markham *Faulkner* WBk	British	778	169·4	34·3	19·4	Kouchibou- [guac,N.B.	1878	J.Tennant	[illegible]	G.K.McLeod
71049	479	V	Bessie May *Chambers -88* WBn	British	342	119·1	29·3	12·9	DorchesterNB	1875	E.Chambers	[illegible]	J.Hickman
60894	480		Bessie Morris *Thomas -84* WBk	British	486	154·2	29·1	15·5	Devonport	1860	H M Government	[illegible]	S.N.Castle
50533	481	R	Bessie Parker (WRECKED) *Baker -87* WBk	British	671	149·0	33·5	19·6	St.Martin's,NB	1864	Parker	[illegible]	Troop & Son
28155	482		Bessie Rowe WBn	British	107	83·3	21·2	11·0	Gloucester	1862	Pickersgill		J.Jones
74531	483		Bessie Whineray WBn	British	153	93·0	23·2	12·5	Ulverston	1865	Wilson		D.T.Sharper
3187	484	R	Bessie Whiting *Dayton* W3mSr	American	532	119·0	32·2	16·8	Pt.JeffersnNY	1882	J.R.Mather		G.D.Dayton
81029	485	✳	Bessie Williams *Crocker* WK	British	98	83·8	20·2	9·6	Plymouth	1879	Hill		C.F.Williams
35044	486		Bessie Young *Duff* WBkn	British	289	123·1	27·3	15·2	St.John	1856	S.Rowan		J.Monkhouse
	487		Bestsabé *Reos* WSr	Spanish	137	81·0	24·2	11·8	Villajoyosa	1876			José Reos
51109	488	✳	Beta *Beynon 86-86* WBk	British	466	142·6	28·8	17·8	Bristol	1864	Hill	[illegible]	G.Shepherd
68556	489	✳	Beta *Mabbot 86-86* WSr	British	97	81·0	20·0	9·9	Portsmouth	1873	Read		G.Bazeley & Son
91391	490	✳	Beta I Sk	British	90	79·4	20·3	10·1	Beverley	1884	Cochrane, Ha- [milton&Cooper		B.W.Hellyer
2886	491		Beta *Clark* WSr	American	144	98·0	26·2	8·5	E.Machias,Me.	1868			L.M.Collerth
	492	N	Beta (ex Ionnevkristian, ex Sir John Lawrence) *Wille* WBk	Norweg'n	845	187·8	31·8	20·4	Aberdeen	1864	J.Duthie,Sons & Co.	[illegible]	Winge & Co.
	493	DV	Beta *Olsen* WBg	Norweg'n	253	99·8	24·9	10·7	Brake	1868	F.Oltmann's [Wwe.	[illegible]	L.J.Olsen
	494		Betel *Maveglia* WBn		295	112·5	24·2	16.0	Lussinpiccolo	1874	G.S.Cattarinich		
	495	D	Betha (ex Johann Friedrich) *Osman Bazeo Edan* WBg	Dutch	257	111·5	23·9	12·6	Fünthausen	1869	J.F.Strenge	[illegible]	Said AbdulbinAbdul
	496	DV	Betha *Krageuborg -88* WSr	Dutch	113	82·3	20·7	10·3	Pekela	1867	DeWyk		W.J.Kuiper
197	497	N	Bethania *Johansen -85* WSr	Norweg'n	123	88·1	22·2	10·9	Stavanger	1875			Jörgensen & Roe
	498		Bethleem WSr	Greek	407	124·3	29·1	..	Syra	1877			D.Vafiadakis
48	499		Bethleem WBn	Greek	230	88·9	26·2	..	Skiathos	1883	G.G.Mytilinaios		C.Riggs
	500		Bethlehem *Knuk -88* WS.	Russian	255	103·0	23·0	12·0	Gutmanns-belt	1876	J.Martinsohn		J.Martinsohn

✳=Lloyd's Register; D=Nederlandsche Vereeniging; G=Germanischer Lloyd; I=Registro Italiano; R=Amer[ican]
A=Veritas Austro-Ungarico; H=Veritas Ellenico; N=Norske Veritas; V=Bure[au]

In the 4th Column, C=Composite (iron frame, wood planked); I=Iron; Stl=Steel; W=Wood.

The Dates of Metalling (**Col. 11**) are based on the records of the Books in which Vessels are respectively classed, or on the Reports of the Surveyors of Ll[oyd's]

Further particulars of Vessels classed by Lloyd's Register will be found in Lloyd's Register Book and Lloyd's Yacht Register.

No.	Cote	Nom, Capitaine, Ancien Nom	Matériaux et Gréement	Pavillon	Tonnage	Longueur	Largeur	Creux	Construit où	Quand	Par qui	Doublage	Armateurs ou Directeurs-Gérants	Port d'Armement
501		Bethlehem	W Sr	[illegible]	[illegible]	74·	[illegible]	[illegible]	[illegible]	1875			P. Akkermann	Haynasch
502	✱	Betsey Bevan	I Bn	[illegible]	[illegible]	125·3	23·9	14·4	Llanelly	1865	C.W.Nevill&Co.		E. Samuel	Llanelly
503	✱	Betsey Roberts	W Sr	[illegible]	91	74·0	20·2	10·1	Conway	1864	Roberts		E.Williams & Co.	Carnarvon
504		Better Luck	W Sr	[illegible]	[illegible]	83·9	18·6	9·6	Singapore	1876	H.A.Choug		De Bordes & Co.	Banda
505	I	Bettina Bonifacio	W Bn	[illegible]	[illegible]	88·0	22·7	10·4	TorredelGreco	1883	G.Bonifacio	Y 1.88	G. Albanese	Naples
506		Betty Ingebrigtsen -88 (ex [illegible])	W 3m Sr	[illegible]	372	121·9	25·1	16·5	New Glasgow	1857			W. Klitgaard	Frederikshavn
507	DG	Betty Hege 74-88 (ex [illegible])	W Bk	German	798	178·4	32·3	21·3	Harrington	1860	R.Williamson&Son	Y 1.88	E. Burchard	Rostock
508		Betty (ex [illegible])	W Bg	German	267	104·5	26·6	16·6	Durham	1865	Gibbon		A. Bendrat	Danzig
509	N	Betty Haugenstein 76-86 (ex [illegible])	W Bk	Norweg'n	678	[illegible]	[illegible]	19·6	Rotterdam	1855	J.Otto		F. Breda	Christiania
510	DN	Betty Hansen	W Bg	Norweg'n	222	87·2	24·1	15·5	Westerwik	1819		Y 11.88	E. Hansen	Fredrikstad
511		Betty Bengtsson 81-84	W Bg	Swedish	209	112·5	25·2	13·4	Hammar	1856			N. Olsson	Höganäs
512	V	Betty Pettersson	W 3m Sr	Swedish	252	111·2	24·1	11·8	Timm'rnabb'n	1874	C. Nilsson	Y 1.88	T. Pettersson	Wiken
513		Betty (CONDEMNED) Holm	W Bn	Swedish	157	95·1	22·2	10·6	Geestemünde	1862	Schau&Oltmanns		L. Nilsson	Brantevik
514		Betty Thode -85	W Sr	Swedish	120	92·2	21·7	9·6	Oscarshamn	1875	S.Wickberg		O. Wingren	Oscarshamn
515		Betty	W Sr	Swedish	102	71·8	21·4	11·3	Garnsvarf				J.O. Friberg	Granvik
516		Betty Russell Davies 70-88	W Sr	British	146	97·0	22·5	12·6	Barrow	1864	Reay		J. Fisher & Sons	Lancaster
517	N	Betzy Gude Due	W Bk	Norweg'n	450	129·3	30·3	17·1	Grimstad	1869	J.Bang	Y 1.88	J. Bang	Grimstad
518	N	Betzy & Arnold Thorsen -70 (ex Ambassador Tancred)	W Bk	Norweg'n	955	173·3	35·6	22·8	Medford, U.S.	1862	J.F.Foster	Y 1.88	S. Hansen	Christiania
519		Beulah Lohnes -83	W Sr	British	103	81·3	23·6	9·2	LaHave, N.S.	1883	S.Leary		E. Lohnes	Lunenburg, N.S.
520		Beulah Wilson	W 3m Sr	American	339	146·6	35·6	10·9	San Francisco	1882	Dickie Bros.		J. Knowland	San Francisco
521		Bhagurrutta	I Lr	British	547	220·0	32·0	9·5	Calcutta	1883	India General Stm.Nav.Co.Ld.		Irrawaddy Flotilla Co. Ld	Glasgow
522		Bhau	Stl S	British	480	220·0	30·0	8·4	Dumbarton	1866	W.Denny&Bros.		Irrawaddy Flotilla Co. Ld	Glasgow
523		Bhaulay	Stl Lr Barge	British	480	220·2	30·1	8·3	Dumbarton	1884	W.Denny&Bros.		Irrawaddy Flotilla Co. Ld	Glasgow
524	✱	Bhotan Thomas	I Bk	British	1020	214·7	33·9	20·5	Sunderland	1876	R.Thompson, jun.		E. Bates & Sons	Liverpool
525		Bhownugger	I Sr	British	285	114·9	23·0	12·6					J.E. Geary	Bombay
526	V	Biagino Revello -88	W Bk	Italian	467	135·1	31·0	19·5	Spotorno	1868	G.B.Cadenaccio	Y 4.88	B. Chiesa	Genoa
527		Biagio (ex Amaral Sul) Grignetti (ex [illegible])	W Bk	Italian	855	181·7	34·4	24·0	Savona	1875	G.Guastavino	Y 1.88	L. Degregori	Genoa
528	✱	Bianca Hughes 84-88	I S	British	1396	217·1	36·4	25·3	Liverpool	1862	H.M.Lawrence &Co.		H. Owen & Son	Liverpool
529	V	Bianca Lockhart	W Sr	British	180	94·6	28·6	10·6	Newport, N.S.	1888	T.A.Mosher	Y 10.88	T.A. Mosher	Windsor, N.S.
530	I	Bianca Garrolo -88 (ex Sincero Scerdo, ex Italia)	W Bk	Italian	517	141·0	31·2	19·6	Albenga	1870	G.B.Guastavino	Y 1.88	P. Borzani	Spezia
531		Bianca C. (ex Fetrea Donnarno)	W Bg	Italian	318	108·9	28·6	16·3	Sestri	1876	A.Briasco		L. & E. Cignoni	Leghorn
532	I	Bianca Casanova Casanova	W Bg	Italian	838	175·2	35·1	21·0	Sestri Ponente	1875	G.Casanova	Y 6.88	L. Casanova	Genoa
533	I	Bianco Salvatore Mormino -88	W Bk	Italian	269	108·9	26·3	16·3	Gaeta	1863	P.Gallinaro	Y 1.88	F[lli]. Leboffe	Gaeta
534		Biasimo Russo	W Bg	Italian	246	104·6	27·4	16·4	Procida	1866	V.Scotto La Chiama	Y 9.88	A. Mazzella	Naples
535	✱I	Bice Pave -82	W Bk	Italian	878	182·0	33·5	23·3	Prà	1877	L.Durante	Y 10.87	G. Costa	Genoa
536	V	Biche Gillard	W Bn	French	170	86·9	22·6	11·6	Nantes	1873	Alleau&Aubert	Y 11.88	L. Jacquier	Nantes
537		Bidsie & Bell Higham 84-84	W Bn	British	174	106·4	25·1	12·6	Berwick	1873	Gowan	Y 7.81	J. Fisher & Sons	Barrow
538	✱	Bidston Hill Lambert 84-86	I 4m S	British	2434	301·6	42·1	24·7	Liverpool	1886	T.Royden&Sons		Sailing Ship "Bidston Hill" Co. Ld. (W. Price & Co.)	Liverpool
539	N	Bien Christensen	W Bg	Norweg'n	192	100·3	25·5	12·1	Sandefjord	1869			A.C. Larsen	Sandefjord
540	N	Bien Emanuelsen -83	W Sr	Norweg'n	115	88·2	23·1	8·5	Lillesand	1883	N.J.Emanuelsen		O.C. Emanuelsen	Lillesand
541		Biene (CONDEMNED) Degen 86-	W Bn	German	253	103·9	24·6	12·3	Hammelwrdn.	1869	W.Hagens	Y 1.86	G.H. Lübken	Elsfleth
542	V	Bienfaiteur Hubeau -79	W Bk	French	287	115·4	25·6	11·8	Nantes	1875	T.Dubigeon&Fils	Y 2.81	A. Hubeau	St. Nazaire
543		Bientang Borneo	W Bg	Dutch	165	..	..	..					Abdul Hadathin Abdul Rasab	Passir
544		Bientang Trang	W Bk	Dutch	252	129·6	26·7	11·9					Malangkel & Daeng Manjarone	Wadjo
545	V	Bifrost Olsen 85-85	W Bkn	Swedish	309	128·2	27·5	11·4	Timmernabbn	1885	S.Ohlsson	Y 5.85	C.A. Banck & Co.	Helsingborg
546	V	Big Bonanza Andrews 85-85	W S	American	1366	210·2	40·2	24·0	Newburyport, Mass.	1875	J.Currier, jun.	Y 5.85	J. Currier, jun.	Newburypt. Mass.
547		Big River Mats	W Sr	American	158	136·0	30·3	12·3	Freeport, Wsh.	1872			A. Anderson	San Francisco
548		Bigador Jani Costandi	W Bg	Turkish	253	..	..	..	Pendreek	1869			Bigadoslee Teophilo Jani & Co.	Bigados
549	N	Biland Carlsen (ex [illegible])	W Bk	Norweg'n	488	115·8	28·8	18·3	Sunderland	1865	W.Briggs&Sons		L.B. Larsen	Farsund
550	V	Bilda (ex [illegible]) Ernst -81	W Sr	German	108	79·7	15·4	8·0	Mundsandam	1865	Drenth&Sinnige		H.J. Ernst	Oldersum

✱ = Lloyd's Register; D = Nederlandsche Vereeniging; G = Germanischer Lloyd; I = Registro Italiano; R = American Record; A = Veritas Austro-Ungarico; H = Veritas Ellenico; N = Norske Veritas; V = Bureau Veritas.

Dans la 4me Colonne, C = Composite (membrure en fer, bordé en bois); I = Fer; Stl = Acier; W = Bois.

Les Dates du doublage en métal (**Col. 11**) ressortent des livres dans lesquels les navires sont respectivement classés ou des rapports des Inspecteurs du Lloyd's Register.

Pour plus amples renseignements sur les navires classés par le Lloyd's Register, voir le Lloyd's Register Book et le Lloyd's Yacht Register.

SAILING VESSELS.

Official No.	No. in Book.	Registry in which classed.	Ship's Name. Material and Rig. Master, and Dates of Appointment to present Owner's Service and to Ship. Ship's Late Name.	Flag.	Net Tonnage.	Length.	Breadth.	Depth.	Built. Where.	When.	By whom.	Date of Metalling.	Owners or Managers.
	551		Bilgair WBk	Dutch	741	149·6	31·7	..					Mahomad bin Allo bin Nahaboedin
2675	552		Bill Stowe *Phinney* W3mSr	American	437	139·0	32·5	15·3	RockyHillCon	1872			J.A.Rumrill
2467	553	R	Billow WSr	American	135	79·9	23·0	8·6	Brunswick,Me	18 59/86			Cobb, Wight & Co.
2517	554		Billow WSr	American	132	88·0	24·0	8·8	Salem,Mass.	18 55/62			D.B.Gardner
86513	555	✳V	Billow Crest *Mathews* WSr	British	197	91·7	21·3	10·9	Plymouth	1883	Watson&Fox	Y 1·86	S.B.Harvey
49213	556		BillyButts (ex LeatherChild) WSr	British	113	86·0	23·2	9·3	DartmouthUS	1848			John Baker
	557		Bintang *Said Achmet Alkaf* (ex Star) W3mSr	Dutch	404	122·6	29·9	15·4	Boston,U.S.	1865			Said Achmat bin Aloei Alkaf
	558		Bintang Slamat *Daeng Pergoni* WBn	Dutch	147	111·5	21·0	..					Daeng Sahida bin Mastaber
	559		Bintang Timor WBk	Dutch	143	..	..	..					Sultan of Bima
	560		Bio Bio (ex Arestruz, ex Victoria) WBk	Chilian	750	151·2	34·1	19·0	New York	1843			F.Peede
	561	✳	Bio Bio (MISSING) *Turon* 83-83 IBk	French	682	185·7	30·3	19·0	Glasgow	1870	Aitken&Mansel		Ant.Dom.Bordes & fils
	562		Biola *Bergland* W3mSr	Swedish	283	104·9	25·9	13·9	Finland	1873			G.Ahlbom
13780	563		Birch Grove (LIGHTER) WBk	British	543	138·8	28·6	18·7	Sunderland	1856	Austin		James Grice
63772	564	✳	Birdston *Galloway* 81-81 IBk	British	700	189·2	31·0	18·6	Dumbarton	1870	A.M'Millan&Sn		Singleton,Dunn & Co.
	565	V	Birger *Laurin* WBk	Russian	742	167·0	31·2	17·5	Uleåborg	1870	C.Pettersson	Y 12·86	G.&C.Berghom
	566	V	Birger *Lundqvist* -86 WSr	Swedish	191	104·3	23·2	10·9	Figeholm	1883	A.P.Andersson		A.Ljungstedt
	567		Birger Jarl *Wicklund* WSr	Swedish	119	82·7	23·1	10·1	Roslagen	18 57/74	E.Söderland		M.Wicklund
	568		Birgit (ex Empress) *Jörgensen* -88 WBk	Norweg'n	419	130·8	29·0	16·6	P.E.Island	1864	Ellis		J.M.Wiborg
	569	DV	Birgitte *Ipsen* 79-79 WBn	Danish	178	111·4	22·2	10·6	Odense	1875	H.L.Hansen	Y 6.86	Aktieselskabet (J. Hansen)
	570	VN	Birgitte (ex Annie Lorway) *Johannessen* WBk	Norweg'n	760	163·6	34·1	19·3	MaitlandN.S.	1872	J.Trahey	Y 1.88	J.Börresen
	571	NV	Birgitte (ex ... Crest) *Gregertsen* -87 WBk	Norweg'n	584	140·1	31·7	18·1	Rothesay,N.B.	1870	R.Titus	Y 3.87	N.Gregertsen
	572		Birgitte Gjoe *Anthonisen* 82-82 WBg	Danish	296	119·0	26·4	12·5	St.Johns,NFL	1863			AktieSelskab (Th.Dahl)
	573		Birinji Yorghi WBg	Turkish	120	..	..	..	Samos	1883			Kerrasoudloe Panayotti Surmelee
	574	N	Birkebeneren *Gundersen* [81-86] WBg	Norweg'n	224	107·1	27·3	12·6	Tvedestrand	1865	Chr.Olsen		And.B.Olsen
70475	575	✳	Birker *Lightbody* IBk	British	1008	209·0	34·1	20·9	Barrow	1875	BarrowS.B.Co.		Shaw, Bushby & Co.
	576	DV	Birma *Hallmann* 85-85 IBk	German	1430	227·3	38·1	21·4	Vegesack	1885	J.Lange		D.Cordes & Co.
	577	DNV	Birma *Waage* WS	Norweg'n	1002	187·8	38·2	21·2	Drammen	1875	J.Jörgensen	Y 8.86	E.Berentsen
60076	578	✳	Birmah *Garner* 88-88 IBk	British	826	192·5	32·1	19·8	Glasgow	1868	Dobie&Co.	Y 8.88	P.Denniston & Co.
72276	579	V	Birnam Wood *Smith* WBk	British	1333	198·4	38·4	23·5	St.John,N.B.	1876	J.Mahoney	Y 11.87	W.Thomson & Co.
81517	580	✳	Birthday WK	British	65	82·0	19·0	6·9	Charlton	1879	W.Cory & Sons		G.A.Rooke
	581	N	Biskop Brun *Johansen* WBk	Norweg'n	532	148·0	32·3	17·5	Drammen	1867	Jörgensen& Knudsen	Y 8.88	Johan Thorne
72578	582		Bismarck WSr	British	302	131·7	26·0	11·5	P.Dalh'sieOnt.	1871	S.D.Andrews& Sons		Kingston & Montreal Forwarding Co.
56639	583		Bispham WSr	British	119	92·7	21·7	11·6	Applestore	1867	Cook		J.Chauncey
85604	584	V	Bittern *Bell* 86 W3mSr	British	399	139·5	29·2	2·13	Tynem'th,N.B.	1883			James Lilly
	585		Bjarke *Gotherstrom* WBg	Swedish	280	96·1	25·4	15·0	Norway				C.M.Carlsson
	586		Bjarkö WSr	Norweg'n	105	73·6	31·7	9·8	Namsos	1870	Salvesen		D.W.Lund
	587	N	Bjorgvin *Nielsen* 80-83 WBg	Norweg'n	245	103·7	26·1	14·8	Bergen	1849		Y 8.86	L.M.Iversen
	588	D	Bjorn (ex Fenwick) *Björnstad* WBk	Norweg'n	416	133·8	30·6	18·1	Horton,N.S.	1867	J.B.North	Y 8.88	Th.Björnstad
	589	G	Bjorn (ex Graf Bohn-Neuendank) *Johansson* -88 WBkn	Swedish	395	111·2	21·8	15·0	Damgarten	1869	B.Dierling		O.F.Johansson
	590	N	Bjornetraa *Olsen* -87 WBk	Norweg'n	348	112·8	30·0	15·7	Grimstad	1867	KnudAndersen	Y 1.8.	T.Olsen
	591	N	Björviken *Sorensen* WBk	Norweg'n	508	143·2	32·2	17·6	Christiania	1868		Y 8.88	Interessentsk (A Winge)
91711	592	✳	Black Diamond *Boy* 85-85 WK	British	59	71·2	19·1	9·0	Lowestoft	1885	H.Reynolds		J.S.Sterry
	593		Black Diamond (ex Maggie Taylor) WBn	Sarawak	347	130·4	30·9	12·6	Moncton, N.B.	1881	R.Cochrane	Y 8.81	SarawakGovernment
	594	GD	BlackHawk *Wackenrath* 87-88 WS	German	1059	182·3	38·3	23·1	NewYork	1856	W.H.Webb	Y 8.87	Gildemeister & Ries
2930	595		Black Prince *Dixon* 86-86 WBg	British	258	99·6	25·5	16·8	Maryport	1838			A.Storm
63573	596	✳	Blackadder *Cratram* 81-88 IBk	British	917	216·6	35·2	20·5	London	1870	Maudslay&Sons		J.Willis & Son
49316	597		Blackwall WBk	British	307	107·7	27·6	14·5	L'Orient				Gopathy Verasawmy Chetty
71650	598	✳D	Blair Athole *Taylor* IS	British	1697	259·1	40·0	23·2	Glasgow	1874	J.Elder&Co.		W.T.MacLellan
71687	599	✳	Blair Drummond *Holt* IS	British	[illegible]	246·0	37·4	22·3	Greenock	1874	Caird&Co.		W.T.MacLellan
7171	600	✳D	Blairgowrie *Dark* 71-75 IS	British	[illegible]	252·0	38·2	22·3	Glasgow	1875	J.&G.Thomson		Thomson & Gray

✳ = **Lloyd's Register**; D = Nederlandsche Vereeniging; G = Germanischer Lloyd; I = Registro Italiano; R = Ameri[can]; Λ = Veritas Austro-Ungarico; Π = Veritas Ellenico; N = Norske Veritas; V = Bure[au Veritas]

In the 4th Column, C = Composite (iron frame, wood planked); I = Iron; Stl = Steel; W = Wood

The Dates of Metalling (**Col. 11**) are based on the records of the Books in which Vessels are respectively classed, or on the Reports of the Surveyors ...

Further particulars of Vessels classed by Lloyd's Register will be found in Lloyd's Register Book and Lloyd's Yacht Register.

No.	Reg.	Nom, Matelot et Gréement	Pavillon	Tonnage	Longueur	Largeur	Creux	Construit: Où	Quand	Par qui	Doublage	Armateurs ou Directeurs-Gérants	Port d'Armement
601	✻	Blairhoyle Hawkins I Bk	British	1291	233·7	35·8	22·1	Glasgow	1875	C.Connell & Co.		Thomson,Dickie & Co.	Glasgow
602	DRV	Blanche Foster 86-88 W Bk	British	822	167·0	35·5	19·6	River John,N.S.	1878	J.Kitchin	11.88	D.Sutherland	Pictou,N.S.
603	R	Blanche Owens -81 W Bg	British	273	121·1	26·0	14·9	P.E.Island	1874	McDougall		T.O.Davies	Swansea
604	R	Blanche [illegible] -85 W Sr	British	22[illegible]	113·0	28·0	11·0	Horton,N.S.	1882	J.B.North	[illegible]	H.J.B.Woods	Windsor,N.S.
605	✻	Blanche Palfrey 15-85 W Bkn	British	195	120·7	25·1	13·0	Barnstaple	1875	Westacott	[illegible]	Job Bros.	Liverpool
606	✻	Blanche [illegible] W Sr	British	74	76·2	19·5	9·8	Plymouth	1873	Banks		J.Westcott	Plymouth
607	✻	Blanche Granatt -88 W Bk	Swedish	313	119·8	26·7	15·0	Littlehampton	1868	Harvey	[illegible]	J.M.Abrandt	Calmar
608	✻	Blanche Currey W Bg	British	180	103·8	24·2	12·6	Portmadoc	1875	Jones	[illegible]	W.Prichard & Co.	Carnarvon
609	R	Blanche H. King W 4m Sr	American	1098	192·0	42·0	17·0	Bath, Me.	1887	New England S.B. Co.		T.T.Anderson	New York
610	R	Blanche Hopkins Tassell W 3m Sr	American	602	152·5	36·1	16·4	Camden,N.J.	1880	S.W.Tilton		W.Irelan	Philadelphia
611	R	Blanche O. Whitney W Sr	British	126	86·0	24·4	10·3	Chester Bsn NS	1885	L.Oxner		George McMinn	Lunenburg,N.S.
612	V	Blanche & Louis W Bn	French	12[illegible]	81·0	21·6	11·1	Verton	1875	P.Doussain		J.Pichaud	Nantes
613	✻	Blanche & Louise I Bk	French	572	171·5	28·8	18·2	Liverpool	1868	Bowdler,Chaffer & Co.		Ant.Dom.Bordes & fils	Bordeaux
614	✻	Blanco Tucker W Bk	British	344	122·2	29·5	12·8	Alma,N.B.	1875	N.H.Foster	[illegible]	James Kirk	St.John,N.B.
615	A	Blandina P. Polinovich W Bk	Austrian	791	166·0	31·9	20·2	Fiume	1873	Flli Schiavoni	[illegible]	A.Poicurich fu M.	Fiume
616		Blast Maxey W Sr	American	118	99·6	17·6	8·1	Hamburg,Pa.	1844			J.A.Andersen	Trenton,N.J.
617	V	Blenda Trapp W Bk	Swedish	597	159·1	30·5	16·3	Söderhamn	1877	J.Bäst	[illegible]	J.F.Schöning	Söderhamn
618		Blenda W 3m Sr	Swedish	120	97·1	20·9	8·2	Sjötorp	1883	Groth		F.G.Sandvall	Jönköping
619		Bleng W Bk	British	215	91·6	22·2	14·8	Whitehaven	1841	L.Kennedy & Co.		Athmaccory Authenarraina Chetty	Madras
620	✻	Blengfell Irving 81-82 I S	British	1154	232·1	35·2	21·1	Whitehaven	1876	Whthvn.S.B.Co.		I.Coleisank & Co.	Liverpool
621	✻	Blenheim Colville I S (ex Wanganui)	British	1077	215·3	34·0	20·4	Glasgow	1877	A.Stephen & Sons		J.Leslie	Aberdeen
622	✻	Blenheim Harris -88 I Bk	British	574	170·1	28·3	17·3	Harrington	1869	R.Williamson & S.		D.Jones	Liverpool
623	V	Blitz [illegible] (ex Johann Gottlieb) W 3m Sr	German	206	108·5	23·7	11·6	Burg a.d.Weser	18 61/79	H.Bosse	[illegible]	H.Bauer	Rostock
624	R	Blizzard Rudd W Sr	British	132	85·0	24·5	10·1	Volger's Cove,N.S.	1883	Reinhardt		A.J.Wolff	Lunenburg,N.S.
625		Blonde Kemp 85-88 W Sr	British	208	115·1	25·1	13·1	Newhaven	1874	Tolman		G.H.Goldfinch & Co.	Faversham
626	V	Blonde Lecorre W Sr	French	129	92·8	20·3	11·3	St.Malo	1883	Gautier		Vve.L.Bahot-Launay	Paimpol
627		Blonde Lesant -87 W Sr	French	118	72·5	21·0	11·3	Nantes	1855	Chauvelon		Coyac	Lorient
628	D	Blondine Lelin W Bn	French	129	78·1	20·3	10·8	Nantes	1869	Lefrançois		Lelin	Nantes
629	DG	Blücher Leaf 83-87 (ex Kinnoulland e Prince Royal) W Bk	German	1209	192·5	39·1	24·1	N.Brunswick	1865	Stewart	[illegible]	Von Riegen & Stindt	Bremerhaven
630	✻	Blue Bell W Sk	British	90	86·7	21·0	10·1	Hull	1886	Hunt & Fowler		R.Roach	Hull
631		Blue Jacket Jones W Sr	British	100	79·5	20·9	10·9	Aberystwith	1860	Jones		Mrs.E.Rees	Aberystwith
632		Blue Jacket Storrs 81-87 W S	American	1338	193·8	40·5	24·1	Green Point,(N.Y.)	1865			P. B. Cornwall	San Francisco
633		Blue Jacket W Sr	American	110	86·6	23·5	9·3	Essex,Mass.	1887			A.Leighton	Gloucester
634	✻	Blue & White Black -87 W Sr	British	143	96·8	21·3	10·7	Kingston	1887	W.Kinloch		R.Davies	Ayr
635		Bluebird Hitchener -88 W Bk	British	393	132·7	30·0	14·1	Cornwallis,NS	1870		[illegible]	A.C.Ells	Windsor,N.S.
636	✻	Blythswood Farmer 81-87 I S	British	1697	259·8	39·0	23·3	Port Glasgow	1875	R.Duncan & Co.		R.Shankland & Co.	Greenock
637		Boa Nova Da Silva W Sr	Brazilian	100	..	..	..	Brazil			[illegible]	M.L.Martins	Laguna
638	V	Boa Sorte Serra -87 W Bg	Portug'se	183	89·9	22·5	10·0	Oporto	1867	J.F.da Lapa	[illegible]	Soares & Co.	Oporto
639		Boa Viagem W Sr	Portug'se	117	83·0	20·3	9·1	Villa do Conde	1876	A.L.Duarte		Alkaim & Co.	Lisbon
640	✻	Boadicea Davies 81-87 Stl I S	British	1896	264·0	39·0	23·6	Workington	1887	R.Williamson & S.		"Boadicea" Ship Co. Ld. (Brown, Jenkinson & Co.)	London
641	✻	Boadicea Francis W Sr	British	100	83·8	20·9	10·8	Perth	1859	D.Smeaton		H.Francis & Co.	Carnarvon
642		Boatswain Parsons W Bg	British	228	99·6	26·2	15·1	Sunderland	1863	Dunstan		A. & F.Manuelle	Guernsey
643		Boavista Frixone W Bn	Italian	121	88·9	21·6	10·5	Elsfleth	1858			G.B.Frixone fu M.	Genoa
644		Bobolink [illegible] W Sr	American	162	104·5	29·[illegible]	8·9	Oakland,Cal.	1868			O.Aufindsen	San Francisco
645		Bobsien Kaegsdorf W Bk	German	347	112·9	28·0	16·3	Rostock	1864	O.Ludewig		A.Burchard	Rostock
646		Bohycito [illegible] W Bk	British	[illegible]	12[illegible]·8	29·3	16·2	P.E.Island	1866			Balmain Steam Ferry Co.	Sydney,N.S.W.
647	✻	Bodrhyddan [illegible] 86-87 I Bk	British	524	167·1	27·2	18·1	Hull	1858	M.Samuelson & C.		Cotesworth, Lyne & Co.	Liverpool
648	N	Bödvar [illegible] 83-86 W Bk	Norweg'n	454	150·1	27·9	16·7	St.Heliers,Jsy.	1855	F.C.Clarke		S.C.Larsen	Brevig
649	I	Bogliasco Risso W Bk	Italian	735	160·1	33·4	21·5	Sestri Ponente	1874	S.Bozano	[illegible]	F.Risso	Genoa
650		Bogota [illegible] W Bn	American	279	99·5	26·8	16·0	Fairhavn Conn	1862		[illegible]	L.Snow	N.Bedford,Mass.

✻ = Lloyd's Register; D = Nederlandsche Vereeniging; G = Germanischer Lloyd; I = Registro Italiano; R = American Record; A = Veritas Austro-Ungarico; H = Veritas Ellenico; N = Norske Veritas; V = Bureau Veritas.

Dans la 4me Colonne, C = Composite (membrure en fer, bordé en bois); I = Fer; Stl = Acier; W = Bois.

✻ Dates du doublage en métal (Col. 11) ressortent des livres dans lesquels les navires sont respectivement classés ou des rapports des Inspecteurs du Lloyd's Register.

Pour plus amples renseignements sur les navires classés par le Lloyd's Register, voir le Lloyd's Register Book et le Lloyd's Yacht Register.

1 Official No.	2 No. in Book	3 Society in which Classed	4 Ship's Name. Material and Rig. Master, and Dates of Appointment to present Owner's Service and to Ship. Ship's Late Name.	5 Flag.	6 Net Tonnage	7 Dimensions. Length.	Breadth.	Depth.	8 Built. Where.	9 When	10 By whom.	11 Date of Metalling	12 Owners or Managers.
	651	V	Bohallard *Colard* W Bn	French	126	76·1	21·0	10·6	Nantes	1857	Viaud		L.Jacquier
[illegible]	652	RV	Bohemia *Trask* W S	American	1633	221·7	40·2	25·5	Bath,Me	1875	Houghton Bros.	[illegible]	Houghton Brothers
	653		Bohemia *Colbeaux* -87 W Bn	French	149	93·4	24·1	10·1	Barringtn,NS	1881	W.Doane	[illegible]	Folquet & fils
[illegible]	654		Bohemian Girl W3mSr	British	175	103·9	22·5	11·8	Waterford	1843	White		S.McGiffin
	655		Bohuslan *Johansson* W Bk	Swedish	345	133·8	24·8	15·8	Sunderland	1853			C.Bruce
	656	V	Boieldieu *Tual* -88 W Bn	French	142	85·3	22·6	10·5	Rouen	1864	J.B.Lemire		Billy
	657	V	Bois Rosé *Tolain* W3mSr	French	265	117·7	29·6	12·5	Fécamp	1884	Benoist Capon	[illegible]	H.Chédru
[illegible]	658	✳	Bolan *Hughes* 87-88 I S	British	1993	273·2	40·0	24·3	Southampton	1882	Oswald, Mord'nt & Co.		T. & J. Brocklebank
	659		Bolette (WRECKED) (ex Bolette Taggaard) *Jorgensen* -87 W Bg	Danish	188	90·8	23·3	13·6	Nakskov	1840	K.Jørgensen		J.P.Simonsen
	660	V	Bølgen *Schrøder* -87 W Sr	Danish	119	89·1	20·1	10·0	Nakskov	18[illegible]	Ridersborg & Troehmann		H.P.Ridersborg
	661	N	Bölgen *Evensen* 83-83 W Bk	Norweg'n	363	133·6	28·6	12·3	Grimstad	1883	S.Johnsen	[illegible]	Sch.P.Messel
	662		Bolivar *Basele* W Bk	French	292	114·8	24·9	13·0	Bordeaux	1868	Raymond frères & fils	[illegible]	R.Aradel
[illegible]	663		Bolivar (ex Voltri) *Schiaffino* W Bk	Italian	511	137·8	32·0	20·2	Voltri	1871	L.Briasco fu GB	[illegible]	N.Schiaffino
[illegible]	664	V	Bolivia *Hawthorne* 84-84 W Bk	British	922	179·4	35·2	21·4	NovaScotia	1873	F.Weston	[illegible]	T.Dixon & Sons
[illegible]	665	✳	Bolivia *Purley* 71-81 Stl Bk	British	878	202·3	33·1	19·9	PortGlasgow	1881	J.Reid&Co.		Nicholson & McGill
[illegible]	666		Bolivia *Venning* 76-84 W Bk	British	378	117·8	25·9	17·3	Whitehaven	1857	L.Kennedy&Co.	[illegible]	R.& W.King
	667		Bolke *Henriksson* -87 W Bn	Swedish	257	105·4	23·2	12·2	Bardenfleth	1868	Gebr.Focke		E.Julin
	668		Bollilidi W Bg	Turkish	162	..	..	..	Greece	1866			TrebizondleeConstantin Bollilidi
	669		Bom Fin W Sr	Brazilian	153	..	..	..					M.J. de Sant'Anna
	670		Bom Jesus de Iguape W Bn	Brazilian	214	..	..	..					A.J.D. do Prado
	671		Bom Jesus dos Navegantes *de Miranda* W Sr	Brazilian	130	..	..	..	Pernambuco	1879		[illegible]	J.I.Pereira
43719	672	✳	Bombay (ex City of Bombay) *Eldridge* 86-86 I Bk	British	991	212·0	31·6	21·6	Glasgow	1862	A.Stephen&Sns.		J.N.Ward & Co.
[illegible]	673	I	Bombay (ex Furioso) *Puglisi* -88 W Bk	Italian	446	130·5	30·8	19·2	Voltri	1871	G.Fava	[illegible]	N.Villa
48217	674		Bomerang W Bn	British	167	104·8	25·8	10·5	St.John, N.B.	1864			S.Allen
	675		Bomfim (ex Zaida) *Mattos* W Sr	Portug'se	80	87·5	22·3	10·4	Figueira	1872	J.Cochulo	[illegible]	M.Joaquin
	676	✳	Bon Accord *Coutsoudakis* W Bk	Greek	398	141·0	27·4	16·0	Sunderland	1876	Crown	[illegible]	G.Coutsoudakis
	677		Bon Pasteur W Sr	French	102	66·9	19·3	10·7	Nantes	1858	Lemerle		L.Lemerle
	678		Bon Père	French	118	..	..	..					F.A.Moriani
	679		Bon Père W Sr	French	117	75·1	22·0	10·8	Auray	1849			Jouano
	680		Bon Pére (CONDEMNED) *Dalest* W Bg	French	105	70·5	20·3	10·2	Martigues	1865	Vidal		Dalest frères
	681	N	Bona Fide (ex Jacob Lecy) *Sørlie* -88 W Bk	Norweg'n	411	117·7	27·3	18·6	Yarmouth,Me.	1851			A.Jacobsen
	682	NV	Bona Fides *Nielsen* W Bk	Norweg'n	1002	175·7	35·9	20·8	Sestri,Italy	1875	A.Briasco	[illegible]	A.& F.Smith
	683		Bonafous *da Silveira* W Sr	Brazilian	162	..	..	..					A.daCunhaVasconcellos
71017	684	DV	Bonanza *O'Brien* -87 W S	British	1079	189·0	36·8	22·0	Plympton,NS	1875	E.Everett	[illegible]	J.H.Crosby
[illegible]	685		Bonanza *Keller* 59-87 W Bk	American	1356	207·8	39·2	24·1	Bath,Me	1875	W.Rogers	[illegible]	PugetSoundCommercial Co.
[illegible]	686		Bonanza *Miller* W Sr	American	130	96·0	23·5	9·0	Danverspoint, Mass.	1875			G.H.Collins
	687	N	Bonanza *Knudsen* -82 W Bk	Norweg'n	534	150·5	31·7	16·9	Lillesand	1881	O.&P.Knudsen	[illegible]	O.& P.Knudsen
	688		Bonapartis *Manocharizis* W Bg		195	84·0	25·0	14·6	Piraeus	1865		[illegible]	
	689	N	Bonden *Tellefsen* W3mSr	Norweg'n	343	120·1	29·7	14·2	Grimstad	1888	N.O.Gjømle		A.Tellefsen
	690	V	Bonden *Lundh* 85-85 W3mSr	Swedish	299	127·9	25·2	12·4	Framnäs	1876	E.Ostman	[illegible]	C.F.Neess
	691	N	Bondevennen *Ludwigsen* -87 W Bk	Norweg'n	436	131·9	30·6	17·1	Kragerö	1875	T.G.Litangen	[illegible]	Peder Rinde
	692	N	Bondevennen *Henriksen* 86-86 W3mSr	Norweg'n	278	126·8	27·7	13·2	Haugesund	1875	L.Armes	[illegible]	Johan Thorsen
	693	NV	Bonheur *Frivold* 84-84 W Bk	Norweg'n	927	184·0	34·7	20·3	Grimstad	1884	S.Johnsen	[illegible]	C.Klöcker
[illegible]	694	R	Boniform *Potter* -88 W Sr	British	148	99·4	37·1	9·8	Cornwallis,N.S.	1888	J.E.Bigelow		A.Potter
	695		Bonita	Norweg'n		..	..	..					E.Christensen
	696	N	Bonita (ex Bonito) *Henricksen* 86-86 W Bk	Norweg'n	599	127·9	35·2	12·4	RiverJohn,NS	1871	J.Kitchen	[illegible]	Th.Thronsen
[illegible]	697	✳	Bonito *Holland* I Sk	British	90	79·2	20·3	10·1	Beverley	1884	Cochrane&Co.		J.W.Hellyer
	698	GV	Bonito *Kock* 75-83 W Bn	German	223	112·5	24·9	11·2	Brake	1863	J.Oltmann Wwe.	[illegible]	H.A.Hellmers
	699	NV	Bonito *Danielsen* W Bk	Norweg'n	520	145·3	30·5	16·9	Vegesack	1868	H.F.Ulrichs	[illegible]	R.Josephsen
	700												

✳ = Lloyd's Register; D = Nederlandsche Vereeniging; G = Germanischer Lloyd; I = Registro Italiano; R = Amer[illegible]
A = Veritas Austro-Ungarico; H = Veritas Ellenico; N = Norske Veritas; V = [illegible]

In the 4th Column, C = Composite (iron frame, wood planked); I = Iron; Stl = Steel; W = Wood.

The Dates of Metalling (Col. 11) are based on the records of the Books in which Vessels are respectively classed, or on the Reports of the Surveyors of [illegible]

Further particulars of Vessels classed by Lloyd's Register will be found in Lloyd's Register Book and Lloyd's Yacht Register.

1	Nº dans le livre (2)	Cote (3)	Noms, Matériaux et Gréement. Capitaines et dates d'entrée en service des Capitaines actuels à bord du navire. Ancien Nom. (4)	Pavillon (5)	Tonnage net (6)	Longueur (7)	Largeur	Creux	Où (8)	Quand (9)	Par qui (10)	Doublage (11)	Armateurs ou Directeurs Gérants (12)	Port d'Armement (13)
	701	V	Bonne Joséphine, Tyrement 86, W Bg	French	224	102·6	22·0	13·4	St. Malo	1867	Carnet		Feray	Granville
	702		Bonne Julienne, Tréhuidic, W Bn	French	117	70·9	19·7	9·4	Lorient	1858	Lacroix		V. Tréhuidic	Lorient
	703	V	Bonne Mère, Boudrault 86, W Bk	French	551	135·1	27·2	19·1	Honfleur	1886	Leviels		D. Auger	Havre
	704		Bonnibel, W Sr	British	141	90·5	21·5	10·8	Isaacs Har., NS	1873			R. H. McMillan	Halifax, N.S.
	705	V	Bonny Doon, Burgess, W Bkn	American	542	145·0	30·6	17·3	Machias, Me.	1876	John Shaw		Miller & Houghton	New York
	706		Bonny Mary, Luke, W Bn	British	149	96·6	20·7	12·0	Jersey	1864	Bellot		Clement & Co.	Jersey
	707	N	Bonus, Hansen, W Sr	Norweg'n	199	100·9	23·0	12·1	Kragerö	1878	T. O. Litangen		C. Börresen	Kragerö
	708	V	Boomerang, Pettersen, W Bk	Norweg'n	372	138·0	29·5	12·9	Pictou, N.S.	1865			H. Helland	Fredrikstad
	709	R	Booth Brothers, W 3m Sr	American	330	136·8	31·0	10·2	Madison, Ct.	1887	Wm. C. Miner		Carlos Barry	N. London
	710	✱	Bootle, Lewis 76-81, I Bn	British	298	133·1	23·6	12·8	Whitehaven	1873	Whithvn S.B. Co.		G. Nelson & Sons	Whitehaven
	711	✱	Border Chief, I Bk	German	1011	212·0	35·0	20·3	South Shields	1870	Redhead Softley & Co.		Conr. Paulsen	Elsfleth
	712	✱	Border Queen, Sinclair, W Bn	Portug'se	183	108·0	23·2	12·7	Berwick	1877	Gowan		Bensaude & Co.	Lisbon
	713	V	Bore, Adolfsson, W Bk	Swedish	382	123·6	25·6	13·9	Westervik	1877	Fridelt		C. D. Westermark	Westervik
	714		Bore, Johansson, W Bg	Swedish	283	115·4	25·3	13·4	Sundsvall	1868	J. A. Nordström		J. Johansson	Stenkyrka, Gotl.
	715	V	Boréal, Mesnard 79-86, W Bg	French	239	104·0	23·3	13·0	St. Vaast	1876	E. Lévêque		De Boismenu	St. Malo
	716	G	Borealis, W. Kruse, W Bn	British	131	87·0	25·0	10·5	Mahurangi NZ	1880	J. Darrach		K. McKenzie	Auckland, N.Z.
	717	DV	Boreas, Gorter 86, W Sr	Dutch	100	79·1	20·0	9·5	Sappemeer	1864	J. Berg		W. Bos	Groningen
	718	N	Boreas, Poulsen, W Bk	Norweg'n	460	132·6	29·8	16·9	Grimstad	1882	J. N. Ribe		J. Jörgensen	Grimstad
	719	N	Boreas, Svendsen 78, W Bk	Norweg'n	437	136·0	29·8	16·3	Risör	1876	T. Olsen		L. Alsing	Kragerö
	720		Borgn, W Bk	Norweg'n	1190	174·5	35·1	24·3	Medfort	1855			F. Olsen	Hvidsteen
	721		Borges I., da Rocha, W Bn	Brazilian	142	..	..	..	Campos	1850			A. J. da Rocha	Itajahy
	722	NV	Borghild, Haugeland, W Bk	Norweg'n	757	162·4	35·1	19·0	Porsgrund	1876	H. J. Gregersen		P. Anker	Fredrikshald
	723	N	Borghild, Olsen, W Bk	Norweg'n	589	148·0	22·3	17·9	Tusket, N.S.	1872			C. E. Holter	Sarpsborg
	724		Borneo, Tjiong Tjen Kwey, W Bk	Dutch	435	124·0	29·1	..	Grisace, Java	1857	Sieman		Ang Lam Thaion, Tho Tjhiang	Banjermassin
	725	G	Bornholm, Grønnegaard 73-73, W Bg	Danish	218	114·1	22·7	12·6	P. E. Island	1864/78			M. Christiansen	Rønne
	726	✱	Boroma, Thomas 83-87, W Bk	British	793	171·3	33·7	19·6	Kingston, N.B.	1879	J. & T. Jardine		J. & T. Jardine	Chatham, N.B.
	727	G	Borough Belle, W Bn	British	210	103·8	24·1	11·8	Pyrmont NSW	1875	Banfield & Allan		Mackay Labour Co., Ld.	Sydney, N.S.W.
	728	N	Borre, Nass 79-79, W Bk	Norweg'n	278	93·5	27·7	15·7	Hamburg	1833			E. Rasmussen	Porsgrund
	729	✱	Borrowdale, Cathrie 87-87, I S	British	1197	226·4	36·4	22·0	Liverpool	1868	W. H. Potter & C.		J. D. Newton & Co.	Liverpool
	730	V	Borsenhalle, Hubert, I Sr	German	150	102·3	22·0	7·5	Rostock	1886	Rostock S. B. & E. Co.		Zürbruggen & Hubert	Hamburg
	731	G	Borussia, Zessin 84-86, W Bk	German	278	101·6	25·6	15·4	Königsberg	1847	J. F. Fischer		Schoenberg & Domansky	Danzig
	732	R	Borzone, Desmond 87, W Bk	British	540	130·5	31·6	18·9	Varazze	1870	D. Craviolo		D. J. Ritchie & Co.	Chatham, N.B.
	733	N	Bos, Pedersen, W 3m Sr	Norweg'n	145	104·4	23·5	9·9	Tvedestrand	1874	L. Tellefsen		M. H. Smith	Tvedestrand
	734	A	Bosana, W Bk	Austrian	473	130·5	29·2	18·7	Trieste	1869	G. Spadon		Cristofors Miletich	Trieste
	735	A	Boschetto M., Ocsolini, W Bg	Italian	384	121·3	30·2	17·6	Varazze	1865	G. Scala		G. Schiaffino	Genoa
	736		Bosphorus, W Bk	British	299	107·6	26·3	16·3	Dundee	1851			J. L. & W. A. Simpson & Elder, Smith & Co.	Adelaide, S.A.
	737		Boston, Adams, W Sr	American	162	113·6	25·0	9·0	Asht'bula, Ohio	1861			W. C. Crosby	Barnstable, Mass.
	738	R	Boston Marine, Le Blanc 87, W Bn	British	172	92·6	26·0	11·0	Tusket Wedge, N.S.	1883	A. O. Porter		A. O. Porter	Yarmouth, N.S.
	739		Botatoung, I Barge	British	355	195·0	25·2	8·4	Port Glasgow	1879	R. Duncan & Co.		Irrawaddy Flotilla Co. Ld.	Glasgow
	740		Bothilda, Andersson, W Sr	Swedish	174	104·6	22·0	11·8	Danzig	1848	Rosalsky		M. Hansson	Nyhamn
	741	✱	Bothnia, Jones 84-84, I Bk	British	753	187·0	31·0	18·7	Dumbarton	1878	Birrell, Stenhouse & Co.		W. Nicol & Co.	Liverpool
	742	DG	Bothnia, W 3m Sr	Dutch	283	122·7	28·4	12·6	Halifax, N.S.	1868	McPherson		S. Tromp Meesters	Steenwijk
	743	N	Bothnia, W Bk	Norweg'n	919	169·9	34·0	21·5	St. Martin's NB	1871	H. McQuiston		L. Rafen	Holmestrand
	744	✱	Bothwell, Stl S	British	1986	283·5	41·2	24·9	Port Glasgow	1886	W. Hamilton & C.		W. Hamilton & Co.	Port Glasgow
	745		Bounding Billow, Lee, W Bk	American	228	119·0	27·0	22·0	Chelsea, Mass.	1854			E. H. Hanson	San Francisco
	746		Bounty, W Bg	British	166	78·8	22·9	13·4	Whitby	1849			R. Storm	Whitby
	747		Boussole, W Bg	French	188	87·3	23·3	12·0	Fécamp	1877	F. Capon		Dr. Slawecki	Bordeaux
	748	✱	Bowdon, Hegarty, I S	British	1624	250·1	38·8	23·6	Sunderland	1875	T. R. Oswald		Hargrove & Hellon	Liverpool
	749	✱ D	Bowman B. Law, Abbott, I Bk	British	1359	231·4	37·0	21·7	Dumbarton	1885	A. McMillan & Sn		W. Law	Yarmouth, N.S.
	750		Boxer, W Bn	British	192	108·0	23·4	12·8	N. Brunswick	1865	Davies		Holden & Gann	Faversham

✱ = Lloyd's Register; D = Nederlandsche Vereeniging; G = Germanischer Lloyd; I = Registro Italiano; R = American Record; A = Veritas Austro-Ungarico; H = Veritas Ellenico; N = Norske Veritas; V = Bureau Veritas.

Dans la 4me Colonne, C = Composite (membrure en fer, bordé en bois); I = Fer; Stl = Acier; W = Bois.

Les Dates du doublage en métal (Col. 11) ressortent des livres dans lesquels les navires sont respectivement classés ou des rapports des Inspecteurs du Lloyd's Register.

Pour plus amples renseignements sur les navires classés par le Lloyd's Register, voir le Lloyd's Register Book et le Lloyd's Yacht Register.

1 Official No.	2 No. in Book	3 Registry in which classed	4 Ship's Name. Material and Rig. Master, and Dates of Appointment to present Owner's Service and to Ship. Ship's Late Name.	5 Flag.	6 Net Tonnage	7 Dimensions. Length.	Breadth.	Depth.	8 Built. Where.	9 When.	10 By whom.	11 Date of Metalling.	12 Owners or Managers.
	751	N	Boylestad *Lydersen* WBk	Norweg'n	336	120·1	29·3	16·0	Arendal	1872	A.Aaronsen		O.Lydersen,sen.
3085	752	RV	Boylston *Small 72-86* WBk	American	540	135·6	31·6	18·3	Bucksport,Me.	1879	Genn&Hill	Y 11.87	J.S.Emery & Co.
	753	✻	Brabant *Baumann 87-87* IBk (ex Martha Jackson)	Belgian	502	161·4	26·3	17·4	Sunderland	1863	J.Laing		F.Speth & Co.
	754	AD	Braca R. *Racich -88* WBn (ex Paolina T. ex Pitma)	Austrian	227	102·3	23·6	13·1	Porto Re	1874	J.Archianin	Y [illegible]	E.Racich
93387	755	✻	Bracadale *Peebles 82-87* I4mBk	British	1862	285·2	40·5	23·8	Glasgow	1887	A.Stephen&Sns.		J.&A.Roxburgh
3282	756		Bradford C. French W3mSr *Hassell -88*	American	929	184·3	37·5	19·2	KennebunkMe	1884	D.Clark		J.B.Phillips
	757	N	Brækka *Jensen* WBk	Norweg'n	368	121·7	29·0	15·8	Arendal	1877	E.Guttormsen	Y 8.88	H.H.Pettersen
90003	758	✻	Braemar *Cato 85-85* IBk	British	1036	216·0	31·0	19·9	Dumbarton	1884	A.McMillan&Sn		Thomson,Dickie & Co.
	759		Braemar *Larsen 86-86* WBk	Norweg'n	309	110·2	27·4	16·1	Moulmein	1838			E.Garstad (Blockstad, Holm & Co.)
	760	V	Brage *Jørgensen 81-74* WSr	Danish	144	94·8	21·2	10·8	Svendborg	1874	C.C.Hansen& H.P.Hansen,jr		J.Anderskouv
	761		Brage *Larsen* WGlt	Danish	114	85·1	19·6	10·0	Stavanger	1864			P.C.Larsen
	762	N	Brage *Jörgensen* WBk	Norweg'n	444	135·8	30·3	16·1	Kragerö	1865	J.Brandi		O.Lyng
	763		Brage *Garneij -86* WBk	Swedish	282	129·2	24·3	11·6	Westby	1871	P.Fröberg		E.Norberg
	764	V	Brage *Strombom 88-88* CSr	Swedish	241	118·3	23·1	10·0	Gefle	1886	O.A.Brodin	C 11.86	O.A.Brodin
	765		Brage *Thomaton* WSr	Swedish	115	90·2	22·1	9·1	Oscarshamn	1858			F.Johansson
72113	766	✻	Brahmin *McKenzie 62-85* IS	British	1264	231·1	36·3	21·9	Greenock	1876	R.Steele&Co.		D.McGillivray
80733	767	✻	Braidwood *Rowlands86-86* WBk	British	950	177·5	35·5	21·4	Quebec	1880	Charland,jun.	Y 3.88	J.Roberts
73827	768	✻	Brambletye *Fisher 87-87* IS	British	1513	241·5	39·0	22·9	Barrow	1876	Barrow S.B.Co.		W. & J.Crawford
2506	769		Bramhall *Drinkwater87-87* WSr	American	135	94·2	23·9	8·2	Westbrook,Me.	1863			W.S.Jordan & Co.
17380	770		Branch WSr	British	120	82·5	20·0	11·5	Yarmouth	1857	Fellows & Son		J.Norman
83335	771	V	Brandon *McCleave 87-87* WBk	British	1250	197·5	38·5	23·4	Quebec	1882	W.Charland,Sen	Y 8.88	John Herron & Co.
92835	772		Brandon (WRECKED) W3mSr	British	517	167·0	32·8	10·8	Kingston,Ont.	1887	J.Gaskin		Montreal Transportation Co.
29880	773		Brandon *Jones 85-87* WSr	British	110	88·3	21·3	19·8	Portsmouth	1862	White		J.Fisher & Sons
	774	V	Brani *Belbeach 85-86* WBg	French	251	102·6	23·0	14·3	Nantes	1867	Jollet&Babin	Y [illegible]	F.Vermeil&A.Vincent
169	775	A	Brasile *Gentili* WBn	Italian	218	131·8	22·6	13·9	Fiume	1869		Y 3.86	L.Micoli Coscano
	776		Brasilianeren *Jurcker* WBg	Danish	144	96·3	23·8	11·2	Oscarshamn	18 45/86			B.J.Jans
2874	777		Brave WSr	American	108	85·8	25·2	7·9	Hancock,Me.	1872			A.P.Foss
	778		Brave (ABANDONED) *Gosse 82-84* WBk	French	396	143·5	26·1	14·7	Jersey	1867		Y 11.81	E.Delalun
	779	N	Bravo *Christopherson* WBk	Norweg'n	452	137·5	30·9	16·6	Langesund	1869	J.Abrahamsen	Y 11.88	B.A.Olsen
	780	N	Bravo *Johnsen* WBk	Norweg'n	451	138·7	30·1	16·3	Kragerö	1876	T.G.Litangen	Y 10.88	Anton Natvig
81	781	V	Bravo *Kontermowitz* WBk	Russian	314	120·0	29·0	13·2	Riga	1868	G.Lindner		Kontermowitz,Arnold & Knokling
	782	G	Bravo *Torngren 81* WBg	Swedish	281	114·5	24·7	13·8	Carlshamn	1861	D.R.Andersson	Y 2.82	C.Elg
	783	V	Bravo *Froberg 81-81* WBk	Swedish	259	120·4	24·6	12·0	Timm'rnabb'n	1872	C.Bring	Y [illegible]	E.T.Killman
83217	784	V	Brazil *Davidson* WBkn	British	344	126·0	32·0	12·0	Hantsport,NS	1882	E.Churchill& Sons	Y [illegible]	Geo.W.Churchill
90498	785	V	Brazil (MISSING) *Le Marchant* WBn	British	266	113·6	27·9	12·0	Newport,N.S.	1885	T.A.Mosher	Y [illegible]	R.I.Hart
	786		Brazil (ex Trople) WBn	Brazilian	141	93·0	23·1	11·7	P.E.Island	1875	White		J.M.da Silveira
	787	NV	Brazil *Evensen 86* WSr	Norweg'n	194	102·5	27·0	9·9	Kragerö	1886		Y 8.86	A.O.Lindvig
93381	788		Brazileira *Stirling -88* WBk	British	306	134·5	28·0	12·3	Vegesack	1866	H.F.Ulrichs	Y 12.86	John Henderson
	789	✻	Brazileiro *Le Berre 84-84* WBk	French	309	118·1	27·9	11·1	Nantes	1878	Dubigeon & fils	C 11.83	Denis Crouan fils
62106	790		Brazilian (STRANDED) *Gasston* WSr	British	239	114·0	24·4	14·0	Lossiemouth	1869	Jack	Y 3.83	J. Gasston (J. Ingram, Manager)
24110	791		Brazilian *Winder* WBg	British	190	86·1	23·5	15·0	Stockwith	1842	Bomforth		R.Cory
	792	N	Brazilian *Ugland* WBk	Norweg'n	415	131·9	29·4	13·3	Grimstad	1882	Jens Tellefsen	Y [illegible]	P.Pedersen
	793		Brazo fuerte (ex Catalina) WSr	Spanish	115	79·7	25·8	9·5	United States	1864			Pascual Franco
[illegible]	794	RV	Brazos *Edgett -88* WBk	American	871	168·0	37·4	20·8	Portland, Conn.	1870	S.Gildersleeve	Y 1.87	F.E.Scammel
[illegible]	795		Breadalbane WBk	British	215	104·7	23·0	13·1	Baddeck,C.B.	1854			L.Saabye
[illegible]	796		Breeze *Arlington -85* WSw	British	216	94·0	24·5	14·7	Sunderland	1859	G.Short		W.Milburn
[illegible]	797		Breeze WSr	American	115	99·4	17·6	7·9	Hamburg,Pa.	1864			J.A.Anderson
	798	N	Breidablik *Halvorsen* WBk	Norweg'n	364	122·4	28·5	16·2	Grimstad	1872	N.Osmundsen	Y 9.88	J.E.Einersen
98447	799	✻	Breiz Izel *Malcolm -88* IBk	British	405	148·0	25·5	15·0	Nantes	1869	Jollet&Babin		J.de V.Le Couteur
	800	V	Brema *Lutterbrodt 82-82* WBk	German	360	123·0	27·1	15·0	Burg	1864	Gebr.Bosse	Y 8.87	Franzius,Henschen&

✻ = Lloyd's Register; D = Nederlandsche Vereeniging; G = Germanischer Lloyd; I = Registro Italiano; R = Ame[rican]
A = Veritas Austro-Ungarico; H = Veritas Ellenico; N = Norske Veritas; V = Bu[reau]

In the 4th Column, C = Composite (iron frame, wood planked); I = Iron; Stl = Steel; W = Wood.

The Dates of Metalling (Col. 11) are based on the records of the Books in which Vessels are respectively classed, or on the Reports of the Surveyors of [illegible]

Further particulars of Vessels classed by Lloyd's Register will be found in Lloyd's Register Book and Lloyd's Yacht Register.

N° dans le livre	Registre	Nom. Matériaux et Gréement. Capitaine et dates d'entrée au service des Armateurs actuels et à bord du navire. Ancien Nom.		Pavillon	Tonnage	Dimensions. Longueur	Largeur	Creux	Construit. Où.	Quand.	Par qui.	Date du doublage en métal	Armateurs ou Directeurs-Gérants.	Port d'Armement.
801	V	Bremen *Kugler*	W Bk	British	852	165·0	35·0	19·8	Cornwallis, N.S.	1876	A. C. Ellis	V 5.88	E. Churchill & Sons	Windsor, N.S.
802	G	Bremen *Hoffschildt 77-87*	W Bk	German	949	178·5	35·1	21·2	Vegesack	1854	H. Ulrichs	V 6.87	J. H. Meyer	Pillau
803	GV	Bromerhaven *Pathe -87* (ex James Montgomery)	W Bk	German	1042	172·5	36·2	2[illegible]	Portsmouth, N.H.	1852		V 5.88	D. Heinrichs	Bremerhaven
804		Bremner	W Bn	British	198	103·7	21·4	12·[illegible]	P.E. Island	1865	M'Millan		J. Watson	N. Shields
805		Brémontier *Bajos*	W Bk	Brazilian	456	128·9	29·8	17·6	Bordeaux	1863	Chantiers de l'Océan		F. A. X. Pintreiro	RioJaneiro
806	✻	Brenda *Hosiery -82-88*	I S	British	1260	228·0	36·8	21·1	Glasgow	1877	C. Connell & Co.		Sandbach, Tinne & Co.	Liverpool
807		Brenda *Simpson -82*	W Bn	British	249	116·4	26·6	12·9	P.E. Island	1870	Peake		Holden & Gaun	Faversham
808	✻	Brenhilda *Morris 75-83*	I S	British	1321	240·4	36·8	22·3	Glasgow	1874	Barclay, Curle & Co.		W. Letham	Greenock
809		Breslau *Peters -86*	W Bg	German	317	114·0	27·7	18·1	Danzig	1857	Kraer & Devrient		H. Peters	Stolpe
810	V	Bretagne *Lecomte*	W Bk	French	374	121·7	24·9	15·7	St. Malo	1872	Hercent	V 2.88	Guibert & fils	St. Malo
811	✻	Bretagne *David* (ex Albert Victor)	W Bkn	French	257	120·0	24·9	13·0	Dartmouth	1878	Philip & Son	V 12.88	Em. Gilles & A. Bellet	Fécamp
812		Bretagne *Nerman*	W Bg	French	221	86·3	23·6	11·7	Le Legué	1847			Perrot	Paimpol
813		Bride	W Bk	British	3[illegible]	118·7	27·7	14·6	P.E. Island	1870	W. T. Ellis		J. H. Davis	Sydney, N.S.W.
814	✻	Bride *Poppitz -86-88*	C S	German	8[illegible]	1[illegible]·4	32·6	19·0	Sunderland	1870	Briggs	V 12.88	C. Rahtkens	Rostock
815	✻	Bridesmaid *Davies 77-88*	W 3m Sr	British	497	157·0	29·7	17·3	P.E. Island	1875	Richards	V 2.87	H. Hoskin	Swansea
816		Bridget Annie *Taylor*	I 3m Sr	British	109	95·2	19·6	8·6	Barrow	1877	D. Noble & Co.		W. H. Walton	Barrow
817		Bridget Smith *(Robinson -81)*	W Sr	British	115	85·0	21·6	11·6	Berwick	1860	Gowan		J. Fisher & Sons	Lancaster
818		Bridgitt	W Sr	British	120	84·0	20·1	9·6	ThreeRivers	1854	L. Beaulieu		H. Brownlee	Quebec
819	R	Bridgewater *Abergh -88*	W S	American	1482	1[illegible]9·8	41·5	28·3	Philadelphia	1855	W. Cramp & Sons	V 6.85	Jno. H. Allen	New York
820	✻	Brier Holme *Jameson 72-76*	I Bk	British	894	2[illegible]·4	33·6	19·0	Sunderland	1876	J. L. Thompson & Sons		Hine Bros.	Maryport
821	R	Brigadier *Cousins*	W 3m Sr	American	295	119·0	29·0	12·6	Stockton, Me.	1873	Hichborn	V 2.88	J. C. Cousin	Rockland, Me.
822	N	Brigida *Olsen 83-88*	W Bk	Norweg'n	456	133·7	29·9	17·6	St. Martin's, N.B.	1861			A. F. Klaveness	Sandefjord
823	V	Brigneti *Simonetti -88*	W Bk	Italian	450	124·6	30·2	19·5	Sestri	1868	Cademnacio	V 1.88	G. B. Brigneti	Genoa
824	DV	Brilhante *Schmidt -87*	W Bn	German	233	114·0	26·8	11·2	Elmshorn	1879	D. W. Kremer	V 8.87	H. N. A. Meyer	Blankenese
825	✻	Brillant *Haffie -88* (ex Marcus)	I Bk	German	1151	2[illegible]7·8	34·3	22·9	Hull	1864	M. Samuelson & Co.		C. M. D. Jörgensen	Hamburg
826	G	Brillant *Abrahamsen -86* (ex Meroë, ex Bombay, ex Baron de Hertslaer)	W Bk	Norweg'n	740	140·4	35·8	21·3	Middelburg	1862			L. Vauvert	Porsgrund
827	✻	Brilliant *Bowden 75*	I S	British	1613	254·8	39·7	24·2	Aberdeen	1877	J. Duthie, Sns & Co.		J. Duthie, Sons & Co.	Aberdeen
828		Brilliant	W Bg	British	19[illegible]	88·0	21·2	11·0	St. Mary's, N.S.	1855			I. Nickerson	Halifax, N.S.
829		Brilliant	W Bg	British	181	84·1	23·2	14·6	Montrose	1839	W. Dunn		G. Hardy	W. Hartlepool
830		Brilliant *Rosevear*	W Sr	British	124	74·3	19·6	12·2	Fowey	1848	G. Nickels		J. C. Willcocks	Fowey
831	NV	Brilliant *Isaks*	W Bk	Norweg'n	764	165·5	35·3	19·8	Porsgrund	1878	C. A. Hansen	V 8.86	H. G. Kirsebom	Christiansand
832		Brilliant *Rasmussen 86-86* (ex Annie Laurie)	W Bk	Norweg'n	626	153·8	31·3	18·7	New Glasgow, N.S.	1857	G. McKenzie		T. H. Nördaas	Egersund
833	N	Brilliant *Abrahamsen -88*	W Bk	Norweg'n	465	126·7	30·4	17·2	Grimstad	1867	C. Paulsen	V 8.88	Bernt Einersen	Grimstad
834	V	Brimiga (WRECKED) *Gelsie*	W Bk	British	603	155·0	32·1	17·9	Maitland, N.S.	1873	Anth'ny & Br'wn	V 2.87	A. Smith	Windsor, N.S.
835		Brinkburn Priory *Davies*	W Bk	British	431	134·0	28·7	17·7	Sunderland	1868	Hardie	V 8.86	D. Davies	Newport
836	NV	Brio *Gabrielsen 78-86*	W Bn	Norweg'n	167	99·2	25·2	10·5	Christiansand	1876	Torstensen	V 8.87	Joh. Boysen	Christiansand
837		Brisk	W Bg	British	128	85·3	22·6	10·4					Hajee Abdul Ramen Mamoojee	Port Louis
838	DV	Bristol *Lawrence*	W Bk	British	1305	196·9	39·7	23·6	Hantsport, NS	1878	J. Davison	V 8.88	Geo. W. Churchill	Windsor, N.S.
839	✻	Bristow *Johns 84-87*	C Bk	British	351	133·3	25·0	15·3	Liverpool	1854	Jones, Getty & C.	V 8.88	J. Loveluck & Son	London
840	✻	Britannia *Scott -87*	I S	British	2160	289·5	40·1	24·5	Southampton	1883	Oswald, Mordaunt & Co		Ship Britannia Co. Ld. (J. Campbell)	London
841		Britannia	W 3m Bn	British	199	106·8	24·3	13·5	Garmouth	1866	W. Kinloch		T. G. Robins & Co.	Guernsey
842	✻	Britannia *Curly 80-88*	W K	British	73	76·0	16·1	9·1	Stromness	1883	F. Stanger		T. J. O'Keefe	Wexford
843		Britannia *Hotchkiss*	W Bn	American	266	128·2	30·3	12·5	Madison, Conn	1874	Miner & Son	V 11.83	H. Trowbridge's Sons	NewHaven, Conn
844	G	Britannia *Greifing -86*	I Bk	German	841	174·0	31·5	19·7	Glasgow	1874	A. Stephen & Sns		D. H. Wätjen & Co.	Bremen
845	I	Britannia *Valle -87*	W Bk	Italian	431	124·6	30·6	19·9	Savona	1868	F. Sirello	V 2.88	A. Demartino	Genoa
846	DNV	Britannia *Christensen*	W Bk	Norweg'n	652	159·4	33·8	19·5	Risör	1881	A. Andersen	V 1.87	M. Danielsen	Risör
847	N	Britannia *Andersen*	W Bk	Norweg'n	647	154·5	34·3	18·4	Gefle	1863	P. Ellfstrand	V 1.88	J. Gamborgs Bo.	Christiania
848		Britannia (WRECKED) *Langlie -87*	W Bk	Norweg'n	416	127·9	31·9	16·8	Jacobstad	1857			H. Langlie	Soon
849	✻	British Ambassador *Jones 1888*	I S	British	1794	262·0	41·9	23·8	Liverpool	1873	Potter & Hodgkinson		British Ship Owners' Co. Ld.	Liverpool
850	DV	British America *Lawrence*	W S	British	1050	184·0	35·2	21·7	Hantsport, NS	1869		V 4.88	British American Ship Co. Ld. (J. S. Wright & Co.)	Belfast

✻ = Lloyd's Register; D = Nederlandsche Vereeniging; G = Germanischer Lloyd; I = Registro Italiano; R = American Record; A = Veritas Austro-Ungarico; H = Veritas Ellenico; N = Norske Veritas; V = Bureau Veritas.

Dans la 4me Colonne, C = Composite (membrure en fer, bordé en bois); I = Fer; Stl = Acier; W = Bois.

Les Dates du doublage en métal (Col. 11) ressortent des livres dans lesquels les navires sont respectivement classés ou des rapports des Inspecteurs du Lloyd's Register.

Pour plus amples renseignements sur les navires classés par le Lloyd's Register, voir le Lloyd's Register Book et le Lloyd's Yacht Register.

Official No.	No. in Book	Registry in which classed	Ship's Name. Material and Rig. Master, and Dates of Appointment to present Owner's Service and to Ship. Ship's Late Name.	Flag.	Net Tonnage	Dimensions. Length.	Breadth.	Depth.	Built. Where.	When.	By whom.	Date of Metalling.	Owners or Managers.
[illegible]	851	R	British American *Flurl* W S	British	1208	186·5	38·7	24·0	N.Brunswick	1863		[illegible]	British American Ship Co., Ld. (J. S. Wing)
[illegible]	852	✻	British Army (ex Centre) I Bk	British	1289	221·1	36·9	22·6	Sunderland	1869	Oswald & Co.		Gillison & Chadwick
[illegible]	853	✻	British Commodore (ex Knight Bachelor) *Owen* 85-86 I S	British	1390	221·6	36·1	24·7	Newcastle	1868	Palmers' Co.		Parry Jones & Co.
[illegible]	854		British Constitution *Robinson* W Bk	British	427	127·5	29·5	18·0	SouthShields	1863	Young		J.T. Davison
[illegible]	855	✻	British Duke (WRECKED) *Fofield* 80-86 I Bk	British	1429	232·8	38·0	23·5	Barrow	1875	BarrowSBC.,Ld		British Ship Owners Co. Ld.
[illegible]	856	✻	British Empire *Kyle* 79-84 I S	British	1499	245·1	39·7	23·4	Sunderland	1869	W.Pile&Co.		G. Duncan & Co.
[illegible]	857	✻	British Empire *Jones* 88-88 I Bk	British	484	155·6	27·2	17·0	Hartlepool	1868	Denton,Gray&Co.		J. Rees
54098	858	✻	British Envoy *Crowest* 85-87 I S	British	1265	214·5	35·9	22·9	Liverpool	1866	T.Royden&Sons		J. Coupland
60[illegible]	859	✻	British General *Tulloch* 67-74 I S	British	1754	257·6	41·9	23·9	Liverpool	1874	T. Royden&Son		British Ship Owners Co. Ld.
31433	860	✻ D	British India (ex Surabjee Jamsetjee Jeejeebhoy) *Cleeve* 83-85 I Bk	British	1199	207·2	35·1	22·8	Birkenhead	1864	LairdBros.		R. W. Leyland & Co.
87[illegible]	861	✻	British Isles *Stoddard* 75-84 Stl S	British	2394	308·9	43·9	24·8	Pt.Glasgow	1884	J.Reid&Co.		British Ship Owners Co.
81[illegible]	862	✻	British Merchant *Mulray* 77-80 I S	British	1696	262·9	38·3	23·7	Belfast	1880	Harland&Wolff		British Ship Owners Co.
51482	863	✻	British Monarch *Morrow* 73-79 I Bk	British	1262	212·1	36·4	23·2	Birkenhead	1865	G.R.Clover&Co.		H. Fernie & Sons
51452	864	✻	British Peer *Irons* 86-86 I S	British	1428	247·5	36·4	22·5	Belfast	1865	Harland&Wolff		J. Nourse
51435	865	✻ D	British Princess *Pierson* 84-84 I Bk	British	1480	246·7	36·7	22·8	Birkenhead	1865	G.R.Clover&Co.		Gracie, Beazley & Co.
50471	866	R	British Princess (WRECKED) *McFadyen* W Bk	British	1346	199·7	38·8	23·9	N.Brunswick	1864	Nevins		W. Clark
51083	867	✻	British Queen *Sharples* I Sr	British	145	106·0	21·2	10·3	Preston	1864	W.Allsup&Co.		W.H. Walton
52033	868		British Queen W Bg	British	113	82·3	23·2	9·4	Granville,N.S.	1865			Edward Foster
83429	869		British Queen W Sr	British	100	90·3	23·9	7·9	SmithBay	1863	R.J.M.Morgan		J.T. Young
60068	870	✻	British Sceptre *Neville* 85-86 I S	British	1436	224·3	37·8	23·5	Liverpool	1868	T.Royden&Son		British Ship Owners Co. Ld.
81391	871	✻	British Yeoman *White* 84-84 I S	British	1898	269·2	39·8	24·2	Southampton	1880	Oswald, Mordaunt & Co.		British Ship Owners Co. Ld.
	872		BritoJunior (ex [illegible]) *Nascimento* [illegible]-82 W Sr	Brazilian	148	..	..	..	VilladoConde	1868		Y [illegible]	J.M.R. Brito, jr.
48416	873		Briton W Bk	British	452	140·0	28·6	18·0	Invercargill, [N.Z.	1864	J.Nicol		W.H. Smith & Sons
76992	874	✻	Briton *Beckett* 87-87 W Bkn	British	240	117·2	25·1	12·8	Portsmouth	1876	Crampton		J.T. Crampton
65657	875		Brocklesby *Rix* 85-85 W Bn	British	195	98·8	24·3	12·1	Lowestoft	1871	Westaway	Y 11.82	J. Hall, jun., & Co.
	876		Broder *Johansson* W Sr	Russian	127	..	..	..		1872			J.F. Sundberg
	877	D N V	Broderfolket *Falck-Dessen* W Bk	Norweg'n	667	169·1	33·9	18·3	Christiania	1878	H.Hansen	Y [illegible]	Chr. Brinch
85207	878		Brodeur Demers W Sp	British	142	106·1	22·8	7·7	Sorel,Quebec	1882	O.Frechette		Damase Leroux
78682	879	✻	BrodickBay *Wakeham* 88-88 I Bk	British	753	187·0	31·1	18·5	Dumbarton	1878	Birrell, Stenhouse & Co.		Hatfield, Cameron & Co.
71743	880	✻	Brodick Castle *Fraser* 85-86 I S	British	1785	258·6	40·3	23·0	Glasgow	1875	T.Wingate&Co.		T. Skinner & Co.
	881	V	Brødrene *Baumann* -82 W Bg	Danish	214	111·9	23·7	11·7	Rudkjobing	1873	S.Boas	Y [illegible]	C.C. Nielsen
	882	V	Brødrene *Jørgensen* 81-81 W Sr	Danish	150	96·9	22·1	11·1	Thurø	1881	J.Ph.Jørgensen		J.P. Jørgensen
	883	N	Brödrene (ex Sølve) *Halvorsen* [illegible]-83 W Bk	Norweg'n	679	153·4	32·9	20·5	[illegible]	1858		Y 11.85	G. Gundersen
	884	D N	Brödrene *Kildahl* W Bk	Norweg'n	665	150·7	33·9	18·4	Risör	1879	E.Knudsen	Y [illegible]	J. Kildahl
	885		Brödrene *Thorsen* -85 W Bk	Norweg'n	464	151·9	31·3	15·6	Finland	1864			Ths.S. Falck
	886		Brödrene (ex St. Philbert) *Evensen* -87 W Bk	Norweg'n	454	135·0	30·3	17·4	Nantes	1861	Guibert		Brödr. Jorgensen & Brødr. Brok
	887	N	Brödrene *Bie* -87 W Bk	Norweg'n	411	119·5	29·7	17·0	Arendal	1867	A.Bie		J. Bie
	888	N	Brödrene *Evensen* -71 W Bk	Norweg'n	389	131·4	30·5	16·4	Tvedestrand	1874	H.Evensen	Y 11.[illegible]	Brödrene Thorvildsen
	889	N	Brödrene *Olsen* 72-86 W Bg	Norweg'n	293	101·6	27·5	15·7	Tvedestrand	1863	C.A.Smith		P. Pedersen
	890	N	Brödrene (ex [illegible]) *Olsen* W Bk	Norweg'n	276	112·0	26·2	14·4	Stavanger	1854		Y [illegible]	L. Larsen
	891	N	BrödreneClausen *Paulsen* W Bg	Norweg'n	288	107·6	26·9	14·8	Christiansand	1851			H.M. Paulsen
	892		Brödres Minde *Hansen* W Bg	Norweg'n	210	83·1	26·3	15·9	Tönsberg	1829			Chr. Johansen
	893	✻	Broederschap (ex [illegible]) *Meiboom* W 3m Sr	Dutch	295	108·5	22·3	15·7	Rye	1852	Wassell&H'lmes		W.S. van de Wetering
	894	✻	Broedertrouw *Kuppen* 85-86 W St	Dutch	183	105·9	21·9	11·9	Hoogezand	1872	J.A.Hooites	Y [illegible]	J.A. Hooites
	895		Bromarf *Nordström* -81 W Bn	Russian	272	108·3	29·4	13·0	Tenala	1875	K.Henriksson		J.F. Strandberg
682[illegible]	896	✻	Broomhall *Dundas* 86-86 I Bk	British	1389	243·7	37·2	22·3	Sunderland	1874	W.Doxford&Ss.		W.S. Croudace
3238	897	R	Brooxxe B. Rokes *Scrivener* -88 W 3m Sr	American	319	133·4	33·2	9·5	Bath,Me.	1883	Geo. Sawyer & [Packard		B.B. Rokes
88102	898		Brothers *Gibson* W Sr	British	125	78·6	26·0	10·4	Eleuthera	1884	A.E.Bethel		R. Gibson Bros.
36812	899		Brothers W Sr	British	111	106·5	23·3	12·1	P.E.Island	1860	McMillan		P. Donnelly
[illegible]	900		Brothers (STRANDED) *Thwaites* 70-70 W Bn	British	108	..	..	..	Maryport	1786			J. Thwaites

✻ = **Lloyd's Register;** D = Nederlandsche Vereeniging; G = Germanischer Lloyd; I = Registro Italiano; R = Ameri[can]
A = Veritas Austro-Ungarico; H = Veritas Ellenico; N = Norske Veritas; V = Bureau [Veritas]

In the 4th Column, C = Composite (iron frame, wood planked); I = Iron; Stl = Steel; W = Wood.

The Dates of Metalling (Col. 11) are based on the records of the Books in which Vessels are respectively classed, or on the Reports of the Surveyors of Lloyd's [Register].

Further particulars of Vessels classed by Lloyd's Register will be found in Lloyd's Register Book and Lloyd's Yacht Register.

No.	Cote	Nom. Matériaux et Gréement. Capitaine	Mat.	Pavillon	Tonnage	Longueur	Largeur	Creux	Où	Quand	Par qui	Doublage	Armateurs ou Intéressés-Gérants	Port d'Armement
901	✱	Brothers [illegible]	W Sk	British	46	[illegible]	[illegible]	[illegible]	[illegible]	1848	D.Griffiths		J.Thomas	Cardigan
902		Brothers [illegible] (ex La Bonne Vierge)	W Sr	[illegible]	14[illegible]	[illegible]	17·3	[illegible]	[illegible]	1844				
903	V	Brothers Apap [illegible]	W Bk	British	[illegible]	[illegible]	[illegible]	[illegible]	Malta	187[illegible]		Y 1.8[illegible]	P.Apap & figli	Malta
904		Brothers German (N.Pomenti)	W Bk	British	[illegible]	[illegible]	[illegible]	[illegible]	Malta	1862	German Bros.	Y 9.8[illegible]	A.Bennett	Malta
905	✱	Broughton [illegible] 70-79	I Bk	British	571	[illegible]	27·[illegible]	1[illegible]	[illegible]	1868	Harland & Wolff		Thomas Watkins & Jenkins	Swansea
906		Browney	I Sp	British	168	[illegible]	27·[illegible]	7·[illegible]	[illegible]	1874	R.Dixon & Co.		Bell Bros.,Ld.	Middlesbro'
907	✱	Brownrigg Williamson (ex [illegible]) [illegible]	I [illegible]	British	[illegible]	[illegible]	[illegible]	24·3	Greenock	1884	Russell & Co.		J.Houston	Liverpool
908	✱	Bruce Macking [illegible]	W S	British	1[illegible]	[illegible]	[illegible]	22·[illegible]	[illegible]	1857	W G. Russell	Y 9.8[illegible]	Bruce Sailing Ship Co., Ld. (J. L. Murray)	Port Glasgow
909	✱	Bruckley Castle [illegible]	C Bk	British	1[illegible]	[illegible]	31·5	22·[illegible]	[illegible]	1867	[illegible]	[illegible]	Bk."Bruckley Castle"Co. Ld. (J. S. Brown & Co.)	London
910		Bruin	W	[illegible]	1[illegible]	87·[illegible]	[illegible]	5·3	Gravenhurst	1881	J.Gerrard		Muskoka & Nipissing Nav. Co.	Toronto, Ont.
911	V	Bruns Canadian	W Sr	[illegible]	142	[illegible]	[illegible]	11·[illegible]	[illegible]	1860	G.Mal[illegible] & Co.		Buhot-Launay aîné	Paimpol
912	✱	Brunel	Stl S	British	[illegible]	[illegible]	[illegible]	22·7	Pt.Glasgow	1889	R.Duncan & Co.		Leitch & Muir	Greenock
913	✱	Brunette (ex Pride of the West)	I Bk	British	37[illegible]	11[illegible]	25·4	15·8	[illegible]	1859	C.Langley		J.C.Ellis	Sydney,N.S.W.
914	✱	Brunette Edwards 75-85	I Bar[illegible]	British	308	1[illegible]	2[illegible]	14·7	[illegible]	1866	C.W.Nevill & Co.		Llanelly Iron Ship Co.Ld.	Llanelly
915	✱D	Brunette Strike	W Sr	British	213	115·[illegible]	2[illegible]	12·9	Newhaven	1872	[illegible]	Y 8.8[illegible]	H. & J.Strike	London
916	V	Brünnow Gossland 14-83 (ex [illegible])	W Bk	German	482	12[illegible]	30·[illegible]	17·1	[illegible]	1867	W.R.Groth	Z 4.8[illegible]	Eduard. Hempden-macher'sche Erben	Rügenwalde
917	N	Br[illegible] Grande [illegible]	W Bk	Norwegian	3[illegible]	108·1	27·7	11·[illegible]	[illegible]	1844		Y 1.8[illegible]	Johan Haussen	Arendal
918	V	Bruno & Marie [illegible] (ex [illegible])	W Bg	German	297	118·2	27·2	16·0	Altona	1862	[illegible]	Y 9.8[illegible]	Franz Schultze	Rostock
919		Brush-Gedeh	W	[illegible]	18[illegible]	..	..	..	Meboli	1871				Meboli
920	✱	Brussels Fleison (ex City of Brussels)	I Bk	British	991	211·3	31·9	21·5	Glasgow	1863	C.Connell & Co.		H.Hogarth	Ardrossan
921	✱D	Bryn Gwyn Roberts -85 (ex [illegible])	I Bk	British	1226	217·[illegible]	36·7	22·0	Liverpool	1883	W.H.Potter & Ss		W.H.Owen	Liverpool
922	✱	Bryn Ivor Mathias 83-83	W Bk	British	368	129·8	27·6	17·0	Sunderland	1866	[illegible]	Y 5.8[illegible]	W.Pickford	Newport
923	✱	Brynhilda Meikle 74-85	I S	British	1409	2[illegible]	[illegible]	21·7	[illegible]	1885	A.Stephen & Sns		J.W.Carmichael & Co	Glasgow
924	N	Brynild Nielsen	W Sr	Norweg'n	151	91·6	24·3	11·0	Sandefjord	1877			S.Larsen	Sandefjord
925		Bua Pan Sentorjch	W S	Siamese	544	140·9	32·4	17·3	Bangkok				Chessua Poh Hee	Bangkok
926	A	Bubulina (ex [illegible]) Economo	W Sr	Austrian	262	..	..	..	[illegible]	1884	E.[illegible]	Y 2.87	D.Economo	Cassos
927		Bubulina	W Bn	Greek	171	81·3	24·9	13·1	Syra	1874	Andrico	Y 7.8[illegible]	P.Svoronos	Cephalonia
928	✱	Buccleuch Johansen -87	I S	British	1991	269·0	41·6	24·3	Liverpool	1885	T. Royden & Sns.		W.H.Ross & Co.	Liverpool
929		Bucco Gough -88	W Sr	British	145	87·9	26·2	9·5	St.John,N.B.	1866			Scammell Bros.	St.John,N.B.
930		Bucephalus Nicholson	W S	British	1036	18[illegible]	[illegible]	21·9	[illegible]	1854	J.Lemelin,jun.		R.S.Allen	St.John,N.B.
931	✱	Buckhurst Adamson -88	I S	British	1877	275·0	40·3	23·7	Dumbarton	1880	M.M'Millan & Son		"Buckhurst" Ship Co.,Ld. (W. R. Price & Co.)	London
932	✱	Buckingham Lyalias -88	Stl 4m S	British	2613	[illegible]7·7	45·1	24·2	Liverpool	1888	T. Royden & Sns.		Steel Sail Ship Buckingham Co. Ld. (M'Vicar, Marshall & Co.)	Liverpool
933	N	Bud Larsen -88	W Bg	Norwegian	146	86·1	22·1	11·7	Sydney (Rebuilt)	1874			B.A.Ellingsen	Christiania
934	V	Buda Corbett -83	W Bn	British	346	127·7	29·6	13·0	MossGlen, NB	1881	G.Merritt	Y 11.83	H.J.Olive	St.John,N.B.
935	N	Budstikken Ettesk	W Bk	Norwegian	240	117·6	26·2	12·2	Porsgrund	1870		Y 3.8[illegible]	P.M.Petersen & Sön	Porsgund
936	✱	Buenos Ayres Ahrens	W Bk	German	643	16[illegible]·2	31·5	18·9	Bordeaux	1876	Rochetta Lar-Igetau	Y 8.8[illegible]	E.Burchard & Co.	Rostock
937		Bugeldi (ex [illegible]) Aslami	W Bk	Montenegrin	390	147·4	27·2	18·6	[illegible]	18[illegible]	Poli	Y 1.8[illegible]	D.Aslatni	Dulcigno
938		Bullion	W S	American	1243	19[illegible]	38·6	24·0	Bath, Me.	1877	Br'wn & Stantial		J.B.Torrey & Co.	Bath,Me.
939		Bundaleer Williams	W S	[illegible]	921	196·4	34·2	20·2	Aberdeen	1869	Humphrey	Y 8.82	Robert Moore	Antofagasta
940	A	Buon Giuseppe Trevia (ex [illegible])	W Bg	[illegible]	42[illegible]	121·7	29·6	20·4	[illegible]	1869	[illegible]	Y 8.8[illegible]	F.Dall' Orso di D.	Spezia
941	I	Buon Pastore Bertolotto	W Bg	Italian	266	115·4	24·6	15·9	[illegible]	1878	[illegible]	Y 7.8[illegible]	P.Lavarello	Genoa
942	I	Buona Nuova Esposito (ex [illegible])	W Bg	Italian	271	105·9	27·7	16·[illegible]	[illegible]	1863	F.S.Mauro	Y 1.8[illegible]	F[lli]. Massa	Naples
943		Buona Sorte Veli Aga (ex [illegible])	W Bg	Montegrn	270	10[illegible]	27·7	16·6	[illegible]	1868	[illegible]	Y 1.8[illegible]	Selim Veli Aga	Dulcigno
944		Buona Sorte Heggia	W Bn	Montegrn	125	73·8	18·9	10·2	Venice	1855			Selim Veli & Co.	Dulcigno
945		Buona Sorte (ex Colombus)	W Bk	Turkish	363	118·7	26·9	15·7	[illegible]	1858	Aurelli	Y 4.89	Ibrahim Bilali	S.Giovanni di Medua
946		Buona Sorte S.	W Bn	Montegrn	144	73·5	23·3	13·5	Chioggia	1859			Asan Aga & Co.	Dulcigno
947		Buoni Amici Piana	W Sr	Italian	197	10[illegible]·6	25·8	13·5	Savona	1885	G.B.Gustavino	Y 6.8[illegible]	F.Cassini	Porto Maurizio
948		Burdett Hart	W Sr	American	319	126·3	30·0	10·0	New Haven, Conn.	1857			E.G.Ostenberg	New York
949	✱	Burgedin Griffiths 83-77	W Sr	British	93	79·1	21·1	9·1	Rhyl	1877	Jones		W.Aaron & Co.	Liverpool
950														

✱ = Lloyd's Register; D = Nederlandsche Vereeniging; G = Germanischer Lloyd; I = Registro Italiano; R = American Record; A = Veritas Austro-Ungarico; H = Veritas Ellenico; N = Norske Veritas; V = Bureau Veritas.

Dans la 4me Colonne, C = Composite (membrure en fer, bordé en bois); I = Fer; Stl = Acier; W = Bois.

Les Dates du doublage en métal (Col. 11) ressortent des livres dans lesquels les navires sont respectivement classés ou des rapports des Inspecteurs du Lloyd's Register.

Pour plus amples renseignements sur les navires classés par le Lloyd's Register, voir le Lloyd's Register Book et le Lloyd's Yacht Register.

Official No.	No. in Book.	Registry in which classed.	Ship's Name. Material and Rig. Master, and Dates of Appointment to present Owner's Service and to Ship. Ship's Late Name.	Flag.	Net Tonnage.	Dimensions. Length.	Breadth.	Depth.	Built. Where.	When.	By whom.	Date of Metalling.	Owners or Managers.
	951	✳ D	Burgemeester Schorer *Zoverdt* WBk	Dutch	1236	193·5	41·0	24·8	Middelburg	1878	Middelburg's Wharf	Y 7.87	G. van de Tievel & van Vooren
	952		Bürgermeister Bauer *Mussel 79-* WBk	German	328	108·0	26·8	14·6	Rostock	1854	J.Möller		F.W.Fischer
	953	V	Bürgermstr. Kirstein *Ebrecsrich* WBk	German	356	124·0	28·2	16·2	Loitz	1860	J.Peuss	Y 9.87	Louis Klein
	954		Burgermeister Oom (RULE) *Heuts 84-* WSr	German	151	78·7	21·8	12·6	Barth	1847	F.W.Liegnitz		Wilh.A.Sarnow
	955	G	Burgermstr.Petersen WBg	Swedish	285	105·3	26·0	16·3	Rostock	1850	A.Schönemann		A.Leffler & Sohn
	956	G	Bürgermeister Stuve (ABANDONED) *Meermann -79* WBg	German	232	105·2	24·5	11·5	Papenburg	1865	H.W.Meyer		W.Hoveller
792	957	A	Burlamacchi *Francesconi* (ex Anna Giovanna) WBn	Italian	113	86·6	23·4	10·8	Leghorn	1858	Bartiani	Z 9.86	Fili. Francesconi fu G.C.
7947	958		Burns *Hacher 69-76* WBk	British	385	123·5	25·8	17·2	Annan	1852	J.Nicholson&Co		R.&W.King
	959	V	Buron *Bourdonnais 65-81* WBg	French	265	101·0	26·6	12·0	St.Malo	1875	J.Parnet	Z 5.85	Vve Pringué
220	960		Burra Norfol WBk	Siamese	340	130·0	24·5	15·0	Bangkok	1862			Goh Ah Chiam
	961	V	Bury St. Edmunds WBk	Argentine	656	153·0	28·4	21·0	Sunderland	1853	W.Wilkinson	Y 12.88	J.Morté
2832	962	R	Bushrod W. Hill *Higgins* W3mSr	American	605	152·2	34·1	16·0	Bath,Me.	1872	A.Hathorn		C.E.Morrison
66970	963	R	Busiris *Morris -86* WSr	British	249	108·0	28·7	10·5	Alma,N.B.	1874	P.C.Copeland		J.D.Seely
85620	964		Buster *Kefkegle* WBkn	British	305	129·2	31·4	12·0	Parrsboro,N.S	1884	N.Cathoum	Y 12.85	J.Williams
	965		Buston Vale WBk	British	412	133·3	28·0	17·7	Newcastle	1863	Hutchinson		Indian Governme[illegible]
37594	966		Busy WBn	British	139	93·8	24·1	11·7	NovaScotia	1857	Young		H.Craig & Co.
2762	967		Busy WSr	American	313	128·1	28·2	13·2	Norfolk,Va.	1869			F.H.Chappell
91804	968	✳	Busy Bee *Dyer 85-85* WSr	British	132	92·3	21·8	11·8	Plymouth	1885	C.Gent	Y 10.85	C.Gent
95084	969	✳	Buteshire *McGibbon 79-88* Stl4mS	British	1854	266·5	40·0	23·7	Dumbarton	1888	Birrell, Stenhouse & Co.		T.Law & Co.
72281	970	RV	Buteshire *Steele -87* WBk	British	967	179·5	34·8	21·0	Black Riv.,N.B	1876	J.&R.McLeod	Y 9.87	D.Carmichael
87847	971	✳	Buttercup *Burnett 84-84* WK	British	93	78·3	20·3	10·2	Goole	1884	G.W.Outwin		G.W.Outwin
92820	972	✳	Buttercup *Neurse* WSk	British	84	73·0	20·1	10·3	Elmshorn	1886	J.Kremer		H.Mudd
88191	973	✳	Buttercup *Johnson* WSk	British	76	74·6	20·4	10·3	Rye	1884	G.&T.Smith		Capt.W.Theobald
76414	974	✳	Buttermere *Owens 86-86* IBk	British	993	214·0	34·1	20·2	Whitehaven	1877	WhthvnS.B.Co.		Fisher & Sprott
96301	975	✳	Byculla *Raffles -87* Wk	British	77	85·4	20·7	6·9	Ipswich	1887	W.Bayley&Sons		E.Garnham
	976	DN	Byfoged Christensen *Iversen 81-81* W3mBk	Norweg'n	339	132·8	26·9	14·3	Christiansund (rebuilt)	18 67/77	P.Holm	Y 1.88	O.Reinertsen & Sønner
	977		Byfoged Christie (ABANDONED) *Iversen* WSr	Norweg'n	257	121·1	25·7	11·7	Kragerö	1881	A.Juliusen	Y 5.88	A.Terkelsen
	978	N	Byfoged Lindahl *Stabell 82-87* WBk	Norweg'n	291	137·8	27·1	13·4	Christiansund	1858	O.Muhle		N.H.Knudtzon
	979	NV	Bygdö *Henschien -88* WBk	Norweg'n	513	141·4	31·2	17·1	Christiania	1871	Mathiesen	Y 1.88	I.A.Henschien
	980	GV	Bylgia *Fock -87* WBk	German	333	121·0	26·6	13·0	Korsör	1877	A.Peterson	Y 3.88	J.G.Lund
85540	981		Byron WSr	British	124	91·0	24·3	9·5	Shelburne,N.S	1883	A.Hood		B.Hines
	982	NV	Byron *Johanessen 85-85* WBk	Norweg'n	346	125·3	28·6	15·3	Tvedestrand	1875	M.Eldrup	Y 8.88	L.Larssen
83218	983	R	Byron M. *Lecain -87* WSr	British	128	84·0	25·0	9·1	Cornwallis NS	1882	J.Steele,jun.		J.Stech,Jr.
240	984	A	Byzantion *Csichlis* WSr	Greek	245	103·0	25·6	..	Syra	1879		Y 11.87	L.Brilis
	985	✳	Byzantium *Fanni* (ex Marietta W., ex Giuseppe Baccarcich) WS	Turkish	781	163·3	32·5	21·4	Fiume	1863	G.Brazzoduro	Y 4.87	G.Chryssaphidis

Additional Ships are inserted in the Supplement at the end of the Register.

✳ = Lloyd's Register; D = Nederlandsche Vereeniging; G = Germanischer Lloyd; I = Registro Italiano; R = [illegible]
A = Veritas Austro-Ungarico; H = Veritas Ellenico; N = Norske Veritas; V = Bu[illegible]

In the 4th Column, C = Composite (iron frame, wood planked); I = Iron; Stl = Steel; W = Wood.

The Dates of Metalling (Col. 11) are based on the records of the Books in which Vessels are respectively classed, or on the Reports of the Surveyors of [illegible]

Further particulars of Vessels classed by Lloyd's Register will be found in Lloyd's Register Book and Lloyd's Yacht Register.

No.	Class.	Nom, Maître et Gréement (Ancien Nom)	Pavillon	Tonnage	Longueur	Largeur	Creux	Construit Où	Quand	Par qui	Doublage	Armateurs ou Directeurs-Gérants	Port d'Armement
1		C. *Phillips* W3mSr	American	139	123·0	23·8	6·1	Bridgeton, N.J	1881	Phillips		L.B.Phillips	Bridgeton, N.J
2	✱	C.A. *Harvey* I 3mSr	British	116	98·8	19·1	9·8	Llanelly	1872	Nevill Bros.		Sailing Ship C. A. Co., Ld. (E. F. Hickman)	London
3		C. A. White *Fogg* W3mSr	American	791	178·5	38·0	15·6	Bath, Me.	1884	New Eng.S.B.Co		W.B.Church	Fall River, Mass
4	R	C. B. Church *Anderson* W3mSr	American	797	175·2	32·2	18·0	Bath, Me	1884	B.W.&H.F.Morse		B.W. & H.F.Morse	Bath, Me.
5	✱	C.B.K.F. *Hutchings* W K	British	64	74·0	19·9	9·0	Falmouth	1881	C.Burt		C.B.Kelway	Falmouth
6		C. B. Paine *Hilyard* WSr	American	196	103·0	28·2	9·9	Eastport, Me.	1875	Huston		C.A.Paine	Eastport, Me.
7		C. B. Wood *Stanley* WSr	American	225	110·6	29·6	9·2	Dennisville N.J	1867	R.S.Leeming		B.D.Wood	Boston, Mass.
8		C. Bernier WSr	British	117	86·0	26·0	10·3	Cap St. Ignace	1867			A.Belanger	Quebec
9	D	C. Boschetto (Giuseppe B.) *Marini* W Bk	Italian	723	166·0	33·5	21·8	Varazze	1875	D.Craviotto	[illegible]	G.B.Razeto fu F.	Genoa
10	V	C. C. Chapman *Hichborn* WS	American	1587	222·3	39·9	25·2	Bath, Me.	1877	W.Rogers	[illegible]	J.S.Winslow & Co.	Portland, Me.
11		C. C. Davidson W3mSr	American	110	97·0	27·4	6·0	Milton, Del.	1887			J.Ponder	Wilmington
12	V	C. C. Funk *Glaser* W Bkn	American	513	163·0	36·8	13·9	Coos Bay, Oreg	1882	H.Reed		A.C.Glaser	San Francisco
13	V	C. G. Hornung (Anna ...) W Bn	Danish	195	109·0	21·9	11·7	Ekensø	1871	H.L.Hanssen	Y 7.86	Actieselskab (S. & B. Horbsen)	Skjelskør
14		C. C. Lane *Bergman* W3mSr	American	306	130·0	32·0	11·0	E.Haven, Conn	1873	W.O.Nettleton		F.W.Benedict	New Haven, Conn
15	R	C. C. Sweeney *Robertson* W Bn	American	596	135·0	33·3	17·8	Harrington Me	1871		[illegible]	A.Robertson	New York
16		C. C. Van Horn *Fraser* W Bn	British	377	116·7	29·5	12·6	Windsor, N.S.	1868	S.Dimock		G.W.Churchill	Windsor, N.S.
17	R	C.C.Wehrum *Coulter* W Sr	American	376	139·0	36·0	10·10	Wilmingtn Del	1887			C.W.Alcott & Co.	New York
18	RV	C. D. Bryant *Hopkins* W Bk	American	887	176·0	37·0	20·6	Searsport, Me.	1878	Pendleton & Co.	[illegible]	J.C.Gilmore	Searsport
19		C. D. Dos Hermanos (San José Felipe) W Sr	Spanish	105	68·6	20·1	8·5	Vinaros	1856			P.Ballester	Alicante
20	✱	C. E. C. G. *Cavanagh* W K	British	56	68·2	17·7	8·1	Jersey	1869	C.W.Aubin		J.F.Picot	Jersey
21	V	C. E. Robinson *Green* W Bk	British	531	113·4	30·0	17·6	Gardner's Creek, N.B	1876	W.&R.Wallace	[illegible]	G.H.Nixon	St. John, N.B.
22	✱	C. E. Spooner *Jones* W3mSr	British	172	103·2	24·0	12·7	Portmadoc	1878	Jones	Y 2.88	J.Jones & Co.	Carnarvon
23	V	C. E. White *St. Croix* W3mSr	British	228	111·4	27·9	10·5	St. John, N.B	1886	F.S.Bonnell	[illegible]	F.S.Bonnell	St. John, N.B.
24	GV	C. F. Ivers *Lenz* W Bk	German	303	121·9	27·6	13·9	Stettin	1865	C.A.Domcke		F.Ivers	Stettin
25	G	C. F. Maass *Zeplien* W Bg	German	267	104·3	27·0	14·1	Rostock	1865	O.Ludewig		H.Mentz	Rostock
26	RV	C. F. Sargent *Meyer* WS	American	1638	220·3	41·3	26·0	Yarmouth, Me.	1874	C.F.Sargent	[illegible]	George Plummer	San Francisco
27		C. G. Cranmer *Walton* W3mSr	American	211	115·0	30·2	10·3	Tuckerton, N.J	1865			J.Mair	Philadelphia
28		C. H. Eaton *Sinclair* W Sr	American	161	101·0	27·9	9·0	Calais, Me.	1869	O.B.Rideout		Rideout & Lord	Calais, Me.
29	R	C. H. Foster *Coombs* W3mSr	American	366	130·0	30·2	10·0 14·6	Duxbury, Mass	1872		[illegible]	C.Foster	Boston, Mass.
30	R	C.H.S. *Thomas* I Bn	British	198	112·8	24·2	12·8	Hayle	1863	Harvey & Co.		J.Samuel	Llanelly
31		C. H. Sampson W Bn	British	104	83·1	22·4	7·6	Bath, U.S.	1851			Thomas Winniett	Halifax, N.S.
32		C. Henry Kirk *Curtis* W Sr	American	175	99·7	28·8	8·7	Belleville, N.J	18 65/80			E.R.Kirk	New York
33		C.J. W Bg	French	124	74·8	20·7	10·8	Agde	1868	Contreste		Laurent & Ferriere	Philippeville
34	G	C. Karl (ex Emilie Nayport) *Jansson* W Bg	Swedish	290	108·5	25·2	14·2	Ribnitz	1867	J.H.Wilken		A.G.Jansson	Väddö, Rsign.
35	GV	C. L. Weyer *Frank* W Bk	German	418	121·6	29·0	17·1	Greifswald	1863	J.W.Sadelkow		J.N.Rodbertus	Barth
36		C. Lovett W Bn	British	197	91·6	25·6	12·0	Weymouth N.S	1866			R.Morrow	Digby, N.S.
37		C. M. Reynolds *Magill* W Bn	British	144	100·3	26·6	10·8	Lepreaux, N.B.	1864	W.Reynolds		J.Neill	Belfast
38		C. Mathews *Miller* W Sr	American	131	94·0	24·8	8·3	Searsport, Me	1854			Curran & Hook	Bangor, Me.
39	R	C. O. Whitmore *Thompson* W Bk	American	855	164·1	35·3	22·6	Bath, Me.	1868	Goss & Sawyer	[illegible]	C.Holmes	San Francisco
40	RV	C. P. Dixon *Rhodes* W Bk	American	692	152·0	33·5	19·7	Belfast, Me.	1881	J.T.Cottrell	[illegible]	G.Gregory	New York
41	DV	C. Paulsen (ex Beaumanoir) *Thiele* I Bk	German	680	169·2	28·5	19·2	Nantes	1870			C.Paulsen	Elsfleth
42	GV	C. R. Bishop (ex Hebe ...) *Jacobsen* W Bk	German	880	164·7	30·9	20·8	Bremerhaven	1872	R.C.Rickmers	[illegible]	J.C. Pflüger & Co.	Bremen
43		C.R.C. *Le Couteur* W Bg	British	248	118·3	26·0	13·7	Quebec	1871	Renout		G.Le Gros	Jersey
44	R	C. R. Flint *Brown* W3mSr	American	253	130·3	30·7	8·3	Bath, Me.	1871		Y 9.82	B.Flint	New York
45	✱	C. Rice *Mogarry* W Sk	British	88	80·0	21·0	10·6	Hull	1880	Alex.Dk.Ship-[wright Co.		J.Munby	Hull
46	✱	C. S. Atkinson *Crellin* W Sr	British	105	93·0	21·7	9·9	Carrickfergus	1878	Rodgers	[illegible]	J.Atkinson & Co.	Belfast
47	R	C. S. Bushnell *Lewis* W Bkn	American	645	150·5	32·7	19·3	E.Haven, Conn	1867		[illegible]	S.Gildersleeve & Co.	New Haven, Conn.
48	R	C. S. Packard (Lord) *Nausloie* W Bn	American	305	115·5	28·2	14·7	Rockport, Me.	1869		Y 1.86	Raynald Thomas	Philadelphia
49	R	C. S. Parnell *Healey* W Bk	British	744	164·0	34·2	18·0	Richibucto, N.B	1880	R.O'Leary	Y 9.87	E.Lantalum	Richibucto, N.B
50	RV	C. Southard Hulbert *Wallis* W Bk	American	1036	178·6	35·7	21·2	Richmond, Me	1881	Southard & Son	[illegible]	T.J.Southard & Son	Bath, Me.

✱ = Lloyd's Register; D = Nederlandsche Vereeniging; G = Germanischer Lloyd; I = Registro Italiano; R = American Record;
A = Veritas Austro-Ungarico; H = Veritas Ellenico; N = Norske Veritas; V = Bureau Veritas.

Dans la 4me Colonne, C = Composite (membrure en fer, bordé en bois); I = Fer; Stl = Acier; W = Bois.

Les Dates du doublage en métal (**Col. 11**) ressortent des livres dans lesquels les navires sont respectivement classés ou des rapports des Inspecteurs du Lloyd's Register.

Pour plus amples renseignements sur les navires classés par le Lloyd's Register, voir le Lloyd's Register Book et le Lloyd's Yacht Register.

Official No.	No.	Classed	Ship's Name. Material and Rig. / Ship's Late Name.	Rig	Flag	Net Tonnage	Length	Breadth	Depth	Built: Where	When	By whom	Metalling	Owners or Managers
	51	N V	C. Tobias *Knudtzen*	W Bk	Norweg'n	830	165·4	35·7	18·8	Arendal	1886	N. Evensen	[illegible]	Hans H. Pettersen
[illegible]	52		C. U. Chandler *Buck* (STRANGERS)	W Sr	British	100	81·3	22·1	8·3	Rockland, N.B.	1881	H. A. Chapman		Geo. W. Chandler
[illegible]	53		C. W. Bangs	W Brgn	British	152	105·0	25·0	7·4	Ottawa	1886	J. Neville		T. H. Kirby
[illegible]	54	R V	C. W. Janes *Crosby*	W Bkn	British	825	165·3	36·0	20·5	Meteghan [River, N.S.	1884	W. D. B. James	[illegible]	J. G. Ross
[illegible]	55	R	C. W. Lewis *Fehring -87*	W 3mSr	American	306	135·3	31·5	10·7	Thomaston Me	1875		[illegible]	Henry Bros.
4316	56		C. W. Locke *Chase* (OCTAVIA [illegible])	W Sr	American	219	109·0	27·1	8·9	Port Republic, [N.J.	1862			A. J. Chase
	57	G	C. von Platen *Lembke*	W Bg	German	203	96·8	24·2	12·8	Stralsund	1868	J. Pense		W. A. Sarnow
94073	58	✻	C. & M. Slater *Beecham -91*	W Sk	British	86	78·8	20·5	10·5	Rye	1887	G. & T. Smith		W. Empson
[illegible]	59	✻	Cabar Feidh *Sandison 77-81*	W Sr	British	86	74·0	20·5	9·8	Stornoway	1877	Cook		B. Swanson
[illegible]	60		Cabarfeidh *Hughes 84-84*	W Bk	British	333	125·7	29·6	12·9	N. Glasgow NS	1874	J. W. Carmichael	[illegible]	C. B. Stone
	61		Cabieces	W Sr	Spanish	448	132·8	27·9	16·7	Bilbao	1870			P. A. San Martin
	62	✻	Cabinda (ex Challenger) *Evans*	C Bk	Portug'se	599	158·0	28·8	18·2	Middlesboro'	1865	Riddle	[illegible]	Portuguese Government
	63		Cabo Menorca *Fernenias* (ex [illegible])	W Sr	Spanish	164	87·6	25·8	12·3	Calella	1875	A. Amargos		B. B. Capella
	64		Cabocla	W Bn	Brazilian	101	..	..	..					J. A. P. Porto
	65		Cabral *Ommundsen 80-80*	W Bk	Norweg'n	222	123·0	27·2	11·3	Arendal	1865	S. Johannesen		T. Thomassen & Smith
	66		Cabral	W Sr	Brazilian	111	..	..	..					J. A. Cabral
	67		Cabral I. *Bainha*	W Sr	Brazilian	133	..	..	..	S. Joan da Barr	1857		[illegible]	M. M. Cabral
[illegible]	68	✻	Cabul *Rowe*	I S	British	1397	223·2	37·3	22·7	Southampton	1877	T. R. Oswald		E. Bates & Sons
	69	V	Cacilda (ex [illegible]) *Palo -87*	W Bg	Portug'se	237	110·2	24·3	11·7	Villa do Conde	1871	M. G. Rodrigues	[illegible]	J. Nogueira Pinto
	70		Cacilie *Schlichting*	I Cr	German	105	..	..	..		1884			J. H. Ewald
[illegible]	71		Cacique	W Bk	British	141	84·0	19·4	10·5	Isle of Man	1831			Henry Moss
[illegible]	72		Cader Box	W Bg	British	159	85·1	20·8	11·3	Cochin	1868			Ana Rosa S. V. Chitty
[illegible]	73	D	Cadetto *Peragallo*	W Bk	Italian	394	127·6	30·2	19·2	Sestri	1869	G. B. Cadenaccio	[illegible]	G. B. Peragallo
[illegible]	74	✻	Cadwalader Jones *Cadwaladr 78-78*	W Sr	British	95	81·5	21·8	10·9	Portmadoc	1878	Owen		D. Jones & Co.
[illegible]	75	✻	Cadwgan *Jones 87-87*	Stl Bk	British	1264	222·0	36·5	21·4	Sunderland	1885	W. Doxford & Ss		R. Thomas & Co.
[illegible]	76	✻	Cadwgan *James 83-83*	W Dy	British	69	72·1	20·1	7·9	Aberayron	1883	D. Jones		Davies & Evans
[illegible]	77	✻ D	Cadzow Forest *Smith 80-80*	I Bk	British	1068	207·7	34·0	21·0	Port Glasgow	1878	Russell & Co.		J. C. King
	78	V	Cæsar *Lauritzen 55-70*	W Sr	Danish	112	95·6	19·5	9·6	Troense	18[55/81]	C. R. Møller		W. Petersen
	79	✻	Cæsar *Sundqvist 80-83*	W Bk	Swedish	463	142·7	26·6	16·3	Frannäs	1881	E. Ostman	[illegible]	Frannäs Warfs Aktiebol
	80		Cæsar *Norman*	W Sr	Swedish	180	100·0	20·6	11·6	P. E. Island	1862	C. J. Riis		J. Norman
	81		Cæsar *Nilsson*	W Sr	Swedish	129	88·9	20·6	9·5	Frannäs	1874	Frannäs Aktbol		R. Fogelmarck
[illegible]	82	✻	Cæsarea *Hansford 86-86*	I S	British	1257	229·5	36·2	22·1	Sunderland	1877	W. Doxford & Ss		W. Pellier
[illegible]	83		Cagliari (ex Panchito) *Rozzo*	W Bk	Italian	578	143·7	32·8	20·2	Recco	1865	F. Rolla	[illegible]	B. Fravega
[illegible]	84	R	Caibarien *Perkins* (ex Jens [illegible])	W Bk	American	383	122·2	29·8	16·5	Arendal	1869		[illegible]	E. L. G. Steele & Co.
[illegible]	85	✻	Cairnbulg *Murray*	I Bk	British	1567	261·3	39·0	23·0	Aberdeen	1874	J. Duthie & Co.		W. Duthie, jun.
[illegible]	86	✻	Cairnsmore *Scott 89-84*	S Bk	British	878	199·1	32·8	20·3	Port Glasgow	1884	J. Reid & Co.		Nicholson & McGill
[illegible]	87	✻ D	Caitloch *Dornan 73-86*	I S	British	1264	235·0	37·0	21·8	Port Glasgow	1874	R. Duncan & Co.		Hunter, Brown & Co.
[illegible]	88	✻ D	Calabar *Phillips 81-83*	I Bkn	British	305	151·6	25·1	12·0	Liverpool	1877	W. H. Potter & Co.		T. Harrison & Co.
[illegible]	89	R V	Calabria *Palmer*	W 3mSr	British	530	154·4	35·6	16·2	Parrsboro, NS	1881	G. E. Pettis	[illegible]	J. Beretta
	90		Calalina (ex Sofia, ex Teresa A. Keene)	W Bm	Hawaiian	316	130·0	32·3	10·1	Waldoboro	1875			J. J. Lorca
[illegible]	91	✻	Calcutta *Cathro 85-85* (ex City of Calcutta)	I S	British	984	212·0	32·0	21·7	Glasgow	1860	A. Stephen & Sons		C. Barrie
[illegible]	92		Calcutta	W S	American	843	163·0	33·0	21·0	Portsm'uth NH	1868			E. Lawrence, jun.
	93		Calcutta *La Varde -87*	W Bk	French	454	135·1	29·4	18·8	Bordeaux	1868	Raymond frères	[illegible]	A. Hubeau
[illegible]	94	✻	Caldbeck *Bevre 79-79*	I Bk	British	747	184·0	30·4	19·2	Dublin	1864	Walpole & Webb		R. Nicholson & Sons
[illegible]	95	V	Caldera *McKenzie 76-76*	W S	British	1575	230·5	39·5	24·1	River John, NS	1884	A. McKenzie	[illegible]	D. Sutherland
[illegible]	96	✻	Caleb *Ashton*	W K	British	78	75·2	19·6	9·0	Barton Stather	1884	J. Garside		T. Ashton & Son
	97	N	Caleb *Johnsen -83*	W Bk	Norweg'n	242	114·3	26·4	12·9	Porsgrund	1866			A. L. Bache
	98		Caleb *Killberg 81-81*	W Sr	Swedish	135	90·5	21·3	9·8	Timm'rnabb'n	1876	C. Nilsson		J. Larsson
[illegible]	99	R V	Caleb S. Ridgway *Townsend 77-84*	W 3mSr	American	384	141·8	34·2	12·0	Wilmington Del	1879	Jackson & Sharp		J. E. Ridgway
[illegible]	100	✻	Caledonia *Hoffmeyer 73-75*	W Bkn	British	312	136·0	27·5	12·9	Sunderland	1878	Pickersgill	[illegible]	J. Duncan

✻=**Lloyd's Register;** D=Nederlandsche Vereeniging; G=Germanischer Lloyd; I=Registro Italiano; R=Am[illegible]
A=Veritas Austro-Ungarico; H=Veritas Ellenico; N=Norske Veritas; V=Bur[illegible]

In the 4th Column, C=Composite (iron frame, wood planked); I=Iron; Stl=Steel; W=Wood.

The Dates of Metalling (Col. 11) are based on the records of the Books in which Vessels are respectively classed, or on the Reports of the Surveyors of [illegible]

Further particulars of Vessels classed by Lloyd's Register will be found in Lloyd's Register Book and Lloyd's Yacht Register.

No.	Cotes	Nom. Capitaine. Ancien Nom.	Gréement	Pavillon	Tonnage	Longueur	Largeur	Creux	Construit Où	Quand	Par qui	Doublage	Armateurs ou Propriétaires	Port d'Armement
101		Caledonia *Osborn*	WSr	British	191	[illegible]	23·8	13·2	Garmouth	1864	Geddie		H.P.Coleman	Faversham
102		Caledonia *Cundy 79-83*	ISr	British	160	102·6	22·0	12·3	Dumbarton	1860	W.Denny & Bros		W.Dusting	Plymouth
103	N	Caledonia	WBk	Norweg'n	711	159·9	34·4	18·1	Pugwash,N.S.	1869	J.Oulton	Y.s.86	J.S.Lorentzen	Drammen
104	DN	Caledonia *Christiansen*	WBg	Norweg'n	447	105·7	28·9	15·8	Brakestad	1850			H.Chr.Erichsen	Fredrikstad
105	A	Cali Tichi *Skinopoli*	WBg	Turkish	396	[illegible]	[illegible]	..	[illegible]	1871	P.Buni		S.Kurubasot	Agatapoli
106		California *Davis*	WBk	American	799	133·5	33·9	20·0	New York	18[illegible]	Webb		N.Bichard	San Francisco
107		California *Brightman*	WBk	American	397	114·4	28·4	18·0	N.Bedford,M.	1842			J.F.Tucker & Co.	New Bedford
108		California *Olsen*	W3mSr	American	114	95·5	22·8	8·0	SanFrancisco	1869			S.B.Peterson	San Francisco
109	✻D	California *Kipsch*	IBk	Dutch	719	153·5	32·1	23·[illegible]	[illegible]	1853	Fop Smit		A.Köpcke	Rotterdam
110	✻	Calistoga *Massey*	IS	British	[illegible]	268·1	42·1	24·0	Liverpool	1884	T.Royden & Sons		W.Nevins	Liverpool
111	DN	Calla *Hellerslossen*	WBk	Norweg'n	377	135·6	30·4	10·3	PortElizabeth	1853			A.Schive	Tönsberg
112	✻	Callao *James*	StlBk	British	978	2[illegible]·7	33·1	20·0	Belfast	1885	Harland & Wolff		NorthWesternShipping Co. Ld.	Liverpool
113		Callao *Brannan 90-76*	WBg	British	127	87·5	21·7	13·6	Whitehaven	1842	& J.Brocklebank		J.Hodgson	Whitehaven
114	✻	Caller Ou *McDonald*	CBk	British	668	172·6	28·8	18·1	Grangemouth	1866	Adamson	Y.s.85	Anderson,Webster & Co.	London
115	R	Calliope *Nickerson 87-87*	WBk	British	1202	184·8	37·1	23·9	S.John,N.B.	1868	Gass	Y.s.86	B.Smith & Sons	Windsor,N.S.
116	✻	Callirrhoe *Robson*	IS	British	1249	213·2	36·2	23·7	[illegible]	1865	Humber I.Wks.		E.Bates & Sons	Liverpool
117	DV	Callixene *Salter*	WS	British	1387	195·2	38·5	24·1	N.Brunswick	1869	Nevins & Fraser		J.R.Suiter	Maryport
118	A	Calomira *Vlassopulo*	WBn	Greek	244	..	..	..	Galaxidi	1876	C.Canata		E.Vlassopulo	Zante
119	R	Calvin B. Orcutt *Foster*	W4mSr	American	905	183·3	40·1	18·6	Bath,Me.	1888	New England S.B.Co.		Geo.Bailey	Manasqu'n
120	R	Calvin F. Baker *Baker*	W3mSr	American	538	144·5	32·9	15·4	Bath,Me.	1873	Robinson		Doane & Crowell	Dennis,Mass.
121		Calvin P. Harris *Higgins*	W3mSr	American	442	138·5	33·2	15·2	Bath,Me.	1871			A.S.Higgins	[illegible]
122		Calvin S. Edwards *Scull*	WSr	American	215	109·7	28·7	9·3	Dorchester,NJ	1860			F.Smith & Co.	Philadelphia
123	V	Calypso *Papademitri*	WBk	Russian	182	82·0	23·0	13·9	Cassos	1871			V.F.Demergi	Odessa
124		Calypso	WBg	Turkish	175	..	..	..	Cashot	1872			Costi Bala Dimitri	Cashot
125		Camagüeyana *Rivera*	WPol	Spanish	191	84·0	25·0	11·1	Voltri,Italy	1858			IgnacioBotet	Barcelona
126	✻	Camana *Worrell*	IBk	British	562	171·7	27·9	18·0	Liverpool	1865	T.Vernon & Son		Nicholson & McGill	Liverpool
127	IV	Camano *Chiapella*	WBk	Italian	497	148·1	31·0	20·1	Varazze	1870	Baglietto		B.Prato	Genoa
128	✻	Cambalu *Buck*	WBkn	British	485	141·5	27·6	19·1	Workington	1852	C.Lamport		Thomas,Watkins & Jenkins	Swansea
129	R	Cambay *Humphreys 83-83* (ex Arizona)	WBk	British	796	156·3	32·8	22·8	UnitedStates	1857			Trayes & Co.	Liverpool
130		Cambois *Gray*	WSw	British	270	112·3	26·8	16·0	Blyth	1870	Un.Co-op.Sh.C.		Miss M.M.Gray	North Shields
131	✻	Camborne *Owens*	WSr	British	99	83·2	22·5	10·1	Amlwch	1884	W.C.Paynter & Co.		T.Morgan & Co.	Beaumaris
132	RV	Cambria *Dunbar*	WBk	British	1299	201·6	38·0	23·5	Quebec	1884	E.Samson		E.Samson	Quebec
133		Cambria *Walters*	WBk	British	211	110·2	24·2	13·2	Newquay	1853	T.Thomas		D.T.Sharper	Aberystwith
134	✻	Cambria	WSrYt	British	105	102·6	21·1	11·6	Cowes	1868	M.Ratsey		F.Swanzy	Cowes
135	✻	Cambrian (Coll...)	IBk	British	629	176·0	29·0	18·0	Liverpool	1867	Bowdler, Chaffer & Co.		J.B.Walmsley & Co.	Liverpool
136	✻	Cambrian *Phillips*	WBg	British	182	100·5	24·3	13·0	Swansea	1872	Harris		Burgess & Co.	Swansea
137	✻D	Cambrian Chieftain *Thomas*	IBk	British	1458	2[illegible]	37·5	22·5	Sunderland	1885	Osbourne, Graham & Co.		T.Williams & Co.	Liverpool
138	✻	Cambrian Duchess *Jones*	IS	British	1266	221·0	36·0	21·7	Sunderland	1876	Mounsey & Foster		T.Williams & Co.	Liverpool
139	✻	Cambrian Monarch *Williams*	IS	British	1306	216·6	37·0	23·0	Southampton	1876	T.R.Oswald		T.Williams & Co.	Liverpool
140	✻	Cambrian Prince *Davies*	IS	British	1349	224·7	37·1	22·6	Southampton	1876	T.R.Oswald		T.Williams & Co.	Liverpool
141	✻	Cambrian Princess *Evans*	IS	British	1350	221·7	37·1	22·6	Southampton	1877	T.R.Oswald		T.Williams & Co.	Liverpool
142	✻	Cambrian Queen *Thomas*	IS	British	1300	221·7	36·8	22·8	Sunderland	1868	Oswald & Co.		T.Williams & Co.	Liverpool
143	D	Cambridge *Stanley*	WS	Chilian	1185	189·8	38·2	22·5	Meteghan,NS	1874	D.Saulnier		W.S.Stanley & Co.	Valparaiso
144		Cambridge	WBn	British	121	95·0	25·5	10·5	Cambridge, N.B.	1864	Wilson		R.Neill & Sons	Belfast
145	✻	Cambridgeshire *Long*	WK	British	79	71·9	20·2	10·6	Brixham	1883	S.Dewdney & Son		R.Long	Grimsby
146	V	Cambronne *Melvan*	WBk	French	296	109·5	23·9	13·0	Nantes	1875	T.Dubigeon & fils		Perthuis	Nantes
147	V	Cambusdoon *Borland*	WBk	British	1198	194·0	38·4	22·5	Gardiner's Creek,N.B.	1881	W.&R.Wallace		Jas.Kennedy	St.John,N.B.
148		Camden *Gotter*	WBk	American	501	132·0	29·3	17·2	Derby,Conn.	1847			Puget Sound Commercial Co.	PrtTwnsnd
149		Camelia	WBn	Brazilian	193	..	..	..					P.J.V. de Lima	Pernambuco
150	GV	Camelia *Arians*	WS	German	1301	188·0	37·4	23·8	Bath,Me.	1869	Moses & Sons		D.Schilling	Bremen

✻ = Lloyd's Register; D = Nederlandsche Vereeniging; G = Germanischer Lloyd; I = Registro Italiano; R = American Record; A = Veritas Austro-Ungarico; H = Veritas Ellenico; N = Norske Veritas; V = Bureau Veritas.

Dans la 4me Colonne, C = Composite (membrure en fer, bordé en bois); I = Fer; Stl = Acier; W = Bois.

Les Dates du doublage en métal (Col. 11) ressortent des livres dans lesquels les navires sont respectivement classés ou des rapports des Inspecteurs du Lloyd's Register.

Pour plus amples renseignements sur les navires classés par le Lloyd's Register, voir le Lloyd's Register Book et le Lloyd's Yacht Register.

Official No.	No. in Book		Ship's Name. Master, and Dates of Appointment to present Owner's Service and to Ship. Ship's Late Name.	Material and Rig	Flag.		Dimensions. Length	Breadth	Depth	Built. Where.	When.	By whom.		Owners or Managers.
56811	151		Camellia *Hassey*	W Bk	British	184	116·0	24·0	12·1	Bideford	1873	Cox	[illegible]	P.G.Tessier
75402	152		Camellia *Hands*	W Sr	British	101	89·3	21·1	9·7	Portsmouth	1876	Crampton		J.T.Crampton
62831	153	❊	Camelot *Murphy*	W Bk	British	369	129·2	26·5	16·6	Pt.Dinorwic	1868	Jones	[illegible]	M.Denning
125688	154		Cameo *Mahlman*	W Bn	American	231	111·5	27·1	10·5	Bath,Me.	1878	Goss&Sawyer		G.Wessels
	155		Camila 1a de Tossa	W Bn		171	95·5	26·0	12·7	Tosa,Spain	1859			Viuda de Moret
32808	156		Camilla *Garth*	W Bn	British	190	85·9	23·6	14·5	Leith	1834			Belbin&Co.
17068	157		Camilla	W Bn	British	161	96·5	22·5	12·5	P.E.Island	1856			G.Sutton & Son
	158	V	Camilla *Jensen 84-84*	W Sr	Danish	141	94·2	20·6	10·8	Troense	1873	C.J.Jensen		R.G.Thomsen
	159	G	Camilla *Beckmann 68-88*	W Bk	German	447	127·4	29·0	17·1	Greifswald	1868	L.Doodt		C.Graedener
64.	160		Camilla *Giribaldi -88*	W Bg	Italian	272	97·4	27·0	17·4	Varazze	1859	F.Tassara	[illegible]	G.Traverso
	161	N	Camilla *Sørensen*	W Bk	Norweg'n	443	124·6	29·9	17·0	Arendal	1874	C.A.Halvorsen	[illegible]	HansH.Pettersen
	162	N	Camilla *Zachariassen* (ex [illegible])	W Bk	Norweg'n	418	131·4	32·0	15·7	Björneborg	1875	G.Gustaffson	[illegible]	C.H.Sörensen
	163	N	Camilla *Hansen 76-86*	W Bg	Norweg'n	344	109·5	29·1	17·2	Baccari	1857			J.E.Jacobsen
	164	N	Camilla *Andersen*	W Bg	Norweg'n	188	80·9	25·4	14·3	Arendal	1831			N.Thorsen
	165		Camilla *Thoresen*	W Sr	Norweg'n	128	90·9	21·5	10·2	Malmö	1870	J.Bonnesen		A.Taraldsen
30834	166		Camille *Williams*	W Bk	British	365	112·0	27·6	16·5	Nantes	1855			Edwd.Stafford
57370	167		Camille	W Bg	British	221	100·0	26·0	12·0	Nantes	1867		[illegible]	Henry Guthrie
70967	168	❊	Campana *Rutchart*	I Bk	British	772	192·5	31·7	19·2	Liverpool	1875	R.&J.Evans		G.Bell & Co.
44661	169		Campanero *Springsteen 1 -88*	W Bk	British	271	116·8	27·8	11·2	Baltimore	1859	J.J.Abrahams (& Sons)	[illegible]	J.H.S.Millard (J.M.Randall & Sons)
	170	V	Campbell *Simonsen 86-86*	W Bk	Norweg'n	1132	190·9	38·0	21·8	WeymouthNS	1879	C.Campbell	[illegible]	Johan Thorsen
	171	N	Campbell *Wilhelmsen* (ex John Campbell)	W Bk	Norweg'n	715	152·5	32·8	19·3	WeymouthNS	1868	R.Hankinson		Chr.Eyde & Hein
	172	V	Campêche *Leprince*	W Bk	French	395	138·4	26·6	15·3	Honfleur	1866	E.Cardon	[illegible]	Vve.G.Oriot
72[illegible]	173	❊	Camperdown *Ferguson (88-)*	I S	British	1487	245·9	38·2	22·9	Glasgow	1875	A.Stephen&Ss		D.Bruce & Co.
	174		Campidoglio (HULK)	W Bk	French	549	137·7	32·1	19·6	Varazze	1868	G.Aschieri	[illegible]	
	175		Camponez *Chaves*	W Bn	Brazilian	178	..	..	..	S.João da Barra	18 73		[illegible]	F.J.Pinto,jr.,& Co.
	176		Camponeza	W Sr	Brazilian	152	..	..	..					J.F.S.Vieira
70281	177	❊	Campsie Glen (CONDEMNED) *Webster [86-86]*	W Bk	British	490	145·4	30·0	17·9	Sunderland	1874	Gardiner	[illegible]	W.Davison & Co.
	178	❊	Camtoos *Holm 82-82*	W Bn	Danish	210	98·9	20·7	14·0	Alloa	1862	T.Adamson		M.J.Ipsen
2046	179		Camuelin	W Bk	Italian	537	145·2	31·9	19·5	Sestri Ponente	1870	Marcenaro		G.N.Storace
75516	180	❊	Camwood *Coad*	W Sr	British	125	85·5	23·6	10·8	P.E.Island	1877	White		J.G.Coad
	181		Canabarro	W Bg	Brazilian	131	..	..	..					J.V.S.Canabarro
5308	182		Canada *Warner*	W S	American	1145	176·6	36·8	23·9	Bath Me.	1859			G.E.Atkinson
	183	V	Canada *Herman -87*	W Bk	German	467	123·9	29·5	17·9	Memel	1865	B.Pieper		F.W.Ogilvie
	184		Canada *Nilsson 78-78*	W Bk	Swedish	362	121·3	27·6	12·5	Windsor,N.S.	1868			N.Olsson
47738	185	R	Canadian *Welham*	W Sw	British	293	108·5	26·7	16·6	Sunderland	1864	S.P.Austin&Sn.	[illegible]	J.Welham
77921	186	R	Canadian *Landry -88*	W Bn	British	232	102·9	25·1	12·5	P.E.Island	1878	J.E.McDonald		W.Le Blanc
6063	187		Canadienne *Rivard*	W Sr	British	128	93·0	23·7	9·6	Grondines	1874			A.Rivard
73099	188		Canadienne	W Sp	British	113	102·0	22·2	7·0	Yamaska	1875			A.Robillard
46691	189		Canadienne	W Sr	British	104	87·2	22·0	8·3	Lotbinière	1863			Samolé Parent
88674	190	RV	Canara *Dexter 87-87*	W S	British	1549	217·6	40·4	24·3	St.John,N.B.	1884	Stew'rt&Ritchie	[illegible]	W.Thomson & Co.
	191		Canario (ex Carea) *Espinoza*	W Bk	Chilian	417	136·0	26·1	16·2	Aberdeen	1859	Hood		Sanchez & Cardem
54985	192	❊	Candahar *Peterkin*	I S	British	1418	239·4	37·2	23·6	Belfast	1866	Harland&Wolff		T.& J.Brocklebank
	193	N	Candase *Johannesen -83*	W Bk	Norweg'n	383	129·3	28·5	15·5	Lillesand	1882	G.Vatlesvord	[illegible]	O.& P.Knudsen
51103	194		Candaswamy Puravy	W Bg	British	107	80·1	21·0	9·7	Quilon	1873			KoonaPanaCA MChitty
	195	N	Candeur *Nielsen 71-86*	W Bk	Norweg'n	451	133·5	31·5	16·1	Arendal	1874	J.Aanonsen	[illegible]	J.E.Hanssen
76290	196	❊	Candida *Mathewson 75-75*	I S	British	1222	226·0	36·1	22·1	Whitehaven	1875	WhithvnS.B.Co.		Lowden,Edgar&Co.
452	197		Candida *Tonietti*	W Bn	Italian	108	77·1	22·8	12·3	Prà	1851	Tixi		P.Tonietti fu B.
60065	198	❊	Candidate *Sharpe*	I Bk	British	765	189·5	31·9	19·7	Glasgow	1868	Bobie&Co.		J.Gambles
30566	199		Candidate *Pullen -87*	W Bk	British	583	139·3	28·6	20·3	London	1842	E.Hawkes		T.Daniel & Sons
82849	200	❊	Candidate *Pullen*	W K	British	69	72·0	19·2	10·0	Lowestoft	1886	S.Richards		T.&J.Leleu,Ld.

❊ = Lloyd's Register; D = Nederlandsche Vereeniging; G = Germanischer Lloyd; I = Registro Italiano; R = Am[erican]; A = Veritas Austro-Ungarico; H = Veritas Ellenico; N = Norske Veritas; V = Bu[reau Veritas]

In the 4th Column, C = Composite (iron frame, wood planked); I = Iron; Stl = Steel; W = Wood.

The Dates of Metalling (Col. 11) are based on the records of the Books in which Vessels are respectively classed, or on the Reports of the Surveyors of [...]

Further particulars of Vessels classed by Lloyd's Register will be found in Lloyd's Register Book and Lloyd's Yacht Register.

No.	Classe	Nom / Capitaine	Gréement	Pavillon	Tonnage	Longueur	Largeur	Creux	Construit où	Année	Par qui	Doublage	Armateurs ou Directeurs Gérants	Port d'Armement
	V	Candidezza-Lauro	W Bg	Italian		116·8	28·3	19·2	Cassano	1872	A. Castellano		F. Lauro	Castell.di Stab.
202	V	Candido	W Bk			126·9	29·5	18·0		1867	G. Mauro		V. Trapani & C.	Castell.di Stab.
203	✱	Canmore	I Bk	British	729	193·3	31·5	17·9	Dundee	1876	Brown & Simpson		Albany Ship Co. Ld.	Dundee
204	R V	Canning	W Bk		857	151·0	31·9	18·3	Cornwallis, N.S.	1874	E. Bigelow		C. A. Palmer	Windsor, N.S.
205	✱	Canopus	I Bk			144·4	28·1	17·2	Sunderland	1866			H. F. A. Evers	Hamburg
		Canova	W Bk	British	1359	201·4	37·1	23·0	N. Brunswick	1859	Rodbeck		T. Emerson	Newcastle
	N V	Canova	W Bk	Norweg'n	219	103·5	24·6	12·4	Arendal	1877	A. Henriksen		D. Herlofson & Axel	Arendal
	✱	Canterbury	I S	British	1245	230·7	36·0	20·8	Port Glasgow	1874	R. Duncan & Co.		Shaw, Savill & Albion Co. Ld.	Glasgow
		Canton	W		313	117·0	29·3	10·7	Saco, Me.	1874	Merrill		C. A. Whittier	Portland, Me.
		Canton	W Bk	American	227	164·4	24·8	15·4		18[illegible]			J. F. Tucker & Co.	New Bedfrd. Mass.
		Canton	W Bk	Swedish	364	131·8	25·8	15·0		1863	C. H. Kraus		G. Sjöberg	Malmö
	V	Canute	W S	British	1337	200·8	40·1	24·8	St. John	1863	Hilyard		J. Edgett	London
		Canute	W S	British	1215	197·5	38·4	23·1	Quebec	1869	M'Kay		Trayes & Co.	Cardiff
	✱	Caonabo	W Bk	Norweg'n	666	163·7	32·6	17·9	P. E. Island	1882	A. McMillan		Eger & Sorensen	Christiania
	N V	Cap	W Bk	Norweg'n		158·8	33·1	17·5	Brevig	1875	P. Knudsen		M. H. Smith	Tvedestrand
	✱ V	Cap Horn	I Bk	French	2037	305·7	44·6	24·2	Pt. Glasgow	1888	Russell & Co.		Ant. Dom. Bordes & fils	Bordeaux
217	✱	Cape Breton	I S	British	1421	239·3	37·8	22·1	Glasgow	1877	J. & G. Thomson		A. Lyle & Sons	Greenock
		Cape City	I Bk	British		112·0	26·0	16·0	Sunderland	1861	W. Pile & Co.		G. Lewis	London
	✱	Cape Clear	I Bk	British		189·7	31·2	20·5	Liverpool	1869	Bowdler, Chaffer & Co.		W. J. Myers, Son & Co.	Liverpool
	✱	Cape Finisterre	I Bk	British	882	198·5	33·3	18·6	Glasgow	1874	T. Wingate & Co.		A. Lyle & Sons	Greenock
		Cape Horn Pigeon	W Bk	American	201	100·0	21·9	14·4	Dartm'th Mass.	1854			E. F. Potter	New Bedfrd. Mass.
	✱	Cape Race	I S	British		197·2	34·2	18·3	Greenock	1874	Caird & Co.		A. Lyle & Sons	Greenock
	✱	Cape St. Vincent	I S	British	1422	239·1	37·8	22·2	Glasgow	1877	J. & G. Thomson		A. Lyle & Sons	Greenock
	✱	Cape Verde	I S	British	1711	249·0	40·3	23·1	Glasgow	1874	T. Wingate & Co.		A. Lyle & Sons	Greenock
	✱	Cape Wrath	I B	British	1199	231·0	37·0	23·4	Port Glasgow	1871	J. Reid & Co.		A. Lyle & Sons	Greenock
	✱	Cape of Good Hope	I S	British	1339	238·8	37·7	22·5	Glasgow	1876	J. & G. Thomson		A. Lyle & Sons	Greenock
		Capella	W Bn	British	123	85·0	21·4	10·2	P. E. Island	1856	Pinett		J. R. Lobb	Fowey
	D V	Capella	I Bk	German	915	185·7	32·4	19·3	Vegesack	1875	H. F. Ulrichs		W. A. Fritze & Co.	Bremen
	✱	Capella	I Bk	German	504	155·7	27·7	17·1	Sunderland	1863	Oswald & Co.		Oldrey, Williamson & Pier	Hamburg
		Capella	W Bk	German	501	143·2	28·6	16·4	Bremerhaven	1858	C. Lange		A. Burchard	Rostock
	N	Capella	W Bk	Norweg'n		136·7	35·1	20·8	Alblasserdam	1855			F. Breda	Christiania
	N	Capella	W Bk	Norweg'n	617	149·3	32·9	17·1	Nystad, Finland	1864			A. F. Klaveness	Sandefjord
	N	Capella	W Bk	Norweg'n	296	119·1	27·2	13·5	Brevig	1869	N. Aslesen		J. E. Jacobsen	Farsund
	N	Capella	W Sr	Norweg'n	107	88·3	23·8	10·1	Rosendal	1866	L. Skåle		C. Mehus	Bergen
	V	Capenhurst	W Bk	British		148·0	31·9	17·9	Alma, N.B.	1878	Foster		J. Ritchie, jun.	Liverpool
		Cappelen	W Sr	German	112	81·3	19·0	9·5	Papenburg	1861	J. F. Veen		Chr. Reuter	Iheringsfehn
	N	Caprice	W Sr	Norweg'n	311	111·0	28·2	16·0	Syra	1879			Carl Dahl	Fredrikshald
		Capricho	W Bn	Spanish	127	79·1	23·5	12·1	Nantes	1867			F. San Juan y Souza	Havana
	N	Capriciosa	W Bg	Norweg'n	182	96·0	25·9	11·2	America				Ths. W. Schwartz	Fredriksstad
	A	Capricorno	W Bk	Austrian		139·1	31·1	20·6	Fiume	1883	G. G. B. Bellaz		L. Cosulich di C.	Fiume
	N	Captain Dan	W Bk	Norweg'n	469	140·0	30·4	14·9	Pictou, N.S.	1872	Primrose		T. Th. Ironsen	Grimstad
		Captain Levi Young	W	Canadian	153	110·6	22·8	6·9		1886	A. Miller		D. Murphy	Ottawa, Ont.
	N	Captain Peter Dahl	W Bk	Norweg'n		151·7	32·2	17·6	Grimstad	1875	J. Bang		J. Bang	Grimstad
	A	Capurro	W Bg	Italian	299	112·8	27·9	17·1	Meta	1854	Frat. Mauro		P. Capurro	Genoa
	A	Carattere	W Sr	Austrian	168	84·3	23·4	13·2	Lussinpiccolo	1884	N. Martinolich		N. Martinolich	Lussinpiccolo
	R V	Caravan	W S	American	1395	190·2	39·5	24·0	Bath, Me.	18 55/72	Hall, Snow & Co.		T. Dunham's Nephew & Co.	New York
	✱	Carbet Castle	I S	British	1585	248·4	38·1	22·9	Sunderland	1875	Mounsey & Foster		L. H. Macintyre & Co.	Liverpool
		Carbone	W Bk	Italian	532	137·1	31·8	19·7	Voltri	1870	B. Fabiano		N. Ansaldo	Genoa
		Carboniere	W Bg	Italian	183	89·9	24·6	13·6	St. Malo	1847			C. Dapelo fu G.	Genoa

✱ = Lloyd's Register; D = Nederlandsche Vereeniging; G = Germanischer Lloyd; I = Registro Italiano; R = American Record; A = Veritas Austro-Ungarico; H = Veritas Ellenico; N = Norske Veritas; V = Bureau Veritas.

Dans la 4me Colonne, C = Composite (membrure en fer, bordé en bois); I = Fer; Stl = Acier; W = Bois.

Les Dates du doublage en métal (Col. 11) ressortent des livres dans lesquels les navires sont respectivement classés ou des rapports des Inspecteurs du Lloyd's Register.

Pour plus amples renseignements sur les navires classés par le Lloyd's Register, voir le Lloyd's Register Book et le Lloyd's Yacht Register.

Official No.	No. in Book	Registry in which Classed	Ship's Name. Master, and Dates of Appointment to present Owner's Service and to Ship. Ship's Late Name.	Material and Rig	Flag	Net Tonnage	Length	Breadth	Depth	Built: Where	When	By whom	Date of Metalling	Owners or Managers
[illegible]	251	R	Cardenas *Horton -88*	W Bk	American	369	131·8	30·3	12·0	Calais, Me.	1867		[illegible]	C.T.Geyer
	252		Cardenas Packet *Jaski -80* (MISSING) (ex Casilda)	W Bn	Dutch	126	96·8	24·6	9·4	[illegible]	1853		[illegible]	F.Stuelt & Zonen
[illegible]	253	R	Cardiff (ex Emmanuele) *Haggia*	W Bk	Austrian	1084	177·4	35·8	25·0	Bath, Me.	1858		[illegible]	Stefano Kraus fu G.
[illegible]	254		Cardigan *Driscol*	W Sr	British	121	81·0	22·3	10·1	P.E.Island	1875	M'Donald		J.Coohig
[illegible]	255	❋	Cardigan Castle *Lewis 83-88*	I S	British	1200	228·0	35·7	21·6	Liverpool	1870	R.&J.Evans		L.Davies & Co.
70514	256	❋	Cardiganshire *Taylor [80-8*	I S	British	1400	235·4	38·1	23·0	Liverpool	1877	T.Royden & Sons		Hughes & Co.
[illegible]	257	❋	Cardiganshire *Jones*	W Bkn	British	365	124·0	26·4	17·0	Neyland	1864	Gaddern	[illegible]	R.Jones
[illegible]	258		Cardillo *D'Ambrosio* (ex Annetta)	W Bg	Italian	322	117·7	27·6	17·6	SestriPonente	1860		[illegible]	G.Cardillo
	259	V	Cardinal *Zange 79-77*	W Bk	German	358	114·8	27·6	16·0	Damgarten	1879	H.Dierling	[illegible]	C.A.Beug
[illegible]	260		Carhue (ex Meredith)	W Bn	Argentine	127	..	..	..	Germany				J.Bozzano & Co.
[illegible]	261	R	Carib *Stahl*	W Bk	American	280	116·6	28·4	14·5	P.Jefferson, NY	1868		[illegible]	J.Williams
[illegible]	262	R	Carib *Montgomery*	W Bn	American	164	97·3	25·0	10·0	Boston, Mass.	1880	R.Crosbie	[illegible]	Eggers & Heinlein
[illegible]	263	R	Caribbean *Blackstock -87* (WRECKED)	W Sr	American	162	102·0	26·4	8·9	Stonybr'k, NY	1875		[illegible]	W.R.Knighton
[illegible]	264	R	Caribou *Fegan -87*	W Bk	British	697	155·0	32·0	20·0	Quebec	1870	Valin	[illegible]	S.Fegan
	265	N	Carin (ex Rachel) *Wilhelmsen*	W S	Norweg'n	957	180·4	38·3	20·3	Calais, Me.	1856		[illegible]	Johan Simonsen
46865	266		Carioca *Swan 79-87*	W Bk	British	299	129·6	24·3	14·9	Liverpool	1864	R.&J.Evans		R.&W.King
	267	❋	Carioca *Engrand -84*	I S	French	775	172·2	33·4	19·4	Nantes	1866	E.Gouin		Ant. Dom. Bordes & fils
60872	268	❋	Carisbrooke Castle *Hamlton 09-85*	I S	British	1415	239·4	37·6	22·9	Glasgow	1868	Barclay, Curle & Co.		D.Currie & Co.
80930	269	❋	Carita *Patrick -87*	W Bn	British	275	115·5	26·4	13·0	P.E.Island	1880	PeakeBros.&Co.	[illegible]	J.Loveluck & Son
	270	G	Carl *Michielli -88*	W S	Austrian	1099	174·8	32·1	23·0	Vegesack	1857	J.Lange	[illegible]	A.Ménard
	271		Carl (ex Elbe) *Bang*	W Bg	Danish	233	99·5	23·0	12·0	Ronnebeck	1867	C.Diercks & Co.	[illegible]	C.S.H.Praetorius
	272	V	Carl *Jørgensen 73-80*	W Sr	Danish	128	93·2	21·1	10·3	Svendborg	1880	P.Troensegaard		C.Jensen
	273	V	Carl (ex Tabor) *Backhaus 83-*	W S	German	1393	195·5	36·8	24·6	Bath, Maine	1869	E.&A.Sewall	[illegible]	Schilling & Co.
	274	D❋	Carl *Koehler 84-84*	I Bk	German	964	205·8	33·5	19·5	Sunderland	1884	R.Thompson & Sons		C.H.H.Winters
	275	GV	Carl *Lassow 81-*	W Bg	German	174	90·5	24·4	12·7	Greifswald	1861	J.W.Sadelkow		C.A.Beug
	276	G	Carl	W Sr	German	144	85·6	23·4	10·9	Wollin	1856	C.H.Brasewitz		H.Langhoff
	277	V	Carl (ex Nuova Angiolina)	W Bg	Norweg'n	379	116·8	30·2	18·9	SestriPonente	1865	D.Tixi	[illegible]	C.A.E.Carlsen
	278	N	Carl *Christensen 85-85*	W Bg	Norweg'n	240	107·5	25·4	13·3	Medemblik, Holland	1857	Tinkelenburg & Zonen		C.F.Seeberg
	279	N	Carl *Enge*	W Bg	Norweg'n	237	103·5	25·7	12·6	Drammen	18 [illegible]		[illegible]	O.Enge
	280	DN	Carl (ex Antverpia) *Bentzen*	W Bg	Norweg'n	210	99·1	25·3	12·0	Antwerp (rebuilt)	1840			B.Bentzen & J.Höegh
	281		Carl (STRANDED) *Fredriksson*	W Bg	Russian	221	..	..	..	Germany	1874			O.Fredriksson
	282	V	Carl August *Bonn 72-74*	W 3m Sr	German	293	116·8	26·2	14·1	Stralsund	1874	J.Peuss	[illegible]	C.A.Beug
	283		Carl August *Lüder 82-82*	W Bg	German	241	107·5	25·9	13·3	Ziegenort	1863	H.Brasewitz		J.C.Holtz
	284		Carl Both *Niemann 68-73*	W Bk	German	528	115·3	30·7	18·5	Rostock	1872	O.Ludewig	[illegible]	Franz Crull
	285	DGV	Carl Burchard *Harder [-88*	W Bk	German	543	112·6	31·3	18·5	Rostock	1883	E.Burchard & Co.	[illegible]	H.Janentzky
63	286		Carl Christoph *Rohrig [-85* (ex Helene)	W Sr	Russian	162	95·3	21·1	11·7	Papenburg	1868	F.Beckmann		Wittkowsky, Goerichl & Co.
125151	287	R	Carl D. Lothrop *Gill*	W Sr	American	278	110·0	29·0	13·8	Lubec, Me.	1873	McBride		Marcus M.Clark
	288		Carl Emil *Petersen 83-84* (ex Catharina & Hermanna)	W Sr	Danish	133	81·6	21·7	11·9	Amsterdam	1854	A.van der Hoog		Martin Jensen
	289		Carl Franz *Steinoff 82-* (LOST)	W Bg	German	178	85·5	24·0	12·7	Ueckermünde	18[43/67]	L.Wittenberg		Edwin Schütz
	290	V	Carl Friedrich *Fröhlich* (ex Red Cloud)	W S	German	2208	239·5	43·2	21·2	Quincy, Mass.	1878	G.Thomas	[illegible]	Siedenburg, Wendt & Co.
	291	GV	Carl Friedrich *Bolton*	W Bk	German	460	130·2	30·3	17·6	Damgarten	1860	H.Dierling	[illegible]	W.A.Sarnow
	292		Carl Friedrich	W Sr	German	129	74·8	22·3	10·5	Stolpmünde	1846	G.Werner		J.Fritz
	293	DG	Carl Friedrich *Haefke* (WRECKED) (ex Colberg) *-88-*	W Sr	German	108	80·9	21·1	11·2	Colberg	1853	G.Mögenburg		C.F.Hackbarth
	294	DGV	Carl Gerhard	W Bk		331	115·1	26·9	13·8	Brake	1870	J.F.Nicolai	[illegible]	
	295	DN	Carl Gustaf *Basberg 76-83*	W Bk	Norweg'n	671	167·4	35·4	18·2	Wasa, Finland	1868	C.C.Wolff		H.J.Basberg
	296		Carl Gustaf *Lindqvist [81-83* (STRANDED)	W Bg	Russian	388	131·5	27·2	15·0	Christinestad	1875	K.J.Lundgren	[illegible]	C.G.Hydén
	297	VN	Carl Haasted *Petersen [-86*	W Bk	Norweg'n	880	183·1	35·6	19·8	Aalesund	1877	C.J.Haasted	[illegible]	J.P.Pedersen
	298		Carl Heinrich *Mandt -85*	W Bk	German	224	93·2	27·3	14·1	Barth	1849			E.Mundt
	299	N	Carl Imenæs *Johnsen 81-87*	W Bk	Norweg'n	365	127·5	28·4	15·4	Grimstad	1873	O.&L.Dømmesmoen	[illegible]	N.Thue Johnsen
	300	N	Carl Johan *Gautesen* (ex Waljekost) (ABANDONED) [65-82	W Bk	Norweg'n	577	155·0	33·4	18·6	Nystad	1866	H.Kjaldström		Carl Foss

❋ = Lloyd's Register; D = Nederlandsche Vereeniging; G = Germanischer Lloyd; I = Registro Italiano; R = Am[erican]
A = Veritas Austro-Ungarico: H = Veritas Ellenico; N = Norske Veritas; V = Bur[eau Veritas]

In the 4th Column, C = Composite (iron frame, wood planked); I = Iron; Stl = Steel; W = Wood.

The Dates of Metalling (Col. 11) are based on the records of the Books in which Vessels are respectively classed, or on the Reports of the Surveyors of [illegible]

Further particulars of Vessels classed by Lloyd's Register will be found in Lloyd's Register Book and Lloyd's Yacht Register.

No.	Cote	Nom, Capitaine, Matériaux et Gréement	Pavillon	Tonnage	Longueur	Largeur	Creux	Construit: Où	Quand	Par qui	Doublage	Armateur ou Principaux Gérants	Port d'Armement
301		Carl Johan WBg	[illegible]	[illegible]	88·0	23·5	12·1	Kimito	1867	[illegible]		K.Gefterleng	Abo
302	V	Carl Johann WBg	German	[illegible]	97·8	21·6	13·8	Ueckermünde	1856	J.C.Pretzer		Carl Wiehards	Stettin
303	G	Carl Linck WBk	[illegible]	[illegible]	131·5	33·2	22·1	Danzig	1873	J.W.Klawitter		H.Janentzky	Rostock
304		Carl Max WBkn	German	204	112·0	27·2	13·5	Rostock	1873	F.Burchard&Co.		F.Grampp	Rostock
305		Carl Milberg Jorgensen WSr	Danish	11[illegible]	78·6	19·9	10·1	Litau	1853			Chr.F.Schmit	Assens
306	NV	Carl Pihl WBk	Norweg'n	720	170·0	33·9	19·1	Aalesund	1875	C.Unsgaard		Ths.S.Falck	Stavanger
307	N	Carl Ritter WBg	Norweg'n	243	108·7	28·1	13·5	Hamburg	1850			S.C.Knudsen	Tönsberg
308	N	Carl Rosenius WBn	Norweg'n	222	121·8	25·8	11·9	Stavanger	1874	Paul Otöen		H.M.Wrangell	Haugesund
309	V	Carl W. Boman WBk	German	480	130·1	29·6	17·9	Rostock	1877	F.Burchard&Co.		Beckmann & Co.	Rostock
310	❋	Carl & Louise WBn	British	160	100·[illegible]	21·2	12·4	Pwllheli	1878	Williams		Mrs.G.Owen	Carnarvon
311	N	Carl XV. Nielsen WBk	Norweg'n	21[illegible]	111·5	27·4	12·1	Arendal	1866	J.Aanonsen		O.S.Nielsen	Arendal
312		Carla Persich WSr	Austrian	[illegible]	118·7	26·8	16·8	Fiume	1874	G.G.Bacich [Be...]		A.Baccarcich fu A.	Fiume
313	N	Carla Torjusen WBk	Norweg'n	[illegible]	115·3	28·4	18·1	Trieste	1857			O.A.Strömme	Christiansand
314	❋	Carleton StlBk	British	[illegible]	231·0	37·5	23·4	PortGlasgow	1881	J.Reid&Co.		J.Kerr & Co.	Greenock
315		Carleton Bell WSr	American	13[illegible]	93·7	21·5	9·[illegible]	Bath. Me.	1866	Deering&Donn'l		S.Nickerson & Sons	Boothbay Me.
316	❋	Carlisle Armstrong IS	British	1121	[illegible]	33·8	21·8	Hartlepool	1864	Denton,Gray&C°		R.Nicholson & Sons	Liverpool
317	❋	Carlisle Castle IS	British	[illegible]	22[illegible]	37·[illegible]	22·8	London	1868	R.&H.Green		R.& H.Green	London
318	IV	Carlo Trapani WBk	Italian	7[illegible]1	171·[illegible]	[illegible]·5	20·8	Castellam. di Stab.	1879	G.Bonifacio		M.Cacace	Castell. di Stab.
319	V	Carlo Andrea F. WBn	Italian	163	98·8	24·3	13·0	Viareggio	1885	A.Raffaelli		C.Fabricotti	Spezia
320	IV	Carlo Mainetto WBk	Italian	[illegible]	167·6	29·5	22·6	Sestri	1879	A.Briasco fu L.		C.Mainetto	Genon
321	V	Carlos Irmãos WBn	Portug'se	[illegible]	9[illegible]·2	22·0	12·5	Lagoaña	1871			C.M.Gonçalves	Lisbon
322	V	Carlos Strugmar W3mSr	Russian	[illegible]	107·6	23·3	13·0	Dagö	1882	J.Soel		Th.Laisberg & Co.	Dagö
323		Carlos Vincenzi WSr	Brazilian	17[illegible]	..	..	..					J.N.de Vincenzi& Filhos	Rio Janeiro
324		Carlos & Carlota WBk	Brazilian	[illegible]	..	..	..					F.S.Villa Lobos	Rio Janeiro
325	V	Carlota Sampera WBk	Spanish	[illegible]	10[illegible]	27·5	12·8	Arenys,Spain	1858			J.Sampere&M.Llauat	Barcelona
326		Carlota & Abreu WSr	Brazilian	1[illegible]	..	..	..					A.F.M.d'Abreu	Itapemerim
327		Carlotta WBk	British	57[illegible]	157·9	30·0	17·9	Sunderland	1875	Richardson		R.N.Smith&Co.,Ld	Newcastle
328		Carlotta Reid WSr	British	24[illegible]	104·9	27·7	11·3	Hopewell Cape, N.B.	1882	J.Cook		J.Cook	St.John,N.B.
329		Carlotta WSrYt	Italian	16[illegible]	88·2	19·2	10·6	Gosport	1851	Camper & [Nicholson]		ContePieroMasetti	Leghorn
330		Carlottina WBg	Italian	2[illegible]	10[illegible]	28·7	16·8	Voltri	1852	L.Briasco		G.Tonietti fu R.	Porto Ferrato
331		Carlskrona WBk	Norweg'n	261	10[illegible]	28·5	15·1	Carlskrona	1853			Lorck & Co.	Drontheim
332	I	Carmagnola WSr	Italian	18[illegible]	102·3	24·2	13·7	SestriPonente	1872	Celesia & Debarbieri		Filippo Calli	Catania
333		Carmel WSr	British	13[illegible]	88·3	20·5	11·3	Exmouth	1858			P.Curran	Exeter
334	NV	Carmel WBk	Norweg'n	757	161·1	33·9	19·7	St.Martins, N.B.	1870	H.McQueston		Ole.Andersen	Sandefjord
335	NV	Carmel Thorsen WBk	Norweg'n	60[illegible]	154·5	32·6	17·5	Porsgrund	1881	L.Pedersen		G.& A.Berg	Stavanger
336	N	Carmel WBk	Norweg'n	[illegible]	152·2	30·5	16·7	Risör	1875	G.Pedersen		J.F.Juell	Risör
337	I	Carmela C. Romano WBk	Italian	597	151·2	31·2	20·1	Alinsuri	1879	G.Starita		F.Cacace	Castellam.
338	I	Carmela Massa Massa WBk	Italian	493	129·6	26·5	19·0	Chioggia	1869			Fratelli Massa	Naples
339	IV	Carmelina WBk	Italian	871	170·5	33·1	22·2	SestriPonente	1874	G.Casanova		G.Olivari	Genon
340		Carmelina Orefice WBk	Italian	334	115·1	32·6	17·6	SestriPonente	1864	B.Debarbieri		G.Vassallo	Catania
341	V	Carmelina WSr	Italian	11[illegible]	91·2	23·6	10·6	Livorno	1882	F.lli Picchiotti		S.Campisi	Catania
342		Carmelina S. Spampinato WBn	Italian	205	101·6	22·6	14·0	Lussinpiccolo	1870			F.Spampinato	Catania
343	V	Carmelina V. WBk	Italian	561	143·5	31·4	20·3	SestriPonente	1867			G.Viacava	Genoa
344	❋	Carmelita CBk	British	588	161·7	28·6	17·7	Glasgow	1865	A.Stephen&Sns.		J.Jose & Co.	Falmouth
345		Carmelita WSr	Dutch	115	96·0	24·3	7·8	Maracaibo	1862	F.Cureta		J.A.Jesurun & Zoon	Curaçao
346	V	Carmelita WBk	Mexican	291	109·9	25·9	12·1	Lillesand	1881			A.Zaldivar	Progresso
347		Carmelita WBn	Portug'se	396	95·8	23·6	11·6	NovaScotia	1865			J.J.R.Leitaõ & Filhos	Funchal
348	R	Carmelita WBk	Swedish	428	134·8	26·1	15·5	Hammelwarden	1864	Gebr. Gross		L.Pählsson	Lerberget
349	❋	Carmelita Rocca WBk	Italian	603	158·4	32·0	20·0	Lavagna	1868	D.Briasco		F.Salvini	Spezia
350	❋	Carmelo WBk	British	68[illegible]	177·4	30·7	18·5	Sunderland	1876	S.P.Austin& Hunter		H.Gourlay & Co.	Glasgow

❋=Lloyd's Register; D=Nederlandsche Vereeniging; G=Germanischer Lloyd; I=Registro Italiano; R=American Record;
A=Veritas Austro-Ungarico; H=Veritas Ellenico; N=Norske Veritas; V=Bureau Veritas.

Dans la 4me Colonne, C=Composite (membrure en fer, bordé en bois); I=Fer; Stl=Acier; W=Bois.

Les Dates du doublage en métal (Col. 11) ressortent des livres dans lesquels les navires sont respectivement classés ou des rapports des Inspecteurs du Lloyd's Register.

Pour plus amples renseignements sur les navires classés par le Lloyd's Register, voir le Lloyd's Register Book et le Lloyd's Yacht Register.

Official No.	No. in Book	Society	Ship's Name, Material and Rig. Master, and Dates of Appointment to present Owner's Service and to Ship. Ship's Late Name.		Flag	Net Tons	Length	Breadth	Depth	Where Built	When	By whom	Metalling	Owners or Manager
[illegible]	351		Carmelo (ABANDONED) [illegible]	W Bk	Italian	322	112·5	27·8	17·6	Fiume	1861			G. Vassallo fu C.
[illegible]	352		Carmelo *Napoli*	W Bn	Italian	197	106·6	23·1	14·4	Servola	1871	G. Schiavoni	[illegible]	S. Munzone fu V.
	353		Carmen *Fabregas* (ex Ann [illegible])	W Sr	Mexican	139	88·8	23·1	12·0	Portmadoc	1869	Hugh Williams	[illegible]	Gonzales Bros. & C
	354		Carmen	W Bg		276	109·5	28·9	14·3	Blanes	1859		[illegible]	
	355		Carmen (ex [illegible])	W Bn	Spanish	111	95·8	19·1	8·6	Tayabas, Luzon	1864			E. Alcantara
	356		Carmen Cobian	W Bn	Spanish	114	76·4	22·9	11·7	Villagarcia	1865			Viuda de Cobian
[illegible]	357		Carmenta	W Bn	British	199	103·0	24·3	12·7	P.E.Island	1875	McMillan		W. Kemp
[illegible]	358	✻	CarnTual *Sweetland* -87	I Bk	British	498	148·9	27·1	17·5	Sunderland	1859	T. R. Oswald		J. C. Ellis
[illegible]	359	✻	Carnarvon Castle *Collins* 86-88	I Bk	British	1200	229·0	36·0	22·1	Glasgow	1867	Barclay, Curle & Co.		[illegible]
[illegible]	360	✻	Carnarvon Castle *Ellis* 87-88	I Bk	British	729	186·0	31·3	19·0	Liverpool	1870	R. & J. Evans		W. Thomas & Co.
[illegible]	361	V	Carnarvonshire (ex Eliza) *Williams* 87-88	W S	British	1336	193·6	38·1	24·2	St. John, N.B.	1870	Fisher	[illegible]	T. Williams & Co
[illegible]	362	✻	Carnarvonshire *Roberts* [illegible]	I S	British	1227	224·1	36·8	21·8	Liverpool	1876	T. Royden & Sons		Hughes & Co.
[illegible]	363	✻	Carnarvonshire *Johansen* 86-86	W 3m Sr	British	378	132·2	26·2	16·5	Neyland	1866	J. Gaddaen	[illegible]	G. B. Wadsworth, j
78023	364	✻ D	Carnegie *Thompson* 82-84	I Bk	British	1020	208·0	34·9	20·7	Port Glasgow	1876	Reid		Tomlinson Hodgetts
	365		Carneiro I°	W Sr	Brazilian	165	..	..	..					Carneiro & Irmão
	366		Carneiro II°	W Sr	Brazilian	171	..	..	..					Carneiro & Irmão
	367	✻	Carniehill	Stl 4m Bk	British	2564	313·0	42·0	24·5	Pt. Glasgow	1889	Russell & Co.		W. T. Dixon & Son
77[illegible]	368	V	Carniola *Wright* 81-87	W Bk	British	732	155·6	33·5	19·2	Bear River, NS	1879	Rice & Clark	[illegible]	Wm. Thomson & C
90110	369	✻	Carnmoney *Patey* 87-87	I Bk	British	1255	235·5	36·2	21·0	Belfast	1884	Workman, Clark & Co.		Carnmoney Ship Co. J (W. Porter & Sons)
	370		Carola (ex Ana[illegible]) *Teagrott*	W Sr	Brazilian	227	103·6	24·0	13·3	P.E.Island	1876		[illegible]	
	371	✻	Carola (ABANDONED) (ex British Queen) *Dorneth* -87	W Bk	German	404	139·8	27·0	16·4	Aberdeen	1858	J. Duthie & Co.	[illegible]	Robert Dorneth
	372		Carola (ex Flora) *Carlmark*	W Bk	Norweg'n	305	111·9	27·2	15·5	Rostock	1861			C. J. & E. G. Carlma
	373		Carola	W Bg	Swedish	214	105·6	23·6	11·8	Värnamo	1855			C. J. Olsson
125980	374		Carolina *O'Neal*	W Sr	American	111	101·0	23·8	8·7	Beaufort, N.C.	1874	J. A. Farrow		J. L. Fowle
3165	375	A	Carolina *Negovetich* -88	W Bk	Austrian	559	142·7	29·3	18·5	Fiume	1875	F. Schiavoni	[illegible]	A. E. Descovich
	376		Carolina	W Bn	Brazilian	219	..	..	..					J. A. Ribeiro
	377		Carolina	W 3m Sr	Brazilian	219	..	..	..					F. T. A. & J. H. da Si
	378		Carolina *Holstius*	W Bk	Russian	489	134·8	31·8	18·9	Christinestad	1866			C. E. Carlström
[illegible]	379	A V	Carolina G. *Frangini* -88 (ex John, ex George)	W Bg	Austrian	375	121·7	26·7	16·7	Fiume	1869	C. Giacovcich	[illegible]	M. F. Gelletich
	380		Carolina G. (ex Abdul[illegible])	W Bg	Turkish	273	..	..	..	Fiume	1848	A. Zanon		
1466	381		Carolina Galatola	W Bk	Italian	542	147·3	28·2	19·5	Sunderland	1858			L. Galatola
10872	382	✻	Caroline *Davis* -84 (ex Athen[illegible])	I S	British	984	213·3	30·4	19·7	Glasgow	1854	Smith & Rodger		A. E. Kinnear & Co
89197	383		Caroline *Rushmore* 86-86	W Bn	British	173	95·1	23·1	12·7	P.E.Island	1857	Coffin & Son		C. Hoult & Co.
	384		Caroline	W Bkn	British	165	118·4	25·3	8·7	Cleveland, Ohio	1848			J. Neill
68173	385	✻	Caroline *Baddeley*	W Bn	British	158	107·1	22·5	11·6	Bideford	1874	Johnson	[illegible]	E. Baddeley
84081	386		Caroline *Goodall* -81	W	British	106	93·0	23·6	9·6	Sweden				W. Malcolm
5774	387		Caroline *Hutchings*	W Sr	American	106	81·0	27·6	7·6	Cherryfield Me	1869			S. N. Campbell
	388	V	Caroline *Ugelbjerg* 73-73	W Bg	Danish	181	101·6	24·4	11·0	Kalmar	18 57/81	H. Olsen		M. Rasmussen
	389	V	Caroline *Pfeffer* -87 (ex Ane Kathrine) (ABANDONED)	W Bn	Danish	156	97·5	22·7	11·5	Sønderho Fanø	1875	S. Sonnichsen		J. C. Pfeffer
	390	V	Caroline *Lauritzen* 78-78	W Bn	Danish	147	89·2	21·6	11·7	Marstal	1871	J. J. Bager		H. C. Christensens Enk
	391	V	Caroline *Petersen* (ex Frederik[illegible])	W Sr	Danish	144	90·9	21·9	10·4	Papenburg	1884	B. Thelen	[illegible]	C. S. Sorensen
	392	V	Caroline *Andersen* 73-73	W Sr	Danish	122	91·4	20·7	10·3	Kiøge	1873	J. W. L. Flindt		Aktie Selskab "Carolin (A. A. Henrichsen)
	393	V	Caroline *Begge*	W Sr	Danish	112	88·4	20·4	10·3	Holbæk	1869	P. Mortensen		H. L. Smiths Enke
	394	V	Caroline *Nielsen* 74-74	W Sr	Danish	101	85·0	19·7	10·2	Kiøge	1874	J. W. L. Flindt		J. Nielsen
	395	✻	Caroline *Martin* 84-86	I Bk	French	680	181·8	29·1	18·5	Glasgow	1869	A. Stephen & Sons		Ant. Dom. Bordes & fils
	396	V	Caroline *Le Pivert* 84-85	W Bk	French	395	151·5	27·7	14·0	St. Vaast	1875	E. Costey	[illegible]	H. Auger aîné
	397		Caroline (ex Result)	W Bn	French	101	72·5	21·6	10·7	Nantes	1858			L. H. Merton & Trel (Laval)
	398	V	Caroline *Schlüter* 83-83 (ex Valhalla)	W S	German	1255	191·6	37·1	24·1	Newburyport, Mass.	1863	J. Currier	[illegible]	Fr. Roters
	399	V	Caroline *Kruse* 74-74	W Glt	German	110	81·5	22·6	10·4	Neustadt	1863	J. Tornoe		N. Hiss
	400	D N V	Caroline *Jørgensen* (ex Madlen, ex [illegible])	W S	Norweg'n	1175	183·3	36·2	23·3	Quebec	1869	J. E. Gingras	[illegible]	J. M. Bryde

✻ = Lloyd's Register; D = Nederlandsche Vereeniging; G = Germanischer Lloyd; I = Registro Italiano; R = Am
A = Veritas Austro-Ungarico; H = Veritas Ellenico; N = Norske Veritas; V = Bu

In the 4th Column, C = Composite (iron frame, wood planked); I = Iron; Stl = Steel; W = Wood.

The Dates of Metalling (Col. 11) are based on the records of the Books in which Vessels are respectively classed, or on the Reports of the Surveyors of

Further particulars of Vessels classed by Lloyd's Register will be found in Lloyd's Register Book and Lloyd's Yacht Register.

No.	Cote	Nom du navire	Gréement	Pavillon	Tonnage	Longueur	Largeur	Creux	Lieu de construction	Année	Constructeur	Armateurs ou Directeurs-Gérants	Port d'Armement
401	N	[illegible]	W Bk	[illegible]	[illegible]	[illegible]	[illegible]	[illegible]	[illegible]	1851		J.H.[illegible]	[illegible]
402	GV	[illegible]	W Bg	[illegible]	[illegible]	[illegible]	[illegible]	[illegible]	[illegible]	1869	[illegible]	S.[illegible]	[illegible]
403	GNV	[illegible]	W Bk	[illegible]	[illegible]	[illegible]	[illegible]	[illegible]	[illegible]	1861	[illegible]	N.H.[illegible]	[illegible]
404	N	Caroline	W Bk	[illegible]	[illegible]	[illegible]	[illegible]	[illegible]	[illegible]	1866	[illegible]	C.Wrobleen	[illegible]
405		Caroline	[illegible]	[illegible]	119	82.6	25.7	9.7	[illegible]	1875	[illegible]	J.Wefang	[illegible]
406		Caroline	[illegible]	[illegible]	313	121.4	27.0	[illegible]				[illegible]	[illegible]
407	V	Caroline Augusta	[illegible]	[illegible]	165	[illegible]	[illegible]	[illegible]	[illegible]	1871	[illegible]	O.E.Olsen	[illegible]
408	GV	Caroline Behn	[illegible]	[illegible]	679	167.8	[illegible]	[illegible]	[illegible]	1874	[illegible]	H.Folsch & Co.	Hamburg
409	V	Caroline Elizabeth	[illegible]	[illegible]	141	[illegible]	21.6	[illegible]	[illegible]	1870	[illegible]	[illegible]	Horsens
410	R	Caroline Gray	W Bk	American	315	141.[illegible]	27.[illegible]	12.[illegible]	[illegible]	1869	[illegible]	J.L.Snow & Co.	Rockland, Me.
411		Caroline Hall	[illegible]	American	[illegible]	107.[illegible]	[illegible]	[illegible]	[illegible]	1879	[illegible]	J.W.Hall	Wilmington, Del.
412	※	Caroline Morris	[illegible]	British	932	205.[illegible]	32.7	19.6	[illegible]	1877	[illegible]	T.Beynon & Co.	Newport
413		Caroline Smith	[illegible]	Danish	284	103.9	25.4	11.1	[illegible]	1850	[illegible]	Chr.Simoni	Aalborg
414	※	Caroline Spooner	W Bk	British	663	162.0	33.2	19.5	Aberystwith	1878	[illegible]	E.Jones & Co.	Aberystwith
415	V	Caroline & Trine	W S	Danish	142	102.6	21.6	10.6	[illegible]	1864	C.Möller	K.H.Mogensen	Faaborg
416	V	Carondelet	[illegible]	American	1376	202.5	40.3	21.6	[illegible], Me.	1872	[illegible]	Pugetsound [illegible]	San Francisco
417	※	Carpasian	W Bk	British	288	131.6	26.1	12.9	[illegible]	1875	[illegible]	Thorburn & Tessier	St.John's, N.F.
418	※	Carpathian	I S	British	1444	210.1	36.6	22.6	Hull	1874	Humphrys & Pearson	McDiarmid Greenshield	Liverpool
419	V	Carpincho	W S	British	192	120.6	28.0	9.0	[illegible]	1885	J.B.North	W.Folker	Windsor, N.S.
420	※	Carr Rock	[illegible]	British	1386	219.0	39.0	22.3	Glasgow	1889	A.Stephen & Sons	J.Cornfoot & Co.	Glasgow
421	I	Carrara	W Bk	Italian	434	133.8	29.1	19.1	[illegible]	1875	Briasco	B.Fabricotti di D.	Leghorn
422	I	Carrarese	W Bk	Italian	195	84.3	22.6	10.7	[illegible]	1887	A.Terizzani	G.Fiaschi	Spezia
423		Carreiros	W S	Brazilian	257	..	..	..				A.J.Villaça	Rio Grande
424	V	Carrick	W Bk	British	314	121.0	28.5	12.7	Tynemouth	1877	J.A.Lovatt	James Kennedy	St.John, N.B.
425	N	Carricks	W S	Norwegian	913	186.1	34.0	21.9	Workington	1867	J.F.Fell & Co.	A.J.Jacobsen & O.R.Rød	Tönsberg
426		Carrie	W S	British	94	85.6	21.8	10.4	Truro	1878	Dyer	G.Hugoe	Truro
427	※	Carrie	W S	British	99	81.6	20.8	10.6	[illegible]	1885	E.Burchard & Co.	W.R.Challis	Hull
428	R	Carrie A. Bucknam	W Bk	American	273	138.0	29.3	10.6	Bucksport, Me.	1883	W.Beazley & Co.	M.Gray	Bucksport, Me.
429	R	Carrie A. Lane	W 3m Sr	American	760	178.2	35.9	17.3	Bath, Me.	1887	A.Sewall & Co.	A.Sewall & Co.	Bath, Me.
430	R	Carrie A. Norton	W [illegible]	American	537	158.9	33.3	12.0	Bath, Me.	1882	Adams & Hitchcock	Adams & Hitchcock	Bath, Me.
431		Carrie Bell	W Sr	British	121	95.3	21.4	11.3	Glasson Dock	1862	Simpson	J.H.Barrow	Lancaster
432	R	Carrie Bell	W 3m Sr	American	260	123.1	29.5	10.1	Kennebunk, Me.	1873		M.P.Emery	Kennebunk, Me.
433		Carrie C. Miles	W Sr	American	101	89.8	25.2	8.8	East Boston	1871		D.W.Dowling	Provincet'n, Mass.
434	R	Carrie C. Ware	W 3m Sr	American	176	105.4	28.7	8.4	Cherryfield, Me.	1884	J.Strout	C.Keene	Machias, Me.
435		Carrie D. Knowles	W Sr	American	115	86.0	22.9	8.8	Essex, Mass.	1887	Jno.James	George D.Knowles	Provincetown
436	V	Carrie Delap	W Bk	British	1109	186.1	38.5	20.7	Annapolis, N.S.	1874		Troop & Son	St.John, N.B.
437	※	Carrie Dingle	W Bn	British	183	105.6	21.5	12.9	Plymouth	1873	Williams	Mrs.E.Roberts	Liverpool
438	R	Carrie E. Long	W Bk	American	549	138.6	30.2	17.5	Stockton, Me.	1866	Fletcher	Ryan & Kelsey	New York
439		Carrie E. Phillips	W Sr	American	110	..	..	..	Essex, Mass.	1887	Story	Phillips	Boston, Mass.
440		Carrie E. Pickering	W [illegible]	American	264	119.0	29.3	10.2	Bangor, Me.	1869		J.P.Ellicott	Boston, Mass.
441	R	Carrie E. Woodbury	W 3m Sr	American	247	127.3	29.5	10.2	Brewer, Me.	1871	J.Oakes	M.E.Strickland	Bangor, Me.
442	R	Carrie Easler	W S	British	154	93.8	26.8	9.8	Parrsboro, N.S.	1883	W.Moore	Wm.Brown	Parrsboro, N.S.
443		Carrie Hall Lister	W Sr	American	142	103.0	27.5	7.4	Milford, Del.	1878	J.W.Abbott & Co.	John W Hall	Wilmingtn, Del.
444	※	Carrie Harvey	W S	British	99	86.7	21.8	10.7	Plymouth	1881	D.Banks & Co.	S.B.Harvey	Plymouth
445	RV	Carrie Heckle	W Bk	American	473	135.7	31.1	16.5	Cape Elizabeth, Me.	1877	J.F.Randall	M.E.Colcord	Portland, Me.
446		Carrie Holmes	W 3m Sr	American	326	129.0	32.0	10.0	[illegible], N.Y.	1866		J.Holmes	Perth Amboy
447	R	Carrie L. Godfrey	W 3m Sr	American	414	126.0	34.8	9.9	Dorchester, N.J.	1874		A.H.Jarman	Bridgeton
448		Carrie L. Hix	W Sr	American	140	93.3	26.3	8.9	Rockland, Me.	1871		H.Hix	Rockland, Me.
449	RV	Carrie L. Tyler	W Bk	American	586	143.8	32.5	17.3	[illegible], N.Y.	1874	J.Carll	G.M.Smith	New York
450	R	Carrie S. Bailey	W [illegible]	American	396	137.6	32.1	11.7	Bath, Me.	1878	[illegible]	A.Sewall & Co.	Bath, Me.

※ = Lloyd's Register; D = Nederlandsche Vereeniging; G = Germanischer Lloyd; I = Registro Italiano; R = American Record; A = Veritas Austro-Ungarico; H = Veritas Ellenico; N = Norske Veritas; V = Bureau Veritas.

Dans la 4me Colonne, C = Composite (membrure en fer, bordé en bois); I = Fer; Stl = Acier; W = Bois.

L. Dates du doublage en métal (Col. 11) ressortant des livres dans lesquels les navires sont respectivement classés ou des rapports des Inspecteurs du Lloyd's Register.

Pour plus ample renseignement sur les navires classés par le Lloyd's Register, voir le Lloyd's Register Book et le Lloyd's Yacht Register.

Official No.	No. in Book	Registry in which classed	Ship's Name. Material and Rig. Master, and Dates of Appointment to present Owner's Service and to Ship. Ship's Late Name.	Flag.	Net Tonnage.	Dimensions. Length.	Breadth.	Depth.	Built. Where.	When.	By whom.	Date of Metalling.	Owners or Managers.
[illegible]	451		Carrie S. Hart *Southard* W3mSr	American	503	132·4	35·5	15·3	FallRiver, Mass.	1873	JosephTerry		John H.Cady
[illegible]	452		Carrie Sanford WSr	British	121	107·5	23·9	7·0	WesterfordU'S	1855			R.Alleyn
[illegible]	453	R	Carrie Strong *Strong 77-82* W3mSr	American	450	149·0	35·0	12·6	ThomastonMe	1882	Dunn& Elliot		Dunn & Elliot
[illegible]	454		Carrie Walker *Starkey* WSr	American	165	100·0	28·8	9·5	ThomastonMe	1867	Walker& Dunn		S.T.King
[illegible]	455	RV	Carrie Winslow *Barrett* WBk	American	944	173·7	34·5	20·7	E.Deering Me	1880	G.W.Russell		J.S.Winslow& Co.
[illegible]	456	V	Carrier Dove *Wright* WBk	British	1026	178·4	36·8	22·2	St.John,N.B.	1879	Stew'rt& Ritchie	[illegible]	Wm.Thomson& Co.
[illegible]	457		Carrington *Strickland* W3mBn	British	296	112·2	25·4	12·9	Gloucester	1867	Miller		G.F.Carrington& Co.
[illegible]	458	✻	Carrizal *Organ 74-74* IBk	British	599	166·5	29·1	18·9	Sunderland	1864	Oswald& Co.		Shallcross & Higham
[illegible]	459		Carroll WSr	American	100	82·5	24·7	7·6	E.Machias,Me	1867			Geo.Sawyer
[illegible]	460		Carrollton *Lewis* WS	American	1390	198·2	36·6	24·6	Bath,Me.	1872	E.& A.Sewall	[illegible]	A.M.Simpson
	461	NV	Carsten Boe *Dannevig 85-86* WBk (ex Hasselhorst)	Norweg'n	851	166·9	35·2	20·4	Annaplis,N.S.	1876	Pickles & Mills	[illegible]	N.Olsen
	462	NV	Carte Blanche *Johannessen 88-88* WBk	Norweg'n	829	170·2	35·0	19·8	Arendal	1885	Henricksen		H.P.Pettersen
	463	V	Carvoeira *[illegible] -87* IBk (ex [illegible])	British	1293	227·0	37·8	22·3	Liverpool	1869	T.Royden& Son		R.Singlehurst& Co.
	464		Casa Blanca WBn (ex Velocidor, ex San Antonio)	Spanish	144	89·2	22·1	11·5	Santander	1863			B.Borras Pujol
[illegible]	465	✻	Casablanca *Bray 87-87* IBk	British	533	172·1	28·0	17·5	Liverpool	1868	T.Royden& Sons		Nicholson & McGill
[illegible]	466	DV	Cashier *Telfier* WS	British	1380	198·0	37·8	22·5 / 27·0	L'ndnderryNS	1876	R.Lewis	[illegible]	F.C.Mahon
[illegible]	467		Cashier *Reinhardt -88* WSr	British	122	88·5	24·0	9·4	ConquerallNS	1888	J.N.Rafuse		T.A.Wilson
[illegible]	468	✻	Cashmere *Thomson 85-85* IBk	British	1245	222·8	36·7	22·7	Sunderland	1868	Oswald& Co.		H.Fernie & Sons
[illegible]	469	V	Cashmere *Sutherland* WS	British	1238	200·5	37·3	23·1	S.Stephen,NB	1873	J.& C.Short	[illegible]	D.Sutherland
	470	V	Casilda (ex Carmen) *Alsina* WSr	Spanish	212	94·2	25·3	13·8	Blanes	1858		[illegible]	Silverio Millet
	471		Casilda *Baines* WSr	Spanish	175	92·8	26·0	12·1	Vinaros	1856			Puig & Carbo
	472		Casimir Périer *Hessy* WSr	French	163	..	..	..	Fécamp	1879			P.Casmir Périer
[illegible]	473	✻	Casket *Coath 77-84* WBk	British	330	130·6	26·1	15·7	Bridport	1869	Cox	[illegible]	J.Marquand & Co.
	474	✻	Casma *Sönniksen -88* IBk	Danish	649	181·1	29·1	18·5	Liverpool	1869	W.R.Potter& Co		P.N.Winther
	475	DG	Caspar de Robles *Bu 85-* WBk (ex Maria & Helena)	Dutch	300	128·2	27·3	12·5	Motzen	1864	D.Oltmann's Wwe.	[illegible]	J.Fontein
	476	V	Casper *Hansen 75-74* WSr	Danish	136	90·8	21·1	10·3	Thurø	1875	P.Bom		P.Bom
[illegible]	477		Casper Heft *Owens* WSr	American	102	74·0	22·0	6·6	WilmingtnDel	1851			J.P.Owens
[illegible]	478	✻	Cassandra *Cromarty 83-83* IBk	British	711	174·5	30·8	19·6	Sunderland	1868	Oswald& Co		"Cassandra" S.S. Co. (J. Steel & Son)
	479	G	Cassandra *Perliet 84-84* WBg	German	288	104·0	27·5	16·0	Greifswald	1857	L.Wittenberg		C.A.Beug
	480	V	Cassandra *Schwerwinsky 1 -87* WBg	German	242	94·5	24·0	13·6	Ribnitz	1849			Retzlaff & Schober
[illegible]	481		Cassandra Adams (WRECKED) *Whelan* WBk	American	1083	196·5	40·3	22·2	Seabec,Wash.	1876	H.Doncaster		Hansen& Co.
	482	A	Cassiani *Georgalis* WSr	Russian	135	..	..	..	Cassos	1881	E.G.Diagusi		S.Mussuri
[illegible]	483	R	Cassie F. Bronson *Anderson* W4mSr	American	1068	193·2	40·1	17·8	Bath,Me.	1886	NewEnglandSB Co.		W.T.Anderson
[illegible]	484		Cassie Hayward *Brown 86-88* W3mSr	American	188	112·0	31·0	9·6	P.Ludl'wWash	1875			Pollard & Dodge
[illegible]	485		Cassie Jameson *Colvin* W3mSr	American	379	143·9	32·1	12·8	ThomastonMe	1874		[illegible]	W.C.Haskins& Sons
[illegible]	486		Castalia *Hammond 81-81* WSrYt	British	120	103·4	23·2	12·1	Lymington	1874	Inman& Son	[illegible]	Earl of Shrewsbury & Talbot
[illegible]	487	R	Castalia *Jackson* WBn	American	499	131·6	30·6	17·0	RichmondMe	1874	G.Farrin	[illegible]	P.M.Whitmore
	488	N	Castelar *Quale 81-82* WBk	Norweg'n	563	159·6	33·0	18·1	Bergen	1871	J.Gran	[illegible]	T.Samuelsen
[illegible]	489	I	Castello Dragone *Schiaffino* WBk	Italian	649	149·9	32·2	21·9	Lavagna	1878	G.Brignole	[illegible]	A.Schiaffino
[illegible]	490		Castilian *Means* WSr	American	119	87·6	24·3	8·3	Ellsworth,Me	1852			E.S.Means
	491		Castilla *Yusti* WBk	Spanish	340	110·8	31·1	16·1	Blanes	1853			Joan Pombo
	492	DGRV	Castine (CONDEMNED) *Lohmann* WS	German	1068	170·9	36·4	23·4	Castine,Maine	1857	J.Fischer	[illegible]	Schilling & Brüning
[illegible]	493	✻	Castle Holme *Bryce 73-82* IBk	British	996	213·9	34·5	20·7	Sunderland	1875	Bartram,Haswell & Co		Hine Bros.
[illegible]	494	✻	Castle Roy *Smith 83-83* IS	British	1664	259·1	40·0	23·1	Glasgow	1874	J.Elder& Co.		G.Gilroy Sons & Co.
[illegible]	495	✻D	Castlehead *Fearon* IS	British	825	196·0	32·2	19·4	Liverpool	1869	R.& J.Evans		North Western Shipping Co., Ld.
[illegible]	496		Castlehow *Williamson* WBk	British	243	103·6	24·7	16·1	Harrington	1860	Williamson		J.Davidson & Sons
[illegible]	497	✻	Castleton *Diore* IBk (ex Liberté [illegible])	British	499	166·4	26·2	18·5	Stockton	1862	M.Pearse& Co.		John P.Ellis
[illegible]	498	✻	Castor *Croudace 84-88* StlBk	British	2005	278·1	41·2	23·9	Leith	1886	Ramages Ferguson		W.S.Croudace
	499	N	Castor *Olsen* WBk	Norweg'n	404	118·6	30·8	17·0	Arendal	18 50/72		[illegible]	S.Olsen
	500												

✻ = Lloyd's Register; D = Nederlandsche Vereeniging; G = Germanischer Lloyd; I = Registro Italiano; R = American [illegible]
A = Veritas Austro-Ungarico; H = Veritas Ellenico; N = Norske Veritas; V = Bureau [illegible]

In the 4th Column, C = Composite (iron frame, wood planked); I = Iron; Stl = Steel; W = Wood.

The Dates of Metalling (Col. 11) are based on the records of the Books in which Vessels are respectively classed, or on the Reports of the Surveyors of L[illegible]

Further particulars of Vessels classed by Lloyd's Register will be found in Lloyd's Register Book and Lloyd's Yacht Register.

No.	Registre	Nom. Capitaine. Ancien Nom	Matériaux et Gréement	Pavillon	Tonnage	Longueur	Largeur	Creux	Construit: Où	Quand	Par qui	Doublage en métal	Armateurs ou Directeurs-Gérants	Port d'Armement
501	N	Castor *Hansen*	W Bg	Norweg'n	[illegible]	103·6	26·6	12·6	Skien	1872			O. Halvorsen	Sandefjord
502	N	Castro *Ock*	W Bk	Norweg'n	275	112·4	25·1	8·9	Philadelphia	1847			H. Cappelens Enke	Skien
503		Castro III.	W Bk	Brazilian	[illegible]	..	..	..					A.D.S. Castro	Rio Janeiro
504		Casualita *Simile*	W Bg	Italian	307	110·2	27·8	18·1	Voltri	1868	L. Briasco		V. Messina	Palermo
505	✻	Caswell *Lewis*	I Bk	British	499	156·5	28·0	16·8	Dumbarton	1875	Mc Kellar, [Mc Millan & Co.		Simpson Bros.	Swansea
506		Catalano *Spirito*	W Bg	Italian	16[illegible]	88·3	23·8	13·7	Elba	1855			F. Giannoni fu Carlo	Porto Ferraio
507	✻	Catalina *Gardiner*	I Bk	British	48[illegible]	151·1	28·1	16·8	Sunderland	1868	W. Doxford		W. O. Gilchrist	Sydney
508	V	Catalina *Garriga*	W Bk	Spanish	49[illegible]	139·7	28·8	18·6	Sunderland	1864	Briggs & Sons		Romulo Bosch & Co.	Barcelona
509		Catalina B. *Vianelle* 87-87	W Bn	Peruvian	1[illegible]	77·7	22·3	8·8					M. Badell	Callao
510		Catania *Costanzo*	W Bg	Italian	1[illegible]	..	..	..	Trieste	1869	S. Bartelo		G. Costanzo	Catania
511		Cataract	W Sr	British	193	105·5	24·0	10·5	Sophiasbg, Cnt	1874	John Tait		S. Shepard	Pictou, Ont.
512		Catawamteak *Perry*	W Sr	American	141	81·0	23·0	9·0	Rockland, Me.	18 64/83			Snow & Pearsons	Rockland, Me
513	V	Caterina *Coloria*	W Bg	Italian	302	122·7	26·8	16·8	Capodistria	1872	Piscitello & Poli		S. Grassi	Catania
514		Caterina *Barbero*	W Bg	Italian	214	..	..	..	Varazze	1854			Flli De-Lieto	Reggio
515	I	Caterina *Virgilio*	W Bg	Italian	216	107·9	25·3	14·2	Castellam. di [Sta.	1876	M. Trombarolo		A. Messina & V. Virgilio	Trapani
516	G	Caterina *Pietro* 85-	W Bn	Italian	134	87·9	22·3	13·5	Sira	1849			R. Prestopino	Messina
517		Caterina	W Bg	Turkish	16[illegible]	..	..	..	Barten	1871			Samaico Nicola Constanti	Samos
518	I	Caterina Accame	W Bk	Italian	776	159·1	30·2	20·7	Loano	1883	N. Cerruti		P. Accame & fi.	Savona
519		Caterina B. *Said* 81-84	W Bn	British	1[illegible]	81·0	21·9	10·0	Malta	1872			Lorenzo Said	Malta
520		Caterina C. *Olivari*	W Bg	Italian	185	88·9	25·0	12·8	Nantes	1868	Dousain		P. Crovari	Genoa
521		Caterina C. *Cocurullo* -88	W Bn	Italian	159	98·4	25·6	9·5	[illegible]	1888			F. Attanazio	Castellam.
522	I	Caterina Cacace *D'Amoro*	W Bk	Italian	664	155·8	30·8	20·0	Cassano	1878	A. Castellano		Eredi di A. Cacace	Castellam.
523		Caterina Camilla (WRECKED) *Leoni*	W Bg	Italian	240	106·9	24·8	17·3	Varazze	1857	G. Cerruti		G. B. Leoni	Genoa
524	I	Caterina Capurro *Rebarbero*	W Bk	Italian	550	150·9	31·6	19·1	Sestri	1879	A. Carlevaro		Frat. Capurro	Genoa
525	I	Caterina G. *D'Aste* 76-76	W Bk	Italian	627	143·7	32·3	21·0	Sestri	1874	L. Tixi		G. Maggiolo	Genoa
526		Caterina G. *Mari* (WRECKED)	W Bg	Italian	351	136·4	23·1	13·7	Fiume	1871	G. Brazzoduro		E. Giacopini	Spezia
527		Caterina Guida *Parv*	W Bn	Italian	287	107·6	26·3	16·5	Procida	1862	Luigi Guida		L. Guida	Naples
528		Caterina Madre (NOT SEAGOING) *Bozzano*	W Bk	Italian	887	176·8	33·8	23·0	Sestri	1870	A. Briasco fu L.		C. Mainetto	Genoa
529		Caterina Pietra (HULK)	W Bk	Italian	681	148·9	29·7	21·1	Sestri Ponente	1868	G. Casanova		E. Accame fu L.	Genoa
530	IV	Caterina R. *Bertolo*	W Bk	Italian	748	180·7	32·1	20·2	La Roque	1873	Roy		A. Rainusso	Genoa
531	I	Caterina S. *Simonetti*	W Bk	Italian	619	153·2	32·2	21·3	Savona	1873	F. Sivello		A. Simonetti	Genoa
532	I	Caterina S.	W Bg	Italian	228	114·5	24·4	13·8	Sampierdar'na	1873	Frasinetti		G. Scandurra & Flli	Catania
533		Caterina S. (WRECKED) *Salomone*	W Bn	Italian	125	91·5	22·2	11·7	Savona	1883	G. B. Pongiglione		G. B. Salomone	Savona
534		Caterina Stuch (WRECKED) *Stuch*	W Bk	Austrian	378	114·8	27·8	17·7	Fiume	1863	Flli Schiavoni		F. Stuch fu S.	Orebich
535		Catharina *Ohlsen*	W Bn	Danish	156	96·6	22·2	11·0	Nordby, Fanø	1870	J. W. Christensen		F. Ohlsen	Nordby, Fanø
536	V	Catharina V *Vaccaro*	W Bk	Italian	599	139·1	31·7	20·8	Sestri	1868	G. Calcagno		A. Vaccaro	Spezia
537		Catharina *Rasmussen*	W Bn	Danish	112	69·1	21·8	12·1	Svendborg	1841	L. N. Møller		G. R. Berg	Svendborg
538	D V	Catharina *Roon* 83-83	C Bk	Dutch	1014	187·0	36·1	20·9	Amsterdam	1883	Huygens & Van [Gelder		J. Veth	Amsterdam
539		Catharina *Dinkla*	W Sr	German	159	97·1	21·6	11·5	Hoogezand	1856	W. C. Wildervank		W. J. Schuster	Dornumersiel
540	V	Catharina *Campen* -81	W Sr	German	121	86·6	20·0	9·8	Edewecht	1868	H. Kramer		S. Wreesmann	Leer
541		Catharina *Pedersen*	W Bk	Norweg'n	378	119·7	27·8	15·7	Vegesack	1852	C. Sager		John Christensen	Drammen
542	N	Catharina *Gundersen*	W Bk	Norweg'n	288	120·7	27·3	12·8	Virginia	1849			H. Hansen	Lillesand
543	N	Catharina *Kjørvig* 87-7	W Bk	Norweg'n	261	123·8	26·3	13·1	Christiansund	1868	P. Wallum		N. H. Knudtzon	Christiansund
544		Catharina	W Sr	Russian	228	102·1	23·5	12·0	Pernigel	1875			G. Schnore	Riga
545		Catharina	W Cr	Russian	116	69·7	23·8	9·1	Eckendorf, [Kurland	1874			J. Gïrtbahn & P. Jaeger	Riga
546		Catharina *Rassmus* -83	W Sr	Russian	102	66·0	21·3	9·6	Dondangen	1878	Lachs		F. E. Rassmus	Riga
547	D	Catharina Helena	W Bk	Dutch	361	121·3	30·3	13·9	St. Mary's Bay	1869	A. J. Raymond		Canne & Balwé	Amsterdam
548		Catharina Suzanna *Meesters*	W Bn	Dutch	194	107·9	22·7	11·7	Wildervank	1868	J. M. Meesters & [Son		J. J. van den Berg	The Hague
549	G	Catharina Wilhelmina *Zeeu*	W Bn	Dutch	218	101·0	26·8	11·7	Windsor, N.S.	1862			F. W. van der Elst & Co.	Schiedam
550		Catharine *Wallace* 79-79	W Sr	British	100	82·0	20·7	10·0	Newburgh	1860	Fenton		D. Sutherland	Montrose

✻ = Lloyd's Register; D = Nederlandsche Vereeniging; G = Germanischer Lloyd; I = Registro Italiano; R = American Record; A = Veritas Austro-Ungarico; H = Veritas Ellenico; N = Norske Veritas; V = Bureau Veritas.

Dans la 4me Colonne, C = Composite (membrure en fer, bordé en bois); I = Fer; Stl = Acier; W = Bois.

Les Dates du doublage en métal (Col. 11) ressortent des livres dans lesquels les navires sont respectivement classés ou des rapports des Inspecteurs du Lloyd's Register.

Pour plus amples renseignements sur les navires classés par le Lloyd's Register, voir le Lloyd's Register Book et le Lloyd's Yacht Register.

Official No.	No. in Book	Registry in which classed	Ship's Name. Material and Rig. Master, and Dates of Appointment to present Owner's Service and to Ship. Ship's Late Name.	Flag.	Net Tonnage	Dimensions: Length	Breadth	Depth	Built: Where	When	By whom	Date of Metalling	Owners or Managers
	551	V	Catharine *Meier 74-74* W Sr	German	175	97·8	23·0	11·8	Elsfleth	18 77/79	G. Hagens		J. J. Wallis & Sohn
[illegible]	552		Catharine Sudden *Lund -87* W Bkn	American	368	142·0	34·2	11·1	Port Ludlow, [Wash.	1878	Hall Bros.		S. H. Harmon
	553		Catharinense W S[illegible]	Brazilian	166	..	..	..					J. P. L. da Silva
[illegible]	554	✱	Cathaya *Morgan 82-83* W Bk	British	790	178·1	32·2	20·7	Workington	1863	C. Lamport	[illegible]	J. I. Jacobs & Co.
[illegible]	555	✱	Cathcart *McChesney 76-86* I S	British	1387	232·0	36·7	23·4	Greenock	1869	R. Steele		Adam, Hamilton & Co.
[illegible]	556	✱	Catherine *Reeve 77-77* W Bkn	British	316	118·2	26·3	15·0	Kingsbridge	1877	Date	[illegible]	W. W. Steer
[illegible]	557		Catherine *Dorrell 85-85* W Bg	British	229	112·0	25·3	14·3	P. E. Island	1866	White		T. H. Franks
[illegible]	558		Caterine (WRECKED) *[illegible]* W Bg	British	194	105·5	25·1	13·5	[illegible]	1866	M'Pherson		C. Marwood
[illegible]	559		Catherine *Dobson 83-83* W Bn	British	123	90·0	20·0	10·9	Sunderland	1855	W. Wilkinson		R. H. Williams
[illegible]	560	✱	Catherine *Richards 73-73* W Sr	British	96	72·2	21·1	11·3	Portmadoc	1856	D. Richards		J. Richards
[illegible]	561	✱	Catherine *Harvey* W Sk	British	84	81·0	21·3	10·3	Hull	1887	Albert Dk. Ship-[wright Co.		H. Toozes
[illegible]	562	✱	Catherine Auld *Evans 88-88* W Sr	British	164	96·4	23·9	11·7	Inverkeithing	1881	R. & D. Ross	[illegible]	W. C. Phillips
[illegible]	563	✱	Catherine Richards *Morgan* W Sr	British	155	95·5	22·6	12·8	Portmadoc	1874	Williams	[illegible]	J. Hughes & Co.
[illegible]	564	✱	Catherine & Margaret *Roberts* W Sr	British	91	70·4	20·5	10·8	Portmadoc	1858	H. Williams		J. Roberts
[illegible]	565	✱	Catherine & Mary *Nicholas* W Sr	British	84	69·2	20·4	10·5	Newquay	1858	D. Evans		R. Nicholas
75402	566		Cathie C. Berry *Foster* W 3m Sr	American	304	132·0	31·1	11·6	Thomaston Me	1873			
	567	N	Cathinca *Hammerberg -87* W Bk	Norweg'n	374	124·0	30·0	15·9	Flensburg	1856			I. A. Berggrav
	568	V	Cathrine *Winther 76-76* W Bg	Danish	251	119·2	24·5	12·4	Fredericia	1876	J. R. Schröder	[illegible]	Actieselskab. ([illegible])
	569	GV	Cathrine *Poulsen 86-88* W Bg	Danish	225	114·5	23·7	12·3	[illegible]	1869	C. A. Hahn	[illegible]	Acktieselskab. (M. S. Mathiesen)
	570	DV	Cathrine *Dam -82* W Bn	Danish	163	104·4	21·6	10·8	Veile	1876	Gylding & [Lindtner	[illegible]	J. N. Jessen
	571		Catina (ex [illegible]) W Bn	Turkish	130	..	..	..	Samos	1864			N. Clini
	572	V	Cato (ex Cato-Annette) *Klitzing -88* W Bn	German	160	102·5	22·6	10·6	Nieuwendam	1864	W. & H. [illegible]	[illegible]	H. v. Ehren
	573	N	Cato *Lange 81-87* W Bk	Norweg'n	476	142·7	31·6	16·7	Stavanger	1869	M. Magnus	[illegible]	Ths. S. Falck
	574	N	Cato *Johannesen -86* W Bk	Norweg'n	342	117·5	28·8	15·8	Grimstad	1869	S. Johnsen	[illegible]	C. H. Johnsen
	575		Cato *Sjögren -88* W Bn	Swedish	169	87·9	26·2	10·7	Finland	1849			P. A. Nelsson
	576	NV	Catrine *Ohlsen -88* W Sr	Norweg'n	155	96·8	23·6	10·8	Nordby, Fanø	1870	J. W. Christensen	[illegible]	F. Ohlsen
[illegible]	577	A	Catterina B. (ex [illegible]) *Babich -84* W Bk	Austrian	422	133·8	28·2	17·0	Sunderland	1865	Taylor & Scoular	[illegible]	Ca. Babich
[illegible]	578		Cattina (ex [illegible]) *Valentin* W Sr	Austrian	175	93·1	21·5	17·8	[illegible]	1868	G. Giurkotich	[illegible]	G. Marussich
[illegible]	579	A	Cattina R. *Sforzina* W Bk	Austrian	381	122·7	27·2	16·9	Lussinpiccolo	1868	G. S. Cattarinich	[illegible]	A. Sforzina
[illegible]	580	IV	Cattolico *Simonetti* W Bk	Italian	981	181·4	33·1	25·3	Varazze	1875	G. Ascheri	[illegible]	P. Lavarello
[illegible]	581		Cautious W Sr	British	153	86·0	23·3	12·6	Rye	1854	W. Mills		W. W. Foreman
[illegible]	582	A	Caval Marino W Bg	Turkish	140	72·2	21·3	10·8	Dulcigno	1873	Salih	[illegible]	Mehmet Giuli
[illegible]	583	R	Cavalier *Parkhill -88* W S	British	1089	176·8	37·3	23·7	Portland, N.B.	1866	Sorrell	[illegible]	W. H. Ross & Co.
[illegible]	584		Cavalier W S	British	299	137·0	26·2	11·7	Quebec	1867			J. (jr.), A. & P. McArthur
	585	N	Cavalier *Christophersen* W Bk	Norweg'n	687	167·1	32·5	17·6	Forhine Brd. [P. E. I.	1874	Duncan	[illegible]	A. Larsen-Naur
[illegible]	586	V	Cavaliere Ivanissevich (Lost) (ex [illegible]) *Sincevich -88* W Bk	Austrian	906	162·4	34·7	22·7	Trieste	1871	Stabilimento [Tecnico	[illegible]	G. Brailli & Co.
[illegible]	587		Cavaliere Krapf (ex [illegible]) *Lucovich 81-81* (CONDEMNED) W Bk	Austrian	358	123·0	27·9	17·2	Sunderland	1869	G. Short	[illegible]	A. T. Lucovich
[illegible]	588	R	Cavour *Brison -88* W Bk	British	1299	193·6	38·4	24·7	New Brunsw'k	1862	Hilyard	[illegible]	[illegible] Shipping Co. Ld
	589	N	Cavour *Boy* W Bk	Norweg'n	377	125·4	30·3	16·0	Harburg	1855			J. O. Boy
[illegible]	590	✱	Cawdor *Jardalla 87-87* I 4m S	British	2355	302·3	41·1	25·1	Southampton	1884	Oswald, Mord'nt [& Co.		J. Houston
	591		Cazador W Bn	Spanish	214	108·9	23·9	12·3	Albay, Luzon	1875			F. Escaño
	592		Cecil (ex [illegible]) *Olsen -88* W S	Norweg'n	1070	175·3	36·3	22·8	St John, N.B.	1857	J. M'Donald	[illegible]	Actieselskabet "Achilles"
[illegible]	593	✱	Cecil Brindley *Davies* W Sr	British	94	80·1	21·0	10·5	Aberdovey	1871	Richards		D. Jones & Co.
[illegible]	594	R	Cecile *Steele* W Sr	American	165	104·8	27·2	9·4	Newburyport, [Mass.	1873		[illegible]	J. L. Mott
	595	V	Cécile (WRECKED) *Lacoste -87* I Bk	French	683	171·2	29·8	18·6	La Seyne	1876	Forges & C. de la [Médit.		C. F. Brunellière frères
	596	V	Cécile Auger *Lemaître -88* W Bk	French	495	131·2	27·2	19·5	Honfleur	1874	Leviels	[illegible]	A. Auger
[illegible]	597	V	Cecilia *Agnello -87* W Bk	British	466	139·5	30·9	19·4	[illegible]	1865	G. Tixi	[illegible]	S. Gauci
[illegible]	598		Cecilia *Macarier* W Bn	British	171	104·0	24·0	12·0	P. E. Island	1864	Robinson		J. Oakes
	599		Cecilia *Trindade* W Bk	Brazilian	332	..	..	..	Villa do Conde			[illegible]	O. A. de Moura
	600		Cecilia W Sr	Brazilian	179	..	..	..					F. P. d'Almeida

✱ = **Lloyd's Register;** D = Nederlandsche Vereeniging; G = Germanischer Lloyd; I = Registro Italiano; R = [illegible]
A = Veritas Austro-Ungarico; H = Veritas Ellenico; N = Norske Veritas; V = Bureau [illegible]

In the 4th Column, C = Composite (iron frame, wood planked); I = Iron; Stl = Steel; W = Wood.

The Dates of Metalling (**Col. 11**) are based on the records of the Books in which Vessels are respectively classed, or on the Reports of the Surveyors of [illegible]

Further particulars of Vessels classed by Lloyd's Register will be found in Lloyd's Register Book and Lloyd's Yacht Register.

	Nos dans le livre	Classification	Nom. Capitaine	Matériaux et Gréement	Pavillon	Tonnage	Longueur	Largeur	Creux	Construit Où	Quand	Par qui	Doublage	Armateurs ou Directeurs-Gérants	Port d'Armement
	601	V	Cecilia *Gilbernau*	W Bg	Spanish	191	90·9	26·8	13·1	Masnou	1855		V 5.84	J.Gilbernau y Maristany	Barcelona
	602	DV	Cecilia *Nilsson* ·87	W Bk	Swedish	335	130·5	26·5	13·0	Sundsvall	1873	N.Hägglund	V 1.82	A.Åkerbladh	Sundsvall
	603		Cecilia K. *Kettina* 87-87 (ex Eliza)	W Bk	Peruvian	272	104·9	26·4	14·2					V.Kettina	Callao
	604	DV	Cecilie *Petersen* 71-76	W Bk	Danish	237	120·2	24·1	12·3	Rudkjöbing	1875	S.Boas	V 2.83	E.Petersen	Rudkjöbing
	605	V	Cecilie *Pettersen* ·86	W Bg	Danish	224	110·2	24·5	12·1	Nordby,Fanö	1875	S.Abrahamsen	V 5.87	Akt.Selsk (S.B.Hansen)	Nordby,Fanö
	606		Cecilie *Lund Nielsen* 78-86 (ex Arcadia)	W Bg	Danish	189	79·3	22·1	14·1	Sunderland	18$^{46}_{52}$			H.P.Lund	Rønne
	607	V	Cecilie *Rasmussen* 76-76	W Sr	Danish	138	96·1	22·9	10·7	Æroeskjöbing	1876	R.Mortensen		P.E.Jensen	Æroeskjöbing
	608	N	Cecilie *Sørensen* 72-	W Sw	Norweg'n	434	125·8	31·7	17·3	Svartvik	1847			Oscar Hansen	Christiania
	609	RV	Cedar Croft *Moss*	W Bk	British	1099	186·6	37·5	22·2	Portland,N.B.	1877	D.Lynch	C 11.88	H.D.Troop & Son	St.John,N.B.
	610	G	Ceder *Zehl* [illegible] (ex Joseph [illegible])	W Bk	German	718	119·6	33·1	19·6	Bremerhaven	1866	F.Tecklenborg	C 12.85	H.Danelsberg	Bremen
	611		Cedia *Lignell*	W Sr	Russian	228	..	..	..	Lappfjerd	1874			K.A.Lundman	Lemland
	612	✻	Cedric *Rook* ·88	W Sr	British	99	85·6	21·2	10·7	Kingsbridge	1888	W.Date & Sons		J.Gray & Sons	Falmouth
	613	✻	Cedric the Saxon *Vaughan* ·88	I S	British	1619	259·8	40·1	23·6	PortGlasgow	1875	J.Reid&Co		Williamson,Milligan&Co	Liverpool
	614		Cefalu *Rizzo* ·85	W Bn	Italian	120	88·6	24·0	9·7	Halifax,N.S.				G.B.Cefalu di Deo.	Palermo
	615		Ceferino *Majo*	W Bn		183	100·0	25·5	11·1	Arens	1872				
	616		Cekrops *Braun* 80-80	W Sr	Danish	118	81·0	19·6	11·2	Troense	1852	H.R.Møller		H.T.Helweg	Odense
	617	N	Celer *Rosenkild* ·88	W Bk	Norweg'n	633	151·3	32·4	18·0	Grimstad	1880	N.Osmundsen	V 2.87	Joh Klöcker	Arendal
	618	DN	Celeritas *Hansen*	W Bk	Norweg'n	248	101·6	25·2	14·2	Libau	1846			C.A.Hansen	Fredrikstad
	619		Celeritas *Bengtsson*	W Bk	Swedish	299	120·4	25·3	13·9	Ekenäs	1858	C.J.Bring		C.J.Bengtsson	Fiskebäckskil
	620		Celerity *Fuggle* ·81	W Bg	British	200	103·3	25·2	13·8	Middlesbro'	1867	R.Craggs&Sons		J.S.Sterry	Lowestoft
	621	V	Celeste Burrill *Robertson*	W S	British	1764	226·0	41·2	24·0	Clare,N.S.	1886	J.L.Burrill	V 2.86	Wm.Burrill&Co.	Yarmouth,N.S.
	622		Celestia	W Bk	British	225	102·6	25·0	10·3	Dixbury,U.S.	1852			A.J.S.Gibbs	Dunedin,N.Z.
	623	✻	Celestial Empire *Duncan* 87-88	I S	British	1757	267·2	40·1	23·9	Sunderland	1877	Osbourne,Graham&Co		G.Duncan & Co.	London
	624	A	Celestina *Vlassich* ·88	W Bk	Austrian	468	132·5	27·7	18·1	Fiume	1874	G.Brazzoduro	Y 12.85	Ma.Battagliarini	Fiume
	625	I	Celestina *Carniglia* (ex Maria Bellamy)	W Bk	Italian	481	131·8	30·5	19·7	Sampierdarena	1870	G.Casanova	Y 12.87	Vincenzo Carniglia	Spezia
	626		Celestina *Paoli*	W Bg	Italian	248	104·0	26·2	15·9	Varazze	1854	B.Camogli		G.Tonietti fu R.	Porto Ferraio
	627		Celestina O. *Olivari* (ex Hortensia [illegible])	W Bkn	Italian	239	114·8	25·1	14·2	SestriPonente	1876	A.Briasco	V 6.88	G. & M.Olivari	Genoa
	628	V	Celestina Rocca *Olivari* (ex Emilio Bordasano (W. 85R E.P.))	W Bn	Italian	613	156·1	32·7	21·5	SestriPonente	1867	B.Celesia	V 1.81	A.Rocca fu B.	Spezia
	629	V	Celestine *Luco* 88-88	W Sr	French	132	81·0	21·0	11·0	Vannes	1860	LeFel		P.Besnard	Brest
	630	✻D	Celia *Lewis* ·88	W Bkn	British	326	136·9	27·7	12·9	Sunderland	1876	Pickersgill	V 2.88	W.C.Jarvis	Liverpool
	631		Celia *Pell* (ex Fola... Mercedes Renom)	W Bn	Spanish	249	164·9	26·2	9·8	Blanes	1856			Carlos Roses	Barcelona
	632		Celia *Nilsson* ·87	W Bk	Swedish	358	124·0	24·9	16·2	Halifax,N.S.	18$^{56}_{64}$			J.Z.Wennerberg	Uddevalla
	633	R	Celina *Murray*	W 3mSr	American	628	161·5	35·4	13·0	Bath,Me.	1882	Adams&Hitchck		Adams & Hitchcock	Bath,Me.
	634	R	Celina *Mountfort* 82-87	W Bk	American	549	152·7	32·7	14·5	Deering,Me.	1873	R.Lewis&Co.	11.87	R.Lewis & Co.	Portland,Me.
	635		Celina	W Sr	Brazilian	158	..	..	..					J.M.Moreira	Rio Grande
	636		Celine *Lundberg* ·87 (ex M... [illegible])	W Bg	Swedish	306	108·6	26·6	16·5	Sunderland	1859	Alcock		E.Jansson	Malmö
	637		Celorio *Renteria*	W Bk	British	248	117·1	25·4	14·6	Sunderland	1870	Gill	V 3.83	J.F.Lucas	Bristol
	638	✻D	Celtic Chief *Pugmire* 83-85	I S	British	1746	266·8	39·5	22·8	Dumbarton	1885	A.M'Millan&Son		Celtic Chief Ship Co. Ld. (Parry, Jones & Co.)	Liverpool
	639	✻	CelticMonarch *Morrison* 86-86	I S	British	2075	277·3	42·5	24·2	Liverpool	1884	T.Royden&Sons		Celtic Monarch Ship Co. Ld. (Parry, Jones & Co.)	Liverpool
	640	G	Celurca *Whittingham* ·88	W Bk	British	356	123·8	27·8	15·6	Montrose	1870	Birnie	V 2.88	H.Campbell	Montrose
	641	✻	Centaur *Flinn* ·88	I Bk	British	1590	232·0	37·2	25·4	London	1864	MillwallIronCo.		Johnston, Sproule & Co.	Liverpool
	642	GV	Centaur *Jansen* ·84	W Bk	German	469	135·2	31·0	17·1	Apenrade	1875	J.Paulsen	Y 11.87	H.Mathiesen	Apenrade
	643	N	Centauri *Olsen*	W Bn	Norweg'n	139	93·6	23·2	11·3	Stavanger	1874			S.T.Dahle	Skudesnæs
	644		Centennial *Phipps* ·82	W Sr	British	124	92·0	26·3	7·6	Clifton,N.B.	1883	G.W.Flewelling		J.M.Taylor	St.John,N.B.
	645		Centennial *Bonnell*	W Sr	British	118	87·5	24·2	10·5	George's River [C.B.	1876	R.Young		Edward Bonnell	Sydney,C.B.
	646	V	Centennial *Colcord* ·88	W S	American	1287	190·1	38·0	24·0	E.Boston	1875	Smith&Twnsnd	V 8.87	Smith & Townsend	Boston,Mass.
	647	R	Centennial *Willets* 83-88	W 3mSr	American	527	149·7	36·0	16·4	Camden,N.J.	1876	Risley&Co.		A.Birdsall	PerthAmboy,NJ
	648		Centennial *Williams*	W Sr	American	116	88·0	23·5	8·8	Essex,Mass.	1876			D.C.& H.Babson	Gloucester,Mass
	649		Centennial *Barrett*	W Sr	American	108	85·4	28·4	5·9	May'sLndg,NJ	1876			W.Barrett	SomersPoint,NJ
	650		Centre Amérique *Rose* 81-85	W Bk	French	633	147·6	31·4	19·0	Bordeaux	1876	Chaigneau [frères	V 1.81	Ant.Dom.Bordes & fils	Bordeaux

✻ = Lloyd's Register; D = Nederlandsche Vereeniging; G = Germanischer Lloyd; I = Registro Italiano; R = American Record;
A = Veritas Austro-Ungarico; H = Veritas Ellenico; N = Norske Veritas; V = Bureau Veritas.

Dans la 4me Colonne, C = Composite (membrure en fer, bordé en bois); I = Fer; Stl = Acier; W = Bois.

Les Dates du doublage en métal (Col. 11) ressortent des livres dans lesquels les navires sont respectivement classés ou des rapports des Inspecteurs du Lloyd's Register.

Pour plus amples renseignements sur les navires classés par le Lloyd's Register, voir le Lloyd's Register Book et le Lloyd's Yacht Register.

Official No.	No. in Book.	Registry in which classed.	Ship's Name. Material and Rig. Master, and Dates of Appointment to present Owner's Service and to Ship. Ship's Late Name.	Flag.	Net Tonnage.	Dimensions. Length.	Breadth.	Depth.	Built. Where.	When	By whom.	Date of Metalling.	Owners or Managers.
60388	651	✳	Centurion *Widdicombe 87-71* I S	British	1194	231·2	35·1	22·0	Glasgow	1868	A.Stephen&Son		J.&A.Roxburgh
4287	652		Centurion *Blodgett* W Sr	American	114	84·4	23·6	8·0	Searsport,Me.	1840			F.A.Linnell
55267	653		Century *Romeril 83-88* W Bn	British	181	107·3	21·2	12·6	Jersey	1866	LeVesconte		G.Le Gros
11	654		Cephalonia W Bg	Greek	219	96·5	27·2	..					A.Vagliano
520	655		Cephalus W S	Greek	118	77·8	22·6	..	[illegible]	1888			E.Terentinos
41	656		Cephissus I	Greek	312	131·2	23·0	13·6	Austria	1876			Theophilatos&Stathatos
61586	657		Cepola W Sr	British	101	83·0	23·0	8·5	PortleBurt	1875			ThomasJohnson
	658		Ceralia *Grönqvist 82-82* W S	Russian	169	92·7	24·2	11·3	Sideby	1877			B.F.Westerling
	659	✳	Cerastes *Brumund 85-88* I Bk	German	573	177·0	29·2	17·9	Sunderland	1868	W.Pile&Co.		D.W.Stege
92358	660		Cerdic *J. J. Shields* W Sr	British	120	82·9	26·2	7·0	Alma,N.B.	1886	J.J.Shields		F.A.French
53037	661	✳	Ceredig *Evans 83-83* W Bg	British	193	103·1	24·3	12·9	Newquay	1865	Davies	Y 1.86	J.Evans
16786	662		Ceres *Sharman 82-82* W Bk	British	124	82·8	20·6	10·5	Prussia	1840			W.P.Brown
42583	663	✳	Ceres *Humphreys* W S	British	90	77·2	20·6	10·5	Aberdovey	1862	Richards		E.Humphreys
	664		Ceres *Meikelsen 85-88* (ex Van Brakel) W3mSr	Danish	316	104·0	26·9	17·6	Amsterdam	1849	J.Meyjes&Sons		William Bille
	665		Ceres *Anderson 78-87* W Bk	Danish	307	111·6	25·3	14·6	Colberg	1865	Most&F.Richter		M.Christiansen
	666		Ceres *Nielsen 79-83* W Bk	Danish	285	121·5	25·5	14·4	Troense	1867	T.Petersen		Den Kongelige Grønlandske Handel (H.Bórnse)
	667	V	Ceres *Nielsen -88* W Bn	Danish	118	77·7	21·6	10·4	Assens	1869	C.Sørensen		N.F.Nielsen
	668	D	Ceres *Visser -87* W3mSr	Dutch	260	113·5	24·6	14·8	Sunderland	1855	Austin& Mills	Z.88	M.E.Kuipers
	669	✳V	Ceres (LOST) (ex Kirkham) *Kaphamel 85-85* I Bk	German	1043	201·9	34·2	21·7	Birkenhead	1856	J.Laird		M.G.Amsinck
	670	GV	Ceres *Bülow 86-* W Bk	German	526	136·0	31·0	17·6	Rostock	1876	E.Burchard&Co	Y 6.86	H.Bauer
	671	V	Ceres *Horck* W Bk	German	384	120·8	29·5	17·8	Memel	1868	B.Pieper		Graff & Bannitz
	672	GV	Ceres *Socken 81-* W Bg	German	288	118·8	25·6	13·1	Elsfleth	1872	J.Ahlers	Y 9.86	J.D.Borgstede
	673	G	Ceres *Schleinhege 83-* W Bn	German	135	85·0	20·0	10·0	Papenburg	1862	L.R.Röttjers		G.J.Schulte Wwe.
	674	N	Ceres *Olsen* W Bk	Norweg'n	515	139·6	31·8	17·2	Bergen	1866	J.Trampy	Y 5.88	Julius Olsen
	675	N	Ceres *Jörgensen* W Bk	Norweg'n	375	125·3	28·1	15·1	Grimstad	1872	J.Tellefsen	Y 6.88	J.E.Einersen
	676	N	Ceres *Larsen 83-83* W Bg	Norweg'n	219	85·3	25·5	14·9	Sweden	1793/1842			T.Ihlen
	677	N	Ceres *Börresen* W Bg	Norweg'n	195	98·1	24·8	11·9	Kragerö	1873		Y 3.88	H.Biörn,jun.
	678	N	Ceres *Christiansen 82-82* W Bg	Norweg'n	178	96·0	25·3	11·4	Haderstleben	1847	Kieldriep		S.Mörch
	679	N	Ceres *Stol -87* W Bn	Norweg'n	171	108·4	22·7	11·2	Skien	1870	A.Olsen	Y 6.88	L.Larsen
	680	N	Ceres *Pedersen* W Sr	Norweg'n	138	87·3	21·9	10·8	Libau	1845			N.Pedersen
	681	N	Ceres *Öveland* W Sr	Norweg'n	108	68·7	21·9	11·3	Kragerö	1843			C.A.Oveland
	682	V	Ceres *Rodrigues* W Bk	Portug'se	289	112·2	27·2	12·1	Villa do Conde	1868	J.M.deAraujo	Y 10.87	J.Nogueira Pinto
	683		Ceres *Jansson* W3mSr	Russian	273	..	..	..		1873			P.Eriksson
	684		Ceres W Bg	Russian	179	..	..	..		1840			F.T.Lindroth
	685	V	Ceres *Öhmann* W Bk	Swedish	305	125·6	25·5	12·6	Hernösand	1871	J.O.Strandberg	Y 2.86	A.Nätterqvist
	686	D	Ceres II. (ex Atilla, ex Dorothea) *Christensen 85-88* W3mSr	Danish	334	116·5	24·1	16·6	Sunderland	1862	Naisby		W.Bille & Co.
	687	✳	Cerro Alègre *Gostin -88* W Bk		608	165·6	31·3	18·4	Bordeaux	1873	Rochet&Largeteau	Y 3.86	
4616	688		Cerro Gordo *Chapman* W Sr	American	121	93·5	26·0	8·2	Westford N.Y.	1848			H.G.Chapman
	689		César W Bk	French	228	100·6	24·3	14·2	Fécamp	1867			G.Vasse & Du Hamel
	690		Cesar (ex [illegible] & Matosinho) *Da Corte Biala* W Bn	Portug'se	168	94·5	21·0	10·8	NovaScotia	18 67/81			J. dos Santos & Silva
	691		César & Jean *Robion -87* W3mSr	French	194	94·2	23·3	11·8	Bastia	1867		Y 6.84	Dupius Robial
487	692		Cesare Augusto *Mori* W Bn	Italian	150	91·2	25·9	14·7	[illegible]	1848	Bastiani		G.Tonietti fu R.
470	693	I	Cesare Secondo *Indelicato* W Bn	Italian	114	82·0	22·6	10·2	[illegible]	1868	Pichiotti	Z 2.84	M. & G.Catalano
1163	694	IV	Cesarina *Romano -91* W Bg	Italian	420	125·0	27·2	18·2	[illegible]	1878	G.Mauro	Y 6.87	C.A.Romano
3307	695	✳I	Cesira (ex Velta, ex Nostra Signora Genova) *Dodero -88* W Bk	Italian	643	152·6	32·4	20·7	[illegible]	1876	Traverso	Y 1.88	F.Dodero
631	696	I	Cesira Madre (ex Antonietta Vernici, ex Dariena) *Giampello -88* W Bn	Italian	106	77·4	24·2	10·8	Limite	1869	Picchiotti		C.Giannoni & Figli
	697	V	C'est Moi (WRECKED) *Couëdel* W Bg	French	136	80·4	20·3	11·3	Martigues	1865	Mauros&Finastre		Couëdel
67769	698		Cetonia *Guinness* W Sr Yt	British	113	107·0	21·1	12·4	Cowes	1873	M.Ratsey		Sir E.C.Guinness,Bt.
1339	699	I	Ceylan *Lauro* W Bk	Italian	497	140·4	30·2	19·0	Alinuri	1882	Mauro	Y 6.88	P.Lauro
66519	700	V	Ceylon *Owens -88* W S	British	944	179·5	36·5	22·1	Newport,N.S.	1873	N.Mosher	Y 1.88	F.Curry

✳ = **Lloyd's Register**; D = Nederlandsche Vereeniging; G = Germanischer Lloyd; I = Registro Italiano; R = [illegible]
A = Veritas Austro-Ungarico; H = Veritas Ellenico; N = Norske Veritas; V = [illegible]

In the 4th Column, C = Composite (iron frame, wood planked); I = Iron; Stl = Steel; W = Wood.

The Dates of Metalling (**Col. 11**) are based on the records of the Books in which Vessels are respectively classed, or on the Reports of the Surveyors of L[illegible]

Further particulars of Vessels classed by Lloyd's Register will be found in Lloyd's Register Book and Lloyd's Yacht Register.

No. du Registre	Registres dans lesquels le navire est classé	Nom, Matériel et Gréement. Capitaine et Année d'entrée au service des Armateurs associés et à bord du navire. Ancien Nom.	Pavillon	Tonnage	Longueur	Largeur	Creux	Construit Où	Quand	Par qui	Date du doublage	Armateurs ou Directeurs-Gérants.	Port d'Armement.
701	R	Ceylon *[illegible]* WBk	American	647	150·1	33·6	20·2	Boston,Mass.	18[illegible]	Wm.Hall	[illegible]	R.Calhoun	Pt.Twnsnd,Wash.
702	DG	Ceylon *[illegible]* WBk	German	561	146·8	30·4	19·7	Preston	1857			Lichenheim & Pincus	Rostock
703	N	Ceylon *[illegible] -72* WBk	Norweg'n	565	151·6	34·8	16·9	Miramichi	1858			Soren Christensen	Drammen
704	V	Ceylon *[illegible]* WBk	Swedish	370	133·8	26·7	12·9	Gefle	1838	O.A.Brodin	[illegible]	F.Rettig	Gefle
705	❋	Chacma *[illegible]* I Bk	British	580	174·0	29·2	17·9	Sunderland	1863	W.Pile&Co.		R.Jones	London
706		Chakkir Hahgus W	Turkish	1[illegible]	..	..	..	Chakkir Hahgus	1868			Tusiji-bashi Mustapha & Co.	ChakkirHahgus
707		Chakraz WBg	Turkish	1[illegible]	..	..	..	Chakraz	1876			Catalada [illegible]	Chakraz
708	❋	Chala *Park 87-87* I Bk	British	1026	213·0	[illegible]	20·1	[illegible]	1885	W.Pickersgill& Sons		S.Wakeham & Sons	Liverpool
709	R	Chalgrove *Frances* WBk	British	513	157·2	28·0	19·1	[illegible]	1862	Lampert	[illegible]	W.Barrett	London
710	❋	Challenge *Davey 82-82* WSr	British	79	80·1	20·2	9·7	[illegible]	1870	Luke	[illegible]	A.Luke	Fowey
711		Challenge *[illegible]* WSr	American	112	90·0	24·0	8·1	St.George,Me.	1844			R.Long	St.George,Me.
712	RV	Challenger *Thompson 77-77* WS	American	1456	212·0	39·7	24·[illegible]	Bath,Me.	1877	E.A.Sewall	[illegible]	A.Sewall & Co.	Bath,Me.
713		Challenger *Hamilton [illegible]* W3mSr	American	266	127·9	32·5	9·2	[illegible]	1883	A.Cookson		W.J.Sweasey	Eureka,Cal.
714	RV	Chalmette *Taylor -87* WBk	American	806	151·6	33·0	21·8	[illegible]	1865	G.&C.Bliss	[illegible]	G.S.Dearborn	New York
715	❋V	Champenois *Chevriot 84* WBk	French	430	126·1	28·6	16·0	[illegible]	1883	Sevestre frères	[illegible]	J.B.Demange	Nantes
716	V	Champenoise *Berticat* WSr	French	131	112·5	20·3	10·7	Dunkirk	1851			Vve.Meleart	Paimpol
717		Champenoise *Flocher 83-83* WSr	French	107	83·7	22·1	10·7	Dunkirk	1848			F.Levee	Brest
718	❋	Champion *Noel 88-88* IS	British	1441	210·4	37·2	23·5	Sunderland	1875	T.R.Oswald		H.Fernie & Sons	Liverpool
719		Champion *Haddie* WSr	British	149	90·1	21·0	11·7	Southampton	1827	Watson		T.&S.Oates	Lowestoft
720		Champion *Verge 82-82* WBn	British	146	97·1	25·0	10·3	PetiteRivière	1866		[illegible]	A.W.West	Halifax,N.S.
721	R	Champion (WRECKED) *Gifford -88* WBn	British	139	83·2	27·0	10·2	RiverJohn,N.S.	1884	G.Heighton		H.R.Troop	Pictou,N.S.
722	❋	Chanaral *Thomas 88-88* I Bk	British	58[illegible]	168·7	27·6	19·3	Sunderland	1862	T.R.Oswald		Shallcross&Higham	Liverpool
723	❋	Chanaral WBk	Chilian	72[illegible]	170·5	31·4	20·0	Bordeaux	1874	Roy	[illegible]	P.S.Codes	Valparaiso
724		Chance (ex North West) WBk	British	28[illegible]	100·3	24·5	17·6	Salem,U.S.	1846			S.Nichol	Invercargill,NZ
725	NV	Chance *Pedersen* W3mSr	Norweg'n	235	113·7	27·6	10·8	Tvedestrand	1885	E.Gundersen	[illegible]	Nikolai Nielsen	Christiansand
726	V	Chanito *Berdagier* WBg	Spanish	184	92·2	27·5	12·4	Barcelona	1850		[illegible]	J.Felpa & Mir	Barcelona
727		Chanticleer WBk	British	301	115·1	26·6	12·0	Williamsburg [U.S.]	1854			Henry Thackray	Turks Islands
728		Chanticleer *Simpson 55-85* WBg	British	221	107·3	20·2	11·1	Rye	1842			Newman,Hunt&Co.	London
729		Chapman WSr	French	145	80·4	19·7	11·9	Nantes	1857	Dubigeon		Jego	Nantes
730	N	Chapman *Larsen 85-85* WBk	Norweg'n	486	130·8	30·3	17·7	Grimstad	1877	Nils Osmundsen	[illegible]	N.Thue Johnsen	Grimstad
731	N	Chapman *Christensen -87* WBk	Norweg'n	425	143·0	30·9	15·9	Westerwik	1856	Fredell		W.Wilhelmsen	Tönsberg
732		Charger *Gundersen -88* WBk	British	1041	174·5	37·7	22·6	N.Brunswick	1856	J.T.Smith		T.Dixon & Sons	Belfast
733	RV	Charger *Goodell -88* WS	American	1444	205·2	39·8	24·0	E.Boston	1874	Smith&Twnsnd	[illegible]	H. Hastings & Co.	Boston,Mass.
734		Charibdis WBn	Greek	118	..	..	..	Venice	1854			A.Tzacassianos	Zante
735	A	Chariclea (ex Milan) *Russo -88* WBk	Greek	622	139·7	29·5	20·5	Lussinpiccolo	1875	N.Martinolich	[illegible]	N.P.Russo	Patras
736	❋	Chariclia (ex Raymond) WBn	Italian	199	104·9	25·3	12·9	P.E.Island	1884	D.Egan		S.Stathopulo	Naples
737	V	Charité (WRECKED) *Mépy -87* WBk	French	255	104·9	22·6	12·8	St.Malo	1876	L.Mallard	[illegible]	Irasque	Havre
738	❋	Charity *Williams* WBk	British	432	110·7	28·8	17·5	Sunderland	1872	Gardner	[illegible]	Eills & Co.	Liverpool
739		Charity *Bell* WBg	British	199	97·2	23·6	13·9	Arbroath	1857	W.Stephen		W.Milburn	Montrose
740	V	Charles *Perry* WS	British	1500	212·0	39·4	23·8	Tusket,N.S.	1879	C.W.Murphy	[illegible]	John Murphy	Yarmouth,N.S.
741		Charles WBn	British	178	96·0	23·4	12·8	P.E.Island	1862			G.Peters	Faversham
742		Charles *Roberts -86* WSrFlat	British	48	64·6	18·2	6·6	Amlwch	1871	Paynter		J.Carson	Beaumaris
743		Charles WBk	French	596	..	..	..	Granville	1871		[illegible]	C.Leclere	Granville
744		Charles WBk	French	213	92·8	23·3	13·8	St.Serran	1850	Lenormand	[illegible]	Lesant	Vannes
745	DV	Charles *Bandou* WBn	French	210	102·0	23·3	12·8	Pr.EdwardIs.	1867		[illegible]	A.Postel & ses fils	Cherbourg
746	V	Charles *Suquet* WBk	French	144	85·3	22·3	11·6	LaSeyne	1871	Curet,Fils	[illegible]	Ch.Suquet	Cette
747	G	Charles *Langhoff 74* W3mSr	German	306	122·3	27·0	14·3	Ueckermünde	1864	C.Pretzer		H.Langhoff	Stettin
748	G	Charles (ex [illegible]) *Sonesson [illegible]* WBg	Swedish	220	114·5	24·8	11·9	Tillingenabben	1867	L.O.Nordström		N.Sonesson	Wiken
749		Charles A. Briggs *Howard -82* W3mSr	American	720	163·9	40·2	16·1	Bath,Me.	1879	Goss&Sawyer		J.B.Phillips	Fall River,Mass.
750													

Lloyd's Register; D=Nederlandsche Vereeniging; G=Germanischer Lloyd; I=Registro Italiano; R=American Record; …=Veritas Austro-Ungarico; H=Veritas Ellenico; N=Norske Veritas; V=Bureau Veritas.

Dans la 4me Colonne, C=Composite (membrure en fer, bordé en bois); I=Fer; Stl=Acier; W=Bois.

…Dates du doublage en métal (Col. 11) ressortent des livres dans lesquels les navires sont respectivement classés ou des rapports des Inspecteurs du Lloyd's Register.

Pour plus amples renseignements sur les navires classés par le Lloyd's Register, voir le Lloyd's Register Book et le Lloyd's Yacht Register.

SAILING VESSELS.

1 Official No.	2 No. in Book	3 Registry & Class	4 Ship's Name. Master, and Dates of Appointment to present Owner's Service and to Ship. Ship's Late Name.	Material and Rig	5 Flag	6 Net Tonnage	7 Length	Breadth	Depth	8 Built: Where	9 When	10 By whom	11 Date of Metalling	12 Owners or Managers
125115	751	R	Charles A. Coulomb *Moore*	W3mSr	American	421	128·5	32·0	16·0	Milton, Del.	1873	N.W. Megee		G.W. Fenimore
67925	752		Charles A. Hoard *Boyd*	W Bo	British	179	102·4	26·1	12·1	Pubnico, N.S.	1865			J. Kingsberry
125063	753	R	Charles A. Sparks *Harris*	W Bn	American	423	130·7	30·1	15·8	Damariscotta, Me.	1872	Merry	Y 10.86	E.C. Harris
125691	754	RV	Charles B. Kenney *Rice*	W Bk	American	1073	179·3	37·2	20·8	Bath, Me.	1878	Goss, Sawyer & Packard	[illegible]	J. Jensen
64480	755	V	Charles Bal *Monk 82-86* (WRECKED)	W S	British	1432	202·0	39·6	24·5	St. John, N.B.	1870	Nevins & Fraser	[illegible]	"Charles Bal" Ship Co.
	756	DN	Charles Bal *Nielsen*	W Bk	Norweg'n	724	175·6	33·9	19·6	Porsgrund	1870	H.J. Gregersen	[illegible]	Steen & Strom
126083	757	R	Charles C. Dame *Daniels*	W3mSr	American	568	161·0	35·0	15·0	Newburyport, Mass.	1882	Atkinson & Fillmore		F.L. Atkinson
126160	758		Charles C. Lister *Truitt -84*	W3mSr	American	251	119·0	30·8	9·0	Milford, Del.	1883	Simpler & Revill		J.W. Hall
126042	759		Charles C. Lister, jr. *Palmer 82*	W Sr	American	152	90·0	27·0	7·0	Frederica, Del.	1882	Nathaniel Lank		J.W. Hall
4170	760		Charles C. Warren *Ratcliffe*	W Sr	American	103	85·5	24·4	8·0	Essex, Mass.	1866			A.N. Rackliffe
65104	761		Charles Chalmers *Patrick*	C Bg	British	177	102·4	23·1	12·6	Aberdeen	1874	Hood		Aberdeen Commercial
5187	762		Charles Colgate *Rogers*	W Sr	American	170	97·7	27·1	10·2	Patchogue NY	1850			S.D. Lawrence
	763		Charles Collet *Lamblotte 74-81*	W Bk	French	406	131·5	25·9	15·8	Dunkirk	1870	Vandezande	[illegible]	C.L. & P. Collet-Taverne fils
74526	764	✱	Charles Cotesworth *Kitchen 75-84*	I Bk	British	1031	205·5	34·7	21·5	Liverpool	1876	T. Royden & Sns.		Cotesworth, Lynn & Co.
73140	765	V	Charles Cox *Nielsen 86.86*	W Bk	British	677	155·2	34·0	18·2	Maitland, N.S.	1875	Charles Cox	[illegible]	John A. Ledward
125384	766	RV	Charles Dennis *Edgett -88*	W S	American	1710	215·7	39·8	24·8	Richmond, Me.	1875	Southard & Son	[illegible]	T.J. Southard & Son
125175	767		Charles Dennis *Haskell -83*	W Bn	American	373	138·5	31·0	11·9	Richmond, Me.	1873	Southard	[illegible]	T.J. Southard & Son
	768	✱V	Charles Dickens *Petersen* (ex Steamer Danubio) I -88	I S	German	1329	250·0	31·6	24·2	Glasgow	1856	Smith & Rodger		R.M. Sloman & Co.
83304	769	R	Charles Duncan *M'Culloch 83-87*	W Bn	British	149	94·5	24·0	10·8	Pt Medway NS	1883	J. Butt	[illegible]	A.G. Jones & Co.
126044	770	R	Charles E. Balch *Crocker -88*	W4mSr	American	802	181·6	36·6	14·3	Bath, Me.	1882	B.W. & H.F. Morse		B.W. & H.F. Morse
88658	771	✱	Charles E. Lefurgey *Kemp*	W Bk	British	950	190·9	36·0	20·9	P.E. Island	1884	J. Lefurgey	[illegible]	John Lefurgey
126070	772	RV	Charles E. Moody *Leonard 83-84*	W S	American	1915	233·9	43·4	26·6	Bath, Me.	1882	Goss & Sawyer	[illegible]	John R. Kelley
5702	773		Charles E. Moody *Leonard -88*	W Sr	American	120	97·0	25·4	8·5	Frankfort, Me.	1867		[illegible]	
5783	774	R	Charles E. Morrison *Smith 78-*	W3mSr	American	506	139·0	32·0	16·0	Prtsm'th, NH.	1869	W.F. Fernald		E.H. Atwood
4215	775		Charles E. Raymond *Kelley*	W Sr	American	309	116·0	29·5	9·8	Wilmingtn Del	1864			P. Thacher
126407	776	R	Charles E. Babbitt (ex Bertha A. Watt) *Gray -87*	W Sr	American	138	90·2	24·9	10·7	Lunenburg NS	1881	D. Smith	[illegible]	Dix & Wilkins
126177	777	R	Charles E. Schmidt *Sharp*	W3mSr	American	514	139·0	35·0	14·9	Mauricetn, N.J.	1883	J.W. Vanneman & Bros.		I. Sharp
125303	778		Charles E. Sears *Allen*	W Sr	American	152	96·0	26·6	9·5	Lubec, Me.	1874	Mufgad		C.J. Staples
5682	779		Charles E. Smith *Ketchum 83-84*	W Sr	American	203	105·0	28·0	8·5	Philadelphia	1869	Philadelphia & Reading Ry Co.		Philadelphia & Reading Ry. Co.
126032	780	R	Charles E. Young *Corson*	W3mSr	American	534	147·5	35·3	15·5	Tuckahoe, N.J.	1882	Sheppard		George May
	781	V	Charles Elisa *Le Saer -88*	W Sr	French	102	78·7	20·7	9·3	Dieppe	1872	Vve. P. Coree	[illegible]	A. Buhot-Launay
125201	782		Charles F. Sampson *Norris 86-88*	W3mSr	American	507	151·7	35·0	15·8	Bath, Me.	1874	Goss & Sawyer		Doane & Crowell
	783		Charles F. Tuttle *Jennings*	W3mSr	American	738	177·3	36·5	..	West HavnCt	1886	Gesner & Marr		Henry Sutton
125453	784	R	Charles F. Ward *Coombs 84-89*	W Bk	American	515	138·4	32·9	17·4	Millbridge, Me	1875		[illegible]	J.S. Emery & Co.
5748	785	R	Charles Fobes *Price -88*	W Bk	American	511	132·1	29·6	17·6	Yarmouth, Me.	1869	Hutchings & Stubbs	[illegible]	J.B. Drake
126078	786	R	Charles G. Joyce *Wheatley*	W Sr	American	117	97·9	26·9	7·6	Baltimore	1882	W.E. Woodall		T.W. Wheatley
125758	787	RV	Charles G. Rice *Lund -88*	W Bk	American	680	151·3	32·4	18·3	Yarmouth, Me.	1879	Hutchings & Stubbs	[illegible]	J. Bain & Co.
	788		Charles Gustave *Leduc*	W Bk	French	268	..	..	..	Fécamp	1869			Veuve Valin
125295	789		Charles H. Fabens *Curtis 83-89*	W3mSr	American	286	131·5	30·9	10·3	Bucksport, Me	1874	W. Beazley & Co.	[illegible]	B.H. Fabens
125897	790	R	Charles H. Haskell *Sibley*	W3mSr	American	452	115·8	34·0	12·4	Bath, Me.	1882	Deering & Donnell		Deering & Donnell
4122	791		Charles H. Hodgdon *Campbell -86*	W Sr	American	107	81·2	22·9	8·7	Essex, Mass.	1864			J.D. Hilliard
125622	792		Charles H. Merchant *Olsen -88*	W3mSr	American	269	120·0	33·2	10·5	Coos Bay, Oreg	1877			E.B. Dean
64501	793		Charles H. Oulton (ex Bavaria) (BROKEN UP)	W S	British	1160	161·4	36·5	21·0	New York, US	1846			W.W. Willink
126056	794	R	Charles H. Sprague *Nickerson 81-83*	W3mSr	American	303	128·0	30·8	12·9	Madison, Con.	1882	C.M. Miner		C.H. Sprague & Son
125719	795	R	Charles H. Trickey *Kelley -8*	W3mSr	American	267	120·1	32·2	8·1	Bath, Me.	1879	Goss & Sawyer		B.F. Nealley
126266	796	R	Charles H. Valentine *Curtis*	W3mSr	American	611	159·0	35·0	15·0	New Havn, Cn.	1884	H.H. Hanscom		A. Curtis
126113	797	R	Charles H. Wolston *Dunton -88*	W3mSr	American	333	133·0	30·3	10·8	Bath, Me.	1883	Goss & Sawyer	[illegible]	F.C. Curtis
125893	798		Charles Hanson *Sprague*	W Sr	American	183	107·0	30·0	8·0	Eureka, Cal.	1881			W. Brown
4186	799		Charles Heath *Pendleton*	W Sr	American	159	93·4	23·9	9·4	Searsport, Me.	1855			W.S. Pendleton
[illegible]	800	R	Charles J. Willard *Wallace*	W3mSr	American	241	106·1	29·2	12·8	Yarmouth, Me.	1874			A.B. Hall

✱ = **Lloyd's Register**; D = Nederlandsche Vereeniging; G = Germanischer Lloyd; I = Registro Italiano; R = [illegible]
A = Veritas Austro-Ungarico; H = Veritas Ellenico; N = Norske Veritas; V = [illegible]

In the 4th Column, C = Composite (iron frame, wood planked); I = Iron; Stl = Steel; W = Wood.

The Dates of Metalling (**Col. 11**) are based on the records of the Books in which Vessels are respectively classed, or on the Reports of the Surveyors [illegible]

Further particulars of Vessels classed by Lloyd's Register will be found in Lloyd's Register Book and Lloyd's Yacht Register.

No.	Registre	Nom. Capitaine et dates d'entrée au service des Armateurs actuels et à bord du navire. Ancien Nom.	Matériaux et Gréement	Pavillon	Tonnage net	Longueur	Largeur	Creux	Construit Où	Quand	Par qui	Date du doublage en métal	Armateurs ou Directeurs-Gérants	Port d'Armement
803	*	Charles James *Jones 77-77*	W3mS	British	138	93·0	24·0	11·7	Pwllheli	1877	D. Williams		W. Johnson & Co.	Carnarvon
804	V	Charles Kahl *E. Ferde 85-85*	W Bk	German	393	114·5	28·5	15·6	Stralsund	1869	C. W. Mohr		C. A. Berg	Stralsund
805	R	Charles L. Jeffrey *Russell*	W3mSr	American	351	129·5	30·5	11·7	E. Boston	1881	J. M. Brooks		C. Lovell & Son	Boston, Mass.
806	R	Charles L. Mitchell *Flood*	W3mSr	American	567	150·0	35·0	15·2	Fair Haven	1881	H. H. Hanscom		C. & J. F. Baker	Boston, Mass.
807		Charles Lawrence *Taylor*	W3mSr	American	436	130·0	33·0	14·0	May's Ldg N.J.	1874			E. Taylor	Somers Pnt. N.J.
808	R	Charles Loring *Thestrup*	W Bk	American	525	144·0	31·2	17·4	Yarmouth, Me	1878	G. Loring	Y 5.86	J. Bain	Portland, Me
809	I·G	Charles Luling *Wicke 61-72*	W Bk	German	1231	194·4	39·0	23·2	Quincy, Mass.	1865	G. Thomas	1.87	D. H. Watjen & Co.	Bremen
810	R	Charles Morford *Herbert 88*	W3mSr	American	346	131·0	31·0	14·0	Chester, Pa.	1871			W. T. Schultz	New York
811	*	Charles Napier *Bray 88*	W Sr	British	96	79·6	17·4	10·4	Southampton	1841	F. Andrews	Y 4.86	J. Dible	Southampton
812		Chas. Noble Simmons *Babbitt 81*	W3mSr	American	716	168·0	35·5	17·7	Bath, Me.	1881			H. L. Babbitt	Fall River, Mass
813	N	Charles Northcote *Henriksen*	W Bk	Norweg'n	332	106·2	28·8	16·9	Risör	1853			Stian Finne	Risör
814		Charles P. Knight *Babcock*	W Bn	British	189	101·7	24·2	12·9	P.E. Island	1874	Knight		W. J. Lawson	Newport
815		Chas. P Sinnickson *Collins*	W3mSr	American	181	113·8	23·5	8·7	Chester, Pa.	1863			G. A. Twibill	Philadelphia
816		Charles P. Stickney *Hawes*	W Sr	American	179	108·6	27·8	9·0	Philadelphia	1859			G. W. Shepherd, jun.	Philadelphia
817	R	Charles Purves *Small*	W Bg	American	523	131·7	32·0	17·6	Baltimore	1871	Hardy & Wood'll	Y	A. Purves & Son	Philadelphia
818		Charles R. Campbell (1-9-T) *McFarland*	W3mSr	American	416	140·0	34·0	10·5	Bath, Me.	1883	B W & H F Morse	Y 7.83	B. W. & H. F. Morse	Bath, Me.
819	V	Charles R. Lewis *Reed, Skew*	W Bk	American	727	165·4	33·4	17·7	Deering, Me.	1875	R. Lewis & Co.	Y 7.86	R. Lewis & Co.	Portland, Me.
820		Charles S. Baylis *Hall*	W3mSr	American	444	122·0	33·0	16·0	S Rondout, NY	1866			Morse & Co.	New York
821	R	Charles S. Davis *Sooy*	W3mSr	American	508	149·8	34·8	15·0	Cooper's Pt N.J	1883	S. W. Tilton		W. C. Sooy	Somers Pnt. N.J.
822	V	Charles S. Whitney *Spicer*	W S	British	1651	236·7	42·0	23·8	Parrsboro' NS	1885	Spencer's Is. Co.		G. D. Spicer	Parrsboro', N.S.
823		Charles Thompson *Odell*	W Sr	American	145	100·7	25·6	9·5	Kenneb'nk, Me	1865			G. O. Odell	New York
824	V	Charles Tottie *Andersson*	W Bk	Swedish	448	123·3	28·0	16·1	Svartvik	18 46/64		Y 12.83	G. H. Hegardt & Co.	Gothenburg
825		Charles Tucker	W Sr	British	106	80·6	20·4	10·3	Swansea	1839	W. Meager		W. Escott	Penzance
826		Charles W. Alcott *French 69-13*	W3mSr	American	282	132·3	36·9	9·0	Greenprt. N.Y.	1873	O. H. Bishop		C. C. Wehrum	New York
827		Charles W. Church *Lent*	W3mSr	American	802	179·3	37·9	15·5	Bath, Me.	1884	N England SB Co		J. H. Church	Fall River, Mass.
828		Charles W. Morgan *Keith*	W Bk	American	314	105·6	27·7	17·6	N. Bedford Mas	1841	Hillman Bros.		J. & W. R. Wing	New Bedfrd. Mass.
829		Charles W. Morse *Rose*	W Sr	American	106	85·7	23·8	9·2	Shelburne, NS	1873		Y 10.81	J. McCullough	New Bedfrd. Mass.
830	*	Charles W. Oulton *Fitzgerald 87-87*	W Bn	British	178	102·0	25·9	11·0	Botsford, N.B.	1874	Oulton	Y 8.84	W. P. Munn	St. Johns, N.F.L.
831		Charles & Arthur	W Bn	British	170	108·8	24·4	10·0	Tasmania	1870	G. Lapthorne		S. Rossiter	Melbourne
832	*	Charles & Ellen *Vickerman*	I Sr	British	145	106·6	22·6	10·5	Barrow	1878	D. Noble & Co.		W. Postlethwaite	Barrow
833	V	Charles & Jeanne *Lejeunen*	W Bk	French	391	127·3	25·6	15·4	Dunkirk	1871	D. Van Cauwenberghe		D. Van Cauwenberghe	Dunkirk
834	V	Charles & Raoul (ex Pompeia; Le Clapdoin 86)	W Bk	French	396	118·7	26·1	17·2	Nantes	1866	Sevestre	Y 7.88	A. Postel & ses fils	Havre
835	N	Charleston *Tuft 88*	W Bk	Norweg'n	448	131·8	30·9	17·1	Arendal	1875	Johnsen	Y 8.86	Hans. H. Pettersen	Arendal
836		Charley (ex Siam) *Gehrk 85-86*	W Bk	British	360	125·8	26·0	15·5	Hammel-Warden	1864	Menke & Botter		P. V. Grant	Shanghai
837	*	Charley *Howard*	W Sk	British	90	81·6	20·8	10·6	Bostock	1885	F. Burchard & Co		Mrs. A. E. White	Hull
838	R	Charley Bucki *French 87*	W3mSr	American	221	121·0	30·2	9·5	Belfast, Me.	1873		Y 11.87	B. Conant	Belfast, Me.
839		Charley Woolsey *Law*	W Sr	American	197	107·7	29·0	9·7	Deep River, Cn	1865			E. H. Weaver	Newhaven, Con.
840		Charlie Baker *Cain 88-88*	W S	British	1063	183·6	36·7	22·1	Brookville, NS	1874	R. K. Beveridge	Y 9.86	L. E. Baker	Yarmouth, N.S.
841		Charlie Blackwood *Palmer*	W Bkn	British	317	120·2	27·1	14·9	P.E. Island	1876	Longworth		The Hull Ice Co. Ld.	Hull
842		Charlie & Willie *Philbrook*	W Sr	American	117	86·9	24·0	8·2	Damariscotta, Me.	18 49/83			Cobb, Wight & Co.	Rockland, Me.
843	R	Charlotta *Borrman*	W Bk	Swedish	423	145·6	27·7	15·4	Svartvik	1854	Kyrkagard	Y 6.83	G. F. Gadda	Gothenburg
844		Charlotta *Kihlberg 74-82*	W Sr	Swedish	185	92·2	23·0	12·8	Bastholmen	1870	J. Johansson		C. J. Ödman	Sälvik
845		Charlotta *Edman*	W Sr	Swedish	158	86·6	22·5	11·8	Oroust	1868	S. Abrahamsson		J. Larsson	Hellevikstrand
846		Charlotta *Nilsson*	W Sr	Swedish	151	85·0	21·2	12·2	Dramsvik	1857	Olof Johansson		J. Andersson	Fiskebäckskil
847		Charlotta *Tegg*	W Sr	Swedish	137	80·0	22·3	10·9	Uddevalla	1876	Andersson		J. Hasselgren	Uddevalla
848		Charlotta Maria *Hansson*	W Sr	Swedish	113	76·4	21·3	10·2	Kristinehamn	1850			A. Halfvordsen	Hogar Bohuslän
849	*	Charlotte *Nicholas*	I Bk	British	378	145·0	27·1	16·1	Sunderland	1868	Iliff & Mounsey		C. Hooper	Liverpool
850		Charlotte *Flood*	W Sp	British	137	91·5	23·5	9·2	Bayonne	1837			Tyne Wherry Co. Ld.	Newcastle
851	*	Charlotte *Hansford*	W Sr	British	97	88·9	20·2	9·9	Jersey	1874	Bellot	Y 1.86	J. Cantell & Co.	Jersey
852	R	Charlotte *Davis 87*	W Bn	American	319	117·8	29·0	19·6	Brewer, Me.	1866		Y 1.84	J. H. Saunders	Key West

*=Lloyd's Register; D=Nederlandsche Vereeniging; G=Germanischer Lloyd; I=Registro Italiano; R=American Record;
A=Veritas Austro-Ungarico; H=Veritas Ellenico; N=Norske Veritas; V=Bureau Veritas.

Dans la 4me Colonne, C=Composite (membrure en fer, bordé en bois); I=Fer; Stl=Acier; W=Bois.

Les Dates du doublage en métal (Col. 11) ressortent des livres dans lesquels les navires sont respectivement classés ou des rapports des Inspecteurs du Lloyd's Register.

Pour plus amples renseignements sur les navires classés par le Lloyd's Register, voir le Lloyd's Register Book et le Lloyd's Yacht Register.

1 Official No.	2 No. in Book	3 Registry in which classed	4 Ship's Name. Material and Rig. Master, and Dates of Appointment to present Owner's Service and to Ship. Ship's Late Name.	5 Flag.	6 Net Tonnage	7 Dimensions. Length.	Breadth.	Depth.	8 Built. Where.	9 When.	10 By whom.	11 Date of Metalling.	12 Owners or Managers.
	851		Charlotte (BROKEN UP) Falp WBk	Danish	300	98·5	26·8	15·4	Finland	1815			K.H.Walloe
	852	GV	Charlotte Stern WBk	German	820	173·2	31·8	20·7	Bremerhaven	1874	C.Lange	[illegible]	[illegible]enburg, Wendt & [illegible]
	853	GV	Charlotte Bohndorf 81-86 WBn	German	354	118·4	24·3	14·8	Elsfleth	1872	J.Ahlers	[illegible]	A.Schiff
	854		Charlotte Lund [illegible] WBk	Norweg'n	159	97·0	23·5	11·6	Stavanger	1875			H.Thorsen
[illegible]	855		Charlotte WCr	Russian	108	67·8	23·9	8·9	Kesterzeem	1875			G.Beckmann
	856	V	Charlotte A. Littlefield WBk (ex Charlotte A. Littlefield) [illegible]	Norweg'n	513	139·8	26·8	20·4	Bangor,Me.	1864	J.Littlefield	[illegible]	P.L.Möller
[illegible]	857		Charlotte Buck WSr Pendleton -87	American	221	100·0	28·3	10·8	Orland,Me.	1865	Newcombe		F.H.Smith
[illegible]	858	✻	Charlotte Croom IS Bowden 86-88	British	1661	262·3	38·3	23·2	Glasgow	1874	C.Connell&Co.		W.Mackinnon (A. Mackay & Co. Ma[illegible])
[illegible]	859		Charlotte Fish Gardner WSr	American	222	115·0	28·0	10·0	Waldoboro,Me	1864	Fish		C.W.Stimpson
	860	N	Charlotte Helberg WSr (ex J. Bickmore) (WRECKED) [illegible]	Norweg'n	289	107·1	29·2	9·6	St.George, Me	1862			H.H.Pettersen
[illegible]	861	✻	Charlotte Kilner WK	British	75	75·4	19·6	9·3	Goole	1882	Ouswin		G.Kilner & Co.
	862	V	Charlotte Lange WBk (ABANDONED) Zepl [illegible] 73	German	406	124·5	28·9	17·3	Rostock	1874	F.Burchard&Co	[illegible]	W.Lange
	863	N	Charlotte Marie WBg (ex Sofie) Melsom	Norweg'n	180	99·2	24·8	11·4	Söderköping	1851			P.Michelsen
[illegible]	864	✻	Charlotte Padbury WBk Bisson 85-88	British	636	161·5	31·5	18·0	Falmouth	1874	Trethowan	[illegible]	F.A.Edelsten & Co.
	865	N	Charlotte Stringer WBk Tonnesen	Norweg'n	374	109·7	30·6	17·1	Grimstad	1857	N.Olsen Gjömle		G.Holst
[illegible]	866	R	Charlotte T. Sibley W3mSr Bartlett	American	358	140·0	32·8	11·0	Belfast,Me.	1882	Carter Bros		E.Sibley
[illegible]	867	✻	Charlotte Young IBk Berryman 84-84	British	302	133·3	24·2	13·9	Hartlepool	1867	Denton Gray & C.		R.& W.King
	868		Charlotte of Derby WBk (STRANDED) Lundegaard	Norweg'n	385	117·8	27·6	17·9	Sunderland	1852	Thompson& [Sons		P.A.Pettersen
	869	V	Charlotte & Anna WBk Kröger 82-	German	434	123·0	28·0	16·6	Danzig	1867	J.W.Klawitter		R.Seeger & Co.
[illegible]	870	✻	Charlwood Hiscocks IBk	British	837	197·7	32·2	19·6	Sunderland	1877	W.Doxford&Ss		G.H.Fletcher & Co
[illegible]	871		Charmer WSr	British	163	96·0	26·0	9·2	Massachusetts [U.S.	1856			J.B.Brown.
[illegible]	872	RV	Charmer Holmes 87-87 WS	American	1796	221·7	42·4	26·3	Bath,Me.	1881	Wm.Rogers	[illegible]	J.Rosenfeld'sSons
[illegible]	873	R	Charmer Dabolt WSr	American	395	136·0	30·3	12·3	Noank,Conn.	1872			F.Dabole
	874	V	Charmeuse Ruffez WSr	French	125	94·8	23·0	11·0	Dunkirk	1886	Pyotte Baynert		Ambrosius&Dinoir [illegible]
[illegible]	875	✻	Charmian Wissman -87 CBk (ex Rosalbahad)	British	555	150·4	28·8	18·7	Sunderland	1866	Moore	[illegible]	J.M.Grant
[illegible]	876		Charon Wattara WSr	Siamese	657	159·0	33·4	16·6	Bangkok	1875			Lace Mah Teng
[illegible]	877	DR	Charter Oak (BARGE) Laffin WS	American	941	168·2	35·6	22·0	Searsport,Me.	1864		[illegible]	Benj.Carver
[illegible]	878		Charter Oak (BARGE) Hook WSr	American	106	80·0	25·0	8·0	Saybrook,Con	18[illegible]			R.H.Tucker
[illegible]	879	✻	Chasca Scott 83-85 IBk	British	638	182·0	29·1	18·4	Liverpool	1869	W.H.Potter&Co		S.Wakeham & Sons
[illegible]	880		Chase Pollard WSr	British	112	81·1	21·6	12·5	P.E.Island	1850			W.Pollard
[illegible]	881	✻	Chase Hughes WSr	British	98	85·7	22·0	10·0	Connah's Quay	1878	Ferguson		R.Ferguson & Co.
	882	V	Chasseur Tillard IBk	French	284	127·8	24·8	14·8	Nantes	1869	Jollet&Babin		A.Giffaut
	883	NV	Chasseur Olsen WBk	Norweg'n	352	121·3	28·6	15·4	Grimstad	1870	A.Bentsen	[illegible]	J.E.Einersen
	884	V	Chateau Laffitte WBk Gotthard -88	French	320	117·1	26·7	15·2	Nantes	1873	T.Dubigeon&fils	V 11.87	Pannerel
[illegible]	885	GV	Chateaubriand Edmunds WBk [illegible] -87	British	409	129·5	26·6	15·9	St.Malo	1868		[illegible]	C.A.L.Dunn
	886	V	Chateaubriand WBk Lab[illegible] -88	French	278	114·5	26·9	14·8	Nantes	1868	T.Dubigeon	[illegible]	E.Labour
	887	N	Chatham Tollefsen WBk	Norweg'n	348	128·6	30·0	12·8	Miramachi	1864	Johnston		Ths.Melsom
	888		Chattal Zeitoon WSr	Turkish	125	..	..	..	ChattalZeito'n	1874			Uzum Mehemed oglou Mehemed & Sons
[illegible]	889	R	Chattanooga Lewis WBk (WRECKED)	American	500	135·0	30·0	18·0	Calais,Me.	1864		[illegible]	J.J.Matthias
[illegible]	890		Chattanooga WBg	American	193	94·0	27·0	10·6	Baltimore	1852	J.J.Abraham		J.E.Burtt
[illegible]	891		Chattanooga Lawson WSr	American	148	93·0	26·0	8·9	Searsport,Me.	1865			H.Lord
	892		Chaves II. WBn	Brazilian	142	..	..	..	S.João da Barra	1865		V 11.84	A.F.Guimarães
[illegible]	893	✻	Chazalie WK	British	90	79·7	20·9	10·6	Brixham	1885	J.W.&A.Upham		F.Hellyer
[illegible]	894	✻	Chazalie Kent WSr	British	79	79·4	19·1	9·5	Jersey	1878	LeSueur	V [illegible]87	T.de Faye & Co.
[illegible]	895	I	Checchina (ex Alfeo) Lipari WBk	Italian	561	146·3	29·8	17·3	Palermo	1873	N.Santocanale	[illegible]	P. Adragna
[illegible]	896		Cheduba StlLr	British	546	250·3	30·1	8·3	Dumbarton	1888	W.Denny & Brs.		Irrawaddy Flotilla Co, Ld.
62	897		Chelidon WSr	Greek	122	87·6	20·5	10·5	England	1872			F.Svoronos
75073	898	V	Chelmsford Collins WBkn	British	381	142·1	27·5	15·3	Quebec	1877	H.Warner	[illegible]	Mackay & Dix
	899	N	Chelone Kage WBg	Norweg'n	316	98·5	27·6	17·2	Aasgaardst'nd	1839			S.A.S[illegible]
91251	900	✻	Chepica Hughes 85-85 IBk	British	1025	213·6	34·7	20·1	Sunderland	1885	W.Pickersgill [& Sons		S.Wakeham & S[illegible]

✻ = **Lloyd's Register**; D = Nederlandsche Vereeniging; G = Germanischer Lloyd; I = Registro Italiano; R = Am[illegible]; A = Veritas Austro-Ungarico; H = Veritas Ellenico; N = Norske Veritas; V = [illegible]

In the 4th Column, C = Composite (iron frame, wood planked); I = Iron; Stl = Steel; W = Wood.

The Dates of Metalling (**Col. 11**) are based on the records of the Books in which Vessels are respectively classed, or on the Reports of the Surveyors of [illegible]

Further particulars of Vessels classed by Lloyd's Register will be found in Lloyd's Register Book and Lloyd's Yacht Register.

Numéro officiel	N° dans le livre	Registre dans lequel le navire est classé	Nom, Matériaux et Gréement. Capitaine et dates d'entrée au service des Armateurs actuels et à bord du navire. Ancien Nom.	Pavillon.	Tonnage net.	Longueur	Largeur	Creux	Construit: Où.	Quand.	Par qui.	Date du doublage en métal	Armateurs ou Directeurs Gérants.	Port d'Armement.
	901		Cherassous *Tsilinghiris* W Bn	Turkish	175	..	..	..	Cherasounda	1885			T.Tsilinghiris	Cherasounda
	902	V	Cherbourg *Gauvin* W Bk	French	738	167·0	34·1	19·1	St.Stephen	1871	J.&C.Short	[illegible]	A.Postel & ses fils	Cherbourg
	903		Cherokee *Fenik 75-82* W Bg (BROKEN UP)	German	283	98·7	23·0	17·6	Falmouth,NS	1829			A.Schaffenorth	Memel
	904	R	Cherub *Svensson* W Bg	Swedish	212	94·8	21·5	14·0	Arbroath	1861			N.Olsson	Höganäs
	905	✱	Cherwell *Tait* I Bk	British	1130	210·0	34·1	22·5	Middlesbro'	1864	Backhouse& [Dixon		E.Bates & Sons	Liverpool
	906		Chesapeake *Street* W Bn	British	194	107·3	26·6	9·6	Baltimore,US	1858			W.H.Perot	Georget'wn,Dem.
	907	RV	Cheseborough *Erickson 84-84* W S	American	1461	212·4	40·0	24·1	Bath,Me.	1878	E.&A.Sewall	[illegible]	A.Sewall & Co.	Bath,Me.
	908	V	Cheshire *Crosby -87* W Bk	British	1307	197·8	38·7	23·8	Quebec	1864	W.Charland	[illegible]	J.G.Ross & Co.	Quebec
	909	V	Cheshire *Snow -88* W Bk	American	701	152·5	33·2	20·1	East Boston	1875	J.Taylor	[illegible]	J.H.Flitner	Boston,Mass.
	910	R	Chestina Redman *Watts -84* W Bkn	American	559	142·6	32·4	16·4	Boston,Mass.	1873	A.Cove	[illegible]	J.H.Conant & Co.	Boston,Mass.
	911	V	Chevreuil *Delamee* W Bg	French	171	92·2	22·6	11·6	St.Malo	1874	J.Paruet	[illegible]	Dupuis Robial	Saint Brieuc
	912	✱	Chevy Chase *Goulden* I Bk (RIGGING)	British	678	176·7	28·6	19·2	Liverpool	1863	T.Vernon&Son		G.H.Fletcher & Co.	Liverpool
	913	A	Chiaini Bahri *Hassan*	Turkish	100	..	..	..	Carındag				Mechmet Jasigi	Constantinople
	914	I	Chiara (LOST) *Saglietto* W Sr	Italian	126	80·7	24·2	12·3	Savona	1866	G.Tixi	[illegible]	G.B.Saglietto	Porto Maurizio
	915	I	Chiara Penco *Penco 87* W Bn (ex Bianca Gavarrede)	Italian	489	141·0	31·0	18·8	Sestri	1876	G.Peragallo	[illegible]	Chiara Penco	Genoa
	916	I	Chiarina *Cuomo* W Bk	Italian	601	180·4	31·8	20·3	Cassano	1880	A.Castellano	[illegible]	F.S.Ciampa	Castellamare
	917		Chica W S	Spanish	1467	197·5	41·7	23·8	England	1869			Joaquin Gurri	Corunna
	918		Chiclayo W Bk	British	647	153·1	34·8	18·0	Sweden	1865			E.Vickery	Sydney,N.S.W.
	919		Chief *Jane* W Sr	American	104	80·5	25·7	8·3	Rockland,Me.	1837			C.C.Jayne	Pt.Jefferson,N.Y.
	920	V	Chieftain *Felton* W Bk	British	908	176·1	36·2	20·6	Londond'ry NS.	1876	McLennan	[illegible]	A.W.McLelan	Halifax,N.S.
	921		Chieftain W Bg	British	290	118·0	26·7	11·6	Shoreham	1867	May	[illegible]	J.E.Mitchell	Sydney,N.S.W.
	922		Chieftain *Gellatly* W 3m Sr	British	169	106·9	23·9	13·1	Lossiemouth	1868	Jack	[illegible]	J.F.Birrell	Dundee
	923		Chieftain *Christensen* W Bn	Danish	231	102·7	22·3	15·2	Scilly	1846			H.Christensen	Marstal
	924	V	Chignecto *Knowlton -87* W Bk	British	1032	168·7	36·0	19·3	ThreeSstrs.NS	1874	I.J.Olive,jun.	[illegible]	John Black	Windsor,N.S.
	925		Chihaya Maru W Bk	Japanese	433	144·6	26·5	14·7	Scotland	1875			Hajo Kwaisha	Nagasaki
	926		Chilas W Bg	Greek	229	..	..	..	Syra	1866			N.Chilas	Syra
	927	R	Childe Harold *Manson* W 3m Sr	American	742	179·6	36·0	18·1	Bath,Me.	1886	N.EnglandSBC		G.Manson	Newhaven
	928	✱	Childers *M'Carry -88* I Bk	British	896	179·9	32·4	21·0	Sunderland	1868	W.Doxford&Ss.		R.J.Swyny	Liverpool
	929	✱	Childwall *Rinder -87* I Bk (ex Edith Watson)	British	1286	213·0	37·1	23·3	Birkenhead	1868	G.R.Clover&Co.		C.W.Kellock & Co.	Liverpool
	930	V	Chilé *Lewis 87-87* I Bk	British	739	182·0	28·4	21·1	London	1856	C.Langley		R.W.Leyland & Co	Liverpool
	931	V	Chile *Coll* W Bg	Spanish	173	93·8	26·2	12·3	Blanes	1858		[illegible]	A.Llompart	Masnou
	932	✱	Chilena *Davies 76-76* I Bk	British	680	181·0	29·9	19·2	Sunderland	1876	W.Doxford&Ss.		Tomlinson,Hodgetts & Co.	Liverpool
	933	R	Chilena *Hansley -87* W Bk (BROKEN UP)	British	445	121·0	21·0	18·4	Liverpool	1854	T.Royden&Sns.	[illegible]	C.W.Polkinghorne	Greenock
	934	✱	Chili *McKenzie 88-88* I Bk	British	678	184·4	30·1	17·6	Kinghorn	1877	J.Key&Sons		J.Stewart & Co.	Dundee
	935	✱	Chili *Leguen 77-88* I Bk	French	1278	230·0	36·6	21·0	Glasgow	1885	W.B.Thompson		Ant.Dom.Bordes & fils	Bordeaux
	936	DV	Chili *Sägström* W Bk	Swedish	713	156·9	31·0	18·5	Gefle	1877	O.A.Brodin	[illegible]	J.Rettig	Gefle
	937	✱	Chilian *McMurchy 86-86* I Bk	British	601	176·7	30·1	17·4	Glasgow	1875	D&W Henders'n		Donaldson Bros.	Glasgow
	938	N	Chimea *Nygaard -88* W Bk	Norweg'n	435	137·5	29·0	17·0	Stavanger	1866	K.Kaisen	[illegible]	S.Bergesen	Stavanger
	939		Chin Dwin (ex Steamer) I Flat	British		190·7	22·1	7·5	PortGlasgow	1875	Robt.Duncan & Co.		Irrawaddy Flotilla Co., Ld.	Glasgow
	940		Chin San W Lorcha	Chinese	241	125·3	28·3	10·1	Hankow	1877				
	941		China W Flat	British	133	85·9	20·7	8·7	Northwich	1870	H.Gibson		R.Verdin	Liverpool
	942	R	China *Dunphy* W Sr	British	113	78·0	23·8	9·5	SheetHarbor, [N.S.	1880	W.Rutledge	[illegible]	Emma M.Richter	Liverpool,N.S.
	943	G	China *Selberg 86-87* W Bk	German	889	160·0	32·7	22·1	Brake	1857	J.Oltmann's [Wwe.	[illegible]	O.Thyen	Bremen
	944	V	China *Thorgersen* W Bk	Norweg'n	730	156·1	31·2	20·4	Gefle	1871	O.A.Brodin	[illegible]	O.Erikssen	Christiania
	945	V	Chinampas *McQuarrie* W Bk	British	674	156·8	33·5	17·8	RiverJohn,NS	1878	McKenzie& [Lander	[illegible]	Chas.McLennan	Pictou,N.S.
	946		Chinchinha W Bg	Brazilian	233	..	..	..	Barcelona	1845			A.M.B.Pereira	Rio Janeiro
	947		Chindaro *Serra* W Bn	Italian	108	82·0	22·4	11·4	Spezia	1877	G.Faggioni		L.Costa fu A.	Spezia
	948	✱	Chinese Gordon *Alp 85-85* W K	British	99	80·0	21·1	10·3	Galmpton	1885	W.A.Gibbs		J.Alp	Grimsby
	949	✱	Chinsura *Williams 79-83* I S	British	1266	215·5	37·1	22·8	Birkenhead	1868	G.R.Clover&Co.		Hughes & Co.	Liverpool
	950	IV	Chioggia *Scarpa -88* W Bk	Italian	483	121·7	30·2	19·8	Chioggia	1872	A.Scarpa	[illegible]	P.Ardisson & O.Scarpa	Venice

✱ = Lloyd's Register; D = Nederlandsche Vereeniging; **G = Germanischer Lloyd;** I = Registro Italiano; **R = American Record;**
A = Veritas Austro-Ungarico; H = Veritas Ellenico; N = Norske Veritas; V = Bureau Veritas.

Dans la 4me Colonne, C = Composite (membrure en fer, bordé en bois); I = Fer; Stl = Acier; W = Bois.

Les Dates du doublage en métal (Col. 11) ressortent des livres dans lesquels les navires sont respectivement classés ou des rapports des Inspecteurs du Lloyd's Register.

Pour plus amples renseignements sur les navires classés par le Lloyd's Register, voir le Lloyd's Register Book et le Lloyd's Yacht Register.

Official No. (1)	No. in Book (2)	Registry in which classed (3)	Ship's Name. Master, and Dates of Appointment to present Owner's Service and to Ship. Ship's Late Name. (4)	Material and Rig	Flag (5)	Net Tonnage (6)	Length (7)	Breadth	Depth	Built: Where (8)	When (9)	By whom (10)	Date of Metalling (11)	Owners or Managers (12)
9	951	A	**Chios** *Pittas*	W Bk	Greek	576	147·3	32·3	..	Chios	1876		[illegible]	D.F.Pittas
436	952	AH	**Chios** *Conjopandeli* (ex J. F. Sollierstrom)	W Bk	Greek	574	138·3	32·4	17·7	Parrsboro N.S.	1874		[illegible]	N.Conjopandeli
76	953		**Chios**	W Bk	Greek	230	90·9	27·5	..	Syra	1872			C.Moschos
44	954		**Chios**	W Sp	Greek	212	131·2	20·7	10·5	Vienna	1874			T.Stathatos
241	955		**Chios**	W Bg	Greek	174	98·1	22·0	..	England	1860			I.Boumboulis
50381	956	V	**Chipman** *Belyea*	W S	British	1083	191·2	36·6	22·5	St Stephen N.B.	1877	J.&C. Short	[illegible]	J.R.DeWolf & Son
40170	957	V	**Chippewa** *Lynch* 80-80	W Bk	British	1072	187·3	35·6	22·5	Quebec	1863	Gingras	[illegible]	R.Burns
71215	958		**Chippewa**	W Sp	British	132	94·0	23·8	7·8	Muskota, Ont.	1874	Alfred Merrill		Muskota Mill & Lumber Co.
74607	959		**Chira** (ex Trinita)	W Sr	British	333	128·6	27·0	13·7	Savona	1869			Australasian S.N. Co. Ld.
60912	960	✻	**Chiselhurst** *Maitland* (ex Charles Marceau) 1885-	I Bk	British	353	151·2	28·1	12·3	Dumbarton	1869	A.M'Millan&S.		Taylor, Sons & Co.
194	961		**Chishima Maru**	W Bkn	Japanese	254	110·6	23·4	13·2	Osaka	1879	Nagata Sanjiro		Nazaki Sebei
85787	962	✻	**Chislehurst** *Griffith* [84-84	W Bkn	British	439	152·5	29·2	16·3	P.E.Island	1883	J.Yeo	[illegible]	Wyllie & Gordon (T.F.Richards & Co., Managers)
51029	963	✻D	**Chittagong** *Warner* 85-85 (ex Arundel Castle)	I Bk	British	1042	210·0	33·6	21·7	Greenock	1864	R.Steele&Co.		C.Barrie
70803	964	✻D	**Chittagong** *Davies* -87	W 3m Sr	British	323	135·0	27·5	12·9	Sunderland	1874	Richardson	[illegible]	W.C.Jarvis
	965		**Chittagong** *Bourcier*	W Bk	French	255	107·9	26·2	13·0	Verton	1873	Doussain	[illegible]	G.Baratoux
62806	966	✻	**Chittoor** *Gerlley* 69-75	W 3m Bn	British	217	112·0	24·5	13·0	Dartmouth	1875	Philip	[illegible]	W.H.Punchard&Co
207	967		**Choan Maru**	W Sr	Japanese	111	86·3	18·7	9·1	Tokio	1878	Hirano Ship-[building yard		S.Seizayemon
71806	968		**Chocola** *Krauell* (ex Notre Dame de la Providence)	W Bk	British	284	111·6	25·5	14·1					J.R.Jones
63514	969	✻	**Chocolate Girl** *Alexander*	W Sr	British	244	126·6	25·2	13·0	Littlehampton	1869	Harvey	[illegible]	DePutron&Routh
641	970		**Choho Maru**	W Sr	Japanese	164	89·2	21·6	11·4	Germany	1871			Kubo Kuisaku
50885	971	R	**Choice** *McMurtry* 84-86	W Bk	British	1102	184·5	36·2	23·0	Miramichi	1865	Harley	[illegible]	J.Carswell
84	972		**Chomei Maru**	W Sr	Japanese	113	87·0	18·7	8·7	Singapore	1864			Kikuchi Choshichi
	973	N	**Chondor** *Danielsen* -87	W 3m Sr	Norweg'n	319	125·5	28·9	12·3	Grimstad	1887	S. Johnsen Vig	[illegible]	O.T.Tonnevold
386	974		**Chosei Maru**	W Sr	Japanese	220	109·5	22·1	11·6	Osaka	1878	Nagata Sanjiro		Masamura Heibei
	975		**Chosovitissa**	W Sr	Greek	131	..	..	..	Syra	1880		[illegible]	C.Mavromatis
872	976		**Choyei Maru**	W Bk	Japanese	293	113·5	23·6	16·5	Sunderland	1866	W.Adamson		Yamagawa Matsugoro
141	977		**Choyetsu Maru**	W Bk	Japanese	467	145·9	27·5	13·3	United States	1865			Tatsuma Yohei
56946	978		**Christabel** *Hickman* 74-74	W Sr	British	175	98·8	23·1	12·9	Rye	1867	Hessell		T.J.Gann & Co.
47007	979	✻	**Christabel** *Lawrey*	W K	British	159	103·6	21·3	12·3	Teignmouth	1863	Owen	[illegible]	M.Lawrey
	980	GV	**Christel** *Bräunings* 81-81	W Bk	German	879	163·3	33·8	21·5	Vegesack	1864	Gebr.Bosse	V 1.86	E.C.Schramm & Co.
	981	G	**Christel** *Johannesen* 65-85 (ex Alberta, ex Daniel Elliott)	W Bk	Norweg'n	670	141·1	31·5	21·4	Brunswick	1854			H.B.Johannesen
	982	N	**Christen Knudsen** *Thomsen*	W Bk	Norweg'n	569	151·2	33·1	17·4	Porsgrund	1875	G.Knudsen	[illegible]	J.C.&G.Knudsen
	983	V	**Christian** *Friis* 72-72	W Sr	Danish	100	77·9	20·7	10·3	Marstal	1872	H.J.Bager		C.H.Kromann, jun.
	984	DGV	**Christian** *Thuesen* -88	W Bn	German	250	114·5	25·7	13·2	Apenrade	1877	G.Raben	V 1.87	G.Raben & Co.
	985		**Christian** *Lübcke* 80-	W Bg	German	171	86·8	24·6	13·0	Barth	1862	A.Witte		J.J.Wallis & Sohn
	986	N	**Christian** *Christiansen* [-76	W Bk	Norweg'n	551	150·9	31·9	16·9	Perth Amboy	1854		V 1.86	C.J.Gullichsen
	987	V	**Christian** *Wegle* -88 (ex Graf-Klot-Trautvetter)	W 3m Sr	Norweg'n	256	108·6	25·0	13·6	Greifswald	1873	A.Drossel	[illegible]	A.M.Flörnaes
317	988		**Christian** *Andersohn*	W Sr	Russian	231	100·0	27·3	12·7	Ugmuzeem	1878	Muzeneek		K.Grüneisen
	989	NV	**Christian Scriver** *Rosberg* -86	W Bk	Norweg'n	608	153·5	33·1	18·5	Grimstad	1879	J.Bang	V 10.86	J.Bang
	990	N	**Christian Wilhelm** (ex Rothsay) *Kvernold* -82	W Bk	Norweg'n	386	140·0	29·2	13·9	River John, N.S.	1868	Kitchen		C.Wroldsen
	991	V	**Christian IX.** *Knudsen* -84	W 3m Sr	Danish	206	115·4	25·4	11·8	Marstal	1878	J.J.Bager		H.C.Christensens Enke
	992		**Christiane** *Mortensen*	W Bg	Danish	195	109·2	25·3	11·2	Varnaes	1867	C.A.Hahn		Actieselskab (P.Jørgensen)
	993		**Christiane** *Kramer* 85-85 (ex Malmo)	W Bn	Danish	138	92·5	18·3	11·6	Hull	1856			P.Kramer
	994	N	**Christiane** *Jensen* 85-85 (ex Riccardo)	W Bk	Norweg'n	402	131·5	29·6	16·6	Hamburg	1864	H.C.Stülcken	[illegible]	S.Eyde
	995	N	**Christiane** *Olsen*	W Bk	Norweg'n	293	103·1	27·1	16·0	Arendal	1847	N.Evensen		N.Rognaas
	996	N	**Christiane** *Christoffersen* 80-82	W 3m Sr	Norweg'n	222	123·1	25·0	11·4	Hamburg	1868	D.Oltmann Wwe		Westergaard & Co.
	997		**Christiane** *Olsson*	W Bn	Swedish	136	95·3	23·1	10·7	Kiel	1866	Reuter&Ihms		H.Olsson
	998	N	**Christiane Sophie** *Andersen*	W Bk	Norweg'n	429	103·5	32·5	19·2	Tönsberg	1839			E.A.Andersen
	999		**Christiane & Marie** (ex Alton) *Lund* 78-79	W Bn	Danish	242	113·7	25·4	12·6	P.E.Island	1873	Owen		M.Christiansen
	1000	N	**Christiania** *Danielsen* (ex Nautilus, ex Ottone)	W Bk	Norweg'n	508	141·9	30·2	18·3	Medford, Mass.	1858		[illegible]	L.Fuyen

✻ = **Lloyd's Register**; D = Nederlandsche Vereeniging; G = Germanischer Lloyd; I = Registro Italiano; R = Am[…]
A = Veritas Austro-Ungarico; H = Veritas Ellenico; N = Norske Veritas; V = Bur[…]

In the 4th Column, C = Composite (iron frame, wood planked); I = Iron; Stl = Steel; W = Wood.

The Dates of Metalling (Col. 11) are based on the records of the Books in which Vessels are respectively classed, or on the Reports of the Surveyors of Ll[oyd's …]

Further particulars of Vessels classed by Lloyd's Register will be found in Lloyd's Register Book and Lloyd's Yacht Register.

No.	Cote	Nom, Matériaux et Gréement. Capitaine et dates d'entrée en service. Ancien Nom.			Pavillon	Tonnage net	Longueur	Largeur	Creux	Construit Où	Quand	Par qui	Doublage	Armateurs ou Directeurs-Gérants	Port d'Armement
1001	N	Christianne		W Bg	Norweg'n	179	96·3	25·9	13·3	Drobak	1856			Act.Selsk. "Ilusa"	Tönsberg
1002	N	Christiansand		W Bg	Norweg'n	254	103·1	26·8	15·3	Christiansand	1853			C.A.Leschbrandt	Mandal
1003	V	Christina (BURNT)		W Bk	British	1056	189·7	37·0	22·3	Maitland,N.S.	1882	A.A.McDougall		A.A.McDougall	Maitland,N.S.
1004	R	Christina (ABANDONED)	Wilkie	W Bk	British	564	140·5	32·0	18·2	St.Andrews,N.B.	18 75/76			John Watson	St.Andrews,N.B.
1005		Christina	Reynolds	W Bn	British	201	110·2	26·5	12·5	P.E.Island	1876	Poole		J.N.Rogers	Folkestone
1006	R	Christina (ABANDONED)	Fanning	W Bn	British	154	86·2	23·5	11·0	Moser R.,N.S.	1875			William Muir	Halifax,N.S.
1007		Christina		W Sr	Brazilian	124	..	..	..	Riodejaneiro	18 81			F.S.Ferreira	Rio de Janeiro
1008		Christina		W Sr	Brazilian	124	..	..	..					J.R.Baptista	Macury
1009	V	Christina	Ferrichs -73	W Sr	German	139	81·3	19·1	9·1	Varel	1873	Chr.Schwoon		H.H.Cassens	Timmel
1010	A	Christina	Jerojanni -87	W Bg	Greek	383	125·9	26·9	17·7	Cassos	1887	EliaSofo		D.M.Jerojanni	Syra
1011	RV	Christina (ex J. K. Pettis)		W Sr	Mexican	292	119·9	29·1	12·1	Parrsboro,N.S.	1879	W.Pettis		Mendez & Co.	Vera-Cruz
1012		Christina	Hansen 81-85	W Bk	Norweg'n	529	143·6	29·8	18·1	Quebec	1864	Dubord		Chr.Anker	Fredrikshald
1013	N	Christina	Olsen	W Bk	Norweg'n	358	100·9	27·8	18·8	Tönsberg (rebuilt)	18 49			I.H.Christiansen	Tönsberg
1014	V	Christina	Cardoso	W Sr	Portug'se	179	94·5	22·3	10·8	Villa do Conde	1871	J.M.de Araujo		M.da Graça de Souza [illegible]	Oporto
1015	V	Christina	Simonsen 78-78	W3m Sr	Russian	262	109·4	29·1	13·3	Haynasch	1877	J.Brackmann		A.Weide	Haynasch
1016		Christina Charlotta		W Bk	Russian	389	115·0	30·0	16·0	Lovisa	1848	J.Kjeld		H.V.Söderlund	Åbo
1017	V	Christina Moore	Card	W Sr	British	216	103·0	27·5	11·9	Kempt,N.S.	1881	R.Rose		R.Rose	Windsor,N.S.
1018		Christina Navarro		W Bk	Chilian	623	158·1	31·4	18·3	Miramichi,N.B.	1864			M.Velasquez	Valparaiso
1019		Christine	LOST	W Bk	Danish	379	131·3	25·8	15·2	Sonderburg	1866	H.H.Hanschild		Peter Paulsen	Veile
1020	DV	Christine	Troensgaard	W Bn	Danish	192	106·7	22·3	11·0	Thurö	1869	N.Troensegaard		E.P.Hoffgaard	Svendborg
1021		Christine	Mortensen 76-76	W Bn	Danish	161	81·0	22·3	12·6	Rostock	1830			H.Iversen	Svendborg
1022		Christine	Petersen 74-87	W Bn	Danish	139	93·1	21·5	11·3	Faaborg	1867	D.Dyreborg		N.Petersen	Marstal
1023	GV	Christine	Schulze 77-79	W Bk	German	1045	173·5	36·1	23·6	Medford	1868	J.O.Curtis	V 2.86	Boyes & Ruyter	Bremen
1024	DV	Christine	Ries 85-85	W Bk	German	542	146·6	31·7	18·0	Apenrade	1875	N.Jacobsen	V 1.88	C.P.Schütt	Apenrade
1025	GV	Christine	Schriefer 88-88	W Bn	German	274	105·3	26·5	11·3	Brake	1873	J.Oltmann's Wwe.	V 1.87	D.Haye	Brake
1026	V	Christine (MISSING)	Hangke 87-87	W Sr	German	230	101·9	27·5	10·8	Hammelwarden	1872	W.G.Hagens		C.G.Beermann	Elsfleth
1027	N	Christine	Jacobsen 72-72	W Bk	Norweg'n	741	140·1	35·0	21·9	Dordrecht	1836			I.H.Christiansen	Tönsberg
1028	N	Christine	Halvorsen	W Bk	Norweg'n	372	127·9	27·5	16·8	Sunderland	1863	Mills		Aktieselskabet "Achilles" (Jens Olsen)	Kragerö
1029	N	Christine	Thorsen	W Bk	Norweg'n	313	129·1	28·6	12·4	Drammen	1876	Svensen		M.Aasaker	Drammen
1030	N	Christine	Fredriksen -82	W Bk	Norweg'n	293	118·7	26·2	13·8	Havre	1849			N.Nielsen	Kragerö
1031	DN	Christine	Berg	W Bk	Norweg'n	252	108·7	25·8	14·1	Weymouth N.S.	1850			H.Berg	Fredrikstad
1032		Christine Edela (ex Goodell)	Jaensten -88	W Bk	Norweg'n	840	160·0	38·8	22·2	Searsport,Me.	1866	D.S.Goodall		K.O.Aanonsen	Lillesand
1033	ND	Christine Elisabeth	Lundegaard	W3m Sr	Norweg'n	290	129·1	26·8	12·3	Stavanger	1876	F.Magnus		Brödrene Jacobsen	Haugesund
1034	N	Christine Elisabeth	Thorsen 79-79	W3m Sr	Norweg'n	289	118·9	26·8	13·7	Stavanger	1876	J.H.Jondahl		E.Rönneberg	Stavanger
1035		Christoforos		W Bk	Greek	272	..	..	..	Trieste	1845			C.Stamopoulos	Piræus
1036		Christoph		W Sr	Russian	113	80·0	26·8	10·4	Kaleten	1884	D.Smilge		D.& J.Smilge	Riga
1037	DV	Christoph Kasten	Vierow	W3m Sr	German	185	93·5	23·2	12·4	Barth	1874	J.Holzerland		E.Vierow	Stralsund
1038		Christophe			French	138	..	..	..					Benerettini & Co.	Bastia
1039		Christophe Colomb	Hubello	W Bk	French	250	104·3	24·9	14·1	Fécamp	1866	Brument & Capon		Em. Gilles & A.Bellet	Fécamp
1040	G	Christopher (ex Christopher Hage)	Schmidt -82	W Bn	Danish	100	82·5	20·2	9·8	Sköge?	1852	A.Peterson		N.Petersen	Marstal
1041	A	Christos Nicolaidis	Calogerogianis	W Bk	Greek	380	123·0	28·4	..	Syra	1876	P.Baunis		C.Calogerogianis	Syra
1042		Chromo	Bunker -82	W Sr	American	122	95·0	26·4	8·0	Franklin,Me.	1874			F.R.Bunker	Ellsworth
1043	A	Chrussakis (ex Afghan)	Tsalatrons -88	W Sr	Greek	230	101·1	27·0	12·0	P.E.Island	1878	A.J.&J.R.Macdonald		Chrussakis Bros.	Galaxidi
1044	A	Chrysallis (ex Crinolis?)	Correntis	W Sr	Greek	198	92·8	19·2	14·0	Galaxidi	1879	C.Papapetros		G.Corentis	Galaxidi
1045		Chrysolite	James	W3m Sr	British	159	95·6	21·6	12·0	Whitehaven	1869	Shepherd		J.Denley	Whitehaven
1046	V	Chrysolite	Rödseth 85-85	W Bk	Norweg'n	1273	190·6	38·0	24·5	St.John,N.B.	1868	Nevins & Fraser		Act.Selsk.(J.Thorsen)	Haugesund
1047	R	Chrysolite	Hansen -88	W Bk	Norweg'n	1098	182·6	36·5	22·9	N.Brunswick	1858	W.& A.Boultenhouse		E.Schwartz	Drammen
1048	✻	Chrysomene	Hogan 83-83	I S	British	1778	261·6	41·8	23·8	Liverpool	1873	Potter & Hodgkinson		H.Fernie & Sons	Liverpool
1049	A	Chrysoula		W Sr	Greek	161	87·6	22·6	13·1	Galaxidi	1885	J.Kanatis		N.C.Costelos	Galaxidi
1050															

✻ = Lloyd's Register; D = Nederlandsche Vereeniging; G = Germanischer Lloyd; I = Registro Italiano; R = American Record; A = Veritas Austro-Ungarico; H = Veritas Ellenico; N = Norske Veritas; V = Bureau Veritas.

Dans la 4me Colonne, C = Composite (membrure en fer, bordé en bois); I = Fer; Stl = Acier; W = Bois.

Les Dates du doublage en métal (Col. 11) ressortent des livres dans lesquels les navires sont respectivement classés ou des rapports des Inspecteurs du Lloyd's Register.

Pour plus amples renseignements sur les navires classés par le Lloyd's Register, voir le Lloyd's Register Book et le Lloyd's Yacht Register.

Official No.	No. in Book	Registry in which classed	Ship's Name. Material and Rig. Master, and Dates of Appointment to present Owner's Service and to Ship. Ship's Late Name.	Flag	Net Tonnage	Length	Breadth	Depth	Built: Where	When	By whom	Dates of Metalling	Owners or Managers
1	2	3	4	5	6	7			8	9	10	11	12
	1051		Chungching W Lorcha	Chinese	135	90·4	26·0	9·8	Hankow	1883	Chinese		Shee Ching San
73415	1052	✱	Churchstow *Adams* W Bk	British	473	145·6	30·1	16·5	Falmouth	1876	Trethowan	[illegible]	R.Jarvis
51470	1053		Chusan *Morrison* C Bk	British	786	185·8	31·6	20·0	Birkenhead	1865	WoodsideGrav-ingDockCo.	[illegible]	TheVictorianGovernm. (The Minister of Defence)
[illegible]	1054	IV	Ciampa Emilia *Cocurullo* W S	Italian	861	169·6	32·0	23·6	Cassano	1874	A.Castellano	[illegible]	F.S.Ciampa
55286	1055	✱	Cicelia *Le Masurier* W K	British	74	78·3	19·0	8·8	Jersey	1867	LeSueur		T.Le Masurier
42629	1056	✱	Cicero *Nixon* 85-85 I Bk	British	1115	209·0	34·0	22·6	Liverpool	1861	T.Vernon&Son		Lowden,Edgar&Co.
58048	1057	✱	Cicero *Hunter* -88 I Bk	British	468	148·5	27·7	17·3	Sunderland	1867	W.Doxford		J.Tully
43500	1058	✱	Cicerone *Prout* 3mSr	British	144	99·5	22·0	12·5	Dartmouth	1861	Kelly		W.Endean
	1059	V	Cicilie *Hermann* 79-79 W Sr	Danish	113	89·4	21·0	10·3	Veile	1870	P.M.Illum		Aktie Selskab "Cicilie" [illegible]
	1060	V	Cid *Cavalin* -87 W Bk	French	247	99·1	25·9	13·0	Nantes	1869	Clairgeau	[illegible]	P.Rozier
	1061		CidCampeador *Theus* -87 W Bn	Spanish	148	87·6	23·1	11·8	Bilbao	1864		[illegible]	F.Carreras
	1062	DV	Cigana *Petersen* -88 W Bk	Russian	252	108·2	23·9	12·5	Gloucester	1867		[illegible]	Witt & Co.
	1063	V	Cilaos *Robard* 82-82 W Bk	French	536	147·9	28·5	17·1	St.Malo	1869	Vve Dandin&fils	[illegible]	V.Vincent
3579	1064	AV	Cileno *Dedussi* -88 W Bk	Greek	399	122·0	27·9	19·1	Sestri	1869	G.Traverso	[illegible]	G.Dedussi
	1065		Cimandef I Bk	French	425	149·2	28·5	16·3	Sunderland	1868	Pile		Lebeaud père, fils & Co.
77444	1066	✱	Cimba *Fenwister* 69-78 I S	British	1117	223·0	34·6	21·7	Aberdeen	1878	W.Hood&Co.		A.Nicol & Co.
69916	1067	✱	Cimbri *Allen* 84-86 W Bn	British	132	99·3	22·3	11·5	Barnstaple	1877	Westacott	[illegible]	Isaac Allen
	1068	DG	Cimbria *Nohr* 86- W Bk	Russian	304	110·5	26·8	15·5	Libau	1868	G.Buchhoff	[illegible]	J.C.Schneider & Co.
	1069		Cimbria *Wendell* 87 W Sr	Swedish	141	95·8	21·7	9·3	Timm'rnabb'n	1872	CarlNilsson		H.Hallberg
	1070		Cimbrishamn (ex Hongkong) *Olson* -71 W Bkn	Swedish	217	117·5	22·2	11·2	Hamburg	1863			O.Wingren
	1071		Cincora W Sr	Brazilian	176	..	..	..					A.J.S.Beleu
63732	1072	✱	Cingalese *Stannard* I Bk	British	698	196·9	32·3	17·1	Glasgow	1869	London&Glasg. ShipCo.		C.W.Turner
	1073		Cinq Frères (ex Charles Thompson) *Godard* W Bn	French	158	100·7	25·6	9·5	Kennebunk	18 65/83			Ch.Landry
3199	1074	I	Cinque *Sussone* W Bk	Italian	1060	199·8	35·8	24·1	Sestri	1873	Debarbieri& Celesia	[illegible]	G.Volpe
205	1075	A	Cinta (ex Tempo) *Petricevich* -88 W Bg	Austrian	321	119·7	25·1	15·7	Trieste	1867	Stabilimento Tecnico	[illegible]	N.Petricevich
	1076	N	Circassia (ex Lady Muriel May) *Bjorgolfsen* [-88 W Bk	Norweg'n	467	144·5	31·7	13·2	Quebec	1873	Oliver	[illegible]	Chr.Hannevig
85530	1077		Circassian *Gayton* W Sr	British	99	87·0	23·9	10·0	Shelburne,N.S.	1883	J.A.McGowan, jun.		Wm.Ryder
59542	1078		Circe *Priest* W Bn	British	145	102·0	23·4	9·0	Brisbane Water,N.S.W.	1869			W.S.Waterston
	1079	✱	Circé *Sevrain* 75-88 I S	French	1610	249·0	39·5	22·5	Glasgow	1885	A.Stephen&Sns		A.C.Le Quellec
	1080		Circe de Suffo	Spanish		..	..	..					F.Riera
	1081		Ciscar *Sobrida* W Bk	Spanish	292	113·5	28·2	13·4	Majorca	1869			A.Canoves y Coll
	1082	G	Cisneros *Noms* 81- W Bg	Spanish	215	104·0	29·0	14·0	Azens de Mar	1857	S.Busquets	[illegible]	Rosa Maristany
75685	1083	V	Citadel *Yuill* -86 W Bk	British	1363	207·3	38·3	24·4	Quebec	1877	Samson	[illegible]	W.H.Ross & Co.
53867	1084		Cite W Sr	British	123	96·0	23·5	8·5	Grondines	1866			Thos.Frenette
	1085	V	Cité d'Aleth *Pelletan* W Bk	French	273	108·6	23·6	13·4	St.Malo	1870	Tranchemer	[illegible]	Danguillen aîné
54367	1086		Citizen *Loughlin* 84-84 W Bn	British	197	97·7	27·1	11·1	NovaScotia	1869	Lent		M.Fleming
	1087	DV	Cito *Hansen* -88 W Bg	Danish	290	116·9	25·6	13·8	Nordby,Fanø	1880	S.Abrahamsen	[illegible]	Aktie Selskab "Cito" (H.Kallesen)
	1088	G	Cito *Boye* 84-84 W Bn	Danish	157	90·7	21·1	11·4	Elbing	1867	Broderene Mutzstaff		Chr.Ohlsen
	1089		Cito *Ahrens* 74-74 W Sr	German	102	73·1	21·1	10·6	Damgarten	1853	H.Dierling		H.Scharmberg
	1090	N	Cito (ex Tropic) *Bergh* -88 W Bk	Norweg'n	524	142·1	31·2	17·6	Gefle	1868	O.A.Brodin	[illegible]	H.Pettersen
	1091	N	Cito *Höyer* -89 W Bk	Norweg'n	494	134·6	31·5	16·8	Arendal	1879	Guttormsen	[illegible]	Chr.Th.Boe
	1092	N	Cito *Holm* 77-87 W Bk	Norweg'n	317	110·8	29·2	15·2	Drammen	1857	Jörgensen		H.A.Lohre
	1093	N	Cito *Amundsen* -86 W Bg	Norweg'n	257	110·5	27·2	13·8	Mandal	1861	Geralisen		T.Jacobsen Nass
59298	1094	V	City Camp *Knox* -88 W S	British	990	176·8	34·6	22·0	St.John,N.B.	1870		[illegible]	Wm.Thomson&Co.
[illegible]	1095	✱	City Merchant *Bond* W K	British	78	80·1	20·0	9·0	Harwich	1874	Vaux		Walker, Howard & Co.
50836	1096	✱	City of Adelaide *McMurtry* 88-88 C Bk	British	719	176·8	33·3	18·8	Sunderland	1864	W.Pile&Co.	[illegible]	T.Dixon & Sons
29148	1097	✱	City of Agra *Young* I S	British	1074	213·6	34·7	20·6	W.Hartlepool	1860	Pile,Spence&Co.		J.Blyth & Co.
56166	1098	✱	City of Athens *Morrison* 84-84 I S	British	1199	222·9	34·2	22·8	Greenock	1866	R.Steele		G.Smith & Sons
125816	1099	R	Cityof Augusta *Meady* W 3mSr	American	552	149·6	35·3	15·6	Bath,Me.	1880	Goss,Sawyer& Packard		J.M.Haynes
126190	1100	R	City of Baltimore *Towne* -84 W 3mSr	American	339	138·0	33·4	9·9	Bath,Me.	1884	Goss&Sawyer		L.S.Taws

✱ = Lloyd's Register; D = Nederlandsche Vereeniging; G = Germanischer Lloyd; I = Registro Italiano; R = Ame[rican]
A = Veritas Austro-Ungarico; H = Veritas Ellenico; N = Norske Veritas; V = Bur[eau]

In the 4th Column, C = Composite (iron frame, wood planked); I = Iron; Stl = Steel; W = Wood.

The Dates of Metalling (Col. 11) are based on the records of the Books in which Vessels are respectively classed, or on the Reports of the Surveyors of [illegible]

Further particulars of Vessels classed by Lloyd's Register will be found in Lloyd's Register Book and Lloyd's Yacht Register.

Numéro officiel	No dans le livre	Registre dans lequel le navire est coté	Nom. Materiaux et Gréement. Capitaine et dates d'entrée en service des Armateurs actuels et à bord du navire. Ancien Nom.	Pavillon.	Tonnage net.	Longueur	Largeur	Creux	Où.	Quand	Par qui.	Date du doublage en métal	Armateurs ou Directeurs-Gérants.	Port d'Armement.
[illegible]	1101	*	City of Benares *Adair 85-92* I S	British	1574	259·0	38·2	23·0	Glasgow	1882	Barclay, Curle [& Co.		G. Smith & Sons	Glasgow
[illegible]	1102	*	City of Bristol W Sk	British	86	77·8	20·4	10·7	Hull	1885	W. & J. McCann		G. Bowman	Hull
[illegible]	1103	* D	City of Canton *Carnochan* [illegible] I Bk	British	909	197·4	31·2	20·8	Greenock	1857	R. Steele & Co.		J. Wilson	Glasgow
[illegible]	1104	* D	City of Carlisle *Moore* [illegible] I Bk	British	823	195·9	32·0	19·6	Glasgow	1875	Dobie & Co.		P. Iredale & Son	Liverpool
[illegible]	1105	*	City of Corinth *Taylor* [illegible] I S (CORLEASTON)	British	1228	235·2	35·0	22·3	[illegible]	1870	Barclay, Curle [& Co.		G. Smith & Sons	Glasgow
[illegible]	1106	*	City of Delhi *Drysdale 75-92* I S	British	1198	227·7	34·8	22·7	[illegible]	1867	Barclay, Curle [& Co.		G. Smith & Sons	Glasgow
[illegible]	1107	*	City of Edinboro' W Sk	British	84	76·2	20·1	10·6	Hull	1884	W. & J. McCann		G. Bowman	Hull
[illegible]	1108	*	City of Florence *Leask 82-92* I S	British	1209	226·9	34·4	22·1	[illegible]	1867	C. Connell & Co.		G. Smith & Sons	Glasgow
[illegible]	1109	*	City of Glasgow *Ingram* [illegible] I S	British	1198	228·0	34·0	22·2	[illegible]	1867	Barclay, Curle [& Co.		G. Smith & Sons	Glasgow
[illegible]	1110	*	City of Glasgow W Sk	British	84	77·5	20·5	10·6	Hull	1885	W. & J. McCann		G. Bowman	Hull
[illegible]	1111	*	City of Hankow *King* C S	British	1193	224·0	35·2	22·3	[illegible]	1869	A. Stephen & Sns.		G. Smith & Sons	Glasgow
[illegible]	1112	R	City of Jacksonville W 3m Sr *Sibbald*	American	387	138·5	35·2	9·6	Bath, Me.	1883	Goss, Sawyer & [Packard		E. Rokes	Baltimore
[illegible]	1113	*	City of Lincoln W Sk	British	85	80·7	21·5	10·7	N. Holland	1885	R. Day		G. Bowman	Hull
[illegible]	1114	R V	City of Liverpool *McCarthy* [illegible] W Sr	British	1379	193·0	39·0	24·0	Quebec	1864	Ross	11.86	W. H. Ross & Co.	Liverpool
[illegible]	1115	*	City of Liverpool W Sk	British	81	76·0	20·3	10·6	Hull	1884	W. & J. McCann		G. Bowman	Hull
[illegible]	1116	*	City of London *Gurney* W K	British	88	78·2	20·9	10·5	Brixham	1885	R. Jackman		J. Randall	Grimsby
[illegible]	1117	*	City of London W Sk	British	71	73·9	20·0	10·0	Rye	1883	J. C. Head		G. Bowman	Hull
[illegible]	1118	*	City of Lucknow *Halley* [84-94] I S	British	1196	231·1	35·5	22·2	Glasgow	1869	Barclay, Curle [& Co.		G. Smith & Sons	Glasgow
[illegible]	1119	*	City of Madras *Halley 76-92* I S	British	1577	260·2	38·4	23·0	Glasgow	1882	Bcly, Curle & Co.		G. Smith & Sons	Glasgow
[illegible]	1120	*	City of Madrid I S (now Agnes Sibbald)	British	1191	231·6	35·3	22·2	Glasgow	1869	C. Connell & Co.		S. Goldberg	Glasgow
[illegible]	1121	*	City of Manchester *Coppill* W Sk	British	79	77·6	20·6	10·6	Hull	1884	T. R. Westerman		G. Bowman	Hull
[illegible]	1122		City of Melbourne *Brown* W	British	1397	217·3	40·8	15·0	Newcastle, U. S.	1853			The Victorian Governmt	Melbourne
[illegible]	1123	R	City of Montreal *Coss* W S	American	1117	177·4	34·5	23·8	Portsm'th, N. H.	1861	Tobey & Co.	[illegible]	T. Dunham's Nephew & C.	New York
[illegible]	1124		City of New York W S	American	1815	214·8	42·8	22·1	New York	1854			J. H. Thomas	New York
[illegible]	1125	*	City of Norwich *Johnson* W K	British	82	78·9	20·9	10·8	Brixham	1885	R. Jackman		J. Randall	Grimsby
[illegible]	1126		City of Ottawa *Bentyman* W S	British	884	169·7	33·0	21·3	Quebec	1860	Gingras	[illegible]	City of Ottawa Sailing Ship Co. Ld. (J. P. Rogers)	Plymouth
[illegible]	1127	V	City of Papeete *Rernde* W Bkn	American	379	145·3	32·1	12·4	Fairhaven, Cal	1883	H. Bendixon		A. L. Schmidt	San Francisco
[illegible]	1128	R V	City of Philadelphia *Winn* W S	American	1384	202·3	40·3	24·3	Bath, Me.	1875	Goss & Sawyer	[illegible]	P. Fitzpatrick	Philadelphia
[illegible]	1129	R	City of Philadelphia *Dodson* W 3m Sr	American	367	140·0	34·0	10·0	Bath, Me.	1885	Goss & Sawyer		P. Fitzpatrick	Philadelphia
[illegible]	1130	*	City of Quebec *Carnegie* [83-85] I S	British	708	184·4	30·6	20·1	Greenock	1855	R. Steele & Co.		Montgomerie & Workman	London
[illegible]	1131	* D	City of Sparta *Bogle 88-88* I S	British	1193	234·2	35·0	22·4	Glasgow	1870	A. Stephen & Sns		G. Smith & Sons	Glasgow
[illegible]	1132	*	City of Tanjore *Thomson* [illegible] I Bk	British	768	194·1	28·7	20·7	Greenock	1856	Steele		T. C. Guthrie	Glasgow
[illegible]	1133	*	City of York *Baxter 85-85* I S	British	1195	222·7	35·8	21·7	Glasgow	1869	J. Elder & Co.		G. Smith & Sons	Glasgow
[illegible]	1134	*	City of York *J. Dishman* W Sk	British	88	78·5	20·5	10·5	Rye	1886	G. & T. Smith		J. Eskriet	Rye
[illegible]	1135	*	City of York *Jones* W Sk	British	81	76·4	20·2	10·5	Hull	1884	W. & J. McCann	[illegible]	G. Bowman	Hull
	1136		Ciudad La Plata *Edilon* W Bn (La Georgiana)	Argentine	345	113·6	28·0	11·1	Damariscotta, [Me.	18 [illegible] 72			C. F. Kellam	Buenos Ayres
[illegible]	1137	*	Civil Lord *Alborough* W K	British	69	71·2	19·2	10·2	Lowestoft	1885	S. Richards		T. & J. Lelen, Ld.	Yarmouth
[illegible]	1138	*	Clachnacuddin *Ford* W Bkn	British	225	113·2	25·4	13·7	Inverness	1871	Stewart	[illegible]	R. Allix	Guernsey
[illegible]	1139	* D	Clackmannanshire *Cooper* [illegible] I S	British	1482	250·0	38·3	22·9	Greenock	1884	Russell & Co.		T. Law & Co.	Glasgow
[illegible]	1140		Clacton *Sycell 78-79* W 3m Bn (ex Jacob Lindstrom)	British	192	104·5	21·2	13·6	Venersborg	1875			J. H. Vaux	Harwich
	1141		Claire & Marie *Gudo* W Bg	French	243	95·8	23·0	14·6	Dunkirk	1869	Pyotte Baynert		H. Monnier & fils	Fécamp
[illegible]	1142	*	Clairmont *Leathwaite* [illegible] I Bk	British	755	193·0	31·2	18·9	Glasgow	1876	Dobie & Co.		N. British Shipping Co.	Glasgow
[illegible]	1143	*	Clan Buchanan *Harris 82-88* I S	British	2072	283·5	40·5	24·5	Port Glasgow	1887	Russell & Co.		T. Dunlop & Sons	Glasgow
[illegible]	1144	*	Clan Ferguson *Michi* [illegible] I Bk	British	739	199·2	31·6	19·3	Port Glasgow	1876	J. Reid & Co.		T. Dunlop & Sons	Glasgow
[illegible]	1145	*	Clan Grant *Moody 87-97* I Bk	British	1095	211·0	34·1	21·0	Glasgow	1878	D. & W. Hender- [son & Co.		T. Dunlop & Sons	Glasgow
[illegible]	1146	*	Clan Macfarlane *Johnson* [illegible] I S	British	1484	249·1	38·3	22·8	Greenock	1881	Russell & Co.		T. Dunlop & Sons	Glasgow
[illegible]	1147	*	Clan Mackenzie *Simpson* [illegible] I S	British	1597	259·5	38·2	23·1	Port Glasgow	1882	R. Duncan & Co.		T. Dunlop & Sons	Glasgow
[illegible]	1148	* D	Clan Macleod *Macleod* I Bk	British	646	179·5	31·3	17·5	Sunderland	1874	Bartram, Has- [well & Co.		Sir R. W. Cameron	Glasgow
[illegible]	1149	*	Clan Macpherson *Wilson* [76-85] I S	British	1601	260·5	38·2	23·2	Port Glasgow	1885	Russell & Co.		T. Dunlop & Sons	Glasgow
[illegible]	1150	*	Clan Robertson *Wilson* [79-86] I S	British	1625	259·5	38·3	23·3	Port Glasgow	1884	R. Duncan & Co.		T. Dunlop & Sons	Glasgow

* = Lloyd's Register; D = Nederlandsche Vereeniging; G = Germanischer Lloyd; I = Registro Italiano; R = American Record; A = Veritas Austro-Ungarico; H = Veritas Ellenico; N = Norske Veritas; V = Bureau Veritas.

Dans la 4me Colonne, C = Composite (membrure en fer, bordé en bois); I = Fer; Stl = Acier; W = Bois.

Les Dates du doublage en métal (Col. 11) ressortent des livres dans lesquels les navires sont respectivement classés ou des rapports des Inspecteurs du Lloyd's Register

Pour plus amples renseignements sur les navires classés par le Lloyd's Register, voir le Lloyd's Register Book et le Lloyd's Yacht Register.

Official No.	No. in Book	Registry in which classed	Ship's Name. Material and Rig. Master, and Dates of Appointment to present Owner's Service and to Ship. Ship's Late Name.	Flag	Net Tonnage	Length	Breadth	Depth	Where	When	By whom	Date of Metalling	Owners or Managers
78373	1151		Clansman *Robertson* -86 WBn	British	157	96·0	23·5	11·6	Mechanics Bay N.Z.	1879	McQuarrie & McCallum		W.S.Laurie
64954	1152		Clara WBg	British	190	96·0	24·9	12·8					W.Kelley
59826	1153		Clara (BROKEN UP) *Randell* WBn	British	182	102·4	23·9	12·7	P.E.Island	1870	Bourke		J.Powell
53970	1154		Clara (ex Sarah Abigail) WBg	British	179	95·7	20·6	10·9	Massachusetts U.S.	1834			Henry Moss
51944	1155	R	Clara *Collins* 88-88 WBn	British	178	94·5	24·0	13·5	P.E.Island	1865	Mellish		J.Jenkins
50790	1156		Clara *Sheffield* 85-85 WBn	British	136	90·0	23·2	11·2	NovaScotia	1865			B.Sheffield
50380	1157		Clara *Kilpatrick* WSr	British	132	95·0	24·3	9·1	Baltimore, U.S	1851			J.A.Wheaton
77402	1158	*	Clara *Trembeth* 83-86 WSr	British	101	84·8	22·3	10·6	Padstow	1877	Stribley	Y 10.85	J.P.Dunstan
4972	1159		Clara *Arnold* WSr	American	281	115·0	29·3	12·7	Philadelphia	1860			J.P.Dole
	1160		Clara WBn	Chilian	305	114·8	25·7	..	Shoreham	1865			P.Schmidt & E.Mahn
	1161	V	Clara *Thögersen* -88 WBn	Danish	162	100·1	21·9	10·9	Faaborg	1874	R.Møller	Y 11.80	Aktie Selskab [illegible]
	1162	V	Clara *Kromann* 78-78 WSr	Danish	156	96·0	22·6	11·3	Marstal	1878	H.J.Bager		C.H.Kromann, jun.
	1163		Clara (ex Ane Cecilie) *Schönning* 81-85 WSr	Danish	118	100·4	21·4	10·3	Troense	1865	C.R.Møller	Y 4.85	Aktie Selskab "Clara" (S.C.Kjærgaard)
	1164	D	Clara (ex Bah, ex Rothley) *Waterborg* -87 WS	Dutch	1057	176·8	35·8	22·6	Kingston, Can.	1865			S.Vander Hei
	1165	DV	Clara (ex Success) *Harding* -87 WBn	Dutch	183	104·0	22·6	11·9	Pekela	1875	C.Maathuis	Y 4.87	P.P.Kolk
	1166	G	Clara *Kuhlmann* 77-85 WS	German	1637	207·0	41·2	26·7	Bremerhaven	1882	N.F.Ulrichs	Y 7.86	D.H.Wätjen & Co.
	1167	DGV	Clara (ex Betty, ex Marianne) *Voss* 76-76 WSr	German	364	122·4	28·4	15·6	Greenpoint U.S.	1861	Wecknow		Lichenheim & Pineas
	1168		Clara *Prohn* 80- WBg	German	288	102·2	27·9	14·7	Barth	1852	A.Schlör		C.Siebe
	1169	V	Clara *Röttgers* 84- WBg	German	271	114·0	26·2	14·0	Brake	1865	I.Oltmann's Wwe	Y 3.86	H.W.Meyer
	1170	V	Clara *Röttige* 74- WBn	German	140	95·0	23·0	12·0	Halte	1871	A.Klasen	Y 3.85	A.Klasen
	1171	N	Clara (ABANDONED) (ex Clara Killam) *Hansen* -88 WBk	Norweg'n	802	168·0	34·5	19·7	Tusket, N.S.	1868	N.Gardner	Y 5.87	L.Dahl (in liquidation)
	1172	*V	Clara (ex Clara B., ex Sumatra) *Langlie* -88 WS	Norweg'n	643	172·4	32·0	20·8	Whitehaven	1858	T. & J. Brocklebank	Y 9.88	Actieselskabet "Clara" (C. Møller)
	1173	N	Clara (ex Claribel) *Jacobsen* -87 WBk	Norweg'n	445	137·9	28·8	16·9	Miramichi	1857		Y 12.83	Actieselskabet "Clara" (Chr. Nielsen)
	1174	V	Clara *Matamata* WBg	Spanish	284	108·2	28·4	14·8	San Feliu de Guixols	1870		C 2.88	M Compañy y Verg[illegible]
	1175	GV	Clara Andrea *Riber* 86-88 WBg	Danish	217	106·1	22·7	11·8	Nyborg	1876	P.Petersen	Y 10.86	N.H.Mathiasen
4516	1176		Clara Davidson *Booye* WSr	American	123	89·0	27·0	7·9	Great Egg Harbour, N.J.	1865			T.J.Todd
	1177	V	Clara Diekelmann *Schroder* 79-82 WSr	German	150	86·0	22·4	11·7	Stralsund	1863	O.Kirchhoff		H.Schröder
125287	1178		Clara E. Bergen *Burroughs* -87 W3mSr	American	457	141·0	34·9	17·5	Port Jefferson, N.Y.	1874			Evans, Ball & Co.
126181	1179	R	Clara E. Colcord *Colcord* W3mSr	American	489	146·6	34·0	12·4	Belfast, Me.	1883	J.Y.Cottrall		J.S.Case
125167	1180		Clara E. McGilvery *Treat* -85 WBkn	American	382	136·6	30·3	10·5 / 16·8	Searsport, Me.	1873	W.McGilvery & Co.	Y 1.85	J.G.Pendleton
125048	1181		Clara E. Rogers *Rogers* -83 WSr	American	137	101·0	26·8	8·9	Jonesport, Me	1872			J.E.Rogers
125261	1182		Clara E. Simpson *Hawtrel* W3mSr	American	401	136·0	32·5	14·7	Portsm'th, N.H	1874	W.F.Fernald		J.C.Hontrel
126461	1183	R	Clara Goodwin *Wyman* W4mSr	American	902	191·2	38·8	18·5	Bath, Me.	1887	Kelley & Spear		John R.Kelley
47565	1184		Clara Hargrave WBk	British	338	135·5	30·4	14·0	Boston, U.S.	1856			New Zealand Government (Postmaster-General)
79014	1185	*	Clara J. Wilbur *Haley* W3mSr	British	204	109·3	28·1	10·5	Dorchester N.B	1883	T.B.Wilbur	Y 4.86	T.B.Wilbur
5106	1186		Clara Jane *Allen* WSr	American	119	86·5	26·5	8·1	Pembroke, Me.	1864			S.Staples & Son
125292	1187	R	Clara Leavitt *Lombard* W3mSr	American	433	127·7	31·1	11·4	Cape Elizabeth Me	1874		Y 9.86	B.F.Lombard
29403	1188	*	Clara Manning *Rowancefell* WBn	British	145	91·1	22·8	12·4	Swansea	1862	Meager		W.H.Lean
	1189		Clara Maria *Freimann* 83 WSr	Russian	187	..	..	..	Gutmannsbach	1874			J.Behrsing
25597	1190		Clara Novello (SUNK) *Edmunds* 86-86 WBn	British	145	98·1	23·4	12·0	Llanelly	1858	J.Bevans	Y 11.84	A.Stone
5312	1191	R	Clara Pickens *Eddy* WBn	American	491	135·0	31·8	18·8	Warren, R.I.	1861	H.D.Chase	Y 1.87	J.H.Cady
4521	1192		Clara Rankin *Myers* WSr	American	144	95·1	27·0	8·0	Kennebunk Me	1867			W.E.Cordrey
	1193	G	Clara & Cathinka *Albertsen* 85-87 WSr	Danish	110	79·7	20·5	10·8	Altona	1847	O.Dolck		J.Albertsen
	1194		Clara & Hermann (STRANDED) *Viemann* -85 WBk	German	247	94·6	26·5	16·3	Greifswald	1844	J.D.Gaede		C.Wallis
90372	1195	V	Clare *Heany* 87-87 WBn	British	229	111·0	27·0	11·3	Clare, N.S.	1886	W.D.Lovitt	Y 6.86	W.D.Lovitt
65375	1196	*	Clare *Williams* WSr	British	88	75·0	20·8	10·8	Carnarvon	1868	Williams		W.Williams
86521	1197	*	Clareen *Davey* WK	British	95	81·9	20·5	9·4	Plymouth	1884	R.Hill & Sons		C.F.Williams
20848	1198	*	Clarence *Webb* 72-82 WBk	British	1032	198·0	36·5	22·5	Sunderland	1858	W.Pile	Y 1.86	Pickford, Ford & Co.
17324	1199		Clarence (CONDEMNED) *Wood* 84-84 WBg	British	187	102·3	24·8	10·9	N.Brunswick	1853			T.H.Franks
83071	1200		Clarence WBarge	British	139	110·0	22·3	6·4	Rockland	1881	J.Wood & Co.		James Wood

✳=Lloyd's Register; D=Nederlandsche Vereeniging; G=Germanischer Lloyd; I=Registro Italiano; R=Amer[ican]
A=Veritas Austro-Ungarico; H=Veritas Ellenico; N=Norske Veritas; Y=Bure[au]

In the 4th Column, C=Composite (iron frame, wood planked); I=Iron; Stl=Steel; W=Wood.

The Dates of Metalling (Col. 11) are based on the records of the Books in which Vessels are respectively classed, or on the Reports of the Surveyors of Ll[oyd's]

Further particulars of Vessels classed by Lloyd's Register will be found in Lloyd's Register Book and Lloyd's Yacht Register.

1	2 N° d'ordre	3 Registre dans lequel le navire est classé	4 Nom. Matériaux et Gréement. Capitaine et année d'entrée au service des Armateurs actuels et à bord du navire. Ancien Nom.	5 Pavillon.	6 Tonnage net	7 Dimensions. Longueur	Largeur	Creux	8 Construit. Où.	9 Quand.	10 Par qui.	11 Date du doublage en métal	12 Armateurs ou Directeurs-Gérants.	13 Port d'Armement.
[illegible]	1201		Clarence G. Sinclair WSr	British	81	81·4	19·9	9·9	Bridgwater	1878	Gough		G. Stark	Wick
[illegible]	1202	*	Clarence S. Bemont IS	American	1889	259·9	40·6	23·6	Philadelphia	1884	American S.B.C.		J.E. Ridgway	Philadelphia
[illegible]	1203		Claribel *Murchison* WBk	British	421	133·4	28·3	16·6	P.E.Island	1876	A.W. Owen	Y 12.82	L.C. Owen	Charlottetown, P.E.I.
[illegible]	1204		Clarice *Gonnelli* WBnSr (ex Antonietta)	Italian	187	90·9	24·6	13·5	Sorrento	1858			V. Del Buono	Porto Ferraio
[illegible]	1205		Clarissa *Kerr* WBg	British	254	105·7	26·4	15·7	Sunderland	1867	Wheatly		A. Buckwell	Shoreham
[illegible]	1206		Clarissa Allen *Wakeley* WSr	American	103	91·1	25·7	7·4	Stratford Conn	1867			Cornelia A. Tisdale	Fall River, Mass.
	1207		Clarisse Covacevich WBk	Russian	349	..	..	..	Trieste	1868			G. Antunadis	Marianopoli
	1208		Clarita *Johnson 81-89* WBk (ex Orient)	Chilian	308	118·0	27·5	9·9	Willmsbg., NY	1864	Webb		E.H. Langdon	Ancud
[illegible]	1209		Claro Babuyan *Brown* WBk	British	359	124·6	27·0	17·1	Sunderland	1862	W. Adamson		F.E. White	HongKong
	1210		Claude *Nouault* WBg	French	139	..	..	..	St. Malo	1857			Vve E. Thomazeau & Co.	St. Malo
[illegible]	1211	*	Claudia *Williams* WBk (WRECKED)	British	408	129·5	28·5	17·3	Sunderland	1871	Douglas	Y 1.87	T. Morris & Co.	Aberystwith
	1212	V	Claudina *Correia* WBk	Portug'se	374	122·7	27·9	14·0	Oporto	1863	C.M. da Santos	Y 2.82	Cie Alliança Maritima Portuense	Oporto
[illegible]	1213	V	Claudine *Robertson -88* WBk	British	576	146·8	32·0	19·9	Recco	1871	Rollo	Y 1.88	W.C. Rafarel	Barnstaple
	1214	N	Claudine *Olsen 84-87* WSr	Norweg'n	218	105·1	25·0	12·8	Hammelwarden	1873	A.P. Botter	Y 3.85	S.C. Larsen	Brevig
	1215		Claudio Vincenzi WSr	Brazilian	172	..	..	..					J.N. de Vincenzi & Sons	Rio de Janeiro
	1216	DV	Claus *Sonnichsen 81-87* WBg	Danish	240	119·2	24·4	12·3	Varnaults	1880	A.P. Bring	Y 6.86	Akt. Selsk. (J.N. Hansen)	Nordby, Fanø
	1217	N	Claus Thomeson WBk	Norweg'n	271	96·5	27·2	15·1	Kragerö	1849			L. Larsen	Kragerö
	1218		Clavileno WBn	Spanish	150	89·6	21·8	11·8	Pangasinan	1846			C. Arévalo	Manila
[illegible]	1219		Claymore *Baldon* WBk	British	273	114·0	26·4	14·0	Leith	1858	R. James	Y 11.87	John Bickers	Adelaide, S.A.
[illegible]	1220		Cleddu Belle *Connell -88* WSr	British	115	85·7	21·5	11·3	Milford	1871	Francis Bros.	Y 2.84	John Rorke	St. Johns, N.F.
	1221		Clémence Marie WBk	French	271	109·5	23·3	13·9	St. Malo	1865	L. Lhotelier		L. Demalvilain	St. Malo
[illegible]	1222	I	Clementina *Porzio 76-76* WBn	Italian	745	176·5	31·9	20·8	Equa	1876	A. Gargiulo	Y 6.83	R. Ajello	Naples
	1223	A	Clementina *Count -88* WBmSr	Spanish	410	115·1	29·3	17·3	Denia	1857	A. Puig	Y 8.88	J. Ballester	Majorca
[illegible]	1224	*	Clementine *Ball 83-85* WBn	British	144	99·0	23·7	11·0	Ipswich	1885	W Bayley & Sons	Y 2.84	H. Harrington & Co.	London
	1225		Clémentine *Boringen* WSr	French	131	81·4	22·0	10·6	Nantes	1866	Dubigeon		Robert	Nantes
	1226	F	Clementine *Sjöland* WBk	Russian	368	122·5	28·8	16·4	Nantes	1862	Olivier		Chaulnier	Mariehamn
[illegible]	1227	*	Cleomene *Mays 73-78* IS	British	1739	251·0	38·9	24·0	Southampton	1877	T.R. Oswald		H. Fernie & Sons	Liverpool
[illegible]	1228		Cleopatra *Evans 81-81* WBg	British	187	88·0	24·5	14·2	Coxgreen	1848			J. Rogers	Maldon
	1229		Cleopatra	American	100	86·0	23·0	9·0	Essex, Mass.	1885			D.C. & B. Babson	Gloucester, Mass.
	1230	G	Cleopatra *Bremers 88-88* WS	German	1233	187·0	39·9	23·7	Williamsburg, [L.I.	1867	T. Stock	Y 1.86	D. Hinrichs	Bremerhaven
316	1231	A	Cleopatra WBn	Greek	337	115·4	27·7	..	Syra	1880		[illegible]	G. Possidon & Co.	Syra
130	1232		Cleopatra (ex Giovanni) *Romeo* WBg	Italian	203	101·0	27·8	14·5	Castellam.	1858		[illegible]	G. De Natale	Messina
	1233	V	Cléta *Marc -88* WBk	French	260	102·6	23·9	15·1	Nantes	1872	T. Dubigeon & fils	[illegible]	G. Craquelin	Havre
	1234		Cleto (ex Giovine Carlo) (LOST) WBk		304	118·1	27·8	15·9	Rostock	1867	W. Zeltz			
	1235	DN	Cleveland *Wiesnes 80-85* WS	Norweg'n	1230	206·3	37·8	22·9	Quebec	1873	W. Charland	[illegible]	Nils P. Höyer	Skien
[illegible]	1236		Clifford *Hamon* WSr	British	132	88·0	25·0	10·1	Lunenburg	1885	L. Young		L. Anderson	Lunenburg, N.S.
[illegible]	1237	R	Clifford *Crabtree* WBmSr	American	262	121·6	29·5	10·3	Lamoine, Me.	1875		[illegible]	W. King & Co.	Lamoine, Me.
[illegible]	1238		Clifford C. *Dykeman* WSr	British	107	81·0	26·5	7·5	Kingston, N.B.	1882	S.J. Holder		D.J. Purdy	St. John, N.B.
[illegible]	1239	V	Clifton *Curry 88-* WSr	British	473	149·0	32·6	16·0	Windsor, N.S.	1883	S. Dimock	Y 11.83	S. Dimock	Windsor, N.S.
[illegible]	1240		Clifton *Cowper* WBk	British	380	128·5	28·2	16·9	Sunderland	1869	Gibbon	Y 10.81	J.M. Jones & F. Proctor	Dunedin, N.Z.
[illegible]	1241	V	Clifton *McLean* WSr	British	218	99·5	28·0	10·6	Moss Glen, [King's Co. N.B.	1885	G. Merritt	[illegible]	F.S. Bonnell	St. John, N.B.
	1242		Clinton *Rengtsson* WBg	Swedish	299	113·5	24·1	14·0	New York	1838			J.H. Gren	Fiskebäckskil
[illegible]	1243		Clio WBn	British	190	109·1	27·0	11·7					H. Boyd	Ardrossan
[illegible]	1244		Clio *Gerhardt* WBn	British	163	94·6	25·0	10·1	Chester, N.S.	1884	J.F. Park		J. Eisenhauer	Lunenburg, N.S.
[illegible]	1245		Clio WBn	British	119	87·7	22·9	10·4	Halifax, N.S.	1862			E.S. Delisle	Halifax, N.S.
	1246		Clio *Eilertsen -83* WBk (WRECKED)	Norweg'n	310	105·2	29·1	15·6	Grimstad	1852	N. Olsen Gjömle		Gj. Olsen	Arendal
	1247	N	Clio *Christiansen* WBk	Norweg'n	271	117·2	26·8	13·6	Kragerö	1870	A. Johnsen	Y 12.87	B. Knoph	Kragerö
[illegible]	1248	A	Clita *Crivellari* WSr	Austrian	241	104·6	24·2	14·5	Fiume	1868	G.G. Baccich [Bellaz	Y 6.88	P. Crivellari & Co.	Cherso
[illegible]	1249	*	Clodian *Thompson* WBk	British	480	154·2	27·9	17·3	Dundee	1863	Tay Ship Co.	Y 11.84	R. Jolly & Son	London
[illegible]	1250	*	Cloncaird *Mould ch 72-82* Stl Bk	British	1300	234·0	37·5	23·5	Pt. Glasgow	1882	J. Reid & Co.		J. Kerr & Co.	Greenock

*=Lloyd's Register; D=Nederlandsche Vereeniging; G=Germanischer Lloyd; I=Registro Italiano; R=American Record; A=Veritas Austro-Ungarico; H=Veritas Ellenico; N=Norske Veritas; V=Bureau Veritas.

Dans la 4me Colonne, C=Composite (membrure en fer, bordé en bois); I=Fer; Stl=Acier; W=Bois.

Les Dates du doublage en métal (Col. 11) ressortent des livres dans lesquels les navires sont respectivement classés ou des rapports des Inspecteurs du Lloyd's Register.

Pour plus amples renseignements sur les navires classés par le Lloyd's Register, voir le Lloyd's Register Book et le Lloyd's Yacht Register.

Official No.	No. in Book.	Registry in which classed.	Ship's Name. Material and Rig. Master, and Dates of Appointment to present Owner's Service and to Ship. Ship's Late Name.	Flag.	Net Tonnage.	Dimensions. Length.	Breadth.	Depth.	Built. Where.	When.	By whom.	Date of Metalling.	Owners or Managers.
	1251		Clorinda (ex Admiral Grenfell) Stiglich W Bk	Chilian	283	127·9	24·4	12·3	Bideford	1851			Sanchez & Cardenas
85859	1252	*	Closeburn Johnston 84-84 I Bk	British	897	201·0	32·7	19·5	Pt.Glasgow	1881	Russell&Co.		T.C.Guthrie
5672	1253	R	Clotilde Bowden 86-87 W Bk	American	443	129·6	30·6	16·6	Portsmouth, N.H.	1869	Tobey&Littlefield	[illegible]	J.S.Emery & Co.
177	1254	·	Clotilde Paoli W Pol	Italian	137	90·9	25·1	10·4					A.Giannoni fu G.
9	1255		Clotilde (ex Franklin) Portella 85-86 W Sr	Peruvian	198	107·2	26·3	10·5					J.Portella & hijo
	1256	G	Clotilde Tomas 78- W Bn	Spanish	279	110·8	28·3	14·3	Blanes	1869	J.Baguer	V 7.88	Balsells & Saldran
	1257		Clotilde B (ex Emilia) Zino W 3m Sr		437	132·5	29·6	19·1	Sestri	1867	A.Briasco fu G B	V 1.88	
987	1258	I	Clotilde Solari Spano W Bn	Italian	235	114·1	26·4	14·3	Sestri	1873	G.Casanova	V 1.87	F.Berfe di G.
86500	1259	*	Cluny Castle Pierce 83-83 I 1m S	British	1934	276·5	41·2	24·0	Glasgow	1883	Barclay,Curle&Co.		D.Currie & Co.
79420	1260	*	Clutha Joyce 79-85 Bkn	British	244	125·0	24·8	12·8	Ardrossan	1881	Baird&Shearer	V 3.87	W.P.Munn
28848	1261	*	Clyde Johnson I Bk	British	1140	212·1	33·1	22·6	Glasgow	1860	A.Stephen&Sns.		A.Gibson & Co.
88020	1262	V	Clyde Strum 84- W Bn	British	237	109·9	27·0	11·9	MahoneBy.NS	1884	T.Langville		A.Chisholm
63752	1263		Clyde Frena 87-87 W Bg	British	194	107·9	24·1	12·9	Teignmouth	1871	Mansfield		Whitstable Shipping Co.,Ld.
45483	1264		Clyde Rive 85-85 W Sw	British	186	101·0	21·5	13·3	Scarborough	1863	Tindall	V 12.86	Ockenden & Son
60381	1265	*	Clyde Vale Jones 77-77 I Bk	British	474	157·6	27·6	16·8	Glasgow	1868	A.Stephen&Sns.		Browne & Watson
78564	1266	*	Clydebank Greig 85-85 I Bk	British	863	195·2	32·2	19·2	Dumbarton	1877	Birrell,Stenhouse&Co.		G.G.Macfarlane & Co.
85874	1267	* D	Clynder Roach 87-87 I Bk	British	1117	215·2	35·3	21·1	Greenock	1881	Russell&Co.		Thom & Cameron
91874	1268	*	Clytie Rickaby 87-87 Stl Bkn	British	320	144·2	27·1	11·5	Grangemouth	1885	Grangemouth Dockyard Co.		Clytie Sailing Ship Co Ld. (Townsend & Spearing)
125472	1269		Clytie Laughton W 3m Sr	American	416	141·0	32·3	15·0	Pembroke,Me.	1875	J.B.Nutt.	V 11.86	J.B.Nutt & Co.
	1270		Clytie Fraser W Bk	Argentine	1079	175·5	36·3	23·5	St.John,N.B.	1864	Gass	V 1.88	
	1271	V	Cobija Fatout -88 W Bk	French	682	154·2	30·2	21·0	Honfleur	1866	Cardon	V 2.88	D.Auger
63088	1272	*	Cochin Oudney 77-79 I S	British	1200	229·0	36·0	22·2	Greenock	1869	Robertson&Co.		J.Kerr & Co.
72522	1273	* D	Cochrina Leighton 76-76 I Bk	British	989	203·8	33·7	20·5	Glasgow	1876	Aitken&Mansel		J.Guthrie
76364	1274	*	Cock o' the North Fellis W Bkn	British	235	116·3	24·1	13·4	Kingston	1877	Geddie,sen.	V 11.82	Anderson,Webster&Co.
63913	1275	D	Cock o' the Walk Bloomfield 76-76 W K	British	100	96·8	23·0	8·0	London	1876	Surridge		R.Cox
74544	1276	*	Cockermouth Little 79-86 I S	British	1297	238·8	36·7	21·4	Glasgow	1876	A.Stephen&Sns.		P.Iredale & Son
	1277	N	Codan Johansen 86-86 W Bk	Norweg'n	406	133·5	28·8	16·5	Apenrade	1863	A.P.Damm	V 11.88	Alexander Krog
51160	1278	*	Cognac (LIGHTER) Drake -88 W Sw	British	294	107·7	27·0	16·2	Sunderland	1864	Gibbon	V 3.82	R.B.Wright
22480	1279		Cognac Packet Button [84-85] W Sw	British	169	86·4	23·0	14·6	Bursledon	1792			J.H.Vaux
	1280	V	Cohnheim (CONDEMNED) (ex Dunmore) Zephen 77- W Bk	German	435	138·7	28·2	17·8	Alloa	1861	Adamson	V 7.80	W.Maack
60056	1281		Coila Court 83-83 W Bg	British	164	97·8	22·5	12·7	Broadhaven	1871	Whitehead		G.H.Goldfinch
29020	1282	*	Coimbatore Jones 85-85 I S	British	1193	215·8	35·1	22·5	Glasgow	1865	Barcl'y Curle&C		Eyre,Evans & Co.
	1283		Coimbra W Sr	Brazilian	237	..	..	..					J.G.de Barros
	1284	V	Colbert I Bk	French	512	172·2	29·4	13·6	Bordeaux	1872	Forges & C. de l'Océan		J.H.Tandonnet frères
69130	1285	V	Colchester Starling 86-88 W S	British	1385	216·0	40·0	23·7	Princeport NS.	1875	Saunderson	V 4.81	Martin Dickie
29963	1286	*	Coldinghame (ex Gavin Steel) Stevens 7-87 I Bk	British	1059	211·8	33·7	21·4	Liverpool	1866	Hart&Sinnott		J.Willis & Son
311	1287		Coldstream Woodgett W Bk	British	684	134·7	29·3	20·9	Moulmein	1849	Cockerell&Co.		Tatham,Bromage & Co.
45385	1288	*	Coldstream Luen I Bk	British	546	170·3	27·3	17·9	Sunderland	1862	J.Laing		E.C.Friend & Co.
60601	1289	*	Coleridge Tregaskes -87 W Sr	British	193	103·7	23·6	12·8	Topsham	1866	Holman	V 1.87	C.T.Bennett
	1290	N V	Colin Archer Andersen -86 W Bk	Norweg'n	657	156·6	35·2	17·3	Arendal	1880	Olsen	V 3.81	Peder Olsen
125093	1291		Colin C. Baker Baker W 3m Sr	American	388	132·0	31·6	14·7	Bath,Me.	1873			J.Baker
32279	1292		Collaroy (ex Steamer) Ball I 1m Sr	British	402	180·9	23·3	11·1	Liverpool	1859	Fawcett,Preston&Co		A.Burns
	1293	N V	Collector (ex Peter Cremar) Andersen 84-84 W Bk	Norweg'n	634	160·1	32·8	17·6	RiverJohnNS.	1868	C.M.Sellon	[illegible]	Tellef Jensen
29065	1294		Collina W Bn	British	168	100·3	22·4	11·5	P.E.Island	1860	Coffin&Son		A.Nicholls
60930	1295	*	Collingrove Angel C Bk	British	861	181·4	33·5	19·6	Sunderland	1869	J.Laing	[illegible]	A.L.Elder & Co.
65731	1296	*	Collingwood Forbes 78-78 I S	British	1011	211·1	34·8	21·0	Aberdeen	1872	W.Hood & Co.		Devitt & Moore
	1297	G V	Collmar Bösaker -86 W Bg	German	223	103·6	23·3	12·2	Brake	18 52/78	D.Denker	[illegible]	Frau B.W.P.Seek
5706	1298	V	Coloma Waape W Bk	American	814	168·0	35·0	20·0	Warren,R.I.	1869	J.J.Cally	[illegible]	C.M.Noyes
2668	1299	D I	Colomba Stagno W Bk	Italian	503	146·6	31·1	19·6	Varazze	1876	G.B.Baglietto	[illegible]	Em. Stagno
	1300												

* = Lloyd's Register; D = Nederlandsche Vereeniging; G = Germanischer Lloyd; I = Registro Italiano; R = A…
A = Veritas Austro-Ungarico; H = Veritas Ellenico; N = Norske Veritas; V = Bur…

In the 4th Column, C = Composite (iron frame, wood planked); I = Iron; Stl = Steel; W = Wood.

The Dates of Metalling (Col. 11) are based on the records of the Books in which Vessels are respectively classed, or on the Reports of the Surveyors of …

Further particulars of Vessels classed by Lloyd's Register will be found in Lloyd's Register Book and Lloyd's Yacht Register.

Numéro d'ordre	Registre dans lequel le navire est coté	Nom. Capitaine et dates d'entrée au service des Armateurs actuels et à bord du navire. Ancien Nom.	Matériaux et Gréement	Pavillon	Tonnage	Longueur	Largeur	Creux	Construit: Où.	Quand.	Par qui.	Date du doublage en métal	Armateurs ou Directeurs-Gérants.	Port d'Armement.
1301		Colomba *Agen*	W Sr	Italian	124	88·9	23·0	10·8	Lionite	1876	N. Pichiotti		G. Agen	Porto Maurizio
1302		Colombe *Meslen 85-85*	W Sr	French	122	92·5	22·3	10·8	Gravelines	1884	Collin		H. Torris	Gravelines
1303	I	Colombina *Merello* (ex Marian B.)	W Bk	Italian	648	148·6	30·7	22·4	Lavagna	1877	Brignole	[illegible]	M. Merello fu G. B.	Genoa
1304	✻	Colombo *Sprosic 85-88*	I Bk	British	1199	230·5	36·3	22·4	PortGlasgow	1868	R. Duncan & Co.		G. Petrie	London
1305		Colombo *Di Milia -87*	W Bn	Italian	482	144·3	30·1	19·3	Gaeta	1876	P. Contreras	[illegible]	B. Capodanno & Co.	Gaeta
1306		Colombo *Pastorino*	W Bk	Italian	279	107·9	27·7	17·4	Savona	1859	A. Gandolfo		F.lli Poldi	Genoa
1307		Colon	W Bn	Spanish	213	100·7	23·7	12·7	Albay	1867			José Muñoz	Manila
1308	✻	Colonel Burnaby	W Sk	British	92	80·4	20·7	10·6	Goole	1886	Cottingham Brs.		Pickering & Haldane	Hull
1309		Colonel Denfert *Valeau* (ex Constance Wilson) [illegible]	W Sr	French	178	99·0	23·0	12·5	Gloucester	1867	Barnes		A. Verdier & Co.	La Rochelle
1310		Colonel G. M. Totten	W Sr	American	178	100·0	31·0	7·1	Brooklyn	1883			C. Colne	NewYork
1311		Colonisador	W Bn	Brazilian	306	..	..	..					J. R. Ferreira	Rio de Janeiro
1312		Colonist	W Sr	British	109	78·6	20·5	10·4	Dumbarton	1861	Denny & Rankin		E. F. Robinson	Sydney, N.S.W.
1313	N	Colonist *Tollefsen*	W Bk	Norweg'n	442	128·4	29·7	16·8	N. Brunswick	1863	Anderson		A. Tollefsen	Tönsberg
1314	✻ D	Colony *J. Owen*	I S	British	1694	258·0	39·5	23·1	Sunderland	1886	W. Doxford & Sons		Ogwen Ship Co., Ld. (W. Thomas & Co.)	Liverpool
1315		Colorado *Marchinnie* (MISSING)	W Bk	British	660	155·0	31·7	19·1	Tusket, N.S.	1870	N. Churchill		J. Cochran	Belfast
1316	R V	Colorado *Dunn -88*	W Bk	American	1036	170·9	35·5	22·7	SouthBoston	1864	H. O. & F. Briggs	[illegible]	J. S. Emery & Co.	Boston
1317	✻	Columba *Rossiter 87-88*	I S	British	1748	263·0	39·1	24·1	Liverpool	1881	W. H. Potter & Sn		Sandbach, Tinne & Co.	Liverpool
1318	N V	Columba *Okegren 86-86*	W Bk	Norweg'n	558	148·4	31·3	18·0	Christiansand	1876	M. Knudsen	Y 7.86	M. F. Stray	Christiansand
1319	N	Columba *Hansen*	W Bk	Norweg'n	391	124·0	30·0	16·3	Lillesand	1870	A. Stainsen	[illegible]	Sören Nielsen	Arendal
1320	V	Columbia *Hogan*	W S	American	1472	205·8	40·0	24·0	Bath, Me.	1871	Houghton Bros.	[illegible]	Houghton Bros.	Bath, Me.
1321		Columbia *Davidson -87*	W Bk	American	940	169·5	37·0	21·4	New York	1846	Webb		C. L. Dingley	San Francisco
1322		Columbia	W Sr Yt	American	106	99·5	25·1	8·3	Chester, Pa.	18 71/84	J. B. VanDeusen		H. M. Flagler	New York
1323		Columbia *Donovan*	W Sr	American	105	76·0	27·3	7·3	Coxsbackie, N.Y.	1847			D. Donovan	New York
1324	I	Columbia *Esposito -87*	W Sr	Argentine	126	89·6	22·1	9·0	Cassano	1886	A. Castellano	Y 11.86	L. Zaccola	BuenosAyres
1325	✻	Columbine *Mitchell 86-86*	W Sk	British	84	75·0	21·0	10·8	Goole	1885	G. W. Outwin		Mrs. E. Jipson	Hull
1326		Columbine *Smith 86-86*	W Cr	British	35	54·7	15·7	7·6	Jersey	1877	Aubin	[illegible]	R. H. Isbister	Lerwick
1327	✻	Columbus *Griffiths 88-88*	C Bk	British	744	182·4	31·0	18·2	Glasgow	1868	Randolph, Elder & Co	[illegible]	J. Lidgett & Sons	London
1328	✻	Columbus *Jonghans* (ex Robert Anderson, ex Princesse)	I Sr	British	221	102·3	27·5	12·9	W. Hartlepool	1865	Irvine & Co.		Aga Abdoul Rassoul & Co.	Port Louis
1329	✻	Columbus	W Sk	British	87	78·0	20·5	10·3	Goole	1882	Cottingham Brs		Pickering & Haldane	Hull
1330	V	Columbus *Spieske -87*	W Bu	Ecuador	257	107·6	21·3	12·7	Brake	1870	F. Oltmanns Wwe	[illegible]	Ecuadorian Lloyd	Guayaquil
1331	V	Columbus *Steuermitch 80-83*	W S	German	1731	218·7	41·3	28·5	Kennebunk	1870	N. L. Thompson	Y 10.88	J. F. Arens	Bremen
1332	V	Columbus *Haesloop 78-86* (ex Brown Brothers)	W S	German	1429	210·6	40·3	24·1	Newburyport, Mass.	1876	Atkinson & Fillmore	Y 10.86	Wm. Stisser & Co.	Bremen
1333	✻	Columbus *Niemann 88-88* (ex Santique, ex Warrior)	I Bk	German	484	155·8	26·1	16·3	Stockton	1858	M. Pearse & Co.		J. C. M. Block	Hamburg
1334	A V	Columbus *Mangano* (ex Dragomeri Dubrovacki)	W Bk	Italian	1255	194·5	34·4	27·5	Gravosa	1875	N. S. Brudicich	[illegible]	G. Starace	Castellamare di Stabia
1335	N	Columbus *Niemann -88*	W Bk	Norweg'n	894	159·2	39·9	22·6	Gefle	1872	O. A. Brodin	[illegible]	A. F. Klaveness	Sandefjord
1336		Columbus *Olsen*	W Bk	Norweg'n	366	107·9	28·8	17·1	Kragerö	18 62/68			J. I. Duus	Kragerö
1337	N	Columbus *Christoffersen* [illegible] -77	W Bk	Norweg'n	348	116·8	28·4	16·2	Arendal	1861	A. Hendriksen		S. T. Eeg	Brevig
1338		Columbus *Clausen*	W Bg	Norweg'n	273	100·7	26·2	15·8	Laurvig	1853	E. Olsen		T. Thommassen	Laurvig
1339		Columbus *Larsen -75* (WRECKED)	W Sr	Norweg'n	239	93·6	21·5	15·3	Poole	1809			Actieselskabet "Columbus" (F. W. Werner)	Drymmen
1340	D N	Columbus *Halvorsen*	W Bk	Norweg'n	251	97·1	25·7	13·7	Mandal	1844			Rederiselskabet "Columbus" (H. Holland)	Fredrikstad
1341	N	Columbus *Wilhelmsen -88*	W Bg	Norweg'n	202	91·9	23·9	13·9	Norrköping	1840			O. Wilhelmsen	Tonsberg
1342	V	Columbus *Bauer -84*	W Bk	Russian	595	150·2	31·2	18·2	Vegesack	1861	F. Ulrichs	Y 9.88	Navigation Co. "Austra"	Riga
1343		Colusa *Berdine 84-87*	W Bk	American	1129	187·3	38·3	22·5	Bath, Me	1873	W. Rogers	Y 3.86	G. F. Smith & Co.	SanFrancisco
1344	✻ D	Colwyn *Robertson 84-84*	I Bk	British	1160	228·4	35·0	21·3	Sunderland	1877	W. Doxford & Sons		G. Traill & Sons	London
1345	✻	Comadre *Doran 84-86*	I Bk	British	772	195·3	31·7	19·4	Glasgow	1868	A. Stephen & Sons		Tomlinson, Hodgetts & Co.	Liverpool
1346		Comarasamypuravy	W Dy	British	118	91·3	22·7	14·0	Valvertyturie	1884	T. Muttovalo		Arunasalem Vairamuttu	Jaffna
1347	✻	Combermere *Jenkins 88-88*	I S	British	1686	251·3	39·1	24·1	Stockton	1881	Richardson, Duck & Co.		C. W. Kellock & Co.	Liverpool
1348		Comet	W Bk	British	337	122·3	25·0	15·7	Quebec	1857			Ameer Jan Saib	Rangoon
1349		Comet *Allan 86-86*	W Sr	British	262	115·8	25·7	14·7	Montrose	1865	Petrie		Montrose Commercial Co.	Montrose
1350		Comet	I Sr	British	122	87·4	18·7	..	Liverpool	1867	W. H. Potter & Co.		Mersey Docks & Harbour Board	Liverpool

✻ = Lloyd's Register; D = Nederlandsche Vereeniging; G = Germanischer Lloyd; I = Registro Italiano; R = American Record;
A = Veritas Austro-Ungarico; H = Veritas Ellenico; N = Norske Veritas; V = Bureau Veritas.

Dans la 4me Colonne, C = Composite (membrure en fer, bordé en bois); I = Fer; Stl = Acier; W = Bois.

Les Dates du doublage en métal (Col. 11) ressortent des livres dans lesquels les navires sont respectivement classés ou des rapports des Inspecteurs du Lloyd's Register.

Pour plus amples renseignements sur les navires classés par le Lloyd's Register, voir le Lloyd's Register Book et le Lloyd's Yacht Register.

Official No. (1)	No. in Book (2)	Registry in which classed (3)	Ship's Name. Master, and Dates of Appointment to present Owner's Service and to ship. Ship's Late Name. (4)	Material and Rig	Flag (5)	Net Tonnage (6)	Dimensions: Length (7)	Breadth	Depth	Built: Where (8)	When (9)	By whom (10)	Date of Metalling (11)	Owners or Managers (12)
126870	1351		Comet *Lass -86*	W3mSr	American	408	144·6	35·2	11·4	Port Blakeley	1886	W.G.Hall		J.A.&F.P.Hooper
125[illegible]	1352	R	Comet *Aldrich*	W3mSr	American	286	117·8	30·0	13·2	Port Jefferson, N.Y.	1877	J.M.Bayles&Son	Y 3.86	A.Dayton & Co.
	1353	V	Comet *Krippner 84-87*	WS	German	1083	191·2	37·1	23·6	Chelsea, Mass.	1869	Pierce & McMichael	Y 2.86	C.Bolken
	1354	V	Comet *Krull 76-76*	WBg	German	222	100·2	25·2	12·8	Ueckermünde	1859	Brusewitz		W.Ludendorff,jun.
	1355		Cometa *Vieira* (ex Ivory, ex Jatobolao)	WBn	Brazilian	150	..	..	..	Holland	1862		Y 9.86	F.Dexheimer
	1356	DV	Cometen *Tillstrom -86*	WBk	Swedish	495	135·5	27·5	16·5	Carlshamn	1874	D.R.Andersson	Y 2.86	F.A.Hurtus
	1357	N	Comforter *Gjertsen 75-81*	WBg	Norweg'n	210	101·4	26·2	12·3	Tvedestrand	1867	B.Halvorsen		A.Speilberg
	1358	N	Commandör Sölling *Aralsen*	WBg	Norweg'n	270	106·5	27·5	12·7	Grimstad	1860	T&T.Gundersen		J.Halvorsen
974	1359	I	Commendatore Alfredo *Mori -88*	WBn	Italian	123	94·2	23·2	11·6	Sestri	1884	A.Carlevaro	Y 7.86	W.Walton & Nipote
	1360	V	Commerçant *Dagorne -78*	WBk	French	281	110·8	23·3	14·6	St.Malo	1872	Gautier	Y 1.86	Debry
29579	1361		Commerce *Halvorsen 84-84*	WSw	British	254	117·5	26·0	14·5	Shoreham	1862	Balley	Y 9.81	J. & N.C.Bull
5303	1362		Commerce *Chase* (DISMANTLED)	WBk	American	453	127·8	30·0	17·0	Newcastle,Me.	1861	Gautier	Y 6.88	D.W.McLean
4621	1363	R	Commerce *Nash*	WSr	American	102	84·5	24·4	8·2	Newburyport, Mass.	1867			J.J.Perry
	1364	N	Commerce *Backer*	WBk	Norweg'n	250	111·4	27·0	13·0	Skien	1860			O.Bjærtness
	1365		Commerce *Nielsen*	WBg	Norweg'n	182	82·4	23·0	14·6	Scotland	1825			C.C.Knudsen
	1366	N	Commerce d'Amiens *Nielsen*	WBg	Norweg'n	177	89·4	24·9	11·9	Dieppe	1845			H.Tambs
	1367	V	Commercio *De Carvalho* (ex Marianna IV., ex Rialto)	WBn	Portug'se	437	125·3	27·2	16·4	E.Machias,Me	1874		Y 5.84	Viuva Macieira & Sons
	1368	V	Commercio *Mattos* (ex Rio Lima)	WBn	Portug'se	163	87·9	22·3	10·8	Vianna	1874	J.G.Martins Branco		Magalhaes & Filho
	1369	V	CommerzienrathRodbertus *Waldmann 81-87*	WBk	German	567	143·5	30·5	17·7	Barth	1878	J.S.Kraeft	Y 7.86	J.N.Rodbertus
	1370		Commewyne	WBk	French	315	129·1	27·4	12·7	Alloa	1875	Roy	Y 8.81	P.Tougard
125728	1371	RV	Commodore *Jordan*	WS	American	1909	226·9	41·8	27·6	Yarmouth,Me.	1879	BlanchardBros.	Y 1.85	C.F.Crooker
4689	1372		Commodore Morris *Winslow*	WBk	American	338	107·0	27·7	17·9	Falm'th,Mass.	1841			Aiken & Swift
126216	1373	RV	CommodoreT.H.Allen *Oates*	WS	American	2271	245·3	41·7	28·3	Richmond,Me.	1884	TJSouthard&Sn	Y 10.88	R.L.Merriman
73548	1374	✳	Commonwealth *Cooper 75-75*	IS	British	1345	237·5	36·2	21·6	PortGlasgow	1875	Russell&Co.		C.H.Stewart
	1375	N	Como *Evensen 80-80*	WBg	Norweg'n	270	116·3	28·1	12·8	Cornwallis,NS	1866	B.&J.Bigelow	Y 3.86	A.C.Evensen
	1376	N	Comorin *Pedersen -85*	WBk	Norweg'n	835	187·0	32·4	21·5	Whitehaven	1855	T.&J.Brocklebank	Y 1.85	O.L.Rød
55090	1377	✳	Compadre *Jones 86-86*	IBk	British	800	190·6	30·4	19·5	Liverpool	1867	R.&J.Evans&Co		Tomlinson,Hodgetts&Co
	1378		Companus *Larsson 88-88*	WSr	Swedish	144	93·8	21·4	10·2	Christinopel	1878			O.Pettersson
125985	1379		Compeer *Peterson 83-86*	W3mSr	American	330	136·0	33·0	11·0	Humboldt,Cal	1877			S.B.Peterson
	1380		Competidor *Santos Lauro*	WBn	Brazilian	195	..	..	..	SJoãodaBarra	1884		Y 8.86	Um. Cabotagem de S. João da Barra
	1381	GV	Competitor *Rosenbaum* (ex Lorelyi) [87-87]	WBk	German	678	165·6	33·5	20·7	Boston,U.S.	1852	J.Curtis	Y 7.88	J.H.Meyer
80400	1382	✳D	Comus *Lamont 88-88*	IBk	British	715	186·7	30·6	18·0	Glasgow	1879	C.Connell&Co.		J.Thomson,jun.
	1383		Conceição	WBk	Brazilian	546	..	..	..					A.J.D.Ferreira
	1384		Concepcion *Raygada* (ex Geneva) [87-87]	WBk	Peruvian	157	101·3	24·9	10·3	UnitedStates				M.Raygada
	1385	V	Concepción *Canto -87* (ex Silvia)	WBk	Spanish	556	137·8	31·2	19·6	SestriPonente	1869	A.Briasco	Y 4.85	Civilis & Ravell
	1386	V	Concepción *Estape*	WBk	Spanish	248	102·6	27·8	14·1	Blanes	1867		Y 10.83	M.Sunol
	1387		Concepción (ex Lepanto)	WBg	Spanish	199	103·3	21·8	12·1	Ilo Ilo	1858			Marcaida & Granades
	1388		Concepción (ex Bosquerna)	WBn	Spanish	110	88·9	20·0	9·1	Mindoro	1875			José M.Garcia
	1389		Concepcion (ex TamO'Shanter)	WSr	Ven'zuel'n	108	90·0	25·0	7·0	New York	1876			Braulio Sifontes
1496	1390		Concetta *Tissone* (ex Marianna Ferro)	WBk	Italian	460	131·5	30·4	19·4	Savona	1868	A.Calcagno	Y 10.82	G.Caflero
3310	1391		Concetta A. (ex Rouzou)	WBk	Italian	140	82·3	24·6	12·5	LaSeyne	1856	Jongler		P.Avegno
2531	1392	I	Concettina *Lubrano* (ex Giuseppino)	WBn	Italian	380	117·7	28·9	17·9	Cassano	1873	F.deRosa	Y 9.87	Fllı. Nugues
2555	1393		Concettina M. *Mazzella* (ex Luisa Isonnio)	WBg	Italian	277	102·6	26·6	15·2	Agde	1865	Cavallier		G.Mazzella
323	1394	A	Concettina T. *Papele* (ex Nuovo Francesco)	WBn	Italian	186	101·7	22·3	13·4	Trieste	1863		Y 11.85	C.Todisco
32	1395	I	Concettina Z. *Ajello -88* (ex Tre Sorelle)	WBn	Italian	199	..	..	..	Capo d'Istria	1850		Y 9.83	I.Zappola
2070	1396	I	Concezione *Chiesa*	WBk	Italian	455	126·9	30·9	19·9	Sestri	1870	G.Bruzzone	Y 7.86	M.Chiappella & C.Ravello
1228	1397	IV	Concezione *Russo -88*	WBk	Italian	406	122·3	27·5	18·2	Alimari	1879	S.Mauro	Y 5.82	F.S.Russo
365	1398	V	Concezione *Leboffe*	WBk	Italian	362	120·7	29·7	17·0	Gaeta	1871	F.Contreras	Y 9.87	Flli. Leboffe
958	1399	I	Concezione Immacolata *Siano*	WBg	Italian	386	126·6	29·4	18·7	Cassano	1873	F.deRosa	Y 12.82	F.Trombetta
2000	1400	I	Concezione S. *Ferrara* (WRECKED)	WBk	Italian	739	163·0	33·5	22·4	Savona	1878	G.Calamaro	Y 1.86	G.Stagno fu G.

✳ = Lloyd's Register; D = Nederlandsche Vereeniging; G = Germanischer Lloyd; I = Registro Italiano; R = Ame[rican]; A = Veritas Austro-Ungarico; H = Veritas Ellenico; N = Norske Veritas; V = Bur[eau]

In the 4th Column, C = Composite (iron frame, wood planked); I = Iron; Stl = Steel; W = Wood.

The Dates of Metalling (Col. 11) are based on the records of the Books in which Vessels are respectively classed, or on the Reports of the Surveyors of L[...]

Further particulars of Vessels classed by Lloyd's Register will be found in Lloyd's Register Book and Lloyd's Yacht Register.

1	2 N° dans la liste	3 Registre dans lequel le navire est coté	4 Nom. Materiaux et Grèement. Capitaine et dates d'entrée au service des Armateurs actuels et à bord du navire. Ancien Nom.		5 Pavillon	6 Tonnage net	7 Dimensions. Longueur	Largeur	Creux	8 Construit. Où.	9 Quand	10 Par qui	11 Date du doublage en métal	12 Armateurs ou Directeurs Gérants.	13 Port d'Armement.
	1401	V	Concha Ballester (ex Javen Joaquin ...) (ABANDONED)	W Bg	Spanish	246	101·7	28·2	13·8	Barcelona	1856			M. Sendon y Garcia	Manila
	1402		Conchita	W Bn	Spanish	106	89·6	18·9	8·2	Tayabas, Luz'n	1874			Manuel Rosario	Manila
	1403		Concord	W Sn	British	195	86·9	24·1	14·2	Middlesbro'	1851	Scruffton & Co.		R. Race	Middlesbro'
	1404	V	Concord *Heyer -84*	W Bk	German	376	128·1	28·1	17·0	Sunderland	1868	Hodgson	Y 3.83	A. H. Volcker	Stralsund
	1405	NV	Concord *Gudersen*	W 3m Sr	Norweg'n	374	126·5	29·1	15·5	Stavanger	1875	C. Knudson	Y 2.86	A. Bryne	Stavanger
	1406	FV	Concorde *Osmonts -87* (WRECKED)	W Sr	French	105	..	..	..	Fécamp	1878			G. Dinoir	Dunkirk
	1407	V	Concordia *Andersen 86-87* (ex Margar Christine)	W Bk	Danish	364	123·3	27·6	17·0	Blyth	1871	Geo. Robinson	Y 5.86	J. Harboe	Skjelskor
	1408	DV	Concordia *Sweet 85-87*	I Bk	Dutch	1002	193·9	35·4	20·6	Amsterdam	1886	Huygens & Van [Gelder		W. A. Huijgens	Amsterdam
	1409		Concordia *Brouwer*	W Bk	Dutch	627	136·1	34·1	20·5	Alblasserdam	1855	C. Smit		W. Smit	Purmerend
	1410	D	Concordia *Jager*	W Bg	Dutch	197	97·4	24·1	12·8	Dordrecht	1858	J. Schouten		A. J. Jager	Farmsum
	1411	D	Concordia *Huisman* (CONDEMNED)	W Sr	Dutch	179	95·8	23·7	11·9	Elsfleth	1852	G. Helmers		J. Hofman & Co.	Bierum
	1412	V	Concordia *Certin 83-*	I Bk	French	463	150·9	27·2	15·9	Sunderland	1869	Oswald & Co.		L. de Baecque	Dunkirk
	1413	V	Concordia *Clausen 83-83*	W Bk	German	652	158·8	33·5	19·6	Apenrade	1883	J. Paulsen & Sohn	Y 1.87	Aktien-Gesellsch. "Concordia" (C. Dollum)	Apenrade
	1414	DV	Concordia *Destmann 85*	W 3m Sr	German	475	130·2	27·6	14·9	Cappeln	1882	C. A. Steen	Y 5.86	H. v. Ehren	Blankenese
	1415		Concordia *Wilke 86-88*	W Bk	German	437	128·3	30·4	16·9	Danzig	1858	F. Devrient		A. Gibsone	Danzig
	1416	V	Concordia *Closter*	W Bn	German	104	86·0	19·7	9·3	Assens	1876	C. Sorensen		N. Closter	Apenrade
	1417	I	Concordia *Cille*	W Bk	Italian	1079	189·9	35·8	21·4	Chiavari	1877	M. Tappani	Y 1.89	G. Bertollo	Genoa
	1418	N	Concordia *Nielsen 78-79*	W Bk	Norweg'n	436	134·7	30·7	17·4	Porsgrund	1867	Kirkegaard	Y 2.86	J. C. & G. Knudsen	Porsgrund
	1419	N	Concordia *Jensen*	W Bk	Norweg'n	268	127·3	27·2	12·4	Drammen	1873			K. Erichsen	Drammen
	1420	N	Concordia *Petersen -85*	W Bg	Norweg'n	253	110·8	26·5	12·5	Arendal	1863	H. Stinusen		P. C. Pedersen	Grimstad
	1421	N	Concordia *Kolderup*	W Bk	Norweg'n	248	126·2	27·4	11·5	Drammen	1870	J. Jörgensen	Y 4.81	J. C. Colderup	Drammen
	1422	N	Concordia *Harristeff*	W Bg	Norweg'n	217	86·1	25·5	14·9	Arendal	1838			Rederiet Concordia (Wilh. Klein)	Fredrikshald
	1423		Concordia *Schlossmann 71-71*	W Bk	Russian	279	120·5	27·2	14·0	Windau	1862	H. Grube		Gutschmidt Bros.	Windau
	1424	V	Concordia *Andersson*	W Bk	Swedish	415	143·3	27·7	15·9	Nynäs	1867	J. G. Truelson	Y 1.87	J. P. Andersson	Höganäs
	1425		Concordia *Tamba -88*	W Bg	Swedish	172	86·6	22·2	12·8	Wisby	18$^{38}_{74}$			L. P. Cronberg	Nyhamn
	1426		Concorrente *Da Silva*	W Bn	Brazilian	171	..	..	..	S. João da Barra	1884	M. D. O. Cintra	Y 5.84	Cia. Cabotagem de S. João da Barra	S. João da Barra
	1427	DV	Concurrent *Leveling 79-86*	W Bn	Dutch	144	89·2	20·0	10·8	Veendam	1875	K. & J. Wilkens	Y 7.85	H. W. Homveld	Hoogezand
	1428	N	Concurrent *Sackner 86-88*	W Bk	Norweg'n	547	129·5	32·9	16·9	Rotterdam	1857	Gebrüder Visser		F. Breda	Christiania
	1429		Condesa de Vergara (ex Maria de Vergara 81-87)	W Bk	Peruvian	396	129·9	26·9	12·8	Brake	1873	T. Nicolai		J. Revoredo	Callao
	1430	✻	Condor *Roberts 88-88*	I Bk	British	1322	228·8	37·2	21·5	Southampton	1885	Oswald, Mor- [daunt & Co.		G. Petrie	London
	1431	V	Condor *M. Null -88*	W Bk	British	975	169·0	34·5	19·2	Prt Phillip, NS.	1876	P. Mockler	Y 8.86	P. Mockler	Pictou, N.S.
	1432		Condor *Nilsson*	W Bk	American	392	131·4	28·7	15·8	Sandefjord	1875	S. Olsen	Y 4.80		
	1433	✻	Condor	W Yawl Yt	American	102	93·9	20·1	11·8	Fairlie	1878	Fife & Son	C 1.88	J. M. Macdonough	New York
	1434	✻	Condor *Curtiss -88*	W Bk	French	425	129·0	27·5	17·5	Bordeaux	1870	Raymond fr. & [fils	Y 12.88	E. Labalette	Bordeaux
	1435	N	Condor *Sycertsen*	W Bk	Norweg'n	489	155·7	30·6	15·7	Westervik	1868	C. R. Fridell	Y 1.87	S. L. Movig	Fredriksstad
	1436	NV	Condor *Nielsen 85-86*	W Bk	Norweg'n	381	131·9	29·3	13·8	Drammen	1876	Jorgensen & [Knudsen	Y 1.87	Ths. S. Falck	Stavanger
	1437	N	Condor (ex Lapwing) *Pedersen*	W Bk	Norweg'n	334	140·8	28·0	14·3	Baltimore	1853		Y 10.84	O. A. Föyen	Tönsberg
	1438	N	Condor *Christiansen -87*	W Bk	Norweg'n	445	128·9	30·2	17·6	Gefle	1850			Th. Olrog	Fredriksstad
	1439		Condor *Svensson 86-88* (ex G. C. Luttens Meyer)	W Bk	Swedish	316	124·6	24·0	13·7	Hamburg	1863	J. P. Dircks		C. A. Banck & Co.	Helsingborg
	1440		Condoren *Nielsen 82-82* (ex Familiens Haab, ex Colibri)	W Bn	Danish	198	101·9	22·1	12·4	Apenrade	1866	A. Mailing		Akt Selsk. M. Michaelsen	Randers
	1441	V	Condoren *Törnström -87* (ex Mary Emma)	W S	Swedish	1174	182·4	32·0	23·4	Brunswick	1864	Pennell Bros.	Y 12.88	Rederibolaget Örnen (J. F. Stromberg)	Gothenburg
	1442	DV	Conductor *Lee 86-88*	W Bk	British	1063	176·0	37·5	22·8	Cornwallis, NS	1880	G. Bigeton	Y 3.88	C. R. Burgess	Windsor, N.S.
	1443		Conductor	W Sr	British	114	81·3	24·2	9·5	Lunenb'rg, NS.	1883	D. Smith		S. W. Oxner	Lunenburg, N.S.
	1444	RV	Conecuh *Southard -83*	W 3m Sr	American	781	175·0	35·6	17·9	Richmond, Me.	1885	T J Southard & Sn	Y 2.87	T. J. Southard & Son	Bath, Me.
	1445		Confederate	W Bg	British	171	109·0	24·0	12·2	P. E. Island	1873			Thomas Green	St. Johns, N.F.
	1446		Conference *Pratt*	I Bk	British	399	164·8	26·3	15·9	Warrington	1855	Tayleur & Co.		J. Dransfield	Geelong
	1447	D	Confiance *Grasdijk* (ex Petronella)	W Bg	Dutch	225	85·0	22·1	10·4	Alkmaar	1863			H. Grasdijk	Groningen
	1448	DV	Confiance *De Groot -88* (WRECKED)	W Sr	Dutch	125	..	..	..	Hoogezand	1860	J. A. Hooites		J. J. Koerts & Zoon	Oude-Pekela
	1449	FV	Confiance *Lazzarel 86-88*	W 3m Sr	French	312	117·4	26·3	16·2	Bordeaux	1876	Tranchand		J. H. Tandonnet Frères	Bordeaux
	1450	V	Confiance *Merlen*	W Sr	French	128	90·5	22·6	11·1	Dunkirk	1885	P. Meuwisse		V. Régnez	Dunkirk

✻ = Lloyd's Register; D = Nederlandsche Vereeniging; G = Germanischer Lloyd; I = Registro Italiano; R = American Record; A = Veritas Austro-Ungarico; H = Veritas Ellenico; N = Norske Veritas; V = Bureau Veritas.

Dans la 4me Colonne, C = Composite (membrure en fer, bordé en bois); I = Fer; Stl = Acier; W = Bois.

Les Dates du doublage en métal (Col. 11) ressortent des livres dans lesquels les navires sont respectivement classés ou des rapports des Inspecteurs du Lloyd's Register.

Pour plus amples renseignements sur les navires classés par le Lloyd's Register, voir le Lloyd's Register Book et le Lloyd's Yacht Register.

Official No.	No. in Book.	Registry in which classed.	Ship's Name. Material and Rig. Master, and Dates of Appointment to present Owner's Service and to Ship. Ship's Late Name.	Flag.	Net Tonnage.	Length.	Breadth.	Depth.	Built. Where.	When.	By whom.	Date of Metalling.	Owners or Managers.
42588	1451	*	Confidence *Jones -81* W Sr	British	75	68·0	19·2	9·5	Aberystwith	1862	Jones		J.Jones
	1452	V	Confidence *Mourritzen 76-76* W Bn	Danish	141	96·9	21·2	10·7	Thuro	1876	J.Ph.Jorgensen		J.Ph.Jorgensens
	1453		Confidentia *Schultz* W Sr	Swedish	115	78·7	24·9	10·1	Visby	1842			C.Molander
1860	1454	IV	Confidenza *Repetto* W Bk	Italian	542	143·0	29·1	19·8	Varazze	1869	D.Craviotto	[illegible]	L.Schiaffino
375	1455	A	Confidenza *Parodi 83-83* W Bk	Italian	534	136·1	29·3	19·7	Sestri	1872	E.Celesia	[illegible]	M.A.Dall'Orso fu G.
91548	1456	*	Confido *Davis 86-86* W K	British	90	80·0	21·1	10·3	G'mptn	1885	W.A.Gibbs		A.R.Roberts
45738	1457		Conflict *Stevens* W Sw	British	227	99·5	24·5	14·8	Whitby	1864	Barrick		W.Pannett
206	1458		Confucius W Sr	Siamese	258	102·4	26·7	13·3	Bangkok				Poh Pan
	1459	N	Congal *Danielsen 84-84* W Bg	Norweg'n	332	112·3	29·2	15·6	Grimstad	1869	O.Kittelsen		C.Stiansen
	1460	*D	Congo *Hauenstein 85-85* W Bn (ex Alexander)	Dutch	187	100·0	24·9	12·0	Stavanger	1873	Knudsen	[illegible]	Nieuwe Afrikaansche Handels Vennootschap
	1461	N	Congo *Simonsen 87-87* W Bk	Norweg'n	442	146·2	31·2	14·2	Grimstad	1886		[illegible]	J.Jörgensen
	1462	NV	Congo *Ellefsen* W 3mSr	Norweg'n	293	125·3	28·9	10·9	Tvedestrand	1886	J.Gundersen	[illegible]	A. & F.Smith
	1463	N	Congo (ex Fido) *Møller 87-* W Bg	Norweg'n	288	105·6	25·7	16·0	Sunderland	1866	J.Crown	[illegible]	L.G.S.Larsen
5423	1464		Congress *Willard* W Sr	American	157	98·8	26·7	9·1	Bath,Me.	1868	Palmer&Hawley		Palmer & Hawley
67787	1465	*	Connaught Ranger I S *Jackson*	British	1153	229·7	34·6	21·3	Belfast	1875	Harland& Wolff		SwanseaShippingCo,L
50402	1466		Conoid *Hubbard 86-86* W Sr	British	168	100·0	23·7	11·8	Peterhead	1865	Carnegie		Whitstable ShippingCo. Ltd.
125276	1467	V	Conqueror *Lothrop -87* W S	American	1621	215·3	41·3	24·8	East Boston	1874	Smith&T'wnsnd	[illegible]	B.F.Wells
	1468	*	Conqueror *White* W Sr	French	60	75·2	19·0	9·0	Jersey	1876	Picot		A.Postel & Co.
94048	1469	*	Conquest W Sk	British	85	76·7	20·7	10·5	Elmshorn	1887	J.Kremer		T.Baskcomb
4238	1470	RV	Conquest *Dixon* W Bk	American	490	133·0	29·4	18·2	East Boston	1862	P.Curtis	[illegible]	C.N.Dixon
922	1471	IV	Conquistatore *Orsi* W Bk	Italian	599	136·4	29·5	20·5	Cassano	1872	F.deRosa	[illegible]	A.Lauro
	1472		Conrad W Bg	Russian	352	..	..	..		1876			Nystedt & Wiberg
50922	1473		Conrade *Hawksfield 80-75* W Bn	British	198	101·9	26·3	12·6	P.E.Island	1864	Green		J.H.Hawksfield
	1474		Conselheiro *Valle -87* W 3mSr	Brazilian	320	..	..	..	S.Joao da Barra	1886			Cia.CabotagemS.Joao da Barra
88129	1475	*	Conservative *Challis* W Sk	British	83	76·3	20·1	10·5	Hull	1883	W.&J.McCann		W.Weife
2073	1476	I	Consiglio Galatola W Bn *Iaccino -87*	Italian	531	110·7	31·5	20·1	Procida	1876	G.Bonifacio	[illegible]	A.Galatola
88673	1477	DV	Constance *Tingley* W S	British	1592	221·3	41·1	24·3	Harvey,N.B.	1884	G.S.Turner	[illegible]	G.F.Smith
29237	1478	*DR	Constance *Trimmer* W Bk	British	234	121·0	23·2	13·9	Greenock	1860	Steele	[illegible]	S.W.Herbert
56383	1479	*	Constance *Langlais 85-* W Sr	British	157	94·6	23·8	12·5	Ipswich	1865	Lambert	[illegible]	E.Rillston
9937	1480		Constance *Hanlsen* W Sr Yt	British	146	103·2	21·6	11·7	East Cowes	1850	J.S.White		P.C.Lovett
44285	1481	*	Constance *Roberts 79-79* W Sr	British	142	90·1	22·0	12·2	Portmadoc	1862	Roberts		H.Roberts & Co.
19346	1482		Constance W Sr	British	129	85·0	21·8	9·0	Bombay	1838	Govt.Dockyard		Abdool Kader Adamjee
58290	1483	*	Constance *McKellar 81-81* W Sr	British	93	80·0	21·6	10·3	Banff	1868	Geddie		J.W.Simpson
	1484		Constance *Bonde 84-85* W Bg	Danish	216	110·2	26·8	14·9	Flensburg	1856	H.J.Nielsen		Den Kongelige Grønl. Handel (R.Hørring)
	1485	V	Constance *Pettersen -88* W Bk	Norweg'n	632	118·3	32·6	19·5	Chiavari	1870	A.Bonifacio	[illegible]	M.Brunsgaard
	1486	N	Constance *Meland 86-86* W Sr	Norweg'n	114	90·6	22·3	10·5	Bergen	1882	S.Sandvig		E.Lindoen
	1487		Constance *Pihl -86* W Bk	Swedish	296	125·0	24·9	13·6	NovaScotia	1859			A.Lysell
	1488		Constance *Wallin 84-* W Bn	Swedish	163	95·5	22·8	10·5	Luleå	1868	E.A.Brandlof		J.C.Hulthen
63523	1489		Constance Ellen W Bn [illegible]	British	171	103·2	24·0	13·0	Porthcawl	1871	Martin		P.Smith
30545	1490		Constance Emma W S	British	509	122·0	29·0	..					Hajee Jacoob
70175	1491	*	Constance Mary W 3mSr *Rowan*	British	167	104·0	22·4	12·0	GlassonDock	1875	Simpson		Porter'sShippingCo.Ld
	1492		Constancia *Echevarria* W Bk	Spanish	453	142·0	28·1	16·1	Bilbao	1869	S.de Arana		Augustin Gonzalez
58080	1493		Constant W Bn	British	191	98·0	24·1	12·9	P.E.Island	1865	T.Simpson		W.Pascoe
	1494	NV	Constant *Øverland 82-86* W Bk	Norweg'n	279	133·3	27·6	11·0	Stavanger	1886		[illegible]	K.S.Bertelsen
	1495	R	Constante W Bn	Mexican	152	..	..	..	Campeche	1866		[illegible]	A.Y.Ramirez
	1496	DV	Constantia *Schall 80-83* W Bk (LOST) (ex Cornelis Smit)	Dutch	655	137·8	35·4	22·0	Kinderdyk	1865	L.Smit&Son	[illegible]	A.Köpcke
	1497	N	Constantia *Jacobsen 82-82* W Bk	Norweg'n	447	136·9	30·0	16·7	Risör	1872	K.Gunnersen	[illegible]	J.W.Probensen
	1498	DN	Constantia *Fredriksen -87* W Bg	Norweg'n	228	85·6	25·5	15·7	Christiansand	1827			N.S.Andersen
115	1499	G	Constantin *Maydmann 85-* W Bkn	Russian	368	123·0	30·2	15·1	Haynasch	1885	Otto Weide		Otto Weide Wwe
	1500		Constantin *Öhman 64-75* W Sr	Swedish	143	82·7	23·2	10·5	Pernau	1871			S.P.Forssberg

*=Lloyd's Register; D=Nederlandsche Vereeniging; G=Germanischer Lloyd; I=Registro Italiano; R=Ame[rican]
A=Veritas Austro-Ungarico; H=Veritas Ellenico; N=Norske Veritas; V=Bure[au]

In the 4th Column, C=Composite (iron frame, wood planked); I=Iron; Stl=Steel; W=Wood.

The Dates of Metalling (Col. 11) are based on the records of the Books in which Vessels are respectively classed, or on the Reports of the Surveyors of Ll[oyd's]

Further particulars of Vessels classed by Lloyd's Register will be found in Lloyd's Register Book and Lloyd's Yacht Register.

No.	Cote	Nom. Capitaine et date d'entrée au service des Armateurs actuels et à bord du navire. Ancien Nom.	Matériaux et Gréement	Pavillon	Tonnage net	Longueur	Largeur	Creux	Construit: Où	Quand	Par qui	Date du doublage en métal	Armateurs ou Directeurs-Gérants	Port d'Armement
1501		Constantin *Andersson*	W 3mSr	Swedish	117	93·8	21·7	8·3	Carlstad				O. Hallengren	Venersborg
1502	V	Constantin v. Reinecke	W Bk	German	323	112·0	27·6	16·2	Rostock	1858	W. Zeltz		W. Maack	Rostock
1503		Constantinos	W Bg	Greek	247	96·5	25·9	..	Castellorizon	1878			E. Aizanis	Syra
1504		Constantinos	W Bg	Turkish	144	..	..	..	Russia	1882			Yalles Pandeli Nikolopulo & Co.	Constantinople
1505		Constantinos Hydras Ermoupolis	W Bg	Greek	245	98·8	25·9	..	Syra	1878			A. Hatzijannis	Syra
1506	V	Constanza Barnuevo	W Bn	Spanish	334	132·5	28·8	15·1	Palma	1877			G. Fuster y Fuster	Palma
1507	V	Constanze *Clausen*	I S	German	978	204·0	33·1	20·9	Flensburg	1877	Flensbg S.B.Co.		Gebr. Hustede	Elsfleth
1508		Constitution	W Sr	American	284	110·0	29·0	8·0	Glenwood, NY	1862			G. W. Taylor	New York
1509		Constitution *Colly*	W Bkn	American	277	157·5	25·3	11·9	Philadelphia	1856 / 1877			N. Bichard	San Francisco
1510	✱	Consuelo *Bannear*	W Sr	British	85	75·7	20·9	10·0	Devoran	1878	Stephens	V	W. H. Mare	Plymouth
1511	V	Consuelo *Robertson 86-88*	W Bn	American	279	133·0	31·0	11·9	San Francisco	1880	M. Turner	V	Oceanic S.S. Co.	San Francisco
1512	V	Consuelo *Roig*	W Bk	Spanish	443	128·9	31·8	16·8	Arens, Spain	1863		V	J. Isern y Garcia	Barcelona
1513	A V	Conte Arturo L.	W Bk	Austrian	518	135·1	29·4	19·2	Capod'Istria	1872	L. Polli	V	A. T. Lucovich	Trieste
1514		Conte Romas	W Bn	Greek	483	139·7	30·5	..	Genoa	1874			Maria N. Hadji Nicolau	Thera
1515	A	Conte Géza Szápáry	W Bk	Austrian	593	118·6	29·3	19·7	Fiume	1874	G. Brazzoduro	V	C. cav. Cosulich	Fiume
1516	✱A	Conte Iginio L.	W Bk	Austrian	463	137·1	27·9	17·6	Sunderland	1864	W. Naizby	V	A. T. Lucovich	Trieste
1517	A V	Conte Oscar L.	W Bk	Austrian	925	165·0	32·8	22·1	Spotorno	1871	C. Cochella	V	A. T. Lucovich	Trieste
1518		Contender *Trindell 87-87*	W Sr	British	100	83·4	18·9	10·9	Rye	1856	Hoad Bros.		Miss C. L. Allen	Lowestoft
1519	✱	Contest *Goffin 88-88*	C Bk	British	486	138·0	28·4	17·7	Sunderland	1867	J. Blumer & Co.	V	J. L. Walsh	Sunderland
1520		Contest *Dawson*	W Bg	British	254	113·4	25·6	14·9	Guernsey	1866	Thom		T. Dawson	Guernsey
1521		Continent	W S	American	1221	167·2	36·5	21·2	Bath, Me.	1850			H. Benner	New York
1522	R	Continental *Gilkey*	W S	American	1668	220·0	42·2	25·1	Bath, Me.	1875	E. & A. Sewall	V	De Groot & Peck	New York
1523		Convivial *Collins*	W Bn	British	94	69·9	20·4	12·4	Yarmouth	1824			J. Davies	Whitehaven
1524	R	Cook Borden *Lunt*	W 3mSr	American	170	109·0	29·6	8·9	Calais, Me.	1874	W. Robinson		C. Borden	Fall River, Mass.
1525	✱	Cooleen *Hall 87-87*	I Bk	British	629	180·5	30·2	17·5	Glasgow	1876	J. Elder & Co.		J. M. Barkley	Belfast
1526	G	Coorong *Mills 85-85*	W Bk	British	344	130·0	26·3	13·1	Rhode Is., U.S.	1864	P. Chase	V	W. Hamilton	Adelaide, S.A.
1527		Copenhagen *Miller*	W S	British	876	163·5	31·1	21·5	India	1855	T. H. Gladstone		S. N. Castle	London
1528	✱V	Copernicus *Lassen 82-82*	I S	German	1212	221·0	34·1	23·3	Hamburg	1878	Reiherstieg S. Works		R. M. Sloman & Co.	Hamburg
1529	✱	Copley *Trumble 86-86*	I S	British	1741	263·0	39·1	24·1	Liverpool	1881	W. H. Potter & Sn		North Western Shipping Co. Ltd.	Liverpool
1530	✱D	Copsefield *Chichester 86-86*	I Bk	British	451	151·1	28·2	16·2	Renfrew	1868	Henderson Coulborn & Co.		J. Graham	London
1531		Coq *Chatelard*	W Bk	French	281	116·8	29·1	12·4	Bordeaux	1858	Chaigneau frères	V	Mme. F. Chatelard	Bordeaux
1532		Coquette	W Sr	British	214	124·0	24·6	10·6	Eaglet'n, NSW	1883	J. Roderick		John Henderson	Newcastle, N.S.W.
1533		Coquette	W Bn	British	157	90·3	25·0	10·5	Nassau, N.P.	1848			J. Wright, jun.	Turk's Island
1534		Coquette *McMaster 70-70*	W Sr	British	131	85·0	23·6	11·6	Nw Brunswick	1851			J. M. Barkley	Belfast
1535		Coquette	W	French	170	..	..	..						Arles
1536	V	Coquette *Neukelmaeca*	W Sr	French	111	93·2	21·3	11·0	Dunkirk	1882	Hecquet Van Rapenbusche		Hecquet V. Rapenbusche	Dunkirk
1537	N	Coquimbo *Andersen*	W Bk	Norweg'n	425	137·0	29·5	17·3	Honfleur	1854	Viel	V 12.81	O. A. Johansen	Frederiksstad
1538	R V	Cora *Roy*	W S	American	1491	200·2	39·0	23·8	Belfast, Me.	1869		V	P. R. Hazeltine & Co.	Belfast, Me.
1539		Cora *Devig 87*	W 3mSr	American	336	134·2	31·3	11·0	Phipsb'gh Me	1873	C. V. Minott		Doane & Crowell	Boston
1540		Cora *Gaskill*	W Sr	American	166	86·0	24·8	12·3	Washingtn NC	1881	J. A. Farrow		J. L. Fowle	New Berne, N.C.
1541	N	Cora *Ofiedahl*	W 3mSr	Norweg'n	356	131·7	29·1	14·2	Stavanger	1876		V	J. Berner, Jr.	Stavanger
1542	R	Cora Dunn *Harrington*	W 3mSr	American	527	169·9	35·5	12·7	Thomaston, Me.	1888	Simmons		Dunn & Elliot	Thomaston
1543		Cora Etta *Fales*	W 3mSr	American	231	114·8	28·5	11·0	S Thomast'n Me	1867			J. A. Fales	Rockland, Me.
1544	R	Cora Green *Philbrook*	W Bn	American	236	117·0	28·0	10·0	Bangor, Me.	1875		V 10.85	E. A. Green	Bangor, Me.
1545		Cora S. Mackay *Matheson*	W 3mSr	American	189	..	..	..	Essex, Mass	1888			A. McKay	Provincetown
1546		Cora S. Van Gilder *Lee*	W 3mSr	American	176	117·0	23·0	8·8	Dennisville NJ	1876			T. Van Gilder	Bridgeton, N.J.
1547		Coral *Owen*	W Bk	American	362	109·1	27·9	18·2	N'wburypt Mass	1819			H. C. Wright	San Francisco
1548		Coralline *Avery 81-86*	W Bn	British	174	89·2	23·9	13·8	Hartlepool	1857	Denton & Co.		I. Mills	Whitby
1549	✱	Coranna *Jessen*	I Bk	Danish	820	189·2	31·0	18·6	Dumbarton	1876	Birrell, Stenhouse & Co.		Et Actieselskab (P. C. Saarest)	Nordby, Fanø
1550	✱	Corby *Stewart 86-86*	I S	British	1399	235·6	38·0	22·5	Liverpool	1876	Bowdler, Chaffer & Co.		R. Nicholson & Sons	Liverpool

= Lloyd's Register; D = Nederlandsche Vereeniging; G = Germanischer Lloyd; I = Registro Italiano; R = American Record;
= Veritas Austro-Ungarico; H = Veritas Ellenico; N = Norske Veritas; V = Bureau Veritas.

Dans la 4me Colonne, C = Composite (membrure en fer, bordé en bois); I = Fer; Stl = Acier; W = Bois.

Les Dates du doublage en métal (Col. 11) ressortent des livres dans lesquels les navires sont respectivement classés ou des rapports des Inspecteurs du Lloyd's Register.

Pour plus amples renseignements sur les navires classés par le Lloyd's Register, voir le Lloyd's Register Book et le Lloyd's Yacht Register.

Official No.	No. in Book	Registry in which classed	Ship's Name. Material and Rig. Master, and Dates of Appointment to present Owner's Service and to Ship. Ship's Late Name.	Flag.	Net Tonnage.	Dimensions. Length.	Dimensions. Breadth.	Dimensions. Depth.	Built. Where.	Built. When	Built. By whom.	Date of Metalling	Owners or Managers
20008	1551		**Corcyra** (BROKEN UP) *Sharp -78* W Bg	British	185	88·4	21·5	11·5	Sunderland	1850			G. Brown
60010	1552	✱	**Cordelia** *Owen 85-85* I Bk	British	598	172·4	28·2	18·5	PortGlasgow	1867	J. Reid&Co.		C.T. Bowring & C
62358	1553		**Cordelia** (*ex* Wakendaw) W Sr	British	142	98·9	20·0	11·0					L. Hart
16645	1554		**Cordelia** (BROKEN UP) W Bg	British	131	82·5	22·2	13·4	Bristol	1844			R.J. Kennedy
47754	1555	✱	**Cordelia** *Price 77-77* W Sr	British	99	79·4	21·8	10·9	Portmadoc	1863	Williams		S. Richards & Co.
4962	1556		**Cordelia Newkirk** *Huntley -83* W Sr	American	191	109·3	28·9	8·5	NacatCr'k N.J	1860			G.W. Shepherd, jun
	1557		**Cordialidade** W Bg	Portug'se	223	92·8	24·9	12·4	Caminha	1855		V 10.81	Rodrigues&Ribeir
54095	1558	✱	**Cordillera** *Higgins 85-88* I Bk	British	852	186·4	30·3	20·5	Liverpool	1866	T. Vernon&Son		W.J. Myers, Son & Co.
45377	1559	✱	**Cordova** (LOST) *Jones 87-87* I Bk	British	521	152·1	27·2	17·7	Stockton	1862	Richardson, Duck [&Co.		Parry Jones & Co.
21333	1560		**Corea** *Ahier* W Bk	British	581	167·2	29·0	19·3	Workington	1859	J. Fell		J.E. Sassoon
5448	1561		**Corea** *Gould* W Bk	American	565	133·4	31·5	18·4	East Boston	1868	C. Sampson&Co.		Arctic Fishing Co.
94	1562		**Corebo II.** *Schiaffino* W Bk	Italian	217	108·6	26·3	15·1	Varazze	1843	Celesia & Die- [barbieri	V 5.88	Eredi F. Schiaffino
	1563		**Corgaz 3d.** *Baralo* (*ex* Visconde de Livramento, *ex* Rousingo, *ex* Henrietta) W Bk	Brazilian	227	106·9	25·5	12·5	Zoutkamp	1868			A.F. Corga
	1564		**Corina** W Bn	Brazilian	282	..	..	..					O.A. da Silva
76990	1565	V	**Coringa** *Bodge 88-88* W S	British	1343	193·0	38·7	23·5	Harvey, N.B.	1879	G.T. Turner	V 4.88	B. Smith & Sons
	1566	NV	**Corinna** *Oftedahl -87* W Bk	Norweg'n	475	143·3	30·7	16·5	Stavanger	1877	Chr. Knudsen	V 4.87	Holdt & Isachsen
61165	1567		**Corinne** (LOST) *Hassel* W Bn	British	124	91·8	23·3	12·6	Montreal	1873		V 11.85	J. Byrene
63602	1568	✱	**Corinth** *Littler 77-77* C Bk	British	614	165·0	31·1	18·3	Sunderland	1870	W. Pile&Co.	V 10.87	T.B. Walker
	1569	N	**Coriolan** *Dahl 86-86* W Bk	Norweg'n	469	131·7	29·4	17·7	Lübeck	1849			Nils P. Hoyer
	1570		**Coriolano** *RibeirodosSantos* (*ex* Rosina) W Sr	Portug'se	174	..	..	..	Villa doConde	1869	J.M. Araujo		A.L. Duarte
73665	1571	✱D	**Coriolanus** *Torkilsen 87-87* I S	British	1046	217·4	35·2	20·1	Dumbarton	1876	A. M'Millan & S		J. Stewart & Co.
71163	1572		**Corisande** (*ex* Star) W Sr	British	276	137·0	26·0	11·1	MarineCity US	1873	P. Rice		Huron & Ont. Trans.Co
67912	1573	✱	**Corisande** *McGairn 87-87* W Sr	British	187	104·7	23·9	12·9	Dartmouth	1873	Philip	V 3.87	J. Grieve, jun.
20972	1574	✱	**Cormorant** *Hume 83-83* I Bk	British	1073	211·4	33·6	21·6	Liverpool	1865	T. Vernon & Son		W.J. Myers, Son & Co
	1575	N	**Cormorant** *Wroldsen -83* W Bg	Norweg'n	238	111·9	28·3	12·0	Tvedestrand	1872	M. Eldrup		S. Guttormsen
5188	1576		**Cornelia** *Lyon* W Sr	American	148	85·0	25·7	10·0	Chatham, Conn	1841			M. Lyon
	1577	DV	**Cornelia** *De Boer -88* I Bk	Dutch	654	149·2	34·4	19·5	Kinderdyk	1863	FopSmit		A. Köpeke
	1578	V	**Cornelia** *Voordewind -87* W Sr	Dutch	163	101·3	18·4	7·8	Nieuwendam	1857	W.&A.H. Meur- [sing		K.H. Voordewind
	1579	D	**Cornelia** *Teerling* (TO BE BROKEN UP) W Sr	Dutch	138	87·9	21·5	9·9	Pekela	1850	B. Kuiper		F.J. Teerling
	1580	G	**Cornelia** *Schon 83-* (*ex* Cornelis Bakker) W Sr	German	153	96·4	21·7	11·0	Sappemeer	1867	E. Bakker		C. Cassens
	1581	D	**Cornelia** *Heerma* W Sr	German	106	76·0	18·0	8·0	Jemgum	1868	G.H. Aden		H. Heerma
	1582	D	**Cornelia Elisabeth** (*ex* Anne Royden) *de Greeve 80-81* W Bk	Dutch	1119	194·7	35·6	22·7	Liverpool	1856	T. Royden	V 11.87	P. Landberg&Zoon
126324	1583	R	**Cornelia Soule** (STRANDED) *Francis* W 3mSr	American	291	123·6	31·0	10·0	Madison	1885	Miner&Son		C. Barry
3480	1584	IV	**Cornelio Zino** *Patrone* (*ex* Calpe) [87 W Bk	Italian	975	205·3	30·7	25·5	Gefle	1873		V 6.86	F. Zino
	1585	D	**Cornelis** *Dalmeijer -87* (*ex* Clara, *ex* Springbok) W Bk	Dutch	293	107·9	27·2	16·1	Sunderland	1862	Pickersgill	V 9.86	J.J. van den Berg
4695	1586		**Cornelius** *Horner* W Sr	American	147	94·3	26·8	8·4	Tom's Riv'r NJ	$18\frac{61}{71}$	J.A. Emmins		W. Booth
	1587	V	**Cornelius** *Windhorst 83-83* (*ex* Saint Fillans) (CONDEMNED) W Bk	German	1105	171·6	35·5	26·7	Cardiff	1863	Batchelor	V 9.86	Gildemeister & Ries
4830	1588		**Cornelius Grinnell** W S	American	980	172·5	37·2	22·8	Boston, Mass.	1850			C.A. Williams
	1589		**Cornubia** *Thorsell* W Sr	Swedish	115	80·4	21·2	10·1	Carlstad				O. Hallongren
52502	1590		**Cornucopia** *Pitt* W Sr	British	207	104·0	24·5	13·3	Findhorn	1865	Anderson		R.L. Wheeler
27869	1591		**Cornucopia** *Robson -88* W Bg	British	155	98·3	21·0	12·0	Shippegan, NB	1859	Henry		J.&E. Collas
63520	1592	✱	**Cornuvia** *Baker 84-69* I Bk	British	799	191·3	32·1	20·0	SouthShields	1869	Readhead, Soft- [ley&Co.		Tomlinson, Hodgetts & Co.
70989	1593	✱	**Corolla** *Messer 73-75* I S	British	1264	228·0	36·7	21·8	Liverpool	1875	W.H. Potter&Co.		J. Bell & Son
71745	1594	✱	**Coromandel** *Curd -87* I Bk	British	849	204·6	32·2	19·9	Glasgow	1875	Lon.&Glas.Co.		R. Gilchrist
52575	1595	✱	**Corona** *Milne* C Bk	British	1193	209·6	35·0	22·0	Dundee	1866	A. Stephen&Sns	V 1.86	A. Stephen & Sons
126168	1596		**Corona** *McAllep -83* W 3mSr	American	372	142·1	34·0	16·3	PortBlakely, [Wash	1883	HallBros.		W.G. Hall
	1597	NV	**Coronæ** *Larsen* W Bk	Norweg'n	456	141·2	31·1	16·9	Stavanger	1881	M. Magnus	V 3.86	J.P. Brodahl & Co
55036	1598	✱	**Coronel** *Williams 86-88* W Bk	British	387	127·5	27·4	17·3	Sunderland	1866	Hardie	V 1.86	Thomas, Watkins & Jenkins
42843	1599		**Coronella** *Craine* W Bn	British	108	91·8	24·5	10·4	Walton, N.S.	1862	Miller		W. Wallace
92622	1600		**Coronet** *Walters -87* W Sr	British	115	83·0	24·2	9·7	Lunenburg NS	1887	L. Young		W.N. Zwicker

✱ = **Lloyd's Register**; D = Nederlandsche Vereeniging; G = Germanischer Lloyd; I = Registro Italiano; R = Ame[rican]
A = Veritas Austro-Ungarico; H = Veritas Ellenico; N = Norske Veritas; V = Bur[eau]

In the 4th Column, C = Composite (iron frame, wood planked); I = Iron; Stl = Steel; W = Wood.

The Dates of Metalling (Col. 11) are based on the records of the Books in which Vessels are respectively classed, or on the Reports of the Surveyors of ...

Further particulars of Vessels classed by Lloyd's Register will be found in Lloyd's Register Book and Lloyd's Yacht Register.

Numéro	Registre	Nom. Capitaine et dates d'entrée au service des Armateurs actuels et à bord du navire. Ancien Nom.	Matériaux et Grèement	Pavillon	Tonnage net	Longueur	Largeur	Creux	Construit Où	Quand	Par qui	Date du doublage	Armateurs ou Directeurs-Gérants.	Port d'Armement.
1601	*	Coronet *Challis -87*	WK	British	96	79·3	21·1	10·2	Galmpton	1887	W.A.[illegible]		W.T.Roberts	Grimsby
1602	*	Coronilla *Sand 81-87*	I Bk	British	524	155·9	26·7	17·6	Liverpool	186[illegible]	[illegible]		J.Gambles	Liverpool
1603	*	Coronilla	WK	British	6[illegible]	75·9	19·6	8·5	Plymouth	188[illegible]	[illegible]		J.Barraclough	Plymouth
1604	N	Correo *Sebastian 81-87*	W3mSr	Norweg'n	29[illegible]	126·[illegible]	30·0	12·2	Christiansand	1884	[illegible]	Y.3.88	S.Otto & Co.	Christiansand
1605	*	Corryvrechan *[illegible]*	Stl Bk	British	1299	240·0	36·2	21·5	Port Glasgow	1885	R.Duncan&Co.		H.Hogarth	Ardrossan
1606	*	Corsair *Messervey*	W Bk	British	34[illegible]	128·0	26·5	16·0	Plymouth	1861	[illegible]	4.88	DePutron & Rooth	London
1607	R	Corsica *Farrington*	WS	American	1270	191·0	38·1	23·7	Thomastn,Me.	1865	Stetson&Gerry	Y.3.88	Snow & Burgess	New York
1608	R	Corsican *Musterd -88*	W Bn	British	167	96·0	26·8	10·9	Cardigan, P.E.I.	1888	J.E.Macdonald		J.Musterd	Charlottetown, P.E.I.
1609	N	Cort Adelaer *Hansen 72-74*	W Bk	Norweg'n	264	121·8	26·9	12·5	Brevig	1868	P.[illegible]	Y.2.88	H.E.Münster	Brevig
1610	N	Cort Adelaer *Knudsen -86*	W3mSr	Norweg'n	135	95·1	24·5	10·5	Laurvig	1861	[illegible]		M.Johannesen	Fredriksværn
1611		Cortés *(ex Palestina)*	W Pol	Spanish	15[illegible]	86·9	26·6	12·0	Blanes	1855			B.C.Vallariola	Majorca
1612	A	Cortese *Grimaldo -88*	W Bk	Italian	34[illegible]	116·1	29·0	17·6	Sestri	1865	G.B.Cadenaccio	Y.4.88	V.Greco fu A.	Trapani
1613	I	Cortesia *Brignet* (ex Lauren) (ABANDONED)	W Bk	Italian	72[illegible]	156·1	31·6	21·4	Pra	1872	L.[illegible]	Y 12.85	B.Chiesa	Genoa
1614	*	Cortez *Sharp 78-86*	I S	British	2233	281·0	40·0	24·2	Southampton	1886	Oswald Mordaunt & Co.		G.Petrie	London
1615	N	Cortez *Hansen*	W Bk	Norweg'n	33[illegible]	130·9	28·2	12·2	Grimstad	1877	S.Johnsen	Y.1.87	T.G.Gundersen	Grimstad
1616		Corunez *Mendiola 87-87* (ex [illegible] Cesar)	W Bg	Peruvian	184	101·7	26·6	11·7	Spain				J.E.Benites	Callao
1617	*	Corwena *Ueens*	W Bg	British	192	102·5	21·2	13·2	Portmadoc	1869	Pierce	Y.4.88	E.Roberts & Co.	Carnarvon
1618	RV	Coryphene *Grosse*	W Bk	American	771	160·2	34·2	20·4	Millbridge, Me.	1878	J.Sawyer	Y.2.88	G.Hume	San Francisco
1619		Cosma & Damiano (ex Talon) *Pessindji*	W Bk	Russian	164	82·3	23·0	11·5	Paros	1862			V.Demirdji	Eupatoria
1620		Cosme *(ex Pepa)*	W Bn	Spanish	145	91·2	23·9	11·8	Mataro	1858			Pedro Gibernau	Barcelona
1621	V	Cosmo *Blackmore -88*	WS	British	1220	200·0	37·3	23·5	Quebec	1877	H.Dinning	Y.3.88	J.G.Ross & Co.	Quebec
1622		Cosmo G. *Gamulin -88* (ex Achille)	W Sr	Austrian	11[illegible]	76·1	20·1	11·2	Segna	1860	[illegible]	Y.8.88	M.Gamulin	Curzola
1623	*	Cosmopolita *Griffiths -88*	I Bk	British	495	151·6	28·2	16·8	Sunderland	1869	W.Doxford&Ss.	Y.1.88	M.Tutton	Swansea
1624	V	Costa Lobo *dos Santos*	W3mSr	Portug'se	331	130·5	25·3	11·7	Villa do Conde	1884	J.F.da Laje	Y.1.88	A.L.Duarte	Oporto
1625		Costa Rica (BROKEN UP)	W Bk	French	401	130·5	27·9	17·8	Bordeaux	1863	P.Chaironje.		L.LeQuellec	Bordeaux
1626		Costa Rica Packet *Donaldson*	W Bk	British	531	162·0	28·6	17·7	Guernsey	1861	Sebire	Y.A.88	Burns, Philp, & Co.	Sydney
1627		Costanti	W Bg	Turkish	300	..	..	..	Heraclea	1882			Keresteh Tijari Avram Christophor	Heraclea
1628		Costanti *Jassis Velidsh Costanti*	W Bg	Turkish	300	..	..	..	Heraclea	1885			Keresteh Tijari Avram Christophor	Heraclea
1629		Costantia	W Bg	Turkish	300	..	..	..	Antaria	1874			Maiz Adahlee Marten oglou Kirsco	MaizAdah
1630		Costantino	W Bg	Portug'se	384	115·1	26·0	18·0	Savona	1866			A.F.Fernandes	
1631		Costantino	W Bg	Turkish	350	..	..	..	Maiz Adah	1871			Maiz AdahleeCostanti	MaizAdah
1632		Costantinos	W Bg	Turkish	184	..	..	..	Greece	1872			Dimitri Sarandi	Constantinople
1633		Costantinos	W Bg	Turkish	184	..	..	..	Greece	1874			Alatjee Dimitri Vassil	Constantinople
1634		Costanza *Capurro -86* (WRECKED) (ex Cesar Etienne)	W Bk	Italian	313	111·6	28·6	10·5	Savona	1872	G.B.Tessitore	Y 10.86	N.Mortola	Genoa
1635	I	Cottardin *Bojetto -88*	W Bk	Italian	811	161·6	29·6	22·1	Varazze	1872	G.Seda	Y.1.88	P.Lavarello	Genoa
1636	*	Cottica *McWicker -88*	W Bk	British	319	133·0	26·6	13·0	Perth	1869	Wallace	Y.2.88	H.McWhirter	Glasgow
1637	NV	Coulant *Melsom*	W Sr	Norweg'n	169	100·3	24·7	11·4	Moss	1880	Vogt&Holst	Y.1.88	H.L.Baltzersen	Fredrikstad
1638	*D	Coulnakyle *Carmichael -88*	W Bk	British	589	168·5	30·4	18·0	Aberdeen	1862	Hall	Y.5.88	Elliott	Sydney
1639	*	Countess of Derby *Hunter 77-77*	I Bk	British	750	186·8	34·0	18·6	Dumbarton	1877	Birrell,Stenhouse&Co.		J.Allison & Son	Dundee
1640	*	Countess of Devon *Adams -88*	W Bkn	British	233	117·5	23·7	12·7	Plymouth	1873	Shilston	Y.1.87	W.H.Shilston	Plymouth
1641		Countess of Dufferin *Pobb 84-88*	W Bk	British	540	143·0	33·5	17·1	Quebec	1872	Rosa		B.McCorkell	Londonderry
1642		Countess of Errol *Taylor*	W Sr	British	218	113·8	24·2	12·9	Peterhead	1873	Stephen&Forbes	Y.6.88	C.G.Warburton	Sydney,N.S.W.
1643	*	Countess of Lonsdale *Hughes 84-88*	W3mSr	British	183	113·5	25·3	12·6	Millom	1878	Thomas	Y.8.87	W.Thomas	Amlwch
1644	*	Countess of Rothes *Mullen*	I Bk	British	738	181·4	31·0	18·6	Dumbarton	1876	Birrell,Stenhouse&Co.		J.Allison & Sons	Dundee
1645		Countess of Seafield	W Bn	British	150	95·5	23·0	12·9	Banff	1856	Watson		A.Mackenzie	Inverness
1646	*	County of Anglesea *Wynne*	I Bk	British	1067	211·1	35·2	21·2	Liverpool	1877	R&JEvans&Co		W.Thomas & Co.	Liverpool
1647	*	County of Ayr *Isles -87*	I Bk	British	499	170·0	26·5	17·[illegible]	Glasgow	1863	C.Connell&Co.		J.G.Isles (Robertson Bros., Managers)	Glasgow
1648	*	County of Bute *Nicola* (WRECKED)	CS	British	793	187·0	31·9	19·4	Glasgow	1866	C.Connell&Co.	Y.2.88	Hajee Selim Khonjee	Glasgow
1649	*	County of Caithness *Dunnamps*	I 4mS	British	1636	266·1	38·8	23·6	Glasgow	1876	Barclay,Curle &Co.		R.& J.Craig	Glasgow
1650	*	County of Cardigan *Williams*	I S	British	1292	229·1	37·1	22·7	Liverpool	1877	R&JEvans&Co.		W.Thomas & Co.	Liverpool

*=Lloyd's Register; D=Nederlandsche Vereeniging; G=Germanischer Lloyd; I=Registro Italiano; R=American Record;
A=Veritas Austro-Ungarico; H=Veritas Ellenico; N=Norske Veritas; V=Bureau Veritas.

Dans la 4me Colonne, C=Composite (membrure en fer, bordé en bois); I=Fer; Stl=Acier; W=Bois.

Les Dates du doublage en métal (Col. 11) ressortent des livres dans lesquels les navires sont respectivement classés ou des rapports des Inspecteurs du Lloyd's Register.

Pour plus amples renseignements sur les navires classés par le Lloyd's Register, voir le Lloyd's Register Book et le Lloyd's Yacht Register.

Official No.	No. in Book.	Registry in which classed.	Ship's Name. Master, and Dates of Appointment to present Owner's Service and to Ship. Ship's Late Name.	Material and Rig	Flag.	Net Tonnage.	Length.	Breadth.	Depth.	Built. Where.	When.	By whom.	Date of Metalling.	Owners or Managers.
[illegible]	1651	✻	County of Carnarvon *Roberts 82-82*	I S	British	1267	227·6	37·0	22·5	[illegible]	1877	R.&J.Evans&Co.		W.Thomas & Co.
[illegible]	1652	✻ D	County of Clare *M'Neill 86-86*	I S	British	1498	235·6	38·3	23·2	[illegible]	1884	Oswald, Mordaunt&Co.		R.J.Swyny
[illegible]	1653	✻	County of Dumfries *Cameron*	I 4mS	British	1640	266·2	38·8	23·6	[illegible]	1878	Barclay,Curle [&Co.		R.& J.Craig
[illegible]	1654	✻	County of Edinburgh *Forsyth -85*	I 4mS	British	2078	285·6	42·5	24·6	[illegible]	1885	Barclay,Curle [&Co.		R.& J.Craig
[illegible]	1655	✻	County of Flint *Jones -88*	I Bk	British	1083	219·5	35·1	21·3	Sunderland	1877	W.Doxford&Sns.		W.Thomas & Co.
[illegible]	1656	✻	County of Forfar *McCalpin*	C S	British	997	208·3	34·6	20·4	Glasgow	1869	C.Connell&Co.	11.87	Ship "County of Forfar" Co. (Woodside & Workman)
80432	1657	✻	County of Haddington *Petrey*	I 4mS	British	1865	281·0	40·4	24·1	Glasgow	1878	Barclay,Curle [&Co.		R.& J.Craig
76733	1658	✻	County of Inverness *Young*	I 4mS	British	1636	266·6	38·8	23·6	Glasgow	1877	Barclay,Curle [&Co.		R.& J.Craig
78570	1659	✻	County of Kinross *Lloyd -88*	I 4mS	British	1641	267·0	38·8	23·6	Glasgow	1878	Barclay,Curle [&Co.		R.& J.Craig
[illegible]	1660	✻	County of Linlithgow *Mackenzie 87-87*	I S	British	2122	286·9	43·5	24·1	Glasgow	1887	Barclay Curle [&Co.(Ltd.)		R.& J.Craig
81341	1661	✻ D	County of Merioneth *Meredith 84-88*	I Bk	British	1065	221·5	35·1	20·9	Sunderland	1880	W.Doxford&Ss.		W.Thomas & Co.
71742	1662	✻	County of Peebles *Murdoch 84-87*	I 4mS	British	1614	266·6	38·7	23·4	Glasgow	1875	Barclay,Curle [&Co.		R. & J.Craig
84171	1663	✻	County of Pembroke *Jones*	I Bk	British	1065	221·5	35·0	20·9	Sunderland	1881	W.Doxford&Ss.		W.Thomas & Co.
83277	1664	✻	County of Roxburgh *J. Shreak*	I S	British	2121	285·6	43·5	24·0	Glasgow	1886	Barclay,Curle [&Co.		R. & J.Craig
80419	1665	✻	County of Selkirk *Stegoni*	I 4mS	British	1865	281·0	40·1	24·1	Glasgow	1878	Barclay,Curle [&Co.		R.& J.Craig
88582	1666	R V	County of Yarmouth *Scott 85-85*	W S	British	2154	243·0	44·5	24·0	Belleveau's [Cove,N.S.	1884	H.P.Boudreau	12.87	W.D.Lovitt
	1667		Courant *(ex Bacon)*	W Bk	Norweg'n	1189	174·5	35·1	24·3	Medford	1855		Y 8.86	H.Pettersen
	1668	N	Coureren *Olsen*	W Bk	Norweg'n	266	120·3	29·3	13·1	Gefle	1841			A.C.Olsen
2570	1669		Courier *Whitehead 85-85*	W Bg	British	115	80·6	20·8	12·3	Whitehaven	1845	T.&J.Brocklebank		P.Whitehead
54898	1670		Courier *Rockett 73-73*	W Sr	British	100	82·0	20·4	10·0	Gloucester	1866	Barnes		J.Bannister
	1671	G	Courier *Sandkop 80-*	W Bk	German	361	115·6	28·1	16·1	Barth	1856	J.S.Krafft		J.N.Rodbertus
	1672	V	Courier *Weideman 78-*	W Bk	German	254	110·4	26·8	12·6	Neustadt	1866	J.H.W.Bernitt		Ed.Volcker
	1673	G	Courier *Prohl 82-82*	W Bn	German	183	103·6	23·8	11·9	Danzig	1858	F.Devrient		Ernst Wendt
	1674	✻	Courrier du Pacifique *Angeron* *(ex Phocéen)*	W Bk	French	731	161·9	31·2	17·5	Marseilles	1856		Y 10.88	L.Le Quellec
125706	1675		Courser *Colby 82-*	W 3mSr	American	340	149·0	36·0	10·0	P.Ludlow,Wsh	1876	Hall Bros.		Renton, Holmes & Co.
126173	1676		Courtney Ford *Miller*	W Bn	American	381	146·2	34·1	12·4	Benicia,Cal.	1883	M.Turner		J.J.McKinnon
125963	1677	R	Cowlitz *Gammanes 81-81*	W Bk	American	779	172·1	40·0	15·2	Bath,Me.	1881	Goss&Sawyer	11.81	Puget Sound Comm. Co.
125930	1678	R	Cox & Green *Thompson -83*	W 3mSr	American	590	164·3	35·8	15·7	Newburyport, [Mass.	1881	Atkinson&Fillmore		C.B.King
71277	1679		Craftsman *Bates*	W Sr	British	266	135·3	26·1	11·8	Port Burwell	1873	H.McDernard		J.Cunningham
7140	1680		Craig Alvah *Keene 74-83* *(ex J. T. Fraser)*	W Bn	British	232	118·0	27·3	12·7	P.E.Island	1874	Vickerson		J.A.Beeching
	1681	N	Craigallion *Ball -88* *(ex Maria Greenleaf)*	W S	Norweg'n	1253	194·9	38·3	23·5	Brunswick,Me	1855	J.S.Given	[illegible]	J.L.Johanson
87409	1682	✻	Craigburn *Kerr 87-87*	I 4mS	British	1997	289·7	42·2	23·9	Glasgow	1884	W.B.Thompson		R. Shankland & Co.
	1683	✻	Craigerne	Stl 4mBk	British	1826	270·0	40·0	23·7	Pt.Glasgow	1889	R.Duncan & Co.		R.R.Paterson
95684	1684	✻	Craigisla *Griffiths 88-88*	I Bk	British	1040	215·0	33·8	20·2	Sunderland	1886	S.P.Austin&Son		W.Nicol & Co.
74474	1685	✻	Craigmullen *Kelly -85*	I Bk	British	761	189·4	31·5	19·4	Liverpool	1875	R.&J.Evans		Staveley, Taylor & Co.
70909	1686	✻	Craignair *Melvin 87-87*	I Bk	British	601	177·1	30·1	17·3	Port Glasgow	1875	R.Duncan&Co.		A.Rae
78812	1687	✻	Craigwhinnie *Bradford*	I Bk	British	834	192·4	33·1	20·2	Liverpool	1878	R.&J.Evans&C.		Staveley, Taylor & Co.
	1688	N	Credo *Christophersen 83-86*	W Bk	Norweg'n	371	123·1	29·6	15·8	Arendal	1867			B.J.Christoffersen
88676	1689	V	Creedmoor *Kennedy -85*	W S	British	1448	214·1	40·2	24·0	St.John,N.B.	1884	W.&R.Wallace	10.88	James Kennedy
2924	1690	I	Creola *(ex Ida P.)* *Gibelli*	W Bk	Italian	673	160·7	31·9	20·3	Sestri Ponente	1873	A.Briasco fu G.B.	Y 1.81	G.Piaggio fu G.B.
	1691	V	Creoula *(ex Hyacintha)* *Silva*	W Sr	Portug'se	180	95·8	20·4	11·8	Brixham	1862		[illegible]	V.A.d'Oliveira
57197	1692		Crescent *Burns 84-86*	W Bn	British	116	86·0	23·0	10·8	Nova Scotia	1868	Moody		J.M'Cann
125009	1693	R	Crescent *Bartlett -87*	W Bk	American	613	137·8	33·1	17·0	Bath,Me.	1877	Goss&Sawyer	[illegible]	N.M.Bartlett
[illegible]	1694		Crescent *Mehaffey*	W 3mSr	American	440	150·8	34·8	12·4	New London, [Conn.	1883	Crocker, Davidson&Co.		J.Luther
1953	1695		Crescenzo *Malinconico*	W Bn	Italian	115	86·6	21·8	10·7	Torre del Greco	1874	L.Trambarulo	[illegible]	G. Mazza
87309	1696	✻	Cressington *Bromley 82-83*	I S	British	2160	289·5	40·4	24·5	Southampton	1883	Oswald, Mordaunt&Co.		Sailing Ship Cressington Co. Ltd.
74939	1697		Crest of the Wave	W Sr	British	112	96·0	21·0	8·6	Lake Macquarie N.S.W.	1877	S.W.Batten		Wm.Speer
	1698		Crested Wave *Frenn -88*	W Bk	Danish	361	137·1	23·3	17·7	Guernsey	1867	Domaille	[illegible]	G.Beretta
80954	1699		Crestline *Ambrose 86-86*	W Sr	British	117	85·9	27·8	7·6	St.Martin's N.B.	1881	D.DeLong		W.J.Davidson
79806	1700	✻	Cricket *Crouch 78-78*	W Bk	British	321	137·4	27·9	12·8	Sunderland	1878	Gibbon	[illegible]	W.C.Jarvis

✻ = Lloyd's Register; D = Nederlandsche Vereeniging; G = Germanischer Lloyd; I = Registro Italiano; R = [illegible]
A = Veritas Austro-Ungarico; H = Veritas Ellenico; N = Norske Veritas; V = [illegible]

In the 4th Column, C = Composite (iron frame, wood planked); I = Iron; Stl = Steel; W = Wood.

The Dates of Metalling (Col. 11) are based on the records of the Books in which Vessels are respectively classed, or on the Reports of the Surveyors of [illegible]

Further particulars of Vessels classed by Lloyd's Register will be found in Lloyd's Register Book and Lloyd's Yacht Register.

N° dans le livre	Registre dans lequel le navire est classé	Nom. Capitaine et dates d'entrée. Matériaux et Gréement. Ancien Nom.			Pavillon.	Tonnage net	Longueur	Largeur	Creux	Construit. Où	Quand	Par qui	Date du doublage en métal	Armateurs ou Directeurs-Gérants.	Port d'Armement.
1701		Cricket	*Kennedy*	W.Sr	British	128	89·1	27·0	7·9	St.John,N.B.	1883	H.J.Oliver		H.J.Oliver	St.John,N.B.
1702	V	Criemhilde	*[illegible]*	W Bg	Danish	326	120·7	25·8	15·1	Nordby,Fanø	1877	S.Abrahamsen	[illegible]	Akt.Selsk.(S.B.Hansen)	NordbyFanø
1703		Crishna		W Bk	British	254	96·3	24·2	14·8	Douglas, Isle of Man	1843			David Munroe	Melbourne
1704	✱	Cristal	*Renouf*	W Bg	British	32	60·8	15·8	6·8	Jersey	1863	Cantell		R.Watson	Jersey
1705	A	Cristina Spaño (ex Tom)	*Spano*	W B	Italian	266	105·2	16·3	..	Buccari	1865	G.Giacovich	[illegible]	G.Spano	Catania
1706	IV	Cristoforo Colombo (Odero)		W Bk	Italian	825	172·5	32·8	21·2	Sestri Ponente	1875	A.Briasco fu L.	[illegible]	P.Antola	Genoa
1707		Crocodile	*Kimber*	W Bk	British	139	113·0	23·1	13·5	Dartmouth	1856			T.G.Robins & Co.	Dartmouth
1708	✱	Crofton Hall	*Lucas*	I Bk	British	2074	291·7	39·1	23·6	Liverpool	1883	W.H.Potter&S		Globe Shipping Co. Ld.	Liverpool
1709	✱	Cromartyshire	*Douglas*	I S	British	1462	248·8	38·1	22·8	PortGlasgow	1879	Russell&Co.		T.Law & Co.	Glasgow
1710	NR	Cromwell	*Jacobsen*	W Bk	Norweg'n	878	157·2	34·8	23·2	Medford,Mass	1848	P.Curtis	[illegible]	A.Jacobsen	Horten
1711	A	Crono	*[illegible]*	W B	Austrian	512	128·2	27·4	19·8	Fiume	1871	G.G.Baccich [Bellaz	[illegible]	F.Minach fu G.	Fiume
1712		Cronómetro	*Hulet*	W Sr	Spanish	120	82·3	23·7	10·9	Barcelona	1840			J.S.Bennasar	Majorca
1713	✱	Crossowen (ex Gourka)	*[illegible]*	W Bn	British	234	114·7	22·6	12·8	Grangemouth	1878	Adamson	[illegible]	Thom & Cameron	Glasgow
1714	DNV	Crown	*Danielsen*	W Bk	Norweg'n	655	158·1	32·3	18·6	[illegible]	1878	A.Andersen	[illegible]	M.Danielsen	Risör
1715	R	Crown Jewel	*Crofton*	W Bk	British	716	151·8	32·9	19·2	Granville,N.S.	1868		[illegible]	H.D.Troop	St.John,N.B.
1716		Crown Point		W Sr	British	124	87·0	22·6	7·6	Newburyport				A.H.Browne	MontegoBay
1717	NV	Crown Prince	*Wickstrøm*	W Bk	Norweg'n	987	174·6	35·6	22·0	St.Martin's N.B.	1871	J.H.Moran	[illegible]	Act. S. "Crown Prince" (C.Hansen & Co.)	Christiania
1718	✱D	Crown of Denmark	*Jackson*	I S	British	2029	283·5	41·3	24·1	Dumbarton	1882	A.M'Millan&Sn		Robertson, Cruickshank & Co.	Liverpool
1719	✱	Crown of England	*Wilson*	I S	British	1799	267·0	39·1	23·6	Workington	1883	R.Williamson&Son		Robertson, Cruickshank & Co.	Liverpool
1720	✱	Crown of India	*Howe*	I [illegible] Bk	British	1975	276·7	41·3	24·1	Leith	1885	Ramage&Ferguson		Robertson, Cruickshank & Co.	Liverpool
1721	✱	Crown of Italy	*[illegible]*	I S	British	1551	251·3	38·2	21·9	Leith	1885	Ramage&Ferguson		Robertson, Cruickshank & Co.	Liverpool
1722	✱	Crown of Scotland	*Smith*	I S	British	1550	249·7	38·3	22·0	Liverpool	1883	T.Royden&Sns		Robertson, Cruickshank & Co.	Liverpool
1723		Crowpill (ex John W...)	*Phillips*	W K	British	149	93·6	24·1	8·8	Harwich	1873			J.G.Sully	Bridgewater
1724	R	Croydon	*Brown*	W Bk	British	376	120·5	27·8	17·0	Shields	1862	Marshall	[illegible]	Thomas, Watkins & Jenkins	Swansea
1725	✱	Cruiser	*Irwin*	W K	British	50	52·7	17·8	8·2	Watermouth	1866	Symons		G.Irwin	Barnstaple
1726	✱	Crummock Water	*Amos*	I Bk	British	995	216·0	34·4	20·4	Whitehaven	1878	WhitehavenS.[B.Co.,Ld.		Jackson, Metcalf & Co.	Liverpool
1727	RV	Crusader	*Hutchinson 87-87*	W S	British	1125	185·6	37·8	22·6	Shelburne, NS	1876	W.T.Kelley	[illegible]	R.Bell	Liverpool
1728	✱	Crusader	*Perriam 87-87*	I S	British	1058	210·7	35·1	21·4	Glasgow	1865	C.Connell&Co.		Shaw, Savill & Albion Shipping Co. Ld.	Glasgow
1729	✱	Crusader	*Williams 77-83*	W Bg	British	183	101·5	23·6	12·7	Aberystwith	1865	Evans		W.Morris	Aberystwith
1730		Crusader	*Scott*	W Bk	American	643	145·5	33·4	21·5	Medford,Mass.	1848	W.W.Goddard	[illegible]	L.V.Sone	New York
1731		Crusoe	*Bateman*	W Sr	American	109	86·0	24·7	7·8	E.Machias,Me	1856			W.H.Leighton	Machias,Me.
1732	V	Cruzeiro	*Vinod -87*	W Bk	French	268	108·9	25·9	13·0	Nantes	1875	Dubigeon	[illegible]	Menier	Havre
1733		Crystal	*Webber*	W Bk	British	263	127·1	25·7	13·1	Aberdeen	1872	J.Humphrey&C	[illegible]	A.Dauban	Port Louis
1734	✱	Crystal Stream	*Bay -87*	W Sr	British	120	85·5	22·3	11·1	Par	1877	Moss	[illegible]	W.Stephens	Fowey
1735		Crystaline		I Bk	British	249	115·5	23·5	13·3	Chester	1856	Cram		W.M'Master	Liverpool
1736		Cuatro Amigos	*(ex F. C.)*	W Bn	Spanish	112	75·8	23·1	11·8	Carril	1849			José Sierra	Villagarcia
1737	V	Cuba	*Davidson*	W Bkn	British	453	140·0	34·0	13·8	Hantsport,NS	1883	E.Churchill&Ss.	V 11.84	G.W.Churchill	Windsor,N.S.
1738		Cuba	*Mahy -87*	W Bk	British	250	120·3	25·7	12·8	Dysart	1875	Watt	V 11.83	J.Helman & Co.	Guernsey
1739		Cuba		W Sr	American	697	209·0	35·0	17·8					W.B.Harnod	New York
1740	V	Cuba	*Springer 86-81*	W S	German	1128	180·1	35·1	22·8	Richmond,Me.	1872	Harward&[Theobald	[illegible]	H.von Riegen & Stindt	Bremerhaven
1741		Cuba		W Bg		199	114·7	25·3	12·3	Bergen	1867	A.Dekke	[illegible]		
1742	V	Cuba (ex [illegible])	*Petterson*	W Bk	Swedish	570	144·6	29·0	18·5	Annapolis,NS	1866	S.Hall	[illegible]	T.Pettersson	Wiken
1743	N	Cuba	*Söblé -88*	W Bg	Swedish	202	111·5	23·1	11·8	Porsgrund	1868			A.Bergman	Halsingborg
1744		Cuba & Canarias (ex Cuba)	*Balaguer*	W Bg	Spanish	295	113·8	26·6	12·4	Cornwallis,NS	1864			J.R.Amador	Havana
1745	✱	Cubana	*Malgrof 74-88*	I Bk	British	499	145·0	27·0	18·1	Sunderland	1862	W.Pile&Co.		J.L.Jacobs & Co.	London
1746	✱	Cuerero	*Williams 88-88*	W Bk	British	239	122·0	24·1	12·9	Bideford	1870	Cox	V 12.88	M.Jones & Co.	Bristol
1747	AI	Cugini (Due Cugini)	*Di Paolo*	W Sr	Italian	179	98·4	24·8	13·6	Fiume	1870	G.G.Baccich [Bellaz	[illegible]	R.A.di Paolo	Bari
1748	V	Culdoon	*Branscomb -88*	W Bkn	British	392	138·1	34·4	13·0	Gardner'sCreek [St.John	1885	W.&R.Wallace	V 11.85	J.Kennedy	St.John,N.B.
1749		Culebra		W Sr	American	158	..	..	..	Camden,N.J.	1883			H.Moody	New York
1750	✱	Culzean Castle	*Jeffery*	W [illegible] Sr	British	168	113·8	22·6	11·6	Troon	1882	TroonS.B.Co.	V 10.87	Daly& Foley	Cork

✱=Lloyd's Register; D=Nederlandsche Vereeniging; G=Germanischer Lloyd; I=Registro Italiano; R=American Record; A=Veritas Austro-Ungarico; H=Veritas Ellenico; N=Norske Veritas; V=Bur[eau Veritas]

Dans la 4me Colonne, C=Composite (membrure en fer, bordé en bois); I=Fer; Stl=Acier; W=Bois.

Les Dates du doublage en métal (Col. 11) ressortent des livres dans lesquels les navires sont respectivement classés ou des rapports des Inspecteurs du Lloyd's Register.

Pour plus amples renseignements sur les navires classés par le Lloyd's Register, voir le Lloyd's Register Book et le Lloyd's Yacht Register.

Official No.	No. in Book	Registry in which classed	Ship's Name. Material and Rig. Master, and Dates of Appointment to present Owner's Service and to Ship. Ship's Late Name.	Flag.	Net Tonnage.	Dimensions. Length.	Breadth.	Depth.	Built. Where.	When	By whom.	Date of Metalling	Owners or Managers
[illegible]	1751	✻	Cumberland *Jones 79-83* I&StlS	British	1798	267·1	39·1	23·6	Workington	1884	R.Williamson & Son		T.Williamson
[illegible]	1752		Cumberland *Sharer 79-82* WBk	British	332	123·4	27·2	16·7	Sunderland	1879	Chilton	Y 4.89	W.Robertson
[illegible]	1753	R	Cumberland *Webber* W3mSr	American	393	129·1	30·0	15·4	Freeport, Me.	1874		Y 10.88	A.B.Hall
[illegible]	1754		Cumberland Lassie *Thomas 77-85* WBn	British	208	114·6	24·0	13·8	Amlwch	1874	Thomas	Y 4.86	W.Postlethwaite
[illegible]	1755	✻	Cumbria *Milnes* I Bk	British	641	177·0	31·3	17·5	Sunderland	1874	Bartram, Haswell&Co.		P.Hick & Co.
[illegible]	1756		Cumbria (ex Glaucus) WBgYt	British	227	111·5	23·3	14·8	Greenock	1851	R.Steele&Co.		Directors of H.M.Training Ship "Cumberland."
[illegible]	1757	✻	Cumbrian *Bartlett 88-88* IBk	British	1053	214·5	34·4	21·4	Sunderland	1879	R.Thompson, jn.		P.Iredale & Son
[illegible]	1758	✻	Cumeria (MISSING) *Hilton 84-84* I S	British	1284	218·6	37·1	22·8	Birkenhead	1869	G.R.Clover&Co.		R.Nicholson & Sons
[illegible]	1759		Cumminger *Hayes* WSr	British	155	92·6	26·3	10·4	St.Mary's River, N.S.	1876		Y 2.81	W.Hayes
	1760		Cunha WSr	Brazilian	142	..	..	..					F.C.L. e Silva
[illegible]	1761	I	Cuomo I. *Cuomo* WBk	Italian	443	130·5	29·1	18·2	Castellamm're	1875	M.Trambarulo	Y 1.88	L. & G.Cuomo
[illegible]	1762	I	Cuore *Gargiulo* WBg	Italian	444	119·7	26·9	18·4	Cassano	1875	A.Castellano	Y 11.88	F[lli]. Paturzo
[illegible]	1763	✻	Cupica StlBk	British	1169	226·0	36·4	21·9	Londonderry	1889	C.J.Bigger		SteelBk."Cupica"Co.Ld (Mac-lean, Marshall & Co.)
[illegible]	1764	✻	Cupid *Pratt -88* WBk	British	652	156·2	32·9	18·5	Maitland, N.S.	1871	Smith	Y 7.89	D.Anderson
[illegible]	1765		Cupido WBn	British	172	98·2	23·3	11·8	Sonderborg	1873	H.Hauschild		Casimir Nairac
	1766	G	Cupido *Hage 87-87* WBk	Norweg'n	397	115·7	30·1	17·4	Lussino	1858			J.H.Hage
[illegible]	1767	V	Curacao *Preist 82-88* WBkn	British	369	134·0	29·5	11·2	Hantsport, NS	1872	J.B.North	Y 10.89	J.B.North
52910	1768		Curlew *Smiley -88* WS	British	1237	188·5	36·8	23·9	Quebec	1867	Gilmour	Y 6.87	Curlew Ship Co. Ld. (J. S. Wright & Co.)
[illegible]	1769	RV	Curlew *Winchester 83-* WBn	British	331	125·8	29·5	13·0	Gardner's Creek, N.B.	1881	Mallery& McKenzie	Y 9.88	Schofield & Co., Ld
[illegible]	1770		Curlew WSr	American	124	79·3	22·4	9·9	P.E.Island	1862			M.H.Roberts
	1771	DV	Curonia (ex Brightest Star) *Wibroe* WBk	Russian	386	124·3	26·6	17·1	Sunderland	1868	Parkes	Y 12.83	J.C.Bruchmann
[illegible]	1772	..	Curreem Bux WBk	British	345	103·5	24·0	15·4	Coringa	1846			Moula Sab
41048	1773		Curreen Bux WBk	British	240	84·0	23·0	15·3	Coringa	1846			Booran Saib
	1774	V	Curry Treffenberg *Edlund* WBk	Swedish	418	141·0	26·6	14·6	Sundsvall	1880	J.Sjölen	Y 10.86	Sundsvall Skeppsaktiebolag (O. L. Widholm)
125412	1775	R	Curtis Akerly (WRECKED) *Bonlo* W3mSr	American	359	122·4	30·1	11·0	Claremont, Va	1875			Paul A.Bonlo
[illegible]	1776		Curtis Goodwin *Abdill* WSr	American	109	101·0	17·0	7·0	Hamburg, Pa.	1863			J.E.Abdill
	1777		Custodio de Abreu WSr	Brazilian	135	..	..	..					C.T.L.Sobrinho
30573	1778		Cutch Merchant WS	British	640	143·1	32·1	20·6	Cutch	1859	Ballajee	Y 1.89	Thacker Govindjee Veerjee
2414	1779		Cutelin Dall'Orso (ex Gostanda) *Questa* WBk	Italian	649	154·2	31·1	21·2	Sestri	1873	A.Briasco fu L.	Y 6.87	Maria Dagnino ved Dall'Orso & figli
[illegible]	1780	✻	Cuthona *Dutchburn 59-87* I S	British	1702	250·8	39·0	24·0	Stockton	1885	Richardson, Duck&Co.		J.H.Worthington & Co.
65063	1781	✻	Cuthona *Gifford 78-78* WBkn	British	201	102·3	24·1	12·9	Dysart	1878	Watt	Y 9.84	J.Gifford
63557	1782	✻	Cutty Sark *Woodget 80-85* C S	British	921	212·5	36·0	21·0	Dumbarton	1869	Scott&Linton	Y 4.89	J.Willis & Son
993	1783	V	Cvietnidan (ex Marietta B.) *Jellusich -88* WBg	Austrian	352	112·2	27·9	26·7	Fiume	1857		Y 11.84	Raimondo Bachich
[illegible]	1784	✻	Cwm Avon *Hughes 83-87* WSr	British	74	71·0	19·1	12·0	Appledore	1868	Cook		R.Jones & Co.
74193	1785	✻	Cwm Donkin *Watkins -88* WBkn	British	571	169·0	29·4	17·9	P.E.Island	1876	Richards	Y 2.88	H.Hoskin
4536	1786		Cygnus *Cole* WSr	American	177	101·5	26·3	9·5	E.Machias, Me.	1865			Ellen M.Mitchell
47327	1787		Cymbeline (BROKEN UP) *Roach* WBk	British	391	130·0	26·2	16·8	Bridport	1863	Cox	Y 4.86	Smart & Simpson
88348	1788		Cymbeline *Croser -86* WSr	British	103	80·2	23·9	9·2	ConquerallNS	1884	N.Rafuse		N.Rafuse
93432	1789	✻	Cymro *Williams -88* WSk	British	28	56·4	16·8	6·4	Milford	1888	J. & W.Francis		W.Francis
[illegible]	1790	✻	Cynisca *Kerr 88-88* I Bk	British	846	200·4	32·6	19·4	Glasgow	1878	Dobie&Co.		H.McEwan
68639	1791	✻	Cynosure *Macfarlane -86* I Bk	British	764	192·7	31·2	19·0	Glasgow	1875	Dobie&Co.		H.McEwan
[illegible]	1792		Cynthia Jane *Gardner* WSr	American	109	99·2	22·0	7·6	Derby, Conn.	1858			Estate of T. B. Gardner
[illegible]	1793	✻	Cypher *Davies 84-88* W3mBkn	British	399	151·6	29·0	16·0	P.E.Island	1882	J.Yeo.	Y 12.85	Wylie & Gordon (T. P. Richards & Co., Managers)
[illegible]	1794		Cypress *Bambra -78* WSw	British	236	101·0	24·5	14·7	Sunderland	1860	Austin		R.G.Sanders
[illegible]	1795	✻	Cypromene *Robertson 83-83* I S	British	1834	260·0	39·5	24·0	Southampton	1878	Oswald, Mordaunt&Co.		H.Fernie & Sons
[illegible]	1796	RV	Cyprus *Steeves -88* WBk	British	1091	186·9	36·6	20·9	Bridget'wnNS	1878	A.Young	Y 11.89	H.D.Troop
[illegible]	1797		Cyrene WSr	British	107	81·2	23·9	9·2	Bridgewater, N.S.	1883	L.Hewitt		John Shankle
[illegible]	1798	✻	Cyril *Cooper -87* W Yawl	British	73	77·0	17·2	10·0	Falmouth	1870	Trethowan	Y 1.86	E.D.Anderton
[illegible]	1799		Cyrus Chamberlain *Steel* WSr	American	121	83·0	24·0	8·0	Freetown, Pa.	1846			J.O.Cushing & Co.
[illegible]	1800	R	Cyrus Hall *Coombs -88* W3mSr	American	384	140·4	32·0	11·4	East Boston	1874	A.Gove		J.H.Conant & Co.

✻=Lloyd's Register; D=Nederlandsche Vereeniging; G=Germanischer Lloyd; I=Registro Italiano; R=A[illegible]
[illegible]=Veritas Austro-Ungarico; H=Veritas Ellenico; N=Norske Veritas; V=Bu[illegible]

In the 4th Column, C=Composite (iron frame, wood planked); I=Iron; Stl=Steel; W=Wood.

The [illegible]s of Metalling (**Col. 11**) are based on the records of the Books in which Vessels are respectively classed, or on the Reports of the Surveyors of L[illegible]

Further particulars of Vessels classed by Lloyd's Register will be found in Lloyd's Register Book and Lloyd's Yacht Register.

1	2	3	4 Nom, Matériaux et Gréement. Capitaine et date d'entrée au service des Armateurs actuels et à bord du navire. Ancien Nom.	5 Pavillon.	6 Tonnage	7 Dimensions Longueur	Largeur	Creux	8 Construit Où.	9 Quand.	10 Par qui.	11	12 Armateurs ou Directeurs-Gérants.	13 Port d'Armement.
	1801	V	Cyrus Wakefield W S	American	2013	217·0	43·7	28·6	Thomaston Me	1882	S.Watts	[illegible]	S.Watts	Thomaston Me
	1802		Cysne W Bn	Brazilian	187	..	..	..					P.Vianna & Co.	Rio Grande
	1803	G	Czaar Peter W Bk (… Registers)	Dutch	309	118·0	27·4	14·9	P.E.Island	1865	M'Donald	[illegible]	P.H.Creandijk & Zoon	Amsterdam
	1804		Czar W Sr	American	137	98·0	26·0	9·7	San Francisco	1881	M.Turner		McCollam Fishing & Trading Co	San Francisco
	1805		Czarowitz W Bn	British	196	106·6	23·3	12·9	Brixham	1866	Upham	[illegible]	W.Polkinghorne	Plymouth

ADDITIONAL SHIPS ARE INSERTED IN THE SUPPLEMENT AT THE END OF THE REGISTER.

✱ = Lloyd's Register; A = Veritas Austro-Ungarico; D = Nederlandsche Vereeniging; G = Germanischer Lloyd; H = Veritas Ellenico; I = Registro Italiano; N = Norske Veritas; R = American Record; V = Bureau Veritas.

Dans la 4me Colonne, C = Composite (membrure en fer, bordé en bois); I = Fer; Stl = Acier; W = Bois.

Les Dates du doublage en métal (Col. 11) ressortent des livres dans lesquels les navires sont respectivement classés ou des rapports des Inspecteurs du Lloyd's Register.

Pour plus amples renseignements sur les navires classés par le Lloyd's Register, voir le Lloyd's Register Book et le Lloyd's Yacht Register.

Official No.	No. in Book	Registry in which classed	Ship's Name. Material and Rig. Master, and Dates of Appointment to present Owner's Service and to Ship. Ship's Late Name.	Flag	Net Tonnage	Length	Breadth	Depth	Built: Where	Built: When	Built: By whom	Date of Metalling	Owners or Managers
6782	1	R	D. A. Brayton *Huntley* [-86] WBkn	American	530	138·6	32·0	17·4	FallRiver, Mass.	1873	J.C.Terry	[illegible]	H.O.Gray
86986	2		D. A. Huntley (ABANDONED) WSr	British	108	78·6	24·6	8·4	Parrsboro'N.S.	1882	A.Huntley		M.Tobin
58825	3		D. Alverty WBn	British	129	90·0	25·0	8·0					C.Collins
6185	4		D. B. Steelman *Patterson* WSr	American	103	85·0	26·0	6·0	EnglishCreek, N.J.	1861			A.A.McCullough
6419	5		D. C. Murray (WRECKED) *Clifford* WBk	American	432	130·6	31·0	12·1	Greenp'ntN.Y.	1860			C.M.Bradshaw
6833	6	R	D. Chapin (FOUNDERED) *Hall* 69-86 WBk	American	514	128·2	31·0	17·8	Portsm'thNH.	1869	Tobey&Littlefield	[illegible]	J.S.Emery & Co.
157047	7	R	D. D. Haskell *Haskell* W3mSr	American	302	130·2	31·1	11·0	Belfast,Me.	1881		[illegible]	S.G.Haskell
42359	8		D. Grant WSr	British	117	82·7	23·4	10·0	Antigonish, N.S.	1863			W.Boudroit
72185	9	V	D. H. Morris *Morris* WS	British	1197	193·6	39·2	23·5	Walton,N.S.	1876	D.H.Morris	[illegible]	D.H.Morris
	10	G	D. H. Watjen *Moll* 84-84 WBk	German	1215	186·7	37·6	23·1	[illegible]	1867	J.Lange	[illegible]	D.H.Wätjen & Co.
157102	11	R	D. K. Baker *Rulon* -88 W3mSr	American	493	143·8	33·0	14·7	[illegible] N.Y.	1883	Mather		J.L.Merrill
6293	12		D. L. Sturgis *Prichard* WSr	American	129	90·0	25·0	7·6	Bridgeport, Conn.	1852			S.Burchard,Jun.
6708	13		D. M. Anthony *Cactus* -86 W3mSr	American	528	146·7	36·4	15·5	FallRiv'r Mass	1873	J.C.Terry		W.Chase
91	14	A	D. Petrocochinos WBn	Greek	318	108·2	28·2	..	[illegible]	1871			C.Isichlis
6539	15		D. Sawyer *Beal* -86 WSr	American	114	89·8	24·7	7·8	Jonesport,Me.	1868			D.J.Sawyer
88418	16		D. W. B. *McLean* WSr	British	121	88·3	26·8	7·7	Westfield, King'sCo.NB	1884	D.W.Belyea		W.B.McLeod
157085	17	R	D. W. McLean *Hudson* W3mSr	American	281	128·2	29·5	10·5	Norwalk,Conn	1882	G.S.Bell		G.C.Hallock
	18	V	D'Artagnan *Louis* WBk	French	317	116·4	25·9	13·0	Means	1875	E.Ollivaud	[illegible]	P.Legal
	19	V	Da Capo *Suerken* -87 WBn	German	294	118·6	25·0	12·7	Papenburg	1876	H.W.Meyer	[illegible]	H.W.Meyer
	20		Da Capo WBg	Norweg'n	299	125·3	27·9	14·3	Björneborg	1857			K.Soiland
	21		Da Capo *Soma* 87-87 WBg	Norweg'n	162	101·8	23·6	11·4	Stavanger	1855	T.Michelsen		Brödrene Berner
	22	N	Da Capo *Dahler* -88 WSr	Norweg'n	149	97·8	24·1	10·6	Aasgaard-strand	1868			H.&O.Dahler
	23	N	Dacapo *Olsen* WSr	Norweg'n	125	89·3	22·7	10·3	Sandefjord	1878			A. Olsen
51458	24	✳	Dacca (ex Tasmanian Settler) *Baird* 86-86 IBk	British	1057	210·3	33·9	22·0	Greenock	1865	R.Steele		C.Barrie
	25		Dadin *Pélicier* WBg	French	100	..	..	..	St.Servan	1855			Guibert & fils
	26	NV	Dæmring *Hansen* -86 WBk	Norweg'n	287	131·5	25·6	10·8	Bergen	1886	Brunchorst & Dekke	[illegible]	K.S.Bertelsen
	27		Dagen *Jurgensen* WBg	Norweg'n	174	98·5	24·5	11·6	Porsgrund	1867			Svend Föyn
	28	N	Daggry *Svendsen* WBk	Norweg'n	876	175·7	36·0	20·0	Drammen	1875	B.O.Hærem	[illegible]	E.Berentsen
	29	N	Daggry (ex Telegraph) *Augensen* -87 WBk	Norweg'n	354	128·6	26·8	16·2	Quebec	1855			Rederiselskabet "Daggry" (C.A. [illegible])
	30	N	Daggry *Thorsen* W3mSr	Norweg'n	185	99·6	24·7	11·8	Sandefjord	1875			S.Larsen Lunnely
85884	31		Daghine IBarge	British	789	195·5	25·6	8·5	PortGlasgow	1879	Robb,Duncan & Co.		IrrawaddyFlotillaCo.
	32	N	Dagmal *Harboe* WBk	Norweg'n	435	135·3	29·1	17·1	Elbing	1867	Gebrüder Mitz[laff]	[illegible]	Brödrene Berner
53461	33		Dagmar *Mahler* WBg	British	192	101·9	23·7	13·0	Hartlepool	1866	Blumer		H.W.Gunn
	34		Dagmar (ex [illegible]) *Andersen* -88 WBk	Danish	383	110·4	25·1	16·5	Østerhallsvik	18 51/72	H.D.Carstensen	[illegible]	Rasmussen & Fische[r]
	35	V	Dagmar *Paulsen* 78-78 WBg	Danish	222	115·8	23·6	11·4	Veile	1874	P.H.Uldum	[illegible]	Akt Selsk.(C.Andersen)
	36		Dagmar (ex Alma) *Christensen* -87 WBg	Danish	181	95·9	22·0	11·8	Flensburg	18 56/71	J.C.Christensen		A.H.B.Kromann
	37	DV	Dagmar *Jensen* -87 WBn	Danish	177	109·1	22·9	10·7	Malmö	1864	J.Bonnesen		C.M.Christiansen
	38		Dagmar (ex Glen Tilt) *Solberg* WS	Norweg'n	1013	172·5	35·3	22·8	Quebec	1865	Valin	[illegible]	I.M.Bryde
	39	N	Dagmar *Nielsen* -86 WBk	Norweg'n	259	115·3	26·8	12·4	[illegible]	1871	O.&P.Knudsen	[illegible]	O. & P. Knudsen
	40	N	Dagmar *Öhre* -88 WBg	Norweg'n	164	93·4	23·4	11·5	[illegible]	1874			M.N.Ohre
	41	N	Dagmar *Nielsen* WSr	Norweg'n	100	83·5	21·9	8·8	Thisted,Denmk	1868	P.Hald		N.T.Nielsen
	42	V	Dagmar (ABANDONED) (ex Thesaurus) *Snellman* WS	Russian	743	155·0	31·0	21·0	Richibucto	1865	Jardine	[illegible]	G.W.Snellman
	43	V	Dagny (ex A. H. Willis, ex Sea Rover) *Olsen* WS	Norweg'n	1046	137·0	35·0	21·0	NewburyPort, Mass.U.S.	1859		[illegible]	A.Weyergang
	44	N	Dagny *Gjertsen* WBk	Norweg'n	549	121·2	33·4	20·2	Rotterdam	1839		[illegible]	L.Vanvert
	45	N	Dagny *Johannessen* WBg	Norweg'n	151	96·8	21·6	11·4	Lovisa,Finlnd	1860			Chr.Johannesen
	46	N	Dagny *Ellingsen* WBn	Norweg'n	140	93·8	24·0	11·2	Christiansund	1872		[illegible]	Parelius & Lossius
	47	D	Dagny (ex Chapman) *Göransson* -88 WBk	Swedish	453	123·6	32·5	16·7	Brahestad	1862	A.Westerlund		B.Göransson
	48		Dagopithi WBg	Turkish	225	..	..	..	Sakarieh	1881			Eceeefehji Karamus-oglou Sotiri
322	49		Daiho Maru WSr	Japanese	160	97·2	21·6	11·8	Osaka	1879	TodaShojiro		Suzuki Yohei
93346	50		Daing StlLr	British	484	220·0	30·0	8·3	Dumbarton	1887	W.Denny&Bros		IrrawaddyFlotillaCo.

✳=Lloyd's Register; A=Veritas Austro-Ungarico; D=Nederlandsche Vereeniging; G=Germanischer Lloyd; H=Veritas Ellenico; I=Registro Italiano; N=Norske Veritas; R=Ame[rican]; V=Bure[au]

In the 4th Column, C=Composite (iron frame, wood planked); I=Iron; Stl=Steel; W=Wood.

The Dates of Metalling (Col. 11) are based on the records of the Books in which Vessels are respectively classed, or on the Reports of the Surveyors of Ll[oyd's]

Further particulars of Vessels classed by Lloyd's Register will be found in Lloyd's Register Book and Lloyd's Yacht Register.

1	2	3	4	5	6	7			8	9	10	11	12	13
	Nos dans le livre	Registre dans lequel classé	Nom. Matériaux et Gréement. Capitaine et dates d'entrée en service des Armateurs actuels et à bord du navire. Ancien Nom.	Pavillon	Tonnage net	Dimensions			Construit			Dates de doublage en métal	Armateurs ou Directeurs-Gérants.	Port d'Armement.
						Longueur	Largeur	Creux	Où	Quand	Par qui			
	51	I	Daino *Profumo* W Bk	Italian	200	119·1	25·3	10·9	Sestri	1873	A. Bozzano fu G.B.	[illegible]	Flli Bozzano	Genoa
	52		Dairymaid *Wilson 89-* W Sr	British	155	101·6	22·5	11·5	Glasson Dock	1867	Simpson		Porter's Shipping Co. Ld.	Fleetwood
	53	✻	Daisy *Lewis 74-* W Bk	British	489	133·8	30·2	17·8	P. E. Island	1872	Carroll	[illegible]	T. Pugsley	Newport
	54		Daisy *[illegible]* W Bn	British	124	81·0	22·3	10·0	Sheet Harbor, N.S.	1870			A. W. West	Halifax, N.S.
	55	✻	Daisy (WRECKED) *[illegible]* W Sr	British	91	82·1	20·5	10·6	Chester	1865	Rishton		C. Davison	Chester
	56	✻	Daisy *Johnson* W K	British	91	82·2	21·1	10·7	Falmouth	1885	E. Martin		S. F. Bartlett	Hull
	57	✻	Daisy *Cawthorne* W K	British	75	76·2	19·6	9·1	Goole	1883	G. W. Gutwin		J. & D. Cawthorne	Goole
	58	R	Daisy *Nash* W Bn	American	418	123·0	31·6	17·1	Setauket, L.I.	1872	N. Hand	[illegible]	Swan & Son	New York, N.Y.
	59	R	Daisy Boynton *Harding* W Bn	American	390	117·5	29·8	16·6	Eastport, Me.	1870	Huston	[illegible]	Louis Chemidlin	New York
	60	R	Daisy E. Parkhurst *Hopper* W 3m Sr	American	342	138·0	32·0	11·9	St. George, Me	1874	Bean	[illegible]	P. O. Hooper	Providence, R.I.
	61	R	Daisy Read *Mitchell* W Bkn	American	445	140·0	33·0	13·0	Harringt'n, Me	1882	A. Sproul	[illegible]	A. M. Nash	Machias, Me.
	62		Daisy Rowe *Petersen* W Sr	American	117	94·5	29·0	7·5	Humboldt Bay [Cal.	1879			E. Higgins	San Francisco
	63		Daiyei Maru W Bg	Japanese	125	91·9	19·2	9·3	Tokio	1879	Kawasaki Ship [Bdg. Yd.		S. Ginsaku	Nariwa, Owari
	64		Daiyei Maru W Sr	Japanese	125	78·7	20·0	11·0	Osaka	1879	Toda Shojita		Uno Kichiou	Hakodote
	65		Daiyei Maru No. 2 W Bg	Japanese	159	97·3	23·9	11·1	Owari	1885	Okamoto Zen- [ki-Chi		S. Ginsaku	Nariwa, Owari
	66		Daiyo Maru W Bkn	Japanese	422	110·0	26·3	18·0	Hiogo	1878	J. W. Barry		Tatsuma	Nishinomiya
	67		Dakia W Sr	Greek	120	..	..	..	Galaxidi	1869			G. Foca	Galaxidi
	68	R	Dakota (STRANDED) *Schiffer 77-88* W S	German	1203	196·8	38·1	23·6	Bath	1881	W. Rogers	[illegible]	G. Hilder	Bremerhaven
	69	V	Dalhanna *Jones -87* W Bk	British	972	187·1	36·2	22·1	Maitland, N.S.	1880	A. A. McDougall		J. M. Campbell	Glasgow
	70	G	Dalkarlså *Johanson -87* W Bg	Swedish	301	112·2	24·3	15·2	Dalkarlsâ	1863	Sundström		A. Nilsson	Wäsby, [illegible]
	71	✻	Dallam Tower *Williamson 87-87* I S	British	1464	243·7	38·3	23·9	Birkenhead	1866	G. R. Clover & Co.		J. Forrester J. [illegible] & Co.	Lancaster
	72		Dallas Hill (LOST) *Jones* W Sr	British	110	82·2	26·6	7·2	Alma, N.B.	1883	P. C. Copeland		B. J. Driscoll	St. John, N.B.
	73	V	Dalmazia *Vidovich 78-86* W Bk ex Ellen [illegible]	Austrian	520	137·4	29·6	20·3	Malta	1869	Westermann	[illegible]	M. Orebich fu G.	Orebich
	74	✻	Dalswinton Stl S	British	1600	255·0	39·0	22·7	Port Glasgow	1889	R. Duncan & Co.		T. C. Guthrie	Glasgow
	75		Dame Blanche (WRECKED) *Koost* W Sr	French	107	84·3	21·6	11·0	Dunkirk	1882			H. Durin	Dunkirk
	76		Damiao *Dos Santos* W Sr	Brazilian	164	..	..	..					A. J. Vilhaça	Rio Grando
	77		Damon *Torrey* W Sr	American	158	96·0	26·3	8·0	Waldoboro, Me.	1856			S. G. Haskell	Deer Isle, Me.
	78	A	Dampier *Prischich* W Bk	Austrian	411	127·3	26·3	16·7	Fiume	1862	Flli. Schiavoni	[illegible]	A. F. Poschich	Fiume
	79		Dan *[illegible] Olsen 85-85* W Bk	Danish	267	119·2	25·6	11·9	Bucksport, Me	1855			Et. Act. Selsk. (P. Berg)	Nexo
	80		Dan *Berg* W Bg	Norweg'n	186	107·4	26·0	12·1	Westervik	1856			Carl Dahl	Fredrikshald
	81		Dan *Månsson -73* W Bg	Swedish	255	122·0	24·7	13·0	Wärnanäs	1863	Hahn		C. Henckel	Helsingborg
	82	V	Dana *Bönnelycke 85-88* W Bn	Danish	180	107·1	22·9	10·8	Kjöge	1870	J. C. Sorensen	[illegible]	C. Bonnelycke	Næstved
	83	V	Dana *Petersen* W Bn	Danish	177	103·0	21·6	11·8	Horsens	1874	Chr. Schrøder	[illegible]	Aktie Selskab "Dana" (R. Petersen)	Horsens
	84		Danæ *Holckhamen* W	Russian	105	..	..	..		1883			A. Waitlinen	Impilaks
	85	✻ D	Danaé *Henoil 85-85* I Bk ex James Bald	French	705	182·8	30·8	18·4	Newcastle	1876	Wigham [Richardson & Co		A. C. Le Quellec	Bordeaux
	86		Danebod *Danielsen 85-86* W Bk (LOST) [illegible]	Danish	450	137·6	27·8	16·6	River John, NS	18 63/74		V 12.81	M. Christiansen	Ronne
	87		Dania *Simonsen 85-87* W Bg ex [illegible]	Danish	225	101·3	22·5	12·3	Hammelwrdn	1864	Meneke, Botter & [Co.		Kingo Pedersen	Odense
	88	G V	Daniel (BROKEN UP) W Bk	German	416	111·8	28·8	15·7	Altona	1865	Ernst Dreyer	[illegible]		Hamburg
	89		Daniel *Grant 49-* W Sr	Russian	216	100·0	27·0	12·4	Koltzen	1876	S. Grewe		S. Grewe	Riga
	90		Daniel *Busch* W Sr	Russian	210	92·0	23·0	11·0	Orrenhof	1877	Em. Titan		Rein Grant	Haynasch
	91	R V	Daniel Barnes *Storer* W S	American	1485	210·3	39·9	24·1	Bath, Me.	1877	W. Rogers	[illegible]	R. P. Buck & Co.	New York
	92		Daniel Brittain *Colemate* W 3m Sr	American	236	111·0	29·0	9·6	P. Republic NJ	1863			E. P. Boggs	Boston, Mass.
	93		Daniel Brown *Curtis* W Sr	American	194	109·0	30·0	10·0	Poughkeepsie, [N.Y.	1853	Daniel Brown			Fall River, Mass.
	94	V	Daniel I. Tenny *Wilson 79-80* W S	American	1633	212·4	40·4	28·8	Newb'ryprt Ms	1875	J. Currier, jun.	V 11.87	Eben Sumner	Newburypt, Mass.
	95		Daniel M. French *Childs* W Sr	American	182	111·0	30·0	7·0	Bass River, NJ	1858			A. F. Childs	Barnstable, Mass.
	96		Daniel Marcy *Nickerson* W Sr	American	110	88·4	23·0	9·9	Essex, Mass.	1874			Howard Sargent	Boston, Mass.
	97		Daniel Morris *Meekin* W Sr	American	100	89·8	22·5	6·7	Brandywine Dl.	1855			W. C. Reynolds	Hartford, Con.
	98		Daniel Pierson *Harris* W Sr	American	252	114·0	29·0	10·6	Waldoboro Me	1866			Geo. Rowley	St. George, Me.
	99		Daniel S. Miller W Sr	American	125	96·0	27·6	8·4	Belleville, N.J.	1852			J. Steelman	New York
	100	R	Daniel S Williams, Jr. *Bennett* W 3m Sr	American	629	161·5	30·5	16·5	Wilmingt'n Dl	1877	Jackson & Sharp [Co.		T. M. Perot	Philadelphia

✻ = Lloyd's Register; D = Nederlandsche Vereeniging; G = Germanischer Lloyd; I = Registro Italiano; R = American Record;
A = Veritas Austro-Ungarico; H = Veritas Ellenico; N = Norske Veritas; V = Bureau Veritas.

Dans la 4me Colonne, C = Composite (membrure en fer, bordé en bois); I = Fer; Stl = Acier; W = Bois.

Les Dates du doublage en métal (Col. 11) ressortent des livres dans lesquels les navires sont respectivement classés ou des rapports des Inspecteurs du Lloyd's Register.
Pour plus amples renseignements sur les navires classés par le Lloyd's Register, voir le Lloyd's Register Book et le Lloyd's Yacht Register.

Official No.	No. in Book	Registry in which classed	Ship's Name, Material and Rig. Master, and Dates of Appointment to present Owner's Service and to Ship. Ship's Late Name.	Flag.	Net Tonnage	Dimensions: Length	Breadth	Depth	Built: Where	When	By whom	Date of Metalling	Owners or Managers
[illegible]	101		Daniel Webster *Smith* WSr	American	115	77·0	25·1	8·2	[illegible] NY	1838	Hadlyme Ct.		W.S.Hungerford
[illegible]	102		Daniele Lavaggi (ex [illegible], ex Thomas [illegible]) WBn	Italian	186	101·6	24·1	13·8	Voltri	1859	G.B.Briasco		D.Lavaggi
	103		Danmark *Jans 82-82* (ex Nimbe, ex Neenden) WBk	Danish	398	127·6	32·3	17·2	Copenhagen	18 53/85	R.Danish Dkyrd	[illegible]	P.P.Schmidt
	104	V	Danmark *Clausen 83-83* W3mSr	Danish	209	111·7	24·4	12·2	Faaborg	1884	R.Møller	[illegible]	H.C.Christensens Enke
	105	V	Danmark *Rasmussen* WBn	Danish	171	97·8	23·2	12·0	Faaborg	1871	R.Dyreborg	[illegible]	N.R.Storm
	106	V	Danmark *Ørsted 74-74* WSr	Danish	122	92·7	20·6	10·1	Nakskov	1874	H P Ridersborg [&Trochmann		F.E.Jorgensen
	107	V	Dannebrog *Nielsen 56-56* (ex Christian den Ottende) WBg	Danish	194	82·1	21·2	13·4	Svendborg	18 40/68	S.Møller		C.P.Buch
	108		Dannebrog *Martensen* [82-82 W3mSr	Danish	194	94·8	24·7	12·4	Marstal	1882	H.J.Bager		H.J.Bager
	109	V	Dannebrog *Thurøe 83-83* WSr	Danish	131	85·5	22·3	10·8	Thurø	1883	C.Bom		C.Bom
	110	G	Dannebrog *Nielsen -78* WSr	Danish	115	89·6	20·8	8·8	Thisted	1866	J.Thomsen		A.L.Nielsen
	111	G	Dannevirke *Clausen* (ex Albert) [-85 WBn	Danish	141	84·5	23·3	12·4	Korsør	1849	Hensche		A.H.Carlsen
218	112	A	Dante *Cosulich* WBk	Austrian	535	136·1	28·8	18·4	Lussinpiccolo	1877	G.S.Cattarinich	[illegible]	Simeone L.Cosulich
444	113		Dante (PONTOON) *Giannoni* WBg	Italian	161	86·9	25·1	11·9	Voltri	1851	L.Briasco		E.Muti fu G.
[illegible]	114	V	Dante B. *Capitanio* WBg	Italian	184	97·8	24·1	14·8	Venice	1865		[illegible]	V.Ballarin
780	115	I	Dante C. *Pozzo -88* (ex Gilda Dazzoni) WBk	Italian	349	112·5	28·9	18·2	Pra	1861	D.Tixi	[illegible]	L.Costa
13061	116		Danube *Harrison -87* WBg	British	192	92·6	23·8	14·2	Whitby	1855			G.Milburn
41154	117		Danube *King* WBn	British	150	100·2	24·9	11·0	Wallace,NS.	1856			T.S.Owen
	118	V	Danube *Gagnerot* I Bk	French	642	178·1	29·2	17·8	Bordeaux	1870	Forges&C.de [l'Océan		Maurel frères
275	119		Danubio *Gari* W3mSr	Italian	245	106·6	27·9	15·2	Voltri	1867			Gius.Pertuso
	120		Danzig *Isendyck 83-84* WBk	German	484	118·6	30·7	19·1	Danzig	1845	J.W.Klawitter		Schoenberg&Domansky
63180	121	✳	Daphne *Morrish 84-84* I Bk	British	955	204·0	34·0	21·7	PortGlasgow	1869	J.Reid&Co.		McDiarmid,Greenshields & Co.
90024	122	R	Daphne *Swain -87* WSr	British	137	84·4	25·8	8·7	IndiantwnNB	1880	J.Rowan		A.Mills
29097	123		Daphne WSrYt	British	120	99·6	19·2	10·1	Cowes	1861	M.Ratsey		SirS.M.Lockhart Bt.
	124	V	Daphné *Rozzel -87* WSr	French	136	88·9	21·9	11·1	Nantes	1878	J.Sevestre	C [illegible]	Vve Gustave Gautier
525	125		Daphne WBg	Greek	291	104·3	27·9	..	Syra	1888			J.Siskos
	126	N	Daphne *Madsen* WBk	Norweg'n	528	135·4	31·5	18·4	Arendal	1867	J.Smith	[illegible]	T.Thommesen & Smith
	127	N	Daphne *Larsen 82-82* WBk	Norweg'n	480	122·5	30·9	18·3	Tvedestrand	1855	C.A.Schmidt		H.E.Jansen
	128		Daphne (CONDEMNED) *Hansen* WBk	Norweg'n	307	114·0	28·6	14·7	Kragerö	1868	J.Brandi	[illegible]	C.Næss
	129	V	Daphne *Pattersson 80-82* WBk	Swedish	717	157·4	31·6	18·3	Gefle	1877	O.A.Brodin	[illegible]	V.T.Engvall
78370	130		Darcy Pratt *Smith -84* WBn	British	149	96·6	23·1	10·5	Jarvis Bay, [N.S.W.	1876	W.Pererley		B.Gilmer
157063	131		Dare *Perry* W3mSr	American	247	130·8	35·8	9·2	N.Bend,Oreg	1882			A.M.Simpson
53022	132		Daring *Banloft* WSw	British	191	98·6	24·8	12·9	Sunderland	1865	Barkes		T.Jefferson
27209	133		Daring *Moore* WSr	British	149	100·0	22·1	12·2	Salcombe	1859	Vivian		W.S.Allport & Co.
3593	134	I	Dario *Giusto -88* WBk	Italian	449	134·5	30·2	19·0	Sestri	1869	A.Briasco fu G B	[illegible]	Mart. Fravega & Eredi di G. Fravega
	135		Darja *Koishmutinen 87* WSr	Russian	105	..	..	..	Salmis	1883			A.Waittinen
88618	136		Darling *Sarty* WSr	British	113	82·5	24·3	9·6	Conquerall NS	1884	J.N.Rafuse		J.N.Rafuse
72297	137	V	Darpa *Gilmore -87* (ex Mary A. Kersten) WBg	British	344	131·5	29·0	12·9	Castleton,N.B.	1877	B.H.Appleby	[illegible]	W.W.Turnbull
53729	138	✳	Darra *Telfer* CBk	British	999	190·7	33·0	21·4	Aberdeen	1865	Hall	[illegible]	G.F.Mason
76075	139	✳	Dartford *Sampson -88* I S	British	1274	221·5	36·0	21·9	Sunderland	1877	Mounsey& [Foster		G.Traill & Sons
27788	140	✳	Dartmouth *Moore -88* WBk	British	915	185·4	34·3	21·6	Dundee	1859	A.Stephen&Ss.	[illegible]	G.Brownell
88643	141	✳	Darwin *Jago -88* W4mSr	British	500	161·9	29·9	17·3	P.E.Island	1884	J.W.Richards	[illegible]	T.P.Richards & Co.
94310	142	✳	Darwin *Sanders 87-* WSr	British	91	78·3	20·2	10·0	Littlehampton	1887	J.&W.B.Harvey	[illegible]	S.Manode
88135	143	✳	Dasher WK	British	84	76·0	20·8	10·6	Burt'nStather	1883	J.Wray&Son		W.L.Robins
56459	144	✳	Dashing Wave *Cundy 69-* W3mSr	British	159	96·4	22·8	12·1	Newburgh	1868	Scrimgeour	[illegible]	N.Hocken
6583	145		Dashing Wave *Conner* WS	American	1012	181·8	39·6	21·3	Portsm'thNH	1853			J.W.Ackerson
6546	146		Dashing Wave *Gaffney* [-83 WSr	American	141	106·0	27·0	9·7	Eureka,Cal.	1867			W.C.Lynde
78452	147	✳	Dashwood *Smith 85-85* WSr	British	98	86·0	22·0	10·5	Bridgwater	1878	Gough		T.Allen
	148	N	Dato *Hansen -88* (ex [illegible]) WBg	Norweg'n	493	133·2	31·7	16·8	Ekenäs,Finland	1872			Actieselskabet "Dato" ([illegible])
83021	149		Dauntless WSr	British	156	105·5	23·4	9·1	Oakville	1867	J.Potter		Mrs.I.Stark
71139	150		Dauntless WSr	British	144	101·6	22·0	8·8	PortDover	1874	W.Allan		W.Allan

✳=**Lloyd's Register**; D=Nederlandsche Vereeniging; G=Germanischer Lloyd; I=Registro Italiano; R=Americ[an]
A=Veritas Austro-Ungarico; H=Veritas Ellenico; N=Norske Veritas; V=Bureau

In the 4th Column, C=Composite (iron frame, wood planked); I=Iron; Stl=Steel; W=Wood.

The Dates of Metalling (**Col. 11**) are based on the records of the Books in which Vessels are respectively classed, or are the Reports of the Surveyors of Lloyd's

Further particulars of Vessels classed by Lloyd's Register will be found in Lloyd's Register Book and Lloyd's Yacht Register.

1	2 N° dans le livre	3 Registre dans lequel le navire est classé	4 Nom. Materiaux et Gréement. Capitaine et dates d'entrée au service des Armateurs actuels et à ceux du navire. Ancien Nom.	5 Pavillon	6 Tonnage net	7 Longueur	Largeur	Creux	8 Construit: Où	9 Quand	10 Par qui	11 Date du doublage en métal	12 Armateurs ou Directeurs-Gérants.	13 Port d'Armement.
	151		Dauntless *Hengford* W K	British	143	98·8	23·2	8·1	Sittingbourne	1873	Spencely		J.Smith	Maldon
	152		Dauntless *Wade* -85 W Bn	British	108	79·6	21·3	11·8	Harrington	1860	Williamson		T.Owens	Waterford
	153		Dauntless *Carroll* W Sr Yt	American	120	130·3	24·6	10·0	Mystic,Conn.	1866			C.H.Colt	Newport,R.I.
	154	✻ V	Dauphin *Bréhaut* W Bk	French	404	122·7	25·7	14·4	St.Malo	1867	Parnet		A.Hubeau	St.Nazaire
	155		David W 3m Sr	Brazilian	221	..	..	..					D.F.Grillo & Co.	Rio Grande
	156	I	David *Bellino* W Bk	Italian	809	158·1	33·1	23·2	Chiavari	1876	M.Tappani		D.Dall'Orso fu G.	Spezia
	157	D V	David *Behrsing* W 3m Sr	Russian	315	124·0	29·1	13·9	Magnushof	1874	H.Grewe		M.Grewe	Riga
	158		David A. Small *Walden* W Bn	American	114	87·6	22·9	9·1	Provincetown, Mass.	1868	J.A.Whitcomb		W.Curren	Provincetn.Mass.
	159		David Auterson W Bk	Salvador	256	105·8	24·1	15·6	Scilly	1870	Edwards		Angulo & Salazar	San Salvador
	160	R	David Bugbee *Stowers* W Bn	American	417	128·6	28·1	16·8	Stockton,Me.	1868			Pendleton & Carver	Portland
	161		David Carll *Somers* W Sr	American	119	103·0	27·0	8·2	CityIslandNY	1876	D.Carll		Jacob van Name	New York
	162	R	David Clarkson W 3m Sr	American	279	129·0	32·2	9·2	Millville,N.J.	1873			D.Ireland	Camden,N.J.
	163	R	David Crockett *Anderson* W S	American	1483	218·8	41·0	27·0	Mystic,Conn.	18 53/68			J Dunham's Nephew & Co.	New York
	164		David Currie *Pease* W Sr	American	144	100·0	26·0	7·0	Portland,Conn	1866	Gildersleeve		W.A.Pease	Hartford.Con.
	165		David Faust *Smith* W Sr	American	206	102·0	28·1	9·5	Philadelphia	1854			Isaiah Blaisdell	Ellsworth.Me.
	166	R	David Rees *Williams* W Bn	British	209	103·8	25·4	12·8	P.E.Island	1879	J.R.Bourke		J.Williams	Charlottetn.P.E.I.
	167		David S. Siner *Guthrie* W Sr	American	221	115·6	28·9	9·3	Milford,Del.	1861	H.&J.Deputy		J.May	Philadelphia
	168	✻	David Sinclair *Jones* W Sr	British	122	87·0	21·5	11·9	Brixham	1872	Ferneaux		H.Jones	Brixham
	169		David Stewart *Davison* W S	Argentine	673	148·0	31·7	20·3	Baltimore	1857	J.A.Robb		C.Morton Stewart	Buenos Ayres
	170	R	David Taylor *Grafton* -88 W Bk	British	599	146·4	33·2	18·7	Rockland,NB	1869	R.E.Chapman		C.M.A.Godfrey	St.John,N.B.
	171		David Torrey *Orne* 86-86 W Sr	American	158	96·3	27·6	8·2	Brunswick,Me	1873	DanielBrewer		W.S.Jordan & Co.	Portland,Me.
	172	R	David W. Hunt *Merritt* -87 W 3m Sr	American	331	135·4	32·3	10·4	Bath,Me.	1881	Donnell& Deering		B.W.&H.F.Morse	Bath.Me.
	173		Davida *Roberts* W Sr	British	377	133·5	28·8	16·5	Parrsboro',NS	1881	T.Trahey		D.S.Howard	Parrsboro',N.S.
	174	D	Dawn *Smart* -87 W Bn	British	154	102·2	22·5	11·9	Gaspé	1874	Gaudin		J. & E. Collas	Gaspé
	175		Dawn W Bn	British	113	92·5	22·3	10·6	P.E.Island	1870			M.Tobin	St.Johns.N.F.
	176		Dawn *Williams* W Bk	American	247	126·0	27·8	12·1	New York	1857			H.C.Wright	San Francisco
	177	✻	Dawpool *Eaton* I S	British	1697	262·9	38·3	23·3	Belfast	1880	Harland & Wolff		North Western Shipping Co., Ld.	Liverpool
	178		Dawstone *Hosen* 85-86 W Bk	British	496	137·8	21·5	17·6	Arbroath	1853			R.&W.King	Bristol
	179		Day Spring *Braithwaite* W Sr	British	159	110·4	20·6	10·4	Johnson's Bay, N.S.W.	1873	Duncan&Sorrie		James Cosh	Sydney.N.S.W.
	180	✻	Daybreak *Horn* W Bk	British	289	110·6	27·3	16·9	Sunderland	1861	Rawson&W'ts'n		W.Yeoman	North Shields
	181		Daybreak *Duffy* 86-86 W Sr	American	166	99·5	26·6	8·0	Stockton,Me.	1865			Turner & Oates	Mobile.Ala.
	182		Daylight *Hogdon* W 3m Sr	American	628	158·0	35·6	15·8	Belfast,Me.	1882	W.Cotterell		Armstrong & Co.	Newhaven.Con.
	183		Daylight *Foster* W Bn	American	227	118·0	28·0	10·0	E.Haven,Con.	1872			L.W. & P.Armstrong	New York
	184	N	De fire Søskende *Agerup* W Bg	Norweg'n	203	84·5	24·3	15·7	Kragerö	1806			C.N.Agerup	Tönsberg
	185	N	De fire Søskende *Gulliksen* W Bg	Norweg'n	187	81·7	30·7	14·0	Kragerö	1848			G. Davidsen & G. A. Gulliksen	Tönsberg
	186		De Goede Hoop W Bn	Dutch	186	104·9	23·5	11·5	Nieuwediep	1868			Nio Ing Sing	Macassar
	187	I	DeMari Marcello *Gazzolo* -88 W Bk	Italian	892	173·8	31·7	24·1	Varazze	1884	G.Scala		P.Lavarello	Genoa
	188		DeMoryGray *Thomas* -87 W 3m Sr	American	381	131·3	32·3	9·9	P.JeffersonNY	1873	J.M.Bayles&Son		S.H.Miller	New York
	189		De Tvende Brødre W Sr	Danish	107	71·5	22·7	11·6	Svendborg	1831	L.N.Møller		L.Clausen	Svendborg
	190		De Twee Vrienden W Bk	Dutch	568	141·7	29·8	17·5	Elsfleth	1874	J.Jürgens		DeWitt,Schippers & Co.	Sourabaya
	191	G	De Zwaan *De Vries* W Sr	German	158	92·6	21·5	10·9	Middelburg	1858	DeCommercie		Haye de Vries	Leer
	192	A	Dea *Gavagnin* W Bk	Italian	416	137·8	30·5	20·4	Chioggia	1874	G.Poli	Y 11.88	A.Gavagnin	Venice
	193	N	Dea (ex Olive) *Hansen* 80-82 W Bk	Norweg'n	558	140·5	29·6	19·5	North Shields	1854	T.&W.Smith		M.Mulvad	Fredriksstad
	194		Deacon W Bg	American	402	131·0	34·0	12·6	Bath,Me.	1863			O.Merrill	San Francisco
	195	✻	Deanfield *Irvine* -81 I Bk	British	1072	208·0	34·0	21·0	PortGlasgow	1878	Russell&Co		W.H.Ross & Co.	Liverpool
	196	✻	Deborah *Owen* 87-87 W Sr	British	94	79·0	20·5	10·1	Wales	1869	Jones		J.Owen	Caernarvon
	197	G	Deborah *Haack* -84 W Sr	German	143	94·8	22·2	10·1	Leer	1879	C.Schwoon		F.Janssen Haak	Karolinensiel
	198	✻	Decapolis *Candler* 80-80 I Bk	British	632	187·5	30·0	18·0	Sunderland	1868	W.Pile&Co.		T.B.Walker	London
	199		Decatur Oaks *Hoker* 70-76 W Sr	American	147	97·0	27·2	9·0	Setauket,N.Y.	1856			J.Sayer	Hartford.Conn.
	200		Decidée *Devonainck* W Sr	French	112	86·3	19·7	11·1	Dunkirk	1868	E.Sagary	Y 12.88	Gustave Dinoir	Dunkirk

✻ = Lloyd's Register; D = Nederlandsche Vereeniging; G = Germanischer Lloyd; I = Registro Italiano; R = American Record; A = Veritas Austro-Ungarico; H = Veritas Ellenico; N = Norske Veritas; V = Bureau Veritas.

Dans la 4me Colonne, C = Composite (membrure en fer, bordé en bois); I = Fer; Stl = Acier; W = Bois.

Les Dates du doublage en métal (Col. 11) ressortent des livres dans lesquels les navires sont respectivement classés ou des rapports des Inspecteurs du Lloyd's Register.

Pour plus amples renseignements sur les navires classés par le Lloyd's Register, voir le Lloyd's Register Book et le Lloyd's Yacht Register.

Official No.	No. in Book	Registry in which classed	Ship's Name. Material and Rig. Master, and Dates of Appointment to present Owner's Service and to Ship. Ship's Late Name.	Flag	Net Tonnage	Length	Breadth	Depth	Built Where	When	By whom	Date of Metalling	Owners or Managers
	201		Decidido (ex Ajace) de Sousa W Bg	Portug'se	228	93·2	22·6	7·3	Castellamare	1853	G. Bonifacio		F. Ricca
	202	N V	Decima Johnson W Bk	Norweg'n	770	175·3	33·8	19·4	[illegible]	1884	J. Bang	[illegible]	J. Bang
6195	203		Decorra Berry 76-76 W Sr	American	151	96·8	25·1	9·2	Machias, Me.	1866	Machias Port		J. W. Berry
89846	204	✱	Dee Pape 85-85 I Bk	British	1115	215·2	35·1	21·2	PortGlasgow	1885	Russell & Co.		P. Macfarlane
	205	N	Defensor Olsen 75-84 W Bk	Norweg'n	870	175·4	36·3	21·2	Porsgrund	1884	J. Jeremiassen	[illegible]	Johan Jeremiassen
78394	206		Defiance Bystantson 86-86 W Bn	British	199	102·8	24·9	12·2	Freeman's Bay (N.Z.	1880	R. Mackay	[illegible]	R. Mackay
3176	207	✱ I	Degregori A. Righetti (ex Giorgio Washington) -88 W Bk	Italian	834	171·2	34·1	21·9	Pra	1875	Parodi	[illegible]	F. Degregori
64014	208		Dei Gratia Donovan 77-77 W Bn	British	296	111·0	27·5	13·0	BearRiver, N.S	1871	L. Bogart		M. Fleming
73381	209		Delaware W Bk	British	249	106·0	26·5	11·9	Damariscotta, [U.S.	1848			Government of New Zealand
83152	210		Delaware W Sr	British	152	105·0	22·0	8·0	Oswego, U.S.	1880			N. Allen
	211	R	Delfin Alander -87 (ex S.S. Björneborg) W Bk	Russian	463	118·1	28·9	16·7	Björneborg	1873	J. Bandorff		J. Lindqvist
	212		Delfin Turnroos W Sr	Russian	280	..	..	..	Kimito	1875			J. A. Ekblom
	213	R	Delfin Holmén -77 W Sr	Russian	219	92·5	27·2	13·4	Tenala	1877	K. Henriksson		A. Aminoff
	214		Delfin Årholm W Sr	Russian	171	..	..	..					J. Arholm
	215		Delfin W Sr	Spanish	118	84·0	24·1	10·2	Blanes	1858			[illegible]
	216		Delfin (SCNK) Gorin W Bg		210	111·2	23·8	11·1	Hernösand	1864	J. O. Strandberg		
	217	G	Delfin Møller 68-79 W Sr	Swedish	154	93·8	21·4	11·5	Sundsvall	1847			P. Möller
39267	218		Delhi Allen 81-84 W Bg	British	186	94·0	24·5	13·8	P.E. Island	1857	R. Orr		P. Ward
6739	219	R	Delhi Green W Sr	American	253	113·1	28·0	12·5	Saco, Me.	1872	W. C. Merrill		R. F. C. Hartley
	220	D V	Deli Van Duin (ex Gijsen en Priesterer) W Bk	Dutch	871	164·8	33·5	21·4	Rotterdam	1876	Smit	[illegible]	Von Lindern & Van Vo[illegible]
	221		Deli El Hair Said Mohd. Basfoun (ex Delfin) C Bk	Dutch	373	144·0	27·2	..	Amsterdam	1873		[illegible]	Said Agil bin Aljdroos bin Agil bin Salim
	222	D V	Deliane Von Lindern 76-76 C Bk	Dutch	1316	208·3	39·7	24·8	Alblasserdam	1876	F W Von Lindern	[illegible]	J. Veth
3158	223	I	Delicato Oanone -88 W Bk	Italian	439	118·4	27·9	19·4	Varazze	1868	A. Bello	[illegible]	P. Lavarello
72971	224	R	Delma C. (WRECKED) Spurr W Bn	British	236	105·9	28·7	11·2	BearRiver, NS	1875			G. F. Miller
48828	225	F	Delmira (ex Guarany) Sonne I Bk	British	338	134·7	24·3	15·6	Chester	1864	RhoodeeIronW.		J. Benn
6026	226		Delmont Locke W Sr	American	133	..	..	..	Bucksport, Me	1861			A. P. Veazie
	227	V	Delphin Janssen 87-87 W SmSr	Dutch	230	99·7	24·1	11·3	Cappeln	1871	C. A. Stoen	[illegible]	P. H. Craandijk & Zn
	228	V	Delphin Wehmeyer W Bg	German	231	102·6	21·3	12·0	Brake	1865	J. D. Behrens		B. F. de Haan
	229	V	Delphin Stahl 64-84 W Sr	German	135	91·5	23·1	10·5	Kiel	1866	Reuters & Dums		H. Lafrenz
	230	N	Delphin Olsen W Bk	Norweg'n	351	127·1	29·2	15·2	Arendal	1871	C. Halvorsen	[illegible]	A. D. Westergaard
	231	N	Delphin Bjönness W Bg	Norweg'n	326	125·5	28·8	14·6	Umeå	1842			N N Bjönness's Enke
	232	N	Delphin Simensen W Sr	Norweg'n	110	79·2	22·7	9·1	Kragerö	1871			H. Biörn, jr.
	233	D	Delphin Olender W Bk	Russian	616	145·6	32·5	17·1	Wasa	18 64/72			C. G. Wolff
	234	G	Delphin Barron 86-85 W Bg	Russian	236	89·6	21·6	11·0	Libau	1865	G. Buchhoff		J. C. Schneider & Co.
	235		Delphin W Sr	Russian	165	..	..	..		1841			O. A. Färdig
	236	V	Delphine Avelles W Bg	French	151	79·1	23·0	10·7	Paimbœuf	1865	Baudet & Co.		A. Cardinal fils aîné
	237	V	Delphine Anna Guillard W Bn	French	148	86·9	20·0	11·4	LeLegué	1868	Frelicot		Ravoux Blévin
	238	V	Delphine Mélanie Gesnager -88 I Bk	French	444	129·9	27·2	16·4	Nantes	1867	T. H. Dubigeon [& fils		J. & C. Baratoux
70654	239	✱	Delscey Leleau 80-80 I Bk	British	670	181·0	29·8	19·2	Sunderland	1875	W. Doxford & Sn		Shallcross & Higham
51107	240	✱	Delta Finch -88 I Bk	British	537	162·6	28·7	18·2	Cardiff	1865	C. Hill & Son		Swansea Merchant Shipowners Co. Ld.
	241	N	Delta Johansen W Bk	Norweg'n	524	141·8	30·3	18·0	Granville, N.S.	1866	L. Delap	[illegible]	O. A. Johansen
	242		Delta Jansohn 75- (ex Et-ten, ex Enjetto) W Sr	Russian	116	82·8	21·8	10·1	Pekela	1860	B. Drenth		Count N. Stenbock
1867	243	I	Demarchi Cappano -88 W Bk	Italian	469	139·7	30·7	19·9	Sestri	1869	G. Calcagno	[illegible]	G. Demarchi
	244	G	Demetra Schramm 86- W Bk	German	428	134·3	29·5	17·2	Elbing	1863	GebrMitzlaff		A. Hohorst
93318	245		Demostenes Menzies Stl Sr	British	151	115·0	24·0	8·1	Dumbarton	1886	W Denny & Brs		La Platense Flotilla Ld.
92380	246		Demozelle Martin -88 W Sr	British	163	94·0	28·1	9·2	HopewellCape, NB	1888	S. F. Rose		S. F. Rose
	247	N	Den gode Hensigt Nissen W Bg	Norweg'n	195	82·5	24·1	14·4	Svelvig	18[illegible]			O. Olsen's Enke
74497	248	✱	Denbighshire Atkinson 85-85 I S	British	1367	231·4	38·9	23·6	Liverpool	1876	T. Royden & Sns.		Hughes & Co.
	249		Deniense W Sr	Spanish	136	76·4	22·4	10·0	Benidorme	1852			Antonio Ferrer
	250												

✱ = Lloyd's Register ; D = Nederlandsche Vereeniging ; G = Germanischer Lloyd ; I = Registro Italiano ; R = Ame[illegible]
A = Veritas Austro-Ungarico ; H = Veritas Ellenico ; N = Norske Veritas ; V = Bure[illegible]

In the 4th Column, C = Composite (iron frame, wood planked) ; I = Iron ; Stl = Steel ; W = Wood.

The Dates of Metalling (Col. 11) are based on the records of the Books in which Vessels are respectively classed, or on the Reports of the Surveyors [illegible]

Further particulars of Vessels classed by Lloyd's Register will be found in Lloyd's Register Book and Lloyd's Yacht Register.

1	2 N° dans le livre	3 Registre	4 Nom. Matériaux et gréement. Capitaines et dates d'entrée au service des Armateurs actuels et à bord du navire. Ancien Nom.		5 Pavillon.	6 Tonneaux	7 Dimensions. Longueur	Largeur	Creux	8 Construit. Où.	9 Quand.	10 Par qui.	11 Date du doublage	12 Armateurs ou Directeurs-Gérants.	13 Port d'Armement.
[illegible]	251		Denmark	W Sr	British	[illegible]	134·8	25·6	10·9	Garden I. Ont	1867	H. Roney		The Calvin Co., Ld.	Kingston, Ont.
[illegible]	252		Denmark	W Bg	British	221	94·0	21·2	11·[illegible]	[illegible]				J.B. Bell	Kingston, Jam.
[illegible]	253		Dennis Brundrit *Thomas*	W Bk	British	462	144·1	27·[illegible]	18·[illegible]	[illegible]	1856		[illegible]	Geo. M. Dean	Stanley, F.I.
	254	N	Deo Gloria *[illegible]*	W Sr	Norweg'n	162	85·8	22·4	11·7	Stralsund	1859			H.C.P. Bakkeskoug	Tönsberg
[illegible]	255		Deodar *Hammond -87*	W Bk	British	406	132·3	27·5	17·6	Dundee	1858	Dundee S.B.Co.		H. Campbell	North Shields
	256	V	Deodata *Arntzen [illegible]*	W Bk	Danish	340	114·9	26·7	16·4	Stavanger	18 55/78		[illegible]	Th.J.E. Arntzen	Dragör
	257	N	Deodata *[illegible]*	W Bk	Norweg'n	462	120·4	32·1	18·6	Brahestad	1857			C.M. Haavaldsen	Drammen
	258	N D	Deodata *Andersen -87*	W Bk	Norweg'n	459	139·5	28·7	17·4	Sunderland	1869	Richard Thompson	[illegible]	Chr. Hannerig	Christiania
	259	N	Deodata *Nicolaysen -87*	W Bk	Norweg'n	369	114·6	29·4	16·4	Drammen	1851			H.T. Kiær	Drammen
	260	N	Deodata *Andersen*	W Bk	Norweg'n	361	114·8	28·7	15·9	Tvedestrand	1868	O. Gundersen		A. Gundersen	Tvedestrand
	261	N	Deodata *Hansen*	W Bg	Norweg'n	360	105·5	28·9	17·5	Tvedestrand	1851			H.A. Thorbjörnsen	Tönsberg
	262	N	Deodata *Abrahamsen*	W Bg	Norweg'n	234	99·4	27·3	13·0	Grimstad	1866	A. Osmundsen		P.B. Einersen	Grimstad
	263	V	Deodata *Nielsen -87*	W Sr	Norweg'n	153	96·2	23·8	11·7	Stavanger	1866	R.O. Berrum		R. Sernaes	Flekkefjord
	264		Deolinda	W Bn	Brazilian	182	..	..	..					Caroline J. Neves	Rio de Janeiro
	265		Deolinda	W Sr	Brazilian	125	..	..	..					A.M. Grant	Rio de Janeiro
7013	266		Dependence *Beale 74-74*	W Sr	British	114	85·5	21·2	10·8	Itchenor	1874	Beale		E.J. Beale, jun.	Littlehampton
	267	V	Der Freischütz (ABANDONED) *Wolter [illegible]*	W Bg	German	177	90·5	21·6	12·3	Wollin	1858	C. Brüsewitz		J. Wiesenewsky	Stettin
	268	G	Der Friede *Biedenweg 80-80*	W Bk	German	329	108·0	27·8	16·1	Greifswald	1862	L. Wittenberg		C. Graedener	Greifswald
	269	V	Derjunge Prinz *Sassen -76*	W Sr	German	105	86·0	20·1	9·0	Nordloh	1868	E. Reil		J.B. Sassen	Varelerhafen
	270	V	Der Kurlander *[illegible] -88*	W 3m Sr	Russian	372	134·1	26·9	14·2	Windau	1888	Redicktum		Gebr. Gutschmidt	Windau
	271	V	Der Nord *Werner*	W Bk	German	499	133·3	31·0	18·1	Stettin	1858	E. Zieske	Y 8.86	W. Schlutow	Stettin
	272	V	Der Nordpol *Schalt -77*	W Bk	German	366	118·4	28·6	16·5	Zingst	1863	Drossel		W.A. Sarnow	Barth
	273	G V	Der Pommer *Bohm -73*	W Bg	German	234	99·0	26·2	14·3	Ueckermünde	1845	L. Wittenberg		H. Bohm	Swinemünde
	274		Der Versuch (LIGHTER) *Prischpenk* [illegible]	W Bk	Russian	426	134·5	31·7	15·9	Nova Scotia	1863	McLennan		J. Johannsen	Riga
	275		Der Versuch *Kudiet*	W 3m Sr	Russian	287	111·2	26·2	14·1	Kesterzeem	1875	C.J. Silinsky		C. Giel	Kesterzeem
	276	V	Der Wanderer *Beyer 86-88*	W Bk	German	600	148·5	33·1	18·0	Danzig	1862	F. Devrient		A. Gibsone	Danzig
[illegible]	277		Der Zehnte Juni *Edwards -86*	W Jary	British	263	102·8	27·8	15·8	Germany	1838			J.B. Bushell	S. Shields
	278		Der 27 Juni *Muller* [illegible]	W Bg	Russian	231	92·0	21·0	11·3	Fünfhausen	1860	J. Strenge		M. Grant	Libau
	279	G	Derby *Schierhorst 86-88*	W S	Norweg'n	1087	177·5	33·1	23·6	Chelsea, Mass.	1856		[illegible]		
[illegible]	280	✻	Derby Park (WRECKED) *[illegible] 86-86*	Stl Bk	British	1292	240·7	36·1	21·3	Belfast	1887	Workman, Clark & Co.		P. Iredale & Son	Liverpool
[illegible]	281	✻	Derbyshire *Williams -87*	I S	British	1163	207·2	33·9	23·5	Liverpool	1863	Jones, Quiggin & Co.		C.H. Rees	Liverpool
	282		Dervisch	W Bg	Turkish	137	..	..	..	Greece	1876			Tijar Abdullah [illegible]	Abahnah
	283		Dervisch	W Bg	Turkish	137	..	..	..	Greece	1870			Kilindar Kiayani Hussein Effendi	Constantinople
	284		Dervisch	W Sr	Turkish	130	..	..	..	Russia	1879			Makarchji Esnaf Rejeb Ousta	Constantinople
	285		Dervisch	W Sr	Turkish	125	..	..	..	Aidos	1876			Osman Chaoush oglou Ibrahim	Aidos
	286		Dervisch	W Bg	Turkish	100	..	..	..	Barten	1871			Bartenlee Mehemed Hadjah	Barten
	287		Dervisch	W Bg	Turkish	100	..	..	..	Greece	1871			Kushah oglou Teodori & Bros.	Constantinople
	288		Dervisch Bahri *Abdul [illegible]*	W Bg	Turkish	182	..	..	..	Greece	1885			[illegible] Balikji Ahmed Zadeh Abdul Rahman	Heraclea
	289		Dervisch Bahri	W Sr	Turkish	175	..	..	..	Barten	1874			Hadji Ali Bey zadeh Hadji Hussein Agha & Co.	Barten
	290		Dervisch Bahri	W Bg	Turkish	120	..	..	..	Tereboli	1885			Terebolielee Hadji Feizi & Co.	Terebolie
	291		Dervisch Bahri	W Bg	Turkish	112	..	..	..	Barten	1876			Bartenlee Tijar Hadji Ali Bey zadeh Daoud Effendi	Barten
	292		Dervisch Bahri	W Sr	Turkish	100	..	..	..	Chakraz	1869			Mollah Halil oglou Mehemet & Co.	Chakraz
	293		Dervisch Bahri	W Bg	Turkish	100	..	..	..	Barten	1876			Bartenlee Choujour oglou Ali	Barten
	294		Dervisch Bahri	W Sr	Turkish	100	..	..	..	Shillee	1884			Bichakji Hadji Halil Effendi	Shillee
	295		Dervisch Bahri	W Sr	Turkish	100	..	..	..	Allaplee	1876			Allaplee Pehlevan oglou Hussein & Co.	Allaplee
[illegible]	296	✻	Derwent *Andrew 81-84*	I S	British	1890	275·0	40·2	23·7	Dumbarton	1884	McMillan & Son		Devitt & Moore	London
[illegible]	297		Derwent *Westlake 69-83*	W Bk	British	372	128·0	28·1	17·5	Sunderland	1868	J. Crown	Y 8.82	Great Grimsby Ice Co. Ld.	Grimsby
[illegible]	298		Derwent	W Bg	British	221	87·5	24·0	16·6	Workington	1834			C.W. Turner	Auckland, N.Z.
[illegible]	299		Derwent Hunter (ex North [illegible])	W Bk	British	368	106·2	29·0	18·6	Kensingt'n U S	1811			A. McGregor	Hobart Town
[illegible]	300	A	Descovich A. *Negovetich -87*	W Bk	Austrian	440	123·3	27·8	18·3	Fiume	1862	G.G. Baccich [Bellan]	Y 2.83	A.E. Descovich	Fiume

✻ = Lloyd's Register; D = Nederlandsche Vereeniging; G = Germanischer Lloyd; I = Registro Italiano; R = American Record; A = Veritas Austro-Ungarico; H = Veritas Ellenico; N = Norske Veritas; V = Bureau Veritas.

Dans la 4me Colonne, C = Composite (membrure en fer, bordé en bois); I = Fer; Stl = Acier; W = Bois.

Les Dates du doublage en métal (Col. 11) ressortent des livres dans lesquels les navires sont respectivement classés ou des rapports des Inspecteurs du Lloyd's Register.

Pour plus amples renseignements sur les navires classés par le Lloyd's Register, voir le Lloyd's Register Book et le Lloyd's Yacht Register.

Official No.	No. in Book.	Registry in which classed	Ship's Name. Material and Rig. Master, and Dates of Appointment to present Owner's Service and to Ship. Ship's Late Name.	Flag.	Net Tonnage.	Dimensions. Length.	Breadth.	Depth.	Built. Where.	When.	By whom.	Date of Metalling	Owners or Managers
1	2	3	4	5	6	7			8	9	10	11	12
76011	301	✳	Desdemona *Baron 83-83* I S	British	1490	242·0	37·7	22·9	Liverpool	1875	W.H.Potter&Co.		C.T.Bowring & Co.
67620	302	✳	Desdemona *Evans* W Sr	British	88	79·3	20·8	10·2	Aberystwith	1874	Evans		Miss M.Evans
6218	303		Desdemona *Davis* W Bk	American	237	98·6	26·6	14·7	Middletown, [Conn.	18 23/45			Aiken & Swift
	304		Desempenho W Bn	Brazilian	150	..	..	..					J.A.V. de Amorim
	305		Desengano W Sr	Brazilian	189	..	..	..					A.G. da Silva
	306	DNV	Desideria *Jörgensen* W Bk	Norweg'n	724	159·0	34·9	19·4	Tvedestrand	1880	A.&F.Smith	Y 1.86	A. & F.Smith
46100	307		Despatch W Bn	British	226	105·3	26·0	12·6	Shelburne,NS.	1863			B.Crawley
83516	308	✳	Despatch *Roberts 86-86* W3mSr	British	145	103·0	24·4	11·1	Runcorn	1886	Brundrit & Co.		J.D.Davies
	309		Despatch (W.R.D.K.E.) *Gulbitzen* W Bk	Norweg'n	293	114·6	26·9	13·9	Grimstad	1883	N.O.Gjömle	Y 6.86	O.M.Pedersen
	310		Despejada W Pol	Spanish	133	80·4	23·5	12·4	Arens	1840			F.Montaner y R…
	311	A	Despina *Lazzaroni* W Bg	Russian	285	..	..	..	Chios	1878	G.DeMaria	Y 11.84	O.Lazzaroni
301	312	✳	Despina (ex Professor Airy) W Bk	Greek	403	139·0	28·4	17·6	Barnstaple	1866	Westacott	Y 8.86	I.Mamoussis
136	313		Despina *Papalas* W Bn	Greek	196	84·0	24·9	..	Syra	1869			D.Mikakoulis
453	314		Despina W3mSr	Greek	194	99·1	23·3	..	Scotland	1870			Dem.Milokos
92	315		Despina W Bg	Greek	161	85·6	24·6	..	Syra	1867			C.Cavouras
127	316	A	Despina *Papas -88* W Bg	Greek	159	86·0	26·2	15·1	Syra	1868			A.Papas
129	317		Despina W Sr	Greek	150	80·7	23·8	..	Syra	1875			N.Benjamin
	318		Despina Prashko W Bg	Turkish	190	..	..	..	Evranyieh	1881			Kerestchjii Christopher oglou Christopher
	319	GV	Detmar *Ruckens 87-87* W Bn	German	261	106·3	25·0	12·6	Bardenfleth	1869	FockeBros.	Y 3.85	Seetzen Gebrüder
6547	320		Detroit *McAllup -86* W Bk	American	1438	198·3	38·5	27·0	Yarmouth,Me	1864			J.J.McKinnon
53105	321	✳D	Deucalion *Baker 87-87* C Bk	British	399	122·9	27·3	17·0	Sunderland	1865	R.Thompson, jr	Y 12.86	W.A.Hornbuckle
	322	V	Deus regit *Knudsen 86-86* W Bn	Danish	142	92·3	21·3	11·8	Korsør	1868	C.Lund		E.C.Jorgensen
	323		Deus te ajude W Sr	Brazilian	176	..	..	..					F.P. de Barros
	324		Deus te guarde W Sr	Brazilian	120	..	..	..	Pernambuco			Y 80	B.Lourenço
	325		Deus te guie W Sr	Brazilian	120	..	..	..	Pernambuco				B.Lourenço
	326	V	Deutschland (ex Tennyson) *Steffen -88* W S	German	1251	193·9	37·1	23·6	Barth	1864	Sewall	Y 6.86	R.S.Carr
	327	V	Deutschland *Bosse 87-87* I S	German	833	176·9	32·1	20·4	Hamburg	1858	ReiherstiegShip [Works		J.C.Pflüger & Co.
	328	V	Deutschland *Lüders [81-81* W3mSr	German	381	130·5	27·6	14·2	Cappeln	1881	C.A.Steen	Y 12.86	C.Brandt
	329	G	Deutschland *Rehberg 78-78* W Bg	German	289	107·8	26·8	15·6	Rostock	1857	W.Zeltz		A.Burchard
	330	✳	Deux Amis *Armagnac 82-86* W Bg	French	443	131·9	27·9	14·7	SestriPonente	1873		Y 12.87	Vve.A.Cabrol jeune&…
	331		Deux Amis W	French	211	110·8	21·3	11·5	Ajaccio	1864	L.Arman		Fabiani
	332		Deux Amis	French	100	..	..	..					Mattei frères
	333	V	Deux Cousines *Villeland -87* W Bg	French	104	82·0	20·0	11·0	Le Legué	1876	Minic		Floury
	334	V	Deux Empereurs *Langlinay -87* W Bk	French	195	90·5	23·6	13·0	Fécamp	1864	Brument&Cap'n		Beust père & fils
	335	✳	Deux Frères *Bussanich* W Bk	Austrian	352	126·1	29·0	15·7	Bordeaux	1867	Bichon frères	Y 3.86	Dr.Robt.Schlesinger
	336	V	Deux Frères (ex Catherine) *Pelton -88* W Bg	French	194	101·0	23·5	13·8	Sunderland	1865	A.Simey		Peunors & Simon
	337		Deux Pierre *Huet* W Bg	French	141	77·8	21·6	12·3	Cayeux	1866	P.Danois	Z 2.85	E.Pourpoint & fils
	338	V	Deux Sœurs *Beveys -87* W Bg	French	242	91·2	23·9	14·9	Nantes	1869	Tillé	Y 1.86	Quintin
60288	339	✳	Deva *Davis 88-88* I Bk	British	761	187·2	31·7	19·8	Liverpool	1873	T.Royden & Son		W.Just & Co.
	340		Deva W Bk	Peruvian	350	..	..	..	Sestri	1868			Serdio & Hermanos
72400	341	✳D	Deveron *Patterson 75-75* I S	British	1256	231·3	36·2	22·0	Greenock	1875	R.Steele & Co.		J.Russell
62449	342	R	Deveron *Angus 86-86* W Sr	British	133	89·0	20·8	11·4	Banff	1871	Geddie		John W.Simpson
47730	343	R	Devon *Hodgson -88* W Bk	British	355	119·3	27·3	17·0	Sunderland	1864	Barker	Y 3.86	J.Tully
12710	344		Devon *White 63-* W Sr	British	107	78·2	20·0	11·3	Bideford,Eng.	1850			H.J.Stabb
47883	345		Devonia *Newman* W Sr	British	161	98·1	24·0	12·3	Ipswich	1865	Read		E.B.Styring
77089	346		Devonport *Campbell -86* W Bk	British	291	119·0	27·8	12·5	P.E.Island	1878	J.Fairchild	Y 12.87	John Bigelow
491	347		Devonport *Tope* W Sr	British	134	89·7	21·0	12·4	Appledore	1853	Chibbett		T.Rhys
13159	348		Devonshire Lass (H.T.L.K.) *Bailey* W Bg	British	133	72·4	18·9	13·1	Appledore	1840	T.Green		H.Seown
91540	349	✳	Devotion *Buckingham* W Sk	British	82	76·3	20·6	10·4	Rye	1885	G.&T.Smith		S.Buckingham
67609	350	✳	Dewdrop *Gill -80* W Sr	British	84	77·0	20·3	9·7	Goole	1874	Pearson		G.B.Wadsworth, jr

✳ = **Lloyd's Register**; D = Nederlandsche Vereeniging; G = Germanischer Lloyd; I = Registro Italiano; R = Amer…
A = Veritas Austro-Ungarico; H = Veritas Ellenico; N = Norske Veritas; V = Burea…

In the 4th Column, C = Composite (iron frame, wood planked); I = Iron; Stl = Steel; W = Wood.

The Dates of Metalling (**Col. 11**) are based on the records of the Books in which Vessels are respectively classed, or on the Reports of the Surveyors of Ll…

Further particulars of Vessels classed by Lloyd's Register will be found in Lloyd's Register Book and Lloyd's Yacht Register.

1	2 N° dans le livre	3 Registre dans lequel le navire est coté	4 Nom. Capitaine et dates d'entrée en service des Armateurs actuels et à bord du navire. Ancien Nom	Matériaux et Gréement	5 Pavillon	6 Tonnage	7 Longueur	Largeur	Creux	8 Construit. Où.	9 Quand	10 Par qui.	11 Date du doublage en métal	12 Armateurs ou Directeurs-Gérants.	13 Port d'Armement.
	351		Dewi Lass *Spriggs -82*	WSr	British	160	78·0	20·8	10·7	Aberayron	1862	Jones		J.Spriggs	Aberystwith
	352		Dexter *Crooks*	WSr	British	149	94·0	26·0	10·3	Brooklyn,N.S.	1878			J.V. & J.H.Dexter	Liverpool,N.S.
	353		Dexter Clark *Theall*	WSr	American	135	104·4	28·5	7·3	Ellsworth,Me	1873			H.E.Wellman	Providence,R.I.
	354		Dezenove de Junho *Serra*	WSr	Portug'se	165	84·6	23·0	10·9	Caminha	1863/72	A.J.Gonçalves		L.Barboza e Silva	Caminha
	355		Dhalia	WSr	Brazilian	175	..	..	..					J.M.Ribeiro	RioJaneiro
	356	✱	Dharwar *Hutchins 62-89*	I S	British	1300	226·2	37·2	23·3	Belfast	1864	Harland&Wolff		J.Willis & Son	London
	357		Diadem *Crasso*	WBn	British	162	89·6	24·0	10·4	LunenburgNS	1885	D.Smith		W.N.Zwicker	Lunenburg,N.S.
	358		Diadem *Morrison* (ex Lizzie)	WBn	British	121	95·8	23·7	11·2	P.E.Island	1869	Lefurgey		P.Wright & Co.	Belfast
	359	N	Diadem *Johnsen*	WBg	Norweg'n	281	107·7	26·6	14·5	Arendal	1856	N.R.Jensen		G.Tellefsen	Grimstad
	360		Diadem	WBg	Swedish	233	106·3	24·1	14·0	Gothenburg	1849	Gamla varfvet		J.P.Johnsson	Kärra,Bhln
	361	I	Diadema *Chiesa*	WBg	Italian	419	125·3	33·5	19·6	Savona	1868	F.Sirello	V 10,84	EredidiG.B.Avegno	Genoa
	362	G	Diamant *Bégoic -7-87* (ex Sondermann)	WBg	French	286	115·1	23·0	13·5	Geestemünde	1864	Schau&Oltmans	V 1,82	Charles Simon	Nantes
	363	✱	Diamant *Steffens -88* (ex Fairchild Steamer John Bell)	I S	German	992	231·4	32·0	20·6	Glasgow	1854	A.Stephen & Sons		C.M.D.Jörgensen	Hamburg
	364	NGV	Diamant *Larsen*	WBk	Norweg'n	282	117·9	27·1	12·4	Rönnebeck	1869	D.Oltmann	V 5,89	S.A.Larsen	Tvedestrand
	365	N	Diamant *Jacobsen 78-84*	WBk	Norweg'n	270	125·0	27·3	12·4	Porsgrund	1867	Dekke	V 1,82	W.D.Shepherd	Christiania
	366		Diamant *Andersen 69-80*	WBk	Norweg'n	220	107·5	26·1	12·5	Burg,Germ'ny	1843			L.Lömsland	Christiansand
	367		Diamant	WBn	Uruguay	172	98·1	25·1	12·0	Blankenese	1866	J.Finck		A.Borzone	Monte Video
	368		Diamante	WSr	Brazilian	209	..	..	..					A.J.S.Belem	Bahia
	369	N	Diamanten *Torkildsen76-76* (ex Ringleader)	WBg	Norweg'n	303	103·5	27·0	16·6	Sunderland	1862	J.Robinson		E.Brönlund	Arendal
	370	N	Diamanten *Thorsteinsen -88*	WBg	Norweg'n	155	91·3	24·7	10·8	Porsgrund	1868			H.E.Münster	Skien
	371		Diamanten *Olsson 69-74*	WSr	Swedish	188	93·8	23·4	12·6	Bastholmen	1874	J.Johansson		J.Olsson	Gåsö
	372		Diamantina *Church* (ex Sanmer)	I Sr	British	285	158·1	23·6	8·7	Newcastle-on-[Tyne	1861	C.Mitchell&Co.		Burns,Philp & Co.	Sydney,N.S.W.
	373	✱D	Diamond *Walker* (WRECKED)	WBk	British	391	123·9	28·1	17·6	Sunderland	1863	W.Doxford	V 3,86	Wright & Orr	Melbourne
	374		Diamond (ex Steamer)	I 3mSr	British	194	139·9	21·0	12·4	Liverpool	1846	P.Cato		J.Tyrrell	Dublin
	375		Diamond *Marr*	WSr	British	101	80·6	24·0	9·4	Guysboro,N.S.	1882	J.W.Hadley		F.Broudrot	Halifax,N.S.
	376		Diamond City	WBk	Siamese	261	114·0	24·3	12·6	Siam	1864			Poh Chin Soo	Bangkok
	377	✱D	Diana *Stephens*	I Bk	British	715	186·7	30·6	18·0	Glasgow	1880	C.Connell&Co.		J.Thomson, jun.	Glasgow
	378		Diana *McCalpin*	WBg	British	132	99·5	22·3	11·9	P.E.Island	1862			W.P.McCalpin	Belfast
	379		Diana (MISSING)	WSr	British	118	86·0	22·0	11·1	Garmouth	1868	Duncan		C.A.Forslind	Aberdeen
	380		Diana	WBk	Brazilian	189	..	..	..					J.F.Falçao	Rio Janeiro
	381	V	Diana *Fallentin*	WSr	Danish	139	93·4	20·7	10·7	Troense	1871	C.R.Møller		C.R.Møller	Svendborg
	382	GV	Diana *Forbrodt*	W3mSr	German	370	126·6	28·2	15·3	Geestemünde	1878	Schau&Oltmans	V 3,86	F.Westermann	Geestemünde
	383	DV	Diana *Schröder 78-*	WBg	German	314	108·0	27·0	14·0	Rostock	1869	E.Padderatz		H.Bauer	Rostock
	384	G	Diana *Mann 83-83*	WBg	German	247	104·3	24·9	13·8	Stralsund	1866	O.Kirchhoff		C.A.Beug	Stralsund
	385		Diana *Palmieri*	WSr	Italian	100	..	..	..	Capo d'Istria	1862	L.Pelli		N.Palmieri	Trani
	386	N	Diana *Thomassen -85*	WBk	Norweg'n	543	146·4	29·6	17·4	P.E.Island	1862			M.Schjöth	Mandal
	387		Diana *Tainis -87*	WBk	Norweg'n	409	136·3	29·0	16·1	Sikea	1871	O.P.Åberg	V 10,82	R.G.Ramsland	Stavanger
	388	N	Diana *Simonsen*	WBg	Norweg'n	342	117·3	29·0	14·9	Arendal	1870	H.Stiansen		Chr.Eyde & Hein	Arendal
	389		Diana (LOST) *Jonassen -71*	WBg	Norweg'n	240	110·6	27·8	12·6	Grimstad	1869	T.&T.Gundersen	V 10,84	H.Isachsen	Grimstad
	390	R	Diana (ex Tale) *Hehrent*	WBk	Swedish	390	134·1	25·5	15·7	Värnanäs	1862	Haman		C.J.Nordström	Gothenburg
	391	V	Diana	WBg	Swedish	287	113·0	24·4	12·7	Elsfleth	1868	G.Wempe	V 4,83	D.W.Pride	Höganäs
	392		Diana *Andersson -78*	WBk	Swedish	285	126·3	24·1	13·1	Bergqvara	1858	O.Ohlsson		J.Bengtsson	Fiskebäckskil
	393	V	Diana *Andersson*	WSr	Swedish	114	85·3	20·0	10·0	Sved	1878	Jansson		C.Hasse	Gothenburg
	394	N	Diaz *Agerholt*	WBk	Norweg'n	645	154·0	35·1	17·3	Arendal	1882	O.M.Olsen	V 3,88	H.H.Pettersen	Arendal
	395		Dichosa	WSr	Spanish	154	80·7	21·1	10·6	Villajoyosa	1874			Gabriel Paeres	Alicante
	396		Dick Williams *White -87*	WSr	American	176	105·8	26·8	8·5	Camden,N.J.	1866			A.White	Dennis,Mass.
	397		Dictator *McSparren*	WBn	British	190	104·9	24·6	12·8	P.E.Island	1874	M.Williams		J.Harvey	Newport
	398	N	Dictator *Tidlund -88* (ex Vangelsaas)	WBk	Norweg'n	1242	191·5	37·8	23·8	St.John,N.B.	1868	Nevins	V 9,87	C.M.Holst	Moss
	399	N	Dictator *Jäger 86-86*	WS	Norweg'n	831	172·7	35·8	20·4	Stavanger	1874	M.A.Magnus	V 4,86	G.Gundersen	Stavanger
	400	NV	Dictator *Bernhdsen*	WBk	Norweg'n	552	147·9	32·5	18·1	Tvedestrand	1874	M.Eldrup	V 4,88	Lars Larssen	Tvedestrand

✱ = Lloyd's Register; D = Nederlandsche Vereeniging; G = Germanischer Lloyd; I = Registro Italiano; R = American Record; A = Veritas Austro-Ungarico; H = Veritas Ellenico; N = Norske Veritas; V = Bureau Veritas.

Dans la 4me Colonne, C = Composite (membrure en fer, bordé en bois); I = Fer; Stl = Acier; W = Bois.

Les Dates du doublage en métal (Col. 11) ressortent des livres dans lesquels les navires sont respectivement classés ou des rapports des Inspecteurs du Lloyd's Register.

Pour plus amples renseignements sur les navires classés par le Lloyd's Register, voir le Lloyd's Register Book et le Lloyd's Yacht Register.

Official No.	No. in Book	Registry in which classed	Ship's Name. Material and Rig. Master, and Dates of Appointment to present Owner's Service and to Ship. Ship's Late Name.	Flag	Net Tonnage	Length	Breadth	Depth	Built. Where	When	By whom	Date of Metalling	Owners or Managers
	401	D	Dido (ex Paul Friedrich Pogge) *Gundersen* W3mSr	Norweg'n	372	121·5	28·5	16·1	Wismar	1854	C.Barmann		P.Knaff
	402		Dido (ex Williamina) *Gundersen* W Bn	Norweg'n	173	97·8	23·8	12·2	New Brunsw'k	1861			Chr.Jakhelln
	403	N	Dido *Olsen 82-82* W Sr	Norweg'n	117	84·8	20·7	10·6	Papenburg	1852			C.Zernichow
	404	G	Dido *Stromberg -87* W3mSr	Swedish	196	98·1	21·9	12·5	Wenersborg	1875	A.Eliasson	V 7.86	A.Pejlitz
80	405	H	Didymi (ex Theo. Botassis) W Bg	Greek	191	86·0	22·7	15·8	Syra	1867			Th.Ghinis
	406	V	Die Gartenlaube *Kruger 71-83* W Bk	German	438	125·6	26·7	17·1	Stralsund	1872	O.Kirchhoff	I 11.83	Gust.Siewert
	407	G V	Die Geschwister (ex Hinderika) *Froschmeister -87* W Sr	Russian	175	95·6	23·3	12·1	Windau	1846	Maschewitz		Gutschmidt Bros
	408		Die Krone *Wiechm 75-* W Bk	German	338	109·3	28·2	16·0	Barth	1859	J.S.Kraeft		J.N.Rodbertus
	409		Die Krone *Wiencke -86* W Bk	German	276	110·7	27·0	13·9	Ribnitz	1864	J.Mielbradt		Albr.Kossel
	410		Die Ostsee (BROKEN UP) *Schultz* W Bg	German	170	85·6	23·9	12·8	Grabow	1837	J.C.Nüske		Retzlaff & Schober
	411	V	Dieguito (ex Soledad) *Ibern -88* W S	Spanish	200	97·8	26·9	12·5	Blanes	1865		V 1.84	T.Fabregas
	412	✳	Diem Selamet *Ben Hajji Mehmet* W Bk	Turkish	480	113·0	26·9	17·0	Bulmont	1879	D.P.Repeten		Mustapha Bey
	413	V	Dieudonné *Simoni* W Bg	French	263	100·0	24·6	[illegible]	Nantes	1881	Sevestre freres	V 8.87	P.Boery
	414		Dikran W Bk	Turkish	425	..	..	..	Sinope	1874			Trebizondlee Tjar Gerikino Stephan & 1
52705	415	✳	Dilbhur *Norris 72-75* C Bk	British	1281	226·0	34·0	22·8	London	1865	C.Langley	V 1.86	J.Lidgett & Sons
	416		Diligence (STRANDED) *Rasmussen* W Bg	Norweg'n	193	82·4	25·9	14·0	Grimstad	1847			P.B.Rasmussen
	417		Diligencia W Sr	Spanish	114	78·4	22·4	11·6	Palma	1863			Doña Isabel Moll
	418	V	Diligente *Pierrin* W Bn	French	116	92·2	18·4	10·6	Cherbourg	1866	Laidier		Pierrin
	419	V	Diligentia (ex Robina) *Uffen 76-76* W Bk	German	350	116·5	27·5	15·9	Sunderland	1865	W.Pile&Co.	V 1.86	R.Reemtsma
	420		Dilzari-Bahri W Bg	Turkish	169	..	..	..	Greece	1883			Ismeh Bey Keremesi Zabik Hannum
	421	✳	Dimitra *Nicolazos* W Sr	Greek	95	72·0	20·0	10·8	Syra	1887	Andrico	V 1.88	N.Necolezze
155	422	A	Dimitrios *Pneumatios -88* W Bk	Greek	348	113·8	27·2	..	Syra	1876		V 3.86	H.P.Misto & Sons
	423	H	Dimitrios *Marromatis* W Bk	Greek	284	..	..	..	Syra	1876			E.Constanditis
23	424	H	Dimitrios (ex Mali Simun) W Bg	Greek	248	89·2	..	..	Segna	1863			Helen Caromlini
62	425		Dimitrios W Bg	Greek	248	98·4	26·6	..	Syra	1876			Styliano Zannos
	426	A	Dimitrios *Avgerino -88* W Bn	Greek	227	..	..	..	Syra	1878	P.Bumis		B.Avgerino
2	427	A	Dimitrios *Economic* W Bn	Greek	211	93·2	24·6	15·4	Galaxidi	1878			S.Caralambo
16	428		Dimitrios W Bg	Greek	198	84·6	26·1	15·8	Syra	1868		V 7.83	R.Barous
431	429	V	Dimitrios *Koufopantelis* W Bn	Greek	197	96·1	24·9	17·0	Syra	1885	A.Livadaros		E.Koufopantelis
10	430		Dimitrios (ex Amalia) W Bg	Greek	148	..	..	..	Naples	1868			G.H.Karavias
144	431		Dimitrios W Sr	Greek	126	77·4	22·0	..	Syra	1876			C.Vicadis
	432		Dimitrios W Bg	Turkish	214	..	..	..	Russia	1874			Vangeli Scorroleo
	433		Dimitrios W Bg	Turkish	170	..	..	..	Greece	1883			Mureitablee Joachim Sari-Aleko &
	434		Dimitrios W Bg	Turkish	142	..	..	..	Russia	1883			Midrieclee Apostolos Grimanis
93	435		Dimitrios Petrokcohino: W Bk	Greek	318	108·2	28·2	..	Syra	1871			C.Tsichlis
1	436	H	Dimitrios Voulgaris W Bg	Greek	220	90·9	25·4	15·0	Syra	1869	P.Boumi		A.S.Lichnos
34	437	A	Dimocritos (ex Agios Georgios) W Bg	Greek	146	79·7	23·0	..	Syra	1869			T.G.Lemos
	438	G	Dina (ex St. Jean Baptiste) *Kruize* W3mSr	Dutch	415	133·1	29·1	16·5	Bordeaux	1868	Raymond Bros.		H.B.Kruize
	439		Dina (ex Jonge Eveli) (ABANDONED) *Wieringa* W Bg	Dutch	210	107·2	24·2	12·3	Nieuwendam	1868	A.H.Meursing	V 4.85	F.J.Wieringa
	440		Dina *Mørch -77* W Bk	Norweg'n	447	129·4	30·0	16·5	Fredriksstad	1857			O.Hansen
47612	441	✳	Dinapore *Titmarsh 88-88* I Bk	British	1047	205·3	33·7	21·9	W.Hartlepool	1863	Pile,Spence&Co		Jackson,Metcalf&Co
28145	442		Dinorah (ex Louis & Albert) W Bg	British	105	83·4	22·2	10·3	Libourne	1853			Mrs.A.Stitt
	443		Dinorah *Olsen -82* W Bk	Norweg'n	458	128·5	30·8	17·8	Grimstad	1864	O.& L.Olsen		T.Meling
	444	N	Dinorah *Salvesen* W Bk	Norweg'n	317	123·2	28·8	15·0	Tvedestrand	1872		V 9.86	T.A.Morland
44234	445		Dinorwic *Ellis 72-72* W Sr	British	124	80·3	21·0	11·5	Port Dinorwic	1862	Jones		O.Morgan
	446	A	Dio Fili *Mina* W Sr	Greek	131	..	..	..	Syra	1887	C.Cristali		G.A.Mina & Co.
	447		Diogenes *Walker* Stl Sr	British	151	115·0	24·0	8·1	Dumbarton	1886	W.Denny&Bros		La Platense Flotilla C
	448		Diomede *Costanti Kaloni* W Bg	Turkish	134	..	..	..	Greece	1883			Dimitri Costanti
	449		Diomede *Christo Costanti* W Bg	Turkish	134	..	..	..	Greece	1880			Syroslee Christo Yorghi & Co.
54791	450		Dione *Berohldsen 84-87* W Bg	British	278	118·8	26·6	14·7	Shoreham	1867	Wm.May	[illegible]	J.& N.C.Bull

✳ = Lloyd's Register; D = Nederlandsche Vereeniging; G = Germanischer Lloyd; I = Registro Italiano; R = Ame
A = Veritas Austro-Ungarico; H = Veritas Ellenico; N = Norske Veritas; V = Bure

In the 4th Column, C = Composite (iron frame, wood planked); I = Iron; Stl = Steel; W = Wood.

The Dates of Metalling (Col. 11) are based on the records of the Books in which Vessels are respectively classed, or on the Reports of the Surveyors of

Further particulars of Vessels classed by Lloyd's Register will be found in Lloyd's Register Book and Lloyd's Yacht Register.

N° dans le livre	Registre dans lequel le navire est classé	Nom. Capitaine et dates d'entrée au service des Armateurs actuels et à bord du navire. Ancien Nom.	Matériaux et Gréement	Pavillon	Tonnage net.	Longueur	Largeur	Creux	Construit Où	Quand	Par qui	Date du doublage en métal	Armateurs ou Directeurs-Gérants.	Port d'Armement.
451		Dione *Haynes*	W Sr	American	[illegible]	11[illegible]·5	27·6	10·8	Stockton, Me.	1871			J.H.Futner	Boston, Mass.
452	V	Dione *Israel 82-83*	I Bk	German	642	160·0	29·0	18·9	Hamburg	1877	Reiherstieg Ship (W.[illegible]		Wachsmuth & Krogmann	Hamburg
453	N	Dione *Pettersen 76-84*	W Bk	Norweg'n	764	174·3	[illegible]	19·6	Porsgrund	1881	E.Coalstad	[illegible]	Johan Jeremiassen	Porsgrund
454	A	Dionyssios *Zannos*	W Bg	Greek	416	122·7	[illegible]	..	Syra	1879		[illegible]	D.Zannos	Santorin
455	AH	Dionyssios	W Bn	Greek	220	93·5	25·2	..	Syra				B.Lebessis	Piraeus
456	H	Dionyssios	W Bg	Greek	18[illegible]	..	..	..	Galaxidi	1873		[illegible]	N.P.Bougas	Galaxidi
457		Diplomata		Brazilian	201	..	..	..					J.A.C.Vianna	RioJaneiro
458		Dipton *Larsen*	W Bk	Norweg'n	346	116·7	27·3	17·0	[illegible]	1865	Rebd.Thompson		Chr.Dahl	Drøbak
459	DV	Director Barrow *Borgonil 84-84* (ex [illegible])	W 3mSr	Chinese	326	122·7	27·1	14·4	[illegible]	1876	J.[illegible]	[illegible]		
460		Dirigo	W Bn	American	367	113·2	28·2	11·5	Harrington, Me.	1862			W.S.Baker	New York
461	V	Diritto *Anaglio*	W 3mSr	Italian	496	112·1	29·5	19·7	SestriPonente	1864	A.Brisasco fu G.B.	[illegible]	A.Revello	Genoa
462		Discovery *Lee*	W Bkn	American	395	150·0	37·0	12·0	PortDiscovery, Wash	1874			E.E.Kentfield	San Francisco
463		Dispatch *MacKinney 82-84*	W Bn	British	172	102·5	24·1	12·5	Newcastle	1846	J.Wilkinson		J.Kingsberry	Belfast
464	✻	Dispatch *Mair 88-88*	W Sr	British	199	90·1	21·5	10·3	Portsoy(?)	1888	J.Geddie		J.Mill	Banff
465	N	Disponent *Jensen -88*	W Bk	Norweg'n	508	147·9	31·3	17·2	Porsgrund	1876		[illegible]	Chr.Th.Boe	Arendal
466		Diva *Scibigtia*	W Bg	Italian	164	90·2	25·3	13·0	Varazze	1861	B.Camogli	[illegible]	Di Battista & Bellassai	Prto Empedocle
467	NV	Divico *Møller 81-81*	W Bk	Norweg'n	501	141·2	30·5	17·5	Bergen	1864	A.Dekke	[illegible]	S.Otto & Co.	Christiansand
468	V	Divine Providence *Audoin*	W Bn	French	108	74·8	20·7	10·3	Martigues	1862	Vidal & Michel	[illegible]	Suchard	Cette
469		Divo	W Bn	Brazilian	158	..	..	..	S.João da Barra	1850		[illegible]	Fernando & Cabral	Laguna
470	✻	Dizzy Dunlop *Williams [illegible]*	W Sr	British	99	84·5	22·2	10·7	Portmadoc	1878	Jones		T.E.Williams & Co.	Carnarvon
471		Djadul Gair (ex Alaska)	W Bk	Dutch	538	136·2	31·3	..	Madison	1868	Miner & Son	[illegible]	Sech Achmat bin Oemar Fagie	Sourabaya
472	AV	Djehlan Bahri *Vent* (ex [illegible])	W Bn	Turkish	428	118·7	27·7	16·8	Fiume	1876	F[lli] Schiavon	[illegible]	Islam Effendi	Constantinople
473		Djeilanie *Sech Asea bin* (Mohamed [illegible])	W Bk	Dutch	262	103·6	22·7	..	Grissee	1845		[illegible]	Said Hassan bin Pangeran Sjarief Ali bin Sech Aboebakar	Palembang
474		Djihander Bir	W Sr	Turkish	140	..	..	..	Barton	1880			Kerestehdjilari Dawud Agha	Barton
475		Djoedoel Karim (ex Diana)	W Bk	Dutch	327	122·6	25·0	15·5	Bristol	1857		[illegible]	Said Hassan bin Abdul Rachman Mola Hadis	Sourabaya
476		Djoedoel Karim *Said Abdul Hamid*	W Sr	Dutch	135	97·1	19·9	..					Said Abdul bin Abdul Rachman Hamdan	Grissee
477		Dmitrei *Dmitrejeff*	W Glt	Russian	143	..	..	..	Impilaks	1878			Pitkaronda Bruksbolag	Impilaks
478	✻D	Dochra *Taylor 87-87*	I Bk	British	966	204·9	33·2	20·6	Greenock	1877	Scott & Co.		Macneilage, Adam & Co.	Greenock
479	V	Dockan *Isaksson*	W 3mSr	Swedish	197	111·2	21·3	11·1	Oscarshamn	1876	O.Olsson	[illegible]	F.Johansson	Oscarshamn
480	V	Docteur Ollive *Mahéo*	W Bg	French	223	98·4	25·6	12·7	Nantes	1870	Tillé		Mahéo	Nantes
481		Doctor *Worth 88-88*	W 3mSr	British	148	103·0	22·5	12·9	BurtonStather	1864	Wray		J.Fisher & Sons	Barrow
482	DGV	Doctor Lasker *[illegible] 76-77*	W Bn	German	266	112·7	26·8	13·[illegible]	Geestemünde	1874	Schau & Oltmann	[illegible]	F.Westermann	Geestemünde
483		Doctor Livingstone	W Sr	British	123	89·9	23·5	10·0	CapeBreton	1872			J.H.Christie	Sydney, C.B.
484	N	Doctor Mezger *Engelstad*	W Bk	Norweg'n	652	165·6	33·5	17·9	Porsgrund	1873	K.Haugland	[illegible]	P.M.Petersen & Søn	Porsgrund
485	G	Doctor Witte *Zeplien 65-74*	W Bg	German	280	110·3	27·1	14·5	Rostock	1874	O.Ludewig		W.Maack	Rostock
486		Dofuku Maru No. 4	W Sr	Japanese	189	110·3	26·0	12·2	Scotland	1869	J.Tacking		Nishikawa Teijiro	Osaka
487	I	Dogali (ex Veloce) *Clarlo -87*	W Bk	Italian	576	143·0	31·8	20·2	Varazze	1870	Craviotto	[illegible]	ConstantinoAnselmo	Savona
488		Dogali *Mazzdia -87* (ex Giuseppe Patrellussi)	W Bg	Italian	219	95·8	26·8	16·2	Alimuri	1856	Mauri	[illegible]	F.Scotto di D.	CivitaVecchia
489		Dogonai (ex Conrad) *Kroeft*	W Bg	Russian	184	100·0	25·8	11·0	Kiel	1863	A.Conradi		J.Seebeck	St.Petersburg
490	V	Dois Amigos *Milhomens* (ex Marat)	W Bn	Portug'se	217	96·8	24·6	12·4	Figueira	1856	Mendes & Irmão	[illegible]	C.J.Rodrigues	Oporto
491	V	Dois Irmaos *Correa -87* (ex Maria da Gloria, ex Vanda) (ABANDONED)	W Bn	Portug'se	185	88·6	23·0	11·5	Figueira	1856	M.G.Mendes & Irmão	[illegible]	C.J.Rodrigues	Oporto
492	✻	Dolbadern Castle *Le Maistre [illegible]*	I S	British	989	198·9	34·1	20·9	Sunderland	1863	T.R.Oswald		Hughes & Co.	Liverpool
493		Dolen	W Bg	Norweg'n	362	117·4	28·3	15·6	Arendal	1872	J.H.Ondedt		E.Stenersen	Arendal
494		Dolfijn (WRECKED) *Grilk* (ex Anna Catharina Wilhelmina)	W Glt	Dutch	132	89·3	22·2	9·8	Hoogezand	1862	J.A.Hoattes		H.J.Plant & Co.	Schiedam
495		Dolly Varden *Allen 76-76* (ex [illegible] Rose)	W Bn	British	186	115·9	23·4	12·4	Bideford	1871	Cox		W.G.Lawson	London
496	AV	Dolores *Marzella di Bosco* (ex Maddalena)	W Bg	Italian	331	111·5	26·1	18·0	Cassano	1870	Castellano	[illegible]	N.De Martino	Naples
497		Dolores *Villamil* (ex Silver Lining)	W 3mSr	Spanish	358	130·6	27·3	14·7	Brixham	1881	J.W & A.Upham	[illegible]	M.Romano	Barcelona
498		Dolores *Cisa*	W Bn	Spanish	249	172·5	27·5	12·5	Blanes	1866			S.Cisa y Olano	Barcelona
499		Dolores (ex Dalia)	W Sr	Spanish	104	85·9	22·0	9·1	SanFeliu de Guixols	1860			F.San Juan y Sotsa	Nuevitas, Cuba
500														

=Lloyd's Register; D=Nederlandsche Vereeniging; G=Germanischer Lloyd; I=Registro Italiano; R=American Record; =Veritas Austro-Ungarico; H=Veritas Ellenico; N=Norske Veritas; V=Bureau Veritas.

Dans la 4me Colonne, C=Composite (membrure en fer, bordé en bois); I=Fer; Stl=Acier; W=Bois.

Les Dates du doublage en métal (Col. 11) ressortent des livres dans lesquels les navires sont respectivement classés ou des rapports des Inspecteurs du Lloyd's Register.

Pour plus amples renseignements sur les navires classés par le Lloyd's Register, voir le Lloyd's Register Book et le Lloyd's Yacht Register.

1 Official No.	2 No. in Book.	3 Registry in which classed.	4 Ship's Name. Material and Rig. Master, and Dates of Appointment to present Owner's Service and to Ship. Ship's Late Name.	5 Flag.	6 Net Tonnage.	7 Dimensions. Length.	Breadth.	Depth.	8 Built. Where.	9 When	10 By whom.	11 Date of Metalling.	12 Owners or Managers.
	501	G	Dolphijn *Leslaiver -88* W Bn (ex Skjold)	Dutch	168	81·2	25·4	12·2	Drammen	1869			M. van Neck
28024	502		Dolphin *Llewellin* W Bk	British	344	121·0	24·8	16·9	LymeRegis	1851		Y 5.86	N. McLeod
50392	503	R	Dolphin *Annis* W Bk (STRANDED)	British	294	118·4	26·7	14·6	Shoreham	1864	Shuttleworth	Y 6.84	Mrs. Mary Annis
6192	504		Dolphin *Grover* W Sr	American	113	87·5	24·2	7·9	E. Machias, Me	1853			Todd & Co.
33211	505		Dolphine W Sr	British	110	80·4	23·2	8·8	Gentilly	1856			T. Chartier
	506		Dom Aurelio W Bn	Brazilian	228	..	..	..					J. J. A. e Silva
	507		Dom Luiz W Sr	Brazilian	100	..	..	..					D. L. de Azevedo
6950	508	RV	Dom Pedro II *Johnson* W Bk	American	465	136·0	30·6	15·6	Baltimore	1878	W. Skinner&Sns	Y 9.87	T. Rollins
	509		Dom Pedro II. W Bn	Brazilian	620	..	..	..					R. B. Montalvão
635	510	V	Domenico *Lauro -88* W Bk	Italian	383	121·3	27·0	17·1	Cassano	1866	F. de Rosa	Y 10.88	D. Ciampa fu M. & D. Jacarino
2665	511		Domenico (ex Felix) *DelGatto* W Bk	Italian	315	126·9	26·9	12·8	Bathurst	1866		Y 8.85	G. Del Gatto
92	512		Domenico *Fasce* W 3m Sr (ex Edoardo)	Italian	270	105·9	26·5	12·8	Leghorn	1858	G. Micheli	Y 10.85	D. Fascia
	513	A	Domenico (ex Fido) W Bn		242	119·4	24·8	12·9	Savona	1873	G. B. Tessitore	Y 1.87	
334	514	A	Domenico Boso *Boso* W Bn (ex Catania)	Italian	154	101·7	23·6	10·2	Trieste	1869		Y 8.85	G. Boso fu D.
2398	515		DomenicoLanata *Farina* W Bk	Italian	559	155·1	31·4	17·7	Sestri Ponente	1873	Fli Cadennacio	Y 5.85	G. B. Farina
2861	516	I	Domenico M. *Cavallo* W Bk (ex Compagnia Genio) [87-88	Italian	864	176·0	35·4	22·9	Spezia	1874	Faggioni	Y 11.87	A. Cavallo
630	517	A	Domenico P. *G. Fama* W Bk (ex Figlio)	Italian	523	144·3	31·8	20·3	Volosca	1864	Fll. Schiavoni	Y 10.87	B. Fama & D. Piazza
	518		Dominga W Bk (ex Queen of the Seas)	Spanish	506	135·5	31·2	17·4	UnitedStates	1858			L. R. Yangco
	519	N	Domingo *Larke* W Bg	Norweg'n	171	90·5	24·6	11·6	Bremerhaven	1844			A. Rasmussen
	520		Domingo Santa Maria W S (ex Chancellor) *Pardo*	Chilian	1813	226·7	43·0	28·6	Newcastle, Me	1856			F. Capurro
	521		Domingo Solari W Bn (ex Little Susie) (now Kosmos)	Dutch	184	98·7	22·3	12·9	Plymouth	1875	Williams	Y 6.86	H. Rietveld
69867	522	V	Dominion *Rowland -87* W S	British	1256	198·7	38·0	23·9	Quebec	1875	Baldwin	Y 8.86	W. E. Jones
89847	523	✻	Don *Haslewood 87-87* I Bk	British	1115	215·2	35·1	21·2	PortGlasgow	1885	Russell & Co.		P. Macfarlane
	524	RV	Don Carlos *Jacobson -87* W Bk (ex Quang Tung ex Principe di Lucedio)	Bolivian	694	159·1	30·0	20·8	Sestri Ponente	1875	Calcagno	Y 8.87	A. Bichard
3275	525		Don Diego *Francis* I S	British	320	145·6	24·2	13·9	Greenock	1855	Scott & Co.		W. H. Smith & Sons Ld.
70984	526	V	Don Enrique *Corning -83* W S	British	1345	201·6	37·8	24·1	Newbury Port, Mass.	1872		Y 4.87	B. Cremor
	527		Don Fernando *Johnsen* W Bk (ex Atacama) [86-87	Chilian	481	140·5	28·2	18·0	Hrtlepool	1861	Denton	Y 11.89	F. Lopez Jofré
6676	528	RV	Don Jacinto *Gillett, Jr.* [-88 W Bn	American	490	131·9	29·0	17·0	Freeport, Me.	1870		Y 2.84	S. B. Kelsey
	529		Don Juan (FOUNDERED) W Bk (ex Alster, ex Polly Lewis) *Petersen*	Chilian	561	143·2	30·4	18·9	Calais, Me.	1869			E. Peters & E. M...
	530	DGV	Don Juan *Wenhold 86-86* W Bk	German	994	182·4	34·8	22·1	Brake	1883	J. Oltman's Wwe.	Y 2.86	J. M. Gildemeister
	531		Don Juan *Gangoitia* W S	Spanish	919	176·8	35·4	20·7	Santander	1871	E. Garris		Juan Pombo
	532	✻	Don Juan *Thompson* I Sr	Urug'yan	72	78·0	25·0	5·9	Sunderland	1884	J. Knox & Co.		Fynn, Butler & Co.
6649	533	V	Don Justo *Neilson 87-87* W Bk	American	709	145·6	32·1	19·1	Damariscotta, Me	1869	G. W. Lawrence		D. T. Morton
	534	R	Don Leon (FOUNDERED) W 3m Sr (ex Lottie Fairfield) *Wagner -87*	Bolivian	165	103·8	23·5	10·0	Pt. Gilbert, NS	1883	J. Bingay	Y 12.83	A. Block
	535		Don Pedro *Schneider* W Bn (ex Don Pedro V.)	German	182	96·1	23·3	10·4	Villa do Conde	1877	J. G. da Silva Rodrigues	Y 6.81	A. Beckmann
	536	GV	Don Quixote *Hilmer 84-84* W Bk	German	1168	179·4	37·4	23·9	Medford, Mass	1868	J. T. Foster	Y 1.87	L. F. Kalkmann & Co.
	537		Don Roberto I	Argentine	158	120·0	24·9	8·0	San Nicolas	1886	C. Geddes		Green & Johnson
	538		Dona Anna I Sr	Brazilian	100	100·1	20·2	5·9	PortGlasgow	1884	Blackwood & Gordon		Amazon Steam Navigation Co. Ld.
	539		Dona Anna (ex Nora) W Bg (BROKEN UP)	Portug'se	257	113·5	23·3	12·5	St. Johns, NFL	1862			J. M. Bettencourt
	540		Dona Antonia W Sr	Brazilian	146	..	..	..					A. A. P. da Silva
	541		Dona Clara (ex Josephina) W Bk	Brazilian	302	..	..	..	P. E. Island	1868			J. G. da Fonte
	542	V	Doña Dolores *Wenke 83-83* W Bk	German	335	118·0	26·0	11·0	Geestemünde	1883	D. Oltmann	Y 4.86	Seetzen Gebrüder
	543	V	Dona Eliza *Quaresma* W Bn (ex Dona Antonia, ex Brother's Pride)	Portug'se	151	85·3	21·6	10·5	Villa do Conde	1878	J. M. de Araujo	Y 2.87	Elisa P. Porto
	544		Dona Francisca W Bg (ex Anaïs)	Brazilian	231	109·2	26·2	12·3	Tvedestrand	1872	T. Gundersen	Y 6.81	A. d'Oliveira Maia
	545		Dona Francisca W Sr	Brazilian	162	..	..	..	Villa do Conde	1870		Y [illegible]	D. M. G. C. Corrêa
	546		Dona Guilhermina W Bn *da Piedade*	Brazilian	185	..	..	..	Portugal			Y 84	Maria Guilhermina Chaby Corrêa
	547		Dona Julia W Sr	Brazilian	120	..	..	..	Pernambuco	1861		Y [illegible]	B. Lourenço
	548		Dona Luisa *Durand -88* W 3m Sr	French	263	131·2	26·5	14·0	Irvine	1871	IrvineShipCo.	Y 1.86	P. Villars
	549	V	Dona Luisa *Nommensen* W 3m Sr (ex Ernst Bruyer) [86-86	German	227	113·8	23·9	12·6	Hamburg	1869	E. Dreyer	Y [illegible]	A. H. Wappäus
	550		Dona Margarida W 3m Sr	Brazilian	196	..	..	..					V. I. da Nova

✻ = Lloyd's Register; D = Nederlandsche Vereeniging; G = Germanischer Lloyd; I = Registro Italiano; R = ...
A = Veritas Austro-Ungarico; H = Veritas Ellenico; N = Norske Veritas; V = ...

In the 4th Column, C = Composite (iron frame, wood planked); I = Iron; Stl = Steel; W = Wood.

The Dates of Metalling (Col. 11) are based on the records of the Books in which Vessels are respectively classed, or on the Reports of the Surveyors of ...

Further particulars of Vessels classed by Lloyd's Register will be found in Lloyd's Register Book and Lloyd's Yacht Register.

No.	Cote	Nom. Capitaine et Gréement. Ancien Nom.			Pavillon.	Tonneaux net.	Longueur	Largeur	Creux	Construit. Où	Quand.	Par qui.	Metallage	Armateurs ou Directeurs-Gérants.	Port d'Armement.
[illegible]		Dona Maria		W Sr	Brazilian	324	..	..	..					P.J. de Azevedo	Rio Janeiro
[illegible]		Doña Maria		I Sr	Brazilian	100	100·1	20·2	5·9	PortGlasgow	1884	Blackwood & [Gordon		Amazon Steam Navign. Co. Ld.	Para
553	DV	DoñaZoyla	Schacht 76-88	W3mSr	German	359	129·5	25·9	12·9	Brake	1878	J.Oltmanns	[illegible]	A.H.Wappäus	Hamburg
554	V	Dona Zulmira	[illegible]	W Bn	Brazilian	205	105·3	22·9	9·8	Gefle	1884	O.A.Brodin [Wwe.		Successors of B.Carneiro Lemos	PortoAlegre
555		Donabew		I Flat	British	777	197·0	25·7	8·3	Port Glasgow	1876	R.Duncan&Co.		IrrawaddyFlotillaCo.Ld	Glasgow
556	R	Donald Cann	[illegible]	W Sr	British	124	93·0	24·0	9·6	Shelburne,NS	1888	A.Hood		H.B.Cann	Yarmouth,N.S.
557		Donald Ferguson	Campbell [illegible]	W Bk	British	584	147·6	31·3	16·5	NGlasgow, Ns	1873	Graham	[illegible]	Robb Bros.	Pictou,N.S.
558		Donatus	Eriksson	Bk	Russian	319	..	..	..	Lemland	1874			G.Mattson	Lemland
559	D	Donegal	[illegible]	W Bk	British	693	160·0	31·8	19·9	Nieuwendam	1869	W&H.Menrgals	[illegible]	G.Milne & Co.	Aberdeen
560	V	Donna Evelina	[illegible]	W3mSr	German	350	129·2	25·6	12·9	Brake	1877	C.Oltmanns [Wwe.	[illegible]	A.H.Wappäus	Hamburg
561		Donna Sofia	Campbell	W Bk	British	133	91·4	21·3	10·9	P.E.Island	1868	M'Kenzie		J.Stewart	Belfast
562		Donnerstag	Greve 76-76	W Bg	German	219	97·9	25·9	13·0	Ueckermünde	1862	Preiser		C.G.Runge	Ueckermünde
563		Donzella	Lalle	W Sr	British	119	82·4	24·0	10·1	Vogler'sCove	1882	S.D.McKean		L.Andersen	Lunenburg,NS
564	✱D	Doon	Angus	I Bk	British	817	195·7	32·9	19·1	Glasgow	1877	Lon.&Glas.Co.		D.Hunter	Ayr
565	✱	Dora	May 82-88	W3mSr	British	212	118·6	24·3	13·0	Bridport	1875	Cox	[illegible]	Job Bros.	Liverpool
566		Dora	Hewett 77-88	W Bk	British	198	114·6	19·5	12·6	Bristol	1854	W.Patterson	[illegible]	G.H.Bridges	Bristol
567	DV	Dora	Hansen	W Bk	Danish	361	126·2	26·6	15·1	Yarmouth	1876	F.Sunde	[illegible]	Akt.Selsk. (N.Hansen)	Nordby,Fanø
568	V	Dora	Meyer [illegible]	W S	German	1259	200·1	37·4	21·5	Boston	1868	Townsend	[illegible]	Köper,Docke & Co	Bremen
569		Dora	Ardoino 87	W Sr	Italian	134	90·2	21·4	10·3	Spotorno	1886	F^lli Cadenaccio	[illegible]	F^lli Cadenaccio	Genoa
570	N	Dora (ex Dallas)	Hansen	W Bk	Norweg'n	955	180·8	35·6	21·3	Quebec	1872	Marquis	[illegible]	M.Pedersen	Mandal
571	G	Dora Ahrens	Peters 87-87	W Bk	German	429	129·0	29·8	17·7	Rostock	1870	J.Müller	[illegible]	W.Maack	Rostock
572	R	Dora Allison	Ross	W3mSr	American	322	134·5	33·7	10·1	TottenvlleNY	1884	J.C.Ellis		J.A.Curtis	Richmond,Va.
573	✱	Dora Ann	Lloyd 75-76	I Bk	British	589	171·5	28·9	17·7	Sunderland	1876	J.Laing		R.James	Aberystwith
574	R	Dora Bluhm	Moore	W3mSr	American	316	127·0	33·0	10·0	PortBlakely, [Wash.	1883	W.G.Hall		W.H.Smith	San Francisco
575	R	Dora M. French	French 81-81	W3mSr	American	205	113·0	29·2	9·1	Brewer,Me.	1874	Jos.Oakes	[illegible]	J.S.Jenness	Bangor,Me.
576	R	Dora Mathews	Bralne	W3mSr	American	373	136·7	33·0	12·0	Camden,N.J.	1882	H.M.Bean	[illegible]	W.H.Parsons	New York
577	DV	Dorane	Rasmussen 83-83	W Bg	Danish	299	120·5	25·1	14·0	Nordby,Fanø	1883	S.Abrahamsen	[illegible]	Akt.Selsk.(C.H.Nielsen)	Nordby,Fanø
578		Dorchester		W Barge	British	375	148·5	27·1	11·1	Quebec	1876	F.Cantin		MontrealTransprtn.Co.	Montreal
579		Dores	Grandström	W Bn	Russian	327	121·0	27·9	12·5	Björneborg	1874	Lindgren		J.H.Ståhlberg	Raumo
580		Doretta (WRECKED)		W Bg	Siamese	300	118·3	27·4	14·9	P.E.Island	1872	Lefurgey		Phra Wichet Saratsye	Bangkok
581		Dorette	Nyström 65-81	W Bg	Swedish	201	88·9	23·2	13·6	Wismar	1838	J.C.Marén		F.Nyström	Slite,Gotland
582		Dori	Stefano 88	W Bg	Italian	309	108·9	28·5	17·4	Savona	1862	Calamaro	Y 6.81	F.G.Peragallo	Genoa
583		Dorina	Mikovich	W Bk	Austrian	498	140·0	27·0	19·0	Fiume	1870	F^lli Schiavoni	[illegible]	A.Luppis	Fiume
584	✱	Doris	McGlashan 87-87	Stl & I Bk	British	1296	248·6	35·3	21·6	Dundee	1887	A.Stephen & Ss.		A.Stephen & Sons	Dundee
585	✱	Doris	Hill 85-88	W Bkn	British	293	128·9	27·0	12·9	Sunderland	1873	J.Crown	[illegible]	W.Brooks	Newport
586		Doris	Gerhardt 87	W Bn	British	187	96·2	26·4	10·5	Chester Basin, [N.S.	1887	L.Oxner		J.Eisenhauser	Lunenburg,NS.
587	✱	Doris	Richards 85-85	W Sr	British	94	80·0	20·4	9·9	Salcombe	1880	Harnden	Y 6.86	W.W.Steer	Salcombe
588	V	Doris	Jørgensen 84-84	W Sr	Danish	157	103·0	21·9	11·3	Thurø	1884	J.Ph.Jørgensen	Y 5.87	J.Jørgensen	Svendborg
589	V	Doris	[illegible]	W S	German	1160	190·9	40·3	23·7	Kennebunk	1869	N.L.Thompson	Y 3.87	Boyes & Ruyter	Bremen
590	V	Doris	[illegible] 83-87	W Bk	Swedish	358	130·2	26·8	14·4	Helsingborg	1887	S.Nordström	Y 1.87	C.Sylvan	Helsingborg
591	✱	Doris Brodersen	Nielsen	I S	Danish	647	177·7	30·2	17·3	Flensburg	1875	Flensburger. [S.B.Co.		H.C.Brodersen	St.Thomas
592	RV	Doris Eckhoff	Todd	W Bk	American	530	142·3	30·8	17·1	Millbridge, Me	1874		Y 8.87	H.P.Booth	New York
593	G	Doris Mentz	Mahn 81-73	W Bn	German	178	86·9	24.1	13.3	Ribnitz	1861	J.H.Wilcken		W. Maack	Rostock
594		Dorotea	Perez	W Sr	Spanish	143	83·3	24·3	12·4	Masnou	1847			Viuda de A. Pagès y Cia	Barcelona
595		Dorotea	Penero	W Bn	Spanish	134	86·9	23·9	11·2	SanSebastian	1857			P. de la Torriente	Santander
596	IV	Dorotea M.	Barrese	W Bn	Italian	436	136·8	29·3	17·9	Cassano	1877	F.DeRosa	[illegible]	G.Maresca	Castell. di Stab.
597	DV	Dorothea	Jakobs [illegible]	W Bg	Dutch	290	123·8	26·3	14·3	Fanø	1866	F.Schönau	[illegible]	A.Köpeke	Rotterdam
598	V	Dorothea	Biel 88	I S	German	1016	185·0	33·0	20·9	Hamburg	1870	ReiherstiegShip [Works		I.G.Lübeken	Elsfleth
599	V	Dorothea	Moller 73-73	W Bk	German	620	152·5	30·5	19·0	Hamburg	1872	H.C.Stülcken	[illegible]	H.C.StülckenWwe.	Hamburg
600	DN	Dorothea	Andersen 87	W Bk	Norweg'n	390	110·0	28·9	17·1	Sestri Ponente	1862	G.Casanova		A.M.Thoresen	Fredrikstad

✱ = Lloyd's Register; D = Nederlandsche Vereeniging; G = Germanischer Lloyd; I = Registro Italiano; R = American Record; A = Veritas Austro-Ungarico; H = Veritas Ellenico; N = Norske Veritas; V = Bureau Veritas.

In the 4th Column, C = Composite (iron frame, wood planked); I = Iron; Stl = Steel; W = Wood.

The Dates of Metalling (Col. 11) are based on the records of the Books in which Vessels are respectively classed, or on the Reports of the Surveyors of Lloyd's Register.

Further particulars of Vessels classed by Lloyd's Register will be found in Lloyd's Register Book and Lloyd's Yacht Register.

Official No.	No. in Book.	Registry in which classed.	Ship's Name. Master, and Dates of Appointment to present Owner's Service and to Ship. Ship's Late Name.	Material and Rig	Flag.	Net Tonnage.	Length.	Breadth.	Depth.	Built. Where.	When	By whom.	Date of Metalling.	Owners or Managers.
	601	V	Dorothea *Andrez -86*	W Sr	Russian	123	75·1	22·0	10·9	Ugezeem	1886	Surting		R. Seeberg
[illegible]	602	*	Dorothy *Riordan*	W Bk	British	442	134·5	28·0	17·6	Sunderland	1861	W. Briggs&Sons	n.se	J. W. Evans
[illegible]	603		Dorothy *Croal*	W Bk	British	310	117·9	25·8	15·6	Gloucester	1864	Miller	[illegible]	R. Smith
[illegible]	604		Dorothy *Robinson*	W Sw	British	219	91·0	25·4	15·2	Sunderland	1856	[illegible]		W. Milburn
	605		Dorothy *Sellman -88*	W Bk	Swedish	346	115·0	27·5	17·5	Sunderland	1860	Harkess	[illegible]	T. Flensburg
[illegible]	606		Dorothy Watson *Graves -88*	W K	British	99	86·7	21·1	9·1	Blyth	1881	UnionCo-op. S.B. (Soc.Ld.)		J. E. Cutter
[illegible]	607		Dorset *Vickers*	W Sr	British	125	86·6	22·2	10·8	Sunderland	1868	Liddle		W. Vickers
	608	V	Dorthea *Hansen 76-83*	W Bn	Danish	200	107·7	23·5	11·7	Svendborg	1883	T. R. Andersen	[illegible]	H. Christensen
	609	DV	Dorthea *Christensen 81-81* (WRECKED)	W Bg	Danish	171	98·5	21·2	11·4	Nordby, Fanø	1870	J. N. Christensen	[illegible]	Actieselskabet J. H. Christensen
	610	V	Dorthea *Brinck 76-86*	W Bn	Danish	142	91·7	21·2	11·0	Nordby, Fanø	1871	J. N. Christensen	[illegible]	Actieselskabet "Dorthea" (P. J. Brinck)
	611	V	Dorthea *Mortensen*	W Sr	Danish	1?8	84·5	20·2	10·2	Nordby, Fanø	1869	J. Th. Hansen	[illegible]	Akt. Selsk. (M. J. Mortensen)
	612	DV	Dorthea Marie *Brinck*	W Bg	Danish	156	91·7	22·6	11·4	Esbjerg	1875	T. Dahl	[illegible]	Akt. Selsk. (T. H. Brinck)
	613		Dos Hermanas	W Bn	Spanish	148	101·0	20·9	9·8	Cavite	1884	M. Inocencio		M. Inocencio
	614	V	Dos Hermanas (ex Calamus Pobre ?)	W Sr	Spanish	10?	78·7	21·7	10·1	Villajoyosa	1877			J. Urios y M. Lopez
	615	V	Dos Hermanos *Castejo*	W Bk	Spanish	345	120·1	26·5	15·2	Manila	1864		[illegible]	N. Gelato & Co.
	616		Dos de Mayo *San Juan*	W Bn	Spanish	219	101·0	27·1	13·4	Masnou	1867	A. Escarpentes		G. Pla
48345	617		Douglas (ex Alfonso de Albuquerque)	W S	British	540	135·0	31·0	19·0	Rhode Island, U.S.	1847			W. Morrison
92351	618	V	Douglas *Crosby -86*	W Bk	British	509	161·7	33·1	13·0	St. John, N.B.	1886	E. McGuiggin		H. D. Troop
14852	619	*	Douglas *Sidler -88*	W Bn	British	168	100·0	23·0	12·1	Guernsey	1841	P. Brouard	[illegible]	N. Bate
	620	*	Douglas Castle *Brandi -87*	C Bk	Danish	678	176·6	30·1	18·7	Glasgow	1864	C. Connell & Co.	[illegible]	G. Brandi & Co.
6734	621		Douglas Haynes *Boston*	W 3m Sr	American	165	104·2	28·1	7·7	Bath, Me.	1872			J. F. Wright
157107	622	R	Douglas Hovey (STRANDED) *W. ? -83*	W 3m Sr	American	468	145·1	34·9	12·8	Bath, Me.	1883	N. Rogers	[illegible]	J. B. Drake
312	623	A	Dounavis *Levantis*	W Bg	Greek	296	105·6	27·1	16·0	Syra	1880	P. Bouml	[illegible]	D. Levantis
60415	624	*	Doune Castle *Jones 78-88*	C Bk	British	8?7	197·1	34·0	19·8	Glasgow	1869	J. Elder & Co.	[illegible]	Foley & Co.
	625		Douro	Stl Lr	British	3?	83·0	14·5	4·5	London	1883	J. S. Watson		J. Watson
	626		Dous Anjos	W Bn	Brazilian	129	..	..	..					J. P. de Figueiredo
	627		Dous Irmãos	W Sr	Brazilian	174	..	..	..					J. F. Coelho
	628		Dous Irmãos	W Sr	Brazilian	116	..	..	..					A. A. e Silva
	629		Dous Irmãos *dos Santos*	W Sr	Brazilian	106	..	..	..	S. João da Barra	1882		[illegible]	F. J. dos Santos Dias
59653	630		Douse *Allen 88-88*	W Bk	British	1??	102·6	24·0	12·9	P. E. Island	1868	Douse		H. Andrews
	631		Douze Juillet *Rochard*		French	151	..	..	..					Guibert & fils
63334	632	*	Dovenby *Fraser 88-88*	I Bk	British	833	197·0	32·0	19·9	Harrington	1871	R. Williamson & Son		P. Iredale & Son
91234	633	*	Dovenby Hall *Kilbert 87-88*	I S	British	1998	280·0	40·0	24·2	Newcastle	1885	Palmers' Co., Ld.		Globe Shipping Co., Ld.
58398	634	*	Dovey Belle *James*	W Sr	British	8?	73·5	20·0	10·0	Aberdovey	1867	Richards		J. James & Co.
	635	NV	Dovre *Pedersen -85*	W Bk	Norweg'n	42?	131·3	29·0	15·5	Tvedestrand	1885	O. Gundersen	[illegible]	A. Biering Olsen
	636	N	Dovre *Torkildsen -88*	W Bg	Norweg'n	348	113·2	28·5	16·2	Grimstad	1862	O. & L. Dannmos ?sen		J. Halvorsen
	637	N	Dovre *Andreassen*	W Bg	Norweg'n	312	107·2	27·8	15·3	Arendal	1858			G. Berthelsen
	638	N	Dovre *Christoffersen -86* (WRECKED)	W Bg	Norweg'n	21?	83·3	26·8	15·1	Grimstad	1837	T. Gundersen		C. Anderssen
60078	639	*	Doxford *Oakley 85-85*	I Bk	British	68?	178·0	29·8	19·1	Sunderland	1868	W. Doxford & Sons		R. W. Leyland & Co.
	640	N	Drafn *Syvertsen*	W Bk	Norweg'n	336	130·1	28·8	14·8	Nova Scotia	1863			S. Olsen & E. Syvertsen
	641	N	Drafna *Hansen 86-86*	W Bk	Norweg'n	41?	11?·9	30·1	18·4	Christiansand	1836	J. T. T[illegible]		S. Andersen
	642		Dragen	W Lighter	Danish	137	86·5	20·5	11·4	Tvedestrand	1843			Flindt, Drost & Co.
50122	643	*	Dragon (ABANDONED) *Roberts*	I Bk	British	696	182·7	28·9	18·5	Stockton	1864	M. Pearse & Co.		Ship "Dragon" Co. (F. M. Tucker)
3268	644	DI	Draguette *Galbiati -88*	W Bk	Italian	728	159·1	33·1	22·0	Lavagna	1852	Brignole	[illegible]	G. B. Schiaffino
	645	N	Drammenseren (ex Jason) *Syvertsen -88*	W Bk	Norweg'n	778	170·5	36·1	20·2	Rotterdam	1856	C. Lehmann	[illegible]	O. Eliasen
	646	N	Draupner *Werge 73-80*	W Bk	Norweg'n	44?	133·3	31·3	17·6	Kragerö	1864	T. Pedersen		Actieselskabet (A. C. Andersen)
6 ?42	647		Dreadnaught *Nickerson*	W 3m Sr	American	32?	134·5	32·5	10·0	N. Haven, Con.	1877	Gessner & Marr		L. Armstrong
53049	648	*	Dreadnought (WRECKED) *Davies*	W Bn	British	14?	94·5	23·1	12·2	Newquay	1866	Davies	[illegible]	J. Davies & Co.
2?8?6	649	*	Dreadnought *[illegible] 78*	W Sr	British	[illegible]	76·6	21·4	10·5	Liverpool	1859	J. D. ?wen		J. Thomas
[illegible]	650	R	Dreadnought *Smith*	W Sr Yt	American	128	120·0	25·0	10·0	Brooklyn	1871	Poillon	[illegible]	A. W. Nickerson

* = Lloyd's Register; D = Nederlandsche Vereeniging; G = Germanischer Lloyd; I = Registro Italiano; [illegible]
A = Veritas Austro-Ungarico; H = Veritas Ellenico; N = Norske Veritas; V[illegible]

In the 4th Column, C = Composite (iron frame, wood planked); I = Iron; Stl = Steel; W = Wood.

The Dates of Metalling (Col. 11) are based on the records of the Books in which Vessels are respectively classed, or on the Reports of the Surveyors [illegible]

Further particulars of Vessels classed by Lloyd's Register will be found in Lloyd's Register Book and Lloyd's Yacht Register.

No.	Cote	Nom. Matériaux et Gréement. (Capitaine et dates d'entrée; Ancien Nom)		Pavillon	Tonnage	Longueur	Largeur	Creux	Construit: Où.	Quand	Par qui.	Doublage	Armateurs ou Directeurs-Gérants.	Port d'Armement.
651	G	Drenna	IS	German	1461	247·0	37·0	23·0	Vegesack	1886	[illegible]		D.H. Wätjen & Co.	Bremen
652	V	Drei Brunies	W Bk	Belgian	6?7	[illegible]	30·8	19·0	St.John,N.B.	1865	Nevins		H. Rich & Co.	Antwerp
653	G	Drei Gebruder	W Sr	German	[illegible]	[illegible]	[illegible]	[illegible]	[illegible]	1856	W. Hagen		W. Freij	Papenburg
654	G	Drei Gebruder	W Sr	German	116	[illegible]	[illegible]	9·7	Grossefehn	1866	F.J.Cassens & [Sohn		[illegible] J. Cassens & Sohn	Grossefehn
655		Drei Geschwister	W Bg	German	2?4	[illegible]	[illegible]	[illegible]	Rostock	1862	J. Möller		P. Grampp	Rostock
656	D	Drie Gezusters	W Sr	Dutch	152	89·3	22·0	10·2	Hoogezand	1859	Niehuis		W. Velthuis	Hoogezand
657	G	Drie Zusters	W Sr	German	111	[illegible]	18·?	8·?	[illegible]	1851	H.J.Schroeder		Abr. Beckman	Papenburg
658		Drisco	W 3m Sr	American	2?8	115·0	31·5	9·5	Millbridge Me	1888	J.Sawyer & Sons		D.J. Sawyer	[illegible]
659	E	Dristig	W Bk	Swedish	4?3	132·5	25·3	16·3	Archangel	18^{54}_{76}			J.E. Olsson	Gothenburg
660		Driving Mist	W Bn	British	145	88·8	23·5	11·9	Sunderland	1854	W. Pile		E. Joiner	Faversham
661		Drogden	W Bk	Danish	247	108·7	22·4	14·4	Arbroath	1858			Ludvig Olsen	Copenhagen
662	V	Dronning Louise	W Bg	Danish	268	119·6	23·7	13·3	Kjøge	1866	J.P.Johansen		J. Skrike	Copenhagen
663	V	Dronning Louise	W Sr	Danish	198	111·2	25·3	12·4	Marstal	1879	J.J. Bager		H.C.Christensens Enke	Svendborg
664	N	Dronning Louise	W Bk	Norweg'n	4??	127·1	32·1	17·0	Tvedestrand	1871	A. Speilberg		A. & F. Smith	Tvedestrand
665	N	Dronning Sophie	W Bk	Norweg'n	7??	154·7	32·2	19·9	Quebec	1873	P. Baldwin		J. Valentinsen	Haugesund
666	N	Dronning Sophie	W Bk	Norweg'n	5?8	133·6	30·2	19·6	Moulmein	1847	Moulds		B. Nielsen	Fredrikstad
667	N	Dronning Sophie	W 3m Sr	Norweg'n	4??	137·2	29·0	16·3	Stavanger	1876			E. Berentsen	Stavanger
668	N	Dronningen	W Bk	Norweg'n	1093	182·1	37·1	22·9	Belleveau's [Cove N.S.	1873			P. Semb	Laurvig
669	V	Drot	W Bk	Norweg'n	1164	186·2	37·3	23·7	Bath, Me.	1874	A. Lemont		E. Berentsen	Stavanger
670	V	Drott	W Sr	Swedish	139	86·6	21·1	10·6	Söderköping	1884	J. Hagglund		J. Johnsson	Söderköping
671	V	Drottning Sofia	W Bk	Swedish	648	155·1	30·3	18·0	Sundsvall	1873	E. Östman		J.C. Hulthen	Helsingborg
672		Drover	W Bg	British	17?	1??·1	22·4	13·0	Aberdeen	1845			J. Allen	Sydney, N.S.W.
673		Druid	W Sr	British	102	?1·3	27·2	7·5	Portland, N.B.	1882	J. Calhoun		J.A. Gregory	St.John, N.B.
674	V	Druide	W Bn	French	138	82·7	23·6	11·3	Redon	1875	Mahon Fils		Calbourdin	Auray
675	❋	Drumblair	I 4m S	British	1848	257·4	40·2	24·1	Greenock	1883	Russell & Co.		Gillison & Chadwick	Liverpool
676	❋	Drumburton	I 4m S	British	1840	256·7	40·2	23·9	Port Glasgow	1881	Russell & Co.		Gillison & Chadwick	Liverpool
677	❋	Drumcliff	I 4m S	British	2408	211·3	45·2	24·2	Greenock	1887	Russell & Co.		Gillison & Chadwick	Liverpool
678	❋	Drumcraig	I 4m Bk	British	1919	289·3	41·1	23·5	Barrow	1885	Barrow S.B.Co.		Gillison & Chadwick	Liverpool
679	❋	Drumeltan	I 4m S	British	1848	257·3	40·3	24·1	Greenock	1883	Russell & Co.		Gillison & Chadwick	Liverpool
680	❋ D	Drumlanrig	I S	British	1425	241·5	37·7	22·6	Port Glasgow	1876	Russell & Co.		Gillison & Chadwick	Liverpool
681	❋	Drummoir	I S	British	17?8	270·5	39·2	24·0	Liverpool	1882	W.H. Potter & Sn		Gillison & Chadwick	Liverpool
682	❋	Drumpark	I S	British	1470	245·6	37·9	22·6	Liverpool	1877	W.H. Potter & So		Gillison & Chadwick	Liverpool
683	❋	Dryad	I S	British	1035	2?3·4	34·7	21·2	Liverpool	1874	T. Royden & Sns		J.B. Walmsley & Co.	Liverpool
684		Dryad	W Bg	British	199	96·2	2?·5	16·0	Whitehaven	1837	L.&J. Brocklebank		W.R. Kelly	Whitehaven
685	V	Dub	W Bk	Austrian	805	158·1	30·5	24·2	Gravosa	1873	N.S. Bradicich		Ciuarlizza & Co.	Ragusa
686		Duc	W Bg	French	192	93·8	23·0	14·7	St. Malo	1858	L. Mallard		Auguste Lemoine	St. Malo
687	V	Duc	W Bk	French	15?	91·8	24·9	11·7	St. Denis	1863	Joineau père		Riotteau & fils	Granville
688	V	Duc d'Aumale	W Bk	French	367	128·6	24·9	16·1	Lübeck	1871	J.A. Meyer		J.B. Demange	Nantes
689	I V	Duca di Galliera	W Bk	Italian	838	168·2	3?·8	22·3	Varazze	1875	C. Cerutti		G. Viganigo & figli	Cagliari
690	❋	Duchalburn	Stl 4m S	British	1998	287·6	42·5	24·0	Glasgow	1887	Barclay, Curle [& Co.		R. Shankland & Co.	Greenock
691	❋	Duchess	W Sr	British	99	85·8	22·7	10·4	Connah's Quay	1878	S. Owen		C. Davison	Chester
692	❋	Duchess of Albany	I S	British	1746	253·0	40·3	23·0	Liverpool	1884	T. Royden & Sns		W. & R. Wright (R. Hamilton, Trustee)	Liverpool
693	❋	Duchess of Edinburgh	I S	British	18?3	255·2	40·0	23·5	Sunderland	1874	Mounsey & Foster		E.F. Powell & Co.	Liverpool
694		Due Amici	W Bn	Italian	119	89·6	23·2	11·2	[illegible]	1878	F. Casanuro		Bronzi & Carpena	Spezia
695		Due Amici L.	W Bn		111	83·6	22·4	10·8	[illegible]	1884	Flli. Fava			
696	I	Due Cognati	W Bk	Italian	488	137·4	30·5	19·5	[illegible]	1869			F. Bozzo	Genoa
697		Due Cognati	W Bn	Italian	146	78·7	23·8	14·3	[illegible]	1857	M. Tappani		Ansaldo & Olivari	Genoa
698	A	Due Figli	W Bk	Italian	394	128·6	28·4	17·9	[illegible]	1868			G. Zecchini fu C.	Leghorn
699		Due Fratelli	W Bg	Italian	346	111·5	28·9	18·0	Chiavari	1860	F. Gotuzzo		Flli. Mortola	Genoa
700	I	Due Fratelli	W Bk	Italian	360	125·0	28·0	16·4	Gaeta	1876	L. Contreras		E. Acceta	Gaeta

❋ = Lloyd's Register; D = Nederlandsche Vereeniging; G = Germanischer Lloyd; I = Registro Italiano; R = American Record;
A = Veritas Austro-Ungarico; H = Veritas Ellenico; N = Norske Veritas; V = Bureau Veritas.

Dans la 4me Colonne, C = Composite (membrure en fer, bordé en bois); I = Fer; Stl = Acier; W = Bois.

Les Dates du doublage en métal (Col. 11) ressortent des livres dans lesquels les navires sont respectivement classés ou des rapports des Inspecteurs du Lloyd's Register.

Pour plus amples renseignements sur les navires classés par le Lloyd's Register, voir le Lloyd's Register Book et le Lloyd's Yacht Register.

Official No.	No. in Book	Registry by which Classed	Ship's Name. Master, and Dates of Appointment to present Owner's Service and to Ship. Ship's Late Name.	Material and Rig	Flag	Net Tonnage	Length	Breadth	Depth	Where Built	When	By whom	Date of Metalling	Owners or Managers
[illegible]	701	A	Due Fratelli *Cannavo* (ex Mille Fr.)	W Bg	Italian	252	113·1	26·0	16·2	[illegible]	1868	[illegible]	[illegible]	F. Cannavo fu C.
[illegible]	702	I	Due Fratelli *Carpena*	W Bn	Italian	123	92·2	21·8	11·5	[illegible]	1882	G. B. Consigliere	[illegible]	A. Carpena fu G.
	703	I	DueFratelliMassone	W Bark		341	121·3	27·7	16·7	Sestri	1879	A. Carlevaro	[illegible]	
[illegible]	704		Due Sorelle *Tancredi* (ex Rosa T.)	W Bk	Italian	147	83·3	23·4	14·6	Piano di Sorrento	1856			E. Tonietti fu L.
[illegible]	705	I	Due Virginie *Jacona*	W Bk	Italian	299	121·9	26·9	16·5	Castel. di Stab.	1878	E. D'Amico	[illegible]	S. Munzone fu V.
	706	N	Duen *Olsen -83*	W Bk	Norweg'n	262	92·8	26·8	15·4	Arendal	18?8			H. H. Pettersen
	707	N	Duen *Andersen 84-77*	W Bk	Norweg'n	246	118·4	26·2	11·6	Kragerö	1874		[illegible]	H. Biorn, jr.
	708	N	Duero *Johansen -88*	W Bk	Norweg'n	265	115·9	27·3	11·7	Risör	1877		[illegible]	P. Pedersen
	709		Duero	W Bk	Peruvian	355	..	..	..	Quebec	1864			Serdio & Hermanos
	710		Duguay Trouin *Hantebert*	W Bk	French	387	127·9	27·2	14·8	St. Malo	1873	O. Hervot	[illegible]	Ch. Leclère
	711	V	Duisburg *Hastom 87-87*	W Bk	Norweg'n	997	183·1	35·4	21·7	Vegesack	1856	Johann Lange	[illegible]	Westergaard & Co.
[illegible]	712		Duke *Gripps 75-83*	W Bk	British	387	122·6	27·9	17·8	Sunderland	1867	T. Metcalf	[illegible]	J. T. Davison
[illegible]	713		Duke	W Sr	British	106	81·0	21·0	7·8	Penang	1874	N. H. Eng		Tan Kay Beng
[illegible]	714	✻	Duke of Abercorn *Jaxcamme -88*	C S	British	1050	212·0	35·1	20·5	Glasgow	1869	C. Connell & Co.	[illegible]	Montgomerie & Workman
[illegible]	715	✻ D	Duke of Argyll *M. Donald 85-84*	I S	British	960	199·7	33·2	20·9	Dumbarton	1865	Denny & Rankin		Montgomerie & Workman
[illegible]	716	✻	Duke of Athole *Nicholson* [illegible]	I S	British	963	199·2	33·2	29·0	Dumbarton	1865	Denny & Rankin		Devitt & Moore
[illegible]	717	✻	Duke of Newcastle [illegible]	W Sr	British	87	77·5	21·4	10·4	Barton Stather	1886	J. Garside		W. Brown
	718		Duldul Bahri	W S	Turkish	175	..	..	..	Jeddah	1871			Yeri Bahan Zadeh Stephanaki & Co.
[illegible]	719	✻	Duleep Singh *Lower 82-82*	I S	British	1198	210·7	34·6	22·7	Liverpool	1863	T. Vernon & Son		J. H. Worthington & Co.
[illegible]	720	✻ D	Dumbartonshire *Norgrove 84-85*	I Bk	British	915	204·4	33·6	19·9	Glasgow	1874	Dobie & Co.		T. Law & Co.
[illegible]	721	✻	Dumfriesshire *Johnson 85-82*	I Bk	British	1221	226·7	36·2	21·5	Port Glasgow	1877	Russell & Co.		J. & W. Goffey
	722	N	Duna *Grepne 88-87*	W Sr	Norweg'n	128	91·3	23·3	10·8	Stavanger	1865	R. & T. Rasmussen		S. Grepne
	723	N	Dunaborg *Omsted 83-83*	W Bn	Norweg'n	187	106·4	25·4	11·0	Bergen	1866	J. Frumphy & Son		G. H. Omsted
104	724	V	Dunahof *Kalning -86*	W S	Russian	171	86·0	23·0	11·1	Haynasch	1881	C. Ohsoling		J. Mikkelsohn
[illegible]	725	✻	Dunard *Quayle -84*	I Bk	British	707	186·2	31·2	18·4	Glasgow	1877	Dobie & Co.		J. Dunn & Sons
[illegible]	726	✻	Dunblane *Manson -86*	I Bk	British	720	187·8	31·0	18·5	Glasgow	1878	D. & W. Henderson		J. Dunn & Sons
[illegible]	727	✻	Dunboyne *O'Neill -88*	I S	British	1380	243·1	37·5	21·5	Whitehaven	1888	[illegible]		R. Martin & Co.
[illegible]	728	✻	Dunbritton *Emmett*	I S	British	1471	237·4	38·4	22·5	Dumbarton	1875	A. M'Millan & Son		J. & A. Allan
[illegible]	729	✻	Duncow *Large 78-78*	I S	British	1635	252·0	40·0	23·5	Sunderland	1878	W. Doxford & Sons		J. Houston
[illegible]	730	✻	Duncrag *Pirie 84-84*	I Bk	British	893	200·1	32·6	19·6	Port Glasgow	1881	Russell & Co.		J. Dunn & Sons
	731	✻	Duncraig *Freimuth 87-87*	I Bk	German	899	188·5	31·0	18·5	Dumbarton	1870	A. M'Millan & Son		W. Maack
[illegible]	732	✻	Dundale *Trevena -85*	I Bk	British	1114	215·2	35·1	21·2	Port Glasgow	1885	Russell & Co.	[illegible]	J. Dunn & Sons
[illegible]	733	✻	Dundarg *Gibb -84*	W Bn	British	133	96·5	23·7	10·7	Fraserburgh	1875	J. Webster		R. Smith
[illegible]	734	✻	Dundee *Low 82-88*	4m S	British	1998	291·5	43·2	23·5	Dundee	1882	W. B. Thompson		C. Barrie
[illegible]	735	✻	Dundonald *Goodall 87-87*	I Bk	British	591	173·2	27·8	18·0	Liverpool	1861	T. Vernon & Son		Nicholson & McGill
[illegible]	736	✻	Dundrennan *Palmer 88-88*	I S	British	1891	268·0	39·7	24·2	Southampton	1880	Oswald, Mordaunt & Co.		J. Houston
[illegible]	737		Dunedaw	I Bg	British	352	195·1	25·0	8·4	Port Glasgow	1879	R. Duncan & Co.		Irrawaddy Flotilla Co. Ld.
[illegible]	738	✻	Dunedin *Roberts 77-86*	I S	British	1250	241·0	36·1	20·9	Port Glasgow	1874	R. Duncan & Co.		Shaw, Savill & Albion Co. Ld.
[illegible]	739	✻	Dunfillan *Ritchie 87-78*	I S	British	1244	228·2	36·3	21·3	Glasgow	1878	Dobie & Co.		W. Ross & Co.
[illegible]	740	✻	Dunkeld *White*	I Bk	British	1005	219·4	34·0	20·0	Dumbarton	1877	A. M'Millan & Son		D. Finlayson
	741		Dnnkerque	Stl 4m Bk	French	3100	330·0	46·5	25·1	Port Glasgow	1889	Russell & Co.		A. D. Bordes & fils
[illegible]	742	R	Dunleary *Kavanagh* (ex Glasin fall Asssoon) (STRANDED)	W Sr	British	265	127·2	27·8	14·1	Trieste	1865	[illegible]		W. Wallace
[illegible]	743	✻	Dunloe *McNeily 85-87*	I S	British	674	180·9	30·4	18·4	Glasgow	1870	Dobie & Co.		J. B. Walmsley & Co.
[illegible]	744	✻	Dunnerdale *Williamson 75-81* (ex Cholula)	I Bk	British	1066	210·2	35·1	21·6	Birkenhead	1867	G. R. Clover & Co.		Hughes & Co.
[illegible]	745	✻	Dunolly *Handyside 84-84*	I S	British	1478	255·6	37·6	22·6	Glasgow	1884	C. Connell & Co.		J. Brown
43873	746	✻	Dunphaile Castle *Johnson 85-86*	W S	British	720	160·0	32·2	20·0	Sunderland	1861	J. Laing	[illegible]	F. S. Whitehead & [illegible]
72142	747	V	Dunrobin *Travis 88-88*	W S	British	1375	198·2	39·2	24·0	[illegible]	1874	J. [illegible]	[illegible]	B. Smith & Co.
[illegible]	748		Dunrobin Castle *Comer* [illegible]	W S	British	545	131·1	25·9	18·7	Aberdeen	1851	A. Hall & Co.	[illegible]	G. Milne & Co.
[illegible]	749	✻	Dunscore *Hunt 83-83* (ex City of Vienna)	I S	British	1006	213·0	31·8	21·1	Glasgow	1868	C. Connell & Co.		T. C. Guthrie
[illegible]	750		Dunsinane *Maisy 85-85*	W Bn	British	271	130·9	25·5	14·0	Dundee	1874	Dundee S. B. Co.	[illegible]	J. Helman & Co.

✻ = Lloyd's Register; D = Nederlandsche Vereeniging; G = Germanischer Lloyd; I = Registro Italiano; A = Veritas Austro-Ungarico; H = Veritas Ellenico; N = Norske Veritas;

In the 4th Column, C = Composite (iron frame, wood planked); I = Iron; Stl = Steel; W = Wood.

The Dates of Metalling (Col. 11) are based on the records of the Books in which Vessels are respectively classed, or on the Reports of the Surveyors

Further particulars of Vessels classed by Lloyd's Register will be found in Lloyd's Register Book and Lloyd's Yacht Register

1	2	4 Nom, Matériaux et Gréement. Capitaines et dates d'entrée en service. Armateurs actuels et à bord de navire. Anciens Noms.	5 Pavillon.	6 Tonnage net.	7 Dimensions. Longueur.	Largeur.	Creux.	8 Construit. Où.	9 Quand.	10 Par qui.	11	12 Armateurs ou Directeurs Gérants.	13 Port d'Armement.
751	V	Dunstaffnage *Jones 80* W Bk	British	950	179·5	35·4	19·5	Kouchibouguac, N.B.	1882	Geo. McLeod	[illegible]	W. H. McLeod	Chatham, N.B.
752	✻	Duntrune *[illegible]* I S	British	1488	245·2	38·3	23·0	Dundee	1875	A. Stephen & Sons		D. Bruce & Co.	Dundee
753	✻	Dunure *[illegible]* W Bgtn	British	186	112·6	22·9	11·8	Troon	1881	Troon S. B. Co.	[illegible]	John Steer	St. Johns, N.F.
754	✻	Dunvegan *[illegible]* W Bk	British	786	172·7	35·0	19·6	Kouchibouguac, N.B.	1881	Geo. McLeod	[illegible]	G. K. McLeod	Chatham, N.B.
755	R	Duo *[illegible]* W Bg	Swedish	254	101·9	24·4	13·0	Ribnitz	1858	J. Mittelstadt		R. Fogelmarck	Westervik
756	V	Duppel *[illegible]* W Bk	German	480	128·1	30·6	18·0	Danzig	1864	F. Devrient		A. Gibsone	Danzig
757	V	Duquesne *Le Mateled 86-84* W Bk	French	156	84·6	21·6	12·7	Dieppe	1866	J. Carpentier		Amaury Simon	Redon
758		Duque da Victoria W Bn	Brazilian	222	..	..	..					J. M. C. Sol	Rio Janeiro
759	V	Durango *Wahler 85-86* W 3m Sr	German	290	129·9	28·0	11·8	Reiherstieg	1876	C. C. A. Dreyer	[illegible]	E. B. Crasemann	Hamburg
760	N	Durango *Berg* W Bk	Norweg'n	328	164·5	25·9	17·3	N. Brunswick	1841	Cunard & Co.		O. A. Berg	Fredrikstad
761	✻	Durham *Bailey* I S	British	1997	286·2	39·1	26·4	London	1874	M. Wigram & Sns		Sailing Ship "Durham" Co. (J. Herron & Co.)	Liverpool
762	✻	Durisdeer *[illegible]* I S	British	989	202·0	34·8	21·4	Glasgow	1864	A. Stephen & Sons		T. C. Guthrie	Glasgow
763	✻	Dussau *Farochine* W Bk	Austrian	438	140·0	29·5	17·8	Sunderland	1869	Towers & Nesbitt	[illegible]	C. Andrianich fu A.	Buccari
764	V	Dusty Miller *Hughes 84-85* W Bk	British	596	142·5	32·3	19·5	N. Brunswick	1862	McLachlan	[illegible]	H. Owen & Son	Liverpool
765	N	Dux *Pedersen* W Bk	Norweg'n	381	121·8	28·5	15·5	Tvedestrand	1874	O. Gundersen	[illegible]	M. H. Smith	Tvedestrand
766	N	Dux *[illegible]* W 3m Sp	Norweg'n	347	116·9	30·6	14·6	Kaskö, Finl'nd	1875			P. Hansen	Sarpsborg
767		Dygden *Axman* W Bk	Russian	499	..	..	..	Raumo	1865			J. H. Stahlberg	Raumo
768	V	Dygve *[illegible]* W Bg	Swedish	240	106·6	22·6	11·2	Perth	1865	Perth S. B. Co.	[illegible]	A. Brolin	Söderhamn
769	✻	Dynomene *Williams 80-85* I S	British	1900	270·0	39·8	24·1	Liverpool	1882	W. H. Potter & Sn		H. Fernie & Sons	Liverpool
770	A H	Dyo Adelphi *Nicolaki* W Bg	Greek	322	117·4	25·9	..	Sunderland	1869		[illegible]	D. Nicolaki & Co.	Syra
771	A	Dyo Adelphi W Bg	Greek	293	126·6	27·7	15·4	Syra	1880		[illegible]	J. Goulandris	Andros
772	H	Dyo Adelphi W Bg	Greek	153	..	..	..	Piræus	1861			E. Caravanghelis	Trikera
773		Dyo Adelphi W Bg	Greek	138	..	..	..	Syra	1863	A. Poulzos		C. Coufopantelis	Zante
774	R	Dyring *Nilsson* W Bn	Swedish	175	87·3	23·5	11·5	Westervik	1877	C. R. Fridell		G. Idestrom	Westervik
775	V	Dzaoudzi *Cumoray* W Bk	French	480	141·0	30·2	17·2	Nantes	1886	Sevestre	[illegible]	A. Viot	Nantes

ADDITIONAL SHIPS ARE INSERTED IN THE SUPPLEMENT AT THE END OF THE REGISTER.

✻ = Lloyd's Register; D = Nederlandsche Vereeniging; G = Germanischer Lloyd; I = Registro Italiano; R = American Record; A = Veritas Austro-Ungarico; H = Veritas Ellenico; N = Norske Veritas; V = Bureau Veritas.

Dans la 4me Colonne, C = Composite (membrure en fer, bordé en bois); I = Fer; Stl = Acier; W = Bois.

Les Dates du doublage en métal (Col. 11) ressortent des livres dans lesquels les navires sont respectivement classés ou des rapports des Inspecteurs du Lloyd's Register.

Pour plus amples renseignements sur les navires classés par le Lloyd's Register, voir le Lloyd's Register Book et le Lloyd's Yacht Register.

1 Official No.	2 No. in Book	3 Registry in which classed	4 Ship's Name	Master	Material and Rig	5 Flag	6 Net Tonnage	7 Length	Breadth	Depth	8 Built: Where	9 When	10 By whom	11 Date of Metalling	12 Managers
[illegible]	1	RV	E. B. Sutton	Carter	WS	American	1758	227·8	48·1	25·6	Bath, Me.	1881	I. F. Chapman	[illegible]	A. G. Ropes
[illegible]	2	✱	E. Birkbeck	Vine	WSk	British	87	77·4	20·6	10·5	Goole	1883	Cottingham Brs.		W. Agerskow
[illegible]	3	R	E. C. Allen	Meady	W3mSr	American	475	141·2	34·6	15·1	Bath, Me.	1879	Goss & Sawyer		S. B. Meaby
[illegible]	4		E. C. Gates	Freeman	WSr	American	127	100·0	26·2	8·0	Perry, Me.	1867	Frost		E. C. Gates
[illegible]	5		E. C. Knight, jun. (Heming 79-83)		W3mSr	American	235	121·3	32·4	9·0	Port Republic, N.J.	1874			G. W. Shepherd, jun.
[illegible]	6		E. C. T.	Toms	WSr	British	105	83·1	20·6	10·7	Falmouth	1870	W. H. Lean		T. Guyatt
[illegible]	7	R	E. Chambers (WRECKED)	Hattendorf	WSr	British	215	103·4	28·6	10·2	Rockport, N.B.	1872			James D. Seely
[illegible]	8		E. Closson	Hardy	WSr	American	136	95·5	26·3	8·7	Sedgwick, Me.	1860			E. C. Arey
	9		E. F. Thompson (ABANDONED)	Jas -80	WBn	Norwegn	230	107·2	25·8	11·2	Baltimore	1860			Bt. Halvorsen
[illegible]	10		E. Flower	Brooks	WSr	American	124	91·0	25·0	7·0	E. Haddam, Conn.	1863			W. J. Mack
[illegible]	11		E. G. Benedict		WSr	British	155	104·5	26·2	8·3	Ontario	1869	Rathbun & Son		J. H. Walker
[illegible]	12	R	E. G. Irwin	Dodd	WSr	American	179	106·0	28·5	8·5	Milton, Del.	1865			Silas H. Dodd
[illegible]	13		E. H. Bronson		W Barge	British	158	109·0	22·0	7·0	Hull, Quebec	1878	A. Miller		Margaret Burll
[illegible]	14		E. H. Cornell	Wass	W3mSr	American	338	137·4	33·0	10·3	Bath, Me.	1881	Deering & Donnell	[illegible]	B. W. & H. F. Morse
85510	15		E. H. Foster	Tufts	WSr	British	124	86·0	27·6	7·7	St. Martin's N.B.	1883	W. J. Davidson		W. J. Davidson
135021	16	R	E. H. Harriman	Wood	W3mSr	American	226	121·0	28·9	9·2	Bucksport, Me.	1874		[illegible]	H. Lord
8661	17		E. H. King		WSr	American	101	85·0	24·6	7·8	Trenton, Me.	1869			T. L. Holmes
77694	18		E. H. Rutherford		WSr	British	286	133·6	22·2	10·7	Port Dalhousie	1881	J. Shickaluna		E. Browne
135582	19	R	E. H. Weaver (Weaver 86-88)		W3mSr	American	653	153·6	35·4	15·5	Bath, Me.	1882	W. Rogers		E. H. Weaver
8028	20		E. J. Hamilton	Mills	WSr	American	179	111·0	30·0	8·0	Nyack, N.Y.	1873	J. B. Voris		S. E. Birdsall
71250	21		E J. Peters		WSr	British	120	96·9	21·6	7·3	Port Dalhousie	1867	S. D. Andrews		E. Winton
64706	22	✱	E. J. Spence	Gill	I Bk	British	519	159·7	28·1	17·0	Sunderland	1871	J. Blumer & Co.		A. Brown
80369	23	DV	E. J. Spicer	Spicer 72-84	WS	British	1318	203·5	39·7	23·8	Parrsboro, N.S.	1880	A. Loomer	[illegible]	G. D. Spicer
135995	24		E. K. Wood		W4mSr	American	495	170·2	37·5	11·7	P. Blakeley W.T.	1888	Hall Bros.		N. C. Lassen
6778	25		E. L. Higgins	Cunningham	WSr	American	100	76·7	23·4	8·0	Eden, Me.	1863/71			H. B. Eaton
62854	26	✱	E. L. Margaret	Williams -88	WBg	British	216	107·1	24·9	14·2	Pwllheli	1873	Prichard	[illegible]	J. Williams
[illegible]	27	RV	E. L. Pettengill	White 79-81	WBk	American	842	161·8	34·3	20·3	Deering, Me.	1878	G. W. Russell	[illegible]	J. S. Winslow & Co.
43484	28		E. Lacourciere		WSr	British	109	91·2	22·0	7·4	Batiscan	1862			James Myles
135077	29	R	E. M. Bacon	McKay 86-6	WSr	American	176	125·0	26·0	10·4	Essex, Mass.	1883	Story		Geo. T. Littlefield
8593	30		E. M. Sawyer	Watts	WSr	American	125	98·0	26·5	8·0	Jonesport, Me.	1869			D. J. Sawyer
80395	31		E. Merriam	Merriam	WSr	British	331	132·1	31·3	12·9	Parrsboro, N.S.	1882	E. Merriam, jun.		E. Merriam, jun.
135729	32	R	E. O. Clark	Stahl	WBkn	American	656	152·0	35·2	17·6	Waldobo'ro, Me.	1883	E. O. Clark	[illegible]	E. O. Clark
71367	33		E. R. C. Proctor		WSr	British	163	109·4	25·5	9·2	Brighton, Ont.	1878	John Allen		J. E. Proctor
[illegible]	34	✱	E. S. Hocken	Hocken 85-88	WBkn	British	283	126·2	26·0	13·9	Fowey	1879	J. Slade & Sons	[illegible]	J. H. Hocken
8870	35	R	E. S. Newman	Lowell	W3mSr	American	373	133·1	31·4	10·9	Bucksport, Me.	1872		[illegible]	R. P. Buck & Co.
8969	36	R	E. S. Powell	McCormick	WBkn	American	558	147·0	36·7	15·0	Tottenville	1873	E. S. Powell	[illegible]	L. E. Davis
[illegible]	37	V	E. Sutton	Vaughan 86-86	WBk	British	545	139·0	34·8	16·7	Grand Bay, N.B.	1876	E. Sutton	[illegible]	[illegible]
[illegible]	38		E. T. Campbell (ex Shamrock)	Doolan -88	WBn	American	243	111·6	25·8	12·2	Cardigan, P.E.I.	1877	McDonald		H. B. Mason
74287	39	V	E. T. G.	Steinen -88	WBk	British	950	167·5	36·5	21·9	Quebec	1877	Rosa	[illegible]	J. P. Evans
	40	V	E. V. Almqvist	Myra	WBk	Swedish	529	138·4	28·9	17·7	Umeå	1873	E. A. Löfroth	[illegible]	C. O. Scharin
7356	41		E. V. Glover	Ingersoll	W3mSr	American	274	116·3	29·8	9·7	Bridgeport, Conn.	1867	J. H. Moore		J. H. Moore
80058	42	V	E. W. Gale	Smith -88	WBn	British	298	118·0	28·4	13·0	Rothesay, N.B.	1880	H. Titus	[illegible]	V. S. White
7770	43	RV	E. W. Stetson	Knight -88	WS	American	1106	173·2	38·2	23·6	Damariscotta, Me.	1862		[illegible]	T. Dunham's Nephew & Co.
64021	44	R	E. Walsh (LOST)	Anderson	WSr	British	146	96·0	21·9	10·9	Bear River N.S.	1872	J. H. Benson		W. W. Clarke
7008	45		E. Waterman	Hinckley	WSr	American	102	91·1	23·6	7·8	Washington, N.J.	1857			W. Hinckley
135043	46		E. & G. W. Hinds	Hill	WSr	American	109	89·0	26·0	8·3	Calais, Me.	1874	Hinds		E. A. Barnard
35150	47		E. & J. Oakley	Newman	WSr	American	163	113·0	28·0	7·9	Calais, Me.	1871			H. S. Newman
71445	48	✱	Eagle	Williams	WBn	British	221	110·0	24·6	13·0	P. E. Island	1874	Davies	[illegible]	R. Williams
51010	49		Eagle	Murphy	WBn	British	144	95·4	23·3	11·5	P. E. Island	1865	McMillan		W. H. Farrell
89136	50		Eagle		WBg	British	140	95·0	26·1	11·0	Bangkok	1875			Chuah Yu Pung

✱ = Lloyd's Register; D = Nederlandsche Vereeniging; G = Germanischer Lloyd; I = Registro Italiano; R = [illegible]
A = Veritas Austro-Ungarico; H = Veritas Ellenico; N = Norske Veritas; V = [illegible]

In the 4th Column, C = Composite (iron frame, wood planked); I = Iron; Stl = Steel; W = Wood.

The Dates of Metalling (Col. 11) are based on the records of the Books in which Vessels are respectively classed, or on the Reports of the Surveyors [illegible]

Further particulars of Vessels classed by Lloyd's Register will be found in Lloyd's Register Book and Lloyd's Yacht Register.

No.	Cote	Nom, Capitaine, Ancien Nom	Matériel et Gréement	Pavillon	Tonnage net	Long.	Larg.	Creux	Construit où	Quand	Par qui	Armateurs ou Directeurs Gérants	Port d'Armement
51		Eagle *Royle*	W Bn	British	[illegible]	89·6	23·0	11·0	Shelburne, N.S.	1872		W. A. Conrod	Halifax, N.S.
52		Eagle *Bell*	W Bn	British	[illegible]	93·4	22·1	9·9	[illegible]	1863	Chatfield	A. King	Belfast
53		Eagle *Meade*	W Sr	British	[illegible]	86·0	21·9	11·3	Montrose	1864	J. Brown & Co.	Bessey & Palmer	London
54		Eagle	W Sr	British	114	73·3	21·4	11·3				Kather	Penang
55		Eagle *Walters*	W Sr	American	177	100·8	27·0	8·1	Kennebunk Me	1869	R. Oakes	Hayford & Stetson	Bangor, Me.
56		Eaglet	W Bn	British	198	103·0	26·7	11·0	Jefferson, U.S.	1859		C. Nickelson	Tortola
57		Eaglet *Cruickshank*	W Sr	British	[illegible]	88·0	22·2	11·0	Fraserburgh	1863	J. Webster	J. Vickers	Fraserburgh
58	R	Eaglet *Holmes*	W Sr	American	124	90·0	26·9	7·3	Millbridge, Me	1882	E. Dyer	J. L. Dickinson	Port Jefferson
59	V	Earl Burgess *Cogill*	W S	British	1749	231·0	42·3	24·1	Cornwallis, N.S.	1885	E. Cox	C. R. Burgess	Windsor, N.S.
60	✻	Earl Cairns	W Sr	British	[illegible]	93·1	23·0	10·4	Connah's Quay	1883	Ferguson & Baird	G. T. Raynes	Liverpool
61	✻	Earl Derby	I Bk	British	961	208·2	33·5	20·0	Port Glasgow	1876	R. Duncan & Co.	T. C. Guthrie	Glasgow
62	✻	Earl Dunraven *Foreman*	I Bk	British	1310	250·8	36·2	21·7	Port Glasgow	1885	Russell & Co.	Stuart Brothers	Glasgow
63	V	Earl Granville *Copp*	W S	British	1193	188·1	37·7	23·1	Harvey, N.B.	1877	G. S. Turner	T. A. Peters	St. John, N.B.
64	✻	Earl Granville *Teston*	I Bk	British	956	207·7	33·4	20·0	Port Glasgow	1877	R. Duncan & Co.	F. M. Tucker	London
65		Earl H. Potter	W 3m Sr	American	332	137·0	31·5	11·0	St. George, Me.	1872		F. Sheerer	St. George, Me.
66		Earl Mulgrave	W Sr	[illegible]	143	..	..	..	[illegible]	1838	J. Young	J. Battershill	London
67	R	Earl P. Mason *Nickerson*	W 3m Sr	American	[illegible]	[illegible]	33·8	[illegible]	[illegible]	1872		H. Nickerson & Co.	Providence, R.I.
68	✻	Earl Rosebery	I Bk	British	1121	[illegible]	35·1	21·2	Port Glasgow	1885	Russell & Co.	W. & J. Crawford	Glasgow
69	✻	Earl Wemyss	I S	British	1411	217·0	37·8	21·4	Port Glasgow	1884	R. Duncan & Co.	Crawford & Rowat	Glasgow
70	✻	Earl of Aberdeen	I 3m Bk	British	2133	271·2	42·6	24·0	Glasgow	1886	C. Connell & Co.	Earl Shipping Ship Line, Ld. (D. Brown & Sons)	London
71	✻	Earl of Beaconsfield	W Sr	British	[illegible]	86·0	22·6	10·9	Connah's Quay	1881	Ferguson & Baird	G. T. Raynes	Liverpool
72	✻	Earl of Dalhousie	Stl S	British	1677	261·0	38·7	23·4	Dundee	1884	A. Stephen & Sns	A. Stephen & Sons	Dundee
73	✻	Earl of Devon *Short*	W Bk	British	441	146·0	28·[illegible]	16·6	Plymouth	1876	Shilston	W. H. Shilston	Plymouth
74	✻	Earl of Lathom	W 3m Sr	British	[illegible]	94·5	23·2	10·4	Connah's Quay	1885	Ferguson & Baird	G. T. Raynes	Liverpool
75	✻	Earl of Shaftesbury	I 4m S	British	[illegible]	289·6	42·1	24·0	Leith	1883	Ramage & Ferguson	"Earl of Shaftesbury" S. S. Co. Ld. (Brown & Sons)	London
76	✻	Earl of Zetland *Cunningham*	I S	British	1461	246·3	38·1	22·7	Dumbarton	1875	A. McMillan & Sn	J. R. Menzies	Glasgow
77	✻	Earlscourt *Anton*	I Bk	British	1113	214·6	35·1	21·2	Port Glasgow	1885	Russell & Co.	Stevenson & Ferguson	Greenock
78	✻	Earlshall *Stephen*	I Bk	British	396	142·0	26·9	15·8	Dundee	1876	W. B. Thompson	Robertson Bros.	Dundee
79	✻	Earn *Anderson*	W Bk	British	312	129·9	26·0	14·4	Dundee	1880	Fay S. B. Co.	Robertson Bros.	Dundee
80	✻	Earnest *Heale*	W 3m Sr	British	161	115·7	22·2	10·9	Barnstaple	1873	W. Westacott	Stephan Bros.	Cape Town
81	✻	Earnest *Taylor*	W 3m Sr	British	96	92·0	21·5	9·9	Tarlton	1884	[illegible]	J. Taylor	Preston
82	✻D	Earnock *Parsons*	I S	British	[illegible]	227·7	35·7	21·5	Dumbarton	1877	Birrell, Stenhouse & Co.	J. Shepherd & Co.	Belfast
83	✻	East Croft *Kimmer*	I S	British	1312	[illegible]	35·[illegible]	22·3	Belfast	1875	Harland & Wolff	J. Gambles	Workington
84	✻	East Lothian *Ross*	I S	British	[illegible]	258·1	36·7	22·2	[illegible]	1876	C. Connell & Co.	J. Boyd	Glasgow
85		East Lothian *Davies*	W Bk	British	349	129·3	26·0	16·6	[illegible]	1853	D. Wright	C. Clark	Melbourne
86	✻	Eastern Belle	W K Barge	British	89	86·0	21·6	7·4	[illegible]	1883	W. Colchester	Hawkins, Holman & Co.	London
87	✻	Eastern Chief	W Bk	British	401	128·3	26·9	17·9	[illegible]	1861	R. Williamson & C	J. Carr	Liverpool
88	V	Eastern Light	W S	Norweg'n	1245	[illegible]	38·1	21·0	[illegible]	1869	J. Fisher & Son	J. Thorsen	Haugesund
89	✻	Eastern Monarch *Jones*	I S	British	1706	256·5	40·3	23·5	[illegible]	1874	Mounsey & Foster	R. Thomas & Co.	London
90	✻	Eastern Queen	W Bk	British	431	140·2	28·2	18·2	[illegible]	1864	R. Thompson, jr.	R. Thomas & Co.	Newcastle
91		Eastern Star	W Bn	American	227	104·0	27·0	[illegible]	[illegible]	18 56/77		H. Trowbridge's Sons	N. Haven, Conn.
92		Eastern Star	W Bk	Argentine	1081	179·3	36·3	23·0	[illegible]	1856		Argentine Government	
93	✻D	Eastminster *Rees*	I S	British	[illegible]	226·0	35·3	20·5	Port Glasgow	1876	J. Reid & Co.	Berryman & Turnbull	London
94	✻	Eastward *McKenzie*	I Sr	British	137	98·2	22·[illegible]	10·5	[illegible]	1875	Scott & Macgill	D. McArthur	Greenock
95	✻	Eastward *Jones*	W Sr	British	67	72·5	19·1	8·7	[illegible]	1871	McLea	D. Jones	Aberystwith
96	✻	Eaton Hall *Lawrence*	I S	British	1779	257·0	42·0	23·9	[illegible]	1870	T. Royden & Son	Balfour, Williamson & Co.	Liverpool
97	V	Ebba *Svensen*	W S	Norweg'n	1410	199·1	39·4	21·4	Baltimore	1853		Actieselskabet "Ebba" (Jbs. Aschehoug)	Arendal
98		Ebba	W Bk	Swedish	816	158·8	31·2	20·5	Yarmouth, N.S.	1862	J. Lovitt	C. A. Banck & Co.	Helsingborg
99		Eben Fisher *Reynolds*	W 3m Sr	American	283	108·5	30·0	9·5	Addison, Me.	1871		W. W. Lord	Boston, Mass.
100	DV	Eben Haezer *Forma*	W Sr	[illegible]	[illegible]	80·5	22·5	10·6	Sappemeer	1867	H. Bakker	A. Forma	Groningen

✻ = Lloyd's Register; D = Nederlandsche Vereeniging; G = Germanischer Lloyd; I = Registro Italiano; R = American Record; A = Veritas Austro-Ungarico; H = Veritas Ellenico; N = Norske Veritas; V = Bureau Veritas.

Dans la 4me Colonne, C = Composite (membrure en fer, bordé en bois); I = Fer; Stl = Acier; W = Bois.

Les Dates du doublage en métal (Col. 11) ressortent des livres dans lesquels les navires sont respectivement classés ou des rapports des Inspecteurs du Lloyd's Register.

Pour plus amples renseignements sur les navires classés par le Lloyd's Register, voir le Lloyd's Register Book et le Lloyd's Yacht Register.

Official No.	No. in Book	Registry in which classed	Ship's Name	Master, and Dates of Appointment to present Owner's Service and to Ship; Ship's Late Name	Material and Rig	Flag	Net Tonnage	Length	Breadth	Depth	Built: Where	Built: When	Built: By whom	Date of Metalling	Owners or Managers
63352	101	✻	Ebenezer	Meese -77	W Bkn	British	317	132·0	26·4	14·1	Peterhead	1876	Stephen	[illegible]	J. Wilson
54085	102		Ebenezer	Lloyd 88-88	W Bg	British	197	106·6	23·9	12·8	Montrose	1865	J. Birnie & Co.		W. R. Holden
27430	103		Ebenezer	Allen 66-68	W Bg	British	177	108·7	22·5	11·4	Shoreham	1860			T. P. Cattley
21328	104	✻	Ebenezer	Roberts 80-81	W Sr	British	96	75·0	20·5	11·0	Portmadoc	1858	H. Jones		P. Jones & Co.
[illegible]	105	✻	Ebenezer	Woodford 83-83	W Sr	British	93	88·5	20·7	8·7	Sunderland	1875	G. Dunn	[illegible]	W. Harty & Co.
	106	DN	Ebenezer	Nielsen	W Bk	Norweg'n	518	129·6	32·2	18·3	Tvedestrand	1869	A. & F. Smith		A. & F. Smith
	107	N	Ebenezer	Gundersen 73-73	W Bk	Norweg'n	443	122·2	30·8	17·6	Grimstad	1867	N. O. Gjömle		N. Olsen Gjömle
	108	N	Ebenezer	Tonnessen -88	W Bk	Norweg'n	356	106·3	26·2	18·2	America	1824			Th. Nordaas
	109	N	Ebenezer	Westermark 83-83	W Bk	Norweg'n	344	119·1	29·0	14·5	Grimstad	1874	J. Boe	[illegible]	Christen Boe
	110	N	Ebenezer	Salvesen 83-83 (ex Tro y Sobrino, ex Lopez pelayo)	W 3mSr	Norweg'n	315	114·2	26·2	15·0	Asturias Spain	1865		[illegible]	Th. Salvesen (D. Backlund)
	111	N	Ebenezer	Gjessen 79-79	W 3mSr	Norweg'n	176	101·5	24·1	11·2	Porsgrund	1866	K. Hanland		C. J. Gjessen
77412	112	✻	Ebenezer Parry	Parry	W 3mBkn	British	182	100·5	24·6	13·2	Nevin	1877	G. Owen	[illegible]	H. Parry & Co.
76513	113	✻D	Eboe	Nelson -89	I Bkn	British	805	151·6	25·1	12·0	Liverpool	1877	W H Potter & Son		T. Harrison & Co.
33100	114		Ebouloise		W Sr	British	120	105·0	22·0	6·6	Les Eboulemts	1855			T. Forsyth
	115		Ecaterini	Constantino	W Bk	Russian	182	92·5	23·6	14·1	Syra	18 59/60			L. Condilato
[illegible]	116	✻V	Ecclefechan	Dow 80-82	I S	British	2058	290·7	42·2	23·8	Port Glasgow	1882	R. Duncan & Co.		T. C. Guthrie
11	117	H	Echmalotos	Zissimos	W Bg	Greek	255	99·4	26·2	..	Galaxidi	1869		[illegible]	Zissimos Bros.
17	118	H	Echmalotos		W Bg	Greek	220	92·5	21·5	..	Galaxidi	1871			N. A. Katsoulis
[illegible]	119	V	Echo	Bogart	W Bn	British	354	132·3	29·7	13·0	Moss Glen, N.B.	1883	G. Merritt	[illegible]	R. C. Elkin
27918	120		Echo	Irving	W Bk	British	243	118·0	24·0	15·0	Sunderland	1859	Peverall & Co.	[illegible]	Lucas Bros. & Co.
38201	121		Echo		W Bg	British	209	117·4	27·0	13·1	Nova Scotia	1862			A. Hall
56630	122	·✻	Echo	Rees 87-87	W Sr	British	170	101·9	23·1	12·6	Dartmouth	1867	Philip & Son	[illegible]	J. Rees
	123	DNV	Echo	Andreasen 88-88	W Bk	Norweg'n	464	132·3	29·7	17·2	Lillesand	1877	G. Olsen	[illegible]	S. N. Hansen
	124		Eclair	Ohlesson 78-83	W Bg	Swedish	218	100·4	24·8	13·2	Gefle	1838			J. C. Hulthen
	125	V	Eclaireur	Pinto -83	W Bn	French	189	93·8	23·9	12·2	Nantes	1878	T. Dubigeon & fils	[illegible]	T. Dubigeon & fils
31960	126		Eclipse		W Sr	British	180	108·7	23·3	10·6	Hobart Town	1853			J. Johnson
94053	127	✻	Eclipse	Davis -87	W K	British	91	79·0	21·1	10·6	Galmpton	1887	T. Mathews		J. W. Haylock
135348	128	RV	Eclipse	Shillaber -83	W S	American	1536	221·7	40·3	24·3	Bath, Me.	1878	Goss & Sawyer	[illegible]	G. S. Dearborn
7991	129		Eclipse	Nelson -87	W Sr	American	222	120·0	29·3	9·3	New York	1852			C. Nelson
	130		Eclipse	Schwede 77-	W Sr	German	106	85·9	18·5	10·2	Gent	18 [illegible]			G. J. Schulte's Wwe.
	131	GV	Eclipse	Franco 85-	W Bn	Spanish	161	100·4	21·9	11·9	Bremerhaven	1875	F. W. Wencke	[illegible]	F. G. Pott
	132		Eclipse (ABANDONED)	Pehrsson 80-83	W Bk	Swedish	290	113·8	23·5	16·1	Whitby	1858			J. C. Hulthen
	133	V	Ecliptic	Johannsen -88	W 3mSr	Russian	316	113·8	26·2	13·5	Orrenhof	1871	A. Ferle		J. Johannsen
71343	134	R	Economy	Morris -88	W Bn	British	416	136·8	31·0	13·5	Economy N.S.	1874	D. P. Soley		Randall A. Morris
2871	135		Economy	Renson 81-81	W Bg	British	137	72·9	20·1	11·8	Sunderland	1848	J. Robinson		W. Renson
7191	136	R	Economy	Read	W Bn	British	133	72·4	20·1	12·8	Peterhead	1831	H. Thompson		W. G. Cole
71008	137		Ecuador	Hughes	W Bk	British	1059	186·0	36·5	22·0	St. John, N.B.	1875	Hilyard Bros.	[illegible]	F. Killam
	138	N	Eda (ex Express)	Andresen 85-88	W Bk	Norweg'n	582	150·1	32·8	17·1	Gefle	1875	Bergforss & Son	[illegible]	Petter Olsen
	139	N	Edda (ex Garland)	Larsen	W Bk	Norweg'n	555	171·1	31·4	16·6	St. John, N.B.	1854			E. T. Nielsen
	140	N	Edda (ex Moselle)	Larke 86-86	W Bk	Norweg'n	322	114·2	27·0	16·0	P. E. Island	1870	Peake		J. Franklin
	141	V	Edda (WRECKED)	Andersen	W Bk	Norweg'n	155	98·4	21·6	11·1	Stavanger	1867	T. Nielsen	[illegible]	B. Bergesen
78834	142	✻	Edderside	Evans	I S	British	1306	231·0	36·7	21·8	Whitehaven	1878	Whitehaven S.B.Co.		F. Bell (R. Taylor & Co.)
[illegible]	143		Eddie	Harrison	W Sr	American	114	104·0	26·6	7·0	Milton, Del.	1874			J. H. Cannon
28501	144		Eddystone (WRECKED)	Wood	W Bk	British	400	121·0	28·6	17·5	Shields	1860	Young		Great Grimsby Ice Co. Ld.
	145		Edellene (ex Lydia)		W Bn	British	218	97·2	26·0	12·6	Ekenäs, Finlnd	1871			
[illegible]	146	D	Eden	James 87-87	W Bk	British	313	129·7	26·0	14·3	Dundee	1878	Tay Ship Co.	[illegible]	Robertson Bros.
	147	V	Eden	Petersen 84-84	W 3mSr	Danish	215	111·6	23·8	12·4	Svendborg	1884	P. Tronsegaard		N. Petersen
	148		Eden (ex Johanne Louise)	Schmit	W Sr	Danish	156	96·3	21·1	11·4	Elsfleth	1857	G. Hagens & Co.		N. P. Schmit
	149	✻D	Eden	Schoepman 88-88	I Bk	German	493	156·0	26·8	17·6	Stockton	1863	M. Pearse & Co.		F. Hilken
62035	150	✻	Eden Holme	Randall 73-76	I Bk	British	794	201·8	32·2	18·5	Sunderland	1875	Bartram, Haswell & Co.		Hine Bros.

✻ = Lloyd's Register; D = Nederlandsche Vereeniging; G = Germanischer Lloyd; I = Registro Italiano; R = [illegible]
A = Veritas Austro-Ungarico; H = Veritas Ellenico; N = Norske Veritas; V = [illegible]

In the 4th Column, C = Composite (iron frame, wood planked); I = Iron; Stl = Steel; W = Wood.

The Dates of Metalling (Col. 11) are based on the records of the Books in which Vessels are respectively classed, or on the Reports of the Surveyors [illegible]

Further particulars of Vessels classed by Lloyd's Register will be found in Lloyd's Register Book and Lloyd's Yacht Regist[er]

No.	Cote	Nom — Capitaine — Gréement	Pavillon	Tonnage	Longueur	Largeur	Creux	Construction — Où	Quand	Par qui	Doublage	Armateurs ou Directeurs Gérants	Port d'Armement
151	✻	Edgar *Nutter* W Bk	British	435	138·0	28·0	18·6	Yarmouth	1865	Fellows	[illegible]	J.H.Fellows	Yarmouth
152	R	Edgar C. Ross *Quillen* W3mSr	American	380	144·2	31·3	10·2	Bath,Me.	1887	[illegible]		W.J.Quillen	Seaford
153	✻	Edina *Johnson* W Bn	British	249	116·8	[illegible]	12·3	P.E.Island	1877	Peake	[illegible]	Goodyear & Co.	Liverpool
154	✻ D	Edinburgh *Hardie* I Bk	British	1406	241·[illegible]	[illegible]	21·5	Glasgow	1885	C.Connell&Co.		Bramwell & Gardiner	Glasgow
155	H V	Edinburgh *Holder* W Bk	British	1299	205·[illegible]	[illegible]	23·9	Quebec	1883	W.Charland,jun	[illegible]	J.Herron & Co.	Quebec
156	✻	Edinburghshire *Peer* I Bk	British	1277	258·[illegible]	[illegible]	21·0	Glasgow	1884	A.Stephen&Sons		T.Law & Co.	Glasgow
157		Edissa *Guille* W Bn	British	161	106·7	[illegible]	13·7	Guernsey	1861	R.Stoneiake		N.P.Barbenson	Guernsey
158	[illegible]	Edith *Merritt* W S	British	893	170·0	35·1	21·3	Tusket,N.S.	1870			T.C.Jones & Co.	Yarmouth,N.S.
159		Edith W Bk	British	449	131·2	[illegible]	18·2	Maitland,N.S.	1872			Geo.Douglas	Halifax,N.S.
160	✻ H	Edith *McRae* W Bn	British	312	125·5	[illegible]	12·9	M'nt.Stewart P.E.I.	1863	D.Egan	[illegible]	W.Welsh	Charlottetown,P.E.I.
161	H	Edith *Campbell* W Bn	British	172	[illegible]	[illegible]	11·6	Sheet Harb,NS	1873		[illegible]	Geo.R.Taylor	Halifax,N.S.
162		Edith *Sennot* W Bg	British	129	[illegible]	[illegible]	9·0	New Haven,US	1844			W.F.O'Gilvie	Bridgetown,Barbados
163		Edith W Sr	British	100	[illegible]	[illegible]	11·5	Montrose	1847			D.Sutherland	Montrose
164	R	Edith *Foster* W Bn	American	188	110·0	28·6	8·5	ColumbiaF,Me.	1866	J.Crandon	[illegible]	J.S.Emery & Co.	Boston,Mass.
165	N	Edith *Fugeseth* W Bg	Norweg'n	246	112·2	28·3	12·1	St.Martin's,NB	1854			J.I.Asker	Christiania
166		Edith B. Coombs W Sr	American	109	83·4	24·0	9·[illegible]	[illegible]	1878	E.Young		F.D.Hodgkins	Ellsworth,Me.
167	✻	Edith Crossfield W3mSr	British	100	94·2	22·5	9·6	[illegible] Fergus	1885	P.Rodgers		J.Fisher & Sons	Barrow
168	✻	Edith Eleanor *Jones* W Sr	British	96	89·4	21·5	10·4	[illegible]	1881	J.Warrell&Co.	[illegible]	D.C.Roberts & Co.	Aberystwith
169	R	Edith McIntyre *Hodgson* W Sr	American	152	[illegible]	21·[illegible]	[illegible]	[illegible]	1883	T.M.Hagan&Sn		P.A.Whorf	P'vincetown,Mass.
170	✻	Edith Mary *Cormack* I Bkn	British	319	144·3	25·7	12·4	Beverley	1885	Cochrane&Co.		J.C.Peacock	Glasgow
171	✻	Edith Mary *Strachan* W Sr	British	228	118·5	24·7	12·2	Dundee	1872	Tay S.B.Co.	[illegible]	Robertson Bros.	Dundee
172	D G V	Edith Mary *Schultz* W Bg	German	259	100·4	26·2	15·4	Sunderland	1862	[illegible]	[illegible]	H.Wulff	Hamburg
173	✻	EdithMaud *Thomas* W3mSr	British	125	92·0	21·8	10·8	Neyland	1888	J.Mills	[illegible]	J.Mills	Milford
174		Edith May *Anderson* W Sr	British	213	111·6	24·0	10·5	ManningRiver N.S.W.	1875	McCulloch& Sinclair		J.Taylor	Sydney,N.S.W.
175	L	Edith R. Seward W3mSr	American	232	116·2	[illegible]	9·3	Dorchester Countyd.	1880	Seward		T.J.Seward	Baltimore
176	H	Edith Sheraton W Bkn	British	314	122·0	29·0	11·0	Shelburne,NS	1887	S.Muir,Sr.		SamuelMuir,Sr.	Shelburn,N.S.
177		Edith T. Gandy *Gandy* W3mSr	American	238	123·5	37·1	9·0	Den'isville,NJ	1883	J.W.Gandy		S.W.Stiles	Camden,N.J.
178		Edith & May *Kelley* W Sr	American	110	90·9	26·1	7·2	Millbridge,Me	1887	G.Kelley		G.Kelley	Machias,Me.
179		Edla *Slingberg* W S	Russian	239	..	..	..	[illegible]	1872			P.Sittkoff	Hammarland
180		Elda *Kristiansson* W Bg	Swedish	269	98·1	24·5	14·8	Vifsta Varf	1838			C.A.Svahn	Kongsbacka
181		Edmond W Bn	French	256	100·0	23·3	11·1	Noël,N.S.	1863	J.B.Elliot		Contentin Gimer	Rouen
182	V	Edmond Alonse W Bg	French	138	78·1	20·3	12·1	Cherbourg	1858	Hamel	[illegible]	A.Goday	Cherbourg
183	V	Edmond Gressier W Bk	French	346	116·1	23·9	14·1	St.Malo	1869	VveDandin&fils	[illegible]	Cie de Transport Marit. & Fafret.	St.Malo
184	H V	Edmonton W Bk	British	1298	200·5	38·5	23·7	Quebec	1882	W.Charland,jun	[illegible]	J. Herron & Co.	Quebec
185	H	Edmund W Bn	British	149	96·7	25·0	10·5	Digby,N.S.	1879	E.Burnham	[illegible]	G.W.Merritt	Digby,N.S.
186		Edmund *Tower* W Sr	British	107	74·0	23·[illegible]	9·0	Alma,N.B.	1879	O.Martin		E.L.Read	Amherst,N.S.
187		Edmund W Bk	Danish	412	120·7	26·4	17·5	Greifswald	1858	L.Wittenberg		N.S.Prast	Dragør
188		Edmund *Eriksen* W Bn	Danish	197	101·6	25·1	11·8	Teswer	1870	J.Jansen		H.C.ChristensensEnke	Svendborg
189		Edmund Kaye *Thompson* W S	British	1091	191·3	37·2	22·4	St.John,N.B.	1855			H.Milvain (per R.S.Allan)	Newcastle
190	H	Edmund Phinney W Bk	American	751	156·5	33·6	18·7	Deering,Me.	1873		[illegible]	J.S.Winslow & Co.	Portland,Me.
191		Edmund Richardson W S	British	292	129·0	26·8	12·9	Sunderland	1873	W.Pickersgill		R. & W.King	Bristol
192		Edna *Pilcher* W Bn	British	190	104·0	24·5	12·8	P.E.Island	1871	Lefurgey	[illegible]	F.J.Jarvis	Folkestone
193		Edna A. Pogue *Dutton* W3mSr	American	154	121·0	23·8	8·2	Cedarville,NJ	1883	Claypole&Parsn		D.P.Mulford	Bridgeton,N.J.
194	✻	Ednyfed *Jones* I Bk	British	1081	221·0	35·0	21·3	Sunderland	1862	W.Doxford&Ss		R.Thomas & Co.	London
195	V	Edouard *Richaud* W Bk	French	550	146·3	28·5	18·0	Nantes	1875	Jollet&Babin	[illegible]	Briandeau fils & Co.	Nantes
196	V	Edouard *L'Hotellier* W Bk	French	233	100·4	23·9	12·7	Nantes	1873	Barban	[illegible]	Thiriot Neumager	Nantes
197		Edouard *Coudel* W Bg	French	159	85·3	22·6	12·8	St.Malo	1856			A.Simon	Redon
198		Edouard W Bg		151	78·8	22·3	12·8	Syra	1868				
199	V	Edouard *Poulier* W S	French	129	[illegible]·2	20·3	11·1	Granville	1872	Gombert		Lacombe & Latté	Binic
200	V	Edouard Héloïse W Bg	French	179	[illegible]	21·[illegible]	11·9	Nantes	1855	Barban		F.Rousseau	Nantes

✻ = Lloyd's Register; D = Nederlandsche Vereeniging; G = Germanischer Lloyd; I = Registro Italiano; R = American Record; A = Veritas Austro-Ungarico; H = Veritas Ellenico; N = Norske Veritas; V = Bureau Veritas.

Dans la 4me Colonne, C = Composite (membrure en fer, bordé en bois); I = Fer; Stl = Acier; W = Bois.

Les Dates du doublage en métal (Col. 11) ressortent des livres dans lesquels les navires sont respectivement classés ou des rapports des Inspecteurs du Lloyd's Register.

Pour plus amples renseignements sur les navires classés par le Lloyd's Register, voir le Lloyd's Register Book et le Lloyd's Yacht Register.

Official No.	No. in Book	Registry in which classed	Ship's Name. Material and Rig. Master, and Dates of Appointment to present Owner's Service and to Ship. Ship's Late Name.	Flag.	Net Tonnage.	Dimensions. Length.	Breadth.	Depth.	Built. Where.	When.	By whom.	Date of Metalling	Owners or Managers.
	201	G	Eduard *Ulech* -86 W Bk	German	602	148·6	30·9	18·4	Vegesack	1857	H. F. Ulrichs	[illegible]	H. Wessels
	202	G V	Eduard *Warnke* 87-74 W Bk	German	345	124·0	25·3	13·0	Brake	1836	J. Oltmann's {Wwe	[illegible]	Seetzen Gebrüder
	203		Eduard (STRANDED) W Sr		111	79·5	25·7	10·7	Arnis	1858	J. P. Jörgensen		
	204	V	Eduard Pens *Kipp* 68-68 W Bk	German	357	121·7	28·1	16·3	Stralsund	1868	J. P. Juhl		Carl Lobeck
	205		Eduard & Laura W 3m Sr *[illegible]*	Russian	214	106·0	25·8	12·5	Riga	1873	A. Kalks		Pisciuti
	206		Eduviges W Sr	Brazilian	172	..	..	..					J. C. S. Guimarães
	207		Eduvigis *Roqueta* W Pol	Spanish	149	85·6	26·3	11·7	Palamos	1851		[illegible]	J. Maristany
	208	N	Edvard *Hegstedt* W Sr	Norweg'n	104	91·0	21·7	9·4	Bergen	1882			J. Tungesvig
	209	D V	Edvard (ex Helvetia) *Akermark* W S (LOST)	Swedish	1196	189·9	32·8	23·8	Newburyport, {Mass.	1854		[illegible]	P. Ahrison
	210	G	Edvard *Ericksson* 87-87 W Bk (ex James Rogers)	Swedish	266	103·6	23·4	15·0	Sunderland	1862	J. Baker		C. Söderbergh
	211		Edvard *Hansson* 82 W Bk (ex Egeria)	Swedish	253	102·3	24·4	13·4	Ribnitz	1858	J. H. Wilken		L. Petersson
	212		Edvige (ABANDONED) *Nollo* W Bk (ex Padre, ex Bortolomeo Calvi)	Italian	627	117·6	33·4	19·8	Varazze	1870		[illegible]	E. Sordino & G. B. Olivari
	213	V	Edvin *Hermansson* W Sr	Swedish	151	90·9	21·6	11·1	Kilanda	1883	A. Olsson		O. Olsson
	214	N	Edvina *Johannesen* -84 W Sr	Norweg'n	118	83·6	23·4	11·2	Amål, Sweden	1861			H. Johannesen
64699	215		Edward W Bn	British	193	101·0	26·0	10·4	Freeport, N.S.	1870			A. Foulis
30383	216		Edward W Bn	British	183	105·7	25·2	8·9	Stempden, U.S.	1854			N. K. Clements
58430	217	V	Edward *Gregory* W Sr	British	131	93·8	23·0	11·4	Arklow	1871	Tracy		J. Kearon
74863	218	✳	Edward *Ellis* 76-76 W Sr	British	92	79·5	21·8	10·8	Portmadoc	1876	Jones		E. Roberts & Co.
	219	D N	Edward *Hegstad* -82 W Bg	Norweg'n	267	122·7	26·9	12·5	Drontheim	1874	Trondhjems {Skibswærft	[illegible]	J. Larsen
135050	220	R	Edward A. Sanchez W Bkn *Foote*	American	468	135·6	33·0	15·7	Baltimore, Md.	1874		[illegible]	L. Ehrman
74878	221	✳	Edward Arthur *Ellis* W Sr (1885)	British	141	95·5	22·9	12·4	Portmadoc	1877	Williams		Mrs. M. Davies
135576	222	R	Edward B Leisenring W 3m Sr *Davidson*	American	410	138·2	34·0	11·9	Camden, N.J.	1882	Morris & Mathis		N. P. Davidson
73912	223		Edward Blake W Sr	British	329	136·0	23·6	17·6	Port Burwell	1872	S. McDermand		B. R. Clarkson
89057	224		Edward Blake *Matthews* -87 W Sr	British	109	83·0	23·9	9·9	Lewis Head N.S.	1882	S. Griffin		S. Griffin
135800	225	R	Edward C. Allen True W Sr *McLaughlin* 85-85	American	783	168·2	35·2	17·8	Bath, Me.	1885	W. Rogers	[illegible]	J. B. Drake
135163	226		Edward Cushing W Bkn *Cassidy* 1885	American	472	138·0	31·3	16·4	Camden, Me.	1875		[illegible]	J. Coombs
83130	227	R	Edward D. *Doty* -87 W Bn	British	163	97·4	26·8	10·0	Meteghan {River, N.S.	1884	C. LeBlanc		C. Doucett
61551	228	✳	Edward D. Jewett W Bk *Lockhart* 85-85	British	880	175·2	34·3	20·0	Portland, N.B.	1871	Scammell	[illegible]	Scammell Bros.
7850	229		Edward Ewing *Walker* W Sr	American	105	99·0	17·6	7·9	Hamburg, Pa.	1862			C. McCoy
135848	230	R	Edward G. Taulane W Sr (ABANDONED)	American	418	141·4	36·4	13·4	May's Lndg, N.J.	1885	S. Gaskill		J. J. Barrett
7197	231		Edward H. Furber W Sr *Nickerson*	American	302	116·0	29·0	9·0	Portsmth, N.H.	1864			E. E. Rackett
8933	232	R	Edward H. Williams W Bn *Gould* 80-80	American	387	126·0	30·1	12·1	Yarmouth, Me.	1873	G. Loring	[illegible]	J. S. Winslow & Co.
135026	233	R	Edward Johnson W 3m Sr *Warren* 85-85	American	380	129·4	29·6	15·0	Belfast, Me.	1874	C. Carter	[illegible]	J. S. Emery & Co.
135054	234	R V	Edward Kidder *Staples* -87 W Bk	American	1015	179·0	35·5	22·2	Newburyport, {Mass.	1874	Atkinson {& Fillmore	[illegible]	E. L. Griffen
135323	235	R V	Edward L. Mayberry W Bk *Knight*	American	624	145·9	32·7	18·5	Yarmouth, Me.	1878	G. Loring	[illegible]	Chas. Leavitt & Co.
135649	236		Edward L. Warren *Colson* W Sr	American	124	92·4	27·1	8·0	Searsport, Me.	1882	H. H. Grant		H. H. Grant
7548	237		Edward Lameyer *Reed* W Sr	American	138	94·4	24·9	8·2	Newburyport, {Mass.	1859			Cobb, Wight & Co.
135175	238		Edward M. Reed *Parker* W 3m Sr	American	400	133·6	33·8	13·0	E. Haven, Conn.	1875			H. Parker
135042	239	R V	Edward May *Mahaney* 83-83 W Bk	American	890	169·3	35·5	21·3	Boston, Mass.	1874	Smith {& Townsend	[illegible]	C. Brewer & Co.
135629	240	R V	Edward O'Brien *Oliver* W S	American	2157	259·1	42·5	28·9	Thomaston, Me.	1882	Ed. E. O'Brien	[illegible]	Edwd. E. O'Brien
8108	241	V	Edward O'Brien *Taylor* W S	American	1725	211·7	39·6	30·5	Thomaston, Me.	1863	E. O'Brien	[illegible]	David Oliver
135708	242	R	Edward P. Avery W 3m Sr *Hawley* -84	American	547	116·0	35·0	12·3	Bath, Me.	1884	G. Hawley		H. Prescott
135263	243		Edward Parke *Bolt* 84 W Sr	American	140	107·0	29·7	8·7	Eureka, Cal.	1876	Vernon Bros.		J. Dollner
42622	244	✳	Edward Percy *Cart* -87 I Bk	British	862	183·7	31·0	20·9	Birkenhead	1861	J. Laird Sons & Co.		A. Blaney (D. & Brabazon, Managers)
8652	245		Edward R. Emerson W 3m Sr *Child*	American	323	121·9	29·0	11·4	Waldoboro', Me.	1871			S. G. Rogers
74866	246	✳	Edward Seymour W Bkn *Hughes* 72-88	British	172	106·9	24·4	13·2	Portmadoc	1876	Williams	[illegible]	L. Hughes & Son
7755	247		Edward Slade *Vars* -87 W Sr	American	188	110·0	30·5	9·1	Brooklyn, N.Y.	1855			J. N. Vars
135714	248	R	Edward Stewart W 3m Sr *Harlow* 1885	American	378	140·9	33·1	12·1	Brewer, Me.	1883	J. Hathorn	[illegible]	F. J. Stewart & Co.
571	249		Edward Thomson W Bg	British	136	74·0	21·0	12·0	Scilly	1850			C. Collins
	250												

✳ = Lloyd's Register; D = Nederlandsche Vereeniging; G = Germanischer Lloyd; I = Registro Italiano; R = Ame[rican]
A = Veritas Austro-Ungarico; H = Veritas Ellenico; N = Norske Veritas; V = Bure[au]

In the 4th Column, C = Composite (iron frame, wood planked); I = Iron; Stl = Steel; W = Wood.

The Dates of Metalling (**Col. 11**) are based on the records of the Books in which Vessels are respectively classed, or on the Reports of the Surveyors of [illegible]

Further particulars of Vessels classed by Lloyd's Register will be found in Lloyd's Register Book and Lloyd's Yacht Register.

No.	Classe	Nom, Capitaine et dates d'entrée au service des Armateurs actuels et à bord du navire. Ancien Nom.	Gréement	Pavillon	Tonnage net	Longueur	Largeur	Creux	Construit: Où.	Quand.	Par qui.	Doublage	Armateurs ou Directeurs-Gérants.	Port d'Armement.
251		Edward W. Perry	W3mSr	American	234	122·0	27·0	9·0	DennisvilleN.J	1852	L.Diverty		W.Depuy	Camden,N.J.
252	R	Edward W. Schmidt	W3mSr	American	420	111·6	31·1	12·2	Bath,Me.	1881	Goss&Sawyer		J.W.Cramer	Philadelphia
253		Edward W Young	W3mSr	American	488	..	..	..	E.Boston,Ma	1888	J.M.Brooks		G.B.Reynolds	Newport
254		Edward Waenerlund	W Bk	German	386	122·0	28·2	16·3	Barth	1873	A.Witte	V [illegible]	J.J.Wallis & Sohn	Barth
255	R	Edward Waite	W3mSr	American	456	128·9	32·6	14·8	Bath,Me.	1872		V 11.85	H.M.Sargent	Portland,Me.
256	❄	Edward Windus	W Bn	British	156	90·3	23·3	12·5	Portmadoc	1864	Griffiths	[illegible]	L.Griffiths	Carnarvon
257	V	Edwin *Dickie*	W Bk	British	636	113·6	33·5	18·5	Cornwallis,NS	1874	A.Loomer		J.F.Dickie	Windsor,N.S.
258	❄	Edwin *Williams 13-83*	W Sr	British	157	97·2	22·6	12·5	Pwllheli	1873	Williams	[illegible]	E.Williams	Carnarvon
259		Edwin *Nelson*	W Sr	American	111	94·2	26·1	8·6	Norwalk,Conn	1864			F.Borden	Fall River,Mass
260	R	Edwin A. Gaskill	W3mSr	American	339	131·9	35·2	11·0	Leesburg,N.J.	1881			J.Russell	Bridgeton,N.J.
261		Edwin A. Stevens *Kendall*	W Sr	American	108	80·0	25·6	8·6	New Brunswick,N.J.	1838			F.Hight	Bangor,Me.
262		Edwin Fox *Patterson*	W Bk	British	838	144·8	29·8	23·6	Calcutta	1853		[illegible]	Shaw Savill & Albion Co.Ld.	London
263		Edwin I. Morrison	W3mSr	American	508	137·8	32·1	16·0	Portsm'th,NH	1873	W.F.Fernald		S.S.Lavender	Boston,Mass.
264		Edwin J. Palmer	W Sr	American	187	102·0	28·2	8·3	Baltimore	1864			R.B.Little	Providence,R.I.
265		Edwin R. Kirk *Barnett*	W3mSr	American	252	126·0	31·0	10·0	Keyport,N.J.	1864			C.Burnett	New York
266	RV	Edwin Reed *Fulton*	W Bk	American	1164	190·8	37·2	23·5	Bath,Me.	1874	Ad'ms&Hitchck	V 11.87	A.Fulton	New York
267	❄	Effendi *Jones -88*	W Bkn	British	469	151·1	28·9	16·8	Summerside,[P.E.Island	1882	J.W.Richards	V 1.86	J. W. Richards (T. P. Richards & Co. Managers)	Swansea
268	R	Effie J. Simmons	W3mSr	American	203	108·4	28·4	9·3	Thomaston Me	1871	Walker&Dunn		Dunn & Elliot	Thomaston,Me.
269		Effie Sweet *King*	W Sr	British	179	96·9	26·9	11·1	Cy.H'rbourNS	1879		V 1.85	S.Sweet & Co.	Halifax,N.S.
270		Effort *Curson*	W Sw	British	167	82·4	23·5	13·5	Sunderland	1849			J.S.Sterry	Lowestoft
271		Effort *Turner -88*	W Sr	British	110	77·1	23·2	10·0	Mahon,N.S.	1879	S.McDonnell		G.E.Forsyth	PtH'wk'sbry,NS
272	N	Efraim *Knudsen*	W Bn	Norweg'n	182	101·3	23·8	12·1	Stavanger	1875		[illegible]	T.Bryne	Stavanger
273		Eftichia *Chatugeo*	W Sr	Turkish	256	106·9	25·3	13·1	Trieste	1870	Stabilimento [Tecnico	V 8.85	C.Taglieo	Alexandria
274	V	Egalité *Heuten*	W Bn	French	183	91·2	23·3	11·5	Nantes	1876	E.Clergeau	V 12.83	F.Rousseau	Nantes
275	G	Egbert Jan *Schoon 84-*	W Bg	German	195	101·6	24·1	11·9	Amsterdam	1864	E.J.Bok & Sons		T.J.Cassens & Sohn	Grossfehn
276	D	Egbert & Nicolaas	W Bk	Dutch	607	143·2	24·9	16·2	Papendrecht	1856		V 3.86	J. & J.Vinke	Amsterdam
277	N	Egden *Berthelsen*	W Bg	Norweg'n	346	116·8	29·0	15·5	Arendal	1867	Dekke	V 3.81	G.Berthelsen	Arendal
278	RV	Egeria *Kerr*	W Bk	British	897	173·1	35·9	19·5	Harvey,N.B.	1879	G.S.Turner	V 11.87	G.F.Smith	St.John,N.B.
279		Egeria *Wolf 87-*	W Sr	British	108	79·3	24·5	9·8	Petite Rivers,[N.S.	1884	J.D.Sperry		J.D.Sperry	Lunenburg,N.S.
280	N	Egero *Gerontsen*	W Bk	Norweg'n	1072	181·8	36·7	22·2	Levis,Quebec	1877	W.Charland, jr.	V 10.86	Chr.Hansen Löning	Egersund
281	N	Egersund *Staalberg*	W Bg	Norweg'n	358	124·6	28·9	14·9	Amsterdam	1863		V 2.83	A.Birkeland	Egersund
282	N	Egia *Gundersen*	W Bg	Norweg'n	254	96·1	27·8	14·2	Grimstad	1856	A.Omundsen		G.Thorbjörnsen	Arendal
283	I	Egilda B. *Matrocco -87*	W Bk	Italian	712	156·5	29·6	19·3	Sestri	1869	B.Celesia	V 9.86	G.B.Beverino	Spezia
284		Egine	W Bn		171	99·8	24·6	11·8	Pekela	1869	J.Kuiper	Z 10.82		
285	V	Eglantine *Agernaes -88*	W Bk	Norweg'n	515	141·0	28·2	13·4	LaSeyne	1872	Cuzet	V 11.87	H.Helliksen	Christiansand
286		Eglantine *Dahl-Blop*	W Bk	Norweg'n	321	121·5	28·8	14·4	Quebec	1866	M'Kay		A.C.Nielsen	Kragerö
287	N	Egmont *Andersen -87*	W Bg	Norweg'n	187	102·0	25·5	11·3	Laurvig	1872	Dahl		E.B.Lindseth	Drammen
288		Egor	W Sr	Russian	269	..	..	..		1872			E.Kutujeff	Salmis
289	A	Egyptos *Constantinidis*	W Bk	Greek	375	119·1	28·9	..	Syra	1879	L.Manassis	V 7.84	G.Constantinidis	Syra
290	V	Ehe *de Vries 13-73*	W Sr	German	104	83·0	19·9	8·7	Leer	1873	B.D.Middendorf	Z 8.83	J.H.Waterborg	Leer
291	❄	Ehen *Pray 83-88*	W Bk	French	301	124·0	24·9	15·6	Whitehaven	1863	Kennedy	V 1.86	A.E.Pray	Havre
292		Eiche	W Bn		222	107·2	24·4	11·8	Brake	1867				
293	N	Eidbo *Henriksen -88*	W Bk	Norweg'n	297	109·2	27·9	15·3	Tvedestrand	1860	O.Gundersen		C.P.Andersen	Tvedestrand
294	NV	Eidsiva *Christensen 87-87*	W3mSr	Norweg'n	454	136·9	29·5	16·4	Drammen	1876	R.V.Haerem	V 7.82	H.T.Kiær	Drammen
295		Eidsvold *Lycke*	W Bk	Norweg'n	564	127·5	31·2	20·4	Alblasserdam	1841			V.Fürst	Christiania
296	N	Eidsvold *Egeberg 77-77*	W Bk	Norweg'n	331	122·8	28·2	14·6	Skien	1867	Hauland	V 8.81	C.Egeberg	Tönsberg
297	VN	Eidsvold *Nielsen 82-*	W3mSr	Norweg'n	272	121·7	26·7	12·3	Kragerö	1882	J.Nielsen	V 8.87	T.Jensen	Kragerö
298		Eighty-Five *Le Gros 83-86*	W Bn	British	140	93·2	20·8	12·4	Canada	1860		V 8.85	G.Le Gros	Jersey
299	❄N	Eikundasund	W Bk	Norweg'n	818	182·8	32·6	20·4	Pembroke	1877	Co-operative [S.B.Co.Ld.	V 1.88	A.Bör	Egersund
300		Eiland Banka *Pastor*	W Bk	Dutch	197	..	..	..					Sultan van Bima	Bima

❄ = Lloyd's Register; D = Nederlandsche Vereeniging; G = Germanischer Lloyd; I = Registro Italiano; R = American Record; A = Veritas Austro-Ungarico; H = Veritas Ellenico; N = Norske Veritas; V = Bureau Veritas.

Dans la 4me Colonne, C = Composite (membrure en fer, bordé en bois); I = Fer; Stl = Acier; W = Bois.

Les Dates de doublage en métal (Col. 11) ressortent des livres dans lesquels les navires sont respectivement classés ou des rapports des Inspecteurs du Lloyd's Register.

Pour plus amples renseignements sur les navires classés par le Lloyd's Register, voir le Lloyd's Register Book et le Lloyd's Yacht Register.

Official No.	No. in Book	Registry in which classed	Ship's Name. Master, and Dates of Appointment to present Owner's Service and to Ship. Ship's Late Name	Material and Rig	Flag	Net Tonnage	Length	Breadth	Depth	Built: Where	Built: When	Built: By whom	Date of Metalling	Owners or Managers
[illegible]	301	✻	Eildenhope *Garrick 85-86*	I Bk	British	1498	240·0	37·3	22·9	W.Hartlepool	1876	W.Gray&Co.		Adam, Hamilton & [illegible]
[illegible]	302		Eillan Donan *McKenzie [-87*	W Bn	British	270	115·5	26·8	12·8	PortClyde,NS	1888	T.Coffin&Co	[illegible]	D.Ross
	303		Eimund (PONDER) *Hauen*	W Sr	Norweg'n	273	127·0	28·1	12·8	Bergen	1876	Erichsen	[illegible]	R.F.Olsen
	304	N V	Einar (ex [illegible] Hinrich) *Tschudi -88*	W Bk	Norweg'n	639	154·3	33·0	19·1	Rostock	1876	E.Burchard&Co	[illegible]	G.C.Hansen
	305		Einar (LOST) *Falsen*	W Bk	Norweg'n	222	115·9	26·0	12·1	Drontheim	1866	Klingenberg	[illegible]	A.Jenssen & Co.
	306		Einar (ex Nil Desperandum) *Jorgensen*	W 3m Sr	Norweg'n	198	115·5	24·0	9·8	Bremen	1863			H.C.Jörgensen
	307	N	Einar *Hagen*	W Sr	Norweg'n	156	97·9	24·3	10·6	Christianssand	1879	H.Larsen	[illegible]	G.T.Jörgensen
	308		Einar *Hämäläinen*	W Sr	Russian	142	..	..	..	[illegible]	1877			E.Hubbard & Co.
	309	N	Einar Tambarskjelver (ex Petit Pourrier, ex Wilberforce) *Christophersen 86-86*	W S	Norweg'n	1145	183·7	35·8	23·3	Chelsea, Mass.	1855		[illegible]	Nils P.Höyer
	310	V	Einar Thambarskelver *[illegible]*	W Bk	Norweg'n	935	169·0	34·6	22·3	Bremerhaven	1873	J.C.Tecklenborg	[illegible]	Ths.S.Falck
	311		Einingkeit (BROKEN UP) *Schmidt 70-*	W Bk	German	439	100·7	30·1	17·4	Greifswald	1858	J.W.Sadelkow		C.Graedener
	312		Einigkeit *Kantel -68*	W Bk	German	397	120·8	27·3	16·9	Elbing	1860	Gebr.Mitzlaff		J.G.Kantel
	313	V	Einigkeit *Wallis 65-*	W Bg	German	195	87·0	25·0	13·2	Barth	1856	J.Holzerland		J.C.Beug
	314	G	Eintracht *Schliee 86-86*	W Bk	German	399	120·4	28·6	16·8	Greifswald	1868	Aug.Spruth		Ferd.Muswieck
	315	G	Eintracht *Ratzki 86-*	W Bk	German	377	112·2	28·5	17·2	Elbing	1848	M.Mitzlaff		Albert Koehne
	316	G	Eintracht (ex Rugenwalde) *Tode 83-*	W Bg	German	272	103·3	28·0	14·8	Swinemünde	1857	Liebnitzky		W.Lüdke & Co.
	317	N	Eira *Larsen 86-84*	W Bk	Norweg'n	1006	188·4	38·0	20·9	Sandefjord	1886	A.Andersen	10.88	A.Andersen
80229	318	✻	Eivion *Davis 78-81*	I Bk	British	1133	211·1	35·0	21·8	Sunderland	1879	Osbourne, Graham&Co.		Ship "Eivion" Co, Ld (W.Thomas & Son)
	319		Ejdern (ex Kinma) *Berin -83*	W Bk	Swedish	615	155·8	29·8	17·8	[illegible],NS	1869		[illegible]	Rederibolaget Ørnen (J.P.[illegible])
434	320	A	Ekaterini *Giourgas*	W Bg	Greek	387	125·0	29·4	11·9	Syra	1874		[illegible]	J.Giourgas
516	321	A	Ekaterini *Calojera*	W Bg	Greek	286	112·8	26·2	..	[illegible]	1888			M.P.Calojera&Co.
64	322	A	Ekaterini *Grigoriou*	W Bk	Greek	227	101·7	27·6	16·1	[illegible]	1881	C.Papapetros		M.Grigoriou
58	323		Ekaterini	W Bg	Greek	178	86·6	25·6	13·4	Syra	1870			R.P.Manzavinos
	324		Ekaterini *Lynguris*	W Bg	Greek	177	..	..	..	[illegible]	1860		[illegible]	J.Lascaratos
	325	H	Ekaterini *Pastos*	W Bk	Greek	177	..	..	..	Syra	1870			J.Pastos
	326		Ekaterini *Carvounis*	W Sr	Greek	103	..	..	..	Syra	1873			S. & K.Carvounis
	327	A	Ekaterini Coupa *Spanaris*	W Bn	Greek	354	..	..	..	[illegible]	1866			G.E.Marulla
168	328	H	Ekaterini Sevastopoulou	W Bn	Greek	319	115·1	22·7	16·9	England	1872			G.Panoutzos
	329	V	Ekaterini Sevastopoulou (ex Fritz von Gadow) *Metaxas*	W Bn	Greek	170	94·0	23·6	11·7	Barth	1879		[illegible]	C.S.Sevastopoulou (D.N.Metaxas)
	330		Eklipsis *Prokopio*	W Bg	Greek	148	78·1	23·6	..	Black Sea	1879			P.Prokopiou
8949	331	V	El Capitan *Humphrey -87*	W S	American	1418	205·3	37·2	25·5	Bath,Me.	1873	E.&A.Sewall	[illegible]	De Groot & Peck
740	332		El Cometa (ex [illegible]) *Valle*	W Bg	Italian	281	..	..	..	Varazze	1861			S.Anselmi
	333		El Kerim Atta	Bg	Turkish	174	..	..	..					D.Khatab
	334		El Kerim Atta	Bg	Turkish	127	..	..	..					Karakiri
	335		Elanchove	W Bn	Spanish	260	112·9	25·5	12·7	Magallanes	1887			Aldecoa & Co.
60185	336	✻ R	Elba *Swan 85-86*	W Bn	British	253	117·5	28·3	12·7	NovaScotia	1873		[illegible]	A.King
7761	337	R	Elba *Tilton -87*	W Bk	American	424	120·0	28·0	17·0	Pembroke,Me.	1863		[illegible]	D.S.Risley
486	338		Elba (ex Leopoldo Petre [illegible]) *Giordano*	W Bg	Italian	160	89·9	25·0	14·0	Castellamare	1858			V.Giordani
84299	339	✻	Elbe *Hunt 85-87*	I S	British	1616	257·0	38·2	23·1	PortGlasgow	1887	Russell&Co.		J.Nourse
	340	V	Elbe (ex [illegible]) *Tranitzen*	I Bk	German	469	149·9	25·9	16·7	Greenock	1862	Steele		J.C.M.Block
135698	341	R	ElbridgeSouther *Fales*	W 3m S	American	658	156·6	36·2	17·4	ThomastonMe	1883	S.Watts	[illegible]	S.Watts
	342		Elcano	W Bn	Spanish	169	104·9	23·4	10·2	Albay	1885			Aldecoa & Co.
132	343	A	Elce *Dominich*	W Bk	Austrian	480	124·3	28·4	19·1	Fiume	1869	P.Buranello	[illegible]	D.Zorovich fu D.
83225	344	✻	Elcho *Oxelgren 83-84*	W Bkn	British	267	118·0	25·9	12·9	P.E.Island	1881	J.R.Bourke jun.	[illegible]	Geo.Little & Co.
	345		Elcira (ex [illegible]) *Waldbahm -83*	W Bn	Chilian	260	105·6	26·2	12·3	Porsgrund	1876			G.R.Jepp
	346	N	Elder (ex Osnabrück) *Jacobsen*	W Bk	Norweg'n	411	138·1	28·1	16·2	Rönnebeck	1856		[illegible]	O.Gjeruldsen
89881	347		Eldorado *Staples -88*	W Bg	British	228	100·0	28·0	8·5	Quebec	1884	Samson		E.Hutchings
	348	N	Eldorado *Halvorsen*	W Bk	Norweg'n	315	115·9	27·9	15·2	Tvedestrand	1867	Eldrup		L.Larsen
[illegible]	349	✻	Eldra *Willis 82-2*	W Sr	British	215	111·0	24·6	12·1	Plymouth	1873	Hill&Son	[illegible]	J.B.Pinkham
	350		Elé *Eriksson 84-84*	W Sr	Russian	227	..	..	..	Åbo	1874			A.Gustafsson

✻ = Lloyd's Register; D = Nederlandsche Vereeniging; G = Germanischer Lloyd; I = Registro Italiano; R = [illegible]
A = Veritas Austro-Ungarico; H = Veritas Ellenico; N = Norske Veritas; V = [illegible]

In the 4th Column, C = Composite (iron frame, wood planked); I = Iron; Stl = Steel; W = Wood.

The Dates of Metalling (Col. 11) are based on the records of the Books in which Vessels are respectively classed, or on the Reports of the Surveyors [illegible]

Further particulars of Vessels classed by Lloyd's Register will be found in Lloyd's Register Book and Lloyd's Yacht Register

No.		Nom, Matériaux et Gréement; Noms des Capitaines		Pavillon	Tonnage	Long.	Larg.	Creux	Construit: Où	Année	Par qui		Armateurs	Port d'Armement
351		Eleanor	W Bk	British	714	161·7	[illegible]	[illegible]	[illegible]	[illegible]	[illegible]		W.D.Douglas	Dorchester,N.B.
352		Eleanor	W Bk	British	429	136·1	27·5	17·1	Milford	1867	R.Lewis		W.Lund	London
353		Eleanor	W Bn	British	176	89·7	22·7	12·8	Yarmouth	1860	BrechingBros.		J.Reeves	Yarmouth
354	R	Eleanor	W 3m Sr	American	353	140·0	32·9	10·0	Newcastle,Me.	1883	E.Haggett		E.Haggett	Damariscotta
355	R	Eleanor	W 3m Sr	American	342	125·0	31·0	13·6	PortJefferson, N.Y.	1876	J.M.Bayles&Son		Floyd & Newins	New York
356	N	Eleanor *Nielsen*	W 3m Sr	Norweg'n	210	104·9	21·7	12·8	P.E.Island	1876	Bourke		C.Thiis	Fredrikestad
357	R	Eleanor *Reinhardt*	W Sr	Norweg'n	163	87·5	22·9	13·8	Redbridge	1837			O.A.Bastiansen	Drammen
358	✱	Eleanor Margaret	I in Bk	German	327	148·8	39·1	28·3	London	1860	ThamesS.B.Co. Ld.		J.D.Bischoff	Bremen
359		Eleanor Thomas	W Sr	British	120	84·5	21·0	11·5	Bridport	1858	Cox		Mrs.Anne Jones	Aberystwith
360	✱	Eleanor & Jane *Jones*	W Sr	British	110	86·2	22·2	11·1	Padstow	1884	C.Rawle		D.Davies	Aberystwith
361		Eleanora	W Bg	British	277	110·0	28·6	10·0	Port Jefferson N.Y.	1856			John Powers	Bridget'n,Barbados
362		ElectaBailey	W 3m Sr	American	327	118·0	31·1	9·2	PetersburgN.J.	1866	D.Bailey		J.C.Darrah	Philadelphia
363	✱	Electra *Fisher*	C S	British	663	176·0	30·6	18·0	Aberdeen	1866	A.Hall&Co.		J.Stewart & Co.	London
364		Electra *McNutt*	W Bk	British	380	124·0	26·6	16·3	Kennebecasis, N.B.	1859	Milledge		Clarence & Richmond River Steam Nav. Co.	Sydney,N.S.W.
365	✱	Electra *Le Sueur*	W Bkn	British	158	118·6	20·6	11·7	Jersey	1854			F.Le Sueur & Son	Jersey
366	V	Electra *Evans*	W Bk	American	940	168·0	34·8	21·9	EastBoston	1868	Sampson&Co.		L.Sloss,junr.	SanFrancisco
367	R	ElectricLight *Nelson*	W Bn	British	249	108·0	28·1	11·7	Shulee,N.S.	1879	J.T.Smith		J.R.Lord	Montreal
368		Electric Light	W 3m Sr	American	537	146·6	33·6	11·6	Bath,Me.	1880	Deering & Donnell		Deering & Donnell	Bath,Me.
369		Electris *Jansson*	W Sr	Swedish	132	..	..	..	Orousť	1850			A.Jansson	Torp,Orousť
370		Eledona *Brown*	W Bn	British	250	112·5	28·2	12·7	[illegible]derry, N.S.	1865			Miss M.A.Finnie	Ardrossan
371	V	Elegant	W Bn	German	182	105·9	24·3	12·5	Blankenese	1868	J.Fieck			Hamburg
372	N	Elektra *Sørrig*	W Bk	Norweg'n	597	160·3	32·9	17·8	Laurvig	1871	Dahl		M.Oppen	Laurvig
373	DN	Elektra	W Sr	Norweg'n	113	87·2	23·7	10·1	Stordöen, Norway	1876	G.Bengtson		P.Kjellesvig	Flekkefjord
374	✱	Elena	W Bk	Austrian	352	115·8	31·3	15·1	Fiume	1854	Zanon		S.Millinovich	Trieste
375		Elena	W Bk	Chilian	828	172·5	33·9	21·8	SestriPonente	1873	A.Briasco fu L.		G.B.Caffarena	Valparaiso
376	DG	Elena *Gerber*	W Bk	German	791	154·0	33·5	21·1	Bremerhaven	1859	J.C.Tecklenborg		A.Burchard	Rostock
377		Elena *Sperno*	W Bg	Italian	306	127·0	27·9	19·8	Lussinpiccolo	1863	Martinolich		Guerrera & Co.	Catania
378	A	Elena *Del Re*	W 3m Sr	Italian	294	115·4	26·3	15·1	Bilbao	1867			B.Barrera	Civita Vecchia
379	V	Elena	W Bg	Spanish	209	92·5	27·1	13·6	Blanes	1847			M.Paques	Barcelona
380	I	Elena B.	W Bk	Greek	435	125·9	27·7	18·9	Savona	1869	G.Tixi		G.Rediadis	Zante
381		Elena M. *Maralli*	W Bg	Italian	141	95·5	23·1	10·7	Varazze	1881	S.Craviotto		P.G.Maralli	Genoa
382	A	Elena Nicolich *Nicolich*	W Sr	Austrian	165	..	..	..	Fiume	1862	Belaz		M.Nicolich fu A.	Trieste
383		Elena Prima	W 3m Sr	Italian	153	91·9	24·0	12·8	Sestri	1857	L.Briasco		C.Sagorara fu B.	Genoa
384	I	Elena Rodocanachi	W Bk	Greek	492	136·1	31·1	19·6	Savona	1868	G.Tixi		M.Culucundis	Zante
385	A	Elenko *Psalti*	W Sr	Russian	137	..	..	..	Sinope	1886			D.G.Psalti	Eupatoria
386	R	Eleonor *Johansson*	W Bk	Swedish	651	150·8	32·5	17·6	St.John,N.B.	1866			T.Nilsson	Gothenburg
387		Eleonora *Åberg*	W Sr	Russian	259	..	..	..	Lappjerd	1872			N.Sittkoff	Mariehamn
388		Eleonore	W Bk	British	680	154·2	31·0	19·6	Quebec	1868	P.V.Valin		P.V.Valin	Quebec
389		Eleonore	W Sr	British	135	93·6	23·0	9·2	St.Pierre des Becquets	1854			L.Gariepy	Quebec
390	V	Eléonore *Lehegarat*	W Sr	French	124	95·5	20·0	11·1	Paimpol	1883	Laboureur		Lehégarat	Paimpol
391	N	Eleonore *Andersen*	W Sr	Norweg'n	137	85·1	21·3	10·8	Hernösand	1859			E.A.Ellingsen	Christiania
392		Elephant	W	British	114	85·6	21·0	11·0	Quebec	1856			J.U.Laird	Quebec
393	A	Eleussa	W Sr	Turkish	141	..	..	..	Sisopoli	1878			Elia Pkinitis	Chios
394	H	Eleutheria	W Bk	Greek	331	113·5	28·5	..	Syra	1864			Ar. Papanikolis	Syra
395	H	Eleutheria	W Bg		223	91·9	22·6	15·9	Spezzia	1869				
396	H	Eleutherios	W Bg	Greek	295	108·9	25·6	..	Spezzia	1870			E.Caffos	Piraeus
397		Eleuthero Tris *Agalon*	W Bn	Greek	208	88·6	24·1	..	Scopelos	1869	P.Barelas		E.K.Agalon	Scopelos
398	✱ D	Elfin	W 3m Sr	Russian	272	112·7	25·7	15·6	Plymouth	1859	Moore		T.Renny & Co.	Riga
399		Elfrida *Dunsford*	W Sr	British	137	90·0	21·3	11·5	Plymouth	1869	Gent		J.Taylor	Plymouth
400	✱	Elgin	W Bk	British	549	138·8	31·3	18·2	Hopewell,N.B.	1877	E.Pye		A.R.McClelan	St.John,N.B.

✱ = Lloyd's Register; D = Nederlandsche Vereeniging; G = Germanischer Lloyd; I = Registro Italiano; R = American Record; A = Veritas Austro-Ungarico; H = Veritas Ellenico; N = Norske Veritas; V = Bureau Veritas.

Dans la 4me Colonne, C = Composite (membrure en fer, bordé en bois); I = Fer; Stl = Acier; W = Bois.

Les Dates du doublage en métal (Col. 11) ressortent des livres dans lesquels les navires sont respectivement classés ou des rapports des Inspecteurs du Lloyd's Register.

Pour plus amples renseignements sur les navires classés par le Lloyd's Register, voir le Lloyd's Register Book et le Lloyd's Yacht Register.

Official No.	No. in Book	Registry in which classed	Ship's Name. Material and Rig. Master, and Dates of Appointment to present Owner's Service and to Ship. Ship's Late Name.	Flag	Net Tonnage	Length	Breadth	Depth	Built. Where.	When	By whom	Date of Metalling	Owners or Managers
51985	401		Elgin *Johnston 66-65* W Bk	British	261	126·1	27·0	12·8	Argyle, N.S.	1866	Hogg		J.J.Carroll
	402		Eli *Gustofsson -87* W Sr	Russian	190	91·9	28·4	12·3	SalebyKapoli	1874	J.Jossfolk		R.A.Mattsen
1340	403	I	Elia (ex Maria) *Corso -88* W Bn	Italian	138	79.7	23·8	13·4	Gaeta	18^{63}_{82}	P.Gallinaro	[illegible]	O.Volpe
141	404		Elias W Bg	Greek	307	..	..	..	Galaxidi	1877			A.Antoniou
166	405		Elias *Katikas* W Bg	Greek	302	103·6	26·2	17·1	Syra	1875		[illegible]	S.Katikas
	406		Elias *A. Mattson -87* W Sr	Russian	355	..	..	..		1873			M.Mattson
	407	A	Elias *Ibrahim -86* W Bg	Turkish	320	111·5	27·4	..	Scopeles	1879		[illegible]	Ibrahim
8366	408		Elias Moore *Falkenbury* W3mSr	American	349	123·0	13·6	9·0	Baltimore	1867			W.De Groot
8008	409		Elias Ross *Lewis* W3mSr	American	178	116·5	22·6	9·0	NBrunswick N.J. (N.Y.)	1873			W.Mershon
	410	V	Elida *Jørgensen 86-87* W Bk	Danish	253	113·3	24·6	13·5	Kragerö	1862	C.Jensen		Actieselskab (M.Lund)
	411	N V	Elida *Mortensen 76-85* W Bk	Norweg'n	757	164·8	31·2	19·2	Tvedestrand	1885	J.Brandi	[illegible]	Carl Bech
	412	N	Elida (ex Weonah) *Waess -88* W Bk	Norweg'n	664	160·5	33·1	18·4	Falmouth, N.S.	1870		[illegible]	A.F.Klaveness
	413	N	Elida *Krog* W Bk	Norweg'n	537	125·6	29·2	20·6	United States	1839			Enkefru F.Moldenhauer
	414	N	Elida *Pettersen 87-87* W Bk	Norweg'n	537	116·6	31·8	17·3	Drammen	1867	Jörgensen	[illegible]	P.G.Pettersen
	415	N	Elida *Hofgaard* W Bk	Norweg'n	380	111·4	29·1	17·7	Piteå, Sweden	1814			J.Hofgaard
	416	N	Elida *Andersen* W Bk	Norweg'n	354	111·8	29·6	17·8	Horten	1848	Coske		Chr.Olsen
	417		Elida (WRECKED) *Christoffersen -85* W Bk	Norweg'n	316	120·3	27·1	13·9	Grimstad	1885	N.OlsenGjÆrale	[illegible]	J.D.Havstad
	418	N	Elida *Gundersen -88* W Bg	Norweg'n	267	100·5	28·0	14·6	Tönsberg	1859			O.Eliasen
	419	N	Elida *Stern -86* W Bk	Norweg'n	233	118·0	24·2	13·3	Aberdeen	1857		[illegible]	O.Bjertness
	420	N	Elida *Olausen -87* W Bg	Norweg'n	221	112·5	26·3	12·1	Kragerö	1870	A.Johnsen	[illegible]	C.Gjerlöff
	421	N	Elida (WRECKED) *Hoff* W Bk	Norweg'n	207	113·2	24·5	11·5	Drontheim	18^{66}_{83}	Jensen	[illegible]	H.W Finne&Sönner
	422	N	Elida *Christensen -89* W Sr	Norweg'n	152	90·1	23·3	11·4	Sarpsborg	18^{48}_{85}	Singdahlsen		A.P.Ljungberg
	423	N	Elida *Maibøe* W Sr	Norweg'n	119	98·1	21·2	10·6	Stavanger	18^{59}_{84}			L.Riisdahl
	424		Elie *Lediabat* W Bn	French	145	84·3	23·0	11·3	Méans	1873	Mahé	[illegible]	P.G.Lediabat
	425	N V	Elieser *Knudsen* W Bk	Norweg'n	346	133·6	29·0	12·2	Kragerö	1884	A.Thomesen	[illegible]	J.A.Larsen & Co.
	426	N	Elieser *Halvorsen 85-85* W3mSr	Norweg'n	299	119·4	29·0	12·1	Tvedestrand	1885	C.Wroldsen	[illegible]	C.Wroldsen
	427	N	Eliezer (ex Thor) *Abrahamsen 84-87* W Bk	Norweg'n	865	173·1	33·7	20·9	Lübeck	1873	J.A.Meyer	[illegible]	A.Nilsen
	428	N V	Eliezer *Andreassen 79-85* W Bk	Norweg'n	593	144·2	32·3	19·0	Grimstad	1875	J.T.Tryde	[illegible]	Krog & Henschien
	429	N	Eliezer *Christensen -87* W Bk	Norweg'n	448	131·4	31·8	17·2	Sandefjord	1865	Chr.Skalberg		B.Christensen
	430	D N	Eliezer *Jensen 85-85* W Bk	Norweg'n	439	135·0	30·1	16·7	Grimstad	1878	A.Bentsen	[illegible]	N.Thue Johnsen
	431	N V	Eliezer *Gausland -84* W Bk	Norweg'n	294	133·2	27·2	12·8	Bergen	1864	J.Tramphy&Sn	[illegible]	N.S.Gausland
	432	N	Eliezer *Terjesen -86* W Bg	Norweg'n	211	103·8	26·3	12·5	Tvedestrand	1864	O.Gundersen		G.Terjesen
	433	N	Eliezer *Melsom* W Sr	Norweg'n	135	89·9	23·2	9·9	Tönsberg	1875			Chr.Melsom
	434	N	Eliezer *Reversen* W Sr	Norweg'n	133	90·3	22·9	11·0	Stavanger	1875		[illegible]	H.E.Norem
	435	N	Eliezer *Theisen* W Sr	Norweg'n	131	91·8	23·8	10·8	Bergen	1867	P.Hendriksen [&O.Knutsen		P.Rasmussen
	436	V	Eliezer *Håkansson 81-86* W Sr	Swedish	154	98.4	21·5	9·9	Ljunghusen	1875	Olsson		P.Hall
2399	437		Elijah Packer *Crawford [83-86* W Sr	British	158	97·4	21·9	10·9	Sunderland	1848			J.Hart
	438	N	Elim *Tengelsen 84-84* W Bk	Norweg'n	348	131·6	29·3	12·3	Tvedestrand	1884	T.A.Mörland	[illegible]	C.Wroldsen
	439	N	Elim (WRECKED) *Jacobsen* W Sr	Norweg'n	146	95·9	23·3	11·6	Stavanger	1872	Magnus	[illegible]	A.Jansen
	440		Elin *Karlsson -87* W Gls	Russian	121	..	..	..	Saleby	1885			J.E.Johansson
	441		Elin (ex Baltic) *Löfgren -86* W Bk	Swedish	438	126·6	26·8	16·5	GamlaKarl'by	1860			J.J.Ekman
	442		Elin (LIGHTER) *Jansson -88* W Sr	Swedish	130	101·7	18·3	7·7	Finland				G.Förselius
	443		Eline (ex Familie) *Lauritzen 85-88* W Sr	Danish	114	96·1	20·1	9·7	Troense	1863	C.R.Møller		A.H.Petersen
15208	444		Elinor *Ryder* W Sr	British	114	94·0	21·8	11·7	Kingsbridge	1856			W.R.Beer
	445	V	Elinor (ex Bliss) *Bøye -88* W Bn	Danish	174	100·1	22·2	11·1	Blankenese	1872	J.Finck		R.O.Boye
135266	446	R	Elinor Vernon *Berry* W Bkn	American	545	150·0	32·5	17·0	Pembroke, Me.	1876	J.B.Nott	[illegible]	Arnold,Cheney&Co.
228	447		Elios (WRECKED) *Gladulich -88* W Bk	Austrian	655	139·7	31·9	20·8	Lussinpiccolo	1871	N.Martinolich	[illegible]	Nicolo Zar
	448		Elioth *Mattson 81-* W Sr	Russian	122	82·7	27·3	10·6	Saltvik	1875	A.A.Söderström		K.E.Janson
	449		Elisa (ex [illegible]) *Holmsen -81* W Bg	[illegible]	138	87·9	19·7	9·2	New Zealand	1864			Doll & Co.
	450		Elisa (ex [illegible]) (WRECKED) W Bg	[illegible]	221	95·5	23·6	15·1	Almuri	1859			D.Martinetti

✱ = Lloyd's Register; D = Nederlandsche Vereeniging; G = Germanischer Lloyd; I = Registro Italiano; R = [illegible]
A = Veritas Austro-Ungarico; H = Veritas Ellenico; N = Norske Veritas; V = [illegible]

In the 4th Column, C = Composite (iron frame, wood planked); I = Iron; Stl = Steel; W = Wood.

The Dates of Metalling (**Col. 11**) are based on the records of the Books in which Vessels are respectively classed, or on the Reports of the Surveyors of [illegible]

Further particulars of Vessels classed by Lloyd's Register will be found in Lloyd's Register Book and Lloyd's Yacht Register.

No.	Cote	Nom. Matériaux et Gréement. Capitaine et dates d'entrée en service avec Armateurs actuels et à bord du navire. Ancien Nom.	Pavillon.	Tonnage net.	Longueur	Largeur	Creux	Construit Où	Quand.	Par qui	Doublage	Armateurs ou Directeurs-Gérants.	Port d'Armement.
451	V	Elisa Denis ..88 W Bk	French	138	95·1	20·3	11·0	Granville	1837	Epron	[illegible]	Benst père & fils	Granville
452		Elisa W Bg	French	114	73·8	21·3	10·5	St. Tropez	1857	Aurignon		Mallet	St. Tropez
453	I	Elisa (Giulio) Vassallo W Bk	Italian	621	147·9	32·8	21·6	Chiavari	1870	M.Tappani	[illegible]	V.Vassallo	Genoa
454		Elisa Ranieri W Bg	Italian	224	106·9	25·2	16·1	Savona	1861	F.Calamaro	[illegible]	Flli Drago	Genoa
455		Elisa Esposito W Bn	Italian	127	91·8	22·3	11·2	TorredelGreco	1881	A.Speranza	[illegible]	Ciro Jengo	Naples
456	N	Elisa Kjeldsen W Bk	Norweg'n	335	131·7	27·8	14·7	Skien	1868		[illegible]	L.Riisdahl	Skudesnæs
457		Elisa W Bg	Norweg'n	188	90·3	21·5	12·9	Risör	1847			A.Jensen	Tvedestrand
458	N	Elisa Abrahamsen W Bn	Norweg'n	165	102·9	23·6	11·0	Porsgrund	1878			O.J.Abrahamsen	Porsgrund
459	N	Elisa Christensen ..80 W Sr	Norweg'n	151	93·2	22·9	10·3	Mandal	1881	J.S.Hansen		G.Christensen	Mandal
460	V	Elisa da Silva ..87 W Bn	Portug'se	179	98·1	21·0	11·1	Papenburg	1879	A.Beckmann	[illegible]	J.J.R.de Magalhaes	Oporto
461		Elisa W Bk	Spanish	405	123·3	26·5	17·3	Sunderland				F. de Paula Reyes	Corunna
462		Elisa W Sr	Spanish	100	73·5	21·9	10·5	Noya,Spain	1860			Pablo Rotra	Villagarcia
463		Elisa E. [illegible] W Bn (ex [illegible])	Italian	264	105·6	26·8	15·3	Savona	1862	G.B.Gustavino		AntonioGaribaldiVedova [illegible]	Genoa
464		Elisa Marie Maréchal W Sr	French	140	83·8	20·0	11·3	Dunkirk	1857	G.Mal		Manrois	Dunkirk
465		Elisa Villegas W Pg	Spanish	130	78·4	22·5	12·8	Tarragona	1865			J.B.Puellano	Majorca
466	V	Elisabet Ingvardsson ..81 W Bk (ex [illegible])	Swedish	803	163·8	34·8	19·5	Maitland, N.S.	1874	Pratt&Co.	[illegible]	J.D.Björkegren	Cimbrishamn
467		Elisabet Nilsson ..87 W Bk (ex [illegible])	Swedish	809	167·8	30·6	17·9	Guernsey	1864	DelaMare	[illegible]	H.Nilsson	Elleholm
468	V	Elisabet Strandberg W Bk	Swedish	429	125·0	27·6	16·7	Svartvik	1852		[illegible]	J.M.Moller	Gothenburg
469	V	Elisabet Jansson 88-89 W Bk (ex [illegible])	Swedish	406	129·9	27·4	14·9	Tatamagouche [N.S.	1868			E.Jansson	Cimbrishamn
470	V	Elisabet [illegible] ..87 W Bk	Swedish	321	129·2	25·7	12·5	Hernösand	1872	J.O.Strandberg	[illegible]	[illegible] Skeppsr. [illegible]	Hernösand
471		Elisabet Anderson W Bk	Swedish	228	101·0	22·6	11·5	Arbroath	1864	ArbroathS.Co.		S.P.Anderson	Höganäs
472		Elisabet Olsson W Bkn	Swedish	223	100·7	22·0	13·3	France	1853			J.A.Nilsson	Forshälla, Bkslo.
473		Elisabet Herlitz ..80 W Sr	Swedish	134	80·7	23·8	10·5	Pernau	1871			D.Herlitz	Helvi,Gotland
474		Elisabet [illegible] ..86 W Bn&Sr	Swedish	122	84·6	21·7	9·5	Wenersborg	1866			C.J.Andersson	Wenersborg
475		Elisabeth [illegible] 86-88 W Bk (AMANDA)	Danish	269	106·1	25·9	13·9	Pitea	1850			H.P.Lund	Rønne
476	V	Elisabeth Nielsen ..85 W Bn	Danish	120	89·8	20·1	10·4	Rudkjöbing	1866	T.Boas		Brödrene (C. & F. Nielsen)	Rudkjöbing
477	G	Elisabeth Petersen 73- W Sn	Danish	100	79·2	17·6	9·8	Elsfleth	1853	H.Ahlers		Chr.Simoni	Aalborg
478	D	Elisabeth Wind ..86 W Sr (ex [illegible])	Dutch	157	93·8	21·0	11·0	Pekela	1863	F.L.Dreuth		J.K.Wind	Termunterzijl
479		Elisabeth Arents W Sr		150	93·1	21·3	10·4	Pekela	1841	F.L.Wortelboer			
480	DGV	Elisabeth van der Veen ..W Sr	Dutch	148	96·4	22·1	10·9	Delfzyl	1857	B.Kuiper		J.J.Sijdzes	Farmsum
481	V	Elisabeth Mory ..88 W Bg	French	152	86·0	22·3	11·5	Nantes	1875	E.Clergeau	[illegible]	Laffetay	Caen
482	V	Elisabeth Hesry ..86 W Bn	French	142	76·8	20·7	11·5	St.Servan	1875	H.Hamon	[illegible]	Anatoll Lemoine	St.Malo
483	DG	Elisabeth Kramer 73-88 W Bk	German	1170	190·2	37·1	24·3	Bremerhaven	1874	UlrichsDock	[illegible]	D.H.Wätjen & Co.	Bremen
484	GV	Elisabeth Groth 76-84 I Bk	German	834	180·0	31·9	18·6	Kiel	1884	GermaniaWrks		C.M.Matzen	Hamburg
485	GV	Elisabeth Trost 86-88 W Bk	German	447	141·7	30·2	17·2	Kiel	1877	C.Ilms	[illegible]	Sartori & Berger	Kiel
486	G	Elisabeth Zulligers ..84 W Bk (ex [illegible])	German	337	112·8	25·8	16·3	Memel	1861	B.Pieper		Lamb.Röttgers	Papenburg
487	G	Elisabeth Hamoro ..79 W Bg (ex [illegible] ARANDONEG)	German	330	108·0	28·3	16·3	Memel	1853	B.Pieper		Frau H.F.Carsjens	Memel
488	G	Elisabeth Wabbel 75-77 W Bk	German	271	114·4	27·5	14·2	Greifswald	1864	J.W.Sadelkow		C.Graedener	Greifswald
489	G	Elisabeth Schleinkoye ..86 W Bg	German	156	98·4	21·5	9·9	Papenburg	1860	H.J.C.Hilling		H.H.Mamme'sWwe	Papenburg
490	G	Elisabeth Olthau ..80 W Bn (ex [illegible])	German	136	85·0	20·0	11·1	Papenburg	1857	B.Tholen		Herm.Heyen	Papenburg
491	G	Elisabeth [illegible]Brauer ..82 W Sr	German	114	83·3	20·7	10·4	Bardenfleth	1851			Hermann Ernst	Oldersum
492		Elisabeth Boltzer ..78 W Sr (WELCKER)	German	111	84·2	20·0	9·9	Ribnitz	1878	C.H.Staben	[illegible]	Liebenheim & Plaass	Rostock
493	N	Elisabeth Torjusen W Bk	Norweg'n	561	149·5	31·6	16·8	Westervik	1873	C.R.Fridell	[illegible]	H.C.Torjusen	Kragerö
494	N	Elisabeth Gundersen 77-77 W Bk	Norweg'n	324	115·5	27·5	15·7	Vegesack	1852			C.Huslund	Arendal
495	N	Elisabeth Andersen W Bk	Norweg'n	239	98·7	28·5	16·6	Hamburg	1836			H.Andersen	Tönsberg
496		Elisabeth Penin W Bg	Russian	322	112·1	28·6	14·3	Fredrikshamn	1873	G.Häkela		N.Sapetoff	Björkö
497		Elisabeth Andersson ..81 W Bn (ex [illegible])	Russian	144	87·1	24·2	11·4	Finkenwärder	1859	C.Wriede		A.Karonius	Lumparland
498		Elisabeth W Sr	Russian	141	..	..	..		1870			N.Siitoin	Sordavala
499		Elisabeth W Ct	Russian	100	76·2	25·0	9·7	Riga	1879	M.Andersohn		G.M.Beckmann	Riga
500													

= Lloyd's Register; D = Nederlandsche Vereeniging; G = Germanischer Lloyd; I = Registro Italiano; R = American Record;
A = Veritas Austro-Ungarico; H = Veritas Ellenico; N = Norske Veritas; V = Bureau Veritas.

Dans la 4me Colonne, C = Composite (membrure en fer, bordé en bois); I = Fer; Stl = Acier; W = Bois.

L = Dates du doublage en métal (Col. 11) ressortent des livres dans lesquels les navires sont respectivement classés ou des rapports des Inspecteurs du Lloyd's Register.

Pour plus amples renseignements sur les navires classés par le Lloyd's Register, voir le Lloyd's Register Book et le Lloyd's Yacht Register.

Official No.	No. in Book	Society in which classed	Ship's Name. Master, and Dates of Appointment to present Owner's Service and to Ship. Ship's Late Name.	Material and Rig	Flag	Net Tonnage	Length	Breadth	Depth	Where	When	By whom	Date of Metalling	Owners or Managers
	501	GV	Elisabeth Ahrens *Cz 8-[illegible]*	W Bk	German	396	126·0	28·[illegible]	17·[illegible]	Rostock	1875	J. Möller	[illegible]	A. Burchard
	502		Elisabeth Catharine *Dybdahl*	W Bk	Norweg'n	366	[illegible]	2[illegible]·2	17·6	Mandal	[illegible]			A. Neumann
	503	GV	Elisabeth Mentz *Johnke [illegible]*	W Bk	German	[illegible]	[illegible]	[illegible]	18·0	Rostock	1871	J. Möller	[illegible]	W. Maack
	504	N	Elisabeth Morton (ex Elisabeth Morgan) *Nielsen [illegible]*	W Sr	Norweg'n	[illegible]	[illegible]	2[illegible]·7	12·0	Ipswich	1870	Robertson	[illegible]	O. H. Holta
	505	V	Elisabeth Rickmers *Hencke [illegible]*	W Bk	German	[illegible]	[illegible]	[illegible]·7	24·2	Geestemünde	1877	R. C. Rickmers	[illegible]	R. C. Rickmers
	506		Elisabeth Wiemann (ex Wilhelm) *Neumann -88*	W Bn	Dutch	181	97·4	[illegible]·7	11·8	Elsfleth	1856	Geert Hagens		P. Stratingh & Co.
[illegible]	507		Elisabetta *Omero*	W Bg	[illegible]	214	92·2	25·3	15·[illegible]	[illegible]	1852			E. Intersich
[illegible]	508	I	Elisabetta C. *Cataldo -88*	W Bn	Italian	187	101·0	24·8	13·9	Castellamare [di Stab	1875	B. Bonifacio	[illegible]	G. Cussisa & Co.
[illegible]	509	✻	Elise *Christian*	W Bkn	British	396	110·0	28·5	15·9	P. E. Island	1878	Ellis	[illegible]	J. M. Campbell
	510	V	Elise *Hansen 79-79*	W Sr	Danish	141	98·7	22·2	10·0	Odense	1875	H. L. Hansen		Kingo Pedersen
	511	DV	Elise *Hansen 86-86*	W Sr	Danish	13[illegible]	93·1	20·7	10·5	Faaborg	1877	R. H. Møller		P. Nielsen
	512	G	Elise (ex Poseidon) *Lodewijks*	W Sr	Dutch	16[illegible]	92·8	23·3	11·9	Fünfhausen	1853	B. H. Steenken		J. J. Lodewijks
	513	V	Elise (ex Fratel) *Rowehl 83-85*	W S	German	[illegible]	194·2	39·0	24·6	Mystic Bridge, [Conn.	1869	Greenman Bros.	[illegible]	Fr. Roters
	514	GV	Elise (ex Fortuna) *Winters 88-*	W Bk	German	984	171·8	31·8	22·9	Boston	1868	Curtis Smith & [Co.	[illegible]	Joh. Fr. Wessels
	515	G	Elise *Steinhagen -79*	W S[illegible]	German	278	108·9	27·0	14·6	Wismar	1861	J. H. D. Barmann		J. C. Thormann
	516	V	Elise *Schwartz -83*	W [illegible]	German	18[illegible]	97·7	23·8	10·5	Weener	1866	R. Borg		Friedr. Tulp
	517		Elise (LIGHTER)	W Sr		14[illegible]	88·[illegible]	22·7	11·0	Weener	1859	R. Borg	[illegible]	
	518	V	Elise *De Freese 86-*	W Sr	German	13[illegible]	88·9	21·0	10·1	Varel	1880	A. Schwoon	[illegible]	Meine de Freese
	519		Elise *Meijer 86*	W [illegible]	German	127	91·2	20·5	9·8	[illegible]	1877	B. D. Muhlendorf		J. G. Ukena
	520	N	Elise *Kjeldsen*	W Bk	Norweg'n	350	111·2	28·5	16·9	Risör	1854			A. C. Selvig
	521	N	Elise *Christensen -87*	W Bg	Norweg'n	146	93·0	22·5	10·8	Laurvig	1857			R. Olsen Rød
	522	R	Elise *Hermansson 83-84*	W S	Swedish	341	101·5	24·1	18·0	England	1845			J. E. Olsen
	523	V	Elise *Gren 72-84*	W Sr	Swedish	152	92·5	21·8	11·1	Calmar	1855	T. A. Svensson		J. H. Gren
	524		Elise *Strålbach*	W Sr	Swedish	151	91·9	20·5	11·8	Scotland	1851			H. Pettersson
	525	V	Elise Bay *Petersen 74-74*	W Sr	Danish	11[illegible]	82·3	20·3	10·0	Thurø	1874	L. Kaas		N. W. Kaas
	526	GV	Elise Both *Ahrens -81*	W B[illegible]	German	[illegible]	127·2	28·5	17·9	Rostock	1870	Em. Padderatz	[illegible]	Franz Crull
	527	V	Elise Høy *Hansen -88*	W [illegible]	Danish	195	104·4	23·0	12·4	Horsens	1876	Chr. Schrøder	[illegible]	Akt. Selsk. "Elise Høy" [illegible]
	528	G	Elise Linck *Falcke 83-83*	W B[illegible]	German	513	126·[illegible]	30·8	19·2	Danzig	1879	J. W. Klawitter		Schoenberg & Doe[illegible]
	529	DV	Elise Susanne *Van Gelder -79*	W B[illegible]	Dutch	228	110·8	25·2	12·9	Rotterdam	1867	Gebr. Visser	[illegible]	Hr. Muller & Co.
[illegible]	530	I	Elisetta *Chiappella -88*	W B[illegible]	Italian	363	120·0	29·9	17·4	Prà	1871	M. & D. Tixi	[illegible]	Fll. Croce fu Ant.
[illegible]	531	R	Elisha Gibbs *Woods [illegible]*	W 3m Sr	American	485	142·2	33·2	15·4	Bath, Me.	188[illegible]	[illegible] Sawyer		E. Gibbs
[illegible]	532	✻	Elissa *Chaddock 85-88*	I Bk	British	409	14[illegible]·6	28·1	14·6	Aberdeen	1877	A. Hall & Co.		H. F. Watt
[illegible]	533	✻	Elite *Hargrave*	W Sr	British	11[illegible]	89·6	23·8	11·4	Howden Dyke	1883	W. Cawley	[illegible]	R. Hargrave & Co.
[illegible]	534	✻	Eliza *Wheaton 76-92*	I Bk	British	2[illegible]	1[illegible]·6	25·8	12·5	Greenock	1876	J. E. Scott		Goodyear & Co.
[illegible]	535		Eliza *Gerrior*	W Bn	British	205	101·0	24·2	12·9	P. E. Island	1875	J. Duncan & Co.		D. Le Blanc
[illegible]	536		Eliza	W Bn	British	197	101·2	24·8	12·3	Nova Scotia	1861	Merritt		Mrs. M. Donelly
[illegible]	537		Eliza *Tyrrell 78-81*	W Sr	British	192	108·0	24·8	13·3	Lossiemouth	1867	Geddie		H. Kearon
[illegible]	538		Eliza *Fennis*	W Bg	British	183	101·2	24·0	13·2	France	1801			Clement & Co.
[illegible]	539		Eliza	W Bn	British	140	94·9	23·3	10·5	Tidnish, N.S.	1865			S. A. Taylor
[illegible]	540		Eliza	W Bn	British	131	84·8	22·4	13·0	Peterhead	1844			J. Rippett
[illegible]	541		Eliza (LIGHTER)	W Sr	British	129	73·0	21·0	12·9	Peterhead	1828			E. Bannister
[illegible]	542		Eliza	W Bn	British	109	91·7	25·0	6·7	Belize, B.H.	1849	J. Savery		Belize Estate & [illegible] Co.
[illegible]	543		Eliza *Lowry*	W Bg	British	1[illegible]	68·7	21·4	12·0	Whitehaven	1792			J. Curran
[illegible]	544	✻	Eliza *Williams*	W Sr	British	94	76·0	20·0	10·3	Bangor	1866	Parry		W. Williams
[illegible]	545		Eliza *Murray*	W Bk	American	29[illegible]	10[illegible]·6	27·2	16·2	Rochester, Mass.	1856			H. C. Wright
	546		Eliza (ex Diamantes) *Cavalcante ira*	W Bn	Brazilian	209	102·3	25·2	11·1	Pokrija	1877		[illegible]	J. R. D. Cruz
	547		Eliza (ex [illegible]) *Campos*	W Bn	Brazilian	141	..	..	..	United States	1860		[illegible]	J. C. Babo
	548	✻	Eliza (ex [illegible]) *Knudsen -88*	I S	Danish	489	170·5	26·4	16·2	Stockton	1861	M. Pearse & Co.		J. M. Knudsen
	549	N	Eliza *Larsen*	W Bk	Norweg'n	257	106·7	26·4	11·[illegible]	Drontheim	1832			L. Eliassen
[illegible]	550		Eliza A. Anderson *Carroll*	W Sr	American	132	106·0	26·4	7·0	Hartford, Conn.	1869			J. Rankin

✻ = Lloyd's Register; D = Nederlandsche Vereeniging; G = Germanischer Lloyd; I = Registro Italiano; R = [illegible]
A = Veritas Austro-Ungarico; H = Veritas Ellenico; N = Norske Veritas; V = [illegible]

In the 4th Column, C = Composite (iron frame, wood planked); I = Iron; Stl = Steel; W = Wood.

The Dates of Metalling (Col. 11) are based on the records of the Books in which Vessels are respectively classed, or on the Reports of the [illegible]

Further particulars of Vessels classed by Lloyd's Register will be found in Lloyd's Register Book and Lloyd's Yacht [illegible]

No.	Reg.	Nom, Matériaux et Gréement; Capitaine; Ancien Nom	Pavillon	Tonnage	Longueur	Largeur	Creux	Construit: Où	Quand	Par qui	Date du doublage	Armateurs ou Directeurs-Gérants	Port d'Armement
551		Eliza Adams *Hoxfsted* W S	American	388	111·8	28·5	19·5	Fairh'en Mass	1835			Tabor & Gordon	N.Bedford,Mas.
552		Eliza Allan W Sr	British	119	84·7	21·4	7·7	Port Dover	1875	H.A.Dakin		John Allan	Port Dover,Ont
553		Eliza Allan W Bn	British	126	88·4	21·4	8·2	Nambuccra River,N.S.W.	1879	M.Holm		J.H.Allan	Sydney,N.S.W.
554		Eliza Ann *Jamieson* W Sr	American	187	113·0	22·6	7·8	Haverhill, Mas	1875	Manson		S.D.Keen	New York
555		Eliza Ann Hooper *Huntley* W Sr	American	159	106·1	28·1	7·1	Smyrna,Del.	1868			C.W.Vood	New York
556	*	Eliza Anne *Evans* W Sk	British	32	54·9	15·5	7·1	Newquay	1877	Daniel		D.Evans	Cardigan
557		Eliza B. Coffin *Mann* W Sr	American	126	87·8	26·6	8·5	Harringt'n,Me	1868			H.W.Cook	Machias,Me.
558		Eliza B. Emory *Tegner* W 3m Sr	American	316	115·8	29·7	9·3	May'sLandng. [N.J.	1866			J.Taylor	New York
559		Eliza Bain *Couch* -87 W Sr	British	150	107·0	24·0	12·4	Hayle	1864	Harvey		D.W.Bain & Co.	Hayle
560		Eliza Bares *Young* -87 W Bk	British	222	102·2	25·5	9·9	Maine,U.S.	1848			G.S.Tucker	Hamilton Bermuda
561		Eliza Blanche *(RULES)* W Bg	British	170	104·0	23·1	12·5	Sunderland	1861			W.Owston	Fremantle, W.A.
562		Eliza Blaxland *Blaxland* W Sr	British	174	98·0	23·7	12·4	Rye	1863	Hoad		A.Nicholls	Faversham
563		Eliza Emma W Bn	British	156	82·0	20·4	13·2	Hartlepool	1845	Denton		W.Watson	Shoreham
564		Eliza Firth *Stevens* W Bn	British	143	97·8	22·9	10·3	JervisBay,NSW	1869			E.H.Castle	Melbourne
565		Eliza Francis *Edwards* W Sr	British	63	65·2	16·5	9·5	Aberystwith	1852			L.S.Edwards	Aberystwith
566	R	Eliza J. McManemy *Dixon* -83 W Bkn	American	746	152·5	34·5	18·6	Milton,Del.	1878	D.H.Atkins	Y 1.88	J.W.Hall	Philadelphia
567	R	Eliza J. Scribner *Jimerson* -84 W 3m Sr	American	378	128·4	30·0	9·2	Milford,Del.	1872			J.W.Hall	Philadelphia
568		Eliza Jane *Chidgey* W Sr	British	107	89·5	22·6	10·1	Bridgwater	1873	Gough		J.W.Sully	Bridgwater
569	R	Eliza Levensaler *Kellar* (82-83) W Sr	American	152	88·0	25·0	8·0	ThomastonMe	1882	Cushing		A.Levensaler	Thomaston,Me.
570	N	Eliza Lines *Andreassen* (ex Lady Hobart) [-88] W Bk	Norweg'n	461	148·4	29·6	17·5	Southampton	1864		Y 1.88	D.Davidsen	Farsund
571		Eliza Mary W Bn	British	161	100·8	20·9	11·1	Lake Macquarie [N.S.W.	1873	C.Moynahan		C.R.Powell	Maryboro'
572		Eliza Miller *Anderson* W Sr	American	149	103·5	28·8	8·6	SanFrancisco	1883	White		J.Jensen	San Francisco
573		Eliza O'Keefe *Clifford* W Bn	British	112	85·6	22·1	11·1	Youghal	1856	P.Kidney		Clifford & Co.	Waterford
574	*	Eliza Patience *Cooper* W Barge	British	67	81·4	19·8	6·6	Ipswich	1883	W.J.Curtis		N.Orvis	Ipswich
575		Eliza S. Potter *Hammond* W Sr	American	321	109·0	28·6	10·4	Mystic,Conn.	18$\frac{57}{77}$			H.F.Champlin	Bristol,R.I.
576		Eliza Shairp W Bk	British	364	102·6	26·7	18·8	Sunderland	1844			C.Parbury	Melbourne
577		Eliza White *Rooney* W Sr	British	106	92·0	23·2	8·3	Pt.Burwell	18$\frac{57}{87}$	G.W.Patten		D.Rooney	Port Burwell
578		Eliza & Jane *Curry* W Sw	British	170	84·6	24·0	13·0	Sunderland	1848			J.C.Butterwick	Whitby
579		Elizabeth W Bk	British	277	120·0	25·5	14·5	P.E.Island	1856			L.Kickham	P.E.Island
580		Elizabeth *Lakie* 64-86 W Bk	British	263	118·9	24·4	14·4	Sunderland	1863	W.Briggs&Sons	Y 2.88	John Bickers	Adelaide,S.A.
581		Elizabeth *Brett* W Bk	British	245	119·8	27·0	12·4	Shelburne,NS.	1868	John Muir		N.McLeod	Melbourne
582		Elizabeth *Hammel* W Bn	British	131	93·6	21·2	11·2	Teignmouth	1856	Owen		J.Lang & Co.	Drogheda
583		Elizabeth W Barge	British	131	104·0	22·6	8·1	Point Levis	1872			John Baile	Quebec
584		Elizabeth *Watts* W Bn	British	122	82·1	23·1	10·7	Russia	18$\frac{47}{75}$			W.Jewitt	Bideford
585		Elizabeth *Humphreys* W Bn	British	122	82·0	21·5	11·6	Sunderland	1847	Lister&Co.		Dawber & Co.	Hull
586		Elizabeth *Roy* W Sr	British	106	82·2	23·1	9·1	Margaretville, [N.S	1876	J.Roy		John Roy	Annapolis,N.S.
587		Elizabeth W Sr	British	100	88·1	20·5	11·7	Hayle	1836	Harvey		T.N.Gudge	Hayle
588	*	Elizabeth *Sinclair* 75-77 W Sr	British	80	79·0	20·0	9·2	Banff	1877	Watson		J.Stephen	Wick
589	RV	Elizabeth *Colcord* -87 W S	American	1773	231·5	41·8	27·5	Newcastle,Me.	1882	Haggett&Co.	Y 4.87	Haggett & Co.	Searsport,Me.
590	NR	Elizabeth *Schildhuis* (ex Elske) W Bg	Dutch	285	109·0	26·7	14·9	Arendal	18$\frac{54}{73}$			J.Schildhuis	Farmsum
591	R	Elizabeth A. Baizley *Townshend* W 3m Sr	American	354	126·0	33·5	9·2	Coop'rsPnt N.J	1874			G.W.Bailey	Camden
592		Elizabeth Alice *Kearns* (83-87) W Sr	British	142	96·6	22·2	12·1	Bowling	1866	Scott & Macgill		J.Fisher & Sons	Barrow
593		Elizabeth Ann *De Costa* W Sr	British	173	94·0	28·3	10·6	Harbor au Bouche,N.S.	1866	Fiddell&deCosta		Wm.Gerroir	Pictou,N.S.
594	*	Elizabeth Bennett *Bennett* W 3m Sr	British	150	103·8	23·4	11·0	Runcorn	1884	Brundrit&Co.		A.Green & Co.	Liverpool
595	*	Elizabeth Charlotte *Jones* W Sr	British	141	93·8	22·4	11·8	Holyhead	1860	Roberts		G.Williams	Carnarvon
596	*	Elizabeth Childs *Christophera* W Bk	Turkish	400	129·9	28·2	17·3	Sunderland	1870	T.Metcalf	Y 2.85	L.Vendopulos	Smyrna
597		Elizabeth Cowman *Blenkinsop* W Bn	British	116	84·3	20·4	11·7	Bristol	1855			A.Cullen	Whitehaven
598		Elizabeth De Hart (WRECKED) *Burke* -87 W Sr	American	218	107·0	28·7	9·1	Essex,Mass.	1860			W.D.McIntyre	New York
599		Elizabeth Drew *Wilcocks* W Sr	British	100	82·5	22·7	10·8	Padstow	1871	Stribley		T.S.Grenfell	Plymouth
600		Elizabeth Graham C Bk		598	165·6	29·2	18·1	Middlesbro'	1869	W.Harkess	Y 9.84	C.S.Hodge	Dunedin

* = Lloyd's Register; D = Nederlandsche Vereeniging; G = Germanischer Lloyd; I = Registro Italiano; R = American Record; A = Veritas Austro-Ungarico; H = Veritas Ellenico; N = Norske Veritas; V = Bureau Veritas.

Dans la 4me Colonne, C = Composite (membrure en fer, bordé en bois); I = Fer; Stl = Acier; W = Bois.

Les Dates du doublage en métal (Col. 11) ressortent des livres dans lesquels les navires sont respectivement classés ou des rapports des Inspecteurs du Lloyd's Register.

Pour plus amples renseignements sur les navires classés par le Lloyd's Register, voir le Lloyd's Register Book et le Lloyd's Yacht Register.

Official No.	No. in Book	Registry in which classed	Ship's Name. Material and Rig. Master, and Dates of Appointment to present Owner's Service and to Ship. Ship's Late Name.	Flag.	Net Tonnage.	Length.	Breadth.	Depth.	Built. Where.	When.	By whom.	Date of Metalling.	Owners or Managers.
1	2	3	4	5	6	7			8	9	10	11	12
[illegible]	601	✻ D	Elizabeth Jane *Hooper* W Bg	British	143	100·4	22·6	11·4	Ipswich	1860	W. Read		J. Hutchings
53235	602		Elizabeth Kelly *Wilson* W Bn	British	116	83·9	21·2	12·3	Whitehaven	1866	Shepherd		John Kearns
65570	603	✻	Elizabeth Kilner *Ashton 73-78* W Sr	British	88	80·1	20·1	10·0	Thorne	1871	Staniland		G. Kilner & Co.
135068	604		Elizabeth M. Cook *Clark* W 3m Sr	American	268	104·8	32·4	10·9	Calais, Me.	1874	Robinson Bros	[illegible]	J.G. Murchie & S…
27620	605	✻	Elizabeth M'Lea *Bai…* [76-…] W Bn	British	128	99·7	20·2	10·8	Greenock	1860	Steele	[illegible]	R. Cock
54086	606	✻	Elizabeth Mary (WRECKED) *Col* W Bk	British	250	129·0	26·4	13·0	Sunderland	1866	W. Watson	[illegible]	E.T. Miles
47226	607	✻	Elizabeth Nicholson *Spilsman 77-83* W Bk	British	904	192·5	32·5	22·2	Annan	1863	Nicholson	[illegible]	Montgomerie & Workman
62082	608	✻	Elizabeth Ostle *Rennie* W Bk	British	740	165·3	33·0	21·7	Maryport	1869	Ritson	[illegible]	Hayton & Simpson
83098	609	✻	Elizabeth Peers *Peers [84-85]* I 3m Sr	British	165	107·1	23·6	10·4	Amlwch	1885	W. Thomas & Sons		W. Postlethwaite
27212	610		Elizabeth Ray *Moon 76-76* W Bn	British	199	100·0	24·5	13·0	Sunderland	1859	Doxford		A.E. Flowers
67601	611		Elizabeth Roy *King* W Bkn	British	223	116·1	25·1	12·1	Dysart	1873	Watt		J. Harvey
36653	612		Elizabeth Ryan W Sr	British	123	88·0	23·0	11·1	Gabarus, N.S.	1864			P. Bagnell
135107	613		Elizabeth S. Lee *Lee* W 3m Sr	American	467	113·4	34·5	9·0	May's Landing [N.J.	1875			J.S. Lee
36660	614		Elizabeth Seldon W Sr	British	100	81·8	20·6	7·7	Tweed River, [N.S.W.	1861			R. Seldon
63369	615	✻	Elizabeth Stevens (WRECKED) *Strike* W 3m Bkn	British	198	114·0	24·2	12·6	Ipswich	1871	Bayley	[illegible]	H. & J. Strike
8071	616		Elizabeth T. Cottingham *Smith* W 3m Sr	American	252	120·4	31·8	9·4	English Crk N.J.	1873			J. Smith
31766	617		Elizabeth Thompson W Sq	British	289	90·4	22·4	16·0	Sunderland	1842			Geo. Tully
8806	618	R	Elizabeth Winslow *Oakes 84-86* W Bn	American	372	126·3	30·1	11·5	Yarmouth, Me.	1873	Loring	[illegible]	J.S. Winslow & Co.
62571	619	✻	Elizabeths *Roberts 87-87* W Bk	British	373	132·3	28·1	16·8	Sunderland	1870	B. Hodgson	[illegible]	R. Thomas & Co.
15098	620		Elizabeths *Peterson -88* W Bk	British	329	109·0	27·0	17·0	Newcastle	1857	Gaddy & Lamb		G. Walker
60969	621		Ella W Bk	British	985	178·0	36·0	21·2	Quebec	1873	P.V. Valin	[illegible]	W. Laud
28359	622		Ella (ex Humphrey Purinton) *Thomas* W Bk	British	672	137·8	30·6	21·7	Maine, U.S.	1857			T.S. Blues
62512	623	✻	Ella *Jones 76-87* W Bk	British	382	122·3	28·2	17·4	Sunderland	1869	T. Metcalf	[illegible]	J. Davies
[illegible]	624		Ella *Hewitt* W Bn	British	194	104·1	24·4	12·6	P.E. Island	1864	Owen		J.W. Hawkes
46003	625		Ella W Bn	British	168	97·0	25·0	11·2	Riv. Hebert NS	1868			J.B. Read
8572	626	R	Ella *Lewis 61-87* W Bk	American	622	149·5	32·1	17·2	Westbrook, Me	1869	R. Lewis & Co.	[illegible]	R. Lewis & Co.
135036	627		Ella *Hansen -88* W Bkn	American	249	131·8	28·7	10·8	Freep'rt, Wash	1874	W. Begaut	[illegible]	Welsh & Co.
7190	628		Ella *Coombes 81-* W Sr	American	184	105·6	28·8	8·6	Petty's Isld N.J.	18 69/70			T.J. Stewart & Co.
8130	629		Ella *McVane* W Sr	American	160	104·0	28·0	8·0	Newcastle, Me.	1864			E. Haggett
8867	630		Ella *Gupill* W Sr	American	106	85·3	27·7	7·5	Harringt'n, Me	1872			R. Mitchell
	631		Ella (ex Prince Alfred) *Folkmann 80-88* W Sr	Danish	136	91·7	21·4	11·3	Perth	1867	Wallace		P. Berg
	632		Ella (ex Scotia Queen) *Nilsholm -88* W Bk	Swedish	420	125·0	26·7	12·6	Maitland	1867	Nova Scotia		Anna E. Wancke
135410	633	R	Ella A. Warner *Hallbrook* W Sr	American	141	100·0	25·0	9·0	Essex, Mass.	1879	J. Story		E.L. Atwood
8196	634		Ella Brown *Keene* W Sr	American	156	101·5	26·3	8·8	E. Machias, Me.	1869			O.B. Hall
135082	635		Ella Clifton *Mansfield* W Sr	American	103	89·0	23·9	7·3	St. John, N.B.	1872	Sutton		J.B. Mansfield
7180	636		Ella F. Crowell *Freeman* W Sr	American	157	93·8	26·4	8·0	E. Haven, Conn	1864			A.C. Freeman
8080	637	R	Ella Francis *Hall* W Sr	American	119	93·0	26·0	8·0	Mnt Desert, Me	1870	W.S. Newman		G.L. Farrand
7558	638		Ella H. Barnes *Hardcastle* W Sr	American	180	101·5	28·5	9·6	Madison, Conn	1865	C.M. Miner & Son		E.K. Johnson
135544	639	R	Ella M. Hawes *Farington* W 3m Sr	American	282	128·0	32·8	9·0	Bath, Me.	1881	B.W. & H.F. Morse		B.W. & H.F. Mors…
8946	640	R	Ella M. Storer *Wade [-88]* W 3m Sr	American	427	135·7	32·0	14·9	Waldoboro' Me	1873	H. Kennedy & Co.	[illegible]	J. Baker & Co.
8910	641	R	Ella M. Watts *Steven [81-]* W 3m Sr	American	443	137·8	31·2	16·0	Thomaston Me	1873	Stetson, Gerry & [Co.	[illegible]	S.S. Gerry & Co.
59709	642		Ella Mae (BROKEN UP) *Chadwick* W Bn	British	145	94·4	24·8	11·3	P.E. Island	1868	M'Lure		C. Neill
80867	643		Ella Maud Wood *Dixon* W Sr	British	160	93·4	27·7	10·6	Joggin Mines, [N.S.	1885	A.F. Melanson		Fred W. Sumner
57191	644	V	Ella Moore *Byer* W Bk	British	391	136·0	30·1	14·6	Cornwallis, NS	1867	J. Buckman		Henry A. Eaton
71067	645		Ella Murton W Sr	British	229	118·3	26·2	10·1	Mill Point	1875	W. Jamieson		J.W. Murton
	646	✻	Ella Nicolai (ex Dalton) *Meisterfeld [82-87]* I Bk	German	572	174·4	27·7	17·9	Liverpool	1868	Bowdler Chaffer [& Co.		J.H. Nicolai
8827	647		Ella Powell *Bishop* W Sr	American	134	99·4	22·5	7·8	Bridgeton N.J.	1872			Mrs. Jane Bishop
8846	648		Ella Pressey *Pressey* W Sr	American	157	98·8	27·3	8·7	Thomaston Me	1872			A. Levensaler
48811	649		Ella Reed W Bg	British	240	105·0	25·5	11·5	Maine, U.S.	1853			E. Bayley
135009	650		Ella T. Little *Frambes -87* W 3m S	American	236	114·3	30·2	9·7	Chester, Pa.	1874			T.G. Crawford

✻ = Lloyd's Register; D = Nederlandsche Vereeniging; G = Germanischer Lloyd; I = Registro Italiano; R = [illegible]
A = Veritas Austro-Ungarico; H = Veritas Ellenico; N = Norske Veritas; V = [illegible]

In the 4th Column, C = Composite (iron frame, wood planked); I = Iron; Stl = Steel; W = Wood.

The Dates of Metalling (Col. 11) are based on the records of the Books in which Vessels are respectively classed, or on the Reports of the Surveyors [illegible]

Further particulars of Vessels classed by Lloyd's Register will be found in Lloyd's Register Book and Lloyd's Yacht Register

[illegible]	No. dans la liste	Cote	Nom. Matériaux et Gréement. Capitaine et dates d'entrée au service des Armateurs actuels et à bord du navire. Ancien Nom.		Pavillon.	Tonnage	Longueur	Largeur	Creux	Construit. Où.	Quand.	Par qui.	11	Armateurs ou Directeurs-Gérants.	Port d'Armement.
[illegible]	651	V	Ella Vose *Doody* 75 (ex-MANTLER)	W Bk	British	[illegible]	[illegible]	[illegible]	[illegible]	[illegible]	1875	D. McPherson	[illegible]	G. J. Troop	Windsor, N.S.
[illegible]	652		EllanoravanDeusen *Gulpen*	W 3m Sr	American	[illegible]	[illegible]	[illegible]	11·4	[illegible]	1874			T. W. Godfrey	Camden, N.J.
[illegible]	653		Ellen *Stephens*	W Bk	British	[illegible]	[illegible]	[illegible]	17·7	[illegible]	1865	J. & S. W. Olive		J. Henderson	Newcastle, N.S.W.
[illegible]	654	✻	Ellen *Snelling*	W 3m Sr	British	[illegible]	[illegible]·8	24·0	12·4	[illegible]	1867	Harland	[illegible]	T. J. Snelling	Chester
[illegible]	655		Ellen	W Bn	British	[illegible]	94·1	24·0	11·4	Lond'd'rry [illegible]	1864	[illegible]Grath		E. Murray	Dublin
[illegible]	656	✻	Ellen *Owens*	W Sr	British	129	94·2	22·4	11·[illegible]	[illegible]	1858			J. Owens	Carnarvon
[illegible]	657		Ellen	W	British	109	76·8	26·1	9·6	Sunderland	1860			C. W. Marshall	Maldon
[illegible]	658	✻	Ellen *Johnston*	W K	British	78	75·6	20·[illegible]	10·[illegible]	[illegible]	1884	S. Dowdney & Sns		W. Acaster	Grimsby
	659	G	Ellen *Brag* 66-66	W Bn	Danish	126	89·7	20·3	10·5	Odense	1866	H. L. Hansen		H. T. Helweg	Odense
	660	N	Ellen *Taarvig*	W Bg	Norweg'n	324	[illegible]	29·1	16·8	[illegible]	1854			C. Taarvig	Fredrikstad
	661	V	Ellen *Jurvelius* (ex Sterling Star, re 1861)	W Bk	Russian	745	157·[illegible]	35·3	18·9	[illegible]	1850		[illegible]	A. Carlström	Nikolaistad
	662	G	Ellen *Backe*	W Sr	Swedish	112	84·0	19·8	9·9	Nordby	1866	T. Schönnan		N. N. Backe	Råå
[illegible]	663	V	Ellen A. Read *Hatfield*	W S	British	1756	2[illegible]·5	41·6	24·[illegible]	[illegible], N.S.	1884	Hatfields Bros.	[illegible]	William Law	Yarmouth, N.S.
[illegible]	664		Ellen A. Swift *Brown*	W Sr	American	125	88·0	24·0	9·0	Essex, Mass.	1882	J. James		S. S. Swift	Provincet'n, Mass.
[illegible]	665	✻	Ellen Anne *Jones* 79-88	W Sr	British	96	75·7	20·7	10·1	[illegible]	1866	Daniel		T. Jones	Aberystwith
[illegible]	666		Ellen Ashcroft *H. McMurtry*	W Bg	British	176	92·8	23·8	13·[illegible]	W[illegible]	1866	Williamson		H. McMurtry	Belfast
[illegible]	667	✻	Ellen Beatrice *Jones*	W Sr	British	88	[illegible]	20·0	10·[illegible]	[illegible]	1865	Evans		W. Owens	Aberystwith
[illegible]	668	E	Ellen Catherine *Savage*	W Bn	British	129	81·0	21·8	12·4	[illegible]	1850			E. Timothy	Aberystwith
[illegible]	669	R	EllenCrusoe *Tupley* 83-88	W 3m Sr	American	383	1[illegible]·8	26·1	13·[illegible]	[illegible], Me.	1883	R. Lewis & Co.	[illegible]	R. Lewis & Co.	Portland, Me.
	670	N	Ellen Grant *Isaksen* 87	W Bk	Norweg'n	353	115·6	31·9	17·6	[illegible], N.B.	1857		[illegible]	P. H. Sörensen	Christiania
[illegible]	671	✻	Ellen Greaves *[illegible]*	W Bg	British	254	114·0	25·0	15·0	Portmadoc	1876	Jones	[illegible]	M. Jones & Co.	Carnarvon
[illegible]	672		Ellen H. *Olsen*	W Bn	British	179	96·0	26·2	11·0	St. John, N.B.	1867			R. W. Saul	Lowestoft
[illegible]	673		Ellen Holgate *Betts* 85-	W Sr	American	160	105·4	28·7	7·5	Philadelphia	1866			J. A. J. Sheets	Philadelphia
[illegible]	674	✻	Ellen Holt *Coulson* 88	W 3m Sr	British	310	135·3	27·5	12·9	Sunderland	1875	W. [illegible]	[illegible]	J. Sutcliffe & Son	Grimsby
	675	✻	Ellen Isabelle *Arnaud* [83-83]	W Bk	French	663	151·9	32·0	19·5	Bordeaux	1875	[illegible]	[illegible]	A. Salle & Co.	Bordeaux
[illegible]	676	✻	Ellen Kerr *Walsh*	W Sr	British	65	68·2	19·4	9·3	Perth	1860	PerthShipB. Co.		W. Armstrong	Wexford
	677	D V	Ellen Kirstine *Petersen* [79-79]	W Bn	Danish	167	106·7	22·1	10·9	Thurø	1879	L. Kaas		L. Kaas	Svendborg
	678		Ellen Lines *Nielsen* (ex Fiery Cross)	W Bk	Norweg'n	695	186·6	31·7	19·0	Liverpool	1861	Chaloner	[illegible]	[illegible] "Ellen Lines" (Christensen & Co.)	Fredrikstad
[illegible]	679		Ellen M. Baxter *Gray* 79-82	W Sr	American	206	121·6	25·2	9·7	Wellington Sq. [Ont.	18[81/82]			W. B. Floyd	Boston, Mass.
[illegible]	680	R	Ellen M. Golder *Johnstone* 87-87	W 3m Sr	American	619	1[illegible]2·5	35·0	17·1	Bath, Me.	1874	Adams & Hitch'k		S. R. Crowell	Boston, Mass.
[illegible]	681	R	Ellen M. Mitchell *[illegible]* 88	W Bn	American	376	142·0	33·0	12·9	Machias, Me.	1875	H. Seavey		A. Eaton	Machias, Me.
[illegible]	682		Ellen Maria *Clay* 82-89	W Bn	American	192	110·0	27·0	9·5	Brewer, Me.	1867			H. Lord	Bangor, Me.
	683		Ellen Maria *[illegible]* (ex Zaotpa)	W Bn	Swedish	144	92·2	21·9	11·4	Bandholm	1856	H. J. Möller		V. Elfversson	Calmar
[illegible]	684		Ellen Masters	W Bk	British	498	137·0	26·5	16·0	N. Brunswick	1855			Aga Mah. bin Abdool Russoul	Bombay
[illegible]	685		Ellen Morrison *Fernald*	W Sr	American	163	100·0	24·4	9·7	New Haven, [Conn.	1850			Swett & Co.	Bangor, Me.
[illegible]	686		Ellen Perkins *Tenball*	W Sr	American	163	100·8	25·7	7·3	Bangor, Me.	1864			H. C. Kendall	Bangor, Me.
	687	V	Ellen Rickmers *Nielsen* 87-87	W Bk	German	1369	203·4	37·4	25·2	Geestemünde	1884	R. C. Rickmers	[illegible]	R. C. Rickmers	Geestemünde
[illegible]	688	✻	Ellen Roberts *Roberts*	W Sr	British	87	75·0	20·0	10·6	Pwllheli	1868	Roberts		J. Roberts	Carnarvon
[illegible]	689		Ellen Smeed *Weaight*	W Sr	British	134	107·8	22·6	8·8	Sittingbourne	1872	Smeed		C. W. Marshall	Rochester
[illegible]	690		Ellen Sutton *Walsh* (ex [illegible])	W Bn	British	181	105·3	23·5	12·7	P. E. Island	1868	Duncan		G. Sutton, jun.	Cork
[illegible]	691	R	Ellen Tobin *Hopkins*	W 3m Sr	American	414	136·7	32·0	14·5	Kenneb'nk Me.	1874			C. Burnett	New York
[illegible]	692		Ellen Vair *Richards* 88	W Sr	British	107	92·2	22·0	11·4	Hayle	1866	Harvey		D. W. Bain & Co.	Hayle
[illegible]	693	✻	Ellenbank *Frew*	I S	British	1426	242·5	37·5	21·6	Maryport	1885	Ritson & Co.		McDiarmid, Greenshields & Co.	Liverpool
	694		Elleni	W Sr	Turkish	175	..	..	..	Scio	1874			Sakyziee Dimitri Rangani	Scio
	695		Elleni	W Bg	Turkish	175	..	..	..	England	1876			Vassil Ikoslee Vassil Pappa Yorghi	Vassilikos
	696		Elleni	W Bg	Turkish	150	..	..	..	Greece	1880			Orkanleeki Tijar	Constantinople
	697		Elleni	W Bg	Turkish	140	..	..	..	Greece	1873			LimnileeNicolaYanni	Lemnos
[illegible]	698	V	Ellerslie *Marsh*	W S	British	1346	199·4	38·8	24·0	N. Brunswick	1872	Nevins	[illegible]	A. Gibson & Co.	Liverpool
[illegible]	699	✻	Ellesmere *Summers* 86	I S	British	2645	308·0	45·3	24·7	Southampton	1886	Oswald, Mordaunt & Co.		Sailing Ship "Ellesmere" Co. Ld. ([illegible])	London
[illegible]	700		Ellias *Allapitchy*	W Bg	British	176	87·5	19·1	11·1	[illegible]	1878			Moona [illegible] Vavana Nalla Abubaker Maricar	Colombo

✻ = Lloyd's Register; D = Nederlandsche Vereeniging; G = Germanischer Lloyd; I = Registro Italiano; R = American Record; A = Veritas Austro-Ungarico; H = Veritas Ellenico; N = Norske Veritas; V = Bureau Veritas.

Dans la 4me Colonne, C = Composite (membrure en fer, bordé en bois); I = Fer; Stl = Acier; W = Bois.

Les Dates du doublage en métal (**Col. 11**) ressortent des livres dans lesquels les navires sont respectivement classés ou des rapports des Inspecteurs du Lloyd's Register.

Pour plus amples renseignements sur les navires classés par le Lloyd's Register, voir le Lloyd's Register Book et le Lloyd's Yacht Register.

Official No.	No. in Book	Registry in which classed	Ship's Name. Material and Rig. Master, and Dates of Appointment to present Owner's Service and to Ship. Ship's Late Name.	Flag	Net Tonnage	Length	Breadth	Depth	Where	When	By whom	Date of Metalling	Owners or Managers
	701		**Ellida** WBg		256	99·0	25·2	14·6					
	702		**Ellida** *Pedersen 87-87* WSr	Danish	109	72·1	19·8	11·6	Nykjøbing	18 42/61	L.Dresler		R.Petersen
	703		**Ellida** *Krog* W3mSr	Norweg'n	200	100·0	22·0	12·2	[illegible]	1879			
	704		**Ellida** *Teir* W3mSr	Russian	293	111·6	29·2	14·1	[illegible]	1874	M.Hannus		A.Teir & Co.
	705		**Ellida** *Lundberg -87* W3mSr	Russian	287	..	..	..		1871			S.G.Jansson
	706		**Ellida** WBg	Russian	211	..	..	..		1875			K.K.Kynberg
	707		**Ellida** (CONDEMNED) *Anderson* WBk	Swedish	602	141·4	31·7	18·4	Brahestad	1866			J.Ingmansson
	708	V	**Ellida** *Gren 56-76* WSr	Swedish	199	101·3	23·4	12·3	Eckerna	1876			J.H.Gren
	709		**Ellida** *Lycke -86* WSr	Swedish	189	92·2	23·2	12·8	Bastholmen	1872	A.Ahlberg		J.Johansson's Estate
	710		**Ellida** *Johansson* WSr	Swedish	140	85·0	20·7	11·8	Dramsvik	1868	Dramsviks Varf		J.A.Samuelsson
74074	711		**Ellie** *Mundy -87* WSr	British	117	84·0	24·0	9·7	W.Dublin,NS.	1876			D.Cronan
95554	712	R	**EllieCarter** *Morrissey -88* WBn	British	178	92·0	24·6	10·9	Humbr.Sound [N.F.L.	1886	W.Mckay		J.Butler
67904	713	✻	**Elliot** *Mackay 73-77* IBk	British	1065	219·0	35·1	21·0	Aberdeen	1877	A.Hall&Co.		J.Muir
8080	714		**Elliott L.Dow** *Lamphere* W3mSr	American	381	126·0	37·7	14·9	Noank,Conn.	1874	R.Palmer		H.L.Crandall
	715	N	**Ellisif** (ex Ghol Tidings) *Abrahamsen -78* WBk	Norweg'n	944	169·6	36·3	22·0	New York	1856	Rosefelt&Joyce	Y 10.88	S.Rumi
	716	NV	**Ellisif** *Nielsen 86-86* WBk	Norweg'n	419	138·5	30·5	16·7	Drammen	1868	Jörgensen & Knudsen	Y 2.85	Ths.S.Falck
87984	717	✻	**Ellisland** *Hind 87-87* I4mS	British	2355	302·3	41·4	25·0	Southampton	1884	Oswald, Mordaunt &Co.		J.Houston
27200	718	✻	**Ellora** (ex Steamer, ex Cadiz) *Beale* IS	British	1727	261·2	36·2	25·8	Liverpool	1855	J.Laird		John Blyth & Co.
135597	719	R	**Ellwood Harlow** *King* W3mSr	American	794	168·6	36·0	18·6	Alex'ndria,Va	1882	W.H.Crawford		A.J.Newbury
30201	720		**Elm** WBk	British	158	93·0	23·5	10·2	United States				T.Chown
135430	721		**Elm City** *Etheridge* W3mSr	American	597	152·2	35·5	15·9	Bath,Me.	1880	Goss,Sawyer& Packard		F.M.Etheridge
58780	722		**Elma** *Gregory 85-85* WBn	British	136	93·6	24·0	11·3	Quebec	1871	Commeaux		R.Kearon
	723		**Elma** *Tönnesen -88* WBk	Norweg'n	771	163·9	34·0	20·3	Quebec	1876	Clairland	[illegible]	J.C.Tönnesen
	724		**Elma** *Eriksson -82* WSr	Swedish	171	92·8	22·8	11·7	[illegible]	1882	L.Wåhnlund		P.Andersson
	725	R	**Elma** *Backman* W3mSr	Swedish	159	94·5	21·1	10·8	Söderhamn	1867			C.Backman
	726		**Elma** (ex Gyda, ex Karolina & Sara) WSr	Swedish	110	80·4	21·4	9·3	Brevig	1852			L.Larsson
87393	727	✻D	**Elmhurst** *McKenzie* IS	British	1712	264·9	39·8	23·4	PortGlasgow	1883	R.Duncan & Co.		C.S.Caird
68572	728		**Elmina** WSrYt	British	179	128·6	25·0	13·0	Gosport	1874	Camper & Nicholson	[illegible]	Sir R.F.Sutton,Bart.
	729		**Elmina** (CONDEMNED) *Jaski 65-80* CBg	Dutch	290	123·6	27·3	12·7	Amsterdam	1865	W.& A.H. Meursing	[illegible]	F.C.Jaski
135050	730	R	**Elmiranda** *Moberly* WBk	American	623	150·0	30·0	18·2	Stockton,Me.	1874	Staples	[illegible]	N.B.Mansfield
225	731	✻	**Elmstone** *Cossovich -88* CS	Austrian	698	166·6	31·5	18·6	Sunderland	1866	G.S.Moore&Co.	[illegible]	T.Cossovich
	732	V	**Elna** (ex Elis) *Madsen -88* IBk	Danish	464	145·2	26·7	17·1	Hamburg	1875	Reiherstieg Ship Works		Et Actieselskab (H.C.Svarter)
	733		**Elna** *Tork -87* W3mSr	Russian	206	..	..	..	Sakkijärvi	1876			J.W.Söderlund
8832	734		**Elnorah** *Yunggren* W3mSr	American	138	106·5	31·1	7·8	Newport,Oreg.	18 71/72			J.A.Hooper
441	735	I	**Eloisa** *Rallo -87* WBg	Italian	169	95·2	24·9	14·9	Alimuri	1848		Y 5.86	S.Malato
	736		**Eloisa** WBn	Spanish	125	84·0	20·1	10·9	Bilbao	1852			R.Momenez
	737	N	**Elphinstone** (ex H. B. Maidussy) *Olsen 74-82* WS	Norweg'n	902	167·6	34·1	21·2	Saco,Me.	1856		Y 2.84	C.Dahl
	738		**Elpida** WBg	Turkish	150	..	..	..	[illegible]	1878			Aktarpaleelze Yorgki Bellasa
21	739	A	**Elpis** (ex Gipsia, or Pa[illegible]) *Sanoudis* WBk	Greek	437	136·1	29·3	18·0	[illegible]	1868	W.Pickersgill		E.Sanoudis
	740		**Elpis** (ex Encasis) *Mauzuris* WBk	Greek	433	..	..	..	Syra	1872	C.Cristali	[illegible]	F.Pieros
58	741	H	**Elpis** WBg	Greek	295	..	..	..	Galaxidi	1877	C.Papapetros		E.Moscholios
423	742	A	**Elpis** *Coulapoutis* WBn	Greek	245	97·1	26·7	..	Syra	1884	A.Livadara		M.Coulapoutis
2	743		**Elpis** *Byzautios* WBn	Greek	218	90·2	25·6	15·2	Syra	1868			A.Byzautios
160	744		**Elpis** WBg	Greek	206	92·8	24·6	..	Syra	1868			G.Cosmos
	745	H	**Elpis** *Broulidakis* WSr	Greek	149	..	..	..	Syra	1880		[illegible]	M.Bourakis
70	746	H	**Elpis** WBg	Greek	148	88·3	23·3	..	Cassos	1877			Paul Nicalatos
411	747	A	**Elpis** *Lemos* WBn	Greek	131	77·8	23·3	..	Syra	1883	P.Bum		C.Loros
31	748		**Elpis** WSr	Greek	119	71·8	22·1	..	Skiathos	1881	C.G.Mytilinaus		N.D.Diniscos
38	749	A	**Elpis** *Sirasturi -88* WSr	Greek	102	74·5	21·0	..	Syra	1881	C.Cristali		T.G.Syrastouris
	750		**Elpis** *Lindholm -87* WBg	Russian	244	..	..	..	Korpo	1868			K.Ehrman

✻ = **Lloyd's Register**; D = Nederlandsche Vereeniging; G = Germanischer Lloyd; I = Registro Italiano; R = [illegible]
A = Veritas Austro-Ungarico; H = Veritas Ellenico; N = Norske Veritas; V = [illegible]

In the 4th Column, C = Composite (iron frame, wood planked); I = Iron; Stl = Steel; W = Wood.

The Dates of Metalling (Col. 11) are based on the records of the Books in which Vessels are respectively classed, or on the Reports of the Surveyors [illegible]

Further particulars of Vessels classed by Lloyd's Register will be found in Lloyd's Register Book and Lloyd's Yacht Register

Numéro officiel	N° dans le livre	Registre dans lequel le navire est coté	Nom, Ancien Nom	Capitaine	Matériaux et Gréement	Pavillon	Tonnage	Longueur	Largeur	Creux	Construit Où	Quand	Par qui	Date du doublage	Armateurs ou Directeurs-Gérants	Port d'Armement
	751	DV	Elsa (ex ...)	Schull	? Bk	Dutch	851	177·1	34·1	19·1	Kinderdijk	1857	Kop.Smit		Aug. Köpke	Rotterdam
	752	V	Elsa (ex ...)	Pfeiffer -86	? Bk	German	474	155·5	27·8	16·9	Glasgow	1869	A.Stephen&Son		C.M.Matzen	Hamburg
	753		Elsa (ex ...)		W Sr	German	108	82·1	20·3	9·7	Oldenburg	1857	H.C.Brand		C.Mentzen	Brake
	754		Elsa	Lindblom -87	W 3mSr	Russian	249	..	..	..		1867			M.H.Mårtensson	Korpo
	755	DV	Elsbeth (ex Eliza E. Milligan)		W Bk	Norweg'n	885	168·5	31·9	19·7	Rothesay N.B.	1874		V 2.88	A.Monsen	Tönsberg
	756	V	Else (ex Ida & Renny)	Brokers 78-84	W S	German	1375	205·7	38·1	24·0	Searsport, Me	1875	W.McGilvery	V 3.86	Joh.Fr.Arens	Bremen
	757	G	Else (ex Lady ...)		W Bg	German	287	110·1	27·2	16·0	Sunderland	1868	C.Wheatly	V 8.85	Aug.Piehl	Danzig
	758	N	Else Kathrine	Olsen	W Sr	Norweg'n	131	94·8	21·6	10·6	Tönsberg	1871			M.Larsen Röd	Sandefjord
	759		Elsie	Murphy	W Bn	Columbia	112	82·0	23·0	10·1	Lunenburg NS	18 61/62		V 8.85	A.Wolff&Co.	Baranquilla
	760	V	Elsie (ex ...)	Christiansen	W S	Norweg'n	1271	191·0	39·0	23·5	Quebec	1866	Baldwin	V 8.87	Actieselskabet "Elsie"	Christiania
	761	R	Elsie A. Bayles	Benson	W 3mSr	American	281	108·0	30·0	13·6	Port Jefferson, N.Y.	1883	J.M.Bayles&Son	V 11.84	C.A.Bayles	New York
	762		Elsie Fay	Renner -87	W 3mSr	American	163	107·0	26·2	9·0	Essex, Mass.	1884	J.Story		C.R.Keefe	Boston, Mass.
	763	R	Elsie Iversen	Iversen -87	W Sr	American	117	86·5	27·2	7·5	Benicia, Cal.	1887	M.Turner		N.Iversen	San Francisco
	764		Elsie M. Smith	Jewett	W Sr	American	107	93·5	24·0	9·1	Kennebunk Me	1882	Christianson		A.M.Smith	Portland, Me.
	765	V	Elsinore	Simonson	W Bk	American	658	117·6	30·5	17·4	Newburyport, Mass.	1866		V 4.85	A.M.Simpson	San Francisco
	766		Elter Water	Kneh 85-88	W Sr	British	100	80·0	20·3	12·3	Maryport	1857			R.Kearon	Dublin
	767		Eltie	Comeau	W Sr	British	124	92·9	26·6	7·6	Moss Glen, N.B.	1887	G.Merritt		W.H.Merritt	St.John, N.B.
	768	R	Elva E. Pettengill	Dodge	W 3mSr	American	319	118·2	29·0	9·8	Deering, Me.	1874	G.Russell		J.S.Winslow & Co.	Portland, Me.
	769		Elvenia	Gee	W Sr	American	141	98·0	29·7	8·2	Humboldt, Cal.	1872			F.Gee	San Francisco
	770	N	Elverhoi	Nielsen -88	W Bk	Norweg'n	434	133·3	29·7	17·6	Königsberg	1868	L.Eggert	V 6.88	L.Gundersen	Porsgrund
	771		Elverton		W Sr	American	272	109·6	25·7	10·3	Roxbury, Mass	1854			A.M.Jewett	Portland, Me.
	772	R	Elvina	Durfee -88	W Bkn	American	336	127·2	30·7	12·4	Bath, Me.	1874		V 11.88	B.F.Metcalf & Co.	New York
	773		Elvina	Cornelius -79	W Sr	Swedish	153	90·5	21·1	11·1	Lübeck				S.R.Söderbom	Strömstad
	774	*	Elvira	Watson 79-82	I Bk	British	464	138·7	28·1	14·7	Aberdeen	1882	A.Hall&Co.		H.F.Watt	Liverpool
	775		Elvira	Watts	W Sr	American	113	89·0	23·9	8·1	E.Machias	18 50/68			G.E.Watts	Machias, Me.
	776		Elvira		W Sr	Brazilian	297	..	..	..					A.S.Pereira	Itapemirim
	777		Elvira		W Bg	Brazilian	186	..	..	..					J.E.Ribeiro	Pernambuco
	778		Elvira		W Bk	Brazilian	183	..	..	..					J.A.C.Vianna	Rio Janeiro
	779	V	Elvira (ex Sovereign of the Seas)	Evers 81-84	W Bk	German	1502	200·1	50·8	24·0	Boston	1868	D.McKay&Co.	V 10.87	J.D.Bischoff	Bremen
	780	A	Elvira	Vianello	W Bg	Italian	172	92·5	24·1	13·8	Venice	1859	G.Grasso	V 8.82	Fdo Vianello	Venice
	781	V	Elvira	Sus	W Pol	Spanish	179	88·9	27·0	13·0	Arens	1854		V 12.83	Juan Sus	Barcelona
	782	V	Elvira	Curell	W Bn	Spanish	129	85·3	25·3	12·2	Blanes	1850		V 9.85	J.Curell Canals	Barcelona
	783		Elvira	Sjögren -86	W Bg	Swedish	231	110·8	21·8	12·1	Söderhamn	1870			A.P.Sjögren	Väddö, Rslgn.
	784		Elvira (ex ...)	Wetterberg 87-87	W Bg	Swedish	220	106·6	24·0	12·2	Ny Karleby	1857			N.Jonsson	Brannby, Malmöhln.
	785		Elvira A.	Saglietto	W Bn	Italian	179	99·1	24·9	12·1	Savona	1874	G.B.Ciarlo	V 8.86	G.B.Ameglio & figli	Porto Maurizio
	786	IV	Elvira Ardisson	Airaldi	W Bn	Italian	250	121·3	26·6	12·0	Viareggio	1886	A.Raffaelli	V 8.86	G.Ardisson	Leghorn
	787	*D	Elvira Camino (ex Doña Juana)	Simmonds 87-87	C Bk	British	294	126·5	26·1	13·8	Glasgow	1872	C.Connell&Co.	V 1.87	R.& W.King	Bristol
	788		Elvira D. (ex Pilade)	Durazzano	W Bn	Italian	117	84·3	22·9	11·5	Castell.di Stab.	1858	A.Bonifacio		N.Durazzano	Gaeta
	789		Elvira Owen		W S	American	981	167·6	34·0	23·2	Bath, Me.	1852			A.B.Helmboe	New York
	790	RV	Elwell	Barston -83	W S	American	1461	212·3	39·1	24·0	Damariscotta, Me.	1875	Norris&Co.	V 8.86	D.W.Chapman	Damariscotta, Me.
	791		Elwine & Friedericke		W Sr	German	143	78·1	22·3	12·2	Ueckermünde	1841	L.Wittenberg		Carl Wichards	Stettin
	792	R	Elwood Burton	Hutchins -88	W 3mSr	American	375	128·0	31·7	11·7	Bath, Me.	1872	Hotchkiss & Soule		L.Jarvis	Philadelphia
	793		Elwood Doran		W Sr	American	224	117·0	29·7	9·6	Milford, Del.	1862			J.W.Knight	Philadelphia
	794	*	Elwy	Owen 87-88	I Bk	British	769	185·8	32·0	20·0	Liverpool	1876	T.Royden&Son		D.Jones	Liverpool
	795	RV	Elze	Hellgren -84	W S	Swedish	357	122·7	25·4	14·5	Ueckermünde	1868	C.Wittenberg	V 3.86	S.Hörberg	Hellevik
	796		Emana	Lindqvist -87	W 3mSr	Russian	325	..	..	..	Lumparland	1866			J.Svibergson	Lemland
	797	V	Emanuel	Clausen	W Bn	Danish	140	88·4	20·6	11·4	Nordby, Fanø	1866	J.Th.Hansen		Actieselskab. (C.J. Clausen)	Nordby, Fanø
	798	V	Emanuel	Petersen 78-80	W S	Danish	134	97·4	21·6	10·1	Svendborg	1880	N.P.Pedersen		R.P.Pedersen	Svendborg
	799		Emanuel		W S	Norweg'n	595	168·4	34·2	18·0	Tvedestrand	1861	O.Jörgensen		O.Jörgensen	Risör
	800															

*=Lloyd's Register; D=Nederlandsche Vereeniging; G=Germanischer Lloyd; I=Registro Italiano; R=American Record; A=Veritas Austro-Ungarico; H=Veritas Ellenico; N=Norske Veritas; V=Bureau Veritas.

Dans la 4me Colonne, C=Composite (membrure en fer, bordé en bois); I=Fer; Stl=Acier; W=Bois.

Les Dates du doublage en métal (Col. 11) ressortent des livres dans lesquels les navires sont respectivement classés ou des rapports des Inspecteurs du Lloyd's Register.

Pour plus amples renseignements sur les navires classés par le Lloyd's Register, voir le Lloyd's Register Book et le Lloyd's Yacht Register.

Official No.	No. in Book	Registry in which classed	Ship's Name. Master, and Dates of Appointment to present Owner's Service and to Ship. Ship's Late Name.	Material and Rig	Flag.	Net Tonnage	Length	Breadth	Depth	Where built	Year	By whom	Date of Metalling	Owners or Managers
[illegible]	601	✻ D	Elizabeth Jane [illegible]	W Bg	British	143	100·4	22·6	11·4	Ipswich	1860	W. Read		J. Hutchings
53185	602		Elizabeth Kelly [illegible]	W Bk	British	116	83·9	21·2	12·3	Whitehaven	1866	Shepherd		John Kearns
[illegible]	603	✻	Elizabeth Kilner *Ashton*	W Sr	British	88	80·1	20·1	10·0	Thorne	1871	Staniland		G. Kilner & Co.
[illegible]	604		Elizabeth M. Cook *Clark*	W 3m Sr	American	268	104·8	32·4	10·9	Calais, Me.	1874	Robinson Bros.	Y [illegible]	J.G. Murchie & S[illegible]
[illegible]	605	✻	Elizabeth M'Lea *Bate 76-77*	W Bn	British	128	99·7	20·2	10·8	Greenock	1860	Steele	Y [illegible]	R. Cock
54086	606	✻	Elizabeth Mary (WRECKED) *Cot*	W Bk	British	250	129·0	26·4	13·0	Sunderland	1866	W. Watson	Y [illegible]	E.T. Miles
47220	607	✻	Elizabeth Nicholson *Falconer 77-83*	W Bk	British	904	132·5	32·5	22·2	[illegible]	1863	Nicholson	Y [illegible]	Montgomerie & Workman
[illegible]	608	✻	Elizabeth Ostle *Rennie*	W Bk	British	740	165·3	33·0	21·7	Maryport	1869	Ritson	Y [illegible]	Hayton & Simpson
[illegible]	609	✻	Elizabeth Peers *Peers 84-85*	I 3m Sr	British	165	107·1	23·6	10·4	Amlwch	1885	W. Thomas & Sns		W. Postlethwaite
27212	610		Elizabeth Ray *Moon 76-76*	W Bn	British	199	100·0	24·5	13·0	Sunderland	1859	Doxford		A.E. Flowers
67001	611		Elizabeth Roy *King*	W Bkn	British	223	116·1	25·1	12·1	Dysart	1873	Watt		J. Harvey
[illegible]	612		Elizabeth Ryan	W Sr	British	123	88·0	23·0	11·1	Gabarus, N.S.	1864			P. Bagnell
[illegible]	613		Elizabeth S. Lee *Lee*	W 3m Sr	American	467	143·4	34·5	9·0	May's Landing [N.J.	1875			J.S. Lee
[illegible]	614		Elizabeth Seldon	W Sr	British	100	81·8	20·6	7·7	Tweed River, [N.S.W.	1861			R. Seldon
[illegible]	615	✻	Elizabeth Stevens (WRECKED) *Strike*	W 3m Bkn	British	198	114·0	24·2	12·6	Ipswich	1871	Bayley	Y 7.87	H. & J. Strike
[illegible]	616		Elizabeth T. Cottingham *Smith*	W 3m Sr	American	252	120·4	31·8	9·4	English Crk, N.J.	1873			J. Smith
[illegible]	617		Elizabeth Thompson	W Sq	British	289	90·4	22·4	16·0	Sunderland	1842			Geo. Tully
[illegible]	618	R	Elizabeth Winslow *Oakes 84-85*	W Bn	American	372	126·3	30·1	11·5	Yarmouth, Me.	1873	Loring	Y 8.87	J.S. Winslow & Co.
62771	619	✻	Elizabeths *Roberts 87-87*	W Bk	British	373	132·3	28·1	16·8	Sunderland	1870	B. Hodgson	Y 10.87	R. Thomas & Co.
[illegible]	620		Elizabeths *Peterson -88*	W Bk	British	329	109·0	27·0	17·0	Newcastle	1857	Gaddy & Lamb		G. Walker
[illegible]	621		Ella	W Bk	British	985	178·0	36·0	21·2	Quebec	1873	P.V. Valin	Y 5.81	W. Lund
[illegible]	622		Ella *Thomas* (ex Humphrey Purinton)	W Bk	British	672	137·8	30·6	21·7	Maine, U.S.	1857			T.S. Blues
[illegible]	623	✻	Ella *Jones 76-87*	W Bk	British	382	122·3	28·2	17·4	Sunderland	1869	T. Metcalf	Y 9.88	J. Davies
[illegible]	624		Ella *Hewitt*	W Bn	British	194	104·1	24·4	12·6	P.E. Island	1864	Owen		J.W. Hawkes
[illegible]	625		Ella	W Bn	British	168	97·0	25·0	11·2	Riv. Hebert N.S.	1863			J.B. Read
[illegible]	626	R	Ella *Lewis 61-87*	W Bk	American	622	149·5	32·1	17·2	Westbrook, Me.	1869	R. Lewis & Co.	Y [illegible]	R. Lewis & Co.
[illegible]	627		Ella *Hansen -88*	W Bkn	American	249	131·8	28·7	10·8	Freep'rt, Wash.	1874	W. Begaut	Y 5.88	Welsh & Co.
[illegible]	628		Ella *Coombes 81-*	W Sr	American	184	105·6	28·8	8·6	Petty's Isl'd N.J.	18 53/70			T.J. Stewart & Co.
[illegible]	629		Ella *McVane*	W Sr	American	160	104·0	28·0	8·0	Newcastle, Me.	1864			E. Haggett
[illegible]	630		Ella *Guptil*	W Sr	American	106	85·3	27·7	7·5	Harringt'n, Me.	1872			R. Mitchell
	631		Ella *Folkmann 80-88* (ex Prince Alfred)	W Sr	Danish	136	91·7	21·4	11·3	Perth	1867	Wallace		P. Berg
	632		Ella *Nikholm -88* (ex Scotia Queen)	W Bk	Swedish	420	125·0	26·7	12·6	Maitland	1867	Nova Scotia		Anna E. Wancke
[illegible]	633	R	Ella A. Warner *Hallbrook*	W Sr	American	141	100·0	25·0	9·0	Essex, Mass.	1879	J. Story		E.L. Atwood
[illegible]	634		Ella Brown *Keene*	W Sr	American	156	101·5	26·3	8·8	E. Machias, Me.	1869			O.B. Hall
[illegible]	635		Ella Clifton *Mansfield*	W Sr	American	103	89·0	23·9	7·3	St. John, N.B.	1872	Sutton		J.B. Mansfield
7180	636		Ella F. Crowell *Freeman*	W Sr	American	157	93·8	26·1	8·0	E. Haven, Conn.	1864			A.C. Freeman
[illegible]	637	R	Ella Francis *Hall*	W Sr	American	119	93·0	26·0	8·0	Mt Desert, Me.	1870	W.S. Newman		G.L. Farrand
[illegible]	638		Ella H. Barnes *Hardcastle*	W Sr	American	180	101·5	28·5	9·6	Madison, Conn.	1865	G.M. Miner & Son		E.K. Johnson
[illegible]	639	R	Ella M. Hawes *Purington*	W 3m Sr	American	282	128·0	32·8	9·0	Bath, Me.	1881	B.W. & H.F. Morse		B.W. & H.F. Mo[illegible]
[illegible]	640	R	Ella M. Storer *Wade -88*	W 3m Sr	American	427	135·7	32·0	14·9	Waldoboro, Me.	1873	H. Kennedy & Co.	Y 11.83	J. Baker & Co.
[illegible]	641	R	Ella M. Watts *Stevens 84-*	W 3m Sr	American	443	137·8	31·2	16·0	Thomaston, Me.	1873	[illegible], Gerry & Co.	Y 10.8[illegible]	S.S. Gerry & Co.
[illegible]	642		Ella Mae (BROKEN UP) *Chadwick*	W Bn	British	145	94·4	24·8	11·3	P.E. Island	1868	McLure		C. Neill
[illegible]	643		Ella Maud Wood *Dixon*	W Sr	British	160	93·4	27·7	10·0	Joggin Mines, [N.S.	1885	A.F. Melanson		Fred W. Sumner
[illegible]	644	V	Ella Moore *Byer*	W Bk	British	391	136·0	30·1	14·6	Cornwallis, N.S.	1867	J. Buckman		Henry A. Eaton
71067	645		Ella Murton	W Sr	British	229	118·3	26·2	10·1	Mill Point	1875	W. Jamieson		J.W. Murton
	646	✻	Ella Nicolai *Meisterfeld 83-87* (ex Dalton)	I Bk	German	572	174·4	27·7	17·9	Liverpool	1868	Bowdler Chaffer & Co.		J.H. Nicolai
[illegible]	647		Ella Powell *Bishop*	W Sr	American	134	99·4	22·5	7·8	Bridgeton, N.J.	1872			Mrs. Jane Bishop
[illegible]	648		Ella Pressey *Pressey*	W Sr	American	157	98·8	27·3	8·7	Thomaston, Me.	1872			A. Levensaler
[illegible]	649		Ella Reed	W Bg	British	240	105·0	25·5	11·5	Maine, U.S.	1853			E. Bayley
[illegible]	650		Ella T. Little *Frambes -87*	W 3m S	American	236	114·3	30·2	9·7	Chester, Pa.	1874			T.G. Crawford

✻ = Lloyd's Register; D = Nederlandsche Vereeniging; G = Germanischer Lloyd; I = Registro Italiano; R = [illegible]
A = Veritas Austro-Ungarico; H = Veritas Ellenico; N = Norske Veritas; V = [illegible]

In the 4th Column, C = Composite (iron frame, wood planked); I = Iron; Stl = Steel; W = Wood.

The Dates of Metalling (Col. 11) are based on the records of the Books in which Vessels are respectively classed, or on the Reports of the Surveyors [illegible]

Further particulars of Vessels classed by Lloyd's Register will be found in Lloyd's Register Book and Lloyd's Yacht [illegible]

1	2 N° dans le livre	3 Registry	4 Nom. Matériaux et Gréement. Capitaine et année d'entrée au service des Armateurs actuels et à bord du navire. Ancien Nom.	5 Pavillon.	6 Tonnage net.	7 Dimensions. Longueur.	Largeur.	Creux.	8 Construit. Où.	9 Quand.	10 Par qui.	11	12 Armateurs ou Directeurs Gérants.	13 Port d'Armement.
[illegible]	651	V	Ella Vose *Davis 75-* W Bk (ex [illegible])	British	884	173·0	34·5	21·1	Halifax, N.S.	1875	D. McPherson	[illegible]	T. J. Troop	Windsor, N.S.
[illegible]	652		Ellanora van Deusen *Godfrey* W 3m Sr	American	294	126·0	32·2	11·4	Leesburg, N.J.	1874			T. W. Godfrey	Camden, N.J.
[illegible]	653		Ellen *Stephens* W Bk	British	499	135·1	30·2	17·7	St. John, N.B.	1865	J. & S. W. Olive		J. Henderson	Newcastle, N.S.W.
[illegible]	654	✱	Ellen *Snelling* W 3m Sr	British	245	110·8	24·9	12·4	Liverpool	1867	Barland	[illegible]	T. J. Snelling	Chester
[illegible]	655		Ellen W Bn	British	136	91·1	24·9	11·4	Londonderry, N.S.	1864	Grath		E. Murray	Dublin
[illegible]	656	✱	Ellen *Owens* W Sr	British	129	93·2	22·4	11·2	Conway	1858			J. Owens	Carnarvon
[illegible]	657		Ellen W	British	160	76·8	26·4	9·6	Sunderland	1860			C. W. Marshall	Maldon
[illegible]	658	✱	Ellen *Johnston* W K	British	78	75·6	20·3	10·6	Brixham	1884	S. Dewdney & Sns		W. Acaster	Grimsby
[illegible]	659	G	Ellen *Hang 66-66* W Bn	Danish	126	89·7	20·3	10·5	Odense	1866	H. L. Hansen		H. T. Helweg	Odense
	660	N	Ellen *Taarvig 85-85* W Bg	Norweg'n	324	105·5	29·1	16·3	Grimstad	1854			C. Taarvig	Fredrikstad
	661	V	Ellen *Jerrelius* W Bk (ex Shooting Star, ex [illegible])	Russian	745	157·6	35·9	18·9	Williamsbg L.I.	1850		[illegible]	A. Carlström	Nikolaistad
	662	G	Ellen *Backe 86* W Sr	Swedish	112	84·0	19·8	9·9	Nordby	1866	T. Schiönau		N. N. Backe	Råå
[illegible]	663	V	Ellen A. Read *Hatfield* W S	British	1750	230·5	41·6	24·0	Tusket, N.S.	1884	Hatfield & Bros.	[illegible]	William Law	Yarmouth, N.S.
[illegible]	664		Ellen A. Swift *Brown* W Sr	American	126	88·0	24·0	9·0	Essex, Mass.	1882	J. James		S. S. Swift	Provincet'n, Mass.
[illegible]	665	✱	Ellen Anne *Jones 79-76* W Sr	British	95	75·7	20·7	10·1	Quay Bach	1866	Daniel		T. Jones	Aberystwith
[illegible]	666		Ellen Ashcroft *H. McMurtry* W Bg	British	176	102·8	23·8	13·0	Whitehaven	1866	Williamson		H. McMurtry	Belfast
[illegible]	667	✱	Ellen Beatrice *Jones* W Sr	British	88	73·5	20·0	10·3	Aberystwith	1865	Evans		W. Owens	Aberystwith
[illegible]	668	R	Ellen Catherine *[illegible]* W Bn	British	129	81·0	21·8	12·1	Newquay	1850			E. Timothy	Aberystwith
[illegible]	669	R	Ellen Crusoe *Tapley 8-88* W 3m Sr	American	283	125·8	26·1	13·3	E. Deering, Me.	1883	R. Lewis & Co.	[illegible]	R. Lewis & Co.	Portland, Me.
	670	N	Ellen Grant *Andersen -87* W Bk	Norweg'n	553	145·6	31·9	17·6	St. John, N.B.	1857		[illegible]	P. H. Sörensen	Christiania
[illegible]	671	✱	Ellen Greaves *[illegible]* W Bg	British	254	114·0	25·0	15·0	Portmadoc	1876	Jones	[illegible]	M. Jones & Co.	Carnarvon
[illegible]	672		Ellen H. *Otten* W Bn	British	179	96·0	26·2	11·0	St. John, N.B.	1867			R. W. Saul	Lowestoft
[illegible]	673		Ellen Holgate *Betts 85-* W Sr	American	160	105·4	28·7	7·3	Philadelphia	1866			J. A. J. Sheets	Philadelphia
[illegible]	674	✱	Ellen Holt *Camber 86-86* W 3m Sr	British	310	135·3	27·5	12·9	Sunderland	1875	W. Richardson	[illegible]	J. Sutcliffe & Son	Grimsby
	675	✱	Ellen Isabelle *Arnaud 63-83* W Bk	French	663	151·9	32·0	19·5	Bordeaux	1875	Chaigneau frères	[illegible]	A. Salle & Co.	Bordeaux
[illegible]	676	✱	Ellen Kerr *Walsh* W Sr	British	65	68·2	19·4	9·3	Perth	1860	Perth Ship B. Co.		W. Armstrong	Wexford
	677	D V	Ellen Kirstine *Pedersen 79-79* W Bn	Danish	167	106·7	22·1	10·9	Thurø	1879	L. Kaas		L. Kaas	Svendborg
	678		Ellen Lines *Nielsen -87* W Bk (ex Forty-Nine)	Norweg'n	695	186·6	31·7	19·0	Liverpool	1861	Chaloner	[illegible]	Actieselskabt. "Ellen Lines" (Christensen & Co.)	Fredrikstad
[illegible]	679		Ellen M. Baxter *Gray 79-82* W Sr	American	206	121·6	25·2	9·7	Wellington Sq. [Ont.	18[illegible]			W. B. Floyd	Boston, Mass.
[illegible]	680	R	Ellen M. Golder *Johnthan 87-87* W 3m Sr	American	619	152·5	35·0	17·1	Bath, Me.	1874	Adams & Hitch'k		S. R. Crowell	Boston, Mass.
[illegible]	681	R	Ellen M. Mitchell *[illegible] -88* W Bn	American	376	142·0	33·0	12·9	Machias, Me.	1875	H. Seavey		A. Eaton	Machias, Me.
[illegible]	682		Ellen Maria *Clay 82-80* W Bn	American	192	110·0	27·0	9·5	Brewer, Me.	1867			H. Lord	Bangor, Me.
	683		Ellen Maria *Ideal -88* W Bn (ex Emanuel)	Swedish	144	92·2	21·9	11·4	Bandholm	1856	H. J. Müller		V. Elfversson	Calmar
[illegible]	684		Ellen Masters W Bk	British	498	137·0	26·5	16·0	N. Brunswick	1855			Aga Mah. bin Abdool Russool	Bombay
[illegible]	685		Ellen Morrison *Fernald* W Sr	American	163	100·0	24·4	9·7	New Haven, [Conn.	1850			Swett & Co.	Bangor, Me.
[illegible]	686		Ellen Perkins *Tendall* W Sr	American	163	100·8	25·7	7·3	Bangor, Me.	1864			H. C. Kendall	Bangor, Me.
	687	V	Ellen Rickmers *Noden 87-87* W Bk	German	1369	203·4	37·4	25·2	Geestemünde	1884	R. C. Rickmers	[illegible]	R. C. Rickmers	Geestemünde
[illegible]	688	✱	Ellen Roberts *Roberts* W Sr	British	87	75·0	20·0	10·6	Pwllheli	1868	Roberts		J. Roberts	Carnarvon
[illegible]	689		Ellen Smeed *Wraight* W Sr	British	134	107·8	22·6	8·8	Sittingbourne	1872	Smeed		C. W. Marshall	Rochester
[illegible]	690		Ellen Sutton *Walsh* W Bn (ex [illegible])	British	181	105·3	23·5	12·7	P. E. Island	1868	Duncan		G. Sutton, jun.	Cork
[illegible]	691	R	Ellen Tobin *Hawkins* W 3m Sr	American	414	136·7	32·0	11·5	Kenneb'nk Me.	1874			C. Burnett	New York
[illegible]	692		Ellen Vair *Richards -88* W Sr	British	107	92·2	22·0	11·4	Hayle	1866	Harvey		D. W. Bain & Co.	Hayle
[illegible]	693	✱	Ellenbank *Pres 88-88* I S	British	1426	242·5	37·5	21·6	Maryport	1885	Ritson & Co.		McDiarmid, Greenshields & Co.	Liverpool
	694		Elleni W Sr	Turkish	175	..	..	..	Scio	1874			[illegible] Dimitri Rappani	Scio
	695		Elleni W Bg	Turkish	175	..	..	..	England	1876			Vassilikoslee Vassil Papa Yorghi	Vassilikos
	696		Elleni W Bg	Turkish	150	..	..	..	Greece	1880			Orkanlecki Tijar	Constantinople
	697		Elleni W Bg	Turkish	140	..	..	..	Greece	1873			Limnilee Nicola Yanni	Lemnos
[illegible]	698	V	Ellerslie *Marsh* W S	British	1346	199·4	38·8	24·0	N. Brunswick	1872	Nevins	[illegible]	A. Gibson & Co.	Liverpool
[illegible]	699	✱	Ellesmere *Summers -86* I S	British	2645	308·0	45·3	24·7	Southampton	1886	Oswald, Mordaunt & Co.		[illegible]	London
[illegible]	700		Ellias *Allagaitchay* W Bg	British	176	87·5	19·1	11·1	[illegible]	1878			[illegible]	Colombo

✱ = Lloyd's Register; D = Nederlandsche Vereeniging; G = Germanischer Lloyd; I = Registro Italiano; R = American Record; A = Veritas Austro-Ungarico; H = Veritas Ellenico; N = Norske Veritas; V = Bureau Veritas.

Dans la 4me Colonne, C = Composite (membrure en fer, bordé en bois); I = Fer; Stl = Acier; W = Bois.

Les Dates du doublage en métal (Col. 11) ressortent des livres dans lesquels les navires sont respectivement classés ou des rapports des Inspecteurs du Lloyd's Register.

Pour plus amples renseignements sur les navires classés par le Lloyd's Register, voir le Lloyd's Register Book et le Lloyd's Yacht Register.

Official No.	No. in Book	Registry in which classed	Ship's Name. Material and Rig. Master, and Dates of Appointment to present Owner's Service and to Ship. Ship's Late Name.	Flag	Net Tonnage	Length	Breadth	Depth	Built Where	When	By whom	Date of Metalling	Owners or Mana[gers]
	701		**Ellida** WBg		256	99·0	25·2	11·6					
	702		**Ellida** *Pedersen 87-87* WSr	Danish	100	72·1	19·8	11·6	[illegible]	18$^{42}_{61}$	L.Dresler		R.Petersen
	703		**Ellida** *Krog* W3mSr	Norweg'n	200	100·0	22·0	12·2	[illegible]	1879			
	704		**Ellida** *Teir* W3mSr	Russian	293	111·6	29·2	14·1	[illegible]	1874	M.Hannus		A.Teir & Co.
	705		**Ellida** *Lundberg -87* W3mSr	Russian	287	..	..	..		1871			S.G.Jansson
	706		**Ellida** WBg	Russian	211	..	..	..		1875			K.K.Kynberg
	707		**Ellida** (CONDEMNED) *Anderson* WBk	Swedish	602	141·4	31·7	18·4	Brahestad	1866			J.Ingmansson
	708	V	**Ellida** *Gren 66-76* WSr	Swedish	199	101·3	23·4	12·3	Eckerna	1878			J.H.Gren
	709		**Ellida** *Lycke -86* WSr	Swedish	189	92·2	23·2	12·8	Bastholmen	1872	A.Ahlberg		J.Johansson'sEsta[te]
	710		**Ellida** *Johansson* WSr	Swedish	140	85·0	20·7	11·8	Dramsvik	1868	Dramsviks Varf		J.A.Samuelsson
74074	711		**Ellie** *Mundy -87* WSr	British	117	84·0	24·0	9·7	W.Dublin,NS.	1876			D.Cronan
93554	712	R	**EllieCarter** *Morrissey -88* WBn	British	178	92·0	24·6	10·9	Humbr.Sound (N.F.L.)	1886	W.Mckay		J.Butler
67604	713	✳	**Elliot** *Mackay 73-77* I Bk	British	1065	219·0	35·1	21·0	Aberdeen	1877	A.Hall&Co.		J.Muir
8900	714		**Elliott L.Dow** *Lamphere* W3mSr	American	381	126·0	37·7	14·9	Noank,Conn.	1874	R.Palmer		H.L.Crandall
	715	N	**Ellisif** (ex Glad Tidings) *Abrahamsen -78* WBk	Norweg'n	944	169·6	36·3	22·0	New York	1856	Rosefelt&Joyce	Y 10.88	S.Rumi
	716	NV	**Ellisif** *Nielsen 85-86* WBk	Norweg'n	419	138·5	30·5	16·7	Drammen	1868	Jörgensen & Knudsen	Y 12.85	Ths.S.Falck
87084	717	✳	**Ellisland** *Hind 81-87* I 4mS	British	2355	302·3	41·4	25·0	Southampton	1884	Oswald, Mord'nt &Co.		J.Houston
27200	718	✳	**Ellora** (ex Steamer, ex Cadiz) *Beale* I S	British	1727	261·2	36·2	25·8	Liverpool	1855	J.Laird		John Blyth & Co.
135597	719	R	**Ellwood Harlow** *King* W3mSr	American	794	168·6	36·0	18·6	Alex'ndria,Va	1882	W.H.Crawford		A.J.Newbury
30291	720		**Elm** WBk	British	158	93·0	23·5	10·2	United States				T.Chown
135489	721		**Elm City** *Etheridge* W3mSr	American	597	152·2	35·5	15·9	Bath,Me.	1880	Goss,Sawyer& Packard		F.M.Etheridge
38769	722		**Elma** *Gregory 85-85* WBn	British	136	93·6	21·0	11·3	Quebec	1871	Commeaux		R.Kearon
	723		**Elma** *Tönnesen -88* WBk	Norweg'n	771	163·9	34·0	20·3	Quebec	1876	Charland	[illegible]	J.C.Tönnesen
	724		**Elma** *Eriksson -82* WSr	Swedish	171	92·8	22·8	11·7	Väte	1882	L.Wännlund		P.Andersson
	725	R	**Elma** *Backman* W3mSr	Swedish	159	94·5	21·1	10·8	Söderhamn	1867			C.Backman
	726		**Elma** (ex Gyda, ex Karolina & Syra) WSr	Swedish	110	80·4	21·4	9·3	Brevig	1852			L.Larsson
87393	727	✳D	**Elmhurst** *McKenzie* I S	British	1712	261·9	39·8	23·4	PortGlasgow	1883	R.Duncan & Co.		C.S.Caird
68572	728		**Elmina** WSrYt	British	179	128·6	25·0	13·0	Gosport	1874	Camper & Nicholson	Y 5.88	Sir R.F.Sutton Bart
	729		**Elmina** (CONDEMNED) *Jaski 65-80* CBg	Dutch	290	123·6	27·3	12·7	Amsterdam	1865	W.&A.H.Meursing	Y 6.86	F.C.Jaski
135059	730	R	**Elmiranda** *Moberly* WBk	American	623	150·0	30·0	18·2	Stockton,Me.	1874	Staples	Y 8.88	N.B.Mansfield
225	731	✳	**Elmstone** *Cossovich -88* CS	Austrian	698	166·6	31·5	18·6	Sunderland	1866	G.S.Moore&Co.	[illegible]	T.Cossovich
	732	V	**Elna** (ex Ella) *Madsen -88* I Bk	Danish	464	115·2	26·7	17·1	Hamburg	1875	Reiherstieg Ship Works		Et Actieselskab (F. C. Svarter)
	733		**Elna** *Tork -87* W3mSr	Russian	206	..	..	..	Sakkijärvi	1876			J.W.Söderlund
8832	734		**Elnorah** *Yunggren* W3mSr	American	138	105·5	31·1	7·8	Newport,Oreg.	18$^{71}_{72}$			J.A.Hooper
441	735	I	**Eloisa** *Rallo -87* WBg	Italian	169	95·2	24·9	11·9	Alimuri	1848		Y 3.86	S.Malato
	736		**Eloisa** WBn	Spanish	125	84·0	20·1	10·9	Bilbao	1852			R.Momenez
	737	N	**Elphinstone** (ex H. B. Mildmay) *Olsen 74-82* WS	Norweg'n	902	167·6	34·1	21·2	Saco,Me.	1856		Y 2.86	C.Dahl
	738		**Elpida** WBg	Turkish	150	..	..	..	Russia	1873			Aktarpoleelee Yorghi Bakash
21	739	A	**Elpis** (ex Gitana, ex Passover) *Simoudis* WBk	Greek	437	136·1	29·3	18·0	Sunderland	1868	W.Pickersgill		E.Simoudis
	740		**Elpis** (ex Ennosis) *Manzuris* WBk	Greek	433	..	..	..	Syra	1872	C.Cristali	Z 3.84	F.Pieros
58	741	H	**Elpis** WBg	Greek	295	..	..	..	Galaxidi	1877	C.Papapetros		E.Moscholios
423	742	A	**Elpis** *Coulapoutis* WBn	Greek	245	97·1	26·7	..	Syra	1884	A.Livadara		M.Coulapoutis
[illegible]	743		**Elpis** *Byzantios* WBn	Greek	218	90·2	25·6	15·2	Syra	1868			A.Byzantios
168	744		**Elpis** WBg	Greek	206	92·8	24·6	..	Syra	1868			G.Cosmos
	745	H	**Elpis** *Brouilidakis* WSr	Greek	149	..	..	..	Syra	1880		Y 4.86	M.Bourakis
70	746	H	**Elpis** WBg	Greek	148	88·3	23·3	..	Cassos	1877			Paul Nicalatos
411	747	A	**Elpis** *Lemos* WBn	Greek	131	77·8	23·3	..	Syra	1883	P.Bami		C.Loros
31	748		**Elpis** WSr	Greek	119	71·8	22·1	..	[illegible]	1881	C.G.Mytilinaus		N.D.Diniscos
38	749	A	**Elpis** *Sirasturi -88* WSr	Greek	102	74·5	21·0	..	Syra	1881	C.Cristali		T.G.Syrastouris
	750		**Elpis** *Lindholm -87* WBg	Russian	244	..	..	..	[illegible]	1868			K.Ehrman

✳ = Lloyd's Register; D = Nederlandsche Vereeniging; G = Germanischer Lloyd; I = Registro Italiano; R = Ame[rican]
A = Veritas Austro-Ungarico; H = Veritas Ellenico; N = Norske Veritas; V = Bu[reau Veritas]

In the 4th Column, C=Composite (iron frame, wood planked); I=Iron; Stl=Steel; W=Wood.

The Dates of Metalling (Col. 11) are based on the records of the Books in which Vessels are respectively classed, or on the Reports of the Surveyors of the ...

Further particulars of Vessels classed by Lloyd's Register will be found in Lloyd's Register Book and Lloyd's Yacht Register

Ancien Numéro	Numéro du livre	Registre	Nom. Matériaux et Gréement. Capitaine et dates d'entrée au service des Armateurs actuels et à bord du navire. Ancien Nom.	Pavillon	Tonnage net	Longueur	Largeur	Creux	Construit Où	Quand	Par qui	Date du doublage	Armateurs ou Directeurs Gérants.	Port d'Armement.
	751	DV	Elsa *Schell* -88 I Bk (ex [illegible])	Dutch	8[illegible]	177·[illegible]	[illegible]	19·[illegible]	[illegible]	1857	Fop Smit		Aug. Köpeke	R[illegible]
	752	V	Elsa *Pfeiffe* 86-86 I Bk (ex Raney Nouvel)	German	47[illegible]	1[illegible]	27·8	1[illegible]	[illegible]	1869	A.Stephen&Son		C.M.Matzen	Hamburg
	753		Elsa [illegible] WSr	German	1[illegible]	82·1	[illegible]	9·7	[illegible]	1857	H.C.Brand		C.Mentzen	[illegible]
	754		Elsa *Lindblom* 80-81 W3mSr	Russian	24	..	..	..		1887			M.H.Martenson	[illegible]
	755	DV	Elsboth *Haavland* 85-88 W Bk (ex Frau E. Hildebr)	Norweg'n	8[illegible]	168·[illegible]	[illegible]	19·7	[illegible]	1874	[illegible]	[illegible]	A.Monsen	[illegible]
	756	V	Else *Bremer* 76-84 WS (ex [illegible])	German	147[illegible]	[illegible]	[illegible]	24·0	Searsport,Me	1875	W.McGilvery	[illegible]	Joh.Fr.Arens	Bremen
	757	G	Else [illegible] W Bg (ex Lady Aberdeen)	German	28[illegible]	110·1	27·2	16·[illegible]	[illegible]	1868	L.Wheatly	[illegible]	Aug.Piehl	Danzig
	758	N	Else Kathrine *Olsen* WSr	Norweg'n	13[illegible]	91·8	21·[illegible]	1[illegible]	[illegible]	1871			M.Larsen Rød	[illegible]
	759		Elsie *Murphy* W Bn	Columbia	11[illegible]	[illegible]	23·0	10·1	LunenburgNS	18[illegible]		[illegible]	A.W[illegible]&Co.	Baranquilla
	760	V	Elsie *Christiansen* -88 WS (ex [illegible])	Norweg'n	1271	191·0	39·0	23·5	Quebec	1866	[illegible]	[illegible]	Actieselskabet "Elsie"	Christiania
	761	R	Elsie A. Bayles *Benson* W3mSr	American	281	1[illegible]	30·0	13·6	PortJefferson, [N.Y.	1883	J.M.Bayles&Son	[illegible]	C.A.Bayles	New York
	762		Elsie Fay *Beener* -87 W3mSr	American	168	107·0	26·2	9·0	Essex,Mass.	1884	[illegible]		C.R.Keefe	Boston,Mass.
	763	R	ElsieIversen *Iversen* -87 WSr (WRECKED)	American	117	93·5	27·2	7·5	Benicia,Cal	1887	M.[illegible]		N.Iversen	SanFrancisco
	764		Elsie M. Smith *Jewell* WSr	American	107	93·5	24·0	9·1	Kennebunk Me	1882	Christiansen		A.M.Smith	Portland,Me.
	765	V	Elsinore *Simonton* W Bk	American	658	137·0	31·5	17·1	Newburyport, [Mass.	1866		[illegible]	A.M.Simpson	San Francisco
	766		Elter Water *Kinch* 85-85 WSr	British	190	89·0	20·3	12·3	Maryport	1857			R.Kearon	Dublin
	767		Eltie *Comeau* -88 WSr	British	124	92·9	26·6	7·6	MossGlen,NB.	1887	G.Merritt		W.H.Merritt	St.John,N.B.
	768	R	Elva E. Pettengill W3mSr (WRECKED) *Dodge*	American	319	118·2	29·0	9·8	Deering,Me.	1874	G.Russell		J.S.Winslow & Co.	Portland,Me.
	769		Elvenia *Gee* WSr	American	141	98·0	29·7	8·2	Humboldt,Cal.	1872			F.Gee	San Francisco
	770	N	Elverhoi *Nielsen* -88 W Bk	Norweg'n	434	123·3	29·7	17·6	Königsberg	1868	L.Eggert	[illegible]	L.Gundersen	Porsgrund
	771		Elverton WSr	American	272	109·6	25·7	10·5	Roxbury,Mass	1854			A.M.Jewett	Portland,Me.
	772	R	Elvina *Durfee* -85 W Bkn	American	336	127·2	30·7	12·4	Bath,Me.	1874		V 11,88	B.F.Metcalf & Co.	New York
	773		Elvina *Carlsson* -79 WSr	Swedish	153	96·5	21·1	11·1	Lübeck				S.R.Söderbom	Strömstad
	774	*	Elvira *Watson* 79-82 I Bk	British	464	168·7	28·1	14·7	Aberdeen	1882	A.Hall&Co.		H.F.Watt	Liverpool
7479	775		Elvira *Watts* WSr	American	113	89·0	25·3	8·4	E.Machias	18[illegible]			G.E.Watts	Machias,Me.
	776		Elvira WSr	Brazilian	297	..	..	..					A.S.Pereira	Itapemirim
	777		Elvira W Bg	Brazilian	186	..	..	..					J.E.Ribeiro	Pernambuco
	778		Elvira W Bk	Brazilian	183	..	..	..					J.A.C.Vianna	Rio Janeiro
	779	V	Elvira *Evers* 84-84 W Bk (ex Sovereign of the Seas)	German	1502	200·1	50·8	24·0	Boston	1868	D.McKay&Co.	V 10,87	J.D.Bischoff	Bremen
	780	A	Elvira *Vianello* W Bg	Italian	172	92·5	24·1	13·8	Venice	1859	G.Grasso	[illegible]	F[lli] Vianello	Venice
	781	V	Elvira *Sus* W Pol	Spanish	179	88·9	27·0	13·0	Arens	1854		V 12,84	Juan Sus	Barcelona
	782	V	Elvira *Curell* W Bn	Spanish	129	85·3	25·3	12·2	Blanes	1850		[illegible]	J.Curell Canals	Barcelona
	783		Elvira *Sjögren* -86 W Bg	Swedish	231	110·8	24·8	12·1	Söderhamn	1870			A.P.Sjogren	Väddö,Rslgn.
	784		Elvira *Westberg* 87-87 W Bg (ex Dorotea)	Swedish	220	106·6	21·0	12·2	NyKarleby	1857			N.Jonsson	Brommby, Malmöhln.
	785		Elvira A. *Saglietto* W Bn	Italian	179	99·1	24·9	12·1	Savona	1874	G.B.Ciarlo	[illegible]	G.B.Ameglio&figli	Porto Maurizio
	786	IV	ElviraArdisson *Airaldi* W Bn	Italian	250	124·3	26·6	12·0	Viareggio	1886	A.Raffaelli	[illegible]	G.Ardisson	Leghorn
	787	*D	Elvira Camino (ex Doña Juana) *Salmond* 87-87 C Bk	British	294	126·5	26·1	13·8	Glasgow	1872	C.Connell&Co.	[illegible]	R. & W.King	Bristol
	788		Elvira D. *Durazzano* W Bn (ex Pilade)	Italian	117	84·3	22·9	11·5	Castell.di Stab.	1858	A.Bonifacio		N.Durazzano	Gaeta
	789		Elvira Owen WS	American	951	167·6	34·0	23·2	Bath,Me.	1852			A.B.Helmboe	New York
	790	RV	Elwell *Barston* -83 WS	American	1461	212·3	39·1	24·0	Damariscotta, [Me.	1875	Norris&Co.	[illegible]	D.W.Chapman	Damariscotta,Me.
	791		Elwine & Friedericke WSr	German	143	78·1	22·3	12·2	Ueckermünde	1841	L.Wittenberg		Carl Wichards	Stettin
	792	R	Elwood Burton *Hitchens* -88 W3mSr	American	375	128·0	31·7	11·7	Bath,Me.	1872	Hotchkiss& [Sons		L.Jarvis	Philadelphia
	793		Elwood Doran WSr	American	224	117·0	29·7	9·6	Milford,Del.	1862			J.W.Knight	Philadelphia
	794	*	Elwy *Owen* 87-88 I Bk	British	769	185·8	32·0	20·0	Liverpool	1876	T.Royden&Son		D.Jones	Liverpool
	795	RV	Elze *Bellgren* -84 WS	Swedish	357	122·7	25·4	14·5	Ueckermünde	1868	C.Wittenberg	[illegible]	S.Hörberg	Hellevik
	796		Emana *Lindqvist* -87 W3mSr	Russian	325	..	..	..	Lamparland	1866			J.Svibergson	Lemland
	797	V	Emanuel *Clausen* W Bn	Danish	140	88·1	20·6	11·1	Nordby,Fanø	1866	J.Th.Hansen		Actieselskab. (C.J. Clausen)	Nordby,Fanø
	798	V	Emanuel *Pedersen* 78-80 WSr	Danish	134	97·1	21·6	10·1	Svendborg	1880	N.P.Pedersen		R.P.Pedersen	Svendborg
	799		Emanuel WS	Norweg'n	595	168·1	34·2	18·0	Tvedestrand	1861	O.Jörgensen		O.Jörgensen	Risör
	800													

*=Lloyd's Register; D=Nederlandsche Vereeniging; G=Germanischer Lloyd; I=Registro Italiano; R=American Record;
A=Veritas Austro-Ungarico; H=Veritas Ellenico; N=Norske Veritas; V=Bureau Veritas.

Dans la 4me Colonne, C=Composite (membrure en fer, bordé en bois); I=Fer; Stl=Acier; W=Bois.

Les Dates du doublage en métal (Col. 11) ressortent des livres dans lesquels les navires sont respectivement classés ou des rapports des Inspecteurs du Lloyd's Register.

Pour plus amples renseignements sur les navires classés par le Lloyd's Register, voir le Lloyd's Register Book et le Lloyd's Yacht Register.

Official No.	No. in Book	Registry in which classed	Ship's Name. Master, and Dates of Appointment to present Owner's Service and to Ship. Ship's Late Name.	Material and Rig	Flag	Net Tonnage	Length	Breadth	Depth	Built: Where	When	By whom	Date of Metalling	Owners or Managers
	801	N	Emanuel *Bremer 80-86*	W Bk	Norweg'n	389	111·2	28·8	18·1	Sunderland	1850			A.N.Walloe
	802	N	Emanuel *Jensen*	W Bk	Norweg'n	301	112·7	27·6	14·7	Arendal	1870	C.A.Halvorsen		O.Lydersen,sen,
	803	N	Emanuel *Erichsen*	W Bg	Norweg'n	293	108·4	27·7	13·8	Fredrikshald	1857	Chr.Wiig		Jul Erichsen
	804		Emanuel *Gundersen*	W Bg	Norweg'n	275	93·3	27·1	15·8	Tvedestrand	1838			E.Jörgensen
	805	N	Emanuel *Einersen*	W Bk	Norweg'n	239	105·5	27·6	12·5	Grimstad	1864	JensTellefsen		J.E.Einersen
	806	G	Emanuel *Nielsen 80-*	W Bn	Norweg'n	225	110·0	26·4	12·0	Sundsvall	1866	J.A.Nordström	Y [illegible]	Th.Nielsen
	807		Emanuel *Smedsvig*	W Sr	Norweg'n	152	96·7	23·8	11·4	Stavanger	1867	R.Harem	Y [illegible]	O.E.Smedsvig
[illegible]	808		Emanuel *Paulschen -86*	W Sr	Russian	142	81·6	22·6	11·9	Orrenhof	1880	Peter Weide		Georg Weide
	809		Emanuel *A. Kääriä -87*	W Sr	Russian	114	..	..	..		1865			E.Kääriä
	810		Emanuel *Gavilla*	W Bn	Spanish	138	81·4	21·0	11·9	Damgarten	1866	H.Dierling		A.P.Pedri
	811		Emanuel (M1884NG) *Stephansson*	W Bk	Swedish	425	134·1	23·4	18·6	Sunderland	1856			L.P.Stephansson
	812	V	Emanuel *Andersson*	W 3m Sr	Swedish	295	122·3	24·4	13·5	Apenrade	1871	H.Jürgensen	[illegible]	J.Wennerberg
	813	V	Emanuel *Hoglund*	W 3m Sr	Swedish	198	100·4	21·3	12·4	Wenersborg	1876		Y [illegible]	C.J.Andersson
	814		Emanuel *Eriksson*	W Sr	Swedish	110	71·8	23·3	9·8	Roslagen	1875			J.E.Eriksson
75739	815	V	EmanuelSwedenborg *Keach*	W Bk	British	799	161·6	33·6	21·0	PortGilbert, [N.S.	1878	L.McDonald	Y 12.88	Z.Goudey
39737	816		Emanuel d'Hely	W Bg	British	115	..	..	..					Adele M.C.Marle
1332	817	I	Emanuele *Lachiana -83*	W Bg	Italian	394	117·1	29·7	18·1	Procida	1867		Y 4.86	E.D'Abbundo
1205	818	I	Emanuele *Marly -85*	W 3m Sr	Italian	366	109·9	26·7	16·7	Cassano	1878	F.DeRosa	Y 10.87	T. & S.Maresca
791	819	I	Emanuele *Schiaffino*	W Bk	Italian	310	105·6	29·2	17·6	Savona	1862	F.Sirello	Y 11.84	F.Schiaffino
607	820	I	Emanuele (ex Amor) *Benvenuto*	W Bg	Italian	281	107·2	27·7	16·5	Savona	1859	G.Tixi	Y 3.85	G.Benvenuto
2707	821	IV	Emanuele Mainetto *Bixi*	W Bk	Italian	1008	192·9	35·1	23·4	SestriPonente	1876	A.Briasco fu L.	Y 11.88	C.Mainetto
1989	822	IV	Emanuele Massone (COLLISION) *Pinco*	W Bg	Italian	385	122·3	29·4	18·4	Varazze	1870	S.Craviotto	Y 7.88	E.Massone
1870	823	I	Emanuele P. *Risso*	W Bg	Italian	363	113·8	30·1	18·1	SestriPonente	1869	G.Borzone	Y 3.84	C.Marchese
	824	V	Embla *Winther 80-87*	W Bk	Danish	384	128·9	26·6	14·1	Nordby	1885	S.Abrahamsen	Y 11.87	Actieselskab. (C.H. Nielsen)
	825	V	Embla *Mikkelsen 75-75*	W Sr	Danish	149	96·0	21·2	10·8	Troense	1875	C.J.Jensen		M.R.Gammelgaard
	826	N	Embla *Möller*	W Bk	Norweg'n	538	137·6	30·7	18·2	Arendal	1868		Y [illegible]	T.Thommesen & Smith
	827	N	Embla *Espersen -88*	W Bk	Norweg'n	379	117·8	29·4	16·1	Grimstad	1871	JensTellefsen	Y [illegible]	P.Pedersen
	828	N	Embla *Olausen*	W Bg	Norweg'n	139	88·7	22·6	11·1	Kragerö	1855	J.Andreassen		J.Dahll
	829	DN	Embla *Enes*	W Sr	Norweg'n	116	89·4	23·3	10·2	Bergen	1877	L.Sandvik	Y [illegible]	H.T.Enes
	830		Embla (ex Prudentia) *Svensson*	W Bg	Swedish	251	102·6	23·7	13·7	Wollin	1858			S.B.Svensson
80749	831	RV	Emblem *Lewis -88*	W S	British	1152	192·5	36·5	22·6	St. Joseph de [Levis	1880	E.Samson	Y [illegible]	E.Samson
	832		Emblem	W Bk	Chilian	519	135·9	29·9	18·5	St.John's,N.B.	1862	R.Bryson		Cia. Ferro-Carril de Copiapo
	833		Emblem *Espersen*	W Bg	Danish	266	99·4	25·9	16·1	Sunderland	1857	Hodgson& [Gardner		Chro.Ipsen
84880	834	✻	Embleton *Bennett 88-88*	I Bk	British	1196	226·5	36·0	21·5	Sunderland	1881	R.Thomps'n&S.		P.Iredale & Son
47886	835	✻	Emblyn *Fishley 84-88*	W 3m Bkn	British	172	117·0	24·0	12·7	Ipswich	1864	Read		R. Williams & J. Hutchings
	836	A	Emboriani *Lahanas*	W Sr	Greek	105	..	..	..	[illegible]	1882			G.Lahanas
67	837		Embros	W Bg	Greek	177	82·0	24·6	..	Spezzia	1866			J.G.Kaftopoulos
	838	G	Emden *Dirks 82-86*	W Sr	German	112	77·1	19·8	12·0	Leer	1882	Chr.Schwoon	Y [illegible]	EmsLootsenGesellschaft
63537	839	✻G	Eme *Summers 85-85*	C Bk	British	774	199·7	32·6	19·0	Glasgow	1869	C.Connell&Co.	Y [illegible]	Robertson & Co.
	840		Emeli *Blomqvist*	W Bk	Russian	609	156·0	31·9	17·2	Nerpes	1877	A.Grauskog		L.Olofsson
	841		Emelia *Karlson -87*	W 3m Sr	Russian	303	..	..	..		1874			K.G.Erikson
	842		Emelia *Petterson -87*	W 3m Sr	Russian	270	..	..	..		1870			E.Eriksson
	843	V	Emelie (ex Fort Regent) *Olsson -88*	W S	Swedish	655	164·7	27·6	18·8	Jersey	1863	LeVesconte	Y [illegible]	J.C.Hulthén
	844		Emelie (ex Emilie, ex Excelsior) *Eriksson -85*	W Bk	Swedish	290	114·9	24·2	14·7	P.E.Island	1865	Carroll		M.Lundqvist
	845		Emelie *Andersson*	W Bg	Swedish	238	113·8	23·5	12·3	Gefle	1850			J.Johansson's estate
	846		Emelie (ex Rhoda) *Solberg -88*	W Bg	Swedish	228	100·4	23·4	13·1	Ueckermünde	1858	L.Wittenberg		C.J.Löfgren
8616	847		Emelie E. Birdsall *Chambers -87*	W 3m Sr	American	400	134·8	33·8	12·4	Wilmington, [Del.	1874	Jackson, Sharp [& Co.		Jacob H.Birdsall
72501	848	V	Emeline (MISSING) *Daniels*	W Bn	British	304	117·4	27·9	12·7	Maitland,N.S.	1875	J.Monteith	[illegible]	P.E.Kenny
61332	849		Emeline	W Sr	British	120	82·0	22·6	8·5	NewBandon, [N.B.	1868			R.Dawson
7483	850		Emeline G. Sawyer *Rogers*	W Sr	American	138	90·0	26·2	8·8	Jonesport,Me.	1860			D.J.Sawyer

✻=Lloyd's Register; D=Nederlandsche Vereeniging; G=Germanischer Lloyd; I=Registro Italiano; R=A[illegible]
A=Veritas Austro-Ungarico; H=Veritas Ellenico; N=Norske Veritas; V=[illegible]

In the 4th Column, C=Composite (iron frame, wood planked); I=Iron; Stl=Steel; W=Wood.

The Dates of Metalling (Col. 11) are based on the records of the Books in which Vessels are respectively classed, or on the Reports of the Surveyors[illegible]

Further particulars of Vessels classed by Lloyd's Register will be found in Lloyd's Register Book and Lloyd's Yacht Register.

Ancien No.	No.	Classif.	Nom, Matériaux et Gréement; Capitaine; Ancien Nom	Pavillon	Tonnage	Longueur	Largeur	Creux	Où	Quand	Par qui	Date du doublage	Armateurs ou Directeurs-Gérants	Port d'Armement
	850	✱	Emeline Jenkin *Cann* W Sr	British	136	91·7	23·0	11·6	Fowey	1878	Stephens		J.Tregaskes & Co.	Fowey
	852		Emerald W Bk	British	322	139·0	25·6	11·6	Port Colborne, [Ont.]	1872	Geo. Hardison		J.Matthews	St.Cath'rines
	853		Emerald (ex Steamer) *Tyrrell* I Bkn	British	198	139·2	20·8	12·3	Liverpool	1846	P.Cato		W.McMaster	Dublin
	854		Emerald *Ford* 87-87 W Bk	American	1134	181·0	36·0	23·0	New York	1856			PugetSoundCommercial Co.	Pt. Townshend
	855	N	Emerald *Pedersen* 80-86 W Bk	Norweg'n	798	162·6	33·5	23·3	Laurvig	1880	G.Brown		M.Oppen	Laurvig
	856		Emeraude *Mory* -86 W Bg	French	160	99·1	22·3	9·6	Nantes	1874	T.Dubigeon		Anatole Lemoine	St. Malo
	857		Emerentia *Westerberg* W Sr (BROKEN UP)	Swedish	128	83·0	21·2	11·2	Orust	1863				
	858	A	Emerico (ex Gost) *Valentin* W Bg	Austrian	259	100·8	25·3	15·2	Ura	1856	S.Seppich		G.Marussich fu G.	Fiume
	859		EmersonRokes *Marston* W 3m Sr	American	419	134·5	31·8	14·4	Rockland, Me.	1873			S.M.Bird	Rockland, Me.
	860	N	Emigrant *Aalholm* W Bk	Norweg'n	579	131·6	32·1	20·1	Bremen	18 45/77			H.Aalholm	Arendal
	861		Emil *Holm* -84 W Bk (ex Nossa Mairo, ex Norma)	Danish	348	132·5	25·7	14·9	Vegesack	1859			N.P.Petersen	Marstal
	862		Emil *Danielsen* 1884 W 3m Sr	Norweg'n	240	93·6	25·5	15·6	America	1823			H.C.Sunde	Flekkefjord
	863	N	Emil (ex Flora) *Halvorsen* 82-89 W Bg	Norweg'n	193	87·1	26·2	13·0	Grimstad	1850			Blakstad, Holta & Co.	Skien
	864		Emil *Widemark* -87 W Bk	Russian	565	150·0	36·0	17·8	Nystad	1876			J.Mallén	Nystad
	865		Emil *Ivanoff-Fadjukoff* W Sr	Russian	171	..	..	..	Hiitola	1879			J.Pösö	Hiitola
	866		Emil *Jeppson* 80-86 W Bk (ex Gerhard & Bauch)	Swedish	406	121·7	27·7	17·1	Greifswald	1856	L.Wittenberg		L.Bauer	Malmö
	867	D	Emil *Carlsson* -86 W Bg	Swedish	264	112·5	25·3	12·4	Sundsvall	1867	J.A.Nordström		E.Sandelin	Sundsvall
	868		Emil *Sonesson* 71-83 W Bg	Swedish	218	101·0	25·8	12·9	Finland	1871			A.Sonesson	Wiken
	869		Emil Devrient *Siebert* -76 W Bg	German	244	105·8	25·9	13·3	Wolgast	1861	W.Lübke		F.Neumann	Wolgast
	870	✱V	Emil Julius *Borgwardt* I Bk	German	481	150·0	26·0	17·1	Sunderland	1862	J.Laing		C.A.Beug	Stralsund
	871		Emile (BARGE)	French	526	..	..	..						Marseilles
	872	V	Emile *Lemerle* W Bg	French	241	97·1	24·3	13·9	Nantes	1872	Barban		A.Lemerle	Nantes
	873	V	Emile *Mériel* W 3m Sr	French	227	101·7	22·8	11·6/15·6	Honfleur	1873	Lefoulon		E.de Baupte	Rouen
	874	V	Emile *Gallien* W Bg	French	185	91·9	21·3	13·4	Cherbourg	1875	E.Lecerf		Godey	Cherbourg
	875	V	Emile *Houzé* -87 W Bg	French	151	89·9	21·3	11·9	Dunkirk	1859	L.Verharne		Comolet Frères & les fils de Gauss	Cette
	876	V	Emile Menier W Bk	French	368	123·6	27·4	16·8	Nantes	1883	E.Clergeau	Y 12.83	Denis Crouan fils	Nantes
	877	V	Emile Postel *Guérie* -88 W S (ex Dep Jnoni)	French	1089	174·3	29·8	24·0	Bath, Me.	1861			A.Postel & ses fils	Cherbourg
	878	V	Emile & Louise *Benard* W Sr (WRECKED)	French	120	86·6	21·6	11·5	Dunkirk	1880			Cortier	Dunkirk
63917	879		Emilia *Dye* W Bk (ex Susan H.) (BROKEN UP)	British	885	159·5	34·1	23·0	Medford, N.J.	1861			J.Grieveson	South Shields
	880		Emilia W Sr	Brazilian	138	..	..	..					J.F.Stael	Bahia
	881	IV	Emilia *De Genaro* W 3m Sr	Italian	306	125·3	24·6	15·2	Cassano	1877	A.Castellano	Y 11.85	F.S.Ciampa	Castellam. di Stab.
	882		Emilia (WRECKED) *Orsolino* W Bg	Italian	152	84·0	23·6	14·1	Chiavari	1854	F.Gotuzzo		G.Schiaffino	Genoa
	883		Emilia *Ferraro* -88 W Bn (ex San Nino) (WRECKED)	Italian	107	77·7	22·0	12·1	Catania	1865		2 11.83	F.Consoli in Jacono	Catania
	884	V	Emilia *Lundqvist* -86 W Bk	Russian	635	150·0	32·7	20·7	Uleaborg	18 67/78	G.Pettersson		C.Nylund	Nykarleby
	885		Emilia *Johansson* 63- W 3m Sr	Russian	186	90·8	28·2	11·2	Björneborg	1873	J.Sandberg		A.Stephanson	Åland
	886		Emilia *Pettersson* -87 W 3m Sr	Russian	168	..	..	..		1873			M.Mottson	Wardö
	887		Emilia *Mellanen* W Sr	Russian	143	..	..	..	St.Johannes	1869			T.Mellanen	St.Johannes
	888		Emilia *Christensen* -83 W 3m Sr	Swedish	120	92·5	20·1	9·3	Lilla Edet	1874			C.Andersson	Strömstad
	889		Emilia Adenecar W Sr (ex Catrina)	Argentine	111	78·0	22·8	10·5	Arras	18 81/83	F.Johansen		J.B.Camossi	Buenos Ayres
	890	V	EmiliaC. (ex Regia) *Terricilo* W Bk	Italian	398	137·1	26·6	17·2	Bristol	1862	Hill		B.Castaldi	Naples
	891	V	Emilia Ciampa *Caffiero* W Bk	Italian	426	131·2	28·0	17·5	Sorrento	1865	A.Castellano		F.S.Ciampa	Castellamare
	892	I	Emilia F. *Ferrari* W Bk (ex Francesco Piccaluga)	Italian	484	141·0	31·5	19·3	Savona	1869	F.Sirello		G.B.Ferrari fu G.	Genoa
	893	I	EmiliaMignano *Mazzella* W Bg	Italian	247	96·5	27·6	16·1	Procida	1863	A.Mazzella		A.Mignano	Naples
	894		Emilia R. (ex Ida Lilly) W S	Chilian	757	145·4	30·8	22·0	Richmond, Me.	1864	J.M.Hagar		A. von der Heyde	Valparaiso
	895	I	Emilia Revello *Dellacasa* -88 W Bk	Italian	488	126·9	31·7	20·3	Varazze	1868	G.Scala		Flli. Schiaffino di D.	Genoa
	896	✱I	Emilia S. *Schiaffino* W Bk (ex Maggie)	Italian	279	123·5	26·5	15·3	P.E.Island	1867			B.Schiaffino	Genoa
	897	V	Emilia T. *Radonicich* -87 W Bk	Austrian	712	165·3	26·5	20·9	Trieste	1875	G.Spadon		M.Terni fu A.	Trieste
	898		Emilie *Schlich* -87 W Bk	Austrian	591	142·7	32·3	19·6	Fiume	1877	Flli. Schiavon		I.Bonetich	Fiume
	899		Emilie *Smith* -88 W Bk (ex Sheriff)	Bolivian	729	163·2	32·8	19·1	Quebec	1875	Charland		A.Bichaud	Cobija
	900	V	Emilie *Hansen* 83-84 W 3m Sr (ex Kris)	Danish	235	114·5	24·3	11·5	Wisby	1870	C.G.Scherner		H.P.Christiansen	Copenhagen

✱ = Lloyd's Register; D = Nederlandsche Vereeniging; G = Germanischer Lloyd; I = Registro Italiano; R = American Record; A = Veritas Austro-Ungarico; H = Veritas Ellenico; N = Norske Veritas; V = Bureau Veritas.

Dans la 4me Colonne, C = Composite (membrure en fer, bordé en bois); I = Fer; Stl = Acier; W = Bois.

Les Dates du doublage en métal (Col. 11) ressortent des livres dans lesquels les navires sont respectivement classés ou des rapports des Inspecteurs du Lloyd's Register.

Pour plus amples renseignements sur les navires classés par le Lloyd's Register, voir le Lloyd's Register Book et le Lloyd's Yacht Register.

Official No.	No. in Book	Registry in which classed	Ship's Name. Material and Rig. Master, and Dates of Appointment to present Owner's Service and to Ship. Ship's Late Name.	Flag	Net Tonnage	Length	Breadth	Depth	Built: Where	Built: When	Built: By whom	Date of Metalling	Owners or Managers
	901	V	Emilie (ex Reine [illegible]) *Maitan 84-84* I Bk	French	358	190·9	32·5	20·9	Glasgow	1860	W. Simons & Co.		Ant. Dom. Bordes & [illegible]
	902		Emilie (ex Flore [illegible] Margaret) W3m Bn	French	302	129·2	27·0	12·9	Sunderland	1874	W. Pickersgill	Y 1.83	G. Vasse & J. Duhamel
	903	V	Emilie *Ferrodo -84* W Bg	French	181	86·6	23·6	13·4	Redon	1867	F. Coyac		Vve. Bruand
	904	V	Emilie *Ruello -87* W Bn	French	131	82·3	20·0	10·4	Granville	1873	Lace	Z 2.87	Prenveille
	905	GV	Emilie *Ohling 88-88* W Bk	German	872	159·3	33·5	21·7	Bremerhaven	1861	J C Tecklenborg	Y 10.86	Jul Wieting
	906		Emilie (ex Betty) *Burkhardt -83* W Bk	German	521	138·1	31·2	18·2	Brake	1856			C. Kriegel
	907	V	Emilie *Shultz 84-84* W Bk	German	419	126·6	29·1	16·4	Ueckermünde	1858	C. Pretzer		Wwe. Emilie F. C. Lau[illegible]
	908		Emilie (ex Ida Maria de Baas[illegible]) *Le Coutre 83-* W Bkn	German	284	123·6	26·5	13·6	Rotterdam	1856			A. Hohorst
	909	V	Emilie *Martens -86* W Bg	German	215	101·3	22·3	12·6	Stralsund	1875	A. Juhl	Y 6.86	H. L. Juhl
	910	✱N	Emilie (ex Parnass) *Halvorsen -84* W Bk	Norweg'n	442	141·7	27·4	17·8	Hartlepool	1857	Pile, Spence & Co.	Y 10.81	Actieselskabet "Ad[illegible]"
	911	N	Emilie *Haagensen -58* W Bk	Norweg'n	427	127·3	30·0	16·6	Mandal	1876	A. Jacobsen	Y 2.85	A. Jacobsen
	912	N	Emilie *Torp 85-84* W Bk	Norweg'n	306	119·1	29·6	12·1	Drammen	1884	J. Jörgensen	Y 10.84	O. Caspersen
407	913		Emilie *Treu -85* W Sr	Russian	165	92·6	26·0	12·8	Augern	1882	Schall		J. Massapsche
73263	914	✱	Emilie Dingle *Roose 187-88* W Bk	British	278	126·7	25·3	13·6	Bridport	1876	Cox	Y 3.86	W. W. Dingle
	915	DGV	Emilie Hessenmüller *Gerdau 84-84* W3m Sr	German	319	117·4	26·3	12·8	Harburg	1881	G. Renik jun.	Y 10.86	F. J. Hubert
	916	G	Emilie Kahl *Franck 75-75* W3m Sr	German	290	112·2	26·6	13·9	Stralsund	1872	C. W. Mohr		J. J. Wallis & Sohn
80616	917	V	Emilie L. Boyd *Killam* W Bk	British	1241	201·0	38·0	23·0	Green Cove, NS	1881	J. W. Raymond	Y 1.86	William Law
	918	NV	Emilie Marie *Tonnesen* W Bk	Norweg'n	601	156·5	32·0	18·4	Christiansand	1876	N. Christensen	Y 4.88	J. A. Salmundsen
	919	V	Emilien Marie *Bobehier* W Bk	French	244	102·0	23·9	13·0	Nantes	1874	Alleau & Aubert	Y 6.81	Leguales de Mesmi[illegible]
3019	920	V	Emilio M. (ex Parquerette) *Mortola* W Bk	Italian	638	167·0	28·6	18·5	La Seyne	1874	Curet fils	Y 12.83	F[lli]. Mortola
1970	921		Emily *Baker 77-85* W Bn	British	196	109·4	23·6	14·0	South Shields	1850	Edwards & Shi'ld		F. W. Fairbrass
43990	922		Emily *W. Johnson -86* W Bg	British	190	115·5	22·7	12·6	Aberdeen	1862	A. Hall & Sons		J. Sanderson
33878	923		Emily *Cory* W Bn	British	166	101·3	23·4	12·9	St. Johns, NFL	1854			J. Menzies
29590	924	✱	Emily (COLLISION) *Remick 84-84* W Bn	British	145	96·6	22·6	12·6	Rye	1861	Hoad		W. C. Phillips
83510	925	✱	Emily *Anderson 84-84* W3m Sr	British	100	94·1	22·4	9·5	Carrickfergus	1884	P. Rodgers		T. Rayner
68325	926	✱	Emily *Ley* W Sr	British	90	77·2	22·1	9·7	Plymouth	1873	Marshall		J. Westacott
	927	G	Emily *Crichton 77-* W Bg	American	296	119·9	27·0	14·7	Shoreham	1868	W. May	Y 1.83	Pacific S.S. Co.
7709	928		Emily *Watt* W Bk	American	231	102·3	26·4	11·5					W. Ray
	929	N	Emily *Larsen* W Bk	Norweg'n	368	119·0	27·6	17·5	Italy	1857			L. Larssen
	930		Emily W Bg	Norweg'n	222	95·3	24·4	14·8	Cork	1844			Th. Nielsen
70507	931	✱	Emily A. Davies *Evans 86-86* I Bk	British	396	163·0	27·3	13·6	Sunderland	1876	J. L. Thompson [& Sons		J. Davies
48878	932		Emily Anne *Cole* W Bn	British	121	91·0	21·5	9·7	Sunderland	1867	Mills		T. Smart
73161	933	✱	Emily Chaplin *Smith* I Bk	British	733	185·0	31·6	18·4	Sunderland	1875	Austin & Hunter		R. N. Smith
32047	934		Emily Downing W Bk	British	269	90·0	26·3	17·8	Tasmania	1841			A. McGregor
135712	935		Emily E. Johnson *Paul* W Sr	American	116	95·4	26·7	7·9	Matthews Co., [Va.	1883	F. R. Haynes		C. C. Paul
48890	936		Emily Eliza *Green* W Bn	British	157	90·4	23·1	12·5	Sunderland	1865	Douglas		T. Smart
135732	937	R	Emily F. Northam *Sauer -88* W3m Sr	American	316	131·2	33·0	10·5	Portland, Con.	1883	Gildersleeve & S.		H. H. Stetson
135468	938	RV	Emily F. Whitney *Rollins* W S	American	1249	193·0	37·6	23·0	East Boston	1879	Gove	Y 8.84	J. H. Flitner
7884	939		Emily H. Naylor *Wilson* W Sr	American	266	117·0	30·2	8·8	Camden, N.J.	1866	Taylor & Mathis		R. B. Fisher
68366	940		Emily Lloyd (ex Wm. Lovett) *Spencer* W K Barge	British	126	95·8	24·0	9·4	Sittingbourne	1872			J. Smith
67160	941	✱	Emily Millington *Harris* W Sr	British	99	88·9	22·2	10·3	Selby	1876	Banks	Y 1.86	G. Millington
135472	942	V	Emily Reed *Shillidas* W S	American	1489	215·0	40·6	24·1	Waldobro', Me	1880	Reed	Y 7.86	Yates & Porterfield
135800	943		Emily S. Baymore *Thompson* W Sr	American	243	124·0	27·0	9·0	S. Dennis, N.J.	1884	J. H. Diverty		W. C. Stiles
135630	944		Emily Shepard *Coggins* W Sr	American	171	102·5	28·0	8·5	Portland, Con	1882	Gildersleeve & S.		F. Gildersleeve
67034	945		Emily Smeed *Bennett 82-82* W Sr	British	286	133·3	25·8	13·3	Sittingbourne	1872	Smeed		J. & N. C. Ball
36245	946		Emily Stewart (ex Graf Adolf) *Thompson 85-85* W Sr	British	124	95·0	21·8	11·0	Holland	1855			Hilton, Anderson & Co.
67128	947		Emily Warbrick *Bennett 85-85* W Bn	British	159	105·6	23·3	12·0	Fleetwood	1872	Gibson		Porter's Shipping Co.
29597	948	✱	Emily Waters *Langhorne -86* I Bk	British	446	155·8	25·5	17·1	Liverpool	1861	W. C. Miller		W. B. Jones
64464	949		Emily Watters *Perry* W Bn	British	310	112·9	28·2	13·0	St. John, N.B.	1870	Jenkins		J. C. Woodin
8392	950		Emily & Jenny *Douglas* W3m S	American	328	119·2	29·7	9·8	May's Landing, [N.J.	1867	J. Clark		Bartlett & She[illegible]

✱ = Lloyd's Register; D = Nederlandsche Vereeniging; G = Germanischer Lloyd; I = Registro Italiano; R = Amer[ican]; A = Veritas Austro-Ungarico; H = Veritas Ellenico; N = Norske Veritas; V = [Bureau Veritas]

In the 4th Column, C = Composite (iron frame, wood planked); I = Iron; Stl = Steel; W = Wood.

The Dates of Metalling (Col. 11) are based on the records of the Books in which Vessels are respectively classed, or on the Reports of the Surveyors [illegible]

Further particulars of Vessels classed by Lloyd's Register will be found in Lloyd's Register Book and Lloyd's Yacht Register.

No.	Cote	Nom. Changements de nom; Capitaines et dates d'entrée en service des Armateurs actuels et à bord du navire. Année. Nom.	Constr.	Pavillon	Tonnage net.	Dimensions. Longueur	Largeur	Creux.	Construit Où.	Quand.	Par qui.	Date du doublage	Armateurs ou Directeurs-Gérants.	Port d'Armement.
951	[illegible]	Emita *Nash*	W Bk	American	449	139·5	32·0	12·6	Harrington Me	1883	A. Nash	[illegible]	A. Nash	Machias, Me.
952		Emma *Borrett 77-77*	W Bg	British	184	91·0	23·7	13·4	Hylton	1865	Briggs		C. Paldon	Sunderland
953		Emma (ex . . . Chapman)	W Bn	British	170	95·8	24·6	8·8	Talbot, U.S.	1856			Chas. Hickman	Nassau, N.P.
954	[illegible]	Emma *Jenkin*	W Sr	British	138	97·0	23·0	12·0	Padstow	1877	Cowl		Mrs. C. Jenkin	Padstow
955		Emma *Smith 81-83*	W Bn	American	409	119·1	28·3	16·6	Yarmouth, Me.	1866	Hutchings & Stubbs	[illegible]	B. Webster	Portland, Me.
956	R	Emma (STRANDED) *Lillegahs*	W 3m Sr	American	337	127·7	31·0	11·7	Portland, Me.	1882	D. Brewer		C. H. Chase	Portland, Me.
957		Emma *Thick*	W Bk	Austrian	577	146·6	29·5	17·7	Lussinpiccolo	1869	N. Martinolich	[illegible]	G. Strudthoff	Trieste
958	G	Emma *Nielsen*	W Bg	Danish	244	102·6	26·6	14·1	Greifswald	1862	J. W. Sadelkow		L. v. d. Hude	Svendborg
959		Emma (ex . . .) *Trauber, Schadt* (BARKEN CP)	W Bt	Danish	156	76·6	21·8	13·1	Rostock	1843	W. Zeltz		M. P. Schmidt	Dragor
960	V	Emma *Nielsen*	W 3m Sr	Danish	127	90·2	21·2	10·5	Greifswald	18 54/58	J. W. Sadelkow		P. Andersen	Thisted
961	D V	Emma *Schott* (ex . . .)	W Bk	Dutch	595	117·6	29·8	17·5	Apenrade	1876	J. J. Paulsen & Sohn	[illegible]	J. Schall	Rotterdam
962	V	Emma *Beauregard*	W Bg	French	178	91·2	22·0	12·1	Dunkirk	1867	L. Verharne	[illegible]	Comolet frères	Cette
963		Emma	W Sr	French	154	86·4	26·6	10·2	Pompquet, N.S.	1864			Executors of F. Cardon	St. Pierre (Miquelon)
964		Emma *Clemence*	W Bg	French	142	84·0	21·0	12·3	Dieppe	1865	P. Corne		E. Pourpoint & fils	Dieppe
965	V	Emma *Boulogne*	W Sr	French	124	86·6	20·7	11·6	Dunkirk	1868	D. VanCauwenberghe		Numa VanCauwenb'rghe	Dunkirk
966	V	Emma (WRECKED) *Lohlsen 87-87*	W Bg	German	349	112·9	25·6	13·4	Elsfleth	1871	J. Ahlers	[illegible]	H. G. Lübken	Elsfleth
967		Emma *Weweltzer 65-65*	W Bg	German	217	90·5	25·5	14·3	Barth	1853	Lignitz		C. A. Beug	Stralsund
968	V	Emma *Focke 85-88*	W Bg	German	215	101·0	25·2	13·4	Stralsund	1861	J. Kasten		W. A. Sarnow	Barth
969	V	Emma *Funck 80-*	W Bg	German	200	94·4	23·6	13·2	Ueckermünde	1856	L. Wittenberg		H. F. Poll	Stettin
970	V	Emma *Bogge 84-73*	W Sr	German	149	99·5	22·1	12·2	Wolgast	1873	J. H. Kräft		Carl Lobeck	Stralsund
971		Emma	W Sr	Argentine	116	79·4	18·7	10·0	Stralsund	1878	A. Juhl			
972		Emma (ex . . .) *Costanzo* (ex . . . Geo. Meade, ex Mary . . . Maria)	W Bg	Italian	400	135·1	30·8	12·8	Br'khaven, L.I.	1860	P. Jefferson		Società Generale Italiana	Palermo
973	N	Emma *Jansen 82*	W Bk	Norweg'n	724	152·9	31·5	21·4	Richmond, Me	1854		[illegible]	N. M. Bugge	Laurvig
974	N R	Emma *Olsen*	W Bg	Norweg'n	305	119·4	27·0	14·9	Littlehampton	1866	J. & W. B. Harvey	[illegible]	A. B. Andersen	Tönsberg
975		Emma (WRECKED) *Isaachsen*	W Bk	Norweg'n	292	114·5	27·6	13·7	Barcelona	1875		[illegible]	H. Kjösterud & Sönner	Drammen
976		Emma *Wirkki 88*	W Bk	Russian	428	122·7	27·6	15·0	Björkö	1872			W. & H. Martelin	Björkö
977		Emma *Kaartinen 87*	W Bg	Russian	371	..	..	..		1872			E. Mortelin	St. Johannis
978	V	Emma *Susling 86-86*	W 3m Sr	Russian	285	119·2	28·6	13·1	Lubb Essern	1885	J. Kohpmann		J. Kirstein	Riga
979		Emma *Lauzeth 73-85*	W 3m	Russian	239	..	..	..	Kimito	1875	G. Sjöström		C. G. Holmberg	Westanfjord, Kimito
980		Emma *Hamberg 87*	W Sr	Russian	195	..	..	..		1864			G. Kollin	Hangö
981		Emma	W	Russian	111	..	..	..		1874			S. Junnola	Säkkjärvi
982		Emma (ex Emma Sophie) *Andersson 87*	W Bg	Swedish	243	99·6	24·7	13·3	Rostock	1862	J. Möller		A. Andersson	Skepsmyra, Rslgn.
983	G	Emma (WRECKED) (ex Lutetia) *Nilsson 83*	W 3m Sr	Swedish	234	106·6	23·4	12·5	Elsfleth	1865	J. Jürgens		A. Nilsson	Arildsläge
984	V	Emma *Johansson*	W Bn	Swedish	199	97·1	23·4	12·3	Eckerna	1875	A. Ahlberg		O. Olsson	Käringön Bhsln
985		Emma *Larsson*	W Sr	Swedish	194	98·1	23·3	12·5	Finland	1865			O. Holm	Sölvesborg
986		Emma *Olsen*	W Sr	Swedish	156	88·6	24·5	10·9	Öfverlöpe	1873			A. Jansson	Öfverlöpe Rsign
987		Emma (ex Alst) *Carlsson 88*	W Sr	Swedish	146	97·2	22·0	10·4	Haugesund	1866			J. A. Carlsson	Calmar
988	I	Emma A. (ex Francesco G.) *Inbernardi*	W Bk	Italian	856	177·1	35·8	22·4	Sestri Ponente	1875	C. Tubino	[illegible]	S. Anselmo	Savona
989		Emma Augusta *Holton*	W Bkn	American	285	115·0	32·0	11·4	San Francisco	1867			S. B. Peterson	San Francisco
990		Emma B. Shaw *Shaw*	W Sr	American	236	110·5	29·2	8·5	Camden, N.J.	1867			F. Smith	Philadelphia
991	V	Emma Beug *Salomon*	W Bg	German	311	117·3	27·2	14·3	Barth	1871	Holzerland	[illegible]	C. A. Beug	Stralsund
992		Emma C. (ex Somers)	W Bn	British	104	73·6	22·2	7·2	Philadelphia	1845			Thos. Colton, junr.	Hamilton, Bermda
993		Emma C.	W Sr	British	100	76·7	25·1	8·3	Dorchester, N.B.	1883	J. Chambers		S. Groves	Annapolis, N.S.
994	R	Emma C. Cotton *Ayres*	W 3m Sr	American	334	138·8	32·7	10·6	Camden, N.J.	1882	Shoe & Chard		T. Ayres	Philadelphia
995		Emma C. Jones	W S	American	307	105·8	27·0	16·9	Fairh'v'n Mass	1849			J. McCullough	New Bedfrd, Mass
996	R	Emma C. Knowles *Mayhew*	W 3m Sr	American	587	147·7	34·9	15·0	Mauricetown, N.J.	1884	Vanneman Bros.		A. W. Mahew	Bridgeton, N.J.
997	R	Emma C. Middleton *Foster*	W 3m Sr	American	501	138·3	35·0	10·4	Philadelphia	1881	Birely, Hillman & Streaker		W. J. Banks	Philadelphia
998		Emma Claudina *Maas*	W 3m Sr	American	186	112·8	31·4	8·6	San Francisco	1882	M. Turner		G. W. Chandler	Eureka
999	R	Emma Crosby *Crosby 81*	W 3m Sr	American	214	120·0	29·3	9·7	Calais, Me	1873	Nickerson		G. G. King	Calais, Me.
1000		Emma D. Endicott *Rowen 86*	W 3m Sr	American	319	120·0	30·0	11·7	[illegible]	1864			G. W. Shepherd, jun.	Philadelphia

[illegible] = Lloyd's Register; D = Nederlandsche Vereeniging; G = Germanischer Lloyd; I = Registro Italiano; R = American Record; A = Veritas Austro-Ungarico; H = Veritas Ellenico; N = Norske Veritas; V = Bureau Veritas.

Dans la 4me Colonne, C = Composite (membrure en fer, bordé en bois); I = Fer; Stl = Acier; W = Bois.

Les Dates du doublage en métal (Col. 11) ressortent des livres dans lesquels les navires sont respectivement classés ou des rapports des Inspecteurs du Lloyd's Register.

Pour plus amples renseignements sur les navires classés par le Lloyd's Register, voir le Lloyd's Register Book et le Lloyd's Yacht Register.

1 Official No.	2 No. in Book.	3 Registry in which classed.	4 Ship's Name. Material and Rig. Master, and Dates of Appointment to present Owner's Service and to Ship. Ship's Late Name.	5 Flag.	6 Net Tonnage.	7 Dimensions. Length.	Breadth.	Depth.	8 Built. Where.	9 When.	10 By whom.	11 Date of Metalling.	12 Owners or Manag
	1001		Emma Dean W Bn		270	117·6	26·9	13·2	Greenpoint U.S.	1860			
57470	1002		Emma E. Potter W Sr	British	147	92·0	24·0	9·4	Clements, N.S.	1870			B. V. Blarcom
70565	1003	✻	Emma Ernest Gill 68-79 W Bg	British	207	109·2	23·7	12·7	Milford	1876	Jones	y s s	Ockenden & Son
135676	1004		Emma F. Angell Trapp -83 W3mSr	American	819	163·0	37·0	15·5	Bath, Me.	1883			W. B. Church
8040	1005		Emma F. Harriman Biber W Bk	American	366	118·8	27·3	16·4	Searsport, Me.	1861			A. W. Moore
7295	1006		Emma F. Hart Keen W3mSr	American	328	133·0	30·0	18·0	Camden, Me.	1866	J. E. Dailey		J. P. Ellicott & Co.
830	1007		Emma Felice Tedesco W Bn	Italian	130	90·5	22·3	11·6	Castellam. di [Stab.	1883	Bonifacio	[illegible]	C. Tedesco
64502	1008	V	Emma G. Scammell Kirkland 83-83 W Bk	British	531	138·5	30·1	16·9	St. John, N.B.	1871	Baizley	y s s	Scammell Brothers
67133	1009		Emma Grace Iddon W Sr	British	102	86·0	20·5	10·7	[illegible]	1872	Barclay		T. Fisher
8564	1010		Emma Green Smith W Sr	American	180	108·4	28·8	8·6	Bangor, Me.	1869			J. Crosby
8759	1011		Emma Heather Lacey W3mSr	American	278	124·4	27·5	11·6	Milton, Del.	1871	Atkins		J. W. Hall
43054	1012		Emma Ives W Bn	British	192	110·3	27·2	11·8	Picton, N.S.	1862	Ives		S. N. Robinson
135790	1013	R	Emma J. Meyer Meyer W3mSr	American	392	140·1	33·6	12·2	Milton, Del.	1884	D. H. Atkins		J. R. Magee
38796	1014		Emma Jane (BROKEN UP) W Bn	British	113	102·9	19·7	8·4	Richmond R., [N.S.W.	1865			William C. Rogers
135057	1015		Emma K. Smalley Johnson W Sr	American	186	101·0	29·7	9·1	Cutler, Me.	1874	Porter		E. M. Mitchell
135177	1016		Emma L. Cottingham Somers W3mSr	American	496	139·8	34·0	15·9	Somers Pnt. N.J.	1875	J. VanSant		J. B. Steelman
72141	1017	V	Emma L. Shaw Porter [-81 W Bn	British	567	149·6	32·4	17·9	Walton, N.S.	1874	W. S. Vaughan	y s s	Mark Shaw
74745	1018	✻	Emma Louise Parsons W Sr	British	80	79·7	21·0	9·8	Emsworth	1877	Apps		L. Parsons
84475	1019	✻	Emma Louise Drake 83-83 W K	British	66	75·4	19·8	8·3	Barnstaple	1883	W. Westacott		F. Drake & Co.
10898	1020		Emma Louise (ex George B. Somes) W Sr	American	112	88·0	25·7	8·0	Mt. Desert, Me.	1867			Boardman, Todd & Chapin
8458	1021		Emma M. Fox Brooks W Sr	American	186	111·4	29·8	8·8	Greenport, N.Y.	1867		y s s	C. Brooks
8917	1022		Emma McAdam Young W Sr	American	158	101·0	27·4	8·6	[illegible], Me.	1872	Rideout & Lord		Rideout & Lord
397	1023	V	Emma Maria Baltzer W 3mSr	Russian	287	123·9	29·8	11·9	[illegible]	1882	J. Baltzer		J. Baltzer
	1024	V	Emma Marie Bichot 73-73 W Bn	French	151	80·0	22·5	11·1	Nantes	1873	J. & A. Lefrançois	y s s	A. Lefrançois
72306	1025		Emma Marr Brewster 82-82 W Bk	British	799	163·9	33·4	19·7	St. Martins, N.B.	1877	J. Marr	Y 10.85	F. Tufts
1919	1026	I	Emma Merello Massa W Bk	Italian	397	121·3	30·3	18·5	[illegible]	1869	B. Debarbieri	y s s	M. Merello
	1027	V	Emma Müller Küding [72- W Bk	German	505	133·6	30·3	17·5	Bath	1873	J. S. Kraeft	y s s	J. N. Rodbertus
	1028	N	Emma Parker Larsen [78-78 W Bk	Norweg'n	498	140·9	29·4	18·0	Tusket, N.S.	1869	D. Ross	y s s	Sch. P. Messel
66513	1029	DV	Emma Payzant Dexter W Bk	British	847	170·9	35·0	21·3	Newport, N.S.	1873	J. Mosher	Y 11.83	G. P. Payzant
85430	1030	V	Emma R. Smith Faulkner W3mSr	British	371	131·0	30·0	15·8	Hantsport, N.S.	1883	J. B. North	y s s	R. D. G. Faulkner
	1031	DGV	Emma Römer Persicke [86- W Bk	German	424	126·3	28·9	17·3	Rostock	1875	E. Burchard & Co.	y s s	H. Bauer
135614	1032	R	Emma S. Briggs Gray W3mSr	American	279	125·0	33·0	9·0	Wiscasset, Me.	1882	Haynes		J. M. Haynes
68885	1033	✻	Emma Sims Wigley -86 W Bk	British	417	135·3	28·5	17·5	Sunderland	1873	W. Richardson	y s s	J. W. Davies
	1034	G	Emma Sofia Jönsson -85 W Sr	Swedish	137	80·7	21·4	11·9	[illegible]	1852	H. Thomsen		C. O. Schéle
135363	1035	RV	Emma T. Crowell Pendleton W Bk	American	1087	181·2	37·3	22·3	Bath, Me.	1878	Goss & Sawyer	Y 10.85	Pendleton, Carver & Nichols
46044	1036		Emma Tuttle W Sr	British	108	73·7	23·4	6·7	N. Havre, U.S.	1859			A. J. Adderley
135206	1037		Emma Utter Allen W3mSr	American	265	127·0	33·0	9·6	Port Ludlow, [Wash.	1875	Hall Bros.		W. S. Browne
58212	1038		Emma & Ester Lloyd W Sr	British	100	83·6	21·8	10·2	Perth	1871	Wallace		Royle & Co.
	1039	V	Emma & Johanna Hradecke 79-79 W Bk	German	346	114·1	27·1	15·6	Memel	1862	B. Pieper		J. G. Gerlach
	1040	V	Emma & Robert Schmidt 83-83 W Bk	German	447	127·3	26·8	18·6	Rostock	1867	Em. Puhlleratz	y s s	A. Burchard
61978	1041		Emmanuel W Sr	British	104	89·8	23·0	7·1	St. Antoine de [Tilly	1871			O. Gignac
91507	1042	✻	Emmanuel Dobson W K	British	83	77·2	20·6	10·3	Galmpton	1885	A. Gibbs		C. H. Dobson
	1043	V	Emmanuel Hemmes (ex Stewart Freeman) W S	Dutch	1485	218·0	39·6	23·7	[illegible], N.S.	1876	S. Freeman & Co.	y s s	S. J. Melchers
	1044	V	Emmanuel Pasquier -87 W Bg	French	231	99·4	23·6	13·0	Nantes	1874	Clergeau	Y 11.82	F. Rousseau
436	1045		Emmanuel (FOUNDERED) W Bg	Greek	115	73·8	21·8	..	Syra	1869			M. S. Franges
	1046	H	Emmanuel Franges W Bg	Turkish	120	..	..	..	Syra	1870		y s s	S. Franges
	1047	V	Emmanuel Auger Josse W Bk	French	452	122·7	26·6	19·0	Honfleur	1875	Leviels	y s s	D. Auger
	1048	N	Emmeline Brygge 87-87 (ex Eleanor Wood) W Bk	Norweg'n	507	148·3	28·5	17·1	Quebec	1863	Samson	Y 9.84	H. C. Torgersen [illegible]
	1049	N	Emmeline Hansen 79-79 W3mSr	Norweg'n	303	100·0	25·3	12·0	Bordeaux	18[illegible]		y s s	C. F. Wetlesen
	1050												

✻ = Lloyd's Register; D = Nederlandsche Vereeniging; G = Germanischer Lloyd; I = Registro Italiano; R = Amer…
A = Veritas Austro-Ungarico; H = Veritas Ellenico; N = Norske Veritas; V = Bure…

In the 4th Column, C = Composite (iron frame, wood planked); I = Iron; Stl = Steel; W = Wood.

The Dates of Metalling (Col. 11) are based on the records of the Books in which Vessels are respectively classed, or on the Reports of the Surveyors of …

Further particulars of Vessels classed by Lloyd's Register will be found in Lloyd's Register Book and Lloyd's Yacht Register.

N° dans ce livre	Registre dans lequel le navire est classé	Nom. Capitaine et dates d'entrée au service des Armateurs actuels et à bord du navire. Ancien Nom.	Matériaux et Gréement	Pavillon	Tonnage	Longueur	Largeur	Creux	Construit: Où.	Quand.	Par qui.	Date du doublage en métal	Armateurs ou Directeurs-Gérants.	Port d'Armement.
1051		Emmeline *Kendtle*	W Sr	Russian	280	117·0	28·6	13·1	Neusalis	1872	R. Grass		G. Ohsoling	Riga
1052	*	Emmie *Nurse 86-86*	I S	British	120	89·4	22·1	8·4	Middlesbro'	1877	R. Craggs & Sons		Champion & Co.	Port Natal
1053		Emmy *Södermann -87*	W 3m Sr	Russian	205	95·8	27·6	13·0	Kitvari	1863	J. Jansson		K. F. Björkroos	Nystad
1054		Emperador	W Sr	American	158	..	..	..	Camden, N.J.	1883			H. Moody	New York
1055	*	Emperor *Lyall 85-83*	W Sr	British	99	85·8	21·8	10·6	Banff	1882	W. Geddie		J. W. Simpson	Banff
1056	*	Emperor *Rawbottom -88*	W K	British	91	80·8	20·8	10·4	Goole	1888	Cottingham Bros.		W. J. Robins	Hull
1057	V	Empire *Snow 85-82*	W S	American	1075	183·5	35·6	23·4	Kennebunk Me	1869	N. L. Thompson	Y [illegible]	Geo. C. Lord	Boston, Mass.
1058	*	Empress *Goldfinch 85-88*	W 3m Sr	British	120	94·5	23·2	10·4	Connah's Quay	1885	Fergus'nd Baird		Whitstable Shipping Co. Ld.	Whitstable
1059	*	Empress *Grant 88-88*	W Sr	British	91	82·9	20·0	9·9	St. Malo	1876	Gautier	Y [illegible]	J. Buhts & Co.	Jersey
1060		Empress *Kendall* (ex [illegible])	W Sr	American	121	86·0	26·0	7·0	Bristol, Me.	1877			L. J. Morse	Bangor, Me.
1061		Empress *Hatch*	W Sp	American	115	74·0	24·0	6·8	Prospect, Me.	1857			J. E. Farnsworth	Rockland, Me.
1062	N	Empress *Bjønness*	W Bg	Norweg'n	281	94·3	26·6	17·0	Sunderland	1836	J. Hutchinson		L. Hofgaard & Co.	Tönsberg
1063		Empress of China *Facy*	W 3m Bn	British	255	120·7	26·1	12·8	Padstow	1874	Stribley		P. Facy	Hobart
1064	*	Empress of India *Birch*	W Sk	British	77	75·9	20·1	10·2	Rye	1883	J. C. Head		T. Birch	Hull
1065	*	Empresa *Allan 76-78*	C Bk	British	236	123·3	25·1	11·3	Glasgow	1865	Hedderwick & Co.	[illegible]	John Bickers	Adelaide, S.A.
1066	*	Emulator *Kendrick 83-84*	W Sr	British	155	98·8	23·3	11·7	Kingston-on-Spey	1878	Kinloch	Y [illegible]	C. T. Bennett	St. Johns, N.F.L.
1067		Emulo *Busetto*	W Bg	Italian	309	106·9	24·6	16·9	Venice	1866	P. Grasso	Y [illegible]	A. Busetto & Co.	Venice
1068		Encarnacion (ex Dama Petronila, ex Lorenzo)	W Bk	Spanish	624	152·5	30·2	17·9	Vinaros	1874	Solomo	Y [illegible]	Jané & Co.	Barcelona
1069	V	Encarnación *Moraga*	W Bn	Spanish	181	99·1	24·4	12·2	Villajoyosa	1877	Vila		V. Galiana & V. Domingo	Alicante
1070		Encarnacion *Marcgas* (ex Casimiro)	W Bn	Spanish	173	72·2	19·2	9·8	Barcelona	$18\frac{37}{72}$			José Agulla	Vigo
1071	*R	Enchanter *Scott*	W Bk	British	461	125·7	24·3	18·4	Dundee	1851		Y 12.88	F. Scott	Liverpool
1072	R	Enchantress *Holwood -88*	W Bk	British	577	112·2	32·1	18·1	Moss Glen, N.B.	1870	Witmore	Y [illegible]	A. Blaney (per D. R. Bannatyne)	Glasgow
1073		Enchantress	W Sr Yt	British	223	130·5	24·6	..	New York	1871	Fish		Maj.-Gen. O. Williams M.P.	Cowes
1074		Enchantress *Rollent*	W Sr	American	353	125·6	30·6	10·0	Hoboken, N.J.	1852			F. Rollent	Port Jefferson, N.Y.
1075		Endeavour	W Bg	British	271	100·5	26·5	16·0	Sunderland	1859	J. Lister		E. Murray	Limerick
1076	V	Endelave *Olsen 76-76*	W Bn	Danish	142	90·6	21·4	10·8	Horsens	1876	Chr. Schroder	Y [illegible]	Actieselskab "Endelave" (M. Pedersen)	Horsens
1077	V	Endrick *Warner -83*	W Bn	British	313	124·1	28·7	13·0	Young's Cove, N.S.	1880	A. Malcolm	Y [illegible]	A. Malcolm	St. John, N.B.
1078	*	Endymion *McKechnie 82-84*	I Bk	British	759	188·7	31·9	18·4	Glasgow	1869	J. G. Lawrie		J. E. Munro	London
1079	*D	Endymion *Ryder -88*	W Bg	British	251	115·8	24·6	14·6	Salcombe	1870	Evans	Y [illegible]	Balkwill & Co.	Salcombe
1080	I	Enea *Descalzo*	W Sr	Italian	108	77·4	22·3	11·7	Chiavari	1869	L. Gotuzzo	Y [illegible]	Flli Questa	Spezia
1081		Eneide (ex Fede, ex C. Cosio)	W Bk	Italian	753	164·0	33·1	21·9	Voltri	1874	A. Canepa	Y [illegible]	Budd, Dellepiane & Co.	Genoa
1082		Energi (ex Johannes, ex Don Juan, ex Johannes) (CONDEMNED) *Hansen*	W Bg	Norweg'n	257	106·3	26·9	14·0	Wolgast	1846	G. Ehrich		J. Börresen	Grimstad
1083	N	Energi *Johnsen*	W Bn	Norweg'n	151	87·3	23·9	11·9	Christiania	1867	Mathisen	Y [illegible]	C. F. Rönneberg & Sönner	Aalesund
1084	V	Energy *McBride*	W Bn	British	193	100·0	26·6	11·1	Parrsboro'	1883	N. W. Barto		John Porter	Parrsboro', N.S.
1085	*	Energy *Strong 86-86*	W Bn	British	146	96·0	22·1	11·0	Rye	1878	Hoad Bros.	Y [illegible]	A. Goodridge & Sons	St. Johns, N.F.L.
1086	DV	Engelbrekt *Eckman*	W Bk	Swedish	659	150·5	30·6	18·1	Umeå	1874	O. Jonsson	Y [illegible]	C. G. Moritz	Umeå
1087		Engelbrekt *Larsson*	W Sr	Swedish	121	84·0	22·5	9·9	Jernavik	1870	Jonasson		M. Magni	Jernavik
1088		Engla *Andersson 79-*	W Sr	Russian	231	99·2	25·5	13·7	Westanfjerd	1874	G. Sjöström		C. M. Eriksson	Westanfjerd
1089	*	England's Hero *Ellis -87*	W K	British	99	84·2	21·2	11·0	Brixham	1887	S. Dowdney & Sons		J. W. Webb	Hull
1090	*	England's Rose *Curr 85-85*	W Sk	British	88	76·5	21·0	10·5	Grimsby	1885	Smith Bros.		W. Carr	Grimsby
1091		English Girl *Peacock 84-86*	W Sr	British	128	92·0	22·8	11·3	Preston	1869	W. Allsup & Sons		Porter's Ship Co., Ld.	Fleetwood
1092		Englishman	W 3m Sr	British	183	112·5	21·7	11·1	Cowes	1854	John White		J. S. Sterry	Lowestoft
1093		Englishman *Green*	W Sr	British	129	94·0	22·7	11·2	Glasson Dock	1864	Simpson		Carter & Co.	Fleetwood
1094	N	Enigheden *Nielsen 87-88*	W Bk	Norweg'n	278	98·2	26·4	15·1	Norrbotten	1812			J. Busch	Risör
1095	N	Enigheden *Knudsen 85-83*	W Sr	Norweg'n	151	78·2	24·3	12·2	Barth	$18\frac{35}{67}$			J. Petersen	Holmestrand
1096	N	Enigheden *Ruenaess -86*	W Sr	Norweg'n	143	78·2	22·6	12·0	Arendal	1856	N. Evensen		O. Ruenaess	Christiansand
1097	RV	Enigma *Bedell*	W Bk	British	649	148·3	32·5	18·6	St. Martins, N.B.	1871		Y 10.86	Geo. Bedell	St. John, N.B.
1098	*D	Enjetta *Buil*	W Bn	Dutch	171	97·1	23·2	11·7	Pekela	1867	J. H. de Boer	[illegible]	J. J. Koerts, jun.	Oude Pekela
1099		Enmore *Hutchinson 83-85* (BROKEN UP)	W Bk	British	561	118·0	30·2	19·0	Bristol	1858			R. L. Hutchinson	North Shields
1100	*	Ennerdale *Dunnan 74-83*	I S	British	1249	226·4	36·3	21·8	Liverpool	1874	W. H. Potter & Co.		J. D. Newton & Co.	Liverpool

*=Lloyd's Register; D=Nederlandsche Vereeniging; G=Germanischer Lloyd; I=Registro Italiano; R=American Record;
A=Veritas Austro-Ungarico; H=Veritas Ellenico; N=Norske Veritas; V=Bureau Veritas.

Dans la 4me Colonne, C=Composite (membrure en fer, bordé en bois); I=Fer; Stl=Acier; W=Bois.

Les Dates du doublage en métal (Col. 11) ressortent des livres dans lesquels les navires sont respectivement classés ou des rapports des Inspecteurs du Lloyd's Register.

Pour plus amples renseignements sur les navires classés par le Lloyd's Register, voir le Lloyd's Register Book et le Lloyd's Yacht Register.

1 Official No.	2 No. in Book	3 Registry in which classed	4 Ship's Name. Master, and Dates of Appointment to present Owner's Service and to Ship. Ship's Late Name.	Material and Rig	5 Flag	6 Net Tonnage	7 Dimensions. Length	Breadth	Depth	8 Built. Where	9 When	10 By whom	11 Date of Metalling	12 Owners or Managers
8227	1101		Enoch Talbot *Stannard* [-88	W Bk	American	1139	178·0	36·0	24·9	Freeport, Me.	1857			E.E. Keutfield
	1102	N	Enok *Mitchelsen 69-69*	W Sr	Norweg'n	113	85·7	21·1	9·9	Tönsberg	1867	SörenOlsen		J.T. Lund
	1103	N	Enos *Petersen 83-83*	W Bn	Norweg'n	186	113·0	25·2	11·6	[illegible]	1870	KarlHansen		A.J.Nielsen & Co.
	1104		Enos	W Bg	Turkish	130	..	..	..	Enos	1863			Madame Christophel Iconvell & Co.
	1105		Enos	W Bg	Turkish	125	..	..	..	Enos	1869			Iskelledili Costanti Angeloni Tredell & Co.
8895	1106		Enos B. Phillips *Baker -75*	W3mSr	American	359	131·0	30·0	14·0	Wilmington,Del	1870			Doane & Co.
8591	1107	V	Enos Soule *Soule*	W Bk	American	1519	198·4	38·1	26·9	Freeport, Me.	1869	E.C.Soule	[illegible]	E.Soule
35	1108		Enosis (ex Stat Iran)	W Bk	Greek	392	117·1	28·9	..	[illegible]	1858			N.Cokinos
	1109	H	Enosis	W Bg	Greek	165	79·7	23·6	..	Spezia	1862			E.Castriotis
738	1110	A	Enrica *Stipanovich*	W Bg	Austrian	432	133.2	27.1	17.8	Fiume	1868	F^lli Schiavon	[illegible]	A.Luppis fu M.
3598	1111	DIV	Enrichetta Accame *Giola*	W Bk	Italian	995	191·9	34·8	23·4	Varazze	1881	B.Cerruti	[illegible]	Em Accame & Figli
2855	1112	I	Enrichetta M. *Massa*	W Bg	Italian	298	114·5	24·8	15,6	Sestri	1873	E.Tixi	[illegible]	Franc. Massa
2777	1113	IV	Enrichino (ex Taormina) *Schiaffino* [-88	W Bk	Italian	955	187·0	33·5	23·0	SestriPonente	1876	O.Casanova	[illegible]	F.Mortola
519	1114		Enrico (COLLISION) *Ventre -87*	W Bk	Italian	399	116·1	30·0	18·7	Chiavari	1870	M.Tappani	[illegible]	Piaraggini
605	1115	A	Enrico (ex Fortunata) *Tonietti*	W3mSr	Italian	376	122·0	29·1	18·5	Savona	1868	F.Sirello	[illegible]	G.Tonietti
33	1116		Enrico *Fravega*	W Bg	Italian	139	91·5	23·5	12·2	[illegible]	1848	N.Pichiotti		S.Tonietti
934	1117	I	Enrico Andrea (ex Enrico Padre) *Arena*	W Bn	Italian	115	87·6	22·3	10·7	Castellamare	1884	G.Bonifacio	[illegible]	A.Arena
2046	1118	✱I	Enrico S. (ex Giovanni B.) *Sajoni -88*	W Bk	Italian	700	164·5	33·1	22·1	Savona	1876	F.Sirello	[illegible]	G.B.Sirombra
	1119	DV	Enrique (STRANDED) *Petersen 68-83*	W Bg	German	213	112·2	24·8	12·5	Hadersleben	1864	P.Kieldrup	[illegible]	H.L.Rochow
	1120	V	Enrique (ex Francisco Aballi) *Martinez -88*	W Bn	Spanish	225	98·1	25·2	14·8	Barcelona	1853		[illegible]	B.Prats
	1121		Enrique (ex Astera)	W Bg	Spanish	211	118·2	24·2	12·9	Bideford	1869	Johnson		Mariano Bertolues
	1122	V	Enriqueta *Alsina*	W Bg	Spanish	187	100·4	24·5	12·7	Blanes	18 43/67			Gusi Hermanos
	1123		Enriqueta (ex Rosario)	W Bn	Spanish	113	82·	22·5	10·4	Villajoyosa	1876	M.Vaello		A.Garriga y Ventura
	1124		Ensayo (ex Steamer)	W Bn	Spanish	209	141·0	22·8	8·9	Tayabas, [Luzon	1884	C.Alcautara		R.Reyes & Co.
45898	1125	✱	Ensign *Phillips 76-82*	W Bk	British	431	142·2	26·7	17·2	Dundee	1864	TayS.B.Co.	Y 11.85	E.Phillips
58039	1126		Ensign (LOST) (ex Banneret) *Simpson 76-84*	W Bk	British	315	129·6	27·5	14·8	Shoreham	1870			James Paterson
79957	1127	✱	Ensign *Pirie 86-86*	W Sr	British	100	87·0	21·3	10·2	Garmouth	1883	W. Anderson	[illegible]	J.Geddes
1706	1128	V	Entella *Mangini -88*	W Bk	Italian	694	153·1	32·4	20·2	Lavagna	1868	G.Saccomano	[illegible]	B.Roncagliolo
	1129	✱	Enterkin	Stl S	British	1600	255·0	39·0	22·7	Pt.Glasgow	1889	R.Duncan & Co.		T.C.Guthrie
33274	1130		Enterprise	W Sr	British	124	86·0	23·0	8·3	Quebec	1856			L. Beaudouin
80254	1131		Enterprise *Carter -88*	W K	British	121	94·5	22·9	8·5	Rye	1888	G. & T.Smith		E.Bayley
1988	1132		Enterprise *Quin 69-78*	W Sw	British	119	81·8	20·3	13·3	Portmadoc	1845			I.Hodgson
36823	1133	D	Enterprise	W Bn	British	113	82·0	20·2	11·4	P.E.Island	1860	Sampson		G.Potter
67961	1134	✱	Enterprise	W Sr	British	95	87·1	21·2	9·6	Bridgwater	1875	Gough		J.Reney
63487	1135	✱	Enterprise	W K	British	71	78.8	19·3	9·3	Shoreham	1872	May		D.W.Foreman
7083	1136		Enterprise	W Sr	American	180	100·6	21·5	9·0	CoosBay,Oreg	1863			H.V.Winslow
8728	1137		Enterprise *Robinson*	W Sr	American	136	88·8	26·0	8·2	Cherryfield,Me	1871			J.Hinckley
	1138	✱	Enterprise (ex Trieste, ex Steamer)	I Bk	Argentine	594	176·5	28·2	16·1	Sunderland	1856	J.Laing		Argentine Govt.
	1139	NV	Enterprise *Gundersen 87-87*	W Bk	Norweg'n	1461	210·0	40·8	24·8	Quebec	1875	Lane	[illegible]	W.Wilhelmsen
21543	1140	✱	Enterprize *Evans*	W Sr	British	102	87·3	19·3	10·6	Newport	1846			T.Morgan & Co.
	1141		Entilla (ex Minnie H. Gerow) *Serra*	W Bk	Chilian	1069	199·9	37·8	23·7	Moncton,N.B.	1874			Solari & Briguardelli
	1142		Entreprenante *Agueray*	W3mSr	French	135	86·9	20·0	11·6	Dunkirk	18 69/70	PyotteBaynert		A.Demarle-Fetel
	1143		Entrepreneur *Bodigust*	W Bn	French	132	79·7	20·0	11·5	St.Malo	1867	Tranchemer	[illegible]	A.Michel & Mars
3501	1144	I	Eolo (ex Albom) *Tassara -8?*	W Bk	Italian	490	139·4	31·2	19·7	Pra	1871	Durante	[illegible]	A.Tassara&Bartol
	1145	N	Eos *Petersen 85-85*	W Bk	Norweg'n	400	122·3	29·1	16·3	Lillesand	1871	T.Aslagsen	[illegible]	O.C.Heldahl
80161	1146		Eothen	W Sr	British	111	82·4	21·5	9·0					A.H.Browne
7187	1147		Eothen *Thomas*	Sr	American	102	83·0	22·0	8·0	Calais, Me.	1860			Alexander Bros.
27	1148	A	Epaminondas (ex Presidente Saruscotto) *Brulotto*	W Bk	Greek	551	146·9	30·9	20·2	Genoa	1868	M.Bossi	[illegible]	N.Calamaris
	1149	✱	Epervier (WRECKED) *Portier*	W Bk	French	650	159·1	28·5	18·7	Bordeaux	1870	Chaigneaufrères	[illegible]	Devé & Co.
	1150	N	Ephialtes *Henrichsen86-86*	W Bk	Norweg'n	669	164·5	32·8	18·4	Grimstad	1879	JacobBoe	[illegible]	Christen Boe

✱=Lloyd's Register; D=Nederlandsche Vereeniging; G=Germanischer Lloyd; I=Registro Italiano; R=Am
A=Veritas Austro-Ungarico; H=Veritas Ellenico; N=Norske Veritas; V=Bu

In the 4th Column, C=Composite (iron frame, wood planked); I=Iron; Stl=Steel; W=Wood.

The Dates of Metalling (Col. 11) are based on the records of the Books in which Vessels are respectively classed, or on the Reports of the Surveyors

Further particulars of Vessels classed by Lloyd's Register will be found in Lloyd's Register Book and Lloyd's Yacht Register

No.	Classe	Nom. Capitaine et dates d'entrée au service des Armateurs actuels et à bord du navire. — Ancien Nom.	Matériaux et Gréement	Pavillon	Tonnage	Dimensions: Longueur	Largeur	Creux	Construit: Où	Quand	Par qui	Date du doublage en métal	Armateurs ou Directeurs-Gérants	Port d'Armement
1151		Ephormia	W Bg	Turkish	136	..	..	..	Greece	1879			Artemisia Hortoradi	Constantinople
1152		Ephraim & Anna *Stewart*	W Sr	American	206	105·9	27·9	9·0	Wilmgtn.Del.	1852			J.E.Reinhart	Philadelphia
1153	✻	Ephratah *Williams*	W Bg	British	230	112·3	25·1	11·4	Portmadoc	1873	Roberts	Y 3.88	J.Owen & Co.	Carnarvon
1154	N	Ephrussi *Samuel 82-87*	W Bk	Norweg'n	346	133·9	29·1	15·2	Bergen	1869	JensGram	Y 3.87	O.Nygaard	Stavanger
1155		Epirus	W Bg	Greek	334	107·6	27·0	..	Scopelos	1867	John Alexander			Scopelos
1156	✻	Epsilon *Jones*	W Bk	British	521	150·9	29·5	18·8	Bideford	1865	Cox	Y 6.88	G.Shepherd	Swansea
1157	RV	Equator *Crosby -86*	W S	British	1273	200·5	38·3	22·6	SalmonRvrNS	1878	J.S.Allen	Y 3.88	H.Lewis	Yarmouth,N.S.
1158	V	Eqvator *Holmqvist -88*	W Bk	Swedish	511	156·5	22·4	16·9	Sundsvall	1877	J.Sjölen	Y 5.89	Sundsvalls Skeppsrederi-Bol (A.C. Wallberg)	Sundsvall
1159		Era *Whilford*	W Bn	British	181	88·3	21·7	14·2	Bristol	1849		Y 4.88	W.F.Ball	London
1160		Era *Spicer -87*	W Sr	American	127	93·0	26·0	9·0	Bostn,Mass.	1847			C.A.Williams	NewLondon,Con.
1161		Erachio *Ag Hassan* — Ape W. & Jamaica	W Bk	Turkish	367	123·6	27·8	15·7	Ancona	1853			F.& M.Christodolu	
1162		Erasmina *Brunetti* — ex Black Wasp	W Sr	Italian	143	102·6	21·3	12·2	Kingstown	1875	Kinloch	Y 2.88	C.Pendibene	Genoa
1163	I	Erasmo *Podestà -87*	W Bk	Italian	568	151·9	31·4	20·9	Chiavari	1867	M.Tappani	Y 10.87	E.Raffo fu Anto.	Savona
1164		Erasmo C. *Castella*	W Bn	Italian	176	105·0	25·1	12·9	Savona	1881	F.Sirello	Y 3.88	E.Costella	Genoa
1165	R	Erastus Wyman *Betts*	W 3m Sr	American	568	148·6	36·3	15·5	Bath,Me.	1880	Goss&Sawyer		T.T.Anderson	New York
1166	✻	Erato *Green 86-88*	I S	British	1205	206·0	35·3	23·1	Isle of Man	1864	RamseySh.Co.		Foley & Co.	London
1167	DV	Erato *Schumacher -88*	W Bk	German	641	150·9	32·5	19·1	Hamburg	1873	BWenckeSöhne	Y 4.88	B.Wencke Söhne	Hamburg
1168	N	Erato *Fruitz*	W Bk	Norweg'n	639	152·9	34·3	18·5	Drammen	1875	Svendsen	Y 2.87	T.Bassöe	Christiania
1169	N	Erato *Jensen 80-88*	W Bk	Norweg'n	544	133·3	33·3	18·4	Brahestad	1859			I.Hesselberg	Laurvig
1170		Erato *Hansen 85-85*	W Bg	Norweg'n	409	123·5	27·8	18·1	Sunderland	1853			Thommesen & Son	Arendal
1171	N	Erato *Salvesen 75-79*	W Bk	Norweg'n	284	109·3	27·1	14·0	Grimstad	1879	N.O.Gjömle	Y 3.88	P.C.Falchenberg	Grimstad
1172	V	Erato *Paterson*	W Bk	Swedish	585	150·5	29·9	18·3	Gothenburg	1869	N.C.Kierkegard	Y 1.88	G.D.Kennedy	Gothenburg
1173	N	Erbrin *Petersen*	W Bk	Norweg'n	599	152·5	33·0	18·2	Arendal	1879	J.Halvorsen	Y 6.88	H.Aalholm	Arendal
1174	V	Ercole *Ruggiero*	W Bk	Italian	517	141·0	33·4	19·9	Cassano	1872	A.Castellano	Y 5.88	Arcangelo Cacace	Castellam. di Stabbia
1175	I	Ercole *Albano*	W Bg	Italian	358	112·2	28·7	17·5	Procida	1868	N.Lubrano	Y 8.88	A.Lubrano	Naples
1176	I	Erede *Acquarone 86-87*	W Bk	Italian	576	139·7	28·6	19·9	Lavagna	1871	Brignole	Y 7.88	L.Bianchi	Genoa
1177	✻	Erema	W Bkn	British	299	122·0	26·9	14·5	P.E.Island	1877	PeakeBros&Co.	Y 6.87	Geo.Peake	Charl'ttetn.P.E.I.
1178	A	Ergo *Giadrossich*	W Bk	Austrian	569	136·1	28·8	18·5	Lussinpiccolo	1875	M.A.Starcich	Y 7.88	L.S.Giadrossich	Trieste
1179		Erhard *Lodwell -86*	W Sr	Russian	119	87·1	25·0	10·3	Kurkis	1875		Y 7.87	W.von Aderkas	Haynasch
1180		Eri *Sherman*	W Sr	American	137	92·6	25·9	9·1	Machias,Me.	1866			A.Peters	New York
1181		Eric *Rekka -85*	W Bn	Russian	187	96·4	25·4	10·9	Dagö	1869	P.Pargo		Graf E.Ungern-Sternberg	Dagö
1182	A	Erica *Czar*	W Bn	Austrian	104	75·7	18·8	10·6	Lussinpiccolo	1880		Y 5.86	G.Bussanich	Lussinpiccolo
1183	I	Erice (ex Caroghi) *Brignati*	W Bk	Italian	407	122·3	30·7	18·1	Sestri Ponente	1864	G.Peragallo	Y 12.85	Cav.D'Alì	Trapani
1184		Ericeirense (ex Vigilante)	W Bn	Portug'se	161	82·7	22·3	11·1	Oporto	1857			A.da Costa Morgado	Lisbon
1185		Ericsson *Maine 87-*	W S	American	1568	248·0	40·4	27·0	New York	1852		Y 11.87	Geo.Plummer	San Francisco
1186		Eridano *Juricevich* (WRECKED)	W Sr	Austrian	127	83·6	20·0	11·1	Lussinpiccolo	1877	M.A.Starcich	Y 4.86	D.G.Luzzatto	Trieste
1187		Erie *Theall*	W Sr	American	219	122·6	26·2	10·4	P.Huron,Mich	1866			F.E.Winslow	Boston,Mass.
1188		Erie Belle	W Sr	British	275	135·5	24·4	9·8	PortBurwell	1873	G.W.Pontine		DonaldMonson	Pt.Burwell,Ont.
1189		Erie Queen	W Sr	British	217	113·0	23·6	10·2	Port Rowan	1874	J.Hudson		O.Barrett	Pt.Rowan
1190		Erie Stewart	W Sr	British	230	117·7	23·0	10·7	PortDover	1874	G.S.Waterberry		E.Hammond	Dover,Ont.
1191	GV	Erik *Rasmussen 80-86*	W 3m Sr	Danish	213	114·9	23·4	11·2	Svendborg	1880	J.R.Andersen	Y 10.87	N.W.Kaas	Svendborg
1192		Erik (WRECKED) *Andersen*	W Bg	Norweg'n	220	90·4	24·9	15·0	Hernösand	1815			C.Taarvig	Fredrikstad
1193		Erik *Nilsson*	W Bg	Swedish	241	93·8	23·8	15·1	Strömstad re-bt.	18 70			L.O.Asberg	Strömstad
1194	NV	Erik Dale *Jacobsen -88*	W 3m Sr	Norweg'n	366	128·3	29·5	13·3	Arendal	1888	H.Stiansen	Y 6.88	E.Dale (A.Evenbye)	Arendal
1195		Erika *Mattsson 69-*	W Sr	Russian	250	107·2	27·3	12·6	Christiansand	1869	Graufrey		G.P.Markström	Oeland
1196		Erika *Henriksson -87*	W Sr	Russian	246	107·0	29·0	13·5	Kimito	1873	G.Sjöström		A.Gustafsson	Westanfjerd
1197		Erika *Sandell*	W Sr	Russian	129	..	..	..		1876			J.F.Strandberg	Bromarf
1198		Erika (ex Sylvia) *Jansson* (LIGHTER)	W Bn	Swedish	233	100·8	26·0	12·4	Finland	1863			Lagerstedt & Johansson	Stockholm
1199		Eriminta *Dunn 83-89*	W S	British	1213	186·2	38·3	23·6	St.John,N.B.	1867	Milledge	Y 4.88	T.J.Dixon	Greenock
1200	N	Erin *Holst -87*	W Bg	Norweg'n	260	94·5	27·4	15·3	Grimstad	1849	JensTellefsen		F.H.Holst	Grimstad

✻ = Lloyd's Register; D = Nederlandsche Vereeniging; G = Germanischer Lloyd; I = Registro Italiano; R = American Record; A = Veritas Austro-Ungarico; H = Veritas Ellenico; N = Norske Veritas; V = Bureau Veritas.

Dans la 4me Colonne, C = Composite (membrure en fer, bordé en bois); I = Fer; Stl = Acier; W = Bois.

Les Dates du doublage en métal (**Col. 11**) ressortent des livres dans lesquels les navires sont respectivement classés ou des rapports des Inspecteurs du Lloyd's Register.

Pour plus amples renseignements sur les navires classés par le Lloyd's Register, voir le Lloyd's Register Book et le Lloyd's Yacht Register.

Official No.	No. in Book	Society in which classed	Ship's Name. Material and Rig. Master, and Dates of Appointment to present Owner's Service and to Ship. Ship's Late Name.	Flag	Net Tonnage	Dimensions: Length	Breadth	Depth	Built: Where	When	By whom	Date of Metalling	Owners or Managers
72311	1201	✻V	Erin's Isle *Waycott 71-87* WS	British	1799	221·5	40·2	25·3	St.John	1877	M.Fee	[illegible]	J. R. de Wolf & S[illegible]
357	1202		Erinio *Parthenopoulos* WBn	Greek	311	106·6	26·9	..	Syra	1875		[illegible]	D.Parthenopoulos
	1203		Erinula *Belivanachi* WBn	Greek	263	..	..	..	Syra	1871		[illegible]	G.Cornilachi
2988	1204	D	Eritreo *Castagnola 89-88* WBk (ex Marinin S., ex Lorenzo)	Italian	779	171·1	30·1	22·6	Varazze	1874	G.Scala	[illegible]	A.Badino
	1205		Erkedo WBg	Turkish	175	..	..	..	Scio	1870			Manolaki Kor[illegible]
	1206	N	Erling *Christiansen* WBk	Norweg'n	400	120·5	30·8	16·5	Carlskrona	1848			Brödr.Christians[illegible]
	1207	NV	Erling *Svendsen -81* WBk	Norweg'n	347	121·9	27·6	14·9	Grimstad	1875	Jörgen Tellefsen	[illegible]	Th.Nielsen
	1208		Erling *Olsen* WBk	Norweg'n	294	121·0	27·4	14·2	Shoreham	1856			Actieselskabet [illegible]
	1209		Erling *Andersen* WSr	Norweg'n	198	99·0	25·9	11·7	NovaScotia	1867			M.Pedersen
	1210	N	Erling *Odland 79-79* WSr	Norweg'n	113	92·3	23·1	9·9	Bergen	1879			Waage & Olsen
	1211	N	Erling Skjalgson *Sandved* WBk	Norweg'n	567	147·2	32·7	17·7	Stavanger	1864	P.Petersen	[illegible]	M.G.Monsen
	1212	V	Erlkönig (HULK) WBk	Chinese	457	129·9	28·9	17·4	Stralsund	1873	A.Juhl	[illegible]	Chinese Govern[illegible]
74627	1213	✻	Erme *Wyatt 83-83* WSr	British	109	92·2	23·3	8·5	Plymouth	1876	Marshall		R.Bayly
	1214		Ermesinda WSr	Spanish	115	77·8	19·8	10·8	Barcelona	1853			F.Fuster y Valls
73507	1215	✻D	Erminia *Davies 74-77* WBk	British	631	158·2	32·3	17·7	P.E.Island	1877	Yeo	[illegible]	E. Davies (T. P. Richards & Co., Mangrs.)
14566	1216		Erminia *Colwell* WSrYt	British	183	107·0	23·3	12·9	Gosport	1849	Camper		Major J.B.Colwell
	1217	N	Erminia *Schanke* WBk (ex Erminia Wilson)	Norweg'n	408	121·8	29·6	17·7	Bordeaux	1854			J.Kraft Pedersen
3076	1218	I	Erminia C. *Chiappe -87* WBk (ex Gius. Bennetti)	Italian	542	112·7	31·3	19·0	Arenzano	1876	Debarbieri	[illegible]	G.Chiappe
3536	1219	I	Ermina V. *Santagata* WBk (ex Olga)	Italian	549	136·4	31·4	19·3	Savona	1872	G.Calamaro	[illegible]	B.Varé di G.
	1220	V	Erna *Poulsen* WBn	Danish	134	93·8	20·8	10·5	Assens	1870	H.Christoffersen		H.M.Poulsen
	1221	V	Erna *Böger* WBk	German	582	141·0	29·5	15·1	Vegesack	1868	J.Lange	[illegible]	E.Burchard
	1222	N	Erna *Asbjörnsen* WBk	Norweg'n	484	133·8	31·8	16·5	Arendal	1867	A.K.Andersen	[illegible]	E.Asbjörnsen
	1223	N	Erna *Thommesen 87-87* WBk	Norweg'n	228	116·7	27·3	14·7	Porsgrund	1866	NilsLarsen		Chr.Schöning
483	1224		Ernast David W3mSr	Russian	289	120·9	29·6	12·9	Livland	1887			S.Grewe
91923	1225	✻	Erne *Lardner 86-86* IS	British	1613	255·9	38·3	23·2	Greenock	1886	Russell & Co.		J.Nourse
56964	1226	✻	Ernest *Rhodes -87* W3mSr	British	183	112·9	23·1	12·5	GlassonDock	1878	Simpson	[illegible]	J.Helme
	1227		Ernest *Allais* WBk	French	291	105·9	24·9	15·0	Dunkirk	1866	Pyotte-Bayaert		Gosselin & Dene[illegible]
	1228		Ernest Marie *Hénry* WBg	French	180	95·2	20·3	11·8	St.Malo	1866	Vve.Dandin& [fils]		Guibert & Fils
3627	1229		Ernesta G. (ex Ledeetta, ex Maria D., ex Maria Bollo) WBk	Italian	563	151·2	30·8	19·9	SestriPonente	1868	A.Briasco fu G.B.	[illegible]	L.Dodero
8990	1230	R	Ernestine *Whittier -87* WBg	American	400	120·2	28·6	16·4	Yarmouth,Me	1868	Hutchins& Stubbs	[illegible]	C.H.Chase
	1231	V	Ernestine *Oger* WSr	French	138	93·2	20·3	11·5	St.Malo	1882	F.Gautier		Sécherie de Morue [illegible] Port de Bouc
	1232	V	Ernestine *Le Person* WBn (ex Nieuwe Maas)	French	101	76·9	22·3	13·6	Vlaardingen	1866	Roelants		V.M.Le Person
	1233	DGV	Ernestine *Schiemann -89* WBk	German	827	168·0	34·1	20·5	Elsfleth	1884	J.Jürgens	[illegible]	C.G.Beermann
	1234	V	Ernestine *Matern 80.* WBn	German	161	74·8	22·3	11·9	Ueckermünde	18 33/75			A.Matern
	1235		Ernestine *Söderberg* WSr	Swedish	151	88·9	21·5	12·0	Kiel	1853	A.Conradi		L.E.Söderberg
	1236		Ernesto WSr	Brazilian	160	..	..	..					M.A.Moreira
531	1237	A	Ernesto *Martelli* WBn	Italian	208	109·5	24·9	13·4	Lussinpiccolo	1874	N.Martinolich	[illegible]	Ant.Martelli di G.
	1238		Ernesto (ex Jaime Millet) *Pla* WBn	Spanish	284	107·9	28·4	13·7	Blanes	1872			Silverio Millet
3335	1239		Ernesto M. (ex Nicaragua) WB	Italian	107	90·2	20·7	10·8	Savona	1871	G.Ferari		Elisa Garibaldi
	1240	G	Ernst *Ahrens* WBk (ex Aardenburg, ex Psyche)	German	659	165·0	31·8	18·8	Quebec	1866	Dinning	[illegible]	G.Kindler
	1241	GV	Ernst *Bulow -88* W3mSr	German	339	131·5	27·6	16·0	Swinemünde	1865	G.Müller	[illegible]	R.Josephy
	1242		Ernst *Wallis -77* WBk	German	323	110·5	27·7	16·7	Barth	1855	Schlör		J.H.Niemann
	1243	V	Ernst *Klickow 83-85* WSr	German	155	83·3	21·2	11·5	Stralsund	1875	O.Kirchhoff		Carl Lobeck
	1244		Ernst (ex Avance) *Grönvall 73-81* WBg	Swedish	213	92·8	24·5	13·7	Grimstad	1857			L.Bauer
	1245	V	Ernst Friedrich *Gering 83.* WBkn	German	250	111·0	27·3	12·4	Colberg	1863	F.Richter		E.F.Hackbarth [illegible]
	1246	GV	Ernst Ludwig Holtz *Epping 76-86* WBk	German	470	130·6	30·4	18·5	Rostock	1876	H.Rickmann	[illegible]	E.Burchard
8001	1247		Ernst T. Lee *Thomas* WSr	American	165	104·0	27·5	9·0	Calais,Me.	1873	Rideout&Lord		Rideout & Lord
	1248	V	Ernst & David *Greave -87* WSr	Russian	288	119·1	26·2	12·5	PetersCapell	1887	S.Greave		SimonGreave
	1249	G	Ernst & Georg *Schoon* WBn (ex Mary Nesch [illegible])	German	113	83·6	20·1	10·4	Papenburg	185?	B.J.Albers		H.Hovene
	1250		Ernst & Maria *Sørensen 83-84* WBk	Norweg'n	283	125·6	25·5	13·7	Bremerhaven	1864	R.C.Rickmers		H.H.Pettersen

✻ = Lloyd's Register; D = Nederlandsche Vereeniging; G = Germanischer Lloyd; I = Registro Italiano; R = [illegible]
A = Veritas Austro-Ungarico; H = Veritas Ellenico; N = Norske Veritas; V = [illegible]

In the 4th Column, C = Composite (iron frame, wood planked); I = Iron; Stl = Steel; W = Wood.

The Dates of Metalling (Col. 11) are based on the records of the Books in which Vessels are respectively classed, or on the Reports of the Surveyors [illegible]

Further particulars of Vessels classed by Lloyd's Register will be found in Lloyd's Register Book and Lloyd's Yacht Register

No.	Cote	Nom / Ancien Nom	Capitaine	Matériaux et Gréement	Pavillon	Tonnage net	Longueur	Largeur	Creux	Construit où	Quand	Par qui	Doublage	Armateurs ou Managing Owners	Port d'Armement
1251	N	Eros	Andersen	WBk	Norweg'n	[illegible]	117·7	[illegible]	16·5	Hernösand	1850			M.Pedersen	Mandal
1252	N	Eros	Pettersen	WBk	Norweg'n	[illegible]	[illegible]	28·3	13·9	Raumo	1868			R.Josephsen	Farsund
1253		Eros	Corneliussen	WBg	Norweg'n	243	92·3	25·[illegible]	14·6	Lillesand	1840			S.Corneliussen	Lillesand
1254	N	Erragon	Weber	WBk	Norweg'n	420	128·5	30·1	16·6	Grimstad	1873	O.Nittelsen	[illegible]	Eskohm C.Pedersen (E.Pedersen)	Grimstad
1255	A	Errante	Nicolich	WBk	Austrian	[illegible]	[illegible]	29·0	19·2	Lussinpiccolo	1876	G.S.Cattarinich	[illegible]	A.[illegible]	Lussinpiccolo
1256	A	Errico M.	Agresta	WBg	Italian	[illegible]	11[illegible]·4	26·6	16·4	Equa	1856		[illegible]	F.Troisi & L.Musso	Naples
1257		Erriestratos		WBg	Turkish	[illegible]	..	..	..	Vassilikos	1883			Vassilikoslee Panayot	Vassilikos
1258	A	Ersilia (ex Nuova Incoronata)		WBg	Italian	117	[illegible]	22·6	12·1	Procida	1871			N.Rosato	Naples
1259		Ersilia	Dos Santos	W3mSr	Portug'se	[illegible]	11[illegible]·8	22·2	10·2	Villa do Conde	1876	J.F.da Loja		A.L.Duarte	Oporto
1260	N	Erstatningen	Nielsen	WBk	Norweg'n	377	12[illegible]·5	29·2	15·9	Arendal	1873		[illegible]	O.Lydersen,jun.	Tvedestrand
1261		Erwin	Lübcke 77-88	WBk	German	[illegible]	121·1	28·5	16·7	Rostock	1857	O.Ludewig		A.Burchard	Rostock
1262	V	Erwin Rickmers	Mahlstede	WBk	German	[illegible]	2[illegible]·3	41·0	25·8	Geestemünde	1882	R.C.Rickmers	[illegible]	R.C.Rickmers	Geestemünde
1263	✻	Erycina	Peachy 81-87	WBk	British	[illegible]	152·3	29·8	19·8	Southwick	1865	W.Nichols'n&Sn	[illegible]	"Erycina" Ship Co. Ld (B.Rosser)	Swansea
1264		EsaiasTegnér	Jonsson -86	WSr	Swedish	121	78·1	19·9	12·0	Råå	1856			N.P.Jönsson	Helsingborg
1265	V	Esbroufe	Soale -82	WBk	French	371	127·3	25·3	15·2	St.Servan	1881	Lenormand frs	[illegible]	G.Lauriol	Nantes
1266		Escort	Flynn	WSr	British	130	87·1	18·8	11·7	TorquayDvon	1849			T.G.Kelly	Sydney N.S.W.
1267		Escort	Waterhouse	WBk	American	695	142·0	32·2	19·2	NewburyptMs	1870	Fillmore& Atkinson	[illegible]	B.F.Butler	Boston,Mass.
1268	✻A	Esempio	Petrina 77-71	WBk	Austrian	658	1[illegible]·1	32·3	20·2	Lussingrande	1877	P.Zucchi	[illegible]	C.Jellersitz	Lussingrande
1269	I	Esempio	Oneto 82-86	WBk	Italian	420	124·6	30·8	19·0	Voltri	1869	D.Salomone	[illegible]	G.B.Degregori	Genoa
1270	I	Esemplare	Serra -88	WBk	Italian	547	147·6	31·1	19·6	SestriPonente	1872	M.Bossi	[illegible]	Federico Vaccaro	Genoa
1271		Esk (ex Mapaelia)		WBk	British	415	1[illegible]·1	26·9	19·5					John Mill	Dunedin,N.Z.
1272	R	Esk (WRECKED)	Smelzer -87	WSr	British	147	98·0	24·0	9·8	BridgwaterNS	1887	T.A.Wilson	[illegible]	T.A.Wilson	Lunenburg. NS
1273	✻	Eskasoni	Burchill	IS	British	1715	2[illegible]·0	39·0	23·5	Stockton	1886	Richardson, Duck&Co.		Eskasoni Ship.Com.Lim. (P.C.Malo)	Windsor,N.S.
1274	✻	Eskdale	Murdock 86-85	IS	British	1225	22[illegible]·5	36·2	22·1	Whitehaven	1873	Whithvn S.B.C.		J.D.Newton & Co.	Liverpool
1275		Eskedal (WRECKED)	Rasmussen 87-87	WBk	Norweg'n	493	134·1	29·0	16·1	Grimstad	1878	J.Terjesen	[illegible]	J.Börresen	Grimstad
1276	N	Eskimo	Petersen -83	WSr	Norweg'n	111	[illegible]	22·6	10·3	Sandefjord	1882	Chr.Christensen		Interessentsk "Fisken" (Alf.Monsen)	Tönsberg
1277	✻	Esmeralda	Blair -87	IS	British	735	18[illegible]·1	31·1	19·0	Liverpool	1866	T.Vernon&Son		North Western Shipping Co., Ld.	Liverpool
1278	✻	Esmeralda	Smith 81-81	WSr	British	145	97·5	21·8	11·7	Brixham	1860	Richardson	[illegible]	J.Bovey & Co.	Brixham
1279	✻	Esmeralda	Williams -88	WSr	British	130	81·9	23·3	9·2	Yokohama	1888	YokohamaEng. & IrnWksCoLd		H.T.Williams	Shanghai
1280		Esmeralda		WSr	Brazilian	142	..	..	..					E.N.Doux	RioJaneiro
1281		Esmeralda (ex Puelche)	Armas 72-83	WBg	Chilian	189	1[illegible]·8	24·8	10·8	Sunderland	1864	Hardie		Cooper & Co.	Constitucion
1282		Esmeralda	Noland	WBk	French	384	1[illegible]	27·2	16·2	Nantes	1869	Jollet&Babin	[illegible]	H.Joyau	Nantes
1283		Esmeralda	Lebris	WSr	French	118	7[illegible]·8	19·0	10·8	Honfleur	1850			Mazure	St.Malo
1284	V	Esmeralda	Weller -88	WBk	German	788	160·1	30·8	20·0	Bremerhaven	1875	H.F.Ulrichs	[illegible]	Neu Guinea Cie.	Hamburg
1285	N	Esmeralda	Kjölner -83	WBk	Norweg'n	[illegible]	1[illegible]·8	27·9	14·8	Vegesack	1860	H.F.Ulrichs		J.C.B.Wang	Tönsberg
1286	N	Esmeralda	Jensen 85-85	WBk	Norweg'n	295	144·1	28·4	12·7	Carlshamn	1863	D.Andersson		Morten Mulvad	Frederikstad
1287	V	Esmeralda (ex Basconi)	Molins -87	WBk	Spanish	644	158·1	32·4	21·0	SestriPonente	1877	A.Carlevaro	[illegible]	Cia Brugués y Gali	Barcelona
1288	V	Esmeralda	Larsson	WBg	Swedish	218	11[illegible]·8	23·3	11·3	Kalmar	1872	M.Deck	[illegible]	J.Jönsson	Brantevik
1289		Espadarte (ex Alvaro ex Pinto)	Lagoa	WSr	Brazilian	115	..	..	..	SJoãodaBarra	1870		[illegible]	M.F.Lagôa	Rio de Janeiro
1290		España (ex Maria and Dolores, ex England)	Cisa	WBk	Spanish	581	144·5	27·4	18.8	Stockton	1857			R.Bosch y Alzina	Barcelona
1291	N	Espeland	Nielsen 75-90	WBk	Norweg'n	495	128·0	30·6	16·0	Risör	1875	A.Langemyhr	[illegible]	A. & F.Smith	Tvedestrand
1292		Esperança		WBk	Brazilian	243	..	..	..					A.A.Guimarães	RioJaneiro
1293		Esperança (ex John C. [illegible])	Valerra	WBk	Portug'se	429	137·9	29·5	17·0	Salem,Mass.	1865		[illegible]	J.S.d'Avellar	Terceira
1294	✻	Esperança (ex [illegible])		WSr	Portug'se	163	101·0	23·8	12·9	Dartmouth	1875	Philip	[illegible]	F.Henriques	Madeira
1295	DV	Esperance		WSr		165	7[illegible]·4	21·2	10·5	Pekela	1873	F.L.Drenth	[illegible]		
1296	A	Esperance (ex Speme Luigi Paleo)	Audoli -88	WBk	French	726	152·8	34·1	18·8	Leghorn	1854		[illegible]	Michel Segaro, père	Philippeville
1297	V	Espérance	Pignonels -87	WBg	French	166	78·1	22·3	14·1	LeLegue	1862	Lhotelier		A.Besnier	Binic
1298	V	Esperance	Benard -88	WSr	French	121	91·2	21·6	11·8	Paimpol	1886	Laboureur		L.Gortghebeur	Dunkirk
1299	N	Esperance	Christophersen (67-67)	WBk	Norweg'n	318	116·1	28·9	14·9	Risör	1867	T.Larsen		J.M.A.Marcussen	Risör
1300															

[illegible]=Lloyd's Register; D=Nederlandsche Vereeniging; G=Germanischer Lloyd; I=Registro Italiano; R=American Record;
[illegible]=Veritas Austro-Ungarico. H=Veritas Ellenico; N=Norske Veritas; V=Bureau Veritas.

Dans la 4me Colonne, C=Composite (membrure en fer, bordé en bois); I=Fer; Stl=Acier; W=Bois.

Les Dates du doublage en métal (Col. 11) ressortent des livres dans lesquels les navires sont respectivement classés ou des rapports des Inspecteurs du Lloyd's Register.

Pour plus amples renseignements sur les navires classés par le Lloyd's Register, voir le Lloyd's Register Book et le Lloyd's Yacht Register.

Official No.	No. in Book	Registry in which classed	Ship's Name. Material and Rig. Master, and Dates of Appointment to present Owner's Service and to Ship. Ship's Late Name.	Flag	Net Tonnage	Dimensions. Length	Breadth	Depth	Built. Where	When	By whom	Date of Metalling	Owners or Managers
	1301	N	Esperance *Kjær* W Bk	Norweg'n	253	107·9	27·5	13·7	Dunkirk	1856		[illegible]	O. Bjertness
	1302		Esperançoso II. *Lourenço* W Bn	Portug'se	193	96·5	23·0	11·1	Caminha	1870	A. J. Gonçalves		L. Barboza e Silva
20272	1303		Esperanza *Winter 72-75* W Bn	British	121	90·8	19·7	11·4	Liverpool	1857			A. Nelson
	1304		Esperanza *Gueglio 86-86* W Bk	Chilian	238	99·1	23·9	18·0	Warren, R.I.	1848			Cie. de Maderas [illegible]
	1305		Esperanza *Bengochea* (ex Louisa) W Bn	Spanish	198	100·4	26·4	12·3	Kyborg	1868		[illegible]	E. Cortada & Co.
	1306		Esperanza W Bn	Spanish	117	82·3	21·1	9·7	Samar	1873			Julian Cabañas
	1307		Esperanza *Jansson* W Bk	Swedish	360	131·3	24·1	16·5	England	1863			J. F. Sjögren
3423	1308	I	Espero (ex Prosperita) *Costa* W Bk	Italian	715	163·5	32·4	22·3	Savona	1875	F. Sirello	[illegible]	Paolo Costa
	1309	V	Espiègle *Miniac -88* W Bg	French	232	101·7	23·0	13·1	St. Servan	1885	Hamond	[illegible]	Guibert & fils
710	1310	V	Esploratore *Longobardi* W Bk	Italian	371	111·8	29·7	18·1	Alimuri	1868	F. Mauro	[illegible]	Flli. Longobardi fu [illegible]
	1311		Espoir *St. Andrea* (ex Malouine) W Sr	French	114	134·8	18·4	10·2	Greenock	1855			M. E. Puygreffier
	1312		Espora I Sr	Argentine	247	134·0	30·3	7·8	Port Glasgow	1884	Murdoch & [illegible] Murray		La Platense Flotilla Co.
625	1313	I	Espresso *Schiaffino -88* W 3m Sr	Italian	431	139·4	28·8	17·8	Rapallo	1873	G. Saccomano	[illegible]	F. Roncallo
3447	1314	I	Espresso T. *Massone* W Bk	Italian	507	131·2	31·8	19·2	Varazze	1868	S. Craviotto	[illegible]	N. Massone di [illegible]
	1315	N	Esra *Jørgensen 74-74* W Bk	Norweg'n	412	125·6	29·9	16·0	Arendal	1874	S. Johnsen	[illegible]	N. Olsen
	1316	N	Esras (MISSING) *Pedersen 86-88* W Bg	Norweg'n	271	104·8	27·8	14·6	Tvedestrand	1869			L. Pedersen
	1317		Essed-i-Bahri W Bg	Turkish	100	..	..	..	Oolooseh	1884			[illegible]
	1318		Esseri Hhair W Bg	Turkish	261	..	..	..	Greece	1879			[illegible] Zadeh [illegible]
57160	1319		Essex *Bradley* W Bn	British	271	110·0	27·0	12·6	Newport, N.S.	1866	Vaughan		F. H. Watt
8672	1320	RV	Essex *Schaube -87* W Bk	American	630	121·4	32·1	19·1	Newburyport, Mass.	1870		[illegible]	Arnold, Cheney & Co.
2887	1321	IV	Est *Tassino* W Bk	Italian	978	170·5	36·0	23·2	Chiavari	1874	M. Tappani	[illegible]	N. Parodi
80516	1322		Estafette *Russell 65-81* W Bn	British	277	127·2	25·9	14·1					F. W. Fairbrass
88406	1323	R	Estella *Spurr -8[illegible]* W Bn	British	270	117·4	27·8	11·6	Bear River, N.S.	1884	J. H. Benson	[illegible]	J. H. Benson
8580	1324		Estella *Avery* W Bk	American	538	133·5	29·4	18·2	Yarmouth, Me.	1869			E. Avery
	1325	V	Estelle *Legrand* W Sr	French	112	87·9	19·7	11·2	Dunkirk	1863	Sagary fils		A. Buhot-Launay
58884	1326		Estelle *Schnieders 79-83* W 3m Sr	British	195	104·3	24·6	12·7	Peterhead	1869	Geo Bernie & Co.	[illegible]	F. Guy
8673	1327		Estelle Day W Sr	American	216	111·0	29·0	10·5	Milton, Del.	1870			E. Tracy
70177	1328		Esther *Stevenson* W 3m Sr	British	151	101·6	23·8	12·2	Fleetwood	1875	Gibson		J. F. Marginson
	1329		Esther (ex Paolina Zino, ex Due Sorelle) W Bk	Chilian	850	166·6	31·1	22·2	Savona	1874	G. Calansaro		F. Vicuña
	1330		Esther W Bk	French	284	104·3	24·6	15·0	St. Malo	1856			J. B. Bosquet
	1331	✻N	Esther (ex Malinche) *Larsen 65-65* W Bk	Norweg'n	749	167·4	33·3	17·9	Summerside, P.E.I.	1879	M'Millan	[illegible]	C. Larsen
	1332	N	Esther *Olsen -87* W Bk	Norweg'n	490	137·1	31·0	16·8	Grimstad	1886		[illegible]	O. Jörgensen
135946	1333		Esther Buhne *Sagelhorst -81* W 3m Sr	American	273	138·6	34·0	9·0	Fairhaven, Cal.	1887	H. D. Bendixon		A. C. Sagelhorst
72625	1334	V	Esther Roy *Roy* W S	British	1534	219·1	41·3	23·7	Maitland, N.S.	1877	A. Roy	[illegible]	A. Roy
	1335		Estland *Forth 77-* W Sr	Russian	238	100·0	23·0	12·0	Gutmannsbch	1874	G. Lorentz		G. Lorenz
	1336	✻	Estoile *Davis -88* W Sk	British	82	76·7	20·6	10·5	Rye	1888	G. & T. Smith		J. Smith
30000	1337	✻	Estrella *Jones 86-86* I Bk	British	499	153·4	27·7	17·5	Glasgow	1866	Hedderwick & Co.		Doward, Dickson & Co.
58353	1338	✻	Estrella de Chile (WRECKED) *Dorward 81-84* I Bk	British	556	171·5	28·1	17·0	Port Glasgow	1867	L. Hill & Co.		A. C. Gow & Co.
	1339	V	Estrella de la Mañana (ex Siete de Noviembre, ex Violantina) *Abril* W Bk	Spanish	661	158·1	33·1	19·3	Varazze	1868	A. Cerutti		L. Albi Ferrandi
	1340		Estrella del Mar W Bn	Spanish	105	84·0	19·8	8·7	Ilocos, Luzon	1884			Mariano Acosta
	1341		Estrella do Norte W Sr	Brazilian	119	..	..	..					V. P. d'Oliveira
	1342	GV	Estrup *Andreasen 79-79* W Sr	Danish	128	90·9	21·3	10·6	Thurø	1879	C. Bom		P. Bom
	1343	V	Eté *Lebas* W Bn	French	119	81·6	18·9	11·0	St. Malo	1882	L. Malliard	[illegible]	Dauphin
	1344	V	Etha Rickmers *Reinken 81-81* W Bk	German	1018	180·4	31·8	22·6	Geestemünde	1871	R. C. Rickmers	[illegible]	R. C. Rickmers
88895	1345	V	Ethandune *Jamieson* W Bkn	British	393	132·9	31·9	13·0	Albert, N.B.	1885	W. B. Jamieson	[illegible]	J. M. Steeves & Co.
54607	1346	✻	Ethel *Ross* C Bk	British	556	164·0	29·4	17·1	Sunderland	1866	W. Pile & Co.	[illegible]	Fenwick & Co.
70817	1347	✻	Ethel *Kidd -88* W Bkn	British	281	129·6	25·5	13·6	Bridport	1874	Cox	[illegible]	Job Bros.
73412	1348		Ethel *Hocken 86-86* W Bn	British	195	103·0	24·4	12·9	Kingsbridge	1876	Date	[illegible]	J. Ennor, Sr.
70328	1349		Ethel W Bn	British	180	97·6	23·6	12·7	Mahurangi, N.Z.	1875	J. Darrick		J. A. Reddell
88883	1350		Ethel *Arenburg* W Sr	British	162	90·2	24·9	10·1	Lunenburg, N.S.	1880	D. Smith		L. Anderson

✻ = Lloyd's Register; D = Nederlandsche Vereeniging; G = Germanischer Lloyd; I = Registro Italiano; R = [illegible]
A = Veritas Austro-Ungarico; H = Veritas Ellenico; N = Norske Veritas; V = [illegible]

In the 4th Column, C = Composite (iron frame, wood planked); I = Iron; Stl = Steel; W = Wood.

The Dates of Metalling (Col. 11) are based on the records of the Books in which Vessels are respectively classed, or on the Reports of the Sur[illegible]

Further particulars of Vessels classed by Lloyd's Register will be found in Lloyd's Register Book and Lloyd's Yacht R[illegible]

Numéro officiel	N° dans le livre	Registre dans lequel classé	Nom. Matériaux et Gréement. Capitaines et dates d'entrée au service des Armateurs actuels et à bord du navire. Ancien Nom.	Pavillon	Tonnage net	Longueur	Largeur	Creux	Construit Où	Quand	Par qui	Date doublage métal	Armateurs ou Directeurs-Gérants	Port d'Armement
[illegible]	1351		Ethel *Phelps* [illegible] W Bn	British	188	93·5	21·8	11·6	[illegible]	1862			G.H.Gann	Faversham
[illegible]	1352	✻	Ethel *Gill* [illegible] W Sr	British	114	[illegible]	[illegible]	11·[illegible]	[illegible]	1872	Gent	[illegible]	J.B.Pinkham	[illegible]
[illegible]	[illegible]		Ethel *Sandreau* [illegible] W Sr	[illegible]	117	88·5	23·6	9·5	TusketWedge [N.S.	1884	A.O.Porter		A.O.Porter	Yarmouth N.S.
[illegible]	1354	✻	Ethel W Sk	British	79	73·2	20·6	10·1	BurtonStather	1884	J.Wray&Son		H.Toozes	Hull
[illegible]	1355	V	Ethel *Thompson* W Bk	American	[illegible]	181·6	35·0	13·2	Portland Me.	1881	Brewer	[illegible]	C.Merrill	Portland,Me.
	1356	N	Ethel (ex Anna) *Hansen* W Sr	Norweg'n	111	81	22·4	10·[illegible]	[illegible]	1865			A. & H.Backer	Drammen
[illegible]	1357		EthelA.Merritt *Cobb* [illegible] W Sr	American	131	90·2	24·5	9·1	[illegible]	1878	Sampson&Co.		C.S.Crowell	Philadelphia
[illegible]	1358		Ethel Emmerson *Tower* [illegible] -81 W Sr	[illegible]	177	[illegible]	27·6	10·7	[illegible] N.B.	1882	Jas.Chambers	[illegible]	E.Tower	Dorchester,N.B.
[illegible]	1359		Ethel Granville *Fuller* W Sr	British	114	97·0	26·5	8·0	Parrsboro'	1881	Fitzgerald		J.W.Smith	Parrsboro',N.S.
[illegible]	[illegible]	R	Ethel M. Davis *Ray* -87 W3mSr (ex [illegible])	American	298	135·4	32·0	9·6	Yarmouth,Me.	1884	Loring&Cobb	[illegible]	Horace F.Davis	Portland
[illegible]	1361		Ethel Swift *Ackerly* W Sr	American	124	95·2	25·4	9·6	Essex,Mass	1883	J.James & Co.		S.S.Swift	Provincet'n,Mass.
[illegible]	1362	✻	Ethels *Piper* W Bkn	British	219	117·5	23·9	14·0	Kincardine	1863	D.Wright&Son	[illegible]	David & Piper	London
[illegible]	1363	V	Ethiopian *Jones* W Bk	British	3[illegible]	11[illegible]	26·5	16·4	Sunderland	1867	J.&F.Lumsden	[illegible]	J.Trattles	Sunderland
	1364	✻	Ethiopian *Evensen* [illegible] W Bk	Norweg'n	[illegible]	[illegible]	33·6	20·3	Aberdeen	1864	Hood	[illegible]	[illegible]	Frederikstadt
[illegible]	1365		Ethiopian *Gillis* 85-86 W Bkn (ex Duke of [illegible])	British	156	[illegible]	26·1	10·5	Newhav'n,Con	1852			R.Clark	Belfast
	1366	V	Etincelle *Dartaguez* 85-86 W Sr	French	111	88·6	19·3	11·1	Paimpol	1880	L.Laboureur		Y.Le Goaster & [illegible]	Paimpol
	1367	V	Etincelle *Hars* W Sr	French	110	90·5	21·3	10·9	Dunkirk	1879	Lefebvre		[illegible].Dalier	Dunkirk
[illegible]	1368	R	Etna *McCumber* -87 W3mSr	American	297	117·5	3[illegible]	12·0	Millbridge,Me	1875			H.M.Sargent	Portland,Me.
	1369		Etoile des Mers W Bg	[illegible]	15[illegible]	..	..	..	Granville	1850		[illegible]	P.Villars	Granville
	1370		Etoile des Mers W Bn	French	105	72·5	20·7	10·3	St.Vaast	1858			Perrodeau	Nantes
	1371	V	Etoile du Marin *Legrand* W Sr	French	113	84·0	20·9	11·0	Dunkirk	1861	Deryeke		[illegible] & O. Legrand	Dunkirk
	1372	V	Etoile du Nord *Brovez* I [illegible] (ex Star of the North)	French	663	177·5	30·7	18·4	Sunderland	1855	Pile,Hay&Co.		H.Prentout Leblond	Rouen
	1373	V	Etoile du Sud [illegible] -87 I Bk (ex Star of the South) (COLLISION)	French	668	177·5	30·7	18·6	Sunderland	1865	Pile,Hay&Co.		M.Thomas	Havre
	1374	V	Etoile du Sud *Nielsen* -87 W Bk (ex Dronerosdal) (WRECKED)	Norweg'n	530	141·0	31·8	20·2	Varazze	1871	G.Aschieri	[illegible]	Actieselskabet "Normand" (A. Kruge)	Tönsberg
[illegible]	1375	A	Etra *D'Arrigo* -84 W Bn	Italian	255	111·5	24·9	14·9	Lussinpiccolo	1876	M.A.Starcich	[illegible]	F.lli Va D'Arrigo & Co.	Catania
[illegible]	1376		Etra (ex Marina) *Figari* W Bg	Italian	134	90·2	23·0	12·5	Chiavari	1854	Gotuzzo		A.Ciechero	Genoa
[illegible]	1377	R	Etta (WRECKED) *Arthur* 86-87 W S	British	1154	181·0	36·4	24·0	Quebec	1863	Dinning	[illegible]	"Etta" Ship Co.,Ld. (J. S. Wright)	Belfast
[illegible]	1378		Etta *Porter* 83-85 W Bn	British	123	88·6	20·6	9·7	Emden	1863			Simon Soutar	Kingston, Jam.
[illegible]	1379		Etta W Sr	British	107	81·6	24·3	9·6	Tatamagouche [N.S.	1878	J.Simple,jun.		Henry E.Oakes	Yarmouth, N.S.
[illegible]	1380	I	Etta (ex Adelina) *Macciò* -87 W Bk	Italian	551	144·6	31·5	20·4	Varazze	1868	A.Cerruti	[illegible]	Antonietta Olivari	Genoa
[illegible]	1381	R	Etta A. Stimpson *Closson* W3mSr	American	293	125·5	31·2	10·9	ThomastonMe	1874	Stets'n,Gerry&C		S.S.Gerry	Thomaston, Me.
[illegible]	1382		Etta A. Watt *Thorren* -88 W Sr	British	108	85·5	24·0	9·2	Lunenburg N.S.	1879		[illegible]	L.L.Fraser	Lunenberg N.S.
[illegible]	1383	R	Etta M. Barter *Barter* -79 W3mSr	American	259	114·3	27·9	10·4	ThomastonMe	1873	Walk'r,Dunn&C		J. Barter	Thomaston, Me.
[illegible]	1384	V	EttaStewart *Maynes* -83 W Bk	British	787	161·2	34·5	20·0	St.John,N.B.	1872	Cruikshank [Pitfield	[illegible]	D.Carmichael	St.John,N.B.
[illegible]	1385		Ettie H. Lister *Mason* -86 W3mSr (ex Wilmington)	American	312	132·1	33·0	9·0	Frederica,Del.	1884	NathanielLank		J.W.Hall	Wilmington,Del.
	1386		Ettine *Feldmann* -83 W Sr (ex [illegible])	German	119	86·7	19·3	8·8	Papenburg	1866	J.T.Veen		WiebertHinrichsFocken	Warsingsfehn
[illegible]	1387	✻	Eudora *Marshall* 88-88 Stl 4m Bk	British	1938	287·5	40·5	23·7	Dundee	1888	A.Stephen & Sons		W.Stephen	Dundee
[illegible]	1388	V	Eudora *Fallon* W Bk	British	1143	185·4	38·9	22·3	Maccan,N.S.	1882	J.O'Brien	[illegible]	J.O'Brien	Halifax, N.S.
[illegible]	1389		Eudora *Marshall* -88 W Bk	British	302	130·4	25·8	14·5	Montrose	1864	Birnie		F.Croft	Grimsby
	1390		Eufemia *Adolfsen* -87 W Bk	Russian	382	..	..	..		1873	Geta		E.Malmström	Geta
	1391	V	Eugen *Beckmann* 85-88 W Bk	German	730	160·2	33·3	20·0	Stralsund	1876	O.Kirchhoff	[illegible]	Knöhr & Burchard	Hamburg
	1392	V	Eugène *Hely* W Bg	French	263	110·5	24·7	14·0	St.Malo	1872	Parnet	[illegible]	Sire fils	St.Malo
	1393		Eugène *Nilsson* 83-83 W Kff	Swedish	135	84·3	19·4	9·9	Blankenese	1853			J.C.Hulthén	Helsingborg
[illegible]	1394		Eugene A. Reed W Bn	British	177	95·3	24·3	8·1	Boothby,U.S.				J.E.Body	Kingston,Jam.
	1395		Eugène André W Bk		292	106·9	24·7	14·8	Nantes	1864	Dubigeon		Minaud	
[illegible]	1396		Eugene Borda *Van Norden* W Sr	American	197	107·2	29·7	8·0	Tuckahoe, N.J.	1870			J.M.Vanvote	PerthAmboy,NJ
[illegible]	1397		Eugene Demers W Sp	British	159	99·8	24·2	8·9	Gentilly	1881	L.Arcand		T.Desrosiers	Montreal
	1398	V	Eugène Gaston *Laniel* W Bg	French	176	89·0	24·5	12·0	Nantes	1875	Alleau&Aubert		Laniel	Vannes
[illegible]	1399	R	Eugene Hale *Harding* W Bn	American	421	124·0	29·0	17·0	Calais,Me.	1878	Rideout&Lord	[illegible]	G.W.Lord	Calais,Me.
	1400	V	Eugène & Raoul *Beven* W Bg	French	216	98·1	22·0	13·0	Nantes	1874	Alleau&Aubert	[illegible]	Beven	Nantes

✻ = Lloyd's Register; D = Nederlandsche Vereeniging; G = Germanischer Lloyd; I = Registro Italiano; R = American Record; A = Veritas Austro-Ungarico; H = Veritas Ellenico; N = Norske Veritas; V = Bureau Veritas.

Dans la 4me Colonne, C = Composite (membrure en fer, bordé en bois); I = Fer; Stl = Acier; W = Bois.

Les Dates du doublage en métal (**Col. 11**) ressortent des livres dans lesquels les navires sont respectivement classés ou des rapports des Inspecteurs du Lloyd's Register.

Pour plus amples renseignements sur les navires classés par le Lloyd's Register, voir le Lloyd's Register Book et le Lloyd's Yacht Register.

SAILING VESSELS.

Official No.	No. in Book	Registry in which classed	Ship's Name. Master, and Dates of Appointment to present Owner's Service and to Ship. Ship's Late Name.	Material and Rig	Flag	Net Tonnage	Length	Breadth	Depth	Built: Where	When	By whom	Date of Metalling	Owners or Managers
327	1401	A	**Eugenia** *Fdaghi* (ex Edoardo [illegible])	W Bk	Italian	273	119·7	26·0	11·[illegible]	[illegible]	1870			F[lli] Zino
	1402		**Eugenia** *Sert* (ex Extern [illegible])	W Bk	Spanish	429	115·8	30·5	17·9	Syra	1871		[illegible]	M. Taltabull
	1403	V	**Eugenia** *Domato* (ex Pucher)	W Bn	Spanish	315	104·3	28·7	14·[illegible]	[illegible]	1861		Y 6.86	A. Esbert
	1404		**Eugenia** *Rahola* (ex Tomas, ex Dolores)	W Bk	Spanish	172	90·9	25·5	12·[illegible]	[illegible]	1855		Y [illegible]	F. Funosas
	1405	V	**Eugenia** *Lyberg* (ex Altmark) (CONDEMNED)	W Bk	Swedish	389	125·6	25·2	15·[illegible]	[illegible]	1866	Paulsen & Co.	Y 4.86	A. Ekman
2730	1406	I	**Eugenia D.** *Delachi* (ex Figaro)	W Bk	Italian	515	135·4	31·4	20·[illegible]	[illegible]	1872	A. Calcagno	Y 12.86	N. Delachi
83231	1407	✱ R	**Eugenie** *Foley*	W Bk	British	696	171·3	33·5	17·[illegible]	P. E. Island	1881	M. Foley	Y 4.86	F. R. Bartling
75652	1408		**Eugenie**	W Sr	British	196	104·9	23·5	11·7	[illegible]	1877	L. Methot		L. P. Methot
77835	1409	R	**Eugenie** *Munro*	W Bn	British	145	91·0	23·5	10·[illegible]	[illegible], N.S.	1879	W. T. Kelley	Y 10.86	D. Munro
15288	1410		**Eugenie** *Greer 88-88*	W Sr	British	136	88·0	21·8	12·0	[illegible]	1856			B. Piper
45084	1411		**Eugenie** (COLLISION) *Corlett*	W Sr	British	127	91·6	22·3	11·8	Whitehaven	1863	Williamson		J. Ashcroft & Co
	1412		**Eugénie** *Sauct 84-88*	W Bg	French	198	105·3	27·2	12·[illegible]	[illegible], NS	1865			H. Lecharpentier
	1413	V	**Eugénie** *Zonnequin*	W Sr	French	133	83·7	20·7	12·3	[illegible]	1867	Lecerf		Cortier
	1414	✱ V	**Eugenie** *Foley -88*	I Bk	German	697	155·6	30·0	18·6	[illegible]	1865	Reiherstieg Ship Works		R. M. Sloman & Co
	1415	N	**Eugenie** *Bjorge*	W Bk	Norweg'n	525	113·5	32·2	17·9	[illegible], N.S.	1860			[illegible] (H. Holland)
	1416	N	**Eugenie** *Marum -8[illegible]*	W Bn	Norweg'n	117	72·5	22·4	10·9	[illegible]	1844			H. Ludvigsen
182	1417		**Eugenie**	W S	Russian	228	96·0	23·0	12·0	Lahti-Essern	1873			Kirstein & Co.
	1418	G	**Eugenie** *Bernstein*	W Sr	Russian	148	91·9	25·9	9·9	[illegible]	1887	A. Sewers		F. Kirschstein
305	1419	V	**Eugenie** *Grünwald -82*	W Sr	Russian	105	73·0	22·2	10·6	[illegible] Essern	1882	K. Oschmunzene's		R. Grünwald
12592	1420		**Eugenie Smith**	W Sr	British	101	80·0	23·5	8·2	[illegible], U.S.	1844			R. [illegible]
	1421		**Eulalia** (ex Maria Augusta, ex Cannach I.)	W Bn	Portug'se	180	96·1	21·6	11·1	S. Martinho do Porto	1871			S. d'A[illegible]
	1422	GV	**Eulalia** *Svensson 83-83*	W Bk	Swedish	528	118·5	30·4	16·3	Wifsta Warf	1863	A. Kjellen	Y 1.86	F. R. H[illegible]
68412	1423	✱	**Eulie** *Johnson 86-88*	W Bk	British	335	136·0	25·9	14·8	Dartmouth	1873	Redway	Y [illegible]	F. W. Brightman
7590	1424		**Eunice H. Adams** *De Costa -8[illegible]*	W Bg	American	102	81·3	23·3	8·0	Bristol, R.I.	1844		Y 1.86	S. Osborn, jun.
7164	1425		**Eunice P. Newcomb** *Eldridge*	W Sr	American	103	86·0	25·0	8·0	Boston, Mass.	1860			S. H. McDonald
2944	1426		**Eunie McKown** *Wells* (ex Bessie) (WRECKED) -87	W Sr	American	136	100·6	25·0	8·0	BoothBay, Me.	1875			H. P. Dyer
135101	1427		**Eunity R. Dyer** *Sabason*	W Sr	American	104	85·0	26·5	6·0	Lebanon, Del.	1875			S. W. Gabriel
	1428	N	**Eunomia** *Haussen 80-86*	W Bk	Norweg'n	340	120·5	29·8	15·9	Arendal	1868		Y [illegible]	J. E. Haussen
	1429		**Eunonia** *Petterson 83-83*	W Sr	Swedish	153	88·6	21·5	11·3	[illegible]	1864			I. V. Petterson
80633	1430	RV	**Euphemia** *Cann -87*	W Sr	British	1368	205·0	39·6	23·5	[illegible], NS.	1882	J. Allen	[illegible]	H. Lewis
92802	1431	✱	**Euphony** *Johnston 86-86*	W Sk	British	83	77·1	20·7	10·1	Elmshorn	1885	J. Kremer		T. Baskcomb
76191	1432	✱	**Euphrates** *Cameron 90-90*	I m Bk	British	1643	268·9	39·2	23·0	PortGlasgow	1879	H. Murray & Co.		R. Hill
87425	1433	✱ D	**Euphrosyne** *Thomson*	Stl S	British	1799	270·6	40·0	23·4	PortGlasgow	1885	R. Duncan & Co.		C. S. Caird
30	1434		**Euphrosyne**	W Bg	Greek	285	77·8	26·2	..	Galaxidi	1871	C. Papapetros		J. N. Mitropoulos
19	1435		**Euphrosyne** *Caradimos*	W Bg	Greek	199	79·7	..	13·[illegible]	Syra	1863		Y 7.86	C. Caradimos
167	1436		**Euphrosyne Vagliano**	W Bg	Greek	282	95·1	25·6	16·2	Syra	1870			A. A. Vagliano
91252	1437	✱	**Eurasia** *Parkes 85-85*	I S	British	1825	262·3	40·3	23·9	Greenock	1885	Russell & Co.		J. & W. Goffey
11327	1438		**Eureka** *Gray 78-80*	W Sr	British	178	103·4	25·8	11·3	Finland				F. W. Fairhurst
51923	1439		**Eureka** *Tyrrell*	W Bn	British	111	89·5	22·3	10·5	P. E. Island	1865	Orr		G. Philpot
135213	1440	RV	**Eureka** *Southard -87*	W S	American	1930	230·9	42·1	26·5	Richmond, Me	1876	T. J. Southard & S[ons]	[illegible]	T. J. Southard & S[ons]
8436	1441		**Eureka** *Meyer*	W Bkn	American	276	134·3	33·0	10·5	Humboldt Bay, Cal.	1868			E. E. Kentfield
135631	1442		**Eureka** *Anderson -87*	W Sr	American	118	89·5	28·5	7·0	Benicia, Cal.	1887	M. Turner		L. A. Maison
271	1443	A	**Eureti**	W Bn	Greek	334	114·1	28·2	..	Syra	1870			S. Patronos
	1444		**Eurico** (ex Europa)	W Bn	Brazilian	104	..	..	..	Mataró	18 4/79		Y 1.86	J. C. Balo
126	1445	A	**Euro** *Cosulich -88*	W Bk	Austrian	794	147·9	33·1	20·2	Lussinpiccolo	1875	S. Cattarinich	Y 12.86	Maria Vedova [illegible]
72941	1446		**Europa**	W Barge	British	244	129·0	23·1	9·6	Quebec	1875	F. Canton		Montreal Transp[ortation]
70830	1447		**Europa**	W Barge	British	167	100·0	27·6	8·5	Middlesbro'	1875	Dixon		Bell Bros. Ld.
7344	1448		**Europa**	W Bk	American	416	119·6	27·6	18·2	Cohasset, Mass	1849			C. G. Chadbourne
	1449	DGV	**Europa** *Brech*	I Bk	German	1231	227·6	36·1	20·7	Hamburg	1884	Blohm & Voss		C. [illegible] & A. [illegible]
	1450	✱	**Europa** *Krohn 79-83* (ex Shipwreck)	C Bk	German	608	171·0	29·1	17·7	London	1864	C. Langley	Y 1.86	C. Brandt

✱ = Lloyd's Register; D = Nederlandsche Vereeniging; G = Germanischer Lloyd; I = Registro Italiano; R = [illegible]
A = Veritas Austro-Ungarico; H = Veritas Ellenico; N = Norske Veritas; V = [illegible]

In the 4th Column, C = Composite (iron frame, wood planked); I = Iron; Stl = Steel; W = Wood.

The Dates of Metalling (Col. 11) are based on the records of the Books in which Vessels are respectively classed, or on the Reports of the Surveyors.

Further particulars of Vessels classed by Lloyd's Register will be found in Lloyd's Register Book and Lloyd's Yacht Register.

No.	Reg.	Nom. Capitaine	Matériaux et Gréement	Pavillon	Tonnage	Longueur	Largeur	Creux	Où	Quand	Par qui	Doublage	Armateurs ou Propriétaires	Port d'Armement
1451		Europa (ex [illegible]) Robajoli	W Bk	Italian	[illegible]	1[illegible]4·6	26·4	17·9	Sestri Ponente	1856	G.Casanova		V.[illegible]	Genoa
1452	V	Europa Mosca	W Bn	Italian	1[illegible]	86·0	24·8	8·2	Marina [illegible]	1885	A.Castellano	[illegible]	D.Ciampa	Castell. di Stab.
1453	✳	Europa [illegible] 77-81	I [illegible]	[illegible]	1142	221·6	35·3	21·1	Glasgow	1875	Dobie&Co.		Harloff & Böe	Bergen
1454	N	Europa [illegible]	W [illegible]	[illegible]	[illegible]	15[illegible]	33·2	17·5	Sik.4.Sweden	1872	O.P.Aberg	[illegible]	Joh.Hanssen	Arendal
1455	N	Europa [illegible] 77-87	W [illegible]	[illegible]	531	147·4	32·4	[illegible]	Kragerö	$18\frac{64}{84}$	E.Taraldsen		P.I. & P.Berg	Laurvig
1456	N	Europa [illegible]	W Bk	Norweg'n	474	[illegible]	31·8	16·4	Wasa, Finland	1863			E.Olsen	Fredrikstad
1457	V	Europa Petersen 68-[illegible]	W [illegible]	[illegible]	1018	190·0	37·0	20·9	Jacobstad	$18\frac{77}{84}$	K.Lofbjum	[illegible]	P.Malm,jun.	Jacobstad
1458		Europa Löw	W Bk	Swedish	869	159·4	30·3	21·5	Vegesack	1854	J.Lange		J.Ingmansson	Elleholm
1459		Europa	W Bg	[illegible]	2[illegible]	..	..	..	[illegible]	1875			Anastos Vassil	Constantinople
1460		Europa	W	British	[illegible]	1[illegible]·4	30·9	13·8	Bath, U.S.	1833			H.Moss	Melbourne
1461	A	Eurota Peranovich [illegible]	W Sr	Austrian	138	81·6	20·0	10·8	Lussinpiccolo	1877		[illegible]	S.G.Penarovich	Lussinpiccolo
1462		Eurotas Barlow	W Sr	American	1[illegible]	108·0	30·0	9·0	Watertown N.J	1865			E.E.Rackett	New York
1463		Eurus Smith 71-87	W Bk	British	327	132·9	27·6	14·9	Shoreham	1870	May	[illegible]	J. & N.C.Bull	Shoreham
1464	✳D	Eurydice Gloss 80-83	I S	British	1465	253·1	38·2	22·8	Greenock	1874	R.Steele&Co.	[illegible]	Baine & Johnston	Greenock
1465	E	Eurydice Olsen -83	W S	British	1247	185·5	38·0	23·9	St.John,N.B.	1864			H.Barrs & Co.	St.John,N.B.
1466	✳	Eurydice Tait 65-84	I S	British	1152	205·2	35·4	24·1	Ramsey,I.M.	1865	Isle of ManSBC		W.H.Ross & Co.	London
1467	✳	Eusemere Windemer -84	I 1mS	British	2651	300·0	45·2	25·0	Liverpool	1884	WHPotter&Sns		Fisher & Sprott	London
1468	A	Eustathios E. Giannagis	W Bn	Greek	217	93·5	25·1	..	Cassos	1877		[illegible]	I.Giannagis	Syra
1469		Eustathios	W Sr	Greek	195	92·2	24·6	16·1	Galaxidi	1882	C.Papapetros	[illegible]	Astero Catzoulis	Galaxidi
1470	✳	Euterpe Bowling 75-83	I S	British	1197	205·5	35·2	23·4	Ramsey	1863	Gibson&Co.		Shaw Savill & Albion Co. Ld.	Southampton
1471		Euterpe	W S	American	1509	220·0	44·0	25·6	Rockland, Me.	1854			Kate Jaquemot	New York
1472	V	Euterpe (Lorr) Pétrel -87	W Sr	French	112	77·1	20·0	11·5	Dunkirk	1872	B.Derycke		Volboudt	Dunkirk
1473	✳	Euterpe Krause 88-88	I 4mBk	German	2052	290·7	42·7	23·9	Barrow	1884	BarrowS.B.Co.		B.Wencke Söhne	Hamburg
1474	✳	Euterpe (ex Knowsley) Ballmüller 84-84	I Bk	German	623	168·3	29·3	19·0	Sunderland	1864	Oswald&Co		J.C.Pflugk	Hamburg
1475	A	Euterpi Pipoulis	W Bk	Greek	415	125·9	29·4	..	Syra	1877		[illegible]	P.I.Pipoulis	Syra
1476		Euterpi (ex Youngarest) Micalopulo	W Sw	Greek	280	100·0	25·5	15·6	Sunderland	1862	J.Lister		✗Micalopulo	Ithaca
1477		Euthymia Levantis	W Bg	Greek	190	..	..	..	Galaxidi	1866			E.Levantis	Zante
1478	AH	Euthymia	W [illegible]	Greek	180	91·5	24·2	15·7	Galaxidi	1880	C.Papapetros		J.Catzoulis	Galaxidi
1479		Euthymios	W [illegible]	Greek	231	[illegible]	25·[illegible]	..	Syra	1884	A.Livadara		E.Levantis	Syra
1480	H	Euthymios G.Lemos	W Bg	Turkish	117	..	..	..	Chios	1874			G.Lemos	Chios
1481	A	Eutychia (ex Ergasia)	W Bg	Greek	185	90·9	24·3	14·1	Syra	1878			S.Kypriotis	Argostoli
1482		Eutychia	W Sr	Greek	175	91·2	23·0	..	Syra	1879			J.A.Tsipis	Piraeus
1483		Eutychia	W [illegible]	Greek	152	79·7	21·6	12·5	Syra	1871	S.Koufoudakis		D.Scarletis	Argostoli
1484	A	Eutychia Batlis -88	W Sr	Greek	114	72·2	21·0	..	Syra	1876			P.G.Batlis	Skiathos
1485		Eutychia	W Sr	Greek	111	68·9	22·0	..	Skiathos	1883	C.G.Mytilinaios		H.G.Koumbis	Skiathos
1486	A	Eutychia (ex "Taxiarchis")	W Bk	Turkish	110	..	..	..	Parthem	1876			C.Valandassi	Scio
1487	N	Euxinus Olsen -81	W Bk	Norweg'n	247	128·2	27·6	11·5	Arendal	1870	S.Johannesen	[illegible]	S.M.Klöcker	Arendal
1488		Eva Wickman	W 3mSr	American	263	128·6	33·7	9·1	Seabeck,Wash	1880	Murray		E.E.Kentfield	San Francisco
1489	V	Eva Mortensen -84-84	W 3mSr	Danish	221	118·6	25·2	12·4	Rudkjøbing	1884	Joh.Boas		Th.Mortensen	Rudkjøbing
1490	DV	Eva Brengers -81	W Bn	Dutch	227	1[illegible]·6	26·6	11·1	Maitland, N.S.	1873	Grant	[illegible]	K. & J.Wilkens	Veendam
1491	H	Eva	W [illegible]	[illegible]	450	1[illegible]·8	30·2	..	Genoa	1869			L.Glycas	Syra
1492	V	Eva Domingo	W Bn	Italian	166	85·3	22·3	11·0	Limite	1885	Picchiotti	[illegible]	Franc.Domingo	Leghorn
1493	N	Eva (ex [illegible]) Zachariassen -79	W Bk	Norweg'n	343	133·6	30·2	12·8	RiverJohn,NS	1866	I.Kitchin	[illegible]	P.A.Christensen	Langesund
1494	V	Eva Lopez	W Bg	Spanish	224	99·1	27·8	14·4	Lloret	1857			O.Cabá & Valls	Barcelona
1495	R	Eva A.Danenhower Johnson	W 3mSr	American	217	116·6	30·2	8·7	Camden,N.J.	1882	Shoe&Chard		D.W.Johnson	Camden,N.J.
1496		Eva Adell Westcott	W Sr	American	163	1[illegible]·2	26·8	8·4	Stockton,Me.	1867	Staples&Griffin		N.Bridge	Boston,Mass.
1497	R	Eva B. Douglass Letts	W 4mSr	American	1040	194·6	40·5	18·3	Bath,Me.	1886	New England S. Bdg. Co.		T.T.Anderson	NewYork
1498		Eva Bell	W Sr	American	187	..	..	..	Abseccom, N.J.	1857			A.A.Lee	Georgetown, DC
1499		Eva D. Rose Branin	W Sr	American	100	89·3	28·6	7·3	Mauricetn,N.J	1880			S.Wood	Philadelphia
1500		Eva Diverty Gandy	W Sr	American	16[illegible]	110·4	23·1	8·5	DennisvilleNJ	1872			J.F.Forepaugh	Philadelphia

[illegible] Lloyd's Register; D = Nederlandsche Vereeniging; G = Germanischer Lloyd; I = Registro Italiano; R = American Record; [illegible] Veritas Austro-Ungarico; H = Veritas Ellenico; N = Norske Veritas; V = Bureau Veritas.

Dans la 4me Colonne, C = Composite (membrure en fer, bordé en bois); I = Fer; Stl = Acier; W = Bois.

[illegible] du doublage en métal (Col. 11) ressortent des livres dans lesquels les navires sont respectivement classés ou des rapports des Inspecteurs du Lloyd's Register.

Pour plus amples renseignements sur les navires classés par le Lloyd's Register, voir le Lloyd's Register Book et le Lloyd's Yacht Register.

Official No.	No. in Book.	Registry in which Classed.	Ship's Name. Material and Rig. Master, and Dates of Appointment to present Owner's Service and to Ship. Ship's Late Name.	Flag	Net Tonnage.	Dimensions. Length.	Breadth.	Depth.	Built. Where.	When.	By whom.	Date of Metalling.	Owners or Managers.
135164	1501		Eva I. Smith *Adams* W3mSr	American	413	126·5	34·0	13·5	Eng.Creek,NJ	1874			R.Adams
135637	1502	R	Eva L. Ferris *St. John* W3mSr	American	561	155·2	35·5	16·0	Newburyport, Mass.	1881	[illegible]		J.H.Ferris
8133	1503		Eva L. Leonard *McMullen* W Sr	American	109	83·2	25·2	8·0	Cherryfield Me	1868			O.H.Perry
60677	1504	V	Eva Lynch *Sunderland* W Bkn	British	462	154·6	32·0	13·2	St.John,N.B.	1864	D.Lynch	[illegible]	H.D.Troop
80847	1505	R	Eva Maud *Holder* -88 W S	British	267	114·0	27·7	13·0	Maitland,N.S.	1880	W.P.Cameron	[illegible]	J.F.Watson
8194	1506		Eva May *M'Duffie* W Sr	American	324	118·5	28·2	15·3	Yarmouth,Me.	1867	Hutchins& Stubbs	[illegible]	S.A.Pierce
83136	1507		Eva Stewart *Mosman* W Sr	British	103	82·6	24·2	9·3	B'dgewaterNS	1881	L.Hewitt		R.Mosman
8129	1508	RV	Evanelle *Colcord* -88 W Bk	American	487	134·0	29·9	17·3	Stockton,Me.	1868	N.G.Hitchborn	[illegible]	N.W.Bridge
58418	1509	V	Evangeline *Eales* W Bk	British	345	116·2	27·2	16·5	[illegible]	1868	W. [illegible]	[illegible]	T.D.Woodhead&Co
61806	1510	R	Evangeline [illegible] W Bn	British	142	90·7	24·3	10·8	[illegible] N.S	1872		[illegible]	D.Cronan
	1511		Evangeline [illegible] -88 W S	French	137	89·2	22·0	8·8	[illegible]	1876	A.Cantin		A.Lemoine
	1512	A	Evangelis *Camenos* -87 W Bg	Greek	285	..	..	..	Syra	1880	P.Bumis		C.D.Bacogeorgis
7	1513	A	Evangelismos *Economides* W Bk	Greek	368	108·2	27·9	..	Syra	1872	S.Konfondakis	[illegible]	G.Œconomides
	1514	H	Evangelismos *Platsoucas* W Bn	Greek	353	..	..	..	Syra	1875		[illegible]	L.G.Platsoucas
54	1515	H	Evangelismos W Bn	Greek	299	125·0	34·4	21·0	Syra	1880			I.Saris
367	1516		Evangelismos W Bk	Greek	295	105·6	26·6	..	Syra	1869			A.Inglessis
	1517		Evangelismos W Bn		190	..	..	..	Skiathos	1866			
32	1518		Evangelismos W Sr	Greek	121	75·8	22·3	..	Skiathos	1881	D.Mbotsanis		M.D.Mitarona
	1519		Evangelismos *Kairi* W Bk	Russian	179	91·5	23·0	16·7	Syra	1873	A.Potie		G.F.Deminga
	1520		Evangelismos *Triandafilo* W Sr	Russian	106	..	..	..	Syra	1868			G.Damulachi
	1521	AH	Evangelistria *Caratza* W Bg	Greek	350	..	..	..	Syra	1874		[illegible]	Piangon Bros.
250	1522		Evangelistria W Bn	Greek	347	110·2	28·5	..	Cassos	1867	N.Mina		M.Gerogianis
26	1523		Evangelistria W Bg	Greek	337	108·6	31·8	18·0	Syra	1875			L.Platsoukas
50	1524		Evangelistria W Bg	Greek	304	108·6	25·7	16·7	Syra	1877			J.A.Kyriazis
15	1525	A	Evangelistria *N.Cokinos* W Bg	Greek	298	105·6	26·9	..	Syra	1874		[illegible]	N.Cokinos
56	1526		Evangelistria *Moussouris* W Bg	Greek	297	110·2	27·4	16·4	Syra	1874	N.Mboumis	[illegible]	A.B.Moussouris
278	1527		Evangelistria W Bg	Greek	285	102·3	27·9	..	Syra	1880			H.Bokogiorgis
	1528	V	Evangelistria *Caradgias* -87 W Bg	Greek	274	100·0	26·9	16·0	Syra	1887	A.Livadaros		A.Caradgias
61	1529	H	Evangelistria *Vombos* -88 W Bg	Greek	262	97·1	26·6	13·8	Spezzia	1871			A.Pappageorgacopoulo
47	1530	H	Evangelistria *Dictas* W Bg	Greek	257	104·3	25·1	..	Syra	1874			A.Papajanis
373	1531		Evangelistria W Bg	Greek	257	98·4	26·2	..	Syra	1868	P.Bumis		Z.Michalopoulos
10	1532	H	Evangelistria *Karatzas* W Bn	Greek	254	96·1	27·2	14·8	Syra	1871	P.Bumis	[illegible]	A.Karatzas
	1533		Evangelistria W Bg	Greek	253	..	..	..	Galaxidi	1865			E.Merintzos
60	1534		Evangelistria *Comnos* W Bg	Greek	250	93·8	25·4	..	Spezzia	1870		[illegible]	J.Comnoupiotis
14	1535	H	Evangelistria *Chymon* W Bg	Greek	245	90·2	26·2	..	Syra	1870	E. Poumpas	[illegible]	C.D.Chymou
	1536	A	Evangelistria *Mneris* W Bg	Greek	244	..	..	..	Syra	1886	A.Livadaros		D.E.Mneris
44	1537	H	Evangelistria W Bg	Greek	243	..	..	..	Galaxidi	1871	E.Kanatis		Dem.Levantis
233	1538		Evangelistria (ex [illegible]) W Bg	Greek	235	92·2	26·2	..	Syra	1869	A.Livadaros		P.Cailas
18	1539	H	Evangelistria *P.Petratzas* W Bg	Greek	229	96·5	25·3	..	Galaxidi	1866			P.Petratzas
118	1540		Evangelistria W Bg	Greek	215	89·9	25·6	..	Syra	1875			N.Cazakas
	1541		Evangelistria *Dedoussis* W Bg	Greek	210	92·5	23·9	14·6	Galaxidi	1868	C.Pappapetros		A.S.Bessiris
79	1542		Evangelistria W Bg	Greek	208	89·2	26·0	11·5	Galaxidi	1863	E.Kanatis		N.I.Catzoulis
370	1543		Evangelistria W Bg	Greek	204	88·6	23·3	..	Syra	1869			E.Markantonis
	1544	H	Evangelistria *Cacaroucas* W Bk	Greek	202	..	..	..	Syra	1875			D.Cacaroucas
327	1545		Evangelistria W Bg	Greek	198	93·2	24·3	..	Castelorizzo	1876			J.Bolonakis
56	1546		Evangelistria [illegible] W Bg	Greek	197	91·5	21·1	..	Galaxidi	1873			Société Alkyon
308	1547		Evangelistria W Bg	Greek	194	86·9	24·4	..		1874			S.G.Gabriel
142	1548		Evangelistria W Bg	Greek	193	91·5	24·6	..	Syra	1873			N.Calamaris
	1549	H	Evangelistria W Bg		191	83·7	22·0	15·1	Spezzia	1867			
23	1550		Evangelistria W Bg	Greek	190	82·0	23·9	..	Scopelos	1866	P.Barclas		

*=Lloyd's Register; D=Nederlandsche Vereeniging; G=Germanischer Lloyd; I=Registro Italiano, R
A=Veritas Austro-Ungarico; H=Veritas Ellenico; N=Norske Veritas; V

In the 4th Column, C=Composite (iron frame, wood planked); I=Iron; Stl=Steel; W=Wood.

The Dates of Metalling (Col. 11) are based on the records of the Books in which Vessels are respectively classed, or on the Reports of the Sur[illegible]

Further particulars of Vessels classed by Lloyd's Register will be found in Lloyd's Register Book and Lloyd's Yacht [illegible]

Numéro	Registre	Nom, Matériaux et Gréement; Capitaine et dates d'entrée	Pavillon	Tonnage	Longueur	Largeur	Creux	Construit: Où	Quand	Par qui	Date du doublage	Armateurs ou Directeurs-Gérants	Port d'Armement
1551	A	Evangelistria W Bn	Greek	186	..	..	..	Galaxidi	1873			M.Caueno	Galaxidi
1552	H	Evangelistria W Bg	Greek	185	86·2	23·3	..	Spezzia	1857			E.B.Caffos	Piræus
1553	A	Evangelistria W Sr	Greek	170	87·6	24·6	13·5	Syra	1869			M.Condis	Syra
1554		Evangelistria *Manolas* W Bg	Greek	169	82·3	21·8	..	Spezzia	1868			C.N.Apostolou	Piræus
1555		Evangelistria W Bg	Greek	166	80·7	21·4	..	Spezzia	1863			G.Glinos	Syra
1556		Evangelistria W Bn	Greek	158	82·3	21·6	12·1	Galaxidi	1863	A.Michopoulos		N.Canelogiannis	Galaxidi
1557		Evangelistria W Bg	Greek	153	81·0	22·3	18·7	Syra				John Liebnos	Hydra
1558		Evangelistria *Kalogiannis* W Bg	Greek	149	81·4	22·1	..	Galaxidi	1868			E.Kalogiannis	Zante
1559	V	Evangelistria *Lorcaso* W Sr	Greek	146	86·6	21·8	13·6	Scopelos	1882	Panacoth (Harvas)		G.Cosmelos	Argostoli
1560	A	Evangelistria *Stamatiu* W Bg	Greek	146	..	..	..	Syra	1881	C.Cristali		F^lli Stamatiu	Cami
1561		Evangelistria *G. Vocos* W Bn	Greek	143	..	..	..	Spezzia	1869			T.Vocos	Piræus
1562		Evangelistria W Bg	Greek	141	76·8	20·5	13·1	Syra	1864			G.E.Pantoleon	Spezzia
1563		Evangelistria W Bg	Greek	139	71·8	22·3	..					A.Tsakonnindos	Syra
1564		Evangelistria *Mitropoulos* W Bg	Greek	138	..	..	..	Spezzia	1862			I.Logothetis	Galaxidi
1565		Evangelistria W Bn	Greek	138	76·1	21·5	14·0	Spezzia	1864	G.Mikeli		G.L.Glinou	Spezzia
1566	A	Evangelistria *Varelas* W Sr	Greek	133	76·8	24·6	..	Syra	1877			D.L.Varelas	Syra
1567		Evangelistria *Varitsis* W Sr	Greek	127	81·0	24·8	13·5	Galaxidi	1876	C.Papapetros		T.Chrissicos	Galaxidi
1568		Evangelistria W Bg	Greek	126	77·4	22·0	..	Chios				J.A.Mathioudis	Syra
1569	A	Evangelistria *Crisanthi* W Sr	Greek	125	..	..	..	Syra	1888	A.Livadaros		A.Crisanthi	Syra
1570	AH	Evangelistria *Roussos* W Sr	Greek	125	70·9	22·8	..	Syra	1875			D.Roussos	Syra
1571	A	Evangelistria W Sr	Greek	121	..	..	..	Skiathos	1881			M.Mataroua	Skiathos
1572		Evangelistria W Bg	Greek	120	74·5	23·8	11·8	Syra	1868			A.A.Tselentis	Argostoli
1573	A	Evangelistria *Zagoras* W Sr	Greek	112	74·1	21·1	..	Scopelos	1876	K.Plakaris		C.Zagoras	Scopelos
1574	H	Evangelistria W Bg	Greek	110	70·5	22·1	..	Syra	1873			P.Voubos	Piræus
1575	H	Evangelistria W Sr	Greek	107	68·6	21·6	..	Syra	1870			Ant.Chrysanthis	Syra
1576	H	Evangelistria *L. Metaxas* W Bmb	Greek	106	68·9	23·3	10·2	Spezzia	1883			N.Metaxas	Spezzia
1577		Evangelistria W Bg	Greek	102	60·8	21·3	..	Cassos	1878			Z.N.Erodiadis	Syra
1578		Evangelistria W Bg	Greek	102	..	..	..	Eubœa	1865			P.G.Pantelis	Chalkis
1579		Evangelistria *Bartolommeo* W Bk	Russian	165	82·7	22·0	14·5	Syra	1874	S.Cristofi		G.Bartolommeo	Casteloriso
1580	A	Evangelistria *Margassiotti* W Sr	Russian	102	..	..	..	Syra	1870			G.Margassiotti	Eupatoria
1581	H	Evangelistria *Hassan bey* W Bk	Turkish	213	..	..	..	Galaxidi	1873			S.Diamandaras	Castelor
1582	A	Evangelos W Sr	Greek	131	79·4	23·1	..	Syra	1886			P.Primides	Hermopolis
1583	V	Evanthia *Emanoel* W Bk	Greek	297	103·3	27·2	17·6	Cassos	1879	A.Emmanuel		M.H.Emmanuel	Syra
1584	IV	Evarista *Rongardi* -88 W Bg	Italian	221	110·8	25·3	14·6	Leghorn	1876	E.Penco		L.Berti di F^co	Leghorn
1585	V	Evelina *Niemann* 85-88 W Bk	German	545	144·7	29·9	17·6	Hamelwarden	1871	K.Gross		A.Burchard	Rostock
1586		Evelina *Mattson* -87 W 3mSr	Russian	289	..	..	..		1882			E.Söderström	Geta
1587		Evelpis *J. Apostolou* W Bg	Greek	275	99·7	26·9	..	Spezzia	1870			G.H.Apostolou	Piræus
1588	✱	Evelyn *Milne* 85-88 I Bk	British	1179	204·0	34·0	23·0	Liverpool	1863	Jones,Quiggin &Co.		J.Cornfoot&Co.	Liverpool
1589	✱	Evelyn *Roberts* -79 W Bg	British	202	104·9	25·6	13·9	Portmadoc	1877	Williams		H.Roberts Sr.&Co.	Carnarvon
1590	✱	Evening Star *T. Bennett* W K	British	92	79·3	21·3	10·6	Galmpton	1886	T.Matthews		W.Haylock	Grimsby
1591		Evenus *McCarthy* W Bn	British	197	105·8	24·3	12·4	P.E.Island	1865	M·Dougall		P.Donovan	Cork
1592	RV	Everest *Hibbert* 87-87 W S	British	1680	218·0	41·3	24·9	Belliveau's Cove, N.S.	1878	C.Belliveau		W.D.Lovitt	Yarmouth,N.S.
1593		Everett *(ex Sea Breeze) Gordon* W Sr	American	143	95·0	25·1	8·9	Bucksport, Me.	1851			A.G.Prentiss	Saco,Me.
1594	R	Everett Webster *Baker* W 3mSr	American	452	140·6	35·0	11·6	Camden,N.J.	1883	S.W.Tilton		E.E.Baker	Philadelphia
1595		Everglade *Turner* W Sr	American	122	83·8	27·5	8·0	Harrington Me	18[illegible]			R.Caulfield	New York
1596		Evert *Erackson* -87 W Sr	Russian	181	..	..	..		1877			E.Söderström	Geta
1597	✱	Evesham Abbey *Crotty* 79-82 I S	British	1613	245·9	40·3	23·5	Liverpool	1876	T.Royden&Son		J.Poole & Co.	Liverpool
1598	AH	Evgenia *Soltiri* -88 W Bn	Russian	151	78·4	21·6	12·8	Syra	1869			G.Demerdgi	Eupatoria
1599	R	Evie B. Hall *Hall* W 3mSr	American	405	148·4	33·9	11·5	Waldoboro,Me	1883	E.O.Clark		E.O.Clark	Rockland
1600	RV	Evie J. Ray *Sawyer* -87 W Bk	American	919	161·2	35·1	21·7	Harrington Me	1878	A.Nash		Simpson,Clap & Co.	New York

✱ = Lloyd's Register; D = Nederlandsche Vereeniging; G = Germanischer Lloyd; I = Registro Italiano; R = American Record; A = Veritas Austro-Ungarico; H = Veritas Ellenico; N = Norske Veritas; V = Bureau Veritas.

Dans la 4me Colonne, C = Composite (membrure en fer, bordé en bois); I = Fer; Stl = Acier; W = Bois.

Les Dates du doublage en métal (Col. 11) ressortent des livres dans lesquels les navires sont respectivement classés ou des rapports des Inspecteurs du Lloyd's Register.

Pour plus amples renseignements sur les navires classés par le Lloyd's Register, voir le Lloyd's Register Book et le Lloyd's Yacht Register.

1 Official No.	2 No. in Book	3 Registry in which classed	4 Ship's Name. Material and Rig. Master, and Dates of Appointment to present Owner's Service and to Ship. Ship's Late Name.		5 Flag.	6 Net Tonnage	7 Dimensions. Length	Breadth	Depth	8 Built. Where.	9 When	10 By whom.	11 Date of Metalling	12 Owners or Managers
[illegible]	1601	RV	Evie Reed *Whittier*	W Bk	American	714	152·8	32·3	18·7	[illegible] Me.	1878	D.O.Blaisdell	[illegible]	Aug.Palmer
[illegible]	1602	AH	Evnia *Coufopantelis*	W Sr	Greek	143	84·3	22·3	13·4	[illegible]	1880	A.Livadaros	[illegible]	A.Nostrakio
	1603		Evogina *Wikström*	W Sr	Russian	117	..	..	..		1875			F.Wikström
	1604		Evora *Campos*	W Sr	Brazilian	118	..	..	..	Groningen	1866			M.M.da Costa
	1605		Evranyieh	W Bg	Turkish	220	..	..	..	Evranyich	1869			Isstam bolda-Tuar-Basji-Ali Effendi & Co.
	1606		Evranyieh	W	Turkish	212	..	..	..	[illegible]	1872			Shekerzadeh-Ibrahim-Bey & Co.
78494	1607	V	Evviva *Stuart -88*	W Bn	British	272	117·7	27·4	12·5	Meteghan, N.S.	1876	W.Ross	[illegible]	S.Terrio
	1608	N	Evviva *Bruhn*	W S	Norweg'n	488	146·6	30·8	16·9	Grimstad	1872	J.T.Tryde	[illegible]	Anton Bruun
	1609		Examiner	W Bk	British	266	139·6	29·7	9·8	Richmond [River, N.S.W.	1870	W.Yabsley		J.M.Brydon
	1610	N	Exaudi *Nielsen 83-83*	W Bk	Norweg'n	332	120·2	28·4	15·7	Risör	1867	A.Olsen		N.E.Nielsen's E[illegible]
21092	1611		Excel	W Bn	British	109	84·7	20·1	12·0	Whitehaven	1859	Kennedy		T.Curran
50507	1612	R	Excelsior *Fraser -82* (WRECKED) (ex Couronnement)	W S	British	1218	194·2	37·2	23·0	Moncton, N.B.	1864	Sumner	[illegible]	W.H.Ross & Co.
60983	1613	*	Excelsior *Edgar -87*	W Bk	British	380	128·5	27·7	16·3	Montrose	1867	Birnie	Y 12.86	J.C.Ellis
53877	1614	*	Excelsior *Hughes 86-85*	W Bg	British	177	93·5	23·7	12·8	Portmadoc	1868	Roberts	Y 10.88	D.Morris
135245	1615		Excelsior *Pitts*	W 3m Sr	American	331	132·8	32·5	10·8	Humboldt, Cal.	1876			D.Steffens
	1616	V	Excelsior *Olsen -87*	W Bk	Danish	456	137·1	26·2	17·7	Sunderland	18⁶⁄₇	G.Barker	[illegible]	Actieselskab. (Chr. Schmidt)
2366	1617	V	Excelsior *Longobardi* (ex Angelina)	W Bk	Italian	452	133·8	30·4	19·8	Castellamare [di Stabbia	1867	A.Bonifacio	Y 11.86	L.Musso
	1618	NV	Excelsior *Kristiensen 88-85*	W Bk	Norweg'n	354	134·3	29·3	12·0	Kragerö	1884	T.G.Litangen	[illegible]	O.P.Olsen
	1619	V	Excelsior *Schersten 87-87*	W Bk	Swedish	397	139·1	26·5	15·4	Hamburg	1867	E.Dreyer	[illegible]	J.G.Scherstén
68041	1620	V	Exile *Pearce*	W Bk	British	934	177·0	35·6	19·8	St.Martin's, N.B.	1873	A.Parks & Son	[illegible]	Herbert W.Northrup
48601	1621		Exile *Storm*	W Sr	British	200	108·0	24·3	13·5	[illegible]	1864	Anderson		Mrs.M.West
54296	1622		Exit *Blair 86-86*	W Sr	British	106	82·2	22·7	9·1	P.E.Island	1866			D.Blair
	1623	N	Expedit *Christensen -88*	W Bk	Norweg'n	319	132·7	28·7	11·4	Porsgrund	1877		[illegible]	H.Biörn,jun.
	1624	N	Expedit *Davidsen -88*	W Bn	Norweg'n	186	101·6	23·6	12·6	Stavanger	1873	M.Magnus	Y 11.88	D.J.Davidsen
66723	1625		Experiment	W Sr	British	225	128·0	32·0	9·6	Shelburne, NS.	1872			J.McKenzie
	1626		Experiment *Nilsson 74-74*	W Bn	Swedish	214	107·2	26·1	11·0	Gefle	1840	C.Thoren		H.Andersson
	1627		Expertise	W Sr	French	138	..	..	..	NovaScotia	1865			Tréhau
	1628		Explorador (ex Alfred Michel)	W Bn	Spanish	202	82·3	24·8	14·6	Vacanze	1855			Vidal Hers, & Co.
54021	1629	*	Explorer *Partridge -87* (COLLISION)	I Bk	British	750	187·2	30·4	19·3	Liverpool	1866	R.&J.Evans & Co		North Western Ship[illegible] Company, Limited.
46809	1630		Export *Mitchell 73-73*	W Bn	British	137	91·5	22·8	10·6	P.E.Island	1863			H.Gibson
8976	1631	RV	Exporter *Howard -88*	W S	American	1312	199·5	38·2	24·0	Newburyport, [Mass.	1874	G.W.Jackson	Y 12.87	Elen Sumner
	1632	V	Expres *Hansen 80-80*	W Sr	Danish	151	96·7	21·5	10·7	Svendborg	1876	J.R.Andersen	[illegible]	L.v.d.Hude
	1633		Expres	W Bk	Norweg'n	322	115·1	27·4	15·1	Risör	1868			M.Danielsen
57047	1634		Express *Rook* (BROKEN UP)	W Bn	British	199	105·2	21·8	12·8	P.E.Island	1867	M'Dougal		E.T.Fraser
80652	1635		Express	W Bg	British	136	80·3	21·3	10·6	Dartmouth NS	18⁵¹⁄₈₉			J.Forde
28401	1636		Express *Kinch 77-82*	W Sr	British	106	86·0	20·6	11·2	Glasson Dock	1860	Simpson		J.Kinch
82972	1637	*	Express *Murphy 85-85*	W Sr	British	89	81·3	20·0	10·0	Milford	1880	J.Rees		L.J.Murphy
	1638	G	Express *Zelck 88-87*	W Bg	German	276	108·5	27·2	14·2	Rostock	1868	W.Zeltz		H.Mentz,jun.
	1639	V	Express *Le Coutre -82*	W Bk	German	250	105·7	25·2	14·5	Königsberg	1861	I.Fischer		F.Stricks
	1640	V	Express *Bäck -87*	W 3m Sr	Russian	357	135·9	27·7	12·3	Poddaisvarf	1870	U.Strömberg	[illegible]	G.A.Petrelius
	1641		Express *Larsen 87-87*	W Bk	Swedish	309	121·7	26·8	13·5	Gefle	1867	P.Bergforst	[illegible]	C.M.Andersson
	1642	DV	Exquis *Knudsen*	W Bk	Norweg'n	258	118·9	27·5	12·2	Risör	1879	A.Andersen	[illegible]	M.Danielsen
	1643	N	Eyer *Bugge*	W Bk	Norweg'n	397	118·3	29·1	17·0	Drammen	1870	A.Tjörn	[illegible]	S.A.Solberg
	1644		Eyrene *Geyken 83-86*	W Sr	German	109	81·0	21·2	9·6	Grossefehn	1872	F.J.Cassens & Sn		H.F.Geyken
3711	1645	I	Ezechiele V. *Villa -88* (ex Luigi C.)	W Bk	Italian	573	143·6	32·1	20·7	Savona	1870	E.Calamaro	[illegible]	E.Villa
68140	1646	*	Ezel *Deacon*	W 3m Sr	British	155	100·0	22·1	12·6	Cardiff	1873	Tonkin		Mrs.M.K.Deacon

ADDITIONAL SHIPS ARE INSERTED IN THE SUPPLEMENT AT THE END OF THE REGISTER.

*=Lloyd's Register; D=Nederlandsche Vereeniging; G=Germanischer Lloyd; I=Registro Italiano; R=[illegible]
A=Veritas Austro-Ungarico; H=Veritas Ellenico; N=Norske Veritas; V=[illegible]

In the 4th Column, C=Composite (iron frame, wood planked); I=Iron; Stl=Steel; W=Wood.

The Dates of Metalling (Col. 11) are based on the records of the Books in which Vessels are respectively classed, or on the Reports of the Surveyors [illegible]

Further particulars of Vessels classed by Lloyd's Register will be found in Lloyd's Register Book and Lloyd's Yacht Register.

Numéro officiel	No. dans le livre	Enregistré dans le	Nom, Matériaux et Gréement. Capitaines et dates d'entrée en service, les Armateurs actuels et à bord du navire. Ancien Nom.	Pavillon	Tonnage net	Longueur	Largeur	Creux	Construit Où	Quand	Par qui	Doublage	Armateurs ou Directeurs-Gérants.	Port d'Armement.
[illegible]	1	✳	F. A. M. E. *Scribbins* WK	British	49	56·3	17·9	7·8	Plymouth	1880	Hill		G.H.Vicary	Barnstaple
[illegible]	2		F. A. Pike *Norwood* WSr	American	116	92·0	26·5	8·0	Perry, Me.	1866	A.Frost,2nd		E.C.Gates	Calais, Me.
[illegible]	3	I	F. B. Risveglio *Bighetto* WSr	[illegible]	107	80·0	20·7	10·8	SestriPonente	1884	A.Carlevaro	[illegible]	V.Bortini	PortoMaurizio
[illegible]	4	R	F. C. Pendleton *Fletcher* W3mSr	American	388	145·2	33·2	12·3	Bangor, Me.	1882	CrosbyBros.		F.C.Pendleton	Bangor, Me.
	5	DV	F. C. Sieben *VanHoorel* WBk	German	641	155·8	33·6	19·1	Apenrade	1880	J.Paulsen	[illegible]	Th.&H.Elmbcke	Hamburg
[illegible]	6	R	F. D. Hodgkins W3mSr	American	169	107·6	25·7	9·7	Ellsworth, Me.	1883	J.H.Grant		F.D.Hodgkins	Ellsworth
[illegible]	7		F. E. Givan *Brien* WSr	British	109	78·1	27·0	7·9	Cambr'dg, N.B.	1886	H.Humphrey		W.R.Givan	Monckton, N.B.
[illegible]	8		F. E. McDonald W3mSr	American	241	127·7	29·6	9·6	Belfast, Me.	1873		[illegible]	F.L.Clayton & Co.	Baltimore
	9	V	F. G. *Gnarillan* WBk	Spanish	537	138·4	31·2	18·0	Sestri Ponente	1868		[illegible]	F.Gusi	Barcelona
[illegible]	10		F.G.French *French* W3mSr	American	175	98·2	28·1	8·3	Bath, Me.	1887	Hagan&Co.		FrankBenedict	N.Haven
[illegible]	11		F. Grenville Russell WSr	American	164	94·0	28·0	8·2	PortlandConn.	1862-84	S.Gildersleeve & Sons		Middlesex Quarry Co.	MiddletownConn
	12	GV	F. H. Drews WBk	German	623	153·5	31·6	18·6	Stralsund	1875	J.Peuss	[illegible]	Fr.Lancken	Stralsund
	13	V	F. H. Lolling *Wilfers* WBk	German	350	118·8	28·9	13·3	Elsfleth	1870	J.Jürgens	[illegible]	E. tom Dieck	Elsfleth
[illegible]	14	R	F. H. Odiorne *Crowell* W3mSr	American	307	121·0	31·2	12·0	Newburyport, Mass.	1871			W. F. Greene & Son	Boston
	15	DV	F. H. von Lindern IBk	Dutch	985	188·6	35·4	20·8	Amsterdam	1884	C&JvonLindern		Von Lindern & Van [illegible]	Rotterdam
[illegible]	16		F. Koupa *Koupas* WBg	Greek	440	135·1	25·1	18·0	Genoa	1874			Nicholas Koupas	Zante
[illegible]	17	R	F. L. Richardson W3mSr	American	488	144·8	32·0	12·7	ThomastonMe	1873	S.Watts	[illegible]	S.Watts	Thomaston, Me.
[illegible]	18		F. Merwin *Slater* W3mSr	American	324	125·7	31·5	11·0	West Haven, Conn.	1870			J.Pierce	New York
[illegible]	19		F. Nelson *Hinckley* WSr	American	123	86·9	24·0	8·0	Warren, Me.	1850			G.W.Rawley	St.George, Me.
[illegible]	20		F. O. Lamarche WSp	British	105	99·6	22·3	6·8	Sorel	1879	D.Richard		L.Tontaul	Montreal
	21		F. Roura *Pringue* WBn	Spanish	144	85·6	22·7	9·3	Blanes	1875			J.F.Portales	Villagarcia
[illegible]	22	✳I	F. S. Ciampa *Cocurullo* StlS	Italian	1527	238·0	39·2	22·6	Glasgow	1886	D. & W.Henderson & Co.		F.S.Ciampa & Figli	Castellammare
[illegible]	23		F. S. Redfield W3mSr	American	453	159·6	37·1	11·4	PortBlakeley, Wash. Ter.	1887	H.K.Hall		C.Birkholm	SanFrancisco
[illegible]	24		F.T.Drisko *Drisko* WSr	American	155	96·6	27·0	7·9	Kennebunk, Me.	1882			T.A.Drisko	Machias, Me.
	25	GV	F. W. Fischer *Abs* WBn	German	213	97·5	25·5	13·9	Rostock	1873	Burchard&Co.		F.W.Fischer	Rostock
	26	R	F. & T. Lupton W4mSr	American	876	188·6	40·0	17·1	Wilmington	1888	Jackson,Sharp&Co.		T.Burrows	Keyport
[illegible]	27	I	Fabbricotti F. *Carpena* WBn	Italian	145	93·2	23·0	12·8	Lerici	1885	F.Medusei	[illegible]	Vedova G.Carpena	Spezia
	28		Fabiné W3mSr	French	103	..	..	..	Manchester, N.H.	1870			G.Devès	St.Louis, Sénégal
[illegible]	29	I	Fabio *Pastorino* WBg	Italian	535	139·4	31·5	20·3	Savona	1871	A.Calcagno	[illegible]	Fili. Pastorino	Genoa
[illegible]	30		Fabiola WSr	British	131	95·0	22·3	9·1	Oakville, Ont.	1858-76			D.W.Allison	Picton, Ont.
[illegible]	31		Fabiola WSr	British	147	100·0	22·0	9·6	Deschambault	1872			C.A.Boucher	Quebec
[illegible]	32	A	Factis *C. Scarpa* WBg	Italian	308	117·4	27·4	17·3	Fiume	1866	G.Buranello	[illegible]	V.Scarpa	Venice
	33		Fäderneslandet WBk	Russian	303	..	..	..	Nerpes	1877			A.V.Häggblom	Wardö
	34		Faders Minde W3mSr	Danish	299	116·8	24·9	14·8	Nyborg	1859	P.Petersen		C.Eberth	Copenhagen
	35	V	Faders Minde WSr	Danish	129	93·1	20·7	9·9	Marstal	1879	J.O.Christensen		N.J.Ohlsen	Svendborg
	36		Fædres Minde *Hansen* WBg	Norweg'n	197	81·4	25·0	14·3	Tönsberg	1827			L.Föyen	Tönsberg
	37	DN	Færder *Stibolt* WBk	Norweg'n	298	121·1	26·0	14·2	Ostend	1857			I.C.Stibolt	Drammen
	38		Fafel *Rodrigues* WBn	Portug'se	218	97·4	23·0	13·5	SanMartinho	1859			F.Ricca	Lisbon
	39		Faidi-mend WBg	Turkish	214	..	..	..	Oonia	1873			Sari-Hadji-Zadeh Ibrahim Effendi&Co.	Oonia
[illegible]	40	✳	Fair Fanny *Carter* WK	British	63	70·6	19·1	9·0	Bideford	1886	H.M.Restarick		T. & J. Lelen, Lim.	Yarmouth
[illegible]	41	✳	Fair Wind WSr	British	147	98·3	23·8	11·1	Banff	1876	Watson	[illegible]	G.Steele	Banff
[illegible]	42		Fairfield (ex Lethbe) *Brown* WBn	British	264	110·8	29·3	13·4	Windsor, N.S.	1873		[illegible]	T.E.Sherwood	Turk's Islands
[illegible]	43	✳	Fairholme (ORIENT) IS	British	1706	251·0	39·0	24·0	Stockton	1885	Richardson, Duck&Co.		Sailing Ship "Fairholme" Co. (Carr & Ashcroft)	Liverpool
[illegible]	44		Fairlie WSr	British	177	101·1	23·7	12·7	Buckie, Banff	1865	YoungBros.		J.Anderson	Sydney N.S.W.
[illegible]	45	DV	Fairmount *King* WBk	British	1099	180·8	38·0	22·7	Windsor, N.S.	1879	S.Dimock	[illegible]	S.Dimock	Windsor, N.S.
[illegible]	46	✳	Fairwater *Neal* WK	British	60	71·4	18·6	8·9	Jersey	1877	Picot		J.M.Kent & Co.	Jersey
[illegible]	47	✳	Fairy *Oren* 87-87 WBk	British	482	141·2	29·9	18·4	Sunderland	1866	W.Pickersgill	[illegible]	J.I.Jacobs & Co.	London
	48		Fairy Belle *Spence* WBk		519	139·0	30·1	17·6	Clifton, N.B.	1862				
[illegible]	49	✳D	Fairy Flower *Williams* WSr	British	100	86·3	22·0	10·7	Padstow	1878	Rawle		W.H.Williams	Padstow
[illegible]	50	✳	Fairy King *Allen* 74-84 WSr	British	100	86·0	22·0	10·9	Padstow	1878	Rawle	[illegible]	W.R.Allen	Padstow

✳=Lloyd's Register; D=Nederlandsche Vereeniging; G=Germanischer Lloyd; I=Registro Italiano; R=American Record;
A=Veritas Austro-Ungarico; H=Veritas Ellenico; N=Norske Veritas; V=Bureau Veritas.

Dans la 4me Colonne, C=Composite (membrure en fer, bordé en bois); I=Fer; Stl=Acier; W=Bois.

Les Dates du doublage en métal (Col. 11) ressortent des livres dans lesquels les navires sont respectivement classés ou des rapports des Inspecteurs du Lloyd's Register.

Pour plus amples renseignements sur les navires classés par le Lloyd's Register, voir le Lloyd's Register Book et le Lloyd's Yacht Register.

Official No.	No. in Book	Registry in which classed	Ship's Name. Material and Rig. Master, and Dates of Appointment to present Owner's Service and to Ship. Ship's Late Name.	Flag.	Net Tonnage	Length	Breadth	Depth	Built. Where.	When	By whom.	Date of Metalling	Owners or Managers
[illegible]	51		Fairy Maid Williams 84-80 WSr	British	100	87·0	22·9	11·0	Newquay	1876	Williams		W.H.Williams
27219	52		Fairy Rock McFie WBg	British	192	101·0	22·6	12·6	Rye	1859			H.F.Armstrong
60290	53	✻	Faith Baker 87-87 WBk	British	481	147·2	29·5	17·6	Sunderland	1873	J.Gardner	Y 4.87	C.McLean
13227	54		Faith WBn	British	132	74·5	21·6	13·6	Poole	1838			W.Waterman,jun.
21778	55	✻	Faith Evans WSr	British	76	80·0	18·5	8·8	Topsham	1860	Holman		H.Evans
88707	56	✻	Faithful Barber -88 W3mSr	British	99	94·5	23·2	10·4	Connah'sQuay	1887	Ferguson & Baird		J.Baird&others
58879	57		Faithlie Sellerfield 85-85 WSr	British	216	116·0	24·6	13·1	Fraserburgh	1868	Webster		G.H.Gann
68894	58		Faiz Rabhan Mittington WBk (ex Tayaby, ex Cy[illegible])	British	608	155·5	31·7	18·4	Sunderland	1873	R.Thompson	Y 1.88	Almabarif[illegible] Kanaa [illegible]
33946	59		Falcon WBk	British	383	104·0	24·7	17·5	P.E.Island	1840			J.P.Bickell
7107	60		Falcon (BROKEN UP) WSr	British	122	92·8	19·6	11·8	Dundee	1854	Fay S.B.Co.		S.Reay
9318	61		Falcon Hardy WBk	American	271	100·5	24·9	16·2	Medford,Mass.	1817			P.Reed
120633	62		Falcon Müller -88 W3mSr	American	196	125·6	29·1	9·1	Eureka,Cal.	1882	H.D.Bendixon	Y 12.87	H.R.Jacobson
	63		Falcon (TO BE BROKEN UP) WBk	Norweg'n	375	106·0	29·2	18·8	Drontheim	1842			J.Wedel Jarlsberg
133	64		Falcon WBk	Siamese	368	130·6	27·4	13·0	Flensburg	1865	F.C.Christiansen		Goh Kim Fan
87108	65	✻	Falconhurst Jones IS	British	1998	281·0	41·2	24·1	Dumbarton	1883	M'Millan&Sons		Falconhurst Ship Co. Ld (W. R. Price & Co.)
	66	N	Falka (exTroubadour) Pande -87 WS	Norweg'n	677	161·0	30·9	19·5	Dumbarton	1867	A.M'Millan&Son	Y 4.87	Pande,Sundbye&Co.
	67	DV	Falke Bauer -79 WBn	German	150	88·9	22·7	11·3	Ribnitz	1878	J.H.Wilken		J.Jacobson
	68		Falke (CONDEMNED) Mathiesen WBn	Norweg'n	164	95·9	23·6	10·8	Elsfleth	1867	J.Ahlers	Y3.86	J.Mathiesen
	69	G	Falken Jensen 72-79 WSr (ex Emily Corlet)	Danish	100	94·5	18·5	10·3	Jersey	1861	Vautier		L.N.Jansen
	70	DN	Falken Christensen 84-84 WBk	Norweg'n	254	118·8	27·3	12·3	Arendal	1869	A.Aanonsen		Christen Boe
89898	71	✻	Falkland Hill Gordon -88 IS	British	1429	245·0	37·6	21·4	PortGlasgow	1884	Russell&Co.		J.R.Dickson
	72	N	Falkon Hansen WBk	Norweg'n	396	130·1	31·3	16·6	Brahestad	18 51/76		Y 11.88	J.Hansen
85916	73	✻	Falls of Afton McGarrity [88-88] I4mS	British	1899	276·7	41·0	23·7	Greenock	1882	Russell&Co.		Ship Falls of Afton Co. Ld (Wright & Breakenridge)
80430	74	✻	Falls of Clyde Harry [88-88] I4mS	British	1741	266·1	40·0	23·5	PortGlasgow	1878	Russell&Co.		Ship Falls of Clyde Co. Ld (Wright & Breakenridge)
85935	75	✻	Falls of Dee Seator 82-85 I4mS	British	1916	276·7	41·0	23·7	Greenock	1882	Russell&Co.		Ship Falls of Dee Co. Ld (Wright & Breakenridge)
89933	76	✻	Falls of Earn [illegible] 88-88 I4mS	British	2292	302·6	42·1	24·5	Greenock	1884	Russell&Co.		Ship Falls of Earn Co. Ld (Wright & Breakenridge)
87653	77	✻	Falls of Foyers Wilson [88-88] I4mS	British	1917	274·1	41·1	24·0	Greenock	1883	Russell&Co.		Ship Falls of Foyers Co. Ld (Wright & Breakenridge)
93287	78	✻	Falls of Garry Lawrence [81-86] IBk	British	2026	275·1	41·6	23·9	P.Glasgow	1886	Russell&Co.		Ship Falls of Garry Co. Ld (Wright & Breakenridge)
93300	79	✻	Falls of Halladale Peters 88-86 I4mS	British	2026	275·2	41·6	23·9	Greenock	1886	Russell&Co.		Ship Falls of Halladale Co. Ld (Wright & Breakenridge)
72184	80	V	Falmouth Merriam -87 WBkn	British	501	138·2	30·8	17·6	Windsor,N.S.	1876	J.Davidson	Y 2.87	G.W.Churchill
120371	81	R	Falmouth Clark 82-83 W3mSr	American	425	139·8	30·0	16·0	EastDeering,Me.	1883	G.Russell		J.S.Winslow & Co.
70023	82	✻	Falstaff Cornell IS	British	1419	238·0	38·0	23·5	Barrow	1875	Barrow S.B.Co.		J.Beazley
	83	DN	Fama Zachariasen 76-82 WS (ex Brewster)	Norweg'n	1091	177·3	36·0	24·9	Newburyport, Mass.	1855		Y3.87	P.A.Christensen
	84	N	Fama Taraldsen -88 WBk	Norweg'n	403	124·5	30·0	16·3	Arendal	1872	Sommundsen	Y 11.86	S.Eriksen
	85	N	Fama Andersen -87 WSr	Norweg'n	156	91·1	24·6	11·1	Sandefjord	1872	S.Olsen		Jac.Christoffersen
	86		Fama Sjöholm -86 WBn	Swedish	173	95·7	23·6	10·5	Finland	1856			O.Persson
	87		Fama de Canarias Marrero WBk	Spanish	415	124·6	28·4	15·9	SantaCruz	1871	L.R.Gonzalez		Rodriguez & Co.
20246	88		Fame Kelly 74-74 WBkn	British	193	108·7	24·5	13·2	St John,N.B.	1858	T.Hilyard		J.Chadwick,jun.
81398	89	✻	Fame Just 81-86 WK	British	75	80·4	20·8	10·6	Lowestoft	1884	H.Reynolds		J.J.Hobbs
70455	90	✻	Famenoth Renaut IBk	British	983	205·9	33·9	20·0	Glasgow	1876	Dobie&Co.		J.Cook
950	91	I	Famiglia Aliotto WBn (ex Risveglio)	Italian	286	106·9	25·9	15·7	Chiavari	1880	M.Tappani	Y 1.87	Piraino & Lopresti
1829	92	I	Famiglia E. Marchio -88 WBk (ex Anna Dente)	Italian	556	139·4	31·5	19·9	Varazze	1869	G.Scala	Y 8.87	Dea. Enrigo & figlie Giacomo
2377	93	I	Famiglia Ferro Ferro[illegible] WBk	Italian	537	142·0	32·3	20·8	SestriPonente	1873	G.Peragallo	Y 5.86	Flli. Ballestrino
2543	94	IV	Famiglia S. Massa WBk (ex Primo Genero)	Italian	799	170·9	33·8	21·8	Varazze	1874	Tassara	Y 4.86	Flli. & Sti. Schiaffino
	95		Familia Troensegaard W3mSr	Danish	219	97·7	24·5	12·8	Troense	18 52/63	N.R.Troensegaard		A.P.F.Troensegaard
	96	✻	Familia Tauregizar 81-86 IBk (ex Lizzie Leslie)	Spanish	371	121·7	25·1	16·4	S.Shields	1866	Readhead, Softley&Co.		J.T.de Garayalde
	97	V	Familie Kähler WBg (ex Conrad Kähler)	German	205	89·3	26·0	13·4	Ueckermünde	1846	C.Radmann		C.Kaehler
	98	V	Familien Bager W3mSr	Danish	192	111·2	25·3	10·8	Marstal	1873	J.J.Bager		B.C.Christensen's [illegible]
	99	N	Familien Næss WBk	Norweg'n	422	124·5	31·7	16·3	Sandefjord	1869			T.Jacobsen Næss
	100	N	Familien Nilsen -87 WBk	Norweg'n	372	105·1	29·9	18·8	Holmestrand	1814		[illegible]	L.Johnsen

✻ = Lloyd's Register; D = Nederlandsche Vereeniging; G = Germanischer Lloyd; I = Registro Italiano; R = Am[illegible]
A = Veritas Austro-Ungarico; H = Veritas Ellenico; N = Norske Veritas; V = Bu[illegible]

In the 4th Column, C = Composite (iron frame, wood planked); I = Iron; Stl = Steel; W = Wood.

The Dates of Metalling (Col. 11) are based on the records of the Books in which Vessels are respectively classed, or on the Reports of the Surveyors [illegible]

Further particulars of Vessels classed by Lloyd's Register will be found in Lloyd's Register Book and Lloyd's Yacht Register.

Numéro officiel	N° dans le Registre	Cote	Nom. Matériaux et Gréement. Capitaines et dates d'entrée au service des Armateurs actuels et à bord du navire. Ancien Nom.	Pavillon	Tonnage	Longueur	Largeur	Creux	Construit. Où.	Quand.	Par qui.	Date du doublage en métal	Armateurs ou Directeurs-Gérants.	Port d'Armement.
	101	✱	Familien *Nielsen* W Bg	Norweg'n	[illegible]	115·0	26·1	12·9	P.E.Island	1875	Gillan	[illegible]	O.J.Skadberg	Egersund
	102		Familien *Andreassen* -89 W Bg	Norweg'n	[illegible]	96·4	25·6	14·8	Seahara	1856			J.T.Ostensen	Farsund
	103		Familien *Hansen* 78-78 W Bg	Norweg'n	[illegible]	90·7	24·7	12·6	Wisby	1827			H.Kierulf	Drammen
	104	N	Familien *Smith* W Bg	[illegible]	170	75·4	23·6	14·5	Brevig	18[illegible]			H.Smith	Drammen
	105	V	Familien W3mSr	Norweg'n	148	96·8	20·6	10·3	Ziegenort	18[illegible]	H.Brüsewitz	[illegible]	Brödrene Stave	Skudesnæs
	106	N	Familien *Melsom* W	[illegible]	127	91·8	21·4	10·0	Brevig	1871			G.B.Melsom	Tönsberg
	107	DV	Familiens Haab *Jürgensen* W	[illegible]	169	95·8	21·5	11·3	SønderhoFanø	1870	S.J.Sønnichsen	[illegible]	Akt.Selsk.(J.Svendsen)	Sønderho Fanø
	108	V	Familiens Haab *Nissen* W Bg	Danish	161	98·3	21·8	10·9	SønderhoFanø	1876	N.S.Nielsen	[illegible]	Akt.Selsk.(H.Nissen)	Sønderho Fanø
	109	V	Familiens Haab *Brandt* W Bg	Danish	111	87·6	20·4	10·6	Ærøskjøbing	1871	R.Møller		N.H.Brandt	Ærøskjøbing
	110		Familiens Haab *Henriksen* W Bg	Norweg'n	188	97·6	23·3	12·6	Tönsberg	1853			Chr.Henriksen	Tönsberg
	111	N	Familiens Minde *Halvorsen* W Bk	Norweg'n	347	119·3	29·4	16·2	Risör	1871	T.J.Vardeen	[illegible]	Ole Johnsen	Risör
	112	G	Familjen *Bengtsson* W3mSr	Swedish	192	106·3	22·6	11·6	Carlshamn	1863	Andersson		P.N.Horndahl	Nyhamn
	113		Familjen *Andersson* -84 WSr	Swedish	135	85·3	21·7	11·2	Värö	1859			A.P.Olsson	Wester Edsvik, Rolgn
	114	R	Familjen *Rosiff* -83 WSr	Swedish	101	75·5	20·3	10·0	Spången	1874	Spångens vari		A.Lycke	Lysekil
[illegible]	115		Fancy *Leys* W Bg	British	128	85·0	21·7	11·4	Fraserburgh	1848	Webster		R.Brown	Aberdeen
[illegible]	116		Faneromeni *Basilis* W Bg	[illegible]	149	79·4	23·8	15·7	Scopelos	1879	P.Barelas		Garoufalis & Kathiniotis	Scopelos
[illegible]	117		Faneromeni (ex Vilasini) WSr	Greek	142	89·6	21·5	..	Lussinpiccolo	1874		[illegible]	G.J.Stamatopulos	Corfu
	118		Fani (ex ...) *G.Zifiroupoulos* W Bk	Greek	185	100·0	23·8	22·3	Galaxidi	1871	C.Papapetros	[illegible]	P.Zifiroupoulos	Galaxidi
	119		Fani W S		134	..	..	..	[illegible]	1886				
	120	N	Fannie *Larsen* 88 W Bk	Norweg'n	299	127·9	27·2	12·1	Sedgwick,Me.	1857			B.A.Ellingsen	Christiania
[illegible]	121	R	Fannie A. Gorham *Wallis* -89 W3mSr	American	308	130·0	32·8	9·5	Belfast,Me.	1881	CarterBros.		E.Sibley	Belfast,Me.
[illegible]	122		Fannie Adele W3mSr	American	222	115·6	30·8	9·6	SanFrancisco	1883	DickieBros.		R.C.Ogilvie	San Francisco
[illegible]	123	R	Fannie B. Tucker *Mountford* -88 W Bn	American	389	126·2	30·0	11·7	Yarmouth,Me.	1874	G.Loring	[illegible]	J.S.Winslow & Co.	Portland,Me.
[illegible]	124	R	Fannie Brown *Sharrett* W3mSr	American	482	150·5	36·5	14·0	Northport,NY	1881	J.Carll		J.A.Curtis	Richmond,Va.
[illegible]	125		Fannie Butler *Randlett* WSr	American	189	101·5	25·9	10·2	Bangor,Me.	1854			G.Crosby	Bangor,Me.
	126	R ✱ A	Fannie C. *Nacari* -81 W Bn	Austrian	316	1[illegible]·8	30·2	12·6	Moncton,N.B.	1880	R.Cochrane	Y 10.83	Teresa Va. Suttora	Lussinpiccolo
[illegible]	127	R	Fannie E. Wolston *Marr* -82 W3mSr	American	282	120·6	30·4	10·3	Bath,Me.	1882	Goss&Sawyer	[illegible]	G.H.Dunton	Bath,Me.
[illegible]	128		Fannie Flint *Warren* WSr	American	167	101·0	28·9	8·7	Cutler, Me.	1872	Townsend		W.H.Hunter	Lubec,Me.
[illegible]	129		Fannie Hanmer *Brooks* WSr	American	122	75·0	27·0	6·0	Essex,Conn.	1864			J.S.Raynor	Greenport,N.Y.
[illegible]	130		Fannie H. Stewart *Lane* W3mSr	American	334	136·0	33·0	11·0	YorkTown,Va.	1882			N.V.Lane	Norfolk
[illegible]	131	R	Fannie J. Bartlett *Breyer* W3mSr	American	789	179·4	35·8	18·2	Bath,Me.	1888	E.S.Crosby		E.S.Crosby	Bath,Me.
[illegible]	132	R	Fannie Kimmey *Wolf* W3mSr	American	365	129·0	32·8	12·0	Milton,Del.	1879	J.L.Black		T.Winsmore	Philadelphia
[illegible]	133	R	Fannie L. Child *Hart* W3mSr	American	404	139·0	34·0	12·0	Waldoboro'Me	1881	A.Welt	[illegible]	Augustus Welt	Boston,Mass.
[illegible]	134		Fannie Skolfield *Whitmore* -88 W Bk	American	1063	173·3	36·7	22·5	Brunswick,Me	1873	SkolfieldBros.	[illegible]	Skolfield Bros.	Portland,Me.
[illegible]	135		Fannie Tracy *Tilton* W3mSr	American	232	117·0	31·2	10·9	ForkedRiver, N.J.	1875			W.Tilton	PerthAmboy,N.J.
[illegible]	136		Fannie Tucker *Greenleaf* WS	American	1457	211·2	40·0	24·0	Wiscasset,Me.	1875	Hodgkins & Brown	[illegible]	J.Tucker	Wiscasset,Me.
[illegible]	137	R	Fannie Whitmore *Whitmore* W3mSr	American	553	154·0	34·7	13·2	Rockland,Me.	1882	Ames	Y 11.83	A.F.Ames	Rockland, Me.
[illegible]	138		Fannie & Fay *Mehaffy* W3mSr	American	222	117·5	30·8	8·8	N.London,Ct.	1885	J.Davidson		J.M.Glover	Boston,Mass.
[illegible]	139	✱	Fanny (WRECKED) *Smith* W Bn	British	301	121·8	26·3	14·9	Dundas,P.E.I.	1877	J.McDougall	[illegible]	W.Teeton	Liverpool
[illegible]	140		Fanny *Baxter* -88 WSr	British	208	120·0	24·9	12·9	Barnstaple	1875	Westacott	Y 10.81	Job Bros.	Liverpool
[illegible]	141		Fanny (BROKEN UP) *Pillar* W Bn	British	175	105·3	24·0	12·2	P.E.Island	1868	McMillan		C.B.Bramwell	Teignmouth
	142	V	Fanny *Le Feuvre* -87 W3mBn	French	340	117·4	26·9	12·4 / 18·7	Nantes	1887	E.Clergeau fils	Y 9.87	F.Batard & Co.	Nantes
	143		Fanny W	French	101	..	..	..						Cannes
	144		Fanny (BROKEN UP) *Focrek* 81 W Bk	German	429	128·3	26·5	18·0	London	1850			Lichenheim & Pincus	Rostock
	145		Fanny (ex Catalina) W Bn		142	..	..	..	[illegible]	1852				
	146	N	Fanny *Christensen* -88 W Bg	Norweg'n	220	100·8	24·0	12·1	[illegible]	1838			H.Molvig	Frederiksstad
	147		Fanny (ex ...) *Jorge* -87 W Bn	Portug'se	142	88·9	21·8	7·4	Aveiro	1874	J.F.da Lapa	[illegible]	J.C.Carrington	Figueira
	148		Fanny *Combothecras* W Bk	Russian	319	103·3	26·2	16·7	Syra	1869			P.Filippachi	Odessa
	149		Fanny *Autio* -84 W Bg	Russian	264	1[illegible]·0	31·6	18·0	Fredrikshamn	1872	F.Wilkke		M.Brunilla	Fredrikshamn
	150	V	Fanny (ex ...) *Marisiany* -87 W Bk	Spanish	402	123·0	32·0	17·0	Blanes	1856		[illegible]	M.Company	Barcelona

✱ = Lloyd's Register; D = Nederlandsche Vereeniging; G = Germanischer Lloyd; I = Registro Italiano; R = American Record;
✱ = Veritas Austro-Ungarico; H = Veritas Ellenico; N = Norske Veritas; V = Bureau Veritas.

Dans la 4me Colonne, C = Composite (membrure en fer, bordé en bois); I = Fer; Stl = Acier; W = Bois.

Les Dates du doublage en métal (Col. 11) ressortent des livres dans lesquels les navires sont respectivement classés ou des rapports des Inspecteurs du Lloyd's Register.

Pour plus amples renseignements sur les navires classés par le Lloyd's Register, voir le Lloyd's Register Book et le Lloyd's Yacht Register.

Official No.	No. in Book	Registry in which classed	Ship's Name. Master, and Dates of Appointment to present Owner's Service and to Ship. Ship's Late Name.	Material and Rig	Flag	Net Tonnage	Length	Breadth	Depth	Built: Where	Built: When	Built: By whom	Date of Metalling	Owners or Managers
	151		Fanny Aurora *Pahlberg*	W Sr	Swedish	217	101·0	22·5	12·4	Finland				H.C.Klockhoff
[illegible]	152		Fanny Breslauer *Jones 85-89*	W Bk	British	274	131·4	27·1	13·2	Plymouth	1871	Ridley		W.Prichard & Co.
[illegible]	153	✻	Fanny Crossfield *Megee 85-88*	W3mSr	British	99	95·6	22·1	9·8	Carrickfergus	1880	Rodgers		J.Fisher & Sons
[illegible]	154		Fanny Dutard *Bessart*	W3mSr	American	252	123·0	31·8	11·0	SanFrancisco	1882	White		H.Madison
[illegible]	155		Fanny Fern *Eaton*	W Sr	American	116	91·5	25·3	7·1	Newport,R.I.	1859	G.Crandall		T.C.Eaton
[illegible]	156		Fanny Fisher	W Bk	British	219	94·6	25·3	14·2	Manning River (N.S.W.	1847	J.Nicholson		A.Campbell
[illegible]	157		Fanny Fothergill *Jones -89*	W3mSr	British	167	97·4	23·3	12·7	Aberystwith	1865	Williams		M.Nicholson
[illegible]	158	RV	Fanny L. Cann *Churchill [illegible]-88*	W Bk	British	797	162·0	35·3	19·8	PortClyde,NS	1878	T.Coffin	[illegible]	L.Cann
	159		Fanny Lélia	W Bg	French	168	80·1	22·3	12·6	St.Tropez	1867	Sevoulle		Vve. Guaitella
[illegible]	160	A	Fanny Mimbelli *Murzi*	W Bk	Italian	531	157·4	32·8	18·2	Trieste	1870	A.Schiavoni	[illegible]	A.Murzi
[illegible]	161		Fanny Rapson	W Bn	British	122	101·0	21·3	8·5	Sittingbourne	1860			C.Burley
[illegible]	162	✻	FannyScott *Rennels 87-88* (ex Santona)	I Bg	British	233	112·3	23·6	13·5	Newcastle	1866	Marshall&Co.		Sailing Ship "Fanny Scott" Co. Ld. [illegible]
	163		Fanny von Schack	W Bg	German	231	99·1	25·1	14·7	Rostock	1849	F.Dethleff		
	164	VD	Fanø *Mathiasen -88*	W Bg	Danish	227	107·9	24·0	12·6	Nordby,Fanø	1878	S.Abrahamsen	[illegible]	Akt.Selsk.(S.B.Hansen)
[illegible]	165	R	Fantee *Chase 85-86* (ex Samuel G. Reed)	W Bk	American	652	140·0	32·3	20·0	Duxbury,Mass	1869	Keen	[illegible]	J.F.Brooks
	166	GV	Farewell *Klein 75-87*	W Bk	German	536	139·2	31·6	18·3	Königsberg	1866	L.Eggert	[illegible]	J.H.Meyer
	167		Farola	W Lorcha	Spanish	101	79·1	20·9	6·2	Manila	1885	G.Rodriguez		Luis R.Yango
[illegible]	168	RV	Farragut *Hardwick 84-84*	W S	American	1476	212·7	40·0	25·2	Newburyport, Mass.	1876	J.Currier,jun.	[illegible]	E.Lawrence
	169	NV	Farvel *Andersen 84-86*	W3mSr	Norweg'n	333	127·2	29·3	12·7	Lillesand	1886		[illegible]	S.N.Hansen
	170	N	Farvel *Tellefsen -88*	W Bg	Norweg'n	134	75·6	23·2	11·9	Christianssand (?)	1854			A.Tönnesen
	171	NV	Fasan *Pallesen 88-88*	W Bk	Norweg'n	294	123·0	29·2	12·7	[illegible]	1875	Schinnerud	[illegible]	Carl Bech
	172		Fasc	W Bg	French	101	78·1	21·0	11·3	[illegible]	1856	Fastre		Gregorj & Co.
	173		Fatal Karim	W Bk	British	376	98·3	25·4	17·9	Cochin	1842			Hajee Mohomed Saddeck
[illegible]	174		Fatha Salam	W Bk	British	224	85·0	24·0	15·4					Anamalag Chitty
	175		Fathahoel Noor	W Sr	Dutch	135	98·4	21·0	..					Hadjee Oemar bin Ali
	176		Fathi-i-Tijaret	W Bg	Turkish	425	..	..	..	[illegible]	1880			Kalafatji Oustaya [illegible] Agmoglo Zekkee [illegible] [illegible] Hassan
	177		Fathool Bab	W Bg	Dutch	167	..	..	..					
	178		Fathool Barie (ex Bintang Timor)	W Bk		132	86·6	19·7	..					
	179		Fathool Haer *Seth Mohamet bin Oesep*	W Bk	Dutch	394	126·6	27·9	..	[illegible]	1869		[illegible]	Said Abdoela bin Aloe Machoer
	180		Fathool Haer (TO BE BROKEN UP)	W Bk	Dutch	261	98·4	26·0	..					Said Mahomad bin Ahmad Alkaf
	181		Fathool Haer (ex Samson)	W Bk	Dutch	253	..	..	..	Rockland,Me.	1854			Said Hassan bin Mohamet Alkaf
	182		Fathool Haer	W Bg	Dutch	113	..	..	..					HadjeeAbdulRachman binHadjee Abdul Rasag
	183		Fathool Kariem (ex Hong Kong)	W Bn	Dutch	103	88·9	14·8	10·0					Daeng Malangket [illegible] Basoko
	184		Fathool Karim *Said Mahamet Alhabsy*	W Bk	Dutch	314	110·8	29·5	..					Said Aloei bin Mohamet Alkaf
	185		Fathool Moebarak	W Bk	Dutch	241	90·9	25·9	..	[illegible]				[illegible]
	186		Fathool Rachman *[illegible]*	W Bk	Dutch	284	103·6	25·7	..	[illegible]	1847		[illegible]	[illegible]
	187		Fathool Rachman	W Bg	Dutch	203	104·6	22·1	..					[illegible]
	188		Fathool Rahim	W Sr	Dutch	123	71·8	19·7	..					Mohamad Ali bin Lako
	189		Fathool Salam	W Bk	Dutch	273	..	..	..					Said Abdul Rachman Alkaf
[illegible]	190		Fathur Rahaman	W Bg	British	100	71·9	19·6	10·7	[illegible]	1855			Koona Mana Soona [illegible] Chitty
	191		Fati Reymany	W Bk	British	180	71·3	20·0	16·2	[illegible]				Chowakarah [illegible] Cauty
	192		Fattle Mobarruck	W Bk	Oman	485	108·0	31·0	17·4	[illegible]	1855			Binissan
	193	N	Fatum *Olsen 85-85*	W Bg	Norweg'n	299	106·3	28·4	15·3	[illegible]	1863	Knud Andersen		O.T.Tönnevold
	194	V	Faucon *Lambert -88*	W Bg	French	135	86·9	21·6	12·1	St.Malo	1868	M.Parnet	[illegible]	Lessard, Lainé & Co.
	195		Faugh Balaugh (ex Glion)	W3mSr		278	122·1	26·3	11·6	[illegible]	1870	J.Jörgensen	[illegible]	
	196	N	Fauna *Gabrielsen* (ex Th[illegible] Petre)	W Bk	Norweg'n	453	126·4	32·6	17·9	[illegible]	1841	LarsAugust'sen.		C.Solberg
	197	V	Faune (ex Lucas) *Roussel*	W Bg	French	192	96·4	23·2	12·8	[illegible]	1869		[illegible]	Danguilhen, aîné
	198	V	Faune *Largouet*	W Bg	French	147	77·4	21·3	12·9	[illegible]	1865	Olliveaud		Largouet père
	199	V	Faust *Pagen*	W Bg	French	217	103·6	21·0	13·4	St. Malo	1873	L.Mallard	[illegible]	Blanchard frères
	200													

✻=Lloyd's Register; D=Nederlandsche Vereeniging; G=Germanischer Lloyd; I=Registro Italiano; R=Am[illegible]
A=Veritas Austro-Ungarico; H=Veritas Ellenico; N=Norske Veritas; V=Bu[illegible]

In the 4th Column, C=Composite (iron frame, wood planked); I=Iron; Stl=Steel; W=Wood.

The Dates of Metalling (Col. 11) are based on the records of the Books in which Vessels are respectively classed, or on the Reports of the Surveyors [illegible]

Further particulars of Vessels classed by Lloyd's Register will be found in Lloyd's Register Book and Lloyd's Yacht Register.

No.	Cote	Nom	Capitaine	Matériaux et Gréement	Pavillon	Tonnage net	Longueur	Largeur	Creux	Construit Où	Quand	Par qui	Doublage	Armateurs ou Directeurs-Gérants	Port d'Armement
201	DNV	Faust	*Otterbeck*	WBk	Norweg'n	811	165·4	36·7	18·7	Arendal	1886	A.Henricksen	·86	H.Aalholm	Arendal
202		Fausto (ex Ellis Quirindo)	*Spinella*	WBk	Italian	881	169·9	32·5	20·3	SestriPonente	1868	A.Briasco fu L.	Y·85	O.Spinella	Messina
203	V	Fauvette	*Henry 81-86*	WSr	French	117	89·6	19·3	10·5	St.Malo	1882	L.Trancheme		Y.Le Gouster fils	Paimpol
204	V	Favonius	*Deckham 77-83*	WS	British	1526	213·6	40·3	24·3	St.John,N.B.	1883	Stew'rt& Ritchie		Wm.Thomson & Co.	St.John,N.B.
205	DN	Favorit	*Johannessen -88*	WS	Norweg'n	933	186·7	35·9	21·2	Bergen	1875	Brunchorst & Dekke	Y·88	Holdt & Isachsen	Stavanger
206	N	Favorit	*Thorsassen*	WBk	Norweg'n	480	141·7	31·4	16·7	Kragerö	1880	C.Knoph	Y·2·87	B.Knoph	Kragerö
207	NV	Favorit	*Hansen*	WBn	Norweg'n	280	121·9	27·6	13·8	Christiansand	1874	M.Knudsen	Y·7·85	H.Natvig	Christiansand
208	DV	Favorita	*Stellmann -87*	WS	German	1211	188·0	37·0	24·0	Mystic,Conn.	1863	C.Greenman	Y·87	John C.Meincke	Geestemünde
209		Favorita (ex Favourite)	*Palmi*	WBn	Italian	195	105·5	24·0	12·9	P.E.Island	1865	Bulls		P.Delfino	Genoa
210		Favorita		WSr	Spanish	136	87·3	22·1	10·5	Gijon	1848			Bonifacio López	Gijon
211		Favorita (ex Atlantic)	*Beverly -87*	WBg	American	151	87·0	24·0	10·0	CherryfieldMe	18		Y·86	H.Trowbridge's Sons	New Haven Conn.
212		Favorite		WBn	American	137	87·0	25·8	8·2	Branford,Con.	18			G.Hotchkiss	New Haven Conn.
213	V	Favorite	*Fery*	WSr	French	113	83·7	20·0	11·2	Dunkirk	1870	G.Malo&Co.		Bonny	Binic
214	V	Favorite	*Cusus*	WSr	French	105	82·7	19·3	10·9	Paimpol	1878	L.Laboureur		Rognez & Vanmaekers-berche	Dunkirk
215		Favoriten	*Hogvall -84*	WSr	Swedish	145	90·5	22·6	10·8	Sundsvall	1862			H.Hogvall	Storugns, Gotland
216	N	Favour	*Olsen*	WBg	Norweg'n	219	108·7	26·3	11·2	Porsgrund	1864	N.A.Larsen		Rederiselskabet Favour (Chr. Ellefsen)	Fredrikstad
217	✻	Favourite (ex Whisker)	*Netherton 83-87*	WBn	British	206	114·5	23·4	13·2	W.Hartlepool	1865	E.Irvine&Co.		W.H.Stott&Co.	Liverpool
218	✻	Favourite	*Fraser -86*	WBk	British	84	78·0	20·9	10·7	Elmshorn	1886	J.Kremer		W.J.Fraser	Grimsby
219		Fawn	*Brazewell*	WBg	British	216	103·0	24·1	14·0	Aberdeen	1862			Edward J.P.Kelly	Lyttelton,N.Z.
220		Fawn	*Langley -79*	WBn	British	192	109·1	23·2	12·9	Dartmouth	1866	Phillips		G.H.Hutchings	Dartmouth
221	✻	Fawn	*Robinson*	WK	British	82	71·4	20·4	10·2	Galmpton	1882	A.Gibbs		J.P.Robinson	Grimsby
222	R	Fawn	*Hopkins*	WS	American	1015	170·2	34·5	23·6	Bath,Me.	1860	Patten&Sons	Y·83	T.Daggett	New York
223		Fawn	*Robinson 86-*	WSr	American	298	108·5	29·6	8·5	Philadelphia	1866			E.H.Atwood	Boston,Mass.
224	✻	Fay (BURNT)	*Evans*	WSr	British	96	81·0	21·1	10·0	Berwick	1878	Gowan		P.Sinnott	Runcorn
225		Fazelcurreem		WBk	Omani	637	120·0	30·0	20·0	Cochin	1856			Rashed Mahomed Van Asia	Muscat
226		Fe	*Bayas*	WBn	Spanish	241	103·6	25·9	13·8	Corunna	1878	M.L.London	Y·85	F.R.Lopez	Corunna
227		Feahmanillah		WBn	Turkish	150	..	..	..	Akjeh-Shehir	1879			Akjeh Shehirlee Hadji Osman	Akjeh Shehir
228		Fearful		WBg	British	221	97·1	24·2	14·7	Portsea	1849			Mrs.C.Byerley	Portsmouth
229		Fearless		WBn	British	255	111·0	26·9	12·4	Granville,N.S.	1861			J.Johnson	Annapolis,N.S.
230		Fearless	*Wheaton*	WBkn	British	230	106·0	24·6	14·5	Brixham	1857			T.G.Robins & Co.	Guernsey
231	✻	Fearless	*Annis 84-84*	WK Barge	British	108	90·4	21·0	8·2	Ipswich	1876	Colchester		F.A.Christie	Ipswich
232		Fearless		WSr	British	100	86·6	21·2	8·7	Riv.Clyde,NSW	1876	G.Coleson		T.O'Dwyer	Maryboro'
233	✻	Fearless	*Watt -86*	WSr	British	80	80·5	18·4	10·1	Guernsey	1876	De la Mare	Y·86	G.De la Mare & Co.	Guernsey
234	R	Fearnought (ex 1885-86)	*Laffredi -88*	WS	British	1292	197·0	37·6	24·1	Boston,Mass.	1859			Cox Bros.	Liverpool
235	I	Fede (ext. Ursula) (batr "Racide")		WBk	Italian	753	164·0	33·1	21·9	Voltri	1874	A.Canepa	Y·86	Badd. Dellepiane & Co.	Genoa
236		Fede	*Stvina*	WBn	Italian	153	93·8	24·7	13·3	SestriPonente	1849	Briasco	Y·84	G.Tonietti fu A.	Genoa
237	✻ DI	Fede & Amore	*Lambruti*	WBk	Italian	1295	212·0	39·0	26·4	Savona	1883	F.Sirello	Y·87	G.B.Bertolotto	Savona
238	IV	Fede & Speranza	*d'Aste*	WBk	Italian	563	146·0	31·8	19·5	Arenzano	1872	A.Salomone	Y 1·86	L.Carbone & V.d'Aste	Genoa
239		Fedele (ex Bartolomeo Simone)	*Ciano -87*	WBk	Italian	517	139·4	30·7	20·1	Loano	1870	O.Amico	Y·88	Flli.Polese&Ciano	Leghorn
240	I	Fedele	*Valle -88*	WBk	Italian	445	133·8	30·2	19·3	SestriPonente	1868	G.Feragallo	Y·85	B.Figari	Genoa
241	I	Fedelta (ex Angelo Fasce)	*Fasce -88*	WS	Italian	628	155·8	32·7	20·3	St.Malo	1868	Mallard	Y·88	Fasce & Gardella	Genoa
242		Federica (ex Valaque & Federica Maria)	*Ohlsen*	WBk	Chilian	746	164·0	32·6	19·5	Quebec	1865	H.Dubord		C.Jessen	Valparaiso
243	IV	Federico	*Vianello*	WBk	Italian	425	131·8	29·7	19·4	Chioggia	1877	G.B.Poli	11·86	G.Gavagnin	Venice
244		Federico	*Cavasso*	W3mSr	Italian	386	123·0	29·5	19·0	Savona	1871	G.Calamaro	Y 11·87	G.Stagno fu G.	Genoa
245	A	Federico	*Ritondo -88*	WBn	Italian	125	77·1	22·0	12·5	Barletta	1856			F.&G.Paronecelli & A.Denovelli	Bari
246		Federico Barreras (ex Nuova Barreras)	*Lorenzo 1-88*	WBn	Spanish	225	96·5	27·8	13·2	Blanes	1870	J.Dieta		J.Barreras y Casellas	Vigo
247		Feizi-Bahri		WBg	Turkish	147	..	..	..		1872			Koorocieh-Shillehlee Fehmilah	Shilleh
248		Feizar Hudar		WBg	Turkish	129	..	..	..	Greece	1878			Mustapha Agha-zevferai Zendjikkha Hashem	Constantinople
249		Feizi Rabani		WBn	Montenegrn	155	78·7	22·1	13·1	Dulcigno	1874	JusOmeri		Selim Alil Aga & Co.	Dulcigno
250		Feizi Rahman		WSr	Turkish	211	..	..	..	Heraclea	1875			Tijar Saghir Oghlou Hailou	Heraclea

✻ = Lloyd's Register; D = Nederlandsche Vereeniging; G = Germanischer Lloyd; I = Registro Italiano; R = American Record; A = Veritas Austro-Ungarico; H = Veritas Ellenico; N = Norske Veritas; V = Bureau Veritas.

Dans la 4me Colonne, C = Composite (membrure en fer, bordé en bois); I = Fer; Stl = Acier; W = Bois.

Les Dates du doublage en métal (Col. 11) ressortent des livres dans lesquels les navires sont respectivement classés ou des rapports des Inspecteurs du Lloyd's Register.

Pour plus amples renseignements sur les navires classés par le Lloyd's Register, voir le Lloyd's Register Book et le Lloyd's Yacht Register.

Official No.	No. in Book.	Registry in which Classed.	Ship's Name. Material and Rig. Master, and Dates of Appointment to present Owner's Service and to Ship. Ship's Late Name.	Flag.	Net Tonnage.	Length.	Breadth.	Depth.	Built Where.	When.	By whom.	Date of Metalling.	Owners or Managers.
	251		**Feizi Rahman** *Halal* W Bmb	Turkish	125	..	..	..	Gallipoli	1885			Alishan Bey
	252		**Feizi-Rahman** W Sr	Turkish	107	..	..	..	Antaria	1874			Antarialee Jaffer Tr[illegible]
	253	G	**Fekeline** *Sweers 79-* W Sr	German	119	82·7	19·0	9·5	Emden	1876	H. Klattenhoff		Gerd Sweers
1336	254		**Felice** *Terrentini* W Bk	Italian	322	116·4	20·4	16·6	Sestri Ponente	1860			R. Caputi
1090	255		**Felice** (ex Lucia C.) *Ciano* W Bg	Italian	265	113·8	25·9	16·4	Sestri	1862			R. Cinno & Co.
654	256		**Felice** *Guarassi* W Bn	Italian	109	80·7	21·5	11·4	Messina	1858		[illegible]	Salv. Pipitone
1101	257	V	**Felice B.** *Bibbolino -87* W Bk (ex Stefano Minbelli)	Italian	513	152·8	33·1	17·8	Trieste	1870	A. Schiavoni	[illegible]	F[lli]. & S[lle]. Bibbolin[illegible]
889	258	I	**Felicia** *Cafiero* W Bg	Italian	298	115·1	26·2	17·8	Alimuri	1871	M. Mauro	[illegible]	Eredi di M. Roman[illegible]
	259		**Felicia** W Pol		172	89·9	26·4	13·0	Arcos	1849			
	260		**Felicia** *Trägårdh -86* W Bk	Swedish	385	134·1	26·4	14·5	Calmar	1875	S. P. Beck	[illegible]	C. Söderbergh
	261	R	**Feliciana** *Gonzales -87* W Bk (ex Esmeralda)	Spanish	301	109·9	27·0	14·8	Bilbao	1860	S. de Arana	[illegible]	A. Genas
	262	N	**Felicié** *Gule* W Bm Sr	Norweg'n	250	128·1	25·9	11·5	St. Magnus	1862	H G. Rischmüller	[illegible]	O. Bjærtness
606	263		**Felicina** *Caffarena* W Bg	Italian	137	80·4	24·0	13·0	Voltri	1846	G. B. Briasco	[illegible]	R. Denaro fu F.
2863	264	D I V	**Felicina Antonio P.** *Brignoli* W Bk	Italian	938	187·3	34·4	23·3	Varazze	1877	Nicolo Cerruti	[illegible]	A. L. Pendibene
3352	265	I	**Felicina Ferrari** *Massone* W Bk (ex Nicolo P.)	Italian	823	167·9	33·8	23·3	Savona	1877	F. Sirello	[illegible]	Vedova F. Ferrari
3440	266		**Felicita S.** *Schiaffino -87* W Bg (ex Colibri)	Italian	248	97·4	26·1	14·5	Agde	1867	Contrasti		G. Schiaffino
69631	267	V	**Felicitas** *Jones 86-86* W Bk	British	754	163·0	32·4	20·0	Quebec	1874	Baldwin	[illegible]	R. Thomas & Co.
	268	V	**Félicité** *Abé* W Bg	French	220	93·8	22·6	13·8	Dieppe	1866	P. Corue		Boismeau frères
	269		**Félicité** *Pilastre -86* W Sr	French	124	88·0	21·4	10·6	Gravelines	1862	Legoulon		Mme. Dermigny
	270		**Felinto** W Sr	Brazilian	189	..	..	..					J. G. Cardia
	271	V	**Felisa** *Laurreta 82-83* I Bn (ex Rosario)	Spanish	185	110·3	22·7	11·8	Middlesbro'	1876	W. Harkess		J. R. de Uriarte & C[illegible]
	272		**Felisa** W Bn	Spanish	106	86·0	18·3	8·4	Tayabas Luzon	1877			F. Llorente
	273	V	**Felisberta** *Pereira* (ex Dois Amigos ex S. Lawrence ex Washington) W Bk	Portug'se	332	150·2	21·8	11·7	Baltimore	18[69/79]		[illegible]	F. Rodrigues
	274		**Felix** (ex Flamengo) *Oliveira* W Bn	Brazilian	127	..	..	..	Campos	18[4?/84]		Y 12.81	J. R. de Azevedo
	275		**Félix** *Foliard -87* W Bg	French	179	..	..	..	St. Servan	1854	Hamon		Guibert & fils
	276	V	**Felix** (LOST) *Mönnich 83-83* W Bn	German	261	105·6	24·6	12·7	Elsfleth	1869	F. Deetjen	[illegible]	J. H. Hustede
	277	G	**Felix** *Steinbrink -83* W Bg	German	234	100·0	24·3	13·4	Stettin	18[illegible]			J. Steinbrink
	278	N	**Felix** *Olsen* W Bg	Norweg'n	241	106·5	27·7	12·9	Tvedestrand	1864	M. Eldrup		A. & F. Smith
	279	N	**Felix** *Larsen -86* W Bg	Norweg'n	177	102·0	24·1	11·4	Tönsberg	1873	P. G. Gundersen		O. Bjærtness
	280	N	**Felix** W Sr	Norweg'n	175	107·0	26·6	9·0	Boston	1866			Winge & Co.
	281	V	**Felix** *Dahlberg* W Bk	Russian	871	168·6	34·1	21·1	Uleaborg	1874	J. Hildanen	[illegible]	G. J. Dahlberg
	282		**Felix** *Lindroos -76* W Sr	Russian	112	..	..	..	Bromarf	1873			J. F. Strandberg
	283		**Felix** *Jarden -82* W Sr	Swedish	126	80·7	21·0	10·7	Hudiksvall	1845			R. Snöbohm
	284		**Felix** *Sjögren* W Sr	Swedish	102	75·5	23·3	8·9	Timm'rnabb'n	1854			N. M. Sjögren
	285		**Felix Brandt** *Brandt* W Bk (ex Eastern Province)	Danish	363	125·5	27·2	16·5	Sunderland	1858	Peverall & D'vison		Akt. Selsk. (H. T. Brandt)
	286	V	**Felix Estivant** *Masson -87* W Bk (ex Italie)	French	879	162·0	30·8	24·6	Quebec	1869	Charl'nd & M'rq's	[illegible]	J. Martin
	287	V	**Félix Marie** *Chenio* W Sr	French	102	67·9	20·7	10·7	Nantes	1862	Chauvelon		Chénio
	288	G	**Felix Mendelssohn** *Freiwurst -87* W Bk	German	923	165·4	34·3	22·6	Bremerhaven	1866	J. C. Tecklenborg	[illegible]	H. Janentsky
	289	V	**Félix Théophile** *Harel* W Sr	French	141	73·5	21·6	10·4	Nantes	1860	Clergeau & Boju		E. Jamont fils
	290	G V	**Felix II.** *Hagens 84-84* W Bk	German	370	125·9	26·8	13·6	Hammel-[warden	1876	E. P. Botter	[illegible]	G. Bolte
	291		**Feliz** W Bk	Spanish	190	92·5	27·3	12·8	Barcelona	1856			Isidro Pla
	292		**Feliz Lembranca** W Sp	Portug'se	134	..	..	..					J. R. Ascencio
	293		**Feliz Uniāo** W Plc	Brazilian	207	..	..	..					J. A. G. Bastos
71275	294		**Fellow Craft** W Bk	British	209	115·7	26·0	10·5	Port Burwell	1873	D. Foster		W. Y. Emery
37307	295		**Fellow Craft** *Kirker* W Sr	British	115	87·8	23·3	9·3	Clements, N.S.	1856			J. Kelly
	296		**Felo** *Riberas* W Bk (ex Castrosol, ex Cyrene)	Spanish	509	128·6	29·6	18·0	Maitland, N.S.	1863	A. Roy		Gimerez & Co.
	297		**Fem Sødskende** *Petersen* W Sr (ex Paul, ex Vestr-möen)	Danish	116	90·7	18·3	9·7	Groningen	1851			J. C. Petersen
	298	✻	**Fénelon** ex Stu[illegible] (now named Victorine) W Bk	British	1138	187·0	36·0	18·8	Honfleur	1864	Leviels	[illegible]	J. Real
	299		**Fenix** *Nyman 85-86* W Bk	Swedish	271	104·0	23·8	14·3	Libau	1839			N. A. Löfgren
	300	V	**Fenja** *Ågren 86-88* C Bn	Swedish	255	118·1	26·0	9·8	Gefle	1888	O. A. Brodin		O. A. Brodin

✻ = Lloyd's Register; D = Nederlandsche Vereeniging; G = Germanischer Lloyd; I = Registro Italiano; R = A[illegible]
A = Veritas Austro-Ungarico; H = Veritas Ellenico; N = Norske Veritas; V = B[illegible]

In the 4th Column, C = Composite (iron frame, wood planked); I = Iron; Stl = Steel; W = Wood.

The Dates of Metalling (**Col. 11**) are based on the records of the Books in which Vessels are respectively classed, or on the Reports of the Surveyors [illegible]

Further particulars of Vessels classed by Lloyd's Register will be found in Lloyd's Register Book and Lloyd's Yacht Register

No.	Cote	Nom, Matériaux et Gréement	Pavillon	Tonnage	Longueur	Largeur	Creux	Construit Où	Quand	Par qui	Doublage	Armateurs ou Directeurs-Gérants	Port d'Armement
301		Fenna & Willemina W Sr	British	115	86·2	22·0	9·7	[illegible]				F.C.Mathews	Penzance
302		Fennahrompni W Bg	[illegible]	120	..	..	..	Greece	1880			Kumardji-Hadji Dimitri Phoni & Co.	Constantinople
303		Fennia *Karlsson* W SmSr	[illegible]	249	..	..	..		1874			G.A.Eriksson	Lemland
304		Fensalè *[illegible]* -71 W Bk	Norweg'n	311	[illegible]	[illegible]	[illegible]	[illegible]	1861			Fr.Kleppe	Stavanger
305		Feodor *Loppinen* W	Russian	307	..	..	..	Impilaks	1876			J.A.R.Lindberg	Impilaks
306		Feodor W Sr	Russian	301	..	..	..		1871			F.Markoff	Sordavala
307	G	Feodor *Salack* -88 W SmSr	Russian	296	123·6	28·3	13·0	Gutorunskirk	1884	J.Saul		F.D.Dubkowsky	Haynaseh
308		Feodor *Hamalainen* W Glt	Russian	285	..	..	..	Impilaks	1871			M.Andrejeff	Impilaks
309		Feodor *Jegoroff* W Glt	Russian	196	..	..	..	Salmis	1883			W.Hosainoff	Salmis
310	V	Ferda *Nielsen* W Bk	Norweg'n	611	145·3	32·1	19·4	Stavanger	1870	K.Kaisen	Y 1.86	S.Berners Sönner	Stavanger
311		Ferdijihan W Bg	Turkish	150	..	..	..	Barten	1871			Trebizondlee Yekendji-zade-Hadji Mustapha	Trebizond
312		Ferdijihan W Bk	Turkish	135	..	..	..	Barten	1875			Azapdar Hammam Jess Hadji Nazif (Effendi & Co.)	Barten
313		Ferdinand *Ruelle* W Sr	French	266	107·9	25·3	14·1	Fecamp	1869	F.& V.Capon		G.Malandain	Fécamp
314	G V	Ferdinand *Lotz 87-87* (ex Hoville) W Bk	German	686	150·9	29·5	20·4	Alloa	1867	Adamson	Y 1.82	Gildemeister & Ries	Bremen
315	V	Ferdinand *Bliesener 85-85* W Bk	German	416	136·8	31·8	16·7	Hamburg	1868	H.C.Stülcken	Y 6.82	H.C.Stülcken Wwe.	Hamburg
316	V	Ferdinand *Döring 86-86* W Bk	German	383	121·2	29·2	16·0	Danzig	1861	F.Devrient		A.Gibsone	Danzig
317	D N	Ferdinand *Larsen 73-83* (ex Jarvis) W S	Norweg'n	1267	194·3	38·5	23·8	St.John,N.B.	1868	Fisher	Y 5.86	F.Breda	Christiania
318		Ferdinand *Vesterberg -88* (ex Erbgrossherzog Friedrich Franz) W Bg	Swedish	251	98·7	24·9	15·1	Rostock	1858	W.Zeltz		O.Vesterberg	HäfveröRslgn.
319	✻ V	Ferdinand Fischer *Kratz 87-88* I S	German	1726	258·5	40·0	23·7	Flensburg	1885	FlensburgSBCo.		J.D.Bischoff	Vegesack
320	N	Fergus *Petersen -88* (ARANDIA DE) W Sr	Norweg'n	170	101·0	24·4	10·6	Grimstad	1869	OleKittelsen		Chr.Th.Boe	Arendal
321	R	Ferland *McWhinnie* W Sr	British	138	87·2	21·6	9·9	Annapolis,NS.	1882	L.Delap		W.McCormick	Annapolis,N.S.
322	N V	Ferm *Just 76-76* W Bn	Norweg'n	305	123·1	27·2	14·3	Christiansand	1874	O.Olsen Hagen	Y 1.86	M.F.Stray	Christiansand
323		Ferm *[illegible] -88* (ex Nordsee Amalia) W Sr	Norweg'n	143	92·8	22·5	10·0	Hoogezand	1862	J.H.Hoeites	Y 85	R.Apenæs	Frederikstad
324	V	Fermanville *Schuby* W Bg	French	138	82·0	21·0	11·8	Cherbourg	1867	Hamel	Z 1.88	A.Postel & ses fils	Cherbourg
325		Fermina (ex Fermina de Vila) W Bk	Spanish	230	109·2	28·6	16·4	Ferrol	1868	M.Lorenzo		A.Suné y Casanovas	Barcelona
326		Fern *Gibbons* W Sr	British	177	103·0	23·2	12·6	Ipswich	1864	W.Read	Y 3.82	B.Byrnes	Sydney
327	✻	Fern *Fennill* W Sr	British	100	79·0	21·5	10·6	Goole	1865	Pearson		T.Ramsey	London
328	✻	Fern *Sleight -88* W Sk	British	89	78·7	20·9	10·5	Grimsby	1888	SmithBros.		SmithBros.	Grimsby
329	N V	Fernando *Jensen* W Bk	Norweg'n	347	130·4	30·1	12·6	Arendal	1888	C.Larsen	Y 3.88	H.Mogensen	Tvedestrand
330		Fernão de Magalhães *Bisia -8* W Sr	Portug'se	171	90·5	24·6	10·2	Villa do Conde	1875	M.R.Gomes		A.I.Serrao	Aveira
331		Fernwood W Sr	American	124	85·4	24·0	9·6	Essex,Mass.	1887			John Pew & Son	GloucesterMass
332	V	Ferona *Nielsen 88-88* W Sr	Danish	137	89·8	20·5	10·7	Troense	1868	C.J.Jensen		N.Nielsen	Svendborg
333		Ferrah-i-Bahri W Bg	Turkish	250	..	..	..	Oonia	1874			Genahji-zadeh-Rifaat Bey & Bro.	Oonia
334		Ferreh-Numa *Omer Israd -86* W Bg	Turkish	195	..	..	..	Karaja-Sinan	1886			Heracleelee Kara Mahmoud-oglou Hadji-Ahmed	Heraclea
335	V	Ferreri *Partalo -88* W Bk	Italian	448	132·2	27·4	18·2	Cassano	1870	F.Derosa	Y 1.81	Flle Cafiero di Ant.	Castellam.di stab
336	R V	Ferris S. Thompson *Paulsen* W Bk	American	531	141·5	32·0	17·0	Setauket,N.Y.	1874		Y 10.86	Lawrence,Giles&Co.	NewYork
337	N	Festinalente *Sammelsen* (ex Henry Jacob [illegible]) [1-83] W Bk	Norweg'n	1032	177·2	36·0	22·6	St.John,N.B.	1860	King	Y 12.88	Johanson,Rötter & Co.	Christiania
338	N	Festina lente *Christensen [-86]* W Bk	Norweg'n	398	126·4	30·1	16·3	Risör	1869	J.Stenersen	Y 2.84	P.D.Prebensen	Risör
339		Fethi Tidjaret W Sr	Turkish	225	..	..	..	Heraclea	1870			Ahmasrahlee Mehemed Bey & Co.	Ahmasrah
340		Fethi Tidjaret W Bg	Turkish	152	..	..	..	Rizeh	1873			Galata Yorghanji Hassan Effendi	Rizeh
341		Fetouh El Khair W Bg	Turkish	218	..	..	..					Mohamed E.Mosa	Alexandria
342		Feu Follet W Sr	British	117	87·8	23·5	8·6	Grondines	1854			L.Murray	Quebec
343		Fiamman-Illah W Bg	Turkish	237	..	..	..	Sakarieh	1875			Nalbour-Hadji Shefki Effendi	Oonia
344		Fiamman-Illah *Feizi* W Bn	Turkish	150	..	..	..	Akjeh-Shehir	1885			Hattab-tijari Keriaco velideh Simplas & Co.	Akjeh Shehir
345	V	Fiancée *Agnieray* W Sr	French	110	84·0	22·0	11·0	Dunkirk	1880	Vanderiele fils		Nissen	Dunkirk
346	R	Fidelia *Blake 84-* W Bn	American	315	112·0	27·6	17·0	Bangor,Me.	1864		Y 9.87	T.J.Stewart & Co.	Bangor,Me.
347	N	Fidelia *Johnsen 86-86* W Bg	Norweg'n	298	104·4	27·4	15·6	Sunderland	1838			T.Eeg	Langesund
348		Fidelidad W Bn	Spanish	128	93·8	20·7	9·3	Albay	1858			Pascual Ledesma	Manila
349	G V	Fidelio *Sänder 88-88* (ex Wim[illegible]) W S	German	1235	194·8	36·7	24·5	Newburyport	1862	J.Currier	Y 7.86	D.Cordes & Co.	Bremen
350		Fidelity *McKinnon 82-82* W 3mSr	American	182	106·5	30·5	8·5	Eureka,Cal.	1882	A.Cookson		W.J.Sweasey	Eureka,Cal.

✻ = Lloyd's Register; D = Nederlandsche Vereeniging; G = Germanischer Lloyd; I = Registro Italiano; R = American Record;
Δ = Veritas Austro-Ungarico; H = Veritas Ellenico; N = Norske Veritas; V = Bureau Veritas.

Dans la 4me Colonne, C = Composite (membrure en fer, bordé en bois); I = Fer; Stl = Acier; W = Bois.

Les Dates du doublage en métal (Col. 11) ressortent des livres dans lesquels les navires sont respectivement classés ou des rapports des Inspecteurs du Lloyd's Register.

Pour plus amples renseignements sur les navires classés par le Lloyd's Register, voir le Lloyd's Register Book et le Lloyd's Yacht Register.

Official No. (1)	No. in Book (2)	Registry in which classed (3)	Ship's Name. Material and Rig. Master, and Dates of Appointment to present Owner's Service and to Ship. Ship's Late Name. (4)	Flag. (5)	Net Tonnage. (6)	Dimensions. Length. (7)	Breadth.	Depth.	Built. Where. (8)	When. (9)	By whom. (10)	Date of Metalling. (11)	Owners or Manag[ers] (12)
	351		Fides (ex Geertruida Strating) W 3mSr	Danish	210	105·6	25·2	12·2	Pekela	1878	Strating		C. Jensen
	352	N	Fides *Olsen* W Bg	Norweg'n	308	104·2	28·3	15·7	Grimstad	1853	A. Omundsen		Gunder Olsen
2556	353	I	Fidia D. *Pagliano* W Bk	Italian	749	169·6	32·4	21·6	Sestri Ponente	1875	B. Canepa	Y 5.85	F. Dall' Orso fu P[illegible]
	354	V	Fido *Jongebloed 76-* W Bn	German	191	100·7	24·0	11·4	Papenburg	1871	J. Beckmann	Y 8.85	J. Beckmann
756	355		Fido (Postiglion) *Campodonico* W Bk	Italian	399	122·0	29·7	18·3	Chiavari	1861	F. Gotuzzo	Y 10.85	F. Figari
	356	N	Fido *Olsen 88-88* W Bk	Norweg'n	520	141·6	30·9	17·4	Arendal	1876	C. H. Christensen	Y 12.85	C. Beck
	357	N	Fido *Steffensen 83-83* W Bg	Norweg'n	214	107·7	26·8	12·3	Tvedestrand	1867	A. & F. Smith		O. T. Steffensen
1723	358	I R	Fiducia (ex Luisa) *Cravari* [-85 W 3mSr	Italian	575	152·5	30·2	19·4	Sestri Ponente	1868	A. Briasco fu G. B.	Y 11.87	Gerolamo Schiaffino
1098	359	A	Fieramosca *Cinquini* W Bn	Italian	103	71·1	22·5	12·3	Castellamare	1849			P. Martinelli
78047	360	*	Fiery Cross *Casey 81-87* I S	British	1399	243·6	37·0	21·9	Glasgow	1878	C. Connell & Co.		J. D. Clink
62457	361	*	Fiery Cross *Junner* W Bn	British	338	133·2	24·3	15·3	Kingston	1872	Geddie	Y 11.85	Anderson, Webster & Co.
90021	362	*	Fifeshire *Greig 75-85* I Bk	British	1318	239·9	36·0	21·6	Port Glasgow	1885	Russell & Co.		T. Law & Co.
	363	G V	Figaro *Heinsel 86-86* W Bk	German	1007	177·1	35·4	23·1	Bremerhaven	1879	J. C. Tecklenborg	Y 12.85	F. Tecklenborg
	364		Figaro *Johansson 78-85* W Sr	Swedish	154	95·8	22·7	10·1	Oscarshamn	1875			J. C. Hulthén
224	365	A	Figlia Penelope *Nichetich* W Bg	Austrian	305	109·9	24,7	15,6	Trieste	1864	F. Poll	Y 5.87	Ved. Maria Morosin[i]
597	366		Figlio d'Italia *Nicolini* W Bk	Italian	407	126·3	31·0	18·8	Lerici	1869	F. Rolla	Y 3.86	F. Giacopello
	367	V	Figueirense (ex Tito, ex Rio Vouga) *Gaspar -87* W Bg	Portug'se	209	93·8	24·3	11·8	Aveiro	18^{68}_{69}	Assenço	Y 6.87	Abilio A. F. Aguas
74478	368	*D	Fiji *Hill 86-86* I S	British	1357	229·4	36·3	23·1	Belfast	1875	Harland & Wolff		W. J. Myers, Son & Co.
180	369	A V	Filadelfia *Zagabria -88* W Bk	Austrian	830	163·7	31·3	21·8	Capo d'Istria	1876	L. Poli	Y 5.86	G. L. Tonetti
	370	A	Filaretos (ex Henoch) *Cochino* W Bk	Greek	430	..	..	..	Trieste	1868	G. Tonello		M. Crussachi
	371	V	Fileuse *Evrard* W Sr	French	109	83·7	20·0	11·3	Dunkirk	1870	G. Malo & Co.		H. Widmer
	372		Filios *Hallil* W Bk	Turkish	535	..	..	..	Austria	1886			Tjari-Djewairdu-Oghlou-Boboski [illegible]
	373		Filios W Bg	Turkish	175	..	..	..	Filios	1869			Abdul Ghani Effendi
	374		Filip *Wilson* W Sr	Swedish	192	103·6	19,6	11·9	Papenburg	1859			P. Mattsson
2565	375	I	Filippino *Chiesa -88* W Bk	Italian	920	175·4	34·4	22·7	Sestri Ponente	1875	A Briasco fu L.	Y 2.86	A. Nattini
2568	376	*I	Filippo (ex Agostino D.) *Bertolotto* W Bk	Italian	749	164·0	32·9	22·3	Varazze	1875	M. Baglietto	Y 9.86	P. Bertolotto
57	377	I	Filippo *Dilietrio* W Bk	Italian	478	137·8	29·5	19·0	Castellamare [di Stab.	1873	L. Trambarulo	Y 5.86	L. Dilietri
982	378	I	Filippo (ex Nuova Chiarina) *Criscuolo -87* W Bk	Italian	425	131·2	26·4	18·4	Alimuri	1873	F. de Rosa	Y 7.87	L. Criscuolo
3388	379	I	Filippo R. (ex Rosario) *Mantabin -88* W Bk	Italian	452	122·3	31·0	19·0	Varazze	1870	A. Cerruti	Y 3.85	A. Rainusso
592	380		Filomena Danovaro *Tonetti* W Bk	Italian	417	125·6	29·8	18·9	Prà	1864	M. Tixi	Y 3.86	L. Tonietti
	381		Fils Croussaki *Mandalakis* W Bg	Greek	214	..	..	..	Syra	1874		C 85	J. Croussakis
	382	V	Fils Unique *Leveux* W Bg	French	139	79·4	22·0	11·5	St. Malo	1853	Vve Dandin & fils		Leveux
	383	N	Fimafenger *Mortensen -88* W Bk	Norweg'n	558	132·9	33·8	19·4	Dordrecht	1847			Chr. Eyde & Hein
	384	N	Fin (ex Lina) *Juul -88* W Bk	Norweg'n	536	158·6	33·0	16·8	Christinestad	1877			L. Vanvert
	385	N	Fin (ex Maria) *Andreassen -84* W Bn	Norweg'n	179	104·7	24·4	11·3	Altona	1867	F. Dreyer	Y 4.88	P. Wiborg
	386	D N	Fin *Økland 83-83* W Sr	Norweg'n	135	96·4	24·4	11·0	Hardanger, [Norway	1883	K. Skaaluren		Frithjof Eide
	387	N	Finboge *Mejlaender -88* W Bg	Norweg'n	216	111·1	26·1	11·9	Porsgrund	1869	P. M. Pettersen	Y 10.84	K. Knudsen
	388		Finella *Jorgensen -88* W Bk	Norweg'n	439	141·3	27·1	17·5	Montrose	1857	J. & D. Birnie		A. A. Thoresen
81442	389	*	Fingal *Fatmore 87-87* Stl 1mS	British	2510	308·2	42·8	25·0	Belfast	1883	Harland & Wolff		R. Martin & Co.
	390	N V	Fingal *Fagiberg -86* W 3mSr	Norweg'n	336	130·0	28·3	14·2	Stavanger	1876	M. A. Magnus	Y 9.86	K. S. Bertelsen
	391	N	Fingal *Salvesen -83* W 3mSr	Norweg'n	243	110·4	27·8	12·2	Grimstad	1867	O. Kittelsen		D. Salvesen
	392	V	Fingal (FOUNDERED) *Ofverberg -86* W Bg	Swedish	260	119·4	23·7	12·6	Hernösand	1868	F. Hägglund	Y 9.86	M. B. Bergström
	393	H	Finix *Calgeras* W Bg	Greek	252	..	..	..	Syra	1869		C 7.81	G. Issaias
	394	N	Finland *Braastad* W Bk	Norweg'n	539	142·0	33·2	17·5	Åbo	1857	Jörgensen	Y 8.81	Joh. Spörck
	395	V	Finland *Lundqvist -83* W Bk	Russian	537	159·4	29·2	17·1	Raumo	1875	J. F. Uhrnberg	Y 1.86	R. Mattsson
	396	D	Finland *Tarlin -87* W Bk	Russian	439	130·0	32·0	17·0	Helsingfors	1870			F. J. F. Sjöblom
	397	N	Finn *Fayen -83* W Bk	Norweg'n	425	136·2	30·2	16·2	Drammen	1870	J. Jörgensen	Y 1.86	H. A. Nielsen
	398	D N	Finn *Anderson -87* W Bg	Norweg'n	261	97·4	26·1	14·8	Sunderland	18^{52}_{76}			K. Johannesen
	399	N V	Finnvid *Sorensen 83-84* W Bn	Norweg'n	258	113·7	26·2	12·8	Christianssand	1879	P. von der Lippe	Y [illegible]	H. Natvig
	400		Finstrom *Isaksson -87* W Bk	Russian	367	..	..	..	Finström	1866			E. O. Woivalin

*=Lloyd's Register; D=Nederlandsche Vereeniging; G=Germanischer Lloyd; I=Registro Italiano; R=[illegible]
A=Veritas Austro-Ungarico; H=Veritas Ellenico; N=Norske Veritas; V=[illegible]

In the 4th Column, C=Composite (iron frame, wood planked); I=Iron; Stl=Steel; W=Wood.

The Dates of Metalling (Col. 11) are based on the records of the Books in which Vessels are respectively classed, or on the Reports of the Surveyors of [illegible]

Further particulars of Vessels classed by Lloyd's Register will be found in Lloyd's Register Book and Lloyd's Yacht Register

No.	Reg.	Nom, Capitaine (Anciens Noms)	Matériaux et Gréement	Pavillon	Tonnage	Longueur	Largeur	Creux	Construit: Où	Quand	Par qui	Doublage	Armateurs ou Directeurs-Gérants	Port d'Armement
401		Finzel *Wittgfening*	W Bk	Chilian	508	150·0	27·9	17·9	Liverpool	1881	Evans		Mathews & Raby	Lota
402	*	Fiona *Le Gros -84*	W K	British	40	62·7	16·6	7·8	Jersey	1884	De le Sueur & Son		J.G. Le Gros	Jersey
403		Fiona *Iversen -81*	W Bk	[illegible]	218	111·5	24·5	11·1	Peterhead	1864	Stephen & F'rbes		H. Iversen	Svendborg
404		Fire Fly	W Bk	British	364	117·8	26·0	16·2	Boston, U.S.				H.E. Meara	Hong Kong
405	N	Fire Söskende *Smith*	W Bk	Norweg'n	272	107·1	28·0	15·1	Luleå, Sweden	1849			M.B. Tvethe	Christiania
406	A	Firenze (ex ...) *R. Napoli*	W Bn	Italian	271	111·8	25·3	17·2	Lussinpiccolo	1868	G. Premuda	V 1·86	J. D'Arrigo ved. Napoli	Catania
407		Firm *Le Couteur*	W Bg	British	127	100·1	22·1	11·1	Jersey	1862	Deslandes		T. Gavey	Guernsey
408	*	First Lancashire *Jones 161-75*	I S	British	1354	232·6	36·9	22·6	Sunderland	1875	Osbourne, Graham & Co.		W.H. Owen	Liverpool
409	* D	Firth of Clyde *Smith 82-82*	I Bk	British	1207	228·1	36·1	21·5	Glasgow	1882	Dobie & Co.		J. Spencer & Co.	Glasgow
410	*	Firth of Cromarty *M'Knight*	Stl S	British	1452	244·7	37·5	21·4	Pt. Glasgow	1888	Russell & Co.		J. Spencer & Co.	Glasgow
411	*	Firth of Dornoch *Milne*	I Bk	British	907	200·4	32·5	19·0	Greenock	1880	Russell & Co.		J. Spencer & Co.	Glasgow
412	*	Firth of Forth *Watt 87-87*	I Bk	British	830	194·8	32·7	19·4	Dumbarton	1875	M'Kellar, M'Millan & Co.		J. Spencer & Co.	Glasgow
413	*	Firth of Lorn *M'Lean 83-83*	I Bk	British	833	196·5	32·6	19·5	Glasgow	1878	Dobie & Co.		J. Spencer & Co.	Glasgow
414	*	Firth of Solway *Craw 86-88*	I Bk	British	1245	228·4	36·5	21·0	Glasgow	1885	W.B. Thompson		J. Spencer & Co.	Glasgow
415	*	Firth of Stronsa	I Bk	British	1251	228·6	36·5	21·0	Glasgow	1885	W.B. Thompson		J. Spencer & Co.	Glasgow
416	*	Firth of Tay *Craig 85-85*	I Bk	British	826	195·3	32·6	19·3	Glasgow	1877	Dobie & Co.		J. Spencer & Co.	Glasgow
417	A	Fischetti *Barrigo*	W Bk	Italian	335	133·8	25·9	15·9	Trieste	1874	S. Piscitelli	V 7·86	V. Fischetti & F[lli]	Catania
418	*	Fitzroy *Jones 81-87*	C Bk	British	573	153·6	28·2	17·8	Sunderland	1863	G. Peverall	V 4·86	R. Davies	London
419	A	Fiume E. *Blecich -87*	W Bk	Austrian	710	150·2	34·1	20·5	Fiume	1881	F[lli] Schiavoni	V 3·86	G. Valcich	Fiume
420		Five Brothers *Worth -87*	W 3m Sr	American	219	114·5	31·0	9·0	Milford, Del.	1878	J.H. Harper		J.C. L'Engle	Jacksonville, Fla
421		Five Sisters	W Bg	British	245	107·5	26·0	12·6	Bathurst, R.G.	1860			F. Paer	Gibraltar
422		Fix *Gundersen*	W Bk	Norweg'n	287	110·7	27·7	12·6	Arendal	1864			O. Johnsen	Arendal
423	N	Fix (ex Ingeman) *Olsen*	W Bk	Norweg'n	237	101·1	27·6	13·8	Sönderborg	1842			H.I. Olsen	Fredrikstad
424	G	Fix (ex Samaritan) *Olsen 86-86*	W Sr	Norweg'n	237	102·2	26·2	12·7	P.E. Island	1865	Muirhead		Emil Olsen	Arendal
425	N	Fix (ex Kemi) *Olsen*	W Bg	Norweg'n	190	108·3	24·8	11·4	Torneå, Swed'n	1858			O.G. Olsen	Christiania
426	R	Fix *Lindholm*	W Sr	Russian	155	97·6	27·5	8·5	Bromarf	1885			G.S. Brodenberg	Tenala
427	N	Fjeldgutten *Hansen 84-84*	W Bg	Norweg'n	162	110·2	22·5	11·9	Skien	1869	J. Brevig	V 8·86	H. Larsen	Kragerö
428	N	Fjellstedt *Gregersen 86-86*	W Bk	Norweg'n	480	133·7	31·2	17·5	Lillesand	1865	P. Knudsen	V 4·86	S.N. Hansen	Lillesand
429	N	Fjerde Novembre *Salvesen*	W Bk	Norweg'n	327	115·1	27·7	15·2	Grimstad	1856	J. Teilefsen		Jacob Holst	Grimstad
430	N	Flamingo (ADANSONIA) *Walker 86-86*	W Bg	Norweg'n	267	120·3	26·9	12·5	Arendal	1869	H. Stiansen		Christen Boe	Arendal
431	V	Flash Light *Knowlton -88*	W Bkn	British	570	143·6	32·6	18·4	Three Sisters, N.S.	1875	L.J. Olive, jun.	V 11·84	Henry A. Eaton	Windsor, N.S.
432	DG	Flocke Jouwer (ex Theophania) *Kroese -88*	W Sr	Dutch	240	93·2	27·2	15·4	Spezia	1865		V 3·86	H.J. Berger	Jeure
433		Fleetwing *Turnell*	W Bk	British	786	161·0	33·0	21·9	Medford, U.S.	1854			Adelaide S.S. Co. Ld.	Melbourne
434	*	Fleetwing *Jones*	W Bg	British	226	107·4	25·0	14·6	Portmadoc	1874	Jones	1·86	W. Prichard & Co.	Carnarvon
435		Fleetwing *Manley*	W Bn	British	185	106·3	23·8	12·9	Brixham	1867	Dewdney		W.G. Kendrick & Co.	Brixham
436		Fleetwing *Mathers -83*	W Sr	British	162	108·0	24·6	7·9	Wilson, N.Y.	1863			J. Cochrane	Windsor, Ont.
437	*	Fleetwing *Simpson*	W Sr	British	110	85·2	21·7	10·3	Banff	1874	Watson		H. Munro	Banff
438	R	Fleetwing (WRECKED) *Nye*	W Bk	American	328	114·1	28·5	17·2	Pt Jefferson NY	1877	J.M. Bayles & Son	V 10·77	J. & W.R. Wing	New Bedfrd Mass
439		Fleetwing *Maddock*	W Sr	American	113	91·1	25·7	7·4	Wilmington, Del.	1865			F. Cobb	Rockland, Me.
440	R	Fleetwing *Elliot*	W Sr Yt	American	106	121·3	23·5	10·5	Pt Jefferson NY	1865	J.M. Bayles & Son	4·86	R.S. Elliott	New York
441		Fleetwing	W Bk	Danish	349	124·4	27·2	17·0	Sunderland	1867	W.H. Pearson, jr	V 11·84	J.J. Bernard	St. Thomas W.I.
442	DR	Fleetwing (WRECKED) *Rule*	W 3m Sr	Norweg'n	335	146·2	31·6	11·4	Setauket, L.I.	1855			Westergaard & Hannevig	Christiania
443	*	Fleetwood *Williams 86-86*	I S	British	630	183·4	28·6	18·4	Greenock	1867	R. Steele		J. Richardson & Co.	Glasgow
444	N	Flekkefjord (BROKEN UP) *Tollaksen 84-*	W Bk	Norweg'n	504	138·1	30·7	18·8	Apenrade	1854		V 8·86	Rederiet "Njord"	Flekkefjord
445		Flensborg *Nielsen*	W Bg	Danish	160	88·2	21·5	11·7	Flensburg	18 52/68	B. Techant		N.W. Thomsen	Nykjøbing, Sjæld.
446	*	Fleur de Lis *Berry 74-87*	I S	British	972	199·7	33·7	21·5	Glasgow	1862	Barclay, Curle & Co.		L. Tulloch	Swansea
447		Fleur de Marie	W Sp	British	113	99·5	23·0	7·0	Quebec	1878	A. McKitchen		Edouard Heynemand	Montreal
448	V	Fleur de Marie (ex Adèle) *Bazin*	W Bn	French	248	97·1	23·3	13·5	Lussinpiccolo	1879	N. Martinolich	V 1·86	D. Anger	Havre
449		Fleur de Marie *Le Rohellec*	W Bn	French	131	82·0	22·0	10·7	Honfleur	1862	Carden		J. Le Rohellec	Vannes
450														

*=Lloyd's Register; D=Nederlandsche Vereeniging; G=Germanischer Lloyd; I=Registro Italiano; R=American Record; A=Veritas Austro-Ungarico; H=Veritas Ellenico; N=Norske Veritas; V=Bureau Veritas.

Dans la 4me Colonne, C=Composite (membrure en fer, bordé en bois); I=Fer; Stl=Acier; W=Bois.

Les Dates du doublage en métal (Col. 11) ressortent des livres dans lesquels les navires sont respectivement classés ou des rapports des Inspecteurs du Lloyd's Register.

Pour plus amples renseignements sur les navires classés par le Lloyd's Register, voir le Lloyd's Register Book et le Lloyd's Yacht Register.

Official No.	No. in Book	Registry in which classed	Ship's Name. Master, and Dates of Appointment to present Owner's Service and to Ship. Ship's Late Name.	Material and Rig	Flag	Net Tonnage	Length	Breadth	Depth	Built: Where	When	By whom	Date of Metalling	Owners or Managers
46038	451	✳	**Fleur de Maurice** *Walker 81*	WBk	British	317	131·0	26·8	15·6	[illegible]	1864	Clarke	Y 10,87	John Walker
	452	GN	**Flid** *Bjørnestad -88* (ex Anna von Klein)	WBk	Norweg'n	340	117·4	27·7	15·5	Bremen	1874	H. Dierling	[illegible]	L. Larsen-Sunde
	453	N	**Flid** *Øgvist -86*	WBk	Norweg'n	339	116·0	28·9	15·4	Arendal	1873	E. Guttormsen	[illegible]	L. L. Sund
	454	N	**Flid** *Jörgensen 88-88*	WBk	Norweg'n	332	123·5	28·6	14·5	Grimstad	1876	J. N. Ribe	[illegible]	J. Jörgensen
	455	N	**Flid** *Ruberg*	WSr	Norweg'n	130	94·6	21·7	10·2	[illegible]	1872	P. Knudsen		L. L. Tangvald
44387	456		**Flimby** *Higgins*	WBn	British	127	89·5	20·7	11·5	Maryport	1862	Wood		A. McConaig
	457		**Flink** *Christensen -88*	WBg	Danish	126	81·9	20·3	11·5	Svaneke	18^{54}_{63}	H. P. Brandt		R. B. Christensen
	458	V	**Flink** *Löffler* (ex Harry C. de Wolf)	WSr	German	175	98·6	27·6	9·2	Parrsboro'	1883	J. W. de Wolf	[illegible]	Jaluit Gesellschaft
	459	N	**Flink** *Olsen-Norsted -87*	WBk	Norweg'n	260	120·7	27·9	12·8	Sarpsborg	1871		[illegible]	Olsen-Norsted
	460		**Flink** (ex Fenna) *Olsen -87*	WGlt	Norweg'n	133	90·5	21·5	9·8	[illegible]	1862	J. A. Hooites		C. Apenes
74557	461	✳	**Flintshire** *Pierce 78-82*	IS	British	1226	221·1	36·8	21·9	Liverpool	1876	T. Royden & Son		Hughes & Co.
59757	462		**Flirt** *Wilce*	WBn	British	187	104·3	24·1	12·7	P. E. Island	1869	Richards		G. B. Joiner
45238	463		**Flirt** *Shilson 85-85*	WSr	British	151	89·0	23·0	11·3	Rye	1862	Head		Holden & Gann
	464	V	**Floke** *Kielland 84-84*	WBk	Norweg'n	334	128·7	28·7	15·2	[illegible]	1870	Jörgensen & Knudsen	Y 8,88	Ths. S. Falck
	465		**Flor d' Angra** *Amancio* (ex Realhina)	W3mSr	Portug'se	310	117·4	22·5	15·0	Sunderland	1858		Y 11,81	A. P. Simoes
	466		**Flor d'Ovar** (ex Flor d'Ilhavo)	WSr	Portug'se	105	74·8	20·5	9·1	Aveiro	1873	J. F. da Lapa		F. da Fonseca Soares
	467		**Flor de Mayo**	WBn	Spanish	128	97·4	21·0	8·1	Tayabas	1879			N. Roqueño
	468		**Flor de Setubal**	WSr	Portug'se	130	82·7	20·5	9·8	Villa do Conde	1872			J. M. Trindade & A. J. [illegible]
	469		**Flor do Rio**	WBn	Brazilian	205	..	..	..					L. J. F. Tinoco
75900	470	R	**Flora** *Thomson 85-86*	WBk	British	977	193·5	37·3	21·6	P. E. Island	1877	W. Richards	Y 12,84 [illegible]	W. Richards (T. P. Richards & Co.)
71323	471	V	**Flora** *Abbott*	WBkn	British	571	172·3	31·7	18·6	Maccan, N.S.	1875	J. O'Brien	Y 8,86	J. O'Brien
79163	472	✳	**Flora** *Pike 74-80* (ex Alice Lyne)	WBkn	British	289	126·0	25·8	14·9	Newport	1878	Newport Dock	Y 5,86	J. Munn & Co.
53258	473		**Flora** *Masterton*	WBkn	British	249	110·5	25·6	14·2	Sunderland	1866	Lumsden		J. Masterton
72118	474	V	**Flora** *Menton*	WSr	British	182	102·4	23·2	10·4	Kiel	1869	Conradi & Son	Y 4,86	Aga Abdool [illegible]
50145	475		**Flora** (WRECKED)	WSr	British	114	94·4	20·1	10·6	London	1863	Langton		J. Hamilton
75160	476	✳	**Flora** *Boutillier 83-83*	WSr	British	99	84·7	21·8	10·8	Truro	1876	Trethowan		J. Bisson
74439	477	✳	**Flora** *Sara 82-82*	WSr	British	73	76·1	20·1	9·2	Malpas	1877	Scoble		T. F. Hitchins
	478	A	**Flora** (ex Henry) *Begna*	WBk	Austrian	549	146·0	27·9	18·2	Nantes	1864	Jollet & Babin	Y 10,86	Pietro Begna
	479	GV	**Flora** (LOST) *Schubert*	WBk	Danish	387	127·3	27·1	14·8	Brake	1874	J. D. Behrens	Y 8,87	A. Andersen
	480		**Flora** *Kopfod 67-77* (ex Flora Woodhouse)	WBn	Danish	324	112·5	26·6	15·4	Fairhaven, U.S.	1862			Chr. Simoni
	481	✳V	**Flora** *Hinge*	IBk	German	970	194·9	33·4	19·9	Hamburg	1880	Blohm & Voss		M. G. Amsinck
	482	G	**Flora** *Bohn 84-*	WSr	German	149	81·5	19·5	11·0	Damgarten	1856	H. Dierling		Carl Siebe
	483	N	**Flora** *Nielsen 87-87* (ex Lord Macaulay)	WBk	Norweg'n	813	171·7	35·0	20·8	Sunderland	1860	Hall	Y 5,86	Alf Monsen
	484	V	**Flora** *Halvorsen 86-86* (ex Flora [illegible])	WBk	Norweg'n	742	172·6	33·0	19·3	Londond'ry, NS	1878	Lewis	Y 10,85	Tellef Jensen
	485	N	**Flora** *Andersen*	WBk	Norweg'n	487	129·7	30·4	17·4	Tvedestrand	1871	R. Pedersen	Y 2,86	J. P. Pedersen
	486	DN	**Flora** (ex Koh-i-noor) *Andersen*	WBk	Norweg'n	440	126·2	27·7	17·4	Perth	1852		Y 11,84	N. S. Andersen
	487	N	**Flora** *Nilsen -86*	WBk	Norweg'n	326	115·8	29·4	15·7	Drontheim	1848		Y 4,86	Thv. B. Olsen
	488	N	**Flora** *Dietrichs*	WBg	Norweg'n	314	93·7	28·3	17·8	Langesund	1840	Jens Nielsen		A. M. Dietrichs
	489	NV	**Flora** *Kloster*	WBk	Norweg'n	296	114·8	28·2	14·6	Grimstad	1873	A. Johannessen	Y 4,8[illegible]	J. Borresen
	490	N	**Flora** *Torgersen 73-73*	WBk	Norweg'n	289	114·7	29·2	15·3	Norrköping	1854			Enkefru A. M. [illegible]
	491		**Flora** (WRECKED) *Ekström -87*	WBk	Russian	619	..	..	..	Tomala	1876			J. G. Sjöström
	492	V	**Flora** *Azcarregui*	WBk	Spanish	498	142·0	30·0	18·4	Sunderland	1866	J. Gill	Y 5,8[illegible]	A. Goytisolo
	493		**Flora**	WBg	Spanish	215	95·5	26·7	13·2	Barcelona	1857			Pablo Sisa
	494		**Flora** (BROKEN UP)	WBn	Spanish	153	78·7	22·5	11·7	[illegible]	1858			Gaspar Carvajal
	495		**Flora** (LIGHTER) *Jansson*	WSr	Swedish	135	85·6	20·8	10·5	Roslagen	1875			J. Sjödin
	496		**Flora** *Lindström*	WSr	Swedish	103	82·0	21·5	8·7	[illegible]	1857			G. F. Burman
71187	497		**Flora Cawith**	WSr	British	190	115·0	23·5	8·9	Mill Point, Ont.	1873	W. Jamieson		J. Cawith
120044	498		**Flora Condon** *French -87*	W3mSr	American	219	123·0	29·3	9·9	Belfast, Me.	1872			J. G. Brooks
57161	499		**Flora del Mar** *MacHugh*	WBn	British	160	88·3	24·9	11·0	Parrsboro', N.S.	1866	Pellow		H. Lane
77539	500	✳	**Flora Emily** *Wright 85-85*	WSr	British	100	87·8	21·1	10·3	Garmouth	1885	W. Anderson		C. F. Morrison

✳ = **Lloyd's Register**; D = Nederlandsche Vereeniging; G = Germanischer Lloyd; I = Registro Italiano; R[illegible]
A = Veritas Austro-Ungarico; H = Veritas Ellenico; N = Norske Veritas; V[illegible]

In the 4th Column, C = Composite (iron frame, wood planked); I = Iron; Stl = Steel; W = Wood.

The Dates of Metalling (Col. 11) are based on the records of the Books in which Vessels are respectively classed, or on the Reports of the Surveyors [illegible]

Further particulars of Vessels classed by Lloyd's Register will be found in Lloyd's Register Book and Lloyd's Yacht Re[illegible]

No.	Cote	Nom, Capitaine	Gréement	Pavillon	Tonnage	Longueur	Largeur	Creux	Où construit	Quand	Par qui	Armateurs	Port d'Armement
501		Flora King *Brett*	W Sr	American	[illegible]	[illegible]	[illegible]	[illegible]	[illegible]	1854		A.H.Sawyer	Calais, Me.
502	V	Flora P. Stafford	W S	British	[illegible]	[illegible]	[illegible]	[illegible]	[illegible]	1883	Vaughan & [Lawrence]	R.M.Rand	Windsor, N.S.
503		Flora Rogers	W 3m S	American	[illegible]	[illegible]	[illegible]	[illegible]	[illegible] Me.	1879	Bean & Robinson	R.Bowers	Camden, Me.
		Flora Ross *Jackson*	W Bn	British	134	[illegible]	[illegible]	[illegible]	[illegible]	1867	Ross	J.Chalmers	Bo'ness
		Flora Woodhouse *Hale*	W 3m Sr	American	[illegible]	[illegible]	[illegible]	[illegible]	[illegible]	1875	Baldwin & Paine	M.D.Lawrence	New York
		Fiorella	W Sr	British	126	87	[illegible]	9·7	Five Islands, N.S.	1875	Roberts	D.J.Green	Londonderry, N.S.
	✻	Florence *Thomson* 81-89	I Bk	British	809	190·2	31·0	19·5	Grangemouth	1869	Adamson & Co.	Thomson, Dickie & Co.	Glasgow
	✻	Florence *Wear*	W Bk	British	436	138·5	28·6	17·4	Plymouth	1870	Moore	G.F.Urling	London
	R	Florence *Thomas*	W Sr	British	278	110·5	27·0	15·8	Sunderland	1867	J.Gill	L.Tulloch	Sunderland
510		Florence	W Bn	British	247	108·2	26·3	[illegible]	Hopewell, N.B.	1867		H.T.Rahming	Nassau, N.P.
511	V	Florence *Furey*	W Bkn	British	212	114·8	24·5	13·1	Brixham	1875	Upham & Sons	E.J.Duder	St. Johns, N.F.
512		Florence *Nichols*	W Bn	British	196	99·3	24·9	12·9	Sunderland	1866	Barker	H.L.Daly	Newport
		Florence *May*	W 3m Sr	British	181	108·1	24·4	12·9	Looe	1869	Shapcott	T.May	Plymouth
		Florence *Price*	W Sr	British	128	92·0	26·7	9·0	St. John, N.B.	1871	Jenkins	J.Armstrong	Belfast
	✻	Florence *Evans*	W Sr	British	100	84·5	22·0	11·1	Portmadoc	1878	Williams	Mrs. M.A.Evans	Carnarvon
	✻	Florence *Parker* -87	W K	British	45	63·4	17·6	9·0	Galmpton	1887	W.A.Gibbs	W.R.Parker	Ramsgate
		Florence *Duncan*	W S	American	1604	223·1	41·0	26·0	Bath, Me.	1877	Goss & Sawyer	C.Davenport	Bath, Me.
		Florence *Carter*	W Bk	American	357	125·5	27·0	12·0	East Boston	1859	[illegible]	D.E.Mayo	Boston, Mass.
		Florence *Flynn*	W Bn	American	318	110·2	28·5	9·8	Noank, Conn.	1865		J.D.Fish & Co.	Stonington
		Florence *Estevs* -87	W Sr	Argentine	114	86·0	23·3	9·5	Shelburne, N.S.	1882	J.Dexter & Co.		Buenos Ayres
		Florence Abbott	W Sr	British	121	84·0	24·6	9·5	Maitland, N.S.	1877	J.Moore	T.Townsend	Maitland, N.S.
		Florence Christene	W Sr	British	122	83·1	24·4	8·8	Bear River, N.S.	1880	[illegible] D.Vroom	Chas. McCormick	Digby, N.S.
523		Florence Danvers *Müller*	W Bk	German	492	140·7	28·5	17·9	Bideford	1865	Johnson	Neu Guinea Cie	Hamburg
524		Florence I. Lockwood *Wing*	W 3m Sr	American	284	103·0	30·5	11·0	Norwalk, Conn.	1867		Levi C.Wing	Boston, Mass.
	R	Florence J. Allen *Drake*	W 3m Sr	American	458	142·6	34·1	12·8	Bath, Me.	1882	Goss, Sawyer & Packard	J.B.Drake	Bath, Me.
526	R	Florence Leland *Adams*	W 3m Sr	American	327	139·0	32·6	11·3	Camden, Me.	1882	Combs & Day	J. & B.C.Adams	Camden, Me.
527		Florence Nightingale	W Bn	British	185	116·7	25·7	11·1	Hopewell, N.B.	1857	Russell	W.Cuthbert	Londonderry
		Florence Nowell *Nickerson*	W Sr	American	212	105·8	28·7	8·0	Chester, Pa.	1867		J.F.Nickerson	Dennis, Mass.
529		Florence Randall *Randall*	W 3m Sr	American	704	159·1	35·4	16·2	Bath, Me.	1882	Goss & Sawyer	H.M.Randall	New Haven, Conn.
	R	Florence Rogers	W 3m Sr	American	396	123·0	30·0	10·0	Bellville, N.J.	1860		L.S.Davis	New York
531		Florence Shay	W 3m Sr	American	385	137·0	30·0	15·0	Pt. Jeffers'n, N.Y.	1867		Evans, Ball & Co.	New York
		Florence Treat *Spring*	W Bk	British	724	152·4	33·3	20·5	Brewer, U.S.	1865	J.Oakes	J.C.Ellis	Sydney, N.S.W.
	✻	Florence & Ann *Beckworth*	W Sk	British	83	76·3	20·2	10·3	Goole	1883	Cottingham Brs	W.Brinham	Hull
534		Florence & Lillian	W 3m Sr	American	240	124·0	30·3	8·9	Leesburg, N.J.	1873	J.Russell	C.S.Smith	Bridgeton, N.J.
	✻	Florencia	I 3m Sr	Argentine	430	168·0	28·2	12·5	Sunderland	1884	Sunderland Ship-Building Co.	B.Denegre	Buenos Ayres
		Flores de Maria	W Bn	Spanish	283	116·4	27·2	13·0	Pantatnan	1855		José Reyes	Manila
		Floresta	W Sr	Brazilian	164	..	..	..				A.R.S.Leite	Rio Janeiro
	V	Floresta *Tur* -86	W Pol	Spanish	176	87·6	26·1	12·7	Barcelona	1855		B.Tur	Ibiza
		Florida -87	W Sr	British	216	112·6	25·0	12·4	Perth	1870	Wallace	J.Williams	Sydney, N.S.W.
	R	Florida *Crosby*	W Bn	British	100	84·5	24·2	9·0	Meteghan, N.S.	1877	B.Robishan	Th.B.Flint	Yarmouth, N.S.
		Florida *Priest*	W 3m Sr	American	272	133·0	32·1	9·4	Belfast, Me.	1872	Carter	M.T.Emery	Portland
	A	Floride *Martinolich*	W S	Austrian	1264	192·5	38·7	23·9	St. John, N.B.	1867	King	A.Tischbein & Co.	Lussinpiccolo
	I	Florida *Repetto* -88	W Bk	Italian	754	152·5	34·1	23·0	Lavagna	1870	F.Rolla	E.Piaggio fu G.B.	Genoa
	I	Florida *Poli*	W Sr	Italian	163	89·4	23·8	9·7	Chioggia	1882	G.Poli	R.Zennari ved. Poli	Venice
	N	Florida 72-83	W Bk	Norweg'n	545	140·7	31·2	17·6	Windsor, N.S.	1876		Et Aktie-Selskab (C.Gjerløff)	Kragerö
	V	Florida *Larsson* -88	W 3m Sr	Swedish	298	112·1	25·2	12·9	Helsingborg	1874	J.Nordström	J.C.Hulthén	Helsingborg
		Florin *Hoskin* 85-88	W Sr	British	158	96·3	24·5	9·5	Newcastle	1864		J.Watson	Newcastle
	V	Floriudo Padre *Tonietti*	W Bg	Italian	246	117·1	24·8	13·2	[illegible]	1873	C.Galli	Egidio Tonietti	Porto Ferraio
		Floritza	W Bg	Turkish	283	..	..	..	[illegible]	1856		I.D.Palamoutis	Galatz
	✻	Florrie 85-88	W K	British	49	[illegible]	[illegible]	7·5	[illegible]	1888	R.Cock	R.Cock	Bideford

✻=Lloyd's Register; D=Nederlandsche Vereeniging; G=Germanischer Lloyd; I=Registro Italiano; R=American Record; A=Veritas Austro-Ungarico; H=Veritas Ellenico; N=Norske Veritas; V=Bureau Veritas.

Dans la 4me Colonne, C=Composite (membrure en fer, bordé en bois); I=Fer; Stl=Acier; W=Bois.

Les Dates du doublage en métal (Col. 11) ressortent des livres dans lesquels les navires sont respectivement classés ou des rapports des Inspecteurs du Lloyd's Register.

Pour plus amples renseignements sur les navires classés par le Lloyd's Register, voir le Lloyd's Register Book et le Lloyd's Yacht Register.

Official No.	No. in Book	Registry in which classed	Ship's Name. Master, and Dates of Appointment to present Owner's Service and to Ship. Ship's Late Name.	Material and Rig	Flag	Net Tonnage	Length	Breadth	Depth	Built: Where	Built: When	Built: By whom	Date of Metalling	Owners or Managers
	551	DN	Flosta *Nielsen 86-85*	WBg	Norweg'n	213	110·9	25·8	12·2	Tvedestrand	1869			John Smith
70367	552	✳	Flower of Olrig *Sinclair 81-84*	WSr	British	81	82·2	20·8	9·1	Stromness	1881	G.&P.Copland		G.Alexander
62048	553	✳	Flower of the Fal *Whitburn*	WSr	British	139	98·0	23·3	12·0	Padstow	1870	Trethwen	[illegible]	J.Whitburn
[illegible]	554	✳	Flower of the Forest *Oxley*	WSk	British	76	72·4	20·2	10·4	Elmshorn	1884	J.Kremer		H.C.Oxley
[illegible]	555	✳	Flower of the Valley	WSk	British	85	78·9	20·5	10·6	Rye	1885	G.&T.Smith		A.Johnson
78065	556		Flowerdale *Hicks*	WBn	British	113	99·3	21·8	8·2	Wynyard,Tas	1878			James Lilly
15589	557		Flown	WSr	British	100	83·5	19·3	10·3	Bideford	1856			J.Leith
	558		Fluminense	WBn	Brazilian	247	..	..	..					J.J.A.Souza
120439	559	R	Fluorine *Dix -88*	WBk	American	298	136·4	29·6	11·4	Bath,Me	1881	Goss&Sawyer		C.R.Dix
9482	560		Fly *Baxter*	WSr	American	108	90·3	26·1	6·9	Camden,N.J.	1856			J.F.Baxter
9064	561		Fly Away *Bingham*	WSr	American	152	94·9	27·5	8·1	ChathamMass	1854			J.P.Dole
76236	562	✳	Flying Fish *Coles*	WSr	British	66	76·5	19·2	9·1	Jersey	1878	Picot	[illegible]	J.F.Picot
76272	563	✳	Flying Foam *Noel*	WSr	British	100	88·4	20·4	10·5	Jersey	1879	LeHuquet	[illegible]	C.Blampied
29418	564	✳	Flying Scud *Corlett*	WBn	British	169	113·1	20·5	12·0	Appledore	1861	Cock	[illegible]	T.J.M.Corlett
[illegible]	565	V	Flying Venus	IS	British	1660	232·0	37·0	31·4	London	1855	C.J.Mare&Co		J.Kerr
[illegible]	566		Foam *Reynolds*	WSr	American	103	100·0	17·5	7·6	Hamburg,Pa.	1864			J.A.Anderson
54274	567		Foh Kien	WSr	British	154	79·0	21·4	12·4	Ipswich	1838			Shaik Madar Sahib [illegible]
	568	V	Foi *Herot*	WBg	French	205	101·0	20·7	12·9	St.Malo	1879	L.Mallard	[illegible]	Vve L.Mallard
	569	N	Folkefesten *Ramsland* (CONDEMNED)	WBk	Norweg'n	383	127·4	29·8	16·3	Lillesand	1874	G.Olsen	[illegible]	F.Smith Petersen
	570	N	Folkvang *Henriksen 75-80*	WBk	Norweg'n	552	141·5	32·9	18·6	Drammen	1864	[illegible] & Knudsen	[illegible]	Ths.S.Falck
126	571	A	Follia *Gramenuda*	WSr	Austrian	375	116·1	25·8	17·3	Cherso	1878	C.Giadrovich	[illegible]	G.Gramenuda fu [illegible]
[illegible]	572	✳	Fomalhaut *Williams 84-*	WSr	British	72	71·4	20·0	9·2	Bangor	1874	Parry		W.Williams
50030	573	✳	Fontenaye *Booth 84-84*	IS	British	635	170·1	28·3	18·6	Sunderland	1864	G.S.Moore&Co.		G.L.Munro & Co
[illegible]	574		Fonthill *Mathews -88*	W3mSr	British	203	105·2	23·6	11·8	Southampton	1878	Ransom	[illegible]	J.Dible
138	575		Foochow	WBg	Siamese	300	116·8	24·5	13·1	Bangkok	1862			Lau Mah Teng
	576	✳	FoohngSuey *Newell 75-88*	StlBk	Hawaiian	1036	211·0	34·5	19·2	Glasgow	1888	Napier, Shanks & Bell		E.Brewer
	577	N	Foraaret *Olsen -75*	WBn	Norweg'n	136	94·6	23·6	11·4	Stavanger	1871	M.Magnus		C.Olsen(JohnC.M[illegible])
	578	N	Forældrene Christie *Jacobsen*	WBg	Norweg'n	149	90·0	24·6	12·1	Christianssund	1855			A.B.Jacobsen
	579	G	Förening *Thore 70-77*	W3mSr	Swedish	213	110·8	23·7	11·9	Timm'rnabben	1869	J.Nordström	[illegible]	B.M.Andersson
	580		Föreningen *Blomberg*	WSr	Russian	133	..	..	..	[illegible]	1870			J.A.Blomberg
	581		Föreningen *Strandström*	WSr	Russian	124	..	..	..		1873			J.F.Strandi[illegible]
	582	V	Föreningen *Olsson -80*	WBkn	Swedish	264	118·4	25·6	11·7	[illegible]	1871	O.A.Brodin	[illegible]	S.Hörberg
	583		Föreningen *Jansson*	WSr	Swedish	115	82·3	21·0	10·0	[illegible]	1851			J.E.Jansson
72310	584	DV	Forest *Perry 87-87*	WBk	British	747	163·9	34·6	19·5	[illegible], N.B.	1877	G.H.Flewelling	[illegible]	Wm.Thomson & Co
63384	585		Forest Belle *Coppack 74-74*	WSr	British	122	96·2	22·1	10·4	Nevin	1870	Huddart		W.Sloane, jun.
9034	586		Forest City *Coulter*	WSr	American	112	87·9	27·0	7·2	Sullivan,Me.	1870	Sullivan,Me		J.Davis
[illegible]	587		Forest Fairy *Davis -87* (ex Josephine)	W3mSr	British	116	101·0	19·1	10·6	Ipswich	1859	Bayley	[illegible]	R.Neville
120489	588	R	Forest Home *Blair -87*	W3mSr	American	116	102·0	25·5	6·5	[illegible],NY	18[illegible]			Poitevant & Favre
73484	589	DV	Forest King *Doyle 86-86*	WS	British	1[illegible]2	215·6	41·0	24·4	[illegible] N.S.	1877	J.B.North	[illegible]	J.B.North
[illegible]	590		Forest King *Wigmore*	WBn	British	158	98·2	22·3	13·1	Plymouth	1868	Gent	[illegible]	J.Williams
[illegible]	591	✳	Forest Princess *Wills*	WSr	British	171	104·0	22·5	12·5	Newport	1866	[illegible]	[illegible]	W.Brooks
[illegible]	592		Forest Queen *Winding*	WBk	American	486	176·0	37·0	13·6	Port Ludlow, [Wash.	[illegible]		[illegible]	C.Nelson
27716	593	R	Forest Rights *Murphy* (ex Edward Everett ex James)	WS	British	1132	181·4	37·3	23·2	Boston,U.S.	1860		[illegible]	Sir J.Arnott
[illegible]	594		Forester *Johns*	WSr	British	115	89·0	20·7	11·1	Rye	1866	Hessel		J.Bowen
[illegible]	595	✳	Forfarshire *Evans 85-87*	IBk	British	1300	239·6	36·0	21·5	Dumbarton	1885	Birrell,Stenhouse&Co		T.Law & Co.
[illegible]	596	✳	Forfarshire *Rolfe 88-88*	CS	British	1238	201·9	37·1	22·1	Sunderland	1867	J.Morison	[illegible]	Shaw Savill & Albion Co Ld.
	597		Forganhall *Olsen -88*	WS	Norweg'n	1070	175·3	36·3	22·8	[illegible] N.B.	1857	J.McDonald	[illegible]	[illegible]
[illegible]	598		Forget-me-not	WBk	British	233	117·0	25·9	11·4	Danzig	1866			T.D.Boyd
[illegible]	599	✳	Forget-me-not	WSk	British	77	75·7	20·2	10·4	Elmshorn	1885	J.Kremer		D.McCarthy
	600	NV	Formica *Mathisen -88*	WBk	Norweg'n	408	143·3	28·5	15·4	Tönsberg	1864	A.B.Bull	[illegible]	[illegible]

✳=Lloyd's Register; D=Nederlandsche Vereeniging; G=Germanischer Lloyd; I=Registro Italiano; R=Ame[rican]
A=Veritas Austro-Ungarico; H=Veritas Ellenico; N=Norske Veritas; V=Bure[au Veritas]

In the 4th Column, C=Composite (iron frame, wood planked); I=Iron; Stl=Steel; W=Wood.

The Dates of Metalling (Col. 11) are based on the records of the Books in which Vessels are respectively classed, or on the Reports of the Surveyors of [illegible]

Further particulars of Vessels classed by Lloyd's Register will be found in Lloyd's Register Book and Lloyd's Yacht Register

No.	Reg.	Nom. Capitaine et Gréement. Ancien Nom.	Rig	Pavillon	Tonn.	Long.	Larg.	Creux	Lieu	Date	Fait par	Doublage	Armateur ou Propriétaire Gérant.	Port d'Attache.
[illegible]	N	Formica	W Bk	Norweg'n	398	116·7	31·5	16·4	Arendal	1857	JørgenMørland		Chr. Klucker	Arendal
602	✻	Formosa	I S	British	[illegible]	[illegible]	38·1	22·7	Port Glasgow	1883	Russell & Co.		P. Denniston & Co.	Glasgow
603	✻	Formosa	I Bk	British	[illegible]	[illegible]	32·8	19·8	Dumbarton	1875	A. McMillan & Son		Ellinor Main & Montgomery	Liverpool
604	RV	Formosa	W Bk	American	[illegible]	[illegible]	31·4	16·9	Brewer, Me.	1876	J. Oakes & Sons		T. H. Pierce	New York
605	N	Fornjot	W Bk	Norweg'n	482	136·0	33·8	16·7	[illegible]	1870	Ole Eriksen		C. H. Mölbach	Krerø
606	✻	Forrest Hall	I S	British	[illegible]	[illegible]	[illegible]	[illegible]	[illegible]	1883	R. Williamson & Son		Globe Shipping Co., Ld.	Liverpool
607	DN	Forsete	W Bk	Norweg'n	[illegible]	[illegible]	[illegible]	[illegible]	[illegible]	1832	H. T. Thryde		O. Christensen	Arendal
608	N	Forsete	W Bk	Norweg'n	184	99·1	24·4	10·2	Sandefjord	1875			P. A. Grøn	Sandefjord
609		Forskandet	W Sr	Swedish	[illegible]	..	..	..	GamlaKarleby				L. Johansson	Oravet
610	N	Forsöget	W Bk	Norweg'n	628	148·2	32·7	20·2	Chiavari	1866	P. Cluggini		H. Rasmussen	Porsgrund
611	N	Forsöget	W Bk	Norweg'n	255	108·1	25·1	14·0	Finland	1857			[illegible]	[illegible]
612	N	Forsøget	W Bg	[illegible]	214	93·8	25·1	11·2	Langesund	1855			M. Borge	Tönsberg
613		Forsoket	W Bg	Swedish	257	104·6	25·1	13·0	Grohn	1863	Angener & Co.		C. Andersson	Strömstad
614		Forsoket	W Sr	Swedish	234	104·7	25·2	12·1					O. F. Hakansson	Landskrona
615	✻	Fort George	I 3mS	British	1686	260·0	40·0	23·1	Belfast	1884	Workman, Clark & Co.		Clark & Service	Glasgow
616	✻	Fort James	I S	British	1687	261·2	38·9	23·3	Belfast	1885	Workman, Clark & Co.		Clark & Service	Glasgow
617		Forte	W S	French	184	91·9	24·8	12·3	Gravelines	1869	Hochart & Lavallée		Winmaert & Co.	Dunkirk
618		Fortis Fortuna	W G	Russian	106	73·0	21·0	8·0	Livland	1876			R. Von Vegesack	Haynasch
619		Fortroendet	W Sr	Swedish	133	78·2	21·4	9·1	(Rebuilt)	18..			J. Johansson	Simmesby
620		Fortuna	W Sr Yt	British	152	[illegible]	25·5	13·7	Lymington	1876	G. Inman		A. E. Hope	Southampton
621	✻	Fortuna	W Sk	British	76	74·5	20·5	10·5	Hamburg	1885	C. C. A. Dreyer		H. Chanter	Grimsby
622	R	Fortuna	W 3mSr	American	[illegible]	151·4	34·5	14·5	Boston, Mass.	1883	Smith & Townsend		W. B. Bowne	Boston, Mass.
623		Fortuna	W Sr	American	138	102·5	29·6	8·2	Fairhaven, Cal.	1886			J. A. Sinclair	San Francisco
624		Fortuna	W Bk	Chilian	[illegible]	116·4	28·0	17·1	Medford, Mass.	1850			[illegible] & Burgos	Valparaiso
625	V	Fortuna	I Bk	German	951	185·0	33·0	20·1	Hamburg	1869	Reiherstieg Ship Works		J. C. Pflugk	Hamburg
626		Fortuna	W Bk	German	[illegible]	129·7	31·3	19·5	Danzig	1868	J. W. Klawitter		O. Münsterberg	Danzig
627	DV	Fortuna	W Bk	German	428	12?·6	29·8	16·6	Elbing	1863	G. Fechter		J. H. Meyer	Pillau
628		Fortuna	W Bk	Italian	128	..	..	..	Catania	18..			Letterio Trapano	Catania
629	I	Fortuna	W Bn	Italian	[illegible]	87·6	21·8	10·5	Vietri	1878	P. Savastano		R. Schiazzano	Messina
630	N	Fortuna	W Bk	Norweg'n	616	126·6	33·2	21·1	Christinestad	1804			Peter Hansen	Christiania
631	N	Fortuna	W Bk	Norweg'n	448	129·4	30·4	16·7	Lillesand	1875	H. Jacobsen		T. Larsen	Lillesand
632	N	Fortuna	W Bk	Norweg'n	434	127·4	29·9	16·8	Grimstad	1874	Jens Tellefsen		P. Pedersen	Grimstad
633	N	Fortuna	W Bg	Norweg'n	320	84·0	26·0	15·0	Arendal	1848			Axel Smith	Arendal
634	N	Fortuna	W Sr	Norweg'n	215	108·5	25·5	12·1	Kragerö	1877	Alf Johnsen		B. Knoph	Kragerö
635	NV	Fortuna	W Bg	Norweg'n	194	10?·?	25·7	11·0	Drammen	1873			M. Andersen	Christiania
636	N	Fortuna	W Bg	Norweg'n	168	92·0	24·3	12·1	Drammen	1859			B. Corneliussen	Tönsberg
637		Fortuna	W Bg	Norweg'n	119	77·7	29·2	10·2					P. Christophersen	Bodo
638	✻	Fortuna	I Bk	Portug'se	518	156·7	27·1	18·3	Sunderland	1865	W. Pile & Co.		J. H. Andresen	Oporto
639		Fortuna	W Bg	Russian	324	92·5	27·5	14·7	Memel	1847	E. Freundt		W. C. Linde	Riga
640	V	Fortuna	W 3mSr	Russian	225	94·0	23·9	11·8	Riga	1879	Mühlberg		C. Grundmann	Riga
641		Fortuna	W Sr	Spanish	137	77·8	23·4	11·1	Arichat, N.B.	1868			B. Cutillas y Alacid	Canary Islands
642		Fortuna	W Bg	Swedish	[illegible]	110·2	23·4	16·0	Sunderland	1853			T. Bolmeer	Bergqvara
643		Fortuna	W 3mSr	Swedish	[illegible]	100·4	23·3	12·4	Skeppsmyra	1883	Skeppsmyra Varf		P. Holmgren	Skeppsmyra
644		Fortuna	W Bg	Swedish	177	93·5	22·7	12·5	GamlaKarleby	1836			N. Olsson	[illegible]
645		Fortuna O.	W Bk	Italian	444	125·3	29·5	18·6	Savona	1862	F. Sivello		B. Revello	Genoa
646	IV	Fortunata	W Bk	Italian	520	135·1	27·9	19·6	Alimuri	1874	G. Maura		F. S. Ciampa & figlio	Castell. di Stab.
647	I	Fortunata	W Bg	Italian	429	139·1	29·2	18·0	Sorrento	1878	A. Paturzo		L. Russo & Co.	Castell. di Stab.
648		Fortunata	W Bk	Norweg'n	277	106·0	27·9	14·7	Bremen	1845	Wencke		J. H. Apeness	Horten
649	A	Fortunata B.	W Bn	Italian	100	80·1	22·6	11·4	Chioggia	1883	L. Menetti		S. Ballarin	Venice
650	✻I	Fortunata Caterina	W Bk	Italian	891	179·7	34·3	23·3	Savona	1876	G. B. Poggi		N. Accame	Savona

✻ Lloyd's Register; D = Nederlandsche Vereeniging; G = Germanischer Lloyd; I = Registro Italiano; R = American Record; A = Registro Austro-Ungarico; H = Veritas Ellenico; N = Norske Veritas; V = Bureau Veritas.

Dans la 4me Colonne, C = Composite (membrure en fer, bordé en bois); I = Fer; Stl = Acier; W = Bois.

Les Dates du doublage en métal (Col. 11) ressortent des livres dans lesquels les navires sont respectivement classés ou des rapports des Inspecteurs du Lloyd's Register.

Pour plus amples renseignements sur les navires classés par le Lloyd's Register, voir le Lloyd's Register Book et le Lloyd's Yacht Register.

1 Official No.	2 No. in Book	3 Registry in which classed	4 Ship's Name. Master, and Dates of Appointment to present Owner's Service and to Ship. Ship's Late Name.	Material and Rig	5 Flag	6 Net Tonnage	Dimensions: Length	Breadth	Depth	9 Built: Where	When	By whom	11 Metalling	12 Owners or Managers
[illegible]	651	✱ I	Fortunata M. *Moriola* (ex Havenne)	W Bk	Italian	682	154·8	27·4	19·9	Sestri Ponente	1873	A. Briasco	[illegible]	S. Marchese
[illegible]	652		Fortunate *Stevens 83-*	W Sr	British	110	95·8	21·8	8·4	Southampton	1849	G. Blaker		J. Dible
[illegible]	653		Fortunate *Cock 77-81*	W Sr	British	199	86·0	22·5	10·9	Padstow	1874	Stribley	[illegible]	R. H. W[illegible]
[illegible]	654	✱	Fortunate *Taylor 84-*	W Sk	British	77	72·4	20·2	10·4	Elmshorn	1884	J. Kremer		H. L. Taylor
[illegible]	655	✱	Fortunatino (ex Metapont) *Pompeo -88*	W Bk	Italian	492	147·6	27·6	21·0	Sunderland	1867	W. Watson	[illegible]	F. [illegible]
2[illegible]	656		Fortunato	W Bn	M'nt'n'grn	128	70·5	22·0	13·1	Venice	1868			Suleiman Zakali [illegible]
	657		Fortunato O. *Olivari* (ex Elisa)	W Bkn	Italian	146	..	..	..	Fiume	1845		[illegible]	A. Olivari di F.
[illegible]	658	D I	Fortunato Repetto (ex Bucefalo) *Schiaffino 86-88*	W Bk	Italian	741	164·7	33·1	22·7	Varazze	1874	C. Scala	[illegible]	G. B. Repetto fu A[illegible]
71478	659		Fortune *Daniel -86*	W Bn	British	199	103·0	24·3	12·9	P.E. Island	1875	Duncan	[illegible]	J. De C. Treffry
61	660		Fortune	W Bk	Siamese	447	127·0	29·4	16·6	Bangkok	1859			Phra Phe Boon[illegible]
[illegible]	661		Forward (ex Arthur Pickering)	W Bk	British	202	100·8	24·2	11·4	Pembroke, US	1847			W. H. Smith & S[illegible]
[illegible]	662	✱	Forward *Coombe -81*	W Sr	British	131	96·7	22·3	11·4	Bideford	1877	Johnson	[illegible]	W. H. Mare
75248	663		Forward *Hughes*	W Sr	British	109	85·3	22·0	10·4	Peterhead	1876	Stephen		R. Edwards
	664	✱	Forward *Solaim -88*	W K	French	71	72·0	19·5	8·8	Peterhead	1885	Stephens & F'bes		J. Solaim
134	665	A	Forza *Percich*	W Bk	Austrian	439	133·2	27·3	18·2	Fiume	1869	Flli. Schiavoni	[illegible]	V. Warglien
	666	N	Fosnæs *Sorensen -87*	W Bk	Norweg'n	299	92·7	27·8	16·0	Tönsberg	1837			Langbryggens Verft (J. Lorange)
	667		Fosterlandet *Lundqvist*	W Bk	Swedish	333	123·6	27·5	14·6	Hernösand	1856			S. Z. Holmberg
120383	668	R	Fostina *Phillbrook 79-79*	W 3m Sr	American	364	137·5	31·5	11·6	Camden, Me.	1879	Coombs & Brown		I. Coombs
67162	669		Four Brothers *Anderson 80-8-*	W Sr	British	83	83·8	20·4	9·3	Northwich	1876	Thompson		T. Anderson
90372	670	✱	Four Sisters	W Sk	British	81	73·3	21·0	10·5	Grimsby	1884	Farmer & Leaver		A. Doust
91200	671	✱	Four Winds *Ritchie 73-83*	I S	British	1799	267·0	39·3	23·7	Newcastle	1885	Palmer's Co. Ld		G. N. Gardiner &[illegible]
	672		Foussingø (WRECKED) *Strang*	W Bkn	Danish	146	91·5	20·9	10·7	Randers	1870	T. Schoman	[illegible]	Actieselskabet [illegible]
83506	673	✱	Fox *Dunhavand 77-83*	W 3m Sr	British	100	96·4	22·5	10·2	Runcorn	1883	Brundrit		T. C. Cooper & Co
	674	N V	Fox *Samuelsen -88*	W Bk	Norweg'n	311	129·9	27·6	13·4	Porsgrund	1870	H. Olsen	[illegible]	Axel Smith
68977	675	✱	Foxglove *Woodward*	I Bk	British	843	193·7	32·3	19·0	Sunderland	1878	J. Laing		Duncan, Fox & C[illegible]
58800	676	✱	Foxhound *Walters -88*	I Bk	British	444	147·4	27·4	17·3	Hartlepool	1867	Denton Gray & C		L. Tulloch
70642	677	✱	Foyle *Cutting 79-82*	I S	British	1598	243·0	38·8	24·2	Sunderland	1874	T. R. Oswald		J. Nourse
	678	N	Föynland (ex Erslina) *Föyen 69-83*	W Bk	Norweg'n	695	156·8	35·1	18·9	[illegible]	1870	P. Bergfors	[illegible]	W. Wilhelmsen
	679		Fraga I	W Bn	Brazilian	190	..	..	..					G. P. de Souza
	680	N	Fram (ex Coxhill) *Christophersen*	W 3m Sr	Norweg'n	339	105·6	27·8	17·5	[illegible]	1847	Wm. Pearson		S. Andersen
	681	N	Fram *Markussen*	W Bg	Norweg'n	239	113·9	27·1	11·9	[illegible]	1872	M. Eldrup	[illegible]	Jörgensen & Jerk[illegible]
	682	N V	Fram *Johnsen*	W Bn	Norweg'n	211	109·9	25·1	12·2	[illegible]	1871	M. Knudsen	[illegible]	H. Natvig
	683	V	Framåt *Åkerman 88-88*	W Bk	Swedish	399	127·3	25·0	12·0	[illegible]	1870	A. N. Liedberg	[illegible]	P. Larsson
	684	V	Framåt *Mannberg*	W Bg	Swedish	252	112·8	24·5	12·3	[illegible]	1870	Jonsson	[illegible]	C. G. Moritz
	685	G N	Framatyr *Danielsen*	W Bn	Norweg'n	185	99·0	21·8	12·8	[illegible]	1870	M. Magnus	[illegible]	J. Haugland
	686	V	Framfart *Marsteen 80-85*	W Bk	Norweg'n	399	128·0	29·6	15·5	[illegible]	1877	Arlesen	[illegible]	Chr. Johnsen
	687	✱	Framnas *Nilsson 84-84*	W Bk	Swedish	324	127·0	28·5	13·0	[illegible]	1880	E. Östman	[illegible]	Framnäs Wharf [illegible]
120210	688	V	Franc Lambirth *Hehdl, -86*	W Bkn	American	465	145·4	29·5	16·4	Bath, Me.	1875	A. Robinson	[illegible]	T. L. Mayo
	689	V	France *Cholet -88*	W Bk	French	499	132·5	26·9	19·1	Quebec	1869	Charland & Marquis	[illegible]	D. Auger
	690	F	France *Maillard*	W Bk	French	270	110·5	27·4	15·0	Fecamp	1877		[illegible]	A. Chancerel
	691	V	France *Vauhille -81*	W S	French	129	93·8	22·6	11·0	Dunkirk	1884	P. Meuwisse		J. Van Cauwenberg[illegible]
	692	✱	France Chérie (ex Beth Shan) *Blondel (78-8-)*	I S	French	657	174·5	29·6	18·7	Sunderland	1863	Pile, Hay & Co.		J. H. Tandonnet [illegible]
[illegible]	693		Frances *Pretty 77-77*	W Bk	British	298	123·0	25·1	15·5	Bristol	1858	Claughton		G. H. Bridges
[illegible]	694	✱	Frances *Pettit*	W Sk	British	90	81·6	20·8	10·6	Rostock	1885	E. Burchard & Co		J. Pettit
[illegible]	695	R	Frances *Benner -87*	W Bkn	American	644	164·0	33·8	17·0	Belfast, Me	1887	G. W. Cottrell	[illegible]	C. Morton Stewart [illegible]
[illegible]	696		Frances *Henrichsen*	W Bk	Norweg'n	490	137·0	28·5	18·5	Sunderland	1860	W. Briggs & Sons	[illegible]	Th. Henrichsen
[illegible]	697		Frances A. Barstow *Macarter -8-*	W Bn	American	121	83·1	23·5	10·1	New Bedford, Mass	1877	Howland	[illegible]	P. H. Reed
[illegible]	698	✱	Frances Fisher *[illegible] 85-85*	Stl Bk	British	1430	241·0	37·5	21·3	Dumbarton	1885	A. McMillan & So[illegible]		[illegible]
[illegible]	699	✱	Frances Milly (CONDEMNED) *Davies [illegible]*	W Bk	British	287	130·0	23·0	14·5	Millom	1858	J. D. Roberts & Co	[illegible]	M. Davies
[illegible]	700	R	Frances Z. *Berry*	W Sr	British	146	97·0	27·0	9·5	[illegible], N.S.	1888	W. A. Gilliatt		W. A. Gilliatt

✱ = Lloyd's Register; D = Nederlandsche Vereeniging; G = Germanischer Lloyd; I = Registro Italiano; A = Veritas Austro-Ungarico; H = Veritas Ellenico; N = Norske Veritas; V[illegible]

In the 4th Column, C = Composite (iron frame, wood planked); I = Iron; Stl = Steel; W = Wood.

The Dates of Metalling (Col. 11) are based on the records of the Books in which Vessels are respectively classed, or on the Reports of the S[illegible]

Further particulars of Vessels classed by Lloyd's Register will be found in Lloyd's Register Book and Lloyd's Yacht [illegible]

	No.		Nom. Matériaux et Gréement. Changements dans l'entrée au registre des Armateurs... Ancien Nom	Pavillon	Tonnage	Longueur	Largeur	Creux	Où	Quand	Par qui		Armateurs ou Directeurs-Gérants	Port d'Armement
[illegible]	701		Frances & [illegible]		[illegible]	119·8	25·7	[illegible]	[illegible]	1878	Vaux		J. M[illegible]	[illegible]
[illegible]	702	[illegible]	Frances & T[illegible]		[illegible]	[illegible]	[illegible]	[illegible]	[illegible]	1874	N. M[illegible]	[illegible]	M[illegible]	[illegible]
[illegible]	703	I	Francesca *Russano* W Bg	Italian	330	109·5	26·1	17·7	Alinari	1875	M. Mauro	[illegible]	F[illegible]	Castellam. di Stab.
[illegible]	704		Franceschino [illegible] W Bk	Italian	521	146·0	31·4	19·6	Varazze	1870	F. Tassara	[illegible]	[illegible]	Venice
[illegible]	705	IV	Francesco *Ruffino* W Bk	Italian	748	166·3	32·2	22·3	Cassano	1875	A. Castellano	V [illegible]	F. & M. Castellano	Castellam. di Stab.
[illegible]	706		Francesco *Ciano* W Bk	Italian	358	112·8	30·4	17·9	Sestri Ponente	1863	G. Casanova	V [illegible]	R. Ciano & Co.	Genoa
[illegible]	707	A	Francesco *Malusa* W Bg	Italian	137	..	..	..		1884			A. Malusa	Venice
	708		Francesco W Bg	Italian	129	80·7	21·5	11·6	Lerici	1861	F. Rolla		F[lli] Faggioni	Spezia
[illegible]	709	IV	Francesco B. *Bonifacio* W Bk	Italian	763	169·9	35·5	21·1	Varazze	1874	N. Imosei	V [illegible]	G. Bonifacio & Co.	Genoa
[illegible]	710	I	Francesco C. *Chiapella* W Bg	Italian	320	116·8	27·6	17·1	Foce	1876	Celesia & Debarbier	V [illegible]	F. Chiapella fu G.	Genoa
[illegible]	711		Francesco C. *Napoli* W Sr	Italian	109	77·4	20·7	12·5	Trieste	1845			G. Decarli fu A.	Porto Maurizio
[illegible]	712	I	Francesco Cilento *Cacace* W Bk	Italian	385	129·6	28·0	18·0	Equa	1870	M. Mauro	V [illegible]	M. & S. Cacace	Castellam. di Stab.
[illegible]	713	I	Francesco Costa *Costa* W Bk	Italian	494	140·7	31·4	20·6	Varazze	1868	N. Imosei	[illegible]	Cassa Marittima	Genoa
	714	V	Francesco Gotusso I Bk	Italian	442	153·8	26·2	16·7	Nantes	1868	Gouin		A. Gotusso	Genoa
[illegible]	715	A	Francesco Padre W Bg	Italian	190	102·3	25·6	13·0	Arenzano	1868	Gerziglia		G. Lenci	Leghorn
[illegible]	716	A	Francesco Paolo W Bg	Italian	244	101·0	27·6	16·1	Castella mare di Stabia	1857	G. Bonifacio	[illegible]	F[lli] Cacciuttolo	Naples
[illegible]	717		Francesco Paolo G. W Bk	Italian	127	89·2	20·8	11·7	Torredel Greco	1886	M. Donnarumma		M. Garofano	Naples
[illegible]	718	III	Francesco R. *Califano* W Bk	Italian	506	135·4	28·4	20·3	Cassano	1875	A. Castellano	V [illegible]	R. Califano & Co.	Castellam. di Stab.
[illegible]	719	I	Francesco Saverio W Bk	Italian	347	118·4	27·5	17·4	Equa	1870	Mauro	[illegible]	C. Starace	Castellamare
[illegible]	720	A	Francesco Saverio Villano W Bg	Italian	127	..	..	..	Torredel Greco	1886	M. Donnarumma	V [illegible]	M. Villano	Naples
[illegible]	721	I	Francesco I. *Mori* -88 W Bg	Italian	131	89·6	23·0	12·3	Lerici	1882	F. Medusei	[illegible]	N. Mori	Spezia
[illegible]	722		Franchise *Donald* -84 W Bg	British	132	86·4	23·4	10·4	Bay de Verte, N.B.	1856			H. M'Dowell	Belfast
[illegible]	723		Francis *Griffiths* W Sr	British	115	89·0	21·2	11·0	Barrow	1861	Hoy		J. Fisher & Sons	Lancaster
[illegible]	724	RV	Francis *Brag* -87 W S	American	1974	231·0	43·4	26·6	Bath, Me.	1885	N England SB Co	[illegible]	W. H. Besse	New Bedford
	725		Francis *Lefèvre* -87 W Bk	French	330	115·1	23·6	15·9	St. Malo	1866	Hervot		J. Bosquet	St. Malo
	726	V	Francis (ex [illegible]) *Sohier* 87-88 W Bg	French	158	94·0	21·0	10·1	Barrington NS	1879	W. Doane	V [illegible]	L. Mazier	St. Pierre, Miq.
	727	N	Francis *Ingebrethsen* -87 W Sr	Norweg'n	145	77·4	23·8	12·2	St. Servan	1834			J. S. Lorentzen	Drammen
[illegible]	728		Francis Alice *Smith* -86 W Sr	American	125	95·5	23·9	10·3	St Helen's Oreg	1881	Stephens		N. Bichard	San Francisco
[illegible]	729	R	Francis Allyn *Fuller* W Sr	American	101	85·4	22·3	8·6	Duxbury, Mass	1869		V [illegible]	C. A. Williams & Co.	N. London Conn.
[illegible]	730		Francis Arthemus *Patterson* -87 W Sr	British	103	85·5	25·6	8·0	Harringtn, US	1875			W. Mitchell, jun.	Amherst, N.S.
[illegible]	731		Francis C. Smith *Miller* W Sr	American	119	87·6	27·8	8·6	New Creek, NJ	1854			E. Miller	New York
[illegible]	732	R	Francis C. Yarnell *Scott* W 3m Sr	American	471	152·3	35·0	12·3	Wilmington, Del.	1881	Jackson & Sharp Co.		F. M. Scott	Philadelphia
[illegible]	733	R	Francis E. Hallock *Bulger* W Sr	American	205	118·8	30·8	7·8	Northport, NY	1869			J. J. Moule	Cold Spring N.Y.
[illegible]	734		Francis E. Waters *Tull* W Sr	American	141	99·4	28·1	7·4	Taylor's Island Ind.	1882			T. J. Seward	Baltimore
[illegible]	735		Francis Edwards *Cookins* W Sr	American	203	104·0	29·0	9·0	Absecom, N.J.	1858			S. R. Hutchings	Castine, Me.
[illegible]	736	✻	Francis Henry *Jones* W Sr (MISSING)	British	100	76·0	20·5	10·5	Pwllheli	1865	Williams	V [illegible]	O. Jones	Carnarvon
[illegible]	737		Francis John *Hawkins* -86 W Bg	British	218	116·7	21·3	13·1	Jersey	1869	Grandin	V [illegible]	F. J. Yarrow	Jersey
[illegible]	738	R	Francis L. Godfrey *Willbank* W 3m Sr	American	338	131·5	38·8	11·1	Dorchester, NJ	1872			C. P. Perry	Philadelphia
	739	V	Francis Marie *Lesquel* -87 W Bk	French	253	106·6	22·6	13·9	St. Malo	1867	L. Mallard		V[ve] Lesquel	Vannes
[illegible]	740		Francis Molison W Bg	British	141	90·6	21·6	11·7	Plymouth	1881	C. Gent		Trustees of Mars Training Ship Institution	Dundee
[illegible]	741		Francis Palmer W Bk	American	200	112·7	25·0	11·0	Robbinston Me	1852	Balksen		T. P. H. Whitelaw	San Francisco
[illegible]	742		Francis R. Baird *Webb* W Sr	American	326	120·8	29·4	10·1	Baltimore	1863			B. Shoe	Philadelphia
[illegible]	743		Francis Schubert *Gifford* W 3m Sr	American	206	105·0	31·0	8·0	May's Landing, N.J.	1872			E. Gifford	Somers' Pt., N.J.
[illegible]	744	✻	Francis Thorpe *Cormack* I S	British	1257	218·0	37·1	22·8	Stockton	1868	Richardson, Duck & Co.		A. Weir & Co.	Glasgow
	745	V	Francisca *Jor[illegible]* W 3m Sr	German	368	138·1	28·6	16·2	Apenrade	1863	N. Jacobsen	V [illegible]	L. P. Michelsen	Flensburg
[illegible]	746		Francisca *Simonetti* W Bg (WRECKED)	Italian	168	85·0	25·2	14·2	Savona	1854	G. Tixi		Gius. Soldi	Genoa
[illegible]	747	I	Francisca R. W Bk	Italian	522	147·6	31·9	20·7	Sestri Ponente	1870	A. Briasco fu G.B.	V [illegible]	G. Rivarola	Genoa
	748	V	Francisca de Vila *Marquez* -87 W Bg	Spanish	426	127·3	29·7	18·1	Ferrol	1871	M. Lorenzo	[illegible]	José Marquez	Bilbao
	749	V	Francisco *Ferrer* W Bg	Spanish	224	98·1	28·3	13·9	Barcelona	1850		[illegible]	J. Ferrer y Casanova	Barcelona
	750	V	Francisco *Sala* W Bg	Spanish	191	90·2	24·6	11·9	Arenys	1867		[illegible]	Francisco Sala	Barcelona

✻ = Lloyd's Register; D = Nederlandsche Vereeniging; G = Germanischer Lloyd; I = Registro Italiano; R = American Record;
A = Veritas Austro-Ungarico; H = Veritas Ellenico; N = Norske Veritas; V = Bureau Veritas.

Dans la 4me Colonne, C = Composite (membrure en fer, bordé en bois); I = Fer; Stl = Acier; W = Bois.

Les Dates du doublage en métal (Col. 11) ressortent des livres dans lesquels les navires sont respectivement classés ou des rapports des Inspecteurs du Lloyd's Register.
Pour plus amples renseignements sur les navires classés par le Lloyd's Register, voir le Lloyd's Register Book et le Lloyd's Yacht Register.

Official No.	No. in Book	Registry	Ship's Name. Material and Rig. Master, and Dates of Appointment to present Owner's Service and to Ship. Ship's late Name.	Flag	Net Tonnage	Dimensions: Length	Breadth	Depth	Built: Where.	When.	By whom.	Date of Metalling	Owners or Managers
[illegible]	751	RV	Francisco Garguilo *Jomez* W Bk	British	489	131·6	29·9	19·8	Vico,Ital.	1875		[illegible]	J. Baretta
	752	V	Francisco Torres *Torres* (ex Maria) W Bn	Spanish	155	87·6	22·8	10·9	Arens	1865	S. Busquets	[illegible]	F. Torres
	753		Francisqueta W Pol	Spanish	156	90·5	25·8	12·3	Barcelona	1858			B. Ragur
[illegible]	754	I	Franco *Questa* W Bg	Italian	284	111·5	26·6	16·3	SestriPonente	1875	A. Brasco fu L.	[illegible]	F. Sanguineti fu [illegible]
[illegible]	755	V	Franceforte *Scotto* [illegible] (ex New York) W Bg	Italian	301	115·8	28·8	14·5	Castellam. di [Stabia	1860	A. Bonifacio	[illegible]	L. Musso
	756	DV	François *Visser* [illegible] (ex Moline) C S	Dutch	1050	165·6	37·4	22·9	Amsterdam	1869	Meursing& [Huijgens	[illegible]	G. Van de Geyel
	757		François Joseph (L'[illegible]) *Hor[illegible]* W Bk		506	141·0	30·3	17·9	Bordeaux	1867	Rochet&Lange- [lenu	[illegible]	
	758		François Joseph W	French	116	..	..	..					Joseph Musse
	759	V	François Joseph *Kergaroun* W Sr	French	110	91·8	19·3	10·8	Granville	1885	Julienne	[illegible]	Benst & fils
	760		Francolin *Ramos* (ex Campbell) W Sr	Brazilian	170	..	..	..	Riolanyero	1862			J.L.P. Ervira & Co
[illegible]	761		Franconia *Priest* -88 W 3m Sr	American	216	116·0	29·7	8·7	Saco,Me.	1868			N. Falker
[illegible]	762		Franconia *Young* W Sr	American	129	92·2	25·5	8·3	Surrey,Me	1862			J.H. Higgins
[illegible]	763		Frank W Bk	British	564	142·3	30·5	18·5	WeymouthNS	1857			J.H. Jones
	764	N	Frank [illegible] *Braun* (ex [illegible]) W 3m Sr	Norweg'n	500	155·9	30·9	15·1	[illegible] [Mass	1866	GeorgeSance!		T. Brunn
	765	N	Frank *Larsen* -88 W Bk	Norweg'n	337	121·2	28·6	15·4	Risör	1872	E. Knudsen	[illegible]	B.A. Olsen
	766	N	Frank *Wenstoff* -87 W Bk	Norweg'n	332	130·8	29·3	13·3	Porsgrund	1874		[illegible]	J. Jeremiassen
	767	N	Frank *Larsen* W Bk	Norweg'n	302	124·7	27·6	13·3	Sarpsborg	1874	T. Birch	[illegible]	A. & O. Larsen
[illegible]	768		Frank A. Magee *Stanwood* W Sr	American	110	95·0	26·9	7·7	Ellsworth,Me.	1873			Hall Bros.
[illegible]	769		Frank Carvill *Chapman* [illegible] W S	British	1489	209·2	39·2	24·0	St. John	1874	McFee	[illegible]	R. Martin & Co.
[illegible]	770		Frank Emett *Pater* W Sr	British	150	101·2	23·6	12·4	Brixham	1867	Ferneaux		C.W. Pater
[illegible]	771		Frank G. Dow *Kelley* 87-91 W 3m Sr	American	391	126·7	31·0	14·2	Newburyport, [Mass	1872	G.E. Currier		P.H. Crowell
[illegible]	772		Frank G. Rich W Sr	American	100	82·0	23·3	8·2	Essex, Mass.	1883	W.A. Burnham		C.A. Gorham
[illegible]	773		Frank Guy W Sr	British	191	126·0	25·0	9·6	Bateman's Bay, [N.S.W.	1875	F. Guy		H. McQuarrie
[illegible]	774		Frank Harrington *Kent* W 3m Sr	American	603	160·0	33·5	15·6	Waldoboro'Me	1880	A. Starer&Son		J. Baker & Co.
[illegible]	775		Frank Herbert *Herrick* W Sr	American	174	110·6	27·8	9·7	E. Haven,Conn	1855			A.P. Veazie & Co
[illegible]	776		Frank L. P. *Lawson* W Sr	British	124	84·2	27·5	7·9	Greenwich, NB	1880	C.H. Peters		C.H. Peters
[illegible]	777		Frank Leaming *M. Seager* -84 W 3m Sr	American	245	120·1	32·0	9·5	DennisvilleNJ	1873	Leaming		G.W. Shepherd jr.
[illegible]	778	R	Frank M. Howes *McKinnon* W 3m Sr	American	170	119·8	25·0	8·4	Baltimore	1882	Beacham & Bro	[illegible]	J.A. Reid
[illegible]	779	R	Frank M. McGear *Sharp* W 3m Sr	American	356	136·4	34·6	10·8	[illegible]	1881	J W Abbott & Co		A. Sharp
[illegible]	780	R	Frank M. Noyes *Mayent* W 3m Sr	American	146	101·6	25·0	9·2	Essex, Mass.	1879	J. James&Co.	[illegible]	Dix & Wilkins
[illegible]	781	R	Frank McDonnell *Cannon* W 3m Sr	American	264	123·0	29·6	9·8	Camden, N.J.	1879	McDonnell Bros		J. & F. McDonnell
[illegible]	782	R	Frank O. Dame *Rogers* [18]82 W 3m Sr	American	654	163·0	39·0	17·2	HadlymeConn	1882	H.T. Comstock		N.T. Rodgers
[illegible]	783	V	Frank Pendleton *Nichols* [illegible] W S	American	1358	200·0	39·0	21·0	Belfast, Me.	1874	H. McGilvery	[illegible]	W.G. Nicholls
[illegible]	784	R	Frank Pratt Lee *Lee* [illegible] W 3m Sr	American	571	144·3	35·0	15·7	Camden, N.J.	1882	Morris&Mathis		R.B. Lee
[illegible]	785		Frank S. Hall *Holland* 86- W Sr	American	152	101·0	29·0	7·0	Wilmington [Del.	1882	J.H. Cahall&Co.		J.W. Hall
	786	R	Frank S. Warren *Falkner* W 3m Sr	American	540	159·6	35·0	13·5	Bath, Me.	1888	Kelley & Spear		J.R. Kelley
[illegible]	787	V	Frank Stafford *Smith* -85 W Bk	British	1169	181·0	37·5	22·5	Cornwallis, NS	1881	Vaughan&Lon- [mon	[illegible]	R.M. Ratal
	788	R	Frank T. Stinson *Reed* W 4m Sr	American	923	186·8	38·2	18·2	Bath, Me.	1887	E.S. Crosby		E.S. Crosby
[illegible]	789	R	Frank Vanderherchen *Chamberlain* -8[illegible] W 3m Sr	American	511	144·5	34·8	15·1	Camden, N.J.	1884	S.W. Tilton		C.A. Pettit
	790		Frank Wilson *Brackmeyer* (M[illegible]) [illegible]-83 I Bk	German	337	134·0	26·0	14·6	Sunderland	1868	Oswald & Co.		C. Paulsen
[illegible]	791		Frankby *Braithwaite* W Bk	British	375	125·1	17·8	9·9	Liverpool	1857	T. Royden&Son		Lucas Bros. & Co.
[illegible]	792	✻	Frankistan *Brown* 79-81 I S	British	1944	270·2	39·5	24·1	Southampton	1882	Oswald Mor- [daunt&Co		[illegible] & Eastern [illegible] Ld
[illegible]	793	R	Franklin *Terry* W Sr	American	280	112·0	30·0	8·0	Baltimore, Md.	1867			W.V. Tupper
	794	V	Franklin *Mourrat* -86 W Bk	French	553	154·5	29·9	18·4	Nantes	1865	Jollet & Babin	[illegible]	[illegible]
	795	N	Franklin *Bonnevitzen* -[illegible] (ex Valparaiso) W Bk	Norweg'n	709	161·3	33·1	19·6	Honfleur	1867	Colombon	[illegible]	A. Abrahamsen
	796	N	Franklin *Nielsen* W Bk	Norweg'n	483	131·6	31·0	17·2	Arendal	1873	C.P. Olsen	[illegible]	[illegible] Olsen
	797	N	Franklin *Torkildsen* (ex [illegible]) W Bk	Norweg'n	408	144·4	31·2	12·9	Quebec	1867	Samson	[illegible]	O. Bjartness
[illegible]	798		Franklin Bell *Smith* W Sr	American	155	96·7	27·0	8·7	Brookhaven, [N.Y.	1853			S.E. Reed
[illegible]	799		Franklin Nickerson *Nash* W Sr	American	202	105·0	28·0	9·0	Greenpoint NY	1854			J.P. Ellicott
[illegible]	800		Franklin Woodruff *Corrigan* W Sr	American	130	93·7	24·5	9·2	Essex, Mass.	1882			S.S. Swift

✻ = **Lloyd's Register**; D = Nederlandsche Vereeniging; G = Germanischer Lloyd; I = Registro Italiano; H = [illegible]
A = Veritas Austro-Ungarico; Π = Veritas Ellenico; N = Norske Veritas; V = [illegible]

In the 4th Column, C = Composite (iron frame, wood planked); I = Iron; Stl = Steel; W = Wood.

The Dates of Metalling (Col. 11) are based on the records of the Books in which Vessels are respectively classed, or on the Reports of the Surveyors [illegible]

Further particulars of Vessels classed by Lloyd's Register will be found in Lloyd's Register Book and Lloyd's Yacht Register.

N° dans la liste	Cotes	Nom. Capitaine et dates d'entrée au service des Armateurs actuels et à bord du navire. Ancien Nom.	Matériaux et Gréement	Pavillon	Tonnage net	Longueur	Largeur	Creux	Construit Où	Année	Par qui	Date du doublage en métal	Armateurs ou Directeurs-Gérants	Port d'Armement
801	V	Franky *[illegible]*	W Bg	Austrian	323	105.3	25.5	16.7	Rostock	1866	O.Ludewig	[illegible]	S.A.Kraus fu G.	Trieste
802	G	Frans *[illegible]*	W Bk	Dutch	308	125.9	27.8	12.8	Port Medway, [N.S	1865		[illegible]	F.W. van der Elst & Co.	Schiedam
803		Frans *Pettersson*	W Bk	Swedish	428	136.4	28.5	16.1	Norrtelje	1877	G.Wenland		F.Lundman	Norrtelje
804	V	Frans *Ostman*	W Bk	Swedish	356	125.9	26.2	14.4	Sundsvall	1871	E.Ostman	[illegible]	E.Sandelin	Sundsvall
805		Fransiskus *Andersson*	W Bk	Swedish	249	118.7	24.8	11.8	Mönsterås	1869	Carstensen		G.Andersson	Landscrona
806	V	Franz *[illegible]*	W Bn	British	282	110.0	21.7	12.0	[illegible]	1863	H.Gross	[illegible]	H.C.Pigott	Melbourne
807		Franz *[illegible]*	W Bg	German	185	81.7	24.3	13.6	Zingst	1863	H.C.Drossel		J.N.Rodbertus	Barth
808		Franz Bottcher	W Sr	German	172	83.3	21.6	13.0	Greifswald	1859	J.W.Sadelkow			Kappeln
809	G V	Franz Fischer *[illegible]*	W Sr	German	162	95.5	23.0	11.9	Rostock	1876	E.Burchard & Co.		F.W.Fischer	Rostock
810	G	Franz von Mathies *Grahl*	W Bg	British	233	104.0	26.2	13.7	Rostock	1859	A.Schonemann	[illegible]	C.Grahl	West Hartlepool
811	V	Franziska *Johansen*	W Gt	Danish	[illegible]	75.3	19.8	10.3	Kiel	1871	Conradi & Sohn		H.C.Johansen	Marstal
812		Franziska *Faas 78-86*	W Bg	German	21[illegible]	94.5	26.1	13.9	Ribnitz	1849	J.Mielbrodt		E.Burchard	Rostock
813	N V	Franziska *Adsen*	W Bk	Norweg'n	781	165.9	[illegible]	[illegible]	[illegible]	1881	H.&J.Gregersen	[illegible]	G.Gundersen	Stavanger
814		Frasquita	W Pol	Spanish	232	114.8	23.8	12.1	Palma	1853			Antonio Morey	Majorca
815	V	Frasquita *Pascual*	W Bn	Spanish	21[illegible]	95.5	27.0	13.3	Blanes	1866		[illegible]	J.P.Arabi	Barcelona
816	V	Frasquito *Saez*	W Bn	Spanish	312	109.9	27.9	15.0	Blanes	1867	Vieta	[illegible]	F.Oliver y Alsina	Barcelona
817	✻D	Fratellanza *Schiaffino* (FORENINGEN)	W Bk	Italian	797	172.7	32.8	23.2	Varazze	1878	V.Baglietto	[illegible]	F.De Gregori fu A.	Genoa
818		Fratelli Benfante	W Bn	Italian	179	89.9	23.4	14.5	Castellam. di [Stabia	1851	Trambarulo		F[lli] Benfante	Palermo
819	I	Fratelli Bianchi	W 3m Sr	Italian	582	143.3	31.3	[illegible]	Sampierdarena	1872	B.Fraissinetti	[illegible]	Fratelli Bianchi	Spezia
820		Fratelli D'Amico *Cannizzo*	W Bk	Italian	436	127.9	29.5	19.7	Castellamare	1866	Bonifazio	[illegible]	V.D'Amico	Milazzo
821	I	Fratelli Doge *Molledo*	W Bk	Italian	618	155.5	33.1	21.4	Voltri	1872	D.Salomone	[illegible]	G.Moltedo di G.B	Genoa
822	A	Fratelli F. *Tomasovich*	W Bk	Austrian	55[illegible]	14[illegible].3	29.8	19.4	Curzola	1875	Foretich	[illegible]	F.Fabris	Curzola
823	I	Fratelli Gazzolo *Gazzolo*	W Bk	Italian	611	14[illegible].6	32.4	20.4	Varazze	1875	D.Craviotto	[illegible]	G.Gazzolo	Genoa
824		Fratelli Isola *Ottone*	I Sr		177	101.7	25.4	10.2	Sampier-[darena	1886	Frat.Roncallo			
825	✻	Fratelli Laurin *Laurin*	W Bk	Italian	616	142.3	30.9	18.7	Trieste	1871	Tonello	[illegible]	L.Laurin	Genoa
826	A	Fratelli M. *Ciano*	W Bn	Italian	124	80.5	21.9	11.1	Chantenay	1856		[illegible]	Ugo Ciano	Leghorn
827		Fratelli Scarsella *Olivari*	W Bk	Italian	8[illegible]6	162.4	31.3	22.5	Chiavari	1874	M.Tappani	[illegible]	A.Scarsella	Spezia
828	I	Fratelli Tixi *Tixi*	W Bk	Italian	624	144.9	32.9	21.5	Prà	1874	B.Tixi	[illegible]	B.Tixi	Genoa
829	I	Fratelli Tubino *Felugo*	W Bk	Italian	596	[illegible]	31.4	20.0	Savona	1869	F.Sirello	[illegible]	F[lli] Tubino fu G.	Genoa
830		Fratelli Vagliano	W Bg	Italian	351	112.2	28.4	17.[illegible]	[illegible]	1863	Barbieri	[illegible]	G.Denatale	Messina
831		Fraternidade	W Bk	Brazilian	322	..	..	..					J.L.dos Reis	Rio de Janeiro
832		Fraternidade *de Mattos* (MISSING)	W Bn	Portug'se	142	86.6	21.3	9.8	Caminha	1883	A.A.Netto	[illegible]	D.J. & A.J.Lourenço	Caminha
833	N	Fraternitas *Gundersen*	W Bk	Norweg'n	427	121.4	30.9	17.2	Grimstad	1871	N.O.Gjömle	[illegible]	J.Jörgensen	Grimstad
834	N	Fraternité *Torgersen*	W Bg	Norweg'n	201	82.3	21.5	14.8	Grimstad	1827			M.Madsen	Tonsberg
835	✻	Frau Minna Petersen	W 3m Sr	British	16[illegible]	102.0	24.3	12.6	Portmadoc	1878	Jones	[illegible]	H.Parry	Carnarvon
836		Frauke *Baumann*	W Gt	German	137	90.4	21.2	10.4	Pekela	1862	B.B.Drenth		C.F.Baumann	Emden
837		Fray Bentos *Olsen*	W Bk	British	511	152.6	28.8	17.0	Lübeck	1863			William Chinham	Pt Adelaide, S.A.
838	N	Fred *Jensen*	W Bk	Norweg'n	451	135.4	29.2	17.5	New Brunswick	1857			L.Iversen (Chr.Nielsen)	Laurvig
839	N	Fred *Rasmussen*	W Bg	Norweg'n	277	109.4	26.7	14.9	Grimstad	1864	Gjerald Larsen		A.Annonsen (H.Tvede ?)	Grimstad
840	N	Fred *Thorsen*	W S	Norweg'n	27[illegible]	120.3	27.6	12.1	Risör	1886	T.Olsen	[illegible]	L.Knudsen	Risör
841	R	Fred A. Small *Thompson*	W 3m Sr	American	588	142.4	32.0	18.2	Machias, Me.	1886	J.Shaw	[illegible]	Miller & Houghton	Newburyport
842	✻	Fred Archer *Beedham*	W Sk	British	51	65.5	18.5	9.1	Burton Stather	1887	J.Garside		J.Beedham	Grimsby
843	R V	Fred B. Taylor *Tilley*	W S	British	1797	2[illegible].8	42.0	24.0	Tusket, N.S.	1883	J.A.Hatfield	[illegible]	William Law	Yarmouth, N.S.
844		Fred C. Holden *[illegible]*	W Sr	American	131	96.8	27.2	7.9	Tremont, Me.	1873			J.G.Wilson	Ellsworth, Me.
845	R	Fred E. Cox *McDonald*	W Sr	British	145	8[illegible].8	26.2	10.4	Isaac's H'rbour [N.S.	1881	L.Hane	[illegible]	S.Street	Guysboro' N.S.
846	R	Fred E. Richards *Prindle*	W Bkn	American	644	[illegible]	35.0	17.6	Camden, Me.	1882	Carleton, Norwood & Co.	[illegible]	Carleton, Norwood & Co	Camden, Me.
847		Fred E. Sander *Ross*	W 3m Sr	American	449	157.7	37.1	11.5	Port Blakely, [Washington	1887	H.K.Hall		W.G.Hall	San Francisco
848	V	Fred E. Scammell *Shaw*	W S	British	13[illegible]	[illegible]	39.2	23.1	Eatonville, N.S.	1880	D.&C.Eaton	[illegible]	Henry A.Eaton	Parrsboro' N.S.
849	R	Fred Jackson	W 3m Sr	American	2[illegible]	11[illegible].8	28.7	9.8	East Deering, Me.	1873	G.Russell	[illegible]	J.S.Winslow & Co.	Portland, Me.
850	V	Fred P. Litchfield *Young*	W Bk	American	1042	177.5	[illegible]	22.3	Bath, Me.	1876	Goss & Sawyer	[illegible]	B.Litchfield & Sons	New York

[A] = Lloyd's Register; D = Nederlandsche Vereeniging; G = Germanischer Lloyd; I = Registro Italiano; R = American Record; [A] = Veritas Austro-Ungarico; H = Veritas Ellenico; N = Norske Veritas; V = Bureau Veritas.

Dans la 4me Colonne, C = Composite (membrure en fer, bordé en bois); I = Fer; Stl = Acier; W = Bois.

Les Dates du doublage en métal (Col. 11) ressortent des livres dans lesquels les navires sont respectivement classés ou des rapports des Inspecteurs du Lloyd's Register.

Pour plus amples renseignements sur les navires classés par le Lloyd's Register, voir le Lloyd's Register Book et le Lloyd's Yacht Register.

Official No.	No. in Book	Register in which Classed	Ship's Name, Material and Rig. Master, and Dates of Appointment to present Owner's Service and to Ship. Ship's Late Name.		Flag	Net Tonnage	Length	Breadth	Depth	Built: Where.	When	By whom.	Metalling	Owners or Managers
[illegible]	851		Fred Smith Whitney -88	W Sr	American	188	109·3	27·0	9·4	Brewer, Me.	1867			H. Lord
1842	852		Fred Snow (ex Stevens)	W Sr	American	[illegible]	78·6	26·0	7·9	Athens, N.Y.	1865			J. M. Tate
60781	853	✻	Fred Thomson Cogan (Lost)	W Bg	British	247	108·8	25·5	14·4	Dundee	1868	Dundee Sh. Co.		W. Foley
83332	854		Fred Toms	W	British	161	110·3	22·2	7·6	Ottawa, Ont.	1881			Thomas H. Ki[illegible]
9477	855		Fred Tyler	W Sr	American	119	98·0	21·2	7·9	Philadelphia, Pa.	1848			E. B. Darling
120190	856	V	Fred W. Carlon Reed	W Bk	American	505	132·2	31·8	17·0	Searsport, Me.	1875	W. McGilvery	[illegible]	W. M. Reed
1893	857		Fred Walton Smith	W 3m Sr	American	441	135·8	31·8	14·7	Bowdoinham, Me.	1869	Purington & Co.		P. Atkins
	858		Freda Langstrom 87-87	W Sr	Russian	221	100·1	31·2	12·3	Nagu	1872	M. Lindquist		G. R. Lindblom
120518	859	R	Freddie Hencken Hendrickson	W 3m Sr	American	326	135·0	25·0	10·6	Greenport, NY	1882	Smith & Terry		H. Hencken
	860	R	Freden Johnsen -88 (ex Duke of Northumberland)	W Bk	Norweg'n	481	127·4	28·7	18·9	Sunderland	18$^{5}_{68}$		y ss	C. Tollesen
	861	NV	Freden Salvesen 79-83	W Bk	Norweg'n	415	128·4	30·1	16·3	Tvedestrand	1869	T. Gundersen	y s s	Carl Bech
	862		Freden Andersen	W Sr	Norweg'n	225	114·4	26·7	11·0	Maine	1855			W. M. Michaelsen
	863		Freden	W Sr	Russian	216	..	..	..		1836			J. Malmgren
2519	864		Freden Fagerlund -83	W Sr	Russian	265	97·4	29·8	13·0	Nagu	1879	Lindqvist		K. L. Lindberg
	865		Fredenborg Karlsson -87	W Bk	Russian	431	..	..	..	Geta				E. Malmström
	866	DV	Fredensborg Jonsson	W Bk	Swedish	322	124·3	25·8	13·6	Elsinore	1874	Lehmann & Barfod	y s s	J. Jönsson
	867	V	Frédéric Suzanne Guyonlon -81	W Bk	French	332	116·9	26·2	14·4	Nantes	1875	J. Sevestre	c s s	J. B. Demange
88411	868	V	Frederica Roberts 85-85	W Bkn	British	429	146·0	31·2	13·0	Portland, St. John, N.B.	1884	F. E. Sayre	y s s	F. E. Sayre
21681	869		Frederica	W Bk	British	376	128·0	25·7	15·6	Yarmouth	1858	T. W. Branford		J. Huddart
48293	870		Frederick	W S	British	380	122·6	28·7	13·2					Charles Key
53786	871		Frederick	W Bk	British	162	93·0	22·8	10·2			F. Fader		F. Fader
	872		Frederick Wirolaynen st.	W Bg	Russian	273	109·5	25·3	12·7	St. Johannis, Lane	1876	M. Reppanen		F. Wirolaynen
45944	873		Frederick Bassil Reed	W Bk	British	342	119·1	27·1	17·1	Sunderland	1862	Taylor		J. Mill
	874	RV	Frederick Billings Sherman -85	W 4m S	American	2497	278·0	44·8	29·0	Rockport, Me.	1885	Carleton, Norwood & Co.	[illegible]	Carleton, Norw[illegible]
9101	875		Frederick Reed Pendleton	W Sr	American	105	85·0	23·5	7·7	Wiscasset, Me.	1851			M. Pendleton
28608	876	✻	Frederick Stonard Robinson 77-86	W Sr	British	123	85·5	21·4	11·2	Ulverston	1861	Schollock		T. Morgan & Co.
135310	877		Fredericka Schepp (ex Emma Ritch) Brown -81	W Sr	American	255	111·7	28·0	13·0	Northport, NY	1877	J. Carll	y s s	L. Schepp
	878	DV	Frederik Carl (WRECKED) (ex Teresa Mulligan) Bolt	W Bg	Dutch	350	119·7	29·2	17·7	Savona	1866		c s s	F. L. Drenth
	879		Frederik den 6te Minde (ex Kirk)	W Sr	Danish	119	87·5	20·3	9·1	England	18$^{24}_{54}$			L. C. Nielsen
	880	V	Frederik & Ane	W S	Danish	121	88·3	20·3	10·2	Faaborg	1876	M. Dyreborg		O. Smith
	881	V	Frederikke Andersen -87 (ex Sophie)	W Sr	Danish	129	97·9	20·9	9·8	Odense	18$^{75}_{??}$	E. Knudsen		N. W. Kaas
	882		Frederikke Wilhelmsen 75-75	W S	Danish	117	85·0	20·4	10·5	Odense	1861	H. L. Hansen		H. T. Helweg
	883	V	Frederikke Madsen 18-78	W Sr	Danish	104	80·0	21·4	9·7	Thurø	1865	L. Kaas		P. O. Petersen
	884	G	Frederikke Louise Kofoed 58-85	W Bk	Danish	290	120·3	23·9	14·1	Rønne	18$^{52}_{74}$	T. P. Bech		C. Thorsen
	885	N	Frederiksteen Jahnsen (1885)	W Bg	Norweg'n	298	125·7	28·1	12·8	Nova Scotia	1868			G. G. Maurud's B[illegible]
	886	V	Frederiksvaag Retdorff (1880) (ex Vigilante)	W Sr	Danish	118	89·8	19·6	11·6	Dunkirk	1866	E. Sagary		M. C. Resdorff & [illegible]
	887	N	Frednæs Petersen 74-74	W Bk	Norweg'n	398	133·4	31·1	17·0	Porsgrund	1863	Chr. Knudsen		J. C. & G. Knudsen
	888		Fredrik Tainer -87	W Bk	Russian	361	..	..	..		1876			F. Richardt
	889		Fredrik Eriksson -87	W 3m Sr	Swedish	104	84·0	21·7	8·1	Bollsby	1885			J. O. Eriksson
	890	N	Fredrik Holst Halvorsen	W Sr	Norweg'n	182	93·8	21·7	12·3	Moss	1867	E. Pedersen		N. W. Sørensen
	891	NV	Fredrik Stang Uckermann	W S	Norweg'n	991	194·8	37·5	22·1	Bergen	1877	Jens Gran	y s s	Rasmus F. Olsen
	892	N	Fredrik William Hansen	W Bg	Norweg'n	264	109·3	29·2	12·7	Arendal	1864	C. Johnsen		L. Andersen
	893	N	Fredriksværn Berg -81	W Bk	Norweg'n	293	131·5	25·4	13·8	St. John, N.B.	1852			E. Rasmussen
	894	N	Fredrique Olsen -87	W Sr	Norweg'n	134	93·7	23·7	9·6	Wenersborg	18[illegible]			H. N. Jensen
	895	N	Fredsæl Zachariassen -88	W B	Norweg'n	494	144·1	31·2	16·3	Porsgrund	1872	J. Kolstad	y s s	B. Condrup
54434	896		Free Trade Hough (ex [illegible] G. Jewett)	W Sr	British	133	82·0	26·0	9·0	Grand Bay, NB	1866			E. Lantalum
9413	897		Free Trade	W Bkn	American	340	125·0	30·8	11·7	Greenport, NY	1866			E. P. Flint
31958	898		Free Trader Bowden	W Bk	British	206	109·8	23·0	11·2	Hobart Town	1856			H. Bowden
120422	899	RV	Freeda A. Willey Hopkins -81	W Bkn	American	507	158·1	33·4	12·4	Thomaston, Me.	1880	Dunn & Elliot	y s s	Dunn & Elliot
81388	900	✻	Freedom Powlett	W K	British	86	79·3	20·8	10·4	Brisbane	1884	W. & A. Upham		J. Dugdale

✻ = Lloyd's Register; D = Nederlandsche Vereeniging; G = Germanischer Lloyd; I = Registro Italiano; A = Veritas Austro-Ungarico; H = Veritas Ellenico; N = Norske Veritas; V = [illegible]

In the 4th Column, C = Composite (iron frame, wood planked); I = Iron; Stl = Steel; W = Wood.

The Dates of Metalling (Col. 11) are based on the records of the Books in which Vessels are respectively classed, or on the Reports of the Surveyors [illegible]

Further particulars of Vessels classed by Lloyd's Register will be found in Lloyd's Register Book and Lloyd's Yacht [illegible]

No.		Nom, Matériaux et Gréement. Capitaine et année d'entrée … des Armateurs actuels … Ancien Nom.												Armateurs ou Directeurs-Gérants.	Port d'Armement.
901	✻	Freedom	Steel 71-71	WK	British	93	75·5	18·0	8·3	Jersey	1847	Lelliquet		W.Steel	Guernsey
902	V	Freeman	Simmer 88-88	WBk	American	1138	184·5	39·0	22·5	E.Boston	1878	J.M.Brooks		R.R.Freeman	Boston,Mass.
903	V	Frégate	Lavallée	WSr	French	[illegible]	92·5	22·0	10·8	Gravelines	1885	Calbo		Gombert Lavallée	Gravelines
904	H	Freia	Frederiksen 86-86	WBn	[illegible]	165	100·5	22·5	11·4	Aalborg	1869	C.T.Vang		H.Iversen	Svendborg
905		Freia (ex-De Brodre …)	Olsen 85-85	WS	Norweg'n	952	161·6	40·2	24·7	Copenhagen	1824	DanskeMarine	Y 11.86	S.C.Larsen	Brevig
906	N	Freia	Salvesen, Mey	WBk	Norweg'n	650	158·1	30·2	20·2	Aberdeen	1852	A.Duthie & Co.	Y 11.86	O.K.Abrahamsen	Porsgrund
907	N	Freia	Johnsen 83-85	WBk	Norweg'n	578	157·5	32·2	18·4	Bergen	1869	Brandhorst & [Dekke	Y 5.88	Bergh & Helland	Bergen
908		Freia	Helliesen 82	WBk	Norweg'n	406	120·3	30·6	17·0	Carlskrona	1845	Petersson		Helliesen & Co.	Stavanger
909	N	Freia	Martinsen 88	WBk	Norweg'n	387	106·5	29·5	18·3	AlterStettin	1814			C.J.Martinsen	Moss
910	N	Freia	Olsen 89	WSr	Norweg'n	152	95·9	23·5	10·3	Tönsberg	1851			Svend Föyen	Tönsberg
911		Freia	Brunning 83-84	WSr	Norweg'n	109	76·1	20·1	10·7	Christiansand	18 69			W.S.Brunning	Christiansand
912	N	Freidig	Berg	W3mSr	Norweg'n	421	132·7	29·8	15·5	Stavanger	1874	C.Knudsen	Y 10.87	L.Larsen	Stavanger
913	N	Freidig	Bie 87-87	WBk	Norweg'n	270	127·6	26·8	12·4	Grimstad	1868	A.Omandsen	Y 7.88	N.Thne Johnsen	Grimstad
914		Freidig (ex-Brødre …)	Halvorsen 89	W3mSr	Norweg'n	243	117·3	26·5	11·8	Sandefjord	1874	Grön		P.A.Grön	Sandefjord
915	DGV	Freihandel	Arkema	WBk	Dutch	771	153·0	32·6	19·2	Bremerhaven	1865	H.F.Ulrichs	Y 3.88	S. van der Hei	Heveskes
916	DG	Freiheit	Wiets 76-76	WSr	German	102	93·4	23·1	10·8	Grossefehn	1876	T.J.Cassens		TönjesJ.Cassens & Sohn	Grossefehn
917	V	Freir	Christensen 77-77	WBn	Danish	158	97·6	21·9	11·0	Svendborg	1874	J.R.Andersen		A.N.Svennevig	Svendborg
918	V	Freitag	Tesch 87	WBk	German	311	114·8	28·4	15·0	Ueckermünde	1864	C.Radmann		C.Tesch	Neuwarp
919	V	Frej	Sandberg 89-89	W3mSr	Swedish	421	133·8	24·4	17·0	Gefle	1877	A.Brodin	Y 11.86	O.A.Brodin	Gefle
920	V	Frej	Wistman 86	WBk	Swedish	313	129·9	25·6	12·3	Sundsvall	1873	E.Östman	Y 7.86	E.Sandelin	Sundsvall
921		Frej	Wimmerström	WSr	Swedish	111	79·7	21·8	8·8	Westervik	1877	H.Fridell		G.Idström	Westervik
922	R	Freja	Nylund 87	W3mSr	Russian	263	115·0	29·5	14·4	Mariehamn	1883			E.P.Karlson	Lemland
923	V	Fremad	Christensen 84	WSr	Danish	101	80·0	20·6	9·8	Marstal	1873	F.Hansen		L.Hansen	Marstal
924	DN	Fremad	Simonsen 82-82	WBk	Norweg'n	499	137·5	29·9	17·1	Grimstad	1880	K.Andersen	Y 1.89	K.Andersen	Grimstad
925		Fremad	Sundene	WBk	Norweg'n	314	121·6	26·2	14·5	Brevig	1857			T.& L.Bache	Tönsberg
926	✻	Fremad (ex-Port …)	Larsen 88	WBg	Norweg'n	246	111·0	26·0	12·9	P.E.Island	1877	Coffin	Y 7.88	Aktieselskabet Fremad	Laurvig
927	N	Fremad	Jensen 77-86	WSr	Norweg'n	134	91·1	22·4	10·4	Christiansand	1868	W.Antzee	Y 9.81	N.T.Langfeldt	Christiansand
928	N	Fremad	Andersen 88	WSr	Norweg'n	118	89·6	21·7	10·0	Sandefjord	1871			Chr.Christensen	Sandefjord
929		Fremad	Johnsen	WSr	Norweg'n	112	83·2	21·1	9·4	Grossefehn	1869			Krogh & Henschien	Lillesand
930	N	Fremad	Jacobsen	WSr	Norweg'n	103	84·2	21·4	9·3	Brevig	1870			A.Jacobsen	Holmestrand
931		Fremgang	Johnsen 89	W3mSr	Norweg'n	418	141·8	29·8	12·6	Grimstad	1887			N.Thne Johnsen	Grimstad
932	N	Fremgang (ex-Johann …)	Andersen 89	WBk	Norweg'n	[illegible]	143·5	28·6	15·0	Bremen	1866	F.W.Wencke	Y 11.84	A.Birkeland	Egersund
933	R	Fremont	Swenson 77-	WBkn	American	328	107·9	25·2	11·8	Philadelphia	18 77		Y 10.86	Lynde & Hough	San Francisco
934	N	Fremtiden	Reinsen	WBk	Norweg'n	388	126·0	29·5	16·4	Tvedestrand	1875	M.Ektrup	Y 7.88	A.Gundersen	Tvedestrand
935		French van Gilder		W3mSr	American	239	120·6	29·6	10·6	DennisvilleNJ	1879			P.M.Wheaton	Bridgeton,N.J.
936	V	Frères & Sœurs	Gaare	WSr	French	113	89·6	19·3	10·5	St Malo	1878	L.Trenchamer		F.Savin	Paimpol
937		Fresno	Hansen 89-89	WBk	American	1187	197·2	38·6	23·1	Bath,Me.	1874	W.Rogers		Puget Sound Commercial Co.	PortTownsend
938	✻	Freuchny	Smith 82-82	WBk	British	324	128·3	26·3	15·0	Garmouth	1875	Geddie	Y 3.88	A.Hendry	Banff
939	H	Freude	De Jonge 88	WBg	German	271	106·8	25·8	14·1	Greifswald	1864	J.W.Sadelkow		H.de Jonge	Emden
940		Freundschaft	Förster	WBg	[illegible]	349	108·6	29·2	17·3	Memel	1846	B.Pieper		J.G.Gerlach	Memel
941		Freundschaft	Abshagen	WBk	German	285	109·9	27·0	13·9	Ribnitz	1862	J.Miebrodt		F.W.Fischer	Rostock
942	N	Frey	Haugen	WBk	Norweg'n	471	124·5	31·8	18·2	Brahestad	1858			O.A.Föyen	Tönsberg
943		Frey	Pettersen	WBk	Norweg'n	312	107·3	28·7	15·6	Arendal	1852		Y 4.84	E.Eyde	Arendal
944	N	Frey	Abrahamsen 87	WBk	Norweg'n	252	109·9	28·5	12·2	Porsgrund	1868	NilsLarsen	Y 10.86	Chr.Schöning	Porsgrund
945	NV	Frey	Petersen 86-86	WBg	Norweg'n	246	113·1	27·5	10·7	Tvedestrand	1886	E.Gundersen	Y 5.86	Brödrene Thorvildsen	Tvedestrand
946	V	Freya	Jeppesen 78-82	WBn	Danish	226	108·1	24·7	12·9	Nyborg	1878	Petersen	Y 6.81	Moses & Søn,G.Melchior	Copenhagen
947	DGV	Freya	Kollis 83-83	WBk	German	659	136·1	33·1	19·1	Hammelwarden	1882	J.T.Strenge & [Sohn	Y 6.85	Gerh.Eilers	Brake
948	V	Freya (ex-St … Bloom)	Eliassen 84-87	WBk	Norweg'n	817	168·0	33·7	20·0	St Martin's NB	1867	J.H.Moran		A.Monsen	Tonsberg
949	N	Freya	Jensen 84-84	WBg	Norweg'n	158	81·3	24·4	11·5	Sweden				Th.& H.Lorentzen	Drammen
950	G	Freya (ex-…)	Eriksson	WSr	[illegible]	136	95·6	19·7	10·6	Flensburg	1863	Westermann		M.Eriksson	Väddö,Rslgn.

✻ = Lloyd's Register; D = Nederlandsche Vereeniging; G = Germanischer Lloyd; I = Registro Italiano; R = American Record; A = Veritas Austro-Ungarico; H = Veritas Ellenico; N = Norske Veritas; V = Bureau Veritas.

Dans la 4me Colonne, C = Composite (membrure en fer, bordé en bois); I = Fer; Stl = Acier; W = Bois.

Les Dates du doublage en métal (**Col. 11**) ressortent des livres dans lesquels les navires sont respectivement classés ou des rapports des Inspecteurs du Lloyd's Register.

Pour plus amples renseignements sur les navires classés par le Lloyd's Register, voir le Lloyd's Register Book et le Lloyd's Yacht Register.

Official No.	No. in Book	Registry in which classed	Ship's Name. Master, and Dates of Appointment to present Owner's Service and to 1??. Ship's Late Name.	Material and Rig	Flag	Net Tonnage	Length	Breadth	Depth	Built Where	When	By whom	Date of Metalling	Owners
	951	V	Freyr *Hansen*	W Sr	Danish	108	86·4	20·2	10·3	Rudkjöbing	1878	S.Bons		N. Hansen
	952	N	Fri *Christiansen -87*	W Bk	Norweg'n	477	138·6	30·8	16·6	Porsgrund	1873	Hansen		A. Amundsen
	953	NV	Fri *Hansen -88*	W 3mSr	Norweg'n	264	121·4	26·5	12·7	Stavanger	1875	O.Solem		G. & A. Berg
	954	GV	Frida *Hansen* (ex Frida Spliedt, ex Joseph)	W Bn	Danish	179	98·4	23·1	11·5	Papenburg	1867	H.W.Meyer		P. Nielsen
	955		Frida (ex Heron) *Olsen*	W Bk	Norweg'n	774	165·0	32·1	20·3	Quebec	1864	Dinning		Aktieselskabet Frida (Chr. Rosted)
	956	NV	Frida *Rasmussen*	W Bk	Norweg'n	382	126·2	29·4	15·3	Grimstad	1875	A.Omundsen		J. Bang
	957		Frida *Jansson -87* (ex Dr. Moeve)	W Bg	Swedish	321	109·3	25·4	15·2	Wismar	1867	M.Maren		A. Jansson
	958	G	Frida *Larsson 84-*	W Sr	Swedish	182	89·2	23·0	12·3	Lubeck	1864			O. Holm
	959	GR	Fride *Tibell -86*	W Bg	Swedish	264	110·8	24·3	13·1	WilstaWarf	1858	F. Hägglund		Bjork & Engström
	960	N	Fridleif *Olsen*	W Bk	Norweg'n	505	131·2	32·5	17·5	Tvedestrand	1870	A.&F.Smith		A.&F. Smith
	961	V	Frieda Grampp (ex Granville, ex Nicolo B.) *Lindeberg -83*	W Bk	German	499	131·2	31·1	19·8	Sestri a Ponente	1868	S.Debarbieri		Paul Grampp
	962	DV	Friede *Krenius 85-88*	W Bg	German	192	96·1	24·1	11·9	Papenburg	1866	B.Mammes Wwe		B. Mammes Wwe
	963		Friede *Jongeblood 79-79*	W Bn	German	171	92·8	23·2	12·2	Weener	1867	R.Borg		J. Huising
84095	964	✻	Friedeburg (MISSING) *Lloyd 85-85*	I S	British	769	183·1	30·0	18·1	Glasgow	1869	A.Stephen&Sns		Tomlinson, Hodgetts & Co.
	965	GV	Friederike (ex Saratoga) *Grote -81*	W S	German	1398	206·0	40·0	24·1	Boston, Mass.	1874	Campbell & Brooks		J. Schween & Co
	966	V	Friederike *Strackmeyer 81-81*	W Bg	German	272	107·3	25·6	11·4	Brake	1872	J.D.Behrens		Gerd. Bolte
	967		Friederike *Stranwitz 64-64*	W Bg	German	181	92·2	24·2	13·4	Barth	1851	Siewert		C.A. Berg
	968	V	Friederike *Hellmiller -76*	W Bn	German	158	87·9	21·6	10·2	Papenburg	1888	B.Tholen		B. Tholen
	969	V	Friederike Weyer (CONDEMNED) *Kniest*	W Bk	German	365	119·4	25·9	16·2	Stralsund	1865	J.P.Juhl		Seitz & Kindt
	970	V	Friederike & Marie *Sprenger 78-8?*	W Bk	German	307	123·0	27·6	13·8	Grabow	1863	C.A.Domcke		Edwin Schütz
	971	GV	Friedlander *Bellmer 80-*	W S	German	1584	215·8	39·4	26·0	K'nnebunk Me	1872	N.L.Thompson		D. Schilling
	972	GV	Friedrich *Korff 87-87*	W Bk	German	672	162·0	34·1	19·1	Apenrade	1876	G.Raben		W. Manck
	973		Friedrich *Gelzin -85*	W Bk	German	354	112·2	28·3	17·3	Colberg	1852	G.Moegenburg		E.F.Hackbarth & Sohn
	974	DG	Friedrich *Korff -87*	W Bn	German	230	110·3	27·0	12·5	Brake	1866	J.Oltmann's Wwe.		T.J. Cassens & Sohn
	975	G	Friedrich *Dreijer 85-85*	W Sr	German	146	90·7	21·4	11·3	Papenburg	1860	H.W.Meyer		W. Hoveler
	976		Friedrich (ex Venture) (LOST) *Mahler 62-*	W Sr	German	101	75·8	19·7	10·9	Garmouth	1855	J.J.Young		Claus Mahler
	977	N	Friedrich (ex Ocean Express) *Hansen*	W S	Norweg'n	1571	227·0	42·4	24·1	Medford, Mass	1854	J.O.Curtis		E. Rasmussen
	978	✻	Friedrich Carl *Langhinrichs*	W Sk	German	90	84·6	20·8	10·6	Rostock	1886	E.Burchard&Co.		F.A. Pust
	979		Friedrich Gelpcke (BROKEN UP) *Borrardt 80-*	W S	German	684	132·0	33·0	22·5	Danzig	1864	J.W.Klawitter		G. Linck
	980	G	Friedrich Hartwig *Svensson -86*	W Bk	Swedish	303	120·6	23·6	13·9	Hammelwarde	1866	K.Gross		J.P. Jonsson
	981		Friedrich Hasselmann *Schmidt*	W S	German	1209	..	..	..		1854			F. Sussekind
	982		Friedrich Wilhelm (ex Oscar Heinrich) *Rothbart*	W Bg	German	236	105·4	26·3	13·1	Damgarten	1861	H.Dierling		F.W. Fleischer
	983	V	Friedrich W. Jebens *Rabow*	W Bk	German	425	119·0	26·3	17·1	Danzig	1864	J.W.Klawitter		H. Gromm
	984	G	Friedrich Wilhelm IV	W Bk		369	111·1	27·9	17·4	Barth	1854	F.S.Kraeft		
	985		Friedrich der Grosse *Bohnsack 80-81*	W Bk	German	633	130·5	33·7	22·1	Danzig	1853	J.W.Klawitter		Schoenberg & Domansky
	986	N	Frier *Sörensen 84-86*	W 3mSr	Norweg'n	274	126·8	28·9	12·1	Porsgrund	1884	Ole Johnsen		H. Thorsen
	987		Friesche Trouw	W Sr	Danish	162	82·4	18·5	10·8	Emden	1850			Randers [illegible] Selskab
	988	D	Friesland *Bopp 87-87*	W Bk	Dutch	631	143·4	32·2	21·6	Harlingen	1858	D.&L.Alta		E.J. Bok & Zonen
	989	V	Frigga (ex [illegible]) *Bang 62-80*	W Bg	Danish	267	105·6	24·2	15·7	Sunderland	1866	Walker & Carr		J. Hansen
	990	V	Frigga *Møller 84-84*	W Sr	Danish	102	83·9	20·2	9·8	Troense	1878	R.V.Moller		R.V. Moller
	991	V	Frigga (WRECKED) *Cheyssen -87*	W Sr	French	139	86·0	21·0	11·8	Dunkirk	1870	F. Sagary		Vve Gysel Sagary
	992	VN	Frigga *Bodensen 88-88*	W Bk	Norweg'n	924	176·4	36·4	21·9	Stavanger	1877	Strömstenens Vert		Ths.S. Falck
	993	DN	Frigga *Naess 83-83*	W Sr	Norweg'n	162	95·1	22·4	11·3	Sappemeer	1861	J.G.Berg		And.& Chr. Ellef…
	994		Frigga	W Sr	Norweg'n	138	..	..	..					C.C. Bjerke
	995	NV	Friheden (ex Bessie Judd) *Tonstad -87*	W Bk	Norweg'n	672	155·6	33·5	17·8	Riv. John, N.S.	1877	McKenzie & Hatcher		O.A. Jolle
	996	V	Frileuse *Chaudelt 86-76*	W Sr	French	109	86·9	20·0	10·3	St. Malo	1878	F.Gautier		Mignot
	997	DG	Frisch *Voss 86-70*	W Bk	German	412	135·6	28·8	17·3	Rostock	1869	Em.Paddenatz		W. Manck
	998		Frisia (ex Albertina Jacoba) *Bossegh*	W Bk	Dutch	309	125·9	28·5	12·0	Bangor, Me.	1855			H. Jans & Co.
	999		Frisk (ex Mathilde Geertruida) *Hendriksen 85-85*	W Sr	Norweg'n	130	93·6	22·0	9·7	Hoogezand	1861	J.van der Werff		C. Apenes & Son
	1000													

✻ = Lloyd's Register; D = Nederlandsche Vereeniging; G = Germanischer Lloyd; I = Registro Italiano;
A = Veritas Austro-Ungarico; H = Veritas Ellenico; N = Norske Veritas;

In the 4th Column, C = Composite (iron frame, wood planked); I = Iron; Stl = Steel; W = Wood.

The Dates of Metalling (Col. 11) are based on the records of the Books in which Vessels are respectively classed, or on the Reports of the Surveyors…

Further particulars of Vessels classed by Lloyd's Register will be found in Lloyd's Register Book and Lloyd's Yacht Register.

No.	Cl.	Nom. Capitaine et Gréement. Ancien Nom.			Pavillon	Tonnage net.	Longueur	Largeur	Creux	Où.	Quand.	Par qui.	Doublage	Armateurs ou Directeurs-gérants.	Port d'Armement.
		Frithiof	Mattson	W Sr	American	231	116·0	32·0	13·8	Coos Bay Oreg	1874			O.Mattson	San Francisco
		Frithiof	Dahlin	I Bk	Swedish	547	166·7	28·8	15·8	Sjöberg	1880	Bachsn		G.F.Cawallin	Frammas
		Frithiof	Pettersson	W Sr	Swedish	216	108·6	22·8	11·9	Västervik	1875	C.Thorén		E.A.Kjellgren	Nyköping
		Frithiof	Kullenberg	W Sr	Swedish	211	101·6	22·8	12·3	America	1872			L.Jönsson	Katsenbergstorp
		Frithiof	Jansson	W Sr	Swedish	151	83·3	25·8	11·9	Vätö	1840			M.Abrader	Vätö,Rolgn.
		Frithiof	Zetterberg	W Sr	Swedish	148	83·3	25·4	11·6	Vätö	1854			J.Zetterberg	Oregrund
		Frithjof	Johannesen	W Bk	Norweg'n	583	125·3	34·7	21·1	Fiume	1855			Sömme & Wattne	Stavanger
		Frithjof	Johnsen	W Bk	Norweg'n	441	135·3	29·6	16·5	Grimstad	1878	J.F.Tryde		S.Salvesen	Tvedestrand
		Frithjof	Christensen	W Bg	Norweg'n	280	111·7	27·9	14·1	Tönsberg	1855			N.Bugge	Tönsberg
1010		Frithjof	Larsen	W Bg	Norweg'n	278	94·6	28·3	16·0	Arendal	1835			F.Lorentzen	Tönsberg
1011		Frithjof	Roth	W Bk	Norweg'n	215	112·9	25·5	11·7	Bergen	1875	R.Tungesvig		J.Hansen	Bergen
1012		Frithjof	Bjönness	W Sr	Norweg'n	153	93·1	22·6	11·8	Seelvig	1852			Ole Bjönness	Tönsberg
1013		Frithiof	Jörgensen	W Bg	Norweg'n	128	83·9	22·9	10·1	Brevig	1854			F.Bentzen	Skien
1014		Frithjof	Christiansen	W Sr	Norweg'n	103	83·1	21·0	10·7	Tvedestrand	1865			M.Olsen Belland	Farsund
1015		Frits Emil	Rasmussen	W Bn	Danish	183	97·3	23·1	12·3	Kjerteminde	1880	Ch.Kraup		R.D.Rasmussen	Svendborg
1016		Fritz	Harlagen	W S	German	1492	192·9	40·7	28·2	Thomaston Me	1865	J.Watts&Co		Sietenburg,Wendt & Co	Bremen
1017		Fritz	Andresen	W Bk	German	412	129·5	28·7	17·8	Rostock	1870	Em Padderatz		Franz Crull	Rostock
1018		Fritz	[illegible]	W Bg	German	229	104·0	24·6	13·4	Barth	1868	J.Holzerland		J.J.Wallis & Sohn	Barth
1019		Fritz	Barth	W Bg	Swedish	243	110·8	24·2	12·6	Lubeck	1873	J.A.Meyer		F.A.Harms	Carlshamn
1020		Fritz Reuter	Rambusch	I S	German	1475	256·0	36·0	22·0	Glasgow	$18\frac{57}{74}$	Smith&Rodger		R.M.Sloman & Co.	Hamburg
1021		Fritz Reuter	Seerin	W Bk	German	298	107·0	26·0	16·0	Sunderland	1853			E.Burchard	Rostock
1022		Fritz Schmidt	[illegible]	W Bk	German	397	127·1	27·9	16·9	Rostock	1874	H.Rickmann		Wilh.Maack	Rostock
1023		Fritz Smith	Larsen	W Bk	Norweg'n	455	137·5	31·0	16·2	Risör	1877	T.Taraldsen		A.&F.Smith	Tvedestrand
1024		Fritz von der Lancken	[illegible]	W Bk	German	323	117·1	27·4	14·2	Seedorf	1877	Kranger		C.A.Berg	Stralsund
1025		Fritzoe	Eckfeet	I Bk	Norweg'n	1135	213·6	35·6	20·8	W.Hartlepool	1884	W.Gray&Co.		Act.Selsk.„Australia"	Laurvig
1026		Frivold	Johnsen	W Bk	Norweg'n	387	113·8	30·3	16·9	Grimstad	1857	A.Omundsen		J.D.Havstad	Grimstad
1027		Frixos	Thermioti	W Sr	Greek	219	..	..	..	Syra	1887	A.Livadara		D.Thermioti	Syra
1028		Frode	Guttormsen	W Bg	Norweg'n	267	104·3	26·8	14·2	Grimstad	1871	N.Olsen Gjömle		A.Rickman	Grimstad
1029		Frode	Gramdorsen	W Sr	Norweg'n	164	103·4	25·8	10·7	Moss	1874	L.Dammes		G.A.Larsen	Fredriksstad
1030		Frode	Jansson	W Bg	Swedish	188	86·0	22·7	14·0	Svaneke	1843			C.Henckel	Helsingborg
1031		Fronia Couppa		W Bg	Greek	528	143·5	27·8	18·8	Sestri Ponente	1874	Bossi		N.Couppa	Cephalonia
1032		Fröy	Christiansen	W Bk	Norweg'n	353	119·9	28·5	15·9	Antwerp	1857	P.Fertin		A.O.Olsen	Stavanger
1033		Frey	Mortensen	W Bk	Norweg'n	324	112·5	28·8	15·3	Lillesand	1860	Th.Aslagsen		H.J.Hammers Enke	Lillesand
1034		Fruen	Bentsen	W Bk	Norweg'n	372	126·5	28·6	15·7	Grimstad	1871	M.Smith Petersen		F.Smith Petersen	Grimstad
1035		Frühling	Albrecht	W Bk	German	370	123·3	27·3	17·6	Sunderland	1865	Barker		Davids & Carow	Wismar
1036		Fruit Girl	[illegible]	W Br	British	125	93·2	22·1	11·2	Ipswich	1866	Read		K.R.Prowse	St Johns,N.F.
1037		Fruiterer	Koffsing	W 3mSr	Dutch	355	119·2	27·8	16·7	Sunderland	1865	Robinson		F.L.Dreuth	Oude Pekela
1038	V	Fuchs	[illegible]	W Sr	Russian	173	89·9	24·9	11·8	Haynasch	1886	F.Rusek		J.Mikkelsohn	Haynasch
1039	V	Fuglen	[illegible]	W 3mSr	Danish	219	112·9	24·2	11·9	Marstal	1884	J.O.Christensen		H.C Christensens Enke	Svendborg
1040		Fuki Maru		W Sr	Japanese	233	110·4	23·1	12·5	Yechizen	1879	Takashiba [Hisakichi		Morita Saburoyemon	Sakai,Yechizen
1041	G	Fulda	[illegible]	I Bk	German	884	186·0	31·8	19·7	Glasgow	1876	A.Stephen&Sons		D.H.Wätjen & Co.	Bremen
1042		Fulwood	McNeal	I S	British	2118	279·0	40·2	24·1	Southampton	1885	Oswald,Mordaunt&Co.		R.W.Leyland & Co.	Liverpool
1043		Furio Camillo	Lagemma	W Bn	Italian	122	..	..	..	Sinigaglia	1856			G.Poli & Co.	Bari
1044		Furman L. Mulford	Little	W 3mSr	American	559	140·0	35·7	17·3	Millville,N.J.	1877	J.D.Bell		J.S.Little	Philadelphia
1045	V	Furness Abbey		W Bk	American	1044	193·5	35·3	12·2	Kennebunk Me	1879	Thompson		J.G.Morse	Boston
1046	G V	Fürst Bismarck		I Bk	German	968	189·9	35·5	20·5	Bremen	1877	Weser Ship B.Co		Schrader & Gaertner	Bremen
1047	V	Fürst Bismarck		W Bk	German	333	127·3	27·9	14·9	Bremen	1867	J.H.Knickmann		Oetling Gebrüder	Hamburg
1048		Fury	Coggin	W Bk	American	294	129·3	28·4	11·5	Portsm'th,NH	1860			H.O.Roberts	Salem,Mass.
1049		Futtanare		W Sr	British	169	82·0	20·8	11·5	Coringa	1870			Jeevanjee Noorbhoy	Galle
1050		Futteh Hosseen		W Bg	British	248	90·0	22·0	15·2	Chittagong	1869			Allee Deedee	Galle

Lloyd's Register; D = Nederlandsche Vereeniging; G = Germanischer Lloyd; I = Registro Italiano; R = American Record; Veritas Austro-Ungarico; H = Veritas Ellenico; N = Norske Veritas; V = Bureau Veritas.

Dans la 4me Colonne, C = Composite (membrure en fer, bordé en bois); I = Fer; Stl = Acier; W = Bois.

Dates du doublage en métal (Col. 11) ressortent des livres dans lesquels les navires sont respectivement classés ou des rapports des Inspecteurs du Lloyd's Register.

Pour plus amples renseignements sur les navires classés par le Lloyd's Register, voir le Lloyd's Register Book et le Lloyd's Yacht Register.

Official No.	No. in Book	Classed in	Ship's Name. Material and Rig. Master, and Dates of Appointment to present Owner's Service and to Ship. Ship's Late Name.		Flag.	Net Tonnage.	Length	Breadth	Depth	Where	When	By whom	Date of Metalling	Owners or Managers
	1051		Futtel Curreem	WBk	British	651	146·0	31·0	22·0	Cochin	1858		Y 12.87	Hajee Roomaol Bin [illegible]
	1052	※ D	FuttelRahaman *Suliman* (ex County of Lancaster)	CS	Arabian	999	208·9	34·7	20·3	Glasgow	1870	C.Connell & Co.	[illegible]	M.B.Goolam & A.R. Sulliman
	1053		Futtel Razac (ex [illegible])	WS	British	392	119·3	29·0	17·6	Denmark	1845			Ebrahim bin Yoosuf
	1054		Fuzzal Curreem	WBk	British	637	146·0	31·0	22·0	Cochin	1858		Y 12.87	Hajee Essoom bin Mahomed
	1055	N	Fylgja *Danielsen*	WBn	Norweg'n	555	136·0	32·5	19·1	Arendal	1870	C.Jensen	Y [illegible]	D.Danielsen
	1056	V	Fylla *Funch*	WBn	Danish	218	107·1	24·7	11·8	Thuro	1873	L.Kaas	N [illegible]	Emil Hansen
	1057		Fylla (ex Queen Emma)	WBg	Danish	193	103·0	28·5	12·9	Quebec	1866	Labbee	Y [illegible]	M.Christiansen
	1058	V	Fylla *Jensen* 78-88	WSr	Danish	121	86·5	22·0	10·3	Assens	1869	H.Christoffersen		F.Winding
	1059	V	Fylla *Hansen* 86-79	WSr	Danish	115	95·0	21·0	10·0	Nakskov	1858	H. P. Rüdersborg & Trochman		Jochum Hansen
	1060	V	Fylla *Eriksen* 73-73	WSr	Danish	108	70·5	21·0	10·5	Marstal	1873	F.Hansen		N.Petersen
	1061		Fylla	WBk	Swedish	239	105·3	23·0	13·9	Leith	1856	R.Innes		Höganäs Bruksb[illegible]
	1062		Fyra Bröder *Wermelin* 88-88	WBk	Russian	333	105·8	28·7	17·0	Sideby	1845			O.Blom

Additional Ships are inserted in the Supplement at the end of the Register.

※ = Lloyd's Register; D = Nederlandsche Vereeniging; G = Germanischer Lloyd; I = Registro Italiano; R = [illegible]
A = Veritas Austro-Ungarico; H = Veritas Ellenico; N = Norske Veritas; V = [illegible]

In the 4th Column, C = Composite (iron frame, wood planked); I = Iron; Stl = Steel; W = Wood.

The Dates of Metalling (Col. 11) are based on the records of the Books in which Vessels are respectively classed, or on the Reports of the Surveyors [illegible]

Further particulars of Vessels classed by Lloyd's Register will be found in Lloyd's Register Book and Lloyd's Yacht Register.

No.	Class.	Nom, Matériaux et Gréement. Capitaine et date d'entrée en fonction des Armateurs actuels et à bord du navire. Ancien Nom.		Pavillon.	Tonnage net.	Dimensions. Longueur.	Largeur.	Creux.	Construit. Où.	Quand.	Par qui.	Date du doublage en métal.	Armateurs ou Directeurs-Gérants.	Port d'Armement.
1		G. A. Hayden *Green*	W Sr	American	103	96·0	26·0	7·0	Essex, Conn.	18[illegible]			E. Brainard Quarry Co.	Hartford, Conn.
2		G. B. [illegible]	W Bk	French	530	125·6	27·9	19·2	Pra	1867	M. Tixi		Gaillard Brian	Marseilles
3		G. B *Ozier* ·87	W Bk	French	275	116·1	25·9	15·2	Dunkirk	1868	Beck		J. F. Merlin	Fécamp
4		G. B. ·87	W Sr	French	1[illegible]	76·8	20·8	11·1	Landerneau	1872	Quefferus		S. cherie du Port de Bonc	St. Malo
5	V	G. B. Lockhart	W Bk	British	296	120·4	29·0	10·7	Horton	1887	J. B. North		G. B. Lockhart	Windsor, N.S.
6	✻ I	G. B. Patrone *Andrea*	W Bk	Italian	[illegible]	[illegible]	33·8	20·4	Sestri Ponente	1874	S. Bruzzone	[illegible]	G. B. Patrone	Spezia
7	V	G. C. No. 12	W Sr	French	14[illegible]	85·6	23·6	11·6	Fécamp	1878	D. Capon	[illegible]	[illegible] Craquelin	Fécamp
8	✻	G. C. Gradwell	W Sr	British	1[illegible]	99·4	22·1	12·1	Appledore	1873	Cock	[illegible]	W. Burnitt	London
9		G. Calvocoressis	W Bn	Greek	336	116·1	28·5	..	Syra	1888			M. J. Pneumatikos	Syra
10		G. D.	W Sr	British	1[illegible]	98·2	21·7	7·4	Yamaska, P.Q.	1885	P. Letendre		Moïse Lesperance	Montreal
11		G. F. Day	W Sr	British	14[illegible]	[illegible]8·3	26·1	10·2	Parrsboro' N.S.	1871			Ingram B. Slocomb	Parrsboro', N.S.
12		G. F. Peniston	W Bn	British	217	[illegible]	26·0	12·0	Walton, N.S.	1864			F. W. J. Hurst	Windsor, N.S.
13		G. G. M.	W Bg	French	14[illegible]	[illegible]	23·3	11·1	Nantes	1868	Sevestre	[illegible]	G. Garand	Lorient
14	R	G. H. Gordon	W Bk	British	47[illegible]	[illegible]	29·8	17·8	P. E. Island	1874	D. Gordon		T. Tanner	Ch'rl'ttn. P.E.I.
15	R	G. H. Holden	W Sr	American	[illegible]	111·3	28·5	10·1	Bath, Me.	1883	T. Hagan & Co.		J. H. Pinkham	Booth Bay, Me.
16		G. H. Wappaus	W Bk	British	533	147·0	29·8	17·3	Lubeck	1871	P. Heitmann		P. V. Grant	Shanghai
17	✻ V	G. H. Wappaus	I Bk	German	1225	218·7	36·2	21·7	Flensberg	1887	Flensberg S.B.Co.		A. H. Wappaus	Hamburg
18		G. L. Waters *Cave*	W Bkn	British	1[illegible]7	103·1	21·6	12·1	Workington	1865	Lamport		R. Kearon	Workington
19		G. M. Brainard	W Sr	American	[illegible]	[illegible]	30·0	[illegible]	[illegible], Me.	1873			Cobb, Wight & Co.	Rockland, Me.
20		G. M. Carins	W Bk	British	1100	175·2	36·8	22·9	St. John, N.B.	1859	M'Donald		W. Heselton	Newcastle
21		G. M. Neelon	W Sr	British	315	138·5	25·8	11·6	Prt. Dalhousie	1873	S. D. Andrews		S. Neelon	S. Cathrns., Ont.
22		G. M. Porter *Johnson*	W Sr	American	137	98·0	26·7	8·[illegible]	[illegible], Me.	1871	Nickerson & [illegible]		C. H. Porter	Calais, Me.
23	R V	G. M. Stanwood	W Bk	American	522	135·8	31·4	16·8	Harpswell, Me.	1879	Bailey Bros.	[illegible]	G. M. Stanwood	Portland, Me.
24	✻ D	G. N. Wilcox *Busch*	W Bk	German	570	152·8	30·7	18·0	[illegible]	1876	R. Thompson	[illegible]	[illegible] Henoch Man [illegible]	Bremen
25	N V	G. P. Harbitz	W Bk	Norweg'n	653	165·8	34·0	17·[illegible]	[illegible]	1876	Chr. Brinch	[illegible]	G. Knap	Tönsberg
26		G. Philaretos	W Bn	Greek	430	135·4	30·2	..	Trieste	1868			N. A. Georgiadis	Piræus
27		G. Reusens *Engel*	W Bkn	American	464	1[illegible]·5	29·5	16·7	Millbridge, Me.	1875			E. B. Risley	Philadelphia
28	R V	G. S. Penry *Foster*	W Bk	British	737	155·2	33·4	19·0	Black Riv. N.B.	1876	J. & R. McLeod	[illegible]	J. W. Penry	St. John, N.B.
29	✻	G. T. D. *Bernier*	W Sr	British	19[illegible]	95·6	27·0	10·6	[illegible]	1883	G. T. Davie		G. T. Davie	Québec
30		G. T. Kemp	W Bk	British	372	129·0	28·7	18·1	Rockland, US	1853	Thomas		W. Hall	Gibraltar
31	D	G. T. Ray *Smith*	W Sr	Dutch	356	119·7	27·0	16·9	Sunderland	1866	W. Naizby	[illegible]	G. G. Albers	Farmsum
32		G. W. Bentley *McLeod*	W Sr	American	1[illegible]	90·0	23·0	8·0	Essex, Mass.	1876			J. Hatch	Provincetn., Mass.
33		G. W. Rawley *Holbrook*	W Sr	American	148	97·6	27·8	8·0	Waldoboro' Me.	1860			G. W. Rawley	St. George, Me.
34	✻	G. W. Wolff *Stevenson*	I Bk	British	1663	257·1	38·3	23·2	Belfast	1878	Harland & Wolff		S. Lawther	Belfast
35	✻	G. & W. Jones	W Sr	British	103	89·5	21·3	11·1	Portmadoc	1872	Griffiths		R. Evans & Co.	Carnarvon
36		G. de Zaldo *Miller*	W Bk	American	4[illegible]3	139·0	31·3	17·0	Br'khaven NY	1870	N. Hand & Sons		F. Waydell	New York
37	N V	Gaa paa *Hansen* ·87	W Bk	Norweg'n	663	157·5	35·8	17·2	[illegible]	1886	O. M. Olsen	[illegible]	H. H. Pettersen	Arendal
38		Gabriela	W Sr	Spanish	148	84·6	21·3	12·[illegible]	[illegible]	1858			G. Ramentol	Barcelona
39		Gabriel *Östlin*	W Sr	Russian	22[illegible]	..	..	..	[illegible]	1883			A. Östlin	Gustaf
40		Gabrielle	W Bk	French	362	111·2	26·6	15·6	[illegible]	1870	Brianne		Fournier Cousins	Marseilles
41	V	Gabrielle *Barier* ·88	W Bg	French	226	101·7	25·9	13·1	St. Malo	1884	Parnet	[illegible]	Sire fils	St. Malo
42		Gabrielle	W Bg	French	187	94·2	26·3	13·5	Fécamp	1866			G. Craquelin	Havre
43	V	Gabriella	W Bn	French	14[illegible]	89·2	23·8	11·1	Varazze	1882	G. Ravecca	[illegible]	R. Poulain de St. Père	Tréguier
44	I	Gabriello Lavaggi	W Bg	Italian	2[illegible]	112·5	25·6	14·0	Sestri Ponente	1872	F[lli] Calcmaccio	[illegible]	G. Lavaggi di O.	Catania
45		Gaditana *Henriques*	W Sr	Brazilian	12[illegible]	..	..	..	United States	1867		[illegible]	H. Rodrigues	Rio de Janeiro
46	✻	Gaerwen	W Bk	British	267	121·6	25·7	12·6	Ulverston	1877	Roberts	[illegible]	Mrs. M. J. Gunn	Salcombe
47	I	Gaeta *Leboffe*	W Bk	Italian	448	132·5	27·8	18·5	Gaeta	1868	F. Contreras	[illegible]	F[lli] Leboffe	Gaeta
48	I	Gaetanin	W Bk	Italian	542	139·4	32·1	20·3	Pra	1870	L. Dunante	[illegible]	G. Schiaffino	Genoa
49		Gaetano *Schiaffino*	W Bk	Italian	421	123·6	30·0	19·0	Lerici	1866	F. Rolla	[illegible]	G. Razeto	Genoa
50	I	Gaetano Cicconardi	W Bk	Italian	591	145·6	30·7	19·2	Sorrento	1875	M. Paturzo	[illegible]	G. Cicconardi	Gaeta

✻ = Lloyd's Register; D = Nederlandsche Vereeniging; G = Germanischer Lloyd; I = Registro Italiano; R = American Record; A = Veritas Austro-Ungarico; H = Veritas Ellenico; N = Norske Veritas; V = Bureau Veritas.

Dans la 4me Colonne, C = Composite (membrure en fer, bordé en bois); I = Fer; Stl = Acier; W = Bois.

Les Dates du doublage en métal (**Col. 11**) ressortent des livres dans lesquels les navires sont respectivement classés ou des rapports des Inspecteurs du Lloyd's Register.

Pour plus amples renseignements sur les navires classés par le Lloyd's Register, voir le Lloyd's Register Book et le Lloyd's Yacht Register.

Official No.	No. in Book	Registry in which classed	Ship's Name. Material and Rig. Master, and Dates of Appointment to present Owner's Service and to Ship. Ship's Late Name.	Flag.	Net Tonnage.	Dimensions: Length.	Breadth.	Depth.	Built: Where.	When.	By whom.	Date of Metalling.	Owners or Managers.
[illegible]	51		Gaetano P. *Assentato -87* (LOST) (ex Ste Germaine) W Bg	Italian	332	104·3	24·6	15·3	Agde	1870	Cavalier		R. Prestopino
[illegible]	52	I	Gaetano Repetto *[illegible]* W Bk	Italian	573	112·8	32·3	20·3	Voltri	1869	D. Fabiano	[illegible]	G. B. Repetto fu A.
[illegible]	53		Gage H. Phillips *Dyer 81-81* W Sr	American	101	80·5	23·2	8·4	Essex, Mass.	1866			S. Cook
[illegible]	54	✻	Gainsborough *McPhail -83* I S	British	974	206·0	31·4	20·4	London	1868	C. Langley		A. McPhail
	55		Gala W Bn	Spanish	111	81·4	22·5	10·5	Corunna	1866	M. Vaamonde		F. Romani
	56	V	Galant *Von Appen -87* (MISSING) W Bn	German	353	126·0	26·0	13·6	Kappeln	1876	Carl Steen	[illegible]	H. v. Appen
[illegible]	57	✻	Galatea *Colville* I S	British	1694	265·0	39·7	23·4	Port Glasgow	1884	R. Duncan & Co.		C. S. Caird
[illegible]	58	DV	Galatea *Campling -88* W Bk	British	1178	194·0	38·3	22·7	Harvey, N.B.	1881	G. S. Turner	[illegible]	G. F. Smith
[illegible]	59	✻	Galatea *Wilkins 84-84* W Bn	British	152	93·7	22·1	11·6	Weymouth	1884	R. A. Ayles	[illegible]	E. J. Duder
[illegible]	60		Galatea *Cock* W Sr	British	133	100·0	22·9	11·4	Truro	1868	Dyer		W. S. Allport & Co.
[illegible]	61	R	Galatea *Coombes -88* W Sr	British	122	83·3	28·2	7·5	Gibson, York [Co., N.B.	1888	J. Gibson		John Gibson
[illegible]	62		Galatea W Sr Yt	British	117	94·0	18·5	9·8	Cowes	1860	C. Hansen & Sons		Frederick W. Wa[illegible]
[illegible]	63	✻	Galatea *Bucktrout* W K	British	66	70·6	17·9	8·2	Castleford	1879	Cliff		J. Bucktrout & Co.
	64	V	Galatea *Lassen 84-84* W Sr	Danish	116	87·3	21·9	10·6	Thuro	1871	P. Troensegaard		H. A. Hansen
	65	V	Galatea *Hornblik -88* W Bg	Swedish	332	121·4	24·0	15·4	Gothenburg	1878	J. E. Hilbe	[illegible]	P. N. Horndahl
	66	N	Galathea *Samuelsen -88* W Bk	Norweg'n	604	140·9	35·2	22·5	Copenhagen	1832		[illegible]	N. Olsen & S. Samuelsen
	67	DN	Galathea *Thomsen 73-86* W Bk	Norweg'n	384	140·9	32·5	17·8	Tvedestrand	1879	A. & F. Smith	[illegible]	A. & F. Smith
	68	✻	Galathée *Hugues -73* I Bk	French	1218	228·7	35·9	20·3	Glasgow	1884	A. Stephen & Sons		A. C. Le Quellec
	69	V	Galathée *Fauchon* W Sr	French	121	88·6	19·7	11·1	Dunkirk	1865	Sagary fils		V[illegible] Gysel Sagary
	70		Galaxidi *Cattario* (ex Angela B. ex Giuseppe Lazarello) W Bk	Greek	596	145·0	31·9	20·4	Varazze	1865		[illegible]	Cost. Cattario e Co.
[illegible]	71		Gale *Riley* W Sr	American	112	93·6	17·5	7·8	Hamburg, Pa.	1864			J. A. Anderson
[illegible]	72	R	Galena *Parks 82-83* W Sr	British	331	145·0	32·4	12·8	Horton, N.B.	1883	J. B. North	[illegible]	A. Haley
	73		Galeno W Bn	Spanish	149	92·8	21·0	11·1	Albay	1860			J. M. Tuazón & Co.
	74		Galeon *Kalsboll 63-77* W 3m Sr	Danish	313	126·7	26·9	13·2	Quebec	1871	W. Charland		C. Langballe & Son
[illegible]	75	✻	Galeota *McMillan 77-86* (ex Ochiltree) W Bk	British	447	146·0	28·8	14·8	Alloa	1878	Roy	[illegible]	G. Turnbull & Co.
[illegible]	76	✻	Galera *Watson 81-81* W Bk	British	436	147·5	27·2	16·4	Grangemouth	1873	Adamson	[illegible]	G. Turnbull & Co.
[illegible]	77	✻	Galgate *Commins 87-88* Stl 3m S	British	2201	293·5	42·8	24·2	Whitehaven	1888	Whitehaven S.B. [Co., Ld.		Galgate Ship Co. Ld. (J. Joyce)
[illegible]	78	✻	Galgate *Commins 87-88* (BURNT) I S	British	1635	252·3	39·0	24·0	Stockton	1884	Richardson, [Duck & Co.		J. Joyce
[illegible]	79	A	Galileo *Papuccio* W Bg	Italian	153	85·6	24·8	13·6	Marciana	1852	Braschi	[illegible]	[illegible] Gherardi
	80		Galileo *Negra -88* W Bg	Spanish	221	102·0	27·7	13·6	Lloret	1846			F. A. Piñeyro
[illegible]	81	DI	Galileo S. *Schiaffino -85* (ex Tre Fratelli) W Bk	Italian	618	147·6	30·5	19·8	Sestri Ponente	1874	G. Peragallo	[illegible]	P. Schiaffino
[illegible]	82	✻	Gallovidian *Price 83-83* I Bk	British	400	141·0	26·5	15·9	Glasgow	1867	Dobie & Co.		A. Rae
	83	V	Galofré *Gelpi* W Bk	Spanish	369	117·7	28·5	15·6	San Feliu	1871		[illegible]	S. Galofré
	84	DGV	Galveston *Stunkel 78-78* W Bk	German	619	143·2	33·4	16·5	Vegesack	1869	H. F. Ulrichs	[illegible]	Boyes & Ruyter
[illegible]	85		Gamaliel *Mason* W K	British	67	75·6	19·0	8·9	Goole	1873	Pearson		G. Kilner & Co.
[illegible]	86	R	Gamaliel *Crochett* (WRECKED) W Bk	American	539	133·0	32·5	18·0	Baltimore	1874		[illegible]	A. Jones
[illegible]	87	A	Gambetta *Coulandris* W Bg	Greek	325	116·1	27·7	..	Syra	1878		[illegible]	A. N. Goulandris
	88	N	Gambetta *Heraldsen -88* W 3m Sr	Norweg'n	299	124·5	29·0	13·2	Porsgrund	1871		[illegible]	J. C. & G. Knudsen
	89	N	Gambetta *Andersen -84* (ex Gloria) W Bg	Norweg'n	236	101·7	26·6	13·5	Risöer	1865	J. H. Wicken		L. Lonsland
	90		Gambia I Sr	French	119	90·7	22·3	9·4	Hoboken	1879	Soc. Cockerill		Maurel frères
	91	V	Gamon *Gillenstedt 79-79* (ex Baron Nandi[illegible]) W Bk	Swedish	917	183·0	31·5	21·6	Varazze	1874	Craviotto	[illegible]	Rederiaktiebol. Gama (J. [illegible])
[illegible]	92	R	Gamma *Pratt -88* W Sr	British	175	100·0	29·3	10·3	Kempt, N.S.	1888	R. Rose		R. Pratt
[illegible]	93		Gamma *Brown* W Sr	American	154	100·5	26·3	8·8	E. Machias, Me.	1869			G. T. Jenkins
	94	N	Gamma *Pedersen -87* W Bk	Norweg'n	553	151·8	29·5	18·9	Bideford	1865	Cox		Gammas Interessenter (H. [illegible])
	95		Gamo (ex Reindeer) *Nunes* W 3m Sr	Portug'se	307	125·0	23·3	13·4	Sunderland	1838			Bensaude & Co.
	96	GV	Gandal *Sorensen 85-87* (ex Wilhelm, ex Louisa) W Bk	Norweg'n	371	124·9	27·6	16·7	Sunderland	1867	G. & J. Mills	[illegible]	Holdt & Isachsen
	97	N	Ganger Rolf *Evensen 84-84* W Bk	Norweg'n	415	132·0	29·6	16·3	Risöer	1870	T. Jensen	[illegible]	J. W. Probensen, [illegible]
	98	N	Ganger Rolf *Olsen -88* W Bk	Norweg'n	377	120·9	30·9	16·7	Drammen	1866			B. Nielsen
	99	N	Ganger Rolf *Flatmark* (ex Engelst) W Bk	Norweg'n	386	117·8	27·2	14·4	Vegesack	1861	J. Lange	[illegible]	C. Wroldsen
	100	N	Ganger Rolf *Dannie -86* W Bk	Norweg'n	312	114·7	28·1	13·4	Risöer	1864	S. Gundersen	[illegible]	T. Larsen & Enke

✻ = Lloyd's Register; D = Nederlandsche Vereeniging; G = Germanischer Lloyd; I = Registro Italiano; R = [illegible]
A = Veritas Austro-Ungarico; H = Veritas Ellenico; N = Norske Veritas; V = [illegible]

In the 4th Column, C = Composite (iron frame, wood planked); I = Iron; Stl = Steel; W = Wood.

The Dates of Metalling (Col. 11) are based on the records of the Books in which Vessels are respectively classed, or on the Reports of the Surveyors [illegible]

Further particulars of Vessels classed by Lloyd's Register will be found in Lloyd's Register Book and Lloyd's Yacht Register.

No.	Classe	Nom, Capitaine	Gréement	Pavillon	Tonnage	Longueur	Largeur	Creux	Lieu de construction	Date	Constructeur	Armateurs ou Directeurs-Gérants	Port d'Armement
101	✻	Ganges *Hook*	I Bk	British	[illegible]	[illegible]	[illegible]	[illegible]	[illegible]	1882	Osbourne, Graham & Co.	J. Nourse	London
102	✻	Ganges *Hamilton*	I Bk	British	[illegible]	[illegible]	[illegible]	[illegible]	[illegible]	1865	G.R. Clover & Co.	E.F. Powell & Co.	Liverpool
103	✻ N	Gann	W Bk	Norweg'n	[illegible]	[illegible]	[illegible]	[illegible]	[illegible]	1856		A.O. Olsen	Stavanger
104	✻	Gannet *Johnson*	I S	British	[illegible]	[illegible]	18·5	9·1	Barton Ferry	1855	Neath Abbey I.C.	Garriock & Co.	Lerwick
105		Gannet *Anderson*	W Bk	Swedish	[illegible]	[illegible]	22·0	[illegible]	[illegible]	1857		F. Petersson	Ystad
106	G	Ganymed *Møller*	W Sr	Danish	[illegible]	[illegible]	[illegible]	[illegible]	[illegible]	1852	E.C. Benzon	C.H. Møller	Odense
107	✻	Ganymede *Alison*	I Bk	British	[illegible]	[illegible]	[illegible]	17·7	[illegible]	1868	W. Doxford	A. White	Lyttelton, N.Z.
108	N	Ganymedes *Olsen*	W Bk	Norweg'n	411	127·7	29·7	16·6	[illegible]	1869		H. Bang	Risör
109	V	Garam *Melalkin*	W 3m Sr	Russian	262	114·0	24·1	[illegible]	[illegible]	1886	O. Kirstein	J. Stein	Riga
110		Garcia	W Sr	American	111	[illegible]	[illegible]	7·	[illegible]	1882	[illegible]	S.B. Peterson	San Francisco
111		Gard, No. 1.		French	24[illegible]	..	..	..					Marseilles
112		Gardiner B. Reynolds	W 3m Sr	American	378	112·4	33·4	12·1	E. Boston, Mass.	1887	J. M. Brooks	Gardiner B. Reynolds	Newport, R.I.
113	R	Gardiner G. Deering *Rogers*	W 3m Sr	American	682	161·2	34·8	18·0	Bath, Me.	1884	Deering & Donnell	Deering & Donnell	Bath, Me.
114	✻	Gareloch	I S	British	1177	227·2	36·1	21·1	Glasgow	1873	Dobie & Co.	P. Rintoul, Son & Co.	Glasgow
115	✻	Garfield	Stl S	British	2290	292·8	41·2	24·8	Belfast	1882	Harland & Wolff	[illegible]	Liverpool
116	✻	Garfield	I Bk	British	297	128·3	25·6	13·9	Glasgow	1869	T. Wingate & Co.	A. Campbell	Glasgow
117	DN	Garfield *Pedersen*	W Bk	Norweg'n	7[illegible]	1[illegible]·1	35·1	17·1	Arendal	1884	E. Gottormsen	H.H. Pettersen	Arendal
118		Garibaldi *Moffet*	W Bn	British	172	100·8	22·4	13·5	Cowes	1860	J. White	D.T. Sharper	W. Hartlepool
119		Garibaldi	W Sr	British	164	103·0	25·5	9·8	Port Rowan	1863		J. McPherson	Hamilton, Ont.
120		Garibaldi	W Sr	British	122	95·2	24·8	8·3	Port Huron U.S.	1861	R. Steel	W.H. Brand	Port Hope, Ont.
121	✻	Garibaldi *Inglety*	W Sr	British	[illegible]	91·0	21·6	10·5	Gainsboro'	1860	Lister	T. Ramsey	London
122	V	Garibaldi	W Sr	Danish	11[illegible]	8[illegible]·6	19·7	10·3	Thurø	1861	N. Troensegaard	J.P. Jorgensen	Svendborg
123	N	Garibaldi *Williamsen*	W Bk	Norweg'n	5[illegible]	12[illegible]·6	32·7	20·0	Alblasserdam	1838		J.C.A. Johannesen	Christiania
124		Garibaldi	W Bk	Norweg'n	53[illegible]	1[illegible]·8	28·0	19·9	Camden, U.S.	1842		M.H. Smith	Arendal
125	V	Garibaldi *Tallaksen*	W Bk	Norweg'n	3[illegible]	128·[illegible]	28·6	12·0	Sundsvall	1870	Ostmann	H. Kallevig	Arendal
126	NV	Garibaldi	W Sr	Norweg'n	27[illegible]	12[illegible]	26·9	12·7	Haugesund	1876	J. Hauge	P.G. Pettersen	Bergen
127		Garibaldi	W Bg	Norweg'n	144	91·5	22·8	10·8	Christianssand	1861	P. Preus	N.T. Langfeldt	Christianssand
128		Garibaldi	W Sr	Swedish	110	86·0	21·4	9·2	Påskallavik	1860		A. Pyk	Helsingborg
129		Garibaldis	W Sr	Greek	149	85·6	24·3	..	[illegible]	1880		M. Vourakis	Syra
130		Garigliano	W Bk	Italian	6[illegible]2	1[illegible]8·1	32·4	20·[illegible]	[illegible]	1875	A. Gargiulo	Scotto di Tello & Co	Naples
131	✻	Garland *Roberts*	I Bk	British	5[illegible]	171·[illegible]	27·7	17·8	Liverpool	1865	R.&J. Evans & Co.	J.B. Walmsley & Co.	Liverpool
132		Garland *Libby*	W Sr	American	158	95·[illegible]	26·6	8·0	E. Machias, Me.	1857		E.I. White	Machias, Me.
133		Garland *Eriksson*	W Bk	Russian	745	155·7	34·1	19·9	Nova Scotia	1865	Young	P.A. Karlsson	Finström
134	✻	Garnett Brothers	W K	British	91	72·7	21·0	10·6	Ponshedu, Cr'k	1884	E. Martin	D. & A. Garnett	Hull
135		Garron *Thompson*	W Sr	British	181	10[illegible]	23·3	12·7	Peterhead	1867	Stephen	C.G. Warburton	Sydney, N.S.W.
136	✻	Garston *Davies*	Stl S	British	1812	2[illegible]	39·1	23·6	Workington	1883	R. Williamson & S.	R.W. Leyland & Co.	Liverpool
137	✻	Gartmore *Ritchie*	Stl Bk	British	1[illegible]	21[illegible]	34·0	19·9	Dumbarton	1885	McMillan & Son	Thomson, Dickie & Co.	Glasgow
138		Gartmorn *Herbert*	W Bk	British	327	12[illegible]·4	27·8	12·8	Alloa	1877	Roy	R.&W. King	Bristol
139		Gartsherrie	W Bk	British	3[illegible]	1[illegible]	26·6	15·4	Quebec	1858	W. Cotman	J. Kelly	Belfast
140	IV	Gaspare *Lauro*	W Bk	Italian	473	125·3	27·6	18·7	Alimuri	1875	M. Mauro	Pietro Cafiero	Castellamare
141	R	Gaspee *Pickey*	W Bk	British	3[illegible]	17[illegible]·8	35·3	22·9	Newbrypt, U.S.	1858		N. Lavoie	Quebec
142	V	Gaston Albert	W Bg	French	173	95·1	24·6	12·8	Gaspé	1858	E. Collas	Daudrun	Brest
143		Gaston P. Hubbard	W Sr	American	1[illegible]	109·5	28·5	8·8	Middletn, Conn.	1861		W.H. Hastings	New York
144		Gaston & Maurice	W 3m Sr	French	[illegible]	1[illegible]2·2	30·0	13·6	[illegible], N.S.	1873	Crow	A. Leblond	Havre
145		Gast[illegible]ula Maru	I Bk	Japanese	9[illegible]	1[illegible]·7	[illegible]·2	21·2	[illegible]	1863	T.R. Oswald & Co.	H.J. Hubenet	Kobé
146	✻	Gateacre	I S	British	1355	231·[illegible]	37·8	22·[illegible]	[illegible]	1877	W. Doxford & Sns	C.W. Kellock & Co.	Liverpool
147	✻	Gateside *Roberts*	I Bk	British	692	184·6	29·6	18·5	Dumbarton	1869	A. McMillan & Sn.	Goldberg & Co.	Swansea
148	V	Gatherer	W S	American	1509	208·1	40·2	21·8	Bath, Me.	1874	A. Hathorn	T. Jensen	San Francisco
149		Gatineau	W S	British	1165	178·9	36·9	23·6	Quebec	1862	Gilmour	Gatineau Ship Co., Ld.	Greenock

Lloyd's Register; D = Nederlandsche Vereeniging; G = Germanischer Lloyd; I = Registro Italiano; R = American Record;
Veritas Austro-Ungarico; H = Veritas Ellenico; N = Norske Veritas; V = Bureau Veritas.

Dans la 4me Colonne, C = Composite (membrure en fer, bordé en bois); I = Fer; Stl = Acier; W = Bois.

Les Dates du doublage en métal (Col. 11) ressortent des livres dans lesquels les navires sont respectivement classés ou des rapports des Inspecteurs du Lloyd's Register.

Pour plus amples renseignements sur les navires classés par le Lloyd's Register, voir le Lloyd's Register Book et le Lloyd's Yacht Register.

Official No.	No.	Registry in which classed	Ship's Name. Material and Rig. Master and Dates of Appointment to present Owner's Service and to Ship. Ship's Late Name.		Flag.	Tonnage	Dimensions. Length	Breadth	Depth	Built. Where.	When.	By whom.	Date of Metalling	Owners or M[illegible]
	151	V	Gaulois *[illegible]*	W Bn	French	[illegible]	8[illegible]	[illegible]	[illegible]	[illegible]	1873	Malo	[illegible]	Ezano
	152		Gauloise (WRECKED) *[illegible]*	W Bk	French	[illegible]	12[illegible]	[illegible]	[illegible]	[illegible]	1866	Raym[illegible]		A. Léon [illegible]
	153	V	Gauloise *Denard*	WSr	French	12[illegible]	84·0	20·3	11·6	Dunkirk	1868	R. Deryckè		E. Didier
	154	V	Gauloise *Molinant*	WSr	French	114	88·3	19·7	10·5	St. Malo	1879	L. Trancheimer		P. Bertho
[illegible]	155		Gauntlet *Jones [illegible]*	WSr	British	100	86·5	21·1	11·8	Glasson Dock	1857	Simpson		J. Fisher & Sons
[illegible]	156	*	Gavenwood *Farquhar* ·87	WSr	British	130	91·8	22·3	11·5	Limehouse	1875	Whitehead	[illegible]	[illegible]
[illegible]	157		Gay Head *Jenny*	WBk	American	252	110·7	26·9	15·1	Mattapoisett, [Mass.	1877	J. Holmes		J. P. Knowles [illegible]
[illegible]	158	R	Gay Thomas (WRECKED) *Sloat*	WSr	American	320	131·7	32·6	10·6	Milford, Del.	1882	T. Carlisle		J. A. Curtis
[illegible]	159	*	Gayton (WRECKED) *McKay*	I Bk	British	936	197·2	32·8	19·1	Bristol	1884	C. Hill & Sons		Duncan, Fox & Co.
[illegible]	160		Gazell	WBg	British	250	98·5	24·0	11·6	Cochin	184[illegible]			Meera Mahomed [illegible] Abdool Rumeh [illegible]
	161	V	Gazella *Leal*	WBn	Portuguese	161	89·2	22·0	10·5	Lisbon	1883	J. A. Sampaio	[illegible]	Bensaude & Co.
[illegible]	162	*	Gazelle (ex George Knox) *Munro*	I Bk	British	336	143·5	27·1	13·4	Glasgow	1877	A. Stephen & Sons		W. S. Waterston
[illegible]	163	V	Gazelle *Starkey* ·87	W3mSr	British	[illegible]	12[illegible]·7	30·0	10·5	St. John, N.B.	1887	E. McGuiggan		G. F. Baird
[illegible]	164		Gazelle	WBg	British	[illegible]	10[illegible]	24·6	17·0	Cochin	1837			Union S.S. Co. [illegible]
	165		Gazelle	WSw	British	149	77·8	20·0	12·4	Shelburne, N.S.	1839			T. Cannell
19442	166		Gazelle *Paterson* 81-84	WSr	British	128	8[illegible]·5	22·2	11·4	Montrose	1857	J. Cochar		W. Morrison
[illegible]	167		Gazelle *Warner*	WSr	British	121	9[illegible]	24·7	9·2	Clare, N.S.	1878	J. H. Munroe		C. T. Warner
6537[illegible]	168	*	Gazelle *Butcher*	W K Barge	British	1[illegible]4	91·8	22·6	8·3	Ipswich	1877	Curtis		A. Beaumont & Co.
[illegible]	169	*	Gazelle	WK	British	87	79·3	20·7	10·4	Galmpton	1884	W. A. Gibbs		J. Thorpe
1035[illegible]	170		Gazelle	WBk	American	260	103·3	27·2	16·3	Mattapoisett, [Mass.	1852			Swift & Allen
	171	V	Gazelle *Barbelot* ·84	WBn	French	131	82·0	22·6	11·0	St. Malo	1874	P. Farnet	[illegible]	Sire fils
	172	V	Gazelle *Brockmoldt*	WBn	German	362	126·0	26·0	14·0	Kappeln	187[illegible]	C. Stoen	[illegible]	H. von Appen
	173	N	Gazelle *Lorentzen*	WBg	Norweg'n	229	107·8	25·4	13·4	Vegesack	1856			H. Heiberg
	174	V	Gazelle *Ehlerdt* ·87	WBk	Russian	403	131·2	26·9	15·7	[illegible]	1866	C. Diercks & Co.	[illegible]	A. J. Ehlerdt
	175	N	Gazelle *Ehlerdt* ·87	W3mSr	Russian	332	125·4	28·2	11·6	Kragerö	1884	J. Nielsen	[illegible]	A. T. Ehlerdt
	176	D	Gazellen ([illegible]) *Olsen*	WBg	Norweg'n	263	108·7	26·5	13·6	Grimstad	1875	Jens Tellefsen	[illegible]	P. Pedersen
	177	*DV	Gebroeders Smit *Mulder* [illegible]	WS	Dutch	1378	191·5	40·0	25·1	Alblasserdam	1875	J. Smit Czn.	[illegible]	J. Smit Czn.
	178	V	Gebruder *Strohschneider*	WSr	German	125	83·3	20·6	9·7	Nordloh	1867	K. H. Reil		J. Strohschneide[illegible]
	179	DV	Gebrüder *Deters*	WSr	German	101	81·3	20·3	9·6	Edewecht	1878	G. Dey		J. & H. Deters
[illegible]	180		Gebrüder Frindlin *[illegible]*	WSr	Russian	16[illegible]	91·5	27·5	12·0	Nogallen	1874	J. Stahl		M. Stahl
	181	D	Gedeh (ex [illegible]) *Sech Achmet Tagie*	CBg	Dutch	417	122·3	30·6	16·9	Amsterdam	1866	W. & A. H. Meursing	[illegible]	Tic Ing Tjai
	182		Geerdina	W Koff		117	84·6	18·5	10·0	Martenshoek	1857	H. Boerema		
	183		Geertruida (ex [illegible])	WS	Arabian	[illegible]	..	..	..	Carleton, N.B.	1862			
	184	DV	Geertruida (ex Agatha) *Hoffman* [illegible]	WBn	Dutch	17[illegible]	95·3	16·5	9·5	Nieuwendam	1866	W. & A. H. Meursing	[illegible]	J. J. van den Be[illegible]
	185		Geertruida Strating (now [illegible])	W3mSr	Danish	2[illegible]	1[illegible]5·6	25·2	12·2	Pekela, Str'ting	187[illegible]	Strating		
	186	GV	Geestemünde (ex Tim Whiffler, ex Madore [illegible]) *Gördes* ·88	WS	German	1098	189·0	37·1	23·1	N. Bedford, Mass.	1862	A. French	[illegible]	J. C. Meinke
	187		Gefina *Ekblom*	WBk	Russian	421	130·0	32·0	20·0	Åland	1862	Olmstrom		N. Sittkoff
	188	G	Gefion *Madsen* [illegible]	WSr	Danish	1[illegible]	75·2	[illegible]	10·2	Nyborg	1852	T. P. Petersen		W. Krusse
	189	DN	Gefion *Sannesen*	WBk	Norweg'n	[illegible]	1[illegible]1·3	31·3	18·5	Libau, Russia	1855			H. P. Sannesen
	190	N	Gefion *Barck*	WBk	Norweg'n	44[illegible]	120·9	31·0	17·9	Grimstad	1866	Ole Kittelsen		Hans Kallevig
	191	N	Gefion *Grimsrud* ·87	WBg	Norweg'n	[illegible]	112·9	29·3	16·2	Svelvig	1858			Enkefru [illegible]
	192	N	Gefion *Bache*	WBk	Norweg'n	[illegible]	1[illegible]4·9	28·0	16·0	Kragerö	1856			T. & L. Bache
	193	N	Gefion *Wig* ·88	WSr	Norweg'n	1[illegible]	99·3	21·7	10·1	Stavanger	1859		[illegible]	L. Larsen
	194	V	Gefion (ex [illegible]) *Borjesson*	WBk	Swedish	465	136·1	26·2	16·9	Brake	1869	D. Rogge	[illegible]	[illegible]
[illegible]	195	A	Gehon *Cosulich*	WBk	Austrian	840	149·9	32·6	23·3	Lussinpiccolo	1881	N. Martinolich		[illegible]
[illegible]	196	I	Gehon *Dodero*	WBk	Italian	498	140·4	30·2	20·9	Sestri Ponente	1866	G. Casanova		[illegible]
	197	DN	Geir *Kristensen*	WBg	Norweg'n	231	112·9	26·5	11·9	Arendal	1867	Dekke		C. A. Christensen
[illegible]	198	*	Gelert *Kneale* 88-88	I 3mSr	British	201	124·6	24·0	10·9	Amlwch	1887	W. Thomas & Sons		Sir R. W. Cam[illegible]
	199		Gelliboplo	W Bmb	Turkish	106	..	..	..	Gallipoli	1880			Marmarabilico [illegible]
[illegible]	200		Gelsomina *Treglia*	WBn	Italian	149	89·2	23·8	12·2	Castellam. di [Stab.	1868	G. Bonifacio	[illegible]	F. Miloloi & V. V[illegible]

*=Lloyd's Register; [1] D=Nederlandsche Vereeniging; G=Germanischer Lloyd; I=Registro Italiano; R=[illegible]
A=Veritas Austro-Ungarico; H=Veritas Ellenico; N=Norske Veritas; V=[illegible]

In the 4th Column, C=Composite (iron frame, wood planked); I=Iron; Stl=Steel; W=Wood.

The Dates of Metalling (Col. 11) are based on the records of the Books in which Vessels are respectively classed, or on the Reports of the S[illegible]

Further particulars of Vessels classed by Lloyd's Register will be found in Lloyd's Register Book and Lloyd's Yacht Reg[illegible]

No. dans la liste	Cote	Nom	Capitaines et dates d'entrée	Matériaux et Gréement	Pavillon	Tonnage	Longueur	Largeur	Creux	Construit: Où	Quand	Par qui	Dates du doublage	Armateurs et Directeurs-Gérants	Port d'Armement
201	I	Gelsomina	Virgilio	W Bn	Italian	184	79·1	22·3	13·	Sorrento	1858	M.Valerio	[illegible]	D.Aula	Trapani
202	✻	Gem	[illegible]	W Bk	British	452	138·8	29·1	17·6	Sunderland	1866	G.Short	[illegible]	T.S.Whitehead & Chambois	London
203	✻	Gem	Smith	W Bk	British	164	102·6	22·5	12·9	Fowey	1871	Butson	[illegible]	Mrs.M.H.Hayes	Fowey
204		Gem	[illegible]	W Bn	British	131	97·7	23·3	11·3	Wallace,N.S.	1862			A.Giovannetti	Sydney,C.B.
205	R	Gem	Lord	W Bkn	American	[illegible]	141·5	31·0	15·0	Bull's F'rry N.J.	186.			J.S.Emery & Co.	Boston,Mass.
206		Gem	[illegible] -72	W Bk	American	[illegible]	101·9	23·5	10·1	N.Haven Con.	1868			Miller,Bull & Co.	New York
207		Gen (ex [illegible])	[illegible] -87	W Bg	Norweg'n	133	84·3	21·6	11·0	Bristol	18 39/64			J.M.Feght	Drammen
208		Gemma (ex [illegible])		W Bg	British	[illegible]	120·4	26·7	13·5	Elsfleth	1868	J.Ahlers		C.Bethell & Co.	London
209	N	Gemma	Olsen	W Bk	Norweg'n	429	133·3	30·8	16·3	Stavanger	1873	R.Nielsen	[illegible]	Joh.Hanssen	Arendal
210	N	Gemma	Jensen 85-86	W Sr	Norweg'n	134	92·3	23·9	10·1	Sandefjord	1880			A.Olsen	Sandefjord
211	DI	Gemma C.	[illegible] -86	W Bk	Italian	749	166·9	33·0	22·7	Voltri	1877	G.Fava	[illegible]	F.Chiozza	Genoa
212	N	Gemsen	Adamsen -87	W Bg	Norweg'n	152	92·2	23·7	11·4	Christianssand	1869	S.I.Omundsen		N.F.Adamsen	Drammen
213	E	Gen. Adelbert Ames	[illegible] -87	W 3m Sr	American	452	141·0	33·6	12·0	Camden,Me.	1881	H.H.Bean		G.G.Jameson	Rockland,Me.
214		Général Aupick	Hattes	W Bg		157	84·3	21·3	11·8	Gravelines	1854	Decarpentry			
215		General Banks (ex Mary & [illegible])	Cole	W Sr	American	102	79·3	23·0	7·8	Damariscotta,Me.	18 38/[illegible]			J.S.Emery & Co.	Boston,Mass.
216		General Banning	Bjorn	W Sr	American	169	112·5	30·0	8·4	NavarroRivCal	1883	T.H.Petersen		R.G.Byxbee	San Francisco
217		General Baquedano (ex [illegible])	[illegible]	W Bk	Chilian	949	..	..	..	Richibucto	1861			A.Bafico	Valparaiso
218	V	Général Bergé	[illegible]	W Bn	French	256	112·8	28·2	10·9	Granville	1876		[illegible]	Ch.Leclère	Granville
219	V	Général Berthaut	[illegible] -88	W Bk	French	599	134·5	27·9	13·3 / 12·5	Honfleur	1884	Leviels	[illegible]	D.Auger	Havre
220	V	General Briaimont (ex [illegible])	[illegible]	W Bk	Chilian	865	167·3	29·8	21·0	Sestri Ponente	1874	Cadenacio Bros.	[illegible]	Meyer,Fonck & Valck	Valparaiso
221		General Butler	[illegible] [55-85]	W Bk	American	1113	182·8	37·0	23·5	Bath,Me.	1862			Puget Sound Commercial Co.	Pt.Townsend, Wash.
222	V	Général Chanzy	[illegible] 77-85	W Bk	French	427	131·2	28·3	13·4	La Seyne	1874	Chantiers de la Seyne	[illegible]	H.Auger ainé	Havre
223	V	General Domville	[illegible] 83-87	W S	British	1570	212·9	40·2	24·1	St.John,N.B.	1876	Cruickshank &	[illegible]	W.Thomson & Co.	St.John,N.B.
224	V	General Fairchild	[illegible] 87-87	W Bk	American	1428	203·1	38·8	24·3	Freeport,Me.	1874	Briggs&Cushing	[illegible]	C.Cushing	Portland,Me.
225	✻	General Gordon	[illegible] [86-88]	I S	British	1615	258·0	38·8	22·7	Liverpool	1886	R.&J.Evans&Co		L.Davies & Co.	Liverpool
226	✻	General Gordon	Russell [84-85]	W K	British	84	78·1	20·5	10·4	Brixham	1884	S.Dewdney&Sns		H.Russell	Hull
227		General Hall	Frisby	W Sr	American	111	87·0	26·0	7·7	Waldoboro'Me	1871			C.H.Simmons	Waldoboro',Me.
228		General Havelock		W Sr	British	121	80·4	21·8	10·9	Sunderland	1858	P.Foster		E.T.Fraser	Portsmouth
229		General Hereaux (ex Berlin)	[illegible] -87	W Bn	Dominic'n	252	118·5	26·0	14·3	Howden	1873	Banks	[illegible]	H.L.Banc	St.Domingo
230		General Howard		W Sr	American	111	90·5	26·1	7·9	Bath,Me.	1865			M.Claudio	Greenport,N.Y.
231		General Knox	Libby 81-86	W S	American	2141	251·1	42·4	29·1	Thomaston Me	1881	E.O'Brien	[illegible]	E.E.O'Brien	Thomaston,Me.
232		General Lee	Leech	W Bn	British	146	93·9	25·8	12·1	Rothsay,N.B.	1865	Maynes		W.Philpot	Dublin
233	RV	General McClellan	[illegible]	W S	American	1518	191·0	39·3	28·6	Thomaston Me	1862	S.Watts	[illegible]	J.W.Elwell	New York
234	✻	General Middleton	[illegible]	W S	British	67	68·5	21·5	8·4	Baie de Vin	1885	D.P.Watts		Sinclair & Co.	Chatham,N.B.
235	D✻	General Nott	James 77-77	I Bk	British	783	187·9	31·5	18·4	Sunderland	1877	Austin&Hunter		T.Morris	Aberystwith
236	✻	General Picton	Morgans	I S	British	1589	258·9	38·2	22·8	Port Glasgow	1883	Russell&Co.		Ship "General Picton" Co., Ld. (L.Davies & Co.)	London
237	✻	General Picton	Williams [illegible]	I Bk	British	1199	226·0	35·8	20·9	Sunderland	1884	S.P.Austin&Son		T.Morris	Aberystwith
238	N	General Prim (ex [illegible])	[illegible] -86	W Sr	Norweg'n	112	89·1	21·6	10·7	Christianssand	1869	Larsen		J.A.Arntzen	Christianssand
239	✻	General Roberts	Bailey 81-81	I m S	British	1914	274·3	40·2	24·2	Greenock	1884	Russell&Co.		Ship "General Roberts" Co., Ld. (L.Davies & Co.)	Liverpool
240	R	General S. B. Merwin	[illegible]	W Sr	American	759	[illegible]	[illegible]	[illegible]	W.Haven Con	1884	H.Sutton		H.Sutton	New Haven,Conn.
241		General Sheridan	[illegible]	W Sr	American	146	100·1	27·0	[illegible]	[illegible]	1865	D.Mack		T.Stewart	Hartford,Conn.
242	✻	General Stewart		W Sk	British	90	79·4	20·8	10·4	Goole	1885	Cottingham Brs.		Pickering & Haldane	Hull
243		General Urquiza	Bell	W Bg	Spanish	211	97·1	26·6	13·0	Blanes	1860			M.Barbera	Barcelona
244	G	Gen. Lieut. von Witzleben	[illegible] -85	W Sr	German	177	91·9	24·2	13·3	Wismar	1858	J.H.D.Barmann		J.C.Thormann	Wismar
245	I	Generosa (ex Arturo)	Corso	W Bg	Italian	112	69·5	19·7	8·9	Leghorn	1850		[illegible]	B.Corso	Messina
246	A	Generoso II.	[illegible] -88	W Bg	Italian	241	99·4	27·7	16·4	Pietra Ligure	1866	G.Acinelli		F.Degregori	Genoa
247	✻	Genesta	Waters 83-85	W K Barge	British	113	96·7	22·8	8·5	Harwich	1886	J.H.Vaux		W.S.Gane & others	Harwich
248	R	Genesta	Stewart -87	W Sr	British	110	80·6	26·8	8·0	Marysville,N.B.	1887	J.Gibson		J.Gibson	St.John,N.B.
249	✻	Genesta	G.S.Smith	W Sr	British	93	90·4	21·1	10·8	Plymouth	1886	W.H.Shilston	[illegible]	Falkland Islands Co.	London
250	✻	Genesta		W K	British	96	82·7	21·2	10·9	Brixham	1887	S.Dewdney&Sns		J.W.Haley	Hull

✻ = Lloyd's Register; D = Nederlandsche Vereeniging; G = Germanischer Lloyd; I = Registro Italiano; R = American Record; A = Veritas Austro-Ungarico; H = Veritas Ellenico; N = Norske Veritas; V = Bureau Veritas.

Dans la 4me Colonne, C = Composite (membrure en fer, bordé en bois); I = Fer; Stl = Acier; W = Bois.

Les Dates du doublage en métal (Col. 11) ressortent des livres dans lesquels les navires sont respectivement classés ou des rapports des Inspecteurs du Lloyd's Register.

Pour plus amples renseignements sur les navires classés par le Lloyd's Register, voir le Lloyd's Register Book et le Lloyd's Yacht Register.

1 Official No.	2 No. in Book	3 Registry in which classed	4 Ship's Name, Material and Rig. Master, and Dates of Appointment to present Owner's Service and to Ship. Ship's Late Name.	5 Flag	6 Net Tons	7 Dimensions: Length	Breadth	Depth	8 Built: Where	9 When	10 By whom	11 Metalling	12 Owners
[illegible]	251	✻	Genesta *Baskcomb 85* WSk	British	88	76·7	20·7	10·5	Elmsborn	1886	J. Kremer		J.H. Baskcomb
[illegible]	252		Geneva *Walter* WSr	British	107	86·4	24·6	9·3	Lunenburg,NS	1884	J. Mayner		W.H. Kaulback
[illegible]	253	R	Geneva *Gregory 81-84* WBkn	American	446	128·0	31·0	16·0	Bucksport,Me.	1873	Genn&Hill	Y 11.86	J.S. Emery & Co.
[illegible]	254		Genevie M. Tucker WBk	British	519	136·4	29·8	16·8	Portland,U.S.	1870			M. Kennedy
[illegible]	255	R	Genevieve *Dutch -87* W3mSr	American	544	146·3	34·4	15·5	Cooper's Point [N.J.	1882	S.W. Tilton		F. Smith & Co.
[illegible]	256		Genio *Claris* WBk	Italian	322	117·7	28·1	17·2	SestriPonente	1860	G. Casanova	Y 4.83	Cesare Claris
[illegible]	257		Genisis Theotokou WSr	Greek	263	100·7	23·9	16·2	Spezzia	1865			G.I. Coutsis
[illegible]	258		Genisis WSr	Greek	131	72·2	23·6	..	Syra				D. Philokas
[illegible]	259	A	Genisis Theotokou WSr	Greek	128	76·9	23·1	..	[illegible]	1886			S.A. Boyazis
[illegible]	260	✻	Genista *Cook 82-88* IS	British	1796	270·5	39·2	24·0	Liverpool	1882	W.H. Potter&Sn		Steinbach, Tinne & Co
	261		Genitore *Hellriegtle* WBk (WRECKED)	German	424	120·4	30·9	18·0	Venice	1854			F.W. Fischer
591	262		Genitore *Scotto* WBk (ex Michelangelo) (BROKEN UP)	Italian	325	106·4	28·5	18·0	Procida	1855		Y 4.83	A. Scotto
1724	263	V	Genitore S. *Scarpa* WBn	Italian	217	97·4	23·3	13·3	Chioggia	1885	G. Poli	Y 1.86	A. Scarpa
712	264		Genitori *Sivori* WBk	Italian	745	166·0	33·5	21·1	SestriPonente	1869	A. Briasco fu L.	Y 6.86	G.B. Ravenna
[illegible]	265	IV	Genitori F. *Olivari* WBk (ex Gentile)	Italian	806	165·0	33·5	23·6	Recco	1875	Rolla & Succo-[mano	Y 7.86	Casabona Vedova Ferrari
[illegible]	266	AV	Genitori Tarabochia WBk *Soppa -88*	Austrian	776	136·1	29·8	24·4	Lussinpiccolo	1878	N. Martinolich	Y 8.86	F.R. Tarabochia
	267	DV	Genius *Albertsen 79-79* WBn	Danish	106	90·6	24·7	11·3	Marstal	1879	H.J. Bager		H.J. Bager
	268	GV	Genius *Assing -87* WBn	German	245	108·9	25·5	13·1	Hammelwarden	1868	W. Hagens	Y 1.86	A.H. Arnold
2717	269	I	Gennaro *Ferrara -87* WBn	Italian	105	83·3	20·1	10·4	[illegible]	1887	M. Donnarumo		S. Ferrara
[illegible]	270		Gennisistis Theotokou WSr	Greek	120	77·1	22·3	..	Syra	1880			B.I. Pappas
72170	271	V	Genoa *Davidson* WBn	British	462	133·0	31·1	17·7	Newport, N.S.	1875	R.H. Burgess	Y 1.86	E.W. Churchill
	272		Genoveva (ex Valentina) WSr	Spanish	105	77·4	22·8	10·0	[illegible]	1852			J.B. Gallard
58092	273		Gensing WBn	British	123	93·4	22·1	10·7	[illegible]	1869			J. Gigina
	274		Gentil Lagunense *Silva e Irmão* WBn	Brazilian	117	..	..	..	Laguna	1868		Y 10.18	F.F. Martins
	275	V	Gentille *Le Breton -87* WSr	French	140	88·6	21·6	12·1	Dunkirk	1869	Vandestande		A. Navet
	276	GV	Georg *Schultz 85-* WS (ex [illegible] Hope, ex Star of Hope)	German	1195	188·6	37·4	23·0	[illegible]	1868	Gove&Choate	Y 7.86	[illegible]tenburg, Wendt & Co
	277	N	Georg *Henricksen -87* WBg (ex Georg Henrich)	Norweg'n	217	102·8	25·2	13·7	Apenrade	1851			Th. Christiansen
	278	G	Georg *Johannsen 87-87* WSr	Russian	126	82·7	25·0	10·0	Oesel	1882	A. Sep		J.M. Johannsen
	279	V	Georg Blohm *Gerloff [87-88* WBk (ex G.D. Wappaus)	German	466	141·1	28·5	16·1	Lübeck	1867	T.H. Evers	Y 8.86	J. Wimmer
	280		Georg Gustav *Kalning [80-83* W3mSr	Russian	270	113·5	29·2	13·0	Orrenhof	1876	J. Saul		G. Kalning
	281	G	Georg Nicolaus *Persson* WBk	Swedish	347	126·9	25·0	11·6	Altona	1863	J.J. Dircks	Y 3.83	L. Jönsson
	282	V	Georg Suppicich *Dale [73-73* WBk	German	443	125·6	29·5	16·5	Rostock	1873	E. Burchard&Co	Y 12.86	H. Bauer
	283		Georg & Adolf *Rang -75* WBg	German	298	108·2	27·4	15·8	Wismar	1857	J.C. Marén		J.C. Thormann
	284		Georg & Louise *Krauel* WBg (WRECKED) (ex Hans Christine) [-77	German	245	100·0	25·3	14·7	Rostock	1848	F.H.C. Dethloff		E. Burchard
411	285	Y	Georg & Luci *Schnohre* WCr	Russian	195	73·2	26·4	10·5	Mühlgraben	1882	A. Ferle		G.M. Beckmann
	286		Georg & Ludvig *Pedersen* WBk	Swedish	278	104·3	24·0	14·4	Burg Weser	1851	J.H. Besse		C.J. [illegible]
[illegible]	287	R	George *McDougall* WBk	British	943	174·4	36·5	22·4	[illegible], N.S.	1872	Putnam	Y 2.86	T.E. Kenny
[illegible]	288		George *Edgett -87* WBn	British	198	105·2	25·6	12·1	Merigomish,N.S	1864	Copeland	Y 5.86	J. Harvey
[illegible]	289		George *Thompson* WBkn (BROKEN UP)	British	192	103·4	24·1	12·6	P.E.Island	1868	Strong		J. Graham
[illegible]	290		George *Edgett -87* WBn	British	179	93·0	24·9	12·0	PtPiswick,N.S	1866			J.M. Huntington & Co
20432	291		George WBg	British	148	75·6	23·5	12·6	Sunderland	1836			Danby & Lennard
[illegible]	292		George *Safford* WSr	American	272	103·0	27·0	13·6	Lynchb'rgTex	1880	H. Gibbs		American Salt Co
	293	DGV	George *Nieuwenhuis -87* WBk (ex [illegible])	Dutch	769	155·1	33·2	20·3	Greenp'nt,NY	1863	E. Williams	Y 3.86	P.T. Nyhoff
	294	D	George *[illegible] Alphen 85-85* WSr (ex Folkje)	Dutch	128	110·2	23·7	11·7	Nieuwendam	1873		Y 1.86	P. Landberg&Zoon
	295		George *Harwardt 85-88* WS	German	653	129·9	33·8	22·4	Danzig	1858	J.W. Klawitter		Schoenberg & Dau[illegible]
[illegible]	296		George *Stahlberg 85-85* WBkn	Russian	266	104·5	23·3	11·0	Haynasch	1869	Stefanor		G. Weide
[illegible]	297		George A. Howes *Clark -84* W3mSr	American	124	95·3	27·4	7·0	May'sLanding, [N.J.	1882			J.A.J. Sheets
	298	R	George A. McFadden *Robinson* W3mSr	American	1047	194·6	40·4	19·5	Bath,Me.	1888	W.T. Donnell		W.T. Donnell
[illegible]	299	R	George A. Laury *Peckham 84* WSr	American	109	85·7	23·8	7·6	Jonesboro',Me.	1886	W.F. Tupper		E.D. Dobbin
[illegible]	300		George Aery *Rodgers* W3mSr	American	307	138·6	34·0	12·2	Jamestown,Va	1878	L. Matthews		S.E. Birdsall

✻ = Lloyd's Register; D = Nederlandsche Vereeniging; G = Germanischer Lloyd; I = Registro Italiano; R = A[illegible]
A = Veritas Austro-Ungarico; H = Veritas Ellenico; N = Norske Veritas; V [illegible]

In the 4th Column, C = Composite (iron frame, wood planked); I = Iron; Stl = Steel; W = Wood.

The Dates of Metalling (Col. 11) are based on the records of the Books in which Vessels are respectively classed, or on the Reports of the Surveyors [illegible]

Further particulars of Vessels classed by Lloyd's Register will be found in Lloyd's Register Book and Lloyd's Yacht [illegible]

No.	Classe	Nom. Capitaine. Ancien Nom.	Matériaux et Gréement	Pavillon	Tonnage	Longueur	Largeur	Creux	Où	Quand	Par qui	Date du doublage	Armateurs ou Directeurs-Gérants	Port d'Armement
301	✱	George B. Balfour	W&I Bk	British	191	1[illegible]	24·1	10·0	Carrickfergus	1885	P. Rodgers		J. Fisher & Sons	Barrow
302	R	George B. Doane	W Bk	British	942	168·0	33·6	22·3	Plymouth, N.S.	1872	J. Sims		A.C. Robbins	Yarmouth, N.S.
303		George B. Ferguson	W Sr	American	1[illegible]	96·1	26·7	7·7	Belfast, Me.	1872			G.B. Ferguson	Belfast, Me.
304	✱	George Bartram	W Bg	British	239	103·1	26·5	16·2	Hylton	1865	Bartram		J. Paul & Co.	[illegible]
305	R	George Bird *Gray*	W 3m Sr	American	[illegible]	[illegible]·8	28·0	9·3	Rockland, Me.	1883	W. Spear		W. Spear	New York
306	✱	George Booth *Forest*	W Sr	British	[illegible]	123·5	26·6	12·0	Sunderland	1872	W. Pickersgill	V [illegible]	W.C. Jarvis	Liverpool
307		George C. A. Travers	W Sr	American	1[illegible]	[illegible]·1	27·0	6·9	Baltimore	1873			R.H. Mitchell	Baltimore
308	V	George C. Perkins	W Bkn	American	[illegible]	142·0	33·5	11·8	Coos Bay, Oreg	1880	Reed		H.A. Ackerman	San Francisco
309	✱	George Casson *Jones*	W Bn	British	134	92·3	22·3	12·3	Portmadoc	1863	Pierce		W. Prichard & Co.	Carnarvon
310	V	George Curtis *Sproul*	W S	American	1746	231·0	42·4	25·2	Waldoboro' Me	1884	A.R. Reed	V [illegible]	A.R. Reed	New York
311		George D. Fullerton	W Sr	British	1?7	102·6	28·5	11·0	Parrsboro', N.S.	1874	Roberts		B. Jeffers	Cork
312		George D. Perry	W Sr	American	11?	88·1	24·5	8·3	Machias, Me.	1873	J. Shaw		J. Inglee	Machias, Me.
313	RV	George Davis *Faulkner*	W Bk	British	643	150·2	31·0	17·5	Windsor, N.S.	1877	S. Dimock	V [illegible]	S. Dimock	Windsor, N.S.
314	V	George E. Corbitt	W Bk	British	4??	14?·7	32·8	13·0	Digby, N.S.	1874	J.H. Hughes	V [illegible]	J.A. McBride	Annapolis, N.S.
315		George E. Dale *Bartlett*	W Bn	American	218	1?1·?	27·6	9·8	Brewer, Me.	1867		V [illegible]	A.D. Pierce	Bangor, Me.
316	R	George E. Prescott	W Sr	American	13?	90·0	24·0	7·9	Camden, Me.	18 48/84			G.M. Brainard	Rockland, Me.
317		George F. *King*	W Bn	British	16?	102·0	25·3	12·1	Bear River, N.S.	1863	Miller		S.N. Robinson	Dublin
318		George F. Edmunds	Sr	American	112	96·0	25·7	10·3	Essex, Mass.	1887			W.G. Pool	Gloucester, Mas.
319	V	George F. Manson *Morse*	W S	American	[illegible]	206·1	39·2	23·9	Bath, Me.	1875	A. Hathorn	V [illegible]		San Francisco
320		George Fox *Parsons*	W Sr	British	1??	93·0	25·0	15·0	Sunderland	1856			W. Milburn	Sunderland
321	R	George G. Green *Burton*	W 3m Sr	American	6??	1?6·8	35·2	16·9	Camden, N.J.	1879	Morris & Mathis		H.L. Benton	Philadelphia
322		George Gilroy *McGill*	W Bk	British	1??1	182·6	36·0	23·5	Medford, U.S.	1862		V [illegible]	Craig, Scott & Co.	Greenock
323		George H. Ames *Marshall*	W 3m Sr	American	421	[illegible]	[illegible]	11·0	Waldoboro' Me	1882	Stover, A.		J. Bliss	Boston, Mass.
324	✱	George H. Douglass	W Bn	Hawaiian	248	115·9	21·1	12·8	P.E. Island	1884	A. McMillan	V [illegible]	J.D. Spreckles & Co.	Honolulu
325		George H. Mills *Tillotson*	W Sr	American	281	117·6	31·0	9·0	Pt Jeff'rson, NY	1867	Bayles & Son		J.M. Cross	Providence, R.I.
326	✱	George Hall	W Sk	British	84	77·0	20·4	10·6	Hull	1884	W. & J. McCann		H. Cook	Hull
327	✱	George Harry Mudd	W Sk	British	77	76·5	20·0	10·3	Elmshorn	1885	J. Kremer		T. Bowne	Grimsby
328		George Henry *Preston*	W Bk	British	48?	134·?	28·5	16·4	Shelburne, N.S.	1865			F.W.L. Popham	Cork
329	R	George K. Hatch *Condon*	W 3m Sr	American	350	120·5	31·1	10·1	Pembroke, Me.	1874	A. Frost	V [illegible]	C.E. & B.H. Fabens	Salem, Mass.
330		George L. Fessenden *Weeks*	W 3m Sr	American	334	133·6	33·2	14·5	Belfast, Me.	1874			P.M. Wheaton	Bridgeton, N.J.
331		George Lamb	W Sr	British	1?8	[illegible]	27·3	9·3	St. Andrews, N.B.	1882	J. Starkie		A.W. Adams	St. Andrew's, N.B.
332		George Linck *Niemann*	W S	German	67?	15?·?	32·8	22·7	Danzig	1870	J.W. Klawitter		[illegible]	Danzig
333		George M. Adams *Saunders*	W 3m Sr	American	6??	167·5	35·3	13·1	Bath, Me.	1881	Adams & Hitchcock		Adams & Hitchcock	Bath, Me.
334	R	George Moon *Wolfe*	W Bk	American	881	172·6	[illegible]	20·1	Mystic, Conn.	1874	Hill & Grinnell	V [illegible]	S. Fish	New York
335	R	George Moulton, Jr.	W 3m S	American	387	138·2	33·3	11·9	Bath, Me.	1881	B.W. & H.F. Morse	V [illegible]	B.W. & H.F. Morse	Bath, Me.
336		George Nevinger *Smith*	W Sr	American	238	117·6	30·0	9·0	Milton, Del.	1865			W. Perkins	Bath, Me.
337		George Noble *Anthony*	W 3m Sr	British	25?	121·?	25·3	14·0	Fraserburgh	1869	Webster		Uh Chong	Sydney, N.S.W.
338	R	George P. Hallock *Edwards*	W 3m Sr	American	307	132·0	33·0	9·6	Port Richmond, N.Y.	1874			B. Edmunds	New York
339		George P. Trigg	W Sr	American	117	83·2	24·7	7·4	E. Boston, Mass.	1866			E.A. Holmes	Gloucester, Mass.
340	✱	George Peake *Matthias*	W Bk	British	554	150·?	30·3	17·5	P.E. Island	1874	Peake	V [illegible]	R. Farrell	Cork
341		George Prescott	W Sr	British	110	77·6	22·7	9·6	Granville, N.S.	1865				Montego By. Jam.
342	R	George R. Congdon *Boyles*	W 3m Sr	American	43?	133·0	32·4	14·8	Pt Jeff'rson, NY	1879	G.R. Mather		G.L. Hammond	New York
343	V	George R. Skolfield	W S	American	1645	232·1	39·9	24·6	Brunswick, Me	1885	G.R. Skolfield	V [illegible]	Skolfield Brothers	Portland, Me.
344		George R. Vreeland *Blake*	W 3m Sr	American	4??	128·2	33·4	16·0	W. Haven, Con	1873			A. Van Horn	New York
345		George Reynolds *Evans*	W Bg	British	188	10?·1	24·1	14·3	Bristol	1851	Peter Symons		J. Hood	West Hartlepool
346	RV	George S. Homer *Wieland*	W Bk	American	1267	204·9	39·6	20·9	Bath, Me.	1882	Goss, Sawyer & Packard	V [illegible]	W.H. Besse	Portland, Conn
347	R	George S. Marts *Adams*	W 3m Sr	American	42?	130·?	33·1	14·8	Dorchester, NJ	1874	Brainer & Champion		Major D. Henderson	Bridgeton, N.J.
348		George S. Tarbell *Higgins*	W 3m Sr	American	4??	13?·1	32·2	15·0	Bath, Me.	1873		V 10.83	L.H. Higgins	Wellfleet, Mass.
349		George Shotton	C Bk	British	549	155·2	28·2	18·1	Sunderland	1866	Moore	V [illegible]	British & Continental African Co. Ld.	Liverpool
350	V	George Skolfield *Ball*	W S	American	1314	187·2	38·2	23·9	Brunswick, Me	1870	Skolfield Bros.	V [illegible]	I.L. Skolfield & Co.	Portland, Me.

✱ = Lloyd's Register; D = Nederlandsche Vereeniging; G = Germanischer Lloyd; I = Registro Italiano; R = American Record; A = Veritas Austro-Ungarico; H = Veritas Ellenico; N = Norske Veritas; V = Bureau Veritas.

Dans la 4me Colonne, C = Composite (membrure en fer, bordé en bois); I = Fer; Stl = Acier; W = Bois.

Les Dates du doublage en métal (Col. 11) ressortent des livres dans lesquels les navires sont respectivement classés ou des rapports des Inspecteurs du Lloyd's Register.

Pour plus amples renseignements sur les navires classés par le Lloyd's Register, voir le Lloyd's Register Book et le Lloyd's Yacht Register.

1 Official No.	2 No. in Book	3 Registry in which classed	4 Ship's Name. Material and Rig. Master, and Dates of Appointment to present Owner's Service and to Ship. Ship's Late Name.	5 Flag.	6 Net Tonnage.	7 Dimensions. Length	Breadth	Depth	8 Built. Where.	9 When	10 By whom.	11 Date of Metalling.	12 Owners or Mana[gers]
85632	351	R V	George Stetson *Wood* -87 W S	American	1780	232·9	41·3	26·3	Bath, Me.	1880	A. Hathorn	[illegible]	G. Stetson
90514	352	V	George T. Hay *Spicer* -88 W S	British	1647	235·6	42·0	23·6	Parrsboro' N.S.	1887	Spencer's Island Co	[illegible]	George D. Spicer
85736	353	R	George Taulane, Jr. *Little* -86 W 3m Sr	American	442	135·0	35·2	12·2	Milton, Del.	1882	D. H. Atkins		J. F. Fisher
48686	354		George Thompson *Benson* W S	British	1128	209·1	36·2	22·5	Aberdeen	1865	W. Hood & Co.	[illegible]	T. Cowlishaw
75056	355		George Thornton (LOST) W Sr	British	182	111·0	25·3	8·8	Jervis Bay, N.S.W.	1880	W. Peverley		J. H. Young
85394	356	V	George V. Jordan *Byrne* W 3m Sr	American	697	158·6	35·0	16·7	Kennebunk Me	1874	D. Clark	[illegible]	B. F. Hamilton
85267	357		George W. Anderson *Cole* W 3m Sr	American	213	115·4	32·2	9·0	Newburgh N Y	1873			J. A. Curtis
85371	358		George W. Churchman *Cole* -87 W 3m Sr	American	268	120·6	31·1	9·7	Gr'nsboro' Md	1874			J. W. Hall, jun.
10079	359		George W. Cummings W Sr	American	131	89·5	25·6	8·2	Leipsic, Del.	1853			M. C. Hurd
85807	360	R	George W. Fenimore *Rogers* -86 W 3m Sr	American	603	157·0	35·0	17·0	Milton, Del.	1883	D. H. Atkins		C. T. Megee
85274	361		George W. Jewett *McKown* -87 W 3m Sr	American	267	121·6	30·2	9·6	Bath, Me.	1873			F. Cushing
91674	362	R	George W. Lochner *Work* -87 W 3m Sr	British	429	136·8	33·8	10·2	Milton, Del.	1877	J. L. Black & Bro.	[illegible]	John Alfred
10311	363		George W. Snow *Plummer* W Sr	American	109	85·0	24·6	6·5	Bangor, Me.	1850			Louisa N. Plum[mer]
85471	364		George W. Sweeney *Hewitt* W Bkn	American	646	150·6	35·5	16·8	May's Landing, N.J.	1876	N. V. Lane	[illegible]	G. Taulane
85165	365		George W. Whitford *Alexander* -85 W Sr	American	182	102·0	26·2	12·6	Kennebunk Me	1871			J. E. Armstrong
85232	366	R	George Walker *Mitchell* -85 W 3m Sr	American	353	121·0	29·3	10·3	Machias, Me.	1872			Miller & Hough[ton]
	367	D G V	George Washington (ex Glendower) *Probst* 72- W Bk	German	1185	188·6	38·4	23·9	Newburyport	1860	T. Currier	[illegible]	Stolenburg, Wendt
68017	368	✻	George Watson *Rear* -88 W Bk	British	770	181·0	31·6	19·0	Sunderland	1875	N. Gibbon	[illegible]	G. Watson & Son
10102	369	R	George & Albert (WRECKED) *Wentworth* W Sr	American	115	88·7	26·2	6·7	Belfast, Me.	1867			S. Stearns & Co.
69789	370	✻	George and Mary *Wheeler* W Sr	British	91	83·2	20·6	10·0	Jersey	1875	LeSueur	[illegible]	J. Wright
10749	371		George & Mary W Bk	American	100	79·0	22·5	10·5	Dartm'th Mass	1852			J. McCullough
67781	372		George & Susan *Kearon* W Sr	British	100	81·9	22·0	10·4	Arklow	1875	Hagan		E. Kearon
68216	373		George 4th *Jones* 76-77 W Sr	British	111	90·2	21·6	10·4	Ulverstone	1873	White		W. Postlethwaite
	374		Georges W Bk	French	376	125·5	28·1	17·5	Sunderland	1868	Reay		Diffon & fils
	375		Georges *Kergalant* W Bk	French	221	97·4	22·3	13·0	St. Servan	1862	Lenormand	[illegible]	E. Famin
	376		Georges *Camus* W Bg	French	152	83·0	21·6	12·2	Dieppe	1866	P. Corne	[illegible]	Pourpoint & fils
	377	V	Georges Paul *Broche* W Bk	French	204	99·4	26·9	13·7	Fecamp	1866			Vve Livois
	378		Georges René *Lepretre* -88 W 3m Sr	French	171	..	..	..	Fecamp	1879			Vve Valin
	379	V	Georges Ville *Ricordel* [illegible] W Bk	French	328	113·8	23·9	14·6	Nantes	1873	Jollet & Babin	[illegible]	Desbois
20391	380	✻	Georgetown *Beales* 83-84 W S	British	414	135·0	26·5	16·6	Sunderland	1858	R. Thompson & Sons	[illegible]	Fawcett & Wau[gh]
85045	381		Georgetta Lawrence *Hawkins* W 3m Sr	American	299	122·6	31·7	15·1	Setauket, N.Y.	1869	Hand		Evans, Ball & Co
85292	382	R	Georgia *Coffin* W 3m Sr	American	333	122·0	30·0	13·0	Harrington, Me	1873		[illegible]	B. T. Thurlow & Sons
487	383		Georgia W Bn	Greek	238	90·2	27·9	..	[illegible]	1885	A. Livadaras		P. Prinzos
85095	384	R	Georgia B. McFarland *Strong* -82 W 3m Sr	American	254	111·8	29·5	10·1	Thomaston Me	1869	Walker & Dunn		Dunn & Elliot
2505	385	R	Georgia Berry (ex Hawthorn) *Ginn* W Sr	American	252	109·3	27·4	13·3	Westbrook Me	1866			A. F. Crockett &
77552	386		Georgiana *Burke* W Sp	British	123	101·8	22·8	7·2	St. Francois du Lac	1877	Desmarais		F. Desmarais
10437	387		Georgiana F. Geery *Conklin* W Bn	American	355	108·5	28·4	16·7	Norwalk, Conn	1861			S. Conklin
83203	388	R	Georgie *Neville* -88 W Bn	British	229	107·0	29·0	11·5	Walton, N.S.	1881	W. S. Vaughan	[illegible]	D. H. Morris
85243	389		Georgie Clark *Bartlett* 69-73 W 3m Sr	American	347	123·9	31·2	13·0	Cooper's Point, N.J.	1872			E. S. Bartlett
85227	390		Georgie D. Loud *Murphy* W Sr	American	167	101·6	28·0	8·0	Waldoboro' Me	1872			J. Whitehouse
85772	391	R	Georgie D. Paine *Paine* [illegible] W Sr	American	166	100·6	25·6	10·0	Bath, Me.	1883	T. Hagan & Co.		W. Matheson
85708	392	R	Georgie L. Dickson *Harding* W 3m Sr	American	556	155·5	34·7	13·6	E. Boston	1883	J. M. Brooks		A. P. Friend
85766	393	R	Georgie L. Drake *Willey* W 3m Sr	American	443	143·3	35·1	12·3	Bath, Me.	1883	W. Rogers	[illegible]	L. S. Wiley
85288	394	R	Georgie Shepard *Rich* -72 W 3m Sr	American	557	147·5	33·4	16·0	Bath, Me.	1873	Deering & D'nn B		T. L. Mayo
85323	395	R	Georgietta *Kasten* W Bk	American	436	138·5	30·5	12·1	Millbridge, Me	1874		[illegible]	G. P. Ockershausen
76302	396	✻	Georgina *Melhuish* 74-81 I Bk	British	628	186·2	29·8	17·6	W. Hartlepool	1875	W. Gray & Co		B. M. Hennessy
57272	397		Georgina W Bn	British	104	84·8	23·5	10·0	Port Hawkesbury, N.S.	1867			R. I. Hart
	398		Georgina W Bk	Brazilian	231	..	..	..					A. S. Ribeiro
	399		Georgine W Bn	Swedish	155	85·4	22·1	11·9	Schulan	1857	J. J. Behrens		J. Jensen
	400	N	Georgio Avassioti *Larsen* -87 W Bg	Norweg'n	319	107·7	29·8	16·9	Fiume	1854		[illegible]	L. A. Larsen

✻ = Lloyd's Register; D = Nederlandsche Vereeniging; G = Germanischer Lloyd; I = Registro Italiano; A = Veritas Austro-Ungarico; H = Veritas Ellenico; N = Norske Veritas;

In the 4th Column, C = Composite (iron frame, wood planked); I = Iron; Stl = Steel; W = Wood.

The Dates of Metalling (Col. 11) are based on the records of the Books in which Vessels are respectively classed, or on the Reports of the …

Further particulars of Vessels classed by Lloyd's Register will be found in Lloyd's Register Book and Lloyd's Yacht …

No.	Cl.	Nom. Capitaines et dates d'entrée au service des Armateurs actuels et à bord du navire. Ancien Nom.	Matériel et Gréement	Pavillon	Tonnage	Longueur	Largeur	Creux	Construit: Où	Quand	Par qui	Date du doublage en métal	Armateurs ou Directeurs-Gérants	Port d'Armement
401	A	Georgios *Livano*	WBr	Greek	[illegible]	..	..	..	Syra	1887	A.Livadara		E.Livanó	Syra
402	AH	Georgios (ex Astoria)	WBk	Greek	[illegible]	126·[illegible]	[illegible]	2·17	[illegible]	1869		Y [illegible]	C.Gandilis	Syra
403		Georgios *Ruben*	WBn	[illegible]	[illegible]	..	..	..	Syra	1866	A.Livaderas		M.Ruben	Syra
404		Georgios	WBg	[illegible]	[illegible]	..	..	..					S.G.Kirizis	Poros
405		Georgios *Russos*	WBn	[illegible]	[illegible]	..	..	..	Syra	1868	P.Bounals	Y [illegible]	N.Russo	Syra
406	H	Georgios *Toilimingras*	WBn	[illegible]	[illegible]	104·7	25·[illegible]	..	[illegible]	1869			M.Toilimingras	Zante
407		Georgios	WBg	[illegible]	[illegible]	95·8	26·6	..	Syra	1869			N.G.Roussos	Syra
408		Georgios (ex Teodoro)	WBk	[illegible]	244	98·4	25·3	..	Leghorn	1856			E.Vlassopulos	Ithaca
409	A	Georgios	WBn		[illegible]	96·1	21·6	..	Casos	1882				
410	H	Georgios *E.Bouconvalas*	WBg	Greek	194	..	..	..	Syra	1871			E.Bouconvalas	Zante
411	A	Georgios	WSr	Greek	183	94·2	24·8	13·[illegible]	[illegible]	1881	J.Canatos		L.Lipardis	Galaxidi
412		Georgios	WBg	Greek	180	[illegible]	24·[illegible]	15·[illegible]	Syra	1869	P.Buoni	Y [illegible]	D.Ch.P.Misthos	Syra
413		Georgios	WBg	Turkish	175	..	..	..	Chios	1867			N.Cacaroucas	Chios
414		Georgios	WBg	Turkish	124	..	..	..		1863			G.Inglessis	Samos
415	GV	Georgios Beltzos (ex [illegible]) *Caldima*	WBk	Greek	656	160·7	33·8	18·7	Apenrade	1874	G.Raben	Y [illegible]	Geo. Beltzos et fils	Syra
416	H	Georgios Filaretos	WBk	Greek	504	..	..	..	Trieste	1868			Filaretos	
417		Georgios I.	WBg	Greek	198	..	..	..	Galaxidi	1865			E.&N.M.Catsouli	Patras
418	✻	Geraldine *Walsh*	WS	British	1168	196·3	38·1	23·0	Quebec	1879	Charland	Y [illegible]	Cox Bros.	Waterford
419	✻	Geraldine *Mowat*	W3mSr	British	254	120·1	24·0	12·9	Salcombe	1873	Vivian	Y [illegible]	W.G.Crinks	Salcombe
420	R	Geraldine *Jamieson 83-86*	WSr	British	140	92·5	25·0	10·2	Shelburne,NS.	1882	S.Muir	Y [illegible]	A.N.Whitman	Halifax,N.S.
421	✻	Geraldine *Pierce -86*	WBn	British	132	89·2	23·0	12·2	Barmouth	1860	Jones		L.Hughes & Son	Carnarvon
422	RV	Gerard C.Tobey *[illegible]*	WS	American	1390	208·7	39·1	23·6	Bath,Me.	1878	Goss,Sawyer, [&Packard	Y [illegible]	W.H.Besse	New Bedford, Mass.
423		Gerazeh	[illegible]	Turkish	160	..	..	..	Gerazeh	1883			Serafenzi zadeh Kadri Yazedji & Co.	Gerazeh
424	GV	Gerd Heye *Ladewig 72-81*	WBk	German	576	140·7	31·6	17·4	Hammelwrdn	1881	J.Deetjen	Y [illegible]	A.Schiff	Elsfleth
425		Gerda *Neill*	WBg	British	138	102·2	22·2	11·9	P.E.Island	1855			J.Neill	Dublin
426		Gerda (ex Lonetar) *Jans*	WBk	Danish	486	111·1	29·7	16·6	Nystad	1865	HH Kjüldström		Ludvig Olsen	Copenhagen
427	V	Gerda *Christensen -86*	WSr	Danish	111	89·7	21·0	8·4	Timmernab-	1886	C.Nilsson		H.Christensen	Marstal
428	DGV	Gerda *Meyer 86-75*	WBk	German	735	161·4	34·6	19·6	Hammelwrdn	1885	J.F.Strenge	Y [illegible]	G.Eilers	Brake
429		Gerda *Johannesen 78-83*	WBk	Norweg'n	489	132·9	33·9	18·6	Gothenburg	1848	Petterson		O.P.Olsen	Dröbak
430	N	Gerda *Olsen -87*	WBk	Norweg'n	407	136·8	30·6	15·1	Drammen	1869	J.Jörgensen	Y [illegible]	N.M.Gram	Drammen
431	G	Gerda *Henrikson -7* (ex Concordia)	WBk	Swedish	375	117·9	26·6	15·6	Rügenwalde	18 64/77	C.F.Zieske		J.D.Björkegren	Cimbrishamn
432		Gerda *Söderlund -86* (ex Betsy ex Evening Star)	WBk	Swedish	297	101·2	24·4	11·3	Harpswell C.S	1859			J.H.Söderlund	Karlshamn
433		Gerda *Mattson -88*	WBg	Swedish	221	110·2	24·9	11·1	Gefle	1869	O.A.Brodin		B.Olsson	Morlanda, Bohsln.
434	G	Gerda *Nilsson 59-66*	WBg	Swedish	219	113·5	24·5	11·8	Tillingenabbn	1867	O.L.Nordström		O.Nilsson	Wiken
435		Gerda *Johansson*	WSr	Swedish	16[illegible]	89·6	22·8	11·9	Krogarne	1870	Samuelsson		J.A.Samuelsson	Mollösund
436		Gerda *Svensson*	WSr	Swedish	138	88·9	21·1	10·0	Finland	1870			C.J.Hellberg	Qville,Bohsln.
437		Gerente *Belen*	WSr	Brazilian	112	..	..	..	[illegible]	1870		Y [illegible]	J.J.de Almeida Silva	Rio de Janeiro
438	G	Gerhard *Läken 87-87*	WSr	German	143	91·7	22·2	11·0	Papenburg	1859	H.W.Meyer		B.Freeriks	Papenburg
439	V	Gerhard Erdwin (WRECKED) *Heerhs -88*	WBn	German	218	99·1	25·2	12·7	Brake	1869	F.Nicolai	Y [illegible]	U.C.Koch	Brake
440		Gerhard & Adolph	WBg		319	10[illegible]	27·7	15·3	Rostock	1869	E.Burchard&Co.			
441	G	Gerhardina *Hemmen 82*	WBn	German	212	102·[illegible]	22·6	12·0	Emden	1858	M.Schmidt		S.Fehr	Oldersum
442	V	Gerhardine (WRECKED) *Thorsen -85*	WBk	German	393	12[illegible]	26·[illegible]	12·7	Neurönnebeck	1869	C.D.Oltmann	Y 12.87	A.Grapengeter&Sohn	Hamburg
443	V	Gerhardus *Bruijs 81-87* (ex Gerhard, ex Albert Edward)	WBk	Dutch	937	165·8	35·4	22·0	Newburyport	1860	E.Manson	Y [illegible]	Ch. de Groof & Co.	Flushing
444	A	Germana Antonio *Lupetich*	WBg	Austrian	371	119·7	26·7	16·[illegible]	Cherso	1871	A.Rizzi	Y 10.86	G.Canarich	Cherso
445		Germania *Kearon, Jr. -87*	W3mSr	British	121	104·5	22·0	11·2	Memel	1871	R.Banisch		G.Kearon	Dublin
446		Germania *Marshall -87*	WSr	British	113	87·8	21·8	10·1	Geestemünde	1869	TCTecklenberg		J.Marshall (W.H.Leask)	Peterhead
447		Germania *Griffin 83-88*	WBk	American	958	180·0	36·0	22·0	Portsm'th,N.H.	1850			P.B.Cornwall	San Francisco
448	G	Germania *Sauerland 82-83*	IBk	German	826	180·1	31·8	19·8	Glasgow	1874	A.Stephen&Sns.		D.H.Watjen & Co.	Bremen
449	G	Germania *Windhorst 82-88* (ex Cumberland)	WBk	German	817	154·5	33·9	21·4	Baltimore	1852		Y [illegible]	D.Heinrichs	Bremerhaven

✻=Lloyd's Register; D=Nederlandsche Vereeniging; G=Germanischer Lloyd; I=Registro Italiano; R=American Record; A=Veritas Austro-Ungarico; H=Veritas Ellenico; N=Norske Veritas; V=Bureau Veritas.

Dans la 4me Colonne, C=Composite (membrure en fer, bordé en bois); I=Fer; Stl=Acier; W=Bois.

Dates du doublage en métal (Col. 11) ressortent des livres dans lesquels les navires sont respectivement classés ou des rapports des Inspecteurs du Lloyd's Register.

Pour plus amples renseignements sur les navires classés par le Lloyd's Register, voir le Lloyd's Register Book et le Lloyd's Yacht Register.

Official No.	No. in Book	Registry in which classed	Ship's Name. Master, and Dates of Appointment to present Owner's Service and to Ship. Ship's Late Name.	Material and Rig	Flag	Net Tonnage	Dimensions: Length	Breadth	Depth	Built: Where	When	By whom	Date of Metalling	Owners or Managers
	451		Germania Aschendorff 79-83 (BROKEN UP)	W Bk	German	360	112·5	28·6	16·1	Danzig	1837	J.W.Klawitter		J.F.Giesebrecht
	452	V	Germania Kühl 86-	W 3m Sr	German	179	103·3	22·0	11·1	Memel	1871	R.Banisch	[illegible]	C.L.E.von M[illegible]
	453	G	Germania Suhrbier -80	W Bn	German	158	94·6	22·6	11·9	Wismar	1874	C.Barmann		J.C.Thormann
	454	V	Germanic Samuelsen -87	W S	Norweg'n	1296	198·9	38·7	23·9	Quebec	1878	P.Baldwin	[illegible]	N.Eyde
50	455		Geroghiannis (ex Adelphoi)	W Bk	Greek	273	98·8	26·6	..	Cossed	1877			E.M.Papa
267	456	A	Gerolama (ex Elisde) Scarpa -86	W Bn	Italian	134	88·6	22·2	12·5	Lussinpiccolo	1881	Martinolich	[illegible]	F[lli] Ballarin
1401	457	I V	Gerolamo Mortola Olivari -86	W Bk	Italian	491	134·8	31·4	19·6	Varazze	1865		[illegible]	G.Mortola
61	458		Geron Gionis	W Bk	Greek	231	..	..	..	Naples	1849			J.Gionis
	459		Geron Nicolaos	W Bg	Greek	211	..	..	..	Syra	1868			N.Stamatopoulos
1084	460	I	Geronima Bianchi (ex Elvetica) Chiesa	W Bk	Italian	646	150·9	31·5	21·1	Sestri Ponente	1876	G.Peragallo	[illegible]	M.Bianchi
2654	461	I	Geronima Madre Lertora	W Bk	Italian	725	156·1	32·8	22·3	Sampierdarena	1875	B.Frassinetti	[illegible]	Eredi di P.Brignati
	462		Gerrassimo	W Bg	Turkish	204	..	..	..	Greece	1870			Costanti Hizwidi
	463	D	Gerredina Wilhelmina (ex [illegible])	W 3m Sr	Dutch	377	111·5	28·5	18·0	Amsterdam	1853	Meyjes & Sons	[illegible]	G.G.Albers
	464		Gerrit (ex Pei Ho) (BROKEN UP)	W Bk	Dutch	262	117·1	25·0	..	Sonderburg	1864	H.H.Hauschildt	[illegible]	Said Aleel bin Ab[illegible]
	465	D	Gerrit & Willem Vliss (ex Hoogly) 68-75	W S	Dutch	641	151·5	35·1	21·8	Kinderdyk	1856			J. & J.Vinke
	466	✻	Gers Thoreux 85-	I Bk	French	723	178·0	30·0	19·0	Stockton	1870	M.Pearse & Co.		Ant.Dom.Bordes & [illegible]
85870	467		Gertie M. Rickerson Andrews	W Sr	American	208	..	..	..	Chatham,Con.	1884	J.H.Woodhouse		J.H.Wood[illegible]
16780	468		Gertrude	W Bk	British	950	171·4	33·6	22·4	Shields	1856	MiddleDockCo.		T.M·Veagh
69410	469	✻D	Gertrude France 77-77	W Bk	British	484	142·5	29·0	18·0	Bristol	1877	Hill	[illegible]	C.Hill & Sons
83038	470		Gertrude Mathews	W Bn	British	293	115·8	26·6	12·4	Mahone Bay, N.S.	1883	J.H.Zwicker	[illegible]	W.A.Conrod
49238	471		Gertrude (ex Amatico)	W Sr	British	199	104·0	26·0	9·4					T.B.Manley
31921	472		Gertrude	W Bg	British	117	76·0	21·4	11·7	Newfoundl'nd	1853			R.R.S.Bowker
85756	473	R	Gertrude Abbott Champion	W 3m Sr	American	565	144·0	35·0	16·5	Dennisville NJ	1882	R.S.Leaming		J.Champion
85824	474	R	Gertrude L. Trundy Parris	W 3m Sr	American	461	151·0	35·3	12·5	Perry, Me.	1883	J.B.Nutt	[illegible]	J.S.Winslow & Co.
85511	475		Gertrude T. Browning Soloman -86	W Sr	American	128	94·9	28·8	6·8	Absecom, N.J.	1877	E.J.Risley		G.A.Howes
88195	476	✻	Gertrude & Ethel Jackson	W Sk	British	84	79·4	20·7	10·9	Hull	1884	Day		F.Jackson
	477	G V	Gesina Oltmann 78-78	W Sr	German	114	87·0	19·5	9·7	Papenburg	1877	F.W.Meyer		F.W.Meyer W[illegible]
	478	G	Gesina De Buhr 73-	W Sr	German	104	79·5	19·5	9·0	Petkum	1862	J.J.Lücht		A.de Buhr
	479	G V	Gesine Müller 85-85	W Bg	German	248	110·2	25·2	12·8	Motzen	1866	Gebr.Focke	[illegible]	J.C.F.Meyer
	480	V	Gesine Cortemalier 85-85	W Sr	German	119	86·4	21·4	10·3	Leer	1883	B.T.Middendorf	[illegible]	J.Weers de Buhr
	481	V	Gesine Brunmann	W Sr	German	116	84·6	20·0	9·2	Jemgum	1869	G.H.Aden		G.& H.Aden
	482	G	Gesine Gortemöller -88	W Sr	German	112	86·3	21·0	9·3	Papenburg	1878	B.Thelen		Bernh.Thelen W[illegible]
	483	V	Gesine Rinderhagen -78	W Sr	German	107	81·1	18·7	9·6	Varel	1874	C.Schwoon		H.K.Buss
	484		Gesoria Galiana	W Bg	Spanish	232	93·2	25·6	14·6	San Feliu	1851		[illegible]	J.Orrios Compañ[illegible]
	485		Gessner (ex Lo[illegible]) Hjulman	W Bk	Russian	712	141·4	34·2	20·7	Schiedam	1854	Gips & Zoonen	[illegible]	L.W.Wendelin
94271	486	R	Gettysburg Stewart 87-87	W Bk	British	1025	173·5	35·5	23·3	Richmond, Me.	1864	H.S.Hagar	[illegible]	G.Milne & Co.
	487		Geutch-Giin	W Bg	Turkish	197	..	..	..	Geutch-Giin	1870			Ahmussrahlen-[illegible]
	488		Geutch-Giin Achmet	W Martico	Turkish	137	..	..	..	Geutch-Giin	1865			Kuorooleh-[illegible]
	489	V	Gevalia Sjöström 77-84	W Bk	Swedish	707	153·6	30·6	19·7	Gefle	1874	Bergforss & Son	[illegible]	S.F.Pousette
	490	D G	Gezina (ex Jantina) Nieuwenhuis 82-	W Sr	Dutch	157	96·8	21·8	11·3	Pekela	1868	B.H.H.Dreuth		H.W.Houwald
	491	D G	Gezusters Plukker 86-	W Sr	Dutch	191	91·9	17·1	10·9	Amsterdam	1857	Meyjes & Sons	[illegible]	G. Plukker [illegible]
	492	D V	Gezusters van Haaften Schulte 85-85	I Bk	Dutch	968	191·2	35·1	20·8	Amsterdam	1885	C & J von Lindern		Von Lindern & [illegible]
	493		Gherzeh	W Bg	Turkish	146	..	..	..	Greece	1879			Hadji Izzet [illegible]
3506	494		Giacinta S. Massa	W Bg	Italian	274	101·7	27·9	16·1	Voltri	1858	D.Briasco	[illegible]	A.Massa
610	495		Giacomino Bonifacio	W Bk	Italian	514	140·1	30·7	20·5	Lerici	1878	G.Saccomanno	[illegible]	[illegible]
2618	496	✻I	Giacomo (ex Ermanno) Brignola 82-86	W Bk	Italian	884	166·9	34·5	23·5	Lavagna	1875	G.Brignole	[illegible]	F.Schiaffino fu G.
1850	497	V	Giacomo Mortola Mortola -86	W Bk	Italian	499	138·1	31·4	19·3	Recco	1868	P.Rolla	[illegible]	G.Mortola
2228	498	I V	Giacomo Tubino Olivari	W 3m Sr	Italian	340	126·3	28·4	18·2	Sestri Ponente	1872	S.Debarbieri	[illegible]	F[lli] Tubino [illegible]
3333	499		Giacomo V. (DISMANTLED) (ex Francesco Antonio)	W Bk	Italian	470	133·2	31·2	19·2	Sestri Ponente	1865	F.Tixi	[illegible]	[illegible]
2168	500	D I	Giaele Fiuola -86	W Bk	Italian	749	159·1	34·5	21·2	Castellamare	1876	G.Bonifacio	[illegible]	K.[illegible]

✻=Lloyd's Register; D=Nederlandsche Vereeniging; G=Germanischer Lloyd; I=Registro Italiano; A=Veritas Austro-Ungarico; H=Veritas Ellenico; N=Norske Veritas;

In the 4th Column, C=Composite (iron frame, wood planked); I=Iron; Stl=Steel; W=Wood.

The Dates of Metalling (Col. 11) are based on the records of the Books in which Vessels are respectively classed, or on the Reports of the [illegible]

Further particulars of Vessels classed by Lloyd's Register will be found in Lloyd's Register Book and Lloyd's Yacht [illegible]

No.	Reg.	Nom. Capitaine	Matériau et Gréement	Pavillon	Tonnage net	Longueur	Largeur	Creux	Où	Quand	Par qui	Doublage	Armateurs ou Propriétaires	Port d'Armement
501		Giambattista C.	W Bk	Italian	458	121·5	30·0	18·7	Prà	1865			A. Bonnevie	Genoa
502		Giambattista Primo *Serra*	W Bg	Italian	128	..	..	..	Sestri	1874			L. Serra fu G. B.	Spezia
503	A	Gian Michel *Merlo*	W Bg	Italian	258	101·7	28·9	17·1	Varazze	1863	A. Cerruti		P. Zigliara	Genoa
504	A	Giannula *Caravia*	W Bg	Greek	115	72·2	22·6	..	Syra	1883			D. Caravias	Syra
505		Giano *Sepich* -81	W Bk	Austrian	585	151·5	29·5	20·3	Lussinpiccolo	1869	N. Martinolich		Andrea Marussich & Co.	Genoa
506	A	Giasone *Padoan*	W Bg	Italian	100	..	..	..	Chioggia	1882	G. Poli		D. Padoan	Chioggia
507		Gibaja	I Bk	Spanish	155	96·5	22·0	12·5	Glasgow	1862	R. Napier		A. G. Gordon	Santander
508		Gibraltar	W Bk	British	270	127·5	25·9	10·7	St Catharines (Ont.)	1854	L. Shickluna		Mrs. Marg. Manning	St Cath'rines, Ont.
509		Gideon *Lundberg* 87	W Bg	Russian	247	110·0	29·0	11·2	Getha	1862	E. A. Mattssen		K. S. Eriksson & Co.	Finström
510		Gideross	W Bg	Turkish	200	..	..	..	Gideross	1883			Kerestеn Hadji Said Hadji Ahmed & Sons	Gideross
511		Gideross	W Bg	Turkish	141	..	..	..	Gideross	1880			Unia Hadji-Hafiz-... Mustapha & Co.	Unia Majorca
512		Giemi I	W Bg	Spanish	191	103·0	24·1	12·6	P. E. Island	1868			B. Canet Ferrer	Majorca
513	*	Gift *Sugars*	W Bk	British	286	112·1	26·5	16·1	Sunderland	1870	J. G. & H. Lister		P. Rowe & Sons	Bristol
514	V	Gil *Da Rocha*	W Bk	Portug'se	270	117·7	23·3	12·6	Gloucester	1867	Miller & Son		A. J. da Rocha	Oporto
515		Gilberte	W Sr	French	171	96·0	22·0	11·6	Newport	1867	Newport S. B. Co.		R. de Fraissinet	Marseilles
516	*	Gilcruix *Cowley* -86	I ShS	British	2239	289·3	42·0	24·3	Whitehaven	1886	Whitehaven S. B. Co.		North Western Shipping Co. Ld	Liverpool
517	I	Gilda M. *Vassallo*	W Bk	Italian	779	167·0	34·1	22·2	Lavagna	1875	G. Brignole		Gaetano Moglia	Genoa
518	N	Gilead *Olsen* -87	W Bk	Norweg'n	435	134·1	30·4	16·2	Risör	1879	O. O. Haugen		N. A. Nielsen	Kragerö
519		Giles Lang *Tonkin*	W Sr	British	124	83·7	21·8	11·5	Padstow	1864	Tredwen		R. Jennings	Haylo
520		Gilpin *Paul*	W Bg	British	187	108·1	25·5	12·4	P. E. Island	1869	Richards		Whitstable Shpng Co. Ld	Faversham
521	*	Gilroy *Jamieson*	I S	British	1678	259·2	39·9	23·2	Glasgow	1875	J. Elder & Co.		Gilroy, Sons & Co.	Dundee
522	*	Gilsland *Lark*	W Bk	British	333	145·3	26·9	16·5	Sunderland	1869	Liddle & Sutcliff		A. McDonald	Stornoway
523	N	Gimle *Jensen*	W Bk	Norweg'n	352	106·1	30·1	16·9	Arendal	18^{47}_{73}	S. Johnnsen		G. M. Klöcker	Arendal
524		Gimso -75	W Bk	Norweg'n	383	115·6	28·4	16·7	Skien	1841			Chr. Th. Boe	Arendal
525	I	Gin *Brunello*	W Bk	Italian	900	181·1	30·8	23·9	Varazze	1882	Stef. Cravioto		G. B. Lavarello di P.	Genoa
526	V	Gine *Sand* -88	W Bg	Norweg'n	265	115·8	28·5	11·3	Altona	1866	J. J. Diercks		C. Sand	Stavanger
527		Gioanin *Paolino*	W Bg	Italian	287	104·9	28·1	17·1	Foce	1857	Debarbieri		V. Firpo	Genoa
528	V	Gioja	W Bg	Italian	314	106·6	26·6	16·9	Meta	1860	F. Mauro		L. Musso	Naples
529	N	Gioja *Ellingsen* 78-86	W Bk	Norweg'n	351	121·6	28·1	15·5	Grimstad	1878	J. N. Ribe		J. Jörgensen	Grimstad
530		Giorgio Boscovich	W Bk	Austrian	537	144·0	28·3	18·9	Leano	1866	Cerruti		M. Boscovich fu G.	Megline
531		Giorgio Washington *Malfino* -87	W Bk	Italian	542	146·9	31·4	19·3	Varazze	1869	F. Tassara		Molfino & Truppia	Genoa
532		Giorgios	W Bg	Greek	156	80·1	22·6	..	Syra	1865			M. Cacuroannias	Syra
533		Giosuè *Di Majo*	W Bg	Italian	237	95·5	25·6	15·7	Sorrento	1861	A. Castellano		Eredi di S. Maresca	Castell. di Stab.
534		Giovanna Rosa	W Sr	British	106	89·6	22·8	11·1	Malta	1875			A. Mercieca	Malta
535	? A	Giovanni	W Bk	British	710	166·5	34·6	19·4	Spezia	1869	C. Faggioni		G. Milne & Co.	Aberdeen
536	I	Giovanni *Trapani*	W BmSr	Italian	493	149·2	29·8	19·0	Castellam. di Stabia	1884	B. Bonifacio		N. Gambardella	Castellamare
537	IV	Giovanni *Gavagnin*	W Bk	Italian	465	137·4	31·2	19·8	Chioggia	1877	G. Poli		F[lli] Gavagnin & Vianello	Venice
538	V	Giovanni *Lauro*	W Bk	Italian	440	129·2	28·5	18·0	Equa	1869	P. Savastano		F. & E. Lauro	Castellamare
539		Giovanni	W Bk	Italian	418	126·9	30·0	19·2	Varazze	1867	N. Iaiosci			Naples
540		Giovanni *Sardi*	W Bg	Italian	198	91·2	16·9	15·1	Savona	1853	Calamora		P. Gherardi	Porto Ferraio
541	A	Giovanni	W Bg	Italian	183	97·4	24·6	15·8	Vietri	1866	P. Savastano		S. Campisi	Catania
542		Giovanni *Succo*	W Bn	Italian	124	75·5	23·4	13·2	Varazze	1849	Craviotto		Dom. Fascia	Civitavecchia
543		Giovanni *Cillari*	W B	Italian	106	..	..	..	Gaeta	1859			E. Riccobono	Naples
544	V	Giovanni A. *Drago*	W Bk	Italian	831	193·9	33·1	21·9	Trieste	1871	G. Tonello		G. Moglia	Genoa
545	I	Giovanni Ansaldo	W Bk	Italian	547	147·6	31·5	20·6	Savona	1872	E. Calamaro		G. P. Ansaldo	Genoa
546		Giovanni Batta	W Bg	Italian	107	77·1	23·3	11·8	Varazze	1850	B. Cerruti		L. M. Olivari	Genoa
547	DI	Giovanni Batta Barabino	W Bk	Italian	825	176·0	34·3	22·6	Sestri Ponente	1877	A. Carlevaro		F[lli] Barabino	Genoa
548		Giovanni Batta Murzi	W Bk	Brazilian	466	127·9	29·5	20·7	Voltri	1869			Martins, Praça & Co.	Bahia
549	IV	Giovanni Batta Repetto	W Bk	Italian	1236	209·6	37·1	25·6	Prà	1883	L. Durante		G. B. Repetto fu A.	Genoa
550	I	Giovanni Battista	W Bg	Italian	144	83·6	21·9	12·1	Gaeta	1867	F. Contreras		N. A. Di Fonzo	Gaeta

* = Lloyd's Register; D = Nederlandsche Vereeniging; G = Germanischer Lloyd; I = Registro Italiano; R = American Record;
A = Veritas Austro-Ungarico; H = Veritas Ellenico; N = Norske Veritas; V = Bureau Veritas.

Dans la 4me Colonne, C = Composite (membrure en fer, bordé en bois); I = Fer; Stl = Acier; W = Bois.

Les Dates du doublage en métal (Col. 11) ressortent des livres dans lesquels les navires sont respectivement classés ou des rapports des Inspecteurs du Lloyd's Register.

Pour plus amples renseignements sur les navires classés par le Lloyd's Register, voir le Lloyd's Register Book et le Lloyd's Yacht Register.

Official No.	No. in Book	Registry in which Classed	Ship's Name. Material and Rig. Master, and Dates of Appointment to present Owner's Service and to Ship. Ship's Late Name.		Flag	Net Tonnage	Dimensions. Length	Breadth	Depth	Built. Where	When	By whom	Date of Metalling	Owners or Managers
[illegible]	551		Giovanni Battista (ex Umbra) *Pacenti*	W Bn	[illegible]	121	73·4	22·5	12·2	Avenza	1859			F^lli Bonfante
[illegible]	552	I	Giovanni Battista O. *Olivari*	W Bk	Italian	439	128·6	31·1	19·5	Savona	1870	F.Sirello	[illegible]	Eredi di B.Olivari
[illegible]	553	I	Giovanni L. *Rolla* -88 (ex Josephine)	W Sr	Italian	172	105·9	24·7	13·2	Leghorn	1874		[illegible]	G.Languasco
[illegible]	554		Giovanni O. *Laura* (ex Due Amici)	W Bn	Italian	140	89·7	24·2	11·1	[illegible]	1876	M.Baglietto		G.B. & M.Olivari
[illegible]	555		Giovanni Padre (ex Anna) *Garassini*	W Bg	Italian	227	95·5	25·5	15·0	[illegible]	1859	F.Sirello		Giannoni & Arecco
[illegible]	556		Giovanni Padre *Bennati*	W Bn	Italian	114	87·9	22·1	11·1	Sestri Ponente	1881	A.Briasco fu L.		Eredi Bennati
[illegible]	557		Giovanni Pietro T. (ex Giuditta Danovaro, *Torrcca* ex Francesco)	W Bk	Italian	294	104·9	28·1	17·8	[illegible]	1859	B.Camogli		F^lli Torucca
[illegible]	558	A	Giovanni S. *Martinolich* (ex Rebus) [-87	W Bk	Austrian	845	158·8	31·7	20·1	[illegible]	1878	M.A.Stareich	[illegible]	Vedova Rosa Scopinich
[illegible]	559	V	Giovannina *Esposito* -88	W Bk	Italian	481	136·1	31·8	19·9	Cassano	1869	A.Castellano	[illegible]	F.Ciampa
[illegible]	560	A	Giovannina *Duria* -87	W Trb	Italian	100	79·4	22·9	9·6	Chioggia	1882	A.Nordio	[illegible]	F^lli Gavagnin
[illegible]	561	A	Giovannina S. *Letterio*	W Sr	Italian	227	108·6	25·8	15·0	Trieste	1869	G.Tonello	[illegible]	F^lli Sieni
[illegible]	562	V	Giovannino *Salese*	W Bg	Italian	159	97·4	23·3	12·5	Torre del Gr'co	1883	G.Bonifacio	[illegible]	L.Musso
[illegible]	563		Giovannino *Puglisi* (ex Paulina)	W Bg	Italian	132	84·7	20·3	13·3	Volosca	1853		[illegible]	T.D'Alessandro
[illegible]	564		Giove (now named "Camelia")	W Bk	Italian	537	145·2	31·9	19·5	Sestri Ponente	1870	Marcenaro		S.Moltedo
[illegible]	565		Gipsy *Swanson* -68	W Sr	British	187	103·0	24·0	13·2	Garmouth	1868	Geddie		G.Swanson
	566	V	Gipsy *Martin*	W Bk	French	424	132·5	29·3	16·9	Nantes	1870	E.Dubigeon [& fils	[illegible]	F.Boissière
[illegible]	567	R	Gipsy Queen *Jordan* [illegible]	W Bn	American	342	117·9	27·7	11·6	Wesbrook, Me.	1866		[illegible]	W.S.Jordan & Co.
	568		Gipsy Queen *Bately*	W Bk	Norweg'n	453	131·9	28·1	17·6	Littlehampt'n	1862	Harvey	[illegible]	C.Taraldsen
	569		Giralda *Le Diabat*	W Bn	French	112	77·1	22·5	10·5	[illegible]	1861	Rimbert & fils		H.E.Prud'homme
	570		Girassimo	W Bk	Russian	205	95·8	23·9	9·5	[illegible]	1872	Gregoire		J.Grisini
[illegible]	571	✳	Girl of Devon *Johnsen* [-88	W 3m Sr	British	136	101·9	21·5	10·9	[illegible]	1880	Shilston	[illegible]	W.H.Shilston
[illegible]	572		Girl of the Period *Goodwin* -87	W Bn	British	211	117·8	24·9	13·8	Hayle	1870	Harvey		Holden & Gann
	573		Giro *Mickelson* -87	W Bg	Russian	250	..	..	..	Sideby	1868			M.A.Jansson
	574	V	Giro Ilardi *Los*	W Bk	Greek	272	108·2	24·6	15·4	Scio	1885		[illegible]	Los Bros.
1088	575	I	Girolmina *Cafiero* -88	W 3m Sr	Italian	348	118·1	26·2	17·1	Alimuri	1876	S.Mauro	[illegible]	M.Jaccarino
90038	576	✳	Girvan *Nisbet*	Stl Bk	British	1277	233·1	36·1	21·1	Glasgow	1885	Lon.& Glasgw Co		D.Hunter
350	577	A	Gisella T. (ex Ettore) *Pitre*	W Bg	Italian	138	90·2	21·6	12·3	Fiume	1863		[illegible]	R.Mineo fu G.
29626	578	✳	Gitana *Byron* 87-87	I S	British	1367	220·5	36·0	24·9	Chester	1861	N.Cox		R.W.Leyland & Co
10084	579		Gitana (WRECKED)	W Sr	British	103	80·5	21·6	10·4	Cowes	1853	J.White		J.Kymbrell
51167	580		Gitanilla Maru (HULK)	W Bk	Japanese	484	135·6	28·5	18·3	Sunderland	1865	R.Thompson jr.	[illegible]	
[illegible]	581	✳ I	Giuditta D. *Dodero* -77	W Bk	Italian	800	164·1	33·3	21·5	Sestri Ponente	1877	G.Pergallo	[illegible]	Franc.Dodero
412	582	I	Giulia *Rocca*	W Bk	Italian	470	146·0	30·0	18·6	Lavagna	1865	D.Briasco fu G.P.	[illegible]	G.B.Repetto & Eredi Repetto
	583	A	Giulia *Zammit* -88 (ex Nicola, ex Shields, ex Ginfiot)	W Bk	British	402	128·6	28·5	17·5	Sunderland	1869	Pickersgill & [Miller	[illegible]	S.Ganci
541	584		Giulia *Inzerillo*	W Bk	Italian	214	111·1	24·8	10·9	Baltimore, Md.	1859			L.Marino
831	585		Giulia (ex Luigi) *Cattabiano*	W Bn	Italian	170	96·8	24·5	13·6	Castellamare	1879	G.Bonifacio	[illegible]	Vita Cassissa & Co
	586		Giulia Anna	W Bk	Spanish	619	..	..	..	Lavagna	1874	G.Brignole	[illegible]	José V.Gallud
3544	587		Giulia Soldi *Soldi* -88 (ex Anna [illegible])	W Bk	Italian	413	150·8	30·0	18·2	[illegible]	1862	D.Bianchi	[illegible]	Soldi
936	588		Giulietta (BROKEN UP) *Gazzolo*	W Bk	Italian	435	132·5	30·1	18·0	Savona	1864	G.Guastavino	[illegible]	Giov.Rissetto
106	589		Giulietta *Cignoni*	W Bg	Italian	123	75·8	22·6	12·6	[illegible]	1852			Giov.Cignoni fu L.
1019	590	I	Giulio *Martingan* -88	W Bk	Italian	492	137·1	29·7	19·6	Castellamare di [Stabia	1874	L.Bonifacio	[illegible]	D.Paolillo
	591		GiulioCesare (ex [illegible])	W Bk	Argentine	401	..	..	..	Chiavari	1864	F.Gotuzzo		J.Catoni
1090	592	I	Giulio R. *Romano*	W Bk	Italian	449	126·6	30·5	18·9	Alimuri	1875	M.Mauro	[illegible]	Francesco Romano fu G.
231	593	A	Giulio R. *Furlan* -88	W Bn	Italian	100	..	..	..	Chioggia	1879	L.Menetto	[illegible]	M.Salvagno & Co.
1301	594	✳	Giulio & Clemenza (ex Cotherstone) *Siena*	W Bk	Italian	355	125·0	26·6	17·3	Sunderland	1860	Austin	[illegible]	M.A.Lubrano
167	595	V	Giurko *Petricevich*	W Bk	Austrian	517	136·4	26·5	16·7	Trieste	1870	O.Spadon	[illegible]	P.L.Tripcovich
	596		Giuseppe *Weymouth* (ex Martin W. Brett)	W Bk	Danish	448	128·9	26·0	17·3	Pembroke, [Maine	1863			G.Copello
[illegible]	597	I	Giuseppe *Lauro*	W Bk	Italian	984	180·1	33·1	21·6	Cassano	1886	M.A.Patuzzo	[illegible]	Conte Mazzarino
1479	598	V	Giuseppe *Enrigo*	W Bk	Italian	756	154·1	33·1	21·8	Sestri Ponente	1866	F.Cadenaccio	[illegible]	D.Enrigo
1278	599	I	Giuseppe *Criscuolo*	W Bk	Italian	498	136·1	30·8	19·2	Castell'diStab.	1880	Trambarulo	[illegible]	L.Criscuolo
594	600		Giuseppe *Sardi*	W Bk	Italian	343	120·1	27·2	18·3	Sorrento	1856			O.Pulis & Caputo

✳ = **Lloyd's Register**; D = Nederlandsche Vereeniging; G = Germanischer Lloyd; I = Registro Italiano; R = [illegible]
A = **Veritas Austro-Ungarico**; H = Veritas Ellenico; N = Norske Veritas; V = [illegible]

In the 4th Column, C = Composite (iron frame, wood planked); I = Iron; Stl = Steel; W = Wood.

The Dates of Metalling (**Col. 11**) are based on the records of the Books in which Vessels are respectively classed, or on the Reports of the Surveyors [illegible]

Further particulars of Vessels classed by Lloyd's Register will be found in Lloyd's Register Book and Lloyd's Yacht Register.

No.	Classe	Nom, Capitaine, Gréement	Pavillon	Tonnage	Longueur	Largeur	Creux	Construit où	Quand	Par qui	Doublage	Armateurs ou Directeurs-Gérants	Port d'Armement
601	A	Giuseppe Gambino W3mSr (ex Maddalena)	Italian	277	1[illegible]3·6	24·0	14·9	Finale	1876	S.Ponsiolo	V.	M. & A.Dall'Orso	Spezia
602		Giuseppe W Bg	Montenegrin	148	77·1	20·5	12·3	Galaxidi	1867			Selim Mambeg & Co.	Dulcigno
603	D	Giuseppe Accame (Chiabrera) W3mSr	Italian	472	138·7	29·6	18·7	Loano	1880	A.Cerruti	V.	Pietro Accame & figli	Savona
604	V	Giuseppe-Anna Bonanno W Bk	Italian	1099	201·1	36·1	21·4	SestriPonente	1878	A.Briasco fu L.	V.	G.Olivari	Genoa
605	I	Giuseppe Boni Raggazzi W Bn	Italian	17[illegible]	[illegible]·1	25·3	13·4	Savona	1879	F.Sirello	V.	G.P.Tori	Spezia
606		Giuseppe Cappadona W3mSr	Italian	18[illegible]	[illegible]·9	23·6	13·1	Castellamare	1877	M.Trombarulo	V.	Cappadona & Marullo	Castellamare
607	V	Giuseppe Costagliola W Bk	Italian	61[illegible]	[illegible]·8	31·7	21·3	Ischia	1877	G.Bonifacio	V.	G.Costagliola	Naples
608		Giuseppe D. W Bg	Italian	34[illegible]	[illegible]·5	25·3	15·4	Savona	1857	F.Sirello	V.	F.Mortola	Genoa
609	V	Giuseppe D'Abundo W Bk	Italian	574	1[illegible]·0	31·5	18·9	Procida	1870	N.Lubrano	V.	E.D'Abundo	Naples
610	I	Giuseppe Emmanuele W Bk	Italian	621	1[illegible]·3	31·3	19·7	Rapallo	1871	G.Saccomanno	V.	F.Denegri	Genoa
611	H	Giuseppe Esposito Esposito W Bn (ex Felice)	Italian	228	9[illegible]·2	27·5	15·4	Trieste	1856			G.Esposito	Bari
612	I	Giuseppe Lanata W3mSr	Italian	462	1[illegible]·8	31·2	20·1	SestriPonente	1869	Casella	V.	E.Bonsignore & G. Vaglieri	Genoa
613		Giuseppe Massone W Bk		49[illegible]	1[illegible]4·5	31·8	19·3	Prà	1868	L.Durante	V.		
614	I	Giuseppe Olivari W Bk	Italian	44[illegible]	13[illegible]·2	30·2	19·0	SestriPonente	1867	S.Debarbieri	V.	Franc.Olivari	Genoa
615		Giuseppe P. W Bk (ex Pietro Antonio)	Italian	381	117·4	29·4	17·9	Pietra Ligure	1864	G.Accinelli		G.Pertuso	Savona
616		Ginseppe Padre (ex ...) W Bk	Italian	3[illegible]	11[illegible]·4	29·5	17·9	Varazze	1862	N.Cerruti		F[lli] Soldi fu G.	Genoa
617	I	Giuseppe Padre W3mSr	Italian	2[illegible]	1[illegible]3·5	27·3	12·7	Sestri	1869	Tubino	V.	Gius.Rabagliati	Savona
618		Giuseppe Paladino W Bk	Italian	28[illegible]	..	..	..	Sestri	1854			Salv.Paladino	Messina
619	I	Giuseppe Pignone W Bk	Italian	613	147·6	31·8	21·2	SestriPonente	1875	A.Carlevaro	V.	Gius.Pignone & figli	Genoa
620		Giuseppe Revello W Bk		431	123·0	31·2	19·7	Varazze	1868	Giacomo Scala	V.		
621	I	Giuseppe S. W Bk	Italian	639	140·0	32·7	22·3	Sestri	1872	A.Briasco fu L.	V.	F[lli] Sanguinetti fu G.	Spezia
622	I	Giuseppe Verdi Ferraro W Bk	Italian	489	135·5	31·6	20·1	Varazze	1869	S.Craviote	V.	Em. Agrenta & E. Ferraro	Genoa
623	A	Giuseppina Lisita -88 W Bg	Italian	2[illegible]	99·7	25·8	15·1	Castellam.di Stabia	1859	A.Bonifacio	V.	L.Musso	Naples
624	V	Giuseppina Calabretta W Bn	Italian	165	1[illegible]·5	25·0	12·6	Savona	1875		V.	T.Calabretta fu V.	Catania
625	V	Giuseppina Tonci W Sr	Italian	118	90·2	22·5	11·4	Viareggio	1878	A.Raffaelli		A.Raffaelli	Leghorn
626		Giuseppina Galletti W Bn	Italian	190	82·7	23·3	10·2	Limite	1869	N.Pichiotti		G.Tonietti fu R.	P.Ferraio
627	❋I	Giuseppina Bertollo I Bk	Italian	1268	2[illegible]·5	36·1	21·0	Dumbarton	1883	A.McMillan & Sn		G.Bertollo	Genoa
628	A	Giuseppina C. (ex Antonietta) W Bn	Italian	104	7[illegible]·4	22·0	11·8	Venice	18[illegible]		V.	A.Conigliaro	Catania
629		GiuseppinaG. -88 W Bk	Argentine	478	134·8	31·9	19·3	SestriPonente	1870	F.Cadennacio	V.	S.Auselmi	Buenos Ayres
630		Giuseppina Naccari	Italian	193	96·8	24·5	13·3	Gaeta	1860	F.Contreras		A.Butera & Co.	Palermo
631		Giuseppina Oneto W Bk (BROKEN UP)	Italian	401	123·3	29·6	18·8	Varazze	1866	N.Cerruti	V.	A.Oneto	Genoa
632		Giuseppina Pizzorni I Bn	Italian	161	106·6	27·2	9·6	Sampierdar'na	1885	A.Carlevaro			
633	I	Giuseppina R. Ruggiero W Bk	Italian	52[illegible]	138·7	29·5	20·7	Alimuri	1876	G.Mauro	V.	N.Ruggiero	Castell.di Stab.
634	I	Giuseppino C. Chiappella W Bk	Italian	595	147·6	32·6	20·1	Foce	1876	Debarb.e Celesia	V.	N.Chiappella	Genoa
635		Giuseppino P. Vergura W Bn (FOUNDERED)	Italian	110	75·8	23·6	12·7	Prà	1841	M.Tixi		F.Peragallo	Genoa
636	I	Giustizia Saglietto W Bn	Italian	138	93·8	22·6	11·9	Savona	1877	G.Pongiglione	V.	Cristof.Astengo	Savona
637	A	Giusto Colombis W Bk	Austrian	793	15[illegible]·2	32·7	17·7	Lussinpiccolo	1875	N.Martinolich	V.	A.D.Purich	Cherso
638	N	Gjallarhorn Leonhardsen W3mSr	Norweg'n	236	114·0	25·9	13·2	Stavanger	1872	M.Magnus	V.	Th.Hamre	Stavanger
639	N	Gjede Johansen -88 W Sr	Norweg'n	211	1[illegible]·3	26·0	12·3	Maitland,N.S.	1862			H.H.Parr	Dröbak
640	N	Gjemle Jorgensen W Bk	Norweg'n	373	114·2	30·2	16·5	Grimstad	1859	N.OlsenGjömle		J.E.Einersen	Grimstad
641	N	Gjendin Valdassen W Bg	Norweg'n	276	126·0	27·6	12·0	Porsgrund	1876		V.	J.C. & G.Knudsen	Porsgrund
642	❋	Gjerterud Severtsen 71-71 W Bk	Norweg'n	555	128·2	28·3	18·2	Rothesay,N.B.	1875	R.Titus	V.	M.Skredsvig	Dröbak
643	V	Gjertrud Jorgensen 70-77 W Sr	Danish	15[illegible]	96·4	22·3	10·8	Marstal	1877	H.J.&L.J.Bager		L.J.Bager	Marstal
644		Glacier (ex Charles Keen) W Bk	British	481	149·0	30·2	16·6	Perth,U.S.	1853			C.Thomson	Halifax,N.S.
645		Glad Lauridsen 82-82 W Bk	Danish	399	128·1	29·0	14·9	P.E.Island	1874	John Gillan	V.	G.Lauridsen	Sonderho,Fano
646	❋	Glad Tidings Williams W Sr	British	95	77·6	20·6	10·6	Aberystwith	1866	Jones		J.Williams & Co.	Aberystwith
647	R	Glad Tidings Roberts W Bkn	American	626	159·2	33·1	17·0	Belfast,Me.	1884	G.W.Cottrell	V.	C.M.Stewart & Co.	Baltimore
648	V	Gladan Andersson -85 W Bk (ex Robert Ritson)	Swedish	579	140·7	27·0	20·6	Maryport	1851		V.	Rederiaktiebolaget (Ömeu (J.F. Strömbeck))	Gothenburg
649		Gladdys McAuley W Bn	British	180	95·9	24·7	11·1	P.E.Island	1882	B.Rogers		P.G.& L.Tessier	St.John's,N.F.L.
650	❋	Gladiateur Keenan -84 W Bk	British	428	137·3	28·0	16·9	Jersey	1866	Clarke	V.	W.Keenan	London

❋=Lloyd's Register; D=Nederlandsche Vereeniging; G=Germanischer Lloyd; I=Registro Italiano; R=American Record; A=Veritas Austro-Ungarico; H=Veritas Ellenico; N=Norske Veritas; V=Bureau Veritas.

Dans la 4me Colonne, C=Composite (membrure en fer, bordé en bois); I=Fer; Stl=Acier; W=Bois.

Les Dates du doublage en métal (Col. 11) ressortent des livres dans lesquels les navires sont respectivement classés ou des rapports des Inspecteurs du Lloyd's Register.

Pour plus amples renseignements sur les navires classés par le Lloyd's Register, voir le Lloyd's Register Book et le Lloyd's Yacht Register.

Official No.	No. in Book	Registry	Ship's Name; Master, and Dates of Appointment to present Owner's Service and to Ship; Ship's Late Name	Material and Rig	Flag	Net Tonnage	Length	Breadth	Depth	Where Built	When	By whom	Date of Metalling	Owners or Managers
	651	N	Gladiator *Helgesen -88*	W Bk	Norweg'n	399	104·0	29·2	17·2	Arendal	1853			R.O.Helgesen
	652	N	Gladiator *Pettersen 87-77*	W Bg	Norweg'n	294	96·4	24·6	14·3	Grimstad	1863	N.Olsen Gjömle		P.Pedersen
[illegible]	653	✻	Gladova *Roberts 88-88*	I Bk	British	1057	216·0	34·5	20·1	Liverpool	1888	R&J.Evans&Co.		S.W[illegible]
	654		Gladovia *Knowlton*	W Bk	Chilian	631	149·8	33·8	18·3	AdvocateHarbour,N.S.	1877	W.Moore	[illegible]	[illegible]
[illegible]	655	✻	Gladstone *Low* (ex Steamer Francisco Calderon)	I S	British	1250	237·3	34·0	20·9	Dumbarton	1873	A.M'Millan&Sn.		W.[illegible]
[illegible]	656	R	Gladstone (ex Louise) *Sutter*	W S	British	1057	182·1	34·5	23·8	Bowdoinham, Me.	1860		[illegible]	J.R.S[illegible]
[illegible]	657		Gladstone *McBain 83-*	W Sr	British	144	95·6	22·7	11·7	Banff	1873	Geddie		R.Morrison
[illegible]	658	R	Gladstone (WRECKED)	W Sr	British	130	85·5	23·9	10·3	Pugwash, N.S.	1887	D.Redmand		D.Redmand
[illegible]	659	A	Gladstone *Calafati*	W Bn	Greek	281	110·2	26·6	. .	Syra	1880	A.Livadara	[illegible]	M.Calafati
[illegible]	660		Gladys *Harding 85-86*	W Sr	British	148	92·2	27·6	8·5	St Martin's N B	1882	W.J.Davidson		W.J.Davidson
[illegible]	661	✻	Glamis *Dalgarno 86-86*	I Bk	British	1150	225·3	34·8	21·9	Dundee	1876	A.Stephen&Sns		D.Bruce & Co.
[illegible]	662	✻D	Glance *Williams 76-88*	I Bk	British	861	197·7	33·1	20·8	Sunderland	1869	W.Watson		G.H.Meager
[illegible]	663	✻	Glance *Noble*	W Sr	British	162	102·0	23·5	12·2	Kingston	1873	Spence	[illegible]	A.Coull
[illegible]	664		Glance *Lewis*	W Sr	British	144	87·3	21·6	12·6	Hull	1852	Pashley&Wood		D.T.Sharper
[illegible]	665	✻	Glandinorwig *Jones 84-84*	I Bk	British	1081	219·0	35·1	21·3	Sunderland	1876	W.Doxford&Ss		Williams,Jones[illegible]
	666	V	Glaneuse *Gorce*	W Sr	French	158	93·5	22·0	11·8	Dunkirk	1878	A.Fyotte		H.Durin
[illegible]	667	✻	Glanivor *Thomas 83-83*	I Bk	British	1084	220·5	35·1	21·3	Sunderland	1878	W.Doxford&Ss		The Eryri Shipping [illegible] Ltd
[illegible]	668	✻	Glanogwen *Parry 86-86*	W Bn	British	119	82·5	20·0	12·6	Bangor	1855	T.T.Parry		T.T.Parry
[illegible]	669	✻	Glanpadarn *Jones 83-83*	I Bk	British	1082	220·6	35·1	21·2	Sunderland	1877	W.Doxford&Ss		Williams,Jones[illegible]
[illegible]	670	✻	Glanperis *Jones 85-85*	I Bk	British	1073	221·0	34·8	21·2	Sunderland	1877	W.Doxford&Ss		Williams,Jones[illegible]
[illegible]	671	✻	Glanrafon *Dewan 79-86*	I Bk	British	472	156·0	26·6	17·1	[illegible]	1866	Brown&Simpson		Richardson & Co.
	672	N	Glarus *Walkvist*	W 3mSr	Norweg'n	294	138·9	29·6	11·7	Tönsberg	1868	Larsen	Y. 1887	C.Walkvist
[illegible]	673	✻D	Glasgow *Singer 81-87* (ex Pembroke Castle)	I Bk	British	1148	208·2	33·9	22·4	Glasgow	1863	R.Napier&Sons		C.Barrie
[illegible]	674		Glasgow *Wilson*	W Bkn	British	148	98·1	22·2	12·0	Newcastle	1816			G.Mitchelson
79111	675	✻	Gleaner *Linklater 76-82*	W Bn	British	171	101·0	23·4	12·4	Scilly	1878	Edwards	Y.[illegible]	J.F.Pemsel & B.L[illegible]
86023	676		Gleaner *Henderson 82-83*	W Sr	British	116	82·7	26·2	8·5	[illegible] N.B.	1880	J.O.Stackhouse		Scammell Brothers
57511	677		Gleaner *Henderson*	W Sr	British	115	89·8	21·2	8·0	[illegible]	1870			J.Sinclair
30214	678		Gleaner	W Bn	British	109	71·8	19·0	11·4	[illegible]	1815			P.Colford
83307	679	✻	Gleaner *Paul 73-83*	W 3mSr	British	98	94·5	22·5	9·0	[illegible]	1883	P.Rodgers		J.Pritchard
75208	680	✻	Gleaner *Jones*	W Sr	British	94	82·8	19·8	10·0	[illegible]	1878	Delamare	Y.[illegible]	O.Jones
	681	N	Gleaner *Olsen -86*	W Bg	Norweg'n	148	90·5	22·0	12·1	[illegible]	1862	Whitehead		H.Hvistendahl
90749	682		Glen *Seely -87*	W Sr	British	125	91·4	26·1	7·6	[illegible] N B	1886	G.Merritt		W.Hamlyn
[illegible]	683		Glen	W Sr	American	121	[illegible]	23·0	8·3	[illegible]	1883			E.P.Nissen
[illegible]	684	✻	Glen Afton *Webster 71-81*	Stl Bk	British	1130	231·2	35·7	20·5	[illegible]	1884	W.Hamilton&C		[illegible] Co.
[illegible]	685	✻	Glen Caladh *Lamont*	I Bk	British	494	158·1	28·0	17·3	[illegible]	1876	Dobie&Co.		R.Lamont
[illegible]	686	V	Glen Grant *Crutchley -87*	W Bk	British	741	166·0	35·7	20·0	SherbrookeNS	1874	A.M'Donald	[illegible]	"Glen Grant" Sailing [illegible] Co. Ld. J. Sutherland
[illegible]	687	✻	Glen Osmond *Barrys 82-82*	C Bk	British	783	172·3	33·5	19·0	Sunderland	1867	J.Laing	[illegible]	A.L.Elder&Co.
[illegible]	688	✻	Glen Ville *Monse 74-86*	W Bkn	British	293	135·0	28·4	12·8	Sunderland	1874	N.Gibbon		A.Meek
[illegible]	689	✻	Glenalvon *Jones 88-88*	I S	British	2072	289·6	42·1	24·0	Greenock	1888	Russell & Co.		J.R.DeWolf & Son [illegible]
[illegible]	690	✻	Glenariff *Jones 71-84*	I Bk	British	488	161·2	28·2	17·0	Dumbarton	1869	A.M'Millan&Sn		M.Tutton
[illegible]	691	✻	Glenavna Park *Mooberley 85-85*	Stl&I Bkn	British	357	156·0	26·3	11·9	Carrickfergus	1885	P.Rodgers		J.W.Valentine[illegible]
[illegible]	692	✻D	Glenavon (MISSING) *Turney 81-81*	I Bk	British	1083	208·1	34·7	20·9	Dumbarton	1868	A.M'Millan&Sn		R.C.Hall
[illegible]	693	✻	Glenbervie *Groundwater 84-*	I Bk	British	896	195·0	31·3	19·7	Glasgow	1866	C.Connell&Co.		Bramwell & G[illegible]
[illegible]	694	✻	Glenboyn *Smith -86*	W Bkn	British	216	111·5	24·7	12·4	Banff	1878	Watson	Y.[illegible]	J.W.Simpson
[illegible]	695	✻	Glenburn *Johnston 87-87*	I S	British	1478	218·3	37·2	23·1	Glasgow	1876	B'clay,Curle&C.		R.Shankland & [illegible]
	696	✻	Glencaird	Stl 4m Bk	British	2564	313·0	42·0	24·5	PortGlasgow	1889	Russell & Co.		W.T.Dixon & Sons
[illegible]	697	✻	Glencairn *Wilson*	I 4m Bk	British	1584	252·1	40·0	22·4	Glasgow	1878	Dobie & Co.		J.& A.Allan
[illegible]	698	✻	Glencairn *McCourt*	W Bn	British	189	105·0	24·4	12·9	P.E.Island	1877	Duncan		T.H.Franks
[illegible]	699		Glencoe *Lane*	W Bg	American	169	95·6	24·0	10·8	Gardiner,Me.	1848	Lon. & Glasgow Co.		P.Lane
	700													

✻=Lloyd's Register; D=Nederlandsche Vereeniging; G=Germanischer Lloyd; I=Registro Italiano; R=[illegible]
A=Veritas Austro-Ungarico; H=Veritas Ellenico; N=Norske Veritas; V=[illegible]

In the 4th Column, C=Composite (iron frame, wood planked); I=Iron; Stl=Steel; W=Wood.

The Dates of Metalling (Col. 11) are based on the records of the Books in which Vessels are respectively classed, or on the Reports of the Surveyors [illegible]

Further particulars of Vessels classed by Lloyd's Register will be found in Lloyd's Register Book and Lloyd's Yacht Register.

Numéro	Registre	Nom. Capitaine et dates d'entrée au service des Armateurs actuels et à bord du navire. Ancien Nom		Matériaux et Gréement	Pavillon	Tonnage	Dimensions: Longueur	Largeur	Creux	Construit: Où	Quand	Par qui	Date de doublage	Armateurs ou Directeurs-Gérants.	Port d'Armement.
701	✻	Glencorse	Hodge 86-87	I S	British	1046	209·9	35·2	20·8	Liverpool	1868	T. Royden & Son		J.B.Walmsley&Co.	Liverpool
702	✻	Glencoyn	Jones 87-87	I Bk	British	426	140·0	26·1	16·8	Glasgow	1863	A. Stephen & Ss.		W.Sherwen	Whitehaven
703		Glendale	Fisk -88	W3mSr	American	221	112·2	34·0	9·0	Fairhaven	1888	Bendixen		G.H.Collins	San Francisco
704		Glendalough	Doyle	W Sr	British	132	94·0	23·0	10·9	Arklow	18			J.Doyle	Dublin
705	✻	Glendaruel	Boyd	I S	British	1761	264·3	40·3	24·1	Glasgow	1876	B'clay,Curle&C.		J. & A.Allan	Glasgow
706	V	Glendon	Fischer -89	W Bk	American	1817	225·0	40·6	28·4	Kennebunk Me	1880	S.Thompson	[illegible]	Edward Lawrence	Boston, Mass.
707		Glendoveer	Brown	W Sr	British	149	100·0	22·2	12·0	Falmouth	1861	Trethowan		Mrs.C.Jenkin	Plymouth
708	✻	Glendovey	Williams	I Bk	British	539	171·8	29·0	17·7	Sunderland	1876	J.Laing		J.Lewis	Aberystwith
709	✻	Glenearn	Johnston 81-81	I Bk	British	601	173·5	28·0	17·4	Dundee	1873	Brown& Simpsn		Dundee Shipowners' Co.	Dundee
710	R V	Gleneida	Cousens -87	W Bk	American	745	162·0	34·8	20·0	Pembroke, Me	1877	J.McVicker	[illegible]	James E. Brett	Calais, Me.
711	✻	Glenericht	Flynn 77-86	I 4mS	British	2341	289·6	44·3	25·2	Liverpool	1885	T.Royden&Sons		L.H.Macintyre&Co	Liverpool
712	✻	Glenesk	Waldner 87-87	I S	British	1298	227·2	37·8	22·3	Liverpool	1869	T.Royden&Sons		Greenshields,Cowie&Co.	Liverpool
713	✻	Glenesslin	Perth 81-85	I S	British	1743	260·0	39·2	23·4	Liverpool	1885	T.Royden&Sons		J.R.de Wolf & Sons	Liverpool
714	✻	Glenfarg	Watt 83-88	I Bk	British	851	203·8	34·1	19·1	Dundee	1881	A.Stephen& Son		Dundee Shipowners' Co.	Dundee
715	✻	Glenfinart	Stirrat	I S	British	1530	254·2	38·5	23·0	Glasgow	1876	B'clay,Curle& C		J.& A.Allan	Glasgow
716	✻	Glenfinlas	Bowles 85-86	I 4mS	British	2148	280·7	41·9	24·8	Sunderland	1882	Sunderland S.B [Co.,Ld		L.H.Macintyre&Co.	Liverpool
717	✻	Glenfyne	Douglas 85-86	Stl Bk	British	913	215·8	34·2	19·1	Dundee	1882	A.Stephen&Sns.		Dundee Shipowners' Co.	Dundee
718	✻	Glengaber	Rollaston -87	I Bk	British	668	172·6	30·1	19·3	Harrington	1868	R.Williamson & [Son		Johnston, Sproule & Co.	Liverpool
719	N	Glengairn	[illegible]		[illegible]	848	188·3	33·0	21·4	Aberdeen	1863	W.Hood&Co.	[illegible]	H.S.Jacobsen	Flekkefjord
720	✻	Glengarry	Wood 83-83	I S	British	1769	257·4	41·8	23·9	Liverpool	1873	T.Royden& Son.		W.Alexander (Shaw Rae, Marshall & Co.)	Liverpool
721	✻D	Glengarry	Murray 85-85	I Bk	British	802	199·8	32·2	19·1	Dundee	1877	A.Stephen& Sns		Dundee Shipowners' Co.	Dundee
722	✻V	Glengarry	McHenry 86-87	W Bk	British	566	156·5	32·5	16·8	Quebec	1876	F.X.Marquis	[illegible]	Maguire & Co.	Quebec
723	✻	Glenhuntley	Abbott 77-77	I Bk	British	547	165·2	28·1	17·5	Port Glasgow	1862	J.Reid & Co.		S.Johnston & Co.	Liverpool
724	✻	Gleniffer	Cumming 88-88	I Bk	British	800	193·0	30·5	20·1	Glasgow	1866	B'clay,Curle& C		Nicholson & McGill	Liverpool
725		Gleniffer		W Sr	British	315	141·0	26·1	11·5	Port Robinson	1873	J.P.Abbey		James Norris	St.Cathrne's,Ont
726	✻	Glenlora (COLLIS...)	Wood 77-85	Stl Bk	British	1220	240·0	36·2	20·7	Port Glasgow	1885	W.Hamilton& C		Dundee Shipowners' Co.	Dundee
727	✻	Glenlora	Sargent 84-84	I Bk	British	774	190·0	31·0	18·5	Liverpool	1864	T.Vernon&Son		Shaw, Savill & Albion Co., Ld.	Southampton
728	✻	Glenlui	Patterson 85-85	I S	British	1847	268·1	42·1	23·9	Liverpool	1884	T.Royden & Sns		W.Alexander (Shaw Rae, Marshall & Co.)	Liverpool
729	✻	Glenlyon	Gale 85-85	I Bk	British	665	179·5	28·6	18·8	Dundee	1868	Brown& Simpsn		Dundee Shipowners' Co.	Dundee
730	✻	Glenmorag	Currie	I S	British	1576	255·1	38·6	22·8	Glasgow	1876	Dobie & Co.		J.& A.Allan	Glasgow
731	✻	Glenmore (LOST)	Lawrence 81-82	I Bk	British	725	190·0	31·4	18·6	Dumbarton	1876	A.M'Millan&Co		W.Porter & Sons	Liverpool
732	✻	Glenogle	Paul 85-88	Stl Bk	British	914	213·8	34·2	19·1	Dundee	1882	A.Stephen& Sns		Dundee Shipowners' Co.	Dundee
733	V	Glenola	Houghton -88	W Bk	British	558	144·7	32·3	17·6	River John, N.S	1875	J.Henry	[illegible]	J.H.Farmer	Pictou, N.S.
734		Glenola	Silver 83-83	W Sr	British	114	85·0	24·4	9·5	Lunenburg N.S	1883	J.Young		C.L.Silver	Lunenburg, N.S
735		Glenora		W Sr	British	628	165·8	31·8	13·4	Kingston, Ont.	1882	R.Robertson		Montreal Transprtn.Co.	Montreal
736	✻	Glenorchy	Taylor 86-87	I 4mS	British	2149	280·7	42·0	24·8	Sunderland	1882	Sunderland Ship [Bdg.Co., Ld.		L.H.Macintyre&Co	Liverpool
737	R	Glenorchy	Ganton -87	W Bn	British	298	122·6	28·1	12·9	Baddeck, N.S.	1883	J.McDonald	[illegible]	J.Carlin	Halifax, N.S.
738	✻	Glenrosa	Macdonald -88	I Bk	British	835	203·7	32·3	20·0	Glasgow	1875	London&Glas- [gow Co.		H.M'Ewan	Glasgow
739		Glenroy	Boon 83-83	W Sr	British	139	92·2	22·4	11·0	P.E.Island	1864	D.Beaton		Starbuck & Rackstraw	London
740	✻D	Glenshee	Watt 77-84	I Bk	British	849	205·8	34·1	19·1	Dundee	1882	A.Stephen& Sns		Dundee Shipowners' Co.	Dundee
741		Glenshee		W Bk	British	317	127·5	24·4	15·6	Arbroath	1855			James Paterson	Melbourne
742	✻	Glentilt (WRECKED)	Paul 82-85	I Bk	British	784	187·0	30·0	18·7	Dundee	1875	Brown&Simpson		Dundee Shipowners' Co.	Dundee
743	R	Glenudal	Hartnell -81	W Bk	British	498	141·6	29·6	18·4	Sunderland	1864	J.Robinson	[illegible]	D.J.Hughes	Swansea
744		Glenwood		W Bk	British	298	133·8	29·4	12·0	Brookhaven, [U.S.	1856			J.E.Alexander	Nassau, N.P.
745	R	Gler (ex Anita e Gler) Shie ds		W Bk	British	485	148·4	30·9	16·3	Porsgrund	1876	Hansen	[illegible]	J.B.Duckworth	Liverpool
746	V	Gler (ex Sappho) Amundsen		W Bk	Norweg'n	707	154·5	34·7	17·3	Mystic, U.S.	1870	R.Palmer	[illegible]	J.Aalborg & Co.	Kragerõe
747		Glide	Pritchard -88	W Bn	British	207	112·3	23·0	12·9	Brixham	1865	Dewdney	[illegible]	S.Deveson & Co.	Ramsgate
748	R	Glide (WRECKED)	Trafalgar -86	W Bk	American	468	129·8	29·2	17·0	Salem, Mass.	1861	J.Bertram	[illegible]	G.R.Emerton	Salem, Mass.
749		Glimpse	Lockyer	W Sr	British	160	103·6	22·8	12·8	Brixham	1864	Dewdney		L.Lockyer	Teignmouth
750	D N V	Glimt	Syvertsen 83-79	W Bk	Norweg'n	714	158·7	34·0	18·6	Lillesand	1879	A.Stiansen	[illegible]	B.Birkeland	Lillesand

✻ = Lloyd's Register; D = Nederlandsche Vereeniging; G = Germanischer Lloyd; I = Registro Italiano; R = American Record; A = Veritas Austro-Ungarico; H = Veritas Ellenico; N = Norske Veritas; V = Bureau Veritas.

Dans la 4me Colonne, C = Composite (membrure en fer, bordé en bois); I = Fer; Stl = Acier; W = Bois.

Les Dates du doublage en métal (Col. 11) ressortent des livres dans lesquels les navires sont respectivement classés ou des rapports des Inspecteurs du Lloyd's Register.

Pour plus amples renseignements sur les navires classés par le Lloyd's Register, voir le Lloyd's Register Book et le Lloyd's Yacht Register.

Official No. (1)	No. in Book (2)	Registry in which classed (3)	Ship's Name. Material and Rig. Master, and Dates of Appointment to present Owner's Service and to Ship. Ship's Late Name. (4)	Flag. (5)	Net Tonnage. (6)	Dimensions (7): Length	Breadth	Depth	Built (9): Where.	When.	By whom. (10)	Date of Metalling (11)	Owners or Managers. (12)
	751	N V	Glimt (ex Mary M. Bird) *Olsen* W Bk	Norweg'n	609	142·2	31·7	18·8	Rockland, Me.	1867	Starrett	[illegible]	Chr. Christensen
	752	N	Glimt *Pedersen -87* W Bk	Norweg'n	411	135·5	30·1	15·5	Drammen	1880	J. Svendsen	[illegible]	K.S. Bertelsen
	753	N	Glitner *Marcussen* W Bk	Norweg'n	421	136·5	30·7	16·0	Tvedestrand	1874	J. Jörgensen & C.	[illegible]	Jörgenson & Jork[illegible]
	754	N	Glitner *Johansen* W Bg	Norweg'n	228	104·3	25·1	14·4	Christiania	1867	L. Mathiesen		A.M. Johansen
55567	755	❋	Gloamin *Norup* W Bk	British	696	158·0	33·1	19·1	Maitland, N.S.	1871	M'Lennan	[illegible]	L.P. Norup
72030	756	D V	Gloaming *Dinsmore* W S	British	1499	222·3	40·6	24·0	Maitland, N.S.	1879	J. Monteath	[illegible]	J. Monteath
52056	757		Globe *Stanbury* W Bkn	British	292	120·0	28·0	13·1	Cornwallis, N.S.	1865	Northup	[illegible]	A. Miller & Co.
1073	758		Globe *McKevitt -81* W Bg	British	204	95·0	23·3	15·9	Whitehaven	1836	T.&J. Brocklebank		F. M'Kevitt
	759		Globo W Sr	Brazilian	149	..	..	..					A.P. Ribeiro
62498	760	❋	Gloria *Newton* I Bk	British	735	172·0	30·1	19·3	Sunderland	1868	W. Doxford		Swansea Sh'pp'ng Co.
23	761	A	Gloria *Premuda* W Bk	Austrian	752	147·9	32·0	20·0	Lussinpiccolo	1880	A. Cattarinich	[illegible]	M.V. Premuda
923	762	I	Gloria *Basile* W Bg	Italian	354	113·8	28·2	18·0	[illegible]	1872	M. Mauro	[illegible]	F. Basile
	763	V	Gloria *Nielsen 84-84* W Bg	Norweg'n	253	121·0	28·0	13·8	Siken, Sweden	1866	O.P. Aberg	[illegible]	Ths. S. Falck
	764	V	Gloria (ex Kintail) *Meyer 76-86* W Bk	Swedish	726	164·7	30·0	18·3	Picton	1880	McKenzie	[illegible]	C.A. Banck
	765		Gloria (ex Melchiorre) *Hessein 87-87* W Bg	Turkish	253	104·9	30·2	14·8	Capo d'Istria	1852	Piscitello		Hussein Pelinen
[illegible]	766		Gloria Carmela R. *Scandura* W Bn	Italian	104	87·6	21·0	10·7	Gaeta	1880	L. Contreras		Salvatore Denaro R[illegible]
28580	767		Gloriana *Ditchburn 79-80* W Sw	British	169	94·0	23·1	12·8	Amble	1860	Sanderson		H. Andrews
67240	768		Gloriana W K	British	116	101·6	22·8	7·3	Harwich	1871	Vaux		W.T. Middleton
85005	769	V	Glory of the Seas *Truman 67-83* W S	American	2009	240·2	44·1	28·3	Boston, Mass.	1869	D. McKay	[illegible]	J.H. Sears & Co.
65542	770	❋	Gloster *Woodruff 80-83* W 3m Bk	British	304	122·6	27·8	14·9	P.E. Island	1873	Richards	[illegible]	R. Moffat
73992	771		Glynwood *Williams 85-85* W Bkn	British	314	122·0	27·0	14·9	P.E. Island	1875	Longworth	[illegible]	C.W. Jones
	772	N	Gna *Andersen* W Bk	Norweg'n	399	131·6	29·0	16·4	Stavanger	1872	F. Magnus	[illegible]	H.E. Dedekam
	773	N	Gnaa *Abrahamsen* W 3m Sr	Norweg'n	216	116·1	26·0	11·4	Grimstad	1876	Terje Eskedal	[illegible]	J.E. Einersen
	774	G	Go Ahead *Vick 88-88* W Bk	German	391	121·3	30·1	17·0	Ribnitz	1857	J.H. Wilken		A. Burchard
	775		Goan Goan (ex Tjiat Bie, ex Goan Goan, ex Soen Hong) *Lie Ba Tjoe* W Bn	Dutch	186	..	..	..	[illegible]	1863	C.H. Kraus	[illegible]	Lie Loo
76150	776	❋	Goatfell *George* I Bk	British	716	179·4	31·4	18·7	Greenock	1876	J.E. Scott		Ship Goatfell Co. Ld (T.B. Lawson & Co.)
	777	N	Gobi *Svenson* W Bk	Norweg'n	261	116·5	27·4	12·4	Grimstad	1865	Fr. Crawfurd	[illegible]	Anton Brunn
23401	778		Goddess *Dossett 66-66* W Bn	British	142	88·5	22·2	12·7	Ipswich	1842	Wm. Bayley		C. Hart
	779	V	Godeffroy (ex Iris) *Stolz 81-81* W Bk	German	532	147·2	31·6	17·6	Gothenburg	1867	Kierkegaard	[illegible]	A. Burchard
93082	780	❋	Godild W K	British	91	79·8	20·9	10·9	Brixham	1886	J.W. & A. Upham		R. Hellyer
86211	781	❋	Godiva *Roberts 82-82* I S	British	1999	269·0	41·6	24·3	Liverpool	1882	T. Royden & Son		Sandbach, Tinne & Co.
	782	V	Goedhart *Oldenburgen* W Bn	Dutch	155	98·1	25·4	11·9	Pekela	1875	Koerts & Kolk	[illegible]	J.J. Koerts & Zoon
	783	G	Goethe *Ahrens 74-74* W Bg	German	239	101·3	25·8	13·6	Rostock	1859	A. Schönemann		W. Mauck
	784		Goethe (ABANDONED) *Johanssen* W Bk	Norweg'n	530	130·4	30·1	20·1	Bremen	1844			A.P. Andersen
	785	G	Goethe (SUNK) *Bohman 84-83* W Sr		115	85·6	21·3	9·8	Westervik	1872	C.R. Fridell		
7442	786		Goffredo W Bk	British	304	101·7	26·7	16·1	Malta	1850	G. Mirabitur	[illegible]	S. Gauci
	787	N	Gogla *Walderup* W Bk	Norweg'n	318	135·1	29·0	12·6	Drammen	1875	Jörgensen & Knudsen	[illegible]	S.S. Svendsen
86722	788	❋	Gogoburn *Thomson* I Bk	British	1068	217·0	34·2	20·0	Dumbarton	1882	Birrell, Stenhouse & Co		M. Carswell
51143	789	❋	Golconda *Salaw 86-86* I Bk	British	499	159·0	26·7	17·3	Dundee	1866	Brown & Simpson		Richardson & Co.
	790	V	Golconda *Thomas -88* W Bn	German	288	114·2	26·5	13·0	[illegible]	1874	F. Johansen	[illegible]	H.R. Lassen
10131	791	R	Golconda *Hall* W Bn	American	479	128·0	30·0	17·4	E. Boston	1866		[illegible]	J.P. Ellicott
88948	792		Gold (ex Daniel S. Williams) W 3m Sr	British	312	137·7	32·7	10·7	Forked River, N.J.	1858			J. Dransfeld, junr.
64791	793		Gold Hunter W Bk	British	206	104·0	24·8	11·8	Bangor, U.S.	1849			Victoria Sugar Co.
54497	794		Gold Hunter *Crowell* W Sr	British	105	82·0	25·6	8·2	Westfield, N.B.	1867	S. Thibodeau		Samuel Potter
[illegible]	795	❋	Golden Fleece *Way* C Bk	British	500	150·0	29·0	17·1	Sunderland	1869	W. Pile & Co.	[illegible]	H. Ellis
42862	796		Golden Fleece *Armstrong 81-86* W Bk	British	225	118·0	24·5	10·0	New Haven U.S.	1863			J.P. Hendy
67272	797	❋	Golden Fleece *[illegible] -88* W Bn	British	176	101·0	23·3	12·8	Salcombe	1873	Harnden & Co.	[illegible]	H.E. Hayward
85420	798		Golden Fleece *Rosealind* W Sr	American	121	94·5	25·0	9·0	Humboldt, Cal.	1875			W.S. Bowne
	799		Golden Fleece *Ullberg* W Bk	Swedish	371	125·0	24·6	16·1	Sunderland	1863	Pile, Hay & Co.		O. Ullberg
[illegible]	800	❋	Golden Gate *McClure 83-84* I Bk	British	899	195·6	33·7	21·0	Liverpool	1869	T. Royden & Son.		Cotesworth, Lynes & Co.

❋ = **Lloyd's Register**; D = Nederlandsche Vereeniging; G = Germanischer Lloyd; I = Registro Italiano; R = Ameri[illegible];
A = Veritas Austro-Ungarico; H = Veritas Ellenico; N = Norske Veritas; V = Bure[illegible]

In the 4th Column, C = Composite (iron frame, wood planked); I = Iron; Stl = Steel; W = Wood.

The Dates of Metalling (Col. 11) are based on the records of the Books in which Vessels are respectively classed, or on the Reports of the Surveyors of L[illegible]

Further particulars of Vessels classed by Lloyd's Register will be found in Lloyd's Register Book and Lloyd's Yacht Register.

Numéro officiel	Numéro du livre	Registre dans lequel le navire est classé	Nom. Matériaux et Gréement. Constructeurs et dates d'entrée en service des Armateurs actuels et à bord du navire. Ancien Nom.	Pavillon	Tonnage net	Longueur	Largeur	Creux	Construit Où	Quand	Par qui	Date du doublage en métal	Armateurs ou Directeurs-Gérants.	Port d'Armement.
[illegible]	[illegible]	✻	Golden Hope WSk	British	84	73·6	20·5	10·0	Elmshorn	1887	T. Kremer		R.B.Baskcomb	Grimsby
[illegible]	[illegible]		Golden Horn *Shannin* WBn	British	195	101·5	25·9	12·8	P.E.Island	1868	Lefurgey		A.Irving	Liverpool
	[illegible]	N	Golden Horn *Tellman* -88 WBk	Norweg'n	1089	191·3	38·2	22·6	Wiscasset,Me.	1854		Y 2.88	A.P.Andersen	Moss
[illegible]	804		Golden Light *Grenfell* WSr	British	103	89·5	21·5	11·3	Truro	1864	W.S.Martyn		W.Calf	Penzance
[illegible]	[illegible]		Golden Rule *Pinkey* WS	British	1169	183·8	37·5	23·4	Newcastle,US.	1855			K.C.Maguire&Co.	Quebec
[illegible]	[illegible]		Golden Rule WSr	American	101	79·0	24·0	7·0	Ellsworth,Me.	1847			G.W.Rawley	St.George,Me.
[illegible]	[illegible]	✻	Golden Russet *Smith* 77-85 IBk	British	408	150·0	27·1	16·1	Sunderland	1868	Hill,Mousey&Co.		W.Brown & Co.	Sunderland
[illegible]	808	R	Golden Sheaf *Chandler* WBkt	American	431	128·2	31·2	11·4	C.Elizabeth,Me.	1874		Y 10.86	W.S.Jordan&Co.	Portland,Me.
[illegible]	809	✻	Golden Sunset *Saltwater* IBk	British	522	159·5	28·1	17·1	Sunderland	1876	J.Blumer&Co.		A.J.T.Goss	Newport
[illegible]	810	✻	Goldenhorn ImBk	British	1842	268·6	40·2	23·7	Greenock	1883	Russell&Co.		J.R.de Wolf & Son	Liverpool
[illegible]	811	V	Golgatha *Lahse* -87 WBk	Russian	369	120·6	30·0	16·2	Riga	1864	F.Witinsky		C.Essman	Riga
[illegible]	812		Goliah WBk	Siamese	543	152·6	33·6	16·9	Bangkok	1862			Chessua Poh Lawn	Bangkolok
[illegible]	813	R	Gondola *Martin* WSr	British	186	100·1	27·9	9·5	Alma,N.B.	1883	N.H.Foster		O.Martin	St.John,N.B.
[illegible]	814	✻	Gondola *Hood* 80-84 WSrYt	British	107	98·5	20·7	12·1	Gosport	1880	Camper&Nicholson		S.Whitbread,M.P.	Portsmouth
[illegible]	815		Gondolier WFlat	British	100	84·4	19·8	8·7	Winsford	1876	Evans		SaltUnion,Ld.	Liverpool
[illegible]	816	R	Good Intent *Forbes* 73-73 WBk	British	399	126·3	28·3	16·6	Montrose	1869	Strachan	Y 10.87	J.Mitchell	Montrose
[illegible]	817	✻	Good Templar *Burnitt* WK	British	74	73·4	19·7	9·3	Goole	1881	Outwin		E.L.Burnitt	Goole
[illegible]	818	✻	Good & Blanchard *Hudder* -86 ISk	British	87	81·2	20·3	10·2	Hull	1885	Cook,Welton&Gemmell		B.W.Hellyer	Grimsby
[illegible]	819		Goodwill *Sweeting* WBg	British	106	70·3	21·5	11·2	Sunderland	1785			W.Earl	London
[illegible]	820	✻	Goodwood *Tudor* -86 WBk	British	583	119·4	29·7	18·3	Littlehampton	1880	Harvey	Y 8.88	A.G.Moffat	Swansea
[illegible]	821	✻	Goonlaze *Coon* 82-83 WSr	British	79	76·2	21·3	10·2	St.Agnes	1874	Hitchins	Y 4.88	M.Hitchins & Co.	Hayle
	822		Göran *Westin* -87 WBk (ex Prid... Wave)	Swedish	382	120·7	26·4	17·6	Sunderland	1859	Pickersgill		A.A.Westin	Tjockö, Roslgn
	823	N	Gordon *Gundersen* WBk (ex William)	Norweg'n	757	158·9	34·0	19·3	P.E.Island	1862	Yeo,Jun.	Y 3.86	M.H.Smiths Bo	Arendal
[illegible]	824		Gordon Jerry WSr	British	110	102·0	24·7	6·1	RiverRascom,Ont.	1884	J.Duprey		Emily Duprey	Windsor,Ont.
[illegible]	825	✻	Gordon'sCharge *Petersen* -87 WSk	British	90	78·6	20·9	10·5	Grimsby	1887	SmithBros.		W.J.S.Hood	Grimsby
	826	N	Gorilla *Gulbrandsen* -88 WBk (ex Aurora)	Norweg'n	387	119·8	31·0	16·5	Gothenberg	1850			L.Jacobsen	Horten
	827	V	Gorm (ex Marie) *Svane* 79-81 WSr	Danish	115	88·3	20·5	10·8	Rendsborg	1866	Hartvig,Holler&Co.		N.Petersen	Marstal
	828		Gorontalo Packet W3mSr	Dutch	133	..	..	..					Lie Boen Yal	Menado
	829	G	Goschen *Hillerns* 76-87 WBk	German	1137	171·7	37·9	23·6	Vegesack	1860	Joh.Lange	Y 12.87	D.H.Wätjen & Co.	Bremen
[illegible]	[illegible]	✻	Goshawk *Starkey* 74-82 WSrYt	British	143	107·8	23·0	12·0	Cowes	1865	C.Hansen&Sons	Y 1.88	J.R.West	Cowes
	831		Gota *Boelin* W3mSr	Swedish	144	93·5	19·3	11·5	Skeen	1865			A.Nilsson	Håseback,Malmöh,Co
	832		Gotfred *Andersson* WSr	Swedish	203	96·5	23·0	12·5	Rostock				G.Sjöberg	Malmö
	833	V	Gotha *Schicken* 84-84 CBk (ex ... ex Bilkh...)	German	810	167·2	32·0	21·1	London	1864	J.Major	Y 4.87	L.F.Mathies & Co.	Hamburg
	834	G	Gotland *Isakson* -88 WBk (ex Hermine ex Mirabilis)	Swedish	269	118·7	22·6	13·7	Brake	1859	L.J.Paulsen		C.G.Idesströn	Oskarshamn
[illegible]	[illegible]		Gotoma *Nelson* WSr	American	189	119·0	32·6	8·6	CoosBay,Oreg.	1872			A.M.Simpson	San Francisco
	836		Gottfrid *Petterson* WBg (ex Th...)	Swedish	253	95·1	24·1	15·6	Sunderland	1851			C.A.Svensson	Bergqvara
	[illegible]		Gouverneur *Wilsen* -88 WBn	Norweg'n	177	107·2	24·8	10·0	Amsterdam	1867	G.Broerse	Y 1.88		
[illegible]	838	✻	Governor *Edmunds* -88 WBk	British	576	114·4	32·5	17·9	P.E.Island	1877	Yeo	Y 2.88	W.Williams & Sons	Newport
[illegible]	839		Governor WSr	British	176	113·0	24·0	9·5	Kingston,Ont.	1847			Wm.Power	Kingston,Ont.
[illegible]	840		Governor *Eaton* WSr	American	121	86·3	25·0	8·2	Seaville,Me.	18 51/75			J.Watson	Sedgwick,Me.
[illegible]	841	RV	Governor Goodwin *Drew* 78-88 WS	American	1414	212·0	40·0	24·0	Boston	1877	Campbell&Brooks	Y 1.87	M.F.Pickering&Co.	Boston,Mass.
[illegible]	842	R	Governor Hall *Cain* W3mSr	American	560	112·3	34·0	16·7	Milford,Del.	1880	J.W.Abbott&Co.		M.Cain	Philadelphia
[illegible]	843		Gov.James Y. Smith *Nickerson* WSr	American	305	114·1	29·0	9·9	Warren,R.I.	1866			A.Nickerson	Providence,R.I.
[illegible]	844	RV	Governor Robie *Blanchard* WS	American	1627	224·1	41·0	23·8	Bath,Me.	1883	W.Rogers	Y 2.88	W.Rogers	Bath,Me.
[illegible]	845	V	Governor Wilmot *Morris* 82-88 WS	British	1611	214·6	41·4	24·7	St.John,N.B.	1873	Cruickshank&Pittfield	Y 9.88	J.W.Scammell	St.John,N.B.
[illegible]	846	✻	Gowan *Taylor* 86-78 WSr	British	135	95·0	22·7	11·1	Banff	1878	Geddie	Y 3.88	J.W.Simpson	Banff
[illegible]	847	✻	Gowanburn *McKay* 86-88 StlImSk	British	1999	289·1	42·2	23·7	Greenock	1886	Scott&Co.		R.Shankland & Co.	Greenock
[illegible]	848		Goward *Tafford* -86 WBn	British	166	93·5	22·9	12·5	Sunderland	1857	Abbay&Co.		J.W.Tutt	Rye
[illegible]	849		Gower WSr	British	105	66·0	18·3	11·2	Ipswich	1824	Wm.Bayley		Brazilian Coal Co., Ld.	St.Ives
[illegible]	[illegible]	✻	Gowerian *Jones* -87 WSr	British	99	79·1	21·2	11·0	Aberayron	1856	EvanJones		J.Jones	Milford

✻ = Lloyd's Register; D = Nederlandsche Vereeniging; G = Germanischer Lloyd; I = Registro Italiano; R = American Record; A = Veritas Austro-Ungarico; H = Veritas Ellenico; N = Norske Veritas; V = Bureau Veritas.

Dans la 4me Colonne, C = Composite (membrure en fer, bordé en bois); I = Fer; Stl = Acier; W = Bois.

Les Dates du doublage en métal (Col. 11) ressortent des livres dans lesquels les navires sont respectivement classés ou des rapports des Inspecteurs du Lloyd's Register.

Pour plus amples renseignements sur les navires classés par le Lloyd's Register, voir le Lloyd's Register Book et le Lloyd's Yacht Register.

Official No.	No. in Book.	Registry in which classed.	Ship's Name. Material and Rig. Master, and Dates of Appointment to present Owner's Service and to Ship. Ship's Late Name.	Flag.	Net Tonnage.	Length	Breadth	Depth	Where.	When	By whom.	Date of Metalling.	Owners or Managers.
1	2	3	4	5	6	7			8	9	10	11	12
	851	✳DV	Graafstroom *Tersma -88* WS	Dutch	1359	200·1	40·0	25·2	Alblasserdam	1879	J.Smit&Co.	[illegible]	J.Smit Czn.
87151	852	✳	Grace *Trussell 84-84* WSr	British	139	97·3	23·3	11·1	Ipswich	1883	W.Bayley&Sons	[illegible]	H.Harrington&Co.
82548	853		Grace *Charles* W3mSr (ex Hendrika Grietje Alzina)	British	122	104·7	20·0	10·6	[illegible]	1867			T.Smart
1175	854		Grace (WRECKED) *Fisher* WBg	British	122	71·1	23·3	13·1	[illegible]	1824			T.Brown
85350	855		Grace Andrews *Watts -87* W3mSr	American	540	148·8	33·3	[illegible]	Thomaston, Me.	1874	Stetson,Gerry& [Co.	[illegible]	S.S.Gerry
85289	856	R	Grace Bradley *Smith -88* W3mSr	American	530	146·0	32·0	13·0	Thomaston,Me	1873	S.Watts	[illegible]	S.Watts
88221	857	R	Grace Butler *Balcom* WBn	British	150	92·3	24·5	10·6	SheetHarbour, [N.S.	1883	W.Routledge	[illegible]	James Butler
84767	858		Grace Carter WSr	British	102	83·3	23·0	8·4	BayofIslands, [N.F.	1881	J.McLeod		J.Baird
85385	859		Grace Cushing *[illegible] -88* WS	American	150	101·3	27·9	7·8	Pittston,Me.	1870			J.E.Stafford
85137	860	R	Grace Davis *Dyer 74-83* W3mSr	American	374	127·7	31·1	11·3	Yarmo'th Mass	1873	Hutchins& [S[illegible]	[illegible]	W.S.Jordan & Co.
85483	861	RV	GraceDeering *Pratt 80-84* WBk	American	694	151·8	33·1	18·3	CElizabeth,Me	1877	J.F.Randall	[illegible]	B.Webster
85834	862	R	Grace F. Littleton *Ross* W3mSr	American	161	98·0	25·5	9·9	Bath,Me.	1884	T.M.Hagan&Co.		S.A.Whorf
60012	863	✳	Grace Gibson *Lloyd* IS	British	540	171·4	27·3	17·5	Glasgow	1867	AStephen&Sons		H.W.Hewitt
85852	864	R	Grace Gower *Wilson* W3mSr	American	241	120·0	34·0	8·5	Calais,Me.	1884	O.B.Rideout	[illegible]	W.D.Gower
	865	✳	Grace Harwar StlS	British	1800	265·3	39·2	23·5	PortGlasgow	1889	W.Hamilton & Co.		W.Montgomery & Co.
85835	866		Grace K.Green *Seaward* W3mSr	American	385	140·0	34·0	12.2	Portsmouth. [N.H.	1884	W.Fernald		C.E.Walker
	867		Grace Marks *Seaward* WBk	Chilian	256	106.3	24·6	..					Cia.Chileña de Balles
85425	868	R	Grace Rice *Marshall* WSr	British	145	93·6	25·0	9·2	WeymouthNS.	1883	T.C.&E.Rice		T.C.Rice
10870	869		Grace Roberts *Larsen* WBkn	American	255	129·5	32·0	9·0	PortOrchard, [Wash.	1868			Glidden & Co.
85998	870	R	GraceSeymour *Holmes* W3mSr	American	602	166·0	34·2	16·0	Fair Haven	1888	H.H.Hanscom		H.H.Hanscom
85336	871		Grace van Dusen *Lee* W3mSr	American	288	125·6	33·0	9·0	Millbridge,Me.	1874	Z.Dyer		L.Matthews
10563	872		Grace Watson *Thomas* WSr	American	106	96·2	21·9	7·1	Philadelphia	1854			J.Ewing
10721	873		Grace Webster *Jewell* WSr	American	321	118·9	29·2	12·4	Portland,Me.	1867	Curtis	[illegible]	L.H.McKenny
18070	874		Graceful WSr	British	125	88·5	20·3	11·3	Wells	1857			E.Shortus
88194	875	✳	Gracie *Beard* WSk	British	84	76·0	21·0	10·4	Goole	1884	CottinghamBrs.		J.Beard
85995	876	R	Gracie D. Buchanan W4mSr	American	1084	194·0	40·0	18·1	Bath,Me.	1888	New England [S.B.Co.		W.A.Anderson
90857	877		GracieD.Chambers W3mSr (ex Mary C. Arnold) *Chambers*	American	379	133·5	33·0	14·0	EastRiv.,Conn	1875			W.H.Kenzie
	878		Gracieuse WSr	French	129	..	..	..	[illegible]	1853			H.Vanrapenbusch
556	879	A	Grad Karlovac *Stipanovich -87* WBk	Austrian	497	141·7	26·4	18·2	Portoré	1869	I.Arcadin	[illegible]	E.Giusti fu V.
1071	880		Grad Zagreb *Prassen -87* WBk	Austrian	537	144·3	26·3	17·9	Portoré	1871	I.Arcadin	[illegible]	G.Mikulicich
	881	G	Graf Bismark *Kleist 79-79* WBg	German	248	95·6	24·2	12·9	Rostock	1867	J.Möller		Lichenheim & Finen
	882	V	Graf Eulenburg *Huisenga* W3mSr	German	271	104·3	26·6	12·8	Jemgum	1878	G.&H.Aden	[illegible]	Huisenga
	883	DV	Graf Otto zu Solms *Vick* W3mSr	German	235	104·3	25·4	13·6	Stralsund	1870	O.Kirchhoff	[illegible]	Seitz & Kindt
	884		Graf von Wrangel (WRECKED) *Tietz 87-* WBk	German	299	121·5	27·0	14·1	Colberg	1866	F.Richter		C.F.Hackbarth
70068	885	V	Grampian *Nicholas 87-87* WS	British	1493	210·4	39·9	24·1	St.John,N.B.	1874	Cruickshank	[illegible]	H.Fernie & Sons
	886	R	Gran Canaria *Circoinas -87* WS	Spanish	385	129·9	31·2	15·0	LasPalmas	1860	S.Arosena	[illegible]	P.Arosena & Co.
85293	887	R	Granada *Carle -88* WBkn	American	376	133·0	31·4	12·1	Damariscotta, [Me.	1873	B.D.Metcalf	[illegible]	B.F.Metcalf & Co.
	888	✳V	Grand Citoyen *Jeray -83* WBk	French	336	113·8	27·2	15·4	Nantes	1873	Tillé	[illegible]	J.B.Demange
90463	889	DV	Grandee *Ellis* WS	British	1578	225·9	42·0	23·7	Maitland,N.S.	1884	J.Monteith	[illegible]	A.Smith
85280	890	RV	Grandee *Winn -83* WS	American	1295	193·0	38·5	23·8	Portsmrth.NH.	1873	Tobey&Little- [field	[illegible]	C.H.Mendum
	891	N	Grane *Knudsen* WBk	Norweg'n	387	127·6	28·5	16·2	[illegible]	1877	A.J.Andreasen		Et Aktieselskabet (T. Jensen)
	892		Granen (CONDEMNED) *Balversen* WBk	Norweg'n	298	128·3	29·0	12·1	Sandefjord	1873		[illegible]	P.A.Grön
	893		Granit (FOUNDERED) *Bach* WBk	Norweg'n	753	170·2	35·5	19·1	Christiansand	1874	H.Olsen	[illegible]	H.G.Kirschon
603	894		Granite (WRECKED) *Lang 85-85* WBg	British	180	109·3	23·5	13·4	Aberdeen	1846			T.Jefferson
85408	895	V	Granite State *Ross -87* WS	American	1684	228·9	41·4	24·0	Kittery,Me.	1877	J.Neal	[illegible]	D.Marcy
73041	896		Grantham WSr	British	325	140·0	26·0	12·0	PortRobinson, [Ont.	1873	J.P.Abbey		H.A.Breck
92890	897	✳	Granville WSk	British	86	78·0	21·0	10·5	Grimsby	1886	SmithBros.		Smith Bros.
69886	898	DV	Granville Belle *Whitney* WBk	British	1129	179·3	38·5	21·9	Granville,N.S.	1874	J.W.Olivier	[illegible]	J.E.Shafner
68776	899	✳	Graphic *Le Seelleur* WSr	British	60	75·1	18·9	8·8	Jersey	1874	LeHuquet	[illegible]	P.LeSeelleur&Co.
74486	900	✳	Grasmere *Carter* IBk	British	1246	226·0	36·1	22·1	Whitehaven	1875	Whitehaven [S.B.Co		Fisher & Sprott

✳=Lloyd's Register; D=Nederlandsche Vereeniging; G=Germanischer Lloyd; I=Registro Italiano; R=Ame[rican]
A=Veritas Austro-Ungarico; H=Veritas Ellenico; N=Norske Veritas; V=Bu[reau]

In the 4th Column, C=Composite (iron frame, wood planked); I=Iron; Stl=Steel; W=Wood.

The Dates of Metalling (**Col. 11**) are based on the records of the Books in which Vessels are respectively classed, or on the Reports of the Surveyors

Further particulars of Vessels classed by Lloyd's Register will be found in Lloyd's Register Book and Lloyd's Yacht Register.

No.	Reg.	Nom. Matériaux et Gréement. Capitaines et dates d'entrée au service des Armateurs actuels et à bord du navire. Ancien Nom.	Pavillon.	Tonnage.	Longueur	Largeur	Creux	Où.	Quand	Par qui.	Doublage	Armateurs ou Directeurs-Gérants.	Port d'Armement.
900		Grasmere *Sharpe* W Bk	British	423	142·7	29·5	17·5	Sunderland	1865	G.Gardner		[illegible]Co.Ld.	Melbourne
901	❋	Grassendale *Forbes 86-87* I S	British	1819	267·0	39·1	23·6	Workington	1885	R.Williamson & Son		[illegible]	Liverpool
902		Grasshopper *Webb* W Bn	British	154	95·1	20·0	12·3	Truro	1837			F.A.Rosamen	Southampton
903	❋	Gratia *Nosworthy 83-85* W Bn	British	157	97·1	21·8	12·4	Scilly	1874	Edwards	[illegible]	A.Goodbridge & Sons	St.John's,N.F.
904	D	Gratia *Andersen* W Bk [illegible]	Norweg'n	484	117·6	33·0	16·5	Uleåborg	1859		[illegible]	J.M.Bryde	Sandefjord
905		Gratidão *Ferreira* W Bk	Portug'se	246	99·1	24·3	13·0	Lisbon	1856	B.J.d'Azevedo		J.P.Borges	Lisbon
907		Gratitud W Bn		194	91·9	26·7	13·9	Barcelona	1855				
908		Gratitude *Holt 81-90* W Bkn	British	298	126·0	26·0	15·2	Whitstable	1875	Dyason		F.W.Fairbrass	Rochester
909		Gratitude *Rogers* W Ketch	British	122	76·0	22·1	9·3	Terrigal	1888	T.Davis		W.Buckingham	Sydney.N.S.W.
910	N	Gratitude *Jacobsen* W Bg	Norweg'n	193	82·2	25·0	13·9	Tönsberg	1827	Abrahamsen		A.C.& A.J.Jacobsen	Tönsberg
911	V	Gravelinoise *Decottier -87* W Sr	French	142	[illegible]	21·2	11·0	Dunkirk	1887	A.Pyotte		Torris	Gravelines
912		Gravia *Mitropulo* W Bg	Greek	[illegible]	124	26·8	..	[illegible]	1876		[illegible]	J.D.Mitropoulos	Galaxidi
913		Gray Eagle *Powell -87* W Bk [illegible]	American	[illegible]	[illegible]	31·4	15·1	Philadelphia	18[illegible]	C.T.Noank		M.L.Bandel	Baltimore
914		Gray Hound *Allen* W Bk	American	[illegible]	95·6	24·4	12·2	New Bedford, Mass.	1851			J.Medina	Boston,Mass.
915	I V	Grazia [illegible] *Trapani -88*	Italian	767	[illegible]	32·0	22·8	[illegible]	1876	G.Mauro		G.Trapani	Castellamare
916	N	Graziella *Andreasen* W Bk	Norweg'n	[illegible]	132·2	29·4	16·0	Sikea, Sweden	1873	O.P.Alberg	[illegible]	A.Kristoffersen Vatne	Farsund
917	V	Great Admiral *Rowell* W S	American	1497	214·2	40·3	25·3	East Boston	1869	R.E.Jackson	[illegible]	W.F.Weld & Co.	Boston,Mass.
918		Great Basses W Sr	British	198	116·[illegible]	21·3	10·8	London	1869			Corporation of Trinity House	London
919	❋	Great Victoria [illegible] I S	British	[illegible]	286·[illegible]	[illegible]	[illegible]	[illegible]	1854	Guibert & fils		W.T.Dixon & Sons	Liverpool
920	❋	Grecian *Trowbridge 84-87* I Bk	British	1272	223·[illegible]	36·2	22·5	Greenock	1868	Scott&Co.		J.Nourse	London
921		Grecian Bend *Rowe* W Bkn	British	21[illegible]	107·[illegible]	23·8	11·8	Tranmere	1873	Harland Brown & Co.		J.Y.Phillips	Sydney.N.S.W.
922		Grecian Bend *Layton* W Sr	British	188	97·7	27·1	11·0	Hantsport,N.S.	1873			G.W.Churchill	Windsor,N.S.
923	❋	Green Olive *Robinson* W Bg	British	244	101·5	26·2	14·8	Hylton	1865	Gray&Young		J.Robinson	Littlehampton
924	V	Greenock *Johnsen 87-87* W S	Norweg'n	1225	198·[illegible]	[illegible]	23·3	Quebec	1873	Marquis	[illegible]	Johan Thorsen	Haugesund
925		Greenwich W Bn	British	214	97·[illegible]	24·5	13·6	[illegible]	1850	Head Bros.		W.W.Foreman	Faversham
926	❋	Greenwood Tree W K	British	64	70·6	19·1	9·0	Bideford	1886	H.M.Restarick		T.&J.Leleu,Ld.	Gt.Yarmouth
927		Gregoria W Bn	Spanish	135	[illegible]	20·8	9·0	Dasol	1884	S.Melliza		Sabas Melissa	Manila
928		Gregorio W Bn	Spanish	117	84·0	19·6	9·4	Pangasinan	1872			J.de Agubita	Manila
929	V	Grenada *Parks -88* W Bkn	British	63[illegible]	16[illegible]	34·6	15·1	Horton,N.S.	1888	J.B.North	10.88	A.Haley	Windsor,N.S.
930		Grenadier *Blondel 85-85* W Bk	French	527	151·5	28·5	18·5	Liverpool	1862	R.& J.Evans & Co.	[illegible]	E.Lavergne	Dunkirk
931	❋	Greta *Cassidy 82-83* I S	British	[illegible]	226·0	36·2	21·9	Whitehaven	1874	Whitehaven S.B.Co.		Lowden,Edgar & Co.	Liverpool
932	G	GretaMaria *Kalning 81-90* W Sr	Russian	176	93·8	24·5	12·7	Haynasch	1880	J.Kalning		J.Kalning	Haynasch
933		Gretchen *Langer 65-* W Sr	German	124	87·[illegible]	20·3	10·5	Papenburg	1854	F.W.Meyer		F.W.Meyer Wwe	Papenburg
934	V	Grethe *Hansen 72-72* W Sr	Danish	146	95·[illegible]	23·6	10·4	[illegible]	1872	L.Kaas		N.W.Kaas	Svendborg
935	N	Grethe [illegible] *-88* W Bk [illegible]	Norweg'n	399	124·6	25·8	16·8	Sunderland	1863	G.&J.Mills	[illegible]	A.C.Nielsen	Kragerö
936	❋	Gretna *Young 81-83* I S	British	1705	264·9	3[illegible]·8	23·4	PortGlasgow	1883	R.Duncan&Co.		T.C.Guthrie	Glasgow
937	V	Greve Frijs *Jansen 88-88* W Bn	Danish	174	100·1	22·1	11·6	Horsens	1872	Chr.Schrøder	[illegible]	Actieselskab "Bjerre-[illegible]"	Horsens
938		Grevinde Karen Wedel [illegible] W Bk	Norweg'n	313	104·0	23·4	16·3	Drammen	1812			J.M.Jacobsen	Arendal
939	❋	Greyhound *Williams 87-87* W 3m Sr	British	167	120·5	22·8	11·[illegible]	[illegible]	1886	W.Thomas&Co.	[illegible]	W.Thomas	Liverpool
940	❋	Greystoke Castle *Clemence 81-86* I S	British	[illegible]	262·0	39·0	23·6	Workington	1886	R.Williamson & Son		GreystokeCastle S.Co.Ld. (J.Chambers & Co.)	Liverpool
941	NV	Grid *Olsen 81-81* W Bk	Norweg'n	[illegible]	[illegible]·8	30·8	16·2	Tvedestrand	1876	M.Eldrup	[illegible]	Carl Bech	Tvedestrand
942	V	Grietje Krythe [illegible] *Gerler -86* W Bkn	Dutch	147	88·6	22·3	10·9	Hoogezand	1867	L.H.Wildervank		W.Bos	Groningen
943		Grigorii W Sc	Russian	181	..	..	..		1873			G.& S.Hermanoff	Impilaks
944	D	Grigorius W Sc	Greek	296	..	..	..	Syra	1876			N.Grigoriadis	Syra
945	❋D	Grijalva *Mitchell 87-87* I Bkn	British	290	150·4	24·1	11·0	Llanelly	1874	Nevill Bros		T.Harrison & Co.	Liverpool
946		Grimaldi W Sr	British	14[illegible]	81·5	23·4	12·7	Liverpool	1841			R.Harris	Fowey
947	I	Grimaldo *Schiaffino* W Bk	Italian	55[illegible]	14[illegible]·0	32·8	19·9	Varazze	1871	B.Camogli	[illegible]	P.Schiaffino	Genoa
948		Grimsby C Sr	British	391	137·8	26·2	11·8	St.Catharine's, Ont.	1874	M.Simpson		James Matthews	St.CatherinesOnt
949	N	Grimstad *Rolleguard* W Bg	Norweg'n	33[illegible]	[illegible]·2	27·0	12·2	Grimstad	1859	M.Smith Petersen		G.Hiuseland	Grimstad
950													

❋=Lloyd's Register; D=Nederlandsche Vereeniging; G=Germanischer Lloyd; I=Registro Italiano; R=American Record; A=Veritas Austro-Ungarico; H=Veritas Ellenico; N=Norske Veritas; V=Bureau Veritas.

Dans la 4me Colonne, C=Composite (membrure en fer, bordé en bois); I=Fer; Stl=Acier; W=Bois.

Les Dates du doublage en métal (Col. 11) ressortent des livres dans lesquels les navires sont respectivement classés ou des rapports des Inspecteurs du Lloyd's Register.

Pour plus amples renseignements sur les navires classés par le Lloyd's Register, voir le Lloyd's Register Book et le Lloyd's Yacht Register.

Official No.	No. in Book	Registry in which classed	Ship's Name. Material and Rig. Master, and Dates of Appointment to present Owner's Service and to Ship. Ship's Late Name.	Flag	Net Tonnage	Length	Breadth	Depth	Built. Where	Built. When	Built. By whom	Date of Metalling	Owners or Managers
	951	V	Gripen Von Matérn 74-74 WS (ex J. F. Foster)	Swedish	1172	179·2	32·8	23·5	Medford, N.J.	1870	J.T.Foster	Y 1.87	Rederiaktbol. Grans (C.F.Strandberg)
	952	❋	Gripen Jansson -86 W Bk (ex Oriole)	Swedish	377	123·3	24·9	17·0	Sunderland	1857	W.H.Pearson	Y 3.86	L.D.Kemner
	953	N	Griqua Pedersen W Bk	Norweg'n	295	129·1	26·8	12·7	Alloa	1873	Roy	Y 11.87	P.Pedersen
70944	954	❋	Grisedale Campbell 84-84 I S	British	1222	226·7	36·1	21·7	Liverpool	1875	W.H.Potter&Co		J.D.Newton & Co.
1546	955	A	Grobnik (ex Pietro) Rubani -88 W Bg	Austrian	362	117·7	29·8	17·5	Fiume	1865	G.Giacovcich	Y 8.87	E.Durbessich fu G.
	956	N	Grönsvœr Bastrup W Bk	Norweg'n	474	131·9	30·5	17·6	Grimstad	1875	F.Smith Petersen	Y 1.86	F.Smith Petersen
60946	957	❋	Grosvenor Bulkeley -87 I Bk (ex Spirit of the South)	British	517	165·2	28·1	17·3	Sunderland	1869	Iliff, Mounsey & Co		P.F.Bulkeley
91397	958	❋	Grovehill I Sk	British	90	79·4	20·3	10·1	Beverley	1884	Cochrane&Co.		J.W.Hellyer
	959		Grumete W3mSr	Brazilian	130	..	..	..					Lindon C.F.Mendens
	960	DN	Grundloven Ellingsen W Bk (ex Independenza) (Thoes	Norweg'n	809	164·3	34·1	21·6	SestriPonente	1874	G.Peragallo	Y 11.81	Ths.S.Falck
53437	961	❋	Gryfe Roberts I S	British	1069	207·1	33·1	21·6	Greenock	1866	R.Steele		J.Russell
	962		Guadalquiver Andrew 86-86 W3mSr	Peruvian	178	110·2	27·2	9·7	Chili	1871	Court Hermanos		J.Portella & F.Andrew
	963		Guadalquivir Linare -88 W Bn	Spanish	328	121·3	27·5	14·1	Stockton, Me.	18 63/84			J.J.Murga
	964		Guadalupe Riera W Sr	Spanish	147	96·1	23·4	10·9	Masnou	1853			D.F.Vives y Alsina
	965		Guadiana (ex Carvalho) W Bg	Portug'se	233	95·5	23·1	10·1	Lisbon	1852			M.R.de Lima
	966	V	Guadiana De Conto W Bkn	Portug'se	232	97·4	24·3	12·6	Lisbon	18 74	Vareta	Y 6.86	J.J.Rodriguez
	967		Guahyra W Bn	Brazilian	174	..	..	..					J.M.de Oliveira
54883	968		Guam Mavns W Sr	Japanese	291	112·8	25·7	16·8	Sunderland	1866	W.Adamson		A.Clark
	969		Guanabara Carvalho W Bg (ex Angel)	Brazilian	155	..	..	..	Bilbao	1867		Y 8.86	F.M.de Carvalho
	970		Guapo W Sr	Brazilian	171	..	..	..					C.E.Meury
	971		Guarany W3mSr	Brazilian	160	..	..	..					J.A.Martins
	972		Guarany Pires W3mSr (ex Lydia B. Rowl)	Brazilian	151	98·6	22·7	10·1	Southwick	1864	Pickersgill	Y 8.84	J.da Rocha Souza
10896	973		Guardian Marsden -87 WS	American	1073	173·0	38·8	23·5	Damariscotta, Me.	1863			W.G.Adams&Co.
	974		Guaresma W Sr	Portug'se	116	80·7	21·0	8·8	Fão	1882	D.D.S.Borda		J.C.Guaresma
	975	V	Guayaquil Calvé -87 W Bk (ex Meta)	French	372	122·7	27·8	16·6	Rostock	1867	Padderatz	Y 1.86	J.L.Tréhondart
	976		Guayaquil Font -87 W Bk	Spanish	313	83·3	29·5	14·6	Blanes	1859			José M.Serra
	977		Guayaquilena W Bk (ex Eliza Schmidt)	Ecuador	396	..	..	..	Altona	1860			A.Monti Verde
	978	❋V	Guaymas Burmester 81-87 I Bk (ex Sancta Elisa)	German	497	149·6	26·6	17·2	Harrington	1863	R.Williamson [&Son		Oetling Gebrüder
	979		Guaymas Krag 86-83 W Bk (CONDEMNED) (ex Esperanza)	German	306	121·6	26·9	13·8	Hamburg	1864	C.Loesby	Y 10.83	H.M.Krag
30254	980		Gudgeon W Sr	British	102	83·2	21·8	10·1	P.E.Island	1857	Pope		N.Roe
86186	981	V	Gudrun W Bn	British	228	115·3	25·1	12·5	Varazze	1874		Y 2.86	G.H.Bate
	982	❋	Gudrun Prehn 83-86 I Bk	German	1424	248·0	37·6	21·3	Rostock	1886	Rostock S.B.Co.		C.M.Matzen
	983		Gudrund Harthun W Sr (WRECKED)	Norweg'n	124	76·2	29·3	10·3	Stenkjær(Ttms)	18 81			L.Haug
	984		Gudvang Schmidt -86 W Bk (WRECKED) (ex Maria Maddalena)	Norweg'n	495	143·2	31·7	16·7	Sundsvall	1864	J.A.Nordström	Y 10.86	K.S.Bertelsen
	985		Guerra I. (ex Conceição) W Bn	Portug'se	136	83·0	20·3	10·3	Esposende	1873			D.da Costa Guerra
3427	986		Guglielmina Tonietti W Bk (ex Amelia)	Italian	229	100·0	26·2	15·9	Sira	1866			Anna Lertora
1082	987	I	Guglielmo C. Stanford M. Maresca W Bk	Italian	434	136·1	29·0	18·7	Cassano	1874	F.De Rosa	Y 12.86	G.Maresca
15	988	DV	Guglielmo D. (CONDEMNED) Pasinovich -87 W Bk	Austrian	522	134·1	26·5	19·0	Capo d'Istria	1877	L.Polli	Y 10.84	G.Druscovich
1332	989	IV	Guglielmo T. Treglia -88 W Bk (ex Lui Matoaka)	Italian	237	105·9	25·5	13·9	Meana	1874		Y 8.86	G.B. & E Treglia & figli
80631	990	RV	Guiana Ossinger -88 W Bk	British	1266	190·0	38·3	22·6	Tusket, N.S.	1882	T.Killam	Y 8.86	T.Killam
40893	991		Guiana W Bg	British	172	79·0	20·8	13·0	Quebec	1838			Henry Moss
	992	D	Guiana Gerdes -88 W Bk (ex Zanzibar)	Dutch	352	132·2	28·2	15·1	Reiherstieg	1866	E.Dreyer	Y 10.88	P.H.Craandijk & Zonen
454	993	V	Guida Celeste Giannoni W Bg	Italian	169	98·4	24·3	13·6	Limite	1857	N.Pichcotti	Y 8.87	Fili Giannoni fu P.
59779	994		Guide Allen W Bg	British	180	101·0	24·8	12·5	P.E.Island	1870	Ellis		W.Harnett
25893	995		Guide Elliott W Sr	British	122	98·8	17·7	11·2	Dartmouth	1854	Holman& Kelly		F.Nurse
85970	996		Guide Thorsen -87 W3mSr	American	137	126·0	26·7	7·4	San Francisco	1887	M.R.Roberts		G.W.Chandler
58144	997	❋	Guiding Star Skin 84-84 W Bk	British	349	121·8	27·6	16·7	Sunderland	1868	W.H.Pearson, jr	Y 8.87	J.Winship & Co.
62435	998		Guiding Star Skin W Bkn	British	249	117·0	25·5	13·4	Banff	1869	Watson	Y 8.85	P.Facy
80406	999	❋	Guiding Star Trethewey (WRECKED) (75-79 W Sr	British	100	88·5	23·0	10·9	Newquay	1879	Clemens		N.Lewarne
1966	1000	A	Guido Gelletich W Bg	Austrian	327	114·5	25·1	16·4	Bra	1869	Schiavon	Y 8.87	A.Gelletich

❋ = **Lloyd's Register**; D = Nederlandsche Vereeniging; G = Germanischer Lloyd; I = Registro Italiano; R = ...
A = Veritas Austro-Ungarico; H = Veritas Ellenico; N = Norske Veritas; V = ...

In the 4th Column, C = Composite (iron frame, wood planked); I = Iron; Stl = Steel; W = Wood.

The Dates of Metalling (**Col. 11**) are based on the records of the Books in which Vessels are respectively classed, or on the Reports of the Surveyors ...

Further particulars of Vessels classed by Lloyd's Register will be found in Lloyd's Register Book and Lloyd's Yacht R...

N°	Classe	Nom. Matériaux et Gréement. Capitaine et dates d'entrée. Ancien Nom.		Pavillon	Tonnage net.	Longueur	Largeur	Creux	Où.	Quand	Par qui.	Doublage	Armateurs ou Directeurs-Gérants.	Port d'Armement.
1001		Guido	W Bk	Italian	436	126·3	30·6	19·1	Varazze	1864	N.Cerruti			Leghorn
1002		Guido *Lindqvist*	W Bk		353	124·6	30·2	16·1	Tenala	1876	C.Jansson		G.Kullin	Bromarf
1003		Guido-Rosa	W F	Italian	525	137·8	31·2	19·7	Sestri Ponente	1867	A.Briasco fu L.		Ant. Arcangelo Vianello	Venice
1004	❋	Guild Mayor *F. Fallon*	W Sr	British	96	93·0	22·6	9·9	Preston	1862	Smith		P.Terman	Drogheda
1005		Guilherme	W Sr	Portug'se	116	77·1	18·7	10·3	Aveiro				V.I.d'Arruda & Co.	Ponta Delgada
1006		Guilhermina	W Bn	Brazilian	250	..	..	..					J.V.Barreto	Rio de Janeiro
1007	V	Guilhermino&Irmaãs	W Bn	Portug'se	138	96·0	24·4	10·0	Liverpool,N.S.	1877			G.R.deGouvea	Lisbon
1008	V	Guillaume Tell	W Bk	French	143	80·1	20·7	11·1	Means	1874	Ollivande		Pitre Legal	Nantes
1009		Guillerma	I Sr	Argentine	133	108·0	26·3	7·9	Paisley	1883	J.McArthur&Co		C.T.Getting	Buenos Ayres
1010	V	Guillermina	W 3mSr	German	241	108·2	23·0	12·3	Arnis	1871	F.Johansen		A.H.Wappäus	Hamburg
1011		Guinevere	W Sr Yt	British	180	125·2	23·6	13·3	Gosport	1868	Camper& (Nicholson		J.D.Gardner	Portsmouth
1012		Guinevere *Whiffen -87*	W Bk	Chilian	960	172·5	36·0	21·8	Digby,N.S.	1870			A.Sutherland	Valparaiso
1013	❋ D	Guinevere	C F	German	857	197·0	34·0	19·7	Glasgow	1868	Randolph, Elder (&Co		C.Rahtkens	Rostock
1014	V	Guldborg *Rasmussen 82-87*	W Sr	Danish	130	93·9	21·8	10·3	Nysted	1877	F.Sparre		L.C.Rasmussen	Svendborg
1015	N V	Guldbringa *Hansen -88*	W Bk	Norweg'n	694	156·7	33·0	19·7	Grimstad	1874	J.Tryde		A.Bjering Olsen	Tvedestrand
1016	N	Guldfaxe *Nielsen*	W Bk	Norweg'n	623	162·3	33·3	19·1	Arendal	1862	G.Knudsen		E.A.Thondes Enke	Arendal
1017	N	Guldregn *Ericksen -87*	W Bk	Norweg'n	871	168·0	33·5	21·7	Miramichi,NB	1864	Johnston		T.M.Wiborg	Kragerö
1018	N	Guldregn *Jacobsen*	W S	Norweg'n	714	153·3	33·4	20·2	Grimstad	1872	M.SmithPetersn		M.Smith Petersen	Grimstad
1019	N	Guldregn *Hansen*	W Bk	Norweg'n	652	167·1	34·4	18·6	Sandefjord	1879	S.Olsen		Anton Andersen	Sandefjord
1020	❋	Gulf Stream *Barstead 81-84*	I Bk	British	1378	234·2	38·1	21·8	Port Glasgow	1884	Russell & Co.		A.L.Polson	Glasgow
1021		Gulnair	W Sr	British	325	142·5	25·5	12·3	Port Robinson	1873	J.P.Abbey		T.Myles	Hamilton,Ont.
1022		Gulnare (ex Aldeganda, ex Gulnare)	W Sr Yt	British	110	112·3	22·0	10·9	Deptford	1874	Walker&Co.		H.Morgan	Cowes
1023	N	Gulnare *Hansen -88*	W Bk	Norweg'n	549	147·5	30·6	17·9	Cardigan,P.E.I.	1873	H.L.McDonald		H.J.Hansen	Porsgrund
1024	N V	Gulnare *Sivertsen -88*	W Bk	Norweg'n	456	135·8	31·4	16·3	Risör	1874	J.Stenersen		D.Herlofson (Axel Smith)	Arendal
1025		Guluessa	W Bg	Turkish	250	..	..	..	Guluessa	1876			Oonialee Drlaver oglou Hassan	Oonia
1026		Gulzari-Bahri *Mehmed Ali*	W Bg	Turkish	169	..	..	..	Greece	1883			Saraf-Vassilaki Kosmo	Constantinople
1027		Gulzari-Oonia	W Bg	Turkish	312	..	..	..	Oonia	1871			Iskenderieh tijari Seid Ali bin Ali Yalnashi	Oonia
1028		Gulzari-Tannah	W Schleppe	Turkish	187	..	..	..	Rustchuk	1868			Rustchuklee Hafiz Bey	Rustchuk
1029		Gunesa *Lemon 79-88* (ex Ginetta, ex Stella Sinope)	W Bk	Swedish	418	132·8	30·1	19·3	Loano	1868	N.Cerruti		J.F.Isleben	Stockholm
1030	N	Gungner *Nilsen*	W Bg	Norweg'n	173	100·1	23·3	11·4	Kragerö	1875			Joh.Dahll	Kragerö
1031		Gunhild *Andersen 86-88* (ex Bremerlehe ex Sawely Chindow) [WRECKED]	W Bk	Norweg'n	955	179·4	35·5	21·9	Bremerhaven	1857	R.C.Rickmers		A.C.Andersen	Laurvig
1032	V	Gunhild *Abrahamsson -87*	W Bk	Swedish	290	119·1	24·9	13·4	Krogarne	1880	B.Samuelsson		J.N.Sanne	Uddevalla
1033	❋	Gunhilde *Pettersen -88* (ex Alexander Duthie)	W Bk	Norweg'n	1143	213·1	35·2	22·1	Aberdeen	1867	W.Duthie,Jr.		Chr.Anker	Fredrikshald
1034		Gunilla *Jonsson 82-82*	W 3mSr	Swedish	251	104·0	25·9	13·1	Väse	1875	C.Andersson		C.Andersson	Harg,Roslagen
1035	D N	Gunn *Bellmann 85-86*	W Bk	Norweg'n	468	143·1	30·7	16·0	Porsgrund	1875			L.Vauvert	Porsgrund
1036	N	Gunvor *Andersen 87-88* (ex Coldstream)	W Bg	Norweg'n	269	102·3	27·0	15·1	Sunderland	1867	J.Barkes		A.F.Kolderup	Christiania
1037		Gurgio	W Bg	Turkish	150	..	..	..	Russia	1871			Limnilee Yorghi Strati	Lemnos
1038	V	Gurli *Knudsen -88*	W Bk	Swedish	721	167·6	29·5	19·0	Gothenburg	1879	J.E.Hube		G.D.Kennedy	Gothenburg
1039	❋	Gurli *Strömberg* (ex Crest of the Wave)	W Bk	Swedish	686	175·8	30·0	19·5	Sunderland	1853			A.F.Strömberg	Gothenburg
1040		Gurnet *Jefferson*	W Sr	American	101	84·9	25·2	6·2	Bath,Me.	1883	C.B.Harrington		Standard Fertilizing Co.	Duxbury,Mass.
1041		Gust	W Sr	American	113	99·6	17·6	8·0	Hamburg,Pa.	1864			J.A.Anderson	Trenton,N.J.
1042	N	Gusta Helene	W Bk	Norweg'n	429	126·2	30·5	17·0	Lillesand	1869	O.&P.Knudsen		P.N.Jahnsen	Lillesand
1043		Gustaf *Andersson*	W 3mSr	Swedish	501	129·2	30·2	18·2	Finland				G.Sjöberg	Malmö
1044		Gustaf *Olsson*	W Sr	Swedish	203	104·6	23·0	11·8	Timmernabben	1869	M.Gudmundsson		N.M.Pahlsson	Helsingborg
1045	N	Gustaf *Andersson*	W Sr	Swedish	189	101·3	23·9	11·5	Björneborg	1874	J.Bustorff		A.Ödman	Fiskebäckskil
1046	V	Gustaf *Jonsson -82* (ex St. Pierre)	W Bg	Swedish	187	83·3	23·0	13·1	Les Sables	1869	Chaigneau		O.Eckerman	Helsingborg
1047	G	Gustaf *Rhodin 77-78*	W Sr	Swedish	130	89·9	23·0	8·8	Westervik	1868	C.R.Fridell		A.Rhodin	Fiskebäckskil
1048		Gustaf *Johansson -81*	W Sr	Swedish	121	78·1	25·1	10·2	Valdö	1853			H.Högvall	Storugns, Gotland
1049														
1050														

❋ = Lloyd's Register; D = Nederlandsche Vereeniging; G = Germanischer Lloyd; I = Registro Italiano; R = American Record; A = Veritas Austro-Ungarico; H = Veritas Ellenico; N = Norske Veritas; V = Bureau Veritas.

Dans la 4me Colonne, C = Composite (membrure en fer, bordé en bois); I = Fer; Stl = Acier; W = Bois.

Les Dates du doublage en métal (**Col. 11**) ressortent des livres dans lesquels les navires sont respectivement classés ou des rapports des Inspecteurs du Lloyd's Register.

Pour plus amples renseignements sur les navires classés par le Lloyd's Register, voir le Lloyd's Register Book et le Lloyd's Yacht Register.

Official No.	No. in Book	Registry in which classed	Ship's Name. Material and Rig. Master, and Dates of Appointment to present Owner's Service and to Ship. Ship's Late Name.	Flag	Net Tonnage	Length	Breadth	Depth	Where	When	By whom	Date of Metalling	Owners or Managers
	1051	V	Gustaf Adolf *Hellgren* W Bk	Swedish	334	127·9	25·9	12·1	Sundsvall	1871	W. Hägglund	[illegible]	A. Akerblad[illegible]
	1052	R	Gustaf Adolf *Lind 82-87* W Bg (ex Kinguwa)	Swedish	252	114·8	24·6	12·8	P.E.Island	1875	Duncan		Aug. Nilsson
	1053		Gustaf Vasa *Nilsson 88-88* W Bk	Swedish	305	106·6	24·3	16·1	Westervik	1832	P. Blomquist		F. A. Harms
	1054		Gustafva W 3m Sr	Russian	310	118·0	27·8	12·6	Hitis	1877	Vanstrum		E. J. Westerlund
	1055	R	Gustafva *Lundberg 87-87* W Bk (ex Fortuna)	Swedish	525	140·0	27·2	18·4	St Martin's N B	1852			Joh. Bengtsson
	1056	V	Gustafva *Andersson 88-88* W Bk	Swedish	313	130·2	25·6	12·1	Sundsvall	1872	W. Hägglund	[illegible]	A. Andersson
	1057	R	Gustafva *Larsson -78* W Bg	Swedish	232	108·2	23·1	12·6	[illegible]	1878	A. Samuelsson		G. Mellin
	1058	V	Gustav *Nielsen* W Bn (ex Gustav Lange)	German	181	99·7	21·7	13·3	[illegible]	1869		[illegible]	J. C. A. F. Zerrener
	1059	G	Gustav *Olsen* W Bg	German	18[illegible]	96·6	25·1	12·6	Wolgast	1860	W. Lübke		W. Krüger
	1060	N	Gustav *Roer* W Bk	Norweg'n	417	113·7	30·1	15·1	[illegible]	1858			Saxlund & Roer
	1061	N	Gustav *Helgesen* W Bg	Norweg'n	218	101·5	26·5	12·5	Kiel	1862	Convalt&Sohn		H. Nielsen
	1062	N V	Gustav Adolph *Jæger* W Bk	Norweg'n	738	160·5	33·0	17·8	[illegible]	1879		[illegible]	O. B. Sorensen
	1063	N	Gustav Adolph *Sorensen* W Bk (STRANDED)	Norweg'n	707	162·8	33·3	20·3	Bergen	1866	A. Bekke	[illegible]	T. Svanöe
	1064	N	Gustav Adolph *Olsen* W Bk (ex Gustaf Adolf)	Norweg'n	376	136·7	29·6	14·1	Svartvik	1862	O. Dickson	[illegible]	A. O. Lindvig
	1065	V	Gustav Friedrich W Bk [illegible] *Ballmann -81*	German	354	123·9	26·7	15·9	Greifswald	1839	J. N. Juhl		J. H. Rhetz
	1066	G V	Gustav Fried Focking *Mukoeff -82* W Bk	German	59[illegible]	131·7	30·0	17·8	Danzig	1866	J. W. Klawitter		G. F. Focking
	1067	V	Gustav Friedrich *[illegible] -81* W Sr	Russian	113	71·5	19·2	10·1	Dunamünde	1887			Chr. Berg
	1068	G V	Gustav Metzler *Torck -74* W Bk	German	352	115·5	27·8	16·3	Rostock	1874	E. Borchardt&Co.	Y 3.88	Wilhelm Munck
[illegible]	1069	D R	Gustav Molien *Maltzahl -84* W Bk (ex Aurora)	Russian	672	152·0	31·3	20·0	Yarmouth, N.S.	1867	Robins&Co.		A. Wirckau
	1070	G	Gustav Wilhelm *Niemann -89* W Sr	German	135	87·3	22·8	11·0	Ribnitz	1874	J. H. Wilkin		G. W. Löwe
	1071	G V	Gustav & Oscar *Scrmann -82* W S (ex Carrie Reed)	German	1352	201·1	39·4	25·1	Kennebunk Me	1870	W. Thompson	Y 1.88	E. C. Schramm & Co.
	1072	V	Gustava *Ramsberg* W Bg	German	212	97·4	25·1	13·2	Wolgast	1860	W. Lübke		H. Ballschmiede
	1073		Gustava W Bkn	Russian	314	118·0	27·8	12·6	Kimito	1877	Vanstrum		Westerlund & [illegible]
	1074		Gustava Egner *Erdbecher 83-83* W Sr	German	126	82·0	21·5	10·2	Zingst	1879	A. Dressel		C. A. Beug
	1075		Gustave Adolphe W Bk	French	233	..	..	..	Fécamp	1865	Brunets & Capon		Veuve Craquelin
2746	1076		Gustavo (PONTOON) *Caliero* W Bk	Italian	332	118·1	30·2	18·8	Equa	1868	S. Mauro	[illegible]	F[lli] Trapani fu Stefano
	1077		Gustavo *Marti* W Bg (ex Maria Antonia Sensat)	Spanish	239	103·0	26·9	13·8	Arens	1869	Busquet	[illegible]	Successors of J.M.Serra & Co.
	1078	✻	Gustavo Adolfo *Lorentzen -88* I Bk (ex Embleo, ex City of Ningpo)	German	986	204·0	32·1	21·3	Glasgow	1862	Barclay, Curle & Co.		Bahr, Gebruder
	1079	✻ V	Gutenberg *Meyer 88-88* I Bk (ex Edinburgh)	German	627	172·6	29·3	18·8	Newcastle	1862	Wigham, Richardson&Co.		R. M. Slomann & Co.
	1080		Guttmeer W Bk	British	321	102·2	25·1	16·5	Cochin	1847			Hadjee Furragee Bem Tousour
35602	1081	R V	Guy C. Goss *[illegible]* W Bk	American	1524	213·8	39·8	24·0	Bath, Me.	1879	Goss&Sawyer	[illegible]	W. H. Besse
78580	1082	✻ D	Guy Mannering *McLaren 79-79* I Bk	British	807	194·8	32·1	18·5	Dumbarton	1878	A. McMillan&So.		J. Hunter
63745	1083	✻	Guyana *Perrin* W Bk	British	598	150·8	30·4	18·5	Ayr	1870	Fullarton	Y 1.88	Guyana Sailing Ship Co. L.L. (Cox [illegible])
	1084		Guzmàn W Sr (ex Dorado M[illegible], ex Edith May)	Spanish	12[illegible]	85·9	21·9	8·2	Chelsea Mass	1866			C. Miranda
41[illegible]	1085		Gwain Maid *Llewellin* W Bn	British	130	73·2	21·5	12·3	Fishguard	1846			Miss E. Donoghue
1151[illegible]	1086		Gwalia *Foster 81-81* W Sr	British	110	89·2	22·4	11·5	Aberayron	1853	E. Jones		J. Callaghan, jun.
87297	1087	✻	Gwen *Roberts -82* W 3m Sr	British	145	100·3	21·6	11·4	Conway	1886	P.&H. Lewis		H. Lewis
62183	1088	✻	Gwendolin W Sr Yt	British	109	104·7	21·2	12·3	Gosport	1870	Camper& Nicholson	[illegible]	W. Greenwell
62743	1089	✻	Gwendoline *Matthews 86-88* W Bk	British	346	128·0	25·9	16·1	Bristol	1869	Hill	[illegible]	Turner, Edwards & [illegible]
93431	1090	✻	Gwendoline *Jones -88* W K	British	33	55·1	15·2	6·5	Milford	1888	F. Jones		F. Jones
47068	1091	✻	Gwladys *Jones* W Bn	British	157	95·3	22·7	12·5	Cardigan	1863	Harries	[illegible]	R. Morgan
70315	1092	✻	Gwrtheyrn Castle *Parry 82-85* I Bk	British	778	191·7	31·8	19·2	Sunderland	1876	Osbourne, Graham&Co.		R. Thomas & Co.
74873	1093	✻	Gwynedd *Davies* I Bk	British	1053	204·6	34·6	21·2	Sunderland	1877	Osbourne, Graham&Co.		"Gwynedd" Ship Co. L. (W. Thomas & Son)
	1094	N	Gyda *Eriksen* W Bk	Norweg'n	658	149·5	34·2	18·3	Arendal	1883	J. Anonsondsen	Y 3.88	Stian Eriksen
	1095	N	Gyda *Sørensen 89-90* W Bk (ex [illegible])	Norweg'n	521	125·5	30·0	20·2	Greenock	1849		[illegible]	T. M. Nielsen & Sø[illegible]
	1096	N	Gyda *Falch* W Bk	Norweg'n	307	120·2	28·7	12·7	Porsgrund	1877	J. Olsen	[illegible]	R. N. Parelius
	1097	N	Gyldenlove *Hansen 78-7[illegible]* W Bk	Norweg'n	371	125·5	29·1	16·3	Laurvig	1855		[illegible]	J. Petersen
83357	1098	D	Gylfe *Le Guardais -88* W Bk	British	984	168·1	33·1	23·5					H. Dinning
	1099	D N	Gyller *Jacobsen -88* W Bk	Norweg'n	489	144·4	30·7	16·9	Porsgrund	1876	Knudsen	[illegible]	Interessentskab i Aalborg
	1100												

✻ = Lloyd's Register; D = Nederlandsche Vereeniging; G = Germanischer Lloyd; I = Registro Italiano; R = [illegible]
A = Veritas Austro-Ungarico; H = Veritas Ellenico; N = Norske Veritas; V = [illegible]

In the 4th Column, C = Composite (iron frame, wood planked); I = Iron; Stl = Steel; W = Wood.

The Dates of Metalling (Col. 11) are based on the records of the Books in which Vessels are respectively classed, or on the Reports of the Surveyors [illegible]

Further particulars of Vessels classed by Lloyd's Register will be found in Lloyd's Register Book and Lloyd's Yacht Register.

No.	Classification	Nom. Matériaux et Gréement. Capitaine et dates d'entrée au service des Armateurs actuels et à bord du navire. Ancien Nom.	Pavillon	Tonnage	Longueur	Largeur	Creux	Construit: Où.	Quand	Par qui	Date du doublage en métal	Armateurs ou Directeurs-Gérants.	Port d'Armement.
1101	N	Gyller Johnsen 89 W.S[illegible]	Norweg'n	202	130·2	22·4	11·4	Hoorn	1856			A.P.Ljungberg	Fredrikstad
1102	N	Gyller Hansen 88 W.Bg	Norweg'n	197	114·4	25·4	12·2	Farsund	1857			N.Engebretsen	Fredrikstad
1103		Gymkhana 80 Lr	British	480	220·0	30·0	8·4	Dumbarton	1887	W. Denny & Bros.		Irrawaddy Flotilla Co. Ld.	Glasgow

Additional Ships are inserted in the Supplement at the end of the Register.

✠ = Lloyd's Register; D = Nederlandsche Vereeniging; G = Germanischer Lloyd; I = Registro Italiano; R = American Record; A = Veritas Austro-Ungarico; H = Veritas Ellenico; N = Norske Veritas; V = Bureau Veritas.

Dans la 4me Colonne, C = Composite (membrure en fer, bordé en bois); I = Fer; Stl = Acier; W = Bois.

Les Dates du doublage en métal (Col. 11) ressortent des livres dans lesquels les navires sont respectivement classés ou des rapports des Inspecteurs du Lloyd's Register.

Pour plus amples renseignements sur les navires classés par le Lloyd's Register, voir le Lloyd's Register Book et le Lloyd's Yacht Register.

Official No.	No. in Book	Registry in which classed	Ship's Name. Material and Rig. Master, and Dates of Appointment to present Owner's Service and to Ship. Ship's Late Name.	Flag	Net Tonnage	Dimensions: Length	Breadth	Depth	Built: Where	When	By whom	Metalling	Owners or Managers
1	2	3	4	5	6	7			8	9	10	11	12
95205	1		H.A.DeWitt *Delahanty* W3mSr	American	228	118·3	30·1	8·6	Bath,Me.	1873	Goss&Sawyer		B.W.&H.F.Morse
11296	2		H. A. Deming *Boardman* WSr	American	131	93·0	25·0	7·0	E.Haddam,Ct.	1864	J.Lord		F.Beardman
	3		H.A.Walter *Möller* WBk	German	394	119·5	28·4	17·2	Rostock	1864	W.Zeltz		E.Burchard
80026	4	RV	H. B. Cann *Cann 81-81* WBk	British	1299	201·0	37·7	22·9	Yarmouth,NS	1882	J.N.Gardner	[illegible]	H.Cann
95381	5		H. B. Griffin *Nelson* WSr	American	118	89·0	24·0	10·0	Essex,Mass.	1875			D.Allen,jun.
94745	6	V	H.B.Homan *Hasson* W3mSr	British	299	131·6	31·2	10·6	Portland,N.B.	1883	E.McGuiggin		R.C.Elkin
95730	7	V	H.B.Hussey *Hodgdon -87* WBn	American	518	160·6	36·2	12·3	Richmond,Me	1883	T.J.Southard&Son	[illegible]	T.J.Southard & Son
	8	✳	H. Bischoff (ex steamer Ville de Paris) IIm.S	German	2600	369·1	43·3	36·7	Glasgow	1865	R.Napier&Sons		H.Bischoff & Co.
	9	GV	H. Bremer *Stehr -88* W3mSr	German	332	121·7	26·9	13·4	Lübeck	1876	J.Steffen	[illegible]	P.Bremer
49402	10		H. Brooks *Risly* WSr	British	107	78·0	23·0	9·4	WeymouthNS	1865			H.Brooks
95357	11	R	H. Buddig *Voss* WSr	American	142	105·0	26·2	7·2	Pascagoula, [Mis.	1874		[illegible]	H.Buddig
	12	GV	H. C. Berg (ex Wilhelmine) *Paulsen -87* WBk	Danish	360	126·2	26·1	15·3	Burg, Bremen	1868	J.H.Bosse	[illegible]	Akt.Selsk (N.J.Os...
	13		H. C. Gram (ex Sparfrid) *Ure -86* WSr	Danish	171	93·3	23·4	12·5	Jocensaw	1874	S.Lokas		C.Hassager
95769	14	R	H. C. Higginson *Ross* W3mSr	American	493	145·0	35·0	12·8	Bath,Me.	1883	E.Soule&Co.		S.M.Bird
95061	15	R	H. C. Sibley *Hichborn* WBn	American	481	131·0	30·0	17·0	Belfast,Me.	1869	Carter&Co.	[illegible]	N.W.Bridge
11156	16		H. Curtis *Spofford* WSr	American	118	84·4	25·5	8·3	HarringtonMe	1850			S.G.Haskell
48275	17		H. G. Berry WBk	British	290	109·5	25·8	11·6	Boothbay,U.S.	1855			H.Allison
95483	18	RV	H. G. Johnson *Colley 76-* WBk	American	1081	199·5	36·8	21·8	N'wbury p't Ms	1877	G.E.Currier	[illegible]	W.A.Johnson
95018	19	RV	H. H Wright *Meyers* WBn	American	387	127·0	29·5	15·1	Searsport,Me.	1869	W.McGilvery	[illegible]	J.H.Lane
	20	✳	H.Hackfeld (ex British Nation) *Wolters 72-88* IBk	German	1302	216·2	35·3	22·9	Dublin	1865	Walpole&Webb		C.Hencock,jun. (J.C.Pflüger & Co.
	21	V	H.J.Baagøe *Hansen 75-75* WSr	Danish	136	94·2	21·4	10·8	Thurø	1875	P.Troensegaard		J.P.Baagøe
130231	22	R	H. J. Cottrell *Haskins* W3mSr	American	337	134·0	32·1	11·5	Belfast,Me.	1882	J.Y.Cottrell	[illegible]	S.G.Haskell
	23		H.J.Hammer (MISSING) *Juell -84* WBk	Norweg'n	325	130·0	28·3	13·1	Lillesand	1884	A.Reinertsen	[illegible]	Krog & Henschien
95336	24	V	H.J.Libby *Richardson 70-84* WBk	American	621	143·0	32·2	16·9	Bath,Me.	1873	Hagen&Thurl'w	[illegible]	B.Webster
	25		H. Jäde (ex Widziwa) *Carlsson -88* WBkn	Russian	274	110·9	25·7	14·0	Rostock	1863	E.Padderatz		Carlsson
	26		H. L. C. *Lambert 87-87* WBn	French	120	82·7	23·2	10·9	St.Pierre- [Miquelon	1880	P.Beronet		H.Lecharpentier
11841	27	RV	H. L. Routh *Kirby* WBk	American	972	167·0	36·6	22·5	New York	1864	Webb&Bell	[illegible]	T.Dunham's Nephew&Co.
11271	28		H. L. Slaight *Blackburn* WSr	American	225	111·0	29·6	9·4	Tuckerton,NJ	1866			W.Cox
11396	29		H. L. Tierman *Melander* WSr	American	145	96·0	27·5	9·1	SanFrancisco	1867			A.Crawford
	30	G	H. Mulder (ex Peter) *Reimers* W3mSr	Dutch	383	120·1	26·0	17·2	Sunderland	1860		10.85	H.Mulder
80631	31		H. N. Todman WSr	British	110	92·2	22·6	8·3	Wellington	1867	D.Tait		James Clark
	32	GV	H. Printzenberg *Ahrens* WBk	German	553	139·3	30·8	18·7	Rostock	1876	C.Schönemann	[illegible]	F.Behnk & Even
95702	33	R	H. S. Lanfair *Woodland* W3mSr	American	298	138·0	32·7	10·0	Baltimore,Md	1884	J.S.Beacham& [Brother		F.L.Clayton & Co.
95622	34	R	H. S. Marlor *Wines* W3mSr	American	324	121·4	32·2	14·1	P.Jeffers'n NY	1869			J.W.Wines
11564	35		H. T. Hedges *Hagner 73-86* WSr	American	227	104·7	28·1	8·1	WilmingtnDel	1868			W.H.Crawley
95071	36	R	H. T. Townsend *Smith* WSr	American	175	104·4	29·0	8·8	Pembroke,Me	1870	J.Wilder & Co.		L.S.Smith
	37		H. Upmann *Weber 77-* WBk	German	427	132·8	28·4	16·6	Brake	1866	D.Rogge		H.Upmann & Co.
59404	38		H. V. Crandall *Britt* WSr	British	148	93·2	27·0	9·9	Parrsboro',NS	1869			R.Ross
11365	39		H.W.Godfrey *Corson -87* WSr	American	187	109·0	28·1	8·2	Tuckerton,NJ	1859			F.L.Godfrey
78911	40	✳	H.W.Palmer *Boyd 83-83* WBkn	British	471	138·3	32·7	12·8	Dorchester NB	1882	P.J.Palmer	Y 1883	P.J.Palmer
95570	41	R	H. & J. Blendermann *Crowley* W3mSr	American	399	139·4	34·0	12·1	P.Jefferson NY	1879	J.M.Bayles&Son		Samuel E.Birdsall
	42	V	H. & L. *Lebeaux -86* WBg	French	246	103·6	26·2	12·9	Nantes	1872	Clairgeau	[illegible]	Serpette & Co.
	43	NV	Haab *Møller -88* WBk	Norweg'n	870	176·1	36·1	19·8	Moss	1877	G.Berentsen	[illegible]	M.Pedersen
	44	V	Haabet *Duhn 84-84* WBg	Danish	228	113·8	23·1	12·0	Svendborg	1876	T.R.Andersen	[illegible]	O.E.Olsen
	45	V	Haabet *Jacobsen 81-81* WBn	Danish	138	89·9	20·5	10·9	Troense	1867	C.J.Jensen		P.Christensen
	46	V	Haabet *Zachariassen 83-83* WSr	Danish	115	80·5	20·4	11·4	Nyborg	1856	P.Petersen		M.Jensen
	47		Haabet (BROKEN UP) *Møller 80-83* WBn	Danish	107	67·8	21·1	12·3	Svendborg	1842	L.Møller		H.L.Møller's Exe.
	48		Haabet (ex Edmond, ex ... Gips) *Paulsen* WBk	Norweg'n	599	132·0	32·3	19·8	Dordrecht	1854	Gips and Zoon		Th.Helgesen
	49	N	Haabet *Halvorsen* WBk	Norweg'n	417	117·5	29·1	18·5	Newcastle	1851			W.Wilhelmsen
	50	N	Haabet *Johnsen* WBk	Norweg'n	404	110·5	29·3	18·2	Finland	18..			Rederiet "Haabet" (T.Klem)

✳ = **Lloyd's Register**; D = Nederlandsche Vereeniging; G = Germanischer Lloyd; I = Registro Italiano; R = ...
A = Veritas Austro-Ungarico; H = Veritas Ellenico; N = Norske Veritas; V = ...

In the 4th Column, C = Composite (iron frame, wood planked); I = Iron; Stl = Steel; W = Wood.

The Dates of Metalling (Col. 11) are based on the records of the Books in which Vessels are respectively classed, or on the Reports of the ...

Further particulars of Vessels classed by Lloyd's Register will be found in Lloyd's Register Book and Lloyd's Yacht Register.

N° Lloyd's	N°	Classe	Nom, Matériau et Gréement. Capitaine et dates d'entrée en service des Armateurs actuels et à bord du navire. Ancien Nom	Matériau / Gréement	Pavillon	Tonnage	Longueur	Largeur	Creux	Construit Où	Quand	Par qui	Date du doublage en métal	Armateurs ou Directeurs-Gérants.	Port d'Armement.
	51	N	Haabet *Nielsen -88*	W Bk	Norweg'n	[illegible]	[illegible]	[illegible]	[illegible]	Grimstad	1881	Jörgen Tellefsen	[illegible]	Jörgen Tellefsen	Grimstad
	52	N	Haabet *Gundersen -88*	W Bk	Norweg'n	346	[illegible]	[illegible]	13·2	Drammen	1865			S.Christensen	Drammen
	53	N	Haabet *Olsen*	W Bg	Norweg'n	[illegible]	[illegible]	[illegible]	[illegible]	Tönsberg	1856			Chr.Melsom	Tönsberg
	54	N	Haabet *Johannessen -87*	W Bk	Norweg'n	[illegible]	91·3	[illegible]	15·7	Tönsberg	1817			K.Gui	Holmestrand
	55	N	Haabet *Johnsen -81*	W Bk	Norweg'n	[illegible]	115·8	27·2	14·5	Porsgrund	1867	E.Rasmussen	[illegible]	M.Morck	Fredrikstad
	56	N	Haabet *Olsen*	W Bm Sr	Norweg'n	[illegible]	[illegible]	[illegible]	12·6	Stavanger	1875		[illegible]	M.G.Mottsen	Stavanger
	57	N	Haabet *Jensen*	W Bk	Norweg'n	[illegible]	94·5	[illegible]	15·4	Bordeaux	184[illegible]			M.Pedersen	Skien
	58		Haabet *(ex Gideon à Jens)*	W Sr	Norweg'n	122	80·7	23·4	11·2	Greece	186[illegible]			H.M.Slettebö	Egersund
	59		Haabets Anker *Christensen*	W Bk	Norweg'n	304	91·6	27·5	17·3	Tönsberg	1847			W.Wilhelmsen	Tönsberg
	60	N	Haabets Anker *Larsen*	W Bk	Norweg'n	289	92·3	26·5	15·8	Sandefjord	1848			Ths.Arbo Hoegh	Laurvik
	61		Haabets Anker	W Bg	Norweg'n	188	92·3	24·4	14·8	Tvedestrand	1854			Jacobsen	Christiansand
	62	N	Haabets Anker *Pedersen 83-89*	W Bg	Norweg'n	143	81·5	23·7	12·2	Arendal	1854			J.F.Juell	Risör
	63	N V	Haakon Haakonsen *Jansen*	W Bk	Norweg'n	619	158·5	33·4	18·8	[illegible]	1873	R.Havem	[illegible]	S.Salvesen	Tvedestrand
	64	N	Haavund *Olsen -88*	W Bk	Norweg'n	534	117·0	32·5	17·2	Porsgrund	1877	N.J.Myren	[illegible]	B.Coudrup	Porsgrund
	65	V	Habana *Mayubei (ex Joaquin Torrents)*	W Bk	Spanish	684	139·1	31·7	20·1	[illegible]	1868	J.Vieta	[illegible]	Balsells & Subiran	Barcelona
[illegible]	66	V	Habitant *Potter 81-83*	W S	British	1319	225·0	42·7	21·0	[illegible]	1885	Sheffield & Wickwire	[illegible]	Sheffield& Wickwire	Windsor,N.S.
	67	N	Habor *Thorsen 82-87*	W Bg	Norweg'n	391	107·1	27·1	16·3	[illegible]	1853			G.Nilsen	Sandefjord
	68		Hadari Bahri	W Bn	Turkish	182	..	..	..	Sinope	1876			Sinoplee Kalionjee-zadeh Aiasti Teboboji & Co.	Sinope
[illegible]	69	❋	Haddington *Belpaire (BURST) 85-86*	I Bk	British	1206	217·0	33·8	22·5	Liverpool	1846	T.Vernon		A.E.Kinnear	London
[illegible]	70	❋	Haddon Hall *Andrews 85-88*	I S	British	1416	225·1	37·9	23·6	Liverpool	1868	T.Royden&Sons		J.R.deWolf & Son	Liverpool
2901	71	❋	Hafren *Brooking*	W Bn	British	143	93·5	22·7	11·9	Gloucester	1875	Hipwood	[illegible]	S.P.Brooking	Brixham
	72	N	Hafrsfjord *Dahl 80-86*	W Bk	Norweg'n	482	140·6	31·2	17·3	Christiania	1869	Chr.Brinch	[illegible]	F.Breda	Christiania
	73		Hahlin	W Bn	Turkish	150	..	..	..	Barten	1871			Oonji-zadeh Apik Effendi	Constantinople
[illegible]	74		Hahnemann *Jefferson -87*	W Bk	British	1313	192·2	39·0	25·0	St.John,N.B.	1861	M'Donald	[illegible]	J.Hall & Co.	Newcastle
	75		Haick *Hussein*	W Bal	Turkish	100	..	..	..	Barten	1885			Ekmek[illegible]ari-Zaffer-Sliden	Barten
	76		Haidar Bahri	W Bn	Turkish	225	..	..	..	Yarda Skelessi	1877			Alibabonblee Ali-Baba zadeh-Hassan	Ahiabanah
	77	❋	Haidee *Moure 73-88*	I Bk	French	755	187·6	30·8	19·0	Newcastle	1869	C.Mitchell&Co.		J.J.Bordes	Bordeaux
	78		Hainan	W 3m Sr	Norweg'n	262	119·3	25·7	11·7	Motzen	1860	Oltmanns Wwe		A.S.Ohlsen	Farsund
	79	V	Hainaut *DeVries 86-87*	Stl S	Belgian	1709	248·8	40·3	22·2	Barrow	1887	Barrow S.B.Co.		F.Speth & Co.	Antwerp
	80		Hairlié	W Bg	Montenegrn.	130	65·6	22·3	9·8	Dulcigno	1865	Jus Omeri		Suleiman Ahmet&C.	Dulcigno
	81		Hairoer Rachman *(ex [illegible]) Said Abd bin Abdulrachman Albas*	W Bk	Dutch	218	104·3	29·9	..				[illegible]	Said Abdulrachman bin Abdulla Albas	Palembang
	82		Hairool Bari *(ex Concordia) Gulje Horst -86*	W Bg	Dutch	265	..	..	..	Bangkok	1863			Said Mohammed bin Osman Arlan	Cheribon
	83	V	Haiti *Revez*	W Bk	French	218	96·1	22·3	13·6	Cherbourg	1864	Mahieu		H.Chedru	Fécamp
	84		Hajee Murad	W Bg	Turkish	183	..	..	..	Greece	1871			Killijiler Mudiri Hajee Hussein Effendi	Constantinople
	85	V	Hak Lee *(ex Roderic Hay)*	W Bk	Chinese	290	127·1	27·0	12·9	[illegible]	1873	N.Gibbon	[illegible]		
	86	N	Hakon Adelsteen *Andersen -86*	W Bg	Norweg'n	217	106·6	24·9	13·5	Bergen	1854	Brunchorst & Dekke		Jörgensen & Röe	Brevig
	87	N	Hakon Jarl *Korsrup 83-89*	W Bk	Norweg'n	491	136·8	31·5	17·1	Grimstad	1868	Jacob Röe	[illegible]	Christen Röe	Arendal
[illegible]	88		Hakutsû Maru	W Sr	Japanese	180	100·3	22·0	10·9	Osaka	1880	Heshiki Hikozo		Hayashi Seiichi	Osaka
[illegible]	89	❋	Halcione *Kelly -87*	I S	British	843	191·7	29·1	18·8	Greenock	1869	R.Steele		Shaw,Savill & Co.	Southampton
[illegible]	90		Halcyon *Rice -87*	W 3m Sr Yt	American	278	133·2	32·0	10·0	Eureka,Cal.	1881	A.Cookson		J.Dolbeer	San Francisco
	91	N	Halden *Dahl -88 (ex Kate Covert)*	W Bk	Norweg'n	843	165·3	33·7	21·1	Oromocto,N.B.	1869	J.S.Covert	[illegible]	Actieselskabet" Halden" (Aug.Olsen)	Fredrikshald
	92	D N	Haldis *Alexandersen -88 (ex Cassi)*	W Bk	Norweg'n	357	122·4	28·2	16·0	St.Servan	1855			A.H.Kiaer & Co.	Fredrikstad
[illegible]	93	❋	Halewood *Stap 87-87*	I S	British	2106	274·3	40·1	24·9	Southampton	1885	Oswald,Mordaunt&Co.		[illegible] Ship "Halewood" Co.(W. Wayland & Co.)	Liverpool
	94		Halfdan *(ex [illegible]) Hansen 86-86*	W Bk	Norweg'n	347	123·6	25·7	14·5	Brake	1864	D.Dencker	[illegible]	A.Meling	Stavanger
	95		Halfdan *Brevig 86-89*	W Bg	Norweg'n	293	107·3	27·5	14·3	Dröbak	1854			M.H.Grönsand	Dröbak
	96		Haliss Tidjaret	W Bn	Turkish	400	..	..	..	Heraclea	1871			Col.Hussein Bey	Heraclea
[illegible]	97	❋	Hallgerda *Hayden -88*	W S	British	1088	190·7	36·0	22·9	N.Glasgow N.S.	1874	Carmichael	[illegible]	J.W.Carmichael & Co.	Pictou,N.S.
3736	98		Hally Bayley	W S	British	113	96·7	22·0	9·0	Hobart	1869	J.McGregor		T.G.Kelly	Sydney,N.S.W.
	99	N	Halvmaanen *Olsen*	W Sr	Norweg'n	160	98·1	25·7	11·4	Grimstad	1856	Chr. Paulsen		N.W.Sörensen	Sandefjord
	100														

❋ = Lloyd's Register; D = Nederlandsche Vereeniging; G = Germanischer Lloyd; I = Registro Italiano; R = American Record; A = Veritas Austro Ungarico; H = Veritas Ellenico; N = Norske Veritas; V = Bureau Veritas.

Dans la 4me Colonne, C = Composite (membrure en fer, bordé en bois); I = Fer; Stl = Acier; W = Bois.

Les Dates de doublage en métal (Col. 11) ressortent des livres dans lesquels les navires sont respectivement classés ou des rapports des Inspecteurs du Lloyd's Register.

Pour plus amples renseignements sur les navires classés par le Lloyd's Register, voir le Lloyd's Register Book et le Lloyd's Yacht Register.

1 Official No.	2 No. in Book.	3 Registry in which classed.	4 Ship's Name. Material and Rig. Master, and Dates of Appointment to present Owner's Service and to Ship. Ship's Late Name.	5 Flag.	6 Net Tonnage.	7 Dimensions. Length.	Breadth.	Depth.	8 Built. Where.	9 When.	10 By whom.	11 Date of Metalling	12 Owners or Managers.
	101	N	**Hama** *Jorgensen 72-86* W Bk	Norweg'n	634	146·5	33·7	19·1	Arendal	1879		[illegible]	J.E. Hanssen
	102		**Hamady** W S	British	308	89·5	23·0	16·0	Cochin	1848			Chowacara Kelsto[illegible]
	103		**Hamalainen** W 3m Sr	Russian	206	98·5	22·6	13·8	Björkö	1876			M. Hämalainen
93482	104	V	**Hamburg** *Caldwell -86* W Bk	British	1699	216·2	43·0	24·0	Hantsport	1886	E. Churchill & Sons	[illegible]	G.W. Churchill
11275	105		**Hamburg** *Small* W Sr	American	174	102·1	26·1	9·0	E. Machias, Me.	1865			J.R. Talbot
26477	106	RV	**Hamilton Fish** *Mitchell* W S (ex Wm. H. Stover)	American	1582	200·0	40·0	28·0	Waldoboro, Me	1856		[illegible]	T. Dunham's Nephew & Co.
	107	NV	**Hamingja** *Johnsen -88* W Bk	Norweg'n	472	139·3	31·7	16·4	Drammen	1883	P. Svendsen	[illegible]	O.&P. Knudsen
	108	V	**Hamlet** *Quist* W Bn	Danish	156	95·5	21·7	11·1	Thurø	1871	P. Bom		P. Bom
	109	N	**Hamlet** *Hansen* W Bg	Norweg'n	264	117·5	27·3	12·8	Laurvig	1870			Joh. Hansen
	110		**Hamody** W Bk	British	438	107·3	25·3	17·8	Cochin	1850			Shaik Abdool L[illegible]
71531	111		**Hamperan Perak** W Sr	British	186	122·0	23·0	8·2	Singapore	1876			Lee Pee Chuan
63580	112	✻	**Hampshire** *Mathias 83-83* C S	British	1164	209·2	34·3	22·0	Greenock	1870	R. Steele	[illegible]	J. Hossack, sen.
50514	113		**Hampton** W Sr	British	294	132·3	30·5	12·7	Lower Hills-(boro'), N.B.	1863	Bennett		R. Tedcastle & Co.
	114	GV	**Hampton Court** *Kruse 76-* W Bk	German	956	168·9	35·1	22·6	Kennebunk	1860	C.&T. Ward	[illegible]	Siedenburg, Wendt & [illegible]
	115	V	**Hananger** *Olsen -87* (ex Hilda) W 3m Sr	Norweg'n	200	100·0	22·0	12·2	Wenersborg	1879		[illegible]	L. Larsen Sunde
95008	116	R	**Hancock** *Gupdill 82-86* W Bkn	American	392	121·5	29·5	15·8	Bucksport, Me	1869	Grant & Hill	[illegible]	J.S. Emery & Co.
82695	117	✻	**Handa Isle** *Johns 85-87* W 3m Sr	British	261	125·6	26·8	12·1	Auckland, N.Z.	1881	J. Bigelow & Son	[illegible]	D.H. McKenzie
143	118		**Handa Maru** W Sr	Japanese	222	104·2	22·2	12·3	Ominato, Ise	1879	Haniu Tamotsu		Takenouchi Seakichi
297	119		**Handa Maru No. 2** W Sr	Japanese	127	81·8	19·4	10·5	Shima	1880	Fukuzawa S.B.Y.		Uchinni Hansosa Kwaisha
52455	120		**Handy** W Sr	British	126	84·8	20·1	8·9	Groninesd	1854	C. Areand		C. Areand
	121	N	**Handy** *Narloff* W 3m Sr	Norweg'n	270	122·0	27·9	12·4	Sarpsborg	1884	C. Olsen	[illegible]	Aug. Olsen
	122	R	**Hango** *Skog 81-81* W Bg	Russian	270	116·3	28·3	14·0	Bromarf	1877			G.S. Bredenberg
80952	123		**Hanlan** W Sr	British	107	86·0	21·0	8·6	Picton, Ont.	1880	W. Redmond		James Williams
	124	G	**Hanna** *Schiebe 63-67* W Bg	German	293	112·1	27·2	14·1	Greifswald	1867	J.W. Sadelkow		C. Gruedener
	125	N	**Hanna** (ex Chloe) *Maus 83-83* W S	Norweg'n	935	179·5	35·7	21·3	Shelburne, N.S.	1872	J. McAlpine		F. Breda
	126	N	**Hanna** *Wiengel 71-71* W Bk	Norweg'n	552	128·3	32·1	19·5	Lekkerkerk	1856			C.H. Möllbach
	127	N	**Hanna** *Christiansen* W Bk	Norweg'n	285	121·5	28·1	12·4	Tatmagouch N.S.	1863			T. Christiansen
	128	N	**Hanna** *Christensen -88* (ex Annie) W Bk	Norweg'n	277	108·0	24·7	15·3	Hull	1857			I. Christensen
	129	N	**Hanna** *Paulsen -88* W Bk	Norweg'n	269	125·5	28·1	12·4	Risör	1865	R. Tellefsen	[illegible]	Chr. Swane
	130	DN	**Hanna** *Stomgaard -87* (ex [illegible]) W Sr	Norweg'n	219	97·3	27·3	13·7	Sweden	18[illegible]			S. Baltzersen's Enke
	131		**Hanna** *Nordberg* W Bg	Russian	385	123·6	27·9	15·1	Kantvik	1874	C. Fredriksson		J. Nordberg
	132		**Hanna** *Janson* W 3m Sr	Russian	318	..	..	..	Lappfjerd				E.P. Akerlund
	133		**Hanna** *Johansson* W 3m Sr	Russian	313	165·0	38·0	18·0	Borgå	1871	Anderson		J.E. Johansson
	134		**Hanna** *Söderlund* W Galeas	Russian	159	..	..	..		1873			E. Söderlund
	135		**Hanna** *Fox 71-74* W Bk	Swedish	355	127·9	25·6	13·9	Quebec	1866			N.P. Andersson
	136		**Hanna** *Andersson -78* W Bg	Swedish	313	116·1	26·0	15·0	Wiborg	1866			J. Gudmundsson
	137		**Hanna** *Andersson 71-84* W Sr	Swedish	150	83·3	22·0	11·8	Orousт	1874	J.O.J. Collén		J.Z. Vennerberg
	138		**Hanna Maria** *Ekqvist* W Bk	Russian	661	160·7	36·0	18·7	Borgå	1866		[illegible]	A. Simolin
42587	139		**Hannah** *Jones* W Swr	British	243	99·0	25·8	15·8	Sunderland	1862	J. Robson		W. Milburn
73080	140	R	**Hannah** *Edwards* W Bn	British	190	103·0	24·5	12·9	P.E. Island	1875	Duncan	[illegible]	J. Powell
29411	141		**Hannah** *Williams* W Bn	British	145	98·0	23·1	12·2	Padstow	1865	Treiwen		M.R. Waddle
11484	142		**Hannah** *Mutters* W Sr	British	121	70·5	21·5	11·0	Sunderland	1852	G.W. & W.J. Hall		S. Mutters
	143	N	**Hannah** *Nielsen -87* (ex United) W Bg	Norweg'n	312	98·2	27·6	17·3	Hartlepool	1846	J.P. Denton & Co		L. Schübeler
	144	N	**Hannah** (WRECKED) *Andreasen 83-85* (ex Malinche) W Bk	Norweg'n	284	119·7	26·0	14·0	Ueckermünde	1864	L. Wittenberg	[illegible]	A.S. Olsen
	145		**Hannah** *Scheller -83* W Bg	Swedish	280	96·8	24·4	16·0	Memel	1840	B. Pieper		J. Berntsson
11536	146		**Hannah Blackman** *Bradick* W Sr	American	135	91·0	26·5	7·2	Tuckahoe, N.J.	1858			W. Booth
78994	147	DV	**Hannah Blanchard** *Smith -88* W Bk	British	991	173·9	35·5	22·3	Newport, N.S.	1878	J. Mosher	[illegible]	G.P. Payzant
80958	148		**Hannah Butler** (SUNK) *O'Hagan -87* W 2m	British	126	89·4	20·0	8·1	Cobourg, Ont.	1856	J. Taylor		W. Ostrander
54550	149		**Hannah Croasdell** *Lewthwaite* W Sr	British	141	89·6	23·1	11·6	Ulverston	1866	Wilson		W. Postlethwaite
95810	150		**Hannah F. Carleton** *Bryant* W Sr	American	214	103·3	30·0	10·6	Pembroke, Me.	1884	W. Welch		W. Welch

✻ = **Lloyd's Register**; D = Nederlandsche Vereeniging; G = Germanischer Lloyd; I = Registro Italiano; R = [illegible]
A = Veritas Austro-Ungarico; H = Veritas Ellenico; N = Norske Veritas; V = [illegible]

In the 4th Column, C = Composite (iron frame, wood planked); I = Iron; Stl = Steel; W = Wood.

The Dates of Metalling (Col. 11) are based on the records of the Books in which Vessels are respectively classed, or on the Reports of the Surveyor[illegible]

Further particulars of Vessels classed by Lloyd's Register will be found in Lloyd's Register Book and Lloyd's Yacht Register.

[illegible]	No. dans le livre	Registre	Nom. Matériaux et Gréement. Capitaines et dates d'entrée en fonction des Armateurs actuels et à bord du navire. Ancien Nom.		Pavillon	Tonnage	Dimensions			Construit. Où	Quand	Par qui	Date du doublage en métal	Armateurs ou Directeurs-Gérants.	Port d'Armement.
[illegible]	151	✻D	Hannah Landles *Smith*	I Bk	British	1293	237·6	36·1	29·9	Glasgow	1878	A.Stephen&Sns		T.Law & Co.	Glasgow
[illegible]	152	R	HannahMcLoon *Reed*	W Bkn	American	518	137·0	31·5	14·5	Rockland,Me.	1874	Starrat&Co.	[illegible]	George Gregory	Rockland,Me.
[illegible]	153		Hannah Nicholson	W Bk	British	252	116·4	23·6	14·4	Whitehaven	1853	L.Kennedy&Co.		John Bickers	Adelaide,S.A.
[illegible]	154	✻	Hannah Ranson *Williams*	W Sr	British	104	89·2	19·2	9·9	Southampton	1876	Ransom	Y [illegible]	J.Dible	Southampton
[illegible]	155	✻	Hannah & Jane *Roberts*	W Sr	British	92	74·0	20·1	11·0	Pwllheli	1867	Jones		E.Jones & Co.	Carnarvon
[illegible]	156		Hannah & Mary	W Bk	British	330	122·8	28·1	17·5	Sunderland	1868	Barker	Y [illegible]	J.Paul & Co.	Lynn
	157	V	Hanne *Jørgensen 75-76*	W Bn	Danish	135	95·1	21·2	10·1	Thurø	1875	P.Troensegaard		Carl Jensen	Svendborg
	158		Hanne Mansoor	W Bn	Turkish	187	..	..	..	Oonia	1883			Oundalee Kotchzadeh (Mousapha & Co.)	Oonia
	159	D	Hannechiena (ex Lone Star) *Engelsman*	W Bkn	Dutch	245	106·9	25·5	12·6	Shelburne	1863	Carnel		H.Engelsman	Stads Kanaal
[illegible]	160	✻	Hannibal	I S	British	1204	219·0	34·8	22·8	Glasgow	1862	Smith&Rodger		W.R.Davies	Liverpool
[illegible]	161		Hannibal *Kearon 76-82*	W Bn	British	189	114·0	21·4	13·6	Hayle	1868	Harvey		R.Kearon	Dublin
[illegible]	162		Hannibal *Kimball*	W Sr	American	136	93·5	26·7	8·7	Bangor,Me.	1856			F.C.Pendleton	Belfast,Me.
	163	✻V	Hannover *Gehrcke 85-86*	C Bk	German	540	161·0	27·9	17·5	Dumbarton	1866	A.McMillan&Sn	Y [illegible]	L.F.Mathies & Co.	Hamburg
	164		Hans *Folmer*	W 3m Sr	Danish	219	114·4	24·7	12·5	Faaborg	1888	R.Møller		H.Christensen	Svendborg
	165		Hans (ex Arctic) *Hensen 85-*	W Sr	German	123	88·7	21·3	8·9	Sundsvall	1879	E.Ostmann	Y [illegible]	Hernsheim&Co.	Hamburg
	166		Hans *Ericksen*	W Bk	Norweg'n	423	133·1	30·8	16·2	Hamburg	1856		Y 10.87	L.Björnstad	Sarpsborg
	167		Hans *Lindholm*	W Galeas	Russian	127	..	..	..	Finby	1886			H.V.Söderlund	Finby
	168	V	Hans	W Sr	Russian	125	81·0	21·9	9·7	Dinamünde	18..			R.Ehmann	Riga
	169	G	Hans Andreas *Espersen 68-88*	W 3m Sr	Danish	103	84·5	18·7	9·8	Nexø	1864/78	H.L.Blem		P.Berg	Nexø
	170	V	Hans Christensens Minde *Nielsen*	W Sr	Danish	131	88·0	21·7	10·6	[illegible]	1866	P.Nielsen		P.Nielsen	Faaborg
	171	N	Hans Gude *Gregersen*	W Bg	Norweg'n	295	106·8	27·7	14·9	Grimstad	1862	J.Tellefsen		A.N.Grönn	Lillesand
	172	V	Hans Kragh *Hansen 75-78*	W Sr	Danish	102	86·0	18·5	9·5	Troense	1874	C.J.Jensen		L.v.d.Hude	Svendborg
	173	DN	Hans Nielsen Hauge *Christensen*	W Bk	Norweg'n	472	142·4	32·0	16·9	Drammen	1872	A.Caspersen	Y 7.86	H.T.Kiaer	Drammen
	174	N	Hans Olsen *Malboer*	W Bn	Norweg'n	178	101·8	23·6	12·7	Stavanger	1874		Y 12.87	A.Johnsen	Skudesnaes
	175	N	Hans Thiis *Thiis*	W Bk	Norweg'n	389	126·9	27·1	16·8	St.John,N.B.	1862		Y 10.84	C.H.Møllerch	Risör
	176	DN	Hans Thode *Josephsen*	W 3m Sr	Norweg'n	259	116·0	27·1	13·9	Stavanger	1872		Y [illegible]	E.H.Jondahl	Stavanger
	177	DG	Hansa (ex Baron von ...) *Hidde*	W Sr	Dutch	155	86·0	23·3	11·9	Barth	1876	R.Schlör	Y [illegible]	A.Köpcke	Rotterdam
	178	V	Hansa *Lorenzen*	W Bk	German	499	148·3	30·9	17·4	Harburg	1869	H.Kraus	Y 10.86	H.H.Pott Nachf.	Flensburg
	179	V	Hansa	W Bn	German	251	106·3	24·0	12·6	Ham'elwarden	1876	K.Lühring	Y [illegible]	Joh.de Harde	Brake
	180	V	Hansa *Gerdt 75-*	W Bn	German	238	105·6	29·9	11·0	Elmshorn	1873	D.Kremer	Y 12.86	M.H.Schlichting	Hamburg
	181		Hansa *Peters 87-*	I Sr	German	140	..	..	..		1872			H.Peters	Hamburg
	182		Hansine *Rasmussen 77-87*	W Bn	Danish	160	101·1	22·1	11·4	Aarhuus	1868	N.Ulstrup & E.Meisner	Y [illegible]	J.P.Jorgensen	Svendborg
	183	DV	Hansine *Larsen 84-*	W Bn	Danish	132	92·5	21·4	10·5	Faaborg	1874	R.H.Møller		J.F.Hoffmann	Faaborg
	184	V	Hansine Marie *Rasmussen*	W Bn	Danish	134	94·3	21·0	10·1	NordBy,Fanø	1878	N.Nielsen	Y [illegible]	Aktieselskab "Hansine Marie" (J.N.Jensen)	Nordby,Fanø
[illegible]	185		Happy Return (ex Volunteer)	W Bg	British	185	88·6	24·5	14·6	St.Malo	1838			F.Picot	Guernsey
[illegible]	186		Happy-go-Lucky (ex Garland)	W Bn	British	164	96·0	25·0	11·7	Eastport,U.S.	1847			J.M.Barkley	Belfast
	187		Harald *Jensen 75-78*	W Sr	Danish	143	95·6	21·7	11·1	Holbæk	1867	H.P.Mortensen		H.L.Smith's Enke	Holbæk
	188	NV	Harald *Larsen -87*	W S	Norweg'n	208	116·1	26·5	10·3	Christianssand	1887	Brödr.Larsen	Y [illegible]	B.Balken	Christianssand
	189	R	Harald *Johansson -77*	W Bk	Swedish	505	141·0	29·5	15·7	Framnäs	1871	J.A.Strandberg	Y [illegible]	G.F.Cawallin	Sundsvall
	190	V	Harald *Bruce -86*	W Bk	Swedish	483	146·3	27·5	16·6	Quebec	1855		Y 12.81	Rederiaktiebolaget Orion	Gothenburg
	191		Harald Haarfager *Henriksen*	W Bk	Norweg'n	1057	176·0	35·9	23·9	Newburyport, Mass.	1850		Y [illegible]	Helliesen & Co.	Stavanger
	192		Harald Haarfager (ex Carrie Wright) *Lauritzen*	W Bk	Norweg'n	532	145·6	31·3	18·0	Hillsboro',N.S.	1864	Marshall		Chr.Hannevig	Christiania
	193	N	Harald Haarfager *Bergh*	W Bk	Norweg'n	512	139·6	31·7	16·9	Bergen	1864	J.Trumphy&Sn.		W.A.Steffens	Fredrikstad
[illegible]	194	R	HarbesonHickman *Robinson -71*	W 3m Sr	American	419	140·5	32·0	15·0	Milford,Del.	1874			J.W.Hall	Wilmington,Del.
[illegible]	195	✻	Harbinger *Bell 78-84*	I S	British	1506	252·5	37·6	22·4	Greenock	1876	R.Steele&Co.		Anderson,Anderson&Co	Aberdeen
[illegible]	196		Harbinger *Hayes*	W Sr	American	103	84·0	23·0	7·0	Bucksport,Me	1848			A.P.Veazie	Bangor,Me.
	197	N	Hardangeren	W Sr	Norweg'n	136	92·6	23·9	11·3	Stavanger	1872			T.Mangor	Haugesund
	198		Hardi *Pedersen*	W Bg	Norweg'n	331	122·1	29·8	14·0	Åbo	1856			Nicolay Olsen	Arendal
	199	N	Hardi *Nielsen*	W Bg	Norweg'n	279	107·0	26·4	14·3	Grimstad	1875	N.OlsenGjömle	Y [illegible]	G.Huseland	Grimstad
[illegible]	200		Hare Bell *Gibb*	W Bg	British	237	102·2	26·0	14·6	Newcastle	1864	Hutchinson		Bessey & Palmer	Newcastle

✻ = Lloyd's Register; D = Nederlandsche Vereeniging; G = Germanischer Lloyd; I = Registro Italiano; R = American Record; A = Veritas Austro-Ungarico; H = Veritas Ellenico; N = Norske Veritas; V = Bureau Veritas.

Dans la 4me Colonne, C = Composite (membrure en fer, bordé en bois); I = Fer; Stl = Acier; W = Bois.

Les Dates du doublage en métal (Col. 11) ressortent des livres dans lesquels les navires sont respectivement classés ou des rapports des Inspecteurs du Lloyd's Register.

Pour plus amples renseignements sur les navires classés par le Lloyd's Register, voir le Lloyd's Register Book et le Lloyd's Yacht Register.

Official No.	No. in Book.	Registry in which classed.	Ship's Name. Material and Rig. Master, and Dates of Appointment to present Owner's Service and to Ship. Ship's Late Name.	Flag.	Net Tonnage.	Dimensions. Length.	Breadth.	Depth.	Built. Where.	When	By whom.	Date of Metalling.	Owners or Managers.
[illegible]	201		Haricleia Roussou W Bn	Greek	517	145·0	30·8	..	[illegible]	1875			N.P. Roussos
	202	✳	Harland *Bidwell 79-85* I S	British	1695	250·8	39·0	24·1	Stockton	1885	Richardson [Duck & Co		W. Lund
	203		Harlingen *Fluckmacher* (ex Sind Leander) 71-83 W Bk	Dutch	372	121·5	30·5	15·8	Holland	1854	Kater		H. Jans & Co.
	204		Harmine *Buss* W Sr	German	129	83·3	21·0	9·6	Grossefehn	1874	Schapp& Meyer		H.L. Buss
	205	✳ V	Harmodius *Spitzen 86-86* I Bk	German	496	151·1	29·0	16·2	Sunderland	1867	W. Pile& Co.		L.F. Mathies & Co.
11069	206		Harmona *Sylvester* W Sr	American	145	85·2	24·2	8·1	Bristol, Me.	1846			H. Perkins
	207	N	Harmoni *Abrahamsen -88* W Bk	Norweg'n	462	117·1	27·2	18·3	Medford, N.J.	1846			J.L. Larsen
	208		Harmonia W Bk	Brazilian	526	..	..	..					Maria C.S. Castro
	209		Harmonia W Bn	Brazilian	122	..	..	..					A.T. de Magalhães
	210	G V	Harmonia *Karselohm 83-86* W S	German	1454	208·7	40·0	24·3	Newburyport	1875	J. Currier	[illegible]	J.C. Meinke
	211	N	Harmonia *Jensen -85* W Bk	Norweg'n	256	111·0	27·5	12·1	Grimstad	1866	John Terjesen		Aktieselskabet "Harmonia" (T. Thorsen)
	212	✳ D	Harmonie *Bakker 73-73* (ex Redine, ex [illegible]) W Sr	Dutch	152	90·9	22·2	10·7	Hoogezand	1868	J.A. Hooites	[illegible]	J.A. Hooites
	213	D G	Harmonie *Buss 80-80* W Sr	German	129	88·9	20·9	10·4	Veendam	1860	L. van Lingn Ezn		Maertens
	214	G	Harmonie *Oldenburg* W Sr	German	103	72·4	22·8	8·7	Bardenfleth	1874	H. Focke		Joh Burch Hellmers
	215	N V	Harmonie *Grunden 81-79* W Bk	Norweg'n	697	141·4	36·3	20·9	Björneborg	18[illegible]			M. Oppen
	216	N	Harmonie *Olsen* (ex Carl August) W Bk	Norweg'n	513	131·8	30·3	18·3	Königsberg	1863	L. Eggert		Chr. Hansen
	217		Harmonie *Wettre* (WILDFIRE) W Bk	Norweg'n	441	131·5	31·6	17·5	Helsingfors	1858			T. Wettre
	218	N	Harmonie *Jensen* W Bk	Norweg'n	391	123·1	29·6	16·5	Arendal	1873		[illegible]	H.P.R. Mathiesen
	219	N	Harmonie *Christiansen* W Bk	Norweg'n	322	..	..	..	Raumo	1860			J. Christophersen
	220	V	Harmonie *Lindström -87* W Bk	Swedish	494	151·9	26·5	16·4	Sundsvall	1872	E. Östmann	[illegible]	G.L. Widholm
	221	D	Harmonie *Arnesen -88* W Bg	Norweg'n	220	92·8	25·6	13·7	Arendal	18[illegible]			Edw. Stangebye
29772	222		Harmony *Linklater* W Bk	British	251	102·2	25·5	15·5	Yarmouth	1861	Fellows		J.F. Pemsel & B. La Trobe
58237	223	✳	Harmony *Haynes* W Bkn	British	167	101·4	23·2	11·1	Newburg	1868	Scrimgeour		J. Haynes
95419	224	✳	Harold *King 88-88* Stl Bk	British	1299	240·0	36·5	21·3	Port Glasgow	1888	R. Duncan & Co.		T. Stephens & Sons
64997	225	D V	Harold *Earl 86-86* W Bk	British	882	175·5	35·8	21·6	Princeport, NS	1872			M. Dickie
59352	226		Harold W Bn	British	142	105·0	28·0	10·5	St. Patrick, [N.B.	1873	J.J. McMullin		W. Wallace
88133	227	✳	Harold *Agenskow* W Sk	British	75	74·1	20·0	10·4	Rye	1883	J.C. Hood		H. Cook
	228		Harold (BROKEN UP) *Jean* W Bk	French	653	136·8	26·6	21·1	Whitehaven	1849	Brocklebank	[illegible]	E. Corblet
95730	229	R	Harold B. Cousens *Davies 82-81* W 3m Sr	American	361	139·0	34·0	11·6	Boston, Mass.	1882	Crosbie& Son	[illegible]	C.A. Cousens
95708	230	R	Harold C. Beecher *Nickerson -88* W 3m Sr	American	346	127·0	33·0	12·4	Madison, Conn	1883	C.M. Miner& Son		E.H. Weaver
95802	231	R V	Haroldine *Hussey* W 4m Sr	American	1294	193·0	40·5	18·7	North Weymouth, Mass.	1884	N.P. Keen	[illegible]	C.B. Farnsworth
93575	232		Harrier *Hennessy* W Sy	British	120	93·4	18·6	12·0	Cowes	1876			J.P. Sunderland
20038	233		Harriet W Sp	British	104	90·4	23·6	8·6	Shoreham	1858			T.F. Ward
20265	234	✳	Harriet (ex Commodore) W K	British	54	67·7	16·8	8·9	Brixham	1857			W. Burgess
11400	235		Harriet W Bk	American	751	154·8	32·9	12·1	Brooklyn	1853			I. Taylor
	236	V	Harriet *Hansen* W Bn	Danish	175	102·6	23·0	11·3	Svendborg	1875	P. Troensegaard	[illegible]	J.P. Rasmussen
53567	237		Harriet Amelia *Reason* W Bn	British	185	97·3	26·3	11·9	Truro, N.S.	1865	Alexander		J. Ronaldson
95832	238		Harriet C. Kerlin *Martz* W 3m Sr	American	491	143·0	35·0	14·0	Leesburg, N.J.	1884	John Russell		G.S. Martz
64036	239	R V	Harriet Campbell *Pearson -88* W Bk	British	649	148·0	31·5	19·2	Weymouth NS	1873	C. Campbell	[illegible]	S. Edgett
95504	240	R	Harriet G. *Stanley* W Bn	American	247	130·0	28·0	10·5	Norfolk, Va.	1877	W.A. Beach	[illegible]	L.Y. Stanley
66925	241	V	Harriet Hickman *Dwyer* W Bk	British	851	165·7	35·0	20·7	Dorchester NB	1873	W. Hickman	[illegible]	W. Hickman
20727	242		Harriet Julia *Dempsey* W Bn	British	138	76·3	24·8	13·0	Inverness	1855			J. Fitzpatrick
57514	243		Harriet McGregor *[illegible]* W Bk	British	331	134·2	27·6	15·9	Hobart Town	1871	McGregor	[illegible]	A. McGregor
11801	244		Harriet S. Brooks *Smith -86* W Sr	American	220	108·0	29·5	14·0	Smyrna, Del.	1867			G.W. Shepherd, jun.
95265	245	R	Harriet S. Jackson *Brown* W Bkn	American	472	140·0	32·4	16·2	Yarmouth, Me	1874	Hutchings & [Stubbs	[illegible]	A.H. White
71349	246		Harriet Upham *Spicer -87* (ARANDOSE[illegible]) W Bkn	British	352	135·0	29·8	12·9	Londonderry, NS	1874		[illegible]	T.E. Kenny
58353	247		Harriet Williams *[illegible]* W Sr	British	144	96·6	21·9	12·3	Burton Stathr	1866	Wray		G.H. Bate
19016	248		Harriett *Smith* W Bn	British	182	91·6	22·4	13·1	Lewes	1848	J. Berry		T. Fieldgate
32497	249		Harriett *Fieldgate* W Bg	British	166	81·6	19·5	13·1	Aberdeen	1844			H. Dunsford
44401	250		Harriett Wardle *Hancock* W Bg	British	229	104·1	25·6	14·9	Sunderland	1862	W. Briggs& Sons		T.G. Robins & Co.

✳ = Lloyd's Register; D = Nederlandsche Vereeniging; G = Germanischer Lloyd; I = Registro Italiano; R = [illegible]
A = Veritas Austro-Ungarico; H = Veritas Ellenico; N = Norske Veritas; V = [illegible]

In the 4th Column, C = Composite (iron frame, wood planked); I = Iron; Stl = Steel; W = Wood.

The Dates of Metalling (**Col. 11**) are based on the records of the Books in which Vessels are respectively classed, or on the Reports of the Surveyor[illegible]

Further particulars of Vessels classed by Lloyd's Register will be found in Lloyd's Register Book and Lloyd's Yacht Register.

No.	Classe	Nom. Matériaux et Gréement. Capitaine et dates d'entrée en service. Ancien Nom.	Nationalité	Tonnage	Longueur	Largeur	Creux	Construit: Où	Quand	Par qui	Date de doublage en métal	Armateurs ou Directeurs-Gérants.	Port d'Armement.
251	✱ I	Harrington		572	161·6	28·4	17·4	Harrington	1868	R. Williamson & S.		R. Connon & Co.	Arbroath
252	✱	Harry *Nairne 75-85* W3mSr	British	525	165·5	28·1	17·4	P.E.Island	1881	W. Richards		H. Hoskin	Swansea
253		Harry W Bn	British	170	93·4	22·0	9·3	Prospect, U.S.	1853			J. B. Bell	Trinidad
254		Harry B. Diverty *Nickerson* W Sr	American	167	110·3	23·0	8·6	Dennisville NJ	1873			J. H. Gifford	Camden, N.J.
255	R	Harry B. Ritter W3mSr	American	612	141·0	35·4	17·5	Mauricet'n, NJ	1873	Vannaman Bros.		J. W. Sunaman	Bridgeton, N.J.
256	R	Harry Bailey *Moore* W Bk	British	686	152·2	34·3	19·4	Hopewell, NB	1874	J. Calhoun		L. H. Vaughan	St. John, N.B.
257	R	Harry Buschman W Bk	British	528	145·6	32·5	16·6	Clifton, N.S.	1873	J. Crowe		A. Haley	Halifax, N.S.
258	R	Harry C. Shepherd *Bunge* W3mSr	American	203	108·5	29·8	8·7	May's Ldg NJ	1870	J. Clark		Charles W. Booye	Somers Pt., N.J.
259		Harry Herbert *Jones* W Sr	British	112	83·9	23·7	11·6	Rhyl	1860	Jones		D. Murray	Liverpool
260	R	Harry L. Whiton *Rich* W3mSr	American	457	134·7	32·9	15·2	Duxbury Mass.	1873	Merritt Bros.		J. Whiton	Boston, Mass.
261		Harry Landell *Crowley 87* W Sr	American	227	127·0	30·0	9·0	Belleville, N.J.	1866			W. A. Ripley	Newark, N.J.
262	R	Harry Messer *Swett 89-90* W3mSr	American	596	153·7	35·5	15·8	Bath, Me.	1880	Goss & Sawyer		Jas. F. Sears	Boston, Mass.
263		Harry Morse *Murphy 87* W Bk	American	1313	198·2	37·2	23·8	Bath, Me.	1871	J. P. Morse & Co.		J. Rosenfeld	San Francisco
264		Harry Mundahl *Jansen 85-86* W Dy	British	122	84·0	20·8	11·1	Hamburg	1882	C. C. A. Drever		C. M. Mundahl	Grimsby
265	R	Harry Prescott *Doane 87* W3mSr	American	412	140·0	34·0	11·6	Bath, Me.	1862	W. Rogers		H. Prescott	New Haven, Conn.
266		Harry Russell *Davies 69-83* W Sr	British	140	95·0	22·5	12·6	Barrow	1863	Reay		J. Fisher & Sons	Lancaster
267	R	Harry Smith *Hutchins* W Bn	American	520	131·1	30·9	17·2	Bangor, Me.	1875	Crosby Bros.		B. S. Crosby	Bangor, Me.
268	R	Harry Stewart *Blake 87* W Bn	American	285	113·5	28·0	11·6	Bangor, Me.	1867			T. J. Stewart & Co.	Bangor, Me.
269	R	Harry White *Sopher 75-76* W3mSr	American	180	102·4	27·5	7·9	Bucksport, Me.	1870	W. Beazley & Co.		E. Hopkins	Bucksport, Me.
270	R	Harry & Aubrey *Tobin* W Bn	British	223	117·2	26·0	11·5	Pugwash, N.S.	1864			Hugh Kelly	Pugwash, N.S.
271	✱	Hartfell *Harman 74-76* I S	British	1224	215·0	35·5	22·4	Glasgow	1864	Barclay Curle & C.		Eyre, Evans & Co.	Liverpool
272	✱	Hartfield *Thomson 85-86* I S	British	1815	261·7	39·3	24·1	Whitehaven	1884	Whitehaven S.B. Co. Ld.		J. B. Walmsley & Co.	Liverpool
273	✱	Hartfield *Long 85-88* I S	British	815	195·1	31·3	19·8	Greenock	1868	R. Steele		J. Richardson & Co.	Glasgow
274	R V	Harvard *Thompson 85-88* W Bk	American	982	181·3	35·7	21·2	Newb'ryp't Ms.	1878	Atkinson & Fillmore		M. F. Pickering & Co.	Boston, Mass.
275	✱	Harvest Home *Rimmer* W Sr	British	88	88·1	21·1	9·6	Tarleton	1882	T. Land		J. Taylor	Preston
276		Harvest Home *Barilari* I Bk	Argentine	544	173·2	28·3	17·1	Liverpool	1855	T. Vernon & Son		Argentine Govt.	
277	✱ D	Harvest King *Boden 80-* W Sr	British	100	91·3	21·7	10·5	Runcorn	1879	Brundrit		T. Rayner	Runcorn
278		Harvest Maid *Swann* W Sr	British	100	87·0	19·0	10·5	Tarleton	1860	Banister		James Taylor	Preston
279		Harvest Man *Walters* W Sr	British	133	100·8	20·9	11·4	Jersey	1865	Valpy		F. Searle	Plymouth
280	✱	Harvest Moon *Roiall* W Sr	British	87	78·0	21·3	10·6	Burton Stather	1886	J. Garside		M. Roiall	Grimsby
281	V	Harvest Queen *Forsyth* W S	British	2020	257·0	45·0	24·6	Cornwallis NS	1887	E. Cox		C. R. Burgess	Windsor, N.S.
282	✱	Harvest Queen *Rob-n 78* W Sr	British	99	90·7	21·3	10·3	Ardrossan	1870	Barr & Shearer		T. Rayner	Runcorn
283	R V	Harvester (STRANDED) *Alwood* W Bk	American	700	156·6	32·7	19·9	Newburyport N.S.	1871	Atkinson & Fillmore		E. G. Stoddart & Co.	New Haven, Conn.
284	R	Harvester *McLennan 87-87* W Sr	British	113	82·1	26·2	8·4	Carleton, N.B.	1880	J. O. Stackhouse		Scammell Bros.	St. John, N.B.
285	R V	Harvester *Allen 87* W S	American	1494	210·1	39·7	24·0	Bath, Me.	1875	E. & A. Sewall		A. Sewall & Co.	Bath, Me.
286		Harvey *Moloney* W Sr	British	146	98·7	22·2	9·3	Harvey, N.B.	1873	Brewster		M. Moloney	Waterford
287		Harvey W. Anderson *Endicott 87* W3mSr	American	394	132·0	32·0	9·5	Greenbank NJ	1877	J. Vansant & Sns.		A. E. Babcock	Somers Pt., N.J.
288	D N V	Hassel *Ellertsen 85-85* (ex Maria Magdalena) W Bk	Norweg'n	569	155·2	31·6	18·1	Hamburg	1878	J. H. Marks		J. A. Samuelsen	Farsund
289	D N	Hasselnodder *Nielsen* W Bk	Norweg'n	544	151·6	31·8	17·5	Grimstad	1880	F. Smith Petersen		F. Smith Petersen	Grimstad
290		Hassir Bahri W Bn	Turkish	295	..	..	..	Unia	1879			Balinoom li-Hafiz Hamid Shakir Effendi & Co.	Unia
291	N	Hast *Henriksen* W Sr	Norweg'n	264	106·1	26·2	13·3	Calais, Me.	1861			A. Berntdsen	Farsund
292	R	Hastings *Lawrence 88* W Bn	British	339	122·0	29·0	12·9	Jordan Rv. N.S.	1882	R. W. Freeman		R. W. Freeman	Shelburne, N.S.
293		Hatteras *Botz* W Sr	American	730	171·5	35·5	19·3	Brooklyn	1865			F. F. Chambers	New York
294	R	Hattie *Coombs 81-81* W Bn	American	328	109·4	27·4	10·5	Bangor, Me.	1864	Thurston & Crosby		J. S. Emery & Co.	Boston, Mass.
295	R	Hattie *Low* W Sr	American	162	100·0	26·8	9·0	Belfast, Me.	1866			S. G. Haskell	Deer Isle, Me.
296		Hattie A. Butler *Carroll* W Sr	American	146	94·0	27·0	7·0	Essex, Conn.	1868			J. Carroll	Hartford, Conn.
297		Hattie A. Marsh *Barker* W3mSr	American	414	140·8	33·6	11·4	Madison, Ct.	1883	M. B. McDonald		C. Barry	New London
298		Hattie Baker *Parker* W Sr	American	328	116·4	30·4	9·5	Boston, Mass.	1864	A. & G. T. Sampson		J. Baker & Co.	Boston, Mass.
299		Hattie C. *Stewart 83* W Sr	British	182	97·5	28·4	10·1	Hopew'll Cape, N.B.	1883	C. Dowling		C. Dowling	St. John, N.B.
300		Hattie Dell *Cook* W Sr	British	111	85·0	23·0	9·7	Barringt'n N.S.	1884	R. B. Stoddart		W. W. Kenney	Barrington, N.S.

✱ = Lloyd's Register; ✱ = Veritas Austro-Ungarico; D = Nederlandsche Vereeniging; G = Germanischer Lloyd; H = Veritas Ellenico; I = Registro Italiano; N = Norske Veritas; R = American Record; V = Bureau Veritas.

Dans la 4me Colonne, C = Composite (membrure en fer, bordé en bois); I = Fer; Stl = Acier; W = Bois.

Les Dates de doublage en métal (Col. 11) ressortent des livres dans lesquels les navires sont respectivement classés ou des rapports des Inspecteurs du Lloyd's Register.

Pour plus amples renseignements sur les navires classés par le Lloyd's Register, voir le Lloyd's Register Book et le Lloyd's Yacht Register.

Official No.	No. in Book	Registry in which classed	Ship's Name, Material and Rig. Master, and Dates of Appointment to present Owner's Service and to Ship. Ship's Late Name.	Flag	Net Tonnage	Dimensions: Length	Dimensions: Breadth	Dimensions: Depth	Built: Where	Built: When	Built: By whom	Date of Metalling	Owners or Managers
95815	301	R	Hattie Dunn *Poland* W3mSr	American	414	115·1	35·1	11·6	Thomaston Me	1884	Dunn&Elliot	[illegible]	Dunn & Elliot
[illegible]	302		Hattie E. Giles *Griffith* W3mSr	American	128	110·5	23·7	7·0	Lewesville Del	1874			T.B.Giles
95238	303		Hattie E. King *Hines* W3mSr	American	189	115·0	28·3	9·0	Calais, Me.	1873	Hinds		C.W.King
[illegible]	304		Hattie E. Smith *Evans* WSr	American	100	87·3	23·0	8·2	Newbrypt.Mass	1869			S.Osborn, jun
[illegible]	305		Hattie E. Tapley *Telfer* WS	British	907	164·0	34·3	22·0	Brewer U.S.	1865	J.Dunning	[illegible]	G.J.Finlayson
[illegible]	306		Hattie F. Rich *Chadsey* WBn	British	118	85·0	24·0	9·4	Pt.le Bert, N.S.	1876	S.Harlow		Thomas Johnson
[illegible]	307	R	Hattie G. Dixon *Dean* -88 W Bkn	American	503	140·0	31·7	16·4	Damariscotta, Me.	1876	G.Merry	[illegible]	J.E.Yates
[illegible]	308	R	Hattie G. McFarland *Dodge* W Bkn	American	519	142·4	32·8	16·5	Thomaston Me	1872	Dunn&Elliot	[illegible]	A.Dodge
[illegible]	309	V	Hattie H. *Dill* -81 W Bkn	British	403	139·6	31·0	12·9	Newport, N.S.	1872	J.A.Harvie&Co.	[illegible]	J.A.Harvie
[illegible]	310	R	Hattie H. Barbour *Fletcher* W3mSr	American	287	121·5	32·5	11·6	Brewer, Me.	1883	Hopkins		P.McConville
95230	311		Hattie L. Curtis (ABANDONED) WSr	American	106	90·0	26·6	7·6	Surry, Me.	1873			George Hook
95707	312	R	Hattie L. Sheets *Dale* W3mSr	American	175	107·8	28·0	8·0	Milford, Del.	1882	J.W.Ablott&Co.		L.Matthews
95436	313		Hattie Lollis *Sharp* -84 W3mSr	American	152	100·0	28·8	8·0	Milton, Del.	1876	D.Atkins		J.W.Hall
[illegible]	314	R	Hattie Louise *Barnard* W Bn	British	200	101·4	28·5	11·5	Sherbrooke, NS.	1886	C.S.McIntosh	[illegible]	A.M.Wright
95228	315	R	Hattie M. Bain *McDonald* -81 W Bn	American	372	125·8	30·1	11·5	Yarmouth, Me	1873		[illegible]	J.Bain
95231	316		Hattie M. Crowell *Chase* W3mSr	American	410	135·0	32·0	15·0	Greenport, N.Y.	1873			J.H.Cady
11959	317	R	Hattie M. Howes *Hobbs* W3mSr	American	188	113·0	29·1	8·5	Providence, R.I.	1869	Chase		
95216	318	R	Hattie McG. Buck *Putnam* W3mSr	American	212	115·5	28·8	10·2	Bucksville, SC	1873	W.McGilvery		J.C.Nickels
95261	319		Hattie N. Bangs *Bangs* W Bkn	American	542	140·0	32·6	16·2	Essex, Mass.	1874	J.McKenzie	[illegible]	Jas. E. Brett
95199	320		Hattie N. Gove *Chase* (LOST) W3mSr	American	330	134·6	30·6	14·4	E.Boston	1873			B.D.Kelly, jun
95635	321		Hattie Nickerson *Nickerson* WSr	American	164	112·0	23·0	9·0	Dennisville N.J.	1881			T.W.Garretson
11811	322		Hattie Paige *Tryon* WSr	American	282	115·1	29·5	9·1	Bridgeton, N.J.	1867			W.B.Trewhells
95182	323		Hattie Perry *Chase* -73 WSr	American	165	103·3	30·1	7·2	Bath, Me.	1872			S.C.Hart
11341	324		Hattie S. Collins *Mapes* WSr	American	117	97·2	26·7	7·2	E.Haven, Conn	1867	G.W.Baldwin		F.W.Benedict
95561	325	R	Hattie S. Williams *Bray* W3mSr	American	853	181·2	39·9	18·0	Bath, Me.	1879	Goss&Sawyer		S.N.Staples
95171	326	R	Hattie Turner *Glass* -88 W3mSr	American	280	129·2	31·9	10·1	Thomaston Me	1872			R.Glass
95324	327		Hattie V. Kelsey *Chase* -88 W3mSr	American	393	140·2	34·0	16·0	W.Haven, Conn	1874	A.Thomas		W.F.Armstrong
82272	328	✻	Hattonburn (ex Our Boys) *Cadier* W Bk	British	235	114·7	28·6	12·8	Grangemouth	1878	Adamson&Co.	[illegible]	A.Cadier
	329	N	Haugar *Omundsen* 84-84 W Bn	Norweg'n	187	107·7	26·2	12·1	Bergen	1874			S.Christiansen
	330	N V	Haugesund *Fagerland* WS	Norweg'n	1035	195·8	37·1	21·9	Haugesund	1876	J.Hauge	[illegible]	A.Lothe
	331		Haugesund (ex Rambler) *Gabrielsen* W Bk	Norweg'n	483	125·0	28·3	19·9	Newbrypt.Mass	1844			H.L.Gabrielsen
	332	V	Haumet *Perrigault* W Bk	French	218	98·4	23·3	12·8	St.Malo	1876	J.Parnet	[illegible]	Vve Le Mével & [illegible]
[illegible]	333	R V	Havana *Simmons* -87 W Bk	American	617	159·1	23·4	15·3	Bath, Me.	1879	W.Rogers	[illegible]	W.D.Manson
	334	N V	Havelock *Jacobsen* -84 W Bk	Norweg'n	1079	181·3	37·5	22·4	Quebec	1875	Charland	[illegible]	A.J.Freberg
	335		Havfruen (ex Effort) *Hansen* -87 W Bg	Danish	390	119·5	26·2	17·4	Parrsbro', N.S.	1878	W.Moore		H.Iversen
	336	N	Havfruen *Gulliksen* 82-82 W Bk	Norweg'n	643	163·3	34·5	18·2	Sandefjord	1880	Chr.Christensen	[illegible]	A.F.Klaveness
	337	N	Havfruen *Hansen* W Bk	Norweg'n	287	100·9	27·8	15·5	Grimstad	1853	T.Gundersen		O.Christensen&Co.
95494	338	R	Havilah *Richardson* -88 W Bk	American	507	136·5	31·3	17·3	Bangor, Me.	1877	Crosby Bros.	[illegible]	Swan & Son
	339	✻ V	Havilah *Mortensen* 88-88 I Bk	Danish	489	156·0	26·8	17·2	Dundee	1868	Brown&Simpson		Act.Selsk Havilah (P.C.Svarrer)
78994	340	V	Havre *Davidson* W Bk	British	642	155·2	32·2	18·0	Hantsp'rt, N.S.	1879	I.Davison	[illegible]	G.W.Churchill
	341	✻	Havre *Laffrentz* W Bk	Chilian	639	165·9	31·8	18·7	Bordeaux	1875	Chaigneaufreres	[illegible]	H.Battle&Laffrentz
	342		Havre (CONDEMNED) *Sire* 78-81 W Bk	French	488	115·6	29·5	15·5	La Mailleraie	1871	E.Lefranc	[illegible]	H.Auger ainé
	343	N	Havre *Gundersen* -88 WS	Norweg'n	826	160·8	36·0	20·6	New York	1845	W.H.Webb		T.Thomassen & Sn
64105	344		Hawaii W Bg	British	212	101·5	24·0	12·5	Oldenburg	1862			J.McLeod
63186	345	✻	Hawarden Castle *Hughes* -85 I S	British	1101	215·0	35·1	21·0	Liverpool	1869	R.&J.Evans&Co.		Richards, Mills & Co.
	346	✻	Hawarden Castle *Hunter* -88 WSk	British	92	81·7	21·1	10·3	Hull	1888	Albert Dock Shipwright Co.		Widdowson & Sp[illegible]
	347		Hawk (ex Mattos Segundo) WSr	British	121	85·3	21·2	10·4	Portugal	1845			R.Carr
82684	348	✻	Hawk *Press* WSr	British	84	86·1	20·3	8·3	Aratapu, N.Z.	1881	J.Barbour		Melbourne Pilot Co.
	349	✻	Hawkesbury *Norfelt* -88 C S	Swedish	1120	203·0	36·2	21·5	Sunderland	1868	W.Pile&Co.	[illegible]	J.Ingemansson
	350												

✻ = Lloyd's Register; D = Nederlandsche Vereeniging; G = Germanischer Lloyd; I = Registro Italiano; A = Veritas Austro-Ungarico; H = Veritas Ellenico; N = Norske Veritas;

In the 4th Column, C = Composite (iron frame, wood planked); I = Iron; Stl = Steel; W = Wood.

The Dates of Metalling (Col. 11) are based on the records of the Books in which Vessels are respectively classed, or on the Reports of the Surveyors [illegible]

Further particulars of Vessels classed by Lloyd's Register will be found in Lloyd's Register Book and Lloyd's Yacht Register.

Nombre	No	Société	Nom. Matériaux et Gréement. Capitaines et dates d'entrée au service des Armateurs actuels et à bord du navire. Ancien Nom.	Pavillon	Tonnage	Longueur	Largeur	Creux	Où	Quand	Par qui	Doublage	Armateurs ou Directeurs-Gérants.	Port d'Armement.
	351	✻	Hawkeye *Embericos* WBk	Greek	505	143·0	29·0	19·0	Bideford	1864	Cox	[illegible]	S.M.Embericos	Andros
	352	DR	Hawthorn *Colquhoun 85-86* WBkn	British	621	158·6	32·4	17·5	St.John,N.B.	1874	J.Rowan	Y	T.S.Adams	St.John,N.B.
	353	✻	Hawthorn *Andrew -87* WSk	British	82	76·8	20·5	10·4	Rye	1887	G.&T.Smith		W.R.O.Oxley	Grimsby
	354		Hayashi Maru WSr	Japanese	209	106·4	21·0	11·2	Fushiki	1884	TodaShojiro		Hayashi Shojiro	Kagoshima
	355	V	Haydée *Gottolin* WSr	French	113	81·4	21·6	10·5	Fécamp	1878			P.Crétolle	Binic
	356	V	Haydn Brown *Havener* WBk	American	821	162·0	34·9	19·8	Newburyport,Mass	1876	Atkinson&Fillmore	Y	C.H.Havener	Boston,Mass.
	357		Hayes *Hansen* WSr	American	194	102·0	33·0	9·1	Umpqua Oreg.	1876			B.H.Madison	San Francisco
	358		Hayle WBn	British	109	72·8	20·7	12·5	Hayle	1835	Harvey & Co.		T.Jones	Llanelly
	359	V	Haynasch *Kraukle -87* WSr	Russian	383	131·4	26·6	13·0	Haynasch	1887	F.Basak		J.Mikkelsohn&J.Kraukle	Haynasch
	360		Hazard *Goodman 82-85* WBg	Hawaian	306	120·0	30·0	15·6	E.Boston,Mass	1865		Y 12.81	Pacific Navigation Co.	Honolulu
	361	N	Hazard *ex Norwood, Thorsen* WBg	Norweg'n	269	108·4	26·2	13·6	P.E.Island	1868	Ramsay		H.A.Thorsen	Tönsberg
	362	R	Hazard *Swensson -87* (ex Tord Bonde) WSr	Swedish	127	71·8	21·3	11·7	Faaborg	1853			J.R.Svensson	Råå
	363		Hazard *Wallén 70-81* (ex Eduard & Emma) WSr	Swedish	100	66·9	20·1	10·2	Rostock	1838			C.F.Wallén	Kylley,Gotland
1115	364		Haze WSr	American	112	99·6	17·6	8·0	Hamburg,Pa.	1864			J.A.Anderson	Trenton,N.J.
	365		Hazelwoode *Cleveland -88* WSr	British	121	93·3	26·6	7·6	Moss Glen,N.B.	1888	G.Merritt		W.H.Merritt	St.John,N.B.
	366		Heath *Campbell* W3mSr	British	187	118·9	21·0	10·5	Aberdeen	1873	J.Humphr'y&Co		B.Byrnes	Sydney,N.S.W.
	367		Heather Bell *Ross* WBk	British	479	153·9	28·3	17·8	Aberdeen	1855	Duthie&Co.	Y	H.Cuthbertson	Newc'stle,NSW
	368	R	HeatherBell *Harding -80* WBk	British	258	119·6	24·0	14·1	Bangor	1859	Ellis	Y	J.Owens	Aberystwith
	369		Heather Belle WSr	British	144	99·0	22·4	8·8	Picton,Ont.	1868	W.Redmond		D.C.Strachan	Goderich,Ont.
	370	✻	Hebe *Rowe* WBkn	British	243	120·5	24·9	13·4	Littlehampt'n	1875	Harvey	Y	Y.Bisson	Falmouth
	371	✻	Hebe *Kennedy 85-87* WSr	British	231	125·5	24·9	12·9	Ardrossan	1873	Barr & Shearer	Y	Baine & Johnston	Greenock
	372		Hebe WBg	British	214	105·5	24·0	12·8	P.E.Island	1862			A.Campbell	Sydney,N.S.W.
	373		Hebe WBg	British	176	85·5	20·2	14·9	Poole	1839			Thomas Jones	Melbourne
	374		Hebe WBn	British	122	92·0	22·0	10·7	Wallace,N.S.	1857			J.J.Adam	St.John's,N.F.
	375		Hebe *Mercer -88* WSr	British	121	87·4	23·8	10·0	BayRoberts,N.F.I.	1888	S.Dawe		Ch.Dawe	St.John's,N.F.
	376	DV	Hebe *Frederiksen 87-87* WSr	Danish	117	93·6	21·3	10·2	Nakskov	1862	H.P.Ridersborg & Troelman		P.S.Winchells Enke	Nakskov
	377		Hebe *Claassen 74-80* (ex ...) WBk	German	311	103·3	23·9	16·7	Sunderland	1851			Herm.Behrent	Danzig
	378		Hebe *Meyn* WSr	German	236	126·0	22·4	12·1	South Shields	1845			Alsen'sche Portland Cement Fabriken	Hamburg
47	379		Hebe WSr	Greek	269	111·8	26·6	..	Syra	1886	A.Livadaras		J.D.Papas	Syra
	380	N	Hebe *Taarvig* WBk	Norweg'n	622	127·4	34·6	21·6	Amsterdam	1842			Bernt Einersen	Grimstad
	381	N	Hebe *Andersen* WBk	Norweg'n	541	151·5	30·9	17·2	Lillesand	1876	J.A.Henschien	Y	I.A.Henschien	Lillesand
	382	N	Hebe *Christen -87* WBk	Norweg'n	518	147·3	31·0	17·7	Kragerö	1876	A.Johnsen	Y 10.82	A.Natvig	Kragerö
	383		Hebe *Holm* WBk	Norweg'n	475	136·0	31·1	16·9	Stavanger	18 52/76	P.Gjosdahl		B.Middelthon	Stavanger
	384	N	Hebe *Sinnes -88* WBk	Norweg'n	458	136·7	30·9	16·5	Tönsberg	1869	T.Hansen	Y 10.81	L.Dahl, in liquidation	Tönsberg
	385	N	Hebe *Hansen* WBg	Norweg'n	239	88·9	25·9	15·4	Kragerö	1846			O.Bastiansen	Tönsberg
	386		Hebe *Johansén* WSr	Russian	186	98·3	27·2	10·7	Kimito	1876	E.Lindroos		G.F.Isaksson	Dragsfjerd
	387		Hebe *Gustafsson -88* WSr	Russian	140	..	..	..		1876			C.Fjäder	Karuna
	388		Hebe *Enroth* WBk	Swedish	336	127·3	26·9	14·3	Gefle	1857			C.G.Lundqvist	Gefle
	389		Hebe *Johansson* W3mSr	Swedish	133	93·5	21·6	9·5	Wenersborg	18 54/62			C.J.Andersson	Wenersborg
	390	N	Hebron *Jensen -88* (ex Asker) WBk	Norweg'n	261	115·4	21·6	14·5	Dundee	1866	TayShipB.Co.	Y	Th.Lövold	Arendal
	391	DN	Hebron *Rodne* WSr	Norweg'n	120	92·3	22·5	9·7	Stavanger	1878		Y	J.Knudsen	Stavanger
	392	DV	Hecht *Plötz 79* W3mSr	German	358	131·1	28·3	14·4	Stralsund	1879	A.Juhl	Y	W.Plötz	Stralsund
	393	V	Hecla *McDougall -84* WBk	British	872	174·6	35·5	20·2	Maitland,N.S.	1875	A.McDougall	Y 10.80	McDougall & Co.	Maitland,N.S.
	394	RV	Hecla *Reese 81-* WS	American	1529	210·5	40·2	24·0	Bath,Me.	1877	Goss&Sawyer	Y	J.W.Marr	Bath,Me.
	395	V	Hectanooga *Cann -87* WS	British	1066	192·0	37·0	22·1	PortClyde,N.S.	1875	T.Coffin	Y	A.C.Robbins	Yarmouth,N.S.
	396	V	Hector *Parker -88* WBkn	British	498	156·6	32·6	13·0	St.John,N.B.	1885	E.McGuggin	Y	H.D.Troop	St.John,N.B.
	397		Hector * *Campbell -86* WBn	British	199	99·1	27·1	10·9	P.E.Island	1878	J.M.Stewart		A.Holland	Brisbane
	398	N	Hector *Johannesen 87-87* WBk	Norweg'n	400	118·9	30·0	16·3	Grimstad	1873	O.Jörgensen	Y	C.Klöcker	Arendal
	399	G	Heda *(ex Ane Christine)* WBn	Danish	136	85·1	21·2	11·4	Ribe	1867	J.N.Nielsen		H.C.Dreier	Svendborg
	400	✻	Hedessa *Williams* WSr	British	92	64·8	18·6	10·8	Newquay	1853			T.H.Williams	Aberystwith

✻ = Lloyd's Register; A = Veritas Austro-Ungarico; D = Nederlandsche Vereeniging; G = Germanischer Lloyd; H = Veritas Ellenico; I = Registro Italiano; N = Norske Veritas; R = American Record; V = Bureau Veritas.

Dans la 4me Colonne, C = Composite (membrure en fer, bordé en bois); I = Fer; Stl = Acier; W = Bois.

Les Dates du doublage en métal (Col. 11) ressortent des livres dans lesquels les navires sont respectivement classés ou des rapports des Inspecteurs du Lloyd's Register.

Pour plus amples renseignements sur les navires classés par le Lloyd's Register, voir le Lloyd's Register Book et le Lloyd's Yacht Register.

Official No.	No. in Book	Registry in which Classed	Ship's Name. Master, and Dates of Appointment to present Owner's Service and to Ship. Ship's Late Name.	Material and Rig	Flag	Net Tonnage	Length	Breadth	Depth	Where	When	By whom	Metalling	Owners or Managers
	401	N	Hedevig *Ellingsen -88*	W Bk	Norweg'n	487	133·3	32·8	16·8	Moss	1853			R.M.Peterson
72739	402		Hedvig *Kalweit 86-88*	W Bk	British	376	140·3	28·0	18·6	Helsingborg	1870			P.V.Grant
	403		Hedvig	W Sr	Russian	122	92·7	22·7	10·8	Fanö	1868			A.Antutjef
	404		Hedvig	W Bk	Swedish	343	122·0	24·0	16·6	Sunderland	1853			J.M.Möller
	405		Hedvig *Nilsson -86*	W3mSr	Swedish	125	86·6	21·7	9·6	Wenersborg	1867			C.J.Andersson
	406		Hedvig Sofia *Åström -86*	W Bg	Swedish	258	112·5	25·2	12·1	Sundsvall	1868	A.Nordström		C.G.Vickberg
	407	G	Hedwig *Warnken 72-72*	W Bk	German	808	162·1	37·8	21·9	Bremerhaven	1872	J.C.Tecklenborg	Y 7.86	D.H.Wätjen & Co.
	408	✳	Hedwig *Burmister 87-87*	Stl3mSr	German	322	130·9	28·0	11·8	Kiel	1885	G.Howaldt		C.Sodermann
	409	G	Hedwig *Haefke 73-73*	W Bk	German	316	123·3	26·8	14·2	Stralsund	1873	G.Dienstbach	Y 8.85	J.J.Wallis & Sohn
	410	DG	Hedwig *Jäger 87-87*	W Bg	German	288	100·0	23·6	14·1	Wismar	1868	M.Maren	Y 8.86	Fr.Jäger
	411		Hedwig (ex Clara Wilsnack)	W3mSr		277	99·4	26·3	14·0	Rostock	1856	O.Ludewig		
	412	V	Hedwig *Bodecowes -77*	W3mSr	German	237	101·7	23·0	12·0	Emden	1877	C.Cassens	[illegible]	Reinhard van Hoorn
	413		Hedwig *Zabel*	W Bn	German	144	82·5	24·6	12·0	Barth	1875	Schlor		J.C.Beug
	414	G	Hedwig Siebe *Knaack*	W Bk	German	356	125·0	28·0	14·9	Stralsund	1869	J.F.Juhl		Carl Siebe
	415	D	Heemse *Visser 66-82* (ex India Packet)	W Bk	Dutch	710	130·9	34·8	22·5	Alblasserdam	1861	Smit	C [illegible]	J.& J.Vinke
	416	N	Heffhi *Blix -87* (ex Johan Feilman)	W Bk	Norweg'n	827	181·0	35·4	20·7	Brahestad	1875	J.Feilman	Y 12.86	Joh.Thorne
71587	417		Hei Cheong (BROKEN UP)	W Bk	British	325	126·3	27·0	13·2	Bangkok	1876	M.T.Nahwab		Chong Beong
1030	418		Heian Maru	W Bk	Japanese	350	130·3	25·6	14·2	Germany	1871			Abo Hikotaro
	419	N	Heidi (CONDEMNED) *Hansen*	W Bk	Norweg'n	459	139·3	32·4	17·2	NyCarleby	1857		Z 12.87	H.U.Singdahlsen
	420	R	Heidi *Kullsen -87*	W Bk	Swedish	432	140·4	28·4	16·4	Gothenburg	1864	C.Kierkegaard	Y 10.86	G.D.Kennedy
	421	V	Heidrun *Gundahl 87-87* (ex Cyprus)	W Bk	Norweg'n	916	173·4	35·5	21·4	Quebec	1878	Charland,jun.	Y 2.86	Actieselskabet "Heidrun" (Th.S.Falck)
	422	N	Heidrun *Berg -88*	W Bk	Norweg'n	277	107·8	26·7	15·0	Flekkefjord	1850			L.Berg
	423		Heimath	W Bg	German	231	96·4	25·1	13·4	Stettin	1854	A.Nüske		G.Radmann
	424	V	Heimath *Ewert 82-*	W Sr	German	107	74·5	20·8	10·0	Stralsund	1875	A.Juhl		H.L.Juhl
	425	V	Heimdal *Andersen 84-84*	W Sr	Danish	123	91·3	21·2	10·6	Faaborg	1873	R.H.Møller		P.Nielsen
	426		Heimdal *Frantzen 88-88*	W Sr	Danish	107	76·3	20·1	11·3	Middelfart	1856	R.Illum		E.P.W.Geisnar
	427	N	Heimdal *Kruge 84-84*	W Bk	Norweg'n	372	127·0	30·3	15·2	Drammen	1869	J.Jörgensen		Ole Naess
	428	N	Heimdal *Hansen 77-77*	W3mSr	Norweg'n	216	112·6	25·9	11·5	Holmestrand	1873	A.Pedersen	Y 2.86	T.Ihlen
	429	N	Heimdal *Pederson 81*	W Sr	Norweg'n	140	90·8	23·7	10·5	Sandefjord	1871			P.A.Grön
	430	N	Heimdal *Knudsen -84*	W Sr	Norweg'n	106	88·2	22·3	9·7	Bergen	1878	H. Hillestad		Carsten Hansen
	431		Heimdal	W Sr	Swedish	117	87·3	21·6	11·0	Nykjöbing, [Falster	1854	E.C.Benzon		H.A.Bergan & Co.
	432		Heimo *Åkerros*	W Bk	Russian	487	140·0	30·7	17·8	Åbo	1867	Stromberg		J.A.Leven
	433	GV	Heinrich *Weyhausen 75-80* (ex Trojan)	W S	German	1484	212·9	41·0	24·2	Kennebunk, [Maine	1875	Titcomb& [Thompson	Y 1.87	Fr.Roters
	434	RV	Heinrich *Baunau 77-86*	W Bk	German	923	172·5	31·2	20·6	Hamburg	1886	Stulcken	Y 10.86	H.C.Stulcken's Wwe
	435	G	Heinrich *Rohde*	W3mSr	German	228	106·5	25·2	12·3	Barth	1860	J.Kraeft		W.A.Sarnow
	436	V	Heinrich *Janssen*	W Bn	German	141	93·3	21·2	11·5	Papenburg	1861	B.Mamme's Wwe		B.Mamme's Wwe
	437	V	Heinrich Bauer *Frehse 85-85*	W Bk	German	571	140·0	31·0	17·0	Rostock	1885	E.Burchard&Co	Y 1.86	H.Bauer
	438	✳	Heinrich Bötel *Sichr -88* (ex Donau, ex Constancia)	I Bk	German	507	152·3	27·0	17·5	Middlesbro'	1864	Candlish,Fox & Co.		J.H.L.Wilckens
488	439	V	Heinrich Emma	W Sr	Russian	251	106·2	28·3	13·9	Korland	1887			H.& J.Damkala
	440		Heinrich Gerdes *Pieplow 87-87*	W Bg	German	242	102·4	27·1	13·7	Rostock	1863	J.Möller		W.Maack
	441		Heinrich Jade *Jansson [-88* (ex Ernst E. ...)	W Bk	Swedish	392	123·6	27·9	15·8	Ribnitz	1854	J.H.Wilken		A.P.Jansson
	442	V	Heinrich Moll *Meyer 79-79*	W Bg	German	273	108·0	26·9	13·9	Rostock	1867	W.Zeltz		A.Burchard
	443	V	Heinrich Ramien *Kregmann 78-83*	W Bk	German	619	143·7	32·0	17·9	Elsfleth	1880	J.Jurgens	Y 8.87	E. tom Dieck
	444	V	Heinrich von Schroeder	W Bk	Belgian	545	112·6	30,6	18·5	Memel	1873	B.Pieper		Belgian Gov[illegible]
	445	G	Heinrich&Catharina *Wack -86* (ex Sanning)	W Bk	Russian	327	123·2	27·5	14·8	Ueckermünde	1867	C.Radmann		C.Essmann
	446	G	Heinrich & Tonio *Thöhl 86-86* (ex Franklin)	W Bk	Norweg'n	1091	171·9	37·1	24·0	Boston	1859	P.Curtis	Y 1.86	
	447		Hekla *Nielsen 86-86*	W Bn	Danish	132	87·0	23·0	11·3	Piteå	18 41/57	C.J.Jörgensen		N.W.Thomsen
	448	N	Heldine *Hansen -88*	W Bg	Norweg'n	156	84·4	23·9	12·3	Sandefjord	1854			M.Larsen Red
	449	N	Heldos *Lydersen*	W Bk	Norweg'n	355	124·1	28·8	15·5	Tvedestrand	1873	E.Ellingsen	Y 5.86	E.Ellingsen
	450	N	Heldos *Askilsen*	W Bg	Norweg'n	256	106·8	26·8	14·2	Grimstad	1865	N.O.Gjömle		J.E.Einersen

✳ = Lloyd's Register; D = Nederlandsche Vereeniging; G = Germanischer Lloyd; I = Registro Italiano; A = Veritas Austro-Ungarico; H = Veritas Ellenico; N = Norske Veritas;

In the 4th Column, C=Composite (iron frame, wood planked); I=Iron; Stl=Steel; W=Wood.

The Dates of Metalling (Col. 11) are based on the records of the Books in which Vessels are respectively classed, or on the Reports of the Surveyors

Further particulars of Vessels classed by Lloyd's Register will be found in Lloyd's Register Book and Lloyd's Yacht Register

No.	Registre	Nom, Matériaux et Gréement. Capitaine et dates d'entrée au service des Armateurs actuels et à bord du navire. Anciens Noms.		Pavillon	Tonneaux	Longueur	Largeur	Creux	Construit: Où.	Quand.	Par qui.	Doublage	Armateurs ou Directeurs-Gérants.	Port d'Armement.
451	*	Helen *Locken*	W Bk	British	378	128·0	26·6	16·9	Sunderland	1862	G. & J. Mills	[illegible]	W. W. Dingle	Plymouth
452		Helen	W Bk	British	343	135·3	25·3	15·8	Greenock	1864	R. Steele	V 8.86	A. McGregor	Hobart
453		Helen	W Sc Yt	British	128	115·9	23·7	11·9	Cowes	1864	J. S. White		Earl of Feversham	Cowes
454		Helen	W Bn	British	190	72·8	22·7	10·0	Clare, N.S.	1859			M. Fitzgerald	Dublin
455		Helen	W Sr	American	185	104·0	28·9	9·1	Harrington, Me.	1874			H. W. Ramsdell	New York
456		Helen *Chase*	W Sr	American	138	97·9	22·7	8·0	New Bedford, Mass.	1864	S. Dammon		H. Taber	Edgartown
457		Helen	W Sr	American	115	81·0	24·0	8·0	Thomaston, Me.	18[illegible]			C. T. Williams	Rockland, Me.
458		Helen	W Bn	Greek	322	..	..	..	P. E. Island	1869			P. Stathopoulos	Poros
459	R	Helen A. Chase	W 3m Sr	American	569	147·0	34·0	13·0	Wiscasset, Me.	1881			J. M. Haynes	Bath, Me.
460		Helen Augusta	W Bk	British	204	105·4	24·1	10·9					H. Thackray	Turks Islands
461		Helen Augusta *Pratt*	W Sr	American	134	91·0	27·0	7·0	Portland, Conn.	1865	Gildersleeve		Brainard Quarry Co.	Hartford, Conn.
462		Helen Burns	W Bn	British	139	97·5	22·6	11·6	Cardiff	1862	Hodge		W. J. Jones	Teignmouth
463		Helen Churchill	W Bn	British	112	86·0	21·0	9·3	Port Le Bear, N.S.	1879	S. Harlow		T. Johnson	Shelburne, N.S.
464	*	Helen Denny	I Bk	British	728	187·5	31·2	19·1	Port Glasgow	1866	R. Duncan & Co.		Shaw, Savill & Albion Co. Ld.	Southampton
465		Helen G. King *Brown*	W Sr	American	146	93·0	25·0	9·0	Calais, Me.	1867	O. B. Rideout		A. H. Sawyer	Calais, Me.
466	R	Helen G. Moseley *Holt*	W 3m Sr	American	447	146·1	34·3	12·5	Belfast, Me.	1883	Brown & McDonald	V 12.88	A. B. Perry & Co.	Boston, Mass.
467		Helen H. Benedict	W 3m Sr	American	720	161·6	35·1	16·8	Bath, Me.	1881	Goss, Sawyer & Packard		Armstrong & Co.	New Haven, Conn.
468		Helen Hasbrouck	W 3m Sr	American	310	130·0	32·5	10·7	E. Haven, Conn.	1870			Ezra Soper	New York
469	*	Helen Isabel	I Bk	British	250	132·2	24·3	13·9	Greenock	1869	Robertson		Baine & Johnston	Greenock
470		Helen J. Holway *Libby*	W Sr	American	212	100·8	26·3	10·0	Machias, Me.	1867			W. Holway	Machias, Me.
471		Helen Jane	W Bg	British	176	90·1	25·0	9·5					J. T. Hogg	Nassau, N.P.
472	R	Helen L. Martin	W 3m Sr	American	403	140·5	33·8	12·3	Thomaston, Me.	1882	S. S. Gerry & Co.	V 5.88	S. S. Gerry	Thomaston, Me.
473	R	Helen M. Rowley	W Bn	American	379	122·8	30·2	15·2	Port Jefferson, N.Y.	1867		V 8.87	S. H. Rowley	New York
474	*	Helen McGregor *Abram*	W Sr	British	77	86·0	20·0	9·0	Ayr	1866	Fullerton		S. Ravenscroft & Co.	Chester
475		Helen Mar *Ellis*	W Bk	American	308	109·7	27·2	17·1	New Bedford, Mass.	1855			E. H. Hansen	San Francisco
476		Helen Mar *Curree*	W Sr	American	226	108·0	28·0	10·0	Stratford, Conn.	1855			G. H. Newhall	Providence, R.I.
477		Helen Mar	W Sr	American	129	96·2	27·2	8·6	Port Jefferson, N.Y.	18$^{52}_{61}$			Chas. Jennings	Bridgeport, Conn.
478	R	Helen Marion *Malcolm*	W Bn	British	335	130·4	30·5	15·4	East Machias, Me.	1871	S. H. Whittemore	V 12.88	E. Lantalum	Yarmouth, N.S.
479		Helen Marshall *Hubbard*	W Bn	British	117	89·2	21·2	11·5	Garlieston	1863	Hannah		Bessey & Palmer	Yarmouth
480	R	Helen Montague	W 3m Sr	American	387	131·0	32·0	12·0	Rockland, Me.	1881	Ames		A. F. Ames	Rockland, Me.
481		Helen N. Kimball *Madsen*	W Sr	American	183	112·5	30·8	9·0	Cuffey's Cove, Cal.	1881			E. L. Allen	San Francisco
482		Helen P. *Jones*	W Sr	American	138	104·0	28·0	7·0	East Haddam, Conn.	1864			Frank Brainard	Hartford, Conn.
483		Helen R. Cooper	W S	American	743	145·4	32·6	21·5	Pittston, Me.	1854			B. J. H. Trask, jun.	New York
484	*	Helen Scott	I Bk	British	1118	204·2	31·4	22·9	Liverpool	1863	Jones, Quiggin & Co.		J. Coupland	Liverpool
485		Helen W. Almy	W Bk	American	299	116·4	29·0	12·5	Fairhaven, Mass.	1859			F. H. Miner	San Francisco
486	*	Helen West	W Sr	British	99	87·0	21·6	10·6	Banff	1878	Watson		A. Wilson	Banff
487		Helena	W Bn	British	126	83·7	23·1	10·1	Mahurangi, N.Z.	1874			F. Buss	Maryboro'
488		Helena *Potter*	W Sr	British	118	82·0	24·0	10·1	Wilmot, N.S.	1864			H. V. Crandall	Annapolis, N.S.
489		Helena	W Bk	American	603	141·7	31·7	18·2	Brewer, Me.	1866			N. C. Ayer	Bangor, Me.
490		Helena	W Bk	Brazilian	335	..	..	..					J. A. Ribino	Rio Janeiro
491		Helena	W S	Brazilian	112	..	..	..					Mendes & Garcia	Rio Janeiro
492	V	Helena *Verbeel* (ex Warren)	W Bk	Dutch	1170	183·7	35·4	23·5	Boston, U.S.	1863	P. Curtis	V [illegible]	W. A. J. Wittkampf	Schiedam
493	N	Helena *Gjertsen*	W Bk	Norweg'n	965	173·9	35·4	22·4	Quebec	1870	Dunn	V 12.86	L. Hofgaard & Co.	Tönsberg
494	N	Helena *Hansen*	W Bk	Norweg'n	682	151·1	32·1	18·9	Kennebunk, Me.	1863		V [illegible]	A. F. Klaveness	Sandefjord
495	N	Helena *Dahl*	W Bk	Norweg'n	660	135·2	33·6	21·1	Bolnes	1856	J. Pot		F. Breda	Christiania
496	N	Helena	W 3m Sr	Norweg'n	298	125·7	29·8	11·9	New York	1857		[illegible]	H. Molvig	Fredrikshald
497	N	Helena	W Bg	Norweg'n	276	118·4	27·3	13·5	Tvedestrand	1874	J. P. Pedersen	[illegible]	N. A. Salvesen	Tvedestrand
498	N V	Helena *Gjeruldsen*	W Bk	Norweg'n	266	111·4	25·5	14·7	Honfleur	1861	E. Cardon	[illegible]	K. Gjeruldsen	Grimstad
499	V	Helena	W 3m Sr	Norweg'n	199	103·5	23·9	12·1	Wenersborg	1878	G. V. Plate	[illegible]	C. A. Jörgensen	Dröbak
500		Helena	W Bg	Swedish	237	117·4	24·3	12·4	Karlshamn	1868	D. R. Andersson	V 1.89	N. A. Liedqvist	Helsingborg

* = Lloyd's Register; D = Nederlandsche Vereeniging; G = Germanischer Lloyd; I = Registro Italiano; R = American Record; A = Veritas Austro-Ungarico; H = Veritas Ellenico; N = Norske Veritas; V = Bureau Veritas.

Dans la 4me Colonne, C = Composite (membrure en fer, bordé en bois); I = Fer; Stl = Acier; W = Bois.

Les Dates de doublage en métal (Col. 11) ressortent des livres dans lesquels les navires sont respectivement classés ou des rapports des Inspecteurs du Lloyd's Register.

Pour plus amples renseignements sur les navires classés par le Lloyd's Register, voir le Lloyd's Register Book et le Lloyd's Yacht Register.

1 Official No.	2 No. in Book	3 Registry in which classed	4 Ship's Name. Material and Rig. Master, and Dates of Appointment to present Owner's Service and to Ship. Ship's Late Name.	5 Flag	6 Net Tonnage	7 Dimensions. Length	Breadth	Depth	8 Built. Where	9 When	10 By whom	11 Date of Metalling	12 Owners or Managers
65357	501	✳	Helena Anna (WRECKED) *Nicholls* W3mSr	British	162	106·2	25·2	11·7	Pekela	1870	Drenth	[illegible]	F.C.Mathews
11293	502		Helena E. Russell *McNaffy* WSr	American	158	101·0	27·6	7·8	Hartford,Conn	1867	S. & E.S. Belden		S.H.Butler
21	503		Helena Marie *Brackmann* WSr	Russian	199	90·0	24·0	11·0	Orrenhof	1875			P.Jürgens & F.Tiedke
85568	504		Helena Maude *Lyons* WSr	British	112	86·0	24·0	9·7	Shelburne, NS.	1883	S.Muir,sen.		J.H.Lyons
73654	505	✳	Helena Mena *Abrahamson 86-6* CBk	British	673	165·0	30·3	18·8	Sunderland	1876	R.Thompson,jr	Y 6.88	Trinder,Anderson
58675	506		Helena Tregenza *Paull 82-82* WSr	British	136	103·0	23·0	11·5	Portreath	1869	Massey		J.L.Tregenza
59092	507		Helene (LOST) *Hartell* WSr	British	111	81·3	21·6	9·6	L'Islet	1871			C.Archibald
	508	DGV	Helene (ex Geo. M. Adams) *Bambach 83-* WS	German	1241	196·5	36·7	23·6	Bath,Maine	1868	J.P.Morse&Co.	Y 11.88	Siedenburg,W[illegible]
	509	GV	Helene *Dunker 72-80* WBk	German	703	158·1	33·3	19·3	Hammelwarden	1880	C.Lühring	Y 6.89	J.C.Warns
	510	V	Helene *Kolln -86* W3mSr	German	430	136·0	29·0	15·4	Brake	1881	J.Oltmann's Wwe.	Y 4.87	H.P.Breckwoldt,jr.
	511	GV	Helene *Behrmann 83-83* W3mSr	German	426	141·0	28·8	14·1	Lübeck	1883	J.Steffen	Y 10.86	C.Brandt
	512	GV	Helene *Ufken 80-* WBk	German	372	136·1	28·7	15·3	Kiel	1875	A.Conradi&Sohn	Y 8.88	R.Ufken
	513	GV	Helene *Gelpke* WBg	German	358	118·0	25·9	13·9	Hammelwarden	1871	W.G.Hagens	Y 3.88	G.H.Lübken
	514	DV	Helene (FOUNDERED) (Ancient Briton) *Burmeister* WBg	German	291	104·1	26·3	16·4	Sunderland	1870	J.Hutchinson	Y 11.88	Carl Rahtkens
	515	GV	Helene *Strenge 87-87* WSr	German	149	90·9	23·2	11·2	Kiel	1874	Chr.Ihms	Y 7.87	J.G.Ukena
	516	V	Helene *Duis 80-* WSr	German	141	94·2	19·2	9·6	Emden	1868	M.J.Schulte		Diedr.Höcker
	517	V	Helene *Baumann -84* WSr	German	129	81·3	18·7	9·7	Edewecht	1883	G.Deye	Y 11.91	H.Setje
	518	V	Helene *Willenborg 77-77* WSr	German	112	83·0	19·4	9·7	Edewecht	1877	H.Kremer		Georg H.Willenborg
	519	V	Helene *Lafrenz 69-* WSr	German	111	81·7	20·8	10·1	Lübeck	1857	T.H.Evers		C.Mackeprang
	520	DGV	Helene *Gowers 69-* WSr	German	106	77·6	19·6	8·6	Weener	1865	B.Berg		L.J.Gowers
	521		Helene *Vollmer -88* WSr	German	106	81·3	19·7	9·6	Papenburg	1854	H.W.Meyer		P.Becker
	522	G	Helene Christiane *Jacobs* WSr	German	106	82·0	20·0	10·0	Emden	1857	M.Smidt		C.H.Langenberg
	523	N	Helene Elisabeth *Olsen -87* WSr	Norweg'n	105	76·8	23·0	9·1	Gothenburg	1849			O.A.Olsen
	524	V	Hélène Georgina *Dolo -87* WBk	French	313	110·2	21·6	13·4	Honfleur	1866	Leviels	Y 1.89	D.Auger
144	525	A	Heleni (ex Antonio S.) *Mavrojalli* WBn	Greek	468	103·3	30·5	..	Fiume	1864	Fm.Schiavoni		G.Mavrojalli
	526	H	Heleni (ex Hellas) *Perranos* WBk	Greek	335	..	..	..	Spezzia	1863			Passac Zayian
	527		Heleni *Tsavouris* WBk	Greek	293	..	..	..	Lormont	1861			E.Argyros
21	528		Heleni WBn	Greek	280	110·2	27·4	..	France	1861			E.G.Argyros
	529		Heleni *Lissatos* WBn	Greek	240	..	..	..	Trieste	1880	G.Spadon		Lissatos & Vlachos
61	530		Heleni (ex Helen) WBg	Greek	209	88·9	25·3	..	Syra	1866			S.Canalas
	531	A	Heleni *Poliora* WSr	Greek	192	..	..	..	Syra	1886			G.Poliora
391	532		Heleni *Andreades* WSr	Greek	187	92·8	23·6	..	Chios	1881	Jean Mar[illegible]		Ant.Bachas
183	533		Heleni WSr	Greek	157	81·0	23·9	..	Chios				A.Mavroudis
293	534	A	Heleni WSr	Greek	147	82·7	24·6	..	Syra	1880	S.Cristali		J.Polymeropoulos
417	535	A	Heleni *Oeconomos* WBn	Greek	144	86·9	21·8	..	Syra	1883		Y 8.83	C.S.Oeconomos
18	536		Heleni WSr	Greek	124	79·7	22·4	12·8	Scopelos	1879	G.Papadimitrion		Sons of P.Barb[illegible]
	537	H	Heleni *Garofalo* WSr	Greek	119	..	..	..	Scopelos	1880			D.Garofalo
66	538		Heleni Coupa WBg	Greek	276	100·7	27·6	..	Syra	1878			E.Couloucountas
33	539		Heleni Proti (ex Feros) WBg	Greek	246	..	..	..	Syra	1879			
65017	540	✳	Helens *Oliver 75-85* WBk	British	433	110·0	28·9	17·2	Sunderland	1872	J.Gardner	Y 11.84	Grimsby Co-op. (re Co.Ld.)
87727	541	✳	Helensburgh *Denver* IS	British	1699	264·9	39·8	23·4	PortGlasgow	1883	R.Duncan&Co.		Thom & Cameron
86366	542	✳	Helenslea *Reid 82-82* StlBk	British	1315	249·8	35·4	21·6	Dundee	1882	A.Stephen&Sons		A.Stephen & Sons
	543	N	Helga *Schildstok 76* WS	Norweg'n	598	153·3	34·2	18·6	Tönsberg	1867	Andersen	[illegible]	C.G.Brunn
	544	N	Helga *Frodesen* WBg	Norweg'n	177	102·8	24·8	10·7	Oscarshamn	1855			O.G.Frodesen
	545	R	Helge *Lundqvist 81-86* WSr	Swedish	157	88·6	21·0	12·1	Copenhagen	1858			G.Lundqvist
	546	V	Helgesen *Nielsen 87-87* WBk	Danish	282	102·9	25·2	14·6	Korsör	1852	C.C.Hansen		Ad. Harboe
	547	✳	Helicon *Herrenbrodt 87-87* StlS	German	1612	256·5	38·6	22·7	Glasgow	1887	C.Connell&Co.		B.Wencke Söhne
	548	✳	Helios (ex Strathardle) *Niejahr 87-87* IS	German	998	202·5	34·6	21·1	PortGlasgow	1864	R.Duncan&Co.		Oldrey, Williams & Day
	549	V	Helios *Maass -86* WBg	German	356	114·8	28·0	17·3	Rostock	1867	O.Ludewig	[illegible]	Alw.K[illegible]
	550	NV	Helios *Johansen -88* WBk	Norweg'n	967	176·2	36·2	22·1	Grimstad	1888	I.Tryde	[illegible]	J.A.H[illegible]

✳ = Lloyd's Register; D = Nederlandsche Vereeniging; G = Germanischer Lloyd; I = Registro Italiano; A = Veritas Austro-Ungarico; H = Veritas Ellenico; N = Norske Veritas;

In the 4th Column, C = Composite (iron frame, wood planked); I = Iron; Stl = Steel; W = Wood.

The Dates of Metalling (Col. 11) are based on the records of the Books in which Vessels are respectively classed, or on the Reports of the Surveyors...

Further particulars of Vessels classed by Lloyd's Register will be found in Lloyd's Register Book and Lloyd's Yacht Register.

	No.	Classe	Nom. Capitaines et dates d'entrée en service des Armateurs actuels et Ancien Nom	Matériaux et Gréement	Pavillon	Tonnage net	Longueur	Largeur	Creux	Où	Quand	Par qui	Dates du doublage en métal	Armateurs ou Directeurs-Gérants	Port d'Armement
	551	V	Helios *Olfedahl*	W Bk	Norweg'n	437	127·3	26·9	16·7	Elbing	1867	Gebr Mitzlaff	Y 4·85	E.H.Jondal	Stavanger
	552	N	Helios *Halvorsen*	W Bk	Norweg'n	384	121·4	28·6	16·8	Grimstad	1865	P.&L.Lemmines [Tausa	Y 10.88	J.Borresen	Grimstad
	553	N	Helios *Wold*	W 3mSr	Norweg'n	294	119·5	28·0	11·9	Tvedestrand	1879		Y 4.84	W.Wilhelmsen	Tönsberg
	554	N	Helios (ex Wille) *Henriksen*	W 3mSr	Norweg'n	233	116·6	25·1	12·3	Dundee	1851			S.Andersen	Tönsberg
	555	DG	Helios *Elfving 86-*	W Bk	Russian	620	164·5	33·5	17·9	Uleåborg	18 64/65	Pettersen	Y 5.86	H.W.Snellman	Uleåborg
	556		Helios *Åkerlund*	W Bk	Russian	324	..	..	..	Lovisa	18 47/64			A.Björksten	Lovisa
	557	GV	Hellas *Bradhering*	W Bk	German	416	121·0	29·3	18·3	Rostock	1867	O.Ludewig	Y 6.88	G.Kindler	Rostock
	558		Hellas (ex Anna Beniborn) *Apostolo*	W Bk	Greek	338	116·4	24·2	17·6	South Shields	1860	Metcalfe	H 7.84	N.P.Goudis	Spezzia
	559		Helleno	W Sn	Brazilian	108	..	..	..					E.V.de Freitas	Itapemerim
	560	DN	Hellespont (ex Scuffingsane) *Hysten 87-83*	W Bk	Norweg'n	560	149·9	30·0	18·6	Alloa	1863		Y 1.86	Et Aktieselskab (C.Herntsbow)	Christiania
	561		Helmi *Pettersson*	W Sr	Russian	144	..	..	..		1875			F.Norrgård	Björneborg
	562		Helmi *Svensson*	W Bg	Swedish	324	107·2	26·6	16·0	Finland	1866			N.Olsson	Höganäs
	563		Helmich	W Bg	Norweg'n	252	105·3	26·0	15·2	Stavanger	1848			A.Hansen	Mandal
	564	V	Héloïse *David*	W Bn	French	214	79·0	26·2	12·6	Caen	1874		Y 5.86	A.Laffetay & Co.	Caen
	565		Helvetia (Willian Dick) *Syvdsen*	W Bk	Norweg'n	438	135·5	27·8	17·2	Bremerhaven	1858		Y 5.86	L.Jacobsen	Horten
	566		Helvetia *Grasmus 85-86*	W Bk	Norweg'n	395	125·6	29·2	16·3	Christiansand	18 54/63		Y 7.86	S.Haug	Drammen
	567		Helvetia	W Sr	Russian	181	102·3	24·3	9·2	Kimito	1875	A.Ölander		J.Stünzi	Kimito
	568	DV	Hendrik Jan. (ex M. Stewart) *Burghoud -85*	W Sr	Dutch	216	106·9	25·5	11·9	BelleCreekP.E.I	1870	J.M.Stewart	Y 3.86	H.J.Plaat & Co.	Schiedam
	569	D	Hendrik Onnos (ex Bengale) *Middel*	W Bk	Dutch	338	126·0	26·9	15·7	Honfleur	1858			N.Kranenborg	Oostwold
[illegible]	570	✱D	Hengist *Jacobs 84-87*	I S	British	1125	218·8	34·1	21·6	Greenock	1860	Scott&Co.		Star Navigation Co.,Ld.	Liverpool
	571	✱	Henny (ex ...) *Nielsen 85-89*	I Bk	Danish	498	155·0	26·7	17·2	Liverpool	1866	R.&J.Evans&Co		N.H.Brinck	Sonderho
	572		Henny *Spångberg*	W Sr	Swedish	129	89·9	21·7	9·3	Gamleby	1876			H.Cornelius	Gamleby
	573		Henri	W Sr	French	120	90·2	20·7	11·5	Paimpol	1884	Toudic		L. Huchet du Guermeur	Paimpol
	574		Henri *Gradnais*	W Bo	French	111	76·8	20·7	11·2	St.Valery en [Caux	1868	Duchemin		A.Lemoine	St.Malo
	575	V	Henri & Leontine *Le Rebeller*	W Bn	French	154	80·4	23·0	11·6	Means	1873	Lamort		Le Bihan	Nantes
	576	N	Henrich Biorn *Nystern*	W Bk	Norweg'n	363	119·5	29·9	16·0	Tvedestrand	1873	T.Gunderson	Y 10.87	Anton Natvig	Kragerö
[illegible]	577		Henrietta	W Sr	British	107	74·0	19·3	11·5	Sunderland	1845			P.McIntosh	Sydney N.S.W.
[illegible]	578	✱	Henrietta *White 82-82*	W Sr	British	74	74·9	19·9	9·4	Malpas	1878	Scoble		T.F.Hitchins	Truro
[illegible]	579	RV	Henrietta *Gilmore -88*	W S	American	1203	201·0	39·0	24·0	Bucksville,SC	1875	Nickels	Y 1.88	J.C.Nickels	Searsport.Me.
[illegible]	580		Henrietta *Stedman*	W Sr	American	111	100·5	17·5	8·3	Hamburg,Pa	1865			J.McIlinden	PerthAmboy,NJ
	581	DG	Henrietta (SUNK) *Schuitema*	W 3mSr	Dutch	378	119·7	28·5	17·5	Amsterdam	1855	Meyjes&Sons		J.Schuitema	Delfzijl
[illegible]	582	R	Henrietta J.Powell *Mason*	W 3mSr	American	300	132·3	30·3	9·4	Bath,Me.	1883	Goss&Sawyer	Y 3.86	J.L.Mason	Philadelphia
[illegible]	583		Henrietta Simmons *Smith*	W Sr	American	218	112·4	28·7	9·1	Leesburg,N.J.	1865			J.May & Sons	Philadelphia
[illegible]	584		Henrietta & Leonard Preston	W Bg	British	146	89·3	21·7	11·4	Stockton	1856			J.W.Ellenor	Sunderland
	585	V	Henriette (ex C.M.L. ... Henriette Melchior) *Nielsen 76-88*	W Bn	Danish	210	104·0	24·9	13·4	P.E.Island	1875	A.W.Owen	Y 3.84	P.Nielsen	Faaborg
	586	V	Henriette *André -87*	I Bk	French	690	169·6	29·2	18·6	LaSeyne	1874	Forges&Ch de [la Medit.		C.F.Branellere frères	Nantes
	587	V	Henriette *Lesnol*	W Bg	French	157	80·0	21·0	12·6	Cherbourg	1857	Hamel		Guillard	Granville
	588	V	Henriette (ex Sarah Higgens) *Buckelmann -87*	W S	German	1419	200·0	38·6	24·0	Boston,Mass.	1869	CurtisSmith&Co	Y 10.8	E.C.Schramm & Co.	Bremen
	589	DG	Henriette *Wobcken 75-75*	W Bk	German	947	174·8	31·5	21·6	Bremerhaven	1873	H.F.Ulrichs	Y 8.87	D.H.Wätjen & Co.	Bremen
	590	V	Henriette *Brahms -86*	W Sr	German	102	83·0	20·8	9·2	Varel	1877	A.Schwoon		J.H.Brahms	Boekzetelerfehn
	591	N	Henriette (ex J. Steele) *Bull 66-75*	W Bk	Norweg'n	583	149·2	31·6	18·3	Cornwallis	1863	E.Steele		C.Langaard	Christiania
	592		Henriette	W Bg	Norweg'n	140	..	..	..					E.Deger	Christiania
	593	G	Henriette *Anderson 86-96*	W S	Swedish	664	165·0	30·3	19·0	Hudiksvall	1857	J.H.Fought		P.A.Lindberg	Gothenburg
	594		Henriette *Petersson -82*	W Sr	Swedish	147	91·5	21·5	10·9	Oscarshamn	1841	P.Pettersson		S.E.Petersson	Oscarshamn
	595		Henriette Elisabeth	W Bk	Dutch	357	..	..	..	Siam	1870			Oei Bok Sing & Oei Bôk Liva	Riouw
	596		Henriette Steinorth (WRECKED) *Kraeft*	W Bk	German	321	109·3	27·3	16·5	Barth	1861	A.Witte		J.N.Rodbertus	Barth
	597		Henriette Susanne (ex Alexandra Zwolau) *Schotte -84*	W Bk	Dutch	590	136·1	33·1	20·5	Capelle	1857	VanHoogendyk [&Co.		J.Vriesendorp & Zoonen	Dordrecht
	598		Henrik (ex ...) *Johansson -85*	W Bg	Swedish	273	103·6	24·9	15·5	N.Shields	1854			E.Andersson	Edsby.Bhln.
	599		Henrik *Nilsson*	W Sr	Swedish	163	95·1	21·6	12·0	Hult	1857			T.Nilsson	Arildsläge
	600														

✱ = Lloyd's Register; D = Nederlandsche Vereeniging; G = Germanischer Lloyd; I = Registro Italiano; R = American Record; A = Veritas Austro-Ungarico; H = Veritas Ellenico; N = Norske Veritas; V = Bureau Veritas.

Dans la 4me Colonne, C = Composite (membrure en fer, bordé en bois); I = Fer; Stl = Acier; W = Bois.

Les Dates du doublage en métal (**Col. 11**) ressortent des livres dans lesquels les navires sont respectivement classés ou des rapports des Inspecteurs du Lloyd's Register.

Pour plus amples renseignements sur les navires classés par le Lloyd's Register, voir le Lloyd's Register Book et le Lloyd's Yacht Register.

Official No.	No. in Book	Register in which classed	Ship's Name. Material and Rig. Master, and Dates of Appointment to present Owner's Service and to Ship. Ship's Late Name.	Flag	Net Tonnage	Dimensions. Length	Breadth	Depth	Built. Where	When	By whom	Date of Metalling	Owners or Managers
1	2	3	4	5	6	7			8	9	10	11	12
	601	V	Henrik *Nystrom 62-72* WSr	Swedish	157	96·8	21·9	10·1	Kolboda	1857	Högland		F.Nyström
	602	NV	Henrik Ibsen *[illegible]* WBk	Norweg'n	319	177·9	35·6	20·3	Bergen	1876	Jens Gran	[illegible]	P.G.Pettersen
	603	NV	Henrik Ibsen *Vogt 86-86* WBk	Norweg'n	495	141·1	32·0	17·2	Moss	1874	Vogt & Holst	[illegible]	A.&F.Smith
	604		Henrik Wergeland *Johnsen* WBk	Norweg'n	383	105·0	30·0	17·2	Bisör	1846			J.W.Prelensen
	605	N	Henrik Wergeland *Hansen* WBg	Norweg'n	208	90·3	25·1	13·5	Rostock	1861			R.Hansen
	606	N	Henrik Wergeland *Nyman* WBn	Norweg'n	142	93·9	21·6	10·9	Stavanger	1877	L.L.Aas	[illegible]	Chr.Aadnesen
	607	V	Henrika *Eekhof 86-86* WBn	Dutch	175	97·1	23·0	11·6	Pekela	1876	J.W.Kuiper	[illegible]	P.P.Kolk
	608		Henrique WSr	Brazilian	142	..	..	..					A.J.do Amaral
66288	609		Henry *Stone* WBn	British	186	106·9	21·7	12·8	P.E.Island	1872	McMillan		J.Reeves
11270	610		Henry *Faulkingham* WSr	American	145	95·0	26·0	9·5	Jonesport,Me.	18[illegible]			W.H.Falkinham
11334	611		Henry WSr	American	133	90·4	26·0	8·9	Addison,Me	1868			H.Nash
	612	V	Henry *Bager -88* WSr	Danish	108	86·7	19·4	10·2	Ærøskjøbing	1872	R.Møller		C.H.Bager
	613		Henry *Fischer 80-80* (ex Kossuth Flad) (BURNT) WS	German	1266	185·3	38·7	23·9	ThomastonMe	1867	S.Watts & Co.	[illegible]	J.D.Bischoff&Co
	614	N	Henry *Haugneland -87* (ex Olaf) (BURNT) WBk	Norweg'n	1062	190·3	37·8	21·2	Gefle	1873	J.A.Bang	[illegible]	Interessentskab (A. Amundsen)
	615	N	Henry *Martinussen 75-75* WBk	Norweg'n	583	133·0	31·4	20·6	Sunderland	1851		[illegible]	I.Beylegaard
	616		Henry WSr	Norweg'n	197	104·0	26·8	10·1	Maine	1848			N.Olsen
	617	V	Henry *Knudsson -85* W3mSr	Swedish	292	127·9	25·5	12·0	Gefle	1874	O.A.Brodin	[illegible]	A.Tornsten
95207	618	R	Henry A. Burnham *Ross -86* WBkn	American	473	140·6	31·9	11·8	E.Boston	1873	A.Gove	[illegible]	J.H.Conant & Co.
95014	619	V	Henry A. Litchfield *Dunn* WBk	American	638	147·0	31·0	19·0	Rockland,Me.	1867	B.Litchfield	[illegible]	B.Litchfield
11[illegible]	620		Henry A. Tabor *Cash* WSr	American	122	91·9	27·5	7·3	P.Jefferson,NY	1860			Tabor & Gordon
	621	V	Henri Adrien *Laroque* (ex Plebiscite) [-76 WBn	French	135	85·0	20·3	11·3	St.Vaast	1870	C.Lévêque		Const.Lévêque
95302	622	R	Henry B. Cleaves *Charleson* WBn	American	370	125·8	30·2	11·7	Bath,Me.	1874		[illegible]	O.Charleson
95828	623	RV	Henry B. Hyde *Pendleton* WS	American	2423	267·9	45·0	28·8	Bath,Me.	1884	J.McDonald	[illegible]	B.Pendleton
11[illegible]	624		Henry B. Metcalf *Fossett 81-8[illegible]* WSr	American	152	105·0	28·0	8·0	Bristol,Me	1856			J.E.Nichols
96027	625		Henry B. Wright WS	American	703	143·1	31·6	19·2	Stockton,Me.	1858			J.H.Lane
11695	626		Henry Buck *Miller -88* WBk	American	651	143·3	31·5	21·5	Searsport,Me	1852	J.Carver		A.Packsher
95140	627	R	Henry C. Winship *Poone* W3mSr	American	472	138·3	32·2	15·0	Bath,Me.	1873	Jewell Bros.		E.H.Atwood
95814	628	R	Henry Crosby *Sweet 83-81* W3mSr	American	391	144·6	33·5	12·7	Brewer,Me.	1884	J.Hathorn		Henry Lord
11570	629		Henry Croskey WSr	American	266	116·0	29·0	8·6	Wilmington, [Del.	1864			E.I.Rackett
25671	630		Henry Curwen *Robinson* WBn	British	130	79·0	21·7	12·5	Workington	1842			R.M'Ferran
95106	631	R	Henry D. May *Morris* W3mSr	American	264	121·0	29·5	11·6	Milford,Del.	1871			J.May & Sons
60763	632		Henry Edmonds WBn	British	162	101·9	24·8	10·9	Aarhuus	1871	Ulstrup and [Meisner		J.Reney
	633	V	Henry Evelina *Levisage* WBg	French	143	78·1	21·6	11·4	Bayonne	1866	Descandé fils [ainé	[illegible]	G.Le Roux
95689	634	RV	Henry Failing *Merriman* WS	American	1899	230·6	43·1	26·3	Bath,Me.	1882	Goss & Sawyer	[illegible]	W.H.Besse
11373	635		Henry Finch *Hickman* WSr	American	198	116·6	30·6	9·6	Poughkeepsie [N.Y.	1854			J.S.Lounsbery
11539	636		Henry G. Ely *Moore* WSr	American	111	100·3	17·5	7·9	Philadelphia	1864			Macdonald Bros.
46016	637		Henry Gilbert *Mooney [85-85]* WBn	British	157	94·1	26·0	11·3	Parrsboro' [N.S.	1863			W.Hinde
95397	638		Henry H. Grant *Snyder* W3mSr	American	236	123·5	29·8	8·6	RockyHillCon	1875	Belden		S. & E.S.Belden
95024	639		Henry H. Olds *Sneed* (WRECKED) [-87 W3mSr	American	829	180·0	37·0	18·0	West HavenCt	1887	Gesner & Marr		Henry Sutton
95251	640		Henry H. Pitts *Goodbee 84-84* WS	American	218	103·0	30·5	9·0	Rondout,N.Y.	1873	J.Snyder		C.W.Alcott & Co.
17508	641		Henry Harvey *Gregory [50-85]* WBn	British	148	93·2	21·8	11·2	Hayle	1857	Harvey & Co.		W.Harvey
11797	642		Henry J. Raymond *Grammer* WSr	American	181	113·0	29·5	9·0	Newburg,N.Y.	1864			C.L.McLean
85920	643	✱	Henry James *Rathmore* (WRECKED) [-82 IBk	British	945	203·3	33·6	19·7	Glasgow	1882	Dobie & Co.		N.BritishShipping Co.
95245	644	V	Henry L. Gregg *Dow -88* WBkn	American	470	131·6	32·0	15·9	ColumbiaFalls [Me.	1873	J.H.Crandon	[illegible]	MarcusHunter&Co.
95281	645	R	Henry Lippitt *Howes* W3mSr	American	622	149·0	35·0	16·7	S.Dennis,Mass	1874	P.Keen		J.H.Conant & Co.
95476	646	R	Henry M. Clark *Thompson -87* WSr	American	165	109·0	25·0	10·0	Essex,Mass.	1877		[illegible]	G.B.Howard
11360	647		Henry May *Hallowell* WSr	American	165	109·7	27·2	9·0	Milford,Del.	1854			E.L.Hallowell
11742	648		Henry N. Squire *Megathlin* WSr	American	292	115·0	29·0	9·0	P.Jefferson,N.Y.	1867			W.B.Fisk
95248	649	R	Henry Norwell *[illegible] -86* WBkn	American	597	141·6	33·4	15·6	Bath,Me.	1873	D.O.Blaisdell	[illegible]	Deane & Crow[illegible]
95219	650		Henry P. Hallock *Tisdale* WSr	American	131	90·8	27·8	6·5	S.Norwalk,Con	1873			C.H.Paull

✱=Lloyd's Register; D=Nederlandsche Vereeniging; G=Germanischer Lloyd; I=Registro Italiano;
A=Veritas Austro-Ungarico; H=Veritas Ellenico; N=Norske Veritas;

In the 4th Column, C=Composite (iron frame, wood planked); I=Iron; Stl=Steel; W=Wood

The Dates of Metalling (Col. 11) are based on the records of the Books in which Vessels are respectively classed, or on the Reports of the Sur[veyors]...

Further particulars of Vessels classed by Lloyd's Register will be found in Lloyd's Register Book and Lloyd's Yacht R[egister]...

No.	Cote	Nom. Matériaux et Gréement. Capitaines et dates d'entrée en fonction. Ancien Nom.	Pavillon.	Tonnage net.	Dimensions			Construit Où	Quand	Par qui		Armateurs ou Directeurs-Gérants.	Port d'Armement.
651		Henry P. Havens *Coffin* W3mSr	American	286	130·0	31·2	9·6	Newburg, NY	1874	Bulman&Brown		F.C.Parker	New York
652	R	Henry P. Mason *Perry* W3mSr	American	684	166·2	35·1	13·6	Bath, Me.	1884	Adams&Hitchcock	Y ss	Adams & Hitchcock	Bath, Me.
653	R	Henry P. Simmons *Grace* W3mSr	American	615	140·0	35·0	18·0	Cooper'sPt. NJ	1884	Morris&Mathis		R.C.Grace	Philadelphia
654		Henry Parker *Green* WSr	American	215	112·0	29·0	9·0	Black Rock, Co.	1868			W.H.Green	New York
655	N	Henry Parr *Larsen* W Bk (ex ...)	Norweg'n	608	143·8	35·0	18·6	Brahestad	1865	Möller		H.H.Parr	Dröbak
656		Henry R. Congdon *Ryan* W3mSr	American	356	131·7	32·5	13·5	RockyHill, Co.	1874			H.F.Withersell	Boston, Mass.
657		Henry R. Tilton *Harris* W3mSr	American	468	152·0	35·0	16·0	Wilmington, Del.	1874	Jackson&Sharp Co.		S.E.Birdsall	New York
658	R	Henry S. Culver *Crowley* W3mSr	American	716	178·0	35·0	19·2	Alexandria, Va.	1883	J.P.Agnew & Co.		J.G.Crowley	Alexandria, Va.
659	R	Henry S. Sandford *Sewall* W S	American	1101	180·6	35·8	23·8	Bowdoinham, Me.	1869	Harvard			New York
660		Henry S. Woodruff W3mSr	American	210	110·9	27·1	10·5	..., Mass.	1886			F.D.Hodgkins	Ellsworth, Me.
661	R	Henry Souther *Rogers* W3mSr	American	680	162·3	35·7	17·4	Thomaston, Me.	1884	S.Watts&Co.		S.Watts	Thomaston, Me.
662		Henry Sutton *Manson* W3mSr	American	572	118·7	35·8	15·0	New Haven, Co.	1879			G.Manson	New Haven, Conn.
663		Henry T. Wood *Roblin* W Sr	American	128	96·0	28·4	8·0	Belleville, N.J.	18$\frac{54}{71}$			P.Phinny	New York
664	RV	Henry Villard *Perkins* W S	American	1553	219·0	39·0	24·0	Bath, Me.	1882	A.Sewall	Y M	A.Sewall & Co.	New York
665	R	Henry Waddington *Mayee* W3mSr	American	470	139·0	35·0	12·0	Milton, Del.	1881	D.H.Atkins		G.Taulane	Philadelphia
666	R	Henry Warner *Paine* W Bkn	American	331	129·8	27·4	14·8	Deering, Me.	1879	R.Lewis&Co.	Y ss	J.H.Flitner	Portland, Me.
667		Henry Whitney *Tapley* W Sr	American	139	96·2	25·5	8·6	Brooksville, Me.	1869			J.H.Gardner	Castine, Me.
668		Henry Withington *Fox* W3mSr	American	502	146·2	34·2	19·1	Newburyport, Mass.	1874	G.E.Currier		Doane & Crowell	Boston, Mass.
669		Henzadah I Flat	British	777	197·0	25·7	8·3	Port Glasgow	1876	R.Duncan & Co.		Irrawaddy Flotilla Co. Ld.	Glasgow
670	✱	Hephzibah *Thomas* W Sr	British	98	87·6	20·3	9·9	Portsmouth	1875	Read		J.Taylor	Plymouth
671		Hepta Adelphi *Broxlios* W Bg	Greek	148	78·7	20·7	14·3	Spezzia	1863			A.T.Broxlios	Spezzia
672		Hera *Clausen* W3mSr	American	339	128·4	29·2	11·8	E.Boston	1869			Metcalfham Fishing and Towing Co.	San Francisco
673	DGV	Hera *Wempe* W Bk	German	1037	176·1	36·3	23·3	Bremerhaven	1875	F.W.Wencke	Y ss	Diedrich Schilling	Bremen
674	N	Hera *Christensen* W Bg	Norweg'n	292	120·0	26·8	13·8	Arendal	1877		Y ss	T.Thommesen & Smith	Arendal
675	DN	Hera *Danielsen* W Bk	Norweg'n	244	116·8	27·6	11·6	Risör	1874	B.Tellefsen	Y ss	Stian Finne	Risör
676		Heraclee W Bg	Turkish	287	..	..	..	Heraclea	1869			Heracleelee Hadji Mustapha & Co.	Heraclea
677		Heraclee W Bg	Turkish	275	..	..	..	Heraclea	1874			Heracleelee Balukjee Ahmed oglou Ahmed	Heraclea
678		Heraclee *Mehemed Halil* W Bg	Turkish	250	..	..	..	Heraclea	1866			Heracleelee Chakkir Abdullah oglou & Co.	Heraclea
679		Heraclee W Bg	Turkish	234	..	..	..	Heraclea	1869			Paramalade Chileckli ... zevjesi Paruda Hanum & Co.	Heraclea
680		Heraclee *Ahmed* W Bg	Turkish	225	..	..	..	Heraclea	1869			Heracleelee Hadji Mustapha & Co.	Heraclea
681		Heraclee W Bg	Turkish	211	..	..	..	Greece	1875			Kerastelli Tüjar Mikkhai Hadji Panayotti	Constantinople
682		Heraclee W Bg	Turkish	200	..	..	..	Heraclea	1870			Heracleelee Balikji Ahmed oglou Hassan	Heraclea
683		Heraclee W Bg	Turkish	175	..	..	..	Heraclea	1869			Heracleelee Balikji oglou ... Ali & Co.	Heraclea
684	R	Herald *Hogan* W3mSr	American	475	144·0	33·3	16·2	Belfast, Me.	1883	Brown&Donald	Y ss	R.P.Buck & Co.	New York
685	✱	Herald of Mercy *Mashfield* W K	British	68	73·6	19·5	9·6	Troon	1887	AilsaS.B.Co.	Y ss	H.Cook	Portsmouth
686	✱	Herat *Nicholson* I S	British	1400	222·6	37·5	22·8	Southampton	1877	Oswald, Mordaunt &Co.		E.Bates & Sons	Liverpool
687	R	Herbert *Towse* W Bn	British	198	114·7	27·5	10·8	Sackville, N.B.	1883	E.W.Ogden	Y ss	J.Wood	Sackville, N.B.
688	✱	Herbert *Williams* W Sr	British	81	67·1	18·6	10·7	Nevin	1848			R.Griffiths	Carnarvon
689		Herbert *Perkins* W Sr	American	109	91·6	26·1	6·5	E.Boston, Mass.	1883	M.Campbell		Rockport Granite Co.	Gloucester, Mass.
690	RV	Herbert W Bk	Norweg'n	1367	182·8	38·4	26·1	Boston, Mass.	18$\frac{69}{75}$	Brewer, Me.		N.Jobson	Arendal
691	RV	Herbert Black *Hoyt* W Bk	American	545	140·0	31·0	18·9	Searsport, Me.	1873	W.McGilvery & Co.	Y ss	A.Nichols	Searsport, Me.
692	V	Herbert C. Hall *Perry* W Bk	British	622	152·4	33·5	19·8	Meteghan, N.S.	1873	C.Le Blanc	Y ss	S.Killam	Yarmouth, N.S.
693		Herbert Dudley *Parsons* W Sr	British	225	114·0	24·3	10·1	Portsmouth, Canada	1875	D.Ainslie		D.Rooney	Kingston, Ont.
694		Herbert E. *Fisher* W3mSr	American	373	129·5	31·0	13·7	Northport, N.Y.	1874			E.H.Atwood	Boston, Mass.
695	R	Herbert Rice *Crandall* W Sr	British	149	99·4	25·5	9·0	Weymouth, N.S.	1888	E.Rice	Y ss	E.Rice	Weymouth, N.S.
696		Hercules W Bn	British	240	131·0	26·0	10·3	Port Burwell	1875	G.W.Poutine		W.B.Martin	Port Burwell, Ont.
697		Hercules W S	American	1215	194·1	38·1	24·0	Bath, Me.	1868	W.Rogers			Boston, Mass.
698		Hercules *McInnes* W Bk	[illegible]	311	102·6	27·2	17·9	New York	18$\frac{15}{30}$			Aiken & Swift	New Bedford, Mass.
699		Hercules W Sr	[illegible]	117	90·0	28·7	7·3	New Haven	1882			Poitevent & Favre	Pearlington, Miss.
700	D	Hercules *Koenes Jr.* W Bk (ex ...)	Dutch	328	115·8	29·8	14·7	Christinestad	1866			C.Koenes Jr.	Delfzijl

✱ = Lloyd's Register; D = Nederlandsche Vereeniging; G = Germanischer Lloyd; I = Registro Italiano; R = American Record; A = Veritas Austro-Ungarico; H = Veritas Ellenico; N = Norske Veritas; V = Bureau Veritas.

Dans la 4me Colonne, C = Composite (membrure en fer, bordé en bois); I = Fer; Stl = Acier; W = Bois.

Les dates du doublage en métal (Col. 11) ressortent des livres dans lesquels les navires sont respectivement classés ou des rapports des Inspecteurs du Lloyd's Register.

Pour plus amples renseignements sur les navires classés par le Lloyd's Register, voir le Lloyd's Register Book et le Lloyd's Yacht Register.

Official No.	No. in Book	Registry in which classed	Ship's Name. Master, and Dates of Appointment to present Owner's Service and to Ship. Ship's Late Name.	Material and Rig	Flag	Net Tonnage	Dimensions: Length	Breadth	Depth	Built: Where.	When	By whom.	Date of Metalling.	Owners or Managers
	701	V	Hercules *Bless 81-81*	I Bk	German	561	150·1	29·1	16·9	Hamburg	1871	C.H.Krause		Conr.Paulsen
	702	N	Hercules *Mathsen 87-87*	W Bk	Norweg'n	471	120·3	32·4	18·8	Sweden	1850			G.I.Jörgensen
	703	N	Hercules *Larsen*	W Bk	Norweg'n	254	115·9	26·6	12·3	Bordeaux	1863			T.Larsen & Sönner
	704	N	Hercules *Greaager*	W Bg	Norweg'n	173	95·3	25·2	11·0	Sandefjord	1853			Carl Dahl
	705		Hercules *Kortman*	W Bk	Russian	601	..	..	..	Helsingfors	1864			C.M.Lindroth
	706	N	Herdis *Rasmussen 87-87*	W Sr	Norweg'n	102	78·9	21·4	9·9	Tvedestrand	1862	A.Smith		N.Chr.Olsen
	707		Here *Andersson*	W Sr	Swedish	129	80·4	21·2	10·6	Denmark	1864			G.A.Osterman
60978	708	❋	Hereford *Wichman 87-87*	I S	British	1469	241·1	37·2	23·2	Glasgow	1869	J.Elder&Co.		J.Nourse
77010	709	❋	Hereward *Inglis 77-77*	I S	British	1513	254·0	39·0	23·2	PortGlasgow	1877	R.Duncan&Co.		J.Potter
	710	N	Hereward (ex Kubana) *Riegen*	W Bk	Norweg'n	760	167·1	32·9	19·6	Richibucto	1876	Jardine	[illegible]	A.Larsen-Naar
48	711	A	Heribea *Velcich -83*	W Bk	Austrian	540	130·2	29·5	21·0	Lussinpiccolo	1880	N.Martinolich	[illegible]	[illegible]
	712		Heribert *Hansson*	W Sr	Swedish	229	100·7	25·0	13·5	Roslagen	1861			P.Hansson
	713	NV	Herlof Herlofson *Krüger*	W Bk	Norweg'n	766	162·7	34·0	19·7	Arendal	1875	O.J.Herlofson	[illegible]	H.Herlofson's Enk
	714	V	Herman *Christensen Isbeg*	W Sn Sr	Danish	341	129·8	25·5	15·2	Bjørneborg	1872	A.Lindgren	[illegible]	J.Harboe
	715	N	Herman *Wellesen -84*	W Bk	Norweg'n	746	151·6	34·7	20·6	Bremerhaven	1853			C.U.Wetlesen [illegible]
	716	N	Herman *Mortensen 82-83*	W Bn	Norweg'n	244	128·7	25·6	11·6	Bergen	1873	S.K.Plage	[illegible]	E.Schmidt,jun.
	717		Herman *Tiainen -83*	W Sr	Russian	140	95·8	23·1	8·8	Joensuu	1875	J.Pettersson		H. Cederberg & [illegible]
95823	718	R	Herman B. Ogden *Harrington*	W 3m Sr	American	609	146·6	35·0	16·6	Millbridge, Me	1884	J.W.Sawyer		A.J.Newbury
	719		Herman F. Kimball *Kimball*	W Sr	American	119	185·0	26·5	7·1	Bath,Me.	1888	C.Hodgkins		F.Kimball
	720	N	Herman Lehmkuhl *Paulsen 8-81*	W S	Norweg'n	1309	219·9	39·7	22·7	Vestnæs	1875	Havanerans	[illegible]	J.E.Lehmkuhl
	721	DN	Herman & Eugen *Gulbrandsen*	W Sr	Norweg'n	134	99·6	24·1	10·5	Bergen	1861			O.Guthormsen
	722	G	Hermann *Hansen 83-*	W Bg	Danish	184	101·0	22·0	12·9	Rönne	18 42/57	S.P.Bech		J.P.Hansen
	723	V	Hermann *Knudsen*	W Sr	Danish	101	81·7	20·0	9·5	Thurø	1872	P. Troensegaard		P.Knudsen
	724	GV	Hermann (ex Jairus B Lincoln) *Willenberg -88*	W S	German	1739	207·7	40·5	28·1	Freeport	1869	Briggs&C'shing	[illegible]	H.Wessels
	725	V	Hermann (ex Wester Schelde, ex California) *Schütte 81-87*	W S	German	1318	193·5	38·1	24·5	Boston	1864	S.Hall	[illegible]	H.Bischoff & Co.
	726	GV	Hermann (LOST) *Binge 85-88*	I S	German	849	182·0	32·1	19·5	Bremen	1875	Act.Ges."Weser"		M.G.Amsinck
	727	V	Hermann *Traulsen -81*	W Bk	German	444	137·5	27·2	16·1	Heiligenhafen	1875	J.Tornoe	[illegible]	P.Kock
	728	G	Hermann (ex Herman) *Ernst 80-82*	W Sr	German	151	89·9	15·6	9·2	Sappemeer	1865	K.Bakker		Hermann Ernst
	729	G	Hermann *Sassen 84-*	W Sr	German	147	93·4	21·3	10·5	Papenburg	1864	J.A.Klasen		A.Klasen
	730	G	Hermann *Schneiders 66*	W Sr	German	132	89·6	21·2	10·2	Papenburg	1852	F.W.Meyer		J.Dieckhaus
	731		Hermann (WRECKED) *Carls 88-88*	W Sr	German	127	85·3	22·5	10·9	Stralsund	1863	A.Kirchhoff		J.J.Wallis & Sohn
	732		Hermann *Onken 87-87*	W Sr	German	113	85·0	21·3	9·7	Leer	1879	B.D.Middendorf	[illegible]	J.J.Onken
	733		Hermann *Mohnsen 80-*	G Sr	German	103	100·4	21·3	6·1	Hamburg	1876	F.H.Schmidt		J.J.H.Mohnsen
	734	V	Hermann *Deters 86*	W Sr	German	100	76·5	17·1	8·8	Edewecht	1875	Kramer		H.Deters
	735	GV	Hermann (ex Hermana) *Leonardsen*	W Bk	Norweg'n	446	139·0	32·2	17·0	Apenrade	1874	J.Paulsen & Sohn	[illegible]	P.A. [illegible]
	736	V	Hermann Hinrich (WRECKED) *Held*	W Sr	German	112	77·0	21·9	7·5	Grossefehn	1869	T.J.Cassens & Sons		F.J. [illegible]
	737	N	Hermanos (ex Gustav) *Araldsen*	W Bk	Norweg'n	579	132·6	31·7	19·4	Bremen	1836			[illegible]
	738	N	Hermanos *Nielsen -87*	W Bk	Norweg'n	379	124·7	30·1	16·3	Arendal	1868	O.P.Olsen	[illegible]	[illegible]
	739	NV	Hermanos *Rasmussen*	W Bn	Norweg'n	221	107·7	26·2	12·7	Christiansand	1873	H.Larsen	[illegible]	[illegible]
	740	R	Hermelin *Andersson*	W Bk	Swedish	286	111·8	26·0	14·3	Töreforss	18[illegible]		[illegible]	J.F.O[illegible]
59912	741		Hermena	W Bg	British	207	91·0	27·0	13·0					F.H.O.Tross
95083	742		Hermes	Stl Sr	British	219	115·2	28·0	9·2	Dumbarton	1888	W.Denny & Brs		La Platense Flotilla Co. Ld.
6	743	A	Hermes *Gounaris 83-*	W Bg	Greek	369	121·3	29·0	..	Syra	1873			[illegible]
42	744	H	Hermes	W Bg	Greek	294	..	..	..	Syra	1869	[illegible]Koufoudakis		[illegible]
	745	H	Hermes *Pappapetrou*	W Bg	Greek	285	117·4	27·1	15·7	Galaxidi	1872	C.Papapetrou	[illegible]	T.Papap[illegible]
	746		Hermes	W Bg	Greek	274	100·7	27·1	..	Syra	1867	F.Cafieleschi		[illegible]
	747	N	Hermes *Teilefsen -87*	W Bk	Norweg'n	549	138·8	32·0	17·1	Grimstad	1887	A.Johannessen	[illegible]	N.A.Grönn
	748		Hermes (ABANDONED) *Abrahamsen*	W Bk	Norweg'n	289	98·8	27·4	15·5	Lillesand	1845			Edv.Abrahamsen
	749	N	Hermes *Olsen -88*	W Bg	Norweg'n	259	113·4	27·6	12·3	Tvedestrand	1877		[illegible]	A.Olsen
	750		Hermes	W Bk	Russian	511	125·6	29·5	16·6	Brahestad	1863	F.Danielson		J.S.Landberg & [illegible]

❋ = Lloyd's Register; D = Nederlandsche Vereeniging; G = Germanischer Lloyd; I = Registro Italiano; A = Veritas Austro-Ungarico; H = Veritas Ellenico; N = Norske Veritas; [illegible]

In the 4th Column, C = Composite (iron frame, wood planked); I = Iron; Stl = Steel; W = Wood.

The Dates of Metalling (Col. 11) are based on the records of the Books in which Vessels are respectively classed, or on the Reports of the [illegible]

Further particulars of Vessels classed by Lloyd's Register will be found in Lloyd's Register Book and Lloyd's Yacht [illegible]

No.	Cote	Nom, Capitaine et dates d'entrée au service des Armateurs actuels et à bord du navire. Ancien Nom.	Matériaux et Gréement	Pavillon	Tonnage	Longueur	Largeur	Creux	Construit: Où	Quand	Par qui	Doublage	Armateurs ou Directeurs-Gérants	Port d'Armement
751	GV	Hermes (ex W…) *Holm -83*	WBk	Swedish	475	112·0	27·8	14·8	Tatamagouche, N.S.	1870	D.E.Campbell	[illegible]	L.Jönsson	Brantevik
752	V	Hermes *Tandel 86-86*	W3mBn	Swedish	317	115·1	26·7	11·1	Sikea varf	1886	O.P.Åberg	[illegible]	Fr.Jacobsen	Robertsfors
753	❊	Hermine *Jones*	I Bk	British	538	159·7	27·4	18·0	Liverpool	1865	J.Evans&Co.		G.H.Fletcher & Co.	Liverpool
754		Hermine *Page*	W3TwgL	British	[illegible]	105·8	26·1	12·6	[illegible]				TyneWherryCo.,Ld.	S.Shields
755	V	Hermine *Schröder -81*	WSr	German	[illegible]	77·1	[illegible]	10·1	[illegible]	1871	A.Juhl		H.Koch	Stralsund
756	GV	Hermine *Fehmiken -83*	WBk	Swedish	[illegible]	118·7	25·1	14·[illegible]	[illegible]	1858		[illegible]	A.G.Forselius	Stockholm
757		Herminia	WBk	Chilian	[illegible]	125·9	27·1	17·8					J.A.Verdugo	Valparaiso
758	❊	Hermione *Parker*	I S	British	11[illegible]	219·4	[illegible]	21·[illegible]	[illegible]	1876	A.Hall&Co.		Shaw, Savill & Co.	Southampton
759		Hermo *Nyman*	WBg	Russian	[illegible]	110·0	31·2	14·[illegible]	[illegible]	1876	A.Tengström		R.Strömberg	Korpo
760	V	Hermod *Hermansen 79-84*	W3mSr	Danish	150	93·2	22·8	10·8	[illegible]	1878	F.Hansen		H.J.Hansen (J.Kromann)	Marstal
761	N	Hermod (ex Elma?) *Fredriksen 80-79*	WBk	Norweg'n	516	138·2	30·8	18·4	[illegible]	1872	Geo.Mitzlaff	[illegible]	H.J.HammersEnke	Lillesand
762	N	Hermod *Olsen*	WBk	Norweg'n	391	131·1	29·6	16·4	[illegible]	1864	Andreasen		J.Dahll	Kragerö
763		Hermod (CONDEMNED?) *Gundersen -85*	WSr	Norweg'n	176	101·6	24·6	11·2	[illegible]	1874	J.Melkevig	[illegible]	K.S.Bertelsen	Stavanger
764	❊	Hermon (WRECKED) *Dutkie*	WBk	British	407	132·3	28·0	17·0	[illegible]	1869	Hodgson	V 1886	T.S.Howitt	North Shields
765	DNV	Hermon *Olsen*	WS	Norweg'n	1077	178·0	37·2	23·1	[illegible], N.B.	1865	Ring	[illegible]	A.J.Myhre	Porsgrund
766		Hermosa *Duthie 58-63*	WSr	British	136	87·2	23·5	6·4	Baltimore,US	1854			Geo.Wood	Nassau,N.P.
767	V	Hermoupolis	WBk	Greek	40[illegible]	139·4	30·0	..	[illegible]	1882	A.Livadaros		M.& J.Pissias	Syra
768		Hermoupolis	WBn	Greek	218	104·2	26·1	..	[illegible]	1866			G.I.Chadoulas	Syra
769	D	Hernösand (ex H…) *Janssens*	WBk	Dutch	5[illegible]	140·4	32·7	19·4	[illegible]haven	1855	De Hoog	[illegible]	W.Porr	Edam
770	V	Hernosand *Pettersson*	WBk	Swedish	335	131·2	26·1	12·5	Hernösand	1875	Fr.Hägglund	[illegible]	Arbetare Skeppsb.-Aktb. (A.[illegible])	Hernösand
771		Hero *McDonald*	WBn	British	178	105·3	24·9	11·6	Five Islands, N.S.	1876	Thompson		A.McDonald	Glasgow
772	❊	Hero *Cordy*	WK	British	8[illegible]	77·8	19·9	9·9	Jersey	1872	Picot		Mrs.R.J.Chambers &Co.	Jersey
773		Hero *Nickerson 82-82*	WSr	American	139	86·5	25·7	7·9	Ipswich, Mass.	1859			C.H.Sprague	Providence,R.I.
774	GV	Hero *Hayessen 82-82*	WBg	German	264	112·0	25·6	13·0	Brake	1868	J.F.Nicolai	V 18[illegible]	J.H.Nicolai	Brake
775	N	Hero *Olsen -83*	WBk	Norweg'n	613	148·1	[illegible]·6	20·4	Sunderland	1855	Geo.Booth		J.S.Lorentzen	Drammen
776	❊	Hero *Olsen -88*	WBk	Norweg'n	374	129·0	28·1	16·0	P.E.Island	1874	Harkness	[illegible]	Actieselskabet "Hero" (P.Dorman&Co.)	Laurvig
777	N	Hero *Olsen 81-87*	WSr	Norweg'n	136	90·7	23·4	10·5	Kragerö	1870	J.Hamborg	V 18[illegible]	H.Biörn,jun.	Kragerö
778		Hero *Lindeman*	WBk	Russian	647	167·9	36·4	19·2	Wasa	1868			L.Olofsson	Mariehamn
779		Hero *Eriksson*	W3mSr	Russian	18[illegible]	92·6	27·6	10·9	Sideby	1879			K.Karlsson	Lemland
780		Hero *Åsberg*	WSr	Swedish	111	78·7	20·7	9·5	Ekenäs	1883	Ekenäs Skepps-varf		G.Pettersson	Ekenäs
781	❊	Hero of Khartoum *Wellsby -87*	WK	British	88	79·7	21·1	10·5	Galmpton	1886	T.Mathews		J.W.Haylock	Grimsby
782	N	Heroen *Hauff -86*	WBg	Norweg'n	272	121·8	27·5	12·9	Porsgrund	1877	A.Olsen		H.Thorsen	Porsgrund
783	V	Héroine *Magdelain -88*	WSr	French	120	84·6	18·7	10·6	St.Vaast	1877	E.Costey	V 18[illegible]	Beust père & fils	Granville
784		Herolden *Hansen 86-86*	WSr	Norweg'n	119	85·8	21·2	9·3	Kaskö,Finland	1863			Carl Evensen	Christiania
785	N	Heros *Thorsen -85*	WS	Norweg'n	599	168·5	32·2	18·0	Sikeå,Sweden	1867	Fr.Petersen	V 18[illegible]	G.Gundersen	Stavanger
786	❊A	Herradura *Scopinich*	CBk	Austrian	613	169·6	30·6	17·6	Aberdeen	1868	A.Hall & Co.	[illegible]	A.G.Scopinich	Lussinpiccolo
787	V	Herrmann Becker *Matzell -87*	W3mSr	German	3[illegible]	135·8	26·7	12·1	Grabow	1869	C.Domcke	V 18[illegible]	C.A.Domcke	Stettin
788	❊D	Herschel *Detlef -88*	I S	German	787	164·0	30·2	21·3	Birkenhead	1857	Canada Works		R.M.Sloman & Co.	Hamburg
789		Hersilia	I Sr	Argentine	199	130·0	27·2	9·0	Glasgow	1885	AbercornSBCo		V.Casares & Son	Buenos Ayres
790		Hersilia	WBk	Brazilian	272	..	..	..						Bahia
791		Hersilia	WBk	Portug'se	290	118·4	24·5	14·4	Villa do Conde				L.P.da S.Moreira	Oporto
792	❊	Hertfordshire *Smith 1888*	I Bk	British	855	180·0	30·0	20·5	Liverpool	1863	T.Vernon&Sons		W.J.Myers,Son &Co.	Liverpool
793	V	Hertha *Schmidt 86-86*	WBn	Danish	177	101·1	22·9	11·3	Marstal	1875	H.J.Bager		H.Hansen	Marstal
794	DV	Hertha (ex Bolette (?Levy)) *Knudsen*	WSr	Danish	140	95·3	21·0	10·5	Thurø	1868	L.J.Kaas		H.A.Hansen	Svendborg
795		Hertha (LOST) *Petersen 70-76*	WSr	Danish	103	71·0	21·0	10·4	Rudkjøbing	1852	N.H.Rasmussen		J.P.Petersen	Rudkjøbing
796		Hertha (WRECKED) *Bruwen -87*	WBg	German	197	96·8	22·7	12·6	Stettin	18 4?/6?	A.E.Nüske		G.Metzler	Stettin
797	DNV	Hertha *Nielsen -81*	WBk	Norweg'n	558	151·1	31·5	17·3	Drammen	1877	Jürgensen & Knudsen	[illegible]	O.C.Reinhardt	Christiansand
798	N	Hertha *Gjeruldsen -88*	WBk	Norweg'n	416	120·6	30·7	17·1	Grimstad	1863	N.O.Gjömle		A.M.Gjeruldsen	Grimstad
799	V	Hertig Oskar Fredrik *Sellström*	WS	Swedish	557	150·5	28·8	17·5	Sundsvall	1870	Fr.Östman	V 18[illegible]	Sundsvalls Skeppsrederi Bolag (G.A.[illegible])	Sundsvall
800		HerzogGeorg *Andreas -7-*	WBk	German	310	111·9	28·0	16·1	Rostock	1853	W.Zeltz		E.Burchard	Rostock

❊ = Lloyd's Register; D = Nederlandsche Vereeniging; G = Germanischer Lloyd; I = Registro Italiano; R = American Record; A = Veritas Austro-Ungarico; H = Veritas Ellenico; N = Norske Veritas; V = Bureau Veritas.

Dans la 4me Colonne, C = Composite (membrure en fer, bordé en bois); I = Fer; Stl = Acier; W = Bois.

Les Dates du doublage en métal (Col. 11) ressortent des livres dans lesquels les navires sont respectivement classés ou des rapports des Inspecteurs du Lloyd's Register.

Pour plus amples renseignements sur les navires classés par le Lloyd's Register, voir le Lloyd's Register Book et le Lloyd's Yacht Register.

Official No.	No. in Book	Registry in which classed	Ship's Name. Material and Rig. Master, and Dates of Appointment to present Owner's Service and to Ship. Ship's Late Name.	Flag	Net Tonnage	Length	Breadth	Depth	Where	When	By whom	Patent Metalling	Owners or Managers
83779	801		Hesione W Bn	British	158	97·9	19·0	9·2	Montreal	1828			T. Hart
	802		Hesperas (M[illegible]) Andersen W3mSr	Norweg'n	238	111·2	26·7	11·3	Grimstad	1886	J. Tellefsen	Y [illegible]	T. G. Gundersen
81162	803		Hesper Jennings 88-88 W K Bge	British	114	101·0	22·6	8·0	Harwich	1879	Vaux		E. Whitmore
85717	804	V	Hesper Ryder W Bk	American	664	163·0	38·0	15·0	Port Blakely, [Wash.	1882	Hall Bros.	Y [illegible]	W. G. Hall
	805	N	Hesperia Nielsen 88-88 W Bk	Norweg'n	499	134·7	30·2	18·6	Castellamare	1875	Longobardi	Y [illegible]	Gunder Olsen
70706	806	✻	Hesperides Wrenn 87-87 I S	British	1338	236·7	37·8	22·6	Sunderland	1875	Short Bros.		J. Stewart & Co.
	807		Hesperos [illegible] W Sr	Norweg'n	171	97·4	23·8	11·6	Stavanger	1868	[illegible]	Y [illegible]	J. C. Knivsberg
68090	808	✻	Hesperus Harry 77-77 I S	British	1777	262·2	39·7	23·5	Greenock	1873	R. Steele & Co.		Anderson, [illegible]
	809		Hesperus Reiter -76 W Sr	German	105	81·8	19·7	10·2	[illegible]	1840	D. Oltmanns [Wwe.		K. E. Aukes
	810		Hesperus Persson W Sr	Swedish	137	80·7	21·6	10·9	[illegible]	1851			B. Olsson
95854	811		Hester A. Seward Thomas W Sr	American	150	108·8	28·8	7·6	Taylor's Island [Md.	1885			T. J. Seward
	812	V	Hestia Bottcher 79-79 W Bk	German	550	143·0	30·5	18·7	Memel	1872	B. Pieper		Graff & Bannis
95916	813		Hettie J. Dorman Tomlinson W3mSr	American	119	100·0	27·0	6·1	Milton, Del.	1887	D. H. Atkins		J. B. Dorman
74428	814	✻	Hetty Stibbins W Sr	British	100	87·2	26·7	11·2	Point	1877	Stephens	Y [illegible]	W. Chellew
	815		Hevelius Bussert 88-88 W Bk	German	374	110·9	29·0	16·8	Danzig	1842	J. W. Klawitter		A. Gibsone
	816	✻	Heybel (ex Elwe) Hussein Effendi W Bk	Turkish	620	134·0	34·0	20·0	Trieste	1870	Tonello	[illegible]	[illegible] Bey
73180	817		Hiawatha (ex Kingsport) Little 63-77 W Bk	British	934	172·0	36·5	20·9	Cornwallis, NS	1876	Crichton		B. McCorkell
71320	818	V	Hiawatha McKay W Bk	British	390	147·0	30·5	12·9	Port Elgin, NB	1878	Copp Bros.	Y [illegible]	W. H. Copp & Co.
65536	819		Hiawatha (DISMANTLED) Jordan W Bg	British	280	116·1	26·5	14·9	P. E. Island	1871	Richards		W. Wright
48623	820		Hibernica Turner W Bn	British	166	104·5	22·0	12·4	Shippegan NB.	1863	J. Henry		J. P. Tocque
11693	821		Hidalgo Hardwick W Bg	American	175	101·0	27·0	8·5	E. Machias Me.	1855			L. W. Williams
	822		Hidayetti Bahri W Bg	Turkish	149	..	..	..	[illegible]	1871			Nouri Bey & Co.
	823		Hidayetti Rahmin W Bn	Turkish	474	..	..	..	[illegible]	1872			Her[illegible]
673	824		Hidesato Maru W Bk	Japanese	332	113·3	24·2	14·2	[illegible]	1872			Nippon Yusen [illegible]
	825		Hiemayoon Shah W Bk	British	872	162·0	35·0	24·0	[illegible]	1852			Ammum
	826	GV	Hieronymus Ohle 65-87 W Bk	German	425	141·7	29·2	16·5	Apenrade	1875	G. Raben	[illegible]	A. Burchard
	827		Hiftzi Huda W Bg	Turkish	125	..	..	..	Roumania	1879			[illegible]
	828		Hiftzi Rahmin W Bg	Turkish	237	..	..	..	[illegible]	1870			[illegible]
	829		Hiftzi Rahmin Ismael W Bn	Turkish	212	..	..	..	[illegible]	1884			Heracleeloo Bali [illegible] Ahmed zadeh [illegible]
	830		Hiftzi Rahmin W Bn	Turkish	212	..	..	..	[illegible]	1877			Trebizondlee [illegible]
	831		Hiftzi Rahmin W Bn	Turkish	200	..	..	..		1873			Trebizondlee [illegible] Kutchuk Osman & Co.
	832		Hiftzi Rahmin W Bg	Turkish	170	..	..	..	[illegible]	1879			Heracleelee Djevdet [illegible]
	833		Hiftzi Rahmin Ferrad W	Turkish	157	..	..	..	[illegible]	1885			Boodighli zadeh [illegible]
	834		Hiftzi Rahmin W Bg	Turkish	150	..	..	..	Roumania	1881			Hattab [illegible] Hajee Yusuf
	835		Hiftzi Rahmin W Bn	Turkish	149	..	..	..	Heraclea	1882			Cotch zadeh [illegible]
	836		Hiftzi Rahmin W Bg	Turkish	105	..	..	..	Jeddah	1886			Haliblecloe [illegible] Ahmed
	837		Hiftzi Rahmin Ahmed W Bg	Turkish	105	..	..	..	Jeddah	1881			Altinverrakji [illegible] & Co.
	838		Hiftzi Rahmin W Bn	Turkish	100	..	..	..	Ebioul Ilhair [Skelessi	1877			Oonoapanindli [illegible]
	839		Hiftzi Rahmin Omer W Bg	Turkish	100	..	..	..	Abdul Khair [Skelessi	1885			[illegible]
79	840		Higashi Maru W Sr	Japanese	489	113·6	29·1	15·2	Hiogo	1883	Sumiyoshi [Yasunoske		Tatsuma Yohel
	841	✻	Highflyer Sieuer 88-88 W Bk	German	960	193·7	35·5	20·0	London	1861	R. & H. Green	Y [illegible]	E. tom Dieck
89908	842	✻D	Highland Forest McWhirl I Bk	British	995	209·6	34·2	19·7	Leith	1884	Ramage [& Ferguson		J. R. Cuthbertson & Co.
86724	843	✻	Highland Glen Dawson I Bk	British	982	211·3	34·0	19·6	Leith	1882	Ramage [& Ferguson		Crane, Colvil & Co.
93256	844	✻	Highland Home Ferguson I Bk	British	1298	234·0	37·0	21·2	Leith	1886	Ramage [& Ferguson		J. R. Cuthbertson & Co.
95318	845		Highland Light Sayer -87 W S	American	1265	194·9	38·1	24·3	Bath, Me.	1874	W. Rogers	[illegible]	R. G. Bybee
90738	846		Highlander Morrison W Bn	British	170	92·0	26·1	10·9	Pt. Hawkesbury NS	1888	D. Morrison		D. A. Morrison
11956	847	RV	Highlander Edgett -88 W S	American	1286	190·3	38·8	24·0	East Boston	1868	T. Hall	[illegible]	[illegible]
86256	848	V	Highlands Owen W Bk	British	1234	199·6	38·2	22·6	Portland, N.B.	1883	D. Lynch	[illegible]	H. D. T[illegible]
62906	849	✻	Highmoor (ex [illegible]) Hailey 87-87 I Bk	British	1113	207·0	35·1	22·0	Sunderland	1866	R. Thompson, jr.		[illegible]
	850		Hii Chrysakis W Bg	Greek	215	89·9	26·8	16·7	Syra	1874			M. Chrysakis

✻ = **Lloyd's Register**; D = Nederlandsche Vereeniging; G = Germanischer Lloyd; I = Registro Italiano;
A = Veritas Austro-Ungarico; H = Veritas Ellenico; N = Norske Veritas;

In the 4th Column, C = Composite (iron frame, wood planked); I = Iron; Stl = Steel; W = Wood.

The Dates of Metalling (Col. 11) are based on the records of the Books in which Vessels are respectively classed, or on the Reports of the [illegible]

Further particulars of Vessels classed by Lloyd's Register will be found in Lloyd's Register Book and Lloyd's Yacht [illegible]

No.	Classe	Nom, Capitaine et Gréement	Pavillon	Tonnage	Longueur	Largeur	Creux	Construit où	Quand	Par qui	Doublage	Armateurs ou Directeurs Gérants	Port d'Armement
851	N	Hilaöo (ex ...) *Jørgensen* W Bk	Norweg'n	357	106·7	27·3	17·8	Baltimore, Md.	1828			S.Eriksen	Arendal
852		Hikaku Maru W S		201	93·0	22·0	12·0	Tokio	1880			Nakamura Heijiro	Tokio
853	V	Hilaria *Buck 66-8* W S	British	1675	225·0	41·5	24·3	St.John,N.B.	1886	J.K.Dunlop	[illegible]	W.Thomson & Co.	St.John,N.B.
854	N	Hild *Hansson* W Gt	Norweg'n	136	95·1	23·4	11·3	Stavanger	1872	M.A.Magnus		B.Hansen	Stavanger
855	V	Hilda *Dawson -82* W Bk	British	719	160·6	34·6	18·7	N.Glasgow,N.S.	1878	J.McColl	[illegible]	J.D.McGregor	Pictou,N.S.
856		Hilda (ex ...) *Thompson* W Bg	British	306	122·0	24·6	14·0	Sweden	1865			P.V.Grant	Shanghai
857	*	Hilda *Langdon 82-83* W Bg	British	238	116·5	24·9	12·9	Appledore	1879	Pickard	[illegible]	J.W.Pring	Brixham
858	*	Hilda *Paterson* W Sr	British	91	79·6	22·0	9·8	Inverness	1879	Stewart		W.Paterson	Inverness
859	N	Hilda (ex Regina) *Banowsen* W Bk	Norweg'n	567	143·3	32·8	17·7	Sherbrook N.S.	1866	J.Cumminger		K.Hansen	Christiansand
860		Hilda W Bk	Norweg'n	361	126·9	29·1	15·4	Borgå	1857			A.Olsen	Fredrikshald
861	N	Hilda (ex Katharina) *Christoffersen* W Bg	Norweg'n	228	100·0	28·2	12·4	Hamburg	1852			J.H.Christoffersen	Tönsberg
862		Hilda W Bg	Norweg'n	195	91·8	26·8	12·6	Åbo	1856	C.J.F.Jörgensen		Erik Gjessen	Skudesnæs
863		Hilda *Olsen 86-86* W Sr	Norweg'n	107	75·9	21·2	10·3	Risör	1853	B.Rolfsen		J.W.Prebensen,sen.	Risör
864		Hilda *Karlsson* W Bk	Russian	394	115·0	27·0	16·0	Åland	1865	R.E.Lindman		M.Mickelson	Åland
865		Hilda *Sundvik 85-85* W Bk	Russian	386	..	..	..		1876			G.Stenroos	Lavia
866		Hilda *Hett* W Bk	Russian	365	119·7	25·3	16·1	Borgå	1844			J.L.Roth	Borgå
867		Hilda *Jansson* W Bg	Russian	282	110·0	30·0	15·0	Sastmola	1868	A.Ahlsten		M.Jansson	Korpo
868	H	Hilda *Fagerström 84-86* W Sr	Russian	247	103·6	26·0	12·9	Westanfjerd	1879	O.F.Olander		O.F.Olander	Westanfjerd
869		Hilda *Mattsson* W Sr	Russian	201	98·2	27·9	12·8	Biskopsö	1869	O.Alander		J.E.Tengström	Korpo
870	* V	Hilda *Afzelius 83-81* W Bk	Swedish	475	113·0	27·1	16·5	Framnäs	1879	E.Östman	[illegible]	Framnäs Varfs Aktieb	Framnäs
871	V	Hilda *Berlet* W Bk	Swedish	383	126·6	25·4	14·9	Sölvesborg	1874	E.Wiréen	[illegible]	N.M.Pahlsson	Helsingborg
872	V	Hilda *Jacobsen 88-85* W Bk	Swedish	355	135·5	26·3	13·6	Carlskrona	1868	O.Bergström	[illegible]	J.E.Olsson	Gothenburg
873		Hilda (ex Venus) *Pettersson* W Bg	Swedish	223	92·8	23·3	14·3	Demmlorf	18 4/6			A.G.Förselius	Stockholm
874	V	Hilda *Abrahamsson* W Sr	Swedish	149	95·8	21·4	9·6	Oroust	1873	A.Isaksson		B.Johansson	Hesselvial, Bohuslän
875		Hilda *Sjögren 85-85* W Sr	Swedish	106	79·1	20·1	9·8	Fanö				O.Jönsson	Hellevik
876	V	Hilda Maria *Lönner -88* W Bk	Swedish	276	123·0	23·2	13·7	Hamburg	1864	J.Marbs	[illegible]	C.G.Bergfors	Gefle
877	N	Hildegard (ex ...) *Thorkildsen* W Bk	Norweg'n	432	130·1	30·0	16·6	Brevig	1865		[illegible]	Oscar Hansen	Christiania
878		Hildegard *Pettersson -88* W Bg	Swedish	256	111·8	25·3	12·1	Sundsvall	1866	A.Nordström		C.E.Olsson	Väddö,Rslgn
879		Hildegard *Andersson* W3mSr	Swedish	158	97·1	23·2	10·4	Vätö	1883			J.E.Jansson	Kulla,Rslgn
880	*	Hildegarde *Wadley -84* W Sr Yt	American	112	106·5	20·0	11·9	Gosport	1874	Camper & Nicholson	[illegible]	G.J.Gould	New York
881	N	Hilding *Tellefsen 85-85* W Bg	Norweg'n	238	105·4	27·2	12·4	Grimstad	1870	A.Johansen		G.Tellefsen	Arendal
882		Hildred *Merrill 82-86* W Bn	British	244	115·2	26·6	12·9	P.E.Island	1878	Peake		Holden & Gann	Faversham
883	N	Hildur *Olsen -88* W Bk	Norweg'n	279	126·2	27·3	12·6	Brevig	1873		Y 10.86	L.Pedersen	Tönsberg
884	V	Hildur *Larsson -88* W Bk	Swedish	341	123·6	26·5	15·1	Söderhamn	1868	A.O.Wanberg	[illegible]	J.Brolin & Son	Söderhamn
885	V	Hildur *Pettersson -85* W3mSr	Swedish	145	91·5	21·8	10·0	Timmernabben	1881	C.Nilsson	Y 11.84	S.Olsson	Timmernabben
886	DG	Hilkea *Veit 77-* W Sr	German	168	97·5	26·3	11·6	Papenburg	1864	H.W.Meyer		W.Höveler	Papenburg
887	N	Hill *Olsen 81-85* W Bg	Norweg'n	216	107·5	25·8	12·2	Arendal	1872	F.Guttormsen		H.H.Pettersen	Arendal
888	N	Hilla *Ose* W Sr	Norweg'n	102	83·6	22·4	11·3	Tvedestrand	1876			O. & R.Ose	Aalesund
889		Hilma *Karlsson 73-79* W Bg	Russian	272	104·5	25·9	14·8	Föglö	1865			O.A.Lindeman	Wardö
890		Hilma *Henriksson* W Sr	Russian	181	..	..	..	Hiitis	1873			B.Andersson	Hiitis
891	V	Hilma *Cronberg -88* W S	Swedish	901	171·9	33·8	20·4	Luleå	1876	Martensson	[illegible]	F.O.Sandström	Piteå
892	V	Hilma *Frisell* W3mSr	Swedish	219	115·8	22·8	11·0	Oscarshamn	1884	C.Hasselbom	[illegible]	F.Johansson	Oscarshamn
893	G	Hilma *Hofrenberg 76-76* W Bn	Swedish	188	100·0	23·0	11·6	Wiken	1867	P.Jacobsson		L.Jönsson	Wiken
894		Hilma *Persson -87* W Bg	Swedish	166	99·7	22·6	11·0	Piteå	1861	J.C.Christensson		C.M.Lundberg	Cimbrishamn
895		Hilma (Broken ...) *Johansson -84* W Sr	Swedish	163	87·9	22·1	11·3	Calmar	1856			C.Bruce	Fiskebäckskil
896	*	Hilma *Engqvist* W Sr	Swedish	155	96·5	21·4	11·0	Oscarshamn	1881	G.Tillberg	[illegible]	C.J.Ödman	Sälvik
897	*	Hilston *Chivers -88* I S	British	1998	278·1	40·0	24·4	Greenock	1885	Russell&Co		Hilston Ship Co. Ld. (Graham & Co.)	London
898	*	Himalaya *Patterson 86-86* I Bk	British	1008	201·6	33·0	20·5	Sunderland	1863	Pile,Hay&Co.		Shaw,Savill& Albion Co. Ld.	Southampton
899		Hin Goan *Lie Tjay Kheng* W Sr	Dutch	160	92·5	23·6	..	Siam	1872		[illegible]	Lie Boen Kiem	Banjarmassin
900													

* Lloyd's Register; D = Nederlandsche Vereeniging; G = Germanischer Lloyd; I = Registro Italiano; R = American Record; Veritas Austro-Ungarico; H = Veritas Ellenico; N = Norske Veritas; V = Bureau Veritas.

Dans la 4me Colonne, C = Composite (membrure en fer, bordé en bois); I = Fer; Stl = Acier; W = Bois.

Dates du doublage en métal (Col. 11) ressortent des livres dans lesquels les navires sont respectivement classés ou des rapports des Inspecteurs du Lloyd's Register.

Pour plus amples renseignements sur les navires classés par le Lloyd's Register, voir le Lloyd's Register Book et le Lloyd's Yacht Register.

Official No.	No. in Book	Registry in which classed	Ship's Name. Ship's Late Name	Master, and Dates of Appointment to present Owner's Service and to Ship	Material and Rig	Flag	Net Tonnage	Length	Breadth	Depth	Built: Where	Built: When	Built: By whom	Date of Metalling	Owners or Managers
47222	901		Hind	Phillips	W Sr	British	114	92·0	20·0	11·8	Cardiff	1862	[illegible]		N. House
58729	902		Hind	Jones	W Sr	British	109	80·7	21·1	10·8	Burton Stath'r	1868	Wray		E. Jackson
69642	903	✻	Hinda	Jones 82-87	I Bk	British	443	157·0	27·2	16·0	Llanelly	1869	Neville Bros.		[illegible] & Co.
	904		Hinderika		W Bn		165	92·8	21·0	10·1	Emden	1873	H. Klattenhoff		
	905		Hinderika Rolina (ex [illegible])	Ohlsen -86	W K	Dutch	362	108·0	27·0	18·0	Sunderland	1855	Potts, Bros. & Co.		B. Ohlsen
	906	✻	Hinderk Robers	Leemhuis, sr. [illegible]	W 3m Sr	Dutch	237	124·6	23·0	11·6	[illegible]	1879	J. A. Hooites	Y [illegible]	J. A. Hooites
	907	V	Hindoo	Matthiesen 76-86	W Bk	German	541	147·5	31·3	17·9	Apenrade	1877	N. Jacobsen	Y [illegible]	J. Matthiessen
	908		Hindoo	Gustafsson	W Bg	Swedish	215	94·8	23·5	13·4	Norrköping	1840			A. Svensson
43162	909	✻	Hindoostan	Gerner -87	W Bk	British	463	138·9	29·4	18·0	Sunderland	1863	J. Gill	Y [illegible]	G. Lough
47022	910	✻	Hindostan	Allen 87-87	I Bk	British	1262	219·5	35·0	24·3	Jarrow	1863	Palmer Bros.		Shallcross & [illegible]
45208	911	R	Hindostan	Cranch 87-88	W S	British	674	143·3	31·5	..	Medford, Mass	1847		Y 1·87	G. Milne & Co.
	912	V	Hindostan	Welsh	W S	Chilian	1547	217·9	39·3	24·1	St. John, N.B.	1874	Nevins, Fras'r & C.	Y 10·81	Castle Maderas & [illegible]
	913		Hindostan (PONTOON)	Cornu	W Bk	French	444	135·4	29·4	18·2	Bordeaux	1865	Raymond frères & fils.	Y [illegible]	J. P. Lestonnat & Co.
	914	V	Hinrich	Nagel 87-	W Bn	German	263	109·2	23·3	12·1	Bardenfleth	1872	Gebr. Focke	Y [illegible]	J. A. Reinecke
	915	V	Hinrich	Janssen -85	W Bn	German	164	97·4	23·0	10·5	Grossefehn	1885	T. J. Cassens	Y [illegible]	T. J. Cassens
	916	V	Hinrika	Eckhoff	W Sr	German	124	85·0	20·0	9·8	Grossefehn	1877	Schapp & Meyer		B. H. de Freese
	917	V	Hinrike (FOUNDERED)	Fock 80-	W Bg	German	282	111·0	26·3	12·8	Rönnebeck	1871	H. Oltmanns [Wwe	Y [illegible]	Hein Fock
273	918	A D V	Hiperion	Zar -83	W Bk	Austrian	985	156·5	31·9	26·5	Lussinpiccolo	1879	N. Martinolich	Y [illegible]	N. Zar fu P.
10568	919	✻	Hippolyta	Gerrard 80-81	I S	British	852	192·0	32·2	20·0	Port Glasgow	1856	J. Reid & Co.		J. Heap & Sons
42958	920		Hippolyte (FOUNDERED)	Pyet	W Bn	British	189	107·8	23·2	12·5	P. E. Island	1862	Longworth		R. Rudd
	921		Hippolyte Marie		W Bn	French	132	85·6	24·6	11·5	St. Pierre-[Miquelon	1875	P. Beroult		H. Lecharpentier
73943	922		Hiram		W Barge	British	134	104·0	22·0	7·0	Ottawa	1876	G. Gariebois		Simon T. Easton
	923	V	Hiram	Jensen	W Bn	Danish	166	95·8	23·5	10·1	Kolboda	18 65/68	C. Olsen		N. W. Thomsen
	924	N	Hiram	Nielsen	W Bk	Norweg'n	376	111·7	29·8	16·8	Grimstad	1867	A. Gunnulsen		A. Löversen
	925		Hiram	Pålsson -86	W Bg	Swedish	238	116·1	24·9	12·9	Oscarshamn	1858	D. R. Andersson		B. M. Andersson
95491	926	R V	Hiram Emery	Gorham (7)-81	W Bk	American	759	161·2	33·2	21·3	Kenneb'nk Me.	1877	N. L. Thompson	Y [illegible]	J. S. Emery & Co.
95097	927		Hiram Tucker	Knowlton	W Sr	American	136	88·0	25·7	7·2	Dennisville Me	1868	Watt		N. S. Allen
	928		Hirondelle	Garnier	W Sr	French	139	91·2	21·0	11·5	Dunkirk	1873	E. Sagary		L. Verry
	929	V	Hirondelle	Bertho	W Sr	French	120	83·7	23·0	10·9	St. Malo	1884	Tranchemer		Bertho frères
	930		Hirondelle	Agniéray 85-86	W Sr	French	126	95·1	22·3	10·9	Gravelines	1878	H. Collin		H. Torris
	931		Hirondelle (ex [illegible])		W Sr Yt	French	105	103·5	20·6	11·5	Gosport	1862	Camper [& Nicholson		H.S.H. Prince of Monaco
230	932		Hisago Maru		W Bg	Japanese	246	115·1	23·8	12·7	Nova Scotia	1868	Pettes		Sugimura Shotaro
2134	933	A	Hitar	Suich	W Bk	Austrian	506	132·8	29·2	18·9	Baccari	1876	Mattergiian	Y [illegible]	P. Randich fu M.
	934	G	Hjalm	Svensson -87	W Sr	Swedish	122	91·6	20·1	10·2	Thurö	1862	N R Troenseg'rd		C. Svensson
	935	V	Hjalmar	Jørgensen	W Sr	Danish	143	102·0	23·2	11·3	Svendborg	1883	P. Troensegaard		F. H. Jorgensen
	936	N	Hjalmar	Brunn 70-76	W Bk	Norweg'n	360	139·4	28·0	14·9	Sandsvall	1857		Y 12·82	P. Bruun
	937	N	Hjalmar (ex Wilhelmine)	Florenæs -88	W Bg	Norweg'n	252	107·6	25·6	13·4	Wolgast	1865	W. Lübke		C. B. Florenæs
	938		Hjalmar	Andersson -83	W Sr	Swedish	135	82·3	22·3	10·5	Ronneby	1847	A. J. Christenson		C. Andersson
	939	V	Hjemmet (ex Heimath)	Ingemundsen (-	W Bk	Norweg'n	380	135·1	26·5	13·7	Bremerhaven	1865	Joh. Lange	Y [illegible]	T. Meling
	940	N	Hjorleif	Hauge	W Sr	Norweg'n	128	96·3	24·2	10·7	Bergen	1882			C. Brække & S. [illegible]
82257	941		Hline		I Flat	British	789	195·5	25·6	8·7	Port Glasgow	1879	Robt. Duncan & Co.		Irrawaddy Flotilla Co.
	942	N	Hnos	Eliassen 80-80	W Bg	Norweg'n	264	107·7	27·2	13·7	[illegible]	1866	T. Langström		Krog & Henschien
	943		Ho Goan (ex Citadel)	Tan Soen Yjoe 80-	W 3m Sr	Dutch	252	110·0	25·0	11·0	[illegible]	1869	Stewart	Y [illegible]	Oei Hoët King
	944		Ho Soen Goan	Oei Tjong Kie -86	W Bk	Dutch	301	..	..	..	Java	1835			Oei Boen Goan
	945		Hoat Sing Batavia (ex Hab Goan)	Lim Kwee Han	W Bk	Dutch	277	120·3	26·9	..	[illegible]				Oey Thee [illegible]
81154	946	✻	Hobah	Cock	W K	British	71	78·6	19·3	9·0	Trellew	1879	Gray		T. Gray
65047	947	✻	Hodbarrow Miner	Griffiths	W Sr	British	108	89·2	20·6	10·5	Ulverston	1871	Charnley		W. Thomas
547	948		Hodsumi Maru		W Sr	Japanese	131	88·8	19·9	10·5	Kagoshima	1881	Komori [Kamekich		Hamasaki Heiz[illegible]
	949		Hoffnung (MISSING)	Niemann 85-86	W Bk	German	372	120·0	28·5	16·9	Rostock	1868	J. Möller		W. Maack
	950		Hoffnung	Wallis	W Sr	[illegible]	340	124·9	27·3	16·2	Barth	1836	R. Schloer		J. C. Berg

✻ = Lloyd's Register; A = Veritas Austro-Ungarico; D = Nederlandsche Vereeniging; G = Germanischer Lloyd; H = Veritas Ellenico; I = Registro Italiano; N = Norske Veritas; [illegible]

In the 4th Column, C = Composite (iron frame, wood planked); I = Iron; Stl = Steel; W = Wood.

The Dates of Metalling (Col. 11) are based on the records of the Books in which Vessels are respectively classed, or on the Reports of the [illegible]

Further particulars of Vessels classed by Lloyd's Register will be found in Lloyd's Register Book and Lloyd's Yacht [illegible]

No.	Reg.	Nom. Matériaux et Gréement. Capitaine. Ancien Nom.	Pavillon	Tonnage	Longueur	Largeur	Creux	Construit à	Quand	Par qui	Doublage	Armateurs ou Directeurs-Gérants.	Port d'Armement.
951	V	Hoffnung *Rakke* W Bg	German	237	101·3	27·1	13·7	Barth	1875	R.Schlor		W.A.Sarnow	Barth
952	V	Hoganas *Nilsson* W3mSr	Swedish	247	115·4	25·3	12·2	Gefle	1871	P.Bergfors		S.N.Hornuhl	Höganäs
953	❋	Hoghton Tower *Partridge* I S	British	1598	217·0	40·1	23·7	Birkenhead	1869	G.R.Clover&Co.		North Western Shipping Co. Ld.	Liverpool
954		Hohenzollern *Dirks* W Bk (MORGEN ST)	German	372	116·4	28·6	16·8	Greifswald	1854	L.Wittenberg		C.Graedener	Greifswald
955	V	Hoiden *Augustin* W S	Norweg'n	751	173·2	33·6	18·8	Arendal	1878	Th.Ombolt	Y 10.86	H.Biorn,jun.	Kragerö
956		Hoken *Andersson* W Sr	Swedish	106	97·4	20·5	7·4	Karlstad				A.N.Andersson	Jönköping
957	V	Holger *Ibsen* W Sr	Danish	101	86·7	21·3	9·9	Doderhult	1854	H.D.Carstensen		H.A.Clausen	Copenhagen
958	G	Holger Danske W Bn	Danish	117	83·6	21·0	11·5	Faaborg	1855	R.Dyreborg		R.Holbeck & Son	Odense
959	❋	Holkar *Ellery* 79-89 Stl Bk	British	3009	329·3	45·2	26·7	Belfast	1888	Harland & Wolff		T.&J.Brocklebank	Liverpool
960	❋	Hollandia *Seinstra* -87 W Bk (ex Kathe)	Dutch	382	110·0	27·9	14·9	London	1874	Salisbury	Y 10.86	P.H.Craandijk & Zoon	Amsterdam
961		Holly How *Williams* W Sr	British	116	81·6	21·6	11·7	Berwick	1866	Gowan		J.Fisher & Sons	Barrow
962		Holmes *Lauder* W Sr	British	121	87·0	26·8	7·9	Clifton,N.B.	1882	G.H.Flewelling		G.M.Branscombe	St.John,N.B.
963	❋	Holmsdale *Crombie* W Bk	British	1250	206·8	37·7	22·4	Sunderland	1858	J.Reed		G.Milne & Co.	Aberdeen
964		Holsatia *Boje* 75-81 W Sr (ALEXANDER)	German	189	97·1	24·1	13·7	Neustadt	1855	J.H.W.Bernitt		H.C.Boye	Heiligenhafen
965	❋	Holt Hill *Robertson* I(m)Bk	British	2441	300·0	42·2	24·7	Liverpool	1884	W.H.Potter&Ss		Sailing S Holt Hill Co. Ld. (W.Price & Co.)	Liverpool
966	NV	Holtingen *Borge* W Bk	Norweg'n	406	130·0	30·7	16·6	Risör	1874	Jens Jörgensen		A. & F.Smith	Tvedestrand
967	❋	Holyrood *Combie* I S	British	1992	275·0	40·1	24·4	Liverpool	1886	W.H. Potter&Ss		Macvicar,Marshall & Co.	Liverpool
968	❋	Holy Wath *Williams* W Sr	British	99	88·4	22·4	10·5	Amlwch	1872	Thomas		W.Morgan	Whitehaven
969		Homan Maru W Bk	Japanese	402	131·5	25·9	14·6	Hiogo	1879	Hosaku Bun [Kioku		Seiko Kwaisha	Hiogo
970	V	Homborgsund *Tobiesen* W Bk	Norweg'n	404	117·8	29·9	17·4	Grimstad	1865	J.Tellefsen		T.Nielsen	Lillesand
971	A	Homer (ΟΜΗΡΟΣ) *Carvouni* W3mSr	Greek	626	159·4	31·4	20·3	Sestri	1869	A.Briasco fu G.B.		N.Beniamin	Chios
972	V	Homewood *Klaveness* W Bk	Norweg'n	1125	180·0	36·5	22·8	St.John,N.B.	1874	Cruickshank [&Pittfield		Actieselskabet	Christiania
973	V	Honauwar *Lane* W S	British	1619	221·5	41·4	24·4	St John,N.B.	1883	John Fraser		W.Thomson & Co.	St.John,N.B.
974		Hondeklip *Browne* W Bg	British	225	106·0	24·6	14·5	Swansea	1870	Richardson		J.C.& G.S.Richardson	Swansea
975		Hong Liong (BROKEN UP) W Sr	Dutch	104	..	..	..						Batavia
976	GV	Hongkong *Seeger* W Bk (ex St. Magnus)	German	385	135·8	26·8	14·4	St.Magnus	18^{63}_{79}	D.Oltmann'sWwe		M.Seeger	Stettin
977		Honkator *Pöntynen* W Sr	Russian	116	79·3	22·1	10·3	Säkkjärvi	1874	M.Kolkonen		A.Pöntynen	Säkkjärvi
978	V	Honolulu *Leary* W S	British	1558	221·2	41·3	24·3	St.John,N.B.	1882	J.Fraser		W.Thomson & Co.	St.John,N.B.
979	N	Honor *Christensen* W Bk	Norweg'n	351	130·5	28·1	14·5	Arendal	1876	Th.Omholt		Actieselskabet "Honor" (Cas. Lyde & Hein)	Arendal
980		Honorina *Lastres* W Bn (ex Maria Rosa)	Spanish	218	113·8	23·5	12·3	Blanes	1856			Viuda & Hijos de Canosa	Gijon
981	RV	Hoogly *Lewis* W S	American	1101	176·0	37·5	23·7	East Boston	1862	J.A.& G.T.Sampson	Y 12.87	C.Wakefield & Co.	Boston,Mass.
982	DV	Hoop *Lukkien* 66-66 W Sr	Dutch	117	83·3	21·4	12·1	Wildervank	1866	R.G. van de [Werf&Son		G.Herwig	Veendam
983	❋	Hope *Griffiths* W Bk	British	537	152·2	30·2	17·9	Sunderland	1871	J.Crown		G.Brownell	Liverpool
984	❋	Hope *Evans* W Bk	British	276	116·0	26·3	12·9	Quebec	1865	Labbie		J.Evans	Aberystwith
985		Hope *Gardener* 76-86 W Bn	British	143	93·5	22·6	11·9	Bosham	1867	Apps		E.Collins	Portsmouth
986	V	Hope *Spalding* W Bn	British	115	82·0	19·5	11·2	Aberdeen	1848			J.Spalding	Arbroath
987	❋	Hope *Horan* 76-76 W Sr	British	85	71·5	19·7	10·7	Bideford	1849			J.Horan	Newport
988	❋	Hope *Watkins* 65-65 W Sr	British	55	57·5	17·5	9·5	P.E.Island	1821			W.Watkins	Aberystwith
989		Hope *Killeran* -88 W Bk	American	750	153·3	32·1	21·2	Bucksport,Me	1862	J.Dunham		C.L.Taylor	San Francisco
990		Hope Haynes *Sawyer* W3mSr	American	295	108·0	30·0	10·7	Wiscasset, Me.	1880			J.M.Haynes	Bath,Me.
991		Hopeful W Lorcha	Chinese	216	111·4	26·3	12·8	Padstow	1876	Stribley			
992		Hoppet *Vassvik* W Bk (ex SORRENTO)	Norweg'n	522	121·6	33·1	19·8	Åbo	1827			P.Semb	Laurvig
993		Hoppet *Granqvist* W Bk	Russian	566	116·9	29·2	16·2	Raumo	1878	G.Gustafsson		G.Granlund	Raumo
994		Hoppet W Bk	Russian	471	156·1	31·8	16·4	Wasa	1859			C.G.Wolff	Wasa
995		Hoppet *Eriksson* W Sr	Russian	131	..	..	..					E.Söderlund	Finstrom
996	V	Hoppet *Andersson* -85 W3mSr	Swedish	251	116·8	24·6	11·7	Gefle	1878	P.Bergfors&Sn		M.Eriksson	Häfverö,Rslgn.
997		Hoppet *Knutsson* -87 W Bkn (ex SWORD)	Swedish	236	97·1	24·9	12·3	Roslagen	1875			H.T.Knutsson	Strömstad
998		Hoppet *Andersson* -86 W Sr	Swedish	169	89·3	23·7	10·7	Vätö	1878	C.Sundberg		P.Pettersson	Skeppsmyra
999	G	Hoppet *Jonsson* -84 W Sr (ex Rosalie)	Swedish	141	77·1	21·6	12·0	Wismar	1840	J.C.Maren		O.Eckerman	Helsingborg
1000	R	Horace P. Shares *Dobson* W3mSr	American	393	132·2	35·0	11·8	FairHvn,Conn	1882	H.H.Hanscom		J.S.Hemingway	New York

❋ = Lloyd's Register; D = Nederlandsche Vereeniging; G = Germanischer Lloyd; I = Registro Italiano; R = American Record;
A = Veritas Austro-Ungarico; H = Veritas Ellenico; N = Norske Veritas; V = Bureau Veritas.

Dans la 4me Colonne, C = Composite (membrure en fer, bordé en bois); I = Fer; Stl = Acier; W = Bois.

Les Dates de doublage en métal (Col. 11) ressortent des livres dans lesquels les navires sont respectivement classés ou des rapports des Inspecteurs du Lloyd's Register.

Pour plus amples renseignements sur les navires classés par le Lloyd's Register, voir le Lloyd's Register Book et le Lloyd's Yacht Register.

Official No.	No. in Book	Registry in which classed	Ship's Name. Material and Rig. Master, and Dates of Appointment to present Owner's Service and to Ship. Ship's Late Name.	Flag	Net Tonnage	Length	Breadth	Depth	Built Where	When	By whom	Date of Metalling	Owners or Managers
[illegible]	1001		Horatio *Perry* W3mSr	American	388	135·1	32·2	13·7	Bath,Me.	1880	Goss,Sawyer &Packard		S.T.Perry
[illegible]	1002	R	Horatio *Morse* WS	American	332	115·7	28·6	17·5	PJeffersonNY	1877	J.M.Bayles&Son		Tabor & Gordon
[illegible]	1003	R	Horatio L. Baker *Crowell* W3mSr	American	787	180·0	35·8	18·1	Bath,Me.	1888	G.G.Deering		G.G.Deering
[illegible]	1004		Horatio Nichols *Rudd* W3mSr	American	171	117·0	26·0	8·0	NewHvn,Conn	1865			J.H.Cannon
[illegible]	1005		Horizon *Leek* WSr	American	104	102·2	21·7	7·2	Stamford,Conn	1857	Wite		J.Leet
	1006	✱V	Horizon *Gautier* 88-88 IBk	French	1054	209·4	34·0	20·0	Londonderry	1888	C.J.Bigger		Estier frères
	1007	GV	Horizont *Lange* -79 WBn	German	163	98·0	23·0	12·0	Papenburg	1871	H.A.Klasen	[illegible]	H.A.Klasen
[illegible]	1008	✱	Hornby Castle *Smith* IS (ex Duncan Campbell) 188[illegible]	British	1436	233·7	37·9	22·9	Liverpool	1877	T.Royden&Son		Lancaster Shipowners Co.Ld.
[illegible]	1009	V	Hornet *McDonald* WBkn	British	407	140·0	30·0	13·0	Carleton,N.B.	1882	E.McGuiggin	[illegible]	H.D.T[illegible]
[illegible]	1010		Hornet WSr	American	122	84·4	23·8	8·5	HarringtonMe	1858			[illegible]
[illegible]	1011	✱D	Horsa *Wilson* 82-82 IS	British	1128	220·0	34·2	21·7	Greenock	1860	Scott&Co.		StarNavigationCo. [illegible]
	1012		Hortencia WSr	Brazilian	155								F.J.da Costa
	1013		Hortencia WSr	Brazilian	155	..	..	..					A.P.F.Liberato
	1014		Hortencia *Honten* WBn	Brazilian	155	..	..	..	United States	1871		[illegible]	G.Assburg
	1015	V	Hortense *Loreau* WSr	French	151	84·3	23·6	11·2	Means	1867	Olliveaud		Labour
	1016		Hortense W	French	146	..	..	..	Dunkirk	1864			Ledret frères
[illegible]	1017		Hortensia *Sanborn* 68-78 WSr	American	163	102·5	28·1	8·9	E.Machias,Me.	1868	S.W.Pope		W.Pope & Sons
	1018	V	Hortensia *Lindgaard* 63-88 WBn	Danish	139	94·6	21·4	10·6	Wennersborg	1860			L.C.Lindgaard
	1019	N	Hortensia *Simonsen* WBk	Norweg'n	795	150·8	32·6	22·6	Medford,Mass.	1853		[illegible]	Louis Vauvert
[illegible]	1020		Horton *Porrier* 87-87 WSr	British	122	82·0	24·3	10·2	Maitland,N.S.	1864			G.E.Bosk
[illegible]	1021		Hosaki Maru WBg	Japanese	188	99·5	21·6	12·2	Kagoshima	1879	K.Kamekichi		Sawada Giheiji
[illegible]	1022	✱	Hospodar *Lloyd* IS	British	1625	249·8	38·8	23·7	Sunderland	1875	T.R.Oswald		Hargrove & Hel[illegible]
	1023		Hostel WBg	British	102	70·7	17·2	10·2					Ahmed Lebah
	1024		Hösten *Amundsen* -83 WBk	Norweg'n	546	145·6	30·6	17·9	Grimstad	1879	N.Osmundsen		J.D.Havstad
[illegible]	1025		Hotha StlLr	British	480	220·0	30·0	8·4	Dumbarton	1887	W.Denny&Bros.		Irrawaddy [illegible]
	1026	V	Hother *Jensen* 84-84 WSr	Danish	135	88·0	21·6	11·2	Horsens	1875	C.Schröder		Joh.Gaarn
[illegible]	1027	R	Hound *McEachrean* -87 WSr	British	132	93·1	25·0	9·8	Chester,N.S.	1885	A.C.Walker		G.E.deWitt
	1028	N	Hövding (ex Gimson) *Beck* WS	Norweg'n	1192	199·8	36·9	23·2	NewYork	1852		[illegible]	Joh.Thorne
	1029	N	Hövding *Engelstad* 83-83 WS	Norweg'n	1096	199·4	37·6	21·6	[illegible],Me.	1854			Actieselskabet [illegible] (S.[illegible])
	1030	N	Hövding *Reynholds* WBk	Norweg'n	759	168·2	34·3	19·7	[illegible]	1868	Jensen	[illegible]	J.C.& G.K[illegible]
	1031	N	Hövding *Olsen* WBk	Norweg'n	452	135·8	30·9	16·7	[illegible]	1876	M.Ekkrup		[illegible]
[illegible]	1032	✱	Howard *Vance* 78-78 WBk	British	1063	188·8	34·0	21·7	[illegible]NS	1878	W.P.Cameron		John [illegible]
[illegible]	1033		Howard *Pitkeikly* -86 WBk	British	580	148·0	32·2	18·8	[illegible],N.S	1864		[illegible]	J.P[illegible] & Co.
[illegible]	1034	✱	Howard *Rendle* WK	British	50	63·6	17·3	8·3	[illegible]	1884	J.Mills&Co.		M.Tremayne
[illegible]	1035		Howard *Dodd* WSr	American	119	92·5	28·5	7·9	San Francisco	1881			R.G.Byxbee
[illegible]	1036		Howard A. Hunt *Tubbing* W3mSr	American	322	118·5	29·7	9·6	[illegible],NJ	1867	Vansaman		E.L.Ross
[illegible]	1037	V	Howard A. Turner *Putnam* -88 WBk	British	504	158·5	33·5	13·2	Harvey,N.B.	1886	G.S.Turner	[illegible] 11.87	J.N.Smith
[illegible]	1038		HowardH.Hanscom *Holmes* -87 W3mSr	American	740	171·0	37·0	18·0	FairHaven	1887	H.H.Hanscom		H.H.Hanscom
[illegible]	1039	R	Howard W. Middleton *Smith* W3mSr	American	560	145·1	35·0	15·2	Camden,N.J.	1883	S.W.Tilton		David Lee
[illegible]	1040		Howard Williams *Foster* W3mSr	American	362	135·8	33·1	11·4	Wilmngtn,Del	1872	Jackson&Sharp Co.		D.S.Williams
[illegible]	1041	✱	Howden *Evans* 86-86 IBk	British	1163	198·6	34·6	22·0	Hull	1855	C.&W.Earle		Howden Ship [illegible]
[illegible]	1042	✱	Howrah *Fish* 87-87 IS	British	1098	207·7	33·7	21·2	Sunderland	1864	Pile,Hay&Co.		Tyser & Co.
[illegible]	1043		Hoyei Maru WBg	Japanese	232	116·0	24·2	12·0	Denmark	1866			Okada Matabei
[illegible]	1044		Hozumi Maru WSr	Japanese	132	88·8	19·6	10·6	Osumi	1881	KomoriKane-kichi		Imaye Seibie
[illegible]	1045	A	Hrabren *Fiscovich* WBk	Austrian	502	138·1	26·4	16·9	Fiume	1869	Schiavoni	[illegible]	M.Fiscovich
[illegible]	1046	A	Hrvat *Sverljuga* WBk	Austrian	574	140·7	31·5	19·2	Buccari	1875	Materljan	[illegible]	M.Sverljuga
	1047	N	Huano *Höyer* 82-82 WBk	Norweg'n	840	163·6	33·2	21·8	ThomastonMe	1850			T.Höyer
	1048		Huasco *Gillmore* -83 WBk	Chilian	371	131·3	27·8	17·6	Bideford	1860	Cox		J.Condon & [illegible]
[illegible]	1049	✱	Huasquina *Anderson* WBk	British	428	139·8	28·2	17·3	Workington	1864	Fell	[illegible]	Lowden,Edgar[illegible]
	1050		Huda Kerim WBn	Turkish	250	..	..	..	Oonia	1877			[illegible]

✱=Lloyd's Register; D=Nederlandsche Vereeniging; G=Germanischer Lloyd; I=Registro Italiano; R=Am[illegible]
A=Veritas Austro-Ungarico; H=Veritas Ellenico; N=Norske Veritas; V=[illegible]

In the 4th Column, C=Composite (iron frame, wood planked); I=Iron; Stl=Steel; W=Wood.

The Dates of Metalling (Col. 11) are based on the records of the Books in which Vessels are respectively classed, or on the Reports of the Surveyors [illegible]

Further particulars of Vessels classed by Lloyd's Register will be found in Lloyd's Register Book and Lloyd's Yacht Register.

No.	Cote	Nom. Matériaux et Gréement. Capitaine. Anciens Noms.	Pavillon	Tonnage net	Longueur	Largeur	Creux	Lieu de construction	Année	Constructeurs	Doublage	Armateurs ou Directeurs-Gérants.	Port d'Armement.
1051	✻	Huda Verdi *Halil* WBk	Turkish	702	160·7	[illegible]	21·7	Sestri Ponente	1870	[illegible]	[illegible]	Coramanian Shakin-Oglu & Co.	Constantinople
1052		Huda Verdi WBg	Turkish	378	..	..	..	[illegible]	1866			Mehemed Habib zadeh Noureddin & Co.	Constantinople
1053	A	Huda Verdi *Kiani*	Turkish	260	..	..	..	[illegible]	1882			Haggi Sefki	Constantinople
1054		Huda Verdi *Ismael* WBn	Turkish	224	..	..	..	[illegible]	1882			Hajee Shefki Effendi	Sakarieh
1055		Huda Verdi *Hussein* WBmb	Turkish	220	..	..	..	[illegible]	1885			Ahbahnahlee Ahmed and Brother	Ahbahnah
1056	A	Huda Verdi *Hassan* WBk	Turkish	217	..	..	..	[illegible]	1872			Mehmet Effendi	Constantinople
1057		Huda Verdi *Halil* WBg	Turkish	210	..	..	..	[illegible]	1884			Hassan Pasha zadeh Hassan Bey	Constantinople
1058	H	Huda Verdi *Houssein* WBk	Turkish	210	..	..	..	[illegible]	1874			Halil Houssein	Constantinople
1059		Huda Verdi *Ismael* WBg	Turkish	198	..	..	..	[illegible]	1884			Kerentin Mehemed bin Yusuf Effendi	Constantinople
1060		Huda Verdi WBn	Turkish	178	..	..	..	[illegible]	1878			Ahmasgrahlee Hajee Pire Oglou Mehemed Ali & Co.	Ahmassarah
1061		Huda Verdi WSr	Turkish	162	..	..	..	[illegible]	1873			Enwizadeh Hajee Ahmed Bros.	Aidos
1062		Huda Verdi WBg	Turkish	160	..	..	..	Ilisk	1870			Tekkehaishui zadeh Mehemed Ali & Co.	Ilish
1063		Huda Verdi *Osman Ibrahim* WBn	Turkish	160	..	..	..	Ilisk	1884			Tekkimshut Oglou Mehemed & Co.	Ilish
1064		Huda Verdi *Mehemed Ali* WBg	Turkish	145	..	..	..	[illegible]	1885			Hussein Ahmed & Co.	Constantinople
1065		Huda Verdi *Hussein* WBg	Turkish	144	..	..	..	[illegible]	1884			Ahbahnahlee Hajee Ahmed & Bros.	Ahbahnah
1066		Huda Verdi WSr	Turkish	125	..	..	..	Guetch Güin	1882			Ooneapaninda Tujar Hussein Effendi & Co.	Guetch Giin
1067		Huda Verdi *Mehemed Ali* WSr	Turkish	125	..	..	..	Guetch Güin	1885			Hattab Tijari Mehemed Effendi & Co.	Guetch Giin
1068		Huda Verdi *Ismael* WBg	Turkish	125	..	..	..	[illegible]	1886			Hettab Tijari Varnalee Ismael & Co.	Varna
1069		Huda Verdi WBg	Turkish	119	..	..	..	[illegible]	1884			Cappandakkik Tijari Ibrahim Hassan Effendi	Constantinople
1070		Huda Verdi WBg	Turkish	116	..	..	..	[illegible]	1883			Mokamdji Rejeb Oustah & Co.	Constantinople
1071		Huda Verdi *Yanni* WBg	Turkish	110	..	..	..	[illegible]	1885			Yeikenji Hajee Ibrahim bin Moussa	Constantinople
1072		Hudavendighiar *Kya'mil Reis* WSr	Turkish	100	..	..	..	Barten	1881			[illegible] Hasan Suleiman Effendi	Barten
1073		Hudavendighiar *Kyamil* WSr	Turkish	100	..	..	..	Barten	1884			[illegible] Tijari [illegible]	Barten
1074	DV	Hudiksvall *Bjönnd* WBk (ex François I.)	Swedish	1363	212·2	32·6	24·9	[illegible]	1867	Lucas frères	V 2·87	C. Almgren	Hudiksvall
1075	✻	Hudson *Kemp* 85-85 IBk	British	797	194·2	30·6	19·4	Stockton	1869	M. Pearse & Co.		Shaw, Savill & Albion Co. Ld.	Southampton
1076		Hudson *McRae* WSr	British	100	75·6	21·0	9·8	New Richm'nd	1872			P. McRea	New Carlisle
1077	RV	Hudson *Curtis* -87 WBk	American	637	149·0	32·5	17·7	Bucksport Me	1867	S.B. Swazey & Co.	Y 1885	R.P. Buck & Co.	New York
1078		Hueneme *Elliot* -77 W3mSr	American	347	142·3	34·2	10·7	Pt Ludlow Wsh	1877	Hall Bros.		W.G. Hall	San Francisco
1079	DV	Hugh Cann *Bent* -86 WBk	British	1074	184·0	37·0	22·0	Salmon Riv NS	1876	J.T. Allen	V 2·87	N.B. Lewis	Yarmouth, N.S.
1080	R	Hugh Fortescue *Gundersen* -87 WBk	Norweg'n	480	139·0	28·6	18·9	Topsham	1865	Holman	Y C 85	A.M. Geelmuyden	Arendal
1081	✻	Hugh Roberts *Edwards* 1-1887 WBg	British	181	104·7	23·6	13·0	Portmadoc	1875	Williams	Y C 88	H. Roberts & Son	Carnarvon
1082	N	Hugin *Beckeman* WBk	Norweg'n	396	130·3	29·7	16·9	Kragerö	1867		Y 7·84	J. Dahll	Kragerö
1083	DV	Hugo *Köhler* -88 WS (ex W. [illegible])	German	1515	214·9	41·3	24·3	Portsmouth	1875	D. Marcy	Y 88	G. Hilder	Bremen
1084	V	Hugo *Bärhofen* -87 WSr	German	111	78·4	21·6	10·6	Stralsund	1864	C.W. Mohr		Franz Böttcher	Stralsund
1085	V	Hugo *Sala* WBn	Spanish	163	83·2	26·2	12·9	Vinaroz	1840		Y 84	A. Pujol & Co.	Ibiza
1086		Hugo *Ohlson* 75-85 WBg	Swedish	251	113·1	21·6	12·2	Wärnanäs	1873	C.A. Hahn	Y 86	A.R. Andersson	Calmar
1087	GV	Hugo Georg *Wilde* WBg	German	194	100·5	24·1	12·3	Stettin	1862	F. Zieske		W.F. Wilde	Stettin
1088		Hulda *Sandell* WSr	Russian	192	90·8	26·7	11·1	Norrlångsvik	1874	P. Eeklund		A. Jansson	Kimito
1089	V	Hulda *Olsson* -88 W3mSr (ex San José)	Swedish	229	106·1	22·6	12·5	Libau	1858	G. Backhoff		P. Olsson	Viken Malmöhusän
1090		Hulda WSr	Swedish	125	76·1	21·8	11·0	Denmark				J. Svensson	Alem, Calmar
1091		Hulda Erika *Andersson* -88 WSr	Swedish	123	81·0	20·4	11·2	N. America	18..			E. Åhrlund	Öregrund
1092	N	Huldren *Stangeby* -88 WBk	Norweg'n	367	123·5	29·3	15·9	Guadaloupe	1856			R.M. Peterson	Moss
1093		Humaeyee-Bahri WSr	Turkish	100	..	..	..	Gherzeh	1874			Gherzehlee Ali & Co.	Gherzeh
1094	R	Humarock *Thatcher* W3mSr (ex Isaac Carlton)	American	432	133·5	33·9	11·3	Columbia Falls, M.	18 83/87	I. Carlton	Y 1887	J. Baker & Co.	Boston, Mass.
1095		Humayoon Shah WS	British	872	142·0	32·0	23·0	Cochin	1852		[illegible]	E.D.J. Ezra	Calcutta
1096	✻V	Humboldt *Jensen* 83-88 IS	German	719	174·6	30·2	18·4	Glasgow	1867	A. Stephen & Sns		R.M. Sloman & Co.	Hamburg
1097	DGV	Humboldt *Meyer* -87 WBk	German	329	108·7	26·1	12·6	Elsfleth	1869	J. Albers	Y 88	A.H. Arnold	Brake
1098	V	Humildade *Frzeira, jun.* WBk	Portug'se	297	117·4	24·3	14·4	Oporto	1866	C.J.A. Vicente	Y 84	Cia Alliança Portuense	Maritimo
1099	✻V	Humor *Behn* I3mSr (ex La Bretagne, ex The Gem)	German	200	110·3	23·8	11·3	Inverkeithing	1878	Inverkeithing Foundry Co.		Wilken Stehr	Hamburg
1100		Humpen *Maybrun* 75- WS	Russian	114	85·0	21·0	10·0	[illegible]	1875			R. von Stähl-Holstein	Pernau

✻ = Lloyd's Register; D = Nederlandsche Vereeniging; G = Germanischer Lloyd; I = Registro Italiano; R = American Record; A = Veritas Austro-Ungarico; H = Veritas Ellenico; N = Norske Veritas; V = Bureau Veritas.

Dans la 4me Colonne, C = Composite (membrure en fer, bordé en bois); I = Fer; Stl = Acier; W = Bois.

Les Dates du doublage en métal (Col. 11) ressortent des livres dans lesquels les navires sont respectivement classés ou des rapports des Inspecteurs du Lloyd's Register.

Pour plus amples renseignements sur les navires classés par le Lloyd's Register, voir le Lloyd's Register Book et le Lloyd's Yacht Register.

Official No.	No. in Book.	Registry in which classed.	Ship's Name. Master, and Dates of Appointment to present Owner's Service and to Ship. Ship's Late Name.	Material and Rig.	Flag.	Net Tonnage.	Dimensions. Length	Breadth	Depth	Built. Where.	When.	By whom.	Date of Metalling.	Owners or Managers.
	1101	H	Hundi Verdi	WBk	Turkish	370	..	..	..	England	1855			M.Lazaron
11317	1102		Hunter	WBk	American	337	124·0	28·2	17·8	Gardiner,Me.	1854			F.M.Green
[illegible]	1103	R	Hunter *Whitten -83*	WSr	American	187	105·2	28·1	9·2	Orland,Me.	1876	J.A.Buck		W.S.White
[illegible]	1104		Huntleys *Allen 88-85*	WBn	British	185	92·4	24·1	13·4	Sunderland	1865	J.Robinson		G.H.Gann
53882	1105	✻	Huntly Castle *Nicola*	CBk	British	624	172·0	29·0	18·0	Glasgow	1866	C.Connell&Co.	[illegible]	Hajee Mahomed … & Co.
45172	1106		Huntress *Gunn*	WBkn	British	227	132·8	24·5	12·4	Salcombe	1862	Vivian	[illegible]	Brazilian Coal Co.Ld.
37059	1107		Huntress *Hasley*	WBn	British	107	75·1	21·6	11·0	Fox Harbour, (N.S.	1847			C.Marrow
71202	1108		Huron	WSr	British	234	123·2	23·5	10·3	Ontario	1874	D.McMurchie		C.Lobb
73384	1109	✻	Hurunui *Sinclair 86-87*	I Bk	British	1012	204·1	34·2	20·0	Newcastle	1875	Palmers Co.,Ld.		New Zealand S.Co.…
	1110	N	Huseland *Jorgensen*	W3mSr	Norweg'n	232	112·2	27·5	11·6	Grimstad	1866	PederHuseland		H.Omundsen
87854	1111	✻	Hutton Hall *Slater 81-84*	I S	British	1989	277·0	40·1	24·2	Liverpool	1884	W.H.Potter&Ss		Globe Shipping Co.Ld.
	1112		Hvalfisken *Bang 84-85*	WBg	Danish	195	88·8	21·2	14·7	Calmar	1801			Den [illegible]
	1113	N	Hvide Stjerne *Andersen -88*	WSr	Norweg'n	165	94·5	23·0	12·0	P.E.Island	1857			A.Kröger
	1114	N	Hvideörn *Andersen -88*	WS	Norweg'n	562	148·6	33·2	17·8	Christiania	1869	C.Brinch	V 11.88	A.M.&T.M.Wiel
26155	1115		Hyacinth	WBk	British	344	128·6	24·0	15·0	Troon	1859			Asgar Ally
83481	1116	R	Hyaline *McLeod -81*	WBn	British	187	101·7	27·0	10·8	Brooklyn,N.S.	1881	J.Gardner	V 10.88	J.V. & J.H.Dexter
41527	1117		Hyder Hoosain (WRECKED)	WSr	British	228	95·6	24·0	12·6	Moulmein	1885			Fakeer Mahomed
87420	1118	✻	Hydorabad *Malcolm 86-85*	I Bk	British	1120	215·2	35·1	21·2	PortGlasgow	1885	Russell&Co.		J.Russell (W.&J.Crawford)
72574	1119		Hyderabad	WSr	British	290	124·4	26·2	11·7	Kingston,Ont.	1876	W.Power&Co.		Wm.Power
	1120		Hydra	WSr		155	82·0	17·1	8·5	Sappemeer	1863	J.G.Bodewes		
	1121	V	Hydra *Christensen 74-88*	WBk	Danish	786	161·0	30·8	20·0	Bremerhaven	1876	H.Ulrichs	[illegible]	J.Harboe
	1122	V	Hydra	WBg		292	100·7	28·3	16·4	Syra	1874			
	1123	N	Hydra *Henriksen*	WBg	Norweg'n	198	105·6	25·2	11·9	Arendal	1872	A.Aamundsen		H.H.Pettersen
89886	1124	✻	Hydrology *Morgan*	WSk	British	83	77·5	20·5	10·5	Elmshorn	1886	J.Kremer		M.Carter
41035	1125		Hydrose (ex Phœnix)	WBg	British	214	94·1	21·1	13·1	Bowmen				S.V.Mayandee Chetty …
81074	1126		Hydrose *Noor Mohamedo -87*	WBg	British	163	80·5	20·0	12·2	Negapatam	1859			Chana Allapitchey …
	1127		Hydross	WBg	British	134	73·0	18·0	12·0	Cochin	1845			Cootty Assen Cooty …
11621	1128		Hyena *Gardner*	WSr	American	117	90·0	25·0	8·0	St.George,Me	1851			R.Long
	1129		Hygæa *Nielsen*	WSr	Danish	162	95·0	22·2	11·3	Nexo	1859			J.A.Larsen
	1130	N	Hygea *Eilertsen* (ex Fort de France)	W3mSr	Norweg'n	297	103·9	27·2	15·6	St.Tropez	1855	Blaignière		J.Eilertsen
	1131	N	Hygiea *Johnsen 68-76*	WBg	Norweg'n	244	100·0	27·5	13·4	Grimstad	1868	S.Johnsen		J.M.A.Marcussen
54487	1132	R	Hypatia *Bradley -87*	WBk	British	730	151·8	33·2	20·6	Kingston,N.B.	1867	Titus		Francis Tufts
	1133	N	Hypatia *Thomassen 80-85* (ex Florence) (WRECKED)	WBk	Norweg'n	896	169·9	36·1	22·1	Boston,U.S.	1856		V 12.86	Joh.Schanning
67814	1134	✻	Hyperion *Perfect -80*	WSr	British	77	73·2	19·0	10·0	Knottingley	1874	Garlick		G.Perfect
11425	1135		Hyperion *Hadley -87* (COLLISION)	WBn	American	291	110·3	27·1	14·1	Wiscasset,Me	1864		[illegible]	B.J.Willard
	1136		Hypolite B. (ex Jena) (CONDEMNED)	WS	French	548	130·8	27·4	19·3	Savona	1868	Guastavino	[illegible]	H.Bonvard
11775	1137		Hyue *Fletcher*	WSr	American	148	99·9	26·8	7·8	Phippsb'rg,Me	1868			C.V.Minott

Additional Ships are inserted in the Supplement at the end of the Register.

✻ = Lloyd's Register; D = Nederlandsche Vereeniging; G = Germanischer Lloyd; I = Registro Italiano; R = [illegible]; Λ = Veritas Austro-Ungarico; H = Veritas Ellenico; N = Norske Veritas; V = [illegible]

In the 4th Column, C = Composite (iron frame, wood planked); I = Iron; Stl = Steel; W = Wood.

The Dates of Metalling (Col. 11) are based on the records of the Books in which Vessels are respectively classed, or on the Reports of the Surveyors …

Further particulars of Vessels classed by Lloyd's Register will be found in Lloyd's Register Book and Lloyd's Yacht Register.

N°	Registre	Nom. Matériaux et Gréement. Capitaine et dates d'entrée au service des Armateurs actuels et à bord du navire. Ancien Nom.		Pavillon.	Tonnage net	Dimensions. Longueur	Largeur	Creux	Construit. Où.	Quand.	Par qui.	Date du doublage en métal	Armateurs ou Directeurs Gérants.	Port d'Armement.
1		I. A. Horsey	W Bn	[illegible]	1[illegible]	96·5	26·0	11·3	Cornwallis,NS	1874	J.H.Pena		W.Le Gallais	New Carlisle
2		I. Benham *Klaveness* (ex Augustin ex Antonello) (BOOTHS)	W Bk	Norweg'n	[illegible]	[illegible]	[illegible]	[illegible]	[illegible]	1851	H.A.Cooper	Y 10.86	H.Normann	Lillesand
3	*	I. C. U.	W Sr	British	[illegible]	79·5	18·0	8·5	Guernsey	1884	P.Ogier		P.Ogier	Guernsey
4	I	I Due Fratelli *Cerbone* (ex Pasquale)	W Bk	Italian	47[illegible]	136·1	29·4	18·7	Alicante	1874	F.SaverioMauro	Y [illegible]	F.lli Trapani	Castellammare
5		I Due Fratelli Jacono *Jacono*	W Bn	[illegible]	14[illegible]	[illegible]	22·[illegible]	12·[illegible]	Castellamm. di Stabia	1873	A.Bonifacio		G.Libroja	Naples
6	RV	I. F. Chapman *Chapman*	W S	American	2038	237·5	42·7	27·5	Bath,Me.	1882	I.F.Chapman	Y 12.87	I.F.Chapman	New York
7	V	I. F. Whitney *Morris*	W Bk	[illegible]	731	1[illegible]	3[illegible]	19·[illegible]	Parrsboro'NS	1872			G.D.Spicer	Parrsboro,'N.S.
8	N	I. H. Schwonesen *Gunnaulsen*	W Bk	Norweg'n	37[illegible]	[illegible]	[illegible]	[illegible]	[illegible]	1866	A.Johnsen	Y 11.83	Anton Natvig	Kragerö
9	V	I. L. Skolfield *Skolfield*	W Bk	American	1[illegible]	[illegible]	[illegible]	23·4	Brunswick,Me	1878	Skolfield	Y [illegible]	Skolfield Bros.	Portland,Me.
10		I. M. T. Hannum	W Barge	British	15[illegible]	1[illegible]·7	21·9	7·0	Quebec	1876	A.Miller		D.Murphy	Ottawa,Ont.
11		I. N. Oswell	W Barge	British	11[illegible]	1[illegible]	21·7	6·0	Ontario	1875	J.Chaffey		R.Waters	Ottawa,Ont.
12	N	I. P. Berg *Nielsen*	W Bk	Norweg'n	467	137·[illegible]	31·0	16·[illegible]	[illegible]	1874	T.Gundersen	Y [illegible]	B.A.Olsen	Risör
13		I. P. Blake	W Sr	British	116	76·8	25·4	9·3	Parrsboro'NS	1875			Henry A.Eaton	Parrsboro,N.S.
14		I. Savard	W Sr	British	13[illegible]	87·[illegible]	24·5	1[illegible]·5	LesEboulemnts	1875	L.Savard		F.A.Ménard	Quebec
15	I	I Tre Giuseppe *Schiaffino*	W Bk	Italian	3[illegible]	1[illegible]·[illegible]	31·1	18·3	Savona	1869	G.Tixi	Y 10.86	R.Schiaffino	Genoa
16		I. W. Hine *Babbidge*	W Sr	American	128	101·[illegible]	23·7	7·4	N.Haven,Con.	1862	G.W.Baldwin		F.W.Benedict	Newhaven,Con.
17	V	I. W. Oliver *Cogswell*	W Bk	British	752	152·3	34·7	19·5	Granville,N.S.	1873	J.W.Oliver	Y [illegible]	J.E.Shafner	Annapolis,N.S.
18	RV	I. W. Parker *Kane*	W Bn	American	361	124·[illegible]	29·5	14·7	Belfast,Me.	1874	C.Curtis&Co.	Y [illegible]	J.W.Kane	New York
19	N	I. Williams *Madsen -87*	W Sr	Norweg'n	3[illegible]	121·2	29·5	12·8	Advocate,N.S.	1871	W.Moore		P.A.Grön	Sandefjord
20		Iaconos	W Bn	Greek	308	136·4	23·9	14·1	Scotland	1878			H.S.Stathopoulou	
21		Ianira	W Sr Yt	British	105	99·1	22·2	11·3	Lymington	1874	Inman&Son		Capt.T.Hargreaves	Southampton
22	*	Ianthe *Davis 82-82*	W Bk	British	380	135·6	26·0	16·6	Dundee	1858	A.Stephen&Son	Y [illegible]	L.Tulloch	Sunderland
23		Ianthe *Howe*	W Sr	British	174	83·3	20·0	14·3	Yarmouth	1846	J.HFellows&Sn		J.Nellist	Yarmouth
24	A	Iason *Fogari*	W Bk	Greek	467	134·5	29·8	..	Syra	1874	A.Lividaros	Y [illegible]	G.Zografafis	Syra
25	A	Ibar *Lazianovich -89*	W Bg	Austrian	676	134·5	29·8	19·8	Fiume	1878	G.BacciaBellaz	Y [illegible]	G.Papi fu M.	Ombla
26		Ibex	W Sr	British	135	87·5	24·0	10·2	Maitland,N.S.	1865			John Graham	Halifax,N.S.
27		Ibis *Adams*	W Bg	British	357	107·5	29·4	17·5					A.Stuart	Sydney,N.S.W.
28		Ibis *Orsell 84-86*	W Bk	British	249	114·0	24·0	14·5	[illegible]	1861	Peverall		Hy.Rider	Bristol
29		Ibis *Carden 81-81*	W Bn	British	295	110·2	24·4	12·9	P.E.Island	1875	Davies		Holden & Gann	Faversham
30	*	Ibis *Shebden -81*	W Sr	British	89	80·0	20·2	10·0	London	1887	W.Walker	C [illegible]	C.T.Winter	London
31	NV	Ibis *Christensen -87*	W Bk	Norweg'n	456	128·4	31·0	16·9	Arendal	1874	E.Guthormsen	Y 7.88	N.A.Flöystad	Arendal
32		Ibraila	W Bg	Greek	200	..	..	..	Syra	1863			I.Panas	Syra
33		Icamiaba	W Bk	Brazilian	153	..	..	..					Barãode Maccio	Amazonas
34	RV	Icarus *Toye -86*	W Bkn	British	4[illegible]	141·0	30·0	17·6	Windsor,N.S.	1875	E.E.Fuller	Y 10.87	John Keith	Windsor,N.S.
35	RV	Iceberg *Treat -87*	W S	American	1136	175·6	37·0	24·0	E.Boston	1877	J.E.Taylor	Y [illegible]	W.E.Barnes	New York
36	V	Iconia *Janssen 85-82*	W Bg	German	227	108·9	24·9	8·6	Leer	1867	O.Wiemann	Y [illegible]	H.Wiemann	Leer
37	R	Ida *Morrison -88*	W Bn	British	2[illegible]	109·0	28·2	12·4	Liverpool,NS	1872		Y [illegible]	J.D.McClearn	Liverpool,N.S.
38	*	Ida *Resnick*	W Bn	British	199	105·4	24·1	12·9	P.E.Island	1873	McLaughlin	Y 7.80	M.Guy	Padstow
39	*	Ida *Williams*	W Bkn	British	179	103·5	24·2	13·4	Portmadoc	1877	Jones	Y 5.87	H.Williams & Co.	Carnarvon
40	*	Ida *Stone*	W K Barge	British	136	94·2	23·3	9·2	Ipswich	1881	W Bayley&Sons		R.S.&W.F.Paul	Ipswich
41	V	Ida (ex Wave of Life) *Machado*	W Bk	Brazilian	811	187·0	34·5	21·0	Aberdeen	1856	Hood	Y 1.87	Jos.J.Reis,jun.	Rio de Janeiro
42		Ida *de Campos*	W Sr	Brazilian	197	..	..	..	Fairhaven,U.S	1885		Y 10.85	J.Pinto de Campos	Rio de Janeiro
43		Ida (LOST) *Nielsen -87*	W Bg	Danish	267	104·7	24·3	16·1	Christinestad	1859			M.Christiansen	Rønne
44		Ida (ex Polarnaja Swesda)	W Bn	Danish	127	87·2	21·3	10·5	Soma	1856			H.C.Petersen	Æröskjöbing
45	G	Ida (ex Margaretha 1880 Juli) *Lund 80-82*	W Sr	Danish	110	71·6	21·5	10·1	Carentan	18 44/74			Hans Friis	Nyborg
46	DV	Ida (ex Drie Gezusters) *Hazewinkel -88*	W Bk	Dutch	465	125·3	20·0	16·3	Amsterdam	1856	Abbema		K.&J.Wilkens	Veendam
47	D	Ida (ex Elenor) *Wolkammer*	W Bk	Dutch	393	131·1	28·9	16·4	P.E.Island	1866	Ellis		A.H.Wolkammer	Farmsum
48		Ida *Hazelwinkel 64-64* (TO BE BROKEN UP)	W Bn	Dutch	154	90·9	22·2	11·1	Wildervank	1864	J.M.Meihuizen & Zoon		De Groof & Co.	Flushing
49	V	Ida (ex D. H. Watson) *Winters -85*	W S	German	1298	195·8	37·7	24·2	Boston,U.S.	1865	P.Curtis	Y [illegible]	Joh.Friedr.Wessels	Bremen
50	V	Ida *Smith*	W Bn Sr	German	337	119·7	23·6	12·7	Papenburg	1883	H.W.Meyer	Y [illegible]	H.W.Meyer	Papenburg

*=Lloyd's Register; D=Nederlandsche Vereeniging; G=Germanischer Lloyd; I=Registro Italiano; R=American Record; A=Veritas Austro-Ungarico; H=Veritas Ellenico; N=Norske Veritas; V=Bureau Veritas.

Dans la 4me Colonne, C=Composite (membrure en fer, bordé en bois); I=Fer; Stl=Acier; W=Bois.

Les Dates du doublage en métal (Col. 11) ressortent des livres dans lesquels les navires sont respectivement classés ou des rapports des Inspecteurs du Lloyd's Register.

Pour plus amples renseignements sur les navires classés par le Lloyd's Register, voir le Lloyd's Register Book et le Lloyd's Yacht Register.

Official No.	No. in Book	Society in which classed	Ship's Name. Material and Rig. Master, and Dates of Appointment to present Owner's Service and to Ship. Ship's Late Name.	Flag.	Net Tonnage	Dimensions. Length	Breadth	Depth	Built. Where.	When	By whom.	Date of Metalling.	Owners or Managers.
	51		Ida (ex Sta Regina) *Hermann* 81- W Bn	German	161	93·0	21·1	11·6	Papenburg	1856	J. Beckmann		A. Klasen
	52	G	Ida *Blankmann* 81-82 W Sr	German	106	97·4	21·6	9·7	Troense	1866	J. Jensen	[illegible]	Carl Franz Sien[illegible]
1087	53	I	Ida *Gazzolo* W Bk	Italian	495	131·2	31·8	19·3	Varazze	1870	M. Craviotto	[illegible]	G. Gazzolo
621	54	V	Ida *Travega* -88 W Bk	Italian	231	113·8	24·5	15·0	Leghorn	1863	P. Micheli	[illegible]	G. Tonietti
[illegible]	55		Ida *Razzetti* W Sr	Italian	127	82·0	23·0	12·0	Ancona	1849			S. Gargano
	56	DN	Ida *Gundersen* -88 W Bk	Norwegian	656	169·0	32·2	18·2	Christianssand	1879	O.O. Hagen	[illegible]	Thr. B. Olsen
	57		Ida (CONDEMNED) W Bk	Norwegian	596	134·0	29·0	20·8	Medford, Mass.	1835			T. W. Hauff
	58	NV	Ida *Andersen* W Bk	Norwegian	329	132·5	28·1	14·4	Christianssand	1875	Thorstensen	[illegible]	Et Interessentskab (E.J. Jorgensen)
	59	N	Ida *Nielsen* W Bg	Norwegian	298	114·6	26·7	14·6	Buctouche, NB	1864			N. Friis
	60		Ida (LOST) *Markoff* W Bg	Russian	315	..	..	..		1867			K. Sundholm
73	61		Ida W Bk	Russian	314	119·0	27·0	12·8	Finland	1867	Loewe		W. Querfeldt & Co.
	62		Ida *Dahlberg* 76-86 W 3m Sr	Russian	300	109·0	28·8	14·2	Lappfjerd	1876	R. Karlsson		J. Janson
	63		Ida W Bg	Russian	290	..	..	..	Christinestad	1859			S. A. Vendelin
	64		Ida *Hientz* -67 W Bk	Russian	264	111·8	25·6	12·5	Jacobstad	1867	J. Lof	[illegible]	Wittkowsky, Questen[illegible] & Co.
	65		Ida *Celander* 81-86 W Sr	Russian	247	..	..	..		1877			C. G. Holmberg
	66		Ida *Dahlberg* W Sr	Russian	238	102·4	28·7	12·7	Brändö	1876	J. Elg		F. W. Gustafsson & C[illegible]
	67		Ida W Sr	Russian	223	..	..	..		1870			J. E. Helin
	68		Ida W 3m Sr	Russian	196	..	..	..		1865			C. Salström
	69	DGV	Ida (ex Celestial Empire) *Carlsson* -86 W Bk	Swedish	1267	196·8	38·4	22·6	[illegible]	1861	Baldwin	[illegible]	E. Frisk
	70	V	Ida (ex Josephine) *Olsson* 83-88 W Bk	Swedish	627	155·8	30·5	19·2	Sunderland	1861	Doxford	[illegible]	C. A. Banck
	71	V	Ida (ex Tynedale) *Tornberg* -85 W Bk	Swedish	569	149·5	30·5	18·9	Sunderland	1865	W. Pickersgill	[illegible]	Rederiaktiebolaget (L. Rostlund)
	72		Ida (ex Hannover) *Jönsson* W Bk	Swedish	351	123·3	25·2	15·0	Harburg	1856			P. Jönsson
	73	V	Ida (ex Fame) *Rosenqvist* -87 W Bk	Swedish	338	114·2	26·9	16·8	Sunderland	1861	Ratcliff & Co.		E. Jansson
	74		Ida *Petersson* -86 W Bk	Swedish	308	111·2	24·2	16·0	Apenrade	1853			B. Olsson
	75	V	Ida *Lindström* 87-87 W Bg	Swedish	282	122·3	25·1	12·0	Hernösand	1870	Fr. Hägglund	[illegible]	H. N[illegible]
	76		Ida W Bg	Swedish	259	100·4	23·3	15·1	Sunderland	1856			A. G. Johansen
	77		Ida *Olsson* W Bg	Swedish	222	103·0	25·6	13·2	Påskallavik	1855			M. Johnsson
	78	V	Ida (ex Courier) *Jonasson* 83-83 W Bg	Swedish	179	103·6	20·7	12·9	P.E. Island	18 54/74			P. Reinhold
	79		Ida *Önnberg* W Sr	Swedish	165	87·9	22·9	11·8	Oroust	1870	A. Samuelsson		J. N. Sanne
	80		Ida *Ingversson* 78-89 W Sr	Swedish	156	97·1	21·3	10·3	Oscarshamn	1880	C. Thoren		C. A. Ideström
	81		Ida *Petersen* 83-83 W Sr	Swedish	144	86·6	21·4	11·0	Blankenese	1859			K. Mattsson
	82		Ida (DISMANTLED) *Soderlund* W Sr		139	86·6	24·4	9·3	Finland	1871			
	83		Ida *Nyström* -88 W Sr	Swedish	128	80·3	21·7	11·2	Roslagen	1859			E. B. Wallerström
	84	G	Ida Amalia *Lossmann* [illegible] W 3m Sr	Russian	171	106·2	21·7	11·8	Libau	18 69/81	Dargewitz		C. Lossmann
21	85		Ida B. (ex Sora) *Laus* 83-85 W Bk	Austrian	767	148·6	29·8	19·7	Fiume	1865	G. Brazzoduro	Y 11.88	G. Brailli
1230	86	A	Ida B. *Ballarin* W Trb	Italian	160	72·8	24·3	10·5	Chioggia	1878	G. Menetto	[illegible]	Ant. Ballarin
100111	87		Ida B. Gibson *Scoffinberg* W 3m Sr	American	224	121·0	32·6	9·0	Chester, Pa.	1874			C. Bowker
52036	88		Ida C. *Bowker* W Bn	British	183	95·0	24·6	11·2	Nova Scotia	1865			D. G. Trefry
100350	89	R	Ida C. Schoolcraft *Booyee* W 3m Sr	American	304	134·3	32·6	9·6	Camden, N.J.	1883	Morris & Mathis		J. Booye
	90		Ida Charlotta *Samuelsson* W Sr	Swedish	154	86·4	22·4	11·8	Oroust	1873			J. J. Svensson
100125	91		Ida E. Latham *Thacher* W 3m Sr	American	440	136·0	33·[illegible]	15·0	Belfast, Me.	1874			E. Trubee
100395	92		Ida H. Mathis *Gifford* W 3m Sr	American	482	142·2	33·5	14·4	Camden, N.J.	1886	Morris & Mathis		D. Gifford
100312	93	R	Ida L. Hull *Hull* W 3m Sr	American	473	148·1	35·3	14·3	Newburyport, Mass.	1882	G.E. Currier		C. A. Hull
12064	94		Ida L. Ray *Marshall* W Sr	American	149	92·3	20·1	8·1	Harringt'n, Me	1864			J. Walter
100217	95	R	Ida Lawrence *Young* W 3m Sr	American	489	147·5	33·6	16·8	Camden, N.J.	1878	D.S. Risley & Co.		F. Smith & Co.
1238	96		Ida M. *Macera* W Bg	Italian	106	79·0	22·9	11·2	Malta	1867			Od. Macera
100262	97		Ida McKay *Brandt* W 3m Sr	American	179	113·0	30·0	8·6	Fairhaven, Cal	1880	H. Bendixon		A. McKay
	98	V	Ida Mathilde *Schütt* 77- W Bk	German	384	123·0	28·9	17·4	Stralsund	1857	F.P. Juhl		A. Schütt
75777	99	R	Ida Maud *Purdy* W Bn	British	172	98·8	26·4	10·0	Liverpool, N.S.	1880	A. Gardner		W. McFatridge
[illegible]4150	100		Ida May W Sr	British	100	78·5	24·3	9·4	Sherbrook, N.S.	1867			J. Cumminger

※ = **Lloyd's Register**; D = Nederlandsche Vereeniging; G = Germanischer Lloyd; I = Registro Italiano; R = [illegible]
A = Veritas Austro-Ungarico; H = Veritas Ellenico; N = Norske Veritas; V = [illegible]

In the 4th Column, C = Composite (iron frame, wood planked); I = Iron; Stl = Steel; W = Wood.

The Dates of Metalling (**Col. 11**) are based on the records of the Books in which Vessels are respectively classed, or on the Reports of the Surveyors [illegible]

Further particulars of Vessels classed by Lloyd's Register will be found in Lloyd's Register Book and Lloyd's Yacht Register.

	1	2	3 Nom. Capitaine et dates d'entrée successives des Armateurs actuels et à bord du navire. Ancien Nom.	Matériaux et Gréement	Pavillon.	Tonnage net.	Dimensions. Longueur.	Largeur.	Creux.	Construit Où.	Quand.	Par qui.	11	12 Armateurs ou Directeurs-Gérants.	13 Port d'Armement.
	101		Ida S. *Pavletich* (ex Radovan)	WSr	Austrian	130	85·3	20·1	11·2	Lussinpiccolo	1880	M.A.Starcich	V 10.83	G.Seraval di G.	Trieste
93	102	A	Ida S. *Beltramino*	WBk	Italian	328	126·9	23·2	17·0	Trieste	1856	Strudhoff	V 1.86	G.A.Gavagnin	Venice
[illegible]	103		Ida Schnauer *Beck* -87	WSr	American	295	117·6	31·3	9·4	PortLudlow, (Wash.	1875	Hall Bros.	V 1.86	J.D.Spreckels & Bros.	San Francisco
	104		Ida Sofia *Hansen* -82	WSr	Swedish	170	87·3	22·5	12·1	Sabbön	1871	A.Olsson		J.Berntsson	Kallands,Bhlen.
[illegible]	105		Ida Walker (ex New Hampshire)	WSr	British	187	109·0	26·0	9·1	Sophiasburg	1867			J.McLeod	Whitby,Ont.
	106		Ida da Conceição	WBn	Portug'se	122	78·7	20·3	10·5	Setubal	1868			J.A.Freire	Lisbon
	107	D	Ida & Johanna *Doewes*	WBk	Dutch	150	91·5	21·1	10·8	Muntendam	1871	Drenth&Sinnige	V 10.81	A.H.Doewes	Veendam
[illegible]	108		Idaho *Richardson*	WBk	American	390	138·0	29·6	12·0	Fairhav'n,Con.	1864			R.A.Tucker	New York
	109		Idalina	WBk	Brazilian	382	..	..	..					J.J.S.Imenes	Rio de Janeiro
[illegible]	110	❋	Iddesleigh *Gibson* -87	WBn	British	312	128·0	28·8	12·8	P.E.Island	1886	J.Yeo	V 3.88	S.B.Harvey	CharlotteTn,PEI
[illegible]	111	❋	Idea *Williams* -87	WSr	British	79	61·1	15·9	10·0	Bristol	1847			J.Williams	Aberystwith
[illegible]	112	I	Idea *Castelletto*	WBk	Italian	575	151·2	33·5	18·7	Lerici	1870	G.Faggioni	V 2.86	Flli. Borzone fu G.	Spezia
[illegible]	113	V	Idea *Gilanas* -87	WBk	Italian	494	147·3	30·8	19·9	Castellam. di Stabia	1869	A.Bonifazio	V 9.85	M.Cacace	Castellam. di Stabia
	114	N	Ideal *Gjesvre*	W3mSr	Norweg'n	323	133·1	28·0	14·0	Stavanger	1877		V 8.86	G. & A.Berg	Stavanger
	115		Idesbalde *Wulff* -88 (ex Anna ...)	WSr	German	210	97·8	24·2	12·7	Ghent	1838	Landsheer		Joach.Johannsen	Lubeck
	116	G	Idog *Larsson*	WBg	Swedish	271	116·8	24·5	13·6	Sundsvall	1868	E.Ostman		N.Tufvesson	Cimbrishamn
	117		Idor *Kurppinnen*	WSr	Russian	166	..	..	..	Joensuu	1863			P.Parviainen	Joensuu
[illegible]	118	❋	Idris *Parry* 60-80	WSr	British	91	70·8	20·0	10·2	Aberdovey	1860	Owen		W.Parry & Co.	Aberystwith
	119	N	Idun *Andersen* -83	WBk	Norweg'n	341	127·3	28·6	15·2	Kragerö	1873	J.Andreasen	V 11.84	H.Biörn,jr.	Kragerö
	120		Idun	Bk	Norweg'n		..	..	..					A.Garboe	Laurvig
	121	V	Idun *Gren* 71-74	WSr	Swedish	207	101·3	23·5	12·4	Eckerna	1877	J.Ahlberg		J.H.Gren	Fiskebäckskil
[illegible]	122	❋	Iduna *Hoen*	WBk	British	334	118·7	27·5	17·1	Sunderland	1866	R.Thompson&S	V 3.86	G.Lough	Newcastle
	123		Iduna *Bradhering* 79-86 (BROKEN UP)	WBg	German	223	92·5	26·0	14·1	Rostock	1847	A.Schönemann		E.Burchard	Rostock
	124	N	Iduna *Petersen*	WBk	Norweg'n	554	126·1	32·0	19·8	Maasluys	1842			A.C.Petersen	Porsgrund
	125	V	Iduna *Arvidsson*	WBn	Swedish	194	103·3	21·1	11·8	Kolboda	1875	H.Olsen	V 3.86	M.Johnsson	Gäsö
	126	DV	Ignatz Breum *Consbruch* 86	WSmSr	Danish	221	114·3	22·9	11·3	Svendborg	1884	J.R.Andersen	V 3.84	Actieselskab. "Calcbar" (F.P.H.Jensen)	Svendborg
[illegible]	127	V	Ignazio *Di Gennaro*	WBk	Italian	368	126·3	26·9	17·9	Cassano	1870	De Rosa	V 6.86	G.Parlato	Castellamare
	128	V	Ihana *Sjöblom*	WBk	Russian	412	125·3	28·9	16·0	Nystad	1872	H.H.Kjaldström		J.A.Forsström	Nystad
	129		Ihana	WSr	Russian	274	..	..	..		1874			C.F.Holmberg	Luvia
	130	G	Ihn & Sohn *Rohde* 88-88	WBg	German	279	100·5	28·2	14·4	Wismar	1849	H.Barmann		A.Burchard	Rostock
	131		Ika *Färdig*	WBk	Russian	563	..	..	..	Gamlakarleby	1861			C.Blom	Nystad
	132		Ikhsan Huda	WBg	Turkish	270	..	..	..	Austria	1879			Heracleelee Hosan Oglou Hajee Hassan & Co.	Heraclea
	133		Ikhsan Huda *Mehemed*	WBg	Turkish	260	..	..	..	Austria	1884			Mustapha Bey & Co	Constantinople
	134		Ikhsan Huda	WBn	Turkish	237	..	..	..	Greece	1865			Bartenlee Hajee Bekr Oglou Ahmet	Barten
	135		Ikhsan Huda	WBg	Turkish	218	..	..	..	Greece	1885			Ahbahmahlee Yuzbashi Oglou Hajee Ahmet & Co.	Ahbahmah
	136		Ikhsan Huda *Suleiman*	WBg	Turkish	184	..	..	..	Greece	1883			Mustapha Pasha Zadeh Ahmed Bey	Constantinople
	137		Ikhsan Huda *Hussein*	WBg	Turkish	151	..	..	..	Greece	1884			Hajee Emin & Co.	Constantinople
	138		Ikhsan Huda	WSr	Turkish	125	..	..	..	Ahyandon	1874			Richakiee Hajee Halil Effendi & others	Ahyandon
	139		Ikhsan Huda *Yelmmel Ali*	WBnb	Turkish	100	..	..	..	Barten	1886			Beshiktashlee Ahmed bin Ali	Beshiktash
	140		Ikhsanni Huda	WSr	Turkish	199	..	..	..	Russia	1882			Toolaji Hajee Mustapha Effendi	Constantinople
	141		Iki Kardash	WBg	Turkish	150	..	..	..	Focchia	1868			Tijar Hajee Sprohti	Focchia
	142		Iki Kardash	WBg	Turkish	125	..	..	..	Enos	1869			Edvinehlee Hajee Todoraki Djumawar & Co.	Edvineh
	143	I	Il Cacciatore *Romano*	WBg	Italian	351	..	..	..	Alimuri	1876		V 1.86	T.Rivara	Castellamare
[illegible]	144		Il Campidoglio *Cacciottolo* -88	WBg	Italian	311	111·5	28·7	14·7	Procida	1870	N.Lubrano	V 8.86	N.Dimartino	Naples
[illegible]	145		Il Carmine *Masiello* (ex ...)	WBn	Italian	148	86·9	23·7	13·7	Gaeta	1869			Gaetano Masiello	Naples
[illegible]	146		Il Francesco *Gentile*	WBg	Italian	116	87·2	21·8	11·1	Castellam. di Stabia	1882	C.Bonifacio	I 11.83	F.Marmino & Co.	Palermo
[illegible]	147		Il Francesco *De Cesare*	WBn	Italian	112	88·9	21·3	10·6	Vietri	1875	M.Donnarumma	V 8.85	P.Avallone	Castellam. di Stabia
[illegible]	148		Il Guerriero *Costa*	WBg	Italian	266	106·9	25·9	15·4	Venice	1862	A.Francesco Amadi	V 7.85	A.Filippini	Venice
[illegible]	149	A	Il Monarca del Mare *Spano*	WBg	Italian	243	101·7	25·5	16·4	Cassano	1859			G.Amato fu B.	Catania
[illegible]	150		Il Mugnaio *Greco*	WBk	Italian	243	113·8	26·3	13·5	Spezia	1869			G.Bonacorsi	Messina

❋=Lloyd's Register; A=Veritas Austro-Ungarico; D=Nederlandsche Vereeniging; G=Germanischer Lloyd; H=Veritas Ellenico; I=Registro Italiano; N=Norske Veritas; R=American Record; V=Bureau Veritas.

Dans la 4me Colonne, C=Composite (membrure en fer, bordé en bois); I=Fer; Stl=Acier; W=Bois.

Les Dates du doublage en métal (Col. 11) ressortent des livres dans lesquels les navires sont respectivement classés ou des rapports des Inspecteurs du Lloyd's Register.

Pour plus amples renseignements sur les navires classés par le Lloyd's Register, voir le Lloyd's Register Book et le Lloyd's Yacht Register.

Official No.	No. in Book.	Registry in which classed.	Ship's Name. Material and Rig. Master, and Dates of Appointment to present Owner's Service and to Ship. Ship's Late Name.	Flag.	Net Tonnage.	Length.	Breadth.	Depth.	Built. Where.	When.	By whom.	Date of Metalling.	Owners or Managers.
1993	151		Il Nuovo Giacomino *Treglia* W Bn	Italian	184	85·6	24·1	16·0	Meta	1859	Mauro		G.Treglia
524	152		Il Nuovo Michele *Vassalo -87* W Bn	Italian	152	96·4	24·2	12·6	Castellam. di [Stabia	1875	M.Trambarulo	[illegible]	M.Pace
274	153		Il Risorto *Carletti* W Bn	Italian	105	72·2	22·5	12·2	Savona	1852	Tixi		F^lli Carletti fu T.
1550	154		Il Salvatore W Bg	Italian	287	115·4	27·7	17·2	Gaeta	1870	L.Contreras	[illegible]	B.Zeccolilla
334	155		Il Salvatore *Miano* W Bg	Italian	141	87·9	23·3	9·3	Limite	1856	N.Pichiotti	[illegible]	Gius.Salvo
602	156	A	Il Terrible *Magliocca -88* W Bn	Italian	149	97·1	22·3	12·8	Gaeta	1858			F^lli Matarazzo fu A.
1325	157		Il Tomasino W Bn	Italian	100	80·0	22·0	10·9	TorredelGreco	1867			F.Oscione
780	158		Il Vero *Cafiero* W Bk	Italian	475	129·6	30·8	19·1	Cassono	1869	M.Patarzo	[illegible]	DeMartino F^lli & [illegible]
	159		Il Vincenzo *Cacciattolo* W Bg	Italian	165	..	..	..	Sira	1863			M.Cacciuttolo
86254	160	✻	Ilala *Parsell 83-95* I S	British	1329	235·0	37·5	21·5	Whitehaven	1882	WhitehvnSBCo		Lowden,Edgar & Co.
271	161	AD	Ilario (ex Mirto) *Persich -88* W Bg	Austrian	260	106·9	24·7	14·7	Fiume	1869	C.Giacovicich		G.Persich
	162	V	Ile de Terre Neuve *Lyon* W Bk	French	229	100·0	23·0	13·1	St.Servan	1873	Hamon	[illegible]	Guibert & Fils
121	163	✻A	Ilia (ex William D.Seed) *Tomicich -87* W Bk	Austrian	746	173·5	33·5	19·7	Sunderland	1874	W.Pickersgill	[illegible]	G.Milossevich & Co.
6	164		Ilias *Castriotis* W Bg	Greek	267	102.0	..	..	Syra	1874		[illegible]	A.& G.Thermisiot[illegible]
	165		Ilish W Sr	Turkish	125	..	..	..	Ilish	1865			Ineboli Hajee Fe[illegible] Oglou & Co.
46	166		Ilissus W	Greek	167	124·6	20·0	5·9	Austria	1876			Theophilatos Stathatos
45210	167		I'll Try *Henderson* W Bg	British	152	97·0	21·9	12·1	Aberdeen	1863	Smith		J.Allan
85076	168	✻	Illawarra *Carvosso 68-82* I S	British	1887	269·1	40·6	24·0	Glasgow	1881	Dobie & Co.		Devitt & Moore
	169	N	Illeri *Iverson -85* W Bn	Norweg'n	228	109·7	26·5	11·8	Sandefjord	1878			P.Tobiassen
70500	170	✻	Illimani *Popham* I Bk	British	590	174·0	29·2	17·4	Glasgow	1875	Dobie & Co.		J.E.Burgess
	171	N	Ilma *Albrethsen* W Bk	Norweg'n	437	129·0	29·7	15·8	Arendal	1870	O.B.Sörensen	[illegible]	O.B.Sörensen
	172	D✻	Ilma *Jacobsen* Stl 3mSr	Norweg'n	332	145·8	27·0	11·1	Grangemouth	1885	Grangemouth [DockyardCo.		Akt. Selsk. "Ilma" (A. Klem)
	173		Ilma *Eklund 80-83* W Bk	Russian	522	134·6	33·5	18·0	Åland	1849	M.Bergendahl		H.H.Mårtensen
	174		Ilma *Andersson* W Sr	Russian	307	111·4	28·7	14·2	Karis	1877	G.F.Sandell		K.F.Hjelt
	175		Ilma *Salminen* W Sr	Russian	220	102·2	27·7	13·0	Töfsala	1876	G.Urnber		B.Gustafson
	176		Ilma *Jansson* W Sr	Russian	110	..	..	..	Lappfjerd	1873			J.P.Mattson
	177		Ilmarinen *Westergren 83-85* W Bg	Russian	258	101·0	25·9	14·0	Fredrikshamn	1874	T.Wilkki		Leonard Kraki
	178	N	Ilmatar *Bonde 64-76* W Bk	Norweg'n	579	139·5	34·3	18·6	Brahestad	1869	A.Westerlund		V.Lerche
	179		Ilmatar *Carlsson 86-36* W Bg	Russian	224	108·0	25·5	14·3	Wyborg	1875	M.Repo		K.Nilsson
	180		Ilo W Sr	Norweg'n	185	99·3	25·2	10·8	Christinestad	1859			H.H.Parr
	181	N	Ilos (ex Julius) *Olsen 82-82* W Bk	Norweg'n	729	145·0	33·5	21·5	Vegesack	1848	P.Sager	[illegible]	Nicolai Olsen (A.S[illegible])
	182	NV	Ilos *Tharaldsen -88* W S	Norweg'n	720	163·0	34·5	18·8	Arendal	1878	O.J.Herlofson	[illegible]	O.J.Herlofson
	183	V	Ilos (ex Jenny) *Junnelius -86* W Bk	Russian	533	140·0	31·8	20·7	Uleaborg	1867	J.Kjeldström	[illegible]	H.O.Store
	184	V	Ilpotar *Johansen* W Bk	Norweg'n	586	144·1	34·8	18·6	Brahestad	18 64/84			H.N.Gjersöe
	185	DN	Imacos *Olsen 88-88* W Bk	Norweg'n	577	146·1	33·5	18·2	Tvedestrand	1880	C.Beck	[illegible]	Carl Beck
	186		Imanta *Ansack* W Sr	Russian	180	89·0	23·0	11·0	NeuSalis	1876	C.Obseling		M.[illegible]
	187	V	Imatra *Snellman -87* W S	Russian	1024	192·9	37·4	23·0	Christinestad	1868	M.Hannus	[illegible]	C.G.Hyden
	188		Imatra *Hamberg* W 3mSr	Russian	394	..	..	..	Korpo	1875			J.V.Anderson
87033	189	✻	Imberhorne *Hansen* I S	British	7199	284·1	41·2	24·1	Dumbarton	1882	M'Millan & Son		"Imberhorne" Ship [illegible] Ld. (W. R. Price & Co.)
43	190		Imbros W	Greek	179	121·3	19·7	9·8	Austria	1871			Theofilatos Stathatos
	191		Imenes W Sr	Brazilian	189	..	..	..					J.J.S.I.Junior
	192	V	Imes *Hogstrom* W 3mSr	Swedish	252	112·8	23·2	12·1	Westervik	1880	G.R.Fridell	[illegible]	G.Idestrom
2861	193		Immacolato (ex Pietro G.) *Muro* W Bk	Italian	637	146·6	33·0	21·3	Varazze	1871	G.Scala	[illegible]	M.V.Muro
1171	194	I	Immacolata *Longobardi* W 3mSr	Italian	353	119·7	26·4	17·2	Alimuri	1877	G.Mauro	[illegible]	[illegible]
2202	195	V	Immacolata Concezione (ex [illegible]) *[illegible]* W Bg	Italian	234	103·6	25·3	15·7	SestriPonente	1871	D.Tixi	[illegible]	[illegible]
354	196	I	Immacolata Concezione *Pasieri* W Bg	Italian	132	83·9	22·3	12·7	SestriPonente	1855	D.Tixi	[illegible]	F^lli Pace
	197	V	Immanuel *Kisby 78-* W Sr	Danish	137	84·7	21·5	11·3	Marstal	1867	J.J.Bager		H.C.Christensen[illegible]
	198	G	Immanuel *Strömpel 87-87* W Bg	German	213	96·4	24·9	13·4	Greifswald	1861	L.Wittenberg		J.J.Wallis & Sohn
	199	N	Immanuel *Bruun -88* W Bk	Norweg'n	465	126·0	30·8	17·3	Skien	1853	L.Jörgensen		A.Bruun
	200	N	Immanuel *Ribe* W Bk	Norweg'n	460	138·5	30·0	16·4	Grimstad	1876	Hans Ribe	[illegible]	Hans Ribe

✻ = Lloyd's Register; D = Nederlandsche Vereeniging; G = Germanischer Lloyd; I = Registro Italiano; [illegible]
A = Veritas Austro-Ungarico; H = Veritas Ellenico; N = Norske Veritas; [illegible]

In the 4th Column, C = Composite (iron frame, wood planked); I = Iron; Stl = Steel; W = Wood.

The Dates of Metalling (**Col. 11**) are based on the records of the Books in which Vessels are respectively classed, or on the Reports of the Surveyors [illegible]

Further particulars of Vessels classed by Lloyd's Register will be found in Lloyd's Register Book and Lloyd's Yacht Register.

No.	Régistre	Nom, Capitaine et Gréement	Pavillon	Tonnage	Longueur	Largeur	Creux	Construit à	Quand	Par qui	Doublage	Armateurs ou Directeurs-Gérants	Port d'Armement
201	N	Immanuel *Molbech* W Bk	Norweg'n	424	129·0	30·1	16·6	Porsgrund	1868		Y 5.84	P.A.Stoeren	Brevig
202	N	Immanuel *Naestden 77-77* W Bg	Norweg'n	[illegible]	83·8	25·3	14·8	Sunderland	1832			C.A.Berg	Holmestrand
203	N	Immanuel *Kjölner 75-75* W	Norweg'n	10[illegible]	91·8	22·1	11·0	Skien	18[illegible]			F.C.& K.Kjölner	Tönsberg
204	N	Immanuel W Bk	Norweg'n	126	91·8	22·[illegible]	9·1	[illegible]	1860	[illegible]		M.E.Schilbred	Brevig
205	DN	Immanuel *Gulbrandsen* W	Norweg'n	111	7[illegible]	[illegible]1·7	10·3	Stavanger	18[illegible]			I.T.Rove	Sandefjord
206		Impatlay	British	652	250·1	34·0	8·8	Dumbarton	1888	W.Denny&Bros		IrrawaddyFlotillaCoLd.	Glasgow
207		Imperador W Bg	Brazilian	11[illegible]	..	..	..					J.R. e Souza	Rio de Janeiro
208		Imperador do Brazil W Bg	Brazilian	21[illegible]	..	..	..					C.V.Vianna	Rio de Janeiro
209	NV	Imperator W	Norweg'n	1082	199·7	[illegible]	21·5	Stavanger	1876	[illegible]	Y 5.84	Holdt & Isachsen	Stavanger
210	N	Imperator *Nielsen* W	Norweg'n	[illegible]	150·[illegible]	[illegible]	17·[illegible]	[illegible]	1879	[illegible]	Y 5.84	Joh Hanssen	Arendal
211		Imperator (CONDEMNED) *Sundström* W Bk	Russian	3[illegible]	..	..	..	[illegible]	1857			F.E.Nyfors	Åbo
212	A	Imperatore Francesco Giuseppe I. *Premuda* W Bk	Austrian	667	117·6	29·6	17·[illegible]	[illegible]	1875	[illegible]	Y 6.87	M.V.Premuda	Lussinpiccolo
213	V	Imperatrice *Wagnder* W	French	141	88·[illegible]	21·3	12·4	Boulogne	1868	[illegible]		Noël	Paimpol
214	I	Imperatrice W Bg	Italian	16[illegible]	88·[illegible]	20·6	10·7	TorredelGreco	1880	[illegible]	Y 4.87	A.Brancamo	Naples
215		Imperial W	British	100	82·[illegible]	20·6	10·4	[illegible]	1867	[illegible]		C.Davison	Chester
216	RV	Imperial W S	American	1288	188·7	38·0	23·5	Quincy,Mass	1869	[illegible]	11.86	J.E.Crosby	Boston,Mass.
217		Impérial	French	13[illegible]	..	..	..						Arles
218	V	Impérieuse *Brinck 87-87* W Bk	Danish	354	132·8	27·1	14·4	Hamburg	1863	[illegible]	Y 6.86	N.H.Brinck	Sønderho Fanø
219		Impetuous (WRECKED) *Mader* W	British	162	94·7	23·1	12·7	[illegible]	1854			W.W.Foreman	Guernsey
220		Impi *Nystrom 75-87* W Bk	Russian	637	143·0	31·1	20·5	Brahestad	1868	[illegible]		J.Lang	Brahestad
221		Impi W	Russian	2[illegible]	97·1	26·9	13·4	[illegible]	1873	[illegible]		F.A.Söderström	Åbo
222		Impi W Bg	Russian	237	114·6	26·9	14·1	Fredrikshamn	1874	[illegible]		V.Söderlund	Åbo
223	V	Importer W	American	12[illegible]	136·6	32·2	23·9	Newburyport, Mass.	1870	J.Currier	Y 6.86	H.Koop & Co.	New York
224		Improviso *De Lellis* W Bg	Brazilian	105	..	..	..	[illegible]	1873		Y 4.84	Donna C. M. da Silva & Filhos	Paranaguá
225		Imsland W Bg	Norweg'n	237	110·2	26·1	12·6	Stavanger	1875	Torjus Imsland	[illegible]	B.Hansen	Stavanger
226	✻	Imuncina *Brown* I Bk	British	583	165·6	30·2	17·4	[illegible]	1867	[illegible]		Donaldson Brothers	Glasgow
227	G	Ina (ex Wm. Hansen) W Bk	German	389	136·4	27·5	17·1	Whitby	1867	T.Turnbull&Son	12.86	C.F.Maass	Rostock
228	V	Ina W & Sr	Norweg'n	252	115·0	25·0	13·9	Sestri Ponente	1874	[illegible]	11.85	Nils R.Parelius	Christiansund
229		Inca *Pook* W Bk	British	398	132·6	26·6	18·[illegible]	Liverpool	1862	[illegible]		Lucas Bros. & Co.	Bristol
230		Inca W Bk	German	720	143·9	34·3	18·6	[illegible]	1882	[illegible]	Y 8.88	J.H.Hustede	Elsfleth
231		Incantatrice (ex Far Madness) W	Italian	146	105·4	21·1	11·6	Ipswich	1874	Wm.Bayley & Sons	[illegible]	D.G.Garuffi	Catania
232	✻D	Inch Keith *Grainger 81-81* I Bk	British	12[illegible]	236·3	35·5	21·6	[illegible]	1878	Barclay,Curle&Co.		J.Hardie	Glasgow
233	✻D	Inch Murren (BURNT) *Petrie 79-81* I	British	1254	231·1	35·8	21·6	Dumbarton	1878	Birrell, Stenhouse&Co.		J.Hardie	Glasgow
234	✻	Inchcape Rock *Armour* I S	British	1493	250·5	38·3	23·0	Port Glasgow	1886	Russell&Co.		J.Cornfoot & Co.	Glasgow
235	✻	Inchgreen *Miller* I Bk	British	1091	219·5	34·1	20·9	Greenock	1876	[illegible]		D.MacDougall	Greenock
236		Indécis W Bg	French	166	..	..	..	St Malo	1840			Guibert & fils	St.Malo
237		Indefatigable W	British	139	71·5	[illegible]	12·2	[illegible]	1801			S.Boustead	Drogheda
238		Independence *Farrell* W S	American	[illegible]	16[illegible]	34·2	22·7	E.Boston,Mass	1871	[illegible]& Smith		C.P.Hemenway	Boston,Mass.
239		Independencia *Tavanja* W Sr	Brazilian	115	..	..	..	S.João da Barr	1861		Y 8.85	Cia. Cabotagem S. João da Barra	Rio de Janeiro
240		Index *Garrison* W 3m Sr	American	318	1[illegible]	[illegible]	8·[illegible]	Camden,N.J.	1870	Taylor&Mathis		L.Garrison	Bridgeton,N.J.
241		India W S	British	4[illegible]	121·2	2[illegible]·2	20·8	Portsm'th,U.S	1833			J.V.Hall	Sydney,N.S.W.
242	V	India *Clapp* W S	American	1295	132·1	[illegible]	23·9	Bath,Me.	1868	J.P[illegible] Son	Y 9.84	J.Rosenfeld's Sons	Bath,Me.
243	G	India I Bk	German	974	1[illegible]	[illegible]·3	19·5	Glasgow	1876	A.Stephen&Sns	Y 1.86	D.H.Wätjen & Co.	Bremen
244		India (WRECKED) *Saff* W Bk	German	574	148·3	31·6	17·9	Elbing	1856	[illegible]		J.Paleske	Danzig
245	V	India *Pundt* W Bn	German	28[illegible]	103·[illegible]	25·3	10·[illegible]	[illegible]	1872	[illegible] Wwe.	[illegible]	D.Haye	Brake
246	NE	India (ex British India) W Bk	Norweg'n	1216	1[illegible]·2	[illegible]	23·6	St John N.B.	1862	[illegible]	[illegible]	Akt. Selsk. "India" (A. Brauer)	Arendal
247	DV	India (CONDEMNED) W Bk	Norweg'n	1141	201·0	[illegible]	23·3	[illegible]	1874	J.[illegible]	Y 8.87	M.Beyer	Bergen
248	DNV	India W Bk	Norweg'n	7[illegible]	168·2	[illegible]	20·0	Bremerhaven	1875	[illegible]	[illegible]	O.C.Reinhardt	Christiansand
249	N	India W Bk	Norweg'n	7[illegible]	1[illegible]	[illegible]·9	21·1	Drammen	1871	[illegible]	Y 3.84	A.P.Andersen	Moss
250													

[illegible] = Lloyd's Register; D = Nederlandsche Vereeniging; G = Germanischer Lloyd; I = Registro Italiano; R = American Record; [illegible] = Veritas Austro-Ungarico; H = Veritas Ellenico; N = Norske Veritas; V = Bureau Veritas.

Dans la 4me Colonne, C = Composite (membrure en fer, bordé en bois); I = Fer; Stl = Acier; W = Bois.

Les Dates de doublage en métal (Col. 11) ressortent des livres dans lesquels les navires sont respectivement classés ou des rapports des Inspecteurs du Lloyd's Register.

Pour plus amples renseignements sur les navires classés par le Lloyd's Register, voir le Lloyd's Register Book et le Lloyd's Yacht Register.

1 Official No.	2 No. in Book	3 Registry in which classed	4 Ship's Name. Material and Rig. Master, and Dates of Appointment to present Owner's Service and to Ship. Ship's Late Name.	5 Flag.	6 Net Tonnage	7 Dimensions: Length	Breadth	Depth	8 Built: Where.	9 When	10 By whom.	11	12 Owners or Managers
	251	N	India *Jorgensen* W3mSr	Norweg'n	350	131·3	27·6	12·0	Grimstad	1885	JohanJörgensen	[illegible]	J. Jorgensen
	252		India *Freixas* WBk	Spanish	570	150·5	34·1	22·0	Barcelona	1869		[illegible]	J. Guerri
	253	V	India *Pla* WPol	Spanish	182	91·2	26·7	12·1	Blanes	1856		[illegible]	Lorenzo Plá
	254		India WBn	Spanish	110	88·3	19·7	9·0	Albay	1871			José Cardell
	255		India *Petterson* -85 WBk	Swedish	492	129·6	29·7	18·1	Törefors	1857			P. Persson
79038	256	✻	Indian Chief *Weir* 81-81 WBkn	British	339	125·7	26·2	14·7	Kingston	1878	Kinloch	[illegible]	W. Kinloch
52705	257	✻	Indian Empire *Watson* [illegible] IS	British	1515	230·0	38·8	23·5	W.Hartlepool	1865	Pile,Spence&Co.		G. Duncan & Co.
35308	258		Indian Queen *Armstrong* WBn	British	113	91·4	22·6	8·7	Calais,Me.	1849			J. Brown
68028	259	✻	Indiana *Penwill* 85-85 W3mSr	British	289	130·1	26·7	12·2	Sunderland	1873	W. Pickersgill	[illegible]	W. C. Jarvis
100189	260	V	Indiana *Morrison* 81-81 WS	American	1414	208·9	40·0	23·9	Bath,Me.	1876	A.Sewall&Co.	[illegible]	A. Sewall & Co.
	261	N	Indiana *Bjönness* 80-83 WBk	Norweg'n	521	130·2	30·7	18·1	Montrose	1852			H. Zernichow
	262	V	Indien *Roussen* -87 WBk	French	422	118·1	28·2	18·5	Marseilles	1867	V. Jouvin	[illegible]	Lallement
	263	V	Indien *Frumerie* 65-85 WBk	Swedish	532	158·8	29·6	17·4	Westervik	1863	C.R. Fridell		C. Carlbom
952	264		Indipendente *Simile* (ex Giuseppe) WBg	Italian	224	100·0	24·9	15·4	Castellamare	1874			V. Culotta
597	265	I	Indipendenza *Regesta* [86-86] WBk	Italian	598	147·6	32·1	20·9	Chiavari	1872	Gotuzzo	[illegible]	F. Dall'Orso & [illegible]
	266		Indo *Soderman* WBk	Swedish	254	106·6	25·8	13·0	Visby	1875			M. Soderman
91276	267	✻	Indore *Jenney* 81-85 IS	British	2088	279·0	40·7	24·6	Stockton	1885	Richardson, [Duck&Co		Eyre, Evans & Co.
	268	V	Indra *Gieseke* 84-84 IBk	German	695	166·0	30·0	18·9	Hamburg	1877	ReiherstiegS.W		Wachsmuth&Krogmann
13884	269		Indus (ex steamer) *King* IBk	British	1319	242·1	38·5	21·1	London	1847			Eastern&S.African Tele-graph Co.Ld.
46713	270		Indus WSr	British	126	94·0	24·4	7·8	Dorchester,US	1857			J. Eneas
2484	271	I	Indus *Righetti* WS	Italian	1111	196·5	31·7	24·0	SestriPonente	1874	Flli Cadenacio	[illegible]	G. Lavarello di P.
1374	272	A	Industre *Sodich* WBk	Austrian	563	149·2	28·9	19·5	Fiume	1868	V. Brazzoduro	[illegible]	C. cav. Cosulich
	273	N	Industri *Sampson* -89 (ex Industry) WSr	Norweg'n	212	107·3	23·0	12·5	P.E.Island	1863			M. Andreasen
	274	N	Industri *Andresen* W3mSr	Norweg'n	201	103·7	25·6	12·3	Sandefjord	1870			M. A. Andresen
2336	275	I	Industria *Olivari* -89 WBk	Italian	600	141·7	29·2	19·2	Pegli	1873	B. Canepa	[illegible]	F. Schiaffino
3263	276		Industria *Di Meo* WBg	Italian	210	104·9	25·5	16·2	Savona	1853	G. Tixi		Paolo Ciccheró
	277	V	Industria *Duarte* (ex Marianna III) WBk	Portug'se	316	122·0	27·9	12·8	SanMartinhoPort	1862	J.R. Assinço	[illegible]	Viuva Macieira & [illegible]
	278		Industria *Auroja* WSr	Portug'se	155	88·9	21·3	10·5	Vianna	1874	M.P. Netto		J.A. de Magelhaes
	279	V	Industrie *Esnouf* WBk	French	187	94·8	22·6	13·3	Granville	1863	Duvard	[illegible]	J. Panier
	280		Industrie See "Industry," No. 289 WSr	French	148	..	..	..	Jersey	1867			
	281	V	Industrie *Kirchhoff* 82-82 IS	German	1642	228·0	43·0	24·2	Kinderdyk	1872	L.Smit&Zoon		C. H. H. Winters
	282	V	Industrie *Fretwurst* 77-77 WBk	German	333	119·4	26·9	15·5	Danzig	1859	W.R. Grott		Lichenheim & [illegible]
	283		Industrie *Hillel* WBg	German	248	109·9	25·1	12·5	Brake	1866	F. Nicolai		H. Hillel
	284		Industrie *Meislahn* 70-74 WBg	German	188	87·3	24·3	12·9	Lubeck	1848	J. Landmann		J. Rahlff & Sohn
	285	V	Industrie *Kleen* 82-82 WGlt	German	100	83·0	19·8	9·2	Grossefehn	1868	C. Meyer		E. C. Aden
17573	286		Industry *Tyrrell* 85-85 WBkn	British	189	108·2	21·4	13·0	St.John,N.B.	1857			E. Kearon
75915	287		Industry W Barge	British	105	95·8	21·7	7·5	MillPoint	1879	W. Jamieson		H. B. Rathbun & Son
60577	288	✻	Industry *Jones* 86-88 WSr	British	84	79·0	18·6	9·1	Gloucester	1867	Davies		J. Jones
	289		Industry (BROKEN UP) *Make* 88-88 WBn	French	144	98·9	20·1	12·1	Jersey	1867	Le Vesconte		Adam
	290		Ines *Warmis* WBn	Russian	222	104·6	25·4	12·6	Strömaly	1874	K. Makela		A. Michelsson
	291		Inés (ex Araucano) *Coll* WBn	Spanish	386	105·9	28·1	17·1	Savona	1861	Calamaro		Doña Rita Mas
	292	G	Inés *Mas Espinas* WBk	Spanish	200	105·6	25·4	12·3	Palma	1858	M. Planes	[illegible]	M. Mas Espinas
	293		Inés (ex Dolores, ex Elena) WBn	Spanish	142	87·6	24·2	10·9	Blanes	1852			J. M. Roca
	294		Inés (ex Vestal) WSr	Spanish	104	75·8	22·9	10·4	Arens	1851			F. Bellver
	295	DV	Inés Røhl *Andersen* WBg	Danish	248	115·1	24·3	12·7	Varnanäs	1881	A.F. Bring	[illegible]	Aktie Selskab [illegible]
	296		Inez *Voisin* WSr	French	107	95·8	19·9	11·0	Bristol	1860	Patterson	[illegible]	Danguilhen frères
	297	N	Infatigable *Falck* 84-84 WBk	Norweg'n	843	180·5	34·8	20·0	Grimstad	1884	A. Knudsen	[illegible]	M. Falck
	298	G	Infatigable *Stephansson* 67-87 WBk	Swedish	390	116·4	28·1	16·6	Grimstad	1860			L. P. Stephansson
80490	299	✻	Inflexible *Race* 85-85 WK	British	90	82·0	20·8	10·9	Brixham	1884	S. Dewdney & [Sons		C. Osborn
	300	RV	Inga *Graam* 78-88 (ex Rosa Rainbow) WBk	Norweg'n	1391	188·0	40·1	23·6	St.John,N.B.	1867	J.&R. Reed	[illegible]	T. M. Nielsen & [illegible]

✻ = Lloyd's Register; D = Nederlandsche Vereeniging; G = Germanischer Lloyd; I = Registro Italiano; R[illegible]
A = Veritas Austro-Ungarico; H = Veritas Ellenico; N = Norske Veritas; V = [illegible]

In the 4th Column, C = Composite (iron frame, wood planked); I = Iron; Stl = Steel; W = Wood.

The Dates of Metalling (Col. 11) are based on the records of the Books in which Vessels are respectively classed, or on the Reports of the Sur[illegible]

Further particulars of Vessels classed by Lloyd's Register will be found in Lloyd's Register Book and Lloyd's Yacht R[illegible]

No.	Cote	Nom. Matériaux et Gréement. Capitaine et dates d'entrée. Ancien Nom.	Pavillon	Tonnage	Longueur	Largeur	Creux	Où	Quand	Par qui	Doublage	Armateurs ou Directeurs Gérants.	Port d'Armement.
301	✱	Inga [illegible] W Bk	Norweg'n	287	192·2	31·6	20·5	Dundee	1862	A.Doublie Sons & Co		L.O.& Larsen	Sandefjord
302	N	Inga [illegible] W Bk	Norweg'n	752	141·7	31·3	21·0	Portsm'th,NH	1836			F.J.Holst'sEnke	Moss
303	DN	Inga [illegible] W Bk	Norweg'n	480	137·7	32·1	13·0	Economy,N.S.	1873	Fraser		Chr.Swane	Christiania
304	NV	Inga *Larsen* -86 W Bk	Norweg'n	457	136·9	30·6	17·1	Christiansand	1877	Brsk.Larsen		T.A.Olsen	Christiansand
305	N	Inga *Heffermehl* 80-86 W Bg	Norweg'n	214	102·3	26·8	12·1	Drammen	1883	Jorgensen& [Knudsen		L.Heffermehl	Drammen
306	NV	Inga *Örbeck* -86 W Sr	Norweg'n	156	95·5	23·5	10·8	Sönderho	1886	P.Sonnichsen		A.N.Örbeck	Lillesand
307	V	Ingeborg *Jensen* -87 W Bk	Danish	195	110·2	22·3	11·7	Copenhagen	1868	Popp & Michel- [sen		C.J.Hoepfner	Copenhagen
308	N	Ingeborg *Roer* -86 W Bk	Norweg'n	397	130·1	27·7	15·9	Bathurst	1861	O'Brien		M.Röer	Christiania
309	NV	Ingeborg *Knudsen* -86 W Bg	Norweg'n	242	101·9	26·1	15·5	Sunderland	1866			Chr.Hannevig	Christiania
310		Ingeborg *Mannerstrom* -86 W Bk	Russian	537	..	..	..	Geta	1874			E.Montell	Geta
311	V	Ingeborg *Fex* W Bk	Swedish	315	132·2	25·6	12·5	Thamernabba	1878	C.Nilsson		H.Fex	Lerhamn
312	V	Ingeborg *Abrahamsson* W Sr	Swedish	202	100·0	23·2	12·0	Oberg	1880	J.H.Selden		B.Johansson	Uddevalla
313		Ingeborg *Nilsson* -86 W Sr	Swedish	125	84·3	19·8	10·0	Brake	1858	Oltmanns		N.Nilsson	Brantevik
314	G	Ingeborg *Sandberg* W Sr	Swedish	115	86·6	22·4	8·7	Ström	1858			A.Andersson	Utsund Rslgn.
315	GV	Ingeburg *Lenks* -86 W Bk	German	397	119·7	25·9	13·6	Apenrade	1871	G.Haben		Th.& F.Eimbeke	Hamburg
316	V	Inger Margrethe W Bk	Danish	160	96·9	26·5	11·5	Veendam	1864			C.Clemmensen	Skive
317	V	Ingleside *Hayes* -85 W Bk	British	926	172·1	34·6	21·0	MossGlen,NB.	1877	G.Merritt		TheBarkDirectorCo.	St.John,N.B.
318	✱	Inglewood *Beeching* -86 I Bk	British	1043	215·0	34·1	21·1	Harrington	1875	R.Williamson & [Son		R.Williamson&Son	Workington
319	DV	Ingmar *Jörgensen* W Bg	Danish	204	103·8	23·5	11·6	Nordby,Fanø	1874	C.Graff	Y 7.8	Akt.Selsk.(J.N.Hansen)	Nordby,Fanø
320		Ingo ex Alt *Maybaum* W Bg	Russian	254	95·0	23·9	13·6	Moss	1866	E.Pedersen		Maybaum	Reval
321	N	Ingolf *Knudsen* 80-86 W Bk	Norweg'n	367	128·2	30·1	15·9	Tvedestrand	1874	A.L.Holm		S.A.Knudsen (A.& F. Smith)	Tvedestrand
322	V	Ingomar *Sinclair* W S	British	1175	192·7	37·8	24·0	Newport,N.S.	1875	Harvey		A.Gibson& Co.	Liverpool
323	✱	Ingrid *Moore* W Sr	British	99	83·8	22·7	10·8	Padstow	1873	Stribley		Mrs.C.Jenkin	Padstow
324		Ingrid *Österling* W 3mSr	Russian	262	..	..	..	Bjerno	1865			W.Akerblom	Nystad
325		Ingrid *Sardin* 72-83 W Sr	Swedish	136	89·6	26·6	10·7	LerwkScotlnd	1859			G.E.Ekström	Oscarshamn
326	N	Inheritance *Smith* -86 W Bk	Norweg'n	406	131·6	28·2	17·2	Sunderland	1869	Pearson		Actieselskabet [illegible] (Chr.Möller)	Christiania
327	V	Innerwick *Waters* 82-83 W Bk	British	1238	194·7	38·6	22·8	RiverJohn,NS	1882	J.Kitchin		A.C.MacDonald	Picton,N.S.
328	✱	Innisfail *Clark* 66-86 W Bn	British	200	99·2	25·1	13·1	Dublin	1859	Good		Goodyear & Co.	Liverpool
329	✱	Innisfallen *Lodge* -88 W Sw	British	232	112·2	26·7	16·2	Blyth	1868	Robinson		Mrs.S.D.Stonehouse	SouthShields
330	I	Innocenta *Schiaffino* W Bk	Italian	751	170·2	33·7	20·9	SestriPonente	1871	A.Briasco fu L.		Alberto Nattini	Genoa
331	D	Innocenti *Canepa* W Bk	Italian	902	173·8	31·2	23·8	Varazze	1883	G.Siala		P.Lavarello	Genoa
332	I	Innocenza Rosa *Napoli* W Bk [illegible] -86	Italian	407	116·4	30·5	18·9	Savona	1866	G.Tixi		S.Napoli	Catania
333	V	Ino *Nagel* 75-75 W Bn	Danish	147	93·7	21·4	11·3	Nysted	1871	F.Sparre		A.H.Friis	Marstal
334	V	Ino W Bk		215	104·6	25·0	11·2	Brake	1867	F.Nicolai			
335	N	Ino *Jonassen* 73-73 W Bk	Norweg'n	337	114·9	28·8	16·1	Grimstad	1856	JörgenOlsen		Christen Boe	Arendal
336	N	Ino *Sterk* -83 W Bn	Norweg'n	160	111·3	25·5	9·8	Kragerö	1868	Dekke		H.Biörn,jr.	Kragerö
337	GV	Ino *Pyk* -86 W Bk	Swedish	326	129·5	26·3	13·6	Hamburg	1864	M.G.Amsinck		G.Sjöberg	Malmö
338	DV	Ino *Bogmberg* 83-85 W Bk	Swedish	325	131·5	25·6	12·9	Kolbada	1874	H.Olsen		C.Sylvan	Helsingborg
339	H	Insani Hounda *Syleiman* W Bk	Turkish	154	..	..	..	Spezzia	1868			Ibraim Effendi	Constantinople
340	GV	Insel *Holtgers* 79- W Bg	German	192	94·7	23·2	11·0	Papenburg	1867	A.Beckmann		Abr.Beckmann	Papenburg
341	I	Insonne *Sponza* W Bn	Italian	139	86·6	23·7	11·4	Chioggia	1883	G.Poli		Ant.Sponza	Venice
342	N	Insula Capri *Larsen* -86 W Bk	Norweg'n	394	125·3	29·7	16·7	Arendal	1869	S.Johansen		S.M.Klocker	Arendal
343	N	Insulan *Pedersen* W Bg	Norweg'n	216	106·1	25·3	11·9	Risör	1866	O.A.Olsen		B.A.Olsen	Risör
344		Insulaneren *Hansen* 80-86 W Bg	Danish	231	103·5	25·7	13·9	Svendborg	1857	H.P.Hansen		Thorvald Hansen	Svendborg
345	DV	Insulinde *Spaanderman* 81-86 W S	Dutch	1326	202·1	39·7	24·8	Alblasserdam	1879	F.H.vonLindern		Von Lindern & Van Vessem	Rotterdam
346		Integridad *Solpi* W Bg	Spanish	282	101·7	29·5	15·1	Mataro	1854	F.Segarto		P.Sust y Mes	Barcelona
347		International I Sp	British	382	120·0	31·0	11·9	London	1879	Fletcher,Son& [Fearnall		London Patent Elevator Co., Ld.	London
348		International *Court* W Bn	British	192	108·0	23·3	12·8	P.E.Island	1862	Coffin		W.Goldfinch	London
349	✱	Intrepid *Le Lacheur* W K	British	59	79·4	20·3	9·2	Falmouth	1882	Burt		S.Taylor	Guernsey

✱ = Lloyd's Register; D = Nederlandsche Vereeniging; G = Germanischer Lloyd; I = Registro Italiano; R = American Record; [illegible] Austro-Ungarico; H = Veritas Ellenico; N = Norske Veritas; V = Bureau Veritas.

Dans la 4me Colonne, C = Composite (membrure en fer, bordé en bois); I = Fer; Stl = Acier; W = Bois.

Les Dates du doublage en métal (Col. 11) ressortent des livres dans lesquels les navires sont respectivement classés ou des rapports des Inspecteurs du Lloyd's Register.

Pour plus amples renseignements sur les navires classés par le Lloyd's Register, voir le Lloyd's Register Book et le Lloyd's Yacht Register.

Official No.	No. in Book	Registry in Society's classes	Ship's Name. Master, and Dates of Appointment to present Owner's Service and to Ship. Ship's Late Name.	Material and Rig	Flag	Net Tonnage	Length	Breadth	Depth	Where	When	By whom	Date of Metalling	Owners or Managers
00314	351	R	Intrepid *Clausen -87*	WSrYt	American	118	105·5	25·3	11·2	Brooklyn	1878	C.& R.Poillon	[illegible]	Lloyd Phœnix
	352	V	Intrépide *Rivard*	WSr	French	115	88·3	21·6	11·2	Dunkirk	1885	E.Baryske		A.Govard
	353		Intrépide *David*	WBn	French	105	67·6	19·7	10·6	Bayonne	1860	Descande fils		J.A.K.David
	354	V	Intrépide Corse *Elliot -87*	WBk	French	276	110·5	23·3	12·6	Honfleur	1866	Leviels frères	V 8.87	D. Auger
68417	355	*	Invercargill *Muir 74-7*	IS	British	1246	239·7	36·0	20·7	PortGlasgow	1874	R.Duncan&Co.		Shaw,Savill & Albion Co. Ld.
	356	N	Inverdruie *Johannesen*	WBk	Norweg'n	595	162·2	30·5	17·9	Kincardine	1867	Wright	V 12.8[illegible]	J.C.A.[illegible]
03527	357	V	Inveresk *Bray*	WBk	British	799	173·6	35·1	19·2	Maitland,N.S.	1874	Frahey	V 8.86	A.Gibson & Co.
28083	358		Invermark *Cocking*	WSr	British	104	81·8	21·5	11·4	Montrose	1862	Petrie		H.Miller
63513	359	*	Inverness *Mudder 79-87*	CBk	British	722	177·6	32·0	18·9	Aberdeen	1869	Hall,Russell&Co	[illegible]	J.C.&G.S.Richardson
52092	360		Invicta *Allen 85-89*	WSr	British	178	98·8	23·7	12·8	Rye	1865	Head		Whitstable Shipping Co. Ld.
89143	361	*	Invincible *Jackson*	WSk	British	80	76·3	21·3	10·2	Newhaven	1883	J.C.Tolman		W.C.Edwards
00096	362		Invincible *Lewis -87*	WS	American	1460	202·4	40·3	21·0	Bath,Me.	1873	W.V.Moses&Sn.	[illegible]	[illegible]
12316	363		Invincible *Strickland*	WS	American	1325	220·5	42·5	17·7	New York	1861			C.Metzger
524	364		Ioanna	WBk	Greek	430	125·3	29·3	..	Cattaro	1873			N.Voudakis
400	365		Ioannis (ex Mathilde) *Mathiodis*	WBk	Greek	626	119·6	30·3	..	Quebec	1866		V 11.80	P.J.Calontes
133	366	V	Ioannis (ex [illegible]) *Gialis*	WBk	Greek	578	148·3	30·8	19·7	[illegible]	1872	N.Martinolich	V 8.86	D.Neamonitachi
89	367		Ioannis *Bouccoucalis*	WBg	Greek	276	128·6	25·9	..	Piræus	1868			G.Malokinis
14	368	H	Ioannis (ex Elpis)	WBg	Greek	204	87·9	24·8	..	Spezzia	1869		[illegible]	E.B.Cosafos
386	369	A	Ioannis *Batzevis*	WBn	Greek	171	81·4	23·6	..	Chios	1875			G.P.Frangakis
	370	H	Ioannis (ex Nicolaos) *Lirus*	WBk	Turkish	264	..	..	..	Syra	1869			Coulonmoutros
	371		Ioannis	WBn	Turkish	217	..	..	..	Chios	1871			I.G.Brilis
	372		Ioannis	WBg	Turkish	170	..	..	..	Syra	1858			Miliade frères
	373		Ioannis Dedoussis *Gazalis*	WBk	Turkish	327	..	..	..	Bordeaux	1867		V 8.83	G.Dedoussis
	374		Ioannis Prodromos	WBg	Russian	120	..	..	..	Syra	1871			G.G.Deligeroglou
85630	375	V	Iodine *Smith*	WBk	British	295	136·0	31·9	11·9	Eatonville,NS	1885	J.Parsons		J.G. Ross & Co.
87898	376	*V	Iolanthe *Stanton 85-85*	IS	British	1593	255·3	38·2	22·5	Dumbarton	1883	A.M'Millan&Sn		Cruickshank,[illegible]
83447	377	V	Iolanthe *Card 83-83*	WSr	British	393	144·0	33·0	12·9	Newport,N.S.	1883	J.Mosher	V 8.86	A. Haley
88150	378	*	Iolanthe *Holder*	WK	British	83	79·1	20·7	10·5	Brixham	1883	J.W.& A.Upham		C. Hellyer
41246	379		Iona	WBk	British	550	144·0	29·0	..					Tehwa Cheang [illegible]
67966	380	*	Iona *Trail 82-82*	WSr	British	99	88·1	22·1	10·1	Fraserburg	1874	Webster		A. Stephen
	381	N	Iona (ex Jona) *Klem -85*	WBk	Norweg'n	796	160·6	35·0	21·1	Quebec	1857	Russell&Co.	V 1.86	F. Klem
	382	DN	Iona *Stromberg -86*	WBk	Norweg'n	579	140·1	32·6	18·9	Arendal	1876	T.Lassen	V 12.87	Tellef Lassen
	383	*	Iona *Pehrsson 88-88*	WBk	Swedish	378	131·9	27·5	16·3	Dundee	1870	Tay Ship. Co.	V 11.88	J. Hasselgren
62564	384	*D	Ione *Edwards -87*	IBk	British	520	165·0	28·1	17·3	Sunderland	1869	Iliff, Mounsey Co		H.M'Ewan
83236	385	*	Ione *Jordan -87*	WBkn	British	319	121·5	27·8	14·9	P.E.Island	1882	J.Yeo	V 8.86	A.G.Moffatt(T.P.Rowland & Co., Managers)
81605	386	*	Ione *Jordan*	WBg	British	238	109·9	24·7	14·1	Cork	1877	J.Wheeler	V 12.86	W.B.Thoyts(Matthews & [illegible], Managers)
	387		Ionia	WBg	Turkish	186	..	..	..	Greece	1881			[illegible]
51117	388	*	Iota *Leonard 76-86*	WBk	British	551	151·2	29·6	19·1	Bideford	1866	Cox	V 11.86	Swansea Merchant Shipowners Co. Ld.
	389		Ioulia *Saratos*	WBg	Greek	124	..	..	..	Galixidi	1866			M.Sorotos
	390		Ipapandi *Vassilachi*		Greek	198	..	..	..	Syra	1880	E.Finui		N.Vassilachi
88724	391		Iphigenia	WSr	British	248	115·0	25·7	14·0	Christiania	1876	A.Olsen		C.C.Wyse
	392		Iphigenia *Kofoed 82-87*	WBn	Danish	251	99·9	24·6	15·1	Pillau	1861	H.O.Becker		M.Christensen
	393	G	Iphigenia *Söderlund 84-84*	WBk	Russian	464	146·5	29·3	17·2	Hamburg	1868	J.Marle		K.G.Söderlund
	394	H	Ipiros *Jeanoulis -88*	WSr	Greek	107	..	..	..	Syra	1879			N.D.Gialouris
12191	395		Ira Bliss *Hudson -75*	WSr	American	166	90·4	23·7	9·1	New York	18[illegible]			J.A. Hudson
100002	396		Ira D.Sturgis *Blaisdell*	W3mSr	American	223	118·3	30·0	8·5	Bath,Me.	1873			J.M.Haynes
12063	397		Ira Laffrinier *Crowell*	WSr	American	189	123·0	25·0	9·0	Clevel'd.,Ohio	1863			C. Lovell & Son
105	398	A	Iraclis *Candilaki*	WBg	Greek	240	93·2	26·1	..	Syra	1865			Andrew H.[illegible]
31121	399		Iravady Soundra Nayaghee	WSr	British	126	84·5	21·8	10·0	Keelicarre	1877			AnaCoos[illegible]
	400		Iraype *Rodrigues*	WB	Brazilian	154	..	..	..	P.Seguro	1875		V 7.88	J.P.Marques

*=Lloyd's Register; D=Nederlandsche Vereeniging; G=Germanischer Lloyd; I=Registro Italiano; [illegible]
A=Veritas Austro-Ungarico; H=Veritas Ellenico; N=Norske Veritas; V=[illegible]

In the 4th Column, C=Composite (iron frame, wood planked); I=Iron; Stl=Steel; W=Wood.

The Dates of Metalling (Col. 11) are based on the records of the Books in which Vessels are respectively classed, or on the Reports of the Sur[illegible]

Further particulars of Vessels classed by Lloyd's Register will be found in Lloyd's Register Book and Lloyd's Yacht [illegible]

Numéro officiel	No. dans la liste	Classification	Nom. Matériaux et Gréement. Capitaines et dates d'entrée au service des Armateurs actuels et à bord du navire. Ancien Nom.	Pavillon.	Tonnage net.	Longueur.	Largeur.	Creux.	Où.	Quand.	Par qui.	Date du doublage en métal	Armateurs ou Directeurs-Gérants.	Port d'Armement.
	401	*	Irby Shaw 85-81 I S	British	1480	237·6	38·0	23·3	Liverpool	1881	R.&J.Evans&Co.		[illegible]	Liverpool
	402		Irena Karlson W ImSr	Russian	189	..	..	..					N.Sittkoff	Mariehamn
	403		Irene Molree W Bg	British	260	116·0	26·4	12·9	P.E.Island	1877	Egan	Y.8.86	Mrs. M.Caughy	Liverpool
	404	*	Irene Rickards 15-84 W K	British	72	79·6	20·1	9·3	Falmouth	1884	C.Burt		J.Gray & Sons	Falmouth
	405		Irene Harding W Bn	American	468	131·6	31·3	16·5	Setauket,N.Y.	1876	G.E.Hand		Yates & Porterfield	New York
	406	V	Irene Pergoon W Bk	Austrian	532	131·5	26·4	18·9	Fiume	1874	Schiavoni		A.Luppis	Fiume
	407	*	Irène Nobé 77 I3m Bk	French	722	184·5	30·1	19·0	Stockton	186	M.Pearse&Co.		Ant. Dom.Bordes & fils	Bordeaux
	408		Irene [illegible] W Bn	German	267	106·6	26·1	14·2	Hamburg	1868	M.G.Amsinck		W.O'Swald & Co.	Hamburg
	409		Irene [illegible] Chrchot W Bk	Greek	464	137·8	27·9	17·8	Cherbourg	1859	Fl Letellier		P.Collakis	Syra
	410		Irene Sirigos W Bg	Greek	262	101·0	27·1	..	Syra	1877			A.Alafouzos	Santorin
	411		Irene Patronas W Bg	Greek	248	53·1	26·2	..	Mytilene	1870	Dem.Vasitis		S.Demetroulaki	Syra
	412		Irene W Bg	Greek	248	96·8	26·7	..	Syra	1868			E.A.Revidis	Piræus
	413		Irene W Sr		169	86·9	23·8	..	Syra	1877				
	414	H	Irene Dambasis W Sr	Greek	102	75·8	21·0	..	Syra	1886			L.Vroutisis	Andros
	415		Irene [illegible] W Bg	Italian	270	115·4	27·7	17·2	Gaeta	1870	L.Contreras		B.Zeccolilla	Naples
	416		Irene [illegible] W Bk	Norweg'n	471	133·5	30·3	20·6	Pra	1871	D.Tixi		S.B.Johnsen	Christiansand
	417	DN	Irene [illegible] Boirenen W Bk	Norweg'n	409	116·9	30·7	17·3	Brevig	1855			S.Johansen&N.B.Olsen	Drammen
	418		Irene Olsen -88 W Bk	Norweg'n	385	119·2	30·1	16·5	Grimstad	1868	H.Jörgensen		Osuld Jörgensen	Grimstad
	419		Irene W Sr	Russian	115	..	..	..		1869			K.Sopetoff	Sordavala
	420		Irene W Bg	Turkish	166	..	..	..	Cashot	1872			Yanni Manyarli	Cashot
	421		Irene E. Meservey W Sr	American	233	114·0	29·0	10·0	Waldob'ro'Me	1866	W.Welt		G.W.Rawley	St.George,Me.
	422		Irene Isigonis W Br	Greek	373	130·9	..	..	Liverpool	1862			Dem.Isigonis	Skiathos
	423		Ireneo [illegible] W Bk	Austrian	531	142·7	26·5	17·5	Fiume	1870	Schiavoni		G.Schnautz fu A.	Buccari
	424	H	Irenoula W Bg	Greek	263	98·8	27·2	..	Syra	1871			N.Z.Goulandris	Andros
	425	*	Irex W Sk	British	83	77·2	20·5	10·5	Elmshorn	1888	J.Kremer		T.Baskcomb	Grimsby
	426		Iris Griffiths -87 W S	British	967	173·2	36·2	22·6	Newport,N.S.	1870	Mosher		R.Griffiths	Liverpool
	427		Iris Williamson -88 W Bk	British	340	120·0	27·7	17·0	Sunderland	1869	T.Metcalfe		Gillan & Hay	Newcastle,N.S.W.
	428		Iris Shaw 83-79 W Sr	British	206	116·7	22·7	10·3	FremantleW.A.	1879			J.Bateman	Fremantle,W.A.
	429		Iris (NOT IN EXISTENCE) W Bn	British	127	85·0	25·2	10·6	Wallace,N.S.	1855			R.B.Huestis	Picton,N.S.
	430	*V	Iris [illegible] Ohlsen 73-79 I 3m Sr	German	507	156·5	27·2	17·3	Harrington	1863	R.Williamson		M.G.Amsinck	Hamburg
	431	N	Iris Andersen 65-78 W Bk	Norweg'n	480	123·7	31·1	18·6	Uleaborg	1840	Langesund		A.Andersen	Grimstad
	432	*	Iris Hevmann -88 W Bk	Norweg'n	458	122·0	29·6	15·7	Vegesack	1856	J.Lange		Actieselskabet "Iris"	
	433	N	Iris Ulsted 82-87 W Bg	Norweg'n	222	110·0	27·4	12·1	Arendal	1867	J.Aanonsen		S.Ulsted	Arendal
	434	N	Iris Gulliksen W Bg	Norweg'n	13[illegible]	93·6	23·7	11·9	Porsgrund	1853			I.C.Lund	Tönsberg
	435	N	Iris Cornelinsen W Bn	Norweg'n	159	91·8	23·9	10·2	Molde	1856			O.A.Föyen	Tönsberg
	436	DV	Iris Strom W Bk	Russian	767	173·1	36·7	19·6	Brahestad	1876	Z.Durchman		Z.Durchman	Brahestad
	437		Iris W Bk	Spanish	222	104·3	22·8	14·6	Bilbao	1850			Ramon Ramos	Barcelona
	438	V	Iris Hansson -78 W Bk	Swedish	3[illegible]	121·3	25·7	16·[illegible]	Nantes	1863	Pelletier & Guibert	V 10.81	C.Molander	Wisby
	439		Iris Z.auerström -88 W Bk	Swedish	3[illegible]	120·0	25·5	11·5	[illegible]	1868	J.Jörnvik		A.Söderström	Calmar
	440	*	Irish Girl McKern's 76-76 W Sr	British	125	93·0	[illegible]	11·1	Dundalk	1876	Connick		R.McClenahan	Dundalk
	441	*	Irish Minstrel [illegible] W Bn	British	136	99·1	25·0	11·1	Dundalk	1879	Connick		H.C.Edwards	Dundalk
	442		Irma W Bk	British	265	108·0	24·0	9·0	[illegible]	1847			N.M.Mills	Bridgetown
	443	V	Irma Escarcel W Sr	French	109	82·7	[illegible]	[illegible]	Dunkirk	1861	Sagary fils		Boone	Dunkirk
	444		Irmak Ali Osman W Bn	Turkish	156	..	..	..	Irmak	1868			Hajee Elias & Sons & Co.	Irmak
	445	*	Iron Crag Janet 78-78 I Bk	British	827	201·0	32·3	19·6	Whitehaven	1877	WhitehavenS.B. Co.,Ld.		T.Connell	Liverpool
	446		Iron Cross Jacob 86-8 I S	British	1558	247·0	35·2	18·7	Greenock	1854	Scott&Co.		H.Fernie & Sons	Liverpool
	447	*V	Iron Duke Bunge 78-86 I S	German	1435	217·0	36·2	25·4	Liverpool	1863	H.M.Lawrence &Co.		H.Bischoff & Co.	Bremen
	448	*	Iron Queen Freeman I Bk	British	796	185·0	29·8	21·1	Liverpool	1862	W.H.Potter&Co.		G.H.Fletcher & Co.	Liverpool
	449	*D	Iron Queen Williams 85-87 I Bk	British	234	113·0	23·4	13·7	Newport	1865	Huddart		T.Pugsley & Co.	Newport
	450	*	Ironside Ford -87 I Bk	British	899	185·0	32·7	20·4	Sunderland	1861	T.R.Oswald		Temperleys, Carter & Darke	London

* = Lloyd's Register; D = Nederlandsche Vereeniging; G = Germanischer Lloyd; I = Registro Italiano; R = American Record; A = Veritas Austro-Ungarico; H = Veritas Ellenico; N = Norske Veritas; V = Bureau Veritas.

Dans la 4me Colonne, C = Composite (membrure en fer, bordé en bois); I = Fer; Stl = Acier; W = Bois.

Les Dates du doublage en métal (Col. 11) ressortent des livres dans lesquels les navires sont respectivement classés ou des rapports des Inspecteurs du Lloyd's Register.

Pour plus amples renseignements sur les navires classés par le Lloyd's Register, voir le Lloyd's Register Book et le Lloyd's Yacht Register.

Official No.	No.	Classed	Ship's Name. Material and Rig. Master, and Dates of Appointment to present Owner's Service and to Ship. Ship's Late Name.	Flag	Net Tonnage	Length	Breadth	Depth	Built: Where	When	By whom	Date of Metalling	Owners
60296	451	RV	Iroquois *Nichols* 81-81 WS	American	2005	239·1	43·6	27·9	Bath, Me.	1881	A. Sewall	[illegible]	A. Sewall & Co.
	452		Irskatt *Pettersson* WSr	Swedish	104	76·1	24·3	9·7	Roslagen	1862	J. Ostensson		J. E. Ersson
38561	453	※D	Irvine *Boyd* I Bk	British	655	170·6	29·5	19·1	Sunderland	1867	W. Doxford		D. Hunter
62129	454	※	Isa Reid *Winchester* 83-88 WSr	British	99	88·7	21·3	10·1	Garmouth	1876	Geddie, jun.		A. Coull
[illegible]	455	R	Isaac Burpee *Matthews* -88 WSr	British	180	95·0	28·1	9·0	Westfield, N.B.	1880	S. J. Holder		E. C. Elkin
100398	456	R	Isaac Dodge *Gaul* WBkn (WRECKED)	American	515	118·6	34·0	13·0	Newcastle, Me.	1885	Gay & Haggott	11.85	T. E. Gay
100359	457	R	Isaac H. Tillyer *Risley* -88 W3mSr	American	568	144·6	35·0	16·8	Dennisville, N.J.	1884	R. S. Leaming		W. H. Smith
12130	458		Isaac Hall *Stone* W Bk	American	580	140·0	32·0	17·0	Stockton, Me.	1869	Colcord & Berry		J. H. Winchester & Co.
100123	459	R	Isaac Jackson *Foss* WBk	American	586	145·3	32·6	17·8	Deering, Me.	1874		[illegible]	G. A. Goodwin
100345	460	R	Isaac N. Kerlin *Steelman* W3mSr	American	349	136·7	35·0	10·0	Leesburg, N.J.	1883	J. Russell		J. Russell
12402	461	Y	Isaac Orbeton *Trim* W3mSr	American	367	120·0	29·4	12·0	Newcastle, Me.	1870	J. Clark		John Cassiday
100156	462	RV	Isaac Reed *Wallace* -87 WS	American	1489	212·0	40·0	24·2	Waldoboro', Me.	1875	A. Reed & Co.	[illegible]	Yates & Porterfield
100245	463		Isaac T. Campbell *Mathews* W3mSr	American	557	157·0	33·9	15·3	Waldoboro', Me.	1879	A. Storer & Son		J. Baker & Co.
6	464		Isaak *Lauter* -83 W3mSr	Russian	336	110·0	26·5	14·0	Haynasch	1873	J. Brackmann		A. Weide
358	465	DG	Isaak *Stahl* -86 W3mSr	Russian	262	113·0	28·3	13·1	Nogallen	1880	J. Stahl		J. & M. Stahl
86282	466	※	Isabel *Drew* 76-88 WBk	British	579	152·0	30·8	18·9	Sunderland	1868	J. Gill	12.87	W. C. Jarvis
80918	467	R	Isabel *McClure* WBk	British	350	119·2	23·3	15·6	P. E. Island	1880	C. McLure	[illegible]	C. McLure
49867	468		Isabel *Fishley* WSr	British	177	102·1	23·8	12·7	Newquay	1865	Davies		J. M. Fishley, jun.
74736	469	※	Isabel *Tupman* I 3mSr	British	149	99·0	21·5	11·6	Whitehaven	1880	Whitehaven S. B. Co., Ld.		B. Tupman
7160	470		Isabel *Stott* 85-85 WBn	British	127	86·5	22·9	11·9	P. E. Island	1850			G. Sinclair
12263	471		Isabel (ABANDONED) *Rainford* WSr	American	176	103·2	29·8	9·1	Coos Bay, Oreg.	1864			A. Anderson
	472		Isabel I Sr	Argentine	199	130·0	27·2	9·0	Glasgow	1884	Abercorn S. B. Co.		V. Casares & Son
	473		Isabel WSr	Brazilian	146	..	..	..					A. F. P. da Cruz
	474	※V	Isabel WBk	Norweg'n	1264	202·8	39·2	22·3	S. Maitland, N.S.	1880	W. P. Cameron	[illegible]	Johan Thorsen
	475		Isabel WBk	Norweg'n	324	123·0	27·8	13·3	Rhode Island	1854			G. G. Maarud
	476		Isabel (ex Carmencita) WBk	Spanish	215	106·9	25·1	14·1	Viavelez	1866	J. Rom		M. Bonet Coll
	477		Isabel WBn	Spanish	149	92·5	23·1	10·7	Villajoyosa	1852			V. M. Galiana
12365	478		Isabel Alberto *Bishop* WSr	American	218	103·4	28·7	10·0	Port Jefferson, N.Y.	18[illegible]			S. L. Keene
90077	479	※	Isabel Browne *James* 78-85 I Bk	British	1289	230·5	36·1	21·1	Port Glasgow	1885	Russell & Co.		Browne & Watson
	480		Isabel Peral (ex Louis) WBk	Peruvian	297	117·0	25·0	14·0	France	1869			J. V. Peral
	481	DV	Isabelita *Nikera* 83-83 WBn	Spanish	249	108·6	21·5	13·6	Bilbao	1871	A. de Arana	[illegible]	M. de Astigurraga
	482		Isabelita *San Juan* WPol	Spanish	180	89·2	26·4	10·9	Mahon	1854		[illegible]	P. Maristany
	483		Isabelita WSr	Spanish	102	78·7	22·5	9·9	Adra	1871			Jose Cuenca
[illegible]	484		Isabella (STRANDED) *Baker* 84-91 WBg	British	270	103·7	26·2	16·1	Sunderland	1861	G. Bartram		J. W. Woods
[illegible]	485		Isabella *Fraser* WSw	British	110	77·8	25·0	9·3	Bay St. George (N.F.)	1877			J. McFatridge
[illegible]	486	※	Isabella *Coppack* 81-86 WSr	British	85	88·5	21·2	9·0	Barrow	1878	W. Ashburner		T. Ashburner & Co.
[illegible]	487	※	Isabella *Jones* -89 WSr	British	79	75·0	18·6	9·9	Fraserburg	1855			O. Jones
[illegible]	488	V	Isabella Balcolm *Armstrong* WBn	British	301	126·0	28·7	12·6	Parrsboro', N.S.	1881	J. Balcolm	[illegible]	E. E. Hutchings
	489	DN	Isabella Blyth *Berner* WBk	Norweg'n	633	139·0	30·9	20·8	London	1847		11.86	Carl Berner
[illegible]	490		Isabella Diston *Tucker* 85-85 WSr	British	155	93·5	22·5	12·7	Inverkeithing	1870	R. & D. Ross		J. Helme
[illegible]	491		Isabella Hall (WRECKED) WSr	British	140	95·5	22·2	12·3	Glasgow	1865	Scott & Co.		J. Fisher & Sons
[illegible]	492	※	Isabella Helen *Lewis* WSr	British	98	79·5	21·6	10·6	Plymouth	1877	C. Gent	[illegible]	J. Prizeman
[illegible]	493		Isabella Leith (FOUNDERED) WSr	British	163	91·4	23·6	13·3	Sunderland	1848			C. Barkaway
	494	V	Isabella V. (ex Domenica II.) *Marciani* WBk	Italian	457	136·8	29·4	19·7	Sestri	1868	Peragallo	11.86	B. Varé
[illegible]	495	※	Isabella Walker *Masson* WBn	British	179	103·7	22·3	12·8	Inverkeithing	1870	R. and D. Ross		J. Chalmers
[illegible]	496		Isabella Wilson *Ellis* -88 WSr	British	169	102·0	23·1	13·0	Garmouth	1865	W. Kinloch		W. G. Crinks
[illegible]	497	※	Isabelle *Williams* 76-86 WBkn	British	289	122·6	27·3	14·9	P. E. Island	1875	Richards	11.86	T. P. Richards & Co.
[illegible]	498		Isabelle WBn	British	129	97·5	23·2	10·0	George River, N.S.W.	1863			B. Byrnes
[illegible]	499		Isabelle (ex Sabrina) WB	British	123	83·5	22·5	11·5	Sunderland	1862	W. Barkley & Co.	[illegible]	I. Baird
	500		Isabelle *Harr* 86-87 WSr	French	122	89·6	22·6	11·1	Dunkirk	1880	G. Beck		G. Beck

※ = Lloyd's Register; D = Nederlandsche Vereeniging; G = Germanischer Lloyd; I = Registro Italiano;
A = Veritas Austro-Ungarico; H = Veritas Ellenico; N = Norske Veritas;

In the 4th Column, C = Composite (iron frame, wood planked); I = Iron; Stl = Steel; W = Wood.

The Dates of Metalling (Col. 11) are based on the records of the Books in which Vessels are respectively classed, or on the Reports of the [illegible]

Further particulars of Vessels classed by Lloyd's Register will be found in Lloyd's Register Book and Lloyd's Yacht [illegible]

No.	Cote	Nom. Matériaux et Gréement. Capitaine et dates d'entrée en service des Armateurs actuels et à bord du navire. Ancien Nom.	Pavillon	Tonnage	Longueur	Largeur	Creux	Construit où	Quand	Par qui	11	Armateurs ou Directeurs-Gérants.	Port d'Armement.
[illegible]		Isabelle Marie Padron WSr	French	[illegible]	[illegible]			[illegible]				Verry Carantan	Binic
502	R	Isaiah Hart Nickerson WBmSr	American	853	183·1	35·1	[illegible]	[illegible]				W.H.Nickerson	Boston, Mass.
503	R	Isaiah K. Stetson WBk	American	398	130·9	31·3	10·[illegible]	[illegible], Me.				G.Stetson	Bangor, Me.
504		Isaline Peixoto WBg	Brazilian	180	..	..	..	Pelotas				C.Dubeaux	Pernambuco
505	N	Isbaadan Jacobsen WBn	Norweg'n	393	123·3	26·1	11·[illegible]	Dundas, P.E.I.	1875	M'Kean		Act.Selsk."Ispilen" (J.Olsen)	Kragerö
506	N	Isbjörnen Larsen WBk	Norweg'n	342	11[illegible]	27·8	16·5	Sunderland	1863	J.Robinson		Act.Selsk."Ispilen" (J.Olsen)	Kragerö
507		Isca Colins WBg	British	293	125·3	25·5	14·6	Exmouth	1868	Redway		Ockenden & Son	London
508	N	Isfareren Olsen WBk	Norweg'n	392	133·8	30·3	12·7	Newport, N.S.	1869	L.Harvey		Act.Selsk."Ispilen" (J.Olsen)	Kragerö
509		Ishiki Maru WBg	Japanese	162	99·9	22·0	10·8	Ominato, Ise	1882	Matsuzaki Yohei		Natsume Ichirobei	Owari
510	✱I	Iside Schiaffino WBk	Italian	735	164·0	[illegible]·4	19·4	Rapallo	1873	G.Merello		R.Schiaffino	Genoa
511	I	Iside B. Foni WBk (ex Tommaso di Savoia)	Italian	1004	183·0	29·5	23·1	Varazze	1879	A.Cerruti		G.B.Beverino	Spezia
512	✱	Isidora I Sr	Uruguay'n	64	75·5	25·5	6·2	London	1884	Forrest & Son		W.B.Cranwell & Co.	Monte Video
513		Isidoros Caloyani WBg	Turkish	238	99·1	27·2	17·4	Spezzia	1867			Nicola Yorghiadi	Vassilikos
514	V	Isidra Fabregas WPol	Spanish	185	88·3	27·1	13·0	Arens	1858			Maristany hermanos.	Barcelona
515		Isidro (ex Signo de Oro) WBn	Spanish	175	95·5	23·4	11·7	Alkey	1859			V.Balbas & Co.	Manila
516	✱	Isipingo Phillips WBk	British	310	135·2	27·1	12·9	London	1872	Salisbury	V 3.87	J.Loveluck & Son	London
517	A	Isis Vidulich WSr	Austrian	134	83·3	20·2	11·2	Lussinpiccolo	1883	G.Giuricich	V 9.83	A.D.Vidulich fu G.	Lussinpiccolo
518	✱	Iskender Mehemed WBk	Turkish	548	133·0	29·3	14·5	Trieste	1867	Strendoff	C 7.86	Sadyk Bey	Constantinople
519		Iskender Mehemed WBk	Turkish	512	..	..	..	Austria	1886			Artin Caramanian Effendi	Constantinople
520		Iskenderieh WBn	Turkish	332	..	..	..	France	1867			Iskenderieh Tijaretkhane	Iskenderieh
521		Isla de Cuba WSr	Spanish	104	83·0	22·8	16·2	Blanes	1861			I.Carbonell y Juncosa	Mataro
522	N	Island Gjertsen WBk	Norweg'n	640	156·7	32·8	19·1	Horton, N.S.	1864	J.B.North		I.F. & G.F.Gjertsen	Tönsberg
523	N	Island Christiansen WBn	Norweg'n	112	85·0	21·6	9·7	Mandal	1879			J.Fredriksen	Mandal
524	N	Island Sørensen WSr	Norweg'n	105	81·0	22·8	10·4	Bergen	1874			N.B.Dahl	Christiansund
525		Island City Warner WBk	British	421	133·6	30·4	16·3	Farmingham, U.S.	1852			J.Gillan	Newcastle, N.S.W.
526		Island City Yorkis WBmSr	American	406	137·1	28·8	15·0	E.Boston, Mass.	1871	T.Nickerson		L.J.Dale & Co.	Savannah, Ga.
527	R	Island Lass Bradling WSr	British	176	102·2	24·5	12·9	Fowey	1870	Batson		H.B.Varwell	Exeter
528	✱	Island Maid Beal WSr	British	150	89·5	23·3	12·1	Salcombe	1865	Harnden	V 8.86	E.Rillston	Fowey
529		Island Maid Johnston WBn	British	135	91·8	22·6	11·6	Whitehaven	1865	Williamson		J.Storey	Belfast
530		Island Maid Bowes WSr	British	124	96·2	20·4	11·2	Plymouth	1863	Shilston		W.H.Cox	Plymouth
531	✱	Island Queen Brabazon WBk	British	353	123·0	28·8	14·8	P.E.Island	1866	Green		Sk."Island Queen" Co.	Newport
532		Islander WBk	British	277	104·0	26·7	15·1	Fairhaven U.S.	1856			W.J.Gillam	Fremantle, W.A.
533		Islander Boucher WBn	British	186	103·1	21·2	12·9	P.E.Island	1876	Duncan		N.Potter & Co.	Guernsey
534		Islay Brian WBn	British	134	94·6	22·8	12·0	Wallace, N.S.	1860			John Munn & Co.	St. John's, N.F.
535	✱	Isle of Anglesea Bennett IBk	British	725	187·3	31·2	18·5	Glasgow	1872	Dobie & Co.		J.Bushby	Liverpool
536	✱D	Isle of Bute Whalley IBk	British	929	209·6	34·3	20·7	Port Glasgow	1875	J.Reid & Co.		J.Bushby	Liverpool
537	✱	Isle of Erin IBk	British	889	202·5	32·5	19·5	Glasgow	1875	D.& W.Henderson & Co		J.Bushby	Liverpool
538		Isle of Pines Legg WSr	American	162	100·0	30·0	7·9	Bass River, NJ.	1856			W.H.Legg	New York
539	V	Ismaele (ex Maresca) Caputo WBk	Italian	410	136·8	27·7	17·8	Cassano	1867	F.De Rosa		T.Maresca	Castellamaro
540	✱V	Ismir Sanders 81-87 WS	British	1260	200·3	36·8	22·6	St. Mary's Bay, N.S.	1878	J.Bingay		J.Bingay	Yarmouth, N.S.
541		Ismyr Crisp WBn	British	210	102·6	23·2	15·1	Liverpool	1856	Royden		W.Milburn	North Shields
542	✱	Isobel Davies 83-86 WBk	British	349	120·8	28·1	14·9	P.E.Island	1875	H.McDonald		D.Davies	Aberystwith
543	A	Isodia Mouromatis WSr	Greek	199	98·1	25·1	..	Syra	1885	A.Lividaros		N.Mouromatis	Syra
544	A	Isodia Makrisis WBg	Russian	181	..	..	..	Spezia	1867			H.Filitoff	Kertsch
545	✱	Isola Gulley IBk	British	653	180·0	30·2	17·7	Glasgow	1876	J.Elder & Co.		J.Kelso	Glasgow
546		Isolina WSr	Brazilian	119	..	..	..					J.A.O.Maia	Porto Alegre
547		Isolina Tonietti WBmSr	Italian	163	86·0	25·8	14·4	Pt.Longone	1859			V.Soldani fu D.	Porto Ferraio
548		Ispilen Knudsen 87-87 WBg	Norweg'n	294	108·3	26·3	16·4	Sunderland	1865	J.Crown		Act.Selsk."Ispilen" (J.Olsen)	Kragerö
549	N	Isploven Knudsen WBk	Norweg'n	369	118·8	27·1	16·7	Sunderland	1865	W.H.Pearson		Act.Selsk."Ispilen" (J.Olsen)	Kragerö
550	N	Ispolen Bee WBg	Norweg'n	234	108·1	26·0	12·1	St.Magnus	1865	D.Olausen		Act.Selsk."Ispilen" (J.Olsen)	Kragerö

✱ = Lloyd's Register; D = Nederlandsche Vereeniging; G = Germanischer Lloyd; I = Registro Italiano; R = American Record; A = Veritas Austro-Ungarico; H = Veritas Ellenico; N = Norske Veritas; V = Bureau Veritas.

Dans la 4me Colonne, C = Composite (membrure en fer, bordé en bois); I = Fer; Stl = Acier; W = Bois.

Les Dates de doublage en métal (Col. 11) ressortent des livres dans lesquels les navires sont respectivement classés ou des rapports des Inspecteurs du Lloyd's Register.

Pour plus amples renseignements sur les navires classés par le Lloyd's Register, voir le Lloyd's Register Book et le Lloyd's Yacht Register.

Official No.	No. in Book	Registry in which Classed	Ship's Name. Material and Rig. Master, and Dates of Appointment to present Owner's Service and to Ship. Ship's Late Name.	Flag.	Tons	Dimensions: Length	Breadth	Depth	Built: Where.	When	By whom.	Date of Metalling	Owners or Managers
	551		Issavlah *Angelidis* W Bg	Turkish	152	..	..	..	[illegible]	1869			Pavlikos, Nakkos & Co.
	552	✻	Issidoros *Founies* W Bg	Turkish	320	..	..	..		1868		[illegible]	Mrs. M Caloyanni
	553		Issim Sis W S	Turkish	406	..	..	..	[illegible]	1883			Saraf Vassilaki Emme Effendi
	554		Issim Sis W Bg	Turkish	275	..	..	..		1873			Rahtvalee-tijar Hassan Agha
	555		Issim Sis W Bg	Turkish	255	..	..	..	[illegible]	1870			Hajee Mustapha Eff.
	556		Issim Sis W Bn	Turkish	250	..	..	..	[illegible]	1869			Isskenderieh-tijario-den Ibrahim Tahsene
	557		Issim Sis W Bn	Turkish	206	..	..	..	Sakarieh	1869			Stambolloe Nicolaki-Velideh-Ermak
	558		Issim Sis *Omer* W Bn	Turkish	199	..	..	..	Greece	1868			Hairie-tijar Osman Yazidji & Co.
	559		Issim Sis W Sr	Turkish	199	..	..	..	Barten	1882			Mustapha Vehbi & Co.
	560		Issim Sis W Bg	Turkish	191	..	..	..	Italy	1878			Helthi Corahy
	561		Issim Sis *Chroni Reiss* W Bg	Turkish	184	..	..	..	Greece	1869			Nanno Petro & Co.
	562		Issim Sis W Bn	Turkish	178	..	..	..	Kara Agatch	1869			Mahaujtler-Kaharand Mehemed Son's Wife Hanoum
	563		Issim Sis W Bn	Turkish	175	..	..	..					Yazidjeh-zadeh Elias Osman
	564		Issim Sis W Sr	Turkish	174	..	..	..	Roumania	1878			Aktarpopulu Nicholas Dimitri Kaheh
	565		Issim Sis W Bg	Turkish	162	..	..	..	Greece	1869			Andon Salidi
	566		Issim Sis W Bn	Turkish	162	..	..	..	Greece	1869			Zishtolee-Vezin-oglou Ibrahim
	567		Issim Sis W Sr	Turkish	158	..	..	..	Vassilikos	1868			[illegible] & Co.
	568		Issim Sis W Bn	Turkish	152	..	..	..	Greece	1868			[illegible]
	569		Issim Sis *Seid Reiss* W S	Turkish	150	..	..	..	Rustchuk	1869			[illegible] & Co.
	570		Issim Sis *Ismael* W S	Turkish	150	..	..	..	Rustchuk	1869			[illegible] Ibrahim & [illegible]
	571		Issim Sis *Mehemed Numan* W Bn	Turkish	150	..	..	..	Barten	1884			Bomtakki Tijari Reiss Zadeh Mehemed [illegible]
	572		Issim Sis W Bg	Turkish	149	..	..	..	England	1883			Hajee Ali Effendi Zeyoe Enneeh Hallah Hassan
	573		Issim Sis *Ismael* W Sr	Turkish	149	..	..	..	Haramson [Skelleast	1884			[illegible]
	574		Issim Sis W Bn	Turkish	146	..	..	..	Greece	1869			Iskendarloe Pappas [illegible] Hotelsador
	575		Issim Sis W Bg	Turkish	142	..	..	..	Russia	1869			Alatlar Dimitri & Co.
	576		Issim Sis W Bg	Turkish	140	..	..	..	Greece	1869			Sakyzlee Yanni Velo Marmora
	577		Issim Sis W Bg	Turkish	137	..	..	..	Roumania	1868			Kiriazi Yanni
	578		Issim Sis W Bn	Turkish	119	..	..	..	Greece	1869			Limnilee Yamandi Mikkiai
	579		Issim Sis W Bg	Turkish	118	..	..	..	Greece	1869			Petro Anestas
	580		Issim Sis W Bn	Turkish	115	..	..	..	Barten	1869			Stambollee Panayoti Kallitopolo
	581		Issim Sis W	Turkish	106	..	..	..	Kooroofeh-Shi	1869			Vassilikealee Costi F. amb
	582		Issim Sis W S	Turkish	100	..	..	..	Greece	1882			[illegible]
	583		Issim Sis W Sr	Turkish	100	..	..	..	Barten	1869			[illegible] Sirdurloe Gaze Bekir & Bros.
	584		Istephan *Saleh* W Bn	Turkish	102	..	..	..	Istephan	1868			Abtahnablee Ali Tehe Iebi Oglou Faut
	585		Italia *McCollan* W Bg	British	116	92·0	25·7	10·0	Lunenburg, NS	1864			John F. Morris
255	586	I	Italia *Ferra* W Bk	Italian	899	183·0	35·0	23·0	Loano	1876	A. Cerruti	[illegible]	Flli. Coxe fu F.
1210	587	IV	Italia (ex Giuseppina) *Esposito* W Bk	Italian	535	147·6	30·0	19·4	Sorrento	1879	M. Paturzo	Y 12.87	G.P. & L. Lauro fra.
	588	N	Italia (ex Valparaiso) *Isaacson* -88 W Bk	Norwegian	829	169·1	33·9	21·1	Genoa	1871	A. Briasco	Y 10.83	Enkefru H. Blom
1716	589		Italo *Panese* W Bk	Italian	480	136·4	31·4	19·5	Sestri Ponente	1868	Frat. Cadenaccio	Y 10.88	Giulio D'Aste
87910	590	✻	Itata *Goddard* Stl Bk	British	927	202·0	33·0	19·9	Liverpool	1883	R&J Evans & Co.		R. & J. Evans & Co.
7	591		Ithaka W Sr	Greek	148	88·6	23·6	..	Braila	1863			D. Vlassopoulos
518	592	A	Ithaka *Chandos* W Sr	Greek	146	86·0	23·4	..	Syra	1888	A. Livadana		P. A. Chandos
66273	593	D	Ithuriel *Clements* 85-86 W Bk	British	319	121·2	26·5	14·9	P. E. Island	1871	Knight	[illegible]	Bevan, Evans & Co.
	594		Itiuba W Sr	Brazilian	162	..	..	..					J. S. Belem
	595	✻	Iton (ex Lady Louisa) *Regnier* -88 C Bk	French	542	152·0	28·3	17·9	Sunderland	1865	R. Thompson Jr.	[illegible]	F. Regnier
90017	596	RV	Itonus *Buckman* 75-80 W Bk	American	818	162·4	34·8	21·1	Yarmouth, Me.	1876	Hutchins & [Stubbs	[illegible]	B. Webster
29	597	A	Ivan (ex Zaro) *Bielich* W Bk	Austrian	609	144·0	33·5	18·6	Trieste	1871	Strudhoff	Y 12.87	G. Bielich
	598	✻	Ivan *Andersen* 83-88 W Bn	Danish	213	103·4	21·6	12·5	P. E. Island	1876	McLaren		C. P. Bentzen
	599		Ivan *Sullinen* W Glt	Russian	285	..	..	..	[illegible]	1872			J. A. R. Lindberg
	600		Ivan *Prokofejeff* W Glt	Russian	229	..	..	..	[illegible]	1878			I. Sidoroff

✻ = Lloyd's Register; D = Nederlandsche Vereeniging; G = Germanischer Lloyd; I = Registro Italiano;
A = Veritas Austro-Ungarico; H = Veritas Ellenico; N = Norske Veritas;

In the 4th Column, C = Composite (iron frame, wood planked); I = Iron; Stl = Steel; W = Wood.

The Dates of Metalling (Col. 11) are based on the records of the Books in which Vessels are respectively classed, or on the Reports of the [illegible]

Further particulars of Vessels classed by Lloyd's Register will be found in Lloyd's Register Book and Lloyd's Yacht Register.

No.		Nom. Matériaux et Gréement. Capitaine et dates d'entrée au service des Armateurs actuels et à bord du navire.			Pavillon	Tonnage	Dimensions.			Construit.			Doublage	Armateurs ou Directeurs-Gérants.	Port d'Armement.
		Nom	Capitaine				Longueur	Largeur	Creux	Où.	Quand.	Par qui.			
601		Ivan	Baksejeff	WG[illegible]	Russian	2[illegible]	..	..	..	Sordavala	1884			I.Spiridonoff	Sordavala
602		Ivan	Safronoff	WGR	Russian	139	..	..	..	Sordavala	1875			H.N.Moldakoff	Sordavala
603	✻	Ivanhoe	George -87	I Bk	British	1332	235·2	37·1	23·7	PortGlasgow	1868	JohnReid&Co.		J.Whittingham	Melbourne
604	✻	Ivanhoe	Williams	WBk	British	347	130·7	26·8	16·3	Aberystwith	1871	Evans	1888	W.Morris & Co.	Aberystwith
605		Ivanhoe	Carter 85-85	WS	American	1563	202·3	39·3	27·6	Belfast,Me.	1865	C.P.Carter	[illegible]	A.W.Rose, junr.	San Francisco
606		Ivanhoe		WS	American	114	86·0	28·2	8·5	CoosBay,Oreg.	1869			T.Pollard	San Francisco
607	N	Ivar	Henriksen	W3mSr	Norweg'n	314	118·3	28·2	14·8	Finland	1865			W.M.Michaelsen	Moss
608	V	Ivar	Larsson -23	WSr	Swedish	102	81·0	26·8	8·4	Carlshamn	1880	D.R Andersson		N.Tufvesson	Brantevik
609		IvarHvitfeldt	Hansen	WBk	Danish	290	111·2	24·7	16·0	Wismar	1883	J.H.D.Barmann		H.P.Christiansen	Copenhagen
610	V	Ivigtut	Andersen --	WBk	British	331	122·3	28·7	15·7	Quebec	1877	McKay&Dix		McKay & Dix	Quebec
611	A	Ivis	Pappa	WSr	Greek	268	..	..	..	Syra	1886	M.Cafudachi		G.D.Pappa	Syra
612	V	Ivy	Caron	WBk	British	580	157·8	28·8	17·7	Quebec	1879	P.V.Valin	[illegible]	P.V.Valin	Quebec
613	✻	Ivy	Bone 85-85	WK	British	65	74·6	19·2	9·2	Point	1878	Stephens		W.Hosken & Co.	Falmouth
614	RV	Ivy	Gardiner -87	WS	American	1181	184·3	37·2	23·8	Phippsburgh, Me.	1876	C.V.Minott	[illegible]	C.V.Minott	Bath,Me.
615		Ivy	Gillis	WSr	American	135	102·5	28·8	8·8	SanFrancisco	1882			J.J.McKinnon	San Francisco
616		Ivy Belle	Lord	WSr	American	117	88·3	25·2	7·8	Bristol,Me.	1870			E.Lord	Damariscotta,Me.
617		Iwan	Johannsen	WSr	Russian	100	65·0	21·0	9·0	Orrenhof	1879	A.Dreymann		M.Birk	Haynasch
618		Iwata Maru		WSr	Japanese	195	100·7	21·2	12·9					UkonGonsayemon	Osaka
619	GV	Izabel (ex Adele)	D'Araujo	WBk	Portug'se	1143	185·8	36·9	23·6	Bremerhaven	1881	F.W.Wencke	[illegible]	C.da Silva Ferreira	Oporto
620		Izetta	Armstrong	WSr	American	190	106·9	27·8	9·0	Bangor,Me.	1866			S.Hincks	Bangor,Me.
621		Izolina (ex [illegible])	Silva Barbosa	WBk	Portug'se	242	118·7	25·6	10·5	Oporto	1867	C.J.d'Azevedo Varella	[illegible]	L.Alves,Rodrigues&Co.	Oporto
622	V	Izvor (ex [illegible])	Radonicich	WBk	Austrian	705	153·2	30·9	20·6	Capo d' Istria	1875	L.Polli	[illegible]	A.Tripcovich fu N.	Meglino
623		Izzedin		WSr	Turkish	112	..	..	..	SudaBay,Crete	1869			SudaleeIzraki Hussein	Suda Bay

ADDITIONAL SHIPS ARE INSERTED IN THE SUPPLEMENT AT THE END OF THE REGISTER.

✻ = Lloyd's Register; [illegible] = Veritas Austro-Ungarico; D = Nederlandsche Vereeniging; G = Germanischer Lloyd; H = Veritas Ellenico; I = Registro Italiano; N = Norske Veritas; R = American Record; V = Bureau Veritas.

Dans la 4me Colonne, C = Composite (membrure en fer, bordé en bois); I = Fer; Stl = Acier; W = Bois.

Les Dates du doublage en métal (Col. 11) ressortent des livres dans lesquels les navires sont respectivement classés ou des rapports des Inspecteurs du Lloyd's Register.

Pour plus amples renseignements sur les navires classés par le Lloyd's Register, voir le Lloyd's Register Book et le Lloyd's Yacht Register.

Official No.	No. in Book	Registry in which classed	Ship's Name. Master, and Dates of Appointment to present Owner's Service and to Ship. Ship's Late Name.	Material and Rig	Flag	Net Tonnage	Length	Breadth	Depth	Built: Where	When	By whom	Date of Metalling	Owners or Managers
1900	151		Il Nuovo Giacomino *Treglia*	W Bn	Italian	184	85·6	24·1	16·0	Meta	1859	Mauro		G. Treglia
[illegible]	152		Il Nuovo Michele *Vuono -83*	W Bn	Italian	152	96·4	24·2	12·6	Castellam. di Stabia	1875	M. Trambarulo	Y 7.87	M. Pace
[illegible]	153		Il Risorto *Carletti*	W Bn	Italian	105	72·2	22·5	12·2	Savona	1852	Tixi		Flli Carletti fu T.
[illegible]	154		Il Salvatore	W Bg	Italian	287	115·4	27·7	17·2	Gaeta	1870	L. Contreras	Y 8.84	B. Zeccolilla
[illegible]	155		Il Salvatore *Miano*	W Bg	Italian	141	87·9	23·3	9·3	Limite	1856	N. Pichiotti	[illegible]	Gius. Salvo
[illegible]	156	A	Il Terrible *Magliocca -88*	W Bn	Italian	149	97·1	22·3	12·8	Gaeta	1858			Flli Mutarazzo fu A.
1325	157		Il Tomasino	W Bn	Italian	100	80·0	22·0	10·9	Torre del Greco	1867			F. Osciono
789	158		Il Vero *Cafiero*	W Bk	Italian	475	129·6	30·8	19·1	Cassano	1869	M. Paturzo	Y 12.87	De Martino Flli & Sarell
	159		Il Vincenzo *Cacciuttolo*	W Bg	Italian	165	..	..	..	Sira	1863			M. Cacciuttolo
86254	160	✻	Ilala *Parsell 83-86*	I S	British	1329	235·0	37·5	21·5	Whitehaven	1882	WhitehvnSBCo		Lowden, Edgar & Co.
271	161	A D	Ilario (ex Mirto) *Persich -88*	W Bg	Austrian	260	106·9	24·7	14·7	Fiume	1869	C. Giacovicich		G. Persich
	162	V	Ile de Terre Neuve *Lyon*	W Bk	French	229	100·0	23·0	13·1	St. Servan	1873	Hamon	Y 4.87	Guibert & Fils
121	163	✻ A	Ilia (ex William D. Seed) *Tomicich -87*	W Bk	Austrian	746	173·5	33·5	19·7	Sunderland	1874	W. Pickersgill	Y 5.86	G. Milossevich & Co.
6	164		Ilias *Castrisiotis*	W Bg	Greek	267	102.0	..	..	Syra	1874		C 75	A. & G. Thermisiotis
	165		Ilish	W Sr	Turkish	125	..	..	..	Ilish	1865			Ineboli Hatce Fam Oglou & Co.
46	166		Ilissus	W	Greek	167	124·6	20·0	5·9	Austria	1876			[illegible]
45210	167		I'll Try *Henderson*	W Bg	British	152	97·0	21·9	12·1	Aberdeen	1863	Smith		J. Allan
85076	168	✻	Illawarra *Carnosso 66-83*	I S	British	1887	269·1	40·6	24·0	Glasgow	1881	Dobie & Co.		Devitt & Moore
	169	N	Illeri *Iverson -85*	W Bn	Norweg'n	228	109·7	26·5	11·8	Sandefjord	1878			P. Tobiassen
70500	170	✻	Illimani *Popham*	I Bk	British	590	174·0	29·2	17·4	Glasgow	1875	Dobie & Co.		J. E. Burgess
	171	N	Ilma *Albrethsen*	W Bk	Norweg'n	437	129·0	29·7	15·8	Arendal	1870	O. B. Sörensen	Y 10.85	O. B. Sörensen
	172	D ✻	Ilma *Jacobsen*	Stl 3m Sr	Norweg'n	332	145·8	27·0	11·1	Grangemouth	1885	Grangemouth Dockyard Co		Akt. Selsk. "Ilma" (E. Klaus)
	173		Ilma *Eklund 80-83*	W Bk	Russian	522	134·6	33·5	18·0	Åland	1849	M. Bergendahl		H. H. Märtenson
	174		Ilma *Andersson*	W Sr	Russian	307	111·4	28·7	14·2	Karis	1877	G. F. Sandell		K. F. Hjelt
	175		Ilma *Salminen*	W Sr	Russian	220	102·2	27·7	13·0	Töfsala	1876	G. Craber		B. Gustafson
	176		Ilma *Jansson*	W Sr	Russian	110	..	..	..	Lappfjerd	1873			J. P. Mattson
	177		Ilmarinen *Westergren 83-85*	W Bg	Russian	258	101·0	25·9	14·0	Fredrikshamn	1874	T. Wilkki		Leonard Kaski
	178	N	Ilmatar *Bonde 64-76*	W Bk	Norweg'n	579	139·5	34·3	18·6	Brahestad	1869	A. Westerlund		V. Lerche
	179		Ilmatar *Carlsson 86-86*	W Bg	Russian	224	108·0	25·5	14·3	Wyborg	1875	M. Repo		K. Nilsson
	180		Ilo	W Sr	Norweg'n	185	99·3	25·2	10·8	Christinestad	1859			H. H. Parr
	181	N	Ilos (ex Julius) *Olsen 82-82*	W Bk	Norweg'n	729	145·0	33·5	21·5	Vegesack	1848	P. Sager	[illegible]	[illegible]
	182	N V	Ilos *Tharaldsen -88*	W S	Norweg'n	720	163·0	34·5	18·8	Arendal	1878	O. J. Heriofson	Y 3.87	O. J. Heriofson
	183	V	Ilos (ex Jenny) *Junnelius -86*	W Bk	Russian	533	140·0	31·8	20·7	Uleaborg	1867	J. Kjeldström	Y 1.86	H. O. Store
	184	V	Ilpotar *Johansen*	W Bk	Norweg'n	586	144·4	34·8	18·6	Brahestad	18[illegible]			H. N. Gjersöe
	185	D N	Imacos *Olsen 88-88*	W Bk	Norweg'n	577	146·1	33·5	18·2	Tvedestrand	1880	C. Beck	Y 7.86	Carl Beck
	186		Imanta *Ansack*	W Sr	Russian	180	89·0	23·0	11·0	Neu Salis	1876	C. Ohsoling		M. Behrsing
	187	V	Imatra *Snellman -87*	W S	Russian	1024	192·9	37·4	23·0	Christinestad	1868	M. Hannus	Y 12.87	C. G. Hyden
	188		Imatra *Hamberg*	W 3m Sr	Russian	394	..	..	..	Korpo	1875			J. V. Anderson
87033	189	✻	Imberhorne *Hansen*	I S	British	7199	284·1	41·2	24·1	Dumbarton	1882	M'Millan & Son		"Imberhorne" Ship Co. Ld. (W. R. Price & Co.)
43	190		Imbros	W	Greek	179	121·3	19·7	9·8	Austria	1871			Theofilatos Stathatos
	191		Imenes	W Sr	Brazilian	189	..	..	..					J. J. S. I. Junior
	192	V	Imes *Hogstrom*	W 3m Sr	Swedish	252	112·8	23·2	12·1	Westervik	1880	G. R. Fridell	[illegible]	G. Ideström
2661	193		Immacolato (ex Pietro G.) *Muro*	W Bk	Italian	637	146·6	33·0	21·3	Varazze	1871	G. Scala	Y 4.86	M. V. Muro
1171	194	I	Immacolata *Longobardi*	W 3m Sr	Italian	353	119·7	26·4	17·2	Alimuri	1877	G. Mauro	Y 7.86	G. B. Longobardi & Figli
2202	195	V	Immacolata Concezione (ex [illegible]) *Demauria*	W Bg	Italian	234	103·6	25·3	15·7	Sestri Ponente	1871	D. Tixi	Y 10.87	Giov. Demauria
354	196	I	Immacolata Concezione *Pavone*	W Bg	Italian	132	83·9	22·3	12·7	Sestri Ponente	1855	D. Tixi	[illegible]	Flli Pace
	197	V	Immanuel *Kisby 78-*	W Sr	Danish	137	84·7	21·5	11·3	Marstal	1867	J. J. Bager		H. C. Christensen [illegible]
	198	G	Immanuel *Strömpel 81-87*	W Bg	German	218	96·4	24·9	13·4	Greifswald	1861	L. Wittenberg		J. J. Wallis & Sohn
	199	N	Immanuel *Bruun -88*	W Bk	Norweg'n	465	126·0	30·8	17·3	Skien	1853	L. Jörgensen		A. Bruun
	200	N	Immanuel *Ribe*	W Bk	Norweg'n	460	138·5	30·0	16·4	Grimstad	1876	Hans Ribe	[illegible]	Hans Ribe

✻ = **Lloyd's Register**; D = Nederlandsche Vereeniging; G = Germanischer Lloyd; I = Registro Italiano; R = [illegible]
A = Veritas Austro-Ungarico; H = Veritas Ellenico; N = Norske Veritas; V = [illegible]

In the 4th Column, C = Composite (iron frame, wood planked); I = Iron; Stl = Steel; W = Wood.

The Dates of Metalling (**Col. 11**) are based on the records of the Books in which Vessels are respectively classed, or on the Reports of the Surveyors [illegible]

Further particulars of Vessels classed by Lloyd's Register will be found in Lloyd's Register Book and Lloyd's Yacht [illegible]

No.	Cote	Nom. Matériaux et Gréement. Capitaine et dates d'entrée au service des Armateurs actuels et à bord du navire. Ancien Nom.		Pavillon.	Tonnage net.	Longueur	Largeur	Creux	Construit. Où.	Quand.	Par qui.	11	Armateurs ou Directeurs-Gérants.	Port d'Armement.
201	[illegible]	Immanuel *Mikkelsen -83*	W Bk	Norweg'n	424	129·0	30·1	16·6	Porsgrund	1868		Y [illegible]	P.A.Steesen	Brevig
202	N	Immanuel *Svendsen 77-77*	W Bg	Norweg'n	208	83·8	25·3	14·8	Sunderland	1832			C.A.Berg	Holmestrand
203	N	Immanuel *Kjölner 75-75*	W Sr	Norweg'n	129	91·8	22·1	11·0	Skien	1865			F.C.& K.Kjölner	Tönsberg
204	N	Immanuel *Olsen*	W Bn	Norweg'n	126	91·9	22·6	9·4	Brevig	1869	P.Knudsen		M.E.Schilbred	Brevig
205	DN	Immanuel *Gulbrandsen* ex [illegible]	W Sr	Norweg'n	111	73·3	21·7	10·3	Stavanger	1825			I.T.Rove	Sandefjord
206		Impallay	Stl Bg	British	652	250·1	34·0	8·8	Dumbarton	1888	W.Denny & Bros		Irrawaddy Flotilla Co Ld.	Glasgow
207		Imperador	W Bn	Brazilian	149	..	..	..					J.R. e Souza	Rio de Janeiro
208		Imperador do Brazil	W Bn	Brazilian	249	..	..	..					C.V.Vianna	Rio de Janeiro
209	NV	Imperator *Hove 84-84*	W S	Norweg'n	1082	199·7	38·0	21·5	Stavanger	1876	C.Knudsen	Y [illegible]	Holdt & Isachsen	Stavanger
210	N	Imperator *Nielsen -87*	W Bk	Norweg'n	566	156·9	33·1	17·9	Laurvig	1870	O.Dahl	Y [illegible]	Joh Hanssen	Arendal
211		Imperator *Sonnstrom* ([illegible])	W Bk	Russian	303	..	..	..	Baltimore	1857			F.E.Nyfors	Åbo
212	A	Imperatore Francesco Giuseppe I. *Premuda*	W Bk	Austrian	667	147·6	29·6	17·5	Trieste	1875	A.Schiavoni	Y [illegible]	M.V.Premuda	Lussinpiccolo
213	V	Impératrice *Gueder -87*	W Sr	French	141	88·9	21·3	12·4	Boulogne	1868	Lecerf		Noël	Paimpol
214	I	Imperatrice *[illegible]*	W Bg	Italian	103	88·2	20·6	10·7	TorredelGreco	1886	M.Frambarelo	Y [illegible]	A.Brancamo	Naples
215		Imperial *Ratcliffe*	W S	British	100	82·0	20·6	10·4	Chester	1867	Bishton		C.Davison	Chester
216	RV	Imperial *Crosby -84*	W S	American	1288	188·7	38·0	23·5	Quincy,Mass.	1869	G.Thomas	Y 11.86	J.E.Crosby	Boston,Mass.
217		Impérial		French	123	..	..	..						Arles
218	V	Impérieuse *Brinck 81-87*	W Bk	Danish	354	132·8	27·1	14·4	Hamburg	1865	E.Dreyer	Y [illegible]	N.H.Brinck	Sønderho Fanø
219		Impetuous *Hunter -61* ([illegible])	W Bn	British	163	91·7	23·1	12·7	Rye	1854			W.W.Foreman	Guernsey
220		Impi *Bystrom 71-77*	W Bk	Russian	637	143·6	31·1	20·5	Brahestad	1868	E.Klubb		J.Lang	Brahestad
221		Impi	W Bk	Russian	250	97·1	25·9	13·4	Pyterlax	1873	D.Hietala		F.A.Söderström	Åbo
222		Impi	W Bg	Russian	237	114·6	26·9	14·1	Fredrikshamn	1874	D.Hietala		V.Söderlund	Åbo
223	V	Importer *Wessels*	W Bk	American	1206	196·6	32·2	23·9	Newburyport, Mass.	1870	J.Currier	Y [illegible]	H.Koop & Co.	New York
224		Improviso *De Lellis*	W Bn	Brazilian	105	..	..	..	Fão	1870		Y [illegible]	Donna C. M. da Silva & Filhos	Paranaguá
225		Imsland *Nygaard*	W Bn	Norweg'n	237	110·2	26·1	12·6	Stavanger	1875	Torjus Imsland	Y [illegible]	B.Hansen	Stavanger
226	✻	Imuncina *Brown 88-88*	I Bk	British	583	165·6	30·2	17·4	Glasgow	1867	Dobie & Co.		Donaldson Brothers	Glasgow
227	G	Ina *Voss 72-72* (ex Wm. Hunter)	W Bk	German	389	136·4	27·5	17·1	Whitby	1867	T.Turnbull & Son	Y 10.88	C.F.Maass	Rostock
228	V	Ina *Sandvold -88* (ex Due Sorelle)	W 3m Sr	Norweg'n	252	115·0	25·0	13·9	Sestri Ponente	1874	B.Canepa	Y [illegible]	Nils R.Parelius	Christiansand
229		Inca *Troth*	W Bk	British	398	132·6	26·6	18·3	Liverpool	1862	Royden		Lucas Bros. & Co.	Bristol
230		Inca *Gilli 81-86*	W Bk	German	729	149·9	31·3	18·6	Elsfleth	1882	F.Deetjen	Y [illegible]	J.H.Hustede	Elsfleth
231		Incantatrice *Citrano* (ex Enchantress)	W Sr	Italian	146	105·4	21·1	11·6	Ipswich	1874	Wm.Bayley & Sons	Y [illegible]	D.G.Garufli	Catania
232	✻D	Inch Keith *Crawford 84-88*	I Bk	British	1238	236·3	35·5	21·6	Glasgow	1878	Barclay,Curle & Co.		J.Hardie	Glasgow
233	✻D	Inch Murren *Petrie 79-81* ([illegible])	I S	British	1254	231·1	35·8	21·6	Dumbarton	1878	Birrell,Stenhouse & Co.		J.Hardie	Glasgow
234	✻	Inchcape Rock *[illegible] 88-88*	I S	British	1493	250·5	38·3	23·0	Port Glasgow	1886	Russell & Co.		J.Cornfoot & Co.	Glasgow
235	✻	Inchgreen *Miller 80-86*	I Bk	British	1091	219·5	34·1	20·9	Greenock	1876	Caird & Co.		D.MacDougall	Greenock
236		Indécis *Delalande -87*	W Bg	French	106	..	..	..	St.Malo	1840			Guibert & fils	St.Malo
237		Indefatigable *Duff*	W Sr	British	139	71·5	20·4	12·2	Trefrne.	1801			S.Boustead	Drogheda
238		Independence *Farrell -88*	W S	American	895	165·6	34·2	22·7	E.Boston,Mass	1871	Curtis & Smith		C.P.Hemenway	Boston,Mass.
239		Independencia *d'Araujo*	W Sr	Brazilian	145	..	..	..	S.João da Barr	1864		Y [illegible]	Cia Cabotagem S. João da Barra	Rio de Janeiro
240		Index *Garrison*	W 3m Sr	American	318	116·5	31·0	8·5	Camden,N.J.	1870	Taylor & Mathis		L.Garrison	Bridgeton,N.J.
241		India	W S	British	436	121·2	26·2	20·8	Portsm'th,U.S	1833			J.V.Hall	Sydney,N.S.W.
242	V	India *Clapp 83-85*	W S	American	1295	192·1	36·5	23·9	Bath,Me.	1868	J.Patten & Son	Y [illegible]	J.Rosenfeld's Sons	Bath,Me.
243	G	India *Regener 71-81*	C Bk	German	974	185·0	34·3	19·5	Glasgow	1876	A.Stephen & Sns	Y [illegible]	D.H.Wätjen & Co.	Bremen
244		India ([illegible]) *Kalff -80*	W Bk	German	574	118·3	31·6	17·9	Elbing	1856	G.A.Mitzlaff		J.Paleske	Danzig
245	V	India *Pundt 88-88*	W Bk	German	284	103·5	25·5	10·5	[illegible]	1872	J.Oltmann's Wwe.	Y [illegible]	D.Haye	Brake
246	NR	India *[illegible] -86* ([illegible])	W Bk	Norweg'n	1216	189·2	38·8	23·6	St.John,N.B.	1862	Gass	Y 10.86	Akt. Selsk. "India" (A. Bruun)	Arendal
247	DV	India ([illegible]) *Imsland*	W Bk	Norweg'n	1141	201·0	38·9	23·3	Bergen	1874	J.Gran	Y [illegible]	M.Beyer	Bergen
248	DNV	India *Gundersen -85*	W Bk	Norweg'n	798	168·2	33·8	20·0	[illegible]	1875	H.F.Ulrichs	Y [illegible]	O.C.Reinhardt	Christiansand
249	N	India *Andersen -88*	W Bk	Norweg'n	796	190·0	35·9	21·1	Drammen	1871	T.Lorentzen	Y [illegible]	A.P.Andersen	Moss
250														

✻ = Lloyd's Register; D = Nederlandsche Vereeniging; G = Germanischer Lloyd; I = Registro Italiano; R = American Record; A = Veritas Austro-Ungarico; H = Veritas Ellenico; N = Norske Veritas; V = Bureau Veritas.

Dans la 4me Colonne, C = Composite (membrure en fer, bordé en bois); I = Fer; Stl = Acier; W = Bois.

Les Dates du doublage en métal (Col. 11) ressortent des livres dans lesquels les navires sont respectivement classés ou des rapports des Inspecteurs du Lloyd's Register.

Pour plus amples renseignements sur les navires classés par le Lloyd's Register, voir le Lloyd's Register Book et le Lloyd's Yacht Register.

1 Official No.	2 No. in Book	3 Registry in which classed	4 Ship's Name. Material and Rig. Master, and Dates of Appointment to present Owner's Service and to Ship. Ship's Late Name.	5 Flag.	6 Net Tonnage.	7 Dimensions. Length	Breadth	Depth	8 Built. Where	9 When	10 By whom	11 Dates of Metalling	12 Owners or Managers
	251	N	India *Jorgensen* W3mSr	Norweg'n	350	131·3	27·6	12·0	Grimstad	1885	JohanJörgensen		[illegible]
	252		India *Freixas* WBk	Spanish	570	150·5	34·1	22·0	Barcelona	1869			J.Ferri
	253	V	India *Pla* WPol	Spanish	182	91·2	26·7	12·1	Blanes	1856			[illegible]
	254		India WBn	Spanish	110	88·3	19·7	9·0	Albay	1871			José Cardell
	255		India *Petterson* -85 WBk	Swedish	492	129·6	29·7	18·1	Töreforo	1857			P.Persson
79933	256	✱	Indian Chief *Weir* 81-81 WBkn	British	339	125·7	26·2	14·7	Kingston	1878	Kinloch		W.K[illegible]
5579	257	✱	Indian Empire *Watson* 75-75 IS	British	1515	230·0	38·8	23·5	W.Hartlepon	1865	Pile,Spence&Co.		G.Duncan & Co.
35288	258		Indian Queen *Armstrong* WBn	British	113	91·4	22·6	8·7	Calais,Me.	1849			J.Brown
66028	259	✱	Indiana *Pennill* 85-85 W3mSr	British	289	130·1	26·7	12·9	Sunderland	1873	W.Pickersgill		W.C.[illegible]
100186	260	V	Indiana *Morrison* 81-81 WS	American	1414	208·9	40·0	23·9	Bath,Me.	1876	A.Sewall&Co.		A.Sewall & Co.
	261	N	Indiana *Björness* 80-83 WBk	Norweg'n	521	139·2	30·7	18·1	Montrose	1852			H.Zernichow
	262	V	Indien *Roussen* -87 WBk	French	422	118·1	28·2	18·5	Marseilles	1867	V.Jeavin		Lallement
	263	V	Indien *Frumerie* 65-85 WBk	Swedish	592	158·8	29·6	17·4	Westervik	1863	C.R.Fridell		C.Carlbom
352	264		Indipendente (ex Giuseppe) *Sicaile* WBg	Italian	224	100·0	24·9	15·4	Castellamare	1874			V.Culotta
597	265	I	Indipendenza *Regesta* [84-84] WBk	Italian	598	147·6	32·1	20·9	Chiavari	1872	Gotuzzo		F.Dall'O[illegible]
	266		Indo *Soderman* WBk	Swedish	254	106·6	25·8	13·0	Vätö	1875			M.Soderman
91276	267	✱	Indore *Jeaney* 81-85 IS	British	2088	279·0	40·7	24·6	Stockton	1885	Richardson, [Duck&Co.		Eyre,Evans & Co.
	268	V	Indra *Gieseke* 84-84 IBk	German	695	166·0	30·0	18·9	Hamburg	1877	ReiherstiegS.W.		[illegible]
13884	269		Indus (ex steamer) *King* IBk	British	1319	242·1	38·5	21·1	London	1847			[illegible] Telegraph Co.,Ld.
46713	270		Indus WSr	British	126	94·0	24·4	7·8	Dorchester,US	1857			J.Eneas
2484	271	I	Indus *Righetti* WS	Italian	1111	196·5	31·7	24·0	SestriPonente	1874	Flli Cadenacio	Y 1.86	G.Lavarello di P.
1374	272	A	Industre *Sodich* WBk	Austrian	563	119·2	28·9	19·5	Fiume	1868	V.Brazzoduro		C.cav.Cosulich
	273	N	Industri (ex Industry) *Sampson* -89 WSr	Norweg'n	212	107·3	23·0	12·5	P.E.Island	1863			M.Andreasen
	274	N	Industri *Andresen* W3mSr	Norweg'n	201	103·7	25·6	12·3	Sandefjord	1870			M.A.Andresen
2336	275	I	Industria *Olivari* -88 WBk	Italian	600	141·7	29·2	19·2	Prà	1873	B.Canepa		F.Schiaffino
3263	276		Industria *Di Meo* WBg	Italian	210	104·9	25·5	16·2	Savona	1853	G.Tixi		Paolo Cicchero
	277	V	Industria (ex Marianna III.) *Duarte* WBk	Portug'se	316	122·0	27·9	12·8	SanMartinhoPort	1862	J.R.Assinço	Y 4.85	ViuvaMacieira & [illegible]
	278		Industria *Arroja* WSr	Portug'se	155	88·9	21·3	10·5	Vianna	1874	M.P.Netto		J.A. de Magelhaes
	279	V	Industrie *Esnouf* WBk	French	187	94·8	22·6	13·3	Granville	1863	Duvanl	Y 1.86	J.Pattier
	280		Industrie See "Industry," No. 289. WSr	French	148	..	..	..	Jersey	1867			
	281	V	Industrie *Kirchhoff* 82-82 IS	German	1642	228·0	43·0	24·2	Kinderdyk	1872	L.Smit&Zoon		C.H.H.Winters
	282	V	Industrie *Pretwurst* 77-77 WBk	German	333	119·4	26·9	15·5	Danzig	1859	W.R.Grott		Lichenheim & [illegible]
	283		Industrie *Hillet* WBg	German	248	109·9	25·1	12·5	Brake	1866	F.Nicolai		H.Hillel
	284		Industrie *Meislahn* 70-74 WBg	German	188	87·3	24·3	12·9	Lubeck	1848	J.Landmann		J.Rahlff & Sohn
	285	V	Industrie *Kleen* 82-82 WGlt	German	100	83·0	19·8	9·2	Grossefehn	1868	C.Meyer		E.C.Aden
17573	286		Industry *Tyrrell* 85-85 WBkn	British	189	108·2	24·4	13·0	St.John,N.B.	1857			E.Kearon
73915	287		Industry W Barge	British	105	95·8	21·7	7·5	MillPoint	1879	W.Jamieson		H.B.Rathbun & [illegible]
60577	288	✱	Industry *Jones* 86-88 WSr	British	84	79·0	18·6	9·1	Gloucester	1867	Davies		J.Jones
	289		Industry (BROKEN UP) *Moke* 88-88 WBn	French	144	98·9	20·1	12·1	Jersey	1867	Le Vesconte		Adam
	290		Ines *Worms* WBn	Russian	222	104·6	25·4	12·6	Strömly	1874	K.Mäkela		A.Michelsson
	291		Inés (ex Arangelo) *Coll* WBn	Spanish	336	105·9	28·1	17·1	Savona	1861	Calamaro		Doña Rita [illegible]
	292	G	Inés *Mas Espinas* WBk	Spanish	200	105·6	25·4	12·3	Palma	1858	M.Planes	Y 10.86	M.Mas Espinas
	293		Inés (ex Dolores, ex Elena) WBn	Spanish	142	87·6	24·2	10·9	Blanes	1852			J.M.Roca
	294		Inés (ex Vestal) WSr	Spanish	104	75·8	22·9	10·1	Arens	1851			F.Bellver
	295	DV	Inés Røhl *Andersen* WBg	Danish	248	115·1	24·3	12·7	Varnanäs	1881	A.F.Bring		Aktie Selskab [illegible]
	296		Inez *Voisin* WSr	French	197	95·8	19·9	11·0	Bristol	1860	Patterson		[illegible]
	297	N	Infatigable *Falck* 84-84 WBk	Norweg'n	843	180·5	34·8	20·0	Grimstad	1884	A.Knudsen		M.Falck
	298	G	Infatigable *Stephansson* 87-87 WBk	Swedish	330	116·4	28·1	16·6	Grimstad	1860			L.P.Stephansson
90406	299	✱	Inflexible *Hace* 85-85 WK	British	90	82·0	20·8	10·9	Brixham	1884	S.Dewdney & [Sons		[illegible]
	300	RV	Inga (ex Rosa Ronhanri) *Grann* 78-88 WBk	Norweg'n	1391	188·0	40·1	23·6	St.John,N.B.	1867	J.&R.Reed		T.M.N[illegible]

✱=Lloyd's Register; D=Nederlandsche Vereeniging; G=Germanischer Lloyd; I=Registro Italiano;
A=Veritas Austro-Ungarico; H=Veritas Ellenico; N=Norske Veritas;

In the 4th Column, C=Composite (iron frame, wood planked); I=Iron; Stl=Steel; W=Wood.

The Dates of Metalling (Col. 11) are based on the records of the Books in which Vessels are respectively classed, or on the Reports of the [illegible]

Further particulars of Vessels classed by Lloyd's Register will be found in Lloyd's Register Book and Lloyd's Yacht [illegible]

No.	Classif.	Nom. Capitaine et dates d'entrée. Ancien Nom.	Matériaux et Gréement	Pavillon	Tonnage	Longueur	Largeur	Creux	Construit Où	Quand	Par qui	Date doublage	Armateurs ou Directeurs Gérants.	Port d'Armement.
301	[illegible]	Inga [illegible]	W Bk	Norweg'n	537	132·2	31·6	20·3	Aberdeen	1865	J. Duthie Sons [& Co]	[illegible]	[illegible] Larsen	Sandefiord
302	[illegible]	[illegible]	W Bk	Norweg'n	732	141·7	31·3	21·0	Portsm'th, N.B.	1838			F.J. Holst's Enke	Moss
303	DV	[illegible]	W Bk	Norweg'n	436	137·7	32·1	15·0	Economy, N.S.	1873	Fraser	[illegible]	[illegible]	Christiania
304	NV	Inga *Isaksen -86*	W Bk	Norweg'n	457	136·9	30·6	17·4	Christiansand	1877	Br&Sr. Larsen	[illegible]	T. A. O[illegible]	Christiansand
305	N	Inga *Heffermehl 86-88*	W Bg	Norweg'n	214	102·3	26·8	12·1	Drammen	1868	Jorgensen [&] Knudsen		L. Heffermeld	Drammen
306	NV	Inga *Orbeck -88*	W Sr	Norweg'n	156	[illegible]	[illegible]	10·8	Sönderho	1868	F. [illegible]	[illegible]	A.N. Ørbeck	Lillesand
307	V	Ingeborg *Jensen -87* [illegible]	W Bk	Danish	195	110·2	22·3	11·7	Copenhagen	1868	Trap & Michelsen	[illegible]	C.J. Hœpfner	Copenhagen
308	N	Ingeborg *Roer* (ex Sparkling Wave)	W Bk	Norweg'n	397	130·1	27·[illegible]	15·9	Bathurst	1861	O'Brien		M. Röer	Christiania
309	NV	Ingeborg *Knudsen*	W Bg	Norweg'n	242	101·9	26·4	15·9	Sunderland	1866			Chr. Hannevig	Christiania
310		Ingeborg *Hammerström -86*	W Bk	Russian	537	..	..	..	Geta	1874			E. Montell	Geta
311	V	Ingeborg *Fex*	W Bk	Swedish	315	132·2	25·6	12·5	Timmernabben	1878	C. Nilsson	[illegible]	G. Fex	Lerhamn
312	V	Ingeborg *Abrahamsson*	W Sr	Swedish	202	100·0	23·[illegible]	12·0	Oberg	1880	J.H. Selden		B. Johansson	Uddevalla
313		Ingeborg *Nilsson -87*	W Sr	Swedish	120	84·5	19·8	10·0	Brake	1868	Oltmann		N. Nilsson	Brantevik
314	B	Ingeborg *Smedberg*	W Sr	Swedish	115	86·4	22·1	8·7	Ström	1858			A. Andersson	Utsand, Relgn.
315	GV	Ingeburg *Lueks -88*	W Bk	German	397	119·7	25·9	15·6	Apenrade	1871	C. Raben	[illegible]	Th. & F. Eimbcke	Hamburg
316	V	Inger Margrethe [illegible]	W Bn	Danish	160	86·2	20·5	11·5	Veenlam	1864			C. Clemmensen	Skive
317	V	Ingleside *Hayes -88*	W Bk	British	926	172·1	34·6	21·0	MossGlen, N.B.	1877	G. Merritt	V 1.88	The Bark Director Co.	St John, N.B.
318	✻	Inglewood *Beeching -88*	I Bk	British	1043	215·0	34·1	21·1	Harrington	1875	R. Williamson & Son		R. Williamson & Son	Workington
319	DV	Ingmar *Jørgensen*	W Bg	Danish	204	103·8	23·5	14·6	Nordby, Fanø	1874	C. Graff	V 7.8	Akt. Selsk. (J.S.N. Hansen)	Nordby, Fanø
320		Ingo (ex Alf) *Maybaum*	W Bg	Russian	254	95·0	23·9	13·6	Moss	1866	E. Pedersen		Maybaum	Reval
321	N	Ingolf *Knudsen 86-86*	W Bk	Norweg'n	367	128·2	30·1	15·9	Tvedestrand	1874	A.L. Helm	V 7.8	R.A. Knudsen (A. & F. Smith)	Tvedestrand
322	V	Ingomar *Sinclair*	W S	British	1175	192·7	37·8	24·0	Newport, N.S.	1875	Harvey	[illegible]	A. Gilison & Co.	Liverpool
323	✻	Ingrid *Moore*	W Sr	British	99	83·8	22·7	10·8	Padstow	1873	Stribley		Mrs. C. Jenkin	Padstow
324		Ingrid *Österling*	W Sm Sr	Russian	262	..	..	..	Bjerno	1865			W. Akerblom	Nystad
325		Ingrid *Sundin 72-81*	W Sr	Swedish	136	89·6	20·6	10·7	Lerwick, Scotland	1859			G.E. Ekström	Oscarshamn
326	N	Inheritance *Smith -88*	W Bk	Norweg'n	406	131·6	28·2	17·2	Sunderland	1860	Pearson	[illegible]	Actieselskabet "Inheritance" [illegible]	Christiania
327	V	Innerwick *Waters 82-82*	W Bk	British	1238	194·7	36·6	22·8	River John, N.S.	1882	J. Kitchin	[illegible]	A.C. MacDonald	Pictou, N.S.
328	✻	Innisfail *Core 86-88*	W Bn	British	200	99·2	25·1	13·1	Dublin	1859	Good	1.88	Goodyear & Co.	Liverpool
329	✻	Innisfallen *Lodge -88*	W Sw	British	232	112·2	26·7	16·2	Blyth	1868	Robinson	[illegible]	Mrs. S.D. Stonehouse	South Shields
330	I	Innocenta *Schiaffino*	W Bk	Italian	751	170·2	33·7	20·9	Sestri Ponente	1871	A. Briasco fu L.	V 7.87	Alberto Nattini	Genoa
331	II	Innocenti *Canepa*	W Bk	Italian	902	173·8	31·2	23·8	Varazze	1883	G. Scala	V 2.88	P. Lavarello	Genoa
332	I	Innocenza Rosa *Napoli* [illegible]	W Bk	Italian	407	116·4	30·5	18·9	Savona	1866	G. Tixi	V 2.88	S. Napoli	Catania
333	V	Ino *Søgel 75-75*	W Bn	Danish	147	93·7	21·4	11·3	Nysted	1871	F. Sparre		A.H. Friis	Marstal
334	V	Ino	W Bn		215	104·6	25·0	14·2	Brake	1867	F. Nicolai	V 1.8		
335	N	Ino *Jonassen 72-73*	W Bk	Norweg'n	337	111·9	28·8	16·1	Grimstad	1856	Jörgen Olsen	V 1.8	Christen Boe	Arendal
336	N	Ino *Stark -88*	W Bn	Norweg'n	160	111·3	25·5	9·8	Kragerö	1868	Dekke	V 3.8	H. Biörn, jr.	Kragerö
337	GV	Ino *Pyk 86-*	W Bk	Swedish	326	129·5	26·3	13·6	Hamburg	1864	M.G. Amsinck	V 8.8	G. Sjöberg	Malmö
338	DV	Ino *Roemberg 82-85*	W Bk	Swedish	325	131·5	25·6	12·9	Kolboda	1874	H. Olsen	V 2.8	C. Sylvan	Helsingborg
339	H	Insani Hounda [illegible]	W Bk	Turkish	154	..	..	..	Spezzia	1838			Ibraim Effendi	Constantinople
340	GV	Insel *Rettgers 78-*	W Bg	German	192	94·7	23·2	11·0	Papenburg	1867	A. Beckmann	V 12.85	Abr. Beckmann	Papenburg
341	I	Insonne *Sponza*	W Bn	Italian	139	86·6	25·7	11·4	Chioggia	1833	G. Poli	V 8.8	Ant. Sponza	Venice
342	N	Insula Capri *Larsen -87*	W Bk	Norweg'n	394	125·3	29·7	16·7	Arendal	1869	S. Johansen	V 9.8	S.M. Klocker	Arendal
343	N	Insulan *Pedersen*	W Bg	Norweg'n	216	106·1	25·3	11·9	Risör	1866	O.A. Olsen		B.A. Olsen	Risör
344		Insulaneren *Hansen 86-*	W Bg	Danish	231	103·5	23·7	13·9	Svendborg	1857	H.P. Hansen		Thorvald Hansen	Svendborg
345	DV	Insulinde *Spaanderman 81-86*	W S	Dutch	1326	202·1	39·7	24·8	Alblasserdam	1879	F.H. van Lindern	V 1.87	Von Lindern & Van Vessem	Rotterdam
346		Integridad *Gelpi*	W Bg	Spanish	282	101·7	29·5	15·1	Mataro	1854	F. Segario		P. Sust y Mes	Barcelona
347		International	I Sp	British	382	120·0	31·0	11·9	London	1879	Fletcher, Son & Fearnall		London Patent Elevator Co. Ld.	London
348		International *Court*	W Bn	British	192	108·0	23·3	12·8	P.E. Island	1862	Coffin		W. Goldfinch	London
349	✻	Intrepid *Le Lacheur*	W K	British	59	79·1	20·3	9·2	Falmouth	1882	Burt		S. Taylor	Guernsey
350														

✻ = Lloyd's Register; D = Nederlandsche Vereeniging; G = Germanischer Lloyd; I = Registro Italiano; R = American Record; [illegible] = Veritas Austro-Ungarico; H = Veritas Ellenico; N = Norske Veritas; V = Bureau Veritas.

Dans la 4me Colonne, C = Composite (membrure en fer, bordé en bois); I = Fer; Stl = Acier; W = Bois.

Les Dates de doublage en métal (Col. 11) ressortent des livres dans lesquels les navires sont respectivement classés ou des rapports des Inspecteurs du Lloyd's Register.

Pour plus amples renseignements sur les navires classés par le Lloyd's Register, voir le Lloyd's Register Book et le Lloyd's Yacht Register.

1 Official No.	2 No. in Book	3 Registry in which classed	4 Ship's Name, Master, and Dates of Appointment to present Owner's Service and to Ship. Ship's Late Name. Material and Rig.	5 Flag.	6 Net Tonnage.	7 Dimensions: Length	Breadth	Depth	8 Built: Where.	9 When.	10 By whom.	11 Date of Metalling.	12 Owners or Managers.
68214	351	R	Intrepid *Clausen -87* W Sr Yt	American	118	105·5	25·3	11·2	Brooklyn	1878	C.& R.Poillon	[illegible]	Lloyd Phœnix
	352	V	Intrépide *Evrard* W Sr	French	115	88·3	21·6	11·2	Dunkirk	1885	C.Derycke		A.Govard
	353		Intrépide *David* W Bn	French	105	67·6	19·7	10·6	Bayonne	1860	Descande fils aîné		J.A.K.David
	354	V	Intrépide Corse *Eliot -* W Bk	French	276	110·5	23·3	12·6	Honfleur	1866	Leviels frères	[illegible]	D.Auger
68117	355	✻	Invercargill *Mair 74-77* I S	British	1246	239·7	36·0	20·7	PortGlasgow	1874	R.Duncan&Co.		Shaw, Savill & [illegible]
	356	N	Inverdruie *Johannesen* W Bk	Norweg'n	595	162·2	30·5	17·9	Kincardine	1867	Wright	[illegible]	J.C.A.Johannesen
63527	357	V	Inveresk *Brag* W Bk	British	799	173·6	35·1	19·2	Maitland,N.S.	1874	Frahey	[illegible]	A.Gibson & Co.
29082	358		Invermark *Cocking* W Sr	British	104	81·8	21·5	11·4	Montrose	1862	Petrie		H.Miller
63713	359	✻	Inverness *Mathber 79-87* C Bk	British	722	177·6	32·0	18·9	Aberdeen	1869	Hall,Russell & Co.	Y 1.88	J.C.&G.S.Richardson
52892	360		Invicta *Allen 83-83* W Sr	British	178	98·8	23·7	12·8	Rye	1866	Hoad		[illegible]
88443	361	✻	Invincible *Jackson* W Sk	British	80	76·3	21·3	10·2	Newhaven	1883	J.C.Tolman		W.C.Edwards
88880	362		Invincible *Lewis -81* W S	American	1460	202·4	40·3	24·0	Bath,Me.	1873	W.V.Moses&Sn.	[illegible]	Renton,Holmes & Co.
12310	363		Invincible *Strickland* W S	American	1325	220·5	42·5	17·7	NewYork	1861			C.Metzger
324	364		Ioanna W Bk	Greek	430	125·3	29·3	..	Chiavari	1873			N.Voudakis
400	365		Ioannis *(ex Mathilde) Mathioudes* W Bk	Greek	626	119·6	30·3	..	Quebec	1866		[illegible]	P.J.Calestas
133	366	V	Ioannis *(ex [illegible]) Gialis* W Bk	Greek	578	118·3	30·8	19·7	Lussinpiccolo	1872	N.Martinolich	[illegible]	D.Neamonitachi
83	367		Ioannis *Bonconealis* W Bg	Greek	276	128·6	25·9	..	Piræus	1868			G.Malokinis
14	368	H	Ioannis *(ex [illegible])* W Bg	Greek	204	87·9	24·8	..	Spezzia	1869		[illegible]	E.B.C[illegible]
886	369	A	Ioannis *Dazeris* W Bn	Greek	171	81·4	23·6	..	Chios	1875			G.P.Frangakes
	370	H	Ioannis *(ex Nicolaos) Liras* W Bk	Turkish	264	..	..	..	Syra	1869			Coulomnoutres
	371		Ioannis W Bn	Turkish	217	..	..	..	Chios	1871			L.G.Brilis
	372		Ioannis W Bg	Turkish	170	..	..	..	Syra	1858			Millude frères
	373		Ioannis Dedoussis *Cassalis* W Bk	Turkish	327	..	..	..	Bordeaux	1867		[illegible]	G.Dedoussis
	374		Ioannis Prodromos W Bg	Russian	120	..	..	..	Syra	1871			G.G.Deligeroglou
85630	375	V	Iodine *Smith* W Bk	British	295	136·0	31·9	11·9	Eatonville,N.S.	1885	J.Parsons		J.G.Ross & Co.
87808	376	✻V	Iolanthe *Stanton 85-85* I S	British	1593	255·8	38·2	22·5	Dumbarton	1883	A.M'Milland&Sn.		Cruickshank,[illegible]
85447	377	V	Iolanthe *Card 83-83* W Sr	British	393	144·0	33·0	12·9	Newport,N.S.	1883	J.Mosher	[illegible]	A.Haley
88150	378	✻	Iolanthe *Hellyer* W K	British	83	79·1	20·7	10·5	Brixham	1883	J.W.&A.Upham		C.Hellyer
51246	379		Iona W Bk	British	550	144·0	29·0	..					Tchwa Cheang S[illegible]
57960	380	✻	Iona *Trail 82-82* W Sr	British	93	88·1	22·1	10·1	Fraserburg	1874	Wester		A.Stephen
	381	N	Iona *(ex Jones) Klem -83* W Bk	Norweg'n	796	160·6	35·0	21·1	Quebec	1857	Russell&Co.	[illegible]	F.Klem
	382	DN	Iona *Strömberg -85* W Bk	Norweg'n	579	140·4	32·6	18·9	Arendal	1876	T.Lassen	Y 12.87	Tellef Lassen
	383	✻	Iona *Pehrsson 88-88* W Bk	Swedish	378	131·9	27·5	16·3	Dundee	1870	Tay Ship Co.	Y 11.88	J.Husselgren
62564	384	✻D	Ione *Edwards -87* I Bk	British	520	165·0	28·1	17·3	Sunderland	1869	Iliff,Mounsey Co.		H.M'Ewan
83236	385	✻	Ione *Jordan -87* W Bkn	British	319	121·5	27·8	14·9	P.E.Island	1882	J.Yeo	Y [illegible]	A.G.Moffatt,T.P.[illegible] & Co. Managers
81005	386	✻	Ione *Jordan* W Bg	British	238	109·9	24·7	14·1	Cork	1877	J.Wheeler	Y 12.88	W.B.Th[illegible] & [illegible] Managers
	387		Ionia W Bg	Turkish	186	..	..	..	Greece	1881			Bigadoglou-Levedi Yanni
51117	388	✻	Iota *Leonard 76-86* W Bk	British	551	151·2	29·6	19·1	Bideford	1866	Cox	[illegible]	Swansea Mar[illegible] owners Co. Ld.
	389		Ioulia *Sorotos* W Bg	Greek	124	..	..	..	Galaxidi	1866			M.Sorotos
	390		Ipapandi *Vassilachi*	Greek	198	..	..	..	Syra	1880	E.Fiorni		N.Vassilachi
58724	391		Iphigenia W Sr	British	248	115·0	25·7	14·0	Christiania	1876	A.Olsen		C.C.Wyse
	392		Iphigenia *Kofoed 82-87* W Bn	Danish	251	99·9	24·6	15·1	Pillau	1861	H.O.Becker		M.Christiansen
	393	G	Iphigenia *Söderlund 84-84* W Bk	Russian	464	146·5	29·3	17·2	Hamburg	1868	J.Marks		K.G.Soderlund
	394	H	Ipiros *Jeanaslis -84* W Sr	Greek	107	..	..	..	Syra	1879			N.D.Gialouris
12101	395		Ira Bliss *Hudson -78* W Sr	American	166	90·4	23·7	9·1	NewYork	18 44/78			J.A.Hudson
100062	396		Ira D.Sturgis *Blaisdell* W 3m Sr	American	223	118·3	30·0	8·5	Bath,Me.	1873			J.M.Haynes
12663	397		Ira Laffrinier *Crowell* W Sr	American	189	123·0	25·0	9·0	Clevel'd,Ohio	1863			C.Lovell & S[illegible]
105	398	A	Iraclis *Candilaki* W Bg	Greek	240	93·2	26·1	..	Syra	1865			Andrew H.[illegible]
31121	399		Iravady Soundra Nayaghee W Sr	British	126	84·5	21·8	10·0	Keelicarre	1877			[illegible]
	400		Iraype *Rodrigues* W B	Brazilian	154	..	..	..	P.Seguro	1875		[illegible]	J.P.Mar[illegible]

✻ = Lloyd's Register; D = Nederlandsche Vereeniging; G = Germanischer Lloyd; I = Registro Italiano; A = Veritas Austro-Ungarico; H = Veritas Ellenico; N = Norske Veritas;

In the 4th Column, C = Composite (iron frame, wood planked); I = Iron; Stl = Steel; W = Wood.

The Dates of Metalling (**Col. 11**) are based on the records of the Books in which Vessels are respectively classed, or on the Reports of the Sur[illegible]

Further particulars of Vessels classed by Lloyd's Register will be found in Lloyd's Register Book and Lloyd's Yacht [illegible]

Numéro officiel	No. dans le livre	Classification	Nom. Matériaux et Gréement. Capitaine et dates d'entrée en fonction des Armateurs actuels et à bord du navire. Ancien Nom.	Pavillon	Tonnage net	Dimensions: Longueur	Largeur	Creux	Construit: Où	Quand	Par qui	Date du doublage en métal	Armateurs ou Directeurs-Gérants.	Port d'Armement.
[illegible]	401	✻	Irby Shaw 85-89 I S	British	1480	237·6	38·9	23·8	Liverpool	1881	R&J.Evans&Co		Irby Ship Co.,Ld.(J.Joyce)	Liverpool
	402		Irena Karlson W Sr	[illegible]	1[illegible]	..	..	..					N.Sittkoff	Mariehamn
[illegible]	403		Irene Mulere W Bg	British	260	116·0	26·4	12·9	P.E.Island	1877	Egan	Y.8.81	Mrs.M.Caughy	Liverpool
[illegible]	404	✻	Irene Richards 78-81 W K	British	72	79·6	20·1	9·3	Falmouth	1884	C.Burt		J.Gray & Sons	Falmouth
[illegible]	405	R	Irene Harding W Bn	American	468	131·6	31·3	16·5	Setauket, N.Y.	1876	G.E.Hand	Y.7.8[illegible]	Yates & Porterfield	New York
[illegible]	406	V	Irene Pasquan W Bk	Austrian	532	131·5	26·4	18·9	Fiume	1874	Schiavoni	Y.[illegible]	A.Luppis	Fiume
	407	✻	Irène Mahé 77- I&Sm Bk	French	722	184·5	30·1	19·0	Stockton	186	M.Pearse&Co.		Ant. Dom. Bordes & fils	Bordeaux
	408		Irene (ex [illegible]) Bendixen W Bn	German	267	106·0	26·1	14·2	Hamburg	1868	M.G.Amsinck	Y.[illegible]	W.O'Swald & Co.	Hamburg
	409	AI	Irene (ex [illegible]) [illegible] W Bk	Greek	464	137·8	27·9	17·8	Cherbourg	1859	Fi Letutier	[illegible]	P.Collakis	Syra
[illegible]	410	A	Irene Syrigos W Bg	Greek	262	101·0	27·1	..	Syra	1877		Y.[illegible]	A.Alafouzos	Santorin
[illegible]	411	H	Irene Patrounas W Bg	Greek	248	53·1	26·2	..	Mytilene	1870	Dem.Vasitis		S.Demetroulaki	Syra
	412		Irene W Bg	Greek	248	96·8	26·7	..	Syra	1868			E.A.Revidis	Piræus
	413		Irene W Sr		169	86·9	23·8	..	Syra	1877				
	414	H	Irene Dombettis W Sr	Greek	102	75·8	21·0	..	Syra	1880			L.Vrontisis	Andros
[illegible]	415	I	Irene Zeccolilla -88 (ex [illegible]) W Bg	Italian	270	115·4	27·7	17·2	Gaeta	1870	L.Contreras	Y.[illegible]	B.Zeccolilla	Naples
	416	R	Irene Johnsen -87 (ex [illegible]) W Bk	Norweg'n	471	133·5	30·3	20·0	Pra	1871	D.Tixi	Y.[illegible]	S.B.Johnsen	Christiansand
	417	DN	Irene (ex [illegible]) Hovrsten W Bk	Norweg'n	409	116·9	30·7	17·3	Brevig	1855			S.Johansen&N.B.Olsen	Drammen
	418	N	Irene Olsen -88 W Bk	Norweg'n	385	119·2	30·1	16·5	Grimstad	1868	H.Jörgensen		Osuld Jörgensen	Grimstad
	419		Irene W Sr	Russian	115	..	..	..		1869			K.Sopetoff	Sordavala
	420		Irene W Bg	Turkish	166	..	..	..	Cashot	1872			Yanni Manyarli	Cashot
[illegible]	421		Irene E. Meservey [illegible] W Sr	American	283	114·0	29·0	10·0	Waldob'ro'Me	1866	W.Welt		G.W.Rawley	St.George,Me.
[illegible]	422		Irene Isigonis [illegible] W Br	Greek	373	130·9	..	..	Liverpool	1862			Dem.Isigonis	Skiathos
[illegible]	423		Ireneo Schnautz (ex [illegible]) W Bk	Austrian	531	142·7	26·5	17·5	Fiume	1870	Schiavoni	Y.[illegible]	G.Schnautz fu A.	Buccari
[illegible]	424	H	Irenoula W Bg	Greek	263	98·8	27·2	..	Syra	1871			N.Z.Goulandris	Andros
[illegible]	425	✻	Irex W Sl	British	83	77·2	20·5	10·5	Elmshorn	1888	J.Kremer		T.Baskcomb	Grimsby
[illegible]	426		Iris Griffiths -87 W S	British	967	173·2	36·2	22·6	Newport,N.S.	1870	Mosher	Y.[illegible]	R.Griffiths	Liverpool
[illegible]	427		Iris Williamson -88 W Bk	British	340	120·0	27·7	17·0	Sunderland	1869	T.Metcalfe	Y.[illegible]	Gillan & Hay	Newcastle,N.S.W.
[illegible]	428		Iris Stone 88-72 W Sr	British	206	116·7	22·7	10·3	FremantleW.A.	1879			J.Bateman	Fremantle,W.A.
[illegible]	429		Iris (NOT IN EXISTENCE) W Bn	British	127	85·0	23·2	10·6	Wallace,N.S.	1855			R.B.Huestis	Picton,N.S.
	430	✻V	Iris (ex [illegible]) Ohlsen 73-79 I Sm Sr	German	507	156·5	27·2	17·3	Harrington	1863	R.Williamson		M.G.Amsinck	Hamburg
	431	N	Iris Andersen 85-78 W Bk	Norweg'n	480	123·7	31·1	18·6	Uleåborg	1840	Langeamal		A.Andersen	Grimstad
	432	✻	Iris Hansen -88 W Bk	Norweg'n	458	1[illegible]2·0	29·6	15·7	Vegesack	1856	J.Lange	Y.[illegible]	Actieselskabet "Iris"	
	433	N	Iris Ulsted 85-85 W Bg	Norweg'n	222	110·0	27·4	12·1	Arendal	1867	J.Annonsen		S.Ulsted	Arendal
	434	N	Iris Gulliksen W Bg	Norweg'n	15[illegible]	93·6	23·7	11·9	Porsgrund	1853			I.C.Lund	Tönsberg
	435	N	Iris Cornelliussen W Br	Norweg'n	159	91·8	23·9	10·2	Molde	1856			O.A.Föyen	Tönsberg
	436	DV	Iris Strom W Bk	Russian	767	173·1	30·7	19·6	Brahestad	1876	Z.Durchman	Y.7.8[illegible]	Z.Durchman	Brahestad
	437		Iris W Bk	Spanish	222	104·3	22·8	14·6	Bilbao	1850			Ramon Ramos	Barcelona
	438	V	Iris Thorgaen -72 W Bk	Swedish	389	121·3	25·7	16·3	Nantes	1863	Pelletier & Guibert	Y 10.8[illegible]	C.Molander	Wisby
	439	G	Iris Z [illegible] -88 W Bk	Swedish	330	120·0	25·5	14·3	Vasa	1868	J.Jörnvik		A.Söderström	Calmar
	440	✻	Irish Girl McElroy 76-[illegible] W Sr	British	125	93·0	23·3	11·1	Dundalk	1876	Connick		R.McClenahan	Dundalk
	441	✻	Irish Minstrel Edwards W Bn	British	136	99·1	25·0	11·4	Dundalk	1879	Connick		H.C.Edwards	Dundalk
	442		Irma W Bk	British	265	108·0	24·0	9·0	Philadelphia	1847			N.M.Mills	Bridgetown
	443	V	Irma Fauvart W S[illegible]	French	100	82·7	19·3	10·9	Dunkirk	1861	Sagary fils		Boone	Dunkirk
	444		Irmak Ali Dennra W Bn	Turkish	150	..	..	..	Irmak	1868			Hajee Elias & Sons & Co.	Irmak
	445	✻	Iron Crag Jones 14-72 I Bk	British	827	201·0	32·3	19·6	Whitehaven	1877	WhitehavenS.B. Co.,Ld.		T.Connell	Liverpool
	446		Iron Cross Lamb 86-6 (ex [illegible] Robert Lowe) I S	British	1508	247·0	35·2	18·7	Greenock	1854	Scott&Co.		H.Fernie & Sons	Liverpool
	447	✻V	Iron Duke Ban[illegible] 78-80 I S	German	1435	217·0	36·2	25·4	Liverpool	1863	H.M.Lawrence &Co.		H.Bischoff & Co.	Bremen
	448	✻	Iron Queen Thomson (ex [illegible]) I Bk	British	796	185·0	29·8	21·1	Liverpool	1863	W.H.Potter&Co		G.H.Fletcher & Co.	Liverpool
	449	✻D	Iron Queen Withenige 85-87 I Bk	British	234	113·0	23·4	13·7	Newport	1865	Huddart		T.Pugsley & Co.	Newport
	450	✻	Ironside Ford -87 I Bk	British	899	185·0	32·7	20·4	Sunderland	1861	T.R.Oswald		Temperleys, Carter & Darke	London

✻ = Lloyd's Register; A = Veritas Austro-Ungarico; D = Nederlandsche Vereeniging; G = Germanischer Lloyd; H = Veritas Ellenico; I = Registro Italiano; N = Norske Veritas; R = American Record; V = Bureau Veritas.

Dans la 4me Colonne, C = Composite (membrure en fer, bordé en bois); I = Fer; Stl = Acier; W = Bois.

Les Dates du doublage en métal (Col. 11) ressortent des livres dans lesquels les navires sont respectivement classés ou des rapports des Inspecteurs du Lloyd's Register.

Pour plus amples renseignements sur les navires classés par le Lloyd's Register, voir le Lloyd's Register Book et le Lloyd's Yacht Register.

Official No.	No. in Book	Society by which classed	Ship's Name. Material and Rig. Master, and Dates of Appointment to present Owner's Service and to Ship. Ship's Late Name.	Flag.	Net Tonnage.	Length.	Breadth.	Depth.	Built. Where.	When.	By whom.	Date of Metalling.	Owners.
100396	451	RV	Iroquois *Nichols 81-83* WS	American	2005	259·1	43·6	27·0	Bath, Me.	1881	A. Sewall	[illegible]	A. Sewall & Co.
	452		Irskatt *Pettersson* WSr	Swedish	104	76·1	21·3	9·7	Roslagen	1862	J. Östensson		J. E. Ersson
56561	453	✱D	Irvine *Boyd* I Bk	British	655	170·6	29·5	19·1	Sunderland	1867	W. Doxford		D. Hunter
76129	454	✱	Isa Reid *Winchester 83-85* WSr	British	99	88·7	21·3	10·4	Garmouth	1876	G. Geddie, jun.		A. Coull
80651	455	R	Isaac Burpee *Matthews -85* WSr	British	180	95·0	28·1	9·0	Westfield, N.B.	1880	S. J. Holder		E. C. Elkin
100390	456	R	Isaac Dodge (WRECKED) *Good* WBkn	American	515	148·0	34·0	13·0	Newcastle, Me.	1885	Gay & Haggett	Y. [illegible]	T. E. Gay
100350	457	R	Isaac H. Tillyer *[illegible] -88* W3mSr	American	568	144·6	35·0	16·8	Dennisville, N.J.	1884	R. S. Leaming		W. H. Smith
12430	458		Isaac Hall *Stone* WBk	American	580	140·0	32·9	17·0	Stockton, Me.	1869	Colcord & Berry		J. H. Winchester & Co.
100123	459	R	Isaac Jackson *Berry* WBk	American	586	145·3	32·6	17·8	Deering, Me.	1874		Y. [illegible]	G. A. Goodwin
100345	460	R	Isaac N. Kerlin *Steelman* W3mSr	American	349	136·7	35·0	10·0	Leesburg, N.J.	1883	J. Russell		J. Russell
12492	461	Y	Isaac Orbeton *Trim* W3mSr	American	367	120·0	29·4	12·0	Newcastle, Me.	1870	J. Clark		John Cassiday
100150	462	RV	Isaac Reed *Waldo -87* WS	American	1489	212·0	40·0	24·2	Waldoboro', Me.	1875	A. Reed & Co.	Y. [illegible]	Yates & Porterfield
100245	463		Isaac T. Campbell *Matthews* W3mSr	American	557	157·0	33·9	15·3	Waldoboro', Me.	1879	A. Storer & Son		J. Baker & Co.
[illegible]	464		Isaak *Lauter -85* W3mSr	Russian	336	110·0	26·5	14·0	Haynasch	1873	J. Brackmann		A. Weide
358	465	DG	Isaak *Stahl -86* W3mSr	Russian	262	113·0	28·3	13·1	Nogallen	1880	J. Stahl		J. & M. Stahl
56282	466	✱	Isabel *Drew 76-86* WBk	British	579	152·0	30·8	18·9	Sunderland	1868	J. Gill	Y. 1887	W. C. Jarvis
80918	467	R	Isabel *McClure* WBk	British	350	119·2	23·3	15·6	P. E. Island	1880	C. McLure	Y. [illegible]	C. McLure
49667	468		Isabel *Fishley* WSr	British	177	102·4	23·8	12·7	Newquay	1865	Davies		J. M. Fishley, jun.
74756	469	✱	Isabel *Tupman* I 3mSr	British	149	99·0	21·5	11·6	Whitehaven	1880	Whitehaven S. B. Co., L.		B. Tupman
7100	470		Isabel *Stott 85-85* WBn	British	127	86·5	22·9	11·9	P. E. Island	1850			G. Sinclair
12265	471		Isabel (ABANDONED) *Rainford* WSr	American	176	103·2	29·8	9·4	Coos Bay, Oreg.	1864			A. Anderson
	472		Isabel I Sr	Argentine	199	130·0	27·2	9·0	Glasgow	1834	Abercorn S. B. Co.		V. Casares & Son
	473		Isabel WSr	Brazilian	146	..	..	..					A. F. P. da Cruz
	474	✱V	Isabel WBk	Norweg'n	1264	202·8	39·2	22·3	S. Maitland N.S.	1880	W. P. Cameron	Y. [illegible]	Johan Thorsen
	475		Isabel WBk	Norweg'n	324	123·0	27·8	13·3	Rhode Island	1854			G. G. Maarud
	476		Isabel (ex [illegible]) WBk	Spanish	215	106·9	25·4	14·4	Vizcaya	1866	J. Roca		M. Bonet Coll
	477		Isabel WBn	Spanish	149	92·5	23·1	10·7	Villajoyosa	1852			V. M. Galiana
12363	478		Isabel Alberto *Bishop* WSr	American	218	103·4	28·7	10·0	Port Jefferson, N.Y.	18[illegible]			S. L. Keene
90077	479	✱	Isabel Browne *James 75-85* I Bk	British	1289	230·5	36·1	21·4	Port Glasgow	1885	Russell & Co.		Browne & Watson
	480		Isabel Peral (ex Louis) WBk	Peruvian	297	117·0	25·0	14·0	France	1869			J. V. Peral
	481	DV	Isabelita *Nikera 83-83* WBn	Spanish	249	108.6	24.5	13.6	Bilbao	1871	A. de Arana	Y. [illegible]	M. de Astigarraga
	482		Isabelita *San Juan* WPol	Spanish	180	89·2	26·4	10·9	Masnou	1854		[illegible]	P. Maristany
	483		Isabelita WSr	Spanish	102	78·7	22·5	9·[illegible]	Adra	1871			José Cuenca
28689	484		Isabella (STRANDED) *Baker 84-84* WBg	British	270	103·7	26·2	16·1	Sunderland	1861	G. Bartram		J. W. Woods
75812	485		Isabella *Fraser* WSw	British	110	77·8	25·0	9·2	Bay St. George (N.F.)	1877			J. McFatridge
76880	486	✱	Isabella *Coppack 81-86* WSr	British	85	88·5	21·2	9·0	Barrow	1878	W. Ashburner		T. Ashburner & Co.
19581	487	✱	Isabella *Jones 88-88* WSr	British	79	75·0	18·6	9·9	Fraserburg	1855			O. Jones
80385	488	V	Isabella Balcolm *Armstrong* WBn	British	301	126·0	28·7	12·6	Parrsboro', N.S.	1881	J. Balcolm	Y. [illegible]	E. E. Hutchings
	489	DN	Isabella Blyth *Berner* WBk	Norweg'n	633	139·0	30·9	20·8	London	1847		Y. [illegible]	Carl Berner
65036	490		Isabella Diston *Tucker 83-85* WSr	British	155	93·5	22·5	12·7	Inverkeithing	1870	R. & D. Rose		J. Helme
51039	491		Isabella Hall (WRECKED) WSr	British	140	95·5	22·2	12·3	Glasgow	1865	Scott & Co.		J. Fisher & Sons
74628	492	✱	Isabella Helen *Lewis* WSr	British	98	79·5	21·6	10·6	Plymouth	1877	C. Gent	Y. [illegible]	J. Prizeman
6899	493		Isabella Leith (FOUNDERED) WSr	British	163	91·4	23·6	13·3	Sunderland	1848			C. Barkaway
	494	V	Isabella V. (ex Domenica M.) *Marcidni* WBk	Italian	457	136·8	29·4	19·7	Sestri	1868	Peragallo	Y. [illegible]	B. Varr
56450	495	✱	Isabella Walker *Maxton* WBn	British	179	103·7	22·9	12·8	Inverkeithing	1870	R. and D. Rose		J. Chalmers
51322	496		Isabella Wilson *Ellis -86* WSr	British	169	102·0	23·1	13·0	Garmouth	1865	W. Kinloch		W. G. Crinks
73028	497	✱	Isabelle *Williams 76-86* WBkn	British	289	122·6	27·3	14·9	P. E. Island	1875	Richards	Y. [illegible]	T. P. Richards & Co.
58543	498		Isabelle WBn	British	129	97·5	23·2	10·0	George River, N.S.W.	1869			B. Byrnes
93167	499		Isabelle (ex [illegible]) WB	British	123	83·5	22·5	11·5	Sunderland	1862	W. Barkley & Co.	Y. [illegible]	I. Baird
	500		Isabelle *Harg 80-87* WSr	French	122	89·6	22·6	11·1	Dunkirk	1880	G. Beck		G. Beck

✱ = Lloyd's Register; D = Nederlandsche Vereeniging; G = Germanischer Lloyd; I = Registro Italiano; A = Veritas Austro-Ungarico; H = Veritas Ellenico; N = Norske Veritas;

In the 4th Column, C = Composite (iron frame, wood planked); I = Iron; Stl = Steel; W = Wood.

The Dates of Metalling (Col. 11) are based on the records of the Books in which Vessels are respectively classed, or on the Reports of the [illegible]

Further particulars of Vessels classed by Lloyd's Register will be found in Lloyd's Register Book and Lloyd's Yacht [illegible]

No.	Cote	Nom (Anciens Noms)	Capitaine	Matériaux et Gréement	Pavillon	Tonnage	Longueur	Largeur	Creux	Où construit	Quand	Par qui	Doublage en métal	Armateurs ou Directeurs-Gérants	Port d'Armement
501	V	Isabelle Marie	[illegible]	W Sr	French	416	[illegible]	[illegible]	[illegible]	Boulogne	186[illegible]	[illegible]		Verry Carmantan	Binic
502	V	Isaiah Hart	Nickerson	W Bn	American	353	130·4	35·1	13·3	Waldoboro,Me.	1880	G.L.Welt & Co.		W.H.Nickerson	Boston,Mass.
503	R	Isaiah K.Stetson	Bond	W Bn	American	298	136·3	31·3	10·[illegible]	Brewer,Me.	1882	E.&I.K.Stetson		[illegible]	Bangor,Me.
504		Isaline	Pedrosa	W Bg	Brazilian	180	..	..	..	Pekela				[illegible]	Pernambuco
505	N	Isbanden (ex Sigurd)	[illegible]	W Bn	Norweg'n	393	123·3	26·1	11·5	[illegible] P.E.I.	1873	M'Kean		[illegible]	Kragerö
506	N	Isbjornen (ex Beaufort)	Larsen	W Bk	Norweg'n	342	113·3	27·[illegible]	[illegible]	[illegible]	1862	J.Robinson		[illegible]	Kragerö
507		Isca	Callie	W Bk	British	293	125·3	25·5	14·6	[illegible]	1868	Redway		[illegible] & Son	London
508	N	Isfareran (ex Richard Pearse)	Olsen	W Bk	Norweg'n	392	133·8	30·3	12·7	[illegible]	1866	T.A.Harvey		Act Selsk. "Ispilen"	Kragerö
509		Ishiki Maru		W Bg	Japanese	162	99·9	22·6	11·7	[illegible]	1882	Matsuzaki [Yoke		Natsume Ichirobei	Owari
510	✱ I	Iside	Schiaffino	W Bk	Italian	735	164·9	33·4	19·4	[illegible]	1873	G.Merello		R.Schiaffino	Genoa
511	I	Iside B. (ex Vittorio Di Savoia)	Vitali	W Bk	Italian	1004	183·0	29·5	24·1	[illegible]	1879	A.Cerruti	V 5.6	G.B.Beverino	Spezia
512	✱	Isidora		I Sr	Uruguay'n	64	75·5	25·5	6·2	[illegible]	1884	Forrest & Son		W.B.Cranwell & Co.	Monte Video
513		Isidoros (ex Andreas)	Coluguan	W Bg	Turkish	238	99·1	27·2	17·4	[illegible]	1867			Nicola Yorghiadi	Vassilikos
514	V	Isidra	Fabregas	W Pol	Spanish	185	88·3	27·1	15·4	[illegible]	1859			Maristany hermanos.	Barcelona
515		Isidro (ex Sigle de Oro)		W Bn	Spanish	175	95·5	23·4	11·7	[illegible]	1859			V.Balbas & Co.	Manila
516	✱	Isipingo	Phillips	W Bk	British	310	135·2	27·1	12·9	[illegible]	1872	Salisbury	V 5.87	J.Lovelock & Son	London
517	A	Isis	Vidulich	W Sr	Austrian	134	83·3	20·2	11·2	[illegible]	1883	G.Giuricich	V 8.87	A.D.Vidulich fu G.	Lussinpiccolo
518	✱	Iskender	[illegible] Issak	W Bk	Turkish	548	133·0	29·3	14·5	[illegible]	1867	Stromloff	C 7.85	Sadyk Bey	Constantinople
519		Iskender	Mehemed	W Bk	Turkish	512	..	..	..	[illegible]	1886			Artin Caramanian Effendi	Constantinople
520		Iskenderieh		W Bn	Turkish	332	..	..	..	[illegible]	1867			[illegible]	Iskenderieh
521		Isla de Cuba		W Sr	Spanish	104	83·0	22·8	16·2	[illegible]	1861			I.Carbonell y Juncosa	Mataro
522	N	Island	[illegible] 83-84	W Bk	Norweg'n	640	156·7	32·8	19·4	[illegible] N.S.	1884	J.B.North		I.F. & G.F.Gjertsen	Tönsberg
523	N	Island	Christiansen	W Bn	Norweg'n	112	85·0	21·6	9·7	[illegible]	1879			J.Fredriksen	Mandal
524	N	Island	Sörensen	W Sr	Norweg'n	105	84·0	22·8	19·4	[illegible]	1874			N.B.Dahl	Christiansund
525		Island City	Warner	W Bk	British	421	133·6	30·4	16·3	Farmingham,[U.S.	1852			J.Gillan	Newcastle,N.S.W.
526		Island City	Tourkis	W 3m Sr	American	406	137·1	28·8	15·0	E.Boston,Mass.	1871	T.Nickerson		L.J.Dale & Co.	Savannah,Ga.
527	R	Island Lass	Breeding	W Sr	British	176	102·2	21·5	12·9	[illegible]	1870	Batson		H.B.Varwell	Exeter
528	✱	Island Maid	Beal	W Sr	British	150	89·5	23·3	12·1	Salcombe	1865	Harnden	V 8.86	E.Rillston	Fowey
529		Island Maid	Johnston	W Bn	British	135	91·8	22·6	11·6	Whitehaven	1865	Williamson		J.Storey	Belfast
530		Island Maid	Dawe	W Sr	British	124	96·2	20·4	11·2	Plymouth	1863	Shilston		W.H.Cox	Plymouth
531	✱	Island Queen	Brabazon	W Bk	British	353	123·0	28·8	14·8	P.E.Island	1866	Green	Z 4.88	[illegible] "Island Queen" Co.	Newport
532		Islander		W Bk	British	277	104·0	26·7	15·1	FairhavenU.S.	1856			W.J.Gillam	Fremantle,W.A.
533		Islander	Boucher	W Bn	British	186	103·4	24·2	12·9	P.E.Island	1876	Duncan	V 11.83	T.Potter & Co.	Guernsey
534		Islay	Brien	W Bn	British	134	94·6	22·8	12·0	Wallace N.S.	1860			John Munn & Co.	St.John's,N.F.
535	✱	Isle of Anglesea	Bennett	I Bk	British	725	187·3	31·2	18·5	Glasgow	1872	Dobie & Co.		J.Bushby	Liverpool
536	✱ D	Isle of Bute	[illegible]	I Bk	British	929	209·6	34·3	20·7	PortGlasgow	1875	J.Reid & Co.		J.Bushby	Liverpool
537	✱	Isle of Erin	Inkster	I Bk	British	889	202·5	32·5	19·5	Glasgow	1875	D.&W.Henderson & Co.		J.Bushby	Liverpool
538		Isle of Pines	Legg	W Sr	American	162	100·0	30·0	7·9	Bass River,N.J.	1856			W.H.Legg	New York
539	V	Ismaele (ex Marcsea)	Capelli	W Bk	Italian	410	136·8	27·7	17·8	Cassano	1867	F.DeRosa	V 3.86	T.Marcsea	Castellamare
540	R V	Ismir	Sanders 77-87	W S	British	1269	200·3	36·8	22·6	St.Mary's Bay,[N.S.	1878	J.Bingay	V 11.87	J.Bingay	Yarmouth,N.S.
541		Ismyr	Crisp	W Bn	British	210	102·6	23·2	15·1	Liverpool	1850	Royden	V 8.85	W.Milburn	North Shields
542	✱	Isobel (ARAN)	Davies 80-86	W Bk	British	349	120·8	28·1	11·9	P.E.Island	1875	H.McDonald	V 11.85	D.Davies	Aberystwith
543	A	Isodia	Mouroummatis	W Sr	Greek	199	98·4	27·1	..	Syra	188[illegible]	A.Lividaros	V 7.86	N.Mouroummatis	Syra
544	A	Isodia	Mekoussi	W Bg	Russian	181	..	..	..	Spezia	1867		Z 1.86	H.Filitoff	Kertsch
545	✱	Isola	Oakley	I Bk	British	653	180·0	30·2	17·7	Glasgow	1876	J.Elder & Co.		J.Kelso	Glasgow
546		Isolina		W Sr	Brazilian	118	..	..	..					J.A.O.Maia	Porto Alegre
547		Isolina	Benetti	W 3m Sr	Italian	163	86·0	25·8	14·4	Pt.Longone	1859			V.Soldani fu D.	Porto Ferraio
548		Ispilen (ex Wilson Thurlow) (STRANDEN)	Knudsen 87-87	W Bg	Norweg'n	294	108·3	26·3	16·4	Sunderland	1865	J.Crown	V 7.83	Act.Selsk. "Ispilen"	Kragerö
549	N	Ispilen[illegible]	Knudsen	W Bk	Norweg'n	363	118·8	27·1	16·7	Sunderland	1865	W.H.Pearson	V 3.83	Act.Selsk. "Ispilen"	Kragerö
550	N	Ispo[illegible]	Hae[illegible]	W Bg	Norweg'n	234	108·1	26·0	12·1	St.Magnus	1865	D.Oltmanns wwe	V 3.8	Act.Selsk. "Ispilen"	Kragerö

✱ = Lloyd's Register; D = Nederlandsche Vereeniging; G = Germanischer Lloyd; I = Registro Italiano; R = American Record;
A = Veritas Austro-Ungarico; H = Veritas Ellenico; N = Norske Veritas; V = Bureau Veritas.

Dans la 4me Colonne, C = Composite (membrure en fer, bordé en bois); I = Fer; Stl = Acier; W = Bois.

Les Dates du doublage en métal (Col. 11) ressortent des livres dans lesquels les navires sont respectivement classés ou des rapports des Inspecteurs du Lloyd's Register.

Pour plus amples renseignements sur les navires classés par le Lloyd's Register, voir le Lloyd's Register Book et le Lloyd's Yacht Register.

Official No.	No. in Book	Registry	Ship's Name. Master. Ship's Late Name.	Material and Rig.	Flag.	Net Tonnage.	Length.	Breadth.	Depth.	Built: Where.	When.	By whom.	Date of Metalling.	Owners or Managers.
	551		Issavlah *Angelidis*	W Bg	Turkish	152	..	..	..	Greece	1869			Pavlikos, Nakkas & Co.
	552	❋	Issidoros *Vourios*	W Bg	Turkish	329	..	..	..		1868		[illegible]	Mrs. M. Caloyanni
	553		Issim Sis	W [illegible]	Turkish	496	..	..	..	Greece	1883			Saraf Vassiliki Kosmo Effendi
	554		Issim Sis	W Bg	Turkish	275	..	..	..		1873			Rahtvalee tijar Ilham Agha
	555		Issim Sis	W Bg	Turkish	255	..	..	..	Greece	1870			Hajee Mustapha Eff.
	556		Issim Sis	W Bn	Turkish	250	..	..	..	Tereboli	1869			Iskenderieh-Hurin-zen Ibrahim Tahuree
	557		Issim Sis	W Bn	Turkish	206	..	..	..	Sakarieh	1869			Stambolice Nicolaki Velideh-Ermak
	558		Issim Sis *Omer*	W Bn	Turkish	199	..	..	..	Greece	1868			Hairie-tijar Osman Yussuf & Co.
	559		Issim Sis	W Sr	Turkish	196	..	..	..	Barten	1882			Mustapha Vehbi Effendi & Co.
	560		Issim Sis	W Bg	Turkish	191	..	..	..	Italy	1878			Helthi Corahy
	561		Issim Sis *Chroni Reiss*	W Bg	Turkish	184	..	..	..	Greece	1869			Nanno Petro & C[illegible]
	562		Issim Sis	W Bn	Turkish	178	..	..	..	Kara Agatch	1869			Madsaujdee-Kekoyazadeh Mehemed Son's Wife Hanoum
	563		Issim Sis	W Bn	Turkish	175	..	..	..					Yazidjeh-zadeh Elias Osman
	564		Issim Sis	W Sr	Turkish	174	..	..	..	Roumania	1878			Aktarpopulo Nicholas Dimitri Kakos
	565		Issim Sis	W Bg	Turkish	162	..	..	..	Greece	1869			Andon Salidi
	566		Issim Sis	W Bn	Turkish	162	..	..	..	Greece	1869			[illegible]
	567		Issim Sis	W Sr	Turkish	158	..	..	..	Vassilikos	1868			[illegible]
	568		Issim Sis	W Bn	Turkish	152	..	..	..	Greece	1868			[illegible] & Co.
	569		Issim Sis *Seid Reiss*	W S	Turkish	150	..	..	..	Rustchuk	1869			BulbuljileeStartiYorg[illegible]
	570		Issim Sis *Ismael*	W S	Turkish	150	..	..	..	Rustchuk	1869			Widdinlee Dimitraki & Co.
	571		Issim Sis *Mehemed Numan*	W Bn	Turkish	150	..	..	..	Barten	1884			Rustchukleo[illegible] Zadeh-Ibrahim & Co.
	572		Issim Sis	W Bg	Turkish	149	..	..	..	England	1883			[illegible] Tijari Bahri Zadeh Mehemed Eff.
	573		Issim Sis *Ismael*	W Sr	Turkish	149	..	..	..	Haramson [Skellessi	1884			Hajee Ali Effendi [illegible]
	574		Issim Sis	W Bn	Turkish	146	..	..	..	Greece	1869			Kerestehjee K[illegible]
	575		Issim Sis	W Bg	Turkish	142	..	..	..	Russia	1869			Iskudarlee Pappas [illegible]
	576		Issim Sis	W Bg	Turkish	140	..	..	..	Greece	1869			Alatjee Dimitri [illegible] & Co.
	577		Issim Sis	W Bg	Turkish	137	..	..	..	Roumania	1868			Sakyzlee Yanni [illegible]
	578		Issim Sis	W Bn	Turkish	119	..	..	..	Greece	1869			Kiriazi Yanni
	579		Issim Sis	W Bg	Turkish	118	..	..	..	Greece	1869			Limnilee Yamandi Mikhail
	580		Issim Sis	W Bn	Turkish	115	..	..	..	Barten	1869			Petro Anasta[illegible]
	581		Issim Sis	W	Turkish	108	..	..	..	Kooroofeh-Shi	1869			[illegible]
	582		Issim Sis	W Sr	Turkish	106	..	..	..	Greece	1882			[illegible]
	583		Issim Sis	W Sr	Turkish	100	..	..	..	Barten	1869			[illegible]
	584		Istephan *Saleh*	W Bn	Turkish	102	..	..	..	Istephan	1866			[illegible] Ali [illegible]
	585		Italia *McCallum*	W Bg	British	116	92·0	25·7	10·0	Lunenburg, NS	1864			John F. Morris
255	586	I	Italia *Ferro*	W Bk	Italian	899	183·0	35·0	23·0	Loano	1876	A. Cerruti	[illegible]	Flli Coxe fu F.
1210	587	IV	Italia (ex Giuseppina) *Demartino*	W Bk	Italian	535	147·6	30·0	19·4	Sorrento	1879	M. Paturzo	Y 12.87	G.P. & L. Lauro fra[illegible]
	588	N	Italia (ex Valparaiso) *Isaacson* ·88	W Bk	Norweg'n	829	169·1	33·9	21·1	Genoa	1871	A. Briasco	Y 10.83	Enkefru H. Blom
1716	589		Italo *Fenese*	W Bk	Italian	480	136·4	31·4	19·5	SestriPonente	1868	Frat. Cadenaccio	Y 10.89	Giulio D'Aste
87916	590	❋	Itata *Goddard*	Stl Bk	British	927	202·0	33·0	19·9	Liverpool	1883	R&J Evans&Co.		R. & J. Evans & C[illegible]
7	591		Ithaka	W Sr	Greek	143	88·6	23·6	..	Braila	1868			D. Vlassopoulos
515	592	A	Ithaka *Chandos*	W Sr	Greek	146	86·0	23·4	..	Syra	1888	A. Livadana		P. A. Chandos
66273	593	D	Ithuriel *Clements* 86-88	W Bk	British	319	121·2	26·5	14·9	P.E. Island	1871	Knight	[illegible]	Bevan, Evans & C[illegible]
	594		Itiuba	W Sr	Brazilian	162	..	..	..					J. S. Belem
	595	❋	Iton (ex [illegible] Lonsdale) *Regnier* ·88	C Bk	French	542	152·0	28·3	17·9	Sunderland	1865	R. Thompson Jr.	[illegible]	F. Regnier
[illegible]	596	RV	Itonus *Buckman* 75-86	W Bk	American	818	162·4	34·8	21·1	Yarmouth, Me	1876	Hutchins& [Stubbs	[illegible]	B. Webster
[illegible]	597	A	Ivan (ex [illegible]) *Bielich*	W Bk	Austrian	609	144·0	33·5	18·6	Trieste	1871	Strudthoff	[illegible]	G. Bielich
	598	❋	Ivan *Andersen* 83-[illegible]	W Bn	Danish	213	103·4	21·6	12·5	P.E. Island	1876	McLaren		C. P. Bentzen
	599		Ivan *Sallinen*	W Gt	Russian	235	..	..	..	Impilaks	1872			J. A. R. Lindberg
	600		Ivan *Prokofejeff*	W Gt	Russian	229	..	..	..	Salmis	1878			I. Sidoroff

❋ = Lloyd's Register; D = Nederlandsche Vereeniging; G = Germanischer Lloyd; I = Registro Italiano; R[illegible]
A = Veritas Austro-Ungarico; H = Veritas Ellenico; N = Norske Veritas; V[illegible]

In the 4th Column, C = Composite (iron frame, wood planked); I = Iron; Stl = Steel; W = Wood.

The Dates of Metalling (Col. 11) are based on the records of the Books in which Vessels are respectively classed, or on the Reports of the [illegible]

Further particulars of Vessels classed by Lloyd's Register will be found in Lloyd's Register Book and Lloyd's Yacht R[illegible]

No.		Nom. Matériaux et Gréement. Capitaines et dates d'entrée au service des Armateurs actuels et à bord du navire. Ancien Nom.		Pavillon	Tonnage	Longueur	Largeur	Creux	Où.	Quand.	Par qui.	Doublage	Armateurs ou Directeurs-Gérants.	Port d'Armement.
601		Ivan *Matveyeff*	W G'lt	Russian	210	..	..	..	Sordavala	1884			I.Spiridonoff	Sordavala
602		Ivan *Softonoff*	W G'lt	Russian	139	..	..	..	Sordavala	1875			H.N.Moldakoff	Sordavala
603	✻	Ivanhoe *George* -87	I Bk	British	1383	235·2	37·4	23·7	PortGlasgow	1868	John Reid & Co.		J.Whittingham	Melbourne
604	✻	Ivanhoe *Williams*	W Bk	British	347	130·7	26·8	16·3	Aberystwith	1871	Evans		W.Morris & Co.	Aberystwith
605		Ivanhoe *Carter* 85-5	W S	American	1563	202·3	39·3	27·6	Belfast.Me.	1865	C.P.Carter		A.W.Rose, junr.	San Francisco
606		Ivanhoe	W Sr	American	114	86·0	28·2	8·5	CoosBay,Oreg.	1869			T.Pollard	San Francisco
607	N	Ivar *Henriksen*	W Bn Sr	Norweg'n	314	118·3	28·2	14·8	Finland	1865			W.M.Michaelsen	Moss
608	V	Ivar *Larsson* -83	W Sr	Swedish	102	81·0	20·8	8.4	Carlshamn	1880	D.R.Andersson		N.Tufvesson	Brantevik
609		IverHvitfeldt *Hansen*	W Bk	Danish	290	111·2	24·7	16·0	Wismar	1853	J.H.D.Barmann		H.P.Christiansen	Copenhagen
610	V	Ivigtut *Andersen* -88	W Bk	British	331	122·3	28·7	15·7	Quebec	1877	McKay&Dix		McKay & Dix	Quebec
611	A	Ivis *Pappa*	W Sr	Greek	268	..	..	..	Syra	1886	M.Cufudachi		G.D.Pappa	Syra
612	V	Ivy *Caron*	W Bk	British	580	157·8	28·8	17·7	Quebec	1879	P.V.Valin		P.V.Valin	Quebec
613	✻	Ivy *Snow* 85-85	W K	British	65	74·6	19·2	9·2	Point	1878	Stephens		W.Hosken & Co.	Falmouth
614	RV	Ivy *Gardiner* -87	W S	American	1181	184·3	37·2	23·8	Phippsburgh. [Me.	1876	C.V.Minott		C.V.Minott	Bath.Me.
615		Ivy *Gillis*	W Sr	American	135	102·5	28·8	8·8	SanFrancisco	1882			J.J.McKinnon	San Francisco
616		Ivy Belle *Loud*	W Sr	American	117	88·3	25·2	7·8	Bristol.Me.	1870			E.Loud	Damariscotta,Me.
617		Iwan *Johansson*	W S	Russian	100	65·0	21·0	9·0	Orrenhof	1879	A.Dreymann		M.Birk	Haynasch
618		Iwata Maru	W S	Japanese	195	100·7	21·2	12·9					UkonGonsayemon	Osaka
619	GV	Izabel (ex Adele) *D'Araujo*	W Bk	Portug'se	1143	185·8	36·9	23·6	Bremerhaven	1881	F.W.Wencke		C.da Silva Ferreira	Oporto
620		Izetta *Armstrong*	W Sr	American	190	106·9	27·8	9·0	Bangor.Me.	1865			S.Hincks	Bangor.Me.
621		Izolina (ex Vigilia) *Silva Barbosa*	W Bk	Portug'se	242	118·7	25·6	10·5	Oporto	1867	J.J.d'Avereda [Vareta		L.Alves,Rodrigues&Co.	Oporto
622	V	Izvor (ex ...) (Ex SDERAD) *Radonicich*	W Bk	Austrian	705	153·2	30·9	20·6	Capo d' Istria	1875	L.Polli		A.Tripcovich fu N.	Meglino
623		Izzedin	W S	Turkish	112	..	..	..	SudaBay,Crete	1869			SudaleeIzraki Hussein	Suda Bay

ADDITIONAL SHIPS ARE INSERTED IN THE SUPPLEMENT AT THE END OF THE REGISTER.

✻=Lloyd's Register; B=Nederlandsche Vereeniging; G=Germanischer Lloyd; I=Registro Italiano; R=American Record; A=Veritas Austro-Ungarico; H=Veritas Ellenico; N=Norske Veritas; V=Bureau Veritas.

Dans la 4me Colonne, C=Composite (membrure en fer, bordé en bois); I=Fer; Stl=Acier; W=Bois.

Les Dates du doublage en metal (Col. 11) ressortent des livres dans lesquels les navires sont respectivement classés ou des rapports des Inspecteurs du Lloyd's Register.

Pour plus amples renseignements sur les navires classés par le Lloyd's Register, voir le Lloyd's Register Book et le Lloyd's Yacht Register.

Official No.	No. in Book	Registry in which classed	Ship's Name. Material and Rig. Master, and Dates of Appointment to present Owner's Service and to Ship. Ship's Late Name.	Flag.	Net Tonnage	Dimensions: Length	Breadth	Depth	Built: Where.	When	By whom	Date of Metalling	Owners or Managers
76180	1	R	J. A. Hatry *Green* W3mSr	American	164	113·8	28·0	5·8	Portland, Me.	1881	Brewer	[illegible]	[illegible]
	2	V	J. A. U. *Ingoria 81-82* W Bn	Spanish	188	104·9	23·5	12·1	Bilbao	1878	D. Mendiguren	[illegible]	J. A. [illegible]
75815	3	R	J. B. Atkinson *Donahoe 81-8?* W3mSr	American	403	126·5	33·0	9·7	Milford, Del.	1875	D. Lank		H. D. May
53330	4		J. B. Blanchard W Barge	British	150	110·7	22·6	7·0	Quebec	1882	A. Millar		J. B. Blanchard
75068	5	V	J. B. Brown *Cameron* W Bk	American	1551	207·5	40·5	24·0	Kennebunk Me	1874	Titcomb & Thompson	[illegible]	W. Mighell
12907	6		J. B. Clayton *Parker* W Sr	American	229	103·7	29·4	9·0	May's L'dng, N.J.	1863			E. Clayton & Co.
	7	V	J. B. D. *Horn* W Bk	Norweg'n	399	127·0	28·8	17·1	Dormont	1862		[illegible]	Actieselskabet J. B. D. (N. M. Bos)
76438	8	R	J. B. Holden *Look* W3mSr	American	399	148·0	34·3	10·9	Rockland, Me.	1888	F. Cobb & Co.		Cobb, Wight & Co.
75832	9		J. B. Leeds *Moore* W Sr	American	218	123·0	33·1	9·3	Coquille, Oreg	1876	H. Doncaster		J. Knowland
71047	10	V	J. B. Newcomb *Rutherford* W Bk	British	891	160·8	34·7	21·0	Rockland, N.B.	1875	R. A. Chapman	[illegible]	J. B. Newcomb
75568	11	R	J. B. Rabel *Sawyer* W Bkn	American	428	145·0	33·0	12·6	Columbia Falls, Me.	1882	J. H. Crandon	[illegible]	Swan & Son
13788	12	R	J. B. Van Dusen *Booye* W3mSr	American	211	115·2	29·4	10·5	Dolphi Crk. Md	1867			J. B. Van Dusen
76092	13		J. B. Walker *Hodgman -87* W S	American	2105	247·1	42·2	29·8	Thomaston Me	1879	E. O'Brien	[illegible]	E. E. O'Brien
96411	14		J. B. Wood W Bn	British	126	90·3	23·8	10·5	P. E. Island	1873	Longworth		J. Pollard
	15	G	J. Bergman Olson *Nomen 87-87* W Bk	Swedish	505	133·8	29·2	17·5	Dalkarlsà	1867	Löfgren	[illegible]	Nordmalings Skepps Red. Bol. (C. M. Bolen)
80982	16		J. Burstall W Barge	British	152	110·7	21·7	7·0	Quebec	1880	J. Bolduc		D. Murphy
76312	17		J. C. Ford *Griffiths* W3mSr	American	231	123·0	31·8	10·0	San Francisco	1882	Turner		T. H. Griffiths
	18	GV	J. C. Julius *Christensen -87* I Bk	German	1076	205·0	34·5	19·9	Rostock	1885	Rostocker Act. Ges.		J. Hintze
75305	19		J. C. Nash (LOST) *Crowley* W Sr	American	129	91·8	25·8	9·0	Addison, Me.	1870			A. L. Crowley
	20	✻	J. C. Pfluger (ex W[illegible]) *Kruse 81-88* I S	German	977	210·5	34·1	19·2	Sunderland	1874	J. Blumer & Co.		J. C. Pfluger & Co.
46142	21		J. C. Rahming W Bk	British	199	98·0	24·0	11·7	Maretta, C. S.	1845			J. C. Rahming
	22	V	J. C. Warns *Warns 79-82* W Bk	German	795	154·8	34·1	19·9	Ha'melwarden	1882	C. Lühring	Y 12.88	J. C. Warns
61575	23	RV	J. C. Williams *Smith 83-88* W Bk	British	860	173·0	34·5	20·4	Jordan, N. S.	1874	S. Freeman & Sons	Y 11.88	A. N. Smith
13461	24		J. Clark *Cousins* W Sr	American	191	107·5	28·3	8·5	Great Egg Harbour	1857			G. E. Martin
36846	25		J. Congdon W Bn	British	126	84·7	23·9	10·2	Truro, N. S.	1861			J. Alexander
	26	G	J. D. Auffort (ex Hanna) *Cajfitsen 81-* W Bk	Swedish	390	119·7	29·4	16·1	Jacobstad	1860	L. Lof		A. B. Elfversen
	27	V	J. D. Bischoff (ex Hagerstown) *Meyer 80-88* W S	German	1808	223·7	42·3	26·1	Richmond, Me.	1874	J. M. Hagar	Y 3.88	J. D. Bischoff
12267	28		J. D. Ingraham *Brockway* W Sr	American	145	105·0	25·3	7·6	Hartf'rd, Conn	1869			E. P. Brockway
75802	29	V	J. D. Peters *Lowe -85* W Bk	American	1031	182·2	36·1	21·0	Bath, Me	1875	Goss & Sawyer	[illegible]	C. L. Taylor & Co.
75088	30	R	J. D. Robinson *Hagan* W3mSr	American	447	135·9	30·9	15·1	Phippsb'rg Me	1874		[illegible]	C. V. Minott
88204	31	DV	J. E. Graham *Cochrane* W Bk	British	1385	206·0	40·5	23·9	Newport, N. S.	1881	J. A. Harvie	Y [illegible]	W. H. Mosher
76716	32		J. Eppinger W Sr	American	107	89·0	26·6	7·5	San Francisco	1888	C. G. White		Thomson Collier
	33		J. F. *Febeiro* W Bg	Spanish	186	93·8	25·6	13·3	Carril	1866			P. Ferreiros
12848	24		J. F. Carver W Sr	American	146	96·5	23·3	9·1	Mt. Desert, Me.	1857			W. Carver
	35		J. F. Heydtmann *Rege 74-?* W Bg	German	247	100·0	27·0	14·0	Damgarten	1863	H. Dierling		F. Schultze
75093	36	R	J. F. Merry *Bradley* W Bn	American	448	137·0	30·0	16·9	Damariscotta, Me.	1874		[illegible]	I. Genthner
	37	DG	J. F. Pust *Niejahr 80-* W Bk	German	401	112·0	29·0	16·5	Rostock	1865	O. Ludewig	[illegible]	H. Bauer
42307	38		J. F. Spencer (ex [illegible] Abbott) W Bk	British	313	129·0	28·5	12·7	Londonderry, N. S.	1863			J. S. Langthorn
	39	R	J. Frank Seavey *Kelley* W3mSr	American	392	144·0	33·9	10·2	Bath, Me.	1888	[illegible]		B. F. Neally
88929	40	✻	J. G. C. Ross *Ross* W S	British	92	92·8	22·7	8·5	Cocos Island	1882	G. C. Ross	[illegible]	G. C. Ross
	41	GV	J. G. Fichte *Gronhoff 79-79* W Bg	German	231	106·2	26·0	11·9	Elsfleth	1863	M. Willers	10.88	F. J. Hulett
75148	42		J. G. Norwood W Bk	American	419	121·0	31·0	16·0	Camden, Me.	1869			G. T. Harkness
45083	43		J. H. Barrow *Bowes* W S	British	130	93·5	22·6	11·2	Glasson Dock	1863	Simpson		J. H. Barrow
75748	44	V	J. H. Bowers *Plum* W Bk	American	704	163·5	33·0	18·5	Camden, Me.	1875	Carleton & Norwood	[illegible]	E. W. Harkness
75633	45	R	J. H. Chadwick *Hurlbut* W Bk	American	455	129·9	30·7	11·5	Deering, Me.	1874	W. Sargent	[illegible]	J. L. Howe
	46	G	J. H. Epping *Kuhl -87* W Bg	German	259	101·3	27·0	14·3	Rostock	1865	G. W. Tesdorf		C. Kühl
	47	DV	J. H. Henkes (ex Jacob den Breejen) *Van Heusden* W Bn	Dutch	179	98·8	22·4	11·7	Hoogezand	1868	J. A. Hoentes	[illegible]	J. H. Henkes Hz.
	48	✻	J. H. Hustede *Reinars* I Bk	German	1092	214·7	34·9	21·6	Port Glasgow	1889	Russell & Co.		J. H. Hustede
	49	GV	J. H. Lübcken W Bg	German	340	117·2	28·2	14·2	Geestemünde	1872	Schau & Oltm'nn	[illegible]	C. G. Beermann
66804	50	RV	J. H. McLarren *Wyman* W Bk	British	764	165·5	34·1	18·2	Blk. River, N.B.	1874	J. McLeod	[illegible]	D. Carmichael

✻ = Lloyd's Register; D = Nederlandsche Vereeniging; G = Germanischer Lloyd; I = Registro Italiano; R[illegible]
A = Veritas Austro-Ungarico; H = Veritas Ellenico; N = Norske Veritas; V[illegible]

In the 4th Column, C = Composite (iron frame, wood planked); I = Iron; Stl = Steel; W = Wood.

The Dates of Metalling (Col. 11) are based on the records of the Books in which Vessels are respectively classed, or on the Reports of the [illegible]

Further particulars of Vessels classed by Lloyd's Register will be found in Lloyd's Register Book and Lloyd's Yacht R[illegible]

No.	Cote	Nom. Matériaux et Gréement. Capitaine et dates d'entrée au service des Armateurs actuels et à bord du navire. Ancien Nom.			Pavillon	Tonnage	Dimensions			Construit			Date du doublage	Armateurs ou Directeurs Gérants.	Port d'Armement.
		Nom	Capitaine	Gréement			Longueur	Largeur	Creux	Où	Quand	Par qui			
51	V	J. H. Marsters	Crossley (71-79)	W Bk	[illegible]	632	148·1	33·0	18·1	Kempt,N.S.	1878	R. Rose	V 8.88	R. Rose	Windsor,N.S.
52	V	J. H. Nicolai	Grotz 81-84	W Bk	German	1033	182·0	37·1	21·8	Brake	1884	F.F. Nicolai	V 12.84	J.H. Nicolai	Brake
53		J. H. Parker	Howe, 84-86	W3mSr	American	521	146·7	32·2	18·2	P.Jefferson,NY	1884	J.R. Mather		W.B. Dayton	P.Jefferson,NY
54	GV	J. H. Ramien	Pfeiffer 81-87	W Bk	German	631	142·3	30·5	18·0	Elsfleth	1875	J. Jurgens	V 6.86	E. tom Dieck	Elsfleth
55		J. H. Wainwright		W Sr	American	190	107·1	23·8	9·1	Cape May, N.J.	1869			Swett & Co.	Bangor,Me.
56	R	J. Howell Leeds	Clough	W3mSr	American	394	134·6	34·8	11·5	Mauricet'n,N.J.	1881	Vanneman		A.W. Cranmer	Philadelphia
57	V	J. I. Smith	Tingley 88	W S	British	1277	204·3	38·5	22·1	Dorch'ster,N.B.	1881	W. Hickman	V 8.88	W. Hickman	Dorchester,N.B.
58		J. I. Worthington (STRANGER)	Pitch	W Sr	American	165	108·0	28·0	8·5	Portland,Con.	1864	Gildersleeve		Scammell Bros.	N.London Conn.
59		J. J. Dussumier		W Bk	French	298	106·9	26·9	15·3	Sables d'Olonne	1860		V 12.87	J. Gronhel	Bordeaux
60		J. J. Little	Gandy	W Sr	American	142	110·3	22·8	8·1	Chester, Pa.	1867			H. Matthews	Philadelphia
61		J. J. Moore	Arnold	W3mSr	American	390	136·0	32·0	15·9	Kennebunk, Me.	1873			D.B. Arnold	N.London,Conn.
62	✻	J.K.Allport	Burns 84-84	W Sr	British	100	85·6	22·7	10·8	Padstow	1876	Cowl		C. Allport	Plymouth
63		J. K. Manning	Gandy	W Sr	American	233	111·0	29·2	9·7	DennisvilleNJ	1869			H. Matthews	Philadelphia
64	R	J. L. Bowen	Kay 88	W Bk	American	478	122·0	29·6	16·7	Quincy,Mass.	1869	G.W. Thomas	V 8.86	J.M. Phillips	Boston,Mass.
65		J. L. Cotter	[illegible]	W Sr	British	139	93·5	26·7	8·8	Portland,N.B.	1872			Scammell Brothers	St.John,N.B.
66	R	J. L. Crossley	McGilvery	W Sr	British	170	100·1	26·8	10·1	Kempt,N.S.	1874	R. Rose		Wm.H. McGilvery	Windsor,N.S.
67	✻	J. L. Hall	Owen 75-86	I Bk	British	682	166·8	30·6	19·6	SouthShields	1859	Marshall Bros.		J.L. & W.A. Simpson, & Elder, Smith & Co.	Adelaide,S.A.
68	✻R	J. L. Pendergast	Pendergast 88	W Bk	British	558	156·8	30·7	17·7	Quebec	1871	Dunn	V 7.87	J.H. McKinnell	Quebec
69		J. L. Stewart		W Bn	British	201	97·8	28·4	11·2	Parrsboro'NS	1875	T. Kelly		James Crarie	Parrsboro'N.S.
70	R	J. M. B.	Mon	W Bk	Spanish	289	107·6	26·2	14·5	Riavedeo	1873	J.A. Reguera	V 9.87	J.M.M. de Bengoechea	Rivadeo
71	V	J.M.Buuck	Wickhardt 88	W Sr	German	179	93·5	23·1	12·1	Barth	1876	R.C. Schber	V 4.86	J.J. Wallis & Sohn	Barth
72		J. M. Coleman	Ure, 88-89	W3mSr	American	448	157·0	37·1	11·3	PortBlakley,W.T.	1888	Hall Bros.		J.J. McKinnon & Co.	San Francisco
73		J. M. Griffith	Grey 88	W Bk	American	575	152·0	40·0	14·0	Seabeck,Wash	1882			R. Holyoke	Seabeck,Wash.
74		J. M. Harlow	Soper	W3mSr	American	221	116·0	32·7	9·0	E.HavenConn.	1874	W.O. Nettleton		R. Kendrick	Port Townshend
75		J. M. Morales	Jordan	W Sr	American	169	97·5	27·0	8·8	Bucksport,Me.	1867			M. Hume	Bath,Me.
76	V	J.M.Nielson.	Jorgensen	W Bn	Danish	144	96·5	21·3	10·3	Svendborg	1878	J.R. Andersen		N.W. Kaas	Svendborg
77	R	J. Manchester Haynes	Mathews	W3mSr	American	756	179·6	35·6	20·2	Waldobro'Me.	1885	L. Storer		J. Baker & Co.	Boston,Mass.
78		J. Martin	Jacobsen	W Sr	Russian	193	96·2	27·2	12·5	Nogallen	1883	J. Stahl		J. Stahl	Riga
79		J. N. Carter		W Sr	British	187	112·4	25·0	9·0	Picton,Ont.	1875	W. Reimond		N. Dyment	Picton,Ont.
80	R	J. Nickerson	Nickerson	W Sr	American	171	98·0	26·0	9·5	Brunswick,Me	1879	R. Given		H.M. Sargent	Portland,Me.
81	DRV	J. P. A. (ex Wassenaer, ex Pholde)	Smits 81-89	W Bk	Dutch	1173	197·1	36·0	22·8	Quebec	1867	Dunn	V 8.86	G. Mauritz Hzn	Dordrecht
82	R	J. P.Auger	Brown	W Sr	American	195	94·6	25·3	8·7	Madison,Conn	1862			J. Brown	New York
83	V	J. P. Lassen	Jensen	W Bn	Danish	152	90·6	22·7	10·5	Aalborg	1872	S.L. Sorensen	V 10.83	L.C. Nielsen	Aalborg
84		J. P. Wyman	Bearill	W Sr	American	157	100·4	26·2	8·7	Franklin,Me.	1867		V 8.86	E.B. Wiley	Boston,Mass.
85		J. Paine	Rich	W Sr	American	232	108·5	29·4	8·7	Philadelphia	1863			H.F. Smith	P'vincetwn.Mass.
86		J. Ponder, jun.	Bulfork	W Sr	American	143	98·0	28·0	7·0	Milton,Del.	1857			W. Pickering, jun.	Salem,Mass.
87	R	J. R. Bodwell	Meteif	W Sr	American	161	100·0	26·0	8·0	Stockton,Me.	1874			Cobb, Wight & Co.	Rockland,Me.
88	R	J.S.Hoskins	Rich 86	W3mSr	American	391	193·9	34·2	11·2	Bath, Me.	1886	New England [V.B.& E.Co.	V 8.86	B.B. Rokes	Baltimore
89	GV	J.Schoentjes	Ohle 88	W Bk	German	363	116·4	28·9	17·3	Rostock	1878	O. Ludewig	V 7.88	E.C. Winter	Rostock
90		J. S. Lamprey	Harding	W3mSr	American	291	130·0	31·0	12·0	Salem,Mass.	1872			S. Harding	Gloucester,Mass.
91		J. S. Terry	Raynor	W Sr	American	118	104·4	22·4	7·2	EHaddamCon.	1864			D. Raynor	P.Jefferson,NY
92	V	J. Steffen	Brahn 87-87	W Bk	German	522	118·9	30·5	16·5	Lübeck	1887	J. Steffen	[illegible]	C. Brandt	Blankenese
93		J. T.	Le Bail	W Bg	French	168	88·0	21·0	11·5	Leghorn	18^{56}_{74}	Basmont		Amaury Simon	Redon
94	✻	J. T. North	Hammond 83-87	I Bk	British	803	196·4	31·8	18·7	PortGlasgow	1883	Russell&Co.		W. & J. Lockett	Liverpool
95		J. T. S.	Robbins 83-84	W Sr	British	129	89·0	21·7	12·0	Sunderland	1862	Watson		J. Fisher & Sons	Barrow
96	N	J. T. Smith	Fougley 88	W Bk	Norweg'n	434	126·7	30·5	17·2	FoxRiver,N.S.	1870	Chs.Smith	V 8.86	H.O. Hansen	Skien
97		J. T. Townsend		W3mSr	American	604	154·6	35·0	17·0	Camden,N.J.	1885	S.W. Tilton		G. Taulane	Philadelphia
98	RV	J. V. Troop	Pratt 88	W S	British	1295	200·0	39·0	23·0	Tynemo'thN.B.	1879	J.S. Parker	V 10.85	H.D. Troop	St.John,N.B.
99		J.V.Wellington	Philpot	W Sr	American	244	112·0	29·0	9·0	Philadelphia	1861			B.H. Chipman	Boston,Mass.
100	✻	J. W. Bebell	Evans 82-82	W Sr	British	99	82·3	21·0	10·6	Newport	1875	Newport Dry [Dock Co.		O. Hughes & Co.	Beaumaris

✻ = Lloyd's Register; D = Nederlandsche Vereeniging; G = Germanischer Lloyd; I = Registro Italiano; R = American Record;
A = Veritas Austro-Ungarico; H = Veritas Ellenico; N = Norske Veritas; V = Bureau Veritas.

Dans la 4me Colonne, C = Composite (membrure en fer, bordé en bois); I = Fer; Stl = Acier; W = Bois.

Les Dates de doublage en métal (Col. 11) ressortent des livres dans lesquels les navires sont respectivement classés ou des rapports des Inspecteurs du Lloyd's Register.

Pour plus amples renseignements sur les navires classés par le Lloyd's Register, voir le Lloyd's Register Book et le Lloyd's Yacht Register.

1 Official No.	2 No. in Book	3 Registry in which classed	4 Ship's Name. Master, and Dates of Appointment to present Owner's Service and to Ship. Ship's Late Name.	Material and Rig	5 Flag	6 Net Tonnage	7 Dimensions: Length	Breadth	Depth	8 Built: Where	9 When	10 By whom	11 Date of Metalling	12 Owners or Managers
75857	101	R	J.W.Dresser *Parker* -83	W Bk	American	634	137·9	31·8	17·8	Castine, Me.	1877	J.A.Webster	[illegible]	J.W.Dresser
76370	102	R	J. W. Gaskill *Godfrey*	W3mSr	American	402	143·9	35·4	12·3	Leesburg, N.J.	1882	J.Russell		A.S.Godfrey
	103	V	J. W. Gildemeister (ex Monaltrie) *Meyer*	W S	German	1006	211·9	36·1	20·6	Aberdeen	1872	Humphrey&Co.	[illegible]	Gildemeister & Ries
61500	104	DV	J.W.Holmes *Newcomb* 81-81	W Bk	British	918	172·0	35·0	22·4	Horton, N.S.	1871	North	[illegible]	J.B.North
37498	105		J.W. Johnston	W Bg	British	187	104·3	24·2	12·1	ColeHarb'r N.S.	1856			G.A.Frink
75814	106	V	J.W. Marr *Cotton* 76	W S	American	1296	186·2	39·6	24·0	Bath, Me.	1875	D.O.Blaisdell	[illegible]	J.W.Marr
74420	107		J. W. T. *Cock* 86-86	W Sr	British	100	87·4	22·0	10·9	Pill	1877	Hitchins		W.Webber
13402	108		J. W. Vannaman *Champy*	W3mSr	American	339	133·0	29·0	14·8	Mauricetown, N.J.	1863			F.W.Benedict
	109	G	J. W. Wendt (ex Morning Light) *Lass*	W S	German	2287	263·7	44·3	28·5	St.John, N.B.	1855	W.&R.Wright	[illegible]	Siedenburg, Wendt&Co.
13696	110		J. W. Woodruff *Smith*	W Sr	American	139	89·2	24·3	9·0	Machias, Me.	1859			S.G.Haskell
64590	111	✳	J. Walter Scammell *M'Farlane* -86	W Bk	British	910	176·7	34·7	20·2	Portland, N.B.	1872		[illegible]	Scammell Brothers
71205	112		J. Walters	W Sr	British	176	108·5	23·5	8·3	Picton, Ont.	1874	W.Redmond		H.A.Patterson
	113	GV	J. Weissenhorn (ex Samuel Watts) *Van Treeten* 84-	W S	German	2024	223·7	40·0	30·5	Thomaston	1870	S.Watts&Co.	[illegible]	Siedenburg, Wendt&Co.
	114	D	J. de Jong (ex Ann, ex Breonero) *Reinders* 80-83	W Bk	Dutch	767	171·2	32·8	..	Sestri Ponente	1874		[illegible]	P.Landberg & Zoon
12603	115		J. & H. Crowley	W Sr	American	180	104·8	28·7	8·7	Addison, Me.	1857	J.C.Nash		Rideout & Lord
76530	116		J. & H. Scull *Ingersoll*	W Sr	American	106	87·0	28·3	6·2	Eng.Creek, N.J.	1884	D.H.Gaudy		J.Scull
75606	117		J.& L.Bryan *Risley* 84-	W3mSr	American	365	135·0	31·8	9·9	Absecom, N.J.	1874	E.S.Risley		R.Babcock
88702	118	✳	J. & M. Garratt *Garratt* 84-84	W Sr	British	87	92·4	21·5	9·2	Barrow	1884	W.Ashburner & Sons		Coppack, Carter & Co.
76522	119		J. & W. Gurney *Gurney*	W Sr	American	114	86·8	28·6	6·9	Rondout, N.Y.	1882			W.B.Gurney
75906	120	RV	Jabez Howes *Henry* 81-84	W S	American	1581	218·8	40·1	26·0	Newburyport, Mass.	1877	J.Currier	[illegible]	J.Rosenfeld's Sons
	121		Jacinta	W Bg	Spanish	115	73·1	23·5	12·3	Arens	1845			Salvador Soler
76743	122	✳	Jacinth *Heath* 87-87	W Sr	British	100	86·3	21·7	10·5	Kingston	1877	Duncan		Colhoun Bros.
	123		Jacinto Goday (ex Pedro Tarull, ex Joaquin) *Fusconde*	W Bn	Spanish	198	88·9	26·5	14·3	Blanes	1848			M.G.Gual
	124	V	Jacmel *Ameline* -86	W Sr	French	207	117·1	22·3	12·6	Havre	1867	B.Normand fils	[illegible]	H.Guillebot
1032	125	I	Jacmel (ex Rodolfo Danovaro) *Bennati*	W Bk	Italian	411	126·3	30·2	18·7	Prà	1864	M.Tixi	[illegible]	G.Costa
43266	126		Jacmel Packet (ex Hastico)	W Bn	British	115	79·7	21·2	9·0	Salem, U.S.	1844			H.Okill
	127		Jacob *Rønne* 85-88	W Bk	Danish	302	108·0	26·3	15·6	Copenhagen	18 38/62	P.Brandt		A.P.Ellebye
	128		Jacob *Nymann* 80-86	W Bg	Norweg'n	171	95·2	23·7	12·0	Åbo, Finland	1850			H.L.Gabrielsen
	129	V	Jacob *Johansson* -86	W Bk	Swedish	556	147·3	28·5	17·8	Frammäs	1876	E.Östmann	[illegible]	Rederi-Aktiebol. Unica (G.O.Wallenberg)
	130	R	Jacob	W Bn		239	107·6	25·1	13·4	Uddevalla	1882	J.Johansson	[illegible]	
75949	131	R	Jacob A. Howland (WHALER) *M'Innes* 75-86	W Bk	American	355	119·0	28·0	17·0	NewBedford, Mass.	1877	S.Dammon	[illegible]	Aiken & Swift
13118	132	G	Jacob A. Stamler *Crystal* 75-75	W S	American	955	175·0	39·0	23·3	Greenpoint N.Y.	1856	J.Stack & Co.	[illegible]	Boyd & Hincken
	133	N	Jacob Aall *Beruldsen*	W Bk	Norweg'n	499	141·2	31·4	17·1	Tvedestrand	1876	M.Eldrup	[illegible]	Jørgensen & [illegible]
	134	V	Jacob Arendt *Mestermann* 73-86	W Bk	German	436	124·3	28·8	17·5	Danzig	1861	J.W.Klawitter		H.Behrent
76237	135	RV	Jacob E. Ridgway *Colt*	W S	American	1720	219·1	41·0	25·4	Bath, Me.	1881	Goss & Sawyer	[illegible]	J.E.Ridgway
75511	136	R	Jacob M. Haskell *Crowell*	W3mSr	American	461	141·8	32·1	15·4	East Boston	1874		[illegible]	J.H.Conant & Co.
	137	V	Jacob Rauers (ex James Ives) *Carlson* 88-88	W Bk	Swedish	518	144·0	28·7	17·1	RiverJohn, N.S.	1869		[illegible]	J.F.Olsson
76350	138	R	Jacob Reed *Nickerson*	W3mSr	American	457	146·0	35·6	13·4	N'wH'v'nCon.	1882	Hanscom		J.H.Conant & Co.
13409	139		Jacob S. Welden *Dow*	W Sr	American	138	99·0	27·0	8·0	Wilmington, Del.	1853			G.S.Dow
27[illegible]	140	DG	Jacob & Catharina *Wiedner* 77-	W Bkn	Russian	332	121·5	28·7	14·2	Riga	1878	K.Thire		J.Dombrowsky
2[illegible]	141		Jacoba (ex Mary Ann Holman) *Andraca* 87-87	W Bk	Peruvian	2671	117·0	24·5	14·8	Topsham	1859	Holman		J.B.Andraca
	142	DGV	Jacobine (ex Rehbel) *Janssen* 82-83	W Bk	German	565	147·1	30·9	18·4	Rostock	1874	E.Borchard	[illegible]	Joh.de Harde
	143	G	Jacobine *Wessels* 79-	W Bn	German	163	95·6	22·3	11·5	Papenburg	1864	H.R.Zeemeyer		Lamb, Jongebloed Wwe.
	144	A	Jacobos (ex [illegible]) *Zaldaras* -86	W Bk	Greek	338	139·9	26·4	14·7	Dundee	1875	Dundee S.B.Co.	[illegible]	S.Stathopulo
	145	DV	Jacobus Johannes *Jacobsen* 86-86	W Bk	Dutch	379	127·9	30·1	14·5	Middelburg	1879	Borsius & Van der Leyer	[illegible]	Den Bouwmeester Sterkebus & v. d. Leije
	146	V	Jacques *Cornelle* -88	W Bg	French	272	111·2	27·6	14·8	Fecamp	1867		[illegible]	C.Sureau
	147	V	Jacques *Leboulanger* -85	W Bg	French	179	102·6	22·3	11·5	Granville	1876	Lecarpentier	[illegible]	Riotteau & fils
	148	DV	Jacques Cartier *Villeneuve*	W Bg	French	222	93·2	23·0	14·4	St.Malo	1862	L.Mallard	[illegible]	J.Blaize
	149	V	Jacques Cœur *Renaud* -85	W Bk	French	436	131·5	26·6	16·1	St.Malo	1875	P.Hervot	[illegible]	A.Postel & ses fils
	150													

✳ = Lloyd's Register; D = Nederlandsche Vereeniging; G = Germanischer Lloyd; I = Registro Italiano; R = [illegible]
A = Veritas Austro-Ungarico; H = Veritas Ellenico; N = Norske Veritas; V = [illegible]

In the 4th Column, C = Composite (iron frame, wood planked); I = Iron; Stl = Steel; W = Wood.

The Dates of Metalling (Col. 11) are based on the records of the Books in which Vessels are respectively classed, or on the Reports of the Surveyors [illegible]

Further particulars of Vessels classed by Lloyd's Register will be found in Lloyd's Register Book and Lloyd's Yacht Register.

No.	Cote	Nom. Matériaux et Gréement. Capitaine et dates d'entrée en service des Armateurs actuels et à bord du navire. Ancien Nom.	Pavillon	Tonnage net	Longueur	Largeur	Creux	Construit Où	Quand	Par qui	11	Armateurs ou Directeurs-Gérants	Port d'Armement
151		Jadar (ex Kate) W3mSr	Norweg'n	253	98·8	25·2	15·1	Peterhead	1841			Thomas Nielsen	Stavanger
152	N	Jaebez *Dahl 78-85* W Bg	Norweg'n	187	99·7	25·4	12·5	Sandefjord	1878	M. Bergan		J.Christophersen	Sandefjord
153	N	Jæderen *Andreassen 77-86* W Bk (WBS-&c.)	Norweg'n	247	103·8	27·4	14·0	Pirth	1848	Petterson		A.Andreasen	Farsund
154	N	Jael *Hansen 74-74* W Sr	Norweg'n	119	82·2	23·4	11·1	Sandefjord	1875			S.Schjulstok	Sandefjord
155	A	Jafet II. *Stuck* W Bk	Austrian	790	159·1	34·8	20·9	Orebich	1879	A.Villovich	Y [illegible]	G.Bielich fu M.	Cuciste
156	N	Jafnahr *Sorensen* W Bk	Norweg'n	483	138·6	30·9	16·1	Porsgrund	1877	Vaugbart& [Høegh	Y [illegible]	H.Sorensen	Laurvik
157		Jahue *Snoling* W Bk	Russian	229	100·0	25·6	12·8	Nogallen	1883	J.Stahl		J.Stahl	Nogarren
158		Jaimito W Bn	Spanish	124	78·7	22·2	11·6	Santander	1852			Jaime Falmer	Barcelona
159	V	Jalapa *Boot 85-85* W Bg	German	218	112·7	25·2	13·9	Reiherstieg	1860	E.Dreyer	Y [illegible]	E.B.Crasemann	Hamburg
160		Jalo W S	Russian	1366	210·5	36·1	23·6	Nystad	1876	H.H.Kjildström	Y [illegible]	V.A.Reims	Åbo
161	✻	Jamaica *Siebert 86-86* I Bk	British	699	178·5	29·9	19·3	Sunderland	1875	W.Doxford & Sons		P.Denniston & Co.	Glasgow
162	R	Jamapa *Gonzalez* W Sr	Mexican	137	98·6	27·0	7·4	Rockland,Me.	1873				
163		James W Barge	British	153	108·6	22·6	6·4	Montreal	1882	P.Lamothe		D.Murphy	Ottawa,Ont.
164		James *Sonnenstein 76-76* W Bk	Swedish	495	148·9	28·5	17·1	Altona	1859	C.Voerman		P.G.Mellander	Lysekil
165	R	James *Larsson* W Bg	Swedish	149	94·8	21·2	12·1	Bermuda	1849			H.Larsson	Mörby
166	R	James A.Borland *Doole* W Bk	American	637	110·4	32·6	19·2	New York	1869	Webb	Y [illegible]	L.Sloss,jun.	San Francisco
167		James A.Brown *Thorndike* W Sr	American	164	95·6	25·8	9·0	Fairhav'n Con.	1859			Burgess & O'Brien	Thomaston,Me.
168	R	James A. Garfield *Woodbury* W3mSr	American	485	112·0	32·4	12·0	Brewer,Me.	1881	Oakes & Son	Y [illegible]	Caleb Holyoke	Bangor,Me.
169		James A. Garfield *Morse* W3mSr	American	300	140·0	37·5	10·2	North Bend, [Oreg.	1881	Kruse		A.M.Simpson	San Francisco
170	R	James A. Gray *Coleman* W Sr	American	124	96·4	27·0	7·0	Bath,Me	1838	Kelley & Spear		R.F.C.Hartley	Saco
171	V	James A. King *Berry* (ex Huron) W Bk	Hawaiian	609	151·5	32·2	18·5	Nova Scotia	1870	Roy	Y [illegible]	Wilder & Co.	Honolulu
172	✻	James A. Mark *Sikey 76-82* W Bk	British	346	131·5	27·0	15·7	Cremyll	1874	Ridley	Y [illegible]	W.W.Dingle	Cardiff
173	RV	James A. Wright *Baker* W Bk	American	867	190·0	36·0	20·0	Wilmington, [Del.	1877	Jackson&Sharp [Co.	Y 3.86	C.&J.F.Baker	Philadelphia
174	✻	James Aikin *Morrison 82-82* I S	British	958	197·5	34·1	21·5	PortGlasgow	1867	J.Reid&Co.		M'Diarmid, Greenshield & Co.	Liverpool
175		James Allen *Lake* W Bk (WHALER)	American	331	116·7	27·8	17·1	Bath,Me.	1877	Goss,Sawyer& [Packard		J.McKenna	N.Bedford,Mass
176		James Arnold *Sherman* W S (WHALER)	American	329	115·2	27·6	17·6	NewBedford, [Mass.	1852			Tabor & Gordon	N.Bedford,Mass
177		James B.Anderson *Bradley* W S	American	155	101·0	27·6	8·2	Frederica,Del.	1871			J.W.Hall	Frederica
178		James B.Johnson *Shaw* W Sr	American	140	91·9	27·3	8·2	EggHarb'r,NJ	1858		Y 11.87	J.B.Van Dusen	Philadelphia,Pa.
179	R	James B. Jordan *Erskine* W3mSr	American	686	162·7	35·5	17·7	ThomastonMe	1884	S.S.Gerry&Co.	Y 11.87	S.S.Gerry	Thomaston,Me.
180	R	James B. Ogden *Newbury* W3mSr	American	648	162·6	36·0	17·5	Alexandria,Va	1880	AlexandriaS.Co.		A.J.Newbury	New York
181	R	James B. Pace *Thrasher 83* W3mSr	American	609	161·4	35·3	18·3	Newburyport, [Mass	1883	G.Carrier		J.W.Thrasher	Richmond,Va.
182		James Barrett *Percy* W Sr	American	148	95·0	26·0	9·0	Milford,Del.	1847			E.Milliken	Bath,Me.
183		James Boyce *Crossley* W3mSr	American	431	142·5	32·5	16·2	E.Haven,Conn	1877	W.O.Nettleton		F.W.Benedict & Co.	New Haven,Con.
184	R	James Boyce, Jr. *Duncan* W3mSr	American	693	166·0	35·4	15·6	Alexandria,Va	1882	Goss&Sawyer		A.A.Duncan	Alexandria,Va.
185	R	James C. Clifford *Clifford* W3mSr	American	359	136·0	32·0	12·0	E.Boston	1886	J.M.Brooks		G.B.Reynolds	Newport,R.I.
186		James Cheston *Plumb 87-87* W Bk	American	948	170·1	36·2	22·3	Baltimore	1854			Puget Sound Commercial Co.	Port Townshend,Wash
187		James Cunningham W Barge	British	176	110·3	22·7	8·0	Quebec	1877	A.Miller		D.Murphy	Ottawa,Ont.
188	R	James D. Dewell *Nelson 88-88* W3mSr	American	573	151·2	34·5	16·0	West Haven, [Conn.	1882	Gesner& Marr		H.Sutton	New Haven,Con.
189		James Davidson *Evans* W Bkn	Brazilian	198	107·1	23·2	14·0	Whitehaven	1865	Shepherd		Brazilian Coal Co.,Ld.	Rio de Janeiro
190		James Diverty *Taylor* W Sr	American	123	101·0	20·0	7·0	DennisvilleNJ	1857			W Taylor	Dennisville,NJ.
191	RV	James Drummond *Carter* W S	American	1479	216·0	40·0	24·2	PhippsbrgMe.	1880	C.V.Minott	Y 2.86	C.V.Minott	Bath.Me.
192		James Duncan *Wahlstrom* W3mSr	Danish	338	113·0	27·5	16·5	Sunderland	1859	J.Hardie		Actieselskab (P.Wilhelmsen)	Copenhagen
193		James E. Bayles *Sterling* W3mSr	American	392	140·4	32·6	10·2	PJeffersonNY	1874	J.M.Bayles	Y 1.87	C.A.Bayles	P.Jefferson,NY
194		James E. Woodhouse *Dawson* W3mSr	American	311	118·5	30·9	12·4	N.Haven,Con.	1878	J.H.Woodhouse	Y [illegible]	A.M.Douglass	New York
195		James English *Perkins* W Sr	American	121	91·2	25·0	8·2	N.Haven,Con.	1850			J.B.Perkins	NewLondon,Con.
196		James Ford *Garfield* W3mSr (ABANDONED)	American	597	146·4	33·9	17·2	Bath,Me.	1871			L.Fisk	Dennis,Mass.
197	V	James G.Bain *McNutt* W Bk	British	597	159·0	34·1	17·8	PortPhilip,NS	1880	P.Meckler	Y [illegible]	P.Meckler	Pictou,N.S.
198	RV	James G. Pendleton *Lancaster* W Bk	American	891	165·6	36·6	22·5	N'buryportMs	1871	Gilman& [Kingsbury	Y 7.87	S.Lancaster	Searsport
199		James G. Worts W Sr	British	309	136·0	26·3	11·3	MillPoint,Ont	1874	W.Jamieson		S.Sylvester	Toronto,Ont.
200	✻	James Garfield W K Barge	British	81	87·5	21·3	7·5	Ipswich	1881	W.J.Curtis		T.G.Wellum & Co.	Ipswich

✻ = Lloyd's Register; D = Nederlandsche Vereeniging; G = Germanischer Lloyd; I = Registro Italiano; R = American Record; A = Veritas Austro-Ungarico; H = Veritas Ellenico; N = Norske Veritas; V = Bureau Veritas.

Dans la 4me Colonne, C = Composite (membrure en fer, bordé en bois); I = Fer; Stl = Acier; W = Bois.

Les Dates de doublage en métal (Col. 11) ressortent des livres dans lesquels les navires sont respectivement classés ou des rapports des Inspecteurs du Lloyd's Register.

Pour plus amples renseignements sur les navires classés par le Lloyd's Register, voir le Lloyd's Register Book et le Lloyd's Yacht Register.

Official No.	No. in Book	Registry in which classed	Ship's Name. Material and Rig. Master, and Dates of Appointment to present Owner's Service and to Ship. Ship's Late Name.	Flag.	Net Tonnage	Dimensions: Length	Breadth	Depth	Built: Where	When	By whom	Date of Metalling	Owners or Managers
85906	201		James Gordon *Pedrick* W	American	147	108·6	22·6	7·[illegible]	[illegible]	1881	A.M[illegible]		Geo.A.Harris
12575	202		James H. Deputy *Macey* WSr	American	12[illegible]	91·7	24·2	7·1	Milford,Del.	1848			J.Varney
13377	203		James H. Gallagher *Lynch* WSr	American	10[illegible]	[illegible]	17·[illegible]	7·[illegible]	[illegible]	186[illegible]			C.Lynch
75335	204		James H. Gordon *Lovell* W3mSr	American	[illegible]	126·[illegible]	31·4	16·[illegible]	[illegible] Landing, N.J.	1871	J.Clark		W.Irelan
7566[illegible]	205		James H. Hoyt *Pomeroy* W3mSr	American	[illegible]	[illegible]	[illegible]	15·[illegible]	N.Haven,Con.	1874			A.C.Lyons
76461	206	R	James H. Woodhouse *Anderson -87* W3mSr	American	27[illegible]	[illegible]·7	31·[illegible]	10·[illegible]	[illegible]	1882	H.W[illegible]	[illegible]	J.H.Woodhouse
75625	207		James Holmes *Ryan 79-83* WSr	American	127	97·[illegible]	25·[illegible]	6·9	[illegible],Me.	1874	B.[illegible]		G.B.Ferguson
76474	208	R	James Ives *Blake 87-87* W3mSr	American	48[illegible]	117·5	35·0	11·[illegible]	[illegible] Haven[illegible]	1883	[illegible]		H.Sutton
54152	209		James Kitchin (STRANDED) *Mosby -88* W Bk	British	500	112·4	31·3	16·9	RiverJohn,NS	1867	Kitchin	[illegible]	W.C.Job
75885	210	RV	James L. Harway *Grant -88* W Bk	British	953	180·3	36·7	20·8	RiverJohn,NS	1880	J.Kitchin	[illegible]	[illegible]
13398	211		James L. Maloy *Gray* WSr	American	166	102·0	28·5	7·9	Milford,Del.	1864			S.L.Keen
87793	212	✻	James Letten *Richards* WSk	British	75	75·4	20·3	10·3	Rye	1881	[illegible]		G.Richards
20763	213	✻	James Livesey *Hughes 79-84* I Bk	British	1072	188·6	34·0	22·1	Liverpool	1857	T.Vernon&Son		"James Livesey" Ship Co.,Ltd.(A.Fyfe)
13066	214		James M. Bayles *Snow* WSr	American	133	90·8	27·1	8·3	Br'kh'ven,NY	1851			G.B.Reynolds
13404	215	✻	James M. Flanagan *Lynch* WSr	American	259	111·6	29·5	9·2	WilmingtnDel	1864			W.F.Greene & Son
75799	216		James McLaren Barge W	British	169	110·0	22·0	7·0	Quebec	1879	A.Mailer		D.Murphy
83190	217	DV	James Martin *Fleming 75-* W Bk	British	1367	206·4	39·3	23·9	RiverJohn,NS	1883	A.McKenzie	[illegible]	D.Sutherland
58111	218	✻	JamesMason *Priestley -87* W Bg	British	276	109·0	26·5	16·4	Sunderland	1868	[illegible]	[illegible]	J.Tully
75963	219	V	James Nesmith *Jackson -87* W S	American	1736	228·4	40·0	26·3	Bath,Me.	1877	A.Hathorn	[illegible]	P.M.Whitmore
12572	220		James O'Donohue *Newton* WSr	American	151	97·0	26·6	9·0	Bangor,Me.	1860			A.P.Veazie
13454	221		James Parker, Sr. *Benjamin* WSr	American	101	94·0	24·0	7·[illegible]	Perth Amboy, N.J.	1856			W.M.Hines
46435	222		JamesPaterson (exSteamer) I Sr	British	579	186·3	23·2	11·6	Glasgow	1862	Smith&Rodger		Australasian United S.N.Co.,Ld.
13733	223	R	James Ponder *Lynch -87* W3mSr	American	258	117·0	30·0	19·0	Milton,Del.	18[illegible]	[illegible].W.Atkins		J.Ponder
83970	224	✻	James Postlethwaite *Williams 82-87* WSr	British	121	90·7	23·0	10·1	Barrow	1881	W.Ashburner		T.Ashburner & Co.
85413	225		James R. Benson WSr	British	325	139·0	26·1	11·4	St.Catharine's, Ont.	1873	L.Shickluna		S.Neelon
75632	226	R	James R. Talbot *Crocker* W3mSr	American	3[illegible]6	117·0	27·0	13·6	[illegible],Me.	1874			Jas.R.Talbot
76514	227	R	James Rothwell *Lowell* W3mSr	American	472	151·1	31·0	14·3	Yarmouth,Me.	1884	[illegible] Smith		James E.[illegible]
76472	228	R	James S. Lowell *Reed -83* W3mSr	American	69[illegible]	168·1	35·8	13·0	Bath,Me.	1882	[illegible].W.Johnson		G.W.Johnson
13855	229	V	James S. Stone *[illegible]* W Bk	American	672	148·0	31·5	19·4	E.Boston	1868	Curtis,Smith&Co.	[illegible]	E.M.Brewer
13401	230	✻R	James S. Watson *Marshall* WSr	American	135	101·0	22·4	8·0	Chester,Pa.	1863			R.F.Hickling
72907	231	✻	James Shearer *Stewart 83-8[illegible]* WSr	British	90	82·0	20·6	10·0	Ayr	1875	Ayr		J.Stewart
26246	232		James Shepherd W Bk	British	272	96·0	24·0	17·[illegible]					Lim Hoean Chee
12533	233		James Simpson *Dunn 83-83* WSr	British	17[illegible]	91·5	25·2	13·[illegible]	[illegible]	1857	W.[illegible]		J.A.Beeching
75627	234	R	James Slater *Conkling* W3mSr	American	295	118·1	30·5	10·2	PJeffersonNY	1874			J.L.Tooker
93992	235	✻	James Smith *Smith -87* WSk	British	87	77·5	21·5	10·8	BurtonStather	1886	[illegible]		S.G.Smith
91521	236	✻	James Spurgeon *[illegible] 86* WSk	British	82	76·3	20·5	10·5	Rye	1885	[illegible]&T.Smith		W.Empson
77955	237	V	James Stafford *Reynolds* W Bk	British	1117	187·2	38·7	22·7	[illegible]	1875	[illegible]	[illegible]	S.J.Bogart
27763	238		James Stonard *[illegible]* W Bk	British	149	83·6	25·0	11·0	Ulverston	1858	[illegible]		J.F.Waldie
13889	239		James Veldren *McKean* WSr	American	194	111·0	29·7	8·[illegible]	[illegible],N.J.	1866			A.R.Pharo
75384	240		James W.Drury *Baker* W3mSr	American	42[illegible]	141·0	32·8	11·[illegible]	[illegible]	1872	Campbell & Brooks		C.Lovell & Son
70470	241	✻	James W. Fisher *Carrey 85-87* W Bkn	British	173	[illegible]	[illegible]	12·[illegible]	[illegible]	1874	[illegible]	[illegible]	J.Fisher & Sons
76379	242	R	Jas.Waples Ponder *Dodge -87* W3mSr	American	[illegible]	121·0	32·4	8·[illegible]	[illegible],Del.	1882	C.C.Davidson		Jas.Ponder
13924	243		JamesWarren *Falkingham* WSr	American	111	88·3	25·8	7·7	[illegible],Me.	1867			W.H.Falkingham
85351	244		James Williamson W Barge	British	155	111·0	22·2	7·[illegible]	[illegible]	1882	J.R.Booth		D.Murphy
84951	245	✻	James Williamson *Close 85-8[illegible]* W3mSr	British	[illegible]	91·0	22·7	10·2	GlassonDock	1882	[illegible] & Marsh		J.Helme
51084	246	✻	James Wishart *[illegible]* I Bk	British	778	[illegible]	[illegible]	[illegible]	[illegible]	1864	[illegible]		J.L.Jacobs & Co.
13605	247	R	James Young *Lankin -82* WSr	American	213	111·6	29·5	11·0	Thomast'n,Me	1866			Dunn & Elliot
51073	248		James & Agnes *Williams 85-8[illegible]* WSr	British	116	92·5	21·9	11·8	Barrow	1864	Ashburner		T.Ashburner & Co.
21466	249		James & Eleanor *Brown* W Sw	British	13[illegible]	[illegible]	24·9	14·[illegible]	[illegible]	1858			J.T.Murray
56700	250		James&Mary *Deakin 85-87* WSr	British	13[illegible]	[illegible]	[illegible]·5	11·[illegible]	GlassonDock	1867	Simpson		E.Porter

✻=Lloyd's Register; D=Nederlandsche Vereeniging; G=Germanischer Lloyd; I=Registro Italiano; R=[illegible]
A=Veritas Austro-Ungarico; H=Veritas Ellenico; N=Norske Veritas; V=[illegible]

In the 4th Column, C=Composite (iron frame, wood planked); I=Iron; Stl=Steel; W=Wood.

The Dates of Metalling (Col. 11) are based on the records of the Books in which Vessels are respectively classed, or on the Reports of the Surveyors.

Further particulars of Vessels classed by Lloyd's Register will be found in Lloyd's Register Book and Lloyd's Yacht Register.

No.	Registre	Nom. Materiaux et Gréement. Capitaine et dates d'entrée au service des Armateurs actuels et à bord du navire. Ancien Nom.	Pavillon.	Tonnage net.	Longueur	Largeur	Creux	Construit Où.	Quand	Par qui.	Date du doublage	Armateurs ou Directeurs Gérants.	Port d'Armement.
251		Jan Frederik (ABANDONED) W.S.	Dutch	183	[illegible]	24.1	11.7	Nieuwendam	1858	W.&A.H.Mouravy		C.de Groof & Co.	Flushing
252	N	Jan Mayn Bødtker W.Bg	Norweg'n	261	107.9	28.0	11.0	Tønsberg	1853			H.A.Bödtker	Svelvig
253	DRV	Jan Melchers W.Bk	Dutch	1417	189.9	38.3	25.5	Belfast Me.	1869	Carter & Co.		S.J.Melchers	Schiedam
254		Jan Pietersnoon Koen W.Bk	Dutch	673	[illegible]	[illegible]	21.5	Alblasserdam	1861	F.Kloos		C.de Groof & Co.	Flushing
255		Jan Smit 83-85 W.Bn	Dutch	153	88.3	23.1	10.9	Hoogezand	1863	J.A.Hooites		J.A.Hooites	Hoogezand
256		Jan & Anton Wekenberg W.Bk	Dutch	194	96.6	23.9	12.1	Harlingen	1861	D.&L.Alta		J.H.Nijhuis	Sappemeer
257		Jan & Carel Steffens W.B.	Dutch	735	153.2	31.8	21.1	Rotterdam	1865	Pot Bros.		Johs.Meijjes,Jer'zn	Amsterdam
258	DGV	Janbaas 84-84 W.Bk	German	[illegible]	177.8	35.4	21.3	Elsfleth	1884	J.Ahlers		Joh.Diedr.Ahlers	Elsfleth
259		Jane Saglund W.Bk	British	612	150.5	32.7	19.3	Sherbrooke, N.S.	1871	McDonald		J.Sutherland	Liverpool
260	*	Jane Edward W.Sw	British	282	109.7	26.4	16.3	Sunderland	1869	J.Lister		W.Morris & Co.	Aberystwith
261	*	Jane (RILEY SUN) 80-81 W.Bkn	British	256	125.2	[illegible]	12.8	Sunderland	1878	W.Pickersgill		J.&J.Denholm	Greenock
262		Jane W. Lighter	British	216	105.9	25.5	10.7	Bathurst,N.B.	1839			Henry Moss	Melbourne
263		Jane Smith W.Bkn	British	213	112.0	25.6	14.7	Littlehampton	1858			M.Storm & Co.	Whitby
264		Jane W.Bn	British	141	88.0	24.2	10.1	Colchester,N.S.	1861			G.B.McLellan	Halifax,N.S.
265		Jane Gillen W.Bn	British	129	94.0	23.0	11.1	Bathurst,N.B.	1857			A.Maclean	Inverness
266		Jane W.Sr	British	108	71.3	20.2	10.8	P.E.Island	1847			M.Sullivan	Belfast
267	*	Jane Jones 79-85 W.Sr	British	94	78.2	23.3	10.6	Pwllheli	1827			R.Jones	Carnarvon
268	A	Jane Viggo W.Bg	Austrian	318	112.2	25.7	16.5	Fiume	1861	Schiavoni		A.Luppis fu M.	Fiume
269	R	Jane A. Falkenburg Lutzen W.Bkn	American	295	131.0	29.6	11.6	N.Bedford Mas.	1854			W.C.Lynde	San Francisco
270		Jane Adeline Oates 80-86 W.Bk	American	375	119.8	28.5	15.1	Westbrook Cn.	1868			J.S.Emery & Co.	Boston,Mass.
271		Jane Ainslie Park 81-82 W.Sr	British	108	82.4	21.9	9.5	P.E.Island	1869	J.Whitman		John Pearce	St.John's,N.F.
272	*	Jane Brown Pedersen W.Sw	Norweg'n	298	107.8	25.7	16.5	Sunderland	1864	J.&J.Gibbon		A.P.Pedersen	Arendal
273		Jane C. Woodruff W.Sr	British	228	139.0	25.5	10.7	St.Catharines	1866	L.Shickluna		J.B.Miller	S.Catharines,Ont.
274	*	Jane Cochrane Foster -78 W.Sr	British	133	91.8	23.9	10.5	Moncton,N.B.	1856			J.W.Hankin	Truro
275	*	Jane Davies Davies W.Sr	British	84	74.6	19.0	10.3	Bideford	1858			D.Davies	Aberystwith
276		Jane Duff Turner 69-69 W.Sr	British	107	84.5	17.0	10.2	Greenock	1847	J.Duncan		Starbuck & Rackstraw	London
277		Jane E. Foster W.Bn	British	174	109.5	23.4	11.7	Portsmouth	1861			J.Danby	Portsmouth
278		Jane E. Hala Hoots W.Bn	British	107	82.5	23.0	10.4	Jordan Riv.N.S.	1870			A.W.West	Halifax,N.S.
279	*	Jane Fairlie Craigen W.Bk	British	377	130.1	25.8	16.7	Scilly	1872	Edwards		A.Stevenson	Greenock
280		Jane Francis May W.Bn	British	174	107.1	24.2	12.3	Padstow	1871	Tredwen		T.May	Plymouth
281	*	Jane Gray Parry 85-85 W.Sr	British	124	85.0	21.5	11.5	Amlwch	1865	Paynter		T.Morgan & Co.	Beaumaris
282	R	Jane Gray Cleveland W.Sr	American	107	82.7	22.0	9.2	Bath,Me.	1887	Kelley & Spear		Wm.H.Kelley	San Francisco
283	*	Jane Harvey (WRECKED) Stevens 74-79 W.Bk	British	347	135.4	27.3	15.0	Littlehampton	1872	Harvey		A.Stevens	Falmouth
284	*	Jane Herbert W.Bn	British	135	86.4	22.0	12.0	Aberystwith	1860	Evans		D.Jones	Aberystwith
285	H	Jane Ingram Balmer W.Sr	British	155	95.1	27.1	9.2	St.Martin's,N.B.	1886	Wm.Ingram		Matthew Balmer	St.John,N.B.
286	*	Jane Kilgour Dewar W.Bk	British	251	126.8	24.1	12.8	Dundee	1874	Tay Ship Bdg.Co.		Robertson Bros.	Dundee
287		Jane L. Newton Stover W.Sr	American	264	107.8	29.0	9.5	Quincy,Mass.	1864			N.Stover	Boston,Mass.
288	V	Jane Lamb Willey 87-87 W.Bk	British	285	117.5	26.2	15.8	Sunderland	1866	G.Gardner	V 10.87	G.Brownell	Liverpool
289		Jane Law Scranton W.Bk	British	1273	196.0	39.3	23.9	Quebec	1869	Baldwin	V 10.87	J.P.Stevenson	Glasgow
290		Jane McLeod W.Sr	British	181	117.6	25.4	9.3	St.Catharines	1868	M.Simpson		T.N.Dancey	S.Catharines,Ont.
291	H	Jane Maria (ABANDONED) W.Bk	British	341	129.0	27.0	16.5	Gloucester	1865	Miller		W.H.Williams (Bevan, [illegible] & Co., Managers)	Ayr
292		Jane Maria (WRECKED) W.Sr	British	182	91.6	23.1	11.0	Southampton	1840			W.J.Sievey	London
293		Jane Martin W.Bk	Chilian	349	..	..	..					Cia Chileña de Vallen	Valparaiso
294		Jane Owens (STRANDED) W.Sr	British	97	78.6	20.2	10.5	Aberdovey	1858			J.Williams	Aberdovey
295	*	Jane Porter Jack 83-88 I.Bk	British	953	200.6	34.8	21.1	Belfast	1860	Harland & Wolff		J.P.Corry & Co.	Belfast
296	*	Jane Rennie Coppack W.Bk	British	198	103.7	21.1	12.7	Arbroath	1876	Roney		S.Coppack	Chester
297	*	Jane Richardson W.Bk	British	257	118.8	25.0	14.2	Bridport	1870	Cox	V 12.87	J.W.Pring	Brixham
298		Jane Slade Slade W.Sr	British	149	97.7	22.9	12.3	Fowey	1870	Slade		W.Geake & Co.	Fowey
299	*	Jane Sprott Ayre 83-84 I.Bk	British	670	176.1	30.0	19.1	Harrington	1868	R.Williamson & S.		G.M.Bushby	Liverpool
300		Jane Williamson W.Bg	British	183	105.0	24.2	13.0	Whitehaven	1870	H.&J.Williamson		R.Kearon	Belfast

* = Lloyd's Register; D = Nederlandsche Vereeniging; G = Germanischer Lloyd; I = Registro Italiano; R = American Record; A = Veritas Austro-Ungarico; H = Veritas Ellenico; N = Norske Veritas; V = Bureau Veritas.

Dans la 4me Colonne, C = Composite (membrure en fer, bordé en bois); I = Fer; Stl = Acier; W = Bois.

Les Dates du doublage en métal (Col. 11) ressortent des livres dans lesquels les navires sont respectivement classés ou des rapports des Inspecteurs du Lloyd's Register.

Pour plus amples renseignements sur les navires classés par le Lloyd's Register, voir le Lloyd's Register Book et le Lloyd's Yacht Register.

Official No.	No. in Book	Class	Ship's Name. Material and Rig. Master, and Dates of Appointment. Ship's Late Name.		Flag	Net Tonnage	Length	Breadth	Depth	Where Built	When	By whom	Metalled	Owners or Managers
[illegible]	301		Jane Woodburn (HULK) Bradborough	W Bg	British	290	103·1	26·8	16·6	Sunderland	1864	Blumer		W.L.R.Ogg
[illegible]	302	✳	Jane & Alice Morris (71-82)	W Sr	British	131	82·0	23·1	11·3	Portmadoc	1868	Williams		H. Morris
[illegible]	303		Jane & Ann Welch 67-	W Bn	British	169	93·9	22·0	11·8	Newburgh	1864	Scringeour		N. Hocken
[illegible]	304		Janes	W Barge	British	158	118·6	23·0	8·0	Quebec	1869			Montreal Transport[illegible]
4144	305		Janet	W Bk	British	475	118·0	28·5	20·0	Quebec	1830			R. Holden
[illegible]	306		Janet	W Bg	British	216	110·0	23·3	13·9	Aberdeen	1867	Hood	v 85	A. Scroggie
73302	307		Janet (WRECKED) Miles -87	W Sr	British	211	120·0	23·9	10·8	Fremantle W.A.	1878			D.J. Avery
43565	308	✳	Janet Jones 85-85	W Sr	British	120	87·0	20·0	11·0	Pwllheli	1861	Jones		J. Jones & Co.
89021	309	✳	Janet	W Sr	British	76	79·8	20·7	9·0	Stromness	1883	G.& P. Copland		J.A. Shearer
60373	310	✳ D	Janet Court Hewer 67-87	I S	British	996	215·0	34·3	20·3	Glasgow	1868	C. Connell & Co.		T. Clavering, Son & [illegible]
55433	311	✳	Janet Cowan Livingstone [87-88]	I S	British	1278	217·4	35·1	22·8	Greenock	1866	R. Steele		[illegible] "Janet Cowan" [illegible] Ld. (Simpson Bro.)
48922	312	✳	Janet Ferguson McGibbon 85-86	C Bk	British	581	158·3	28·6	17·7	Glasgow	1864	A. Stephen & Sons	v 85	W. Ross & Co.
82345	313	✳	Janet M'Niel Jones 84-86	I Bk	British	899	200·6	32·8	19·6	Port Glasgow	1880	Russell & Co.		Browne & Watson
68628	314		Janet Worthington Millmore 8[illegible]-85	W Sr	British	100	87·1	21·4	10·6	Ardrossan	1873	Barclay		T. Rayner
67940	315	✳	Janets & Margaret Moor	W Sr	British	110	92·0	21·5	10·6	Rothesay	1874	Fife	v 86	F.A. Smith
59430	316		Janette T. Hibbard Lundele	W Sr	British	146	95·3	28·3	9·0	Parrsboro' N.S.	1873			John Landels
74422	317	✳ D	Janie Good 84-84	W Sr	British	121	92·9	22·7	11·6	Malpas	1876	Scoble		J.W. Haukin
77400	318	✳	Janie Kernick	W Sr	British	100	86·2	22·9	10·8	Padstow	1878	Cowl		Mrs. C. Jenkin
81951	319	✳	Janie Gough [illegible]	W Bkn	British	212	114·0	24·6	12·5	Carrickfergus	1879	Rodgers	v 83	R. Gray
21762	320	✳	Jannett & Jane Evans	W Sr	British	81	66·0	20·1	10·6	Portmadoc	1858			M. Evans
	321		Janota	W Sr	Brazilian	106	..	..	..					F.J. da Silva
	322	✳	Jantje Lever -71	W Sr	Argentine	150	90·2	22·9	11·5	Hoogezand	1871	J.A. Hoottes	v 85	Juan Senoraus
	323	G D	Jantje (ex Olga) Bloompot [illegible]	W Bk	Dutch	492	133·5	27·5	15·7	Geestemünde	1861	R.C. Rickmers	12.84	M.E. Kuipers
63902	324	✳	Japan George	W Bk	British	394	129·0	27·0	16·6	Dysart	1872	Watt	v 85	G. Sharp (Haines, Batchelor & Co., Mang.)
430	325	G V	Japan Goerke 81-87	W Bk	Russian	669	155·5	33·2	18·8	Gefle	1869	O.A. Breslin	11.87	J. Grell
	326		Jaragua	W Sr	Brazilian	120	..	..	..					S. Leao & Filho
	327	N	Jarl Kjelsen -87	W Sr	Norweg'n	155	95·8	23·7	10·8	Djursvik	1857			Ths. A. Herg
	328	N	Jarlen (ex A.F. Stoneman) Beck	W Bk	Norweg'n	1024	181·3	36·0	22·2	Green Cove, N.S.	1874	E. Raymond	v 85	Joh. Thorne
	329	N	Jarlen (BROKEN UP) Nielsen	W Bg	Norweg'n	237	95·7	25·7	14·9	Sunderland	1855	J. Davison & Co.	85	P. Nielsen
	330	D N V	Jarlsberg (ex Algonquin) Bjønness	W S	Norweg'n	1265	190·0	36·0	23·7	Maitland, N.S.	1870	Ch. Cox	12.86	W. Wilhelmsen
63441	331	✳	Jason Henderson 85-85	I S	British	1512	253·0	38·6	23·2	Glasgow	1870	Barclay, Curle & Co.		A.& J.H. Carmichael
	332		Jason Schumacher -83 (ABANDONED)	W Bk	German	362	113·7	27·1	11·2	Elsfleth	186[illegible]	G. Wempe	10.85	C. Paulsen
	333	N	Jason (ex C.R. Jason ex J.E. Kahn ex Parrsboro) Olsen 72-78	W Bk	Norweg'n	720	154·6	36·7	18·6	Portland, Ct.	1859		v 85	Chr. Ihlen
	334	D N	Jason Olsen 88-88	W Bk	Norweg'n	497	141·3	30·5	17·3	Grimstad	1881	Johan Ribe	10.88	A. Tellefsen
	335	N	Jason Eevnsen -81	W Bk	Norweg'n	374	115·4	30·1	16·3	Kragerö	1851			I. Stub
58338	336		Jasper Lovett	I Sr	British	252	129·4	24·5	13·8	Greenock	1867	Robertson & Co.		C.W. Turner
85825	337	✳	Jasper Hocking	W 3m Sr	British	129	96·0	23·0	11·3	Falmouth	1884	C. Burt		A. Jenkyns
	338	N	Jaspis Einertsen -88	W Bk	Norweg'n	216	103·2	25·8	12·2	Kragerö	1880	T.G. Lüangen	12.85	J.A. Larsen & Co.
	339		Jassimula	W Bk	Greek	307	..	..	..	Galaxidi	1880	G. Canata	11.86	E. Kazulli
88371	340		Java (ex Noach II.) Beynon 87-87 (MISSING)	W Bk	British	948	170·4	38·1	22·0	Ridderkerk	1867	J. Smit	v 85	G. Milne & Co.
84344	341	✳	Java Roff 86-88	I Bk	British	890	202·6	32·7	19·4	Greenock	1881	Russell & Co.		P. Denniston & Co.
48037	342		Java Murray	W Bn	British	135	96·8	23·4	11·4	Granville, N.S.	1864			J. Murray
	343	V	Java Wulff 70-81	W Bk	Swedish	574	151·2	29·6	18·1	Gothenburg	1863	Hinke	v 86	J.A. Kjellberg & S[illegible]
129	344	✳ A	Javor (ex Stanley) Camenarovich	W Bk	Austrian	384	127·0	27·0	17·2	Sunderland	1863	G.& J. Mills		A. Camenarovich [illegible]
	345		Jaylan-i-Bahri	W Bk	Turkish	500	..	..	..	Greece	1874			Ali Riza Bey & Co.
	346		Jaylan-i-Bahri	W Bg	Turkish	290	..	..	..	Oonia	1883			Nafiz Hajee Sheik Effendi
	347		Jaylan-i-Bahri	W Bn	Turkish	220	..	..	..	Vassilikos	1873			Kerestab [illegible]
	348		Jaylan-i-Bahri	W Bn	Turkish	207	..	..	..	Oonia	1873			Sari-Hajee-Zadeh [illegible]
	349		Jaylan-i-Bahri	W Bg	Turkish	200	..	..	..	Oonia	1884			[illegible]
	350		Jaylan-i-Bahri	W Bn	Turkish	168	..	..	..	Heraclea	1882			[illegible]

✳=Lloyd's Register; D=Nederlandsche Vereeniging; G=Germanischer Lloyd; I=Registro Italiano; V=[illegible]
A=Veritas Austro-Ungarico; H=Veritas Ellenico; N=Norske Veritas;

In the 4th Column, C=Composite (iron frame, wood planked); I=Iron; Stl=Steel; W=Wood.

The Dates of Metalling (Col. 11) are based on the records of the Books in which Vessels are respectively classed, or on the Reports of the Surveyors [illegible]

Further particulars of Vessels classed by Lloyd's Register will be found in Lloyd's Register Book and Lloyd's Yacht Register.

Lloyd's	No.	Cote	Nom. (Ancien Nom.)	Capitaine et dates d'entrée au service des Armateurs actuels et à bord du navire	Matériaux et Gréement	Pavillon	Tonnage net	Longueur	Largeur	Creux	Construit: Où	Quand	Par qui	Date du doublage en métal	Armateurs ou Directeurs-Gérants	Port d'Armement
	351		Jaylan i Bahri		W Sr	Turkish	150	..	..	..	Hartensen-zelli	1874			Tchukuji-oglou-Teodo- [illegible]	[illegible]
	352		Jaylan-i Bahri		W Sr	Turkish	145	..	..	..	Greece	18[illegible]			Tujar Imam-Zadeh [illegible]	Constantinople
	353		Jaylan-i Bahri		W Bg	Turkish	137	..	..	..	Barten	1875			[illegible]	Sunnus
	354		Jaylan-i-Bahri		W Bmb	Turkish	125	..	..	..	Kara Simnu	1871			[illegible] & Co	Heraclea
	355		Jaylan i Bahri		W Bn	Turkish	125	..	..	..	Guetch Gihi	1876			[illegible]	Ahmaserah
	356		Jaylan-i Bahri		W Bn	Turkish	125	..	..	..	Barten	1884			[illegible] Tijer Bostanji [illegible] Osman & Bros. & Co.	Barten
	357		Jean		W Sr	Russian	235	..	..	..		1872			I. Judins Arfv.	Impilaks
	358	V	Jean Agathe	Gislon	W Bk	French	206	101·7	26·3	13·9	St. Valery en Caux	1868	A. Coquais		A. Olivier	S. Valery en Caux
[illegible]	359		Jean Anderson	Stewell	W Bn	British	185	102·0	24·5	12·9	P.E. Island	1877	Duncan		T. & J. Connick & Co.	Dundalk
	360		Jean André	Ficq	W Bk	French	357	115·8	26·2	15·5	Martigues	1866	Fiastre, Mauras		Rieoux	Marseilles
[illegible]	361		Jean Baptiste		W Sr	British	114	101·0	22·5	7·7	Batiscan [Quebec	1864			J. Piché	Quebec
	362	DV	Jean Baptiste (ex L. B. [illegible] Bosjaou)	Bruijn -87	W Bk	Dutch	796	166·3	34·1	21·6	Maassluis	1866	Mit & Bogaard	V 9.85	Ch. de Groof & Co.	Flushing
	363		Jean Baptiste ([illegible])	Sparito -87	W Bg	French	232	98·4	25·6	14·4	Agde	1869	Contrasty	Z 5.81	Mme. Mages	Agde
	364		Jean Baptiste	Fafin -85	W Bg	French	132	77·8	20·7	11·8	Cherbourg	1856			J.B. Laronche	Cherbourg
	365	DN	Jean Baptiste	Marcussen	W 3m Sr	Norweg'n	174	109·7	22·9	10·2	Bordeaux	1862	Moulinie frères		C. Marcussen	Horten
	366		Jean Bart		W Bk	French	181	93·2	24·7	13·4	La Reuille	1853		Z 11.81	F.D. Robial.	St. Brieuc
	367	N	Jean Bart	Jespersen -88	W Bk	Norweg'n	309	112·7	27·6	14·9	Nantes	1861	Postec	V 12.84	H.O. Björndahl	Tönsberg
	368	N	Jean Maurice	Toraldsen	W Bk	Norweg'n	230	110·5	27·7	11·6	Bordeaux	1857			H.C. Andersen	Christianssand
[illegible]	369		Jean Pierre	Smith 75-84	W Bk	British	614	161·9	33·1	17·1	Quebec	1865			John Bickers	Adelaide
	370	✻V	Jean Pierre	Serial	W Bk	French	455	132·8	30·0	17·8	Bordeaux	1865	Raymond frères	V [illegible]	Jean Jouin	Bordeaux
	371	D	Jeane (ex Valiny)	Seth Salim	W Sr	Dutch	191	104·3	22·0	..	Wildervank	1874	J.M. Meyhuizen [& Zoon		Seeh Said bin Salim Bahasoan	Grissee
	372	N	Jeanette	Falck -82	W Bg	Norweg'n	211	100·8	26·3	13·1	Drontheim	1855	Throndhjems [Skibswarft		P.V. Pay	Drammen
	373		Jeanette	Johnsen 85-82	W Sr	Norweg'n	167	92·6	25·5	10·5	Baltimore	1856			C. Apenes & Sön	Fredriksstad
	374		Jeanette	Johansson -80	W Sr	Swedish	102	71·8	21·7	9·4	Wenersborg	1855			C.J. Andersson	Wenersborg
[illegible]	375	✻	Jeanie	Fervie	W Sk	British	30	51·3	13·0	9·0	Fairlie	1876	Fife		Nobel's Explosives Co.	Glasgow
[illegible]	376	R	Jeanie-Lippitt	Vinter 84-86	W 3m Sr	American	706	163·5	35·5	17·3	Boston, Mass.	1880	J.M. Brooks		P.H. Crowell	Boston, Mass.
	377	N	Jeanie Louttit	Knudsen -87	W Bk	Norweg'n	466	148·8	27·6	17·1	Kingston	1869	Kinloch	V 12.83	Actieselskabet "Achilles" (Jens Olsen)	Kragerö
[illegible]	378		Jeanie Oswald (ex Hance)		W Bk	British	194	104·6	24·0	14·2	Cherbourg	1853			Henry Moss	Melbourne
	379		Jeanne	Loren -87	W Bk	French	376	121·3	28·7	11·2 / 17·0	St. Malo	1869	L. Mallard		P. Layet	Marseilles
	380		Jeanne (ex Rosalie)	Savalle 86-	W Bkn	French	224	116·0	26·1	12·9	Padstow	1879	C. Rawle	V 9.81	Hermel & Buisson	Fécamp
	381		Jeanne	Bernard	W Sr	French	147	80·4	20·7	11·9	Boulogne	1865	Lecerf		N. Van Cauwenberghe	Dunkirk
	382	V	Jeanne	Morvan	W Sr	French	120	91·2	19·3	11·2	St. Malo	1878	F. Gautier		H. Pasquiou	Paimpol
	383	V	Jeanne	Lemeur	W Sr	French	118	75·1	19·3	11·0	LeLegue	1869	Bouillet		Lemeur	Paimpol
	384		Jeanne (WRECKED)	Beaunnick 87-87	W Sr	French	115	88·6	20·5	11·1	Dunkirk	1870	G. Beck		G. Beck	Dunkirk
	385		Jeanne (ex Kirkwood)	Boimare	W Bk	Haytian	325	114·0	26·6	17·0	Sunderland	1861	Haswell		H. Etienne	C. Haitien, Hayti
	386		Jeanne d'Arc		W 3m Sr	French	186	83·0	21·6	12·3	La Nouvelle	1854			A. Sarvine & fis	Narbonne
	387		Jeanne d'Arc	Cagniard -81	W Bg	French	181	90·9	23·0	12·5	Sunderland	1864			T.A. Cagniard	Coursellcs
	388		Jeanne d'Arc	Hermenk -87	W Sr	French	137	84·6	23·6	11·0	Fecamp	1879	Capon		Hermel Buisson	Fécamp
	389	V	Jeanne d'Arc	Hars -87	W Sr	French	122	87·6	20·3	12·8	Dunkirk	1872	E. Sagary		Woutermartaens	Dunkirk
	390	V	Jeanne d'Arc	Lequiniercr 87	W Bn	French	102	82·3	21·0	10·0	Cherbourg	1872	E. Lecerf		A. Postel & fils	Cherbourg
	391		Jeanne d'Arc	Sideropulo	W Bk	Greek	256	..	..	..	Nantes	1864			G. Marlas	Marsiglia
	392	V	Jeanne & Eugénie (ex Brüs-Johanna)	Lequellec	W Sr	French	100	73·9	20·0	10·9	Wyk	1867	D.L. Freundt		Lequellec	Brest
574	393	R	Jeannette	McGough	W Sr	British	198	100·0	26·5	11·9	Petite Riviere	1883	T. Bouchard			
	394	V	Jeannette	Aden 77-82	W Sr	German	169	95·2	23·8	11·5	Stralsund	1874	C.W. Mohr		Duis J. Aden	Boekzeteler Fehn
	395	D	Jeannette Agnes ([illegible])	Franssen -88	W Sr	Dutch	254	97·1	26·8	14·6	Galaxidi	1871		Z 5.81	J.G. Van Gelder	Amsterdam
[illegible]	396		Jed F. Duren	Noble	W Sr	American	107	87·0	26·4	7·9	Calais, Me.	1874			J.H. Nickerson	Calais, Me.
[illegible]	397		Jed Frye	Langley	W Sr	American	140	100·0	26·2	8·0	Perry, Me.	1867	A. Frost		J.P. Langley	Ellsworth
	398		Jeddah		W Bg	Turkish	287	..	..	..	Jeddah	1869			Zviholee Strati	Zvibo
	399		Jeddah		W Bmb	Turkish	136	..	..	..	Jeddah	1868			Harien Tijari Haice Ahmed Effendi	Jeddah
	400															

[illegible]=Lloyd's Register; D=Nederlandsche Vereeniging; G=Germanischer Lloyd; I=Registro Italiano; R=American Record;
[illegible]=Veritas Austro-Ungarico; H=Veritas Ellenico; N=Norske Veritas; V=Bureau Veritas.

Dans la 4me Colonne, C=Composite (membrure en fer, bordé en bois); I=Fer; Stl=Acier; W=Bois.

Les Dates du doublage en métal (**Col. 11**) ressortent des livres dans lesquels les navires sont respectivement classés ou des rapports des Inspecteurs du Lloyd's Register.

Pour plus amples renseignements sur les navires classés par le Lloyd's Register, voir le Lloyd's Register Book et le Lloyd's Yacht Register.

Official No.	No. in Book	Registry	Ship's Name. Material and Rig. Master, and Dates of Appointment to present Owner's Service and to Ship. Ship's Late Name.	Flag.	Net Tonnage	Dimensions. Length	Breadth	Depth	Built Where.	When	By whom.	Metalling	Owners or Managers
	401		Jeddah WBr	Turkish	112	..	..	..	Jeddah	1872			Hajee Mohemed Sugar & Co.
[illegible]	402		Jeddo *Kogh* WSr	British	103	82·5	25·5	8·0	St. Martins N B	1872			Thomas F. Dixon
[illegible]	403		Jefferson *Griffin* W3mSr	American	309	124·6	30·0	11·5	Brooklyn	1874	C.&R.Poillon	[illegible]	Tupper & Beattie
	404		Jefim WSr	Russian	158	..	..	..		1860			I.Judins arfv.
[illegible]	405		Jehu *Willocks* WBn	British	198	127·0	26·3	11·3	Newburyport, Mass.	1859			W.R.Beer
[illegible]	406		Jehu *Patterson* WSr	British	150	95·1	22·7	12·0	Bideford	1860	Johnson		T.W.Hutchings
	407		Jejuy ISr	Argentine	102	88·3	23·0	7·1	Paisley	1884	J.Fullerton & Co.		La Platense Flotilla Co.
	408		Jelena WSr	Russian	174	..	..	..		1866			I.Judins arfv.
	409		Jenkin WSr	Norweg'n	118	86·1	22·9	9·9	Stavanger	1878			O.H.Baadsvig
[illegible]	410		Jennie *J. Murray* WSr	British	438	168·0	26·6	11·8	Garden Is. Ont.	1871	H.Roney		Montreal Trans. Co.
75251	411	R	Jennie A. Cheney *Arey* WSr	American	270	110·0	27·9	9·3	Thomaston Me	1870	S.Caru		Talbot & Co.
75905	412	R	Jennie A. Stubbs W3mSr *Farnham -88*	American	199	111·8	27·6	10·7	Lamoine, Me.	1877	D.Hodgkins	Y 11.86	F D Hodgkins & Sons
60114	413	V	Jennie B. *Champion -88* WBkn	British	474	135·0	29·8	17·8	Halifax, N.S.	1874	T.J.Bentley	[illegible]	G.J.Waterhouse
75849	414		Jennie Beasley *Dow* WSr	American	204	106·3	28·5	9·2	Bucksport, Me.	1876	W.Beasley	[illegible]	W.Beasley
13690	415	R	Jennie Cushman WBk *Titcomb -88*	American	274	119·2	29·0	13·0	New Bedford, Mass.	1865	J.Howland	[illegible]	F.C.Butman
76483	416		Jennie E. Marshall WSr *Marshall -84*	American	188	103·5	23·8	9·0	Bethel, Del.	1884	M.Griffiths		R.F.Hickling
76707	417		Jennie E. Righter W3mSr *Crowley -87*	American	615	117·0	35·4	14·6	Madison, Ct.	1887	W.C.Miner		F.Benedict
75389	418	R	Jennie E. Simmons WSr *Williams -87*	American	271	114·5	29·6	9·1	Camden, N.J.	1871	Taylor & Mathis		F.Smith
75252	419		Jennie F. Willey W3mSr *Chadwick -83*	American	365	128·2	29·9	10·6	Thomaston Me	1870	Walker & Dunn		Dunn & Elliot
76365	420	R	Jennie G. Pillsbury WSr *Pillsbury*	American	147	93·9	25·8	7·2	Rockland, Me.	1882	C.Starret		White & Case
76454	421	R	Jennie Hall *Hall* W3mSr	American	392	121·8	32·1	10·5	Columbia Falls, Me.	1883	J.H.Crandon		John H.Crandon
76105	422	RV	Jennie Harkness WBk *Amesbury*	American	1373	206·5	38·3	23·2	Rockport, Me.	1879	J.Pascal	[illegible]	Carleton, Norwood & Co.
76147	423	R	Jennie Hulbert *Norton -87* WBn	American	419	143·7	32·6	12·2	Richmond, Me.	1880	Southard	[illegible]	T.J.Southard & Sons
76470	424	R	Jennie Lind *Leighton* WSr	American	117	90·8	25·9	7·5	Cherryfield Me	1883	J.Hinckley		J.Hinckley
76344	425	R	Jennie Lockwood W3mSr *Poland &-*	American	412	142·8	34·1	12·2	Thomaston Me	1882	S.S.Gerry & Co.	[illegible]	S.S.Gerry
75700	426		Jennie M. Carter W3mSr *Bates*	American	281	130·3	33·0	9·2	Newton, Del.	1874			N.G.Carpenter
75338	427		Jennie Middleton W3mSr *Parsell -87*	American	288	122·0	30·0	11·0	Camden, N.J.	1871			G.W.Shepherd
75194	428		Jennie N. Huddell W3mSr *Haines*	American	265	122·0	32·0	10·3	Tuckerton, N.J.	1870	Lane & Son		A.W.Crammer
79981	429	DRV	Jennie Parker *Barker -87* WBk	British	998	177·2	35·4	22·0	Black River N.B.	1878	J.&R.McLeod	[illegible]	J.W.Perry
92370	430	R	Jennie Parker *Corbett -87* W3mSr	British	211	113·0	29·2	10·2	Tynem'th, N.B.	1887	R.L.Parker		R.C.Elkin
75624	431	R	Jennie Phinney WBn *Morris -83*	American	416	129·8	30·0	16·8	Yarmouth, Me.	1874	Loring & Brewster	[illegible]	J.W.Musgrave
75982	432		Jennie R. Morse W3mSr	American	388	137·8	32·0	12·1	Bath, Me.	1878	B.W.& H.F.Morse		B.W. & H.F.Morse
75228	433	R	Jennie Rosaline *Saxton* W3mSr	American	331	125·6	31·0	13·8	P.Jefferson N.Y.	1870	J.M.Bayles & Son		W.W.Saxton
76940	434	R	Jennie S. *Sinclair* WSr *(ex Beta)*	American	176	96·0	26·0	10·3	S'mm'rs'de N.S.	1883	B.D.Bradshaw		C.Sinclair
[illegible]	435	R	Jennie S. Hall W3mSr *Hodgins -85*	American	428	143·0	33·2	12·0	Waldob'ro' Me.	1881	W.Fish	[illegible]	J.P.Ellicott
76118	436	R	Jennie Seaverns *Lightbody* WSr	American	107	87·6	24·0	8·0	Newburyport, Mass.	1880	Colby & Lunt		J.I.Tupper
[illegible]	437		Jennie Stella *Kreis* W3mSr	American	278	132·0	34·4	9·8	Coos Bay, Oreg.	1876			H.Winkelman
75810	438		Jennie Sweeney WBkn *Moss -86*	American	611	147·0	35·4	18·0	Mays L'dng N.J.	1875	Lane & Bro.	[illegible]	S.S.Hudson
[illegible]	439		Jennie Thelin *Forest* WSr	American	139	91·0	29·5	8·2	Davenport Landing, Cal.	1869			H.C.Groeper
[illegible]	440		Jennie Wand W3mSr	American	163	124·0	32·8	9·0	Marshfield, Or.	1883	Hans Reed		H.Winkelman
	441	GV	Jenny (MISSING) *Clampen -87* WBk	German	680	144·3	29·8	20·1	Bremerhaven	1864	R.C.Rickmers	[illegible]	H.R.Lassen
	442	V	Jenny [illegible] *Larsen* WBk	Norweg'n	569	148·7	32·0	17·0	New Glasgow	1874	Little & Meikle	[illegible]	J.M.Bryde
	443	N	Jenny *Kolderup* WBk	Norweg'n	492	135·4	32·4	17·3	Drammen	1865	J.Jörgensen	[illegible]	A.F.Kolderup
	444	N	Jenny *Jansen 79-79* WBg	Norweg'n	255	101·1	26·1	14·7	Gefle	1842		[illegible]	H.E.Jansen
	445	DN	Jenny *Christiansen* WBk	Norweg'n	253	107·9	27·0	13·6	Brahestad	1857			J.N.Jacobsen & Co.
	446	N	Jenny *Gundersen* WBg	Norweg'n	217	95·7	25·4	14·0	Bremen	1844			B. & M.Gundersen
	447	N	Jenny *Nielsen* WSr	Norweg'n	142	89·6	22·5	10·7	Helsingfors	18[illegible]			J.H.Nielsen
	448		Jenny *Janssen -85* WBk	Russian	352	..	..	..		1867			C.Engblom
	449		Jenny *Ivanoff* WGet	Russian	1[illegible]	..	..	..	Björneborg	1882			J.Päsö
	450		Jenny *Eriksson* WSr	Swedish	114	84·0	21·6	9·0	Gothenburg				J.E.Eriksson

✻ = Lloyd's Register; D = Nederlandsche Vereeniging; G = Germanischer Lloyd; I = Registro Italiano; R = [illegible]
A = Veritas Austro-Ungarico; H = Veritas Ellenico; N = Norske Veritas; V = [illegible]

In the 4th Column, C = Composite (iron frame, wood planked); I = Iron; Stl = Steel; W = Wood.

The Dates of Metalling (Col. 11) are based on the records of the Books in which Vessels are respectively classed, or on the Reports of the Surveyors [illegible]

Further particulars of Vessels classed by Lloyd's Register will be found in Lloyd's Register Book and Lloyd's Yacht Register.

No. d'ordre	Signe	Nom. Capitaine et dates d'entrée au service des Armateurs actuels et à bord du navire. Ancien Nom.	Matériaux et Gréement	Pavillon	Tonnage	Longueur	Largeur	Creux	Construit où	Quand	Par qui	Date du doublage en métal	Armateurs ou Directeurs-Gérants	Port d'Armement
451	✻	Jenny Lind [illegible]	W S	Bremen	137	82·0	19·1	13·2	Yarmouth	1860		[illegible]	P. Peterson	London
452		Jenny Lind	W Sr	British	125	93·4	23·0	8·9	[illegible]	1855			N. Hamelin	Quebec
453		Jenny Lind	W Sp	British	101	102·6	22·7	6·9	St. Thomas de [Pierreville	1885	F.X. Manseau		F.X. Mausrau	Montreal
454		Jencer	W Bg	Dutch	228	92·2	23·0	..	Grissee	1844		[illegible]	Said Kasim C. Abet C. [illegible]	Palembang
455	V	Jens Eriksen 87	W Sr	Danish	129	92·1	22·1	10·4	Thoro	1875	P. Trønnegaard		R.P. Nielsen	Svendborg
456	V	Jens Andersson 87-87 (ex [illegible])	W Bg	Swedish	400	111·2	28·7	17·2	Lussinpiccolo	18 56/77		10.86	C.J. Kronker	Gothenburg
457	N	Jens Nielsen Andersen 87	W Bk	Norweg'n	354	111·8	29·6	16·1	Grimstad	1866	Jens Tellefsen	[illegible]	Jacob Holst	Grimstad
458	DN	Jens Ruffen Pedersen	W Sr	Norweg'n	118	72·5	22·6	11·5	Grimstad	1828	Chr. Pharo		Saxlund & Röer	Christiania
459	V	Jerbuen Larsen	W 3m Sr	Norweg'n	272	116·6	28·0	12·9	Drammen	1874	R.O. Hæsem	[illegible]	C.H. Gabrielsen	Stavanger
460	R	Jeremiah Smith Tyler	W 3m Sr	American	389	143·7	35·0	10·3	New London, Ct.	1888	J. Davidson		J. Smith	New Haven, Ct.
461		Jeremiah Thompson Jerry	W S	American	1836	216·0	43·0	28·0	Williamsburg, [N.Y.	1854		[illegible]	C. Knudsen & Co.	San Francisco
462		Jerfalcon Murdock	I 3m Sr	British	287	128·5	25·8	12·8	Port Glasgow	1868	L. Hill & Co.		J.V. Bentley	Melbourne
463	N	Jernæs Jørgensen 85-85	W Bk	Norweg'n	289	110·6	26·9	15·0	Risör	1870	N. Olsen Gjernæs	[illegible]	L. Alsing	Kragerö
464		Jernavik Nilsson 57-81	W Sr	Swedish	158	98·1	22·1	10·7	Jernavik	1872	M. Jonnasson		M. Magni	Jernavik
465	N	Jernbyrd Pedersen 88	W Bk	Norweg'n	252	114·7	26·7	12·4	Grimstad	1870	Smith-Petersen	[illegible]	G.T. Jörgensen	Christiansand
466	V	Jeroianni Jeroianni	W Bg	Greek	273	98·7	25·5	17·9	Cassos	1878	E. Diangussi		Jeroianni	Cassos
467		Jerso Svedman 86	Bkn	Swedish	232	100·7	25·7	13·0	Jersö	1879			P. Persson	Ruddö, Rslgn.
468	A	Jerusalem Tratzaroni	W Bk	Greek	405	123·0	29·5	..	Syra	1877		[illegible]	P.G. Caluta	Syra
469		Jerusalem [illegible]	W Bg	Greek	319	103·3	23·8	16·9	Sunderland	1864	Lister		S. Mangonas	Zante
470	✻	Jerusalem Carlsen	W Bk	Norweg'n	901	196·5	33·7	20·7	Aberdeen	1867	Hood	[illegible]	C.A.E. Carlsen	Lillesand
471		Jesse Hart, 2d. Wall	W Sr	American	242	116·0	29·0	10·0	Waldob'ro' Me	1866			G.W. Rawley	St. George, Me.
472		Jesse L. Leach Pridmore	W Sr	American	236	110·0	30·0	9·7	Baltimore	1863			S. Inman	Tuckerton, N.J.
473		Jesse Murdock Fleming	W 3m Sr	American	342	125·0	30·4	13·0	Bath, Me.	1871	Goss & Sawyer		J.W. Gammons	N. Bedford, Mass.
474		Jesse W. Knight Edwards (S N K)	W Sr	American	181	115·2	26·2	8·3	Str'tford, Conn	1866			H. Fisher	Philadelphia
475	R	Jesse W. Starr Warren	W 3m Sr	American	292	128·0	32·5	10·4	Milford, Del.	1874			P. Fitzpatrick	Philadelphia
476		Jessie Ruxton	W Bkn	British	247	124·0	25·1	12·8	Sunderland	1879	W. Pickersgill	[illegible]	J. Ruxton	Auckland
477	N	Jessie Rasmussen (ex Emilie & Josefine, Marie Emilie)	W Bk	Norweg'n	431	121·5	28·5	19·0	New York	1840			E. Rasmussen	Porsgrund
478		Jessie A. Woodhouse	W Sr	British	158	93·0	24·1	8·0	E. Haddam U.S.	1859			C.F. Feez	Trinidad
479		Jessie Annandale	W Bk	British	123	88·7	21·3	11·1	Sunderland	1857	Brown & [Johnson		A. Harvey	Colchester
480		Jessie Anning Morton (CONDEMNED) [90-90]	W Bk	British	275	122·0	24·9	14·9	Bridport	1865	Cox	Y 10.85	R. Morton	St. Ives
481		Jessie Bennet Martin	W Bn	British	135	90·9	23·3	11·5	Perth	1868	Wallace		J. Samuel	Perth
482		Jessie Drummond	W Bk	British	292	134·5	23·0	11·9	St. Catharines, [Ont.	1864	M. Simpson		William Walsh	Montreal
483		Jessie H. Breck	W Sr	British	305	132·3	25·6	11·7	Pt. Dalhousie	1873	G.D. Andrews		E.A. Booth	Kingston, Ont.
484		Jessie Kelly	W Sr	British	145	103·0	22·2	11·0	Auckland, N.Z.	1866	H. Nicoll		B. Byrnes	Sydney, N.S.W.
485	R	Jessie Lena Hunter	W 3m Sr	American	330	136·5	32·5	11·0	Belfast, Me.	1883	J.Y. Cottrell	Y 12.88	B.O. Knowlton	Boston, Mass.
486	R	Jessie McGregor McFadden 85-89	W Bkn	American	579	151·1	33·5	12·4	Deering, Me.	1882	G. Russell	Y 11.88	J.S. Winslow & Co.	Portland, Me.
487		Jessie Minor Whitney [85-85]	W 3m Sr	American	248	129·0	32·5	9·2	Fairhaven, Cal	1883	Bendixen		W. Whitney	Eureka, Cal.
488		Jessie Morris Fairbairn [illegible]	W Bk	British	819	164·2	31·1	21·3	Walton, N.S.	1872	Vaughan	Y 1.89	J.R. Suiter	Maryport
489		Jessie Nickerson Hughes	W Sr	American	175	112·6	30·2	9·1	Pt. Ludlow, [Wash.	1874			J.J. McKinnon	San Francisco
490	✻ RV	Jessie Osborne Wilson (ex [illegible]) [79-84]	I S	British	1058	221·0	34·3	20·7	Dumbarton	1874	J. & R. Swan		J. Hossack, sen.	Liverpool
491	✻	Jessie Readman Gibson [79-79]	I S	British	962	204·0	32·6	20·8	Greenock	1869	Scott & Co.		Shaw, Savill & Albion Co., Ld.	Glasgow
492		Jessie Scarth	W Sr	British	306	135·0	25·9	12·7	Hamilton, Ont.	1871	A.M. Robertson		J. Conlon	Toronto, Ont.
493	✻	Jessie Stowe Hunter	I Bk	British	645	191·0	27·8	17·2	Sunderland	1863	Oswald & Co.		J.S. Stowe	Liverpool
494		Jessie T. Matheson Campbell	W Sr	American	131	93·0	24·6	9·1	Essex, Mass.	1881	J. James & Co.		W. Matheson	Provincet'n, Mass.
495	✻	Jessomene Hawkins 85-85	I S	British	1898	270·0	39·8	24·1	Liverpool	1881	W.H. Potter & Sn.		H. Fernie & Sons	Liverpool
496	✻	Jessonda Oltmanns 85 (ex [illegible])	W Bk	German	883	168·3	33·0	21·0	Bangkok	1858	Baxter	[illegible]	Gildemeister & Ries	Bremen
497		Jessy Morgan 86-86	W Bk	British	215	115·6	23·7	13·5	Bristol	1859	Symonds		P. Morgan	Drogheda
498		Jessy Nickelsen 89-89 (ex [illegible]) (WRECKED)	W Bk	Norweg'n	383	132·4	29·2	14·6	Tatamagush, [N.S.	1867	Campbell	[illegible]	N.M. Gram	Drammen
499		Jesuina 1. Ramos (ex [illegible])	W Bn	Portug'se	123	85·0	20·5	9·7	Fão	1873	M.D. dos Santos [Borda		Faro & Lima	Lisbon
500		Jeudokia	W	Russian	115	..	..	..		1872			I. Judins arfv.	Hapilaks

✻ = Lloyd's Register; D = Nederlandsche Vereeniging; G = Germanischer Lloyd; I = Registro Italiano; R = American Record;
A = Veritas Austro-Ungarico; H = Veritas Ellenico; N = Norske Veritas; V = Bureau Veritas.

Dans la 4me Colonne, C = Composite (membrure en fer, bordé en bois); I = Fer; Stl = Acier; W = Bois.

Dates du doublage en métal (Col. 11) ressortent des livres dans lesquels les navires sont respectivement classés ou des rapports des Inspecteurs du Lloyd's Register.

Pour plus amples renseignements sur les navires classés par le Lloyd's Register, voir le Lloyd's Register Book et le Lloyd's Yacht Register.

Official No.	No. in List	Registry in which classed	Ship's Name. Master, and Dates of Appointment. Ship's Late Name.	Material and Rig	Flag	Net Tonnage	Length	Breadth	Depth	Where built	When	By whom	Metalling	Owners or Managers
	501		Jeune Actif	W Bg	French	140	79·7	22·3	11·9	Dieppe	1843			F. F[illegible]
	502		Jeune Adolphe [illegible]	W Bg	French	156	..	..	..	St. Servan	1848			Guilbert & fils
	503		Jeune Alexandre	W Bg	French	148	79·7	21·0	12·4	Vannes	1867	Lef[illegible]		Pc[illegible]rodean
	504		Jeune André [illegible]	W Bn	French	145	76·8	23·0	11·4	Nantes	1867	Debonleaux	[illegible]	Cormerais frères
	505		Jeune Anna *Pedron*	W Bn	French	138	85·0	20·3	11·2	St. Malo	1869	Tranchemer		L. Verry
	506		Jeune Anne Marie	W	French	103	65·0	18·7	8·8	Vannes	1851			F. Dupont
	507		Jeune Baptistin	W	French	102	..	..	..					Mattei frères
	508		Jeune Bénoni *Luco*	W Bg	French	133	77·4	21·3	11·2	Mesnes	1870	Lamort		Cardinal ainé
	509	V	Jeune Berthe *Barré* -87 (WRECKED)	W Sr	French	137	83·0	20·0	11·8	Dunkirk	1865	Derycke fils		Macré & Co.
	510	V	JeuneCharles *Raoult* -88	W Bg	French	138	82·0	20·7	11·9	Cherbourg	1857	Mahieu	z 6 8[illegible]	J. B. Laronche
	511		Jeune Felix *Lecreux*	W Sr	French	112	67·6	21·0	11·6	Vannes	1857	Ridel		A. Mahen
	512		Jeune Florestine *Tilleau*	W Sr	French	112	76·4	21·0	10·6	Nantes	1865	Tille		Quimener
	513	V	Jeune France *Le Blay*	W Bn	French	118	82·9	20·3	9·8	Cherbourg	1877	E. Lecerf	z [illegible]	A. Postel & [illegible]
	514		Jeune Henriette	W Sr	French	107	74·5	19·3	10·0	LeLegue	1858			P. Carfanton [illegible]
	515	V	Jeune Léonie *[illegible]* -87	W Sr	French	107	83·7	20·3	11·2	Dunkirk	1879			C. Louvert fils
	516	V	Jeune Marie *Lemasse* -88	W Bg	French	232	102·6	23·6	12·8	St. Malo	1874	Gautier	z 2 6	P. Villars
	517		Jeune Saint Vincent (CONDEMNED)	W Sr	French	137	80·0	21·3	10·7	Carnac	1866	RioPe[illegible]		Le Rouzic
	518		Jew Le	W3mLr	Chinese	163	109·5	25·3	9·6	Shanghai	1877	S.C. Farnham & C[o.]		
51215	519		Jewel *Robinson*	W Bg	British	195	100·0	22·9	13·0	Workington	1867	Fell		J. T. Fell & Co.
98023	520		Jhansi	I Lr	British	547	220·0	32·0	9·5	Calcutta	1879	IndiaGenl. Stm. [Nav. Co. Ld.		Irrawaddy Flotilla [illegible]
	521		Jinko Maru No. 1	W Bg	Japanese	129	91·7	19·1	9·6	Tokio	1874			Takeko Ichibei
	522		Jiquia (ex Isabella Ridley)	W Bk	Brazilian	534	139·6	29·3	18·7	Newcastle	1860	Gaddy		H. Hayward
13738	523		Jireh Perry *Chace*	W S	American	316	121·1	28·8	16·0	Newburyport, [Mass.	1852			Aiken & Swift
79015	524	✻	Jka Vuka *Robertson*	W Bn	British	149	91·0	26·6	9·0	Dorchester N.B.	1884	P. J. Palmer	[illegible]	H. Macfarlane
	525	G V	Joachim Christian *Dilhein* 84	W Bk	German	457	144·5	29·0	16·7	Hamburg	1865	E. Dreyer	Y 6 8[illegible]	H. Bauer
40959	526		Joad el Manaan	W Bk	British	369	107·5	24·0	16·0	Cochin				Adam Ali Sultan [illegible]
	527	N	Joanchas *Nilsen*	W Bg	Norweg'n	135	93·1	23·5	10·4	Sandefjord	1873	L. Engebretsen		A. Andersen
	528	V	Joanna *Vudachi* (ex Giovanni D.)	W Bk	Greek	474	136·4	28·0	19·1	Curzola	1873	A. Villovich	[illegible]	N. Vudachi & Co.
	529		Joannis	W Bg	Turkish	144				Syra	1883			Sokyzise Pera [illegible]
	530	V	Joaquina *De Silva* (ex Conceicao A. M[illegible])	W Sr	Portug'se	151	84·0	22·3	10·2	V[illegible] de Conde	1868	J. M. de Araujo	[illegible]	J. A. R. da Silva
	531		Joaquina *Parada* 85-85 (ex Joaquina Barreras)	W Bn	Spanish	197	86·3	22·7	10·9	Blanes	1871	José Vieta		M. Barreras
	532		Joaquina	W Bn	Spanish	155	89·2	24·6	11·3	Corunna	1869	Baamonde		W. M. Gomez
	533		Job *Deltonen* 87- [illegible]	W Bk	Russian	412	123·9	30·5	21·5	New York				D. Brumila
76306	534	R	Job H. Jackson *Williams* -87	W3mSr	American	494	152·5	35·0	13·8	Wilmington Del	1882	Jackson & Sharp [Co.		D. S. Williams
51830	535		Joe Kelley *Plummer* (ex Exmouth)	W Sr	British	110	84·2	23·4	8·7	Essex, U.S.	1863			A. Montgomery
76450	536	R	Joe Read (LOST) *Allen*	W Bkn	American	468	141·6	33·1	12·9	Harrington Me	1883	D. Downey	[illegible]	A. M. Nash
	537	G	Joel *Svensson* 86	W Sr	Swedish	162	84·3	21·4	11·0	Åland	1867			O. Holm
66748	538	R	Joequinna *Gardiner*	W Bkn	British	356	127·2	29·5	12·9	Liverpool, N.S.	1875	M. McLeod	[illegible]	A. McNutt
	539		Joesserin *[illegible]* (ex Martha [illegible])	I Bk	Dutch	341	126·9	26·5	..	Sonderburg	1868	Matsen		Said Mohamed bin [illegible]
278	540		[illegible]foo Maru	W Sr	Japanese	247	114·5	23·0	13·1	Tokio	1876	YasuiSadayasu		Teishin Sho
	541	N	Jofur *Larsen* 68-72	W Bg	Norweg'n	202	98·8	24·8	12·8	Grimstad	1872	Jens Tellefsen		J. C. Larsen
	542	V	Johan *Pieper* 83-	W Sr	German	145	94·9	20·8	10·2	Papenburg	1876	A. Beckmann	[illegible]	Abr. Beckmann
	543		Johan *Hjid*	W	Russian	126	..	..	..		1877	Salmis		J. Posé
	544	V	Johan *From* -84	W Bn	Swedish	201	108·5	22·7	10·9	Kyrkön	1884	Fr. Höglund	[illegible]	[illegible]
	545		Johan *Gelander*	W Bg	Swedish	167	84·9	20·8	12·6	Skellefteå	1834			C. M. Viklund
	546	N	Johan Dahll *Knudsen* [illegible]	W Sr	Norweg'n	197	102·8	25·5	11·8	Kragerö	1879	A. Thomassen	[illegible]	T. Jensen
	547	N V	Johan Hanssen *Andersen*	W Bk	Norweg'n	595	153·3	31·6	17·6	Risör	1874	Berentsen	[illegible]	J. Hanssen
	548	N V	Johan Irgens *Dahl* -88	W Bk	Norweg'n	774	166·6	35·2	20·1	Bergen	1873	J. Trampe & S[illegible]	[illegible]	Matthias H[illegible]
	549		Johan Smidt *[illegible]*	W Bk	Norweg'n	765	142·3	33·6	23·0	Boston	1834			E. A. [illegible]
	550	V	Johan Theodor (CONDEMNED)	W Bk	Dutch	586	139·7	28·5	17·9	Bo[illegible]	1877	D. R[illegible]	[illegible]	Y. Feenstra

✻=Lloyd's Register; D=Nederlandsche Vereeniging; G=Germanischer Lloyd; I=Registro Italiano; R=[illegible]
A=Veritas Austro-Ungarico; H=Veritas Ellenico; N=Norske Veritas; V=[illegible]

In the 4th Column, C=Composite (iron frame, wood planked); I=Iron; Stl=Steel; W=Wood.

The Dates of Metalling (Col. 11) are based on the records of the Books in which Vessels are respectively classed, or on the Reports of the Surveyors [illegible]

Further particulars of Vessels classed by Lloyd's Register will be found in Lloyd's Register Book and Lloyd's Yacht Register.

No.	Class.	Nom. Capitaine et dates d'entrée. Ancien Nom.	Matériaux et Gréement	Pavillon	Tonnage net	Longueur	Largeur	Creux	Construit Où	Quand	Par qui	Date du doublage	Armateurs ou Directeurs-Gérants	Port d'Armement
551		Johan Vik	W Sr	Swedish	183	88·9	23·8	12·2	Roslagen	1877			J. Mattsson	Lervik, Relgo.
552		Johann	W S		238	100·0	24·8	15·8	Elsfleth	1864	J. Ahlers			
553	G	Johann *Fulfs*	W Bg	German	191	102·6	20·0	12·8	Papenburg	1859	F. W. Meyer		B. V. Fulfs	Dornumersiel
554	V	Johann	W Sr	German	136	87·9	13·4	10·0	Halte	1884	H. Klasen		A. & H. Klasen	Papenburg
555	V	Johann	W Sr	German	127	86·0	21·3	9·5	Grossefehn	1873	Schapp & Meyer		J. W. Eggen Wwe.	Grossefehn
556	V	Johann Adolph	W 3mS	German	310	128·0	27·2	14·0	Heiligenhafen	1882	J. Tormie		P. Kock	Heiligenhafen
557	V	Johann Brodersen	W Bk	Danish	324	122·9	24·0	13·1	Vegesack	1864	P. Sayer		H. C. Brodersen	St. Thomas
558	V	Johann Carl	W Bk	German	500	131·2	30·8	16·1	Elsfleth	1878	G. Wempe		J. C. Warns	Elsfleth
559	G	Johann Daniel	W Bk	German	351	113·2	27·8	17·7	Damgarten	1857	H. Dierling		Liebenheim & Pincus	Rostock
560	GV	Johann Friedrich *ex Wm. M. Reed*	W S	German	1236	192·9	37·1	24·3	Bath, Me.	1867	J. & E. Reed		Siedenburg, Wendt & Co.	Bremen
561		Johann Friedrich	W Bk	German	337	119·7	28·5	16·2	Rostock	1853	W. Zeltz		A. Burchard	Rostock
562	V	Johann Friedrich	W Bg	German	245	106·1	25·9	13·5	Wolgast	1862	W. Lübke		J. Kretzmer	Wolgast
563		Johann Friedrich	W Bg	German	200	85·0	25·0	12·7	Barth	1861	A. Witte		C. Bruhn	Barth
564	G	Johann Gustav	W Sr	German	161	87·2	23·7	12·4	Barth	1866	C. Schröder		J. J. Wallis & Sohn	Barth
565	V	Johann Hinrich	W 3mSr	German	411	139·7	27·9	15·5	Hamburg	1877	P. Petersen		Aug. Bolten	Hamburg
566	V	Johann Kepler	W Bk	German	727	119·9	32·1	18·4	Brake	1861	J. Oltmann's [Wwe		Boyes & Ruyter	Bremen
567	DV	Johann Ludwig	W Bk	Norweg'n	749	150·9	33·1	22·4	Bath, Maine	1870	Goss & Sawyer		Ch. Thorbjörnsen	Christiania
568		Johann Michael	W Sr	Russian	215	95·0	24·0	12·0	Orrenhof	1877	J. Brackmann		M. Eichen	Haynasch
569	V	Johann Wilhelm	W Bk	German	779	155·5	32·7	22·8	Danzig	1863	J. W. Klawitter		A. Gibsone	Danzig
570	G	Johanna *Jensen 72-72*	W Bg	Danish	200	102·3	23·7	11·8	Rönne	1856	S. P. Bech		J. Anderskouv	Svendborg
571	✻DV	Johanna *Holly*	W S	Dutch	1379	190·2	39·0	25·1	Alblasserdam	1870	J. Smit, Czn.	Y 3.87	J. Smit & Co.	Alblasserdam
572		Johanna	W Bk	Dutch	819	161·0	35·4	22·7	Slikkerveer	1866	J. Smit		A. Kopeke	Rotterdam
573		Johanna	W Bk	Dutch	637	..	..	..		1863			J. Meijjes Jerzn.	Amsterdam
574	✻D	Johanna *De Groot*	W Bn	Dutch	205	100·7	23·9	12·8	Pekela	1872	L. H. de Wijk		J. J. Koerts & Zoon	Oude Pekela
575	✻DG	Johanna	W Bn	Dutch	200	99·7	25·6	12·4	Elsfleth	1863	J. Ahlers		K. B. Slotema	Ulrum
576		Johanna *Meyer*	W Bk	German	721	149·7	35·2	21·2	Biddeker	1860	Gebr. Bolle		W. Lübke & Co.	Stettin
577	GV	Johanna	W 3mSr	German	439	131·2	28·5	13·8	Lübeck	1884	J. Steffen		C. Brandt	Blankenese
578	G	Johanna	W Bk	German	354	108·6	27·7	18·1	Sunderland	$18\frac{42}{63}$	Peper		Ernst. Wendt	Danzig
579	DV	Johanna *Meyer*	W 3mSr	German	294	114·8	25·0	12·7	Papenburg	1878	H. W. Meyer		H. W. Meyer	Papenburg
580	V	Johanna *Heurichs*	W Bn	German	179	100·4	24·3	10·3	Papenburg	1886	H. W. Meyer		H. W. Meyer	Papenburg
581	G	Johanna *Hansen 76-76*	W Sr	German	104	76·4	20·1	10·1	Zingst	1876	A. Dressel		J. J. Wallis & Sohn	Barth
582	N	Johanna	W Bk	Norweg'n	885	194·6	36·4	21·7	Boston	1853	Sampson		A. Brodersen	Mandal
583	N	Johanna	W Bk	Norweg'n	558	150·7	30·5	17·2	Richibucto	1867			A. D. Westergaard	Arendal
584	N	Johanna	W Bk	Norweg'n	490	138·5	30·9	16·7	Porsgrund	1873	C. A. Hansen		O. A. Strömme	Christiansand
585		Johanna	W Sr	Russian	150	88·2	26·3	11·2	Wandsen	1885	Schull		Kapkat, Jacke & Behrsing	Riga
586	G	Johanna *Kystein*	W Bn	Russian	133	95·9	23·2	11·1	Björkö	1879	J. Korvenato		G. Kystein	Björkö
587		Johanna *Blomqvist 86-71*	W Sr	Russian	117	80·1	25·1	9·1	Kimito	1877	E. Söderström		K. E. Mattsson	Sund
588	G	Johanna *Hallberg 72-77*	W 3mSr	Swedish	193	105·6	23·6	10·5	Bergqvara	1874	O. Ohlson		J. D. Björkegren	Cimbrishamn
589		Johanna *Thorsson 78-74*	W Sr	Swedish	164	98·4	21·8	10·4	Timmernabb'n	1877	T. Ohlson		M. Jönsson	Kivik
590	✻D	Johanna Geertruida	W Bk	Dutch	322	119·0	26·6	16·2	Southwick	1864	G. & J. Mills		D. Nanninga, Czn.	Winneweer
591		Johanna Karolina	W Sr	Swedish	150	90·5	22·9	10·6	Nova Scotia	1866			C. A. Fahlgren	Örnsköldsvik
592	V	Johanna Kremer *Eggers*	W Bn	German	280	120·0	28·9	13·2	Elmshorn	1876	D. W. Kremer	Y 1.87	Wm. Meyer	Hamburg
593	DV	Johanna Maria	W 3mSr	Dutch	352	118·1	27·1	16·9	Sunderland	1866	J. Thompson		A. Köpeke	Rotterdam
594		Johanna Maria	W Sr	Swedish	121	82·0	26·1	9·3	Björneborg				C. Vigren	Westervik
595	G	Johanna Matilda	W Sr	Swedish	167	95·8	24·8	11·5	Carlshamn	1869	D. R. Andersson		N. Olsson	Raa
596	G	Johanna Matilda	W Sr	Swedish	111	77·8	20·4	10·4	Landamn	1856	L. Olsson		A. Pehrsson	Brantevik
597	V	Johanna Sophia	W Sr	German	1331	196·2	38·4	24·0	Bremerhaven	1886	F. W. Wencke		F. W. Wencke	Bremerhaven
598		Johanna & Lina	W Bg	German	258	109·5	25·9	13·8	Wolgast	1866	W. Lübke		Carl Stüdemann	Greifswald
599	DV	Johanne *Albertsen 01-81*	W 3mSr	Danish	218	112·6	23·7	12·2	Svendborg	1884	P. Troensegaard		H. Hansen	Marstal
600		Johanne *Harris*	W Bg	Danish	191	101·0	23·7	11·9	Rönne	$18\frac{48}{64}$	S. P. Bech		W. A. Harris	Skjelskor

✻ = Lloyd's Register; A = Veritas Austro-Ungarico; D = Nederlandsche Vereeniging; G = Germanischer Lloyd; H = Veritas Ellenico; I = Registro Italiano; N = Norske Veritas; R = American Record; V = Bureau Veritas.

Dans la 4me Colonne, C = Composite (membrure en fer, bordé en bois); I = Fer; Stl = Acier; W = Bois.

Les Dates du doublage en métal (Col. 11) ressortent des livres dans lesquels les navires sont respectivement classés ou des rapports des Inspecteurs du Lloyd's Register.

Pour plus amples renseignements sur les navires classés par le Lloyd's Register, voir le Lloyd's Register Book et le Lloyd's Yacht Register.

Official No.	No. in Book	Registry in which classed	Ship's Name. Material and Rig. Master, and Dates of Appointment to present Owner's Service and to Ship. Ship's Late Name.	Flag	Net Tonnage	Length	Breadth	Depth	Built: Where	Built: When	Built: By whom	Date of Metalling	Owners or Managers
	601	V	Johanne *Jessen 85-85* W Bn	Danish	145	85·5	21·2	10·3	Varnanäs	1871	C. A. Hahn	[illegible]	Act. Selsk. (J. F. Jen[illegible]
	602	V	Johanne *Hjorth 82-82* W S.	Danish	123	80·4	20·8	11·3	Aarhus	1862	O. A. Høyer		C. Hassager
	603	GV	Johanne *Lemcke 86-87* W Bk (ex Hercules, ex Edith)	German	1121	182·7	38·4	23·9	Newburyport, Mass	1864	Manson&Davis	[illegible]	Louis F. Kalkmann[illegible]
	604	GV	Johanne *Saathoff 84-* W Bn	German	121	83·4	20·8	9·8	Brake	1872	J. Oltmann's (Wwe.		Joh. Saathof (D. C[illegible]
	605	N	Johanne *Rafen 77-81* W S (ex Queen of the Mersey)	Norweg'n	1168	190·8	38·0	23·2	St. John, N.B.	1860	Macdonald	[illegible]	L. Rafen
	606	N	Johanne Augusta *Christophersen -74* W Bk	Norweg'n	428	124·7	27·5	17·9	Williamsburg	1864			Enkefru C. Hansen
	607	✳V	Johanne Auguste *[illegible]* W Bk (ex [illegible])	German	895	186·0	35·0	20·0	Aberdeen	1868	A. Hall&Co.	[illegible]	Fr. Hilken
	608		Johanne Emilie *Blaes* W Bg	Danish	219	101·3	27·4	11·8	Colberg	18 54/75	Lincke		J. O. Blaes
	609		Johanne Margaritha W Bk	Norweg'n	340	129·8	27·6	15·3	Shoreham	1868	Shuttleworth		T. Kleiberg
	610		Johanne Marie *Lauritzen -88* W Bk	Danish	349	129·8	25·9	15·9	Apenrade	1863	S. Andersen		P. Mahler
	611	DV	Johanne Marie *Petersen 84-87* W Bk	Norweg'n	667	155·3	31·8	18·9	Bremerhaven	1862	H. F. Ulrichs	[illegible]	Westergaard & Co.
	612		Johannes *Jans 80-80* W 3m Sr (ex Johannes Hage)	Danish	282	105·9	23·9	15·4	Nakskov	18 41/55	K. Jørgensen	[illegible]	P. J. Greisen
	613	V	Johannes *Davidsen 80-83* W Bg	Danish	157	101·4	22·0	11·1	Rudkjøbing	1871	S. Boas	[illegible]	E. Petersen
	614	DG	Johannes *Oosting 85-85* W Bg (ex Johannes, ex Marie, ex Peter, ex Mary)	Dutch	256	108·9	27·6	13·4	NovaScotia	1857			C. Kuhlman, Jzn.
	615	V	Johannes *Stöver 81-81* W Bk	German	979	165·0	37·1	21·0	Baltimore	1862	Cooper&Slicer	[illegible]	W. Stisser & Co.
	616	DV	Johannes *Geyken 85-88* W Sr	German	124	82·7	21·2	10·2	Weener	1867	R. Berg		H. F. Geyken
	617	N	Johannes *Christensen* W Sr (ex Heron)	Norweg'n	156	85·5	21·3	12·5	Dundee	1856			G. Jacobsen
	618		Johannes W Bk	Russian	383	..	..	..		1857			N. Sittkoff
	619	GV	Johannes *Behrting 77-* W 3m Sr	Russian	256	114·7	26·9	13·0	Guttman'sbach	1881	M. Magnus		J. Martinsohn
330	620		Johannes *Martinsson (77-80)* W Sr	Russian	229	95·0	24·8	12·8	Kuerbis	1880			M. Trautmann
	621		Johannes W Sr	Russian	178	..	..	..		1866			G. Sundblom
	622	V	Johannes *Wickstrand 86-86* W Sr (ex Johanna)	Russian	134	92·8	21·6	11·8	Fanö	18 67/80			P. W. Snellman
	623		Johannes *Johansson* W Sr	Russian	100	..	..	..		1865			L. Olofsson
	624	DN	Johannes Foss *Josephson* W Bk	Norweg'n	374	122·5	29·9	15·5	Lillesand	1872	H. Jacobsen	[illegible]	P. Josephsen
	625	N	Johannes Röd *Gjertsen* W Bk	Norweg'n	444	138·3	30·6	16·3	Tönsberg	1866	J. Röds Sönner	[illegible]	Johs. Röds Sönner
68914	626	✳	John *Hossack 84-87* I 3m Sr	British	355	147·5	26·9	14·0	Newcastle	1874	Cole Bros.		J. Tully
	627	N	John *Knudtsen* W Bg	Norweg'n	286	115·8	27·4	15·0	Abo	1857	C. Jørgensen		A. Knudsen
	628	N	John *Olsen* W Bg	Norweg'n	198	106·1	26·0	12·9	Gothenburg	1847			O. Olsen
	629		John *Jansson* W Bk	Russian	519	153·8	31·5	17·9	Nystad	1873			E. Nordström
	630		John *Jansson* W Bk	Russian	452	130·0	30·0	16·5	Nerpes	1866			N. Sittkoff
	631		John *Westerlund 74-86* W Bg	Russian	379	123·0	22·0	16·0	Kimito	1875	Sandell		O. Jansson
	632		John *Karlsson* W 3m Sr	Russian	309	111·6	29·1	14·1	Kimito	1870	C. Söderström		E. A. Karlsson
334	633		John W Sr	Russian	184	89·6	26·3	11·8	Ruthern	1879	J. Wickman		J. Putning
	634		John *Mikkonen* W Sr	Russian	169	..	..	..	Joensuu	1878			P. Parviainen
321	635		John W Cr	Russian	108	68·9	22·0	9·6	Ralnoziam	1878			J. Purmal
	636		John *Norberg 75-83* W Bk	Swedish	394	107·2	24·8	16·6	Finland	1862			A. Norberg
	637		John *Pettersson -83* W Bk	Swedish	298	115·8	25·5	14·5	Nystad	1865			S. Pettersson
	638	V	John *Hillman 86-88* W Bn	Swedish	254	118·0	25·7	9·7	Gefle	1868	O. A. Brodin		O. A. Brodin
	639	G	John *Pettersson* W Bg	Swedish	212	107·9	24·3	12·1	Edingenbben	1866	O. L. Nordström		T. Pettersson
	640		John *Ingemansson -88* W Sr	Swedish	181	99·4	23·1	11·3	Maitland, N.S.	1858			N. Ingemansson
75856	641		John A. Beckerman *Goodman -81* W 3m Sr	American	461	140·7	34·2	16·1	E. Boston, Mass	1876	R. Crosbie		A. Besse
[illegible]	642	V	John A Briggs *Balch -83* W S	American	2033	234·2	41·1	28·2	Freeport, Me.	1878	Briggs&Cushing	[illegible]	G. Plummer
75368	643		John A. Curtis *Williams* W 3m Sr	American	148	111·2	23·6	8·5	Laurel, Del.	1874			G. K. Phillips
13728	644	R	John A. Griffin *Nordbury* W Sr	American	296	117·6	29·6	11·0	Wilmington Del	18 5?/84			Wilmington, [illegible]
76307	645		John A. Matheson *Totten -87* W Sr	American	147	98·4	24·4	9·5	Essex, Mass	1883	J. James		W. Matheson
[illegible]	646	✳	John Abel *Capron* W K	British	77	76·0	20·0	10·5	Brixham	1885	S. Dewdney & Son		H. Capron
12	647		John Anna *Tomson -6* W Sr	Russian	169	98·0	21·0	10·0	Havnasch	1874	Ch. Thomson		J. Mikkelsohn
73675	648		John B. Carrington *Hervey -86* W 3m Sr	American	369	133·5	33·0	9·0	E. River, Conn	1874	E. L. Goodale		E. Barnett & Co.
75?68	649		John B. Hamel, Jun. *Wilmot -86* W 3m Sr	American	305	135·4	32·4	17·6	Milton, Del.	1877	D. H. Atkins		I. R. Fennimore
[illegible]	650	R	John Baizley *Shepard* W Bkn	American	586	154·7	34·0	15·5	Milford, Del.	1880	W. A. Scribner	[illegible]	F. P. Shepherd

✳ = **Lloyd's Register**; D = Nederlandsche Vereeniging; G = Germanischer Lloyd; I = Registro Italiano; A = Veritas Austro-Ungarico; H = Veritas Ellenico; N = Norske Veritas;

In the 4th Column, C = Composite (iron frame, wood planked); I = Iron; Stl = Steel; W = Wood.

The Dates of Metalling (Col. 11) are based on the records of the Books in which Vessels are respectively classed, or on the Reports of the S[illegible]

Further particulars of Vessels classed by Lloyd's Register will be found in Lloyd's Register Book and Lloyd's Yacht [illegible]

N° dans le livre	Classification	Navires — Gréement (Capitaine, Ancien Nom)	Gréement	Pavillon	Tonnage net	Longueur	Largeur	Creux	Construit — Où	Quand	Par qui	Date du doublage en métal	Armateurs ou Directeurs-Gérants	Port d'Armement
651		John Banfield	W Bk	British	528	141·0	29·0	18·6	Scilly	1859	Edwards	V	H.Campbell	Hull
652		John Barnard	W Bn	British	400	121·0	27·4	11·7	Maine,U.S.	1862				Turk's Island
653		John Bird	W3mSr	American	337	118·4	29·5	11·6	Rockport,Me.	1870	Tolman&Fells		S.M.Bird	Rockland,Me.
654	V	John Black	W Bk	British	545	139·0	32·9	17·8	Port Elgin,N.B.	1871			John Black	St.John,N.B.
655		John Boyd	W Bn	British	193	105·0	26·8	12·0	Port Elgin,N.B.	1867			J.F.Norris	Halifax,N.S.
656	R	John Bracewell	W3mSr	American	190	113·6	29·2	9·1	Bath,Me.	1878	Goss&Sawyer		B.F.Neally	Portsmouth,N.H.
657		John Brightman	W Bn	American	352	110·8	29·3	10·7	Calais,Me.	1864			A.W.Hanson	Fall River,Mass.
658	V	John Bryce *Murphy 80-88*	W S	American	1895	217·3	18·0	30·0	Thomaston,Me.	1869	E.O'Brien	V	E.E.O'Brien	Thomaston,Me.
659		John Bull	W Bk	British	248	104·0	26·0	14·0	Dunkirk	1852			T.Roberts	Hong Kong
660	RV	John Bunyan	W S	British	1194	191·1	38·2	22·0	Meteghan,N.S.	1875	B.Robeshean	V	L.E.Baker	Yarmouth,N.S.
661		John C.Cottingham *Bishop*	W3mSr	American	215	111·0	28·5	8·8	Camden,N.J.	1872			Thomas Ayres	Philadelphia
662	R	John C.Gregory *Killen*	W3mSr	American	360	140·0	28·2	10·1	Newburyport,Mass.	1884	Atkinson&Fillmore	V	F.L.Atkinson	Newburyport,Mass.
663	R	John C. Haynes	W3mSr	American	720	172·5	35·1	17·8	Bath,Me.	1887	G.G.Deering		Gardiner G.Deering	Bath
664	*D	John C.Munro *Herbert*	I S	British	613	169·2	28·2	18·3	Sunderland	1862	J.Laing		T.D.Woodhead&Co	Hull
665	RV	John C. Noyes	W Bg	American	344	126·0	30·0	12·9	Calais,Me.	1875	Rideout	V	Miller,Bull & Co.	Newburyport,Mass.
666	V	John C.Potter *Pendleton*	W S	American	1244	186·3	36·8	21·8	Searsport,Me.	1869	W.McGilvery&Co.	V	Charles Nelson	San Francisco
667	R	John C. Smith	W Bkn	American	429	132·8	32·8	15·5	Belfast,Me.	1874		V	E.Sibley	Belfast,Me.
668		John C.Sweeney	W3mSr	American	377	130·0	33·2	14·0	Cooper's Point,N.J.	1874			R.J.Winsmore	Philadelphia
669		John Cadwallader	W Sr	American	118	91·1	26·4	8·5	Tuckahoe,N.J.	1853			J.C.Wall	New Bedford,Mass.
670	HV	John Casper *Lindemann*	W Bk	German	337	110·8	25·1	17·2	Sunderland	1859	P.Foster		Aug.Lindenau	Memel
671		John Clarke	W Bg	British	220	102·1	25·1	15·1	S.Shields	1863	Metcalf		W.Dixon	S.Shields
672		John Coggin	W Sw	British	170	77·8	19·3	12·9	Whitby	1846	H.Barrick		W.Watson	Whitby
673		John Crockford	W Sr	American	113	86·7	25·7	7·9	Belleville,N.J.	1849			R.Bush	New York
674	RV	John Currier *Lawrence*	W S	American	1847	235·8	42·8	26·0	Newburyport,Mass.	1882	J.Currier,jun.	V	Thayer & Lincoln	Boston,Mass.
675	RV	John D. Brewer *Brooklyn 81-82*	W Bk	American	926	174·6	35·5	21·6	Boston,Mass.	1882	Smith&Townsend	V	E.M.Brewer	Boston,Mass.
676	R	John D. Paige *Haley*	W3mSr	American	378	127·0	34·0	9·6	Camden,N.J.	1872	Taylor&Mathis		A.C.Haley	Bridgeton,N.J.
677	V	John D. Spreckels	W Bn	American	254	124·6	31·8	10·1	San Francisco	1880	Turner		Oceanic S.S.Co.	San Francisco
678		John D. Williams	W [illegible]	American	515	140·6	35·0	16·3	Kennebunk,Me.	1873			W.V.Pearce	New York
679	*V	John Davie *Croker*	I Bk	British	1013	208·0	34·9	20·7	Port Glasgow	1876	J.Reid&Co.		Hajee Hossen Esmile	Calcutta
680		John Day	W Sp	British	136	106·3	22·8	8·0	St.Aimé	1870			F.Morin	Quebec
681		John Douglass *Jordan*	W Sr	American	180	103·6	26·9	8·9	Bucksport,Me.	1873			S.H.Barbour	Bangor,Me.
682	R	John E. Chase *Clifford-87*	W Bk	American	634	152·0	32·0	18·0	Calais,Me.	1868	Rideout&Co.	V	Ryan & Kelsey	Portland,Me.
683	V	John E. Sayre	W S	British	1548	220·0	40·2	24·0	Portland,N.B.	1885	F.E.Sayre	V	F.E.Sayre	St.John,N.B.
684		John Elliotson	W Bg	British	196	87·9	25·5	14·5	Southwick	1845	Doxford&Crown		G.A.Cox	Rochester
685		John Emil *Landmann*	W Sr	Russian	157	92·7	25·9	11·1	Angern	1880	Ohschmuzeneek		C.Munder	Riga
686	*	John Evans	W Sr	British	126	85·6	22·9	11·6	Portmadoc	1873	Williams		J.Jones & Co.	Carnarvon
687	*	John Ewing *Williams*	W Sr	British	105	79·6	21·5	10·3	Whitehaven	1867	Williamson		G.Parry	Aberystwith
688	R	John F. Kranz	W3mSr	American	519	140·9	35·0	16·8	Boston,Mass.	1871	Campbell&Brooks	V	R.R.Freeman	Boston,Mass.
689	R	John F. Merrow	W3mSr	American	668	157·0	35·0	19·0	Camden,Me.	1882	H.M.Bean		C. & J.F.Baker	Boston,Mass.
690		John F. Miller	W Sr	American	176	107·0	30·6	9·0	San Francisco	1882			B.H.Hanson	San Francisco
691	R	John F. Rottman	W Bk	American	605	138·8	32·3	17·8	Harrington,Me.	1874	A.Nash	V	Swan & Son	New York
692	GV	John Fougt *Björkman*	W S	Swedish	518	148·9	29·1	16·6	Hudiksvall	1869	J.H.Fougt	V	Björkman	Hudiksvall
693		John G. Fell	W Sr	American	157	106·4	22·5	8·2	Dennisville,N.J.	1871			W.Mershon	Perth Amboy,N.J.
694		John G. North	W3mSr	American	320	143·5	34·2	11·2	Marshfield,Oreg.	1881			R.Nelson	San Francisco
695		John G. Schmitt	W3mSr	American	495	135·6	35·6	9·8	Leesburg,N.J.	1880	J.Russell		J.Russell	Bridgeton,N.J.
696	*	John Gambles	I Bk	British	1027	215·9	34·1	21·0	Harrington	1874	R.Williamson&Son		J.Gambles	Workington
697		John Gaskin	W Sc	British	487	165·8	31·4	12·1	Kingston	1881	R.Robinson		Montreal Transportation Co.	Montreal
698		John Gibson	W Sc	British	100	87·4	20·6	9·[illegible]	Fleetwood	1878	Gibson		Porter's Shipping Co.,Ld	Fleetwood
699	V	John Gill *McKenzie 84*	W Bk	British	1011	180·9	36·8	21·5	River John,N.S.	1881	A.McKenzie	V	D.Sutherland	Pictou,N.S.
700														

* = Lloyd's Register; D = Nederlandsche Vereeniging; G = Germanischer Lloyd; I = Registro Italiano; R = American Record; A = Veritas Austro-Ungarico; H = Veritas Ellenico; N = Norske Veritas; V = Bureau Veritas.

Dans la 4me Colonne, C = Composite (membrure en fer, bordé en bois); I = Fer; Stl = Acier; W = Bois.

Dates du doublage en métal (Col. 11) ressortant des livres dans lesquels les navires sont respectivement classés ou des rapports des Inspecteurs du Lloyd's Register.

Pour plus amples renseignements sur les navires classés par le Lloyd's Register, voir le Lloyd's Register Book et le Lloyd's Yacht Register.

Official No.	No. in Book	Registry in which classed	Ship's Name. Material and Rig. Master, and Dates of Appointment to present Owner's Service and to Ship. Ship's Late Name.	Flag	Net Tonnage	Length	Breadth	Depth	Built: Where	Built: When	Built: By whom	Date of Metalling	Owners or Managers
37027	701		John Gilpin *Whitering* W Bk	British	283	118·3	27·2	12·4	Pembroke, U.S.	1854			T.J.Curtis
83346	702		John Gray W	British	159	110·0	22·0	7·4	Quebec	1881	J.R.Booth		D.Murphy
75064	703	R	John H. Converse *Leighton* W3mSr	American	315	124·0	30·3	12·1	Harrington Me	1874		[illegible]	O.P.Rumball
75702	704		John H. Crandon *Nowell* W Bn	American	495	129·6	31·2	16·9	Columbia Falls, Me	1875	J.H.Crandon	[illegible]	Swan & Son
76327	705	R	John H. Cross *Sweeney* W3mSr	American	385	145·6	34·0	11·5	Waldoboro Me	1882	Kennedy	[illegible]	Wm. H. Gladding
75508	706	V	John H. Kranz *Fiske* W3mSr	American	653	158·0	34·0	17·0	E. Boston	1873	J.M.Brooks	[illegible]	R.R.Freeman
76572	707		John H. McManus W Sr	American	106	88·0	24·0	10·0	Essex, Mass	1885	A.D.Story		M.Leonard
76420	708	R	John H. May *Riggs* W3mSr	American	345	136·2	34·0	10·0	Milford, Del.	1883	T.W.Abbott & Bros.		J.May & Sons
12691	709	R	John H. Pearson *Mahoney* W Bk	American	402	126·0	28·6	12·3	Newburyport, Mass.	1865	Currier	[illegible]	Bates & Barnett
75045	710		John H. Perry *Kelley* (STRANDED) W Sr	American	171	99·0	28·0	8·5	Bath, Me.	1867	Goss & Sawyer		E.Gibbs
13077	711		John H. Rapp *Curtis* W Sr	American	175	106·6	29·5	9·7	Tuckerton, N.J	1866			E.R.Kirk
76525	712	R	John H. Tingue *Burdge* W3mSr	American	525	151·0	35·0	12·2	Fairhaven Conn	1884	H.Hanscom		H.H.Hanscom
76578	713		John Haggerty Sr	American	161	99·5	29·0	9·0	[illegible] N.Y.	1885			I.J.Merritt, jnr.
686	714	I	John Hamnet *Pedone* W Bk	Italian	639	147·6	32·1	20·0	Palermo	1875	V.Conigliaro	[illegible]	Giov. Hamnet
13571	715		John Hancock *Chapman* W3mSr	American	168	143·7	23·5	8·5	Boston	1850			H.H.Buhne
92201	716		John Hannah *Harris* W Bn	British	135	91·6	22·0	11·5	Charlestown	1868	Hannah		Ramsgate Smack Owners Ice Co. Ld.
12988	717	RV	John Harvey *Stewart* W S	American	700	146·0	33·0	20·6	Newburyport, Mass.	1866	W.B.Coffin	[illegible]	J.H.Stewart
79004	718	DV	John Hickman *Thomson* W Bk	British	1241	192·3	38·2	22·5	Dorchester N.B.	1876	W.Hickman	[illegible]	W.Hickman
76549	719	R	John J. Hanson *Stevens* W3mSr	American	651	150·1	34·8	18·7	Bath, Me.	1884	New Eng. S.B.Co.		B.F.Nealley
75527	720	R	John J. Marsh *Whittier* W Bkn	American	390	129·0	29·0	16·8	Newburyport, Mass.	1873		[illegible]	Chase, Leavitt & Co.
	721	R	John J. Snow *Snow* W3mSr	American	186	110·0	27·0	10·0	Rockland, Me.	1888			J.M.Snow & Co.
13627	722		John J. Ward *Vancleaf* W3mSr	American	281	127·0	30·2	10·0	Newark, N.J.	1867			A.J.Newbury
27001	723	✻	John James *Clayton* W Sr	British	78	73·0	18·6	9·5	Aberystwith	1859	James		J.Clayton
69067	724	V	John Johnson *Henderson* W Bk	British	696	119·5	33·8	18·8	Annapolis, N.S.	1874	R.Young	[illegible]	J.Johnson
13414	725	R	John Johnson *Heavon* W Sr	American	299	114·5	31·0	9·5	Milford, Del.	1866	Black		J.Slusman
	726	R	John K. Souther *Thompson* W4mSr	American	944	182·5	38·7	18·5	Thomaston Me	1888	N.E.Vinal		Washburne Bros.
76322	727	R	John K. Souther (WRECKED) *Thompson* W3mSr	American	737	176·6	34·9	17·6	Thomaston Me	1884	S.Watts	[illegible]	Washburne Bros.
24502	728		John Knox (HULK) *Sopwith* W Bk	British	247	126·4	23·8	14·0	Aberdeen	1852	W.Hood & Co.		[illegible] of N.Z.
75880	729	R	John L. Hasbrouck *Thompson* W Bkn	American	772	150·4	33·1	22·4	Kennebunk Me	1876	N.L.Thompson	[illegible]	E.Cobb
76484	730	R	John L. Treat *McLure* W3mSr	American	478	155·0	35·0	11·5	Madison, Conn.	1884	E.M.Miner & Son		J.A.McLure
13358	731		John Lenthall *Clark* W Sr	American	136	100·2	23·3	7·8	N. Brunswick, [N.J.]	18 55/79			G.W.Clark
73511	732		John Leslie Griffiths W Bkn	British	192	120·6	25·2	10·5	Tasmania	1876	J.Griffiths		J.Griffiths
87097	733	✻	John Lockett *Kitchen* I Bk	British	810	200·9	32·0	18·8	Liverpool	1884	R&J.Evans&Co		W. & J.Lockett
90500	734	V	John M. Blaikie *Forbes* W4mS	British	1778	245·2	43·6	24·0	Londonderry, N.S.	1885	J.B.Geddes	[illegible]	J.M.Blaikie
13380	735		John M. Broomall *Gandy* W Sr	American	210	109·2	29·0	9·2	Chester, Pa.	1862			J.Wallace
76425	736	R	John M. Brown *Brown* W3mSr	American	430	147·0	35·0	12·0	N.Haven, Conn	1883	H.H.Hanscom		N.Easterbrook
75801	737	V	John M. Clerk *Pendleton* W Bk	American	719	147·6	33·3	19·4	Camden, Me.	1876	Bean & Coombs	[illegible]	I.Coombs
76001	738	RV	John McDermott *Davis* W Bn	American	536	130·2	33·7	17·6	Pt.Jefferson NY	1878	J.R.Mather	[illegible]	A.Abbot
76081	739	RV	John McDonald *Stores* W S	American	2172	249·4	43·1	28·1	Bath, Me.	1882	B.Flint	[illegible]	R.W.Griffiths
75430	740		John McGinnis *Riggan* W Sr	American	124	98·8	20·5	8·2	Laurel, Del.	1872			G.R.Phillips
92406	741	✻	John McLaggan W3mSr	British	238	103·6	27·8	11·8	Newcastle N.B.	1887	E.Sinclair		E.Sinclair
90742	742	V	John McLeod *Stewart* W S	British	1595	225·4	41·1	24·3	Black Riv. N.B.	1885	J.&R.McLeod	[illegible]	Howard D.Troop
81060	743		John Mary W Bg	British	128	87·5	22·0	10·1	Colasegaram-[patnam]	1876			Moona Kavana [illegible]
67774	744		John Morrison *Kinch* W Bkn	British	184	105·5	25·0	12·1	Arklow	1874	Tracy		E.Kearon
73734	745		John N. Parker *Phillips* W3mSr	American	157	118·3	29·5	8·2	Laurel, Del.	1874			D.Hearn
76443	746		John N. Sherwood W Sr	American	105	87·8	27·2	6·7	Athens, N.Y.	1883	Vandoon & Magee		J.Hong
16284	747	✻	John Nicholson W Bk	British	685	177·6	29·4	20·4	Annan	1858	Nicholson	[illegible]	W.Quine
56907	748	✻	John O'Gaunt *Bowles* I S	British	1275	219·5	37·2	22·8	Birkenhead	1868	G.R.Clover & Co.		J.Beazley
91211	749	✻	John O'Gaunt *Newman* Stl Bk	British	1197	222·4	36·2	21·7	Port Glasgow	1885	J.Reid & Co.		T.Bell
	750	G	John Olof *Norberg* W Bg	Swedish	240	102·9	22·9	13·5	Brake	1860	Behrens Sohn		P.Norberg

✻ = Lloyd's Register; D = Nederlandsche Vereeniging; G = Germanischer Lloyd; I = Registro Italiano; R = [illegible]
Â = Veritas Austro-Ungarico; H = Veritas Ellenico; N = Norske Veritas; V = [illegible]

In the 4th Column, C = Composite (iron frame, wood planked); I = Iron; Stl = Steel; W = Wood.

The Dates of Metalling (**Col. 11**) are based on the records of the Books in which Vessels are respectively classed, or on the Reports of the Surveyors.

Further particulars of Vessels classed by Lloyd's Register will be found in Lloyd's Register Book and Lloyd's Yacht Register.

No.	Classe	Nom. Matériaux et Gréement. Capitaine et dates d'entrée au service des Armateurs actuels et à bord du navire. Ancien Nom	Pavillon	Tonnement	Longueur	Largeur	Creux	Où.	Quand.	Par qui.	Date du doublage en métal	Armateurs ou Directeurs-Gérants.	Port d'Armement.
751		John P. Kelsey *Steelman* W Sr	American	171	100·7	29·5	8·6	Eastteyville N.J	1872			H.L.Steelman	Camden, N.J.
752		John P. West *Smith* W Bk (WHALER)	American	336	117·3	28·6	17·5	N.Bedford Mass	1857	S.Hammon		S.N.West	N.w Bedford Mass
753	*	John Paterson *Rankin* I Bk	British	1256	223·1	36·2	22·1	Sunderland	1877	P.Hardcastle&Co.		T.Clark	Irvine
754	*	John Peirce *Rogers* W 3m Sr	American	385	117·0	34·3	10·0	Bath, Me	1886	New England S.B.Co		John Peirce	New York
755		John Peirce W Sr	American	201	93·0	24·0	9·0	Belfast, Me.	1853			R.S.Bradley	Boston, Mass.
756	*	John Pierce *Hughes 7-78* W Sr	British	97	75·0	21·0	10·4	Aberayron	1860	Jones		J.Hughes	Aberystwith
757		John Pitcairn *Whitwell* W Bg	British	257	119·8	25·2	11·2	P.E.Island	1864	Davies		F.W.Fairbrass	London
758	*	John Potts *Correll* W Bk	British	374	129·0	27·9	16·8	Sunderland	1870	R. Thompson	[illegible]	China Shipowners Association (Morris)	Shanghai
759		John Price *Jones* W Sr	American	194	105·0	27·6	10·1	Patchogue N.Y	1853	H.Gerard		W.F.Greene & Son	Boston, Mass.
760		John Proctor *Robinson* W 3m Sr	American	474	140·0	32·0	15·6	Waldoboro Me	1873	A.Storer&Sara		C.&J.F.Baker	Boston, Mass.
761	R	John R. Bergen *Parsons* W 3m Sr	American	615	152·0	35·0	17·0	Wilmngtn. Del	1882	Jackson& Sharp Co.	[illegible]	C.V.Parsons	New York
762	R	John R. Fell *Deane* W 3m Sr	American	336	131·8	34·1	10·4	Bath, Me	1880	W.Rogers		J.R.Fell	Philadelphia
763		John R. Halladay *Knudsen* W 3m Sr	American	336	125·0	32·0	12·0	Wilmngtn. Del	1872			J.Falkenburg	New York
764	RV	John R. Kelley *Gibbons* W S	American	2255	256·9	45·0	27·8	Bath, Me.	1883	Goss&Sawyer	[illegible]	J.R.Kelley	Bath, Me.
765	R	John R. Penrose *Smith* W 3m Sr	American	433	148·0	34·6	12·2	Milford, Del.	1882	T.Carlisle		S.R.Smith	Philadelphia
766	R	John R. Stanhope *Martin* W Bk	American	391	123·1	28·5	15·9	Rockland, Me.	1877	Snow&Pearsons	[illegible]	J.D.Winter	Rockland, Me.
767	*	John R. Worcester *Smith* C Bk	British	844	191·5	32·4	19·9	Port Glasgow	1865	Marine Investment Co. Ltd.	Y 11.87	John Stewart & Co.	London
768	*	John Rennie *Shelley* I Bk	British	848	177·0	32·6	20·7	London	1863	J.& G. Rennie		Devitt & Moore	London
769		John Richards *McLester* W Bg	British	256	106·0	27·3	13·0	Yarmouth, N.S.	1855			T.McVeagh	Belfast
770	*	John Ritson W Bg	British	513	143·6	29·5	18·2	Sunderland	1868	G.& J.Mills	[illegible]	F.& W.Ritson	Sunderland
771	*	John Roberts *Thomas* W Bg	British	207	105·6	25·3	13·8	Pwllheli	1875	Roberts	[illegible]	E.Thomas	Carnarvon
772	R	John Rose *Allen* W 3m Sr	American	595	161·0	35·0	15·0	Wilmngtn. Del	1883	Jackson&Sharp Co.		B.F.Allen	Philadelphia
773		John Rose W Sr	American	223	..	..	..	Setauket, N.Y.	1861			J.Howell	New York
774	R	John S. Beacham W 3m Sr	American	176	107·4	28·6	8·4	Baltimore	1874	J.S.Beach'm&Bro		W.Moore	Richmond, Va.
775		John S. Case *Frothingham* W 3m Sr	American	193	112·5	28·4	9·2	Brewer, Me.	1878	Wyman		D.J.Sawyer	Machias, Me.
776	R	John S. Davis *Green* W 3m Sr	American	345	133·0	33·6	10·0	Bath, Me.	1883	Goss&Sawyer		P.Fitzpatrick	Philadelphia
777		John S. Gilmore *Smith* W Sr	American	111	87·9	27·2	7·0	May's Lndg N.J	1870			A.W.Hall	Albany, N.Y.
778		John S. Moulton *Sawyer* W Sr	American	152	102·0	26·7	7·0	Addison, Me.	1865			J.D.Sawyer	Machias, Me.
779	R	John S. Wood W 3m Sr	American	521	139·0	34·0	15·3	Somers Pnt N.J	1875	I.B.Smith		J.B.Adams	Camden, N.J.
780		John Shay *Hargrove* W Bn	American	377	118·9	30·2	10·0	Patchogue NY	1864			R.P.Conk	Port Jefferson
781		John Shay *Tilton* W 3m Sr	American	290	116·3	30·0	10·0	Gt Egg Hbr. N.J	1864	S.Gaskell		Mason & May	Philadelphia
782		John Shelly *Fishpool* W Sr	British	175	101·3	24·8	12·5	P.E.Island	1875	Duncan		W.J.Sievey	London
783	*	John Sims *Pestreath* W Sr	British	98	89·9	21·1	10·8	Falmouth	1873	Trethowan		J.D.Perry	Plymouth
784		John Skaltzounis W Bg	Greek	291	89·9	25·3	16·1	Syra				C.P.Tzortzou	Poros
785	*	John Smith *Hendry* W 3m Sr	British	133	101·5	22·4	10·9	Ardrossan	1881	Barr&Shearer	[illegible]	C.Hendry	Ardrossan
786		John Smith *Kustel* W Bkn	American	565	159·3	37·2	14·4	Port Blakey, Wash.	1882	Hall Bros.		O.Kustel	San Francisco
787		John Somes *Robbins* W Sr	American	130	81·7	26·2	8·4	Mnt. Desert Me	1868			J.W.Somes	Ellsworth, Me.
788		John Stockham *Searle* W Sr	American	203	110·1	29·0	8·6	Unionville N.J	1861			G.T.Searl	Providence, R.I.
789		John Stroup *Sturgis* W Sr	American	271	113·7	29·7	9·3	Bridgeton, N.J.	1861			J.Baker	Boston, Mass.
790		John Swan *Huey* W Bn (WHALER)	American	508	129·0	31·5	17·5	Columbia Falls Me.	1872	J.H.Crandon		Swan & Son	New York
791	V	John T. Ives W Bn	British	358	131·6	30·5	12·9	Tatamagouche N.S.	1876	D.Campbell	[illegible]	J.D.McGregor	P.H'wksb'ry, N.S.
792		John T. Williams *Lock* W Sr	American	230	127·0	31·3	9·4	Brooklyn	1857			J.T.Williams	New York
793		John Tway *Wheatley* W Sr	American	151	103·0	29·4	7·8	Allow'ystn N.J	1853			H.C.Wheatley	Crisfield, Md.
794		John W. Hall, Jun. *Fleming* W 3m Sr	American	183	105·0	28·8	8·0	Frederica, Del.	1873	Nathaniel Hank		J.W.Hall	Wilmington, Del.
795		John Walters *Rendell* W Sr	British	162	97·0	21·4	12·4	Exmouth	1866	Walters		G.Ellett & Co.	Exmouth
796		John Ward *Shapter* W K	British	123	103·8	22·6	8·1	Sittingbourne	1871	Smeed		Burham Brick Lime & Cement Co. Ld.	Rochester
797		John Warren *Gardner* W Sr	American	128	89·0	27·0	8·0	Newburg, N.Y.	1850			F.B.Gardner	Cold Spring, N.Y.
798	R	John Watt (BARGE) *Sweetser* W S	American	1108	175·0	35·0	24·0	Bath, Me.	1859		[illegible]	J.Weston	Boston, Mass.
799		John Welsh, jun. (WHALER) *Mahoney* W Bn	American	300	115·0	28·6	15·6	Philadelphia	1862	C.E.Williams	[illegible]	J.J.Matthias	New Haven, Conn.
800		John Wesley *Schnieders* C Bg	German	238	118·0	23·9	13·5	Aberdeen	1867	A.Hall&Sons		Jahnt Gesellschaft	Hamburg

* = Lloyd's Register; D = Nederlandsche Vereeniging; G = Germanischer Lloyd; I = Registro Italiano; R = American Record; A = Veritas Austro-Ungarico; H = Veritas Ellenico; N = Norske Veritas; V = Bureau Veritas.

Dans la 4me Colonne, C = Composite (membrure en fer, bordé en bois); I = Fer; Stl = Acier; W = Bois.

Les Dates du doublage en métal (Col. 11) ressortent des livres dans lesquels les navires sont respectivement classés ou des rapports des Inspecteurs du Lloyd's Register.

Pour plus amples renseignements sur les navires classés par le Lloyd's Register, voir le Lloyd's Register Book et le Lloyd's Yacht Register.

Official No.	No. in Book	Registry in which classed	Ship's Name. Material and Rig. Master, and Dates of Appointment to present Owner's Service and to Ship. Ship's Late Name.	Flag	Net Tonnage	Length	Breadth	Depth	Where	When	By whom	Metalling	Owners or Managers
[illegible]	801	R	John Wesley *Van Gilder -87* W Bn	American	435	104·1	30·6	16·0	Baltimore	1868	J.S.Beacham & B[illegible]		[illegible]
	802	N	John Wesley *Nielsen 85-86* W Bk	Norweg'n	359	120·5	28·6	15·9	Nantes	1855			H.Svendsen
[illegible]	803		John White *Paul* W Bn	British	188	96·1	23·2	13·6	Rye	1855	Hoad Bros.		Whitstable Shipping Co. Ld.
[illegible]	804		John Wignall *Thornton* W 3mSr	British	141	87·7	23·1	12·1	Preston	1870	Allsup		M.Nicholson
[illegible]	805	✻	John Williams *Turpie* W Bk	British	186	106·0	24·5	13·5	Aberdeen	1868	Hall	[illegible]	London Missionary Soc.
[illegible]	806	✻	John Williams *Dellerass-86* W Sr	British	71	59·0	17·6	10·1	Portmadoc	1848	W.Griffith		H.Davies
[illegible]	807	R	John Worster *Johnson 187-87* W Bkn	American	581	144·4	31·0	17·8	Medford, Mass	1867	J.O.Curtis	10.85	J.Jensen
[illegible]	808	V	John & Ann *Mikkelsohn* W Bn	Russian	220	101·2	26·3	12·6	Riga	1868	Lindner		J.Beckmann
	809		John & Gustav *Meyerdiercks* W Sr	German	197	87·3	23·3	12·1	Hammelw'd'n	1855			D.Meyerdiercks
[illegible]	810		John & Jane W Bn	British	174	85·8	24·1	14·6	Sunderland	1850	J.Rogerson		Z.Granger
[illegible]	811		John & Julie *Legsdin* W 3mSr	Russian	282	118·8	27·8	13·0		1882	J.Ehkist		J.K.Ehkist
[illegible]	812		John & Mary W Bk	British	244	106·5	25·0	11·1	America				Bavaria Mash Anarus Ararine Sendemanys Co.
[illegible]	813	✻	John & Sarah Ann W Sk	British	83	76·7	20·5	10·7	Elmshorn	1886	J.Kremer		Letten Bros.
[illegible]	814		John & Winthrop W Bk	American	321	114·4	28·5	16·2	Bath, Me.	1876	Goss Sawyer & Packard		J.P.Knowles (2nd)
	815		Johnny *Jönsson 83-85* W Sr	Swedish	182	100·1	22·8	11·2	Mönsterås	1865	Nordstrom		J.Gudmundsson
[illegible]	816	✻	Johnny Toole W k	British	64	71·3	19·4	9·0	Bideford	1886	H.M.Restarick		T. & J.Lelen, Ld.
	817	V	Joinville *Morgan* W Bk	French	497	143·0	30·2	18·5	Sunderland	1872	John Gill	[illegible]	Briandeau fils & Co.
	818	DV	Jolani *Lilja* W Bk	Swedish	842	171·2	31·3	20·6	Elsfleth	1876	J.Ahlers	[illegible]	T.Pettersson
	819	N	Jolund *Martinsen -86* W Bg	Norweg'n	270	111·7	27·1	14·1	Moss	1869	J.Hansen	[illegible]	R.M.Peterson
	820	N	Jomsborg *ex Avon Svendsen* W Bk	Norweg'n	598	150·4	30·4	18·5	Scarboro'	1856	Tindall	[illegible]	O.Herseth
	821	V	Jon Sjödin *Dahlbom 85-* W Bk	Swedish	302	125·6	25·5	12·3	Hernösand	1872	F.Höggland	[illegible]	H.Nordlander
1778	822	A	Jona *Radalich* W Bk	Austrian	548	142·0	29·4	19·6	Lussinpiccolo	1869	N.Martinolich	[illegible]	E.Giusti
	823	DV	Jonas Lie *Nielsen 75-75* W 3mSr	Norweg'n	285	115·6	26·5	11·5	Stavanger	1874	K.Jacobsen	[illegible]	Ths.S.Falck
	824	N	Jonas Rein *Rothje* W Bn	Norweg'n	158	96·1	25·5	11·1	Bergen	1880	J.Hovershólm	[illegible]	F.C.Svendsen
[illegible]	825	R	Jonathan Bourne *Thompson* W 4mSr	American	673	169·0	35·5	17·0	Camden, Me.	1883	H.M.Bean		W.C.Thompson
[illegible]	826		Jonathan Cone *Coggins* W Sr	American	116	101·5	29·0	8·0	Camden, Me.	18 55/81	Williamson		G.M.Goodwin
[illegible]	827	R	Jonathan May *Cook -88* W 3mSr	American	379	132·2	34·8	11·9	Milford, Del.	1879	J.W.Abbott & Co		J.May
[illegible]	828	R	Jonathan Sawyer *Reynolds 86-86* W 3mSr	American	379	140·0	33·3	10·3	Bath, Me.	1886	New England S. Bdg. Co.		B.F.Nealley
	829	V	Joneheld *Wulff -87* W 3mSr	Swedish	247	110·8	24·7	12·3	Koltuoda	1871	H.Olsen	[illegible]	C.Wulff
	830		Jonstorp *Anderson 86-85* W Bk	Swedish	457	132·5	27·9	15·8	Nova Scotia	1863			B.M.Andersson
	831	R	Jordan L. Mott *Tobin* W Sr	American	138	95·0	23·0	8·9	Newburyport, Mass.	1874	Colby & Laut		R.E.Gregory
[illegible]	832		Jørgen *Friis 83-80* W 3mSr	Danish	206	107·9	23·3	12·7	Korsør	1867	Chr.Lund		E.E.Petersen
	833	DV	Jørgen *Nielsen 86-86* W Sr	Danish	122	88·4	20·9	10·6	Thurø	1868	P.Bom		P.Bom
	834	V	Jørgen J. Lotz *Andersen 71-82* W Bk	Danish	567	152·8	32·1	17·8	Carlskrona	1882	C.P.Mellin	[illegible]	Ad. Harlez
	835		Jorgen Larsen *Jorgensen 82-82* W Bn	Danish	155	100·1	22·1	10·9	Thurø	1879	J.Ph.Jörgensen		L.Jorgensens Enke
	836	N	Jörgen Lorentzen *Wessel* W Bk (ex Hildor, ex Jorgen Lorentsen)	Norweg'n	583	147·8	33·6	18·1	Drammen	1867		12.86	C.M.Haavaldsen
	837	V	Jørgen Olsen *Christensen* W Sr	Danish	292	123·8	26·2	12·6	Marstal	1887	[illegible]	[illegible]	[illegible]
	838	V	Jørgen Ring *Fisker* W Sr	Danish	131	93·5	20·6	9·9	Svendborg	1875	J.R.Andersen		J.R.Andersen
	849	V	Jørgiane & Anne *Petersen 85-87* W Bn	Danish	164	95·3	21·6	10·4	Lohr	1886	Chr.Schwoon	[illegible]	Actieselskab (E.N.Madsen)
	840		Jørgine *(ex Heinrich) Hansen* W Sr	Danish	129	83·1	20·7	9·4	Fakshol	1864			N.P.J.Hansen
	841		José Barreras *Reixis 86-86* W Bg	Spanish	125	84·6	24·3	10·9	Blanes	1868	J.Vieta		J.Barreras y Casella
75784	842		Jose D. Bueno *McCall* W Bk	American	398	136·5	30·5	14·6	Damariscotta, Me.	1875	C.G.Merry		F.Waydell
75883	843		Jose E. More *Lemhard* W Bkn	American	421	134·0	30·7	15·7	Damariscotta, Me.	1875	C.G.Merry	[illegible]	F.Waydell
	844	V	José Estavão *Santo Ilarde -87* W 3mSr	Portug'se	259	113·5	23·9	12·1	Aveiro	1872	J.F.da Lapa	[illegible]	A.Pereiro, Jun. & Co.
	845	V	Jose Ginebra *Albrecht 78-88* W Bg	German	361	120·7	28·3	13·3	Brake	1878	J.F.Nicolai	[illegible]	J.H.Nicolai
	846		José Lauro (ex Theodor, ex [illegible] Adelaide) W Bn	Argentine	145	87·2	22·3	11·4	Hamburg	1861	H.C.Stulcken		F.Ferrari & J.Lauro
	847		José Maria W Bn	Chilian	147	92·2	21·6	9·0	Curanipe, Chili	1887			J.Diaz
1478	848	V	Jose Maria *Jaccarino -88* W Bk	Italian	472	137·8	30·6	19·2	Sestri Ponente	1867	G.Paragallo	[illegible]	Giacinto Demente fu Andrea
	849	V	José Maria *Sisa* W Bn	Spanish	214	96·1	26·9	13·0	Blanes	1869		[illegible]	P.Cisa y Moreno
	850		José Maria W Bn	Spanish	156	82·0	24·9	13·0	Sestri	1856			J.B.Freixas

✻ = Lloyd's Register; D = Nederlandsche Vereeniging; G = Germanischer Lloyd; I = Registro Italiano; R [illegible]
A = Veritas Austro-Ungarico; H = Veritas Ellenico; N = Norske Veritas; V [illegible]

In the 4th Column, C = Composite (iron frame, wood planked); I = Iron; Stl = Steel; W = Wood.

The Dates of Metalling (Col. 11) are based on the records of the Books in which Vessels are respectively classed, or on the Reports of the Surveyors [illegible]

Further particulars of Vessels classed by Lloyd's Register will be found in Lloyd's Register Book and Lloyd's Yacht Register.

Nos dans ce livre	Registre	Nom. Matériaux et Gréement. Capitaine et dates d'entrée au service des Armateurs actuels et à bord du navire. Ancien Nom.	Pavillon.	Tonnage net.	Longueur	Largeur	Creux	Construit: Où.	Quand.	Par qui.	Date de doublage métal	Armateurs ou Directeurs-Gérants.	Port d'Armement.
851	R	Jose Olaverri *Acee* W3mSr	American	[illegible]	144·0	35·0	16·5	Rockland, Me.	18[illegible]	Cobb, Wight & Co	[illegible]	Cobb, Wight & Co.	Rockland, Me.
852	V	José Orts *Nogueroles* ·81 WBn	Spanish	[illegible]	[illegible]	26·7	13·6	Villajoyosa	1878		[illegible]	M. Orts y Ortuño	Alicante
853		Jose Pozzo *Todo* [illegible] WSr	Peruvian	[illegible]	[illegible]	23·4	10·6	San Francisco [illegible]	186?			Isola & Yerboline	Callao
854		Jose Romero *Arenas* ·85-86 WBn	Peruvian	412?	[illegible]	[illegible]	[illegible]					[illegible]	Callao
855		José Ruiz Arteaga WSr	Spanish	112	88·0	21·?	[illegible]	[illegible]	1876	J. Aleina		J.R. Arteaga	Barcelona
856		Jose Zanelli [illegible] WBn	Peruvian	17?	82·7	[illegible]	..		1848?			[illegible]	Callao
857		José I. WSr	Brazilian	111	..	..	..					J.C. Oliveira	Porto Alegre
858	A	Josef *Ivancich* WBk	Austrian	[illegible]	131·7	27·4	18·5	Lussinpiccolo	1875	G.S. Cattarinich	[illegible]	M.A. Ivancich	Lussinpiccolo
859	N	Josef *Andersen* WSr	Norweg'n	139	98·6	23·1	9·8	Lillesand	1882	A. Reinertsen	3.88	O. Andersen	Lillesand
860	R	Josefa [illegible] ·87 WBn	American	[illegible]	[illegible]	[illegible]	17·?	[illegible]	1874	G. Russell		J.F. Bent?	New York
861		Josefa *Berg* 77-? WBk (ex [illegible])	Danish	289?	119·?	[illegible]	[illegible]	[illegible]	18 58	J.C. Smidt		P. Berg	Nexø
862	G	Josefa *Brentano* ·88 IBk	German	815	178·1	32·?	[illegible]	[illegible]	1872	A. Stephen & Sons		D.H. Wätjen & Co.	Bremen
863		Josefa (ex [illegible]) *Font* WBk	Spanish	36?	119·1	29·9	15·0	Barcelona	1866	Font		Solo, Villarrasa & Co.	Barcelona
864		Josefa WPol	Spanish	137	82·3	23·8	12·8	San Feliu de Guixols	1858			Ignacio Fuster	Majorca
865		Josefa (ex [illegible]) WSr	Spanish	114	80·0	23·7	10·7	[illegible]				M. Gonzalez	Gijon
866		Josefa Duvall (ex Nueva Villa de Tossa) WBn	Spanish	246	106·9	27·?	12·?	[illegible]	1872			N. Gelart & Duvall	Habana
867		Josefa Formosa (ex Valparaiso) WS		[illegible]	178·4	34·1	22·2	Newburyport, Mass.	1856				
868		Josefina *Oblom* ·81 W3mSr	Russian	301	..	..	..	[illegible]	18??			C.G. Andersson	Åbo
869		Josefina *Carcon* ·88 WBk (ex [illegible])	Spanish	691	144·4	30·2	18·6	Newburyport, Mass.	1868	C. Currier		[illegible] & Co.	Barcelona
870		Josefina WPol	Spanish	129	88·3	23·2	11·6	Palma	1851			Doña I. Moll	Majorca
871		Josefina *Hagland* WBk	Swedish	287	98·1	22·?	16·9	Whitehaven	1839			M. Lind	Grebbestad
872		Josefina *Andersson* WBg (ex Elina?)	Swedish	206	105·6	22·3	12·2	Canada	1865	Kickham		J.A. Samuelsson	Mollösund
873		Josefina *Larsson* 87-? WBn (ex Josephina)	Swedish	160	86·2	22·2	12·0	Oskarshamn	18			C.G. Larsson	Helsingborg
874		Josefina *Andersson* W3mSr	Swedish	142	98·1	20·8	10·0	Karlstad	1839			C.J. Andersson	Wenersborg
875		Josefina WSr	Swedish	131	77·3	25·3	11·0	Roslagen				J.E. Jansson	Djurhamn
876		Josefita *Jaurreguizar* IBk (ex [illegible])	Spanish	276	[illegible]	25·3	11·1	Newcastle	1856	C. Mitchell & Co.		J.T. de Garayalde	San Sebastian
877		Joseph *Nicholas* WS	British	1541	216·0	41·0	25·0	Maitland, N.S.	1876	A. McCallum	[illegible]	F. Curry	Maitland, N.S.
878		Joseph *Park* ·86 WBn	British	123	104·6	24·1	12·9	P.E. Island	1877	Duncan		W.H. Park	Llanelly
879		Joseph *Harris* ·85 WBn	British	195	103·3	24·8	12·9	P.E. Island	1879	Egan	[illegible]	Whitstable Shipping Co.	Faversham
880		Joseph *Allen* 87-87 WSr	British	100	94·0	21·0	11·2	Barrow	1858	Reay		J. Fisher & Sons	Lancaster
881		Joseph [illegible] ·86 WBn	French	170	89·2	21·6	11·5	Nantes	1876	J.J. Cossard		Anatole Lemoine	Saint Malo
882	G	Joseph *Graver* WSr	German	110	79·6	20·2	9·5	Papenburg	1863	J.F. Veen		B. Frericks	Papenburg
883	NV	Joseph *Christophersen* ·88 WBk	Norweg'n	687	157·3	34·4	18·5	Tvedestrand	1882	T.A. Mörland	[illegible]	C. Wroldsen	Tvedestrand
884	RV	Joseph A. Ropes [illegible] ·86 WBk	American	681	152·8	32·2	19·2	N. Wymth, Me.	1877	N.P. Kean	[illegible]	J.A. Ropes	New York
885	V	Joseph Amédée *Pierre* WSr	French	135	93·1	20·3	11·3	St. Malo	1882	F. Gautier		Sécherie de Morue de [illegible]	Binic
886	V	Joseph André *Berni* ·87 WBk (ex [illegible] Padre)	French	504	134·5	27·2	18·6	[illegible]	1875	D. Briasco	[illegible]	L. de Toppin	Marseilles
887	V	Joseph B. Thomas *Lermond* [illegible] WS	American	1938	234·5	42·3	27·2	Thomaston, Me.	1881	S. Watts	[illegible]	S. Watts & Co.	Thomaston, Me.
888	R	Joseph Baker *Eaton* ·86 WBkn	American	379	138·7	30·8	13·7	[illegible], Me.	1873		[illegible]	H. Lord	Bangor, Me.
889	✻	Joseph Chapman *G. Bruce* WSk	British	88	79·1	21·?	10·8	Elmshorn	1886	J. Kremer		Letten Bros.	Great Grimsby
890	V	Joseph Charles [illegible] ·88 WBn	French	146	82·3	23·7	11·5	St. Servan	1864	Lenormand	[illegible]	Vve. Gouriou	Pontrieux
891		Joseph Eaton, jun. *Chatfield* W3mSr	American	182	106·0	27·?	9·?	Mystic, Conn.	1874			T. Chatfield	Barnstable, Mass.
892		Joseph F. Baker (ex [illegible]) W3mSr	American	44?	140·2	35·?	15·3	Kennebunk, Me.	1874			J.F. Davis	Sag Harbour, N.Y.
893		Joseph G. Dean *Chase* 82-82 W3mSr	American	154	109·5	28·?	9·1	Kennebunk, Me.	1882	D. Clark		Z.H. Chase	Nw. B'dford, Mass.
894		Joseph Grice WSr	British	226	106·0	26·6	9·7	New Jersey, US	1832			H. Thackray	Turks Islands
895	DRV	Joseph H. Scammell [illegible] ·88 WS	British	1411	223·0	39·2	22·5	Eatonville, N.S.	1884	D. & C. Eaton	[illegible]	Scammell Brothers	St. John, N.B.
896		Joseph Hay *Phinney* ·83 WSr	American	179	105·0	29·2	9·8	Cohasset, Mass.	1864			A. Besse	Nw. B'dford, Mass.
897	[illegible]V	Joseph Haydn [illegible] WBk	German	809	161·7	35·8	21·7	Bremerhaven	1874	J.C. Tecklenborg	[illegible]	F. Tecklenborg	Bremen
898		Joseph J. Pharo *Soper* W3mSr	American	218	120·0	32·0	10·0	Allow'ystn, N.J.	1867	Van Zandt		C. Soper	Tuckerton, N.J.
899	R	Joseph M. Fitzpatrick [illegible] ·87 WSr	American	292	108·5	30·0	9·2	Philadelphia	1863	Birely, Hillman & Streaker	[illegible]	Poitevant & Favre	Pearlington
900	R	Joseph M. Hayes (ex [illegible]) *Cooke* W3mSr	American	288	126·0	32·1	9·4	Bath, Me.	1881	B.W. & H.F. Morse	[illegible]	B.W. & H.F. Morse	Bath, Me.

✻ = Lloyd's Register; D = Nederlandsche Vereeniging; G = Germanischer Lloyd; I = Registro Italiano; R = American Record;
A = Veritas Austro-Ungarico; H = Veritas Ellenico; N = Norske Veritas; V = Bureau Veritas.

Dans la 4me Colonne, C = Composite (membrure en fer, bordé en bois); I = Fer; Stl = Acier; W = Bois.

Les Dates de doublage en métal (Col. 11) ressortent des livres dans lesquels les navires sont respectivement classés ou des rapports des Inspecteurs du Lloyd's Register.

Pour plus amples renseignements sur les navires classés par le Lloyd's Register, voir le Lloyd's Register Book et le Lloyd's Yacht Register.

Official No.	No. in Book	Registry in which classed	Ship's Name. Material and Rig. Master, and Dates of Appointment to present Owner's Service and to Ship. Ship's Late Name.	Flag	Net Tonnage	Dimensions: Length	Breadth	Depth	Built: Where	When	By whom	Date of Metalling	Owners or Managers
[illegible]	901	✻	Joseph Nicholson Davies W Sr	British	99	81·0	20·9	10·6	Nevin	1868	Thomson		D. Davies
[illegible]	902		Joseph Oakes Gray W Sr	American	195	103·5	27·7	9·0	Brewer, Me.	1869			A. P. Veazie
[illegible]	903		Joseph P. Macheca W Sr (Francisco ·84)	American	108	93·6	25·5	8·6	Bath, Me.	1877			J. Macheca
[illegible]	904		Joseph Rudd Halleck W 3m Sr	American	351	130·5	31·5	13·2	Northport, NY	1873			Edgar Halleck
[illegible]	905		Joseph Russ Hendricks W 3m Sr	American	235	124·0	30·0	9·6	Eureka, Cal.	1881	H. D. Bendixsen		E. Henrix
[illegible]	906	V	Joseph S. Spinney Curling W S	American	1895	230·9	42·6	27·5	Thomaston Me.	1874	Creighton & Mills	[illegible]	F. F. Curling
[illegible]	907	R	Joseph Souther W 3m Sr (Eaton ·80)	American	362	141·0	31·5	11·0	Thomaston Me.	1874	S. Watts	[illegible]	S. Watts
[illegible]	908		Joseph W. Fish Hart W Sr	American	194	105·0	27·5	9·0	Waldoboro Me.	1860			G. W. Rawley
[illegible]	909	R	Joseph W. Foster W 3m Sr (Robinson)	American	558	142·6	35·5	16·3	Milford, Del.	1879	T. Carlisle	[illegible]	B. W. Robinson
[illegible]	910	R	Joseph Wilde Look ·88 W 3m Sr	American	275	119·0	28·5	9·0	Tremont, Me.	1875		[illegible]	A. B. Perry & Co.
	911		Joseph Xavier Lecoque W Bn	French	149	89·6	21·0	11·6	Caen	1868	Lecornec [illegible]	[illegible]	A. Mesnil frères
	912		Joseph & Marie Thomas W Bn	French	143	75·7	21·9	11·6	Vannes	1860			Troqueit & Co.
61164	913		Josephine W Sr (ex Rosie Dean) (SUNK)	British	200	120·0	22·5	10·4					L. McCallum
52465	914		Josephine W Sr	British	116	90·5	22·5	8·9	St. Jean (D'Echaillons)	1865			H. Dinning
83058	915		Josephine McKay W Bn	British	111	87·0	23·0	9·9	Lockeport, NS	1882	S. Harlow		T. Johnson
[illegible]	916		Josephine W Sr	British	106	82·4	22·5	8·6	Grondines	1861			J. Lavoie
75645	917	V	Josephine Brown W Bkn	American	569	117·0	32·3	17·4	Wald'boro' Me.	1874	J. Clark & Son	[illegible]	H. B. Rawson
[illegible]	918		Josephine (WHALER) Smith 87·87 W Bk	American	385	120·7	28·9	17·2	[illegible]	1877			Aiken & Swift
76126	919	R	Josephine Parker W 3m Sr	American	347	[illegible]	34·0	10·[illegible]	[illegible]	188[illegible]	[illegible]		J. Middleton
13542	920		Josephine Ogilvie W Bg	American	207	104·0	28·5	9·7	[illegible] Me.	1861			P. [illegible]
	921	D	Josephine (ex Grace) Fenenga W Bk	Dutch	392	134·5	29·9	13·4	Windsor, N.S.	1866	W. H. Mosher	[illegible]	W. Stuit
	922	✻	Josephine (ex Aloes) Guiboux W Bkn	French	213	111·7	23·6	12·8	Gloucester	1879	F. C. Hipwood	[illegible]	Guiboux
	923		Joséphine Rossignol W	French	147	74·5	21·6	11·4	La Seyne	1856	Jouglas		Vve Giran
	924	N	Josephine (ex Lucy Jewett) Roger [illegible] W S	Norweg'n	873	169·1	34·6	21·7	Boston	1856		[illegible]	T. M. Nielsen & Son
	925	N	Josephine Olsen W Bg	Norweg'n	260	90·7	27·1	15·2	Sandefjord	1848			J. Christophersen
	926	N	Josephine (ex Cygnet) Johansen W S	Norweg'n	186	101·8	24·2	11·5	P. E. Island	1870	Ramsey		Carl Dahl
	927		Josephine Eg	Norweg'n		..	..	..					A. Johansen
	928		Joséphine Anna W Bk	French	248	114·8	26·1	14·1	Granville	1877			Aldric Basile
[illegible]	929	R	Josephine B. Knowles W Sr (Nickerson ·87)	American	127	97·6	26·3	9·9	Addison, Me.	1865			G. P. Ockershausen
12655	930		Josephine G. Collyer W Sr (Knox)	American	141	101·0	27·0	8·0	New York	1853			J. Knox
	931	I	Josephine Henriette (ex Silvio) Lefegre ·86 W 3m Sr	French	456	148·9	29·4	16·8	Sestri Ponente	1871	A. Briasco	[illegible]	C. Clement
75898	932	R V	Josephus Rogers ·76 W S	American	1397	213·0	39·2	24·4	Newcastle, Me.	1876	E. Haggett	[illegible]	A. Austin
76462	933	R	Joshua Baker Kelly W 3m Sr	American	555	157·0	35·0	17·0	Wald'boro' Me.	1883	A. Stover & Son		J. Baker & Co.
[illegible]	934		Joshua King Breen W Bkn	British	253	120·3	28·1	12·8	Dorchester, N.B.	1869	Chambers		W. W. Robinson
13793	935	R	Joshua Loring (WRECKED) Hayes ·87 W Bk	American	483	127·4	32·0	16·1	Chelsea, Mass.	1866		[illegible]	T. Dunham's Nephew & Co.
76423	936		Josiah R. Smith Case W 3m Sr	American	662	166·8	35·3	15·0	Bath, Me.	1883	Deering & Donnell		Deering & Donnell
[illegible]	937		Josiah Whitehouse Hart W Sr	American	263	123·6	29·0	10·3	W[illegible] Me.	1864			G. W. Rawley
[illegible]	938		Josie R. Burt Burt W 3m Sr	American	722	167·8	35·3	15·2	Bath, Me.	1882	[illegible]		A. E. Burt
[illegible]	939	R V	Josie Troop Cook ·· W Bk	British	1099	187·5	37·3	22·3	[illegible] N. B.	1881	D. Lynch	[illegible]	H. D. Troop
	940	D	Josine Louise (ex Anna & Sophia) Visser 1·87 W Bk	Dutch	656	..	..	..	Amsterdam	1862	P. Haverkamp	[illegible]	J. Meijjes Jerzn
174	941	✻ A D	Josip Luchetich ·88 W Bk	Austrian	596	138·1	32·4	19·9	Fiume	1878	Fr. Schiavon	[illegible]	G. Richich
[illegible]	942	A	Josipa M Medanich ·88 W Bk (ex Mary ex Joseph Broomfield)	Austrian	344	121·0	27·6	17·3	Sunderland	1868	R. Thompson, jr	[illegible]	S. Mariassevich
	943	N	Josva Jacobsen ·87 W Bk	Norweg'n	472	131·8	30·7	16·5	Grimstad	1879	N. Olsen Gjerulde	[illegible]	G. Huseland
	944	V	Josva (ex Minerva) Soiland W Bn	Norweg'n	211	96·5	24·5	11·8	Nantes	1875	J. Sevestre	[illegible]	J. Soiland
	945		Joutsen Hella W 3m Sr	Russian	245	113·8	24·9	12·0	Pyhamaa	1884			J. Hella
	946		Joutsen Kaukiaru 89·89 W Sr (ex Claus Wilhelmine, ex Johanna)	Russian	136	82·5	24·3	10·2	Emden	1885	C. Freericks		A. Santaholma
	947	V	Joutsen (ex Tang[illegible]) Enstrom W Sr	Russian	198	88·6	20·4	11·5	Brixham	1865	Fermoux		V. Mannelin
[illegible]	948	H	Jov [illegible] W Bn	Greek	172	..	..	..	Syra	1868			S. E. Eustathiou
	949		Joven W Bn	Spanish	129	79·7	21·6	9·8	[illegible]	1859			Pedro Marcelo
	950		Joven Adela Isern W S	Spanish	110	78·7	23·8	10·[illegible]	[illegible]	1862			José Isern

✻ = Lloyd's Register; D = Nederlandsche Vereeniging; G = Germanischer Lloyd; I = Registro Italiano;
A = Veritas Austro-Ungarico; H = Veritas Ellenico; N = Norske Veritas;

In the 4th Column, C = Composite (iron frame, wood planked); I = Iron; Stl = Steel; W = Wood.

The Dates of Metalling (Col. 11) are based on the records of the Books in which Vessels are respectively classed, or on the Reports of the Surve[illegible]

Further particulars of Vessels classed by Lloyd's Register will be found in Lloyd's Register Book and Lloyd's Yacht R[illegible]

No.	Registre	Nom. Matériaux et Gréement. Capitaines et dates d'entrée en service des Armateurs actuels et à bord du navire. Ancien Nom.	Pavillon	Tonnage net	Longueur	Largeur	Creux	Construit Où	Quand	Par qui	Date du doublage	Armateurs ou Directeurs-Gérants	Port d'Armement
951	V	Joven Alberto WIm Sr	Portug'se	384	114·8	25·3	17·7	Chiavari	1873			L. Alves Rodrigues & Co.	Oporto
952		Joven Alfredo W Sr	Brazilian	180	..	..	..					A.A. Fernandes	Bahia
953	V	Joven Ana W Bg	Spanish	314	110·2	28·8	14·5	Blanes	1869			M. Domenech Furio	Barcelona
954		Joven Antonio W Bn	Spanish	237	98·4	27·7	15·4	Blanes	1863			R.M. Flechas	Majorca
955		Joven Beatriz W Sr	Spanish	196	102·3	23·0	13·1	Palma	1857			I. Fuster y Forteza	Majorca
956		Joven Correia W Sr	Brazilian	160	..	..	..	Villa do Conde	1873			Maria G. ...	Rio Grande
957		Joven Dolores W Bn	Spanish	239	108·2	25·8	13·8	Bilbao	1859			P.N. Echevarria	San Sebastian
958		Joven Elvira W Sr	Brazilian	145	..	..	..	Villa do Conde	1873			C. Dourado & Co.	Bahia
959		Joven Elvira W Sr	Brazilian	145	..	..	..					S. da Costa	Pernambuco
960		Joven Gabriel W Bn	Spanish	202	97·8	31·8	12·9	Vinaros	1870	Viñas		V. Viñas & Co.	Valencia
961		Joven Guillermo W Bg	Spanish	174	81·7	22·4	10·9	Torrevieja	1854			J. Talavera	Valencia
962		Joven Ida W Bg	Spanish	164	84·0	21·9	10·9	Mahon	1835			Vicente Puig	Barcelona
963		Joven Julian W Bn	Spanish	1231	86·3	22·7	10·2	Orio	1888	A. Mutiozabal		E. Gurruchaga	San Sebastian
964		Joven Laura W Sr	Portug'se	111	75·5	20·3	9·2	Villa Nova	1871	J.F. Lapa		B. Mortenegro	Oporto
965		Joven Mariana W Bn	British	99	72·0	21·1	10·7	Villajoyosa	1845	M. de Costante		L.H. Fara	Gibraltar
966		Joven Paulistana W Sr	Brazilian	110	..	..	..	S. João da Barra	1858			A.L.F. de Moura	Rio de Janeiro
967		Joven Pepita W Bn	Spanish	120	95·1	23·9	12·2	Arens	1852	S. Busquet		José Ferrer	Villagarcia
968	V	Joven Pura W Bn	Spanish	131	88·9	23·0	11·0	Villajoyosa	1877	J. Martinez		L. Lopez	Carthagena
969		Joven Rosalia Bn	Spanish	162	83·7	26·0	12·4	Masnou	1854			M. Garcia Moll	Barcelona
970		Joven Temerario W Bn	Spanish	104	83·3	21·3	10·2	La Palma	1843			P.J. Fuster	Majorca
971		Joven Umberto W Bn	Argentine	105	79·1	19·7	10·0	Finalmarina	1883	F. Ponziolo		G.V. Gagliolo	Buenos Ayres
972	V	Jowina *Rogge* 74-67 W Sr	German	133	84·7	21·3	11·1	Stralsund	1867	O. Kirchhoff		Carl Lobeck	Stralsund
973		Joya del Llobregat W Bn	Spanish	219	114·8	25·6	14·2	Genoa	1872	F. Rolla		Miguel Company	Barcelona
974	A	Joyeuse	French	181	84·3	23·6	..	Lussinpiccolo	1878			Daguerre & Co.	Bayonne
975	A	Jozic *Pasely* W Bk	Austrian	385	125·9	25·7	16·6	Nantes	1865	P. Jellet et Babin		Mattea Jelencic	Buccari
976		Jozic W Bk	French	427	116·4	31·2	20·6	Curzola	1857	Forelich			Marseilles
977		Juan *Core* 84-84 W Sr	British	131	99·8	21·2	10·8	Salcombe	1848	Bonker		Mrs. S.J. Hill	Salcombe
978		Juan *Wicks* W Bn (TO BE BROKEN UP)	British	122	86·0	21·8	11·2	P.E. Island	1866	McLure		J.J. Blight	Guernsey
979		Juan W Bn	Spanish	113	80·4	21·8	9·8	Arens	1857			M. Pous y Pons	Barcelona
980		Juan J. Murga *Fleming* W Bk	Spanish	296	116·4	24·7	14·3	Alloa	1869	Adamson		Juan J. Murga	Seville
981		Juan José de Latorre W Bk	Chilian	404	122·0	27·5	17·5	Sunderland	1861	Davison		R. Barazarte	Valparaiso
982	V	Juana *Lauren* W Bk	Spanish	316	118·4	28·9	15·5	Risör	1874			M.A. de Maruri	Havana
983		Juanita *Thomas* W Bg	British	238	103·5	25·0	14·0	Hartlepool	1860	Denton		E. Jopling	Hartlepool
984		Juanita W Bk	Spanish	425	129·9	30·8	13·8	Blanes	1872			Mir. Marti & Co.	Barcelona
985		Juanita W Pol	Spanish	179	87·9	25·8	12·7	Masnou	1849			José Pagés	Barcelona
986		Juanita W Sr	Spanish	148	75·5	20·7	10·9	Vinaros	1852			Juan Llorca	Vinaros
987	V	Juanito W Bn	Spanish	247	105·3	27·3	13·6	Arens	1868	Busquet		Doña Dolores & J. Fontanills	Barcelona
988		Juanito W Bn	Spanish	150	90·2	24·9	10·5	Lisbon	1853			Juan Boeta y Real	Canary Islands
989		Juanito W Bg	Spanish	105	75·8	21·9	10·9	Santander	1851			Miguel Baldó	Corunna
990	✱	Juanpore *Evans* W Bk	British	460	144·4	26·7	18·6	Whitehaven	1859	T.&J. Brocklbank		Mrs. M. Davies	London
991	✱	Jubilee *Hawkes* 88-6 I S	British	2239	284·0	40·1	24·1	Southampton	1887	Oswald, Mordaunt & Co.		R.W. Leyland & Co.	Liverpool
992		Jubilee *Evans* W S	British	741	192·4	31·0	21·3	Workington	1857	Workington S.B. Co.		Gear Meat Preserving Co.	Wellington, N.Z.
993		Jubilee W Im Sr	British	174	121·0	26·1	9·4	Harwich	1887	J.H. Vaux		J.H. Vaux	Harwich
994		Juellerne *Nielsen* 77-77 W Bn	Danish	113	69·8	21·2	12·1	Troense	1849	N.D. Nielsen		P. Jensen	Svendborg
995	V	Jugor *Jensen* W Bn	Danish	153	90·9	21·2	11·2	Nordby, Fanø	1868	T. Schoman		Akt. Selsk. (J. Svendsen)	Sonderho, Fanø
996		Juhachi Maru W Sr	Japanese	354	117·6	27·6	14·9	Osaka	1877	Nagata Sanjuro		Katano Kisuke	Hiogo
997		Juhachi Maru No. 2 W Sr	Japanese	374	123·8	28·3	15·4	Tokio	1879	Hirano S.B. Yard		Matsuzawa Magohachi	Tokio
998		Juhre Anna W Sr	Russian	110	90·8	20·5	9·8	Dondangen	1875	J. Kiel		Smith & Witt	Liban
999		Jules W Bg	French	129	75·8	19·7	11·6	Charente	1858	Allain		Prado frères	Havre
1000		Jules Bertrand *Ollivier* W Bn (CONDEMNED)	French	153	93·2	21·3	11·6	Rotterdam	1857			A. Postel & ses fils	Cherbourg

✱ = Lloyd's Register; D = Nederlandsche Vereeniging; G = Germanischer Lloyd; I = Registro Italiano; R = American Record;
A = Veritas Austro-Ungarico; H = Veritas Ellenico; N = Norske Veritas; V = Bureau Veritas.

Dans la 4me Colonne, C = Composite (membrure en fer, bordé en bois); I = Fer; Stl = Acier; W = Bois.

Les Dates du doublage en métal (Col. 11) ressortent des livres dans lesquels les navires sont respectivement classés ou des rapports des Inspecteurs du Lloyd's Register.

Pour plus amples renseignements sur les navires classés par le Lloyd's Register, voir le Lloyd's Register Book et le Lloyd's Yacht Register.

1	2	3	4 Ship's Name, Material and Rig. Master, and Dates of Appointment to present Owner's Service and to Ship. Ship's Late Name.		5 Flag.	6 Net Tonnage.	7 Dimensions. Length.	Breadth.	Depth.	8 Built. Where.	9 When.	10 By whom.	11 Date of Metalling.	12 Owners or Managers
	1001	V	Jules Estelle *Ledagne -83*	W Bg	[illegible]	122	76·1	20·0	11·9	Cherbourg	1856			Godet
70251	1002		Jules Marie	W Bk	British	257	107·6	26·6	13·3	Chantenay	1868			Henry Moss
	1003		Jules Marie *Richard 71-81*	W Sr	French	100	70·0	21·0	11·0	St. Nazaire	1867	Maire		Paul Richard
	1004	V	Jules Noémie *Guinot*	W Lr	French	117	72·8	20·3	10·7	Nantes	1861	Bojn&Clairgeau		Guinot
	1005	G	Jules & Ernest *Giequel* (ex Mamans)	W Bo	French	144	..	..	..	Sunderland	1837			Giequel
	1006	DNV	Juletrœ *Eriksen*	W S	Norweg'n	730	156·0	34·1	20·1	Grimstad	1877	F.Smith Petersen		F.Smith Petersen
[illegible]	1007	RV	Julia *Bremner -81*	W S	British	1076	177·3	36·7	23·7	Newport, N.S.	1868			A.Gibson
[illegible]	1008		Julia (ex Grand Turk)	W Bk	British	265	107·3	23·7	11·0	BlueHill, U.S.	1852			W.Morrice
42833	1009		Julia	W Bk	British	186	103·8	25·6	9·7	Baltimore. US.	1857			E.Key
[illegible]	1010	✻	Julia *Hocking 76-79*	W3mSr	British	169	106·5	25·5	12·4	Padstow	1879	Stribley		J.P.Dunstan
58531	1011		Julia *Worth 82-88*	W Sr	British	148	101·5	22·1	11·4	GlassonDock	1865	Simpson		J.Fisher & Sons
36296	1012		Julia	W Sr	British	113	75·0	22·9	10·4	OldBarns, N.S.	1860			R.Smith
72550	1013		Julia	W Sr	British	108	89·0	19·8	7·9	SmithFalls, Ont.	1875	W.O'Mara		T.Donnelly
51831	1014		Julia (ex Jas. C. Russell)	W Sr	British	102	81·0	24·0	7·5	Rockport, U.S.	1854			D.McDonald
58380	1015	✻	Julia *Avery*	W Sr	British	90	85·8	19·9	10·5	Par	1870	Moss		S.Truscott
75002	1016	RV	Julia *Smith -81*	W Bk	American	759	155·0	34·0	19·5	Ellsworth, Me.	1877	I.M.Grant		J.M.Higgins
12938	1017		Julia	W Sr	American	102	79·8	23·8	8·4	Harringt'n, Me	1864			A.B.Strout
	1018		Julia	W Sr	Brazilian	157	..	..	..					J.F.Capella
	1019		Julia (ex Anita ex Richard Kelly)	W S	Chilian	1159	179·1	36·7	24·4	Boston, Mass.	1858			F.Peede
	1020		Julia *Robine -81*	W Bu	French	113	78·7	23·0	10·8	Granville	1875	Lecharpentier		J.M.Panier
	1021		Julia *Blomqvist*	W Sr	Russian	132	83·5	25·0	11·7	Mariehamn	1873	M.Sittkoff		N.Sittkoff
	1022		Julia	W Bn	Spanish	151	89·6	22·5	9·5	Tayabas	1882			Rafael Calvo
	1023		Julia	W Bn	Spanish	116	85·6	24·5	10·4	Palma	1855			M.R.Figueroda
	1024		Julia *Isidor 81-87*	W Bg	Swedish	204	91·5	23·2	13·4	Wisby	1839			M.Hansson
	1025		Julia *Gustafson*	W Sr	Swedish	107	73·2	22·8	9·8	Vätö	1881			C.P.Mattsson
15025	1026		Julia A. Berkele	W Sr	American	160	105·5	25·0	7·9	E.Haven, Conn.	1866	G.W.Baldwin		F.W.Benedict
64688	1027		Julia A. Merritt *Cook -88*	W Bn	British	178	94·2	27·4	10·0	MossGlen, NB.	1872	Wesmore& Merritt		W.Lander
75636	1028	R	Julia A. Ward *Rich -88*	W3mSr	American	561	146·2	33·0	15·1	KennebunkMe	1874			Joseph Stevens
	1029	DG	Julia Alida *Martinson -82*	W3mS	Russian	260	114·1	27·7	13·8	Guttmannsb'ch	1873	G.Martenson		G.Martenson
	1030	V	Julia Alida *Kiel*	W Glt	Russian	179	83·0	23·9	11·7	Karal	1883			Redick Thom
59116	1031	R	Julia Blake *Patterson*	W Bn	British	200	97·6	25·3	10·4	Parrsboro, N.S.	1869			Sarah Grimes
157	1032		Julia Catharina *Trautmann*	W Sr	Russian	206	106·0	26·7	12·2	Livland	1872	G.Gross		Trautmann&Gross
	1033	✻	Julia F. *Gallitee*	I Sr	Argentine	67	79·1	21·3	7·0	Newcastle	1884	SirW.G.Armstrong&Co.Ld.		Borra, Baile[illegible]
76503	1034		Julia Foard *Bergman 84-86* (WRECKED)	W Bk	American	446	136·7	29·4	17·2	Bordeaux	18[illegible]			A.P.Lorentzen
76401	1035		Julia Fowler *Parker -88*	W3mSr	American	213	114·0	27·0	11·0	Essex, Mass.	1884	J.James&Co.		W.H.Ryder
[illegible]	1036	R	Julia H. *Master -87* (MISSING)	W Bk	British	579	146·5	31·2	17·1	Burton, N.B.	1874	J.S.Covert		J.R.Haws & Co.
	1037		Julia Katharina *Tobias 79-*	W Sr	Russian	142	75·0	21·8	10·4	Takkerort	1879	Eisenschmidt		J.Petersohn
63625	1038		Julia M. Avery	W Bg	British	127	95·3	23·5	9·6	Baltimore, U.S.				R.H.D.White
251	1039		Julia Maria *Muzneek 83-88*	W Sr	Russian	232	96·4	26·7	12·2	Wandsen	1877	K.Oschmann		M.Muzneek
76550	1040	V	Julia Rollins *Johnsen -87*	W Bk	American	586	146·4	34·9	16·9	Baltimore	1884	W.Skinner&Son		T.Rollins
76235	1041	R	Julia S. Bailey *Otis -86*	W3mSr	American	306	124·6	30·3	12·2	Bath, Me.	1881	Goss, Sawyer& Packard		A.Bailey
	1042		Julia II. (ex Carrie D. Allen)	W3mSr	Portug'se	151	98·0	25·0	9·0	Essex, Mass.	1874			A.Marianna do [illegible]
13317	1043		Julia & Martha *Hopps*	W Sr	American	112	84·8	25·0	8·0	BoothBay, Me.	18[illegible]			H.B.Eaton
75133	1044		Julian Nelson *Kelley -81*	W Sr	American	200	112·0	29·7	8·9	Tuckerton, NJ.	1867			E.Gibbs
	1045		Juliana *Österman*	W3mSr	Swedish	129	87·6	20·9	9·2	Svartvik	1857			C.Ekman
	1046		Juliana Gonzalez *Martinez*	W Bk	Spanish	352	118·4	29·3	15·4	Oviedo	1867			J.M.Martinez
	1047		Juliane *Krüge 75-75* ([illegible])	W Bu	Norweg'n	168	92·6	23·7	11·6	Sandefjord	1875			Ole Krüge
	1048		Julianehaab *Schmidt* (ex Gustaf)	W Bk	Danish	224	96·5	23·6	14·2	Cleaborg	1838	C.Lund		J.J.Schmidt
92	1049		Julianna *Johnson -85*	W Sr	Russian	161	97·7	26·4	12·1	Koltsen	18[illegible]	M.Breede		M.Breede
89487	1050		Julie	W Sr	British	100	86·8	20·6	9·8	Neustadt	1875	J.H.H.Bernitt		W.Coward

✻ = Lloyd's Register; D = Nederlandsche Vereeniging; G = Germanischer Lloyd; I = Registro Italiano; R = [illegible]
A = Veritas Austro-Ungarico; H = Veritas Ellenico; N = Norske Veritas; V = [illegible]

In the 4th Column, C = Composite (iron frame, wood planked); I = Iron; Stl = Steel; W = Wood.

The Dates of Metalling (Col. 11) are based on the records of the Books in which Vessels are respectively classed, or on the Reports of the Surveyors [illegible]

Further particulars of Vessels classed by Lloyd's Register will be found in Lloyd's Register Book and Lloyd's Yacht Register.

No.	Registre	Nom. Capitaine (et année d'entrée). Ancien Nom.	Matériaux et Gréement	Pavillon	Tonnage net	Longueur	Largeur	Creux	Où	Quand	Par qui	Date du doublage	Armateurs ou Directeurs-Gérants	Port d'Armement
1051		Julie *Schou*	W Bg	Danish	180	85·8	21·8	12·8	Demmin	1842	Wacke		Akt.Selsk. (J.P.Folkmann)	Rønne
1052	G	Julie *Petersen 83-86* (ex Margaret)	W Sr	Danish	110	71·6	21·5	10·1	Carentan	18^{44}_{74}			I.S.Wiberg	Nibe
1053	V	Julie *Fleury*	I Bk	French	504	161·7	26·6	16·0	Sunderland	1870	Iliff, Mounsey & Co.		F.Foerster	Havre
1054		Julie *Besnard 86-84*	W Sr	French	119	89·2	22·7	10·8	Dunkirk	1877	G.Beck		G.Beck	Dunkirk
1055		Julie *Müller*	W Bk	German	390	110·3	29·7	17·7	Alblasserdam	1853	C.Smit		J.H.Kraeft	Wolgast
1056	DV	Julie (ex Cornelia) *Steen*	W S	Norweg'n	1284	192·5	38·3	24·6	Alblasserdam	1867	J.Smit, Czn.	V 11.84	Actieselskabet "Julie" (L.Meldahl)	Laurvig
1057	NV	Julie *Jörgensen 85-85* (ex Georgia Campbell)	W Bk	Norweg'n	845	167·0	34·6	20·3	Weymouth, N.S.	1876	C.Campbell, jr.	V 11.86	F.Smith Petersen	Grimstad
1058		Julie	W Bg	Norweg'n	371	105·1	28·3	17·7	Trieste	1857			L.O.Röed	Christiania
1059	N	Julie *Jensen*	W Bg	Norweg'n	257	114·8	28·3	12·1	Arendal	1875	E.Gottormsen	V 9.83	H.H.Pettersen	Arendal
1060	G	Julie *Freymann*	W Sr	Russian	107	77·0	25·1	9·5	Wandsen	1871	M.Motzneck		Lembert & Freymann	Baltisch
1061	G	Julie & Auguste *Zeplien 81-87*	W Bg	Danish	198	95·8	25·3	12·8	Wolgast	18^{51}_{72}	W.Lübke			Rønne
1062	R	Julien (ex White Swan)	W Sr	Dominic'n	142	93·0	23·8	8·5	Calais, Me.	1865			New York & St.Domingo Lumber & Mining Co.	St.Domingo
1063		Julien *Bernard*	W Bk	French	293	114·8	23·8	16·1	Whitby	1864			J.Demarle	Boulogne
1064	V	Julien Gabrielle	W Bn	French	123	84·3	23·3	9·6	Granville	1866	Clément	Z 1.88	H.Lecharpentier	Granville
1065		Julien Marie *Lefranc -88*	W Sr	French	137	73·2	22·5	10·3	Nantes	1860	Chauvelon		Lefranc	Vannes
1066		Julieta (ex La rata) *Secalini*	W Bn	Brazilian	178	..	..	..	Germany			Y 13.88	J.M.Chaves	Rio Grande
1067	N	Juliette *Svendsen* (ex Jeanne d'Arc)	W Sr	Norweg'n	115	82·4	21·5	11·0	Boulogne	1861	Lecerf		H.Tambs	Fredrikshald
1068		Julio	W Bn	Brazilian	178	..	..	..					Cia. E. F. Bahia & Minas	Rio Janeiro
1069		Julio	W Bn	Spanish	132	86·9	22·1	8·7	Tayabas	1883	C.Borromeo		N.R.Bautista	Manila
1070	V	Julio Teodoro *Schultz -83*	I Bk	German	1223	225·0	35·8	21·1	Flensburg	1885	Flensburg Schiffs.Ges.		Bahr Gebrüder	Hamburg
1071		Julita (ex C.H.Knowles)	W Sr	Spanish	172	95·8	26·7	9·0	Addison, Me.	1868	C.H.Knowles		J.B.Frexas	Barcelona
1072		Julito	W Bg	Spanish	184	86·6	25·2	13·8	Blanes	18^{48}_{67}			J.B.Freixas	Barcelona
1073	V	Julius *Jacobsen 79-75* (ex ... Holland)	I S	German	968	193·5	34·8	19·9	Kinderdyk	1864	F.Smit		H.Bischoff & Co.	Bremen
1074		Julius *Wallis 83-82*	W Bg	German	235	95·1	26·5	14·4	Barth	1849	J.Bohl		J.J.Wallis & Sohn	Barth
1075		Julius (ex Fanny)	W Sr	Norweg'n	162	103·5	22·1	11·0	P.E. Island	1859	Creswell		H.A.Haraldsen	Tönsberg
1076	V	Julius (ex Sønnøveren) *Vieira 1863*	W Bk	Portug'se	568	138·1	28·2	18·4	Risör	1875	O.Jürgensen	V 5.86	J.H.Andresen	Oporto
1077		Julius *Olofsson*	W Sr	Russian	155	87·9	27·0	11·1	Sideby	1875	J.F.Johansson		K.H.Olofsson	Wårdö
1078		Julius *Pettersson -76*	W Sr	Swedish	176	91·5	24·4	12·1	Roslagen	1872			M.Lundqvist	Östhammar
1079	G	Julius *Idman*	W Sr	Swedish	141	88·9	21·3	10·4	Norrtelje	1876			F.Lundman	Norrtelje
1080		Julius *Jönsson -86*	W S	Swedish	110	80·0	19·3	9·5	Elsfleth	1851	J.Ahlers		S.Andersson	Höganäs
1081		Julius Jäde *Sjöberg -88*	W Bk	Swedish	310	111·7	25·2	14·9	Wismar	1863	H.D.Barmann		A.F.Sjöberg	Norrtelje
1082	V	Julius Skrike *Rønne*	W Bk	Danish	283	121·6	24·7	13·5	Elsinore	1870	C.Roman & Barfoed	V 5.88	J.Skrike	Copenhagen
1083		July Fourth	W Sr	American	114	89·9	25·2	8·0	Brewer, Me.	1863			H.Lord	Bangor, Me.
1084	V	Jumbo *Welch 88-88* (ex Stranger)	W Bn	British	346	127·6	24·4	13·0	Moss Glen, N.B.	1882	G.Merritt	V 7.88	Scammell Bros.	St.John, N.B.
1085		Jumbo	W	British	146	92·0	24·5	7·8	Quebec	1883	T.N.Tweddell		W.M.Macpherson	Quebec
1086	*	Jumna *Robinson 87-88*	I S	British	1048	208·6	34·1	20·9	Sunderland	1867	W.Pile & Co.		J.Nourse	London
1087	DV	Junak *Clerarton -87*	W Bk	British	500	136·0	32·0	17·7	Trieste	1869	Stabilimento Tecnico		T.H.Jenkins	Liverpool
1088		June Bright *Bartlett*	W 3m Sr	American	329	141·0	31·0	10·0	Bath, Me.	1882	W.Rogers	Y 7.87	S.J.Watts	Boston, Mass.
1089	GH	Jungfrau *Glycas -86*	C Bk	Greek	578	158·3	30·0	18·6	Sunderland	1867	W.Doxford	Y 10.82	L.Glycas	Syra
1090	N	Juni *Nielsen*	W Sr	Norweg'n	194	91·6	25·3	12·8	Grimstad	1862	A.Osmundsen		G.Huseland	Grimstad
1091	N	Junior *Olsen*	W Bk	Norweg'n	367	116·3	27·6	16·3	Haddam, Conn.	1831			Actieselskabet "Junior" (H.M.…)	Drammen
1092	V	Juno *Hansen*	W Sr	Danish	126	80·6	21·1	10·8	Svendborg	18^{46}_{72}	H.P.Hansen		M.Hansen	Svendborg
1093	*	Juno *Schwarting* (ex …)	I S	German	1325	240·0	36·0	22·6	Liverpool	1861	T.Vernon & Son		D.Schilling	Bremen
1094	V	Juno *Müller 87-87*	W Bk	German	517	139·1	29·8	18·0	Hamburg	1874	M.G.Amsinck	V 8.88	M.G.Amsinck	Hamburg
1095	V	Juno *Lindt 81-67*	W Bk	German	409	115·2	28·2	17·7	Pillau	1864	H.O.Becker		J.G.Gerlach	Memel
1096		Juno	W Bk	[illegible]	266	122·6	25·6	13·8	Sonderburg	1865	H.Jensen		Kugai Tajemon	Kobé
1097	N	Juno *Svendsen -83*	W Bk	Norweg'n	995	174·2	36·1	23·2	Newport, N.S.	1865	J.A.Harvey	V 8.88	Louis Vauvert	Porsgrund
1098	N	Juno *Halvorsen -62*	W Bk	Norweg'n	305	106·4	28·5	15·9	Tvedestrand	1862	L.Holm		A.Biering Olsen	Tvedestrand
1099	N	Juno *Svendsen 18-81*	W Bk	Norweg'n	287	105·7	27·0	15·6	Sunderland	1862	Gibbon		C.Svendsen	Christiania
1100		Juno *Gelberg*	W Sr	Swedish	196	95·1	27·7	12·6					O.Jonsson	Calmar

* = Lloyd's Register; D = Nederlandsche Vereeniging; G = Germanischer Lloyd; I = Registro Italiano; R = American Record; A = Veritas Austro-Hongrois; H = Veritas Ellenico; N = Norske Veritas; V = Bureau Veritas.

Dans la 4me Colonne, C = Composite (membrure en fer, bordé en bois); I = Fer; Stl = Acier; W = Bois.

Les Dates du doublage en métal (**Col. 11**) ressortent des livres dans lesquels les navires sont respectivement classés ou des rapports des Inspecteurs du Lloyd's Register.

Pour plus amples renseignements sur les navires classés par le Lloyd's Register, voir le Lloyd's Register Book et le Lloyd's Yacht Register.

Official No.	No. in Book	Registry in which classed	Ship's Name. Material and Rig. Master, and Dates of Appointment to present Owner's Service and to Ship. Ship's Late Name.	Flag.	Net Tonnage	Dimensions. Length.	Breadth.	Depth.	Built. Where.	When	By whom.	Date of Metalling	Owners or Managers.
	1101	V	Junon *Fee* -83 W Bg	French	171	90·5	22·3	11·1	Granville	1868	Clement		Riottean & fils
62373	1102	✱	Jupiter *Jones* 86-86 I Bk	British	720	186·5	30·7	19·2	Sunderland	1870	Iliff, Mounsey & Co.		J. Steel & Son
	1103		Jupiter (ABANDONED) *Jones* 87-87 W Bk	British	686	151·8	31·7	20·4	Alblasserdam	1865	Kloos	[illegible]	G. Milne & Co.
	1104	V	Jupiter *Le Rodet* W Bg	French	138	78·1	22·6	13·0	Meaus	1864	Ollivaud		Pitre Legal
	1105		Jupiter *Loczau* 86-73 W Bk	German	664	131·1	33·6	22·6	Danzig	1872	J. W. Klawitter		Schoenberg & Domansky
	1106	V	Jupiter *Zigahn* 81-81 W Bk	German	442	143·7	29·3	17·1	Elbing	1870	G. Fechter	[illegible]	J. H. Meyer
	1107		Jupiter *Klenow* 85-85 W Bg	German	244	102·8	27·2	14·0	Rostock	1858	W. Zeltz		H. Mentz
	1108	N	Jupiter *Nielsen* -87 W Bk	Norweg'n	592	140·0	33·5	19·6	Arendal	1858		[illegible]	M. B. Nielsen
	1109	N	Jupiter *Tellefsen* 87-87 W Bk	Norweg'n	313	119·1	27·3	14·0	Grimstad	1887	N. Olsen Gjömle	[illegible]	N. Thue Johnsen
456	1110	V	Jupiter *Kodritky* -87 W Bk	Russian	677	145·0	31·5	18·8	Vegesack	1864	J. Lange	[illegible]	Navigation Co. [illegible]
	1111		Jupiter *Snellman* -83 W Bk	Russian	618	157·8	29·2	17·5	Wasa	18 63/81	A. Lönnbäck		A. Carlstrom
	1112	G	Jupiter *Grant* 81-87 3m Sr	Russian	317	129·9	29·0	13·1	Orrenhof	1887	Las Busch		J. Markson
	1113	V	Jupiter *Bjarke* -87 W Bk	Swedish	727	157·9	31·3	18·4	Gefle	1880	O. A. Brodin	[illegible]	Rederiaktie Bolaget "[illegible]" (O. d. Wallenberg)
	1114		Jupiter *Nilsson* W Bk	Swedish	692	132·5	30·7	22·1	Danzig	1863	Link		H. Pettersson
	1115	D V	Jupiter *Natterqvist* -86 W Bk	Swedish	446	138·1	27·0	16·1	Hernösand	1874	J. Strandberg	[illegible]	L. Nisholm
	1116	G	Jupiter *Hall* 75-86 W Bg	Swedish	295	112·2	24·6	13·9	Hamburg	1855	J. v. Somm		A. Bergman
76081	1117	✱	Jura *Russell* 83-85 I S	British	1199	230·3	37·1	20·8	Glasgow	1875	C. Connell & Co.		Sandbach, Tinne & Co.
	1118	V	Jurgen *Beckmann* 76-78 W Bk	German	296	123·0	27·9	13·8	Kiel	1867	A. Cornelius & Son	[illegible]	Wwe Marg. Beckmann
	1119	N	Jury *Adsem* W Bk	Norweg'n	357	127·7	28·6	15·1	Stavanger	1860	R. O. Haerem	[illegible]	C. H. Gabrielsen
	1120		Juserin W Sr	Dutch	204	..	..	..					
8934	1121	N	Just in Time *Hansen* W Sr	British	102	91·4	20·7	7·2	Stavanger	1884	Haerem	[illegible]	E. Davies
75891	1122	R V	Justine H. Ingersoll *Peterson* W Bk	American	578	142·6	31·5	17·0	Columbia Falls (Me	1876	J. H. Crandon	[illegible]	Swan & Son
	1123		Justizrath von Paepke *Carlsson* -87 W Bk	Swedish	322	108·2	25·7	16·1	Rostock	1858	O. Ludwig		C. Carlsson
	1124		Juvenal (ex [illegible]) *Netto* W 3m Sr	Brazilian	204	114·8	22·6	12·5	Hoogezand	1871	J. A. Boites		A. d'Oliveira Maia

ADDITIONAL SHIPS ARE INSERTED IN THE SUPPLEMENT AT THE END OF THE REGISTER.

✱ = Lloyd's Register; D = Nederlandsche Vereeniging; G = Germanischer Lloyd; I = Registro Italiano; A = Veritas Austro-Ungarico; H = Veritas Ellenico; N = Norske Veritas;

In the 4th Column, C = Composite (iron frame, wood planked); I = Iron; Stl = Steel; W = Wood.

The Dates of Metalling (Col. 11) are based on the records of the Books in which Vessels are respectively classed, or on the Reports of the Surveyors[illegible]

Further particulars of Vessels classed by Lloyd's Register will be found in Lloyd's Register Book and Lloyd's Yacht Register.

No.	Classé	Nom	Capitaine	Matériaux et Gréement	Pavillon	Tonnage net	Longueur	Largeur	Creux	Construit Où	Quand	Par qui	Doublage	Armateurs ou Directeurs Gérants	Port d'Armement
1		Kache		StlLg	British	480	220·0	30·0	8·4	Dumbarton	1887	W. Denny & Brs.		Irrawaddy Flotilla Co. Ld.	Glasgow
2		Kachyen		StlLg	British	480	220·0	30·0	8·1	Dumbarton	1886	W. Denny & Brs.		Irrawaddy Flotilla Co. Ld.	Glasgow
3		Kadai Maru		WSr	Japanese	187	103·7	19·9	13·0	Hiogo	1879	Kita Hanbei		Hayashi Seiichi	Hiogo
4	✱H	Kaffir Chief	Thomas	WBkn	British	336	127·0	25·9	14·8	Kingston	1878	Kinloch	V 3.88	W. Kinloch	Banff
5		Kafuku Maru		WBg	Japanese	211	97·3	25·2	14·	Osaka	1881	Nishioka [Tyemon		Masuya Heisaburo	Nagato
6		Kahlamba	Doherty	Bk	British	309	126·6	24·8	14·1	[illegible]	1856	I. Reid & Co.		J. O'Conner	Greenock
7		Kahnee		Barge	British	353	196·3	24·	8·	[illegible]	1882	R. Duncan & Co.		Irrawaddy Flotilla Co. Ld.	Glasgow
8		Kaifoo Maru		WSr	Japanese	250	108·5	21·3	13·7	Hiogo	1881	Kawasaki [Shojiro		Kamigawa Yasaburo	Hiogo
9		Kaisei Maru		WBg	Japanese	490	141·8	27·7	17·1	Osaka	1882	Nagata Sanjiro		Chiashi Riyemon	Hiogo
10		Kaisei Maru		WSr	Japanese	344	120·6	25·5	15·9	Nagasaki	1882	Kosaku Bun [Kioku		Tsukuda Motokiyo	Tokio
11		Kaisen Maru		WSr	Japanese	115	81·0	19·0	11·0	Osaka	1882			Kondo Kiroku	Hiogo
12	G	Kaiser	[illegible]	WS	German	1240	192·8	35·7	24·0	Bremerhaven	1877	H.F. Ulrichs	V	D.H. Wätjen & Co.	Bremen
13	✱	Kaisow	Davies	CBk	British	795	193·2	33·0	20·3	Greenock	1863	R. Steele	V	W. Bowen, jun.	London
14		Kaitais	Eriksson	WBk	Russian	400	..	..	..	Tetsala	1866			F.E. Lindberg	Gustafs Kapell
15	N	Kaja	Amundsen 85	WBk	Norweg'n	414	122·2	28·1	12·5	Bordeaux	1847			C.C. Wetlesen	Christiania
16		Kalagouk		StlLg	British	546	250·3	30·1	8·3	Dumbarton	1888	W. Denny & Brs.		Irrawaddy Flotilla Co. Ld.	Glasgow
17	V	Kalakaua	Mora	WBk	French	448	132·8	26·2	16·5	St. Malo	1881	F. Gautier	V	Lemoine	St. Malo
18		Kalakaua (ex Mary E. ...)	Anderson		Hawaian	369	..	..	..	Parkersburg	1864			John Merrill	Honolulu
19		Kalander Bux		WBg	British	138	74·6	21·4	12·8	Porto Novo	1856			Mahomed Alli Meat / Ahmed Khan	Penang
20		Kalay		Sr	British	392	199·5	34·1	6·0	Port Glasgow	1883	R. Duncan & Co.		Irrawaddy Flotilla Co. Ld.	Glasgow
21	G	Kale	Witt 86	WBk	German	778	158·4	33·1	20·3	[illegible]	1872	H.F. Ulrichs	V	J.H. Hustede	Elsfleth
22	N	Kaleb	Christophersen	WBgSr	Norweg'n	179	101·5	25·1	12·	Stavanger	1867	M. Magnus	V	S.A. Ashjörnsen	Flekkefjord
23	✱	Kaleda	Edmonds	WBkn	British	202	118·4	22·5	11·8	Troon	1884	Troon Ship B. Co.	V	T. Steele	Ayr
24	DN	Kaleva	Olsen	WBk	Norweg'n	450	134·9	32·0	16·8	Gamla Carleby	1863	J. Bjornvik		I. Olsen	Brevig
25		Kaleva	Dahlberg	WBk	Russian	510	155·6	35·2	18·0	Uleaborg	18 61/79	Berghom	V	H.J. Dahlberg	Uleaborg
26	N	Kali Maas	Hansen	WBgSr	Norweg'n	289	117·1	26·8	15·1	[illegible]	1860			S.A. Hansen	Tönsberg
27		Kali Tychi	Apostolou	WSr	Greek	104	..	..	..	Spezzia			1.80	A. Apostolon	Spezzia
28	A	Kalk	Cerlich 88	WBk	Austrian	444	134·1	27·7	18·0	Fiume	1871	V. Brazzoduro	V 6.88	Flli Cosulich	Fiume
29		Kalla	Balkenhausen	WBk	Swedish	329	105·3	24·0	17·3	England				A. Balkenhausen	Sölvesborg
30	✱	Kalliope	Hilgerloh 88	StlS	German	1622	258·8	38·7	22·7	Glasgow	1888	C. Connell & Co.		B. Wencke Söhne	Hamburg
31	N	Kalliope	Severtsen 87	WBk	Norweg'n	954	175·1	31·2	21·6	St. John, N.B.	1856		V 8.88	L.G.S. Larsen	Sandefjord
32	A	Kalliopi	Mavris	WBk	Greek	528	143·3	31·5	19·8	Sestri	1871	Fr. Cardenaccio	V	G. Mavris	Syra
33	A	Kalliopi	G. Paraskeftiou	WBk	Greek	314	107·2	26·9	..	[illegible]	1868	N. Catramados	Z	M. Pneumaticos	Syra
34	A	Kalliopi	Tsargis	WBn	Greek	262	102·6	26·2	..	Castellamare	1867			Stamatopoulos Bros.	Corfu
35		Kalliopi	Broutskos	WBn	Greek	249	92·5	26·2	16·1	Syra	1867			K.D. Broutskos	Hydra
36		Kalliopi		WBn		239	94·5	27·9	..	Syra	1866				
37	A	Kalliopi	Manialis	WBn	Greek	239	97·1	25·1	..	Casos	1883	A. Emmanuel		G.E. Manialis	Syra
38	A	Kalliopi		WSr		229	93·5	25·6	..	Casos	1884	A. Emmanuel			
39		Kalliopi	Kavidis	WBg	Greek	206	89·2	25·6	..	Syra	1876	S. Cafudaki		N. Cavidas	Syra
40		Kalliopi Ariadni		WBg	Greek	186	91·2	21·7	14·0	Naples	1858			J.C. Gamonis	Spezzia
41		Kalliroi		WBg	Greek	309	..	..	..		1847			G. Ioannidis	Syra
42		Kalliroi	S. Catzourakis	WBn	Greek	282	106·9	26·9	..	Syra	1883	C. Cristali		N. Catzourakis	Syra
43	I	Kalmia	Revell	WBn	British	173	111·0	23·2	11·6	Bideford	1872	Cox	V	P.G. Tessier	St. John's, N.F.
44	AH	Kalomira	Vlassopoulos	WBg	Greek	244	92·8	26·9	..	Galaxidi	1876	J. Canatis		E. Vlassopoulos	Zante
45	DN	Kalos	Petter... 88	WBk	Norweg'n	818	161·7	35·5	20·5	St. Martin's, NB	1858		V	Et Interessentskab (Chr. Müller)	Christiania
46	N	Kalstad	Hansen	WBk	Norweg'n	480	135·6	31·5	17·3	Kragerö	1870	J. Nielsen	V	H. Biörn, jun.	Kragerö
47	R	Kaluna	Mills	WBn	American	330	121·0	30·4	12·7	Harrington Me.	1873			J.S. Winslow & Co.	Portland, Me.
48		Kama		Barge	British	791	194·0	25·6	8·2	Port Glasgow	1875	[illegible]		Irrawaddy Flotilla Co. Ld.	Glasgow
49	V	Kambira	Boswell 88	WS	British	1952	237·2	44·0	26·7	Kingsport, NS	1882	P.B. Crichton	V	C.R. Burgess	Windsor, N.S.
50	✱H	Kamehameha IV	Crilton	WBk	British	495	154·7	26·5	18·2	Workington	1857	Peile, Scott & Co.	V	H. Tyrer & Co.	Greenock

= Lloyd's Register; D = Nederlandsche Vereeniging; G = Germanischer Lloyd; I = Registro Italiano; R = American Record;
= Veritas Austro-Ungarico; H = Veritas Ellenico; N = Norske Veritas; V = Bureau Veritas.

Dans la 4me Colonne, C = Composite (membrure en fer, bordé en bois); I = Fer; Stl = Acier; W = Bois.

Les Dates de doublage en métal (Col. 11) ressortent des livres dans lesquels les navires sont respectivement classés ou des rapports des Inspecteurs du Lloyd's Register.

Pour plus amples renseignements sur les navires classés par le Lloyd's Register, voir le Lloyd's Register Book et le Lloyd's Yacht Register.

Official No.	No. in Book	Society in which classed	Ship's Name. Master, and Dates of Appointment to present Owner's Service and to Ship. Ship's Late Name.	Material and Rig	Flag.	Net Tonnage	Length	Breadth	Depth	Built Where.	When.	By whom.	Date of Metalling	Owners or Managers.
	51	DN	Kamfjord *Andersen 71-84*	W Bk	Norweg'n	752	168·2	[illegible]	19·[illegible]	Sandefjord	1881	SørenOlsen	[illegible]	A.C.Gogstad
	52	V	Kammerherre Schütte *Hansen*	W Bn	Danish	192	106·8	2[illegible]	11·5	Horsens	1877	Chr.Schroder	[illegible]	Actieselskab "Kam[illegible]"
	53		Kamp *(ex Gardar)*	W Sr	Norweg'n	100	85·8	19·4	10·4	Hull	1866	Parkinson		H.J.Ball
	54	N	Kamschatka *Floistadt*	W Bk	Norweg'n	515	139·1	32·9	18·2	Hamburg	1853		[illegible]	H.H.Pettersen
[illegible]	55		Kanagawa Maru	I Bk	Japanese	1150	236·0	32·0	20·0	Glasgow	1882	Todd& McGregor		Nippon Yusen Kaisha
[illegible]	56		Kanko Maru *(ex Thomas Brown)*	W Bk	Japanese	279	141·1	22·9	13·7	Australia	1868			Inouye Zenyemon
[illegible]	57		Kapodistrias	W Bg	Greek	182	90·9	21·7	14·5	Spezzia	1859			L.Croutiras
[illegible]	58	✻	Kappa *Smith 74-[illegible]*	C Bk	British	491	156·3	27·2	17·3	Glasgow	1865	A Stephen& Son	[illegible]	Swansea Merchant Ship Owners' Co., Ld.
	59		Kara Agatch	W Bg	Turkish	158	..	..	..	Kara Agatch	1859			[illegible]
	60		Kara Agatch	W Bn	Turkish	159	..	..	..	Kara Agatch	1872			Osmanli Zadeh Apik Effendi
	61		Kara Kush	W Bg	Turkish	143	..	..	..	Greece	1875			Barlenlee Kaloghlou Hussein
	62		Kara Sou	W Bn	Turkish	125	..	..	..	Barten	1869			[illegible] Zoolar Osman Hussein & Co
	63		Kara Suleiman	W Bg	Turkish	160	..	..	..	Syra	1869			Kiritlee Ah.B Popach
	64		Karajah	W Bn	Turkish	137	..	..	..	Karajah	1869			Bende Ereklealee Sandal Zadeh Ahmed & C
[illegible]	65		Karen	Stl La	British	480	220·0	30·0	8·4	Dumbarton	1886	W.Denny& Bros		Irrawaddy Flotilla Co. Ld.
	66	G	Karen *Thomsen* *(ex [illegible]gaard)*	W Bn	Danish	200	100·9	23·6	11·8	Parrsboro'N.S	1866	WellsCole	[illegible]	Actieselskab[illegible]
	67	G	Karen *Jacobsen* *(ex H[illegible])*	W Sr	Danish	122	89·4	22·0	10·4	Nexø	1858	H.L.Blem		N.Jensen
	68	V	Karen *Jacobsen -86*	W Sr	Danish	118	89·6	20·4	10·3	Faaborg	1878	M.Dyreborg		Hans Christensen
	69		Karen Elise *Andersen* *(ex James Primrose, ex TRANDE[illegible])*	W Bk	Norweg'n	336	132·0	29·5	12·9	River John N.S	1865	J.Kitchin		J.B.Rode's Enke
	70		Karen Kirstine Roeskilde *(ex Vincent)* *Larsen*	W Sr	Danish	125	90·1	20·2	9·9	Sappemeer	18[illegible]			Valdemar Frandsen
	71	V	Karikal *Raubaud -87*	I Bk	French	615	192·4	26·3	16·0	South Shields	1868	Readhead Softley & Co.		A.Viale
	72	✻	Karin *Leufstadius 79-80*	W Bk	Swedish	472	148·0	27·0	16·3	Framnas	1877	E.Ostman	[illegible]	Framnäs Warfs Aktiebolag (C.W.Lilljeqvist)
	73	V	Karin *Westerlund -84*	W Bk	Swedish	345	138·1	25·8	12·2	Westby	1876	O.Froberg		E.Norberg
	74	DV	Karin *Tengwall 87-87*	W Bn	Swedish	265	117·4	24·5	11·7	Ortviken	1876	J.A.Strandberg	[illegible]	Ortvikens Angsaga-aktiebol. (C.A.M.[illegible])
	75		Karine (BROKEN UP)	W Sr	Norweg'n	108	77·5	21·1	10·8	Christiansand	1857			Ole Bentsen
	76	V	Karl *Petrell 87-87* *(ex Stirling, ex Admiral)*	W Bk	Russian	766	155·8	32·1	22·0	Boston U.S.	18[illegible]	E.&H.Briggs	[illegible]	K.Pentzin
	77		Karl *Lundqvist*	W Bk	Russian	359	..	..	..	Kaskö	1874			V.A.Reims
	78		Karl *Röman*	W3mSr	Russian	234	114·7	26·0	11·8	Libau	1853			A.T.Hedberg's[illegible]
	79		Karl *Janson*	W3mSr	Russian	231	..	..	..		1874			G.Lindholm
	80	V	Karl *Andersson -86*	W3mSr	Swedish	325	127·6	26·2	12·6	Piteå	1870	C.Brandtöl	[illegible]	C.A.Hedqvist
	81		Karl *[illegible] -88* *(ex Valborg[illegible])*	W Bk	Swedish	284	97·9	25·4	16·2	Hamburg	1846	Rose		G.Ideström
	82		Karl *Pettersson*	W3mSr	Swedish	245	105·6	24·3	13·6	Våto	1873			A.Andersson
	83		Karl *Carlsson -86*	W Sr	Swedish	233	108·6	26·5	10·6	N.America				G.Sjöberg
	84		Karl *Erikson*	W Bg	Swedish	196	95·1	24·0	11·8	Piteå	1863	G.E.Lundqvist		M.Lindstedt
	85		Karl *Olsson*	W Sr	Swedish	169	96·8	22·7	10·4	Figeholm	1880			A.Olsson
	86		Karl *Johansson*	W Sr	Swedish	128	85·3	22·7	9·4	Figeholm	1868			G.& F.Johansson
	87	G	Karl Albert *Petersson -87*	W S	Swedish	127	78·4	20·6	11·4	Wiken	1856	A.Jacobsson		N.Pettersson
	88	G	Karl Emil *Bengtsson 78-* *(ex [illegible])*	W Bg	Swedish	171	95·8	21·6	12·6	Apenrade	1850			N.P.Nilsson
	89	V	Karl Hindric *[illegible] 73-*	W S	Swedish	1067	164·3	31·5	22·2	Framnas	1872	J.A.Strandberg	[illegible]	G.F.Cawallin
	90		Karl Johan *Hoglin -86*	W Bk	Swedish	426	143·3	28·2	15·6	Hernösand	1860			O.R.O.Norlin
	91		Karl Johan *Andersson -84*	W Sr	Swedish	143	78·7	20·8	12·0	Rostock	1842			A.Pyk
	92		Karl Magnus *Johansson* *(ex [illegible])*	W Sr	Swedish	175	91·2	22·8	11·9	Harburg	1854			C.M.Westberg
	93		Karl Olof *Olsson*	W Sr	Swedish	135	87·6	21·4	10·8	Carlskrona	1866	O.Bergström		A.Olsson
	94		Karl Oscar *Kullén -86*	W Sr	Swedish	145	89·6	21·3	11·2	Wenersborg	1855			A.B.Elfversson
	95	G	Karl Peter *Påhlsson -87*	W Bn	Swedish	178	102·0	21·8	10·9	Värnanäs	1868	G.A.Hahn		H.P.Pettersson
	96	V	Karl v. Döbeln *Jonsson 70-77*	W Bk	Swedish	451	129·2	28·2	15·8	Umeå	1866	O.Jonsson	[illegible]	O.Jonsson
	97		Karl & Anna *Strömberg*	W Sr	Swedish	114	75·5	23·8	10·3	Roslagen	1869			C.P.Jansson
	98	G	Karl XI. *Ewald 72-73*	W Bg	Swedish	236	101·6	23·9	12·6	Oscarshamn	1872	P.[illegible]sson		L.Jonsson
	99		Karl XV. *Westerlund 70-76*	W Bk	Swedish	397	133·5	26·5	15·4	[illegible]	1860			N.J.Nyberg
[illegible]	100	RV	Karnak *Marsell*	W Bk	British	899	171·5	35·7	20·7	St.John N.B.	1878	O.[illegible]	[illegible]	

✻ = Lloyd's Register; D = Nederlandsche Vereeniging; G = Germanischer Lloyd; I = Registro Italiano; B[illegible]
A = Veritas Austro-Ungarico; H = Veritas Ellenico; N = Norske Veritas; V[illegible]

In the 4th Column, C = Composite (iron frame, wood planked); I = Iron; Stl = Steel; W = Wood.

The Dates of Metalling (**Col. 11**) are based on the records of the Books in which Vessels are respectively classed, or on the Reports of the Surveyors [illegible]

Further particulars of Vessels classed by Lloyd's Register will be found in Lloyd's Register Book and Lloyd's Yacht [illegible]

No.	Classe	Nom. Matériaux et Gréement. Capitaine et dates d'entrée au service des Armateurs actuels et à bord du navire. Ancien Nom.	Capitaine	Mat./Grém.	Pavillon	Tonnage net	Longueur	Largeur	Creux	Construit où	Quand	Par qui	Date du doublage en métal	Armateurs ou Directeurs-Gérants.	Port d'Armement.
101		Karolina	Mattson	W Bg	Russian	242	..	..	..					B.Suttkoff	Hammarland
102		Karolina	Sjöstrom	W Sr	Russian	145	..	..	..	Sideby	1864			G.R.Ekström	Jomala
103		Karolina (ex Eliza Mary)	Holmström	W Bk	Swedish	580	168·3	27·1	15·8	Kingston	1855			M.Holmström	Helsingborg
104	R	Karolina (ex [illegible])	Pettersson	W Bk	Swedish	578	159·1	30·9	17·0	Svartvik	1856			J.Bengtsson	Gothenburg
105	V	Karolina	Berlet -88	W Bk	Swedish	367	122·3	26·5	16·2	Umea	1859		Y 1.88	J.F.Berlet	Helsingborg
106		Karolina	Andersson	W Bk	Swedish	274	111·5	24·6	13·6	Finland	1867			A.Göransson	Helsingborg
107		Karolina (ex Caroline Helding)	Persson -88	W Bg	Swedish	241	100·3	24·0	13·4	Rostock	1864	G.W.Tesdorpf		O.Persson	Wik.Christinaet. [illegible]
108		Karolina (ex Anna Maria, ex Adolphine)	Svensson -87	W3mSr	Swedish	236	98·8	25·1	13·8	Odense	1854	H.Land		P.Lindroth	Fjellbacka
109		Karolina (ex Ernst, ex [illegible])	[illegible]	W Bg	Swedish	223	98·8	25·6	12·9	Rostock	1860	H.Rickmann		C.M.Olsson	Mönsterås
110		Karolina (ex [illegible])	Persson -88	W Bn	Swedish	200	95·8	23·2	11·0	NovaScotia	1864			P.Persson	Helsingborg
111		Karolina	Göransson	W Sr	Swedish	134	78·7	21·8	11·3	Aarhus	1847	Tomsen		C.Göransson	Arildsläge
112	G	Karolina	Olsson 82-82	W Sr	Swedish	128	86·0	20·4	10·8	Lerhamn	1857	O.Nilsson		J.Holm	Kivik
113	R	Karolina	Jansson --	W Bn	Swedish	125	83·3	20·7	11·2	Sjötorp	1859	Larsson& Blin		A.Jansson	Teguneby.Bhln.
114		Karolina	Karlsson 84-86	W Sr	Swedish	121	78·4	24·7	11·0	Gräsö	1853			M.Jansson	Veda.Roslagen
115		Karolina Mathilda	Kohlman	W Sr	Swedish	105	82·0	20·3	10·0	Sjötorp	1867	A. Andersson	Y 5.88	J.M.Lysell	Fiskebäckskil
116		Karolos Nazos		W Bn	Greek	243	91·2	26·2	..	Syra	1866	S.Cafudachi		P.N.Passari	Syra
117	V	Karoo	Munroe 86-88	W S	British	2031	240·0	44·3	26·6	Cornwallis.NS	1884	C.R.Burgess		C.R.Burgess	Windsor.N.S.
118		Karoola		W Sr	British	201	119·8	23·8	9·9	Nambucera, N.S.W.	1884	J.Stuart		P.F.Fagan	Sydney,N.S.W.
119	R	Karslie	Delong	W Sr	British	125	87·8	27·4	7·7	St.Martin'sNB	1884	D.DeLong		W.De Long	St.John.N.B.
120	N	Karsten Langaard (ex Lima)	Gundersen	W Bk	Norweg'n	645	151·0	32·4	18·0	RiverJohn,NS	1872	C.McLennan	Y 1.87	M.Langaard & Co.	Christiania
121		Kas Balgair (ex General Michiels)		W Bk	Dutch	400	112·2	29·5	..	Java	1849			Said Achmad b. Sech b. Achmad b. Saleh	Palembang
122		Kashiu Maru		W Sr	Japanese	154	91·4	21·5	10·9	Osaka	1879	TemmaSaburoka		Hiromi Nisaburo	Osaka
123		Kassa	Anderson	W Bk	British	307	125·1	25·1	12·6	Jersey	1868	Esnouf& Mauger		P.Facy	Hobart
124		Kassed Kerim		Bg	Turkish	293	..	..	..					Bayumi Effendi	Alexandria
125		Kassed Kerim		Bg	Turkish	289	..	..	..					El Said Ali El Masri	Alexandria
126		Kassed Kerim		Bg	Turkish	195	..	..	..					A.Callagussef	Alexandria
127		Kassed Kerim		Bg	Turkish	140	..	..	..					Mohamed Khlab	Alexandria
128		Kassos	Papakonakis	W Bg	Greek	308	..	..	..	Syra	1864		Z.81	G.Zambonnieris	Zante
129	V	Kata (ex Acasia) (WRECKED)	Haudich 87-87	W Bk	Austrian	488	128·2	30·5	17·9	Trieste	1867	Stabilimento Tecnico	Y 3.87	F.Randich	Buccari
130	V	Katahdin	Marsters -88	W Bk	British	1173	194·8	38·2	22·3	Kingsport,NS	1880	P.R.Crichton	Y 7.88	W.Thomson & Co.	St.John.N.B.
131		Katahdin	Farquharson	W3mSr	British	381	150·0	30·9	10·9	Owen Sound	1888	J.Simpson		Spanish Rvr.LumberCo.	Owen Sound
132		Katahdin	Hays	W Bn	American	203	112·5	27·3	9·3	Brewer,Me.	1864			A.P.Veazie	Bangor,Me.
133		Katarina		W Bk	Russian	487	130·9	28·9	18·3	Wiborg	1874	S.Koln		W. & H.Martelin	Wiborg
134		Katarina	Reponen	W Bk	Russian	476	..	..	..		1874			J.Reponen	Wiborg
135	V	Katarina	Burström -87	W Bk	Swedish	605	140·0	30·7	18·6	Umea	1870	E.A.Löfgren	Y 9.86	C.O.Scharin	Umea
136		Katarina	Svensson 87-87	W3mSr	Swedish	285	121·0	24·2	11·7	Gefle	1865	O.A.Brodin		L.Olsson	Lerhamn
137	V	Katarina	Andersson	W Sr	Swedish	143	84·7	21·4	11·0	Fanö	18 64/85	T.Schönau		R.Fogelmarck	Westervik
138	G	Katarina	Nilsson 70-74	W Sr	Swedish	121	87·3	20·4	8·9	Timmernabben	1874	G.Nilsson		N.Nilsson	Kivik
139		Kate	Honkin	W Bg	British	212	107·5	25·4	13·5	Guernsey	1865	Thom		J.H.Guilbert & Co.	Guernsey
140		Kate	Gatenby	W Bn	British	187	106·3	23·5	12·9	P.E.Island	1864	Duncan		G.Hopper	St.Ives
141		Kate	Parsons	W Bn	British	141	91·0	21·5	12·1	Dartmouth	1868	W.Couch	Y 11.86	J.Rorke	St.John's,N.F.
142		Kate	Watterson	W Sr	British	122	91·0	22·0	11·3	Isle of Man	1872	Graves		Mrs.C.Graves	Peel.I.M.
143		Kate	McIlwaine	W S	British	100	89·4	20·3	11·0	Appledore	1865	W.Pickard		J.McIlwaine	London
144	R	Kate	Crowley -84	W Bkn	American	561	142·7	32·8	18·1	HarringtonMe	1884	A.Nash	Y 3.84	A.P.Nash	Machias.Me.
145	N	Kate	Haraldsen	W Sr	Norweg'n	122	80·1	21·0	11·5	England	1845			H.Haraldsen	Tönsberg
146	R	KateB.Ogden	Newberry	W3mSr	American	594	148·0	35·0	16·0	Millbridge,Me	1883	J.Sawyer		A.J.Newberry	New York
147	V	KateBurrill	Beveridge 87-87	W Bk	British	688	154·0	32·5	19·4	LkBrook,N.S.	1871	A.Blinn	Y 1.87	W.Burrill & Co.	Yarmouth.N.S.
148	R	Kate C. Maguire (ex [illegible])	[illegible] -83	W Bk	Argentine	1457	168·0	38·2	26·5	Yarmouth, Me	1867		Y 7.84	Maguire & Co.	Buenos Ayres
149		Kate Cann	Reid	W Bk	British	952	182·0	35·5	21·4	Yarmouth,NS	1874	J.N.Gardner		H.D.Cann	Yarmouth,N.S.
150		Kate Carlton		W Sr	American	206	102·8	26·4	9·0	Maine,U.S.	1862			BostonTowBoat Co.	Boston

* = Lloyd's Register; D = Nederlandsche Vereeniging; G = Germanischer Lloyd; I = Registro Italiano; R = American Record; A = Veritas Austro-Ungarico; H = Veritas Ellenico; N = Norske Veritas; V = Bureau Veritas.

Dans la 4me Colonne, C = Composite (membrure en fer, bordé en bois); I = Fer; Stl = Acier; W = Bois.

Les Dates du doublage en métal (Col. 11) ressortent des livres dans lesquels les navires sont respectivement classés ou des rapports des Inspecteurs du Lloyd's Registe.

Pour plus amples renseignements sur les navires classés par le Lloyd's Register, voir le Lloyd's Register Book et le Lloyd's Yacht Register.

1 Official No.	2 No. in Book	3 Registry in which classed.	4 Ship's Name, Material and Rig. Master, and Dates of Appointment to present Owner's Service and to Ship. Ship's Late Name.	5 Flag.	6 Net Tonnage.	7 Dimensions. Length.	Breadth.	Depth.	8 Built. Where.	9 When	10 By whom.	11 Date of Metalling	12 Owners or Managers.
	151	DN	Kate Carnie (BANSHEE) *Liverten* W Bk	Norweg'n	540	152·9	28·5	18·8	Greenock	1855	R.Steele&Co.		A.L.Liversen
[illegible]	152	R	Kate Crosby *Hansen* -84 W Bk	British	696	160·8	33·5	19·7	Pleasant Cove, N.S.	1868			R. & D.A.Duncan
[illegible]	153	V	Kate Davenport *Howland* -86 W S	American	1194	187·0	37·1	23·9	Bath.Me.	1866	A.&G.Hathorn		Renton,Holmes & Co.
[illegible]	154	R	Kate E. Gifford *Wright* W3mSr	American	420	126·5	32·8	15·6	May'sLanding N.J.	1874	J.Clark		T.H.Wright
[illegible]	155	R	KateE.Morse *Carlson* -85 W3mSr	American	698	161·5	35·3	18·2	PhippsburgMe	1885	J.G.Morse		J.G.Morse
[illegible]	156		Kate E. Rich *Bowen* WSr	American	286	113·2	30·0	9·8	Absecom,N.J.	1865			B.Doughty
[illegible]	157	DV	Kate F. Troop *Banks*-87 W Bk	British	1097	187·0	37·5	22·1	Tynem'th,NB	1881	J.S.Parker		H.D.Troop
[illegible]	158		Kate Foster *Tate* WSr	American	159	93·3	26·2	8·0	CherryfieldMe	1866			E.J.White
[illegible]	159		KateHarding *McDougall* [88-88 W Bk	British	714	148·9	32·1	20·7	ThomastonMe	1869	Watts&Co.		Maguire & Co.
[illegible]	160		Kate Kallahan *Cropper* WSr	American	101	91·0	21·0	7·1	Philadelphia	1853			G.W.Hart
[illegible]	161	R	Kate M.Hilton *Johnson* 83-84 W3mSr	American	510	140·0	32·9	15·4	Bath.Me.	1873	Adms&Hitchc'k		Saml.R.Crowell
[illegible]	162		Kate Markee *Allen* W3mSr	American	499	142·8	31·2	14·8	Bath,Me.	1880	A.Sewall&Co.		J.F.Allen
[illegible]	163	V	Kate Sancton *Evans* -87 W Bk	British	665	150·7	32·1	19·5	Church Point, N.S.	1868	V.LeBlanc		J.Helme
[illegible]	164		Kate Scranton *Pond* WSr	American	118	95·0	27·2	7·8	Madison,Con.	1864			J.H.Eldridge
[illegible]	165	✻	KateTatham (ex Yah-Shing) I3mSr	British	275	138·7	23·1	12·6	Greenock	1863	R.Steele		Geo.Tait
[illegible]	166	✻	Kate Thomas *Evans* -84 I 4mS	British	1693	258·0	39·5	23·1	Sunderland	1885	W.Doxford& Ss.		Kate Thomas Sailing Ship Co. (W.Thomas & Co.)
[illegible]	167		Kate V. Aitkin *Brower* W3mSr	American	398	128·2	31·5	15·2	Tuckahoe,N.J.	1874	J.C.Sheppard		P.Brower,jun.
[illegible]	168		Kate Walker *Ficket* WSr	American	133	90·8	24·4	8·0	Brewer,Me.	1860			A.P.Veazie
[illegible]	169	D	Kate & Anne *Trembeth* WSr	British	117	85·0	21·5	12·0	Fowey	1858			J.Polkinghorne
	170		Katharina *Martinsohn* -86 W Sr	Russian	238	98·0	24·0	13·0	Kuerbis	1878			A.Martinsohn
	171	V	Katharina *Girtehan* -86 WSr	Russian	134	79·7	23·9	10·5	Angern	1886	Ohrmann		J.Kirtehan&PeterEigel[illegible]
	172	✻	Katharine (ex Baron Aberdare) *Spille* 77-85 I S	German	1630	259·0	39·7	23·4	Sunderland	1874	W.Watson		Boyes & Ruyter
	173		Katharine *Silling* 76- WSr	Russian	231	108·3	26·5	13·1	Pernigel	1876	P.Andreison		G.Schnohre
[illegible]	174	✻	Katherine Ellen *Williams* 79-79 WSr	British	99	79·0	21·0	10·5	Waterford	1862	White		Bon Accord Slate Merchant Co.
[illegible]	175		Kathleen W Bn	British	222	113·5	24·7	14·6	P.E.Island	1864	F.Pauling		J.Munnings
[illegible]	176	R	Kathleen *Cook* 79-83 W Bn	British	213	106·0	26·3	11·2	JordanRiv.NS	1883	C.A.Holden		A.G.Jones & Co.
[illegible]	177		Kathleen *Sandy* 86-86 W Bn	British	155	90·9	22·2	12·2	P.E.Island	1856			C.Honlt & Co.
[illegible]	178		Kathleen (WHALER) W Bk	American	195	101·0	25·2	12·5	Philadelphia	1844			J. & W.R.Wing
[illegible]	179		Katie *Le Blanc* W Bo	British	191	99·9	26·2	12·6	P.E.Island	1870			M.Le Blanc
[illegible]	180		Katie *Swaine* WSr	British	131	91·0	24·6	9·4	IndianRiv.NS	1879	S.Freeman&Co.		R.W.Freeman
[illegible]	181	✻	Katie *Jenkin* WSr	British	98	88·6	23·0	10·9	Padstow	1881	Cowl&Sons		Mrs.C.Jenkin
[illegible]	182	✻	Katie *Pierce* -83 WSr	British	97	87·3	20·3	9·7	Portsmouth	1879	Read		British & Irish Gunpowder Manufacturing Co.
[illegible]	183	✻	Katie Cluett *Harvey* WSr	British	126	94·4	23·1	11·6	Fowey	1876	Stephens		C.Isaac
[illegible]	184		Katie Eccles WSr	British	121	95·0	24·0	8·0	Mill Point	1877	W.Jamieson		D.Eccles
[illegible]	185		KatieFlickinger *Young* 175-86 W Bkn	American	449	145·2	35·9	13·6	Seattle,Wash.	1876	Gibson		S.B.Peterson
[illegible]	186		Katie G. Robinson *Lore* W3mSr	American	299	116·4	29·6	9·3	May'sLanding, N.J.	1873	J.Clark		J.Scull
[illegible]	187	R	Katie J.Barrett *McLeod* -87 W4mSr	American	916	191·2	38·0	19·0	Bath,Me.	1887	Wm.T.Donnell		Deering & Donnell
[illegible]	188		Katie J. Hoyt *Hearse* WSr	American	209	112·0	30·3	9·3	P.JeffersonNY	1865			F.Harrington
[illegible]	189	R	KatieJ.Irelan *Stedman* W3mSr	American	670	157·7	36·6	18·8	Cooper'sPt.NJ	1883	S.W.Tilton		A.Jones
[illegible]	190		Katie Mitchell *Oliver* WSr	American	107	84·1	26·8	7·3	Bath,Me.	1871			A.Robinson
[illegible]	191	R	KatieStuart *Edwards* -87 W Bk	British	368	123·6	29·5	15·7	P.E.Island	1875	D.C.Ramsay		J.Reid
[illegible]	192	✻	Katie Swyny *Whiteside* W3mSr	British	186	111·1	24·4	12·5	Carrickfergus	1878	P.Rodgers		T.Fisher
[illegible]	193	AH	Katina *Cokina* W Bk	Greek	362	120·4	28·5	..	Syra	1876	S.Cristali		Sons of K.Zer[illegible]
[illegible]	194	H	Katina (ex Emmanouil) W Bg	Greek	199	..	..	..	Galaxidi	1869	C.Papapetros		Heirs Kyriakis
[illegible]	195	H	Katina *T.Cammenas* WS	Greek	186	92·5	24·3	13·7	Galaxidi	1880	C.Papapetros		A.Cammenas
[illegible]	196		Katingo *Tzakalis* WSr	Greek	202	107·2	23·9	..	Sunderland	1868			F.Andreadis
	197	✻	Katinka (ex Ardenscallach) *Seghorn* 85-85 I Bk	German	816	196·4	32·2	18·8	PortGlasgow	1874	Macfadyen&Co.		J.D.Ahlers
	198		Katirlee WS	Turkish	112	..	..	..	Unia	1869			Katirleeioo Sheik [illegible]
	199		Katrina *Weledejeff* WS	Russian	143	..	..	..	Impilaks	1871			M.Andrejeff
	200	V	Katrine *Hansen* 70-70 W Bn	Danish	141	95·2	21·7	11·3	Karrebæksminde	1865	L.Hansen		L.Hansen

✻ = **Lloyd's Register**; D = Nederlandsche Vereeniging; G = Germanischer Lloyd; I = Registro Italiano; R = [illegible] Lloyd; A = Veritas Austro-Ungarico; H = Veritas Elleñico; N = Norske Veritas; V = [illegible]

In the 4th Column, C = Composite (iron frame, wood planked); I = Iron; Stl = Steel; W = Wood.

The Dates of Metalling (**Col. 11**) are based on the records of the Books in which Vessels are respectively classed, or on the Reports of the Surveyors [illegible]

Further particulars of Vessels classed by Lloyd's Register will be found in Lloyd's Register Book and Lloyd's Yacht [illegible]

No.	Classe	Nom. Capitaine et dates d'entrée en service des Armateurs actuels et à bord du navire. Ancien Nom.	Matériaux et Gréement	Pavillon	Tonnage net	Longueur	Largeur	Creux	Construit. Où	Quand	Par qui	Date du doublage en métal	Armateurs ou Directeurs-Gérants	Port d'Armement
201	V	Katy & Hanna Anderssen -83 (ex George & Richard)	W Bg	Swedish	316	110·8	24·7	16·6	Sunderland	1862	J. J. B. Gardner	Y [illegible]	E. Jansson	Cimbrishamn
202	N	Kawe Petersen 83-83	W Bk	Norweg'n	508	132·7	32·2	17·5	Uleåborg	1857		Y [illegible]	H. H. Pettersen	Arendal
203	✻	Kebroyd (IRON) Heslop	W Bk	British	363	120·2	27·8	17·0	Sunderland	1866	Robinson	Y [illegible]	Adelaide S.S. Co. Ld.	Adelaide, S.A.
204	DV	Kedron Fraser 82-85	W Bk	British	1160	193·0	38·0	22·0	Cornwallis, NS	1880	P. R. Crichton	Y [illegible]	C. R. Burgess	Windsor, N.S.
205		Keeratarudren	W Bg	British	155	99·5	20·6	..	Beypoor	1851			Rehmanjee Khanttoy	Bombay
206		Keewaydin Robinson -89 (ABANDONED)	W Bk	British	640	153·0	32·5	18·6	Hopewell [N. S.]	1880	C. Dowling	Y [illegible]	C. Dowling	St. John, N.B.
207		Keian Maru	W Sr	Japanese	124	81·6	18·9	10·5	Hiogo	1881	Ohmatsu Jugoro		Masuda Matayemon	Osaka
208		Keian Maru No. 2	W Sr	Japanese	390	125·0	28·7	17·2	Hiogo	1882	Ohmatsu Jugoro		Seiko Kwaisha	Hiogo
209		Keihan Maru	W Sr	Japanese	306	122·7	23·7	11·0	Osaka	1879	Toda Shojiro		Nagoshi Aisuke	Osaka
210	✻	Keir [illegible] 50-85	I Bk	British	986	212·1	32·2	21·4	Glasgow	1859	Barclay, Curle [& Co.]		T. C. Guthrie	Glasgow
211	✻	Kelat M'Kay	I S	British	1822	261·0	41·0	23·5	Stockton	1881	Richardson, [Duck & Co.		E. Bates & Sons	Liverpool
212		Kelpie Dickenson	W Bn	British	190	102·8	25·4	12·9	P. E. Island	1866	Hunt		J. Kelly	Belfast
213		Kelso Doucett -86	W Sr	British	111	80·0	24·0	9·5	Tusket W'dg NS	1875			J. H. Porter	Yarmouth, N.S.
214	V	Kelverdale Grady -88	W Bk	British	1191	193·0	38·0	22·9	Kingsport, NS	1881	P. R. Crichton	Y [illegible]	W. Thomson & Co.	St. John, N.B.
215	V	Kelvin Thompson 83-87	W Bk	British	1099	187·0	37·3	22·6	Kingsport, NS	1879	P. R. Crichton	Y [illegible]	W. Thomson & Co.	St. John, N.B.
216		Kemendine	I Flat	British	777	197·0	25·7	8·3	Port Glasgow	1878	Robt. Duncan [& Co		Irrawaddy Flotilla Co. Ld.	Glasgow
217		Kendrick Fish Clark	W Sr	American	137	91·0	24·0	8·5	St. George, Me.	1851			G. W. Rawley	St. George, Me.
218		Kenduskeag Whitney	W Sr	American	141	95·0	25·6	7·9	Bangor, Me.	1867	Crosley Bros.		E. A. Green	Bangor, Me.
219		Keng Chuan Seng	W Junk	British	112	92·0	23·0	8·0	Keng Chow	1876	F. Moh		Chea Cheng Siang	Bremen
220	✻	Kenilworth H Nairn 89-87	Stl 4m S	British	2243	300·2	43·1	24·2	Port Glasgow	1887	J. Reid & Co.		Williams'n Milligan & Co.	Liverpool
221		Kenilworth	W Sr	British	113	88·0	22·0	9·8	Auckland, N.Z.	1866			J. Sinclair	Auckland, N.Z.
222	✻	Kenilworth Ahlers 83-87	I Bk	German	695	176·2	28·9	18·4	Preston	1866	Preston Iron SC		H. Bischoff & Co.	Bremen
223	✻D	Kenmore Milne 82-82	I Bk	British	931	207·2	33·3	20·2	Port Glasgow	1882	J. Reid & Co.		A. C. Finlay	Greenock
224	G	Kenna De Walt	W Sr	German	107	75·5	19·7	9·5	Brake	1864	J. H. Addicks		J. Munderto	Emden
225	R	Kennard Bettencourt	W Bk	American	592	144·5	33·5	17·6	E. Machias, Me.	1877	W. H. Stevens	Y 10.88	E. A. Adams & Co.	Boston, Mass.
226	V	Kennebec Love 83-88	W S	American	2025	237·7	43·3	27·3	Bath, Me.	1883	W. Rogers	Y [illegible]	Middlemas & Boole	San Francisco
227		Kennebec Wallis (ex Ida Della T.)	W Sr	American	178	103·7	27·4	9·4	Fairhaven	1852			Elias Milliken	Augusta
228		Kenolee Hussein Reiss	W Bn	Turkish	209	..	..	..	Kenolee	1869			Hairie Tiiari Mehemed & Co.	Kenolee
229	R	Kensett Smith	W 3m Sr	American	436	135·0	30·0	16·0	E. Machias, Me.	1873		Y [illegible]	F. Talbot & Co.	Providence, R.I.
230		Kenshin Maru	W Bk	Japanese	304	130·4	25·0	12·8	Copenhagen	1873	A. Petersen		Nippon Yusen Kaisha	Tokio
231		Kensho Maru	W Sr	Japanese	150	94·9	20·1	11·1	Owari	1884	Yamamoto [Shosabro		Hasegawa Katsuichi	Owari
232		Kensington (BROKEN UP)	W Bn	British	142	96·2	25·0	10·8	Walton, N.S.	1860			J. Curwen	Whitehaven
233		Kent	W Sr	British	104	98·2	20·2	7·2	Brisbne. Water	1884	E. G. Beattie		R. Langley	Sydney, N.S.W.
234		Kent Caballero	W Bk	Chilian	381	139·0	28·0	17·5	Sunderland	1862	Marshall		J. Jones	Lota
235	V	Kentigern Dexter -88	W Bk	British	825	164·0	35·0	19·9	Cornwallis, NS	1877	P. R. Crichton	Y 11.87	W. Thomson & Co.	St. John, N.B.
236		Kentish Lass Mosley	W Bk	British	306	133·1	26·1	13·0	Bideford	1872	J. Cox	Y 11.81	James Hutchin	Wellington N.Z.
237	✻	Kentmere Raymond -88	I 4m S	British	2457	300·0	42·2	24·7	Liverpool	1883	W. H. Potter & Sn		Ship Kentmere Co. Ld.	London
238	✻D	Kenton McPherson 86-86	I Bk	British	668	180·9	30·0	17·6	Sunderland	1877	W. Doxford & Ss.		A. Gibson & Co.	London
239	✻	Kenyon Keith 83-83	I Bk	British	1191	208·0	34·6	22·8	Liverpool	1862	T. Vernon & Son		J. Poole & Co.	Liverpool
240	N	Kepha Terjesen	W Bk	Norweg'n	474	139·4	30·0	16·7	Grimstad	1875	John Terjesen	Y [illegible]	O. Terjesen	Lillesand
241	✻V	Kepler Haase 84-84	I S	German	1193	221·0	34·1	21·4	Hamburg	1878	Reiherstieg Ship [Work		R. M. Sloman & Co.	Hamburg
242	RV	Kepler Oelken 81-82	W Bk	German	774	163·0	34·1	19·8	Bath, Me.	1877	Goss & Sawyer	Y [illegible]	Balleer & Marting	Bremen
243	V	Kerchalon Mahé -88	W Bk	French	268	104·9	24·9	13·8	Nantes	1872	Sevestre frères	Y [illegible]	F. Batard & Co.	Nantes
244	N	Kerdalea Halvorsen 86-86 (ex Johanne Mathilde)	W Bk	Norweg'n	415	134·4	29·4	16·5	Hamburg	1863	B. Wencke	Y [illegible]	J. E. Hanssen	Arendal
245		Kerkyra Coutsoufas	W Bg	Greek	227	107·2	24·5	..	Syra	1875			John K. Mazavinos	Corfu
246		Kerrasonde	W Bn	Turkish	175	..	..	..	Kerrasunda	1868			Kerrasoudleo-[illegible]	Kerrasunda
247	DV	Kersbergen Klasen 77-77	C Bk	Dutch	985	183·0	36·4	21·2	Amsterdam	1877	W. H. Meursing	Y [illegible]	W. H. Meursing	Baarn
248		Keshish Daghi	W Schleppe	Turkish	196	..	..	..	Austria	1883			Jiwaneh Orsenaki	Constantinople
249	R	Kesmark Tennyson -86 (LOST)	W Bk		627	146·3	32·3	18·3	Monckton, NB	1879	Robt. Cochrane		H. O'Leary	
250	V	Kestrel Norton 83-85	W Bk	British	467	133·8	30·2	12·6	Horton, N.S.	1871	J. B. North	Y 12.81	J. B. North	Windsor, N.S.

✻ = Lloyd's Register; D = Nederlandsche Vereeniging; G = Germanischer Lloyd; I = Registro Italiano; R = American Record;
A = Veritas Austro-Ungarico; H = Veritas Ellenico; N = Norske Veritas; V = Bureau Veritas.

Dans la 4me Colonne, C = Composite (membrure en fer, bordé en bois); I = Fer; Stl = Acier; W = Bois.

Dates du doublage en métal (Col. 11) ressortent des livres dans lesquels les navires sont respectivement classés ou des rapports des Inspecteurs du Lloyd's Register.

Pour plus amples renseignements sur les navires classés par le Lloyd's Register, voir le Lloyd's Register Book et le Lloyd's Yacht Register.

Official No.	No. in Book	Registry in which classed	Ship's Name. Master, and Dates of Appointment to present Owner's Service and to Ship. Ship's Late Name.	Material and Rig	Flag	Net Tonnage	Length	Breadth	Depth	Built: Where	When	By whom	Date of Metalling	Owners or Managers
85344	251	✱	Kestrel [illegible]	W S	British	152	102·6	22·0	11·9	Falmouth	1874	Trethowan	[illegible]	J. Munn & Co.
80796	252	V	Keswick *Galliatt*	W Bk	British	924	173·5	36·4	21·5	BearRiver,NS	1882	W.F.Marshall	[illegible]	ClarenceA.Vaughan
[illegible]	253		Keystone *Wilder 66-77*	W3mSr	American	178	110·5	29·4	8·5	Pembroke,Me.	1873	G.Frost		E. A. Wilder
[illegible]	254		Kezia [illegible]	W S	British	136	85·0	21·6	10·2	Conquerall NS	1883	Hafuse & [Corkum		W. Sarty
	255		Khadem Hhair	W Bg	Turkish	137	..	..	..	[illegible]	1877			Ahmussahlee Hajee Nari
	256		Khadem Hhair	W Bg	Turkish	125	..	..	..	[illegible]	1868			Tijar Hajee Saleh & Co.
	257		Khadem Hhair	W Bg	Turkish	112	..	..	..	[illegible]	1878			Kumurdji Ethem Zadeh Ahmed Hakki [illegible]
	258		Khair Eddin *Kaisi 73-83*	W	Turkish	209	..	..	..					Ahmed Rostom Chorji Ali Zadeh & Co.
	259		Khania	W Bg	Turkish	138	..	..	..	[illegible]	1874			[illegible]
[illegible]	260		Khanoung	I L	British	355	194·8	25·1	8·5	PortGlasgow	1880	R.Duncan&Co.		[illegible] Ld.
	261		Kharbaneh	W	Turkish	145	..	..	..	[illegible]	1871			Riza Bey-Zadeh Sala[illegible] & Bros.
82291	262		Khat Tee Ya	I L	British	357	195·6	25·1	8·5	PortGlasgow	1889	R.Duncan&Co.		Irrawaddy Flotilla Co. Ld.
62541	263	✱	Khedive *McDonald 85-85*	W Bk	British	394	128·7	28·3	17·4	Sunderland	1870	J.Robinson	[illegible]	J. Tweddell, jun.
8108	264	✱	Khersonese *Sleightholm (88-85)*	I S	British	1711	246·1	38·9	[illegible]	Belfast	1855	R.Hickson&Co.		G. Duncan & Co.
14770	265	✱	Khimjee Oodowjee *Rowe*	I Bk	British	903	212·0	33·5	20·4	Liverpool	1856	J.Jones,jr.		Eastern Telegraph Co. Ld.
80188	266		Kho Khine	I Bsd	British	777	197·0	25·7	8·3	PortGlasgow	1878	R.Duncan&Co.		Irrawaddy Flotilla Co. Ld.
	267	✱	Khorasan [illegible]	I Bk	German	1039	190·5	34·4	22·0	PortGlasgow	1864	J.Reid&Co.		C. H. H. Winters
81300	268	✱	Khyber [illegible]	I S	British	1967	276·6	40·1	24·2	Liverpool	1880	W.H.Potter&Sn.		T. & J. Brocklebank
	269	V	Kiana *Carlsen 74-78*	W Bn	Danish	193	106·3	23·3	11·8	Marstal	1879	H.J.Bager		H. Hansen
44733	270	✱	Kiandra [illegible]	I Bk	British	990	210·5	31·8	21·4	Glasgow	1862	C.Connell&Co.		H. Bischoff & Co.
	271		Kiem Hien Taij *Achmad*	W Bk	Dutch	427	132·5	25·5	..	Siam	1868		[illegible]	Lee Bewn K[illegible]
	272		Kiem Seng Hoeat *Gho Kian Bie*	W S	Dutch	135	88·3	20·1	..	Siam	1864		[illegible]	Go Djin Gang
	273		Kiem Sin Hin [illegible]	W Bg	Dutch	167	94·8	22·6	..					Oe Soen Ong
	274		Kiem Soen Hoeat (ex Henry Melvill, ex Sing [illegible])	W Bk	Dutch	275	109·9	24·8	..	[illegible]	1857	Bas Siman	[illegible]	Goei Boen Giok an Lie[illegible]
	275		Kiem Tiang Hoeat [illegible]	W Bk	Dutch	223	107·9	22·1	..	Java	1850		[illegible]	Goei Boen Giok [illegible] Tjeet Hoen
	276		Kiemiss	W Bmb	Turkish	100	..	..	..	[illegible]	1874			Limnilee Dimitri Mahoria
	277	V	Kierstine Elise *Hansen (69-89)*	W Bn	Danish	146	93·5	21·9	11·1	[illegible]	1869	Chr.Schrøder		J. Hansen
[illegible]	278		Kiho Maru	W S	Japanese	153	88·5	20·5	11·0	[illegible]	1884	Takeda Saizo		Tsukui Chojuro
[illegible]	279		Kiko Maru	W Bk	Japanese	220	115·3	22·9	13·4	[illegible]	1880	NagataSanjiro		Morita Saburoyemon
[illegible]	280	✱	Kilbrannan *McCallum 84-81*	I S	British	1572	256·4	38·2	22·8	PortGlasgow	1882	Russell&Co.		Kerr, Newton & Co.
[illegible]	281	✱	Kildonan *Jones 83-83*	I Bk	British	650	177·3	30·1	18·0	Glasgow	1869	A.Stephen&Sns		W. B. Jones
[illegible]	282		Kildonan	W3mSr	British	439	174·1	33·0	11·1	Kingston,Ont	1888	MontrealTrans- [portationCo.		Montreal Transportation Co.
[illegible]	283	V	Kildonan *O'Brien -83*	W Bn	British	372	126·9	30·4	12·9	RiverJohn,NS	1876	McKenzie& [Sander	[illegible]	J. McNeil
	284		Kilityia	W Bg	Turkish	1100	..	..	..					Tijar Frank Oglou [illegible]
[illegible]	285	✱	Kilkerran *Ferguson 77-81* [illegible]	I S	British	199	231·0	37·0	23·4	PortGlasgow	1871	J.Reid&Co.		J. Kerr & Co.
[illegible]	286		Killarney	W Bk	British	1401	183·5	29·0	17·9	[illegible]	1869	Macfarlane		The Kauri Timber Co.Ld.
[illegible]	287	✱	Killean *Mackinnon 86-86*	I S	British	1660	261·4	39·3	23·8	PortGlasgow	1875	J.Reid&Co.		M'Diarmid, Greenshields & Co.
[illegible]	288	✱ D	Killeena *Blake 85-85*	I Bk	British	795	193·7	31·7	19·2	Glasgow	1875	D.&W.Hender- [son&Co.		Kerr, Newton & Co.
[illegible]	289	✱	Killia Lass *Lloyd 69-83*	W Bg	British	185	99·6	24·0	12·6	Aberayron	1860	DavidJones	[illegible]	M. Jones & Co.
[illegible]	290	✱	Killochan *Munson 73-81*	I S	British	1233	226·5	37·1	21·6	Dumbarton	1874	A.M'Millan&Sn		J. Kerr & Co.
[illegible]	291	✱ D	Kilmory [illegible] *75-85*	I S	British	1569	259·1	38·2	22·8	PortGlasgow	1885	Russell&Co.		Kerr, Newton & Co.
[illegible]	292		Kim Chye Hup	W S	British	127	78·0	21·6	7·8	[illegible]	1884	I Yalya		Ong Kew Ho
[illegible]	293		Kim Fook Lee	W S	British	102	74·0	21·4	10·0	Singapore	1888	VongFookTak		Vung Fook Choy
	294		Kim Guan Huat	W Bg		224	111·0	25·0	12·3					
[illegible]	295		Kim Ho Tong	W S	British	111	85·0	20·0	10·0	Singapore	1880	Tham Ah Hong		Tham Boon Yew
[illegible]	296		Kim Hock Hwat	W Kattoo	British	170	84·5	26·0	9·7	[illegible]	1871			Khoo T[illegible]
[illegible]	297		Kim Kong Hwat	W Kattoo	British	159	83·0	24·5	9·0	[illegible]	1871			Nyo Tsein Pale
[illegible]	298		Kim Lee Watt	W S	British	120	74·0	24·0	9·6	[illegible]	1874			Ooi K[illegible]
[illegible]	299		Kim Oh Chin	W S	British	108	86·1	21·6	10·0	Singapore	1885	Tham Ah Hong		Tham Boon Yew
	300		Kim Sin Lie [illegible] (ex Harald Haarfager)	W Bk	Dutch	336	122·3	27·5	12·2	Christiansund	1878			Tam Djim Seng

✱ = Lloyd's Register; D = Nederlandsche Vereeniging; G = Germanischer Lloyd; I = Registro Italiano; [illegible]
A = Veritas Austro-Ungarico; H = Veritas Ellenico; N = Norske Veritas; V[illegible]

In the 4th Column, C = Composite (iron frame, wood planked); I = Iron; Stl = Steel; W = Wood.

The Dates of Metalling (Col. 11) are based on the records of the Books in which Vessels are respectively classed, or on the Reports of the Surveyors.

Further particulars of Vessels classed by Lloyd's Register will be found in Lloyd's Register Book and Lloyd's Yacht Register.

No.	Cote	1. Nom. Capitaine et dates d'entrée au service des Armateurs actuels et à bord du navire. Ancien Nom.	Matériaux et Gréement	2. Pavillon	3. Tonnage net	4. Dimensions. Longueur	Largeur	Creux	5. Construit. Où	9. Quand	10. Par qui	11. Date du doublage en métal	12. Armateurs ou Directeurs-Gérants.	13. Port d'Armement.
301		Kim Soon Ho	W [illegible]	[illegible]	146	[illegible]	21·7	..	[illegible]	1870		12.83	Khu Liem Yang	Banjarmassin
302		Kim Yap Soon	W Sr	British	112	78·[illegible]	23·7	8·4	Singapore	1884	L.S.Hong		Yeo Yiam Chuah	Singapore
303	V	Kimisis *Karas*	W Bk	Greek	244	105·6	24·8	14·1	[illegible]	1865		10.81	F.Kara	Santorin
304		Kimisis	W Bg	[illegible]	19[illegible]	83·0	24·0	..	Syra	1874			S.Damigo	Santorin
305		Kimisis Theotocou *Tribas* (ex Andrea Vagliano)	W Bk	Greek	26[illegible]	..	..	..	Syra	1871			Vª Revidi	Hydra
306		Kimon *Handras*	W Bk	Greek	3[illegible]	125·[illegible]	28·[illegible]	..	[illegible]	1868	Faggioni	10.83	M.Handras	Syra
307	N	Kimon *Andersen 72-74*	W Bg	Norweg'n	271	101·4	28·2	14·[illegible]	[illegible]	1868	A.Omundsen	V [illegible]	Harald Isachsen	Grimstad
308		Kinburn *Tooker -87*	W S	British	1198	18[illegible]	37·0	22·8	[illegible], N.S.	1875	J.H.Zwicker	V 10.81	J.W.Moody	Yarmouth,N.S.
309	*	Kinclune *Vigget 80-86*	I Bk	British	718	184·7	30·1	18·4	Dundee	1876	A.Stephen&Sns		C.Couper	Dundee
310	DV	Kinderdijk *Dunshard -83*	I Bk	Dutch	948	168·3	36·4	21·0	Kinderdijk	1876	J.K.Smit		M.Lels & Zoon	Alblasserdam
311	*	Kinfauns *Piggott -88*	Stl Bk	British	981	204·0	34·5	19·[illegible]	Dundee	1886	A.Stephen &Sns		C.Couper	Dundee
312	*	King Alfred *Rutherford*	I Bk	British	1349	228·0	36·5	21·[illegible]	Glasgow	1886	Barclay, Curle & Co.		J.A.Walker & Co.	Glasgow
313	*	King Arthur *Drummond -87*	I 3m S	British	1598	257·6	38·2	23·2	Glasgow	1887	W.B.Thompson & Co., Lim.		J.A.Walker & Co.	Glasgow
314	V	King Cenric *Edwards 85-86*	W S	British	1491	208·5	39·5	24·0	St.John, N.B.	1874	J.K.Dunlop	V 8.86	R.W.Leyland & Co.	Liverpool
315	*	King Malcolm *McDonald -87*	I Bk	British	1256	228·0	36·5	21·0	Glasgow	1885	W.B.Thompson		J.A.Walker & Co.	Glasgow
316	R	King Philip *Phillips*	W 4m Sr	American	1164	211·0	42·5	20·4	Camden, Me.	1886	H.M Bean		J.M.Phillips	Belfast,Me.
317	*	King Robert	Stl S	British	1829	258·7	38·2	23·2	Port Glasgow	1889	W.Hamilton & Co.		A.J.Walker & Co.	Glasgow
318	*	Kingdom of Saxony *Howard* (ex [illegible])	I Bk	British	538	161·1	27·1	17·3	Hartlepool	1864	Denton,Gray&C		Trinder,Anderson & Co.	London
319	*	Kingdom of Sweden *Cooper*	I Bk	British	788	191·7	32·0	18·4	Sunderland	1875	Mounsey & Foster		Gosman & Smith	London
320		King of Tyre	W Bk	British	260	108·0	27·4	12·6	Tatamagouche, N.S.	1865			C.A.De Wolf	Halifax,N.S.
321		Kingfisher	I Bk	British	164	126·1	18·8	9·3	Renfrew	1854			P.& O.S.N.Co.	Melbourne
322		Kingo Maru	W Bg	Japanese	168	107·5	19·8	11·5	Osaka	1878	Nagata Sanjiro		Nagoshi Aistike	Osaka
323		Kingo Maru No. 2	W Bg	Japanese	161	91·0	22·0	9·5	Osaka	1879	TodaShojiro		Nishidani Shohachi	Osaka
324	DV	King's County *Kinsman 87-89*	W Bk	British	859	165·0	36·0	21·1	[illegible] N.S.	1871	J.E.Steele	V [illegible]	S.Sheffield	Windsor,N.S.
325	*	King's Oak *Cox*	W K	British	63	73·3	19·1	9·5	Gt.Yarmouth	1884	J.H.Fellows&Ss		T. & J. Leleu,Ld.	Gt.Yarmouth
326	V	Kingsport *Morehouse 85-88*	W S	British	1161	190·0	38·0	22·3	[illegible] N.S.	1878	P.R.Crichton	V 4.88	Wm.Thomson &Co.	St.John,N.B.
327		Kinjaboo *Bernard -87* (ex Elizabeth Taylor)	W Bk	French	261	123·5	26·3	12·9	Sunderland	1872	Pickersgill		A.Verdier	La Rochelle
328	*	Kinkora *Wymark 88-89*	I S	British	1799	267·7	39·7	23·5	Maryport	1888	Ritson & Co.		J.Porter	Belfast
329	*	Kinnaird *Rees*	W Sr	British	138	98·9	23·[illegible]	10·7	Fraserburgh	1874	Webster	V 12.81	G.Kilner	Liverpool
330		Kinokuni Maru No. 1	W Bg	Japanese	168	97·8	20·9	11·7	Osaka	1879	NagataSanjiro		Abe Ichitaro	Osaka
331		Kinokuni Maru No. 2	W Bk	Japanese	217	108·4	21·9	13·3	Osaka	1881	NagataSanjiro		Abe Ichitaro	Osaka
332	*	Kinross *Sinclair*	I S	British	1399	242·5	37·0	22·0	Liverpool	1877	W.H.Potter&Ss		A.Gibson & Co.	Liverpool
333	*V	Kintore *Clement 87-88* (ex Johannes & Margaretha)	W Bk	British	972	173·3	38·2	22·5	Kinderdyk	1870	Rykee	V 1.89	G.Milne & Co.	Aberdeen
334	N	Kintu (ex Excelsior) *Magnussen 86-88*	W Bg	Norweg'n	240	97·0	21·9	15·3	Sunderland	1851	Booth & Blacklock		A.Magnussen	Sarpsborg
335		Kirkconnell	W Bkn	British	227	109·8	25·3	11·5	Glencaple	1855			P.M'Fee	Ardrossan
336	*	Kirkcudbrightshire *[illegible]*	I S	British	1482	250·0	38·3	22·9	Greenock	1884	Russell&Co.		T.Law & Co.	Glasgow
337	*D	Kirkdale *Davies*	I Bk	British	760	191·7	31·1	19·3	Sunderland	1877	W.Doxford&Ss.		J.Steel & Son	Liverpool
338	*	Kirkloch *Brunel*	I Bk	British	755	191·8	31·0	19·2	Sunderland	1876	W.Doxford&Ss.		J.Steel & Son	Liverpool
339	*	Kirkmichael *Pritchard 82-85*	Stl Bk	British	903	202·0	33·1	20·1	Sunderland	1882	W.Doxford&Ss.		J.Steel & Son	Liverpool
340		Kirlangitch *[illegible]*	W Bmb	Turkish	125	..	..	..	Gallipoli	1883			Mureftahlee Yorghaki & Co.	Mureftah
341		Kirlangitch Bahri	W Bg	Turkish	120	..	..	..	Barten	1871			Galatadah Kumwidji Hajee Mustapha	Barten
342	V	Kirstine (ex Emery) *[illegible]*	W Sr	Danish	129	88·3	20·1	10·8	Newport,N.S.	18 53/74			G.R.Berg	Svendborg
343		Kirstine *Brunn*	W Bn	Danish	128	84·3	21·4	11·1	Aalborg	1857	L.Bonnesen		Chr.Simoni	Aalborg
344	*	Kishon *May 87-89*	W Bk	British	491	143·0	29·3	17·9	Sunderland	1872	J.Gardner	V [illegible]	T.S.Howitt	North Shields
345	N	Kismet *[illegible]* (ex [illegible])	W [illegible]	Norweg'n	672	146·0	30·0	21·4	Jersey	1850			B.M.Paulsen	Drammen
346	*	Kistna *Smith*	I S	British	2149	289·6	40·5	24·5	Southampton	1883	Oswald,Mordaunt&Co.		E.Bates & Sons	Liverpool
347	R	Kit Carson *Smith 86*	W Sr	American	201	108·0	27·8	9·4	Bucksport,Me	1877	W.Beasley	V 10.86	T.J.Stewart & Co.	Bucksport,Me.
348		Kite *Davey*	W Bg	British	177	100·4	21·9	12·1	Jersey	1859	Deslandes	V 1.8[illegible]	P.Dreland	Guernsey
349		Kithos *Christo Yorghi*	W Bg	Turkish	240	..	..	..	Greece	1880			Muttah-bashi-zadeh Hajee Bitaat Effendi	Barten
350														

* = Lloyd's Register; D = Noderlandsche Vereeniging; G = Germanischer Lloyd; I = Registro Italiano; R = American Record; A = Veritas Austro-Ungarico; H = Veritas Ellenico; N = Norske Veritas; V = Bureau Veritas.

Dans la 4me Colonne, C = Composite (membrure en fer, bordé en bois); I = Fer; Stl = Acier; W = Bois.

Les Dates du doublage en métal (Col. 11) ressortent des livres dans lesquels les navires sont respectivement classés ou des rapports des Inspecteurs du Lloyd's Register.

Pour plus amples renseignements sur les navires classés par le Lloyd's Register, voir le Lloyd's Register Book et le Lloyd's Yacht Register.

1 Official No.	2 No. in Book	3 Registry in which classed	4 Ship's Name. Material and Rig. Master, and Dates of Appointment to present Owner's Service and to Ship. Ship's Late Name.	5 Flag	6 Net Tonnage	7 Dimensions: Length	Breadth	Depth	8 Built: Where	9 When	10 By whom	11 Date of Metalling	12 Owners or Managers
[illegible]	351		Kitsap *Tibbetts 86-87* W4mSr	American	756	188·2	40·5	16·0	Port Ludlow	1887	Middlemas & Boole		Puget Sound Commercial Co.
	352	N	Kittil Motzfeldt *Gunnulsen -87* WSr	Norweg'n	105	88·4	23·6	9·6	Fjeldb'rg Nwy	1881			G.T.Jörgensen
[illegible]	353		Kitty (ex Stad Enschede) *Laird* I Bk	British	803	177·7	31·4	19·6	Amsterdam	1856	Konin Fabrick		W.Dunbar
[illegible]	354		Kitty Clyde *Burns* WBn	British	120	93·6	22·6	10·8	Tatamagouche N.S	18[illegible]			T.Burns
	355	✳	Kitty Meadows *Hardy -87* WK	British	60	67·6	18·6	9·6	Galmpton	1887	T.Mathews		Hardy & Long
	356	✳	Kitty Read *Greaves* WSk	British	89	78·5	21·0	10·4	Grimsby	1886	Smith Bros.		J.Greaves
[illegible]78	357		Kiuho Maru WSr	Japanese	135	79·9	20·5	10·1	Osaka	1877	Mihara Jintaro		Fujimoto Tashiro
	358	N V	Kjartan *Johannesen -88* WBn	Norweg'n	341	126·9	28·2	14·0	Christiansand	1875	O.P.Moe&Søn	Y [illegible]	J. Johannessen
	359	N V	Kjelland *Larsen -88* WSr	Norweg'n	301	117·1	28·4	14·4	Christiansand	1877	O.Hagen	Y 8.88	O.Hagen & S.[illegible]
	360	N	Kjellestad *Haagensen 81-81* WBk	Norweg'n	362	113·6	28·8	16·4	Brevig	1858	A.Blehr		H.A.Haagensen
	361	V	Kjerteminde *Olsen* WSr	Danish	115	91·9	19·9	10·0	Korsør	1867	Chr.Lund		P.Olsen
	362		Kjøge *Pii 86-86* WBk	Danish	270	121·2	25·0	13·2	Kolboda	1869	B.Olsen	[illegible]	C.Eberth
	363		Klaistotimos WBn	Turkish	175	..	..	..	Sakarieh	1876			Kara Agob Oglou [illegible]
	364	N	Klara (ex Clara) *Haslum -87* WBk	Norweg'n	897	179·7	35·0	21·9	Sunderland	1859	J.Laing	Y [illegible]	H.D.Movig
	365		Klara *Karlsson-85* WSr	Russian	121	..	..	..		1868			J.G.Österlund
	366	V	Klara *Berntsson 84-84* WBn	Swedish	161	89·9	22·8	11·6	Eckerna	1874	A.Ahlberg		C.J.Olsson
	367	G	Klara (ex Engelina Hermanna, ex Dido) *Andersson 86-* WSr	Swedish	149	92·5	20·0	10·9	Sappemeer	1860	C.Bakker		J.Andersson
	368		Klara *Johansson* W3mSr	Swedish	146	98·4	21·5	9·6	[illegible]				O.Hallongren
	369		Klara *Persson* WSr	Swedish	130	82·0	20·5	11·4	[illegible]	1849	H.P.Carstensen		N.Pehrsson
	370		Klara *Schmidt* WSr	Swedish	114	80·0	22·1	9·[illegible]	[illegible]	1856			Iggesund's Bruks Intressenter (A.Lindner)
	371		Klara Josefina *Östermann 85-85* WSr	Swedish	231	105·9	25·3	11·8	[illegible]	1871	J.P.Elfving		J.Norman
	372		KlaraMaria *Granberg -86* WBk	Swedish	821	143·5	31·9	22·0	[illegible]	1850			E.Frisk
14395	373		Klikitat *Cutler 81-82* WBkn	American	469	163·0	38·8	12·7	[illegible] Bay, Ora	1881			Puget Sound Commercial Co.
	374	✳	Klio *Bramslow 87-88* StlS	German	1621	258·8	38·7	22·7	Glasgow	1888	C.Connell & Co.		B.Wencke Söhne
	375		Klio *Sultanitchi* WSr	Greek	262	..	..	..	[illegible]	1884	E.Soffos		D.Economo
48802	376	✳	Knight Commander *Bell 83-83* I S	British	1435	213·5	36·5	25·1	Dublin	1864	Walpole&Webb		H.Fernie & Sons
52845	377	✳	Knight Companion *Davies 83-83* I S	British	1470	221·4	36·2	24·7	Liverpool	1865	T.Royden&Son		Davison,Bissett & Co.
	378		KnightTemplar *Heinsohn* W Bk	Chilian	449	135·9	27·4	17·4	Liverpool	1861	Chaloner		J.Heinsohn
86205	379	✳	Knight of St. Michael *Williams 75-81* Stl4mS	British	2221	294·7	42·2	24·1	Glasgow	1883	W.B.Thompson		Greenshields,Cowie & Co.
78724	380	✳	Knight of the Garter *Gordon 75-82* I S	British	1434	233·0	37·9	22·8	Liverpool	1877	T.Royden&Son		Greenshields,Cowie & Co.
	381	✳	Knight of the Garter WSk	British	87	78·0	21·4	10·6	BurtonStath'r	1886	J.Garside		G.Stannard
78762	382	✳	Knight of the Thistle *[illegible] 87-87* I S	British	1428	232·6	37·9	22·7	Liverpool	1878	T.Royden&Son		Greenshields,Cowie & Co.
	383	V	Knud *Jensen 78-78* WSr	Danish	122	91·0	22·1	10·3	Nysted	1875	F.Sparre		T.N.Brendstrup
	384	N	Knudsvig *Svensen* WBk	Norweg'n	349	119·6	28·8	15·2	Tvedestrand	1870	O.Gundersen	Y [illegible]	A.Gundersen
	385		Knut Åkerhjelm *Larsson -86* WBg	Swedish	244	101·5	24·1	13·0	Luleå	1855			C.Henckel
	386	D N	Knut Alfson *Kvernelaud* W3mSr	Norweg'n	292	126·7	26·1	13·9	Stavanger	1877		Y [illegible]	E.Berentsen
	387		Knut Bonde *Stakre* W3mSr	Swedish	156	93·9	21·6	10·9	Kilane	1872	A.Olsson		O.M.Carlsson
	388	N	Kobe (ex [illegible]) *Andresen -88* WBk	Norweg'n	334	133·1	28·0	12·9	Martigues	1866	Flastre&Mauras	Y [illegible]	Chr.Th.Boe
14273	389	R	Kocheko *Morehead -87* W3mSr	American	353	125·0	29·7	10·1	E.Machias,Me	1872		Y [illegible]	H.T.Knowlton
	390	G	Kodan *Christensen 82-82* WBn	Danish	136	97·2	21·8	11·8	Svendborg	1868	J.R.Andersen		N.Christensen
	391	V	Kodan *Hansen 76-76* WSr	Danish	133	88·8	22·0	11·3	Marstal	1876	F.Hansen		A.H.B.Kromann (B.Kromann, Jun.)
14390	392		Kodiak *Wessels* WSr	American	103	93·0	25·0	8·2	SanFrancisco	1881	M.Turner		AlaskaCommercialCo.
	393	V	Koerier (ex [illegible] Solari, ex [illegible]) *Wilkenstijer* WBn	Dutch	192	101·2	24·4	12·9	Plymouth	1875	Williams	Y [illegible]	H.Rietveld
	394		Koetus *Nyström* WBg	Russian	251	..	..	..		1869			N.Sittkoff
[illegible]	395		Kofuku Maru WBg	Japanese	211	97·3	25·2	11·5	Osaka	1881	Nishioka Nyemon		Abe Ikotaro
[illegible]	396		Koh-i-noor WBk	British	281	107·0	25·7	15·8	Glanyd'aCarn	1852			R.Gardner
[illegible]	397	✳	Koh-i-noor *Dingle 77-77* WBkn	British	237	120·2	25·3	12·9	Polruan	1877	Slade	[illegible]	Mrs.M.H.Hayes
[illegible]	398	✳	Koh-i-noor *Klingsick* WK	British	92	77·2	21·1	10·5	Whitby	1886	Whitby & Robin Hood's Bay S.B.C		J.W.Haylock
[illegible]	399	✳	Koh-i-noor *Edwards* WSr	British	53	..	14·6	8·0	Aberdovey	1853	R.Lewis		G.Edwards
	402	V	Koldinghuus *Sørensen -83* WBg	Danish	248	112·3	24·4	13·7	Kolding	1870	C.Hansen	[illegible]	M.R.B[illegible]k

✳=Lloyd's Register; D=Nederlandsche Vereeniging; G=Germanischer Lloyd; I=Registro Italiano; R=[illegible]
A=Veritas Austro-Ungarico; H=Veritas Ellenico; N=Norske Veritas; V=[illegible]

In the 4th Column, C=Composite (iron frame, wood planked); I=Iron; Stl=Steel; W=Wood.

The Dates of Metalling (Col. 11) are based on the records of the Books in which Vessels are respectively classed, or on the Reports of the Surveyors of [illegible]

Further particulars of Vessels classed by Lloyd's Register will be found in Lloyd's Register Book and Lloyd's Yacht Register.

		Registre dans lequel le navire est classé	Nom. Capitaine et dates d'entrée au service des Armateurs actuels et à bord du navire. Ancien Nom.		Matériaux et Gréement	Pavillon	Tonnage net	Dimensions Longueur	Largeur	Creux	Construit Où	Quand	Par qui	Date du doublage en métal	Armateurs ou Directeurs-Gérants	Port d'Armement
	401		Kolon	Libby	WSr	American	143	93·3	23·3	8·5	E.Machias,Me	1866			E.I.White	Machias,Me.
	402	G	Koltrasten (ex Thrush)	Nilsson 81-87	WBk	Swedish	283	107·5	23·9	15·5	Sunderland	1866	R.H.Pott Bros.		L.Jeansson	Calmar
	403		Kolumbus	Eriksson	WGlt	Russian	126	..	..	..	Wardö	18^{57}_{6}			A.England	Wardö
	404		Komena		WBg	Turkish	136	..	..	..	Greece	1866			Limnilee Panayotti Oglou Kemena	Lemnos
	405	N	Kommandör Svend Föyn		WS	Norweg'n	2493	259·0	17·8	27·4	Maitland,N.S.	1874	Lawrence	Y 12.88	Chr.Christensen	Sandefjord
	406	G	Kommendör (ARMENIEN)	Pålsson	WBg	Swedish	205	97·4	23·3	12·5	Hamburg	1855	M.Toosbuy	Y 6.88	C.R.Pahlsson	Trelleborg
	407		Komna	D. Comna	WBg	Greek	231	..	..	..	Syra	1867		Y 10.82	Comna frères	Syra
	408		Kondylo	Nautas	WSr	Greek	134	..	..	..	Syra	1887			Callitgas	Ydra
	409	N	Kong Carl	Knudsen 71-73	WBk	Norweg'n	504	143·1	29·5	17·2	Bremen	1865	P.Sager	Y 10.85	Sömme & Wattne	Stavanger
	410	N	Kong Carl	Olsen	WBk	Norweg'n	502	136·0	31·2	18·5	Risör	1868	T.Jensen		Jacob Prebensen	Risör
	411	N	Kong Carl	Olsen	WBk	Norweg'n	447	126·7	31·3	16·3	Tvedestrand	1871	A.&F.Smith	Y 11.87	A.& F.Smith	Tvedestrand
	412	N	Kong Carl	Hansen 80-82	WBg	Norweg'n	436	128·2	32·1	16·8	Aasg'rdstrand	1865	C.Christensen		J.Wesmann	Aasgaardstrand
	413	N	Kong Carl XV.	Eriksen	WBg	Norweg'n	236	115·3	26·2	12·4	Drammen	1867	J.Jörgensen		E.Eriksen&L.Tobiasen	Horten
	414	N	Kong Eystein	Larsen	WBk	Norweg'n	463	138·9	29·1	17·2	Kragerö	1868	Ths.Thoinesen	Y 11.83	A.Larsen-Nuur	Kragerö
	415	N	Kong Kaare (ABANDONED)	Torkildsen 85-88	WBn	Norweg'n	201	105·9	26·8	12·1	Finna,Norway	1875		Y 11.86	Holdt & Isachsen	Stavanger
	416	N	Kong Oscar	Nilsen 80-86	WBk	Norweg'n	357	121·6	29·8	15·1	Sandefjord	1874	J.B.Hotvedt		S.Larsen	Sandefjord
	417	N	Kong Oscar II. (ex May Dundas)	Kjorbe	WS	Norweg'n	790	156·5	34·6	20·2	P.E.Island	1859	Ellis	Y 2.89	Chr.Winsnes	Skien
	418	V	Kong Sverre	Salvesen -87	WBk	Norweg'n	526	163·6	32·1	16·5	Bergen	1871	J.Trampy&Sön	Y 6.87	Rasmus F.Olsen	Bergen
	419	DN	Kong Sverre	Andersen -85	WBk	Norweg'n	474	140·8	31·2	17·4	Sandefjord	1870	S.Hansen	Y 4.86	Ole Nielsen	Sandefjord
	420	N	KongSverre	Gittleson 86-86	WBk	Norweg'n	457	133·0	30·8	16·3	Grimstad	1877	N.OlsenGjömle	Y 5.88	Krog & Henschien	Lillesand
	421	DN	Kong Sverre	Hansen -86	WBk	Norweg'n	336	110·4	29·1	17·1	Lübeck	1841		Y 6.86	C.Christensen	Laurvig
	422	N	Kong Sverre	Thorsen 82-85	WBn	Norweg'n	219	113·6	25·9	11·0	Kragerö	1879		Y 11.87	L.Larsen	Kragerö
	423		Kongsberg	Berg	WBg	Norweg'n	238	103·4	27·1	13·4	Porsgrund	1857			L.W.Berg	Fredrikstad
	424	N	Kongsbyrd	Nielsen -83	WBk	Norweg'n	267	120·1	26·9	12·4	Grimstad	1874	F.SmithPeter-[sen	Y 10.86	M.Smith Petersen	Grimstad
	425		Kongsek (WRECKED)	Tönnessen	WSmSr	Norweg'n	228	113·5	26·2	12·3	Tönsberg	1877	Hansen	Y 2.81	Simonsen & Bjertness	Holmestrand
	426	V	Königin Augusta	Wilde 78-78	WBk	German	460	145·0	29·4	16·3	Stettin	1865	C.Domcke		C.A.Domcke	Stettin
	427		Konigin Elizabeth Louise	Meyer 88-89	WBk	German	485	118·9	31·0	19·5	Danzig	1845	J.W.Klawitter		Ernst Wendt	Danzig
1626	428	✻	Konigsberg	Jones -87	WSr	British	99	87·4	21·4	10·5	Kingston	1877	Duncan	Y 1.87	R.Reid	Banff
	429		KoninginAtiatoolRachman (ex Steamer Koninginder Nederlanden) (BROKEN UP)		WBk	Dutch	381	153·5	26·2	..	Java	1836	J.Brown		Sech Assan bin Aboot bin Abdul Alkaterie	Sourabaya
7134	430	✻	Konoma	Thompson 79-82	WBk	British	800	169·0	33·5	20·0	Kingston,N.B.	1881	J.&T.Jardine	Y 1.87	J.&T.Jardine	Richibucto,N.B.
	431		Konrad		WBg	Russian	323	122·8	28·8	14·6	Swartbäck	1876	J.Kjellman		A.F.Brendenberg	Ingo
	432		Konstantin		WSr	Russian	220	..	..	..		1874			I.Judins,Arfv.	Impilaks
	433		Konstantin		WBk	Russian	120	85·6	19·7	9·8	Cherson	1869			T.P.Rodocanachi	Odessa
	434	A	Konstantinos (ex Andreas)	Cirli	WBk	Greek	516	141·7	30·7	..	Syra	1875	C.Cristali	Y 8.83	Sons of K. Zerboudaki	Syra
	435		Konstantinos	Coutsoudakis	WBk	Greek	393	..	..	..	Syra	1887			G.Coutsoudakis	Syra
	436		Konstantinos (ex Panayhia Portaitissa)	G. Filinas	WBk	Greek	391	123·3	28·9	..	Castellorosso	1877	Nicolettos	Y 6.83	K.Filinas Bros.	Syra
	437	N	Konstantinos	Frangoulis	WBg	Greek	288	106·9	25·9	..	Sunderland	1871		Y 12.85	N.Frangoulis	Piræus
	438	H	Konstantinos	Musuris	WBg	Greek	250	93·8	25·3	14·9	Syra	1871			E.Musuris	Zante
	439		Konstantinos	Stefos	WBg	Greek	237	97·8	24·1	..	Galaxidi	1866	C.Papapetros		G.Stefos	Galaxidi
	440		Konstantinos	A. Currentis	WBn	Greek	200	..	..	..	Galaxidi	1869		Y 2.83	A.Currentis	Galaxidi
	441	V	Konstantinos	Argentinis	WBn	Greek	160	80·4	23·0	14·3	Syra	18^{67}_{82}			G.N.Dedes	Spezzia
	442		Konstantinos		WSr	Greek	114	76·8	21·5	..	Syra	1882			L.Koufopendelis	Syra
	443	A	KonstantinosKriaras		WBk	Greek	476	146·1	29·9	18·4	Cardiff	1872	Batchelor	Y 11.83	P.Kriaras	Syra
	444		Konte Mieñoris		WBg	Greek	278	92·8	25·3	15·4	Galaxidi	1870			P.Amelopoulos	Poros
	445		Koogee		Stl Lr	British	484	220·0	30·0	8·3	Dumbarton	1887	W Denny & Bros		IrrawaddyFlotillaCo Ld	Glasgow
	446		Koonywah		I Flat	British	777	197·0	25·7	8·8	PortGlasgow	1878	Robt. Duncan [&Co.		Irrawaddy Flotilla Co., Ld.	Glasgow
	447		Koophandel (ex Elisabeth Maria, ex Elisabeth)	Ekamp	WBk	Dutch	613	135·8	33·1	20·5	Krimpen	1856	J.&C.Smith&Zn.		N.Brantjes	Purmerend
	448	✻D	Koophandel	Haldeman 83-87	WSmSr	Dutch	216	104·6	23·4	12·8	Bridgwater	1865	Gough	Y 11.83	van Harwegen & den Breeum	Vlaardingen
	449	✻	Kooringa (ex Invermore)	Gibbons 84-86	I Bk	British	1175	226·0	35·2	21·6	S.Shields	1874	Softley&Co.		Trinder, Anderson & Co.	London
	450		Kooroojeh Shilleh		WSr	Turkish	125	..	..	..	Kooroojeh Sh.	1869			Russmnallee-Aghassin ben Suleiman Agha	Russumat

✻=Lloyd's Register; D=Nederlandsche Vereeniging; G=Germanischer Lloyd; I=Registro Italiano; R=American Record; A=Veritas Austro-Ungarico; H=Veritas Ellenico; N=Norske Veritas; V=Bureau Veritas.

Dans la 4me Colonne, C=Composite (membrure en fer, bordé en bois); I=Fer; Stl=Acier; W=Bois.

Les Dates du doublage en métal (Col. 11) ressortent des livres dans lesquels les navires sont respectivement classés ou des rapports des Inspecteurs du Lloyd's Register.

Pour plus amples renseignements sur les navires classés par le Lloyd's Register, voir le Lloyd's Register Book et le Lloyd's Yacht Register.

Official No.	No. in Book	Society in which classed	Ship's Name. Material and Rig. Master, and Date of Appointment to present Owner's Service and to Ship. Ship's Late Name.	Flag	Net Tonnage	Length	Breadth	Depth	Built: Where	When	By whom	Date of Metalling	Owners or Managers	Port
[illegible]	451	A	Korais *Saliaris* W Bk	Greek	651	152·2	32·3	..	Syra	1879		Y 6.86	H.Saliaris	[illegible]
296	452		Korais W S	Greek	184	91·9	23·8	..	Syra	1884			A.N.Pithis	[illegible]
354	453		Korio Maru (ex Culoria) W Bg	Japanese	149	83·2	20·3	10·1	England	1865			Tokuda Yataro	[illegible]
40	454		Korlos Nayos W Bg	Greek	243	94·2	26·2	..	Syra	1865			P.Pasaras	Syra
	455	N	Kornmo *Korsen -88* W Bk	Norweg'n	568	119·9	31·9	17·9	Grimstad	1883	F.SmithP'tersen	Y 7.86	F.Smith Petersen	[illegible]
	456	N	Korsvei *Nilsen* W 3mSr	Norweg'n	377	135·2	29·8	14·2	Porsgrund	1879		[illegible]	J.C.& G.Knudsen	[illegible]
	457	V	Kosak *Tork -88* W Bk	Russian	244	130·2	28·4	11·8	Raumo	1884	Ehrnberg	Y 1.87	P.J.Sallmen	Raumo
47290	458		Kosciusko *Logan* W Bk	British	1192	212·5	36·3	22·3	Aberdeen	1862	W.Hood&Co.	Y 6.85	M.C.Cowlishaw	[illegible]
	459	N	Kosmos *Jorgensen* (ex Sylvia, ex Lizzie Bunstat) W Bk	Norweg'n	709	155·0	31·4	20·1	Portland,Me.	1854		Y 3.84	Tygesen & Aalholm	[illegible]
31058	460		Kosoomogaidin W Bg	British	136	87·0	22·3	10·4	Negapatam	1858			Vana Moona Ravana Mana Narsinapa Chaty	[illegible]
14180	461		Kossack *Sproule* W Bn	American	328	114·1	28·2	15·8	E.Machias,Me	1866			Carleton & Moffat	[illegible]
	462		Kotka W	Russian	204	..	..	..		1865			W.Jansson	[illegible]
74	463		Kotonowo Maru W S	Japanese	776	161·8	32·1	17·6	New York				Sato Jubei	[illegible]
658	464		Koun Maru W Bg	Japanese	120	87·4	18·8	10·2	Yetchin	1881	Kobayashi [Otojiro		Abe Yechitaro	[illegible]
210	465		Koupas *Podara -87* W Bg	Greek	347	115·8	28·4	..	Syra	1876		Y 10.85	N.Podaras	Syra
629	466		Koyei Maru W Sr	Japanese	135	93·0	24·0	11·5	Hakodate	1879	Tsuzaki Toyoji		Naka Yeisuke	[illegible]
864	467		Koyei Maru W S	Japanese	100	51·4	20·8	13·7	Owari	1882	ItoKinpei		Takamatsu Chobei	[illegible]
	468	N	Kragerö *Halvorsen -87* W Bk	Norweg'n	531	156·3	32·2	17·4	Kragerö	1872	J.Hamborg	[illegible]	H.Biörn,jun.	[illegible]
	469	N	Kreon *Jensen 78-78* W Bk	Norweg'n	447	129·0	31·0	17·0	Arendal	1878	Gabrsen	[illegible]	Chr.Th.Boe	[illegible]
	470		Krey-Woggersin *Kramer -88* W Bk	German	376	120·0	28·5	16·0	Rostock	1862	W.Zeitz		H.Bauer	[illegible]
	471	✻G	Kriemhild *Meyer 84-84* I Bk	German	807	177·0	31·7	18·6	Kiel	1884	GermaniaWrks.		C.M.Matzen & Cie.	[illegible]
89607	472	✻	Kriemhilda *Toogood* W K	British	87	79·0	20·8	10·6	Brixham	1885	S.Dewdney&Sn.		J.Guzzwell	[illegible]
	473	V	Krimpen aan de Lek *Van der Vegte 82-85* I Bk	Dutch	1070	214·9	35·1	20·7	Krimpen	1885	J.&K.Smit		P.van der Hoog	[illegible]
	474	A	Krina *Catzudachi -88* W Sr	Greek	214	..	..	..	Syra	1888	A.Evdemon		L.Vafiadachi	Syra
17	475		Krini *Catzudachi* W Bk	Greek	291	103·3	27·5	..	Syra	1867	S.Cufalachi	[illegible]	D.Vafiadakis	Syra
	476		Kristina *Lindholm 79-* W Sr	Russian	195	93·5	26·6	12·4	Norrnäs	1875	G.Skomars		H.Snickars	[illegible]
	477		Kristina W Sr	Russian	182	..	..	..		1875			K.A.Karsten	[illegible]
	478	G	Kristina *Jansson -84* W Sr	Russian	182	91·6	25·3	13·5	Lappfjärd	1876			C.A.Hallman	[illegible]
	479	V	Kristina *Anderson* W Bk	Swedish	530	138·7	29·1	18·1	Dalkarlså	1872	E.A.Lofgren	Y 2.88	F.R.Hullström	[illegible]
	480	V	Kristina *Hansen* W Bk	Swedish	314	130·5	25·5	13·3	Wärmnäs	1875	C.A.Hahn	[illegible]	H.Fex	[illegible]
	481		Kristina *Thurin -88* W Bk	Swedish	289	105·6	24·7	13·6	Greifswald	1865			J.Nilsson	[illegible]
	482		Kristina *Kindberg 74-88* W S	Swedish	216	105·8	24·2	12·1	N.America				F.Nyström	[illegible]
	483		Kristina *Backman -87* W Sr	Swedish	174	107·5	22·5	11·1	Vestervik	1857			E.Andersson	[illegible]
	484	GV	Kristina Nilsson *Gostard -88* W Bg	German	280	117·4	25·0	12·7	Poskallavik	1877	N.Nilsson	Y 12.86	J.Thomsen	[illegible]
	485		Kristo W Bn	Turkish	188	..	..	..	Greece	1866			Kerrabolee Kristo	[illegible]
	486		Krona *Jansson* W Glt	Russian	113	..	..	..					K.A.Lundman	[illegible]
	487	V	Krona *Jonsson* W Bk	Swedish	204	119·4	24·8	12·8	Landscrona	1873	Avendskoug	[illegible]	N.Olsson	[illegible]
	488	N	Krone *Tonning* W 3mSr	Norweg'n	303	129·8	27·9	13·7	Stavanger	1877		[illegible]	S.Rasmussen	[illegible]
	489	N	Kronos *Bentzen 85-85* W Bk	Norweg'n	353	121·8	28·5	15·0	Grimstad	1874	A.Bentsen	[illegible]	N.Thue Johnsen	[illegible]
	490	N	Kronprinds Carl *Christensen* W Bg	Norweg'n	257	110·9	26·6	13·3	Molde	1858		[illegible]	T.Christensen	[illegible]
	491		Kronprindsesse Caroline *Hansen 86-88* W Bn	Danish	167	81·7	23·3	14·1	Svendborg	1842	L.Møller		H.C.Christensens Enke	[illegible]
34	492		Kubli W Cr	Russian	100	78·0	21·0	10·0	Ubla,Russia	1875			R. von Stahl-Holstein	[illegible]
	493		Kuhiriwan W Sr	Turkish	154	..	..	..	Heraclea	1882			Hattab Tujari Kurekli Haje Rashid	[illegible]
746	494		Kumasaka Maru (ex Coomassie) W Bk	Japanese	448	136·0	27·2	17·2	Sunderland	1874	Gibbon	[illegible]	Tsukada M[illegible]	[illegible]
	495		Kung Tai *Smith -88* W Lorcha	British	316	130·0	32·0	9·7	Shanghai	1888	Boyd & Co.		C.A.Sm[illegible]	[illegible]
	496		Kungsborg *Nilsson* W Bkn	Swedish	121	84·0	21·3	10·0	Borgvik	1856			A.B.Nilsson	[illegible]
	497		Kungsskogen *Carlsson -87* W 3mSr	Swedish	135	91·9	21·2	10·1	Borgvik	1856			H.E.Fuhlroth	[illegible]
	498		Kuopio *Jansson* W Sr	Russian	105	94·0	22·0	9·5	Kuopio	1858	P.Myohausen		V.Janson	[illegible]
	499		Kushadi Bahri W Bg	Turkish	100	..	..	..	Terebolie	1883			Nakib zadeh Ali Effendi	[illegible]
[illegible]	500		Kutana Maru No. 1 W Sr	Japanese	285	112·4	25·1	12·5	Tokio	1878	HiranoS.B.Y'rd		Uchimura Hansaku Kwaisha	[illegible]

✻ = **Lloyd's Register**; D = Nederlandsche Vereeniging; G = Germanischer Lloyd; I = Registro Italiano; [illegible]
A = Veritas Austro-Ungarico; H = Veritas Ellenico; N = Norske Veritas; V[illegible]

In the 4th Column, C = Composite (iron frame, wood planked); I = Iron; Stl = Steel; W = Wood.

The Dates of Metalling (Col. 11) are based on the records of the Books in which Vessels are respectively classed, or on the Reports of the Surveyors of [illegible]

Further particulars of Vessels classed by Lloyd's Register will be found in Lloyd's Register Book and Lloyd's Yacht Register.

No.		Nom. Matériaux et Gréement. Capitaine et dates d'entrée au service des Armateurs actuels et à bord du navire. Ancien Nom.		Pavillon	Tonnage net.	Longueur	Largeur	Creux	Construit. Où.	Quand.	Par qui.	Date du doublage en métal	Armateurs ou Directeurs-Gérants.	Port d'Armement.
501		Kutana Maru No. 2	WSr	Japanese	286	113·0	26·0	14·1	Tokio	1878	HiraneS.B.Y'nl		Seiko Kwaisha	Hiogo
502	XV	Kvik *Sørensen 87-89*	WBk	Norweg'n	416	150·6	39·8	14·5	Drammen	1874	J.Jörgensen	[illegible]	S.C.Larsen	Brevig
503		Kvik *Halvorsen -87* (CONDEMNED)	WSr	Norweg'n	175	98·2	23·4	11·1	Oldenburg	1860		[illegible]	C.Thiis	Fredriksstad
504	X	Kvik *Christiansen*	WSr	Norweg'n	153	93·4	24·6	11·0	Porsgrund	1872			B.I.Larsen	Sandefjord
505		Kvik *Carlsen*	WSr	Norweg'n	147	96·2	23·4	11·9	Stavanger	1868	Iverem	[illegible]	C.E.Rönneberg&Sönner	Aalesund
506	X	Kvik *Eriksen 73-73*	WSr	Norweg'n	112	84·6	22·5	10·0	Krageró	1873	J.Nilsen		P.Bredsdorff	Kragerö
507		Kwaika Maru	WBg	Japanese	262	125·3	23·2	13·2	Germany	1876			Kamada Katsutaro	Sanuki
508		Kwaiso Maru	WBkn	Japanese	315	115·0	25·8	15·2	Hiogo	1883	OhmatsuJugoro		Tatsuma Kiyo	Hiogo
509		Kwaiso Maru No. 1	WSr	Japanese	300	120·0	25·0	12·0	Tokio	1880	KawasakiS.B.Y.		Kaigun Zosenjo	Tokio
510		Kwaiso Maru No. 2	WSr	Japanese	200	101·5	20·4	9·0	Tokio	1880	KawasakiS.B.Y.		Kaigun Zosenjo	Tokio
511		Kwanko Maru	WBk	Japanese	278	141·1	22·9	13·6	Tasmania	1868			Inouye Zenyomon	Hiogo
512		Kwanko Maru	WBk	Japanese	103	77·9	18·6	9·0	Osaka	1880	TodaShojiro		Nagoshi Aisuke	Osaka
513		Kwansho Maru	WBg	Japanese	182	97·7	20·2	12·0	Settsu	1887	NagataSanjiro		Abe Ichitaro	Osaka
514	*	Kylemore *Thompson*	StlBk	British	1198	226·2	36·4	22·1	Port Glasgow	1880	J.Reid&Co.		Nicholson & M·Gill	Liverpool
515		Kymatoessa	WBg	Greek	315	..	..	..	Castellamare	1864			G.Micaro	Syra
516		Kyprus	WSr	Greek	145	84·0	24·9	12·7	Cyprus	1881	C.Hatzikostis		K.Kanakis	Zante
517		Kyriakoula	WBg	Greek	199	84·3	25·6	..	Syra	1868			M.H.D.Xenou	Syra
518	H	Kyriakoula *Fassili*	WSr	Greek	172	..	..	..	Syra	1877		C 92	D.Fassili	Syra
519		Kyzicos *Pneumaticos*	WSr	Greek	100	..	..	..	Syra	1879			N.Vassiliadi	Syra

ADDITIONAL SHIPS ARE INSERTED IN THE SUPPLEMENT AT THE END OF THE REGISTER.

* = Lloyd's Register; D = Nederlandsche Vereeniging; G = Germanischer Lloyd; I = Registro Italiano; R = American Record; A = Veritas Austro-Ungarico; H = Veritas Ellenico; N = Norske Veritas; V = Bureau Veritas.

Dans la 4me Colonne, C = Composite (membrure en fer, bordé en bois); I = Fer; Stl = Acier; W = Bois.

Les Dates de doublage en métal (**Col. 11**) ressortent des livres dans lesquels les navires sont respectivement classés ou des rapports des Inspecteurs du Lloyd's Register.

Pour plus amples renseignements sur les navires classés par le Lloyd's Register, voir le Lloyd's Register Book et le Lloyd's Yacht Register.

Official No.	No. in Book.	Registry in which classed.	Ship's Name. Material and Rig. Master, and Dates of Appointment to present Owner's Service and to Ship. Ship's Late Name.		Flag.	Net Tonnage.	Dimensions. Length.	Breadth.	Depth.	Built. Where.	When.	By whom.	Date of Metalling.	Owners or Managers.
[illegible]	1	R	**L. A. Burnham** *Watts -87*	W3mSr	American	370	138·8	32·0	11·0	E.Boston	1873	J.M.Brooks	[illegible]	J.H.Conant & Co.
[illegible]	2		**L. A. Plummer**	W3mSr	American	375	129·7	32·2	12·4	Bath,Me	1882	Goss, Sawyer, [& Packard		I.W.Benjamin
[illegible]	3		**L. A. Rose** *Cox*	WSr	American	138	98·0	27·5	7·4	Patcouck [Creek,N.J.	1868	J.Risley		G.M.Vaulant
[illegible]	4		**L. B. Curtis**	WSp	American	171	102·0	26·6	9·6	Boston, Mass.	1880	Crosby		R.S.Bradley
[illegible]	5	DV	**L.B.Gilchrist** *Wood* (DISMANTLED)	WS	American	1100	183·0	37·0	24·3	ThomastonMe	1886	S.Watts&Co.	[illegible]	W.B.Wood
[illegible]	6	V	**L.C.K.** *Brebier* (CONDEMNED) (ex[illegible])	WBk	British	530	118·6	28·6	17·[illegible]	[illegible]	1879	S.Gautier	[illegible]	Irlapoulle Valaydon[illegible]
[illegible]	7		**L. C. Owen**	WBn	British	189	101·2	24·8	12·8	P.E.Island	1868	McDougall		J.Reeves
[illegible]	8	R	**L. F. Munson** *McKown*	WBn	American	392	127·0	30·4	15·2	[illegible]	1871		[illegible]	F.Talbot & Co.
	9	V	**L. Hagen** *Busch 79-79*	WBk	German	503	133·5	30·2	17·7	[illegible]	1879	R.Schlor	[illegible]	J.J.W[illegible]
[illegible]	10		**L. Holway** *Bryant*	WSr	American	115	88·0	26·0	7·7	[illegible], Me.	1869			J.C.Bryant
[illegible]	11		**L. L. Hamlin** *Harvey*	WSr	American	140	105·0	26·6	8·8	Calais,Me.	18[illegible]	Nickerson &		J.M.Velsor
[illegible]	12	V	**L. M. Smith** *Smith -84*	WBkn	British	379	130·9	32·0	12·4	Kempt,N.S.	1884	R.Rose [Ridcout	[illegible]	R.Rose
[illegible]	13		**L. M. Strout** (NOT NOW EXISTING)	WSr	American	100	84·5	25·1	8·9	Harringt'n,Me	1854			E.C.Arey
[illegible]	14		**L. P. Churchill** *Murray*	WSr	British	113	84·0	21·0	9·[illegible]	Shelburne,NS.	1874			J.G.Murray
[illegible]	15		**L. Q. C. Wishart** *Carlin*	W3mSr	American	227	115·2	29·5	9·1	GreenwichNJ.	1867			J.Kiengle
[illegible]	16		**L. S. Levering** *Carson*	WSr	American	284	119·1	29·3	9·9	WilmingtnDel	1854			Wilmington Nav.Co.
[illegible]	17	RV	**L. Schepp** *Gates*	WS	American	1776	224·3	42·1	27·1	Kennebu'nkMe.	1878	Titcomb [&Thompson	Y 7.86	J.F.Chapman
[illegible]	18	R	**L. Staples** *Stowers*	WBn	American	369	123·0	27·5	16·3	Stockton,Me.	1865		11.87	N.W.Bridge
[illegible]	19		**L. T. Whitmore** *Blackington*	W3mSr	American	280	120·9	30·1	10·6	Rockland,Me.	1874			Amos & Rankin
[illegible]	20		**L. & D. Bullock**	WSr	British	208	110·0	26·3	8·5	MillPoint	1874	W.Jameson		A.Eccles
[illegible]	21		**L. & W. Armstrong** *Alexander*	WBg	American	159	92·0	20·0	10·0	N.Havn,Conn	18[illegible]			L.W.&P.Armstron[illegible]
[illegible]	22		**L'Amico** *Esposito* (WRECKED)	WBg	Italian	350	110·8	28·6	18·8	[illegible]	1865	S.Mauro	Y 7.86	V.Esposito
	23		**L'Amitié** *Liem Djoen Kie*	WBk	Dutch	136	88·6	21·3	10·9	Grisses	18[illegible]	P.A.Berest	[illegible]	Liem Djoen Tik
[illegible]	24		**L'Angelina**	WBn	Italian	138	95·2	26·6	9·8		18[illegible]			B.Mione di V.
[illegible]	25	A	**L'Arcangelo** *Romeo*	WBn	Italian	270	112·2	26·3	17·6	Castellamare [di Stabbia	1857		Y [illegible]	Maria Vedª Castag[illegible]
[illegible]	26	I	**L'Armonia** *Tori -87*	WSr	Italian	129	87·3	22·9	12·3	Savona	1878	G.B.Magnano	Y 6.84	F.Salvini di P[illegible]
	27		**L'Aurora** *Malinconico*	WSr	Italian	124	84·3	21·9	10·9	Castellamare	1886	G.Bonifacio		G.Malinconico
[illegible]	28	✻	**L'Avenir** *Green*	WSk	British	76	75·3	20·3	10·3	Rye	1884	J.C.Hoad		S.Green
[illegible]	29		**L'Avvenire** *Rogers*	WBn	British	351	127·6	26·8	15·4	[illegible]				C.Rogers
[illegible]	30		**L'Esperance** *Jenkins 82-82*	WBg	British	288	110·6	26·9	16·6	Sunderland	1868	J.J.Gibbon	Y 12.86	W.Robertson
	31	V	**L'Esperance** *Errard*	WSr	French	111	85·3	20·3	10·8	Dunkirk	1876	Vandewiele		E.Muchery
	32		**L'Esperance** *Knick 83-73*	WBg	German	212	97·0	26·0	13·2	Anclam	1857	E.Lober		E.Schütz
[illegible]	33	A	**L'Immacolata** *Sagliobene*	WBkn	Italian	360	118·4	29·5	18·0	Riposto	1855		Y 4.86	S.Scuto d'Allio
[illegible]	34	I	**L'Immacolata** *Dolciame*	WBn	Italian	114	82·0	23·0	11·0	TorredelGr'co	1862			S.Langella
[illegible]	35		**L'Indifferente** (ex "Salvatore Naccari")	WBg	Italian	222	96·8	26·6	16·0	Procida	1858			T.V.N.Naccaro
[illegible]	36		**L'Italia** *Langella -87* (ex Catello Ciro)	WBn	Italian	242	103·6	25·1	15·9	Castellamare [di Stabbia	1870	G.B.Bonifazio	Y [illegible]	Saglimbene fratelli
	37	I	**L'Italia** *Sassella -88*	WBn	Italian	115	..	..	..	Castellamare [di Stabbia	1881	Bonifacio	Y [illegible]	Flli. Pandolfi
[illegible]	38	VI	**L'Onesto** *Tomei*	WSr	Italian	118	90·9	23·0	10·5	Viareggio	1885	C.A.Raffaelli	[illegible]	Cav. A.Raffaelli
	39	DN	**L'Union** *Andersen*	WBg	Norweg'n	205	98·7	25·6	13·5	Bergen	1850	Berendsen		Paul Hansen
[illegible]	40	IV	**L'Unione** *Rizzo* (ex Antonietta P.)	W3mSr	Italian	308	117·4	26·6	14·8	Sestri Ponente	1882	G.Peragallo	Y [illegible]	F.Piraino fu G.
	41	✻	**La Araucana** *Leech -88* (ex Glamorgan)	WBk	Chilian	600	151·2	2[illegible]	19·2	Bideford	1859	Cox	Y 6.88	Cia Esplotadora de Lota [illegible]
	42	N	**La Bella** *Olsen*	W3mSr	Norweg'n	256	118·8	27·9	13·2	Tvedestrand	1870	M.Ekhrup	[illegible]	Johan Smith
[illegible]	43		**LaChalaca** *Maniani 87-87* (ex Porcia)	WBk	Peruvian	232	113·5	25·3	11·4					F.Maniani
[illegible]	44	✻	**La Escocesa** *Evans 62-68*	IS	British	946	202·0	34·2	21·3	Dundee	1868	Gourlay Bros.		Balfour, Williamson & [illegible]
[illegible]	45	I	**La Fiducia** *Fevola*	WBk	Italian	631	147·3	31·1	20·5	Procida	1884	N.Lubrano	Y 5.86	V.Demartino & Co.
[illegible]	46		**La Flor del Plata** *Salkon 73-[illegible]*	W3mBn	British	322	121·8	24·1	16·9	Maryport	1856	R.Ritson&Co.		H.J.Rider
	47	V	**La Foi** *Agnieray -87*	WSr	French	111	84·6	20·3	10·8	Dunkirk	1875	Vandewiele		Bouvarlet frères
[illegible]	48		**LaFortuna** (ex Rolfsen) *Nigga* (BROKEN UP)	IBk	Austrian	1022	227·0	30·4	18·4	Amsterdam	1859			BancaGeneraled'[illegible]
[illegible]	49		**La Fortuna** *Treglia* (ex Andolia)	WBn	Italian	133	91·1	23·1	12·0	Taranto	1863			G.La Porta
	50													

✻ = **Lloyd's Register**; D = Nederlandsche Vereeniging; G = Germanischer Lloyd; I = Registro Italiano; R = [illegible]
A = Veritas Austro-Ungarico; H = Veritas Ellenico; N = Norske Veritas; V = [illegible]

In the 4th Column, C = Composite (iron frame, wood planked); I = Iron; Stl = Steel; W = Wood.

The Dates of Metalling (**Col. 11**) are based on the records of the Books in which Vessels are respectively classed, or on the Reports of the Surveyors [illegible]

Further particulars of Vessels classed by Lloyd's Register will be found in Lloyd's Register Book and Lloyd's Yacht Register.

No.	Registre	Nom. Matériaux et Gréement. Capitaine et dates d'entrée au service des Armateurs actuels et à bord des navires. Ancien Nom.	Pavillon	Tonnage	Longueur	Largeur	Creux	Construit Où	Quand	Par qui	Date du doublage	Armateurs ou Directeurs-Gérants	Port d'Armement
51	V	La Gaité *Durand* WSr	French	124	80·7	23·0	10·5	Nantes	1865	Guibert&Blond'l		A.Venard	Nantes
52		La Gironde *Dill* WSr	American	195	105·0	31·0	9·0	PortLudlow, [Wash	1875	[illegible] Bros.		A.Rowe	San Francisco
53	N	La Gitana *Myhre 85-88* WBk	Norweg'n	285	116·0	28·5	12·5	Tönsberg	1875	[illegible]	[illegible]	Morten Magnus	Christiania
54		La Gloria *[illegible]* WBn	Italian	169	83·3	24·9	14·8	Procida	1857			V.Lubrano	Naples
55	✻	La Hogue *Wagstaff* WS	British	1331	226·0	35·0	22·9	Sunderland	1855	J.Laing	Y 3.85	T.Hick	London
56	✻	La Manche *Charles 84-87* WBg (ex [illegible])	French	249	114·1	24·6	14·2	Salcombe	1863	Whiddin	Y 2.86	H.Lecharpentier	Granville
57	I	La Maria *Trapani* WBk	Italian	412	122·0	28·7	18·0	Castellamare [di Stabbia	1874	A.[illegible]	Y 10.86	G.Calcagno	Naples
58		La Maria *Sorrentino* WBn	Italian	151	..	..	..	Castellamare	1857			F.Starace & Co.	Naples
59	✻	La Mascotte *Kerwood* WSk	British	82	76·8	20·5	10·6	Rye	1885	G.&T.Smith		S.Green	Grimsby
60		La Ninfa *Williams 89* WSr	American	120	86·0	25·0	7·7	PJeffersonNY	1877	J.J.Harris		T.P.H.Whitelaw	San Francisco
61	I	La Nuova Maria Liberi *Parmense* WBg (ex Twenty)	Italian	326	102·6	29·3	18·0	Varazze	1864	N.Cerruti	Y 3.85	V.Mazzella	Naples
62	V	La Nuova Provvidenza *Mangiapane* WBg	Italian	252	104·0	25·5	15·9	Castellamare [di Stabbia	1854	L.Trambarulo [fu M.	Y 6.88	F.Starace	Castell. di Stab.
63		La Perla WBk	Chilian	167	96·8	23·0	12·0		1886			Cia.Chilena de Vallen	Valparaiso
64		La Plata WSr	Argentine	136	100·0	26·1	9·8	Sestri Ponente	1884	F^lli Cadenaccio	Y 3.84	Ortelli & Luca	Buenos Ayres
65	I	La Plata *Moglia* WBk (ex Ronoletta)	Italian	433	129·9	30·8	19·6	Sestri Ponente	1871	G.Peragallo	Y 4.87	Gaet. Moglia	Genoa
66	N	La Plata *Hansen -87* WBk ([illegible])	Norweg'n	629	143·5	30·2	19·0	Miramichi	1866	Harley	[illegible]	N.Olsen	Arendal
67	N	La Plata *Andreasen* W3mSr	Norweg'n	260	122·4	26·0	12·5	Stavanger	1874	M.Magnus	[illegible]	Aarenes & Dalens Interessentskab	Flekkefjord
68		La Plata (ex Lanno) WSr	Uruguayn	131	89·3	22·1	10·5	Elsfleth	1862	J.Ahlers		L.Spiguo	Monte Video
69	✻	La Querida *Evans 86-86* I Bk	British	679	178·3	29·9	19·3	Sunderland	1876	W.Doxford&Sns		Tomlinson, Hodgetts & Co.	Liverpool
70	I	La Saletta *Lubrano -88* WBn	Italian	426	116·1	30·8	19·9	Procida	1873	N.Lubrano	Y 7.8[illegible]	A.Scotto	Naples
71		La Speranza *Amabile* WBn (1885)	Italian	104	79·4	19·7	10·4	Vietrie sul [Mare	1886	P.Savastano		P.Savastano	Castellamaro
72	✻	La Tour d'Auvergne [illegible] WBg	French	178	101·2	23·6	12·1	St. Malo	1878	Parnet	Y 2.87	L.Hovins	Saint Malo
73		La Ville d'Ostende *Horn* WBk	Belgian	559	135·0	28·8	18·6	Loano	1870		Y 10.83	Assoc. Internationale du Congo	Antwerp
74		La Zingara *Sprowles 84-* WBk (BRISING)	British	264	123·6	23·5	14·1	Liverpool	1860	T.Royden&Son		R. & W.King	Bristol
75	N	Labora *Kallberg -88* WBk (ex Aleppo)	Norweg'n	662	153·9	31·2	21·7	Quebec	1867	Gingra	Y 6.86	Carl Tollefsen	Egersund
76	V	Laborieuse *Fontaine* WSr	French	102	85·3	18·1	10·7	Paimpol	1878	T.Pilvin	Z 10.83	A.Buhot Launay	Paimpol
77	V	Laborieux *Philippe 73-84* WBg	French	188	92·5	23·3	15·0	St.Malo	1875	L.Mallard		J. Le Pomellec	Binic
78		Labrador (ex [illegible]) W3mSr	Portug'se	233	102·6	22·8	12·8					Bensande & Co.	Lisbon
79		Lac Superior WSp	British	105	98·0	22·8	6·9	St.Thomas de [Pierreville	1879	A.Lapurierre		D.Alie	Montreal
80	I	Lacaruna *Simonetti* WB (ex Maria [illegible])	Italian	786	171·8	33·1	21·5	Varazze	1875	S.Craviotto	Y 4.89	F^lli Mortola	Genoa
81	R	Lackawanna *Closson* WSr	American	157	103·0	28·0	8·5	Searsport.Me.	1881	Dunbar	Y 10.85	A.G.Closson	Searsport,Me.
82		Lada (WRECKED) *Medanich* WBk	Austrian	544	137·8	29·4	19·7	Fiume	1871	L.Rosa	Y 12.86	G.Paicurich fu M.	Fiume
83	✻	Ladakh *Thomas* IS	British	1942	269·0	39·5	24·2	Southampton	1883	Oswald, Mord'nt [&Co.		E.Bates & Sons	Liverpool
84	✻	Ladstock *Whitten -87* I Bk	British	816	201·0	32·2	19·5	Whitehaven	1876	Whiteh'vnSBCo		T.Connell	Liverpool
85		Lady WSr	Russian	225	..	..	..		1874			J.Karlsson	Sund
86	✻	Lady Agnes *Friend* WBk	British	357	144·0	26·3	14·3	Salcombe	1867	Harnden	Y 1.87	J.Friend	Salcombe
87	✻	Lady Agnes *Piper 83-83* WSr	British	84	79·1	21·5	10·1	S.Agnes	1877	Hitchins	Y 6.8[illegible]	M.T.Hitchins	Hayle
88	✻	Lady Bertha *Lewis -88* WSr	British	98	85·7	20·7	10·2	Salcombe	1878	Harnden	Y 1.88	R.H.Sladen & Co.	Salcombe
89		Lady Bird *Brien* WBn	British	173	94·0	25·0	11·9	Quebec	1872	Samson		J.Bourgard	Quebec
90		Lady Bird WSrYt	British	130	111·7	23·1	12·3	Cowes	1865	J.S.White		M.Cope	Cowes
91	DN	Lady Blessington *[illegible] -88* W-	Norweg'n	1123	178·3	37·1	23·4	Belfast,Me.	1855	Ring	Y 2.87	Act [illegible] "Magnum" (Nathan)	Tvedestrand
92	✻	Lady Cairns *Morrissey 87-87* IS	British	1265	216·4	35·4	22·5	Belfast	1869	Harland&Wolff		R.Martin & Co.	Dublin
93	✻	Lady Cecilia Hay *Ladds* WSr	British	74	76·6	19·7	9·1	Peterhead	1878	Stephen		J.Ladds	Peterhead
94		Lady Chandos (ex [illegible]) WSr	British	100	72·8	21·2	9·9	St.Malo	1861			Powley, Thomas & Co.	Yarmouth
95		Lady Constance *Abbott* WBg	British	195	90·0	24·5	18·6	Sunderland	1858	Taylor&Co.		J.G.Hill & Co.	Southampton
96		Lady Denison WBg	British	129	93·3	18·5	10·0	Sydney,NSW	1859	J.Cuthbert		W.Walton	Sydney,N.S.W.
97	✻	Lady Douglas *Keen -87* I Bk	British	594	166·5	27·3	17·9	Greenock	1856	Scott&Co.		Taylor Sons & Co.	London
98	✻	Lady Dufferin *Marshall 85-88* WBk (STRANDED)	British	894	178·4	36·1	20·0	P.E.Island	1874	Yeo	Y 9.88	W.W.Dingle	Liverpool
99	✻	Lady Elibank *Buxton 86-87* W3mBn	British	219	113·2	24·5	12·9	Plymouth	1880	Hill	Y 9.88	N.B.Thoyts (Matthews & Luff)	London
100	✻	Lady Elizabeth *Karran 85-87* I Bk	British	1155	223·0	35·0	24·4	Sunderland	1879	R.Thompson,jr.		G.C.Karran	Castletown,I.M

✻=Lloyd's Register; D=Nederlandsche Vereeniging; G=Germanischer Lloyd; I=Registro Italiano; R=American Record;
A=Veritas Austro-Ungarico; H=Veritas Ellenico; N=Norske Veritas; V=Bureau Veritas.

Dans la 4me Colonne, C=Composite (membrure en fer, bordé en bois); I=Fer; Stl=Acier; W=Bois.

Les Dates du doublage en métal (Col. 11) ressortent des livres dans lesquels les navires sont respectivement classés ou des rapports des Inspecteurs du Lloyd's Register

Pour plus amples renseignements sur les navires classés par le Lloyd's Register, voir le Lloyd's Register Book et le Lloyd's Yacht Register.

Official No.	No. in Book	Registry in which classed	Ship's Name, Material and Rig. Master, and Dates of Appointment to present Owner's Service and to Ship. Ship's Late Name.	Flag	Net Tonnage	Length	Breadth	Depth	Built: Where	When	By whom	Metalled	Owners or Managers
14741	101	R	Lady Ellen *Clark* W Sr	American	207	104·7	29·1	8·9	Dorchester N.J.	1856			Obed Clark
5242	102		Lady Flora Hastings *Joiner 85-85* W Bn	British	155	102·9	20·5	12·1	Dundee	1839			Whitstable Shipping Co.
75076	103		Lady Franklin W Bn	British	166	109·8	24·8	8·7	Jervis Bay, N.S.W.	1880	A. Coulon		W. Yabsley
60576	104	✻ R	Lady Gertrude *Mush 85-88* (ex Paci) W Bk	British	500	136·3	31·1	18·9	Fiume	1867		y.s.s.	G. Milne & Co.
	105	N V	Lady Gladys *Hansen -87* (ex Maggie R. Scott) W S	Norweg'n	1365	203·2	41·0	23·8	Maitland, N.S.	1874	Putnam	y.s.s.	A. C. Hansen
	106		Lady Grace W Bk	Chilian	545	160·4	29·6	15·7	Bordeaux	1866			Hazen & Compton
[illegible]	107		Lady Harewood *Williams* W Bk	British	382	131·8	27·5	16·4	Plymouth	1868	Williams	y.s.s.	T. H. Williams
22074	108	✻	Lady Head *Bishop* W Bk	British	457	138·9	28·2	17·5	Sunderland	1865	Mills	y.s.s.	Hudson's Bay Co.
86090	109	✻	Lady Isabella *Congdon 75-82* I S	British	1462	255·7	38·3	20·6	Dumbarton	1882	A. M'Millan & Son		North British Shipping Co.
80460	110	✻	Lady Jane *Toms* W Sr	British	96	84·6	23·3	10·7	Porth	1880	J. Osborne		R. E. Bennett
11923	111	✻	Lady Jocelyn *Watt 83-83* (ex steamer Brazil) I S	British	2138	251·0	39·0	24·9 / 32·6	London	1852	C. J. Mare & Co.		Shaw, Savill & Albion Co., Ld.
	112		Lady Lampson *Soldergren -88* W Bk	Hawaian	412	131·5	28·5	17·4	Sunderland	1868	G. & J. Mills		C. Brewer & Co., Ld.
58300	113	✻	Lady Lawrence *McAvoy 83-88* I S	British	1467	225·0	37·7	23·0	Liverpool	1868	T. Royden & Son		J. T. Cowley
[illegible]	114	✻	Lady Lisgar *Hughes 83-88* W Bk	British	1207	202·0	38·0	23·0	[illegible]	1871	Baldwin	y.s.s.	T. O. Jones
18715	115		Lady Louisa Pennant *Hughes -87* W Sr	British	57	62·3	16·5	8·9	Bangor	1847	John Parry		J. Ellis & Co.
80412	116		Lady Mabel *Corbett* W Bn	British	215	118·3	26·8	10·7	Parrsboro' N.S.	1883	W. B. Connabell	y.s.s.	A. Hayes
71238	117		Lady McDonald W Sr	British	284	137·5	25·7	12·2	P. Burwell, Ont.	1873	G. W. Pontine		W. Yonell
[illegible]	118		Lady Milne *Bonneville -87* W Bk	British	304	114·7	26·1	15·2	Bermuda	1864			J. Bonneville
31878	119		Lady Monck W Bg	British	179	92·1	25·0	11·4					P. R. Middlemiss
[illegible]	120		Lady Nairn *Richards 83-83* W Bk	British	446	135·0	30·0	16·2	Georgtwn P.E.I.	1882	D. Gordon	y.s.s.	D. Gordon
[illegible]	121	✻	Lady Neave *Jones* W Sr	British	89	87·3	21·0	9·7	Amlwch	1876	W. Thomas		W. Thomas
6104	122	✻	Lady Octavia *McMurtry 82-82* I Bk	British	1132	200·6	36·4	22·2	Warrington	1854			G. G. Macfarlane & Co.
[illegible]	123		Lady Palmerston *Le Feuvre 85-88* I Bk	British	1247	219·4	35·4	22·5	Greenock	1864	R. Steele		Lowden, Edgar & Co.
[illegible]	124	✻	Lady Pelham *Whitworth* I Sk	British	89	80·6	20·0	9·6	Beverley	1885	Cochrane & Co.		J. W. Haylock
74464	125	✻ D	Lady Penrhyn *Thomas 85-85* I Bk	British	815	197·7	31·6	19·1	Glasgow	1875	A. Stephen & Sns.		J. Richards & Co.
47474	126		Lady Rowena *Grady 87-87* (ex Robert Cushman) (ABANDONED) W Bk	British	1181	179·1	37·5	24·4	Bath, Me.	1858		y.s.s.	G. T. Soley & Co.
[illegible]	127		Lady Russell *Laurenson* (HULK) W Bk	British	920	164·0	30·2	21·8	Moncton, N.B.	1851	Salter	10.83	Francia Bros.
72887	128	✻	Lady Ruthven *Kavanagh* I S	British	1591	258·3	39·1	23·6	Greenock	1875	R. Steele		Adam, Hamilton & Co.
[illegible]	129		Lady St. Aubyn *Dalziel* W Sr	British	150	101·0	25·6	11·7	Penzance	1870	M. Matthews		T. Heselton
[illegible]	130		Lady Stanley *Greenwood 85-86* (WRECKED) W Bg	British	331	161·0	24·4	16·0	Shields	1851	T. Metcalfe & Sns.		J. Greenwood
[illegible]	131	✻	Lady Stuart *Brown -87* (LOST) W Sr	British	96	81·5	21·6	10·0	Perth	1869	M'Pherson	y.s.s.	W. Knowles
91251	132	✻	Lady Wolseley *Evans 84-88* I Bk	British	1238	230·0	35·2	21·3	Sunderland	1884	North of Englnd. S.B. Co., Ld.		T. Beynon & Co.
[illegible]	133	✻	Lady of Avenel *Richards* W Sr	British	152	99·2	23·0	12·2	Falmouth	1874	Trethowan	y.s.s.	E. D. Anderton
[illegible]	134		Lady of the Lake *Mosher* W Bk	British	552	146·0	30·8	18·6	Economy, N.S.	1873	McKenzie		S. Mosher
[illegible]	135	✻	Lady of the Lake *Lake* W K	British	88	80·0	21·0	10·6	Galmpton	1886	T. Mathews		J. W. Haylock
[illegible]	136		Lafreniere et St. Onge W S	British	131	103·2	22·5	8·0	Yamaska	1874			James Williamson
56757	137		Lafrowda *Harvey 62-67* W Sk	British	122	93·5	22·1	11·3	Plymouth	1867	Ridley		A. H. James
	138		Lage W Bn	Brazilian	255	..	..	..					A. M. Lage
	139	N	Lagertha *Melvig* (ex Water Lily) W Bg	Norwegian	341	122·0	28·8	15·0	Clifton, N.B.	1864	W. Flewelling		J. N. Jacobsen
	140	N	Lagertha *Apenes* W Bg	Norwegian	281	114·5	27·7	14·6	Drammen	1857	Holmens Värft		H. A. Apenes
[illegible]	141	I V	Laghetto *Vitterbo* W Bk	Italian	900	177·8	34·8	23·2	Sestri Ponente	1882	G. Bozzano	y.s.s.	Luigi Penco
[illegible]	142		Lagoda *Parker 87-87* W Bk	American	271	107·5	26·8	18·3	Scituate, Mass.	18 26/52			W. Lewis
	143	V	Lagos W Bk		350	131·5	28·9	15·3	Hamburg	1867	E. Dreyer	V 11.85	
	144	N	Lagos *Olsen* W Bn	Norweg'n	125	92·7	21·8	11·1	Stavanger	1880	F. Magnus	y.s.s.	E. Berentsen
	145		Laguna *Nicolai* (ex Impératrice Charlotte) W Bk	French	375	142·7	28·5	13·7	Bordeaux	1865	E. Cardon	y.s.s.	Vve G. Oriot
[illegible]	146	R	Lahaina *Allen 77-86* W Bn	American	417	126·5	32·0	12·0	Millbridge, Me.	1878	E. Dyer	y.s.s.	J. S. Emery & Co.
	147		Laine *Hallonen 78-85* W S	Russian	231	102·7	29·1	12·3	[illegible]	1877	P. Sandström		J. Esko
	148		Lainetar *Nyross* W Bk	Russian	534	140·0	32·1	18·5	Nykarleby	1865	H. Kjäldström		C. G. Sandback
	149		Lainetar *Nyross* W Bk	Russian	408	133·0	29·2	17·5	Raumo	1868	E. Justin		G. Granlund
[illegible]	150	✻	Laira *Hughes* I Bk	British	492	156·1	27·3	16·6	Sunderland	1870	W. Pile & Co.		C. B. Stone

✻ = Lloyd's Register; D = Nederlandsche Vereeniging; G = Germanischer Lloyd; I = Registro Italiano; R = [illegible]
A = Veritas Austro-Ungarico; H = Veritas Ellenico; N = Norske Veritas; V = Bureau [illegible]

In the 4th Column, C = Composite (iron frame, wood planked); I = Iron; Stl = Steel; W = Wood.

The Dates of Metalling (Col. 11) are based on the records of the Books in which Vessels are respectively classed, or on the Reports of the Surveyors of [illegible]

Further particulars of Vessels classed by Lloyd's Register will be found in Lloyd's Register Book and Lloyd's Yacht Register.

No. dans le livre	Inscrits dans registre	Nom. Matériaux et Gréement. Capitaine et dates d'entrée au service des Armateurs actuels et à bord du navire. Ancien Nom.	Pavillon	Tonnage	Longueur	Largeur	Creux	Construit: Où	Quand	Par qui	Date du doublage en métal	Armateurs ou Directeurs-Gérants.	Port d'Armement.
151	* D	Lake Erie *Mallam 86-86* I Bk	British	935	202·4	33·6	20·3	Glasgow	1868	Barcl'y,Curle&C		CanadaShippingCo.,Ld.	Montreal
152	*	Lake Leman *Magey* C S	Chilian	1035	202·4	34·9	21·6	Renfrew	1867	W.Simons&Co.	V 10.86	F.Caparro	Valparaiso
153	*	Lake Ontario *McKie 86-86* I S	British	1061	217·7	34·7	21·0	Glasgow	1868	Barcl'y,Curle&C		CanadaShippingCo.,Ld.	Montreal
154	*	Lake Patos *Callan -87* W3mSr	British	107	101·9	22·9	9·1	Dundalk	1884	DundalkP.S. [&S.B.Co.,Ld	V 2.86	W.Larmour	Liverpool
155	*	Lake Simcoe *Susson -86* I Bk	British	384	152·2	24·1	14·3	Greenock	1871	R.Steele		F.W.Brightman	London
156	*	Lake Superior *Friston 84-84* I S	British	1274	231·9	35·9	22·3	Greenock	1868	R.Steele		J.Shepherd & Co.	London
157	V	Lakefield *Griffiths 87-87* W Bk	British	997	176·6	35·6	22·7	Quebec	1875	Dunn	V 7.87	R.Thomas & Co	Liverpool
158		Lakoeadjan W Bg	Dutch	106	..	..	..						Dongala (Celebes)
159		Laleah W Sr	British	109	83·3	22·8	9·0	Hillsboro',N.S.	1855			F.P.Payson	Digby,N.S.
[illegible]		Lalia (ABANDONED) *Lyon* W Bk	British	396	136·0	29·3	14·5	GreenCoveNS	1867	Goudy		J.C.Rogers	St.John,N.B.
161		Laliberté W Sp	British	135	110·0	24·0	8·0	St.Jean [Deschaillons	1875	J.Laliberte		C.Pageau	Quebec
162	V	Lalla *Nielsen* W Bk	Norweg'n	1099	184·0	37·1	21·3	Maccan.N.S.	1874	J.O'Brien	V 8.86	Chr.Hannevig	Christiania
163	*	Lalla Rookh *Kinnear 86-86* I Bk	British	814	196·3	31·9	19·7	Liverpool	1876	R.&J.Evans&Co		E.C.Friend & Co.	Liverpool
164		Lalla Rookh W Bg	British	147	98·3	20·8	12·2	Peterhead	1848			Bernard Byrnes	Sydney,N.S.W.
165	N	Lamartine *Hansen* W Bg	Norweg'n	192	96·2	24·2	12·5	Wenersborg	1848			L.Dahl (in liquidation)	Tönsberg
166	DG	Lambertha *Bos -82* W Glt (1st N.R.)	Dutch	137	92·0	22·3	10·0	Hoogezand	1862	J.A.Hoites		J.W.Bos	Groningen
167	A	Lamek *Marcovich -87* W Bk	Austrian	531	141·7	28·8	18·8	Trieste	1868	Stabilimento [Tecnico	V 6.83	V.Marincovich	Cucisto
168		Lamplighter W Bk	American	248	110·0	29·3	9·6	Calais,Me.	1859			R.W.Burch	New York
169	*	Lampo I Sr	Spanish	59	78·4	51·0	8·4	Liverpool	1878	W.H.Potter&Sn		Woodhouse & Co.	Marsala
170	R	Lanarkshire *Curnew -88* W S	British	1422	211·0	39·0	23·7	Quebec	1864	Gingras	V 8.87	W.H.Ross & Co.	Glasgow
171	*	Lanarkshire *Fairlie 86-86* I Bk	British	794	195·0	32·2	19·7	Glasgow	1873	Dobie&Co.		T.Law & Co.	Glasgow
172	*	Lancashire *Evans 84-84* W S	British	1159	190·5	38·0	23·1	Quebec	1880	Gingras	V 8.87	Parry,Jones & Co.	Liverpool
173		LancashireLad *Brown [Davies* W3mSr	British	142	92·8	22·9	11·7	GlassonDock	1870	Simpson		Porter's ShippingCo,Ld.	Fleetwood
174	*	Lancashire Lass *Sloggett* W3mSt	British	135	110·3	23·2	10·6	Newport	1863	Wilmott		Mrs.H.Sloggett	Fleetwood
175		Lancashire Witch *Jackson* W S	Peruvian	1574	210·0	35·5	22·5 / 33·6	Quebec	1854	Lomas&Sewell		Grace Bros.& Co.	Lima
176	*	Lancaster *Walker* W K	British	82	76·5	20·4	10·3	Galmpton	1883	A.Gibbs		T.Fieldgate	Grimsby
177	*	Lancaster Castle *Thorpe [1884* I S	British	2095	278·8	40·4	24·4	Workington	1884	RWilliamsn&Sn		LancasterCastleShipCo. Ld.(J.Chambers & Co.)	Liverpool
178		Lancaster W Barge	British	220	125·0	22·8	9·0	Lancaster	1873			Montreal Transprtn. Co.	Montreal
179	V	Lancefield *Delap* W Bk	British	994	179·0	36·5	21·5	Monckton.NB	1881	Crandall [&Croudace	V 11.81	A.E.Killam	Monckton,N.B.
180		Lancer (WHALER) *Whiteside 86-86* W Bk	American	295	115·7	28·0	16·2	Newburyport, [Mass.	1852			W.Lewis	New Bdfrd,Mass.
181		Lancet *Lewis -83* W Bg	British	214	97·5	24·0	15·0	Sunderland	1855	W.Taylor		W.Milburn	Whitby
182		Lancing *Holfield -88* I 4mSr	British	2600	356·0	43·8	27·3	Glasgow	1865	R.Napier & Sons		A.E.Kinnear & Co.	London
183	G	Landbouw *Ekker -87* W Sr	Dutch	134	89·6	21·1	10·3	Pekela	1863	B.B.Drenth		B.H.Kraus	Veendam
184	N	Landsblomsten *Ellefsen [87-83* W Bg	Norweg'n	215	102·9	25·9	12·6	Sandefjord	1875	J.Brandi		M.Oppen	Laurvig
185	RV	Landseer *Smith 83-83* W S	American	1348	200·0	39·6	24·0	Newburyport, [Mass.	1874	G.W.Jackman,jr	V 3.87	G.C.Lord & Co.	Boston,Mass.
186	V	Landskrona *Sinclair 70-86* W Bk	British	1370	206·8	39·7	23·5	Gardner's [Creek, N.B.	1886	W.&R.Wallace	V 8.86	B.Smith & Sons	Windsor,N.S.
187	N	Lanercost *Mörch* W Bk	Norweg'n	545	151·8	30·2	17·6	Sunderland	1874	N.Gibbon	V 1.88	C.Mörch	Fredriksstad
[illegible]	*	Langdale *McAllister 81-87* I S	British	1966	275·0	40·1	24·2	Liverpool	1885	W.H.Potter&Ss		J.D.Newton & Co.	Liverpool
189	*	Langland *Morgan 85-86* I Bk	British	639	173·6	29·6	17·6	Middlesbro'	1875	R.Dixon&Co.		Simpson Bros.	Swansea
190	*	Langstone *Tonkin 83-83* I Bk	British	746	186·4	31·2	19·6	Sunderland	1869	W.Pile&Co.		Shaw, Savill & Albion Co. Ld.	Southampton
191	R	Lanie Cobb *Cobb 80-80* W3mSr	American	231	114·0	30·0	8·6	Brewer,Me.	1874	J.Oakes		H.Lord	Bangor,Me.
192	*	Lanisley *Beckerleg -87* I Sr	British	148	104·0	23·1	11·2	Hayle	1887	Harvey&CoLim		T.Bolitho & Sons	Penzance
193		Lanko *Mattsson* W Sr	Russian	232	100·0	27·4	13·1	Halliko	1867			J.E.Andersson	Eckerö
194		Lannen W Bn	British	103	70·9	18·5	9·8	WeymouthNS	1846			Henry Moss	Melbourne
195	*	Lanoma (WRECKED) *Whittingham 64-79* I Bk	British	665	184·6	31·1	17·9	Sunderland	1876	S.P.Austin& [Hunter		T.B.Walker	London
196	*	Lanoy *Richards -88* W Bk	British	229	120·3	23·6	12·7	Bideford	1869	Cox	V 10.85	Rider,Son & Andrew	Bristol
197		Lansdowne *Lockhart 83-84* W S	British	1486	219·0	40·2	23·7	Horton,N.S.	1884	J.B.North		J.B.North	Windsor,N.S.
198	*	Lantana *Murchison -88* W Bn	British	246	111·0	26·2	11·8	Georgetown, [P.E.I.	1885	McDonald& [Westaway	V 6.86	M.McDonald	Charlottetn,P.E.I.
199	*	Laomene *Ramsall 74-87* I S	British	1746	263·0	39·4	24·1	Liverpool	1880	W.H.Potter&Son		H.Fernie & Sons	Liverpool
200	RV	Lapland *Carlisle 57-80* W Bk	American	617	147·6	32·9	18·7	Bucksport,Me	1880	Genn&Hill	V 3.88	J.S.Emery & Co.	Boston,Mass.

*=Lloyd's Register; D=Nederlandsche Vereeniging; G=Germanischer Lloyd; I=Registro Italiano; R=American Record;
A=Veritas Austro-Ungarico; H=Veritas Ellenico; N=Norske Veritas; V=Bureau Veritas.

Dans la 4me Colonne, C=Composite (membrure en fer, bordé en bois); I=Fer; Stl=Acier; W=Bois.

Les Dates du doublage en métal (Col. 11) ressortent des livres dans lesquels les navires sont respectivement classés ou des rapports des Inspecteurs du Lloyd's Register.

Pour plus amples renseignements sur les navires classés par le Lloyd's Register, voir le Lloyd's Register Book et le Lloyd's Yacht Register.

Official No.	No. in Book	Registry in which classed	Ship's Name. Material and Rig. Master, and Dates of Appointment to present Owner's Service and to Ship. Ship's Late Name.	Flag	Net Tonnage	Dimensions: Length	Breadth	Depth	Built: Where	When	By whom	Date of Metalling	Owners or Managers
[illegible]	201	✻	Lapwing (ex Jane) *De Gruchy 80-85* I Bk	British	747	189·0	30·8	13·1	Sunderland	1870	[illegible] & Co.		L.W.Evans
	202		Laraquete W S	[illegible]	626	142·3	28·0	19·7	[illegible]	1865			A.von der Heyde
	203		Lareño W Sr	Spanish	105	82·7	22·1	9·2	Palma	1870			P.Mayol
[illegible]	204	✻D	Largo Bay *Smith 88-88* I Bk	British	1178	221·9	35·8	21·2	Kinghorn	1878	J.Key&Sons		Hatfield,Cameron & Co.
[illegible]	205	✻	Largo Law *Niven 76-81* I S	British	1587	253·8	38·7	22·8	Glasgow	1881	Napier,Shanks [& Bell		T.Law & Co.
[illegible]	206		Larissa *Wilkinson* W Bk	British	316	116·0	25·2	16·1	Sunderland	1855	L.Gales	[illegible]	J.Davison
[illegible]	207		Lark (ex Early Bird) W Bk	British	442	132·0	29·0	17·0	Newcastle,U.S.	1856			H.B.Crum
[illegible]	208		Lark *Osborne -87* W Sr	British	197	108·2	23·1	11·2	Barnstaple	1880	Westacott&Sns		G.Thompson
	209	V	Larkan *Övergaard 88-88* W Sr	Norweg'n	228	122·6	25·2	11·8	Gothenburg	1875	F.O.Bergström	[illegible]	Rederiet 'Lärkan' (W.Klein)
[illegible]	210	✻	Larnaca *Giles -88* I S	British	1435	234·5	38·0	23·0	Liverpool	1878	T.Royden&Son		[illegible]
[illegible]	211	V	Larnica *Boyd 88-88* W S	British	1458	209·1	39·4	24·0	Gardner's Creek, N.B.	1878	W.&R.Wallace	[illegible]	B.Smith & Sons
	212		Lars W	Russian	134	..	..	..		1872			L.Hagan
	213	V	Lars Jørgensen *Jørgensen* W Bn	Danish	120	93·6	21·2	9·4	Svendborg	1871	J.R.Andersen		J.Jørgensen
	214	N	Larus *Ohlsen -88* W Bg	Norweg'n	204	102·7	26·5	12·6	Tvedestrand	1864	O.Gundersen		J.W.Prebensen
	215		Las Palmas W Bg		207	92·2	25·8	13·4	LasPalmas	1853			
43135	216		Lass of Gawler *Dixon* W Bg	British	224	119·3	23·5	13·8	Poole	1862	Meadus		Tasmanian S.N.Co.
490	217	V	Latava *Ohsoling 87-87* W 3m Sr	Russian	528	151·2	32·0	15·9	Majorenhof	1886	Ohrmann&Kies		Navigation Co."Austra"
91868	218	✻	Latimer *Robertson 87-87* I S	British	1734	260·0	38·7	23·3	W.Hartlepool	1885	W.Gray&Co.		J.Lidgett & Sons
36194	219	R	Latona (ex Squando) *Marconini* W Bk	British	1032	176·6	36·2	24·8	Rockland,Me.	1858		Y 2.87	T.Hughes & Son
[illegible]	220	RV	Latona *Lipsett -88* W Bk	British	948	180·1	36·3	20·1	Harvey,N.B.	1880	G.S.Turner	[illegible]	G.F.Smith
[illegible]	221	✻	Latona *Bowden 79-85* W Bkn	British	282	115·0	25·0	14·8	Port Dinorwic	1866	Jones	[illegible]	F.C.Hipwood
[illegible]	222		Latona *Romkey -87* W Sr	British	107	79·9	24·0	9·4	Lunenburg NS	1886	D.Smith		L.Anderson
71879	223		Latona *Seaboyer* W Yl Yt	British	101	95·7	20·3	12·5	Cowes	1875	J.S.White		[illegible] A.W [illegible]
	224	V	Latona *Brandt 77-77* W S	Danish	133	93·0	20·5	10·4	Thurø	1866	L.Kaas		Sm.J.Møller
	225	V	Latona *Petersen 88-88* W Sr	Danish	112	82·4	20·1	10·7	Middelfart	1865	H.Illum		F.Winding
	226	N	Latona *Gahrsen -88* W Bk	Norweg'n	558	143·2	31·1	17·5	Risör	1874	O.Eriksen	[illegible]	J.Prebensen
	227	N	Latona *Jensen -87* W Bk	Norweg'n	342	113·4	28·6	15·9	Arendal	1856		[illegible]	A.Arvesen
457	228		Latwija *Martinson -86* W 3m Sr	Russian	343	133·5	29·5	13·7	Wandsen	1885	J.Ohrmann		H. & J.[illegible]
	229	N	Laugen *Pettersen* W 3m Sr	Norweg'n	265	118·9	26·9	13·2	Stavanger	1874		[illegible]	R.Elfsvig
	230	NV	Lauget *Sivertsen* W Bk	Norweg'n	538	143·2	31·8	17·9	Risör	1870		[illegible]	O.Jørgensen
[illegible]	231		Laughing Wave *Miles 85* W Bg	British	161	105·7	21·8	11·0	Fre'm'ntle W A	1868			J.Bateman
870	232		Laujei Maru No. 2 W Bg	Japanese	159	97·3	23·8	11·1	Owari	1885	Okamoto Zen-[kichi		Sakakibara [illegible]
[illegible]	233		Laura *Olsen* W Bk	British	349	144·8	27·8	16·8	Sunderland	1858		Y 2.87	E.Lantalum
[illegible]	234	V	Laura *McArthur -87* W Bkn	British	329	126·1	30·3	12·9	N.GlasgowNS	1876	J.McColl	[illegible]	H.D.McArthur
[illegible]	235		Laura *Fowle 67-72* W Bn	British	186	100·0	24·0	13·0	Newquay	1861	Thomas		C.C.Fowle
[illegible]	236		Laura *Attwood 86-86* W Bn	British	174	105·9	23·1	11·9	Ipswich	1861	Read		J.R.Daniels
[illegible]	237		Laura *Lemon* W Bg	British	162	94·8	20·7	13·3	S.Shields	1855	Bushell		R.S.Pearce
[illegible]	238		Laura *Quinlan* W Sr	British	125	86·0	26·0	8·0	St.John,N.B.	1871			James F.Raymond
[illegible]	239	R	Laura *Lamson 79-79* W 3m Sr	American	523	138·9	32·7	14·7	Millbridge,Me.	1879	J.W.Sawyer	[illegible]	J.S.Emery & Co.
	240	V	Laura *Mortensen 72-88* W Bn	Danish	136	96·1	21·3	10·3	Thurø	1870	L.Kaas		J.M.Nielsens Enke
	241	DG	Laura (ex [illegible], ex Golden West) *Kester 64-* W Bk	Dutch	342	129·2	29·3	14·4	Meteghen,N.S.	1867	J.Thibido	[illegible]	H.Jans & Co.
	242		Laura *Strey 79-* W Bk	German	311	106·3	24·0	16·1	Colberg	1853	G.Mogenburg		E.F.Hackbarth
	243		Laura (ex A.Philippe) *Jancke -66* W Sr	German	140	96·1	23·2	10·2	Elbing	1865	Fechter		G.W.M.Jancke
1241	244	I	Laura *Castellano* W Bk	Italian	425	140·7	29·3	17·8	Cassano	1879	F.De Rosa	[illegible]	Eredi di P.A.Castellano
[illegible]	245		Laura *Domenica* W Bg	Italian	351	107·2	30·0	18·2	Venice	1856		[illegible]	Paolo Adragna
	246		Laura W [illegible]		334	125·9	28·4	15·4	Savona	1876	G.B.Pongiglione		
[illegible]	247		Laura (ex Emma) *Villa* W Bk	Italian	265	124·0	25·7	14·2	Apenrade	1867		[illegible]	P.Villa
[illegible]	248		Laura *Gentile* W Bk	Italian	222	98·4	24·9	15·7	Castellammare	1862	A.Bonifacio	[illegible]	Sarcinelli & Gentile
[illegible]	249	A	Laura *Arcucci* W Bk	Italian	114	79·1	22·6	12·7	Lerici	1858	P.Rolla		Slli. Arcucci
	250	N	Laura *Hansen* W Bk	Norweg'n	355	130·5	28·3	16·5	Bergen	1853	J.Trumpy	[illegible]	P.Semb

✻ = Lloyd's Register; D = Nederlandsche Vereeniging; G = Germanischer Lloyd; I = Registro Italiano; R = Amer[ican] [illegible]
A = Veritas Austro-Ungarico; H = Veritas Ellenico; N = Norske Veritas; V = Bureau [Veritas]

In the 4th Column, C = Composite (iron frame, wood planked); I = Iron; Stl = Steel; W = Wood.

The Dates of Metalling (Col. 11) are based on the records of the Books in which Vessels are respectively classed, or on the Reports of the Surveyors of [illegible]

Further particulars of Vessels classed by Lloyd's Register will be found in Lloyd's Register Book and Lloyd's Yacht Register.

No.	Cl.	Nom. Capitaine et dates d'entrée successives des Armateurs actuels et à bord du navire. Ancien Nom.	Matériaux et Gréement	Pavillon	Tonnage net	Longueur	Largeur	Creux	Construit: Où	Quand	Par qui	Date du doublage en métal	Armateurs ou Directeurs-Gérants	Port d'Armement
	¥	Laura *Olsen*	W Bk	Norweg'n	335	103·4	26·4	17·4	Liverpool	1829	Bland & Co.		Joh. Spörck	Fredrikshald
		Laura *Gazzolo 86-86* (ex L...)	W Bk	Peruvian	192	104·3	26·2	10·9					G. Artadi	Callao
	V	Laura *Cesario* (ex Rodrigo Nortem)	W Bn	Portug'se	507	128·6	27·6	17·2	Economy	1870		Y 1.87	J. J. M. Feijoo	Brava, CapeV'rd
		Laura *Protzing*	W3mSr	Russian	137	66·9	23·3	8·8	Windau	1872			Gebr. Gutschmidt	Windau
	¥V	Laura *Lindgren 81-86*	W Bk	Swedish	333	128·9	24·5	14·1	Harburg	1864	G. H. Kraus	Y 5.84	L. V. Akerson	Gefle
		Laura *Soderstrom 71-76*	W Sr	Swedish	102	78·7	18·5	9·4	Hanover	1856			C. Smitterberg	Klintehamn
	V	Laura	W Sr	Uruguay'n	208	109·9	24·0	10·7	Sundsvall	1871	W. Hägglund	Y 4.85		
		Laura Ann *Balls*	W Sr	British	110	85·4	26·0	11·5	Yarmouth	1855	Fellows		W. T. Balls	Yarmouth
	I	Laura B. *Santiago -88* (ex Emilia N.)	W Bk	Italian	647	152·5	32·1	21·8	Sestri Ponente	1876	P. Cadenaccio	Y 8.86	Laura Bagnasco-Boero	Genoa
		Laura C. Wolff *Huston*	W Sr	American	111	100·0	17·5	8·3	Hamburg, Pa.	1874			D. McDevitt	Philadelphia
	R	Laura E. Mosser *Grogan*	W3mSr	American	405	132·0	32·0	12·0	Belfast, Me.	18 77/83			A. J. Bird	Rockland, Me.
	V	Laura Emily *Spicer -88*	W Bk	British	768	162·8	34·6	19·9	Maitland, N.S.	1875	B. Smith	Y 2.89	T. E. Kenny	Maitland, N.S.
	✻	Laura Emma *Blamey 84-87*	W Sr	British	99	88·2	23·1	10·9	Padstow	1880	Cowl & Sons	Y 6.87	W. Philp	Padstow
	✻	Laura Griffith *Williams*	W Sr	British	95	74·8	20·4	10·6	Pwllheli	1865	Roberts		E. Williams	Carnarvon
		Laura Madsen *Madsen*	W3mSr	American	329	140·0	38·0	10·0	CoosBay, Oreg	1882			B. H. Madison	San Francisco
	V	Laura Maria *Schroeder*	W Bk	German	486	130·3	30·6	18·2	Danzig	1868	F. Devrient		A. Gibsone	Danzig
		Laura May *Dieling*	W3mSr	American	234	121·6	32·6	9·5	CoosBay, Oreg	1875			W. S. Browne	San Francisco
		Laura Pike *Baas*	W Sr	American	139	96·0	29·0	17·7	Eureka, Cal.	1875			C. A. Baas	San Francisco
		Laura Robinson *Mathews*	W Sr	American	124	92·8	27·6	7·3	Mattapoisett, [Mass.	1869	L. C. Holmes		E. Harding	Edgartown, Mass
		Laura Rosa	W Bk	Chilian	224	106·6	24·3	..	Massachusetts	1852			D. Fuentes & B. Squella	Talcahuano
		Laura Rosalia *Newman 18-88* (ex Pitindaro)	W S	Chilian	1121	186·6	36·6	23·3	Bath, Me.	1864	J. P. Morse		Charles Lyon	Valparaiso
		Laura V. Rose *Leeds*	W3mSr	American	466	134·6	35·0	15·0	Bridgeton, NJ	1875			J. Allen	Philadelphia
	V	Laura & Gertude (ex Susie M. Jones) *Hilla*	W Bk	German	991	168·6	35·1	23·3	Baltimore	1863	Cooper & Slicer	Y 1.89	D. Cordes & Co.	Bremen
		Laura & Isabel	W Bkn	British	147	88·9	22·7	12·6	Barry Island	1861	Crawshay		R. Fell, jun., & Co.	Troon
		Laure *Lhermite*	W Bg	French	154	80·1	21·3	12·1	Granville	1843			Riotteau & fils	Granville
	V	Laureano *Barcelo*	W Bn	Spanish	152	87·6	25·0	11·5	Blanes	1858		Y 1.88	J. B. Barcelo	Barcelona
		Laureano *Mirambell*	W Plc	Spanish	148	117·7	23·6	11·4	Mataro	1885		Y 2.86	Mirambell	Barcelona
		Laurentius *Lundberg 84-86*	W Sr	Swedish	102	71·5	23·5	8·7	Roslagen	1878	A. Lundberg		L. Lundberg	Norrtelje
	DN	Laurenze *Hansen*	W Bg	Norweg'n	261	105·9	27·6	13·8	Drammen	1863	J. Jörgensen		Chr. Swane	Christiania
	V	Lauretta G. *Giannelli*	W Bg	Italian	152	93·2	24·7	13·1	Sestri Ponente	1878	G. Pezzani	Y 1.86	C. Giannelli	Porto Ferraio
		Lauria	W Bn	Spanish	125	153·2	19·1	8·4	Manila	1884	J. Rodriguez		Muñoz hers. & sobrinos	Manila
		Lauria (ex Leila)	W Bn	Spanish	120	87·3	19·1	8·2	Tayabas	1876			M. Garcia	Manila
	G	Laurita *Beyen -88*	W3mSr	German	235	113·8	24·6	12·6	Geestemünde	1868	R. C. Rickmers	Y 6.81	Alsen'sche Portland Cem. Fabrik	Hamburg
		Lauritz *Mortensen -67*	W Sr	Danish	147	92·9	23·2	11·3	Karrebæks- [minde	1867	L. Hansen		P. A. Mortensen	Næstved
		Lautaro (ex Sparkling Dew)	W Bk	Chilian	314	114·0	26·9	16·8	Sunderland	1863	Lister		G. Jenkins	Valparaiso
	AV	Lav *Lakmasica -88* (ex Jedanesti Dubrovacki)	W Bk	Austrian	742	157·4	29·3	21·3	Gravosa	1873	N. S. Bradicich	Y 7.88	L. Ciurlizza fu A.	Ragusa
	I	Lavarello *Martolo -86* (ex Credito)	W Bk	Italian	461	131·2	31·4	18·8	Varazze	1869	B. Camogli	Y 11.87	F. Schiappacasse & Co.	Genoa
		Lavina	W Bn	British	119	98·6	20·8	9·1	Port Stephens, [N.S.W.	1876	Roderick & [McPherson		R. McBurney	Brisbane
		Lavinia *Valentine 76-76*	W Bk	British	349	125·8	27·7	15·4	P. E. Island	1873	Mason	Y 11.86	W. Hamilton	Adelaide, S.A.
	✻	Lavinia *Mitchell 69-84*	C Bk	British	251	125·1	24·1	13·6	Greenock	1869	R. Steele	Y 2.87	J. & W. Stewart	Greenock
	G	Lavinia *Griwan -88*	W3mSr	Russian	162	89·0	23·0	11·5	Libau	1857	D. Dargewitz		L. Griwan	Rohjen
		Lavinia Bell *Delaney*	W Sr	American	147	97·5	28·2	8·0	Br'khaven, NY	1864			P. McConville	Bangor, Me.
	R	Lavinia Campbell (Col. Lib...) *Poil Sbbo*	W3mSr	American	697	175·9	37·0	17·1	Kennebunk- [port, Me.	1883	D. Clark		C. N. Franklin	Greenport
		Lavolta *Whittemore*	W Sr	American	160	104·0	27·6	9·0	Ellsworth, Me.	1870	D. P. Jordan		J. M. Higgins	Ellsworth, Me.
	A	Lavoro *Nicolich -87*	W Sr	Austrian	166	85·3	23·4	13·1	Lussinpiccolo	1884	N. Martinolich	Y 9.86	D. Sablich	Lussinpiccolo
296		Lavoro *Catalcrese*	W Bk	Italian	252	91·2	24·4	15·5	Spezia	1868	G. Calabrese	Y 5.86	C. Lavaguino	Genoa
297		Lavoro	W Bg	Spanish	147	109·2	26·8	9·6	Savona	1884	B. Calcagno	Y 5.8		
298	R	Lawrence Haynes *Tibbetts*	W3mSr	American	244	124·9	30·0	8·6	Wiscasset, Me.	1883	J. M. Haynes		J. M. Haynes	Bath, Me.
299		Lawrence N. McKenzie *Allen*	W Sr	American	147	98·0	24·6	9·6	Essex, Mass.	1882			N. McKenzie	Provincetn. Mass

=Lloyd's Register; D=Nederlandsche Vereeniging; G=Germanischer Lloyd; I=Registro Italiano; R=American Record;
=Veritas Austro-Ungarico; H=Veritas Ellenico; N=Norske Veritas; V=Bureau Veritas.

Dans la 4me Colonne, C=Composite (membrure en fer, bordé en bois); I=Fer; Stl=Acier; W=Bois.

Les Dates du doublage en métal (Col. 11) ressortent des livres dans lesquels les navires sont respectivement classés ou des rapports des Inspecteurs du Lloyd's Register.

Pour plus amples renseignements sur les navires classés par le Lloyd's Register, voir le Lloyd's Register Book et le Lloyd's Yacht Register.

Official No.	No. in Book	Registry in which classed	Ship's Name. Material and Rig. Master, and Dates of Appointment to present Owner's Service and to Ship. Ship's Late Name.	Flag.	Net Tonnage	Dimensions: Length	Breadth	Depth	Built: Where	When	By whom	Date of Metalling	Owners or Managers
3421	301	I	Lazzaro Anna (ex Hector, & Athalia) *Maggiolo* W Bn	Italian	389	118·7	27·6	16·1	Chiavari	1878	L.Gotuzzo	Y 3.86	L. Maggiolo
992	302	I	Lazzaro Bianchi (ex Rudolph, Dall'Orso) *Dall'Orso* W Bk	Italian	557	144·6	32·2	20·6	Savona	1869	G.Tixi	Y 3.86	F^lli Esauchi fu L.
1846	303		Le Grazie (BROKEN UP) *Zulio* W Bk	Italian	512	131·2	32·5	21·6	Savona	1869	G.B.Guastavino	Y 3.86	Costant. Fulle
	304	V	Le Locquirec *Finisternegol* -88 W Bk	French	256	105·0	25·6	12·5	Wallace, N.S.	1867	J.Batty	Z 4.88	M. Francastel
	305	V	Le Méléart *Scoarnec* -86 W Bn	French	113	86·0	19·3	10·5	St. Malo	1877	L.Tranchemer		Scoarnec
	306		Le Pellorin (ex Charles Henry) *Bertho* W Bn	French	300	115·8	25·6	12·6	Richmond, Me.	1865	G.McLellan		Le Kanaff
2705	307	I	Le Sirene *Accetta* -87 W Sr	Italian	118	90·5	20·9	11·1	Torre del [Grecco	1887	M.Trambarolo		V.Magliolo
	308		Lea W3mSr	Norweg'n	173	104·9	23·8	12·2	Stavanger	1865	J.Mikkelson		A.Lindtner
	309		Lea *Jansson* W3mSr	Russian	280	..	..	..		1870			K.Karlsson
60660	310	✻	Leader *Jones* 86-86 W Bn	British	196	113·0	24·2	12·9	Salcombe	1869	Evans	Y 12.88	J.Jones (Mary A. Jones)
863	311		Leader (FOUNDERED) *Davies* W Sr	British	124	86·8	20·2	11·6	Plymouth	1852	J.Banks		Mrs. L. Davies
73006	312	✻	Leading Chief *Richards* 80-80 W Bkn	British	315	124·0	26·0	14·6	Kingston	1876	Kinloch	Y 9.85	W. Kinloch
140020	313	V	Leading Wind *Hinckley* -86 W S	American	1159	186·6	37·3	22·7	Bath, Me.	1874	Goss&Sawyer	Y 9.84	W. A. Rust
2147	314	I	Lealta (ex Rosa B.) *Gandolfo* -86 W Bk	Italian	488	136·8	31·1	19·8	Sestri Ponente	1871	G.Traverso	Y 3.86	L.Cichero&G.Gandolfo
	315		Lealtad *Barcelo* W Bg	Spanish	291	109·2	26·2	15·7	Palma	1849	J. & T. Mateo		M. Estela y Calafat
56878	316	✻	Leander *Hamilton* 86-86 C S	British	848	215·5	35·2	20·7	Glasgow	1867	J.G.Lawrie	Y 11.87	W. Ross & Co.
56460	317	✻	Leander *Congdon* W3mSr	British	227	114·8	20·5	12·7	Newburgh	1869	Scrimgeour	Y 7.84	Thorburn&Tessier
27309	318		Leander *Jones* W Sr	British	72	65·5	19·0	10·0	Wales	1859	Harris		J.Jones
140602	319		Leander V. Beebe *Beebe* W3mSr	American	712	160·9	35·7	16·3	Bath, Me.	1883	Goss&Sawyer		L.V. Beebe
41370	320		Leandre W Bn	British	117	80·0	22·2	10·4	Cap. St. Ignace	1858			L. Methot
204	321	A	Leandro *Martinolich* -88 W Bk	Austrian	651	137·4	30·8	20·3	Lussinpiccolo	1879	R.Scopinich	Y 1.89	V.L. Martinolich fu M.
60005	322	✻D	Lebu *Lewis* 86-86 I Bk	British	726	184·0	31·3	19·6	Dundee	1868	Gourlay Bros.		Nicholson&McGill
	323	N	Lector Schweigaard *Davidsen* W Bk	Norweg'n	328	103·1	29·3	16·7	Grimstad	1840	G.Thorsen		E.Davidsen
49617	324		Leda *Sword* I Sp	British	158	96·0	26·0	9·6	Newcastle	1864	Rogerson		Bell Bros.
199	325	V	Leda *Budinich* W Bk	Austrian	533	129·2	29·5	16·7	Lussinpiccolo	1876	Tarabocchia	Y 10.88	Emma Tarabocchia
	326		Leda W3mSr		323	118·6	26·3	16·2	Königsberg	18 53/63	L.Eggert		
	327	DV	Leda *Alberts* W Bn	German	202	105·0	24·0	10·9	Leer	1883	C.Schwoon	Y 6.85	S. Wreesmann Wwe.
	328	G	Leda (ex Marianne) *Kruger* 75- W Bn	German	158	90·0	20·9	11·7	Papenburg	1857	H.W. Meyer		W. Lüdke & Co.
	329	N	Leda *Brinck* -88 W Bk	Norweg'n	531	133·3	33·6	18·9	Arendal	1854			S.Jörgensen
	330	NV	Leetch (ex Little Harry) *Sundegaard* -88 W Bg	Norweg'n	371	129·5	29·9	12·9	St. Martin's N.B.	1873	W.H.&J.Rourk	Y 11.86	P.A. Petersen
	331	A	Lefki Peristeris *Savas* W Lg	Greek	120	..	..	..	Castelrosso	1887	A.Fitos		S.N. Savas
	332		Lefve *Sundgren* -87 W Sr	Russian	229	..	..	..	Raumo	1884			G.Granlund
15024	333		Legal Tender *Avery* W Bk	American	200	101·0	24·0	12·0	Port Madison, [Wash.	1863			E.S. Rowe
	334		Legiona W Sr	Russian	213	..	..	..		1870			A. Waldonen
15184	335		Lehman Blew *Parsons* W Sr	American	261	112·0	29·4	9·0	Bridgeton N.J.	1867			Jeremiah Smith & Brother
85031	336	✻	Leicester Castle *Tewksbury* 84-86 I S	British	2009	273·4	40·3	24·2	Southampton	1882	Oswald, Mor- [daunt&Co.		"Leicester Castle" Ship Co., Ld. (J. Coupland)
	337	V	Leif (ex Kho) *Endresen* 88-88 W Bk	Norweg'n	820	168·6	33·5	21·5	Savona	1873	Guastavino	Y 1.88	T. Nordaas
	338	NV	Leif *Ericksen* 69-78 W Bk	Norweg'n	455	142·0	29·9	16·3	Stavanger	1877	Strömstenens [Verft	Y 6.87	Ths. S. Falck
	339	N	Leif *Aass* -87 W Bk	Norweg'n	260	125·7	27·0	12·0	Skien	1868	Lanfoss	Y 9.82	P. Ltz. Aas
	340	N	Leif *Evensen* -88 W Bn	Norweg'n	179	106·0	24·4	11·6	Thurö	1866	N.Troensegaard		T.Torkildsen
	341		Leif (ex Helena) *Olsen* W Sr	Norweg'n	158	100·8	21·2	11·0	Newport, N.S.	1852			J.Prebensen
	342	NV	Leif Erikson *Kroger* -88 W Bk	Norweg'n	624	161·5	33·0	18·1	Laurvig	1874	Treschow	Y 3.87	G.Olsen
31617	343		Leipzig W Lighter	British	159	106·8	24·1	10·0	Holderness	1814			Henry Moss
	344		Leivo *Anckar* W Sr	Russian	162	..	..	..	Pyhämaa	1873			D.Lundmark
	345		Lekna *Leknaes* -88 W3mSr	Norweg'n	197	113·0	25·0	11·5	Bergen	1877		Y 6.88	Joachim Hansen
	346		Lemnos W Bk	French	268	102·0	26·3	13·5	Nantes	1870	Sevestre	Y 11.87	Lampoignard
	347		Lemnos W Bg	Turkish	105	..	..	..	Lemnos	1869			Lemnocles Mascandaran & Co.
	348		Lemnos W Bg	Turkish	100	..	..	..	Lemnos	1874			Alatjes Costanti Ioles
14816	349	R	Lemuel H. Hopkins *Crammer* W3mSr	American	221	109·5	30·6	10·2	Barnegat, N.J.	1865	Pridmore& [Hadine		D.Appleton
77087	350	✻	Lena *Richards* 78-78 W Bkn	British	277	123·4	25·6	14·9	P.E. Island	1878	Carrol	Y 4.87	I.E. Ashmole (T.P. Richards & [illegible])

✻ = **Lloyd's Register**; D = Nederlandsche Vereeniging; G = Germanischer Lloyd; I = Registro Italiano;
A = Veritas Austro-Ungarico; H = Veritas Ellenico; N = Norske Veritas;

In the 4th Column, C = Composite (iron frame, wood planked); I = Iron; Stl = Steel; W = Wood.

The Dates of Metalling (Col. 11) are based on the records of the Books in which Vessels are respectively classed, or on the Reports of the Surveyors.

Further particulars of Vessels classed by Lloyd's Register will be found in Lloyd's Register Book and Lloyd's Yacht Register.

No.	Cote	Nom. Capitaine et dates d'entrée au service des Armateurs actuels et à bord du navire. Ancien Nom.	Matériaux et Gréement	Pavillon	Tonnage net	Dimensions: Longueur	Largeur	Creux	Construit: Où	Quand	Par qui	Doublage	Armateurs ou Directeurs-Gérants	Port d'Armement
351		Lena *Rennels*	W Sr	British	164	99·7	23·2	12·7	Dartmouth	1867	Follett	[illegible]	J. F. Ferris	Brixham
352	✻	Lena *Marlow*	W K	British	82	76·8	20·7	10·7	BurtonStather	1883	J. Wray & Sons		W. J. Robins	Hull
353	R	Lena *Persson*	W Bk	Swedish	999	180·0	35·4	22·6	Quebec	1865	Dubord	[illegible]	J. Ingmansson	Elleholm
354		Lena Breed *Jones -88*	W Sr	American	216	111·0	29·7	8·8	Milton, Del.	1872			L. Mathews	Philadelphia
355	R	Lena R. Storer *Lee -88*	W 3m Sr	American	238	114·6	29·6	10·2	Waldob'ro' Me	1873	G. W. Caldwell	[illegible]	A. Soderstrom	Shieldsbro', Miss.
356		Lena Sweasey *Ray -88*	W 3m Sr	American	244	130·2	32·0	9·0	Eureka, Cal.	1882	A. Cookson		W. J. Sweasey	Eureka, Cal.
357	D	Lennatar *Schwela*	W Bk	Russian	600	154·8	32·5	19·8	Åbo	1868	L. P. Kjäldström	[illegible]	G. A. Lindblom	Åbo
358	V	Lennatin *Bergkoff -84*	W Bk	Russian	707	163·0	34·1	20·3	Nystad	1868	N. H. Kjäldström	[illegible]	J. Rivell	Nystad
359		Lenni *Mattson*	W Sr	Russian	224	..	..	..	Sideby	1876			M. Eriksson	Lemland
360	V	Lennie *Swanson 80-86*	W Bk	British	989	175·0	35·0	22·0	Belliveaus Cove, N.S.	1871	W. D. Lovitt	[illegible]	W. D. Lovitt	Yarmouth, N.S.
361		Lennie *Porter 80-84*	W Sr	British	100	84·0	24·0	8·4	Yarmouth, N.S.	1882	H. Jenkins		Geo. Crosby	Yarmouth, N.S.
362	V	Lennie Burrill *Bell -84-88*	W S	British	1328	198·5	38·7	23·7	Little Br'k, N.S.	1877	I. L. Burrill	[illegible]	W. Burrill & Co.	Yarmouth, N.S.
363	G	Lennok *Klein 78-85*	W Bn	Russian	365	135·4	29·0	13·3	Haynasch	1885	L. Lepp		M. Klein	Haynasch
364		Lenore *Pont 88-88*	W Bn	British	192	105·2	24·2	12·9	P. E. Island	1873	McMillan	[illegible]	H. Pont	Faversham
365	✻	Lenore	Stl & I Sr Yt	British	102	97·8	17·2	14·4	Belfast	1882	Workman, Clark & Co.		George Smith	Glasgow
366	V	Lénore *Bruneau -88*	W Bk	French	392	128·1	28·2	17·2	Sunderland	1869	Metcalfe	[illegible]	L. Mabon fils	Nantes
367	I	Lenuccia *Guida -87* (ex [illegible])	W Bg	Italian	386	114·5	28·9	18·4	Equa	1874	M. Maresca	[illegible]	A. Guida	Naples
368		Leo *Hammett -84*	W Bn	British	165	97·2	24·6	10·8	Mahone By. N.S.	1882	T. Langille		J. Eisenhauer	Lunenburg N.S.
369		Leo (ex Rev. Cut. Reliance) *Pillz*	W Sr	American	152	102·0	26·6	9·8	Baltimore	1865			J. G. Brady	Sitka, Alaska
370	✻	Leo (ex Medora)	W Sr Yt	French	90	96·9	20·3	11·5	Gosport	1875	Camper	[illegible]	A. Bouard	Boulogne
371	N	Leo (ex Undine) *Nielsen*	W Bg	Norweg'n	326	112·0	27·4	16·3	Sunderland	1860	Rawson & Watson		Jens Nielsen	Tvedestrand
372		Leo *Eriksson*	W 3m Sr	Russian	289	..	..	..		1870			K. Karlsson	Lemland
373		Leo *Cederlund*	W Bg	Russian	281	120·1	26·5	12·3	Strömme	1866	E. Sandelin		F. W. Wahlberg	Nystad
374		Leo *Lindblom*	W Sr	Russian	133	82·8	26·4	10·3	Kimito	1875	Sjöberg		S. Isaksson	Kimito
375	V	Leocadia *Stöff 81-81*	W Bk	German	799	150·9	32·1	20·7	Motzen	1862	D. Oltmann's Wwe.	[illegible]	Schaer & Heitmann	Bremen
376		Leocadia Francisca	W Sr	Brazilian	100	..	..	..					J. S. Caldas	Espirito Santo
377	V	Léon *Boutin -87*	W 3m Sr	French	247	106·9	22·6	14·2	St. Malo	1868	Vve Dandin & fils	[illegible]	Mentomais & Pannier	Granville
378	H	Leon *Focca*	W Bk	Greek	353	118·1	27·9	..	Cassos	1872		[illegible]	A. Vagliano	Argostoli
379	N	Leon (ex C. Hooper) *Sahlqvist*	W Sr	Norweg'n	323	122·6	31·0	13·0	Cape Breton	1876	A. Hooper		L. Sahlqvist	Sarpsborg
380	N	Leon *Andersen -87*	W Bn	Norweg'n	281	110·7	28·0	15·2	Laurvig	1880		[illegible]	D. Herlofson (A. Smith)	Arendal
381		Leon Gambetas	W Bk	Greek	430	129·6	27·9	..	Sunderland	1868			I. Mamounis	Syra
382		Leon Jeanne *Murray*	W Sr	British	129	88·8	23·2	12·1	Dunkirk	1866	Derycke		J. Murray	Dublin
383	I	Leon Pancaldo *Mortola*	W Bk	Italian	729	161·0	33·5	22·0	Savona	1876	G. Calamaro	[illegible]	Gerol. Mortola	Genoa
384		Leon Raymundo *Stephens -88*	W Sr	British	100	87·5	19·4	10·8	Rye	1856	Hessel & Holmes		H. & E. Crundall	Dover
385		Leon S. Swift *McDonald -88*	W Sr	American	147	95·7	25·4	9·8	Glouc'st'r, Mass.	1881			S. S. Swift	Provincetn, Mass.
386		Léon & Toni *Roubiex -88*	W Bn	French	115	74·5	20·8	10·2	Antibes	1855			Vogley & Cochois	Antibes
387		Leona *Hanlan 86-86*	W Bn	British	297	113·5	26·5	11·9	Dorchester, N.B.	1866	Chapman		J. J. Carroll	Dublin
388		Leona *Nickle*	W Sr	British	116	84·7	24·0	9·6	Granville, N.S.	1868	Delap		R. Curran	Belfast
389	V	Léona *Talleux -88*	W Sr	French	107	89·9	20·7	10·6	Dunkirk	1879	Ogez		L. Vincent	Dunkirk
390	I	Leona *Drago* (ex [illegible])	W Bk	Italian	598	149·3	31·7	20·6	Varazze	1873	Baglietto	[illegible]	Flli. Drago fu F.	Genoa
391	A	Leonardo *Vatini*	W Bg	Greek	273	..	..	..	Syra	1887	A. Livadara	[illegible]	L. Vati	Syra
392	I	Leonardo *Cafiero* (ex Arcangelo)	W Bg	Italian	433	123·3	25·3	18·7	Cassano	1874	A. Castellano	[illegible]	G. Califano	Castell. di Stab.
393		Léoncie *Aubert*	W Bg	French	135	..	..	..	St. Servan	1844			Guibert & fils	St. Malo
394	IV	Leone *Esposito 68-87*	W Bk	Italian	605	150·9	29·2	20·7	Cassano	1879	A. Castellano	[illegible]	F. S. Ciampa	Castell. di Stab.
395	V	Leone (ex [illegible]) *[illegible] -88*	W Bk	Italian	408	131·2	29·2	18·2	Gaeta	1870	P. Gallinaro	[illegible]	G. Cicconardi	Gaeta
396	I	Leone S. (ex Persone)	W 3m Sr	Italian	380	131·8	29·9	17·2	Voltri	1880	Flli. Fava	[illegible]	Santo Valle	Genoa
397		Leonessa *Fullerschell*	W 3m Sr	American	203	102·0	25·0	8·9	Rockland, Me.	1860			T. N. Mayer	Rockland, Me.
398	V	Leonhard *Brahms -88* (ex [illegible])	W Bn	German	146	91·2	21·7	10·3	Esbjerg	1876	Th. Dahl		A. Brahms	Beckzeteler-Fhn.
399		Leonhard	W Sr	Russian	180	95·0	25·0	13·2	France				M. Grant	Riga
400	IV	Leonida *Trapani* (ex [illegible])	W Bk	Italian	617	150·2	33·5	20·4	Lussinpiccolo	1875	G. S. Cattarinich	[illegible]	M. Cacace	Castell. di Stab.

✻ = Lloyd's Register; A = Veritas Austro-Ungarico; D = Nederlandsche Vereeniging; G = Germanischer Lloyd; H = Veritas Ellenico; I = Registro Italiano; N = Norske Veritas; R = American Record; V = Bureau Veritas.

Dans la 4me Colonne, C = Composite (membrure en fer, bordé en bois); I = Fer; Stl = Acier; W = Bois.

Les Dates du doublage en métal (Col. 11) ressortent des livres dans lesquels les navires sont respectivement classés ou des rapports des Inspecteurs du Lloyd's Register.

Pour plus amples renseignements sur les navires classés par le Lloyd's Register, voir le Lloyd's Register Book et le Lloyd's Yacht Register.

Official No.	No. in Book	Registry in which classed	Ship's Name. Master, and Dates of Appointment to present Owner's Service and to Ship. Ship's Late Name.	Material and Rig	Flag	Net Tonnage	Length	Breadth	Depth	Built: Where	When	By whom	Date of Metalling	Owners or Managers
1532	401		**Leonida** *Risso*	W Bk	Italian	409	124·0	30·2	19·2	Sestri	1867	F. Tixi	[illegible]	Eredi di E. Risso
1075	402		**Leonidas** *Dic*	W Bk	British	580	138·5	31·3	20·0	Sunderland	1853	W. Wilkinson		J. L. Robson
	403	V	**Léonidas** *Coffournic*	W Bn	French	139	80·4	21·6	10·4	Nantes	1861	Sevestre		J. Coffournic
13	404		**Leonidas**	W Bk	Greek	498	142·7	27·9	19·7	Sunderland	1868			L. A. Glykas
333	405	A	**Leonidas** *Abatzalos -88*	W Bk	Greek	316	105·6	27·1	..	Syra	1850	S. Cufudachi		A. Babas
26	406	H	**Leonidas** *Dascalopoulos*	W Bg	Greek	296	104·3	27·2	..	Galaxidi	1875			N. Katsoulis
31	407	H	**Leonidas**	W Bg	Greek	291	106·9	27·1	..	Syra	1874		[illegible]	S. Tsamizis
58864	408		**Leonie**	W Bk	British	459	127·0	29·0	12·7	Quebec	1870			N. Rosa
	409		**Léonie** *Guillerme 82-82*	W Bn	French	132	76·4	20·3	11·2	Méans	1873	Ollivaud		P. Ch. Grenet
	410	IV	**LeonildaGauci** *Cassoar 1 -88* (ex Trento)	W Bk	British	653	141·0	32·8	20·7	SestriPonente	1868	A. Briasco		A. Gauci
	411		**Leonor** *De Souza Jorge*	W Sr	Brazilian	129	..	..	..					A. J. de Souza Belles
	412		**Leonor** (ex Laura McLennan)	W Bk	Chilian	801	161·0	30·8	20·0	RiverJohn, NS	1874	C. McLennan		
	413		**Leonor** *Fonseca* (ex Nuevo Ignacio, ex Zulmira)	W Bg	Portug'se	440	115·8	26·9	16·4	Ferrol	1865		[illegible]	A. T. R[illegible]
	414		**Leonor**	Bn	Spanish	188	105·6	24·9	10·7	Albay	1875			D. Echevarria
76451	415	* D	**Leonor Troncoso** *Rudd*	I Bn	British	313	136·7	25·1	14·0	PortGlasgow	1875	W Hamilton&Co		R. & W. King
91529	416	*	**Leonora**	W Sk	British	96	80·0	21·5	10·9	Burt'nStather	1885	J. Gamble		J. Plastow
140015	417	R	**Leonora** *Monroe*	W Bn	American	436	126·5	30·6	15·1	ColumbiaFal[ls], Me.	1874	J. Crandon	Y 8.84	Swan & Son
15266	418		**Leonora** *Bonsey*	W Sr	American	110	94·0	26·0	7·5	Ellsworth, Me.	1873			I. M. Grant
12	419	V	**Leonora** *Moresco -88*	W S	Austrian	1418	198·8	39·0	24·0	Belfast, Me.	1869	J. W. White & Co.	[illegible]	F. Demartini fu G.
	420	DG	**Leonore** *Tobben 74-*	W Bg	German	196	106·4	21·0	11·8	Papenburg	1858	F. W. Meyer		J. Dieckhaus
	421		**Leontin** *Eriksson -86*	W Sr	Swedish	164	89·2	23·6	11·6	Roslagen	1856			C. M. Vicklund
	422		**Léontine** *Hars 84-84*	W Sr	French	118	88·9	22·7	11·0	Dunkirk	1878	G. Beck		G. Beck
	423		**Leontine** *Eriksson*	W Glt	Russian	113	..	..	..					K. A. Lundman
	424		**Léontine Amélie** *Loche 85-85*	W Bg	French	189	85·3	22·6	12·8	Girard	1863			Lebourdiec
	425		**Leontius** *Eriksson*	W Sr	Swedish	116	78·7	22·5	10·1	Axmar	1856			J. E. Törnqvist
	426	N	**Leopard** *Olsen 88-88*	W Bk	Norweg'n	274	116·7	28·6	12·1	Grimstad	1867	K. Andersen	[illegible]	J. Prebensen
	427	V	**Leopold** (ex Sharpsea) *Gautier*	W Bk	French	331	143·3	26·5	15·3	Liverpool	1857			P. Villars
	428		**Leopold** *Norberg 87-76*	W Sr	Swedish	148	88·6	23·8	10·3	Finland	1874			P. Norberg
	429	GV	**Leopold von Vangerow** (ex Stafford) *Schilling 86-89*	W S	German	1056	184·5	36·6	22·3	Quebec	1877	Dunn&Samson	[illegible]	H. Addicks
1108	430	IV	**Leopoldo** *Pesce*	W Bk	Italian	708	170·2	34·7	19·8	Cassano	1878	F. De Rosa	[illegible]	F. Petrelluzzi
	431	V	**Leopoldshall** *Brandt* (ex Prince Rudolph)	W S	German	1372	210·1	38·0	23·6	S. Martins, NB	1874	J. H. Moran		R. S. Carr
3194	432	I	**Lepanto** *Cremonini -87*	W Bk	Italian	863	173·8	31·1	23·7	Varazze	1881	G. Scala	[illegible]	P. Lavarello
	433		**Lepanto**	W Bn	Spanish	187	89·6	26·2	12·6	San Felin	1854			Juan Santa
87728	434		**Leppadan**	I Barge	British	353	197·3	25·3	8·5	PortGlasgow	1883	R. Duncan&Co.		Irrawaddy Flotilla Co. Ld.
72254	435	V	**Lepreaux** *Brown 72-76*	W Bk	British	769	157·4	33·2	19·7	St Martin's, NB	1876	J. Marr	[illegible]	F. Tufts
	436		**Les Bons Fils** *Horseau*	W Bk	French	287	106·3	26·6	14·1	Bordeaux	1855			Daniel Riviere
	437	R	**Les Trois Soeurs** *Oswald* (ex The Charley, ex Edward Slatton)	W Bn	Haytian	223	95·8	25·2	12·8	Parrsboro', NS	1875	J. W. Durant	[illegible]	V. Thomas
	438	N	**Lesseps** *Pedersen 67-86*	W Bk	Norweg'n	364	127·1	27·5	15·6	Risör	1870	Stian Finne	[illegible]	Stian Finne
	439		**Lesseps** *Jonasson 87-87* (ex Bonetsang)	W3mSr	Swedish	274	113·8	24·3	14·0	Pictou, N.S.	1854			S. Jonasson
140256	440	R	**Lester A. Lewis** *Hatch*	W3mSr	American	237	123·9	29·5	9·9	Bangor, Me.	1877	Crosby Bros.		Crosby Bros.
	441	DN	**Let** *Nielsen*	W Sr	Norweg'n	119	84·3	22·4	10·3	Sarpsborg	1868	Birch		O. A. Nielsen
39132	442		**Letitia** *Pearce*	W Sr	British	115	81·0	20·2	11·2	Sunderland	1861	Metcalfe		T. Smart
14613	443		**Letitia** *Moller*	W Sr	American	234	114·6	28·6	10·0	Boston, Mass.	1867			J. Jensen
898	444	V	**Letizia** *Maresca*	W Bk	Italian	427	127·3	28·9	18·3	Sorrento	1867	M. Paturzo	[illegible]	MoseCastellano&Co
	445	NV	**Leto** *Wegle -88*	W Bk	Norweg'n	580	147·8	31·6	19·0	Arendal	1880	K. Larsen	[illegible]	T. Thoumoesen & Smith
76294	446	*	**Letterewe** *Gifford 85-85*	I Bk	British	798	191·0	33·3	19·0	Liverpool	1876	W. H. Potter&Co		J. Dowie & Co.
43	447	G	**Lettonia** *Greve 77-*	W3mSr	Russian	365	131·5	30·0	14·4	Adamünde	1882	J. Kopmann		J. Grewe
60702	448	*	**Leucadia** *Mill 67-83*	I Bk	British	896	194·2	33·8	20·4	Aberdeen	1870	Hood		A. Nicol & Co.
	449	V	**Leufsta** (ex [illegible]) *Lindström*	W Sr	Swedish	194	100·0	22·6	11·7	Ängskär	1873	E. Vireen		G. Förselius
	450	N	**Levang** *Salvesen -77*	W Sr	Norweg'n	174	96·4	23·7	11·4	Kragerö	1877	Alf Johnsen		A. Knudsen

* = **Lloyd's Register**; D = Nederlandsche Vereeniging; G = Germanischer Lloyd; I = Registro Italiano; R = [illegible]
A = Veritas Austro-Ungarico; H = Veritas Ellenico; N = Norske Veritas; V = [illegible]

In the 4th Column, C = Composite (iron frame, wood planked); I = Iron; St = Steel; W = Wood.

The Dates of Metalling (Col. 11) are based on the records of the Books in which Vessels are respectively classed, or on the Reports of the Surveyors of

Further particulars of Vessels classed by Lloyd's Register will be found in Lloyd's Register Book and Lloyd's Yacht Register.

No.	Cote	Nom. Capitaine. Ancien Nom.	Matériaux et Gréement	Pavillon	Tonnage net	Longueur	Largeur	Creux	Construit Où	Quand	Par qui	Doublage	Armateurs ou Directeurs-Gérants	Port d'Armement
451	✠	Levant *Holton*	WSr	British	168	102·1	23·4	12·9	Brixham	1868	Upham		W.C.Phillips	Brixham
452	V	Levant *Van Kyo -86*	WSr	Dutch	142	102·6	24·4	10·4	Piteå	1856	F.Pettersson		A.Hoogendijk,Jzn	Vlaardingen
453	NV	Levant *Gjeteer -87*	WBk	Norweg'n	365	110·0	28·3	14·4	Gefle	1875	O.A.Brodin		F.G.Thorsen	Skudesnaes
454	V	Levante *Janzen*	WBn	German	206	109·9	25·6	12·5	Harburg	1868	G.Renck,jun.		Frau B.W.P.Seele	Hamburg
455	R	Levanter *Gerry 82-86*	WBkn	American	612	[illegible]	32·9	17·1	ThomastonMe	1874	Stetson&Gerry		S.S.Gerry	Thomaston,Me.
456	RV	Levi G. Burgess *Rock 76-87*	WS	American	1536	217·5	41·2	24·9	ThomastonMe	1877	S.Watts		J.Jansson	San Francisco
457	R	Levi Hart *Giles*	W3mSr	American	387	130·0	31·0	11·5	St.George,Me.	1873	H.M.Bean		J.P.Ellicott & Co.	Boston,Mass.
458	V	Levi S. Andrews *Watts 79-81*	WBkn	American	636	163·1	34·0	17·6	ThomastonMe	1881	S.S.Gerry&Co.		S.S.Gerry	Thomaston,Me.
459	N	Leviathan *Mosfjeld 82-76*	WS	Norweg'n	1062	186·7	36·2	20·4	Grimstad	1876	Jacob Boe		Christen Boe	Arendal
460	N	Leviathan *Pedersen 84-86*	WBk	Norweg'n	381	117·5	29·1	16·1	Grimstad	1868	John Terjesen		S.Eyde	Arendal
461	R	Levinia F. Warren *Jenson*	W3mSr	American	284	112·0	28·6	13·9	Machias,Me.	1871			Z.R.Thompson	New York
462	V	Levrette *Legoff -87*	WBn	French	108	80·4	20·6	10·8	St.Malo	1875	J.Parnet		Daugnet	St.Malo
463	DV	Levuka *Harris 76-84*	WBk	British	1426	199·6	39·7	24·3	St.John,N.B.	1884	John Fraser		B.Smith & Sons	Windsor,N.S.
464		Levuka *Hanst 77-78*	IBk	German	437	145·3	27·1	15·9	Hamburg	1878	C.H.Kranse		H.F.A.Evers	Hamburg
465		Lewis A. Edwards *Peterson -88*	WSr	American	231	102·0	28·2	8·9	New York	1851			C.F.Peterson	New York
466		Lewis A. Hogdon (W.K.F.K.B.D.) *Lagore*	WSr	American	158	109·0	25·7	8·3	BoothBay,Me.	1884	J.G.Fuller		J.F.Fuller	Booth Bay,Me.
467		Lewis Clark (W.R.B.S.G.D.) *Bartlett*	WSr	American	192	108·1	27·9	9·2	Bangor,Me.	1863			J.H.Crosby	Bangor,Me.
468		Lewis Ehrman *Collison*	W3mSr	American	386	127·1	31·6	15·2	Baltimore	1872	Wolf'nd'r&Buck		L.Ehrman	Baltimore
469		Lewis K. Cottingham *Nilson -87*	W3mSr	American	498	142·5	34·0	13·1	Camden,N.J.	1883	Morris&Mathis		J.D.Whittaker	Camden,N.J.
470	R	Lewis King *Farnham*	WSr	American	142	99·9	25·9	8·7	Ellsworth,Me.	1883	I.H.Grant		L.King	Ellsworth,Me.
471		Lewis L. Squire *Nilsson -87*	WBn	American	424	122·8	31·6	17·4	PtJeffersonNY	1868			S.C.Evans & Co.	New York
472	✠	Lewis M. Lamb *Williams 80-87*	WBk	Chilian	404	124·7	27·6	16·6	Newcastle	1871	Gaddy		E.Koegel	Valparaiso
473		Lewis Ross	WSr	British	253	124·9	23·7	10·3	PtBurwellOnt	1874	L.McDermand		A.Cochrane	Port Hope,Ont.
474	R	Lewis Smith *Wright*	WBk	British	1093	186·2	36·2	21·4	Rockland,N.B	1876	R.A.Chapman		R.A.Chapman	Dorchester,N.B.
475		Lexington *Priest*	WSr	American	142	102·0	28·2	7·4	CherryfieldMe	1875			E.I.White	Machias,Me.
476	✠V	Leyland Brothers *Richardson 86-88*	IS	British	2238	284·0	40·0	24·2	Southampton	1886	Oswald, Mordaunt & Co.		ShipSh "Leyland Brothers" Co.Ld.(R.W.Leyland & Co.)	Liverpool
477	N	Lhadnoi *Aske -81*	WSr	Norweg'n	199	108·7	25·0	12·1	Stavanger	1875	Nielsen		A.Aske	Stavanger
478	V	Lia *Consigliere*	WBg	Italian	172	108·2	22·2	9·9	SestriPonente	1885	C.Bozzano	Y.3.85	Consigliere & Crovetto	Genoa
479	V	Liana *Gower -88*	W3mSr	Swedish	295	125·9	25·9	11·9	Kolbacks	1876	H.Olssen		L.Pahlsson	Lerberget
480	V	Liane *Hars*	WSr	French	138	85·3	21·0	12·3	Boulogne	1866	Lecerf		Macré & Co.	Dunkirk
481		Libanon *Karlsson*	WBk	Russian	269	..	..	..	Geta	1873			K.E.Eriksson	Finström
482	R	Libbie H. *Pye -88*	WBn	British	361	127·0	30·6	12·9	St.Mary's,N.S.	1875	A.McDonald	11.88	John McNab	Halifax,N.S.
483		Liberal	WSr	British	185	107·0	22·8	10·9	[illegible]	1883	F.Grenier		T.Marchesseault	Montreal
484		Liberal	WSr	Brazilian	117	..	..	..					H.A.Joanny	PortoAlegre
485		Liberal	WSr	Brazilian	133	..	..	..					J.C. Pinto & I. A. da Silva	PortoAlegre
486		Liberator *Pennock*	WSr	British	124	87·7	18·5	12·0	Waterford	1833	Smith		T.Toole	London
487		Liberia *Page -87*	WBk	American	317	120·0	29·7	11·5	Newb'ryptMs.	1872			Yates & Porterfield	New York
488		Liberine *G. Scotto Lavina* (ex Alessandro (I.))	WBg	Italian	269	111·8	25·4	15·1	SestriPonente	1857		11.88	M.Scotto Lavina	Naples
489		Libero *Boglio's 87-87*	WBk	Italian	480	133·5	30·2	19·2	Chiavari	1864	F.Gotuzzo		M. & A.Dall'Orso	Spezia
490		Libertad *Landieta*	WBn	Spanish	136	91·2	24·2	10·7	Sorsogon	1868			Ignacio Boncan	Manila
491	V	Libertas (ex Rose of Trump?)	WB	French	596	137·8	31·2	19·5	St.John,N.B.	1863	McLachlan		Gastaldi	Marseilles
492	V	Libertas *Schell 86-88*	WBk	German	418	125·7	30·0	16·8	Danzig	1863	F.Devrient		A.Gibsone	Danzig
493	N	Libertas *Danielsen*	WBg	Norweg'n	240	111·6	27·0	11·7	Laugesund	1858	Tonning		H.A.Danielsen	Tönsberg
494	V	Libertas *Eklund*	WBk	Russian	514	137·8	30·5	17·9	Ekholm	1865/76	A.Lindgren		J.M.Eklund	Abo
495	V	Libertas (ex Fremont)	WBk	Russian	403	135·4	29·7	16·2	Brake	1866	J.Oltmann's Wwe.			Riga
496	V	Liberté *Drouot -87*	WSr	French	159	104·9	23·5	9·8	Nieuwendam	1862	W. & A. H. Meursing	Y.7.83	Drouet	St.Vaast
497	N	Liberte *Tonnesen -87* (ex Charlotte Goodall)	WBk	Norweg'n	516	139·7	30·5	16·6	NwGlasg'w,N.S.	1865	Carmichael		T.Samuelsen	Farsund
498		Liberty	WBg	American	291	115·6	26·6	11·9	Stockton,Me.	1866			R.Devereaux	Searsport,Me.
499		Libra *Krijnen -81*	WBk	Dutch	453	129·6	28·5	17·7	Searsport	1862			C.Krijnen	Nieuwediep
500		Libra	WSr	French	122	99·7	26·7	9·5	Rotterdam	1860			J. E. Buhan père, fils & A. [illegible]	St.Louis,Sénégal

A = Lloyd's Register; D = Nederlandsche Vereeniging; G = Germanischer Lloyd; I = Registro Italiano; R = American Record; [illegible] = Veritas Austro-Ungarico; H = Veritas Ellenico; N = Norske Veritas; V = Bureau Veritas.

Dans la 4me Colonne, C = Composite (membrure en fer, bordé en bois); I = Fer; Stl = Acier; W = Bois.

Les Dates du doublage en métal (Col. 11) ressortent des livres dans lesquels les navires sont respectivement classés ou des rapports des Inspecteurs du Lloyd's Register.

Pour plus amples renseignements sur les navires classés par le Lloyd's Register, voir le Lloyd's Register Book et le Lloyd's Yacht Register.

Official No.	No. in Book	Registry in which classed	Ship's Name, Material and Rig. Master, and Dates of Appointment to present Owner's Service and to Ship. Ship's Late Name.	Flag	Net Tonnage	Dimensions: Length	Breadth	Depth	Built: Where	When	By whom	Date of Metalling	Owners or Managers
	501	N	Libra *Nielsen* W Bk	Norwegian	419	134·5	31·7	14·6	Miramichi	1867		[illegible]	P.H.Roer
	502		Libra *Johanssen* W Bk	Russian	325	..	..	..		1863			M.A.Michelson
	503		Librada (ex Generosa) W Bn	Spanish	116	81·4	21·3	10·4	[illegible]	1859			P.Valenzuela
	504	DN	Liburna *Jacobsen 84-84* W Bk	Norweg'n	483	137·8	31·4	17·3	Grimstad	1873	Jacob Boe	v 7 86	Christen Boe
	505	✱V	Libussa *Mangels 82-87* I S	German	1730	259·1	39·4	23·8	Flensburg	1884	Flensburger Schiffsb. Ges.		J.C.Pflugk
2124	506	I	Liceti Padre *Liceti* W Bk	Italian	481	151·3	31·5	18·5	Lavagna	1871	L.Carbone	v 8 86	F^lli Liceti
15891	507		Lida Babcock W3mSr	American	247	118·0	29·6	9·4	Goshen,N.J.	1872			R.B.Dare
140650	508	R	Lida Fowler *Higgins* W Sr	American	186	105·5	25·8	10·7	Bath,Me.	1883	T.Hagan&Co.	[illegible]	I.A.Wharf
140476	509	R	LidaJ.Lewis *Townshend* W3mSr	American	536	139·2	35·0	15·6	Camden,N.J.	1881	S.W.Tilton		J.Lewis & Co.
	510	V	Lidador *Dos Reis* W Bn (ex Arzé)	Portug'se	159	84·0	22·6	10·5	Oporto	1864	J.F.daLapa	[illegible]	J.L.Fernandes
	511		Lidoina (ex Margarita Antonia) W Sr	Portug'se	122	79·1	21·3	9·5	Herpatha				L.V.Machado
	512	N	Lidskjalf *Nielsen* W Bk	Norweg'n	497	138·6	30·5	17·4	Grimstad	1880	Terje Eskedal	v 2 86	J.E.Einersen
	513	N	Lidskjalf *Gundersen -82* W Bk	Norweg'n	282	107·8	28·6	15·4	Tvedestrand	1866	L.Holm	v 8 86	A.Biering Olsen
	514		Liefjeld *Eeensen 70-73* W Bk (WRECKED) (ex Fresh Breeze)	Norweg'n	605	152·6	31·8	17·8	Maitland,N.S.	1868	W.McDougall	v 3 86	P.Knudsen
	515	DGV	Lienen *Aden -82* W Bg	German	254	110·2	25·5	12·7	Rönnebeck	1869	H.Oltmann's [Wwe.	[illegible]	H.J.Aden
51100	516	✱	Lieutenant Maury *Thomson* C F	British	350	135·0	27·2	17·0	Barnstaple	1865	Westacott	v 10 85	J.Thompson
89344	517		Liffey W Bn	British	102	96·2	22·0	7·2	B'ckwall NSW	1885	R.Davis		B.Corrigan
	518		Ligeira *Silva -87* W Bk	Portug'se	275	113·1	24·3	13·6	Lisbon	1864	C.J.d'Azevedo [Vareta		I.F.Padinhas F^os D.R.[illegible] C.Mascarenhas e Silva
	519		Ligeiro *Eta* W Bg (ex [illegible]) (PONTOON)	Portug'se	212	95·1	24·3	11·7	Figueira	1850	M.G.Mendes		J. J. das Neves & M.[illegible] S. Vasco
	520	G	Ligero *D'Almeira 82* W Bg (ex San Miguel Archangel, ex San Michele Archangelo)	Argentine	323	118·4	24·6	15·2	Fiume	1865	G.Spandoni		T.Burone
	521		Ligero W Bn	Spanish	120	86·0	21·8	8·5	Albay	1882			A.Aranlico
17389	522		Light of the Age *Woolcer* W Bn	British	208	95·5	24·8	13·9	Shoreham	1857	May		T.Chilton
	523		Light of the Age *Welch -87* W Bk	Chilian	491	134·3	29·0	18·5	Sunderland	1863	R.Thompson,jr.		E.H.Langdon
15080	524		Light of the East *Coulter* W Sr	American	130	90·0	27·0	8·0	Tremont,Me.	1869			J.M.Higgins
16706	525		Lightning *Parish 84-84* W Bg	British	154	91·9	23·8	12·6	Newcastle	1857	R.Fulton	[illegible]	T.Jefferson
2461	526	DI	Liguri *Sarzani* W Bk	Italian	744	162·4	33·8	20·7	Voltri	1874	A.Briasco fu G.B.	[illegible]	Murzi & Ajcardi
3344	527	I	Liguria *Schiappacasse -88* W Bk	Italian	417	150·2	30·6	15·9	Sestri Ponente	1884	G.Peragallo	[illegible]	G.Peragallo & Co.
26528	528	✱	Lile *Delbridge* W Bkn	British	228	..	28·8	13·9	Plymouth	1844	Hocking		A.C.P.Willyams
	529	V	Lilia *Theisaul* W Sr	French	116	72·5	19·3	11·1	Dunkirk	1866	B.Derycke		F.Ruellan
60737	530	R	Lilian *Shiere* W Bn	British	300	119·6	28·8	12·8	JordanRiv.NS	1873	Pence	[illegible]	D.S.Bremner
60452	531		Lilian *Tupman* W Bg (ABANDONED)	British	265	121·5	26·2	14·8	P.E.Island	1874	Owen	[illegible]	B.Tupman
81261	532	✱	Lilian *May 85-87* W Bn	British	146	102·0	22·4	11·5	Bridport	1879	Cox	v 1 86	Job Bros.
56294	533	✱	Lilian *Davis -86* W K	British	80	78·7	20·0	9·0	Bridport	1879	Cox		H.Good & Co.
	534	N	Lilian *Henriksen* W Bg	Norweg'n	245	110·2	26·2	13·7	Picton,N.S.	1857	A.Patterson		L.O.Dahl
	535	✱	Lilian&Evelyn *R. Thomas* W K	British	90	79·5	21·1	10·2	Glampton	1886	W.A.Gibbs		J.Thorpe
58870	536	✱	Lilla *Speight 78-78* W Sr	British	84	80·0	21·3	9·9	Middlesbro'	1870	Craggs	v 7 85	T.Shephard & Co.
	537	GV	Lilla *Ipland 80-86* I Bk	German	1076	203·0	34·4	19·9	Rostock	1886	Rostocker Act [Ges fur Sch		Jean Hintze
140610	538		Lillebonne *Hansen 85-85* W3mSr	American	208	122·7	31·6	9·2	Eureka,Cal.	1883	Cousins		Pollard & Dodge
	539	N	Lillesand *Nilsen 72-86* W Bg	Norweg'n	214	112·3	26·4	11·4	Drammen	1870	Jörgensen	[illegible]	H.J.HammersEak[illegible]
	540		Lilli W Sr	Russian	132	..	..	..		1873			E.Hämälain
	541		Lilli *Nykänen* W Sr	Russian	124	..	..	..	Impilaks	1889			N.G.Siitoin
42367	542		Lillian W Bk	British	440	138·0	29·8	16·0	Truro,N.S.	1863			James Crowe
88285	543	V	Lillian *Dexter -88* W Bkn	British	395	134·4	32·1	12·6	Kempt	1886	R.Rose	[illegible]	R.Rose
77737	544		Lillian *Hirt* W Sr	British	134	89·0	24·3	9·2	Plympton,NS	1879	C.E.Everitt		Nelson J.Williams
15043	545	R	Lillian *Willey* W Bk	American	587	135·0	33·0	19·9	Harringt'n,Me	1873		v 7 86	G.W.Coffin
83056	546	R	Lillian B. Jones *Petrie* W Bk (WRECKED) (ex [illegible] E.H.Harrington)	British	274	115·0	26·1	12·1	Duxburg,U.S.	1853		v 1 86	Nelson & Williams
120025	547	R	Lillian Hyatt W Sr (ex Capt. M. [illegible])	American	117	90·0	28·7	7·3	NHaven,Conn	1882		[illegible]	Poillevant & Farre
64881	548	V	Lillian M. Vigus *Morse* W Bk	British	872	173·3	34·5	19·7	Maitland,N.S	1873	Frakey	v 3 86	G.J.Troop
91414	549	✱	Lillian Maud *Owers* W K	British	88	79·5	20·7	10·8	Brixham	1885	S.Dewdney&Sn.		J.Marr
73131	550	✱	Lillian Morris *Davies 77-87* I Bk	British	769	195·3	31·3	19·1	Sunderland	1875	W.Doxford&Sns		T.Beynon & Co.

✱ = Lloyd's Register; D = Nederlandsche Vereeniging; G = Germanischer Lloyd; I = Registro Italiano; R = [illegible]
A = Veritas Austro-Ungarico; H = Veritas Ellenico; N = Norske Veritas; V = [illegible]

In the 4th Column, C = Composite (iron frame, wood planked); I = Iron; Stl = Steel; W = Wood.

The Dates of Metalling (Col. 11) are based on the records of the Books in which Vessels are respectively classed, or on the Reports of the [illegible]

Further particulars of Vessels classed by Lloyd's Register will be found in Lloyd's Register Book and Lloyd's Yacht Register.

No.	Reg.	Nom. Capitaines et dates d'entrée au service des Armateurs actuels et à bord du navire. Ancien Nom.	Matériaux et Gréement	Pavillon	Tonnage net	Longueur	Largeur	Creux	Construit: Où	Quand	Par qui	Doublage	Armateurs ou Directeurs-Gérants	Port d'Armement
551		Lillie (ex Ava)	W Bk	British	467	127·3	24·6	18·6	Yarmouth, U.S.	1861			J. Darling, jun.	Adelaide, S.A.
552		Lillie A. Warford *Pierce*	W Sr	American	163	96·2	27·3	7·0	Westfield, N.Y.	1874			F.O. Bailey	Perth Amboy, N.J.
553	R	Lillie F. Schmidt *Henderson*	W3mSr	American	548	146·3	35·0	15·8	Camden, N.J.	1882	S.W. Tilton		J.G. Schmidt	Philadelphia
554		Lillie Falkinburg *Leeland*	W3mSr	American	206	131·4	30·8	13·1	Greenbak, N.J.	1873			J. Kennedy	Tuckerton, N.J.
555		Lillie Keith (ex Fierian)	W Bk	British	225	98·0	27·5	13·0	Bordeaux	1851			W. Morrice	Nassau, N.P.
556		Lillie O. Wells *Wells*	W Sr	American	166	100·6	28·6	8·6	Stony Br'k N.Y.	1872	T.S. Wells		R.S. Wells	Pt. Jefferson, N.Y.
557	R	Lillie Soullard *Dennis*	W Bk	British	998	175·0	35·5	21·6	Pt. Gilbert, N.S.	1872	Everett		E.C. Dennis	Yarmouth, N.S.
558		Lilly (ex W. Alford, ex Arendt) *Karlsen 88-*	W Bk	Norweg'n	245	115·6	26·7	11·5	Baltimore	1855			B.A. Ellingsen	Christiania
559	V	Lilly *Trautmann -87*	W3mSr	Russian	324	131·4	30·8	13·1	Korbis	1887	M. Trauttmann		A. Trautmann	Riga
560	V	Lily *Dill*	W Bn	British	368	136·5	28·5	12·5	Hantsport, N.S.	1871	T. Tralsy	Y 7.5	G.W. Churchill	Windsor, N.S.
561		Lily *Davison 82-83*	W Bn	British	235	108·9	27·2	12·2	Kempt., N.S.	1866	Starratt		H. Andrews	North Shields
562		Lily *Broadstock*	W Bn	British	195	103·7	24·0	12·7	P.E. Island	1873	McLaine	Y 1.87	W. Brooks	Newport
563		Lily *Walton*	W3mSr	British	147	102·8	22·7	11·3	Ipswich	1858	W. Bayley & Co.	Y 4.81	G. Walton	Ipswich
564		Lily *Williams*	W Bn	British	121	86·0	22·9	10·2	P.E. Island	1863	McMillan		W. Wright	London
565	✻	Lily *Jones*	W Sr	British	93	80·0	20·0	10·5	Chester	1863	Parkes		C. Davison	Chester
566	✻	Lily *Crabb -88*	W Sr	British	76	79·5	21·5	8·3	Bridgwater	1879	Gough		R.S. Arnold	Bridgwater
567		Lily *Noyes*	W Sr	American	135	102·5	28·8	8·8	San Francisco	1882			J.J. McKinnon	San Francisco
568		Lily	W Bn	Spanish	185	95·5	23·9	10·6	Oldenburg	1861			J.D. Oñate	Manila
569		Lily Dale *Driman 79-82*	W Bn	British	197	103·0	23·2	11·9	P.E. Island	1865	Cox		J.H. Scott	Cork
570	R	Lily E. (LOST) *Finlay*	W Sr	British	118	79·1	26·4	7·7	St. Martin's N.B.	1885	W.H. Rourke		J.R. Blauvelt	St. John, N.B.
571	✻	Lily of Devon (FOUNDERED) *Aas -88*	W Bk	Norweg'n	389	134·4	27·7	17·0	Plymouth	1868	Moore	Y 10.86	H.E. Aas	Christiania
572	V	Lima *Carver 81-81*	W Bk	British	892	172·0	34·0	21·3	Yarmouth, N.S.	1872	F. Weston	Y 8.88	T. Killam	Yarmouth, N.S.
573	DV	Lima *Elfstadius 85-85*	W S	Swedish	692	157·8	30·9	18·4	Gefle	1870	O.A. Brodin	Y 8.86	V.T. Engvall	Gefle
574		Lima Faria	W Sr	Brazilian	143	..	..	..					F.A.P. Lima	S. Joas de Barra
575	✻	Limache *Simpson 85-85*	I Bk	British	799	190·8	32·0	18·5	W. Hartlepool	1885	W. Gray & Co.		Bk "Limache," Ld. (F.H. Vaughan)	Liverpool
576	✻	Limari *Monk 84-84*	I Bk	Chilian	600	175·4	27·8	18·2	Glasgow	1867	A. Stephen & Sns.		Lever Murphy & Co.	Valparaiso
577	✻	Limena *Collier 85-85*	I Bk	British	1043	215·0	33·8	20·1	Sunderland	1885	S.P. Austin & Son		W. Nicol & Co.	Liverpool
578		Limonia	W Sr	Turkish	100	..	..	..	Maiz Ada	1872			Stavrimatos	Maiz Ada
579	✻	Lina (ex Britannia) *Schonnandt 88-88*	I Bk	German	1543	236·1	37·1	22·5	Sunderland	1875	Osbourne, Graham & Co.		C.M. Matzen	Hamburg
580	G	Lina *Suhr 67-67*	W Bk	German	433	121·0	29·6	17·3	Barth	1867	J.S. Kraeft	Y 12.88	Joh. Struck & Sohn	Barth
581	V	Lina *Witte*	W Bg	German	238	102·0	26·4	12·0	Papenburg	1868	F.W. Meyer	Y 2.88	G.J. Schulte's Wwe.	Papenburg
582	V	Lina (CONDEMNED) *Gerdts 77-87*	W Bn	German	207	103·3	24·7	12·5	Kippkaune	1876	J.H. Meyer	Y 3.86	E.F.C. Tobias	Brake
583		Lina (ex Henriette Wester) *Racke*	W Sr	German	142	87·3	19·0	9·8	Pekela	1865	H.F. Drent		H.L. Juhl	Stralsund
584	N	Lina (ex Theodor) *Müller 87-87*	W Bk	Norweg'n	802	174·2	34·9	21·7	Bergen	1867	A. Dekke	Y 4.88	P.G. Pettersen	Bergen
585	N	Lina (ex Wave Crest) *Hansen -88*	W Bk	Norweg'n	685	154·5	32·9	19·4	Pleasant Cove, N.S.	1866	Crosby & Co.		L.G.S. Larsen	Sandefjord
586	N	Lina (ex Simplon) *Torgersen*	W Bg	Norweg'n	260	96·9	26·1	15·1	Sunderland	1859	Lister		Th. Terjesen	Kragerö
587		Lina *Andersson 81-*	W Bk	Russian	545	153·4	33·0	17·4	Swiskarwarf	1877	A. Thor		H.J. Knus	Lappfjerd
588		Lina *Eriksson*	W Bg	Russian	399	..	..	..		1866			E. Göransson	Hammarland
589		Lina *Helmberg 80-86*	W Bg	Russian	286	119·1	29·7	13·4	Kimito	1873	O. Oelander		C.G. Holmberg	Westanfjerd
590	V	Lina *Freimann -87*	W Sl	Russian	101	66·6	23·9	10·5	Ilgezeem	1887	C. Ohrmann		R. Seeberg	Riga
591	D	Lina & Johanna (ex Gazelle) *Visser 87-87*	W Bk	Dutch	636	..	..	..	Capelle	1857		Y 1.89	E.J. Bok & Zonen	Amsterdam
592	R	Linah C. Kaminski *Woodbury*	W3mSr	American	421	132·3	33·2	14·4	Bath, Me.	1882	Goss & Sawyer		H.H. Grant	Georgetown, S.C.
593		Linaria *Falt 80-84*	W Sr	British	107	79·3	24·5	9·7	Nova Scotia	1884	U. Falt		William Falt	Lunenberg, N.S.
594	V	Lince (ex Tomas III) *Maristany*	W Bn	Spanish	307	109·9	28·4	13·9	Barcelona	1866		Y 10.86	J. Maristany	Barcelona
595	✻	Lincelles *Razeto -88*	W Bk	Italian	773	172·0	30·6	23·0	Moulmein	1858	Dunbar	Y 2.88	A. Razeto	Genoa
596		Lincluden *Reilly 84-84*	W Bk	British	286	124·2	25·1	14·9	Newport	1855	J. Thomas	Y 6.87	T. Lyons	Belfast
597	A	Lincoln *Cattarinich -87*	W Bk	Austrian	793	159·1	31·3	19·8	Lussinpiccolo	1875	M.A. Starcich	Y 3.88	A.G. Cattarinich	Lussinpiccolo
598	N	Lincoln *Seensen 86-86*	W Sr	Norweg'n	164	84·0	24·5	11·6	Kragerö	1880		Y 12.81	J. Wright	Langesund
599	✻	Linda *Drake -88*	W K	British	34	57·3	17·1	7·6	Falmouth	1878	Burt		F. Drake	Barnstable
600														

✻ = Lloyd's Register; D = Nederlandsche Vereeniging; G = Germanischer Lloyd; I = Registro Italiano; R = American Record; A = Veritas Austro-Ungarico; H = Veritas Ellenico; N = Norske Veritas; V = Bureau Veritas.

Dans la 4me Colonne, C = Composite (membrure en fer, bordé en bois); I = Fer; Stl = Acier; W = Bois.

Les Dates du doublage en métal (**Col. 11**) ressortent des livres dans lesquels les navires sont respectivement classés ou des rapports des Inspecteurs du Lloyd's Register.

Pour plus amples renseignements sur les navires classés par le Lloyd's Register, voir le Lloyd's Register Book et le Lloyd's Yacht Register.

Official No.	No. in Book	Registry in which classed	Ship's Name. Material and Rig. Master, and Dates of Appointment to present Owner's Service and to Ship. Ship's Late Name.	Flag	Net Tonnage	Dimensions: Length	Breadth	Depth	Built: Where	When	By whom	Date of Metalling	Owners or Managers
[illegible]	601		Linda *Boyd* W3mSr	American	225	126·3	31·2	8·7	Mobile, Ala.	1871			W. J. Boyd
	602		Linda *Rey* -[illegible] W Bn	French	109	79·4	22·9	11·2	Chiavari	1877	L. Gotuzzo		François Rey
2[illegible]	603	I	Linda *Nossardi* 83-88 W Bg	Italian	425	125·6	30·4	19·[illegible]	Finale	1871	F. Fonzolo	[illegible]	Gerolamo Nossardi
	604	N V	Linda *Ericksen* W Bk	Norweg'n	332	128·3	29·9	12·2	Kragerö	1883	T. G. Litangen	[illegible]	A. Terkelsen
	605		Linda *Ek* W Sr	Russian	273	112·0	28·3	13·6	Sastmola	1873	J. Ek		L. Kräki
	606		Linda *Troberg* 81- W Sr	Russian	220	96·8	25·3	12·0	Kimito	1869	G. Sjöström		J. Johansson
	607	V	Linda *Ferrer* W Bk	Spanish	361	122·7	29·6	15·1	Arens	18[illegible]		[illegible]	P. Ferrer & Casals
	608		Linda (ex Nuestra Señora del Carmen) W Sr	Spanish	135	80·7	22·8	11·1	Palma	1852			B. Pieras y Florest
	609		Linda Flor *Marques* W Bn	Brazilian	174	..	..	..	S. Joao da Barra	1870		[illegible]	Cia. Cabotagem de S. João da Barra & Campos
	610	I V	Linda G. W Bk		710	154·9	30·8	22·5	Savona	1873	G. Clario	[illegible]	
	611	V	Linda Morgenröden (ex Morgenröden) *Rissen* -86 W Bk	Russian	448	138·6	30·7	17·2	Grimstad	1878	J. Bang	[illegible]	Shipping Co. [illegible]
	612		Linda Palmeira (ex Amelia) *Ins Rocha Cruz* W Bk	Portuguese	364	117·7	26·2	15·1	Oporto	1862	[illegible]	[illegible]	A. J. da Rocha
90124	613	✻	Linda Park *Ryan* -8[illegible] I&Stl Bkn	British	343	155·5	26·5	11·7	[illegible]	1885	S. M. Knight & Co.		J. W. Valentine & Co.
75138	614		Linda Weber *Bushell* -86 W Bn	British	114	86·4	22·5	9·3	Mahurangi, N.Z.	1877	M. Darrach		Donald H. Ross
83128	615	D V	Linden *Crowe* 84-86 W Bk	British	913	175·6	34·4	20·4	Clifton, N.S.	1881	J. Crow	[illegible]	Martin Dickie
	616	N	Lindesnæs (ex Eugenie) *Johnsen* -88 W Bk	Norweg'n	674	157·9	31·3	19·4	Quebec	1869	P. V. Valin	[illegible]	G. C. Hansen
	617	R	Lindesnæs (ex Admiral de Ruyter) *Ingvardsen* -87 W Bk	Norweg'n	518	138·2	29·6	18·7	Amsterdam	1860		[illegible]	J. M. Jonasen
	618	N	Lindheim *Nielsen* 71-71 W Bg	Norweg'n	258	114·8	24·7	13·9	P. E. Island	1866		[illegible]	T. & A. Nielsen
[illegible]	619	✻	Lindisfarne *Exon* 87-88 I S	British	1724	250·8	39·0	24·0	Stockton	1886	Richardson, Duck & Co.		J. Lidgett & Sons
	620	N V	Lindola *Larsen* -87 W Bk	Norweg'n	384	121·4	29·5	16·5	Arendal	1873	A. Aanonsen	[illegible]	G. Jensen
[illegible]	621	✻	Lindores Abbey *Kinnaird* 83-85 I Bk	British	863	197·5	32·6	19·5	Port Glasgow	1877	R. Duncan & Co.		Clark & Service
	622	N V	Lindsay *Bull* W Bk	Norweg'n	559	144·0	33·1	18·5	Tönsberg	1863	A. B. Bull	[illegible]	A. B. Bull's Enke
	623	V	Line *Rasmussen* 75-88 W Bn	Danish	142	91·7	22·5	10·8	Aalborg	1874	C. Andersen		R. H. Rasmussen
	624		Lines Brothers *Svendsen* [illegible] -87 W Bk	Norweg'n	617	161·7	28·3	19·4	Whitehaven	1860			Actieselskabet "Ellen Lines" (Christensen & Co., London)
	625	V	Linga *Lebour* -88 W Bk	French	386	121·7	25·6	16·8	Nantes	1869	Jollet & Babin		P. Ch. Grenet
70[illegible]	626	✻	Linlithgowshire (ex Jessie Landles) *Stott* [illegible]-83 I S	British	1357	241·7	37·4	21·7	Port Glasgow	1877	R. Murray & Co.		T. Law & Co.
	627		Linné *Eriksson* W Sr	Russian	260	101·6	28·6	12·9	Sideby	1872	K. J. Jansson		C. Carlsson
	628		Linnea *Eriksson* W3mSr	Russian	300	..	..	..		1872			A. J. Erikson
	629	N	Linnea *Hanssen* 79-84 W Bk	Norweg'n	349	119·0	28·8	15·5	Risör	1876	Jacob Hansen	[illegible]	O. Caspersen
[illegible]	630	V	Linnet *Graham* -87 W Bkn	British	928	166·0	35·0	22·0	Maitland, N.S.	1884	W. P. Cameron	[illegible]	W. P. Cameron
751[illegible]	631	✻ D	Lintrathen *Alexander* I Bk	British	664	185·6	30·0	17·9	Dundee	1876	Thompson		J. Gibson
	632		Lintu *Berglund* W3mSr	Russian	313	115·5	30·9	13·2	Ekkysby	1874	J. H. Johansson		K. Puukki
19[illegible]	633	✻	Linus *Humphreys* -8[illegible] W Bg	British	170	105·6	22·7	13·5	Nevin	1857	R. Thomas	[illegible]	D. Morris & Co.
	634	V	Linwood *McKenzie* -87 W Bk	British	1233	201·6	38·6	23·0	Maitland, N.S.	1887	E. Roy	[illegible]	A. Roy
33277	635		Linwood *Boyles* -84 W Bg	British	157	92·0	24·7	11·1	Moncton, N.B.	1858	Jones		A. McMurtry
88[illegible]	636	R	Lion *Peters* -87 W Sr	British	146	89·5	24·8	10·4	Mahone Bay, N.S.	1884	J. H. Zwicker	[illegible]	A. Ernst
	637	✻	Lion *Gosse* -88 W Bk	French	470	141·3	29·4	17·3	Littlehampton	1873	Harvey	[illegible]	A. Delaunay
	638		Lionel (ex Ocean Child) W Bk	Norweg'n	299	121·1	29·5	12·7	River John, N.S.	1863	McLennan		Actieselskabet [illegible] (Jens Olsen)
70133	639	✻	Lioness *Robinson* -83 W Bk	British	512	148·2	29·6	17·7	Littlehampton	1875	Harvey	[illegible]	G. & J. Robinson
	640		Lionne *Schreiber* W Bg	French	217	93·5	23·0	13·6	Dunkirk	1847	Malo		F. Denis & Co.
[illegible]	641		Lipari (ex Vincenzo) *Arena* W Bg	Italian	158	97·8	23·0	12·9	Limite	1858	N. Pichiotti	[illegible]	Patino & Arena
	642		Liquidateur *Leroy* -86 W Bg	French	164	84·2	22·1	13·2	St. Malo	1851			Lemoine
	643		Liro de Saffo (ex Aurora, ex Uruguay) W Sr	Spanish	142	84·0	25·8	11·4	Masnou	1855	B. & M. Carreras Hermanos		P. Riera y Suarez
[illegible]	644	I V	Lisa *Semidei* 80-80 W Bk	Italian	745	158·1	33·8	22·3	Sestri Ponente	1880	G. Bozzano	[illegible]	A. Semidei & L. Pesce
[illegible]	645	I	Lisa *Mondaini* W Bg	Italian	146	91·2	23·0	12·8	Venice	1864	O. Zanon		A. Mondaini
	646	V	Lisboa *Håkansson* -81 W Sr	Swedish	146	91·8	21·8	10·7	Timmernabben	1875	C. Bring	[illegible]	A. Arfvidsson
[illegible]	647		Liscio (BROKEN UP) W Bk	Italian	478	132·8	31·7	19·3	Savona	1866	F. Sirello	[illegible]	[illegible]
	648	R	Lisette *Holm* -88 W Bg	Swedish	293	103·6	24·7	15·8	Bollsta	1839			R. Fogelmarck
[illegible]	649		Lisette Ilse Catharina *Petersholz* -8[illegible] W Sr	Russian	178	95·0	26·0	12·0	Angern	1875	M. Berg		A. W. Grünberg
9[illegible]	650	✻	Lismore *Sutton* I S	British	1598	255·0	39·0	22·9	Glasgow	1885	C. Connell & Co.		J. Gardiner & Co.

✻ = Lloyd's Register; D = Nederlandsche Vereeniging; G = Germanischer Lloyd; I = Registro Italiano; R = A[illegible]
A = Veritas Austro-Ungarico; H = Veritas Ellenico; N = Norske Veritas; V = Bureau [illegible]

In the 4th Column, C = Composite (iron frame, wood planked); I = Iron; Stl = Steel; W = Wood.

The Dates of Metalling (Col. 11) are based on the records of the Books in which Vessels are respectively classed, or on the Reports of the Surveyors of L[illegible]

Further particulars of Vessels classed by Lloyd's Register will be found in Lloyd's Register Book and Lloyd's Yacht Register.

		Noms, Matériaux et Gréement. Capitaine et dates d'entrée au service des Armateurs actuels et à bord du navire. Ancien Nom.	Pavillon.	Tonnage	Dimensions. Longueur	Largeur	Creux	Construit. Où.	Quand.	Par qui.	Date du doublage en métal	Armateurs ou Directeurs-Gérants.	Port d'Armement.
651		Lismore WSr	British	181	119·3	23·0	9·9	PortStephens, N.S.W.	1878	J.Rodrick		Emily O.Pratt	Sydney,N.S.W.
652	V	Lista *Coll* WSr	Spanish	201	91·9	26·1	13·1	Arenys	1866		Y 12·81	Viuda de J.Millet	Barcelona
653		Lister (BROKEN UP) WBk	Norweg'n	255	109·7	26·2	13·9	Flekkefjord	1856	Abrahamsen		S.Fedde	Farsund
654		Little Annie *Perrier* -87 WBn	British	228	106·0	23·0	12·9	Quebec	1868	C.W.Wilson	Y 6·87	G.E.Perrier	PortLouis
655	❋	Little Beauty *Braddon* WSr	British	90	86·2	20·2	10·1	Pelrum	1875	Stephens	Y 3·85	J.Stephens	Fowey
656		Little Dorrit *Toms* -84 WSr	British	149	100·0	22·0	12·0	Brixham	1862	Upham		W.A.Hanson (G.H. Eaton)	Fowey
657		Little Jessie WSk	British	71	72·4	20·2	10·4	Elmshorn	1885	J.Kremer		H.Mudd	Great Grimsby
658	❋	Little Mystery *Bowden* -86 WSr	British	100	84·5	21·1	10·8	Kingsbridge	1887	W.Date&Sons	Y 87	J.Stephens	Fowey
659	❋	Little Pet *Bartlett* -87 WSr	British	75	76·0	19·8	9·5	Fowey	1880	Dingle	Y 9·86	J.Stephens	Fowey
660		Little Reaper *Sherlock* (ex Zingara) 1863 W3mSr	British	100	95·6	19·8	9·7	Falmouth	1863	Lean		J.Pritchard	Runcorn
661	❋	Little Secret *Couch* -87 WSr	British	89	88·0	20·4	9·4	Portsmouth	1884	J.Read,jun.	Y 9·87	J.Stephens & Co.	Fowey
662	❋	Little Vixen *Lapage* -88 WSr	British	93	90·0	20·2	10·1	Poole	1877	Meadus	Y 4·89	J.Taylor	Plymouth
663	❋	Little Willie Coath WSr	British	97	88·0	22·0	10·8	Plymouth	1875	Banks	Y 6·85	W.Stephens	Plymouth
664	❋	Little Wonder *Hancock* WSr	British	98	84·3	21·3	10·7	Kingsburg	1886	W.Date&Sons	Y 6·86	J.Stephens	Fowey
665	N	Liv *Bolleaven* WBk	Norweg'n	256	114·0	27·4	12·8	Porsgrund	18 59/80		Y 1·88	J.Harbitz	Tönsberg
666	A	Livadia *Radessich* -88 WSr	Austrian	157	..	..	..	Lussinpiccolo	1884	N.Martinolich	Y 8·84	Guiricich di A.	Lussinpiccolo
667	V	Livadia *Greene* 87-87 W3mSr	Russian	282	117·9	29·0	13·5	Dondangen	1884	RedikTuum		E.Moritz	Riga
668	❋	Liverpool Stl 4mS	British	3330	333·2	47·9	26·5	PortGlasgow	1889	Russell & Co.		R.W.Leyland & Co.	Liverpool
669		Liverpool (ex Sir Robt. Peel) WSr	British	117	86·4	21·1	7·9	Washingtn,US	1853			C.Graveby	Nassau,N.P.
670		Livia (WRECKED) *Mazella* WBn	Italian	240	96·1	28·2	16·5	Procida	1865	A.Lubrano		Flli Mazzella	Naples
671	DV	Livia *Raudsep* -86 WSr	Russian	207	97·8	27·2	13·3	Wandsen	1881	J.Kaimann		J.Jansohn	Riga
672		Livia Costa *Montarso* WBk	Italian	228	115·1	23·6	13·6	Sampierdarena	1874	Frasinetti	C 5·82	G.B.Scandura	Catania
673	A	Livia Olivari (ex Alwina) *Olivari* WLr	Italian	205	..	..	..	Hoogezand	1867	J.A.Hooites	Y 2·86	G.Olivari	Genoa
674	❋	Livietta (ex Adelaide M.) *Scala* 70-86 WBg	Italian	367	116·8	26·9	16·5	SestriPonente	1875	Ag.Carlevari	Y 1·85	M.& A.Dall'Orso	Spezia
675	❋	Livingstone *Iddon* 86-87 W3mSr	British	170	104·1	23·2	12·0	GlassonDock	1877	Simpson	Y 5·83	Porter's ShippingCo.Ld.	Lancaster
676	V	Livingstone *Wenferl* 82-82 WBn	Danish	226	111·7	23·7	12·5	Middelfart	1874	H.Illum	Y 2·86	Aktie-Selskabet " Livingstone" (O.Hansen)	Veile
677	V	Livingstone *Asmus* -87 WBk	German	531	139·1	31·4	17·3	Kiel	1874	Conradi&Sohn	Y 6·87	W.Stehr	Hamburg
678	N	Livingstone (ex Melbourne) *Sukkestad* WBk	Norweg'n	638	166·8	29·7	17·8	Leith	1860	Menzies	Y 2·86	Brödrene Knudsen	Haugesund
679	N	Livingstone *Bowitz* WBk	Norweg'n	436	122·9	31·3	17·3	Grimstad	1872	N.O.Gjömle		N.Jacobsen	Grimstad
680	N	Livingstone *Johnsen* 73-81 WBn	Norweg'n	229	112·0	25·5	12·2	Stavanger	1872	C.Knudsen	Y 5·83	H.Wigsnœs	Stavanger
681		Livlig (ex Spörrelet) *Lobes* W3mSr	Norweg'n	164	114·2	23·1	10·6	Calmar	1864	J.G.Roos		P.Grenager	Fredrikshald
682	❋	Livonia *Finlay* 83-83 WBn	British	268	115·5	27·3	13·9	P.E.Island	1874	Clow		J.Tyrrell	Newport
683		Livonia *Caws* 85-85 WSrYt	British	128	115·8	23·3	11·9	Cowes	1871	M.E.Ratsey		Capt.Thomas Wood	Portsmouth
684		Livonia (ex Coral Isle) *Minkewitz* 79- WBg	Russian	311	116·0	25·1	16·0	Sunderland	1861	Naizby		J.Brakmann	Reval
685		Lizette *Hatfield* WBkn	British	249	141·6	27·2	12·4	Gloucester	1869	W.Hunt		J.Neill	Belfast
686		Lizor WBg	Turkish	227	..	..	..	England	1879			Stilyiatostori Bakkhal	Constantinople
687	V	Lizzie *Sebire* -75 WBkn	British	352	146·3	31·2	11·0	Maitland,N.S.	1885	C.Putnam	Y 9·85	J.De V.Le Couteur	Maitland,N.S.
688		Lizzie *Jones* 85-85 WBkn	British	223	104·1	25·4	11·9	Maine,U.S.	1875	T.Jackson		J.Williams	Sydney,N.S.W.
689	❋	Lizzie (ex Catherine & Alice) *Attwooll* 86-86 W3mSr	British	121	87·2	22·3	10·6	Nevin	1850	E.Ellis		R.A.Ayles	Weymouth
690	❋	Lizzie *Jones* WSr	British	100	84·0	21·7	10·6	Milford	1885	W.Francis	Y 2·85	W.Francis	Milford
691	❋	Lizzie *Richards* 80-88 WSr	British	100	81·8	20·9	10·3	Kingsbridge	1872	Date	Y 7·87	D.Richards	Salcombe
692	❋	Lizzie *Hutton* 85-85 WSr	British	94	84·3	21·6	10·1	Perth	1872	McPherson		Mrs.A.Bowyer (J.Bowyer)	Chester
693	❋	Lizzie *Tallack* 85-85 WSr	British	82	75·8	21·0	10·0	Malpas	1881	Scoble&Davies	Y 4·85	Scoble & Davies	Truro
694		Lizzie B. McNichol WSr	American	154	96·5	27·2	9·4	Lubec,Me.	1874	McBride		P.Gillis	Lubec,Me.
695	R	Lizzie B. Willey *Tyler* -86 W3mSr	American	549	150·4	34·1	16·6	Thomast'n,Me	1881	Dunn&Elliot	Y 5·87	Dunn & Elliot	Thomaston,Me.
696	R	Lizzie Babcock *English* W3mSr	American	560	145·0	35·6	17·2	Dennisville,NJ	1883	R.S.Leaming		F.Babcock	SomersPoint,N.J.
697	❋	Lizzie Bell *Hay* 88-88 IBk	British	1036	214·5	34·1	21·3	Sunderland	1877	R.Thompson,jun.		P.Iredale & Son	Liverpool
698	V	Lizzie Burrill *Johnson* 86-86 WS	British	1185	190·0	37·4	22·2	LittleBrookNS	1875	A.Blinn	Y 8·86	W.Burrill & Co.	Yarmouth,N.S.
699	V	Lizzie C. Troop *Crosby* -87 WS	British	1391	198·0	39·1	24·0	St.John,N.B.	1873	Troop & Son	Y 6·85	H.D.Troop	St.John,N.B.
700	R	Lizzie Carr *Browne* 82-84 W3mSr	American	272	138·9	29·6	10·9	Thomast'n,Me	1868	Walker&Dunn		Dunn & Elliot	Thomaston,Me.

❋=Lloyd's Register; A=Veritas Austro-Ungarico; D=Nederlandsche Vereeniging; G=Germanischer Lloyd; H=Veritas Ellenico; I=Registro Italiano; N=Norske Veritas; R=American Record; V=Bureau Veritas.

Dans la 4me Colonne, C=Composite (membrure en fer, bordé en bois); I=Fer; Stl=Acier; W=Bois.

Les Dates du doublage en métal (Col. 11) ressortent des livres dans lesquels les navires sont respectivement classés ou des rapports des Inspecteurs du Lloyd's Register.

Pour plus amples renseignements sur les navires classés par le Lloyd's Register, voir le Lloyd's Register Book et le Lloyd's Yacht Register.

Official No.	No. in Book	Registry in which classed	Ship's Name. Master, and Dates of Appointment to present Owner's Service and to Ship. Ship's Late Name.	Material and Rig	Flag	Net Tonnage	Length	Breadth	Depth	Where	When	By whom	Date of Metalling	Owners or Managers
140811	701	R	**Lizzie Carter** *Goodwin -87*	W Bkn	American	750	168·9	35·0	18·0	Bath, Me.	1885	N. England SBC	[illegible]	J. R. Kelley
[illegible]	702	R	**Lizzie Chadwick** *Chadwick 70-85*	W 3m Sr	American	449	155·0	35·0	11·5	Thomast'n, Me	1883	Dunn & Elliot	[illegible]	Dunn & Elliot
[illegible]	703		**Lizzie Cochran** *Hopkins*	W Sr	American	175	108·0	27·3	8·8	Brewer, Me.	1870	J. Oakes		J. A. Hopkins
[illegible]	704		**Lizzie Colby** *Allen 86-86*	W Sr	American	143	96·6	24·0	9·3	Essex, Mass.	1882	J. James & Co.		J. A. Mathieson
[illegible]	705	V	**Lizzie Curry** *McCulloch 83-84*	W Bk	British	498	144·4	31·0	16·0	Horton, N.S.	1875	J. B. North		J. B. North
[illegible]	706		**Lizzie D. Small** *Warwick*	W Sr	American	181	109·0	27·2	8·2	Camden, N.J.	1866	Day & Carter		G. A. Warwick
[illegible]	707		**Lizzie Dewey** *Burgi -88*	W 3m Sr	American	356	124·0	30·6	15·0	Lubec, Me.	1872	Langmaid & Mugford	[illegible]	J. Schroder
[illegible]	708		**Lizzie Durkee**	W Bk	British	1021	171·3	36·0	22·8	Bath, Me.	1864			J. Niles
[illegible]	709		**Lizzie Griffin**	W Sr	American	101	82·4	23·0	8·9	Essex, Mass.	1884	M. Adams		S. G. Pool
[illegible]	710	R	**Lizzie H. Patrick** *Meyer*	W 3m Sr	American	449	146·5	34·0	14·0	Milford, Del.	1883	D. Lank		J. Meyer
[illegible]	711	R	**Lizzie Heyer** *Harrington 71-78*	W 3m Sr	American	342	137·0	31·0	11·5	Thomaston, Me	1873	Walker & Dunn	[illegible]	Dunn & Elliot
140854	712		**Lizzie J. Call** *Call -86*	W 3m Sr	American	186	..	..	..	Portsmouth, N.H.				G. J. Bufford
[illegible]	713		**Lizzie L. Mills** *Armstrong* (ex General Scott)	W Sr	American	112	*81·2	24·3	8·5	Brewer, Me.	18[illegible]			J. Littlefield
[illegible]	714	R	**Lizzie Lane** *Herrick*	W 3m Sr	American	226	115·8	29·8	9·2	Searsport, Me.	1874		[illegible]	J. H. Lane
[illegible]	715		**Lizzie Lee** *Wanstall 83-83*	W Sr	British	129	85·4	20·4	11·3	Yarmouth	1856	J. Powell		W. Burnitt
[illegible]	716		**Lizzie M. Eels** (ex Charles Russell) *Greenlaw*	W 3m Sr	American	104	71·9	21·6	7·2	Baltimore, Md.	1860			A. F. Eells
[illegible]	717	✻	**Lizzie Mary** *Gill 86-*	W Sr	British	97	84·0	21·2	10·8	Milford	1874	Morgan		W. C. Phillips
[illegible]	718	R	**Lizzie May** *Hutchinson* (ex May Morn)	W 3m Sr	American	191	104·5	27·4	9·0	Bowdoinham, Me.	1869			E. G. Russell
75135	719	✻	**Lizzie R. Wilce** *Beale*	W 3m Sr	British	155	102·9	23·2	11·7	Point	1876	Stephens	[illegible]	W. Chellew
[illegible]	720		**Lizzie Raymond** *Lord*	W Sr	American	113	98·0	26·6	8·0	Essex, Conn.	1863			R. M. Lord
[illegible]	721	DV	**Lizzie Ross** *Vaughan 80-80*	W S	British	1151	187·0	37·0	23·1	Windsor, N.S.	1873	S. Dimock	[illegible]	S. Dimock
[illegible]	722	R	**Lizzie S. Haynes** *Sawyer*	W 3m Sr	American	415	140·0	32·9	12·9	Wiscasset, Me.	1881	J. M. Haynes		J. M. Haynes
140674	723	R	**Lizzie S. James**	W Sr	American	173	100·6	30·0	7·0	Milford, Del.	1883	J W Abbott & Co.		P. J. Johnsen
[illegible]	724	✻	**Lizzie Thorpe** *Hill 86-86*	W Sk	British	85	75·0	21·0	10·5	Grimsby	1884	Smith Bros.		G. Dobson
58288	725		**Lizzie Trenberth** *Jenkins 81-81*	W Sr	British	100	92·6	22·2	11·5	Par	1867	Moss		W. E. Pearce
140332	726		**Lizzie V. Hall** *Clouting 76-80*	W 3m Sr	American	186	117·0	30·8	9·0	Frederica, Del.	1878	Nathaniel Lank		J. W. Hall
[illegible]	727		**Lizzie W. Matheson** *McKoy 81-81*	W 3m Sr	American	184	98·8	26·5	10·4	Essex, Mass.	1875			H. Cook
[illegible]	728		**Lizzie Waters** *Barsted (-83* (CONDEMNED)	W Sw	British	256	99·0	26·4	15·2	Sunderland	1864	Cray		C. Weatherburn
43246	729		**Lizzie Webster**	W Sr	British	108	78·2	22·6	9·4					S. Parsons, jun.
[illegible]	730	R	**Lizzie Wharton** *Nelson*	W Sr	British	120	85·6	25·8	9·6	Brooklyn, N.S.	1884	A. Gardner		Chas. H. Clements
[illegible]	731		**Lizzie Williams** *Colson -88*	W Bk	American	790	158·1	33·4	20·0	Rockland, Me.	1868			Alaska Packing Co.
[illegible]	732	R	**Lizzie Wilson** *Chadwick* (COLLISION)	W 3m Sr	American	303	124·6	29·5	11·8	Thomaston, Me	1872	Walker, Dunn & Co.	[illegible]	Dunn & Elliot
[illegible]	733		**Lizzie Wright** *Wells*	W Bk	British	933	176·4	35·3	20·7	Rockland, N.B.	1872	R. A. Chapman	[illegible]	J. N. Smith
[illegible]	734		**Lizzie Young** *Pierce*	W 3m Sr	American	361	128·0	32·8	14·6	Seaford, Del.	1873	W. F. Lamb		R. C. Young
	735	V	**Lizzie & Eugenie** *Geerdes* (ex Ellen Goodspeed)	W S	German	1262	196·4	37·6	24·0	Bath, Me.	1868	F. & E. Reed	[illegible]	C. A. E. Solscher
[illegible]	736	✻	**Lizzy Fox** *Vale*	W Bn	British	263	119·0	24·7	14·5	Bridport	1864	Cox	[illegible]	W. C. Rafarel
[illegible]	737		**Ljubezna** *Bonetich* (ex Saint Helena)	W Bk	Austrian	444	138·4	28·7	17·5	Sunderland	1868	C. Short		Antonia Bonetich
	738		**Ljubezni Otac** *Gherterd*	W Bk	Russian	268	..	..	..	Buccari	1868			G. Suttora
	739	D	**Ljubirod** (ex P. F. Webster)	I Bk	British	1250	202·0	37·0	22·0	Liverpool	1856			Burns, Philp & Co. Ld.
[illegible]	740		**Llama** *Layzell*	W K	British	149	106·4	25·5	9·4	Harwich	1875	Vaux		A. Layzell
[illegible]	741	✻	**Llewellyn** *Jones 81-81*	I Bk	British	498	162·1	27·6	17·1	Glasgow	1875	A. Stephen & Sons		J. Rosser
[illegible]	742	RV	**Llewellyn J. Morse** [illegible]	W S	American	1393	198·2	36·6	24·0	Brewer, Me.	1877	J. Oakes & Son	[illegible]	J. Renfield's Sons
	743		**Lloret** *[illegible]*	W Bk	Spanish	184	90·2	26·8	13·1	Lloret	1859			José Cabruja y Roca
	744	R	**Lloret de Mar** *Carreras* (ex Sorata)	W [illegible]	Spanish	348	124·0	31·8	12·8	Tracadie, N.S.	1876	P. Landry	[illegible]	Garriga hers. & Co.
	745	N	**Lloyd** *Jacobsen 76-7[illegible]*	W Bk	Norweg'n	454	132·6	29·7	16·8	Lillesand	1876	Speilberg & Reiersen	[illegible]	B. Birkeland
[illegible]	746	V	**Loanda** *Scott 79-81*	W S	British	1525	209·2	39·8	24·2	St. John, N.B.	1881	J. Fraser	[illegible]	B. Smith & Sons
[illegible]	747	✻	**Lobo** *Sambrook -88*	I Bk	British	899	209·0	32·4	19·9	Sunderland	1877	Osbourne, Graham & Co.		Shallcross & Higham
[illegible]	748	✻	**Loch Awe** *Nicol 78-78*	I S	British	1053	217·7	34·5	21·0	Glasgow	1869	B'clay Curle & Co		J. Wilson
[illegible]	749	✻	**Loch Bredan** *Cumming 75-83*	I Bk	British	956	263·4	33·6	19·9	Glasgow	1882	Dobie & Co.		J. Sproatt
[illegible]	750	✻	**Loch Broom** *Martin 87-88*	I 4m S	British	2075	287·7	42·5	24·1	Glasgow	1885	B'clay Curle & Co		General Shipping Co.

✻ = **Lloyd's Register**; D = Nederlandsche Vereeniging; G = Germanischer Lloyd; I = Registro Italiano; R = Ame[…]
A = Veritas Austro-Ungarico; H = Veritas Ellenico; N = Norske Veritas; V = Bure[…]

In the 4th Column, C = Composite (iron frame, wood planked); I = Iron; Stl = Steel; W = Wood.

The Dates of Metalling (Col. 11) are based on the records of the Books in which Vessels are respectively classed, or on the Reports of the Surveyors of […]

Further particulars of Vessels classed by Lloyd's Register will be found in Lloyd's Register Book and Lloyd's Yacht Register.

1	2	4 Noms. Matériaux et Gréement. Capitaine et dates d'entrée au service des Armateurs actuels et à bord du navire. Ancien Nom.	5 Pavillon	6 Tonnage net.	7 Dimensions. Longueur	Largeur	Creux	8 Construit. Où	9 Quand	10 Par qui	11	12 Armateurs ou Directeurs-Gérants.	13 Port d'Armement.
751	*	Loch Carron *Clarke 85-86* I 4mS	British	2075	287·7	42·5	24·1	Glasgow	1885	B'clayCurle&Co		General Shipping Co.	Glasgow
752	*	Loch Cree *Jones 77-77* I Bk	British	791	197·6	32·1	18·4	PortGlasgow	1874	Macfadyen&Co.		J.Sproat	Liverpool
753	*	Loch Doon *Reblic 84-84* I Bk	British	781	195·6	31·0	19·0	PortGlasgow	1872	Hamilton		J.Sproat	Liverpool
754	*	Loch Eck *Pittendreigh 79-84* I S	British	1641	262·9	38·3	23·1	Glasgow	1874	C.Connell&Co.		J.Wilson	Glasgow
755	*	Loch Etive *Stuart 77-77* I S	British	1235	226·9	35·9	21·6	Glasgow	1877	A.&J.Inglis		General Shipping Co.	Glasgow
756	*	Loch Fergus *Clackrie 80-85* I Bk	British	846	202·1	33·2	18·7	Glasgow	1875	D.&W.Henderson&Co.		J.Sproat	Liverpool
757	*	Loch Fleet *Milbeack 86.86* I Bk	British	713	188·1	31·2	18·4	Glasgow	1872	Dobie&Co.		J.Sproat	Liverpool
758	*	Loch Garry *Horne 77-85* I S	British	1493	250·5	38·4	22·6	Glasgow	1875	J.&G.Thomson		Glasgow Shipping Co.	Glasgow
759	*	Loch Katrine *Mckenna 85-85* I S	British	1200	226·0	35·8	21·5	Glasgow	1869	J.G.Lawrie		Glasgow Shipping Co.	Glasgow
760	*	Loch Ken *Wilson 81-81* I Bk	British	590	179·7	28·6	17·7	PortGlasgow	1869	L.Hill&Co.		J.Sproat	Liverpool
761	*	Loch Linnhe *Vaughan 83-83* I S	British	1382	234·7	37·2	22·5	Glasgow	1876	J.&G.Thomson		J.Wilson	Glasgow
762	*	Loch Lomond *Davies 83.83* I S	British	1200	226·3	35·8	21·5	Glasgow	1870	J.G.Lawrie		Glasgow Shipping Co.	Glasgow
763	*	Loch Long *Armit 85-88* I S	British	1203	228·5	35·8	21·3	Glasgow	1876	J.&G.Thomson		General Shipping Co.	Glasgow
764	*	Loch Moidart *Auld 88-88* I 4mBk	British	2000	287·1	42·6	24·0	Glasgow	1881	B'clayCurle&Co		General Shipping Co.	Glasgow
765		Loch Naw *Pritchard* W Bk	British	277	115·0	25·2	14·5	Isl of Whithorn	1867	W.McWilliam		J.C.Ellis	Sydney,N.S.W.
766	*	Loch Ness *Fernie 83-83* I S	British	1203	225·5	35·6	21·6	Glasgow	1869	B'clayCurle&Co		Glasgow Shipping Co.	Glasgow
767	*	Loch Rannoch *Watt 88-88* (ex Clanranald) I S	British	1185	217·8	35·5	21·0	Glasgow	1868	J.&G.Thomson		Glasgow Shipping Co.	Glasgow
768	* D	Loch Ranza *Boyd 80-86* I Bk	British	1075	219·9	34·7	20·7	Glasgow	1875	C.Connell&Co.	Y 1.83	J.Wilson	Glasgow
769	*	Loch Ryan *Ozanne 77-86* I S	British	1207	228·5	35·8	21·3	Glasgow	1877	J.&G.Thomson		General Shipping Co.	Glasgow
770	*	Loch Shiel *Radford 84-84* I S	British	1218	225·3	35·6	21·1	Glasgow	1877	D&WHenderson		Glasgow Shipping Co.	Glasgow
771	*	Loch Sloy *McLean 85-85* I S	British	1225	225·5	35·6	21·2	Glasgow	1877	D&WHenderson		Glasgow Shipping Co.	Glasgow
772	*	Loch Tay *Gibson 84.84* I S	British	1191	225·4	35·5	21·6	Glasgow	1869	B'clayCurle&Co		Glasgow Shipping Co.	Glasgow
773	*	Loch Torridon *Pattman 82-88* I 4mS	British	2000	287·1	42·6	24·0	Glasgow	1881	B'clayCurle&Co		General Shipping Co.	Glasgow
774	*	Loch Trool *Mainland 82-84* I Bk	British	1410	244·3	37·2	21·5	Glasgow	1884	GovanS.B.Co.		J.Sproat	Liverpool
775	*	Loch Urr *Murdoch 89-90* I Bk	British	716	189·8	30·1	18·6	PortGlasgow	1870	McCulloch, Patterson&Co.		J.Sproat	Liverpool
776	*	Loch Vennachar *Fannell 85-84* I S	British	1485	250·1	38·3	22·4	Glasgow	1875	J.&G.Thomson		Glasgow Shipping Co.	Glasgow
777	*	Lochee *Paton 74-86* I S	British	1728	264·2	39·0	23·1	Dundee	1874	A.Stephen&Sns.		D.Bruce & Co.	Dundee
778		Lochiel *Kerrison* W3mSr	British	216	113·4	24·1	12·6	Peterhead	1875	Steph'n&Forbes	Y 6.83	A.P.Barton	Brisbane
779		Lochiel *Ham -71* W Bn	British	108	89·0	23·0	10·0	Shelburne,NS	1870			J.Pertus,jun.	Halifax,N.S.
780	*	Lochinvar *Hillington 83-87* I Bk	British	718	184·9	30·1	18·3	Glasgow	1876	A.Stephen& Sns		Nicholson & McGill	Liverpool
781		Loch-lee *Scapishorn* W3mSr	British	248	116·1	23·9	13·9	Inverness	1865	Badenoch& Stewart		W.Johnson	Sydney,N.S.W.
782	N	Lochlyn *Olsen -83* W3mSr	Norweg'n	239	114·5	27·6	11·7	Grimstad	1866	OleKittelsen		Chr.Th.Boe	Arendal
783	*	Loch-na-gar *Sinclair* I Bk	British	464	163·0	26·0	16·0	W.Hartlepool	1861	W.Pile,Spence&Co.		Park Bros.	London
784	*	Lochranza Castle *Kerr* W Sr	British	88	84·6	20·5	9·5	Ardrossan	1876	P.Barclay & Co.		R.Kerr	Ardrossan
785	*	Lodestar *Norian 74-84* I S	British	1690	259·3	39·2	22·9	W.Hartlepool	1884	W.Gray&Co.		J.Lidgett & Sons	London
786	N	Lodsen *Jacobsen* W Bn	Norweg'n	287	123·9	27·9	13·4	Stavanger	1874		Y 1.86	N.W.Olsen	Stavanger
787		Loduskia *Treworgy* W Sr	American	103	86·1	24·1	8·0	Freeport,Me.	18 51/68			J.Davis	Ellsworth,Me.
788	V	Lætitia *Denée 71-86* W Bg	French	210	94·8	23·0	12·5	Méans	1874	E.Ollivaud	Z 1.83	P.Ch.Grenet	Nantes
789	*	Loftus (BARGE) *Harris 82-70* I	British	236	130·0	26·6	8·9	Newcastle	1882	WallsendSlipwayCo.		Palmers' Co.,Ld.	Newcastle
790	* D	Lohengrin *Burchert 82-82* (ex Parthenon) I Bk	German	587	175·9	26·6	17·1	Liverpool	1862	T.Vernon&Son		C.M.Matzen	Hamburg
791	N	Löining *Neilsen -88* (ex Isle of the South) W Bk	Norweg'n	850	175·4	31·9	19·9	Sunderland	1859	J.Laing	Y 12.87	O.G.Skadberg	Egersund
792		Lois *Lewis* W Bk	British	393	131·3	29·0	17·8	Sunderland	1868	W.Watson	Y 12.81	J.Gillan	Newcastle,N.S.W.
793		Lois V. Chaples *Hoes 86-86* W3mSr	American	219	125·0	31·0	8·8	Belfast,Me.	1875			J.C.L'Engle	Jacksonville,Fla.
794		Lojo W Sr	Russian	251	..	..	..		1876			C.Finnberg	Karis
795		Lokkha W Sr	Turkish	255	..	..	..	Italy	1884			Bigadoslee Athanas&Co	Bigados
796	*	Lola *Prowse 84-84* W Sk	British	49	73·6	16·9	9·6	Plymouth	1884	Watson&Fox		H.B.Watson	Plymouth
797		Lola W Bg	Spanish	238	116·1	25·7	13·1	Westerby	1859			F.Perez & others	Gijon
798		Lola W Bn	Spanish	112	80·7	22·4	10·9	Oviedo	1886			Fructuoso Perez	Gijon
799		Lola G. *Tenago -87* (ex ... Natividad) W Bk	Spanish	385	134·1	31·6	12·6	St.John,N.B.	1863	Milledge		M.Goday Gual	Villagarcia
800	IV	Lombardia *Ungarro* W Bn	Italian	143	108·9	26·4	9·8	SestriPonente	1886	G.Peragallo	Y 1.86	G.Peragallo	Genoa

*=Lloyd's Register; D=Nederlandsche Vereeniging; G=Germanischer Lloyd; I=Registro Italiano; R=American Record; A=Veritas Austro-Ungarico; H=Veritas Ellenico; N=Norske Veritas; V=Bureau Veritas.

Dans la 4me Colonne, C=Composite (membrure en fer, bordé en bois); I=Fer; Stl=Acier; W=Bois.

Les Dates du doublage en métal (Col. 11) ressortent des livres dans lesquels les navires sont respectivement classés ou des rapports des Inspecteurs du Lloyd's Register.

Pour plus amples renseignements sur les navires classés par le Lloyd's Register, voir le Lloyd's Register Book et le Lloyd's Yacht Register.

Official No.	No. in Book	Registry in which classed	Ship's Name. Material and Rig. Master, and Dates of Appointment to present Owner's Service and to Ship. Ship's Late Name.	Flag.	Net Tonnage.	Dimensions. Length	Breadth	Depth	Built. Where.	When	By whom.	Date of Metalling.	Owners or Managers.
[illegible]	801	*	Lombardian McDonald 85-88 I Bk	British	718	188·1	31·4	17·8	Sunderland	1875	Austin & Hunter		J. Tweddell, jun.
[illegible]	802		Londesborough W Sw	British	165	92·1	23·5	13·1	Cox Green	1856	W. Johnson		W. Watson
[illegible]	803	* D	London (ex Beatrice [illegible]) Craigie 84-84 I Bk	British	1152	208·1	33·9	22·5	Glasgow	1863	R. Napier & Sons		C. Barrie
[illegible]	804		London Steer W Sr	British	134	96·6	21·3	11·7	Bideford	1858	G. Cox		J. Steer
	805	V	London Mikkelsen 85-88 W Sr	Danish	138	89·5	21·6	10·9	Thuro	1885	C. Bom		C. Bom
[illegible]	806	V	Longfellow Faraday 84-84 W Bk	British	812	173·0	33·5	19·6	St Martin's N B	1874	W H & J Rourke	[illegible]	W. T. Dixon & Sons
[illegible]	807	R V	Loochoo Bell 78-84 W Bk	British	495	141·1	29·7	17·9	Sunderland	1867	W. Adamson	[illegible]	J. Adamson
[illegible]	808		Look Out Williams 79-86 W Bn	British	197	112·7	22·4	14·2	Workington	1858	J. T. Fell & Co.		W. H. Williams
[illegible]	809		Look Out W Bn	British	108	88·2	20·0	9·3	Manning River N.S.W.	1866	A. Newton		J. F. Mason
[illegible]	810		Lookout Crane W 3m Sr	American	483	156·0	32·0	16·0	Providence, RI	1870			N. Edwards
[illegible]	811	*	Loong Wha Lord -84 W Bkn	British	374	135·2	29·1	14·9	P. E. Island	1882	John Yeo	[illegible]	John Morris
[illegible]	812	*	Loongana Evans W Bk	British	279	127·0	26·0	13·1	Hobart Town	1879	McGregor		A. McGregor
	813	V	Lopes Duarte Dos Santos W Bk	Portug'se	288	125·3	24·3	11·8	Villa do Conde	1884	J. Fda Lapa	[illegible]	A. L. Duarte
	814	N	Lord Baltimore (ex Fanny) Eriksen 78-[illegible] W Bk	Norweg'n	329	133·0	28·2	13·5	Baltimore	1854			N. W. Coch
[illegible]	815	*	Lord Beaconsfield Goodman 83-88 W Barge	British	73	85·0	21·1	6·8	Littlehampt'n	1888	J. & W. B. Harvey		Hawkins, Holman & Co.
	816		Lord Byron (ex Andreolis) W Bg	Russian	241	..	..	..	[illegible]	1867			B. Mangube
[illegible]	817	*	Lord Cairns Davies 75-82 I S	British	1311	230·8	36·0	22·2	Belfast	1877	Harland & Wolff		Hughes & Co.
[illegible]	818	*	Lord Canning Stevenson 81-87 I S	British	1465	224·2	37·8	23·5	Liverpool	1867	T. Royden & Son		C. W. Kellock & Co.
[illegible]	819	*	Lord Churchill King -88 W K Barge	British	72	84·5	21·1	6·8	Littlehampt'n	1888	J. & W. B. Harvey		Hawkins, Holman & Co.
[illegible]	820	*	Lord Clyde Macbeth 83-86 I Bk	British	560	173·4	28·0	17·6	Glasgow	1875	A. Stephen & Sns.		L. Tulloch
[illegible]	821		Lord Clyde Smith 82-82 W Sr	British	115	86·5	22·3	11·0	Banff	1869	Geddie		J. W. Simpson
[illegible]	822	*	Lord Devon Partridge W Sr	British	98	84·0	21·2	10·7	Salcombe	1885	T. Sanders	[illegible]	Salcombe Shipowning Co.
[illegible]	823	*	Lord Downshire Newcomen 79-88 Stl 4m S	British	2263	299·8	41·3	24·8	Belfast	1882	Harland & Wolff		T. Dixon & Sons
[illegible]	824	*	Lord Dufferin Gilmore 85-88 I S	British	1697	262·9	38·3	23·3	Belfast	1879	Harland & Wolff		T. Dixon & Sons
	825		Lord Dufferin Jorgensen W Bk	Norweg'n	661	138·2	32·2	21·3	Quebec	1846			M. B. Nielsen & N. Olsen
[illegible]	826	*	Lord Duffus Masson 76-77 W Sr	British	152	98·0	23·0	11·5	Kingston	1877	Duncan	[illegible]	A. Marshall
[illegible]	827		Lord Exmouth Sherry 85-85 W Bn	British	105	81·7	19·4	10·3	Garlieston	1827			Mrs. H. Sealer
[illegible]	828	*	Lord Hartington Corbell W	British	80	85·6	20·8	6·8	Ipswich	1887	G. P. Gildea & Co.		Hawkins, Holman & Co.
[illegible]	829	*	Lord Iddesleigh Pope W Barge	British	75	84·4	20·4	6·8	Ipswich	1888	G. P. Gildea & Co.		Hawkins, Holman & Co.
[illegible]	830	*	Lord Kinnaird Simpson 85-88 I Bk	British	844	195·7	32·7	19·2	Dundee	1876	Brown & Simpson		Dundee Shipowners Co. (W. O. Taylor & Co.)
[illegible]	831		Lord Londesborough Johnson -88 W Bn	British	145	90·3	21·2	12·2	Selby	1856			C. Thompson
[illegible]	832	* D	Lord Lyndhurst O'Neill -89 I S	British	1187	212·5	34·4	22·6	Liverpool	1861	T. Vernon & Son		W. H. Stott & Co.
72271	833		Lord Lytton Grierson -87 W S	British	1264	197·4	38·3	23·9	St. Martin's N B	1876	A. Parks & Sons	[illegible]	H. McC. Grierson
73900	834	*	Lord March M[illegible] 73-81 W 3m Sr	British	156	98·9	23·3	12·1	Kingston	1876	Duncan	[illegible]	G. M'Robie
	835		Lord Nelson Grönberg -86 W Bkn	Swedish	238	96·8	22·9	16·0	Cowes	1850	T. & J. White		J. E. Hagström
	836	G V	Lord Palmerston Larsson W Bk	Swedish	417	130·5	28·0	16·8	Hernösand	18[illegible]		[illegible]	G. F. Gädda
93676	837	*	Lord Raglan Campbell 16-86 I S	British	2078	276·0	42·4	24·1	Liverpool	1888	T. Royden & Sns		Sailing Ship "Lord Raglan" Co., Ld. (J. Herron & Co.)
60893	838	*	Lord Reidhaven Nanese 75-[illegible] W Sr	British	146	96·1	22·7	11·8	Kingston	1875	Duncan	[illegible]	J. Thomson
19435	839		Lord Rollo Cowley -84 W Sr	British	129	85·8	20·5	11·0	Perth	1856	Fenton		J. Clancy
91423	840	*	Lord Salisbury Ricketts W Sk	British	92	80·6	20·7	10·5	Hull	1885	Cottingham Bros		Pickering & Haldane
91989	841	*	Lord Salisbury Goodmann -87 W K Barge	British	78	85·2	20·6	6·9	Littlehampt'n	1887	J. & W. B. Harvey		Hawkins, Holman & Co.
93830	842	*	Lord Shaftesbury Johnston 88-88 Stl 4m S	British	2273	293·3	42·8	24·0	Whitehaven	1888	Whitehvn. Shipbuilding Co., Ld.		Sailing Ship "Lord Shaftesbury" Co., Ld. (J. Herron & Co.)
91911	843	*	Lord Shaftesbury Wright 85-86 W K	British	78	84·0	20·1	6·9	Ipswich	1886	G. P. Gildea		Hawkins, Holman & Co.
84082	844		Lord Stanley C Flat	British	139	89·0	20·6	9·4	Northwich	1880	J. Thompson & S		Salt Union, Ld.
14834	845		Lord Strangford Churchill W Bn	British	178	100·8	23·6	12·8	Isle of Wight	1825			T. G. Robins & Co.
63156	846	*	Lord Templetown Hawthorne 85-88 Stl S	British	2093	282·9	40·1	23·7	Belfast	1886	Harland & Wolff		Irish Shipowners' Co.
89540	847	*	Lord Wolseley McVicker 86-86 I 4m S	British	2518	308·2	42·9	25·1	Belfast	1883	Harland & Wolff		Irish Shipowners' Co., Ld.
51651	848	* V	Lord of the Isles Gawne 84-82 I Bk	British	630	183·3	28·7	18·5	Greenock	1864	R. Steele & Co.		W. P. Colchoro & Co.
[illegible]	849	*	Lord of the Isles Meerbach W 3m Sr	British	208	116·4	22·4	11·0	Balmain, N.S.W.	1881	H. Beattie	[illegible]	H. Beattie
	850												

* = Lloyd's Register; D = Nederlandsche Vereeniging; G = Germanischer Lloyd; I = Registro Italiano; A = Veritas Austro-Ungarico; H = Veritas Ellenico; N = Norske Veritas;

In the 4th Column, C = Composite (iron frame, wood planked); I = Iron; Stl = Steel; W = Wood

The Dates of Metalling (Col. 11) are based on the records of the Books in which Vessels are respectively classed, or on the Reports of the Surveyors of [illegible]

Further particulars of Vessels classed by Lloyd's Register will be found in Lloyd's Register Book and Lloyd's Yacht Register.

No.	Cote	Nom. Matériaux et Gréement. Capitaine et dates d'entrée au service des Armateurs actuels et à bord du navire. Ancien Nom.			Pavillon.	Tonnage net.	Dimensions.			Construit.			Date du doublage en métal.	Armateurs ou Directeurs-Gérants.	Port d'Armement.
		Nom	Capitaine	Mat.			Longueur	Largeur	Creux	Où.	Quand.	Par qui.			
851	A	Lordos Byron		W Bg	Greek	320	108·2	29·0	16·4	Scopelos	1880	P. Barelas	Y 2.83	G. D. Praidis	Scopelos
852	V	Loreley	Hausen -88	W Bk	Norweg'n	544	150·5	32·3	19·6	Arendal	1864	A. R. Andersen	Y 2.87	R. M. Petersen	Arendal
853	V	Loreley (ex Samuel [illegible])	Olsen	W Sr	Russian	109	76·1	17·4	9·5	Veendam	1869	K. & J. Wilkens		R. Danzberg	Riga
854	V	Loreley	Arvor -88	W Sr	Swedish	135	73·7	22·1	10·1	Ratteregården	1879	E. Abrahamsson	Y 7.86	C. J. Kronker	Gothenburg
855		Lorencita		W Bn	Spanish	300	165·9	24·7	13·7	Arens	1856			F. Casellas	Barcelona
856	V	Lorenz Hansen	Dillwitz (87-73)	W Bk	German	397	113·1	28·8	17·3	Damgarten	1873	H. Dierling	Y 1.88	W. Maack	Rostock
857	DIV	Lorenzino	Albano 86-88	W Bk	Italian	922	173·1	34·8	23·9	Prà	1882	L. Durante	Y 1.87	M. Amoroso	Genoa
858	V	Lorenzo	Williamson	W S	British	1135	193·5	37·7	23·7	Quebec	1877	Dinning	Y [illegible]	L. T. Merrow	Glasgow
859	IV	Lorenzo (ex [illegible] Bambino)	Olcese -87	W Bk	Italian	479	138·4	31·5	19·3	Sestri Ponente	1867	G. Debarbieri	Y [illegible]	G. A. Brignetti	Genoa
860	I	Lorenzo (ex Celes)	Gregoretti	W Bg	Italian	238	120·7	26·9	15·4	Sestri Ponente	1878		Y [illegible]	F[lli] Vianello Moro	Venice
861	V	Lorenzo	Ørsland	W Bk	Norweg'n	482	141·0	31·0	17·5	Arendal	1870	L. Waaland	Y [illegible]	S. Eyde	Arendal
862	V	Lorenzo (ex [illegible])	Mir	W 3m Sr	Spanish	302	134·5	29·9	15·7	Sestri Ponente	1877	L. Marcenaro	Y 9.86	N. Mir & Co.	Barcelona
863		Lorenzo Daw		W Bn	British	126	94·5	23·1	10·5	P. E. Island	1863			P. McInnis	P. E. Island
864		Lorenzo Padre (ex [illegible])	Bozzo	W Bk	Italian	430	127·9	30·0	19·3	Savona	1868	F. Sirello	Y 7.82	F[lli] Schiaffino	Genoa
865		Loreto (ex [illegible])	Nielsen	W Bk	Chilian	694	152·0	29·0	19·5	Sunderland	1853	McPherson		Cia Esplotadora de Lota y Coronel	Valparaiso
866	I	Loreto (ex Angela F.)	Antola	W Bk	Italian	646	156·8	31·4	19·9	Sampierd'rena	1872	B. Frassinetti	Y 2.87	Gius. Massone	Genoa
867		Loring (ex Anthem)		W Sr	British	194	101·0	26·6	7·7	Mystic, U.S.	1851			D. B. Lemon	St. John, N.B.
868		Loring C. Ballard	Bowen '83	W 3m Sr	American	628	149·6	35·1	18·4	Bath, Me.	1884	N England S B Co		L. C. Ballard	Bath, Me.
869	N	Lorma	Andersen -81	W 3m Sr	Norweg'n	279	122·5	27·3	12·1	Grimstad	1876	Ole Kittelsen	Y 12.87	Chr. Th. Boe	Arendal
870	*	Lorna Doone	Reid	W Bk	British	357	118·2	26·1	15·4	Dundee	1876	Tay Ship B. Co.	Y 5.87	Robertson Bros.	Dundee
871	*	Lorne	Hughes	W Sr	British	74	77·1	19·1	9·1	Chester	1871	Ferguson		J. Hughes	Aberystwith
872	*	Loro	Ascough	I Sr	Argentine	67	79·2	21·3	7·0	Newcastle	1884	Sir W. G. Armstrong, Mitchell & Co. Ld.		Borra, Bailerini & Co.	Buenos Ayres
873	*	Lorraine	Johnson 77-83	I Bk	British	828	196·8	32·8	19·2	Sunderland	1876	W. Doxford & Ss.		J. Lidgett & Sons	London
874	*	Lorton	Rathery 88-89	Stl Bk	British	1380	245·8	37·4	21·6	Belfast	1888	Workman, Clark & Co.		P. Iredale & Son	Liverpool
875		Los Carreras (ex Montreal)		W Bk	Chilian	382	126·3	25·5	16·4	Boston, Mass.	1847			P. Alvarez	Valparaiso
876		Los Emilios	Arendaño	W Bn	Spanish	135	88·9	23·8	10·4	Mataro	1857	L. Zegarra		Heredia hermanos	Malaga
877		Lota (ex Robert Seddon, ex Maria Trinidad)	Johnsen -88	W S	Chilian	1047	196·7	39·4	24·9	St. John, N.B.	1863	King		F. Peede	Valparaiso
878	*	Lothair	Evans 74-86	C Bk	British	794	191·8	33·5	19·0	London	1870	Walker	Y 1.87	W. Bowen, jun.	London
879		Lothair (ex Agnes M. Lovett)	Hawthorne	W Bk	British	657	150·9	33·3	20·0	St. John	1863	Rowan		R. A. Lowther	Belfast
880		Lothair	Mann	W Bn	British	180	102·6	24·6	12·4	P. E. Island	1875	Knight		J. Mann	Harwich
881		Lothair		W K	British	129	101·8	23·2	7·5	Ipswich	1872	Robertson		John Watts	Harwich
882		Lothair	Joyce	W Sr	British	100	83·0	24·8	8·8	Shelburne, NS	1870			W. Joyce	St. John's, N.F.
883	* DV	Lotos	Klinkhamer -87	I S	Dutch	1212	206·8	35·9	20·9	Slikkerveer	1883	H. Bothof & W. Grauesheim		J. Koning & Van Delden	Rotterdam
884	NV	Lotos	Gunnufsen	W Bk	Norweg'n	718	163·9	33·1	19·8	Lillesand	1879	C. Osmundsen	Y 2.88	I. A. Henschien	Lillesand
885		Lotta	Abrahamson 75	W Sr	Swedish	150	86·3	21·7	11·9	Kilanda	1873			J. Berntsson	Kallanda
886	I	Lottero Bertollo (ex [illegible])	[illegible]	W Bk	Italian	499	137·1	31·7	19·9	Spotorno	1870	G. B. Cadennacio	Y 10.89	G. Schiaffino	Genoa
887	R	Lottie	Mills	W Bkn	British	492	110·4	31·0	18·1	Londonderry, N.S.	1876	W. Cameron	Y 10.82	D. L. Mills	Halifax, N.S.
888		Lottie	Brien	W Sr	British	118	87·4	25·8	10·3	Trinity Bay NF	1882	J. Gulliford		J. G. Joy	St. John's, N.F.
889		Lottie	Speed	W Sr	American	266	94·7	29·2	9·3	Cutler, Me.	1866			J. F. Wyckoff	New York
890		Lottie	Somers -74	W Sr	American	295	109·8	28·5	8·4	Milford, Del.	1871			G. W. Shepherd, jun.	Philadelphia
891	R	Lottie	Marshall	W 3m Sr	American	176	98·2	26·3	10·0	Boston	18 55/66			J. A. Creighton	Thomaston, Me.
892	R	Lottie Beard	Burgwood -87	W 3m Sr	American	388	115·0	29·0	14·6	Fr Haven, Mass	1866	W. G. Blackler	Y 6.86	J. W. Perry	N. Bedford, Mass
893	*	Lottie Belle	Fortin -87	W Bn	British	199	103·0	24·7	12·9	P. E. Island	1880	J. R. Bourke, jun.	Y 5.86	F. Carbray	Quebec
894		Lottie Carson	Bonifield 81-87	W 3m S	American	272	127·3	32·8	8·8	Port Blakely, Wash.	1881	Hall Bros.		S. Bonnifield	San Francisco
895	V	Lottie E.	Wyman	W Bn	British	167	95·3	25·8	10·8	Green Cove, NS	1882	J. W. Raymond	Y 2.86	W. Crosby	Yarmouth, N.S.
896		Lottie K. Friend	Gould	W 3m St	American	349	131·6	32·5	13·2	Essex, Mass.	1873			J. A. Gould	Gloucester, Mass.
897	RV	Lottie Moore	Watts -87	W Bk	American	837	161·0	34·2	20·6	Setauket, N.Y.	1877	G. E. Hand	Y 11.86	Yates & Porterfield	New York
898	V	Lottie Stewart	Kenny	W Bk	British	742	159·5	34·1	18·5	St. John, N.B.	1873	J. & R. McLeod	Y 6.86	D. Carmichael	St. John, N.B.
899		Lotus	Webster -88	I Lr	British	263	115·0	26·0	9·2	Shanghai	1888	Boyd & Co.		Taku Tug & Lighter Co.	Shanghai
900	*	Loudon Hill	Holt 84-87	I 4m Bk	British	2066	283·1	40·4	24·5	Port Glasgow	1887	Russell & Co.		J. R. Dickson & Co.	Glasgow

* = Lloyd's Register; D = Nederlandsche Vereeniging; G = Germanischer Lloyd; I = Registro Italiano; R = American Record;
A = Veritas Austro-Ungarico; H = Veritas Ellenico; N = Norske Veritas; V = Bureau Veritas.

Dans la 4me Colonne, C = Composite (membrure en fer, bordé en bois); I = Fer; Stl = Acier; W = Bois.

Les Dates du doublage en métal (Col. 11) ressortent des livres dans lesquels les navires sont respectivement classés ou des rapports des Inspecteurs du Lloyd's Register.

Pour plus amples renseignements sur les navires classés par le Lloyd's Register, voir le Lloyd's Register Book et le Lloyd's Yacht Register.

Official No.	No. in Book.	Registry in which classed.	Ship's Name. Material and Rig. Master, and Dates of Appointment to present owner's Service and to Ship. Ship's Late Name.	Flag	Net Tonnage.	Length.	Breadth.	Depth.	Built. Where.	When.	By whom.	Date of Metalling.	Owners or Managers.
83998	901	✳	Louie Bell *Aldon* 77-84 W3mSr	British	99	94·3	22·6	9·6	Carrickfergus	1882	P.Rodgers		J.Fisher & Sons
140949	902		Louis *Hatch* -88 W3mSr	American	820	193·8	36·0	18·0	NorthBend,Or.	1888	J.Kruse		A.M.Simpson
	903	V	Louis *Vessal* WBk	French	194	94·2	23·6	13·9	St.Valery en Caux	1858	J.Lehoc		Brache & Vessal
	904	V	Louis (ex Roault) *Lamusse* -87 WBg	French	169	106·2	23·3	12·1	Jersey	1863	Clarke		Rioult
	905	GV	Louis *Schmieler* 83- WBk	German	607	145·7	31·4	18·3	Vegesack	1863	J.Lange	[illegible]	G.Kindler
	906	DV	Louis *Freericks* WBg	German	220	101·9	25·5	12·4	Papenburg	1866	B.Thoien	[illegible]	R.H.Freericks
	907		Louis *Philipp* 64-64 WBg	German	193	89·6	25·3	13·7	Greifswald	1863	J.Spörck		C.A.Beug
	908	✳	Louis *Corneliussen* -88 (ex City of Aberdeen) WBk	Norweg'n	569	157·0	29·1	18·0	Aberdeen	1862	Duthie	[illegible]	Actieselskabet "Louis" (H.Corneliussen)
	909	G	Louis *Asbjörnsen* 84- (ex Bréda Lourtel) (TO BE BROKEN UP) WBk	Norweg'n	458	151·4	30·8	16·2	Bordeaux	1856	L.Arman	[illegible]	Akt. S. "Louis" (Eyde & Rolo)
	910	N	Louis *Hansen* 87-87 WBk	Norweg'n	449	133·5	29·5	16·8	Elbing	1854			J.D.Gjersöe
70613	911		Louis Alfred WBn	British	117	90·2	21·0	10·2	Bideford	1874	Cox		A.Ohlsson
	912		Louis Elisa *Granger* 71-81 WBg	French	180	88·9	24·3	11·8	Nantes	1868	Debordeaux	[illegible]	P.Ch.Grenet
	913	V	Louis Eugène *Le Normant* WBk	French	438	132·8	26·2	16·4	St.Malo	1876	L.Mallard	[illegible]	L.Mallard
	914	V	Louis Marie *Hugues* WBg	French	183	90·9	24·6	12·1	Nantes	1873	Clergeau	[illegible]	Hugues
	915	V	Louis Marie *Croisil* -75 WBn	French	140	92·8	22·3	10·5	St.Vaast	1874			C.Lévêque
140858	916		Louis Monette Sr	American	110	97·0	26·9	6·9	Huntington,N.Y.	1886	J.C.Hawkins		J.C.Hawkins
80820	917	✳	Louis Montgomery *Parkman* [-87 WBn	British	209	105·9	25·5	12·1	P.E.Island	1880	D.Montgomery	[illegible]	R.Roberts
	918	V	Louis Pierre Marie *Erkolu* -87 WBg	French	169	87·6	20·0	12·8	St.Malo	1868	Tranchemer fes.	[illegible]	P.Debry
15341	919		Louis Walsh *Pendleton* [63- WS	American	1497	187·4	40·0	27·1	Belfast,Me.	1861	J.P.White	[illegible]	J.G.Pendleton
	920	N	Louis de Geer *Bogasen* -87 WBg	Norweg'n	264	111·7	28·0	14·9	Norrköping	1845			J.Danielsen
	921	V	Louis de Geer *Anderson* WS	Swedish	665	155·5	31·0	18·5	Hernösand	1866	F.Hägglund	[illegible]	J.F.Olson
	922	✳	Louis IX. *Labour* -82 WS	French	405	139·1	28·7	15·3	St.Malo	1877	Gautier	[illegible]	A.Hubeau
32808	923		Louisa *Rees* WS	British	781	147·0	30·5	21·7	P.E.Island	1851		[illegible]	J.Rees
48252	924		Louisa WBn	British	253	110·5	26·6	12·7	Barrington NS	1863			P.Doyle
88036	925		Louisa WBk	British	233	129·2	23·0	10·7	St.Catharine's	1856	L.Shickluna		J.Shickluna
30901	926		Louisa WBk	British	192	96·0	23·0	13·0	Philadelphia				Eyoob Aboo Tauleb
46336	927		Louisa (ex Lillian) WBn	British	164	95·6	25·3	8·9	Eastport,U.S.	1853			A.Johnson
50903	928		Louisa *Wheeler* WBn	British	153	108·1	24·8	10·2	P.E.Island	1864	Bell		W.Holden
35406	929		Louisa WBg	British	148	96·0	24·6	9·6	Cincinnatti US	1850			J.N.O'Viatt
45078	930		Louisa *Williams* WSr	British	130	90·1	21·6	11·6	Berwick	1863	Gowan		J.W.Bussey
57982	931		Louisa WBn	British	123	87·5	23·5	9·7	P.E.Island	1866	Cox		Victorian Government (Minister of Defence)
15237	932		Louisa WSr	American	100	78·1	28·4	7·9	Pittston,Me.	1832			W.Coombs
15986	933		Louisa A. Boardman *Tinker* WSr	American	107	87·0	26·8	7·0	Calais,Me.	1872	Boardman Bros.		G.H.Boardman
15176	934		Louisa A. Danenhower (ABANDONED) *Lee* 86-86 WSr	American	232	111·2	29·0	9·4	Newcastle Del	1857			G.W.Shepherd,jun.
140577	935		Louisa A. Grout *McLeod* 84 WSr	American	148	97·6	25·4	9·1	Essex,Mass.	1882	J.Story		J.Hatch
74318	936	R	Louisa Coipel *Parker* 74-79 WBn	British	144	87·3	23·7	10·8	ChurchPnt NS	1877	J.D.Resemond		E.F.Parker
89224	937		Louisa Messell WSr	British	103	90·0	22·6	7·4	CockleCreek, N.S.W.	1883	W.Woodward		C.F.Messell
15193	938		Louisa P. Pharo *Long* WSr	American	192	109·5	29·5	8·2	Tuckerton,N.J.	18^{55}_{72}			L.N.Long
15163	939	R	Louisa Smith *Webber* WSr	American	137	93·3	25·4	8·6	Brooksville Me	1868			J.Jones
	940		Louisa Thérésa *Denost* WBg	French	201	89·9	24·6	13·6	Redon	1868	F.Coyac		A.Coffre
	941	V	Louisa & Augusta (ex Eliphalet Greeley) *Hansen* -81 WBk	German	1102	173·6	35·3	24·0	Bath,Me.	1856		[illegible]	C.A.E.Solscher
38517	942		Louise (ex Geo. Hughes) *Boudreaut* WSr	British	186	98·5	27·4	10·8	Sherbrooke NS	1872			D.Landry
77940	943	R	Louise *Le Blanc* 71-86 WBn	British	149	88·8	23·0	11·1	P.E.Island	1879			B.LeBlanc
43263	944		Louise WSr	British	125	96·2	25·7	7·5	NewJersey,US	1853			G.D.Harris
69440	945		Louise *Chegwidden* 84-84 WSr	British	100	87·4	22·9	11·0	Newquay	1877	Chegwdden		R.Hockin
	946	V	Louise (ex Delaval) *Hansen* 80-80 WBg	Danish	272	107·2	21·2	15·8	Sunderland	1866	W.Richardson		F.Jensen
	947	DV	Louise *Oldefar* 84-88 WBg	Danish	243	112·8	24·0	12·6	Värmanäs	1872	C.A.Hahn	[illegible]	G.H.E.Wielandt
	948	V	Louise *Holst* WSr	Danish	115	88·7	20·5	10·3	Holbæk	1872	O.C.Mortensen		H.L.Smith's Enke
	949	✳	Louise *Bremand* IBk	French	426	150·0	27·0	16·0	Sunderland	1868	[illegible] Mounsey & C		G.Viale
	950	V	Louise *Mace* W3mSr	French	299	110·2	24·6	13·9	St.Servan	1877	Hamon fils	[illegible]	A.Macé

✳ = Lloyd's Register; A = Veritas Austro-Ungarico; D = Nederlandsche Vereeniging; G = Germanischer Lloyd; H = Veritas Ellenico; I = Registro Italiano; N = Norske Veritas; R = Am[illegible]; V = Bureau [illegible]

In the 4th Column, C = Composite (iron frame, wood planked); I = Iron; Stl = Steel; W = Wood.

The Dates of Metalling (Col. 11) are based on the records of the Books in which Vessels are respectively classed, or on the Reports of the Surveyors of L[illegible]

Further particulars of Vessels classed by Lloyd's Register will be found in Lloyd's Register Book and Lloyd's Yacht Register.

Numéro	Registre	Nom. Capitaine et dates d'entrée au service des Armateurs actuels et à bord du navire. Ancien Nom.	Matériaux et Gréement	Pavillon	Tonnage net	Dimensions. Longueur	Largeur	Creux	Construit. Où	Quand	Par qui	Date du doublage en métal	Armateurs ou Directeurs-Gérants	Port d'Armement
951		Louise *Hubert*	WBk	French	285	128·6	27·1	12·7	Alloa	1875			P.O.Tougard	Fécamp
952		Louise *Sounes 79-79*	WSr	French	125	87·9	20·3	11·6	Binic	1879	Y.Minier		J.Le Pomellec & fils	Binic
953	DGV	Louise *Köhne 75-90*	WBk	German	653	150·2	33·0	18·5	Elsfleth	1876	J.Jürgens	Y 10.87	E.tom Dieck	Elsfleth
954		Louise (STRANDED) *Bohl... 72*	WBk	German	390	121·9	28·8	16·6	Barth	1872	H.Witte	Y 11.85	J.J.Wallis & Sohn	Barth
955		Louise (COLLISION) *Holm 83-*	WBn	German	280	115·4	26·2	12·9	P.E.Island	1874	Peake	Y 1.84	H.C.L.Holm	Flensburg
956	D	Louise *Haack 79-*	WSr	German	218	99·8	24·1	13·3	Barth	1874	H.T.Bracken [wagen		J.C.Beug	Barth
957		Louise *Permien 88*	WSr	German	179	98·1	23·0	11·6	Geestemünde	1862	Schau&Oltmans		J.J.Wallis & Sohn	Barth
958	DN	Louise (ex Joseph S. Barker) *Jacobsen -86*	WBk	Norweg'n	1040	173·2	34·6	23·0	Freeport,Me.	1869	G.& C.Bliss	Y 7.88	Ole Nielsen	Sandefjord
959	DN	Louise (ex ...) *Cägor*	WBk	Norweg'n	949	166·6	35·1	22·4	Quebec	1863	Rosa	Y 6.84	H.C.Mathiesen	Christiania
960		Louise *Hansen 81-81*	WBk	Norweg'n	459	128·5	30·8	19·0	Amsterdam	1854		Y 4.86	T.M.Nielsen & Son	Laurvig
961	N	Louise *Dahl -74*	WBk	Norweg'n	455	139·4	30·2	16·5	Sandsvall	1873	J.A.Strandberg	Y 9.88	Chr.Th.Boe	Arendal
962	N	Louise *Christoffersen 75-76*	WBk	Norweg'n	328	119·3	28·1	15·2	Rönnebeck	18 54/75		Y 12.87	H.A.Lohre	Drammen
963	N	Louise (ex Martha) *Nilsen*	WBg	Norweg'n	213	108·6	24·5	12·7	Geestemünde	1858			H.Andersen	Tönsberg
964	N	Louise *Olsen*	WBg	Norweg'n	179	90·8	23·7	12·2	Sandefjord	18 57/75	M.Larsen		M.Larsen Röd	Sandefjord
965		Louise *Salming 85-85*	WBk	Russian	558	115·4	32·0	18·1	Fredrikshamn	1865	C.Backmann		H.Salming	Riga
966	R	Louise Adelaide *Orr*	WBkn	American	638	151·9	34·2	12·6	Yarmouth,Me	1882	G.Loring	Y 10.88	E.Orr	Portland,Me.
967		Louise Bachmann	WBg	German	285	98·0	26·5	14·7	Rostock	1857	W.Zeltz		Gustav Kindler	Rostock
968		Louise Brockelmann (LOST)	W3mSr	German	240	103·4	26·0	13·8	Ribnitz	1856	J.H.Wilken		A.Burchard	Rostock
969	V	Louise Collet *Huet 85-85*	WBk	French	468	136·8	24·9	16·4	Dunkirk	1876	A.Lefebvre	Y 3.85	C.L.&P.Collet-Taverne fils	Dunkirk
970	R	Louise Hastings *Gray*	WSr	American	117	88·0	23·3	7·7	Brewer,Me.	1885	J.Hathorn		E. & J.K.Stetson	Bangor
971		Louise Henry	WBn		141	89·6	25·3	11·1	Courseulles	1862				
972	GV	Louise Lübeke	W3mSr	German	317	120·0	27·4	15·0	Wismar	1875	M.C.Maren	Y 5.87	G.W.Löwe	Wismar
973	V	Louise Marie *Bon*	WBk	French	334	118·1	24·9	14·4	St.Malo	1875	P.Hervot	Y 10.84	H.Monnier & ses fils	Fécamp
974		Louise Marie (WRECKED) *Desmolt*	WSr	French	145	84·3	20·7	11·2	Dunkirk	1865	D.VanCauwen- [berghe		C.Van Cauwenberghe & fils	Dunkirk
975		Louise Marie	WSr	French	132	94·0	19·7	8·6	LaSeyne	1868	Curet		Grégory	Marseilles
976		Louise Marie	WSr	French	114	..	..	..	Martigues	1868		Y 3.83	M.Michel	Martigues
977	GV	Louise Meyer *Eickmann 81-88*	WBg	German	297	115·8	26·7	14·5	Rostock	1875	J.Müller	Y 3.87	W.Maack	Rostock
978		Louise Otto Warbelow W.	WBg	German	246	102·2	25·2	14·5	Damgarten	1851	H.Dierling		A.Burchard	Rostock
979	V	Louise Wichards (ABANDONED)	WBk	German	350	117·1	27·9	15·4	Ueckermünde	1861	T.Prelzer		C.Wichards	Stettin
980		Louisiana	WSr	British	106	84·3	23·5	9·5	Grondines	1874			J.V.R.Thibaudeau	Quebec
981	RV	Louisiana *Oliver*	WS	American	1365	202·0	40·0	24·2	Bath,Me.	1873	HoughtonBros.	Y 5.88	Houghton Bros.	Bath,Me.
982	DI	Louisiana (TO BE BROKEN UP) *Pedesta*	WBk		639	160·7	33·1	19·6	Sestri	1872	A.Briasco fuG.B.	Y 8.83		
983	V	Louisiane *Cevelan*	WSr	French	128	83·7	21·3	11·4	Dunkirk	1865	Van de Zande		Vve.Buhot Launay	Paimpol
984	AH	Loula Calvocoressi	WBk	Greek	402	125·0	28·4	..	Syra	1878		Y 3.87	G.Calvocoressis	Syra
985	I	Lourdes (ex-Alice) *-87*	WBk	Italian	562	145·9	31·3	21·6	Sestri Ponente	1875	Bozzano	Y 9.87	Luigi Peneo	Genoa
986		Louvain *Harnaes 84-84*	WSr	Norweg'n	144	101·0	25·7	9·0	Arendal	1871			A.B.Bull Larsen	Christiania
987	V	Louvois *Barbu -83*	WBk	French	267	111·2	24·6	13·0	Meaux	1876	E.Ollivaud	Y 1.86	A.Lemoine	St.Malo
988	V	Lovdahl *Thorsen*	WBk	Norweg'n	321	123·4	27·2	13·8	Stralsund	1865	G.Dienstbach	Y 7.82	F.Kleiberg	Stavanger
989	N	Lovehjerte *Michaelsen*	WBk	Norweg'n	498	140·4	31·9	17·7	Grimstad	1878	FSmithPetersen	Y 11.88	F.Smith Petersen	Grimstad
990	N	Loveid *Paus*	WBk	Norweg'n	465	136·0	31·6	17·1	Skien	1868	Dekke	Y 2.85	Chr.Winsnes	Skien
991	N	Löven (STRANDED) *-87*	WBg	Norweg'n	238	90·0	20·8	15·6	Grimstad	1842			E.Dedekam	Arendal
992	N	Lovetand *Kurize*	WBk	Norweg'n	404	120·3	23·7	15·7	Grimstad	18 57/76	MSmithPetersen	Y 2.85	M.Smith Petersen	Grimstad
993	N	Lovfald *Amundsen -87*	WBk	Norweg'n	656	151·8	34·0	19·0	Grimstad	1867	MSmthPetersen	Y 6.88	F.Smith Petersen	Grimstad
994		Lovisa *Larsen -81*	WBk	Danish	240	117·6	23·5	12·4	Aabenraa	1862	N.Jacobsen		P.Berg	Nexo
995	V	Lovisa *Eriksen 85-75*	WBn	Danish	240	110·2	24·7	11·8	Veile	1875	P.H.Illum	Y 3.89	J.P.Eriksen	Veile
996		Lovisa *Jansson -84*	WSr	Russian	128	80·1	27·3	10·2	Hvittisbofjerd	1874	J.Landberg		J.G.Österlund	Wärdö
997	DV	Lovisa (ex Estes) *Olsen*	W3mSr	Swedish	249	118·4	25·1	12·1	Grimstad	1866	A.Geelmuyden	Y 8.86	R.C.Jobson	Warberg
998		Lovisa (LIGHTER) *Jansson*	WSr	Swedish	171	90·2	24·2	11·9	Finland	1862			A.Fröberg	Häfverö
999	V	Lovisa *Bengtsson*	WSr	Swedish	168	91·5	22·6	11·8	Eckerna	1872	A.Ahlberg		B.Bengtsson	Hellevikstrand
1000	N	Lovise (ex R.W.Weisner) *Jacobsen*	WBk	Norweg'n	1054	181·8	35·9	22·0	ClydeRiv.NS	1873	R.B.Stoddart	Y 7.88	M.Langaard & Co.	Christiania

*=Lloyd's Register; D=Nederlandsche Vereeniging; G=Germanischer Lloyd; I=Registro Italiano; R=American Record;
A=Veritas Austro-Ungarico; H=Veritas Ellenico; N=Norske Veritas; V=Bureau Veritas.

Dans la 4me Colonne, C=Composite (membrure en fer, bordé en bois); I=Fer; Stl=Acier; W=Bois.

Les Dates du doublage en métal (**Col. 11**) ressortent des livres dans lesquels les navires sont respectivement classés ou des rapports des Inspecteurs du Lloyd's Register.

Pour plus amples renseignements sur les navires classés par le Lloyd's Register, voir le Lloyd's Register Book et le Lloyd's Yacht Register.

Official No.	No. in Book	Society in which classed	Ship's Name. Material and Rig. Master, and Dates of Appointment to present Owner's Service and to Ship. Ship's Late Name.	Flag.	Net Tonnage	Length	Breadth	Depth	Built: Where.	When	By whom.	Date of Metalling.	Owners or Managers
	1001	DNV	Lovspring *Thomsen* WBk	Norweg'n	547	149·4	30·9	17·8	Christiansand	1873	H.Olsen	[illegible]	H.G.Kirschbom
	1002	N	Lovspring *Schiet -89* WBk	Norweg'n	489	136·4	31·0	16·9	Grimstad	1885	N.O.Gjömle	[illegible]	J.Jörgensen
	1003	N	Lovspring *Thimnesen* WBg	Norweg'n	225	107·5	26·9	12·2	Porsgrund	1872			P.M.Petersen & Sön
72335	1004	RV	Low Wood *Tharber -87* WBk	British	1091	186·7	37·6	22·4	Portland,N.B.	1878	D.Lynch	[illegible]	H.D.Troop
36478	1005		Lowestoft *Hook* WBk	British	293	112·7	26·5	16·5	Sunderland	1866	J.Dennison		R.L.Manners
76360	1006	✱	Loweswater *Smith -88* I Bk	British	603	174·0	30·2	17·5	Whitehaven	1876	Whitehvn SB Co		Jackson,Metcalf & Co.
91201	1007	✱	Lowther Castle *Scarborough -86* I S	British	1824	262·1	39·0	23·3	Workington	1886	R.Williamson & Son		Lancaster Shipowners Co. Ld.
	1008	N	Loyal *Fritz -87* (ex Dreadnought) (WRECKED) WS	Norweg'n	1427	220·2	38·9	24·7	Quebec	1863	Ross	[illegible]	Rederiselskabet "Loyal" (Chr. Sliefsen)
	1009	D	Loyal *Knudsen -88* (WRECKED) (ex Skotte) WBk	Norweg'n	663	159·9	31·9	18·4	Grimstad	1883	S.Johnsen	[illegible]	M.F.Stray
65594	1010	V	Loyalist *Vroom 88-89* WBn	British	348	130·5	29·4	13·0	Portland,N.B.	1863	F.E.Sayre	[illegible]	F.E.Sayre
	1011		Loyo (ex A[illegible]) *San Thome* W3mSr	Brazilian	242	100·0	24·3	11·5	Fao	1876	Borda	[illegible]	J.daSilvaLoyo & Filho
	1012		Lua WSr	Brazilian	150	..	..	..					V.A.e Magalhaes
	1013	N	Luca *Jansen 58-75* WSr	Norweg'n	399	116·3	28·3	18·4	Buccari	1857			H.E.Jansen
88031	1014	V	Lucania *Gibson 73-88* WS	British	1774	229·8	42·1	24·4	St.John,N.B.	1886	J.F.Fraser	[illegible]	B.Smith & Sons
	1015	G	Lucas (ex Lina) *Pocker 85-85* WSr	German	159	91·9	21·1	11·1	Papenburg	1863	F.W.Meyer		G.& H.Aden
	1016	✱	Lucayas *Johansen* I Bk	Danish	445	150·4	25·5	18·1	Greenock	1869	Robertson & Co		Th.Meinertz's Enke
1528	1017	V	Lucco *Degregori -87* (BROKEN UP) WBk	Italian	498	134·5	30·8	20·0	Sestri Ponente	1866	L.Briasco	[illegible]	B.Olivari
1979	1018	I	Luchina C. *Repetto* WBk	Italian	506	134·5	31·2	19·8	Lavagna	1869	G.Saccomano	[illegible]	A.Cichero
70686	1019		Lucia (ex Madre Fidele) *Wood 83-86* C Bk	British	640	162·3	31·1	18·3	Sunderland	1868	J.Blumer & Co.	[illegible]	T.Roberts
532	1020		Lucia (ex Angiolina) *Carletta* WBn	Italian	169	100·0	23·1	14·2	Leghorn	1863	G.Gori	[illegible]	Oreste Del Buono
101	1021		Lucia WSr	Russian	126	89·7	23·6	10·4	Kurland	1876			Oskar von Stryk
15102	1022		Lucia B. Ives (ex E. H. Parker) *Mott* WSr	American	177	96·8	27·2	8·5	Belleville,N.J	18 53/84			A.H.Mott
140498	1023	R	Lucia Porter *Grindle -81* W3mSr	American	316	132·0	32·0	10·5	Belfast,Me.	1881		[illegible]	J.F.Grindel
	1024		Lucie WSr	French	143	88·3	23·3	12·1	Dunkirk	1846			Comolet frères
	1025	N	Lucie *Bidsaay 75-87* WSr	Norweg'n	116	95·2	24·0	10·0	Christiansund	1877		[illegible]	Parelius & Lossius
246	1026		Lucie *Johnsohn 82-82* WSr	Russian	149	89·4	25·5	10·1	Kürbis	1877	Grass		K.Berg & J.Jansohn
	1027		Lucie (ex Dagmar) *Hedberg -88* WBg	Swedish	214	104·0	22·8	12·0	Miramichi	1858			J.Pettersson
	1028		Lucie Gabrielle (DISMANTLED) (ex Peppino P.) WBg	French	417	107·9	26·9	18·7	Sestri Ponente	1866	G.B.Cadenaccio		C.Guérin
	1029	V	Lucie Marie (ex Ayna) *Denis -88* WBn	French	163	95·8	20·0	12·1	Finland	1858			Dupuis-Robiel
	1030		Lucie Marie (BROKEN UP) *Bertho* WSr	French	141	84·3	23·0	11·1	Nantes	1866	Tillé		D.Bertho
	1031	G	LucieRadmann (ex Susanne Johanna, ex Kon og Vaderland) *Gerth -88* WBk	German	447	117·6	29·9	18·9	Vlaardingen	1854	Von Dam		W.Lüdke & Co.
140069	1032		Lucie Wheatley *Pettengill -88* W3mSr	American	180	113·8	29·5	8·2	Frederica,Del.	1874	NathanielLank		J.W.Hall
	1033	V	Lucien & Valentine (ex Anna) *Verdeis* WBg	French	274	110·2	27·6	14·8	Arendal	1877		[illegible]	K.Morvant
522	1034	A	Lucietta (ex Giuseppe) *Musana* WSr	Italian	124	80·7	22·4	12·1	Gaeta	1858		[illegible]	Ant.Burgio & Co.
66855	1035	V	Lucile *Williams* WBkn	British	498	128·8	31·1	17·0	Tynem'th,NB	1872	R.Lowitt	[illegible]	A.J.Williams
76399	1036		Lucile WSrYt	British	157	117·0	23·1	13·4	Gosport	1876	Campr&Nichlsn		N.Rudge
140031	1037	V	Lucile *Serman* WS	American	1323	200·2	40·0	23·9	Freeport,Me.	1874	E.C.Soule	[illegible]	E.C.Soule
71444	1038	✱	Lucille *Kirby -88* WBg	British	278	117·6	26·1	14·9	P.E.Island	1874	Yeo	[illegible]	W.Williams & Sons
57022	1039		Lucinda *White 88-88* WBn	British	199	105·8	24·1	12·9	P.E.Island	1867	Tuplin		H.W.Brightman
140539	1040		Lucinda G. Potter *Potter -88* W3mSr	American	613	151·8	35·8	15·9	Bath,Me.	1882	Goss&Sawyer		W.T.Potter
	1041		Lucinde *Jensen 87-87* WBg	Danish	150	86·5	23·7	13·4	Holbæk	1842	V.Lagoni		Den Kgl.Grønl.Handel
	1042	DV	Lucinde *Krakauschke 86-* WBk	Russian	319	117·0	28·0	14·0	Libau	1866	J.D.Dargewitz		C.Essmann
	1043		Lucinde *Hansson* WSr	Swedish	162	76·8	20·0	10·8	Flensburg	1861			M.Hansson
251	1044		Lucio (ex [illegible]) *[illegible] -87* WBg	Austrian	254	86·0	26·5	16·1	Prà,Italy	1870		[illegible]	Ivcevich & Co.
	1045	V	Lucio I. *Voz* WSn	Portug'se	15[illegible]	89·6	21·5	10·0	Villa do Conde	1884	C.M.de Araujo	[illegible]	V.L.de Almeida
90006	1046	✱	Lucipara *Harder 86-88* I 4mS	British	1563	267·1	40·2	23·8	Greenock	1885	Russell & Co.		P.Denniston & Co.
87113	1047	✱	Lucknow *Lidstone 81-87* StlS	British	1440	245·3	37·1	22·5	Sunderland	1883	Doxford&Sons		Tyser & Co.
79485	1048	✱	Lucy *Davies* WBkn	British	192	112·8	24·1	12·5	Ipswich	1879	Bayley	[illegible]	J.C.Peacock
140315	1049	R	Lucy *Cook* WSr	American	297	130·0	30·5	8·0	Berwick,Me.	1882			American Salt Co.
15046	1050		Lucy *Woosier* WSr	American	129	87·5	27·8	9·1	Eastport,Me.	1864	Buston		L.A.Godfrey

✱=**Lloyd's Register;** D=Nederlandsche Vereeniging; G=Germanischer Lloyd; I=Registro Italiano; R=American [Record];
A=Veritas Austro-Ungarico; H=Veritas Ellenico; N=Norske Veritas; V=Bureau [Veritas].

In the 4th Column, C=Composite (iron frame, wood planked); I=Iron; Stl=Steel; W=Wood.

The Dates of Metalling (Col. 11) are based on the records of the Books in which Vessels are respectively classed, or on the Reports of the Surveyors of [...]

Further particulars of Vessels classed by Lloyd's Register will be found in Lloyd's Register Book and Lloyd's Yacht Register.

		Nom. Matériaux et Gréement. Capitaine et dates d'entrée au service des Armateurs actuels et à bord du navire. Ancien Nom.	Pavillon.	Tonnage net.	Dimensions. Longueur	Largeur	Creux	Construit. Où.	Quand.	Par qui.	Date du doublage en métal	Armateurs ou Directeurs-Gérants.	Port d'Armement.
1	2, 3	4	5	6	7			8	9	10	11	12	13
1051		Lucy *Lirman* -90 WSr (ex Eva Parker)	Dutch	355	120·0	27·9	12·5	Harb'rvilleNS	1873	J.Givan		Kali Anam Maatschappy	Batavia
1052	✱	Lucy *Klausen* -87 WBk (ex Lucy Lloyd)	Norweg'n	739	176·2	33·0	19·5	P.E.Island	1875	J.C.Pope	Y 4.88	Actieselskabet "Lucy" (C.Moller)	Christiania
1053	DN	Lucy *Röer* 72-73 WBg	Norweg'n	312	120·5	27·9	13·2	Cornwall,Can.	1865			Martin Röer	Dröbak
1054		Lucy A. Blossom W3mSr	American	179	108·0	24·0	9·0	ConneautOhio	1846			J.Philbrook	Winterport,Me.
1055	R	Lucy A. Davis W3mSr *Jenning* -88	American	596	118·5	34·2	12·6	Yarmouth,Me	1882	Htchns&Stubbs	Y 9.87	J.W.Davis	Portland,Me.
1056	R	Lucy A. Nickels *Nickels* WS	American	1395	199·0	37·9	23·8	Brewer,Me.	1875	McGilvery	Y 7.87	J.C.Nickels	Searsport,Me.
1057	R	Lucy E. Friend *Collins* W3mSr	American	447	147·0	34·6	12·6	Nwbrypt Mass	1882	G.E.Currier		W.H.Collins	Gloucester,Mass
1058	N	Lucy Frances *Stubbs* -85 W3mSr	Norweg'n	232	114·6	27·0	11·2	Brookville,Me	1856		Y 11.89	Ludwigsen & Schjelderup	Christiania
1059	R	Lucy H. Russell W4mSr *Randall* -87	American	1108	194·3	39·9	18·0	Bath,Me.	1887	NewEng.S.B.Co		H.M.Randall	Pt.Jefferson
1060		Lucy Hammond WSr *Simmons*	American	150	96·5	26·3	8·9	E.Machias,Me	1869			J.R.Talbot	E.Machias,Me.
1061		Lucy M. Collins *Newcomb* WSr	American	159	100·3	27·8	8·3	WinterportMe	1867			G.B.Hook	Bangor,Me.
1062	✱	Lucy March *Griffiths* -88 WBkn	British	244	118·0	26·6	14·1	Pwllheli	1876	Williams	Y 8.88	Morgan,Jones & Co.	Carnarvon
1063	✱	LucyReppen *Bach* -88 WBk (ex Josep Reppen)	Norweg'n	892	180·6	35·3	20·5	N.Bruns.	1877	Brown	Y 12.87	O.C.Johansen	Christiansand
1064		Lucy Richmond WSrBarge *Richmond*	British	137	106·7	24·4	8·6	Ipswich	1875	Robertson		W.T.Richmond & Co.	Maldon
1065	✱ DI	Lucy W. *Viglio* WBk	Italian	683	159·1	33·1	21·6	SestriPonente	1876	A.Calcagno	Y 4.86	P.Zenoglio	Genoa
1066		Lucy W. Snow WBn *Bargen* -85	American	299	112·8	26·9	14·0	Rockland,Me.	1869	Snow&Farwell		F.C.Butman	Boston,Mass.
1067	V	Lucy and Paul *Andreeis* 75-88 WBk	German	377	127·9	28·7	15·9	Elbing	1861	C.G.Fechter		E.Burchard	Rostock
1068		Ludolfina Jantina WBn *Poppen* -86	Dutch	156	95·0	23·0	10·6	Pekela	1874	L.H.DeWijk		H.J.Kuiper (H.L.deWyk,Manager)	Oude Pekela
1069		Ludovic WSr	French	120	89·6	19·7	10·9	St.Malo	1878	Tranchemer frs.		Vve H.Ruellan	Paimpol
1070	N	Ludvig *Larsen* 82-83 WBk (ABANDONED)	Norweg'n	302	124·1	27·4	13·5	Brevig	1870	G.Moe		G.J.Sörensen	Sandefjord
1071	N	Ludvig *Mathisen* WSr	Norweg'n	110	83·6	20·9	9·6	Brake	1851			Chr.& John Mathisen	Horten
1072	DNV	Ludvig Holberg WBk *Hansen* 81-86	Norweg'n	1019	193·1	37·5	21·7	Bergen	1878	Brunchorst& Dekke	Y 8.88	A.F.Klaveness	Sandefjord
1073		Ludvig (MISSING) *Schaver* WBk	German	368	128·6	26·6	13·9	Greifswald	1865	C.Dodt		J.Rudolph	Stettin
1074		Ludwig *Eichen* -80 WSr	Russian	126	82·0	21·0	9·5	Haynasch	1872	M.Eichen		A.Eichen & F.Weide	Haynasch
1075		LudwigCapobus,sen. WBg *Permien* 81-81	German	242	98·3	26·2	14·8	Rostock	1863	W.Zeltz		A.Burchard	Rostock
1076	R	Luella A. Snow *Rose* WSr	American	178	106·7	27·0	8·4	Rockland,Me.	1875	Snow&Farwell	Y 7.87	Snow & Pearsons	Rockland,Me.
1077	✱	Lufra *Rogers* CBk	British	672	179·7	31·1	17·9	Aberdeen	1870	Hall	Y 6.87	A.McGregor	Hobart
1078		Lufra WKYt	British	131	108·1	22·0	12·2	Cowes	1865	M.Ratsey		Exrs. of the Duke of Rutland,K.G.	Cowes
1079	N	Lufra *Evensen* WBk	Norweg'n	483	133·6	31·3	17·6	Grimstad	1874	O&L.Dömmesmoen	Y 8.88	J.E.Einersen	Grimstad
1080	R	Lugano *Clark* WSr	American	166	104·5	28·1	9·0	E.Machias,Me	1867			J.G.Murchie & Sons	New York
1081	✱I	Luigi (ex Caterina B.) *Cuneo* WBk	Italian	828	168·7	33·7	23·6	Finale	1877	F.Ponziolo	Y 7.88	F.Cuneo fu A.	Genoa
1082	A	Luigi *Scattiola* WBk	Italian	511	..	..	..	Brooklyn	1865	Th.Stach	Y 8.88	L.Pirandello	Messina
1083		Luigi *Corradini* WBg	Italian	193	111·8	22·1	14·1	Capo d'Istria	1864			A.Bini & Co.	Genoa
1084	A	Luigi *Di Pace* -88 WBn	Italian	141	78·7	23·3	13·3	Castellamare di Stabia	1863		Z 1.88	F.Troise	Naples
1085	I	Luigi Accame WBk *Olivar* -88	Italian	939	176·8	35·4	22·6	SestriPonente	1870	G.Casanova	Y 1.85	Em.Accame fu L.	Genoa
1086		Luigi C. *Tonerelli* WBk (ex Vittorino)	Italian	249	107·6	26·6	16·3	Leghorn	1853	G.Micheli		E.Pieruzzini	Leghorn
1087	A	Luigi Fortunato WTrb *Ghezzo* -88	Italian	117	83·0	24·4	10·6	Chioggia	1881	L.Menetto	Z 8.86	Flli Ghezzo & Co.	Venice
1088	I	Luigi P. *Moglia* -87 WBk	Italian	657	143·5	32·4	21·0	Prà	1873	B.Canepa	Y 3.88	G.Moglia	Genoa
1089	I	Luigi Padre *Olivari* -88 WBk (ex Olivari Pietro)	Italian	529	140·7	31·2	20·7	Savona	1875	F.Calamaro	Y 12.87	Flli Scinecaluga	Genoa
1090		Luigi Padre (WRECKED) WBn (ex Maria ed Angiolina)	Italian	121	71·2	23·2	13·1	Varazze	1855			C.A.Giorgetti	Leghorn
1091		Luigi Repetto *Repetto* WBk	Italian	220	101·7	25·6	14·8	St.Valery	1852			Fort Repetto	Genoa
1092	I	Luigi Ruggero WBk *Mercuri* -88	Italian	490	131·2	29·2	18·7	Castellamare	1873	L.Tramblarulo	Y 3.86	L.Ruggero	Castellamare
1093		Luigia *Leonardi* WBk	Italian	402	122·3	29·7	18·4	SestriPonente	1866	G.Peragallo		Flli Pendibene	Genoa
1094		Luigia (ex Rosa) *Calabrese* WBg	Italian	231	93·5	26·1	16·2	Piraeus	1867			G.B.Borzone fu G.	Leghorn
1095		Luigia F. (ex Consiglio) *Villa* W	Italian	491	147·6	30·7	19·7	SestriPonente	1866	L.Casella		B.Fravega fu G.B.	Genoa
1096	I	Luigia M. *Massone* -88 WBk (ex Ermenegilda Dasovaro)	Italian	421	128·9	30·2	17·9	SestriPonente	1880	A.Briasco fu G.B.	Y 3.85	E.Massone	Genoa
1097	I	Luigia R. *Daregna* -88 WBk (ex Portofino)	Italian	498	137·8	30·0	18·7	Chiavari	1867	M.Tappani	Y 3.88	F.Rainusso	Genoa
1098	IV	Luigia Raffo *Gattia* WBk	Italian	972	183·4	38·1	23·1	Chiavari	1875	M.Tappani	Y 3.86	Erasmo Raffo fu A.	Spezzia
1099	I	Luigia Rocca *Botaffo* WBk	Italian	546	150·2	32·3	19·7	Lavagna	1870	D.Briasco	Y 8.85	Rocca & Lavello	Genoa
1100													

✱=Lloyd's Register; D=Nederlandsche Vereeniging; G=Germanischer Lloyd; I=Registro Italiano; R=American Record;
A=Veritas Austro-Ungarico; H=Veritas Ellenico; N=Norske Veritas; V=Bureau Veritas.

Dans la 4me Colonne, C=Composite (membrure en fer, bordé en bois); I=Fer; Stl=Acier; W=Bois.

Les Dates du doublage en métal (Col. 11) ressortent des livres dans lesquels les navires sont respectivement classés ou des rapports des Inspecteurs du Lloyd's Register.

Pour plus amples renseignements sur les navires classés par le Lloyd's Register, voir le Lloyd's Register Book et le Lloyd's Yacht Register.

Official No.	No. in Book.	Registry	Ship's Name. Material and Rig. Master, and Dates of Appointment to present Owner's Service and to Ship. Ship's Late Name.	Flag.	Tonnage	Length.	Breadth.	Depth.	Built. Where.	When.	By whom.	Date of Metalling.	Owners or Managers.	Port
	1001	DNV	**Lovspring** *Thomsen* WBk	Norweg'n	547	149·4	30·9	17·8	Christiansand	1873	H.Olsen	Y 2.87	H.G.Kirschom	[illegible]
	1002	N	**Lövspring** *Sakse* -88 WBk	Norweg'n	489	136·4	31·0	16·9	Grimstad	1885	N.O.Gjömle	Y 11.85	J.Jörgensen	Grimsta[illegible]
	1003	N	**Lovspring** *Thorvertsen* WBg	Norweg'n	225	107·5	26·9	12·2	Porsgrund	1872			P.M.Petersen & Sön	[illegible]
72333	1004	RV	**Low Wood** *Tharber* -87 WBk	British	1091	186·7	37·6	22·4	Portland,N.B.	1878	D.Lynch	Y 2.88	H.D.Troop	St.John
[illegible]78	1005		**Lowestoft** *Hook* WBk	British	293	112·7	26·5	16·5	Sunderland	1866	J.Dennison		R.L.Manners	[illegible]
76389	1006	✻	**Loweswater** *Smith* -86 IBk	British	603	174·0	30·2	17·5	Whitehaven	1876	WhitehvnSBC		Jackson,Metcalf & Co.	[illegible]
[illegible]91	1007	✻	**Lowther Castle** *Scarborough* 81-86 IS	British	1824	262·1	39·0	23·3	Workington	1886	R.Williamson & [Son]		Lancaster Shipowners Co.Ld	[illegible]
	1008	N	**Loyal** *Friis* -87 (ex Dundonald) (WRECKED) WS	Norweg'n	1427	220·2	38·9	24·7	Quebec	1868	Rosa	Y 4.87	Rederiselskab "Loyal" (Chr.Ellefsen)	[illegible]
	1009	D	**Loyal** *Knudsen* -88 (WRECKED) (ex Susan) WBk	Norweg'n	663	159·9	31·9	18·4	Grimstad	1883	S.Johnsen	Y 4.86	M.F.Stray	[illegible]
83594	1010	V	**Loyalist** *Fraser* 83-85 WBn	British	348	130·5	29·4	13·0	Portland,N.B.	1883	E.E.Sayre	Y 8.86	E.E.Sayre	St.John
	1011		**Loyo** *San Thome* (ex Alcester) W3mSr	Brazilian	242	100·0	24·3	11·5	Fao	1876	Borda	Y 2.81	J.daSilvaLoyo & Filho	[illegible]
	1012		**Lua** WSr	Brazilian	150	..	..	..					V.A.c Magalhaes	[illegible]
	1013	N	**Luca** *Jansen* 56-75 WSr	Norweg'n	399	116·3	28·3	18·4	Buccari	1857			H.E.Jansen	[illegible]
[illegible]	1014	V	**Lucania** *Gibson* 72-86 WS	British	1774	229·3	42·1	24·4	St.John,N.B.	1886	J.F.Fraser	Y 11.86	B.Smith & Sons	[illegible]
	1015	G	**Lucas** (ex Lina) *Pecker* 85-85 WSr	German	159	94·9	21·1	11·1	Papenburg	1863	F.W.Meyer		G.& H.Aden	[illegible]
	1016	✻	**Lucayas** *Johansen* IBk	Danish	445	150·4	25·5	18·1	Greenock	1869	Robertson & Co		Th.Meinertz's Enke	[illegible]
15[illegible]	1017	V	**Lucco** *Degregori* -87 (BROKEN UP) WBk	Italian	498	134·5	30·8	20·0	Sestri Ponente	1866	L.Briasco	Y 1.86	B.Olivari	[illegible]
10[illegible]	1018	I	**Luchina C.** *Repetto* WBk	Italian	506	134·5	31·2	19·8	Lavagna	1869	G.Saccomano	Y 2.86	A.Cichero	[illegible]
70083	1019		**Lucia** *Wood* 83-86 (ex Mona Fidele) CBk	British	640	162·3	31·1	18·3	Sunderland	1868	J.Blumer & Co.	Y 10.85	T.Roberts	[illegible]
552	1020		**Lucia** (ex Angelina) *Carletta* WBg	Italian	169	100·0	23·1	14·2	Leghorn	1863	G.Gori	Y 9.83	Oreste Del Buono	[illegible]
101	1021		**Lucia** WSr	Russian	126	89·7	23·6	10·4	Kurland	1876			Oskar von Stryk	[illegible]
1510[illegible]	1022		**Lucia B. Ives** *Mott* (ex E. H. Parker) WSr	American	177	96·8	27·2	8·5	Belleville,N.J	18 53/64			A.H.Mott	[illegible]
140498	1023	R	**Lucia Porter** *Grindle* [-81 W3mSr	American	316	132·0	32·0	10·5	Belfast,Me.	1881		Y 11.85	J.F.Grindel	[illegible]
	1024		**Lucie** WSr	French	143	88·3	23·3	12·1	Dunkirk	1846			Compolet frères	[illegible]
	1025	N	**Lucie** *Eidsaay* 75-87 WSr	Norweg'n	116	95·2	24·0	10·0	[illegible]	1877		Y 1.87	Parelius & Lessius	[illegible]
34[illegible]	1026		**Lucie** *Johnsohn* 82-82 WSr	Russian	149	80·4	25·5	10·1	Kürbis	1877	Grass		K.Berg & J.Jansohn	Riga
	1027		**Lucie** *Hedberg* -83 (ex Dagmar) WBg	Swedish	214	104·0	22·8	12·0	Miramichi	1858			J.Petterson	[illegible]
	1028		**Lucie Gabrielle** (DISMANTLED) (ex Pippin P.) WBg	French	417	107·9	26·9	18·7	Sestri Ponente	1866	G.B.Cadenaccio		C.Guérin	[illegible]
	1029	V	**Lucie Marie** *Denis* -88 (ex Anna) WBn	French	163	95·8	20·0	12·1	Finland	1858			Dupuis-Robial	[illegible]
	1030		**Lucie Marie** *Bertho* (BROKEN UP) WSr	French	141	84·3	23·0	11·1	Nantes	1866	Tille		D.Bertho	Nantes
	1031	G	**LucieRadmann** *Gerth* [-88 (ex Susanne Johanne, ex Koning Vaderland) WBk	German	447	117·6	29·9	18·9	Vlaardingen	1854	Von Dam		W.Lüdke & Co.	[illegible]
140063	1032		**Lucie Wheatley** *Pettengill* -88 W3mSr	American	180	113·8	29·5	8·2	Frederica,Del.	1874	NathanielLank		J.W.Hall	[illegible]
	1033	V	**Lucien & Valentine** *Verdon* (ex Anna) WBg	French	274	110·2	27·6	14·8	Arendal	1877		Z 1.86	K.Morvant	[illegible]
53[illegible]	1034	A	**Lucietta** (ex Giuseppe) *Musana* WSr	Italian	124	80·7	22·4	12·1	Gaeta	1858		Z 4.86	Ant.Burgio & Co.	[illegible]
66855	1035	V	**Lucile** *Williams* WBkn	British	438	128·8	31·1	17·0	Tynemo'th,NB	1872	R.Lowitt	Y 1.86	A.J.Williams	[illegible]
76390	1036		**Lucile** WScYt	British	157	117·0	23·1	13·1	Gosport	1876	Camper&Nicholson		N.Rudge	[illegible]
140031	1037	V	**Lucile** *Serman* WS	American	1329	200·2	40·0	23·9	Freeport,Me.	1874	E.C.Soule	Y 11.87	E.C.Soule	[illegible]
7144[illegible]	1038	✻	**Lucille** *Kirby* -88 WBg	British	278	117·6	26·1	14·9	P.E.Island	1874	Yeo	Y 6.87	W.Williams & Sons	[illegible]
5762[illegible]	1039		**Lucinda** *White* 88-88 WBn	British	199	105·8	24·1	12·9	P.E.Island	1867	Tuplin		H.W.Brightman	[illegible]
1403[illegible]	1040		**Lucinda G. Potter** *Potter* -88 W3mSr	American	613	151·8	35·8	15·9	Bath,Me.	1882	Goss&Sawyer		W.T.Potter	[illegible]
	1041		**Lucinde** *Jensen* 87-87 WBg	Danish	150	86·5	23·7	13·4	Holbæk	1842	V.Lagoni		Den.Kgl.Grønl.Handel	[illegible]
	1042	DV	**Lucinde** *Kraknnschke* 86- WBk	Russian	319	117·0	28·0	14·0	Libau	1866	J.D.Dargewitz		C.Essmann	Riga
	1043		**Lucinde** *Hansson* WSr	Swedish	103	76·8	20·0	10·8	Flensburg	1861			M.Hansson	[illegible]
251	1044		**Lucio** *Martinolich* -87 (ex Maria Regina) WBg	Austrian	254	86·0	26·5	16·1	Pra,Italy	1870		Y 11.83	Ivecvich & Co.	[illegible]
	1045	V	**Lucio I.** *Vaz* WSr	Portug'se	150	89·6	21·5	10·0	Villa do Conde	1884	C.M.deAraujo	Y 10.85	V.L.de Almeida	[illegible]
90053	1046	✻	**Lucipara** *Harder* 85-88 I 4mS	British	[illegible]	267·1	40·2	23·8	Greenock	1885	Russell&Co.		P.Denniston & Co.	[illegible]
87113	1047	✻	**Lucknow** *Lidstone* 81-87 StlS	British	1440	245·3	37·1	22·5	Sunderland	1883	Doxford&Sons		Tyser & Co.	[illegible]
79483	1048	✻	**Lucy** *Davies* WBkn	British	192	112·8	24·1	12·5	Ipswich	1879	Bayley	Y 6.87	J.C.Peacock	[illegible]
140515	1049	R	**Lucy** *Cook* WSr	American	297	130·0	30·0	8·0	Barwick,Me.	1882			American Salt Co.	[illegible]
15048	1050		**Lucy** *Webster* WSr	American	129	87·5	27·8	9·1	[illegible],Me.	1864	[illegible]		L.A.Godfrey	[illegible]

✻=Lloyd's Register; D=Nederlandsche Vereeniging; G=Germanischer Lloyd; I=Registro Italiano; R=Ameri[can]…
A=Veritas Austro-Ungarico; H=Veritas Ellenico; N=Norske Veritas; V=Bureau [Veritas]…

In the 4th Column, C=Composite (iron frame, wood planked); I=Iron; Stl=Steel; W=Wood.

The Dates of **Metalling (Col. 11)** are based on the records of the Books in which Vessels are respectively classed, or on the Reports of the Surveyors of …

Further particulars of Vessels classed by Lloyd's Register will be found in Lloyd's Register Book and Lloyd's Yacht Register.

No.	Classe	Nom. Matériaux et Gréement. Capitaine et dates d'entrée au service des Armateurs actuels et à bord du navire. Ancien Nom.	Pavillon.	Tonnage net.	Dimensions. Longueur.	Largeur.	Creux.	Construit. Où.	Quand.	Par qui.	Date du doublage en métal.	Armateurs ou Directeurs-Gérants.	Port d'Armement.
		Lucy *Lüsink* -86 WSr	Dutch	355	120·0	27·9	12·5	Harb'rvilleNS	1873	J.Givan		Kali Anam Maatschappy	Batavia
	✱	Lucy [illegible] -88 WBk (ex Lucy Bond)	Norweg'n	739	176·2	35·0	19·5	P.E.Island	1875	J.C.Pope	Y 4.88	Actieselskabet "Lucy" (C.Mølle)	Christiania
	N	Lucy *Röer* 12-73 WBg	Norweg'n	312	120·5	27·9	13·2	Cornwall,Can.	1865			Martin Röer	Dröbak
		Lucy A. Blossom W3mSr	American	179	108·0	24·0	9·0	ConneautOhio	1846			J.Philbrook	Winterport.Me.
	R	Lucy A. Davis W3mSr *Lyons* -87	American	596	148·5	34·2	12·6	Yarmouth,Me	1882	Htchns&Stubbs	Y 9.87	J.W.Davis	Portland,Me.
	R	Lucy A. Nickels *Nickels* WS	American	1395	199·0	37·9	23·8	Brewer,Me.	1875	McGilvery	Y 7.82	J.C.Nickels	Searsport,Me.
		Lucy E. Friend *Collins* W3mSr	American	447	147·0	34·6	12·6	Nwbrypt Mass	1882	G.E.Currier		W.H.Collins	Gloucester,Mass.
	N	Lucy Frances *Stubbe* W3mSr (-85	Norweg'n	232	114·6	27·0	11·2	Brookville,Me	1856		Y 3.85	Ludwigsen & Schjelderup	Christiania
		Lucy H. Russell W4mSr *Russell* -87	American	1108	194·3	39·9	18·0	Bath,Me.	1887	NewEng.S.B.Co		H.M.Randall	Pt.Jefferson
1060		Lucy Hammond WSr *Robinson*	American	150	96·5	26·3	8·9	E.Machias,Me.	1869			J.R.Talbot	E.Machias,Me.
1061		Lucy M. Collins *Newcomb* WSr	American	159	100·3	27·8	8·3	WinterportMe	1867			G.B.Hook	Bangor,Me.
1062	✱	Lucy March *Griffiths* WBkt (-82	British	244	118·0	26·6	14·1	Pwllheli	1876	Williams	Y 2.86	Morgan,Jones & Co.	Carnarvon
1063	✱	LucyReppen *Hansen* -88 WBk (ex [illegible] Brunswick)	Norweg'n	892	180·6	35·3	20·5	N.Bruns.	1877	Brown	Y 12.87	O.C.Johansen	Christiansand
1064		Lucy Richmond WSrBarge *Richmond*	British	137	106·7	24·4	8·6	Ipswich	1875	Robertson		W.T.Richmond & Co.	Maldon
1065	✱ DI	Lucy W. *Fatta* WBk	Italian	683	159·1	33·1	21·6	SestriPonente	1876	A.Calcagno	Y 4.86	P.Zenoglio	Genoa
1066		Lucy W. Snow WB *Burgess*	American	299	112·8	26·9	14·0	Rockland,Me.	1869	Snow&Farwell		F.C.Butman	Boston,Mass.
1067	V	Lucy and Paul *Andreis* WBk 10-86	German	377	127·9	28·7	15·9	Elbing	1861	C.G.Fechter		E.Burchard	Rostock
1068		Ludolfina Jantina WBn *Pepper* -86	Dutch	156	95·0	23·0	10·6	Pekela	1874	L.H.DeWijk		H.J.Kuiper (O.L.de Wyk,Manager)	Oude Pekela
1069		Ludovic WSr	French	120	89·6	19·7	10·9	St.Malo	1878	Tranchemer frs.		Vve. H.Ruellan	Paimpol
1070	N	Ludvig (ABANDONED) *Larsen* 82-82 WBk	Norweg'n	302	124·1	27·4	13·5	Brevig	1870	G.Moe		G.J.Sörensen	Sandefjord
1071	N	Ludvig *Mathisen* WSr	Norweg'n	110	83·6	20·9	9·6	Broke	1851			Chr.& John Mathisen	Horten
1072	DNV	Ludvig Holberg *Bøhmer* 1-86 WBk	Norweg'n	1019	193·1	37·5	21·7	Bergen	1878	Brunchorst& [Dekke	Y 8.86	A.F.Klaveness	Sandefjord
1073		Ludvig (MISSING) *Schauer* WBk	German	368	128·6	26·5	13·9	Greifswald	1865	C.Dadt		J.Rudolph	Stettin
1074		Ludwig *Eichen* -80 WSr	Russian	126	82·0	21·0	9·5	Haynasch	1872	M.Eichen		A.Eichen & F.Weide	Haynasch
1075		LudwigCapobus,sen. WBg *Permien* 8-84	German	242	98·3	26·2	14·8	Rostock	1863	W.Zeltz		A.Burchard	Rostock
1076	R	Luella A. Snow *Rose* WSr	American	178	106·7	27·0	8·4	Rockland,Me.	1875	Snow&Farwell	Y 7.81	Snow & Pearsons	Rockland,Me.
1077	✱	Lufra *Bagent* CBk	British	672	179·7	31·1	17·9	Aberdeen	1870	Hall	Y 6.81	A.McGregor	Hobart
1078		Lufra WKYt	British	131	108·1	22·0	12·2	Cowes	1865	M.Ratsey		Exrs.of the Duke of Rutland,K.G.	Cowes
1079	N	Lufra *Evensen* WBk	Norweg'n	483	133·6	31·3	17·6	Grimstad	1874	O&I.Dimmesmoen	Y 9.84	J.E.Einersen	Grimstad
1080	R	Lugano *Clark* WSr	American	166	104·5	28·1	9·0	E.Machias,Me	1867			J.G.Murchie & Sons	New York
1081	✱I	Luigi (ex Caterina B.) *Cuneo* WBk	Italian	828	168·7	33·7	23·6	Finale	1877	F.Ponzialo	Y 1.89	F.Cuneo fu A.	Genoa
1082	A	Luigi *Sgroi* WBk	Italian	511	..	..	..	Brooklyn	1865	Th.Stach	Y 8.83	L.Pirandello	Messina
1083		Luigi *Corradini* WBg	Italian	193	111·8	22·7	14·1	Capo d'Istria	1864			A.Bini & Co.	Genoa
1084	A	Luigi *Di Face* -89 WBn	Italian	141	78·7	23·3	13·3	Castellamare [di Stabia	1863		Z 1.88	F.Troise	Naples
1085	I	Luigi Accame *Olivari* -88 WBk	Italian	939	176·8	35·4	22·6	SestriPonente	1870	G.Casanova	Y 7.86	Em.Accame fu L.	Genoa
1086		Luigi C. *Tancredi* WBk (ex Vittoriosa)	Italian	249	107·6	26·6	16·3	Leghorn	1853	G.Micheli		E.Pieruzzini	Leghorn
1087	A	Luigi Fortunato WTrb *Bacco* -89	Italian	117	83·0	24·4	10·6	Chioggia	1881	L.Menetto	Z 8.86	F^lli. Ghezzo & Co.	Venice
1088	I	Luigi P. *Moglia* -87 WBk	Italian	657	143·5	32·4	21·0	Prà	1873	B.Canepa	Y 8.86	G.Moglia	Genoa
1089	I	Luigi Padre *Olivari* WBk (ex Ducati Fiores) (-88	Italian	529	140·7	31·2	20·7	Savona	1875	F.Calamaro	Y 12.87	F^lli. Sciaccaluga	Genoa
1090		Luigi Padre (WRECKED) WBn (ex Garbato, ex Angiolina)	Italian	121	71·2	23·2	13·1	Varazze	1855			C.A.Giorgetti	Leghorn
1091		Luigi Repetto *Repetto* WBk	Italian	220	101·7	25·6	14·8	St.Valery	1853			Fort Repetto	Genoa
1092	I	Luigi Ruggero WBk *Mazzon* -89	Italian	490	131·2	29·2	18·7	Castellamare	1873	L.Trambarulo	Y 3.86	L.Ruggero	Castellamare
1093		Luigia *Bonanati* WBk	Italian	402	122·3	29·7	18·4	SestriPonente	1866	G.Peragallo		F^lli. Pendibene	Genoa
1094		Luigia (ex Gené) *Calabrese* WBg	Italian	231	93·5	26·1	16·2	Piraeus	1867			G.B.Borzone fu G.	Leghorn
1095		Luigia F. (ex Consiglio) *Villa* W	Italian	491	147·6	30·7	19·7	SestriPonente	1866	L.Casella		B.Fravega fu G.B.	Genoa
1096	I	Luigia M. *Bassoul* -88 WBk (ex Emmeregilda Dassano)	Italian	421	128·9	30·2	17·9	SestriPonente	1880	A.Briasco fu G.B.	Y 3.85	E.Massone	Genoa
1097	I	Luigia R. *Daccomi* -88 WBn (ex Portofino)	Italian	498	137·8	30·0	18·7	Chiavari	1867	M.Tappani	Y 4.88	F.Rainusso	Genoa
1098	IV	Luigia Raffo *Gaffio* WBk	Italian	972	183·4	38·1	23·1	Chiavari	1875	M.Tappani	Y 3.86	Erasmo Raffo fu A.	Spezzia
1099	I	Luigia Rocca *Rebaffo* WBk	Italian	546	150·2	32·3	19·7	Lavagna	1870	D.Briasco	Y 8.85	Rocca & Lavello	Genoa
1100													

✱=Lloyd's Register; D=Nederlandsche Vereeniging; G=Germanischer Lloyd; I=Registro Italiano; R=American Record;
A=Veritas Austro-Ungarico; H=Veritas Ellenico; N=Norske Veritas; V=Bureau Veritas.

Dans la 4me Colonne, C=Composite (membrure en fer, bordé en bois); I=Fer; Stl=Acier; W=Bois.

Les Dates du doublage en métal (Col. 11) ressortent des livres dans lesquels les navires sont respectivement classés ou des rapports des Inspecteurs du Lloyd's Register.

Pour plus amples renseignements sur les navires classés par le Lloyd's Register, voir le Lloyd's Register Book et le Lloyd's Yacht Register.

Official No.	No. in Book	Society in which Classed	Ship's Name. Master, and Dates of Appointment. Ship's Late Name.	Material and Rig	Flag	Net Tonnage	Length	Breadth	Depth	Built Where	When	By whom	Date of Metalling	Owners or Managers
	1001	DNV	**Lovspring** *Thomsen*	WBk	Norweg'n	547	149·4	30·9	17·8	Christiansand	1873	H.Olsen	[illegible]	H.G.Kirschom
	1002	N	**Lovspring** *Sakive -88*	WBk	Norweg'n	489	136·4	31·0	16·9	Grimstad	1885	N.O.Gjömle	[illegible]	J.Jörgensen
	1003	N	**Lovspring** *Thonnesen*	WBg	Norweg'n	225	107·5	26·9	12·2	Porsgrund	1872			P.M.Petersen & Søn
72935	1004	RV	**Low Wood** *Thurber -87*	WBk	British	1091	186·7	37·6	22·4	Portland,N.B.	1878	D.Lynch	[illegible]	H.D.Troop
56478	1005		**Lowestoft** *Hook*	WBk	British	293	112·7	26·5	16·5	Sunderland	1866	J.Dennison		R.L.Manners
76389	1006	✻	**Loweswater** *Smith -88*	IBk	British	603	174·0	30·2	17·5	Whitehaven	1876	WhitehvnSBCo		Jackson,Metcalf & Co.
91201	1007	✻	**Lowther Castle** *Scarborough -88*	IS	British	1824	262·1	39·0	23·8	Workington	1886	R.Williamson & Son		Lancaster Shipowners Co.Ld.
	1008	N	**Loyal** *Friis -88* (ex Dreadnought) (WRECKED)	WS	Norweg'n	1427	220·2	38·9	24·7	Quebec	1863	Rosa	[illegible]	[illegible] "Loyal"
	1009	D	**Loyal** *Knudsen -88* (WRECKED) (ex Skotta)	WBk	Norweg'n	663	159·9	31·9	18·4	Grimstad	1883	S.Johnsen	[illegible]	M.F.S[illegible]
85591	1010	V	**Loyalist** *Vroom 88-88*	WBn	British	348	130·5	29·4	13·0	Portland,N.B.	1883	F.E.Sayre	[illegible]	F.E.Sayre
	1011		**Loyo** *San Thome* (ex Arsenea)	W3mSr	Brazilian	242	100·0	24·3	11·5	F[illegible]	1876	Borda	[illegible]	J.daSilvaLoyo & Filho
	1012		**Lua**	WSr	Brazilian	150	..	..	..					V.A.e Magalhaes
	1013	N	**Luca** *Jansen 58-73*	WSr	Norweg'n	399	116·3	28·3	18·4	Baccari	1857			H.E.Jansen
89364	1014	V	**Lucania** *Gibson 72-86*	WS	British	1774	229·3	42·1	24·4	St.John,N.B.	1886	J.F.Fraser	[illegible]	B.Smith & Sons
	1015	G	**Lucas** (ex Lien) *Pocker 85-85*	WSr	German	159	94·9	24·1	11·1	Papenburg	1863	F.W.Meyer		G.& H.Aden
	1016	✻	**Lucayas** *Johansen*	IBk	Danish	445	150·1	25·5	18·1	Greenock	1869	Robertson & Co		Th.Meinertz's Enke
1528	1017	V	**Lucco** *Degregori -81* (BROKEN UP)	WBk	Italian	498	134·5	30·8	20·0	Sestri Ponente	1866	L.Briasco	[illegible]	B.Olivari
1974	1018	I	**Luchina C.** *Repetto*	WBk	Italian	506	134·5	31·2	19·8	Lavagna	1869	G.Saccomano	[illegible]	A.Cichero
70980	1019		**Lucia** *Wood 83-88* (ex Maria Fidele)	CBk	British	640	162·3	31·1	18·3	Sunderland	1868	J.Blumer & Co.	[illegible]	T.Roberts
532	1020		**Lucia** (ex Angiolina) *Carletta*	WBg	Italian	169	100·0	23·1	14·2	Leghorn	1863	G.Geri	[illegible]	Oreste Del Buono
101	1021		**Lucia**	WSr	Russian	126	89·7	23·6	10·4	Kurland	1876			Oskar von Stryk
15102	1022		**Lucia B. Ives** *Mott* (ex B. B. Parker)	WSr	American	177	96·8	27·2	8·5	Belleville,N.J	18 53/64			A.H.Mott
140498	1023	R	**Lucia Porter** *Grindle -86*	W3mSr	American	316	132·0	32·0	10·5	Belfast,Me.	1881		[illegible]	J.F.Grindel
	1024		**Lucie**	WSr	French	143	88·3	23·3	12·1	Dunkirk	1846			Connolet frères
	1025	N	**Lucie** *Eidsaag 75-87*	WSr	Norweg'n	116	95·2	24·0	10·0	Christiansand	1877		[illegible]	Parchus & Lossius
246	1026		**Lucie** *Johnsohn 82-82*	WSr	Russian	149	80·4	25·5	10·1	Kurbis	1877	Grass		K.Berg & J.Jansohn
	1027		**Lucie** *Hedberg -88* (ex Dagmar)	WBg	Swedish	214	104·0	22·8	12·0	Miramichi	1858			J.Pettersson
	1028		**Lucie Gabrielle** (DISMANTLED) (ex Peppino P.)	WBg	French	417	107·9	26·9	18·7	Sestri Ponente	1866	G.B.Cadenaccio		C.Guérin
	1029	V	**Lucie Marie** *Denis -88* (ex Ayna)	WBn	French	163	95·8	20·0	12·1	Finland	1858			Dupuis-Robial
	1030		**Lucie Marie** *Bertho* (BROKEN UP)	WSr	French	141	84·3	23·0	11·1	Nantes	1866	Tillo		D.Bertho
	1031	G	**LucieRadmann** *Gerth -8* (ex Susanne Johanna, ex Koning Vaderland)	WBk	German	447	117·6	29·9	18·9	Vlaardingen	1854	Von Dam		W.Lüdke & Co.
140060	1032		**Lucie Wheatley** *Pillsbury -88*	W3mSr	American	180	113·8	29·5	8·2	Frederica,Del.	1874	NathanielLank		J.W.Hall
	1033	V	**Lucien & Valentine** *Verdois* (ex Anna)	WBg	French	274	110·2	27·6	14·8	Arendal	1877		[illegible]	K.Morvant
522	1034	A	**Lucietta** (ex Giuseppe) *Musana*	WSr	Italian	124	80·7	22·4	12·1	Gaeta	1858		[illegible]	Ant.Burgio & Co.
66835	1035	V	**Lucile** *Williams*	WBkn	British	498	128·8	31·1	17·0	Tynem'th,NB	1872	R.Lowitt	[illegible]	A.J.Williams
75393	1036		**Lucile**	WScYt	British	157	117·0	23·1	13·4	Gosport	1876	Camper&Nicholson		N.Rudge
140031	1037	V	**Lucile** *Sermon*	WS	American	1329	200·2	40·0	23·9	Freeport,Me.	1874	E.C.Soule	[illegible]	E.C.Soule
71144	1038	✻	**Lucille** *Kirby -86*	WBg	British	278	117·6	26·1	14·9	P.E.Island	1874	Yeo	[illegible]	W.Williams & Sons
57022	1039		**Lucinda** *White 88-88*	WBn	British	199	105·8	24·1	12·9	P.E.Island	1867	Tuplin		H.W.Brightman
140550	1040		**Lucinda G. Potter** *Potter -88*	W3mSr	American	613	151·8	35·8	15·9	Bath,Me.	1882	Goss&Sawyer		W.T.Potter
	1041		**Lucinde** *Jensen 87-87*	WBg	Danish	150	86·5	23·7	13·4	Holbæk	1842	V.Lageni		Den.Kgl.Grønl.Handel
	1042	DV	**Lucinde** *Krakauschke 86-*	WBk	Russian	319	117·0	28·0	14·0	Libau	1866	J.D.Dargewitz		C.Essmann
	1043		**Lucinde** *Hansson*	WSr	Swedish	102	76·8	20·0	10·8	[illegible]	1861			M.Hansson
231	1044		**Lucio** *Martinolich -8* (ex Maria Regina)	WBg	Austrian	254	86·0	26·5	16·1	[illegible]	1870		[illegible]	Ivcevich & Co.
	1045	V	**Lucio I.** *Vaz*	WSr	Portug'se	150	89·6	21·5	10·[illegible]	[illegible]	1884	C.M.deAraujo	[illegible]	V.L.de Almeida
90065	1046	✻	**Lucipara** *Harder 86-88*	I4mS	British	1863	267·1	40·2	23·8	[illegible]	1885	Russell&Co.		P.Denniston & Co.
87113	1047	✻	**Lucknow** *Lidstone 81-87*	StlS	British	1440	245·3	37·1	22·[illegible]	[illegible]	1883	Doxford&Sons		Tyser & Co.
70485	1048	✻	**Lucy** *Davies*	WBkn	British	192	112·8	24·1	12·5	Ipswich	1879	Bayley	[illegible]	J.C.Peacock
140315	1049	R	**Lucy** *Cook*	WSr	American	297	130·0	30·0	8·0	Berwick,Me.	1882			American Salt Co.
15046	1050		**Lucy** *Wooster*	WSr	American	129	87·5	27·8	9·1	Eastport,Me.	1864	Huston		L.A.Godfrey

✻ = **Lloyd's Register**; D = Nederlandsche Vereeniging; G = Germanischer Lloyd; I = Registro Italiano; R = Ameri[can]; A = Veritas Austro-Ungarico; H = Veritas Ellenico; N = Norske Veritas; V = Bureau [Veritas]

In the 4th Column, C = Composite (iron frame, wood planked); I = Iron; Stl = Steel; W = Wood.

The Dates of Metalling (**Col. 11**) are based on the records of the Books in which Vessels are respectively classed, or on the Reports of the Surveyors of [illegible]

Further particulars of Vessels classed by Lloyd's Register will be found in Lloyd's Register Book and Lloyd's Yacht Register.

No.	Cote	Nom, Matériaux et Gréement. Capitaine et dates d'entrée en service des Armateurs actuels et à bord du navire. Ancien Nom.	Mat. et Gréem.	Pavillon.	Tonnage net.	Longueur	Largeur	Creux	Où.	Quand.	Par qui.	Date du doublage en métal	Armateurs ou Directeurs-Gérants.	Port d'Armement.
1051		Lucy *Lilord* (ex Tan Pieker)	WSr	Dutch	355	120·0	27·9	12·5	Harb'rvilleNS	1873	J.Givan		Kali Anam Maat-schappy	Batavia
1052	*	Lucy *Klausen* (ex Lucy Rose)	WBk	Norweg'n	739	176·2	33·0	19·5	P.E.Island	1875	J.C.Pope	Y 4.88	Actieselskabet"Lucy" (C.Møller)	Christiania
1053	DN	Lucy *Röer 79-79*	WBg	Norweg'n	312	120·5	27·9	13·2	Cornwall,Can.	1865			Martin Röer	Dröbak
1054		Lucy A. Blossom	W3mSr	American	179	108·0	24·0	9·0	ConneautOhio	1846			J.Philbrook	Winterport,Me.
	R	Lucy A. Davis *Lewis -87*	W3mSr	American	596	148·5	34·2	12·6	Yarmouth,Me	1882	Htchns&Stubbs	Y 9.87	J.W.Davis	Portland,Me.
	R	Lucy A. Nickels *Nichols*	WS	American	1395	199·0	37·9	23·8	Brewer,Me.	1875	McGilvery	Y 7.87	J.C.Nickels	Searsport,Me.
	R	Lucy E. Friend *Rollins*	W3mSr	American	447	147·0	34·6	12·6	Nwbrypt Mass	1882	G.E.Currier		W.H.Collins	Gloucester,Mass.
	N	Lucy Frances *Eskild 1 -85*	W3mSr	Norweg'n	232	114·6	27·0	11·2	Brookville,Me	1856		Y 3.89	Ludwigsen & Schjel-derup	Christiania
1059	R	Lucy H. Russell *Russell -8*	W4mSr	American	1108	194·3	39·9	18·0	Bath,Me.	1887	NewEng.S.B.Co		H.M.Randall	Pt.Jefferson
1060		Lucy Hammond *Hammond*	WSr	American	150	96·5	26·3	8·7	E.Machias,Me.	1869			J.R.Talbot	E.Machias,Me.
1061		Lucy M. Collins *Newcomb*	WSr	American	159	100·3	27·8	8·3	WinterportMe	1867			G.B.Hook	Bangor,Me.
	*	Lucy March *Griffiths -84*	WBk	British	244	118·0	26·6	14·1	Pwllheli	1876	Williams	Y 9.88	Morgan,Jones & Co.	Carnarvon
	*	LucyReppen *Bach -88* (ex Josef Brunick)	WBk	Norweg'n	892	180·6	35·3	20·5	[illegible]	1877	Brown	Y 12.87	O.C.Johansen	Christiansand
1064		Lucy Richmond *Eastwood*	WSrBarge	British	137	106·7	24·4	8·6	[illegible]	1875	Robertson		W.T.Richmond & Co.	Maldon
1065	*DI	Lucy W. *Vallo*	WBk	Italian	683	159·1	33·1	21·6	SestriPonente	1876	A.Calcagno	Y 4.89	P.Zenoglio	Genoa
1066		Lucy W. Snow *Simpson*	WBn	American	299	112·8	26·9	11·0	Rockland,Me.	1869	Snow&Farwell		F.C.Butman	Boston,Mass.
	*	Lucy and Paul *Andreas 79-89*	WBk	German	377	127·9	28·7	15·9	Elbing	1861	C.G.Fechter		E.Burchard	Rostock
1068		Ludolfina Jantina *Pygma*	WBn	Dutch	156	95·0	23·9	10·6	Pekela	1874	L.H.DeWijk		H.J.Kuiper (H.L.deWijk, Manager)	Oude Pekela
1069		Ludovic	WSr	French	120	89·6	19·7	10·9	St.Malo	1878	Tranchemer frs.		Vve. H.Ruellan	Paimpol
1070	N	Ludvig *Larsen 82-82* (ABANDONED)	WBk	Norweg'n	302	121·1	27·4	13·5	Brevig	1870	G.Moe		G.J.Sörensen	Sandefjord
1071	N	Ludvig *Mathisen*	WSr	Norweg'n	110	83·6	20·9	9·6	Brake	1851			Chr.& John Mathisen	Horten
1072	DNV	Ludvig Holberg *Hansen 81-86*	WBk	Norweg'n	1019	193·1	37·5	21·7	Bergen	1878	Brunchorst& Dekke	Y 8.86	A.F.Klaveness	Sandefjord
1073		Ludvig (MISSING) *Schöuer*	WBk	German	368	128·6	26·6	13·9	Greifswald	1866	C.Dolt		J.Rudolph	Stettin
1074		Ludwig *Eichen -80*	WSr	Russian	126	82·0	21·0	9·5	Haynasch	1872	M.Eichen		A.Eichen & F.Weide	Haynasch
1075		LudwigCapobus,sen. *Permen 8-81*	WBg	German	242	98·3	26·2	11·8	[illegible]	1863	W.Zeltz		A.Burchard	Rostock
1076	R	Luella A. Snow *Hoss*	WSr	American	178	106·7	27·0	8·4	Rockland,Me.	1875	Snow&Farwell	Y 7.86	Snow & Pearsons	Rockland,Me.
1077	*	Lufra *Rogers*	CBk	British	672	179·7	31·1	17·9	Aberdeen	1870	Hall	Y 8.87	A.McGregor	Hobart
1078		Lufra	WKYt	British	131	108·1	22·0	12·2	Cowes	1865	M.Ratsey		Exrs. of the Duke of Rutland, K.G.	Cowes
1079	N	Lufra *Evensen*	WBk	Norweg'n	459	133·6	31·5	17·6	Grimstad	1874	O&L Dömmes-moen	Y 8.88	J.E.Einersen	Grimstad
1080	R	Lugano *Clark*	WSr	American	166	104·5	28·1	9·0	E.Machias,Me	1867			J.G.Murchie & Sons	New York
1081	*I	Luigi (ex Caterina B.) *Cuneo*	WBk	Italian	828	168·7	33·7	23·6	Finale	1877	F.Ponziolo	Y 1.89	F.Cuneo fu A.	Genoa
1082	A	Luigi *Scotticia*	WBk	Italian	511	..	..	..	Brooklyn	1865	Th.Stach	Y 6.83	L.Pirandello	Messina
1083		Luigi *Corradini*	WBg	Italian	193	111·8	22·1	14·1	Capo d'Istria	1864			A.Bini & Co.	Genoa
	A	Luigi *Di Pace -88*	WBn	Italian	141	78·7	23·8	13·3	Castellamare di Stabia	1863		I 1.89	F.Troise	Naples
1085	I	Luigi Accame *Olivari -86*	WBk	Italian	939	176·8	35·4	22·6	SestriPonente	1870	G.Casanova	Y 7.83	Em.Accame fu L.	Genoa
1086		Luigi C. *Tancredi* (ex Vittoria)	WBk	Italian	249	107·6	26·6	16·3	Leghorn	1853	G.Micheli		E.Pieruzzini	Leghorn
1087	A	Luigi Fortunato *Ghezzo -86*	WTrb	Italian	117	83·0	24·4	10·6	Chioggia	1881	L.Menetto	Z 8.81	F^lli Ghezzo & Co.	Venice
	I	Luigi P. *Moglia -87*	WBk	Italian	657	143·5	32·4	21·0	Foce	1873	B.Canepa	Y 8.88	G.Moglia	Genoa
	I	Luigi Padre *Olivari 1 -86* (ex Olivari Padre)	WBk	Italian	529	140·7	31·2	20·7	Savona	1875	F.Calamaro	Y 12.87	F^lli Sciaccaluga	Genoa
		Luigi Padre (WRECKED) (ex M.Padre ex Angiolina)	WBn	Italian	121	71·2	23·2	13·1	Varazze	1855			C.A.Giorgetti	Leghorn
		Luigi Repetto *Repetto*	WBk	Italian	220	101·7	25·6	14·8	St.Valery	1853			Fort Repetto	Genoa
	I	Luigi Ruggero *Russo -88*	WBk	Italian	490	131·2	29·2	18·7	Castellamare	1873	L.Trambarulo	Y 3.88	L.Ruggero	Castellamare
		Luigia *Bennati*	WBk	Italian	402	122·3	29·7	18·4	SestriPonente	1866	G.Peragallo		F^lli Pendibene	Genoa
		Luigia (ex Luigi) *Calabrese*	WBg	Italian	231	93·5	26·1	16·2	Piraeus	1867			G.B.Borzone fu G.	Leghorn
		Luigia F. (ex Costantina) *Villa*	W	Italian	491	147·6	30·7	19·7	SestriPonente	1866	L.Casella		B.Fravega fu G.B.	Genoa
	I	Luigia M. *Raimondi -88* (ex Emilia ex Bianca)	WBk	Italian	421	128·9	30·2	17·9	SestriPonente	1880	A.Briasco fu GB	Y 8.88	E.Massone	Genoa
	I	Luigia R. *Dareggio -88* (ex Orfeo)	WBn	Italian	498	137·8	30·0	18·7	Chiavari	1867	M.Tappani	Y 8.88	F.Rainusso	Genoa
	IV	Luigia Raffo *Guilio*	WBk	Italian	972	183·4	38·1	23·1	Chiavari	1875	M.Tappani	Y 3.86	Erasmo Raffo fu A.	Spezzia
	I	Luigia Rocca *Rebaffo*	WBk	Italian	546	150·2	32·3	19·7	Lavagna	1870	D.Briasco	Y 8.88	Rocca & Lavello	Genoa

* = Lloyd's Register; A = Veritas Austro-Ungarico; D = Nederlandsche Vereeniging; G = Germanischer Lloyd; H = Veritas Ellenico; I = Registro Italiano; N = Norske Veritas; R = American Record; V = Bureau Veritas.

Dans la 4me Colonne, C = Composite (membrure en fer, bordé en bois); I = Fer; Stl = Acier; W = Bois.

Les Dates de doublage en métal (Col. 11) ressortent des livres dans lesquels les navires sont respectivement classés ou des rapports des Inspecteurs du Lloyd's Register.

Pour plus amples renseignements sur les navires classés par le Lloyd's Register, voir le Lloyd's Register Book et le Lloyd's Yacht Register.

Official No.	No. in Book	Classed	Ship's Name. Master, and Dates of Appointment to present Owner's Service and to Ship. Ship's Late Name.	Material and Rig	Flag	Net Tonnage	Length	Breadth	Depth	Built: Where	Built: When	Built: By whom	Date of Metalling	Owners or Managers
	1001	DNV	Lovspring *Thomsen*	WBk	Norweg'n	547	149·4	30·9	17·8	Christiansand	1873	H.Olsen	V 2.88	H.G.Kirsebom
	1002	N	Lovspring *Schive -88*	WBk	Norweg'n	489	136·4	31·0	16·9	Grimstad	1885	N.O.Gjömle	11.85	J.Jörgensen
	1003	N	Lovspring *Thinnesen*	WBg	Norweg'n	225	107·5	26·9	12·2	Porsgrund	1872			P.M.Petersen & Sön
72385	1004	RV	Low Wood *Thurber -81*	WBk	British	1091	186·7	37·6	22·4	Portland,N.B.	1878	D.Lynch	V 2.88	H.D.Troop
60478	1005		Lowestoft *Hook*	WBk	British	293	112·7	26·5	16·5	Sunderland	1866	J.Dennison		R.L.Manners
76369	1006	✳	Loweswater *Smith -88*	I Bk	British	603	174·0	30·2	17·5	Whitehaven	1876	Whitehvn SBCo		Jackson,Metcalf & Co
91291	1007	✳	Lowther Castle *Southborough 84-86*	I S	British	1824	262·1	39·0	23·3	Workington	1886	R.Williamson & [Son		Lancaster Shipowners Co.,Ld.
	1008	N	Loyal *Friis -87* (ex Drammensdahl) (WRECKED)	WS	Norweg'n	1427	220·2	38·9	24·7	[illegible]	1863	Rosa		Rederselskab "Loyal" (Chr.Ellefsen)
	1009	D	Loyal *Knudsen -88* (WRECKED) (ex Skaalo)	WBk	Norweg'n	663	159·9	31·9	18·4	Grimstad	1883	S.Johnsen	V 3.86	M.F.Stray
85394	1010	V	Loyalist *Froom 83-88*	WBn	British	348	130·5	29·4	13·0	Portland,N.B.	1883	F.E.Sayre	V 8.88	F.E.Sayre
	1011		Loyo *San Thome* (ex Alberto)	W3mSr	Brazilian	242	100·0	24·3	11·5	[illegible]	1876	Borda		[illegible]
	1012		Lua	WSr	Brazilian	150	..	..	..					V.A.e Magalhaes
	1013	N	Luca *Jansen 58-75*	WSr	Norweg'n	399	116·3	28·3	18·4	Buccari	1857			H.E.Jansen
88394	1014	V	Lucania *Gibson 72-86*	WS	British	1774	229·3	42·1	24·4	St.John,N.B.	1886	J.F.Fraser	11.86	B.Smith & Sons
	1015	G	Lucas (ex Lina) *Pecker 85-88*	WSr	German	159	91·9	21·1	11·1	Papenburg	1863	F.W.Meyer		G.& H.Aden
	1016	✳	Lucayas *Johansen*	I Bk	Danish	445	150·4	25·5	18·1	Greenock	1869	Robertson & Co		Th.Meinertz's Enke
1328	1017	V	Lucco *Degregori -87* (BROKEN UP)	WBk	Italian	498	131·5	30·8	20·0	Sestri Ponente	1866	L.Briasco	V 1.86	B.Olivari
1974	1018	I	Luchina C. *Repetto*	WBk	Italian	506	134·5	31·2	19·8	Lavagna	1869	G.Saccomanno	V 7.86	A.Cichero
70080	1019		Lucia *Wood 82-86* (ex Maria Fidelo)	C Bk	British	640	162·3	31·1	18·3	Sunderland	1868	J.Blumer & Co.	10.85	T.Roberts
552	1020		Lucia (ex Angiolina) *Carletta*	WBg	Italian	169	100·0	23·1	14·2	Leghorn	1868	G.Gori	V 8.85	Oreste Del Buono
101	1021		Lucia	WSr	Russian	126	89·7	23·6	10·4	Kurland	1876			Oskar von Strÿk
15102	1022		Lucia B. Ives *Mott* (ex E. H. Parker)	WSr	American	177	96·8	27·2	8·5	Belleville,N.J	18 63/64			A.H.Mott
140108	1023	R	Lucia Porter *Grindle [-81*	W3mSr	American	316	132·0	32·0	10·5	Belfast,Me.	1881		11.88	J.F.Grindel
	1024		Lucie	WSr	French	143	88·3	23·3	12·1	Dunkirk	1846			Comolet frères
	1025	N	Lucie *Eidsaay 75-87*	WSr	Norweg'n	116	95·2	24·0	10·0	Christiansand	1877		V 1.87	Parelius & Lossius
246	1026		Lucie *Johnsohn 82-83*	WSr	Russian	149	80·4	25·5	10·1	Kürbis	1877	Grass		K.Berg & J.Jansohn
	1027		Lucie *Hedberg -88* (ex Dagmar)	WBg	Swedish	214	104·0	22·8	12·0	Miramichi	1858			J.Pettersson
	1028		Lucie Gabrielle (DISMANTLED) (ex Peppino P.)	WBg	French	417	107·9	26·9	18·7	Sestri Ponente	1866	G.B.Cadenaccio		C.Guérin
	1029	V	Lucie Marie *Denis -88* (ex Aylia)	WBn	French	163	95·8	20·0	12·1	Finland	1858			Dupuis-Robial
	1030		Lucie Marie *Bertho* (BROKEN UP)	WSr	French	141	84·3	23·0	11·1	Nantes	1866	Tillé		D.Bertho
	1031	G	LucieRadmann *Gerth [-88* (ex Susanne Johanna, ex Koning Vaderland)	WBk	German	447	117·6	29·9	18·9	Vlaardingen	1854	Von Dam		W.Lüdke & Co.
140063	1032		Lucie Wheatley *Pettingill -82*	W3mSr	American	180	113·8	29·5	8·2	Frederica,Del.	1874	NathanielLank		J.W.Hall
	1033	V	Lucien & Valentine *Verdus* (ex Anna)	WBg	French	274	110·2	27·6	14·8	Arendal	1877		2.88	K.Morvant
922	1034	A	Lucietta (ex Giuseppe) *Mutana*	WSr	Italian	124	80·7	22·4	12·1	Gaeta	1858		4.84	Ant.Burgio & Co.
[illegible]	1035	V	Lucile *Williams*	WBkn	British	498	128·8	31·1	17·0	Tynem'th,NB	1872	R.Lowitt	V 1.88	A.J.Williams
[illegible]	1036		Lucile	WSrYt	British	157	117·0	23·1	13·1	Gosport	1876	Camp&Nichlsn		N.Rudge
[illegible]	1037	V	Lucile *Serman*	WS	American	1329	200·2	40·0	23·9	Freeport,Me.	1874	E.C.Soule	1.85	E.C.Soule
[illegible]	1038	✳	Lucille *Kirby -86*	WBg	British	278	117·6	26·1	14·9	P.E.Island	1874	Yeo	V 6.87	W.Williams & Sons
[illegible]	1039		Lucinda *White 86-88*	WBn	British	199	105·8	24·1	12·9	P.E.Island	1867	Tuplin		H.W.Brightman
[illegible]	1040		Lucinda G. Potter *Potter -88*	W3mSr	American	613	151·8	35·8	15·9	Bath,Me.	1882	Goss&Sawyer		W.T.Potter
	1041		Lucinde *Jensen 87-87*	WBg	Danish	150	86·5	23·7	13·4	Holbæk	1842	V.Lagoni		Den.Kgl.Grønl.Handel
	1042	DV	Lucinde *Krakauschke 85-*	WBk	Russian	319	117·0	28·0	14·0	Libau	1866	J.D.Dargewitz		C.Essmann
	1043		Lucinde *Hansson*	WSr	Swedish	102	76·8	26·0	10·8	Flensburg	1861			M.Hansson
251	1044		Lucio *Martinolich -87* (ex Mare Regina)	WBk	Austrian	254	86·0	26·5	16·1	Prà,Italy	1870		11.87	Ivcevich & Co.
	1045	V	Lucio I. *Vaz*	WSr	Portug'se	150	89·6	21·5	10·0	Villa do Conde	1884	C.M.de Araujo	1885	V.L.de Almeida
50065	1046	✳	Lucipara *Harder 86-88*	I 4mS	British	1863	267·1	40·2	23·8	Greenock	1885	Russell&Co.		P.Denniston & Co.
87113	1047	✳	Lucknow *Lidstone 81-87*	StlS	British	1440	245·3	37·1	22·5	Sunderland	1883	Doxford&Sons		Tyser & Co.
79483	1048	✳	Lucy *Davies*	WBkn	British	192	112·8	24·1	12·5	Ipswich	1879	Bayley	10.87	J.C.Peacock
140515	1049	R	Lucy *Cook*	WSr	American	297	130·0	30·0	8·0	Berwick,Me.	1882			American Salt Co.
15046	1050		Lucy *Wooster*	WSr	American	129	87·5	27·8	9·1	Eastport,Me.	1864	Huston		L.A.Godfrey

✳=Lloyd's Register; D=Nederlandsche Vereeniging; G=Germanischer Lloyd; I=Registro Italiano; R=American [illegible]
A=Veritas Austro-Ungarico; H=Veritas Ellenico; N=Norske Veritas; V=Bureau [illegible]

In the 4th Column, C=Composite (iron frame, wood planked); I=Iron; Stl=Steel; W=Wood.

The Dates of Metalling (**Col. 11**) are based on the records of the Books in which Vessels are respectively classed, or on the Reports of the Surveyors of [illegible]

Further particulars of Vessels classed by Lloyd's Register will be found in Lloyd's Register Book and Lloyd's Yacht Register.

		Nom. Matériaux et Gréement. Capitaine et dates d'entrée en service des Armateurs actuels et à bord du navire. Ancien Nom.			Pavillon.		Dimensions.			Construit.				Armateurs ou Directeurs-Gérants.	Port d'Armement.
		4			5	6	7			8	9	10	11	12	13
						Tonnage	Longueur	Largeur	Creux	Où.	Quand.	Par qui.	Date du doublage métal.		
1051		Lucy	Lliegt -86	W Sr	Dutch	355	120·0	27·9	12·5	Harb'rvilleNS	1873	J.Givan		Kali Anaia Maatschappy	Batavia
1052	✻	Lucy	Klemn	W Bk	Norweg'n	739	176·2	33·0	19·5	P.E.Island	1875	J.C.Pope	V 4.86	Actieselskabet "Lucy" (C. Møller)	Christiania
1053	DN	Lucy	Röer 79-79	W Bg	Norweg'n	312	120·5	27·9	13·2	Cornwall,Can.	1865			Martin Röer	Dröbak
1054		Lucy A. Blossom		W3mSr	American	179	108·0	24·0	9·0	ConneautOhio	1846			J.Philbrook	Winterport,Me.
1055	R	Lucy A. Davis	Davis -81	W3mSr	American	596	148·5	34·2	12·6	Yarmouth,Me	1882	Htchns&Stubbs	V 9.87	J.W.Davis	Portland,Me.
1056	R	Lucy A. Nickels	Nichols	W S	American	1395	199·0	37·9	23·8	Brewer,Me.	1875	McGilvery	V 7.87	J.C.Nickels	Searsport,Me.
1057	R	Lucy E. Friend	Collins	W3mSr	American	447	147·0	34·6	12·6	Nwbrypt Mass	1882	G.E.Currier		W.H.Collins	Gloucester,Mass.
1058	N	Lucy Frances	Stoll	W3mSr	Norweg'n	232	114·6	27·0	11·2	Brooksville,Me	1856		V 2.88	Ludwigsen & Schjelderup	Christiania
1059	R	Lucy H. Russell	Russell	W 3mSr	American	1108	194·3	39·9	18·0	Bath,Me.	1887	NewEng.S.B.Co		H.M.Randall	Pt.Jefferson
1060		Lucy Hammond		W Sr	American	150	96·5	26·3	8·9	E.Machias,Me.	1869			J.R.Talbot	E.Machias,Me.
1061		Lucy M. Collins	Newcomb	W Sr	American	159	100·3	27·8	8·3	WinterportMe	1867			G.B.Hook	Bangor,Me.
1062	✻	Lucy March	Griffiths	W Bkn	British	244	118·0	26·6	14·1	Pwllheli	1876	Williams	V 9.8	Morgan,Jones & Co.	Carnarvon
1063	✻	LucyReppen	Koch	W Bk	Norweg'n	892	180·6	35·3	20·5	N.Bruns.	1877	Brown	V 12.87	O.C.Johansen	Christiansand
1064		Lucy Richmond	Richmond	W SrBarge	British	137	106·7	24·4	8·6	Ipswich	1875	Robertson		W.T.Richmond & Co.	Maldon
1065	✻ DI	Lucy W.	Vallo	W Bk	Italian	683	159·1	33·1	21·6	SestriPonente	1876	A.Calcagno	V 4.86	P.Zenoglio	Genoa
1066		Lucy W. Snow		W Bn	American	299	112·8	26·9	14·0	Rockland,Me.	1869	Snow&Farwell		F.C.Butman	Boston,Mass.
1067	V	Lucy and Paul	Andreis 1985	W Bk	German	377	127·9	28·7	15·9	Elbing	1861	C.G.Fechter		E.Burchard	Rostock
1068		Ludolfina Jantina	Poppen -88	W Bn	Dutch	156	95·0	23·0	10·6	Pekela	1874	L.H.DeWijk		H.J.Kuiper (H.L.de Wyk, Manager)	Oude Pekela
1069		Ludovic		W Sr	French	120	89·6	19·7	10·9	St.Malo	1878	Tranchemer frs.		Vve. H.Ruellan	Paimpol
1070	N	Ludvig (ABANDONED)	Larsen 82-88	W Bk	Norweg'n	302	124·1	27·4	13·5	Brevig	1870	G.Moe		G.J.Sörensen	Sandefjord
1071	N	Ludvig	Mathisen	W Sr	Norweg'n	110	83·6	20·9	9·6	Brake	1851			Chr.& John Mathisen	Horten
1072	DNV	Ludvig Holberg	Hansen 81-86	W Bk	Norweg'n	1019	193·1	37·5	21·7	Bergen	1878	Brunchorst& Dekke	V 8.88	A.F.Klaveness	Sandefjord
1073		Ludvig (russisch)	Schaver	W Bk	German	368	128·6	26·6	13·9	Greifswald	1865	C.Dodt		J.Rudolph	Stettin
1074		Ludwig	Eichen -89	W Sr	Russian	126	82·0	21·6	9·5	Haynasch	1872	M.Eichen		A.Eichen & F.Weide	Haynasch
1075		LudwigCapobus,sen.	Persian 81-81	W Bg	German	242	98·3	26·2	14·8	Rostock	1863	W.Zeltz		A.Burchard	Rostock
1076	R	Luella A. Snow	Pose	W Sr	American	178	106·7	27·0	8·4	Rockland,Me.	1875	Snow&Farwell	V 7.81	Snow & Pearsons	Rockland,Me.
1077	✻	Lufra	Rogers	C Bk	British	672	179·7	31·1	17·9	Aberdeen	1870	Hall	V 6.87	A.McGregor	Hobart
1078		Lufra		W K Yt	British	131	108·1	22·0	12·2	Cowes	1865	M.Ratsey		Exrs. of the Duke of Rutland, K.G.	Cowes
1079	N	Lufra	Evensen	W Bk	Norweg'n	489	133·6	31·3	17·6	Grimstad	1874	O&L Dömmesnoen	V 3.88	J.E.Einersen	Grimstad
1080	R	Lugano	Clark	W Sr	American	166	104·5	28·1	9·0	E.Machias,Me	1867			J.G.Murchie & Sons	New York
1081	✻ I	Luigi (ex Capelan B.)	Cuneo	W Bk	Italian	828	168·7	33·7	23·6	Finale	1877	F.Pouziolo	V 1.86	F.Cuneo fu A.	Genoa
1082	A	Luigi	Scattiola	W Bk	Italian	511	..	..	..	Rosskelen	1865	Th.Stach	V 6.83	L.Pirandello	Messina
1083		Luigi	Corradini	W Bg	Italian	193	111·8	22·1	14·1	Capo d'Istria	1864			A.Bini & Co.	Genoa
1084	A	Luigi	Di Pace -88	W Bn	Italian	141	78·7	23·3	13·3	Castellamare di Stabia	1863		Z 1.88	F.Troise	Naples
1085	I	Luigi Accame		W Bk	Italian	939	176·8	35·1	22·6	SestriPonente	1870	G.Casanova	V 7.85	Em.Accame fu L.	Genoa
1086		Luigi C.	Tancredi	W Bk	Italian	249	167·6	26·6	16·3	Leghorn	1853	G.Micheli		E.Pieruzzini	Leghorn
1087	A	Luigi Fortunato	Ghezzi -88	W Trb	Italian	117	83·0	24·4	10·6	Chioggia	1881	L.Menetto	Z 8.81	Flli Ghezzo & Co.	Venice
1088	I	Luigi P.	Magi'lo -87	W Bk	Italian	657	143·5	32·4	21·0	Prà	1873	B.Canepa	V 8.88	G.Moglia	Genoa
1089	I	Luigi Padre	Olivari -88	W Bk	Italian	529	140·7	31·2	20·7	Savona	1875	F.Calamaro	V 12.87	Flli Sciaccaluga	Genoa
1090		Luigi Padre		W Bn	Italian	121	71·2	23·2	13·4	Varazze	1855			C.A.Giorgetti	Leghorn
1091		Luigi Repetto	Repetto	W Bk	Italian	220	101·7	25·6	14·8	St.Valery	1853			Fort Repetto	Genoa
1092	I	Luigi Ruggero	Marino -88	W Bk	Italian	490	131·2	29·2	18·7	Castellamare	1873	L.Trambarulo	V 3.88	L.Ruggero	Castellamare
1093		Luigia	Bennati	W Bk	Italian	402	122·3	29·7	18·4	SestriPonente	1866	G.Peragallo		Flli Pendibene	Genoa
1094		Luigia (ex Zeus)	Chiabrese	W Bg	Italian	231	93·5	26·1	16·2	Piræus	1867			G.B.Borzone fu G.	Leghorn
1095		Luigia F.	Consiglio Viola	W	Italian	491	147·6	30·7	19·7	SestriPonente	1866	L.Casella		B.Fravega fu G.B.	Genoa
1096	I	Luigia M.	Bassano -88	W Bk	Italian	421	128·9	30·2	17·9	SestriPonente	1880	A.Briasco fu G.B.	V 5.88	E.Massone	Genoa
1097	I	Luigia R. (ex Port Said)	-88	W Bn	Italian	498	137·8	30·0	18·7	Chiavari	1867	M.Tappani	V 5.88	F.Raimusso	Genoa
1098	IV	Luigia Raffo	Gallio	W Bk	Italian	972	183·4	38·1	23·1	Chiavari	1875	M.Tappani	V 3.86	Erasmo Raffo fu A.	Spezzia
1099	I	Luigia Rocca	Rebuffo	W Bk	Italian	546	150·2	32·3	19·7	Lavagna	1870	D.Briasco	V 5.88	Rocca & Lavello	Genoa
1100															

= Lloyd's Register; D = Nederlandsche Vereeniging; G = Germanischer Lloyd; I = Registro Italiano; R = American Record;
= Veritas Austro-Ungarico; H = Veritas Ellenico; N = Norske Veritas; V = Bureau Veritas.

Dans la 4me Colonne, C = Composite (membrure en fer, bordé en bois); I = Fer; Stl = Acier; W = Bois.

Les Dates du doublage en métal (Col. 11) ressortent des livres dans lesquels les navires sont respectivement classés ou des rapports des Inspecteurs du Lloyd's Register.

Pour plus amples renseignements sur les navires classés par le Lloyd's Register, voir le Lloyd's Register Book et le Lloyd's Yacht Register.

Official No.	No. in Book	Society in which classed	Ship's Name (Ship's Late Name)	Master, and Dates of Appointment	Material and Rig	Flag	Net Tonnage	Length	Breadth	Depth	Where built	When built	By whom	Date of Metalling	Owners or Managers	Port of Registry
	1001	DNV	Lovspring	Thomsen	W Bk	Norweg'n	547	149·4	30·9	17·8	Christiansand	1873	H.Olsen	Y [illegible]	H.G.Kirschom	[illegible]
	1002	N	Lovspring	Schare -88	W Bk	Norweg'n	489	136·4	31·0	16·9	Grimstad	1885	N.O.Gjessdale	Y 11.85	J.Jorgensen	[illegible]
	1003	N	Lovspring	Thomessen	W Bg	Norweg'n	225	107·5	26·9	12·2	Porsgrund	1872			P.M.Petersen & Son	Porsgrund
72835	1004	RV	Low Wood	Thurber -87	W Bk	British	1091	186·7	37·6	22·4	Portland,N.B.	1878	D.Lynch	Y [illegible]	H.D.Troop	St.John
58478	1005		Lowestoft	Hoy	W Bk	British	293	112·7	26·5	16·5	Sunderland	1866	J.Denniston		R.L.Manners	North S[illegible]
76380	1006	✳	Loweswater	Smith -88	I Bk	British	603	174·0	30·2	17·5	Whitehaven	1876	Whitehvn S.B.Co.		Jackson,Metcalf & Co.	Liverpool
91291	1007	✳	Lowther Castle	Scarborough 84-86	I S	British	1824	262·1	39·0	23·3	Workington	1886	R.Williamson & Son		Lancaster Shipowners' Co.Ld.	L[illegible]
	1008	N	Loyal (ex Greenland) (WRECKED)	Prits -87	W S	Norweg'n	1427	220·2	38·9	24·7	Quebec	1863	Rosa	Y [illegible]	Rederiselskab "Loyal" (Chr.Ellefsen)	[illegible]
	1009	D	Loyal (WRECKED) (ex Shasta)	Knudsen -88	W Bk	Norweg'n	663	159·9	31·9	18·4	Grimstad	1883	S.Johnsen	Y [illegible]	M.F.Stray	Chr[illegible]
85584	1010	V	Loyalist	Troop 83-88	W Bn	British	348	130·5	29·4	13·0	Portland,N.B.	1883	F.E.Sayre	Y [illegible]	F.E.Sayre	St.John
	1011		Loyo (ex Alferes)	San Thome	W3mSr	Brazilian	242	100·0	24·3	11·5	Fao	1876	Borda	Y [illegible]	I.daSilvaLoyo & Filh	Pe[illegible]
	1012		Lua		W Sr	Brazilian	150	..	..	..					V.A.e Magalhaes	Parnahyba
	1013	N	Luca	Jansen 58-75	W Sr	Norweg'n	399	116·3	28·3	18·4	Buccari	1857			H.E.Jansen	Pl[illegible]
[illegible]	1014	V	Lucania	Gibson 72-86	W S	British	1774	229·3	42·1	24·4	St.John,N.B.	1886	J.F.Fraser	Y 11.86	B.Smith & Sons	[illegible]
	1015	G	Lucas (ex Lina)	Packer 85-86	W Sr	German	159	91·9	21·1	11·1	Papenburg	1863	F.W.Meyer		G.& H.Aden	[illegible]
	1016	✳	Lucayas	Johansen	I Bk	Danish	445	150·4	25·5	18·1	Greenock	1869	Robertson & Co.		Th.Meinertz's Enke	Fanø
1528	1017	V	Lucco (BROKEN UP)	Degregori -87	W Bk	Italian	498	131·5	30·8	20·0	Sestri Ponente	1866	L.Briasco	Y [illegible]	B.Olivari	Genoa
1974	1018	I	Luchina C.	Repetto	W Bk	Italian	506	134·5	31·2	19·8	Lavagna	1869	G.Saccomanno	Y [illegible]	A.Cichero	Genoa
70680	1019		Lucin (ex Maria Fidele)	Wood 82-86	C Bk	British	640	162·3	31·1	18·3	Sunderland	1868	J.Blumer & Co.	Y 10.85	T.Roberts	[illegible]
552	1020		Lucia (ex Angiolina)	Carletta	W Bg	Italian	169	100·0	23·1	14·2	Leghorn	1863	G.Gori	Y [illegible]	Oreste Del Buono	Porto F[illegible]
101	1021		Lucia		W Sr	Russian	126	89·7	23·6	10·4	Kurland	1876			Oskar von Stryk	[illegible]
15102	1022		Lucia B. Ives (ex E. H. Parker)	Mott	W Sr	American	177	96·8	27·2	8·5	Belleville,N.J	1853/64			A.H.Mott	Pt.Jefferson
140498	1023	R	Lucia Porter	Grindle I -81	W3mSr	American	316	132·0	32·0	10·5	Belfast,Me.	1881		Y 11.86	J.F.Grindel	Belfast
	1024		Lucie		W Sr	French	143	88·3	23·3	12·1	Dunkirk	1846			Connolet frères	St.Servan
	1025	N	Lucie	Bidsaay 75-87	W Sr	Norweg'n	116	95·2	24·0	10·0	Christiansund	1877		Y [illegible]	Parelius & Lossius	Christiansund
246	1026		Lucie	Johnsohn 82-82	W Sr	Russian	149	80·4	25·5	10·1	Kürbis	1877	Grass		K.Berg & J.Jansohn	Riga
	1027		Lucie (ex Dagmar)	Hedberg -88	W Bg	Swedish	214	104·0	22·8	12·0	Miramichi	1858			J.Pettersson	[illegible]
	1028		Lucie Gabrielle (DISMANTLED) (ex Peppino P.)		W Bg	French	417	107·9	26·9	18·7	Sestri Ponente	1866	G.B.Cadenaccio		C.Guérin	Marseilles
	1029	V	Lucie Marie (ex Ayna)	Denis -88	W Bn	French	163	95·8	20·0	12·1	Finland	1858			Dupuis-Robial	St.Brieuc
	1030		Lucie Marie (BROKEN UP)	Bertho	W Sr	French	141	84·3	23·0	11·1	Nantes	1866	Tillé		D.Bertho	Nantes
	1031	G	LucieRadmann (ex Susanne Johanna ex Koning Vaderland)	Gerth I -88	W Bk	German	447	117·6	29·9	18·9	Vlaardingen	1854	Von Dam		W.Lüdke & Co.	Stettin
140069	1032		Lucie Wheatley	Pettingell -88	W3mSr	American	180	113·8	29·5	8·2	Frederica,Del.	1874	NathanielLank		J.W.Hall	W[illegible]
	1033	V	Lucien & Valentine (ex Anna)	Verdoux	W Bg	French	274	110·2	27·6	14·8	Arendal	1877		Y [illegible]	K.Morvant	[illegible]
522	1034	A	Lucietta (ex Giuseppe)	Musana	W Sr	Italian	124	80·7	22·4	12·1	Gaeta	1858		Y 4.84	Ant.Burgio & Co.	[illegible]
66835	1035	V	Lucile	Williams	W Bkn	British	498	128·8	31·1	17·0	Tynem'th,NB	1872	R.Lowitt	Y [illegible]	A.J.Williams	St.John
75389	1036		Lucile		W Sr Yt	British	157	117·0	23·1	13·4	Gosport	1876	Campr&Nichlsn		N.Rudge	[illegible]
140031	1037	V	Lucile	Sermen	W S	American	1329	200·2	40·0	23·9	Freeport,Me.	1874	E.C.Soule	Y 11.87	E.C.Soule	Portland
71441	1038	✳	Lucille	Kirby -88	W Bg	British	278	117·6	26·1	14·9	P.E.Island	1874	Yeo	Y [illegible]	W.Williams & Sons	Newport
57032	1039		Lucinda	White 86-88	W Bn	British	199	105·8	24·1	12·9	P.E.Island	1867	Puplin		H.W.Brightman	North S[illegible]
140530	1040		Lucinda G. Potter	Potter -88	W3mSr	American	613	151·8	35·8	15·9	Bath,Me.	1882	Goss&Sawyer		W.T.Potter	[illegible]
	1041		Lucinde	Jensen 87-87	W Bg	Danish	150	86·5	23·7	13·4	Holback	1842	V.Lageni		Den.Kgl.Gronl.Handel	Copenhagen
	1042	DV	Lucinde	Krakausch[illegible] 86-	W Bk	Russian	319	117·0	28·0	14·0	Libau	1866	J.D.Dargewitz		C.Essmann	Riga
	1043		Lucinde	Hansson	W Sr	Swedish	102	76·8	20·0	10·8	Flensburg	1861			M.Hansson	[illegible]
251	1044		Lucio (ex Maria Regina)	Martinolich -87	W Bg	Austrian	254	86·0	26·5	16·1	Pra,Italy	1870		Y [illegible]	Ivcevich & Co.	[illegible]
	1045	V	Lucio I.	Faz	W Sr	Portug'se	150	89·6	21·5	10·4	Villa do Conde	1884	C.M.de Araujo	Y [illegible]	V.L.de Almeida	Oporto
90003	1046	✳	Lucipara	Harder 86-88	I 4mS	British	1863	267·1	40·2	23·8	Greenock	1885	Russell&Co.		P.Denniston & Co.	Glasgow
87411	1047	✳	Lucknow	Lidstone 81-87	Stl S	British	1440	215·3	37·1	22·5	Sunderland	1883	Doxford&Sons		Tyser & Co.	[illegible]
[illegible]	1048	✳	Lucy	Davies	W Bkn	British	192	112·8	24·1	12·5	Ipswich	1879	Bayley	Y 10.87	J.C.Peacock	[illegible]
140513	1049	R	Lucy	Cook	W Sr	American	297	130·0	30·0	8·0	Berwick,Me.	1882			American Salt Co.	[illegible]
15046	1050		Lucy	Wooster	W Sr	American	129	87·5	27·8	9·1	Eastport,Me.	1864	Huston		L.A.Godfrey	[illegible]

✳ = **Lloyd's Register;** D = Nederlandsche Vereeniging; G = Germanischer Lloyd; I = Registro Italiano; R = American [illegible]; A = Veritas Austro-Ungarico; H = Veritas Ellenico; N = Norske Veritas; V = [illegible]

In the 4th Column, C = Composite (iron frame, wood planked); I = Iron; Stl = Steel; W = Wood.

The Dates of Metalling (Col. 11) are based on the records of the Books in which Vessels are respectively classed, or on the Reports of the Surveyors of [illegible]

Further particulars of Vessels classed by Lloyd's Register will be found in Lloyd's Register Book and Lloyd's Yacht Register.

No.		Nom. Capitaine et dates d'entrée au service des Armateurs actuels et à bord du navire. Ancien Nom.		Pavillon.	Tonnage net.	Dimensions.			Construit. Où.	Quand.	Par qui.	Date du doublage en métal	Armateurs ou Directeurs-Gérants.	Port d'Armement.
1051		Lucy *Litzau -86* (ex Sea Bird)	WSr	Dutch	353	120·0	27·9	12·5	Harb'rvilleNS	1873	[illegible]		Kali Anam Maatschappy	Batavia
1052	✻	Lucy *Klausen -88* (ex [illegible])	WBk	Norweg'n	732	176·2	35·0	19·5	P.E.Island	1875	J.C.Pope	Y 3.88	Actieselskabet "Lucy" (C.Mollerl)	Christiania
1053	DN	Lucy *Roer 79-79*	WBg	Norweg'n	312	120·5	27·9	13·2	Cornwall,Can.	1865			Martin Röer	Dröbak
1054		Lucy A. Blossom	W3mSr	American	179	108·6	24·0	9·0	ConneautOhio	1846			J.Phillbrook	Winterport,Me.
1055	R	Lucy A. Davis *Loring -87*	W3mSr	American	596	148·5	34·2	12·6	Yarmouth,Me	1882	Htchns&Stubbs	Y 2.87	J.W.Davis	Portland,Me.
1056	R	Lucy A. Nickels *Nichols*	WS	American	1385	199·0	37·9	23·8	Brewer,Me.	1875	McGilvery	Y 7.87	J.C.Nickels	Searsport,Me.
1057	R	Lucy E. Friend *Collins*	W3mSr	American	447	147·0	34·6	12·6	Nwbrypt Mass	1882	G.E.Currier		W.H.Collins	Gloucester,Mass
1058	N	Lucy Frances *Stabbe -85*	W3mSr	Norweg'n	232	111·6	27·0	11·2	Brookville,Me	1856		Y 3.82	Ludwigsen & Schjelderup	Christiania
1059	R	Lucy H. Russell *Russell -87*	W4mSr	American	1108	194·2	39·0	18·0	Bath,Me.	1887	NewEng.S.B.Co		H.M.Randall	Pt.Jefferson
1060		Lucy Hammond *Robinson*	WSr	American	159	96·5	26·3	8·9	E.Machias,Me.	1869			J.R.Talbot	E.Machias,Me.
1061		Lucy M. Collins *Newcomb*	WSr	American	159	100·3	27·8	8·3	WinterportMe	1867			G.B.Hook	Bangor,Me.
1062	✻	Lucy March *Griffiths -86*	WBkn	British	244	118·0	26·6	14·1	Pwllheli	1876	Williams	Y 8.8-	Morgan,Jones & Co.	Carnarvon
1063	✻	LucyReppen *Bach -88* (ex [illegible])	WBk	Norweg'n	892	180·6	35·3	20·5	N.Bruns.	1877	Brown	Y 12.87	O.C.Johansen	Christiansand
1064		Lucy Richmond *Richmond*	WSrBarge	British	137	106·7	24·4	8·6	Ipswich	1875	Robertson		W.T.Richmond & Co.	Maldon
1065	✻DI	Lucy W. *Pollo*	WBk	Italian	683	153·1	33·1	21·6	SestriPonente	1876	A.Cabagno	Y 4.86	P.Zenoglio	Genoa
1066		Lucy W. Snow *Sargent*	WBn	American	286	112·8	26·9	14·0	Rockland,Me.	1859	Snow&Farwell		F.C.Batman	Boston,Mass.
1067	V	Lucy and Paul *Andres 79-86*	WBk	German	377	127·9	28·7	15·9	Elbing	1861	C.G.Fechter		E.Burchard	Rostock
1068		Ludolfina Jantina *Poppen -80*	WBn	Dutch	156	95·0	23·0	10·6	Pekela	1874	L.H.De Wijk		H.J.Kuiper (H.L.deWyk,Manager)	Oude Pekela
1069		Ludovic	WSr	French	120	89·6	19·7	10·9	St.Malo	1878	Tranchemer frs.		Vve. H.Ruellan	Paimpol
1070	N	Ludvig *Larsen 85-88* (ABANDONED)	WBk	Norweg'n	392	124·1	27·4	13·5	Brevig	1870	G.Moe		G.J.Sörensen	Sandefjord
1071	N	Ludvig *Mathisen*	WSr	Norweg'n	110	83·6	20·9	9·6	Brake	1851			Chr.& John Mathisen	Horten
1072	DNV	Ludvig Holberg *Hansen 85-88*	WBk	Norweg'n	1019	193·1	37·5	21·7	Bergen	1878	Brauchorst& Dekke	Y 8.88	A.F.Klaveness	Sandefjord
1073		Ludvig (MISSING) *Schauer*	WBk	German	368	128·6	26·6	13·9	Greifswald	1855	C.Dodt		J.Rudolph	Stettin
1074		Ludwig *Eichen -80*	WSr	Russian	126	82·0	21·0	9·5	Haynasch	1872	M.Eichen		A.Eichen & F.Weide	Haynasch
1075		LudwigCapobus,sen. *Premon 84-84*	WBg	German	242	98·3	26·2	14·8	Rostock	1865	W.Zeltz		A.Burchard	Rostock
1076	R	Luella A. Snow *Foss*	WSr	American	178	106·7	27·0	8·4	Rockland,Me.	1875	Snow&Farwell	Y 7.8	Snow & Pearsons	Rockland,Me.
1077	✻	Lufra *Rogers*	CBk	British	672	179·7	31·1	17·9	Aberdeen	1870	Hall	Y 6.87	A.McGregor	Hobart
1078		Lufra	WKYt	British	131	108·1	22·0	12·2	Cowes	1865	M.Ratsey		Exrs. of the Duke of Rutland,K.G.	Cowes
1079	N	Lufra *Evensen*	WBk	Norweg'n	493	133·6	31·3	17·6	Grimstad	1874	O&I.Dimmesluoen	Y 1.86	J.E.Einersen	Grimstad
1080	R	Lugano *Clark*	WSr	American	166	104·5	28·1	9·0	E.Machias,Me	1887			J.G.Murchie & Sons	New York
1081	✻I	Luigi (ex Caterina B.) *Cuneo*	WBk	Italian	828	168·7	33·7	23·6	Finale	1877	F.Potenziolo	Y 1.89	F.Cuneo fu A.	Genoa
1082	A	Luigi *Scattola*	WBk	Italian	511	..	..	..	Brooklyn	1865	Th.Stach	Y 6.83	L.Pirandello	Messina
1083		Luigi *Corradosi*	WBg	Italian	193	111·8	22·1	14·1	Capod'Istria	1864			A.Bini & Co.	Genoa
1084	A	Luigi *Di Paco -85*	WBn	Italian	141	78·7	23·3	13·3	Castellamare [di Stabia	1863		Z 1.88	F.Troise	Naples
1085	I	Luigi Accame *Imbrie -88*	WBk	Italian	938	176·8	35·4	22·6	SestriPonente	1875	G.Cresanova	Y 7.85	Em.Accame fu L.	Genoa
1086		Luigi C. *Tancredi* (ex [illegible])	WBk	Italian	248	107·6	26·6	16·3	Leghorn	1853	G.Michelli		E.Pieruzzini	Leghorn
1087	A	Luigi Fortunato *Ghezzo -86*	WTrb	Italian	117	83·0	21·4	10·6	Chioggia	1851	L.Moretto	Z 8.86	Flli Ghezzo & Co.	Venice
1088	I	Luigi P. *Maglio -87*	WBk	Italian	657	143·5	32·4	21·0	Prà	1873	B.Canepa	Y 8.88	G.Moglia	Genoa
1089	I	Luigi Padre *Olivari -88* (ex Olivari Pietro)	WBk	Italian	529	140·7	31·2	20·7	Savona	1875	F.Calamaro	Y 12.87	Flli Sciaccaluga	Genoa
1090		Luigi Padre (WRECKED) (ex Maria G., ex Angiolina)	WBn	Italian	121	71·2	23·2	13·1	Varazze	1855			C.A.Giorgetti	Leghorn
1091		Luigi Repetto *Repetto*	WBk	Italian	229	101·7	25·6	14·8	St.Valery	1853			Fort. Repetto	Genoa
1092	I	Luigi Ruggero *Mazzeo -88*	WBk	Italian	490	131·2	29·2	18·7	Castellamare	1873	L.Tramborulo	Y 3.88	L.Ruggero	Castellamare
1093		Luigia *Bennati*	WBk	Italian	492	122·3	29·7	18·4	SestriPonente	1866	G.Peragallo		Flli Pendibene	Genoa
1094		Luigia (ex Gessy) *Calabrese*	WBg	Italian	231	93·5	26·1	16·2	Piræus	1867			G.B.Borzone fu G.	Leghorn
1095		Luigia F. (ex Consiglio) *Villa*	W	Italian	491	147·6	30·7	19·7	SestriPonente	1866	L.Casella		B.Fravega fu G.B.	Genoa
1096	I	Luigia M. *Rossoni -88* (ex Sturnegilda Gandvaso)	WBk	Italian	424	128·9	30·2	17·9	SestriPonente	1889	A.Briasco fu G.B.	Y 3.8-	E.Massone	Genoa
1097	I	Luigia R. *Bavegno -88* (ex Costantino)	WBn	Italian	498	137·8	30·0	18·7	Chiavari	1867	M.Tappani	Y 3.88	F.Rainusso	Genoa
1098	IV	Luigia Raffo *Cuneo*	WBk	Italian	972	183·4	38·1	23·1	Chiavari	1875	M.Tappani	Y 3.88	Erasmo Raffo fu A.	Spezzia
1099	I	Luigia Rocca *Rebuffo*	WBk	Italian	546	150·2	32·3	19·7	Lavagna	1870	P.Briasco	Y 8.86	Rocca & Lavello	Genoa

[illegible] Lloyd's Register; [illegible] Veritas Austro-Ungarico; D = Nederlandsche Vereeniging; G = Germanischer Lloyd; H = Veritas Ellenico; I = Registro Italiano; N = Norske Veritas; R = American Record; V = Bureau Veritas.

Dans la 4me Colonne, C = Composite (membrure en fer, bordé en bois); I = Fer; Stl = Acier; W = Bois.

Les Dates du doublage en métal (Col. 11) ressortent des livres dans lesquels les navires sont respectivement classés ou des rapports des Inspecteurs du Lloyd's Register.
Pour plus amples renseignements sur les navires classés par le Lloyd's Register, voir le Lloyd's Register Book et le Lloyd's Yacht Register.

Official No.	No. in Book	Society	Ship's Name. Master, and Dates of Appointment to present Owner's Service and to Ship. Ship's Late Name.	Material and Rig	Flag	Net Tonnage	Length	Breadth	Depth	Built Where	When	By whom	Date of Metalling	Owners or Managers
	1001	DNV	Lovspring *Thomsen*	WBk	Norweg'n	547	119·4	30·9	17·8	Christiansand	1873	H.Olsen	[illegible]	H.G.Kirsebom
	1002	N	Lövspring *Schive -88*	WBk	Norweg'n	489	136·4	31·0	16·9	Grimstad	1885	N.O.Gjömle	[illegible]	J.Jörgensen
	1003	N	Lovspring *Thönnesen*	WBg	Norweg'n	225	107·5	26·9	12·2	Porsgrund	1872			P.M.Petersen & Sön
7233[illegible]	1004	RV	Low Wood *Thurber -87*	WBk	British	1091	186·7	37·6	22·4	Portland,N.B.	1878	D.Lynch	[illegible]	H.D.Troop
[illegible]	1005		Lowestoft *Hook*	WBk	British	293	112·7	26·5	16·5	Sunderland	1866	J.Dennison		R.L.Manners
7038[illegible]	1006	✻	Loweswater *Smith -88*	IBk	British	603	174·0	30·2	17·5	Whitehaven	1876	WhitehvnSBCo		Jackson,Metcalf & Co
91[illegible]	1007	✻	Lowther Castle *Scarborough 81-86*	IS	British	1824	262·1	39·0	23·3	Workington	1886	R.Williamson & [Son		Lancaster Shipowners' Co. Ld.
	1008	N	Loyal *Friis -87* (ex [illegible]) WRECKED	WS	Norweg'n	1427	220·2	38·9	24·7	Quebec	1863	Rosa	[illegible]	Rederiselskab "Loyal" (Chr. Ellefsen)
	1009	D	Loyal *Knudsen -88* (WRECKED) (ex Skotte)	WBk	Norweg'n	663	159·9	31·9	18·4	Grimstad	1888	S.Johnsen	[illegible]	M.F.Stray
85594	1010	V	Loyalist *Froom 88-88*	WBn	British	348	130·5	29·4	13·0	Portland,N.B.	1888	F.L.Sayre	[illegible]	F.E.Sayre
	1011		Loyo (ex Alberto) *San Thome*	W3mSr	Brazilian	242	100·0	24·3	11·5	Fao	1876	Borda	[illegible]	J.daSilvaLoye & Filho
	1012		Lua	WSr	Brazilian	150	..	..	..					V.A.e Magalhaes
	1013	N	Luca *Jansen 58-75*	WSr	Norweg'n	399	116·3	28·3	18·4	Baccari	1857			H.E.Jansen
8[illegible]	1014	V	Lucania *Gibson 72-86*	WS	British	1774	229·3	42·1	24·4	St.John,N.B.	1886	J.F.Fraser	[illegible]	B.Smith & Sons
	1015	G	Lucas (ex Lina) *Poeker 85-85*	WSr	German	159	94·9	21·1	11·1	Papenburg	1863	F.W.Meyer		G.& H.Aden
	1016	✻	Lucayas *Johansen*	IBk	Danish	445	150·4	25·5	18·1	Greenock	1869	Robertson & Co		Th.Meinertz's Enke
152[illegible]	1017	V	Lucco *Degregori -87* (BROKEN UP)	WBk	Italian	498	134·5	30·8	20·0	Sestri Ponente	1866	L.Briasco	[illegible]	R.Olivari
[illegible]	1018	I	Luchina C. *Repetto*	WBk	Italian	506	134·5	31·2	19·8	Lavagna	1869	G.Saccomano	[illegible]	A.Ciehero
70660	1019		Lucia (ex Maria Fedele) *Wood 83-86*	CBk	British	640	162·3	31·1	18·3	Sunderland	1868	J.Blumer & Co.	[illegible]	T.Roberts
562	1020		Lucia (ex Angiolina) *Carletta*	WBg	Italian	169	100·0	23·1	14·2	Leghorn	1863	G.Gori	[illegible]	Oreste Del Buono
101	1021		Lucia	WSr	Russian	126	83·7	23·6	10·4	Kurland	1876			Oskar von Stryk
15102	1022		Lucia B. Ives (ex E. H. Parker) *Mott*	WSr	American	177	96·8	27·2	8·5	[illegible],N.J.	18[illegible]			A.H.Mott
140498	1023	R	Lucia Porter *Grindle -81*	W3mSr	American	316	132·0	32·0	10·5	Belfast,Me.	1881		[illegible]	J.F.Grindel
	1024		Lucie	WSr	French	143	88·3	23·3	12·1	Dunkirk	1846			Comolet fréres
	1025	N	Lucie *Eidsaay 75-87*	WSr	Norweg'n	116	95·2	24·0	10·0	Christiansund	1877		[illegible]	Parelius & Lossius
246	1026		Lucie *Johnsohn 82-82*	WSr	Russian	149	80·4	25·5	10·1	Kürbis	1877	Grass		K.Berg & J.Jansohn
	1027		Lucie (ex Dagmar) *Hedberg -88*	WBg	Swedish	214	104·0	22·8	12·0	Miramichi	1858			J.Pettersson
	1028		Lucie Gabrielle (DISMANTLED) (ex Peppino P.)	WBg	French	417	107·9	26·9	18·7	Sestri Ponente	1866	G.B.Calemaccio		C.Guérin
	1029	V	Lucie Marie (ex Anna) *Denis -88*	WBn	French	163	95·8	20·0	12·1	Finland	1858			Dupuis-Robial
	1030		Lucie Marie (BROKEN UP) *Bertho*	WSr	French	141	81·3	23·0	11·1	Nantes	1866	Tillé		D.Bertho
	1031	G	LucieRadmann (ex Susanna Johanna, ex Koning Vaderland) *Gerth -88*	WBk	German	447	117·6	29·9	18·9	Vlaardingen	1854	Von Dam		W.Lüdke & Co.
140069	1032		Lucie Wheatley *Pettingill -88*	W3mSr	American	180	113·8	29·5	8·2	Frederica,Del.	1874	NathanielLank		J.W.Hall
	1033	V	Lucien & Valentine (ex Anna) *Verdois*	WBg	French	274	110·2	27·6	14·8	Arendal	1877		[illegible]	K.Morvant
522	1034	A	Lucietta (ex Giuseppe) *Musana*	WSr	Italian	124	80·7	22·1	12·1	Gaeta	1858		[illegible]	Ant.Burgio & Co.
66855	1035	V	Lucile *Williams*	WBkn	British	498	128·8	31·1	17·0	Tynem'th,NB	1872	R.Lewitt	[illegible]	A.J.Williams
75309	1036		Lucile	WSrYt	British	157	117·0	23·1	13·4	Gosport	1876	Camp&Nichlsn		N.Rudge
140031	1037	V	Lucile *Serman*	WS	American	1329	200·2	40·0	23·9	Freeport,Me.	1874	E.C.Soule	[illegible]	E.C.Soule
71444	1038	✻	Lucille *Kirby -88*	WBg	British	278	117·6	26·1	14·9	P.E.Island	1874	Yeo	[illegible]	W.Williams & Sons
570[illegible]	1039		Lucinda *White 88-88*	WBn	British	199	105·8	24·1	12·9	P.E.Island	1867	Tuplin		H.W.Brightman
140339	1040		Lucinda G. Potter *Potter -88*	W3mSr	American	613	151·8	35·8	15·9	Bath,Me.	1882	Goss&Sawyer		W.T.Potter
	1041		Lucinde *Jensen 87-87*	WBg	Danish	150	86·5	23·7	13·4	Holbæk	1842	V.Lageni		Den.Kgl.Grønl.Handel
	1042	DV	Lucinde *Krakanschke 86-*	WBk	Russian	319	117·0	28·0	14·0	Libau	1866	J.D.Dargewitz		C.Essmann
	1043		Lucinde *Hansson*	WSr	Swedish	102	76·8	20·0	10·8	Flensburg	1861			M.Hansson
251	1044		Lucio (ex Maria Regina) *Martinolich -87*	WBg	Austrian	254	86·0	26·5	16·1	Peà,Italy	1870		[illegible]	Ivecvich & Co.
	1045	V	Lucio I. *Faz*	WSr	Portug'se	19[illegible]	89·6	21·5	10·0	Villa do Conde	1884	C.M.deAraujo	[illegible]	V.L.de Almeida
90005	1046	✻	Lucipara *Harder 86-88*	I 4mS	British	1863	267·1	40·2	23·8	Greenock	1885	Russell&Co.		P.Denniston & Co.
87113	1047	✻	Lucknow *Lidstone 81-87*	StlS	British	1446	245·3	37·1	22·5	Sunderland	1883	Doxford&Sons		Tyser & Co.
70485	1048	✻	Lucy *Davies*	WBkn	British	192	112·8	24·1	12·5	Ipswich	1879	Bayley	[illegible]	J.C.Peacock
140517	1049	R	Lucy *Cook*	WSr	American	297	130·0	30·0	8·0	Berwick,Me.	1882			American Salt Co.
15046	1050		Lucy *Wooster*	WSr	American	129	87·5	27·8	9·1	Eastport,Me.	1864	Huston		L.A.Godfrey

✻=Lloyd's Register; D=Nederlandsche Vereeniging; G=Germanischer Lloyd; I=Registro Italiano; R=[illegible]
A=Veritas Austro-Ungarico; H=Veritas Ellenico; N=Norske Veritas; V=[illegible]

In the 4th Column, C=Composite (iron frame, wood planked); I=Iron; Stl=Steel; W=Wood.

The Dates of Metalling (Col. 11) are based on the records of the Books in which Vessels are respectively classed, or on the Reports of the Surveyors of [illegible]

Further particulars of Vessels classed by Lloyd's Register will be found in Lloyd's Register Book and Lloyd's Yacht Register.

No.	Cl.	Nom. Capitaine et dates d'entrée au service des Armateurs actuels et à bord du navire. Ancien Nom.	Matériaux et Grèement	Pavillon	Tonnage net	Dimensions: Longueur	Largeur	Creux	Construit: Où	Quand	Par qui	Date de doublage en métal	Armateurs ou Directeurs-Gérants	Port d'Armement
1051		Lucy (ex Jan Dekker) Litcan -86	WSr	Dutch	355	120·0	27·9	12·5	Harb'rvilleNS	1873	J.Givan		Kedi Anam Maatschappy	Batavia
1052	✻	Lucy (ex Lone Pine) Kleinstuh -88	WBk	Norweg'n	739	176·2	33·0	19·5	P.E.Island	1875	J.C.Pope	Y 4.88	Aktieselskabet "Lucy" (C.Moller)	Christiania
1053	DN	Lucy Röer 73-79	WBg	Norweg'n	312	120·5	27·9	13·2	Cornwall,Can.	1865			Martin Röer	Dröbak
1054		Lucy A. Blossom	W3mSr	American	179	108·0	24·0	9·0	ConneautOhio	1846			J.Philbrook	Winterport,Me.
1055	R	Lucy A. Davis Lor ... -87	W3mSr	American	596	148·5	34·2	12·6	Yarmouth,Me	1882	Htchns&Stubbs	Y 9.87	J.W.Davis	Portland.Me.
1056	R	Lucy A. Nickels Nickels	WS	American	1395	199·0	37·9	23·8	Brewer,Me.	1875	McGilvery	Y 7.87	J.C.Nickels	Searsport,Me.
1057	R	Lucy E. Friend Collins	W3mSr	American	447	147·0	34·6	12·6	Nwbrypt Mass	1882	G.E.Currier		W.H.Collins	Gloucester,Mass.
1058	N	Lucy Frances Stokke -8	W3mSr	Norweg'n	232	114·6	27·9	11·2	Brookville,Me	1856		Y 3.82	Ludwigsen & Schjelderup	Christiania
1059	R	Lucy H. Russell Russell -87	W4mSr	American	1108	194·3	39·1	18·0	Bath,Me.	1887	NewEng.S.B.Co		H.M.Randall	Pt.Jefferson
1060		Lucy Hammond	WSr	American	150	96·5	26·3	8·9	E.Machias,Me.	1869			J.R.Talbot	E.Machias,Me.
1061		Lucy M. Collins Newcomb	WSr	American	159	100·3	27·8	8·3	WinterportMe	1867			G.B.Hook	Bangor,Me.
1062	✻	Lucy March Griffiths -86	WBkn	British	244	118·0	26·6	14·1	Pwllheli	1876	Williams	Y 8.86	Morgan,Jones & Co.	Carnarvon
1063	✻	LucyReppen (ex Jessie Scammell) Bach -88	WBk	Norweg'n	892	180·6	35·3	20·5	N.Bruns.	1877	Brown	Y 12.87	O.C.Johansen	Christiansand
1064		Lucy Richmond Richmond	WSrBarge	British	137	106·7	24·4	8·6	Ipswich	1875	Robertson		W.T.Richmond & Co.	Maldon
1065	✻DI	Lucy W. Bello	WBk	Italian	683	159·1	33·1	21·6	SestriPonente	1876	A.Calcagno	Y 4.86	P.Zenoglio	Genoa
1066		Lucy W. Snow	WBn	American	299	112·8	26·9	14·0	Rockland,Me.	1869	Snow&Farwell		F.C.Butman	Boston,Mass.
1067	V	Lucy and Paul Andreis	WBk	German	377	127·9	28·7	15·9	Elbing	1861	C.G.Fechter		E.Burchard	Rostock
1068		Ludolfina Jantina Poppen -86	WBn	Dutch	156	95·0	23·0	10·6	Pekela	1874	L.H.DeWijk		H.J.Kuiper (H.Lale Wijk Manager)	Oude Pekela
1069		Ludovic	WSr	French	120	89·6	19·7	10·9	St.Malo	1878	Tranchemer frs.		Vve H.Ruellan	Paimpol
1070	N	Ludvig (ABANDONED) Larsen 82-83	WBk	Norweg'n	302	124·1	27·4	13·5	Brevig	1870	G.Moe		G.J.Sörensen	Sandefjord
1071	N	Ludvig Mathisen	WSr	Norweg'n	110	83·6	20·9	9·6	Brake	1851			Chr.& John Mathisen	Horten
1072	DNV	Ludvig Holberg Rasmussen 8-86	WBk	Norweg'n	1019	193·1	37·5	21·7	Bergen	1878	Brunchorst& Dekke	Y 8.88	A.F.Klaveness	Sandefjord
1073		Ludvig (MISSING) Schaaer	WBk	German	368	128·6	26·6	13·9	Greifswald	1865	C.Dodt		J.Rudolph	Stettin
1074		Ludwig Eichen -86	WSr	Russian	126	82·0	21·0	9·5	Haynasch	1872	M.Eichen		A.Eichen & F.Weide	Haynasch
1075		LudwigCapobus,sen. Prosern 81-84	WBg	German	242	98·3	26·2	14·8	Rostock	1863	W.Zeltz		A.Burchard	Rostock
1076	R	Luella A. Snow Rose	WSr	American	178	106·7	27·0	8·4	Rockland,Me.	1875	Snow&Farwell	Y 7.81	Snow & Pearsons	Rockland,Me.
1077	✻	Lufra Rogers	CBk	British	672	179·7	31·1	17·9	Aberdeen	1870	Hall	Y 6.87	A.McGregor	Hobart
1078		Lufra	WKYt	British	131	108·1	22·0	12·2	Cowes	1865	M.Ratsey		Exrs. of the Duke of Rutland,K.G.	Cowes
1079	N	Lufra Evensen	WBk	Norweg'n	489	133·6	31·3	17·6	Grimstad	1874	O&I.Dömmesmoen	Y 8.88	J.E.Einersen	Grimstad
1080	R	Lugano Clark	WSr	American	166	104·5	28·1	9·0	E.Machias,Me	1867			J.G.Murchie & Sons	New York
1081	✻I	Luigi (ex Caterina B.) Cuneo	WBk	Italian	823	168·7	33·7	23·6	Finale	1877	F.Ponziolo	Y 1.86	F.Cuneo fu A.	Genoa
1082	A	Luigi Scalfiola	WBk	Italian	511	..	..	..	Brooklyn	1865	Th.Stach	Y 8.83	L.Pirandello	Messina
1083		Luigi Corradini	WBg	Italian	193	111·8	22·1	14·1	Capod'Istria	1864			A.Bini & Co.	Genoa
1084	A	Luigi Di Pace -88	WBn	Italian	141	78·7	23·3	13·3	Castellamare di Stabia	1863		Z 1.88	F.Troise	Naples
1085	I	Luigi Accame Olivari	WBk	Italian	939	176·8	35·4	22·6	SestriPonente	1870	G.Casanova	Y 7.83	Em.Accame fu L.	Genoa
1086		Luigi C. (ex Vittorioso) Tancredi	WBk	Italian	249	107·6	26·6	16·3	Leghorn	1853	G.Micheli		E.Pieruzzini	Leghorn
1087	A	Luigi Fortunato Chiara	WTrb	Italian	117	83·0	24·4	10·6	Chioggia	1881	L.Menetto	Z 8.81	Flli Ghezzo & Co.	Venice
1088	I	Luigi P. Maglio -87	WBk	Italian	637	143·5	32·4	21·0	Prà	1873	B.Canepa	Y 8.88	G.Moglia	Genoa
1089	I	Luigi Padre (ex Bruni Pietro) Olivari -88	WBk	Italian	529	140·7	31·2	20·7	Savona	1875	F.Calamaro	Y 12.87	Flli Sciaccaluga	Genoa
1090		Luigi Padre (WRECKED)	WBn	Italian	121	71·2	23·2	13·1	Varazze	1855			C.A.Giorgetti	Leghorn
1091		Luigi Repetto Repetto	WBk	Italian	220	101·7	25·6	14·8	St.Valery	1853			Fort Repetto	Genoa
1092	I	Luigi Ruggero Moresco -86	WBk	Italian	490	131·2	29·2	18·7	Castellamare	1873	L.Tramborino	Y 8.86	L.Ruggero	Castellamare
1093		Luigia Rinaldi	WBk	Italian	402	122·3	29·7	18·4	SestriPonente	1866	G.Peragallo		Flli Pendibene	Genoa
1094		Luigia (ex Zeno) Calabrese	WBg	Italian	231	93·5	26·1	16·2	Piræus	1867			G.B.Borzone fu G.	Leghorn
1095		Luigia F. (ex Consiglio) Villa	W	Italian	491	147·6	30·7	19·7	SestriPonente	1866	L.Casella		B.Fravega fu G.B.	Genoa
1096	I	Luigia M. Rossoni -88	WBk	Italian	421	128·9	30·2	17·9	SestriPonente	1880	A.Briasco fu G.B.	Y 5.83	E.Massone	Genoa
1097	I	Luigia R. (ex Pasdasso) Pavegno -88	WBn	Italian	498	137·8	30·0	18·7	Chiavari	1867	M.Tappani	Y 5.88	F.Rainusso	Genoa
1098	IV	Luigia Raffo Gaibie	WBk	Italian	972	183·1	38·1	23·1	Chiavari	1875	M.Tappani	Y 3.86	Erasmo Raffo fu A.	Spezzia
1099	I	Luigia Rocca Rebuffo	WBk	Italian	546	150·2	32·3	19·7	Lavagna	1870	D.Briasco	Y 5.83	Rocca & Lavello	Genoa
1100														

=Lloyd's Register; D=Nederlandsche Vereeniging; G=Germanischer Lloyd; I=Registro Italiano; R=American Record;
=Veritas Austro-Ungarico; H=Veritas Ellenico; N=Norske Veritas; V=Bureau Veritas.

Dans la 4me Colonne, C=Composite (membrure en fer, bordé en bois); I=Fer; Stl=Acier; W=Bois.

Les Dates du doublage en métal (Col. 11) ressortent des livres dans lesquels les navires sont respectivement classés ou des rapports des Inspecteurs du Lloyd's Register.

Pour plus amples renseignements sur les navires classés par le Lloyd's Register, voir le Lloyd's Register Book et le Lloyd's Yacht Register.

Official No.	No. in Book	Registry in which classed	Ship's Name, Material and Rig. Master, and Dates of Appointment to present Owner's Service and to Ship. Ship's Late Name.	Rig	Flag	Net Tonnage	Length	Breadth	Depth	Built: Where	When	By whom	Date of Metalling	Owners or Managers	Port
	1001	DNV	Lovspring *Thomsen*	WBk	Norweg'n	547	149·4	30·9	17·8	Christiansand	1873	O.Olsen	[illegible]	H.G.Kirsebom	Chris[illegible]
	1002	N	Lovspring *Schive -88*	WBk	Norweg'n	489	136·4	31·0	16·9	Grimstad	1885	N.O.Gjömle	[illegible]	J.Jörgensen	Grim[illegible]
	1003	N	Lovspring *Thomesen*	WBg	Norweg'n	225	107·5	26·9	12·2	Porsgrund	1872			P.M.Petersen & Son	[illegible]
73935	1004	RV	Low Wood *Thurber -87*	WBk	British	1091	186·7	37·0	22·4	Portland,N.B.	1878	D.Lynch	[illegible]	H.D.Troop	St.[illegible]
56478	1005		Lowestoft *Hook*	WBk	British	293	112·7	26·5	16·5	Sunderland	1866	J.Dennison		R.L.Manners	North[illegible]
76389	1006	✳	Loweswater *Smith -88*	IBk	British	603	174·0	30·2	17·5	Whitehaven	1876	WhitehvnSBCo		Jackson,Metcalf & Co	Liverp[illegible]
91291	1007	✳	Lowther Castle *Scarborough 81-86*	IS	British	1824	262·1	39·0	23·3	Workington	1886	R.Williamson & [Son		Lancaster Shipowners Co.Ld.	Liverp[illegible]
	1008	N	Loyal *Fritz -87* (ex Dreadnought) (WRECKED)	WS	Norweg'n	1427	220·2	38·9	24·7	[illegible]	1869	Rosa	[illegible]	Rederiselskab "Loyal" (Chr.Ellefsen)	[illegible]
	1009	D	Loyal *Knudsen -88* WRECKED (ex Skotta)	WBk	Norweg'n	663	159·9	31·9	18·[illegible]	[illegible]	1889	S.Johnson	[illegible]	M.F.Stray	Chris[illegible]
85304	1010	V	Loyalist *Troop 88-88*	WBn	British	348	130·5	29·4	13·0	[illegible],N.B.	1883	F.E.Sayre	[illegible]	F.E.Sayre	St.Joh[illegible]
	1011		Loyo *San Thomé* (ex Alberto)	W3mSr	Brazilian	242	100·0	24·3	11·5	[illegible]	1876	Borda	[illegible]	J daSilvaLoye & Filho	Pern[illegible]
	1012		Lua	WSr	Brazilian	150	..	..	..					V.A.e Magalhaes	[illegible]
	1013	N	Luca *Jansen 56-75*	WSr	Norweg'n	399	116·3	28·3	18·4	Buccari	1857			H.E.Jansen	Flekkef[illegible]
88391	1014	V	Lucania *Gibson 72-88*	WS	British	1774	229·3	42·1	24·4	St.John,N.B.	1886	J.F.Fraser	[illegible]	B.Smith & Sons	[illegible]
	1015	G	Lucas (ex Lisa) *Pocker 85-85*	WSr	German	159	94·9	21·1	11·1	Papenburg	1863	F.W.Meyer		G.& H.Aden	Jemgum
	1016	✳	Lucayas *Johansen*	IBk	Danish	445	150·4	25·5	18·1	Greenock	1869	Robertson & Co		Th.Meinertz's Enke	Fanø
1528	1017	V	Lucco *Degregori -87* (BROKEN UP)	WBk	Italian	498	131·5	30·3	20·0	Sestri Ponente	1866	L.Briasco	[illegible]	B.Olivari	Genoa
1974	1018	I	Luchina C. *Repetto*	WBk	Italian	506	134·5	31·2	19·8	Lavagna	1869	G.Saccomano	[illegible]	A.Ciebero	Genoa
70699	1019		Lucia *Wood 83-86* (ex Maria Fidela)	CBk	British	640	162·3	31·1	18·3	Sunderland	1868	J.Blumer & Co.	[illegible]	T.Roberts	London
552	1020		Lucia (ex Angiolina) *Carletta*	WBg	Italian	169	100·0	23·1	14·2	Leghorn	1863	G.Gori	[illegible]	Oreste Del Buono	Porto [illegible]
101	1021		Lucia	WSr	Russian	126	89·7	23·6	10·4	Kurland	1876			Oskar von Strÿk	[illegible]
15102	1022		Lucia B. Ives *Mott* (ex E. M. Packer)	WSr	American	177	96·8	27·2	8·5	Belleville,N.J	18 53/64			A.H.Mott	Pt.Jeff[illegible]
140498	1023	R	Lucia Porter *Grindle -84*	W3mSr	American	316	132·0	32·0	10·5	Belfast,Me.	1881		[illegible]	J.F.Grindel	Belfast[illegible]
	1024		Lucie	WSr	French	143	88·3	23·3	12·1	Dunkirk	1846			Comolet frères	St.Serv[illegible]
	1025	N	Lucie *Eidsaay 75-87*	WSr	Norweg'n	116	95·2	24·0	10·0	Christiansund	1877		[illegible]	Parelius & Lessius	Christi[illegible]
246	1026		Lucie *Johnsohn 82-82*	WSr	Russian	149	80·4	25·5	10·1	Kürbis	1877	Grass		K.Berg & J.Jansohn	Riga
	1027		Lucie *Hedberg -88* (ex Dagmar)	WBg	Swedish	214	104·0	22·8	12·0	Miramichi	1858			J.Pettersson	Gotheb[illegible]
	1028		Lucie Gabrielle (DISMANTLED) (ex Peppino P.)	WBg	French	417	107·9	26·9	18·7	Sestri Ponente	1866	G.B.Cadenaccio		C.Guérin	Martigues
	1029	V	Lucie Marie *Denis -88* (ex Anna)	WBn	French	163	95·8	20·0	12·1	Finland	1858			Dupais-Robial	St.[illegible]
	1030		Lucie Marie *Bertho* (BROKEN UP)	WSr	French	141	84·3	23·0	11·1	Nantes	1866	Tillé		D.Bertho	Nantes
	1031	G	LucieRadmann *Gerth -88* (ex Susanne Johanna, ex Koenig Vaderland)	WBk	German	447	117·6	29·9	18·9	Vlaardingen	1854	Von Dam		W.Lüdke & Co.	Stettin
140069	1032		Lucie Wheatley *Pettingill -88*	W3mSr	American	180	113·8	29·5	8·2	Frederica,Del.	1874	NathanielLank		J.W.Hall	[illegible]
	1033	V	Lucien & Valentine *Verdois* (ex Anna)	WBg	French	274	110·2	27·6	14·8	Arendal	1877		[illegible]	K.Morvant	[illegible]
522	1034	A	Lucietta (ex Giuseppe) *Musana*	WSr	Italian	124	80·7	22·4	12·1	Gaeta	1858		[illegible]	Ant.Burgio & Co.	[illegible]
66853	1035	V	Lucile *Williams*	WBkn	British	498	128·8	31·1	17·0	Tynem'th,NB	1872	R.Lowitt	[illegible]	A.J.Williams	St.Joh[illegible]
75309	1036		Lucile	WSrYt	British	157	117·0	23·1	13·4	Gosport	1876	Camper&Nichlsn		N.Rudge	[illegible]
140031	1037	V	Lucile *Serman*	WS	American	1329	200·2	40·0	23·9	Freeport,Me.	1874	E.C.Soule	[illegible]	E.C.Soule	[illegible]
71441	1038	✳	Lucille *Kirby -88*	WBg	British	278	117·6	26·1	14·9	P.E.Island	1874	Yeo	[illegible]	W.Williams & Sons	[illegible]
57022	1039		Lucinda *White 88-88*	WBn	British	199	105·8	24·1	12·9	P.E.Island	1867	Duplin		H.W.Brightman	[illegible]
140530	1040		Lucinda G. Potter *Potter -88*	W3mSr	American	613	151·8	35·8	15·9	Bath,Me.	1882	Goss&Sawyer		W.T.Potter	[illegible]
	1041		Lucinde *Jensen 87-87*	WBg	Danish	150	86·5	23·7	13·4	Holbæk	1842	V.Lagoni		Den.Kgl.Grønl.Handel	Copenh[illegible]
	1042	DV	Lucinde *Krakauschke 86-*	WBk	Russian	319	117·0	28·0	14·0	Libau	1866	J.D.Dargewitz		C.Essmann	[illegible]
	1043		Lucinde *Hansson*	WSr	Swedish	102	76·8	20·0	10·8	Flensburg	1861			M.Hansson	[illegible]
251	1044		Lucio *Martinolich -87* (ex Maria Regina)	WBg	Austrian	254	86·0	26·5	16·1	Prà,Italy	1870		[illegible]	Ivecvich & Co.	[illegible]
	1045	V	Lucio I. *Foz*	WSr	Portug'se	150	89·6	21·5	10·0	Villa do Conde	1884	C.M.de Araujo	[illegible]	V.L.de Almeida	[illegible]
90905	1046	✳	Lucipara *Harder 86-88*	I 4mS	British	1863	267·1	40·2	23·8	Greenock	1885	Russell&Co.		P.Denniston & Co.	[illegible]
87113	1047	✳	Lucknow *Liddione 81-87*	StlS	British	1440	215·3	37·1	22·5	[illegible]	1883	Doxford&Sons		Tyser & Co.	[illegible]
[illegible]	1048	✳	Lucy *Davies*	WBkn	British	192	112·8	24·1	12·5	Ipswich	1879	Bayley	[illegible]	J.C.Peacock	[illegible]
14051	1049	R	Lucy *Cook*	WSr	American	297	130·0	30·0	8·0	Berwick,Me.	1882			American Salt Co.	[illegible]
15046	1050		Lucy *Wooster*	WSr	American	129	87·5	27·8	9·1	Eastport,Me.	1864	Huston		L.A.Godfrey	[illegible]

✳=Lloyd's Register; D=Nederlandsche Vereeniging; G=Germanischer Lloyd; I=Registro Italiano; R=[illegible]
A=Veritas Austro-Ungarico; H=Veritas Ellenico; N=Norske Veritas; V=[illegible]

In the 4th Column, C=Composite (iron frame, wood planked); I=Iron; Stl=Steel; W=Wood.

The Dates of Metalling (Col. 11) are based on the records of the Books in which Vessels are respectively classed, or on the Reports of the Surveyors of [illegible]

Further particulars of Vessels classed by Lloyd's Register will be found in Lloyd's Register Book and Lloyd's Yacht Register.

No.	Reg.	Nom, Capitaine, Ancien Nom	Matériaux et Gréement	Pavillon	Tonnage net	Longueur	Largeur	Creux	Où	Quand	Par qui	Date du doublage en métal	Armateurs ou Directeurs-Gérants	Port d'Armement
1051		Lucy	WSr	Dutch	355	120·0	27·9	12·5	Harb'rvilleNS	1873	J.Givan		Kah Adam Maatschappy	Batavia
1052	✻	Lucy	WBk	Norweg'n	739	176·2	33·0	19·5	P.E.Island	1875	J.C.Pope		Actieselskabet "Lucy" (C.Møller)	Christiania
1053	DN	Lucy Röer 79-79	WBg	Norweg'n	312	120·5	27·9	13·2	Cornwall,Can.	1865			Martin Röer	Dröbak
1054		Lucy A. Blossom	W3mSr	American	179	108·0	24·0	9·0	ConneautOhio	1846			J.Philbrook	Winterport,Me.
1055	R	Lucy A. Davis	W3mSr	American	596	148·5	34·2	12·6	Yarmouth,Me	1882	Htchns&Stubbs		J.W.Davis	Portland,Me.
1056	R	Lucy A. Nickels Nickels	WS	American	1395	199·0	37·9	23·8	Brewer,Me.	1875	McGilvery		J.C.Nickels	Searsport,Me.
1057	R	Lucy E. Friend	W3mSr	American	447	147·0	34·8	12·6	Nwbrypt Mass	1882	G.E.Currier		W.H.Collins	Gloucester,Mass.
1058	N	Lucy Frances Stokke	W3mSr	Norweg'n	232	114·6	27·0	11·2	Brookville,Me	1856			Ludwigsen & Schjelderup	Christiania
1059	R	Lucy H. Russell Rendall	W4mSr	American	1108	194·3	39·0	18·0	Bath,Me.	1887	NewEng.S.B.Co		H.M.Randall	Pt.Jefferson
1060		Lucy Hammond	WSr	American	150	96·5	26·3	8·9	E.Machias,Me.	1869			J.R.Talbot	E.Machias,Me.
1061		Lucy M. Collins	WSr	American	159	100·3	27·8	8·3	Winterport Me	1867			G.B.Hook	Bangor,Me.
1062	✻	Lucy March Griffith	WBkn	British	244	118·0	26·6	14·1	Pwllheli	1876	Williams		Morgan,Jones & Co.	Carnarvon
1063	✻	LucyReppen Bach	WBk	Norweg'n	892	180·6	35·3	20·5	N.Bruns.	1877	Brown	12.87	O.C.Johansen	Christiansand
1064		Lucy Richmond Richmond	WSrBarge	British	137	106·7	24·4	8·6	Ipswich	1875	Robertson		W.T.Richmond & Co.	Maldon
1065	✻DI	Lucy W. Tallo	WBk	Italian	683	159·1	33·1	21·6	SestriPonente	1876	A.Calcagno		P.Zenoglio	Genoa
1066		Lucy W. Snow	WBn	American	299	112·8	26·9	14·0	Rockland,Me.	1869	Snow&Farwell		F.C.Butman	Boston,Mass.
1067	V	Lucy and Paul Andreis	WBk	German	377	127·9	28·7	15·9	Elbing	1861	C.G.Fechter		E.Burchard	Rostock
1068		Ludolfina Jantine Poppen	WBn	Dutch	156	95·0	23·0	10·6	Pekela	1874	L.H.DeWijk		H.J.Kuiper (H.L.deWyk,Manager)	Oude Pekela
1069		Ludovic	WSr	French	120	89·6	19·7	10·9	St.Malo	1878	Trauchemer frs.		Vve. H.Ruellan	Paimpol
1070	N	Ludvig Larsen 82-82 (ABANDONED)	WBk	Norweg'n	302	124·1	27·4	13·5	Brevig	1870	G.Moe		G.J.Sörensen	Sandefjord
1071	N	Ludvig Mathisen	WSr	Norweg'n	110	83·6	20·9	9·6	Brake	1851			Chr.& John Mathisen	Horten
1072	DNV	Ludvig Holberg Hansen	WBk	Norweg'n	1019	193·1	37·5	21·7	Bergen	1878	Bruuchorst& Dekke	Y 8.88	A.F.Klaveness	Sandefjord
1073		Ludvig (MISSING) Schauer	WBk	German	368	128·6	26·6	13·9	Greifswald	1865	C.Dodt		J.Rudolph	Stettin
1074		Ludwig Eichen	WSr	Russian	126	82·0	21·0	9·5	Haynasch	1872	M.Eichen		A.Eichen & F.Weide	Haynasch
1075		LudwigCapobus,sen.	WBg	German	242	98·8	26·2	14·8	Rostock	1863	W.Zeltz		A.Burchard	Rostock
1076	R	Luella A. Snow Rose	WSr	American	178	106·7	27·0	8·4	Rockland,Me.	1875	Snow&Farwell		Snow & Pearsons	Rockland,Me.
1077	✻	Lufra Rogers	CBk	British	672	179·7	31·1	17·9	Aberdeen	1870	Hall	Y 6.87	A.McGregor	Hobart
1078		Lufra	WKYt	British	131	108·1	22·0	12·2	Cowes	1865	M.Ratsey		Exrs. of the Duke of Rutland,K.G.	Cowes
1079	N	Lufra Evensen	WBk	Norweg'n	489	133·6	31·3	17·6	[illegible]	1874	O&L.Dömmesmoen		J.E.Einersen	Grimstad
1080	R	Lugano Clark	WSr	American	166	101·5	28·1	9·0	E.Machias,Me	1867			J.G.Murchie & Sons	New York
1081	✻I	Luigi (ex Caterina B.) Canevaro	WBk	Italian	828	168·7	33·7	23·0	Finale	1877	F.Ponziolo		F.Cuneo fu A.	Genoa
1082	I	Luigi Scattiola	WBk	Italian	511	..	..	..	Brooklyn	1865	Th.Stach		L.Pirandello	Messina
1083		Luigi Corradini	WBg	Italian	193	111·8	22·1	14·1	Capo d'Istria	1864			A.Bini & Co.	Genoa
1084	A	Luigi Di Pace	WBn	Italian	141	78·7	23·3	13·3	Castellamare di Stabia	1863		Z 1.88	F.Troise	Naples
1085	I	Luigi Accame Olivari	WBk	Italian	939	176·8	35·4	22·6	SestriPonente	1870	G.Casanova	Y 7.83	Em.Accame fu L.	Genoa
1086		Luigi C. Tarantini	WBk	Italian	249	107·6	26·6	16·3	Leghorn	1853	G.Micheli		E.Pieruzzini	Leghorn
1087	A	Luigi Fortunato Ghezzo	WTr	Italian	117	83·0	24·4	10·6	Chioggia	1881	L.Menetto	Z 8.81	Flli. Ghezzo & Co.	Venice
1088	I	Luigi P. Maglio	WBk	Italian	657	143·5	32·4	21·0	Pegli	1873	B.Canepa	Y 8.88	G.Moglia	Genoa
1089	I	Luigi Padre Olivari	WBk	Italian	529	140·7	31·2	20·7	Savona	1875	F.Calamaro	Y 12.87	Flli. Sciaccaluga	Genoa
1090		Luigi Padre	WBn	Italian	121	71·2	23·2	13·1	Varazze	1855			C.A.Giorgetti	Leghorn
1091		Luigi Repetto Repetto	WBk	Italian	220	101·7	25·6	14·8	St.Valery	1853			Fort Repetto	Genoa
1092	I	Luigi Ruggero	WBk	Italian	490	131·2	29·2	18·7	Castellamare	1873	L.Trambarulo		L.Ruggero	Castellamare
1093		Luigia	WBk	Italian	402	122·3	29·7	18·4	SestriPonente	1866	G.Peragallo		Flli. Pendibene	Genoa
1094		Luigia (ex Leny) Calabrese	WBg	Italian	231	93·5	26·1	16·2	Piraeus	1867			G.B.Borzone fu G.	Leghorn
1095		Luigia F. Villa	W	Italian	491	147·6	30·7	19·7	SestriPonente	1866	L.Casella		B.Fravega fu G.B.	Genoa
1096	I	Luigia M.	WBk	Italian	421	128·9	30·2	17·9	SestriPonente	1866	A.Briasco fu G.B		E.Massone	Genoa
1097	I	Luigia R. Bacigalupo	WBn	Italian	498	137·8	30·0	18·7	Chiavari	1867	M.Tappani		F.Raimusso	Genoa
1098	IV	Luigia Raffo Guisio	WBk	Italian	972	183·4	38·1	23·1	Chiavari	1875	M.Tappani		Erasmo Raffo fu A.	Spezzia
1099	I	Luigia Rocca Rebuffo	WBk	Italian	546	150·2	32·3	19·7	Lavagna	1870	D.Briasco		Rocca & Lavello	Genoa
1100														

✻=Lloyd's Register; A=Veritas Austro-Ungarico; D=Nederlandsche Vereeniging; G=Germanischer Lloyd; H=Veritas Ellenico; I=Registro Italiano; N=Norske Veritas; R=American Record; V=Bureau Veritas.

Dans la 4me Colonne, C=Composite (membrure en fer, bordé en bois); I=Fer; Stl=Acier; W=Bois.

Les Dates du doublage en métal (Col. 11) ressortent des livres dans lesquels les navires sont respectivement classés ou des rapports des Inspecteurs du Lloyd's Register.

Pour plus amples renseignements sur les navires classés par le Lloyd's Register, voir le Lloyd's Register Book et le Lloyd's Yacht Register.

Official No.	No. in Book	Registry in which classed	Ship's Name. Master, and Dates of Appointment to present Owner's Service and to Ship. Ship's Late Name.	Material and Rig	Flag	Net Tonnage	Length	Breadth	Depth	Built: Where	When	By whom	Date of Metalling	Owners or Managers	Port
	1001	DNV	Lovspring *Thomsen*	W Bk	Norweg'n	547	149·4	30·9	17·8	Christiansand	1873	H.Olsen	[illegible]	H.G.Kiese [illegible]	[illegible]
	1002	N	Lövspring *Sehive* -88	W Bk	Norweg'n	489	136·4	31·0	16·9	Grimstad	1885	N.O.Gjömle	[illegible]	J.Jörgensen	[illegible]
	1003	N	Lovspring *Thönnesen*	W Bg	Norweg'n	225	107·5	26·9	12·2	Porsgrund	1872			P.M.Petersen & Sön	Porsgrund
72335	1004	RV	Low Wood *Thurber* -87	W Bk	British	1091	186·7	37·6	22·4	Portland,N.B.	1876	D.Lynch	Y 8.88	H.D.Troop	St.John
[illegible]178	1005		Lowestoft *Hook*	W Bk	British	293	112·7	26·5	16·5	Sunderland	1866	J.Dennison		R.L.Manners	[illegible]
76389	1006	✻	Loweswater *Smith* -88	I Bk	British	603	174·0	30·2	17·5	Whitehaven	1876	WhitehvnSBC.		Jackson,Metcalf & Co	Liverpool
91291	1007	✻	Lowther Castle *[illegible]*	I S	British	1824	262·1	39·0	23·3	Workington	1886	R.Williamson & [Son]		Lancaster Shipowners' Co.Ld.	Liverpool
	1008	N	Loyal *Fritz* -87 (ex Dreadnought) (WRECKED)	W S	Norweg'n	1427	220·2	38·9	24·7	Quebec	1863	Rosa	[illegible]	Rederselskab "Loyal" (Chr.Ellefsen)	[illegible]
	1009	D	Loyal *Knudsen* -88 (WRECKED) (ex Skotta)	W Bk	Norweg'n	663	159·9	31·9	18·4	Grimstad	1883	S.Johnsen	Y 3.86	M.F.Stray	Christ[iansand]
83594	1010	V	Loyalist *Froom* 88-88	W Bn	British	348	130·5	29·4	13·0	Portland,N.B.	1883	F.E.Sayre	Y 8.88	F.E.Sayre	St.J[ohn]
	1011		Loyo (ex Alberto) *San Thome*	W 3mSr	Brazilian	242	100·0	24·3	11·5	Foz	1876	Borda	Y 3.87	J.daSilvaLoyo & Filho	Pern[ambuco]
	1012		Lua	W Sr	Brazilian	150	..	..	..					V.A.e Magalhaes	Parnahyba
	1013	N	Luca *Jansen* 58-75	W Sr	Norweg'n	399	116·3	28·3	18·4	Buccari	1857			H.E.Jansen	Flekke[fjord]
88329	1014	V	Lucania *Gibson* 72-86	W S	British	1774	229·3	42·1	24·4	St.John,N.B.	1886	J.F.Fraser	Y 11.86	B.Smith & Sons	Windsor
	1015	G	Lucas (ex Lina) *Pocker* 85-85	W Sr	German	159	94·9	21·1	11·1	Papenburg	1863	F.W.Meyer		G.& H.Aden	Jemgum
	1016	✻	Lucayas *Johansen*	I Bk	Danish	445	150·4	25·5	18·1	Greenock	1869	Robertson & Co		Th.Meinertz's Enke	Fanø
152[illegible]	1017	V	Lucco *Degregari* -87 (BROKEN UP)	W Bk	Italian	498	134·5	30·8	20·0	Sestri Ponente	1866	L.Briasco	Y 4.86	B.Olivari	Genoa
1071	1018	I	Luchina C. *Repetto*	W Bk	Italian	596	134·5	31·2	19·8	Lavagna	1869	G.Saccomano	[illegible]	A.Cichero	Genoa
70682	1019		Lucia (ex Maria Fidelio) *Wood* 83-86	C Bk	British	640	162·3	31·1	18·3	Sunderland	1868	J.Blumer & Co.	Y 10.86	T.Roberts	London
532	1020		Lucia (ex Angiolina) *Carletta*	W Bg	Italian	169	100·0	23·1	14·2	Leghorn	1863	G.Gori	[illegible]	Oreste Del Buono	Porto Fe[rrajo]
101	1021		Lucia	W Sr	Russian	126	89·7	23·6	10·4	Kurland	1876			Oskar von Stryk	[illegible]
15102	1022		Lucia B. Ives (ex E. H. Parker) *Mott*	W Sr	American	177	96·8	27·2	8·5	Belleville,N.J	18 63/64			A.H.Mott	Pt.Jeff[erson]
140498	1023	R	Lucia Porter *Grindle* [-81	W 3mSr	American	316	132·0	32·0	10·5	Belfast,Me.	1881		Y 11.88	J.F.Grindel	Belfast,Me.
	1024		Lucie	W Sr	French	143	88·3	23·3	12·1	Dunkirk	1846			Comolet frères	St.Nazaire
	1025	N	Lucie *Eidsaay* 75-87	W Sr	Norweg'n	116	95·2	24·0	10·0	Christiansund	1877		[illegible]	Parelius & Lossius	Christiansund
246	1026		Lucie *Johnsohn* 82-83	W Sr	Russian	149	80·4	25·5	10·1	Küchia	1877	Grass		K.Berg & J.Jansohn	Riga
	1027		Lucie (ex Dagmar) *Hedberg* -88	W Bg	Swedish	214	104·0	22·8	12·0	Miramichi	1858			J.Pettersson	Gothenburg
	1028		Lucie Gabrielle (DISMANTLED) (ex Peppino P.)	W Bg	French	417	107·9	26·3	18·7	Sestri Ponente	1866	G.B.Cadenaccio		C.Guérin	Marseilles
	1029	V	Lucie Marie (ex Agnes) *Denis* -88	W Bn	French	163	95·8	20·0	12·1	Finland	1858			Dupuis-Robial	St.Brieuc
	1030		Lucie Marie (BROKEN UP) *Bertho*	W Sr	French	141	84·3	23·0	11·1	Nantes	1866	Tillé		D.Bertho	Nantes
	1031	G	LucieRadmann (ex Susanne Johanna, ex Koning Vaderland) *Gerth* [-88	W Bk	German	447	117·6	29·9	18·9	Vlaardingen	1854	Von Dam		W.Lüdke & Co.	Stettin
140062	1032		Lucie Wheatley *Pettingell* -88	W 3mSr	American	180	113·8	29·5	8·2	Frederica,Del.	1874	NathanielLank		J.W.Hall	[illegible]
	1033	V	Lucien & Valentine (ex Anna) *Verdois*	W Bg	French	274	110·2	27·6	11·8	Arendal	1877		[illegible]	K.Morvant	Bordeaux
522	1034	A	Lucietta (ex Giuseppe) *Musana*	W Sr	Italian	124	80·7	22·1	12·1	[illegible]	1858		[illegible]	Ant.Burgio & Co.	Catania
66855	1035	V	Lucile *Williams*	W Bkn	British	408	128·8	31·1	17·0	Yarmouth,N.S.	1872	R.Lowitt	[illegible]	A.J.Williams	St.J[ohn]
75399	1036		Lucile	W Sr Yt	British	157	117·0	25·1	13·1	[illegible]	1876	Camper& Nicholsn		N.Rudge	[illegible]
140031	1037	V	Lucile *Serman*	W S	American	1329	200·2	40·0	23·9	Freeport,Me.	1874	E.C.Soule	[illegible]	E.C.Soule	Freeport
71441	1038	✻	Lucille *Kirby* -88	W Bg	British	278	117·6	26·1	14·9	P.E.Island	1874	Yeo	[illegible]	W.Williams & Sons	Newport
57022	1039		Lucinda *White* 88-88	W Bn	British	199	105·8	24·1	12·9	P.E.Island	1867	Tuplin		H.W.Brightman	[illegible]
140336	1040		Lucinda G. Potter *Potter* -88	W 3mSr	American	613	154·8	35·8	15·9	Bath,Me.	1882	Goss&Sawyer		W.T.Potter	[illegible]
	1041		Lucinde *Jensen* 87-87	W Bg	Danish	150	86·5	23·7	13·4	Holbæk	1842	V.Lagoni		Den.Kgl.Grønl.Handel	Copenhagen
	1042	DV	Lucinde *Krakauschke* 86-	W Bk	Russian	319	117·0	28·0	14·0	Libau	1866	J.D.Dargewitz		C.Essmann	Riga
	1043		Lucinde *Hansson*	W Sr	Swedish	102	76·8	20·0	10·8	Flensburg	1861			M.Hansson	[illegible]
251	1044		Lucio (ex Maria Regina) *Martinolich* -87	W Bg	Austrian	254	86·0	26·5	16·1	Prà,Italy	1870		[illegible]	Ivcevich & Co.	Trieste
	1045	V	Lucio I. *Faz*	W Sr	Portug'se	15[illegible]	89·6	21·5	10·0	Villa do Conde	1884	C.M.de Araujo	[illegible]	V.L.de Almeida	Oporto
99005	1046	✻	Lucipara *Harder* 86-88	I 4mS	British	1863	267·1	40·2	23·8	Greenock	1885	Russell& Co.		P.Denniston & Co.	Glasgow
87113	1047	✻	Lucknow *Lidstone* 81-87	Stl S	British	1440	245·3	37·1	22·5	Sunderland	1883	Doxford&Sons		Tyser & Co.	London
79485	1048	✻	Lucy *Davies*	W Bkn	British	192	112·8	24·1	12·5	[illegible]	1879	Bayley	[illegible]	J.C.Peacock	Glasgow
140317	1049	R	Lucy *Cook*	W Sr	American	297	130·0	30·0	8·0	Berwick,Me.	1882			American Salt Co.	[illegible]
15046	1050		Lucy *Wooster*	W Sr	American	129	87·5	27·8	9·1	Eastport,Me.	1864	Huston		L.A.Godfrey	New [illegible]

✻=Lloyd's Register; D=Nederlandsche Vereeniging; G=Germanischer Lloyd; I=Registro Italiano; R=American [Record]; A=Veritas Austro-Ungarico; H=Veritas Ellenico; N=Norske Veritas; V=Bureau Veritas.

In the 4th Column, C=Composite (iron frame, wood planked); I=Iron; Stl=Steel; W=Wood.

The Dates of Metalling (Col. 11) are based on the records of the Books in which Vessels are respectively classed, or on the Reports of the Surveyors of [illegible]

Further particulars of Vessels classed by Lloyd's Register will be found in Lloyd's Register Book and Lloyd's Yacht Register.

No.		Nom. Capitaine. Ancien Nom	Matériaux et Gréement	Pavillon	Tonnage net	Longueur	Largeur	Creux	Où	Quand	Par qui	Date du classement	Armateurs ou Directeurs-Gérants	Port d'Armement
1051		Lucy	WSr	Dutch	355	120·0	27·9	12·5	HarbrvilleNS	1873	J.Givan		Kali Anam Maatschappy	Batavia
1052	✱	Lucy *Klouman*	WBk	Norweg'n	739	176·2	33·0	19·5	P.E.Island	1875	J.C.Pope	Y 4.88	Actieselskabet "Lucy" (C.Møller)	Christiania
1053	DN	Lucy *Röer 79-89*	WBg	Norweg'n	312	120·5	27·9	13·2	Cornwall,Can.	1865			Martin Röer	Dröbak
1054		Lucy A. Blossom	W3mSr	American	179	108·0	24·0	9·0	ConneautOhio	1846			J.Philbrook	Winterport,Me.
1055	R	Lucy A. Davis *Loring -87*	W3mSr	American	596	148·5	34·2	12·6	Yarmouth,Me	1882	Htchns&Stubbs	Y 8.87	J.W.Davis	Portland,Me.
1056	✱	Lucy A. Nickels *Nichols*	WS	American	1395	199·0	37·9	23·8	Brewer,Me.	1875	McGilvery	Y 7.87	J.C.Nickels	Searsport,Me.
1057	R	Lucy E. Friend	W3mSr	American	447	147·0	34·6	12·6	Nwbrypt Mass	1882	G.E.Currier		W.H.Collins	Gloucester,Mass.
1058	N	Lucy Frances *Stokke -85*	W3mSr	Norweg'n	232	114·6	27·0	11·2	Brookville,Me	1856		Y 3.87	Ludwigsen & Schjelderup	Christiania
1059	✱	Lucy H. Russell *Randall -87*	W4mSr	American	1108	194·3	39·9	18·0	Bath,Me.	1887	NewEng.S.B.Co		H.M.Randall	Pt.Jefferson
1060		Lucy Hammond *Robinson*	WSr	American	150	96·5	26·3	8·9	E.Machias,Me	1869			J.R.Talbot	E.Machias,Me.
1061		Lucy M. Collins *Newcomb*	WSr	American	159	100·3	27·8	8·3	WinterportMe	1867			G.B.Hook	Bangor,Me.
1062	✱	Lucy March *Griffiths -84*	WBkn	British	244	118·0	26·6	14·1	Pwllheli	1876	Williams	Y 9.86	Morgan,Jones & Co.	Carnarvon
1063	✱	LucyReppen *Bach -86*	WBk	Norweg'n	892	180·6	35·3	20·5	N.Bruns.	1877	Brown	Y 12.87	O.C.Johansen	Christiansand
1064		Lucy Richmond *Richmond*	WSrBarge	British	137	106·7	24·4	8·6	Ipswich	1875	Robertson		W.T.Richmond & Co.	Maldon
1065	✱DI	Lucy W. *Cello*	WBk	Italian	683	159·1	33·1	21·6	SestriPonente	1876	A.Calcagno	Y 4.86	P.Zenoglio	Genoa
1066		Lucy W. Snow *Burgess*	WBn	American	299	112·8	26·9	14·0	Rockland,Me.	1869	Snow&Farwell		F.C.Butman	Boston,Mass.
1067	V	Lucy and Paul *Andreas*	WBk	German	377	127·9	28·7	15·9	Elbing	1861	C.G.Fechter		E.Burchard	Rostock
1068		Ludolfina Jantina *Crispen -86*	WBn	Dutch	156	95·0	23·0	10·6	Pekela	1874	L.H.DeWijk		H.J.Kuiper (H.L.de Wyk Manager)	Oude Pekela
1069		Ludovic	WSr	French	120	89·6	19·7	10·9	St.Malo	1878	Tranchemer frs.		Vve H.Ruellan	Paimpol
1070	N	ludvig *Larsen 82-83* (ALEXANDER)	WBk	Norweg'n	302	121·1	27·4	13·5	Brevig	1870	G.Moe		G.J.Sörensen	Sandefjord
1071	N	Ludvig *Mathisen*	WSr	Norweg'n	110	83·6	20·9	9·6	Brake	1851			Chr.& John Mathisen	Horten
1072	DNV	Ludvig Holberg *Hansen 84-86*	WBk	Norweg'n	1019	193·1	37·5	21·7	Bergen	1878	Brunchorst&Dekke	Y 8.88	A.F.Klaveness	Sandefjord
1073		Ludvig (MUSSINE) *Schauer*	WBk	German	368	128·6	26·6	13·9	Greifswald	1865	C.Dodt		J.Rudolph	Stettin
1074		Ludwig *Eichen -89*	WSr	Russian	126	82·0	21·0	9·5	Haynasch	1872	M.Eichen		A.Eichen & F.Weide	Haynasch
1075		LudwigCapobus,sen. *Permien 81-84*	WBg	German	242	98·3	26·2	14·8	Rostock	1863	W.Zeltz		A.Burchard	Rostock
1076	R	Luella A. Snow *Rose*	WSr	American	178	106·7	27·0	8·1	Rockland,Me.	1875	Snow&Farwell	Y 7.87	Snow & Pearsons	Rockland,Me.
1077	✱	Lufra *Rogers*	CBk	British	672	179·7	31·1	17·9	Aberdeen	1870	Hall	Y 6.87	A.McGregor	Hobart
1078		Lufra	WKYt	British	131	108·1	22·0	12·2	Cowes	1865	M.Ratsey		Exrs. of the Duke of Rutland,K.G.	Cowes
1079	N	Lufra *Evensen*	WBk	Norweg'n	489	133·6	31·3	17·6	Grimstad	1874	O&L.Dömmesmoen	Y 9.88	J.E.Einersen	Grimstad
1080	R	Lugano *Clark*	WSr	American	166	101·5	28·1	9·0	E.Machias,Me	1867			J.G.Murchie & Sons	New York
1081	✱I	Luigi (ex Caterina B.) *Cuneo*	WBk	Italian	828	168·7	33·7	23·6	Finale	1877	F.Ponziolo	Y 1.86	F.Cuneo fu A.	Genoa
1082	A	Luigi *Scottinia*	WBk	Italian	511	..	..	..	Brooklyn	1865	Th.Stach	Y 6.88	L.Pirandello	Messina
1083		Luigi *Corradini*	WBg	Italian	193	111·8	22·1	14·1	Capo d'Istria	1864			A.Bini & Co.	Genoa
1084	A	Luigi *Di Pace -88*	WBn	Italian	141	78·7	23·3	13·3	Castellamare di Stabia	1863		Z 1.88	F.Troise	Naples
1085	I	Luigi Accame *Olivari*	WBk	Italian	939	176·8	35·4	22·6	SestriPonente	1870	G.Casanova	Y 7.85	Em.Accame fu L.	Genoa
1086		Luigi C. *Tancredi* (ex Vittoria)	WBk	Italian	249	107·6	26·6	16·3	Leghorn	1853	G.Micheli		E.Pieruzzini	Leghorn
1087	A	Luigi Fortunato *Ghezzo*	WTrb	Italian	117	83·0	24·4	10·6	Chioggia	1881	L.Menetto	Z 8.81	Fili Ghezzo & Co.	Venice
1088	I	Luigi P. *Maglia -87*	WBk	Italian	657	143·5	32·4	21·0	Prà	1873	B.Canepa	Y 8.88	G.Moglia	Genoa
1089	I	Luigi Padre *Olivari*	WBk	Italian	529	140·7	31·2	20·7	Savona	1875	F.Calamaro	Y 12.87	Fili Sciaccaluga	Genoa
1090		Luigi Padre	WBn	Italian	121	71·2	23·2	13·1	Varazze	1855			C.A.Giorgetti	Leghorn
1091		Luigi Repetto *Repetto*	WBk	Italian	220	101·7	25·6	14·8	St.Valery	1853			Fort. Repetto	Genoa
1092	I	Luigi Ruggero *Maresca -88*	WBk	Italian	490	131·2	29·2	18·7	Castellamare	1873	L.Trambarulo	Y 3.86	L.Ruggero	Castellamare
1093		Luigia *Bessati*	WBk	Italian	402	122·3	29·7	18·4	SestriPonente	1866	G.Peragallo		Fili Pendibene	Genoa
1094		Luigia *Calabrese*	WBg	Italian	231	93·5	26·1	16·2	Piraeus	1867			G.B.Borzone fu G.	Leghorn
1095		Luigia F. *Villa*	W	Italian	491	147·6	30·7	19·7	SestriPonente	1866	L.Casella		B.Fravega fu G.B.	Genoa
1096	I	Luigia M. *Massone -88*	WBk	Italian	421	128·9	30·2	17·9	SestriPonente	1860	A.Briasco fu G.B	Y 5.88	E.Massone	Genoa
1097	I	Luigia R. *Dorogna -88*	WBn	Italian	498	137·8	30·0	18·7	Chiavari	1867	M.Tappani	Y 5.88	F.Rainusso	Genoa
1098	IV	Luigia Raffo *Gallo*	WBk	Italian	972	183·4	38·1	23·1	Chiavari	1875	M.Tappani	Y 3.86	Erasmo Raffo fu A.	Spezzia
1099	I	Luigia Rocca *Rebuffo*	WBk	Italian	546	150·2	32·3	19·7	Lavagna	1870	D.Briasco	Y 5.85	Rocca & Lavello	Genoa
1100														

✱=Lloyd's Register; D=Nederlandsche Vereeniging; G=Germanischer Lloyd; I=Registro Italiano; R=American Record;
A=Veritas Austro-Ungarico; H=Veritas Ellenico; N=Norske Veritas; V=Bureau Veritas.

Dans la 4me Colonne, C=Composite (membrure en fer, bordé en bois); I=Fer; Stl=Acier; W=Bois.

Les dates de doublage en métal (Col. 11) ressortent des livres dans lesquels les navires sont respectivement classés ou des rapports des Inspecteurs du Lloyd's Register.

Pour plus amples renseignements sur les navires classés par le Lloyd's Register, voir le Lloyd's Register Book et le Lloyd's Yacht Register.

Official No.	No. in Book	Registry in which classed	Ship's Name	Master, and Dates of Appointment / Ship's Late Name	Material and Rig	Flag	Net Tonnage	Length	Breadth	Depth	Built: Where	When	By whom	Date of Metalling	Owners or Managers
	1001	DNV	Lovspring	Thomsen	W Bk	Norweg'n	547	149·4	30·9	17·8	Christianssand	1873	H. Olsen	[illegible]	H. G. Kirsebom
	1002	N	Lovspring	Schive -88	W Bk	Norweg'n	489	136·4	31·0	16·9	Grimstad	1885	N. O. Gjömle	[illegible]	J. Jörgensen
	1003	N	Lovspring	Thönnesen	W Bg	Norweg'n	225	107·5	26·9	12·2	Porsgrund	1872			P. M. Petersen & Sön
72335	1004	RV	Low Wood	Thurber -87	W Bk	British	1091	186·7	37·6	22·4	Portland, N.B.	1878	D. Lynch	[illegible]	H. D. Troop
[illegible]	1005		Lowestoft	Hook	W Bk	British	293	112·7	26·5	16·5	Sunderland	1866	J. Dennison		R. L. Manners
76389	1006	✳	Loweswater	Smith -88	I Bk	British	603	174·0	30·2	17·5	Whitehaven	1876	Whitehvn S B Co		Jackson, Metcalf & Co
91291	1007	✳	Lowther Castle	Scarborough 85-86	I S	British	1824	262·1	39·0	23·3	Workington	1886	R. Williamson & Son		Lancaster Shipowners Co. Ld.
	1008	N	Loyal	Friis -87 (ex Dreadnought) (WRECKED)	W S	Norweg'n	1427	220·2	38·9	24·7	Quebec	1869	Ross	[illegible]	Rederiaktieselskab "Loyal" (Chr. Eilefsen)
	1009	D	Loyal	[illegible] -88 (ex Wilhelm & Elise [illegible])	W Bk	Norweg'n	663	159·9	31·9	18·4	Grimstad	1883	S. Johnsen	[illegible]	M. F. [illegible]
85531	1010	V	Loyalist	Fitson 83-88	W Bn	British	348	130·5	29·4	13·0	Portland, N.B.	1883	F. E. Sayre	[illegible]	F. E. Sayre
	1011		Loyo (ex A [illegible])	San Thome	W 3m Sr	Brazilian	242	100·0	24·3	11·5	[illegible]	1876	Borda	[illegible]	J. da Silva Loyo & Filho
	1012		Lua		W Sr	Brazilian	159	..	..	..					V. A. e Magalhaes
	1013	N	Luca	Jansen 58-75	W Sr	Norweg'n	399	116·3	28·3	18·4	Buccari	1857			H. E. Jansen
88394	1014	V	Lucania	Gibson 72-86	W S	British	1774	229·3	42·1	24·4	St. John, N.B.	1886	J. F. Fraser	[illegible]	R. Smith & Sons
	1015	G	Lucas (ex Lina)	Packer 85-88	W Sr	German	159	91·9	21·1	11·1	Papenburg	1863	F. W. Meyer		G. & H. Aden
	1016	✳	Lucayas	Johansen	I Bk	Danish	445	150·4	25·5	18·1	Greenock	1869	Robertson & Co		Th. Meinertz's Enke
1528	1017	V	Lucco	Degregori -87 (BROKEN UP)	W Bk	Italian	498	134·5	30·8	20·0	Sestri Ponente	1866	L. Brisseo	[illegible]	B. Olivari
1974	1018	I	Luchina C.	Repetto	W Bk	Italian	506	134·5	31·2	19·8	Lavagna	1869	G. Saccomanno	[illegible]	A. Cichero
70689	1019		Lucia (ex Mossa Fidele)	Wood 83-86	C Bk	British	640	162·3	31·1	18·3	Sunderland	1868	J. Blumer & Co.	[illegible]	T. Roberts
552	1020		Lucia (ex Angiolina)	Carletta	W Bg	Italian	169	100·0	23·1	14·2	Leghorn	1863	G. Gori	[illegible]	Oreste Del Buono
101	1021		Lucia		W Sr	Russian	126	89·7	23·6	10·4	Kurland	1876			Oskar von Stryk
15102	1022		Lucia B. Ives (ex E. H. Parker)	Mott	W Sr	American	177	96·8	27·2	8·5	Belleville, N.J.	1853/64			A. H. Mott
140498	1023	R	Lucia Porter	Grindle -81	W 3m Sr	American	316	132·0	32·0	10·5	Belfast, Me.	1881		[illegible]	J. F. Grindel
	1024		Lucie		W Sr	French	143	88·3	23·3	12·1	Dunkirk	1846			Connolet frères
	1025	N	Lucie	Bidsaay 75-87	W Sr	Norweg'n	116	95·2	24·0	10·0	Christiansund	1877		[illegible]	Parelius & Lossius
246	1026		Lucie	Johnsohn 82-82	W Sr	Russian	149	80·4	25·5	10·1	Kürbis	1877	Grass		K. Berg & J. Jansohn
	1027		Lucie (ex Dagny)	Hedberg -88	W Bg	Swedish	214	104·0	22·8	12·0	Miramichi	1858			J. Pettersson
	1028		Lucie Gabrielle	(DISMANTLED) (ex Peppino P.)	W Bg	French	417	107·9	26·9	18·7	Sestri Ponente	1866	G. B. Cadenaccio		C. Guérin
	1029	V	Lucie Marie (ex Ayna)	Denis -88	W Bn	French	163	95·8	20·0	12·1	Finland	1858			Dupuis-Robial
	1030		Lucie Marie	Bertho (BROKEN UP)	W Sr	French	141	81·3	23·0	11·1	Nantes	1866	Tille		D. Bertho
	1031	G	Lucie Radmann (ex Susanne Johanna, ex Koening Vaderland)	Gerth -88	W Bk	German	447	117·6	29·9	18·9	Vlaardingen	1854	Von Dam		W. Lüdke & Co.
140060	1032		Lucie Wheatley	Tellinger -88	W 3m Sr	American	180	113·8	29·5	8·2	Frederica, Del.	1874	Nathaniel Lank		J. W. Hall
	1033	V	Lucien & Valentine (ex Anna)	Verdeis	W Bg	French	274	110·2	27·6	14·8	Arendal	1877		[illegible]	K. Morvant
622	1034	A	Lucietta (ex Giuseppe)	Musana	W Sr	Italian	124	80·7	22·4	12·1	Gaeta	1858		[illegible]	Ant. Burgio & Co.
66855	1035	V	Lucile	Williams	W Bkn	British	438	128·8	31·1	17·6	Tynemo'th, N.B.	1872	R. Lowitt	[illegible]	A. J. Williams
75389	1036		Lucile		W Sr Yt	British	157	117·0	23·1	13·4	Gosport	1876	Camp'r & Nichlsn		N. Rudge
140031	1037	V	Lucile	Serman	W S	American	1329	200·2	40·0	23·9	Freeport, Me.	1874	E. C. Soule	[illegible]	E. C. Soule
71444	1038	✳	Lucille	Kirby -88	W Bg	British	278	117·6	26·1	14·9	P. E. Island	1874	Yeo	[illegible]	W. Williams & Sons
57022	1039		Lucinda	White 88-88	W Bn	British	199	105·8	24·1	12·9	P. E. Island	1867	Tuplin		H. W. Brightman
140536	1040		Lucinda G. Potter	Potter -88	W 3m Sr	American	613	151·8	35·8	15·9	Bath, Me.	1882	Goss & Sawyer		W. T. Potter
	1041		Lucinde	Jensen 81-87	W Bg	Danish	150	86·5	23·7	13·4	Holbæk	1842	V. Lagoni		Den Kgl. Grønl. Handel
	1042	DV	Lucinde	Krakauschke 86-	W Bk	Russian	319	117·0	28·0	14·0	Libau	1868	J. D. Dargewitz		C. Essmann
	1043		Lucinde	Hansson	W Sr	Swedish	192	76·8	20·0	10·8	[illegible]	1861			M. Hansson
231	1044		Lucio (ex Maria Regina)	Martinolich -87	W Bg	Austrian	254	86·0	26·5	16·1	Prà, Italy	1870		[illegible]	Ivecvich & Co.
	1045	V	Lucio I.	Faz	W Sr	Portug'se	150	89·6	21·5	10·6	Villa do Conde	1884	C. M. de Araujo	[illegible]	V. L. de Almeida
90005	1046	✳	Lucipara	Harder 86-88	I 4m S	British	1863	267·1	40·2	23·8	Greenock	1885	Russell & Co.		P. Denniston & Co.
87113	1047	✳	Lucknow	Lidstone 81-87	Stl S	British	1440	245·3	37·1	22·5	Sunderland	1883	Doxford & Sons		Tyser & Co.
79485	1048	✳	Lucy	Davies	W Bkn	British	192	112·8	24·1	12·5	Ipswich	1879	Bayley	[illegible]	J. C. Peacock
140515	1049	R	Lucy	Cook	W Sr	American	297	130·0	30·0	8·0	Berwick, Me.	1882			American Salt Co.
15046	1050		Lucy	Webster	W Sr	American	129	87·5	27·8	9·1	Eastport, Me.	1864	Huston		L. A. Godfrey

✳=Lloyd's Register; D=Nederlandsche Vereeniging; G=Germanischer Lloyd; I=Registro Italiano; R=[illegible]
A=Veritas Austro-Ungarico; H=Veritas Ellenico; N=Norske Veritas; V=[illegible]

In the 4th Column, C=Composite (iron frame, wood planked); I=Iron; Stl=Steel; W=Wood.

The Dates of Metalling (Col. 11) are based on the records of the Books in which Vessels are respectively classed, or on the Reports of the Surveyors of [illegible]

Further particulars of Vessels classed by Lloyd's Register will be found in Lloyd's Register Book and Lloyd's Yacht Register.

No.		Nom. Capitaine et dates d'entrée. Ancien Nom.	Matériaux et Gréement	Pavillon	Tonnage net	Long.	Larg.	Creux	Construit. Où.	Quand.	Par qui.	Date du doublage en métal	Armateurs ou Directeurs-Gérants.	Port d'Armement.
1051		Lucy *Lilzan* (ex Rosa Parker)	W Sr	Dutch	355	120·0	27·9	12·5	Harb'rvilleN.S.	1873	J.Givan		Kali Anam Maatschappy	Batavia
1052	✻	Lucy *Kleinsve* (ex Lucy Perry)	W Bk	Norweg'n	739	176·2	33·0	19·5	P.E.Island	1875	J.C.Pope	Y 4.88	Actieselskabet "Lucy" (C.Møller)	Christiania
1053	DN	Lucy *Röer 73-79*	W Bg	Norweg'n	312	120·5	27·9	13·2	Cornwall,Can.	1865			Martin Röer	Dröbak
1054		Lucy A. Blossom	W3mSr	American	179	108·0	24·0	9·6	Conneaut Ohio	1846			J.Philbrook	Winterport,Me.
1055	R	Lucy A. Davis *Davis -87*	W3mSr	American	596	148·5	34·2	12·6	Yarmouth,Me	1882	Hitchins&Stubbs	Y 9.87	J.W.Davis	Portland,Me.
1056	R	Lucy A. Nickels *Nichols*	W S	American	1395	199·0	37·9	23·8	Brewer,Me.	1875	McGilvery	Y 7.87	J.C.Nickels	Searsport,Me.
1057	R	Lucy E. Friend *Collins*	W3mSr	American	447	147·0	34·6	12·0	[illegible]	1882	G.E.Currier		W.H.Collins	Gloucester,Mass.
1058	N	Lucy Frances *Stokke*	W3mSr	Norweg'n	232	111·6	27·0	11·2	[illegible]	1856		Y 3.82	Ludwigsen & Schjelderup	Christiania
1059	R	Lucy H. Russell *Russell -87*	W 3mSr	American	1108	194·3	39·9	18·0	Bath,Me.	1887	New Eng.S.B.Co		H.M.Randall	Pt.Jefferson
1060		Lucy Hammond *Johnson*	W Sr	American	150	96·5	26·3	8·9	E.Machias,Me	1869			J.R.Talbot	E.Machias,Me.
1061		Lucy M. Collins *Newcomb*	W Sr	American	159	100·3	27·8	8·3	Winterport Me	1867			G.B.Hook	Bangor,Me.
1062	✻	Lucy March *Griffiths -89*	W Bk	British	244	118·0	26·6	14·1	Pwllheli	1876	Williams	Y 9.89	Morgan,Jones & Co.	Carnarvon
1063	✻	Lucy Reppen *-88* (ex Josef Reppen)	W Bk	Norweg'n	892	180·6	35·3	20·5	N.Bruns.	1877	Brown	Y 12.87	O.C.Johansen	Christiansand
1064		Lucy Richmond *Richmond*	W Sr Barge	British	137	106·7	24·4	8·6	Ipswich	1875	Robertson		W.T.Richmond & Co.	Maldon
1065	✻DI	Lucy W. *Pollo*	W Bk	Italian	683	159·1	33·1	21·6	SestriPonente	1876	A.Calcagno	Y 4.85	P.Zenoglio	Genoa
1066		Lucy W. Snow *Barrett*	W Bk	American	299	112·8	26·9	14·0	Rockland,Me.	1869	Snow&Farwell		F.C.Butman	Boston,Mass.
1067	V	Lucy and Paul *Andreis 73-86*	W Bk	German	377	127·9	28·7	15·9	Elbing	1861	C.G.Fechter		E.Burchard	Rostock
1068		Ludolfina Jantina *Peppen*	W Bn	Dutch	156	95·0	23·0	10·6	Pekela	1874	L.H.DeWijk		H.J.Kuiper (H.L.de Wijk, Manager)	Oude Pekela
1069		Ludovic	W Sr	French	120	89·6	19·7	10·9	St.Malo	1878	Tranchemer frs.		Vve H.Ruellan	Paimpol
1070	N	Ludvig *Larsen* (ABANDONED)	W Bk	Norweg'n	302	124·1	27·4	13·5	Brevig	1870	G.Moe		G.J.Sörensen	Sandefjord
1071	N	Ludvig *Mathisen*	W Sr	Norweg'n	110	83·6	20·9	9·6	Brake	1851			Chr.& John Mathisen	Horten
1072	DNV	Ludvig Holberg *Hansen 81-86*	W Bk	Norweg'n	1019	193·1	37·5	21·7	Bergen	1878	Branchorst & Dekke	Y 8.88	A.F.Klaveness	Sandefjord
1073		Ludvig (MISSING) *Schauer*	W Bk	German	368	128·6	26·6	13·9	Greifswald	1865	C.Dodt		J.Rudolph	Stettin
1074		Ludwig *Eichen -86*	W Sr	Russian	126	82·0	21·0	9·5	Haynasch	1872	M.Eichen		A.Eichen & F.Weide	Haynasch
1075		Ludwig Capobus, sen. *Prowien*	W Bg	German	242	98·3	26·2	14·8	Rostock	1863	W.Zeltz		A.Burchard	Rostock
1076	R	Luella A. Snow *Rose*	W Sr	American	178	106·7	27·0	8·4	Rockland,Me.	1875	Snow&Farwell	Y 7.8	Snow & Pearsons	Rockland,Me.
1077	✻	Lufra *Rogers*	C Bk	British	672	179·7	31·1	17·9	Aberdeen	1870	Hall	Y 4.87	A.McGregor	Hobart
1078		Lufra	W K Yt	British	131	108·1	22·0	12·2	Cowes	1865	M.Ratsey		Exrs. of the Duke of Rutland,K.G.	Cowes
1079	N	Lufra *Evensen*	W Bk	Norweg'n	489	135·6	31·5	17·6	Grimstad	1874	O&L Dömmesmoen	Y 8.88	J.E.Einersen	Grimstad
1080	R	Lugano *Clark*	W Sr	American	166	104·5	28·1	9·0	E.Machias,Me	1867			J.G.Murchie & Sons	New York
1081	✻I	Luigi (ex Caterina R.) *Cuneo*	W Bk	Italian	828	168·7	33·7	23·6	[illegible]	1877	F.Pouriolo	Y 1.89	F.Cuneo fu A.	Genoa
1082	A	Luigi *Scaffiola*	W Bk	Italian	511	..	..	..	Brooklyn	1865	Th.Stach	Y 8.83	L.Pirandello	Messina
1083		Luigi *Corradini*	W Bg	Italian	193	111·8	22·1	14·1	Capod'Istria	1864			A.Bini & Co.	Genoa
1084	A	Luigi *Di Pace -88*	W Bn	Italian	141	78·7	23·3	13·3	Castellamare [di Stabia	1863		Z 1.88	F.Troise	Naples
1085		Luigi Accame	W Bk	Italian	939	176·8	35·4	22·6	SestriPonente	1870	G.Casanova	Y 7.88	Em.Accame fu L.	Genoa
1086		Luigi C. (ex Vittoria) *Tancredi*	W Bk	Italian	249	107·6	26·6	16·3	Leghorn	1853	G.Micheli		E.Pieruzzini	Leghorn
1087	A	Luigi Fortunato	W Trb	Italian	117	83·0	24·4	10·6	Chioggia	1881	L.Menetto	Z 8.84	Fli Ghezzo & Co.	Venice
1088	I	Luigi P. *Moglia -87*	W Bk	Italian	657	143·5	32·4	21·0	[illegible]	1873	B.Canepa	Y 8.88	G.Moglia	Genoa
1089		Luigi Padre *Olivari -88* (ex Orazio Piccone)	W Bk	Italian	529	140·7	31·2	20·7	Savona	1875	F.Calamaro	Y 12.87	Fli Sciaccaluga	Genoa
1090		Luigi Padre (ex Marco, ex Angiolina)	W Bn	Italian	121	71·2	23·2	13·1	Varazze	1855			C.A.Giorgetti	Leghorn
1091		Luigi Repetto *Repetto*	W Bk	Italian	220	101·7	25·6	14·8	St.Valery	1853			Fort Repetto	Genoa
1092	I	Luigi Ruggero *Marasco -88*	W Bk	Italian	490	131·2	29·2	18·7	Castellamare	1873	L.Trambarulo	Y 3.88	L.Ruggero	Castellamare
1093		Luigia *Bennati*	W Bk	Italian	402	122·3	29·7	18·4	SestriPonente	1866	G.Peragallo		Fli Pendibene	Genoa
1094		Luigia (ex Zeny) *Calabrese*	W Bg	Italian	231	93·5	26·1	16·2	Piræus	1867			G.B.Borzone fu G.	Leghorn
1095		Luigia F. (ex Consiglio) *Villa*	W	Italian	491	147·6	30·7	19·7	SestriPonente	1866	L.Casella		B.Fravega fu G.B.	Genoa
1096	I	Luigia M. *Rassoni -88* (ex Ermenegilda Danovaro)	W Bk	Italian	421	128·9	30·2	17·9	SestriPonente	1880	A.Briasco fu G.B.	Y 5.8	E.Massone	Genoa
1097	I	Luigia R. *Dasogno -88*	W Bn	Italian	498	137·8	30·0	18·7	Chiavari	1867	M.Tappani	Y 3.88	F.Raimusso	Genoa
1098	IV	Luigia Raffo *Guido*	W Bk	Italian	972	183·4	38·1	23·1	Chiavari	1875	M.Tappani	Y 3.86	Erasmo Raffo fu A.	Spezzia
1099	I	Luigia Rocca *Rebuffo*	W Bk	Italian	546	150·2	32·3	19·7	Lavagna	1870	D.Briasco	Y 8.8	Rocca & Lavello	Genoa
1100														

✻=Lloyd's Register; D=Nederlandsche Vereeniging; G=Germanischer Lloyd; I=Registro Italiano; R=American Record;
A=Veritas Austro-Ungarico; H=Veritas Ellenico; N=Norske Veritas; V=Bureau Veritas.

Dans la 4me Colonne, C=Composite (membrure en fer, bordé en bois); I=Fer; Stl=Acier; W=Bois.

Les Dates du doublage en métal (Col. 11) ressortent des livres dans lesquels les navires sont respectivement classés ou des rapports des Inspecteurs du Lloyd's Register.

Pour plus amples renseignements sur les navires classés par le Lloyd's Register, voir le Lloyd's Register Book et le Lloyd's Yacht Register.

Official No.	No. in Book	Society classed	Ship's Name (Ship's Late Name)	Master, and Dates of Appointment	Material and Rig	Flag	Net Tonnage	Length	Breadth	Depth	Built: Where	When	By whom	Date of Metalling	Owners or Managers	Port
	1001	DNV	Lovspring	Thomsen	WBk	Norweg'n	547	143·4	[illegible]	17·8	[illegible]	1873	[illegible]	[illegible]	H.G.Kirsebom	
	1002	N	Lovspring	Sekere	WBk	Norweg'n	489	136·4	31·0	16·9	Grimstad	1865	N.O.Gjössle	[illegible]	J.Jörgensen	
	1003	N	Lovspring	Thomassen	WBg	Norweg'n	225	107·5	26·9	12·2	[illegible]	1872			P.M.Petersen & Sön	
72335	1004	RV	Low Wood	Thurber -87	WBk	British	1091	186·7	37·6	22·4	Portland,N.B.	1878	D.Lynch	[illegible]	H.D.Troup	St.J
56478	1005		Lowestoft	Hook	WBk	British	293	112·7	26·5	16·5	[illegible]	1866	J.Dennison		R.L.Manners	
76388	1006	✳	Loweswater	Smith -88	IBk	British	603	171·0	30·2	17·5	[illegible]	1876	WhitehvnSBCo		Jackson,Metcalf & Co.	Liverpool
91391	1007	✳	Lowther Castle	[illegible]	IS	British	1824	262·1	39·0	23·3	[illegible]	1886	R.Williamson & Son		Lancaster Shipowners' Co.Ld.	
	1008	N	Loyal (ex Broadsound) (WRECKED)	Friis -87	WS	Norweg'n	1427	220·2	38·9	24·7	Quebec	1863	Rosa	[illegible]	Rederiselskab "Loyal" (Chr.Ellefsen)	
	1009	D	Loyal (WRECKED)	[illegible] -88	WBk	Norweg'n	663	159·9	31·9	18·4	[illegible]	1888	S.Johnsen	[illegible]	M.F.Stray	
83504	1010	V	Loyalist	Troop 88-88	WBn	British	348	130·5	29·4	13·0	[illegible],N.B.	1883	F.E.Sayre	[illegible]	F.E.Sayre	St.John
	1011		Loyo (ex Alberto)	San Thame	W3mSr	Brazilian	242	100·0	24·3	11·5	[illegible]	1876	Borda	[illegible]	J.daSilvaLoyo & Filho	
	1012		Lua		WSr	Brazilian	150	..	..	..					V.A.e Magalhaes	Parnahyba
	1013	N	Luca	Jansen 68-75	WSr	Norweg'n	399	116·3	28·3	18·4	[illegible]	1857			H.E.Jansen	
88304	1014	V	Lucania	Gibson 72-86	WS	British	1774	229·3	42·4	24·4	St.John,N.B.	1886	J.F.Fraser	[illegible]	B.Smith & Sons	Windsor
	1015	G	Lucas (ex Lina)	Pocker 63-85	WSr	German	159	91·9	21·1	11·1	Papenburg	1863	F.W.Meyer		G.& H.Aden	
	1016	✳	Lucayas	Johansen	IBk	Danish	445	150·4	25·5	18·1	Greenock	1869	Robertson & Co		Th.Meinertz's Enke	
1526	1017	V	Lucco (BROKEN UP)	Degregori -87	WBk	Italian	468	134·5	30·8	20·0	Sestri Ponente	1866	L.Briasco	[illegible]	B.Olivari	Genoa
1974	1018	I	Luchina C.	Repetto	WBk	Italian	506	134·5	31·2	19·8	Lavagna	1869	O.Saccomanno	[illegible]	A.Cichero	Genoa
70689	1019		Lucia (ex Maria Estella)	Wood 83-86	CBk	British	640	162·3	31·1	18·3	Sunderland	1868	J.Blumer & Co.	[illegible]	T.Roberts	London
657	1020		Lucia (ex Angiolina)	Carlotta	WBg	Italian	169	100·0	23·1	14·2	Leghorn	1863	G.Gori	[illegible]	Oreste Del Buono	
101	1021		Lucia		WSr	Russian	126	89·7	23·6	10·4	Kurland	1876			Oskar von Stryk	
15102	1022		Lucia B. Ives (ex E. H. Packer)	Mott	WSr	American	177	96·8	27·2	8·5	[illegible],N.J.	1853 64			A.H.Mott	Pt.Jefferson
140498	1023	R	Lucia Porter	Grindle -81	W3mSr	American	316	132·0	32·0	10·5	Belfast,Me.	1881		[illegible]	J.F.Grindel	Belfast,Me.
	1024		Lucie		WSr	French	143	88·3	23·3	12·1	Dunkirk	1846			Comolet frères	
	1025	N	Lucie	Eidsaay 75-87	WSr	Norweg'n	116	95·2	24·0	10·0	Christianssund	1877		[illegible]	Parelius & Lossius	
246	1026		Lucie	Johnsohn 82-83	WSr	Russian	149	89·4	25·5	10·1	Kürbis	1877	Grass		K.Berg & J.Jansohn	Riga
	1027		Lucie (ex Dagmar)	Hedberg -88	WBg	Swedish	214	104·0	22·8	12·0	Miramichi	1858			J.Pettersson	Gothenburg
	1028		Lucie Gabrielle (DISMANTLED) (ex Peppino P.)		WBg	French	417	107·9	26·9	18·7	Sestri Ponente	1866	G.B.Cadenaccio		C.Guérin	Martigues
	1029	V	Lucie Marie (ex Anna)	Denis -88	WBn	French	163	95·8	20·0	12·1	Finland	1858			Dupuis-Robial	St.Brieuc
	1030		Lucie Marie (BROKEN UP)	Bertho	WSr	French	141	84·3	23·0	11·1	Nantes	1866	Tillé		D.Bertho	Nantes
	1031	G	LucieRadmann (ex Susanne Johanna, ex Konig og Vaderland)	Gerth -88	WBk	German	447	117·6	29·9	18·9	Vlaardingen	1854	Von Dam		W.Lüdke & Co.	Stettin
140059	1032		Lucie Wheatley	Pettingell -88	W3mSr	American	180	113·8	29·5	8·2	Frederica,Del.	1874	NathanielLank		J.W.Hall	Wilmington
	1033	V	Lucien & Valentine (ex Anna)	Verdois	WBg	French	274	110·2	27·6	14·8	Arendal	1877		[illegible]	K.Morvant	
522	1034	A	Lucietta (ex Giuseppe)	Musana	WSr	Italian	124	80·7	22·4	12·1	[illegible]	1858		[illegible]	Ant.Burgio & Co.	Catania
60865	1035	V	Lucile	Williams	WBkn	British	498	128·8	31·1	17·0	Tynem'th,NB	1872	R.Lowitt	[illegible]	A.J.Williams	
75309	1036		Lucile		WSrYt	British	157	117·0	23·1	13·1	Gosport	1876	Camper&Nicholson		N.Rudge	Cowes
140031	1037	V	Lucile	Serman	WS	American	1329	200·2	40·0	23·9	Freeport,Me.	1874	E.C.Soule	[illegible]	E.C.Soule	
71441	1038	✳	Lucille	Kirby -88	WBg	British	278	117·6	26·1	14·9	P.E.Island	1874	Yeo	[illegible]	W.Williams & Sons	Newport
57023	1039		Lucinda	White 88-88	WBn	British	199	105·8	24·1	12·9	P.E.Island	1867	Tuplin		H.W.Brightman	
140336	1040		Lucinda G. Potter	Potter -88	W3mSr	American	613	151·8	35·8	15·9	Bath,Me.	1882	Goss&Sawyer		W.T.Potter	
	1041		Lucinde	Jensen 87-87	WBg	Danish	150	86·5	23·7	13·4	Holbæk	1842	V.Lagoni		Den.Kgl.Grønl.Handel	Copenhagen
	1042	DV	Lucinde	Krubauschke 86-	WBk	Russian	319	117·0	28·0	14·0	Libau	1866	J.D.Dargewitz		C.Essmann	Riga
	1043		Lucinde	Hansson	WSr	Swedish	102	76·8	20·0	10·8	Flensburg	1861			M.Hansson	
251	1044		Lucio (ex Maria Rosa)	Marinolich -87	WBg	Austrian	254	86·0	26·5	16·1	Prà,Italy	1870		[illegible]	Ivcevich & Co.	
	1045	V	Lucio I.	Vaz	WSr	Portug'se	153	89·6	21·5	10·0	Villa do Conde	1884	C.M.deAraujo	[illegible]	V.L.de Almeida	
89002	1046	✳	Lucipara	Harder 86-88	I4mS	British	1882	267·1	40·2	23·8	Greenock	1885	Russell&Co.		P.Denniston & Co.	Glasgow
87113	1047	✳	Lucknow	Lidstone 81-87	StlS	British	1440	245·3	37·1	22·5	Sunderland	1883	Doxford&Sons		Tyser & Co.	London
79485	1048	✳	Lucy	Davies	WBkn	British	192	112·8	24·1	12·5	Ipswich	1879	Bayley	[illegible]	J.C.Peacock	
140515	1049	R	Lucy	Cook	WSr	American	297	130·0	30·0	8·0	Berwick,Me.	1882			American Salt Co.	
15046	1050		Lucy	Wooster	WSr	American	129	87·5	27·8	9·1	Eastport,Me.	1864	Huston		L.A.Godfrey	

✳=Lloyd's Register; D=Nederlandsche Vereeniging; G=Germanischer Lloyd; I=Registro Italiano; R=American [illegible];
A=Veritas Austro-Ungarico; H=Veritas Ellenico; N=Norske Veritas; V=Bureau Veritas.

In the 4th Column, C=Composite (iron frame, wood planked); I=Iron; Stl=Steel; W=Wood.

The Dates of Metalling (Col. 11) are based on the records of the Books in which Vessels are respectively classed, or on the Reports of the Surveyors of [illegible]

Further particulars of Vessels classed by Lloyd's Register will be found in Lloyd's Register Book and Lloyd's Yacht Register.

No.	Classe	Nom. Capitaine et dates d'entrée au service des Armateurs actuels et à bord du navire. Ancien Nom.	Matériaux et Gréement	Pavillon	Tonnage net	Longueur	Largeur	Creux	Construit: Où	Quand	Par qui	Date du doublage en métal	Armateurs ou Directeurs-Gérants	Port d'Armement
1051		Lucy [illegible] -86	W Sr	Dutch	355	120·0	27·9	12·5	Harb'rville Nr	1873	J.Givan		Kali Anam Maatschappy	Batavia
1052	※	Lucy *Klouman* -88 (ex [illegible])	W Bk	Norweg'n	739	176·2	33·0	19·5	P.E.Island	1875	J.C.Pope	Y 4.88	Actieselskabet "Lucy" (C.Møller)	Christiania
1053	DN	Lucy *Roer* 79-19	W Bg	Norweg'n	312	120·5	27·9	13·2	Cornwall,Can.	1865			Martin Röer	Dröbak
1054		Lucy A. Blossom	W 3m Sr	American	179	108·0	24·0	9·0	Conneaut Ohio	1846			J.Philbrook	Winterport,Me.
1055	R	Lucy A. Davis *Jordan* -84	W 3m Sr	American	596	148·5	34·2	12·6	Yarmouth,Me	1882	Htchns&Stubbs	Y 9.87	J.W.Davis	Portland,Me.
1056	R	Lucy A. Nickels *Nichols*	W S	American	1395	199·0	37·9	23·8	Brewer,Me.	1875	McGilvery	Y 7.87	J.C.Nickels	Searsport,Me.
1057	R	Lucy E. Friend *Collins*	W 3m Sr	American	447	147·0	34·6	12·6	Nwbrypt Mass	1882	G.E.Currier		W.H.Collins	Gloucester, Mass
1058	N	Lucy Frances *Stokke* -86	W 3m Sr	Norweg'n	232	114·6	27·0	11·2	[illegible]	1856		Y 3.85	Ludw. gaen & Schjelderup	Christiania
1059	R	Lucy H. Russell *Russell* -87	W 4m Sr	American	1108	194·3	39·9	18·6	Bath, Me.	1887	New Eng.S.B.Co		H.M.Randall	Pt.Jefferson
1060		Lucy Hammond *[illegible]*	W Sr	American	150	96·5	26·3	8·9	E.Machias,Me	1869			J.R.Talbot	E.Machias,Me.
1061		Lucy M. Collins *Newcomb*	W Sr	American	159	100·3	27·8	8·3	Winterport Me	1867			G.B.Hook	Bangor,Me.
1062	※	Lucy March *Griffiths* -85	W Bk	British	244	118·0	26·6	14·1	Pwllheli	1870	Williams	Y 9.86	Morgan,Jones & Co.	Carnarvon
1063	※	Lucy Reppen *Bach* -87 (ex [illegible])	W Bk	Norweg'n	892	180·6	35·3	20·5	N.Bruns.	1877	Brown	Y 12.87	O.C.Johansen	Christiansand
1064		Lucy Richmond *Robinson*	W Sr Barge	British	137	106·7	24·4	8·6	Ipswich	1875	Robertson		W.T.Richmond & Co.	Maldon
1065	※ D I	Lucy W. *Vallo*	W Bk	Italian	683	159·1	33·1	21·6	SestriPonente	1876	A.Calcagno	Y 4.86	P.Zenoglio	Genoa
1066		Lucy W. Snow *[illegible]*	W Bn	American	299	112·8	26·9	11·0	Rockland,Me.	1869	Snow&Farwell		F.C.Butman	Boston,Mass.
1067	V	Lucy and Paul *Andree* [illegible]	W Bk	German	377	127·9	28·7	15·9	Elbing	1861	C.G.Fechter		E.Burchard	Rostock
1068		Ludolfina Jantina *Pyppen* -85	W Bn	Dutch	156	95·0	23·0	10·6	Pekela	1874	L.H.DeWijk		H.J.Kuiper (H.L.deWijk, Manager)	Oude Pekela
1069		Ludovic	W Sr	French	120	89·6	19·7	10·9	St.Malo	1878	Tranchemer frs.		Vve. H.Ruellan	Paimpol
1070	N	Ludvig *Larsen* 82-83 (ex [illegible])	W Bk	Norweg'n	302	124·1	27·4	13·5	[illegible]	1870	G.Moe		G.J.Sörensen	Sandefjord
1071	N	Ludvig *Mathisen*	W Sr	Norweg'n	110	83·6	20·9	9·6	[illegible]	1851			Chr.& John Mathisen	Horten
1072	DNV	Ludvig Holberg *Hansen* 81-86	W Bk	Norweg'n	1019	193·1	37·5	21·7	Bergen	1878	Brunchorst& Dekke	Y 8.88	A.F.Klaveness	Sandefjord
1073		Ludvig (ex [illegible]) *Schaner*	W Bk	German	368	128·6	26·6	13·9	Greifswald	1865	C.Dodt		J.Rudolph	Stettin
1074		Ludwig *Eichen* -80	W Sr	Russian	126	82·0	21·0	9·5	Haynasch	1872	M.Eichen		A.Eichen & F.Weide	Haynasch
1075		Ludwig Capobus, sen. *Fermien* 81-85	W Bg	German	242	98·3	26·2	14·8	Rostock	1863	W.Zeltz		A.Burchard	Rostock
1076	R	Luella A. Snow *Rose*	W Sr	American	178	106·7	27·0	8·4	Rockland,Me.	1875	Snow&Farwell	Y 7.84	Snow & Pearsons	Rockland,Me.
1077	※	Lufra *Rogers*	C Bk	British	672	179·7	31·1	17·9	Aberdeen	1870	Hall	Y 6.87	A.McGregor	Hobart
1078		Lufra	W K Yt	British	131	108·1	22·0	12·2	Cowes	1865	M.Ratsey		Exrs. of the Duke of Rutland, K.G.	Cowes
1079	N	Lufra *Evensen*	W Bk	Norweg'n	489	133·6	31·3	17·6	Grimstad	1874	O&L Dönnemoen	Y 8.86	J.E.Einersen	Grimstad
1080		Lugano *Clark*	W Sr	American	166	101·5	28·1	9·0	E.Machias,Me	1867			J.G.Murchie & Sons	New York
1081	※ I	Luigi (ex Caterina B.) *Cuneo*	W Bk	Italian	828	168·7	33·7	23·6	Finale	1877	F.Ponzialo	Y 1.89	F.Cuneo fu A.	Genoa
1082	A	Luigi *Sottiletta*	W Bk	Italian	511	..	..	..	Brooklyn	1865	Th.Stach	Y 6.88	L.Pirandello	Messina
1083		Luigi *Corradini*	W Bg	Italian	193	111·8	22·1	14·1	Capo d'Istria	1864			A.Bini & Co.	Genoa
1084	A	Luigi *Di Pace* -88	W Bn	Italian	141	78·7	23·8	13·3	Castellamare di Stabia	1863		Z 1.88	F.Troise	Naples
1085	I	Luigi Accame *[illegible]*	W Bk	Italian	930	176·8	35·4	22·6	SestriPonente	1870	G.Casanova	Y 7.85	Em.Accame fu L.	Genoa
1086		Luigi C. (ex Vittoria) *Tancredi*	W Bk	Italian	249	107·6	26·6	16·3	Leghorn	1853	G.Micheli		E.Pieruzzini	Leghorn
1087	A	Luigi Fortunato *Ghezzo* -88	W Trb	Italian	117	83·0	24·4	10·6	Chioggia	1881	L.Menetto	Z 8.81	Flli. Ghezzo & Co.	Venice
1088	I	Luigi P. *Maggio* -87	W Bk	Italian	657	143·5	32·4	21·9	Foce	1873	B.Canepa	Y 8.86	G.Moglia	Genoa
1089	I	Luigi Padre (ex [illegible]) *Olivari* -88	W Bk	Italian	529	140·7	31·2	20·7	Savona	1875	F.Calamaro	Y 12.87	Flli. Sciaccaluga	Genoa
1090		Luigi Padre (ex Maria C. ex Angelina) *[illegible]*	W Bn	Italian	121	71·2	23·2	13·1	Varazze	1855			C.A.Giorgetti	Leghorn
1091		Luigi Repetto *Repetto*	W Bk	Italian	220	101·7	25·6	14·8	St.Valery	1853			Fort Repetto	Genoa
1092	I	Luigi Ruggaro *[illegible]*	W Bk	Italian	490	131·2	29·2	18·7	Castellamare	1873	L.Trambarulo	Y 3.86	L.Ruggero	Castellamare
1093		Luigia *Brunati*	W Bk	Italian	402	122·3	29·7	18·4	SestriPonente	1866	G.Peragallo		Flli. Pendibene	Genoa
1094		Luigia (ex [illegible]) *Calabrese*	W Bg	Italian	231	93·5	26·1	16·2	Piraeus	1867			G.B.Borzone fu G.	Leghorn
1095		Luigia F. (ex Consiglio) *Villa*	W	Italian	491	147·6	30·7	19·7	SestriPonente	1866	L.Casella		B.Fravega fu G.B.	Genoa
1096	I	Luigia M. *Massone* (ex [illegible])	W Bk	Italian	421	128·9	30·2	17·9	SestriPonente	1880	A.Briasco fu G.B.	Y 5.85	E.Massone	Genoa
1097	I	Luigia R. (ex [illegible]) *[illegible]* -88	W Bn	Italian	498	137·8	30·0	18·7	Chiavari	1867	M.Tappani	Y 5.88	F.Rainusso	Genoa
1098	IV	Luigia Raffo *Gallo*	W Bk	Italian	972	183·4	38·1	23·1	Chiavari	1875	M.Tappani	Y 8.86	Erasmo Raffo fu A.	Spezzia
1099	I	Luigia Rocca *Rebuffo*	W Bk	Italian	546	150·2	32·3	19·7	Lavagna	1870	D.Briasco	Y 5.85	Rocca & Lavello	Genoa
1100														

[illegible] Lloyd's Register; [illegible] Austro-Ungarico; D = Nederlandsche Vereeniging; G = Germanischer Lloyd; H = Veritas Ellenico; I = Registro Italiano; N = Norske Veritas; R = American Record; V = Bureau Veritas.

Dans la 4me Colonne, C = Composite (membrure en fer, bordé en bois); I = Fer; Stl = Acier; W = Bois.

Les Dates du doublage en métal (Col. 11) ressortent des livres dans lesquels les navires sont respectivement classés ou des rapports des Inspecteurs du Lloyd's Register.

Pour plus amples renseignements sur les navires classés par le Lloyd's Register, voir le Lloyd's Register Book et le Lloyd's Yacht Register.

Official No.	No. in Book.	Society in which classed.	Ship's Name. Material and Rig. Master, and Dates of Appointments to present Owner's Service and to Ship. Ship's Late Name.	Flag.	Net Tonnage.	Length.	Breadth.	Depth.	Built Where.	When.	By whom.	Metalling.	Owners or Managers.
3490	1101		Luigia V. (ex Rosa) Villa W Bk	Italian	562	146·9	32·4	19·7	Sestri Ponente	1871	L.Marcenaro	Y 1.88	E.Villa
2082	1102	A	Luigina (ex Catterina) Repetto W Bk	Italian	396	122·3	29·5	18·4	Recco	1865	P.Rolla	Y 4.88	A.Repetto
1098	1103	I	Luigina (ex Eva) Mando W Bn	Italian	267	102·6	27·4	16·6	Procida	1864	A.Lubrano	[illegible]	A.Galatola
2548	1104		Luigino Guida Scotto W Bk (ex Sebastiano Gattorno)	Italian	456	131·5	31·8	19·4	Sestri Ponente	1867	FlliCadenacio	Y 1.88	L.Guida
60414	1105	V	Luis A. Martinez Scott W Bk	British	540	144·9	31·7	15·9	Merigomish	1873	D.Patterson	[illegible]	D.Patterson
	1106		Luis Alberto W Bk (ex Kilsyth) (Spallanzani)	Chilian	572	144·1	30·5	17·5	St.George,N.B	1872	A.Young		Cia.Esplotadora de Lota & Coronel
140630	1107	R	Luis G. Rabel Murphy [7-88] W3mSr	American	552	152·4	33·5	12·8	E. Deering,Me.	1883	G.Russell	[illegible]	J.S.Winslow & Co.
40073	1108		Luis Mariani W Bn	British	177	89·6	26·0	12·0	Waldob'ro,US	1855			J.Nibs
2523	1109	I	Luisa Narizzano W3mSr	Italian	490	118·0	26·3	15·0	Sestri Ponente	1872	A.Briasco fu G.B	Y 6.87	P.Antola
702	1110		Luisa Carolissimo W Bk (ex G. Paladino, ex Invinta) (PONTOON)	Italian	289	97·6	27·5	17·4	SestriPonente	1854		Y 4.80	F.Cucurullo
334	1111		Luisa L.Casella W Bkn	Italian	278	113·5	25·1	12·1	Sestri Ponente	1886	A.Briasco	Y 4.86	Fran.Casella
	1112		Luisa W P	Spanish	183	88·9	25·9	11·9	Blanes	1854			J.Pastor Marco
	1113		Luisa W Bn	Spanish	129	93·2	20·7	9·4	Tayabas	1886			C.Lagdameo
	1114	V	Luisa Marta (ex Luisa, ex Bismarck, ex Frederica, ex F.U.) W Bk	Chilian	865	167·3	29·8	21·0	SestriPonente	1874	CadenacioBros.	Y 9.84	Meyer,Fonck & Yelck
	1115	✻	Luise Bohne 87-87 I Bk (ex Odalis, ex Great Western)	German	418	138·7	28·6	16·1	Newcastle	1863	Rogerson		C.Dieckhoff & Co.
58	1116		Luise W Cr	Russian	142	80·0	23·0	10·0	Livland	1876			Baroness L.Mengden
	1117		Luisita Gabien W Bg (ex Wilhelmine)	Chilian	250	101·0	26·0	12·0	Elsfleth	1864	G.F.Stindt		J.Saez
	1118		Luiz I. (ex Lizzie Garrow) W Sr	Portug'se	209	94·2	19·3	13·1	Dodbrooke	1854	Date		A.J.Baptista
	1119		Luiza de Vincenzi W Bn (ex Dos Albinos)	Brazilian	242	106·0	26·0	10·0	Carleton,N.B.	1869	J.M.Laughlan		J.N.de Vincenzi & Filhos
58042	1120	✻	Luke Bruce Lott 73-86 I Bk	British	310	138·5	25·2	14·0	Aberdeen	1868	Hall,Russell&C.		T.Harrison & Co.
	1121		Lukianos Frangoulis W Bg (ex Aurora M.)	Greek	334	..	..	..	Lussinpiccolo	1866	G.S.Cattarinich	C.80	G.Frangoulis
140765	1122		Lula E Wilbur Robbins Sr	American	100	87·6	23·7	8·6	Essex,Mass.	1885			J.P.Edwards
21730	1123		Lula M. Quillin W3mSr ex Benning R.R. No. 31 Insley -87	American	122	107·0	18·0	9·0	Hamburg, Pa.	1863			Caleb R.Insley
	1124	V	Luleå Johansson W3mSr	Swedish	250	117·7	23·8	11·8	Luleå	1868	Brandoff		C.O.Olsson
15920	1125		Lulu Briggs -87 W3mSr	American	450	135·5	32·0	14·8	Waldoboro Me	1873	A.Storer&Son		C.& J.F.Baker
140066	1126		Lulu Ammerman W3mSr Predmore -86	American	285	135·0	34·0	11·2	ForkedRivN.J	1873			A.Birdsall
60898	1127	✻	Lumberman's Lassie WS (ex Windsor Castle) Mason 86-87	British	979	198·0	35·3	20·6	Aberdeen	1869	J.Duthie,Ss&Co.	Y 1.89	J.Rust & Son
53877	1128		Lumina WS	British	143	110·0	23·8	8·0	Quebec	1866			T.A.Howard
	1129		Lumparland Mattson W3mSr	Russian	271	..	..	..		1877			E.P.Jansson
6865	1130		Luna Christie 86-86 W Bn	British	171	103·5	23·1	13·0	Aberdeen	1848			A.Farquhar
	1131		Luna Hansen 76-76 W Sr	Danish	115	85·2	20·6	11·5	Troense	1858	C.J.Jensen		P.Jensen
	1132	GV	Luna Hinrichsen 80-85 I Bk	German	820	180·0	31·9	18·6	Kiel	1885	GermaniaWrks.		Wachsmuth&Krogmann
	1133		Luneborg W Bk	Norweg'n	284	111·6	27·7	14·6	Harburg	1853			J.Hansen
83095	1134	✻	Lunesdale Jones 76-86 W3mSr	British	141	102·2	22·9	10·7	GlassonDock	1884	Nichlsn.&Mrsh.		J.Fisher & Sons
15866	1135		Lunet Swain W Sr	American	171	103·0	28·0	8·0	Calais,Me.	1872	Hinds		E.A.Barnard
1528	1136		Luno W Bk	Italian	488	134·9	30·9	20·0	SestriPonente	1867			O.Biagio fu Prospero
74486	1137	✻	Lurlei Hetherington I Bk	British	803	185·0	31·7	19·8	Liverpool	1875	T.Royden&Son		J.B.Walmsley&Co.
76744	1138	✻	Lurline Lawes I Bk	British	761	189·5	31·0	18·0	Glasgow	1877	A.Stephen&Sns.		C.W.Turner
60864	1139		Lurline Searle W Bn	British	189	101·6	23·6	12·9	Salcombe	1870	Hamlen	[illegible]	W.R.Beer
140831	1140	V	Lurline Matson -87 W Bn	American	341	135·0	34·0	13·0	Benicia,Cal.	1887	M.Turner		C.A.Spreckels
	1141	GV	Lusitania Da Costa 84- W Bk (ex Sebastiao Guanabara)	Portug'se	523	151·5	29·2	18·4	Palma	1875			Cia.Alliança Fabril
52454	1142		Lusteria Purrior W Sr (BROKEN UP)	British	104	79·0	22·5	10·4	LesEboulmts.	1865			A.Purrior
83963	1143		Luta Price Copp -88 W Sr	British	121	85·0	27·2	7·9	Portland,N.B.	1882	J.Calhoun		R.B.Price
	1144		Lutfi Bahri W Bg	Turkish	219	..	..	..	Heraclea	1869			[illegible]
	1145		Lutfi Hûda W Sr	Turkish	295	..	..	..	Aliyandon	1885			[illegible]
	1146		Lutfi Huda W Sr	Turkish	295	..	..	..	Ahyandon	1880			[illegible]
	1147		Lutfiyeh W Bg	Turkish	169	..	..	..	Greece	1883			[illegible]
	1148	R	Luther Jansson W Sr	Swedish	139	92·2	21·5	10·2	Figeholm	1876	G.Pettersen		S.J.Johansson
140621	1149	R	Luther A. Roby Blake W3mSr	American	608	150·0	35·5	15·0	E.Boston,Mas.	1884	J.M.Brooks		J.M.Brooks
140591	1150	R	Luther T. Garretson W3mSr Crawford	American	543	146·0	34·6	15·6	Millville,N.J.	1884	J.D.Bell		H.C.Matthews

✻ = Lloyd's Register; D = Nederlandsche Vereeniging; G = Germanischer Lloyd; I = Registro Italiano; K [illegible]
A = Veritas Austro-Ungarico; H = Veritas Ellenico; N = Norske Veritas; V [illegible]

In the 4th Column, C = Composite (iron frame, wood planked); I = Iron; Stl = Steel; W = Wood.

The Dates of Metalling (**Col. 11**) are based on the records of the Books in which Vessels are respectively classed, or on the Reports of the Surveyors of [illegible]

Further particulars of Vessels classed by Lloyd's Register will be found in Lloyd's Register Book and Lloyd's Yacht Register.

No.	Classe	Nom. Capitaine et dates d'entrée au service des Armateurs actuels et à bord du navire. Ancien Nom.	Matériaux et Gréement	Pavillon.	Tonnage net	Dimensions: Longueur	Largeur	Creux	Construit: Où	Quand	Par qui	Date du doublage en métal	Armateurs ou Directeurs Gérants.	Port d'Armement.
1151		Lutin (CONDEMNED) Lerest	W Bg	French	285	108·6	26·9	16·4	[illegible]	1875		V 1.88	Pierre Viton	Marseilles
1152	*	Lutterworth Streater 87-87	I Bk	British	887	209·2	33·4	19·4	[illegible]	1868	DentonGray&C.		Shaw, Savill & Albion Co. Ld.	Southampton
1153	V	Lutzburg Foote -88	W Bn	British	232	106·5	23·7	12·0	[illegible]	1866	J.F.Strange	Y 9.86	A.S.Archer	St.John's, Antigua
1154		Luvia Hallman 80-86	W Sr	Russian	327	..	..	..		1873			G.Stenroos	Luvia
1155		Luxor McLeod	W Bk	British	811	163·3	34·4	20·0	S.Martin's N.B.	1877	DeLongBros.	Y 11.86	J.L.Dunn	St.John.N.B.
1156	DRV	Luz Mudge	W 3mSr	British	190	104·2	23·9	12·9	Dartmouth	1869	Philip	Y 1.88	R.Johnson	Bristol
1157	*	Luz (ex Lizzie Berry)	W Bk	Chilian	498	146·6	31·0	18·1	Sunderland	1875	J.Gardner	Y 1.88	Cia Esplotadora de Lota y Coronel	Coronel
1158		Luz Fardon	W Bk	Spanish	313	108·9	27·1	15·1	Viavelez	1870	JoséRon	Y 1.83	A.Arias	Gijon
1159		Luzitano (ex Ceuba I.) de Oliveira Lima	W Sr	Brazilian	206	..	..	..	Canada	1874			J.C.Babo	Rio de Janeiro
1160	RV	Luzon Park	W S	American	1339	205·8	40·7	24·0	Boston, Mass.	1881	Smith&Townsd.	Y 11.87	De Groot & Peck	New York.N.Y.
1161		Lybeck Hansson	W Sr	Swedish	122	84·7	20·2	10·4	Björneborg	1855			A.Peilitz	Strömstad
1162		Lyckan Högström	W Sr	Russian	141	86·1	26·7	10·6	Luvia	18 65/72	C.Röstelt		G.W.Wahlroos	Raumo
1163		Lycke (ex Blessing) Hansson 77-87	W Bg	Swedish	239	99·4	23·4	15·1	Sunderland	1854			A.Nilsson	Gothenburg
1164	V	Lydia Perry	W S	British	1242	185·8	38·6	23·0	Tusket, N.S.	1874	J.Jeffery		A.C.Robbins	Yarmouth.N.S.
1165	V	Lydia Sprague -87	W Bkn	British	424	122·8	29·5	17·3	Digby, N.S.	1871	M.L.Oliver	Y 11.86	S.Schofield	St.John.N.B.
1166		Lydia (WHALER) McLane 84-86	W Bk	American	313	104·0	27·4	18·0	Mattapoiset, Ms	1841	Holmes		W.Baylies	NewBedfd., Mass.
1167	A	Lydia (ex Nova Hollandia) Da Silva	W Sr	Brazilian	173	..	..	..	Veendam	1874	K. & J.Wilkens	Y 3.85	J.P.Gaspar & Co.	Rio Grande
1168	DV	Lydia Bager -87	W Sr	Danish	118	92·3	20·5	9·8	Marstal	1874	J.J.Bager		H.C.Christensens Enke	Marstal
1169	N	Lydia Cuthbertson	W Bk	Norweg'n	366	131·1	29·6	14·7	Tönsberg	1868	S.Olsen	Y 11.85	N.Aker	Tönsberg
1170		Lydia (TO BE BROKEN UP) Nordberg -87	W Bg	Norweg'n	198	99·9	26·0	11·9	Tvedestrand	1858			C.Speilberg	Arendal
1171		Lydia Nordlund	W 3mSr	Russian	321	..	..	..		1874			A.A.Holmberg	Hammarland
1172		Lydia Selin 79-	W 3mSr	Russian	294	120·5	30·2	12·8	Laitakari	1874	B.Sandström		C.F.Holmberg&Co.	Luvia
1173		Lydia Öblom	W Sr	Russian	242	98·1	28·2	12·8	Sagu	1876	W.Sjöström		M.F.Laurén	Kimito
1174		Lydia Stackelberg 82-82	W Sr	Russian	194	93·3	28·0	12·4	Nerpissocken	1875	M.Finne		Baron W.Stackelberg	Raumo
1175		Lydia Lind	W Sr	Russian	161	89·9	27·8	10·5	Sandvik	1874	K.Henrikson		K.Mattson	Kimito
1176		Lydia B Cowperthwaite Caswell	W 3mSr	American	349	120·0	31·0	10·6	ForkedR., N.J	1856			S.T.Caswell, jr.	Newport.R.I.
1177	*	Lydia Cardell Todd	W 3mBn	British	225	118·0	24·8	13·6	Appledore	1873	Cook	F & C 6.88	W.Geake & Co.	Fowey
1178	R	Lydia H. Roper Thrasher	W 3mSr	American	305	125·0	34·0	10·0	Norfolk, Va.	1875	Sawyer&Taylor		J.L.Roper	Norfolk, Va.
1179		Lydia Middleton Powell	W 3mSr	American	187	120·0	33·3	9·9	Barnegat, N.J.	1875			C.P.Bonnell	PerthAmboy, N.J.
1180	GR	Lydia Peschau (ex Elise) Bernstein -88	W Bk	German	368	120·0	29·0	16·0	Viavelez	1868			F.Westermann	Geestemünde
1181	R	Lydia Skolfield Dunning [-87	W S	American	1201	187·0	37·0	23·9	Brunswick, Me	1860	GeoR.S.Rolfield	Y 7.86	Skolfield Bros.	Portland, Me.
1182		Lye Seng Ngwan	W Sr	British	115	76·0	20·6	8·1	Singapore	1887	WeeAhTak		Goh Ah Tee	Singapore
1183	R	Lygonia (ex Anna Bell) Fullerton 86-86	W Sr	American	172	103·0	29·0	8·5	ColumbiaFalls, Me	1874	IsaacCarlton	Y 5.86	J.S.Emery & Co.	Boston, Mass.
1184		Lymington Leonard 84-84	W Sr	British	163	116·1	24·8	9·0	Harwich	1880	J.H.Vaux		J.H.Vaux	Harwich
1185	DN	Lympha Pettersen 82-82	W Bg	Norweg'n	296	110·5	28·5	15·4	Drammen	1850	Jörgensen& [Knudsen		R.Grauer	Drammen
1186	N	Lyn Knaase -87	W Bk	Norweg'n	252	112·3	27·8	12·5	Arendal	1870	P.Vindholmen	Y 4.88	A.B.Sævig	Christiansand
1187	N	Lyn (ex William & Catherine) Olsen -88	W Bg	Norweg'n	251	100·2	26·0	15·6	Sunderland	1861	B.Thompson & Sons		H.Helland	Fredrikstad
1188	N	Lyn Herland -86	W Sr	Norweg'n	137	102·6	25·5	10·4	Moss	1871	Schmieriel	Y 1.88	A.Basberg	Christiania
1189	N	Lyna (ex John) Wahl 77-85	W Bk	Norweg'n	554	147·5	29·5	18·7	Richibucto	1856	Jardine& [Jardine	Y 4.88	H.A.Lobre	Drammen
1190	N	Lyna (ex Tyrian) Larsen 80-80	W Bk	Norweg'n	233	110·0	26·8	12·3	Barrington, [N.S.	1858		Y 3.88	P.A.Christensen	Langesund
1191	*	Lyndhurst Martin -88	Stl 4mS	British	2249	295·0	42·1	24·1	Dumbarton	1886	A.M'Milland&Sn		Lyndhurst Ship Co. Ld. (W.R.Price & Co.)	London
1192	N	Lyngoer Anderson -87	W Bk	Norweg'n	455	132·4	30·9	16·3	Risör	1876		Y 9.85	P.A.Olsen	Risör
1193		Lynher (ABANDONED) Bailey	W Bn	British	137	75·0	22·0	13·5	Plymouth	1835	W.Kerswill		T.Bailey	Lynn
1194	V	Lynnwood Sinclair 85-87	W Bk	British	1150	190·2	38·2	22·8	St.John, N.B.	1880	J.Fraser	Y 1.87	B.Smith & Sons	Windsor, N.S.
1195		Lynwood Harper 78-83	W Bn	British	184	104·5	24·5	12·9	P.E.Island	1875	Longworth		A.Nelson	Aberystwith
1196		Lynx Finley	W Sr	British	124	88·7	27·0	8·0	Portland, N.B.	1883	J.H.D.Eagles		J.H.D.Eagles	St.John.N.B.
1197	*	Lyra Goldsmith 88-88	I Bk	British	380	140·8	27·2	16·7	Sunderland	1867	Diff, Mounsey&C		H.C.Marsh (Mauvelle, Marsh&Co. Mgrs.)	Shoreham
1198	*	Lyra Fraser	W Trawler	British	81	77·4	20·3	10·1	Glampton	1883	W.A.Gibbs		W.H.Fraser	Grimsby
1199		Lyra Leonard	W Sr	American	106	82·5	25·0	8·5	Bucksport, Me.	1843			A.L.Hopkins	Bangor, Me.
1200	V	Lyra Gautier	W Bn	French	100	83·3	23·1	9·0	Shelburne N.S.	1877		Y 1.88	H.Mignot	St.Malo

*=Lloyd's Register; D=Nederlandsche Vereeniging; G=Germanischer Lloyd; I=Registro Italiano; R=American Record; A=Veritas Austro-Ungarico; H=Veritas Ellenico; N=Norske Veritas; V=Bureau Veritas.

Dans la 4me Colonne, C=Composite (membrure en fer, bordé en bois); I=Fer; Stl=Acier; W=Bois.

Les Dates du doublage en métal (Col. 11) ressortent des livres dans lesquels les navires sont respectivement classés ou des rapports des Inspecteurs du Lloyd's Register.

Pour plus amples renseignements sur les navires classés par le Lloyd's Register, voir le Lloyd's Register Book et le Lloyd's Yacht Register.

Official No.	No. in Book	Registry	Ship's Name. Material and Rig. Master, and Dates of Appointment to present Owner's Service and to Ship. Ship's late Name.	Flag.	Net Tonnage	Length	Breadth	Depth	Built Where	Built When	Built By whom	Date of Metalling	Owners or Managers
	1201		Lyta [illegible]	Russian	388	13[illegible]·3	31·9	15·3	Christinestad	1874	H.Granskoy		N.Sittkoff
59131	1202	✱	Lythemore Stasset W Bn	British	176	103·0	23·0	12·9	Berwick	1865	Gowan	[illegible]	W.Samuel
	1203	R	Lyyli Axplund W[illegible]	Russian	339	124·6	28·4	14·7	[illegible]	1880	J.Dahlroth		F.W.Gustafsson&Co.

Additional Ships are inserted in the Supplement at the end of the Register.

✱=Lloyd's Register; D=Nederlandsche Vereeniging; G=Germanischer Lloyd; I=Registro Italiano R=[illegible]
A=Veritas Austro-Ungarico; H=Veritas Ellenico; N=Norske Veritas; V=Bureau Veritas

In the 4th Column, C=Composite (iron frame, wood planked); I=Iron; Stl=Steel; W=Wood.

The Dates of Metalling (Col. 11) are based on the records of the Books in which Vessels are respectively classed, or on the Reports of the Surveyors of [illegible]

Further particulars of Vessels classed by Lloyd's Register will be found in Lloyd's Register Book and Lloyd's Yacht Register.

		Nom. Matériaux et Grément. Capitaine et dates d'entrée au service des Armateurs actuels et à bord du navire. Ancien Nom.	Pavillon.	Tonnage net.	Dimensions. Longueur.	Largeur.	Creux.	Construit. Où.	Quand.	Par qui.	Date du doublage en métal.	Armateurs ou Directeurs-Gérants.	Port d'Armement.
1	R	M. A. Achorn *Achorn* WBnSr	American	293	127·0	34·0	11·0	Rockland,Me.	1879	A.F.Ames		E.A.Butler	Rockland.Me.
2		M. A. Coombs WSr	American	190	..	..	..	Jennysville, Me.	1868			C.E.Turner	Boston.Mass.
3		M. A. Doran *Moran* WBn	British	316	117·6	29·3	12·9	Walton,N.S.	1875	J.A.Stabatt		The English, Scottish & Australasian Chartered Bank Incorporated	Sydney,N.S.W.
4		M. A. Nutter *Jewett* WSr	British	291	120·1	28·2	11·8	Portland,N.B.	1881	J.Napier	Y 3.85	R.C.Elkin	St.John,N.B.
5	❋	M. A. Wilkinson *Sims 1881* WSr	British	86	82·6	21·2	9·8	Appledore	1879	Cock		J.Parson	Plymouth
6		M. B. Cohn *Gelle 79* WBg	German	232	100·0	26·6	14·0	Rostock	1866	G.W.Tesdorpf		E.Burchard	Rostock
7	R	M. B. Daly *McNeil 88* WBn	British	135	95·0	24·5	9·5	Medway,N.S.	1885	JohnHutt		A.B.Crosby	Medway.N.S.
8	R	M. B. Mahoney *Hodgkins 87* WSr	American	117	84·5	25·2	8·5	Wilmingtn.Dl	1843			P.McConville	Bangor.Me.
9	RV	M. B. Millen *Dwyer* WBmSr	American	319	134·7	32·2	10·9	Chelsea,Mass.	1874	Pierce [Montgomery	Y [illegible]	D.Wadley	Savannah.Ga.
10		M. C. Cameron WSr	British	309	138·0	25·8	12·5	Oswego,U.S.	1862			William Lee	Montreal
11		M. C. Haskell *Perry 88* WBn	American	334	115·0	28·0	16·8	Rockland,Me.	18$^{76}_{78}$	Snow&Farwell	Y 7.80	Snow & Pearsons	Rockland.Me.
12	R	M. C. Moseley *Torrey* WSr	American	189	130·0	28·5	9·2	Franklin,Me.	1866		Y 3.87	J.G.Moseley	Boston.Mass.
13	V	M. Caridad *Zaragoza* WBn	Spanish	107	74·5	22·0	10·9	Villajoyosa	18$^{55}_{60}$		Y 3.88	M.Zaragoza	Alicante
14	AF	M. Cariddi *Caudi* WBk	Italian	627	131·2	28·2	20·6	Rockland,Me	1858		Y 9.88	M.Cariddi	Taranto
15	G	M. D. Rucker *Rehberg 75-76* WBk	German	397	120·8	28·8	17·5	Rostock	1869	G.W.Tesdorpf	Y 1.88	G.Kindler	Rostock
16		M. E. Coipel *Lordly 88* WBn	British	115	89·0	23·5	9·5	Shelburne,N.S	1871			A.F.Stoneman	Yarmouth,N.S.
17	R	M. E. Eldridge *Kelly* WBmSr	American	240	108·0	30·0	12·6	Portsmth.N.H	18$^{78}_{80}$	W.E.Fernald		C. & J.F.Baker	Dennis.Mass.
18	❋	M. E. Johnson *Johnson 175-76* WSr	British	100	98·3	23·0	9·9	Barrow	1879	Ashburner		T.Ashburner & Co.	Barrow
19	❋	M. E. Watson *Pratt 87-88* IS	British	1670	251·4	39·1	24·0	Stockton	1883	Richardson. [Duck&Co.		Gracie,Beazley& Co.	Liverpool
20		M. H. Read *Phinney* WSr	American	152	104·0	28·8	8·7	Brookhvn.NY	18$^{54}_{66}$			Tabor & Gordon	Nw.Bedfrd.Mass
21	R	M. J. Foley *Robinson* WBkn	British	446	139·0	29·3	14·8	P.E.Island	1878	M.Foley	Y 9.86	J.Rodenhiser	Bridgewater,N.S
22		M. K. Rawley *Allen* WBmSr	American	288	125·0	34·0	12·5	Tenants' [Harbour,N.J.	1874			G.W.Rawley	St.George.Me.
23	R	M. Luella Wood *Spofford* WBmSr	American	529	153·5	34·4	13·5	Rockland. Me.	1882	A.F.Ames		A.F.Ames	Rockland.Me.
24		M. M. Hamilton WSp	American	106	84·8	26·2	7·2	Harpswell,Me.	1869			J.F.Hamilton	Portland.Me.
25		M. M. Merriman *Lamphere* WSr	American	138	106·0	28·0	7·5	DeepRivr.Con	1865			T.L.Church	FallRiver.Mass
26	RV	M. P. Grace *Williams* WS	American	1863	229·9	42·1	27·8	Bath,Me.	1875	Chapm'n&Flint	Y 12.[illegible]	B.Flint	New York
27	V	M. Roosval *Nilsson 85* WBk	Swedish	291	121·3	25·8	13·0	Oscarshamn	1874	P.Pettersson	Y 3.[illegible]	C.J.Löfgren	Oscarshamn
28	DN	M. Smith Petersen *Gran* WBk	Norweg'n	453	132·0	30·4	17·1	Grimstad	1873	M.Smith [Petersen	Y 3.86	F.Smith Petersen	Grimstad
29	R	M. V. B. Chase *Blake* WBmSr	American	435	143·3	35·9	12·1	Bath.Me.	1882	Rogers	Y [illegible]	J.B.Drake	Bath.Me.
30		M. V. Cook *Falkenburgh* (ABANDONED) WSr	American	191	104·0	28·0	9·5	Tuckertn.N.J	1865			A.R.Pharo	Tuckerton,N.J.
31		M. W. Griffing *Sterling* WSr	American	124	96·1	27·2	7·5	CityIslandNY	1865	DavidCarll		F.W.Benedict	NewHavn.Conn.
32		M. & E. Cann *Poole 87* WBk	British	920	182·0	34·5	21·1	Yarmouth,N.S	1872			H.D.Cann	Yarmouth.N.S
33	V	M. & E. Cox *Robinson 85-86* WS	British	1185	188·2	38·0	23·7	Maitland,N.S.	1869	G.Cox	Y 2.86	W.T.Dixon & Sons	Halifax.N.S.
34	N	Maagen *Gundersen 88* WSr	Norweg'n	143	88·4	23·2	11·0	Arendal	1873		[illegible]	H.Gunderson	Fredrikstad
35	❋	Mabel *Johns 85-86* WBk	British	719	178·5	34·2	18·6	P.E.Island	1877	G.Bellon	Y 7.8[illegible]	T.P.Richards & Co.	Charlottetown
36		Mabel *Hughes 74-82* WBn	British	140	87·7	22·4	12·3	Portmadoc	1864	Williams		G.Williams & Co.	Carnarvon
37	❋	Mabel *Taylor 87-87* WSr	British	99	88·1	21·5	10·3	Garmouth	1882	Anderson& [Duncan	Y 7.87	H.R.Taylor	Inverness
38	RV	Mabel *Snow* WBk	American	750	150·3	34·3	20·0	Bath,Me.	1874	O.Blaisdell	Y 6.8[illegible]	S.D.Hinckley	Boston,Mass.
39		Mabel Darling *Bonner* WSr	British	111	86·2	24·8	9·8	DunmoreTwn.	1880	J.W.Roberts		John S.Darling	Nassau.N.P.
40		Mabel Gray *Simmit* WBmSr	American	196	116·0	31·4	8·7	Eureka,Cal.	1882	H.D.Bendixon		Geo.D.Gray	San Francisco
41		Mabel Hall *Eardley 2-87* WSr	American	162	92·0	25·9	11·0	Addison.Me.	18[illegible]			R.C.Hall	Rockland.Me.
42	R	Mabel Hooper *Hooper* WBmSr	American	545	155·0	34·4	12·9	Rockland.Me.	1883	A.F.Ames		J.Hooper	Rockland.Me.
43	R	Mabel Howard *Hopkins 88* WSr	British	143	92·0	24·0	10·0	Barrington NS	1882	W.Doane	Y 3.86	R.I.Hart	Halifax.N.S.
44		Mabel L. Phillips *Bowe* WBmSr	American	790	172·8	40·1	17·0	Bath,Me.	1879	Goss&Sawyer		J.B.Phillips	FallRiver,Mass
45		Mabel Rose *Barrett* WBmSr	American	369	127·0	33·2	8·7	Bridgeton,N.J	1874			D.C.Barrett	ColdSpring.NY
46	R	Mabel Stoddard *Wallace* (ex Pet... T.D. Marshall 1887) IBk	American	693	171·5	30·5	20·4	South Shields	1857	T.D.Marshall		Baker Salvage Co.	Norfolk,Va.
47	V	Mabel Taylor *Durkee* WS	British	1298	205·0	37·7	23·0	Tusket,N.S.	1878	J.A.Hatfield	Y 7.82	William Law	Yarmouth.N.S.
48		Mabel Thomas *McKenzie* WBmS	American	319	148·0	32·7	11·7	W.Haven,Con	1874	A.Thomas		C.Lovell & Son	NewHaven,Conn.
49		Mabrook *Said Achmad bin Osman Barkba* WBk	Dutch	177	85·6	23·7	..	Lassem,Java	1845		Y 3.56	Said Achmad bin Osman Barkba	Palembang
50		Mabrook Bg	Turkish	207	..	..	..					Ahmed Lashal	Alexandria

[illegible]=Lloyd's Register; D=Nederlandsche Vereeniging; G=Germanischer Lloyd; I=Registro Italiano; R=American Record;
[illegible]=Veritas Austro-Ungarico; H=Veritas Ellenico; N=Norske Veritas; V=Bureau Veritas.

Dans la 4me Colonne, C=Composite (membrure en fer, bordé en bois); I=Fer; Stl=Acier; W=Bois.

Les Dates du doublage en métal (Col. 11) ressortent des livres dans lesquels les navires sont respectivement classés ou des rapports des Inspecteurs du Lloyd's Register.

Pour plus amples renseignements sur les navires classés par le Lloyd's Register, voir le Lloyd's Register Book et le Lloyd's Yacht Register.

Official No.	No. in Book	Registry in which classed	Ship's Name. Material and Rig. Master, and Dates of Appointment to present Owner's Service and to Ship. Ship's Late Name.	Flag.	Net Tonnage	Length	Breadth	Depth	Where	When	By whom	11	Owners or Managers
	1201		Lyta *Erickson 64-[illegible]*	Russian	388	133·3	31·9	15·[illegible]	Christinestad	1874	H.Granskoy		N. [illegible]
53131	1202	✻	Lythemore *Samuel* W Bo	British	176	103·0	25·0	12·9	Berwick	1865	Gowan	Y	W.Samuel
	1203	R	Lyyli *Asplund* W[illegible]	Russian	339	124·6	28·1	14·7	[illegible]	1880	J.[illegible]		F.W.Gustafsson&Co.

ADDITIONAL SHIPS ARE INSERTED IN THE SUPPLEMENT AT THE END OF THE REGISTER.

✻ = Lloyd's Register; D = Nederlandsche Vereeniging; G = Germanischer Lloyd; I = Registro Italiano; [illegible]
A = Veritas Austro-Ungarico; H = Veritas Ellenico; N = Norske Veritas; [illegible]

In the 4th Column, C = Composite (iron frame, wood planked); I = Iron; Stl = Steel; W = Wood.

The Dates of Metalling (Col. 11) are based on the records of the Books in which Vessels are respectively classed, or on the Reports of the Surveyors of [illegible]

Further particulars of Vessels classed by Lloyd's Register will be found in Lloyd's Register Book and Lloyd's Yacht Register.

No.	Classe	Nom	Capitaine / Ancien Nom	Matériaux et Gréement	Pavillon	Tonnage net	Longueur	Largeur	Creux	Où	Quand	Par qui	Date du doublage en métal	Armateurs ou Directeurs-Gérants	Port d'Armement
1	*	M. A. Achorn	Achorn	W3mSr	American	293	127·0	31·0	11·0	Rockland, Me.	1879	A.F.Ames		E.A.Butler	Rockland, Me.
2		M. A. Coombs		WSr	American	190	..	..	..	Dennysville, Me.	1868			C.E.Turner	Boston, Mass.
3		M. A. Doran	Mahan	WBn	British	316	117·6	29·3	12·9	Walton, N.S.	1875	J.A.Stanatt		Eng. English, Scottish, & Australian Chartered Bank Incorporated	Sydney, N.S.W.
4		M. A. Nutter	Beatt	WSr	British	291	120·1	28·2	11·8	Portland, N.B.	1881	J. Napier	Y 2.88	R.C.Elkin	St.John, N.B.
5	*	M. A. Wilkinson		WSr	British	86	82·6	21·2	9·8	Appledore	1879	Cock		J.Parson	Plymouth
6		M. B. Cohn	Colle -79	WBg	German	232	100·0	26·6	14·0	Rostock	1866	G.W.Tesdorpf		E.Burchard	Rostock
7	R	M. B. Daly	McNeil -88	WBa	British	135	95·0	24·5	9·5	Medway, N.S.	1885	JohnHutt		A.B.Crosby	Medway, N.S.
8	R	M. B. Mahoney	Hodgson -87	WS	American	117	84·5	25·2	8·5	Wilmingtn, Dl	1843			P.McConville	Bangor, Me.
9	RV	M. B. Millen	Boyer	W3mSr	American	319	134·7	32·2	10·9	Chelsea, Mass.	1874	Pierce [Montgomery	Y 11.88	D.Wadley	Savannah, Ga.
10		M. C. Cameron		WSr	British	309	138·0	25·8	12·5	Oswego, U.S.	1862			William Lee	Montreal
11		M. C. Haskell	Perry -88	WBn	American	334	115·0	28·0	16·8	Rockland, Me.	18 65/78	Snow & Farwell	Y 1.86	Snow & Pearsons	Rockland, Me.
12	R	M. C. Moseley	Torrey	WSr	American	189	130·0	28·5	9·2	Franklin, Me.	1866		Y 5.87	J.G.Moseley	Boston, Mass.
13	V	M. Caridad	Zaragoza	WBn	Spanish	107	74·5	22·0	10·9	Villajoyosa	18 ?/?		Y 7.86	M.Zaragoza	Alicante
14	AF	M. Cariddi	Cilona -88	WBk	Italian	627	131·2	28·2	20·6	Rockland, Me	1858		Y 8.82	M.Cariddi	Taranto
15	G	M. D. Rucker	Rehberg 73-74	WBk	German	397	120·8	28·8	17·5	Rostock	1869	G.W.Tesdorpf	Y 1.88	G.Kindler	Rostock
16		M. E. Coipel	Larkin -88	WBn	British	115	89·0	23·5	9·5	Shelburne, N.S	1871			A.F.Stoneman	Yarmouth, N.S.
17		M. E. Eldridge	Kelly	W3mSr	American	240	108·0	30·0	12·6	Portsmth, N.H	18 75/80	W.E.Fernald		C. & J.F.Baker	Dennis, Mass.
18	*	M. E. Johnson	Johnson Davis	WSr	British	109	98·3	23·0	9·9	Barrow	1879	Ashburner		T.Ashburner & Co.	Barrow
19	*	M. E. Watson	Coll 87-89	IS	British	1670	251·1	39·1	24·0	Stockton	1883	Richardson, [Duck & Co.		Gracie, Beazley & Co.	Liverpool
20		M. H. Read	Pikaney	WSr	American	152	104·0	28·8	8·7	Brooklyn, NY	18 54/86			Tabor & Gordon	Nw. Bedfrd, Mass
21	R	M. J. Foley	Rodenhiser -88	WBkn	British	446	139·0	29·3	14·8	P.E.Island	1872	M.Foley	Y 9.86	J.Rodenhiser	Bridgewater, N.S
22		M. K. Rawley	Alley	W3mSr	American	288	125·0	34·0	12·5	Tenants [Harbour, N.J.	1874			G.W.Rawley	St.George, Me.
23	R	M. Luella Wood	Spalding	W3mSr	American	529	153·5	34·1	13·5	Rockland, Me.	1882	A.F.Ames		A.F.Ames	Rockland, Me.
24		M. M. Hamilton		WSp	American	106	84·8	26·2	7·2	Harpswell, Me.	1869			J.F.Hamilton	Portland, Me.
25		M. M. Merriman	Tongue -86	WSr	American	138	106·0	28·0	7·5	DeepRivr, Con	1865			T.L.Church	FallRiver, Mass
26	RV	M. P. Grace	Williams	WS	American	1863	229·9	42·1	27·8	Bath, Me.	1875	Chapman & Flint	Y 12.8[illegible]	B.Flint	New York
27	V	M. Roosval	Nilsson -83	WBk	Swedish	291	121·3	25·8	13·0	Oscarshamn	1874	P.Pettersson	Y 3.86	C.J.Löfgren	Oscarshamn
28	DN	M. Smith Petersen	Grøn	WBk	Norweg'n	458	132·0	30·1	17·1	Grimstad	1873	M.Smith [Petersen	Y 5.87	F.Smith Petersen	Grimstad
29	R	M. V. B. Chase	Blake	W3mSr	American	435	143·3	35·0	12·1	Bath, Me	1882	Rogers	Y 5.8[illegible]	J.B.Drake	Bath, Me.
30		M. V. Cook	Falkenburgh (ABANDONED)	WSr	American	191	101·0	28·0	9·5	Tuckerton, N.J	1866			A.R.Pharo	Tuckerton, N.J.
31		M. W. Griffing	Stocking	WSr	American	124	96·1	27·2	7·5	CityIsland NY	1865	David Carll		F.W.Benedict	NewHavn, Conn
32		M. & E. Cann	Foote -87	WBk	British	920	182·0	34·5	21·1	Yarmouth, N.S	1872			H.D.Cann	Yarmouth, N.S
33	V	M. & E. Cox	Robinson 85-86	WS	British	1185	188·2	38·0	23·7	Maitland, N.S.	1869	C.Cox	Y 3.86	W.T.Dixon & Sons	Halifax, N.S.
34	N	Maagen	Gundersen	WSr	Norweg'n	143	88·4	23·2	11·0	Arendal	1873		Y 10.8[illegible]	H.Gundersen	Fredrikstad
35	*	Mabel	Johns 83-86	WBk	British	719	178·5	34·2	18·6	P.E.Island	1877	G.Bellem	Y 7.82	T.P.Richards & Co.	Charlottetown
36		Mabel	Hughes 74-82	WBn	British	140	87·7	22·4	12·3	Portmadoc	1864	Williams		G.Williams & Co.	Carnarvon
37	*	Mabel	Taylor 81-87	WSr	British	99	88·1	21·5	10·3	Garmouth	1882	Anderson & [Duncan	Y 1.87	H.R.Taylor	Inverness
38	RV	Mabel	Faber	WBk	American	750	150·3	34·3	20·0	Bath, Me.	1874	O.Blaisdell	Y 12.88	S.D.Hinckley	Boston, Mass.
39		Mabel Darling	Ranger	WSr	British	111	86·2	24·8	9·8	DunmoreTwn.	1880	J.W.Roberts		John S.Darling	Nassau, N.P.
40		Mabel Gray	Small	W3mSr	American	196	116·0	31·4	8·7	Eureka, Cal.	1882	H.D.Bendixon		Geo.D.Gray	San Francisco
41		Mabel Hall	Gurley -87	WSr	American	162	92·0	25·9	11·0	Addison, Me	18 64/79			R.C.Hall	Rockland, Me.
42	R	Mabel Hooper	Hooper	W3mS	American	545	155·0	31·4	12·9	Rockland, Me.	1883	A.F.Ames		J.Hooper	Rockland, Me.
43	R	Mabel Howard	Hopkins	WSr	British	143	92·0	24·0	10·0	Barrington NS	1882	W.Doane	Y 3.87	R.I.Hart	Halifax, N.S.
44		Mabel L. Phillips	Heines -88	W3mSr	American	790	172·8	40·1	17·0	Bath, Me.	1879	Goss & Sawyer		J.B.Phillips	FallRiver, Mass
45		Mabel Rose	Everett	W3mSr	American	369	127·0	33·2	8·7	Bridgeton, N.J	1874			D.C.Barrett	ColdSpring, NY
46	R	Mabel Stoddard	Wolf (Ex T. D. Marshall (LOST))	WBk	American	693	171·5	30·3	20·4	South Shields	1857	T.D.Marshall		Baker Salvage Co.	Norfolk, Va.
47	V	Mabel Taylor		WS	British	1298	205·0	37·7	23·0	Tusket, N.S.	1878	J.A.Hatfield	Y 7.83	William Law	Yarmouth, N.S.
48		Mabel Thomas	McKenzie	W3mSr	American	319	148·0	32·7	11·7	W.Haven, Con	1874	A.Thomas		C.Lovell & Son	NewHaven, Conn.
49		Mabrook	Ex Baraka Seria	WBk	Dutch	177	85·6	23·7	..	Lassem, Java	1845		Y 2.86	Said Achmad bin Osman Barakba	Palembang
50		Mabrook		Bg	Turkish	207	..	..	..					Ahmed Lashal	Alexandria

★=Lloyd Register; D=Nederlandsche Vereeniging; G=Germanischer Lloyd; I=Registro Italiano; R=American Record;
A=Veritas Austro-Ungarico; H=Veritas Ellenico; N=Norske Veritas; V=Bureau Veritas.

Dans la 4me Colonne, C=Composite (membrure en fer, bordé en bois); I=Fer; Stl=Acier; W=Bois.

Les Dates du doublage en métal (Col. 11) ressortent des livres dans lesquels les navires sont respectivement classés ou des rapports des Inspecteurs du Lloyd's Register.

Pour plus amples renseignements sur les navires classés par le Lloyd's Register, voir le Lloyd's Register Book et le Lloyd's Yacht Register.

Official No.	No. in Book	Societies	Ship's Name. Material and Rig. Master, and Dates of Appointment to present Owner's Service and to Ship. Ship's Late Name.	Flag.	Net Tonnage.	Dimensions. Length.	Breadth.	Depth.	Built. Where.	When.	By whom.	Date of Metalling.	Owners or Managers.	Port
	51		Mabrook Bg	Turkish	176	..	..	..					Ahmed Eff. Rounstem	Alexandria
	52		Mabrook Bg	Turkish	172	..	..	..					Abdul M. Eff. Khatab	Alexandria
	53		Macahé *Da Costa* WSr (ex Nossa Catalina B.)	Brazilian	147	..	..	..	Malta	1874		Y 5·88	R. da Silveira Gusmão	Rio de Janeiro
88304	54	✱	MacCallum More *Smith* IS	British	1669	265·3	39·5	23·4	PortGlasgow	1873	R.Duncan&Co.		J.Potter	London
	55	✱I	Macdiarmid *Scotto 87-87* IS	Italian	1560	255·3	38·2	22·6	Dundee	1882	A.M'Millan&Sn		M.Amoroso	Palermo
84010	56	DRV	McDougall *Davis* WS	British	1194	197·0	38·2	22·9	Five Mile River, N.S.	1873	McDougall		McDougall	Halifax N.S.
78571	57	✱	Macduff *Jones -87* IS	British	1235	232·4	36·0	21·0	Dumbarton	1877	A.M'Millan&Sn		NorthBritishShippg.Co.	Glasgow
31770	58	✱	Macduff *Pettersen -86* WS	Norweg'n	1122	219·6	35·7	22·1	Aberdeen	1859	Hall	Y 5·86	Chr. Anker	Fredrikstad
91116	59	V	McLaurin *Oak 86-88* WS	American	1330	200·6	39·0	24·0	Newburyport, Mass.	1878	Atkinson&Filmore	Y 5·88	M.F.Pickering&Co.	Boston
86467	60	✱	MacLeod *Henderson -88* IS	British	1556	255·5	38·3	22·5	Dumbarton	1879	A.M'Millan&Sn		A.M'Millan & Son	Glasgow
66584	61	R	Macleod *Cottam -88* WBk	British	633	148·8	34·1	17·8	St.John,N.B.	1873	D.Lynch		W.H.McLeod	St.John N.B.
82205	62	✱D	MacMillan *Grey -79* IS	British	1450	255·5	38·3	20·6	Dumbarton	1879	A.M'Millan&Sn		Macbeth & Gray	Glasgow
90416	63	V	McNear *Dickinson 87-88* WS	American	1245	185·3	37·4	24·0	Belfast,Me.	1872	H.McGilvery	Y 8·86	A.Sewall & Co.	Boston
2976	64	I	Maccabeo *Cerruti 86-88* WBk	Italian	896	174·5	31·5	23·1	Varazze	1879	G.Scain	Y 1·86	Prosp.Lavarello	Genoa
72840	65	DV	Macedon *McDonald -88* WS	British	1486	210·9	39·2	24·4	St.John,N.B.	1878	J.Fraser	Y 1·86	W.Thomson & Co.	St.John N.B.
26772	66		Macedonian *Thompson -85* WS	British	199	95·5	25·0	15·0	Lynn	1854	W.T.Rolin		W.Milburn	Newcastle
86064	67	✱	Machrihanish *Crosby -88* IS	British	1699	264·9	39·8	23·5	PortGlasgow	1883	R.Duncan&Co.		H.Hogarth	Ardrossan
70743	68	✱	Macquarie *Goddard 65-87* IS (ex Melbourne)	British	1867	269·8	40·1	23·7	London	1875	R. & H.Green		Devitt & Moore	London
95037	69	✱	Madagascar *McCrone 88-88* Stl4mBk	British	2077	282·0	40·5	24·6	PortGlasgow	1888	Russell & Co.		J.Boyd	Glasgow
16865	70		Madagascar *Knowlton* WSr	American	107	83·9	23·9	9·9	Ellsworth,Me.	1845			C.H.Potter	Calais
786	71	V	Madawaska *Saracene* WBk	Italian	490	135·5	30·9	17·5	Medford	1866	Hayden&Cudworth	Y 8·86	Eredi Carm.Picciotto	Messina
63403	72		Madcap *Miller* W3mSr	British	199	117·0	24·3	12·9	Brixham	1871	Upham	Y 4·81	G.Spencer	Waterford
	73	G	MaddaSophia *Stahl -88* WSr	Russian	146	87·3	25·2	11·2	Kaleren	1888	J.Kirstein		F.Stahl	Kaleren
1738	74	I	Maddalena *Maggiolo* WBk (ex Cavallo Marino II.) (WRECKED)	Italian	636	152·5	32·3	21·3	SestriPonente	1868	G.Casanova	Y 5·88	P.A.Maggiolo fu G.	Genoa
156	75	V	Maddalena *Cuartero -88* WBk	Italian	453	138·7	30·9	17·8	Sorrento	1862	A.Castellano	Y 5·87	Florio & Giudice	Palermo
2750	76	V	Maddalena *Saltafiro -87* WBk (ex Mooresk Vila) (PONTOON)	Italian	451	134·8	29·5	17·9	Trieste	1868	StabilimentoTecnico	Y 10·82	G.Dalgato	Naples
604	77		MaddalenaT. *Rapisardi* WBg (ex General Garibaldi)	Italian	275	97·8	27·4	17·5	Sestri	1859	G.B.Cadenaccio	Y 10·87	S.Rapisardi fu D.	Catania
78622	78	✱	Madeira *Falck 88-88* IBk	British	845	196·3	32·7	19·2	PortGlasgow	1878	Russell&Co.		P.Denniston & Co.	Glasgow
58008	79		Madeleine *Connell* W3mS (WRECKED)	British	114	93·6	21·5	10·3	Falmouth	1867	H.S.Trethowan		F.H.Palmer	Maryport
	80	V	Madeleine *Dupont -88* WBk (COLLISION)	French	287	119·4	26·9	13·4	Granville	1876		Y 1·88	Beust père & fils	[illegible]
	81	V	Madeleine *Zoonekynd* WSr	French	114	90·2	22·0	10·8	Dunkirk	1885	PyotteBayaert		H.Durin	Dunkirk
	82	V	Madeleine Rickmers *Durls 88-88* WBk	German	1277	198·4	41·0	25·5	Geestemünde	1879	R.C.Rickmers	Y 7·88	R.C.Rickmers	Geestemünde
58230	83		Madeline Ann *McCourt* WBn	British	132	86·4	23·4	12·3	Sunderland	1867	W.Barkley&Co.		R.Brown	Dublin
93358	84		Madeya StlLr	British	215	160·0	22·0	7·3	Dumbarton	1887	WDenny & Bros		IrrawaddyFlotillaCo.L	Glasgow
	85	N	Madonna *Nicolaisen -87* WBk	Norweg'n	463	136·1	28·8	16·2	Grimstad	1880	S.Johnsen	Y 1·89	Bernt Einersen	Grimstad
1329	86	A	Madonna Dell'Arco *Cacace* WBg (ex Sofia)	Italian	382	115·8	28·7	17·7	Procida	1867	Sciuano	Y 7·86	Mod Di Mattia	Naples
1348	87	A	Madonna Della Libera *Scotto* WBn	Italian	112	78·7	22·6	10·8	Limite	1859	N.Pichiotti		A. & V.Scotto	Castellammare
46556	88	✱	Madras *Jones* WBk	British	668	160·5	30·6	19·7	Dumbarton	1865	A.M'Millan&Sn	Y 7·86	E.Jones	Greenock
2483	89	IV	Madre *Costa -87* WBk	Italian	725	162·0	33·1	22·0	Sampierdar'na	1874	B.Frassinelli	Y 5·87	F^lli Casabona	Genoa
1786	90	I	Madre Antola *Ferrari -87* WBk (ex Madre Chiozza)	Italian	436	142·7	30·3	17·8	SestriPonente	1868	G.Casanova	Y 12·86	P.Antola	Genoa
3078	91	I	Madre Maria *Dondero -89* W3mSr	Italian	359	133·5	28·0	15·0	SestriPonente	1880	A.Briasco fu L.	Y 4·88	Fran.De Andreas	Genoa
1964	92	DI	Madre O. *Mezzani* WBk (ex Rosa Eugenio)	Italian	454	128·9	31·4	19·0	SestriPonente	1869	F^lli Cadenaccio	Y 1·87	F.Olivari fu G.	[illegible]
2164	93		Madre Rosa *Della Cava* WBk (ex Rosa Madre)	Italian	713	169·6	33·1	22·2	Savona	1871	A.Calcagno	Y 5·86	F.Ottone	[illegible]
708	94	I	Madre & Figli *Scarpati* WBk	Italian	396	125·9	27·9	18·7	Castellamare	1868	A.Bonifazio	Y 5·87	G.Cafiero	[illegible]
	95		Madrid (ex Craigie Burn) WBk	Urug'yan	1121	194·5	37·6	22·2	Maitland,N.S.	1886	A.A.McDougall			
71282	96	V	Madura *Bernier* WBk	British	970	187·5	36·3	20·7	Quebec	1876	F.X.Marquis	Y 1·88	W.H.Ross & Co.	[illegible]
70065	97		Madura *Andrews* WBk	British	344	123·2	26·8	15·6	Nantes	1866	Poster		J.C.Ellis	[illegible]
	98		Madura Meanatchy WBk (ex Papillon)	British	476	136·1	28·7	16·6	St.Malo	1870	[illegible]	Y 11·88	L.A.R.R.M.Doreiswamy Chetty	Negapatam
90092	99	✱	Maelgwyn *Jones 84-84* IBk	British	1235	220·0	35·6	21·4	Sunderland	1884	W.Doxford &Sn		R.Thomas & Co.	London
	100	N	Mærke To *Henriksen -88* WBk	Norweg'n	449	120·0	30·9	17·5	Porsgrund	1855			M.Johannessen&SJensen	[illegible]

✱ = Lloyd's Register; D = Nederlandsche Vereeniging; G = Germanischer Lloyd; I = Registro Italiano; R = American Record; A = Veritas Austro-Ungarico; H = Veritas Ellenico; N = Norske Veritas; V = Bureau Veritas.

In the 4th Column, C = Composite (iron frame, wood planked); I = Iron; Stl = Steel; W = Wood.

The Dates of Metalling (Col. 11) are based on the records of the Books in which Vessels are respectively classed, or on the Reports of the Surveyors of Lloyd's.

Further particulars of Vessels classed by Lloyd's Register will be found in Lloyd's Register Book and Lloyd's Yacht Register.

1	2	4 Nom. Capitaine et dates d'entrée au service des Armateurs actuels et à bord du navire. Ancien Nom.	Matériaux et Gréement	5 Pavillon	6 Tonnage net	7 Dimensions Longueur	Largeur	Creux	8 Construit Où	9 Quand	10 Par qui	11 Date du doublage en métal	12 Armateurs ou Directeurs-Gérants	13 Port d'Armement
101		Magano	W Bn	Brazilian	224	..	..	..					J.S. Magano	Rio Grande
102	N	Magda *Nielsen*	W Bk	Norweg'n	229	118·0	27·1	11·3	Drammen	1870	Jörgensen	V 3.88	M. Pedersen	Mandal
103		Magda *Wirolainen*	W Bg	Russian	276	110·0	26·7	13·6	Björkö	1873	W. Martelin		M. Reponen	Björkö
104		Magdala	W Sr	British	165	105·0	25·3	9·4	Quebec	1868			Geo. Farewell	Quebec
105	NV	Magdala *Pettersen*	W Bk	Norweg'n	1172	180·4	37·2	23·6	Portland, N.B	1868	Charles Sorrell	V 5.88	R. Ingvardsen	Flekkefjord
106	V	Magdalena *Gundersen -87* (ex ...)	W Bk	Norweg'n	823	177·0	35·2	21·0	Maitland, N.S.	1872	Smith	V 3.87	M. Brunsgaard	Drammen
107	V	Magdalena *Alsina* (ex ... Fortuna)	W Bk	Spanish	235	108·6	28·1	13·3	Blanes	1858		V 1.87	Taltabull & Xiró	Barcelona
108		Magdalena	W Bk	Spanish	203	109·5	22·8	13·7	Arens	1857			E. Alsina y Roca	Barcelona
109		Magdalena Charlotte	W Sr	Russian	106	71·0	21·1	9·0	Kurland	1876			John Mikkelsohn	Riga
110		Magdalene *Meyer 77-*	W S	German	1250	196·8	39·8	22·9	Quincy, Mass.	1856	Thomas	V 7.87	D.H. Wätjen & Co.	Bremen
111	N	Magdalene (ex Agnes) *Jensen*	W Bk	Norweg'n	622	150·2	32·2	18·6	Windsor, N.S.	1864	J. Mosher	V 11.88	M. Pedersen	Mandal
112	N	Magdalene *Johnsen*	W Bk	Norweg'n	267	113·9	26·5	13·8	Hamburg	1854			C.A. Hansen	Fredriksstad
113		Magdalene *Bertling -83*	W Sr	Russian	165	92·4	26·7	11·9	Ilmaten	1883			K. Bertling	Riga
114	A	Magdalini (ex Daisy)	W Bg	Greek	315	112·6	27·3	16·9	Sunderland	1868	Pace		Theofilatos & Stathatos	Ithaca
115		Magellan	W Sr	British	226	103·3	28·0	10·8	Hopewell, N.B	1882	C. Dowling		Geo. H. Irving	St. John, N.B.
116		Magellan *Manson -88*	W S	American	1019	178·3	34·5	22·9	East Boston	1873	Smith & Townsend		C.P. Hemenway	Boston, Mass.
117	V	Magellan *Niss 86-86*	W Bk	German	435	138·5	30·5	17·0	Brake	1870	I. Oltmann's Wwe.	V 7.88	Gebr. Oetling	Hamburg
118		Magellan	W Bk	Norweg'n	614	162·5	28·4	19·3	Whitehaven	1860	Kennedy	V 5.88		
119	RV	Magellan *Srahn -85* (ex ...)	W Bk	Russian	476	141·1	31·6	18·9	Bordeaux	1865		V 1.85	A. Carlström	Nikolaistad
120		Magenta	W Bg	Norweg'n	173	92·7	22·8	12·2	Arendal	1869			A. Aanonsen	Arendal
121	N	Magenta *Baltzersen -87*	W Sr	Norweg'n	133	86·6	23·5	10·7	Hernösand	1852			A. Baltzersen	Dröbak
122		Maggie	W	British	415	166·4	26·4	11·8	Garden Island, Ont.	1869	H. Roney		Montreal Transportation Co.	Kingston, Ont.
123		Maggie	W Sr	British	267	130·0	26·0	10·5	St. Cathrines O.	1881	L. Shickluna		T. Coulon	S. Catharines on.
124		Maggie	W Bn	British	242	111·5	28·3	12·7	Freeport, N.S.	1869			W.W. Robinson	Dublin
125	✻	Maggie *Percy 85-85*	W Sr	British	193	114·0	22·7	11·7	Troon	1883	Troon S.B. Co.	V 6.86	E.J. Duder	St. John, N.F.L.
126	✻D	Maggie *Evans 80-80*	W Bn	British	159	99·8	21·8	12·2	Bideford	1870	Cox	V 6.86	N. Bate	London
127	✻	Maggie *Vickers jun. -88*	W Sr	British	100	88·7	22·1	10·6	Peterhead	1878	Stephen	V 3.8	S. Vickers	Ardrossan
128		Maggie *Findley 83-83*	W Sr	British	100	82·0	21·6	10·5	Banff	1867	Watson		J. Findlay	Wick
129	✻	Maggie *Pennifigan 84-84*	W Sr	British	99	91·9	21·5	10·3	Fraserburgh	1878	Webster		J. Ennor	Fraserburgh
130	✻	Maggie *Leslie*	W Sr	British	74	74·1	20·3	8·9	Stromness	1880	Copland		J.A. Shearer	Kirkwall
131	✻	Maggie *Warner*	W K Bge	British	46	76·8	16·6	5·9	Littlehampt'n	1880	Harvey		J. & W.B. Harvey	Littlehampton
132		Maggie	W Bg	American	152	88·6	22·0	12·5					M. Crumley	New York
133		Maggie A. *Willis 83-83*	W Bt	British	196	103·5	24·6	13·0	St. Peter's Bay, P.E.I.	1877	J.A. M'Laine		Mrs. L.E. Mowlam	Poole
134	R	Maggie Abbott *M'Intosh*	W3mSr	American	213	119·0	28·5	9·8	Bucksport, Me	1874		V 8.8	J. Stevens	Boston, Mass.
135	R	Maggie Andrews *Snow*	W3mSr	American	584	157·5	34·4	17·4	Newburyport, Mass	1884	G.E. Currier	V 10.88	A.F. Snow	Baltimore
136	✻	Maggie Annie *Merse*	W K	British	69	76·5	20·2	9·0	Bridgwater	1881	F.J. Carver		R. Hillman	Gloucester
137		Maggie Bell *Chandler*	W Sr	American	105	80·2	24·2	7·6	Camden, Me.	1859			A.J. Bird & Co.	Rockland, Me.
138	✻	Maggie Browne *Davies -83*	W Bk	British	564	151·5	30·7	17·7	Maitland, N.S.	1871	M'Dougall	V 8.88	J.M. Campbell	Glasgow
139		Maggie C. Russ *Peterson*	W3mSr	American	187	117·0	30·0	9·0	Eureka, Cal.	1881	Cousins		C. Page	San Francisco
140	R	Maggie Cain *Meyer*	W3mSr	American	412	125·6	32·2	14·3	Philadelphia	1870			M. Cain	Philadelphia
141	R	Maggie Dalling *Dalling*	W3mSr	American	333	128·4	30·4	9·4	Portland, Me.	1874	Curtis Bros.	V 8.88	T. Dalling	Philadelphia
142	R	Maggie Dart *Edgett* (ex ...)	W Bk	British	584	145·6	30·0	19·4	Princeport, N.S	1882	R. Dart	V 5.88	R. Dart	Maitland, N.S.
143	✻	Maggie Douglas *Lloyd 1-88*	W Bk	British	630	164·7	32·3	19·2	Ardrossan	1872	Barr & Shearer	V 12.88	J. Wallace	Irvine
144	R	Maggie E. Gray *Peacock*	W3mSr	American	385	127·8	30·2	15·8	Baltimore	1867	Booze & Co.	V 7.83	S.H. Beacham	Baltimore
145		Maggie Ellen *Littlejohn*	W3mSr	American	207	110·0	28·8	8·6	Cape Elizabeth, Me.	1874	N.R. Dyer		H.M. Sargent	Portland, Me.
146	R	Maggie G. Hart *Williams -88*	W3mSr	American	488	150·9	34·0	13·5	Waldoboro' Me	1883	A. West	V 10.88	Yates & Porterfield	New York
147		Maggie J. Chadwick *Dyer*	W Sr	American	272	126·0	29·3	10·2	Calais, Me.	1866			A.B. Hall	Portland, Me.
148		Maggie J. Lawrence *Grace*	W3mSr	American	349	124·0	33·5	14·5	Camden, N.J.	1874			F. Smith & Co.	Philadelphia
149		Maggie Louisa	W Sr	British	133	94·5	26·0	11·6	Tusket, N.S.	1863			A. King	Belfast
150														

✻ = Lloyd's Register; D = Nederlandsche Vereeniging; G = Germanischer Lloyd; I = Registro Italiano; R = American Record;
A = Veritas Austro-Ungarico; H = Veritas Ellenico; N = Norske Veritas; V = Bureau Veritas.

Dans la 4me Colonne, C = Composite (membrure en fer, bordé en bois); I = Fer; Stl = Acier; W = Bois.

Les Dates du doublage en métal (Col. 11) ressortent des livres dans lesquels les navires sont respectivement classés ou des rapports des Inspecteurs du Lloyd's Register.

Pour plus amples renseignements sur les navires classés par le Lloyd's Register, voir le Lloyd's Register Book et le Lloyd's Yacht Register.

Official No.	No. in Book	Society in which classed	Ship's Name	Master, and Dates of Appointment / Ship's Late Name	Material and Rig	Flag	Net Tonnage	Length	Breadth	Depth	Where Built	When	By whom	Date of Metalling	Owners or Managers	Port [of Registry]
	51		Mabrook		Bg	Turkish	176	..	..	..					Ahmed Eff. Roustem	Alexandria
	52		Mabrook		Bg	Turkish	172	..	..	..					Abdul M. Eff. Khatab	Alexandria
	53		Macahé	Da Costa (ex Nossa Senhora B.)	WSr	Brazilian	147	..	..	..	[illegible]	1874		Y 5.86	R. da Silveira Guvenio	Rio de J...
68604	54	✻	MacCallum More	Smith [75-88	IS	British	1669	265·3	39·5	23·4	PortGlasgow	1873	R. Duncan & Co.		J. Potter	London
	55	✻ I	Macdiarmid	Scotto 81-87	IS	Italian	1560	255·3	38·2	22·6	Dumbarton	1883	A. McMillan & Sn		M. Amoroso	Palermo
54919	56	D R V	McDougall	Davis	WS	British	1194	197·0	38·2	22·9	Five Mile River [N.S.	1873	McDougall	Y 11.86	McDougall	Halifax
78571	57	✻	Macduff	Jones -87	IS	British	1235	232·4	36·0	21·0	Dumbarton	1877	A. McMillan & Sn		North British Shippg. Co.	Glasgow
21770	58	✻	Macduff	Pelterson -86	WS	Norweg'n	1122	219·6	35·7	22·1	Aberdeen	1869	Hall	Y 5.86	Chr. Anker	Fredriks...
91116	59	V	McLaurin	Oak. 86-88	WS	American	1330	200·6	39·0	24·0	Newburyport, [Mass.	1878	Atkinson & [Filmore	Y 2.86	M. F. Pickering & Co.	Boston
80467	60	✻	MacLeod	Henderson -88	IS	British	1556	255·5	38·3	22·5	Dumbarton	1879	A. McMillan & Sn.		A. McMillan & Son	Glasgow
66884	61	R	Macleod	Cottam -88	WBk	British	633	148·8	34·1	17·8	St. John, N.B.	1873	D. Lynch	Y 11.87	W. H. McLeod	St. John
82205	62	✻ D	MacMillan	Grey -79	IS	British	1450	255·5	38·3	20·6	Dumbarton	1879	A. McMillan & Sn		Macbeth & Gray	Glasgow
90446	63	V	McNear	Dickinson 87-88	WS	American	1245	185·3	37·4	24·0	Belfast, Me.	1872	H. McGilvery	Y 8.87	A. Sewall & Co.	Boston
2078	64	I	Maccabeo	Cerruti 86-85	WBk	Italian	895	174·5	31·5	23·4	Varazze	1879	G. Scala	Y 4.86	Prosp. Lavarello	Genoa
72840	65	D V	Macedon	McDonald -88	WS	British	1486	210·9	39·2	24·4	St. John, N.B.	1878	J. Fraser	Y 1.86	W. Thomson & Co.	St. John
20772	66		Macedonian	Thompson [-85	WS	British	199	95·5	25·0	15·0	Lynn	1854	W. T. Rolin		W. Milburn	North Sh...
86004	67	✻	Machrihanish	Crosby -88	IS	British	1699	264·9	39·8	23·5	PortGlasgow	1883	R. Duncan & Co.		H. Hogarth	Ardrossan
70740	68	✻	Macquarie	Goddard 65-87 (ex Melbourne)	IS	British	1867	269·8	40·1	23·7	London	1875	R. & H. Green		Devitt & Moore	London
95057	69	✻	Madagascar	McCrone 88-88	Stl 1mBk	British	2077	282·0	40·5	24·6	PortGlasgow	1888	Russell & Co.		J. Boyd	Glasgow
16865	70		Madagascar	Knowlton	WSr	American	107	83·9	23·9	9·9	Ellsworth, Me.	1845			C. H. Potter	Calais
786	71	V	Madawaska	Saracene	WBn	Italian	490	135·5	30·9	17·5	Medford	1866	Hayden & Cud- [worth	Y 2.86	Eredi Carm. Picciotto	Messina
65483	72		Madcap	Miller	W3mSr	British	199	117·0	24·3	12·9	Brixham	1871	Upham	Y 6.86	G. Spencer	Waterford
	73	G	MaddaSophia	Stahl -88	WSr	Russian	146	87·3	25·2	11·2	Kaleten	1888	J. Kirstein		F. Stahl	Kaleten
1738	74	I	Maddalena	Maggiolo (ex Cavallo Marino II.) (WRECKED)	WBk	Italian	636	152·5	32·3	21·3	SestriPonente	1868	G. Casanova	Y 3.86	P. A. Maggiolo fu G.	Genoa
156	75	V	Maddalena	Cuartero -88	WBk	Italian	453	138·7	30·9	17·8	Sorrento	1862	A. Castellano	Y 5.85	Florio & Giudice	Palermo
2736	76	V	Maddalena	Sallejtro -87 (ex Mosorsk Vila) (PONTOON)	WBk	Italian	451	134·8	29·5	17·9	Trieste	1868	Stabilimento Tecnico	Y 10.85	G. Dalgato	Naples
504	77		MaddalenaT.	Rapisardi (ex General Garibaldi)	WBg	Italian	275	97·8	27·4	17·5	Sestri	1859	G. B. Cadenaccio	Y 10.87	S. Rapisardi fu D.	Catania
78622	78	✻	Madeira	Falck 88-88	IBk	British	845	196·3	32·7	19·2	PortGlasgow	1878	Russell & Co.		P. Denniston & Co.	Glasgow
58008	79		Madeleine	Connell (WRECKED)	W3mS	British	114	93·6	21·5	10·3	Falmouth	1867	H. S. Trethowan		F. H. Palmer	Maryport
	80	V	Madeleine	Dupont -88 (COLLISION)	WBk	French	287	119·4	26·9	13·4	Granville	1876		Y 1.88	Beust père & fils	Granville
	81	V	Madeleine	Zoonekynd	WSr	French	114	90·2	22·0	10·8	Dunkirk	1885	Pyotte Bayaert		H. Durin	Dunkirk
	82	V	Madeleine Rickmers	Dierks -88	WBk	German	1277	198·4	41·0	25·5	Geestemünde	1879	R. C. Rickmers	Y 7.88	R. C. Rickmers	Geestemünde
58229	83		Madeline Ann	McCourt	WBn	British	132	86·4	23·4	12·3	Sunderland	1867	W. Barkley & Co.		R. Brown	Dublin
93358	84		Madeya		StlLa	British	215	160·0	22·0	7·3	Dumbarton	1887	W. Denny & Bros		Irrawaddy Flotilla Co. Ld.	Glasgow
	85	N	Madonna	Nicolaisen -87	WBk	Norweg'n	463	136·1	28·8	16·2	Grimstad	1880	S. Johnsen	Y 1.86	Bernt Einersen	Grimstad
1322	86	A	Madonna Dell'Arco	Cestaro (ex Sofia)	WBg	Italian	382	115·8	28·7	17·7	Procida	1867	Schiano	Y 7.86	Mod. Di Mattia	Naples
1343	87	A	Madonna Della Libera	Scamiso	WBn	Italian	112	78·7	22·0	10·8	Limite	1859	N. Pichiotti		A. & V. Scotto	[illegible]
48550	88	✻	Madras	Jones	WBk	British	668	160·5	30·6	19·7	Dumbarton	1865	A. McMillan & Sn	Y 7.86	E. Jones	[illegible]
3480	89	I V	Madre	Costa -87	WBk	Italian	725	162·0	33·1	22·0	Sampierdar'na	1874	B. Frassinelli	Y 6.87	Flli Casalicchio	[illegible]
1786	90	I	Madre Antola	(ex Madre Chiossa) Ferro -87	WBk	Italian	436	142·7	30·3	17·8	SestriPonente	1868	G. Casanova	Y 12.86	P. Antola	[illegible]
3078	91	I	Madre Maria	Dasa... -88	W3mSr	Italian	359	133·5	28·0	15·0	SestriPonente	1880	A. Briasco fu L.	Y 4.86	Fran. De Andreas	[illegible]
1958	92	D I	Madre O.	Marini (ex Rosa Eugenio)	WBk	Italian	454	128·9	31·4	19·0	SestriPonente	1869	[illegible] Cadenaccio	Y 7.86	F. Olivari fu G.	[illegible]
2164	93		Madre Rosa	Della Casa (ex Rosa Madre)	WBk	Italian	713	169·6	33·1	22·2	Savona	1871	A. Calcagno	[illegible]	F. Orione	[illegible]
708	94	I	Madre & Figli	Scarpati	WBk	Italian	396	125·9	27·9	18·7	Castellamare	1868	A. Bonifazio	Y 6.87	G. Cafiero	[illegible]
	95		Madrid	(ex Uraiga Basa)	WBk	Urug'yan	1121	194·5	37·6	22·2	Maitland, N.S.	1885	A. A. McDougall			
74262	96	V	Madura	Bernier	WBk	British	970	187·5	36·3	20·7	Quebec	1876	F. X. Marquis	Y 1.86	W. H. Ross & Co.	Quebec
70802	97		Madura	Andrews	WBk	British	344	123·2	26·8	15·6	Nantes	1866	Pestel		J. C. Ellis	[illegible]
	98		Madura Meanatchy	(ex Papillon)	WBk	British	476	136·1	28·7	16·6	St. Malo	1870	Vve. Bonnet & Fils	Y 11.83	L. A. R. R. M. Doreiswamy Chetty	[illegible]
90023	99	✻	Maelgwyn	Jones 84-84	IBk	British	1235	220·0	35·6	21·4	Sunderland	1884	W. Doxford & Sns		R. Thomas & Co.	[illegible]
	100	N	Mærke To	Henriksen -88	WBk	Norweg'n	449	120·0	30·9	17·5	Porsgrund	1855			M. Johannessen & S. Jensen	[illegible]

✻ = Lloyd's Register; D = Nederlandsche Vereeniging; G = Germanischer Lloyd; I = Registro Italiano; R = Amer[illegible]
A = Veritas Austro-Ungarico; H = Veritas Ellenico; N = Norske Veritas; V = Bureau [illegible]

In the 4th Column, C = Composite (iron frame, wood planked); I = Iron; Stl = Steel; W = Wood.

The Dates of Metalling (Col. 11) are based on the records of the Books in which Vessels are respectively classed, or on the Reports of the Surveyors of [illegible]

Further particulars of Vessels classed by Lloyd's Register will be found in Lloyd's Register Book and Lloyd's Yacht Register.

2 N° dans le livre	3 Registre dans lequel le navire est coté	4 Nom. Capitaine et dates d'entrée au service des Armateurs actuels et à bord du navire. Ancien Nom.	Matériaux et Gréement	5 Pavillon	6 Tonnage net	7 Dimensions: Longueur	Largeur	Creux	8 Construit: Où	9 Quand	10 Par qui	11 Date du doublage en métal	12 Armateurs ou Directeurs-Gérants	13 Port d'Armement
101		Magano	WBn	Brazilian	224	..	..	..					J.S.Magano	Rio Grande
102	N	Magda *Nielsen*	WBk	Norweg'n	229	118·0	27·1	11·3	Drammen	1870	Jörgenson	Y 1.83	M.Pedersen	Mandal
103		Magda *Wirolainen*	WBg	Russian	276	110·0	26·7	13·6	Björkö	1873	W.Martelin		M.Reponen	Björkö
104		Magdala	WSr	British	165	105·0	25·3	9·4	Quebec	1868			Geo.Farewell	Quebec
105	NV	Magdala *Pettersen*	WBk	Norweg'n	1172	180·4	37·2	23·6	Portland, N.B.	1868	CharlesSorrell	[illegible]	R.Ingvardsen	Flekkefjord
106	V	Magdalena *Gundersen -87* (ex Trammete)	WBk	Norweg'n	823	177·0	35·2	21·9	Maitland, N.S.	1872	Smith	Y 6.85	M.Brunsgaard	Drammen
107	V	Magdalena *Alsina* (ex Hermosa Favorita)	WBk	Spanish	235	108·6	28·1	13·3	Blanes	1858		Y 1.82	Taltabull & Xiró	Barcelona
108		Magdalena	WBk	Spanish	203	109·5	22·8	13·7	Arens	1857			E.Alsina y Roca	Barcelona
109		Magdalena Charlotte	WSr	Russian	106	71·0	21·1	9·0	Kurland	1876			John Mikkelsohn	Riga
110	DG	Magdalene *Meyer 77-*	WS	German	1250	196·8	39·8	22·9	Quincy, Mass.	1856	Thomas	Y 7.83	D.H.Wätjen & Co.	Bremen
111	N	Magdalene (ex Agra) *Jensen*	WBk	Norweg'n	622	150·2	32·2	18·6	Windsor, N.S.	1864	J.Mosher	Y 11.86	M.Pedersen	Mandal
112	N	Magdalene *Johnsen*	WBk	Norweg'n	267	113·9	26·5	13·8	Hamburg	1854			C.A.Hansen	Fredriksstad
113		Magdalene *Bertling -83*	WSr	Russian	165	92·4	26·7	11·9	Ilmaten	1883			K.Bertling	Riga
114	A	Magdalini (ex Daisy)	WBg	Greek	315	112·6	27·3	16·9	Sunderland	1868	Pace		Theofilatos & Stathatos	Ithaca
115		Magellan	WSr	British	226	103·3	28·0	10·8	Hopewell, N.B.	1882	C.Dowling		Geo.H.Irving	St.John, N.B.
116		Magellan *Manson -88*	WS	American	1019	178·3	34·5	22·9	EastBoston	1873	Smith&Town-[send		C.P.Hemenway	Boston, Mass.
117	V	Magellan *Nissen 86-86*	WBk	German	435	138·5	30·5	17·0	Brake	1870	I.Oltmann's [Wwe.	Y 7.83	Gebr.Oetling	Hamburg
118		Magellan	WBk	Norweg'n	614	162·5	28·4	19·3	Whitehaven	1860	Kennedy	Y 3.81		
119	RV	Magellan *Stahn -85* (Leon)	WBk	Russian	476	141·1	34·6	18·9	Bordeaux	1865		Y 1.83	A.Carlströn	Nikolaistad
120		Magenta	WBg	Norweg'n	173	92·7	22·8	12·2	Arendal	1869			A.Aanonsen	Arendal
121	N	Magenta *Baltzersen -87*	WSr	Norweg'n	133	86·6	23·5	10·7	Hernösand	1852			A.Baltzersen	Dröbak
122		Maggie	W	British	415	166·1	26·1	11·8	Garden Island, [Ont.	1869	H.Roney		MontrealTransportation Co.	Kingston, Ont.
123		Maggie	WSr	British	267	130·0	26·0	10·5	St.CathrinesO.	1881	L.Shickluna		T.Conlon	S.Catharines Ont.
124		Maggie	WBn	British	242	111·5	28·3	12·7	Freeport, N.S.	1869			W.W.Robinson	Dublin
125	✻	Maggie *Percy 85-85*	WSr	British	193	114·0	22·7	11·7	Troon	1883	TroonS.B.Co.	Y 9.86	E.J.Duder	St.John, N.F.L.
126	✻D	Maggie *Evans 80-80*	WBn	British	159	99·8	21·8	12·2	Bideford	1870	Cox	Y 8.86	N.Bate	London
127	✻	Maggie *Vickers, jun. -83*	WSr	British	100	88·7	22·1	10·6	Peterhead	1878	Stephen	Y 3.8[illegible]	S.Vickers	Ardrossan
128		Maggie *Findlay 83-83*	WSr	British	100	82·0	21·6	10·3	Banff	1867	Watson		J.Findlay	Wick
129	✻	Maggie *Penaligan 84-84*	WSr	British	99	91·9	21·5	10·3	Fraserburgh	1878	Webster		J.Ennor	Fraserburgh
130	✻	Maggie *Leslie*	WSr	British	74	74·1	20·3	8·9	Stromness	1880	Copland		J.A.Shearer	Kirkwall
131	✻	Maggie *Warner*	WKBgc	British	46	76·8	16·6	5·9	Littlehampt'n	1880	Harvey		J.&W.B.Harvey	Littlehampton
132		Maggie	WBg	American	152	88·6	22·0	12·5					M.Crumley	New York
133		Maggie A. *Willis 83-83*	WBn	British	196	103·5	24·6	13·0	St. Peter's [Bay, P.E.I.	1877	J.A.M'Laine		Mrs.L.E.Mowlam	Poole
134	R	Maggie Abbott *M'Intosh*	W3mSr	American	213	119·0	28·5	9·8	Bucksport, Me	1874		Y 3.8[illegible]	J.Stevens	Boston, Mass.
135	R	Maggie Andrews *Snow*	W3mSr	American	584	157·5	34·4	17·4	Newburyport, [Mass	1884	G.E.Currier	Y 10.86	A.F.Snow	Baltimore
136	✻	Maggie Annie *Merse*	WK	British	69	76·5	20·2	9·0	Bridgwater	1881	F.J.Carver		R.Hillman	Gloucester
137		Maggie Bell *Chandler*	WSr	American	105	80·2	24·2	7·6	Camden, Me.	1859			A.J.Bird & Co.	Rockland, Me.
138	✻	Maggie Browne *Daniel -82*	WBk	British	564	151·5	30·7	17·7	Maitland, N.S.	1871	M'Dougall	Y 9.85	J.M.Campbell	Glasgow
139		Maggie C. Russ *Peterson*	W3mSr	American	187	117·0	30·0	9·0	Eureka, Cal.	1881	Cousins		C.Page	San Francisco
140	R	Maggie Cain *Meyer*	W3mSr	American	412	125·6	32·2	14·3	Philadelphia	1870			M.Cain	Philadelphia
141	R	Maggie Dalling *Dalling*	W3mSr	American	333	128·4	30·4	9·4	Portland, Me.	1874	CurtisBros.	Y 8.88	T.Dalling	Philadelphia
142	R	Maggie Dart *Edgett* (CONDEMNED)	WBk	British	584	145·6	30·0	19·1	Princeport, NS	1882	R.Dart	Y 6.85	R.Dart	Maitland, N.S.
143	✻	Maggie Douglas *Boyd [-86*	WBk	British	630	164·7	32·3	19·2	Ardrossan	1872	Barr&Shearer	Y 12.8[illegible]	J.Wallace	Irvine
144	R	Maggie E. Gray *Pedrick*	W3mSr	American	385	127·8	30·2	15·8	Baltimore	1867	Booze&Co.	Y 11.8[illegible]	S.H.Beacham	Baltimore
145		Maggie Ellen *Littlejohn*	W3mSr	American	207	110·0	28·8	8·6	CapeElizabeth [Me.	1874	N.R.Dyer		H.M.Sargent	Portland, Me.
146	R	Maggie G. Hart *Williams -88*	W3mSr	American	488	150·9	34·0	13·5	Waldoboro'Me	1883	A.West	Y 10.88	Yates & Porterfield	New York
147		Maggie J. Chadwick *Orr*	WSr	American	272	126·0	29·3	10·2	Calais, Me.	1866			A.B.Hall	Portland, Me.
148		Maggie J.Lawrence *Orsin*	W3mSr	American	349	124·0	33·5	14·5	Camden, N.J.	1874			F.Smith & Co.	Philadelphia
149		Maggie Louisa	WSr	British	133	94·5	26·0	11·6	Tusket, N.S.	1863			A.King	Belfast
150														

✻ = Lloyd's Register; D = Nederlandsche Vereeniging; G = Germanischer Lloyd; I = Registro Italiano; R = American Record; A = Veritas Austro-Ungarico; H = Veritas Ellenico; N = Norske Veritas; V = Bureau Veritas.

Dans la 4me Colonne, C = Composite (membrure en fer, bordé en bois); I = Fer; Stl = Acier; W = Bois.

Les Dates du doublage en métal (**Col. 11**) ressortent des livres dans lesquels les navires sont respectivement classés ou des rapports des Inspecteurs du Lloyd's Register.

Pour plus amples renseignements sur les navires classés par le Lloyd's Register, voir le Lloyd's Register Book et le Lloyd's Yacht Register.

Official No.	No. in Book	Registry in which classed	Ship's Name. Material and Rig. Master, and Dates of Appointment to present Owner's Service and to Ship. Ship's Late Name.	Rig	Flag	Net Tonnage	Length	Breadth	Depth	Where built	When	By whom		Owners or Managers	Port
[illegible]	151	✻	**Maggie Low** *Melville* 85	WSr	British	100	88·7	21·2	10·1	Kingston	1883	W. Kinloch		G. Stephen, jun.	[illegible]
39287	152	R	**Maggie M.** *[illegible]* (Wm. & Ed.)	WBk	British	430	123·0	29·0	17·0	Rothsay, N.B.	187	R. Titus		E. Lantalum & Co.	[illegible]
93138	153	R	**Maggie Millard** *Ormiston*	WSr	British	120	86·6	24·6	10·3	Liverpool, N.S.	1883	R. McLeod		M. Ormiston	[illegible]
17365	154		**Maggie Mulvey** *[illegible]*	WSr	American	253	114·0	26·0	10·0	Waldoboro Me	1867			F. C. Pendleton	[illegible]
17147	155		**Maggie P. Smith** *Douglass*	WSr	American	184	104·0	26·7	8·8	Camden, N.J.	1866			F. Smith & Co.	[illegible]
77754	156		**Maggie Page** *Dexter*	WBn	British	110	86·0	24·0	9·4	Port Joliett NS	1879	S. Harlow		Th. Johnson	[illegible]
91[illegible]	157	R	**Maggie S. Hart** *[illegible]*	W3mSr	American	645	174·0	35·3	18·3	Waldoboro Me	1885	A. R. Reed		C. T. Geyer	New York
	158		**Maggie Taylor** (now Black Dolphin)	WBg	Sarawak	347	130·4	30·9	12·6	Moncton, N.B.	1881	R. Cochrane	y ss	Govt. of Sarawak	Sarawak
90659	159	R	**Maggie Todd** *Cogswell*	WSr	American	119	102·0	26·8	7·5	Calais, Me.	1873	Boardman Bros.		H. Q. Boardman	Calais
63260	160	✻	**Maggie Trimble** *Morgan* -83	IS	British	780	191·8	31·1	[illegible]	[illegible]	1870	A. Stephen & Sns		R. G. Sharp & Co.	Liverp.
92875	161	✻	**Maggie W. Smith** *Thompson*	W3mSr	British	138	102·1	22·5	11·[illegible]	[illegible]	1886	Barr & Shearer	y ss	A. Thomson	[illegible]
79873	162	R	**Maggie Willett** *Moran* -81	WSr	British	118	84·0	26·1	10·[illegible]	Hopewell Cape, N.B.	1878	J. [illegible]		A. L. Bonnell	St. J.
66882	163		**Maggie Wood** *Williams*	WBn	British	178	103·3	28·4	11·4	Hillsboro', N.B.	1873	Pye		D. Graham	Belfast
52908	164		**Maggie & Helen** *Morison* [illegible]	WS	British	128	88·2	21·0	11·7	Inverkeithing	1866	Ross		W. H. Leask	Peterhead
74191	165	✻	**Magic** *Langlois* -88	WBn	British	182	104·0	24·5	12·9	P.E. Island	1876	Duncan	y ss	E. J. Bellingham & C.	London
77830	166	✻	**Magic** *Lloyd*	WSr	British	136	93·8	22·3	10·5	P.E. Island	1879	M'Dougall	y ss	J. Reney	Chester
	167	N	**Magna** *Christensen* 82-82	WBk	Norweg'n	723	155·7	33·8	19·4	Brooklyn	1863		y ss	P. A. Christensen	Langesund
	168	N	**Magna** *Kjelstad* 82-82 (ex Hanna Maria)	WBk	Norweg'n	529	141·9	32·8	16·5	Maitland, N.S.	1864	Smith		L. A. Morch	Frederiks.
	169	✻	**Magnat** *Ostermann* 85-85	IBk	German	948	206·0	33·5	19·2	Sunderland	1885	S.P. Austin & Son		Gerd. Bolte	Elsfleth
	170	N	**Magnat** *Faust* 77-77	WBk	Norweg'n	630	156·5	33·6	18·3	Sandefjord	1877	Chr. Christensen	y ss	A. F. Klaveness	Sandefj.
	171	N	**Magne** *Sørensen*	WBk	Norweg'n	625	156·9	32·9	18·0	Porsgrund	1875		y ss	J. Harbitz	Tönsberg
[illegible]	172	✻	**Magnet** *Roach*	IBk	British	404	156·0	26·7	17·4	Sunderland	1866	J. Haswell & Son		L. Tulloch	[illegible]
48138	173		**Magnet**	WBn	British	299	115·5	27·9	12·8	Londonderry NS	1864			John Geddes, jun.	Halifax
[illegible]	174		**Magnet**	WK	British	117	91·0	22·6	8·8	Sittingbourne	1884	J. Taylor		J. Smith	Colchester
[illegible]	175		**Magnet** *Fletcher*	WSr	American	137	88·1	24·6	8·8	E. Machias, Me	1860	J. W. Wiswell		A. M. Fletcher	Machias
	176	V	**Magnet** *Jensen* -71	WSr	Danish	123	87·8	20·9	10·8	Thurø	1866	P. Bom		H. F. Nielsen	[illegible]
	177		**Magneten**	WBg	Norweg'n	249	89·6	26·2	15·8	Arendal	1840			O. Johnsen	Arendal
	178	V	**Magnhild** *Tollefsen* (ex Cumberland)	WS	Norweg'n	1555	220·0	41·6	24·0	Maccan, N.S.	1876	J. O'Brien	y ss	G. C. Hansen	[illegible]
47[illegible]	179	RV	**Magnificent** *Merrick* 87-87	WS	British	1283	189·3	37·1	24·8	Quebec	1863	P. Valin	y ss	J. Herron & Co.	[illegible]
[illegible]	180	R	**Magnolia** *Davies*	WS	British	999	177·0	36·1	21·3	Plymouth NS	1874	J. [illegible]	y ss	A. F. Stoneman	Yarmouth
[illegible]	181	R	**Magnolia** *Holle*	WSr	American	117	..	..	..	St. Michaels, Md.	1870				[illegible]
[illegible]	182		**Magnolia**	WSr	American	108	86·0	23·7	9·0	Essex, Mass.	1887			J. G. Farr	[illegible]
	183	N	**Magnolia** *Berg* -86 (ex [illegible])	WBk	Norweg'n	959	206·7	33·8	20·9	Aberdeen	1857	W. Hood & Co.	y ss	O. Svendsen	[illegible]
[illegible]	184	R	**Magnum** *German*	WSr	British	119	82·5	21·0	8·8	Meteghan, N.S	1884	Mark L. Devon		Geo. German	[illegible]
	185	N	**Magnum** *Ness* (ex Sir John [illegible])	WBk	Norweg'n	1360	130·0	38·9	28·0	St. John, N.B.	1859	Smith	y 10.84	A/S [illegible] Magnum	[illegible]
	186		**Magnus** *Petersen* 86-86	WBg	Swedish	230	93·5	23·1	13·3	Dalkarlsä	1846			L. F. Boström	[illegible]
	187	DR	**Magnus** *Skantze* 83-83	WSr	Swedish	179	92·5	23·0	11·9	Krogarne	1875	S. Abrahamsson		C. R. Holst	[illegible]
	188	V	**Magnus Huss** *Hägglun* -88	WBk	Swedish	533	157·1	29·2	17·3	Ortviken	1874	W. Häggl und	y ss	Ortvikens Angsagsaktiebolag (C. A. Forssell)	Ortviken
	189	N	**Magnus Lagaböter** *Nielsen*	WBk	Norweg'n	421	138·7	30·6	16·7	Christiania	1868	H. C. Hansen	y ss	I/S Interessentskab (Klingo & Co.)	[illegible]
	190	R	**Magnus Stenbock** (ex Hakon Adelstein) *Scheel* 85-85	WBk	Swedish	276	114·4	24·0	14·5	Stavanger	18 64/65	K. Kaysen		M. B. Bergström	[illegible]
1768[illegible]	191		**Mahaska** *Fitzgerald*	WSr	American	117	89·0	27·0	7·7	Millbridge Me	18 63/84	Cambridge, Md.		E. L. Palmer & Co.	Baltimore
1663[illegible]	192		**Mahlon Williamson**	WBk	American	461	129·0	28·0	13·0	Wilmington, Del.	1854			W. Bernard	New Yd.
4807[illegible]	193		**Mahmoda** (ex Eucnee)	WBk	British	411	117·5	29·4	17·5	Marseille	1862			Ponnambat Abboobaker	[illegible]
3111[illegible]	194		**Mahomed Samdanee**	WBg	British	189	85·0	21·4	11·4	[illegible]	1875			Mohamed Ismail (Mohamed Ismia)	Colombo
	195	N	**Mai** *Olsen*	WBg	Norweg'n	138	95·5	23·2	16·6	[illegible]	1864	A. Omundsen		H. Hansen	Lillesand
	196		**Maia I.** *Mendes* (ex George Peabody)	WBkn	Brazilian	436	131·6	31·5	18·5	Baltimore, Md.	1871		y ss	A. d'Oliveira Maia	[illegible]
	197		**Maia II.** (ex Barbosa II.) *Almeida*	WBn	Brazilian	165	..	..	..	[illegible]	1875			A. d'Oliveira Maia	[illegible]
	198		**Maid Marian**	WBg		278	106·0	26·7	16·2	Sunderland	1864	Gibbon			
12108	199	✻	**Maid of Anglesea** *Wm. P. Steel*	WSr	British	68	77·0	18·5	9·4	[illegible]	1857	I. Jones		J. W. Honkin	[illegible]
43571	200		**Maid of Honour** *Roffe*	WBkn	British	176	106·0	24·2	12·0	Colchester	1864	Sainty		J. Clancy	

✻ = Lloyd's Register; D = Nederlandsche Vereeniging; G = Germanischer Lloyd; I = Registro Italiano; [illegible]
A = Veritas Austro-Ungarico; H = Veritas Ellenico; N = Norske Veritas; [illegible]

In the 4th Column, C = Composite (iron frame, wood planked); I = Iron; Stl = Steel; W = Wood.

The Dates of Metalling (Col. 11) are based on the records of the Books in which Vessels are respectively classed, or on the Reports of the Surveyors of [illegible]

Further particulars of Vessels classed by Lloyd's Register will be found in Lloyd's Register Book and Lloyd's Yacht Register.

		Nom. Capitaine et dates d'entrée au service des Armateurs actuels et à bord du navire. Ancien Nom.	Matériaux et Gréement	Pavillon	Tonnage net	Dimensions. Longueur	Largeur	Creux	Construit. Où	Quand	Par qui	Date du doublage en métal	Armateurs ou Directeurs-Gérants.	Port d'Armement.
201	*	Maid of Meirion *Jones*	W Sr	British	6[illegible]	65·2	18·4	8·6	Aberystwith	1859	Jacobs		R. Jones	Aberystwith
202		Maid of Mona *Grindley*	W Sr	British	13[illegible]	96·0	22·4	11·5	Ayr	1866	Fullerton		W.E.R. Heskin	Douglas
203	V	Maid of Orleans *Husted*	W Sr	American	172	110·2	29·5	9·6	SanFrancisco	1882	[illegible]kie Bros.		J.J. McKinnon	San Francisco
204		Maid of the Mist (BROKEN UP) *Jordan*	W Sr	British	111	107·7	16·3	9·7	St.John.N.B.	1833			T. Roberts	Dublin
205	*	Maida (HULK) *Graham*	W Bk	British	520	117·4	28·2	19·2	Moulmein	1857		[illegible]	D. M'Gregor	Brisbane
206		Maida *Kearney* -88	W Bn	British	114	76·6	22·0	10·6	Shelburne.N.S	1876	J. Decker		P.G. Tessier	St.Johns,N.F.L.
207	*	Maiden City *Montgomery* [illegible]	Stl Bk	British	1197	22[illegible]	[illegible]	20.7	Londonderry	1887	C.J. Bigger		Thomson,Dickie & Co.	Londonderry
208	V	Maiden City *Humphreys*	W Bk	British	799	165·1	34·1	20·6	St.Martin's NB	1878	J. Osborne	[illegible]	O. Emery	St. John. N.B.
209		Maiee Kisbeyeh	W Bg	Turkish	120	..	..	..	Barten	1872			Altinbilco Ahmed & Co.	Barten
210		Mail *Cowell*	W Sr	American	130	81·9	24·3	8·1	Bristol,R.I.	1850			S.J. Sewall	Boston, Mass.
211	*	Maile *Lane* 77-84	W Sr	British	252	125·7	26·6	11·7	Auckland,N.Z.	1884	Lane & Brown	[illegible]	C.C. McMillan	Auckland.N.Z.
212	V	Maillen (ex Colonel L-Quelle)	W Bk	Chilian	526	135·8	29·8	19·2	Bordeaux	1863	Forges & Co. de l'Océan	[illegible]	T. Moreno	Valparaiso
213	*	Main *Richards* [illegible]	I S	British	1614	256·4	38·3	23·1	Greenock	1884	Russell & Co.		J. Nourse	London
214	R	Maine *Shaw* -86	W Sr	American	181	105·4	28·0	8·5	Bath,Me.	1886	New Eng.S.B.C.		R.D. Wood & Co.	Philadelphia
215		Mainio *Johansson*	W Bk	Russian	339	..	..	..	Lovisa	1857			A. Gratschoff	Lovisa
216		Mainio *Dettlöf*	W Bk	Russian	338	..	..	..	Jacobstad	1854			P.U. Strengberg	Jacobstad
217		Mair & Cranmer *Rudd*	W 3m Sr	American	233	115·6	29·6	9·4	Goshen,N.J.	1871			J. McCauley	Philadelphia
218	*	Mairi Bhan *McIntyre* 7-87	I S	British	1315	239·6	37·0	22·3	Glasgow	1874	Barclay,Curle & Co.		P. M'Intyre	London
219	N	Maitland *Svendsen* -88	W Bk	Norweg'n	736	160·0	33·3	19·3	Maitland,N.S.	1871	G. Smith & Son	[illegible]	J.T. Duus & O.A. Svendsen	Kragerö
220		Maiz	W Bg	Turkish	262	..	..	..	Austria	1874			Boghoz Serkiz & Co.	Constantinople
221	V	Maja *Drejo* 87-88	W Sr	Danish	106	82·4	20·3	10·3	Æroskjobing	1877	R. Mortensen		J. Albertsen	Æroskjobing
222	*	Majestic *Orr* 81-88	I S	British	1884	273·4	40·2	24·3	Belfast	1875	Harland & Wolff		T. & J. Brocklebank	Liverpool
223		Majestic	W Sr	British	120	78·6	21·5	12·1	Newport	1839	J. Young		F.E. Dullin	Lyme
224		Majestic *Bergman*	W Bk	American	1117	177·0	36·9	23·6	Portland,Me.	1866	G. Lawrence	[illegible]	A. Bergman	San Francisco
225	R	Major M. Kock *Kittern*	W Sr	Swedish	149	79·0	22·7	11·8	Uddevalla	1876	M. Hulten		C.J. Backman	Lysekil
226		Makah (LOST) *Larsen*	W Bk	American	667	169·0	38·0	15·0	Port Blakely, [Wash.	1882			H.A. Thompson	SanFrancisco
227		Makboobjian	W Bg	Turkish	287	..	..	..	Antaria	1874			Iskenderieh Tijar [illegible]	Iskenderieh
228		Makedonia	W Bg	Turkish	169	..	..	..	Greece	1883			Costanti Mickal Joklos & Bros.	Constantinople
229		Makesed Khair	W Bg	Turkish	242	..	..	..					El Hag Hassan Gadar	Alexandria
230		Malangen *Abrahamsen* -87	W Bg	Norweg'n	187	100·4	25·0	10·7	Wiborg	1855			O.M. Abrahamsen	Farsund
231		Malay	I Lr	British	362	195·8	25·1	8·5	PortGlasgow	1880	R. Duncan & Co.		Irrawaddy Flotilla Co. Ld.	Glasgow
232		Malay *Crocker* -88	W Bkn	American	706	167·0	33·0	21·6	Chelsea,Mass.	1852			N. Bichard	San Francisco
233	*	Malaysia *Kirkbride* 80-85	I S	British	1827	263·1	40·3	24·0	Greenock	1885	Russell & Co.		J. & W. Goffey	Liverpool
234	*	Malborough *Ircombe*	W Bk	British	297	118·0	25·4	14·7	Salcombe	1870	Vivian	[illegible]	J. Sutcliffe & Son	Grimsby
235		Malcolm	W Bn	British	182	105·2	23·5	12·5	P.E.Island	1862	J. MacDougall		Bernard Byrnes	Sydney,N.S.W.
236		Malecon	W Lorcha	Spanish	101	79·1	20·9	6·2	Manila	1885	G. Rodriquez		Luis.R. Yangco	Manila
237		Malfilatre *Smith*	W Sr	British	116	86·7	21·9	10·2	[illegible]	1859	Guillemette		J.G. Sully	Bridgwater
238		Malik	I Lr	British	700	210·0	35·5	10·3	PortGlasgow	1884	Russell & Co.		Irrawaddy Flotilla Co. Ld.	Glasgow
239	* D	Mallowdale *Davies* 86-88	I Bk	British	1290	211·8	37·1	22·6	Lancaster	1869	Lanc. Ship Co.		Hunter, Brown & Co.	Greenock
240	*	Mallsgate *Spout* 86-86	I Bk	British	1043	215·3	34·1	21·0	Harrington	1877	Williamson		Fisher & Sproit	Workington
241	*	Malpas Belle *Common* [illegible]	W 3m Bn	British	171	109·2	22·7	12·1	Truro	1872	Scoble		W.C. Phillips	Truro
242		Malta (OLD SHIP C.P.) *De Maria*	W S	British	1336	216·7	37·0	28·1	Malta	1864	G. Mirabitur	[illegible]	Marquis J. Scicluna	Malta
243		Malta	W Sr	British	198	137·5	23·5	8·2	St.Catharines		L. Shickluna		Mary M. Blodget	St.Cath'rines.Ont.
244	R	Malta *Lundberg* -88	W S	Swedish	855	157·4	31·9	22·1	NovaScotia	1862		[illegible]	C. Barchmann & Co.	Gothenburg
245		Malvina *Bartram* 85-85 (ex Fredrik)	W Bk	Danish	329	113·2	26·7	15·6	Hernösand	18 45/61	P. Ahlquist		Moses & Søn,G. Melchior	Copenhagen
246	V	Malvina Wendt *Dreyer* -88	W Bk	German	363	116·8	28·5	16·7	Barth	1858	J. Kraeft		J.N. Rodbertus	Barth
247	V	Malvine *Ludwig* 87-87 (ex [illegible])	W Bk	German	479	132·8	28·7	15·5	Bremerhaven	1869	F.W. Wencke	[illegible]	J.B. Reiners & Co.	Brake
248		Malvoisin	W K	British	113	101·0	22·8	8·5	Maldon	1883	J.T. Howard		W.T. Meeson	London
249		Malwina Sophia *Pleik* 85	W Sr	Russian	128	72·0	24·0	11·0	Lab-Essern	1868	Behnhusa		T. Plath	Riga
250		Mam *Ray* -88	C Bk	French	420	115·8	27·2	11·2	Bordeaux	1867	Chaigneau	[illegible]	A. Hubeau	St.Nazaire

* = Lloyd's Register; D = Nederlandsche Vereeniging; G = Germanischer Lloyd; I = Registro Italiano; R = American Record; A = Veritas Austro-Ungarico; H = Veritas Ellenico; N = Norske Veritas; V = Bureau Veritas.

Dans la 4me Colonne, C = Composite (membrure en fer, bordé en bois); I = Fer; Stl = Acier; W = Bois.

Les Dates du doublage en métal (**Col. 11**) ressortent des livres dans lesquels les navires sont respectivement classés ou des rapports des Inspecteurs du Lloyd's Register.

Pour plus amples renseignements sur les navires classés par le Lloyd's Register, voir le Lloyd's Register Book et le Lloyd's Yacht Register.

Official No.	No. in Book	Registry in which classed	Ship's Name. Material and Rig. Master, and Dates of Appointment to present Owner's Service and to Ship. Ship's Late Name.	Flag.	Net Tonnage.	Length.	Breadth.	Depth.	Built. Where.	When.	By whom.	Date of Metalling.	Owners or Managers.	Port
	251		Mamood Samdany WBk	British	215	84·1	22·9	13·6	[illegible]	1850			Mahomed Tamby Marcar	Cochin
[illegible]	252	DV	Man Zu *Stemmer 85* WBn	German	247	121·0	26·2	10·4	Maitland, N.S.	1885	Putnam Bros.	Y 11.86	J.H.L.Wilckens	Hamburg
[illegible]	253		Mana WSr	British	107	87·8	25·7	7·6					T.G.Kelly	[illegible]
	254	I	Manana Savona *Anselmo* WBk (ex Tobia)	Argentine	386	..	..	..	Recco	1864	Rolla	Y 8.88	C.Bettinote	[illegible]
[illegible]	255	✻	Mandalay *Armstrong* IBk	British	904	201·0	32·8	19·7	Greenock	1880	Russell&Co.		Anderson, Webster & Co	Glasgow
[illegible]	256		Mandalay WS	British	745	149·2	32·6	16·5	Millbridge, US	1862			Bombay & Burmah Trading Corporation	Bombay
[illegible]	257	✻	Mandara *Smith* WBkn	British	195	118·6	22·7	11·6	Troon	1885	Troon S.B.Co.	Y 6.88	T.Steele	Ayr
	258	N	Mandarin *Mathiesen 84-84* WBk	Norweg'n	251	131·2	27·6	10·3	Sundsvall	1863	J.A.Ledberg		Morten Mulvad	[illegible]
7745	259		Manekar Theagar Vadyvndiamma WBg	British	154	108·8	22·7	11·0	Cochin	1876			Lana A.R.R.M. Sapramanoopoleo Arumalee	[illegible]
[illegible]	260	✻	Mangalore *Parry 83-83* IS	British	1213	216·0	35·1	22·4	Glasgow	1864	Barclay, Curle &Co.		Eyre, Evans & Co.	Liverpool
	261		Mangerton *Sörensen -88* WBk	Norweg'n	317	120·1	25·6	16·6	Sunderland	1859	J.Laing		H.Andersen	[illegible]
	262	V	Manhem *Jansson* WBkn	Swedish	474	145·0	25·7	17·0	[illegible]	1877	G.Persson	Y 7.88	J.F.Olsson	[illegible]
[illegible]	263	A	Mania WSr	Greek	226	102·3	27·9	..	Chios	1880	C.Scaramanga		A.Lazaros	[illegible]
[illegible]	264	R	Manie Saunders *Hall* W3mSr	American	265	125·0	32·0	8·6	Tuckahoe, N.J.	1883	Still		J.A.Curtis	[illegible]
	265	N	Manifest (ex Adonis) *Paigin* WBn	Russian	184	93·8	23·6	11·1	Kappeln	1867	G.A.Steen		W.Mitrofanoff	[illegible]
[illegible]	266		Manin Cichero WBk *Valpela -88*	Italian	524	140·0	30·6	19·2	Voltri	1867	B.Fabiano	Y 7.88	A.Cichero fu N.	[illegible]
	267	V	Manitoba WBk	Norweg'n	728	158·0	32·7	20·5	Quebec	1870	Gingras	Y 6.88	Th.Nordaas	[illegible]
[illegible]	268		Manitou *Avey* WSr	American	190	98·0	25·3	11·0	Woolwich, Me.	1849			L.Smith	[illegible]
17716	269	R	Manson *Schultz -87* WBn	American	251	125·2	25·5	12·4	Brewer, Me.	1867		Y 5.88	Yates & Porterfield	[illegible]
	270		Manuel WBk (ex Ventnor, ext magna Barreck)	Spanish	991	208·0	34·5	20·7	Aberdeen	1870	Humphry		M.de Iriart	[illegible]
	271	V	Manuel Antoñito *Pose* WBn (ex Victoria Perez, ex Frejar) [81-84]	Spanish	232	115·1	25·9	10·6	Gefle	1866	J.A.Broden	Y 3.88	J.Marquez	[illegible]
	272		Manuel C. Fortunat WBn (ex Unone) *De Gomez*	Mexican	216	103·0	21·7	12·1	Stavanger	1869		Y 8.88	M.C.Fortunat	[illegible]
[illegible]	273	RV	Manuel Llaguno *Smalley* WS	American	1650	221·3	41·5	25·5	Bath, Me.	1879	J.McDonald	[illegible]	B.Flint	[illegible]
[illegible]	274	R	Manuel R. Cuza *Bayles* W3mSr	American	283	119·0	31·5	13·0	Pt Jefferson NY	1875		Y 8.88	S.S.Bayless	[illegible]
	275	DV	Manuelita Ratto WSr (ex Heidelika)	Argentine	126	87·7	21·1	10·5	Veendam	1864	K.&J.Wilkens		B.Ratto	[illegible]
[illegible]	276	✻	Manx King *Karran 84-84* IS	British	1702	251·0	39·0	24·1	Stockton	1884	Richardson, Duck&Co.		R.G.Karran	Castletown
[illegible]	277	✻	Manx Queen *Kee 87-87* IBk	British	303	150·7	25·4	11·6	Barrow	1878	D.Noble&Co.		R.G.Karran	Castletown
[illegible]	278	✻	Manydown *Callender* IS	British	2381	305·0	41·3	25·0	Southampton	1884	Oswald, Mordaunt&Co.		E.Bates & Sons	Liverpool
[illegible]	279	R	Manzanilla *Smith* WSr	British	174	95·1	26·1	10·7	Five Islands, NS	1883	B.Fulmore	Y 12.83	Robert Lewis	[illegible]
[illegible]	280	✻	Maori *Jackson 86-86* WBk	British	656	144·0	32·0	21·2	Sunderland	1851	J.Laing	Z 3.87	A.Major & Co.	London
[illegible]	281	DV	Marabout *Ferguson -87* WS	British	1498	210·8	40·2	24·3	St.John, NB	1882	Stewart&Ritchie	Y 8.86	W.E.Vroom	St.John N.B.
	282	A	Marathon *Colakis* WBg	Greek	212	92·8	23·9	15·2	Syra	1881	A.Livadaros	Y 2.88	P.Colakis	Syra
	283	I	Marathon *Nicolachi* WBg	Greek	187	..	..	..	Syra	1868		Y 2.88	G.C.Nicolakis	[illegible]
[illegible]	284	A	Maratona *Taraboschia -88* WBn	Austrian	258	95·8	25·2	12·9	Lussinpiccolo	1876	N.Martinolich	[illegible]	N.F.Martinolich	[illegible]
[illegible]	285	✻	Maraval *Reith 77-82* IS	British	1257	237·1	36·1	21·1	Dumbarton	1877	A.M'Millan&Sn		W.Kenneth & Co.	Glasgow
[illegible]	286	✻	Maravilla *Smyth* IBk	British	656	165·4	28·4	18·3	Preston	1865	Preston, I.W.		G.H.Fletcher & Co.	Liverpool
[illegible]	287		Marblehead WBk	American	582	167·1	27·0	17·2	Newburyport, Mass.	1862			W.W.Collins	[illegible]
	288	V	Marcel & Emma *Collet* WBg	French	173	67·3	22·0	12·3	Nantes	1869	Clergeau	[illegible] 1.81	E.Collet	[illegible]
	289	V	Marcelina *Pages* WBn	Spanish	240	103·6	27·6	13·6	Arenas	1866		Y 7.82	M.Sitfiel	[illegible]
	290	DV	Marcelino Jané *Amigo -88* WS (ex Sydenham)	Spanish	1061	199·5	36·7	21·9	Christinestad	1873	M.Hannus	X 10.88	M.Jané & Co.	[illegible]
	291		Marcelita *Ferro -88* WPol (ex Mercedita)	Spanish	154	85·6	15·5	12·0	Masnou	1856			T.Fabregado	[illegible]
2551	292	I	Marcello *Dodero* WBk (ex Giambattista)	Italian	792	177·3	33·9	21·8	Savona	1874	V.Dabove	Y 6.88	F.lli Monti	[illegible]
[illegible]	293		Marcello Barbaro WBg (ex Cinque Sorelle) *Barbaro*	Italian	220	109·2	24·2	13·1	Nova Scotia	1862			M.Barbaro	[illegible]
	294	V	Marcelo *Sauns 67-82* WBk	Danish	354	128·5	25·4	14·9	Brake	1865	Menke&Botter	Y 1.87	P.Berg	[illegible]
[illegible]	295	I	Marchesino *Vicini* WBk	Italian	663	144·6	30·2	20·0	Sestri	1871	A.Briasco fu L.	Y 10.86	G.Vicini	[illegible]
1001	296	V	Marchiani *Bratschi* WBk	Italian	287	111·8	25·8	16·5	Sunderland	1865		Y 1.88	F.Marchiani	[illegible]
	297	I	Marchin WBk		741	164·7	33·5	22·2	Varazze	1876	G.Scala	[illegible]		
	298	D	Marchina *Scholtens* WSr	Dutch	149	91·6	22·1	11·1	Hoogezand	1858	J.A.Haaltes		J.S.[illegible]	[illegible]
	299	R	Marcia C. Day *Albretsen -81* WS (ex White Eagle)	German	858	166·0	35·8	22·2	Damariscotta Me	1853		[illegible]	[illegible]	[illegible]
[illegible]	300		Marcia S. Lewis *Kelley* W3mSr	American	330	123·0	29·0	10·0	Portland, Conn.	1867			[illegible]	[illegible]

✻ = Lloyd's Register; D = Nederlandsche Vereeniging; G = Germanischer Lloyd; I = Registro Italiano; R = [illegible]
A = Veritas Austro-Ungarico; H = Veritas Ellenico; N = Norske Veritas; V = [illegible]

In the 4th Column, C = Composite (iron frame, wood planked); I = Iron; Stl = Steel; W = Wood.

The Dates of Metalling (Col. 11) are based on the records of the Books in which Vessels are respectively classed, or on the Reports of the Surveyors of [illegible]

Further particulars of Vessels classed by Lloyd's Register will be found in Lloyd's Register Book and Lloyd's Yacht Register.

No.	Cote	Nom. Capitaine et dates d'entrée au service des Armateurs actuels et à bord du navire. Ancien Nom.	Matériaux et Gréement	Pavillon	Tonnage	Dimensions: Longueur	Largeur	Creux	Construit: Où	Quand	Par qui	Date du doublage en métal	Armateurs ou Directeurs Gérants	Port d'Armement
301		Marcial	WBn	Spanish	115	80·0	21·5	10·6	Bilbao	1865	P.deBilbao		Sevalles	Santander
302	V	Marco (WRECKED) *Mirrochi*	WBk	Italian	415	126·6	30·9	16·8	Malta	1866	A.Busuttil	2.83	M.Miceli	Messina
303		Marco Polo (NOT IN EXISTENCE)	WBk	British	392	106·8	24·6	15·5	Rv St. John N.S	1855	J.Kitchin		James Kitchin	Pictou, N.S.
304	A	Marco Polo *Giada*	WBn	Italian	190	86·1	21·6	10·7	Chioggia	1884	G.Poli	5.87	F[lli] Giada	Venice
305	VN	Marco Polo *Gawiersen* (ex Bengal)	WBk	Norweg'n	754	160·0	32·3	17·1	Vegesack	1870	J.Lange	1.86	B.A.Olsen	Risör
306		Marco Polo *Anderson*	WBk	Norweg'n	374	131·7	28·7	14·0	Bremerhaven	1863	F.W.Wencke	2.81	R.A.Mithassel	Farsund
307		Marcus *Erling*	WSr	Russian	22?	95·0	21·0	11·0	[illegible]	1877	E.Titan		M.Grant	Haynasch
308		Marcus *Jacobsen 88*	WSr	Russian	118	98·9	23·3	9·8	[illegible]	1866	M.Ratix		G.W.Reinicke	Windau
309		Marcus Edwards (italien)	W3mSr	American	274	118·6	27·6	9·0	Bendsville N.J	1875			G.A.Cotton	Philadelphia
310	V	Mardyckoise *Adaoul-87*	WSr	French	113	94·5	22·3	10·9	Dunkirk	1887	Cornemause Brs.		Macré & Co.	Dunkirk
311		Mareego *Patra*	WBg	Turkish	141	..	..	..	Greece	1883			Sakyzlee Margosarzo & Co.	Chios
312		Mareeyah	WBg	Turkish	275	..	..	..	Roumania	1878			Tujar Anastassaki Nicholaidis	Constantinople
313		Mareeyah	WSr	Turkish	247	..	..	..	Greece	1881			Tashi Zadeh Suleiman Effendi	Constantinople
314		Mareeyah	WBg	Turkish	155	..	..	..	Lemnos	1870			Limnilee NicolaManoli	Lemnos
315		Mareeyah	WBg	Turkish	120	..	..	..	Greece	1882			Bulbuljehlee Scotpappa Tesdari	Bulbuljeh
316		Mareeyah	W	Turkish	100	..	..	..	Hasskioi	1880			Tujar Giovanni Kosarinski	Constantinople
317		Mareeyah Vezin	WBg	Turkish	175	..	..	..	Greece	1877			Keresteh Tijari Osraphil Pavlomos	Constantinople
318		Maremma (ex Stefano)		Italian	134	80·4	24·9	13·4	Varazze	1845	B.Cerrutti		A.Dapelo fu A.&Co.	Genoa
319		Maren *Kandes -88*	WBn	British	170	99·8	23·6	10·6	Varmanäs	1876	F.Lhde		J.Westcott	Plymouth
320	V	Maren *Sotrre -88* (ex Adolph)	IBk	Danish	580	159·7	29·8	17·8	Harburg	1873	H.Kraus		C.H.Nielsen	Nordby, Fanø
321		Maren *Christensen 81-81*	W3mSr	Danish	211	112·1	25·9	12·4	Marstal	1883	J.O.Christensen		H.C.Christensens Enke	Marstal
322	V	Maren *Madsen 87-88*	WBn	Danish	150	95·1	21·1	10·7	Nordby, Fanø	1870	J.N.Christensen	7.87	Actieselskab (M.S.Hansen)	Nordby, Fanø
323	V	Maren *Knippel 80-80*	WSr	Danish	130	92·9	21·0	10·6	Faaborg	1880	F.Vang	4.85	P.Nielsen	Faaborg
324		Maren *Jessen*	WSr	Danish	122	86·0	21·0	10·2	SønderhoFanø	1872	P.Sønnichsen		S.A.Sorensen	Sønderho, Fanø
325	V	Maren Kirstine *Jensen*	WSr	Danish	102	82·0	19·6	10·0	Svendborg	1876	J.R.Andersen		J.R.Andersen	Svendborg
326	V	Marena *Moore*	WBn	American	405	125·6	30·5	15·2	ColumbiaFalls [Me	1874	J.Cramden		R.Poillon	New York
[illegible]	I	Maresca *Allegro -88* (ex Luigi Regina)	WBk	Italian	518	138·4	30·9	20·0	Castellamare	1867	B.Bonifacio	8.88	R.Maresca	Castellamare
[illegible]	V	Maresca *Longo* (ex Filomena & Zelda)	WBg	Italian	347	111·8	27·0	18·0	Sorrento	1867	A.Castellano	8.86	Mariangela Maresca	Castellamare
329	R	Margam	WBg	Norweg'n	298	128·0	27·7	14·8	P.E.Island	1872	Richards	10.86	T.M.Tönnesen	Christiansand
330	※	Margaret *Roderick* (COLLISION)	ISr	British	187	127·6	25·7	11·4	Hayle	1866	Poole		D.W.Bain & Co.	Hayle
331		Margaret (ex L'Amiral Rochambeau)	WBg	British	123	74·9	20·8	9·1	Bordeaux	1839			H.Hustin	Belfast
332		Margaret *James* (COLLISION)	WSr	British	116	91·4	22·0	10·1	P.E.Island	1867	McMillan		W.Vickers	Liverpool
333		Margaret *Hough*	WBa	British	109	79·6	22·8	10·5	Argyle, N.S.	1845			J.M.Carnaghan	Maryport
334	※	Margaret *Nicholls*	WSr	British	93	79·9	20·0	10·4	Leith	1867	Menzies		J.H.Nicholls	Falmouth
335	※	Margaret *Luke 85-85*	WSr	British	82	81·0	21·1	9·9	Malpas	1885	Scoble&Davies	5.85	Scoble & Davies	Truro
336	V	Margaret *Nelson* (ex North Britain)	WBk	American	1212	208·0	40·1	23·1	Bergen	1874	Brandhorst & Helder	3.86	C.Nelson	San Francisco
337		Margaret *Leighton*	WSr	American	111	81·8	27·1	7·1	Cherryfield Me	1869			G.R.Campbell	Machias, Me.
338		Margaret	WBarge	British	129	103·4	23·0	7·0	Sorel, Canada	1854			T.Hart	Quebec
339	R	Margaret A. May *Freer*	W3mSr	American	537	134·0	34·0	15·5	Wilmngtn, Del	1884	Jackson& Sharp [Co.		W.N.Jarvis	Philadelphia
340		Margaret Ann *Cubitt 81-81* (BROKEN UP)	WBk	British	718	138·7	28·5	21·2	St.John	1851			W.Cay & Partners, Ld.	Plymouth
341	※	Margaret Galbraith *Forbes 73-83*	IS	British	841	198·5	32·2	19·9	PortGlasgow	1868	R.Duncan&Co.		Shaw, Savill & Albion Co., Ld.	Glasgow
342	※	Margaret Garton *Elliott 80-82*	WSr	British	61	75·6	19·7	8·3	Isle of Man	1876	Qualtrough		J.Qualtrough	Peel, I.M.
343	※	Margaret Gore *Gore*	W3mSr	British	98	94·0	22·3	9·7	Tarlton	1886	Latham & Lund		J.Taylor	Preston
344	※	Margaret Hain *Lowrie 75-86*	WBn	British	170	100·6	23·9	12·9	Rye	1866	Hoad	3.86	E.Hain & Son	Plymouth
[illegible]	※	Margaret Heald *Williams*	IBk	British	605	173·0	29·2	18·1	Liverpool	1877	R.&J.Evans&Co		R.& J.Evans & Co.	Liverpool
[illegible]		Margaret Hobley *Hughes 86-86*	WSr	British	100	86·6	22·2	10·8	PembrokeDck	1868	Warlow		W.Postlethwaite	London
347		Margaret Jane *Owen*	WSr	British	117	82·0	21·3	11·0	Nevin	1857	J.Owen		T.Jones	Carnarvon
[illegible]		Margaret Jane Leeson	W	British	148	113·8	24·9	7·1	Merritton, Ont.	1879	Fred.Sharp		Robert F.Lattimore	St.Catharine, Ont.
[illegible]		Margaret John *Robinson [-85]*	WSr	British	154	90·5	21·5	12·5	Cardiff	1836	T.Jenkins		J.& E.Robinson	Littlehampton
350	R	Margaret L. *Ryan*	Sr	British	169	93·1	25·0	11·8	Big Hrbr, C.B.	1885	A.Fraser		Murdoch & McLeod	Sydney, C.B.

※ = Lloyd's Register; D = Nederlandsche Vereeniging; G = Germanischer Lloyd; I = Registro Italiano; R = American Record;
A = Veritas Austro-Ungarico; H = Veritas Ellenico; N = Norske Veritas; V = Bureau Veritas.

Dans la 4me Colonne, C = Composite (membrure en fer, bordé en bois); I = Fer; Stl = Acier; W = Bois.

Les Dates du doublage en métal (**Col. 11**) ressortent des livres dans lesquels les navires sont respectivement classés ou des rapports des Inspecteurs du Lloyd's Register.

Pour plus amples renseignements sur les navires classés par le Lloyd's Register, voir le Lloyd's Register Book et le Lloyd's Yacht Register.

Official No.	No. in Book	Registers in which classed	Ship's Name. Master and Date of Appointment. Ship's Late Name.	Material and Rig	Nationality	Tonnage	Length	Breadth	Depth	Built Where	When	By whom	Metalling	Owners or Managers	Port
73096	351	A	Margaret Mitchell [illegible]	WBk	British	[illegible]	[illegible]	[illegible]	[illegible]	Maitland, N.S.	1875	McDougall	[illegible]	D.M.Campbell	
[illegible]	352	⁕	Margaret Murray [illegible]	Stl 3mSr	British	176	107·2	23·7	11·7	Grangemouth	1885	Grangemouth [DockyardCo.		A.Murray	London
20113	353		Margaret Nixon [illegible] 75-76	WSw	British	188	[illegible]	25·0	14·2	Newcastle	1861	Fulton		J.Wallis	Whitby
91730	354		Margaret S. Smith [illegible]	WSr	American	139	[illegible]	24·0	9·4	Kennebunk, Me	1885	[illegible]		A.M.Smith	[illegible]
52558	355		MargaretSutton Reynolds	WBn	British	197	104·5	23·7	12·8	Cork	1866	G.Robinson&Co		N.Reynolds	Cork
[illegible]	356	⁕	Margaret & Ann Parry	WSk	British	29	51·6	16·7	7·0	Cardigan	1877	J.Williams		E.Parry	Cardigan
12251	357		Margaret & Elizabeth [illegible]	WSr	British	101	82·0	20·1	10·2	Nairn	1856	Mann		W.Nicolson	Inverness
27434	358		Margaret & Mary [illegible]	WSr	British	100	78·8	20·2	10·7	Pt Dinorwic	1859	R.Jones		W.E.Jones	Carnarvon
55128	359	⁕	Margaret&Mary Smith	WSr	British	83	79·6	21·6	9·7	Perth	1867	[illegible]		J.Smith	Wexford
	360	N	Margareta Dale	WSr	Norweg'n	125	94·0	22·9	10·5	Stavanger	1878			M.Dale	Stavanger
	361		Margareta Cullberg -86 (ex [illegible])	WBk	Swedish	580	146·6	30·4	18·5	Björneborg	1865			A.Cullberg	[illegible]
	362	V	Margareta Carlsson	WBk	Swedish	312	131·2	25·2	12·8	Oscarshamn	1879	C.Hasselbom	[illegible]	F.Johansson	
	363		Margareta Andersson	WBg	Swedish	276	111·5	26·6	13·2	Marum	1874			E.Andersson	
	364	V	Margareta Magnusson	WBk	Swedish	275	119·7	25·1	12·7	Yarmouth NS	1864		[illegible]	A.G.Johansen	
	365	V	Margaretha Potkuis	WBn	Dutch	171	96·5	24·8	12·2	Pekela	1874	F.L.Drenth	[illegible]	F.L.Drenth	[illegible]
	366	G	Margaretha Visser -87 (ex Margaretha) (TO BE BROKEN UP)	W3mSr	Dutch	120	86·6	20·5	10·8	Dundee	1857	Tbay&Co.			
	367	GV	Margaretha Köster 82-87 (ex J. A. Thompson)	WS	German	1287	198·1	38·4	23·7	Bath, Maine	1869	A.Hathorn	[illegible]	Fr.Roters	[illegible]
	368	V	Margaretha DeBoer 78-83 (ex [illegible])	I Bk	German	747	160·8	32·7	16·7	Middlesbro'	1879	R.Craggs&Sons		C.H.H.Winters	[illegible]
	369	GV	Margaretha [illegible]	W3mSr	German	241	108·8	24·4	12·8	Papenburg	1869	H.W.Meyer	[illegible]	H.W.Meyer	[illegible]
	370	V	Margaretha Mahler [illegible]	WSr	German	151	91·9	22·7	11·4	Bardenfleth	1853			Renck & Hemsemüller	[illegible]
	371	G	Margaretha Dethloff Stratten [illegible] (ABANDONED)	WSr	German	129	85·7	21·1	10·4	Rostock	1874	H.Kirkmann		A.Burchard	Rostock
	372	V	Margaretha Gaiser Mathiesen [illegible]	I Bkn	German	387	131·8	28·5	[illegible]	Flensburg	1875	Flensburg [Schiffsb.Ges.		G.L.Gaiser	Hamburg
	373	G	Margaretha Johanna Meiners [illegible] (ex [illegible])	WSr	German	132	85·2	21·1	10·8	Muntendam	1863	R.J.Drenth		J.B.Meiners	[illegible]
	374	GV	Margarethe Pöppe 82-85 (ex Hamilton)	WS	German	1228	186·6	37·1	24·1	Kennebunk US	1871	J.A.Harmsen	[illegible]	Joh.Fiedr.Wessels	Bremen
	375	V	Margarethe [illegible]	WBk	German	475	128·2	30·9	18·2	Memel	1862	B.Pieper		J.E.G.Sternberg	[illegible]
[illegible]	376	G	Margarethe [illegible] (ex Alexandra, ex Elis Libiers)	WSr	Russian	183	96·8	23·4	13·6	Faaborg	18[illegible]	R.Dyreborg		H.D.Schmidt	[illegible]
	377	V	Margarethe Blanca [illegible]	WBk	German	527	131·0	30·8	19·2	Danzig	1865	J.W.Klawitter		A.Gibsone	Danzig
	378	V	Margarida Silva Junior	WBk	Portug'se	351	125·9	25·9	13·8	Villa Nova	1873	J.F.da Lapa	[illegible]	M.Guimarães&Monteiro	Oporto
283	379	A	Margarita Borchi -88	WBk	Austrian	667	154·9	30·3	19·9	Lussinpiccolo	1867	N.Martinolich	[illegible]	M.Terni & Co.	Trieste
	380		Margarita Molina (ex [illegible])	WBk	Chilian	791	176·0	35·3	21·9	Sunderland	1863	Smurthwaite	[illegible]	W.Rothe	[illegible]
174	381		Margarita	WBg	Greek	240	95·8	25·9	..	Galaxidi	1868			T.Nipas	Syra
	382		Margarita (ex Tartar, ex [illegible] Rosenbecker)	WBg	Spanish	265	112·5	26·1	11·9	Geestemunde	1886	Schau&Oltmans		M.Bertolucci	[illegible]
	383	⁕	Margarita	I Sr	Urugu'y'n	83	79·0	25·4	5·6	Sunderland	1884	J.Knox&Co.		J.A.Iliffe	[illegible]
	384		Margaro	WBn	Greek	185	..	..	..	Syra	1865			A.Patrogiannis	Syra
	385		Margem Coneca (ex [illegible] Maria [illegible])	WBg	Brazilian	519	..	..	..	Varazze	1869		[illegible]	F.P.Mayrink	Rio de [illegible]
	386	N	Marget [illegible] -88 (ex Ocean [illegible])	WBk	Norweg'n	345	116·7	27·3	16·4	Sunderland	1862	J.Robinson	[illegible]	G.Uhl	[illegible]
126	387	A	Margherita Bonacich	WSr	Austrian	100	80·7	20·9	10·1	Lussinpiccolo	1886		[illegible]	A.Bonacich Marcheich	[illegible]
1033	388	IV	Margherita Carbone -88 (ex [illegible])	WBk	Italian	498	132·8	29·4	18·7	Alinotri	1874	F.Mauro	[illegible]	Salv.Trapani	[illegible]
305	389	A	Margherita Mondaini -87 (ex [illegible])	WBk	Italian	444	139·7	27·8	17·7	Trieste	1867	G.Spadon	[illegible]	E.Arbib	Venice
2642	390		Margherita Gaglione (ex Maria [illegible])	WSr	Italian	102	78·7	16·1	10·2	Castellamare	1883	G.Bonifacio		Ottavio Gaglione	Naples
3136	391	V	Margherita F. Ferro (ex Elsie)	WBk	Italian	438	138·1	30·8	19·5	Spezia	1869	Faggioni	[illegible]	G.B.Ferro	[illegible]
786	392	I	MargheritaM. [illegible] (ex Regina)	WBn	Italian	144	84·3	23·1	11·9	Alimuri	1877	F.S.Mauro	[illegible]	N.Artese	[illegible]
17589	393		Margie Haskell	WSr	American	131	105·7	27·5	9·3	Eastport, Me.	1867			E.J.Peters	New York
	394	R	Margit Aas -87 (ex Angelina) (STRANDED)	W Bk	Norweg'n	451	129·8	27·8	17·8	Bilbao	1864		[illegible]	H.E.Aas	[illegible]
	395	N	Margith Olsen [illegible] (ex Elvira)	WBg	Norweg'n	178	94·0	23·2	11·8	P.E.Island	1852			A.Olsen	[illegible]
	396		Margot Bernard 86-88	WSr	French	111	88·1	20·7	10·8	Dunkirk	1872	G.Beck		G.Beck	Dunkirk
72468	397	V	Margrethe Riley 77-82	WSr	British	145	104·7	24·2	10·9	Bergen	1874	J.Hauge	[illegible]	G.Robertson	Kirkwall
	398	V	Margrethe [illegible] 83	W3mSr	Danish	186	105·6	22·5	11·3	Nexø	1871	J.H.Benzin	[illegible]	P.[illegible]	[illegible]
	399	V	Margrethe [illegible] (LOST)	WBn	Danish	178	101·1	23·5	11·8	Randers	1869	Th.Schouw	[illegible]	C.Hassager	[illegible]
	400														

⁕=Lloyd's Register; D=Nederlandsche Vereeniging; G=Germanischer Lloyd; I=Registro Italiano; R=American [illegible]; A=Veritas Austro-Ungarico; H=Veritas Ellenico; N=Norske Veritas; V=Bureau [illegible]

In the 4th Column, C=Composite (iron frame, wood planked); I=Iron; Stl=Steel; W=Wood.

The Dates of Metalling (Col. 11) are based on the records of the Books in which Vessels are respectively classed, or on the Reports of the Surveyors of [illegible]

Further particulars of Vessels classed by Lloyd's Register will be found in Lloyd's Register Book and Lloyd's Yacht Register.

		Nom. Capitaine et dates d'entrée au service des Armateurs actuels et à bord du navire. Ancien Nom.	Matériaux et Gréement	Pavillon	Tonnage	Dimensions: Longueur	Largeur	Creux	Construit: Où	Quand	Par qui	Date du doublage en métal	Armateurs ou Directeurs-Gérants.	Port d'Armement.
401	V	Margrethe *Farup 81-88*	W Bn	Danish	159	98·4	21·8	10·6	Värnanäs	1871	C.A.Hahn	Y 8.86	Soren J.Farup	Nordby, Fanø
402		Margrethe	W Bk	Norweg'n	223	91·5	23·3	15·9	England	1835			Oluf Pedersen	Christiania
403		Margrietha *Boerhave* (BROKEN UP)	W Sr	Dutch	144	93·7	22·3	10·3	Sappemeer	1858	A.Maathuis		H.G.Boerhave	Veendam
404		Marguerite *Lear*	W Sr Yt	British	110	88·9	18·7	10·0	Cowes	1849	J.White		Lord Charles F.B.Bruce	Cowes
405		Marguerite *Hollis*	W Sr	American	103	88·0	23·5	8·4	Essex, Mass.	1883	W.A.Burnham		J.F.Wonson & Co.	Gloucester, Mass.
406	V	Marguerite *Ginsset*	W Bk	French	453	132·2	28·5	17·6	St.Malo	1876	Gautier	Y	J.Blaize	Saint Malo
407	V	Marguerite *Laborie*	W Bk	French	409	124·6	28·6	18·8	Nantes	1885	E.Clairgeau	Y	Lassere fils	Bordeaux
408		Marguerite	W Bk		275	118·1	24·5	11·5	Bordeaux	1865	Moulinié frères			
409	V	Marguerite *Germain -81*	W Bn	French	157	85·3	22·3	11·5	St.Malo	1875	J.Parnet		Lemasson	Granville
410	V	Marguerite *Hanselwarker*	W Sr	French	154	95·8	21·3	11·9	Dunkirk	1870	D.VanCauwenberghe		J.N.VanCauwenberghe	Dunkirk
411	V	Marguerite *Regnier -83*	W Bg	French	154	104·0	21·0	11·2	Granville	1876		Y	Beust & fils	Granville
412		Marguerite *Philippe*	W Bg	French	142	78·1	22·3	12·0	Bayonne	1859			Guibert & fils	St.Servan
413	V	Marguerite *Lainé*	W Sr	French	122	91·9	21·0	10·9	St.Vaast	1879	E.Costey		A.Lemoine	St.Malo
414	V	Marguerite *Le Regent*	W Sr	French	111	85·6	19·3	10·9	Paimpol	1878	L.Laboureur		F.Savin	Paimpol
415	V	Marguerite *Doupia*	W Sr	French	110	86·9	19·3	11·0	Paimpol	1883	L.Laboureur		O.Le Goaster	Paimpol
416	R	Marguerite *Åsberg* (ex Scottish Queen)	W Bk	Swedish	1193	185·9	38·5	23·4	Windsor, N.S.	1861	B.Smith	Y 8.86	L.Brolin	Söderhamn
417		Marguerite *Petersson -82*	W Bk	Swedish	341	116·8	25·7	15·8	Bordeaux	18^{50}_{72}			L.Petersson	Gothenburg
418	V	Marguerite Alice *Lebe*	W Bg	French	255	101·0	25·9	13·9	Mesnus	1869	Olliveaud	Y 12.86	Joyaux frères	Nantes
419		Marguerite Louise Regine	W Bk	Dutch	389	134·5	30·1	12·9	Bathurst, N.B.	1863	O'Brien		Jan Foekens	Harlingen
420	*	Marguerite Marie (ex Mary Ellen) *Thébau*	W Sr	French	140	94·1	20·5	12·5	Scilly	1873	Mumford	Y	E. de Baupte	Rouen
421	A	Mari Vagliano *Mazlum*	W Bn	Turkish	260	..	..	..	Syra	1880		Y 8.86	M.Mazlum	Canea
422		Maria (BROKEN UP)	W Bk	British	789	155·0	29·0	23·0					Khoo Tiong Poh	Singapore
423		Maria	W Bk	British	351	125·0	27·0	11·0	Warren, U.S.	1859			Jas.Candler	Hong Kong
424	R	Maria (WRECKED) *Snow*	W Bk	British	228	114·0	25·1	14·5	Dartmouth	1870	Moore	Y 8.8	P. & L.Tessier	St.Johns, N.F.L.
425		Maria	W Bn	British	166	92·4	23·3	9·2	New York				G.A.Lewis	Bermuda
426		Maria *Veale 83-83*	W Bn	British	165	105·6	24·3	12·1	Hayle	1865	Harvey		J. & J.Hooper	Hayle
427		Maria *Harvey 81-86*	W Bn	British	144	90·7	22·7	12·7					R.H.Williams	Plymouth
428		Maria	W Bn	British	136	94·8	23·1	11·6	P.E.Island	1863	Morris		F.H.Watt	Coleraine
429		Maria *Goodwin*	W Sr	British	109	87·6	24·0	9·2	Shelburne, NS.	1883	A.Hood & J. McAlpine		Byron Hines	Yarmouth, N.S.
430	V	Maria *Land*	W Sr	British	109	87·6	19·5	9·2	Assens	1875	H.Christofferssn	Y 4.83	Julius Gau	Cape Town
431		Maria	W Sr	American	118	95·0	28·0	6·6	Westerly, R.I.	1865	N.Rose		N.Rose	New York
432		Maria (ex G. W. Halls)	W Bn	Argentine	457	133·6	30·6	18·0	Economy, N.S.	1870			L.Canale	Buenos Ayres
433	A	Maria (ex Proto) *Stipanovich*	W Bk	Austrian	525	129·6	28·9	18·9	Lussinpiccolo	1878	A.Cattarinich	Y 7.83	Ma Vª Baccarcich & Co.	Fiume
434		Maria (ex Tycoon) *Ricksby*	W Bk	Chilian	346	119·0	25·5	16·5	Sunderland	1863	J.Robinson	Y 7.86	W.Ruthe	Valparaiso
435		Maria	W Bk	Chilian	144	80·7	18·0	12·8	Philadelphia				G.Delano	Concepcion
436		Maria (ex Fama) *Eriksen 82-82*	W Bg	Danish	265	105·8	25·8	14·3	Hudiksvall	1851			T.Eriksen	Drager
437	GV	Maria (ex Nereide) *Larsen*	W Bn	Danish	167	102·5	21·6	11·1	Altona	1863	E.Dreyer		J.A.Larsen	Kjertemindo
438		Maria (ex Jan & Maria, ex Providentia) *Holm 82-83*	W Bn	Danish	135	80·3	19·8	11·9	Ostende	1839			L.Andersen	Odense
439	DV	Maria	W Sr	Danish	125	84·1	21·1	10·4	Veendam	1860	A.de Boer		R.Grönvolds Eftf.	Præstö
440	V	Maria *Thomsen 77-77*	W Sr	Danish	109	81·1	20·5	10·1	Nordby, Fanö	1868	Th.Schonau		Aktieselskab (N.H. Svarrer)	Nordby, Fanø
441	V	Maria *Petersen*	W Sr	Danish	107	81·6	20·4	10·1	Holbæk	1856	H.P.Mortensen		H.L.Smiths Enke	Holbæk
442		Maria (ex Union, ex Lyra) *Meier*	W Bk	Dutch	930	155·0	33·0	22·3	Newburyport	1855			N.Brantjes	Purmerend
443	DV	Maria (ex Ambassador) *Markwardt -85*	W Bk	Dutch	589	146·3	28·9	17·4	Halifax, N.S.	1873		Y 10.87	A.Köpcke	Rotterdam
444	V	Maria (ex Tros) *Schoisema -83*	W Bk	Dutch	401	121·9	27·9	17·1	Sunderland	1866	J.Robinson	Y 10.88	W.A.J.Wittkampf	Schiedam
445	D	Maria *Ketelaar*	W Bg	Dutch	265	..	..	..	Gothenburg	1856			N.Brantjes	Purmerend
446	DV	Maria (ex Trio) *Van Ossen*	W 3m Sr	Dutch	208	115·1	21·3	11·5	Hoogezand	1874	J.A.Hooites	Y 5.86	F.Smelt & Zn	Amsterdam
447		Maria (ABANDONED)	W Bn	Dutch	116	..	..	..					H.Vitalli & Zoon	St.Martin, W.I.
448	F	Maria *Guillerm*	W Bn	French	152	86·0	22·0	11·2	Hoogezand	1868	L.Wilderwank		J.Guillerm	Vannes
449	G	Maria *Zielke 86-86*	W Bk	German	482	123·1	31·7	19·0	Memel	1854	B.Pieper		Schoenberg & Domansky	Danzig
450		Maria	W Bk	German	333	109·2	27·7	16·8	Memel	1861	B.Pieper		K.M.Pitcairn	Memel

* = Lloyd's Register; D = Nederlandsche Vereeniging; G = Germanischer Lloyd; I = Registro Italiano; R = American Record; A = Veritas Austro-Ungarico; H = Veritas Ellenico; N = Norske Veritas; V = Bureau Veritas.

Dans la 4me Colonne, C = Composite (membrure en fer, bordé en bois); I = Fer; Stl = Acier; W = Bois.

Les Dates du doublage en métal (Col. 11) ressortent des livres dans lesquels les navires sont respectivement classés ou des rapports des Inspecteurs du Lloyd's Register.

Pour plus amples renseignements sur les navires classés par le Lloyd's Register, voir le Lloyd's Register Book et le Lloyd's Yacht Register.

Official No.	No. in Book	Registry in which classed	Ship's Name, Material and Rig. Master, and Dates of Appointment to present Owner's Service and to Ship. Ship's Late Name.	Flag	Net Tonnage	Length	Breadth	Depth	Built Where	When	By whom	Metalling	Owners or Managers	Port
	451	V	Maria *Meyer* -88 W3mSr	German	322	127·6	25·0	13·2	[illegible]	1875	[illegible]	[illegible]	R.Lassen	[illegible]
	452	V	Maria *Kugemann* WBk	German	319	117·8	26·3	15·1	La-Grana	1875	M.Lorenzo	[illegible]	W.Hoveler	[illegible]
	453	G	Maria *Ziesemer* -85 WBg	German	234	95·4	25·8	14·4	Ribnitz	1839	J.C.Peters		Ziesemer	[illegible]
	454	G	Maria (ex [illegible]) *Albers* 76- WBg	German	176	99·5	21·3	11·8	Papenburg	1856	H.W.Meyer		H.W.Meyer Ww.	[illegible]
	455	G	Maria *Abrahams* 82-83 WSr	German	165	94·5	21·1	11·7	Papenburg	1863	F.W.Meyer		Joh.Nadorff	[illegible]
	456	G	Maria *Schneiders* 84- WSr	German	152	91·8	21·3	11·5	Papenburg	1861	H.W.Meyer		F.W.Meyer	[illegible]
	457		Maria (WRECKED) *Truggen* 79- WSr	German	150	88·3	20·1	11·3	Papenburg	1847	F.W.Meyer		Schulte & Bruns	[illegible]
	458	V	Maria *Kellwig* WSr	German	143	92·1	21·3	10·7	Grosseiehn	1876	Schapp& Meyer		Dietrich Eggen	[illegible]
	459	V	Maria (CONDEMNED) *DeBuhr* 83-83 WBn	German	135	99·1	22·7	10·5	Veile	1876	P.H.Illum	Y 8.83	J.W.de Buhr	[illegible]
	460	G	Maria (WRECKED) *Ulrichs* WSr	German	134	89·1	21·3	10·5	Papenburg	1856	B.Tholen		T.J.Cassens & Sohn	[illegible]
22	461		Maria (ex Maria) *Vaglianos* WBk	Greek	416	137·8	26·2	17·5	Apenrade	1873	H.Jürgensen	Y [illegible]	G.Vaglianos	Corfu
32	462		Maria WSr	Greek	251	110·5	25·6	..	Naples	1861			H.Limberatos	Zante
404	463	V	Maria *Zouis* WBk	Greek	182	89·2	24·9	..	Syra	1878		Y 8.85	J.E.Georgiadis	Syra
66	464		Maria WSr	Greek	153	82·3	24·3	15·7	Sorrento	1859	A.Castellano		S.Stathopoulos	Argostoli
485	465	A	Maria *Lukas* WBn	Greek	128	82·7	22·0	..	Syra	1885			G.A.Lukas	Syra
51	466		Maria WBg	Greek	106	72·8	21·0	..					S.P.Manzarakis	Volo
1684	467	I	Maria *Ambrosano* WBk	Italian	381	138·5	28·6	17·8	Sunderland	1870	J.Crown	Y 12.85	Ant.Galatola	Naples
308	468	A	Maria *Olivari* -88 WBk	Italian	318	110·8	28·7	18·0	[illegible]	1867	G.Mauro	Y [illegible]	A.Vianello Moro	Venice
110	469		Maria (ex Autonomia) *Bianchi* WBg	Italian	213	111·5	23·0	15·2	Procida	1859	A.Mazzella	Z 8.-6	Francesco Scotto	Naples
3175	470		Maria *Marini* WBg	Italian	168	91·2	25·7	14·8	Venice	1855			G.Marini	Genoa
8	471		Maria *Pezzella* WBg	Italian	152	79·4	24·0	14·8	Varazze	1856	P.Vattino	Y [illegible]	G.Rabagliati	Savona
2029	472	A	Maria *Del Gatto* WSr	Italian	119	..	..	..	TorredelGreco	1885	Domarumma		G.Del Gatto	Naples
1184	473		Maria (WRECKED) *Martinelli* WBg	Italian	105	86·6	21·7	10·9	Trieste	1866			B.Martinelli	[illegible]
	474	N	Maria WBg	Norweg'n	232	108·0	26·5	12·8	Porsgrund	1873		Y 12-87	G.B.Henriksen	[illegible]
	475	N	Maria *Mathisen* WBn	Norweg'n	208	109·3	26·2	11·5	Aalesund	1876		Y 10.86	H.Mathiesen	[illegible]
	476	V	Maria (ex Leinster, ex Lisbon) WBk	Portuguese	500	139·7	29·8	15·4	Villado Conde	1868	M.G.Rodrigues	Y 2.87	Cia. Alliança, Maritimo Portuense	Oporto
	477		Maria *Kiiski* WBk	Russian	439	..	..	..	Wederlaks	1881			A.W.Fronden	[illegible]
	478		Maria (MISSING) *Backman* WBk	Russian	417	136·8	27·9	16·0	Fredrikshamn	1874	J.Mokala		H.L.Bäckmann	Fredrikshamn
	479		Maria (ex Charlotta) *Brown* 73- WBk	Russian	281	111·0	25·0	14·0	Flensburg	1862	VonStemann		F.Savin	[illegible]
	480		Maria *Mattson* WBk	Russian	276	111·5	27·3	15·1	Sage	1859			A.Michelsson	[illegible]
	481		Maria *Sandström* WSr	Russian	210	..	..	..	Tofsala	1866			J.E.Jansson	[illegible]
	482		Maria *Lundstrom* WSr	Russian	180	..	..	..	Joensuu	1874			H.J.Ridderstad	[illegible]
	483		Maria WSr	Russian	187	..	..	..		1873			A.Ostlin	[illegible]
442	484	G	Maria *Jorgenberg* -86 WSr	Russian	162	85·7	26·7	12·0	Kaleten	1884	J.Stahl		J.& M.Stahl	Riga
	485		Maria *Surring* WSr	Russian	141	85·6	26·9	10·9	Domlangen	1879	P.Ranne		A.Geige	[illegible]
	486		Maria *J.Petari* WSr	Russian	140	..	..	..		1876			E.Petari	[illegible]
	487		Maria	Russian	128	..	..	..		1873			G.Andrejeff	[illegible]
	488		Maria *Niemi* WSr	Russian	116	80·0	23·4	10·6	Björkö	1873	J.Koli		M.Hämäläinen & Co	[illegible]
	489	V	Maria (ex [illegible]) *Carran* WBk	Spanish	577	147·6	30·5	20·0	Cagliari	1877	A.Brasco	Y 11.83	Balsells & Sabitac	[illegible]
	490	V	Maria *Careago* -88 WBk	Spanish	493	144·3	30·4	16·9	Bremen	1865	H.F.Ulrichs	Y [illegible]	J.A.Gorocica	[illegible]
	491		Maria (ex Conseller) WBg	Spanish	350	110·8	30·8	16·2	Blanes	1857	Ramon Vieta		Solo, Villarrasa & Co.	[illegible]
	492		Maria (ex Fermina de Vila) *Sust* WBk	Spanish	273	79·4	26·6	15·3	Grana	1868			A.S.Casasnovas	[illegible]
	493	V	Maria (ex Volador) *Sust* WBk	Spanish	273	116·1	27·4	14·0	Sestri	1862		[illegible]	J.Gras y Pérez	[illegible]
	494		María WBk	Spanish	233	109·9	28·9	13·4	Blanes	1859			M.Cutillas	[illegible]
	495		Maria (ex Ramona Suarez, ex Susan) WBn	Spanish	198	97·4	25·6	11·6	Maitland,N.S.	1865			M.G.Carvajal y Pi[illegible]	[illegible]
	496		María (ex Fidels) WSr	Spanish	126	82·7	25·2	10·7	Arens	1857			F.Sacrista y Vidal	[illegible]
	497		Maria WBn	Spanish	106	71·5	21·9	8·6	Villajoyosa	1851			Dona Maria Furio	[illegible]
	498		María (ex Cometa) WSr	Spanish	104	80·7	20·7	10·8	Palma	1847			A.Alberti y Planes	Majorca
	499		Maria WSr	Spanish	103	88·6	24·5	8·0	Palamos				Juan Mir	[illegible]
	500	V	Maria *Carlsen* WS	Swedish	338	135·8	26·3	12·2	Söderhamn	1875	A.O.Vännberg	Y [illegible]	J.F.Schöning	[illegible]

✻ = Lloyd's Register; D = Nederlandsche Vereeniging; G = Germanischer Lloyd; I = Registro Italiano; R = Ameri[illegible]
A = Veritas Austro-Ungarico; H = Veritas Ellenico; N = Norske Veritas; V = Bureau [illegible]

In the 4th Column, C = Composite (iron frame, wood planked); I = Iron; Stl = Steel; W = Wood.

The Dates of Metalling (Col. 11) are based on the records of the Books in which Vessels are respectively classed, or on the Reports of the Surveyors of [illegible]

Further particulars of Vessels classed by Lloyd's Register will be found in Lloyd's Register Book and Lloyd's Yacht Register.

	Nom. Matériaux et Gréement. Capitaine et dates d'entrée en service des Armateurs actuels et à bord du navire. Ancien Nom.	Pavillon	Tonnage net.	Dimensions. Longueur	Largeur	Creux	Construit. Où.	Quand.	Par qui.	Date du doublage en métal	Armateurs ou Directeurs-Gérants.	Port d'Armement.
	Maria *Påhlsson -85* W Bk	Swedish	310	126·9	25·7	12·3	Sundsvall	1871	E.Ostmann	Y 6.83	H.Nordlander	Hernösand
G	Maria (ex Marie) *Westman [76-82]* W Bk	Swedish	289	110·5	25·0	11·8	Rostock	1853	W.Zeltz		R.Snöbohm	Klintehamn
N	Maria (ex Margaret Bleeth) *Petersen -87* W Bg	Swedish	285	122·2	25·6	14·2	Oscarshamn	1860	N.P.Carlsten		G.C.Holst	Visby
	Maria (ex Southern Belle) *Nordgren -88* W Sr	Swedish	274	119·7	27·5	11·9	Islip	1855		Y 11.88	P.Olsson	Sölvesborg
	Maria (ex [illegible]) *Stenström -83* W Bk	Swedish	245	117·4	23·2	12·3	Kiel	1865	Conradi&Sohn	Y 5.88	G.J.Sundberg	Lysekil
	Maria *Olsson -84* W3mSr	Swedish	225	105·3	25·8	12·1	Roslagen	1871	A.Björk		P.Andersson	Björhöfda, Bslgn
	Maria (ex Helene) *Jönsson* W Bg	Swedish	203	90·5	23·4	13·5					J.F.Hallberg	Uddevalla
	Maria *Olsson* W Sr	Swedish	158	86·9	21·8	11·8	Bohuslän	1849			C.A.Svensson	Bergqvara
	Maria *Berntsson -88* W Sr	Swedish	141	92·2	20·1	11·5	Apenrade	1857	Andersson		A.Åkesson	Kivik
	Maria *Pettersson* W Sr	Swedish	140	82·7	21·3	11·9	Orust	1867			O.Olsson	Käringön
	Maria *Pettersson* W Sr	Swedish	117	74·2	19·8	10·3	Wiken	1855	O.Nilsson		C.P.Pettersson	Cimbrishamn
V	Maria (ex Marie) *Bergh -87* W Bn	Swedish	114	85·5	20·5	9·2	Randers	1856	C.Mattat & P. Jörgensen		T.A.G.Bergh	Calmar
	Maria W Sr	Swedish	110	75·6	20·0	10·5	Hamburg	18[illegible]			O.Johansson	Qville, Bohuslän
G	Maria *Götherström* W Sr	Swedish	109	77.4	19.9	10.2	Ladeå	1840		Y 11.81	C.J.Götherström	Timmernabben
A	Maria *Joannis -88* W Bg	Turkish	198	..	..	..	Vasiliko	1867			Nicolaidis frères	Vasiliko
	Maria W Sr	Uruguayn	118	78·1	20·0	11·5	Elmshorn	1863	D.W.Kramer			Monte Video
	Maria *Sanchez* W Sr	Ven'zu'l'n	170	86·6	20·7	12·5	Swinemünde	1873	L.Freundt	Y 6.83	J.Capriles	Maracaibo
	Maria A. *Puisateri* W Bn	Italian	176	102·3	24·7	11·6	Varazze	1883	A.Cerruti		F.Palmisano & Co.	Palermo
I	MariaAccame *Luigi -88* W Bk	Italian	1149	193·5	32·8	24·4	SestriPonente	1883	A.Briasco fu L.	Y 6.87	E.Accame & figli	Genoa
	Maria Adelaide *Jones* W Sr	American	106	96·0	24·9	7·3	Bangor,Me.	1872	J.Oaks		G.B.Hook	Bangor.Me.
B	Maria Adelaide *Cacace* W Bk	Italian	460	138·1	31·5	16·7	Palermo	1872	A.LaRocca	Y 6.83	G.& G. Flli Cacace fu D.	Castell.di Stab.
R	Maria Adelaide *[illegible]* W Bk	Swedish	397	124·0	26·6	16·5	Danzig	1856	A.Gibson		L.H.Sjöström	Cimbrishamn
A	Maria Andrina (ex [illegible]) *Marassich -86* W Bk	Austrian	348	112·8	27·8	16·7	Sunderland	1868		Y 5.88	Maria Vlassich & Co.	Buccari
*	Maria Angelina (ex [illegible]) *Mendes* W Bk	Brazilian	216	117·6	25·2	11·2	Oporto	1873	C.J.A.Vasetalmo	Y 1.88	J.daS.Loyo & Filho	Pernambuco
	Maria Anna *Melkert* W Sr	Russian	192	98·8	26·8	11·2	Raumharja	1875	A.Ferle		A.Melkert	Riga
	Maria Annett W Sr	British	228	125·0	26·2	10·1	Quebec	1867			Robt.Henning	Quebec
R	Maria Antonia *Riera [-88* W Bk	Spanish	423	133·5	28·5	17·1	Cadiz	1864	A.Cabezade Vaca	Y 7.87	S.S.Vergara	Seville
A	Maria Antonietta *[illegible]* W Bn	Italian	178	92·8	24·7	13·9	Gaeta	1862	P.Gallinaro	Y 10.86	S.Munzone fu V.	Catania
	Maria Asumpta W Sr	Spanish	122	86·9	24·0	10·2	Barcelona	1858			A.H.del Rio	Villagarcia
	Maria Augusta (ex [illegible]) W Bk	Dutch	652	136·8	33·8	21·0	Papendrecht	1856	J.Pyle		Sech Abdoela bin Achmat bin Secubia	Tagal
I	Maria B. *Bertolotto -88* (ex Enrichetta) W Bk	Italian	419	122·0	29·3	17·8	Savona	1867	A.Calcagno	Y 8.88	F.Bertolotto	Genoa
	Maria B. (ex [illegible]) W3mSr	Italian	306	126·3	26·5	15·0	Savona	1876	G.Calamaro	Y 3.83	G.Bardi	Genoa
I	Maria Benvenuto W Bn		117	87·9	25·9	9·1	Castellamare	1887	V.Bonifacio	Y 11.87		
V	Maria C. *Ramana* W Bg	Italian	342	122·3	27·9	18·5	Cassano	1871	A.Bonifacio	Y 4.83	P.Cafiero	Castell.di Stab.
	Maria Carolina (ex [illegible]) *[illegible]* W Bk	Portug'se	302	107·2	25·6	14·1	Villa do Conde	18 65/77	J.M.daSilva	Y 8.83	Cia Alliança Maritima Portuense	Oporto
I	MariaCasabona *Olivari [-88* W Bk	Italian	624	149·2	32·6	20·7	Finale	1872	F.Ponziolo	Y 1.88	G.Casabona	Genoa
	Maria Charlotta (ex [illegible]) W3mSr	Swedish	319	124·0	25·0	13·0	Gefle	1863			J.Berglund	Skellefteå
*	MariaCheesman *Withers* W Sk	British	87	78·5	21·0	10·4	Grimsby	1886	Smith Brothers		W.Guy	Grimsby
	Maria Concepción (ex [illegible]) W Bk	Spanish	300	108·2	28·6	15·3	Seville	1854			M.José de la Portilla	Bilbao
I	Maria Copello *Copello* W Bn	Italian	319	119·4	27·9	16·1	Chiavari	1885	L.Gotuzzo	Y 2.85	Flli Copello	Spezia
I	Maria Cristina *Coppola -88* W Bk	Italian	414	126·7	28·9	17·9	Cassano	1870	F.DeBosa	Y 4.83	A.G.Jaccarino	Castellamare
A	Maria D. *Descovich -88* W Bg	Austrian	372	125·0	26·2	15·3	Fiume	1871	Flli Schiavoni	Y 8.83	A.& E.Descovich	Fiume
	Maria D. (ex [illegible]) W Bg	Italian	315	108·2	23·3	16·7	SestriPonente	1859	D.Tixi	Y [illegible]		Genoa
	Maria de las Nieves W Bn	Spanish	252	109·9	28·9	13·4	Blanes				Benito Cutillas Alacid	Malaga
I	Maria Della Catena W Bn	Italian	140	87·6	23·3	13·8	Syra	1880		Y 6.84	N.Artese	Palermo
V	Maria Di Monte Allegro W Bk	Italian	267	108·2	25·6	14·8	Nantes	1869	Jollet & Babin	Y 9.85	Cat.Castagneto Ved Campodonico	Genoa
	Maria Dolores W Bn	Spanish	118	84·3	22·6	8·9	Albay	1860			J.José Reyes	Manila
	Maria E. Smith W3mSr	American	347	141·5	34·6	10·6	Port Blakely, Wash.	1881	HallBros.		J.Tuft	San Francisco
V	Maria Elisa W Bn	Italian	105	83·3	23·0	10·3	Viareggio	1867	A.Raffaelli	Y 12.83	A.Raffaelli fu F.	Leghorn
	Maria Elisabet W Sr	Swedish	107	72·2	24·8	9·0	Finland	1862			P.M.Hansson	Bergqvara

*=Lloyd's Register; D=Nederlandsche Vereeniging; G=Germanischer Lloyd; I=Registro Italiano; R=American Record; A=Veritas Austro-Ungarico; H=Veritas Ellenico; N=Norske Veritas; V=Bureau Veritas.

Dans la 4me Colonne, C=Composite (membrure en fer, bordé en bois); I=Fer; Stl=Acier; W=Bois.

Les Dates du doublage en métal (Col. 11) ressortent des livres dans lesquels les navires sont respectivement classés ou des rapports des Inspecteurs du Lloyd's Register

Pour plus amples renseignements sur les navires classés par le Lloyd's Register, voir le Lloyd's Register Book et le Lloyd's Yacht Register.

Official No.	No. in Book	Character for Hull and Stores	Ship's Name. Material and Rig. Master, and Dates of Appointment to present Owner's Service and to Ship. Ship's Late Name.	Flag	Net Tonnage	Length	Breadth	Depth	Where	When	By whom	Date of Metalling	Owners or Managers
	551		Maria Elisabeth Pommer W Sr (ex Hermann …)	German	112	80·0	18·8	12·3	Middlesbro	1864	R.Craggs&Sons	[illegible]	W.Stindt
50448	552		Maria Elizabeth W Sr (ex Marie Marguerite)	British	123	71·6	21·4	10·4					P.J.Brown
8517	553	I	Maria F. Villa ·87 W Bn	Italian	147	87·9	21·6	11·1	Voltri	1886	F^lli Fava	[illegible]	F^lli Fava
	554		Maria Heydorn W Sr (ex Johanna) Nystrom	Swedish	165	98·8	20·4	10·6	Hanover	1856			A.Persson
	555		Maria Isabel W Bn	Brazilian	236	..	..	..					J.J.S.Breves
	556		Maria Isabel W Sr	Brazilian	123	..	..	..					A.P.C.Carneiro
	557		Maria Isabel W Bk	Spanish	298	106·8	26·0	12·6	[illegible]	1870			Doña M.B.Forteza
18883	558	❋	Maria Jane Jones ·87 W Sr	British	90	75·2	21·2	10·7	[illegible]	1857			J.Ellis & Co.
	559		Maria Josefa (…) W Bn	Spanish	115	84·0	24·3	10·2	[illegible]	1858			J.Bosch & Estape
	560		Maria Katarina W Cr	Russian	101	..	..	..		1874			A.Junnola
	561		Maria Kristina Nilsson W Sr	Swedish	109	73·5	20·8	10·3	Orœust	1864			S.Christiansson
64620	562		Maria Lamb Road 64-83 C 3m Sr	British	189	110·0	23·0	11·8	Milford	1868	Watson		J.Fisher & Sons
1085	563	DI	Maria Laura Gardella W Bk	Italian	440	130·9	30·5	19·2	Savona	1867	F.Sirello	[illegible]	F.Schiaffino
1087	564	IR	Maria Lauretta Olivari ·88 W Bk	Italian	449	133·8	29·8	19·2	Savona	1867	F.Sirello	[illegible]	F.Schiaffino
	565		Maria Libera W Bg (ex Evangelistria) Paramadda	Italian	212	..	..	..	Schiatto	1862			V.Mazzella
435	566	I	Maria Libera Amoretti W Bn	Italian	105	83·0	19·7	10·3	Loano	1886	A.Cerruti	[illegible]	G.Amoretti
	567	G	Maria Louise Nilsson W Bg (WRECKED) [·81	Swedish	238	111·5	24·5	11·9	Sundsvall	1868	C.A.Nordström	[illegible]	S.N.Horndahl
	568		Maria Louise Gustafsson ·86 W Sr	Swedish	116	77·1	21·6	11·3	Edelöse	1860	Lundgren		E.Olsson
63975	569	G	Maria Luigia Hill ·88 W Sr	British	129	89·0	23·0	..	Austria	1867			J.H.Hecken
2285	570	IV	Maria Luigia Massone W Bk	Italian	571	150·2	31·2	20·2	SestriPonente	1872	E.Celesia	[illegible]	Ant.Carbone
1141	571	I	Maria Luigia Savarese W Bk	Italian	458	103·3	29·2	18·8	Equa	1876	A.Gargiulo	[illegible]	F^lli Savarese
	572	V	Maria Luisa Mora ·81 W Bk (ex Maria Macabeo Ramona)	Spanish	641	158·7	32·4	20·5	Varazze	1876	D.Craviotto	[illegible]	R.Taltabull
	573		Maria Luisa Ortega W Bk	Spanish	273	107·2	26·6	13·7	Palma,Santa Cruz	1876	S.Arozena		Viuda & hijos de J.Yanes
	574	V	Maria Luisa Gras ·81 W Bn	Spanish	178	90·2	26·0	13·0	Lloret	1864	Ribas	[illegible]	J.Gran Marti
	575		Maria Luiza de Oliveira W 3m Sr	Brazilian	294	..	..	..	Oporto	1862			Costa & Gallo
	576		Maria Luiza Bolo W Bk	Portug'se	272	105·3	23·7	14·0	Lisbon	1860	C.J.A.Varela		J.J.F.R. & A.C.Lima
	577	V	Maria Lukasopoulos W Bk Lukasopoulos	Greek	297	108·6	27·5	9·7		1877		[illegible]	Lukasopoulos
3229	578	IV	Maria M. Olivari 82-82 W Bk (ex Excelsior, ex Maria)	Italian	393	129·2	27·0	17·9	Sunderland	1868	Davison	[illegible]	G.Olivari
	579	V	Maria M. Di Filippo W Bn (STRANDED)	Italian	154	87·0	20·4	11·5	Castellamare	1885	G.Bonifacio	[illegible]	L.Musso
1366	580	I	Maria Madre Morando ·87 W 3m Sr	Italian	485	137·7	31·3	19·1	Voltri	1873	Fava	[illegible]	F.Degregori
1103	581	I	Maria Madre Mezzanello W Bn (ex S. Benigno, Benvenuto)	Italian	150	82·0	24·4	13·8	Varazze	1869			B.Martinelli fu G.
2507	582		Maria Madre Baglietto W Bn (ex Maria Vittoria)	Italian	140	89·2	23·1	11·8	Arenzano	1875	Salomone		L.Arpe
	583	V	Maria Margretha Jonsson ·88 W Bk	Swedish	740	166·3	31·0	20·0	Framnäs	1877	E.Ostman	[illegible]	Fearnaes Warfs (Aktiebolaget W.Reuterskiöld)
	584		Maria Mathilda Andersson W Sr	Russian	219	101·2	27·3	12·6	Dünamünde	1875	J.Andersson		J.Andersson
	585	❋GV	Maria Mercedes … Bk (ex Marianne Francisca) ·87	German	1090	209·5	35·0	22·1	PortGlasgow	1863	J.Reid&Co.		Bahr Gebr.
	586		Maria Nathalia Schulz W Sr	Russian	218	89·8	20·3	11·1	PetersKapelle	1872	L.Kurtze		J.Schultz
434	587		Maria O. (ex Antonio Olivari) W Bn	Italian	195	90·5	25·3	15·0	SestriPonente	1856			Antonio Olivari
2395	588	IV	Maria Parodi Garibaldi W Bk (ex Maria Gazzino) [·84	Italian	761	168·9	33·5	21·5	Varazze	1873	N.Cerruti	[illegible]	G.Pagliano fu G.S.
17012	589		Maria Pierson Russell W 3m Sr	American	272	126·0	31·0	11·2	Forked River, N.J.	1864			J.Grant
	590		Maria Preciosa W Bn	Brazilian	194	..	..	..					M.S.Pereira
	591	A	Maria Preve W Sr		110	..	..	..	Lussin	1884	G.Picchiotti	[illegible]	
3478	592	A	Maria R. (ex Vera) Randich W Bk	Austrian	498	..	27·0	17·8	Lussinpiccolo	1868	N.Martinolich	[illegible]	N.Randich fu L.
907	593		Maria R. (ex …) (BROKEN UP) Rosa W Bk	Italian	360	117·7	29·8	18·4	SestriPonente	1883	A.Briasco fu G.P.		A.Rebora
946	594	DIV	Maria Raffo Questa 83-86 W S	Italian	1309	213·2	37·7	25·1	Chiavari	1883	M.Tappani	[illegible]	E.Raffo fu A.
3345	595	I	Maria Repetto Arecco ·88 W Bk	Italian	597	..	..	..	Varazze	1872	M.Baglietto	[illegible]	S.Repetto
	596	V	Maria Rickmers Thode W Bk [81-86	German	549	145·0	30·2	18·9	Geestemünde	1867	R.C.Rickmers	[illegible]	R.C.Rickmers
90519	597		Maria S. Knowlton W Sr	American	100	81·0	26·0	7·6	[illegible]	1873			N.S.Allen
27	598	A	Maria S. Scholetich ·87 W Bn	Austrian	102	76·4	19·2	11·4	Lussinpiccolo	1877	A.Cattarinich	[illegible]	G.Racich
[illegible]	599		Maria Stella W Sr	British	138	98·1	23·8	9·6	Grondines	1873			Jos.G.Côté
808	600	IV	Maria Stella [illegible] 83-86 W Bk (ex …)	Italian	437	115·3	28·9	17·1	Sorrento	1865	A.Castellano	[illegible]	D.Aula & Co.

❋ = Lloyd's Register; D = Nederlandsche Vereeniging; G = Germanischer Lloyd; I = Registro Italiano; R = [illegible]
A = Veritas Austro-Ungarico; H = Veritas Ellenico; N = Norske Veritas; V = Bureau [illegible]

In the 4th Column, C = Composite (iron frame, wood planked); I = Iron; Stl = Steel; W = Wood.

The Dates of Metalling (Col. 11) are based on the records of the Books in which Vessels are respectively classed, or on the Reports of the Surveyors of [illegible]

Further particulars of Vessels classed by Lloyd's Register will be found in Lloyd's Register Book and Lloyd's Yacht Register.

		Nom. Capitaine et dates d'entrée. Ancien Nom.	Matériaux et Gréement	Pavillon	Tonnage net.	Dimensions. Longueur	Largeur	Creux	Construit. Où.	Quand.	Par qui.		Armateurs ou Directeurs-Gérants.	Port d'Armement.
601		Maria Stoneman *Pierce*	W Bk	British	959	177·0	35·5	21·7	Shelburne, N.S.	1873	E.P. Young		A.F. Stoneman	Yarmouth, N.S.
602	A	Maria T. *Tomic* (ex Costella Tomic)	W Bk	Austrian	387	118·7	26·1	17·2	Buccari	1870	Mattarglian		T^s Smoquina	Buccari
603		Maria T. *Esposito*	W Bg	Italian	195	..	..	..	Torre del Greco	1875			Dom. Esposito	Naples
604	I	Maria Teresa *Arena*	W Bn	Italian	187	..	..	..	Castellamare	1886	V. Bonifacio		S. Arena	Messina
605		Maria Teresa	W Bn	Montenegro	130	68·9	17·4	13·6	Malta	1852			Dant Suleiman & Co.	Dulcigno
606		Maria Teresa *Estrada*	W Bg	Spanish	[illegible]	114·1	26·0	[illegible]	[illegible]	1868	J. de Arana		Janet & Co.	Barcelona
607		Maria Teresa	W Bn	Spanish	236	95·1	27·7	14·1	St. Felicé	1869			A. Tovia	Havana
608		Maria Thereza	W Bn	Brazilian	243	..	..	..					J.J.S. Imenes	Rio de Janeiro
609		Maria Velasquez *Ybarra* (ex Serenissima)	W S	Chilian	772	166·9	32·7	21·0	Sunderland	1861	Pevcrall		M.G. Velasquez	Valparaiso
610		Maria Virginia *Houlihan* (DISMANTLED)	W Bn	British	283	120·0	29·0	10·0					D.H. McKenzie	Auckland, N.Z.
611		Maria Virginia	W Sr	Brazilian	108	..	..	..					J.F. Stael	Bahia
612		Maria Volpina (ex Gandolfo)	W Bk	Argentine	476	138·1	31·2	19·5	Loano	1861	B. Amicé		J. Bautista Canepa	Buenos Ayres
613	R	Maria W. Norwood *Thompson*	W Bn	American	453	132·6	30·1	18·0	Rockport, Me.	1866	Carlton, Norwood & Co		J.J. Matthias	New Haven, Conn.
614		Maria & Anna *Schnobre*	W 3m Sr	Russian	255	109·0	28·0	13·2	Riga	1866	A. Ferle		P. Schnobre	Riga
615		Maria & Joaquina *Reyes*	W Bn	Spanish	222	101·0	27·0	13·8	Capomanes	1875	A. Perez		J.R. Cierto	Santander
616	G V	Maria & Kathe *Lass*	W Bk	German	419	125·5	27·0	17·0	Rostock	1875	E. Burchard & Co.		H. Jamentzky	Rostock
617	I	Maria & Teresa *Matarazzo*	W S	Italian	980	185·3	34·1	22·8	Gaeta	1879	P. Gallinaro		F^lli Matarazzo	Gaeta
618	✻	Marian *Strath*	W Sr	British	97	90·7	21·8	10·1	Barnstaple	1877	Westacott		R.J. Kellow	Padstow
619	D N	Mariane *Torgersen*	W Bk	Norweg'n	277	106·2	28·7	14·7	Uleaborg	1846			A.Sœberg & Dodsbo (M. Nielsen)	Drammen
620	N	Mariane *Christensen*	W Sr	Norweg'n	150	96·5	23·7	10·7	Aasgaardst'nd	1879	Jensen		J.A. Hansen	Tönsberg
621		Marianinha *Costa*	W Bk	Brazilian	238	114·1	23·3	10·5	Oporto	1876	C.M. da Silva (Santo)		B. Oliveira & Co	Pernambuco
622	V	Marianinha *Paula*	I Bk	Portug'se	869	183·7	28·9	18·9	Glasgow	1867	Connell & Co.		C. Abranches & Co.	Lisbon
623		Marianna	W Bn	Brazilian	176	..	..	..					J.J. da Silveira	Rio de Janeiro
624		Marianna Pepe *Mauro*	W Bg	Italian	279	99·7	28·4	17·2	Alimuri	1861	F^lli Mauro		M. Schiano di Pepe	Naples
625		Marianna II. *De Barros*	W Bn	Portug'se	224	94·5	21·3	13·4	Sunderland	1865	Reay		A.M. Bravo & Filho	Lisbon
626		Marianne V. *Santos*	W S	Portug'se	598	150·2	28·0	18·2	Lisbon	1862	J.A. da Silva		A.M. Bravo & Filho	Lisbon
627		Marianne *Rickards*	W Sr	British	100	87·9	18·8	11·0	Fowey	1857	Nickells		R.T. Michell	Falmouth
628	V	Marianne *Couëdel*	W Lr	French	107	72·5	21·6	10·8	Redon	1868	J. Coyac fils		A. Couëdel	Vannes
629	V	Marianne *Bradhering*	W Bk	German	410	122·4	28·9	17·7	Rostock	1874	E. Burchard & Co.		Beckmann & Co.	Rostock
630	V	Marianne *Zimmermann*	W Bkn	Russian	346	116·9	27·3	16·8	Sunderland	1865	Short Bros.		F. Zimmermann	Riga
631	V	Marianne *Tjernberg*	W Bk	Swedish	518	129·9	28·2	19·0	Danzig	1855			J. Westman	Örnsköldsvik
632	V	Marianne Bertha *Pietsch*	W Bk	German	363	104·3	27·9	16·2	Memel	1847	B. Pieper		Ch. A. Pietsch	Memel
633	✻	Marianne Greaves *Williams*	W Bg	British	185	105·9	25·1	13·0	Portmadoc	1876	Jones		Mrs. A. Williams	Carnarvon
634	R V	Marianne Nottebohm *Negret*	W S	American	1117	186·0	36·6	22·3	Brooklyn	1857			T. Dunham's Nephew & Co.	New York
635	I	Mariannina *Russo*	W Bk	Italian	496	134·1	28·0	19·0	Equa	1873	M. Patturzo		M. Russo	Castell. di Stab.
636	A	Mariannina Galatola	W Bk	Italian	395	129·6	28·1	18·1	Sunderland	1868	John Crown		Galatola & Mancino	Naples
637		Marianno I.	W Sr	Brazilian	101	..	..	..					M.J. do Couto	Porto Alegre
638	✻	Mariano (ex Ida) *Nappo*	W 3m Sr	Italian	322	123·6	27·0	15·0	Pisa	1863	Tixi		M. Messina	Castellamare
639		Mariano *Coll*	W Bn	Spanish	128	86·3	21·5	11·5	Palma	18 59/74			P. Coll y Mas	Majorca
640	H	Marianthi	W Bg	Turkish	138	..	..	..	Syra	1880			D. Maistrali	Cisme
641	A	Marica *Catramados*	W Bk	Greek	596	146·6	31·7	20·3	Recco	1868	E. Ziesko		G.N. Catramados	Syra
642		Marica	W Bg	Turkish	395	..	..	..	[illegible]	1873			Cashotleo Comyanos Yuani	Cashot
643		Marie	W Bk	British	466	145·5	31·1	17·0	Swinemünde	1869			Union S.S. Co. of New Zealand, Ld.	Wellington
644		Marie	W Bg	British	214	104·0	25·4	13·0	Germany	1864			A. Campbell	Sydney, N.S.W.
645		Marie	W Sr	British	163	94·0	19·6	9·0	Bombay	1855			Sultan Ebrohim Noroodin	Maldive Islds.
646	A	Marie *Cimaus*	W Bk	Austrian	430	123·0	27·4	18·3	Volosca	1871	Minach		G. Minach	Fiume
647	V	Marie	W Bn		224	105·4	23·3	12·5	Woltmershausen	1869	J.H. Knickmann			
648	V	Marie *Lausen*	W 3m Sr	Danish	199	110·8	23·1	11·4	Svendborg	1886	J.R. Andersen		J.R. Andersen	Svendborg
649	V	Marie *Nielsen*	W Bg	Danish	198	105·4	23·3	11·8	Nordby, Fanø	1878	C. Graff		Actieselskab (J. Nielsen)	Nordby, Fanø

✻=Lloyd's Register; D=Nederlandsche Vereeniging; G=Germanischer Lloyd; I=Registro Italiano; R=American Record;
A=Veritas Austro-Ungarico; H=Veritas Ellenico; N=Norske Veritas; V=Bureau Veritas.

Dans la 4me Colonne, C=Composite (membrure en fer, bordé en bois); I=Fer; Stl=Acier; W=Bois.

Les dates du doublage en métal (Col. 11) ressortent des livres dans lesquels les navires sont respectivement classés ou des rapports des Inspecteurs du Lloyd's Register.

Pour plus amples renseignements sur les navires classés par le Lloyd's Register, voir le Lloyd's Register Book et le Lloyd's Yacht Register.

Official No.	No. in Book	Registry in which classed	Ship's Name. Material and Rig. Master, and Dates of Appointment to present Owner's Service and to Ship. Ship's Late Name.	Flag	Net Tonnage	Length	Breadth	Depth	Where	When	By whom	Dates of Metalling	Owners or Managers	Port
	651	GV	Marie *Gulderman* 74-74 WBg	Danish	165	100·0	21·8	10·8	Rudkjøbing	1865	Joh.Boas		E.E.Nielsen	
	652	V	Marie (ex Gustav & Adelheid) *Sörensen* -87 WSmSr	Danish	159	93·1	21·7	10·3	Swinemünde	18 41/65	O.Hartwich		J.P.Rasmussen	
	653	V	Marie *Albertsen* 84-84 WSr	Danish	156	93·5	21·5	11·3	Heiligenhafen	1877	J.Tornøe		J.Albertsen	
	654	V	Marie (ex ...) *Jörgensen* 83-83 WBn	Danish	136	93·1	21·9	11·3	Faaborg	1875	M.Dyreborg	Y 8.87	[illegible] (B. Petersen)	
	655	V	Marie *Weber* 74-74 WSr	Danish	124	88·5	22·1	10·7	Marstal	1874	H.J.Bager		S.L.Weber	Marstal
	656	V	Marie *Rasmussen* 85-85 WSr	Danish	123	89·6	23·4	10·3	Nysted	1867	F.Sparre		R.H.Rasmussen	Nakskov
	657		Marie *Lorentsen* 82-86 WSr	Danish	114	74·7	22·3	9·8	Rønne	1856	S.P.Bech		M.Christiansen	Rönne
	658	V	Marie *Nielsen* 75-75 WSr	Danish	107	94·2	18·5	9·5	Troense	1857	C.R.Møller	Y 1.87	Aktie. Selskab. "Marie" (J.P.Møller)	Vejle
	659	V	Marie *Rasmussen* 76-87 WSr	Danish	105	82·8	20·4	9·8	Marstal	1876	J.J.Bager		H.C.Christensens Enke	Marstal
	660	D	Marie (ex Lynx) *Vlieland* (82-86) WBk	Dutch	364	122·7	27·6	14·9	Redon	1874		Y 1.87	P.Landberg & Zoon	Batavia
	661	R	Marie (ex Henrietta, ex Sarah A.Halbe...) (ABANDONED) *Porter* WBn	Dutch	156	93·0	23·6	11·4	Bermuda, W.I.	1857			C.A.Van Romondt & Co.	St. Martin
	662		Marie IBk	French	495	164·2	28·1	17·3	Sunderland	1869	Iliff, Mounsey &Co.		M.Viali	Marseille
	663	✻	Marie *Le Rot* WBk	French	417	129·5	27·8	17·2	Sunderland	1865	J.Robinson	Z 4.87	Le Rot	Havre
	664	DV	Marie *Guguen* WBk	French	330	119·1	24·2	16·1	St.Malo	1873	Gautier	Y 11.87	J.Blaize	St.Malo
	665	V	Marie (WRECKED) *Lateste* WBk	French	234	102·0	23·9	13·0	Nantes	1875	E.Clergeau	Y 4.85	Poulain de Saint Père	Tréguier
	666		Marie *Amour* WBk	French	227	100·0	24·6	13·7	Fécamp	1866	F.Capon		A.Chancerel	Fécamp
	667		Marie *Pedron* 78-78 WBg	French	159	87·3	21·6	13·0	St.Malo	1856	L.Malland		J.Le Pomellec	Binic
	668	V	Marie *Fon* 67- WSr	French	142	87·6	21·0	12·5	Boulogne	1869	Lecerf		Altazin-Petyt	Boulogne
	669	V	Marie *Lego* WSr	French	125	76·4	19·7	12·8	Nantes	1870	Dubigeon&fils	Z 8.85	C.Le Gloahec & fils	Nantes
	670	V	Marie *Carus* -86 WSr	French	119	89·9	20·0	11·3	St.Malo	1884	Trancheuer		Y.Le Goaster & fils	Paimpol
	671		Marie WBn	French	111	..	..	..	Messina	18 47/82		Z 7.82	D.Martinetti	
	672		Marie WBg	French	108	73·8	19·3	10·2	Pô	1865	Roi fils		F.G.De Pietra Corbara	Auray
	673	V	Marie *Denès* WSr	French	104	90·5	21·6	10·5	Granville	1879			J. Le Chapelain-Varin	
	674	V	Marie (ex Charmer) *Warnken* (82-83) WBk	German	1218	197·8	37·7	24·4	Boston	1869	T.L.Townsend	Y 12.86	Köper, Docke & Co.	Bremen
	675	V	Marie *Christians* 72-88 WBk	German	430	128·4	25·6	15·5	Elsfleth	1872	J.Jurgens	Y 6.86	E.tom Dieck	Elsfleth
	676	GV	Marie *Boltzen* -88 WBk	German	429	140·7	29·0	16·7	Apenrade	1876	G.Raben	Y 3.85	F.M.A.Steele	Hamburg
	677		Marie *Kugemann* WBg	German	319	105·1	27·5	16·1	Greifswald	1859	J.W.Sadelkow		F.Böttcher	
	678	V	Marie *Schumacher* 84-86 WBg	German	309	122·0	28·5	13·1	Elsfleth	1879	G.Wempe		J.G.Lubken	Elsfleth
	679		Marie *Redloff* WBk	German	299	113·0	27·0	14·5	Stralsund	1867	J.Peuss		Jas.Stevenson	Stettin
	680	V	Marie *Kreuins* -88 WSr	German	124	85·6	20·7	9·9	Edewacht	1886	Kramer	Y 1.87	S.Creesemann,Wwe	Leer
	681	V	Marie *Hansen* WBk	Norweg'n	714	146·0	34·4	20·9	Alblasserdam	1863	J.Smit		H.C.Hansen	
	682	N	Marie (ex Hendrik Ido Ambacht) *Zachariasen* WBk	Norweg'n	663	137·5	34·9	20·9	Rotterdam	1857			J.Meyer	
	683	N	Marie (ex Morning Star) *Saane* WBk	Norweg'n	662	165·4	33·4	18·9	CornwallisNS.	1856			Steen & Strøm	
	684	GN	Marie (ex Giulio & Anna) *Haugnsen* WBk	Norweg'n	625	149·2	32·5	19·3	Lavagna	1869	F.Rolla		A.G.Samuelsen	
	685	N	Marie *Eliasen* -86 WBk	Norweg'n	453	137·5	30·6	16·7	Bergen	1864			O.Eliasen	
	686	NR	Marie *Krag* -88 WBk	Norweg'n	441	131·5	29·9	16·5	Christiansund	1853	F.Wessel		G.Krag	
	687	N	Marie *Kittelsen* -88 WBk	Norweg'n	430	124·0	30·0	17·3	LaSeyne	1858			Chr.Th.Boe's Bo	
	688	N	Marie (ex H. Upmann) *Borge* WBk	Norweg'n	414	133·7	27·2	16·5	Brake	1866	D.Rogge		J.Dyrdahl	
	689	D	Marie (ex Lancashire Witch) *Helgessen* 87-87 WBk	Norweg'n	379	133·6	26·5	15·2	Sunderland	1852	G.W.&W.J.Hall		Birkeland & Co.	
	690		Marie *Hille* 77-77 WBk	Norweg'n	242	98·4	25·9	14·7	France	1859			J.C.Iversen	
	691	N	Marie *Nilsen* WBk	Norweg'n	242	102·2	26·5	14·4	Agde	1856			P.A.Nielsen	
	692		Marie *Jahre* -88 WSr	Norweg'n	204	110·0	24·3	11·3	Bremen	1858			L.Hansen Jahre	
	693	N	Marie *Iversen* -86 WSr	Norweg'n	141	95·7	24·1	10·9	Bergen	1875			N.W.Olsen	
	694	N	Marie (ex ...) *Syvertsen* 81-81 WSr	Norweg'n	132	83·8	22·1	11·5	Garmouth	1858	J.Geddie & Co.		O.M.Syvertsen	
	695	N	Marie *Egge* 82-82 WSr	Norweg'n	112	92·4	23·4	10·3	Hardanger, Norway	1882	K.Skaalurren		J.J.Egge	
80630	696		Marie Adèle *Lachaud* WSr	British	149	100·8	23·0	9·9	Grondines	1874			Olivier Rivard	
	697	V	Marie Agostini (ex ...) *Savary* WBg	French	279	115·8	26·2	14·7	Trieste	1871			René Gillard de Kerannflec	
	698		Marie Alfred *Jolliest* WBk	French	308	111·5	24·9	13·2	St.Malo	1874	Parnet		Danguet	
	699		Marie Amanda WBn	French	190	..	..	..	Bayonne	1859			Ferdel	
	700		Marie Angélina *Leberigot* WLr	French	122	71·5	21·0	10·9	Meuns	1866	Olivaud		Le Berigot	

✻=Lloyd's Register; D=Nederlandsche Vereeniging; G=Germanischer Lloyd; I=Registro Italiano; R=American; A=Veritas Austro-Ungarico; H=Veritas Ellenico; N=Norske Veritas; V=Bureau

In the 4th Column, C=Composite (iron frame, wood planked); I=Iron; Stl=Steel; W=Wood.

The Dates of Metalling (Col. 11) are based on the records of the Books in which Vessels are respectively classed, or on the Reports of the Surveyors of Lloyd's

Further particulars of Vessels classed by Lloyd's Register will be found in Lloyd's Register Book and Lloyd's Yacht Register.

1	2	4 Nom. Matériaux et Gréement. Capitaine et dates d'entrée au service des Armateurs actuels et à bord du navire. Ancien Nom.		5 Pavillon.	6 Tonneaux net.	7 Dimensions. Longueur	 Largeur	 Creux	8 Construit. Où.	9 Quand.	10 Par qui.	11 Date du doublage en métal.	12 Armateurs ou Directeurs-Gérants.	13 Port d'Armement.
701		Marie Anna	WSr	British	116	82·1	22·0	10·0	St.Irenée	1872			Nap.Lavoie	Quebec
702	V	Marie Anna *Le Bart* -88	WSr	French	132	79·1	20·3	11·7	Nantes	1867	Burhan&Co.		Le Bart	Nantes
703		Marie Anne	WSr	British	139	95·2	23·5	9·1	Batiscan	1874			Pierre Duval	Quebec
704		Marie Anne	WSr	British	142	81·5	23·0	8·1	St.Antoine de [Tilly	1855			Jean Fiset	Quebec
705	GV	Marie B. Korsch	WBk	German	540	136·8	29·9	18·5	Rostock	1882	E.Barchard&Co.	11 88	A.Barchard	Rostock
706	GV	Marie Becker	WBk	German	547	130·1	30·5	15·6	Brake	1871	J.Oltmann's Wwe.	Y 88	H.G.Lübken	Elsfleth
707	DV	Marie Berg *Hendricks*	WBk	German	536	141·2	29·9	17·7	Barth	1879	J.S.Kraeft	Y 1 86	J.N.Rodbertus	Barth
708	NV	Marie Berner *Gundersen* -87	W3mSr	Norweg'n	266	121·0	26·2	13·6	Haugesund	1873	J.Hauge	10.88	S.Berners Sönner	Stavanger
709	V	MarieBerthe *Verdier* -88	WBg	French	237	95·5	23·3	14·0	St.Malo	1867	L.Lhotelier		E.Barrier	Paimpol
710	V	Marie Bertrand *Bertrand*	WBg	French	199	112·2	22·3	12·7	Jersey	1862	Le Viscomte	Y 8.88	C.F.Bertrand	Cherbourg
711		Marie Brockelmann	WBg	British	197	109·3	26·0	13·1	Rostock	1856			A.Murdoch	Drogheda
712	V	Marie Céline *Mauffret*	WSr	French	119	74·1	20·0	10·7	Nantes	1876	E.Clergeau		Mauffret	Nantes
713	V	Marie Charlotte -88	WBk	French	389	123·3	25·3	15·6	St.Malo	1872	Gautier		Caujolle	St.Malo
714		Marie Clémentine 84-84	WBk	French	215	94·2	21·9	14·1	Fécamp	1867	Brument&Capon		Hermel & Buisson	Fécamp
715		Marie Elmire	WSr	British	155	86·2	25·8	10·7	L'sEbo'Tm'nts	1886	J.Bergnon		J.Bergnon	Quebec
716	G	Marie Emilie *Scharmberg*	WBk	German	403	121·3	29·1	16·8	Demmin	1861	Hansen		E.W.Ohrloff Sohn	Barth
717		Marie Erzelie	WSr	British	111	80·9	22·8	10·0	L'Islet	1867			D.Bouchard	Quebec
718		Marie Eugénie (ex Lorenzo Valerio)	W3mSr	French	198	..	..	..	Ancona	1860			De Peretti	Marseilles
719	V	Marie Eugénie *Roussel*	WSr	French	115	86·9	19·3	10·9	St.Malo	1879	F.Gauthier		L.Hubert	St.Servan
720		Marie Eugénie *Rio*	WSr	French	114	76·4	20·7	11·3	Dunkirk	1856			T.Rio	St.Servan
721	V	Marie Fanny *Lesèque* -87	WBn	French	134	87·3	19·3	11·2	St.Malo	1860	L.Mallard	Z 3.88	Dambrung	Brest
722		Marie Georgiana	WSr	British	159	106·0	24·5	9·2	Champlain	1873			L.Sauvageau	Quebec
723		Marie Helene	WSr	British	139	99·5	20·5	6·6	ThreeRivers	1853	L.Beaulieu		J.Normand	Quebec
724	V	Marie Henriette *Thouin*	WSr	French	146	91·5	20·3	11·7	Marseilles	1873			Dupuis-Robial	St.Brieuc
725	V	Marie Heyn *Draheim* -85	WBk	German	323	109·9	26·6	14·7	Stettin	18 61/74	E.Zieske		W.Draheim	Stettin
726		Marie Immaculata	WSp	British	133	97·9	23·0	7·6	Lanoraie	1880	A.McKercher		J.Robillard	Montreal
727	I	MarieJeanne *Beauregard* (ex Giuseppina B.) -88	WBn	French	130	..	..	..	Savona	1879	G.Pongiglione	Y 7.88	E.de Daupte	Rouen
728		Marie Jeanne *Paris*	WSr	French	109	69·9	20·0	10·8	St.Tropez	1864	Blain		Fourcade	Marseilles
729	V	Marie Joseph *Enaut*	WBn	French	107	73·2	20·7	10·4	Caen	1867	Lecorneur		S.Esnol	Caen
730		Marie Joséphine (BROKEN UP) *Mallet* -87	WBg	French	116	80·7	22·0	10·8	Cherbourg	1858	E.Lecerf	10.83	G.Collin	Plouëzec
731	V	Marie Juliette *Bellais* -87	WSr	French	131	92·5	22·0	11·7	Dunkirk	1878	Dorp		Woutermartaens	Dunkirk
732	G	Marie Kuyper *Maass* 87-90	WBk	German	362	119·3	28·2	16·3	Rostock	1868	W.Zeltz	Y 5.84	W.Maack	Rostock
733		Marie Laure	WBk	British	278	103·3	26·0	18·2	Seychelles	1840			J.Pulfer	Hobart
734		Marie Laure *Zoonekindt*	WSr	French	145	95·1	21·3	11·6	Dunkirk	1874	Vanderiele fils		Vanderiele fils	Dunkirk
735	G	Marie Louise *Ertaud* -85	WBk	French	442	132·1	26·6	15·3	Kolding	1876	C.Hansen	Y 3.85	MadameM.Ertaud	Marseilles
736	V	Marie Louise *Pastorel*	WBk	French	426	131·2	25·3	15·3	St.Malo	1875	J.Avril	Y 8.85	L.de Toppia	Marseilles
737		Marie Louise	WSr	French	104	75·5	20·0	9·2	America					Fécamp
738		Marie Louise (BROKEN UP)	W	French	101	..	..	..					Comolet frères	St.Pierre-Miquelon
739	GV	Marie Louise *Behring*	WBk	German	916	168·9	35·4	22·6	Damariscotta, [Maine	1864	A.Hall&Co.	Y 8.87	Siedenburg,Wendt & Co.	Bremen
740		Marie Louise	WS	Norweg'n	711	202·9	34·0	17·9	Havre	1856			Aktie Selskab "Marie Louise"	Christiania
741	N	Marie Louise *Börresen* (CONDEMNED)	WSr	Norweg'n	155	95·1	22·9	10·8	Elsfleth	1855			H.Tambs	Fredrikshald
742		Marie-Louise-Elisabeth (ex Elisa) *Olivier* 85-86	WBn	French	132	78·7	23·4	12·3	Villajuan	1864			P.Pagès & Co.	Cette
743		Marie Marguerite	WBk	French	523	135·5	26·2	19·7	Blyth	1865			Moureau	Chantenay
744	V	Marie Pauline *Jan* -88	WBn	French	159	83·7	24·9	9·5	Canada	1865		12.82	Folquet & fils	St.Malo
745	V	MariePetræus *De Souza* -87	WBn	Brazilian	113	87·5	20·9	10·0	Nordby,Fanö	1873	J.Th.Hansen	Y 2.86	A.J.P.Vianna	Rio de Janeiro
746		Marie Riebeck *Hegland* (BURNT)	WBk	German	504	134·8	27·5	16·8	Brake	1872	D.Rogge		A.W.Lane	Stralsund
747		Marie Rosa	WSr	Brazilian	218	..	..	..					F.R.de Azevedo	Rio de Janeiro
748		Marie Rose *Le Gleahec*	WBn	French	121	73·8	21·0	12·0	Carnac	1857	Rio		M.Le Gleahec	Auray
749	GV	Marie Siedenburg *Stricker* -77	WBk	German	1143	193·2	38·7	23·3	ThomastonMe	1862	Chapman&Flint	Y 3.88	Siedenburg,Wendt & Co.	Bremen
750	GV	Marie Spatz	WBk	German	344	122·0	27·3	14·5	Rostock	1875	E.Barchard&Co.	Y 3.81	W.Lange	Rostock

*=Lloyd's Register; D=Nederlandsche Vereeniging; G=Germanischer Lloyd; I=Registro Italiano; R=American Record;
A=Veritas Austro-Ungarico; H=Veritas Ellenico; N=Norske Veritas; V=Bureau Veritas.

Dans la 4me Colonne, C=Composite (membrure en fer, bordé en bois); I=Fer; Stl=Acier; W=Bois.

Les Dates du doublage en métal (Col. 11) ressortent des livres dans lesquels les navires sont respectivement classés ou des rapports des Inspecteurs du Lloyd's Register.

Pour plus amples renseignements sur les navires classés par le Lloyd's Register, voir le Lloyd's Register Book et le Lloyd's Yacht Register.

Official No.	No. in Book	Registry in which classed	Ship's Name. Material and Rig. Master, and Dates of Appointment to present Owner's Service and to Ship. Ship's Late Name.	Flag	Net Tonnage	Length	Breadth	Depth	Built Where	When	By whom	Date of Metalling	Owners or Managers	Port
	751	DGV	Marie Stahl *Schmidt 79* W IrnSr	German	27[illegible]	[illegible]4·7	26·[illegible]	13·1	[illegible]	1877	[illegible]	[illegible]	F.W.Fischer	[illegible]
	752		Marie Thérèse *Vidal* W Bk	French	22[illegible]	96·8	22·6	13·6	[illegible]	1858			[illegible]	Narbo[illegible]
	753	V	Marie Thérèse *Lecesseur* W Bn	French	195	90·9	24·9	12·1	Nantes	1884	Allenu& Aubert	[illegible]	[illegible] Le Gualès de Mézaubran	[illegible]
	754	V	Marie Thun *Voss 81-78* W Bn	German	171	89·0	24·2	12·7	[illegible]	1878	F.Barchard& Co.		W.M[illegible]	Rostock
	755		Marie Valentine W Bg	French	161	79·4	21·6	12·[illegible]	[illegible]	1856			[illegible]	Auray
	756	V	Marie Valentine *Pollefeyt* W Sr (WRECKED)	French	122	82·7	20·5	11·8	[illegible]	1862	Van[illegible]		[illegible] Van Cauwenberghe	Dunkirk
	757	✻	Marie Victor W Sr	French	100	84·2	20·1	10·1	St.Malo	1877	[illegible]		V.[illegible]	Binic
[illegible]	758		Marie Vigilante W Sc	British	114	76·2	23·0	9·9	BaySt.Paul	1879	[illegible]		M.[illegible]	Quebec
	759	V	Marie von Oldendorp *Voss 73-81* W Bn	German	163	96·4	20·4	11·2	Emden	1881	[illegible]	[illegible]	S.Wreesmann Wwe.	Weener
	760		Marie & Antoinette W Bg		257	98·1	24·[illegible]	15·[illegible]	[illegible]	1868	Contrasty	[illegible]		
	761	✻DV	Marie&Julie *Blankert 77-87* I Bk	Dutch	394	130·7	30·0	14·1	Middelburg	1885	Borsius& Van der Leije		Den Bouwmeester, Borsius & Van der Leije	M[illegible]
	762	V	Mariehamn *Troberg* W Bk (ex Lieutenant)	Russian	494	151·7	27·9	17·6	[illegible]	1866	TayS.B.Co.	[illegible]	K.J.Karlsson	Mariehamn
	763		Marietta *de Souza -87* I Sr (ex Orange)	Brazilian	172	104·2	21·8	11·6	Cork	1867	G.Robinson&Co.		A.de Oliveira Maia	Pern[illegible]
[illegible]	764	I	Marietta *Porta* W Bg	Italian	396	115·8	29·5	18·5	Procida	1867	[illegible]	[illegible]	G.Castagliola	Naples
[illegible]	765	A	Marietta (ex Kabilie) *Sfilio* W Bn	Italian	168	96·4	23·4	13·8	[illegible]	1864	[illegible]	[illegible]	G.B.Bennati	Genoa
[illegible]	766		Marietta *Jacono* W Bn	Italian	164	90·9	23·9	12·6	[illegible]	1863	F.Contreras		Salv.D.Janni	Gaeta
[illegible]	767	GR	Marietta Brailli *Critich 75-84* CS (ex Ravalli)	Austrian	920	205·0	32·0	21·0	London	1864	[illegible]	[illegible]	G.Brailli	Trieste
[illegible]	768	A	Marietta D. *Vusevich -87* W Sr	Austrian	180	90·2	22·2	11·3	Capod'Istria	1870	L.P[illegible]	[illegible]	N.Dubokovic	Gela
[illegible]	769	✻	Marietta D. *Gazzolo -87* W Bk (ex Sir W. F. Williams)	Italian	807	181·1	34·5	26·8	St.John,N.B.	1856		[illegible]	P.Olivari fu G.	Genoa
	770		Marietta S. *Palmento* W Bn	Italian	107	83·9	22·8	10·8	VietrisulMare	1884	P.Savastano		P.Savastano	Castel[illegible]
[illegible]	771	A	Marietta W. *Smojlina* W S (ex Giuseppe Ravarovich; now named "Byzantinus")	Turkish	781	163·3	32·5	21·4	Fiume	1867	[illegible]	[illegible]		
[illegible]	772	A	Mariettina *Baroncelli -88* W Bk (ex Mala Marina)	Italian	357	119·7	29·6	17·1	[illegible]	1868	[illegible]	[illegible]	G.B.Malabotich	Venice
	773		Marigo *Haggi* W Bk	Greek	418	..	..	..	[illegible]	1878			N.N.Haggi & Co.	Cassa[illegible]
	774		Marigo *Nicolaou* W Bk		379	122·0	27·6	..	[illegible]	1878				Syra
[illegible]	775		Marigo W Bn	Greek	376	114·8	24·6	..	[illegible]	1875			M.Daskakis	Syra
[illegible]	776		Marigo W Bn	Greek	319	101·0	25·7	16·9	[illegible]	1876			I.Mavripakis	Syra
[illegible]	777		Marigo *Goulandris* W Bg	Greek	301	106·6	27·1	..	[illegible]	1875		[illegible]	G.Goulandris	Syra
[illegible]	778	A	Marigo W Bkn	Greek	297	117·7	26·1	..	[illegible]	1868	C.Chaludis		J.Y[illegible]	Piraeus
[illegible]	779	H	Marigo *Berdanis* W Bg	Greek	271	95·1	25·9	..	Syra	1868	E.Posnopas	[illegible]	N.Georgiadis	[illegible]
	780		Marigo (ex Tesfrasia) *Alafouzos* W Bg	Greek	252	..	..	..	Capod'Istria	1867			J.Alafouzos	[illegible]
[illegible]	781		Marigo W Bn	Greek	249	98·4	26·2	13·1	Syra	1878	A.L[illegible]	[illegible]	A.[illegible]	[illegible]
[illegible]	782	A	Marigo *Papaggelis -88* W Bg	Greek	240	90·9	16·7	..	Syra	1869		[illegible]	M.Papaggelis	[illegible]
[illegible]	783		Marigo W Bg	Greek	224	86·9	27·2	..	Syra	1869	A.Livadaros	[illegible]	D.T[illegible]	[illegible]
[illegible]	784		Marigo W Bg	Greek	185	81·5	25·6	..	Syra	1865			A.Patrojannis	[illegible]
[illegible]	785	AH	Marigo W 3mSr	Greek	183	91·5	24·5	13·1	[illegible]	1880	E.Papapetros	[illegible]	J.Be[illegible]	[illegible]
[illegible]	786		Marigo *Gregoriadis* W Bg	Greek	182	83·6	25·3	..	[illegible]	1871			G.Gregoriadis	[illegible]
[illegible]	787		Marigo (ex Tessera Adelphi) Bn W	Greek	178	83·7	24·7	..	Syra	1866			E.Drikos	[illegible]
[illegible]	788		Marigo W Bg	Greek	165	82·0	23·9	..	[illegible]	1862			T.Chordavelis	[illegible]
[illegible]	789		Marigo W Sr	Greek	132	73·2	22·6	..	[illegible]	1876			N.Daskalakis	[illegible]
[illegible]	790		Marigo W Sr	Greek	108	72·5	22·1	9·5					N.P[illegible]	[illegible]
	791		Marigo W Bg	Russian	122	..	..	..	[illegible]	1868			Z.Alexandridi	[illegible]
	792		Marigo *Dimitri* W Bg	Turkish	157	..	..	..	[illegible]	1874			Limnilee Manoel [illegible] & Co.	[illegible]
	793		Marigo *Sargo* W Bg	Turkish	141	..	..	..	[illegible]	1885			Madame Marigo Sargo	[illegible]
[illegible]	794		Marigo Calvocoressis *Lentakis* W Sr	Greek	181	86·9	24·6	..	[illegible]	1883	[illegible]		N.Calvocoressi	[illegible]
	795		Marika *Bassis* W Sr	Greek	149	..	..	..	[illegible]	1861		[illegible]	J.Andreates	[illegible]
	796		Marin *Blanlot 86-88* W Bk	French	29[illegible]	1[illegible]8·6	21·3	12·[illegible]	[illegible]	1860	[illegible]		E.[illegible]	Dunkirk
	797		Marin *Jansson* W Bk	Russian	331	..	..	..	England				E.P.Jansson	[illegible]
[illegible]	798		Marina W Sr	Greek	158	..	21·0	11·4	England	1860			A.A.Vaglianos	[illegible]
[illegible]	799	✻	Mariner *Thurmott 65-75* W Bk	British	695	158·0	33·5	18·2	Pt.Elgin,N.B.	1875	[illegible]	[illegible]	Josiah Wood	[illegible]
[illegible]	800		Mariner *Sharp 71-71* W Bg	British	248	119·3	26·6	15·0	[illegible]	1871	[illegible]		[illegible]	[illegible]

✻ = Lloyd's Register; D = Nederlandsche Vereeniging; G = Germanischer Lloyd; I = Registro Italiano; R = A[illegible]; A = Veritas Austro-Ungarico; H = Veritas Ellenico; N = Norske Veritas; V = Bureau [illegible]

In the 4th Column, C = Composite (iron frame, wood planked); I = Iron; Stl = Steel; W = Wood.

The Dates of Metalling (Col. 11) are based on the records of the Books in which Vessels are respectively classed, or on the Reports of the Surveyors of [illegible]

Further particulars of Vessels classed by Lloyd's Register will be found in Lloyd's Register Book and Lloyd's Yacht Register.

1 Numéro d'ordre	2 Registre dans lequel le navire est classé	3 Nom. Matériaux et Gréement. Capitaines et dates d'entrée au service des Armateurs actuels et à bord du navire. Ancien Nom.	4 Pavillon	5 Tonnage net.	6 Dimensions. Longueur	7 Largeur	Creux	8 Construit. Où.	9 Quand.	10 Par qui.	11 Date du doublage en métal.	12 Armateurs ou Directeurs-Gérants.	13 Port d'Armement.
801	I	Marinetta *Olivari* 88 WBk	Italian	411	115·1	27·3	18·1	Savona	1867	Tixi	[illegible]	A.Razetto	Genoa
802		Marinetta *Calcagno* WBn	Italian	113	85·6	22·4	11·2	SestriPonente	1881	A.Carlevaro		Giac.Calcagno	Savona
803	V	MarinhasII. *Marques* 88 WBn	Portug'se	232	116·8	24·3	10·5	Esposende	1883	A.D.dosSantos	[illegible]	M.M.Marinhas	Oporto
804		Marinho I. WSr	Brazilian	268	..	..	..					Marinhos & Co.	Bahia
805	*	Marinho II. *Marques* WBk	Brazilian	267	..	..	..	Sweden	1873		[illegible]	Marinhos & Co.	Bahia
806		Marinho III. WSr	Brazilian	245	..	..	..					Marinhos & Co.	Bahia
807		Marinho V. WBk	Brazilian	240	..	..	..					Marinhos & Co.	Bahia
808		Marinho VI. *Borcs* WSr	Brazilian	186	..	..	..	Oporto			[illegible]	Marinhos & Co.	Bahia
809		Marinho VII. *Da Cunha* WSr	Brazilian	280	..	..	..	Portugal			[illegible]	Marinhos & Co.	Bahia
810		Marinho VIII. WBg	Brazilian	170	..	..	..					Marinhos & Co.	Bahia
811		Marinho IX. WBk	Brazilian	235	..	..	..					Marinhos & Co.	Bahia
812		Marinho XI. *De Souza* WBk	Brazilian	290	..	..	..	Sweden	1873		[illegible]	Marinhos & Co.	Bahia
813		Marinho XII. *Augusini* WBg	Brazilian	185	..	..	..	Oporto	1855			Marinhos & Co.	Bahia
814		Marinho XIII. WBk	Brazilian	244	..	..	..					Marinhos & Co.	Bahia
815		Marinho XIV. WBk	Brazilian	248	..	..	..					Marinhos & Co.	Bahia
816		Marini Padre (MISSING)	Italian	213	107·2	24·8	15·1	Sestri	1873	Celesia		Marini Prospero fu A.	Genoa
817	I	Marinin *Sanguinetti* WBk	Italian	936	184·0	35·4	22·6	Prà	1876	G.Tixi	[illegible]	D.F.Chighizola	Spezia
818	A	Marino *Bassich* WBk	Austrian	537	141·1	29·1	19·4	Fiume	1867	F[lli] Schiavoni	[illegible]	A.Cosulich fu G.	Fiume
819	A	Mario *Coelho* W3mSr	Portug'se	213	114·8	23·6	10·2	Villa do Conde	1878	J.F.da Lapa	[illegible]	A.L.Duarte	Oporto
820	V	Mario *Artigues Munera* 88 WBk	Spanish	258	124·6	26·2	16·1	Quebec	1866	N.Rosa	[illegible]	Jaime Mir	Barcelona
821		Mario B. *Luna* 87 WBk (ex Antonio Fermin)	Italian	515	138·4	30·2	20·3	Varazze	1867	N.Cerruti	[illegible]	A.Bertolotto	Genoa
822	*	Marion *George* 88 WBk	British	315	122·1	27·6	14·9	P.E.Island	1875	Ramsay	[illegible]	Wyllie & Gordon (F.P.Richards & Co.)	Swansea
823		Marion *Payne* 85-85 WBn	British	137	94·0	22·8	12·0	Sunderland	1867	Liddle		H.R.Cornelius	Faversham
824		Marion W3mSr (ex St. ... Santiago de Cuba)	American	977	231·0	38·0	20·2	[illegible] N.Y.	1861			F.H.Chappell	NewLondon,Con.
825	V	Marion *Ericksen* 85-86 WBkn	American	348	125·0	29·0	13·8	Boston,Mass.	1868			I.Gutte	San Francisco
826		Marion *Green* W3mSr	American	224	123·0	32·0	10·0	Boston,Mass	1882	M.Turner		P.K.Genereaux	San Francisco
827	*	Marion Crosbie *Roach* 88 StlBk	British	1026	215·8	34·0	19·9	Dumbarton	1885	A.M'Millan& Sn.		Rogers & Co.	Glasgow
828		Marion Draper *Bailey* WSr	American	174	106·0	27·5	9·1	Pittston,Me.	1867			J.E.Bartlett	Bath,Me.
829		Marion E. Rockhill *Anderson* W3mSr	American	271	119·0	29·7	9·6	Tuckerton,N.J.	1867			W.B.Anderson	Tuckerton,N.J.
830	R	Marion Hill *Armstrong* W3mSr	American	220	116·0	30·0	8·5	E.Boston	1882	Smith& [Townsend	[illegible]	S.Smith	Boston,Mass.
831	*I	Marion Inglis *Patterson* *Inness* StlS	British	1548	250·0	38·1	22·7	Dumbarton	1886	A.McMillan&Sn		Rogers & Co.	Glasgow
832	*	Marion Lightbody *Cardiner* 8-88 StlfmS	British	2131	287·0	42·6	24·1	Glasgow	1888	D. & W. Hender- [son & Co.		Rogers & Co.	Glasgow
833		Marion Ross *Carter* WSr	British	104	80·3	20·9	10·7	Inverkeithing	1863	Ross		J.Chalmers	Bowness
834		Marionga WBk	Greek	384	108·9	27·7	..	Syra	1873			A.Rosolimos	Syra
835		Mariquita WSrYt	British	106	89·9	18·0	10·4	Lymington	1864	G. Inman		S.Grimshawe	Cowes
836	V	Maris Stella *Mauffret* 87 WSr	French	141	87·3	23·9	11·6	Dunkirk	1855	B.Derycke		Mauffret	Vannes
837	A	Maris Vaglianos (ex Marina Doria) *Evangelos* WBk	Greek	485	110·0	31·1	20·0	Genoa	1872		[illegible]	N.Coupas	Zante
838		Maris Vaglianos WBg	Greek	194	102·0	23·8	11·1	Syra	1869			A.Combitsis	Argostoli
839	V	Maristany *Pla* WPol	Spanish	182	88·6	25·9	12·1	Arens	1868		[illegible]	A.S.Primol	Barcelona
840	*	Maritana WTrawler	British	89	80·0	20·9	10·9	Brixham	1886	Dewdney Bros.		J.Randall	Grimsby
841		Marittimo W3mS	Argentine	271	109·5	29·8	17·7	SestriPonente	1860	D.Bianchi		Arseno & Denegri	BuenosAyres
842		Maritza WBg	Greek	163	83·7	24·6	..	Syra	1861	Forges & C.de [l'Océan		C.Scandoulis	Syra
843	*	Maritzburg *Albison* 88 IBk	British	442	157·1	27·3	13·6	Aberdeen	1876	A.Hall & Co.		J.T. Rennie & Son	Aberdeen
844	A	Mariuccia (ex ...) WBk	Italian	582	149·2	31·3	19·1	Bordeaux	1876	Forges & C. de [l'Océan	[illegible]	Guido Aniello	Procida
845	A	Mariuccia Galatola *Ferro* WBg	Italian	267	117·7	27·6	15·1	Trieste	1871	N.Zanon	[illegible]	A.Galatola	Naples
846	I	Mariuccia P. *Di Donna* WBn	Italian	189	92·8	25·6	15·2	Alimuri	1853		[illegible]	L.Scognamiglio	Naples
847	F	Marius (ex ...) *Luce* WBk	French	179	110·8	24·3	11·5	Ajaccio	1865	L. Arnaud	[illegible]	Comolet frères	Cette
848	*	Marjorie WS[...]	British	71	72·1	20·2	10·4	Elmshorn	1885	J.Kremer		Letten Bros.	Grimsby
849	R	Mark Gray *Williams* W3mSr	American	293	115·0	32·0	9·0	N. Weymouth [Mass	1882	Keene		T.F.Cogswell	Newburyport,Ms.
850		MarkPendleton *Pendleton* WSr	American	189	109·2	26·7	8·9	Bangor,Me.	1874	Crosby Bros.		J.H.Crosby	Bangor,Me.

L = Lloyd's Register; D = Nederlandsche Vereeniging; G = Germanischer Lloyd; I = Registro Italiano; R = American Record;
A = Veritas Austro-Ungarico; H = Veritas Ellenico; N = Norske Veritas; V = Bureau Veritas.

Dans la 4me Colonne, C = Composite (membrure en fer, bordé en bois); I = Fer; Stl = Acier; W = Bois.

* Dates du doublage en métal (Col. 11) ressortent des livres dans lesquels les navires sont respectivement classés ou des rapports des Inspecteurs du Lloyd's Register.

Pour plus amples renseignements sur les navires classés par le Lloyd's Register, voir le Lloyd's Register Book et le Lloyd's Yacht Register.

Official No.	No. in Book	Registry	Ship's Name. Material and Rig. Master, and Dates of Appointment to present Owner's Service and to Ship. Ship's Late Name.	Flag	Net Tonnage	Length	Breadth	Depth	Where built	When	By whom	Date of Metalling	Owners or Managers	Port
73885	851	V	Mark Twain *Melvins -85* W Bk	British	765	158·2	34·4	19·3	[illegible]	1878	[illegible]	[illegible]	B.D.Melvin	St.Jo[illegible]
46286	852	RV	Marlborough *Sattler* W S	British	1383	202·2	39·8	25·0	Hantsport,NS	1863	Fuller	[illegible]	G.W.Churchill	Windsor
73638	853	✻	Marlborough *Bird 79-84* I S	British	1124	228·0	35·0	21·0	Port Glasgow	1876	[illegible]		J.Leslie	[illegible]
81170	854	✻	Marlborough Hill *Jones* I 4m Bk	British	2443	300·5	42·2	24·7	Liverpool	1885	[illegible]		Sailing Ship "Marlboro[ugh Hill"] Co. Ld. (W. Price & Co.)	[illegible]
30625	855	✻	Marmion *Piper 75-84* I Bk	British	788	192·8	31·1	20·2	Port Glasgow	1864	[illegible]		Williamson, Milligan & Co.	Li[illegible]
5842	856		Marmion W Bg	British	158	88·5	22·5	14·1	Maryport	1844	Ritson&Co.		J.Graham	Wh[illegible]
	857		Marmorah W Bg	Turkish	106	..	..	..	Barton	1878			Marmorablee Lemba [illegible]	Marmora
	858	D	Marnix (ex Martha Allen) *Kuiper* W Bk	Dutch	231	104·5	24·1	12·2	Boston,U.S.	1849			F.L.Drenth	Oude [illegible]
6047	859	✻D	Maroon (ex Jane Symons) *Macdonald* W Bk	British	362	124·2	25·0	16·6	Bristol	1855	Symons	[illegible]	J.Cummins	Melbourne
54960	860	✻D	Marpesia *Thompson 83-88* I S	British	1453	234·2	38·4	23·9	Port Glasgow	1866	J.Reid&Co.		Australasian Shipping Co. Ld.	Liverpool
30682	861		Marquard Family W Bk	British	338	104·0	23·5	15·0	Chittagong	1851			M[illegible] Cassum Mahomed [illegible]	Bombay
	862		Marques (ex Maria Isabel) *Marques* W Bn	Brazilian	120	..	..	..	S.João d Barra	18 5		[illegible]	M.P.Marques	Rio de J[illegible]
	863		Marquez de Olinda W Bn	Brazilian	208	..	..	..					J.G.de Carvalho	Rio de [illegible]
	864		Marquez de Pombal *Luiz* W Sr	Portug'se	129	84·6	21·6	9·0	[illegible]	1875	[illegible]		D.J.Rebello	Oporto
71240	865		Marquis (ex Tobias Butler) W Sr	British	424	118·0	28·9	11·5	[illegible]	1872	W.[illegible]		A.R.Christie	Toronto
	866	✻	Marquis of Anglesea *Griffiths* W Sr	British	74	67·0	18·5	9·0	Amlwch	1826			O.Jones	Carnarvon
80003	867	V	Marquis of Lorne *Vickery -83* W Bk	British	1143	190·1	37·3	23·5	St.John,N.B.	1879	J.K.Dunlop	[illegible]	Wm.Thomson & Co.	St.John
29593	868		Marquis of Worcester *Evans 83-86* W Bk	British	375	124·0	27·3	18·7	Swansea	1861	Richardson		J.C. & G.S. Richardson	Swansea
1206	869	V	Marrocchita (ex Marachita) *Colasso* W Bg	Italian	229	100·1	27·7	13·8	Arenys	1850		[illegible]	Pasquale Delogu	[illegible]
	870	N	Marryat *Busch* W Bk	Norweg'n	422	132·2	29·9	16·1	Risör	1873		[illegible]	O.Caspersen	Risör
16789	871		Mars *Dyzell 84-85* W Bk	American	243	92·0	27·3	17·0	Boston,Mass.	1825			W.Lewis	[illegible]
	872	N	Mars *Olsen* W Bk	Norweg'n	590	149·0	31·4	18·5	Archangel	1864			Th.W.Bergh	Fredrik[illegible]
	873		Mars (ex Glenthound) W Bk	Norweg'n	237	112·2	27·2	12·6	Picton,N.S.	1855			Martin Pedersen	[illegible]
	874	N	Mars *Johansen* W Sr	Norweg'n	100	79·5	21·8	10·4	Christiansand	1881	J.D.Sörvik		J.D.Sörvik	Christ[illegible]
	875	V	Mars *Berg -85* W Sr	Russian	175	90·8	26·8	11·8	Wandsee	1885			J.Jansohn	Riga
	876	✻	Marseille *Blank 88-88* W Bk	German	697	167·7	31·7	20·2	Bordeaux	1875	Rochet & Largeteau	[illegible]	Ernst Burchard & Co.	Rostock
45133	877		Marshal Keith *Gregory 80-83* W Sr	British	130	94·0	21·7	11·5	Peterhead	1864	Carnegie		R.Johns	Peterhead
16138	878		Marshall Perrin *Packard* W Sr	American	142	103·0	28·0	10·0	Leesburg,N.J.	1859			J.F.Bliss	[illegible]
73118	879		Marshall S. *Craig* W Bn	British	179	96·5	25·7	11·0	St.Marys,N.S.	1876		[illegible]	H.P.Palser,jun.	Sydney,N[illegible]
	880		Marso W Bg	Turkish	174	..	..	..	Greece	1876			Eigasloslee Kueloe Yorghi and his brother	[illegible]
	881		Marta (ex Oromocto) *Sandoval -86* W Bkn	Chilian	406	123·0	29·4	17·6	Oromocto,N.B.	1871	Morrow		G.Mathews&G.C.Rabe	Lota
	882		Marta (UNDERMSNED) *Nordling 85-84* W Bg	Swedish	259	112·8	25·4	12·2	Sundsvall	1867	Nordström	[illegible]	E.Christiernin	Sundsvall
	883		Marta *Magnusson* W Sr	Swedish	131	99·0	21·4	9·3	Westergötland				C.J.Andersson	W[illegible]
15081	884	✻	Martaban *Hodge* I Bk	British	737	172·1	30·3	19·9	Greenock	1853	J.Scott&Sons		W.Clark	Glasgow
147	885	A	Marte *Demiai* W Bk	Austrian	572	143·3	29·7	19·7	Fiume	1866	G.Brazzoduro	[illegible]	C.cav.Cosulich	Fiume
[illegible]	886		Marte *Pandolfi* W 3m Sr	Italian	136	96·1	21·4	10·8	Civita Vecchia	1855			A.Pandolfi	Civita V[illegible]
	887	R	Marten *Carlsson 81-84* W Sr	Swedish	175	94·5	21·4	12·1	Bremen				O.Wingren	[illegible]
72747	888	G	Martha (ex Giacomo Ferdinando ex Battler) *McPherson 75-83* W S	British	853	185·0	35·0	21·0	Rockland,U.S.	1852		[illegible]	China Shipowners' Assn.	Shanghai
83905	889		Martha *Wale -87* W Sr	British	100	86·0	20·5	10·7	Cardiff	1880	J.W.Tonkin		W.J.Lodge	[illegible]
	890	V	Martha *Christiansen 88-88* W Bn	Danish	138	94·4	20·8	10·5	Troense	1872	C.R.Møller		K.Blinkenberg	[illegible]
	891	V	Martha *Carlsen -77* W Sr	Danish	136	90·3	21·9	10·8	Marstal	1876	H.J.Bager		H.J.Bager	Marstal
	892	V	Martha *Andersen 61-74* W Sr	Danish	123	91·0	20·6	10·4	Rudkjöbing	1874	S.Boas		E.Petersen	[illegible]
	893	D	Martha (ex Rolling Wave) *Kremer* W Bg	Dutch	217	94·8	24·0	15·5	Shields	1854	T.Young&Sons		J.Kremer	[illegible]
	894	V	Martha *Vanhille* W Sr	French	132	88·6	21·0	11·0	Dunkirk	1877	D.Van Cauwenbergh		Van Cauwenberghe-Lesaffre	Dunkirk
	895	✻	Martha (ex Remington) I Bk	German	999	198·0	34·9	21·0	Glasgow	1863	Smith&Rodger		Knohr & Burchard	[illegible]
	896	V	Martha *Prignitz 88-88* W Bk	German	415	136·7	30·4	16·4	Brake	1868	J.Oltmann's Wwe	[illegible]	C.A.Beug	[illegible]
	897		Martha *Schroeder 83-85* W Sr	German	149	86·6	23·3	11·3	Ueckermünde	18 62/69	J.M.Pretzen		E.Schroder	[illegible]
	898	N	Martha *Gjertsen* W Bk	Norweg'n	632	152·4	32·3	17·9	St.John,N.B.	1855			C.Zernichow	[illegible]
	899		Martha (ex [illegible]) *Hannevig -87* W 3m Sr	Norweg'n	567	136·6	30·8	19·7	Restigouche, N.B.	1857			Chr.Hannevig	[illegible]
	900	N	Martha (ex [illegible]) *[illegible] -80* W Bk	Norweg'n	530	138·5	30·1	18·2	Maine	1853		[illegible]	Ludwigsen&Schjelderup	[illegible]

✻ = Lloyd's Register; D = Nederlandsche Vereeniging; G = Germanischer Lloyd; I = Registro Italiano; R = American [Record]; A = Veritas Austro-Ungarico; H = Veritas Ellenico; N = Norske Veritas; V = Bureau Veritas.

In the 4th Column, C = Composite (iron frame, wood planked); I = Iron; Stl = Steel; W = Wood.

The Dates of Metalling (**Col. 11**) are based on the records of the Books in which Vessels are respectively classed, or on the Reports of the Surveyors of L[loyd's Register].

Further particulars of Vessels classed by Lloyd's Register will be found in Lloyd's Register Book and Lloyd's Yacht Register.

No.	Reg.	Nom. Matériaux et Gréement. Capitaines et dates d'entrée au service des Armateurs actuels et à bord du navire. Ancien Nom.		Pavillon.	Tonnage	Longueur	Largeur	Creux	Construit. Où.	Quand	Par qui.	Date du doublage en métal	Armateurs ou Directeurs-Gérants.	Port d'Armement.
901	A	Martha Andersen -87	W Bg	Norweg'n	256	96·5	26·1	14·6	Christiania [(rebuilt)	18 63			B.Nielsen	Fredriksstad
902	A	Martha Nielsen (ex [illegible])	W Sr	Norweg'n	124	82·6	23·2	11·5	Calmar	1846			J.Nielsen	Horten
903	R	Martha Birnie White -88	W Bk	British	832	190·0	32·3	19·0	Aberdeen	1863	Duthie	Y 4.86	J.Lugaro	Aberdeen
904	DRV	Martha Brockelmann [illegible]	W Bk	German	488	139·4	30·2	17·4	Rostock	1873	H.Rickmann	Y 1.88	A.Burchard	Rostock
905	✻	Martha C.Craig Adams -87	I Bk	British	449	171·0	28·2	13·6	Belfast	1884	Workman,Clark [&Co.		Barque "Martha C.Craig" Co.,Ld. [illegible]	Belfast
906	RV	Martha Cobb Crosby -87	W S	American	1249	181·9	37·5	25·6	Rockland,Me.	1862	F.Cobb	Y 4.86	C.W.Crosby	New York
907	RV	Martha Davis [illegible] -87	W Bk	American	833	165·3	33·7	21·0	E.Boston Mass	1873	Smith&Towns-[end	Y 11.88	E.M.Brewer	Boston,Mass.
908		Martha E. Bayard	W Sr	American	169	107·0	29·6	7·5	Tuckerton, NJ	1872			A.Camp	Somers Pt.,N.J.
909		Martha Edmonds Galley	W Bg	British	183	102·7	23·0	12·8	Milford	1873	Jones	Y 1.88	E.Rillston	Fowey
910		Martha Ellen	W Sr	British	124	77·0	18·7	11·7	Sunderland	1840			James Allan	Newcastle,N.S.W.
911	✻	Martha Fisher [illegible]	I Bk	British	811	197·8	31·8	18·8	Glasgow	1875	A.Stephen & Sns		P.Iredale & Sons	Liverpool
912	G	Martha Gale [illegible] (CONDEMNED)	W Sr	Chilian	360	122·0	28·0	12·0	Machias,Me.	1872	Willey	Y 11.88	Wehrhahn & Co.	Valparaiso
913		Martha Innis Kelly	W Sr	American	245	106·5	29·0	8·4	Smyrna,Del.	1870			W.A.Lamper	Marblehead,Mass
914	R	Martha J. Brady (WRECKED) [illegible]	W Bn	British	288	111·0	29·4	12·2	Kempt,N.S.	1875	R.Rose	Y 11.87	Henry H.Greene	Windsor,N.S.
915	V	Martha Maria [illegible] -87	W 3m Sr	Russian	301	122·5	28·6	12·8	Gutmannsbach	1887	M.Mangus		M.& J.Martinsohn	Riga
916	R	Martha N. Hall Crowell	W 3m Sr	American	580	159·9	34·6	17·0	Providence R.I.	1873			J.F.Crocker	Boston,Mass.
917		Martha Nickels	W Sr	American	115	95·0	25·5	8·9	Machias,Me.	1858			H.Nash	Machias,Me.
918	V	Martha P. Tucker Smith	W Bk	American	623	140·6	33·1	18·3	Bath,Me.	1874	Goss & Sawyer	Y 7.88	A.W.Smith	New York
919	✻	Martha Percival Jones	W Bkn	British	249	120·1	26·3	14·5	Portmadoc	1877	Roberts	Y 3.87	E.Roberts & Co.	Carnarvon
920	V	Martha Reid Jones 86-86	W Bk	British	631	156·4	31·0	18·0	Welford,N.B.	1873	Brown	Y 10.88	Browne & Watson	Glasgow
921	R	Martha S. Bement [illegible]	W 3m Sr	American	479	146·2	35·0	12·4	Bath,Me.	1881		Y 2.87	J.E.Ridgway	Philadelphia
922		Martha W. Tuft Groth	W Sr	American	165	101·0	34·0	8·5	Humboldt,Cal	1876			J.Jensen	San Francisco
923		Martha Welsh	W 3m Sr	American	305	114·5	23·5	9·0	Chester,Pa.	1872			J.A.Burdge	Philadelphia
924		Martha & Clara Blancke 81-86	W Bg	German	262	100·0	27·0	13·0	Rostock	1863	G.W.Tesdorpf		A.Burchard	Rostock
925	✻	Marti Codolar [illegible]	I Bk	British	382	142·0	27·3	16·0	Port Glasgow	1876	J.Reid & Co.		F.W.Brightman	London
926		Martie A. Holmes [illegible]	W 3m Sr	American	521	153·0	36·0	17·0	Camden,N.J.	1877			J.Holmes & Co.	Perth Amboy,N.J.
927		Martin Nørregaard 75-76	W Bk	Danish	331	113·6	25·9	16·1	Danzig	1847	J.W.Klawitter		Oscar Petersen	Kastrup
928	V	Martin Bruns-Leep -88	W Sr	Russian	242	106·7	28·7	12·2	Wandsen	1883	J.Leepe		M.Bruns-Leep	Wandsen
929		Martin Frederik [illegible]	W Bg	Norweg'n	234	91·3	25·7	14·3	Damgarten	1841			Z.Möller	Fredriksstad
930		Martin L. Smith Smith	W 3m Sr	American	365	128·0	32·4	15·2	Jamestown, [Va.	1875	G.Wheaton		G.Taulane	Philadelphia
931	DNV	Martin Luther Grønneberg 87-88	W Bk	Norweg'n	792	172·2	36·0	20·2	Wiborg	1864	H.L.Schmidt	Y 6.88	H.Kjösterud & Sönner	Drammen
932	DN	Martin Luther Hansen -88	W Bg	Norweg'n	259	98·3	28·4	15·3	Trieste	1859			O.Eliasen	Drammen
933	G	Martin Luther Nilsson (ex [illegible]) -87	W Bk	Swedish	327	120·4	24·6	13·6	Bremen	1864	Botter&Meincke		A.R.Andersson	Calmar
934		Martin Pêcheur Davy	W Bk	French	136	..	..	..	St. Servan	1858			Guibert & fils	St.Malo
935	✻	Martin Scott Bush -88	I S	British	1388	239·4	38·3	21·4	Greenock	1875	J.E.Scott		H.W.Jamieson (Glasgow)	Greenock
936	V	Martin & Michail Jordan 84-85	W Sr	Russian	223	107·0	29·0	13·5	Orrenhof	1882	M.Weide		F.& G.Weide	Orrenhof
937	DRV	Martina Johanna (ex [illegible]) (ABANDONED)	W Bk	Dutch	743	149·9	35·1	22·0	Kinderdyk	1867	B.Pot Bros.	Y 2.87	E.J.Bok & Zonen	Amsterdam
938		Martine	W Sr	British	105	79·5	20·5	7·3	Grondines	1851	C.Heros		Narcisse Rivard	Quebec
939	DIV	Martinin Mezzano	W Bk	Italian	723	164·3	33·1	21·4	Foce	1878	Debarbieri& [Celesia	Y 3.88	S.Razeto	Genoa
940	R	Martinique Lowell	W 3m Sr	American	187	116·0	28·5	8·2	Belfast,Me.	1882	D.W.Dyer	Y 3.87	D.Barnes	New York
941		Martino	W Bk	Chilian	750	169.2	33.3	21.7	Sestri	1873			A.Bafico	Valparaiso
942		Martino	W Sr	Italian	124	84·0	20·7	10·8	Sestri Ponente	1862	A.Briasco fu G.B		Flli Ghio	Genoa
943	I	Martino Cilento Amabrieno -88	W Bk	Italian	520	139·4	30·6	19·6	Equa	1868	A.Gargiulo	Y 12.88	F.Cilento	Naples
944	V	Martinus Nielsen	W Sr	Swedish	132	88·6	21·3	10·0	Timmernab'en	1872	O. Andersson	Z 5.84	J.C.Hulthén	Helsingborg
945	✻	Marwood Williams	W Sr	British	89	75·1	20·6	10·7	New Quay	1860	Evans		W.J.Williams	Carnarvon
946	✻	Mary Jenkins 88-88	W Bk	British	418	136·5	28·1	17·7	Southwick	1864	D.A.Douglas	Y 12.85	J.Jenkins	Greenock
947	R	Mary Milne 79-82	W Bk	British	381	119·4	28·5	16·5	Montrose	1868	Strachan	Y 10.87	J.Mitchell	Montrose
948		Mary	W Bk	British	259	112·0	24·6	11·7	Waldoboro US	1851			B.C.Potter	Halifax,N.S.
949		Mary (ex Thetis) Lee	W Bn	British	196	114·2	19·5	14·2	Jersey	1858	Le Vesconte		Brazilian Coal Co.,Ld.	Bristol
950														

✻=Lloyd's Register; A=Veritas Austro-Ungarico; D=Nederlandsche Vereeniging; G=Germanischer Lloyd; H=Veritas Ellenico; I=Registro Italiano; N=Norske Veritas; R=American Record; V=Bureau Veritas.

Dans la 4me Colonne, C=Composite (membrure en fer, bordé en bois); I=Fer; Stl=Acier; W=Bois.

Les Dates du doublage en métal (Col. 11) ressortent des livres dans lesquels les navires sont respectivement classés ou des rapports des Inspecteurs du Lloyd's Register.

Pour plus amples renseignements sur les navires classés par le Lloyd's Register, voir le Lloyd's Register Book et le Lloyd's Yacht Register.

Official No.	No. in Books	Registry	Ship's Name. Material and Rig. Master, and Dates of Appointment to present Owner's Service and to Ship. Ship's Late Name.	Flag	Net Tonnage	Length	Breadth	Depth	Where built	When	By whom	Date of Metalling	Owners or Managers	Port belonging to
[illegible]	951		Mary *King* W Sr	British	136	86·2	21·2	12·2	Stockton	1826			Bessey & Palmer	[illegible]
[illegible]	952		Mary W Flat	British	111	81·3	20·8	9·0	Tranmere	1873	Anderson		W.B.Hill	Liverpool
[illegible]	953		Mary *Nicoll 86-86* W Sr	British	116	86·4	23·2	8·4					J.Mitchell	Montrose
7674[illegible]	954	❊	Mary *Jackson* W Sr	British	109	89·1	21·4	10·6	Ardrossan	1877	Barclay	Z 88	H.M·Donald	Glasgow
[illegible]	955		Mary (LOST) *Pauck 85-86* W Sr	British	1[illegible]	85·8	21·5	10·4	Garmouth	1873	Geddie		W.C.Phillips	Padstow
7380[illegible]	956	❊	Mary *Muir 86-75* W Sr	British	9[illegible]	86·2	21·5	10·3	Banff	1875	Watson		A.Robertson	Banff
	957	❊	Mary *Young 84-88* W Sr	British	92	77·1	21·1	11·1	Nevin	1859	Thomas		R.E.Williams	Carnarvon
	958	❊	Mary Stl Lighter	British	5[illegible]	70·0	19·0	6·7	London	1887	J.S.Watson		B.Jacob & Sons	London
1788[illegible]	959		Mary *Snow 85-86* W Sr	American	152	92·6	25·6	8·6	Belfast, Me.	1862			J.W.Frederick	Belfast, Me.
	960		Mary ex Reserve, ex Delia, ex ... ex Tres Amigos, Pelican ... W Bk	Chilian	35[illegible]	120·0	27·0	13·7	Quebec	1862	Oliver		G.R.Gepp	Valparaiso
	961		Mary *Mikkelsen 78-78* W Bn	Danish	13[illegible]	87·0	21·0	10·3	Nordby, Fanø	1867	T.Schou[illegible]		J.N.Jensen	Sønderho
	962	D V	Mary *Hasswinkel -84* W S	Dutch	136[illegible]	1[illegible]5·2	39·7	25·0	Alblasserdam	1874	J.Smit & Co.	V 88	F. Suermondt Zoonen & Co.	Rotterdam
	963	V	Mary *Spiesen* W Bn	German	198	106·5	21·0	12·1	Hamburg	1867	E.Dreyer	V 86	C.M.Goedelt	Hamburg
	964	V	Mary *Junge -88* W Bn	German	170	101·3	21·0	11·9	Elmshorn	18$^{64}_{83}$	D.W.Kremer	V 86	M.P.A.Meyer	Hamburg
	965	N	Mary *Andersen* W Bk	Norweg'n	543	150·9	30·5	17·0	Maitland, N.S.	1869	A.McDougall	V 86	Ferd. Melsom	Christiania
	966	N	Mary *Svendsen 74-74* W Bk	Norweg'n	518	141·9	30·3	17·8	Sackville, N.B.	1856			Oscar Hansen	Christiania
	967		Mary *ANDERSEN* ex Ludwig H[illegible] W Bk	Norweg'n	338	135·6	29·0	14·2	Westervik	1858			H.A.Andersen (Chr.Nielsen)	Laurvig
2130[illegible]	968	R	Mary *Skjulestad -88* W Bkn	Norweg'n	292	120·5	25·5	16·2	Aberdeen	1858	Hall	V 86	A.H.A.Davidsen	Stavanger
	969		Mary *Mansson 84-88* W 3m Sr	Swedish	259	120·7	26·2	12·8	Carlshamn	1871	D.R.Andersson	V 86	N.Olsson	Helsingborg
	970		Mary *Bengtsson -79* W Bg	Swedish	245	95·2	23·0	15·7	England	1853			C.J.Olsson	[illegible]
	971		Mary ex Caroline G., ex Adelina G. *Caja* W Bg	Turkish	273	..	..	..	[illegible]	1848	A.Zuna		I.Caja	Scutari
9042[illegible]	972		Mary A. Drury (WRECKED) *Nickerson* W 3m Sr	American	453	130·5	31·7	15·3	E. Boston	1872	Campbell & Brooks		J.H.Conant & Co.	Boston, Mass.
9145[illegible]	973		Mary A. Fisher *Cardes* W Sr	American	111	85·5	26·9	6·6	[illegible], N.J.	1882			E.A.Fisher	Perth Amboy
9110[illegible]	974	R V	Mary A. Greenwood *Tucker* W Bk	American	617	154·1	33·0	17·8	Northport, N.Y.	1878	J.Carll	V 88	Arnold, Cheney & Co.	New York
9144[illegible]	975	R	Mary A. Hall *Curtis -87* W 3m Sr	American	362	143·0	33·0	11·0	Bath, Me.	1882	Brown & McDonald	V 1887	Sloan, Bussell & Co.	New York
1763[illegible]	976		Mary A. Holt W Sr	American	193	101·0	27·9	9·4	[illegible], Me.	1867			R.L.Stanwood	Ellsworth
9056[illegible]	977		Mary A. Hood *Somers* W 3m Sr	American	361	121·0	32·0	8·4	Somers Point, N.J.	1873	Doughty & Chapelle		C.H.Steelman	Philadelphia
[illegible]	978		Mary A. M·Kay *Digby* W Bn	British	192	103·9	24·1	12·5	P.E.Island	1864	M·Kay		Kemp & Co.	Faversham
9013[illegible]	979		Mary A. Power *Keen -87* W 3m Sr	American	472	140·1	31·4	11·6	[illegible], Me.	1872	S.Watts & Co.	V 88	S.Watts	Thomaston
9070[illegible]	980		Mary A. Rice *Lowell* W Sr	American	115	98·0	27·3	6·6	Hampden, Me.	1869			E.B.Gardner	Bucksport
[illegible]	981		Mary A.Trainor *Webster 85-87* W Sr	American	198	112·5	29·5	8·3	[illegible], Del.	1874	Debatz, Maloney & Debatz		John W. Hall	Philadelphia
[illegible]	982	V	Mary A. Troop *Young* W Bk	British	1118	189·4	37·3	22·4	[illegible], N.B.	1882	D.Lynch	V 87	H.D.Troop	St.John, N.B.
[illegible]	983	R	Mary A. Trundy *Dodge* W 3m S	American	405	140·9	33·5	12·0	Newburyport, Mass.	1885	Atkinson & Fillmore	V 88	B.F.Metcalf	New York
1771[illegible]	984		Mary A. Tyler *Curtis* W Sr	American	195	102·2	29·4	9·2	Setauket, N.Y.	1866			H.P.Havens	New York
[illegible]	985	❊	Mary Agnes *Hume* W Sr	British	6[illegible]	76·8	19·5	7·8	Barnstaple	1882	W.Westacott		J.Newall	Dumfries
[illegible]	986	R	Mary Ann (STRANDED) *Williams 87-87* W Bk	British	313	111·1	27·5	15·9	Sunderland	1869	J.Barkes	V 86	R.Williams (Bevan, Evan & Co.)	London
[illegible]	987	❊	Mary Ann *Godfrey 86-86* W Bkn	British	241	119·1	21·3	14·1	Appledore	1863	Cibbles	V 88	T.Pugsley	Newport
[illegible]	988	❊ D	Mary Ann *Woodruff 76-76* W Sr	British	236	111·8	25·7	12·9	Sunderland	1870	W.Adamson	V 86	J.Adamson	Sunderland
[illegible]	989	❊	Mary Ann *Fishley* W Bn	British	14[illegible]	99·0	23·7	12·2	Kingston	1879	A.Spence		J.Hutchings	Fowey
[illegible]	990		Mary Ann W Sr	British	1[illegible]	89·2	22·2	12·[illegible]	[illegible]	1851	Hood Bros.		W.T.Blaxland	Faversham
[illegible]	991		Mary Ann *Mooney* W Sr	British	13[illegible]	93·7	25·6	10·5	Hopewell, N.B.	1857	Bennett		J.Curran	Belfast
[illegible]	992		Mary Ann W Bk	American	341	103·7	26·5	18·0	Fair Haven, Ms.	1838			T.O'Leary	New York
	993	V	Mary Ann *Waleirus* W Bk	Swedish	4[illegible]	130·9	27·8	15·8	Sandsvall	1869	J.A.Nordström	V 86	C.G.Wickberg	Sundsvall
[illegible]	994	❊	Mary Ann Hood *Sandle* (owner) W Sk	British	8[illegible]	75·0	21·0	10·5	Grimsby	1884	Smith Brothers		W. & J.Hood	Grimsby
7110[illegible]	995		Mary Ann Lydon W Sr	British	245	121·0	26·0	9·7	P.Burwell, Ont.	1874	William Yeoell		John Lydon	Port Burwell
[illegible]	996	R	Mary Ann McCann *Bateman* W Sr	American	170	101·8	27·1	8·7	Bangor, Me.	18$^{58}_{84}$	W.Hunnewill		J.Cassidy	Bangor, Me.
[illegible]	997		Mary Ann Predmore *Smith* W Sr	American	12[illegible]	96·0	26·6	7·5	Barnegat, N.J.	1856			A.M.Whipple	N.London
[illegible]	998		Mary Annie *Cuffe* W Bn	British	188	101·2	24·1	12·8	Plymouth	1871	Williams		F.J.Jarvis	Folkestone
[illegible]	999	❊	Mary Armstead *Jones* W 3m Sr	British	93	95·6	22·1	9·8	Carrickfergus	1881	P.Rodgers		J.Fisher & Sons	Barrow
1750[illegible]	1000		Mary Augusta *Holt* W Sr	American	150	98·0	27·2	9·0	Ellsworth, Me.	1867			A.C.Holt	Ellsworth

❊ = Lloyd's Register; D = Nederlandsche Vereeniging; G = Germanischer Lloyd; I = Registro Italiano; R = American Record;
A = Veritas Austro-Ungarico; H = Veritas Ellenico; N = Norske Veritas; V = Bureau Veritas.

In the 4th Column, C=Composite (iron frame, wood planked); I=Iron; Stl=Steel; W=Wood.

The Dates of Metalling (Col. 11) are based on the records of the Books in which Vessels are respectively classed, or on the Reports of the Surveyors of Lloyd's Register.

Further particulars of Vessels classed by Lloyd's Register will be found in Lloyd's Register Book and Lloyd's Yacht Register.

1	2	3 Nom, Matériaux et Gréement. Capitaine et dates d'entrée au service des Armateurs actuels et à bord du navire. Ancien Nom.	4 Pavillon.	5 Tonnage net.	6 Dimensions. Longueur.	7 Largeur.	Creux.	8 Construit. Où.	9 Quand.	10 Par qui.	11 Date du doublage en métal.	12 Armateurs ou Directeurs-Gérants.	13 Port d'Armement.
1001	*	Mary B. Judge *Magee* W3mSr	American	449	144·6	35·3	12·2	Milton,Del.	1884	D.H.Atkins		G.E.Magee	Philadelphia
1002	R	Mary B. Wellington *Bolles* W3mSr	American	294	131·0	32·0	11·0	N.Bedford Mass	1843	J.W.Howland		C.Lovell & Son	N.Bedford, Mass.
1003	R	Mary Bartlett *Holmes* 1-88 W Bn	American	340	114·0	27·6	16·0	Machias,Me.	1888		[illegible]	Holway & Co.	Machias,Me.
1004		Mary Blair *Aslin* -85 W Bk	British	328	111·6	27·0	14·6	Aberdeen	1870	Duthie	[illegible]	R.Honey	Adelaide
1005	*	Mary Blundell *Williams* [85-88] I Bk	British	822	194·3	32·1	19·9	Harrington	1869	R.Williamson &Co.		Tomlinson,Hodgetts & Co.	Liverpool
1006		Mary Boucher W Bg	British	180	102·0	25·1	12·2	Shelburne,N.S	1863			Charles H.Boucher	Shelburne,N.S.
1007	*	Mary Boyes *Frost* 85-81 W K	British	90	80·3	20·8	10·7	Brixham	1885	J.W.&A.Upham		W.Lloyd	Hull
1008		Mary Buhne *Segelhorst* 1-88 WSr	American	140	102·0	29·5	7·5	Eureka,Cal.	1876			Charles Nelson	San Francisco
1009		Mary C. *Bennett* -83 WSr	British	104	79·6	23·7	8·4	Rockport,N.B	1871			George Bennett	Rockland,N.B.
1010		Mary C. Carroll *Carrol* W3mSr	American	165	114·0	23·0	8·7	DennisvilleNJ	1875			J.D.Frambes	Bridgeton, N.J.
1011		Mary C. Comery W Bn	American	348	122·5	29·5	11·0	W'ldoboro'Me	1866			J.K.Terry	Baltimore,Md.
1012	R	Mary C. Hale *Higgins* W Bk	American	539	139·0	34·0	15·8	Ellsworth,Me	1875		11.88	Lewis Friend	Ellsworth,Me.
1013	R	Mary C.Mariner *Wharton* 1-82 W Bn	American	274	110·5	25·9	11·9	Cape Elizabeth,Me.	18 61/76	J.Dyer	11.88	B.Webster	Portland,Me.
1014		Mary Campbell W Bn	British	144	102·4	22·8	10·4	Manning Riv.,N.S.W.	1869	A.Newton		William Glen	Melbourne
1015	*	Mary Claasen *Evans* [87-75] W Bkn	British	182	104·6	25·1	13·5	Portmadoc	1876	Jones	[illegible]	J.Jones & Co.	Carnarvon
1016		Mary Coles *Skinner* W Bn	British	175	103·0	24·6	12·7	P.E.Island	1865	Campbell		P.Wright & Co.	Belfast
1017		Mary Culmer *Bethel* -88 WSr	British	243	119·8	30·4	10·6	Bahamas	1884	J.W.Roberts		R.H.Sawyer	Nassau,N.P.
1018		Mary D. Cranmer *Mathews* WSr	American	203	110·7	29·3	10·0	Tuckerton,NJ	1861			J. & A.Pharo	Tuckerton,N.J.
1019		Mary D. Ireland *Colhoun* WSr	American	252	105·6	28·5	9·5	Tuckahoe,NJ	1865			D.Scull	Philadelphia
1020		Mary Dodge *Gallup* [81-] W3mSr	American	231	120·0	32·5	8·9	Eureka,Cal.	1882	Murray		C.Nelson	San Francisco
1021		Mary E. Amsden *Clark* WSr	American	327	121·0	29·0	14·0	Portsm'th NH	1864			J.G.Murchie	Calais,Me.
1022		Mary E. Bacon *Eskridge* Sr (ex Clara M. Goodman)	American	182	106·5	23·8	9·2	Laurel,Del.	1876			Samuel Bacon	Seaford
1023	R	Mary E. Bliss *Fanning* [78-88] W Bn	British	188	94·5	24·7	10·5	DartmouthNS	1883	E.Williams	[illegible]	Wm.Fanning	Amherst,N.S.
1024	V	Mary E. Chapman *Kuhn* -87 W Bk	British	696	153·6	33·6	18·6	Rockland,N.B	1873	R.A.Chapman	[illegible]	C.A.Palmer	St.John,N.B.
1025		Mary E. Crosby *Brooks* W3mSr (ex Imogene Beverly)	American	179	111·3	23·4	8·6	Dennisville,N.J.	1873			J.L.Brooks	Nantucket, Mass.
1026	R	Mary E. Dana *Burgess* W Bn	American	208	101·0	28·6	9·8	E.Boston	1868			E.Russell	New York
1027		Mary E. Graham *Powell* WSr	American	196	113·8	29·1	8·4	Milford,Del.	1867			T.Winsmore	Philadelphia
1028	R	Mary E. McDougall *Renault* WSr	British	123	88·0	25·0	9·9	M'hone By. NS	1878	Zwicker		Alfred Renault	Halifax,N.S.
1029		Mary E. McLaughlin *Boudrot* WSr	British	168	91·8	26·3	10·6	Parrsboro',NS	1874			Chas.Boudrot	Halifax,N.S.
1030	R	Mary E. Morris *Fitzpatrick* -88 W3mSr	American	417	138·0	32·6	8·8	Camden,N.J.	1884	Morris&Mathis		J.W.Hall	Wilmington,Del.
1031	R	Mary E. Morse *Crocker* W3mSr	American	612	169·7	36·4	13·5	Bath,Me.	1881	BW&HFMorse		B.W.&H.F.Morse	Bath,Me.
1032		Mary E. Nason *Gaspey* WSr	American	103	79·7	22·9	8·9	Essex,Mass.	1864			D.Cornwell	Prvncetwn,Mass.
1033		Mary E. Oliver *Hinckley* WSr	American	140	87·3	26·2	7·7	GeorgetwnMe	1876	Trafton Bros.		E.C.Hinckley	Bath,Me.
1034		Mary E. Pearson *Hall* WSr	American	114	86·5	24·8	8·5	Bucksport,Me	1852			M.Pendleton	Belfast,Me.
1035	R	Mary E. Pennell *Mitchell* -88 W Bn	American	239	110·0	28·2	11·4	Harringtn.Me	1868	Ramsdell & Rumball	[illegible]	R.Mitchell	Machias,Me.
1036	RV	Mary E. Russell *Nickels* W Bk	American	575	117·0	32·0	17·5	Addison,Me.	1875	Nash	[illegible]	J.C.Nash	Machias,Me.
1037		Mary E. Simmons *Manley* W3mSr	American	200	110·3	29·3	8·5	Tuckahoe,N.J.	1866			C.G.Endicott	New York
1038		Mary E. Thayer W Bn	American	263	108·5	27·6	14·8	Brookhaven,N.Y	1868			N.Hand	New York
1039		Mary E. Webber *Holyden* WSr	American	123	94·7	24·7	9·0	Bath,Me	1876	A.Robinson		A.Robinson	Bath,Me.
1040		Mary Eliza *Merriley* 85 WSr	American	107	88·0	25·0	7·5	NewburghNY	1851			Scott & Bridge	Boston
1041	*	Mary Elizabeth *Pearson* -87 W3mBn	British	177	101·8	21·8	13·7	Newport	1862	Cooke	[illegible]	W.Brooks	Newport
1042	*	Mary Elizabeth *Garrett* WSr	British	99	87·0	22·4	10·4	Chester	1877	Ferguson		T.Mathias	Chester
1043	*	Mary Elizabeth *Lockner* [87-87] W Bk (ex Joanna McLeod)	Swedish	750	172·3	29·2	20·0	Southampton	1872	Ransom	[illegible]	C.M.Boden	Nordmaling
1044	R	Mary Ella Mallett *Mallett* W Bn	British	176	90·0	21·5	10·4	Southampton,N.S.	1883	H.S.Mallett	[illegible]	H.S.Mallett	Weymouth,N.S.
1045		Mary Ellen *Clark* WSr	British	113	90·0	24·4	8·8	St.George,N B	1865			Robt.Ross	St.John,N.B.
1046		Mary Ellen (ex Mercy Taylor) W Bn	British	100	87·4	23·6	8·6	UnitedStates	1857			A.King	Belfast
1047		Mary Emily *Jones* 76-76 W Bkn	British	193	110·6	25·1	12·3	Sunderland	1876	W.Pickersgill	[illegible]	J.Jones & Co.	Liverpool
1048	*V	Mary Emily *Eckhardt* 85-88 W Bk	German	461	138·1	29·2	17·3	Littlehamptn	1872	Harvey	[illegible]	Carl Rahtkens	Rostock
1049		Mary Evans *Richards* W Bk	British	306	114·3	26·8	16·3	Newcastle	1861	Hutchinson		J.Gulland	Brisbane
1050		Mary Everitt W3mS	British	198	120·0	26·0	8·1	Oakville, Ont	1867	Paul Dorion		C.Robertson	Belleville,Ont.

*=Lloyd's Register; D=Nederlandsche Vereeniging; G=Germanischer Lloyd; I=Registro Italiano; R=American Record; A=Veritas Austro-Ungarico; H=Veritas Ellenico; N=Norske Veritas; V=Bureau Veritas.

Dans la 4me Colonne, C=Composite (membrure en fer, bordé en bois); I=Fer; Stl=Acier; W=Bois.

Les Dates du doublage en métal (Col. 11) ressortent des livres dans lesquels les navires sont respectivement classés ou des rapports des Inspecteurs du Lloyd's Register.

Pour plus amples renseignements sur les navires classés par le Lloyd's Register, voir le Lloyd's Register Book et le Lloyd's Yacht Register.

Official No.	No. in Book	Register	Ship's Name. Material and Rig. Master, and Dates of Appointment. Ship's Late Name.	Rig	Flag	Net Tonnage	Length	Breadth	Depth	Built Where	When	By whom	Metalling	Owners or Managers
[illegible]	1051		Mary F. Corson *Robinson*	W3mSr	American	[illegible]	[illegible]	[illegible]	[illegible]	[illegible]	1872			J.C.L'Engle
[illegible]	1052	R	Mary F. Godfrey *Godfrey*	W3mSr	American	[illegible]	[illegible]	[illegible]	12·[illegible]	[illegible]	1882	J.B.Myrtetus		
[illegible]	1053		Mary F. Pike *Mitchell*	WSr	American	118	91·[illegible]	[illegible]	7·8	Perry,Me.	1872	Potter		P.Gillis
[illegible]	1054	R	Mary Fink *Myers*	WBg	American	[illegible]	132·[illegible]	[illegible]1·2	16·0	Millbridge,Me	1873		[illegible]	D.B.Darrah
[illegible]	1055		Mary Frances (ex [illegible])	WBk	British	209	[illegible]	[illegible]	11·4	Maine,U.S.	1846			Hy.Moss
[illegible]	1056	D V	Mary Fraser *Card*	WS	British	117[illegible]	186·3	37·[illegible]	23·3	Newport,N.S.	1875	JamesMosher	[illegible]	G.P.Payzant
[illegible]	1057		Mary Frazier *Downs* ([illegible])	WBk	American	[illegible]	[illegible]	[illegible]	15·9	Newburyport, Mass.	18[illegible]			S.Osborn,jun.
[illegible]	1058	✱	Mary Frost *Clark -86*	WSr	British	[illegible]	13[illegible]	27·7	12·8	Sunderland	1874	N.Gibbon	[illegible]	J.Sutcliffe & Son
[illegible]	1059	V	Mary G. Reed *Warren 55-55*	WBk	American	566	144·6	[illegible]·5	17·3	WaldoboroMe	1867	Reed,Welt&Co.	[illegible]	A.Welt
[illegible]	1060	R	Mary Gibbs *Moore 88-88*	WBn	American	[illegible]	122·[illegible]	[illegible]·5	16·8	Wstbrook Con	1869		[illegible]	J.S.Emery & Co.
[illegible]	1061		Mary Glover *Darrah -87*	WBk	American	[illegible]	[illegible]	[illegible]·0	21·9	Boston,Mass.	1849			A.D.Moore
[illegible]	1062	✱	Mary Graham *Lines -88*	WBk	British	67[illegible]	16[illegible]	[illegible]2·6	17·8	Quebec	1877	Oliver	[illegible]	J.Graham & Co.
[illegible]	1063	R V	Mary Hasbrouck *Ludwig*	WBk	American	[illegible]	15[illegible]	[illegible]·3	19·5	Calais,Me.	1878	BoardmanBros	[illegible]	A.J.Newbury
[illegible]	1064		Mary Helen *Baldwin 63-79*	WSr	British	[illegible]	93·[illegible]	24·0	11·5	Salcombe	1861	Evans		W.Sandford
[illegible]	1065	V	Mary Hogarth *Harris 83-83*	WBk	British	[illegible]	[illegible]	[illegible]2·2	18·0	Gardners Creek,N.B.	1875	Wallace	[illegible]	H.Hogarth
[illegible]	1066		Mary Holland	WBn	British	[illegible]	[illegible]	26·0	12·2	Clare,N.S.	1856			J.Milburn
[illegible]	1067		Mary Hounsell	WBn	British	146	94·[illegible]	23·5	12·7	Newfoundl'nd	1842			R.Farrell
[illegible]	1068	V	Mary I. Baker *Scovill 81-83*	WBk	British	844	171·[illegible]	[illegible]·6	20·8	SalmonRr.,NS	1876	R.Beveridge	[illegible]	L.E.Baker
[illegible]	1069	R	Mary J. Castner *Thurber -87*	W3mSr	American	412	127·7	32·5	14·[illegible]	Bath,Me.	1877	Br'wn&Stantial		J.May & Sons
[illegible]	1070	R	Mary J. Cook *Haffses*	W3mSr	American	414	138·[illegible]	32·5	11·[illegible]	[illegible]	1874		[illegible]	W.H.Parsons
[illegible]	1071	R	Mary J. Hubbard *Hubbard 84-84*	W3mSr	American	534	151·[illegible]	[illegible]	14·3	Milford,Del	1884	D.Lank		J.H.Hubbard
[illegible]	1072		Mary James *[illegible] 70-79*	WSr	British	154	99·2	23·[illegible]	12·0	Padstow	1862	Tredwen		A.H.James
[illegible]	1073	✱	Mary Jamieson *Phillips -8[illegible]*	WBkn	British	[illegible]	119·[illegible]	25·[illegible]	12·[illegible]	[illegible]	1875	Redway	[illegible]	F.H[illegible] & Co.
[illegible]	1074	✱	Mary Jane *Albury*	WSk	British	[illegible]	73·8	[illegible]	[illegible]	[illegible]	1835	G.W.Brown		H.Maddock
[illegible]	1075		Mary Jane *Dickson*	WSr	American	117	[illegible]	[illegible]	7·6	Woolwich,Me	1846			J.Arey
	1076	V	Mary Jane *[illegible] -86*	WBk	German	[illegible]	113·1	[illegible]	16·7	Memel	1858	B.Pieper		Siebert & Alexander
[illegible]	1077	R	Mary Jane Lee *Jordan*	WSr	American	132	94·[illegible]	[illegible]	8·6	Hancock,Me.	1874			C.H.Jordon
[illegible]	1078		Mary Jane Wilber *Coffey -76*	WBn	British	[illegible]	120·8	29·2	12·5	Dorch'ster,NB	1866	Wilber		J.J.Carroll
[illegible]	1079	R	Mary Jenness *Jones -85*	WBkn	American	[illegible]	[illegible]	31·3	16·3	Brewer,Me.	1875		[illegible]	[illegible].B.Hook
[illegible]	1080		Mary Johns *Jolly -87*	WBn	British	[illegible]	115·5	24·2	[illegible]	[illegible]	1868	Harvey & Co.	[illegible]	J.Stephens
[illegible]	1081		Mary Johnston *[illegible]*	WBkn	British	[illegible]	[illegible]	[illegible]	[illegible]	Montrose	1862	Birnie		Whicker & Co.
[illegible]	1082	✱	Mary Jose *George*	IBk	British	623	177·[illegible]	[illegible]	17·5	Dumbarton	1876	Birrell,Stenhouse&Co.		J.Jose & Co.
[illegible]	1083		Mary K. Campbell (STRANDED) *[illegible]*	WBk	British	578	[illegible]	31·[illegible]	18·3	WeymouthNS	1872	C.Campbell	[illegible]	D.Anderson
[illegible]	1084	R	Mary L. Allen *[illegible]*	W3mSr	American	[illegible]	127·5	[illegible]	9·7	Bath,Me.	1883		[illegible]	J.B.Drake
[illegible]	1085	V	Mary L. Burrill *Cook*	WS	British	1456	210·[illegible]	[illegible]·5	23·5	Clare,N.S.	1883	J.L.Burrill	[illegible]	W.Burrill & Co.
[illegible]	1086	R V	Mary L. Cushing *Bray -83*	WS	American	[illegible]	[illegible]·7	[illegible]·1	23·7	Newburyport, Mass.	1883	J.Currier,jun.	[illegible]	L.B.Cush[illegible]
[illegible]	1087	R	Mary L. Peters (ex L. P[illegible]) *Williams*	W3mSr	American	[illegible]	140·[illegible]	[illegible]	16·5	EastBoston	1874		[illegible]	J.H.C[illegible] & Co.
[illegible]	1088	V	Mary L. Stone *Carver -83*	WS	American	142[illegible]	198·[illegible]	[illegible]	[illegible]	Bath,Me.	1874	Goss&Sawyer	[illegible]	D[illegible] Goss & P[illegible]
	1089	G	Mary Lee *Oldsson 87-87*	WBk	Swedish	[illegible]	140·7	[illegible]	[illegible]	Maryport	1859	Ritson	[illegible]	J.Nilsson
[illegible]	1090		Mary Lee Newton *Colwell*	WSr	American	137	[illegible]	[illegible]	8·0	Perry,Me.	1868	Potter		F.A.Pike
[illegible]	1091		Mary Lord *Kehoe 59-76*	WBn	British	184	105·[illegible]	22·[illegible]	13·0	Appledore	1859	[illegible]		A.Nelson
[illegible]	1092	R	Mary Lord *Smith -83*	W3mSr	American	347	[illegible]	30·[illegible]	11·9	Ellsworth,Me.	1873	A.Lord	[illegible]	[illegible]
[illegible]	1093		Mary Louisa *[illegible]*	WSr	British	[illegible]	[illegible]·2	23·1	12·8	Brixham	1866	Upham	[illegible]	[illegible]
[illegible]	1094	✱	Mary Low *Robertson -87*	IBk	British	[illegible]	197·[illegible]	32·[illegible]	19·1	PortGlasgow	1876	R.Duncan&Co.		J.Lang
[illegible]	1095		Mary Lowell	WBg	American	[illegible]	[illegible]	25·[illegible]	9·0	E.Machias,Me	1852			C.H.Campbell
[illegible]	1096		Mary Lymburner *Booker*	WSr	American	[illegible]	92·[illegible]	[illegible]	7·[illegible]	Brewer,Me.	1868			E.A.Green & Co.
[illegible]	1097	✱	Mary MacMaster *McLeod 81-8[illegible]*	WBk	British	[illegible]	16[illegible]	30·6	18·[illegible]	PembrokeDk.	1861	T.M'Master&Co.	[illegible]	T.M'Master & Co.
[illegible]	1098		Mary Margaret *[illegible]*	WBn	British	[illegible]	86·5	23·7	9·3	Eastport,U.S.	1851			Wm.[illegible]
[illegible]	1099	R	Mary Mark *Frank 75-75*	WBk	British	271	[illegible]	25·0	12·8	Shoreham	1866	Shuttleworth	[illegible]	F.W.Mark
[illegible]	1100		Mary Means *[illegible]*	WSr	American	125	93·3	24·[illegible]	8·5	Sedgwick,Me.	1855			P.M[illegible]

✱ = Lloyd's Register; D = Nederlandsche Vereeniging; G = Germanischer Lloyd; I = Registro Italiano; R = American [illegible];
A = Veritas Austro-Ungarico; H = Veritas Ellenico; N = Norske Veritas; V = Bureau [illegible]

In the 4th Column, C = Composite (iron frame, wood planked); I = Iron; Stl = Steel; W = Wood.

The Dates of Metalling (Col. 11) are based on the records of the Books in which Vessels are respectively classed, or on the Reports of the Surveyors of [illegible]

Further particulars of Vessels classed by Lloyd's Register will be found in Lloyd's Register Book and Lloyd's Yacht Register.

		Nom. Matériaux et Gréement. Capitaine et dates d'entrée au service des Armateurs actuels et à bord du navire. Ancien Nom.	Pavillon	Tonnage net.	Longueur	Largeur	Creux	Construit. Lieu.	Quand.	Par qui.	Date du doublage en métal	Armateurs ou Directeurs-Gérants.	Port d'Armement.
		Mary Miller *Stewart* WSr	British	123	91·2	22·3	11·1	Irvine	1871	IrvineShipCo.	v 8.88	W. Reney	Kirkcaldy
	*	Mary Miller *Williams 12-88* WSmSr	British	99	93·2	22·1	9·9	Carrickfergus	1881	P. Rodgers		J. Fisher & Sons	Barrow
		Mary Miller *Dayton* WSr	American	159	95·5	28·4	7·8	Philadelphia	18 46/73			G.H. Jenks	Providence, R.I.
	*	Mary Moore *Sewell* C Bk	British	565	164·1	28·1	17·9	Glasgow	1868	A.Stephen&Sns.	v 8.87	W. Montgomery	London
	*	Mary Nish *Stevenson 81-88* WSr	British	100	86·7	21·8	10·3	Banff	1888	W. Geddie		D.B. Shaw	Wigtown
	R	Mary Nowell *Lank* WSmSr	American	377	128·0	32·0	9·5	Marcus Hook, Pa.	1874			W.M. Lank	Philadelphia
	*	Mary Orr *Perry 84-84* C Sr	British	84	81·8	20·7	9·7	Glasgow	1868	C.Connell&Co.		J. Parry & Co.	Glasgow
	*	Mary Owens *Owens 74-74* WSr	British	128	87·5	22·7	11·8	Pwllheli	1874	Pritchard		J. Owens	Carnarvon
	R	Mary P. *Philpot* WSr	British	139	90·8	26·0	10·8	Liverpool, N.S.	1884	R. McLeod		J. Philpot	Liverpool, N.S.
	V	Mary P. Kitchin WBk	Argentine	1073	183·7	35·9	22·1	RiverJohn, N.S.	1874	J. Kitchin	v 9.86	J. Ryan	Buenos Ayres
	*	Mary Peers *Garratt 85-86* WSr	British	132	94·8	23·5	12·0	Gunnel	1875	Clemens		H.C. Carter	Chester
1112		Mary Peverley WSr	British	100	86·7	19·6	9·0	J'rvisByNSW	1874	W. Peverley		John Ritchie	Sydney, N.S.W.
1113		Mary Price *Gandy* WSr	American	119	100·0	28·0	7·0	Wilmngtn, Del	1854			J.M. Applegarth	Baltimore, Del.
1114		Mary Riley *Taylor* WSr	American	227	114·0	28·8	8·6	Bridgeton, NJ	1866			J. Riley	Bridgeton, N.J.
1115	*	Mary Roberts *Lewis 87-87* IBk	British	1363	236·0	37·3	21·8	Sunderland	1887	W. Pickersgill & Sons		R.H. Roberts	Liverpool
1116		Mary Rosanna WSr	British	87	84·1	19·1	10·0	Ardrossan	1860	P. Barclay & Son		M. Knight	Dublin
1117		Mary Rowlands *Parry* WSr	British	109	85·0	22·5	11·3	Carnarvon	1858	M. Rowlands		J. Parry	Carnarvon
1118	V	Mary S. Ames *Crocker* WBk	American	700	147·9	33·1	16·7	EBoston, Mass	1876	G. Gaskill	v 6.86	J.C. Tiereira	Nw. Bedford, Mass.
1119		Mary S. Bradshaw WSmSr *Vanglider*	American	361	120·0	32·0	14·0	Philadelphia	1874			H. Matthews	Philadelphia
1120		Mary S. Tibbitts *McKeen* WSr (ABANDONED)	American	144	97·0	27·9	8·2	StatenIsland, N.Y.	1862	J. Ellis		S.E. Birdsall	New York
1121	R	Mary Sanford WSmSr *Brown*	American	456	136·0	34·0	15·0	EBoston, Mass	1881	J.M. Brooks		R.R. Freeman	Wellfleet, Mass.
1122	D	MarySeymour *Sharpe 77-* WSr	British	138	97·0	23·2	11·6	Padstow	1865	Tredwen		J.T. Crampton	Portsmouth
1123		MarySmethurst *Wheatley* WSk	British	110	91·5	21·0	11·4	Hamburg	1882	C.C.A. Dreyer		W. Whitelaw	Grimsby
		Mary Smith WSr	British	162	94·2	23·5	12·1	Hartlepool	1869			W.E. Marmion	Fremantle, W.A.
1125	R	Mary Sprague *Poland* WSmSr	American	617	158·3	33·1	17·9	Th'maston Me	1880	S.S.Gerry&Co.		G.W. Poland	Thomaston, Me.
1126		Mary Sproat *Walles* WBn	British	125	70·0	20·2	12·0	Maryport	1853			J. Ward	Milford
1127		Mary Standish *Higgins* WSr	American	258	118·0	29·8	9·8	E.Haven, Con.	1860	G.W. Baldwin		E.H. Atwood	Boston, Mass.
1128	V	Mary Stewart WBk *Thompson*	British	484	131·2	29·5	17·0	Black River, N.B.	1871	J. McLeod	v 9.86	A. Thompson	Aberdeen
1129	*	Mary Stewart *Stewart* ISr	British	59	72·5	19·5	8·3	Montrose	1876	Black		J. Stewart, jun.	Greenock
1130		Mary Stewart *Conant* WSr	American	218	111·0	27·4	10·3	Bangor, Me.	1856			Crosby Brothers	Bangor, Me.
1131	*	Mary Stewart *Andersen* WS	Norweg'n	1071	182·8	36·2	22·0	Nappan, N.S.	1874	JFullerton&Co.		C.H. Johnsen	Arendal
1132		Mary Stowe *Herbert* WBg	British	236	98·6	25·2	15·0	Gloucester	1861	Pickersgill		T. Pritchard	Whitby
1133		Mary Swann (LOST) *Olsen* WSr	American	136	95·0	29·0	7·7	Eureka, Cal.	1875			A. Anderson	San Francisco
1134	R	Mary T. Kimball *Gilley* WBn	American	453	134·6	32·0	15·3	E.Machias, Me	1877	C. Fry		J.H. Cady	Providence
1135		Mary W. Hupper WSr	American	264	118·9	29·5	10·6	W'ldoboro' Me	1866			G.W. Rawley	St. George, Me.
1136		Mary Wadley WSr	British	159	112·5	21·3	9·8	MacleayRiver N.S.W.	1874	E. Wadley		Edw.G. Ford	Hobart, Tas.
1137	*	Mary Waters *Waters* WSr	British	100	85·0	22·2	11·0	Padstow	1875	Rawle		J. Stronach	Padstow
1138	*	Mary Watkins *Simon 77-87* WSr (FOUNDERED)	British	169	91·0	23·3	12·7	Nevin	1850	O. Jones		M. Jones & Co.	Carnarvon
1139	*	Mary Watkinson *Elder* WSr	British	149	95·4	22·4	12·1	Berwick	1872	Gowan		J. Fisher & Sons	Barrow
1140		Mary Weaver *Weaver* WSr	American	211	110·0	29·8	9·0	Philadelphia	1867			C. Hillman & Co.	Bridgeton, N.J.
1141		Mary Wighton *Robin* WSr	British	119	85·1	20·5	10·9	Dundee	1859	Tay Ship Co.		J. Bunt	Weymouth
1142	V	Mary Winkelman WBkn	American	496	115·0	36·6	13·0	Seabeck, Wash	1881	Duncater	v 4.83	G.F. Smith & Co.	San Francisco
1143	*	Mary& Eleanor *Jones -84* WSr	British	98	82·0	21·6	10·8	Pwllheli	1874	Williams		W. Jones	Carnarvon
1144	*	Mary & Ellen *Evans 72-88* WSr	British	97	82·9	21·4	10·6	Aberdovey	1872	Richards		J. Evans & Co.	Aberystwith
1145		Mary & Emily WBg	British	203	89·2	21·5	11·5	Middlesbro'	1853			G. Moorsom	Whitby
1146		Mary & Francis *Snell* Sr	American	130	100·0	19·6	7·0	Philadelphia	18 61/70			J. McDonnell	Philadelphia
1147		Mary & Ida *Pitts* WSr	American	175	110·2	29·4	9·7	SanFrancisco	1882	DickieBros.		J.J. McKinnon	San Francisco
1148		Mary & Sus (WRECKED) WBk	American	327	114·7	29·0	16·8	N.Bdfrd. Mass	1857			G.F. Bartlett	NewBedfrd. Mass.
1149		Maryborough *Ald* WBk (Robert Parker)	British	963	181·1	36·2	22·0	Granville, N.S.	1854			T. Emerson	Newcastle
1150		Maryetta WBn	British	171	89·6	23·7	11·2	HillsburghNS	18 65/86	A. Dawe		Chas. Dawe	St. Johns, N.F.L

*=Lloyd's Register; D=Nederlandsche Vereeniging; G=Germanischer Lloyd; I=Registro Italiano; R=American Record;
A=Veritas Austro-Ungarico; H=Veritas Ellenico; N=Norske Veritas; V=Bureau Veritas.

Dans la 4me Colonne, C=Composite (membrure en fer, bordé en bois); I=Fer; Stl=Acier; W=Bois.

Les Dates du doublage en métal (Col. 11) ressortent des livres dans lesquels les navires sont respectivement classés ou des rapports des Inspecteurs du Lloyd's Register.

Pour plus amples renseignements sur les navires classés par le Lloyd's Register, voir le Lloyd's Register Book et le Lloyd's Yacht Register.

1 Official No.	2 No. in Book	3 Society in which classed	4 Ship's Name. Material and Rig. Master, and Dates of Appointment to present Owner's Service and to Ship. Ship's Late Name.			5 Flag.	6 Tons	7 Dimensions. Length	Breadth	Depth	8 Built. Where.	9 When.	10 By whom.	11 Date of Metalling.	12 Owners or Managers.	Port
122	1151		Masaniello	Arenci -87	W Bn	Italian	220	100·0	26·1	15·0	Savona	1854	F.Cabonaro		E.Tonetti fu L.	[illegible]
899	1152		Masashige Maru		W Bg	Japanese	337	111·3	26·9	11·2	Tosio	1882	KawasakiShYd		Oya Hichebei	Tokio
91101	1153	❋	Mascotte	Cramley	W K	British	90	80·0	21·3	10·4	Galupton	1885	T.Matthews		J.Marr	Hull
29129	1154	R	Mascotte (WRECKED)	Barkle	W Bkn	American	594	111·0	22·0	19·0	Greenport,NY	1881		[illegible]	F.H.Smith&Co.	New York
10773	1155		Masonic	Wilson	W Bg	British	156	103·2	20·5	11·7	Jersey	1857	P.Grandin		Whitstable Shipping Co.	London
69895	1156		Masonic	Fauns	W K	British	117	100·1	23·7	7·5	Brightlingsea	1876	Aldous&Co.		H.D.Smith	Ipswich
17199	1157		Masonic		W Bk	American	540	114·0	28·8	17·9	Stockton,Me.	1864			A.P.Blanchard	New York
	1158		Masood		Bg	Turkish	161	..	..	..					Mohamed Goda	Alex[illegible]
	1159	V	Massilia (ex Ayan-Andre)	Kalaris -86	W Bk	Greek	366	115·8	26·2	15·5	Martigues	1866	Fl.stre. Mauras	[illegible]	J.G.Kalaris	Syra
30522	1160		Mastiff		W Bg	British	146	70·4	19·3	13·[illegible]	[illegible]	1848			Koongie Murttayar Ka- ver Marna Murttayar	Bombay
	1161	DGV	Matador	Jordan 80-84	W Bk	German	1438	212·0	38·1	24·1	Bremen	1884	J.Oltmann's Wwe	[illegible]	Gildemeister & Ries	Bremen
114	1162	GV	Matador	Weide -86	W 3mSr	Russian	332	120·0	28·6	13·1	Haynasch	1885	J.Rasuk		Georg Weide	Haynasch
46317	1163		Matamoros		W Sr	British	236	98·0	26·6	8·5	P.JeffersonUS				Wm.Stewart	Nassau N.
	1164		Matanzas	Mateu -88	W Bg	Spanish	243	105·6	25·9	13·1	Blanes	1870	José & Malisa		F.C.Lopez	Canarias
58368	1165	❋	Mataura (ex Dunfillan)	Marlin -87	I S	British	853	199·4	33·3	20·2	Glasgow	1868	Aitken&Mansel		New Zealand Shippg.Co	London
	1166	G	Matautu (ex Mathilde)	Bonnelykke	W Bn	German	107	82·8	21·2	9·6	Auckland,N.Z.	1876	R.Mackay	[illegible]	Wachsmuth & Krogmann	Hamburg
18517	1167		Matchless		W Sr	British	151	95·8	19·9	11·3	Aberdeen	1846	Hall		J.J.R.Meiklejohn	Cape Town
92787	1168	❋	Matchless		W Sk	British	80	75·7	20·2	10·5	Rye	1885	G.&T.Smith		E.Bacon	Grimsby
247	1169	A	Mate	Marussich	W Bk	Austrian	329	105·9	27·2	17·5	Buccari	1853	Mattaglion	[illegible]	A.Corich fu G.	Lus[illegible]
	1170	DN	Mathania (ex Hebe)	Hansen -86	W Bk	Norweg'n	467	139·6	30·3	16·5	Sikeå	1876	O.P.Åberg	[illegible]	B.Bergesen,jun.	Stavanger
67276	1171		Matheran (MISSING)	Bowden	W Sr	British	111	85·4	21·3	10·9	Salcombe	1874	Vivian		W.R.Beer	Salcombe
	1172		Mathias (ex Freinfeld)	Tantra 86-86	W Bk	Russian	360	..	..	..		1876			F.Mickelsson	Wiborg
	1173		Mathias I.		W Sr	Brazilian	217	..	..	..					M.F.Cruz	Rio J[illegible]
	1174		Mathieu & Branas		W Bk	Chilian	232	108·9	27·0	15·7	Massachusetts	1856			Mathieu & Branas	Talcahuano
	1175		Mathieu Orenga	Doli	W Bn	French	146	82·3	23·3	11·6	Bastia	1868		[illegible]	L.Orenga	Marseilles
	1176	D	Mathilda (ex St. (ABANDONED)	Vass -86	W Bk	Dutch	556	135·8	31·8	19·5	Finale	1870	Guastavino	[illegible]	A.Köpcke	Rotterdam
	1177	NV	Mathilda (ex Northern Light)	Hansen	W S	Norweg'n	1795	220·0	43·0	27·5	Quincy,Mass.	1873	G.Thomas	[illegible]	J.C.A.Johannesen	Christiania
	1178		Mathilda	Ahlqvist	W Bk	Russian	401	122·7	26·9	15·9	Fredrikshamn	1874	J.Hietuela		J.Wirkki	Björkö
	1179		Mathilda		W Sr	Russian	165	..	..	..		1873			K.A.Melander	Lemland
	1180		Mathilda	Sinmata	W Sr	Russian	142	..	..	..	Böitis	1874			O.Ginman	Hitis
	1181	V	Mathilda	Rebetsky	W Bk	Swedish	545	159·1	29·7	14·1	Sandsvall	1873	J.A.Nordström	[illegible]	C.G.Vickberg	Sundsvall
	1182		Mathilda	Öfverberg	W Bk	Swedish	390	131·2	28·5	14·7	Gefle	1868	O.A.Brodin		E.T.Killman	Höganäs
	1183		Mathilda		W Bk	Swedish	344	106·3	24·6	18·1	Pomern				E.A.Francke	[illegible]
	1184	R	Mathilda	Helberg 86-86	W Bg	Swedish	226	114·1	23·7	12·9	Uddevalla	1855		[illegible]	Björk & Engström	Gothenburg
	1185		Mathilda	Asmundsson	W Bg	Swedish	203	101·3	23·4	12·5	Styltenäs	1871	A.Isacksson		J.A.Samuelsson	Morlanda
	1186		Mathilda	Jonsson	W Bg	Swedish	140	85·0	20·5	11·1	Lübeck	1847			A.Jonsson	Nyhamn
	1187		Mathilda	Fogelström -86	W Sr	Swedish	138	82·4	21·9	11·3	Orust	1859	J.Hegglund		J.Fogelström	Gefle
	1188		Mathilda (ex Bacchus)	Bengtsson	W Sr	Swedish	104	86·3	18·1	10·1	Westervik	18⁶⁸⁄₆₅			A.B.Elfversson	Gothenburg
85842	1189	V	Mathilde	Dukau -87	W Bn	British	299	..	..	..	Elsfleth	1873	J.Deetjen	[illegible]	J.Smith	Port Louis
51635	1190		Mathilde		W Bn	British	130	98·1	22·9	10·2	Apenrade	1872			J.F.Loroque	Port Louis
	1191		Mathilde (ex Anna)	Wolff	W Bk	Danish	360	134·4	23·2	14·5	Vegesack	1865	Joh.Lange		J.H.Wallöe	Fanö
	1192		Mathilde (LOST)	Larsen 7-87	W Bn	Danish	131	80·1	23·2	11·8	Fredericia	1847	S.Boas		F.W.Valentin	Svendborg
	1193	V	Mathilde	Guiho 81-81	W Bk	French	252	107·2	23·9	12·5	Nantes	1874	Alleaume&Aubert	[illegible]	Me. Guiho	Nantes
	1194		Mathilde	Pen -82	W Bk	French	123	73·5	21·0	10·9	Nantes	1874	E.Clergeau	[illegible]	P.Ch.Grenet	Nantes
	1195	G	Mathilde	Krenzien 86-88	W Bk	German	473	132·4	29·1	17·2	Rostock	1866	O.Ludewig	[illegible]	W.Manck	Rostock
	1196	DV	Mathilde	Boltzen -88	W 3mSr	German	355	139·4	26·6	15·2	Apenrade	1877	G.Baben	[illegible]	Frau B.W.P.Seek	Hamburg
	1197	V	Mathilde	Waack -84	W Bg	German	297	107·0	27·5	13·5	Rostock	1868	E.Paldernitz		A.Burchard	Rostock
	1198	V	Mathilde (ex Matador, ex Lagos)	Hus[illegible] -88	W 3mSr	German	239	101·3	26·0	14·0	Brake	1852	D.Behrens		A.H.Arnold	Brake
	1199		Mathilde (WRECKED)		W Sr		143	90·9	23·1	10·9	Danzig	1838	C.Strey			
	1200															

❋ = Lloyd's Register; D = Nederlandsche Vereeniging; G = Germanischer Lloyd; I = Registro Italiano; R = American [illegible]
A = Veritas Austro-Ungarico; H = Veritas Ellenico; N = Norske Veritas; V = Bureau Veritas

In the 4th Column, C = Composite (iron frame, wood planked); I = Iron; Stl = Steel; W = Wood.

The Dates of Metalling (Col. 11) are based on the records of the Books in which Vessels are respectively classed, or on the Reports of the Surveyors of Lloyd's

Further particulars of Vessels classed by Lloyd's Register will be found in Lloyd's Register Book and Lloyd's Yacht Register.

No.	Classe	Nom. Capitaine et dates d'entrée en service des Armateurs actuels et à bord du navire. Ancien Nom.	Matériaux et Gréement	Pavillon	Tonnage net	Longueur	Largeur	Creux	Construit: Où	Quand	Par qui	Date du doublage en métal	Armateurs ou Directeurs-Gérants	Port d'Armement
	N	Mathilde *Olsen*	W Bk	Norwegian	399	117·7	[illegible]	17·6	Tönsberg	1876	S.Josephsen	Y 10.83	I.H.Christiansen	Tönsberg
1202	N	Mathilde *Egeness* (ex ...)	W Bk	Norwegian	435	135·5	30·1	17·5	St.Stephen, N.B.	1862	J.&C.Short		P.Egeness	Tönsberg
1203	N	Mathilde *Hjalmarsen 1887* (ex Henry ...)	W...	Norwegian	432	135·3	27·9	16·2	Quebec	1865	Sewell		Carl Dahl	Fredrikshald
1204	N	Mathilde *Wergé 75-*	W Sr	Norwegian	136	91·3	21·9	10·6	[illegible]	1872	S.Johansen		M.E.Schildbred	Brevig
1205	N	Mathilde *Bjönness*	W Sr	Norwegian	136	89·7	22·2	10·6	Seldam	1865			N.S.Bjönness	Tönsberg
1206		Mathilde *Frohmuth*	W Sr	Russian	138	81·5	26·4	11·0	Angern	1877	Tellinsk		G.Frohmuth	Riga
1207		Mathilde	W Sr	Russian	125	84·0	26·0	10·7	Angern	1876	Selwinsky		Münder	Riga
1208	I V	Mathilde Hennings *Jude 79-*	W Bk	German	523	140·3	31·1	18·6	Rostock	1879	E.Burchard&Co	Y 9.86	M.Petersen	Rostock
1209		Mathilde Jade *Lindström 1-88* (ex ...)	W Bk	Swedish	257	105·2	24·7	13·2	Ribnitz	1859	J.Miedbrodt		G.W.Lindström	Wätö,Rslgn.
1210	A	Mathusalem *Jacobovich*	W Bk	Austrian	583	140·4	28·6	18·6	[illegible]	1868	G.Tonello	Y 4.86	A.Buntielich fu A.	Rosario
1211	A	Mati Ane *Miletich -87* (ex ...)	W Bk	Austrian	777	160·4	32·7	21·2	[illegible]	1878	A.Villovich	Y 8.88	S.& G.Bielovucich	Ragusa
1212		Matilda (ex Economo)	W Bg	British	224	108·0	23·0	13·5	[illegible]	1855			Wm.Shannon	Hong Kong
1213		Matilda	W Sr	British	124	78·0	23·4	7·7	Essex,U.S.	1860			W.MacDonald	Trinidad
1214		Matilda *Mercer*	W Bg	British	111	93·2	21·4	10·9	TrintyByNFL	1873			John Munn	St.Johns,N.F.L.
1215		Matilda *Potts -87*	W Bk	American	848	158·0	31·4	22·6	Searsport,Me.	1857			A.Spear	San Francisco
1216	R	Matilda Brooks *Jones*	W Sr	American	323	119·9	31·0	15·4	P.Jeff'rsonNY	1867		Y 8.86	E.H.Jones	New York
1217		Matilda Buck *Chisholm*	W Bn	British	271	106·2	24·0	12·0	Rockland,NB	1877	Chapman		M.T.Chisholm	Turks Islands
1218	V	Matilda C. Smith *McKenna*	W Bk	British	670	156·4	32·2	18·8	DorchesterNB	1873	G.Palmer	Y 10.86	J.McCausland	Belfast
1219		Matilda Calder *Joyner 81-86*	W Sr	British	163	101·0	23·7	11·6	Banff	1864	Watson		G.H.Gann	Faversham
1220		Matilda Kristina *Hanström*	W Bk	Swedish	478	126·6	28·1	17·3	[illegible]	18^{41}_{67}			S.J.Hamberg	Hernösand
1221	✱	Matilda Meadows *Ben...*	W K	British	89	79·6	21·2	10·5	Galmpton	1886	T.Mathews		Hardy & Long	Grimsby
1222	✱	Matilda Upton *Dare -87*	W K Bge	British	90	..	..	..	Ipswich	1887	Orvis&Fuller		W.Orvis & Co.	Ipswich
1223		Matilde (ex Sarah Jane)	W Sr	British	100	83·5	22·7	6·1					Thos.H.Tolmé	Nassau.N.P.
1224		Matilde *Dalero* (ex Matilde Dabcovich)	W Bg	Italian	323	116·2	28·7	18·6	[illegible]	1869	M.V.Tixi		C.Geranzini	Genoa
1225		Matilde (PONTOON) (ex Josephine)	W Bk	Italian	251	118·7	28·0	12·5	Liverpool,N.S	1868		Y 3.81	R.Volpicelli	Naples
1226	I	Matilde *Romano*	W Bg	Italian	218	91·5	26·0	15·5	Sorrento	1857	F.DeRosa		Romano	Castellammare
1227		Matilde	W Bn	Spanish	164	90·2	24·1	10·9	[illegible]	1859			D.Perez	Manila
1228		Matilde Encarnación	W Bn	Spanish	223	107·2	24·4	12·6	Albay	1868			C.Aramburu	Manila
1229	I	Matilde Mignano (ex Elisabetta Mignano) *Mosella*	W Bk	Italian	466	135·8	33·5	19·4	Venice	1873	A.Stipitivich	Y 1.88	A.Mignano	Naples
1230		Matsein	Stl Lg	British	480	220·0	30·0	8·4	Dumbarton	1887	W.Denny& Brs.		Irrawaddy Flotilla Co., Ld.	Glasgow
1231		Matsusaka Maru	W Bg	Japanese	179	97·9	20·5	11·9	Ominato,Ise	1878	Masumoto Matsujiro		Hagane Genobei	Osaka
1232		Matsushima Maru	W Bk	Japanese	407	128·7	24·5	16·5	Liverpool	1857			Nagoshi Aisuke	Osaka
1233		Mattapoisett *Reynolds 84*	W Bk	American	104	78·3	22·4	11·0	Mattapoisett, Mass.	18^{35}_{59}			S.Osborn,jun.	Edg'rtown
1234	A	Mattea *Vidulich*	W Bk	Austrian	585	140·7	28·5	20·2	Lussinpiccolo	1872	C.S.Cattarinich	Y 8.88	M.V.Premuda	Lussinpiccolo
1235	✱A	Matteo R. *Ivancich -88* (ex Matteo Mircovich, ex Carlotta)	W Bk	Austrian	373	129·5	28·5	17·3	Sunderland	1870	G.Bartram	Y 1.88	G.Zacevich fu N.	Lussinpiccolo
1236	✱	Matterhorn *Williams 73-87*	I Bk	British	1833	266·5	40·2	23·9	Greenock	1882	Russell & Co.		J.R.de Wolf & Son	Liverpool
1237	G	Matthaeus Rickert *Rieck 86-88*	W Sr	German	114	84·9	20·9	11·0	Rostock	1871	O.Ludewig		H.M.Kühlsen	Fehmarn
1238	R	Matthew Baird *Williams 86-85*	W Bkn	American	412	137·8	29·2	16·8	KennebunkMe	1873	N.L.Thompson	Y 10.85	S.Jordan & Co.	Portland
1239		Matthew Vassar *Chisholm*	W Sr	American	116	91·5	27·5	7·9	Poughkeepsie, N.Y.	1855			S.Brooks	N.Bedford, Mass.
1240	BG	Matthias *Berckmann* (ex Montebello)	W Bk	German	1124	178·8	37·1	24·0	Thomaston	1859	M.Ring	Y 4.85	Gildemeister & Ries	Bremen
1241	V	Mattia *Castellano -88*	W Bk	Italian	382	123·6	27·7	18·5	Alimuri	1870	S.Mauro	Y 1.88	Flli Lauro	Castellammare
1242	R	Mattie A. Franklin *Macdonald 78-81*	W3mSr	American	496	143·2	32·1	15·8	Bath,Me.	1870	Hathorn& Hodgkins		E.H.Atwood	Boston
1243	R	Mattie B. Russell *Lowell*	W Bk	American	370	115·9	28·0	14·5	Westbrook Me	1870	G.Russell	Y 3.86	H.P.Dewey	Portland.Me.
1244	R	Mattie E. Eaton *Dunnage*	W3mS	American	590	160·5	35·5	16·5	ThomastonMe	1887	Washburn Bros		Washburn,Bros. & Co	Thomaston,Me.
1245		Mattie Holmes *Jordan*	W Sr	American	152	108·0	26·5	8·9	Brewer,Me.	1865			H.Lord	Bangor.Me.
1246	R	Mattie J. Alles *Crossett*	W3mS	American	218	120·8	30·0	8·9	Yarmouth,Me.	1883	Hutchns&Stbbs		M.Fickett	Portland,Me.
1247		Mattie May *Richardson*	W3mSr	American	161	108·0	27·2	7·6	Baltimore	1883	Skinner & Sons		W.H. & G.W.Skinner	Baltimore
1248	R	Mattie Newman *Loveland*	W3mSr	American	337	129·6	34·0	9·8	Yorktown,Va.	1883	N.V.Lane		S.E.Birdsall	New York
1249		Mattie T. Dyer *Spurling*	W Sr	American	193	93·5	24·0	9·3	Kenneb'nkMe	1884	G.Christenson		A.M.Smith	Portland Me.
1250		Mattie W. Atwood *Newcomb*	W3mSr	American	620	150·0	33·0	17·0	[illegible], Mass.	1872	James& McKenzie	Y 8.83	E.L.Atwood	Wellfleet, Mass

...= Lloyd's Register; ...= Austro-Ungarico; D = Nederlandsche Vereeniging; G = Germanischer Lloyd; H = Veritas Ellenico; I = Registro Italiano; N = Norske Veritas; R = American Record; V = Bureau Veritas.

Dans la 4me Colonne, C = Composite (membrure en fer, bordé en bois); I = Fer; Stl = Acier; W = Bois.

Les Dates du doublage en métal (Col. 11) ressortent des livres dans lesquels les navires sont respectivement classés ou des rapports des Inspecteurs du Lloyd's Register.

Pour plus amples renseignements sur les navires classés par le Lloyd's Register, voir le Lloyd's Register Book et le Lloyd's Yacht Register.

Official No.	No.	Society in which Classed	Ship's Name. Material and Rig. Master, and Dates of Appointment to present Owner's Service and to Ship. Ship's Late Name.	Flag.	Net Tonnage.	Length.	Breadth.	Depth.	Built: Where.	When.	By whom.	Date of Metalling.	Owners or Managers.	Port of Registry.
	1251	V	**Matts August** *Wirpi* W Bk	Russian	613	119·9	29·2	20·5	Brahestad	1874	N. Möller		Fr. Sovelius & Co.	[illegible]
	1252		**Matwey** *Kabel* -84 W Sr	Russian	152	91·3	24·3	9·2	Narva	1860	A. Prüigger		D. v. Zinovieff & Co.	Narva
41890	1253		**Maud** *Linklater* -87 W Bk	British	1172	185·5	37·5	22·9	Carleton, N.B.	1859	Gilbert		T. Emerson	[illegible]
71164	1254		**Maud** W Barge	British	137	110·0	22·7	7·1	Ottawa, Ont.	1875	J. Bedard		Wm. Owens	Ottawa
67690	1255	✻	**Maud** *Fowler* 81-82 W Sr	British	140	97·0	23·5	11·7	Appledore	1876	Cock	Y 3.84	C. T. Bennett	Bristol
49098	1256		**Maud** *Small* 82-82 W Bn	British	123	98·2	23·3	11·6	Tatamagouche [N.S.	1864			M. Crean	Belfast
90456	1257	R	**Maud** *Robinson* W 3m Sr	American	379	127·0	29·8	13·9	Machias, Me.	1872			Woodbury & Truck	Beverly
90691	1258		**Maud B. Wetherell** *Foster* W Sr	American	102	87·5	23·3	8·5	Essex, Mass.	1877			J. & L. N. Paine	[illegible]
90823	1259	R	**Maud Briggs** *Young* W 3m Sr	American	237	118·4	29·4	10·0	Brewer, Me.	1875		[illegible]	S. H. Barbour	Bangor
68946	1260		**Maud C.** *Gale* -88 W Sr	British	194	103·7	28·1	10·1	Alma, N.B.	1873	N. Foster		Geo. F. Baird	St. J[illegible]
91808	1261	R	**Maud H. Dudley** *Oliver* 84- W 3m Sr	American	293	139·3	33·2	10·5	Bath, Me.	1884	Goss & Sawyer	Y 4.85	F. C. Curtis	Bath, Me.
91814	1262		**Maud McLain** *Craig* I 4m Sr (ex Stn. Frolic)	American	586	250·0	27·0	15·0	England	1885			Williams	New [illegible]
90282	1263		**Maud Malloch** *Coggswell* W Sr	American	111	90·0	26·0	8·0	Perry, Me.	18[66/71]	Wentworth & [Kelley		H. G. Boardman	Calais
	1264	✻	**Maud Mary** W Sr	British	84	77·2	20·6	9·4	Howden Dyke	1889	W. Caisley		G. H. Anderton	Hull
85583	1265		**Maud Pye** *Dixon* 82-83 W Sr	British	100	81·6	26·4	7·1	Hopewell Cape [N.B.	1883	E. W. Lynds		E. W. Lynds	St. John, N.
60872	1266	✻ R V	**Maud Scammell** *Bell* W Bk	British	593	146·0	30·8	18·0	St. John, N.B.	1873	Scammell	Y 4.88	Scammell Bros.	St. John, N.
90302	1267		**Maud Seward** *Fenton* W Sr	American	136	98·2	26·6	7·8	Little Chop- [tank, Md.	1875			T. J. Seward	Baltimore
92951	1268	✻	**Maud of Scotland** W K	British	64	71·4	19·4	9·0	Bideford	1886	H. M. Restarick		T. & J. Lelen, Ld.	Gt. Y[illegible]
78781	1269	✻	**Maude** *Evans* I Bk	British	1077	211·5	35·3	21·3	Liverpool	1873	R & J Evans & Co.		R. & J. Evans & Co.	Liverpool
	1270	N	**Maude** *Hannevig* -88 W Bk (ex Belle Alliance)	Norweg'n	675	150·3	31·5	19·7	Sunderland	1855	A. Leithhead	Y 1.87	E. Hannevig	Arendal
42243	1271		**Maude Annie** *Kelly* W Bn	British	101	81·1	22·0	10·6	Richibucto	1860	Mundle & [Maxwell		T. Hutchinson	Wexford
94655	1272	✻	**Maude Green** *Green* -87 W Bk	British	51	65·5	18·5	9·1	[illegible]	1887	J. Garside		J. H. Knowles	Grimsby
91562	1273	R	**Maude Sherwood** *Kelley* 84-85 W 3m Sr	American	498	152·0	35·0	17·3	Newburyp., Ms	1883	G. E. Currier		Doane & Co.	Boston, Ms.
91590	1274	R	**Maude Snare** *Dow* W 3m Sr	American	265	127·5	31·9	9·2	Bangor, Me.	1883	Crosby Bros.	Y 4.83	F. T. Hall	Bangor, Me.
78395	1275	✻	**Maulesden** *Walker* 84-84 I S	British	1500	245·2	38·3	23·1	Dundee	1875	A. Stephen & Sons		D. Bruce & Co.	Dundee
72640	1276	V	**Mauna Loa** *Putnam* -88 W Bk	British	1071	189·9	37·1	22·0	Maitland, N.S.	1881	C. Putnam	Y 4.86	Chas. Putnam	[illegible]
	1277	V	**Maurice** *Joubert* -88 W Bk	French	254	102·3	24·6	14·1	Nantes	1872	Sevestre	Y 3.88	F. P. Chapron	St. Nazaire
	1278	V	**Maurice** *Robin* -87 W Bg	French	166	82·3	21·6	12·0	Nantes	1865	Bojn & Bertrand		Orain	Nantes
	1279		**Maurice** *Robin* -87 W Bn	French	126	78·7	20·7	11·3	St. Malo	1874	J. Parnet	Z 1.81	Aug. Lemoine	St. Malo
	1280	V	**Maurice Gautier** *Mauger* W Sr	French	113	84·6	21·1	10·5	St. Vaast	1873	Cossey frères		Maurice Gautier	St. Vaast
	1281	R	**Mauritz** (WRECKED) *Clausen* -87 W Bg (ex Friensus)	Swedish	284	109·5	26·7	16·1	Sunderland	1866	Sykes, Talbot [& Sykes	[illegible]	J. Molander	[illegible]
	1282	R	**Maury** (CONDEMNED) *Hansen* 79-82 W Bk	Norweg'n	576	146·4	32·7	18·5	Drammen	1865	Jorgensen & [Knudsen	Y 8.84	Ths. S. Falck	Stavanger
	1283	N	**Maury** *Olsen* W Bk	Norweg'n	496	138·9	31·4	17·3	Arendal	1866	Dekke	Y 3.88	E. Dedekam	Arendal
46084	1284		**Mavilette** (STRANDED) W Sr	British	124	85·6	24·2	9·7	Cape Cove, NS.	1864			J. C. Boudreau	[illegible]
62192	1285		**Mavourneen** W Sr Yt	British	105	95·0	20·0	12·1	Gosport	1871	Camper & [Nicholson		C. E. H. Chadwyck-Healey	[illegible]
296	1286	H	**Mavrogordatos** *Tsiktis* W Bk	Greek	287	102·6	27·4	..	Syra	1878	A. Livadara	Y 3.84	N. Birlis	Syra
	1287	G	**Max** *Bräutigam* 77-71 W Bg	German	233	103·3	27·7	16·0	Rostock	1859	O. Ludewig		W. Maack	Rostock
	1288	V	**Max** *Sehl* 81-84 W 3m Sr	Russian	350	133·5	28·7	13·8	Caspervig	1876	C. Grübner		N. Sehl	Riga
55996	1289		**Maxim** *Owens* 86-86 W Bn	British	126	96·0	22·7	12·0	P. E. Island	1866			H. Smyth	Dro[illegible]
90948	1290		**Maxim** W Sr	American	112	92·0	29·0	7·0	Humboldt By Cl	1876	H. D. Bendixon		J. G. Jackson	San [illegible]
56850	1291	✻	**Maxima** *Maddox* 85-85 I Bk	British	474	153·6	28·4	17·1	Newcastle	1867	Palmers' Co.		G. H. Meager	Swansea
	1292		**Maxima** (WRECKED) *Risso* W Bk (ex Maggie Elliott)	Chilian	799	168·7	35·1	19·0	Maitland, N.S.	1872			Revello & Co.	Valparaiso
	1293		**Maxime David** *Bodiquet* [87-87] W Bk (ex Gratitudine)	French	795	167·3	29·6	21·1	Spotorno	1874	F. Cadenaccio	Y 3.87	Ehrenberg & D. Preston Leblond	[illegible]
	1294		**Maximos** *Berni* W Bk	French	357	116·8	26·9	16·2	Marseilles	1868	Jouvin	Y 7.86	L. de Toppia	Marseilles
	1295		**Maxoudie** *Ibraim* W Bk	Turkish	184	..	..	..	Sinope	1866			Has Moustafa Bey	[illegible]
93714	1296	✻	**Maxwell** *Frazer* -88 I S	British	1800	269·1	39·1	23·5	Liverpool	1887	T. Royden & Sons		Johnston, Sproule & Co.	Liverpool
71420	1297	✻	**May** *McNamara* -88 W Bkn	British	295	121·9	27·5	14·9	P. E. Island	1874	Yeo	Y 4.86	J. Tedford & Co.	Belfast
62536	1298		**May** *Kobe* W 3m Sr	British	237	111·7	25·1	12·8	Sunderland	1869	R. Pace & Sons		A. W. Mackay	Sydney
68637	1299		**May** *Dingle* -88 W Bn	British	156	100·4	22·6	12·0	Teignmouth	1873	Mansfield	Y 7.83	Robt. G. Rendell	[illegible]
69709	1300		**May** *Manning* W Bn	British	148	90·5	23·4	10·5	Lunenburg NS	1874			Lewis Anderson	[illegible]

✻ = Lloyd's Register; D = Nederlandsche Vereeniging; G = Germanischer Lloyd; I = Registro Italiano; R = American [illegible]
A = Veritas Austro-Ungarico; H = Veritas Ellenico; N = Norske Veritas; V = Bureau [illegible]

In the 4th Column, C = Composite (iron frame, wood planked); I = Iron; Stl = Steel; W = Wood.

The Dates of Metalling (Col. 11) are based on the records of the Books in which Vessels are respectively classed, or on the Reports of the Surveyors of [illegible]

Further particulars of Vessels classed by Lloyd's Register will be found in Lloyd's Register Book and Lloyd's Yacht Register.

		Nom. Matériaux et Gréement. Capitaine et dates d'entrée en service des Armateurs actuels et à bord du navire. Ancien Nom.	Pavillon.	Tonnage net.	Dimensions Longueur.	Largeur.	Creux.	Construit. Où.	Quand.	Par qui.	Date du doublage en métal.	Armateurs ou Directeurs-Gérants.	Port d'Armement.
		May (Successo) *Ketonen 83-88* WBk	Norweg'n	322	127·3	25·5	14·9	Lübeck	1855			M.Pedersen	Mandal
		May Bent (Bryn) *King -87* WSr	British	111	87·6	25·8	8·6	Parrsboro,N.S.	1887	D.O'Neil		D.O'Neil	Parrsboro'. N.S.
	*	May Cory *Watt 84-84* W3mBn	British	163	107·8	22·7	11·4	Bideford	1875	Johnson	Y 7.88	C.T.Bowring & Co.	Liverpool
		May Flower *Cruse 84-86* WBn	British	150	96·6	23·0	12·5	Brixham	1874	J.Barter&Son		E.J.Duder	St.Johns N.F.L.
		May Flower *Morrick* WSr	British	148	90·6	26·0	10·0	CornwallisNS	1870			Ezra Churchill	Windsor.N.S.
1306		May Flower *Fraser 76-80* WSr	British	126	91·4	21·2	11·9	Brucehaven	1868	Whitehead		W.Morrison	Inverness
1307		May Flower WSr	American	108	84·6	23·4	8·7	Essex,Mass.	1886			W.Thompson	Gloucester,Mas.
1308	*	May Hulse *Jones 86-86* WBk	British	463	150·9	30·0	17·6	Southampton	1881	J.Ransom	Y 11.88	J.R.Leckie	Glasgow
1309		May M. Dyer *Berry -88* WSr	American	144	93·1	25·5	9·1	Bath,Me.	1879	Hagan&Thrlow	Y 8.81	H.P.Dyer	Baltimore
1310		May McFarland *Montgomery -79* W3mSr	American	456	140·2	32·1	12·2	ThomastonMe	1873		Y 9.82	J.H.Conant & Co.	Thomaston,Me.
1311		May Newton WSr	British	214	144·0	22·8	8·9	ManningRiver [N.S.W.	1878	A.Newton		Alex.Newton	Sydney,N.S.W.
1312	R	May O'Neill *Hart* W3mSr	American	574	150·0	34·0	16·0	Camden,Me.	1883	H.M.Bean		J.P.Ellicott & Co.	Camden,Me.
1313	*	May Queen (Wee Kee) *Colville 80-89* I Bk	British	733	178·6	31·2	19·0	Aberdeen	1869	A.Hall&Co.		J.Leslie	Aberdeen
1314		May Queen *Ryan* WBkn	British	325	130·2	28·9	16·4	Wexford	1867	Devereux		W.W.Robinson	Dublin
1315		May Queen WBk	American	268	120·0	25·3	11·4	Newbury,Mas.	1850			G.E.Teague	New York
1316		May Queen *Esquini* WBk	American	175	101·5	24·2	10·6	Baltimore	1852			Pierce & Co.	Baltimore
1317	R	MayWilliams *Williams -89* W3mSr	American	525	144·5	34·8	15·1	Camden,N.J.	1882	S.W.Tilton		G.B.Fithian	Camden,N.J.
1318	V	Mayer III. (Coelho) *Costa* WBn	Portug'se	222	98·4	22·0	13·1	P.E.Island	1866		Y 1.82	V.A.d'Oliveira	Lisbon
1319		Mayfield *Murdoch 62-81* WBn	British	186	95·5	22·7	13·5	Hylton	1862	Hodgson		J.Murdoch	Liverpool
1320		Mayflower (COLLISION) WBn	British	178	102·3	24·6	10·6	Emsworth	1865			W.Foster	Portsmouth
1321	A	Maylath *Cerich* WBg	Austrian	287	103·3	26·0	15·8	Fiume	1858	Flli Schiavoni	Y 4.83	Car Mohovich	Fiume
1322	R	Maynard Sumner *Dyer* W3mS	American	307	143·3	28·5	10·0	Rockland,Me.	1882	Cobb&Bodwell	Y 5.86	Cobb,Wight & Co.	Rockland,Me.
		Mayo *Equidezu* WBn	Spanish	197	105·6	23·1	12·0	Tayabas	1875			José Reyes	Manila
		Maypó WBk		467	83·7	31·0	17·2	Arcus	18^{49}_{65}		Y 11.83		
		Maysales (ex Lorenzo ex Paula) WBn	Spanish	198	93·2	23·7	11·9	Barcelona	1854			L.Mir	Barcelona
		Mazagán WBn	Spanish	130	88·9	22·8	10·3	Malta	1878			B.Borras	Majorca
	V	Mazatlan *Leerkos 87-87* WBk	German	523	132·0	31·3	17·3	Harburg	1878	C.C.A.Dreyer	Y 8.87	Aug.Schultze	Oldenburg
		Mazeppa *Kinch 88-88* WBg	British	191	113·5	24·9	13·1	P.E.Island	1868	Davies		G.Kinch	Dublin
	*	Mazeppa *Ward* WSk	British	90	81·6	20·8	10·6	Rostock	1885	F.Barchard&Co		Lengfield & Ward	Hull
	*	Mazeppa (BARGE) *Stone* WK	British	82	88·8	20·9	9·7	Harwich	1887	Parsons&Cann		J.Wells & others	Harwich
		Mazeppa (WRECKED) *Hansen* WBk	Norweg'n	475	134·6	31·8	17·4	Stavanger	1854			A.Abrahamsen	Stavanger
		Mead (BROKEN UP) *Lumsden* WSw	British	171	84·0	24·0	14·0	Sunderland	1847	R.Thomps'n&Ss		W.T.Davies	Blyth
		Mechanic *Perrior -86* WBn	British	185	96·4	24·6	9·8	UnitedStates				D.A.Townsend	Halifax,N.S.
	R	Mecklenburg's Hauswirthe *Wiechmann 82-84* WBk	German	314	115·2	28·1	16·2	Rostock	1856	F.Dethloff		W.Maack	Rostock
	*	Meda *Leonard 87-87* WBkn	British	329	136·0	29·6	13·5	P.E.Island	1887	J.Yeo	Y 8.87	J.Yeo	Charlottetown
		Meda WSr	British	124	85·0	20·0	10·5	Barnstaple	1880	Westacott		Government of Western Australia	
		Medari Tefyk WBg	Turkish	370	..	..	..	Heraclea	1875			Hadjie Tilati Trebizondlee [illegible] & Co.	Heraclea
		Medari Tefyk *Ibrahim* WBg	Turkish	170	..	..	..	Greece	1885			Suleiman bin Ahmed & Co.	Constantinople
	N	Medbør *Jensen -88* WBk	Norweg'n	503	138·7	31·0	16·9	Grimstad	1888		Y 7.88	J.Jörgensen	Grimstad
1340	*	Medea *Fraser 84-84* I S	British	1066	217·2	34·5	21·0	Glasgow	1868	Barcly,Curle&C.		A.&J.H.Carmichael&Co.	Greenock
1341		Medea W3mSr	British	203	109·4	24·8	11·5					P.Richmond	London
		Medea *Kücking 75-81* WBk	German	310	116·3	27·3	15·1	Rostock	1869	W.Zeltz		A.Burchard	Rostock
	N	Medéa *Abrahamsen* WBk	Norweg'n	296	104·8	28·8	15·3	Kragerö	1863	T.Pedersen		B.Knoph	Kragerö
	N	Medea *Dalbom* WBg	Norweg'n	272	108·0	27·2	14·0	Risör	1871		Y 10.87	H.Bang	Risör
	V	Medellin *Roullet -83* (COLLISION) (ex Aquila ex Barletta) WBn	French	254	125·0	26·9	10·6	Sestri	1873			H.Guillebot	Granville
	V	Medelpad *Löw -88* WBk	Swedish	333	131·2	26·7	12·5	Sundsvall	1872	J.M.Högglund	Y 6.87	J.Tugmansson	Ellebolm
	N	Mediator *Nielsen -88* WBk	Norweg'n	285	113·9	27·4	14·1	Risör	1873	K.Gundersen	Y 1.88	G.Olsen	Arendal
	G	Medina *Potter -87* WBg	British	341	107·7	26·5	14·0	Granville,N.S.	1873	J.E.Shaffner	Y 1.87	T.E.Sherwood	Turks Island
	V	Méditerranée *Paleard (8.88)* WBg	French	128	80·4	21·3	11·4	St.Malo	1871	Parnet		Vve G.Landel	St.Malo
	V	Medor *Svarrer 84-84* WBg	Danish	266	116·6	24·2	13·0	Varmauäs	1884	C.I.Bring&S. Jönnsson	Y 7.87	Akt. Sels. (N.H.Svarrer)	Nordby,Fanø

*=Lloyd's Register; A=Veritas Austro-Ungarico; D=Nederlandsche Vereeniging; G=Germanischer Lloyd; H=Veritas Ellenico; I=Registro Italiano; N=Norske Veritas; R=American Record; V=Bureau Veritas.

Dans la 4me Colonne, C=Composite (membrure en fer, bordé en bois); I=Fer; Stl=Acier; W=Bois.

Les Dates du doublage en métal (Col. 11) ressortent des livres dans lesquels les navires sont respectivement classés ou des rapports des Inspecteurs du Lloyd's Register.

Pour plus amples renseignements sur les navires classés par le Lloyd's Register, voir le Lloyd's Register Book et le Lloyd's Yacht Register.

SAILING VESSELS.

Official No.	No. in Book	Registry in which classed	Ship's Name (Ship's Late Name)	Master, and Dates of Appointment	Material and Rig	Flag	Net Tonnage	Length	Breadth	Depth	Where Built	When	By whom	Metalled	Owners or Managers	Port
52831	1351		Medora		W Bg	British	232	108·0	26·8	12·3	N[illegible],N.S.	1865			John H. Allen	Halifax
	1352	✳	Medora (ex "Lise")		W Sr Yt	French	90	96·9	20·3	11·5	Gosport	1875	Camper	[illegible]	A. Bonard	Boulogne
1713	1353	V	Medora	Mortola	W Bk	Italian	396	119·1	30·1	19·0	Sestri Ponente	1868	G. Peregallo	[illegible]	G. Mortola	Genoa
	1354	✳	Medora	Lund 87-87	W Bk	Swedish	770	173·4	32·8	20·4	Ardrossan	1867	Barr & Shearer		G. R. Larsergren	Stockholm
	1355		Medusa		W Sr	Brazilian	149	..	..	..					A. L. do Rego	Bahia
	1356	V	Medusa	Krause	W 3m Sr	German	366	120·1	27·6	15·7	Memel	1874	L. Eggert		Siebert & Alexander	Memel
	1357	V	Medusa	Ockolm 86-86	W S	Norweg'n	818	168·9	33·6	21·1	Sunderland	1862	W. Briggs & Sons	[illegible]	T. F. Andorsen	Mandal
	1358	V	Meeden	Mantjes	W Bg	Dutch	194	103·6	25·4	11·7	Pekela	1875	J. H. de Boer	[illegible]	D. S. Detmers	Veendam
81031	1359		Meenatchi Sundram		W Sr	British	130	90·0	19·5	10·0	Cochin	1877			Hettia Kandaga Joseph Fernando	Colombo
	1360	N	Meenstad	Engebretsen	W Bg	Norweg'n	279	110·7	27·3	13·6	Porsgrund	1868			A. Sorensen	Porsgrund
54271	1361		Meera Hussaine (ex Alster) (WRECKED)	Packer Malim -85	W Bk	British	437	124·0	29·2	17·6					M. Othman Lebbei Marcayar	Negapatam
31113	1362		Meera M. Samadany		W Bg	British	112	78·6	19·5	9·7	Quilon	1874			Murkairtamby Abdul-cader	Colombo
31166	1363		Meerasaib		W Bg	British	104	71·9	16·7	11·3	Cochin	1856			Mohidin Abdulcader Meyasan	Colombo
	1364	V	Meerkonig	Carstens -69	W Bkn	German	253	101·2	26·3	12·0	Zingst	1869	F. Zallmer		W. A. Sarnow	Barth
65462	1365	✳	Meggie Dixon	W. Donald [86-86]	W Bk	British	474	114·7	29·5	17·9	Sunderland	1873	J. Gardner	[illegible]	T. S. Hewitt	North [illegible]
17836	1366	RV	Meguntícook	Wallace 8-88	W Bk	American	420	132·0	30·0	17·0	Buc'sport, Me.	1866	W. H. Genn	[illegible]	J. S. Emery & Co.	Boston
81803	1367		Mehrap		W Bg	British	219	95·5	22·6	13·5	Chittagong	1870			Asmut Ally	Rangoon
293	1368		Meikaku Maru No. 2		W Sr	Japanese	202	93·0	22·7	12·4	Tokio	1880	Hirano S. B. Yd.		Tsunoya Jintaro	Osaka
80482	1369		Meingyan		I	British	777	197·0	25·7	8·3	Port Glasgow	1875	Robt. Duncan & Co.		Irrawaddy Flotilla Co. Ld.	Glasgow
	1370	N	Melanchton	Tollaksen	W Bk	Norweg'n	274	113·3	29·0	13·0	Arendal	1867	S. Johnsen		J. E. Hanssen	Arendal
17725	1371		Melancthon	Forrest	W Bk	American	283	133·0	30·5	10·9	N. Bend, Oreg.	1867			A. M. Simpson	San Fr[illegible]
78816	1372	✳D	Melanesia (MISSING)	Woodhouse 85-85	I Bk	British	1224	226·9	36·2	21·5	Port Glasgow	1878	Russell & Co.		J. & W. Goffey	Liverpool
	1373	NV	Melanesia	Klocker 84-84	W Bk	Norweg'n	1281	200·9	41·9	22·4	Arendal	1884	A. Henriksen	[illegible]	C. Klocker	Arendal
	1374	DN	Melanie (ex Regulus)	Olsen	W Bg	Norweg'n	260	102·6	26·1	14·1	Burg, Germany	1846			Wilh. Klein	Fred[illegible]
74550	1375	✳	Melanope	Gray 82-87	I S	British	1608	258·2	40·2	23·8	Liverpool	1876	W. H. Potter & Co.		Australasian Shipping Co. Ld.	Liverpool
	1376	✳D	Melati	Van der Bos	I Bk	Dutch	1071	210·9	34·4	20·1	Newcastle	1885	Campbell, Mackintosh & Bowstead		J. Koning & Van Delden	Rotterdam
40578	1377		Melba (ex Nursery Packet)	Coutanche	W Bg	British	257	110·2	22·7	15·6	London	1858	Wigram		R. Tregaskes	[illegible]
	1378		Melbourne (ABANDONED)	Sandersen [-87]	W 3m Sr	Norweg'n	398	128·0	31·9	15·7	Cruos	1854		[illegible]	Chr. Ellefsen	Fredrikstad
90	1379		Melchior Vidulich (WRECKED)	Tonetti	W Bkn	Austrian	461	121·1	26·6	19·9	Lussinpiccolo	1881	N. Martinolich	[illegible]	M. Tonetti di G.	Fiume
1317	1380	IV	Melchiorre (ex Zarigo, ex Castro)	Isco	W Bk	Italian	791	167·3	33·4	21·5	Sestri Ponente	1875	F. Pezzani	[illegible]	D. Ciampa	Castell[illegible]
33814	1381		Melina	Veale	W Bn	British	132	93·4	21·9	12·9	Newfoundl'nd	1853	W. Knight		L. Veale	Cork
80658	1382		Melinda		W Sp	British	108	95·5	22·8	6·9	Quebec	1880	J. Matton		Eusebe Lussier	Montreal
85773	1383		Melinda		W Sp	British	102	94·0	22·4	6·7	Quebec	1883	H. Gill		Joseph Damran	Montreal
	1384	A	Melissa	Inglese	W Sr	Turkish	195	..	..	..	England	1872		[illegible]	G. Inglese	Samos
90490	1385	R	Melissa A. Willey	Willey	W 3m Sr	American	404	141·7	33·4	12·0	Thomaston, Me.	1874	Dunn & Elliot	[illegible]	Dunn & Elliot	Thomaston
91257	1386	R	Melissa Trask	Trask	W 3m Sr	American	225	115·6	29·0	9·6	Brewer, Me.	1880	Perkins	[illegible]	J. L. Cutter	Bangor, Me.
	1387		Melita		W Sr	Norweg'n	157	95·1	26·1	10·1	England				Smith & Thommesen	Arendal
	1388		Melleekettle Bahir		W Bmb	Turkish	137	..	..	..	Geuteh Gon	1884			Kalafat Oglou [illegible] Agha, Zerjow F. Hassan	Constantinople
	1389	N	Mellikoff (ex Cecilie)	Hansen	W Sr	Norweg'n	248	104·0	27·9	12·8	Åbo	1868			H. H. Pettersen	Arendal
	1390	H	Mellon		W Bn	Turkish	228	..	..	..	Syra	1867			H. A. Jordanidis	Constantinople
40086	1391	R	Melmerby	Olsen -88	W S	British	1487	207·3	40·5	21·0	St. John, N.B.	1863	Nevins	[illegible]	W. H. Ross & Co.	Liverpool
93350	1392		Meloun		Stl Lr	British	215	160·0	22·0	7·3	Dumbarton	1887	W. Denny & Bros.		Irrawaddy Flotilla Co. Ld.	Glasgow
80020	1393	✳	Melpomene	Leach 86-86	I S	British	1211	221·6	36·6	22·1	Sunderland	1876	Osbourne, Graham & Co.	[illegible]	T. Scott	Sunderland
	1394	✳V	Melpomene	Nielsen 82-88	I S	German	1039	205·3	33·1	20·8	Hamburg	1876	Reiherstieg Schiffsb. Ges.		B. Wencke Söhne	Hamburg
54	1395		Melpomeni (ex Sassa)	Kolakis	W Bg	Greek	199	92·5	25·3	..	Naples	1858			M. A. Kolakis	Spezzia
19027	1396		Melrose	Kelly	W Bk	American	944	166·5	35·0	23·1	East Boston	1863	J. Taylor		A. Packsher	[illegible]
	1397	DGV	Melusine	Mehlburger 82-82	W Bk	German	938	172·5	34·8	22·6	Bremerhaven	1876	J. C. Tecklenborg	[illegible]	F. Tecklenborg	[illegible]
90488	1398		Melville (ex Amber)	Hatton	W Sr	American	108	85·2	23·6	7·7	Brewer, Me.	18[illegible]			Swett & Co.	[illegible]
90579	1399		Melvin	Howland	W Sr	American	105	95·0	25·7	8·7	New Bern, N.C.	1873			Mrs. E. Ellis	New [illegible]
	1400	G	Memel (ex Hercules)	Wenzel -87	W Bk	German	376	123·6	28·0	18·0	Sunderland	1857			T. G. Wenzel	Memel

✳ = Lloyd's Register; D = Nederlandsche Vereeniging; G = Germanischer Lloyd; I = Registro Italiano; R = American [illegible]
A = Veritas Austro-Ungarico; H = Veritas Ellenico; N = Norske Veritas; V = Bureau Veritas

In the 4th Column, C = Composite (iron frame, wood planked); I = Iron; Stl = Steel; W = Wood.

The Dates of Metalling (Col. 11) are based on the records of the Books in which Vessels are respectively classed, or on the Reports of the Surveyors of [illegible]

Further particulars of Vessels classed by Lloyd's Register will be found in Lloyd's Register Book and Lloyd's Yacht Register.

No.	Cote	Nom. Capitaine et dates d'entrée. Ancien Nom.	Matériaux et Gréement	Pavillon	Tonnage net	Longueur	Largeur	Creux	Construit: Où.	Quand.	Par qui.	Date du doublage en métal	Armateurs ou Directeurs-Gérants.	Port d'Armement.
1401		Memento *White*	W Sr	American	144	94·0	27·9	9·0	Wareham, N.J.	1855			W. White	New York
1402	N	Memento *Andersen 75-75*	W Bk	Norweg'n	462	122·7	36·6	17·1	Langesund	1867	Kirkegaard	10.81	Chr. Erichsen	Langesund
1403	V	Memlo *Horn*	W Bk	British	954	172·6	35·5	20·8	River John, N.S.	1875	C. McLennan	V 9.87	C. McLennan	Picton, N.S.
1404		Memnon *Fisher -88*	W Bk	American	847	155·5	33·8	22·6	Boston, Mass.	1858	Briggs		L. Blum	San Francisco
1405	N	Memoria *Bolst -83*	W Bk	Norweg'n	383	120·9	30·9	16·2	Drammen	1868	A. Thjomme		Actieselskab "Memoria" (Elias F. Kiær)	Frederikstad
1406	N	Memory *Christiansen -84*	W Bk	Norweg'n	449	136·7	28·8	16·3	Kragerö	1873	Niels Nielsen	V 5.86	J. A. Samuelsen	Farsund
1407	DV	Mena (ex Dan) (CONDEMNED)	W Bg	Venezuel.	200	105·8	25·3	12·5	Kolding	1874	Chr. Hansen	V 6.86		
1408	✻	Menai *Lewis*	I S	British	1377	231·1	36·2	22·0	Sunderland	1880	R. Foster		Menai Ship Co., Ld. (W. Thomas & Co.)	Plymouth
1409		Menawa *Alexander 84-84*	W Sr	American	201	104·8	28·4	10·0	Greenpnt, N.Y.	1864			Henry Lord	Bangor, Me.
1410	V	Mendes (ex Vesuvio) *Garcia -87*	W Bg	Portug'se	257	109·7	23·4	13·9	San Martinho	1870		V 8.86	L. T. Mendes	Lisbon
1411	✻	Mendoza *Symons 87-87*	I S	British	495	170·5	28·0	17·7	Dundee	1861	Gourlay Bros.		T. Symons (S. Stevens)	Plymouth
1412	V	Mendoza *Simmons*	W Bk	American	790	170·7	36·2	17·3	Bath	1882	Goss, Sawyer & Packard	V 1.87	H. P. Booth	New York
1413	A	Menelaos *Palasoff*	W Bn	Russian	106	..	..	..	Chersonesus	1880		V 1.89	Sava Palasoff	Taganrog
1414		Menfi (now "Maria B.")	W Bn	Italian	326	126·3	26·5	15·0	Savona	1876	G. Calamaro	V 5.83	G. Bardi	Genoa
1415	DV	Menno *Schuring 83-*	W Sr	Dutch	133	86·0	22·3	10·4	Hoogezand	1865	L. Wildervank	Z 5.88	H. J. Plant & Co.	Schiedam
1416	✻	Mennock *Chalmers 84-84*	I Bk	British	787	188·4	32·1	19·2	Glasgow	1876	Lond. & Glas. Co.		Bramwell & Gardiner	London
1417		Menshikoff	W Bk	British	223	105·2	23·9	12·3	Baltimore	1847			C. G. Warburton	Sydney, N.S.W.
1418		Mentana *Guguen 83-87*	I S	French	1023	196·2	32·7	21·4	St. Nazaire	1867	John Scott		Ant. Dom Bordes & fils	Bordeaux
1419	N	Mentone *Bain -87*	W Bk	Norweg'n	769	173·9	35·0	19·5	Porsgrund	1878		V 5.87	J. C. & G. Knudsen	Porsgrund
1420		Mentor	W Bn	Brazilian	173	..	..	..					E. J. de Sá	Rio de Janeiro
1421		Mentor	W Bg	French	244	100·0	23·9	13·6	Nantes	1871	Fillé		Thiriot	Nantes
1422	N	Mentor *Amundsen -88*	W Bk	Norweg'n	524	142·8	30·5	17·1	Grimstad	1880	Ole Ribe	10.86	Halvor Ribe	Grimstad
1423		Mentor *Nygaard -82*	W Bg	Norweg'n	220	116·7	25·8	13·0	Flekkefjord	1851			S. Bergesen	Stavanger
1424	N	Mentor *Schöttelvig*	W Bg	Norweg'n	198	102·9	25·4	12·5	Drammen	1864	A. Pedersen		T. Schjöttelvig's Enke	Christiania
1425	V	Mentor *Persson 76-79*	W Bk	Swedish	302	124·0	24·8	15·0	Slesvig	1876	Abrahamsson	V 1.87	J. E. Olsson	Gothenburg
1426		Mera	W Bg	Turkish	270	104·9	25·4	18·0	Fiume	1858			Bet. Giulli	S. Giovanni di Medua
1427	✻	Mercator *White*	W Bk	British	294	116·0	25·5	15·6	Dundee	1864	Dundee Shp. Co.	Y 12.87	White & Co.	Dundee
1428	✻V	Mercator (ex Mercator Saint) *Lange 85-85*	I S	German	1396	216·8	36·1	25·0	Liverpool	1863	T. Royden & Sons		Knöhr & Burchard	Hamburg
1429	✻V	Mercator *Breckwoldt -84*	I Bk	German	807	188·3	31·9	18·6	Kiel	1884	Schiffsb. A.G. "Germania"		H. Hoh & Co.	Blankenese
1430		Mercator *Kittelsen*	W S	Norweg'n	807	143·2	37·4	23·4	Amsterdam	1840			N. Nielsen	Arendal
1431	N	Mercator *Oftedahl*	W Bk	Norweg'n	426	121·6	28·9	17·5	Königsberg	1867	L. Eggert	V 3.86	G. & A. Berg	Stavanger
1432		Mercator (ex Graf Bismarck, Graf Bismarck; Woods -89)	W Bk	Norweg'n	338	127·3	27·9	14·9	Bremen	1867	J. H. Knickmann	12.88	A. Wessels	Drontheim
1433		Merced	W Pol	Spanish	166	86·6	26·0	12·6	Lloret	1851			Juan Verdeguer	Valencia
1434		Mercedes	W Bk	Chilian	216	106·9	23·0	..					Mesa Hermanos	Constitucion
1435	V	Mercedes *Gaza*	W Bn	Spanish	194	85·6	23·3	13·9	Cano	18 58/77		C 7.85	Dª M. Terrer	Majorca
1436		Mercedes (ex Virgen del Mamon) *Millett*	W Sr	Spanish	157	85·0	25·4	12·6	Blanes	1868		C 6.88	F. O. Alsina	Barcelona
1437		Mercedes	W Sr	Spanish	135	81·6	24·1	11·3	Villajoyosa	1868			M. Zaragoza	Alicante
1438		Mercedes (ex Emilio)	W Pol	Spanish	128	78·1	24·1	11·5	Lloret	1854			Vda. y Hijos de J. M. Bonet	Barcelona
1439		Mercedes	W Bk	Swedish	340	125·7	26·4	15·1	Hamburg	1862	H. C. Stülken	V 2.85	J. Sjögren	Malmö
1440		Mercedes Elena ex Sarah (ex ABANDONED) Jackson	W Bn	Chilian	269	108·2	26·8	7·5	United States			V 2.86	P. E. Almeida	Valparaiso
1441		Mercedes Luisa (ex Emilia)	W Sr	Chilian	155	94·5	22·8	11·0	Preston	1866	Smith		J. de D. Peñafiel	Serena
1442		Merchant *Mitchell 69-69*	W Bg	British	282	119·0	26·6	14·1	Shoreham	1869	May		J. Sharp	Shoreham
1443	✻	Mercia *Morey 80-84*	I Bk	British	751	186·0	31·7	18·6	Sunderland	1876	Bartram, Haswell & Co.		P. Hick, jun.	Scarborough
1444		Mercur	W Bn	British	249	110·1	25·7	12·9	Elsfleth	1872	Bretjen		W. Dick	Shanghai
1445	V	Mercur *Hansen*	W Bn	Danish	141	101·0	21·7	10·5	Stege	1866	A. Petersen		C. Thuro	Svendborg
1446	V	Mercur *Gotk 82-82*	W Sr	Danish	118	82·1	20·6	10·5	Rudkjöbing	1868	S. Bois		N. H. Nissen	Rudkjobing
1447	DV	Mercur *Schönrogg*	W Bk	German	661	150·7	32·3	19·4	Bremerhaven	1865	J. C. Tecklenborg	V 1.8	J. N. Rodbertus	Barth
1448	V	Mercur *Heuentier -83*	W Bk	German	481	135·3	30·2	17·4	Barth	1876	J. S. Kraeft	V 6.8	J. N. Rodbertus	Barth
1449	G	Mercur *Lübbers 82-84*	W Sr	German	339	138·8	26·3	11·6	Manitowoc, U.S.	1876	Bond & Co.		Hamburgische Portland Cement Fabrik	Hamburg
1450		Mercur *Ewer 78-86*	W Sr	German	178	89·5	23·0	12·5	Neustadt	1864	J. C. L. Tornoe		C. Jahn	Neustadt

A = Lloyd's Register; D = Nederlandsche Vereeniging; G = Germanischer Lloyd; I = Registro Italiano; R = American Record; A = Veritas Austro-Ungarico; H = Veritas Ellenico; N = Norske Veritas; V = Bureau Veritas.

Dans la 4me Colonne, C = Composite (membrure en fer, bordé en bois); I = Fer; Stl = Acier; W = Bois.

Les Dates du doublage en métal (Col. 11) ressortent des livres dans lesquels les navires sont respectivement classés ou des rapports des Inspecteurs du Lloyd's Register.

Pour plus amples renseignements sur les navires classés par le Lloyd's Register, voir le Lloyd's Register Book et le Lloyd's Yacht Register.

Official No.	No. in Book	Registry in which classed	Ship's Name. Material and Rig. Master, and Dates of Appointment to present Owner's Service and to Ship. Ship's Late Name.	Flag.	Net Tonnage.	Length	Breadth	Depth	Built. Where.	When	By whom	Metalling	Owners or Managers	Port
	1451	G	Mercur *Dornbusch -83* W Sr	German	170	97·1	23·5	13·2	Wismar	1861	C.D.H.Hansen		J.C.Thormann	Wismar
	1452		Mercur *Brauer* I Cr	German	102	81·7	19·3	9·4	Renfrew	1866	Henderson Coul-[born&Co.		A.Sondermann	[illegible]
	1453	NV	Mercur *Frivold* W Bk	Norweg'n	717	160·9	33·4	19·9	Grimstad	1880	J.T.Tryde	V 8-88	J.Klöcker	[illegible]
	1454	N	Mercur (ex Scotland) *Larsen* [-87 W Bk	Norweg'n	536	129·3	32·0	18·1	Amsterdam	1842			Joh.Thorne	Moss
	1455		Mercur (W.... ....) *Zachariassen* W Bk	Norweg'n	464	136·3	32·2	16·2	Wasa	1859	A.Lomback		B.Condrup	[illegible]
	1456	N	Mercur *Brandt -85* W Bk	Norweg'n	339	120·7	29·2	16·1	Drammen	1861			G.C.HellesensEnke	[illegible]
	1457	N	Mercur *Granboe* W3mSr	Norweg'n	396	118·0	26·9	15·9	Baltimore	1858			A.Nielsen	Drammen
	1458		Mercur W Sr	Norweg'n	146	85·7	23·5	11·1	America				Olaves Nilsen	[illegible]
	1459		Mercur *Thomel* W Sr	Russian	194	91·0	24·0	12·5	Kurland	1882			A.Thomel	Riga
	1460		Mercure *Gigot* W Bg	French	173	89·6	23·0	13·8	St.Valery en [Caux	1875	Coquais		N.Pichard	St.Valery
	1461	V	Mercure *Lejardinier 88-87* W Bn	French	116	83·3	20·9	10·6	St.Malo	1869	Nouël		H.Lecharpentier	Granville
	1462	G	Mercurius (ex Mercur) *Kilsma -87* W Bk	Dutch	546	152·6	31·0	18·4	Quebec	1854			M.Van Neck	Nieuwediep
	1463	DN	Mercurius *Isaksen* W Bk	Norweg'n	585	153·0	33·5	17·7	Brevig	1870	P.Knudsen	Y 10·81	H.Ludvigsen	Skien
56610	1464	✻	Mercury (ex Rieve) *Hoare* W S	British	398	139·0	27·0	15·9	Aberdeen	1867	A.Hall&Co.	Y 6·88	C.A.R.Hoare	Aberdeen
63083	1465	✻D	Mercury *Thomas 59-71* W Sr	British	361	144·2	27·1	15·7	Aberdeen	1871	Duthie	Y 3·88	D.Thomas	Aberdeen
32168	1466		Mercury W Bg	British	150	94·5	20·8	11·4	Charleton,N.B.	1849			J.Tulloch	[illegible]
16948	1467		Mercury *Magune -87* W S	American	1156	173·0	39·9	22·5	New York	1851	J.A.Westervelt		Renton Holmes&Co	[illegible]
	1468		Merida (ex Pathfinder) *Andersen 86-86* W3mSr	Danish	270	117·4	26·3	11·2	Somerset	1852			N.R.Storm	Faaborg
	1469	✻	Merida (ex Tynemouth) *Nebendahl 84-86* I Bkn	German	277	137·5	26·2	12·4	Hayle	1881	Harvey & Co.		E.B.Crasemann	Hamburg
45	1470		Meridian *Lange -86* W3mSr	Siamese	294	119·5	25·4	13·0	Bangkok	1863			Poh Chu	Bangkok
	1471	✻	Meridian *Ekelund -88* W Bk	Swedish	389	127·5	28·7	17·6	Sunderland	1866	W.Pickersgill	Y 3·88	G.F.R.Ekelund	Wisby
	1472	V	Méridien *Poirier* W Bk	French	549	135·8	27·9	18·5	Bordeaux	1869	Roy	Y 6·87	Brunellière frères	Nantes
74486	1473	✻	Merioneth *Thomas 74-80* I S	British	1366	231·4	38·9	23·6	Liverpool	1875	T.Royden&Sons		Hughes & Co.	Liverpool
	1474		Merjan W Bg	Turkish	413	..	..	..	Sinope	1872			Trebizondlee Tijar [bash & Co.	Trebizond
	1475		Merkki (ex Merck) *Sundvik 81-* I Bk	Russian	280	102·1	27·9	16·7	Lübeck	1840			K.G.Silander	Nystad
48302	1476	✻	Merle *Davies -87* I Bkn	British	281	141·7	25·3	12·8	London	1863	Westwood		C.T.Bennett	Bristol
16788	1477		Mermaid W Bk	American	260	112·6	24·3	15·4	Westprt.Mass	1855	G.&F.Sisson		E.Gifford	[illegible]
	1478	G	Mermaid *Niemann 85-88* W Bk	German	467	139·7	30·1	16·3	Madison,N.B.	1852			H.Mentz,jun.	Rostock
67904	1479	✻	Mermerus *Coles 78-80* I S	British	1671	264·2	39·8	23·7	Glasgow	1872	Barclay,Curle [&Co.		A.&J.H.Carmichael&Co	Greenock
	1480		Merogat W Bk	Norweg'n	423	126·7	29·0	16·8	Risör	1874			E.J.Ellingsen	[illegible]
90070	1481		Merom *Glover* W S	American	1204	172·3	37·6	24·0	Phippsbrg.Me	1870	C.V.Minott	Y 10·81	C.V.Minott	Bath,Me.
63505	1482	✻	Merope *Thomas 85-85* C Bk	British	1054	203·3	35·1	20·5	Sunderland	1870	Oswald&Co.	Y 8·88	Shaw Savill&AlbionCo. [Ld.	[illegible]
268	1483		Merope W Bk	Greek	346	127·9	26·6	..	Castellamare	1868			N.Vouroecas	Syra
	1484		Merope W Bg		152	77·8	23·9	..	Cassos	1865				
17690	1485		Merrill C. Hart W Sr	American	178	97·9	26·7	9·8	Bristol,Me.	1866			G.W.Rawley	[illegible]
88419	1486	V	Merritt *Johnson* W Bkn	British	449	148·6	32·3	13·0	King'sCo.N.B.	1884	G.Merritt	Y 9·87	R.C.Elkin	[illegible]
27224	1487	RV	Mersey (ex Steamer) *Dann -87* IbnBkn	British	963	260·4	30·2	16·5	London	1859			A.Major	London
83491	1488	R	Mersey Belle *Henson* [-85 W Bn	British	299	117·0	28·0	12·6	Liverpool,N.S.	1883	R.McLeod		S.T.R.B. [illegible]	[illegible]
38762	1489		Mertola *Green* W Bk	British	392	144·2	28·9	14·1	Pugwash,N.S.	1866	Jones		D.R. [illegible]	[illegible]
2257	1490		Merton (LOST) *Sheppard 85-85* W Sr	British	110	71·7	20·9	12·2	Sunderland	1819			R.J.Hoad	Rye
67641	1491	✻	Mervinia *Jones -78* W Sr	British	97	84·0	21·6	10·9	Aberdovey	1878	Richards	Y 3·88	M.Owens & Co.	Aberystwith
	1492	G	Mervyn *Collaki 84-* C Bk	Greek	288	115·1	24·1	15·0	Sunderland	1866	G.Moore & Co.	Y 10·81	P.Collaki	Piraeus
	1493		Mésange *Guerrant -87* W Bk	French	322	130·9	26·6	13·0	La Seyne	1875		Y 9·81	Guerrand & Le Port	[illegible]
	1494	V	Mésange *Hoguet 80-85* W Sr	French	132	93·5	22·3	11·2	Gravelines	1885	Collin		R.Torris	[illegible]
	1495	H	Messaret (ex Dionysios) W Bg	Turkish	198	81·6	26·1	15·8	Syra	1868		Y 1·80	Niaji Bey	[illegible]
	1496	NV	Messel *Nygaard -87* W Bk	Norweg'n	407	131·5	30·0	16·5	Arendal	1870	J.Aanensen	Y 8·88	S.Eyde	Arendal
43041	1497		Messenger *Wilson* W Bg	British	228	101·0	26·2	15·4	Sunderland	1862	Rawson& [Watson		Bessey & Palmer	[illegible]
24726	1498		Messenger *Eyres* W3mBn	British	211	102·6	21·9	13·7	Aberdeen	1839	A.Duthie&Co.		A.[illegible]	Perth
24093	1499	✻	Messenger *Williams* W Sr	British	82	68·8	16·9	10·1	Aberystwith	1841	J.Evans		J.Williams	Aberystwith
91301	1500	R	Messenger *Falkner* W3mSr	American	338	132·1	31·5	10·3	Bath,Me.	1881	W.Rogers	Y 7·83	R.F.C.Hartley	[illegible]

✻ = **Lloyd's Register**; D = Nederlandsche Vereeniging; G = Germanischer Lloyd; I = Registro Italiano; R = American [Record]
A = Veritas Austro-Ungarico; H = Veritas Ellenico; N = Norske Veritas; V = Bureau [Veritas]

In the 4th Column, C = Composite (iron frame, wood planked); I = Iron; Stl = Steel; W = Wood.

The Dates of Metalling (Col. 11) are based on the records of the Books in which Vessels are respectively classed, or on the Reports of the Surveyors of Ll[oyd's]

Further particulars of Vessels classed by Lloyd's Register will be found in Lloyd's Register Book and Lloyd's Yacht Register.

1	2	4 Nom. Capitaine et dates d'entrée au service des Armateurs actuels et à bord du navire. Ancien Nom.	Matériaux et Grèement	5 Pavillon.	6 Tonnage net.	7 Dimensions. Longueur.	Largeur.	Creux.	8 Construit. Où.	9 Quand.	10 Par qui.	11 Date du doublage en métal	12 Armateurs ou Directeurs-Gérants.	13 Port d'Armement.
1500		Messina [illegible] (ex [illegible])	I 3mSr	British	376	[illegible]	21·9	13·5	Glasgow	1862	C.Connell&Co.		J.Luke	Dundee
1501	NV	Messina *Nielsen -88*	WSr	Norweg'n	205	114·4	25·2	11·9	Skien	1871	Hansen	Y 7.8?	J.Nielsen	Porsgrund
1502	V	Messina *Johansson -87*	WBk	Swedish	275	114·8	24·8	12·7	Uddevalla	1875	M.Halten	Y 11.87	A.Lysell	Lysekil
1503		Messoodeh	WBg	Turkish	109	..	..	..	Sakarieh	1872			Hattabji Halil Effendi & Co.	Sakarieh
1504		Messoud	WBg	Turkish	202	..	..	..	Greece	1886			Hassan Pasha Zadeh Hasher & Co.	Constantinople
1505		Messoud	WBg	Turkish	187	..	..	..	Greece	1885			Yorghanjiler Kiyassi Hajee Mustapha Effendi	Constantinople
1507	GV	Meta *De Vries -81*	WSr	German	246	103·0	25·2	11·7	St.Magnus	1869	L.Hilmers&Co.	Y 2.87	B.de Vries	Neermoor
1508		Meta *Rittberger 87-87* (ex [illegible])	WSr	German	132	91·3	22·9	10·8	Brake	1858	J.Oltmanns		M.Rittberger	Geestemünde
1509		Meta *Halvorsen*	WBk	Norweg'n	521	129·5	31·2	18·9	Vegesack	1842			H.Aalholm	Arendal
1510	N	Meta *Torresen 85-85* (ex Explorer)	WBk	Norweg'n	483	133·2	29·5	17·8	Memel	1865	B.Pieper	4.4.8?	D.Davidsen	Farsund
1511	G	Meta *Heinberg -83*	WBk	Russian	428	122·1	29·1	17·5	Barth	1865	A.Witte		C.Frischmuth	Windau
1512		Metamorcos	WBg	Turkish	105	..	..	..	Oonia	1870			Ooniaige Akandol oglou Hajee Yorghi & Bros.	Oonia
1513		Metamorphosis	WBg	Greek	205	82·0	25·0	..	Syra	1867	Theofanis		C.Barberis	Syra
1514		Meteor *Smith 84-88*	WBk	British	248	120·0	24·1	13·5	Greenock	1862	R.Steele	Y 8.88	J.& W.Stewart	Greenock
1515		Meteor *Lamp*	WSr	British	142	88·5	19·3	11·0	Liverpool	1835			Mersey Docks & Harbour Board	Liverpool
1516		Meteor (ex [illegible])	WSr	American	116	86·0	22·0	9·0	Scituate,Mass.	1864			A.F.Tift	Key West.Fla.
1517	G	Meteor *Niemann*	WBk	German	595	145·6	31·3	19·4	Vegesack	1864	C.Lange		E.Burchard	Rostock
1518	N	Meteor *Jensen*	WBk	Norweg'n	440	133·1	29·9	16·2	Porsgrund	1869	P.Grøtnes	Y 4.8?	O.A.Olsen	Arendal
1519	N	Meteor *Breseid -86*	WBk	Norweg'n	330	133·3	28·3	11·7	Carlscrona	1864	F.Bergström		K.Finsdal	Mandal
1520		Methodius *Karlsson*	WBg	Russian	277	..	..	..		1864			J.Hägg blom	Hammarland
1521	*	Metropolis *Williams -87*	I Bk	British	1759	265·7	40·0	22·9	Liverpool	1887	R.&J.Evans		Metropolis Ship Co.,Ld. (W.Thomas & Co.)	Liverpool
1522	DV	Metsola *Larsen -88*	WBk	Norweg'n	585	139·6	34·5	18·9	Brahestad	1870	A.Wikland		J.Gjersöe	Christiania
1523	AV	Metta *Staugher*	WBk	Austrian	617	143·7	33·2	24·9	Fiume	1879	F^lli Schiavoni	Y 7.87	A.Staugher & figli	Volosca
1524	DV	Mette *Paulsen 73*	W3mSr	Danish	217	120·6	21·9	11·8	Randers	1874	T.Shønau	Y 1.86	Actieselskab(J.H Christiansen)	Nordby,Fanø
1525	DV	Mette *Thørgersen 75-75*	WBn	Danish	130	90·6	21·2	10·5	Middelfart	1875	H.Illum	Y 8.88	H.Svarrer & Co.	Fanø
1526	V	Mette *Toft -86*	WBn	Danish	117	86·1	20·2	10·2	Nordby,Fanø	1871	N.Nielsen	Y 11.8?	Akt. Selsk. "Mette" (J. N. Jessen)	Nordby,Fanø
1527	V	Mette Johanne *Christensen*	WBn	Danish	166	99·5	22·2	11·1	Nordby,Fanø	1876	N.Nielsen	Y 4.8?	Akt. Selsk. "Mette Johanne" (C. H. Nielsen)	Nordby,Fanø
1528	V	Mette Kathrine *Fischer*	WBn	Danish	132	91·5	21·8	10·2	Nordby,Fanø	1875	H.J.Bolwig	Y 4.8?	Actieselskab (S.A. Fischer)	Sønderho,Fanø
1529	DV	Mette Katrine *Ankersen*	WBn	Danish	157	97·9	21·6	10·9	Svendborg	1876	Schmidt& Andersen	Y 4.8?	Actieselskab (L.Ankersen)	Nordby,Fanø
1530	G	Metz *Schülke 88-88* (ex Louis Kayser)	WBn	German	228	99·4	25·9	14·1	Rostock	1868	C.Schönemann		Carl Lobeck	Stralsund
1531		Mexican Packet	WBn	British	144	79·7	17·4	11·8	Birkenhead	1840			W.Jackson	Liverpool
1532		Mexico (ex Velasco) *Abel*	WBn	Mexican	222	104·0	23·0	12·2	Arendal	1875	H.Henricksen	Y 3.8?	G.Gonzales	Campeche
1533	V	Mexico *Åkerlund 88-88* (ex Kaiser)	WBg	Swedish	622	143·2	31·8	18·5	Brahestad	1865	E.Danielson		O.Heilborn	Stockholm
1534	R	Meyer & Muller *Perkins -83*	W3mSr	American	400	150·0	35·8	10·5	Belfast,Me.	1883	CarterBros.		J.H.Perkins	Belfast,Me.
1535		Mia Madre *Arena* (ex Di Cagliari)	WBg	Italian	146	86·6	24·8	13·0	Malta	1850		Y 8?	R.Schiaffino	Genoa
1536		Miako *Wood 82-85*	C Bk	British	516	160·1	30·1	17·1	Sunderland	1869	W.Pile&Co.	Y 9.88	T.Roberts	London
1537	V	Miaoulis *Revidis* (ex Angelina)	WBg	Greek	269	100·4	26·9	16·6	Syra	1884	A.Livaláros		T.Revidis	Syra
1538		Miceli *Serra*	WBg	Italian	113	78·7	22·5	11·7	Malta	1851			B.A.Spano di S.	Trapani
1539	G	Michael *Marx 81-*	WBk	German	380	122·0	28·5	17·6	Elbing	1854	M.Mitzlaff		Carl Wichards	Stettin
1540		Michael (ex Calliope)	WBg	Greek	230	88·9	25·0	..	Mykonos	1878			S.Denaxas	Santorin
1541	H	Michael *Roussos*	WBg	Greek	198	88·3	24·6	15·0	Syra	1868	A.Pontzos	Y 10.83	D.I.Roussos	Zante
1542		Michael	WBg	Greek	187	85·6	23·4	11·1	Syra	1870	P.Andrikos		G.Moostakis	Zante
1543	H	Michael	WBg	Greek	183	94·5	24·5	13·4	Galaxidi	1880	C.Papapetros		C.Papapetros	Galaxidi
1544	DV	Michael *Birk*	WSr	Russian	254	111·1	28·7	13·6	Gutman'sbach	1875	A.Frem		M.Klein	Gutmannsbach
1545	V	Michael *Andersohn -86*	WSr	Russian	228	102·8	27·2	12·7	Angern	1884	Kirstein		J.Jansohn	Riga
1546		Michael	WSr	Russian	215	..	..	..		1875			F.Markoff	Sordavala
1547		Michael *Kaasmar -78*	WSr	Russian	130	76·0	20·0	8·0	Haynasch	1877	C.Ohseling		J.Mikkelsohn	Haynasch
1548		Michael *Burack*	WSr	Russian	120	84·4	22·8	11·2	Pernau	1872			H.Raenek	Pernau
1549	*	Michael Hutchinson *Musgrove 87-87?*	I Bk	British	685	175·2	30·1	19·3	Harrington	1867	R.Williamson & Son		J.Bell & Son	Liverpool
1550		Michael Taxiarchis	WBg	Greek	207	90·5	24·0	..	Piræus	1860			G.Fotias	Piræus

* = Lloyd's Register; D = Nederlandsche Vereeniging; G = Germanischer Lloyd; I = Registro Italiano; R = American Record; A = Veritas Austro-Ungarico; H = Veritas Ellenico; N = Norske Veritas; V = Bureau Veritas.

Dans la 4me Colonne, C = Composite (membrure en fer, bordé en bois); I = Fer; Stl = Acier; W = Bois.

Les Dates de doublage en métal (Col. 11) ressortent des livres dans lesquels les navires sont respectivement classés ou des rapports des Inspecteurs du Lloyd's Register.

Pour plus amples renseignements sur les navires classés par le Lloyd's Register, voir le Lloyd's Register Book et le Lloyd's Yacht Register.

Official No.	No. in Book	Registry in which classed	Ship's Name. Master and Dates of Appointment to present Owner's Service and to Ship. Ship's Late Name.	Material and Rig	Flag	Net Tonnage	Length	Breadth	Depth	Where built	When	By whom	Date of Metalling	Owners or Managers	Port
	1551		Michael Treschow *Larsen 70-79*	W Bg	Norweg'n	203	99·6	25·4	13·2	Laurvig	1855	A.L.Hunberg		O.Moe	Fredriks
	1552		Michailos *Diassakis*	W Bn	Greek	178	..	..	..	Cassos	1879			M.Diassakis	Cassos
1080	1553	I	Michelangelo di Pieve *D. Benvenuto*	W Bk	Italian	444	121·3	30·8	19·2	Savona	1867	G.Tisi	[illegible]	Em.Benvenuto	Genoa
1350	1554	I	Michele *Cacace -87*	W Bk	Italian	513	137·8	30·2	19·3	Alimuri	1883	G.Starita	[illegible]	M.S.Cacace	Castellam.
1562	1555		Michele *(ex Solari)*	W Bk		376	129·6	28·8	16·7	Voltri	1866	B.Fabiani	[illegible]		
1394	1556	IV	Michele A. *Trapani -87*	W Bk	Italian	708	158·8	30·5	20·3	Sorrento	1884	A.Castellano	[illegible]	S.& O.Maresca	
3344	1557	I	Michele B. *Bertolotto -88* *(ex Waddell ...)*	W Bk	Italian	655	150·9	32·0	21·3	Recco	1873	P.Rolla	[illegible]	G.Bertolotto	Genoa
1333	1558	I	Michele Dapelo *(ex ...)*	W Bk	Italian	428	128·2	34·4	19·6	Savona	1866	P.Sirello	[illegible]	C.A.Dapelo	Genoa
2021	1559		Michele Padre *Cannavacciuolo*	W Bg	Italian	117	86·6	23·2	11·8	Torre del Greco	1885	F.Iadbrano fu F.		C.Albanese fu M.	Naples
86294	1560	*	Micronesia *Greig 78-82*	I S	British	1577	258·0	38·2	22·9	Greenock	1883	Russell&Co.		J. & W.Goffey	Liverpool
71712	1561	*D	Mid Lothian *Marchie 88-88*	I S	British	1085	220·0	34·6	20·7	Glasgow	1875	C.Connell&Co		J.Boyd	Glasgow
	1562	N	Mida *Tallaksen -87*	W Bk	Norweg'n	439	131·5	28·2	17·6	Trieste	1858		[illegible]	Hans Johnsen	
[illegible]	1563	R	Midas *Whiddon -88*	W Bk	British	839	173·4	35·5	19·8	Maitland,N.S.	1870	A.McAllum	[illegible]	J.Northup	
[illegible]	1564	*	Middlesex *Innes -88*	I S	British	1742	268·4	38·8	23·6	Glasgow	1884	Barclay,Curle [&Co.		G.Marshall & Sons	London
47770	1565		Middleton *H.H*	W Sw	British	284	108·7	26·3	16·5	Sunderland	1863	Pearson		W.Hill	North
71429	1566		Midland Rover *(ex ...)*	W Sr	British	354	140·0	30·0	12·4	Cleveland,U.S	1863	G.H.Notter		T.Marks	
	1567	*DN	Midnatssol *Nielsen*	Stl Bk	Norweg'n	1216	219·4	35·7	21·4	W.Hartlepool	1885	W.Gray&Co.		F.Smith Petersen	
	1568	A	Mie Figlie *Stipanovich* *(ex ...)*	W Bk	Austrian	781	180·0	31·3	20·1	Cardiff	1870	Batchelor	[illegible]	E.Colulovich	
[illegible]	1569	I	Miei Nipoti *Borzone*	W Bk	Italian	865	173·8	31·7	21·5	Varazze	1883	G.Scala	[illegible]	P.Lavarello	Genoa
[illegible]	1570	V	Mignonette *Boulanger -86*	W Sr	British	139	96·5	25·5	11·0	St.Etienne	1886	E.J.Price		E.J.Price	Quebec
	1571	V	Mignonne *Chauvel -88*	W Sr	French	117	88·6	19·7	10·6	St.Malo	1879	L.Tranchemer		Veuve G.Gautier	St.Servan
	1572		Mignonne *Mary -88* *(ex C. A. Meniac)*	W Sr	French	106	83·9	23·6	9·8	PtMedwayNS	1877	J.Meniac		Bichet & Yon	St.Malo
	1573		Miguel Abbatangelo *(ex Emma Zibike)*	W Sr	Argentine	161	88·5	24·0	12·3	Stettin	1869	A.Sietz		D.Abbatangelo	Buenos
	1574		Mihail *Panayottis -87*	W Bk	Greek	242	90·2	24·9	21·0	Syra	1872			Stefanos Panayottis	Syra
	1575		Miikula	W Sr	Russian	111	..	..	..		1872			F.Petroff	
[illegible]	1576	*	Mikado *Rann*	C Bk	British	643	166·3	31·0	18·8	London	1868	Walker&Co.	[illegible]	S.Fegan	London
	1577		Mikkula *Hinsch*	W Sr	Russian	104	..	..	..		1868			A.Nissin	
[illegible]	1578	R	Mildred J. McLean *Murphy -88*	W Sr	British	148	105·0	26·0	10·1	JordanRiver,N.S.	1888	R.McDonald		G.W.McLean	
	1579	N	Miletus *Petersen*	W Bg	Norweg'n	247	113·1	27·5	12·0	Tvedestrand	1871	LarsTellefsen		P.Svarstad& Th.Thorsen	Drammen
[illegible]	1580	R	Milford *Look*	W 3m Sr	American	216	117·0	28·5	9·3	Rockland,Me	1881	Cobb		Cobb,Wight & Co.	
51	1581	A	Milizza *Gairgavich* *(ex ...)*	W Bk	Austrian	390	130·2	27·6	16·5	Capod'Istria	1869	L.Polli	[illegible]	L.Cetcovich fu T.	Trieste
	1582	N	Millia *Sandvig*	W Sr	Norweg'n	128	91·5	23·1	10·4	Bergen	1874			R.Lindöen	
[illegible]	1583	*	Millie Bain *Steer -84*	W Bn	British	232	124·1	25·2	13·1	Hayle	1872	Harvey	[illegible]	D.W.Bain & Co.	
	1584	A	Millosck *Passalisky*	W Bk	Russian	212	95·8	24·9	16·5	Fiume	18[illegible]		[illegible]	L.Damianovich	
[illegible]	1585	R	Millville *Branin -87*	W 3m Sr	American	347	131·0	32·2	11·9	Camden,N.J.	1887	Morris&Mathis		R.D.Wood & Co.	
[illegible]	1586	*	Millwall *Anderson 82-84*	I Bk	British	1165	214·8	34·6	23·0	London	1863	Millwall I.W.C.		T.Law & Co.	
[illegible]	1587		Milo	W Bn	British	155	93·6	23·6	11·6	Londonderry,[N.S.	1857			F.J.Cortissoz	
[illegible]	1588		Milo	W Bg	British	127	89·9	23·6	11·0	Cornwallis,N.S	1854			N.S.Townsend	
[illegible]	1589		Milo *Lewis*	W Bn	British	114	88·6	20·2	10·5	Plymouth	1861	Banks		W.Murton	
[illegible]	1590	*	Miltiades *Ayling 67-83*	I S	British	1452	240·5	39·3	23·3	Aberdeen	1871	W.Hood&Co.		G.Thompson & Co.	Aberdeen
[illegible]	1591	*	Miltiades *Vincent*	Stl Bk	British	1383	230·6	37·0	21·8	Sunderland	1885	J.Laing		T.Scott	
	1592	AH	Miltiadis *Benjamin* *(ex ...)*	W Bk		474	138·4	30·5	..	Gravosa	1871	M.Capor	[illegible]		
[illegible]	1593	R	Milton *Potter*	W Bk	American	373	110·5	28·6	18·5	Milton,Mass	18[illegible]		[illegible]	Tabor & Gordon	
	1594	N	Milton *Krüger -73*	W Bk	Norweg'n	477	111·6	34·3	16·9	Arendal	1874	H.Stiansen	[illegible]	Chr.Th.Boe	Arendal
[illegible]	1595	*	Miltonpark *Hamilton 81-82*	I S	British	1461	255·7	38·3	20·6	Dumbarton	1882	A.McMillan&Son		NorthBritishShippg.Co.	Glasgow
	1596	*	Milufer *Omer*	W Bk	Turkish	574	..	..	..	Lussinpiccolo	1865		[illegible]	Sadyk Bey	
[illegible]	1597	*	Milverton *Adams 78-86*	I S	British	2141	280·0	40·2	25·1	Southampton	1886	Oswald,Mordaunt&Co.	[illegible]	F.Nodin	Liverpool
[illegible]	1598	R	Mima A. Reed *Dixon -87*	W 3m Sr	American	395	121·0	31·4	11·8	Harrington,Me	1874	Ramsdell&[Rumbal	[illegible]	O.P.Rumball	
	1599	NV	Mimer *Hansen 73-85*	W Bk	Norweg'n	419	138·6	29·9	15·9	Drammen	1869	Birgensen&[Knudsen	[illegible]	Ths.S.Falck	
	1600	N	Mimer *Amundsen 87-87*	W Bk	Norweg'n	330	109·9	28·8	15·8	Risör	1859	Knud Hansen		W.Wright	

*=Lloyd's Register; D=Nederlandsche Vereeniging; G=Germanischer Lloyd; I=Registro Italiano; R=American [Record]; A=Veritas Austro-Ungarico; H=Veritas Ellenico; N=Norske Veritas; V=Bureau [Veritas]

In the 4th Column, C=Composite (iron frame, wood planked); I=Iron; Stl=Steel; W=Wood.

The Dates of Metalling (Col. 11) are based on the records of the Books in which Vessels are respectively classed, or on the Reports of the Surveyors of [Lloyd's Register].

Further particulars of Vessels classed by Lloyd's Register will be found in Lloyd's Register Book and Lloyd's Yacht Register.

		Nom. Capitaine et dates d'entrée au service des Armateurs actuels et à bord du navire. Ancien Nom.	Matériaux et Gréement.	Pavillon	Tonnage net.	Dimensions			Construit			Date du doublage en métal	Armateurs ou Directeurs-Gérants.	Port d'Armement.
						Longueur	Largeur	Creux.	Où.	Quand.	Par qui.			
1601	A	Mimi *Teraicich*	W Bk	Austrian	498	135·5	28·0	19·2	Fiume	1865	F[lli] Schiavoni	[illegible]	A.F.Poschich	Fiume
1602	DV	Mimi *Meyer -82*	W Bk	German	789	162·1	34·9	20·2	Elsfleth	1882	J.Jürgens	[illegible]	J.E.Schäffer	Elsfleth
1603		Mimina	W Bk	Greek	410	122·7	23·0	..	Syra	1871			A.M.Nicolius	Syra
1604	V	Mimosa *De Barros*	W Bk	Brazilian	287	122·7	25·9	10·9	Villa do Conde	1870			B.Oliveira & Co.	Pernambuco
1605		Mina *Stirling* (ALEXANDER ... ex Vorwärts & Adelpa ...) *-87*	W Bg	British	258	107·9	25·1	12·1	Ronnebeck, Germany	1867	Ullmans		J.P.Stirling	Sydney, N.S.W.
1606		Mina	W Bn	British	208	112·2	27·4	12·5	Long Reach N.B.	1865			Joseph Graham	St.John, N.B.
1607		Mina *Reinertsen -88*	W Bk	Norweg'n	723	146·9	34·1	20·7	Rotterdam	1867	De Jong, Kortlandt & Co.	[illegible]	A.Berge	Farsund
1608	R	Mina Belle *Thompson*	W 3m Sr	American	445	131·2	32·4	12·8	Machias, Me.	1883	J.Shaw		I.Sargent	Machias, Me.
1609		Minatogawa Maru	W Sr	Japanese	148	88·5	21·0	10·2	Hiogo	1882	Kinoshita Benzo		Iwaki Kikusaburo	Tokio
1610	N	Minde *Andersen -85*	W Bk	Norweg'n	376	125·3	29·4	15·6	Tvedestrand	1877		[illegible]	J.M.A.Marcussen	Tvedestrand
1611	DV	Minden *McDonald -88*	W Bk	British	1287	194·7	39·4	23·5	Horton, N.S.	1879	J.B.North	[illegible]	Allison Smith	Windsor, N.S.
1612	N	Mindet *Sørensen -88*	W Bg	Norweg'n	269	96·2	28·4	14·9	Kragerö	1848			P.Sörensen	Arendal
1613	V	Mindin *Lelouie -82*	I Bk	French	444	135·5	26·4	17·0	Nantes	1868	E.Goüin & Co.		A.Viot	Nantes
1614		Mindora	W S	British	974	174·6	35·4	23·1	St.John, N.B.	1865			The North Illawarra Coal Mining Co., Ld.	Sydney, N.S.W.
1615		Mindora *Billings*	W Sr	American	116	87·6	25·0	8·0	Eden, Me.	1855			J.W.Somers	S.W.Harbor, Me.
1616	N	Mindora *Olsen -88*	W Bk	Norweg'n	295	122·2	28·7	13·9	Richmond, S.I.	1856		[illegible]	Johan Hansen	Brevig
1617	R	Mindoro *Powers -87*	W S	American	971	169·0	38·8	23·5	East Boston	18 64/84	Taylor	[illegible]	W.D.& D.L.Pickman	Salem, Mass.
1618	V	Mine *Møller 85-85*	W Sr	Danish	125	90·8	21·1	10·3	Thurø	1885	J.Ph.Jørgensen		J.Ph.Jorgensen	Svendborg
1619	✻	Minero *Davies 83-83*	I Bk	British	478	150·7	27·5	16·8	Glasgow	1866	Stephen		Trinder, Anderson & Co.	London
1620		Minerva	W Sr	British	135	85·7	21·0	10·0					Thos.Winniett	Hamilton, Ber.
1621		Minerva *Pereira*	W Bn	Brazilian	220	..	..	..	Mundaud	1867		[illegible]	Liberato Irmãos	Itajahy
1622		Minerva	W Bn	Brazilian	185	..	..	..					J.M.do Valle	Sa.Catharina
1623	V	Minerva *Olsen -88*	W Bk	Danish	275	115·4	26·1	13·4	Stettin	18 43/52	F.E.Z.Zieske		R.A.Holst	Svaneke
1624	DG	Minerva *Albers*	W Bk	Dutch	319	114·0	26·6	14·2	Hamelwarden	1869	W.Hagens	[illegible]	Jan G.Albers	Farmsum
1625		Minerva *Schwalfeldt* (ex REB... N.V...)	W Bg	German	417	118·3	29·3	18·1	Memel	1857	B.Pieper		J.N.Rodbertus	Barth
1626	V	Minerva *Tasletta -88*	W Bk	Italian	694	154·1	33·7	21·6	Sampierdarena	1873	G.Gallo	[illegible]	F.Figari	Genoa
1627	I	Minerva *Bronzi*	W Bn	Italian	126	89·2	24·3	11·5	Lerici	187	F.Medusei	[illegible]	C.Fabbricotti fu Dco.	Spezia
1628	N	Minerva *Hansen 81-82*	W Bk	Norweg'n	513	117·0	31·7	17·2	Kragerö	1869	L.Larsen	[illegible]	L.Larsen	Kragerö
1629	N	Minerva *Rød -87*	W Bk	Norweg'n	468	111·0	32·2	19·5	Risör	18 09/62			O.O.Rund	Fredriksstad
1630	N	Minerva *Isaksen -87* (ex Leopold (L))	W Bk	Norweg'n	440	128·9	26·8	17·1	Quebec	1861			O.Chr.Olsen	Tönsberg
1631		Minerva *Gundersen* (ex ...)	W Bk	Norweg'n	362	132·2	27·8	14·9	Bergen	18 51/63	Brunchorst & Dekke		C.Feilberg	Langesund
1632	N	Minerva *Sandersen -87*	W 3m Sr	Norweg'n	266	112·9	26·2	13·9	Harburg	1860		[illegible]	Z.Möller	Fredriksstad
1633		Minerva *Pedersen -82* (BROKEN ...)	W Bk	Norweg'n	164	95·7	21·9	11·5	Drontheim	1856	M.Voigt		J.M.A.Marcussen	Tvedestrand
1634		Minerva *Olsen* (...)	W 3m Sr	Norweg'n	140	89·5	21·5	10·1	Prussia				H.& E.Edvardsen	Drammen
1635		Minerva *Wermolin 76-76*	W Bg	Russian	209	..	..	..	Sideby	18 3?/5?			G.Rivell	Nystad
1636		Minerva	W Bk	Spanish	306	111·5	25·6	13·7	Tayabas	1862			Doña C.Lichanco	Manila
1637		Minerva	W Bn	Spanish	133	86·6	23·7	12·5	Caraminal	1866			S.Soler y Paz	Villagarcia
1638		Minho *Marques -87* (... Admiravel)	W Bk	Portug'se	260	101·0	23·0	14·4	Lisbon	1856	J.A.da Silva	[illegible]	Cia Alliança Maritima Portuense	Oporto
1639		Minihic *Colas -88*	W Bn	French	123	82·3	22·0	10·2	St.Pierre-Miquelon	1877	L.Hubert		L.Hubert	St.Pierre-Miquelon
1640	GD	Minister Camphausen *Ades 73-73*	W Sr	German	119	86·6	21·2	9·7	Grossefehn	1873	T.J.Cassens & Sohn		Tönjes J.Cassens & Sohn	Grossefehn
1641	RV	Minister of Marine *McLaughlan -85*	W S	British	1648	217·1	40·7	25·0	Tynemouth N.B.	1874	J.S.Parker	[illegible]	H.D.Troop	St.John, N.B.
1642		Minmi	W Bn	British	185	120·0	29·0	7·0	Newcastle, N.S.W.	1861	T.Adam		W.H.Smith & Sons	Sydney, N.S.W.
1643		Minna	W Bk	Chinese	435	131·9	30·7	17·0	Apenrade	1871	J.Paulsen		Totai	Foochow
1644	V	Minna *Zeplien 83-83*	W 3m Sr	German	467	135·3	29·8	16·9	Ueckermünde	1862	C.Radmann	[illegible]	H.Maass	Rostock
1645		Minna *Steinorth*	W Bg	German	173	92·2	23·8	12·7	Barth	1863	J.S.Kraeft		J.C.Beug	Barth
1646	✻	Minna Elkan *Griffith* (ex Plas...) *70-85*	W Sr	British	138	91·0	23·3	11·6	Nevin	1872	Owen		T.Griffith	Carnarvon
1647	V	Minna Helene *Mangler -86*	W Bk	German	340	126·0	27·7	15·6	Memel	1876	L.Eggert	[illegible]	Siebert & Alexander	Memel
1648		Minnehaha *Boggs 79-80*	W Bg	British	1072	183·2	37·5	23·0	St.John, N.B.	1860	McMorrin		B.H.McCorkell	Londonderry
1649		Minnesota *Johnson*	W Sr	American	103	96·0	26·0	8·0	May's Landing, N.J.	1852			M.F.Lynch	Boston, Mass.
1650	V	Minnet *Lindstedt*	W Bk	Swedish	539	142·0	29·1	17·3	Dalkarlsa	1875	E.A.Löfgren	[illegible]	P.F.Lindstedt	Landscrona

A = Lloyd's Register; D = Nederlandsche Vereeniging; G = Germanischer Lloyd; I = Registro Italiano; R = American Record; A = Veritas Austro-Ungarico; H = Veritas Ellenico; N = Norske Veritas; V = Bureau Veritas.

Dans la 4me Colonne, C = Composite (membrure en fer, bordé en bois); I = Fer; Stl = Acier; W = Bois.

Les Dates du doublage en métal (**Col. 11**) ressortent des livres dans lesquels les navires sont respectivement classés ou des rapports des Inspecteurs du Lloyd's Register.

Pour plus amples renseignements sur les navires classés par le Lloyd's Register, voir le Lloyd's Register Book et le Lloyd's Yacht Register.

Official No.	No. in Book	Society in which Classed	Ship's Name. Material and Rig. Master, and Dates of Appointment to present Owner's Service and to Ship. Ship's Late Name.	Flag	Net Tonnage	Dimensions: Length	Breadth	Depth	Built: Where	When	By whom	Date of Metalling	Owners or Managers	Port
	1651		**Minnet** (WRECKED) *Lindh* WBk	Swedish	492	119·1	27·5	19·2	Gefle	1836			[illegible]	[illegible]
	1652		**Minnet** *Zyndebadt* -87 WBg	Swedish	238	102·6	23·2	13·0	Dalkarlså	1860		[illegible]	A. Natterqvist	[illegible]
88631	1653	✻R	**Minnia** *Olsen* 79-88 WBkn	British	423	151·9	29·4	15·2	Bideford, P.E.I.	1885	W. Richards	[illegible]	W. Richards	[illegible]
48811	1654		**Minnie** *Brown* WBg (ex [illegible])	British	166	91·4	26·0	12·3					J. Curran	[illegible]
71957	1655	✻	**Minnie** *Mulcahy* 83-83 WBn	British	164	106·0	23·0	11·7	Bideford	1875	Cox	[illegible]	A. Goodridge & Sons	St. John's
35560	1656		**Minnie** (ex Little Aine) WSr	British	100	85·2	21·1	10·0					M. Donoghue	[illegible]
73253	1657	✻	**Minnie** *Collie* 67-85 WSr	British	97	84·2	21·3	10·[illegible]	Peterhead	1878	Carnegie	[illegible]	J. Stewart & Co.	[illegible]
59687	1658		**Minnie** *Row* 79-83 WSr	British	86	83·7	20·0	10·0	Ardrossan	1868	Barr & Shearer	[illegible]	H. Row	[illegible]
98361	1659	✻	**Minnie** *Tomblin* 86- WSk	British	77	72·4	20·2	10·4	Elmshorn	1884	J. Kremer		T. Baskcomb	[illegible]
17702	1660		**Minnie** WSr	American	299	108·4	30·2	9·0	Patchogue N.Y.	1867			A. A. Wicks	[illegible]
91362	1661		**Minnie** (STRANDED) W3mSr	American	273	125·0	30·0	8·0	Berwick City	1881			W. Crooks	[illegible]
16792	1662	R	**Minnie Abbie** *Look* -87 WBn (ex Andrew Johnson)	American	327	128·0	29·1	13·0	N. Bedford Mas.	1866	J. Howland	[illegible]	G. J. Carver	[illegible]
91468	1663	R	**Minnie Anna Bonsall** *Bonsall* W3mS	American	436	153·5	32·0	11·0	Milton, Del.	1882	J. L. Black & Bros		J. L. Bonsall	[illegible]
	1664	✻	**Minnie Bacon** *Bates* -87 Sk	British	82	76·6	20·6	10·4	Rye	1887	G. & T. Smith		E. Bacon	Grimsby
73639	1665	V	**Minnie Browne** *Williams* 87-87 WBk	British	1022	185·3	37·1	22·0	Maitland, N.S.	1881	A. A. McDougall	[illegible]	Browne & Watson	Glasgow
80615	1666	V	**Minnie Burrill** *Cann* -87 WS (WRECKED)	British	1466	206·2	40·1	24·0	Little Brook N.S.	1881	J. L. Burrill	[illegible]	W. Burrill, jun.	[illegible]
90686	1667		**Minnie C. Taylor** *Burns* -87 W3mSr	American	182	104·3	27·2	8·2	Harpswell, Me.	1874			A. Hall	[illegible]
56134	1668	✻	**Minnie Coles** *Jones* 75-87 WSr	British	99	80·6	21·0	11·0	Pt. Dinorwic	1867	Jones		W. E. Jones	[illegible]
94792	1669		**Minnie E. Moody** *Long* -88 WSr	British	112	89·3	24·7	8·9	[illegible], N.B.	1888	G. Long		G. Long	[illegible]
54185	1670		**Minnie Eaton** *Wetherall* [illegible] WBn	British	198	105·5	23·5	12·9	P. E. Island	1866	Longworth		W. E. Wetherall	[illegible]
78473	1671	✻	**Minnie Flossie** *Baltes* [85-85] WK	British	46	60·0	18·0	8·0	Milford	1879	Davies		J. Lang & Co.	[illegible]
85606	1672	V	**Minnie G. Elkin** *Nutter* WBkn	British	429	143·1	31·2	13·1	St. John, N.B.	1883	E. McGuiggin	[illegible]	R. C. Elkin	[illegible]
88386	1673	V	**Minnie G. Whitney** *Foley* -84 WBk	British	1222	192·0	39·8	23·9	Newport, N.S.	1884	T. A. Mosher	[illegible]	Allen Haley	[illegible]
36914	1674		**Minnie Gordon** WBk	British	322	119·8	28·5	13·0	River St. John, N.S.				S. Bethel	[illegible]
91754	1675		**Minnie H. Smithers** (ex Anna M. Edwards) *Insley* WSr	American	130	104·6	21·2	7·8	Bristol, Pa.	18 57/85			W. J. Insley	[illegible]
73074	1676	R	**Minnie Louise** *Boudrot* WSr	British	223	105·7	27·5	11·7	Sherbrook, N.S.	1883	C. S. McIntosh		S. Boudrot, jun.	[illegible]
91657	1677		**Minnie Rowan** *St. John* W3mSr	American	644	168·0	35·5	15·0	E. Boston	1884	Smith & Townsend		C. C. St. John	[illegible]
91440	1678	R	**Minnie Smith** *Rowan* -87 W3mSr	American	371	133·0	32·0	12·0	Bath, Me.	1882	Goss & Sawyer	[illegible]	Cobb, Wight & Co.	[illegible]
74624	1679		**Minnie Sommers** *Eden* WBkn	British	229	117·8	26·4	12·6	Portreath	1876	Massey		T. Fisher	[illegible]
72614	1680	DV	**Minnie Swift** *Liswell* WS	British	1150	187·0	37·7	22·6	Maitland, N.S.	1876	C. Cox	[illegible]	J. W. Liswell	[illegible]
91531	1681	R	**Minnie & Gussie** *French* W3mSr	American	216	112·0	32·2	8·8	Tottenville, N.Y.	1883	J. S. Ellis		J. Blenderman	New York
90940	1682	✻	**Minnyhive** *Webster* [illegible] IBk	British	1296	239·5	36·2	21·5	Port Glasgow	1885	R. Duncan & Co.		T. C. Guthrie	[illegible]
36055	1683		**Minora** *Grambis* WBg	British	249	103·2	26·6	15·0	Sunderland	1867	T. Young		D. Williams	[illegible]
16489	1684		**Minquas** *Cobb* WSr	American	109	91·2	23·5	7·3	Milton, Del.	1866			S. Banks	[illegible]
65030	1685	✻	**Minstrel King** (ABANDONED) *Everingham* 82-86 WBk	British	497	146·6	29·6	17·7	Sunderland	1873	R. Thompson	[illegible]	J. I. Jacobs & Co.	[illegible]
3529	1686	I	**Mio Padre** (ex Ferdinand [illegible]) *Bottino* WBk	Italian	787	165·0	32·8	23·3	[illegible]	1875	P. Rolla	[illegible]	B. Fravega fu G. B.	Genoa
748	1687		**Mioun Maru** WSr	Japanese	126	87·8	19·1	10·3	[illegible]	1882	Higami Gombei		Akagi Shimpoo	[illegible]
	1688	N	**Mira** *Jensen* -82 WBg	Norweg'n	333	107·9	27·6	16·7	[illegible]	1854			M. Jensen	[illegible]
	1689	V	**Mira** *Christensen* -88 WBg	Norweg'n	284	119·6	26·9	14·0	[illegible]	1870		[illegible]	T. Thommesen & Smith	Arendal
90143	1690	R	**Mira A. Pratt** (LOST) *Sherman* -87 WSr	American	139	98·6	28·6	7·[illegible]	[illegible]	1870			Turner & Oates	[illegible]
	1691		**Miraflores** (ex Nuevo Chileno) *Johnsen* (LOST) WBg	Chilian	618	146·6	33·1	..	[illegible]	1867			F. Peede	Valparaiso
62213	1692		**Mirage** WSrYt	British	111	102·7	21·3	11·8	Lymington	1869	Inman & Son			Southampton
	1693		**Miramichi** S	Norweg'n	690	..	..	..					Kr. Hansen	Christiania
62095	1694		**Miranda** *Thomas* 83-83 WSr	British	157	92·6	22·4	12·1	Arbroath	1870	Arbroath Ship C.	[illegible]	W. Balfour	Arbroath
90825	1695	R	**Miranda** *Corbett* WBkn	American	407	131·0	29·8	16·6	Machias, Me.	1874		[illegible]	Brett, Son & Co.	New York
	1696		**Miranda** (ex Gironde) *Eschen* -87 WBn	Danish	330	102·8	24·9	18·5	South Shields	1846			H. C. Christensens Enke	Marstal
	1697		**Miranda** *Jonsson* -86 WSr	Swedish	131	82·0	20·7	10·2	Grundsunda	1858			A. Pyk	[illegible]
	1698	V	**Mireille** (ex [illegible]) *Grondin* W3mSr	French	252	118·8	23·9	13·1	Nordby, Fanö	1876	Chr. Graff	[illegible]	Th. Gaune & F. Martin-Laval	Marseille
54621	1699	✻	**Mirella** *Coulls* 84-84 IBk	British	706	178·8	29·8	19·6	Newcastle	1865	Schlesinger, Davis & Co.		D. B. Lawson & Co.	Liverpool
68247	1700		**Miriam** *Manning* 80-87 WBn	British	127	97·7	22·7	11·0	P. E. Island	1871	White	[illegible]	A. Goodridge & Sons	St. John's

✻ = **Lloyd's Register**; D = Nederlandsche Vereeniging; G = Germanischer Lloyd; I = Registro Italiano; R = American [illegible];
A = Veritas Austro-Ungarico; H = Veritas Ellenico; N = Norske Veritas; V = Bureau [illegible]

In the 4th Column, C = Composite (iron frame, wood planked); I = Iron; Stl = Steel; W = Wood.

The Dates of Metalling (Col. 11) are based on the records of the Books in which Vessels are respectively classed, or on the Reports of the Surveyors of Lloyd's [illegible]

Further particulars of Vessels classed by Lloyd's Register will be found in Lloyd's Register Book and Lloyd's Yacht Register

N°	Cote	Nom. Capitaine et dates d'entrée au service des Armateurs actuels et à bord du navire. Ancien Nom.	Matériaux et Gréement	Pavillon	Tonnage net	Longueur	Largeur	Creux	Construit — Où	Quand	Par qui	Date du doublage en métal	Armateurs ou Directeurs-Gérants	Port d'Armement
1701	Å	Miriam *Lusinovich* (ex Isaac Ledward)	W Bk	Austrian	1111	187·3	35·1	22·0	Maitland, N.S.	1875	I. Trakey	Y 3.87	C. D. Ivancich	Lussinpiccolo
1702		Miriam Duranty (ex Aquila)	W 3m Sr	Colamb'n	129	111·6	26·7	10·0	Dartmouth	1872		I 10.87	C. D. Martinez & Co.	Rio Hache
1703	N	Mirjam *Tomessen 82-85*	W Bk	Norweg'n	522	140·3	31·2	17·3	Gjennesberg	1865			J. M. Pauss	Drammen
1704		Mirliton *Gulez 85-87*	W Bk	French	243	115·1	24·0	13·2	Bouchain	1864	A. Lefebvre		E. Lavergne	Dunkirk
1705	V	Mirtha *Le Hélois* (ex Collision)	W Sr	French	122	74·5	20·3	11·5	Nantes	1858	T. Dubigeon			Nantes
1706	H	Mirtidiotissa *Bletzo*	W Sr	Greek	111	..	..	..	Syra	1868			A. Bletzo	Limni Eboea
1707		Mirto N. *Lucich*	W Bk	Austrian	605	148·3	30·2	18·7	Trieste	1872	Tonello	Y 10.87	P. A. Pazze & Ci.	Trieste
1708	*	Mirzapore *Parker 87-7*	I S	British	1186	209·6	36·4	22·7	W. Hartlepool	1863	Pile, Spence & Co		Mirzapore S. Co. Ld. (W. J. Gibson & Sons)	Liverpool
1709		Misafko	W Sr	Swedish	115	84·4	19·9	10·9	Stavanger	1856			M. Cronlund	Calmar
1710		Mischief (ex Nahua Steeson)	W Sr	British	128	88·4	25·0	8·6					R. Burrowes	Nassau, N.P.
1711		Miss Beck *Jones*	W Sr	British	99	82·1	22·0	11·5	Carnarvon	1857			H. Hughes & Co.	Carnarvon
1712	*	Miss Hughes *Owens 86-88*	W Sr	British	59	72·0	19·2	8·6	Carnarvon	1884	H. Hughes		H. Hughes & Co.	Carnarvon
1713	*	Miss Pritchard *Jones*	W Sr	British	92	75·3	22·0	10·5	Portinllaen	1870	Ellis		J. Pritchard	Carnarvon
1714		Miss Thomas *Humphrey 87-88*	W Sr	British	127	80·0	21·5	11·5	Nevin	1864	Thomas		H. Lloyd & Co.	Carnarvon
1715		Mississippi *Kickne -88*	W Bn	British	228	110·1	25·5	9·0	Baltimore, U.S.	1863		Y 1.86	M. W. Crowell	Georgetn. Dem.
1716	N	Mississippi *Jacobsen*	W Bk	Norweg'n	668	153·2	34·8	18·9	Arendal	1875	L. Waaland	Y 9.8	S. Eyde	Arendal
1717	H	Missolonghion *Lambros*	W Bn	Greek	118	..	..	..	Curzola	1870			C. Lambros	Missolungi
1718		Mist	W Sr	American	114	100·0	17·6	7·7	Hamburgh, Pa	1864			J. A. Anderson	Trenton, N.J.
1719	*	Mistelten *Elfström -87*	W Bg	Swedish	171	100·7	21·8	10·8	Orust	18[illegible]			M. Johansson	Ulmeskärr
1720	V	Misti *Neees*	I Bk	French	486	127·9	26·6	17·1	Nantes	1869	E. Gouin		G. Dor & E. Bouge	Marseilles
1721	RV	Mistletoe *Mers -88*	W Bk	British	821	164·0	34·5	19·7	Young's Cove, N.S.	1875	H. Young	Y 7.84	H. D. Troop	St. John, N.B.
1722	*	Mistletoe *Snow 74-88*	W Bn	British	160	100·3	23·3	12·7	Brixham	1876	Upham	Y 10.88	P. G. & L. Tessier	St. Johns
1723	*	Mistletoe *De Gruchy -88*	W Sr	British	69	73·0	18·3	9·0	Jersey	1876	LeHuquet		J. Le Gros	Jersey
1724	*D	Mistley Hall *Nelson 87-87*	I S	British	1772	260·0	42·3	23·9	Liverpool	1874	Evans		Globe Shipping Co., Ld.	Liverpool
1725		Mistley Park *Rumsey*	W Bn	British	121	72·0	20·6	11·9	Sunderland	1845	B. Hodgson & Co.		W. Brooks	Harwich
1726	*	Mitford *Patt*	W Bk	British	334	121·2	27·4	17·0	Blyth	1868	Floating Dk. Co.	Y 3.83	J. G. Taylor	Newcastle
1727	NV	Mithassel *Andreassen -87* (ex Peter Young)	W Bk	Norweg'n	1024	180·1	35·7	21·5	Shelburne, N.S.	1874	P. Young	Y 10.87	T. Reinertsen	Farsund
1728	*	Mitredale *Davies 86-86*	I S	British	1231	225·8	36·2	21·7	Liverpool	1875	W. H. Potter & Co		J. D. Newton & Co.	Liverpool
1729	H	Mitros	W Bg	Greek	269	92·8	26·1	15·1	Galaxidi	1871	G. Canata		C. Mitropoulos	Galaxidi
1730		Mitsunaka Maru	W Sr	Japanese	408	126·8	27·8	15·1	Tsukiji, Tokio	1883	Kawasaki S. Yd.		Nippon Yusen Kaisha	Tokio
1731	G	Mittwoch *Sprenger 81-*	W Bg	German	280	110·8	27·3	11·1	Ueckermünde	1863	C. Radmann		A. R. Bieback	Stettin
1732		Mixte No. 2 *Guegan* (ex N...) (PONTOON)	C Bk	French	287	121·6	28·3	12·1	Bordeaux	1858	Bichon frères		H. S. Johnston & Co.	Bordeaux
1733	H	Mizela K. *Kalamidas*	W Bn	Greek	294	103·6	27·9	..	Syra	1877	E. Flamis	Y 12.81	S. Calamidas	Syra
1734	N	Mizpa *Bonde 83-83* (ex Bilderm...)	W Bk	Norweg'n	738	165·0	31·2	19·3	S. Stephen, N.B.	1870	J. & C. Short	Y 2.8	G. E. Bonde	Tönsberg
1735	*	Mizpah *Jones 77-87*	W Bk	British	453	138·8	29·5	17·2	P. E. Island	1876	Plestid	Y 12.87	J. M. K. Campbell	Swansea
1736		Mizpah *Gerraert 72-72*	W Sr	British	141	91·2	21·1	12·4	Kingswear	1855	E. Alford		H. & E. Crundall	Dover
1737	*	Mizpah *McKeown*	W K	British	51	65·8	18·1	8·5	Jersey	1883	D. LeSueur & Son		J. C. Renouf & Co.	Jersey
1738	*	Mizpah *Tepp 78-88*	W Bk	German	540	146·7	30·1	18·4	Shoreham	1874	May	Y 9.87	C. Rahtkens	Rostock
1739	NV	Mizpah *Olsen*	W Bk	Norweg'n	406	125·2	29·6	16·3	Grimstad	1876	J. Jörgensen	Y 7.87	Bernt Einersen	Grimstad
1740	N	Mjolner *Nielsen*	W Bk	Norweg'n	361	131·1	28·8	15·6	Sandefjord	1875	R. Ormstad	Y 5.85	J. M. Osmundsen	Sandefjord
1741	N	Mjolner *Andressen*	W Bg	Norweg'n	249	101·5	26·7	12·4	Grimstad	1861	Jens Tellefsen		H. Hansen	Lillesand
1742	N	Mjölner *Ingvoldsen*	W Bn	Norweg'n	206	101·6	25·8	12·1	Kragerö	1879	J. Nielsen	Y 11.83	J. Ingvoldsen	Kragerö
1743	N	Mjosen *Haugersen -86*	W Bg	Norweg'n	296	120·1	28·9	13·5	Wiborg	1864			L. Heffermehl	Drammen
1744		Moa	W Bg	British	219	99·0	25·1	13·1	Auckland	1849			John Mill	Dunedin, N.Z.
1745	V	Mobil	W 3m Bn	Norweg'n	352	122·4	26·5	13·1	Arnis	1876	F. Johansen	Y 8.87	B. J. Nielsen	Arendal
1746	G	Mobil *Kraeft -88*	W Bg	German	156	82·5	23·3	12·5	Zingst	1854	H. C. Drossel		J. G. Kraeft	Prerow
1747	*	Mobile Bay *McPhail 86-86*	I Bk	British	1117	216·9	35·1	21·0	Greenock	1883	Russell & Co.		Hatfield, Cameron & Co.	Glasgow
1748	N	Mod *Birkeland -88*	W Bk	Norweg'n	519	143·6	31·7	17·6	Lillesand	1876	O. & P. Knudsen	Y 1.86	P. Knudsen	Lillesand
1749		Modena *McBride*	W Sr	British	174	92·0	26·3	10·4	Cornwallis N.S.	1874	J. Lockhart		D. M. Dickie	Windsor, N.S.
1750														

* = Lloyd's Register; D = Nederlandsche Vereeniging; G = Germanischer Lloyd; I = Registro Italiano; R = American Record; Å = Veritas Austro-Ungarico; H = Veritas Ellenico; N = Norske Veritas; V = Bureau Veritas.

Dans la 4me Colonne, C = Composite (membrure en fer, bordé en bois); I = Fer; Stl = Acier; W = Bois.

Les Dates du doublage en métal (Col. 11) ressortent des livres dans lesquels les navires sont respectivement classés ou des rapports des Inspecteurs du Lloyd's Register.

Pour plus amples renseignements sur les navires classés par le Lloyd's Register, voir le Lloyd's Register Book et le Lloyd's Yacht Register.

Official No.	No. in Book	Registry in which classed	Ship's Name. Master, and Dates of Appointment to present Owner's Service and to Ship. Ship's Late Name.	Material and Rig	Flag	Net Tonnage	Length	Breadth	Depth	Built: Where	Built: When	Built: By whom	Metalling	Owners or Managers	Port of Registry
	1751	DN	Modesta *Tor Olsen*	W Bn	Norweg'n	255	124·0	26·6	12·7	Bergen	1875	K. Jensenius		A. V. Johannesen	Bergen
	1752	V	Modesta *Gibernau*	W Pol	Spanish	188	90·2	26·1	13·1	Lloret	1854			A. Maristany	Barcelona
2969	1753		Modesto *Fasce*	W Bg	Italian	310	118·1	28·2	16·6	Savona	1870	G. B. Guastavino		E. Piaggio	Genoa
	1754		Modesto *Gibernau* (ex Francesca Mathilde)	W Bn	Spanish	134	81·4	22·6	11·1	Havre	1867			B. Canet	Majorca
16812	1755		Modesty *Nickerson*	W Sr	American	127	94·0	24·0	7·7	Derby, Conn	1866			H. I. Manter	Bangor
90575	1756		Modoc *Busch* -87	W Bkn	American	430	170·0	36·6	12·6	Utsalady Wsh	1873			S. B. Peterson	San Francisco
91519	1757	R	Modoc *Perry*	W Sr	American	180	99·0	29·4	8·5	Columbia Falls Me	1882	I. Carleton		H. W. Buckman	Machias
	1758	N	Modum *Petersen* (ex Gabriel Flater)	W Bk	Norweg'n	294	124·9	28·7	12·4	Drammen	1874	G. Flater		H. L. Baltzersen	Fredrikstad
65869	1759	✻	Modwena	C Sr Yt	British	161	116·1	21·1	13·2	Seacombe	1872	Bowdler, Chaffer & Co.		John Gretton	Liverpool
77418	1760	✻	Moel Eilian *Jones* 84-84	I Bk	British	1081	221·5	35·0	21·2	Sunderland	1877	W. Doxford & Ss.		W. E. Jones	Carnarvon
89234	1761	✻D	Moel Tryvan *Edwards*	I Bk	British	1639	258·0	38·0	23·4	Sunderland	1884	Doxford & Sons		W. E. Jones	Carnarvon
80295	1762		Moel-y-Don *Williams*	I Bk	British	1080	222·0	35·0	21·2	Sunderland	1882	W. Doxford & Ss.		Gwynedd Shipping Co. Ld.	Carnarvon
	1763	D	Moewe *Kr. Lyfeldt* 85-85	I Bk	German	1058	205·6	34·5	20·1	Kiel	1885	Germania Wrks.		Norddeutsche Dampfs. Act. Ges.	Kiel
	1764		Moewe (LOST) *Ahrens* 77-79	W Bk	German	339	113·3	28·1	16·4	Rostock	1869	J. Möller		E. Burchard	Rostock
	1765	DV	Moewo *Rudbach* (ex Fredrikke)	W Sr	Russian	119	78·5	22·7	10·1	Thurö	1871	L. Kaas		N. Rudbach	Wiborg
	1766		Mohamed Agmi	Bg	Turkish	204	..	..	..					Bayumi Effendi	
75095	1767	V	Mohawk *Churchill*	W Bk	British	1338	193·1	39·3	24·0	Quebec	1878	J. E. Gingras		J. G. Ross & Co.	Quebec
	1768		Mohawk *Hansen* (CONDEMNED)	W Bk	Norweg'n	710	132·9	36·1	22·6	Middelburg	1839			J. Augestad	Laurvig
81099	1769		MohayadeenSamdany	W Bg	British	136	80·0	18·8	11·0	Coringa	1856			Chana Meana Haribhoo Mahomed Marikar	Colombo
28631	1770		Mohican *Cowling* -88	W Bk	British	326	130·6	26·6	16·0	Waterford	1859	White		R. & W. King	Bristol
91029	1771	RV	Mohican *Berry* 75-87	W Bk	American	814	160·2	35·2	19·9	Chelsea, Mass	1875	Pierce, Montgomery & Howard		C. S. Gill	Boston
91671	1772		Mohican *Lyceve*	W Sr Yt	American	114	107·4	24·0	10·0	Noank, Conn.	1884	R. Palmer & Sons		H. D. Burnham	[illegible]
31070	1773		Mohidin Cader Hydroos	W Bg	British	105	78·8	19·2	9·8	Ceylon	1870			Ana Coos Mohomado	Colombo
31090	1774		Mohidin Nux	W Bg	British	111	98·5	19·6	12·0	Cochin	1872			Mohomado Meydin	Jaffna
81069	1775		Mohidindux (WRECKED)	W Bg	British	178	84·0	22·0	12·8	Kelecarre	1882	R. Assary		Mana Ena Voona Noora Meerdando	Colombo
41528	1776		Mohoddien Bux	W Bk	British	387	126·0	26·0	14·0	[illegible]	1876			S. Mahomed Gouse Vanya	Moulmein
81073	1777		Mohomed Ebroym-Ally	W Bg	British		77·0	18·5	10·0					Oona Meerasaibo	Colombo
65524	1778	RV	Mohur *Stevens* -87	W S	British	1413	207·0	39·9	23·8	Maitland, N.S.	1874	Monteith		A. Gibson & Co.	Liverpool
90416	1779		Mokmai	Stl Lr	British	652	250·1	34·0	8·8	Dumbarton	1887	W. Denny & Bros		Irrawaddy Flot. Ba Co. Ld.	Glasgow
	1780	N	Moland *Abrahamsen* 81-88	W Bk	Norweg'n	446	140·7	29·9	16·9	Lillesand	1875	O. Nielsen		H. J. HammersEnkla	[illegible]
94686	1781	R	Molega *Lohnes*	W 3m Sr	British	148	91·7	27·0	8·2	[illegible] N.S	1888	J. N. Rafuse		C. H. Davison	Lunenburg
72998	1782	R	Molilamo *Sinclair* -86	W Bk	British	842	172·0	34·1	20·6	Chatham, N.B.	1875	J. Desmond		A. W. Sinclair	Liverpool
30230	1783	I	Molini Callegno *Costa*	W 3m Sr	Italian	295	114·8	26·2	16·1	Savona	1880	G. Calanaro		Gio. Stagno fu G.	Genoa
91532	1784	R	Mollie J. Saunders *Parry* -88	W 3m Sr	American	532	148·0	36·0	14·9	City Island NY	1882	D. Carll		Wm. Ingersoll	New York
91807	1785	R	Mollie Rhodes *Watts* (88-88)	W 3m Sr	American	226	118·0	30·5	9·5	Jonesboro', Me	1885	W. Tupper		E. P. Walker	Thomaston
	1786		Molly *Christensen* 82-82	W 3m Sr	Danish	257	116·6	24·5	14·4	Königsberg	1865	L. Eggers		H. C. Grube	Marstal
91670	1787		Molly Adams *Johnson*	W Sr	American	117	98·2	24·4	8·8	Kennebunk Me	1884	C. Christensen		S. Jacobs	Gloucester
	1788		Molodez *Johansson* -86	W Bg	Swedish	242	97·8	24·9	13·5	Ueckermünde	1850	L. Wittenberg		O. Nilsson	[illegible]
	1789	GV	Moltke *Crantz* 84-84	I S	German	827	191·6	32·2	19·3	Hamburg	1870	Reiherstieg Schiffswerfte		C. H. Donner	Hamburg
	1790	V	Mommsen *Wilms* 71-82 (ex Eliza)	W S	German	1254	195·2	38·4	24·3	Newburyport	1864	J. Currier		D. Cordes & Co.	Bremen
78654	1791	✻	Mona *Kean* 87-87	I Bk	British	1045	205·0	34·5	21·2	Sunderland	1878	Osbourne, Graham & Co.		H. Owen & Son	Liverpool
41296	1792		Mona	W Bkn	British	303	150·0	26·0	9·0	Sweden				D. D. Lewin	[illegible]
	1793		Monarca	W Bg	Spanish	218	89·6	25·6	13·7	Blanes	1850			José Rosell	[illegible]
21087	1794		Monarch *Winterflood*	W Bg	British	285	116·7	27·4	14·6	Guernsey	1859	Sebire		A. & F. Mannelle	[illegible]
[illegible]	1795		Monarch	I Barge	British	191	123·0	25·8	7·9	Goolwa, S.A	1879			A. A. Scott	[illegible]
[illegible]	1796		Monarch	W Sr	British	132	109·0	23·9	7·6	Moreton Bay	1871			A. Heesh	Sydney
	1797		Monarch *Slangebyer* 83-83 (ex [illegible])	W Bk	Norweg'n	756	151·5	32·2	20·2	Sunderland	1856	W. R. Abbay & Co.		Act Selsk. "Monarch"	Fredrikstad
	1798	DV	Monarch *Lindendahl*	W Bk	Swedish	823	170·9	32·0	20·5	Gefle	1880	O. A. Brodin		J. Rettig	Gefle
	1799		Monarchia *de Bessa* -84	W 3m Sr	Brazilian	200	..	..	..	Villa do Conde	1870			[illegible]	[illegible]
	1800	DN	Monark *Abrahamsen* (ex Alber C.)	W Bk	Norweg'n	868	171·1	34·5	21·1	Quebec	1870	McKay		Th. Offenberg	[illegible]

✻ = Lloyd's Register; D = Nederlandsche Vereeniging; G = Germanischer Lloyd; I = Registro Italiano; R = American [illegible]
A = Veritas Austro-Ungarico; H = Veritas Ellenico; N = Norske Veritas; V = Bureau Veritas

In the 4th Column, C = Composite (iron frame, wood planked); I = Iron; Stl = Steel; W = Wood.

The Dates of Metalling **(Col. 11)** are based on the records of the Books in which Vessels are respectively classed, or on the Reports of the Surveyors of [illegible]

Further particulars of Vessels classed by Lloyd's Register will be found in Lloyd's Register Book and Lloyd's Yacht Register.

No.		Nom. Matériaux et Gréement. Capitaine et dates d'entrée au service des Armateurs actuels et à bord du navire. Ancien Nom.		Pavillon.	Tonnage net.	Dimensions. Longueur.	Largeur.	Creux.	Construit. Où.	Quand.	Par qui.	Date du doublage en métal.	Armateurs ou Directeurs-Gérants.	Port d'Armement.
1801	V	Mondego	W Bn	Portug'se	163	91·9	22·3	10·3	Caminha	1880	M.J.Gonçalves		A.T.de Barros	Lisbon
1802		Mondego [illegible]	W Bn	Portug'se	139	68·6	18·8	9·4	Figueira	1848			M.J.da Costa Pedreira	Lisbon
1803		Moneta *Gann*	W Bn	British	158	101·2	22·3	11·4	P.E.Island	1859	Cox		J.Reeves	Faversham
1804		Mongibello *Granata*	W Bn	Italian	218	117·1	25·6	11·0	Sestri	1876	A.Briasco fu L.	V 9.83	Gius.Puglisi fu A.	Catania
1805	N	Monica *Johannesen*	W Bk	Norweg'n	636	158·4	33·5	17·1	Arendal	1883	E.Guttormsen	V 1.87	H.H.Pettersen	Arendal
1806		Monitor *Boyd*	W Bg		247	110·6	27·0	13·6	Carlskrona	1862				
1807		Monitor *Gallop*	W Bkn	American	224	127·0	28·0	9·2	San Francisco	1862			C.Nelson	San Francisco
1808		Monitor	W Sr	American	105	88·0	23·7	9·5	Essex, Mass.	1886	J.James		J.McDonald	Gloucester, Mass
1809	V	Monitor *Ohlsson 83-86*	W Sr	Swedish	220	97·4	26·6	11·0	Gefle	1883	Bergforss & Son	V 11.87	P.Bergforss & Son	Gefle
1810	✳	Monmouthshire *Moulson 87-8*	I Bk	British	1162	207·0	34·1	23·3	Liverpool	1863	B.M.Lawrence		H.Fernie & Sons	Liverpool
1811	V	Monrovia *Jeffrey -88*	W S	British	1493	210·5	39·3	24·4	St.John, N.B.	1878	J.Fraser	V 1.87	Wm.Thomson & Co.	St.John. N.B.
1812	RV	Monrovia *Rogers*	W Bk	American	443	137·5	29·2	17·0	Setauket, N.Y.	1878	G.E.Hand	V 5.86	Yates & Porterfield	New York
1813		Monseratt *ex Jaime Guitarrol*	W Bk	Spanish	527	135·5	28·2	19·1	Sestri Ponente	1868	Flli Cadenaccio		F.Guardiola	Barcelona
1814		Monsita *Wallace 82-87*	W Bkn	British	255	113·5	28·0	11·9	St.John, N.B.	1872			J.P.Hendy	Bridget'n.Bar.
1815		Montana	W Bk	American	597	143·0	31·0	19·6	Stockton, Me.	1864			C.A.Nutsen	San Francisco
1816	V	Montana *Schmidt 84-87*	W Bk	German	481	139·1	27·8	17·4	Kell	1875	A.Conradi & Sohn	V 5.83	Knöhr & Burchard	Hamburg
1817	V	Montbars *Rollando*	W Bg	French	211	97·8	24·9	11·9	Nantes	1870	Tlle	V 1.87	L.Jacquier	Nantes
1818	V	Montbert *Eveno -88*	W Bn	French	132	78·7	21·3	11·0	Nantes	1873	Sevestre		Eveno	Nantes
1819	I	Monte A. *Razetto -87*	W Bk	Italian	812	168·0	33·7	21·3	Varazze	1874	F.Tassara	V 4.88	G.B.Bozzo	Genoa
1820		Monte Alegre	W Sr	Brazilian	169	..	..	..					J.F.Pinto	Rio de Janeiro
1821		Monte Alegre *Duarte*	W Bn	Brazilian	120	..	..	..	S.João da Barra	1872		V 1.83	Cia. Cabotagem de S. João da Barra & Campos	S.João da Barra
1822	I	Monte Allegro M. *Chiesa ex Domenico e Raquilino) I -88*	W Bk	Italian	605	149·2	29·0	20·2	Sestri Ponente	1872	Flli Cadenaccio	V 6.88	Flli Chiesa	Genoa
1823		Monte Bello	W Bn	Brazilian	405	..	..	..					T.C.P.Junior	Rio de Janeiro
1824	IV	Monte Santangelo *Cafiero -87*	W Bk	Italian	466	133·2	29·2	19·3	Cassano	1874	F.DeRosa	V 7.86	Eredi Cafiero	Castellamare
1825	I	Monte Tabor *Razeto*	W Bk	Italian	567	153·2	32·1	19·7	Varazze	1873	D.Craviotto	V 1.86	Ant.Razeto	Genoa
1826		Monteiro II.	W Sr	Brazilian	196	..	..	..					Bastos & Monteiro	Porto Alegre
1827		Monterey *Klocker -87*	W Sr	American	120	96·5	28·0	7·5	Benicia, Cal.	1887	M.Turner		C.L.Dingley	San Francisco
1828	✳	Montezuma *Gittons 85-86*	I Bk	British	326	133·0	22·8	14·8	Port Glasgow	1857	J.Reid & Co.		T.Harrison & Co.	Liverpool
1829	✳	Montgomery Castle *Anderson*	I Bk	British	871	190·7	33·0	20·0	Liverpool	1869	R.& J.Evans & Co		Richards, Mills & Co.	Liverpool
1830	✳D	Montgomeryshire *Hughes 82-83*	I S	British	1401	235·4	38·1	23·0	Liverpool	1877	T.Royden & Sns.		Hughes & Co.	Liverpool
1831		Monticello	W Bg	American	151	88·0	23·0	8·0	Prospect, Me.	1869			I.Taylor	New York
1832	✳	Montmorency *Bernard*	W Bk	French	610	165·7	31·5	18·5	Bordeaux	1872	Rochet & Largeteau	V 1.85	Edmond Boissière	Nantes
1833		Montpeyroux *Buvig*	W BknSr	Norweg'n	229	110·2	26·7	14·3	Fécamp	1847		Z 2.86	Hj.Buvig	Sarpsborg
1834	RV	Montreal *Mitcaucer -87*	W Bk	British	807	165·0	34·0	20·6	Hantsport, N.S.	1868	E.E.Fuller	V 10.87	Ezra Churchill, Jr.	Windsor, N.S.
1835	✳	Montrosa *Linneman -88 ex Montr*	I Bk	German	995	194·8	34·5	21·3	Glasgow	1863	Barclay Curle & Co.		H.P.Breckwoldt, Jr	Hamburg
1836	✳	Montrose *Cooper 87-7*	W Bk	British	365	128·5	27·0	16·0	Jersey	1861	Le Rougetel	V 10.86	W.H.Williams	London
1837		Montrose	W Bn	British	178	97·1	21·9	9·1	Baltimore	1857			Edward Boyce	Trinidad
1838		Montrose *Lindström 76-84*	W Bk	Swedish	467	147·9	29·9	15·7	Svartvik	1853			A.Svensson	Gothenburg
1839		Monyo	I Barge	British	358	196·4	25·2	8·4	Port Glasgow	1881	R.Duncan & Co.		Irrawaddy Flotilla Co. Ld	Glasgow
1840	N	Moody *Tonnesen*	W Bn	Norweg'n	134	94·0	22·7	10·8	Stavanger	1877		V 11.83	Amund Tönnesen	Egersund
1841	✳	Mooltan *Doughty -88*	I S	British	1668	251·0	39·0	24·0	Stockton	1880	Richardson, Duck & Co.		British & Eastern Shipping Co. Ld.	Liverpool
1842	RV	Moonbeam *Dunbar*	W Bk	American	692	140·0	31·9	20·9	Searsport, Me.	1859	McGilvery	V 10.86	J.C.Nickels	Searsport, Me.
1843	✳	Moonlight *Hughes*	W Sr	British	95	83·6	21·4	10·5	Bridgwater	1876	Gough		F.Prince	Liverpool
1844	R	Moonlight *Cranmer*	W Sr	American	189	107·9	30·3	9·3	Belleville, N.J.	1855			S.E.Birdsall	New York, N.Y
1845		Moorarey	W Bg	British	112	85·1	21·0	9·3	Quoylon	1857			Doralzee Rattangee	Colombo
1846	✳	Moorhill *Brown*	W Bk	British	484	111·0	28·8	17·7	Sunderland	1865	A.Simey & Co.	V 3.87	E.Brown	Sunderland
1847		Mooydin Bux	Bg	British	155	..	..	..	Koorikuty	1849			Poodiavittil Kunnatty Veeran	Cochin
1848	✳	Mora *ex Lothair Crosby 86-86*	C Bk	British	620	160·0	34·1	18·6	Sunderland	1871	J.Blumer & Co.	V 7.87	Sandbach, Tinne & Co.	Liverpool
1849	V	Moralidad *Font*	W Bg	Spanish	186	91·5	27·1	13·3	Barcelona	1855		V 1.86	M.Espino	Barcelona
1850	R	Morancy *Wass 83-83*	W Sr	American	188	103·5	28·3	9·5	Sullivan, Me.	1868	G.B.Kingsley	V 9.86	J.S.Emery & Co.	Boston

✳=Lloyd's Register; D=Nederlandsche Vereeniging; G=Germanischer Lloyd; I=Registro Italiano; R=American Record
A=Veritas Austro-Ungarico; H=Veritas Ellenico; N=Norske Veritas; V=Bureau Veritas.

Dans la 4me Colonne, C=Composite (membrure en fer, bordé en bois); I=Fer; Stl=Acier; W=Bois.

Les Dates du doublage en métal (Col. 11) ressortent des livres dans lesquels les navires sont respectivement classés ou des rapports des Inspecteurs du Lloyd's Register.

Pour plus amples renseignements sur les navires classés par le Lloyd's Register, voir le Lloyd's Register Book et le Lloyd's Yacht Register.

Official No.	No. in Book	Registry in which classed	Ship's Name. Material and Rig. Master, and Dates of Appointment to present Owner's Service and to Ship. Ship's Late Name.	Flag	Tons	Length	Breadth	Depth	Where	When	By whom	Date of Metalling	Owners or Managers	Port
72[illegible]	1851		Moravian *Lessing* W Bk	British	968	199·7	33·6	21·2	Aberdeen	1858	W. Hood & Co.		J. E. Ives	
79870	1852	✳	Moray Chief W Bkn	British	314	137·7	26·3	12·9	Kingston	1888	W. Kinloch	[illegible]	W. Kinloch	
73782	1853	✳	Morayshire *Storm 83-83* I S	British	1428	215·7	37·3	21·3	Glasgow	1875	Dobie & Co.		T. Law & Co.	Glasgow
86891	1854	✳	Morecambe Bay *Russell 80-81* I Bk	British	1115	215·1	35·1	21·2	Port Glasgow	1884	Russell & Co.		Hatfield, Cameron & Co.	Glasgow
[illegible]	1855		Morecambe Bay I Lt	British	139	97·0	21·0	..	London	1864			Corporation of Trinity House	London
47785	1856		Morecambe Belle *Farrell* W Sr	British	100	84·8	21·3	12·1	Burton Stather	1864	Wray		J. Farrell	Drogheda
	1857	✳ D	Moree *Mobrig* W Bg	Dutch	265	112·5	25·8	14·1	Perth	1869	Wallace	[illegible]	G. G. Allers, Senr.	Farmsum
60223	1858		Morelight *Webster* W Sr	American	121	89·0	24·8	8·8	Calais, Me.	1868			G. King	Calais, Me.
68363	1859		Morell *Tyrrell* W Bn	British	121	93·0	25·2	9·7	P. E. Island	1871	Hayden		R. McClenahan	Dundalk
86208	1860	✳	Moresby *Williams 86-86* I S	British	1217	223·5	36·1	21·2	Whitehaven	1882	Whitehaven [S. B. Co.]		J. Dodd	Liverpool
	1861		Morey *Rodriguez -88* W Pol	Spanish	248	107·9	25·3	13·6	England	1857			G. C. Gutierrez	Santa...
	1862	N	Morgengryet *Jensen -88* W Bk	Norweg'n	587	145·0	32·9	17·6	Grimstad	1886		[illegible]	J. H. Ugland	Grimstad
55011	1863	✳	Morialta *Phillips 81-81* (ex British Consul) I S	British	1267	211·9	36·0	22·6	Liverpool	1866	T. Royden & Son		Trinder, Anderson & Co.	London
92311	1864	R	Morien *Martell 86-88* W Sr	British	136	91·5	25·1	10·0	Liverpool, N.S.	1886	J. Millard	[illegible]	C. Archibald	Liverpool
	1865		Moringen *Halvorsen* W Bg	Norweg'n	212	98·2	26·2	13·1	Drontheim	1853			Jacob Larsen	Drontheim
76515	1866	✳ D	Morna *Goodwin 81-81* I S	British	1437	233·7	37·9	23·0	Birkenhead	1877	Laird Bros.		J. H. Worthington & Co.	Liverpool
66368	1867	R	Morna (WRECKED) *Bartlett* W Bn	British	255	117·8	26·0	12·9	P. E. Island	1873	J. Duncan		John Woods	S. John's
75744	1868	R V	Morning Light *Ladd* W S	British	1310	212·2	38·0	22·0	Meteghan, N.S.	1878	M. L. Denny	[illegible]	G. H. Perry	Yarmouth
17613	1869		Morning Light *Price* W Bn	American	257	114·0	28·5	10·0	New Haven, Conn.	1867	Lane & Griswold		R. D. Armstrong	New York
60668	1870	✳	Morning Star *Luckham [87-87]* W Bn	British	280	117·0	26·1	14·6	Kingsbridge	1871	Date	[illegible]	W. W. Steer	Salcombe
70562	1871		Morning Star W Bkn	British	258	113·0	24·9	14·7	Kingston	1877	Spence	[illegible]	J. Innes	Banff
42690	1872		Morning Star (WRECKED) *McLester [85-85]* W Bk	British	217	114·9	27·8	12·6	St. John, N.B.	1862	Olive		M. & P. Crean	Belfast
63484	1873	✳	Morning Star *Pritchard -88* W 3m Bn	British	213	119·0	21·4	12·9	Guernsey	1874	De la Mare	[illegible]	J. Marquand & Co.	Guernsey
62164	1874		Morning Star (ex Florida) W Sr Yt	British	126	104·1	20·0	12·6	Greenock	1868	Steele & Co.		Sir James P. Corry, Bart., M.P.	[illegible]
67638	1875	✳	Morning Star *Thomas [77-77]* W Sr	British	100	79·1	21·1	11·1	Aberystwith	1877	Jones Bros.		J. Thomas	Aberystwith
76520	1876	✳	Morning Star *Wallace* W Sr	British	99	88·6	21·7	10·4	Kingston	1878	Duncan		J. McBeath	Thurso
87777	1877	✳	Morning Star *Peters 86-86* W K	British	81	76·4	20·3	10·2	Galmpton	1883	A. Gibbs		W. Padgett	Grimsby
10791	1878		Morning Star *King* W Bk	American	226	104·6	25·6	15·1	Dartmo'th Mas	1853	Matthews [Maskeen & Co.]		J. & W. R. Wing	New Bedford
	1879		Morning Star (ex Van Amerongen, ex Morning Star) (BROKEN UP) W Bk	Dutch	545	159·8	28·1	18·0	Dundee	1858	Tay Shipbuilding Co.	[illegible]	Said Hassan bin Abdul Rachman Moh Badlo	Sourabaya
50183	1880	V	Mornington *Jones 83-83* W S	British	1357	190·8	39·6	24·9	St. John, N.B.	1868	Rudduck	[illegible]	H. Owen & Son	Liverpool
17927	1881	R V	Moro Castle (WRECKED) *Smith* W Bk	American	384	127·0	29·5	17·0	Mystic, Conn.	1868	Hill & Grinnel	[illegible]	D. E. Mayo	[illegible]
91311	1882	R	Morris W. Child *Torrey* W 3m Sr	American	487	145·0	34·0	12·3	Bath, Me.	1881	H. M. Bean	[illegible]	J. P. Ellicott	[illegible]
402	1883		Mortola *Mortola* W Bn	Italian	157	78·4	23·7	13·8	Varazze	1855	G. Camogli		G. Mortola	Genoa
16258	1884		Morton Castle *Monro* W Bk	British	137	87·6	21·8	12·5	Dumfries	1857	G. Thomson		G. Horton	Montrose
	1885	N	Morvig *Olsen* W Bk	Norweg'n	369	127·1	30·1	14·8	Grimstad	1876	A. Bentsen	[illegible]	G. Huseland	Grimstad
260	1886	I	Mosca (ex Tigri) (MISSING) *Molfino* W Bk	Italian	839	171·5	34·1	22·8	Prà	1875	L. Durante	[illegible]	Ag. Schiaffino	Genoa
18	1887	H	Moscha *Empiricos* W Bk	Greek	439	136·1	29·1	18·0	Syra	1873	S. Koufoudakis	[illegible]	J. M. Empiricos	Andros
77680	1888	✳	Moselle *Rendell 81-85* W Bk	British	516	143·8	31·0	16·6	P. E. Island	1878	Peake Bros. & Co.	[illegible]	Peake Bros. & Co.	Charlottetown
79096	1889		Moselle *Gould -87* W Sr	British	139	85·0	23·0	10·3	Cornwallis, N.S.	1879	S. H. Newcomb	[illegible]	W. Y. Fullerton	Windsor
18887	1890		Moses B. Bramhall *Woodhull* W 3m Sr	American	328	122·9	31·3	13·5	Jersey City, N.J.	1860			Evans, Ball & Co.	New York
90638	1891	V	Moses B. Tower *Auerdas -86* W Bkn	American	665	147·2	33·0	17·2	E. Boston, Mass.	1874	Campbell & Brooks	[illegible]	L. Snow & Son	[illegible]
95563	1892	R	Mosquito (ex Henry Johnson) *Johnson -88* W Sr	American	138	93·1	24·5	10·1	Essex, Mass.	1879	J. Story	[illegible]	Gledden & Curtis	[illegible]
83312	1893	R	Moss Glen *Swain -86* W Sr	British	150	98·4	25·6	10·1	Pt. Medway, N.S.	1884	E. Dolliver		T. Forham	[illegible]
	1894	R	Moss Glen (BROKEN UP) *Kriegel -84* W Bk	Swedish	548	135·5	28·6	18·2	Clifton, N.B.	1865	Titus & Whitmore	[illegible]	C. Henckel	[illegible]
88260	1895	V	Moss Rose *Purdy 86-86* W Bn	British	371	135·5	30·1	15·0	St. John, N.B.	1883	J. Mahony	[illegible]	Frederick E. Sayre	St. John
88768	1896	✳	Moss Rose *Hughes -88* W 3m Sr	British	148	108·5	23·0	11·2	[illegible]	1888	Barr & Shearer		Coppack, Carter & Co.	Chester
92383	1897	✳	Moss Rose *Strange [84-84]* W K Barge	British	127	91·9	22·5	9·5	Ipswich	1884	W. Bayley & Sons		R. Cox	[illegible]
88704	1898	✳ D	Mostyn *Evans* W Sr	British	132	103·1	23·1	10·7	[illegible]	1884	Nicholson & Marsh		J. H. Lewis	Chester
	1899		Mota W Pol	Spanish	201	82·3	23·7	12·6	[illegible]	1851			F. Maristany	Palamos
	1900		Motala No. I. *Carlson [-84]* W Sr	Swedish	155	100·0	21·0	10·0	Motala	1875	Motala Co.		S. Sandberg	[illegible]

✳ = Lloyd's Register; D = Nederlandsche Vereeniging; G = Germanischer Lloyd; I = Registro Italiano; R = American [illegible]
A = Veritas Austro-Ungarico; H = Veritas Ellenico; N = Norske Veritas; V = Bureau Veritas

In the 4th Column, C = Composite (iron frame, wood planked); I = Iron; Stl = Steel; W = Wood.

The Dates of Metalling (Col. 11) are based on the records of the Books in which Vessels are respectively classed, or on the Reports of the Surveyors of Lloyd's [illegible]

Further particulars of Vessels classed by Lloyd's Register will be found in Lloyd's Register Book and Lloyd's Yacht Register.

No.	Cote	Nom. Capitaine et dates d'entrée au service des Armateurs actuels et à bord du navire. Ancien Nom.	Matériaux et Gréement	Pavillon	Tonnage net	Longueur	Largeur	Creux	Où	Quand	Par qui	Doublage en métal	Armateurs ou Directeurs-Gérants	Port d'Armement
1901		Motala No. II. Andersson [1884]	W Sr	Swedish	159	100·7	20·9	11·1	Motala	1876	MotalaCo.		Motala Mek. Verkstads Aktiebolag (L. Bromer)	Motala
1902	V	Motia Ferrante -87 (ex Francesco D.)	W Bk	Italian	460	131·2	29·5	18·5	SestriPonente	1876	F.lli Cadenaccio	Y 5.88	Gius. D'Ali	Trapani
1903	*	Motley Harper	W Bn	American	568	145·0	31·5	17·5	Machias, Me.	1877	L. Nash	Y 2.88	H. T. Harper	Machias, Me.
1904	I	Mou (ex Emmanuele) Pezzolo	W Bk	Italian	944	181·3	34·8	24·0	Lavagna	1875	G. Brignole	Y 11.88	Nicolo Bozzo fu G.	Genoa
1905	V	Mouette Boivel	W Sr	French	126	95·8	22·3	11·2	St. Malo	1863	Gautier		Scheult	Binic
1906	V	Mouette Thooris	W Sr	French	111	84·6	19·3	11·8	Dunkirk	1869	A. Lefebvre		E. Didier	Dunkirk
1907		Moulmein	W Sr	British	101	70·0	20·1	8·2	Moulmein	1881			Mackbood Ally	Rangoon
1908	*	Mount Carmel Livingstone [84]	I S	British	1596	263·8	38·7	23·2	Leith	1883	Ramage & Ferguson		Smith & Service	Glasgow
1909		Mount Hope Frowley	W 4m Sr	American	1050	204·0	42·4	19·7	Camden, Me.	1887	H. M. Bean		J. G. Crowley	Taunton
1910		Mount Lebanon Nelson	W Bkn	American	530	134·0	33·0	17·3	St. Martin's NB	1875	J. Carson	Y 3.86	C. A. Nelson	Hong Kong
1911		Mount Macedon Sustin (ex City of Brooklyn)	W S	British	1539	195·0	37·9	21·3 28·7	Brooklyn	18[54] [57]			J. S. Butters	Melbourne
1912	V	Mount Pleasant Fenderson -88	W S	Norweg'n	1493	200·8	40·4	24·9	St. John, N.B.	1864	Reed		Actieselskabet "Mount Pleasant" (J. W. Frost)	Christiania
1913	*	Mount Vernon Jones	I Bk	British	526	171·0	26·8	17·2	Liverpool	1863	T. Vernon & Son		G. Shepherd	Swansea
1914	V	Mount Washington Fox	W S	American	1162	184·6	37·5	23·1	Saco, Me.	1866	C. Hartley	Y 9.86	Leon Blum	San Francisco
1915	*	Mountain Laurel Williams -88	I Bk	British	660	174·4	30·2	19·4	Harrington	1865	R. Williamson & S.		T. Thompson	Liverpool
1916	*	Mountaineer	W Sk	British	90	81·6	20·8	10·6	Rostock	1886	E. Burchard & Co.		Lengfield & Ward	Hull
1917		Mountblairy Steer	W Sr	British	138	95·1	22·3	11·6	Kingston	1874	Geddie		W. H. Steer	Plymouth
1918	*	Mountstuart Wallace 77-85	I S	British	1220	224·1	36·7	21·3	Dumbarton	1878	A. M'Millan & Sn		Hunter, Sheriff & Co.	Glasgow
1919		Moura VIII. (ex Bragança, ex São Paulo)	W	Portug'se	131	74·1	22·0	10·3	VilladoConde	1849	J. M. d'Araujo		J. A. Moura	Funchal
1920	V	Move Hinck -87	W 3m Sr	German	247	111·8	26·7	12·0	Elmshorn	1873	J. Kramer	Y 9.88	C. Hinck	Hamburg
1921	*	Mow Hill Pulford 87-87	I Bkn	British	220	124·5	24·0	11·5	Sunderland	1877	J. Blumer & Co.		W. Price & Co.	Whitehaven
1922	G	Mowe	W Sr	Mexican	177	99·8	21·0	11·0	Motzen	1874	W. Focken	Y 10.83	Zaldivar & Castillo	Campeche
1923	R	Mower Purdy 85-88 (FOUNDERED)	W Sr	British	124	87·1	26·1	8·4	Moss Glen, NB.	1881	G. Merritt		Scammell Bros.	St. John, N.B.
1924	*	Moy Haider 86-86	I S	British	1620	257·0	38·3	23·2	Greenock	1885	Russell & Co.		J. Nourse	London
1925	G	Moya Högberg -88	W Bk	Swedish	318	120·4	25·8	15·1	Grimstad	1869	G. Larsen		C. Molander	Visby
1926	D	Mozambique Pommies (COLLISION)	W Bk	French	451	127·3	25·6	12·5 17·7	LaSeyne	1873	Curet fils	Y 12.85	A. Fabre & fils	Marseilles
1927	DG	Mozart Balle 87-87	W Bk	German	848	161·8	34·1	21·8	Bremerhaven	1865	J. C. Tecklenborg	Y 10.85	H. Janentzky	Rostock
1928		Mozart Christensen 86-86 (CONDEMNED)	W Bk	Norweg'n	320	114·7	29·1	15·3	Tvedestrand	1868	A. & F. Smith	Y 11.84	A. & F. Smith	Tvedestrand
1929	I	Mugnaio Peras	W Bn	Italian	277	..	..	..	Spezzia	1871		Y 2.86	G. Marullo	Milazzo
1930		Muhamedo Mohaidin Bux	W Dhy	British	121	89·4	21·2	12·0	Trincomalee	1873			Madarsaib Rawter (Madarsaib Canny Rawter)	Trincomalee
1931		Muisto Uhlman	W Bk	Russian	643	162·4	33·1	20·5	Raumo	1874	J. H. Justen		G. W. Wahlroos	Raumo
1932		Muisto Pihlman	W Bk	Russian	595	..	..	..	Nystad	1881			J. Malén	Nystad
1933		Mukedamich Khair Mustapha	W Bg	Turkish	125	..	..	..	Oonia	1880			Ooniales Kotch oglou Mustapha	Oonia
1934		Mumbi Tidjaret	W Bg	Turkish	222	..	..	..	Greece	1876			Nalboor Hajee Shefki Effendi	Constantinople
1935	*	Muncaster Castle Valentine 79-82	I 4m Bk	British	2037	300·0	39·2	23·6	Liverpool	1882	W. H. Potter & Sn		Lancaster Shipowners Co. Ltd.	Liverpool
1936		Munsoory	W S	British	1125	188·4	38·2	23·1	Richmond, U.S.	1854			Aga Syed Abdool Hossain	Moulmein
1937	V	Munster Brown 86-86	W S	British	1499	211·5	39·2	24·3	St. John, N.B.	1877	J. Fraser	Y 12.86	Wm. Thomson & Co.	St. John, N.B.
1938		Murad Ali	W Bg	Turkish	225	..	..	..	Alecheeyeh	1874			[illegible]	Alecheeyeh
1939		Murad i Bahri Ali (ex Elisa)	W Bg	Turkish	210	..	..	..	Galaxidi	1866			Receblize Hajee Hassan Effendi & Bros.	Constantinople
1940		Murad i Bahri	W Bg	Turkish	200	..	..	..	Evranyich	1873			[illegible] Oglou	Evranyich
1941		Murad i Bahri	W	Turkish	125	..	..	..	Barten	1869			[illegible] Agha	Barten
1942		Muradie (ex Enrico)	W Bg	M'nt'n'g'o	179	87·3	23·3	17·1	Dulcigno	1875	Jus. Omeri		Zoris Aci Adem Aga	Dulcigno
1943		Muradji Odo Beg (ex Eaton, ex [illegible])	W Bk	Turkish	355	125·5	27·8	17·2	Sunderland	1869	Pickersgill	Y 10.87	Odo Beg	Scutari
1944	I	Murcarolo Balestrino	W Bk	Italian	526	140·3	31·1	21·3	Varazze	1876	F. Tassara	Y 1.85	Stef. Balestrino fu E.	Genoa
1945		Mureftah	W	Turkish	137	..	..	..	Barten	1869			Mureftalidee Strati Hadji Strati	Mureftah
1946		Mureftah	W Bg	Turkish	112	..	..	..	Mureftah	1869			Mureftalidee Andon Hadjeeh	Mureftah
1947	*	Muriel Hale 86-86	W Bk	British	535	158·0	32·3	17·9	P.E. Island	1877	Yeo	Y 1.87	J. M. Campbell	Swansea
1948	*	Muriel Joy	W 3m Sr	British	153	105·9	23·0	12·2	Barnstaple	1878	Westacott & Son	Y 10.87	C. T. Bennett	St. Johns, NFL.
1949	R	Muriel Brinton	W Sr	British	149	89·6	26·0	10·0	BearRiver, NS.	1884	T. H. Rice		T. H. Rice	Annapolis, N.S.
1950	*	Muriel (COLLISION) Rogers	W Sk	British	75	71·8	20·2	10·3	Rye	1884	J. C. Head		R. T. Vivian	Hull

*=Lloyd's Register; D=Nederlandsche Vereeniging; G=Germanischer Lloyd; I=Registro Italiano; R=American Record;
A=Veritas Austro-Ungarico; H=Veritas Ellenico; N=Norske Veritas; V=Bureau Veritas.

Dans la 4me Colonne, C=Composite (membrure en fer, bordé en bois); I=Fer; Stl=Acier; W=Bois.

Les Dates du doublage en métal (Col. 11) ressortent des livres dans lesquels les navires sont respectivement classés ou des rapports des Inspecteurs du Lloyd's Register.

Pour plus amples renseignements sur les navires classés par le Lloyd's Register, voir le Lloyd's Register Book et le Lloyd's Yacht Register.

Official No.	No. in Book	Society	Ship's Name. Master, and Dates of Appointment to present Owner's service and to Ship. Ship's Late Name.	Material and Rig	Flag	Net Tonnage	Length	Breadth	Depth	Built Where	When	By whom	Date of Metalling	Owners or Managers	Port
	1951	V	Muriel	W Bk	Norweg'n	596	142·7	32·7	18·7	St. John, N.B.	1873	Mahony	[illegible]	Actieselskabet "Muriel" ([illegible])	P[illegible]
	1952		Muros	W Bn	Spanish	124	82·0	19·7	9·8	Corunna	1874			Manuel Siava	Corunna
	1953	V	Musca *Gieseke 88-88*	I Bk	German	700	163·4	28·8	16·3	Vegesack	1878	H. F. Ulrichs		J. H. Hustede	Elsfleth
	1954	N	Mustang *Townsen -88*	W Bk	Norweg'n	375	119·3	29·2	16·2	[illegible]	1873	L[illegible]gström	[illegible]	N. S. Gausdal	Lillesand
22290	1955	✱	Mute *Blasby 69-81*	W Sr	British	89	81·6	18·6	9·1	Portsmouth	1853			W. Bayley	Ipswich
77136	1956		Mutto Komarasamypuravy	W Bg	British	114	93·0	21·0	10·0	Valvertyturie	1877			[illegible]	Jaffna
24475	1957		Mutual *Clarkson 70-70*	W Bg	British	197	84·5	25·1	14·2	Sunderland	1846			R. Harrowing	Whitby
41520	1958		Myan Aung	W Sr	British	161	92·0	21·9	10·4	Moulmein	1888			Mah Shive Bevin	Moulmein
86157	1959	✱	Mylomene *Cross 78-83*	I S	British	1900	270·0	39·8	24·0	Liverpool	1882	W. H. Potter & Sn		H. Fernie & Sons	Liverpool
95062	1960		Myohla	Stl Lr	British	480	220·0	30·0	8·4	Dumbarton	1888	W. Denny & Bro.		Irrawaddy Flotilla Co. Ld.	Glasgow
748	1961		Myoun Maru	W Sr	Japanese	143	89·3	21·3	10·5	Hiogo	1882	H. Genbei		Hiroumi Nisabaro	Osaka
54357	1962	✱	Myra *Bower -85*	W Bg	British	218	95·2	25·2	15·0	Sunderland	1866	W. H. Pearson jr.	[illegible]	R. G. Sanders	Folkestone
	1963	R	Myra W. Spear	W 3m Sr	American	148	..	..	..	Bath, Me.	1888	Kelley & Spear		J. R. Kelley	Bath, M.
	1964	G	Myran *Aspenberg -82*	W Sr	Swedish	137	99·4	20·4	9·7	Svanvik	1876	C. Andersson		F. L. Pettersson	Svanvik
17267	1965		Myronus *Chatto -88*	W Bn	American	362	119·5	27·5	11·6	Ellsworth, Me.	1865	D. P. Jordan	[illegible]	J. M. Higgins	Ellsworth
16865	1966		Myrover *Chase -87*	W Sr	American	397	127·0	30·2	10·4	Belleville, N.J.	1854			Evans, Ball & Co.	New York
	1967	V	Myrte (WRECKED) *Gaboret*	W Bk	French	200	121·7	27·0	12·7	Musquash, N.B.	1864		[illegible]	E. Fumin	St. Valery
85002	1968	V	Myrtle *Carter 83-83*	W Bkn	British	387	144·7	30·3	13·1	Black River, N.B.	1883	J. & R. McLeod	[illegible]	J. A. Likely	St. John, N.B.
80792	1969	V	Myrtle *Sherratt*	W Bn	British	169	101·0	25·0	10·0	Bear River, N.S.	1881	G. F. Miller	[illegible]	G. F. Miller	Annapolis
78364	1970		Myrtle *Ross -85*	W Bn	British	167	96·4	23·6	10·3	Auckland, N.Z.	1878	Lane & Brown		A. Tornaros	Sydney
25257	1971		Myrtle (ex Sybarite)	W Sr Yt	British	116	95·6	21·0	11·6	Gosport	1852	Camper & Nicholson	[illegible]	Capt. R. G. Gattsen	Portsmouth
72921	1972	✱	Myrtle Holme *Millican 178-83*	I Bk	British	902	211·0	32·9	19·7	Sunderland	1875	Bartram, Haswell & Co.		Hine Bros.	Maryport
58994	1973	✱	Mysore *Chard 85-87*	I S	British	1594	218·1	40·1	23·8	Birkenhead	1869	G. R. Clover & Co.		Eyre, Evans & Co.	Liverpool
58958	1974	✱	Mysterious Star *Rees*	W Bk	British	397	121·9	28·5	17·5	Sunderland	1867	J. Barkes	[illegible]	R. Davies	Aberystwith
12789	1975		Mystery *Finney 82-82*	W Bn	British	139	79·0	20·0	11·8	Sandwich	1850			D. T. Finney	London
91688	1976		Mystery (LOST) *Martin*	W Sr	American	124	98·2	25·0	9·4	Kennebunk, Me.	1884	G. Christenson		J. F. Wonson & Co.	Gloucester
17930	1977	V	Mystic Belle *Freeman 83-83*	W S	American	717	150·7	33·0	21·1	Medford, Mass.	1867	J. T. Foster	[illegible]	E. Lawrence	Boston, Ms.
63232	1978		Myvanwy *Vincent*	W 3m Sr	British	162	106·0	23·0	12·6	Newport, Mon.	1869	Newport Dry Dock S. B. Co.	[illegible]	W. R. P. Murray	Port Fa[illegible]

ADDITIONAL SHIPS ARE INSERTED IN THE SUPPLEMENT AT THE END OF THE REGISTER.

✱ = Lloyd's Register; D = Nederlandsche Vereeniging; G = Germanischer Lloyd; I = Registro Italiano; R = America[illegible]; A = Veritas Austro-Ungarico; H = Veritas Ellenico; N = Norske Veritas; V = Bureau V[illegible]

In the 4th Column, C = Composite (iron frame, wood planked); I = Iron; Stl = Steel; W = Wood.

The Dates of Metalling (**Col. 11**) are based on the records of the Books in which Vessels are respectively classed, or on the Reports of the Surveyors of Lloyd's [illegible]

Further particulars of Vessels classed by Lloyd's Register will be found in Lloyd's Register Book and Lloyd's Yacht Register.

		4 Nom. Capitaine et dates d'entrée au service des Armateurs actuels et à bord du navire. Ancien Nom.	Matériaux et Gréement.	5 Pavillon.	6 Tonnage net.	7 Dimensions. Longueur	Largeur	Creux	8 Construit. Où.	9 Quand.	10 Par qui.	11 Date du doublage en métal	12 Armateurs ou Directeurs-Gérants.	13 Port d'Armement.
1	RV	N. B. Lewis *Hoffman -88*	WS	British	1326	202·0	39·6	23·2	Salmon River, N.S.	1880	N. B. Lewis	V 2.88	N. B. Lewis	Yarmouth, N.S.
2	V	N. B. Morris *Smith -81*	W Bk	British	699	160·0	33·5	20·2	Advocate, N.S.	1880	W. Moore	V 3.87	W. Moore	Parrsboro', N.S.
3	NV	N. B. Palmer *Sorensen*	W S	Norweg'n	1117	200·0	38·7	21·7	New York	1850	A. Westervelt		T. Thommesen & Sön	Arendal
4		N. Boynton *James 83-88*	W Bk	American	1012	162·7	37·4	23·7	E. Boston Mass	1866	A. Gove	V 3.88	G. C. Lord & Co.	Boston, Mass.
5		N.E.V.A. *Burwood 79-80*	W 3m Bn	British	147	..	21·0	11·8	Newport	1855			E. G. Craske	Lowestoft
6		N. H. Burrows *Taylor*	W 3m Sr	American	240	125·9	32·9	9·5	Green Bank NJ	1874			C. G. Endicott	New York
7		N. H. Skinner *Nickerson*	W Sr	American	284	117·0	30·0	8·0	Wilmington Del	1867	J. Birr		J. Nickerson	Barnstable, Mass.
8		N. J. Miller	W Sr	American	119	81·0	24·0	9·0	Booth Bay, Me.	1873			J. Fairfield	Salem, Mass.
9	G	N.M. Harboe *Andersen 58-88*	W Sr	Danish	141	87·4	21·3	11·7	Nyborg	1858	P. Petersen		S. & R. Harboe	Skjelskør
10	N	N.P. Nielsen *Knudsen -88*	W Bk	Norweg'n	476	137·7	30·5	16·7	Lillesand	1874	O. & P. Knudsen	V 8.87	O. & P. Knudsen	Lillesand
11	DV	Nachtegaal *Beckman 83-*	C Bk	Dutch	1015	189·3	36·7	21·1	Amsterdam	1884	J. F. Meursing	V 5.87	J. F. Meursing	Amsterdam
12		Nachtigal *Ohsoling*	W Sr	Russian	180	95·8	24·3	12·0	Orrenhof	1878	J. Saul		C. Ohsoling & Christiansen	Pernau
13	A	Nada *Percovil 85-88*	W 3m Sr	Austrian	430	127·3	27·6	17·3	Fiume	1878	G. Bacach Bellaz	V 6.86	J. Wimmer & Co.	Fiume
14		Nadeshda (ex Tre ...) *Noak*	W Bk	Russian	345	118·9	27·9	17·2	Sunderland	1866	W. Watson	V 2.86	H. Noak	Riga
15		Nadeshda *Port*	W Sr	Russian	195	97·6	28·9	12·6	Gutmannsbach	1873	J. Jacobsson		J. Port	Gutmannsbach
16	N	Nadia *Thorsen*	W Bk	Norweg'n	590	119·9	31·7	17·8	Kragerö	1879	Alf. Johnsen	V 1.87	Et Interessentskab (J. Aasberg)	Kragerö
17	DV	Nadinka *Schmidt 66-83*	W Bn	Danish	233	115·9	26·3	11·3	Värnanäs	1870	C. A. Hahn	V 1.88	J. H. Schmidt	Ribe
18	G	Naema *Henriksson 78-*	W Sr	Russian	270	87·4	21·3	11·7	Finby	1876	G. Kronstrom		H. V. Söderlund	Finby
19		Naemi (ex Marie Amalia) *Fromén*	W Bk	Swedish	219	109·9	25·7	10·9	Bahia				N. J. Jönsson	Calmar
20	N	Næsheim (ex William Clowes) *Rich*	W Bk	Norweg'n	555	148·7	29·7	19·0	Sunderland	1865	J. Robinson	V 10.87	J. M. Jonasen	Farsund
21		Nagalingapuravy	W Dhy	British	101	85·5	20·4	15·0	Jaffna	1874			Ammiar Vallipuram	Jaffna
22	❋	Nagpore *Fraser 83-85*	I S	British	1209	204·2	35·5	22·1	Glasgow	1863	Barclay, Curle & Co.		Nagpore Ship Co. Ld. (A Fer)	Liverpool
23	R	Nahum Chapin *Arey*	W 3m Sr	American	567	145·0	35·0	15·6	Rockland, Me.	1882	Cobb, Wight & Co	V 11.87	Cobb, Wight & Co.	Rockland, Me.
24	❋	Naiad *Davies 81-87*	I Bk	British	1039	204·8	34·7	21·5	Liverpool	1876	Bowdler, Chaffer & Co.		J. B. Walmsley & Co.	Liverpool
25		Naiad *Samuel*	I S	British	135	93·1	21·6	11·6	Llanelly	1867	Neville Bros.		J. Samuel & Co.	Llanelly
26	❋	Naiad *Corr*	W Sr	British	81	79·7	22·0	8·6	Hopewell N.B.	1862	N. M. Bennet		W. Murphy	Wexford
27		Naim (ex ...)	W Sr	Dutch	249	..	..	..	Värnanäs	1864	G. Hahn	V 2.77	Said Hassan bin Abdul Rachman Mola Hatta	Sourabaya
28		Naima *Åberg*	W Bk	Russian	404	134·0	30·5	17·2	Nagu	1868	T. Lindqvist		A. Anderson	Nagu
29		Naima *Jakobsson 81-*	W Sr	Russian	229	101·7	28·9	13·0	Sideby	1874			K. J. A. Granborg	Åbo
30	❋	Nairnshire *Lowrie 84-84*	I S	British	965	204·0	33·6	19·9	Glasgow	1876	Dobie & Sons		T. Law & Co.	Glasgow
31		Naisnigel *Wennström*	W Gt	Russian	143	..	..	..		1872			G. A. Wennström	Wardö
32	N	Naja *Jacobsen -88*	W Bk	Norweg'n	428	135·7	31·6	16·7	Arendal	1873	J. Aanonsen	V 12.85	Morten Kallevig	Arendal
33	GV	Najade *Fischer 87-88*	Stl S	German	1691	258·8	39·4	23·2	Geestemünde	1888	J. C. Tecklenborg		Gildemeister & Ries	Bremen
34		Najade *Duit 66-*	W Sr	German	113	77·8	21·3	14·3	Elmshorn	1856	D. Kramer		H. D. Duit	Tönning
35		Najaden *Ericksen -86*	W Bg	Danish	288	102·0	28·2	16·1	Tvedestrand	1859	A. & F. Smith		J. C. Sand & C. Holmen	Lönstrup
36	DV	Najaden *Birkholm 76-75*	W Bn	Danish	135	96·1	21·3	10·7	Faaborg	1875	R. H. Møller		P. Nielsen	Faaborg
37		Najaden (Loser) *Edvardsen*	W Bk	Norweg'n	534	143·4	33·5	17·8	Jacobstad	1860			S. A. Solberg	Drammen
38	N	Najaden *Thorsen 72-81*	W 3m Sr	Norweg'n	239	126·7	27·0	10·5	Kragerö	1866	J. Amber	V 12.86	H. Biörn, jun.	Kragerö
39	N	Najaden *Christiansen -87*	W Bg	Norweg'n	203	100·3	25·8	12·6	Grimstad	18 58/72	T. Gundersen		O. Christiansen	Lillesand
40	N	Najaden *Thorgersen*	W Bg	Norweg'n	140	77·0	24·2	12·0	France	1830			C. H. Knudsen	Drammen
41		Najaden	W Bk	Swedish	283	112·8	29·0	15·0	Carlscrona	1834			G. Sjöberg	Malmö
42		Najaden (ex Elise) *Flink*	W Sr	Swedish	130	81·0	21·0	11·2	Holstein	1858			J. D. Björkegren	Cimbrishamn
43		Nalon (ex ...) *Garcia*	W Bk	Spanish	333	116·1	27·5	11·4	Harringtn, Me	1874			Castano & Intriago	Cienfuegos
44		Nameaug	W Bg	American	292	112·0	28·1	14·2	Westerly, R.I.	1854			J. Atkins	New York
45		Nameless	W Bn	British	143	100·0	22·8	12·0	P. E. Island	1857			R. Farrell	Youghal
46	V	Nanaimo *Newby -88*	W Bk	British	397	153·0	34·0	12·5	Nanaimo, B.C.	1882	A. Allen	V 1.87	Robt. P. Rithet	Victoria, B.C.
47	V	Nancy *Heinzelmann 74-83*	W Sr	Danish	115	83·6	20·7	10·4	Stege	18 55/72	A. Petersen	V 1.84	W. C. H. Fischer	Copenhagen
48		Nancy J. Day *Monroe*	W Sr	American	123	97·3	27·1	7·4	Bristol, Me.	1874			J. A. Chamberlain	Damariscotta, Me
49		Nancy M'Sweeney	W Bg	British	197	97·3	24·4	12·6					Mrs. H. Sutton	Cork
50	V	Nancy Pendleton *Pendleton -88*	W S	American	1385	199·6	39·2	24·1	Belfast, Me.	1871	H. McGilvery	V 1.87	W. Beazley	Belfast, Me.

❋ = Lloyd's Register; D = Nederlandsche Vereeniging; G = Germanischer Lloyd; I = Registro Italiano; R = American Record; A = Veritas Austro-Ungarico; H = Veritas Ellenico; N = Norske Veritas; V = Bureau Veritas.

Dans la 4me Colonne, C = Composite (membrure en fer, bordé en bois); I = Fer; Stl = Acier; W = Bois.

Les Dates du doublage en métal (Col. 11) ressortent des livres dans lesquels les navires sont respectivement classés ou des rapports des Inspecteurs du Lloyd's Register.

Pour plus amples renseignements sur les navires classés par le Lloyd's Register, voir le Lloyd's Register Book et le Lloyd's Yacht Register.

1 Official No.	2 No. in Book	3 Registry in which classed	4 Ship's Name. Master, and Dates of Appointment to present Owner's Service and to Ship. Ship's Late Name.	Material and Rig.	5 Flag.	6 Net Tonnage.	7 Dimensions. Length.	Breadth.	Depth.	8 Built. Where.	9 When.	10 By whom.	11 Date of Metalling.	12 Owners or Managers.	Port of Registry.
[illegible]	51	R	Nancy Smith *Hodgkinson 11-78*	WSr	American	419	124·5	31·5	17·6	Stony Brook, N.Y.	1868		[illegible]	L.S.Davis	[illegible]
[illegible]	52		Nanhoron *Williams*	WBn	British	147	92·6	22·7	12·2	Nevis	1859	Thomas		H.Williams	[illegible]
	53	V	Nanine Aglaé *Reanne*	WBk	French	191	90·2	22·6	12·1	Nantes	1875	E.Chergeau	[illegible]	L.Ezanno	[illegible]
	54	V	Nanna *Nielsen -87*	WBk	Danish	364	127·3	26·2	15·5	Nordby	1887	S.Abrahamsen		C.H.Nielsen	[illegible]
	55	N	Nanna *Andersen -87*	WBn	Norweg'n	157	98·2	22·8	11·1	Flekkefjord	18[illegible]		[illegible]	R.Pedersen	[illegible]
	56	D V	Nanna *Grabbström -85*	WS	Swedish	711	165·0	29·6	18·7	Kylorn	1882	Fr.Haggland	[illegible]	[illegible] Skyrederi Aktie-bol. (G. Haggman)	[illegible]
	57	N	Nannette *Nielsen 82-88* (FOUNDERED)	WSr	Norweg'n	139	92·2	22·2	10·4	Elsfleth	1860			H.Sorensen	[illegible]
[illegible]	58	R	Nannie T. Bell *Patterson 1-87*	WBk	American	412	122·8	29·0	16·7	Brooklyn	1865	Williams	[illegible]	W.M.Patterson	[illegible]
	59	*	Nanny *Imethgen* (ex [illegible])	IBk	German	595	172·0	28·6	17·5	Liverpool	1868	T.Royden & Sns.		J.W.Burmester	[illegible]
	60		Nanny *Gronqvist*	WSr	Russian	153	..	..	..	[illegible]	1868			F.W.Ridderstad	[illegible]
	61	G	Nanny *Andersson 87-87*	WBk	Swedish	423	130·9	28·9	16·8	Stockholm	1867	K.Agerskow	[illegible]	D.V.Pride	[illegible]
[illegible]	62	R	Nantasket *Richardson 75-83*	W3mSr	American	496	155·6	34·9	13·0	Waldoboro' Me	1882	E.O.Clark	[illegible]	J.S.Emery & Co.	[illegible]
	63	V	Nantes *Lemerle*	WBk	French	336	123·0	27·1	15·4	Nantes	1874	T.Dubigeon & fils	[illegible]	Rozier & Co.	Nantes
[illegible]	64		Naomi	WBn	British	152	97·2	24·4	12·2	P.E.Island	1874	Peake		M.Morgans	Llanelly
	65		Naomi (ex Harriet White)	WBk	Norweg'n	762	157·0	29·5	22·6	Portsm'th, N.H.	1837			N.Olsen & M.Andreasen	Arendal
	66	*	Naomi *Pettersson -75*	WBk	Swedish	673	155·8	28·9	19·8	Sunderland	1859	Peverall	[illegible]	A.Liljeqvist	Gothenburg
[illegible]	67	D	Napier *Campbell 84-84* (ex [illegible])	CBk	British	993	207·3	31·5	20·1	Port Glasgow	1869	R.Duncan & Co.	[illegible]	J.Hossack, sen.	Liverpool
[illegible]	68		Napoleon	WBk	American	306	105·7	26·4	18·2	Rochester, Mass.	1838			J.Bourne	Edgartown
18240	69		Napoleon *Tyrrel*	W3mSr	American	193	120·0	28·2	8·1	Newburgh, N.Y.	1857			Clarissa A.Terrell	New York
	70	N	Napoleon *Olsen*	WBk	Norweg'n	739	148·3	31·8	21·9	Medford, N.J.	1851		[illegible]	Ex Interessentskab (A. [illegible])	Sarpsborg
130323	71	R	Napoleon Boughton *Sells*	W4mSr	American	786	188·0	37·6	18·7	Fairhaven, Ct.	1888	H.H.Hanscom		H.H.Hanscom	New Haven
	72	N	Napoleon III. *Olsen -86*	WBk	Norweg'n	741	184·5	31·8	18·9	Ardrossan	1855	Barr & Shearer	[illegible]	J.Ihlen	Christiania
	73	I	Napoli *Maresca -88*	WBn		172	115·0	29·5	9·0	[illegible]	1888	A.Castellano	[illegible]		
	74	V	Nara *Perlstein -88*	WSr	Russian	132	80·0	25·2	10·9	Uppesgriew	1888	J.J.Kirstein		F.Lerch	Uppesgriew
[illegible]	75		Naramissie	WBk	British	273	113·0	25·6	11·7	Orland, U.S.	1852			Jas.Huddart	Melbourne
[illegible]	76		Narassa	WSr	British	148	86·4	25·5	10·2	Parrsboro', N.S.	1878			J.A.Hatfield	Parrsboro
	77		Narcisa	WSr	Brazilian	104	..	..	..					J.D.Vianna	Pelotas
[illegible]	78		Narcisa *Sardi*	WBg	Italian	196	96·5	27·6	15·8	C.d'Elba	1849	E.Brischi		G.Tonietti fu Raff	Porto [illegible]
76149	79	*	Narcissus *Eustice*	IS	British	1270	235·0	37·1	22·0	Port Glasgow	1876	R.Duncan & Co.		C.S.Caird	Greenock
[illegible]	80	R	Narcissus *Heisler*	WSr	British	127	85·5	23·7	9·9	Lunenburg, N.S.	1888	D.Smith		S.W.Oxner	Lunenburg
60946	81	*	Nardoo *Campbell -87* (HULK)	WBk	British	379	131·2	27·2	16·4	Dundee	1869	Tay S.B.Co.	[illegible]	Robt.Rae	Melbourne
[illegible]	82	*	Nariva *Reid 78-88*	WBk	British	493	142·3	29·7	18·4	Troon	1871	Troon S.Co.	[illegible]	G.Turnbull & Co.	Glasgow
[illegible]	83		Narragansett *Carson*	W3mSr	American	239	112·2	32·0	8·8	Wilmington, Del.	1874			G.W.Stone	Wilmington
75755	84	R V	Narwhal *Weston 74-79*	WS	British	1328	204·7	39·0	23·7	Saulnierville, N.S.	1879	J.Lovitt	[illegible]	Lovitt & Co.	Yarmouth
	85		Nassir Bahri	WBg	Turkish	300	..	..	..	Oonia	1876			Gonian Zadeh Hafee (Ibrahim Effendi)	Oonia
	86		Nassry	WBg	British	194	78·0	20·9	14·4	Surat	1810			Syud Abdool Rahimon (Baboo [illegible])	Bombay
[illegible]	87		Nassurool Ahmid *[illegible]*	WBg	British	106	77·0	19·6	9·9	Cranganoor	1857			A.C.N.Maricar	Colombo
[illegible]	88	R	Nat Meader *Brown*	W3mSr	American	264	119·8	29·8	10·3	Wiscasset, Me.	1885	J.M.Haynes		J.M.Haynes	Bath, Me.
[illegible]	89	*	Natal *Scorgie 87-87*	IBk	British	446	155·6	27·0	13·6	Aberdeen	1876	A.Hall & Co.		J.T.Rennie & Son	Aberdeen
[illegible]	90	D	Natal *Couch 77-87*	W3mSr	British	189	112·0	23·1	12·7	Alloa	1861	Adamson		W.Stephens	Fowey
	91		Natal (WRECKED) *Hillstedt*	WBn	Swedish	335	118·1	22·8	11·7	Gefle	1878	J.Bergfors & Son	[illegible]	F.Dyback	Gefle
54710	92		Natal Queen *Garth*	WBk	British	230	115·8	23·5	12·7	Grangemouth	1866	Adamson		Bellan & Co.	Hobart
[illegible]	93	V	Natale Gallino *Zino*	WBk	Italian	840	171·9	33·7	22·8	Varazze	1876	N.Cerruti	[illegible]	A.Gallino	Genoa
	94		Natalia *Escauriaza* (ex [illegible])	WBk	Chilian	463	140·0	28·1	17·5	Troon	1861	Portland S.B.Co.		J.Martinez	Valparaiso
[illegible]	95	R	Natalie *McDonald -87*	WBn	British	199	95·3	21·7	12·1	P.E.Island	1878	T.P.Richards	[illegible]	Wm.Fanning	Halifax, N.S.
[illegible]	96	*	Natalla *Davies 86-86*	W3mSr	British	175	110·4	22·7	11·8	Troon	1879	Troon S.B.Co.	[illegible]	J.C.Peacock	Glasgow
71019	97	V	Natant *Porter 83-86*	WBk	British	1064	179·0	36·6	21·9	Saulnierville, N.S.	1875	J.Lovitt	[illegible]	Lovitt & Co.	[illegible]
	98	V	Nathalia *Hager 75-76* (ex Nanny Marie)	WBn	Danish	123	92·9	21·6	10·6	Djursvik	18[illegible]	S.Petersen		Chr.Ohlsen	Skjelskør
	99		Nathalia *Lundberg 86-86*	WBk	Russian	352	..	..	..	[illegible]	1869			K.Eriksson	[illegible]
[illegible]	100	R	Nathan Easterbrook, jun. *Sharp 84-88*	W3mSr	American	677	156·0	35·2	18·1	West Haven, Conn.	1883	[illegible] & Maur	[illegible]	H.Sutton	[illegible]

*=Lloyd's Register; D=Nederlandsche Vereeniging; G=Germanischer Lloyd; I=Registro Italiano; R=American [illegible]
A=Veritas Austro-Ungarico; H=Veritas Ellenico; N=Norske Veritas; V=Bureau Veritas

In the 4th Column, C=Composite (iron frame, wood planked); I=Iron; Stl=Steel; W=Wood.

The Dates of Metalling (Col. 11) are based on the records of the Books in which Vessels are respectively classed, or on the Reports of the Surveyors of Lloyd's Re[gister].

Further particulars of Vessels classed by Lloyd's Register will be found in Lloyd's Register Book and Lloyd's Yacht Register.

No.	Cote	Nom. Capitaine et dates d'entrée au service des Armateurs actuels et à bord du navire. Ancien Nom.	Matériaux et Gréement	Pavillon	Tonnage	Dimensions: Longueur	Largeur	Creux	Construit: Où.	Quand	Par qui.	Date du doublage en métal	Armateurs ou Directeurs Gérants.	Port d'Armement.
101		Nathan Lawrence *Hayes*	W3mSr	American	732	168·4	35·2	15·0	Bath, Me.	1882	Goss&Sawyer		J.B.Phillips	FallRiver,Mass
102	DN	Nathanael *Thue 84-87* (ex Hesperia von Bremen)	WBk	Norweg'n	572	145·7	31·3	17·5	Rönnebeck	1858	Reichmann	Y 11.86	J.Harbitz	Tönsberg
103		Nathaniel *Davis -76*	WSr	British	98	76·6	21·0	11·1	Portmadoc	1858	Wm.Griffiths		A.Davies	Aberystwith
104		Nathaniel Lank *Ferreira -88*	W3mSr	American	268	119·6	33·8	9·0	Frederica,Del.	1883	NathanielLank		J.W.Hall	Wilmington,Del.
105		National	WSp	British	151	108·7	22·8	8·1	St.Thomas de Pierreville PQ	1886	G.Mauseau		V.Gladu	Montreal
106	V	National *Guychague 83-88*	I Bk	French	753	182·4	31·2	19·5	Nantes	1867	T.Dubigeon		Ant.Dom. Bordes & fils	Bordeaux
107		Native Lass	WSr	British	106	73·2	21·1	10·9	Baddeck,C.B.	1851			Thos.G.Kelly	Sydney,N.S.W.
108	R	Natividad (ex Vulture)	WBn	Spanish	174	100·6	23·0	11·5	Newport,R.I.	18		Y 11.85	F.L. del Valle	Trinidad
109		Nativita *Pendibene*	WBn	Italian	142	95·1	22·1	12·9	Varazze	1874	G.B.Baglietto		E.G.B.Pendibene	Genoa
110		Nativité	WBn	French	137	74·1	21·0	12.0	Vertou	1866	Doussin		Doussin	Nantes
111		Nativité *Laurent*	Bn	French	122	..	..	..					Simon Amaury	Redon
112	✻	Natuna *Grahn 85-85*	I Bk	British	1106	214·6	35·1	21·2	Greenock	1885	Russell&Co.		P.Denniston & Co.	Glasgow
113	N	Natvig *Sandblad 84-84*	WBk	Norweg'n	486	135·2	30·9	17·1	Arendal	1875	O.P.Olsen	Y 11.87	Gunder Olsen	Arendal
114	RV	Naupactus *77-84 Lovitt*	WS	British	1399	218·4	39·3	23·0	Digby,N.S.	1884	J.Lovitt	Y 9.86	Lovitt & Co.	Yarmouth,N.S.
115		Nausika *Gahlsen -87* (FOUNDERED)	WBk	Norweg'n	588	149·4	31·6	18·4	Risör	1877	ErikKnudsen	Y 9.84	J.Prebensen	Risör
116	V	Nautik *Mikkelsen -88*	W3mSr	Norweg'n	323	123·4	27·9	13·5	Arnis	1875	F.Johansen	Y 5.86	C.Hansen	Egersund
117		Nautilus *Hawkins*	WBk	British	243	119·4	26·0	14·0	Hobart Town	1872	Mackay		E.W.Knox	Sydney,N.S.W.
118		Nautilus *Toleman*	WSr	American	128	87·6	23·2	8·2	ThomastonMe	18 59/80			N.A.Farwell	Rockland,Me.
119		Nautilus	WBg	Chilian	463	104·9	23·6	..					J.A.Horman	Valparaiso
120		Nautilus *Micheli*	WBk	Chilian	262	110·8	27·6	..	New Bedford	1851			Cia Chileña de Ballens	Valparaiso
121	V	Nautilus *Lilienthal -87*	I Bk	German	725	186·4	31·8	18·0	Hamburg	1878	Reiherstieg [Schiffs Werfte		C.H.Donner	Hamburg
122		Nautilus *Brüning -87*	WSr	German	269	109·4	26·2	15·0	Lübeck	1858			W.Höveler	Papenburg
123	V	Nautilus *Anderson -86*	W3mSr	Swedish	192	107·9	23·0	9·9	Bofvenäs	1886	J.H.Seldén	Y 12.85	J.H.Seldén	Uddevalla
124		Navarch *Wilson 81-81*	WBk	British	995	174·0	35·8	21·8	Saulnierville, [N.S.	1877	J.Lovitt	Y 12.86	Lovitt & Co.	Yarmouth,N.S.
125	R	Navarino *Cole 85-87* (ex Jenny B. Gilkey)	W3mSr	American	321	122·0	30·0	13·7	East Boston	1872	A.Gove	Y 10.85	J.S.Emery & Co.	Boston,Mass.
126		Navarinon	WBg	Greek	211	104·0	24·3	..	Genoa	1869			D.Zouros	Syra
127		Navarro y Burr *Olivier* (ex Lucy, ex Dr. Jones)	WBk	Chilian	485	144·5	28·4	17·3	Sunderland	1866	S.P.Austin&Son		J. de D. Navarro & M. G. Velasquez	Valparaiso
128		Navegador *Eça -88* (ex Harbor Queen)	W3mSr	Portug'se	228	105·0	21·8	12·6					Bensaude & Co.	Lisbon
129		Navegante (ex Loyola, ex Thomas Heron)	WBk	Chilian	568	139·0	31·1	18·4	Manila	1863			A.Steel & Co.	Coquimbo
130	R	Navesink (DISMANTLED)	WBk	American	688	156·9	32·0	20·0	Newcastle,Me.	1866		Y 4.81		
131		Navigator *Trevett 80-80*	WBn	British	194	110·5	23·0	12·9	Cardiff	1865	Batchelor	Y 5.85	R.M.Church	Portsmouth
132		Navigator *Kirk 68-68* (ex Hans)	WSr	British	142	91·6	21·6	10·8	Cuxhaven	1851			J.Saul	Lowestoft
133	DN	Navigator *Gabrielsen*	WBk	Norweg'n	531	141·1	32·8	18·8	Stavanger	1869	R.O.Hœrem	Y 2.87	C.H.Gabrielsen	Stavanger
134	V	Navigator *Petterson 78-81*	WBk	Russian	673	163·0	30·1	18·1	Jakobstad	1866	H.Blassan	Y 8.85	P.Malm,jun.	Jakobstad
135	IV	Navigatore *Ciampa -88*	WBk	Italian	620	148·6	30·0	21·2	Cassano	1877	F.DeRosa	Y 10.86	A.Lauro & Figli	Castellamare
136		Nawan Eljoesoer (ex Abdul Hollandia Hoesan)	WBk	Dutch	318	102·3	26·2	..	Lassem	1843	Brown	Y 12.82	Said Achmat bin Abdul Gader Assegaf	Sourabaya
137	A	Nazaret *Canalas*	WBk	Greek	427	135·1	29·0	..	Syra	1875	A.Livadara	Y 4.88	A.Canalas	Syra
138		Nea Arghiri *C.Culucundi*	WBk	Greek	299	104·3	27·2	..	Syra	1868	C.Cristali	Y 7.85	E.& G.Culucundi	Syra
139	A	Nea Ariadni *Handras*	WBg		264	96·1	29·5	16·8	Syra	1869	N.Paydas	Y 80		
140	A	Nea Emporiani *Spanoudis*	WBn	Greek	254	103·9	26·1	..	Syra	1878			Spanoudis Bros.	Syra
141		Nea Evelpis *Apostolon*	WBn		232	95·5	26·2	..	Syra	1885	M.Cufudachi			
142		Nea Maria	WBk		357	108·2	26·6	17·3	Syra	1868	A.Livadara			
143	II	Nea Marigo	WBg	Greek	225	92·8	25·9	..	Syra	1859			G.A.Stenos	Syra
144		Nea Omonia *Margaris*	WBn	Greek	247	99·4	23·0	..	Syra	1879	S.Cristali		E.Margaris	Syra
145	II	Nea Panagia *Michalizakis* (ex Fleury)	WBg	Greek	218	87·9	26·2	..	Syra	1867		Y 84	A.Michalizakis	Argostoli
146		Nea Rosina *Coutsaucos*	WSr		106	74·1	21·5	..	Syra	1882	A.Livadara			
147	II	Nea Thessalia *Kotzias*	WBg	Greek	248	100·0	23·9	..	Syra	1873	D.Epifanis		G.Kotzias	Volo
148	A	Nea Tychi *Marcantoni*	WBn	Greek	407	128·9	28·7	..	Syra	1876		Y 11.85	K.H.Midoulis	Syra
149	II	Nea Tychi *Tsicanias*	WBg	Greek	229	..	..	..	Syra	1870		C 2.83	N.Catsoulieris	Syra
150		Nea Tychi *C.Macris*	WBn	Greek	203	..	..	..	Syra	1875		Y 1.86	D.E.Macris	Syra

Lloyd's Register; D = Nederlandsche Vereeniging; G = Germanischer Lloyd; I = Registro Italiano; R = American Record; A = Veritas Austro-Ungarico; H = Veritas Ellenico; N = Norske Veritas; V = Bureau Veritas.

Dans la 4me Colonne, C = Composite (membrure en fer, bordé en bois); I = Fer; Stl = Acier; W = Bois.

Les Dates du doublage en métal (Col. 11) ressortent des livres dans lesquels les navires sont respectivement classés ou des rapports des Inspecteurs du Lloyd's Register.

Pour plus amples renseignements sur les navires classés par le Lloyd's Register, voir le Lloyd's Register Book et le Lloyd's Yacht Register.

Official No.	No. in Register	Registry in which classed	Ship's Name. Material and Rig. Master and Dates of Appointment to present Owner's Service and to Ship. Ship's Late Name.	Flag.	Net Tonnage.	Length	Breadth	Depth	Built. Where.	When.	By whom.		Owners or Managers	
10	151		Nea Tychi (ex Pactolos) WBn	Greek	177	81·0	23·6	..	Cephalonia				Const. Spanos	Kyr[illegible]
	152		Nea Tychi [illegible] WBn	Greek	161	90·9	21·6	12·6	Poros	1875	I. Archi[illegible]	[illegible]	J. Elefterion	Cumi
	153		Nea Tychi *Eleutherion* Wg	Greek	152	80·0	22·9	..	Dulcigno [illegible]	1875			J. Eleutherion	Kyrn
	154		Nea Tychi WBg	Greek	144	..	..	..	Syra	1865			G.I. Caravas	Poros
	155		Nea Tychi WBg	Greek	114	69·9	20·7	13·7	Syra	1867			A. Cutculi	Zante
	156		Nea Tychi WBg	Greek	133	70·5	21·5	10·1	Spezzia	1862			Angelos Kontoumapa	Zante
	157	H	Nea Tychi *Kyriazoglou* WBk	Turkish	247	..	..	..	Galaxidi	1867			L. Emmanouelidis	Cons[illegible]
	158		Nea Tychi *Lira* WSr	Turkish	135	..	..	..	Chios	1875			P. Lira	Chios
70	159		Nea Tychi WSr	Greek	104	71·8	21·0	10·7						Spezzia
	160		Nea Virginia [illegible] WBk	Greek	616	152·2	3[illegible]·8	19·3	Trieste	1871	[illegible] [Tecnico		Fratelli Filaretos	Athens
	161		NeaVirginia *Skourdakis* WBk	Greek	593	..	..	..	Syra	1886			N. Nikitas	Syra
78657	162	✳	Nebo *Phizers 83-83* I S	British	1383	216·9	37·1	21·1	Glasgow	1877	Dobie & Co.		Smith & Service	Glasgow
	163	N	Nebo *Olsen* WBk	Norweg'n	590	112·1	34·9	18·7	Tvedestrand	1875	L. Tellefsen	[illegible]	Johan Smith	Tvedes[illegible]
	164	DN	Nebo *Johannessen* WBk	Norweg'n	516	138·4	30·9	17·1	Grimstad	1878	J. Borresen	[illegible]	J. Borresen	Grimstad
	165		Neck *Petersen 84-89* I S	German	1462	217·0	37·0	23·0	Vegesack	1889	Bremer S.B.Ges.		Gildemeister & Ries	Bremen
	166		Neckrosen *Johansson* WBg	Swedish	148	101·3	22·0	9·9	Uddevalla	1859	P. Abrahamson		A.T. Zachau	Uddevalla
	167		Necochea *Tenzo* I Sr	Argentine	238	125·0	29·6	10·6	Genoa	1884	E. Craven & Co.		LaPlatense FlotillaCo.	Buenos [illegible]
18795	168	R	Ned White *Elwell -96* WBn	American	523	130·7	32·5	18·4	Belfast, Me.	1874		[illegible]	A.F. Elwell	Belfast[illegible]
	169		Nedjnie Bahri WBg	Turkish	275	..	..	..	Ooniah	1866			Oontalee Hajee [illegible] Mehemet	[illegible]
	170		Nedjnie Bahri WBg	Turkish	238	..	..	..	Inebolee	1868			Omer Bey	Stamb[illegible]
	171		Nedjnie Bahri WBg	Turkish	106	..	..	..	Aleckreych	1876			[illegible]	Aleck[illegible]
84671	172		Neelayathatchy [illegible] WBg	British	105	77·6	18·1	20·5	Cochin	1873			Akani Koso Mohomado	Colomb[illegible]
	173		Néerlande *Requet -87* WBn	French	106	..	..	..	St. Malo	1873			Demalvilain fils	St. Malo
	174		Neerlands Vlag (ex Pieter Adolph) *Brouwer* WBk	Dutch	771	149·2	33·5	21·5	Newburyport	1864	E.S. Moseley	[illegible]	H.J. Plaat & Co.	Schiedam
85774	175		Negociant WSp	British	129	105·4	22·3	7·7	Quebec	1884	J. Brouillard		J.B. Charland	Montreal
	176		Negraes WSr	Brazilian	286	..	..	..					E.C. Carpinetti	Rio J[illegible]
2419	177	✳I	Negrisuola *Gandolfo -88* WBk	Italian	852	167·9	31·0	21·9	Savona	1873	A. Calcagno	[illegible]	Andrea Schiaffino	Genoa
1288	178	AD	Nehaj *Pesety -88* WBk	Austrian	479	135·1	29·9	18·7	Segna	1874	Societa Navale	[illegible]	G. Vicerina	Tersato
18538	179	V	Nehemiah Gibson *Kelley -87* WBk	American	741	147·0	32·6	19·4	E. Boston	1868	CurtisSmith&C°	[illegible]	S. Smith & Co.	Boston
18791	180		Nehemiah Hand *Hawley -88* WSr	American	185	91·5	26·0	9·8	Setauket, N.Y.	1874			F.G. Phipps	N.B[illegible]
35393	181		Neilly WBk	British	189	100·8	25·0	11·4	Campobello, N.B	1855			J.G. Sully	Bric[illegible]
	182		Neitto *Weisman 70-83* WBg	Russian	276	111·0	27·5	12·5	Hodlesbofjerd	1874	J.S. Sandberg		J. Ahlsten	Nykyrk
85431	183		Nell *Conlon -86* WSr	British	120	85·0	27·1	8·3	Newport, N.S.	1882	J. Mosher	[illegible]	Allen Haley	Wind[illegible]
55483	184	✳	Nellie (WRECKED) *Ellis -87* W3MSr	British	281	123·4	25·0	11·1	Kingsbridge	1867	Date	[illegible]	W.C. Jarvis	Salcom[illegible]
69118	185		Nellie *Reid* WBn	British	281	109·6	29·0	12·5	St. Mary's, N.S.	1874			W.C. Sache	Halifax
18631	186		Nellie *Drinkwater* WSr	American	280	112·2	28·4	11·0	Belfast, Me.	1870			O.W. Pitcher	Belf[illegible]
18782	187		Nellie *Brannock* WSr	American	135	108·0	26·0	7·1	Church Creek, Md.	1873			L. Ehrman	Baltimore
2087	188	V	Nellie *Copezz* WBg	Italian	360	109·5	29·5	18·3	Meta	1864	G. Mauro	[illegible]	F[lli] Cafiero fu V.	Naples
130290	189	R	Nellie A. Drury *Wilson 72-83* W3mSr	American	472	149·0	35·2	12·5	Thomaston, Me.	1883	Dunn & Elliott	[illegible]	Dunn & Elliot	Thoma[illegible]
130331	190	R	Nellie Blanche (ex Cl[illegible]) *Johnson* WSr	American	133	82·4	26·4	10·0	Parrsboro N.S.	1879	M. McLeod	[illegible]	W.H. W[illegible]	Key W[illegible]
18433	191	R	Nelly Bowers (WRECKED) *Magune* WSr	American	282	106·5	26·6	15·0	Camden, Me.	1867		[illegible]	Carleton, Norwood & Co.	Camd[illegible]
130162	192	RV	Nellie Brett *Knight -88* WBk	American	689	157·0	33·1	18·5	Calais, Me.	1877	O.B. Rideout		James E. Brett	New Y[illegible]
75987	193		Nellie Bruce *Sommerville* WSr	British	117	87·0	27·0	7·9	Springfield N.B	1882	J. Somerville		Robert B. Somerville	St.J[illegible]
18864	194		Nellie C. Paine *[illegible]* WSr	American	274	115·0	29·1	8·0	N. Haven Conn	1864	G.W. Baldwin		E.H. Atwood	[illegible]
18750	195		Nellie Clark *Clark 67-73* WSr	American	160	103·0	27·9	9·2	Robbinston Me	1873	J. Starkey		J. Frye	Ne[illegible]
130283	196	R	Nellie Coleman *Hegerthy -88* WSr	American	152	97·0	25·7	9·5	Lamoine, Me	1883	D.D. H[illegible]	[illegible]	N.B. Coolidge	[illegible]
86729	197	V	Nellie Crosby (LIGHTER) *[illegible]* WBn	British	440	132·2	32·6	16·8	Shelburne, N.S.	1874	B.P. King	[illegible]	Benjamin Davis	[illegible]
18477	198		Nellie Doe WSr	American	139	89·2	26·8	8·0	Orrington, Me	1867			T.C. Farris	[illegible]
130241	199	R	Nellie E. Rumball *Jenkins -88* WBkn	American	421	140·0	32·8	13·0	Harrington Me	1882	A. Nash	[illegible]	O.P. Rumball	Mach[illegible]
18397	200		Nellie Eaton *Knowlton -88* WSr	American	112	87·5	26·0	8·1	Calais, Me.	1874	[illegible] & Laud		B.L. Eaton	Cal[illegible]

✳ = Lloyd's Register; D = Nederlandsche Vereeniging; G = Germanischer Lloyd; I = Registro Italiano; R = Americ[an]; A = Veritas Austro-Ungarico; H = Veritas Ellenico; N = Norske Veritas; V = Bureau [illegible]

In the 4th Column, C = Composite (iron frame, wood planked); I = Iron; Stl = Steel; W = Wood.

The Dates of Metalling (Col. 11) are based on the records of the Books in which Vessels are respectively classed, or on the Reports of the Surveyors of [illegible]

Further particulars of Vessels classed by Lloyd's Register will be found in Lloyd's Register Book and Lloyd's Yacht Register.

1	3	4 Nom	Capitaine et dates d'entrée au service des Armateurs actuels et à bord du navire. Ancien Nom	Matériaux et Gréement	5 Pavillon	6 Tonnage net	7 Dimensions Longueur	Largeur	Creux	8 Construit Où	9 Quand	10 Par qui	11	12 Armateurs ou Directeurs-Gérants	13 Port d'Armement
201		Nellie F.	Huntley	W Sr	American	100	86·5	22·5	8·1	E.Machias,Me	1873			C.Huntley	Machias,Me.
202	R	Nellie F. Sawyer		W3mSr	American	270	121·4	31·7	10·0	Brunswick,Me	1873	D.Brewer	Y 10.84	F.A.Avery	Portland,Me.
203	R	Nellie Floyd	Johnson	W3mSr	American	435	136·0	33·0	15·4	P.JeffersonNY	1879	J.M.Bayles&Sn		Floyd & Newins	New York
204		Nellie Grant	Jordan	W Sr	American	113	92·0	26·5	7·5	Ellsworth,Me.	1873			W.Haynes & Co.	Ellsworth,Me.
205		Nellie Hunt		W Bn	British	206	111·3	24·3	9·4	Robbinst'n US	1859			E.Stinson,jun.	Bridgetown,N.S.
206		Nellie Hunter		W Sr	British	172	101·0	25·0	9·7	Dog Lake,Ont.	1874	C.Harris		G.C.Curry	Kingston,Ont.
207		Nellie J. Dinsmore	Parker	W3mSr	American	363	123·0	30·7	15·4	Lubec,Me.	1872	McBride		S.Staples & Son	Lubec,Me.
208		Nellie Lamper	Thompson	W3mSr	American	311	130·0	30·5	10·0	Mystic,Conn.	1873			J.B.Lamper	MarbleheadMass
209	V	NellieM.Slade	Pezzetro-87	W Bk	American	561	143·5	33·4	15·9	E.Boston	1874	R.Crosbie	Y 6.88	E.R.Lewis	Boston,Mass.
210		Nellie May	Austin	W Bk	American	699	148·9	32·0	20·5	Newcastle,Me.	1867	E.Flye	Y 3.88	E.Flye	Damariscotta,Me.
211		Nellie Moody	Stevens -88	W Bk	British	747	163·0	33·5	19·5	GreenCove,NS	1873	Raymond		A.F.Stoneman	Yarmouth,N.S.
212	R	Nellie Parker	Corbett -85	W Sr	British	183	101·0	27·7	10·0	Tynem'th,N.B	1885	R.L.Parker	Y 12.87	Robert L.Parker	Tynemouth,NB
213	R	Nellie Pickup		W Sr	British	150	96·1	26·5	9·9	Granville,NS.	1888	S.Groves		S.Groves	Annapolis,N.S.
214	*	NellieRoberts	Hett	W Sk	British	89	79·0	21·0	10·4	Grimsby	1886	Smith Bros.		W.G.Hett	Grimsby
215	*	Nellie S.	Francis 87-87	W Barge	British	262	131·2	26·0	12·8	Whitstable	1876	Gann	Y 1.88	W.R.Thoyts (Matthews & Luff)	London
216	R	Nellie S. Pickering	Thomas	W3mSr	American	266	128·4	32·8	9·3	Belfast,Me.	1880	Carter& Bro.		T.W.Frederick	Belfast,Me.
217	R	Nellie Smith	Wallace	W Bkn	American	567	146·3	33·0	17·9	Millbridge,Me	1883	J.W.Sawyer	Y 1.86	F.H.Smith & Co.	New York
218		Nellie Starr	Oliv -88	W Sr	American	216	112·8	27·9	9·5	Portland,Me.	1866			I.S.Preble & Co.	Addison,Me.
219		Nellie Swift	Beane	W Sr	American	128	93·2	24·6	9·1	Essex,Mass.	1878	J.James		S.S.Swift	Prvncetwn,Mass
220	R	Nellie T. Guest	Crosby -88	W Bk	British	838	173·0	35·9	20·6	Londonderry, N.S	1873	A.Fraser	Y 1.86	Thos.Guest	Yarmouth, N.S.
221		Nellie T. Morse	Baker	W3mSr	American	438	141·0	33·0	12·2	Bath,Me.	1879	BW& HF Morse		B.W.& H.F.Morse	Bath,Me.
222	R	Nellie V. Rokes	Thompson 7?	W3mSr	American	281	121·0	32·4	9·2	Bath,Me.	1879	Goss,Sawyer& Packard		J.B.Drake	Bath,Me.
223		Nellie W. Craig	Pearce	W3mSr	American	468	146·0	35·4	11·0	P.JeffersonNY	1883	J.M.Bayles& Son		J.Holmes	PerthAmboy,N.J.
224	R	Nellie W. Howlett	Fortune	W3mSr	American	540	148·5	35·1	15·1	Dorchester,N.J	1884	J.W.Vanneman		D.P.Malford	Bridgeton,N.J.
225	R	Nellie Woodbury	Hodgkins -81	W Sr	American	138	96·0	25·5	9·2	Ellsworth,Me.	1883	A.Lord		F.L.Hodgkins	Ellsworth,Me.
226	*	Nelly	Olsen 17-83, ex Nessie	W3mSr	British	184	107·3	23·2	12·6	Berwick	1876	A.B.Gowan&Sns	Y 7.88	N.& S.R.March	St.Johns,N.F.L.
227		Nelly		W Bk	Chilian	421	124·0	27·5	18·0	Sunderland	1860	Todd		N.Antanes	Quicavi
228	V	Nelly	Elia -86	W Bg	Greek	186	90·2	22·6	14·6	Cassos	1886	Emanoil		J.Protopapas & E.H. Elia	Syra
229	IV	Nelly	Mori -88	W Bkn	Italian	234	119·7	26·0	13·9	Viareggio	1886	A.Raffaelli	Y 10.86	F.Nelly	Leghorn
230		Nelly & Mathilda	Norrman 81-87	C Bk	Swedish	529	151·1	27·2	17·2	Sunderland	1866	Gardner	Y 5.82	E.T.Norrman	Malmö
231	*	Nelson	Bannatyne 82-82	I S	British	1247	239·3	36·0	20·7	PortGlasgow	1874	R.Duncan&Co.		Shaw,Savill& AlbionCo. Ld.	Glasgow
232		Nelson	Keating 83-	W Sr	British	144	96·2	26·9	9·1	Coverdale,NB	1870	J.Wright		H.Croskery	Belfast
233	V	Nelson	Johansen 88-88	W S	Norweg'n	943	172·2	35·9	22·1	St.George,NB	1858		Y 9.87	Actieselskabet "Nelson" (Ferd. Melsom)	Christiania
234	R	Nelson Bartlett	Watts	W3mSr	American	637	154·8	34·8	17·8	ThomastonMe	1881	S.Watts	Y 9.86	S.Watts	Thomaston,Me.
235	R	Nelson E. Newbury	King	W3mSr	American	626	148·2	35·4	16·5	Millbridge,Me	1882	J.Sawyer		J.B.Newbury	New York
236		Nelson Harvey	Parker	W Sr	American	104	85·0	25·6	8·6	Barnstaple, Mass.	1854			C.W.Parker	N.Bedford,Mass.
237	*	Nelson Hewertson	Wilson	W Bkn	British	240	118·5	24·9	12·7	Portheawl	1877	Martin	Y 9.85	W.Brooks	Newport
238	V	Némésis	Vinsut	W Bk	French	294	110·5	24·6	14·6	Nantes	1875	Tillé	Y 12.85	Pignot	Nantes
239	*A	Nemirna	Gellati -88	W Bk	Austrian	607	144·6	33·1	19·2	Fiume	1878	V.Brazzoduro	Y A 86	G.Milossevich & Co.	Megline
240		Nemo		W Sr	British	146	90·5	22·2	10·0	Yokohama	1882	A.J.M.Smith		H.J.Snow	Shanghai
241	*	Nenuphar	Jones 83-88	W Bk	British	392	132·5	28·2	17·1	Sunderland	1870	G.Short	Y 5.84	R.Thomas & Co.	Sunderland
242	RV	Neophyte	Goudey 85-88	W S	British	1056	198·4	37·3	19·8	ChurchPnt NS	1882	J.Lovitt	Y 8.85	Lovitt & Co.	Yarmouth,N.S.
243		Neos Cosmos		W Sr	Greek	348	115·4	29·2	18·4	Syra	1877			G.Mousouris	Argostoli
244	V	Neos Eolos	Fardjis	W Bn	Greek	297	102·6	28·0	16·9	Syra	1872	P.Bunis	Y 12.87	Helen D.Fotia	Hydra
245		Neos Evelpis	Apostolou	W Bk	Greek	232	..	..	..	Syra	1885			J.Apostolou	Syra
246	A	Neos Georgios	Polemis	W Bg	Greek	340	114·1	27·9	..	Syra	1874	P.Bunis	Y 1.89	Polemis Bros.	Andros
247		Neos Iraclis		W Bn	Turkish	225	..	..	..	Kerasunda	1869			E.Costantinides	Kerasunda
248		Neos Kimon	G.Marcoris	W Bk	Greek	341	..	..	..	Syra	1877			Focas & Moussaris	Cephalonia
249		Neos Lukianos		W Bn	Greek	375	121·6	27·6	19·6	Germany	1866			J.C.Frangoulis	Corfu
250															

= Lloyd's Register; D = Nederlandsche Vereeniging; G = Germanischer Lloyd; I = Registro Italiano; R = American Record; = Veritas Austro-Ungarico; H = Veritas Ellenico; N = Norske Veritas; V = Bureau Veritas.

Dans la 4me Colonne, C = Composite (membrure en fer, bordé en bois); I = Fer; Stl = Acier; W = Bois.

Les dates du doublage en métal (Col. 11) ressortent des livres dans lesquels les navires sont respectivement classés ou des rapports des Inspecteurs du Lloyd's Register.

Pour plus amples renseignements sur les navires classés par le Lloyd's Register, voir le Lloyd's Register Book et le Lloyd's Yacht Register.

Official No.	No. in Book.	Registry in which classed.	Ship's Name. Material and Rig. Master, and Dates of Appointment to present Owner's Service and to Ship. Ship's Late Name.	Flag.	Net Tonnage.	Length.	Breadth.	Depth.	Built: Where.	When.	By whom.	Date of Metalling.	Owners or Managers.	Port of Registry
	251	N	Nepoter *Nielsen* WBk	Norweg'n	311	114·2	28·6	15·4	Pungmand	1865			G.Nielsen	Arendal
	252	V	Neptun *Thomsen 82-83* WBk	Danish	1173	186·0	37·4	23·7	Bowdoinham [N.A.	1864	R.Purington	[illegible]	O.Hirth	Kolding
	253	V	Neptun *Christensen 86-86* WBn	Danish	145	105·8	21·9	10·2	Malmo	1865	J.Bonnesen		H.Christensen	Næstved
	254	DV	Neptun *Dreyer* W3mSr	German	386	121·4	29·1	13·1	Cappeln	1879	C.A.Steen	[illegible]	Th.Breckwoldt	Blankenese
	255	V	Neptun *Brinkema 80-86* WSr	German	116	77·8	22·2	9·1	Baalenfleth	1884	Aug.Focke		F.Th.Meyer	Vegesack
	256	G	Neptun (ex Wilhelm) *Engel 75* WSr	German	110	82·3	20·1	9·6	Varel	1869	C.Schweers		J.C.Thormann	Wismar
	257	V	Neptun *Gewald* WSr	German	101	83·0	19·4	9·4	Halte	1880	A.&H.Klasen		A. & H.Klasen	Papenburg
202	258	A	Neptun *Gerardi* WBk	Italian	382	125·0	27·7	16·6	Trieste	1862	Stabilimento [Tecnico	[illegible]	A.D.Vianello Moro	Venice
	259		Neptun *Johannesen -83* WBk	Norweg'n	671	135·9	31·9	21·1	Vegesack	1850	C.Lange	[illegible]	H.B.Johannesen	Tönsberg
	260	N	Neptun (ex Dorus) *Johannessen* WBk	Norweg'n	609	150·5	32·9	18·4	Vegesack	1857	J.Lange	[illegible]	O.Bjertnaes	Holmestrand
	261	N	Neptun *Tobiassen* WBk	Norweg'n	429	129·9	28·8	17·0	Grimstad	1874	N.Osmundsen	[illegible]	P.B.Einersen	Grimstad
	262		Neptun (WRECKED) *Andersen* WBk	Norweg'n	359	118·9	28·5	16·2	Svelvig	1855	A.Helgesen		Rederiselskabet Neptun (J. P. Andersen)	Drammen
	263		Neptun *Simonsen* WBk	Norweg'n	312	109·9	28·1	15·4	Grimstad	1864	A.Bentsen		I.S.Knudsen	Drammen
	264	N	Neptun *Borgersen* WBg	Norweg'n	266	105·6	25·8	13·9	Libau	1838			H.A.Borgersen	Drammen
	265	DN	Neptun (ex Neptune) *Sorensen* WBg	Norweg'n	261	100·6	26·0	15·0	Salcombe	1846			T.Mulvad	Fredrikstad
	266		Neptun *Eriksson* WBk	Russian	434	137·8	31·5	15·0	Wasa	18^{69}_{72}	A.Lindbeck		P.W.Holmberg	Lemland
	267		Neptun (ex Skandia) *Wuoti* WBk	Russian	416	116·0	27·1	13·4	Abo	1872	L.P.Kjaldström		G.A.Lindblom	Abo
	268		Neptun WBg	Russian	390	..	..	..		1869			E.Malmberg	Fre[illegible]
	269		Neptun *Dunder* WSr	Russian	229	..	..	..		1876			Joh.Rivell	Mystad
	270		Neptun *Wenschau 69-83* WBg	Russian	201	94·0	23·0	14·3	Dundee	1857			Gebr.Gutschmidt	Windau
	271		Neptun *Ohlson 64-73* WSr	Swedish	387	123·0	27·9	15·5	Milton,N.S.	1864			J.C.Hulthén	Helsingborg
	272		Neptun (ABANDONED) *Swedin 65-83* WBk	Swedish	363	104·9	24·9	17·7	Milton,N.S.	1828			C.Sylvan	Helsingborg
	273	V	Neptun *Hellberg 85-85* WBkn	Swedish	235	107·6	22·9	12·1	Elsfleth	1866	J.Ahlers		Joh.Olsson	[illegible]
54794	274		Neptune *Smith* WBg	British	299	119·5	26·1	15·1	Kingston,Sus.	1867	May&Thwaite		Union SS. Co. of New Zealand	Dunedin
71816	275		Neptune *Rice -75* WSr	British	221	118·5	23·6	11·3	Blackwall, [N.S.W.	1875	R.Davis		W.T.Yeager	Sydney,N.S.W.
	276		Neptune (ex Carrie W. Clark) *Williams* WSr	British	124	88·1	22·7	10·3	Boston,Mass.	1868	Geo.Bode		A.W.R.Dart	Halifax,N.S.
18028	277	R	Neptune *Hill* WBk	American	469	133·1	32·0	16·8	Portsm'th NH	1870	Tobey&Little- [field	[illegible]	E.Atkins & Co.	Boston,Mass.
130297	278		Neptune *Martin* WSr	American	175	106·0	30·0	8·7	FairhavenCal.	1882			H.C.Groeper	San [illegible]
18161	279		Neptune *Sanborn* WSr	American	104	84·0	23·7	7·0	E.Machias Me.	1850			Small Bros.	Machias,Me.
	280		Neptune *Girault* WBg	French	188	88·3	23·0	12·1	Fecamp	1863			Gilbert	Narbonne
	281		Neptuno WBn	Brazilian	193	..	..	..					F.D.S.Junior	Desterro
	282	V	Neptuno (ex Victor) *Rodriguez* WBn	Portug'se	179	108·9	23·0	9·2	Villa do Conde	1873	Rodrigues	[illegible]	M. d'Avellar & A. B. Cardoso	Lisbon
	283	V	Neptunus *Jensen 80-81* WBn	Danish	136	95·9	20·9	10·4	Troense	1876	S.V.Møller		N.Jensen	Svendborg
	284	R	Neptunus *Larsen 81-83* WBn	Danish	107	75·9	19·3	11·1	Copenhagen	1844	K.P.Bonnesen		C.C.Olsen	Copenhagen
	285	V	Neptunus (ex Bene & Benuno) *Zeilinga -86* WBk	Dutch	355	120·1	26·0	15·1	Anclam	1868	F.Bucholtz	[illegible]	J.& S.Wiarda	Harlingen
	286	N	Neptunus *Tobiassen 85-85* WBk	Norweg'n	623	134·4	32·3	20·1	Vegesack	1847		[illegible]	Nils P.Höyer	Skien
	287		Neptunus *Olsen -87* WBk	Norweg'n	320	96·7	29·3	17·1	Kragerö	1831			P. & C.Römer	Kragerö
	288		Neptunus (SUNE) *Andersen 86-86* WBg	Norweg'n	164	82·1	22·8	13·4	Westervik	1799			J.M.Andresen	Fredrikstad
[illegible]	289	G	Neptunus W3mSr	Russian	240	113·5	27·4	12·7	Orrenhof	1881			Otto Grant	Hapsal
	290		Neptunus *Anderson* WBg	Swedish	247	96·1	22·2	14·1	Antwerp	1852			J.P.Johnssen	Grebbestad
27578	291		Nereid *Adam 70-83* WSr	British	183	101·9	22·8	14·2	Aberdeen	1860	Hood		G.Sinclair	Aberdeen
1807?	292	R	Nereid *Clifford* WBk	American	668	118·1	31·4	19·1	EastBoston	1866	Curtis&Smith	[illegible]	H.Clifford	Boston,Ma.
2837	293	IV	Nereo *Gamba 80-80* WBk	Italian	872	184·0	34·4	22·3	Sestri Ponente	1877	A.Briasco fu L.	[illegible]	F.Vaccaro	Genoa
[illegible]	294	✻	Nereus *Laing* IS	British	1272	231·9	37·0	21·8	PortGlasgow	1880	R.Duncan&Co.		C.S.Caird	Greenock
	295	DV	Nereus *Seinstra -88* WBk	Dutch	613	131·2	32·6	19·8	Rotterdam	1863	VisserBros.	[illegible]	P.H.Graandijk & Zoon	Amsterdam
67842	296	✻	Nerissa *Jenkins 78-78* WSr	British	194	84·8	21·8	11·0	Aberystwith	1878	Thomas Jones	[illegible]	D.C.Roberts & Co.	Aberystwith
	297		Nerva (ex Michiel) *Visser 87-87* WBg	Dutch	251	115·8	22·6	14·0	P.E.Island	1866	Duncan		Den Bouwmeester Rotteveel & Van der Leije	Middelburg
	298		Nervion (ex Daisy) *Norraunsescoan (?)* WBk	Spanish	590	150·9	30·1	18·3	Quebec	1874	F.X.Marquis		Fidel de Olenga	Bilbao
	299	N	Nescio *Samuelsen* WBk	Norweg'n	266	130·9	26·0	11·6	Kragerö	1873	A.Johnsen		P. & C.Römer	Kragerö
79191	300	✻	Nesta *Hughes* WSr	British	103	88·5	21·3	10·8	Amlwch	1877	Thomas		S.R.Platt	[illegible]

✻ = Lloyd's Register; D = Nederlandsche Vereeniging; G = Germanischer Lloyd; I = Registro Italiano; R = American [Record]; A = Veritas Austro-Ungarico; H = Veritas Ellenico; N = Norske Veritas; V = Bureau Veritas.

In the 4th Column, C = Composite (iron frame, wood planked); I = Iron; Stl = Steel; W = Wood.

The Dates of Metalling (Col. 11) are based on the records of the Books in which Vessels are respectively classed, or on the Reports of the Surveyors of Lloyd's [Register].

Further particulars of Vessels classed by Lloyd's Register will be found in Lloyd's Register Book and Lloyd's Yacht Register.

1-3		4. Nom. Matériaux et Gréement. Capitaine et dates d'entrée au service des Armateurs actuels et à bord du navire. Ancien Nom.		5. Pavillon.	6. Tonnage net.	7. Dimensions. Longueur	Largeur	Creux	8. Construit. Où.	9. Quand.	10. Par qui.	11. Date du doublage en métal.	12. Armateurs ou Directeurs-Gérants.	13. Port d'Armement.
301		Nestor	W Bn	French	117	70·2	20·7	10·5	La Nouvelle				Lafattut	Narbonne
302	BV	Nestor *Fehr 81-83*	I S	German	1346	207·0	36·8	22·8	Kinderdyk	1867	L. Smit		Knöhr & Burchard	Hamburg
303	G	Nestor *Klanck -88* (ex Ludwig Wittenberg)	W Bk	German	448	129·6	30·0	17·7	Ueckermünde	1859	L. Wittenberg		C. Wichards	Stettin
304		Nestor *Patch 80-80* (WRECKED)	W Bk	German	351	112·1	28·2	17·0	Greifswald	1849	J. D. Gaude		C. Graedener	Greifswald
305		Nestor	W Bk	Spanish	652	115·0	31·1	21·4	Brahestad	1866	N. Möller			
306	✻	Netherby *Chellew*	I S	British	1400	212·6	37·5	21·5	Maryport	1886	Ritson & Co.		J. Dodd	Liverpool
307	✻	Netherton *Jones 79-88*	W Bn	British	188	103·7	23·3	12·9	Salcombe	1872	Vivian	Y 5.89	J. Jones	Salcombe
308		Netherton	W Bn	British	116	95·2	23·8	11·3	Newfoundland	1870			P. Byrne	Dundalk
309	R	Nettie *Larrin*	W Bn	American	477	131·0	32·4	17·2	Tottenville N.Y.	1874	J. Ellis	Y 8.88	G. M. Smith	New York
310	R	Nettie Champion	W 3m Sr	American	423	112·5	35·1	12·2	Mays Ldng. N.J.	1881	J. Clark		J. Champion	Somers Point N.J.
311		Nettie Langdon *Eagley 89-89*	W 3m Sr	American	288	129·0	30·0	10·2	Boston, Mass.	1874			J. C. L'Engle	Boston, Mass.
312		Nettie Merrill *Brownell* (WRECKED)	W Sr	Hawaian	156	86·0	24·9	8·5	Honolulu	1875	D. Foster		P. Isenberg	Honolulu
313	RV	Nettie Murphy *Cosman*	W S	British	1374	205·0	38·0	23·3	Tusket, N.S.	1880	C. W. Murphy	Y 3.86	John Murphy	Yarmouth, N.S.
314	R	Nettie Shipman *Hudson*	W 3m Sr	American	321	118·3	35·0	10·0	P. Jeffers'n N.Y.	1884	J. M. Bayles & Sn.		A. H. Hudson	Pt. Jefferson N.Y.
315		Netto (ex Maggie, ex Agnes L.)	W Bn	Portug'se	169	94·2	23·0	10·2					J. T. Netto & Co.	Ponta Delgada
316	N	Neu Granada *Schildbred* [-89	W Bk	Norweg'n	327	126·9	27·1	13·6	Nw Rönnebeck	1862	C. Diercks & Co.		M. E. Schildbred	Brevig
317	H	Neutral *Andersson 85-85* (ex Columbia)	W 3m Sr	Swedish	373	113·5	24·8	17·6	Genoa	1857		Y 12.85	J. E. Olson	Gothenburg
318		Nev Shehir *Mehemed Ali*	W Bg	Turkish	199	..	..	..	Barten	1884			Nev Shehirlee Haice Mehemed & Co.	Nev Shehir
319		Neva *Black 85-85*	W Bn	British	236	111·7	26·8	11·6	Granville, N.S.	1862	Crosscup		A. Blaney (D. R. Brannigan)	Belfast
320		Neva *Kenwright*	W Sr	British	191	97·5	27·9	10·4	Pompuet, N.S.	1875			A. M. Cunningham	Afton, N.S.
321	R	Neva May *Howard* (ABANDONED)	W 3m Sr	American	217	121·6	30·0	7·8	Milford, Del.	1884	J. W. Abbott & Sn		J. May & Sons	Philadelphia
322	N	Nevada *Thorstensen 82-86*	W Bk	Norweg'n	490	141·7	28·9	17·7	Sunderland	1869	W. Nicholson & Ss	Y 5.86	J. Andorsen	Mandal
323	V	Nevado *Donaldson 87-87*	W Bk	British	674	162·0	33·5	19·5	Colchester, N.S.	1871	J. Geddes	Y 9.87	G. Milne & Co.	Aberdeen
324	✻	New Blessing *Campbell*	W Bn	British	142	93·9	23·1	12·7	Portmadoc	1865	Daniel Griffiths		A. M'Donald	Carnarvon
325		New Brunswick	W Sp	British	105	101·9	22·8	6·4	Canada	1868			F. Desmarais	Montreal
326	N	New Brunswick *Hansen -87*	W Bk	Norweg'n	480	137·4	30·0	17·4	Buctouche, N.B.	1864			N. W. Coch	Brevig
327	BRV	New City *Robinson -88*	W S	British	1442	209·6	39·8	24·1	Black Riv'r N.B.	1878	J. & R. McLeod	Y 10.86	D. Carmichael	St. John, N.B.
328		New Dominion	W Sr	British	196	117·5	25·3	9·2	Port Rowan	1868			Geo. H. Morden	W'llaceburg, Ont
329		New Dominion	W Sr	British	164	115·0	25·7	9·4	Quebec	1867	McKay & Warner		H. W. Allan	Port Dover, Ont
330	V	New Dominion *Lemieux*	W Bn	British	144	87·7	25·0	10·9	Montmagny	18 71/83	J. Joncas	Y 6.87	James King	Quebec
331	N	New Ed *Olsen -87*	W Bg	Norweg'n	229	113·3	23·2	14·4	Sunderland	1853		Y 3.81	Actieselskabet "New Ed" (H. Holland)	Fredriksstad
332		New Era *Fitzgerald -87*	W Bn	British	314	118·0	26·7	14·0	Georgetown P.E.I.	1871	D. Gordon		D. Gordon & Co.	Charlottet'n, P.E.I.
333		New Harmony	W Bn	British	105	72·4	21·5	12·3	Barmouth	1818			E. Lightfoot	Maryport
334	✻	New Leader *Jenne*	W Sr	British	58	72·8	18·2	8·8	Jersey	1873	Bisson		T. Jenne	Jersey
335		New Light *Snow -85*	W Bk	American	451	139·8	29·6	17·1	Baltimore	1855	R. B. Goodwin & C.	Y 1.86		Baltimore
336		New Liverpool	W Sr	British	176	111·6	22·0	9·2	Canada	1856			D. Gaharty	Quebec
337	V	New Pactolus *Nickerson 87-87*	W Bk	British	1035	186·8	37·5	21·0	Noel, N.S.	1885	O. O'Brien	Y 8.85	O. O'Brien	Maitland, N.S.
338		New York (ex Steamer) *Hughes*	I S	British	2699	328·0	40·2	33·5	Greenock	1858	Caird & Co.		E. Bates & Sons	Liverpool
339	I	New York *Messina -87* (ex Angelina Camogli)	W Bk	Italian	503	139·7	31·8	20·6	Varazze	1870	B. Camogli	Y 8.86	Messina & Chiaramonte	Palermo
340	V	New York *Caffiero*	W Bg	Italian	388	125·1	27·2	18·0	Cassano	1871	A. Castellano	Y 2.88	F. S. Ciampa	Castell. di Stab.
341		New Zealand	W Sr	British	112	89·0	22·5	8·0	Quebec	1852			F. S. Boisvert	Quebec
342		New Zealand *Haskell*	W Sr	American	109	81·8	24·9	7·3	Tremont, Me.	1853			J. H. Kelley	Ellsworth, Me.
343	NV	New Zealand *Johnsen 86-88*	W Bk	Norweg'n	660	161·2	33·0	18·3	Grimstad	1883	A. Knudsen	Y 3.87	C. H. Johnsen	Arendal
344	N	Newa *Jensen*	W Bn	German	171	110·1	25·7	10·7	Risör	1869	Stian Finne		Nordischer Bergungs-Verein	Hamburg
345		Newark *Beck -87*	W Sr	American	115	93·5	27·2	7·5	Benicia, Cal.	1887	M. Turner		C. L. Dingley	San Francisco
346	V	Newburgh *Parsons 88-85*	W 3m Sr	British	505	135·9	31·6	16·2	Windsor, N.S.	1888	S. Dimock		S. Dimock	Windsor, N.S.
347	B	Newcastle *Taft 88-88*	W BK	Norweg'n	468	140·2	28·2	18·2	Newcastle	1865	Gaddy		Actieselskabet "Atlas" Krog & Henschien	Lillesand
348		Newfield	Stl & I S	British	1350	243·0	35·2	23·0	Dundee	1889	A. Stephen & Sons		Brownell & Co.	
349	BV	Newman Hall *Dixon -88*	W S	British	1528	215·5	40·8	24·0	St. Martin's N.B.	1875	Bradshaw	Y 8.86	R. Y. Dixon	Liverpool
350		Newsboy	W Bn	British	106	85·9	21·8	10·2	St. John, N.B.	1859	Thompson		J. Stewart	Belfast

B = Lloyd's Register; D = Nederlandsche Vereeniging; G = Germanischer Lloyd; I = Registro Italiano; R = American Record;
A = Veritas Austro-Ungarico; H = Veritas Ellenico; N = Norske Veritas; V = Bureau Veritas.

Dans la 4me **Colonne**, C = Composite (membrure en fer, bordé en bois); I = Fer; Stl = Acier; W = Bois.

Les Dates du doublage en métal (Col. 11) ressortent des livres dans lesquels les navires sont respectivement classés ou des rapports des Inspecteurs du Lloyd's Register.

Pour plus amples renseignements sur les navires classés par le Lloyd's Register, voir le Lloyd's Register Book et le Lloyd's Yacht Register.

Official No.	No. in Book.	Registry in which classed.	Ship's Name. Material and Rig. Master, and Dates of Appointment to present Owner's Service and to Ship. Ship's Late Name.	Flag.	Net Tonnage.	Dimensions. Length.	Breadth.	Depth.	Built. Where.	When	By whom.	Date of Metalling.	Owners or Managers.	Port
	251	N	Nepoter *Nielsen* WBk	Norweg'n	311	114·2	28·6	15·4	Porsgrund	1865			G.Nielsen	
	252	V	Neptun *Thomsen 82-88* WBk	Danish	1173	186·4	37·4	23·7	Bowdoinham [N.A.	1864	R.Purington		O.Hiort	
	253	V	Neptun *Christensen 80-80* WBn	Danish	145	105·8	21·9	10·2	Malmø	1865	J.Bonnesen		H.Christensen	
	254	DV	Neptun *Dreyer* W3mSr	German	386	121·4	29·0	13·3	Cappeln	1879	C.A.Steen		Th.Breckwoldt	
	255	V	Neptun *Brinkema 85-88* WSr	German	116	77·8	22·2	9·1	Bardenfleth	1884	Aug.Focke		F.Th.Meyer	
	256	G	Neptun (ex Wiebke) *Engel 72-* WSr	German	110	82·3	20·1	9·6	Varel	1869	C.Schwoon		J.C.Thormann	
	257	V	Neptun *Gewald* WSr	German	101	83·0	19·1	9·1	Halte	1880	A.&H.Klasen		A. & H.Klasen	
	258	A	Neptun *Girardi* WBk	Italian	382	125·0	27·7	16·6	Trieste	1882	Stabilimento [Tecnico		A.D.Viatello Mor.	Venice
	259		Neptun *Johannesen -83* WBk	Norweg'n	671	135·9	31·9	21·1	Vegesack	1850	C.Lange		H.B.Johannessen	Tönsberg
	260	N	Neptun (ex Susan) *Johannesen* WBk	Norweg'n	609	150·5	32·9	18·4	Vegesack	1857	J.Lange		O.Bjertnæs	Holm.
	261	N	Neptun *Tobiassen* WBk	Norweg'n	429	129·9	28·8	17·0	Grimstad	1874	N.Osmundsen		P.B.Einersen	Grimstad
	262		Neptun (WRECKED) *Andersen* WBk	Norweg'n	359	118·9	28·5	16·2	Svelvig	1855	A.Helgesen		Rederiselskabet Neptun (J.F.Andersen)	Drammen
	263		Neptun *Simonsen* WBk	Norweg'n	312	109·9	28·1	15·4	Grimstad	1864	A.Bentsen		L.S.Knudsen	
	264	N	Neptun *Borgersen* WBg	Norweg'n	266	105·6	25·8	13·9	Libau	1838			H.A.Borgersen	
	265	DN	Neptun (ex Neptunus) *Sörensen* WBg	Norweg'n	261	100·6	26·0	15·0	Salcombe	1846			T.Malvad	
	266		Neptun *Eriksson* WBk	Russian	434	137·8	31·5	15·0	Wasa	18 52/72	A.Lindbeck		P.W.Holmberg	
	267		Neptun (ex Steamer) *Wuoli* WBk	Russian	416	146·0	27·1	13·4		1872	I. P. Kjäldström		G.A.Lindblom	
	268		Neptun WBg	Russian	300	..	..	..		1869			E.Malmberg	
	269		Neptun *Dunder* WSr	Russian	229	..	..	..		1876			Joh.Rivell	
	270		Neptun *Wenschan 63-88* WBg	Russian	201	94·0	23·0	11·3	Dundee	1857			Gebr.Gutschmidt	Windau
	271		Neptun *Ohlson 64-73* WSr	Swedish	387	123·0	27·9	15·5	Milton,N.S.	1864			J.C.Hulthen	Helsingborg
	272		Neptun (ABANDONED) *Sandin 65-83* WBk	Swedish	363	104·9	24·9	17·7	Milton,N.S.	1828			C.Sylvan	Helsingborg
	273	V	Neptun *Hellberg 83-85* WBkn	Swedish	235	107·6	22·9	12·1	Elsfleth	1866	J.Ahlers		Joh.Olsson	
54794	274		Neptune *Smith* WBg	British	299	119·5	26·1	15·1	Kingston, Su-	1867	May&Thwaite		Union SS. Co. of New Zealand	
71816	275		Neptune *Rice -75* WSr	British	221	118·5	23·6	11·3	Blackwall, [N.S.W.	1875	R.Davis		W.T.Yeager	
	276		Neptune (ex Caren W. Clark) *Williams* WSr	British	124	88·1	22·7	10·3	Boston,Mass.	1868	Geo.Boole		A.W.R.Dart	
18023	277	R	Neptune *Hill* WBk	American	469	133·4	32·0	16·8	Portsm'th NH.	1870	Tobey&Little- [field		E.Atkins & Co.	
130227	278		Neptune *Martin* WSr	American	175	106·0	30·0	8·7	FairhavenCal.	1882			H.C.Groeper	
18161	279		Neptune *Sanborn* WSr	American	104	84·0	23·7	7·0	E.Machias,Me.	1850			Small Bros.	
	280		Neptune *Girault* WBg	French	188	88·3	23·0	12·1	Fecamp	1863			Gilbert	
	281		Neptuno WBn	Brazilian	193	..	..	..					F.D.S.Junior	
	282	V	Neptuno (ex Victor) *Rodriguez* WBn	Portug'se	179	108·9	23·0	9·2	Villa do Conde	1873	Rodrigues		M.d'Avellar & A. B. Carneiro	Lisbon
	283	V	Neptunus *Jensen 80-81* WBn	Danish	136	95·9	20·9	10·4	Troense	1876	R.V.Möller		N.Jensen	Svendborg
	284	R	Neptunus *Larsen 81-83* WBn	Danish	107	75·9	19·3	11·1	Copenhagen	1844	E.P.Bonnesen		C.C.Olsen	Copenhagen
	285	V	Neptunus (ex Mina & Benno) *Zeilinga -86* WBk	Dutch	355	120·4	26·0	15·1	Anclam	1868	F. Bucholtz		J.& S.Wiarda	Harlingen
	286	N	Neptunus *Tobiassen 85-85* WBk	Norweg'n	623	134·4	32·3	20·1	Vegesack	1847			Nils P.Hoyer	Skien
	287		Neptunus *Olsen -87* WBk	Norweg'n	320	96·7	29·3	17·1	Kragerö	1831			P. & C.Römer	Kragerö
	288		Neptunus (SUNK) *Andersen 86-89* WBg	Norweg'n	164	82·4	22·8	13·4	Westervik	1799			J.M.Andresen	
	289	G	Neptunus W3mSr	Russian	240	113·5	27·4	12·7		1881			Otto Grant	
	290		Neptunus *Andersson* WBg	Swedish	247	96·1	22·2	14·1		1852			J.P.Johnssen	
	291		Nereid *Adam 70-83* WSr	British	183	101·9	22·8	14·2		1860	Hood		G.Sinclair	Aberdeen
	292	R	Nereid *Clifford* WBk	American	668	148·4	31·4	19·4		1866	Curtis&Smith		H.Clifford	
	293	IV	Nereo *Gamba 80-89* WBk	Italian	872	184·0	31·4	22·3		1877	A.Briasco fu L.		F.Vaccaro	
	294	✻	Nereus *Laing* IS	British	1272	234·9	37·0	21·8	PortGlasgow	1880	R.Duncan&Co.		C.S.Caird	Greenock
	295	DV	Nereus *Schuitema -88* WBk	Dutch	613	131·2	32·6	19·8	Rotterdam	1863	VisserBros.		P.H.Craandijk & Zoon	
67642	296	✻	Nerissa *Jenkins 78-78* WSr	British	104	84·8	21·8	11·0	Aberystwith	1878	Thomas Jones		D.C.Roberts & Co.	
	297		Nerva (ex Eleivera) *Visser 87-87* WBg	Dutch	251	115·8	22·6	11·0	P.E.Island	1866	Duncan		Den Bouwmeester Borsten & Van der Lelie	
	298		Nervion (ex Desiree) *Navarrineuroa 69-88* WBk	Spanish	590	150·9	30·1	18·3	Quebec	1874	F.X.Marquis		Fidel de Olenga	Bilbao
	299	N	Nescio *Sörensen* WBk	Norweg'n	266	130·9	26·6	11·6	Kragerö	1873	A.Johnsen		P. & C.Römer	Kragerö
73101	300	✻	Nesta *Hughes* WSr	British	103	88·5	21·3	10·8	Amlwch	1877	Thomas		S.R.Platt	

✻ = Lloyd's Register; D = Nederlandsche Vereeniging; G = Germanischer Lloyd; I = Registro Italiano; R = American
A = Veritas Austro-Ungarico; H = Veritas Ellenico; N = Norske Veritas; V = Bureau

In the 4th Column, C = Composite (iron frame, wood planked); I = Iron; Stl = Steel; W = Wood.

The Dates of Metalling (**Col. 11**) are based on the records of the Books in which Vessels are respectively classed, or on the Reports of the Surveyors of Lloyd's

Further particulars of Vessels classed by Lloyd's Register will be found in Lloyd's Register Book and Lloyd's Yacht Register.

1	2-3	4 Nom. Materiaux et Greement. Capitaine et dates d'entrée au service des Armateurs actuels et à bord du navire. Ancien Nom.		5 Pavillon.	6 Tonnage net.	7 Dimensions. Longueur	Largeur	Creux	8 Construit. Où.	9 Quand.	10 Par qui.	11 Date du doublage en métal	12 Armateurs ou Directeurs-Gérants.	13 Port d'Armement.
301		Nestor	W Bn	French	117	70·2	20·7	10·5	[illegible]				Lalattut	Narbonne
302	DV	Nestor Edler 81-88	I S	German	1346	207·0	36·8	22·8	Kinderdyk	1867	L.Smit		Knöhr & Burchard	Hamburg
303	G	Nestor Ehnet -88 (ex Ludwig Wilhelm ?)	W Bk	German	448	129·6	30·0	17·7	Ueckermünde	1859	L.Wittenberg		C.Wichards	Stettin
304		Nestor Pasch 80-80 (WRECKED)	W Bk	German	351	112·1	28·2	17·6	Greifswald	1849	J.D.Gaede		C.Graedener	Greifswald
305		Nestor	W Bk	Spanish	652	145·0	34·1	24·4	Brahestad	1866	N.Möller			
306	*	Netherby Chellew	I S	British	1400	212·6	37·5	24·5	Maryport	1886	Ritson& Co.		J.Dodd	Liverpool
307	*	Netherton Jones 79-88	W Bn	British	188	103·7	23·3	12·9	Salcombe	1872	Vivian	Y 2.88	J.Jones	Salcombe
308		Netherton	W Bn	British	116	95·2	23·8	11·3	Newfoundland	1870			P.Byrne	Dundalk
309	R	Nettie Latvia	W Bn	American	477	134·0	32·1	17·2	TottenvilleNY	1874	J.Ellis	Y 8.88	G.M.Smith	New York
310	R	Nettie Champion	W3mSr	American	423	142·5	35·1	12·2	MaysLdng.NJ	1881	J.Clark		J.Champion	SomersPoint,N.J.
311		Nettie Langdon Tapley 82-82	W3mSr	American	288	129·0	30·0	10·2	Boston,Mass.	1874			J.C.L'Engle	Boston,Mass.
312		Nettie Merrill Brownell (WRECKED)	W Sr	Hawaian	156	86·0	21·9	8·5	Honolulu	1875	D.Foster		P.Isenberg	Honolulu
313	RV	Nettie Murphy Cosman	W S	British	1374	205·0	38·0	23·3	Tusket,N.S.	1880	C.W.Murphy	Y 2.88	John Murphy	Yarmouth,N.S.
314	R	Nettie Shipman Hulsen	W3mSr	American	321	118·3	35·0	10·0	P.Jeffers'n NY	1884	J.M.Bayles&Sn.		A.H.Hudson	Pt.Jefferson,N.Y.
315		Netto (ex Maggie, ex Agnes ?)	W Bn	Portug'se	169	91·2	23·0	10·2					J.T.Netto & Co.	Ponta Delgada
316	N	Neu Granada Schildbred I -86	W Bk	Norweg'n	327	126·9	27·1	13·6	NwRönnebeck	1862	C.Diercks&Co.		M.E.Schildbred	Brevig
317	G	Neutral Andersson 85-85 (ex Columbia)	W3mSr	Swedish	373	113·5	24·8	17·6	Genoa	1857		Y 1.88	J.E.Olson	Gothenburg
318		Nev Shehir Mehemed Ali	W Bg	Turkish	199	..	..	..	Barten	1884			Nev Shehirlee Hajee Mehemet & Co.	Nev Shehir
319		Neva Black 85-85	W Bn	British	236	111·7	26·8	11·6	Granville,N.S.	1862	Croscup		A.Blaney (D.R.Brannigan)	Belfast
320		Neva Kenright	W Sr	British	191	97·5	27·9	10·4	Pomquet,N.S.	1875			A.M.Cunningham	Afton,N.S.
321	R	Neva May Howard (ABANDONED)	W3mSr	American	217	121·6	30·0	7·8	Milford,Del.	1884	J.W.Abbott&Sn		J.May & Sons	Philadelphia
322	N	Nevada Thorstensen 82-86	W Bk	Norweg'n	490	141·7	28·9	17·7	Sunderland	1869	W.Nicholson&Ss	Y 3.86	J.Andersen	Mandal
323	V	Nevado Donaldson 87-87	W Bk	British	674	162·0	33·5	19·5	Colchester,N.S	1871	J.Geddes	Y 8.87	G.Milne & Co.	Aberdeen
324	*	New Blessing Campbell	W Bn	British	142	93·9	23·1	12·7	Portmadoc	1865	Daniel Griffiths		A.McDonald	Carnarvon
325		New Brunswick	W Sp	British	105	101·9	22·8	6·4	Canada	1868			F.Desmarais	Montreal
326	N	New Brunswick Hansen -87	W Bk	Norweg'n	480	137·4	33·0	17·4	Buctouche,NB	1864			N.W.Coch	Brevig
327	DRV	New City Robinson -88	W S	British	1442	209·6	39·8	24·1	BlackRiv'rNB	1878	J.&R.McLeod	Y 10.86	D.Carmichael	St.John,N.B.
328		New Dominion	W Sr	British	196	117·5	25·3	9·2	Port Rowan	1868			Geo.H.Morden	W'llaceburg,Ont
329		New Dominion	W Sr	British	164	115·0	25·7	9·4	Quebec	1867	McKay&Warner		H.W.Allan	Port Dover,Ont
330	V	New Dominion Lemieux	W Bn	British	144	87·7	25·0	10·9	Montmagny	18 71/83	J.Joncas	Y 8.87	James King	Quebec
331	N	New Ed Olsen -87	W Bg	Norweg'n	229	113·3	23·2	14·4	Sunderland	1853		Y 7.81	Actieselskabet "New Ed" (H.Helland)	Fredriksstad
332		New Era Fitzgerald -87	W Bn	British	314	118·0	26·7	14·0	Georget'wnPEI	1871	D.Gordon		D.Gordon & Co.	Charlottet'n,PEI
333		New Harmony	W Bn	British	105	72·4	21·5	12·3	Barmouth	1818			E.Lightfoot	Maryport
334	*	New Leader Jenne	W Sr	British	58	72·8	18·2	8·8	Jersey	1873	Bisson		T.Jenne	Jersey
335		New Light Snow -88	W Bk	American	451	139·8	29·6	17·1	Baltimore	1855	R.B.Goodwin&C.	Y 1.86		Baltimore
336		New Liverpool	W Sr	British	176	111·6	22·0	9·2	Canada	1856			D.Gaharty	Quebec
337	V	New Pactolus Nickerson 87-87	W Bk	British	1035	186·8	37·5	21·0	Noel,N.S.	1885	O.O'Brien	Y 9.86	O.O'Brien	Maitland,N.S.
338		New York (ex Steamer) Hughes	I S	British	2699	328·0	40·2	33·5	Greenock	1858	Caird&Co.		E.Bates & Sons	Liverpool
339	I	New York Messina -87 (ex Angelina Camogli)	W Bk	Italian	503	139·7	31·8	20·6	Varazze	1870	B.Camogli	Y 6.88	Messina & Chiaramonte	Palermo
340	V	New York Cafiero	W Bg	Italian	388	125·1	27·2	18·0	Cassano	1871	A.Castellano	Y 2.88	F.S.Ciampa	Castell.di Stab.
341		New Zealand	W Sr	British	112	89·0	22·5	8·0	Quebec	1852			F.S.Boisvert	Quebec
342		New Zealand Haskell	W Sr	American	109	81·8	24·9	7·3	Tremont,Me.	1853			J.H.Kelley	Ellsworth,Me.
343	NV	New Zealand Johnsen 86-83	W Bk	Norweg'n	660	161·2	33·0	18·3	Grimstad	1883	A.Knudsen	Y 5.87	C.H.Johnsen	Arendal
344	N	Newa Jensen	W Bn	German	171	110·4	25·7	10·7	Risör	1869	Stian Finne		Nordischer Bergungs-Verein	Hamburg
345		Newark Beck -87	W Sr	American	115	93·5	27·2	7·5	Benicia,Cal.	1887	M.Turner		C.L.Dingley	San Francisco
346	V	Newburgh Parsons 85-85	W3mSr	British	505	135·9	31·6	16·2	Windsor,N.S.	1888	S.Dimock		S.Dimock	Windsor,N.S.
347	R	Newcastle Taft 88-88	W BK	Norweg'n	468	140·2	28·2	18·2	Newcastle	1865	Gaddy		Actieselskabet "Atlas" Knag & Hanschien	Lillesand
348		Newfield	Stl&I S	British	1350	243·0	35·2	23·0	Dundee	1889	A.Stephen&Sons		Brownell & Co.	
349	DV	Newman Hall Dixon -88	W S	British	1528	215·5	40·8	24·0	St.Martin'sNB	1875	Bradshaw	Y 9.88	R.Y.Dixon	Liverpool
350		Newsboy	W Bn	British	106	85·9	21·8	10·2	St.John,N.B.	1859	Thompson		J.Stewart	Belfast

* = Lloyd's Register; D = Nederlandsche Vereeniging; G = Germanischer Lloyd; I = Registro Italiano; R = American Record; A = Veritas Austro-Ungarico; H = Veritas Ellenico; N = Norske Veritas; V = Bureau Veritas.

Dans la 4me **Colonne**, C = Composite (membrure en fer, bordé en bois); I = Fer; Stl = Acier; W = Bois.

Les Dates du doublage en métal (**Col. 11**) ressortent des livres dans lesquels les navires sont respectivement classés ou des rapports des Inspecteurs du Lloyd's Register.

Pour plus amples renseignements sur les navires classés par le Lloyd's Register, voir le Lloyd's Register Book et le Lloyd's Yacht Register.

Official No.	No. in Book	Society in which classed	Ship's Name. Material and Rig. Master, and Dates of Appointment to present Owner's Service and to Ship. Ship's Late Name.	Flag.	Net Tonnage.	Dimensions. Length.	Breadth.	Depth.	Built. Where.	When.	By whom.	Date of Metalling	Owners or Managers	Port
[illegible]	351		Newsboy *Johnson* W Bk	American	559	168·0	38·0	13·9	San Francisco	1882	Dickey Bros.	[illegible]	J. J. Smith	San Fran
	352	V	Newsky *Nielsen* 84-84 W Sr	Danish	135	89·0	20·8	10·5	Thurø	1875	P. Bom.		P. Bom	Scandinavia
	353		Nexø *Lau* 83-83 W Bg	Danish	162	93·4	23·5	12·4	Nexø	1855	J. H. Brandt		P. Berg	Nexø
18129	354		Ney *Anderson* W Sr	American	117	102·3	25·2	7·5	Baltimore	1854			F. Borden	Fall River
	355		Nezaret W Br	Turkish	307	..	..	..	Russia	1879			Logoperian Husbek & Co.	Constan
85890	356		Ngawoon I Barge	British	791	194·0	25·5	8·2	Dumbarton	1875	Wm. Denny & Bros.		IrrawaddyFlotillaCo.Lt	Glasgow
33025	357		Niagara *Rose* W Bg	British	292	119·3	26·9	14·9	Shoreham	1865	W. Hay		J. Henderson	Sydney N
	358	V	Niagara *Beaulieu* 79-83 W Bn	French	174	103·0	26·0	11·7	Meteghan, N.S.	1884	J. Babine	12.85	Vve. F. Cordon & fils	St Pierre
	359	GV	Niagara *Steenken* 87-87 W Bk	German	920	162·3	34·2	20·3	Bremerhaven	1867	J. Teckienborg	1886	Jul. Wieting	Bremerhaven
	360	V	Niagara *Breckwoldt* -88 I Bk	German	692	166·0	30·0	18·0	Hamburg	1878	Reiherstieg Schiffs Werfte		J. Peters	Altona
18667	361		Niantic *Hall* 83-86 W Sr	American	121	100·0	17·2	8·8	Hamburg, Pa.	1870	J. D. Jarrard		J. N. Hancox	Stonington
90881	362	V	Nicanor *Wolfe* W Bkn	British	453	140·3	31·1	17·0	Mahon Bay NS	1886	J. H. Zwicker		G. J. Tromp	Lunenburg
	363	N	Nicanor *Olsen* W Bk	Norweg'n	424	122·1	29·3	16·9	Skelleftea	1857	N. Sundström		A. Christensen	Drammen
	364		Nicasia *San Juan* W Pol	Spanish	190	99·7	26·6	12·4	Barcelona	1858			J. M. Arribillaga	Barcelona
[illegible]	365		NicholasSmirk (LOST) *Lamidon* W Sw	British	190	85·0	24·3	14·5	Sunderland	1849			M. Harland	Hartlepool
18554	366	V	Nicholas Thayer *Sarason* W Bk	American	585	138·9	31·0	19·3	Thomaston Me	1868	Small	[illegible]	C. C. Rohlffs	San Fran
	367		Niclot (LOST) *Hill* -88 W Bg	German	233	96·5	25·8	14·6	Rostock	1847	Fr. Dethloff		E. H. Utech	Rostock
	368		Nico *Nielsen* 86-88 W Bg (ex Margrethe, ex Margaret)	Norweg'n	241	106·3	22·7	14·5	Montrose	1857	C. Birnie		H. Hoistendahl	Holmestrand
	369		Nicola W Bg	Turkish	210	..	..	..	Cashot	1872			Pappa Nicola	Cashot
	370	G	Nicolaas *Berghout* 76-76 W Bk (ex Ludwig, ex Rivadavia)	Dutch	321	122·0	30·0	11·8	Belfast, Me.	1856			Repko & Co.	Harlingen
	371	DV	Nicolaas Frans *Vegter* W Bn	Dutch	174	100·7	21·2	12·0	Pekela	1867	F. Woortelboer	[illegible]	A. E. Doewes	Veendam
89388	372		Nicolai *P. J. Peterson* -86 W Sr	British	109	93·2	21·6	7·3	Nambucca River, N.S.W.	1885	M. Holm		W. Langley	Sydney N
	373	V	Nicolai *Oestmann* -88 W 3m Sr (WRECKED)	German	289	126·7	27·7	11·1	Hamburg	1870	Petersen	[illegible]	H. v. Ehren	Blankenese
394	374		Nicolai Livland *Bothmann* 87-87 W 3m Sr	Russian	299	120·1	27·6	13·9	Pernigel	1881	A. Spiring		G. Sirnin & J. Sausing	Riga
472	375	A	Nicolaos *Coroniadis* -88 W Bg	Greek	325	119·7	26·6	..	Syra	1886	A. Livadaras		G. Coroniadis	Syra
4	376	H	Nicolaos W Bg	Greek	282	101·5	27·1	15·6	Galaxidi	1875	G. E. Canatas		N. N. Moris & Son	Patras
131	377	H	Nicolaos W Bn	Greek	274	114·5	21·0	15·7	France	1869			C. Vasilakis	Zante
112	378	A	Nicolaos *Caimasius* W Bn	Greek	223	90·5	25·1	..	Syra	1869			P. Caimasius	Syra
127	379		Nicolaos *Platys* W Sr	Greek	168	87·3	23·8	..	Syra	1884	A. Evdomon		C. Platys	Syra
41	380	H	Nicolaos W Bn	Greek	161	83·3	22·2	13·9	Spezzia	1865			Helen D. Gouli	Spezzia
48	381		Nicolaos W Sr	Greek	142	80·1	21·1	13·5	Skiathos	1885			D. N. Mitouras	Skiathos
	382		Nicolaos W Bg	Greek	141	..	..	..	Syra	1858			I. Pavlatos	Zante
48	383	A	Nicolaos *Melisaris* W Bn	Greek	137	83·0	23·4	11·8	Galaxidi	1880	J. Canatas		E. Melisaris	Galaxidi
	384	A	Nicolaos *Pappacostandopolo* W Sr	Greek	121	..	..	..	Galaxidi	1887	G. Canata		A. Baccogiorgi	Galaxidi
	385		Nicolás *Aguirre* W Bn	Spanish	310	122·7	25·6	15·1	Bilbao	1868	S. de Arana	[illegible]	José Morodo y Gallardo	Coruña
	386		Nicolaus *Lammers* 85-87 W Bg (LOST)	German	306	118·8	29·2	13·1	Brake	1875	J. F. Nicolai	[illegible]	J. H. Nicolai	Brake
	387	V	Nicolaus *Jens* -87 W Sr	German	131	79·0	22·6	10·0	Rendsburg	1863	J. Paap & Co.		J. Paap & Co.	Rendsburg
	388		Nicolaus *Jansson* 83-83 W Sr (ex Nicolaos)	Swedish	178	93·5	20·3	11·1	Halmstad	1859			O. Jansson	Calmar
	389		Nicolaus Heinrich *Bade* W Bk	German	307	105·8	27·5	16·1	Rostock	1852	A. Schoenemann			
	390	DN	Nicolay H. Knudtzon *Bull* -87 W Bk	Norweg'n	282	122·8	27·1	13·5	Christiansand	1869	J. Wessell	[illegible]	A. B. Bulls Enke	Tönsberg
	391		Nicoletta F. *Feingo* -88 W 3m Sr		459	133·2	28·9	19·8	Rapallo	1874	G. Merello	[illegible]		
	392	V	Nicolette *Jorgensen* 85-85 W Bk (ex Bannaton)	Dutch	1004	174·2	36·1	22·4	Quebec	1865	Davie		Van Veen & Van der Hoeven	
2179	393	✻I	Nicolina *Penco* W Bk	Italian	515	136·5	31·2	19·9	Varazze	1871	A. Craviotto	[illegible]	Flli Croce fu A.	Genoa
	394		Nicolina *Wesslad* -87 W Bk	Swedish	373	120·1	26·7	16·3	Dalkarlsö	1856	W. Sundström		A. M. S. Wessblad	Stockholm
	395		Nicolina *Jonsson* W Sr	Swedish	133	93·2	21·8	9·0	Baltimore	1852			P. A. Jonsson	Bergqvara
	396	V	Nicoline *Klosborg* W Sr	Danish	116	84·2	20·5	9·5	Aalborg	1874	S. L. Sørensen		Actieselskab "Nicoline" (P. Andersen)	Thisted
	397	V	Nicoline *Mühlmann* -86 W Bn	German	331	122·3	27·6	13·0	Brake	1878	J. Oltmann's Wwe	[illegible]	H. P. Breckwoldt, jun	Blankenese
2550	398	IV	Nicolino *Ferro* 87-87 W Bk	Italian	745	170·6	33·8	21·7	Varazze	1874	S. Craviotto	[illegible]	L. Gazzolo	Genoa
487	399	IV	Nicolino *Laviero* -88 W Bg	Italian	273	104·9	26·2	16·5	Gaeta	1876	A. Di Russo	[illegible]	Luigi Talamo	Gaeta
3704	400	I	Nicolino *D'Amato* -87 W Bg (ex Rinaldo)	Italian	159	82·3	23·3	15·3	Sorrento	1859	De Rosa	[illegible]	S. Ferrara	

✻ = Lloyd's Register; D = Nederlandsche Vereeniging; G = Germanischer Lloyd; I = Registro Italiano; R = American Record;
A = Veritas Austro-Ungarico; H = Veritas Ellenico; N = Norske Veritas; V = Bureau Veritas.

In the 4th Column, C = Composite (iron frame, wood planked); I = Iron; Stl = Steel; W = Wood.

The Dates of Metalling (Col. 11) are based on the records of the Books in which Vessels are respectively classed, or on the Reports of the Surveyors of Lloyd's Register.

Further particulars of Vessels classed by Lloyd's Register will be found in Lloyd's Register Book and Lloyd's Yacht Register.

No.	Classe	Nom. Capitaine et dates d'entrée au service des Armateurs actuels et à bord du navire. Ancien Nom. Matériaux et Gréement.	Pavillon.	Tonnage net.	Longueur	Largeur	Creux.	Où.	Quand.	Par qui.	Date de doublage en métal	Armateurs ou Directeurs-Gérants.	Port d'Armement.
401	V	Nicolino *Del Gatto* W Bg	Italian	115	81·3	22·4	10·9	Vietri	1874	P.Savastano		Flli Pellegrino	Castellamare
402	V	Nicolo (ex Scoglio) *Scarpa* W Bg	Italian	335	119·4	27·2	17·2	Martinsalnizza	1886	C.Giacovcich	[illegible]	A.Scarpa	Venice
403		Nicolo (ex L'Annunziata) *Tedesco* W Bn	Italian	112	83·3	22·9	11·6	Gaeta	1867	F.Contreras	[illegible]	N.Arteso	Palermo
404	I	Nicolo Accame *Boccone* W S	Italian	1371	214·5	33·8	24·9	Sestri Ponente	1884	A.Briasco fu L.	[illegible]	Em.Accame & figli	Genoa
405	A	NicoloDe'Conti *Poli* -88 W Sr	Italian	100	92·5	21·6	12·0	Chioggia	1888	F.Menetto		F.Poli	Chioggia
406	V	Nicolo' P. *Serra* -87 W Bn	Italian	120	92·0	22·0	12·0	Pisa	1880	H.Parodi	[illegible]	Flli Pendibene	Spezia
407	✻	Nicolo Padre *E.Mortola* W Bk (ex [illegible])	Italian	375	127·0	26·0	11·0	St.Malo	1869	L.Mallard	[illegible]	E.Mortola	Genoa
408	V	Nicosia *Macdonald* 81-83 W Bk	British	1094	184·2	36·5	22·3	St.John,N.B.	1878	J.Stewart	[illegible]	Wm.Thomson & Co.	St.John,N.B.
409	✻	Nicoya *Foster* C S	British	594	162·8	30·1	17·6	Aberdeen	1867	A.Hall&Co.	[illegible]	A.McIntyre	Melbourne
410	N	Nidaros *Nielsen* W Bk	Norweg'n	335	105·0	28·0	17·3	Medford,N.J.	1852			O.A.Föyen	Tönsberg
411	N	Nidelven *Norberg* W Bk	Norweg'n	279	100·3	27·8	14·7	Grimstad	1863	A.Omundsen		G.Norberg	Porsgrund
412	V	Niels *Lauritzen* W3mSr	Danish	213	116·0	25·3	11·8	Marstal	1881	J.O.Christensen	[illegible]	H.C.ChristensensEnke	Svendborg
413	DV	Niels Juel *Kruse* -84 W3mSr	Danish	230	113·5	25·1	12·9	Marstal	1883	F.Hansen	[illegible]	N.P.Petersen	Marstal
414	V	Niels Juel *Henriksen* W Bn	Danish	156	98·7	21·3	10·8	Svendborg	1877	P.Troensegaard		J.F.Henriksen	Svendborg
415	V	Niels Juel *Birkholm* -86 W Bn	Danish	143	93·1	21·9	10·8	Faaborg	1873	B.Dyreborg		N.R.Storm	Faaborg
416	N	Niels Koren *Pedersen* -85 W Sr	Norweg'n	156	97·9	24·7	11·2	Bergen	1869	Sandvik		Chr.Pedersen	Egersund
417	V	Nielsine *Brinck* -88 W Bn	Danish	151	96·1	21·9	10·8	Nordby,Fanø	1873	S.Abrahamsen	[illegible]	Actieselskab (N. H. Brinck)	Sonderho,Fanø
418		Niemi *Rosenberg* -88 W3mSr	Russian	305	132·5	27·9	13·0	Peksala	1875			C.J.A.Ahlstedt	Angelniemi
419		Nieuw Guinea (ex W[illegible] Liang Seng) W Sr	Dutch	122	85·3	21·7	10·0	Singapore	1871			C. W. van Renesse van Duijvenbode	Ternate
420		Niger *Hallet* W S	American	391	116·6	29·2	19·2	Rochester,Ma.	1844			Tabor & Gordon	N.Bedford,Mass.
421		Niger W	French	208	115·1	26·9	11·3	Boston	1866			J. E. Buhan père, fils & A. Teisseire	St.Louis,Sénégal
422		NightHawk (BROKEN UP) I Bk (ex Steamer Wm.McCormick) *Dundas*	British	651	213·5	27·9	16·0	Glasgow	1854	A.Stephen&Sns.		J.Bulloch	Calcutta
423	R	Night Watch *Carter* W Bkn	British	331	116·5	26·0	16·2	Sunderland	1859	Naizby		T.Middleton	London
424	✻	Nightingale *Wilson* -87 W K	British	93	80·8	21·3	10·6	Brixham	1887	S.Dewdney&Sn		T.Wilson	Grimsby
425		Nightingale W Sr	American	151	99·0	28·5	8·0	Bristol,R.I.	1854			P.McConville	Bangor,Me.
426	N	Nightingale *Ingebrethsen* W Bk	Norweg'n	657	179·3	35·6	18·4	Portsmouth, N.H.	1851	J.Hanscomb		A.Schweigaard	Kragerö
427	G	Nijenstein *Bussman* 88-88 W Bg (ex Cornelia, ex Roosbergh)	Dutch	304	115·8	26·3	15·0	Rotterdam	1862	Gebr.Visser	[illegible]	S.Barends	Delfzijl
428		Nijssina *Veldman* W Sr (ex Helena)	Dutch	106	84·3	20·2	9·3	Papenburg	18[illegible]			J.K.Veldman	Termunterzijl
429		Nijverheid W Bg	British	182	98·9	24·1	12·0	Slikkerveer	1870		[illegible]	H.C.Myburgh	Cape Town
430	D	Nijverheid *Bakkema* 83-87 W Bk (ex Nagen van Mareen)	Dutch	445	124·0	31·9	17·8	Varazze	1867	G.Scala	[illegible]	G.G.Albers	Farmsum
431		Nikanor *Bagman* W Sr	Russian	277	103·0	29·5	14·5	Kimito	1873	Söderström		E.A.Eriksson	Kimit
432	V	Nikaria *Long* 88-88 W Bk	British	1083	189·0	36·5	21·5	MeteghanN.S.	1888	A.W.Eakins	[illegible]	J.Y.Robbins	Yarmouth,N.S.
433	A	Nikiforos *Marengo* W Bg	Russian	255	94·0	27·5	..	Syra	1865			B.Tongua	Eupatoria
434		Niko (ex Naomi K.) *Bonzio* W Sr (now named "Sophie")	Austrian	400	121·0	26·9	17·5	Fiume	1875	G.B.Baccich [Bellaz	[illegible]	N.Svilocossi fu N.	Ragusa
435	DV	Nikolai *Fredriksson* W Bk	Russian	443	129·9	29·5	17·5	Dragsfjerd	1884	J.E.Danielsen		G.A.Backlund	Dragsfjerd
436		Nikolai *Juselius* W Bk	Russian	414	..	..	..	Jomala	1870			N.Sittkoff	Mariehamn
437		Nikolai *Ivanoff* W Sr	Russian	377	..	..	..	Walamo				WalamoMonastery	Walamo
438		Nikolai *Pavlsi* W Sr	Russian	291	..	..	..	Salmis	1882			W.Hosainoff	Salmis
439		Nikolai *Tschunoff* W Sr	Russian	237	..	..	..	Impilaks	1877			M.Korniloff	Sordavala
440		Nikolai W Sr	Russian	234	..	..	..		1872			N.Mironoff	Sordavala
441		Nikolai *Timofejeff* W Sr	Russian	221	..	..	..	Sordovala	1878			I.Spiridonoff	Sordavala
442		Nikolai *Tallikka* W Sr	Russian	169	..	..	..	Hiitola	1873			A.Lisitzin	Kexholm
443		Nikolai *Carlsson* W Sr	Russian	163	95·1	24·0	11·0	Wiborg	1864	Lane		O.Carlsson	Raumo
444		Nikolai *Lempinen* W Sr	Russian	160	..	..	..	Impilaks	1869			M.Korniloff	Sordavala
445		Nikolai W Sr	Russian	126	..	..	..		1873			I.Spiridonoff	Sordavala
446		Nikolai *Nasarjeff* W	Russian	124	..	..	..	Impilaks	1880			M.Korniloff	Sordavala
447		Nikolai *Wasiljeff* W Sr	Russian	118	..	..	..	Salmis	1874			J.Sidoroff	Salmis
448		Nikolay Lebedeff (CONDEMNED) *Ivanoff* W Bg	Russian	194	97·6	23·3	11·3	Narva	1878	P.Belt		D.v.Zinovieff & Co.	Narva
449	✻	Nil Desperandum *Lamb* W K	British	92	81·3	21·0	10·8	Brixham	1886	S.Dewdney & Ss		J.Lamb	Grimsby
450	✻	Nil Desperandum Stl Bk	Dutch	1165	..	..	..	Amsterdam	1889	F.F.Green		A.Hendrichs & Co.	Amsterdam

✻ = Lloyd's Register; D = Nederlandsche Vereeniging; G = Germanischer Lloyd; I = Registro Italiano; R = American Record;
A = Veritas Austro-Ungarico; H = Veritas Ellenico; N = Norske Veritas; V = Bureau Veritas.

Dans la 4me Colonne, C = Composite (membrure en fer, bordé en bois); I = Fer; Stl = Acier; W = Bois.

Les Dates du doublage en métal (**Col. 11**) ressortent des livres dans lesquels les navires sont respectivement classés ou des rapports des Inspecteurs du Lloyd's Register.

Pour plus amples renseignements sur les navires classés par le Lloyd's Register, voir le Lloyd's Register Book et le Lloyd's Yacht Register.

Official No.	No. in Book	Registry in which classed	Ship's Name. Master, and Dates of Appointment to present Owner's Service and to Ship. Ship's Late Name.	Material and Rig.	Flag.	Net Tonnage	Length	Breadth	Depth	Where.	When	By whom.	Date of Metalling.	Owners or Managers.	
90285	451	*	Nile *Gilmore 86-88*	I 3mS	British	2079	283·5	40·5	24·6	PortGlasgow	1886	Russell & Co.		W.Douglas & A.Brown	Glasgo
10243	452		Nile	W Bk	British	298	124·6	24·7	14·6	Arbroath	1855			Hong Kong, Singapore & Borneo Trading Co.	Hong K
89862	453		Nile *Coffin -88* (ex Robert K. Kirkland)	W Bg	British	164	96·6	24·6	8·2	S.Michaels,Md	1857		[illegible]	H.Mosey	Turks' Is
189	454		Nilo (LOST) *Lallich -87*	W Bg	Austrian	431	122·0	26·1	17·4	Fiume	1876	G.B.B.Bedaz	[illegible]	G.Xigga	Meglin
	455		Nilo *Miguels*	W 3mSr	Brazilian	202	..	..	..	UnitedStates	1856		[illegible]	M.E.Migueis	Bahia
1622	456		Nilo *Ferrari -87*	W Bk	Italian	429	121·7	30·6	19·5	Spotorno	1867	G.B.Cadenaccio	[illegible]	G.B.Ferrari	Genoa
	457		Nils *Nilsson*	W Sr	Swedish	117	87·3	21·4	9·6	Figeholm	1872			N.P.Dahlgren	Figeholm
	458	G V	NilsMagnus *Andersson -87*	W Sr	Swedish	183	110·8	22·3	11·0	Påskallavik	1868			G.Förselius	Stockholm
78866	459	D V	Nimbus *Smith 84-84*	W Bk	British	1222	196·0	38·4	22·7	Cornwallis,NS	1878	G.B.Vaughan	[illegible]	C.R.Burgess	Windsor
40	460	A	Nimrod *Scopinich -88*	W Bk	Austrian	620	136·8	29·6	18·8	Trieste	1867	B.Tonello	[illegible]	C.Gerolimich	Lussinpic
	461		Nimrod *Wieting* (ex Samuel Pascal)	W S	German	695	152·5	34·6	21·5	Medford,N.J.	1855			E.D.B.Wieting	Elsfleth
410	462		Nimrod *Wentzban* (CONDEMNED)	W 3mSr	Russian	266	110·0	29·8	13·1	Picton,N.S.	1860			Gebrüder Gutschmidt	Windau
54244	463		Nina	W Bn	British	186	104·2	24·0	12·9	P.E.Island	1866	Duncan		T.& J.Connick & Co.	Dundalk
71489	464		Nina *Buckham 61-91*	W Bn	British	162	99·5	23·0	12·0	P.E.Island	1875	McMillan		H.Manley	Teignmouth
91303	465	*	Nina	W Sk	British	89	78·4	20·8	10·1	Goole	1884	CuttinghamBr		Pickering & Haldane	Hull
294	466	A	Nina (ex Pietà) *Grando*	W Bg	Austrian	414	115·8	28·4	18·1	Lussinpiccolo	1861	G.S.Cattarinich	[illegible]	L.Grando	Trieste
	467		Nina	W Bk	Norweg'n	655	156·0	32·5	19·7	St.John,N.B.	1856			Brödr.Rosenkilde	Christiania
	468	N	Nina *Melsom 68-86* (ex Tamanki)	W Bk	Norweg'n	544	142·6	30·5	18·4	Quebec	1865	Lee	[illegible]	P.A.Michelsen	Tönsberg
	469	N	Nina *Hansen -88*	W Bk	Norweg'n	496	134·1	31·2	17·2	Tvedestrand	1878	T.Gundersen	[illegible]	A.Biering Olsen	Tvedestrand
1073	470	I	Nina Schiaffino *Gnecco -87*	W Bk	Italian	393	122·0	30·2	18·9	Savona	1867	F.Sirello	[illegible]	L.Perrano	Spezia
	471		Nina Seconda *Nicola*	W Bk	Greek	541	143·7	31·0	20·6	Savona	1872	F.Sirello	1.88	Hajee N.Marigo	Zante
130279	472	R	Nina Tillson *Green*	W 3mSr	American	615	152·0	35·0	12·5	Rockland,Me.	1883	A.F.Ames		A.F.Ames	Rockland
48854	473		Nineveh *Broadfoot -87*	W S	British	1174	209·9	36·3	22·6	Aberdeen	1864	W.Hood & Co.		J.H.Goodlet	Sydney,NS
18784	474	R	Nineveh *Sydney -88*	W Bkn	American	470	126·2	30·5	17·0	East Boston	1874	A.&G.F.Sampson	[illegible]	J.F.Brooks	Boston,Ms
3318	475		Nino Bixio *Nossardi -88*	W Bk	Italian	533	140·0	29·5	16·2	Sestri Ponente	1875	Ag.Briasco fu G.B.	[illegible]	A.Nossardi	Genoa
188	476	V	Nino Bixio *Ravagnan*	W Bn	Italian	199	103·0	22·9	14·3	Ravenna	1875		[illegible]	Flli Vianello-Moro	Venice
5260	477		Nio *Beedle 86-86*	W Bg	British	172	80·0	21·5	13·4	Whitby	1853			I.Mills	Whitby
72378	478	*	Niobe *Ralston 72-75*	I S	British	1469	253·0	38·2	22·8	Greenock	1874	R.Steele		Baine & Johnston	Greenock
	479	N	Niobe *Olsen -88*	W Bk	Norweg'n	458	135·7	30·0	16·9	Grimstad	1874	A.Gelmuyden	[illegible]	P.E.Johannesen	Christiania
	480		Niord *Terkelsen*	W Bk	Norweg'n	440	136·5	29·6	16·7	Risör	1878	Niels Olsen	[illegible]	A.Terkelsen	Kragerö
	481	N	Niord *Petersen 88-*	W Bk	Norweg'n	732	153·4	35·1	19·9	Stavanger	1866	M.A.Magnus	[illegible]	Holdt & Isachsen	Stavanger
	482	N	Niord (ex Anois Emma) *Dahl*	W Bk	Norweg'n	503	136·6	30·8	18·0	Bordeaux	1865	Chaigneau	[illegible]	Johan Thorne	Moss
	483	N	Niord *Lepsöe -87*	W 3mSr	Norweg'n	116	91·5	21·4	10·1	Bergen	1864	S.Fluge		Andr.Olsen	Bergen
	484		Niord *Backmann*	W Sr	Swedish	236	102·3	21·6	11·7	Italy	1869			J.Ohlsson	Smögen
	485		Niord *Pettersen -79*	W Sr	Swedish	206	94·2	27·5	12·6	Vätö	1850	P.Österberg		J.A.Johanson	Östra Sh
	486		Niord *Svennevig 73-86*	W Sr	Swedish	142	96·8	22·1	10·9	Denmark				A.Kastman	Calmar
	487		Niord	W Sr	Swedish	101	79·1	23·5	8·8	Denmark				J.Johansson	Norrköping
2295	488	I	Nipote *Brignati*	W Bk	Italian	403	134·5	29·2	17·4	Voltri	1872	G.Fava	[illegible]	Eredi de Gregori	Genoa
2266	489	D I	Nipoti Accame *Paleste*	W Bk	Italian	898	175·8	34·4	23·1	Leano	1876	A.Cerruti	[illegible]	P.Accame & figli	Savona
2027	490	I	Nipotini *Aste* (ex Gerolamo Figari)	W Bk	Italian	536	131·8	29·2	19·7	Savona	1872	P.Sirello	[illegible]	G.Schiaffino	Genoa
363	491		Nishibetsu Maru	W Sr	Japanese	167	100·3	20·4	10·1	Tokio	1881	Kawasaki Ship Building Yard		Owaki Kikusaburo	Tokio
61	492		Nishiwo Maru	W Sr	Japanese	476	140·0	29·2	16·1	Hiogo	1877	Omatsu Kunimatsu		Hachima Kanesuke	Hiogo
28802	493	*	Nith *Ball 86-88* (ex City of Shanghai)	I Bk	British	990	212·5	32·3	21·5	Glasgow	1860	Brclay,Curle&Co.		Bramwell & Gardiner	Glasgow
	494	G V	Nixe *Feldjuch 83-87*	I S	German	1672	228·0	39·0	24·0	Vegesack	1887	Bremer Schiffbau Ges.		Gildemeister & Ries	Bremen
394	495		Nizzarda *Danesi*	W Bn	Italian	170	84·6	23·8	15·3	Castellamare di Stab.	1853			Eredi di M.Giannoni di B.	Porto Fe
	496	D N V	Njaal *Hansen 87-87*	W Bk	Norweg'n	569	147·0	34·7	16·8	Soon	1881	C.Abrahamsen	[illegible]	Axel.Sch.Hansen	Christiania
	497	V	Njord *Svennevig 73-83*	W Bn	Danish	135	93·9	21·5	10·2	Svendborg	1873	L.B.Andersen		A.N.Svennevig	[illegible]
	498	D V	Noach III. *Kruijt 73-77*	W S	Dutch	1091	189·9	13·0	21·8	Slikkerveer	1869	J.Smit	[illegible]	Fop Smit,jun.	Rotterdam
	499	* D V	Noach IV. *Gouws 76-*	W Bk	Dutch	1303	206·0	41·7	25·0	Slikkerveer	1875	A.Smit	[illegible]	Fop Smit,jun.	Rotterdam
	500	* D	Noach V. *Speenhoff -88*	W Bk	Dutch	1263	207·3	42·3	24·9	Slikkerveer	1877	A.Smit	[illegible]	Fop Smit,jun.	Rotterdam

*=Lloyd's Register; D=Nederlandsche Vereeniging; G=Germanischer Lloyd; I=Registro Italiano; R=American Record; A=Veritas Austro-Ungarico; H=Veritas Ellenico; N=Norske Veritas; V=Bureau Veritas.

In the 4th Column, C=Composite (iron frame, wood planked); I=Iron; Stl=Steel; W=Wood.

The Dates of Metalling (Col. 11) are based on the records of the Books in which Vessels are respectively classed, or on the Reports of the Surveyors of Lloyd's Register.

Further particulars of Vessels classed by Lloyd's Register will be found in Lloyd's Register Book and Lloyd's Yacht Register.

No.	Cote	Nom. Matériaux et Gréement. Capitaine et dates d'entrée au service des Armateurs actuels et à bord du navire. Ancien Nom.	Mat. et Gréem.	Pavillon	Tonnage net	Longueur	Largeur	Creux	Construit: Où	Quand	Par qui	Date du doublage en métal	Armateurs ou Directeurs-Gérants	Port d'Armement
	V	Noach VI. *Pootman*	W Bk	Dutch	1334	204·4	42·0	24·9	Slikkerveer	1879	A. Smit	[illegible]	Pop Smit, jun.	Rotterdam
		Noah	W Sp	British	118	107·5	22·0	7·2	Williamstown [Ont	1873			N. Rousson	Montreal
	X	Noah *Larsen -88*	W Bk	Norweg'n	441	135·5	30·5	16·5	Risör	1876	J. Stenersen	[illegible]	P. D. Prebensen	Risör
	X	Noah *Salvesen*	W Bk	Norweg'n	400	125·1	29·7	16·8	Tvedestrand	1868	J. P. Pedersen	[illegible]	Niels Salvesen	Tvedestrand
	V	Noah *Legzding -88*	W 3m Sr	Russian	333	128·9	26·7	13·4	Adiamünde	1888	P. Krause		J. Ehkis	Adiamünde
		Noahthat	Stl Lr	British	652	250·1	34·0	8·8	Dumbarton	1887	W. Denny & Bros		Irrawaddy Flotilla Co. (Limd.)	Glasgow
507	X	Noatun *Jorgensen*	W Bk	Norweg'n	501	155·9	30·6	17·2	Grimstad	1881	H. A. Weber	[illegible]	Chr. Th. Boe	Arendal
508	X	Noatun *Hansen*	W Sr	Norweg'n	210	104·6	25·9	12·2	Kragerö	1883	I. Nielsen	[illegible]	Chr Sunde	Kragerö
509	X	Noble *Hansen 82-82* (ex [illegible])	W Sr	Norweg'n	146	86·1	24·2	10·9	Blankenese	1851			H. H. Pettersen	Arendal
510	V	Nobreza (ex Adelaid) *S. Jod*	W Bk	Portug'se	372	118·1	24·6	15·3	Searsport Me.	1864		[illegible]	J. N. da Silva & V. Rosa & Co.	Lisbon
511		Nod Coven *Lapie -87*	W Sr	French	126	90·5	21·0	11·4	Bideford	1866			Eugène Lapie	Granville
512	*	Noddleburn *Hall 87-87*	I Bk	British	1053	217·0	34·2	20·0	Dumbarton	1883	Birrell, Stenhouse & Co.		M. Carswell	Glasgow
513	IV	Noè *Cacace 86-86*	W Bk	Italian	513	139·4	26·2	16·7	Castellamare	1886	G. Starita	[illegible]	M. Cacace	Castellamare
514	V	Noel *Knowlton 75-75*	W Bk	British	812	171·2	35·7	20·2	Maitland, N.S.	1875	O. O'Brien	[illegible]	O. O'Brien	Maitland, N.S.
515	NV	Noel *Jörgensen*	W Bk	Norweg'n	489	142·6	30·5	16·9	Lillesand	1879	S. N. Hansen	[illegible]	H. Christoffersen	Arendal
516	V	Noel Henry *Langlois -87* (ex Maid of [illegible])	W Bg	French	250	121·5	25·6	12·8	Newport	1867	Willmott	[illegible]	J. Jamet & fils	Granville
517	A	Noel Tousis *N. Makris*	W Sr	Greek	148	86·6	24·6	12·1	Syra	1882	L. Josipu		G. M. Makris	Syra
518	V	Noemi (WRECKED) *Oger*	W Bk	French	340	119·1	24·6	15·9	St. Malo	1871	Hervot	[illegible]	P. Ch. Grenet	Nantes
519	I	Noemi *Rossi*	W Bn	Italian	362	115·8	29·2	18·2	Varazze	1864	N. Cerruti	[illegible]	N. Oneto	Genoa
520	V	Noemia (PONTOON) *Reis* (ex Mimosa; ex Alegria; ex Monteiro; ex Ermelinda; ex Lydia; ex Julia [illegible])	W Bk	Portug'se	308	99·7	25·6	13·8	Villa Nova	1857	C. M. da Silva [Santos	[illegible]	J. P. de Campos	Oporto
521		Noemis (WRECK [illegible]) *Andouard*	W Sr	French	136	88·6	22·0	8·9	Cape Breton	1868			A. Girardin	St. Malo
522		Nohrborg *Andersson -84*	W Sr	Swedish	146	95·5	21·8	10·3	Figeholm	1880	N. Larsson		N. P. Dahlgren	Figeholm
523	I	Noi (ex Cesira Madre) *Schiaffino*	W Bk	Italian	655	153·5	31·5	20·5	Chiavari	1881	L. Gotuzzo	Y 10.88	A. De Gregori di G. B.	Genoa
524		Noisiel *Savary*	W Bg	French	138	..	..	..	Honfleur	1860				Marseilles
525		Nojet *Eriksson*	W Sr	Russian	236	..	..	..		1870			M. Mattson	Lemland
526	*	Nolan's Charge *Pikens 85*	W Sk	British	86	76·0	21·0	10·0	Grimsby	1885	Smith Bros.		J. H. Bellamy	Grimsby
527		Noli *Strina*	W Bg	Italian	298	89·9	24·9	16·1	Prà	1858	Tixi	[illegible]	G. Cordiglia	Genoa
528	R	Nomad *Sterling*	W Bk	American	476	133·6	30·3	17·2	P. Jeffersn, N.J	1872	Bayles & Son	[illegible]	E. R. Sterling	New York
529		Nomaden *Olsen 86-86*	W Bg	Norweg'n	289	98·5	27·7	15·6	Grimstad	1854	T&T Gundersn.		B. Olsen	Grimstad
530		Nominoé	W Bg	French	134	72·5	21·0	11·3	Nantes	1865	Sevestre		Le Perron	Nantes
531		Nonantum *Fogg -88*	W Bk	American	1100	178·6	36·2	23·7	Nwburypt. Ms	1865	G. W. Jackman		D. R. Sortwell	Boston, Mass.
532		Nonpareil *Finlayson*	W Bk	British	399	125·5	27·4	17·4	Sunderland	1865	J. & E. Lumsden	[illegible]	J. Huddart	Melbourne
533	*	Nonpareil *Cook*	W Sk	British	90	81·6	20·8	10·6	Rostock	1885	F. Burchard & C.		G. Cook	Hull
534		Noord Holland *Blom* (ex Maarten van Rossum)	W Bk	Dutch	617	132·8	33·5	20·7	Alblasserdam	1855	C. Smit		W. Smit	Purmerend
535	DV	Noordster *Smit -88*	I Sr	Dutch	180	104·3	21·7	11·2	Charlois	1879	Rybee & Co.		J. Vroege	Alblasserdam
536		Noorool Hasim (LOST)	W Sr	Dutch	146	..	..	..						Ternate
537	X	Nor *Ekornes* (ex Mashua) (WRECK [illegible])	W S	Norweg'n	796	155·8	32·4	22·7	Medford, Mass	1849		I 12.81	H. C. Seeberg	Tönsberg
538	XV	Nor *Steinart*	W Bk	Norweg'n	581	157·6	31·6	18·5	Christiansand	1878	M. Knudsen	[illegible]	H. Natvig	Christiansand
539	XV	Nor *Buchdal -85*	W Bk	Norweg'n	529	153·5	31·5	16·5	Drontheim	1873	Jensen	[illegible]	Jenssen & Co.	Drontheim
540	X	Nor *Olsen*	W Bk	Norweg'n	450	131·3	29·7	16·8	Nantes	1862		[illegible]	B. Knoph	Kragerö
541	X	Nor (ex Ran) *Hansen 88-88*	W Bk	Norweg'n	409	135·7	29·6	16·3	Carlshamn	1856		[illegible]	A. J. Semb	Laurvig
542	X	Nor *Hansen 81-81* (BANDON [illegible])	W Bk	Norweg'n	330	106·5	29·2	16·2	Arendal	1850		[illegible]	Chr. Klöcker	Arendal
543		Nor *Hansen 87-87* (ex Annie Augusta) (LOST)	W Bk	Norweg'n	280	111·5	27·2	14·3	Bilbao	1854			L. L. Sund	Farsund
544	X	Nor *Sivertsen -87*	W Bg	Norweg'n	253	102·0	27·7	14·3	Grimstad	1859	A. Bentsen		S. Eyde	Arendal
545	X	Nor *Nielsen*	W Bg	Norweg'n	208	97·1	25·8	11·8	Arendal	1857			M. Nielsen	Arendal
546	*	Nor'Wester *Toulson 87-87*	I Bk	British	547	163·5	29·2	16·8	Pt. Glasgow	1864	L. Hill & Co.		J. & A. Roxburgh	Glasgow
547	*	Nora (WRECKED) *Murchison*	W Bn	British	215	107·2	24·5	12·4	Summerside [P.E.I.	1883	A. McMillan	[illegible]	R. T. Holman	Chrlttown, P.E.I.
548		Nora *Wilhelmsen -82* (ex [illegible])	W Bk	Norweg'n	802	156·2	34·0	21·1	Bremen	1851			C. G. Hansen	Christiania
549	X	Nora *Hassel -80* (ex North Wind)	W Bk	Norweg'n	783	169·0	33·9	21·0	Quebec	1865	Richard	[illegible]	J. F. Juell	Risör

✱ = Lloyd's Register; D = Nederlandsche Vereeniging; G = Germanischer Lloyd; I = Registro Italiano; R = American Record; A = Veritas Austro-Ungarico; H = Veritas Ellenico; N = Norske Veritas; V = Bureau Veritas.

Dans la 4me Colonne, C = Composite (membrure en fer, bordé en bois); I = Fer; Stl = Acier; W = Bois.

Les Dates du doublage en métal (Col. 11) ressortent des livres dans lesquels les navires sont respectivement classés ou des rapports des Inspecteurs du Lloyd's Register.

Pour plus amples renseignements sur les navires classés par le Lloyd's Register, voir le Lloyd's Register Book et le Lloyd's Yacht Register.

Official No.	No. in Book	Society in which Classed	Ship's Name. Master, and Dates of Appointment to present Owner's Service and to Ship. Ship's Late Name.	Material and Rig	Flag.	Net Tonnage.	Length.	Breadth.	Depth.	Built: Where.	Built: When.	Built: By whom.	Date of Metalling.	Owners or Managers.
	551		Nora (ex [illegible]) Larsen -86	W Bk	Norweg'n	699	156·0	32·9	16·2	Flensborg	1857			AH Monsen
	552	N	Nora Kriken -84 (ex Lovenzo)	W Bk	Norweg'n	481	132·4	29·7	17·3	Vegesack	1851		11.87	T.M.Wiel
	553		Nora Warner -86 (WRECKED)	W Bk	Norweg'n	371	113·5	30·4	16·7	Amsterdam	1848			Thor Warner
	554	V	Nora Ovregaard (ex Hedwig Eleonora)	W 3m Sr	Norweg'n	212	104·2	24·4	13·1	Wenersborg	1875	S.Oldsson		L.B.Ovregaard
89231	555		Nora Harkins Harkins	W Sr	American	199	124·0	32·4	8·0	PrsrsbrgOreg	1882			G.W.Chandler
90687	556	V	Nora Wiggins Lawrence -88	W Bk	British	484	143·0	33·0	13·1	[illegible] N.S.	1887	R W.Freeman		R.W.Freeman
89702	557	✻	Norcross Reap 72-84	I Bk	British	897	207·0	34·1	18·5	Dumbarton	1884	A.M'Millan&Sn		J.N.Ward & Co.
	558		Nord	W Sr	Russian	167	..	..	..		1847			B.Ekrott
1480	559	I V	Nord America Maresca (ex America)	W Bk	Italian	545	112·3	29·4	19·4	Baccari	1881	M.Martaglian		C.Pollio
	560	V	Nordahl Bruun Henrichsen 70-77	W Bn	Norweg'n	178	104·0	23·3	12·3	Königsberg	1867	L.Eggert		J.G.Myhre
	561	N	Nordboen Olsen -87	W Sr	Norweg'n	247	113·0	26·1	11·3	Luleå	1869	E.Brandlöf		W.B.Jansen
	562		Nordbon Björkman -80	W Sr	Swedish	213	107·9	23·5	11·0	Hernösand	1872	Fr.Häggland		M.Lundqvist
	563	V	Norden Kromann 83-83	W 3m Sr	Danish	229	112·9	25·0	12·7	Marstal	1883	H.J.Bager		N.Petersen
	564		Norden Giverholt 71-81	W Bk	Norweg'n	590	156·5	34·0	17·8	Gefle	1853	P.Elfstrand		J.Eitzen
	565	N	Norden Aanonsen -84	W Bk	Norweg'n	528	151·1	33·3	15·6	Sundsvall	1856			J.L.Ugland
	566	D	Norden Nielsen (CONDEMNED)	W Bk	Norweg'n	356	115·5	29·9	14·6	Stavanger (rebuilt)	18 74			Winge & Co.
	567		Norden Mortensen (WRECKED)	W 3m Sr	Norweg'n	262	118·0	27·4	12·1	Grimstad	1875	O.Kittelsen		J.E.Einersen
	568	N	Norden Gjedsen 85-87	W Bg	Norweg'n	258	94·3	27·3	14·8	Stavanger	1828			S.Nielsen
	569	V	Norden Jensen	W Bn	Norweg'n	236	112·0	25·9	12·7	Stavanger	1880	Chr.Knudsen		B.Hansen
	570		Norden Werge 71-74 (CONDEMNED)	W Bg	Norweg'n	223	103·9	27·6	12·5	Stavanger	1855			O.P.Werge
	571	N	Norden Houlder	W Bg	Norweg'n	209	108·8	25·9	12·1	Ar'ga'rdstrnd	1870			James Houlder
	572	V	Norden Kähr 83-85	W Bk	Swedish	464	138·4	27·7	16·1	Sikeå	1877	O.P.Åberg		Fr.Jacobsen
	573	D	Norden Houlder 86-86	W Bk	Swedish	355	132·5	27·3	12·7	Hernösand	1876	J.A.Strandberg		O.R.O.Nordin
	574	V	Norden Svensson -86	W Bk	Swedish	332	114·8	25·4	11·9	Hernösand	1855			P.A.Lindberg
	575		Norden (ex Barbarossa)	W Bg	Swedish	266	91·0	23·2	16·4	Falmouth	1811			R.Stendahl
	576		Norden Rosenlund	W Sr	Swedish	117	83·1	22·6	10·0	Pukavik	1868	M.Jonasson		B.J.Törngren
	577	N V	Nordenskjöld Christensen -88	W Bk	Norweg'n	725	167·8	34·5	19·5	Bergen	1880	Brunchorst& Dekke		S.Hansen
	578	N	Nordenskjöld Petersen	W Bk	Norweg'n	495	140·4	31·3	17·4	Grimstad	1879	O.Kittelsen		J.Gundersen
	579	N	Nordenskjöld Abrahamsen -88	W 3m Sr	Norweg'n	294	126·7	28·3	12·8	Risöe	1879	H.Jensen		J.Prebensen
	580	N	Nordenskjöld Tveit	W Sr	Norweg'n	111	90·9	23·4	10·2	Bergen	1882			S.Christiansen
	581		Nordenskjold Backman -89	W Bk	Russian	333	123·3	28·2	14·1	Pyhäjoki	1878	H.Nordling		E.Bengts
	582		Nordenskjöld Karlsson		Russian	138	..	..	..					J.A.Nordlund
	583	D V	Nordenskjöld Brandinius -86	W Bk	Swedish	665	161·4	30·8	18·0	Sundsvall	1875	E.Östman		E.Sandelin
73290	584		Nordkap Hellgar 86-86	W Bn	British	151	100·5	23·6	10·8	Stavanger	1866	J.Mikkelsen		T.Anthony
	585	D	Nordkap Reinertsen (ex Marina)	W Bk	Norweg'n	664	162·7	30·3	20·1	Milford	1874	J.D.Warlow		R.Reinertsen
	586		Nordlyset Georg 84-84	W Bk	Danish	248	102·4	27·1	14·4	Elsinore	1852	J.H.Løve& Søn		Kgl.Gronl.Handel
	587	N	Nordlyset Nilsen 84-84	W Bk	Norweg'n	346	114·7	28·9	16·0	Tönsberg	1853			N.Chr.Nielsen
	588	N	Nordlyset Aslaksen 75-86	W Bk	Norweg'n	326	101·1	28·9	16·6	Grimstad	1847	JörgenOlsen		E.Dedekam
	589	V	Nordlyset Insland -86	W Bg	Norweg'n	170	98·0	24·0	11·5	Stavanger	1869	Chr.Knudsen		E.Endresen
	590		Nordmanden Hellicsen -78	W Bg	Norweg'n	246	90·4	24·2	16·1	England	1830			T.Kleiberg
	591		Nordsee (LOST) Ebert -87	W Bk	German	328	121·0	27·4	15·2	Rügenwalde	1866	C.Zielke		Edwin Schutz
	592	N	Nordskov Ellingsen -87	W Bk	Norweg'n	391	115·3	30·6	17·3	Grimstad	1862	N.O.Gjömle		P.Pedersen
	593	G V	Nordsøen (ex [illegible])	W Bk	Dutch	322	119·4	25·2	13·9	Hammelwarden	1863	Gebr.Gross		P.T.Nyhoff
	594	N	Nordsöen Johannessen -87	W 3m Sr	Norweg'n	163	92·5	21·7	10·4	Maine	1848			S.Mikkelsen Ran
	595		Nordstern Haevert 77-79	W Bk	German	323	112·6	27·6	16·1	Damgarten	1855	H.Dierling		W.Maack
	596		Nordstern Burmeister -88	W Bg	German	229	93·8	25·5	12·2	Rostock	1856	O Ludwig		Burmeister
	597	V	Nordstern Clavier 81-	W Bg	German	218	95·1	24·9	14·0	Elbing	18 39/59	Geb.Mitzlaff		Wm.Ludendorff jun.
	598		Nordstern Hollander	W Sr	German	135	87·8	20·9	10·5	Bardenfleth	1852	D.Focke		Joh.Hollander
	599	V	Nordstjernan Olsson	W 3m Sr	Swedish	213	108·2	23·7	11·7	Gefle	1870	O A.Broslin		Edw.Bengtsson
	600		Nordstjernan Nordgren (ex Louisa)	W Bg	Swedish	193	92·8	20·8	13·0	P.E.Island	1859			O.P.Nordgren

✻ = Lloyd's Register; D = Nederlandsche Vereeniging; G = Germanischer Lloyd; I = Registro Italiano; R = American [Record];
A = Veritas Austro-Ungarico; H = Veritas Ellenico; N = Norske Veritas; V = Bureau Veritas.

In the 4th Column, C = Composite (iron frame, wood planked); I = Iron; Stl = Steel; W = Wood.

The Dates of Metalling (Col. 11) are based on the records of the Books in which Vessels are respectively classed, or on the Reports of the Surveyors of Lloyd's [Register].

Further particulars of Vessels classed by Lloyd's Register will be found in Lloyd's Register Book and Lloyd's Yacht Register.

1	2-3	4 Nom. Matériaux et Gréement. Capitaine et dates d'entrée au service des Armateurs actuels et à bord du navire. Ancien Nom.	5 Pavillon.	6 Tonnage net.	7 Dimensions. Longueur.	Largeur.	Creux.	8 Construit. Où.	9 Quand.	10 Par qui.	11 Date du doublage en métal.	12 Armateurs ou Directeurs-Gérants.	13 Port d'Armement.
601		**Nordstjernan** *Mattson* WSr	Swedish	141	87.9	26.7	10.3	Häfverö	1876			O.Ersson	Häfverö
602		**Nordstjernan** *Andersson* WBn (ex Nordstjernen)	Swedish	135	91.7	21.2	10.5	Christiania	1871	W.Ihlen		N.Tufvesson	Brantevik
603	V	**Nordstjernen** *Lehr* -87 WSr (ex Marie Christensen)	Danish	105	85.1	19.9	10.1	Faaborg	1872	R.H.Möller		J.MortensensEnke	TveraaFaroeIsl
604	N	**Nordstjernen** *Olefsen* -83 WBk	Norweg'n	503	143.9	30.1	17.2	Vegesack	1855	C.Pokrantz.	V 9.87	L.G.S.Larsen	Sandefjord
605	N	**Nordstjernen** *Oden* -84 WBk	Norweg'n	352	125.0	28.7	15.5	Grimstad	1871	O.& L.Dönnes-[moen	V 2.86	J.Börresen	Grimstad
606	N	**Nordstjernen** *Jacobsen* WBk	Norweg'n	269	132.9	26.8	11.9	Holmestrand	1874	A.Pedersen	V 3.88	G.Jacobsen	Porsgrund
607		**Nordstjernen** *Hansen* WSr	Norweg'n	137	80.1	23.5	11.2	Bergen	1836			H.Hansen	Tönsberg
608	V	**Nore** *Carlberg* 88-88 WBk	Swedish	469	161.7	24.8	15.0	Helsingborg	1873	J.Nordström	V 1.86	J.C.Hulthén	Helsingborg
609	R	**Norena** *Chase* W3mSr	American	417	129.4	31.1	14.6	Bath,Me.	1874	Goss&Sawyer	V 8.88	G.S.Hunt	Portland,Me.
610		**Norfolk Hero** *Turner* WBn	British	157	83.3	22.0	14.0	Yarmouth	1836			J.Gilmour	Dundee
611	*	**Norfolk Island** *Thomas Stone* StlBk	British	1311	231.1	36.1	21.7	PortGlasgow	1888	Russell & Co.		Stuart Bros.	Glasgow
612	N	**Norge** *Nielsen* -86 WBk	Norweg'n	540	133.1	32.6	19.1	Risör	1859			Chr.Th.Boe	Arendal
613		**Norge** WRECKED *Pedersen* WBk	Norweg'n	446	131.8	30.6	17.1	Risör	1878		V 8.87	O.Jörgensen	Risör
614	*	**Norham Castle** *Holmes* CBk WRECKED *Green*	British	698	177.4	32.1	18.0	Glasgow	1869	A.Stephen&Ss.	V 4.88	Norham Castle Sailing Ship Co.,Ld.(P.H.Cowley&Co.)	Glasgow
615		**Norma** *Lordal* W3mSr	American	310	138.5	34.0	10.5	SanFrancisco	1883	D.Turner		G.S.Hinsdale	San Francisco
616		**Norma** WBk	Brazilian	244	..	..	..					P.P.Nunes	RioGrandedoSul
617	V	**Norma** *Popieul* WSr	French	122	86.9	20.7	11.6	Dunkirk	1869	Van Cauwen-[berghe		J.van Cauwenberghe Lemaire	Dunkirk
618		**Norma** WSr	French	101	73.5	19.7	10.0	LaRochelle	1853			Le Bihan	Vannes
619	V	**Norma** *Grafe* -84 WBk (ex The Richard Green)	German	432	138.4	29.5	16.4	Grimsby	1876	Hadfield	V 5.86	J.H.L.Wilckens	Hamburg
620		**Norma** *Rasch* 70-70 WBk	German	357	114.8	29.5	16.5	Ueckermünde	1849	C.Pretzer		R.C.Gribel	Stettin
621	N	**Norma** *Mathiesen* 66-77 WBk (COLLISION)	Norweg'n	476	149.1	30.1	16.7	Grimstad	1877	JacobBoe	V 8.86	Chr.Th.Boe	Grimstad
622	N	**Norma** *Tralsen* -88 WBk	Norweg'n	358	117.8	28.7	16.0	Lillesand	1861	J.Johannsen	V 12.87	A.J.Semb	Laurvig
623	N	**Norma** *Olsen* -88 WBk	Norweg'n	337	114.7	27.6	15.8	Vegesack	1851			H.C.Foden	Tönsberg
624	NV	**Norma.** *Sivertsen* 85-85 WBn	Norweg'n	218	106.7	25.2	12.3	Stavanger	1879	Chr.Knudsen	V 4.85	Holdt & Isachsen	Stavanger
625		**Norma** W3mSr	Russian	285	..	..	..		1874			K.J.Blomberg	Dragsfjerd
626		**Norma** *Andersson* -86 WSr	Russian	146	..	..	..	Denmark				A.Granskog	Nerpes
627		**Norma** (ex Valia) *Moys* WSr	Spanish	102	78.7	20.4	9.1	Masnou	1851			F.Comerma	Barcelona
628	G	**Norma** *Backman* 81-80 WSr	Swedish	125	82.7	20.9	10.4	Upperud	1850			G.J.Sundberg	Lysekil
629	V	**Norman** *Hawley* -88 WBk	British	870	176.3	35.2	19.6	Maitland,N.S.	1877	A.McDougall	V 12.88	J.Ritchie,jun.	Liverpool
630	R	**Norman** *Holt* -88 W3mSr	American	367	127.0	30.3	14.5	Pembroke,Me.	1874	J.B.Nutt		D.K.Hobart	Eastport,Me.
631	R	**Norman** *Roberts* -88 W3mSr	American	347	136.1	31.0	11.6	Bath,Me.	1877	Adams&[Hitchcock	V 1.85	Adams & Hitchcock	Bath,Me.
632	*	**Norman Macleod** *Lovett* IBk	British	834	202.7	32.2	19.1	PortGlasgow	1876	Mrdch&Murray			Lyttelton
633	DN	**Normand** *Larsen* 74-86 WBg	Norweg'n	281	105.5	27.7	15.7	Grimstad	1854	O.&L.Dönnes-[moen		H.S.Nielsen	Laurvig
634	N	**Normand** *Sörensen* W3mS	Norweg'n	216	97.5	27.0	12.4	Sandefjord	1868	S.Josephsen	Z 4.84	I.H.Christiansen	Tönsberg
635	V	**Normandie** *Ledain* 77-85 WBk	French	710	160.4	32.3	13.0 / 19.8	Honfleur	1885	Leviels	C 8.88	H.Auger aîné	Havre
636		**Normandie** *Lausseud* W3mSr (ex Emma Crosb.)	French	296	131.6	26.9	12.9	Sunderland	1874	W.Pickersgill	V 8.85	E.M.Gilles&A.Bellet	Fécamp
637	RV	**Normandy** *Tukey* WS	American	1166	188.0	38.0	24.2	Damariscotta,[Me.	1877	Clark&Curtis	V 1.87	J.W.Tukey	Damarse'tta,Me
638		**Normandy** *Rivers* -88 W3mSr	American	516	154.0	35.4	13.2	Bath,Me.	1878	Adams&[Hitchcock		Adams & Hitchcock	Bath,Me.
639	N	**Normania** *Thorlifjärnsen* WBk	Norweg'n	358	114.3	29.0	16.3	Arendal	1864	O.Kittelsen		Gunder Olsen	Arendal
640	N	**Normann** *Nielsen* -88 WSr	Norweg'n	165	86.8	29.0	10.9	Namsos(rebuilt)	18			N.Björnm	Namsos
641	N	**Normanvik** *Alfsen* WBk (ex Jacob W. Satterwaith) -88	Norweg'n	711	163.5	35.9	18.9	Sundsvall	1870	Naeström	V 10.86	B.A.Olsen	Risör
642	*D	**Norna** *Reid* 76-81 IBk	British	1050	218.5	34.8	21.0	Glasgow	1868	Barclay,Curle&[Co.		W.Latham	Greenock
643	*N	**Nornen** *Larsen* 87-87 WBk (ex Malta)	Norweg'n	679	154.5	34.6	20.3	Laroque	1876	Roy	V 2.87	S.C.Larsen	Brevig
644	N	**Nornen** *Svendsen* 84-86 WBk	Norweg'n	438	125.5	31.4	18.2	Drontheim	1847		V 3.88	O.P.Olsen	Dröbak
645	N	**Nornen** *Engelbretsen* WBk	Norweg'n	410	109.7	30.0	17.9	Tönsberg	1851	J.O.Röd		Olaus L.Röd	Tönsberg
646	N	**Nornen** *Johnsen* 85-85 WBk	Norweg'n	306	108.7	28.6	15.9	Lillesand	1853	A.Sörensen	V 3.83	H.G.Kirsebom	Christiansand
647	N	**Nornen** *Christensen* WBg	Norweg'n	239	114.1	28.8	12.1	Arendal	1869	J.M.Osmundsen		Blackstad,Holta & Co.	Skien
648	N	**Nornen** *Jensen* WBg	Norweg'n	221	101.1	26.5	11.9	Tvedestrand	1870			Th.Jensen	Tvedestrand
649	N	**Nornen** *Olsen* -88 WBg	Norweg'n	209	112.0	25.1	11.9	Sandefjord	1870	Ch.Christensen		P.A.Grön	Sandefjord
650	R	**Norombega** *Craig* -88 W3mSr	American	295	137.5	32.0	9.5	Verona,Me.	1884	W.Beazley	V 8.88	W.Beazley	Bucksport,Me.

*=Lloyd's Register; D=Nederlandsche Vereeniging; G=Germanischer Lloyd; I=Registro Italiano; R=American Record;
A=Veritas Austro-Ungarico; H=Veritas Ellenico; N=Norske Veritas; V=Bureau Veritas.

Dans la 4me Colonne, C=Composite (membrure en fer, bordé en bois); I=Fer; Stl=Acier; W=Bois.

Les Dates de doublage en métal (Col. 11) ressortent des livres dans lesquels les navires sont respectivement classés ou des rapports des Inspecteurs du Lloyd's Register.

Pour plus amples renseignements sur les navires classés par le Lloyd's Register, voir le Lloyd's Register Book et le Lloyd's Yacht Register.

1 Official No.	2 No. in Book.	3 Registry in which classed.	4 Ship's Name. Material and Rig. Master, and Dates of Appointment to present Owner's Service and to Ship. Ship's Late Name.	5 Flag.	6 Net Tonnage.	7 Dimensions. Length.	Breadth.	Depth.	8 Built Where.	9 When.	10 By whom.	11 Date of Metalling.	12 Owners or Managers.	Port
	651		Norrbotten *Andersson-86* (ex Skenaroch) W Bg	Swedish	207	94·2	23·3	13·1	Grimstad	1859			A.Nystrom	Pitea
132006	652	V	Norris *Weeks* -88 W S	American	1154	181·6	35·5	23·1	Damariscotta, Me	1874	E.Norris&Co.	8.88	D.W.Chapman	Damar...
	653	N	Norrona *Aarø* 81-84 W Bn	Norweg'n	295	126·5	29·0	12·7	Sandefjor	1888	Olsen	Y 2.87	P.A.Grön	Sandefjord
	654		Norrtelje W Sr	Swedish	124	82·1	23·1	10·3	Norrtelje	1873			F.Lundman	Norrtelje
51058	655	✻	Norseman *McClure* 82-87 (BURNT) I Bk	British	599	173·3	28·5	18·9	Hull	1864	[illegible]		Cotesworth,Lynn&Co.	Liverpool
62407	656	✻	Norseman *Main* 79-84 W Sr	British	119	87·7	21·9	10·9	Banff	1873	Watson		J.W.Simpson	Banff
	657	✻	Norseman I Barge	British	98	76·1	18·9	5·9	LongDitton	1885	Phillips&Graves		Phillips & Graves	London
130194	658		Norseman W Sr Yt	American	147	130·0	25·0	11·6	Brooklyn	1881	C&R Poillon		O.Goelet	New York
58807	659	✻ D	North *Newson* 85-86 I S	British	1296	220·1	37·1	22·6	Birkenhead	1868	G.R.Clover&Co		Tyser & Co.	London
59464	660		North America *Landry* (74-86) W Sr	British	175	98·9	26·0	11·4	Nw.Dublin,NS	1870	Langille		Simon Joyce	Halifax,N.S
80901	661	R	North America *Pontret* W Sr	British	138	91·0	24·6	9·8	Yarmouth,N.S.	1880	H.Jenkins		I.Boudrot	Yarmouth
18631	662	D R V	North American *Coleman* W S	American	1584	219·0	40·0	24·0	E.Boston	1873	CurtisSmith	Y 6.87	H.Hastings	Boston,Mass
8817	663		North Ash *Merigan* W Sr	British	108	58·0	20·5	11·8	Brixham	1810			P.Kirwan	Dublin
130108	664		North Bend *Wallace* 77 W Bkn	American	358	153·5	32·8	11·2	N.Bend,Oreg.	1877	J.Kruse		A.M.Simpson	San Francisco
70083	665	✻	North Riding *Rutherford* I S	British	1371	245·8	36·8	21·9	Liverpool	1875	W H Potter&Co.		A.Gibson & Co.	Liverpool
89110	666	✻	North Sea *Loggie* W Dy	British	86	79·3	19·6	9·5	Peterhead	1884	Cruden A Mathew		J.McCombie & Co.	Peterhead
24477	667		North Star (BROKEN UP) W Bg	British	209	81·4	26·3	15·1	Sunderland	1836			D.T.Sharper	Hartlepool
18493	668		North Star *Moorhouse* W Bg	American	389	120·0	29·6	15·6	Boston	1867			J.H.Moorhouse	San Francisco
64944	669	N V	North Star *Scheen* -87 W Bk	Norweg'n	718	156·8	32·5	19·5	Quebec	1871	Angers	Y 1.87	B.T.Scheen	Christiania
	670	✻	Northampton *Pagels* (COLLISION) 86-87 C S	German	1173	211·6	35·6	22·0	Glasgow	1866	C.Connell&Co.	Y 1.88	C.Rahtkens	Rostock
70586	671	✻	Northbrook *Timothy* 82- I S	British	1820	263·0	41·4	23·8	Stockton	1874	Richardson, Duck&Co.		Milton Stuart	Liverpool
76567	672		Northern Belle *Spence* (85-85) W Sr	British	214	114·6	25·0	13·0	Garmouth	1877	Geddie	Y 12.86	W.Whyte & Co.	Banff
91637	673	✻	Northern Belle *Lucas* W K Barge	British	76	84·9	20·2	6·8	Ipswich	1885	G.P.Gibiea&Co.		Hawkins,Holman & Co.	London
87545	674		Northern Chief *Mackay* (-87) W Bk	British	274	125·0	27·4	12·2	Auckland	1886	R.Mackay	Y 12.85	R.Mackay	Auckland
80989	675	V	Northern Empire *Kaneken* -87 W Bk	British	950	188·5	37·0	20·7	Economy,N.S.	1880	D.P.Soley		D.P.Soley	Windsor,N.S
15253	676		Northern Light (BROKEN UP) *Morrison* W	British	429	114·9	25·5	19·2	Shields	1853	J.Young		W.Ross	South Shields
52739	677		Northern Light *Budd* (75-76) W Bn	British	172	101·7	23·0	11·9	P.E.Island	1865			W.Kemp	Faversham
18493	678		Northern Light *Baker* (86) W Bk	American	365	125·4	30·3	17·9	Rochester Mas	1851			J.McKenna	San Francisco
18136	679		Northern Light *Harper* W Sr	American	132	91·0	27·0	8·6	Tremont,Me.	18 65/82			N.A.Harper	Mt Desert
73614	680	✻	Northern Monarch *Stevenson* 85-87 I S	British	1230	227·4	36·6	21·7	Dumbarton	1875	A.McMillan&S.		Lowden, Edgar & Co.	London
73072	681		Northern Star *Lessey* -88 W Bk	British	327	120·0	27·7	14·9	P.E.Island	1875	J.Yeo		J.N.Storey	Sydney,N.S.
64559	682		Northern Star *Sherry* 71- W Bn	British	298	116·0	28·5	12·9	Sackville,N.B.	1871	Anderson	Y 10.84	J.J.Carroll	Dublin
90319	683	✻	Northernhay *Watson* I Bk	British	1221	225·5	35·1	20·9	Sunderland	1885	J.Laing		R.H.Gayner	Sunderland
53894	684	R	Northumbria *Maddison* 88-88 W S	British	1292	196·8	37·9	23·7	Quebec	1866	Samson	Y 2.86	C.E.de Wolf & Co.	Liverpool
54621	685	✻	Northumbria *Curry* -86 W Sw	British	291	110·0	26·5	16·4	Sunderland	1865	G.Bartram	Y 9.87	C.N.Brown	Sunderland
18585	686		Northwest *Maguire* W Bk	American	490	147·9	36·8	13·9	P.Madisn.Wsh	1868			W.P.Sayward	[illegible]
72583	687		Norway W Sr	British	332	135·5	26·0	11·9	GardenIsl.Ont	18 73/91	H.Roney		The Calvin Co.,Ld.	Kingston
160294	688		Norway *Paul* W Sr	American	183	127·0	32·0	9·0	SanFrancisco	1870	G.Boole		A.Anderson	San Francisco
	689	N	Norway *Mortensen* -88 (ex Recipemeoy) W Bk	Norweg'n	667	144·0	32·6	19·9	P.E.Island	1857			Johanson,Rotter & Co.	Christiania
	690	V	Norway *Albrethsen* -88 (ex Anna Florence) W Bk	Norweg'n	641	161·0	31·7	10·2	Varazze	1870		Y 1.87	Th.Albrethsen	Tvedestrand
	691	V	Norwood *Klavenes* 88-88 (ex F.G.Carrill) W S	Norweg'n	1587	221·6	39·3	24·4	St.John,N.B.	1874	McFee	Y 6.88	Homewood Aktieselskab (P.F. Klavenes)	Christiania
897	692	I	Nostra Madre *Fragone* 81-88 W Bk	Italian	915	171·5	33·8	19·9	SestriPonente	1878	G.Peragallo	Y 7.88	Fⁱˡˡⁱ Sanguinetti	Spezia
2426	693	I	Nostra Madre *Palazzo* (ex Banca Natale D.) W 3m Sr	Italian	477	154·8	29·8	16·8	SestriPonente	1873	A.Briasco	Y 10.87	Fⁱˡˡⁱ Drago	Genoa
	694		Nostra Madre (ex Mirra) W Bk	Italian	275	109·5	26·0	17·4	Sorrento	1863			G.Romano	Genoa
2154	695	I	Nostra Signora del Boschetto *Peragallo* -81 W Bk	Italian	572	144·3	32·3	20·7	Voltri	1871	G.Fava	Y 6.86	M.Razeto	Genoa
879	696		Nostra Signora del Boschetto *Pilato* W Bg	Italian	139	76·1	23·1	14·7	Varazze	1853			G.Mattarana	Spezia
2279	697		Nostra Signora del Carmelo (BROKEN UP) *Frecetto* -87 W Bk	Italian	560	134·5	30·8	20·6	Varazze	1872	D.Craviotto	Y 1.87	A.Gazzolo	Genoa
2584	698	I	NostraSignoradelCarmine (ex Agostino Moretti) *Gazzolo* -88 W Bn	Italian	757	170·2	33·8	21·9	Sestri Ponente	1875	L.Marcenaro	Y 4.87	A.Gazzolo	Genoa
1007	699	I R	Nostra Signora del Sacro Cuore di Gesù *Mortola* (ex K. Danovaro) W Bk	Italian	543	147·6	31·1	20·0	Sestri Ponente	1867	G.Casanova	Y 8.86	Gius.Mortola	Genoa
3108	700	V	Nostra Signora DellaSalute (ex Prudentia) *Olivari* W Bk	Italian	421	124·6	30·8	19·0	SestriPonente	1867	Fⁱˡˡⁱ Cadenaccio	Y3.88	G.B.Olivari & R.Bozzo	Genoa

✻ = Lloyd's Register; D = Nederlandsche Vereeniging; G = Germanischer Lloyd; I = Registro Italiano; R = American Record;
A = Veritas Austro-Ungarico; H = Veritas Ellenico; N = Norske Veritas; V = Bureau Veritas.

In the 4th Column, C = Composite (iron frame, wood planked); I = Iron; Stl = Steel; W = Wood.

The Dates of Metalling (Col. 11) are based on the records of the Books in which Vessels are respectively classed, or on the Reports of the Surveyors of Lloyd's Register.

Further particulars of Vessels classed by Lloyd's Register will be found in Lloyd's Register Book and Lloyd's Yacht Register.

		Nom. Matériaux et Gréement. Capitaine et dates d'entrée au service des Armateurs actuels et à bord du navire. Ancien Nom.	Pavillon	Tonnage Net	Longueur	Largeur	Creux	Construit Où.	Quand.	Par qui.		Armateurs ou Directeurs-Gérants.	Port d'Armement.
701		Nostra Zia W.3mSr	Italian	281	119·4	27·8	15·4	SestriPonente	1873	M.Bossi		Anna Palazzo	Genoa
702	✻1	Nostro Padre W.Bk	Italian	527	131·2	29·3	19·7	Pra	1873	L.Durante		F.lli Drago	Genoa
703		Notero *Smith* '83 W.Bk	British	430	142·8	30·4	15·6	Drontheim	1881			Smith & Jakins	Auckland,N.Z.
704		Notre Dame Auxiliatrice *Foccas* W.Bk	Greek	535	147·6	28·2	17·6	St.Malo	1875	Gautier		S.Foccas	Hydra
705		Notre Dame de Bon Secour W.Sp	British	152	108·4	24·3	8·3	Lanoraie,P.Q.	1887	E.Hayneman		Louis E.Bonaventure	Montreal
706		Notre Dame de Bon Secour W.Bn	French	137	81·0	22·0	11·8	Bayonne	1859	Begue & Burgalaire		J.Lafont & Co.	Bayonne
707		Notre Dame d'Espérance W.Bk	French	285	93·8	24·3	15·7	LaLegué	1866	L.Villeferon		P.Rouxel	St.Brieuc
708	V	Notre Dame de la Garde I.Bk	French	457	115·3	27·6	16·4	Stockton	1868	Richardson, Duck&Co.		H.de Peindray	St.Denis
709	V	Notre Dame de la Ronce W.Sr	French	116	90·9	22·3	11·0	Gravelines	1878	Collin		M.F.Ruellan	Binic
710	V	Notre Dame des Dunes W.Sr	French	121	92·8	23·0	10·9	Dunkirk	1886	Pyotte Bayaert		Bonvarlet frères	Dunkirk
711		Notre Dame du Rosaire W.Bg	French	194	88·6	22·3	13·1	La Nouvelle	1856			Azibert	Narbonne
712		Nouri Bahri W.Bg	Turkish	200	..	..	..	Dardanelles	1871			Hajee Ummeh Gulrun Hanoum	Constantinople
713		Nouri Bahri W.Bg	Turkish	119	..	..	..	Tereboilie	1874			Topdjeesada Kumurji Ethem Aga Zadeh Kumurji Hakki	Tereboilie
714		Nouveau Coriolan '86 W.Bk	French	293	111·8	26·6	15·1	Bordeaux	1865	Rocheta Largeteau		Savalle	Fécamp
715		Nova Flora *Avintes* W.Bn	Brazilian	100	..	..	..	Campos	18 72		V 1.81	Cia Cabotagem de S.João da Barra e Campos	S.João da Barra
716		Nova Harnost W.Sr	Austrian	128	79·0	19·4	12·7	Servola	1864	G.Spadon		N.Svilocossi	Ragusa
717		Nova Marianna W.Bk	Brazilian	363	..	..	..					Oliveira & Filhos	Rio Grande
718		Nova S. da Penha W.Sr	Brazilian	162	..	..	..					J.F.Carrico	Rio Doce
719	DV	Nova Scotia *Potter* '82-'85 W.Bk	British	1110	179·0	38·1	22·4	Cornwallis,N.S.	1877	J.E.Steele		S.Sheffield	Windsor,N.S.
720	N	Nova Scotia *Halvorsen* W.Bg	Norweg'n	232	90·9	25·9	14·3	Pictou,N.S.	18 48/73			Halvorsen	Fredrikstad
721	GV	Nova Sympathia W.Bk	Brazilian	361	142·3	25·2	14·7	Oporto	1869	M.Soares		Baltar Oliveira & Co.	Pernambuco
722		Nova Tentativa W.Bk	Brazilian	347	..	..	..					M.A.Cardoso	Rio de Janeiro
723	V	NovaUnião *Conceiçao* '87 W.Bk	Portug'se	399	131·3	27·9	17·3	Sunderland	1865	J.Gill		Jose N.Pinto	Oporto
724		Novelty *Mitchell* W.4mSr	American	450	168·2	39·0	13·0	Coos Bay,Or.	1886	Simpson		A.M.Simpson	San Francisco
725		NoviKlas *Ballon* W.Bk	Austrian	262	89·2	23·1	15·9	Fiume	1856	Wranvich		C.Ballon di C.	Cherso
726	V	Novi Molodez *Peterl* '71-'75 W.Bg	Russian	251	119·1	24·3	11·6	Kjöge	1875	L.Flint		Witt & Co.	St.Petersburg
727		Novo Gloria W.Sr	Portug'se	103	75·5	19·7	8·8	Villa do Conde	1874	J.M.d'Araujo		V.S.A.d'Azevedo	Villa Nova
728		Novo Lima W.Sr	Brazilian	151	..	..	..					A.M.& J.M.de Lima	RioGrandedoSul
729		Novo Machado II. W.Bn	Portug'se	150	87·6	22·1	9·9	Figueira	1883	M.D.dosSantos Borda		M.F.Machado	Aveiro
730		Novo Mathilde '82 W.Bn	Portug'se	165	90·2	21·3	11·6	SanMartinho	1868			A.G.Gomes & Co.	Madeira
731		Novo Momento W.Bn	Brazilian	160	..	..	..	Campos	1867			M.M.S.de Mesquita	Rio de Janeiro
732		Novo Mundo W.Sr	Brazilian	355	..	..	..					F.P.Dantas	Rio de Janeiro
733		Novo Preceito *Preceito* W.Sk	Portug'se	100	72·5	22·0	8·5	Villa do Conde	1885	S.M.d'Araujo		J.J. da CostaGuimarães	Oporto
734	V	Novo Silencio *Ferreira* W.Bk	Portug'se	326	118·7	28·9	15·4	Villa do Conde	1867	M.G.Rodrigues		A.F.M.Guimarães	Oporto
735	D	Nubia *Olsen* W.Bk	Norweg'n	314	128·3	27·0	14·1	Altona	1865	E.Dreyer		S.Olsen	Farsund
736		Nueva Angelita *Lloret* W.Sr	Spanish	121	80·0	23·4	11·7	Villajoyosa	1872			V.L.Linares	Alicante
737		Nueva Carlota W.Sr	Spanish	130	80·0	24·9	10·9	Arens	1858			José Banza	Barcelona
738		Nueva Carmelita W.Sr	Spanish	104	74·1	22·5	10·7	Lloret	1878			B.Uguet&Esteva	Majorca
739		Nueva Concepcion W.Bn	Spanish	103	72·8	22·2	10·8	Villajoyosa	1852			Miguel Lopez	Alicante
740		Nueva Dolores W.Sr	Spanish	116	78·7	22·8	11·3	Palma	1850			Juan Cieluna	Carthagena
741		Nueva Elisa W.Sr	Spanish	113	81·4	23·1	10·0	Blanes	1867			Viuda & hijos de B. Maristany	Barcelona
742		Nueva Fermina W.Bg	Spanish	237	95·1	27·7	14·1	San Feliee	1869			Guma hermanos	Havana
743		Nueva Juliana W.Bn	Spanish	183	93·2	25·9	14·2	Linera	1863			J.M.Martinez	Rivadeo
744	V	Nueva Paula *Maristany* W.3mSr	Spanish	317	119·7	28·0	13·3	Blanes	1877	Vieta		F.Maristany	Barcelona
745		Nueva Providencia W.Pol	Spanish	198	96·8	25·0	12·8	Italy	1871			J.Erquicia	San Sebastian
746	V	Nuevitas *Alsina* W.Pol	Spanish	169	88·3	27·1	5·6	Blanes	1857			J.Gotarra y Guinar	Barcelona
747	V	Nuevo Copernico W.Bg	Spanish	180	102·0	25·9	12·2	Arens	1839		V 10.87	F.Ginebra y Pou	Barcelona
748		Nuevo Corazon Bg	Spanish	110	92·8	24·2	8·8	Palma	1885			A.A.Pons	Majorca
749	V	Nuevo Curritaco *Elias* W.Bn	Spanish	257	100·4	27·4	13·7	Blanes	1864			Miguel Company	Barcelona
750		Nuevo Mal Genio W.Bn	Spanish	165	87·6	23·8	12·6	Corcubion	1851			José Sagristá	Corunna

✻=Lloyd's Register; D=Nederlandsche Vereeniging; G=Germanischer Lloyd; I=Registro Italiano; R=American Record; A=Veritas Austro-Ungarico; H=Veritas Ellenico; N=Norske Veritas; V=Bureau Veritas.

Dans la 4me Colonne, C=Composite (membrure en fer, bordé en bois); I=Fer; Stl=Acier; W=Bois.

Les Dates du doublage en métal (Col. 11) ressortent des livres dans lesquels les navires sont respectivement classés ou des rapports des Inspecteurs du Lloyd's Register.

Pour plus amples renseignements sur les navires classés par le Lloyd's Register, voir le Lloyd's Register Book et le Lloyd's Yacht Register.

Official No.	No. in Book	Registry in which classed	Ship's Name. Material and Rig. Master, and Dates of Appointment to present Owner's Service and to Ship. Ship's Late Name.	Flag	Net Tonnage	Dimensions. Length	Breadth	Depth	Built. Where	When	By whom	Date of Metalling	Owners or Managers	Port
	751		Nuevo Porvenir Bg	Spanish	113	78·0	21·6	8·2	Ilocos-Sur	1864			J. Angeo	Manila
	752		Nuevo Victoria W Bg	Spanish	251	104·0	25·6	13·7	Vigo	1852			F. Kuntz y Amor	Vigo
	753	V	Nuevo Vigilante Corell W Bg	Spanish	191	98·1	25·6	12·2	Palamos	1866		[illegible]	S. Talavera	[illegible]
[illegible]	754		Nulato Sawyer ·88 W Sr	American	123	95·5	25·1	8·2	Machias, Me.	1871			L. Holway	Machias
	755		Numa W Sr	Spanish	201	83·0	23·5	11·9	San Felice	1848			Doña M. Nadal	Palamos
[illegible]	756		No. 1 I Sr	British	100	126·3	19·1	5·0	Hull	1872	Earle's Co.		The Amazon Stm. Nav. Co. Ltd.	London
	757		Nummer Fünf Schmidt ·80·85 W Bk	German	367	119·5	28·3	16·5	Elbing	1863	Gebr. Mitzlaff		Emil Buttkus	Memel
[illegible]	758		Nunobiki Maru W Sr	Japanese	225	103·0	23·5	11·3	Hiogo	1884	Kinoshita Benso		Hirata Jusuke	Hiogo
[illegible]	759		Nuova Amabile Giulietta Sari W Bn	Italian	132	84·3	23·4	12·7	Varazze	1849	B. Craviotto		Luigi Sari fu G.	Porto [illegible]
[illegible]	760	A	Nuova Assunta Rapallo ·88 W Bn	Italian	110	69·2	20·0	12·7	Pra	1850	Tixi	7.85	C. Arado	Savona
1961	761	I	Nuova Caterina (ex Maria Colombo) Costa ·88 W Bk	Italian	354	117·7	29·5	18·0	Arenzano	1869	D. Salomone	[illegible]	Giuseppe Nicolini	Genoa
1423	762	✳	Nuova Eleonora Madre (ex Mofussilite) Quaglia C S	Italian	975	201·1	33·1	20·8	Glasgow	1864	Stephen	[illegible]	Lubrano fratelli di M.	Castellammare
359	763	I	Nuova Galatea Basso W Bk	Italian	515	141·0	30·9	19·9	Sestri Ponente	1870	G. Peragallo	[illegible]	Societa Anonima Metallurgica (Tardy & Benech)	Genoa
1737	764		Nuova Gemma Schiaffino ·88 W Bk	Italian	364	114·8	29·8	18·0	Lavagna	1868	G. Saccomano	[illegible]	P. Schiaffino	Genoa
140	765	A	Nuova Gioconda Bologna ·87 W Bk	Italian	272	..	..	..	Chioggia	1870		[illegible]	A. Antonelli	Venice
132	766		Nuova Giovannina (ex Giovannina Tabuso, ex Aldo D'Arrigo) W Bn	Italian	161	88·9	21·8	13·8	Palermo	1854			G. Tabuso	Catania
981	767	I	Nuova Giulia Manfroni W Bn	Italian	140	105·9	23·4	11·6	Voltri	1884	F[lli] Fava	[illegible]	Bartolomeo Manfroni	Spezia
321	768		Nuova Irene Mirabella W 3m Sr	Italian	159	104·9	20·3	12·6	Trieste	1868		[illegible]	G. Mirabello fu A.	Catania
[illegible]	769		Nuova Isabella (ex Isabella) Accangora ·86 W Sr	Italian	104	75·4	23·2	11·8	Castellamare	1861			C. Altieri	Naples
371	770	A	Nuova Lucia D'Urso W Bg	Italian	172	96·4	22·9	14·1	Castellamare	1862			G. D'Urso	Catania
[illegible]	771	IV	Nuova Margherita Persico W Bk	Italian	566	142·7	29·8	20·3	Ischia	1875	G. Bonifacio	[illegible]	L. Galatola	Naples
104	772	I	Nuova Maria Scarpa W Bg	Italian	247	101·7	27·0	16·0	Chioggia	1873		[illegible]	Pietro Scarpa	Venice
1619	773	A	Nuova Oregina Massa W Bg	Italian	383	122·3	30·2	18·7	Sestri Ponente	1867	F[lli] Cadenaccio	[illegible]	Giacomo Massa	Genoa
3100	774	✳	Nuova Provvidenza (ex Nuova Provvidenza Conte de Lion) Bassetto W S	Italian	799	185·0	33·4	20·8	Jersey	1867	Le Vesconte	[illegible]	G. B. Bozzo	Genoa
	775		Nuova Provvidenza W Bg		108	71·5	22·8	12·0	Savona	1858	Tixi			
[illegible]	776	V	Nuova Rosa Causi W Bk	Italian	557	147·3	30·5	18·8	Sestri Ponente	1871	S. Debarbieri	[illegible]	Ant. Causi	Spezia
[illegible]	777		Nuova Rubinacci (ex Maria Rubinacci) Serpa (CONTON) W Bk	Italian	314	110·8	27·6	15·3	Warren	1857		[illegible]	G. Rubinacci	Naples
[illegible]	778	I	Nuova Sara Lauro ·87 W Bn	Italian	117	85·3	20·7	11·1	Voltri	1887	Frat. Fava	[illegible]	Angelo Lauro	P. Maurizio
224	779	I	Nuova Teresa (ex Augusto) Valenza W Bg	Italian	218	102·0	26·5	15·6	Castellamare	1863	Trambaresto	[illegible]	A. Valenza	Castellammare
[illegible]	780		Nuovo Arturo Del Buono W Bg	Italian	173	94·2	25·2	13·8	Leghorn	1856	Percuco		O. Del Buono di A.	Porto [illegible]
[illegible]	781	A	Nuovo Ciriaco Tomasovich ·88 W Bg	Austrian	270	97·4	25·5	16·1	Fiume	1844		[illegible]	G. Afenduli	Trieste
[illegible]	782	V	Nuovo Cristoforo Colombo Carletti W Bg	Italian	152	91·5	25·5	12·5	Rio	1866	Braschi		F[lli] Cignoni d'Luigi	Porto [illegible]
	783	V	Nuovo Luciano B. (ex Perseverante) Franculli ·87 W Bk	Greek	418	122·3	30·4	19·8	Varazze	1866	Ant. Cerruti	[illegible]	G. Franculli	Corfu
1840	784	IV	Nuovo Manilla Cassallo W Bk	Italian	492	132·8	30·8	19·8	Varazze	1869	G. Aschieri	[illegible]	A. Celle	Genoa
2216	785	I	Nuovo Matteo Ansaldo W Bk	Italian	587	140·0	29·1	20·2	Sestri Ponente	1871	B. Canepa	[illegible]	Eredi L. Ansaldo	[illegible]
651	786		Nuovo Mondo Astarita W Bk	Italian	509	136·1	26·4	19·8	Castellamare di Stabia	1866	L. Trambarulo	[illegible]	F. Starace	[illegible]
[illegible]	787		Nuovo San Giuseppe (ex Vincenzo Maria) Milano W Bg	Italian	282	112·8	27·8	15·5	Savona	1870		[illegible]	Sabia & De Martino	Naples
550	788		Nuovo Sebastiano Cafiero W Bn	Italian	194	105·3	24·7	14·2	Gaeta	1881	L. Contreras	[illegible]	S. Munzone fu V.	[illegible]
[illegible]	789	I	Nuovo Stella Chiozza W Bk	Italian	776	168·9	33·3	22·2	Sestri Ponente	1877	G. Peragallo	[illegible]	G. Ferrari	Genoa
618	790	I	Nuovo Talete Fascia ·88 W Bkn	Italian	215	113·1	25·3	13·1	Savona	1887	E. Calamaro	[illegible]	L. Fascia	Porto [illegible]
[illegible]	791		Nuovo Tre Fratelli Montaldi	Italian	107	87·3	21·8	10·1	Savona	1868	G. B. Gustavino		F[lli] Montaldi di A.	Porto [illegible]
[illegible]	792	I	Nuovo Valoroso Soldani W Bn	Italian	121	90·9	22·1	12·0	Savona	1875	G. B. Magnano	[illegible]	G. Soldani fu A.	Porto [illegible]
[illegible]	793	✳	Nushka Bevan W Bkn	British	417	138·0	29·2	17·0	P. E. Island	1875	Yeo	[illegible]	T. P. Richards & Co.	Swansea
	794		Nusret Bahri W Bg	Turkish	290	..	..	..	Douta	1883			Tashji Zadeh Ahmed Kyamil Effendi	Gallipoli
	795		Nussar W Sr	British	100	76·0	17·4	9·9	Cochin	1850			Shaik Hamed	[illegible]
	796		Nusser Musjeet W Bk	Turkish	667	125·0	30·0	21·0	Cochin	1855			Shawlavinjuor & Son	Jeddah
31112	797		Nusserel Musjeeth W Bg	British	193	94·6	22·3	12·6	Cochin	1874			Meerasaibe Sinne Kadertamby	Colombo
73007	798	✳	Nyanza Milne ·81·84 W 3m Bn	British	232	109·8	24·3	14·2	Kingston	1876	Spence	[illegible]	J. Mill	Banff
56808	799		Nyanza C S Yt	British	131	110·0	21·5	11·8	Greenock	1867	Steele & Co.		J. C. Dewar	Glasgow
	800	D	Nyassa (L.G.37) Solberg W Bk	Norweg'n	799	179·7	32·0	19·5	Montrose	1868	J. & D. Birnie	[illegible]	E. Bertelsen	Grimstad

✳ = Lloyd's Register; D = Nederlandsche Vereeniging; G = Germanischer Lloyd; I = Registro Italiano; R = American [Lloyd's]; A = Veritas Austro-Ungarico; H = Veritas Ellenico; N = Norske Veritas; V = Bureau [Veritas].

In the 4th Column, C = Composite (iron frame, wood planked); I = Iron; Stl = Steel; W = Wood.

The Dates of Metalling (Col. 11) are based on the records of the Books in which Vessels are respectively classed, or on the Reports of the Surveyors of Lloyd's [Register].

Further particulars of Vessels classed by Lloyd's Register will be found in Lloyd's Register Book and Lloyd's Yacht Register.

1	2	4 Nom. Matériaux et Gréement. Capitaine et dates d'entrée au service des Armateurs actuels et à bord du navire. Ancien Nom.	5 Pavillon.	6 Tonnage net.	7 Dimensions. Longueur	Largeur	Creux.	8 Construit. Où.	9 Quand.	10 Par qui.	11	12 Armateurs ou Directeurs-Gérants.	13 Port d'Armement.
[illegible]	DN	**Nyborg** *Jørgensen* -88 W.Bk.Sr	Norweg'n	270	132·3	26·9	12·4	Nyborg	1866	P.Petersen		J.A.Henriksen	Tönsberg
[illegible]		**Nyfart** *Karlsson* 81-84 W.Sr	Russian	269	..	..	..	[illegible]	1881			A.Gustaffson	Westanfjord
[illegible]	V	**Nyhamn** *Cronberg* 72-76 W.Bk	Swedish	291	122·0	25·4	13·4	Nyhamn	1869	P.Cronberg	[illegible]	L.P.Cronberg	Nyhamn
804	V	**Nylghau** *Butler* 84-84 W.S	British	1252	200·2	38·6	23·0	Yarmouth,N.S	1881	J.Lovitt	[illegible]	Lovitt & Co.	Yarmouth.N.S.
805	N	**Nymph** *Pedersen* W.Bk	Norweg'n	420	129·0	29·5	15·9	Granville,N.B.	1866			O.A.Føyen	Tönsberg
806	N	**Nymphe** *Tobiassen* -88 W.Bk	Norweg'n	312	119·0	27·1	14·3	Rönnebeck	1865	H.Oltmann		C.Tobiassen	Horten
807	N	**Nymphen** *ex(Tiow) Eriksen* W.Bk	Norweg'n	699	150·8	31·9	20·8	Freeport,Me.	1855		[illegible]	O.Eriksen&H.N.Gjersöe	Christiania
808	N	**Nymphen** *Wilhelmsen* 66 W.Bk	Norweg'n	312	122·1	27·2	13·9	Lübeck	1873	J.A.Meyer	[illegible]	J.Boysen	Christiansand
[illegible]		**Nymphen** *Falck* -86 W.Sr	Swedish	115	82·1	22·2	8·5	Carlscrona	1886			G.Ernberg & Co.	Carlscrona
[illegible]	N	**Nystad** *Rasmussen* 85-85 W.Bk	Norweg'n	456	142·7	30·3	16·5	Nystad	1865	P.H.Kjaldström		W.Wilhelmsen	Sarpsborg

ADDITIONAL SHIPS ARE INSERTED IN THE SUPPLEMENT AT THE END OF THE REGISTER.

*=Lloyd's Register; D=Nederlandsche Vereeniging; G=Germanischer Lloyd; I=Registro Italiano; R=American Record; A=Veritas Austro-Ungarico; H=Veritas Ellenico; N=Norske Veritas; V=Bureau Veritas.

Dans la 4me Colonne, C=Composite (membrure en fer, bordé en bois); I=Fer; Stl=Acier; W=Bois.

Les Dates du doublage en métal (Col. 11) ressortent des livres dans lesquels les navires sont respectivement classés ou des rapports des Inspecteurs du Lloyd's Register

Pour plus amples renseignements sur les navires classés par le Lloyd's Register, voir le Lloyd's Register Book et le Lloyd's Yacht Register.

1 Official No.	2 No. in Book	3 Registry in which classed	4 Ship's Name. Material and Rig. Master, and Dates of Appointment to present Owner's Service and to Ship. Ship's Late Name.	5 Flag	6 Net Tonnage	7 Dimensions. Length	Breadth	Depth	8 Built. Where	9 When	10 By whom	11 Date of Metalling	12 Owners or Managers	Port
[illegible]	1		O. Blanchard *Le Brun 81-89* W Bg	British	260	123·9	26·0	13·6	Canada	1868			G. Le Gros	Jersey
19409	2	R	O. D. Witherell *Garfield* W 3mSr	American	600	148·0	33·1	15·0	Bath, Me.	1874	Robinson	Y 9.88	Crosby B. Fisk and others	Dee...
	3		O. Ferreos *Caltsiogiani* (ex Eliot Peas...) W Bg	Greek	246	..	..	..	Syra	1879	E. Flamis		S. Duludi	Syra
	4	V	O. Kohan *Witt 86-88* W Bk	German	403	118·9	28·8	15·9	Rostock	1871	E. Burchard & Co.	V 11.81	H. Bauer	Rostock
19312	5		O. M. Marrett *Willis* W Sr	American	164	100·0	26·0	9·0	Booth Bay, M.	18⁶⁸⁄₆₉			C. McKown	Booth Bay
	6	N V	O. P. Olsen *Jacobsen 83-85* W Bg	Norweg'n	308	123·2	28·8	12·1	Kragerö	1885	T. G. Lisangen	? 6.88	O. P. Olsen	Kragerö
10631	7		Oak *Harvard 85-85* W Sp	British	175	79·8	21·1	12·9	Sunderland	1844			J. Watson	Newcastle
75005	8		Oak Hill W Bk	British	590	136·6	28·4	20·4	Scarboro', U.S.	1854			Alex. Stuart	Sydney
18955	9		Oakes Ames *Sheridan* W Sr	American	276	118·0	30·5	8·7	Noank, Conn	1862			L. A. Lewis	New York
79122	10	❋	Oakhurst *Richards 83-* I Bk	British	1032	216·0	34·0	21·0	Harrington	1879	R. Williamson & Son		T. Thompson	Liverpool
19142	11		Oakland *Welfare* W Bk	American	508	150·8	34·5	13·2	Bath, Me.	1865			W. P. Seyward	Pt. Towns...
73645	12	❋	Oaklands *De Stiger* I Bk	British	955	203·8	33·8	19·9	Dundee	1876	Gourlay Bros.		A. M. Lawrence & Co.	London
155017	13	R	Oakwoods *Carr* W Sr	American	131	93·3	27·3	7·8	Kennebunk, Me	1880	Titcomb & Ward		J. C. Tucker	Newport
69332	14	❋	Oakworth *Balker* I S	British	1202	236·7	36·3	21·2	Port Glasgow	1874	W. Hamilton & Co.		Hargrove & Hellon	Liverpool
	15		Oal *Sola* W Sr	Norweg'n	138	92·3	23·7	11·1	Bergen	1868			O. Nygaard	Stavanger
71662	16	❋	Oamaru *Boorman 84-88* I S	British	1306	239·1	36·1	21·0	Greenock	1874	Scott & Co.		Shaw, Savill & Albion Co. Ld.	Glasgow
78358	17		Oamaru *Richards* W Bn	British	156	92·0	23·5	11·5	Mahurangi, N.Z.	1878	J. & J. Darrach		Tho. Richards	Auckland
	18		Oasis (ex Stranger) W Bn	Peruvian	228	109·7	24·0	14·7	Swansea	1859	Meager		F. Imaravoto	Callao
70720	19	❋	Oban Bay *Gorley 68-88* I Bk	British	1068	215·9	34·2	20·7	Glasgow	1876	Dobie & Co.		Hatfield, Cameron & Co.	Glasgow
	20	V	Obdulia *Bonet* W Bk	Spanish	343	123·0	27·3	16·3	Sestri	1857		Y 8.85	Vda e Hijo de C. Tartabull	Barcelona
19456	21	R V	Obed Baxter *Colby* W Bk	American	877	180·2	34·0	19·0	Newburyport, Mass	1876	G. E. Currier	Y 9.85	W. A. Johnson	Boston
	22	G V	Oberbürgermeister v. Winter *Bratz 81-87* W Bk	German	552	143·0	31·2	18·6	Elbing	1866	G. Fechter & F. Schichau		Herm. Gronau	Danzig
60901	23	❋	Oberon (ex Steamer) *Dacy* C S	British	1194	241·1	36·2	21·0	Glasgow	1869	A. & J. Inglis	V 10.87	J. P. McCausland	London
	24	D V	Oberon (ex L. L. Sturges) *Zinke* W S	German	1300	188·5	38·3	24·0	East Boston	1860	S. Hall	Y 8.88	G. Balleer	Bremen
	25	❋	Oberon *Schumacher 84-84* I Bk	German	738	188·0	31·5	18·4	Sunderland	1878	Austin & Hunter		G. Bolte	Elsfleth
	26		Oberon W Bg	German	263	95·1	26·8	16·0	Rügenwalde	18⁶²⁄₆₃	C. F. Zielke		W. Beckmann	Rügenwalde
	27	G	Oberslieutenant von Sulstorff *Bauer 86-87* W Bg	German	241	105·3	26·4	13·9	Ribnitz	1863	J. Miedbrodt		Eduard Burchard	Rostock
22677	28		Obey *Brooks* W Bkn	British	258	118·2	24·2	14·0	Jersey	1858	G. Deslandes & Son	Y 8.88	J. Hamley	Guernsey
183	29	A	Obilic *Chersinus -88* W Bk	Austrian	427	129·6	27·4	17·6	Lussinpiccolo	1870	N. Martinolich	Y 7.86	S. Medin	Meglino
353	30	I	Oblio *Gallo -87* W Bk	Italian	661	154·1	32·9	20·7	Prà	1871	L. Durante	Y 8.87	Sig... Molfino, & F. ...	Genoa
	31	V	Observador *Linares* W Bg	Spanish	201	98·4	25·1	13·8	Almeria	18⁵⁸⁄₇₉		C 3.88	Orozco hermanos	Cartagena
	32	V	Ocata *Pagés* W Bn	Spanish	175	88·9	26·5	12·7	Blanes	18⁵³⁄₃₄		C 9.88	Pedro Pagés	Barcelona
19410	33	V	Occidental *Taylor 83-87* W S	American	1470	210·6	39·8	24·2	Bath, Me.	1874	E. & A. Sewall	V 11.85	A. Sewall & Co.	Bath, Me.
155065	34		Occidental W 3mSr	American	199	119·6	30·5	8·8	Fairhaven, Cal	1884			J. Nelson	San F...
8582	35		Ocean *Curson* W Sw	British	227	93·2	24·8	16·2	Sunderland	1851	W. Pearson		D. T. Sharper	Workington
18083	36		Ocean *Koon* W Bk	American	274	102·4	27·7	16·9	Haddam, Conn	18⁵⁵⁄₅₆			R. W. Bartlett	New Haven
	37	V	Océan *Mauge* W Bg	French	196	100·0	23·6	12·5	St. Vaast	1875		Z 8.88	Eust. Levêque	St. Vaast
	38	❋	Ocean (ex Amazone) *Stuben -88* I S	German	1300	231·1	37·1	23·1	Pt. Glasgow	1873	J. Reid & Co.		A. Evers & Classen	Hamburg
	39	V	Ocean *Klampen 84-84* W Bk	German	465	132·8	30·1	17·1	Hammelwarden	1875	Strenge & Sohn	Y 8.88	Gerh. Eilers	Brake
	40	G	Ocean *Jorck 69-69* W Bg	German	285	103·1	27·1	15·6	Rostock	1853	W. Zeltz		E. Burchard	Rostock
	41	N V	Ocean *Deiss 81-84* W Bk	Norweg'n	984	201·7	38·0	22·2	Bergen	1871	Brunchorst & Dekke	Y 8.86	Harloff & Bøe	Bergen
	42	N	Ocean *Moe -87* W Bk	Norweg'n	664	135·5	32·5	20·6	Vegesack	1850	Joh. Lange		Chr. Hannevig	Christiania
	43	N	Ocean *Toraldsen* W Bk	Norweg'n	530	139·9	32·2	17·1	Tvedestrand	1877	A. & F. Smith	Y 6.88	A. & F. Smith	Tvedestrand
	44	N	Ocean *Wager 77-87* W Bk	Norweg'n	477	129·7	31·1	17·8	Brahestad	1861			H. A. Lohre	Drammen
	45	N	Ocean *Salvesen -88* W Bk	Norweg'n	475	135·1	31·2	16·4	Arendal	1873	O. B. Sörensen	V 12.87	O. B. Sorensen	Arendal
	46	N	Ocean *Andersen* W Bn	Norweg'n	213	111·2	26·5	11·3	Sandefjord	1875	C. Christensen		P. A. Grøn	Sandefjord
	47		Ocean *Haglund* W Sr	Russian	230	..	..	..		1867			K. J. Johansson	Lund
	48	R	Ocean *Hellgren -81* W Bk	Swedish	354	117·7	25·3	16·8	Sunderland	1864	Thompson	Y 8.88	S. Hörberg	Helsingborg
	49		Ocean *Larsson 86-86* W Bk	Swedish	275	99·4	24·5	15·6	Carlshamn	1839			O. Jönsson	Hörby
70370	50	❋	Ocean Belle *Harroway 84-84* W Sw	British	283	118·6	27·5	15·7	Blyth	1874	Union Co-op. S. B. Soc.	Y 8.8?	R. L. Manners	North...

❋ = **Lloyd's Register**; D = Nederlandsche Vereeniging; G = Germanischer Lloyd; I = Registro Italiano; R = American ...; A = Veritas Austro-Ungarico; H = Veritas Ellenico; N = Norske Veritas; V = Bureau Veritas.

In the 4th Column, C = Composite (iron frame, wood planked); I = Iron; Stl = Steel; W = Wood.

The Dates of Metalling (Col. 11) are based on the records of the Books in which Vessels are respectively classed, or on the Reports of the Surveyors of Lloyd's.

Further particulars of Vessels classed by Lloyd's Register will be found in Lloyd's Register Book and Lloyd's Yacht Register.

1 Premier numéro	2 No. dans la liste	3 Registre dans lequel le navire est coté	4 Nom. Matériaux et Gréement. Capitaine et dates d'entrée au service des Armateurs actuels et à bord du navire. Ancien Nom.	5 Pavillon.	6 Tonnage net.	7 Dimensions. Longueur	Largeur	Creux.	8 Construit. Où.	9 Quand.	10 Par qui.	11 Date du doublage en métal	12 Armateurs ou Directeurs-Gérants.	13 Port d'Armement.
	51	*	Ocean Belle *Williams 1887* W Sw	British	183	104·0	23·0	12·8	Sunderland	1859	W. Pile	Y 1.88	R. Williams	Aberystwith
	52		Ocean Belle *Hickman* W Sr	British	104	83·6	25·3	7·9	Waterbro, N.B.	1870	G. Wasson		Dwight Ruggles	Digby, N.S.
	53		Ocean Bride *Dawson* W Bg (BROKEN UP)	British	209	105·0	25·1	13·9	Guernsey	1864	DePutron		J.R. Daniels	Guernsey
	54	*	Ocean Chief *Morgan -88* W Bk	British	386	136·6	23·8	16·1	Kingston	1872	Kinloch	Y 12.88	W. Kinloch	Banff
	55	R	Ocean Home *Carleton* W Sr	British	150	89·2	26·2	10·6	Jeddore, N.S.	1880			J.G. Charlton	Halifax, N.S.
	56		Ocean Lilly *Watt -88* W 3m Sr	British	136	92·0	24·7	10·1	Washabuck	1885	K. McDonald		William Watt	Sydney, C.B.
	57		Ocean Mail *Jorgensen* W Bk	Norweg'n	391	132·2	29·2	15·8	P.E. Island	1866	Ellis		W.M. Michaelsen	Moss
	58		Ocean Monarch *Le Gay* W Bkn	British	215	113·6	25·0	11·1	Guernsey	1866			R.J. Dorey	Guernsey
	59		Ocean Nymph *Cook 84-87* W Bk	British	294	118·1	27·3	12·6	Quebec	1862	Rosa		W. Cook	Lowestoft
	60	R	Ocean Pearl *Staten -87* W Bkn (WRECKED)	American	436	126·9	30·3	16·2	Deering, Me.	1874		Y 12.86	John W. Musgrave	Portland, Me.
	61		Ocean Pearl W Sr	Dutch	125	111·2	28·2	9·3	Essex, Mass.	1880		Y 3.81	A. Laugen	Batavia
	62	*	Ocean Pet *Downey 87* W Sr	British	74	80·9	20·0	9·2	Jersey	1876	Picot	Y 2.87	W. Pickford	Jersey
	63		Ocean Queen *Ireland 1886* W 3m Sr	British	195	121·3	25·2	12·5	Looe	1868	Pengelly		G.F. Treleaven	Plymouth
	64		Ocean Queen *Cantell* W Bkn	British	150	103·9	20·5	12·1	Jersey	1865	LeVesconte		G.T. Bryant	Jersey
	65		Ocean Queen *Chappell 1886* W Bn	British	121	83·2	20·5	11·0	Padstow	1858			A.B. Hutchings	Padstow
	66	*	Ocean Racer *White 87* W Bkn	British	202	109·0	23·8	13·1	Kingston	1875	Spence	Y 3.86	Crinks, Parsons & Co.	Inverness
	67	*	Ocean Ranger *Pearce 75-75* W 3m Bkn	British	268	125·0	25·1	13·9	Appledore	1875	Cook	Y 10.88	J.H. Hocken	Fowey
	68		Ocean Ranger *Gillet* W Bn	British	234	113·0	24·1	14·2	Bridport	1867	Cox	Y 8.82	H. Guthrie	Dunedin, N.Z.
	69	*	Ocean Rover *Davies 88-88* C Bk	British	548	153·0	28·7	18·0	Sunderland	1867	R. Thompson, jr.	Y 1.88	J.C. & G.S. Richardson	Swansea
	70		Ocean Rover W Sr	British	139	96·0	25·2	9·6	Williams River (N.S.W.)	1884	R. Muirhead		D. Munro	Melbourne
	71	*	Ocean Spray *Hocken* W Bkn	British	253	122·8	25·7	13·9	Par	1877	Moss	Y 8.88	N. Hocken	Fowey
	72	*	Ocean Sprite *Moore 81-82* W Bk	British	211	114·0	23·6	12·5	Sunderland	1854		Y 7.86	S.B. Mason	Hull
	73		Ocean Star *Lenman* W Bn	British	176	108·2	28·2	12·0	Parrsboro', NS	1865	Mullon		J. Neill	Belfast
	74		Ocean Star W Sr	British	126	85·5	24·6	9·0	Parrsboro', NS	1882	H. Ross		A. Marvin	Halifax, N.S.
	75		Ocean Swell *Bohann* W Bkn	British	187	111·6	24·4	12·9	Fowey	1875	Stephen	Y 1.88	G.H. Bate	Fowey
	76		Ocean Traveller *Thomasson* W Sr	Danish	112	85·1	22·7	8·7	Nova Scotia	1877		Y 10.84	P. Pentheny	St. Croix, W.I.
	77	*	Ocean Wave *Camden* W Sr	British	157	96·8	22·7	12·2	Milford	1876	Howell	Y 2.88	M. Ellis	Carnarvon
	78	*	Ocean Wave *Toms* W Sr	British	139	99·8	21·8	12·2	Appledore	1870	Cook	Y 10.87	N. Hocken	Fowey
	79		Ocean Wave *Hurd* W Sr	American	120	88·6	24·5	7·2	Camden, N.J.	1853			J.I. Hurd	Bangor, Me.
	80		Ocean Witch W Sr	British	131	91·0	21·8	12·0	Littlehampton	1852	J. & W.B. Harvey		W. Peet	Hull
	81	*	Oceana *Pape 87-87* Stl 4m Bk	British	1827	269·7	40·1	23·7	Port Glasgow	1887	R. Duncan & Co.		R.R. Paterson	Greenock
	82	*	Oceana (ex Thais) W Sr Yt	British	109	105·8	21·4	10·6	Cowes	1880	C. Hansen & Sons	C 6.83	Sir P.F. Shelley, Bt.	Cowes
	83	N	Oceana *Svendsen* W Bk	Norweg'n	467	122·6	30·0	17·8	Arendal	1845			S. Eyde	Arendal
	84		Oceana *Andersson 77-77* W Sr	Russian	265	..	..	..	Hittis	1873			A. Salomonsson	Hittis
	85		Oceania *Vance -88* W 3m Sr	American	410	148·3	36·0	11·3	Port Blakely	1888	Hall Bros.		J. Vance	Eureka
	86		Oceanide *Blanck -75* W Bg	German	204	89·3	25·1	13·6	Elbing	18 38/62	M. Mitzlaff		Schultz & Lübcke	Stettin
	87	*	Oceanie *Saintmartin 76-88* W Bk (ex Margaret Pugh)	French	675	136·0	32·0	20·0	Pwllheli	1862	Edwards	Y 7.87	J.H. Tandonnet frères	Bordeaux
	88		Oceano Bn	Brazilian	313	..	..	..					A.P. de Magalhaes	Bahia
	89		Oceano (ex Oceaan) *Coppa* W Sr	Italian	130	83·0	21·7	11·1	Vlaardingen	1848	L. van Dam		P. Semeria	Genoa
	90		Océano W Bg	Spanish	228	106·3	25·7	11·1	Brake	1854			Rafael Calvo	Manila
	91	A	Oceanos *Trapis -88* W Bn	Greek	438	133·5	29·5	..	Syra	1875	A. Livadara	Y 12.80	A. Giannicoglu	Syra
	92		Oceanus W Sr	American	547	232·0	40·0	10·0	New York	1864			J.C. Haddock	New York
	93		Oceanus *Young* W Sr	American	241	112·8	30·4	8·4	Stony Br'k, NY	1865			Floyd & Newins	Prt. Jefferson, N.Y.
	94	V	Oceola *Smith -88* W B 3m Sr (ex Konig Wilhelm I.)	Norweg'n	227	110·5	25·4	11·8	Reiherstieg	1869	F. Dreyer	Y 2.88	O. Wettre	Christiania
	95	* D	Ochtertyre *Smith 81-85* I Bk	British	1296	239·5	36·2	21·5	Port Glasgow	1885	R. Duncan & Co.		H. Hogarth	Ardrossan
	96		Octavia *Purdy 84-88* W Bk	British	898	164·7	30·3	22·3	Shields	1853	Middle Dock Co.		J. Graham & Co.	South Shields
	97	N	Octavia *Laland -88* W Bk	Norweg'n	307	132·9	27·9	11·5	Baltimore	1857	T. Abrahams		Carl Bentsen	Mandal
	98		Octavie W Bk	French	380	119·1	25·3	16·9	Nantes	1854	M. Viaud		Chaigneau frères	Bordeaux
	99		Octavio *De Barros* W Bg	Brazilian	137	..	..	..	Spain	1864		Y 2.80	C. da Estrada de Ferro da Bahia & Minas	Rio Janeiro
	100		Octo Adelphi W Bg	Greek	212	..	..	..	Syra	1868	A. Pontzos		P.T. Pagoulatos	Zante

*=Lloyd's Register; D=Nederlandsche Vereeniging; G=Germanischer Lloyd; I=Registro Italiano; R=American Record; A=Veritas Austro-Ungarico; H=Veritas Ellenico; N=Norske Veritas; V=Bureau Veritas.

Dans la 4me Colonne, C=Composite (membrure en fer, bordé en bois); I=Fer; Stl=Acier; W=Bois.

Les Dates du doublage en métal (Col. 11) ressortent des livres dans lesquels les navires sont respectivement classés ou des rapports des Inspecteurs du Lloyd's Register

Pour plus amples renseignements sur les navires classés par le Lloyd's Register, voir le Lloyd's Register Book et le Lloyd's Yacht Register.

Official No.	No. in Book	Registry in which classed	Ship's Name. Material and Rig. Master, and Dates of Appointment to present Owner's Service and to Ship. Ship's Late Name.	Flag.	Net Tonnage	Dimensions. Length.	Breadth.	Depth.	Built. Where.	When.	By whom.	Date of Metalling	Owners or Managers.	Port of Registry
[illegible]	101	V	Odabella *Tomietti -83* W Bn	Italian	135	95·5	23·2	12·1	Savona	1882	J.B.Pongiglione		Sorelle Arenoen fu Luigi	Porto
	102	N	Odd *Christensen 82-82* W Bk	Norweg'n	277	121·9	27·9	12·4	Drammen	1872	J.Jörgensen	[illegible]	Carl Bech	Tvede
[illegible]	103		Odell *Wade -84* W Sr	American	134	97·2	26·2	7·5	Bowdoinham, M.	1873			Scott & Bridge	Bo
	104	R	Oden *Stranne* W Bk	Swedish	526	143·7	29·8	17·3	Gefle	1868			N.Christiansson	Ko
	105	V	Oden *Ryfrat 81-81* W Sr	Swedish	183	104·6	22·9	11·0	Oscarshamn	1878	C.Thorén	[illegible]	J.E.Hagström	
	106	G	Odense *Hansen 84-* W Sr	Danish	103	80·9	20·6	10·8	Odense	1856	H.L.Hansen		P.Hansen	Odense
	107		Odila W Bn	Spanish	141	84·6	21·6	11·1	Canary Islands	1866			Ginna hermanos	Havana
	108		Odin *Holm 82-83* W Bg (LOST) (ex Jylland, ex Mads Basell)	Danish	226	104·8	22·2	12·8	Neu Ra'nebeck	1859			Lauritz Andersen	Odense
	109		Odin *Hansen -84* W Bg (ex Excelsior) (BROKEN UP)	Danish	170	87·1	21·8	12·6	Goole	1855	Jas.Wake		H.P.Christiansen	Lemvig
	110	V	Odin *Christiansen 81-81* W Sr	Danish	134	92·3	20·8	10·7	Thurø	1874	P.Bom		Actieselskabet Odin Møller	Svend
	111	V	Odin *Mikkelsen -88* W Sr	Danish	123	89·4	20·0	9·5	Aalborg	1867	S.L.Sørensen		H.A.Hansen	Aalborg
	112	N	Odin (ex Oden) *Hallian -88* W Bk	Norweg'n	510	151·3	31·7	16·0	Wasa	18 53/55		[illegible]	Fr.Allum	Drammen
	113		Odin (ex Thor) *Heel* W Bk	Norweg'n	465	121·0	32·2	18·3	Brahestad	1856			T.Jensen	Kragerö
	114	N	Odin *Skaadal -70* W Bg	Norweg'n	355	104·3	29·0	16·0	Grimstad	1855	J.Telletsen		B.Birkeland	Lillesand
	115	N	Odin *Johnsen 77-77* W Bg	Norweg'n	297	95·4	27·9	16·8	Finland	1802			I.C.Lund	Christiania
	116	N	Odin *Andersen -88* W Sr	Norweg'n	125	81·6	22·8	10·5	Hamburg	1849			Chr.Hannevig	Arendal
463	117	I	Odoardo Virginio *A. Di Janni* W Bk	Italian	324	124·6	29·0	16·8	Gaeta	1873	L.Contreras	[illegible]	Raf.Di Janni & Co.	Gaeta
837	118		Odonsura Maru W Bg	Japanese	284	115·9	25·1	12·8	Rikuzen	1883	S.Chokichis		Yamazawa Chokuro	Yeichi
19394	119	R	Odorilla *Holland* W Bn	American	398	129·1	31·6	14·7	Smyrna, Del	1873	Worden & Evans	[illegible]	Henry D.May	Philada
	120	H	Odysseus *Tagoras* W Bg	Greek	247	..	..	..	Galaxidi	1869			Scouter & Zagoras	Galaxidi
	121		Odysseus W Bg	Greek	192	95·1	22·6	..	England	1867			S.Theofilaton	Ithaca
	122	I	Odysseus *Pericle* W Bgn	Turkish	200	..	..	..	Scio			[illegible]	F[lli] Miliadi	Scio
6	123		Odysseus Ithakis W Bn	Greek	248	101·6	25·4	17·1	Ithaca	1869	P.Lenketsas		D.Pieros	Ithaca
70071	124	✱D	Œnone *Beale 81-84* I S	British	1437	239·7	38·3	23·5	Port Glasgow	1875	J.Reid & Co.		W.& C.Battersby & Co	Liverpool
	125		Oey Sinjo *San Sjior Giok* W Bk (ex Hog Sinjo)	Dutch	266	111·2	25·0	..	Rembang	1856		[illegible]	Said Salim bin Saleh Mohdarta	Pekalonga
	126		Offerton (LOST) *Watson* W Bk	Chilian	494	144·7	30·2	18·5	Sunderland	1870	J.Gill	[illegible]	F.Peede	Valparaiso
	127	N	Ögir *Krogh -76* W Bk	Norweg'n	547	116·0	31·2	17·8	Christiansund	1867	T.Wullum	[illegible]	D.Davidsen	Farsund
	128	N	Ögir *Nielsen* W Bg	Norweg'n	275	110·5	28·1	14·5	Drammen	1854	Jörgensen & Knudsen		Ole Nielsen	Drammen
80400	129	V	Ohio *Crawford 84-* W Bn	British	348	127·6	29·4	13·0	Portland, N.B.	1882	A.B.& W.S. Ruddock	[illegible]	Scammell Bros.	St.John N.B.
18587	130		Ohio *Gifford 87-87* W Bk	American	363	170·9	28·4	18·6	Mattapoisett, Mass.	18 33/53			Aiken & Swift	New Bedford
18960	131		Ohio *Benton* W Bk	American	206	92·6	24·0	14·9	Baltimore	1833			Gilbert Allen	New Bedford
19046	132		Ohio W Sr	American	104	77·8	27·6	7·4	Coxsackie, NY	1824			D.E.Donovan	New York
512	133		Ohkuni Maru W Bk	Japanese	428	136·9	26·8	16·7					Wataru Mohei	Hiogo
71739	134	✱	Oimara *Robertson 84-84* I S	British	1354	241·4	37·1	21·4	Glasgow	1875	C.Connell & Co.		J.D.Clink	Glasgow
55453	135		Oimara *Herbert* C Cr Yt	British	111	96·6	20·0	12·0	Greenock	1867	Steele & Co.		J.Wylie	Greenock
837	136		Oishi Maru W Sr	Japanese	250	99·0	26·4	13·3	Rikuzen	1883	Suzuki Chokichi		Uyeno Yosoji	Rikuzen
63191	137	✱	Oithona *Halliday 87-87* I S (MISSING)	British	1342	212·7	36·2	24·2	Birkenhead	1869	Laird Bros.		J.H.Worthington & Co	Liverpool
	138	DV	Okeia *Nissen 76-79* I Bk	German	688	178·3	31·3	18·0	Flensburg	1879	Flensburger Schiffsb.-Ges.		J.E.M.Bleck	Hamburg
[illegible]	139	✱	Okenbury *Mathew* W Bg	British	232	112·0	24·5	14·9	Salcombe	1869	Harnden	[illegible]	T.Matthews	Salcombe
93314	140		Okkan Stl S	British	480	220·0	30·0	8·4	Dumbarton	1886	W.Denny & Bros		Irrawaddy Flotilla Co.Ld	Glasgow
	141	NV	Okonom *Lassen -83* W3mSr	Norweg'n	319	133·6	27·7	14·0	Stavanger	1878	K.Nielsen	[illegible]	K.S.Bertelsen	Stavanger
512	142		Okuni Maru W Bk	Japanese	428	136·9	26·8	16·7					Hachima Kanesuke	Hiogo
	143	N	Olaf *Grum* W Bk (ex William Tatten)	Norweg'n	667	139·5	30·9	20·8	Richmond, M.	1848		[illegible]	J.Hansen	Tönsberg
	144	N	Olaf *Andersen* W Bg	Norweg'n	244	95·2	26·4	14·8	Montrose	1843			T.J.Lyche	
	145	N	Olaf *Andersson 73-73* W Bg	Norweg'n	205	108·7	25·9	11·5	Sandefjord	1873		[illegible]	C.R.Thoresen	Sand
	146	N	Olaf *Foyen* W Sr	Norweg'n	112	76·7	23·7	10·1	Fredrikshamn	1858			C.A.Föyen	Tönsberg
	147	N	Olaf Kyrre *Hansen 79-80* W Bk	Norweg'n	468	137·0	30·2	17·8	Elbing	1866	Gebrüder Mitzlaff	[illegible]	T.Eeg	Bergen
	148		Olaf Kyrre *Olsen* W Bg (WRECKED)	Norweg'n	291	107·5	28·4	14·9	Drammen	1854			T.W.Bergh	
	149	N	Olaus *Amundsen -76* W Bk	Norweg'n	473	130·9	28·9	19·4	Baltimore	1854			C.Svendsen	Drammen
	150	N	Olaus *Olsen -88* W Bk (ex Minna & Maria)	Norweg'n	233	100·3	25·9	13·1	Rouen	1856	Le Mire		L.A.Torpp	Fr

✱ = Lloyd's Register; D = Nederlandsche Vereeniging; G = Germanischer Lloyd; I = Registro Italiano; R = American [Shipmasters'] …; A = Veritas Austro-Ungarico; H = Veritas Ellenico; N = Norske Veritas; V = Bureau [Veritas].

In the 4th Column, C = Composite (iron frame, wood planked); I = Iron; Stl = Steel; W = Wood.

The Dates of Metalling (Col. 11) are based on the records of the Books in which Vessels are respectively classed, or on the Reports of the Surveyors of L…

Further particulars of Vessels classed by Lloyd's Register will be found in Lloyd's Register Book and Lloyd's Yacht Register

		Nom. Matériaux et Gréement. Capitaine et dates d'entrée au service des Armateurs actuels et à bord du navire. Ancien Nom.	Pavillon	Tonnage net.	Dimensions. Longueur.	Largeur.	Creux.	Construit. Où.	Quand.	Par qui.	Date du doublage en métal	Armateurs ou Directeurs-Gérants.	Port d'Armement.
151	V	Olaus *Robertson* WBn	Swedish	184	96·1	29·9	13·2	Norrsundet	1869	J.Rost	[illegible]	C.A.Robertsson	Helsingborg
152	G	Olbers *Nielsen -88* WBk	Norweg'n	849	157·7	34·0	21·3	Bremerhaven	1863	JCTecklenborg	Y 10.83	J.Pedersen	Christiania
153		Olcese *L. Olcese* WBk	Italian	439	131·2	29·8	19·0	SestriPonente	1868	G.Peragallo	Y 10.83	G.Olcese fu D.	Genoa
154		Old Goody *Ashby 81-81* WSr	British	174	100·4	24·3	12·3	Sandwich	1865	Paine		E.Rigden	Faversham
155	*	Old Head *Cumins* WSr	British	97	89·6	22·0	9·8	Barnstaple	1878	Westacott		T.Crowley & Co.	Cork
156	*	Old Kensington *Jones 81-81* I S	British	1777	262·0	42·1	23·8	Liverpool	1874	W.H.Potter&Co		A.Billbrough & Co.	London
157	*D	Oldambt *H. A. Mulder* W3mSr	Dutch	308	104·8	25·9	17·1	NorthShields	1863	T.&W.Smith	[illegible]	H.R.Toren	Scheemda
158	*	Oldambt II. *Mulder* WBk	Dutch	395	128·4	26·1	17·0	Whitby	1865	Barrick	[illegible]	H.R.Toren	Scheemda
159	N	Ole Boe *Nielsen 82-86* WBk	Norweg'n	770	173·5	33·9	18·6	Grimstad	1882	JacobBoe	[illegible]	Christen Boe	Arendal
160	DN	Ole Bull *Riise 55-78* WBk	Norweg'n	549	155·5	32·8	18·3	Bergen	1870	JensGran	[illegible]	P.G.Pettersen	Bergen
161	DN	Ole Smith Ploug WBk	Norweg'n	651	159·6	33·0	18·9	Stavanger	1871	E.Knudsen	[illegible]	I.A.Henschien	Lillesand
162		Olga *Pennycook* WBg	British	245	117·0	26·1	12·9	P.E.Island	1875	Peake		G.Spencer	Waterford
163		Olga CSrYt	British	108	105·4	22·2	10·0	Renfrew	1874	Henderson, Coulborn&Co.		J.A.Hankey	London
164		Olga *Jung* WBk	Chilian	598	154·2	30·8	18·9	Sestri Ponente	1875	Bossi	[illegible]	Gutschow&Piza	Valparaiso
165		Olga (ex Adlatar) *Kofod* WBg	Danish	211	106·1	25·3	12·1	Viborg	1864	Z.Koli		J.Hintze	Ronne
166	*	Olga (ex Leonine) *Theizen -87* I Bk	German	1340	237·4	36·4	21·9	Glasgow	1875	J.G.Lawrie		J.Hintze	Hamburg
167	*	Olga (ex Amaly) *Wittrock -88* I Bk	German	519	165·0	28·1	17·3	Sunderland	1869	Hill		H.J.Kröncke	Hamburg
168	G	Olga *Fiecke 83-86* WBk	German	349	112·1	27·8	15·6	Ueckermünde	1838			Ernst Wendt	Danzig
169		Olga W3mSr	German	254	..	..	..		1879			W.C.T.Henglin	Hamburg
170		Olga (ex Vittorio) WBn	Greek	497	137·8	29·5	..	Loano	1867	Cerruti		M.Spartali	Syra
171		Olga *Vergotis* WBn	Greek	347	118·1	28·1	..	Syra	1878			A.Vaglianos	Syra
172		Olga WBg	Greek	230	..	..	..	Egina	1860			Catsimbas & Iliopoulos	Zante
173	H	Olga *C. Arvanitis* WBg	Greek	212	85·3	23·0	..	Galaxidi	1867		[illegible]	J.Arvanitis	Poros
174		Olga *Sotirakis* WBg	Greek	210	91·2	23·4	15·5	Syra	1867			G.Sotirakis	Spezzia
175		Olga WSr	Greek	200	89·6	25·3	..	Fiume	1870			D.Phlokas	Syra
176		Olga *Kyriacou* WBg	Greek	199	..	..	..	Spezzia	1867			C.A.Kyriacou	Spezzia
177	H	Olga *Missirlis* WSr	Greek	199	91·5	24·9	..	Fiume	1872			S.Chrysos	Syra
178		Olga WBg	Greek	197	93·5	24·8	..	Syra	1881	C.Cristali		D.Vafiadakis	Syra
179	H	Olga *Tsentzeftis* WBg	Greek	165	76·1	24·0	..	Trieste	1867	Stabilimento Tecnico	[illegible]	S.Moraiti	Corfu
180		Olga *Laimos* WSr	Greek	100	..	..	..	Syra	1877			D.C.Laimos	Syra
181		Olga *Mirabella* WBk	Italian	224	121·7	25·3	16·4	Malta	1863	Fili German	[illegible]	S.Vasta	Catania
182	A	Olga *Ghezzo* WBn	Italian	219	99·0	27·9	15·5	Curzola	1871		[illegible]	D.Ghezzo	Venice
183	N	Olga *Troberg* WBg	Norweg'n	562	138·5	33·1	18·1	Brahestad	1863			J.M.Pauss	Drammen
184	N	Olga (ex Abord) *Thiis -86* WBg	Norweg'n	316	110·5	28·8	15·5	Hernösand	18 43/72			Carl Weigel	Christiania
185	N	Olga *Jörgensen -88* WBg	Norweg'n	220	93·6	24·6	15·2	Sunderland	1852			C.G.Bruun	Tönsberg
186		Olga *Lundqvist* WBk	Russian	561	152·4	32·6	19·3	Mariehamn	1874	Lisman		M.Mattsson	Mariehamn
187		Olga *Wirkki* WBg	Russian	257	..	..	..		1870			J.Wirkki	Björkö
188		Olga WSr	Russian	250	108·3	29·3	12·4	Laitakari	1876	P.Sandström		J.J.Roström	Lavia
189		Olga *Holmberg* WSr	Russian	232	..	..	..	Kimito	1872			A.Anderson	Nagu
190		Olga WBg	Russian	176	..	..	..		1842			L.Kräki	Nystad
191		Olga *Behrsing* WSr	Russian	148	79·5	25·5	9·6	Kesterzeem	1872	Kotzer		Romanow	Riga
192		Olga *Jansson* WSr	Russian	104	..	..	..		1874			Mans A.Jansson	GetaKapell
193		Olga *S.Papandopulo* WBg	Russian	101	..	..	..	Kerasunda	1873	J.Chilinghiri		Z.Papandopulo	Eupatoria
194		Olga WSBn		109	79·4	22·0	8·2	NovaScotia					
195		Olga WSr	Swedish	276	104·6	25·9	13·3	Christinestad	1868			G.T.Sjölin	Jernavik
196		Olga *Hansson* WSr	Swedish	170	93·5	22·5	12·0	Finland	1857			J.B.Jonsson	Lysekil
197		Olga *Andersson* WSr	Swedish	160	92·2	22·5	10·9	Finland	1867			M.Mattsson	Rosättra Rslgn.
198		Olga *Nilsson -88* W3mSr	Swedish	139	89·8	20·7	10·5	Sikeå	1839		[illegible]	J.Balkenhausen	Sölvesborg
199		Olga *Berendtsen 88-88* WSr	Swedish	127	80·7	21·6	11·4	Oronst	1854	O.Johansson		M.Backelin	Fjellbacka
200		Olga *Magnusson* WSr	Swedish	115	72·4	22·9	9·4					C.Ekman	Norrköping

*=Lloyd's Register; D=Nederlandsche Vereeniging; G=Germanischer Lloyd; I=Registro Italiano; R=American Record;
A=Veritas Austro-Ungarico; H=Veritas Ellenico; N=Norske Veritas; V=Bureau Veritas.

Dans la 4me Colonne, C=Composite (membrure en fer, bordé en bois); I=Fer; Stl=Acier; W=Bois.

Les Dates du doublage en métal (Col. 11) ressortent des livres dans lesquels les navires sont respectivement classés ou des rapports des Inspecteurs du Lloyd's Register.

Pour plus amples renseignements sur les navires classés par le Lloyd's Register, voir le Lloyd's Register Book et le Lloyd's Yacht Register.

Official No.	No. in Book	Registry in which classed	Ship's Name. Material and Rig. Master, and Dates of Appointment to present Owner's Service and to Ship. Ship's Late Name.	Flag.	Net Tonnage	Length	Breadth	Depth	Where	When	By whom	Date of Metalling	Owners or Managers	Port
	201		Olga *Jonsson -87* WSr	Swedish	193	73·2	20·4	10·3	Gamla Lödöse	1868	P. Larsson		B. Gustafsson	Merka…
	202	H	Olga *Moustakas* WBk	Turkish	286	..	..	..	Syra	1867			Houssein M. Meleglou	Con…
	203		Olga WBg	Turkish	280	..	..	..	Russia	1883			[illegible] Hassan & Co	Con…
	204		Olga WBg	Turkish	139	..	..	..	Roumania	1870			Fabrikjee Zacharieh	Con…
	205	A	Olga B. (WRECKED) *Bujan* WSr	Austrian	266	101·0	24·7	15·6	Buccari	1871	Matterglian	Y 6.88	A. Bujan fu V.	Bucca…
	206		Olga Emilie WSr	Russian	126	87·1	28·1	10·2	Kesterzeem	1882	Kotzer		D. Romanow	Riga
	207	V	Olga T. *Giurcovich -8* WBk	Austrian	600	138·1	29·8	21·0	Fiume	1877	G. Brazzoduro	Y 1.87	M. Terni & Co	Volos…
	208	A	Olimpia *Carugnalti* W3mSr (ex A. Fedo)	Greek	498	114·3	30·2	18·2	Varazze	1872	G. B. Baglietto	Y 6.88	Th. A. Vuccina & C	Syra
	209		Olinda *Storm -86* WBg	British	234	106·5	24·1	14·2	Arbroath	1869	Arbroath Sh. Co.		Bessey & Palmer	Arbroath
	210		Olinda *de Araujo* WBg	Brazilian	165	..	..	..	Porto Alegre	1874		Y 6.86	Donna M. da Fonseca	Porto…
	211		Olinda *de Lago* WSr	Brazilian	104	..	..	..	Pernambuco	1873		Y 3.81	O. da C. Marcial	Bahia
	212		Olinda *Soares 82-87* WBg (ex Federico Mallfrid)	Portuguese	337	104·3	25·3	16·4	Varazze	1864			J. H. Andresen	Oporto
	213	G	Oline *Hansen -88* WSr	Danish	101	82·2	19·1	9·2	Nordby, Fano	1857	C. R. M. Mecklenborg		H. P. Hansen	Svendb…
	214		Olinto (ex Rivera C.) WBn		180	100·7	23·7	12·7	Rapallo	1875	G. Merello	Y 6.83		
	215		Olio *Martin* WSr	British	105	83·0	26·1	7·7	St. Martin's N.B.	1880	J. Brown		E. G. Dunn	St. Joh…
	216	I	Oliva Speciosa *Dell'Orso -87* WBk	Italian	570	136·3	30·1	18·5	Sestri	1872	Debarbierie [Celesia	Y 2.83	Fllli. Dall'Orso fu A.	Spezia
1830	217		Olivari *Molfino* WBk	Italian	483	138·1	30·9	18·7	Varazze	1868	G. Cerruti	Y 6.84	Andrea Bozzo	Genoa
88346	218		Olive *Getson* WSr	British	103	80·0	24·5	9·0	La Have, N.S.	1883			C. Romkey	Lunen…
87525	219	✻	Olive *Watt 84-86* WSr	British	92	83·4	20·5	8·1	Auckland	1885	Lane & Brown	Y 6.85	C. C. McMillan	Auckl…
19286	220		Olive *Frye* WSr	American	135	96·3	26·3	8·4	Addison, Me.	1869			G. H. Frye	Machi…
	221	✻	Olive *Brehn 80-88* IBk	German	847	197·6	32·2	19·8	Sunderland	1874	W. Pile & Co.		C. M. Matzen & Co.	Hamburg
65350	222		Olive Branch *Rudge* W3mBn	British	180	111·0	23·4	11·4	Peterhead	1872	Carnegie	Y 6.82	H. Good	London
80830	223	R	Olive Branch *Hawthorn -84* WBn	British	149	86·6	25·8	11·0	Mahone Bay N.S.	1880	W. A. Pickels	Y 6.88	W. Pickels	Lunenb…
87640	224	✻	Olive Branch *Griffiths* WSr	British	99	85·4	21·7	11·5	Aberdovey	1880	Richards	Y 6.87	O. Williams	Beaum…
77116	225	✻	Olive Branch *Davies 85-88* WSr	British	97	82·3	20·4	10·5	Shoreham	1878	Dyer	Y 6.88	R. Mitchell	Swansea
[illegible]	226		Olive Mount *Richards -87* WBk	British	941	176·6	36·3	21·2	Maitland, N.S.	1874	Cox	Y 7.87	J. A. Ledward	Halif…
	227		Olive Mount *Williams* WBk	Chilian	604	157·2	30·2	19·3	Quebec	1863	Lee		S. Martin	Valpar…
[illegible]	228	V	Olive S. Southard *Wilson* WS	American	1193	186·2	37·1	23·6	Richmond, Me.	1871	T. J. Southard	Y 6.85	Montgomery & Colwell	Bath, Me.
[illegible]	229	RV	Olive Thurlow *Corbett* WBk	American	634	149·0	32·7	17·6	Calais, Me.	1876	O. B. Rideout	Y 5.87	James E. Brett	New Yo…
	230		Oliveira WBn	Brazilian	143	..	..	..					M. M. Nogueira	Rio de…
	231		Oliveira WBn	Brazilian	112	..	..	..	Guarapary	1859			M. M. Nogueira	Victoria
	232		Oliveira WSr	Brazilian	104	..	..	..					J. F. Bastos	Rio de…
57851	233		Oliver (ex Felix) WBn	British	130	90·0	22·0	10·0	Kiel	1867			H. Thackray	[illegible]
18033	234		Oliver Ames *Handy -84* WSr	American	433	124·0	33·0	16·0	Berkley, Mass.	1866			J. B. Phillips	Fall Riv…
19359	235		Oliver Dyer *Emmerson* WSr	American	198	107·5	28·0	7·8	Saco, Me.	1871			R. F. C. Hartley & Co.	Saco, Me.
66886	236		Oliver Emery *Swabridge* WBk	British	623	142·7	32·3	18·0	St. John, N.B.	1873	W. King		O. Emery	St. Joh…
19030	237		Oliver H. Booth *Cranmer* WSr	American	231	120·0	33·0	9·7	Poughkeepsie, N.Y.	1856			T. S. Parker	New Yo…
	238		Oliver Kromvel *Mattson* WBk	Russian	478	..	..	..	England				M. Eriksson	Fin…
92384	239		Oliver Mowat *Peacock* W3mSr	British	244	131·2	25·9	10·6	Mill Haven, Ont.	1873	Fraser & George		J. O. Guy	Kingst…
155062	240	R	Oliver S. Barrett *Lewis* W3mSr	American	603	153·0	35·4	12·5	Bath, Me.	1884	Deering & Donnell		Deering & Donnell	Bath, Me.
19220	241		Oliver Schofield *Crowell* W3mSr	American	374	130·0	31·0	14·0	Tottenville, S.I.	1867			C. Lovell & Son	New Yo…
26321	242		Olivia *Kennard* WSr	British	134	88·0	22·5	12·0	Sunderland	1860	Woods		T. Smart	[illegible]
85099	243		Olivia *Williams* WSr	British	117	79·5	27·1	7·9	Cambridge N.B.	1882	C. & J. Robinson		John Gibson	St. J…
	244	N	Olivia *Thorsen 87-87* WBg	Norwegian	273	105·0	26·9	14·4	Grimstad	1869	N. Olsen Gj[illegible]		G. Huseland	[illegible]
	245		Olivia *Ginström* WSr	Russian	231	..	..	..	Dragsfjard	1874			A. L. Ginström	[illegible]
80134	246	R	Olivia A. Carrigon *Lowdry* WBn	British	350	127·1	29·4	12·4	Parrsboro', N.S.	1878	J. O. Mullin	[illegible]	P. Doyle	[illegible]
59426	247		Olivia A. Omullin *Rickard* WSr	British	149	88·4	25·1	10·7	Parrsboro', N.S.	1871			Patrick Doyle	[illegible]
	248		Oljona WSr	Russian	229	..	..	..		1860			I. Sidoroff	[illegible]
	249	R	Olof (ex Sofia Charlotta) *Hansson* WBk	Swedish	263	113·8	26·1	12·1	Finland	1849			A. Viklander	[illegible]
	250	V	Olof *Jansson 73-86* WSr	Swedish	223	103·3	26·7	11·8	Vätö	1867			P. Jansson	[illegible]

✻ = **Lloyd's Register**; D = Nederlandsche Vereeniging; G = Germanischer Lloyd; I = Registro Italiano; R = Ame…
A = Veritas Austro-Ungarico; H = Veritas Ellenico; N = Norske Veritas; V = Bureau V…

In the 4th Column, C = Composite (iron frame, wood planked); I = Iron; Stl = Steel; W = Wood.

The Dates of Metalling (Col. 11) are based on the records of the Books in which Vessels are respectively classed, or on the Reports of the Surveyors of Lloyd…

Further particulars of Vessels classed by Lloyd's Register will be found in Lloyd's Register Book and Lloyd's Yacht Register.

1	2	3. Nom. Capitaine et dates d'entrée au service des Armateurs actuels et à bord du navire. Ancien Nom.	4. Matériaux et Gréement	5. Pavillon.	6. Tonnage net.	7. Dimensions. Longueur	Largeur	Creux	8. Construit. Où.	9. Quand.	10. Par qui.	11. Date du doublage en métal	12. Armateurs ou Directeurs-Gérants.	13. Port d'Armement.
251		Olof *Lönngren -88*	WBn	Swedish	125	78·4	25·3	9·6	Hittis	1874	J.L.Isaksson		C.Söderström	Mariefred
252	V	Olof Glas *Anderson*	WBk	Swedish	984	146·9	29·9	17·3	Umeå	1871	O.Jonasson	Y 12.86	C.G.Moritz	Umeå
253	V	Olof Trygvason *Meyer 85-87* (ex Olaf Trygvason)	WBk	Swedish	850	175·6	33·6	20·4	Stavanger	1876	T.Tonning	12.81	C.A.Banck	Helsingborg
254	N	Oluf *Synness -76*	WBk	Norweg'n	342	124·7	26·2	15·3	Bremerhaven	1862	R.C.Rickmers	Y 5.82	I.O.Oyen	Drontheim
255	N	Oluf Nickelson *Olsen* (ex Harlow)	WBk	Norweg'n	348	127·2	29·2	14·5	Dorchester,N.B	1862	Palmer		Thore Nass	Tönsberg
256	V	Olustee *Nickerson* (WRECKED)	WBk	American	470	133·0	29·9	16·6	Machias,Me.	1873	H.Whitmore	Y 3.82	D.F.Hinckley	Boston,Mass.
257	❋	Olymp *Volz 83-88*	StlS	German	1621	258·8	38·7	22·7	Glasgow	1888	C.Connell&Co.		B.Wencke Söhne	Hamburg
258	V	Olympe *Piquot*	WSt	French	110	81·4	18·4	10·1	St.Vaast	1877	E.Costey		Mme.Piquot	La Hogue
259		Olympia *Mavroudis*	WSr	Greek	263	108·2	26·6	..	Chios	1879		Y 2.86	C.Mavroudis	Piræus
260		Olympia	WSr	Turkish	283	..	..	..	Chios	1879			P.Mavroudis	Chios
261	A	Olympos *Xenias*	WBg	Greek	328	109·9	27·6	..	Castellamare	1874			I.Calamaris	Syra
262	V	Oma (ex ...) *Durchman*	WBk	Russian	445	125·6	30·2	14·1	Kaskö	1876	J.Granskog		A.Durchman	Christinestad
263		Omaha *Door -88*	WSr	American	111	89·3	25·3	8·3	Hancock,Me.	1871	J.Wooster		N.H.Powers	Bucksport,Me.
264		Ombros	W	Turkish	301	..	..	..	Roumania	1884			Marmaralee Georgi Frangopolo	Marmara
265		Omega *Brown 82-85*	WBk	British	480	149·8	27·9	17·5	Workington	1869	J.T.Fell & Co.	Y 6.86	ChinaShipowners'Assn	Shanghai
266	A	Omega *Canepa -88* (WRECKED)	WBk	Italian	530	139·1	30·0	19·3	Bath,Me.	1857			F.Bozzo	Genoa
267		Omeo *Anthony* (ex Steamer)	I-mBk	British	789	213·2	30·5	16·7	Newcastle	1858	A.Leslie&Co.		W.H.Smith & Sons	Melbourne
268		Omer *Crowell*	WBn	British	181	96·7	25·0	12·2	Clare,N.S.	1864			T.Williams	Turks Islands
269		Ommelanden *Williams*	WBn	British	116	89·2	21·0	10·2	Groningen	1852	Kater		W.S.Kelly	Falmouth
270	A	Omonia *Pappas* (ex Denbighshire)	WBk	Greek	483	156·5	28·2	17·5	Neyland	1870	Gaddarn	Y 3.86	C.N.Pappadas	Syra
271	H	Omonia	WBn	Greek	269	104·9	26·7	17·7	Scopelos	1876	P.Barelas		G.Polymeropoulo	Scopelos
272	H	Omonia	WBg	Greek	228	93·8	25·8	17·5	Castello-rosso	1869	N.Caralli		G.Stavrikios	Syra
273		Omonia	WBg	Greek	204	86·6	25·6	..	Syra	1871			A.Inglessis	Syra
274		Omonia *Elefterion*	WBg	Greek	202	..	..	..	Syra	1866	P.Mbounnis		N.Elefterion	Zante
275	AH	Omonia *Caparoukas*	WBg	Greek	182	83·3	24·4	..	Syra	1869	S.Cufadaki	Y 12.85	M.Caparoukas	Syra
276	H	Omonia	WBg		175	84·0	24·3	..	Syra	1869				
277		Omonia	WBg	Greek	171	..	..	..	Galaxidi	1865	J.Canatas		G.Mitropoulos	Galaxidi
278	A	Omonia *Ablandis*	WSr	Greek	157	80·7	24·0	..	Cassos	1882	A.G.Flito	Y 2.84	V.Ablanidis	Syra
279	A	Omonia *Tsoukarakis*	WBg	Greek	149	76·8	22·6	..	Skiathos	1875	C.G.Mitylinaios		N.& Y.Tsoukarakis	Skiathos
280	A	Omonia	WSr	Greek	145	75·5	23·5	..	Skiathos	1883	C.G.Mitylinaios		N.D.Tsamasfyris	Skiathos
281		Omonia	WBg	Greek	141	75·5	23·6	..	Syra	1866	S.Koufoudakis		E.Karipis	Andros
282	AH	Omonia *Varlami*	WBg	Greek	134	81·4	23·3	12·3	Galaxidi	1879	C.Papapetros		L.Bisbichi	Galaxidi
283	H	Omonia	WBg		132	76·4	21·9	..	Spezzia	1860				
284		Omonia	WBg	Greek	122	73·8	21·0	..	Syra	1866			D.Kouris	Volo
285		Omonia *Papasmanuele*	WBk	Russian	183	83·0	23·6	15·9	Cassos	1869	M.Marcantonio		G.Fillipovich & Demirge	Odessa
286		Omonia *Franco* (ex ...)	WBk	Turkish	410	..	..	..	Dundee	1860		Y 9.82	G.Franco & Co.	Chios
287		Omonia *Deino Yanni*	WBg	Turkish	113	..	..	..	Greece	1882			Yorghi Velideh Vrioniki	Constantinople
288		Omorkkhia	WBg	Turkish	137	..	..	..	Syra	1869			Ligori Lefter & Co.	Constantinople
289		On Time *Kempton -86*	WSr	British	124	86·7	24·4	9·3	Pt.Medway,NS	1883	J.Hutt		John H.Harlow	Liverpool,N.S.
290	R	Onalaska *Griggs 65-78*	WBn	American	452	124·4	29·1	17·4	Rockland,Me.	1869	S.Starrett	Y 12.87	J.S.Emery & Co.	Boston,Mass.
291	❋	Onaway *Anderson 88-*	WBk	British	420	139·2	31·8	12·9	Sackville,N.B.	1877	E.W.Ogden	Y 1.86	J.Wood	Sackville,N.B.
292	V	Onaway *Crickett 80-81*	WBk	American	886	173·9	36·2	20·0	Yarmouth,Me.	1883	Loring,Chadsey [&Co.	Y 7.88	B.Webster	Portland,Me.
293		Oncle Felix	WS	French	854	164·5	33·2	21·5	Liverpool	1859	Chaloner		H.Ravot & Co.	Havre
294	D	Onderneming *Kuiper* (ex Dania)	WBg	Dutch	280	97·8	28·7	15·4	Pillau	1848			J.Alta Lzn.	Harlingen
295	❋	One *Coates -87* (BROKEN UP)	WBk	British	540	141·0	30·7	18·5	P.E.Island	1870	Richards	Y 5.86	S.J.Roberts	Swansea
296		Oneata *Rousseau -87*	WBk	British	589	136·0	31·6	18·2	Quebec	1869	McKay&Warne	Y 2.83	K.C.Maguire & Co.	Quebec
297	❋	Oneida (ex Steamer) *Clyma*	IS	British	2161	306·8	38·9	26·7	Greenock	1855	Scott&Co.		Ship "Oneida" Co. Ld. (H. Chaplin & Co.)	London
298	RV	Oneida *Erickson*	WS	American	1074	179·0	36·5	23·9	Searsport,Me.	1866	W.McGilvery	Y 5.87	L.Sloss	San Francisco
299	R	Oneota *Hick -88*	WS	British	1047	171·8	36·7	23·1	St.John,N.B.	1868	Milledge	Y 11.88	A.Armstrong	Liverpool
300														

❋=Lloyd's Register; D=Nederlandsche Vereeniging; G=Germanischer Lloyd; I=Registro Italiano; R=American Record; A=Veritas Austro-Ungarico; H=Veritas Ellenico; N=Norske Veritas; V=Bureau Veritas.

Dans la 4me Colonne, C=Composite (membrure en fer, bordé en bois); I=Fer; Stl=Acier; W=Bois.

Les Dates du doublage en métal (**Col. 11**) ressortent des livres dans lesquels les navires sont respectivement classés ou des rapports des Inspecteurs du Lloyd's Register.

Pour plus amples renseignements sur les navires classés par le Lloyd's Register, voir le Lloyd's Register Book et le Lloyd's Yacht Register.

Official No.	No. in Book.	Registry in which classed.	Ship's Name. Master, and Dates of Appointment to present Owner's Service and to Ship. Ship's Late Name.	Material and Rig.	Flag.	Net Tonnage.	Length.	Breadth.	Depth.	Built. Where.	When.	By whom.		Owners or Managers.
	301		Onkahuia *Rasmussen* (ex Caroline Amalie, ex Anne)	W Bg	Danish	192	90·4	21·7	14·0	Blyth	1858			C.H.Carlsen
	302	N	Onkel Christian *Mortensen*	W Bk	Norweg'n	315	105·1	[illegible]	16·6	Fiume	1847			F.J.Holst's Enke
	303		Onnetar	W Sr	Russian	110	..	..	..		1875			C.Söderholm
	304	N	Onni *Abrahamsen*	W Bk	Norweg'n	516	129·1	32·0	18·6	Grimstad	1864	S.Johnsen		Jens Gundersen
	305	V	Onni *Galenius 79-81*	W Bk	Russian	826	167·0	35·1	20·2	Uleaborg	1871	J.Kjeldström		G. & C.Bergbom
	306		Onni *Wirlenen 80-80*	W Bk	Russian	402	..	..	..	Kimito	1857			A.Abrahamsson
	307	R	Onni *Grönblom*	W Bk	Russian	379	127·3	28·5	15·7	Pytrolax	1875	T.Wilk		A.Jusselius
	308		Onni *Anderson -77*	W Bkn	Russian	279	125·8	29·5	12·0	Nagu	1872	A.Henriksson		K.L.Lindberg
	309		Onni *Kohonen*	W Sr	Russian	208	..	..	..	Impilaks	1882			P.Wonninen
	310		Onni *Karvonen*	W Sr	Russian	121	..	..	..		1872			J.Rapell
	311		Onni *Abrahamsson*	W Sp	Russian	108	..	..	..		1874			M.Jansson
18078	312		Onrust *Reed*	W Sr	American	186	109·9	31·2	8·7	Westfield NY.	18[illegible]			C.F.Clark
	313	N	Onsö *Vielsen 82-82* (ex Hal. Athos)	W Bk	Norweg'n	450	139·5	29·5	17·6	Shoreham	1864	Balley		N.L.Nielsen
72100	314	V	Ontario *White -87*	W Bk	British	825	160·7	35·3	20·1	Hantsport NS	1876	J.Davidson		G.W.Churchill
33171	315		Ontario	W Sr	British	338	133·6	23·5	11·9	Quebec	1851	A.Gibson&Co		Hy.Waters
61120	316		Ontario	W Sr	British	161	118·5	24·0	10·0	S'th.Bay,Ont.	1869	J.D.Tait		N.Dyment
77775	317		Ontario	W Sr	British	150	105·0	23·0	9·5	Goderich,Ont.	1867			H.Horton
66351	318		Ontario *Willcock -87*	W Bn	British	129	98·5	23·8	13·8	P.E.Island	1873	Hyndman		S.Slade
75431	319		Ontario	W	British	107	94·6	17·4	7·5	FortAnn,U.S.	1874	M.Farran		R.McLeod
68437	320	❋	Onward *Morris -88* (ex James Smith)	W Bk	British	559	147·1	31·8	16·5	Picton,N.S.	1875	McDonald		W.Whyte & Co.
42743	321		Onward	W	British	220	108·4	20·0	10·4	Quebec	1854			Wm.Jones
7305	322		Onward *Day 69-69*	W Sr	British	119	74·7	21·8	12·1	Bristol	1849			T.Stewart & Co.
79053	323	❋	Onward *Cooper*	W K Barge	British	77	81·8	20·5	6·2	Ipswich	1878	Bayley		R.Peck & Co.
27148	324	❋	Onward *Miles -88*	W K	British	72	71·3	18·1	8·7	Wells	1859	Tyrrell		W.Scott
	325	N	Onward *Frolich 73-88*	W Bk	Norweg'n	558	119·1	29·3	18·7	Dundee	1860	Calman		Ole Baastad
51153	326	❋	Onyx *Paterson*	I Bk	British	403	136·6	25·8	16·5	Sunderland	1864	J.Laing		H.Guthrie
23946	327		Onyx *Stewart* (BROKEN UP)	W Sr	British	190	91·7	22·8	14·5	Greenock	1847			J.Walker
85353	328	R	Onyx *Morrell -87*	W Sr	British	139	93·2	24·4	9·8	Tusket,N.S.	1884	E.F.Parker		E.F.Parker
42985	329		Onyx *Paul 85-85*	W Bn	British	132	95·2	22·7	10·2	P.E.Island	1863	Longworth		G.H.Gann
	330		Oonia	W Bg	Turkish	330	..	..	..	Oonia	1868			Iskenderich Tjiarladza Rustam Effendi & Co.
	331		Oonia	W Bg	Turkish	290	..	..	..	Oonia	1866			[illegible] Ali Oglou Hajee Hassan & Co.
	332		Oonia	W Bg	Turkish	285	..	..	..	Oonia	1866			OonialeeShakir&Co
	333		Oonia	W Bg	Turkish	203	..	..	..	Oonia	1870			[illegible]richlee Adran Oglou Kirkor & Co.
	334		Oonia	W Bg	Turkish	195	..	..	..	Oonia	1867			Oonialee Hajee Hassan & Co.
	335		Oonia	W Bg	Turkish	143	..	..	..	Oonia	1869			Oonialee Ahmed Tefyk & Co.
	336		Oonia	W Bg	Turkish	125	..	..	..	Oonia	1869			Oonialee Mutal Kirkor Oglou [illegible]
	337		Oonia	W Bg	Turkish	125	..	..	..	Oonia	1868			[illegible] Ibrahim Jazid [illegible]
	338		Oonia	W Bg	Turkish	112	..	..	..	Oonia	1868			[illegible] Kji Oglou & Sons
	339		Oonia	W Bg	Turkish	112	..	..	..	Oonia	1868			[illegible] Tappaz Oglou Sawah & Co.
	340		Oonia	W Bg	Turkish	112	..	..	..	Oonia	1871			Saleh Bey & Co.
	341	❋DV	Oostenburg *Visser 86-87*	I Bk	Dutch	1007	182·0	35·0	22·5	Amsterdam	1888	Huijgens & Van Gelder		W.A.Huijgens
	342		Oostzee Packet *Smit* (WRECKED) (ex Josephine)	W Bn	Dutch	237	105·5	27·6	12·2	Truro,N.S.	1861			
	343		Oostzee Packet II (ex A. M. R.)	W Bn	German	262	110·5	28·7	12·8	Arichat,N.S.	1867			
59050	344		Opah *Burns -82*	W Bn	British	113	84·2	23·3	9·9	P.E.Island	1868	E.Alexander		W.Corey
3634	345		Opal *Sherry*	W Bg	British	154	86·7	23·5	13·5	Greenock	1845			G.Hoppet
73718	346	❋	Opawa *Hamon 85-85*	I S	British	1076	215·2	34·0	20·4	Glasgow	1876	A.Stephen&Sns		New Zealand Shipping Co. Ld.
18827	347		Open Sea	W Sr	American	175	91·5	25·1	8·8	Brewer,Me.	1856			E.A.Wyman
47522	348	❋	Ophelia *Fetherstonhaugh*	I Bk	British	1134	217·5	35·7	22·6	PortGlasgow	1863	J.Reid&Co.		C.T.Bowring & Co.
	349		Ophelia *Rosenblad -88*	W Bk	Russian	461	..	..	..	Abo	1876			A.J.Borisoff
30730	350		Ophir	W S	British	847	159·3	35·0	20·0	Bangkok	1858			Hadjee Jeelani Bin [illegible]

❋ = **Lloyd's Register**; D = Nederlandsche Vereeniging; G = Germanischer Lloyd; I = Registro Italiano; R = American [Record]; A = Veritas Austro-Ungarico; H = Veritas Ellenico; N = Norske Veritas; V = Bureau Veritas.

In the 4th Column, C = Composite (iron frame, wood planked); I = Iron; Stl = Steel; W = Wood.

The Dates of Metalling (**Col. 11**) are based on the records of the Books in which Vessels are respectively classed, or on the Reports of the Surveyors of [illegible]

Further particulars of Vessels classed by Lloyd's Register will be found in Lloyd's Register Book and Lloyd's Yacht Register.

		Nom. Capitaine et dates d'entrée en service des Armateurs actuels et à bord du navire. Ancien Nom.	Matériaux et Gréement.	Pavillon.	Tonnage net.	Dimensions. Longueur.	Largeur.	Creux.	Construit. Où.	Quand.	Par qui.	Date du doublage	Armateurs ou Directeurs-Gérants.	Port d'Armement.
351		Ophir *Zacherich*	W Bn	Austrian	268	102·3	25·5	11·7	Lussinpiccolo	1877	M.A.Starcich		G.Giuricich di A	Lussinpiccolo
352		Ophir *Bruquiro -88*	W Bk	Ecuador	778	168·8	32·8	21·2	SestriPonente	1874	Luigi		E.De Sabla & Co.	
353	X	Ophir *Jensen*	W Bk	Norweg'n	439	135·1	30·3	16·1	Risör	1876	O.OlsenHaugen		O.Caspersen	Risör
354	X	Ophir *Sorensen -88*	W Bk	Norweg'n	443	128·5	30·1	16·4	Arendal	1874	[illegible]		O.Christensen	Arendal
355	X	Ophir *Thoresen*	W Bg	Norweg'n	259	116·8	28·1	12·8	Porsgrund	1875			G.Knudsen	Porsgrund
356	X	Ophir *Borgenvin -88*	W Sr	Norweg'n	107	90·9	23·3	9·7	Stordöen, Norway	1882	J. [illegible]		P.Amlie	Haugesund
357		Opit	W Sr	Russian	235	..	..	..	Nicolaieff	1861			Giulia Gellich	Odessa
358	V	Ora *Eckhoff*	W Sr	German	118	87·4	20·7	9·8	Nordloh	1880	[illegible]		J.F.Eckhoff	Iheringsfehn
359	X	Ora et Labora *[illegible] -86*	W Bg	Norweg'n	198	100·9	26·9	12·3	Sandefjord	1868			P.Tobiasen	Sandefjord
360		Oradd *Pettersson*	W Sr	Swedish	214	101·0	25·2	13·6	Hälsvoll	1852			F.Lundman	Norrtelje
361		Orage	W Bg	French	168	..	..	..	Jersey	1859			A.Leroux	St.Malo
362	✻	Orange Grove *Whiting -85*	I Bk	British	385	116·6	25·6	15·9	Glasgow	1861	B'clayCurle&Co		I.J.Whitting (R.Fricker, Manager)	Melbourne
363	A	Orao *Zaccevich*	W Sr	Austrian	171	96·8	23·1	12·1	Lussinpiccolo	1885	O.Richulch		G.Zaccevich	Lussinpiccolo
364	✻	Orari *Miller 85-81*	I S	British	1011	204·1	34·2	20·0	Newcastle	1875	Palmers'S&I Co.		New Zealand Shipping Co. Ld.	London
365		Oratios Coupas	W Bg	Greek	306	108·2	27·9	..	Syra	1885			N.Koupas	Syra
366		Orazio Primo *Cherchi*	W Bg	Italian	194	91·9	25·4	15·9	Varazze	1854			A.Tassara	Genoa
367	✻	Oreadian *Smith 74-74*	W Sr	British	76	75·0	18·6	8·9	Stromness	1874	Copeland		J.Sinclair	Kirkwall
368	✻	Orchid *Dennis 77-87*	W Bk	British	222	129·0	22·6	12·9	Bristol	1861	Hill		G.H.Bridges	Bristol
369	✻	Orchomene *Paxley 83-83*	I S	British	1542	233·3	37·3	22·7	Liverpool	1884	W H Potter&Sns		H.Fernie & Sons	Liverpool
370	✻	Orco (Missano) *Fava*	W Bk	Italian	871	163·9	32·0	22·7	Sampierdar'na	1871	Casanova		Casanova	Genoa
371		Ordon	W Bg	Turkish	130	..	..	..	Ordon	1868			Bulbulchileo Oglou Hatchik & Co.	Bulbuljeh
372	✻	Ordovic *Austin*	W Bk	British	826	168·6	33·8	21·3	Pt.Dinorwic	1877	Jones		J.Austin	Carnarvon
373	X	Ore *Ostensen -88* (ex John Milton)	W Bk	Norweg'n	614	167·3	30·5	18·2	Jersey	1865	F.C.Clarke		J.T.Östensen	Farsund
374	✻	Oread *Anderson 72-82*	I Bk	British	797	189·7	32·0	19·1	Liverpool	1882	T.Royden&Sons		J.Newton & Co.	Liverpool
375	R	Oreala (ex Kong Most)	W Sr	British	187	112·0	23·8	12·4	Gloucester	1863	Pickersgill		J.Kent	London
376	✻	Orealla *Gifford 87-7*	I S	British	1763	250·5	40·2	24·0	Liverpool	1882	T.Royden&Sons		Sandbach,Tinne & Co.	Liverpool
377	I	Oreb *Queirolo -88*	W Bk	Italian	773	164·0	32·3	20·9	Voltri	1875	Frat.Piva		Gerolamo Morice	Genoa
378		Oregina (DISMANTLED)	W Bg	Italian	121	74·5	23·1	13·1	Varazze	1854	G.Canepa		F.Boggiano	Genoa
379	✻	Oregon *Henderson 85-85*	I Bk	British	891	194·4	31·7	18·5	Sunderland	1875	[illegible]		R.Kinnes	Dundee
380	RV	Oregon *Pritchard 87-87*	W S	American	1364	205·6	39·9	24·0	Bath,Me.	1875	W.Rogers		W.E.Mighell	San Francisco
381	DX	Oregon *Thorsen*	W S	Norweg'n	872	187·2	35·8	21·1	Haugesund	1874	J.[illegible]		E.Berentsen	Stavanger
382	✻	Orellana *Gibb 81-85*	Stl Bk	British	866	198·6	33·7	19·8	PortGlasgow	1885	J.Reid & Co.		Nicholson & McGill	Liverpool
383		Orénoque (ex MariaVittoria)	W Sr	British	117	89·0	22·6	10·6	Oneglia	1874			ReneA.J.M.W.Mercier	Port Louis
384	X	Orest *Christoffersen*	W Bg	Norweg'n	278	102·6	27·0	15·5	Kalix,Sweden	1839			O.Christoffersen	Tönsberg
385		Orest	W Sr	Russian	139	89·1	26·5	11·2	Körbis	1880	A.Speak		W.von Aderkas	Riga
386	V	Oreste *Mequa*	W Bk	Italian	363	115·1	28·5	11·1	Alimuri	1865	[illegible]		P.Chirico	Castell.di Stab.
387		Oreste *Paolini*	W Sr	Italian	127	92·8	23·1	12·3	Mariaura	1848	F.[illegible] Baschi		B.Paliani & D.Fascia	Civitta Vecchia
388		Oreste *Donna -87*	W Bn	Italian	116	84·0	22·9	11·5	[illegible]	1858	B.[illegible]		F.lli Goglione	Naples
389	✻	Orestes *Bolton 81-85*	W K	British	57	64·7	19·0	8·4	Plymouth	1885	D.Banks & Co.		D.Banks & Co.	Plymouth
390		Öresund *Petersen -74*	W Sr	Swedish	137	90·8	21·3	10·0	Westervik	1853	A.C.Engström		C.O.Schele	Halmstad
391	✻	Oriana *Mosey 86-86*	I Bk	British	987	201·8	33·2	21·3	[illegible]	1864	Scott & Co.		Trinder,Anderson & Co.	London
392	R	Orient *Sweeny*	W S	British	926	191·0	35·8	21·0	[illegible]	1853	[illegible]		Cox Bros.	Waterford
393		Orient	W Bn	British	124	88·0	20·6	11·1	P.E.Island	1852			B.T.H.Gould	St.Johns,N.F.L.
394		Orient *McDonald*	W Bg	American	296	126·0	30·0	10·6	Brunswick,Me	1865			A.M.Simpson	San Francisco
395	✻	Orient *Koehl -89*	Stl S	German	1619	258·8	38·7	22·7	Glasgow	1889	C.Connell & Co.		A.Evers &Classen	Hamburg
396	V	Orient *Godbert 85-88*	W Bk	German	461	142·4	29·8	17·0	Flensburg	1878	E.Weelemann		H.A.Petersen	Flensburg
397	✻X	Orient *Falkenberg -88*	W Bk	Norweg'n	498	151·9	29·4	16·5	Dundee	1877	DundeeS.B.Co.		N.Nilsen	Arendal
398	X	Orient *Larsen 77-77*	W Bk	Norweg'n	374	133·9	28·4	14·3	St Martin's N B	1856			L.Larsen	Flekkefjord
399	V	Orient *Ahlström -1*	W Bk	Russian	442	132·4	27·0	15·9	Jacobstad	1868	H.Blaese		P.Malm,jun.	Jacobstad
400	V	Orient *Kronall -88*	W Bk	Swedish	367	134·1	26·8	12·8	Gefle	1884	O.A.Brolin		J.Rettig	Gefle

✻=Lloyd's Register; D=Nederlandsche Vereeniging; G=Germanischer Lloyd; I=Registro Italiano; R=American Record; A=Veritas Austro-Ungarico; H=Veritas Ellenico; N=Norske Veritas; V=Bureau Veritas.

Dans la 4me Colonne, C=Composite (membrure en fer, bordé en bois); I=Fer; Stl=Acier; W=Bois.

Les dates de doublage en métal (Col. 11) ressortent des livres dans lesquels les navires sont respectivement classés ou des rapports des Inspecteurs du Lloyd's Register.

Pour plus amples renseignements sur les navires classés par le Lloyd's Register, voir le Lloyd's Register Book et le Lloyd's Yacht Register.

1 Official No.	2 No. in Book	3 Society in which classed	4 Ship's Name. (Ship's Late Name.)	Master, and Dates of Appointment to present Owner's Service and to Ship.	Material and Rig.	5 Flag.	6 Net Tonnage	7 Dimensions. Length	Breadth	Depth	8 Built. Where.	9 When	10 By whom.	11 Date of Metalling	12 Owners or Managers.
78581	401		Oriental (FOUNDERED)		WSr	British	329	136·0	25·6	12·3	Garden Is. Ont.	1864	H. Roney		Wm. Lewis
34116	402		Oriental (BROKEN UP)		W Bn	British	114	92·0	21·2	11·5	Tatamagouche N.S.	1857			G. Spencer
19418	403		Oriental	Slater 80-84	WS	American	1625	220·1	42·2	24·9	Bath. Me.	1874	E. & A. Sewall	Y 12.82	S. Blair
	404	G	Oriental	Waldmann 86-	W Bn	German	214	103·0	24·5	12·7	Elmshorn	1873	D. Kremer	Y 10.86	E. W. Ohrloff & Sohn
54022	405	✻	Oriente	Warn 88-88	I Bk	British	595	171·6	27·9	18·0	Liverpool	1866	T. Vernon & Son		Nicholson & McGill
1866	406		Origine	Gastaldi	W Bk	Italian	444	132·8	30·0	19·3	Voltri	1869	G. Fava	Y 7.86	Carlotta Isnardi
246	407	A	Origine (ex I. Antonio)	Cocconardi	W Bg	Italian	136	83·6	21·7	11·7	Gaeta	1865	P. Gallinaro		Capodanno & Di Campo
87601	408	V	Orinoco	Upham 81-83	W 3m Sr	British	298	124·0	28·7	12·4	Portland, N.B.	1883	A. B. & W. S. Rod [docks	Y 2.86	C. A. Palmer
	409		Orinoco		W Bn	Spanish	284	124·0	25·5	12·1	Grohn	1862			Gama Hermanos
88427	410	R	Oriole	Seacord -86	WSr	British	124	89·0	26·7	7·9	St. John, N.B.	1884	J. C. Ferguson		John Connor
19263	411		Oriole	Crapo 86-	WSr	American	238	105·6	29·8	9·0	Waldsboro', Me	1868			T. Crapo
8251	412		Orion		W Bn	British	116	80·0	21·9	10·8	Tatamagouche N.S.	1846			J. R. Suiter
77764	413		Orion		WSr	British	100	82·2	23·5	9·5	Shelbourne N.S.	1880	J. Brown		Edgar Penny
36025	414	✻	Orion	Ships 84-84	WSr	British	95	82·0	21·0	10·4	Dysart	1868	Watt		W. Ships
19487	415		Orion	Chirdensen	WSr	American	112	92·0	27·7	7·2	Humboldt, Cal.	1878			E. P. Nissen
172	416	A	Orion	Staparich	W Bk	Austrian	323	111·5	26·5	16·7	Lussinpiccolo	1869	G. S. Cattarinich	Y 7.86	G. A. Haracich fu G.
	417	R V	Orion	Clausen -72	W Bk	Danish	333	122·0	26·3	14·0	Hamburg	1862	J. Toosbuy		C. J. Borg
	418	G R	Orion (ex William Leckie)		W Bk	Dutch	416	117·5	27·5	18·5	Sunderland	1855	G. Worthy	Y 7.86	G. G. Albers, sr.
	419	G V	Orion	Forma 87-87	W Bn	Dutch	215	97·1	24·4	12·0	Varel	1868	C. Schwoon	Y 3.86	A. Forma
	420	V	Orion	Grünberg -87	W Bk	German	519	123·0	29·5	17·9	Memel	1866	B. Pieper		Siebert & Alexander
	421	✻ V	Orion (ex Golden Wave)	Fock 85-85	W Bk	German	343	120·6	27·5	17·0	Sunderland	1869	R. Thompson & Sons	Y 8.8	J. O. Williamson
	422	N	Orion (ex Inverauld)	Pettersen	WS	Norweg'n	1237	222·6	36·9	21·3	Aberdeen	1874	Humphrey & Co.	Y 11.86	O. B. Sörensen
	423	N R	Orion (ex F. H. Cockburn)	Mathiesen	W Bk	Norweg'n	1009	170·5	36·7	25·5	Geb.	1864		Y 6.87	A. Mathiesen
	424	N V	Orion	Bonnerie -87	W Bk	Norweg'n	656	152·0	32·1	19·1	Grimstad	1877	S. Johnsen	Y 11.86	Peter L. Lund
	425	N V	Orion (ex Accessory)	Bergersen -88	W 3m Sr	Norweg'n	585	149·9	30·3	18·8	Quebec	1863	Samson		J. Bergersen (J. T. Andersen)
	426		Orion		W Bk	Norweg'n	493	148·3	31·3	16·5	Wasa	1860			S. Solberg
	427	N	Orion	Johannesen 86-86	W Bk	Norweg'n	329	131·2	28·6	12·4	Lillesand	1886		Y 6.86	O. & P. Knudsen
	428	N	Orion	Jorgensen	W Bn	Norweg'n	267	112·2	26·8	13·6	Stavanger	1866	R. O. Bœrem	Y 4.8	E. Berentsen
	429	N	Orion	Jensen	W Bg	Norweg'n	155	97·8	22·3	10·9	Sappemeer	1857			Nils Jensen
	430		Orion	Mattson	W Bg	Russian	276	105·0	26·4	17·4	Sunderland	1861			C. Björklund
	431		Orion	Nordfeldt -84	W Bn	Swedish	160	97·1	23·7	10·5	Sundsvall	1863	J. A. Nordström		B. J. Törngren
2662	432	I	Orione	Schiaffino	W Bk	Italian	647	154·4	32·2	21·2	Savona	1876	Em. Calamaro	Y 6.85	Antola & Schiaffino
	433		Orionis		WSr	Turkish	230	..	..	..	Greece	1878			Yorghi Zaffrapolo
	434		Orisia		WSr	Brazilian		..	..	..					A. J. Dantas
63437	435	✻	Orissa	Penny 83-88	I S	British	1199	225·2	35·3	21·4	Glasgow	1869	J. & G. Thomson		J. Kerr & Co.
78709	436	✻	Orissa (ex Tollie Whitaker)	Winther 86-87	W Bn	British	182	109·9	24·4	11·8	Carrickfergus	1882	P. Rodgers	Y 5.86	J. M. Dean
10504	437		Orixa (LOST)	Lund	W Bk	British	492	..	27·2	18·9	Liverpool	1836			C. C. Lewis, jun.
	438		Orixa (LIGHTER)	Olsen -87	W Bk	Norweg'n	425	134·0	28·9	17·4	Bordeaux	1861	Raymond frères		M. Vauvert
82540	439	✻	Orlando	Foot	W K	British	82	74·5	20·8	10·3	Brixham	1882	J. W. & A. Upham		C. Hellyer
18950	440	R	Orlando	Knaebel	WSr	American	189	109·5	25·7	10·3	Newburyport, Mass.	1856	Greene & Son		Wm. F. Greene & Son
58430	441		Orloff	Boudrot	W Bn	British	235	112·3	28·9	11·3	Parrsboro' N.S.	1872	L. E. Graham		C. A. Boudrot
	44.		Ormelie (CONDEMNED)	Eriksen	W Bk	Norweg'n	516	130·7	29·8	19·6	Glasgow	1847			O. A. Johansen
19065	443		Ormus	Frost -88	W Bk	American	461	133·5	29·3	17·3	Pembroke, Me.	1866	Parrott	Y 4.86	E. A. Shadsford
74443	444		Ornate	Holmes	W K	British	122	101·3	24·1	8·2	Harwich	1875	Vaux		J. Holmes
	445	N	Örne	Johannesen 83-86	WSr	Norweg'n	245	115·1	26·5	11·2	Carlshamn	1878	D. R. Andersson	Y 12.85	Chr. Th. Boe
	446	V	Ørnen	Larsen	WSr	Danish	116	92·1	20·7	11·0	Faaborg	1869	B. Dyreborg		Ørum & Wulff
	447		Ornen (ex Scandinavia ex Majestic)	Nielsen 86-86	W Bk	Norweg'n	781	150·3	33·6	22·2	Bath. Me.	1856		Y 4.80	A. C. Gogstad
	448	N	Ornen	Nielsen -83	W Bk	Norweg'n	534	154·8	31·8	18·4	Bergen	1869	Brunchorst & Dekke	Y 1.87	Chr. Isdahl
	449	N	Örnen	Hansen 86-86	W Bk	Norweg'n	503	131·0	32·5	17·9	Hamburg	1853			E. Rasmussen
	450		Ornen		W Bg	Norweg'n	246	88·0	26·0	15·4	Kragerö	1836			J. Olsen

✻ = Lloyd's Register; D = Nederlandsche Vereeniging; G = Germanischer Lloyd; I = Registro Italiano; R = American
A = Veritas Austro-Ungarico; H = Veritas Ellenico; N = Norske Veritas; V = Bureau

In the 4th Column, C = Composite (iron frame, wood planked); I = Iron; Stl = Steel; W = Wood.

The Dates of Metalling (Col. 11) are based on the records of the Books in which Vessels are respectively classed, or on the Reports of the Surveyors of L...

Further particulars of Vessels classed by Lloyd's Register will be found in Lloyd's Register Book and Lloyd's Yacht Register.

No.	Reg.	Nom. Matériaux et Gréement. Capitaine et dates d'entrée au service des Armateurs actuels et à bord du navire. Ancien Nom.			Pavillon.	Tonnage net.	Dimensions. Longueur.	Largeur.	Creux.	Construit Où.	Quand.	Par qui.	Date du doublage en métal.	Armateurs ou Directeurs-Gérants.	Port d'Armement.
451	N	Ornen	Michaelsen	WSr	Norweg'n	132	101·7	22·2	11·8	Stavanger	1857			H.M.Slettebö	Egersund
452	V	Örnen	Kiserman 73-77	WBk	Swedish	1006	177·1	33·1	22·2	[illegible]	1877	F.O.Bergström	[illegible]	Rederi-Akt.-Bolaget "Örnen" (J.F.Strömberg)	Gothenburg
453		Ornon (ex Aldebaran ex Meteor)	Bakkbom	WBkn	Swedish	300	107·9	21·9	15·6	Dartmouth,N.S.	1856			L.Näsholm	Hernösand
454	✱D	Oronsay	(ex Hogarth) Pinton	IS	British	1569	235·9	37·5	22·3	Liverpool	1859	T.Vernon&Son		J.Shepherd & Co.	London
455	✱	Orontes	Bain 79-81	IS	British	1318	231·8	36·1	22·5	Aberdeen	1881	W.Hood&Co.		G.Thompson & Co.	Aberdeen
456	R	Orontes	Hanson 86-87	WBk	British	740	160·8	33·2	19·3	Dorchester NB	1878	J.Harris	[illegible]	Wm.Thomson & Co.	St.John.N.B.
457		Orozimbo	Gupfill 77-	WSr	American	154	97·0	25·0	9·0	Gouldsbro'Me.	1854			A.H.Sawyer	Calais,Me.
458	✱	Orpheus	Young 79-82	IS	British	1462	253·0	38·1	22·8	Greenock	1874	R.Steele&Co.		Baine & Johnston	Greenock
459	G	Orpheus (ex G.[illegible])	Kramel -87	WBk	German	910	167·3	35·1	22·3	Boston	1860		[illegible]	D.Schilling	Bremen
460	GV	Orpheus	Warns 80-85	WBg	German	261	112·2	25·9	11·6	Brake	1869	J.D.Behrens	[illegible]	Adolph Schiff	Elsfleth
461	A	Orpheus		WBg	Greek	371	105·3	30·0	..	Syra	1878	B.Manassis		E.Anastasion	Syra
462		Orrie V. Drisko	Drisko	W3mSr	American	305	120·0	30·0	13·6	Columbia Falls, Me.	1873		[illegible]	J.Richardson	Machias,Me.
463	V	Orskar	Eastrom	WBk	Swedish	439	136·1	27·9	15·1	Grasön	1875	H.Söderman	[illegible]	A.A.Andersson&Co.	Gefle
464	IV	Orsola	Peturso	WBk	Italian	539	141·0	28·7	19·3	Cassano	1875	F.de Rosa	[illegible]	F[lli]. Caliero	Castell. di Stab.
465	DV	Ortelius	Jansonius -88	WBk	Dutch	636	136·1	31·1	17·1	Alblasserdam	1857	C.Smit	[illegible]	H.J.Plant & Co.	Schiedam
466	✱D	Orthes	Mynose 88-88	IS	British	1206	228·0	35·8	21·4	Glasgow	1877	J.&G.Thomson		J.Hardie & Co.	Glasgow
467		Orthodoxia	Coutzis	WBg	Greek	217	91·8	23·3	16·7	Spezzia	1870	G.Micheli		P.G.Coutzis	Spezzia
468		Orthodoxia		WBg	Turkish	147	..	..	..	Vassilikos	1878			Vassilikosiee Yakbi Nicola	Vassilikos
469	N	Orton	Abrahamsen -88	WSr	Norweg'n	157	83·7	21·6	11·7	Wivenhoe	1842			H.Cappelen's Enke	Skien
470	N	Örvar Odd	Torkelsen	WBk	Norweg'n	512	140·8	31·4	17·8	Stavanger	1865		[illegible]	Brödr.Berner	Stavanger
471	V	Orvarodd (ex St.Petersb)	Krog 87-88		Norweg'n	686	129·7	33·1	22·3	Dramm	1868	C.Linck		A.Krog	Christiania
472		Orwell	Veale -83	W3mSr	British	141	95·1	19·9	10·8	Ipswich	1811			J.Veale	Ipswich
473		Oryx	Cell	WBk	British	270	111·5	25·1	15·2	Dundee	1856			Great Grimsby Ice Co., Ld.	Grimsby
474	✱	Osaka	Jones	CBk	British	517	165·0	30·1	17·2	Sunderland	1869	W.Pile&Co.	[illegible]	T.Roberts	Llanelly
475	I	Osanna	Tanazzi	WBk	Italian	507	134·8	30·8	20·2	Cassano	1871	F.De Rosa	[illegible]	F.& S.Maresca	Castellamare
476	V	Osberga	McKenzie	WBk	British	1158	195·0	37·6	22·5	Maitland,N.S.	1884	A.McDougall	[illegible]	StairsSon&Morrden	Halifax,N.S.
477	✱	Osburgha	Cook 69-75	IBkn	British	346	146·7	26·1	13·0	Glasgow	1875	A.Stephen & Sns		W.Cook	Dundee
478		Oscar (ex Maggie)	Beis	WBn	Brazilian	296	..	..	..	Liverpool	1874		Y 82	A.d'Oliveira Maia	Pernambuco
479	DV	Oscar	le Moult	WBk	German	735	157·4	31·2	19·7	Kiel	1876	A.Conradi&Son	Y 8.85	A.Strantzen	Hamburg
480	N	Oscar (ex Atlantis ex King Dove)	Nielsen -88	WBk	Norweg'n	797	151·1	33·8	20·7	Fairhaven, Conn.	1861		[illegible]	P.Christiansen	Sarpsborg
481	N	Oscar	Larsen 87-87	WBk	Norweg'n	365	111·8	28·0	17·9	Fairhaven, Conn.	1832		[illegible]	L.Dahl (in liquidation) & J.L.Dahl	Tönsberg
482	N	Oscar	Johansen	WBg	Norweg'n	310	105·1	28·1	15·9	Brahestad	1848	J.Danielsen		J.Johansen	Drammen
483	N	Oscar	Knudsen -82	W3mSr	Norweg'n	298	109·9	29·2	14·5	Christinestad	1867			Chr.Gjertsen	Porsgrund
484		Oscar (LIGHTER)	Danielsen -87 (ex Adonis)	WBg	Norweg'n	217	92·1	21·2	14·4	Tönsberg	1834			G.Uhl	Fredriksstad
485	N	Oscar	Henriksen 87-87	WBg	Norweg'n	178	107·8	24·2	11·6	Drontheim	1860	J.Jensen		H.Henriksen	Fredriksstad
486	N	Oscar	Christoffersen	WSr	Norweg'n	164	93·9	21·1	11·8	Brüges	1827			O.C.Christensen	Drammen
487	N	Oscar	Jacobsen 85-85	WSr	Norweg'n	136	81·0	24·7	11·5	Umea	1840			Jacob A.Andorsen	Mandal
488	N	Oscar	Hansen	WBg	Norweg'n	127	86·9	23·7	9·2	New Haven, Conn.	1847			C.Hansen	Horten
489	N	Oscar	Larsen	WSr	Norweg'n	122	77·9	22·6	11·2	Swalen	1852			H.Christensen	Holmestrand
490	N	Oscar	Marsteen	WBn	Norweg'n	113	91·4	22·1	10·4	Molde	18 57/79	O.Hovdenak	[illegible]	L.Rognskaug & Sön	Christiansund
491		Oscar	Schybergson	WBk	Russian	527	134·8	31·5	17·5	Fredrikshamn	1857			O.Simolin	Borga
492		Oscar	Kalnbehrsing	WSr	Russian	229	108·2	27·7	12·9	Pernigel	1875	P.Andersen		M.Kalnbehrsing	Riga
493		Oscar (ex Taggamois)	Tiedemann	WSr	Russian	118	86·0	22·0	10·0	Arensburg	1877			J.Tiedemann	Pernau
494	V	Oscar	Andersson 84-86	WBk	Swedish	802	167·9	32·8	20·4	Christinestad	1857		Y 10.84	P.A.Lindberg	Gothenburg
495		Oscar		WBk	Swedish	209	105·3	21·9	12·5	Elsfleth	1854			O.Engström	Calmar
496		Oscar	Runberg	WBg	Swedish	170	94·5	20·3	12·6	America	1881			C.Runberg	Westervik
497		Oscar	Johansson	WSr	Swedish	107	72·3	21·2	10·9	Dalsland	18 52/71	A.Olson		B.Olsson	Morlanda, Orust
498	R	Oscar C. Schmidt	Bacon	W3mSr	American	513	145·2	34·6	15·2	Mauricetown, N.J.	1882	J.W.Vanneman		A.T.Bacon	Philadelphia
499		Oscar F. Hawley	Tyler	WSr	American	109	86·0	25·8	7·6	Sag Harbor, N.Y.	1853			B.S.Tyler	Pt.Jefferson, N.Y.
500															

✱ = Lloyd's Register; D = Nederlandsche Vereeniging; G = Germanischer Lloyd; I = Registro Italiano; R = American Record;
A = Veritas Austro-Ungarico; H = Veritas Ellenico; N = Norske Veritas; V = Bureau Veritas.

Dans la 4me Colonne, C = Composite (membrure en fer, bordé en bois); I = Fer; Stl = Acier; W = Bois.

Les Dates du doublage en métal (Col. 11) ressortent des livres dans lesquels les navires sont respectivement classés ou des rapports des Inspecteurs du Lloyd's Register.
Pour plus amples renseignements sur les navires classés par le Lloyd's Register, voir le Lloyd's Register Book et le Lloyd's Yacht Register.

Official No.	No. in Book	Society in which Classed	Ship's Name. Material and Rig. Master, and Dates of Appointment to present Owner's Service and to Ship. Ship's Late Name.		Flag.	Net Tonnage.	Dimensions. Length	Breadth	Depth	Built. Where.	When	By whom.	Metalling	Owners or Managers.	Port
	501		Oscar Mooyer (ex Valparaiso) *Rogers -87*	W Bk	German	382	126·6	27·3	15·2	Hamburg	1863	H.H.Stülcken	[illegible]	J.W.Danielsen	[illegible]
	502	V	Oscar Wendt *Schoeneberg*	W Bk	German	477	129·1	29·2	17·1	Barth	1871	J.S.Kraeft	[illegible]	J.N.Rodbertus	Barth
	503		Oscar den Første *Larsen 76-83*	W Bk	Norweg'n	297	97·6	27·5	16·1	[illegible]	1848			T.M.Nielsen	[illegible]
	504		Oscar II. (ex [illegible]) *Ols[illegible]*	W Bk	Norweg'n	576	154·7	31·4	16·8	Quebec	1866	McKay		O.Eliasen	[illegible]
	505		Oscar II. (ex Hampden) *Olsen*	W Bk	Norweg'n	520	151·7	31·7	17·2	St.John,N.B.	1855			H.H.Parr	[illegible]
	506	V	Oscar II. *Lodin 82-82*	W Bk	Swedish	328	130·0	26·3	12·2	Hernösand	1873	F. Hägglund	[illegible]	H.Nordlander	[illegible]
	507	V	Oscar & Georg *Eriksen -85*	W Bk	Russian	526	138·9	33·8	18·7	Brahestad	18[illegible]	J.Lang	[illegible]	A.Erikson	[illegible]
	508	N	Oscarsborg *Henriksen 80-85*	W Sr	Norweg'n	225	107·7	26·6	12·6	Drammen	1871			M.Mulvad	[illegible]
	509	V	Oscarsvarf *Teglund*	W Bk	Swedish	672	151·9	30·5	18·1	Umea	1876	Jonsson	[illegible]	V.Glas	[illegible]
[illegible]	510		Osceola *Pratt*	W S	British	1157	185·0	36·6	23·5	Newburyport, Mass.	1864	J.L.Townsend		James Mills	[illegible]
	511		Osman Ghazi	W Bg	Turkish	101	..	..	..	Ayandon	1872			Hussein Hafiz Effendi & Co.	[illegible]
	512		Osman Ghazi	W	Turkish		..	..	..	Barton	1874			Mehemed Effendi & Co.	[illegible]
77078	513		Osman Pacha *Hellis 89-89*	W Bk	British	497	112·6	29·8	17·9	Shoreham	1878	Dyer		Macbeth & Gray	Glasgow
	514	R	Osmo *Panelius*	W Bk	Russian	793	172·3	34·4	21·0	Raumo	1868	J.Justin	[illegible]	Rederi Aktie Bolag	[illegible]
	515	V	Osmo *Abrahamsen -88*	W Bk	Norweg'n	597	148·8	34·2	18·0	Brahestad	18[illegible]	E.Klubb		H.N.Gjersoe	[illegible]
	516		Osmo *Jonsson 83-86*	W Bm Sr	Russian	257	112·9	28·5	13·1	Sago	1874	J.Holmberg		A.B.Nystrom	[illegible]
73622	517	V	Osmond O'Brien *Moreh[illegible] 83-83*	W Bk	British	878	173·7	35·2	20·2	Noel,N.S.	1877	O.O'Brien	[illegible]	O.O'Brien	[illegible]
[illegible]	518		Ospray *Phelan*	W Bn	British	144	..	..	..	Hastings	1841			Farrell & Son	[illegible]
18978	519		Ospray	W Bk	American	173	101·7	21·1	10·1	Baltimore	1851			G.J.Tarr & Son	[illegible]
19232	520		Ospray	W Sr	American	149	96·2	26·5	8·4	Addison,Me.	1867			Henry Nash & Co.	[illegible]
29452	521		Osprey *Livingston*	W Sr	British	153	97·4	22·0	11·6	Arbroath	1861	Arbroath Sh.Co.		Mrs.M.Livingston	Dundee
66446	522	❋	Osséo *Davies -82*	W Bk	British	568	113·0	32·2	17·9	P.E.Island	1874	G.Carroll	[illegible]	[illegible]	[illegible]
	523		Ossia Prashko *Photaki*	W Bg	Turkish	149	..	..	..	Greece	1884			[illegible]	[illegible]
	524		Ossia Prashko	W Bg	Turkish	149	..	..	..	Greece	1886			[illegible]	[illegible]
	525	N	Ossian *McKay -84*	W Sr	Norweg'n	154	96·1	24·2	10·7	Grimstad	1868	O.Kittelsen		P.[illegible]	[illegible]
	526		Osstorio	W Sr	Turkish	12[illegible]	..	..	..	Lemnos	1876			[illegible]	[illegible]
71388	527	❋	Ossuna *M'Kay 84-84*	W Bk	British	794	171·7	33·5	20·0	Kingston,N.B.	1883	Jardine Bros.	[illegible]	J.& T.Jardine	[illegible]
	528	N V	Osteraa *Larsen -88*	W Bk	Norweg'n	485	138·3	31·6	17·0	Tvedestrand	1873	Eldrup	[illegible]	L.Larssen	[illegible]
	529	N	Österdalen *Hansen -87*	W Sr	Norweg'n	220	102·5	27·3	11·7	St.John,N.B.	1858			C.A.Jensen	[illegible]
	530	N	Östersjöen *Svendsen*	W Sr	Norweg'n	101	85·9	20·1	10·0	Stavanger	1873			L.A.Melkevig	[illegible]
	531		Ostersøen (ex Halcyon) *Hoyer 85-*	W Bg	Danish	281	101·3	23·8	16·2	Sunderland	1852	S.Hodgson		H.C.Nyholm	[illegible]
	532		Ostin *Hofgren*	W Sr	Swedish	121	86·9	21·1	9·7	Brom	1871	O.Ohlsson		M.Peltz	[illegible]
	533		Ostindia *Isackson -86*	W Bk	Swedish	362	137·4	26·7	15·9	Oscarshamn	1859	O.Thoren		G. & F.Johansson	[illegible]
	534	V	Ostsee *Diesner 85-85*	W Bk	German	361	105·3	26·2	18·0	Colberg	1840	Müggenborg		J.Stevenson	[illegible]
	535	V	Osvald *Petersen -87*	W S	Danish	113	88·6	21·3	10·1	Holbæk	1874	R.Mortensen		H.L.Smith's Enke	[illegible]
	536		Osvaldo (ex Pepe Alicantino) *Pujol*	W Bg	Spanish	181	87·3	26·3	13·9	Lloret	1848			J.Anell y Bou	[illegible]
37804	537		Oswego	W Bn	British	128	93·8	24·4	10·3	Windsor,N.S.	1857			S.N.Robinson	[illegible]
2840	538		Oswy *Stevenson 80-80*	W Bg	British	172	84·2	23·2	13·6	Whitby	1849			H.Smith	Whitby
3588	539		Otac (ex Britomart) *Smoich -88*	W Bk	Austrian	42[illegible]	136·1	29·3	18·5	S.Shields	1869	Wilson&Blair	[illegible]	F.Goinbovich fu F.	[illegible]
[illegible]	540	A	Otac Niko (ex [illegible]) *Calafatovich*	W Bk	Austrian	1048	170·9	34·4	21·8	Bremerhaven	1870		[illegible]	S.Bielovucich	[illegible]
68807	541		Otacilius *Donovan 77-77*	W Bn	British	25[illegible]	102·0	26·8	13·0	Sackville,N.B.	1873	A.&W.Ogden		Amos Ogden	[illegible]
70983	542	V	Otago *Gullison -88*	W S	British	1[illegible]	182·0	37·0	22·5	SalmonRiv,N.S.	1874	[illegible] Allen	[illegible]	N.B.Lewis	[illegible]
60449	543	❋ D	Otago *Norman*	C Bk	British	9[illegible]	2[illegible]7·3	34·5	20·1	PortGlasgow	1869	R.Duncan&Co.	[illegible]	J.Hardie	[illegible]
60463	544	❋	Otago *Kornitowski -88*	I Bk	British	346	147·0	26·0	14·0	Glasgow	1869	A.Stephen&Sns		Snadden & Simpson	[illegible]
19315	545		Otago (WRECKED) *Gupitt*	W S	American	85[illegible]	161·3	33·5	22·0	Rockland,Me.	1864	F.Cobb		Renton,Holmes & Co.	[illegible]
73570	546	❋	Otaki *Worsley 83-83*	I S	British	1[illegible]	201·1	34·2	20·0	Newcastle	1875	Palmers S.&I.C.		New Zealand Shipping Co. Ld.	London
[illegible]	547	R	Otello *Bond -87*	W 3m Sr	American	332	139·0	33·0	11·4	East Boston	1882	D.D.Kelley&Son		J.H.Conant & Co.	[illegible]
70806	548	❋	Othello *Baron 83-83*	I S	British	1450	23[illegible]·6	37·8	22·7	Liverpool	1875	T.Royden&Sons		C.T.Bowring & Co.	[illegible]
40281	549		Othello	W Bk	British	314	115·0	27·5	14·0	Fairhaven,U.S.	1857			W.S.Waterston	[illegible]
[illegible]	550	❋	Othello	W K	British	88	80·2	20·9	10·5	Brixham	1884	J.W.&A.Upham		C.Hellyer	[illegible]

❋ = Lloyd's Register; D = Nederlandsche Vereeniging; G = Germanischer Lloyd; I = Registro Italiano; R = America[illegible]
A = Veritas Austro-Ungarico; H = Veritas Ellenico; N = Norske Veritas; V = Bureau [illegible]

In the 4th Column, C = Composite (iron frame, wood planked); I = Iron; Stl = Steel; W = Wood.

The Dates of Metalling (Col. 11) are based on the records of the Books in which Vessels are respectively classed, or on the Reports of the Surveyors of [illegible]

Further particulars of Vessels classed by Lloyd's Register will be found in Lloyd's Register Book and Lloyd's Yacht Register.

No.	Soc.	Nom. Capitaine et dates d'entrée au service des Armateurs actuels et à bord du navire. Ancien Nom.	Matériaux et Gréement.	Pavillon.	Tonnage net.	Longueur	Largeur	Creux	Construit. Où.	Quand.	Par qui.	Date du doublage en métal	Armateurs ou Directeurs-Gérants.	Port d'Armement.
551	✻	Othello *Sempers* -85		British	80	73·5	19·8	12·5	Elmshorn	1886	D.W.Kremer		I.Sempers	Grimsby
552	V	Othello (1883) *Riedbrett* 77-87	W Bk	German	461	129·6	30·4	18·2	Pillau	1866	H.O.Becker		J.G.Gerlach	Memel
553	G	Othello *Butzekat* 89-	W Bk	German	350	101·3	29·5	16·7	Memel	1848	B.Pieper		C.Kröger	Lubeck
554		Othelo *de Oliveira*	W3mSr	Brazilian	204	..	..	..	Villa do Conde	1879		[illegible]	J.da R.Souza	Rio de Janeiro
555	V	Otilde *Suhr* 80-86	W Bk	German	887	172·6	35·8	21·3	Elsfleth	1885	F.Deetjen	[illegible]	H.G.Deetjen	Elsfleth
556		Oto Maru	W Bg	Japanese	234	115·8	24·0	13·4					Koyekishn	Totomi
557	N	Otoen *Steensohn* -85	W3mSr	Norweg'n	294	126·1	27·9	13·5	Stavanger	1875	Paul P.Otten	[illegible]	Johan Bilstad	Egersund
558		Otonabee	W Sr	British	225	125·0	26·2	10·0	Quebec	1867			The Muskoka Mill & Lumber Co.	Toronto
559	NV	Otra *Eliefsen* 86-86	W Bk	Norweg'n	320	128·9	29·0	12·5	Christiansand	1886	Brød. Larsen	[illegible]	S.Otto & Co.	Christiansand
560		Otronto *Lockhart*	W Sr	American	100	79·7	23·3	8·3	Surry, Me.	1833			O.Lockhart	Ellsworth, Me.
561	A	Ottavina *Magnasco* -88	W Bk	Italian	448	129·2	29·8	19·4	Varazze	1866	N.Cerruti	[illegible]	E.Schiaffino	Genoa
562		Ottawa	W Bn	British	143	90·5	24·7	11·0	Parrsboro, N.S.	1867	Kelly		S.Vickers	Chester
563	N	Ottawa *Jorgensen*	W Bk	Norweg'n	615	153·1	30·7	18·5	Kingston, Ont	1866	Desnoyers	[illegible]	N.W.Cech	Brevig
564	R	Otter *Welsh* -88	W Sr	British	199	101·8	28·6	10·0	St.John, N.B.	1884	J.H.D.Eagles	[illegible]	J.H.D.Eagles	St.John, N.B.
565	✻	Otterspool *Evans* 88-88	I S	British	1798	266·5	39·4	23·5	Newcastle	1884	Palmers' Co. Ld		Sailing Ship "Otterspool," Ld. (R.W.Leyland & Co.)	Liverpool
566	R	Ottilia (ex Rosa Bice) *Treglia*	W Bk	Italian	682	143·3	31·8	20·3	Brewer, Me.	1866		[illegible]	M.Somma, Ottilia Ardizzone	Castellamare, Stabia
567	G	Ottilie *Leibauer* 67- (ex Two Brothers)	W Bk	German	567	129·2	31·3	20·8	Boston, U.S.	1848			F.Heyn	Danzig
568	V	Ottilie *Scheel* -88	W Bk	German	323	120·0	27·8	11·3	Ziegenort	1865	H.Brüsewitz	[illegible]		Stettin
569		Otto (Voy. SIEGFRIED) *Wesley* -86 (ex Alphonse Nelie)	W Bk	Chilian	590	137·4	29·2	18·5	Bordeaux	1863	Chaigneau frères		Lund & Co.	Valparaiso
570	V	Otto *Vandel* 89-89	W Bn	Danish	149	96·3	23·0	10·8	Svendborg	1871	J.R.Andersen		I.F.G.Vandel	Svendborg
571	V	Otto *Nielsen* 64-73	W Sr	Danish	114	86·1	20·5	10·7	Rudkjøbing	1868	S.Poas		C.& F.Nielsen	Rudkjøbing
572	V	Otto (ex Matrem) *Fischer* 72-81	W Sr	Danish	103	82·9	20·2	9·7	Thurø	1877	J.Ph.Jørgensen		Akt.Selsk. (N.H.Fischer)	Sønderho
573	G	Otto (ex Tecumseh) *Langen* -86	W S	German	1205	196·5	37·1	24·3	Boston, U.S.	1866	J.L.Townsend	[illegible]	Von Riegen & Stindt	Bremerhaven
574	GV	Otto *Degen* 88-88	W Bn	German	233	97·0	24·9	11·0	Motzen	1875	D.Focke	[illegible]	Gebrüder Hustede	Elsfleth
575	G	Otto *Ewert* -82	W Sr	German	162	86·6	23·5	12·0	Barth	1877	R.Seldör		Wilh.A.Sarnow	Barth
576	G	Otto *Schoon*	W Sr	German	104	85·0	20·5	9·6	Weener	1863	R.Borg	[illegible]	Gerd.M.Schoon	W.Rauderfehn
577	N	Otto *Gundersen* -87 (ex Royal Headed)	W3mSr	Norweg'n	526	137·7	32·0	16·9	Hopewell, N.B.	1872	J.L.Sye	[illegible]	A.C.Gundersen	Tönsberg
578	N	Otto *Knudsen* -78	W Sr	Norweg'n	215	102·5	26·9	11·1	Pth Amboy, N.J.	1849			C.J.Gullichsen	Drammen
579	V	Otto *Granit* 86-86	W Bk	Russian	320	123·6	27·2	13·5	Nagu	1875	J.Jansson		M.Jansson	Nagu, Finland
580		Otto *Andersson*	W Bg	Swedish	368	108·9	27·9	17·5	Trieste	1857			O.M.Schell	Landscrona
581	G	Otto *Olsson* -86	W Sr	Swedish	203	101·7	23·2	11·8	Timmernabben	1868	J.Nordström		L.Pahsson	Lerberget
582	V	Otto Graf zu Stolberg *Wesemann* 71-	W Bn	German	182	97·8	23·2	11·8	Papenburg	1871	A.Beckmann	[illegible]	A.Beckmann	Papenburg
583		Otto Linck *Paetsch* -87	W S	German	668	131·5	33·5	23·4	Danzig	1861	J.W.Klawitter		O.Münsterberg	Danzig
584		Otto von Salis *Teizihl*	W Sr	Russian	196	99·7	24·6	12·5	Salis	1879			O.A.Teizilt	Haynasch
585	GN	Otto & Antonie (ex Hermit) *Gundersen* 77-	W S	Norweg'n	1277	196·7	38·0	23·7	Yarmouth, Me.	1855		[illegible]	L.Dahl (in liquidation)	Tönsberg
586	V	Ottolina *Ternstrom* 83-86	W S	Swedish	1250	193·9	41·0	24·0	Krimpen	1870	J. & K.Smit	[illegible]	G.R.Lagergren	Stockholm
587	N	Ottolina (WRECKED) *Hansen* 88-88	W Bk	Norweg'n	637	135·8	35·6	20·9	Kinderdyk	1853		[illegible]	F.Bredu	Christiania
588	DG	Ottomar *Grant* 81-82	W3mSr	Russian	325	122·0	28·4	15·0	Haynasch	1876	Thomsen		O.Weide, Ww.	Haynasch
589	✻	Ottone *Valle*	W Bk	Italian	620	152·0	31·7	21·0	Savona	1869	A.Calcagno	[illegible]	F.Ottone	Genoa
590	NV	Otus *Ekornaes* -88	W Bk	Norweg'n	567	157·2	31·8	17·4	Grimstad	1875	Jacob Boe	[illegible]	A.Ekornaes	Arendal
591		Ouachita (ex ...) *Keatron* 85-86	W Bn	British	128	92·4	25·7	10·0	New York				J.Hall	Maryport
592	R	Our Annie *Kinond*	W Bkn	British	364	146·0	29·1	12·9	Sunderland	1874	R.Thompson	[illegible]	R.Turner	Quebec
593	✻	Our Boys *Robbens* 84-85	W K	British	63	70·5	19·2	9·1	Lowestoft	1885	Page & Chambrs.		J.S.Sterrys	Lowestoft
594	✻	Our Nellie *Larsen* 78-83	W Sr	British	100	82·3	21·0	10·7	Kingsbridge	1883	W.Date		J.Westcott	Plymouth
595		Ourania (ex Ioannis)	W Bg	Greek	225	..	..	..	Malta	1871			S.N.Focas	Cephalonia
596	✻	Ousuri *Jones*	W Bk	British	357	132·5	25·9	14·8	Arbroath	1863	Armisted	[illegible]	R.Honey	Adelaide, S.A.
597	✻	Overdale *Nicholls* 88-88	I Bk	British	882	203·3	34·2	19·2	Dundee	1877	A.Stephen & Sns		Doward, Dickson & Co.	Liverpool
598		Overman *Olsen* -88	W Bk	Norweg'n	266	112·5	26·1	13·3	Elsfleth	1860			O.N.Hus	Arendal
599		Owari Maru	W Sr	Japanese	209	102·7	23·2	13·0	Kanagawa	1882	Shiranaiye S.B.Yd		Rikni Taemon	Saikai
600														

✻=Lloyd's Register; D=Nederlandsche Vereeniging; G=Germanischer Lloyd; I=Registro Italiano; R=American Record; A=Veritas Austro-Ungarico; H=Veritas Ellenico; N=Norske Veritas; V=Bureau Veritas.

Dans la 4me Colonne, C=Composite (membrure en fer, bordé en bois); I=Fer; Stl=Acier; W=Bois.

Les Dates du doublage en métal (Col. 11) ressortent des livres dans lesquels les navires sont respectivement classés ou des rapports des Inspecteurs du Lloyd's Register.

Pour plus amples renseignements sur les navires classés par le Lloyd's Register, voir le Lloyd's Register Book et le Lloyd's Yacht Register.

Official No.	No. in Book	Registry in which classed	Ship's Name. Material and Rig. / Master, and Dates of Appointment to present Owner's Service and to Ship / Ship's Late Name.	Flag.	Net Tonnage.	Length	Breadth	Depth	Where.	When	By whom.	Date of Metalling	Owners or Managers.	Port of [illegible]
67631	601	✻	**Owny Belle** [illegible] WSr	British	127	89·6	22·7	11·6	Newquay	1875	Daniels	[illegible]	J. Vautier	Jersey
	602	DNV	**Oxo** [illegible] WBk	Norweg'n	604	116·5	31·7	17·7	Tatamagush. (N.S.	1868	A. Campbell	[illegible]	Mathias Hansen	Christia[illegible]
[illegible]	603		**Ozama** (Lost) Pietro [illegible] WBg	Italian	250	111·1	24·4	13·1	Sestri Ponente	1879	A. Briasco fu L.	[illegible]	Gio. Batta Vicini	Genoa

Additional Ships are inserted in the Supplement at the end of the Register.

✻ = **Lloyd's Register**; D = Nederlandsche Vereeniging; G = Germanischer Lloyd; I = Registro Italiano; R = American Re[illegible]
A = Veritas Austro-Ungarico; H = Veritas Ellenico; N = Norske Veritas; V = Bureau Veritas

In the 4th Column, C = Composite (iron frame, wood planked); I = Iron; Stl = Steel; W = Wood.

The Dates of Metalling (**Col. 11**) are based on the records of the Books in which Vessels are respectively classed, or on the Reports of the Surveyors of Lloyd's Re[illegible]

Further particulars of Vessels classed by Lloyd's Register will be found in Lloyd's Register Book and Lloyd's Yacht Register.

No.	Registre	1. Nom. Matériaux et Gréement. Capitaine et dates d'entrée au service des Armateurs actuels et à bord du navire. Ancien Nom.		5. Pavillon	6. Tonnage net.	7. Dimensions. Longueur	Largeur	Creux	8. Construit. Où.	9. Quand.	10. Par qui.	11. Date du doublage en métal.	12. Armateurs ou Directeurs-Gérants.	13. Port d'Armement.
1	DN	**P. A. Munich** Olsen -88	W Bk	Norweg'n	436	128·6	30·6	17·2	Risör	1870	Th.Gundersen		B.A.Olsen	Risör
2	R	**P. C. Merryman** (DISMANTLED) Powers	W Bk	American	876	161·3	33·2	23·0	Freeport, Me.	1866	E.Talbot		Chase, Leavitt & Co.	Portland, Me.
3	NV	**P. C. Peterson** -88	W Bk	Norweg'n	647	166·1	34·3	18·0	Christiania	1874	Chr. Brinch	Y 12.86	M.Sundt	Moss
4		**P.F.**	W Bn	British	198	103·0	21·4	12·4	St. Malo	1878			J.S.Sterry	Lowestoft
5		**P.F.** Cadion	W Bn	French	133	78·4	20·7	11·0	Dieppe	1863	P.Corue		E.Pourpoint fils	Dieppe
6	NV	**P. G. Blanchard** -87	W S	Norweg'n	1250	187·0	37·7	24·8	Yarmouth, Me.	1862	S.C.Blanchard		H.Normann	Lillesand
7	RV	**P. J. Carleton** Crosbie -87	W Bk	American	986	170·4	35·0	21·7	Rockport, Me.	1870	J.Pascal		Carleton, Norwood & Co.	Camden, Me.
8		**P. J. Nevius**	W Bn	British	150	102·0	25·9	10·6	Hopewell, N.B.	1860			J.J.Carroll	Dublin
9	✻	**P. J. Palmer** McKay	W Bkn	British	415	136·8	32·4	12·9	Dorchester	1881	P.J.Palmer	Y 11.88	P.J.Palmer	Dorchester, N.B.
10	V	**P. M. Pederson** Nissen (ex John Trahey) -88	W Bk	Norweg'n	1147	190·6	37·2	21·7	Maitland, N.S.	1876	S.Trahey		A.S.Myhre	Porsgrund
11	V	**P. N. Blanchard** (Blanchard) -86	W S	American	1503	213·1	40·0	25·0	Yarmouth, Me.	1876	Blanchard Bros.	Y 2.87	F.L.Oakes	Boston
12		**P. Pentheny** Thomas	W Bkn	Danish	283	118·4	25·5	12·5	Drammen	1874			P.Pentheny	Christiansted St. Croix, W.I.
13		**Paavo Nuorempi**	W Sr	Russian	218	..	..	..		1878			P.Wonninen	Impilaks
14	V	**Pablo Sensat** Calzada -89	W Bk	Spanish	575	144·3	29·5	16·4	Blanes	1878	J.M.Vieta		J.M.Serra	Barcelona
15		**Paca**	W Sr	Spanish	182	92·2	26·8	12·9	Blanes	1855			J.E.Valles	Barcelona
16		**Pace Schiaffino**	W Bg	Argentine	513	..	..	..	Recco	1868	P.Rolla	11.85		
17	V	**Pachino** Morando	W Bn	Italian	180	104·3	24·4	12·8	Varazze	1885	B.Cerruti	Y 6.88	B.Cerruti di A.	Savona
18	✻	**Pacific** Davies 84-85	I Bk	British	399	146·9	26·2	16·5	Glasgow	1867	A.Stephen & Sns		B.M.Hennessy	Swansea
19		**Pacific** Nash	W Sr	American	121	88·6	26·1	7·8	Addison, Me.	1865			A.B.Nash	Machias, Me.
20	GV	**Pacific** Oltmann 87-87	W Bk	German	460	123·1	28·5	14·2	Brake	1875	I.Oltmann's Wwe.		D.Haye	Brake
21	N	**Pacific** Eriksen 87-87	W Bk	Norweg'n	562	155·5	31·8	17·6	Drammen	1869	Chr.Tofte		F.Breda	Christiania
22		**Pacific Slope** Barnes	W Bk	British	799	187·6	38·9	16·7	San Francisco	1876	Boole & Webster		H.& R.Hudson	Sydney, N.S.W.
23		**Pacifico** Wilson 71-78 (ex Pacific)	W Bk	Chilian	758	146·3	33·5	21·3	Gefle	1867	O.Brodin		F.Peede	Valparaiso
24	N	**Pacifik** Abrahamsen -86	W Bk	Norweg'n	606	167·0	33·0	18·1	Bergen	1872	Jens Gran		C.J.Ludvigsen	Farsund
25		**Paco**	W Bn	Spanish	143	81·0	24·3	12·1	Palma	1856			A.Oliver Bonet	Majorca
26		**Pactole**	W Bk	British	386	124·0	28·8	16·2	Les Sables d'Olonne	1861	Noleau		J.C.Ellis	Sydney, N.S.W.
27		**Pactolus** Richards	W Bk	British	399	132·1	27·7	17·7	Sunderland	1864	J.Gill		T.Bambra	London
28	RV	**Pactolus** Burnham	W S	American	1145	191·4	38·5	23·4	Thomaston, Me.	1865	Chapman & Flint		B.Flint	New York
29	A	**Padosa** Marinovic 87-88	W Bk	Austrian	656	145,6	31,4	19,3	Porto Rè	1878	Arcudin	10.86	T.Lawer fu M.	Fiume
30		**Padoung**	I Barge	British	791	194·0	25·5	8·2	Glasgow	1875	A.& J.Inglis		Irrawaddy Flotilla Co. Ld.	Glasgow
31	A	**Padre** Vellusig	W Bk	Austrian	593	145·0	30·2	18·6	Fiume	1866	Flli Schiavoni		B.Minach	Fiume
32	I	**Padre** Bertolotto	W Bk	Italian	581	143·3	32·4	21·3	Voltri	1871	B.Fabioni		M.L.Bertolotto	Genoa
33	I	**Padre Ageno** Simonetti (ex Padre Chiama) -88	W Bk	Italian	444	134·5	30·2	18·9	Voltri	1869	G.Fava	2.87	G.Ageno	Genoa
34		**Padre Cacique** Guerrire (ex Saladi)	W Sr	Brazilian	159	..	..	..	Villa do Conde			10.85	E.O.T.Ribas	Porto Alegre
35		**Padre Figli** Stuck	W Bk	Austrian	422	125·9	27·9	17·2	Fiume	1874	Flli Schiavoni		A.Stuck	Orebich
36	I	**Padre Francesco** Maggiolo	W Bk	Italian	687	163·0	32·3	21·5	Savona	1874	F.Sirello		S.Mortala ved Valle	Genoa
37	A	**Padre Stefano** Polganik -88	W Bk	Austrian	371	121·3	27·5	..	Trieste	1862	Stabilimento Tecnico		Sorelle Chersa	Orebich
38	I	**Padre & Figlio** Sabia	W Bn	Italian	345	108·2	25·9	16·7	Procida	1885	N.Lubrano		A.& V.Sabia	Procida
39		**Page** (WRECKED) Prescott	W Sr	American	104	81·0	23·6	10·0	Barnstaple, Mass.	1831			A.Center	San Francisco
40		**Pagona** (WRECKED)	W Bg	Greek	127	76·4	23·3	11·7	Galaxidi	1879	J.Canatas		J.Metropoulos	Galaxidi
41	A	**Pagonitsa** Vlanis -88	W Bg	Greek	240	107·2	25·9	..	Galaxidi	1884	J.Canatas		N.D.Metropoulos	Galaxidi
42		**Paing**	Stl Lr	British	484	220·0	30·0	8·3	Dumbarton	1887	W Denny & Bros		Irrawaddy Flotilla Co Ld	Glasgow
43	D	**Pak Wan** Smith -88	C S	British	819	186·2	32·6	19·0	Sunderland	1863	G.Peverall		James Gillan	Newcastle, N.S.W.
44		**Palace** Horne	W Bk	British	339	132·0	28·7	12·9	Miramichi, NB	1866			Thos.Paterson	Dunedin
45	V	**Paladin** Kefting -89	W Bk	Dutch	337	117·6	28·8	16·1	Rostock	1869	G.W.Tesdorpf		K.B.Kuiper	Delfzijl
46	G	**Paladin** Giese 77-80	W Bk	German	229	108·3	26·6	14·7	Swinemünde	1848	C.Ritzerow		Carl F.Eschricht	Swinemünde
47	N	**Palander** Osokildt	W Bk	Norweg'n	467	138·9	31·2	16·8	Arendal	1880			A.Aanonsen	Arendal
48	N	**Palander** Ellefsen -81	W 3m Sr	Norweg'n	394	127·6	30·0	14·9	Porsgrund	1881	Carl Hansen	11.85	O.K.Abrahamsen	Porsgrund
49	N	**Palander** -88	W 3m Sr	Norweg'n	303	121·6	28·8	13·1	Risör	1881			J.Prebensen	Risör
50	R	**Palatka** Chaples	W 3m Sr	American	408	145·0	33·5	12·0	Belfast, Me.	1883	C.P.Carter & Co.		Drew, Hazeltine & Livingstone	Belfast, Me.

✻ = Lloyd's Register; A = Veritas Austro-Ungarico; D = Nederlandsche Vereeniging; G = Germanischer Lloyd; H = Veritas Ellenico; I = Registro Italiano; N = Norske Veritas; R = American Record; V = Bureau Veritas.

Dans la 4me Colonne, C = Composite (membrure en fer, bordé en bois); I = Fer; Stl = Acier; W = Bois.

Les Dates du doublage en métal (Col. 11) ressortent des livres dans lesquels les navires sont respectivement classés ou des rapports des Inspecteurs du Lloyd's Register.

Pour plus amples renseignements sur les navires classés par le Lloyd's Register, voir le Lloyd's Register Book et le Lloyd's Yacht Register.

Official No.	No. in Book	Registry in which classed	Ship's Name. Material and Rig. Master, and Dates of Appointment to present Owner's Service and to Ship. Ship's Late Name.	Flag	Net Tonnage	Length	Breadth	Depth	Built Where	Built When	Built By whom	11	Owners or Managers	Port
[illegible]	51	✳	Palawan *Partridge 85-86* I Bk	British	955	201·0	34·0	21·7	Port Glasgow	1869	J. Reid & Co.		McDiarmid, Greenshields & Co.	Liverpo…
[illegible]	52	V	Palermo *Walker* W Bk	British	799	165·1	34·6	19·8	Granville, N.S.	1872	Olivier	[illegible]	Israel L. Walker	Annapolis
[illegible]	53		Palermo W Bgr	British	133	72·0	21·6	11·4	Nova Scotia	1849			John Greer	Sydney, N…
	54	V	Pales *Thomsen -88* W S (ex Lammerts … Hendericus)	Dutch	131	87·0	20·0	11·0	Veendam	1868	E. van Linden		L. Rozenbeek	Termunt…
1049	55		Palestina *Giannoni* W Bn	Italian	100	82·3	23·4	10·8	B. d'Arno	1867	Picchiotti		Luigi Iacono	Naples
62844	56	✳	Palestine *Williams* W Bg	British	234	109·3	24·8	14·9	Port Dinorwic	1870	Jones	[illegible]	R. Williams	Carnarvo…
150100	57	V	Palestine *Hayden 78-88* W S	American	1470	209·6	40·0	24·0	Bath, Me.	1876	W. V. Moses & Son	[illegible]	S. Blair	San Fran…
19631	58		Palestine *Chadwick -87* W Sr	American	135	109·0	27·0	8·0	Forked Rvr. N.J.	1852			L. M. Godfrey	New York
	59	✳G	Palestine *Kellermann -87* I Bk	German	598	184·3	27·6	16·8	Belfast	1863	Harland & Wolff		J. G. Lund	Hamburg
89613	60	✳D	Palgrave *Higgens -88* I 4m S	British	3078	322·5	49·2	25·4	Port Glasgow	1884	W. Hamilton & Co.		W. Hamilton & Co.	Pt. Glasgow
	61		Palheros *Dos Reis* W Bn	Brazilian	138	..	..	..	Europe	1862		[illegible]	J. Dos Reis	Pernambu…
	62	✳V	Pallas *Nehring 78-88* Stl Bk	German	1407	233·0	37·1	21·7	Sunderland	1888	J. P. Austin & Son		M. G. Amsinck	Hamburg
	63	✳DV	Pallas *Siege 77-77* I Bk	German	597	170·0	29·0	17·7	Sunderland	1877	J. Laing		C. G. Beermann	Elsfleth
	64	H	Pallas (ex Boldon) *Saliari* W Bk	Greek	655	168·1	32·2	19·0	Sunderland	1873	Crown	[illegible]	C. Saliari	Syra
	65	N	Pallas *Pedersen* W Bk	Norweg'n	593	150·9	33·9	18·0	Tvedestrand	1881		[illegible]	J. P. Pedersen	Tvedestr…
	66		Pallas W Bn	Spanish	185	..	..	..					A. M. de Souza	Rio de Ja…
	67	V	Pallas *Elfversen 72-83* W Bk	Swedish	481	137·1	28·7	17·0	Hamburg	1866	M. G. Amsinck	[illegible]	N. Olsson	Höganäs
	68	V	Palme (ex Frederik Tudor) *Bahls 85-83* W S	German	1081	183·0	36·0	28·0	Boston, Mass.	1866	J. Taylor	[illegible]	J. D. Bischoff	Bremen
	69	GV	Palme *Reemts 81-87* W Bg	German	223	100·0	25·0	10·8	Brake	1866	J. Oltmann's Wwe.	[illegible]	Hillrick Arnold	Brake
	70		Palme *Herrmie* W Bn	German	174	95·1	22·8	11·8	Motzen	1859	D. Oltmann's Wwe.		A. T. Duysen	Geestemü…
	71	N	Palme *Finkelsen 75-75* W Bg	Norweg'n	233	108·4	25·3	12·2	Motzen	1866	D. Oltmann's Wwe.		B. A. Finkelsen	Farsund
	72	N	Palmen *Jansen 75-79* W Bg	Norweg'n	335	101·0	27·9	17·3	Arendal	1840			H. E. Jansen	Flekkefjo…
20147	73		Palmer W Sr Yt	American	125	100·0	24·2	10·3	Philadelphia, Pa	18^{65}_{74}	Hy. Steers		R. Stuyvesant	New York
	74	V	Palmerston (ex Herschel, ex Cubana, ex Sumner) *Corolls 86-86* I 4m S	German	1116	246·0	30·0	22·0	Liverpool	1858	Laird & Co.		R. M. Sloman & Co.	Hamburg
20139	75		Palmetto *Tripp* W Bk	American	215	100·0	29·8	12·1	Medford, Mass.	1846			J. F. Tucker & Co.	New Bedf…
	76	V	Palmier W Bk	Greek	736	162·1	32·5	20·2	La Seyne	1874	Curet fils	[illegible]	T. Stamati	Syra
	77		Palmira (ex Virgen del Carmen) W Bk	Spanish	249	112·2	27·7	14·4	Palma	1858	José Koll		Doña M. M. Alena	Majorca
150036	78	RV	Palmyra *Minott 87-87* W S	American	1299	197·0	38·0	24·0	Bath, Me.	1876	Goss & Sawyer	[illegible]	Puget Sound Commercial Co.	Bath, Me.
	79		Paloma *Taylor -88* W Bn	German	223	112·0	27·0	10·0	Humboldt Bay	1875	H. D. Bendixen		Société Commerciale de l'Océanie	Hamburg
[illegible]	80	R	Palos *Haulermann* W Sr	American	246	104·5	27·2	14·0	Eastport, Me.	18^{68}_{73}			Poitvant & Favre	[illegible]
75326	81	✳	Pamela H. Banfield (ex Zeh) *Morgan 87-87* W Bn	British	179	96·0	23·5	11·5	Pekela	1876	Drenth	[illegible]	J. Bateman	[illegible]
	82	✳V	Pamelia *Hellwage 79-88* Stl Bk	German	1403	244·5	38·4	20·8	Hamburg	1888	Blohm & Voss		F. Laeisz	Hamburg
48653	83	✳	Pampero *George 84-84* I Bk	British	588	171·6	27·7	17·5	Liverpool	1864	T. Royden & Sons		Swansea Merchant Shipowners' Co. Ld.	Swansea
	84	N	Pan *Nielsen -87* W Bg	Norweg'n	187	88·8	25·0	13·4	[illegible]	1854			S. Haagensen	Drøbak
	85		Panaghia W Bg	Turkish	137	..	..	..	Greece	1882			Shakkisteo Nicola Anastos & Co.	Shakki…
	86		Panaghia W Bg	Turkish	125	..	..	..	Kerrasonade	1871			Bureckha & Bros.	Kerrasun…
	87		Panaghia W Bmb	Turkish	124	..	..	..	[illegible]	1874			Bulbuljehloo Spiro Dimitri & Co.	Bulbul…
22	88	A	Panagia Akathistou *Alafousos* W Bk	Greek	336	110·8	28·5	..	Syra	1876			M. Alafousos	Santorin
7	89	A	Panagia Akathistou *Zannos* W Bg	Greek	257	104·6	26·9	..	Syra	1870	E. Flumis	[illegible]	D. Zannos	Santorin
	90	A	Panagia Akathistou *Sarris* W Bg	Greek	250	..	..	..	Syra	1877		[illegible]	S. I. Sarris	Santorin
	91		Panagia Akathistou *Demarcaki* W Bg	Russian	226	..	..	..	Syra	1870			G. Zannes	Taganrog
76	92	A	Panagia Erythiani *Zacco* W Sr	Greek	135	84·7	22·5	..	Syra	1875			D. Zacco	Syra
440	93		Panagia Genethlion W Bg	Greek	202	86·9	25·6	..	Syra	1869			A. Livadaras	Syra
	94		Panagia Karoussena *Mitzi* W Bg	Greek	296	103·3	28·4	..	Syra	1868			I. Filippakis	Syra
	95		Panagia Katafiani *Dimezzi* W Bg	Russian	150	..	..	..	Syra	1868	Andrice		P. Plate	Taganrog
[illegible]	96	H	Panagia Malteza W Bg	Greek	167	95·5	23·6	..	Syra	1879	S. Cristali		M. Manolakis	Syra
[illegible]	97		Panagia Malteza W Bn	Greek	152	84·3	20·8	..	Syra	1863			S. Vlouchakis	Volo
[illegible]	98	V	Panagia Mangano *Syvoronos* W Bk	Greek	167	88·6	23·0	13·0	Syra	1885	A. Livadaros		E. I. Syvoronos	Syra
[illegible]	99	A	Panagia Myrtidiotissa W Bn	Greek	293	102·3	26·9	..	Syra	1868	S. Cofudachi	[illegible]	S. Canalas & Co.	Syra
100														

✳ = Lloyd's Register; D = Nederlandsche Vereeniging; G = Germanischer Lloyd; I = Registro Italiano; R = American [illegible];
A = Veritas Austro-Ungarico; H = Veritas Ellenico; N = Norske Veritas; V = Bureau Veritas.

In the 4th Column, C = Composite (iron frame, wood planked); I = Iron; Stl = Steel; W = Wood.

The Dates of Metalling (Col. 11) are based on the records of the Books in which Vessels are respectively classed, or on the Reports of the Surveyors of Lloyd's [illegible]

Further particulars of Vessels classed by Lloyd's Register will be found in Lloyd's Register Book and Lloyd's Yacht Register.

N°	Registre	Nom. Capitaine. Ancien Nom.	Matériaux et Grément	Pavillon	Tonnage net	Longueur	Largeur	Creux	Construit: Où	Quand	Par qui	Date du doublage en métal	Armateurs ou Directeurs-Gérants	Port d'Armement
101	H	Panagia Myrcidiotissa	W Bg	Turkish	131	..	..	..	Syra	1868			Peris Christos	Chios
102		Panagia Vaglianos [illegible]	W Bg	Greek	261	128·9	26·9	..	Syra	1868		V 3.88	A.Vaglianos	Piræus
103		Panagiotis *Nicas*	W Bg	Greek	348	..	..	..	Galaxidi	1877			E.Nicas	Galaxidi
104	A	Panagiotis *Macri*	W Bg	Greek	345	111·5	28·9	16·4	Galaxidi	1880	G.E.Canatas	V 8.88	G.Canatas	Galaxidi
105	V	Panagiotis *Macris -87* (ex Sobat)	W Bg	Greek	188	89·6	22·6	12·5	Nantes	1872	Chairgeau	V 11.86	J.& G.Macris	Galaxidi
106		Panagis	W Bg	Greek	143	75·1	30·8	..	Syra	1868			J.Stathopoulos	Piræus
107		Panagistitsa	W Bg	Greek	158	70·0	21·1	13·7	Limni				S.G.Gavetas	Chalkis
108	A	Panais Cuppa [illegible]	W Bk	Greek	438	139·7	30·9	18·7	Savona	1873	V.Dabove	V 2.87	Nicolo Pappa	Santorin
109		Panalon	W Bg	Turkish	202	..	..	..	Greece	1865			Limnilos Angeli Angelishi	Lemnos
110		Panama	W Sr	American	158	102·0	32·2	6·1	Camden,N.J.	1883			H.Moody	New York
111		Panama [illegible]	W Bk	Ecuador	625	151·7	32·3	20·2	Hampden,Me.	1854				
112	V	Panama *Koster 87-87*	W Bk	German	411	124·6	29·5	16·5	Brake	1869	J.Oltmann's Wwe	V 4.88	A.H.Arnold	Brake
113	RV	Panay *Bray -87*	W S	American	1131	186·7	37·0	23·2	Boston,Mass.	1877	J.Taylor	V 6.88	G.H.Allen	Salem
114		Panayashki	W Bg	Turkish	162	..	..	..	Syra	1870			Aimaros Monastery	Constantinople
115	V	Panchito *Malet*	W Bn	Spanish	228	103·3	26·0	12·8	San Feliu	1866		V 1.87	Roberts & Co.	Barcelona
116	*	Panda *Lewis -88*	W Bk	British	328	140·4	27·2	13·7	Aalborg	1874	Vang	V 10.87	W.C.Jarvis	Liverpool
117	*	Pandora	W K	British	81	77·5	20·6	10·6	Falmouth	1883	Trethowan		F.Hellyer	Hull
118	*	Pandora *Tait 71-76*	W Sr	British	79	79·6	20·7	9·1	Fraserburgh	1876	Webster		J.D.Henderson	Fraserburgh
119	AH	Pandora	W Bg	Greek	279	100·4	28·4	..	Syra	1876			C.Papapetros	Piræus
120		Pandora	W Bn	Greek	243	..	..	..	Galaxidi	1871	E.C.Papapetros		C.D.Cammenos	Zante
121	N	Pandora *Olsen 88-87*	W Bk	Norweg'n	514	140·0	29·8	17·5	Grimstad	1881	JensTellefsen	V 10.88	O.A.Olsen	Arendal
122	V	Pandur *Baysen 85-88*	W Bk	German	595	147·3	29·0	18·6	Geestemünde	1877	Schau&Oltmanns	V 12.86	F.Laeisz	Hamburg
123		Pangwey	Stl Lr	British	450	220·0	30·0	8·4	Dumbarton	1887	W.Denny & Brs.		Irrawaddy Flotilla Co. Ld.	Glasgow
124	A	Panhellenion *Karinos*	W Bk	Greek	470	130·2	28·8	19·0	Syra	1878		V 2.88	M.Karinos	Syra
125		Panhellenion *Mitropoulos*	W Bg	Greek	178	..	..	..	Galaxidi	1867			J.D.Mitropoulos	Galaxidi
126	*	Panmure *Hughes 81-84*	I S	British	1505	245·6	38·3	23·3	Glasgow	1875	A.Stephen&Sns		D.Bruce & Co.	Dundee
127		Pannayotti *Majakliuo*	W Bg	Turkish	134	..	..	..	Russia	1883			Salessiles Yorghi	Chios
128		Pannonia *Higgins*	W Sr	American	196	106·0	30·8	10·0	Coos Bay, Oreg	1875			J.Knowsland	San Francisco
129	R	Panope *Young -86*	W Sr	British	122	82·0	26·0	8·3	St.John,N.B.	1881	J.Mahony		Chas.A.Palmer	St.John,N.B.
130	G	Pansevitz *Wallis -89*	W Bk	German	343	112·8	27·6	16·6	Barth	1860	Witte		Wilh.A.Sarnow	Barth
131	*	Pansy *Coupland*	W Sk	British	92	80·6	21·0	10·6	Goole	1886	Cottingham Brs		S.F.Bartlett	Hull
132		Panteleimon	W Bg	Greek	174	82·7	24·3	..	Syra	1863			G.I.Oeconomos	Cumi
133		Pantoleon	W Bk	British	450	129·4	27·8	18·3	Sunderland	1855			W.B.Neale&B.B.Fisher	Liverpool
134	I	Paola *Pachi -88*	W Bk	Italian	599	148·6	32·0	20·6	Sestri Ponente	1875	E.Tixi	V 9.88	Giovanni Pace	Genoa
135	I	Paola R. *Perco 88-88* (ex Indus)	W Bk	Italian	822	165·0	34·8	22·5	Chiavari	1872	M.Tappani	V 5.88	F.Schiaffino fu G.	Genoa
136	I	Paolina *Fravega*	W Bg	Italian	369	121·0	29·5	18·2	Sestri Ponente	1871	G.Bozzano	V 1.88	Flli Fravega	Genoa
137	I	Paolino *Merabello -88* [illegible]	W Bk	Italian	280	101·7	28·1	16·3	Sestri	1863	A.Briasco fu G.B.	12.86	Munzone	Catania
138	A	Paolino (ex Due Fratelli) *Bruno*	W Bn	Italian	189	91·5	25·0	14·6	Messina	1857	G.Milore	V 4.88	P.Martino & Co.	Messina
139		Paolo *Tancredi*	W Bk	Italian	496	131·2	30·3	20·2	Sorrento	1865	F.Mauro		G.Tonietti fu Raff.	Porto Ferraio
140	A	Paolo Non Cangiò [illegible] *Tonietti*	W Bg	Italian	183	103·3	24·3	13·1	Dunkirk	1864			Paolo Tonietti	Porto Ferraio
141	V	Papa *Nielsen 88-88*	W Bk	Danish	385	128·5	25·5	16·1	Brake	1865	J.Oltmann's Wwe.	V 8.88	Actieselskabet "Papa" (J. Baymann)	Rudkjøbing
142	V	Papa *Heinz 83-85*	W Bk	German	743	170·9	30·2	20·0	Hamburg	1874	H.C.Stülcken Wwe.	V 7.87	H.C.Stülcken Wwe.	Hamburg
143	I	Papà (ex Stefano [illegible]) *Cucurú*	W Bk	Italian	556	135·2	34·5	21·0	Sestri Ponente	1867	Flli Cadenaccio	V 6.88	Ant.Cacace	Castell.di Stab.
144	I	Papà Giacomo *Lauro*	W Bg	Italian	394	124·6	26·2	17·1	Castellamare	1874	Trombarulo	V 11.87	Flli Lauro	Naples
145	I	Papà Giambattista *Carpena*	W Bn	Italian	130	93·8	23·7	12·1	Varazze	1883	S.Craviotto	12.83	G.Carpena	Spezia
146	I	Papà Giovanni *Gimelli*	W Sr	Italian	114	78·1	19·7	12·2	Lavagna	1874	G.Brignole	V 5.86	Eredi di Giovanni Devoto	Cagliari
147	I	Papà Giovanni Batta [illegible]	W Bk	Italian	760	164·0	33·8	21·8	Sestri Ponente	1875	G.Peragallo	V 6.88	S.Anselmo	Savona
148		Papà Rissetto *Zunino* [illegible]	W Bk	Italian	931	189·2	34·6	23·3	Sestri Ponente	1876	Stef.Bozano	V 5.87	Giov.Rissetto	Genoa
149		Papandi	W Bk	Russian	100	57·7	19·3	12·7	Kerescum	1874	Halfosman		G.S.Sclavo	Berdianski
150	DG	Papenburg *Sleur*	W Bk	German	247	113·5	25·4	12·2	Papenburg	1864	B.Mammes's Wwe	V 3.88	A.Wirtjes	Weener

* = Lloyd's Register; D = Nederlandsche Vereeniging; G = Germanischer Lloyd; I = Registro Italiano; R = American Record;
A = Veritas Austro-Ungarico; H = Veritas Ellenico; N = Norske Veritas; V = Bureau Veritas.

Dans la 4me Colonne, C = Composite (membrure en fer, bordé en bois); I = Fer; Stl = Acier; W = Bois.

Les Dates du doublage en métal (**Col. 11**) ressortent des livres dans lesquels les navires sont respectivement classés ou des rapports des Inspecteurs du Lloyd's Register.

Pour plus amples renseignements sur les navires classés par le Lloyd's Register, voir le Lloyd's Register Book et le Lloyd's Yacht Register.

Official No.	No. in Book	Registry	Ship's Name. Master, and Dates of Appointment to present Owner's Service and to Ship. Ship's late Name.	Material and Rig	Flag.	Net Tonnage	Length	Breadth	Depth	Built. Where.	When.	By whom.	Date of Metalling	Owners or Managers.	Port
	151	V	Paposo *Kessler 83-85*	I Bk	German	1098	211·4	34·3	20·0	Hamburg	1885	Blohm & Voss		H. Fölsch & Co.	Hamburg
	152		Pappa Kiriaco	W Bg	Turkish	287	..	..	..	Greece	1874			Ordenico Jeck ogdon [illegible]	Orden
	153		Paquebot de Brest No. 1 *Lepogue*	W Bn	French	112	71·5	19·7	10·3	Girard	1861	Arnaud jeune		Le Gouguec	Auray
	154		Paquete de Cantabria	W Bk	Spanish	300	113·8	27·7	15·8	England	1868			F. Romero Paz	Corunna
	155		Paquete de Itaquahy	W Sr	Brazilian	118	..	..	..					A. D. Pavao	Rio
	156		Paquete de Itajahy *Bedrigo*	W Sr	Brazilian	161	..	..	..	Villa do Conde	18 57/85			Liberato & Irmao	Itajahy
	157		Paquete de Lima *Erquiaga 87-87*	W Bn	Peruvian	215	103·0	23·4	12·3	England	1867			M. Erquiaga	Callao
	158	✻V	Paquita *Moller 88-88* (ex Anna ex Maggie Leslie)	I Bk	German	434	163·7	27·1	16·1	Newcastle	1862	A. Leslie & Co.		A. Burchard	Rostock
	159		Paquita	W Bk	Spanish	398	160·1	27·4	16·4	Liverpool	1855			Pedro Blanco	Santander
	160		Paquita (ex Carlos)	W Bn	Spanish	143	84·6	23·2	11·0	Oviedo	1858			José Orts y Ortuño	Gijon
	161		Paquita	W Pol	Spanish	121	85·0	25·3	10·6	Masnou	1840			J. Villalonga Tomas	Majorca
	162		Paquito	W Bg	Spanish	308	102·6	27·0	15·5	Oviedo	1863			J. M. Martinez	Rivadeo
	163		Paquito	W Bn	Spanish	147	78·7	20·1	10·9	Portosin	1863			Viuda de F. Roura	Villagarcia
[illegible]	164		Para *Casper -87*	W Bg	British	252	115·7	23·4	14·6	Alloa	1862	Adamson		A. Holland	Brisbane
[illegible]	165		Para *Colbert*	W Sr	American	109	92·0	26·0	7·6	E. Machias, Me.	1870			J. K. Ames	Machias, Me.
	166	✻	Para *Lemerle*	W Bk	French	391	131·0	28·8	13·4	Nantes	1888	P. Sevestre		D. Cronan fils	Nantes
	167		Para (ex Success)	W	Portug'se	362	99·7	24·7	17·8					L. T. Mendes	Lisbon
	168		Para *Pik -85*	W Bn	Swedish	180	98·1	22·4	12·0	Törefors	1861	E. A. Löfgren		G. J. Sundberg	Lysekil
[illegible]	169	DI	Paradiso *Prave -86*	W Bk	Italian	756	160·7	29·6	21·9	Varazze	1876	G. Aschieri		P. Lavarello	Genoa
	170		Paradissos *Bouldis*	W Bg	Greek	190	..	..	..	Galaxidi	1869			M. I. Cammenos	Cephalonia
	171	V	Paradox *Larsson -88*	W Bk	Russian	682	168·9	31·9	18·9	Oevelgönne	1876	J. P. Diercks		Wasa Rederi Bolag (K. Lehjonen)	Wasa
[illegible]	172	V	Paragon *Doane*	W Bk	British	820	172·6	35·1	20·0	Maitland, N.S.	1881	S. D. Brown		John Black	Maitland, N.S.
[illegible]	173		Paragon *Newton*	W Bg	British	199	83·9	25·0	15·6	Sunderland	1834			C. Marwood	Whitby
[illegible]	174		Paragon *Shute*	W Sr	American	153	96·6	25·5	8·7	Searsport, Me.	1856	Adamson		H. Shute	Belfast, Me.
	175		Paraguassu	W Bn	Brazilian	218	..	..	..					A. J. S. Pereira	Rio de Janeiro
[illegible]	176	V	Paramatta *Scott 87-90*	W Bk	British	925	175·5	36·3	20·3	Gardner's Creek, N.B.	1880	W. & R. Wallace		F. E. Sayre	St. John, N.B.
[illegible]	177	RV	Paramita *Prescott -88*	W S	American	1573	216·6	41·3	23·1	Freeport, Me.	1879	E. C. Soule & Bros.		E. C. Soule	Portland, Me.
[illegible]	178	V	Paran (CONDEMNED) *Graham*	W Bn	British	240	112·5	27·8	11·8	Londonderry NS	1878	A. Fraser		McLelan & Blackie	Truro, N.S.
	179	✻DG	Parana *Staben 83-86*	W Bk	German	408	129·7	26·9	16·8	Dumbarton	1863	A. McMillan & Sons		E. Burchard	Rostock
	180	N	Parana *Larsen*	W Bk	Norweg'n	707	158·9	32·6	18·9	Miramichi	1865	Mitchell		N. W. Coch	Brevig
	181		Paraskovia	W Sr	Russian	335	..	..	..		1869			I. Indins arfv.	[illegible]
	182		Paratous *Coll -89*	W Pol	Spanish	119	94·5	27·4	14·0	Barcelona	1858			C. F. Roses	Barcelona
[illegible]	183		Pardon G. Thompson *Chase 81-91*	W 3mSr	American	163	110·6	27·5	9·1	Kennebunk, Me.	1884	D. Clark		Z. H. Chase	New Bedford
[illegible]	184	✻	Parejero *Davidson -88*	C Bk	British	345	138·7	25·5	14·7	Greenock	1867	R. Steele		Baine & Johnston	Greenock
[illegible]	185	I	Parenti *Marini -87*	W Bk	Italian	699	159·4	32·8	22·0	Savona	1874	B. Calcagno		Z. Razeto	Genoa
[illegible]	186		Paroora (BROKEN UP) (ex White Eagle) *Rungard*	I S	British	879	203·3	32·8	20·9	Glasgow	1855	A. Stephen & Sons		J. Livingston	London
	187		Pargas *Sjoblom*	W 3mSr	Russian	277	117·7	26·6	12·4	Pargas	1872	G. Sundqvist		K. G. Fredriksson	Pargas
[illegible]	188		Paride (WRECKED) *[illegible]*	W Bg	Italian	108	78·4	22·5	11·8	Limite	1855	N. Pichiotti		Oresto Del Buono Ric	Porto [illegible]
[illegible]	189		Parisian *King*	W Sr	British	107	79·3	24·4	9·2	Mahone B'y NS	1882	T. Langille		Geo. W. Nass	Lunenburg
[illegible]	190		Parissis (ex Giuseppe Antonio) *Russo*	W Bk	Greek	666	158·1	31·0	19·8	Arenzano	1878	A. Debarbieri		N. G. Rousses	Syra
[illegible]	191		Parkend *Sike*	W 3mSr	British	139	101·7	24·0	8·0	Ipswich	1873	Bayley		J. W. Sully	Bridgwater
[illegible]	192	RV	Parker M. Whitmore *Whitmore*	W S	American	2205	239·4	43·6	28·0	Bath, Me.	1881	P. M. Whitmore		P. M. Whitmore	Bath, Me.
[illegible]	193		Parkersburg *Shroester*	W Sr	American	118	100·0	31·0	6·5	Parkersbrg Or.	1883	J. Parker		C. F. Doe	San [illegible]
[illegible]	194	✻D	Parknook *Thompson 76-76*	I Bk	British	793	198·7	32·2	19·5	Whitehaven	1876	Whitehaven S.B.C. Ltd.		R. Conaway & Co.	Whitehaven
[illegible]	195		Parkside	W Sr	British	132	89·9	20·1	12·3	Workington	1860	C. Lamport		R. M'Clenahan	Dundalk
	196	V	Parnass *Fradden 84-84*	W Bk	German	629	165·3	33·1	18·9	Geestemünde	1878	Schau & Oltmanns		F. Laeisz	Hamburg
[illegible]	197	A	Parnassos *Andrikopoulos*	W Sr	Greek	191	89·9	24·7	..	Syra	1885	A. Livadara		P. Andrikopoulis	Syra
[illegible]	198		Parnell *McIsaac 83-83*	W Bn	British	351	127·0	29·1	12·9	P.E. Island	1879	M. Foley		C. B. Stone	Auckland
[illegible]	199	✻N	Parramatta *Hegnander -87*	W S	Norweg'n	1402	231·0	38·2	22·8	Sunderland	1866	J. Laing		J. Simonsen	Mandal
	200	N	Parry *Ellingsen -87*	W Sr	Norweg'n	172	110·5	22·4	11·1	Skien	1871	A. Olsen		E. Ryan	Fredrikstad

✻ = **Lloyd's Register**; D = Nederlandsche Vereeniging; G = Germanischer Lloyd; I = Registro Italiano; R = American Record; A = Veritas Austro-Ungarico; H = Veritas Ellenico; N = Norske Veritas, V = Bureau Veritas

In the 4th Column, C = Composite (iron frame, wood planked); I = Iron; Stl = Steel; W = Wood

The Dates of Metalling (Col. 11) are based on the records of the Books in which Vessels are respectively classed, or on the Reports of the Surveyors of Lloyd's [illegible]

Further particulars of Vessels classed by Lloyd's Register will be found in Lloyd's Register Book and Lloyd's Yacht Register.

No.		Nom. Matériaux et Gréement. Capitaine et dates d'entrée au service des Armateurs actuels et à bord du navire. Ancien Nom.		Pavillon.	Tonnage				Construit. Où.	Quand	Par qui.	Date du doublage en métal	Armateurs ou Directeurs-Gérants.	Port d'Armement.
201	✻D	Parsee *Bailey 74-82*	I Bk	British	1281	227·1	35·9	22·1	Greenock	1868	R.Steele		J. & W.Stewart	Greenock
202	V	Parsee *Cocia -87*	W Bkn	British	875	136·7	32·0	12·8	Newport, N.S.	1887	W.H.Mosher	Y 11.87	R.Curry	Windsor, N.S.
203	DV	Parthena *Tennant*	W S	British	1381	205·0	38·1	24·0	Newport, N.S.	1877		Y 2.88	A. Gibson & Co.	Liverpool
204	✻R	Parthenia *Davies*	W Bk	British	749	179·5	34·7	19·9	P.E.Island	1881	W.Richards	Y 10.88	W. Richards (T.P. Richards & Co.)	Charlottetown, P.E.I.
205	✻	Parthenia *[illegible]*	I Bkn	British	249	138·8	24·1	12·2	Greenock	1875	J.E.Scott		Goodyear & Co.	Liverpool
206		Parthenia *Males*	W Bn	British	172	90·5	23·6	13·1	Lynn	1853			Bessey & Palmer	Wells
207		Parthenia *Simmons -88*	W Sr	British	111	79·8	24·0	9·6	Lunenb'rg N.S.	1882	D.Smith		S.W.Oxner	Lunenburg, N.S.
208	A	Parthenon	W Bg	Greek	232	95·2	25·3	..	Castellamare	1869		Y 8.88	J.Penthis	Syra
209	✻	Parthenope *Heal 86-*	I S	British	1563	250·6	39·9	23·6	Liverpool	1875	R.&J.Evans&Co.		Australasian Shipping Co., Ld.	Liverpool
210	N	Parthenope *Backer -83*	W Sr	Norweg'n	182	105·0	24·5	11·8	Cappeln	1866	Carl A.Staen		I.N.Backer	Drammen
211	✻	Parthia *Ward 87-87*	I Bk	British	1022	215·0	34·1	20·8	Harrington	1874	R.Williamson&S.		W.Nicol & Co.	Liverpool
212	✻	Parton *Christopher 82-82*	W Sr	British	99	81·9	20·7	10·6	Annan	1866	Nicholson		W.Kieley	Whitehaven
213		ParupathaPaykialetchemy *Nagappa Canaathipulle -87*	W Dy	British	133	91·4	23·0	15·2	Point Pedro	1887	Canthuenddy Veloopulle		Sapapathy Panisamutty	Jaffna
214		Pasig *Arriaga*	W Bn	Spanish	216	104·3	25·5	10·4	Albay	1877			I.Boncan	Manila
215		Pasqual	W Bg	British	266	122·0	22·0	13·0	Colombo	1880	T.Appo		Susey Fernando Bastian Appuhamy	Colombo
216	V	Pasqualino *Marcesa -88*	W Bk	Italian	388	116·4	29·5	18·4	Cassano	1864	Castellano	Y 2.88	E.Cacace	Castell.di Stab.
217	✻	Pass-by *Coombes*	W Sr	British	127	100·8	21·7	12·2	Kingsbridge	1862	Date	Y 11.87	J.Stephens	Fowey
218	✻	Pass of Balmaha *Hay 88-88*	St S	British	1498	245·4	38·8	22·5	Port Glasgow	1888	R.Duncan&Co.		Gibson & Clark	Glasgow
219	✻D	Pass of Leny *McDowall 88-88*	I Bk	British	1279	233·5	36·4	21·4	Glasgow	1885	London&Gls.C.		Gibson & Clark	Glasgow
220	N	Passat *Aanonsen -88*	W Bk	Norweg'n	686	158·6	33·4	18·5	Arendal	1878	Tellef Lassen	Y 11.87	J.L.Ugland	Arendal
221	N	Passe partout *Johnson -87*	W Bk	Norweg'n	556	151·8	32·5	17·9	Porsgrund	1877		Y 9.87	H.Biörn, jun.	Kragerö
222	NV	Pasteur *Lydersen -86*	W Bk	Norweg'n	420	136·7	31·7	13·9	Arendal	1886		Y 8.86	J.E.Hanssen	Arendal
223	✻	Pastoriza *(ex [illegible])*	W Bk	Chilian	1138	187·0	36·0	18·8	Honfleur	1864	Leviels	Y 8.83	J.Real	Valparaiso
224	RV	Patagonia *Hibbard -87*	W Bk	British	1199	190·5	38·0	22·5	Tusket, N.S.	1880	T.Killam	Y 12.88	Thos.Killam	Yarmouth, N.S.
225		Patagonia *Margaronis*	W Bk	Greek	493	139·1	29·7	16·9	Brake	1873	J.Oltmann's Wwe.	Y 6.86	P.Margaronis	Syra
226	NV	Patent *Olsen -88*	W Bk	Norweg'n	382	125·8	29·9	16·8	Tvedestrand	1875	H.Nielsen	Y 2.86	H.Nielsen	Tvedestrand
227		Patience *Winney* *(ex [illegible])*	W Bg	British	204	93·6	24·5	12·8	St.Malo	1857			A.M.Sainty	London
228	H	Patir *Oeconomos*	W Brz	Greek	243	..	..	..	Galaxidi	1873	J.Canatas	C 4.86	E.Oeconomos	Galaxidi
229	N	Patmos *Roth*	W Bk	Norweg'n	347	133·2	28·2	13·9	Stavanger	1876	M.Lunde	Y 5.87	J.P.Brodahl & Co.	Stavanger
230	✻	Patra *Hus*	W Bg	British	184	101·4	23·5	12·7	Brixham	1864	Dewdney	Y 1.86	D.Richards & Co.	Brixham
231	V	Patria *Agrell -87*	W Bk	German	391	131·1	28·5	16·4	Brake	1863	J.Oltmann's Wwe.	Y 3.88	G.Kindler	Rostock
232		Patria (BROKEN UP)	W Bk	Norweg'n	368	129·3	28·2	16·5	Stavanger	1857	M. Magnus		Carl Nyman	Stavanger
233	N	Patria *Jensen* *(ex [illegible])*	W Bg	Norweg'n	267	103·6	27·4	14·8	Rostock	1855	O.Ludewig		H.H.Pettersen	Arendal
234	N	Patria *Olsen*	W Bg	Norweg'n	193	98·1	25·4	11·7	Risör	1860			J.P.Pedersen	Tvedestrand
235		Patria *Ulenius 82-* *(ex [illegible])*	W Bk	Russian	508	140·6	33·5	17·6	Uleaborg	1863	G.Pettersson	Y 12.82	J.C.Svendelin & Co.	Uleaborg
236	V	Patria *Hallgren*	W 3mSr	Swedish	234	112·8	23·4	11·4	Westervik	1869	C.R.Fridell	Y 10.83	J.C.Tenger	Westervik
237	✻D	Patriarch *Allan 78-88*	I S	British	1339	221·1	38·1	22·3	Aberdeen	1869	W.Hood&Co.		G.Thompson & Co.	Aberdeen
238	✻	Patrician *Whitehead 85-87*	Stl Sr	British	166	120·5	23·2	10·1	Carrickfergus	1887	P.Rodgers		P.Ternan	Drogheda
239	V	Patrician *Kenney*	W S	American	1254	192·0	36·1	24·0	Damariscotta, Me.	1879	B.D.Metcalf	Y 8.87	W.A.Street	New York
240	N	Patrie *Pedersen*	W Bg	Norweg'n	284	97·6	27·9	15·9	Grimstad	1848	O.Jörgensen		P.C.Pedersen	Arendal
241	✻	Patriot *Saunders 86-86*	W Bn	British	199	107·9	23·6	12·1	Newburg	1866	Scrimgeour		M.W.Pearce	Plymouth
242	V	Patriot *Deutsch*	W Bg	German	234	103·5	25·6	13·1	Ueckermünde	18 49/66	M.Pretzer		W.Deutsch	Ueckermünde
243	NV	Patriot *Tellefsen 82-82*	W Bk	Norweg'n	483	140·9	30·5	17·6	Lillesand	1879	O.Reinertsen	Y 10.87	S.N.Hansen	Lillesand
244		Patriot (WRECKED) *Nergaard*	W Bg	Norweg'n	262	94·7	26·9	15·1	Grimstad	1847	Tellefsen		B.A.Olsen	Risör
245	N	Patriot *Ingvaldsen -76* (BROKEN UP)	W Bn	Norweg'n	207	111·0	26·0	12·9	Haugesund	1876	J.Hauge		K.H.Rönnevig	Haugesund
246	N	Patriot *Jacobsen 80-80*	W Sr	Norweg'n	110	89·2	21·3	10·4	Stavanger	1859			J.Jacobsen	Skudesnæs
247		Patriot *[illegible] Nilsson -83*	W Bk	Swedish	287	107·9	24·7	13·6	Piteå	1852			J.A.Enhörning	Sundsvall
248		Patriot *Pyk -87*	W Bk	Swedish	275	111·5	22·1	15·5	Whitby	1856			N.Pyk	Mölle
249		PatriotQueen *Hepburn -88*	W S	British	510	124·2	25·9	20·2	Whitehaven	1838	T. & J. Brocklebank		J.Sutherland	Liverpool
250	H	Patris *Biovikas*	W Bg	Greek	250	99·4	25·3	15·4	Galaxidi	1873	C.Papapetros	C 4.86	L.Biovikas	Galaxidi

✻=Lloyd's Register; D=Nederlandsche Vereeniging; G=Germanischer Lloyd; I=Registro Italiano; R=American Record; A=Veritas Austro-Ungarico; H=Veritas Ellenico; N=Norske Veritas; V=Bureau Veritas.

Dans la 4me Colonne, C=Composite (membrure en fer, bordé en bois); I=Fer; Stl=Acier; W=Bois.

Les Dates du doublage en métal (**Col. 11**) ressortent des livres dans lesquels les navires sont respectivement classés ou des rapports des Inspecteurs du Lloyd's Register.

Pour plus amples renseignements sur les navires classés par le Lloyd's Register, voir le Lloyd's Register Book et le Lloyd's Yacht Register.

Official No.	No. in List	Register in which classed	Ship's Name. Master, and Dates of Appointment to present Owner's Service and to Ship. Ship's Late Name.	Material and Rig	Flag	Net Tonnage	Length	Breadth	Depth	Where Built	When	By whom	Date of Metalling	Owners or Managers	Port of Registry
[illegible]	251	I	Patrocinio *Olivari*	W Bk	Italian	426	122·7	31·2	14·2	[illegible]	1864	P. Rolla	[illegible]	Maria Degregori	Genoa
[illegible]	252		Patruus *Le Cornu*	W Bg	[illegible]	187	87·1	25·4	14·4	[illegible]	1839			G. Le Gros	Jersey
	253		Patta Hoelabab *Pang Tebudong -84*	W Sr	Dutch	132	84·6	21·5	..					Kalabo	[illegible]
[illegible]	254	✻	Patterdale *Atkinson 83-83*	I S	British	1200	224·3	36·0	21·9	Whitehaven	1871	Whitehvn S B C		J. D. Newton & Co.	Liverpool
10706	255		Paugussett	W S	American	186	108·0	28·0	9·0	WilmingtnDel	1856			H. N. Johnson	Norfolk
	256		Paul *Godichard*	W Bg	French	199	93·5	22·0	12·0	Gravelines	1865	[illegible]		Godichard	Gravelines
	257		Paul	W Bg	French	161	85·3	21·0	12·0	St. Malo	1855			M. Ollivier	St. Malo
	258	V	Paul *Hoverdeck -85*	W Bk	German	744	151·1	34·0	18·7	Kiel	1878	A. Conradi [& Sons]	[illegible]	A. Strautzen	Hamburg
	259	G V	Paul *Klatt 78-85*	W Bk	German	376	120·1	28·4	16·5	Damgarten	18 64/75	H. Dierling		Fr. Lanneken	Stralsund
	260		Paul *Schauer*	W Bn Sr	German	170	95·8	23·8	10·7	Ueckermünde	1860	L. Wittenberg		J. Rudolph	Stettin
	261	G	Paul *Behrens* (ex Jan) (WRECKED)	W Bg	German	135	91·9	24·1	9·8	P. E. Island	1866	Lord		J. & T. Helmers	Rostock
	262		Paul	W Sr	Russian	218	..	..	..		1878			P. Wonninen	Impilahti
[illegible]	263		Paul	W Sr	Russian	104	70·4	22·0	8·8	Mühlgraben	1877	A. Ferle		G. Beckmann	Riga
	264		Paul *Jansson -88*	W Bg	Swedish	232	103·3	24·1	12·9	Luleå	1854			M. Jansson	Häfverö
	265	G	Paul Emil *Rager 84-84*	W Bn	Swedish	157	93·9	22·2	11·8	Kiel	1863	L. Back		R. J. Boye	Svendborg
	266	G	Paul Gerhard *Papist* (ex Mentor) *[188]*	W Bg	German	217	105·2	27·7	14·1	Swinemünde	1841	A. F. Ritzerow		Schoenberg & Domansky	Danzig
	267	✻	Paul Isenberg *Walters 88-88*	Stl Bk	German	1197	226·0	36·4	21·9	Londonderry	1888	C. J. Bigger		J. C. Pflüger & Co.	Bremen
	268		Paul Jones *Langhinrichs*	W Bg	German	303	105·2	27·7	16·0	Damgarten	1852	H. Dierling		Beckmann & Co.	Rostock
	269	V	Paul Marie *Lévêque 86-88*	W Bk	French	324	114·8	23·9	14·3	St. Malo	1868	Vve Dandin & fils	[illegible]	Lole Villers	Saint Malo
	270		Paul Marty *Olsson*	W Sr	Swedish	104	71·2	18·9	10·6	Wismar				N. J. Cedergren	Timrå
[illegible]	271	K V	Paul Revere *Sewall -87*	W S	American	1657	221·0	41·2	24·[illegible]	Boston, Mass	1876	Smith & Townsend	[illegible]	De Groot & Peck	New York
	272	V	Paul Rickmers *Heinsberg 85-87*	W Bk	German	1191	188·6	38·7	25·5	Geestemünde	1876	R. C. Rickmers	[illegible]	R. C. Rickmers	Geestemünde
[illegible]	273		Paul Seavey *Kimble -87*	W Sr	American	225	97·0	26·7	13·0	Brewer, Me.	1866			G. B. Hook	Bangor, Me
[illegible]	274		Paul & Thompson *Weeks*	W Sr	American	195	102·0	29·3	8·3	DorchesterN.J.	1867			T. F. Pollard	Perth Amboy
	275	V	Paula *Diekmann 86-86*	W Bk	German	516	136·1	30·3	16·5	Hammelwarden	1876	C. Lühring	[illegible]	A. Schiff	Elsfleth
	276	V	Paula *Bergmann -87*	W Bn Sr	German	284	120·1	23·6	13·0	Papenburg	1887	Abr. Beckmann	[illegible]	Abr. Beckmann	Papenburg
	277	✻	Pauline *Kylls -88* (ex Little Moreen)	W Bk	Dutch	575	112·5	31·6	18·4	St. John, N. B.	1864	Morrow	[illegible]	S. van der Hei	Harlingen
	278		Pauline *Arnaud* (ex Caroline)	I Bk	French	465	150·0	26·0	17·0	Belfast	1862	Harland & Wolff		Bouge & Dor	Marseille
	279	V	Pauline *Ermand*	W Bg	French	151	83·0	22·0	11·2	Nantes	1867	Title		L. Jacquier	Nantes
	280	V	Pauline *Hars 86-88*	W Sr	French	113	184·6	21·5	10·9	Gravelines	1867	Hochart [Lavallée]		Maillard & T. Claeyssen	Dunkirk
	281		Pauline *Kruger* (CONDEMNED)	W Bk	German	337	106·7	27·9	15·8	Ueckermünde	1862	L. Wittenberg		Retzlaff & Schober	Stettin
	282	N	Pauline *Knudsen -87*	W Bk	Norweg'n	396	112·8	27·8	18·8	N. BedfordMs.	1831			A. Christensen	Drammen
	283	N	Pauline *Lund -82* (ex Norwood)	W Sr	Norweg'n	164	93·5	23·9	11·2	P. E. Island	1873	Lanesworth	[illegible]	B. A. Ellingsen	Christiania
	284		Pauline *Bervahlen* (WRECKED)	W Bn	Norweg'n	146	97·4	24·3	11·1	Bergen	1875		[illegible]	O. G. Skadberg	Egersund
	285	N	Pauline *Mortensen*	W Sr	Norweg'n	137	90·7	22·0	10·1	Elsfleth	1854			Chr. Olsen Holter	Horten
	286		Pauline *Frosch*	W Bg	Russian	252	..	..	..	Finby	1874			G. Signell	Finby
	287		Pauline	W Sr	Russian	190	67·0	22·0	9·0	Kesterwege	1878	Schilinsky		A. W. Grünberg	Riga
	288		Pauline *Östergren -86*	W Sr	Swedish	187	90·5	26·6	12·[illegible]	[illegible]	1869			O. Pettersson	Bro. R[illegible]
	289	V	Pauline Marie *Stephany [-86]*	W Sr	French	127	78·7	20·[illegible]	11·5	Miane	1867	Ollivean		Cardinal	Nantes
	290		Paulita (ex [illegible])	W Sr	Spanish	128	73·2	20·1	5·5	Arens	1851			N. Tur	[illegible]
	291	✻	Paulus *Ludvigsen -86*	I Bk	Norweg'n	587	161·4	29·1	12·2	[illegible]	1885	Stavanger [Stöberie Dok]		Norske Missionsselsk. (T. Melbye)	Stavanger
	292	N	Paulus *Sørensen 78-86*	W Bk	Norweg'n	427	137·0	31·0	16·0	Porsgrund	1875		[illegible]	A. & H. Sörensen	Porsgrund
	293	V	Pauvre Orpheline *Lemaitre -86*	W Bg	French	129	82·3	20·3	11·8	Granville	1860			Clement	Granville
	294		Pavel *Stepanoff*	W Sr	Russian	216	..	..	..	Solom	1883			P. Sidorow	Solom
[illegible]	295		Pavilion	W Sr	American	114	80·2	23·7	8·3	Bucksport, Me	1862			F. C. Pendleton	Belfast
[illegible]	296		Pavilion	W Sr	American	109	84·9	21·8	7·9	Tremont, Me	1849			E. C. Gates	Calais
[illegible]	297		Pavlis	W Bg	Greek	127	71·5	22·6	..	[illegible]	1867			J. Michalinos	Piræus
	298		Pavlos Focas *Caterali*	W Bg	Greek	198	91·1	23·0	11·0	Sunderland	1868			Caterali Bros.	Cephalonia
	299	✻	Pavonia	W Sk	British	86	78·0	21·0	10·4	Grimsby	1886	Smith Bros.		W. Burns	Grimsby
	300	G	Pax *Frankenbusch*	W Sr	German	147	92·5	22·[illegible]	10·6	Elsfleth	1859	G. Wempe		Tobias	Brake

✻ = Lloyd's Register; D = Nederlandsche Vereeniging; G = Germanischer Lloyd; I = Registro Italiano; K = American [Record]; A = Veritas Austro-Ungarico; H = Veritas Ellenico; N = Norske Veritas; V = Bureau V[eritas].

In the 4th Column, C = Composite (iron frame, wood planked); I = Iron; Stl = Steel; W = Wood.

The Dates of Metalling (Col. 11) are based on the records of the Books in which Vessels are respectively classed, or on the Reports of the Surveyors of Lloyd's

Further particulars of Vessels classed by Lloyd's Register will be found in Lloyd's Register Book and Lloyd's Yacht Register.

N°	Registre	Nom. Matériaux et Gréement. Capitaine et dates d'entrée au service des Armateurs actuels et à bord du navire. Ancien Nom.	Pavillon	Tonnage net.	Dimensions. Longueur	Largeur	Creux	Construit. Où.	Quand.	Par qui.	Date du doublage en métal	Armateurs ou Directeurs-Gérants.	Port d'Armement.
301	GR	Pax *Treufeldt* W Bk	Russian	279	122·2	25·8	13·8	Libau	1856	J. Pabre		J. Nicolar	Pirnau
302		Paykéaletchemy W Bky	British	128	96·1	22·6	15·8	Ceylon	1864			Anthony Savorimutto	Jaffna
303	R	Payson Tucker *Foster* W Bkn	American	661	166·9	35·6	17·3	Portland, Me.	1883	J. & D. Brewer	11.88	J. S. Winslow & Co.	Portland, Me.
304	✱	Payta *Alexander 48-81* I Bk	British	563	162·7	29·2	18·1	Sunderland	1867	Oswald & Co.		Turner, Edwards & Co.	Bristol
305	V	Paz [illegible] W Bk	Spanish	278	115·1	27·5	14·1	Corunna	1865			D. L. Malmres	Corunna
306	✱V	Peabody *Allemand -86* W Bk	French	447	132·1	27·6	17·6	St. Malo	1870	Lhotellier	V 7.88	J. B. Demange	Nantes
307	N	Peabody *Christophersen 83-87* W Bk	Norweg'n	457	128·9	30·5	17·3	Arendal	1869	C. P. Olsen	V 2.88	Gunder Olsen	Arendal
308		Peacedale W Sr	American	123	91·7	28·1	7·1	Bath, Me.	1871			J. N. R. Hazard	Newport, R. I.
309	DV	Peacemaker *Clement 64-72* W Bk	British	1364	182·6	36·3	22·4	MossGlen, N. B.	1872	Wetmore & Merritt	V 5.88	The Bark "Peacemaker" (Co.) (W. W. Turnbull)	St. John, N. B.
310	R	Peacemaker [illegible] W Bk	Argentine	682	57·0	33·0	20·0	Shelburne, N.S.	1868		V 2.88	Maguire & Co.	Buenos Ayres
311		Pearl *Hildebrandt -85* W Sr	British	163	102·0	24·2	9·7	Horton, N.S.	1881	J. B. North	[illegible]	Allen Haley	Windsor, N.S.
312	✱	Pearl *Phillips* W Bn	British	159	98·5	23·1	12·2	P. E. Island	1875	Richards	10.88	Edwards Bros. & Co.	Aberystwith
313		Pearl W Bn	British	133	91·6	23·0	11·0	Wallace, N.S.	1860			W. Hennessey	St. Johns, N. F. L.
314	✱	Pearl *Weatherall -87* W Sr	British	129	88·9	22·6	11·5	Salcombe	1887	R. Harnden	10.87	J. H. Weatherall	London
315	✱	Pearl *Murdock -84* W Bn	British	118	87·0	21·1	11·3	Montrose	1861	Petrie		R. Patton	Montrose
316		Pearl *Kelso 82-83* W Bn	British	110	82·6	20·3	11·2	Whitehaven	1867	Shepherd		A. Nelson	Whitehaven
317	✱	Pearl *Lewis 55-80* W Sr	British	100	91·4	21·7	10·0	Amlwch	1880	W. Thomas		T. F. Evans & Co.	Beaumaris
318	✱	Pearl *Lower 85-85* W Sr	British	93	86·0	22·0	10·5	Truro	1875	Dyer	V 2.88	J. Estlick	Truro
319	✱	Pearl *Davison 85-87* W Sr	British	92	85·6	21·7	10·4	Perth	1870	Wallace		J. Reney	Chester
320		Pearl *Murray -87* W Bk	American	509	132·5	29·2	18·5	Medford, Mass.	1862			J. F. Kennedy	San Francisco
321		Pearl *Zimmer -85* W Bg	American	195	105·5	26·0	10·8	Madison, Conn	1872			D. Trowbridge	New York
322		Pearl Nelson *Kemp 79-82* W Sr	American	117	84·0	23·0	9·0	Essex, Mass.	1881	J. James & Co.		R. S. Cook & Co.	Provincetn, Mass.
323	N	Peder Anæus *Knitsen -83* W Bk	Norweg'n	406	127·3	29·8	17·1	Christiansund	1855			S. Christensen	Drammen
324	✱	Pedestrian *Pinches* W Sr	British	132	93·0	22·0	11·0	Barnstaple	1877	Westacott	V 1.88	J. Tregaskes	Fowey
325	I	Pedro *Corsanego -88* W Bk	Italian	407	125·3	29·1	19·0	Sestri Ponente	1867	A. Briasco fu G. B.	V [illegible]	M. Corsanego & P. Dessalesquis	Genoa
326		Pedro Bk	Spanish	502	133·8	27·9	18·1	San Dionisio	1866			F. Rahola	Barcelona
327		Pedro *Ribeiras* W Bn	Spanish	235	101·0	27·1	13·6	Blanes	1870	Vieta		S. Mensua	Valencia
328	V	Pedro Gusi *Reyes* W Bk (ex Rosa Rahola)	Spanish	509	134·2	29·5	..	Prá	1874		V 2.88	Gusi hermanos	Barcelona
329	V	Pedro Maristany *Miradoser* W Bn	Spanish	283	108·2	28·2	14·5	Arens	1870			Doña M. P. Maristany	Barcelona
330		Pedro Peral W Bk (ex Wilhelm Hoffmann)	Peruvian	512	150·5	31·9	16·2	Wolgast	1875	J. H. Kraft		J. V. Peral	Callao
331		Pedro Sueyras *Venegas 87-87* W Bk	Peruvian	689	156·5	28·5	20·7	Jermesey	1853			W. Venegas	Callao
332		Pedro I. W Bn (ex George & Rosalinda Moura I.)	Portug'se	235	95·8	21·0	14·1					J. J. R. Leitao & Filhos	Funchal
333		Pedro II. W Bn	Brazilian	261	..	..	..					M. A. Guimaraes	Paranagua
334		Pedro II. W Bk	Brazilian	167	..	..	..					F. J. Cardoso	Itaguahy
335	✱	Peeblesshire *Webster 83-88* I Bk	British	866	200·0	32·7	19·6	Port Glasgow	1879	Russell & Co.		T. Law & Co.	Glasgow
336		Peep o' Day W Bg (ex Rita Fleming)	British	135	93·5	24·9	9·7	Waldboro', Me.	1848			S. J. Binning	Belfast
337		Peeress *Thompson* W Bk	British	749	158·7	32·5	21·3	Sunderland	1859	Doxford		Salvage Association	North Shields
338		Peeress *McLean -88* W Bn	British	239	101·8	25·0	12·9	Cardigan, P. E. I.	1877	J. Duncan	10.87	C. Outerbridge & Bro.	Hamilton, Bermuda
349	V	Peerless *Morrison* W Bkn	British	317	125·2	30·7	11·9	Lower Selmah, N.S.	1882	J. McK. Blaikie	V 6.88	J. McK. Blaikie	Halifax, N.S.
340		Peerless *Hoogs 82-83* W Bn	British	284	112·0	29·1	12·7	Halifax, N.S.	1873			Chas. B. Stone	Auckland, N. Z.
341		Peerless W Bg	British	147	102·6	20·7	12·6	Greenock	1850	R. Steele & Co.		Samuel Edwards	St. Johns, N. F. L.
342		Peerless *Stockman* W 3m Sr	American	233	127·0	33·0	9·9	Gardiner City, Oreg.	1878	T. Peterson		J. Tuft	San Francisco
343	✱I	Pefetta *Schiaffino* W Bk	Italian	586	112·1	29·1	20·3	Varazze	1872	A. Bollo	V [illegible]	Prosp. Schiaffino	Genoa
344	V	Pégase *Dupont* C Bk	French	471	111·5	29·0	18·8	Bordeaux	1873	Arman	V [illegible]	F. Merle	Bordeaux
345	✱	Pegasus *Ward 75-* I 3m S	British	2564	314·0	42·3	24·9	Liverpool	1884	W. H. Potter & Sns		W. T. Dixon & Sons	Liverpool
346	✱	Peggie Doy *Hill* W Bk	British	248	115·3	23·5	11·7	Leith	1871	T. Adamson	V [illegible]	J. V. Georges	Port Louis
347		Peggy *Prowse 87-87* W 3m Bn	British	247	119·1	25·8	14·8	Salcombe	1863	J. J. Vivian	[illegible]	A. Goodridge & Sons	St. John's
348	NV	Peggy *Zeiffert* W Bg	Norweg'n	229	112·5	25·6	12·3	Christiansund	1859	O. Meble	[illegible]	A. Zeiffert	Arendal
349	V	Pehr Brahe *Jansson -87* W Bk	Russian	577	150·5	31·2	16·6	Brahestad	1877		V [illegible]	N. Sittkoff	Mariehamn
[illegible]	V	Pei-Ho *Lamken 74-* I Bk	German	443	148·9	25·6	13·9	Vegesack	1877	H. F. Ulrichs		Gildemeister & Ries	Bremen

✱ = Lloyd's Register; D = Nederlaandsche Vereeniging; G = Germanischer Lloyd; I = Registro Italiano; R = American Record; [illegible] = Veritas Austro-Ungarico; H = Veritas Ellenico; N = Norske Veritas; V = Bureau Veritas.

Dans la 4me Colonne, C = Composite (membrure en fer, bordé en bois); I = Fer; Stl = Acier; W = Bois.

Les Dates du doublage en métal (Col. 11) ressortent des livres dans lesquels les navires sont respectivement classés ou des rapports des Inspecteurs du Lloyd's Register.

Pour plus amples renseignements sur les navires classés par le Lloyd's Register, voir le Lloyd's Register Book et le Lloyd's Yacht Register.

Official No.	No. in Book	Registry in which classed	Ship's Name. Master, and Dates of Appointment to present Owner's Service and to Ship. Ship's Late Name.	Material and Rig	Flag	Net Tonnage	Length	Breadth	Depth	Built, Where	Built, When	Built, By whom	Metalling	Owners or Managers	Port of Registry
104	251	I	Patrocinio *Olivari*	W Bk	Italian	426	122·7	31·3	19·2	Genoa	1864	P. Rolla	V 1·84	Maria Degregori	Genoa
33641	252		Patruus *Le Cornu*	W Bg	British	167	87·1	23·9	11·4	Gaspé	1839			G. Le Gros	Jersey
	253		Patta Hoelabab *(Pranoe Pabamberg -84)*	W Sr	Dutch	133	84·6	21·9	..					Kalalsi	Palangkabu
65867	254	✱	Patterdale *Allinson* 83-83	I S	British	1266	224·3	36·0	21·9	Whitehaven	1871	Whitehven SBC		J. D. Newton & Co.	Liverpool
19706	255		Paugussett	W Sr	American	186	108·0	28·0	9·0	Wilmington Del	1856			H. N. Johnson	Norfolk, Va.
	256		Paul *Godichard*	W Bg	French	199	93·5	22·0	12·6	Gravelines	1865	Hochart		Godichard	Gravelines
	257		Paul	W Bg	French	161	85·3	21·0	12·0	St. Malo	1855			M. Ollivier	St. Malo
	258	V	Paul *Homerich -85*	W Bk	German	744	151·4	31·0	18·7	Kiel	1878	A. Contain & Sohn	X 10.88	A. Strantzen	Hamburg
	259	G V	Paul *Klatt 78-85*	W Bk	German	379	120·4	28·4	16·5	Damgarten	18 64/75	H. Dierling		Fr. Lancken	Stralsund
	260		Paul *Schauer*	W 3m Sr	German	170	95·8	23·8	10·7	Ueckermünde	1860	L. Wittenberg		J. Rudolph	Stettin
	261	G	Paul *Eckhoff (ex Jane) (WRECKED)*	W Bn	German	136	91·9	24·1	9·8	P. E. Island	1866	Lord		J. & T. Helmers	Barssel
	262		Paul	W Sr	Russian	218	..	..	..		1878			P. Wonninen	Inspilahti
325	263		Paul	W G	Russian	104	70·4	22·6	8·8	Mühlgraben	1877	A. Ferle		G. Beckmann	Riga
	264		Paul *Jansson -88*	W Bg	Swedish	232	103·3	21·1	12·9	Enjer	1854			M. Jansson	Häfverö
	265	G	Paul Emil *Böyer 81-81*	W Bn	Danish	157	93·0	22·2	11·8	Kiel	1863	L. Bock		R. J. Böye	Svendborg
	266	G	Paul Gerhard *Lupist (ex Mentor) 55-88*	W Bg	German	247	105·2	27·7	11·1	Swinemünde	1841	A. F. Ritzerow		Schoenberg & Domansky	Danzig
	267	✱	Paul Isenberg *Walters 86-86*	Stl Bk	German	1197	226·0	36·1	21·9	Londonderry	1888	C. J. Bigger		J. C. Pflüger & Co.	Bremen
	268		Paul Jones *Langhinrichs*	W Bg	German	393	105·2	27·7	16·0	Damgarten	1852	H. Dierling		Beckmann & Co.	Rostock
	269	V	Paul Marie *Lévêque 86-86*	W Bk	French	324	114·8	23·9	14·3	St. Malo	1888	Vve Dandin a fils	X 1.88	L. de Villers	Saint Malo
	270		Paul Marty *Olsson*	W Sr	Swedish	104	71·2	18·9	10·6	Wismar				N. J. Cedergren	Timrå
150000	271	R V	Paul Revere *Small -87*	W S	American	1657	221·0	41·2	24·6	E. Boston, Mass	1876	Smith & Townsend	X 10.85	De Groot & Peck	New York
	272	V	Paul Rickmers *Hesse 85-87*	W Bk	German	1191	188·6	38·7	25·5	Geestemünde	1876	R. C. Rickmers	X 10.85	R. C. Rickmers	Geestemünde
19546	273		Paul Seavey *Kimble -87*	W Sr	American	225	97·0	26·7	13·0	Brewer, Me.	1866			G. B. Hook	Bangor, Me.
20059	274		Paul & Thompson *Wasgatt*	W Sr	American	195	102·0	29·3	8·3	Dorchester N.J.	1867			T. F. Pollard	Perth Amboy
	275	V	Paula *Dickmann 86-86*	W Bk	German	516	136·1	30·3	16·5	Hammelwarden	1876	C. Lühring	V 2.88	A. Schiff	Elsfleth
	276	V	Paula *Breymann -87*	W 3m Sr	German	284	120·1	23·6	13·0	Papenburg	1887	Abr. Beckmann	V 7.87	Abr. Beckmann	Papenburg
	277	✱	Pauline *Eylts -86 (ex Lizzie Morrow)*	W Bk	Dutch	575	142·5	31·6	18·4	St. John, N.B.	1864	Morrow	V 1.86	S. van der Hei	Heveskes
	278		Pauline *Arnaud (ex Caroline, ex Recipe)*	I Bk	French	465	150·0	26·0	17·0	Belfast	1862	Harland & Wolff		Bonge & Dor	Marseille
	279	V	Pauline *Evrard*	W Bg	French	151	85·0	22·0	11·2	Nantes	1867	Tillé		L. Jacquier	Nantes
	280	V	Pauline *Hars 85-85*	W Sr	French	113	84·6	21·5	10·9	Gravelines	1867	Hochart		Maillard & T. Claeyssen	Dunkirk
	281		Pauline *Kruger (CONDEMNED)*	W Bk	German	337	106·7	27·9	15·8	Ueckermünde	1862	L. Wittenberg [Lavallee		Retzlaff & Schober	Stettin
	282	N	Pauline *Knudsen -87*	W Bk	Norweg'n	396	112·8	27·8	18·8	N. Bedford Mass	1831			A. Christensen	Drammen
	283	N	Pauline *Lund -82 (ex Nora)*	W Sr	Norweg'n	164	99·5	23·9	11·2	P. E. Island	1873	Langworth	V 88	R. A. Ellingsen	Christiania
	284		Pauline *Berentsen (WRECKED)*	W Bn	Norweg'n	146	97·4	24·3	11·1	Bergen	1875		X 10.80	O. G. Skalberg	Egersund
	285	N	Pauline *Ma. Nesen*	W Sr	Norweg'n	137	90·7	22·0	10·4	Elsfleth	1854			Chr. Olsen Holter	Horten
	286		Pauline *Fench*	W Bg	Russian	252	..	..	..	Finby	1874			G. Signell	Finby
	287		Pauline	W S	Russian	106	67·0	22·0	9·0	Kesterzeem	1878	Schilinsky		A. W. Grünberg	Riga
	288		Pauline *Östergren -86*	W Sr	Swedish	187	90·5	26·6	12·0	Västö	1869			O. Pettersson	Bro, Sweden
	289	V	Pauline Marie *Stephans [-86*	W Sr	French	127	78·7	20·3	11·5	Means	1867	Olliveau		Cardinal	Nantes
	290		Paulita *(ex Paulita)*	W Sr	Spanish	128	73·2	20·1	5·5	Arenas	1851			N. Tur	Barcelona
	291	✱	Paulus *Ludvigsen -86*	I Bk	Norweg'n	507	161·4	29·1	12·2	Stavanger	1885	Stavanger [Støberi & Dok		Norske Missionsselskab (T. Malhus)	Stavanger
	292	N	Paulus *Sörensen 79-86*	W Bk	Norweg'n	427	137·0	31·0	16·0	Porsgrund	1875		X 1.88	A. & H. Sorensen	Porsgrund
	293	V	Pauvre Orpheline *Lécole -86*	W Bn	French	129	82·3	20·3	11·8	Granville	1860			Clement	Granville
	294		Pavel *Stepanoff*	W Sr	Russian	246	..	..	..	Saimis	1883			P. Sidorow	Salmis
19432	295		Pavilion	W Sr	American	114	80·2	23·7	8·5	Bucksport, Me.	1862			F. C. Pendleton	Belfast, Me.
19951	296		Pavilion	W Sr	American	109	84·9	21·8	7·9	Tremont, Me.	1849			E. C. Gates	Calais, Me.
89	297		Pavlis	W Bg	Greek	127	71·5	22·6	..	Syra	1867			J. Michalinos	Piraeus
	298		Pavlos Focas *Cafevati*	W Bg	Greek	108	91·1	23·0	11·0	Sunderland	1868			Cafevati Bros.	Cephalonia
	299	✱	Pavonia	W Sk	British	86	78·0	21·0	10·4	Grimsby	1886	Smith Bros.		W. Burns	Grimsby
	300	G	Pax *Praukenbusch*	W Sr	German	147	92·5	22·5	10·6	Elsfleth	1859	G. Wempe		Tobias	Brake

✱ = Lloyd's Register; D = Nederlandsche Vereeniging; G = Germanischer Lloyd; I = Registro Italiano; R = American Record; A = Veritas Austro-Ungarico; H = Veritas Ellenico; N = Norske Veritas; V = Bureau Veritas.

In the 4th Column, C = Composite (iron frame, wood planked); I = Iron; Stl = Steel; W = Wood.

The Dates of Metalling (Col. 11) are based on the records of the Books in which Vessels are respectively classed, or on the Reports of the Surveyors of Lloyd's [illegible]

Further particulars of Vessels classed by Lloyd's Register will be found in Lloyd's Register Book and Lloyd's Yacht Register.

		Nom. Matériaux et Gréement.	Capitaine et dates d'entrée au service des Armateurs actuels et à bord du navire. Ancien Nom.		Pavillon	Tonnage net	Dimensions. Longueur	Largeur	Creux	Construit. Où.	Quand	Par qui.	Date du doublage en métal	Armateurs ou Directeurs-Gérants.	Port d'Armement.
301	GR	Pax	Trenfeldt	W Bk	Russian	279	122·2	25·8	13·8	Libau	1856	J.Pubre		J.Nicolar	Pirnan
302		Paykéaletchemy		W Bg	British	128	96·4	22·6	15·8	Ceylon	1864			Anthony Savorimatto	Jaffna
303	R	Payson Tucker		W Bkn	American	661	166·9	35·6	17·3	Portland,Me.	1883	J.&D.Brewer	[illegible]	J.S.Winslow & Co.	Portland,Me.
304	✻	Payta	Alexander 48-87	I Bk	British	569	169·7	29·2	18·1	Sunderland	1867	Oswald & Co.		Turner,Edwards & Co.	Bristol
305	V	Paz	(E Maria Dolores) Peira	W Bk	Spanish	278	115·1	27·5	14·1	Corunna	1865			D.L.Mahures	Corunna
306	✻V	Peabody	Gillirand -85	W Bk	French	467	132·1	27·4	17·6	St.Malo	1870	Lhotellier	[illegible]	J.B.Demange	Nantes
307	N	Peabody	Chri-Andersen 87-87	W Bk	Norweg'n	457	128·9	30·5	17·3	Arendal	1869	O.P.Olsen	[illegible]	Gunder Olsen	Arendal
308		Peacedale		W Sr	American	123	91·7	28·1	7·1	Bath,Me.	1871			J.N.R.Hazard	Newport,R.I.
309	DV	Peacemaker	Clossont 64-72	W Bk	British	1064	182·6	36·3	22·4	MossGlen,N.B	1872	Wetmore & Merritt	[illegible]	The Bark "Peacemaker" (Co.(W.W.Turnbull)	St.John,N.B.
310	R	Peacemaker (ex E. H. Duval)	Gauthier -87	W Bk	Argentine	662	57·0	33·0	20·0	Shelburne,NS.	1868		[illegible]	Maguire & Co.	Buenos Ayres
311		Pearl	Hildebrandt -87	W Sr	British	163	102·0	24·2	9·7	Horton,N.S.	1881	J.B.North	[illegible]	Allen Haley	Windsor,N.S.
312	✻	Pearl	Phillips	W Bn	British	159	98·5	23·1	12·2	P.E.Island	1875	Richards	[illegible]	Edwards Bros.& Co.	Aberystwith
313		Pearl		W Bn	British	133	91·6	23·0	11·0	Wallace,N.S.	1860			W.Hennessey	St.Johns,N.F.L.
314	✻	Pearl	Weatherall -87	W Sr	British	129	88·9	22·6	11·5	Salcombe	1887	H.Harnden	Y 10.87	J.H.Weatherall	London
315	✻	Pearl	Murdock -84	W Bn	British	118	87·0	21·1	11·3	Montrose	1861	Petrie		R.Patton	Montrose
316		Pearl	Kelso 83-83	W Bn	British	110	82·6	20·3	11·2	Whitehaven	1867	Shepherd		A.Nelson	Whitehaven
317	✻	Pearl	Lewis 56-80	W Sr	British	100	91·1	21·7	10·0	Amlwch	1880	W.Thomas		T.F.Evans & Co.	Beaumaris
318	✻	Pearl	Lowes 85-85	W Sr	British	98	86·0	22·0	10·5	Truro	1875	Dyer	Y 4.87	J.Estlick	Truro
319	✻	Pearl	Davison 85-87	W Sr	British	92	85·6	21·7	10·1	Perth	1870	Wallace		J.Reney	Chester
320		Pearl	Murray -87	W Bk	American	509	132·5	29·2	18·5	Medford,Mass.	1862			J.F.Kennedy	San Francisco
321		Pearl	Zimmer -85	W Bg	American	195	105·5	26·0	10·8	Madison,Conn	1872			D.Trowbridge	New York
322		Pearl Nelson	Knapp 73-82	W Sr	American	117	84·0	23·0	9·0	Essex,Mass.	1881	J.James & Co.		R.S.Cook & Co.	Provincetn,Mass.
323	N	Peder Anæus	Eriksen [illegible] -85	W Bk	Norweg'n	406	127·3	29·8	17·1	Christiansund	1855			S.Christensen	Drammen
324	✻	Pedestrian	Punches	W Sr	British	132	93·0	22·0	11·0	Barnstaple	1877	Westacott	[illegible]	J.Tregaskes	Fowey
325	I	Pedro	Carrasa -88	W Bk	Italian	407	125·3	29·1	19·0	SestriPonente	1867	A.Briasco fu G.B.	[illegible]	M.Corsanego & P. Passalacqua	Genoa
326		Pedro		Bk	Spanish	502	133·8	27·9	18·1	SanDionisio	1866			F.Rahola	Barcelona
327		Pedro	Riberas	W Bn	Spanish	235	101·0	27·1	13·6	Blanes	1870	Victa		S.Mensua	Valencia
328	V	Pedro Gusi (ex Rosa Rabel)	Peyes	W Bk	Spanish	593	133·2	29·5	..	Pra	1874		[illegible]	Gusi hermanos	Barcelona
329	V	Pedro Maristany (Maristany)		W Bn	Spanish	283	108·2	28·2	14·5	Arens	1870			Doña M.P.Maristany	Barcelona
330		Pedro Peral (ex Wilhelm Hoyer)		W Bk	Peruvian	512	150·5	31·9	16·2	Wolgast	1875	J.H.Kraft		J.V.Peral	Callao
331		Pedro Sueyras	[illegible] 87-87	W Bk	Peruvian	689	156·5	28·5	20·7	Jermesey	1853			W.Venegas	Callao
332		Pedro I. (ex [illegible] & [illegible] Moura I.)		W Bn	Portug'se	235	95·8	21·0	11·1					J.J.B.Leitne & Filhos	Funchal
333		Pedro II.		W Bn	Brazilian	261	..	..	..					M.A.Guimaraes	Paranagua
334		Pedro II.		W Bk	Brazilian	167	..	..	..					F.J.Cardoso	Itaguahy
335	✻	Peeblesshire	Isbister 83-88	I Bk	British	866	200·0	32·7	19·6	PortGlasgow	1879	Russell & Co.		T.Law & Co.	Glasgow
336		Peep o' Day (ex Ella Fleming)		W Bg	British	135	93·5	21·9	9·7	Waldboro',Me	1848			S.J.Binning	Belfast
337		Peeress	Thompson	W Bk	British	740	158·7	32·5	21·3	Sunderland	1859	Doxford		Salvage Association	North Shields
338		Peeress	McLeod -88	W Bn	British	239	101·8	25·0	12·9	Cardigan,P.E.I	1877	J.Duncan	[illegible]	C.Outerbridge&Bro.	Hamilton, Bermuda
339	V	Peerless	Morrison	W Bkn	British	317	123·2	30·7	11·9	Londond'ry NS	1882	J.McK.Blaikie	[illegible]	J.McK.Blaikie	Halifax,N.S.
340		Peerless	Heays 82-83	W Bn	British	284	112·0	29·4	12·7	Halifax,N.S.	1873			Chas.B.Stone	Auckland,N.Z.
341		Peerless		W Bg	British	147	102·6	20·7	12·6	Greenock	1850	R.Steele & Co.		Samuel Edwards	St.Johns,N.F.L.
342		Peerless	Steckman	W 3mSr	American	233	127·0	33·0	9·9	Gardiner City,Oreg.	1878	T.Peterson		J.Tuft	San Francisco
343	✻I	Pefetta	Schiaffino	W Bk	Italian	586	112·1	29·1	20·3	Varazze	1872	A.Bolla	[illegible]	Prosp.Schiaffino	Genoa
344	V	Pégase	Dupont	C Bk	French	471	111·5	29·0	18·8	Bordeaux	1873	Arman	[illegible]	F.Merle	Bordeaux
345	✻	Pegasus	Ward 75-84	I 4mS	British	2564	311·0	42·3	24·9	Liverpool	1884	W.H.Potter&Sns		W.T.Dixon & Sons	Liverpool
346	✻	Peggie Doy	H.H.	W Bk	British	248	115·3	23·5	11·7	Leith	1871	T.Adamson	[illegible]	J.V.Georges	Port Louis
347		Peggy	Pascoe 7-87	W 3m Bn	British	247	119·1	25·8	14·8	Salcombe	1868	J.J.Vivian	[illegible]	A.Goodridge & Sons	St.John's
348	NV	Peggy	Zeiffert	W Bg	Norweg'n	229	112·5	25·0	12·3	Christiansund	1869	J.Mæhle	[illegible]	A.Zeiffert	Arendal
349	V	Pehr Brahe	Jans'ius -87	W Bk	Russian	577	151·5	31·2	16·6	Haraldholm	1877		[illegible]	N.Sittkoff	Mariehamn
350	V	Pei Ho	Lassen 78-0	I Bk	German	433	148·9	25·3	13·9	Vegesack	1877	H.F.Ulrichs		Gildemeister & Ries	Bremen

✻=Lloyd's Register; D=Nederlandsche Vereeniging; G=Germanischer Lloyd; I=Registro Italiano; R=American Record; A=Veritas Austro-Ungarico; H=Veritas Ellenico; N=Norske Veritas; V=Bureau Veritas.

Dans la 4me Colonne, C=Composite (membrure en fer, bordé en bois); I=Fer; Stl=Acier; W=Bois.

Les Dates du doublage en métal (Col. 11) ressortent des livres dans lesquels les navires sont respectivement classés ou des rapports des Inspecteurs du Lloyd's Register.

Pour plus amples renseignements sur les navires classés par le Lloyd's Register, voir le Lloyd's Register Book et le Lloyd's Yacht Register.

Official No. (1)	No. in Book (2)	Registry in which classed (3)	Ship's Name. Ship's Late Name. (4)	Master, and Dates of Appointment to present Owner's Service and to Ship.	Material and Rig.	Flag. (5)	Net Tonnage (6)	Dimensions (7) Length	Breadth	Depth	Built (8) Where.	(9) When.	(10) By whom.	Date of Metalling. (11)	Owners or Managers. (12)	Port of Registry
10745	351		Peiro	Kelley 75	W Sr	American	113	86·0	25·8	8·4	Jonesport, Me	1867			D.J.Sawyer	Jonesport
303	352	V	Pelagia	Jerefanni	W Sr	Greek	248	96·1	27·9	..	Cassos	1881		V.86	E.Frangoulis	Syra
	353		Pelayo (ex Stomer)		I Bn	Spanish	539	180·1	25·5	17·0	Toulon	18$\frac{5}{7}$			R.de Silva y Ferro	Palma
56	354	A	Pelesac	Buntielich -88	W Bk	Austrian	444	130·5	27·6	17·0	Fiume	1872	F[lli]. Schiavoni	V.88	B.Buntielich fu A.	Cuciste
	355	✳	Pelican	H. Ellis	W K	British	92	79·0	20·8	10·7	Brixham	1886	J.W.&A.Upham		R.Hellyer	Hull
	356		Pelle (ex Tessa) (FOURWEARD)	Åsberg -87	W Bk	Swedish	309	112·0	25·0	14·0	Neckermunde	1854	L.Wittenberg		J.S.Göthe	Söderhamn
3518	357	A	Pellegra	Olivari -87	W Bk	Italian	420	125·6	29·6	19·[illegible]	[illegible]	1860	F[lli]. Tixi	V.88	A.Cichero	Genoa
2730	358	V	Pellegra Madre	Marola -88	W Bk	Italian	755	162·6	32·8	22·9	Sampierdar'na	1876	B.Frassinetti	V. 11.88	G.B.Mortola	Genoa
1070	359	I	Pellegro	Gusto	W Bk	Italian	445	125·3	31·4	19·0	Savona	1865	F.Sirello	V.88	P.Schiaffino	Genoa
973	360	I	Pellerano	Carpena	W Bn	Italian	118	85·6	22·5	12·1	Savona	1884	B.Calcagno	V.88	G.B.Pellerano fu B.	Spezia
531	361	I	Pellicano	Zolezzi	W Bk	Italian	900	174·5	35·8	22·8	Chiavari	1870	M.Tappani	V.86	Dav.Fra.Chighizola	Spezia
216	362		Pelopidas		W Sr	Greek	122	108·9	22·5	..	Syra	1875			A.Makantinas	Syra
	363	V	Pelotense	Castanheira	W Bn	Brazilian	193	94·8	22·3	10·4	[illegible]	1874	J.Brancos	V.88	F.da C.Leite	RioG.
	364		Pelotense III.	Mendes	W Sr	Brazilian	215	..	..	..				V.87	F.da C.Leite	RioG.
44840	365	✳	Pembroke Castle	Lewis -87	I Bkn	British	384	140·2	26·2	16·3	Glasgow	1863	A.Stephen&Sns.		Simpson Bros.	Swansea
	366		Pen-Bas (ex Rebecca)	Allainmat -86	W Sr	French	159	104·2	23·6	12·7	Kingsbridge	1863	Date	V. 10.84	P.Joly	Paimpol
60083	367		Pen Guin		W Sr	British	127	86·0	20·0	10·0	Penang	1875			Lee Kong Boon	Penang
	368		Penalva		W Sr	Brazilian	137	..	..	..					A.J.M.Amarante	Rio Jan.
	369	N	Penda	Christensen -87	W Sr	Norweg'n	162	102·3	25·5	11·8	Arendal	1875	T.Omholdt		Actieselskab "Oceana"	Sandefjord
	370	V	Pende Adelfi	Diakachis 1 -87	W Bk	Greek	372	124·0	28·2	18·4	Cassos	1887			D.Diakachis	Syra
74750	371		Pendle Hill	Evans	I Bkn	British	222	124·6	24·0	11·4	Sunderland	1878	J.Blumer&Co.		Hawkes Bay Shipping Co.,Ld.	Napier,N.Z.
60694	372	✳	Pendragon	Billett 88-88	I S	British	1262	215·6	35·8	22·6	Liverpool	1868	T.Royden&Sns.		M'Diarmid,Greenshields & Co.	Liverpool
	373		Pendreck		W Sr	Turkish	145	..	..	..	Pendreck	1869			Pendrecklee Kior Mahmoud Oghlou Hadji Ahmed & sons	Pendreck
9277	374		Penelope	Moore 85-85	W Bg	British	100	75·5	20·8	12·2	Whitehaven	1836	L.Kennedy&Co.		J.Kingsberry	Belfast
377	375		Penelope		W Bk	Greek	326	113·1	27·6	..	Stockton	1870			S.Garoufalos	Syra
	376		Penelope	Alafousos	W Bg	Greek	305	..	..	..	Trieste	1849			G.G.Alafousos	Syra
6	377		Penelope		W Bg	Greek	131	77·1	23·0	..	Spezzia	1859			D.Nicolau	Skiathos
8	378		Penelope	Chanos	W Bg	Greek	120	77·4	20·0	..	Ithaca	1367	P.Leuketsas	V.87	P.Chanos	Ithaca
	379	N	Penelope	Bjørnstad -88	W Bg	Norweg'n	299	102·5	29·6	14·7	Finland	18$\frac{34}{76}$			H.A.Lehre	[illegible]
	380		Penelope	Johanson 84-84	W Bkn	Swedish	184	95·5	20·8	12·4	Saltkällan	1854	J.Johannsson		A.M.Koch	[illegible]
	381	V	Penerf	Guipo -87	W Bk	French	267	107·9	25·6	13·2	Nantes	1875	Doussain	V. 8.88	K.Mervant	Bordeaux
	382	N V	Penguin	Tjensvold 85-88	W 3m Sr	Norweg'n	214	104·7	26·2	11·9	Arois	1870	F.Johansen	V.88	Ths.S.Falck	Stavanger
86277	383	✳	Pengwern	Davies 74-83	I S	British	1573	256·1	38·2	22·7	Greenock	1882	Russell&Co.		S."Pengwern" Co.,Ld (R.Thomas & Co.)	Liverpool
56488	384		Pennine	Souter 87-87	W Sw	British	296	110·4	27·8	16·7	Sunderland	1867	Austin&Son	V.88	J.& N.C.Bull	Newhaven
19550	385		Pennsylvania (CONDEMNED)		W Sr	American	117	89·0	23·0	8·0	PortRichmond,N.Y.	1848			F.H.Rackett	[illegible]
150131	386	R V	Penobscot	Eatan 81-81	W Bk	American	1067	187·0	37·1	23·0	Bucksport,Me.	1878	Gonn&Hill	V.88	J.S.Emery & Co.	[illegible]
150267	387	R	Penobscot	Carter	W 3m Sr	American	340	137·0	32·0	11·0	Belfast,Me.	1882	CarterBros.		E.Sibley	[illegible]
78079	388	✳	Penpont (ex T. S. B.)	Graham 88-88	I Bn	British	276	137·4	26·2	12·5	Hayle	1877	Harvey&Co.		T.C.Guthrie	Glasgow
83853	389	✳	Penryn	Eaton 73-80	W K	British	69	74·1	20·3	9·2	Ponsharden Creek	1880	E.Martin&Co.		W.Hosken & Co.	Falmouth
332	390	I	Pensa al Domani (ex Clelia Bianca)	Bacco	W Bk	Italian	610	142·0	32·4	20·3	Sestri	1868	G.Casanova	V.86	C.Anselmo	Savona
3542	391	I V	Pensacola (ex Ave Cristoforo)	Parodi -88	W Bk	Italian	982	183·0	34·4	23·6	Varazze	1878	A.Cerruti	V. 12.88	GiuseppePiaggiofuGB	Genoa
	392		Pensamento II. (ex Liverpool)	Ferreira	W Bn	Portug'se	141	84·6	20·3	10·5	Villa do Conde	1874		V.86	A.F.d'Araujo	Lisbon
	393		Pensativo		W Bn	Brazilian	260	..	..	..					M.J.A.Costa	Rio Jan.
	394	V	Pensée	Pillard -88	W Bn	French	211	99·1	25·6	12·6	Bordeaux	1878	Raymond frères	V.86	Mme A.Pillard	Bordeaux
	395	V	Pensée	Aufret -87	W Sr	French	132	83·0	20·0	11·3	Granville	1865	Clément fils		A.Lepésant	Paimpol
	396		Penseur	Morel -87	W Bk	French	473	128·9	28·5	17·8	St.Malo	1869	L.Lhotelier	V. 10.86	J.B.Demange	Nantes
68688	397		Penshaw	Key	W Bk	British	729	175·8	32·6	20·0	Sunderland	1875	J.Crown	V.88	H.G.Brown	Hong Kong
3565	398		Pensiero	Parilli	W Bg	Italian	211	101·8	26·0	15·9	Savona	1849			S.Parilli	Genoa
917	399		Pensiero Italiano	Urbano	W Bn	Italian	277	100·4	28·7	17·4	SestriPonente	1863	D.Tixi	V. 2.88	F.Massa	Genoa
	400	Π	Pente Adelphi	Procopiu	W Bg	Greek	234	92·2	28·9	..	Syra	1867		V.86	G.Panoutsos	Piræus

✳=Lloyd's Register; A=Veritas Austro-Ungarico; D=Nederlandsche Vereeniging; G=Germanischer Lloyd; Π=Veritas Ellenico; I=Registro Italiano; N=Norske Veritas; R=American [illegible]; V=Bureau Veritas

In the 4th Column, C=Composite (iron frame, wood planked); I=Iron; Stl=Steel; W=Wood.

The Dates of Metalling (Col. 11) are based on the records of the Books in which Vessels are respectively classed, or on the Reports of the Surveyors of Lloyd's [illegible]

Further particulars of Vessels classed by Lloyd's Register will be found in Lloyd's Register Book and Lloyd's Yacht Register.

1	2-3	4 Nom. Matériaux et Grément. Capitaine et dates d'entrée au service des Armateurs actuels et à bord du navire. Ancien Nom.	5 Pavillon.	6 Tonnage net.	7 Dimensions. Longueur	Largeur	Creux	8 Construit. Où.	9 Quand.	10 Par qui.	11 Date du doublage	12 Armateurs ou Directeurs-Gérants.	13 Port d'Armement.
		Pente Adelphi	Greek		..	..	..		1887	Andrea Emmanouel			
	✻	Penthesilea *Wilson -88* I S	British	1608	[illegible]	[illegible]	24.6	[illegible]	1869	Oswald&Co.		W.D.Reid	Liverpool
		Pepa (ex Rafaela Galarza, ex Aldershot) W Bk	Spanish	761	163.6	32.9	21.7	[illegible]	1863	Casella		F.Tarongi	Havana
	V	Pepe (ex Sirenita, ex Don Juan) *Font* I Bn	Spanish	449	181.4	25.3	14.1	Newcastle	1853			Alejandro Berge	Barcelona
		Pepe (ex [illegible]) W Bn	Spanish	162	92.5	9.8	10.2	[illegible]	1862	Von Senden		Antonio Forro	Havana
		Pepe Tono *Albi* W Bn Sr	Spanish	241	107.6	28.0	13.0	Javea Alicante	1877	A.Mari		José A.Bolufer	Alicante
		Pepita *Girminole* W Bk	Spanish	693	112.7	31.0	18.3	[illegible]	1867			José M.Lago	Manila
		Pepita W Bk	Spanish	371	108.6	25.4	14.6	[illegible]	1858			Salvio Brillas	Valencia
	G V	Pepita *Johansson -85* W Bg	Swedish	247	121.0	24.4	12.2	Carlshamn	1869	Andersson	V 7.81	L.Pahlsson	Lerberget
		Pepita Piedras (ex Rafael) *Gil -87* W Bk	Spanish	312	112.5	[illegible]	14.6	[illegible]	1856			Pedro F.Rey	Villagarcia
		Pepito W Sr	Argentine	149	103.6	25.9	9.0	Sestri Ponente	1884	C.Bozzano	V 7.84	E.Capurro fu G.	Buenos Ayres
		Pepito *Theus* W Bg	Spanish	141	85.0	21.9	11.3	Valencia	1864			M.T.Pascual & Co.	Barcelona
	V	Pepito (ex Elvira) *Beloña* W Sr	Spanish	104	81.0	21.7	10.3	Corunna	1851			Francisco Roura	Corunna
		Peppina *Ravalich* W Bk	Austrian	542	139.1	29.1	19.1	Fiume	1863	G.Brazzoduro	V 11.86	M.Paicurich	Fiume
	V	Peppina *Paturzo* W Bk	Italian	615	146.0	32.8	19.8	[illegible]	1873	Castellano	V [illegible]	Giudice & Paturzo	Palermo
		Peppina (ex Rosalia [illegible]) *Califano* W Bg	Italian	346	117.4	28.9	17.2	[illegible]	1870	De Rosa	V [illegible]	B.Capiello	Castellamare
		Peppina (ex San Giuseppe) W Bn		127	..	..	..	Galaxidi	1867		V [illegible]		
		Peppina *Rallo* W Bn	Italian	107	..	..	..	Castellamare di Stabia	1876	A.Bonifacio		V. di Maggio	Naples
		Peppina M. (ex [illegible] A.) *Caseo -88* W Bk	Italian	577	137.1	28.2	18.1	Fiume	1871	Flli Schiavoni	V 6.82	G.Maresca	Castell.di Stab
		Peppina S. (ex Alberto) *Spampinato* W Bn	Italian	182	91.8	25.2	15.4	Piano di Sorrento	1858		V 12.80	Sigra.Veda.Spampinato	Catania
	V	Peppino (ex Roberto) *Cafiero* W Bg	Italian	242	107.9	25.1	16.0	[illegible]	1873	F.S.Mauro	V [illegible]	R.Scarpati	Castellamare
		Peppino (ex Leandro) *d'Antoni 86-* W Bn	Italian	233	112.5	25.5	12.8	Brake	1869	F.Nicolai	V 10.83	Mariano Denaro fu Giuseppe	Catania
		Peppino (ex Domenico) *Sallustro* W Bg	Italian	182	92.5	20.7	14.4	Sorrento	1860	F.Mauro		L.Musso	Naples
		Peppino Amodeo *Virzi* W Bn	Italian	111	89.6	21.9	11.6	Trapani	1875		V 10.87	Salv.Amodeo	Trapani
		Peppino Chelli (ex Stefanino) *Minassi* W Bn	Italian	109	81.0	23.7	11.6	Malta	1856			Rizzotti,Chelli & Andrioli	Messina
426	V	Peppino Mignano *De Martino* W Bk	Italian	459	135.0	30.8	19.1	Procida	1868		V 1.83	A.Mignano	Naples
427	✻	Per Ardua *Ramlal 75-86* C Bk	British	788	178.4	32.3	18.7	Sunderland	1871	R.Thompson,jn.	V 1.88	W.Nicol & Co.	Liverpool
428	✻	Percy *Nicoll -88* W Sr	British	127	90.3	22.8	11.3	Neyland	1879	Mills	V 1.87	W.B.Thoyts (Mathews & Luff)	London
429	A	Percy (ex [illegible]) *Ivancich -87* W Bk	Austrian	627	143.7	29.6	20.7	Lussinpiccolo	1871	G.S.Cattarinich	V 6.87	G.Benussi	Trieste
430		Percy Edward *Anderson* W Bn	American	189	105.0	28.0	10.0	Boston,Mass.	1866	Curtis		C.Anderson	San Francisco
431	R	Percy H. Reed W Sr	British	148	91.3	26.0	8.2	Bear River,N.S	1884	Wm.Reed		Wm.Reed	Digby,N.S.
432	✻V	Père Guignard *Chalès -88* W Bk	French	262	108.9	28.7	13.2	Ars,Ré	1880	Guignard	V [illegible]	L.de Villers	St.Malo
433		Père Jacques *Menage* W Bg	French	210	96.5	23.6	12.7	Nantes	1877	Alleau&Aubert	V [illegible]	Guibert & fils	St.Malo
434		Père de Famille *Brunet 83-83* W Bk	French	559	138.7	28.5	18.5	Dunkirk	1866	G.Beck	V 11.86	G.Beck	Dunkirk
435	V	Pereira (ex Bom Jesus de Fao, ex Bayão de Lago) *Gonçalves -87* W Bn	Portug'se	197	95.1	23.6	11.1	Fão	1875	F.D.dos S. Borda,jun.	V [illegible]	J.J.Ribeira de Magalhaes	Oporto
436		Pereira Borges (PONTOON)(ex Maria Carlota, ex Estrella) *Borges* Wbk	Portug'se	313	111.8	28.7	14.5	Lisbon	1854	J.A.da Silva		J.P.Borges	Lisbon
437		Perfect W Sr	British	112	65.8	18.6	12.1	Newport	1836			P.Gorman	Cork
438	✻	Pergamon *Hilgendorf -88* Stl Bk	German	1403	244.1	38.1	20.7	Hamburg	1888	Blohm & Voss		F.Laeisz	Hamburg
439	✻	Peri *Pryde 84-84* I Bk	British	897	212.0	32.1	19.7	South Shields	1868	Readhead, Softley&Co.		J.Stewart & Co.	South Shields
440		Peri (ex Sol[illegible]) W Bg	British	205	96.7	23.9	10.6	Baltimore,U.S	1845			Aboo Buckar	Bombay
441	✻	Pericles *Philip 61-88* I S	British	1598	259.6	39.1	23.6	Aberdeen	1877	W.Hood&Co.		G.Thompson & Co.	Aberdeen
442	V	Périgny *Lainé* W Bk	French	331	116.8	2[illegible].6	11.8	St.Malo	1869	Hervot	V 8.88	H.L.Langevin	Bordeaux
443		Perilla W Bn	British	192	106.9	23.7	12.8	P.E.Island	1863	Gillis		J.Logan	Belfast
444		Perinoo W Bg	Turkish	150	..	..	..	Barten	1882			Hattab tijari Mehemet Ali & Co.	Barten
445		Perla del Océano W Bn	Spanish	215	102.0	23.5	12.6	Albay	1864			J.M.Quiogay & Co.	Manila
446	A	Perla del Plata (ex Michele Pinasco) *Vives -88* W Bk	Argentine	500	145.1	29.5	19.5	Sestri	1866	D.Briasco fu G.B.	V [illegible]	J.Clarammux & Co.	Buenos Ayres
447		Perlan *Carlsson* W Sr	Swedish	100	75.5	21.7	9.3	Ekenäs	1855			G.& F.Johansson	Oscarshamn
448		Perle *Nilsson 83-88* W Bk	British	401	139.9	26.4	16.0	Reiherstieg	1864	F.Dreyer		J. V. Grant.	Shanghai
449		Perle *Fick -87* W Bg	German	271	97.7	26.5	15.7	[illegible]	1836	W.Henberg		J.Stevenson	Stettin
450	D V	Perle *Ackermann* W Bn	German	160	90.9	22.6	10.4	Grossefehn	1870	T.J.Cassens	V 11.86	Tönges J.Cassens&Sohn	Grossefehn

✻ = Lloyd's Register; D = Nederlandsche Vereeniging; G = Germanischer Lloyd; I = Registro Italiano; R = American Record; A = Veritas Austro-Ungarico; H = Veritas Ellenico; N = Norske Veritas; V = Bureau Veritas.

Dans la 4me Colonne, C = Composite (membrure en fer, bordé en bois); I = Fer; Stl = Acier; W = Bois.

Les Dates de doublage en métal (Col. 11) ressortent des livres dans lesquels les navires sont respectivement classés ou des rapports des Inspecteurs du Lloyd's Register.

Pour plus amples renseignements sur les navires classés par le Lloyd's Register, voir le Lloyd's Register Book et le Lloyd's Yacht Register.

Official No.	No. in Book	Society in which classed	Ship's Name. Master, and Dates of Appointment to present Owner's Service and to Ship. Ship's Late Name.	Material and Rig	Flag	Net Tonnage	Length	Breadth	Depth	Where built	When	By whom	Date of Metalling	Owners or Managers	Port
17784	451	❋	Perpetua *Montgomery* (ex Maria M)	W Bk	British	397	125·8	27·3	17·2	Sunderland	1856	R. Wilkinson		J. Montgomery	Liv
	452	V	Perpetua *Hillenberg* -85	W Bn	Swedish	200	97·7	22·1	12·6	Upperud	1877	O. Olsson		I. E. Olsson	
	453	V	Perpignan *Garnier*	W Bk	French	239	100·1	24·3	12·8	Nantes	1874	Bachan		J. Garnier	Nantes
84127	454	❋	Perriton *Lewis* 81-81	W Sr	British	86	71·0	21·1	10·1	Minehead	1881	B. Williams		T. K. Ridler	
	455		Perrosien *Lecalvez*	W Bg	French	195	96·8	21·6	12·5	St. Vaast	1869	E. Costey		L. Lecalvez	
	456		Persante *Clausen* 7-83	W Bn	Danish	204	91·8	21·4	14·1	Dundee	1847			M. Christiansen	
42150	457		Perseus *Machon* 86-86	W Bn	British	126	96·7	21·6	11·3	P.E.Island	1860	Longworth		H. W. Smith	
44894	458	❋	Perseverance *Perry* -86	W Bn	British	165	97·0	22·0	12·0	Fowey	1862	Nickles		W. F. Hannan	Fowey
18548	459		Perseverance *Murray* 85-85	W Bk	British	164	101·0	22·7	13·2	Sunderland	1852			R. J. Kydd	
29470	460		Perseverance *Agnmore* (STRANDED) (ex Claire & Marie)	W Sr	British	117	90·5	22·1	9·9	Nantes	1861			T. Stuart & Co.	
150352	461	R	Perseverance *Willard*	W Sr	American	186	102·9	27·8	9·3	Bath, Me.	1882	A. Palmer		A. Palmer	Bath
	462	❋V	Perseverance *Danion* -74	I m S	French	2511	293·0	41·3	26·1	Glasgow	1886	W. B. Thompson		Ant. Dom. Bordes & fils	Bordeaux
	463	V	Persévérance *Denvilla*	W Sr	French	109	86·3	22·3	10·7	Dunkirk	1880	D. VanCauwenberghe		C. Van Cauwenberghe	Dunk
	464		Persévérant *Bouvie* 85-86 (CONDEMNED)	C Bk	French	496	135·5	27·2	19·9	Nantes	1863	T. Dubigeon		A. Postel & ses fils	Cherb
	465		Persévérant *Durand* -81	W Bk	French	350	108·6	24·3	14·8	Nantes	1871	Tillé		A. Hubeau	St. N
	466		Persévérante *Marcelli* -81	W Sr	French	109	81·0	18·7	10·4	St. Malo	1865	L. Mailard		A. Girardin	Saint
848	467	I	Perseverante *Greco* 85-86 (ex Archimede)	W Bk	Italian	300	115·4	27·5	11·2	Castell di Stab.	1860	A. Bonifacio		G. Greco & P. Siragusa	Messina
66	468	A	Perseveranza *Cosulich* -88	W Sr	Austrian	186	84·9	23·3	13·1	Lussinpiccolo	1884	G. Tarabocchia		G. A. Cosulich	Lussin
92485	469	V	Persia *Malcolm*	W Bkn	British	578	155·7	35·9	17·0	Hantsport N.S.	1886	G. W. Churchill		G. W. Churchill	Windsor
	470	N	Persia *Aarøe*	W Bk	Norweg'n	1770	207·6	40·0	30·7	Quebec	18⁵²/₆₅			Johnson, Ritter & Co.	Christ
77115	471	❋	Persian *Pilkenton* 83-84	I S	British	1387	241·8	37·2	22·5	Greenock	1878	Scott & Co.		A. Mackay & Co.	London
73403	472	❋	Persian *White* -86	W Bk	British	273	126·1	24·6	13·4	Dundee	1876	Tay Ship Co.		J. R. Wilson	Dundee
29807	473	❋	Persian *Abbott* 86-86	W Bn	British	142	81·9	19·2	13·2	Plymouth	1852	C. Gent		C. Gent	Plym
52829	474	❋	Persian Empire *Storey* 88-88	I S	British	1532	230·1	37·0	23·8	W. H.	1865	Pile, Spence & Co.		G. Duncan & Co.	London
	475		Persistant *Cerlain* -86	W Bk	French	552	115·?	30·3	18·2	Nantes	1865	T. Dubigeon		H. Anger aîné	Havre
	476		Pertero Pialel	W Sr	Turkish	113	..	..	..	Gossia	1873	A. H. Pirachi		Ahmet Baggi Pirachi	Gossia
41460	477		Perthshire	W Sr	British	126	68·1	19·0	12·0	Perth	1836			J. Young	Dundee
75168	478	❋	Peru *Orr* 81-80	I Bk	British	683	182·1	30·1	17·8	Kinghorn	1876	J. Key & Sons		J. Stewart & Co.	Dundee
20443	479		Peru *Larrabee*	W S	American	1457	168·0	38·2	26·5	Y'mouth, Me.	1867			S. C. Blanchard	Portland
	480		Peru *Kjöller* 88-88	W Bg	Danish	261	88·5	25·1	17·0	Copenhagen	1846	E. P. Bonnesen		Den. Kgl. Grønl. Handel	Copen
	481	N	Peru *Pedersen* -88	W Bk	Norweg'n	589	139·8	31·6	19·6	Bremen	1851			G. Gundersen	Arendal
	482	N	Peru *Sivertsen*	W Bn	Norweg'n	253	118·3	28·2	12·1	Bergen	1881			H. Svendsen	
	483	V	Peru *Gillberg* 81-81	W Bk	Swedish	583	151·5	29·8	17·2	Gefle	1859			P. J. Hagerstrand	Gefle
	484		Peruana Deriya	W Sr	Turkish	175	..	..	..	Kevken Skell	1880			Hairie Tijari Bessi Mehemed Ali & Co.	Kerken
	485		PeruanaDeriya *Ibrahim*	W Sr	Turkish	175	..	..	..	Kerken Skell	1883			Bezji Zade Mehemed Ali Agha	Kerken
73607	486	❋D	Peruvian *McKinnon* 80-84	I Bk	British	608	176·5	30·1	17·4	Glasgow	1875	D. & W. Henderson		Donaldson Bros.	Glasgow
	487		Peruvienne *Steen* 80-82	W Bk	Norweg'n	510	137·2	30·6	18·6	Bordeaux	1865			A. Steen	Drammen
	488		Pervase Bachri	W Bg	Turkish	188	..	..	..	Dulcigno	1866			Ijet Efentis	
	489		Pervasi Bahri	W Bg	Turkish	130	68·9	21·3	10·2	Dulcigno	1869	Hussein Omer		Mehmet Giuli & Co.	
	490		Pervasi Bahrie (damaged)	W Bn	Montn'g'n	280	*95·5	27·9	19·1	Fiume	1848			Mahir Mehmet Aga	Dulcigno
	491	V	Pervenche *Léban*	W Sr	French	129	93·2	21·3	11·0	St. Vaast	1879	E. Costey		J. Gicquel	Paimpol
	492	V	Pestalozzi *Burgdorff* 74-85	I Bk	German	1039	210·6	34·4	19·9	Hamburg	1884	Blohm & Voss		F. Laeisz	Hamburg
31844	493		Pet *Harrison*	W Bk	British	268	108·1	23·8	15·4	St. Mary's, N.S.	1847			Peter Fony	
45197	494		Pet *Gill* 80	W Sr	British	146	96·8	21·1	12·5	Dundee	1863	Tay Ship Co.		J. Hutet & Co.	Wisbech
70528	495	❋	Pet *Guest*	W Sr	British	98	89·1	21·2	10·4	Kingston	1876	Geddie		R. Jones	Chester
	496		Peter *Schmidt* -79	W Bk	German	301	108·3	26·5	15·6	Barth	1854	J. S. Kraeft		Wilh. A. Sarnow	Barth
	497	V	Peter *Buskell*	W Sr	German	126	82·3	20·7	10·8	Elsfleth	1871	Kramer		Ukena & Janssen	Leer
	498		Peter *Sørensen* (ex ...)	W Bk	Norweg'n	320	111·2	28·0	17·1	Veteri	1859	L. Ellessen		J. P. Sorensen	Laurvig
	499	V	Peter *Christensen* -88	W Bk	Swedish	443	121·6	27·6	16·9		1869	J. H. Wilken		J. Gjersöe	Christ
	500		Peter *Paulsen*	W S	Norweg'n	199	102·7	22·1	13·1	Newcastle	1857			A. Bondé	Christ

❋ = Lloyd's Register; D = Nederlandsche Vereeniging; G = Germanischer Lloyd; I = Registro Italiano; R = American; A = Veritas Austro-Ungarico; H = Veritas Ellenico; N = Norske Veritas; V = Bureau Veritas

In the 4th Column, C = Composite (iron frame, wood planked); I = Iron; Stl = Steel; W = Wood.

The Dates of Metalling (**Col. 11**) are based on the records of the Books in which Vessels are respectively classed, or on the Reports of the Surveyors of Lloyd's

Further particulars of Vessels classed by Lloyd's Register will be found in Lloyd's Register Book and Lloyd's Yacht Register.

No.		Nom. Capitaines et dates d'entrée en service. Ancien Nom.	Matériaux et Gréement	Pavillon	Tonnage net	Longueur	Largeur	Creux	Où	Quand	Par qui	Date du doublage en métal	Armateurs ou Directeurs-Gérants.	Port d'Armement.
501		Peter *Nilsson 79-77*	W Bk	Swedish	515	133·5	26·1	19·1	Boston Ms	1852		v. 3.88	P.A.Lindberg	Gothenburg
502		Peter *Mansson* ex Fredrik	W Bk	Norweg'n	315	107·6	25·4	15·6	Wolgast	1837			C.E.Mansson	Helsingborg
503		Peter *Pettersson*	W Sr	Swedish	109	90·2	21·1	9·1	Christianopel	1876	Sodergren		O.Dahlqvist	Christianopel
504		Peter C. Schultz *Thompson*	W 3m Sr	American	315	138·0	32·0	10·0	Newburg.N.Y	1874	Bulm'n & Brown		A.P.Thompson	PrthAmboy.N.J.
505		Peter Dickson	W Bk	Swedish	382	128·6	27·8	15·4	Hernösand	1859			J.M.Möller	Gothenburg
506		Peter Godeffroy *Möller*	I Bk	German	462	111·0	26·6	17·2	Hamburg	1869	Reiherstieg [Schiffs Werfte		Aug.Bolten	Hamburg
507		Peter H. Crowell	W 3m Sr	American	423	139·0	33·0	14·0	East Boston	1873			P.H.Crowell	Dennis,Mass.
508		Peter Kraeft *Bradin*	W Bk	German	333	115·1	28·0	16·5	Damgarten	1857	H.Diedling		W.A.Sarnow	Barth
509		Peter Landberg	W S	Dutch	1199	183·0	37·7	24·9	Alblasserdam	1867	J.A.von Linden	v. 2.88	P.Landberg & Zoon	Batavia
510		Peter Landberg	W Bk	Dutch	763	160·3	32·5	20·0	Quebec	1875	P.Baldwin	v. 82	Tan Djim Soei	Batavia
511		Peter Rickmers	Stl 4m S	German	2600	331·0	44·5	25·4	PortGlasgow	1889	Russell & Co.		R.C.Rickmers	Bremerhaven
512		Peter Rolt *Jorck 82-*	W 3m Sr	German	500	122·8	31·2	18·8	Danzig	1847	J.W.Klawitter		H.Mauss	Rostock
513		Peter Stuart *Hughes*	I S	British	1447	234·0	38·9	23·0	Dumbarton	1868	A.M'Millan & Sn		T.Stephens & Sons	Liverpool
514		Peter Tordenskjold	W Bk	Danish	249	107·5	26·2	14·4	Rönne	1850	S.P.Bech		H.Wieck & Co.	Copenhagen
515		Peter & Paul	W 3m Sr	Russian	273	111·0	29·4	13·1	Pernigel	1872	C.Grube		Schnore & Andreisoan	Riga
516		Peter & Wilhelm	W Bn	Danish	100	70·4	21·8	10·7	Svendborg	18^{38}_{67}	N.G.Møller		R.Rasmussen	Odense
517	*	Peterborough *Marchie*	I S	British	1680	250·0	41·1	23·8	Stockton	1874	Richardson, [Duck & Co.		Adam Hamilton & Co.	Greenock
518	N	Peterel *Hansen*	W Bg	Norweg'n	260	113·9	27·5	12·4	Tönsberg	1876	Jahnsen	z. 1.88	Otto Thoresen	Tönsberg
519		Peters Kapelle *Ballod*	W Sr	Russian	147	88·8	26·9	10·8	Pabbasch	1872	P.Ballod		Ballod & Zehns	Riga
520	V	Petit Bourgeois	W Bk	French	541	112·0	27·2	18·0	St.Malo	1874	Lhotellier	c. 1.88	J.B.Demange	Nantes
521		Petit Caporal	W Bg	French	187	87·3	22·6	12·8	Nantes	1861	Chauvelon		Garaud	Saint Brieuc
522		Petit Eugène	W Lr	French	106	73·8	21·3	10·9	Vannes	1865	Ridel		Le Moulin	Vannes
523	V	Petit Joseph *Gonëdel*	W Bn	French	135	78·4	17·7	11·9	Méans	1868	Lanort	z. 1.88	Gonëdel	Vannes
524	V	Petitcodiac *Steeve*	W Bk	British	682	146·1	33·6	19·2	Salisbury,N.B	1879	McEwen	v. 9.88	A.L.Wright	Moncton.N.B.
525	A	Petka *Scocetich*	W Sr	Austrian	235	100·7	23·3	..	Gravosa	1875	N.A.Bradirich		G.Racich di N.	Ragusa
526		Petrel	W Bn	British	115	78·5	21·0	11·9	Tatamagouche,N.S.	1850			P.Murphy	Cork
527	*	Petrel *Wallace*	W Sr	British	99	84·6	20·8	11·0	Montrose	1856	C.Barnie		A.Bowman	Montrose
528		Petrel *Reed*	W Bk	American	257	107·4	27·0	15·5	Rochester,Ms	1853		v. 4.88	Philip H.Reed	N.Bedford.Mass.
529		Petri *Eriksson*	W Sr	Swedish	163	86·3	26·6	11·1	Roslagen	1860			A.Österlund	Malnäs Väto
530		Petri *Andersson 85-86*	W Sr	Swedish	123	76·1	21·8	9·8	Väddö				A.Öhman	Lingsläto,Rlsgn
531	G	Petrine *Knudsen 87*	W Bn	Danish	130	86·3	20·1	11·2	Elmshorn	1866	D.W.Kremer		H.F.Knudsen	Svendborg
532		Petrino *Porzio*	W Bn	Italian	209	89·6	25·8	15·5	Cassano	1849			F^lli Lauro	Castellamare
533		Petro	W Bg	Turkish	180	..	..	..		1880			Yazidji Essad Effendi & Co.	Constantinople
534		Petronella *Andersson 85*	W 3m Sr	Swedish	152	92·5	21·7	10·4		1858			A.Peylitz	Strömstrad
535	N	Petropolis *Bull 88-8-*	W Bk	Norweg'n	556	160·5	32·8	17·6	Havre	1854	L.Normand	v. 3.88	J.R.Bull	Christiania
536	V	Petros *Batur*	W Bk	Greek	445	134·2	29·6	18·4	Syra	1875	Confoudakis	v. 7.88	A.Mlahis	Syra
537	H	Petros *Petroulis*	W Bg	Greek	278	..	..	..	Galaxidi	1870			G.M.Petroulis	Galaxidi
538	A	Petros *Petrangis*	W Bg	Greek	222	93·5	25·1	..	Galaxidi	1871	J.Canatas	v. 4.88	G.Petrangis	Galaxidi
539	H	Petros *Volicas*	W Sr	Turkish	145	..	..	..	Syra	1868			E.Volicas	Chios
540		Petros	W Sr	Turkish	129	..	..	..		1866			C.D.Faraclas	Chios
541	N	Petrus *Johnsen 88*	W Bk	Norweg'n	540	143·9	31·4	17·7	Tönsberg	1876	O.Ellefsen	v. 5.88	Sören Hansen	Christiania
542	N	Petrus *Tygesen*	W 3m Sr	Norweg'n	286	112·7	28·5	14·5	Amsterdam	1850			A.Christensen	Drammen
543		Petrus *Pettersson*	W Sr	Swedish	103	78·7	21·9	9·3	Väto	1885			P.Eriksson	Wätö,Roslagen
544		Petter *Rabin*	W Sr	Russian	257	..	..	..	Impilaks	1872			P.Berg	Sordavala
545		Petter *Wanamen*	W Sr	Russian	187	..	..	..	Impilaks	1883			A.Waittinen	Impilaks
546		Petter *Tesofejeff*	W Sr	Russian	179	..	..	..	Impilaks	1882			M.Korniloff	Sordavala
547		Pettu *Brandtkill*	W Sr	Russian	150	83·0	26·0	10·7	Kimito	1865	J.W.Janssen		G.Hafverman	Raumo
548	N	Petty *Larsen* ex Townsend	W Bk	Norweg'n	489	166·5	29·0	13·5	Gosport, Va. [rebuilt	18	Navy Yard		Hans Larsen	Sandefjord
549	*	Petunia *Nichols*	W 3m Bn	British	207	116·6	23·6	12·7	Bideford	1874	Cox	v. 5.87	P.G.Tessier	St.Johns N.F.L.
550	*	Phæton *Pitchers 78-83*	C Bk	British	576	151·4	30·4	18·0	Sunderland	1868	J.Blumer & Co.	v. 3.88	F. & W.Ritson	Sunderland

*=Lloyd's Register; D=Nederlandsche Vereeniging; G=Germanischer Lloyd; I=Registro Italiano; R=American Record; A=Veritas Austro-Ungarico; H=Veritas Ellenico; N=Norske Veritas; V=Bureau Veritas.

Dans la 4me Colonne, C=Composite (membrure en fer, bordé en bois); I=Fer; Stl=Acier; W=Bois.

Dates du doublage en métal (Col. 11) ressortent des livres dans lesquels les navires sont respectivement classés ou des rapports des Inspecteurs du Lloyd's Register.

Pour plus amples renseignements sur les navires classés par le Lloyd's Register, voir le Lloyd's Register Book et le Lloyd's Yacht Register.

Official No.	No. in Book	Registry in which classed	Ship's Name, Material and Rig. Master, and Date of Appointment to present Owner's Service and to Ship. Ship's Late Name.	Flag.	Net Tonnage.	Dimensions: Length.	Breadth.	Depth.	Built: Where.	When.	By whom.	Metalling	Owners or Managers.
[illegible]	551	✳	Phantom *Jones -88* W Bk	British	249	111·2	24·2	11·5	Salcombe	1867	Evans	[illegible]	H. Parry
	552		Pharaó (ex Candida) *Guimaraes* W Pol	Brazilian	231	..	..	..	[illegible]	1869			J. A. A. Monteiro
	553	V	Phare *Signoret* C Bk	French	490	132·8	25·9	[illegible]	[illegible]	1876	J. Tranchard	[illegible]	Lieutaud de Topia
	554	V	Pharo *Suarez -88* W Bk	French	615	147·6	29·8	19·3	Marseilles	1874	V. Jouvin	[illegible]	M. & B. Roletty neveux
[illegible]	555	V	Pharos *Reitzenstein -88* W S	American	2002	235·6	40·7	27·8	Kennebunk Me	1877	N. L. Thompson	[illegible]	T. Nickerson
[illegible]	556	✳	Phasis *Chrystal 82-82* I S	British	1490	248·0	38·0	23·0	Glasgow	1880	Brclay, Curle & Co		A. & J. H. Carmichael & Co.
[illegible]	557		Phebe *Meders* W Sr	American	177	104·5	27·0	9·4	[illegible]	1869			J. G. Meders
[illegible]	558		Phebe J. Woodruff *Kendrick* W 3m Sr	American	499	150·0	35·3	11·4	Booth Bay Me	1882	Fuller		T. B. Baker
	559		Philadelphia (ex Marianna VI, ex North America) W S	Brazilian	1235	194·0	38·4	28·9	[illegible] Me	1852		[illegible]	Cia de Estrada de Ferro de Bahia & Minas
[illegible]	560		Philadelphia *Confopantili* W Bg	Greek	193	86·9	24·9	..	Syra	1874		[illegible]	I. Confopanteli
	561		Philadelphia W Bg	Turkish	225	..	..	..	Greece	1885			Kerem oh [illegible] Emin [illegible] and W[illegible]
[illegible]	562		Philadelphos W Bk	Greek	294	..	..	..	Syra	1868			G. Drakopoulos
	563	F	Philémon *Tribault -67* W Bk	French	310	117·4	25·6	15·1	Honfleur	1866	Leviels	[illegible]	Mounier & Fribouklet
	564	✳	Philip Nelson *Niemann* [87-87] I Bk	German	548	167·1	26·7	17·4	Liverpool	1864	T. Vernon & Son		H. Mentz, jun.
[illegible]	565		Philippine *Boyle* W Bk	British	302	127·8	25·7	14·3	Kincardine	1860	D. Wright		Tasmanian Steam Nav. Co.
[illegible]	566		Philip & Family W Sr	British	108	78·3	17·9	9·0	Mattancheri	1876			Ana Caes Mahamede
[illegible]	567		Phillis *Must 78-78* W Bg	British	240	103·2	23·6	12·9	Miramichi, N B	1859		[illegible]	John Bickers
[illegible]	568		Phillis & Mary *Fletcher* W Bkn	British	233	103·2	25·7	15·0	Sunderland	1865	J. Hardie		T. G. Robins & Co.
[illegible]	569	✳	Philomene *King 84-84* I S	British	1423	240·7	37·1	23·2	Sunderland	1875	T. R. Oswald		H. Fernie & Sons
	570	A	Philopator *Competsis* W Bn		220	89·2	26·2	14·7	Syra	1869			
	571	V	Philosophe *Allançon* W Bk	French	216	91·5	23·6	12·9	Vannes	1877	Malé	[illegible]	Allançon
	572	N	Philotaxe *Stark* W Bk	Norweg'n	388	130·7	28·2	16·3	St. John, N.B.	1852			H. Biörn, jun.
	573	G	Philothea *Schmidt 86-* W 3m Sr	German	198	105·0	24·3	12·0	Papenburg	1866	C. Freericks	[illegible]	A. Klasen
	574		Phineas H. Gay *Bryant* W Sr	American	104	..	..	..	Newcastle, Me	1888			T. E. Gay
[illegible]	575	R	Phineas W. Sprague *Strong -87* W 3m Sr	American	749	175·9	35·1	12·5	Thomaston, Me	1887	A. Simmons	[illegible]	Dunn & Elliot
164	576	A V	Phison *Nicolich* W Bk	Austrian	730	134·5	30·9	21·8	Lussinpiccolo	1879	N. Martinolich	[illegible]	M. A. Cosulich
	577	N	Phobus *Pedersen -84* W Bg	Norweg'n	247	95·9	27·3	13·9	Christiansand	1843			G. Pedersen
[illegible]	578		Phœnix *Pettis -87* W Sr	British	397	144·3	34·0	12·9	Parrsboro, N.S	1883	J. E. Pettis		J. E. Pettis
[illegible]	579		Phœnix W Bk	British	340	125·5	29·9	12·9	Picton, N.S.	1867	Jas. Little		Edward Miles
13	580		Phœnix W Bn	Greek	254	101·7	23·0	..	England	1864			G. Theophilatos
	581		Phœnix W Bg		252	102·3	25·3	15·7	Syra	1869			
	582	V	Phønix (ex Fama) *Hansen 80-80* W Bn	Danish	173	110·8	22·0	10·4	Pasadlavik	18[illegible]			J. F. Andersen
	583	V	Phønix *Eriksen 84-84* W Bn	Danish	165	100·7	22·2	11·3	Marstal	1878	H. J. Bager		N. J. Ohlsen
	584	V	Phønix *Mortensen 73-83* W Sr	Danish	115	87·7	20·6	9·7	Svendborg	1873	I. R. Andersen		R. P. Nielsen
	585	V	Phönix *Diewitz -88* I Bk	German	683	192·0	31·6	17·1	Flensburg	1878	Flensburger Shiffsb. Ges.		P. Diedrichsen
	586		Phönix (ex Flying Dutchman) *Petersen -88* I 3m Sr	German	268	115·3	25·2	15·1	London	1850	Money Wigram		H. L. Rochow
	587	N	Phonix *Nielsen -88* W 3m Sr	Norweg'n	284	127·5	27·5	12·9	Skonevig	1878	Aasheim & Co.	[illegible]	W. Gjertsen
	588	N	Phonix *Jensen* W Bg	Norweg'n	269	98·0	27·6	14·8	Sundsvall	1834	Höglund		J. Svendsen
	589	N	Phönix *Hansen 83-83* W Bg	Norweg'n	138	85·3	22·8	10·7	Drammen (rebuilt)	18[illegible]			A. & H. Backer
	590	N	Phönix *Ferman* W Bg	Norweg'n	133	91·1	23·0	10·6	Kragerö	1863			P. & C. Rönner
	591	G V	Phonizia *Brane 87-82* W Bk	German	764	153·8	33·2	17·9	Brake	1882	J. Oltmann's Wwe.	[illegible]	H. F. C. Tobias
[illegible]	592		Phuttel Barry *Tokur 78-* W S	British	705	130·0	34·5	22·0	Cochin	1833			Hussein Ibrahim bin Johur
	593	✳	Phyllis *Raymond 88-88* I Bk	British	932	204·5	32·6	19·4	Bristol	1888	C. Hill & Sons		C. Hill & Sons
[illegible]	594	✳	Physician *Jones* W Sr	British	90	79·2	20·6	10·7	Pwllheli	1861	Roberts		R. Jones & Co.
	595		Pia W Bn	Brazilian	158	..	..	..					N. Malbourg
[illegible]	596	A	Pia S. *Soldatich* W Sr	Austrian	200	83·3	22·6	13·2	[illegible]	1869	C. Giacovich	[illegible]	G. Soldatich di N.
	597		Piaba W Bn	Brazilian	242	..	..	..					J. J. C. Ferreira
[illegible]	598	✳	Piako *Sutherland 83-85* I S	British	1075	215·3	34·0	20·5	Glasgow	1876	A. Stephen & Sons		New Zealand Shipping Co. Ltd.
[illegible]	599		Picton W Sr	British	160	98·7	22·7	9·6	Picton	1867	T. Walters		R. C. Smith
[illegible]	600	✳	Picton *Jones* W Sr	British	111	81·5	21·4	11·9	Portmadoc	1858			D. Richards & Co.

✳ = Lloyd's Register; D = Nederlandsche Vereeniging; G = Germanischer Lloyd; I = Registro Italiano; R = Ame[illegible]
A = Veritas Austro-Ungarico; H = Veritas Ellenico; N = Norske Veritas; V = Bur[illegible]

In the 4th Column, C = Composite (iron frame, wood planked); I = Iron; Stl = Steel; W = Wood.

The Dates of Metalling (**Col. 11**) are based on the records of the Books in which Vessels are respectively classed, or on the Reports of the Surveyors of [illegible]

Further particulars of Vessels classed by Lloyd's Register will be found in Lloyd's Register Book and Lloyd's Yacht Register.

N°	Registre	Nom. Matériaux et Gréement. Capitaine et dates d'entrée au service des Armateurs actuels et à bord du navire. Ancien Nom.	Pavillon.	Tonnage net.	Longueur.	Largeur.	Creux.	Construit: Où.	Quand.	Par qui.	Date du doublage en métal	Armateurs ou Directeurs-Gérants.	Port d'Armement.
601	*	Picton Castle *Clark 75-79* I Bk	British	567	173·1	28·0	17·6	Glasgow	1875	A.Stephen&Ss.		Simpson Bros.	Swansea
602	V	Pie IX. *Crispel* W Bk	French	380	109·5	26·2	17·3	St.Servan	1869	Hamon	V 4.86	Guibert & fils	Saint Malo
603		Pied Nez W S	British	531	155·7	29·5	17·7	Quebec				H.Dubord	Quebec
604	I	Piemonte *Gambi* W 3mSr	Italian	456	135·1	29·7	19·2	SestriPonente	1864	A.Briasco fu G.B.	V 2.87	Capellino Fratelli	Genoa
605	I	Piemonte *Merlinelli* W Bn	Italian	111	80·3	21·3	9·5	Sestri Ponente	1886	G.Peragallo		G.Peragallo	Genoa
606	V	Pierre *Michel* -88 W Bg	French	179	86·3	21·3	13·1	Bordeaux	1867	Rochete		Th.Michel	Auray
607	V	Pierre Antoine *Chevalier* -87 W Bg	French	154	86·9	22·0	11·5	Dieppe	1867	P.Cerne [Largeteau		Guillebot	Granville
608		Pierre Emilie *Magon* W3mBk	French	245	96·8	26·2	16·1	Fécamp	1877			Hermel & Buisson	Fécamp
609	V	Pierre Emilie *Alix* -80 W Bn	French	144	92·5	22·8	10·8	St.Vaast	1876	C.Levêque		C.Levêque	La Hogue
610		Pierre Philippe *Gosselin* -88 W Bk	French	231	99·1	21·9	13·7	Fécamp	1866	F.Capon	V 11.81	Duboc	Fécamp
611		Pierreville W Sp	British	126	103·0	22·7	7·5	St.Thos.de [Pierreville	1874			A.Charland	Montreal
612	I	Pietrina M. *Mezzano* -88 W Bn	Italian	152	103·6	24·3	11·8	Varazze	1885	G.Aschieri	V 1.88	F.Massa fu F.	Spezia
613	VI	Pietrino *Materazza* -88 W Bk	Italian	650	156·1	31·8	21·0	Gaeta	1878	F[lli] Gallinaro	V 12.87	N.Di Fonzo	Gaeta
614		Pietrino Barbaro *Doria*	Italian	255	98·1	26·8	16·2	Sorrento	1849			P.Barbaro	Palermo
615	I	Pietro *Solari* -87 W Bk	Italian	434	131·5	30·0	19·5	Savona	1867	F.Sirello	V 3.87	E.Isola	Genoa
616	I	Pietro B. *Minetti* 74-84 W Bk	Italian	848	170·6	30·2	22·3	Spezia	1878	G.Faggiani	V 2.87	G.B.Beverino	Spezia
617		Pietro Boscovich W Bk (ex Elisabetta Boscovich)		327	124·0	26·4	17·1	Sunderland	1858		V 9.82		
618		Pietro M. W Bk (ex Bartolomeo Emanuele)	Italian	527	144·3	32·6	19·9	SestriPonente	1867	G.Casanova		A.Danovaro	Genoa
619		Pietro N. *Ferri* W Bn		499	135·0	30·6	18·6	SestriPonente	1867	G.Casanova	V 1.86		
620		Pietro Padre *Labanca*	Italian	243	101·3	24·7	15·1	Agde	1863	Cavallié		P.Tosi & figli	Genoa
621		Pilar W Bg		193	90·5	26·8	14·1	Masnou	1850				
622		Pilar W Bn	Spanish	172	98·8	24·[illegible]	11·3	Albay	1862			James Urrutia	Manila
623		Pilar W Bg	Spanish	109	69·5	20·1	10·0	Asturias	1841			Serra,hermanos	Barcelona
624	*	Pilgrim *Pugh* 87-87 W Sr	British	159	92·1	23·4	12·4	Portmadoc	1866	Richards	V 3.87	D.Richards & Co.	Portmadoc
625	*	Pilgrim *Roberts* W Sk	British	38	60·0	17·0	6·7	Bangor	1879	Parry		T.Roberts & Co.	Beaumaris
626	V	Pilgrim *Freeman* W Bk	American	918	172·4	35·0	21·0	Medford,Mass	1874	J.T.Foster	V 10.86	H.Hastings	Boston.Mass.
627	DV	Pillau *Gerlach* 73-83 W Bk	German	470	128·2	26·9	18·5	Pillau	1863	O.Becker	Z 1.86	J.H.Meyer	Pillau
628		Pilote *Noiseux* W Sr	French	157	76·4	21·3	12·9	Stralsund	1854			J.Noiseux	Belle Ile
629	V	Pimpac *Ferrinho* W Bn	Portug'se	121	85·3	20·6	8·8	Vianna	1885	M.D. dos Santos [Borda	V 8.85	J.M.Mattos & Co.	Vianna
630	*	Pinmore *Maxwell* 82-82 Stl 4m Bk	British	2358	310·1	43·7	24·7	PortGlasgow	1882	J.Reid&Co.		J.Kerr & Co.	Greenock
631		Pinta *Nicholls* 75-78 W Bn	British	199	104·5	24·[illegible]	12·9	P.E.Island	1875	M'Millan		F.W.Fairbrass	Liverpool
632	*	Pinta *Heckils* W Sk	British	86	78·1	20·6	10·5	Goole	1884	CottinghamBs.		Pickering & Haldane	Hull
633		Pinto I. *Freitas* W Sr	Brazilian	122	..	..	..	RioJaneiro	18 75		V 2.86	J.C.Pinto	Porto Alegre
634	A	Pinus *Simunich* W Bg	Austrian	360	118·7	25·5	16·6	Fiume	1870	G.Baccarcich	V 4.80	A.Simunich fu N.	Volosca
635	A	Pio *Persich* 82-82 W Bk (ex Pero Dubrovacki)	Austrian	515	142·7	28·9	17·5	Fiume	1870	F[lli] Schiavoni	V 9.83	Pio de Persich	Lovrana
636	*	Pioneer *Godfrey* W Bg (Colliers)	British	287	119·1	27·0	16·1	Blyth	1873	TheUnionCo.	V 1.84	R.L.Manners	North Shields
637		Pioneer (WRECKED) *Moon* W Bg	British	270	119·0	26·3	14·9	Shoreham	1863	Balley		T.Gates & Co.	Shoreham
638		Pioneer W Bn	British	158	92·0	22·5	10·4	Melbourne	1865			W.H.Smith&Ss.Ld.	Brisbane
639		Pioneer W Sr	British	106	85·2	24·7	9·5	N.Sydney	1886	W.Allan		John Vooght	Sydney,C.B.
640	*	Pioneer *M'Nally* 87-87 I Sr	British	75	75·0	19·1	9·0	Belfast	1887	M'IlwaineLewis [& Co.,Ld.		W.H.M'Donnell	Belfast
641		Pioneer *Hughes* W 3mSr	American	397	138·5	36·3	12·5	Gray's Harbor	1886	A.M.Simpson		A.M.Simpson	San Francisco
642	DN	Pioneer *Nielsen* -88 W Bk (ex Marcia A. McNeil)	Norweg'n	965	178·4	34·8	20·5	ThomastonMe	1868	Burgess& [O'Brien	V 12.87	Act.Selsk.[ab] "Pioneer" (Ths.S.Aschehoug)	Fredrikshald
643	V	Pionier *Petersen* -88 W Bk (ex M. & A. Cox)	German	1123	190·9	34·4	22·5	Maitland,N.S.	1881	C.Cox&Sons	V 7.8	H.P.Breckwoldt,jun.	Hamburg
644	V	Pionnier W Sr	Belgian	110	91·8	22·6	11·1	Antwerp	1883	H.Wappers		D.Mauroy	Antwerp
645	I	Pippo (ex Emilia) *Drago* W Bk	Italian	467	131·8	30·8	19·6	Sestri	1869	S.Borzone	V 3.84	Giu.Raffo fu G.B.	Genoa
646	I	Pippo C. (ex Rosolia) *Costa* W Bk	Italian	516	141·0	31·7	19·3	Voltri	1870	C.Fava	V 12.87	Em.Chiappella	Genoa
647		Pippo Savona *Jurgensen* W Bk	Argentine	449	..	..	..	NovaScotia	1872		V 3.86	Stefano Anselmo	Savona
648	V	Pirat *Kayser* 86-88 I Bk	German	1029	212·6	34·4	20·0	Hamburg	1883	Blohm&Voss		F.Laeisz	Hamburg
649	*	Pirate *Smith* W K	British	48	57·2	19·5	8·0	Stromness	1888	G. & P.Copland		P.Copland	Stromness
650		Piratininga W Bk	Brazilian	230	..	..	..					A.F.C.Cabral	Rio Janeiro

[*] = Lloyd's Register; D = Nederlandsche Vereeniging; G = Germanischer Lloyd; I = Registro Italiano; R = American Record;
[A] = Veritas Austro-Ungarico; H = Veritas Ellenico; N = Norske Veritas; V = Bureau Veritas.

Dans la 4me Colonne, C = Composite (membrure en fer, bordé en bois); I = Fer; Stl = Acier; W = Bois.

[Les da]tes du doublage en métal (Col. 11) ressortent des livres dans lesquels les navires sont respectivement classés ou des rapports des Inspecteurs du Lloyd's Register.

Pour plus amples renseignements sur les navires classés par le Lloyd's Register, voir le Lloyd's Register Book et le Lloyd's Yacht Register.

Official No.	No. in Book	Registry in which classed	Ship's Name. Master, and Dates of Appointment to present Owner's Service and to Ship. Ship's Late Name.	Material and Rig	Flag	Net Tonnage	Length	Breadth	Depth	Built: Where	Built: When	Built: By whom	Dates of Metalling	Owners or Managers
	651		Pires I.	W Sr	Portug'se	132	81·0	20·8	9·6	Rio	1879	M. de Barbosa		A.P. Barbosa
	652		Pirineo	W Bn	Spanish	116	91·8	26·7	9·1	Rosalies	1848			Juan Mir
	653		Pirot	W S	Turkish	432	..	..	..	Greece	1881			Epaminonda Anagnosti
87636	654		Pirro	W Bk	British	251	105·2	28·6	15·0	Leghorn	1856	G. Michelli		O.M. Pulis
	655		Pirros (ex Kepler) Psarali	W Bk	Turkish	432	..	..	..	Varazze	1868		v 88	E. Anagnosti
	656	V	Pisco Eckersberg 88	W Bk	Norweg'n	716	166·2	33·0	20·1	SestriPonente	1870	Cadenaccio	v 88	Westergaard & Co.
68191	657		Piskataqua Faulkner 87-88	W Bk	British	600	151·3	33·2	18·2	Noel, N.S.	1872		v 88	O. O'Brien
4	658		Pistis	W Bk	Greek	417	127·9	28·8	..	Trieste	1866	Stabilimento [Tecnico		D. Mylonas
93346	659		Pitagoras Crediner	Stl Sr	British	151	115·0	24·0	8·1	Dumbarton	1886	W. Denny & Bros		La Platense Flotilla Co Ld
98001	660	*	Pitcairn Island Cornell 88	Stl Bk	British	1320	231·3	36·4	21·7	Greenock	1888	Russell & Co		Stuart, Bros.
	661		Pitea Johansson	W Sr	Swedish	109	81·7	21·3	9·6	Pitea	1880			E.G. Jansson
60936	662	*	Pitho Jones 88	W Bn	British	140	96·0	22·7	12·9	Swansea	1868	Meager	v 88	J.E. Burgess
	663	V	Pitre Alcide Lakgoute	W Bg	French	173	81·7	21·6	12·1	Means	1872	Olliveaud	20 88	Pitre Legal
	664	V	Pitre Anna Lepart	W Bn	French	117	72·8	19·7	10·9	Nantes	1857	T. Dubigeon		L. Jacquier
	665		Pitre Marie Chapron 61-72 (FOUNDERED)	W Bn	French	119	75·1	22·3	10·4	Nantes	1861	Bojo & Clergean	2 9.88	F.P. Chapron
81325	666		Planet	I Sr	British	184	103·0	21·2	11·6	Liverpool	1880	R. & J. Evans & Co		Mersey Docks & Harbour Board
37076	667		Planet Young	W Bn	British	136	86·0	25·0	10·3	Parrsboro, N.S.	1861			F.H.T. Bellew
27809	668		Planet Tubb 73-73	W Bn	British	134	81·7	21·7	12·2	Pwllheli	1860	Roberts		B.M. Harvey
	669	N	Plantagenet Thils 88	W Bk	Norweg'n	1286	193·1	37·9	24·0	St. John, N.B.	1869	Nevins & Fraser	v 88	Act. Selsk. "Plantagenet" (Th. S. Aanensen)
15355	670		Planter	W Bk	British	231	106·9	22·5	15·0	Grimsby	1857			Tasmanian S.N. Co.
150374	671	V	Planter Penhallow 88	W Bn	American	499	158·7	37·5	13·5	Port Ludlow	1886	C. Murray		Edw. E. Kentfield
	672	V	Plata	W Bk	French	657	155·8	30·2	18·7	Honfleur	1862	Cardon	v 88	Ch. Leclère
2566	673		Plata Paolillo 88 (ex Eugenio Accame)	W Bk	Italian	629	154·5	32·2	20·6	Loano	1869	D. Aicardo	v 9.88	G.B. Profumo
90511	674	V	Platina Stevens 87	W Bk	British	297	138·0	33·9	10·0	Eatonville, N.S.	1887	D.R. & C.F. Eaton		H.A. Eaton
19787	675		Platina Gilbert	W Bk	American	203	100·0	24·8	12·1	Rochester, Ms.	1847			J.F. Tucker
	676		Platino Moreira	W Bg	Brazilian	179	..	..	..	Barcelona	18..		v 88	A. Schmidt
	677	N	Platon Andersen	W Bk	Norweg'n	480	133·2	31·1	16·5	Arendal	1875	O. Olsen	v 88	J. Hammer
73437	678	R	Playfair Ludlow 87	W Sr	British	114	86·1	21·3	8·2	Cornwallis, NS	1877	J. Morris	v 88	M. Adams
77622	679		Pleasantville Lauer 86	W Sr	British	103	80·0	23·8	9·4	Pleasantville NS	1878			A. McKean
76355	680	*	Pleiad Oliver 83-83	I Bk	British	617	170·6	29·1	17·9	Liverpool	1876	Bowdler, Chaffer [& Co.		J. Newton & Co.
63500	681	*	Pleiades Morton 88	I S	British	997	209·6	33·1	20·6	Dumbarton	1869	A. M'Millan & Sn		Shaw, Savill & Albion Co. Ld.
35275	682		Pleiades (ex Vesta)	W Bk	British	329	122·2	27·2	15·0	Parrsboro', NS	1858			A.L. Harrold
37817	683		Pleiades	W Sr	British	106	77·5	20·7	10·4	Economy, N.S.	1853			A. Donald
72909	684	*	Pleione Gray 88	I S	British	1092	209·7	34·6	20·3	Glasgow	1876	A. Stephen & Sns		Shaw, Savill & Albion Co. Ld
85641	685		Pleroma Loye	W Sr	British	103	81·5	24·4	9·3	Mahone By. NS	1883	Z. Langille		J. Eisenhauer
269	686	A	Plessa Tsalanghiras	W Bg	Greek	199	90·5	24·4	..	Galaxidi	1875	C.M. Papapetros	v 88	K. Courodelis
	687	N	Plimsoll Olsen 87-87	W Bn Sr	Norweg'n	301	119·6	27·6	13·8	Kobbervig	1876	A. Bentlich	v 88	Holdt & Isachsen
	688	DN	Plimsoll Steen	W Bn	Norweg'n	225	96·2	26·8	13·2	Fredrikshald	1876	R. Lorange	2 88	F.G. Steen
	689	N	Plimsoll Henricksen	W Sr	Norweg'n	201	104·2	26·7	10·6	Riga	1871			H.H. Petersen
2566	690	I	Plinio (ex Gnesea) Corzanego	W S	Italian	1133	180·7	35·8	24·7	SestriPonente	1874	FB Cadenaccio	v 88	Francesco Coxiola
19475	691		Ploughman Fairweather [88	W Bg	British	168	103·0	23·0	13·0	Aberdeen	1867	W. Hood & Co.		G. Milne & Co.
20598	692	*	Plover Poulson	W Bk	British	414	130·5	26·5	16·6	Sunderland	1858	R. Thompson & Co	v 88	A. Poulson
85580	693	V	Plover Lowery 87	W Bn	British	386	132·4	29·9	13·0	St. Martin, NB	1883	J. Marr	v 88	Schofield & Co. Ld.
19652	694		Plow Boy Hallett	W Sr	American	178	118·0	25·7	8·6	Ashtabula, Oh.	1863			J.S. Hallett
	695	* V	Plus Steincke 88-88	I Bk	German	1230	226·6	36·0	20·6	Hamburg	1885	Blohm & Voss		F. Laeisz
	696	V	Plus Jensen 87	W Sr	Norweg'n	158	101·2	23·7	11·3	Christiansand	1868	N. Jørgensen	v 88	Act. Selsk. "Plus"
	697	NV	Plutarch Johnsen	W Bk	Norweg'n	445	136·0	31·5	17·1	Arendal	1877	A. Aanonsen	88	Chr. Hannevig
	698	V	Pluto Evasen 81-88 (ex ...)	I S	German	1133	213·9	35·1	22·8	Liverpool	1862	Jones Quiggin & [Co.		F. Laeisz
	699		Pluton (ex ...)	W Bn		253	108·9	25·0	13·4	Bilbao	1867	J. de Uncueta		
	700	N	Plutus Christiansen	W Bk	Norweg'n	378	110·7	27·1	18·2	P.E. Island	1849			Th. Christiansen

*=Lloyd's Register; D=Nederlandsche Vereeniging; G=Germanischer Lloyd; I=Registro Italiano; R=American ...
A=Veritas Austro-Ungarico; H=Veritas Ellenico; N=Norske Veritas; V=Bureau ...

In the 4th Column, C=Composite (iron frame, wood planked); I=Iron; Stl=Steel; W=Wood.

The Dates of Metalling (**Col. 11**) are based on the records of the Books in which Vessels are respectively classed, or on the Reports of the Surveyors of ...

Further particulars of Vessels classed by Lloyd's Register will be found in Lloyd's Register Book and Lloyd's Yacht Register.

N°	Registre	Nom. Capitaine et dates d'entrée au service des Armateurs actuels et à bord du navire. Ancien Nom.	Matériaux et Gréement	Pavillon	Tonnage net	Longueur	Largeur	Creux	Où	Quand	Par qui	Date du doublage en métal	Armateurs ou Directeurs-Gérants	Port d'Armement
701		Pluvier *Annix*	W Bg	French	135	79·7	20·7	11·7	St. Malo	1857	Hervot	[illegible]	Gautier jeune & fils	St. Malo
702	V	Plymouth *Gadpeel*	W Bk	British	1312	198·0	40·0	23·3	Hantsport, N.S.	1879	J. Davison	V 1.88	G. W. Churchill	Windsor, N.S.
703	V	Plymouth *Gue 85-87*	W Bn	British	199	103·6	26·5	11·3	P. E. Island	1878	R. Westaway	[illegible]	F. C. Wells	St. Johns, N.F.L.
704		Plymouth Rock *Terris* (WRECKED)	W Sr	American	192	97·0	27·0	10·0	Derby, Conn.	1856			G. B. Scammell	Boston, Mass.
705		Plymouth Rock	W Bk	Norweg'n	1017	172·6	36·1	22·2	Boston	1849		V 12.87	C. T. Gogstad	Sandefjord
706		Pocahontas *Phillips*	W 4mSr	American	1313	233·6	46·6	19·0	Camden, Me.	1887	H. M. Bean		J. M. Phillips	Taunton
707	V	Pocahontas	W Sr	German	119	93·1	21·1	10·3	Odense	1875	E. Knudsen		M. de Freese	Warsingsfehn (Hannover)
708		Pochasset *Herrick -85*	W Sr	American	226	120·0	30·6	11·0	Mystic, Conn.	1874			J. M. Cross	Providence, R.I.
709	N	Pocohontas *Hansen (80-86)*	W 3mSr	Norweg'n	197	101·1	25·8	10·9	Lincolnville, Me.	1851			N. Hansen	Tönsberg
710		Pohjola	W Bk	Russian	563	141·0	30·5	16·7	Gamla Carleby	1861			Gamla Carleby Skepps Wasstabagel	Gamla Carleby
711	R	Pohona *Jamieson -87*	W Bk	British	798	170·0	33·7	19·6	Kingston, N.B.	1880	J. & T. Jardine	[illegible]	J. B. Duckworth	London
712		Pola	W Sp	American	108	75·0	29·0	7·0	Deep Riv. Con.	1862			S. F. Shortland	New York
713		Polar Star *Robertson 81-85*	W Bk	British	216	104·9	24·2	14·7	Peterhead	1857	Stephen & Forbes		R. Kinnes	Dundee
714	IV	Polare *Costa*	W Bk	Italian	893	172·9	33·5	22·1	Sestri Ponente	1874	C. Tubino	V 11.88	Tomaso Seghezza	Genoa
715	❋	Polestar *Simeon 87-87*	I Bk	British	625	175·7	28·2	19·0	Hull	1858	M. Samuelson & Co.		Cotesworth, Lyne & Co.	Liverpool
716		Poli	W Bg	Spanish	209	96·8	27·9	11·1	Scotland				P. A. Carvajal	Gijon
717		Policena	W Sr	Brazilian	198	..	..	..					J. J. Moreira	Rio Janeiro
718	V	Polina *Eugenikos*	W Bg	Greek	277	98·8	27·6	..	Syra	1872		[illegible]	G. Eugenikos	Syra
719		Polka	W Sr	British	125	74·9	19·4	10·9	Sunderland	1845			Mrs. M. A. Clarke	Maldon
720		Pollux *Zaage -88*	W Bn	German	241	101·7	23·9	12·4	Leer	1876	C. Schwoon	V 9.88	C. Goedelt	Hamburg
721	V	Pollux *Ackermann*	W Bn	German	146	91·2	20·0	9·9	Varel	1882	A. Schwoon	V 11.88	H. L. Buss	Grossefehn
722	N	Pollux *Svendsen -87*	W Bg	Norweg'n	310	107·7	28·2	15·2	Arendal	18 52/73			H. Svendsen	Arendal
723	N	Pollux *Thorsen -88*	W Bg	Norweg'n	226	110·0	26·4	12·1	Skien	1872			C. R. Thoresen	Sandefjord
724	V	Pollux *Gribwan -86*	W 3mSr	Russian	420	112·0	27·2	14·5	Kurland	1886	R. Tamm	[illegible]	Sandfeldt & Gribwan	Windau
725		Polly *Howland -86*	W Bn	British	173	96·6	23·0	12·4	Quebec	1857			Holden & Gunn	Bristol
726	N	Polly *Sørensen 86-86 (ex Julie Tabham)*	W 3mSr	Norweg'n	321	122·3	26·5	15·3	Cherbourg	1858		[illegible]	S. Sörensen	Brevig
727		Polly Preston *Williams*	W Sr	British	131	89·6	22·6	11·7	Nevin	1863	Thomas		L. Williams, jun.	Carnarvon
728	V	Polly Stott *Harder 86-76*	W Bk	German	432	123·8	29·8	17·7	Damgarten	1876	H. Dierling	V 7.86	Wilh. Maack	Rostock
729	❋	Polly Woodside *Yates*	I Bk	British	648	192·2	30·1	16·0	Belfast	1885	Workman, Clark & Co.		Bk. "Polly Woodside" Co. Ld. (Woodside & Workman)	Belfast
730	❋	Polly & Emily *Whitefield -82*	W 3mSr	British	146	97·7	23·3	12·4	Littlehampton	1873	Harvey	V 9.8	J. Whitefield	Bideford
731	V	Polstjernan *Norman -88*	W Bk	Swedish	327	132·5	26·1	12·3	Hernösand	1876	Fr. Hägglund	V 6.85	Arbetare Akt. Bol. (A. Backlund)	Hernösand
732	N	Polykarp *Olsen -88*	W Bk	Norweg'n	498	142·8	31·7	17·2	Grimstad	1880	J. Börresen	V 7.86	J. Börresen	Grimstad
733	N	Polykarp *Edvardsen -80*	W Bk	Norweg'n	395	128·1	28·8	16·1	Brevig	1868	Hans Pedersen		K. A. Olsen	Kragerö
734	AH	Polykrates *Michalizianos -88 (ex Eliza U. Lawrence)*	W Bg	Greek	296	110·8	26·6	16·7	Sunderland	1871		V 3.88	G. Michalizianos	Argostoli
735	❋V	Polymnia *Ipland 87-88*	I 4mS	German	2052	290·7	42·7	23·9	Hamburg	1886	Blohm & Voss		B. Wencke Söhne	Hamburg
736		Polynemus *Gonzales (ex Lightning)*	W Sr	Spanish	115	107·4	19·5	8·7	Wivenhoe	1852			Heritiers de M. M. Coll	Palma
737	❋	Polynesia *Reitmann 86-87*	I S	German	985	195·0	33·0	20·4	Hamburg	1874	Reiherstieg [Schiffs Werfte		F. Laeisz	Hamburg
738	V	Polynesian *Dumas*	W S	British	1294	199·0	37·6	23·4	Quebec	1878	P. V. Valin	V 6.8	Pierre V. Valin	Quebec
739	❋	Polynesian *Spencer 82-85*	I Bk	British	864	197·5	32·6	19·1	Port Glasgow	1876	R. Duncan & Co.		Thomson, Dickie & Co.	Glasgow
740		Polyxeni *Liras*	W Sr	Turkish	117	..	..	..	Chios	1872			C. Liras	Chios
741	H	Polyxeni Omirou (ex Eleftheria)	W Bg	Greek	229	91·2	26·6	15·4	Syra	1867			N. S. Bouchlas	Cephalonia
742		Pomerania *Fremen* (ABANDONED)	W Bk	German	351	107·2	29·2	16·9	Elbing	1839	M. Metzlaff		C. G. A. Fritze	Stolpemunde
743	A	Pomoc *Tichinz -88*	W Bg	Austrian	342	..	..	..	Lussinpiccolo	1876	G. S. Cattarinich	V 9.87	F. Sussanich	Lussinpiccolo
744	❋	Pomona *Smith 88-88*	I Bk	British	1200	222·9	34·2	22·8	Greenock	1867	R. Steele & Co.		A. Weir & Co.	Glasgow
745	V	Pomona *Nielsen*	W Sr	Danish	122	92·0	20·8	10·0	Svendborg	1868	J. R. Andersen		P. E. Nielsen	Svendborg
746	DV	Pomona *Neumann* (WRECKED)	W Bk	German	434	127·1	29·2	18·1	Memel	1859	R. Pieper		Siebert & Alexander	Memel
747	N	Pomona *Ommundsen 72-84*	W Bk	Norweg'n	417	137·2	29·8	16·1	Risör	1871	Tellefsen	V 6.86	Christen's Boe	Arendal
748		Pomoschnik *Kraeft 85-87 (ex Thor)*	W Bn	Russian	155	97·3	22·2	11·2	Oscarshamn	1866	C. Thoren		Fishing Co. "Ribak"	St. Petersburg
749	V	Poncho *Gropius 85-88 (ex Woosung(S))*	I Bk	German	819	187·0	33·0	21·0	Port Glasgow	1858	John Reid & Co.		F. Laeisz	Hamburg
750	❋	Ponemah *Thornton 85-88*	W Bk	British	796	175·6	34·8	19·9	P. E. Island	1883	John Yeo	V 8.88	J. Yeo (T. P. Richards & Co.)	Charlottetown

❋ = Lloyd's Register; D = Nederlandsche Vereeniging; G = Germanischer Lloyd; I = Registro Italiano; R = American Record; A = Veritas Austro-Ungarico; H = Veritas Ellenico; N = Norske Veritas; V = Bureau Veritas.

Dans la 4me Colonne, C = Composite (membrure en fer, bordé en bois); I = Fer; Stl = Acier; W = Bois.

Les Dates du doublage en métal (**Col. 11**) ressortent des livres dans lesquels les navires sont respectivement classés ou des rapports des Inspecteurs du Lloyd's Register.

Pour plus amples renseignements sur les navires classés par le Lloyd's Register, voir le Lloyd's Register Book et le Lloyd's Yacht Register.

Official No.	No. in Book	Registry in which classed	Ship's Name. Material and Rig. Master, and Dates of Appointment to present Owner's Service and to Ship. Ship's Late Name.	Flag	Net Tonnage	Length	Breadth	Depth	Built: Where	Built: When	Built: By whom	Date of Metalling	Owners or Managers	Port
65397	751		Pons Ælii *Pakenham* W Bk	British	357	120·6	27·1	16·7	Newcastle	1868	[illegible]	[illegible]	E. Nicholson, jun.	Newcastle
	752	N	Pontecorvo *Sørensen* W Bk (ex Frederick W...)	Norweg'n	526	129·3	30·5	19·2	[illegible]	1851		[illegible]	H. Astrup	Christiania
	753	V	Pontiac *Olsen* W Bk	Norweg'n	591	152·5	29·8	18·6	Quebec	1860	[illegible]	[illegible]	H Prentout-Leblond	Christiania
[illegible]	754		Pontomedon W Bg	Greek	138	..	..	..	Galaxidi	1861			M. Zafiropoulos	Patras
95033	755		Poomgi Stl Lr	British	652	250·1	34·0	8·8	Dumbarton	1889	W. Denny & Bros		(Irrawaddy Flotilla Co. Ld)	Glasgow
56880	756	✻	Poonah *Richards 86-86* I S	British	1199	233·5	34·9	21·8	Sunderland	1867	W. Pile & Co.		Tyser & Co.	London
31416	757		Porana Letchemy W Bk	[illegible]	152	95·5	20·3	11·8	Ceylon	1857			Sunasethempavemchet-[illegible]	Jaffna
1356	758	I	Porfirio Di Martino *Esposito* W Bk	Italian	413	122·3	29·9	18·9	Procida	1868	G. Scotto	[illegible]	F. di Martino	Naples
	759	N	Porro *Olsen 78-78* W Bk	Norweg'n	474	135·1	30·1	16·9	[illegible]	1876	T. Jensen	[illegible]	J. Prebensen	Risör
87718	760	✻	Port Adelaide *Kerr 85-85* I Bk	British	1301	237·8	36·1	20·9	Port Glasgow	1883	Russell & Co		Crawford & Rowat	Glasgow
	761	✻	Port Carlisle *McIntyre 88-88* I Bk	British	1328	230·7	36·1	21·7	Greenock	1888	Russell & Co.		Crawford & Rowat	Glasgow
85623	762	✻ D	Port Glasgow *McDonald 88-88* I Bk	British	899	199·0	32·6	19·5	Port Glasgow	1882	Russell & Co.		Crawford & Rowat	Glasgow
86732	763	✻	Port Gordon *Gibb 82-82* I Bk	British	1098	215·0	35·1	21·0	Port Glasgow	1882	Russell & Co.		Crawford & Rowat	Glasgow
84362	764	✻	Port Jackson *Crombie 82-82* I Bk	British	2132	286·2	41·1	25·2	Aberdeen	1882	A. Hall & Co.		Duthie Bros.	Aberdeen
93322	765	✻	Port Ramsay *A. Hamilton* I S (MISSING)	British	1501	250·0	38·1	22·9	Port Glasgow	1886	Russell & Co.		Crawford & Rowat	Glasgow
90080	766	✻ D	Port Sonachan *M. Watson 85-85* I Bk	British	1112	215·4	35·1	21·3	Port Glasgow	1885	Russell & Co.		Crawford & Rowat	Glasgow
93383	767	✻	Port Yarrock *Pattie 86-86* I Bk	British	1318	230·8	36·2	21·7	Port Glasgow	1886	Russell & Co.		Crawford & Rowat	Glasgow
234	768	A	Portaitissa Iviron W Bg	Greek	131	75·5	22·3	..	Syra	1870			D. Mischalinos	Syra
	769	N	Portena *Halvorsen 85-85* W Bk	Norweg'n	321	118·3	25·8	16·0	Sunderland	1852	Austin & Mills		H. Rasmussen and S.B. Halvorsen	Porsgrund
56910	770	✻ D	Portia *Jones 84-87* I S	British	1424	221·6	36·1	21·7	Newcastle	1868	Palmer Bros. & Co		Ship "Portia" Co., Ld (W. Thomas & ...)	London
91413	771	✻	Portia *Ellis* W Sk	British	91	80·1	21·1	10·6	Brixham	1885	J.W. & A. Upham		C. Hellyer	Hull
55898	772	✻ D	Portinscale *Kearon -88* W Bk (ex Beatr... ex Portinscale)	British	447	141·2	27·2	17·3	Shoreham	1866	Shuttleworth	[illegible]	E. Kearon, jun.	London
20457	773		Portland *Donnelly 73-* W Bkn	American	469	161·0	35·5	13·6	Coos Bay, Oreg	1873			A. M. Simpson	San Francisco
150063	774	RV	Portland Lloyds *Hersey 75-81* W S	American	1242	190·6	38·5	22·0	E. Deering, Me.	1875	G. W. Russell	[illegible]	J. S. Winslow & Co.	Portland
	775		Porto Salvo *Cachia -79* W Bk	Greek	308	122·0	27·0	16·2	Malta	1864	A. Bosnich		N. Fermanoglu	Piraeus
	776		Portoense (WRECKED) W Bn (ex Duarte II., ex Florinda) *Dias*	Portug'se	235	97·4	21·3	12·8	Lisbon	1854		[illegible]	A. M. dos Santos	Oporto
	777		Portuense *Ramos* W Bg	Brazilian	175	..	..	..	Villa do Conde	1867			L. J. Ramos	Porto Alegre
	778	V	Porvenir *Lluell -87* W Bg	Spanish	328	115·4	28·9	16·4	Lussino	1862		[illegible]	J. M. Granada	Majorca
1007	779	A	Poschich *Tomsinich* W Bk (ex Bertrand)	Austrian	908	175·5	37·4	22·2	Segna	1874	Tartalich	[illegible]	A. F. Poschich fu A.	Fiume
81815	780	✻	Poseidon *Granger -88* I S	British	1708	264·5	39·8	23·4	Port Glasgow	1881	R. Duncan & Co.		C. S. Caird	Greenock
95062	781		Poseidon *Baird -88* Stl Sr	British	219	115·2	28·0	9·2	Dumbarton	1888	W. Denny & Bros		La Platense Flotilla Co. Ld	Glasgow
	782	G	Poseidon *Christensen 86-86* W Sr	Danish	135	89·2	20·0	11·3	Hoogezand	1862	J. A. H[illegible]		H. C. Christensen's Enke	Svendborg
	783	N	Poseidon *Husted* W Bk	Norweg'n	568	146·1	32·2	17·6	Kragerö	1877	J. Hamborg	[illegible]	H. Biorn, jun.	Kragerö
	784	N	Poseidon *Børresen 80-84* W Bk	Norweg'n	534	141·5	31·8	18·0	Stavanger	1871	M. A. Magnus	[illegible]	G. Gundersen	Stavanger
	785	NV	Poseidon *Terjesen* W Bk	Norweg'n	386	121·8	31·1	16·6	Tvedestrand	1867	A. & F. Smith	[illegible]	A. & F. Smith	Tvedestrand
	786	N	Poseidon *Christiansen -88* W Sr	Norweg'n	178	106·8	21·4	10·8	Grimstad	1869	Jacob Roe		Christen's Roe	Arendal
	787	V	Poseidon *Oegnestad* W Bn Sr	Norweg'n	176	101·3	26·6	10·5	Kolding	1872	C. Hansen	[illegible]	M. Oxnevad	Sandefjord
436	788	A	Posidon *Roussos* W Bk	Greek	351	111·8	28·9	..	Syra	1871		[illegible]	E. Roussos	Syra
	789	A	Posidon *Pappapetro* W Bg	Greek	335	..	..	..	Galaxidi	1887	C. Pappapetro	[illegible]	C. Pappapetro	Galaxidi
86	790		Posidon I Bn	Greek	274	157·2	25·1	..	Glasgow				S. Matkris	Zante
	791		Posidon *Minacoulis* W Bg	Greek	191	..	..	..	Cassos	1859			C. I. Miliarakis	Santorin
	792		Positivo *Coelho* W Bn	Brazilian	170	..	..	..	Villa do Conde				M. F. da Rosa	Rio Grande
	793		Posseidon W Bg	Turkish	362	..	..	..	Ooma	1880			Oozalaoo Kotch Zadeh Mustapha	[illegible]
	794		Posseidon W Bg	Turkish	362	..	..	..	Oonia	1885			Oozalaoo Kotch Zadeh Mustapha Effendi & Sons	Oonia
	795		Posseidon W Bg	Turkish	215	..	..	..	Greece	1874			Enosloe Yanni & Co.	[illegible]
	796		Posseidon W Bg	Turkish	175	..	..	..	Greece	1883			Madame Hernchmisan Sutirisb	[illegible]
20123	797	R	Post Boy *Smith* W Sr	American	163	95·6	28·2	9·0	Bucksport, Me.	1872		[illegible]	S. B. Swasey	Bucksport
	798	G	Posthalter J. C. Wahl *Lassen 67-* W Bg	German	299	109·7	27·0	15·7	Rostock	1867	J. Möller		E. Beckmann & Co.	Rostock
	799	N	Postulant *Johannesen* W Sr	Norweg'n	141	93·3	23·4	11·4	Wiborg	1876			J. Johannessen	Brevik
	800	✻ R	Potengy [illegible] W 3m S (ex [illegible])	Brazilian	351	138·5	26·5	16·3	P.E. Island	1878	Richards	[illegible]	Pineiro	Natal

✻ = **Lloyd's Register**; D = Nederlandsche Vereeniging; G = Germanischer Lloyd; I = Registro Italiano; R = American [...]
A = Veritas Austro-Ungarico; H = Veritas Ellenico; N = Norske Veritas; V = Bureau V[...]

In the 4th Column, C = Composite (iron frame, wood planked); I = Iron; Stl = Steel; W = Wood.

The Dates of Metalling (Col. 11) are based on the records of the Books in which Vessels are respectively classed, or on the Reports of the Surveyors of L[...]

Further particulars of Vessels classed by Lloyd's Register will be found in Lloyd's Register Book and Lloyd's Yacht Register.

No.		Nom. Matériaux et Gréement. Capitaine et dates d'entrée au service des Armateurs actuels et à bord du navire. Ancien Nom.	Pavillon.	Tonnage net.	Longueur	Largeur	Creux	Construit. Où.	Quand	Par qui.	Date du doublage en métal	Armateurs ou Directeurs-Gérants.	Port d'Armement.
		4	5	6	7			8	9	10	11	12	13
801		Potomac *Munsanden* W Bn	British	282	129·1	28·4	10·1	Baltimore	1868			Benjamin V. Abraham	Gibraltar
802		Potomac W	British	161	99·0	24·6	10·7	Melbourne	1857			Henry Moss	Melbourne
803	R	Potosi *Henderson* W3mSr	American	350	126·5	28·5	13·8	CityIslandNY	1871	D. Carll		Walton & Co.	Wilmington
804	✱V	Potrimpos *Rohlke* 78-87 Stl Bk	German	1231	227·7	35·9	20·7	Hamburg	1887	Blohm & Voss		F. Laeisz	Hamburg
805	✱	Potsdam *Müller* -89 Stl Bk	German	1411	241·4	38·1	20·7	Hamburg	1889	Blohm & Voss		F. Laeisz	Hamburg
806		Pouplina W Sr	Greek	216	100·4	24·9	..	Cassos	1884	Elias Sophos		D. E. Oeconomo	Syra
807	R	Powhattan *Oats* W Bk	British	403	130·7	27·8	17·1	Swansea	1866	Meager	Y 1.88	G. H. Meager	Swansea
808		Pozzallo *Scala* W Bn	Italian	107	..	..	..	Castellamare	1875	A. Bonifacio		E. E. V. Giunto	PortoEmpedocle
809	NV	Præcis *Markussen* W3mSr	Norweg'n	290	127·4	27·8	13·5	Stavanger	1876	R&T Rasmussen	Y 8.88	S. Berners Sönner	Stavanger
810		Præsident *Strandberg* -75 W S (ex Kosmos, ex Hiawatha, ex W. Merritt)	Norweg'n	1055	172·7	36·6	26·0	Waldoboro, Me.	1860		Y 12.88	A. Leschbrandt	Mandal
811	N	Præsident *Mikelsen* -87 W Bk	Norweg'n	794	170·4	34·6	20·4	Stavanger	1875	M. Magnus	Y 11.88	E. Berentsen	Stavanger
812	N	Præsident Harbitz *Hansen* W Bk	Norweg'n	562	151·2	34·4	18·3	Sandefjord	1865	S. Andersen		J. Christophersen	Sandefjord
813		Præsident Harbitz (PONTOON) *Flaaand* -88 W3mSr	Norweg'n	339	143·1	28·6	16·2	Bergen	1861	J. Trampy & Sns.	Y 2.86	Georg Wiese	Bergen
814		Præsident Sverdrup (WRECKED) *Bentzen* W Bk	Norweg'n	445	126·5	29·7	16·4	Kragerö	1874	E. O. Rinde	Y 8.87	Peter Rinde	Kragerö
815	V	Praesident von Blumenthal *Scharping* 63-76 W Bk	German	525	124·3	31·6	19·3	Memel	1857	B. Pieper		E. Hauptermacher 'seine Erben	Rügenwalde
816		PrairieGem *Andersson* -78 W Sr	Swedish	248	106·3	21·7	14·6	P. E. Island	1864	Bollen		B. Andersson	Höganäs
817	V	Präsident Trotsche *Ziegule* 76-76 W Bk	German	504	139·7	30·7	18·1	Ribnitz	1876	J. H. Wilken	Y 4.88	A. Burchard	Rostock
818	✱	Prazeres *dos Santos* W Bg	Brazilian	226	128·7	25·0	11·4	Oporto	1873	C. J. A. Vareta	Y 8.88	M. G. da C. Correia	Rio Grande
819	I	Precessore *G. dos Santos* W Bn (ex Geltrude, ex Tammis)	Italian	151	97·4	21·0	11·8	Pekela	1865		I 8.85	Merli & Lugaro	Genoa
820		Preciosa (TO BE BROKEN UP) *Holow* W Bk	German	362	116·2	28·5	16·5	Barth	1860	A. Witte		J. N. Rodbertus	Barth
821	N	Preciosa *Jacobsen* 77-77 W Bk (ex Mikado, ex Virginia Dare)	Norweg'n	797	160·0	33·5	20·7	Baltimore	1860		Y 3.88	A. J. Jacobsen	Tönsberg
822	V	Preciosa *Staudin* -86 W Bk	Russian	737	146·4	34·1	19·1	Bremerhaven	1867	J C Tecklenborg	Y 11.83	C. E. Carlström	Christinestad
823	V	Preciosa *Hager* W Bk	Swedish	326	117·1	23·3	15·5	Gothenburg	1864	A. Ahlberg	Y 7.88	P. A. Lindberg	Gothenburg
824		Preference *Berntson* W Sr	Swedish	205	108·9	23·5	12·1	Westervik	1861			J. S. Edström	Mollösund
825	A	Prelucano *Vladiscovich* W Bg	Austrian	334	117·7	26·5	16·8	Volosca	1870	F[lli] Schiavoni	Y 3.88	G. Minach fu G.	Fiume
826	✱	Premier (HULK) I Bk		486	153·6	26·8	17·2	Dundee	1865	Brown & Simpson			
827	✱	Premier (WRECKED) *Roberts* 87-87 I Sw	British	219	117·4	23·9	12·9	Llanelly	1863	Nevill Bros.		A. L. Jones (Grand Canary Coaling Co.)	Llanelly
828		Premier W Bn	British	176	95·8	21·9	13·4	Rye	1855			J. W. Bussey	Dover
829	R	Premier *Anderson* W3mSr	American	292	137·0	33·4	11·0	P Ludlow, Wsh	1876	Hall Bros.		J. Jensen	San Francisco
830	N	Premier *Stokke* 87-87 W S	Norweg'n	1090	180·4	37·8	23·2	Bangor, Me.	1853		Y 12.87	H. Kjösterud & Sönner	Drammen
831	N	Premier *Gulbrandsen* 71-77 W Bk	Norweg'n	474	131·3	31·2	17·1	Arendal	1877	S. Johnsen	Y 10.85	A. Asbjörnsen	Arendal
832	R	Premier McKenzie *Howard* -87 W Bk	British	398	132·0	29·2	11·4	Newcastle, NB	1878	C. C. Watt	Y 5.88	Miss M. A. G. Watt (Manager, W. E. Boyd)	Chatham, N.B.
833	R	Prescott Hazeltine *Knowlton* -87 W3mSr	American	366	130·0	31·7	14·4	Belfast, Me.	1871			C. Harvey	Belfast, Me.
834	DV	President *Chisholm* -89 W Bk	British	916	186·0	36·3	19·8	Gt. Village, NS	1881	McLean & Blekie	Y 8.87	J. M. Blaikie	Halifax, N.S.
835	V	Président Mabire *Corre* -88 W Bk	French	666	153·8	29·2	20·7	Quebec	1868		Y 4.88	A. Postel & ses fils	Havre
836	V	Président Thiers *Moreau* -88 W Bk	French	399	128·6	28·4	14·9	St. Malo	1872	Parnot	Y 10.86	H. Auger, aîné	Havre
837	D	President van Rijckevorsel *Verkamp* 88-88 W Bk	Dutch	716	149·2	35·8	20·8	Krimpen	1857	Otto	Y 83	E. J. Bok & Zonen	Amsterdam
838	I	Presidente Washington *Feppa* -88 W Bk		453	128·6	27·9	19·0	Lerici	1866	F. Rolla	I 7.86		
839		Presto *Laidman* I Bk	British	384	135·0	28·0	14·5	Amsterdam	1862	W. & A. H. Meursing		H. McQuarrie	Auckland, N.Z.
840	G	Presto *Haase* 86-86 W Bk	German	285	112·0	27·2	14·2	Rostock	1867	H. Rickmann		W. Maack	Rostock
841	N	Presto *Olsen* -82 W Bg	Norweg'n	258	100·4	26·3	15·0	Arendal	1859			Sn. F. Olsen	Arendal
842		Presto *Olsen* -86 W Bg	Norweg'n	173	102·7	24·4	11·9	Stavanger	1858	R. O. Hœrem		T. Meling	Stavanger
843	✱	Preston *Armstrong* 73-75 W Bk	British	336	117·1	27·3	17·0	Sunderland	1865	W. Chilton & Co.	Y 1.88	W. Davison	North Shields
844	V	Providente (ex Daniele) *Via* 86-86 W Bk	Italian	705	167·9	33·1	21·3	Lavagna	1866	D. Briasco	Y 3.86	D. Dellepiane	Savona
845		Pride of America W Sr	British	285	133·7	23·3	12·2	S. Ctherins, Ont	1863	L. Shickluna		M. Ryan	Toronto, Ont.
846		Pride of England *Nash* W S	British	1351	205·5	37·5	22·8	Quebec	18^{63}_{73}	H. Dinning		Government of New South Wales	Sydney, N.S.W.
847	✱D	Pride of Wales *Griffiths* W Bk	British	288	125·3	26·5	14·6	Portmadoc	1869	Jones	Y 1.88	D. Morris	Portmadoc
848	✱	Pride of the Channel *Piper* 75-75 W Bkn	British	176	109·6	24·4	12·8	Charlestown	1872	Lake	Y 6.87	T. Hitchins	Fowey
849	✱	Pride of the Ocean W Sk	British	91	78·2	21·4	10·8	BartonStather	1885	J. Garside		G. W. Parker	Grimsby
850		Pride of the Sea W Bk	British	320	113·0	27·0	11·0	Franklin, U.S.	1854			Wm. Morrice	Nassau, N.P.

✱ = Lloyd's Register; D = Nederlandsche Vereeniging; G = Germanischer Lloyd; I = Registro Italiano; R = American Record;
A = Veritas Austro-Ungarico; H = Veritas Ellenico; N = Norske Veritas; V = Bureau Veritas.

Dans la 4me Colonne, C = Composite (membrure en fer, bordé en bois); I = Fer; Stl = Acier; W = Bois.

Les Dates du doublage en métal (Col. 11) ressortent des livres dans lesquels les navires sont respectivement classés ou des rapports des Inspecteurs du Lloyd's Register.

Pour plus amples renseignements sur les navires classés par le Lloyd's Register, voir le Lloyd's Register Book et le Lloyd's Yacht Register.

Official No.	No. in Book	Society in which classed	Ship's Name. Master and Dates of Appointment to present Owner's Service and to Ship. Ship's Late Name.	Material and Rig	Flag.	Net Tonnage.	Length	Breadth	Depth	Built Where.	When.	By whom.	Dates of Metalling	Owners or Managers.	Port of Registry
[illegible]	851	*	Pride of the South *Pellow*	W Sr	British	130	93·5	23·4	11·7	Padstow	1831	C. Rawle		A. Jenkyns	Padstow
[illegible]	852	* D	Pride of the West *Hicks*	W Sr	British	98	88·6	22·2	11·1	Padstow	1880	C. Rawle		A. Jenkyns	Padstow
	853		Prim	W Bn	Spanish	129	81·1	23·3	10·2					S. Piera	Barcelona
	854		Prima *Wulf*	W Bn	Danish	118	77·1	22·6	11·0	[illegible]	1854	C. J. Bring		P. J. Wulf	Dragør
	855	R	Prima *Christiansen 87-88* (ex Magdala)	W Bg	Norweg'n	275	105·2	26·2	16·0	P. E. Island	1868	Plestid		T. Saeboe	Egersund
	856	V	Prima *Holm*	W Bk	Swedish	504	155·8	22·4	16·9	Sundsvall	1876	J. Sjölin	[illegible]	Sundsvalls Skeppsr. Bolag (R. A. Wildhaling)	Sundsvall
	857		Prima *Rosenberg*	W S	Swedish	254	120·0	24·4	11·9	Timmernabben	1871	C. Bring	[illegible]	N. Olsson	Höganäs
[illegible]	858		Prima Donna	W Sr	British	104	88·4	21·4	11·0	Llanelly	1851			R. T. Holbrook	Penzance
	859	V	Prima Donna *Pedersen*	W Bk	Norweg'n	930	166·0	33·8	22·7	Bideford, Me.	1865	E. Perkins	[illegible]	Helgesen & Co.	Christiania
	860		Primeiro de Janeiro	W Bg	Brazilian	178	..	..	..	Portmadoc	1868			J. B. Bernisson, jun.	Desterro
74484	861	*	Primera *Christian 83-83*	I Bk	British	597	173·7	28·6	17·5	Glasgow	1875	A. Stephen & Sons		W. Sherwen	Liverpool
	862	V	Primevère *Hanis*	W Sr	French	122	90·5	21·0	10·5	St. Vaast	1878	E. Costey		J. Gicquel	Dunkirk
	863	V	Primitiva Maria *Alvado* (ex Ignacia ex Primitiva)	W Sr	Spanish	165	90·8	25·1	11·9	Blanes	1855		[illegible]	Señora M. Orero Zuñiga	Valencia
[illegible]	864	A	Primo *Ciano*	W Bk	Italian	464	131·8	30·5	19·3	Leghorn	1865	L. Patrone	[illegible]	R. Ciano	Leghorn
[illegible]	865	V	Primo *Scarpati*	W Bg	Italian	199	99·1	24·1	15·7	Meta	1860	F. S. Manno	[illegible]	L. Scarpati	Castellammare
	866	N	Primo (ex [illegible]) *Hansen 85-87*	W Bg	Norweg'n	245	117·1	27·1	12·5	Brevig	1876		[illegible]	Jorgensen & Roe	Brevig
[illegible]	867	A	Primo A. *Sparta* -87	W Trb	Italian	139	82·0	22·9	12·0	Pallestrina	1881	G. Vianello	[illegible]	Pietro Ardizzon	Venice
[illegible]	868		Primrose *Loughlan 82-83*	W Sw	British	208	94·8	25·4	15·2	Sunderland	1864	Denniston		T. H. Franks	Folkestone
[illegible]	869	*	Primrose *Hitchins*	W Sr	British	104	91·0	19·7	9·9	Southampton	1886	J. Ransom	[illegible]	I. Dibb	Southampton
	870	*	Primrose *Cook*	W Sk	British	96	82·0	21·2	10·4	Hull	1886	Albt. Dk. Shp. Co.		J. Munby	Hull
	871	G	Primrose *Stromberg 83-83* (ex Primrose of Ireland)	W Sr	Swedish	176	94·5	21·5	11·8	P. E. Island	1864	Dolrant		A. R. Andersson	Calmar
91294	872	*	Primrose Hill *Anderson*	I Bq Sr	British	2436	301·6	42·1	24·7	Liverpool	1886	T. Royden & Sons		[illegible] "Primrose Hill" [illegible] (W. Price & Co.)	Liverpool
	873	N	Primula *Lorentzen 88-88*	W Sr	Norweg'n	257	118·7	28·1	12·9	Nova Scotia	1865		[illegible]	Actieselskabet "Primula" (A. R. [illegible])	Fredrikshald
69762	874	G	Primus *Scott's 85-85* (ex [illegible])	W Bn	British	289	122·6	29·2	13·9	Sastmola	1879		[illegible]	C. S. Rugeroni	Gibraltar
	875	N	Primus *Einarsen*	W Bk	Norweg'n	277	115·1	27·5	13·7	Tvedestrand	1871	G. Berntsen	[illegible]	E. Jorgensen	Tvedestrand
	876	DV	Primus *Heine* -83	W S	Russian	1166	194·8	35·1	21·9	Strömma	1878	L. P. Kjaldström	[illegible]	V. A. Reims	Abo
	877		Primus *Lundell*	W Sr	Russian	237	..	..	..		1874			K. G. Sundberg	Lemland
	878		Primus *Hougelin*	W Sr	Russian	228	..	..	..	Kimito	1875			A. W. Blomqvist	Kimito
	879		Primus *Ohlsson*	W Sr	Swedish	217	98·8	23·6	13·1	[illegible]	1877	A. Ohlsson		J. Abrahamsson	Öregrund
	880	V	Prince Albert *Hanssen 85-87* (ex King [illegible])	W S	Norweg'n	1564	222·2	40·0	24·0	St. John, N.B.	1875	J. K. Dunlop	[illegible]	Actieselskabet Prince Albert (A. [illegible])	[illegible]
44410	881	R	Prince Alfred *Bennett* -87	W Bk	British	531	142·3	30·8	19·4	Queenstown	1864	J. Wheeler	[illegible]	J. Owens	Swansea
43738	882		Prince Alfred *Wilby*	W Sw	British	257	103·0	26·2	16·0	Sunderland	1861	A. Simey & Co.		A. & F. Manuelle	Guernsey
43133	883		Prince Alfred *Rutter*	W Bkn	British	161	111·8	21·8	10·6	Adelaide	18 61/62			M. Kennedy	Lyttelton
72205	884	DV	Prince Amadeo *Nelson* [illegible]	W S	British	1561	224·5	39·9	24·0	St. John, N.B.	1875	J. K. Dunlop	[illegible]	Moran, Galloway & Co.	Liverpool
	885	NV	Prince Arthur *Moeck* [illegible]	W Bk	Norweg'n	990	170·3	34·5	21·6	St. Martins, N.B.	1872	J. H. Moran	[illegible]	Akt. Selsk. "Prince Arthur" (Saxlund & Roer)	Christiania
	886	NV	Prince Charlie *Brevegaard* [illegible]	W Bk	Norweg'n	1386	201·5	40·7	23·9	St. Martins, N.B.	1864	J. H. Moran	[illegible]	H. Kjilsterud & Sönner	Drammen
44435	887		Prince Consort *Pierce* [illegible]	W Bg	British	2[illegible]	105·0	24·0	13·3	Aberdeen	1862	W. Duthie & Co.		J. E. Pearce	Faversham
	888	*	Prince Edward *Cornelissen* (ex [illegible] ex [illegible])	I S	Norweg'n	2180	284·8	41·0	27·7	London	1867	M. Wigram & Sons		Actieselskabet Prince Edward (Saxlund & Roer)	Christiania
	889	NV	Prince Eugène *Nygaard 81-88*	W Bk	Norweg'n	1287	193·5	39·9	23·9	St. Martins, N.B.	1867	J. H. Moran	[illegible]	Johanson, Rötter & Co.	Christiania
66953	890	V	Prince Frederick *Rutter 75-86*	W S	British	1476	220·6	39·1	24·0	St. John, N.B.	1874	J. K. Dunlop	[illegible]	Moran, Galloway & Co.	Liverpool
	891	*	Prince George *Overgaard* (ex Blanche) [illegible]-87	I Bk	Norweg'n	478	157·9	26·8	17·2	Sunderland	1865	W. Doxford		Saxlund & Roer	Christiania
	892		Prince Hassan *Davies*	W Bk	Chilian	387	134·9	28·8	17·3	Littlehampton	1871	H. Harvey & Sons	[illegible]	J. Jones	Coronel
68891	893	V	Prince Henry (ex King [illegible])	W S	British	1288	195·9	38·7	23·8	Hopewell, N.B.	1874	J. Calhoun	[illegible]	Moran, Galloway & Co.	Liverpool
59411	894		Prince Le Boo *Thorne*	W Bn	British	231	101·1	28·6	11·7	Parrsboro, NS	1873	J. Leonard		J. Murray	St. John's N.
28547	895		Prince Llewellyn *Pritchard*	W Sr	British	111	78·3	21·3	11·3	Carnarvon	1860	Barlow		R. Pritchard	Carnarvon
[illegible]	896	V	Prince Louis *Page* 73-84	W S	British	1322	206·3	38·3	23·5	St. Martins, N.B.	1873	J. H. Moran	[illegible]	Moran, Galloway & Co.	Liverpool
72394	897	V	Prince Lucien *Clague* 81-85	W S	British	1549	214·2	40·3	24·2	St. John's, N.B.	1877	J. K. Dunlop	[illegible]	Moran, Galloway & Co.	Liverpool
51403	898	*	Prince Oscar *Henderson* 84-86	I S	British	1292	220·0	37·0	21·7	W. Hartlepool	1864	Pile, Spence & Co.		A. Gibson & Co.	Liverpool
	899	N	Prince Patrick *Nelson* [illegible]	W Bk	Norweg'n	1052	178·3	38·9	22·8	St. Martins, N.B.	1866	J. H. Moran	[illegible]	M. Pedersen	Mandal
	900														

*=Lloyd's Register; A=Veritas Austro-Ungarico; D=Nederlandsche Vereeniging; G=Germanischer Lloyd; H=Veritas Ellenico; I=Registro Italiano; N=Norske Veritas; R=American [illegible]; V=Bureau Veritas.

In the 4th Column, C=Composite (iron frame, wood planked); I=Iron; Stl=Steel; W=Wood.

The Dates of Metalling (**Col. 11**) are based on the records of the Books in which Vessels are respectively classed, or on the Reports of the Surveyors of Lloyd's [illegible]

Further particulars of Vessels classed by Lloyd's Register will be found in Lloyd's Register Book and Lloyd's Yacht Register.

No.	Registre	Nom. Matériaux et Gréement. Capitaine et dates d'entrée au service des Armateurs actuels et à bord du navire. Ancien Nom.		Pavillon.	Tonnage.	Longueur.	Largeur.	Creux.	Où.	Quand.	Par qui.	Doublage	Armateurs ou Directeurs Gérants.	Port d'Armement.
901	V	Prince Regent	W Bk	Norweg'n	1332	197·2	39·6	23·9	St.Martins,N.B	1869	J.H.Moran	V 5.86	Haagensen	Christiania
902	NV	Prince Rupert *O'Malley 75-82*	W Bk	British	1168	191·0	39·2	23·0	St.Martins,N.B	1862	J.H.Moran	V 2.86	A.O'Malley	Liverpool
903	*	Prince Rupert *Hatfield*	W Bk	British	368	139·2	27·4	13·7	Littlehampton	1876	J.&W.B Harvey	V 2.86	Hudson's Bay Co.	London
904	DV	Prince Umberto *Keane*	W Bk	British	1400	215·7	39·1	24·0	St.John,N.B.	1873	J.K.Dunlop	V 5.86	Moran,Galloway & Co.	Liverpool
905	V	Prince Victor *Vande 87-87*	W S	Norweg'n	1134	193·1	36·0	22·0	St Martin'sNB	1875	W.H.Rucke	V 3.87	Aktieselskabet "Prince Victor" (Seeligmann & Berg)	Christiania
906		Prince of Wales	W Bg	British	167	99·5	24·5	12·1	P.E.Island	1860			Wm.Morrice	P.E.Island
907		Princeport *Showab -88*	W Sr	British	122	83·2	24·1	9·7	Princeport	1883	M.C.Norris		A.J.Walker	Truro,N.S.
908		Princesa Dagmar	W Bk	Spanish	431	127·9	28·3	17·9	S.Shields	1865			V.F.Casella	Havana
909		Princess	W Sr	British	137	88·2	25·0	9·4	PtGilbert,N.S.	1879	J.H.Munroe		J.Morehouse	Digby,N.S.
910	*	Princess *Rowbotham*	W K	British	100	78·0	21·4	9·9	Goole	1879	Wake		G.J.King & Co.	Goole
911	R	Princess Alexandra *M'Kinlay*	W Bk	British	1287	192·6	39·7	24·9	St.John,N.B.	1862	M'Donald	V 5.84	Barque "Princess Alexandra" Co.,Ld. (J.&R.Brown & Co.)	Greenock
912	R	Princess Alexandra *Spencer -86*	W S	British	1281	187·7	38·8	24·4	St.John	1863	Ruddock	V 3.87	J.Moore	Belfast
913		Princess Alexandra *Railton 78-83*	W Sw	British	288	107·6	26·8	16·2	Sunderland	1863	Todd Brown		J.Paul & Co.	Lynn
914	*V	Princess Beatrice *Cordes 1-87*	W Bk	German	341	117·5	27·1	16·9	Sunderland	1863	R.Thompson,jr.	V 3.85	C.C.L.Cordes	Hamburg
915	*	Princess Royal *Branstead*	W Bn	British	135	94·6	21·7	10·3	Woodbridge	1843			W.G.Cole	Ipswich
916		Princess Seraphi	W Bk	Siamese	455	145·0	26·3	15·6	Bremen	1862			Siamese Government	Bangkok
917	*	Princess of Thule *Hughes*	W Sr	British	115	88·8	22·6	11·0	Bridgwater	1878	Gough		J.Hughes	Chester
918		Princess of Wales *Williams 72-83*	W Bk	British	332	143·8	26·2	13·9	Harwich	1862	Vaux		J.H.Vaux	Harwich
919		Princess of Wales	W Sr	British	187	105·0	23·0	9·6	Moulmein	1879			Moung Kyoung	Moulmein
920		Princeton *Gray*	W Sr	American	141	95·4	25·2	9·0	Brooklyn	1853			S.G.Haskell	Deer Isle,Me.
921		PrincezaAmelia *(ex Christina)*	W	Brazilian	105	..	..	..					A.J.da Silva	Rio de Janeiro
922	*D	Principality *Jones*	I 4mS	British	1699	258·5	39·6	23·1	Sunderland	1885	W Doxford&Sns		PrincipalityS.Co. (W.Thomas & Co.)	Liverpool
923	V	Principe *Pagés*	W Bn	Spanish	190	92·2	27·0	12·8	Barcelona	1856		V 8.84	Jaime Pagés	Barcelona
924	V	Principe Amedeo Di Savoia *Calabrese*	W Bk	Italian	463	133·2	31·0	19·9	SestriPonente	1868	F^lli^ Cadenaccio	V 7.83	Frat.Calabrese&Lanzow	Bari
925	*	Principe da Beira *(ex Glenisla) Morgan*	W Bkn	Portug'se	283	117·0	24·9	14·2	London	1877	Mills	V 12.87	A.dos S.Gomes	Terceira
926		Principe d. Alfonso	W Bn	Brazilian	142	..	..	..					J.J.S.Breves	Rio de Janeiro
927	IV	Principe Di Napoli *(ex Donna Isabella) Garibaldi -87*	W Bk	Italian	1059	191·2	35·1	24·3	Varazze	1878	N.Cerruti	V 1.84	G.B.Dodero	Genoa
928	I	Principe Di Napoli *Menna*	W Bk	Italian	366	121·3	29·5	17·4	Alimari	1871	F.S.Mauro	V 6.85	P.Menna & F.Megna	Castell. di Stab.
929	V	Principessa Margherita Di Piemonte *Cammarota 87-87*	W Bk	Italian	438	125·6	30·7	18·9	SestriPonente	1869	B.Celesia	V 11.84	V.Cardillo	Naples
930		Principio *Lagomarsino 86-88*	W Bk	Italian	507	131·2	31·4	19·0	Prà	1870	M.Tixi	V 6.82	M. & A.Dall'Orso fuG.	Spezia
931	DN	Prinds Christian August *Andersen*	W 3mSr	Norweg'n	247	95·4	26·2	14·9	Risör	1856			J.K.Andersen	Tönsberg
932	NV	Prinds Leopold *(ex Prince Leopold) Kjønnerberg -88*	W Bk	Norweg'n	1296	195·8	39·6	23·9	StMartins,NB	1868	J.H.Moran	V 6.88	H.Kjösterud & Sönner	Drammen
933	N	Prinds Mauritz *Kreger*	W S	Norweg'n	809	150·7	36·2	21·9	Middelburg	1841		Z 7.81	D.Danielsen	Tönsberg
934	N	Prinds Oscar *Hansen*	W Bk	Norweg'n	662	144·5	35·0	20·2	Rotterdam	1860		V 3.83	R.-Aktieselskab(O.Olsen)	Fredrikshald
935		Prinds Oscar	W Bk	Peruvian	479	133·9	29·9	15·9	Bergen	1865	J.Gran		C.B.Bacon	Callao
936	N	PrindsOscar Frederik *Hansen*	W Bk	Norweg'n	227	96·4	25·9	14·3	Christiansand	1843			Carl Dahl	Fredrikshald
937		Prindsesse Caroline Amalie *Mathiesen 75-77*	W Bg	Danish	177	83·4	25·1	13·6	Svendborg	1830	L.N.Møller		J.Mathiesen	Svendborg
938	NV	Prindsesse Louise *Helvorsen -87*	W Bk	Norweg'n	430	139·6	31·2	15·9	Bergen	1869	J.Trumpy&Sön	V 1.86	O.Reinertsen & Sönner	Farsund
939		Pringle *(ex Serrado)*	I Sr	Argentine	238	125·0	29·6	10·6	Genoa	1884	E.Cravero & Co.		La Platense Flotilla Co.	Buenos Ayres
940		Prins Karl *Jonsson -81*	W Bk	Swedish	382	113·1	27·3	17·3	Stockholm	1843			G.Mellin	Hälle,Bohuslän
941		Prins Karl	W Bk	Swedish	380	115·4	27·6	17·1	Hernösand	1845	Frick		J.M.Möller	Gothenburg
942	G	Prins Oskar Fredrik *Braun -88*	W Bg	Swedish	256	100·7	26·3	14·8	Gothenburg	1852			C.Olsson	Viken
943	DV	Prinses Wilhelmina *Bener 85-85*	W Bk	Dutch	370	133·8	29·8	14·4	Middelburg	1882	Borsius & van der Leye	V 6.85	Den Bouwmeester Borsius & van der Leye	Middelburg
944	NV	Prinsessen *Andersen 70-71*	W Bk	Norweg'n	454	112·3	31·4	17·2	Tvedestrand	1874	T.Tellefsen	V 10.86	J.Smith	Tvedestrand
945	V	Printemps *Carné -84*	W Bn	French	133	82·3	22·5	11·2	Redon	1882	L.Mahon	Z 3.86	L.Mahon fils	Nantes
946	G	Prinz Adalbert *Helander -87*	W Bkn	Swedish	246	106·2	23·6	13·6	Wolgast	1864	E.Schelle		O.F.Helander	Calmar
947	*V	Prinz Albert *Jäger -88*	I Bk	German	570	155·1	28·6	17·6	Hamburg	1861	Reiherstieg Schiffs-Werfte		R.M.Sloman & Co.	Hamburg
948	V	Prinz Friedrich Carl *Fridrichsen 79-87*	W Bk	German	443	119·4	30·2	17·4	Danzig	1866	F.Devrient		F.Hemptenmacher sche Erben	Rügenwalde
949	GV	Prinz Regent *Hennings 66-74*	W Bk	German	467	133·8	29·5	17·2	Memel	18^59^~59~	B.Pieper	V 10.82	F.Braunn	Stettin
950		Prinz von Preussen *Pries -73*	W Bg	German	241	103·1	26·5	13·6	Greifswald	1859	L.Wittenberg			

*=Lloyd's Register; D=Nederlandsche Vereeniging; G=Germanischer Lloyd; I=Registro Italiano; R=American Record; [illegible]=Veritas Austro-Ungarico; H=Veritas Ellenico; N=Norske Veritas; V=Bureau Veritas.

Dans la 4me Colonne, C=Composite (membrure en fer, bordé en bois); I=Fer; Stl=Acier; W=Bois.

Les Dates du doublage en métal (Col. 11) ressortent des livres dans lesquels les navires sont respectivement classés ou des rapports des Inspecteurs du Lloyd's Register

Pour plus amples renseignements sur les navires classés par le Lloyd's Register, voir le Lloyd's Register Book et le Lloyd's Yacht Register.

Official No.	No.	Registry	Ship's Name. Material and Rig. Master, and Dates of Appointment to present Owner's Service and to Ship. Ship's Late Name.	Flag.	Net Tonnage.	Length.	Breadth.	Depth.	Built. Where.	When	By whom.	Date of Metalling.	Owners or Managers	Port of [illegible]
[illegible]	951	✻	Priorhill *Donovan 81-88* I Bk	British	1127	215·3	35·1	21·0	PortGlasgow	1883	Russell&Co.		J.Cornfoot & Co.	Glasgow
[illegible]	952		Priscilla W Bk	British	766	150·0	27·4	21·5	Yarmouth,U.S	1856			King George's Sound Coal Co.	Sydney,N.S.
[illegible]	953	✻	Priscilla *Owen* W Bn	British	163	92·6	23·1	12·9	Aberystwith	1866	Williams		W.Thomas	Aberystwith
[illegible]	954	R	Priscilla *McClain -88* W Bktn	American	612	165·4	34·0	17·1	Belfast,U.S.	1886	G.W.Cottrell	[illegible]	C.M.Stewart & Co.	Baltimore
[illegible]	955		Priscilla Scribner *Dauey [illegible]* W 3mSr	American	432	144·5	33·0	16·5	Milford,Del.	1873	Scribner		J.W.Hall	Wilmington
[illegible]	956	V	Privateer *Landry* W Bk	British	875	168·7	35·3	20·4	Tyn'mouthNB	1872		[illegible]	H.D.V.Troop	St.John,N.B.
	957	V	Probidade (ex W. H. Marsh) *Rocha* W Bk	Portug'se	427	121·0	26·7	16·9	Voltri	1869		[illegible]	Martins,Praça & Co.	Lisbon
[illegible]	958	V	Probo *Cafieri -87* W Bk	Italian	345	114·5	29·6	18·1	Meta	1865	F.Mauro	[illegible]	F^lli De Martino	Castell.di [illegible]
	959		Probus W 3mSr	Russian	187	..	..	..		1852			C.M.Osterlund	Wardo
[illegible]	960	I	Procida (ex [illegible]) *Mazzella [illegible]* W Bn	Italian	307	108·6	29·3	17·3	Procida	1864		[illegible]	M.Mazzella di B.	Naples
	961		Prodomos W Sr	Turkish	174	..	..	..	Greece	1880			Costantinasi & Co.	Constant[illegible]
	962		Prodomos W Bg	Turkish	140	..	..	..	Greece	1880			Salunikiee Christo Dimitri [illegible]	Salanik
[illegible]	963	A	Prodromos *Sanoudas* W Bk	Greek	332	109·2	27·7	..	Syra	1871		[illegible]	A.Sanoudas	Syra
[illegible]	964	H	Prodromos *Parthenopoulos* W Bg	Greek	294	107·3	27·9	..	Syra	1870			J.T.Sachtouris	Piræus
[illegible]	965	A	Prodromos *Dambasis* W Bg	Greek	187	86·6	22·5	..	Spezzia	1865			A.G.Dambasis	Syra
	966	N	Producent *Jensen 80-83* W Bk	Norweg'n	338	116·9	28·6	15·3	Risör	1877	J.Hansen	[illegible]	J.F.Juell	Risör
	967	N	Professo *Andersen* W Bg	Norweg'n	199	93·1	25·8	12·2	Tvedestrand	1867	A.&F.Smith		Ferd Ourdahl	Christiania
	968	V	Professor (ex [illegible]) *Neumann 87-87* I Bk	German	512	143·0	27·6	17·2	Hamburg	1865	Reiherstieg Schiffs Werft		F.Laeisz	Hamburg
	969	V	Professor Cantzler *Buerert [illegible]* W Bg	German	253	99·3	23·5	13·0	Stralsund	1867	O.Kirchhoff		Seitz & Kindt	Stralsund
	970	NV	Professor Johnson *Pedersen [illegible]* W Bk	Norweg'n	1107	187·8	38·6	21·5	Drammen	1877	R.O.Barrem	[illegible]	H.T.Kiær	Drammen
	971	N	Professor Lintner (ex Rachel Black, [illegible]) W Bk	Norweg'n	506	142·2	28·9	17·3	Quebec	1863	P.Valin	[illegible]	M.Langaard & Co.	Christiania
	972	NV	Professor Mohn (CONDEMNED) *Jensen* W Bk	Norweg'n	955	181·7	37·9	21·2	Bergen	1875	J.Trumpy&Sön	[illegible]	Th.Svanöe	Bergen
	973	NV	Professor Nordenskjöld *Aase 87-87* W 3mSr	Norweg'n	453	139·8	29·7	16·4	Stavanger	1876	C.Knudsen	[illegible]	S.Berners Sönner	Stavanger
	974	N	Professor Schweigaard *Didriksen [illegible]* W Bk	Norweg'n	332	122·6	28·2	15·2	Kragerö	1869	J.Andreassen		[illegible] (C. J. Schön-[illegible])	Kragerö
[illegible]	975	I	Profeta *Giucapello* W Bk	Italian	583	143·0	32·8	21·1	Savona	1873	Pongiglione	[illegible]	G.B.Giucapello fu A	Spezia
[illegible]	976	A	Profitis Elias *Michalopulos* W Sr	Greek	213	96·1	25·4	13·0	Galaxidi	1882	J.Canatas		J.Michalopulos	Galaxidi
[illegible]	977		Profitis Elias *Syringos* W Bg	Greek	206	91·5	24·8	..	Syra	1879	A.Livadara		D.Syringos	Syra
[illegible]	978	A	Profitis Elias W Sr	Greek	133	80·1	23·0	..	Syra	1879			Z.Xenios	Syra
	979		Profitis Elias (ex St.Georgio) W Bg	Russian	173	..	..	..	Cassos	1868			B.Mangubi	Eupatoria
[illegible]	980		Progress *Carre -76* W Bktn	British	246	118·8	25·5	11·5	Guernsey	1865	De la Mare		R.M.Lelean & Co.	Guernsey
[illegible]	981		Progress *Innes* W Sr	British	178	100·0	23·8	12·8	Garmouth	1866	Geddie		A.Hendry	Banff
[illegible]	982	✻	Progress *Crimp* W K	British	76	80·2	19·2	9·2	Kingsbridge	1884	W.Date	[illegible]	H.Grant & Co.	Salcombe
[illegible]	983		Progress *Lapham* W Bk	American	358	94·0	25·3	15·2	Westerly,R.I.	18 43/56			G.F.Bartlett	NewBedford
	984	N	Progress (ex [illegible]) *Rafen* W Bk	Norweg'n	929	202·8	38·9	20·4	Portsm'th,NH	1850	G.Raynes	[illegible]	L.Rafen	Holmestrand
	985	N	Progress *Olsen -85* W Bk	Norweg'n	437	132·8	29·7	16·6	Grimstad	1877	J.Tellefsen	[illegible]	A.Tellefsen Ugland	Grimstad
	986	DN	Progress *Nielsen* W Bk	Norweg'n	419	128·4	30·6	16·7	Risör	1875		[illegible]	E.J.Ellingsen	Risör
	987	N	Progress *Gundersen -85* W Bk	Norweg'n	349	119·7	28·8	16·0	Risör	1865	K.Gundersen	[illegible]	Chr.Th.Boe	Arendal
	988		Progress (CONDEMNED) *Larsen* W Bn	Norweg'n	177	97·5	24·0	11·7	P.E.Island	1871			J.E.Larsen	Farsund
	989	N	Progress *Nygaard -87* W Sr	Norweg'n	145	96·2	23·9	10·9	Stavanger	1868	K.K.Oxtre	[illegible]	S.Eriksen	Stavanger
[illegible]	990	I	Progresso (ex [illegible]) *Schiano* W Bk	Italian	479	118·9	30·4	18·7	Venice	1872	Grasso	[illegible]	A.Mignano & Schiano	Naples
[illegible]	991	I	Progresso *Cardella* W Bn	Italian	147	86·6	22·3	12·1	Viareggio	1885	C.Gori	[illegible]	Emilio Bocci	Leghorn
[illegible]	992	I	Progresso Valle *Schiaffino -88* W Bk	Italian	387	126·6	29·7	18·7	Varazze	1866	D.Craviotto	[illegible]	Eredi Franc.Valle	Genoa
[illegible]	993		Prome I Barge	British	370	152·3	25·2	5·7	Dumbarton	1867	W.Denny&Bros		IrrawaddyFlotillaCo.Ld	Glasgow
[illegible]	994		Promise *Huson 72-72* W Bn	British	168	..	22·9	12·6	Howden	1847			F.W.Fairbrass	Exeter
	995	V	Promise *Papiene* W Sr	French	147	93·8	21·6	11·8	Gravelines	1878	Collin		J. Van Cauwenberghe-Lemaire	Dunkirk
	996	✻V	Prompt *Schlüter 81-87* Stl Bk	German	1417	238·8	38·1	20·5	Hamburg	1887	Blohm&Voss		F.Laeisz	Hamburg
	997		Pronta W Bn	Spanish	171	100·7	25·9	12·2	Blanes	1840			Doña Teresa Ros	Barcelona
	998	DN	Prospect *Wold* W Bk	Norweg'n	710	169·7	33·4	19·4	Grimstad	1884	J.T.Tryde	[illegible]	Peder C.Pedersen	Arendal
47016	999		Prospector W Bk	British	235	123·0	24·6	12·1	Peterhead	1863			Government of New South Wales	Newcastle
[illegible]	1000		Prosper *Knights [illegible]* W Sr	British	122	77·9	22·6	11·6	Nevin	1848			J.B.Fletcher	Lowestoft

✻ = Lloyd's Register; D = Nederlandsche Vereeniging; G = Germanischer Lloyd; I = Registro Italiano; R = American Record; A = Veritas Austro-Ungarico; H = Veritas Ellenico; N = Norske Veritas; V = Bureau Veritas

In the 4th Column, C = Composite (iron frame, wood planked); I = Iron; Stl = Steel; W = Wood.

The Dates of Metalling (Col. 11) are based on the records of the Books in which Vessels are respectively classed, or on the Reports of the Surveyors of Lloyd's Re[illegible]

Further particulars of Vessels classed by Lloyd's Register will be found in Lloyd's Register Book and Lloyd's Yacht Register.

No.	Cl.	Nom. Capitaines et dates d'entrée au service des Armateurs actuels et à bord du navire. Ancien Nom.	Matériaux et Gréement	Pavillon	Tonnage net	Dimensions: Longueur	Largeur	Creux	Construit: Où	Quand	Par qui	Date du doublage en métal	Armateurs ou Directeurs-Gérants	Port d'Armement
1001	V	Prosper *Binet*	W Bk	French	234	96·1	23·3	13·7	Dunkirk	1862	D.VanCauwenberghe		N.Van Cauwenberghe	Dunkirk
1002	V	Prosper Corue *Fra*	W Bk	French	179	86·3	22·6	12·4	Dieppe	1870	P.Corue		A.Corue	Dieppe
1003	V	Prosper Jeanne *Bardies*	W Sr	French	120	90·2	20·3	10·7	Nantes	1878	Alleau&Aubert		Carfontan	Saint Brieuc
1004		Prosperina *Magnasco* (BROKEN UP, ex Aslia)	W Bk	Italian	615	..	..	..	Marseilles			Y 1.86	F.Degregori	Genoa
1005	IV	Prosperino *Figari -87*	W Bk	Italian	886	179·5	34·9	24·1	Spotorno	1876	F[lli] Cadenaccio	Y 11.88	F.LavarelloVed.Figari	Genoa
1006		Prosperità *Salazzo*	W Bk	Italian	431	138·4	31·7	17·1	New York	1857		Y 11.81	P.Delogne fu I.	Leghorn
1007		Prosperity	W Bt	British	125	92·3	22·5	9·0	Brisbane Water, NSW.	1868			S.E.Laidley	Sydney
1008	✻	Prosperity *Bew*	W Sr	British	95	79·0	20·6	10·4	Goole	1869	Pearson		P.Farnill	London
1009		Prospero *Skimmins*	W Bg	British	198	108·5	22·4	12·6	Jersey	1862	Clarke		Joseph Graves	Hobart
1010		Prospero *Fuller*	W Bg	British	144	88·7	21·9	12·4	Rye	1856			T.P.Cattley	Shoreham
1011	G	Prospero *Schröder 86-88*	W Bk	German	387	125·3	28·9	16·4	Rostock	1853	W.Zeltz		W.Manck	Rostock
1012	I	Prospero *Repetto*	W Bk	Italian	499	142·0	31·3	20·2	SestriPonente	1872	S.Bozzano	Y 1.86	G.Repetto di G.	Genoa
1013		Prospero *Schiaffino*	W Bg	Italian	150	80·0	24·4	14·3	Varazze	1847	B.Camogli	Y 81	G.Cordiglia	Genoa
1014		Prospero Padre *Rocco* (ex Mio Cognato)	W Bg	Italian	368	114·8	29·9	18·1	Varazze	1864	A.Quartino	Y 10.82	P.Olivari	Genoa
1015	V	Prospero Razeto *Chiesa 1.87-87*	W Bk	Italian	894	120·7	30·5	18·4	SestriPonente	1867	G.Peragallo	Y 6.87	M.Razeto	Genoa
1016	IV	Prospero & Davide *Olivari*	W Bk	Italian	900	180·4	34·8	22·8	SestriPonente	1881	A.Briasco fu L.	Y 6.86	B.Olivari	Genoa
1017	✻	Protector *Hr. Appleton -88*	W Sk	British	94	81·3	21·0	10·8	Hull	1888	Hunt,Fowler & Co.		Hunt,Fowler & Co.	Hull
1018	N	Protector *Pedersen*	W Bk	Norweg'n	662	150·2	33·4	19·4	Arendal	1864	Chr.Stephansen	Y 4.86	Peder C.Pedersen	Arendal
1019		Protégé	W Bg	British	217	113·7	25·8	12·2	Digby,N.S.	1856			Henry Moss	Melbourne
1020	I	Protettore *Riccio -88*	W Bk	Italian	435	122·3	31·2	19·3	Procida	1877	M.Scotto	Y 10.88	V.Lubrano	Naples
1021		Proteus	W Sr	British	151	93·4	25·3	11·4	Java	1815			Henry Moss	Melbourne
1022	R	Proteus *Peterson*	W Bk	American	622	151·0	32·0	19·4	Bucksport,Me	1867	W.H.Genn	Y 12.87	Swan & Son	New York
1023		Proteus *Dale*	W Bn	American	287	107·1	27·0	11·9	Bucksport,Me	1857			W.F.Emery	Portland,Me.
1024	A	Protexia *Mazzuri* (ex Antonio)	W Bk	Russian	222	92·5	25·3	15·2	Syra	1869		Z 2.86	J.Grinstein	Taganrog
1025		Prothesa *Goodyear*	W Bg	British	154	96·0	22·2	12·8	Bideford	1855			T.W.B.Hutchings	Salcombe
1026	N	Prov *Hussebö*	W Sr	Norweg'n	109	83·0	23·6	10·7	Stavanger	1884			T.O.Husebö	Stavanger
1027		Prøven *Eriksen*	W Bg	Danish	140	76·8	20·6	12·6	Nyborg	1827	A.Bonnesen		Wm.Krause	Nyborg
1028		Pröven (ex Adolphine) *Helgesen*	W Bk	Norweg'n	514	125·0	29·3	20·4	Baltimore	1833			O.B.Paulsen	Drammen
1029	V	Provence *Lelasseur -88* (ex Emma?)	W Bk	French	350	125·6	26·6	14·7	Elsfleth	1878	F.Deetjen	Y 10.88	E.Gilles & A.Billet	Fécamp
1030		Providence *Smith* (BURNT)	W Sr	British	137	74·0	22·0	12·0	Sunderland	1846			Humble & Thompson	Newcastle
1031		Providence	W Sp	British	105	99·6	23·0	6·4	Yamaska	1875	M.Beaupre		Mrs.J.Fourquin	Montreal
1032		Providence *Smith*	W 3mSr	American	399	135·5	34·4	14·4	DennisvilleNJ	1874	G.P.Wentzel		L.C.Gandy	Bridgeton,N.J.
1033		Providence *Depagne -88*	W Bk	French	509	136·0	30·0	17·3	Canning,N. S	1867	E.Biqnelow	Y 11.86	A.Postel et ses fils	Havre
1034		Providencia (ex Belle Prescott)	W Bn	Spanish	449	121·3	28·9	16·3	Deering,Me.	1872			J.F.Rosell	Havana
1035	V	Providencia de S. Thomé *Matthias* (ex Providencia)	W Bn	Portug'se	467	121·7	27·1	16·1	Deering,Me.	1872		Y 2.86	Ant. Azancot, Martins & Co.	Lisbon
1036	G	Providentia *Holtz 73-*	W 3mSr	German	235	106·6	25·4	12·8	Greifswald	1873	A.Spruth		F.Muswieck	Greifswald
1037		Providentia & Arnt *Halvorsen*	W Bk	Norweg'n	238	88·0	27·7	15·6	[illegible]	1801			Enkefru H. Blom	Moss
1038	✻	Province *Jones*	I Bk	British	1784	268·5	39·5	23·1	Sunderland	1886	W.Doxford&Sns		Province Ship Co.Ld. (W.Thomas & Co.)	Liverpool
1039	A	Provvidenza *Pesciutti*	W Sr	Austrian	228	105·3	23·5	13·5	Trieste	1865	Stabilim.Tecnco	Y 1.84	A.Pesciutti di G.	Orebich
1040		Provvidenza *Dodero* (BROKEN UP)	W Bk	Italian	370	118·4	30·1	18·2	SestriPonente	1865	D.Tixi	Y 10.82	M.Dodero	Genoa
1041	I	Provvidenza R. *Razeto -86*	W Bk	Italian	896	181·7	34·4	24·1	Savona	1876	G.B.Magnano	Y 1.88	Eman.Razeto	Genoa
1042		Provvidenza R. *Razeto* (ex Dap[illegible] WRECKED)	W Bn	Italian	339	110·2	27·7	17·5	Recco	1864	P.Rolla	Y 7.81	Ant.Razeto	Genoa
1043		Provvido *Sampognaro -87*	W Bn	Italian	150	..	..	..	Lussinpiccolo	1867	G.Giacovich	Y 6.86	Salv.Spina	Catania
1044		Prudent *Dickson -88*	W Sr	British	124	93·9	26·6	7·6	Portland,N.B.	1888	R.Stackhouse		J.M.Taylor	St.John,N.B.
1045	V	Prudent *Moyan*	W Bk	French	591	125·0	28·2	18·9	SestriPonente	1868	G.B.Cadenaccio	Y 12.8[illegible]	E.Bigot	Havre
1046	V	Prudent *Aubin -8[illegible]*	B Wg	French	248	103·3	23·3	12·9	St.Malo	1876	P.Hervot	Y 1.84	L.Lemasson	Granville
1047	I	Prudente *Morlolo* (ex Lussino)	W Bk	Italian	441	128·2	30·8	19·0	Varazze	1869	C.Aschieri	Y 8.88	F.De Gregori	Genoa
1048		Prudente *Ferrer*	W Bn	Spanish	198	92·5	26·0	12·9	Blanes	1864		C 12.83	Pedro Rosés	Barcelona
1049	N	Prudentia *Iversen -78* (ex [illegible])	W Bk	Norweg'n	372	119·5	29·2	16·6	Grimstad	1868	Salve Johnsen	Y 1.83	H.Villadsen	Christiania
1050	I	Prudenza *Gazzolo*	W Bk	Italian	430	128·9	29·8	19·0	Varazze	1865	B.Camogli	Y 7.88	F.Gazzolo & Co.	Genoa

✻ = Lloyd's Register; D = Nederlandsche Vereeniging; G = Germanischer Lloyd; I = Registro Italiano; R = American Record;
A = Veritas Austro-Ungarico; H = Veritas Ellenico; N = Norske Veritas; V = Bureau Veritas.

Dans la 4me Colonne, C = Composite (membrure en fer, bordé en bois); I = Fer; Stl = Acier; W = Bois.

Les Dates du doublage en métal (Col. 11) ressortent des livres dans lesquels les navires sont respectivement classés ou des rapports des Inspecteurs du Lloyd's Register.

Pour plus amples renseignements sur les navires classés par le Lloyd's Register, voir le Lloyd's Register Book et le Lloyd's Yacht Register.

Official No.	No. in Book	Registry classed	Ship's Name. Material and Rig. Master and Dates of Appointment to present Owner's service and to Ship. Ship's Late Name.			Flag.	Net Tonnage	Dimensions. Length.	Breadth.	Depth.	Built. Where.	When.	By whom.	Date of Metalling	Owners or Managers.	Port
	1051	N	Prudhoe	Irgens 85-88	W Bk	Norweg'n	580	152·9	31·2	19·6	Sunderland	1865	J.T.Alcock	[illegible]	J.L.Ugland	Arendal
[illegible]	1052		Prussia		W Sr	British	376	134·0	26·0	13·3	[illegible], Ont.	1879	Calvin & Breck		The Calvin Co.	Kingston
[illegible]	1053	R	Prussia	Reynolds 81	W S	American	1172	186·3	37·1	24·0	Bath, Me	1868	Houghton Bros.		Renton, Holmes & Co.	[illegible]
	1054		Psarino		W Bg	Turkish	250	..	..	..	Kara Agatch	1882			Kotsatch [illegible]	Kara Agatch
[illegible]	1055	✻	Psyche	Lloyd 59-76	W 3mSr	British	167	106·1	23·7	12·1	Bideford	1875	Cox		W.M.Jeffreys	Swansea
	1056	V	Psyche	Hansen 86-86	W Bn	Danish	166	100·6	21·1	10·8	[illegible]	1867	C.R.Møller		Sm.J.Møller	Svendborg
	1057	✻	Psyché	Simon 87-88	I Bk	French	1034	214·6	31·0	20·0	Glasgow	1877	A.Stephen & Sons		A.C.Le Queller	Bordeaux
203	1058		Puan (ex Duilio)	Baracchini	W Bn	Argentine	135	87·9	22·9	12·2	Palermo	1866		[illegible]	J.Bozzano & Co.	Buenos Ayres
	1059	V	Pubilla	Carell	W Bg	Spanish	196	94·8	26·8	13·3	Vilasar	1856		[illegible]	A.C.de Barbara	Barcelona
68715	1060		Pucelle	Stark 81	W Bn	British	177	102·0	24·5	11·2	Port Gilbert, N.S.	1874	Bingay		J.Williams	Leith
	1061	✻V	Puck (ex Puppy [illegible])	[illegible] 88	I Bk	Danish	484	154·0	27·5	17·6	Newcastle	1868	Marshall Bros.		Jernbarkskibet "Puck" Rederi (P.C.Sørensen)	Nordby, F.
86621	1062		Pudda		I La	British	547	220·0	32·0	9·5	Calcutta	1882	India General Stm Nav. Co Ld		Irrawaddy Flotilla Co Ld	Glasgow
	1063	V	Pudel	Paulsen 70-70	W Bk	German	440	111·4	28·7	16·6	Hamburg	1857	H.C.Stülcken	[illegible]	J.H.Meyer	Pillau
[illegible]	1064		PudseyDawson	[illegible]	W Bk	British	699	119·2	31·5	21·0	Whitehaven	1852	S.Kennedy & Co.		C.J.King & Sons	Plymouth
	1065		Puget	Demeron 83	W Bk	French	215	115·1	22·0	12·5	Granville	1857	Gombert		Anatole Lemoine	St.Malo
	1066		Pululo	Coleman	W Bn	Haytian	189	99·5	26·5	10·8	Economy, N.S.	1873			J.Y.Jimenes & Co.	Monte Christi
	1067	N	Punctum	Hansen	W Sr	Norweg'n	228	115·5	26·6	12·4	Moss	1877	Vogt & Holst	[illegible]	Parelius & Lessing	Christiania
	1068		PurisimaConcepcion (ex Nuevo Dromano, ex Astrea)		W Bn	Spanish	304	110·8	27·8	13·0	Advocate, N.S.	1867			Juan J.Murga	Seville
	1069	✻	Puritan	Barber	W Sk	British	84	77·6	20·0	10·8	Grimsby	1885	G.Ontwin		T.Baskcomb	Grimsby
[illegible]	1070	V	Puritan (ex Charles L. Pearson)	Harding 86-86	W 3mSr	American	641	153·5	34·0	17·5	E.Boston	1873	Campbell & Brooks	[illegible]	Peter H.Crowell	Boston
[illegible]	1071		Puritan	Henrix 88	W 4mSr	American	584	172·3	40·2	12·7	Port Madison	1888	Reed		C.N.Hooper & Co.	San Francisco
[illegible]	1072		Puritan	Sargent 86	W S	American	110	85·5	28·7	7·3	Belfast, Me.	1886	Brown & McDown		J.C.Hall	Belfast, Me.
[illegible]	1073	✻	Pursuit	Arnold	W Sr	British	91	78·4	19·3	9·8	Wells	1861	Tyrrell		Mrs.E.Arnold	London
[illegible]	1074	✻	Pursuit		W Sk	British	83	73·0	20·5	10·5	Elmshorn	1887	J.Kremer		T.Baskcomb	Grimsby
	1075	N	Pusnæs	Rommelhoff 88	W Bk	Norweg'n	485	136·2	29·8	17·7	Arendal	1876	H.Johnsen	[illegible]	A.Arvesen	Arendal
87718	1076		Pyaphoon		I La	British	353	197·3	25·2	8·1	Port Glasgow	1883	R.Duncan & Co		Irrawaddy Flotilla Co. Ld.	Glasgow
81338	1077	✻	Pythomene	Archer 85-85	I S	British	1896	265·6	39·7	24·1	Southampton	1880	Oswald, Mordaunt & Co.		H.Fernie & Sons	Liverpool

ADDITIONAL SHIPS ARE INSERTED IN THE SUPPLEMENT AT THE END OF THE REGISTER.

✻ = Lloyd's Register; D = Nederlandsche Vereeniging; G = Germanischer Lloyd; I = Registro Italiano; R = American Record; A = Veritas Austro-Ungarico; H = Veritas Ellenico; N = Norske Veritas; V = Bureau Veritas.

In the 4th Column, C = Composite (iron frame, wood planked); I = Iron; Stl = Steel; W = Wood.

The Dates of Metalling (Col. 11) are based on the records of the Books in which Vessels are respectively classed, or on the Reports of the Surveyors of Lloyd's Register.

Further particulars of Vessels classed by Lloyd's Register will be found in Lloyd's Register Book and Lloyd's Yacht Register.

No.		Nom. Matériaux et Gréement. Capitaine et dates d'entrée au service des Armateurs actuels et à bord du navire. Ancien Nom.	Pavillon.	Tonnage net.	Dimensions. Longueur.	Largeur.	Creux.	Construit. Où.	Quand.	Par qui.	11	Armateurs ou Directeurs Gérants.	Port d'Armement.
1		Quaco Lynch W Bn	British	166	96·0	26·0	11·4	St.Martin's,NB	1871			T.& J.Connick & Co.	Dundalk
2	R	Quaker City W 3mSr	American	485	110·7	31·0	15·0	Bath,Me.	1879	Goss&Sawyer		P.Fitzpatrick	Philadelphia
3	❋	Quarry Man Williams W Sr	British	99	73·8	20·7	10·9	Pwllheli	1840			R.Williams	Carnarvon
4	❋	Quathlamba Early I Bk	British	467	167·9	29·1	13·6	Aberdeen	1879	A.Hall&Co.		J.T.Rennie & Son	Aberdeen
5		Quatorze de Janeiro W	Brazilian	180	..	..	..					F.J.Cardoso	Itaguahy
6	V	Quatre Frères Gouret W Bn	French	204	96·8	23·0	12·3	St.Malo	1876	J.Parnet	[illegible]	Vve.E.Thomazeau&Cie.	St. Malo
7	V	Quatre Frères Oger W 3mSr	French	174	86·0	23·9	10·7	St.Pierre [Miquelon	1881	B.Joret		A.Girardin	St. Malo
8		Quatro Irmaos W	Brazilian	136	..	..	..					M.D.da Cruz	Rio Janeiro
9	I	Quattro Fratelli W Bk	Italian	528	133·5	31·6	20·1	Prà	1869	M.Tixi	[illegible]	M.& A.Dall'Orso fu G.	Spezia
10		Quattro Fratelli S. W Bn	Italian	199	80·3	23·8	9·7	Chioggia	1883	G.Poli	[illegible]	Ang.Bussetto	Venice
11	I	Quattro Sorelle W Bn	Italian	377	121·0	29·4	18·2	Voltri	1871	G.Fava	[illegible]	F.Penco fu Drago	Genoa
12	DR	Quebec Nelson W Bk	British	1462	202·0	41·8	25·0	Hantsport, NS	1867	Churchill	[illegible]	G.W.Churchill	Windsor,N.S.
13	❋	Queen Jones 86-6 W Bk	British	383	128·2	28·4	17·2	Sunderland	1868	L.&J.Mills	[illegible]	R.Thomas & Co.	Sunderland
14	❋	Queen Osborn W S	British	99	86·5	20·4	9·2	Ipswich	1838	Bayley		A.Beaumont	Ipswich
15	❋	Queen Guille 79-79 W Cr	British	33	56·6	16·1	7·3	Jersey	1879	Picot		P.Guille	Jersey
16		Queen W 3mSr	American	234	124·0	31·0	10·0	SanFrancisco	1882			A.P.Lorentzen	San Francisco
17	N	Queen Tallaksen W Bk (ex Queen of Prussia)	Norweg'n	325	131·0	28·0	12·9	Miramichi,N.B	1863	Gough		Tallak Hansen	Arendal
18		Queen Emma W Bk	American	314	126·6	26·2	13·4	Steinwarden	18 69/71	H.C.S.Tönsbu		J.G.Walsh	
19	❋	Queen Mab Smith I Bk	British	1099	205·0	33·5	19·8	Port Glasgow	1887	Russell & Co.		J.Fairlie	Glasgow
20	❋D	Queen Victoria Holmes Stl S	British	1605	256·4	38·6	22·8	Glasgow	1887	C.Connell&Co.		J.Black & Co.	Glasgow
21		Queen Victoria W Sr	British	105	85·5	23·0	8·5	Grosslines	1867			D.Banville	Quebec
22	❋	Queen Victoria W Sr	British	72	80·0	20·2	10·0	Skerries	1880	W.Morris		F.Griffin	Poole
23	❋	Queen Victoria Buckden W Sk	British	75	74·5	20·5	10·5	Hamburg	1885	C.C.A.Dreyer		M.Buckden	Grimsby
24	N	Queen Victoria W Bg	Norweg'n	255	105·9	27·3	14·9	Arendal	1852		[illegible]	O.A.Olsen	Arendal
25	❋D	Queen of Cambria Jones I Bk	British	865	191·2	32·8	20·2	Sunderland	1876	Osbourne, [Graham&Co.		W.Thomas & Son	Carnarvon
26	R	Queen of Clippers Cripps W Sr	British	129	90·5	20·7	11·8	Brixham	1863	Upham	[illegible]	J.W.Upham & Co.	Brixham
27	❋	Queen of England I S	British	2071	276·0	42·4	24·1	Liverpool	1885	T. Royden&Sns.		D.Kennedy & Co.	Liverpool
28		Queen of England W S	Siamese	543	160·0	29·6	15·0	Bangkok	1865			Poh Chin Soo	Bangkok
29		Queen of Mistley W Sr	British	186	110·3	25·2	12·3	Harwich	1872	Vaux		J.Mann	Harwich
30	R	Queen of Nations W S	British	1462	198·9	39·4	25·0	St.John,N.B.	1863		[illegible]	Ship "Queen of Nations" Co. Ld. (D.S.Brown & Co.)	Greenock
31	❋	Queen of Scots McKenzie I S	British	1374	229·4	38·0	22·8	Stockton	1877	Richardson, [Duck&Co.		D.Kennedy & Co.	Liverpool
32	❋	Queen of the Avon I Sw	British	162	101·4	20·4	12·3	Newcastle	1856	J.Vernon		R.Swainston	Newcastle
33	❋	Queen of the Bay W Bk	German	381	133·1	27·6	17·0	Southwick	1864	Hardie	[illegible]	C.Rahtkens	Rostock
34	❋	Queen of the Chase W Sr	British	98	84·0	19·2	10·3	Falmouth	1852			P.A.Jeffery	Falmouth
35		Queen of the Dart W Sr	Norweg'n	129	84·9	21·3	12·1	Dartmouth	1848			O.Mathisen	Tönsberg
36	N	Queen of the East W Bk	Norweg'n	1186	203·5	37·9	22·5	St.John,N.B.	1853		[illegible]	A.P.Andersen	Moss
37	❋	QueenoftheFleet W Bk	British	941	171·6	35·1	21·0	DorchesterNB	1876	G.Palmer	[illegible]	G.Palmer	DorchesterN.B
38	❋	Queen of the Fleet W Bn	British	262	115·0	24·6	14·8	Bridport	1866	Cox	[illegible]	W.Thompson	Bridport
39	❋	Queen of the Fleet W Sk	British	99	81·7	20·9	10·9	Rostock	1884	E.Burchard&C.		T.Hamling	Hull
40		Queen of the Isles Peeps W Sw	British	194	79·6	22·0	15·0	Shields	1837	Stephenson		D.T.Sharper	Lynn
41	❋	Queen of the Isles W Sr	British	149	98·4	22·3	11·2	Newburgh	1867	Scrimgeour	[illegible]	J.Stephen	Wick
42		Queen of the Lakes W Sr	British	253	128·0	23·3	10·5	Portsm'th,Ont	1853	G.N.Ault		TheCongerCoalCo.Ld.	Kingston,Ont.
43		QueenoftheNorth W S	British	1668	202·0	40·3	29·1	St.John,N.B.	1860	M'Morran		R.S.Allen	Newcastle
44	❋	Queen of the North I Bk	British	393	158·0	25·2	15·5	Sunderland	1865	Pile,Hay&Co.		W.Davies	Swansea
45		Queen of the Sea W Sr	British	123	81·2	18·2	11·5	Bideford	1852			J.Short	Penzance
46		Queen of the South W Sr	British	120	82·5	18·6	12·1	Salcombe	1850			H.Pulsford	Salcombe
47		Queen of the Usk W Bn	British	115	80·0	21·5	12·5	Newport	1847			P.Curtis	Dundalk
48	❋	Queen of the Waves W Sk	British	87	78·8	20·5	10·6	Hull	1886	W.&J.McCann		R.W.Rushton	Grimsby
49	❋	Queen of the West I Bk	British	404	117·5	26·6	16·5	Sunderland	1867	Pile,Hay&Co.		W.Price & Co.	London
50		Queen of the West W Sr	British	119	82·6	18·6	12·1	Salcombe	1849		[illegible]	N.Trute	Salcombe

❋=Lloyd's Register; D=Nederlandsche Vereeniging; G=Germanischer Lloyd; I=Registro Italiano; R=American Record;
[illegible]=Veritas Austro-Ungarico; H=Veritas Ellenico; N=Norske Veritas; V=Bureau Veritas.

Dans la 4me Colonne, C=Composite (membrure en fer, bordé en bois); I=Fer; Stl=Acier; W=Bois.

Les Dates du doublage en métal (Col. 11) ressortent des livres dans lesquels les navires sont respectivement classés ou des rapports des Inspecteurs du Lloyd's Register.

Pour plus amples renseignements sur les navires classés par le Lloyd's Register, voir le Lloyd's Register Book et le Lloyd's Yacht Register.

Official No.	No. in Book	Society	Ship's Name. Material and Rig. Master, and Date of Appointment to present Owner's Service and to Ship. Ship's Late Name.	Flag.	Net Tonnage.	Length	Breadth	Depth	Built: Where.	Built: When	Built: By whom.	Dates of Metalling.	Owners or Managers.	Port
[illegible]	51	V	Queen's Cliff *Parsons* -87 W Bk	British	611	153·0	32·0	17·8	Quebec	1873	[illegible]	[illegible]	W. D. Douglas	Abery
[illegible]	52	✻	Queen's Island *Reid* 85-87 Stl&I Bk	British	2038	282·8	40·5	23·6	Belfast	1885	Harland&Wolff		S. Lawther	Belfas
	53	N	Quelle *Bugge* W Bk	Norweg'n	563	116·6	32·0	18·1	Grimstad	1875	S. Johnsen	[illegible]	Julius Stephansen	Arenda
	54		Qui-Qu'en-Grogne *Landgren* -88? W Bg	French	138	80·0	21·6	12·3	St. Malo	1861	Tracheuer		Anatole Lemoine	Saint
20575	55		Quickstep *Jacobsen* 82-82 W Bkn	American	402	117·9	31·1	13·7	Port Ludlow, [Wash.	1876	Hall		S. B. Peterson	San F
74483	56	✻	Quillota *Browne* 81-84 I Bk	British	686	186·8	30·9	18·6	Liverpool	1875	W. H. Potter&Co	[illegible]	S. Wakeham & Sons	Liverp
	57	✻	Quillota *Lopez* 87-87 W Bk	French	723	170·6	32·0	20·4	Bordeaux	1874	Bordeaux [illegible]	[illegible]	Verdaux	Borde
26885	58		Quintus W Bk	British	406	115·0	30·2	18·3	Malta	1853			Alfred Ganci	Malta
	59		Quintus *Boeijin* -84 W Sr (ex Hessel Wiebrens)	Dutch	144	101·0	21·6	9·7	Nieuwendam	1860	W. & A. H. M[illegible]		F. W. van der Elst & Co.	Schied
2196	60	A	Quirinale *Scala* -87 W Bk	Italian	445	131·5	29·8	18·0	Sestri Ponente	1871	F[lli] Cattaneo	[illegible]	F[lli] Mazzella	Naples
	61	V	Quiteria *da Costa* W Bk (ex Luisa Norton, ex Lucia)	Portug'se	315	123·6	24·8	13·5	Norrtelje	1877	[illegible]	[illegible]	L. Marques da Nova	Oporto
20563	62		Quoddy *Sirmann* -76 W Sr (W. R. F. K. [illegible])	American	159	102·5	27·0	9·7	Lubec, Me.	1869	[illegible]		J. W. Lamson	Mach
	63	N	Qvos . . . ! *Olsen* W Bk	Norweg'n	465	139·4	31·0	17·1	Kragerö	1873	A. [illegible]	[illegible]	P. & C. Romer	Krager

ADDITIONAL SHIPS ARE INSERTED IN THE SUPPLEMENT AT THE END OF THE REGISTER.

✻ = Lloyd's Register; D = Nederlandsche Vereeniging; G = Germanischer Lloyd; I = Registro Italiano; R = America[n] ...;
A = Veritas Austro-Ungarico; H = Veritas Hellenico; N = Norske Veritas; V = Bureau [Veritas]

In the 4th Column, C = Composite (iron frame, wood planked); I = Iron; Stl = Steel; W = Wood.

The Dates of Metalling (Col. 11) are based on the records of the Books in which Vessels are respectively classed, or on the Reports of the Surveyors of Lloyd's ...

Further particulars of Vessels classed by Lloyd's Register will be found in Lloyd's Register Book and Lloyd's Yacht Register.

1	2	4 Nom. Matériaux et Gréement. Capitaines et dates d'entrée au service des Armateurs actuels et à bord du navire. Ancien Nom.	5 Pavillon.	6 Tonnage net.	7 Dimensions. Longueur.	Largeur.	Creux.	8 Construit. Où.	9 Quand.	10 Par qui.	11 Date du doublage en métal	12 Armateurs ou Directeurs-Gérants.	13 Port d'Armement.
1	*	R. A. Calderon *Thompson 83-87* Stl Bk	British	815	199·7	31·2	19·5	PortGlasgow	1883	J.Reid & Co.		W. & J. Lockett	Liverpool
2	V	R. Bowers *Thompson -89* W3mSr	American	414	150·5	33·5	12·2	Camden, Me.	1879	H.Bean		J.F.Bliss	Boston, Mass.
3		R. C. A. Ward *Hart* W3mSr	American	216	121·0	31·0	8·0	Belleville, N.J.	1856			C.A.Hart	New York
4	V	R. C. Rickmers *Anderson 81-88* WS	German	1663	239·5	39·4	26·5	Bremerhaven	1888	R.C.Rickmers	V 6.88	R.C.Rickmers	Geestemunde
5		R. Carson *Sweet -88* WSr	British	108	80·8	27·2	7·5	St.Martin's, N.B.	1888	R.Carson		R.Carson	St.John, N.B.
6	R	R. D. Bibber *Pishbon -81* W3mSr	American	728	162·0	35·6	18·6	Bath, Me.	1884	G.W.Johnson	V 6.84	G.W.Johnson	Bath, Me.
7	V	R. D. Rice *Gordon -88* WS	American	2134	252·1	43·7	28·7	ThomastonMe	1883	S.Watts & Co.	V 1.88	S.Watts	Thomaston, Me.
8	R	R.D.Spear *Gilson -88* W3mSr	American	335	133·4	32·9	10·3	Bowdoinham, [Me.	1883	Rideout		J.B.Drake	Bath, Me.
9		R. F. (WRECKED) *Poulter -85* WSr	French	111	85·3	23·6	10·6	St. Valery en [Caux	1879	A.Coquais		A.Gourio	Paimpol
10		R. F. Bell *Edmondson 63-86* WSr	British	175	97·9	24·0	13·0	Barrow	1865	Rawlinson		J.Fisher & Sons	Barrow
11		R. F. Hart *Humphrey* W3mSr	American	328	128·1	30·5	10·3	St.George, Me.	1869			J.H.Conant & Co.	Boston, Mass.
12	R	R. G. Whilldin *Read* WSr	American	179	105·5	28·5	9·1	Milford, Del.	1853			H.F.Champlin	Bristol, R.I.
13		R. H. Wilson WSr	American	147	99·1	27·0	8·0	Pt.JeffersnNY	1853	J.L.Darling		D.G.Floyd	Greenport, N.Y.
14	*	R. Ingersoll WSk	British	78	76·1	20·5	10·5	Rye	1884	G.&T.Smith		S.W.Alward	Grimsby
15		R. K. Ham *Gove 78-* WBk	American	541	173·4	37·0	15·0	PtBlak'lyWsh	1875	W.Bryant		RentonHolmes&Co.	Pt.T'wns'nd, Wsh.
16	*	R. L. T. *Thompson -81* WBn	British	445	134·4	28·7	18·0	Five Islnds NS	1881	R.Lewis	V 1.85	W.Thompson	Halifax, N.S.
17	R	R. L. Tay *Hibbard* WSr	American	144	105·0	27·0	8·7	Milford, Del.	18 50/82			E.C.Gates	Calais, Me.
18		R. M. Brookings *Sawyer -81* WSr	American	161	109·0	21·8	8·6	Wiscasset, Me.	1867	Chism & Brookng		F.C.Curtis	Bath, Me.
19	V	R. Morrow *O'Brien* WBk	British	1156	194·4	37·6	22·2	Maitland, N.S.	1884	E.Roy	V 11.87	Alexander Roy	Maitland, N.S.
20	VR	R. R. Thomas *Nichols 76-76* WS	American	1333	202·0	40·3	24·0	Searsport, Me.	1876	E.Dunbar	V 5.86	J.C.Nickels	Searsport, Me.
21		R. S. Dean *Coleman* WSr	American	129	94·0	24·0	7·2	Taunton, Mass	1855			W.H.Phillips	Fall River, Mass.
22		R.S.Graham *Artis -86* W3mSr	American	325	128·5	30·5	14·1	Milford, Del.	1871	Scribner		J.W.Hall	Wilmington, Del.
23	*	R. T. B. *Hobbs* WK	British	75	79·6	21·2	8·7	Appledore	1884	R.T.Blackmore		R.Blackmore	Bideford
24	*	R. T. K. *Ralden* WSr	British	75	75·4	19·5	9·8	Charlestown	1863	Lake		J.de C.Treffry	Fowey
25		R. & S. Corson *Corson* WSr	American	249	110·0	29·0	8·5	Wilmngtn.Del	1866			H.Barry	Wilmington, Del.
26	N	Rabbi *Einertsen* WBr	Norweg'n	174	96·8	23·4	11·2	Kragerö	1882	T.G.Litangen	V 1.84	T.G.Litangen	Kragerö
27		Rabboni *Craig 83-84* WBg	American	280	114·4	28·0	14·5	Bangor, Me.	1866			T.J.Stewart & Co.	Bangor, Me.
28		Raby Castle *Harris* WS	British	692	151·0	31·5	20·2	Sunderland	1856	G.Short		H.W.White	London
29		Race Horse *Bishop* WSr	American	106	87·8	22·5	8·4	Newb'ryp't Ms	1865			N.Bishop	Rockland, Me.
30		Race Horse WBk	Siamese	387	133·8	25·9	15·2	Bangkok	1858			SiameseGovernmnt.	Bangkok
31	NV	Racehorse *Jensen 73-81* WS	Norweg'n	1011	192·9	37·5	21·3	Bergen	1877	Brunchorst & [Dekke	V 2.84	J.L.Ugland	Arendal
32	*	Racer *Dean -89* WSr	British	100	90·1	21·3	9·9	Garmouth	1889	W.Anderson		G.Smith	Inverness
33	*	Racer *Maddy -87* WSr	British	62	75·0	18·8	9·2	Truro	1866	Stephens	Y 7.87	S.Truscott	Fowey
34		Rachel *Sinclair* WBk	British	282	127·4	24·1	14·0	ManningRiver [N.S.W.	1865			Robert Frith	Sydney, N.S.W.
35		Rachel *Descovich* WBk	Austrian	358	118·7	26·6	17·1	Fiume	1863	F[lli] Schiavoni	V .81	Fratelli Luppis	Fiume
36		Rachel *Cormillet* WSr	French	122	82·0	21·6	8·4	Dunkirk	1873	Dorp		De Grandpré	Dahouët
37		Rachel *Nicolas -87* WSr	French	111	..	..	..	Fécamp	1878			Louis Gaultier	Treguier
38		Rachel A. Collins *Betts* WSr	American	126	91·0	27·0	7·2	Lebanon, Del.	1873			D.C.Betts	Wilmington, Del.
39		Rachel Cohen *Worlege -86* WSr	British	170	105·5	21·0	10·5	ManningRiver [N.S.W.	1871			Henry C.Pigott	Melbourne
40	R	RachelConoy *Bosham -87* WBg	American	277	117·5	27·5	10·1	Brewer, Me.	1866		V 12.86	E.S.Aldrich	Providence, R.I.
41	R	Rachel Emery *Whitemore -88* WBkn	American	640	167·1	35·3	17·4	Waldobro' Me	1883	H.Kennedy&Co.	V 2.87	J.S.Emery & Co.	Boston, Mass.
42		Rachel Lotinga *Robinson -72* WSp	British	242	100·3	26·1	15·9	Sunderland	1855			T.F.Ward	Whitby
43		Rachel S. Miller *Steelman* WSr	American	167	100·0	27·0	9·0	Philadelphia	1852			J.H.Cannon	Norfolk, Va.
44		Rachel Seaman *Smith* WSr	American	298	115·0	30·5	9·7	Philadelphia	18 61/76			J.N.Smith	Fall River, Mass.
45	N	Rackel *Salvesen* WBk	Norweg'n	539	149·0	32·9	16·8	Stavanger	1886		Y 7.86	E.Berentsen	Stavanger
46	I	Radice *Gaglione* WBn	Italian	142	92·1	21·1	11·8	TorredelGreco	1886	G.Guida	Y 10.87	Radice & Gaglione	Naples
47		Radja Laut (ex Cato) WBk		161	117·7	15·6	10·5	Risör	1863			W.Lingard	
48	V	Rafael Pomar *Pino* WBn	Spanish	399	130·2	29·4	15·9	Blanes	1870	Vieta	Y 10.83	R.Pomar y Cortés	Majorca
49	I	Raffaele *Scuarese* WBk	Italian	433	128·9	29·1	18·7	Equa	1868	A.Gargiulo	Y 1.87	F.Cilento	Naples
50	I	RaffaeleD. *Gattinegna -88* (ex Adriatico) WBk	Italian	630	..	..	..	Chiavari	1867	L.Casella	Y 4.88	Dodero Nicolo	Genoa

*=Lloyd's Register; D=Nederlandsche Vereeniging; G=Germanischer Lloyd; I=Registro Italiano; R=American Record; A=Veritas Austro-Ungarico; H=Veritas Ellenico; N=Norske Veritas; V=Bureau Veritas.

Dans la 4me Colonne, C=Composite (membrure en fer, bordé en bois); I=Fer; Stl=Acier; W=Bois.

Les Dates du doublage en métal (Col. 11) ressortent des livres dans lesquels les navires sont respectivement classés ou des rapports des Inspecteurs du Lloyd's Register.

Pour plus amples renseignements sur les navires classés par le Lloyd's Register, voir le Lloyd's Register Book et le Lloyd's Yacht Register.

Official No.	No. in Book	Registry in which classed	Ship's Name. Material and Rig. Master, and Dates of Appointment to present Owner's Service and to Ship. Ship's Late Name.			Flag.	Net Tonnage.	Dimensions. Length	Breadth	Depth	Built. Where.	When.	By whom.	Date of Metalling.	Owners or Managers.	Port
1741	51		Raffaele Liguro (Iaraendonefu)	Consigliere	W Bk	Italian	488	137·1	30·0	19·1	Savona	1868	A. Calcagno	[illegible]	G. Consigliere	Genoa
998	52	I	Raffaele Pandolfi	[illegible]	W Bn	Italian	124	82·6	21·6	10·5	Castellamare	1884	G. Bonifacio		G. C. Pandolfi	[illegible]
2402	53	IV	Raffaelina	Scaparone	W Bk	Italian	805	164·0	34·4	21·4	Sestri	1873	A. Briasco fu L.	[illegible]	A. D. Scaparone	Genoa
74405	54	*	Raglan Castle	Mathews	I Bk	British	1050	215·7	34·8	21·1	Liverpool	1876	R. & J. Evans & Co.		Richards, Mills & Co.	Liverp
	55	NV	Ragna	Siverlsen -86	W S	Norweg'n	1112	189·9	38·0	21·0	Drammen	1878	J. Jörgensen	[illegible]	O. Jörgensen	Risör
	56	RV	Ragna (ex Post Boy... ex Cotswater)	Thomsen -88	W S	Norweg'n	989	176·1	34·1	22·9	Medford	1861	J. O. Curtis	[illegible]	A. Monsen	Tönsb
	57	N	Ragna	Pettersen	W Bk	Norweg'n	487	127·0	31·5	18·2	Arendal	1865	H. Larsen	[illegible]	T. Thoomesen & Smith	Arend
55068	58	DR	Ragnar (ex Calypso)	Robbins 88-88	W S	British	993	180·7	35·2	21·7	NewGlsgwNS	1871	Carmichael	[illegible]	G. T. Soley & Co.	Pictou
	59	V	Ragnar	Carlsson -84	W Sr	Swedish	164	96·5	21·5	11·6	Sundsvall	1874	M. Hägglund		A. Carlsson	Styrsö
	60	N	Ragnhild	Dahl 71-79	W Bk	Norweg'n	518	133·1	32·2	19·3	Drammen	1871		[illegible]	Johs. Harbitz	Tönsb
	61	N	Ragnhild	Rytter -87	W Bn	Norweg'n	197	112·1	25·3	11·9	Soon	1866	Skinnerod	[illegible]	O. Möllbach	Langes
	62		Ragnvald Jarl	Bryhn	W Bn	Norweg'n	140	97·9	24·1	10·5	Aalesund	1873	C. J. Haastel	[illegible]	P. T. Sandborg	Aales
81052	63		Ragoomani		W Bg	British	148	86·6	19·9	11·6	Cochin	1855			Ravana Mana Chavana Carupan Chetty	Colomb
	64	A	Ragusa	Galante	W Sr	Russian	101	..	..	..	Pavloscka	1886	P. Martinolich		A. Duranto	Berdia
	65	*	Rahane	Penfold 79-86	Stl S	British	1649	260·1	39·0	22·7	Port Glasgow	1886	R. Duncan & Co.		Thom & Cameron	Glasgo
	66		Railleuse	Verry 88-88	W Bn	French	105	..	..	..	St. Servan	1873			Anatole Lemoine	Saint S
	67	V	Raimunda (ex [illegible])	Olive	W Bn	Spanish	214	98·8	25·5	12·3	Villajoyosa	1867		[illegible]	G. Olive	Barcel
35791	68		Rainbow		W Bn	British	128	87·6	23·2	12·2	[illegible], N.S.	1854			J. Monkhouse	Whitb
	69	V	Rainbow	Markussen 88-88	W Bk	Norweg'n	796	170·7	34·4	20·1	[illegible]	1876	Dunn	[illegible]	Westergaard & Co.	Christ
	70	*	Rajah (ex [illegible])	Albrecht	I S	German	1230	216·8	34·5	22·0	Liverpool	1864	Hart & Sinnot		D. Schilling	Bremen
86274	71	*	Rajore	Lloyd 77-83	I S	British	2034	278·2	40·3	24·2	Southampton	1882	Oswald, Mordaunt & Co.		Eyre, Evans & Co.	Liverp
	72	*	Rajpoot	Grimm 84-88	I S	German	1261	237·7	33·2	22·7	Liverpool	1864	T. Vernon & Son		H. Folsch & Co.	Hambu
[illegible]	73	*	Rakaia	Banks 87-87	I Bk	British	1022	210·2	34·0	19·2	Sunderland	1873	J. Blumer & Co.		New Zealand Shipping Co. Ltd.	London
	74		Rakkaus	Peltta 82-82	W Bg	Russian	271	120·0	27·0	14·0	Kotka	1873	F. Penttila		J. Penttila	Wiborg
	75		Rakkaus	Lindholm	W Sr	Russian	131	..	..	..	Kuopio	$18\frac{67}{75}$			G. A. Lindholm	Borgå
74176	76	*	Ralph B. Peake	McDonald 67-86	W Bk	British	699	170·0	33·5	18·6	P.E. Island	1876	Peake Bros. & Co.	[illegible]	Peake Bros. & Co.	[illegible]
110315	77		Ralph M. Hayward	Baxter	W Bkn	American	574	152·0	33·6	15·3	Waldoboro Me	1877	A. Starer & Son	[illegible]	J. Baxter	Boston
60357	78	* D	Ralston	Hansen -88	I S	British	815	195·1	31·3	19·8	Greenock	1868	R. Steele		J. Richardson & Co.	Glasgo
	79		Ramazan	Hassan	W Sr	Turkish	150	..	..	..	Ackjee Shehir	1883			Hattat tjar Hajee Shefki Effendi	Aidin
68781	80	*	Rambler	Davey -88	W Sr	British	83	80·7	20·0	9·8	Jersey	1875	Bellot	[illegible]	J. Whitley	Jersey
21909	81		Rambler	Dayton 73-	W Sr Yt	American	130	130·5	25·0	9·4	New London, Conn.	1871		[illegible]	W. H. Thomas	New Y
38224	82		Ramilee		W Bn	British	112	78·6	21·7	10·2	Sandy Cove NS	1862			Robt. McAfee, jun.	[illegible]
110130	83	R	Ramona (ex [illegible])		W Sr Yt	American	115	107·0	25·2	8·4	City Island N.Y.	1871	D. Carll		C. A. Postley	New Yo
	84	NV	Ran	Mahony -83	W Bk	Norweg'n	794	168·4	36·1	18·7	Arendal	1884	O. M. Olsen	[illegible]	H. Natvig	[illegible]
	85	N	Ran	Halvorsen 82-82	W Sr	Norweg'n	134	84·2	22·9	11·4	Drammen	1861			H. A. Lohre	Dramm
	86		Ran	Pihl 86-86	W Bk	Swedish	393	138·1	27·5	15·5	Hudiksvall	1854	P. Elfstrand		A. A. Pihl	Nyhamn
	87	V	Rana	Christiansen -88	W Bg	Norweg'n	263	113·6	26·9	13·5	Arendal	1870	A. Hendriksen	[illegible]	T. Thommesen & Smith	Arend
43801	88	*	Ranavola	Doey	W Bk	British	388	122·7	27·7	17·6	Newcastle	1861	Gaddy	[illegible]	Ranavola Ship Co. Ld. (J. S. Boyd)	Belfast
	89		Rancagua	Kruse	W Bk	Chilian	705	167·0	29·8	21·3	Bordeaux	1874	Bochet & Largeteau	[illegible]	Horacio Battles & Co.	Valpar
	90		Rance	Lebaut	W Bk	French	571	117·9	26·9	19·2	St. Malo	1866	Parnet	[illegible]	Vve. Visé	Saint S
	91	N	Randers	Haraldsen	W Bk	Norweg'n	268	127·9	26·3	13·1	Randers	1867	P. F. Jörgensen		H. A. Haraldsen	Tvedes
48630	92		Randolph	Kasson	W Bk	British	343	122·4	26·2	16·0	Arbroath	1864	Arbroath S. B. Co		Hajee Moosa Hadjee Ahmed Mahomed	Bomba
50282	93	*	Ranee (ex [illegible])	[illegible] 80-82	I Bk	British	1190	217·5	34·5	22·4	Liverpool	1864	Hart & Sinnott		Nicholson & McGill	Liverp
30769	94		Ranee		W Bg	British	159	94·2	23·0	12·2	Cochin	1869			Moosary Kritha Naina Marakair Mahomade Mahoboo Marakair	Cochin
47332	95	*	Rangitiki (ex Scimitar)	Taylor 86-87	I S	British	1188	212·0	35·0	22·8	Hull	1863	Humber Iron Wks Ld		New Zealand Shipping Co.	London
32083	96		Rantipole		W Sr	British	183	99·1	24·6	12·7	Gloucester	1865			John Paterson	[illegible]
	97	V	Raoul & Fernand	Dehilh [illegible]	W Bk	French	376	118·7	28·5	17·4	Nantes	1885	Serestre frères	[illegible]	J. B. Demange	Nantes
	98	V	Raoul & Madeleine	Lequeusnes -88	W 3m Bk	French	236	99·1	24·3	14·9	Nantes	1861	Bojn	[illegible]	Amaury Simon	[illegible]
	99	V	Raoul & Valentine	Bourgeois	W Sr	French	164	90·2	22·3	11·8	St. Malo	1877	J. Parnet	[illegible]	J. Baptiste Bourgeois	[illegible]
	100	N	Rap	Hansen	W Sr	Norweg'n	135	93·8	22·9	9·9	Brevig	1872	N. Oslessen		M. E. Schilbred	Brevig

*=Lloyd's Register; D=Nederlandsche Vereeniging; G=Germanischer Lloyd; I=Registro Italiano; R=American [Lloyd's]; A=Veritas Austro-Ungarico; H=Veritas Ellenico; N=Norske Veritas; V=Bureau [Veritas].

In the 4th Column, C=Composite (iron frame, wood planked); I=Iron; Stl=Steel; W=Wood.

The Dates of Metalling (Col. 11) are based on the records of the Books in which Vessels are respectively classed, or on the Reports of the Surveyors of L[loyd's]

Further particulars of Vessels classed by Lloyd's Register will be found in Lloyd's Register Book and Lloyd's Yacht Register.

Numéro de la page	Registre dans lequel le navire est coté	Nom. Matériaux et Gréement. Capitaine et dates d'entrée au service des Armateurs actuels et à bord du navire. Ancien Nom.		Pavillon.	Tonnage net.	Dimensions: Longueur	Largeur	Creux	Construit: Où	Quand	Par qui	Date du doublage en métal	Armateurs ou Directeurs-Gérants.	Port d'Armement.
101	N	Rap *Eriksen 86-88*	WSr	Norweg'n	122	93·3	23·8	10·7	Bergen	1882	Knud Nces		J. Thorsen	Haugesund
102	N	Rap (ex Stord) *Christoffersen 1-86*	WSr	Norweg'n	118	82·6	22·1	9·9	Elsfleth	1847			A. Christoffersen	Moss
103	N	Rap *Vige 88-88*	WSr	Norweg'n	112	91·8	21·1	10·2	Stavanger	1877	M.A. Magnus		S. Jörgensen	Christiansand
104	N	Rap *Meyer*	WSr	Norweg'n	106	90·9	22·0	10·2	Bergen	1883	S.K. Flage		J.E. Lehmkuhl	Bergen
105	R V	Raphael *Harkness*	WS	American	1543	222·0	46·0	24·3	Camden, Me.	1875	Carleton & N'rwd	Y 2.88	Carleton, Norwood & Co.	Camden, Me.
106	✻	Raphael (ex Copperacue)	WBk	French	307	137·3	27·8	13·6	Sunderland	1880	W. Pickersgill	Y 7.83	J. Malandain	Fécamp
107	R	Rapid *Gagnon 83-87*	WBn	British	325	120·5	28·4	13·0	Rothesay, N.B.	1876	R. Titus	Y 9.86	Maguire & Co.	St. John, N.B.
108		Rapid	WBg	British	191	100·4	24·8	12·4	Shoreham	1861	May		W.H. Digby	Shoreham
109		Rapid *Roney*	WSr	British	126	90·7	22·1	11·5	St. John, N.B.	1856			R. & R. Allen	Wexford
110		Rapid	WSr	British	106	75·5	19·9	11·5	Bristol	1830			W. Wilson	Workington
111	N	Rapid *Pedersen -87*	WBg	Norweg'n	208	101·2	25·9	11·9	Risör	1878	Stenersen		E.J. Ellingsen	Risör
112	N	Rapid *Werge 83-84*	W3mSr	Norweg'n	163	95·6	24·7	11·1	Sandefjord	1869			O.P. Werge	Brevig
113		Rapid	WBk	Siamese	430	135·6	27·6	16·6	Bangkok	1856			Chessna Low Chae	Bangkok
114		Rapid *Lycke*	WSr	Swedish	125	88·2	21·7	8·8	Gefle	1860	O. Brodin		J.B. Jonsson	Lysekil
115		Rapid *Thomasson -86*	WSr	Swedish	101	83·3	21·8	8·7	Pigeholm	1868			O. Thomasson	Cimbrishamn
116	V	Rapide *Marklund 86-86*	WBk	Russian	544	161·1	32·1	18·1	Jakobstad	18^{53}_{55}	Bleszer	Y 9.86	P. Malm, jun.	Jakobstad
117		Rapido	WBn	Brazilian	144	..	..	..					F.D. Machado	Rio de Janeiro
118		Rapido (ex Julia) *Godinhon*	WBn	Portug'se	195	105·9	25·3	9·5					J.T. da Silva Godinhon	Lisbon
119		Rapido	WBg	Spanish	150	94·2	23·7	13·7	Mataro	1835			José Mataró	Barcelona
120		Rare Plant *Tabb 83-83*	WSr	British	128	90·2	21·5	11·7	Dartmouth	1866	Moore	Y 9.81	C. Allport	Padstow
121		Rask (ex Hundredth) *Lindoe 75-87*	WBg	Norweg'n	199	119·8	21·2	11·8	Jersey	1870	Deslandes	Y 3.85	A. Lothe	Haugesund
122	N	Rask *Bie -87*	WBn	Norweg'n	191	109·1	24·6	11·9	Stavanger	1874		Y 7.87	G. & A. Berg	Stavanger
123	N	Rata *Brauer*	WBg	Norweg'n	199	97·5	24·9	13·0	Luleå	1847			F.C. Brauer	Fredriksstad
124	N	Rata *Larsen*	WBg	Norweg'n	167	99·4	24·0	11·4	Drammen	1870			T. Larsen	Tönsberg
125		Ratcliff *Chitham*	WSw	British	186	92·0	25·0	14·5	Sunderland	1860	Barkley		J. Boyd	Ayr
126		Rauha (ex Salmea) *Dowder*	WBn	Russian	229	98·0	29·4	13·4	Sideby	1874			G.N. Söderland	Nystad
127		Rauha *Bosvall*	WSr	Russian	188	..	..	..	Daisbruk	1875			G. Granlund	Raumo
128		Rauha *Palsunoos*	WSr	Russian	183	..	..	..	Wederlako	1873			M.G. Sundblom	Nystad
129	V	Raul *Netto -88*	WSr	Brazilian	229	105·6	25·9	12·7	Rostock	1878	L. Burchard & Co.	Y 8.88	J.R. da Cruz	Marsim
130		Rauma *Söderlund -86*	WBk	Russian	260	116·1	27·9	11·0	Raumo	1881			A.W. Granström	Raumo
131		Raven *Whiting*	WBk	British	344	121·1	27·5	17·0	Sunderland	1864	J. Robinson	Y 6.81	W.R. Cave	Adelaide, S.A.
132	✻	Raven *Card*	WBn	British	211	103·6	24·9	12·9	P.E. Island	1875	Keefe	Y 5.87	J. Sutcliffe & Son	Grimsby
133		Raven *Bentley*	WBn	British	188	106·5	24·2	12·9	P.E. Island	1873	Duncan		G. Holden, jun.	Faversham
134		Raven *Stuart 85-85*	WBn	American	368	117·8	27·4	16·5	Machias, Me.	1865	J.L. Nash	Y 10.82	I. Sargent	Machias, Me.
135		Raven *Pendleton*	WSr	American	120	88·3	25·2	7·8	Bucksport, Me.	1852			M. Pendleton	Belfast, Me.
136	V	Ravenna *Bartholuzzi*	WBn	Italian	224	105·3	25·5	14·6	Ravenna	1874	E. Olivetto	Y 4.88	F.lli Vianello	Venice
137	✻	Ravenscrag *Howells 70-85*	IS	British	1263	219·2	35·0	22·9	Greenock	1866	R. Steele		J.C. & G.S. Richardson	Swansea
138	✻	Ravenswood *McGhie 86-86*	IBk	British	1089	216·2	35·8	20·6	Liverpool	1877	W.H. Potter & Co		Fisher & Sprott	Workington
139	V	Ravissante *Pichon -81*	WSr	French	120	86·0	20·3	11·0	Dunkirk	1878			A. Moret	Dunkirk
140	V	Ravola *Lockhart 86*	WS	British	1493	212·0	41·0	21·0	Newport, N.S.	1886	W.H. Mosher	Y 7.87	W.H. Mosher	Windsor, N.S.
141	✻	Raymond *Evans*	WBn	British	188	101·8	24·3	12·9	P.E. Island	1876	Lefurgey	Y 3.86	E. Davies	Aberystwith
142		Raymond (WRECKED)	WBg		167	81·2	22·3	12·2	Caen	1865	Robine			
143		Raymond T. Maull *Ogilvie*	W3mSr	American	512	140·5	35·0	15·3	Camden, N.J.	1881			J.W. Lee	Somer's P't. N.J.
144	I	Razeto (ex Zio Lorenzo) *Olivari -88*	WBk	Italian	570	143·0	32·8	21·2	Savona	1871	B. Calcagno	Y 9.86	Em. Razeto	Genoa
145	I	Razeto Padre *Bozzo*	WBk	Italian	443	131·5	31·3	19·5	Savona	1869	G. Tixi	Y 8.88	G.B. Razeto	Genoa
146	V	Razoilo I. *Razoilo*	W3mSr	Portug'se	212	98·4	26·2	9·9	San Martinho	1885	J.R. Ascenso	Y 10.85	J. Roilo	Aveiro
147	A	Rea *Adametz*	WSr	Austrian	150	85·3	22·0	11·7	Lussinpiccolo	1884	N. Martinolich	Y 7.84	V. Adametz fu G.	Lussinpiccolo
148		Ready Rhino *Stephens*	WSr	British	128	86·5	22·0	11·0	Sunderland	1860	G.W. & W.J. Hall		J. Stephens	Truro
149		Realidade	WSr	Brazilian	126	..	..	..					Cia. Cabotagem S. Joao da Barra	San Joao da Barra
150		Reaper *McDonald 83-84*	WSr	British	159	93·0	22·5	12·0	Garmouth	1864	Kinloch		W. Morrison	Inverness

...Lloyd's Register; D = Nederlandsche Vereeniging; G = Germanischer Lloyd; I = Registro Italiano; R = American Record;
...Austro-Ungarico; H = Veritas Ellenico; N = Norske Veritas; V = Bureau Veritas.

Dans la 4me Colonne, C = Composite (membrure en fer, bordé en bois); I = Fer; Stl = Acier; W = Bois.

...de doublage en métal (**Col. 11**) ressortent des livres dans lesquels les navires sont respectivement classés ou des rapports des Inspecteurs du Lloyd's Register.

Pour plus amples renseignements sur les navires classés par le Lloyd's Register, voir le Lloyd's Register Book et le Lloyd's Yacht Register.

Official No.	No. in Book	Registry in which classed	Ship's Name, Material and Rig. Master, and Dates of Appointment to present Owner's Service and to Ship. Ship's Late Name.	Flag	Net Tonnage	Length	Breadth	Depth	Built: Where	When	By whom	Date of Metalling	Owners or Managers	Port
51235	151		Reaper *Godfrey 86-86* W Bn	British	137	93·3	20·8	12·3	Gaspé, Canada	1864		Y 1.86	G. LeGros	Jersey
80029	152		Reaper *McLean 86-86* W Sr	British	124	83·5	26·0	8·5	Moss Glen, NB	1880	G. Merritt		Scammell Brothers	St.John's
72553	153	⁂	Reaper *Bustow 47-85* W 3m Sr	British	110	95·6	22·4	10·5	Pill	1875	Hitchens		C. B. Kelway	Truro
110270	154	RV	Reaper *Sawyer 87-87* W S	American	1469	211·6	39·2	24·6	Bath, Me.	1876	E. & A. Sewall	Y [illegible]	A. Sewall & Co.	Bath, Me
87542	155		Rebecca *Hopkins -87* W Bk	British	414	111·0	28·3	16·6	Apenrade	1868	J. Paulsen		T. H. Holdship	Sydney
56391	156	⁂	Rebecca (STRANDED) *May* W Bn	British	184	105·0	24·6	13·0	Aberystwith	1866	Evans		T. May	Plymouth
110308	157		Rebecca *Hughes* W Sr	American	153	105·0	30·6	8·8	Coos Bay, Oreg	1875			R. G. Byxbee	San F...
110400	158		Rebecca *James* W Sr	American	135	95·0	28·0	8·0	Portland, Conn	1879			H. Gildersleeve	Hartford
	159		Rebecca *John 84-* W Sr (ex Gertrude, or Solid)	Russian	162	92·7	23·9	11·5	Heiligenhafen	1871	J. H. Tornoe		F. V. Ilsoff	Amur
	160	V	Rebecca *Ringe -88* W Bk	German	419	126·1	29·5	17·7	Rostock	1862	O. Ludervig		O. Ringe	Danzig
110560	161	R	Rebecca A. Taulane *Champion* W 3m Sr	American	427	138·1	36·4	12·8	May's Landing (N.J.)	1882	S. Gaskell		J. S. Champion	Somers P
21420	162		Rebecca Caruana *Doherty* W Bk	American	444	131·0	29·4	16·8	Brooklyn	1867	C. & R. Soillon	Y [illegible]	F. Waydell	New York
110098	163	RV	Rebecca Crowell *Bray -82* W Bk	American	557	141·0	31·9	16·6	E. Boston, Mas	1873	A. Gore	Y [illegible]	R. W. Ropes	New York
110008	164	R	Rebecca F. Lamdin *Davis* W 3m Sr	American	441	145·0	34·3	13·0	Perry, Me.	1884	J. B. Nutt		J. B. Nutt	Eastport
21117	165	R	Rebecca Goddard *Nelson* W Bk	American	386	134·8	27·5	15·2	Medford, Mass	1860	P. Curtis	Y [illegible]	N. B. Mausfield	Boston, Ms
110402	166		Rebecca J. Moulton *Studley* W Bkn	American	560	113·6	32·5	16·1	East Boston	1873	D. Lank	Y [illegible]	J. H. Conant & Co.	Boston, Ms
110146	167		Rebecca M. Smith *George* W 3m Sr	American	302	116·3	31·1	9·3	Camden, N.J.	1872			F. Smith	Philadel...
110408	168	R	Rebecca M. Walls *Truax* W 3m Sr	American	561	145·6	34·6	16·3	Milton, Del.	1879	D. H. Atkins		J. May	Philadelp...
110530	169		Rebecca R. Nickerson *Tolner -86* W Sr	American	130	95·3	24·5	9·5	Booth Bay, Me.	1882	Hodgdon Bros.		A. McKay	Provincetown
110168	170		Rebecca Shepherd *Smith* W 3m Sr	American	391	116·0	32·8	9·7	Milford, Del.	1873			G. W. Shepherd	Philadelp...
21901	171	R	Rebecca W. Huddell *Depsin* W 3m Sr	American	244	125·0	36·0	7·4	Camden, N.J.	1869	Taylor & Mathis		J. Middleton	Philadelp...
	172	N	Rebekka (ex Union) *Bull -87* W Bk	Norweg'n	569	150·1	30·2	19·0	Quebec	1866	Shipwrights' Association	Y [illegible]	C. Langaard	Christiania
	173	N	Rebekka *Sörensen* W Sr	Norweg'n	104	80·0	20·3	9·6	Mandal	1878		Y 3.87	Albert Jacobsen	Mandal
	174	N	Rebus *Ingest -7* W Bk (ex Queen of the Ocean)	Norweg'n	665	161·2	33·5	19·1	Sunderland	1860	W. Pile, jun.	Y [illegible]	J. L. Ugland	Arendal
3410	175	⁂ I	Recco (ex Gabbio C.) *Gottozzo* W Bk	Italian	759	154·2	30·5	21·4	Sestri Ponente	1876	G. Peragallo	Y [illegible]	S. Ciehero fu G. B.	Genoa
	176		Recife W Sr	Brazilian	909	..	..	..					Barao de Maua	Rio de J.
69644	177		Reciprocity *Jones 75-75* W S	British	1409	205·9	40·7	24·4	Quebec	1874	Oliver	Y [illegible]	J. Thomas	Liverpool
85440	178	V	Record *Forbes* W S	British	1722	230·6	41·5	24·0	Newport, N.S.	1883	J. Mosher	Y [illegible]	A. Gibson & Co.	Liverpool
78987	179	DV	Recovery *Davison 87-87* W Bk	British	1012	180·0	37·0	21·9	Cornwallis, NS	1878	G. Bigelow	Y [illegible]	C. R. Burgess	Windsor
	180		Recurso Secondo W Bg	Spanish	193	88·9	25·3	15·0	Lloret	1847			C. Taltabull	Barcelona
110331	181	R	Red Cross *M'Intyre* W S	American	1301	185·3	38·1	23·1	Richmond, Me.	1877	T. J. Southard & S	Y [illegible]	T. J. Southard & Son	Bath, Me
60633	182	⁂	Red Cross Knight *Mewlee 83-87* W Bk	British	463	143·6	29·3	17·9	Barnstaple	1868	Westacott	Y [illegible]	J. E. Burgess	Swansea
75263	183	⁂	Red Gauntlet *Benest -88* W Sr	British	65	75·8	19·1	9·1	Jersey	1876	Aubin	Y 3.87	W. T. Pugsley	Guernsey
21615	184	R	Red Jacket *Elms* W Sr	American	111	81·3	23·8	9·3	Camden, Me.	18[illegible]			Cobb, Wight & Co.	Rockland
56925	185	⁂	Red Rose *McDonell 80-82* W 3m Sr	British	188	112·2	23·3	12·5	[illegible]	1879	Simpson	Y [illegible]	Porter's Shipping Co. Ld.	Fleetwood
62040	186	⁂	Red Rose *Bushell -88* W Bn	British	145	96·0	21·6	12·3	Bridport	1869	Cox	Y [illegible]	J. Bisson	Falmouth
110624	187	R	Red Wing *Johnson* I 3m Sr	American	415	130·0	33·0	11·6	P. Richmond, N.Y.	1884	American S.B. Co.		D. W. Johnson	Philadelp...
48800	188	⁂	Redgauntlet *Dixon 83-83* I Bk	British	1053	201·8	33·2	21·7	Greenock	1864	R. Steele		Williamson Milligan & Co.	Liverpool
75487	189	⁂	Redwood *Noel -83* W Bg	British	209	112·0	25·8	12·9	P. E. Island	1877	Longworth	Y [illegible]	F. Le Sueur & Son	Guernsey
	190		Reflectant W Bg	Norweg'n	153	..	..	..					H. M. Iversen	Christiania
	191	GV	Reform *Grütz 80-* W Bk	German	252	109·5	24·0	13·0	Stettin	18 46/69	E. Zieske		W. Lüdke & Co.	Stettin
	192	DN	Reform *Andersen* W S	Norweg'n	554	152·0	31·8	18·0	Porsgrund	1861			A. P. Andersen	Fr...
	193	N·	Reform *Sörensen* W S	Norweg'n	399	136·6	30·2	15·3	Satsmola	1866	P. Sandström		H. Grorud	Fr...
	194	N	Reform *Hansen 87-87* W Bg (ex Alders Minde)	Norweg'n	208	81·3	26·0	14·8	Sandefjord	18 46/71			M. Oppen	Laurvig
	195	N	Reform *Larsen 77-86* W Sr	Norweg'n	107	94·7	21·3	10·1	Stavanger	1858	R. O. Hœrem	Y [illegible]	H. L. Gabrielsen	Stavanger
	196		Regaleira *Paris -87* W Sr	Brazilian	155	..	..	..	S. João da Barra	1886			J. J. de Almeida	S. João d...
43648	197		Regalia *Wood* W Bg	British	319	112·0	27·0	16·6	Sunderland	1863	Rawson & Watson		W. Twizell	North S...
26105	198		Regalia (LIGHTER) *Warner -86* W Bn	British	146	90·0	22·0	12·6	Sunderland	1859	Austin		Bessey & Palmer	Yarmouth
88200	199	⁂	Regalia W Sk	British	83	76·1	20·8	10·6	Burton Stather	1884	J. Garside		C. H. Westoby	Hull
	200		Regenerada W Pol	Spanish	107	90·5	24·7	10·9	Blanes	1854			G. Sanjuan	Valencia

⁂ = Lloyd's Register; D = Nederlandsche Vereeniging; G = Germanischer Lloyd; I = Registro Italiano; R = American ...; A = Veritas Austro-Ungarico; H = Veritas Ellenico; N = Norske Veritas; V = Bureau Veritas.

In the 4th Column, C = Composite (iron frame, wood planked); I = Iron; Stl = Steel; W = Wood.

The Dates of Metalling (Col. 11) are based on the records of the Books in which Vessels are respectively classed, or on the Reports of the Surveyors of Lloyd's ...

Further particulars of Vessels classed by Lloyd's Register will be found in Lloyd's Register Book and Lloyd's Yacht Register.

1 No. dans le livre	2 Registre dans lequel le navire est classé	4 Nom. Capitaine et dates d'entrée au service des Armateurs actuels et à bord du navire. Ancien Nom.	Matériaux et Gréement	5 Pavillon.	6 Tonnage net.	7 Dimensions: Longueur	Largeur	Creux	8 Construit: Où.	9 Quand.	10 Par qui.	11 Date du doublage en métal	12 Armateurs ou Directeurs-Gérants.	13 Port d'Armement.
201	V	Regent *Treadwell*	WS	British	1289	199·0	39·3	24·0	Newport, N.S.	1878	Harvie	Y 11.87	A.Gibson & Co.	Liverpool
202	✱	Regent Murray *[illegible] -88*	I Bk	British	818	195·5	32·6	19·3	Glasgow	1876	D&W Henders'n		J.Dickie	Glasgow
203	N	Regia *Buus -83*	W Bk	Norweg'n	510	137·6	29·9	18·2	Grimstad	1883	L.Dahl	Y 1.86	Smith&Thommes'n	Arendal
204		Regina (ex Steamer Europa)	W3mSr	British	411	171·0	25·7	11·3	St.Catharin's (Ont.	18^{79}_{86}	L.Shickluna		Montreal Transportation Co.	Montreal
205		Regina	W Sr	British	100	77·3	19·3	9·8	Coringa	1875			Sheik Tyeb Abdul Ally	Colombo
206	V	Regina *Frederiksen 77-86*	W Bn	Danish	139	95·6	21·1	10·4	Troense	1877	C.J.Jensen		C.J.Jensen	Svendborg
207	NV	Regina *Brånlund 71-76*	W Bk	Norweg'n	826	182·5	33·9	19·9	Laurvig	1876	Dahl	Y 6.88	M.Oppen	Laurvig
208	N	Regina *Petersen -88*	W Bk	Norweg'n	425	131·6	32·1	15·5	Wasa, Finland	1857			E.Rassmussen	Porsgrund
209	DN	Regina *Kruge*	W Bg	Norweg'n	252	111·0	25·8	11·2	Apenrade	1848			A.A.Kruge	Tönsberg
210		Regina *Bjørnstad* (ex Bulgdere)	W Bk	Norweg'n	228	99·3	25·8	13·1	Bayonne	1855			Carl Dahl	Fredrikshald
211		Regina *Johansen -85*	W Bk	Russian	571	116·9	29·2	19·7	Borgå	1874	S.Andersson		A.Eklöf	Borgå
212		Regina *Fabeiro*	W3mSr	Spanish	319	118·4	26·4	11·1	Bilbao	1871	J.De Unzueta	Y 6.81	M.Ferrer de Noya	Villagarcia
213		Regina Cœli *Delacour -88*	W Bk	French	429	121·6	26·6	17·4	Nantes	1855	Viaud			
214		Regina Cœli *Lubrano* (FOXTROT)	W Bk	Italian	338	115·4	28·3	16·5	Sestri Ponente	1863			S.Lubrano	Naples
215		Registro	W Bn	Spanish	214	110·5	23·3	10·6	Pangasinan	1877			D.Noble José	Manila
216	N	Regnbuen *Olsen 88-88*	W Bg	Norweg'n	279	107·2	27·1	14·0	Grimstad	1877	N Olsen Gjämle		N.Jacobsen	Grimstad
217	DG	Regnera *Arents -87*	W Bn	Dutch	151	95·1	22·0	11·1	Muntendam	1865	B.J.Drenth		J.J.Arents	O.Pekela
218	A	Regolo (ex Sveti Krez) *Pollich*	W Bg	Austrian	303	..	..	..	Buccari	1865	M.Matterglian	Y 8.83	Costa.Pollich	Fiume
219		Regolo	W Bn	Italian	221	94·5	23·7	15·8	Savona	1853	Calamaro		G.Tonietti fu R.	
220		Regolo R. *Tonietti* (LOST) (ex Shepstin)	W3mSr	Italian	199	92·5	25·7	15·4	Savona	1854			L.Rabajoli	Porto Ferraio
221	N	Regres *Johannessen*	W Sr	Norweg'n	121	93·6	23·6	9·9	Bergen	1879			Brödr. Knudsen	Haugesund
222	R	Regulator *Drew*	W Bg	British	206	116·8	23·7	12·9	Cardiff	1868	Hodge	Y 1.89	E.F.Angel	Cardiff
223	N	Regulator *Hansen 83-86*	W Sr	Norweg'n	214	108·4	25·6	11·8	Sandefjord	1882	Albert Grön		P.A.Grön	Sandefjord
224		Regulo	W Sr	Brazilian	173	..	..	..					A. de Siqueira	Rio Grande do Sul
225	V	Regulus *Meyer 68-79*	I S	German	1115	206·7	34·4	20·2	Vegesack	1877	H.F.Ulrichs		W.A.Fritze & Co.	Bremen
226	V	Reidar *Olsen -88*	W Sr	Norweg'n	161	96·3	24·0	11·8	Stavanger	1868	K.Kaisen	Y 7.86	G. & A.Berg	Stavanger
227	DN	Reidulf *Knudsen -73*	W3mSr	Norweg'n	257	112·4	28·2	12·2	Risör	1872	G.Pedersen	Y 5.84	J.F.Juell	Risör
228	✱	Reigate *Jones 87-87*	I S	British	1035	210·6	32·1	21·2	London	1862	C.Lungley		W.B.Jones	Swansea
229	✱	Reigate *Welsh 77-85*	W Bkn	British	398	145·8	27·8	15·7	Gnd River, P.E.I.	1884	J.Yeo	Y 6.85	A.G.Moffat (T.P. Richards & Co.)	Swansea
230		Reina del Pacifico (ex Robert Pulsford) *Messina -86*	W Bk	Chilian	450	138·0	28·0	21·0	Whitehaven	1844	Brockelbank		J.Oyarzon	Valparaiso
231		Reindeer *Blyth*	W Sr	British	283	127·9	25·3	13·6	Tayport	1874	Mitchell	Y 11.83	Tatham, Bromage & Co	Dundee
232	✱	Reindeer *Osborne*	W K Barge	British	113	97·1	22·5	8·0	Ipswich	1879	Curtis		A. Beaumont & Co.	Ipswich
233	✱	Reindeer *Browne*	W K	British	91	79·7	21·2	10·4	Galampton	1886	W.A.Gibbs		J.P.Robinson	Grimsby
234	R	Reindeer *Baker 77-77*	W Bk	American	357	119·7	28·4	17·5	N Bedfrd, Mass.	1877	S.Dammon	Y 4.81	Aiken & Swift	N. Bedford, Mass.
235		Reindeer *Strandt*	W Bk	American	215	114·0	27·0	11·5	N.Haven, Conn	18^{53}_{80}		Y 10.81	H.Trowbridge's Sons	N.Haven, Conn.
236	✱	Reine *Hesry 84-88* (ex Teresa G., ex San Michele, ex Clotilde)	W Bkn	French	273	114·9	26·5	15·1	Sestri Ponente	1875	Paragallo	Y 2.89	E.Hondacé	St.Malo
237	V	Reine *Vanpouille*	W Sr	French	142	97·8	21·3	11·9	Dunkirk	1866	G.Malo, Vandericle fils		Herquet Van Rapenbusche	Dunkirk
238		Reine Blanche *Richard*	W Bk	French	213	93·8	23·6	12·9	La Nouvelle	1854		Z 10.83	Lemoine	St.Malo
239		Reine Elisabeth	W Sr	British	109	80·0	22·5	8·5	Portneuf	1871			E.Lacourcerière	Quebec
240	V	Reine Victoire *Ernoul -88*	W Sr	French	169	89·2	23·3	12·9	Dunkirk	1864	G.Beck		A.Govard	Dunkirk
241		Reine Victoria	W Sr	British	116	92·0	22·0	8·4	Champlain	1871			Samuel Perrault	Quebec
242	V	Reine Victoria *Leleillier 1883*	W Sr	French	126	86·9	21·3	10·8	Dunkirk	1856	G.Malo		G.Beck	Dunkirk
243	V	Reine des Mers *Lecheuallier -81*	W Bn	French	176	92·8	21·6	13·1	Dunkirk	1867	Pyotte Bayaert		Winnaert & Co.	Dunkirk
244	GV	Reinhard *Gowers 72-72*	W Sr	German	130	86·3	20·9	10·3	Weener	1869	R.Borg	Y 10.83	J.Cramer & Co.	Weener
245	G	Reinhold *Gerlach 83-83*	W Bk	German	311	113·7	27·9	14·7	Pillau	18^{40}_{58}	C.Eggert		J.H.Meyer	Pillau
246	V	Reinhold *Teter*	W3mSr	Russian	321	128·3	29·6	13·2	Widrisch	1882	P.Andersohn	Z 11.83	P.Andersohn	Riga
247		Reinhold *Pettersson* (ex Diligent)	W Bg	Russian	288	114·1	26·2	15·1	Sunderland	1862			K.L.Lindberg	Nagu
248		Reinosa (ex Johann Anton, ex Eagunia, ex Gellert) *Herrera*	W Bk	Spanish	299	115·1	26·3	13·0	Bardenfleth	1865	Gebruder Focke	Y 1.83	Man.Gonzalez del Corral	Santander
249	DV	Reiziger *Kuiper -87*	W3mSr	Dutch	167	99·1	21·3	11·5	Pekela	1864	F.L.Drenth	Z 10.81	F.L.Drenth	Oude Pekela
250														

✱=Lloyd's Register; D=Nederlandsche Vereeniging; G=Germanischer Lloyd; I=Registro Italiano; R=American Record; A=Veritas Austro-Ungarico; H=Veritas Ellenico; N=Norske Veritas; V=Bureau Veritas.

Dans la 4me Colonne, C=Composite (membrure en fer, bordé en bois); I=Fer; Stl=Acier; W=Bois.

Les Dates du doublage en métal (Col. 11) ressortent des livres dans lesquels les navires sont respectivement classés ou des rapports des Inspecteurs du Lloyd's Register

Pour plus amples renseignements sur les navires classés par le Lloyd's Register, voir le Lloyd's Register Book et le Lloyd's Yacht Register.

Official No. (1)	No. in Book (2)	Registry in which classed (3)	Ship's Name, Material and Rig. Master, and Dates of Appointment to present Owner's Service and to Ship. Ship's Late Name. (4)	Flag (5)	Net Tonnage (6)	Dimensions (7) Length	Breadth	Depth	Built (8) Where	(9) When	(10) By whom	Date of Metalling (11)	Owners or Managers (12)	Port of Registry (13)
87937	251	✳	Reliance *English* 79-84 I 4mS	British	2568	313·7	42·3	24·3	[illegible]	1884	W.H.Potter&Co		W.T.Dixon & Sons	Liverpool
110615	252	R	Relief *Bradley* -87 W3mSr	American	218	104·3	27·1	11·0	Bath, Me.	1884	Wm.Rogers	[illegible]	H.P.Dyer	[illegible]
139	253	I	Religione & Libertà *Lamberti* W Bk	Italian	846	163·5	34·7	23·9	Savona	1873	F.Sirelle	[illegible]	G.B.Bertolotto	[illegible]
	254	N	Relis *Ingebrigtsen* -88 W Galeas	Norweg'n	130	97·3	23·9	10·1	Bergen	1888	J.Sebü		P.Ingebrigtsen	[illegible]
110[illegible]	255	V	Rembrandt *Paine* 77-77 W S	American	1343	220·5	40·3	24·0	Kennebunk Me	1876	Crawford & Perkins	[illegible]	G.C.Lord & Co.	Boston
29467	256		Remembrance *Mennel* -83 W Bg	British	206	97·3	24·7	14·5	Middlesbro'	1862	Craggs		G.Judson	[illegible]
45910	257	✳ D	Remington *Langlois* 85-86 I Bk (now named "Martha")	British	999	198·0	34·9	21·0	Glasgow	1863	Smith&Rodger		Greenshields,Cowie & Co.	Liverpool
	258	N	Remittent *Pedersen* 82-83 W Bk	Norweg'n	532	145·0	31·8	17·2	Porsgrund	1876		[illegible]	J.Smith	[illegible]
78783	259	✳	Remonstrant *Knight* 82-83 I Bk	British	1045	214·0	34·4	21·3	Sunderland	1878	R.Thompson,jun		R.Conaway & Co.	Liverpool
	260		Remu *Sjödahl* W Bkt	Russian	149	88·9	26·9	8·9	Bjernö	1876	Tryg		S.Sjödahl	Toash
65219	261	✳	Remus *Sundholm* 86-86 W Bk	British	737	168·5	32·8	20·0	Bristol	1872	Wapping DkCo.	v 1.88	R.Conaway & Co.	Liverpool
	262	V	René *Macé* 85-85 W Sr	French	102	87·9	18·4	10·9	St.Servan	1878	Hamon fils		P.M.Macé	Pouliguen
	263	V	Renée Adrienne *Jeanno* W Bg	French	220	95·5	23·3	13·4	Nantes	1872	Clairgeau	[illegible]	Jeanno	Vannes
	264	✳ V	Renée Rickmers *Hew* 75-8? I 4mS	German	2064	283·0	40·5	24·6	PortGlasgow	1887	Russell&Co.		R.C.Rickmers	Bremerhaven
	265	N	Renen (ex Black Horse) *Dahl* 76-76 W Sr	Norweg'n	226	109·7	26·8	12·0	Windsor,N.S.	1861			E.Suhrke	Fredrikstad
	276	N	Renen* *Christensen* W Sr	Norweg'n	102	85·0	21·4	10·0	Porsgrund	1866	E.Rasmussen		Chr.Christensen	Sandefjord
71723	277	✳	Renfrewshire (MISSING) *Cummins* [8?88] I S	British	898	202·5	33·1	20·6	PortGlasgow	1875	H.Murray&Co.		T.Law & Co.	Glasgow
	278		Rengo *Bosch* W Sr	Spanish	114	80·0	23·2	11·2	Masnou	1844			Jose M.Serra	Barcelona
	279	V	Renner *Thomsen* W Sr	Danish	155	91·5	21·9	11·3	Arnis	1864	Johansen		P.Thomsen	Sonderborg
21[illegible]	270		Reno *Randall* W Sr	American	135	93·7	25·8	8·2	E.Machias,Me.	1868	S.W.Pope&Co.		J.Inglee	Machias
	271	A	Reno *Gherbaz* W Bk	Austrian	468	131·5	28·2	18·4	Fiume	1871	N.Martinolich	[illegible]	G.Gherbaz fu A	Fiume
	272	V	Renommée *Bryckaert* -87 W Sr	French	121	89·9	19·7	11·5	Dunkirk	1865	E.Sagary fils		A.Maurois	Dunkirk
[illegible]	273		Renovation (ex Thomas & Joseph) *Mallaburn* [86-86] W Sw	British	268	101·4	25·9	15·1	Sunderland	1831	Gates		J.L.Robson	South Shields
56377	274		Renown *Sebire* -88 W Bn	British	216	107·7	25·3	13·9	Guernsey	1867	Thom		H.J.Sebire	Guernsey
76153	275	✳	Renown *McKenzie* 84-81 W3mSr	British	138	98·7	22·3	11·5	Troon	1877	TroonS.B.Co.	[illegible]	D.McArthur	Greenock
63[illegible]	276	✳	Renown *Cruickshank* 35-82 W Sr	British	127	91·7	24·4	10·7	Boddam	1876	Stephen		J.S.Wood	Peterhead
	277	V	Rensche *DeBuhr* 82-82 W Sr	German	107	82·4	20·2	9·6	Papenburg	1872	B.Tholen		B.Tholen Wwe	Papenburg
	278		Renzo (ex Nuovo Dovere) *Marine* W Bk		540	139·1	32·8	19·6	Recco	1870	Rolla	[illegible]		
	279		Reporter *Wood* -87 W Sr	British	149	96·9	25·2	8·5	Bucksport, Me	1852	Genn		C.T.Wood	Moncton
83588	280		Reporter *Gilchrist* 82-82 W Sr	British	122	86·0	26·2	7·8	Wickham, N.B	1883	S.H.Gilchrist		Robert C.Elkin	St.John, N.B.
110196	281	V	Reporter *Spalding* 80-82 W S	American	1286	202·0	38·2	24·2	Newburyport, Mass.	1874	G.W.Jackman&Co	[illegible]	E.Sumner	Newburyport
110[illegible]	282		Reporter *Cook* W3mSr	American	333	141·4	34·0	10·6	PortLudlow Wash.	1876	HallBros.		W.G.Hall	San Francisco
	283		Reptor *Hilchert* W Bg	Russian	264	104·3	26·6	15·1	Libau	1848	Gamper		E.D.Hilchert	Pernau
21885	284	R V	Republic *Holmes* 74-74 W S	American	1293	193·6	39·6	23·7	KennebunkMe	1869	N.L.Thompson	[illegible]	G.C.Lord & Co.	Boston
	285	V	Républicain *David* -87 W Bg	French	186	91·5	22·3	12·2	Nantes	1875	Clergeau	[illegible]	Edmond Boissière	Nantes
	286	V	Republik *Kuiper* -87 W3mSr	German	169	90·0	22·5	10·0	Westerwik	1868	C.R.Fridell		J.Wallenstein	[illegible]
	287	G V	Republik *Nielsen* -88 W Bk	Norweg'n	648	147·3	31·3	19·9	Vegesack	1853			C.K.Wiese	Bergen
[illegible]	288	✳	Rescue *Townsley* 84-85 W Sr	British	132	92·8	21·3	11·5	Gainsboro'	1869	Fidell	[illegible]	J.Wetherall	[illegible]
87[illegible]	289	✳	Rescue *Bates* W K	British	80	76·1	20·3	10·1	Galmpton	1883	W.A.Gibbs		A.Bates	Grimsby
[illegible]	290		Research *Finigan* 80-80 W Bn	British	129	91·1	22·4	11·3	P.E.Island	1855			W.Twibill	[illegible]
[illegible]	291		Residue *Niller* W Sr	British	111	73·5	17·7	10·6	Sunderland	1851	W.Thompson& Sons		J.Galley	[illegible]
	292		Résolu (HULK) *Mouton* W Bk	Italian	532	131·2	30·8	20·3	St.Tropez	1866	Henrigon			
	293		Resolución W Sr	Spanish	129	81·1	25·0	11·5	Villajoyosa	1870			Marcos Lopez	[illegible]
	294	N	Resolut *Jacobsen* 82-86 W Bk	Norweg'n	617	153·7	33·7	20·7	Alblasserdam	1854		[illegible]	H.Aagaard	Tönsberg
	295		Resolut (BROKEN UP) (ex Endymion) W Bk	Norweg'n	543	128·9	30·1	19·7	Liverpool	1847			J.Olsen	Farsund
	296	N	Resolut (ex Der Unfassbar) *Andersen* 87-87 W Bg	Norweg'n	299	114·3	27·3	14·9	Bilbao	1856		[illegible]	Westergaard & Co.	Christiania
38881	297		Resolute *Whitenall* 86-86 W Bkn	British	216	116·6	24·6	13·2	Aberdeen	1869	Duthie	v 1.88	WhitstableShippingCo. Limited	Faversham
73250	298	✳	Resolute *Leask* 84-84 W Sr	British	114	88·4	22·0	10·9	Peterhead	1877	Stephens	v 1.87	W.H.Leask	Peterhead
90986	299	✳	Resolute *Stokes* W K	British	85	79·6	21·3	10·1	Galmpton	1884	T.Matthews		T.Stokes	Grimsby
78262	300	✳	Resolute *Noel, jun.* W Sr	British	76	79·8	19·3	9·3	Jersey	1877	Le Huquet		G.Noel	Jersey

✳=**Lloyd's Register**; D=Nederlandsche Vereeniging; G=Germanischer Lloyd; I=Registro Italiano; R=American [Lloyd's]
A=Veritas Austro-Ungarico; H=Veritas Ellenico; N=Norske Veritas; V=Bureau [Veritas]

In the 4th Column, C=Composite (iron frame, wood planked); I=Iron; Stl=Steel; W=Wood.

The Dates of Metalling (**Col. 11**) are based on the records of the Books in which Vessels are respectively classed, or on the Reports of the Surveyors of Lloyd's [Register].

Further particulars of Vessels classed by Lloyd's Register will be found in Lloyd's Register Book and Lloyd's Yacht Register.

1	2	3 Nom. Matériaux et Gréement. Capitaine et dates d'entrée au service des Armateurs actuels et à bord du navire. Ancien Nom.	4 Pavillon.	5 Tonnage net.	6 Dimensions. Longueur	7 Largeur	Creux	8 Construit. Où.	9 Quand.	10 Par qui.	11 Date du doublage en métal	12 Armateurs ou Directeurs-Gérants.	13 Port d'Armement.
301	D	Resolutie ([illegible]) Cd. Vr[illegible] W Bn	Dutch	266	113·1	26·9	13·0	P.E.Island	1874	McDonald	V 7.88	H.Rietveld	Amsterdam
302	R	Resolution *Bernaux -86* W Sr	British	144	96·0	26·5	9·7	Cornwallis	1886	J.E.Bigelow		W.H.Baxter	Windsor,N.S.
303	✻	Resolution *Carlton* W K	British	77	73·5	20·8	10·3	Brixham	1886	R.Jackman		J.Randall	Grimsby
304		Resolution *Edvardsen* W Bk	Norweg'n	343	106·1	28·8	16·5	Christiansand	1815			A.H.Kiær & Co.	Drammen
305		Resolve W Sr	British	111	79·8	22·0	11·5	Dunkirk	1860			W.Grant	Inverness
306	N	Resolve *Hansen* W Bk	Norweg'n	398	129·4	28·5	16·2	Charlottetown [P.E.I.	1871	Davies	V 1.88	S.Hansen	Sandefjord
307	R	Resolven W Bg	British	142	94·8	22·9	11·3	Bideford,PEI	1872	W.Richards		F.B.Atkinson	Sackville,N.B.
308		Resource *Frederiksen 86-86* W Bk	Norweg'n	528	137·4	30·3	18·2	Memel	1867	B.Pieper		J.A.Andorsen	Mandal
309	✻	Respigadera (STRANDED) *Parsell* I S	British	1629	254·1	38·8	23·6	Sunderland	1874	T.R.Oswald		Hargrove & Hellon	Liverpool
310	N	Respit *Petersen 77-77* W Bk	Norweg'n	458	133·9	30·1	17·1	Kragerö	1877	J.Nielsen	V 2.86	P.A.Petersen	Kragerö
311		Restaurador *Caetano* W Sr	Brazilian	196	..	..	..	S.JoasdeBarra	1885		V 8.85	Cia.Cabotagem de S.Joao da Barra	St.JoasdaBarra
312	I	Restituta Madre *[illegible]* W Bk	Italian	536	113·7	32·2	20·5	Procida	1878	N.Lubrano	V 1.88	A.Luterano	Naples
313		Restless W Bg	British	289	122·8	28·9	12·7	Maitland,N.S.	1866	I.Smith		James C.Ellis	N'weastle,N.S.W.
314		Restless W Bn	British	190	101·0	23·5	12·7	Kingsbridge	1865	Date	V 9.81	W.R.Beer	Salcombe
315	✻	Restless *Cawthorne 88-88* W K	British	75	79·5	19·4	9·4	Dysart	1888	Foster & Co.		R.Hickman	Goole
316		Resuelta (ex Narragansett) *Gran* W Bk	Spanish	594	134·5	30·8	21·0	Warren,R.I.	1849			Jané & Co.	Barcelona
317		Result W Bg	British	242	117·5	25·8	13·3	Guernsey	1864	DePutron		Westergnard & Co.	Guernsey
318		Result *Mawer* W Sr	British	119	81·2	21·0	10·7	Goole	1867	Pearson		W.Green & Co.	London
319	R	Resultado *Frelick -87* W Bn	British	178	103·8	26·0	9·6	Portugal	1871		V 12.81	Ernest M.McLeod	Georgetown,Dem
320		Resumption W Sr	American	294	143·4	29·0	10·3	MilwaukeeWi-	1879			J.D.Ross	Escanaba,Mich
321	R	Retriever (FOUNDERED) *Cornish* W Bk	British	296	116·0	26·7	15·5	Sunderland	1858	J.Laing	V 5.87	F.H.Cornish	London
322		Retriever *Edmonds 76-82* W Bkn	British	204	117·8	22·8	13·6	Dartmouth	1876	Philip	V 1.83	Newman,Hunt&Co.	London
323		Retriever W Bn	British	191	106·0	23·8	12·9	P.E.Island	1863			Josiah Reeves	Faversham
324		Retriever *Gray 83-83* W Bkn	American	520	161·0	37·0	13·0	Seabeck,Wash	1881			R.Kendrick	Port Townsend, U.S.
325	A	Retta *Vidulich -87* W Sr	Austrian	119	83·3	20·0	11·5	Lussinpiccolo	1880	G.Giuricich	V 4.83	P.Scopinich fu A.	Lussinpiccolo
326	✻	Return *Kelso 87-87* I Sr	British	104	91·0	21·0	10·1	Bowling	1877	Scott&Macgill		D.M'Arthur	Greenock
327		Retzia *Sandquist* W Sr	Swedish	119	80·7	22·2	11·4	Visby	1845	H.Hansson		C.Stenberg	Wisby
328		Reuben Eastman *Eastman* W Sr	American	109	81·0	25·1	7·5	Bath,Me.	1878	Blaisdell		R.Eastman	Bath,Me.
329	RV	Reuce *Adams -87* W S	American	1828	229·2	41·0	27·1	KennebunkMe	1881	Thompson	V 11.87	G.H.Theobald	Bath,Me.
330		Reuhas *Proberg 72-79* W Sr	Russian	231	102·7	28·2	12·8	Kromby	1873	Frederiksen		G.W.Sjöblom	Tonala
331		Revenue *Hutchins* W Sr	American	159	98·0	29·0	8·0	CapeMay,N.J.	1854			S.R.Hutchins	Orland,Me.
332	N	Revenue *Green* W Sr	Norweg'n	121	81·1	23·0	10·1	Norway				C.Green	Laurvig
333	V	Réville (ex Heinrich) *Vancostenobel* W Sr	French	100	73·5	18·4	9·7	Stralsund	1872	A.Juhl		A.Postel & ses fils	Havre
334	✻	Revival *Peat* W Sk	British	82	76·6	20·6	10·4	Rye	1886	G.& T.Smith		J.S.Bedford	Grimsby
335		Revnek Bahri W Sr	Turkish	145	..	..	..	Kara Agatch	1877			Doghrah Majeh Zadeh Raif Effendi & Co.	KaraAgatch
336	bV	Revolving Light *Sprague -88* W S	British	1317	196·4	38·3	23·9	Harvey,N.B.	1875	G.S.Turner	V 2.86	W.R.McMann	St.John,N.B.
337	✻	Rewa (ex Hope) *Harbury 83-83* I Bk	British	454	144·8	26·9	16·4	S.Shields	1867	Readhead,Softley&Co.		Ship "Rewa" Co., Ld. (Bethell & Co.)	London
338		Reward W Bn	British	166	89·1	24·2	12·9	Sunderland	1846			J.H.Guilbert & Co.	Guernsey
339	✻	Reward *Wilkie 65-82* W 3m Sr	British	163	110·0	22·5	11·4	Troon	1878	TroonS.B.Co.	V 8.88	D.M'Arthur	Greenock
340	N	Rex *Andersen* W Bk	Norweg'n	508	134·7	30·7	17·3	Grimstad	1881		V 6.88	J.T.Thommesen	Arendal
341	N	Rex *Tholvsen -86* W Bk	Norweg'n	336	118·0	28·7	15·7	Sandefjord	1873		V 1.88	S.Schülstach	Sandefjord
342	N	Rex *Löre -80* W Bg	Norweg'n	131	90·5	23·8	10·0	Kragerö	1857			N.Nielsen	Kragerö
343	R	Reyita (ex Northern Home) *Rogler* W Sr	Colombi'n	173	94·0	26·1	11·0	Cornwallis,NS	1872	J.E.Bigelow	V 8.88	Cano,Dugand & Co.	Rio Hacha
344		Reynard *McGough* W Bk	British	1029	182·6	35·6	23·0	Newburyport, [Mass.	1856	Jackman		K.C.Maguire & Co.	Quebec
345	V	Rhea *Budig -87* W Bk	German	445	136·3	30·0	16·2	Stettin	1866	A.E.Nüske	V 7.87	F.Müller	Stettin
346	V	Rhea *Lundquist* W S	Russian	968	215·8	38·4	21·3	Laitakan	1880		V 5.87	Petrell & Hagner	Björneborg
347	✻	Rheidol *Williams 78-78* W Sr	British	107	88·1	21·5	10·8	Aberystwith	1878	Jones&Bros.	V 2.84	T.H.Jones & Co.	Aberystwith
348	✻	Rhine *Williams 87-87* I S	British	1611	257·2	38·3	23·1	Greenock	1886	Russell&Co.		J.Nourse	London
349	bV	Rhine *Macumb 82-86* W S	British	1093	180·7	36·6	23·0	Windsor,N.S.	1874	S.Dimock	V 6.87	S.Dimock	Windsor,N.S.
350	N	Rhine *Florens* W Bk	Norweg'n	1178	166·7	38·5	26·8	NewYork	1850	J.Westervelt	V 3.89	Act. Selsk. "Ispilon"	Kragerö

✻=Lloyd's Register; D=Nederlandsche Vereeniging; G=Germanischer Lloyd; I=Registro Italiano; R=American Record;
A=Veritas Austro-Ungarico; H=Veritas Ellenico; N=Norske Veritas; V=Bureau Veritas.

Dans la 4me Colonne, C=Composite (membrure en fer, bordé en bois); I=Fer; Stl=Acier; W=Bois.

Les Dates du doublage en métal (**Col. 11**) ressortent des livres dans lesquels les navires sont respectivement classés ou des rapports des Inspecteurs du Lloyd's Register.

Pour plus amples renseignements sur les navires classés par le Lloyd's Register, voir le Lloyd's Register Book et le Lloyd's Yacht Register.

Official No.	No. in Book	Registry in which classed	Ship's Name. Material and Rig. Master, and Dates of Appointment to present Owner's Service and to Ship. Ship's Late Name.	Flag.	Net Tonnage	Dimensions: Length	Breadth	Depth	Built: Where	When	By whom	Date of Metalling	Owners or Managers	Port
21893	351		Rhoda Holmes *Holmes* W3mSr	American	281	128·0	32·0	10·1	Cooper's Pt NJ	1869			J.S.Holmes	Port[illegible]
62036	352	✻	Rhoda Mary *Westlake* W Sr	British	99	101·2	21·9	11·5	Point	1868	Stephens		S.T.Westlake & Co.	Falmouth
21501	353		Rhodella Blew W Sr	American	175	107·3	29·2	8·6	MauricetwnNJ	1859			B.Lyman	New York
	354	V	Rhône *Reboa* I Bk	French	838	193·5	32·0	18·8	Bordeaux	1872	Bichon frères		Maurel frères	Bordeaux
	355		Rhône	French	215	..	..	..						Marseille
[illegible]	356	✻	Rhuddlan Castle *Williams* I Bk (MISSING)	British	600	174·2	28·2	18·0	Liverpool	1867	R&J Evans&Co.		Richards, Mills&Co	Liverpool
[illegible]	357	V	Rialto *Gelson* W Bk	British	1462	206·2	41·7	23·8	Newport, N.S.	1881	J.Masher	V 9.86	A.Gibson & Co.	Liverpool
	358	R	Rialto *Jörgensen 86-86* W Bk	Danish	438	135·3	26·9	17·2	AnnapolisN.S.	1867		[illegible]	M.J.Juul	Svaneke
	359	✻	Rialto *Genters 88-88* I S	German	1182	211·5	36·5	22·0	Sunderland	1869	Oswald&Co.		C.Paulsen	Elsfleth
[illegible]	360	✻	Ribbleton *Francis 85-88* I Bk (ex Matanie)	British	375	138·7	25·3	16·6	Preston	1865	Mackern		R.& W.King	Bristol
	361	A	Ribes *Suttora -88* W Bk	Austrian	271	104·0	24·6	13·5	Lussinpiccolo	1879	N.Martinolich	V 10.85	G.B.Fremuda de Matteo	Lussinpic[illegible]
	362		Ribnitz *Fretwurst 66-86* W3mSp (ex Paul Gotthold)	German	270	106·5	26·7	13·9	Rostock	1862	Geo. Tesdorpf		Liebenheim & Pincus	Rostock
[illegible]	363		Ribston *Neave* W Bk	British	397	127·6	28·1	17·2	Sunderland	1867	J.Robinson	[illegible]	William Owston	Freema[illegible]
	364	N	Ricard *Nielsen* W Sr	Norweg'n	116	78·5	22·9	10·5	Christiansand	1877			I.C.Nielsen	Christ[illegible]
	365		Ricardo W Bn (BROKEN UP) (ex Avillez e Fernando)	Portug'se	182	87·9	21·2	12·5	S.Martinho do Porto	18^{40}_{62}		11.81	M.Baptista	Lisbon
57428	366		Ricardo Barros W Sr	British	111	85·0	26·8	7·8	Long Island	1866			E.P.Masson	Trinidad
21420	367		Ricardo Barros *Newton* W Sr	American	119	85·0	26·8	8·0	PatchoqueNY	1866		11.83	T.Pearsall	New York
26965	368		Riccardo W Bk	British	285	99·3	26·6	15·8	Malta	1851	L.German	[illegible]	Alfred Ganci	Malta
1624	369		Riccardo *Cheli* W Bn	Italian	149	100·7	24·3	10·2	Caletta	1857			E.Crovari	Palermo
74406	370	✻	Richard *Glyka* W Bk	British	531	131·0	30·6	18·6	Richibucto, N.B.	1877	O-Leary	V 6.86	J.E.Stangala	Chatham
	371	V	Richard *Jacobsen -88* W Bk (ex Joseph Hambro)	Danish	290	127·0	25·1	10·9	Rønne	1866	S.P.Bech	10.87	N.W.Thomsen	[illegible]
	372	G	Richard *Paske 77-77* W Bk (ex Sans Souci)	German	463	124·1	28·7	18·3	Swinemünde	1858	Liebnitzky		W.Ludke & Co.	Stettin
	373	G	Richard *Burmeister* W Bg	German	291	104·8	27·2	16·2	Damgarten	1857	H.Dierling		P.Grampp	Rostock
	374	N	Richard *Blichfeldt* W Bk	Norweg'n	254	126·9	25·5	12·7	Risør	1858			Chr.Nielsen	Drammen
	375	DG	Richard *Engblom -88* W Bk	Russian	722	154·2	33·5	20·6	Burg	1860	Gebr. Bosse	[illegible]	C.E.Engblom	Nystad
	376		Richard *Jansson* W Sr	Swedish	164	84·7	22·5	11·5	Stralsund	1848			A.G.Jansson	Frötuna[illegible]
	377	N	Richard Cobden *Thorbjørnsen* W Bk	Norweg'n	421	124·0	29·5	17·4	Burg	1847			Chr.Swane	Christiania
	378		Richard Cobden *Johanssen -85* W Bk	Swedish	295	99·4	25·8	15·8	Brevig	1847			J.Bengtsson	Gotheborg
100701	379	R	Richard F. C. Hartley *Baker* W3mSr	American	446	145·0	31·2	12·0	Bath, Me.	1887	Kelley & Spear		T.W.Penhallow	Boston [illegible]
73752	380		Richard Fisher *Bennett 81-81* W3mSr	British	179	112·0	24·3	12·6	Carrickfergus	1876	Rodgers		T.Fisher	Fleetwood
89243	381	✻	Richard Greaves *Davies* W Sr	British	99	86·0	22·8	11·2	Portmadoc	1885	D.Jones	V 3.86	J.K.McKerrow	Carnarvon
21403	382	R	Richard Hill *Vaux* W Sr	American	152	99·3	26·1	8·4	BrandywineDel	1861			J.A.Creighton	Thomaston
21503	383		Richard Law *Smith 83-86* W Sr	American	130	87·5	24·5	7·9	Stonington, Conn.	18^{48}_{72}	Irons&Grinnell		J.N.Hancox	Stoning[illegible]
2440	384		Richard N. Parker *Paske* W Sr	British	159	80·6	19·9	13·3	Cork	1839	W & H. Browne		Starbuck & Rackstraw	London
110516	385	RV	Richard P. Buck *Carver* W S	American	1491	220·4	40·3	24·0	Bath, Me.	1882	W.Rogers	V 3.87	D.Barnes	New York
110365	386	RV	Richard Parsons *Pendleton* W Bk	American	1160	189·1	37·2	23·2	Camden, Me.	1878	J.Pascal	11.85	Carleton, Norwood & Co.	Camden
21511	387		Richard Peterson *English* W Sr	American	192	106·1	29·1	10·2	Cooper's Pt NJ	1864			D.English	Camden N.[illegible]
	388	G	Richard Porter *Pieplow 79-88* W Bn	German	126	89·9	21·2	11·0	Rostock	1873	O.Ludewig		E.Burchard	Rostock
	389	V	Richard Rickmers *Denker Sternberg* W Bk	German	1345	201·7	40·7	25·5	Geestemünde	1881	R.C.Rickmers	V 7.86	R.C.Rickmers	Geestem[illegible]
110152	390	R	Richard S. Leaming *Shaw* W3mSr	American	463	110·0	35·0	14·6	DennisvilleNJ	1880			R.S.Leaming	Philadel[illegible]
	391	R	Richard T. Green *Kandel -88* W Bn (ex Riberita)	American	289	118·0	29·2	12·2	Rockport, N.B.	1873	A.D.Cole	12.85	R.T.Green	Boston
21510	392		Richard Vaux *Richardson* W Sr	American	234	125·7	30·0	8·8	WilmingtnDel	1855			J.D.Richardson	Bridge[illegible]
110681	393		Richard W. Denham *Terrill* W Sr	American	111	87·7	25·3	7·1	BowdoinhmMe	1873			C.V.Minott	[illegible]
	394	GV	Richard Wagner *Hansen-Sengtz* I 4mS	German	2026	276·0	41·0	23·9	Geestemünde	1886	J.C.Tecklenborg		B.Wencke, Söhne	Hamburg
32408	395		Richard & William *Tinker* W Bg	British	150	82·0	21·2	13·2	London	1845			F.Hillery	London
21660	396		Richard III. *McIntyre* W S	American	955	175·3	33·7	23·5	Portsm'th N.H.	18^{58}_{82}		V 8.86	Calvin Page	P.Townsend
52	397		Richardos W Bn	Greek	377	122·7	25·6	..	England	1878			Alexandra Tsatsaroni	Skiathos
71639	398		Richelieu W Bateau	British	121	105·0	23·0	7·0	Sorel	1874			Joseph Gagnon	Montreal
	399	V	Richelieu *Eudes -87* W Bk	French	275	110·2	24·6	14·8	Fécamp	1885	D.Capon		H.Monnier & fils	Fécamp
	400	R	Richelieu *Ohlsson -86* W Bk (ex Morning Star)	Swedish	296	101·7	24·8	11.3	Harpswell Me.	1859		V 1.88	A.Ohlsson	Åbo

✻ = Lloyd's Register; D = Nederlandsche Vereeniging; G = Germanischer Lloyd; I = Registro Italiano; R = American [illegible]
A = Veritas Austro-Ungarico; H = Veritas Ellenico; N = Norske Veritas; V = Bureau Veritas

In the 4th Column, C = Composite (iron frame, wood planked); I = Iron; Stl = Steel; W = Wood.

The Dates of Metalling (Col. 11) are based on the records of the Books in which Vessels are respectively classed, or on the Reports of the Surveyors of Lloyd's [illegible]

Further particulars of Vessels classed by Lloyd's Register will be found in Lloyd's Register Book and Lloyd's Yacht Register.

No.	Classe	Nom. Matériaux et Grèvement. Capitaine et dates d'entrée au service des Armateurs actuels et à bord du navire. Ancien Nom.	Pavillon	Tonnage net.	Longueur	Largeur	Creux	Où.	Quand.	Par qui.	Date du doublage en métal	Armateurs ou Directeurs-Gérants.	Port d'Armement.
401		Richmond *Stuart* W Sr	British	183	105·8	26·1	11·0	London	1868	J.&W.Dudgeon		Board of Trade	London
402	IV	Riconoscenza *Solari* -84 W Bk (ex Giuseppina B.)	Italian	609	152·8	31·6	20·3	SestriPonente	1872	C.Tubino	Y 6.87	M.Solari	Genoa
403	V	Ricordo *Simonelli* W Bk	Italian	699	167·8	33·6	20·7	SestriPonente	1869	G.Casanova	Y 1.83	A.Degregori fu T.	Genoa
404	I	Ricordo *Consigliere* W 3m Sr	Italian	473	139·4	31·2	19·0	SestriPonente	1872	S.Bozzano	Y 12.88	Eredi di F.Consigliere	Genoa
405	A	Ricordo *Catania* -88 W Bg	Italian	358	108·2	26·6	15·4	Varazze	1863	G.Scala	Y 2.88	A.La Barbera	Trapani
406	R	Ridesdale W Bk	Argentine	374	122·4	26·4	15·6	Sunderland	1859	G W & W J Hall	Y 3.88	W.Barrett	BuenosAyres
407		Ridgewood *Weaver* 85-85 W Sr (RWKN)	American	283	112·0	29·6	14·1	GreenPnt,NY.	1863	GreenPoint,NY		J.C.L'Engle	Jacksonville,Flor.
408		Rifleman *Turner* -79 W Sr	British	132	87·5	21·0	12·0	Teignmouth	1858			J.A.Turner	Teignmouth
409	N	Rifondo *Mathiesen* 77-77 W Bk (ex [illegible])	Norweg'n	246	121·5	28·4	11·5	Baltimore	1855			Peter Höeg	Drammen
410	NV	Riga *Sorensen* -88 W Bk	Norweg'n	335	123·1	29·1	13·8	Middelburg	1869	J.Van Vliet	Y 11.87	A.Jörgensen	Horten
411		Riga *Dansen* -87 W Sr	Norweg'n	117	89·3	22·6	10·8	Stavanger	1879	Lars Haaland		E.Berentsen	Stavanger
412	N	Rigel *Pettersen* W Sr	Norweg'n	207	101·0	25·3	11·8	Arendal	1869	Aanundsen		H.H.Pettersen	Arendal
413	NV	Rigi *Zopfi* 85-85 W Bk	Norweg'n	517	146·7	31·9	17·0	Drammen	1884	P.Svendsen	Y 8.88	S.Otto & Co.	Christiansand
414		Rigmer *Larsen* 71 W Bn (ex J. P. [illegible])	Danish	138	86·9	20·7	11·9	Elsinore	1858	C.Hoffmann & Co.		J.A.Larsen	Kjerteminde
415		Riho Maru W Bk	Japanese	378	126·5	24·8	12·3	Sunderland				Ihashi Chinshichi	Tokio
416	N	Riisoer *Sörensen* -88 W Bk	Norweg'n	381	135·1	28·4	16·1	Risör	1867	O.Jörgensen	Y 1.8[illegible]	O.Caspersen	Risör
417		Rika *Olsson* -87 W Sr	Swedish	166	89·9	22·3	11·7	Oroust	1869	A.Isaksson		J.A.Nilsson	Forshälla
418	R	Rillie S. Derby *Taylor* W 3m Sr	American	398	132·0	34·0	9·3	Camden,N.J.	1874	Taylor&Mathis		G.W.Shepherd,jun.	Philadelphia
419	✻	Rimac *Snelders* 89-89 I Bk	British	593	171·6	27·9	18·0	Liverpool	1865	T.Vernon&Son		Nicholson & McGill	Liverpool
420	NV	Rimfaxe *Salvesen* 86-86 W Bk	Norweg'n	513	135·9	31·2	17·7	Grimstad	1876	J.Tryde	Y 9.87	Axel Smith	Arendal
421	N	Rinde *Holm* -88 W S	Norweg'n	582	144·4	31·8	18·4	Kragerö	1866	P.Pedersen		T.M.Wiborg	Kragerö
422	R	Ring Dove *Marston* W Sr	American	150	100·0	24·0	8·0	Calais,Me.	18 67/79			Cobb,Wight & Co.	Rockland,Me.
423	✻	Ringdove *Walsh* 89-89 W Bk	British	399	130·7	27·4	17·1	Southwick	1864	G.&J.Mills	Y 2.88	R.Farrell	Greenock
424	N	Ringdove *Rund* 86-85 W Bn	Norweg'n	160	95·1	24·4	10·5	P.E.Island	1869	S.Brown		M.Mulvad	Fredrikstad
425	N	Ringhorn *Knudsen* -88 W Bk	Norweg'n	589	147·8	31·5	18·6	Grimstad	1884	N.Osmundsen	Y 12.86	J.E.Einersen	Grimstad
426	DN	Ringhorne *Bentsen* W Bk (ex S. Vaughan)	Norweg'n	951	168·5	35·4	24·5	Cornwallis,NS	1869	Eaton	Y 1.8[illegible]	M.Pedersen	Mandal
427		Ringleader *Hoult* 82-82 W Bn	British	180	96·0	20·4	12·7	Sandwich	1853	Hoad		W.G.Lawson	Faversham
428	✻	Ringleader *Webster* 84-84 W Sr	British	167	104·0	24·6	11·8	Portreath	1884	W.Davies		W.Davies & Co.	Penzance
429	V	Ringleader *Entwistle* 82-84 W S	American	1145	185·0	37·6	22·9	Boston,Mass	1868		Y 9.88	E.Lawrence	Boston,Mass.
430	I	Rinnovato *Sanguinetti* -87 W Bk	Italian	712	152·1	33·3	23·5	Chiavari	1874	L.Gotusso	Y 3.87	A.Dall'OrsoBianchi	Spezzia
431	V	Rinnuschen *Pukse* -85 W 3m Sr	Russian	306	119·5	28·7	14·0	Mühlgraben	1874	A.Ferle		Schwalbe & Rink	Riga
432		Rinula *N. Giorgulis* W Sr	Greek	123	..	..	..	Syra	1878	L.Manassis		F[lli] Giorgulis	Syra
433		Rio W 3m Sr	British	317	129·0	30·6	11·5	Hantsport,NS.	1881	J.Davison		Philip Dolbel	Napier,N.Z.
434		Rio *Johnson* W Bn	British	305	118·9	27·7	13·0	Londonderry, [N.S.	1865	Faulkner		F.J.Ransome	London
435		Rio *Berg* -86 W Sr	Norweg'n	151	110·7	25·4	10·9	Drontheim	1874	Voight		J.Larsen	Drontheim
436		Rio Empedrado I Bk	Argentine	250	126·0	26·5	10·7	Ruhrort	1845		Y 3.87	Delpiano y GotussoHos	BuenosAyres
437	✻	Rio Janeiro *Casareggio* W 3m S	Italian	204	123·3	23·0	12·0	SestriPonente	1878	Bossi		E.Costella	Genoa
438		Rio Loge *Monro* I Bg	British	241	122·3	24·7	12·8	Garston	1869	Miller&Sons	Y 9.87	Chas.W Turner	Lyttelton,N.Z.
439		Riouw W Sr	Dutch	142	..	..	..						Riouw
440	I	Ripamare *Cichero* -88 W Bk	Italian	477	141·0	30·8	18·5	Lavagna	1871	Carbone	Y 12.88	L.Bianchi	Spezia
441		Ripka *Warsiljeff*	Russian	103	..	..	..	Impilaks	1883			M.Korniloff	Sordavala
442		Ripon (ex Steamer) (HULK) I Bg	British	1908	276·8	35·1	28·7	London	1846			G.Turnbull & Co.	Glasgow
443	✻	Rippling Wave *Barnicoat* -87 W Sr	British	117	89·6	22·8	11·5	Fowey	1869	Butson	Z 1.8[illegible]	E.Rillston	Fowey
444		RipplingWave *Anderson* W Sr (ex Odessa)	Chilian	123	93·2	25·4	9·5	New York	1868	Gates&Co.		José Noguiera	Punta Arenas
445	✻	Rising Star *Brough* 86-86 I Bk	British	835	195·9	32·6	18·5	Dumbarton	1877	Birrell,Sten- [house&Co.		H.H.Melmore	Maryport
446		Rising Sun *Becker* W Bn	American	332	121·0	28·0	11·5	N.Haven,Conn	1866			R.D.Armstrong	New York
447	I	Risoluto (ex Luigi F.) *Picasso* W Bk	Italian	550	159·4	31·5	20·4	Trieste	1872	Pico	Y 11.86	F.Picasso	Genoa
448	A	Risorgimento *di Reo* W Bn	Italian	137	86·9	22·3	12·1	Viareggio	1884	Lenzi	Z 1.8[illegible]	E.Bocci	Leghorn
449	V	Risorta *Fascetto* W Bk (ex [illegible])	Italian	515	134·5	31·8	20·0	SestriPonente	1865	F.Cadenaccio	Y 6.8[illegible]	N.Pescetto & Co.	Genoa
450	DI	Rissetto C. *Rittano* W Bk	Italian	889	177·1	34·5	22·7	Chiavari	1875	M.Tappani	Y 2.8[illegible]	S.Cichero fu G.B.	Genoa

✻ = Lloyd's Register; A = Veritas Austro-Ungarico; D = Nederlandsche Vereeniging; G = Germanischer Lloyd; H = Veritas Ellenico; I = Registro Italiano; N = Norske Veritas; R = American Record; V = Bureau Veritas.

Dans la 4me Colonne, C = Composite (membrure en fer, bordé en bois); I = Fer; Stl = Acier; W = Bois.

Les Dates du doublage en métal (Col. 11) ressortent des livres dans lesquels les navires sont respectivement classés ou des rapports des Inspecteurs du Lloyd's Register.

Pour plus amples renseignements sur les navires classés par le Lloyd's Register, voir le Lloyd's Register Book et le Lloyd's Yacht Register.

Official No.	No. in Book	Society in which classed	Ship's Name, Material and Rig. Master, and Dates of Appointment to present Owner's Service and to Ship. Ship's Late Name.	Flag	Net Tonnage	Length	Breadth	Depth	Where	When	By whom	Date of Metallings	Owners or Managers	Port
2054	451	V	Rissoluto P. *Mancini* W Bg (ex Maria Carlotta, ex Mary)	Italian	258	99·4	27·2	16·3	Savona	1862	G.B.Guastavino	[illegible]	A.Mancini & Co.	Naples
[illegible]	452	✱R	Rita *Trainer* -88 W Bkn	British	378	140·0	29·7	14·0	P.E.Island	1868	W.Richards	[illegible]	W.Richards	[illegible]
[illegible]	453		Rita W Bg	British	198	100·3	22·1	12·5	Shoreham	1852			William Robertson	Sydney
	454		Rita (ex Angela) *Cira* -86 W Sr	Spanish	195	87·6	26·6	13·7	Lloret	1854			A.Esbert y Ballés	[illegible]
	455		Rita (ex Valencia) *Castella* W Bn	Spanish	185	95·8	24·9	12·6	Valencia	18$^{55}_{74}$		[illegible]	Antonio Castellar	Barcel
[illegible]	456		Riuscito *Giannoni* W Sr	Italian	170	88·6	25·8	14·4	Sorrento	1857			G.Giannoni fu A.	Porto Fe
76	457		Riusho Maru W Bk	Japanese	342	127·2	27·3	17·1	Germany	1865			Shionoya Seibei	Kugch
54475	458		Rival *McCenney* 85-85 W Bn	British	168	106·0	24·1	12·4	P.E.Island	1865	White		M.Crean	Belfast
98876	459	✱	Rival *McGregor* 85- W Sk	British	71	72·4	20·2	10·4	Elmshorn	1884	J.Kremer		A.E.McGregor	Grimsb
110087	460		Rival *Jacobsen* 82-82 W Sr	American	123	84·2	25·6	7·2	Phippsburg Me	1872			C.V.Minott	Bath Me
	461		Rival W Bn	Brazilian	241	..	..	..	Europe	1871			Amorim & Irmão	Pernam
	462	N	Rival *Pettersen* -88 W Bk	Norweg'n	303	110·6	28·4	14·3	Grimstad	1867	J.Erlandsen		H.C.Thomassen	Christ
	463	N	Rival (STRANDED) *Jacobsen* W Bg	Norweg'n	290	106·6	28·0	15·5	Tvedestrand	1862	L.Holm		A.Speilberg	Tvedes
	464		Rival *Lie* -88 W Bn (ex Gomes de Castro)	Portug'se	133	77·4	21·0	10·5	Fão	1868	M.D.dos Santos [Borda		T. dos Santos Lé	Aveiro
54065	465	✱	River Clyde *Jones* 86-86 I Bk	British	499	153·1	27·8	17·4	Glasgow	1866	Heddrwick & Co.		J.Williams	Liverp
89964	466	✱	River Falloch *Davidson* [82-84] I S	British	1586	257·8	38·2	23·2	Port Glasgow	1884	Russell & Co.		W.D.Denny	Glasgow
55694	467	✱	River Ganges *Budge* 88-88 I Bk	British	642	192·8	28·0	17·5	Port Glasgow	1867	Kirkpatrick & Co		Hargrove & Hellon	Liverp
60680	468	✱	River Indus *Williams* 88-88 I S	British	1045	211·1	31·8	21·2	Glasgow	1868	Dobie & Co.		Hargrove & Hellon	Liverpool
56580	469	✱D	River Nith *Shearer* I S	British	1165	216·3	35·6	22·3	Newcastle	1869	Schlesinger, [Davis & Co.		Hargrove & Hellon	Liverp
21733	470		River Queen *Davis* W Sr	American	123	98·2	26·1	8·8	Haddam, Conn.	1853			J.T.Wilson	[illegible]
50463	471	✱	River Thames *Gould* 87-87 I Bk	British	501	151·2	27·5	17·7	Port Glasgow	1864	Clyde Ship Co.		River Thames Sailing Ship Co. Ltd. (J. L. Moncel)	Liverpool
29953	472	✱V	Riversdale *Farr* 88-88 C S	British	1453	225·7	38·0	23·3	Liverpool	1865	J. Andrews & Co.	[illegible]	Ship "Riversdale" Co. Ld. (C. W. De Wolf & Co.)	Calcutta
	473	✱	Riverside *Hill* -88 I S	British	1623	261·8	38·2	22·9	Port Glasgow	1886	Russell & Co.		T.C.Jones & Co.	Liverpo
69954	474		Riverside (WRECKED) *Clark* W Sr	British	109	81·7	24·5	8·5	Hopewell, N.B.	1874	J.Newcomb		P.Britt	St.And
21715	475	V	Riverside *Langdon* -88 W S	American	1147	183·1	38·1	24·0	Bath, Me.	1868	W V Moses & Sns	[illegible]	W.V.Moses & Sons	Bath Me
27580	476	✱	Riviere (LOST) *Glanville* I S	British	112	87·9	21·1	11·1	Hayle	1860	J.Pool		H.Roach & Co.	Hayle
	477		Riziyee Barri *Rejeb* W Bg	Turkish	110	..	..	..	[illegible]	1885			Rizehlee Rejeb Becker Oglou Rejeb & Co.	Rizeh
	478	V	Rjukan (WRECKED) *Hellisen* -87 W Bn	Norweg'n	209	101·3	25·4	12·8	Stavanger	1865	B.O.Hœrem	[illegible]	J.C.Knivsberg & Co.	Stavang
63940	479	✱	Roanoke *Dowey* W Bk	British	325	129·3	27·5	13·6	[illegible]	1876	[illegible]	[illegible]	S.Bingham	London
62123	480	✱	Rob the Ranter *Coode* W Sr	British	147	96·7	22·5	11·8	Kingston	1874	Geddie		J.G.Coode	Fowey
85994	481		Rob & Harry *Brown* W Sr	British	100	82·3	26·4	7·9	Queen's Co, N.B.	1882	F.Baird		George F.Baird	St.John
79900	482	R	Robbie Godfrey *Cooke* W Sr	British	164	93·8	27·0	9·9	Rockland, N.B.	1882	R.A.Chapman		C.M.A.Godfrey	Dorchest
12655	483		Robert W Bn	British	104	67·6	21·6	13·0	[illegible]	1775			R.Hewitt	Marypo
73249	484	R	Robert *Parsons* -88 W Sr	British	100	85·9	22·0	10·6	[illegible]	1876	Carnegie & [Mathews	[illegible]	Henry E.Hayward	St.John
4184	485	✱	Robert (WRECKED) *Jones* 88 W Sr	British	76	60·3	18·4	10·4	[illegible]	1840	[illegible]		T.Morgan & Co.	Carnarv
4084	486	A	Robert (ex Bolus.) *Cvetini* -88 W Bk	Austrian	423	129·9	25·6	17·7	[illegible]	1876	G.B.B.Bellaz	[illegible]	A. Mariassevich fu Bortolo	Buccari
	487		Robert (LOST) *Lund* -87 W Sr	Danish	104	76·1	19·1	9·5	[illegible]	1852			O.Petersen	Korsør
	488	N	Robert *Aaresold* 85-81 W Bk	Norweg'n	183	91·2	24·7	13·1	[illegible]	1851	Lecorneur père		N.Bugge	Tønsberg
	489	N	Robert (ex [illegible]) *Johnsen* -88 W Sr	Norweg'n	118	71·5	22·2	11·1	[illegible]	1844			P.J.&P.Berg	Laurvik
361	490		Robert *Kalning* -85 W 3m Sr	Russian	254	114·0	27·3	12·9	[illegible]	1880	J.Jacobsohn		P.Sahlit	Riga
	491	V	Robert *Mattsson* 88-88 W Bk	Swedish	331	132·2	26·5	11·8	[illegible]	1875	Fr.Hägglund	[illegible]	Nordmalings Skepps aktiebol. (C.M. Bodén)	[illegible]
	492	V	Robert *Nygren* W Bkn	Swedish	311	126·0	27·5	12·2	[illegible]	1877	A.Helmgren	[illegible]	S.Hedberg	Malmö
	493		Robert *Grönberg* -86 W Bn	Swedish	137	90·2	21·7	10·0	Fischolm	1877	H.P.Andersson		N.J.Carlsson	[illegible]
	494		Robert *Svensson* W Kff	Swedish	119	80·4	21·7	9·6	Holland				J.Svensson	Oscar
	495		Robert *Olsson* -88 W 3m Sr	Swedish	117	95·5	18·5	9·4	Gothenburg	1867	S.Abrahamsson		A.B.Elfversson	Gothen
110717	496	R	Robert A. Snow *Pillsbury* W Sr	American	165	97·8	24·4	7·8	Rockland, Me.	1886	J.L.Snow & Co.		Snow & Pearsons	Rockland
7390	497		Robert Adamson *Bell* W Bn	British	159	86·0	24·1	11·9	Sunderland	1855			C.J.Carter	Maldon
69771	498	✱	Robert Brown *Marrow* 88-88 W Sr	British	100	86·7	21·7	10·8	Warrenpoint	1874	M.Hunter		D.Marrow	Chester
21793	499		Robert Dillon *Leighton* W Bn	American	431	127·4	30·4	17·4	Tottenville NY	1868	J.C.Ellis		E.E.Hawkins	New York
	500													

✱ = Lloyd's Register; D = Nederlandsche Vereeniging; G = Germanischer Lloyd; I = Registro Italiano; R = American [Record]; A = Veritas Austro-Ungarico; H = Veritas Ellenico; N = Norske Veritas; V = Bureau V[eritas].

In the 4th Column, C = Composite (iron frame, wood planked); I = Iron; Stl = Steel; W = Wood.

The Dates of Metalling (Col. 11) are based on the records of the Books in which Vessels are respectively classed, or on the Reports of the Surveyors of Lloyd's [Register].

Further particulars of Vessels classed by Lloyd's Register will be found in Lloyd's Register Book and Lloyd's Yacht Register.

		1 Nom. Matériaux et Grèement. Capitaine et dates d'entrée au service des Armateurs actuels et à bord du navire. Ancien Nom.		5 Pavillon	6 Tonnage net	7 Dimensions. Longueur	Largeur	Creux	8 Construit. Où.	9 Quand.	10 Par qui.	11 Date du doublage en métal	12 Armateurs ou Directeurs-Gérants.	13 Port d'Armement.
501	V	Robert Dixon *Austin -87*	WS	American	1313	194·9	38·7	24·3	Damariscotta, [Me.	1873	A.Hall	Y 2.80	A.Austin	Noblesbro,Me.
502	R	Robert Dority *Lowe 75-80*	WSr	American	124	86·5	27·3	8·0	Sedgwick,Me.	1858	WG Sargent&Sn		H.W.Sargent	Sedgwick,Me.
503		Robert Gaskin	WSr	British	298	132·6	26·3	11·3	Kingston,Ont.	1863	C.W.Jenkins		Samuel Fraser	Kingston,Ont.
504	R	Robert Graham Dun *Crowell 85-88*	W3mSr	American	566	146·1	35·5	15·5	Bath,Me.	1881	Goss&Sawyer		P.H.Crowell	New York
505		Robert H. Mitchell *Mitchell 80-88*	WSr	American	155	104·0	29·6	7·0	Dorchester [County,Md.	1875			F.P.Murphy	Baltimore
506	R	Robert H. Parker *Steelman*	W3mSr	American	361	139·8	33·0	11·5	Milford,Del.	1881	T.Carlisle		J.B.Steelman	Somer's Pt.,N.J.
507	R	Robert H. Rathbun *Crowe*	W3mSr	American	348	131·1	32·0	11·7	Bath,Me.	1881	Goss, Sawyer& [Packard		R.H.Rathbun	Pth.Amboy,N.J.
508	*	Robert Hine *Rees 87-87*	WBk	British	319	119·8	27·3	16·4	Sunderland	1868	Davison	Y 12.87	W.Barrett (Bevan, Bros. & Co.)	Liverpool
509	R	Robert J. Barr *Josephy -88*	W3mSr	American	412	141·8	35·0	12·3	Camden,N.J.	1882	Burton& [Vanneman	Y 10.85	H.C.Galbraith	Philadelphia
510		Robert Kerr (DISMANTLED)	WBk	British	1191	190·6	38·4	23·7	Quebec	1866			W.H.Soule	Vancouver,B.C.
511	RV	Robert L. Belknap *Staples*	WS	American	2251	264·8	43·8	29·3	Rockport,Me.	1884	Carleton, [Norwood&Co.	Y 1.87	Carleton,Norwood & Co.	Camden,Me.
512		Robert Morgan *Crosby*	W3mSr	American	553	151·5	35·0	14·5	Madison,Conn	1883	C.M.Miner&Son		F.W.Benedict	New Haven,Con.
513	*	Robert Morris *Jones 55-76*	WSr	British	135	92·7	22·2	11·8	Portmadoc	1876	Williams	Y 1.87	D.Morris	Carnarvon
514		Robert Mowe *Petersen*	WBn	American	163	97·0	25·4	10·5	Eastport,Me.	18 54/70			H.Trowbridge's Sons	New Haven,Con.
515		Robert Mowe, jun.	WSr	British	122	80·0	21·9	8·2	StGeorge,N.B.	1859			W.B.Newbery	Gibraltar
516		Robert P. King *(or Captain Williams)*	WSr	American	113	90·6	23·6	7·8	Philadelphia	1847			E.Orison	New York
517	*DV	Robert Rickmers *Hesse 79-88*	IBk	German	2211	279·3	42·0	24·4	Greenock	1888	Russell & Co.		R.C.Rickmers	Bremerhaven
518		Robert Ross (STRANDED)	WSr	British	124	102·5	26·0	7·6	S.Andrew's NB	1873	R.Ross		N.C.Corbett	St.Andrew's,N.B.
519	R	Robert Ruff *Stilphen*	W3mSr	American	208	120·6	29·0	8·2	Bath,Me.	1871		Y 4.87	A.H.Voss	Shieldsbro.Miss.
520	V	Robert S. Besnard *Andrews 81-84*	WBk	British	1200	191·0	38·8	23,0	Eatonville, NS	1882	D.R.&C.Eaton	Y 7.83	Scammell Bros.	Parrsboro',N.S.
521		Robert Searles *Petty -88*	W4mSr	American	578	183·8	38·0	13·6	PortBlakely	1888	HallBros.		J.Tuft	SanFrancisco
522	V	Robert Sudden *Petty -88*	WBkn	American	585	167·0	37·7	14·2	PortBlakely	1887	HallBros.		J.J.Smith	SanFrancisco
523		Robert T. Clark *Sanderson*	W3mSr	American	180	118·0	27·9	8·3	PortJefferson, [N.Y.	1873			J.A.Sanderson	SagHarbour
524	R	Robert W. Dasey *Tracy*	W3mSr	American	339	130·2	34·0	10·0	Milton,Del.	1880	D.H.Atkins		Thos.Winsmore	Philadelphia
525	*	Robert the Devil *Germany*	WK	British	86	79·6	20·5	10·6	Brixham	1884	S.Dewdney&Ss.		J.Germany,jun.	Grimsby
526		Roberto	WSr	Spanish	120	96·8	23·9	9·1	France	1885			J.O.Busquets	Majorca
527	V	Robertsfors *Sandborg 75-83*	WBk	Swedish	764	170·6	31·9	19·5	Sikeå	1883	O.P.Åberg	Y 3.87	Fr.Jacobsen	Robertsfors
528		Robie Burns *Livingstone*	WBn	British	201	104·5	24·7	12·9	P.E.Island	1868	Davies		W.H.Turner	Folkestone
529		Robin *Havner*	WBn	British	309	111·0	28·2	10·8	Maine,U.S.	1857			Annie McKidley	St.Andrew's,N.B.
530		Robin *Le Rues 81-86*	WBn	British	151	102·2	21·1	12·3	Paspeliac, [Canada	1866			G.Le Gros	Jersey
531	*	Robin Hood *Ryder -88*	WK	British	94	77·8	21·1	10·7	Whitby	1887	Whitby&Robin- [hdsBaySBCo		W.Ryder	Scarbro'
532		Rochemont	WSr	British	396	126·2	32·5	14·9	Liverpool,N.S.	1888	A.Sponagle		J.S.Sponagle	Liverpool,N.S.
533	N	Rock City *Stangebye*	WBk	Norweg'n	803	174·9	33·1	20·6	Quebec	1868	McKay& [Warner	Y 3.87	O.A.Stangebye	Fredriksstad
534		Rock City *Olsen 79-88*	WBk	Norweg'n	573	161·1	32·0	18·4	Quebec	1853		Y 8.82	W.M.Michaelsen	Moss
535	R	RockTerrace *Atkinson* (ABANDONED)	WS	British	1769	228·2	41·2	25·2	Portland, N.B.	1875	D.Lynch	Y 8.87	Howard D.Troop	St.John,N.B.
536	*	Rock of Ages *Smith*	WK	British	88	76·5	20·5	10·8	Whitby	1885	Whitby&Robin [Hood'sB.SB&Dk.Co		G.M.Smith	Grimsby
537	*	Rockhurst *Jones*	IS	British	1400	229·2	38·3	22·5	Dumbarton	1876	A.M'Millan&Ss.		W.R.Price & Co.	London
538	R	Rocky Glen *Bray 82-82*	WBn	American	380	120·0	30·0	16·1	RobbinstonMe	1866	Don	Y 10.88	J.S.Emery & Co.	Boston,Mass.
539	*	Roderick Dhu *Roddick 83-82*	IS	British	1642	257·1	40·2	22·8	Sunderland	1873	Mounsey& [Foster		Williamson,Milligan & Co.	Liverpool
540		Roderick Dhu *Forsyth 80-86*	WBn	British	163	102·6	24·6	11·4	Auckland,N.Z.	1875			Robert Cran	Maryboro'
541	A	Roditelji Z. *Vicinich*	WSr	Austrian	205	95·1	23·9	13·4	Lussinpiccolo	1886		Y 11.85	G.Zacevich fu N.	Lussinpiccolo
542	*	Rodney *Barrett 72-87*	IS	British	1447	235·6	38·4	22·6	Sunderland	1874	W.Pile&Co.		Devitt & Moore	London
543		RodneyParker *Higgins*	W3mS	American	418	130·0	33·0	10·0	Mystic,Conn.	1874			C.F.Treat	Marblehead,Mass.
544		Rodolphe	WSr	French	121	85·6	23·9	9·9	Ajaccio	1877			W.Vinchaut	Bourbon
545		Roecliff *Barnett 78-78*	WBn	British	147	87·2	22·7	12·5	Sunderland	1860			J.Adamson	Sunderland
546		Rogaciano	WBn	Spanish	122	88·9	20·9	8·4	Tayabas	1872			Joaquin Armada	Manila
547		Rogaland *Marcussen -88*	WBn	Norweg'n	142	92·9	23·6	11·1	Stavanger	18 48/77	K.Kaisen		N.T.Landaas	Stavanger
548	DN	Rogate *Olsen -88*	WBk	Norweg'n	337	116·7	28·1	15·6	Risör	1864	E.Pettersen		Joh.Spörck	Fredrikshald
549	N	Roger Drury *Delay*	W3mSr	American	343	129·2	32·0	11·0	East Boston	1872			J.H.Conant & Co.	Boston,Mass.
550	R	Roger Moore *Gilkey* *(or Isaac L. Clarke)*	WSr	American	318	120·8	32·0	14·2	Brideton,N.J.	1874			W.R.Gilkin	Wilmington

*=Lloyd's Register; D=Nederlandsche Vereeniging; G=Germanischer Lloyd; I=Registro Italiano; R=American Record; A=Veritas Austro-Ungarico; H=Veritas Ellenico; N=Norske Veritas; V=Bureau Veritas.

Dans la 4me Colonne, C=Composite (membrure en fer, bordé en bois); I=Fer; Stl=Acier; W=Bois.

Les Dates du doublage en métal (Col. 11) ressortent des livres dans lesquels les navires sont respectivement classés ou des rapports des Inspecteurs du Lloyd's Register.

Pour plus amples renseignements sur les navires classés par le Lloyd's Register, voir le Lloyd's Register Book et le Lloyd's Yacht Register.

Official No.	No. in Book.	Registry in which classed.	Ship's Name. Material and Rig. Master, and Dates of Appointment to present Owner's Service and to Ship. Ship's Late Name.	Flag.	Net Tonnage.	Dimensions. Length.	Breadth.	Depth.	Built. Where.	When.	By whom.	Date of Metalling.	Owners or Managers.	Port of Registry.
	551		Roger de Flor WBn	Spanish	195	91·5	22·8	10·9	Arens	1857			M.Alsina	Barce…
[illegible]	552		Rogers *Rogers* W3mSr	American	253	117·0	32·0	9·5	Millbridge, Me	1888			D.J.Sawyer	Jonesp…
[illegible]	553	✳	Rohilla *Sunter 88-8* I S	British	1599	258·0	38·2	23·2	PortGlasgow	1885	Russell&Co.		Foley & Co.	London
	554	✳V	Rohilla *Breskwoldt 82-86* C Bk	German	985	201·0	33·2	20·5	Glasgow	1865	A.Stephen&Sns.	[illegible]	L.F.Mathies & Co.	Hambur…
	555	V	Røhl *Mortensen* WBn	Danish	172	97·3	21·4	10·8	Värnamäs	1877	A.T.Bring	[illegible]	Akt.Selsk.(S.B.Hansen)	Nordby…
31120	556		Rohomanee WBg	British	199	86·0	22·3	14·3	Chittagong	1875			Mohomed Ismail Mohomed	Colombo
443	557	V	Roja *Ahl 84-87* W3mSr	Russian	292	117·3	27·4	13·4	Roja	1884	K.Oschmann-[illegible]		C.Wentenberg	Riga
47542	558	✳	Rokeby Hall *Staley 88-88* (ex Lloyd Rayner) I S	British	1044	202·6	33·0	22·5	Liverpool	1863	T.Vernon&Son		Nicholson & McGill	Liverpo…
74161	559	✳	Rokewood *McCafferty 88-88* WBn	British	203	103·8	24·9	12·9	P.E.Island	1876	Longworth	[illegible]	W.J.Woodside&Co.	Belfast
93789	560	✳	Roland *Finlayson* WBkn	British	348	130·4	29·9	13·9	Summerside, P.E.I.	1883	J.Lefurgey	[illegible]	John Lefurgey	[illegible]
	561	G	Roland *Jülicher 63-82* WS	German	1343	197·1	39·0	23·6	Bremerhaven	1879	H.F.Ulrichs	[illegible]	D.H.Watjen & Co.	Bremen
[illegible]	562	I	Rolando Secondo *Rolando* WBk	Italian	559	143·7	32·1	20·5	Savona	1875	G.Calamaro	[illegible]	N.G.Rolando	[illegible]
	563	NV	Rolf *Jörgensen -88* WS	Norweg'n	1211	206·1	39·4	23·0	Bergen	1876	Brunchorst & Dekke	[illegible]	G.C.Hansen	[illegible]
	564	NV	Rolf *Hansen* W3mSr	Norweg'n	362	131·4	29·4	13·6	Arendal	1885	C.Sörensen	[illegible]	C.Sörensen	Arendal
	565		Rolf *Söderberg -84* WSr	Swedish	134	97·4	21·4	9·6	Svanecke	1856			J.P.Jönsson	Landskr…
	566	N	Rolfsöen *Schade* WBk	Norweg'n	289	106·6	38·5	13·7	Finland	1863			S.B.Schade	Sarpsbor…
74663	567	✳	Rollo *Johnson* I S	British	902	216·6	32·7	18·0	Glasgow	1876	Dobie&Co.		Currie & Co.	Melbour…
	568	DN	Rollo *Mangor* WBk	Norweg'n	597	163·4	31·8	18·4	Grimstad	1883	J.T.Tryde	[illegible]	J.Lassen	Arendal
50373	569	RV	Roma *Kennedy -88* I Bk	British	633	205·8	25·0	16·6	Glasgow	1865	A.Stephen&Sns.		Alexander Major	Quebec
59612	570		Roma *Breen* WSr	British	132	95·7	24·3	10·9	P.E.Island	1870			W.Tyrrell	London…
1201	571	I	Roma *Cafiero -88* WBk	Italian	517	138·1	30·6	19·8	Alizauri	1878	G.Starita	[illegible]	M.Izzo & F.Cacace	Castell…
1071	572	I	Roma *Serazzo 88-88* W3mSr	Italian	311	120·0	26·8	15·4	SestriPonente	1880	A.Carlevaro	[illegible]	M. & A.Dall'Orso fu G.	Spezia
633	573		Roma (ex Eole) *Napoli* WBk	Italian	269	105·6	24·3	11·9	Nantes	1871	Tillé	[illegible]	C.Napoli	Catania
	574	NV	Roma *Holberg -88* WBk	Norweg'n	549	138·7	31·9	19·0	Sestri Ponente	1867	A.Briasco di L	[illegible]	H.&C.J.Skougaard	Langesu…
	575	N	Roma *Tønnesen -84* WBk	Norweg'n	495	112·8	30·6	16·7	Porsgrund	1875		[illegible]	J.O.Maberg	Farsund
2589	576		Romagna *Pittoni* WSr	Italian	115	..	..	..	Sestri	1888	G.Peragallo		G.Peragallo	Genoa
53723	577		Roman WBk	British	330	114·7	27·4	15·9	Prussia	1854			William F.Ogilvie	Georget…
	578	✳	Roman I Barge	British	100	75·3	19·5	6·9	London	1884	Phillips& Graves		Phillips & Graves	London
54658	579	✳	Roman Empire *Judson 88-88* I S	British	1542	231·7	38·8	23·4	W.Hartlepool	1866	Pile,Spence&Co		G.Duncan & Co.	London
66584	580	V	Romance *Toye -87* WBk	British	593	151·5	32·0	18·5	Horton,N.S	1873		[illegible]	John W.Barss	Windsor,…
21616	581		Romance *Brown 79-79* WBn	American	204	109·0	24·1	10·9	Baltimore	1853		[illegible]	Navassa Phosphate Co.	New Yo…
	582		Romancito (ex Tres Hermanos) *Valero* WBn	Spanish	130	87·6	22·8	11·1	Villajoyosa	1869			R.B.Guarner	Alicante
70436	583	✳	Romanoff *Mearns 62-84* I S	British	1226	222·1	36·2	22·2	Aberdeen	1874	W.Hood&Co.		A.Nicol & Co.	Aberdeen
74304	584	V	Romanoff *Dotty* WBk	British	1050	175·0	37·4	22·0	Shelburne,N.S	1876	E.P.Young	[illegible]	Augustus F.Stoneman	Yarmo…
	585	V	Romeo *Lubben 83-85* WBg	German	299	109·3	24·3	13·0	Brake	1877	J.F.Nicolai	[illegible]	J.H.Nicolai	Brake
29	586	A	Romolo (ex Nuova Zora) *Soppa* WSr	Austrian	168	88·5	22·5	12·5	Lussinpiccolo	1871	G.S.Cattarinich	[illegible]	R.Soppa fu R.	Lussin…
78356	587	✳	Romsdal *Jarman -88* I 4mS	British	1827	275·9	41·1	23·5	Greenock	1877	R.Steele		J.& A.Allan	Glasgow
1174	588		Romulus *Gundril* WBg	British	164	99·3	22·3	13·1	Bridgwater	1854			H. & E. Crundall	Dover
	589	N	Romulus *Nord -86* WBk	Norweg'n	665	153·8	32·0	19·5	St.John,NB.	1859	Olive	[illegible]	Sophus Erichsen	Porsgru…
60063	590	✳	Rona *Oppenheim 76-4* C S	British	638	156·4	29·6	18·6	Glasgow	1867	A.Stephen&Sns	[illegible]	W.Cliff & Son	Liverpool
85073	591		Rondo *Somerville* WSr	British	123	89·0	26·6	7·7	Belleisle,N.B.	1882	P.McIntyre		Peter McIntyre	St.John,…
	592	N	Rondo (ex Emily Lowther) *Olsen* WBk	Norweg'n	793	161·3	34·8	20·0	Argyle,N.S.	1870	N.W.W.Hogg		M.Olsen	Porsgrund
[illegible]	593	✳	Rookwood *Jewell 71-73* I Bk	British	740	193·9	30·4	18·9	Cardiff	1869	C.Hill&Sons		C.Hill & Sons	Cardiff
	594		Rora *Sundvik 76-* WSr	Russian	228	..	..	..	Lökö	1869			J.Malen	Nykyrk…
62722	595	✳	Rosa *Bale* StlSr	British	52	75·9	18·1	9·7	Belfast	1885	Workm'n Clarke & Co.		Belfast Harbour Commissioners	Belfast
90123	596		Rosa WBn	Brazilian	195	..	..	..					Cia.E.F.Bahia & Minas	Rio…
	597	V	Rosa (ex Rose des Alpes) *Petersen* WSr	Danish	136	91·0	20·6	12·2	Dunkirk	1863	P.Bayaert		T.Gunnarssen	[illegible]
	598		Rosa WSr	French	109	78·1	19·3	11·8	Dunkirk	1866	Pyotte Bayaert		Dupius Robial	St.Brieu…
3834	599	I	Rosa (ex Luigina E.) *Vaccari -87* WBk	Italian	587	140·7	31·4	20·4	Sestri Ponente	1870	G.Traverso	[illegible]	G.Vaccari	Genoa
3651	600	I	Rosa (ex Rosa Danovaro) *Questa -88* WBk	Italian	535	114·0	32·4	19·4	Prà	1866	M.Tixi	[illegible]	F.Sanguinetti	Spezia

✳=Lloyd's Register; D=Nederlandsche Vereeniging; G=Germanischer Lloyd; I=Registro Italiano; R=American Re…
A=Veritas Austro-Ungarico; H=Veritas Ellenico; N=Norske Veritas; V=Bureau Ve…

In the 4th Column, C=Composite (iron frame, wood planked); I=Iron; Stl=Steel; W=Wood.

The Dates of Metalling (Col. 11) are based on the records of the Books in which Vessels are respectively classed, or on the Reports of the Surveyors of Lloyd's Re…

Further particulars of Vessels classed by Lloyd's Register will be found in Lloyd's Register Book and Lloyd's Yacht Register.

No.	Registre dans lequel le navire est classé	Nom. Matériaux et Gréement. Capitaine et dates d'entrée au service des Armateurs actuels et à bord du navire. Ancien Nom.	Pavillon.	Tonnage net.	Dimensions. Longueur	Largeur	Creux	Construit. Où.	Quand.	Par qui.	Date du doublage en métal.	Armateurs ou Directeurs-Gérants.	Port d'Armement.
601	I	**Rosa** (now "Teresinella") W Bk	Italian	484	139·7	27·9	18·4	Cassano	1874	F.DeRosa	Y 3.87	EredìMelch.Ciampa	Castell. di Stab.
602	A	**Rosa** Salvatore -88 W Bk (ex Rosa [illegible])	Italian	417	128·2	30·3	19·0	SestriPonente	1869	F[lli] Cadenaccio	Y 3.86	Salustro & Co.	Naples
603		**Rosa** P. Luisan W Bn (ex [illegible])	Italian	130	87·6	22·3	12·2	Castellamare [di Stabia	1874	M.Paturzo	Y 12.81	G.La Rocca & Oddo	Palermo
604		**Rosa** Sjoman -79 W Bk	Russian	388	..	..	..	Börga	1859			C.W.I.Sundman	Helsingfors
605		**Rosa** W Bk	Spanish	326	120·4	29·3	15·2	Corunna	1869		Y 3.88	José Romani	Villagarcia
606		**Rosa** (ex Juanito) W Bk	Spanish	105	76·4	20·9	10·6	Corunna	1859			Manuel Romero	Corunna
607	V	**Rosa** Ohman W Bk	Swedish	516	136·4	30·3	17·9	Umea	1871	E.A.Löfgren	Y 9.84	A.Scharin	Umeå
608		**Rosa Abbott** W Bn	British	163	92·9	24·2	10·0	Bristol,U.S.	1851			W.D.Campbell	Georget'n. Dem.
609		**Rosa Baker** Howard W Bn	American	103	82·0	23·6	8·9	Scituate,Mass.	1867			H.Smith	Boston,Mass.
610	I	**Rosa C.** Repetto -88 W Bk	Italian	586	160·7	30·4	21·1	Voltri	1873	A.Fabiano	Y 12.84	B.Chiesa	Genoa
611		**Rosa Harriette** Roberts [73-77] W Sr	British	117	84·0	20·6	11·7	Newport	1864	Cooke		J.Lindsay & Son	Montrose
612		**Rosa I.Carmen** Bilbao W Lench	Spanish	261	101·7	28·8	15·2	Barcelona	1854			Zulueta y sobrino	Havana
613	I	**Rosa Maddalena** Scotto W Bk	Italian	729	160·5	33·4	21·7	Savona	1874	C.Ciarlo	Y 9.86	G.Viglianzoni	Savona
614	I	**Rosa Madre** Leboffe -83 W Bk	Italian	911	169·6	34·1	23·6	Gaeta	1876	L.Contreras	Y 7.86	L.Leboffe fu G.	Gaeta
615	I	**Rosa Madre** Rapallo W Bk (ex Buona Famiglia)	Italian	479	130·2	31·5	19·4	Sestri	1868	F.Cadenaccio	Y 11.88	F[lli] Sciaccaluga	Genoa
616	I	**Rosa Madre** Marsella W Bg (ex Nina Emilia)	Italian	257	104·0	27·5	16·5	Procida	1866		Y 6.86	A Mazzella	Naples
617		**Rosa Madre** W Sr	Italian	110	85·3	22·3	10·8	TorredelGreco	1881	F.Lubrano		F[lli] Albanese	Naples
618		**Rosa Maria** (ex Herminea) W Bk	Peruvian	132	88·9	21·2	10·1	UnitedStates				J.Ferrando	Callao
619	❊	**Rosa Meek** Tune -78 W Sr	British	91	84·0	22·0	9·8	Goole	1878	Cottingham	Y 6.86	A.Meek	Goole
620	R	**Rosa Mueller** M'Lean W 3 m Sr	American	268	120·0	30·9	10·3	E.Deering, Me.	1882	G.Russell	Y 8.88	H.M.Sargent	Portland,Me.
621	❊I	**Rosa Rocca** Repetto W Bk	Italian	689	161·7	33·1	21·1	Lavagna	1873	D.Briasco fu G B	Y 6.87	Lavello & Rocca	Genoa
622	V	**Rosa S.** (ex Louis T.) Segaberta W Bk	Italian	374	121·7	28·7	16·8	Nantes	1866	Sevestre	Y 2.81	F.Segaleria	Genoa
623	I	**Rosa V.** (ex Albissola P.) Varaldo W Bk	Italian	526	142·7	31·4	20·2	SestriPonente	1872	F.Cadenaccio	Y 9.87	G.B.Varaldo	Savona
624		**Rosa & Isabel** Dillwitz W Bk (NOT SEAGOING)	German	407	138·0	29·6	16·2	Hamburg	1865	E.Dreyer		A.Burchard	Rostock
625	V	**Rosalba** Sanguinetti W Bk (ex La Bonita)	Italian	403	124·6	30·5	18·6	SestriPonente	1866	G.Peragallo	Y 7.81	F[lli] Sanguinetti fu G	Spezia
626		**Rosales** I Sr	Argentine	247	134·0	30·3	7·8	PortGlasgow	1884	Murdch&Marry		LaPlatenseFlotillaCo.	Buenos Ayres
627	IV	**Rosalia** Gargiulo -87 W Bk	Italian	428	125·0	30·0	19·2	Castellamare [di Stabia	1866	L.Trambarulo	Y 12.86	F[lli] Petrelluzzi	Castell. di Stab.
628	A	**Rosalia** Miglione -88 W Bg	Italian	259	102·0	25·9	16·8	Cassano	1860	A.Maresca	Y 6.82	A.Guida	Naples
629	DV	**Rosalie** (ex B.Hilton) Mörkx W Bk	Swedish	973	175·2	32·7	21·6	Shelburne,N.S	1874	J.L.M'Kelpine		Mörkx	Umeå
630		**Rosalie** Sealey W Bn	British	199	101·6	24·5	15·5	P.E.Island	1848		Y 12.86	G.S.Knight, jun.	London
631		**Rosalie** Oliver W Sr	French	131	85·0	20·7	11·3	Calais	1870	A.Coffre		A.Coffre	Calais
632	V	**Rosalie** Andersen -88 W Sr	German	101	86·3	21·0	9·8	Apenrade	1879	N.Jacobsen		H.D.Andersen	Heiligenhafen
633	V	**Rosalie** Sihlemann W Sr	Russian	238	108·2	26·2	12·9	NeuSalis	1875	G.Grass		G.Schnore	Riga
634		**Rosalie Ahrens** Wiedemann W Bk	German	274	110·5	26·7	13·9	Ribnitz	1862	J.Miedbrodt		J.D.Capell Nachf.	Rostock
635	❊	**Rosalind** Russell 82-85 I Bk	British	352	140·7	27·6	13·0	Aberdeen	1885	A.Hall&Co.		J.Sutcliffe & Son	Grimsby
636	❊	**Rosalind** Holder W K	British	93	80·6	21·1	10·6	Brixham	1885	J.W.&A.Upham		C.Hellyer	Hull
637		**Rosalind** Le Ballister W 3 m Sr	American	274	132·0	32·6	9·5	PtBlakelyWsh	1883	HallBros.		J.E.Le Ballister	San Francisco
638		**Rosario** W Bg		278	114·2	23·4	15·5	Cardiff	1861	Batchelor&Co.	Y 8.86		
639	V	**Rosario** Howard -88 W Sr	American	149	104·9	26·5	9·1	SanFrancisco	1878	M.Turner	Y 1.87	Oceanic S.S.Co.	San Francisco
640	I	**Rosario** Lubrano W Bg (ex Lauro Casnet)	Italian	301	110·2	28·3	17·9	Castellamare	1868	A.Bonifacio	Y 7.88	G.Lubrano&L.Palumbo	Castellamare
641		**Rosario** (ex J.B.V. No.1) Ros W Bk	Spanish	224	98·1	26·5	12·7	Martigues	1863	Fiastre		M.Goday y Gual	Villagarcia
642		**Rosario** Cubas -88 W Bn	Spanish	139	95·1	24·1	11·0	Palma,S.Cruz	1862	S.Arozina& [Lernos		Rodriguez & Co.	Palma
643	V	**Rosario I.** (ex Basilio) Denaro W Bn	Italian	106	84·3	23·2	10·5	Limite	1874	N.Pichiotti		P. Caltabiano Indelicato	Catania
644	❊	**Roscrana** Menmuir 87-87 I Bk (ex Robert Lees)	British	1195	210·5	34·8	22·8	Liverpool	1863	T.Vernon&Son		J.H.Worthington & Co	Liverpool
645		**Rose** D.J.Webster -86 I L	British	202	135·0	22·0	8·8	Shanghai	1886	Boyd & Co.		D.J.Webster	Shanghai
646		**Rose** Harrison 81-81 W Bg	British	186	88·2	24·7	14·8	Sunderland	1848			J.Mills	Whitby
647		**Rose** Pont 76-76 W Bn	British	164	96·9	23·9	12·5	Whithorn	1870	M'William		Holden & Gaun	Sunderland
648	V	**Rose** Darfarich -88 W Bk (ex O'Barris)	Austrian	434	126·3	27·3	17·8	Savona	1868	G.Tixi	Y 3.87	S.Kraus fu G.	Trieste
649	V	**Rose** Vanpouille W Sr	French	143	91·9	22·0	11·3	Gravelines	1877	Collin		A.Maurois	Dunkirk
650	V	**Rose** Pedron W Sr	French	111	69·9	19·7	10·8	Nantes	1856	Chauvelon		F.Pedron	Nantes

❊ = Lloyd's Register; D = Nederlandsche Vereeniging; G = Germanischer Lloyd; I = Registro Italiano; R = American Record;
A = Veritas Austro-Ungarico; H = Veritas Ellenico; N = Norske Veritas; V = Bureau Veritas.

Dans la 4me Colonne, C = Composite (membrure en fer, bordé en bois); I = Fer; Stl = Acier; W = Bois.

Les Dates du doublage en métal (**Col. 11**) ressortent des livres dans lesquels les navires sont respectivement classés ou des rapports des Inspecteurs du Lloyd's Register.

Pour plus amples renseignements sur les navires classés par le Lloyd's Register, voir le Lloyd's Register Book et le Lloyd's Yacht Register.

Official No.	No. in Book	Registry in which classed	Ship's Name. Master, and Dates of Appointment to present Owner's Service and to Ship. Ship's Late Name.	Material and Rig	Flag	Net Tonnage	Length	Breadth	Depth	Where	When	By whom	Date of Metalling	Owners or Managers	Port
	651	V	Rose *Geerds 77-86*	W Bk	German	310	122·5	26·6	11·7	Brake	1877	J.F.Stranger [Sohn	[illegible]	M.Geerds	Emde
	652	RV	Rose C. *Sattis* (ex Louisa Grant)	W Bk	British	419	122·7	28·2	16·0	Swinemunde	1874	O.Hartwich	[illegible]	Judson Edgett	PortS
110621	653	R	Rose Easterbrook *Ross*	W3mSr	American	619	157·5	35·0	20·2	N'b'ryPt Mass	1884	G.E.Currier		B.F.Davis	Boston
82987	654	✻	Rose Hill *Jones 84-86*	I 3mBn	British	196	117·0	22·1	12·0	Harrington	1875	R.Williamson&S		W.Price & Co.	Working
110485	655	RV	Rose Innes *Marwick*	W Bk	American	835	163·2	35·9	19·6	Bath, Me.	1881	W.Rogers	[illegible]	G.S.Hunt	Portlan
59126	656		Rose M. *Crall*	W Bk	British	366	125·4	28·4	12·9	Tatamagouche (N.S.	1865	A.Campbell		James C. Ellis	Newcas
110153	657		Rose McDevitt *Riggings*	W Sr	American	116	100·3	17·4	8·5	Hamburg, Pa.	1873	Wolfe		D.McDevitt	Philade
63147	658	✻	Rose of Devon *Dawe 74-83*	W Bk	British	388	143·0	27·5	16·2	Plymouth	1871	Shilston	[illegible]	W.H.Shilston	Plymou
19140	659		Rose of England *Evans -88*	W Bk	British	395	133·0	27·2	17·5	Hartlepool	1858	Denton Gray & Co.	[illegible]	L.Tulloch	Swanse
82806	660	✻	Rose of Sharon	W Sk	British	86	77·5	20·8	10·3	Gainsboro'	1885	E.Fidell		R.Winterflood	Grimsb
72539	661		Rose of Torridge *Evans 89-86*	W Sr	British	100	88·8	21·4	10·7	Bideford	1875	Cox	[illegible]	R.Hughes	Fowe
10770	662		Roseau *Enchern*	W Bk	British	361	127·8	26·1	16·4	Jersey	1857	T.LeHuguet	[illegible]	W.Morphy	Londo
63113	663	✻	Rosebery *Ellis -87*	W Trawler	British	95	82·8	21·3	10·9	Brixham	1877	S.Dewdney &Sn		J.W.Haley	Hull
73246	664	✻	Rosebud *Collie 77-76* (WRECKED)	W3mSr	British	341	133·7	26·7	14·1	Peterhead	1876	Carnegie	[illegible]	J.Stewart & Co.	Peterhe
47074	665	✻	Rosebud *Beynon -83*	W Sr	British	160	101·5	22·5	12·7	Fowey	1865	Nickels	[illegible]	J.B.Pinkham	Plymou
73140	666	✻	Rosedale *Reid 84-84*	W Bk	British	496	148·3	30·4	17·9	Barnstaple	1876	Westacott	[illegible]	J.Tedford & Co.	Belfast
48491	667		Rosedale (ex Isabella)	W Bk	British	234	107·0	23·6	11·0	Brewer, U.S.	1856			David Lemon	Bridgw
5607	668		Roseland (HULK) *Pirran*	W Sr	British	100	..	..	..	Fowey	1835			J.G.Allen	Falmou
	669	NV	Rosenberg *Thorsen 79-88*	W Bk	Norweg'n	860	180·5	35·5	21·6	Stavanger	1875	C.Knudsen	[illegible]	Ths.S.Falck	Stavan
	670	V	Rosendaël *Everard*	W Bk	French	324	114·5	24·6	14·4	Dunkirk	1867	D. van Cauwen-[berghe		Faucompré	Dunkirk
68220	671	✻	Rosendale *Hughes 85-85*	W Sr	British	134	92·1	22·6	12·1	Chester	1874	Ferguson		W.Postlethwaite	Barrow
20104	672	✻	Roseneath *Browne 88-88*	I Bk	British	735	187·0	31·7	19·4	Glasgow	1857	Barcly,Curle&C		J.Spencer & Co.	Glasgo
17352	673		Roseneath	W Bg	British	236	95·5	24·0	15·0	Cochin	1854			Goolam Mohideen Saibo	Penang
80626	674		Roseneath *Wilson*	W Sr	British	110	84·6	24·0	9·7	Shelburne, NS	1882	B.Hines		Byron Hines	Yarmou
	675	NV	Rosenius *Hoyer -88*	W Bk	Norweg'n	542	142·6	32·2	18·1	Arendal	1880		[illegible]	Hans.H.Pettersen	Arend
	676		Rosenklippe *Swendsen*	W3mSr	Norweg'n	137	95·1	22·3	11·2	Stavanger	1856	K.Kaisen		H.A.Svendsen	Stavan
58225	677	✻	Rosevear *Parsons*	W Bn	British	167	99·6	22·7	12·7	Scilly	1878	W.M.Gluyas	[illegible]	C.T.Bennett	St.John
78358	678	✻	Rosie *Bate 81-87*	W Sr	British	89	83·0	21·0	9·9	Appledore	1885	R.Cock	[illegible]	R.Cock	Bidefor
88823	679	✻	Rosie *Clark*	W K Barge	British	78	85·2	20·8	6·9	Littlehampton	1886	J.&W.B.Harvey		J.&W.B.Harvey	Littleha
10189	680		Rosie Welt *Welt -74*	W S	American	1364	200·0	39·8	24·1	Waldoboro' Me	1874	Reed, Welt&Co.	[illegible]	Yates & Porterfield	New Yo
110107	681		Rosie & Adra *Small*	W Sr	American	112	87·8	27·2	7·0	Bristol, Me.	1873			M.Gardner	Machia
85816	682	✻	Rosina *Stafford 72-84*	W Bn	British	185	110·6	23·6	12·7	Barnstaple	1884	W.Westacott	[illegible]	A.Goodridge & Sons	St.John
875	683	A	Rosina *Evona -88* (ex Susanna M. Jessup)	W Bk	Italian	409	125·9	30·7	18·3	Galaxidi	1872		[illegible]	D.Russo & P.Martino	Messin
911	684		Rosina (WRECKED) (ex Marini B, ex Saterina)	W Bn	Italian	171	89·9	24·6	14·4	Savona	1859	Calamaro		P.Gallo fu P.	Palerm
3227	685	I	Rosina *Viglienzoni -88*	W Bn	Italian	152	94·2	23·0	12·3	Veltri	1882	G.Debarbieri	[illegible]	N.Piccardo	Genoa
	686		Rosina	W Bn	Italian	134	..	..	..	Gaeta	1861	L.Contreras		G.Cocozza	Procid
	687	N	Rosita *Kolberg*	W Bk	Norweg'n	370	128·5	29·0	16·1	Christiania	1875	H.C.Hansen	[illegible]	Chr.Brinch	Christi
	688	✻	Rosita *Sharmann*	I Sr	Urugu'y'n	65	75·5	25·4	6·2	London	1884	Forrestt&Son		W.B.Cranwell & Co.	Monte V
	689	V	Roska *Nielsen 86-88*	W Sr	Danish	150	96·3	20·9	10·6	Troense	1874	C.J.Jensen		N.Nielsen	Svendb
	690	N	Roska *Larsen -88* (ex Helen McGaw)	W Bk	Norweg'n	683	142·2	31·9	21·4	Medford, Mass.	1847		[illegible]	P.A.Christensen	Langes
	691	V	Roska	W Bk	Norweg'n	616	138·1	33·8	21·4	Brahestad	18 67/82	E.Klubb		Actieselskabet "Roska" (Chr.Hansen)	Laurvik
	692	DN	Röskva *Hopp*	W Bg	Norweg'n	243	113·9	26·5	12·4	Kragerö	1876	Danielsen	[illegible]	C.Gjerlöff	Kragerö
	693	DN	Röskva *Johannesen* (ex Santa Cruz)	W Sr	Norweg'n	140	85·5	23·0	11·7	Bilbao	18 54/72			A.H.Kiaer & Co.	Fredrik
56113	694		Rosland *Arundell 84-86*	W Bkn	British	237	115·9	25·0	12·9	Sunderland	1870	J.Crown	[illegible]	J.H.Farmer & Co.	Liverpo
45588	695	✻	Roslin Castle *Symons 85-85*	I Bk	British	644	183·9	29·1	18·7	Glasgow	1868	C.Connell&Co.		Nicholson & McGill	Glasgo
	696	N	Ross *Johannessen*	W Bk	Norweg'n	370	129·8	30·4	14·1	Arendal	1875	T.Larsen	[illegible]	Hans H.Pettersen	Arenda
	697	N	Ross *Nilsen*	W Bn	Norweg'n	181	110·4	22·3	11·4	Skien	1874			N.Nilsen	Brevig
67943	698	✻	Rossdhu *Fleming*	I S	British	1321	232·5	38·2	21·9	Greenock	1874	J.E.Scott		C.S.Caird	Greeno
[illegible]	699	RV	Rossignol *Robbins 68-71*	W S	British	1510	203·6	38·8	27·0	Tusket, N.S.	1872	D.Sperotti	[illegible]	C.T.Soley & Co.	Yarmou
	700	V	Rossini *Crouzel -87*	W Bg	French	275	116·5	26·1	14·0	P.E.Island	1871	Henderson	[illegible]	Olive Cousins	Cette

✻ = Lloyd's Register; D = Nederlandsche Vereeniging; G = Germanischer Lloyd; I = Registro Italiano; R = American
A = Veritas Austro-Ungarico; H = Veritas Ellenico; N = Norske Veritas; V = Bureau Veri

In the 4th Column, C = Composite (iron frame, wood planked); I = Iron; Stl = Steel; W = Wood.

The Dates of Metalling (**Col. 11**) are based on the records of the Books in which Vessels are respectively classed, or on the Reports of the Surveyors of Lloyd's Re

Further particulars of Vessels classed by Lloyd's Register will be found in Lloyd's Register Book and Lloyd's Yacht Register.

No.	Registre	Nom / Ancien Nom	Capitaine et dates d'entrée	Matériaux et Gréement	Pavillon	Tonnage net	Longueur	Largeur	Creux	Construit: Où	Quand	Par qui	Date du doublage en métal	Armateurs ou Directeurs-Gérants	Port d'Armement
701	V	Rota (ex Espérance)	Lorn Jørgensen 84-84	WBg	Danish	246	102·7	23·9	14·7	Sunderland	1866	John Barkes		L.C.Jorgensen	Svendborg
702	V	Rota (WRK 8 Eli)	Nielsen 76-84	WSr	Danish	115	87·8	20·7	10·3	Thoro	1884	J.P.Jørgensen		L.P.Jorgensen	Sønderho
703		Rota		WBk	Norweg'n	988	161·8	39·2	24·0	Copenhagen	18^{22}_{65}	Danish Govt.		Paul B.Lassen	Arendal
704	N	Rota	Sørensen	WBk	Norweg'n	472	130·6	31·6	17·2	Arendal	1874	Hans Christensen	Y 10.88	P.C.Pedersen	Arendal
705	V	Rota	Schnohr	W3mSr	Russian	363	137·0	31·4	14·0	Pernigel	1886	P.Krause	M 10.87	J.Schnohr	Riga
706	N	Rothesay	Feersen -74	WBk	Norweg'n	642	145·4	32·4	19·5	Rothesay,NB	1868	J.Titus	Y [illegible]	N.O.Young	Christiania
707	*D	Rothesay Bay	Cejsay 87-87	I Bk	British	750	186·9	31·1	18·5	Dumbarton	1877	Birrell, Stenhouse & Co.		Hatfield,Cameron&Co.	Glasgow
708	V	Rothiemay	Burns -87	WBk	British	1246	198·7	38·6	23·2	St.John,N.B.	1881	J.K.Dunlop	Y 1.87	Wm.Thomson & Co	St.John,N.B.
709	*	Rothiemay		WK	British	89	79·5	20·9	10·6	Brixham	1884	J.W.&A.Upham		J.Leyman	Hull
710		Rousseau		WBk	American	305	92·9	28·3	18·3	Philadelphia	18^{01}_{45}			Aiken Swift	New Bedfrd Mas.
711	*	Routenbeck	Stitt -88	I Bk	British	930	208·5	32·4	19·6	Whitehaven	1875	Whitehavn SBC		J.Davidson & Son	Whitehaven
712	*	Routenburn	Holmgard [illegible]	I 1mBk	British	1997	289·0	42·2	23·9	Greenock	1881	R.Steele&Co.		R.Shankland & Co.	Greenock
713		Rover	Kennard	WBk	British	273	115·3	28·6	12·0	Wallace,N.S.	1873			Wallace Ship B. Co.	Halifax,N.S.
714		Rover	M. L. Stack -86	WBkn	British	214	104·1	23·1	14·1	Aberdeen	1859			J.F.Norris	Turks Islands
715		Rover		WSr	British	114	83·7	23·0	9·6	P.E.Island	1862			Mrs.Isabella Taggart	Dublin
716		Rover	Smith	WSr	British	94	84·0	23·3	9·3	Shelburne,NS	1878	Thos.Muir		Jonathan Locke	Shelburne,N.S.
717		Rover	McLeod	WSr	American	120	85·0	26·7	8·2	Cambridge Ind	1883	J.H.Johnson		W.S.Hoddinott	Baltimore
718	NV	Royal	Jorgensen -88	WBk	Norweg'n	670	153·8	34·0	19·1	Arendal	1881	A.Henrichsen	Y 2.87	Chr.Th.Boe's Bo	Arendal
719		Royal Adelaide		W3mSr	Norweg'n	198	117·5	21·8	11·3	England	18^{30}_{53}			Dykkercompagniet (N.C.Ring)	Christiania
720	*D	Royal Alexandra	Jones 79-81	I Bk	British	1332	216·7	36·1	23·9	Stockton	1864	Richardson, Duck & Co.		H.Fernie & Sons	Liverpool
721	*	Royal Alfred	James 82-84	I Bk	British	1188	215·5	34·7	22·8	Newcastle	1863	Palmer Brs.&Co.		H.Fernie & Sons	Liverpool
722	*	Royal Alice		I Bk	British	1199	215·8	34·7	22·8	Newcastle	1863	Palmer Brs.&Co		H.Fernie & Sons	Liverpool
723		Royal Arcanum (WRECKED)	Goold [-82	WSr	British	117	84·0	26·3	8·0	Clifton,N.B.	1881	G.H.Flewelling		John M.Taylor	St.John,N.B.
724	V	Royal Arch	Jones -88	WBk	British	345	119·7	27·1	17·0	Sunderland	1868	Sykes	Y 8.82	D.Jenkins	Liverpool
725		Royal Arch	Hamlyn	W3mSr	American	401	132·2	32·2	10·5	E.Haven,Con	1873	H.H.Hanscom		W.H.Lander	New Haven,Con.
726		Royal Duke	Boreham -81	WBn	British	105	86·0	21·0	9·0	McLeay River, N.S.W.	1868	C.Lawson		Kirk,Boreham & Hart	Sydney,N.S.W.
727		Royal Eagle		WBk	British	304	111·5	28·2	13·5					Allohe Chitty	Penang
728	*D	Royal George	Gardner 81-81	I S	British	1404	221·3	38·4	22·9	London	1864	Westwood, Baillie&Co.		M'Diarmid,Greenshields & Co.	Liverpool
729	*	Royal Sovereign	Milne	I S	British	1388	218·8	36·0	24·0	Bristol	1864	W.Patterson		E.Bates & Sons	Liverpool
730		Royal Tar	Walkley	WBk	British	598	171·2	31·4	17·2	Nambucera, N.S.W.	1876	W.Marshall		Arch.McLean	Sydney,N.S.W.
731	*	Royal Tar	Evans	WSr	British	118	92·5	28·8	11·8	Brixham	1859			G.Evans	Brixham
732		Royal Tar	Carlson -81	WBk	Russian	717	158·7	31·6	20·0	Quebec	1864	Lachance	Y 7.88	M.Greve	Riga
733	DN	Royal Visitor	Bull -88	WBk	Norweg'n	1167	189·6	37·1	22·7	Quebec	1860	Oliver	Y 5.86	Skibsactieselskabet "Visitor" (Chr. Rafter)	Christiania
734		Royalist	Bennett	WSr	British	123	88·2	21·4	11·4	Littlehampton	1850			N.House	Padstow
735		Royalist	Eden -88	WSr	British	111	77·0	19·8	8·1	Grand Cayman	1875			Richard H.Rose	Belize,B.H.
736	RV	Rozella Smith	Greene	WBn	British	509	143·7	32·1	17·2	Walton,N.S.	1873	W.S.Vaughan	Y 5.86	R.G.Tootal	London
737		Rozilda		WSp	British	106	95·5	22·0	7·0	Quebec	1865	D.Richard		Joseph E. Lareau	Montreal
738	V	Rubens	Sinsonnet -87	WBk	French	214	107·6	26·5	14·8	Fecamp	1870		V 11.81	Comolet frères	Cette
739	N	Rubens	Knudsen	WBk	Norweg'n	375	110·6	27·1	18·2	Bath,Me.	1831			A.Börresen	Drammen
740	V	Ruby	Robbins	WS	British	1392	202·8	39·6	23·8	Church Pnt NS	1878	J.D.Raymond	Y 6.87	Abel C. Robbins	Yarmouth,N.S.
741		Ruby	Morris	WBk	British	879	181·6	35·6	20·7	Maitland,N.S.	1870	M'Dougall		J.Smyth	Dublin
742		Ruby	Ferguson -88	WBk	British	251	125·4	26·9	12·9	London	1868	Salisbury		East Downshire S.S. Co. Ld.	Belfast
743	*	Ruby	Sinclair	WSr	British	99	91·0	21·1	10·3	Kingston	1884	W.Kinloch	Y 3.85	R.Sinclair	Lerwick
744		Ruby	Alexander -88	WBn	American	222	97·0	26·9	10·6	Brooklyn	1873	C.&R.Poillon		L.W.&P.Armstrong	New York
745		Ruby A. Cousins	Hansen 86-86	W3mSr	American	185	112·5	30·5	8·9	Eureka,Cal.	1882			Pollard & Dodge	San Francisco
746		Rudolf (ex Marquis)	Jacobsson -86	WSr	Swedish	171	90·9	21·5	11·4	France	1853			J.H.Gren	Fiskebäckskil
747		Rudolf	Norberg	WSr	Swedish	101	83·7	21·9	8·8	Sundsvall	1872			M.Norberg	Vermdö
748		Rudolph	Barnard	WSr	British	118	78·1	19·9	9·7	Rønne	1855			D.J.Beatton	Ipswich
749	N	Rudolph	Torjusen -87	WSr	Norweg'n	140	73·2	23·2	11·8	Stettin	1835			R.J.Torjusen	Fredrikshald
750		Rudolph Ebel (MISSING)	Wittenburg 81-84	W3mSr	German	407	138·1	29·7	14·7	Stettin	1864	C.Domcke		C.A.Domcke	Stettin

=Lloyd's Register; D=Nederlandsche Vereeniging; G=Germanischer Lloyd; I=Registro Italiano; R=American Record;
=Veritas Austro-Ungarico; H=Veritas Ellenico; N=Norske Veritas; V=Bureau Veritas.

Dans la 4me Colonne, C=Composite (membrure en fer, bordé en bois); I=Fer; Stl=Acier; W=Bois.

Les dates du doublage en métal (**Col. 11**) ressortent des livres dans lesquels les navires sont respectivement classés ou des rapports des Inspecteurs du Lloyd's Register.

Pour plus amples renseignements sur les navires classés par le Lloyd's Register, voir le Lloyd's Register Book et le Lloyd's Yacht Register.

Official No.	No. in Book.	Registry in which classed.	Ship's Name. Material and Rig. Master, and Dates of Appointment to present Owner's Service and to Ship. Ship's Late Name.	Flag.	Net Tonnage.	Dimensions. Length.	Breadth.	Depth.	Built. Where.	When.	By whom.	Date of Metalling.	Owners or Managers	Port
	751	G V	Rudolph Josephy *Brudhering -83* WBk	German	477	134·2	31·8	18·2	[illegible]	1883	O.Ludwig	[illegible]	Rud.Josephy	Rostock
	752	G	RudolphineBurchard *Dade 73-75* WBg	German	277	108·8	27·1	14·7	Rostock	1873	O.Ludwig	[illegible]	A.Burchard	Rostock
	753	N	Rufus *Andresen* WBg	Norweg'n	199	100·9	26·5	12·1	[illegible]	1871	L.Tellefson	[illegible]	Johan Hansen	Brevig
[illegible]	754	R V	Rufus E. Wood *Farnham* WS	American	1477	200·1	40·4	24·5	E.Deering,Me	1875	Russell	[illegible]	W.E.Mighell	San F...
[illegible]	755		Ruggiero I. (ABANDONED) *Ruggiero* WBk	Italian	463	124·6	28·0	18·7	Alimuri	1872	G.Mauro	[illegible]	S.N.&G.Ruggiero	Castellam...
	756	N	Runeberg *Fersen* WBk	Norweg'n	417	121·7	30·0	16·9	Lillesand	1867	T.Astagzen	[illegible]	D.Hegermann	Christian...
	757		Runeberg *Sjöman -86* (ex Emmanuel) WBk	Russian	331	125·5	30·2	13·7	Ekenäs	1866			F.J.F.Sjöblom	Helsingf...
	758	N	Runer *Langfeldt* WBk	Norweg'n	380	130·8	29·1	15·3	Christiansand	1876	O.Hagen	[illegible]	R.Torstensen	Christian...
	759	✻ V	Runnymede *Jarnion* I Bk	French	700	200·0	30·2	18·0	Sunderland	1869	W.Pile&Co.		H.Joyau	Nantes
59642	760		Rupert *Pout 85-85* WSr	British	163	99·0	23·0	12·5	Gloucester	1867	Barnes		Whitstable Shipg.Co.Ld	Fav...
	761		Rurik *Tengström* WBk	Russian	882	185·0	32·8	20·0	Åbo	18 62/67		[illegible]	G.A.Lindblom	Åbo
	762	N V	Russell *Erichsen 72-87* WBk	Norweg'n	633	151·6	35·0	17·1	Arendal	1882		[illegible]	D.Lohne	Mandal
	763	✻	Russell *Royen -87* WBk	Norweg'n	375	126·1	28·4	17·5	Sunderland	1867	J.Davison	[illegible]	T.Royen	Lillesand
26216	764	G	Russia *Thompson 84-85* (ex Admiral Tromp, ex Russia, ex Mauritius, ex Steamer) I S	British	1547	261·0	39·0	25·7	London	1852	C.J.Mare & Co.		W.J.Fernie (W.T.Dixon & Sons)	Bombay
83043	765	✻	Russia *Nash -87* (ex Richard Hutchinson) WBk	British	768	165·6	33·9	19·9	New Brunswick	1880	C.C.Watt	[illegible]	Troop & Son	Cha...
573	766	V	Russo *Paturzo* WBn	Italian	299	111·5	25·9	16·7	Sorrento	1867	M.Paturzo		Eredi Russo & Co.	Castell...
	767		Rustchuk I. WSchleppe	Turkish	162	..	..	..	Rustchuk	1868			Rustchuklee Hajee Ali & Co.	Rustchuk
	768		Rustchuk II. WSchleppe	Turkish	112	..	..	..	Rustchuk	1868			Rustchuklee Kooros Oushlet & Co.	Rustchuk
	769		Rustchuk III. WSchleppe	Turkish	150	..	..	..	Rustchuk	1868			HajeeTannash & Co	Rustchuk
	770		Rustchuk IV. WSchleppe	Turkish	150	..	..	..	Rustchuk	1868			HajeeTannash & Co	Rustchuk
	771		Rustchuk V. WSchleppe	Turkish	150	..	..	..	Rustchuk	1868			Rustchuklee Gulgher Rosalni & Co.	Rustchuk
	772		Rustchuk VI. WSchleppe	Turkish	150	..	..	..	Rustchuk	1868			Rustchuklee Gulgher Rosalni & Co.	Rustchuk
	773		Rustchuk VII. WSchleppe	Turkish	250	..	..	..	Rustchuk	1868			Rustchuklee Hajee Omer zadeh Hadji Bey	Rustchuk
	774		Rustchuk VIII. WSchleppe	Turkish	250	..	..	..	Rustchuk	1868			Rustchuklee Hajee Omer zadeh Hafiz Bey	Rustchuk
	775		Rustchuk IX. WSchleppe	Turkish	238	..	..	..	Rustchuk	1868			Rustchuklee Hajee Omer zadeh Alish Bey	Rustchuk
	776		Rustchuk X. WSchleppe	Turkish	175	..	..	..	Rustchuk	1868			Rustchuklee Hajee Omer zadeh Alish Bey	Rustchuk
	777		Rustchuk XI. WSchleppe	Turkish	175	..	..	..	Rustchuk	1868			Hajee Omer Keremessi Emingh Gülsun Hannum	Rustchuk
	778		Rustchuk XII. WSchleppe	Turkish	225	..	..	..	Rustchuk	1868			Hajee Omer Keremessi Heywali Hannum	Rustchuk
	779	N	Ruth *Jacobsen -88* WBk	Norweg'n	482	144·5	30·2	17·0	Christiansand	1875	Larsen	[illegible]	H.Hansen	Lillesand
	780	N	Ruth *Grefstad -88* WBk	Norweg'n	469	137·1	30·2	17·6	Grimstad	1879	J.Jörgensen	[illegible]	J.Jörgensen	Grimstad
	781		Ruth (WRECKED) *Röed* WBg	Norweg'n	296	98·1	27·5	16·5	Varazze	1856			J.M.Bryde	Sandefjord
	782	N	Ruth *Torressen 64-74* W3mSr	Norweg'n	294	118·1	26·7	14·3	Stavanger	1874		[illegible]	J.G.Myhre	Stavanger
	783	N	Ruth *Rasmussen 78-83* (ex Electra) WSr	Norweg'n	220	106·0	24·8	12·6	P.E.Island	1869	McDougall		J.F.Juell	Risör
[illegible]	784		Ruth Darling *Lowell* WSr	American	193	107·1	27·9	9·6	Orland,Me.	1874		[illegible]	J.L.Buck	Bucksport
71041	785	✻	Ruth Palmer *Michael* WBk	British	934	169·6	36·1	21·0	D'rchesterNB	1875	G.Palmer	[illegible]	Gideon Palmer	[illegible]
[illegible]	786		Ruth Robinson *Fuller* W3mSr	American	479	139·9	33·0	12·2	Portland,Con.	1874	Gildersleeve		S.H.Mills	New York
21762	787		Ruth Shaw *Carlow* W3mSr	American	366	120·2	30·0	15·0	Millville,N.J.	1867			E.Carlan	Bridgeton
70949	788	✻	Ruth Topping *Wood (Spear)* WSr	British	319	135·2	27·5	12·9	Sunderland	1875	W.Richardson	[illegible]	W.C.Jarvis	Liverpool
52622	789	✻	Ruthin *Cutler -86* (ex City of Benares) I Bk	British	1178	224·4	34·2	22·5	Glasgow	1865	Brclay,Curle&C		G.Traill & Sons	Glasgow
90953	790	✻	Ruthwell *McConnell 81-85* I Bk	British	1296	239·5	36·2	21·5	Pt.Glasgow	1885	R.Duncan&Co.		T.C.Guthrie	Glasgow
8574	791		Rychner (ex Abeonuto) (BROKEN UP) WBk	Italian	381	135·6	28·6	12·9	Pugwash,N.S.	1869			G.B.Figari	Genoa
70908	792	✻	Rydalmere *Boys 87-87* I Bk	British	1246	226·0	36·1	22·1	Whitehaven	1875	WhithavnSBCo		Fisher & Sprott	Liverpool
84964	793	✻	Ryelands *Marrow* W3mSr	British	133	102·2	22·9	10·8	Glasson Dock	1887	Nicholson & Marsh		J.Helme	Lancaster
66364	794	V	Ryerson *Jostin 76-83* WS	British	1423	199·0	39·0	24·1	Salmon River, N.S.	1872	H.K.Richards	[illegible]	W.H.Ross & Co,	Liverpool
78573	795	✻ D	Ryevale *Lewis 86-88* I Bk	British	844	195·4	32·6	18·5	Dumbarton	1878	Birrell,Stenhouse&Co.		Browne & Watson	Glasgow
	796	N	Ryfylke *Knudsen* WBn	Norweg'n	156	100·2	22·8	11·4	Stavanger	1877		[illegible]	N.W.Olsen	Stavanger
	797	N	Ryno *Sörensen* WBg	Norweg'n	282	105·8	27·4	14·3	Grimstad	1873	N.O.Gjësdale	[illegible]	J.Halvorsen	Grimstad
[illegible]	798		Rynsho Maru WBk	Japanese	342	127·2	27·3	13·7	Flensburg	1865			Watanake Shosuki	[illegible]

ADDITIONAL SHIPS ARE INSERTED IN THE SUPPLEMENT AT THE END OF THE REGISTER.

✻ = Lloyd's Register; D = Nederlandsche Vereeniging; G = Germanischer Lloyd; I = Registro Italiano; R = American Record; A = Veritas Austro-Ungarico; H = Veritas Ellenico; N = Norske Veritas; V = Bureau Veritas.

In the 4th Column, C = Composite (iron frame, wood planked); I = Iron; Stl = Steel; W = Wood.

The Dates of Metalling (Col. 11) are based on the records of the Books in which Vessels are respectively classed, or on the Reports of the Surveyors of Lloyd's Register.

Further particulars of Vessels classed by Lloyd's Register will be found in Lloyd's Register Book and Lloyd's Yacht Register.

N° dans la série	Registre dans lequel le navire est classé	Nom. Matériaux et Gréement. Capitaine et dates d'entrée au service des Armateurs actuels et à bord du navire. Ancien Nom.	Pavillon.	Tonnage net.	Dimensions. Longueur	Largeur	Creux.	Construit. Où.	Quand.	Par qui.	Date du doublage en métal	Armateurs ou Directeurs-Gérants.	Port d'Armement.
1	R	S. A. Paine *Bushey -87* WSr	American	137	93·4	24·0	9·5	East Boston	1866			Oliver Bushey	Gloucester
2	R	S. A. Rudolph *Wallis* W3mSr	American	197	121·0	23·6	8·5	Camden, N.J.	1882	McDonnell Bros		J. & F. McDonnell	Philadelphia
3		S. B. Hume WBkn	British	289	117·0	27·5	14·0	Perry, Me.		J. B. Nutt		W. W. Robinson	Dublin
4		S. B. Wheeler *Godfrey* WSr	American	267	118·7	30·0	9·4	Wilmngtn Del	18 $^{88}_{83}$			Carson Harris	Wilmington Del.
5	V	S.C.Allen *Thompson -88* WBk	American	655	177·2	37·0	14·3	Bath, Me.	1888	New England S.B.Co.	Y 5.88	C. S. Holmes	Port Townsend
6		S. C. Tryon *Nickerson* W3mSr	American	402	140·0	32·0	14·0	Greenport NY	1873	O. H. Bishop		C. & J. F. Baker	Boston, Mass.
7		S. F. Hersey *Bennett -88* WS	British	961	168·8	33·0	22·8	Searsport, Me.	1865	W. McGilvery	Y 5.88	C. G. Millar	Melbourne
8		S. F. Maker WSr	American	104	90·0	21·1	9·8	Essex, Mass.	1887			B. F. Payson	Gloucester. Mass.
9	R	S. G. Hart *Smith* W3mSr	American	505	159·6	34·3	12·3	WaldoboroMe	1883	W. Fish	Y 8.88	S. G. Hart	New York
10	V	S. G. Wilder *Paul 85-87* WBkn	American	579	166·8	37·3	15·0	Port Blakely	1887	Hall Brothers		G. F. Smith & Co.	San Francisco
11		S. Guignard WSr	American	178	100·0	31·6	7·1	Brooklyn	1883	J. D. Leary		C. Colne	New York
12		S. H. *Sceeno 85-87* WBk	French	196	98·4	21·7	13·4	Dunkirk	1862	G. Beck		G. Beck	Dunkirk
13	G	S. H. T. *Lundgren -88* WBn	Swedish	125	89·6	20·2	9·3	Calmar	1877	Bäck		J. A. Carlsson	Calmar
14		S. Hale *Coleman 79-79* WSr	American	119	89·6	26·5	8·5	Taunton, Mass	18 $^{53}_{84}$			G. L. Coleman	Fall River. Mass.
15	V	S. J. Bogart *Shaw -87* WBk	British	836	166·0	35·0	20·3	Granville, N.S.	1873	Milner	Y 6.86	Thos. E. Morrison	Annapolis, N.S.
16		S. J. Gilmore *Rose* WSr	American	117	86·4	27·1	7·9	Sullivan, Me.	1868			D. J. Sawyer	Machias. Me.
17		S. J. Lund *Pettersen -72* WBn	Danish	111	79·4	20·2	10·4	Aalborg	1860	B. J. Bonnesen		Actieselskab. (V. Rittschevk)	Aalborg
18		S. J. Musson *Swaine 80-88* WBn	British	128	89·0	23·5	9·6	JordanRvrNS	1878		Y 12.84	Alfred G. Jones	Halifax, N.S.
19	V	S. Louisa *Andersen* WSr	Danish	113	88·4	19·5	9·9	Copenhagen	1872	Popp & Michelsen	Y 8.88	A. Asgeirsson	Isafjørd, Iceland
20	R	S. M. Bird *Muriell [79-71* W3mSr	American	421	130·0	32·0	14·6	Belfast, Me.	1874			S. M. Bird	Rockland, Maine
21		S. M. Thomas W3mSr	American	723	167·0	35·2	15·0	Bath, Me.	1881	Goss & Sawyer		J. B. Phillips	Taunton, Mass.
22	V	S.N.Castle *Hubbard -87* W3mSr	American	489	155·8	35·1	13·7	Port Blakely	1887	Hall Brothers		L. H. Hubbard	San Francisco
23		S. N. Colymore *Davis* WBn		220	105·0	27·5	11·0	ChurchPointNS	1873	Raymond Bros.			
24	N	S. N. Hansen *Salvesen [-80* WBg	Norweg'n	308	114·6	27·7	14·8	Lillesand	1873	O. & P. Knudsen	Y 12.83	O. & P. Knudsen	Lillesand
25	RV	S. P. Hitchcock *Nichols* WS	American	2178	247·4	44·3	28·6	Bath, Maine	1883	I F Chapman & C.	Y 12.88	A. G. Ropes	New York
26		S. P. Hitchcock *Blair* W3mSr	American	574	162·3	35·6	13·9	Bath, Maine	1879	Adms & Hitchck		Adams & Hitchcock	Bath. Me.
27	RV	S. R. Bearce *Thestrup* WBk	American	577	139·7	31·9	17·8	Y'rmouth, Me	1873	G. Loring	Y 11.87	James Bain	Portland. Me.
28	R	S. R. Lyman *Pinkham* WBk	American	569	139·9	32·3	17·2	Y'rmouth, Me.	1873	A. Seabury	Y 1.87	Chase, Leavitt & Co.	Portland. Me.
29	✻	S. R. & H. *More 72-72* WSr	British	140	94·8	22·2	12·8	Teignmouth	1872	Mansfield	Y 6.83	P. Ward	Teignmouth
30		S. S. Bickmore *Wall* WSr	American	214	110·2	29·4	9·4	St. George, Me.	1867			G. A. Gilchrist	Rockland. Me.
31		S. S. Hudson *Vaughan* W3mSr	American	430	127·0	31·9	19·2	May's Landing [N.J.	1869			D. F. Vaughan	Somers Pt. N.J.
32	✻	S. T. *James 88-88* I Bkn	British	298	129·0	25·3	14·6	Llanelly	1880	S. Samuel		Samuel Bros.	Llanelly
33		S. V. W. Simmons WSr	American	184	106·0	29·0	8·6	Camden, N.J.	1869			F. Smith	Philadelphia
34		S. Warren Hall *Douglass Walker [-85* WSr	American	152	90·0	29·0	7·0	Milford, Del.	1882	J. R. Caball & Co.		J. W. Hall	Wilmingtn, Del.
35		S.&J.Collier *Macaulay -87* WSr	British	146	106·0	23·8	8·7	South Marys- [burg, Ont.	1872	J. Tait		Allan Lockhart	Wallaceburg
36		Sa-adan WBk	Dutch	207	94·8	23·0	..	Siam	1881			Isah bin Jasen	Banjarmassin
37	R	Saama *Hangell* WBk	Russian	687	140·0	34·1	20·2	Brahestad	1870	F. Petterson		F. Sovelius	Brahestad
38		Saarbruck *Clark* WSr	American	155	99·5	26·5	8·7	E. Machias Me.	1871			S. H. Talbot	E. Machias. Me.
39		Saarikoski *Wahlroos -85* WBk	Russian	611	151·9	33·1	18·2	Sastmola	1870	M. Hannus		Johan Ahlström	Sastmola
40		Saarila WSr	Russian	182	..	..	..		1872			A. R. Hjelt	Salo
41		Sabao *Flynn* WSr	American	134	87·3	25·8	9·6	Machias, Me.	1866			M. Gardner	Machias. Me.
42		Sabawa *Gunin 79-* W3mSr	Russian	198	116·5	22·5	11·0	Kowda	1878	J. S. Markikoff		J. Klementyf	Kowda
43	V	Sabine *Chambert -88* WBn	French	162	88·5	23·2	10·6	Bayonne	1888	Lacaussade		J. B. Cormier	Bayonne
44	NV	Sabino *Bergersen 86-86* WBk	Norweg'n	1231	185·6	36·3	23·8	Bath, Maine	1862	G. F. Patten & Sn.	Y 3.88	Aktieselskab "Sabino" (J.A.Hansen)	Tönsberg
45	V	Sablais *Provost* WBn	French	132	73·8	20·3	11·5	Sables [d'Olonne	1866	Chaiguean		Vve P. Langlois	Lannion
46	✻	Sabrina *Johnson -88* I Bk	British	767	187·2	31·7	19·9	Liverpool	1874	T. Royden & Sns		J. Newton & Co.	Liverpool
47	✻	Sabrina *Ball 79-86* WBkn	British	317	134·8	27·6	12·9	Sunderland	1876	W. Richardson	Y 7.8	W. W. Steer	Salcombe
48		Sabrina WBn	British	148	95·3	24·3	11·3	P.E.Island	1870	Grady		J. Craig & Co.	Liverpool
49		Sabrina *Trefren* WSr	British	124	86·6	26·6	8·0	Portland, N.B.	1883	C. S. Phipps		Noah H. Urquhart	St. John, N.B.
50		Sabrina WSrYt	British	117	110·5	23·8	11·8	Cowes	1863	J. S. White		M. K. King	Bristol

✻ = Lloyd's Register; D = Nederlandsche Vereeniging; G = Germanischer Lloyd; I = Registro Italiano; R = American Record; A = Veritas Austro-Ungarico; H = Veritas Ellenico; N = Norske Veritas; V = Bureau Veritas.

Dans la 4me Colonne, C = Composite (membrure en fer, bordé en bois); I = Fer; Stl = Acier; W = Bois.

Les Dates du doublage en métal (**Col. 11**) ressortent des livres dans lesquels les navires sont respectivement classés ou des rapports des Inspecteurs du Lloyd's Register.

Pour plus amples renseignements sur les navires classés par le Lloyd's Register, voir le Lloyd's Register Book et le Lloyd's Yacht Register.

Official No.	No. in Book	Registry in which classed	Ship's Name. Material and Rig. Master, and Dates of Appointment to present Owner's Service and to Ship. Ship's Late Name.	Flag.	Net Tonnage	Length	Breadth	Depth	Built. Where.	When	By whom.	Date of Metalling	Owners or Managers.
[illegible]	51	RV	**Sachem** *Bartlett* 83-83 W S	American	1381	194·0	39·1	23·8	Boston, Mass.	1876	C. Sampson		M.F.Pickering&Co.
	52	H	**Sachper** *Dimitrios* W Bg	Turkish	182	..	..	..	Scopelo				Hadji Ibraim Efendi
[illegible]	53		**Sadaleya** (W.REASER) W Bk	British	277	101·8	22·8	13·3	Coringa	1873			Ravana Mana Pana Chana (Moona Chana)
[illegible]	54		**Sadie Corey** *Marshal* W Sr	American	149	101·6	27·0	7·3	Winterport, Me	1880			G. Miller
115102	55		**Sadie F. Caller** *Pebbitts* -88 W 3m Sr	American	393	131·9	31·5	15·1	Essex, Mass.	1873			S.B.Peterson
[illegible]	56	R	**SadieWillcutt** *Hart* -87 W 3m Sr	American	347	129·0	30·0	10·0	St. George, Me.	1874	Pease & Long		G.W.Rawley
	57		**Sadyk Deriyah** W Bg	Turkish	118	..	..	..	Greece	1873			Oomialee Izzet & Co.
	58	RV	**Sæheim** *Monsen* 81-88 W S (ex Ellen ...)	Norweg'n	893	167·4	34·4	21·0	Pt.Gilbert N.S.	1870			A. Monsen
	59		**Sæhrimner** *Olsen* 67-81 W Bk	Norweg'n	321	107·0	27·1	16·5	Newport, N.S.	1845			A. Olsen
	60	DNV	**Særimner** *Hattrig* -87 W S	Norweg'n	1366	216·3	39·6	23·6	Stavanger	1878	L. Magnus		J. Somme
833	61	A	**Saffo** *Richiostri* W Bn	Italian	159	111·8	19·2	11·8	Trieste	1868	G. Spadon		Ant. Barbaro
54050	62		**Saffron** *Mills* W Sw	British	260	107·0	26·1	15·7	Sunderland	1866	J. Denniston		G. Mills
	63	N	**Safi** *Goday* W Bn	Spanish	201	107·2	27·1	10·5	Palma	1873			B. Borras
	64	RV	**Safir** *Keogh* W Bk	Norweg'n	939	172·4	36·3	21·7	St.Stephen N.B.	1876	J. & C. Short		Actieselskabet "Safir"
	65	N	**Safir** *Pedersen* -81 W Sr	Norweg'n	130	91·3	23·5	9·7	Porsgrund	1876	N.J. Klevstrand		J. Johannessen
	66		**Safo** W Bn	Spanish	160	82·3	25·1	12·4	Blanes	1850			F. Montaner
[illegible]	67	V	**Saga** *Williams* 86-87 W Bk	British	499	138·1	30·5	16·8	N.Glasgow N.S.	1869	J.W. Carmichael		J.H. Farmer & Co.
	68	DV	**Saga** *Bang* 70-86 W Bn	Danish	150	96·5	21·1	10·8	Troense	1870	C.J. Jensen		C.J. Jensen
	69	N	**Saga** *Jesmundsen* W Bk	Norweg'n	686	151·1	35·3	19·9	Bergen	1867	Jens Gran	V 9.88	C. Ege
	70	N	**Saga** *Jensen* 70-81 W Bk	Norweg'n	597	146·2	33·4	18·6	Drammen	1866	Jorgensen & Knudsen		F.J. Kjösterud
	71	N	**Saga** *Furop* W Bk	Norweg'n	585	150·7	31·8	18·2	Grimstad	1873	J.T. Tryde		F. Smith Petersen
	72	N	**Saga** *Gundersen* -87 W Bk	Norweg'n	442	131·4	30·3	17·1	Grimstad	1872	S. Johnsen	V 4.88	J. Gundersen
	73	N	**Saga** *Svendsen* W Bk	Norweg'n	435	126·5	30·2	17·1	Lillesand	1869	T. Langström		I.A. Henschien
	74	N	**Saga** *Walberg* 80-84 W Bk	Norweg'n	417	130·4	30·7	16·7	Drammen	1868	J. Jorgensen	V 8.87	Rederiet "Niord"
	75		**Saga** *Mattsson* 81-81 W Bg	Swedish	306	108·9	28·1	16·3	Copenhagen	1848			G. Sjöberg
	76	V	**Saga** *Zinnerstrom* W Bg	Swedish	304	129·6	24·7	12·7	Varnanäs	1875	C.A. Hahn		A.R. Andersen
22170	77		**Sagadahoc** *Sheehan* W Bk	American	506	149·9	29·4	17·8	Bath, Maine	1865			M.D. Sheehan
23414	78		**Sagamore** *Brandenburg* 87-87 W S	American	1274	200·0	38·0	21·0	Ptsmouth, N.H.	1856			Puget Sound Commercial Co.
	79	N	**Sagatun** *Severtsen* W Bk	Norweg'n	366	129·0	29·2	15·0	Porsgrund	1870	J. Olsen		J.A. Hage
15473	80		**Sagitta** *Webber* W Bg	British	283	135·7	25·8	14·6	Guernsey	1856	P. Ogier		T.G. Robins & Co.
	81	V	**Sagitta** *Jorgensen* 87-87 W Bk	German	553	143·2	33·9	17·7	Kempt, N.S.	1875	D. Caldwell		Knöhr & Burchard
406	82	A	**Saggittario** *Sodich* W Bk	Austrian	354	132·8	24·1	16·7	Fiume	1867	G. Brazzoduro		C.cav. Cosulich
71807	83	✻	**Sagona** *Jones* 88-88 W Bk	British	797	169·0	33·7	20·0	Kingston, N.B.	1882	Jardine Bros.		Jardine Bros.
	84	V	**Sagterland** *Becken* -76 W Sr	German	126	86·6	20·9	10·0	Nordloh	1876	E. Reil		Gerh. Lauwer
23867	85		**Sagua** *Munday* W Bn	American	366	127·0	29·0	13·8	Camden, N.J.	1870	Taylor & Mathis		W.G. Munday
	86	A	**Sahim Bahri** W Bn	Turkish	150	..	..	..	Bartan				Halil Ali
	87		**Sahim Bahri** *Suleiman* W Sr	Turkish	125	..	..	..	Caradagh	1882			H.H. Bostanjoglu
	88		**Said** W Bg	Turkish	146	..	..	..	Greece	1877			Tannus Tiliri Massood
	89		**Said Maskoor** W Bk	Dutch	166	90·5	23·9	..	Java	1845			Achmad
[illegible]	90	R	**Saidee** *Dauphney* W Bn	British	155	96·0	25·6	10·5	Jordan R., N.S.	1879	S. Freeman & Co.		Lemuel W. Drew
	91		**Saidoel Aripin** W Bg (ex Wanderer) (Said Achmat Bareiba)	Dutch	136	91·9	21·0	..	Bangkok	1864			Said Achmat bin Osman
308	92		**Saigon** *Langkaker* -87 W Bk	Norweg'n	768	167·0	32·8	20·0	Quebec	1871	Marquis		Actieselskabet "Saigon"
[illegible]	93		**Sailor Boy** *Mitchell* -84 W 3m Sr	American	316	139·0	33·8	12·2	Shoal Water Bay, Cal.	1883	Paterson		A. Crawford & Co.
	94	V	**St. Aignan** *Giraud* -81 W Bk	French	331	120·0	27·1	15·1	Nantes	1869	T. Dubigeon & fils		F. Boissiere
	95		**St. Albans** *Lindhardt* -77 W S (ABANDONED)	Danish	1248	187·8	33·1	23·6	Portsmouth	1856	Marcy		Actieselskab "St. Albans" (C. Dalberg)
	96		**St. Alexei** *Hansen* 58-86 W Bn	Danish	184	108·4	23·4	10·9	Kolbuda	1865	H. Olsen		C. Eberth
64577	97	V	**St. Andrew** *Bergstra* -78 W Bk	British	782	160·6	34·2	19·6	St. John, N.B.	1872	Lovitt		Hall & Fairweather
51516	98		**St. Anne** W S	British	670	147·0	32·5	20·1	Russia	1852			Hans Hagans
80730	99		**St. Anne** W Sr	British	125	83·3	23·9	10·2	Les Eboulments	1880	A. Tremblay		Alexis Tremblay
	100	A	**St. Antoine** *Delaplanche* -85 W Bk	French	432	127·8	28·3	17·3	Sunderland	1870	G. Barker		E. Delalun

✻ = Lloyd's Register; D = Nederlandsche Vereeniging; G = Germanischer Lloyd; I = Registro Italiano; R = American ...; A = Veritas Austro-Ungarico; H = Veritas Ellenico; N = Norske Veritas; V = Bureau Veritas.

In the 4th Column, C = Composite (iron frame, wood planked); I = Iron; Stl = Steel; W = Wood.

The Dates of Metalling (Col. 11) are based on the records of the Books in which Vessels are respectively classed, or on the Reports of the Surveyors of ...

Further particulars of Vessels classed by Lloyd's Register will be found in Lloyd's Register Book and Lloyd's Yacht Register.

		Nom. Matériaux et Gréement. Capitaine et dates d'entrée au service des Armateurs actuels et à bord du navire. Ancien Nom.	Pavillon	Tonnage net.	Dimensions. Longueur	Largeur	Creux.	Construit. Où	Quand	Par qui	Date du doublage en métal	Armateurs ou Directeurs-Gérants.	Port d'Armement.
101	V	St. Augustin *Daboy* WIrnSr (ex Nostre Padre)	French	375	116·4	25·9	15·8	Sestri Ponente	1880	G.Perigallo	V 11.85	A.Fabre & fils	Marseilles
102		St. Bede *Emet 81-91* WBg (WRECKED)	British	289	101·0	25·6	14·5	Sunderland	1858	Pickersgill & Miller		J.R.Watson & Co	Newcastle
103		St. Bernardin WSp	British	101	101·0	22·3	6·7	Quebec	1882	H.Gill		Gibbon Gill	Montreal
104	RV	St. Charles *Burns* WS	American	1662	225·0	41·6	25·0	Phippsbrg.Me	1883	C.V.Minott	V 6.88	C.V.Minott	Bath.Me.
105		St. Christopher WBk	German	610	150·1	33·3	17·9	Danzig	1866	F.Devrient		A.Gibsone	Danzig
106	D	St. Christophorus *Teessen* WBk (ex Singapore)	Dutch	505	162·8	32·6	16·9	Nantes	1858	E.Gouin		N.Brantjes	Purmerend
107		St. Clair *Johnston* WSr	British	142	86·5	21·0	11·6	Stirling	1858			J.Johnston	Inverness
108		St. Clair W	British	101	82·6	21·0	8·0	Sophiasburgh, Ont.	1875	John Tait		James Lamont	Chatham,Ont.
109	D	St. Clemens *Mathiesen* WBn	Danish	176	104·5	25·2	10·8	Värnanäs	1865	Hahn		J.H.Schmidt	Copenhagen
110	V	St. Cloud *Patten 82-83* WS	British	1528	209·5	40·0	24·8	Belliveause Cove,N.S.	1877	N.Belliveau	V 6.88	W.D.Lovitt	Yarmouth,N.S.
111	R	St. Croix *Rosebrook 87-87* WSr	American	210	112·0	28·5	9·8	Calais,Maine	1866	W.Robertson	V 8.86	J.S.Emery & Co.	Boston,Mass.
112	✻	St. Cuthbert *Rae 73-85* IBk	British	999	211·0	33·7	20·0	Glasgow	1884	Barcly,Curle&C°		A.Rae	Liverpool
113		St. Cyprien WSp	British	128	105·0	22·8	7·1	St.Thos.de Pierreville	1872			J.B.Desmarais	Montreal
114	RV	St. David *Peterson -87* WS	American	1536	213·4	40·6	25·2	Bath,Maine	1877	J.McDonald	V 3.88	B.Flint	New York
115		St. Devenick *Lawrence* WSr	British	226	114·8	24·8	14·5	Garmouth	1864	Geddie		R.Allix	Guernsey
116	✻	St. Duthus *Shepherd 70-75* WSr	British	85	81·0	26·8	9·6	Kingston	1875	Duncan		Mackenzie & Ross	Inverness
117		St. Elmo *Watts* WSr	American	115	85·5	25·3	7·6	Round Pnd,Me	1870			A.F.Crockett	Rockland,Me.
118	RV	St. Frances *Scribner* WS	American	1811	231·4	41·8	25·9	Bath,Maine	1882	McDonald	V 3.89	B.Flint	New York
119	R	St. François *Gagnon 87-87* WBn	Argentine	282	111·5	29·6	12·7	L'Islet	1875	R.Normand		A.Spinetto	Buenos Ayres
120		St. George *Brook* WSw	British	182	..	..	..	Sunderland	1848			A.&F.Manuelle	Guernsey
121	✻	St. George *Thorsen -88* WS	Norweg'n	1499	209·1	38·0	24·4	Quebec	1872	Valin	V 1.88	Actieselskabet St.George (F.Melsom)	Christiania
122		St. Germans *John 64-64* WSr	British	100	84·6	20·5	11·1	Bideford	1859	Waters		T.H.Geake	Plymouth
123	✻	St. Halvard *Mann -88* WBkn	British	284	130·4	28·7	12·7	Drammen	1878		V 2.87	Bkn."St.Halvard"Co.Ld. (Wombsell & Workman)	Belfast
124		St. Helena *Karlsson* WBk	Russian	347	..	..	..	Finström	1867			P.Sittkoff	Hammarland
125		St. Hilaire WSp	British	146	102·2	23·0	8·6	Quebec	1882	C.Cinqmars		L.Sauvageau	Montreal
126	V	St. Ideuc *Dehillot -88* WBk	French	369	122·3	25·9	14·9	St.Malo	1876	E.Nonet	V 1.88	G.Anquetil	Fécamp
127		St. James WS	British	1162	185·0	39·7	23·8	Quebec	1866	P.Valin		Edward Hopper	Rangoon
128	RV	St. James *Cook -87* WBk	American	1488	218·5	41·3	23·7	Bath,Maine	1883	J.McDonald	V 7.88	B.Flint	New York
129	V	St. Jean *Rubion -87* WBk	French	339	117·4	25·6	15·1	Fecamp	1887	F.Capon	V 1.87	H.Monnier & fils	Fécamp
130	V	St. Jean (ex Alvington) WBk	French	296	123·6	25·4	15·8	Pallion	1865	Watson	V 5.88	Berst père & fils	Granville
131		St. Jean Baptiste WSp	British	123	99·4	22·5	7·4	Quebec	1872			Jean B.Moin	Montreal
132		St. Johannes *Papp 88-77* WBk	German	676	131·5	34·0	22·5	Danzig	1866	J.W.Klawitter		Ernst Wendt	Danzig
133	N	St. Johannes *Olsen* WBg	Norweg'n	293	115·6	28·4	14·5	Risör	1870	Ellif R.Fidje		J.Prebensen	Risor
134	N	St. Johannes *Sörensen -87* WBg	Norweg'n	248	97·9	26·5	14·9	Pitea	1815			P.Sörensen	Arendal
135	N	St. Johannes *Rohde* WBg	Norweg'n	148	82·9	23·1	11·8	Kragerö	1865	J.Andreassen		E.Larsen & G.A.Rode	Brevig
136	RV	St. John *Harris -88* WIrnSr	British	248	119·7	28·0	11·4	Rothesay,N.B.	1881	R.Titus	V 10.87	Charles A.Palmer	St.John,N.B.
137	RV	St. John *Fales* WS	American	1819	216·3	42·7	28·9	Bath,Maine	1870	Chapman&Flint	V 5.88	I.F.Chapman	New York
138	R	St. John's *Gilmore 70-82* WIrnSr	American	406	111·0	33·0	12·0	Belfast,Me.	1882	Carter Bros.		J.G.Brooks	Belfast,Me.
139		St. Joseph *Thomas -88* WS	British	1138	185·3	36·3	23·4	Bath,Maine	1865	Hitchcock		W.Murphy	Dublin
140	RV	St. Joseph *Langlier* WBn	British	233	111·5	27·1	11·9	Cap St.Ignac	1875	E.Auger	V 12.86	Solymi Gamache	Quebec
141		St. Joseph *Foy* WBn	British	169	100·0	26·0	12·0	Quebec	1867			A.King	Ardrossan
142		St. Joseph WSp	British	116	109·7	22·0	7·1	Willimstn Ont	1873			Noah Rousson	Montreal
143		St. Joseph WSp	British	112	97·6	23·0	6·9	Quebec	1882	N.Lamie		W.Carpentier	Montreal
144		St. Joseph WSp	British	103	97·3	23·0	7·2	Sorel	1884	J.Matton		Louis A.Champagne	Quebec
145		St. Joseph WSp	British	101	104·7	23·0	6·2	Sorel,P.Q.	1886	E.Lussier		E.Lussier	Montreal
146		St. Joseph *Morel 86-87* WBn (ex J. L. Stewart)	French	197	100·4	28·7	10·9	Parrsboro',NS	1875			H.Lecharpentier	Granville
147		St. Joseph *Ouesson* WBg	French	161	90·9	20·0	10·8	St.Malo	1861	L.Mallard		Leroux-Berthevin	St.Malo
148	N	St. Joseph *Christensen 80-86* WIrnBn (ex Dauntless)	Norweg'n	385	137·2	30·1	13·0	Granville,N.S.	1871	W.Oliver	V 3.87	C.H.Johnsen	Arendal
149	RV	St. Julien *Sloan -87* WBk	British	1049	177·7	37·0	22·4	Black Riv.,N.B.	1880	J.&R.McLeod	V 3.88	Howard D. Troop	St.John,N.B.
150		St. Julius *Sjöland 79-* WSr	Russian	205	100·0	26·4	12·1	Nakskö	1874	J.Nordstrom		W.Holmberg	Lemland

✻=Lloyd's Register; D=Nederlandsche Vereeniging; G=Germanischer Lloyd; I=Registro Italiano; R=American Record; A=Veritas Austro-Ungarico; H=Veritas Ellenico; N=Norske Veritas; V=Bureau Veritas.

Dans la 4me Colonne, C=Composite (membrure en fer, bordé en bois); I=Fer; Stl=Acier; W=Bois.

Les Dates du doublage en métal (Col. 11) ressortent des livres dans lesquels les navires sont respectivement classés ou des rapports des Inspecteurs du Lloyd's Register.

Pour plus amples renseignements sur les navires classés par le Lloyd's Register, voir le Lloyd's Register Book et le Lloyd's Yacht Register.

Official No.	No. in Book	Society in which classed	Ship's Name. Master, and Dates of Appointment to present Owner's Service and to Ship. Ship's Late Name.	Material and Rig	Flag	Net Tonnage	Length	Breadth	Depth	Built: Where	When	By whom	Date of Metalling	Owners or Managers
77623	151	V	St. Kilda *Sesley -87*	W Bk	British	1056	186·5	35·4	21·5	Bridgwater N.S	1879	Hines & Co.	[illegible]	Robert Lindsay
[illegible]	152		St. Kilda (BROKEN UP)	C Bk	British	885	188·6	32·7	19·0	Glasgow	1868	A. Stephen & Sns		
[illegible]	153		St. Kilda	I Sr	British	189	127·5	22·0	10·4	[illegible]	1868	J. Fullerton & Co.		Martin Kennedy
29415	154	D	St. Lawrence *Mitchell*	W S	British	1019	179·4	37·4	21·4	[illegible]	1861	Smith	[illegible]	Robey Bros.
71240	155		St. Lawrence	W Sp	British	102	97·6	23·0	7·9	St. Thomas de Pierreville	1872			Charles Mongeau
75836	156		St. Louis	W Bkn	British	361	127·7	26·2	11·9	St. Catharines, Ont.	1877	L. Shickluna		Thos. Williamson
90541	157		St. Louis	W Sl	British	114	97·1	22·4	7·1	Yamaska, P.Q	1886	L. H. Lafleur		A. Thibendeau
	158		St. Louis (*New Olga?*)	W Bk	Chilian	598	154·2	30·8	18·9	Sestri Ponente	1875	Bossi	[illegible]	Gutschow & Piza
	159		St. Louis *Touse*	W Bk	French	404	140·4	26·6	14·3	Toulon	1866	Germain		J. Cathala
	160		St. Louis *Parfait*	W Bg	French	142	89·6	23·3	11·5	St. Valery en Caux	1866	J. Lebot		Parfait
35680	161		St. Luce *Pierre*	W Sr	British	122	79·0	23·6	10·2	St. Luce	1868			P. G. Tesseer
23579	162		St. Lucie *Williams*	W S	American	1264	191·4	37·4	24·0	Bath, Maine	1868	McDonald		I. F. Chapman
115280	163	V	St. Lucie *Erskine*	W Bkn	American	689	150·8	35·6	19·3	[illegible], Me.	1874	Merrow	[illegible]	J. W. Elwell
	164	V	St. Lucien *Rocher*	W Bk	French	293	112·2	27·0	13·7	Nantes	1871	Ollivcaud	V 10·81	F. Boissière
98954	165	*	St. Magnus *Brender 83-88*	I S	British	1262	217·0	35·8	22·8	Liverpool	1867	T. Royden & Sns.		Ship "St. Magnus" Co. Ld. (Rankin, Gilmour & Co.)
	166		St. Marc *Billard -87*	W Sr	French	105	78·4	19·7	10·9	Fécamp	1879	Capen		Aldric Basile
79807	167	*	St. Margaret *Irving 86-86*	I S	British	1368	233·1	37·3	22·5	Liverpool	1877	T. Royden & Sns		S "St. Margaret" Co. Ld. (Rankin, Gilmour & Co.)
77596	168		St. Marie	W Sp	British	117	101·0	22·3	6·5	Quebec	1879	E. Belair		Narcisse Paul
115683	169	R V	St. Mark *Nichols*	W S	American	1892	239·9	42·5	27·5	Bath, Maine	1877	Hitchc'k & Blair	[illegible]	I. F. Chapman
58832	170	*	St. Marnock (Now named [illegible])	I Bk	British	1258	216·8	36·0	22·7	Liverpool	1867	T. Royden & Sns		S "St. Marnock" Co. Ld. (Rankin, Gilmour & Co.)
23640	171	R V	St. Mary *Mears -87*	W Bk	American	680	154·1	33·1	20·3	Eastport, Me.	1869		[illegible]	J. W. Elwell
22865	172		St. Mary	W Bk	American	337	106·0	26·4	7·8					P. McGinnis
91294	173	*	St. Mary's Bay *McKuechie 87-87*	I Bk	British	1115	215·2	35·2	21·2	Port Glasgow	1886	Russell & Co.		Hatfield, Cameron & Co
	174		St. Matheus *Lundqvist*	W Bk	Russian	389	..	..	..	[illegible]	1872			A. Nylund
	175	V	St. Maurice *Delaunay* (ex [illegible])	W Sr	French	106	87·6	21·5	10·4	Peterhead	1873	Stephen		Duglé
75736	176		St. Michel *Foote 86-86*	W Bn	British	121	80·6	23·6	9·6	Tusket Wedge, N.S	1878	F. Porter		Jeremiah H. Porter
66017	177		St. Michel	W Sp	British	103	93·2	23·5	7·2	Quebec	1872			Michel Augé
	178		St. Michel	W	French	109	..	..	..					
76793	179	*	St. Mildred *Scott 86-86*	I S	British	1296	229·6	38·1	22·8	Stockton	1877	Richardson, Duck & Co.		Ship "St. Mildred" Co. Ld. (Rankin, Gilmour & Co.)
60837	180	*	St. Monan *Kane*	I S	British	1453	224·7	37·8	23·6	Liverpool	1868	T. Royden & Sons		R. J. Swyny
23801	181	R V	St. Nicholas *Carver -88*	W S	American	1723	205·9	42·8	29·0	Bath, Maine	1869	Chapman & Flint	[illegible]	B. Flint
62003	182		St. Nicolai	W Sr	British	155	107·0	25·0	10·6	Sweden	1863			T. Murray
	183		St. Nicolai *Koklin*	W Bn	Russian	132	89·0	22·0	10·0	Onega	1870	J. Koklin		J. Koklin
	184		St. Nicolo	W Bg	Russian	169	..	..	..	Cherson	1871			D. Xalaigioglu
	185		St. Nicolo	W Bg	Russian	117	..	..	..	[illegible]	1868			E. Crisanopulo
	186	N	St. Olaf *Saunum -87*	W Bk	Norweg'n	381	130·8	29·2	15·2	[illegible]	1874	J. Olsen	[illegible]	[illegible]
	187	N	St. Olaf *Olsen*	W Bk	Norweg'n	391	105·2	28·1	15·8	Kragerö	1854			W. Wright
	188	N	St. Olaf *Hassel -88*	W Bk	Norweg'n	287	119·1	27·1	13·1	Kragerö	1869	A. Johnson	[illegible]	J. F. Juell
66910	189		St. Olaves *Knox 87-87* (ABANDONED)	W Bk	British	572	145·3	32·0	18·2	Hopewell, N.B	1873	J. L. Pye		W. Thomson & Co.
	190	D	St. Olof *Jansson*	W Bk	Russian	539	131·3	32·3	19·4	Sweden	1857			Olaf Erikson
64538	191	V	St. Patrick *Baker -88*	W Bk	British	708	153·2	33·5	19·0	St. John, N.B.	1871	King	[illegible]	Hall & Fairweather
115300	192	R V	St. Paul *Williams -88*	W S	American	1824	228·2	42·1	27·5	Bath, Maine	1874	Chapman	[illegible]	A. G. Ropes
	193		St. Paul *Jean -85*	W Bn	French	129	75·5	20·7	11·9	Nantes	1865	Boju & Chaigneau		P. Ch. Grenet
	194		St. Paul *Jeppesen*	W Bk	German	351	108·9	25·3	18·0	Hamburg	1841			
	195	N	St. Petersburg *Hansen*	W S	Norweg'n	758	152·8	34·5	20·6	Archangel	1867	S. Schmidt		L. Hofgaard
	196		St. Philip *Hansen 78-*	W Bn	Russian	182	86·0	21·0	11·0	Archangel	1858			N. V. Woronjen
80688	197		St. Pierre	W Sp	British	113	102·2	22·2	6·0	Quebec	1881	G. Mansea		Pierre Mougeau
	198	V	St. Pierre *Delasalle -85*	W Bk	French	306	115·4	25·6	14·9	Fécamp	1885	F. Capen		Lecacheur
	199	V	St. Pierre *Luce* (CONDEMNED)	W Bg	French	227	99·1	23·3	13·0	St. Malo	1877		[illegible]	P. Hervot
18[illegible]	200		St. Sermbos	W Bg	Greek	2[illegible]	87·9	21·9	..	Syra	1868			Miltiades Sirmbis

*=Lloyd's Register; D=Nederlandsche Vereeniging; G=Germanischer Lloyd; I=Registro Italiano, R=American [illegible]
A=Veritas Austro-Ungarico; H=Veritas Ellenico; N=Norske Veritas; V=Bureau Ve[illegible]

In the 4th Column, C=Composite (iron frame, wood planked); I=Iron; Stl=Steel; W=Wood.

The Dates of Metalling (Col. 11) are based on the records of the Books in which Vessels are respectively classed, or on the Reports of the Surveyors of Lloyd's [illegible]

Further particulars of Vessels classed by Lloyd's Register will be found in Lloyd's Register Book and Lloyd's Yacht Register.

Nos dans le livre	Registre dans lequel le navire est coté	Nom. Matériaux et Gréement. Capitaine et dates d'entrée au service des Armateurs actuels et à bord du navire. Ancien Nom.	Pavillon	Tonnage net.	Dimensions. Longueur	Largeur	Creux	Construit. Où.	Quand.	Par qui.	Date du doublage en métal	Armateurs ou Directeurs-Gérants.	Port d'Armement.
1	2	3	5	6	7			8	9	10	11	12	13
201	R	St. Thomas *Kelley* W3mSr	American	705	164·1	35·4	17·7	Phippsb'rg Me	1885	Minott		C.V. Minott	Bath, Maine
202	N	St. Thomas *Ants. Gjersöe* W Sr	Norweg'n	158	93·5	21·8	10·8	Pekela	1865			A.O. Gjersöe	Tönsberg
203		St. Ursula W Sr Yt	British	150	103·9	21·3	11·1	Lymington	1853	G. Inman		Col. E.H. Kennard, M.P.	London
204	V	St. Vincent *Evans -86-86* W Bk	British	1377	200·0	36·5	24·8	Quebec	1872	Dunn	V 11.84	C. Hill & Sons	Bristol
205		St. Yves *Langlois -87* W Sr	French	289	100·0	25·9	14·8	Nantes	1862	Jollet & Babin		H. & L. Sebert	St. Brieuc
206	V	Ste. Adresse *Lemoeuf* W Bk ex Beatrice	French	622	147·9	26·4	19·3	Quebec	1866	Valin	V 12.85	J. Silvain	Havre
207	V	Sainte Albertine W Bk ex Weshaw	French	277	124·2	26·2	12·3	Sestri Ponente	1875	Cideria & Fo'aubieri	V 1.84	Eugénie Herbay	Marseilles
208		Ste. Anne W Sr	British	107	87·4	23·0	8·1	Grondines	1865			Germain Hardy	Quebec
209	V	Ste. Anne *Larguet* W Bn	French	133	76·1	20·7	11·4	Perrieres	1872	Lacroix		Le Tallec	Vannes
210	V	Ste. Espérance *LeBihan* W Sr	French	150	80·4	21·3	11·8	Nantes	1865	Boju & Chargeau		A. Le Bihan	Nantes
211	V	Ste. Eugénie *Mahio* W Bk	French	124	78·7	22·0	11·5	Redon	1867	J. Coyac fils		Hurtaud	Auray
212	V	Ste. Hélène *Guillaume* W Sr	French	119	68·6	20·7	11·5	Carnac	1869	Rio fils		J. Guillaume	Auray
213		Ste. Luce *Perré* W Bk	French	342	108·0	27·2	14·1	Bordeaux	1863	Laudard	Z 1.84	Cardinal fils aîné	Nantes
214	R	Sainte Marie *Painlon* W3mSr	British	168	105·5	26·4	9·1	Church Point, N.S.	1888	L.A. Melançon		L.A. Melançon	Weymouth, N.S.
215	*V	Ste. Marthe *Stevens* I 3mSr ex Resolution	French	420	146·0	29·3	12·7	Inverkeithing	1876	J. Scott & Sons		A. Hubeau	St. Nazaire
216	V	Ste. Nathalie W Sr *Vanderbroeck*	French	118	84·3	20·3	11·5	Gravelines	1877	L. Legonlon		Legonlon-Denève	Gravelines
217	R	Sainval Coipel *Bradan* W Bn (WRECKED)	British	290	115·0	29·0	13·1	Bear River, N.S.	1872	T.H. Rice		Willard G. Clark	Digby, N.S.
218		Sakarieh W Bg	Turkish	200	..	..	..	Sakarieh	1873			Hattabji Tujari Vassil Sgoulo & Co.	Sakarieh
219		Sakarieh W Sr	Turkish	199	..	..	..	Sakarieh	1872			Inlara Kiurekjee Hajee Rashid	Sakarieh
220		Sakarieh W Bg	Turkish	187	..	..	..	Sakarieh	1869			Papuji Bashi Kiniasslee Emin Agha & Sons	Sakarieh
221		Sakarieh W Bg	Turkish	182	..	..	..	Sakarieh	1867			Rizan Bey Zadeh Riffat Bey & Co.	Sakarieh
222		Sakarieh W Bg	Turkish	175	..	..	..	Sakarieh	1879			Kereste Usta Kara Agah oglou Issak Effendi & Co.	Sakarieh
223		Sakarieh W Bg	Turkish	125	..	..	..	Sakarieh	1870			Rizehlee Ahmed Reiss oglou Hussein Agha	Rizeh
224		Sakarieh W Bg	Turkish	125	..	..	..	Sakarieh	1869			Mehemed & Arif Aghas & Co.	Sakarieh
225		Sakarieh W Bg	Turkish	116	..	..	..	Sakarieh	1878			Katirleelee Ligori Niago	Sakarieh
226		Sakkini Deriya W Bg	Turkish	125	..	..	..	Sinope	1883			Tilar Yorghanji zadeh Hajee Mehmed Bey	Sinope
227		Sakyz W Bk	Turkish	450	..	..	..	Scio	1870			Bayirehlee Adirinon	Scio
228	RV	Salacia *Eldredge -87* W Bk	British	797	167·7	35·7	19·1	Harvey, N.B.	1878	G.S. Turner	V 7.86	George F. Smith	St. John, N.B.
229		Salacia *Waiter* W Bn	British	227	104·6	25·4	15·3	Sunderland	1867	W.H. Pearson, jun.		G. Holden, jun.	Faversham
230	*	Salado *Anderson* I Bk	British	442	158·5	26·5	15·7	Port Glasgow	1866	Kirkpatrick		J.M. Macfarlane	Greenock
231	R	Salama *Lezelius* W Bk (CONDEMNED)	Russian	491	165·0	30·0	16·0	Gamla Karl'by	18 67/70	Skepps & Warfs-bolaget		A. Finnilä	Gamla-Karleby
232	*	Salamanca *Bryce 82-82* I Bk	British	1202	227·7	35·9	21·3	Glasgow	1877	J. & G. Thomson		J. Hardie	Glasgow
233	R	Salamanda *Svensson 82-82* W Bg	Swedish	273	107·6	23·2	14·8	[illegible]	1866			M. Svensson	Bergqvara
234		Salamanderen *Abrahamsen -86* W Sr (CONDEMNED)	Norweg'n	166	92·0	28·6	10·6	Hernösand	1843	O. Nyström		G. Abrahamsen	Tönsberg
235		Salamandern *Mattson* W Sr	Swedish	213	99·7	23·4	12·8	Carlscrona	1856			J.A. Samuelsson	Mollösund
236	*	Salamis *Philip 88-88* I S	British	1079	221·6	36·0	21·7	Aberdeen	1875	W. Hood & Co.		G. Thompson & Co.	Aberdeen
237	DV	Salatiga *Caspersen* W S	Norweg'n	1101	171·9	39·7	23·6	Rotterdam	1872	De Jong, Kochland & Anthony	V 1.88	Aktieselskabet "Salatiga" (Chr. M. Beer)	Christiania
238	*V	Salazie *Laffitte* W Bk	French	633	153·0	28·4	19·2	Bordeaux	1872	Chaigneau frères	V 1.88	J. Puy-Bouffat	Bordeaux
239	N	Salcha *Aratsen -88* W3mSr	Norweg'n	274	110·9	28·0	14·6	Arendal	1878		[illegible]	Th. Thorsen	Arendal
240	NV	Salem *Kornsen -88* W Bk	Norweg'n	427	128·1	29·2	17·1	Tvedestrand	1873	S. Tellefsen	[illegible]	Johan Smith	Tvedestrand
241		Salem *Pehrsson* W Bg	Swedish	269	117·5	25·4	12·4	Sundsvall	1869			L. Jonsson	Jonstorp
242		Saleta *Lubraro -87* W Bk (BROKEN UP)	Spanish	406	130·9	27·9	17·4	Sunderland	1865	Tiney	V 12.84	Doña Maria de Escazo y Olazo	Bilbao
243		Salim *Said Salim Alkaf* W Bk	Dutch	205	83·7	22·3	..					Said Hamid Alkaf	Palembang
244	V	Salina *Peterson* W Bk	British	296	138·0	33·9	10·4	Eatonville, N.S.	1886	J.E. Starr		James Gilb. Ross	Parrsboro', N.S.
245	R	Salisbury *Wilbur -87* W Bn (CONDEMNED)	British	343	123·2	30·2	12·9	Salisbury, N.B.	1878	T. McEwen		A.J. Babang	Moncton, N.B.
246		Salisbury *Rottenstein 80-80* W S (ABANDONED)	German	1017	198·0	36·1	22·6	Troon	1866	Portland S.B. Co.	[illegible]	J.D. Bischoff	Bremen
247		Sallie B. *McGlathlin* W Sr	American	273	116·5	30·0	9·9	Philadelphia	1861			R.N. Elwell	Boston, Mass.
248		Sallie Bousall W Bk	British	332	115·5	26·8	13·0	Robbinst'n U.S.	1862			Edward Stinson, jun.	Nassau, N.P.
249		Sallie E. Ludlam W3mSr	American	225	113·7	30·3	9·3	Goshen, N.J.	1873			J. Mair	Philadelphia
250													

*=Lloyd's Register; D=Nederlandsche Vereeniging; G=Germanischer Lloyd; I=Registro Italiano; R=American Record;
=Veritas Austro-Ungarico; H=Veritas Ellenico; N=Norske Veritas; V=Bureau Veritas.

Dans la 4me Colonne, C=Composite (membrure en fer, bordé en bois); I=Fer; Stl=Acier; W=Bois.

Les Dates du doublage en métal (Col. 11) ressortent des livres dans lesquels les navires sont respectivement classés ou des rapports des Inspecteurs du Lloyd's Register.

Pour plus amples renseignements sur les navires classés par le Lloyd's Register, voir le Lloyd's Register Book et le Lloyd's Yacht Register

Official No.	No. in Book	Registry in which classed	Ship's Name (Ship's Late Name)	Master, and Dates of Appointment	Material and Rig	Flag	Net Tonnage	Length	Breadth	Depth	Built: Where	When	By whom	Metalling	Owners or Managers
19028	251	R	Sallie Ton	West	W3mSr	American	523	143·8	34·7	12·8	Searsport, Me.	1884	G. Merrill	[illegible]	W.J. Nickerson
23430	252		Sallie T. Charter	Cumming	WSr	American	116	105·0	21·5	6·8	Lewes, Del.	1858			T.L. Church
	253	DGV	Sally (ex Hedwig, ex Intrepid)	Forsberg -88	WS	Swedish	1126	183·4	36·4	23·5	Bath	1864	E. & A. Sewell	[illegible]	A. Brolin
	254		Sally	Andersson -86	WSr	Swedish	123	86·3	21·3	10·4	Bergqvara	1845	Berggren		A.R. Andersson
22437	255		Sally W. Ponder	Lincoln -76-	WSr	American	102	86·6	25·4	6·4	Milton, Del.	1854			E.F. Lincoln
	256		Salma	Kappelin -8	WBn	Swedish	127	84·7	21·2	10·1	Timmernabben	1871	C. Bring		J.D. Björkegren
	257		Salma		Bg	Turkish	168	..	..	..					Tanous & Wakim
90471	258	V	Salmon	Bradley -88	WBk	British	1163	200·3	38·5	22·2	S'th Maitland	1885	A. McDougall	[illegible]	Sarah McDougall
	259		Salo	Raussi 78-	WBk	Russian	439	143·0	32·0	16·5	Wederlaks	1878	D. Hietala		D. Hietala
	260	N	Salome	Nielsen	WBk	Norweg'n	465	135·4	30·9	16·9	Risör	1874	T. Jensen	[illegible]	Thygesen & Aalholm
1004	261	V	Salomone (ex Templar)	Callabiano -88	WBk	Italian	577	131·5	29·6	18·0	Brewer	1862		[illegible]	D. Salomone
	262	✻	Salsette	T. G. Roulton	IS	British	1614	259·4	38·2	23·2	Dumbarton	1886	Birrell, Stenhouse & Co.		J. Shepherd & Co.
87768	263	✻	Saltaire	Carrington	WSk	British	89	78·8	20·4	11·0	Rye	1883	J.C. Hoad		W.M. Carrington
49501	264		Saltee (WRECKED)	Smith -87	WBk	British	266	111·0	25·4	14·7	Weymouth, N.S.	1868	Savory	[illegible]	Dock Yard Co.
51230	265		Salterbeck	Murray	WBk	British	179	104·8	23·7	12·0	Workington	1869	Beck		T. Murray
78072	266	✻	Saltram	Bishop	WSr	British	129	95·6	22·4	11·5	Plymouth	1877	Marshall		H. Goad
58026	267		Salus	Boylan	WBkn	British	316	129·0	26·6	14·9	Shoreham	1868	Colbran	[illegible]	Miss M.J. Cullen
	268	V	Salus	Bager 81-84	WSr	Danish	144	94·4	21·9	10·8	Marstal	1875	H.J. Bager		L.J. Bager
	269	N	Salus	Hansen	WBg	Norweg'n	247	100·7	26·8	14·6	Sandefjord	1873			K.N. Björndal
	270	N	Salus	Olsen	WBk	Norweg'n	222	106·9	27·6	11·7	Grimstad	1865	K. Andersen		H. Omundsen
	271	N	Saluto	Stray 78-88	WBk	Norweg'n	734	171·5	33·5	18·6	Grimstad	1888	S. Johnsen	[illegible]	M.F. Stray
49194	272		Salvador		WSr	British	162	94·4	26·2	8·6	Baltimore, U.S.	1864			James B. Bell
	273		Salvador		WSr	Brazilian	114	..	..	..					D.B.M. dos Santos
1457	274	V	Salvador	Cordiglia -89	WBk	Italian	406	120·4	31·0	18·9	Varazze	1866	G. Scala	[illegible]	L.F. Valle
	275	✻	Salvador (ex Alice Smith)	Decarata 88-88	IBn	Spanish	239	110·1	23·5	13·2	Sunderland	1866	W. Doxford		F. de Carranza
	276	V	Salvator	Grosse	WBg	French	196	88·3	23·3	12·8	Nantes	1875	J.J. Cassard	[illegible]	P. Bureau
	277	N	Salvator	Schanke	WBk	Norweg'n	356	105·2	29·9	17·0	Porsgrund	1839	J. Gasmann		T. & O. Gurstad
	278		Salvator	Bull -88	W3mSr	Norweg'n	288	110·4	26·2	16·2	Liverpool	1825			T.M. Bull
	279	N	Salvator	Severtsen	WBg	Norweg'n	229	93·1	24·2	14·3	Memel	1826			Et Actieselskab (Chr. Gasmann)
1136	280	IV	Salvatore	Romano	WBk	Italian	487	138·7	28·5	19·3	Sorrento	1877	A. Castellano	[illegible]	F.S. Ciampa & Co.
771	281	I	Salvatore	Olivari	W3mSr	Italian	407	132·8	27·9	17·2	Rapallo	1877	G. Merello	[illegible]	D. Chighizola
517	282	A	Salvatore (ex Anonima)	De Franceschi	WBn	Italian	174	..	..	..	Savona	1872		[illegible]	S. Guarrera
486	283		Salvatore		WBg	Italian	141	77·1	24·1	14·0	Pra	1854	Tixi		G. Cignoni fu L.
3305	284	I	Salvatore Accame	Gavone	WBk	Italian	809	183·0	33·8	22·0	Sestri Ponente	1878	G. Calcagno	[illegible]	E. Accame & F.
224	285		Salvatore Camogli (WRECKED)	[illegible]	WBn	Italian	123	75·5	22·6	12·8	Varazze	1850	G. Cerruti		A. Cichero
3298	286	DI	Salvatore Dall' Orso	Cavagnaro -87	WS	Italian	1198	211·6	37·1	24·8	Sestri	1883	A. Carlevaro	[illegible]	S. Dall'Orso
631	287	V	Salvatore-de-Salvo & figlio (ex F. S. Cimmino ex Raffaello)	[illegible]	WBg	Italian	306	112·2	26·7	17·8	Castellamare di Stab.	1869	A. Bonifacio	[illegible]	Salvatore de Salvo & figlio
1924	288		Salvatore Naccari (ex [illegible])	[illegible] -88	WBn	Italian	222	96·8	26·6	16·0	Procida	1858			N. Naccari
2339	289	A	Salvatore Riccio Figlio (ex Elena Vagliosa)		WBn	Italian	259	123·0	26·6	16·4	Estera			[illegible]	M. Riccio
	290	DN	Salveen	Dalling 72-82	WBk	Norweg'n	271	101·5	25·8	15·0	Moulmein	1851			Emil Salvesen
399	291	I	Salvo	Cirino	WBg	Italian	168	102·3	22·9	12·9	Sestri	1874	Debarbieri & Celesia	[illegible]	F[lli] Salvo
	292	✻	Sam J. Dodson	Rowson	WSk	British	59	67·4	19·0	9·2	Rye	1886	G. & T. Smith		S. Rowson
28288	293	✻	Sam Mendel	Gower -88	IBk	British	1017	215·6	35·0	20·6	W. Hartlepool	1861	Pile, Spence & Co		T. Roberts
115673	294	V	Sam Skolfield, 2d	Bishop -87	WS	American	1514	218·7	39·9	24·2	Brunswick, Me	1883	Skolfield Bros.	[illegible]	Skolfield Bros.
48631	295	✻	Sam Weller	Pennison -88	WBn	British	165	102·5	23·1	12·9	Brixham	1864	Upham	[illegible]	W. Bellingham
91167	296	✻	Samanco	Bell 76-84	IBk	British	815	195·0	32·3	19·1	Sunderland	1884	W. Pickersgill & Sons		S. Wakeham & Son
115426	297	RV	Samar	Emery -88	WS	American	1070	184·0	36·0	22·6	East Boston	1875	Campbell & Brooks	[illegible]	K. Page
115408	298	RV	Samaria	Snow	WS	American	1509	217·6	39·1	24·1	Bath, Maine	1876	Houghton Bros.	[illegible]	Houghton Bros.
76167	299	✻D	Samarkand	Bain 87-87	IBk	British	1104	220·7	35·1	21·1	Glasgow	1877	Barclay, Curle & Co.		W. & J. Crawford
	300	N	Sambo	Wilhelmsen	WBg	Norweg'n	268	98·5	27·5	15·1	Prussia	1853			A. Olsen

✻ = Lloyd's Register; D = Nederlandsche Vereeniging; G = Germanischer Lloyd; I = Registro Italiano; R = American [Record]; A = Veritas Austro-Ungarico; H = Veritas Ellenico; N = Norske Veritas; V = Bureau V[eritas]

In the 4th Column, C = Composite (iron frame, wood planked); I = Iron; Stl = Steel; W = Wood.

The Dates of Metalling (Col. 11) are based on the records of the Books in which Vessels are respectively classed, or on the Reports of the Surveyors of Lloyd's [Register].

Further particulars of Vessels classed by Lloyd's Register will be found in Lloyd's Register Book and Lloyd's Yacht Register.

Nos dans le livre	Registre dans lequel le navire est coté	Nom. / Capitaine et dates d'entrée au service des Armateurs actuels et à bord du navire. / Ancien Nom.	Matériaux et Gréement	Pavillon	Tonnage net	Longueur	Largeur	Creux	Où	Quand	Par qui	Date du doublage en métal	Armateurs ou Directeurs-Gérants	Port d'Armement
		Samboyana	WBk	Spanish	331	98·8	28·5	17·1	Loano	1856			F. Forteza	Barcelona
[illegible]		Samhold	WBk	Norweg'n	287	112·5	28·6	10·8	Bergen	1886			K.S. Bertelsen	Stavanger
[illegible]		Sammy Ford	WSr	American	109	87·0	25·4	7·9	Lubec, Maine	1872	McBride		C.J. Staples	Lubec, Me.
[illegible]	*	Samoa *Jenks 79-86*	IBk	British	1109	214·6	35·1	21·2	Greenock	1886	Russell & Co.		P. Denniston & Co.	Glasgow
[illegible]		Samoa	WSr	British	163	112·3	25·2	11·1	Stockholm	1866			John H. Goodlet	Sydney, N.S.W.
306		Samos *Smith 84-*	W3mSr	American	375	132·5	29·6	15·5	Boston, Mass.	1868	Smith & Townsend		T.L. Mayo	Boston Mass.
307	A	Samos *Arghiros*	WBk	Turkish	331	..	..	..	Syra	1878		[illegible]	G. Nicolan & Co.	Samos
308		Samoset *Schwarting -87*	WBkn	American	601	156·0	30·2	19·1	Portsm'th, NH	1847	Tucker		G.E. Atkinson	P. Townsend, Wash.
309	V	Sampo *Sorensen -87*	WBk	Norweg'n	801	178·5	36·5	19·5	Uleaborg	$18\frac{69}{82}$	G. Petterson	[illegible]	Chr. Möller	Christiania
310		Samson	ISr	British	205	104·3	23·0	12·0	Glasgow	1887	Barclay Curle & Co., Ld.		Rangoon Port Commissioners	Rangoon
311		Samsoom	Bg	Turkish	135	..	..	..					Mohamed Eff. Mosa	Alexandria
312		Samuel *Hervu*	WBg	French	180	86·9	22·6	13·0	St. Malo	1857	Hervé		Amaury Simon	Redon
313	*	Samuel *Ohlrich -88*	WBk	German	427	131·0	27·2	17·3	Sunderland	1861	Mills & Austin	[illegible]	R.S. Carr	Hamburg
314		Samuel (CONDEMNED) *Roman -85*	WSr	Russian	252	96·6	26·5	13·9	Haynasch	1879	J. Brackmann		M. Klein	Haynasch
315	R	Samuel B. Hale *Haven (86-?)*	WBk	American	538	138·9	30·9	16·5	Westbrook, Me	1870	R. Lewis & Co.	[illegible]	R. Lewis & Co.	Portland, Me.
316	R	Samuel B. Hubbard *McLaffey*	W3mSr	American	383	136·0	33·5	13·0	N. London Con.	1882	J. Davidson		Joseph Luther	Nw. Londn., Conn.
317	R	Samuel B. Vrooman *Smith*	W3mSr	American	449	131·2	31·5	15·8	Philadelphia	1882	Doughty & Kapella		G.W. Smith	Philadelphia
318		Samuel C. Hart *Kelly 83-83*	WSr	American	246	112·0	29·2	8·7	Bath, Maine	1870	Goss & Sawyer		E. Gibbs	Nw Bedfrd., Mass.
319		Samuel Castner, jun. *Cook*	WSr	American	228	119·8	29·3	10·0	Milford, Del.	1863			S. Castner, Trustee	Philadelphia
320		Samuel D. Carleton (QUARANTINE SHIP) *Freeman*	WBk	American	843	158·9	34·2	21·0	Camden, Maine	1868	Carlton & Norwood	[illegible]	G.A. Freeman	Boston, Mass.
321	R	Samuel Dillaway *Drinkwater*	W3mSr	American	771	170·4	35·2	17·9	Bath, Me.	1886	Deering & Donnell		Deering & Donnell	Bath, Me.
322	R	Samuel E. Spring (HULK) *Rice -80*	WBk	American	479	126·4	30·0	17·2	Westbrook, Me	1869		[illegible]	H.P. Dewey	Portland, Me.
323	RV	Samuel H. Nickerson *Nason -78*	WBk	American	755	152·0	34·3	20·0	Harringt'n, Me	1878	G. Emery	[illegible]	J. S. Winslow & Co.	Machias, Me.
324		Samuel H. Sharp *Carter 81-84*	WSr	American	236	114·4	28·9	9·9	Mauricetown, N.J.	$18\frac{69}{85}$			C.W. Alcott & Co.	New York
325	R	Samuel H. Walker *Kelly*	W3mSr	American	561	155·5	35·5	15·4	East Boston	1881	R. Crosbie & Son		S.H. Walker	Taunton, Mass.
326		Samuel Hart *Bellbrook*	WSr	American	200	104·0	29·3	9·9	Waldoboro, Me	1871			S. Hart	St. George, Me.
327	*	Samuel Holland *Williams 77-77*	WSr	British	151	92·4	23·1	12·5	Portmadoc	1867	Griffiths		J. Williams	Carnarvon
328		Samuel L. Russell *McClain*	WSr	American	184	109·0	25·5	7·9	Leesburg, N.J.	1869			C.S. Smith	Bridgeton, N.J.
329	*	Samuel Moss *Climore*	WSr	British	141	97·0	23·3	11·1	Par	1878	Moss	[illegible]	S. Truscott	Fowey
330		Samuel P. Adams *Scott*	WSr	American	146	92·0	27·3	9·0	Columbia, Me.	1864			J.W. Green	Deer Isle, Me.
331	*	Samuel Plimsoll *Henderson 75-81*	IS	British	1459	241·3	39·0	23·1	Aberdeen	1873	W. Hood & Co.		G. Thompson & Co.	Aberdeen
332	R	Samuel S. Thorp *Fraser*	W3mSr	American	502	150·0	35·6	15·2	City Islnd, N.Y	1881	D. Carll		F.O. Bailey	Perth Amboy, N.J.
333	R	Samuel Welsh (SUNK) *Irving*	WBkn	American	480	138·0	32·0	17·0	Bord'ntown N.J.	1875	J.S. Lamson		W. Busby	Philadelphia
334	*	Samuel & Ann *Reed 85-85*	WSk	British	86	76·6	20·8	10·4	Goole	1885	Cottnghm Bros.		S. Glansford	Hull
335	*	San Antonio *Troop*	WBk	British	497	143·0	29·5	18·3	Sunderland	1867	J. Hardie	[illegible]	A.B. Troop	Liverpool
336		San Antonio	WPol	Spanish	168	88·3	25·4	13·0	Lloret	1855			A.P. Franch	Majorca
337		San Antonio (ex [illegible], ex E. B. Darling)	WSr	Spanish	150	96·8	27·8	9·2	Port Jefferson	1874			Tomas Bosch	Great Canary
338		San Antonio *Baeza*	WSr	Spanish	146	92·5	26·7	11·4	Villajoyosa	1875			P. Raola	Palamas
339		San Battistino (ex Luigia V.) (BROKEN UP) *Scaparone*	WBk	Italian	558	152·8	31·9	19·2	Sestri Ponente	1870	A. Briasco fu G.B.	[illegible]	G.A. Scaparone	Genoa
340	I	San Beltran *Besà*	WBg	Italian	193	99·4	25·7	14·0				[illegible]	S. Lancella fu F.	Leghorn
341	I	San Biagio *Esposito -87*	WBk	Italian	342	..	..	..	Fiume	1854		[illegible]	Fⁱˡˡ Florentino fu Antonio	Naples
342		San Bruno	WBn	Spanish	174	76·4	27·2	12·5	Villajoyosa	1850			Tomas Uria	Alicante
343		San Buenaventura *Peltz*	WS	American	172	107·0	30·0	8·5	Fairhaven, Cal.	1876			D. Steffens	San Francisco
344		San Carlos *Ebbesson -88*	WBk	Swedish	889	161·6	31·6	22·6	Newbury Mass.	1856		[illegible]	J.P. Ebbesson	Vestra Karup
345	I	San Domenico *Serra*	W3mSr	Italian	467	128·3	27·9	19·0	Savona	1869	G.B. Guastavino	[illegible]	G. Becchi	Savona
346	R	San Domingo (ABANDONED) *[illegible]*	WSr	American	491	131·2	32·4	13·7	Bath, Maine	1884	N. Engl'nd S.B. Co.	[illegible]	E. Rokes	Baltimore
347	I	San Francesco di Paola *Salmone*	WBg	Italian	151	91·5	24·0	12·8	Gaeta	1859	P. Gallinaro		F.S. D'Ottone	Gaeta
348	A	San Francesco di Paola (ex [illegible]) *[illegible]*	WBg	Italian	121	78·7	23·8	12·3	Venice	1856			P. Sforza di F.	Bari
349	G	San Francisco *Arkema (86-*	WBk	Dutch	436	135·5	30·5	16·4	Hamburg	1859	Reiherstieg Schiffs Werfte		S. van der Hei	Heveskes
350		San Francisco *Rosé 83-87*	WBk	French	656	165·6	31·6	19·1	Bordeaux	1876	Roch't & L'rgetau	[illegible]	Ant. Dom. Bordes & fils	Bordeaux

*=Lloyd's Register; D=Nederlandsche Vereeniging; G=Germanischer Lloyd; I=Registro Italiano; R=American Record; A=Veritas Austro-Ungarico; H=Veritas Ellenico; N=Norske Veritas; V=Bureau Veritas.

Dans la 4me Colonne, C=Composite (membrure en fer, bordé en bois); I=Fer; Stl=Acier; W=Bois.

Les Dates du doublage en métal (Col. 11) ressortent des livres dans lesquels les navires sont respectivement classés ou des rapports des Inspecteurs du Lloyd's Register.

Pour plus amples renseignements sur les navires classés par le Lloyd's Register, voir le Lloyd's Register Book et le Lloyd's Yacht Register.

Official No.	No. in Book	Registry	Ship's Name. Master, and Dates of Appointment to present Owner's Service and to Ship. Ship's Late Name.	Material and Rig	Flag	Net Tonnage	Length	Breadth	Depth	Where Built	When	By whom	Date of Metalling	Owners or Managers	Port
[illegible]	351	✻	San Francisco (ex John Allan) Scotto	W Bk	Italian	734	159·1	32·9	20·8	Sunderland	1864	W. Briggs&Sons	[illegible]	Flli Lubrano	Castel…
	352		San Francisco	W Bn	Spanish	108	84·3	19·4	9·0	Tayabas	1885			Rafael Calvo	Manila
	353		San Francisco	W Bn	Spanish	105	86·0	22·0	9·6	Villajoyosa	1850			J. Corbeto y Salinas	Alicante
	354		San Francisco de Paula (ex Lolita)	W Bk	Spanish	574	151·5	30·1	20·1	Barcelona	1866			Balcells & Subirana	Barcelona
	355		San Francisco de Paula (ex Zaragoza)	W Bk	Spanish	158	96·8	24·7	9·1	Buccari	1857			José Riera	Valencia
[illegible]	356	I	San Gaetano Mazzella	W Bg	Italian	411	120·0	31·2	18·8	Procida	1872	D. Fiorentino	[illegible]	A. Mignano	Naples
	357		San George	W Bg	Turkish	119	..	..	..					Tanous & Wakiem	Alexandria
	358		San Giovanni Papacostandopoulo	W Bg	Greek	191	87·9	22·6	13·1	Galaxidi	1870		[illegible]	Papacostandopoulos	Galaxidi
[illegible]	359		San Giovanni (ex Anatolio) Andreoni	W Bg	Italian	296	110·8	27·2	17·8	Varazze	1858			B. Andreoni fu R.	Leghorn
	360		San Giovanni	W Bg	Turkish	270	..	..	..	Italy	1886			Oouili zadeh Apik Effendi	Constant…
[illegible]	361	I	San Giovanni Battista Antonio	W Bg	Italian	238	96·5	27·0	16·1	Procida	1862		[illegible]	M. Mignano	Naples
[illegible]	362		San Giovanni Battista Guerri	W Bg	Italian	200	97·1	24·7	15·5	Sestri Ponente	1854			S. Ancella & Polese	Leghorn
	363		San Giovanni Battista (ex [illegible])	W Bn	Italian	150	..	..	..	Varazze	1869			R. Martinelli	Leghorn
[illegible]	364		San Giovanni Battista (FOUNDERED) Aiolo	W Bg	Italian	120	75·5	23·9	10·1	Sestri Ponente	1847	D. Briasco		S. Carbone	Palermo
[illegible]	365	I	San Giovanni E. Bertolotto -88	W Bk	Italian	831	176·1	33·8	22·7	Savona	1873	G. B. Ponziglione	[illegible]	G. Bertolotto	Genoa
[illegible]	366		San Giovanni Evangelista Ferolie	W Bg	Italian	285	106·6	27·6	17·5	Sestri Ponente	1859	Flli Cadenaccio	[illegible]	S. Schiaffino	Genoa
[illegible]	367	✻	San Giuseppe (ex [illegible]) (WRECKED) Rinaldi	W Bk	Italian	495	137·3	28·7	18·1	Southwick	1864	Mills	[illegible]	M. A. Lubrano	Naples
[illegible]	368	I	San Giuseppe P. (ex Eufemia) Costagliola -87	W Bg	Italian	317	106·9	28·5	17·1	Sorrento	1865	A. Castellano	[illegible]	V. Costagliola	Naples
	369	A	San Giuseppe Maffei	W Bn	Italian	103	..	..	..	Viareggio	1879		[illegible]	Cl. Della Lena	Viareggio
[illegible]	370	I	San Giuseppe B. Benvenuto	W Bg	Italian	229	96·1	25·3	16·0	Savona	1857	Calamaro	[illegible]	E. Benvenuto	Genoa
[illegible]	371	V	San Joaquin Drinkwater	W S	American	1637	219·8	41·6	24·2	Freeport, Me.	1876	E. C. Soule	[illegible]	E. C. Soule	Portland
[illegible]	372		San José Denni -88	W Bk	British	393	130·0	27·7	17·8	Sunderland	1861	Pickersgill		J. C. & G. S. Richardson	Swansea
	373		San José	W Bn	Spanish	228	104·6	27·7	13·4	Palma	1869			Baltasar Cortes	Majorca
	374		San José (ex Antonieta)	W Bn	Spanish	171	86·9	23·7	12·8	France				Juan Nieto	Majorca
	375		San José	W Bn	Spanish	118	80·7	23·5	10·4	Alicante	18 57/80			Romualdo Perez	Alicante
	376	N	San Juan Bache	W Bg	Norweg'n	295	110·5	26·4	16·0	Sunderland	1868	W. Richardson		C. M. Bache	Drammen
[illegible]	377	✻	San Lorenzo Francis	I Bk	British	466	156·2	26·5	17·7	Liverpool	1861	Jones, Quiggin & Co.		J. L. Jacobs & Co.	London
[illegible]	378	V	San Luigi Trapani	W Bk	Italian	491	136·1	31·0	20·0	Equa	1871	G. Mauri	[illegible]	Flli Trapani fu Stef	[illegible]
[illegible]	379	✻	San Luis Lonnen 86-86	I Bk	British	591	171·2	27·8	17·7	Liverpool	1864	T. Royden&Sns		R. W. Leyland & Co.	Liverpool
	380	V	San Luis (ex Clarissa [illegible]) Gundersen	W Bk	Norweg'n	278	124·6	27·6	11·6	Baltimore	1858		[illegible]	S. Larsen	Drammen
	381		San Luis	W	Spanish	461	112·8	28·5	17·0	Havanah	1886			A. Martinez	Havana
	382		San Mariano	W Sr	Spanish	147	91·5	25·8	11·2	Blanes	1864			A. Roca Felañy	Majorca
[illegible]	383	V	San Martin (ex Aquila) ([illegible] Aquila")	W Bk	Italian	321	115·4	27·4	17·8	Sestri Ponente	1860		[illegible]	C. Reali	Leghorn
[illegible]	384		San Martin Onza.n 87-87	W Sr	Peruvian	107	76·7	20·6	10·0					C. Castillo	Callao
[illegible]	385	✻	San Michele (ex Adam Sedgwick) Ambrosino -87	W Bk	Italian	442	142·2	27·5	18·6	Alembon	1861		[illegible]	M. A. Lubrano	Castel…
	386	I	San Michele (ex [illegible], ex Cancon Madre) Calabrese	W Bk	Italian	408	123·3	29·8	18·9	Varazze	1867	A. Cravioto	[illegible]	R. Polese & Co.	Naples
[illegible]	387	A	San Michele (ex Bachelor) Mancino	W Bg	Italian	261	95·8	27·7	16·2	Procida	1872	L. Trambavalo	[illegible]	Flli Mancino	Naples
[illegible]	388	I	San Michele Arcangelo Bitonto	W Bn	Italian	223	95·1	26·8	16·1	Alimuri	1858	M. Mauro	[illegible]	A. di Bitonto	Bari
	389		San Miguel	W	Spanish	103	77·4	23·1	9·4	Barcelona	1866			A. R. Oliver	Valencia
[illegible]	390		San Nicolas (ex Carolina) Risso 87-87	W Bk	Peruvian	350	120·0	27·0	16·5	Sunderland	1869			E. Monteverde	Callao
	391		San Pablo	W Bn	Spanish	114	94·4	19·7	9·2	Pitogo	1886			T. Glinoga	Manila
[illegible]	392	I	San Paolo (ex Emma D.) Scotto -88	W Bk	Italian	585	112·6	31·8	19·8	Voltri	1869	B. Canepa	[illegible]	G. Massone	Genoa
	393		San Pasquale	W E	French	108	74·8	22·0	10·8	Bastia	1876			Leonetti & Co.	Bastia
[illegible]	394	A	San Pasquale Baylon Magliocca	W Bn	Italian	150	86·3	24·2	12·2	Gaeta	1862	F. Contreras		Flli Matarazzo	Gaeta
[illegible]	395		San Pedro	W Bn	British	112	80·0	21·8	10·5					Luigi Psaila	George…
[illegible]	396		San Pietro	W Bk	British	523	142·0	31·3	18·1	Castellamare	1879			James Smith	Port …
[illegible]	397	IV	San Pietro Longello	W Bg	Italian	476	125·6	28·2	13·1	Castellamare	1879	M. Trambarulo	[illegible]	F. Stornce & Co.	Castel…
[illegible]	398	V	San Prisco Bussecu	W Bg	Italian	371	115·8	27·2	18·0	Sorrento	1872	A. Castellano	[illegible]	A. Paturzo	Castel…
[illegible]	399		San Pubblio (ex Roma)	W Bg	British	192	99·8	26·6	13·6					Francesco Guerriera	Malta
	400		San Quintin (ex [illegible]) Maury -87	W 3m Sr	Mexican	200	109·1	23·0	7·4	Bowdoinham, Me.	1873			José Maynus	Campe…

✻ = Lloyd's Register; D = Nederlandsche Vereeniging; G = Germanischer Lloyd; I = Registro Italiano; R = American Re…
A = Veritas Austro-Ungarico; H = Veritas Ellenico; N = Norske Veritas; V = Bureau Veri…

In the 4th Column, C = Composite (iron frame, wood planked); I = Iron; Stl = Steel; W = Wood.

The Dates of Metalling (Col. 11) are based on the records of the Books in which Vessels are respectively classed, or on the Reports of the Surveyors of Lloyd's …

Further particulars of Vessels classed by Lloyd's Register will be found in Lloyd's Register Book and Lloyd's Yacht Register.

No.	Classe	Nom. Matériaux et Gréement. Capitaine et dates d'entrée au service des Armateurs actuels et à bord du navire. Ancien Nom.		Pavillon.	Tonnage net.	Dimensions.			Construit.			Date du doublage en métal.	Armateurs ou Directeurs-Gérants.	Port d'Armement.
						Longueur.	Largeur.	Creux.	Où.	Quand.	Par qui.			
401		San Rafael (ex [illegible])	W Bn	Spanish	237	106·3	26·8	15·[illegible]	Palma	1857			M. Bennasar	Majorca
402	V	San Raffaele *Cacace*	W Bg	Italian	278	109·5	26·2	16·9	[illegible]	1867	G. Mauro	Y 6.88	D. Cacace	Castellamare
403		San Remo *Pera*	W Sr	Italian	240	115·1	25·6	11·8	[illegible]	1849			Stefano Pera	Porto Maurizio
404		San Spiridion *Castroldi*	W Bk	Russian	550	111·4	29·8	17·5	Cherson	1872	Coffidachi		Welikanoff & Carazoli	Odessa
405	RV	San Stefano *Beal* 88-88	W S	British	1197	194·2	38·2	22·5	Beaver Riv. NS	1878	G. E. Jenkins	Y 3.86	Loran E. Baker	Yarmouth, N.S.
406	I	San Tomaso *Vassalo* (ex Tomaso Parodi 2)	W Bk	Italian	430	124·3	30·2	19·3	Varazze	1868	N. Cerruti	Y 11.84	Eredi G. B. Vassallo	Genoa
407		San Vincente (ex Cong-[illegible])	W Bn	Spanish	138	90·2	21·6	9·6	[illegible]	1862			Manuel Trani & Co.	Manila
408		Sanbo Maru	W Bg	Japanese	162	97·9	19·5	11·5	[illegible]	1878	Hann Tunotsu		Sakakibara Kosuke	Handa
409		Sancha	W Bg	Brazilian	141	..	..	..	Rio de Janeiro	1859		Y 3.86	C. P. da Silva	Rio de Janeiro
410	✻	Sancho *Bilgate* 79-15 (MISSIN)	W Sw	British	286	109·3	26·2	16·0	Sunderland	1866	Reay & Naizby	Y 2.86	Sancho Shipping Co. Ld. (Dent & Co.)	North Shields
411		Sancho Panza	W Sr	British	146	87·6	19·7	12·3	Shoreham	1828			J. Spenceclay	Faversham
412	V	Sancta Maria *Lefebvre* 79-88	W 3m Sr	French	283	117·1	23·0	13·5	Dunkirk	18^{51}_{74}	A. Pyotte		G. Winnaert	Dunkirk
413		Sandalphon *Thorburn* -88	W Sr	British	105	87·6	22·3	10·1	Shelburne, NS	1888	G. A. Cox		C. Locke	Shelburne, N.S.
414	N	Sandoen *Olsen*	W 3m Sr	Norweg'n	211	113·2	27·5	11·1	Risör	1867	R. Gundersen		C. A. Olsen's Enke	Tvedestrand
415		Sandra *Söderlund*	W Bk	Russian	213	106·7	26·0	12·3	Björkö	1870	M. Reponen		F. Söderlund	Raumo
416		Sandvik *Petersson* -87 (STRANDED)	W Bk	Swedish	593	146·3	29·9	17·1	Umeå	1873	O. Jonsson	Y 10.87	G. Moritz	Umeå
417	R	Sandy Hook *Barlow* -87	W Sr	American	102	89·1	23·6	7·3	E. Haddam, Ct.	1869			Edw. W. Barlow	Barnstable, Mass.
418		Sanfrid	W Sr	Russian	148	..	..	..		1874			A. J. Mustonen	Joensuu
419	V	Sanglier *Laroque*	W Bg	French	184	93·2	23·0	12·3	St. Malo	1876	J. Parnet	Y 2.89	Sire fils	St. Malo
420		Sango Maru	W Sr	Japanese	264	124·6	24·1	12·6	Aarhuus	1866	Ustrup & Meissner		Miki Mantaro	Harima
421		Sanoma	W Sr	Russian	122	..	..	..		1876			H. Mattila	Sastmola
422	V	Sans Peur *Sicerry*	W Bk	French	284	108·9	23·9	14·1	St. Servan	1884	Hamond	Y 2.8	M. Guibert & fils	St. Malo
423		Sans Souci	W Sr	American	112	71·4	26·3	6·8	Barnegat, N.J.	1876			N. Middleton	Tuckerton, N.J.
424		Sans Souci *Vilala*	W Bg	French	149	93·2	21·3	10·1	St. Malo	1873	Pitrel		Guibert & fils	St. Servan
425		Sans Souci *Foliard* -87	W Bn	French	105	82·7	23·6	9·3	St. Malo	1837			A. Cholet	St. Brieuc
426	A	Sansone *Starcich*	W Bk	Austrian	644	147·9	26·1	20·3	Lussinpiccolo	1869	M. A. Starcich	Y 1.87	M. A. Starcich	Lussinpiccolo
427	A	Sansone *Ostuni*	W Bn	Italian	116	73·8	23·2	12·7	Bari	1858			G. Mandriota	Bari
428		Sanspareil *Bowden*	W Sr	British	132	88·6	20·5	12·2	Bideford	1857			T. W. Hutchings	Teignmouth
429	IV	Sant' Anna *Severino* -88	W Bk	Italian	646	159·4	29·5	19·5	Cassano	1876	F. De Rosa	Y 8.85	F[lli] Cafiero	Castell. di Stab.
430	AI	Sant' Anna & Maria *Tarese* -88	W Bk	Italian	413	128·9	28·7	16·6	Sorrento	1872	M. Paturzo	Y 7.88	A. Guida	Naples
431	AI	Sant'Antonio *Mancino* (ex Ascensione, ex Risurrezione)	W Bk	Italian	497	133·2	30·9	20·2	Procida	1874	A. Bonifacio	Y 9.87	Ant. Galatola	Naples
432	I	Sant' Antonio *Messone* (ex Maria Damonzato) [-89	W Bk	Italian	387	132·2	30·5	16·8	Sestri Ponente	1875	G. Casanovo	Y 7.83	G. Batta Celle	Genoa
433	V	Sant' Antonio *Lizza* (ex Alber[illegible])	W Bn	Italian	147	105·9	22·8	12·6	Cappelli	1867		Y 6.88	N. Cipollina & Co.	Savona
434	I	Sant' Antonio *Cilento* (ex [illegible])	W Bn	Italian	273	115·4	26·1	16·4	Equa	1870	G. Gargiulo	Y 8.85	F. Cilento	Naples
435	✻	Santa *Langford* 78-86	I Bk	British	940	205·8	33·6	19·6	Sunderland	1886	R. Thompson & Sons		Turner, Edwards & Co.	Bristol
436		Santa *Quintavalle* -87 (ex Anna, ex Augusta)	W Bk	Italian	429	109·2	30·4	15·7	Venice	1857	Amadi		L. Quintavalle	Venice
437		Santa Barbara	W Bn	Brazilian	130	..	..	..					J. F. de Barros	Bahia
438	RV	Santa Clara *Dunn*	W S	American	1474	209·5	40·0	25·5	Bath, Maine	1876	Chapman & Flint	Y 3.88	I. F. Chapman	New York
439	✻	Santa Maria *Jackson*	W Sk	British	86	78·4	20·6	10·1	Goole	1884	Cottingham Bro.		C. Pickering & Haldane	Hull
440		Santa Rita *Caetano*	W Sr	Brazilian	110	..	..	..	Campos	1881			Cia. Cabotagem de S. João da Barra	S. João da Barra
441		Santa Rita II.	W Sr	Portug'se	126	77·4	21·7	9·6	Caminha	1879	A. J. Gonçalves		Heirs of A. J. Gonçalves	Caminha
442	✻	Santa Rosa *Thomas* -85	I Bk	British	568	169·5	26·6	17·9	Liverpool	1862	T. Vernon & Son		J. Tweedie	Liverpool
443	IV	Santa Rosa *M. de Rosa* (ex Mary W. Pipes)	W Bk	Italian	572	145·5	31·6	19·0	Nova Scotia	1878	J. R. Pipes	I 11.87	A. Castellano	Castellamare
444		Santa Rosa	W Bn	Spanish	111	77·9	18·5	6·6	Albay	1884			F. Vellones	Manila
445		Santa Thereza	W S	Brazilian	1519	..	..	..					J. R. V. Lobos	Rio de Janeiro
446		Santelmo	W Bg	Brazilian	309	..	..	..					M. Belchior & Co.	Rio Grande
447	✻	Santiago *Gaunson* 67-88	Stl Bk	British	978	207·6	33·1	20·0	Belfast	1885	Harland & Wolff		North Western Shipping Co., Ld.	Liverpool
448		Santiago *Miller* -86 (ex Genoa A. Host)	W Bk	Chilian	1247	185·0	37·7	27·1	St. Martin's NB	1860			Cia. Nacional de Remolcadores	Valparaiso
449	✻DV	Santiago *Hold* 87-87	I Bk	German	460	160·6	25·9	17·1	Methill	1856	Balfour		E. F. C. Tobias	Brake
450		Santiago (ex Escenio de [illegible] Antonio?, ex Strathallan)	W Sr	Spanish	547	162·2	28·3	17·8	Dundee	1857			J. Vidal y Vidal	Majorca

✻ = Lloyd's Register; D = Nederlandsche Vereeniging; G = Germanischer Lloyd; I = Registro Italiano; R = American Record;
A = Veritas Austro-Ungarico; H = Veritas Ellenico; N = Norske Veritas; V = Bureau Veritas.

Dans la 4me Colonne, C = Composite (membrure en fer, bordé en bois); I = Fer; Stl = Acier; W = Bois.

Les Dates du doublage en métal (Col. 11) ressortent des livres dans lesquels les navires sont respectivement classés ou des rapports des Inspecteurs du Lloyd's Register.

Pour plus amples renseignements sur les navires classés par le Lloyd's Register, voir le Lloyd's Register Book et le Lloyd's Yacht Register.

SAILING VESSELS.

Official No.	No. in Book.	Registry in which classed.	Ship's Name. Material and Rig. Master, and Dates of Appointment to present Owner's Service and to Ship. Ship's Late Name.	Flag.	Net Tonnage.	Dimensions. Length.	Breadth.	Depth.	Built. Where.	When.	By whom.	Date of Metalling.	Owners or Managers.	Port belonging to.
	451		Santiago (ex [illegible]) Ferrer W Sr	Spanish	208	103·6	23·3	12·7	Plymouth	1871	Hill		J. Rodriguez	Santiago
	452		Santiago Marti W Sr	Spanish	118	79·1	22·9	11·2	Villajoyosa	18[illegible]			Nadal Marti	Alicante
[illegible]	453		Santiago de Petres (ex Emanuele Mateo) W 3m Sr	Italian	426	127·6	30·1	18·7	SestriPonente	1866	C.Guglielmo		V. Firpo & Co.	Genoa
	454		Santissima Trinitè Malaxa W Sr	Greek	140	..	..	..	Skiathos	1882			G. Mataroua	Skiathos
[illegible]	455	V	Santissima Trinità [illegible] W Bn	Italian	174	103·0	24·9	12·6	Lisaite	1880	S.Pichiotti		F. Carletti fu A.	Porto [illegible]
	456	A	Santissima Trinità W Bn	Turkish	182	..	..	..	Piraeus	1866			B. Patroua	Syra
[illegible]	457	V	Santo Agnello Fienga W Bk	Italian	358	115·1	26·9	17·5	[illegible]	1867	A.Castellano	[illegible]	Fʳˡˡ. Fienga	Castell[illegible]
	458		Santo Antonio W Sr	Brazilian	258	..	..	..					J. J. R. Guimaraes	Rio de J[illegible]
	459		Santo Antonio W 3m Sr	Brazilian	170	..	..	..	Norway	1866			J. P. A. de Carvalho	Bahia
	460		Santo Antonio Dr Antonio W Bg	Brazilian	133	..	..	..	S.JoãodaBarra	1859		[illegible]	C. J. de Bessa	Laguna
115445	461		Santo Oteri [illegible] W Sr	American	162	105·1	26·0	9·7	Bath, Me.	1876	Hagan&Thurlw		S. Oteri	New [illegible]
45[illegible]	462	⁂	Santon Vaughan 81-85 I Bk	British	511	156·0	26·4	17·4	Sunderland	1863	Pile,Hay&Co.		E. Phillips	Liverpool
60447	463	⁂	Santona McLeod 88-88 I Bk	British	855	203·4	32·1	19·3	Glasgow	1869	Dobie&Co.		Donaldson Bros.	Glasgow
	464		Sao Bartholomeu DosReis W Sr	Brazilian	160	..	..	..	Pernambuco			[illegible]	B. Lourenço	P[illegible]
	465		Sao Benedicto Coelho W Bn	Brazilian	156	..	..	..	Campos	1873		[illegible]	F. J. da C. Coelho & Co.	Rio de Ja[illegible]
	466		Sao Joao Baptista W Bg	Brazilian	107	..	..	..	Campos	1879		[illegible]	Cia. Cabotagem de S. João da Barra	S. João da [illegible]
	467		Sao José W Bk	Brazilian	297	..	..	..					J. S. L. Junior	Pernamb[illegible]
	468		Sao José W Bg	Brazilian	202	..	..	..					A. A. Fernandes	Bahia
	469		Sao José W Bg	Brazilian	152	..	..	..					J. R. de Souza	Rio de Ja[illegible]
	470		Sao José W Sr	Brazilian	115	..	..	..					D. A. G. da Silveira	Fortaleza
	471		Sao José Americano W Sr	Brazilian	152	..	..	..					D. M. A. Lessa	Rio de J[illegible]
	472		Sao Lourenço Da Silva (ex Tropico) W Sr	Brazilian	101	..	..	..	PortoAlegre	1870		[illegible]	J. L. R. Ferreira	Pernamb[illegible]
	473		Sao Lourenço Dr [illegible] Mendes W Bg	Portug'se	195	84·6	23·0	11·5	Lisbon	1864	J.d'Azevedo [Vareta		J. J. R. de Magalhaes	Oporto
	474		Sao Manoel W Bn	Brazilian	178	..	..	..					J. F. Porto	Rio de J[illegible]
	475		Sao Manoel Vierra, jr. W Bg	Brazilian	129	..	..	..	[illegible]	1858			F. Dexheimer	Porto [illegible]
	476		Sao Miguel (ex [illegible] Litchfield) W Sr	Portug'se	147	87·9	22·6	8·5	[illegible], Mass.	1877			J. do Rego & Co.	Ponta [illegible]
	477		Sao Thiago W Bg	Portug'se	237	89·2	23·6	14·4					J. C. Serra	Lisbon
	478		Sao Thomé Rodrigues (ex Alejandro, ex Cayenne) W Bg	Portug'se	277	105·6	23·9	14·8	Sestri	1861			Mesquita & Santiago	Lisbon
	479	⁂	Saphir Ollive W Sr	French	75	76·2	12·6	9·0	Kingston	1875	Duncan	[illegible]	C. Hermann	Réunion [illegible]
	480	N	Saphir Langfeldt W Sr	Norweg'n	128	91·8	21·7	10·9	Banff	18[illegible]			L. Lömsland	Christ[illegible]
	481		Sapho W Sr	Brazilian	288	..	..	..					D. F. Grillo	Rio de J[illegible]
	482	D V	Sappemeer II. Huizenga [illegible] (ex Marianne VIII) W Bn	German	202	97·8	26·1	12·1	Quebec	1872	Julien	[illegible]	Tubergen & Daam	Emden
76[illegible]	483	D V	Sapphire Murray W S	British	1273	197·1	38·5	23·9	NewBrun.,US	1864	J.Currie jun.	[illegible]	Louisa S. Murray	Halifax [illegible]
32023	484		Sapphire Sparks W Bk	British	237	95·3	24·7	15·6	Montrose	1845			George Salier	Hobart T[illegible]
88226	485	R	Sapphire Cox -87 W Sr	British	124	89·5	24·0	9·6	PortClyde,NS	1884	T.Coffin&Co.	[illegible]	J. G. Cox	Victor[illegible]
77083	486		Sappho Irving -88 W Bk	British	559	140·9	32·0	18·6	P.E.Island	1878	D.Gordon	[illegible]	Daniel Gordon	Charl[illegible]
[illegible]	487	R	Sappho Taylor 81-81 W Bkn	British	171	111·3	25·3	12·8	Padstow	1869	Stribley	[illegible]	L. A. McGrather	Inverness
80349	488		Sappho (BROKEN UP) W Sr Yt	British	132	128·2	27·0	10·9	New York	1868	Poillon Bros.		G. Marvin	Southa[illegible]
76270	489	⁂	Sappho Gaudin 79-79 W K	British	48	63·9	18·3	8·3	Jersey	1879	LeSueur		J. Gaudin	Jersey
	490	N	Sappho Knudsen W Bk	Norweg'n	515	118·2	31·2	15·3	Grimstad	1885	S.Johnsen	[illegible]	M. Smidt	Arendal
	491	G	Sara Schomacker 76- (ex Victorie Hortense) W Sr	German	129	82·7	23·2	10·8	Cherbourg	1858			J. Wallenstein	Geeste[illegible]
4[illegible]	492		Sara Cignoni W Bg	Italian	168	93·8	21·8	14·5	SestriPonente	1851	L.Briasco		G. Cignoni fu A.	Porto F[illegible]
	493	N	Sara (ex Albion) Host W Bk	Norweg'n	556	136·8	33·8	18·7	Brahestad	1860	J.Danielson		C. M. Host	Arendal
	494	N	Sara (TO BE BROKEN UP) Wright W Bg	Norweg'n	154	82·6	23·8	12·7	Aalborg	1836			E. B. Wright	Porsgrund
	495	V	Sara (ex R. B. Chapman) Wirkke -87 W Bk	Russian	555	142·8	30·5	17·9	Rockland N.B.	1872	R.A. Chapman	[illegible]	J. Wirkkè	Wiborg
	496		Sara Ross W Sr	Spanish	105	78·1	21·8	10·0	Arenys	1851			J. Bosch y Estapé	Barcel[illegible]
[illegible]	497	R	Sara A. Dudmann [illegible] W Bk	Russian	483	131·6	29·7	19·2	Yarmouth,NS	1864			C. Essmann	Riga
	498		Sara Margrethe W Bg	Norweg'n	266	92·6	26·7	15·6	Risör	1829			Knutsen & Herlofsen	Laurvi[illegible]
	499		Sara Mercedes (ex Madge Wildfire) [illegible] W 3m Sr	Peruvian	245	127·0	26·0	12·6	Glasgow	1868			J. Portella	Callao
	500													

⁂ = Lloyd's Register; D = Nederlandsche Vereeniging; G = Germanischer Lloyd; I = Registro Italiano; R = American [illegible]
A = Veritas Austro-Ungarico; H = Veritas Ellenico; N = Norske Veritas; V = Bureau Verit[illegible]

In the 4th Column, C = Composite (iron frame, wood planked); I = Iron; Stl = Steel; W = Wood.

The Dates of Metalling (Col. 11) are based on the records of the Books in which Vessels are respectively classed, or on the Reports of the Surveyors of Lloyd's R[illegible]

Further particulars of Vessels classed by Lloyd's Register will be found in Lloyd's Register Book and Lloyd's Yacht Regist[illegible]

No.	Soc.	Nom. Matériaux et Gréement. Capitaine et dates d'entrée au service des Armateurs actuels et à bord du navire. Ancien Nom.			Pavillon	Tonnage net	Dimensions. Longueur	Largeur	Creux	Construit. Où	Quand	Par qui	Date du doublage en métal	Armateurs ou Directeurs-Gérants.	Port d'Armement.
501	✻	Saraca	Ball 85-85	I Bk	British	836	189·4	31·2	18·5	Dumbarton	1877	Birrell, Stenhouse & Co.		Scotia Shipping Co., Ld.	Glasgow
502	✻	Sarah	Bagwell 88-88	W Bk	British	1142	184·0	38·8	23·0	Tusket, N.S.	1874	D. Ross	12.82	H. Hutton & Co.	Yarmouth, N.S.
503	✻R	Sarah		W Bn		187	104·0	24·5	12·9	P.E. Island	1876	McWilliams	Y 11.83		
504		Sarah	Baker 53-68	W Bg	British	185	110·6	23·0	11·6	Shoreham	1862	May		T.P. Cattley	Shoreham
505		Sarah		W Bg	British	176	95·0	21·6	9·6					John Punton	Turks Island
506		Sarah	Carter 84-84	W Sr	British	149	103·0	22·9	11·9	Kingsbridge	1858	W. Date		J. Ennor	Fowey
507		Sarah	Harper -88	W Sr	British	118	80·0	26·0	8·4	Tynem'th, N.B.	1878	R.L. Parker		Robert L. Parker	St. John, N.B.
508	✻	Sarah	Pritchard 85-85	W Sr	British	100	79·1	23·0	11·6	Portmadoc	1866	Griffiths		J. Ellis & Co.	Carnarvon
509	✻	Sarah	Lile	W Sk	British	27	49·0	16·0	6·6	Milford	1879	Morgan		S.T. Williams	Milford
510	R	Sarah (ABANDONED)	Deane	W Bk	American	628	143·1	30·9	19·2	East Boston	1867	Curtis, Smith & C.	Y 1.87	J.M. Haskell	Boston, Mass.
511	R	Sarah	Hale	W Bk	American	530	135·0	31·2	17·5	Stockton, Me.	1870		Y 5.83	E.A. Adams	Boston, Mass.
512	V	Sarah (ex Bahia-de-Marinheiro)		W 3m Sr	Brazilian	583	160·0	33·5	19·3	Ancona	1878	Cagliesi	Y 10.85	G.A. de Mirande filho	Para
513		Sarah	Mathersen 87-87	W Bk	Norweg'n	569	135·1	29·6	19·5	Bath, Me.	1848			J. Augestad	Laurvig
514		Sarah (ex Vaidor de Mondego)	Francisco	W Bg	Portug'se	267	102·5	23·9	14·8	Rio Mondego				J.F. Gonçalves	Lisbon
515		Sarah A. Falconer	Wilson	W Sr	American	107	87·2	24·4	8·0	Brooklyn N.Y.	1854			B. Wilson	Bridgeport, Conn.
516	R	Sarah A. Fuller	Hart	W 3m Sr	American	468	143·0	35·6	12·3	Camden, Me.	1883	H.M. Bean	Y 3.87	W.H. Gladding	Providence, R.I.
517		Sarah A. Reed	Aylward	W Sr	American	151	100·0	26·0	8·0	Calais, Me.	1868	Kelley		Boardman Bros.	Calais, Me.
518	R	Sarah A. Staples	Gray	W Bk	American	443	126·0	28·8	18·0	Belfast, Me.	1862		Y 6.81	Peter Brown	Philadelphia
519	R	Sarah A. Townsend	Townsend	W Sr	British	149	93·5	25·3	10·8	Pugwash, N.S.	1884	D. Redmond		Wm. H. Townsend	Halifax, N.S.
520		Sarah Amy	Gardner	W Bn	British	158	99·4	23·2	10·7	Emsworth	1874	Foster		J.T. Read	Portsmouth
521		Sarah Ann	Renaut 72-79	W Sw	British	270	105·1	26·6	16·2	Sunderland	1862	Bartram		J. Paul & Co.	Lynn
522	✻	Sarah Ann	George	W Sk	British	24	48·2	15·5	6·4	Milford	1881	J. & W. Frances		W. George	Milford
523		Sarah Ann Widdup		W Sr	British	120	93·5	22·6	11·4	Fleetwood	1869	Gibson		J. Reney	Fleetwood
524		Sarah Anne	Gelling 84-87	W 3m Sr	British	146	95·0	22·6	12·0	Berwick	1859	Gowan		J. Fisher & Sons	Lancaster
525	V	Sarah B. Cann	Whitmore	W Bk	British	757	158·0	35·0	20·0	Digby, N.S.	1870			R.R. Sneden	Yarmouth, N.S.
526	✻	Sarah Bell	Hurst 86-86	I Bk	British	812	194·3	32·1	19·8	Harrington	1870	R. Williamson & S.		J. Bell & Son	Liverpool
527	✻	Sarah Bridget	Jones	W Bn	British	106	86·4	22·5	11·8	Bangor	1858			J.R. Jones	Carnarvon
528		Sarah Burnyeat		W Bg	British	277	104·0	25·0	16·1	Sunderland	1862			Adelaide S.S. Co.	Fremantle, W.A.
529	R	Sarah C. Smith	Henderson -88	W 3m Sr	American	282	118·5	29·5	13·9	Chester, Pa.	1868			C.W. Stevens	Philadelphia
530	R	Sarah Carlin (STRANDED)	Flavian	W Sr	British	149	97·4	27·0	10·8	Sherbrook, N.S.	1881	Sherbrooke		John Carlin	Halifax, N.S.
531	DV	Sarah Chambers	Carter	W Bk	British	1000	175·2	35·6	21·9	Dorchester N.B.	1874	W. Hickman		William Hickman	Dorchester, N.B.
532		Sarah Cullen		W Sr	American	242	112·2	29·6	6·8	Philadelphia	1858			R.P. Vansant	Philadelphia
533	R	Sarah D. Fell	Loveland	W 3m Sr	American	552	145·4	35·3	14·9	Bath, Me.	1882	W. Rogers		J.R. Fell	Philadelphia
534	R	Sarah D. J. Rawson	French 83-83	W 3m Sr	American	368	140·5	31·8	11·1	Camden, Me.	1883	Coombs & Day		Isaac Coombs	Camden, Me.
535	✻	Sarah Davies	Davies 83-83	W Sr	British	90	77·0	20·0	10·2	Aberdovey	1870	Jones		T. Davies & Co.	Aberystwith
536	R	Sarah Doe (WRECKED)	Merryman -83	W Bk	American	616	112·0	32·2	18·2	Cherryfield Me.	1874		Y 5.86	L.S. Davis	New York
537		Sarah E. Kennedy (LOST)	Walters	W Bn	American	402	119·5	28·9	16·9	Waldoboro Me.	1863			Daniel Walters	Somers Pt. N.J.
538	R	Sarah E. Ward	Wixon	W 3m Sr	American	548	176·0	35·5	15·0	Camden, Me.	1884	H.M. Bean		A.R. Wixon	Boston, Mass.
539	R	Sarah Eaton	Hops	W Sr	American	189	100·0	29·0	9·0	Calais, Me.	1874	Nickerson & Holcomb		H.B. Eaton	Calais, Me.
540	✻	Sarah Elizabeth	Coppack	W Sr	British	130	92·8	22·9	11·5	Rhyl	1873	Jones		W. Williams	Chester
541		Sarah Ellen	Hughes 84-	W Sr	British	100	85·0	19·0	11·6	Flint	1855	Ferguson & Baird		J. Hughes	Swansea
542	✻	Sarah Evans	Roberts 76-81	W Sr	British	98	82·5	22·4	10·8	Portmadoc	1877	Jones		R. Evans & Co.	Carnarvon
543	✻	Sarah Fox	Prout	W Sr	British	121	87·5	22·8	11·8	Mevagissey	1868	Lelean		W. Endean	Hayle
544	✻	Sarah Gibson	Benson (75-86)	W 3m Sr	British	188	110·1	23·2	12·4	Berwick	1871	Gowan		J. Fisher & Sons	Barrow
545	R	Sarah Godfrey	Hogan -87	W Sr	British	186	101·2	27·4	10·6	Rockland, N.B.	1883	R.A. Chapman		C.M.A. Godfrey	Dorchester, N.B.
546	N	Sarah Humphreys (ex Belgrave)	Thorbjørnsen	W Bk	Norweg'n	472	127·2	28·8	18·8	Sunderland	1851			Fru M. Ruud	Fredrikstad
547		Sarah Hunter	Mowry	W Sr	British	122	86·4	27·2	7·9	St. John, N.B.	1882	J.H.D. Eagles		W.C. Purves	St. John, N.B.
548		Sarah J. Hoyt	Lake	W Sr	American	158	105·0	28·2	8·8	Barnegat, N.J.	1861			W. Predmore	Tuckerton, N.J.
549		Sarah Jane		W	British	238	131·1	24·5	9·5	Port Robinson	1872	J.S. Abbey		John S. Nesbitt	St. Catherine's
550	✻	Sarah Jane	Foale	W K	British	56	66·0	18·1	7·0	Salcombe	1878	Date		J. Rice	Salcombe

✻ = Lloyd's Register; D = Nederlandsche Vereeniging; G = Germanischer Lloyd; I = Registro Italiano; R = American Record; A = Veritas Austro-Ungarico; H = Veritas Ellenico; N = Norske Veritas; V = Bureau Veritas.

Dans la 4me Colonne, C = Composite (membrure en fer, bordé en bois); I = Fer; Stl = Acier; W = Bois.

Les Dates du doublage en métal (Col. 11) ressortent des livres dans lesquels les navires sont respectivement classés ou des rapports des Inspecteurs du Lloyd's Register.

Pour plus amples renseignements sur les navires classés par le Lloyd's Register, voir le Lloyd's Register Book et le Lloyd's Yacht Register.

Official No.	No. in Book.	Registry in which classed.	Ship's Name. Material and Rig. Master, and Dates of Appointment to present Owner's Service and to Ship. Ship's Late Name.	Flag.	Net Tonnage.	Dimensions. Length.	Breadth.	Depth.	Built. Where.	When.	By whom.	Date of Metalling.	Owners or Managers.	Port of Registry.
22516	551		Sarah Jane Vaughan *Vaughan* W Sr	American	208	111·8	29·7	8·9	May's Landing N.J.	1861			D. Gifford	Somers P.
4987	552		Sarah King *Bennett* W Sr	British	155	94·6	23·1	12·5	Sunderland	1865	Liddle		G. Judson	Portsmouth
113612	553		Sarah L. Davis *Knowland* W Sr	American	185	106·0	25·9	8·9	Belfast, Me.	1871			F. A. Knowles	Belfast, M.
22033	554		Sarah L. Simmons *Hiton* W Sr	American	231	108·6	30·0	9·7	Philadelphia	1864			G. W. Shepherd, jun.	Philadelphia
67135	555		Sarah Lightfoot *Macauley* W Sr	British	145	97·0	23·4	12·3	Littlehampton	1872	Harvey		T. E. Adams	Fleetwood
56453	556	✱	Sarah McDonald *Davison* W Sr	British	90	79·6	21·3	9·8	Perth	1867	Perth S. B. Co.		G. Millington	Chester
23073	557		Sarah Mills *Jary* W	British	162	80·2	23·1	13·5	Newcastle	1838	Taylor		J. B. Bushell	S. Shields
22030	558		Sarah Mills *Howell* W Sr	American	146	98·2	27·2	8·9	Pt Jefferson L.I.	1855			J. H. Howell	Cold Spring
19461	559		Sarah Pile W Bn	British	115	92·6	20·8	10·0	Sydney, N.S.W.	1864	R. Pile		Geo. Tippett	Auckland
115177	560		Sarah Potter *Sumner* W 3m Sr	American	342	129·3	32·0	12·0	E. Boston Mass	1873			G. W. Rawley	St. George
23542	561		Sarah Quinn *Moore* W Sr	American	108	88·0	27·7	7·2	Tuckerton N.J.	1868			I. W. Valentine	Cold Spring
65023	562	✱	Sarah Rowe *Jones* W Sr	British	63	73·8	18·9	9·0	Coston	1872	Morgan		O. Jones	Milford
115341	563	R V	Sarah S. Ridgway *Townsend 75-83* W Bk	American	870	167·5	35·3	23·2	Wilmington Del.	1877	Jackson, Sharp & Co.	[illegible]	J. E. Ridgway	Philadelphia
22568	564		Sarah S Tyler *Williams* W 3m Sr	American	160	104·0	28·0	7·0	Hartford, Ct.	1866	Belden		H. S. Williams	Hartford
69475	565		Sarah Smith *Noel -87* W 3m Sr (BROKEN UP)	British	194	113·0	22·4	12·4	Troon	1869	Troon S. B. Co.		Aboobaker Amod	Port Natal
115432	566	R	Sarah W. Hunt *Miner* W Sr	American	110	88·7	24·7	8·4	Bath, Me.	1876	Hagan & Thurlow	[illegible]	E. Miner & Co.	Hartford
118106	567		Sarah W. Lawrence *Farrow* W 3m Sr	American	1301	217·0	45·2	19·8	Bath, Me.	1886	N England S B Co		J. B. Phillips	Taunton
66912	568		Sarah Wallace *Holderness -88* W Bk	British	216	107·6	27·5	10·9	St. John, N.B.	1873	W. & R. Wallace	[illegible]	F. E. Sayre	St. John, N.B.
22186	569	✱	Sarah Williams *Jones -88* W Sr	British	118	80·5	21·5	11·4	Barmouth	1859	R. Jones		G. Williams & Co.	Carnarvon
23853	570		Sarah Wood *Hickman* W Sr	American	259	110·0	29·9	9·0	Camden, N.J.	1871			Jeremiah Smith & Bro.	Philadelphia
115720	571	R	Sarah & Ellen *Henley* W 3m Sr	American	553	142·5	34·6	12·7	E. Deering, Me.	1880	Russell		J. S. Winslow & Co.	Portland, Me.
29137	572	✱	Sarah & Emma *Lewis 70-79* I Bk	British	1097	199·6	33·8	22·9	Liverpool	1860	T. Vernon & Son		Hughes & Co.	Liverpool
23810	573		Sarah & Emma *Nash* W Bn	American	306	113·0	28·8	14·0	Columbia Falls Me.	1869		[illegible]	Swan & Son	New York
115203	574		Sarah & Lucy *Booker* W 3m Sr	American	240	114·0	33·2	9·0	Holyfork, Va.	1874			S. E. Birdsall, Exetr.	New York
12138	575		Sarah & Margaret W Bg (BROKEN UP)	British	178	87·4	23·7	14·2	Sunderland	1849			G. Moorsom	Whitby
52365	576		Sarah & Mary W Bn	British	145	116·0	24·4	8·4	McLeay River, N.S.W.	1865			Walter S. Waterston	Invercargill
115785	577	R V	Saranac *Shaw* W Bk	American	1081	195·5	34·6	22·5	Kennebunk Me.	1880	N. L. Thompson	[illegible]	W. Simpson, jun.	New York
	578		Saranak *Olsen* W S	Norweg'n	838	153·1	35·0	21·7	Philadelphia	1844			O. B. Sorensen	Arendal
	579	V	Sarcelle *Maillet -87* W Sr (WRECKED)	French	109	85·0	19·3	11·7	Dunkirk	1869	A. Lefebvre		E. Didier	Dunkirk
87421	580	✱ D	Sardhana *Jones 87-87* I Bk	British	1119	215·1	35·1	21·2	Port Glasgow	1885	Russell & Co.		J. Russell (W. & J. Crawford)	Greenock
74294	581	V	Sardinian *Le Blanc -88* W S	British	1543	211·9	40·4	24·7	Quebec	1877	P. V. Valin	[illegible]	F. Ross	Quebec
23525	582		Sardinian *Holbrook* W Sr	American	119	84·6	25·1	8·6	Harrington Me	1859			A. J. Bird	Rockland, Me.
86191	583	✱	Sardomene *Pitt 78-82* I S	British	1943	269·0	39·6	24·1	Southampton	1882	Oswald, Mordaunt & Co.		H. Fernie & Sons	Liverpool
	584		Sargento Cabral *(ex Lola)* I Sr	Argentine	238	125·0	29·5	10·5	Genoa	1883	E. Cravero & Co.		La Platense Flotilla Co.	Buenos Ayres
	585		Saris Bandong Bintang Passir W Sr (ex Hall colonhead)	Dutch	375	123·0	28·6	11·0	Mystic, U.S.	1865			Hadji Kasama bin Hadji Gasa	Batavia
62	586		Sarina Costanzo *Costanzo* W Sr (ex Eolo)	Italian	259	122·0	25·3	15·2	Fiume	1870	Schiavoni	[illegible]	E. Costanzo fu G.	Catania
56	587		Sarina Liardo *Liza* W Bn (ex Maddalena Russo)	Italian	124	86·9	21·0	11·8	Sestri Ponente	1877	M. Bossi		F[lli] Liardo	Porto Empedocle
	588		Sarita Fuentes W Sr	Chilian	124	85·6	24·5	10·2	Noank	1874			D. Fuentes	Talcahuano
7429	589	V	Sarmatian *Morris 78-74* W S	British	1198	190·2	37·2	22·8	Quebec	1877	P. V. Valin	[illegible]	Pierre V. Valin (W. H. Ross & Co.)	Quebec
115057	590	R	Sarmiento *Gould 71-78* W Bk	American	576	150·2	31·6	16·8	E. Deering, Me.	1872	R. Lewis & Co.	[illegible]	R. Lewis & Co.	Portland, Me.
18761	591		Sarnian Gem *Le Lacheur* W Bkn	British	184	115·2	26·0	11·7	Guernsey	1857	P. Ogier		W. Le Lacheur	Guernsey
	592		Sarnos I	Greek	201	131·2	20·4	10·1	Austria	1874			Theofilatos Stathatos	Ithaca
2284	593	I	Saro' Caino *Olivari* W Bk	Italian	613	136·2	30·1	20·5	Savona	1872	B. Calcagno	[illegible]	B. Olivari	Genoa
	594	N	Saron *Olsen 85-85* W Bk	Norweg'n	484	129·9	30·2	16·9	Lillesand	1875	A. Stiansen	[illegible]	Krog & Henschien	Lillesand
	595	N	Saron *Torstensen -86* W Bk	Norweg'n	354	122·1	29·0	15·3	Risör	1875	E. Knudsen	[illegible]	J. W. Prebensen, sen.	Risör
	596	D N	Saron *Johnsen 84-84* W 3m Sr	Norweg'n	345	132·1	28·1	14·1	Stavanger	1877	M. A. Magnus	[illegible]	G. Gundersen	Stavanger
	597	N	Sarpsborg *Klouman -75* W Bk (ex Denmark)	Norweg'n	582	136·2	29·9	19·9	Boston	1835			A. Hesselberg & Co.	Christiania
	598	N	Sarpsborg *Torgersen -86* W Bg	Norweg'n	140	79·8	25·0	11·0	Sarpsborg	1855			O. Bjertness	Holmestrand
8342	599		Sarsfield *Dunne* W Bn	British	129	85·0	21·8	12·4	Cockle Creek, Hillsburgh	1847			J. H. Rutter & Co.	Cork
	600		Sarsati *Daly* I Bg	British	260	121·2	26·1	15·3	Glasgow	1883	Napier, Shanks & Bell		Government of India	Calcutta

✱ = Lloyd's Register; D = Nederlandsche Vereeniging; G = Germanischer Lloyd; I = Registro Italiano; R = American Record; A = Veritas Austro-Ungarico; H = Veritas Ellenico; N = Norske Veritas; V = Bureau Veritas

In the 4th Column, C = Composite (iron frame, wood planked); I = Iron; Stl = Steel; W = Wood.

The Dates of Metalling (Col. 11) are based on the records of the Books in which Vessels are respectively classed, or on the Reports of the Surveyors of Lloyd's Register.

Further particulars of Vessels classed by Lloyd's Register will be found in Lloyd's Register Book and Lloyd's Yacht Register.

		Nom. Materiaux et Grèement. Capitaine et dates d'entrée au service des Armateurs actuels et à bord du navire. Ancien Nom.	Pavillon	Tonnage net.	Dimensions. Longueur	Largeur	Creux	Construit. Où.	Quand	Par qui.	Date du doublage en métal	Armateurs ou Directeurs-Gérants.	Port d'Armement.
601		Satama *Sundberg* -88 W Bk	Russian	519	134·1	29·5	17·4	Brahestad	1863	J. Danielsson		J. Lang	Brahestad
602		Satellite *Keeson* W Bk	British	278	123·6	28·4	12·8	Cocaigne, N.B.	1868	Whittaker		R. Tedcastle & Co.	Dublin
603		Satellite *Penfinson* W Bkn	British	245	120·0	25·0	11·8	Yarmouth	1867	Goldie	V	R. Allix	Guernsey
604		Satilla *Sheffield* -87 W3mSr	American	296	128·9	31·6	10·7	Bath, Me.	1873			A. Sewall & Co.	Bath, Me.
605		Satis *Francis* W Bn	British	192	82·2	21·9	14·4	Sunderland	1845			E. Francis	Rochester
606	V	Satisfaction *Klatus* 71-76 W Bk	German	430	118·8	29·9	18·2	Memel	1854	B. Pieper		F. R. Dittborn	Memel
607		Satisfaction *Kanow* -81 W Bg	German	176	92·2	24·4	13·2	Wolgast	1861	H. Ebrich		J. C. Holtz	Wolgast
608	A	Sator *Rodriguez* -88 W Bk	Austrian	634	..	..	..	Lussinpiccolo	1864	N. Martinolich	V 5.87	A. A. Cosulich & Ce.	Lussinpiccolo
609	XV	Sator *Nielsen* 83-4 W Bk	Norweg'n	329	128·4	28·3	11·5	Lillesand	1884	O.&P. Knudsen	V 1.87	Krog & Henschien	Lillesand
610		Satsuma *Swanson* 85-7 W Bk	British	364	137·9	27·2	15·4	Sunderland	1869	J. Crown	V 8.87	China Shipowners' Association	Shanghai
611		Saturn *Agnew* W Sr	British	115	84·8	24·2	10·6	P.E. Island	1864	Grady		T. M'Donogh	Galway
612	N	Saturn *Carlsen* -87 W Bk	Norweg'n	475	141·5	32·6	17·1	Wasa, Finland	1860	Wulff		I. C. Svendsen	Christiansand
613	DN	Saturn *Olsen* -88 W Bk	Norweg'n	410	125·4	30·2	16·1	Risör	1873	AanonOlsen	V 2.87	I. W. Prebensen	Risör
614	V	Saturne *Mauffret* W Bn	French	126	80·0	20·3	11·5	St. Malo	1863	Carnet		J. Mauffret	Nantes
615		Saturno (ex Saturn) *Slack* -87 W3mSr	Chilian	290	126·0	27·3	12·5	Halsenöen, [Norway	1877		V 2.86	A. Sutherland	Valparaiso
616	V	Saturnus *Hockelmann* 86-88 W Bk	German	608	146·9	29·3	18·6	Hamburg	1871	M. G. Amsinck	V 5.88	M. G. Amsinck	Hamburg
617		Saucy Jack W Bn	British	156	105·6	23·8	9·0	Manning River [N.S.W.	1874	G. R. Nicoll		Lawrence Tulloch	Sydney, N.S.W.
618		Saudade W Sr	Portug'se	156	91·2	21·6	10·8	Vianna	1873	Borda		A. P. & B. J. Barbosa	Vianna
619		Saul (CONSIDERED) *Cacace* W Bg	Italian	343	117·1	25·9	17·1	Cassimo	1868	F. De Rosa	V 1.81	T. Cacace	Castell. di Stab.
620		Savanetta *Purdy* W Bk	British	324	123·2	25·3	16·2	Alloa	1860	Adamson	V 3.82	G. Turnbull & Co.	Glasgow
621	V	Savanna *Collenette* -87 W Bk	British	749	146·2	34·0	20·0	Clifton, N.B.	1868	Hewelling	V 10.85	Doane & Co.	Barbados
622	G	Savannah (ex Winfield Scott) *Haul* -87 W S	German	1399	183·7	37·7	22·3	Newcastle, Me.	1851		V 1.88	Siedenburg, Wendt & Co	Bremen
623	V	Savenas *Thanell* 80-82 W Bk	Swedish	768	158·8	32·1	20·1	Skelleftea	1875	E. A. Löfgren	V 7.87	A. Markstedt & Söner	Skellefteå
624	✻	Savernake *Pritchard* 67-70 I Bk	British	1284	218·0	37·1	22·7	Stockton	1869	Richardson, [Duck & Co.		H. Owen & Son	Liverpool
625	I	Savina (ex Idea G.) *Mazzone* [-87 W Bk	Italian	643	153·0	31·1	20·2	Sestri Ponente	1876	Tixi	V 7.87	C. Marnata & P. Marioni	Genoa
626	IV	Savoia *Gueguina* -88 W S	Italian	1308	200·1	32·8	24·9	Varazze	1884	B. Cerruti	V 2.88	A. Cerruti fu B.	Genoa
627		Saxon *Beaker* -71 W Bk	British	401	128·5	27·6	18·4	Sunderland	1854		V 6.82	J. L. & W. A. Simpson & Elder, Smith & Co.	Adelaide, S.A.
628		Saxon Maid *McCurry* 83-85 W Sr	British	137	94·4	22·3	11·6	Glasson Dock	1871	Simpson		Porter's Shipping Co. Ld.	Fleetwood
629		Say Wan W Bg	Turkish	450	..	..	..	Kara Agatch	1871			Osman Zadeh Apik Effendi	Constantinople
630		Sayah I Barge	British	392	199·5	34·1	5·9	Port Glasgow	1883	R. Duncan & Co.		Irrawaddy Flotilla Co.	Glasgow
631		Sayareh W Bg	Turkish	150	..	..	..	Roumania	1869			Bostanji Zadeh Hasse	Constantinople
632		Sayfan Stl Lr	British	480	220·0	30·0	8·4	Dumbarton	1887	W. Denny & Brs.		Irrawaddy Flotilla Co. Ld.	Glasgow
633	I	Sbrigati (ex Papa Giuseppe Repetto) *Schiaffino* -87 W Bk	Italian	652	154·2	33·1	20·4	Lavagna	1868	D. Briasco	V 4.87	A. Centurini fu I.	Genoa
634	DRV	Scammell Brothers *Lewis* 85-6 W Bk	British	1219	196·5	38·8	22·3	Eatonville, NS	1883	DR. & CF. Eaton	V 6.88	Scammell Brothers	Parrsboro', N.S.
635	N	Scandia *Christensen* -70 W Bk	Norweg'n	415	117·5	31·2	17·3	Sundsvall	1850			M. Christensen	Horten
636	N	Scandia *Nilsen* W Bk	Norweg'n	360	106·0	30·0	16·7	Grimstad	1855	N. OlsenGjömle		H. H. Pettersen	Arendal
637	N	Scandia *Tjøstolsen* W Bk	Norweg'n	335	122·4	28·9	15·2	Gefle	1847			P. Arentsen	Tvedestrand
638	N	Scandia *Hansen* -88 W Bg	Norweg'n	220	107·8	26·5	13·5	Wisby	1847			O. A. Föyen	Tönsberg
639		Scaramanga *Iversen* -86 W Bk	Norweg'n	313	123·1	27·9	15·0	Stavanger	1859			Actieselskabet "Scaramanga"	Fredrikstad
640		Sceptre *Potter* W Bn	British	208	84·6	22·6	15·3	Stockton	1850			W. C. Bray	London
641		Sceptre *King* -87 W Bn	British	142	88·6	24·5	9·7	Lunenburg NS	1887	D. Smith		W. N. Zwicker	Lunenburg, N.S.
642		Scerif Ghildis W Bn	Turkish	232	..	..	..	Unich	1866			Mahmut Maslun	Candia
643	N	Schadberg (ex Arabia) *Salvesen* 86-7 W Bk	Norweg'n	300	128·8	27·0	13·6	Hamburg	1863	E. Dreyer	V 1.81	A. Salvesen	Egersund
644		Schah Jehan W S	British	698	117·4	32·5	20·1	Cochin	1857	Guiezebar		Hajee Essuf Bin Mahomed Hussan Kh'n	Calcutta
645	N	Schamyl *Møller* W Bk	Norweg'n	325	109·8	28·8	15·7	Grimstad	1865	G. Larsen		A. Chr. Hein	Arendal
646		Schelde (CONDEMNED) *B. Grinstad* -77 W Bk	Norweg'n	619	127·3	33·0	21·3	Zierickzee	1851			Asker & Co.	Christiania
647	DG	Schenke *Margaering* 81- W3mSr	German	204	105·0	24·1	11·2	Papenburg	1867	H. J. C. Hilling		W. Hilling	Papenburg
648		Schiaffino W Bk	Italian	528	142·3	31·9	18·7	Varazze	1870	F. Tassara		Eredi P. Schiaffino	Genoa
649	I	Schiaffino Padre *Schiaffino* W Bk	Italian	774	167·0	33·8	22·3	Lavagna	1884	G. Brignole	V 3.83	Flli. Schiaffino	Genoa
650	V	Schiffswerft *Hildebrand* I S	German	867	197·6	33·0	19·0	Flensburg	1877	Flensburg [Schiffsb. Ges.		F. W. Selck	Flensburg

✻ = Lloyd's Register; D = Nederlandsche Vereeniging, G = Germanischer Lloyd; I = Registro Italiano; R = American Record; A = Veritas Austro-Ungarico; H = Veritas Ellenico; N = Norske Veritas; V = Bureau Veritas.

Dans la 4me Colonne, C = Composite (membrure en fer, bordé en bois); I = Fer; Stl = Acier; W = Bois.

Les Dates du doublage en métal (Col. 11) ressortent des livres dans lesquels les navires sont respectivement classés ou des rapports des Inspecteurs du Lloyd's Register.

Pour plus amples renseignements sur les navires classés par le Lloyd's Register, voir le Lloyd's Register Book et le Lloyd's Yacht Register.

Official No.	No. in Book	Registry in which classed	Ship's Name. Master, and Dates of Appointment to present Owner's Service and to Ship. Ship's Late Name.	Material and Rig	Flag	Net Tonnage	Length	Breadth	Depth	Built: Where	Built: When	Built: By whom	Date of Metalling	Owners or Managers	Port
	651	G	**Schiller** *Steinbömer* 83-84	I Bk	German	1227	228·7	34·5	20·5	Vegesack	1879	H.F.Ulrichs		D.H.Wätjen & Co.	Bremen
	652	DGV	**Schiller** *Zeplien* 84-8[illegible]	W Bk	German	590	143·5	31·5	18·4	Motzen	1861	D.Oltmann's [Wwe	[illegible]	W.Maack	Rostock
	653	NV	**Schiller** *Larsen* (ex Jane Wright)	W 3mSr	Norweg'n	435	139·2	30·3	13·4	St.John,N.B	1872	C.Nevins	[illegible]	Brødr.Herlofson's Ro	Arendal
	654	G	**Schlump zu Lulle** *Aang* [illegible]	W Bg	German	179	98·0	21·0	11·5	Papenburg	1858	Veen& Baeren		B.Freericks	Papenb
	655		**Schotter-hoff** [illegible]	W Sr	French	113	91·2	27·9	11·0	Dunkirk	1877	Lefebvre		G.Beck	Dunkirk
	656	GV	**Schwan** *Giese* 88-88	W Bk	German	312	114·0	26·0	11·0	Vegesack	1876	C.Lange	[illegible]	Seetzen Gebrüder	Bremen
	657	NV	**Schwanden** *West*	W Bk	Norweg'n	857	168·1	37·0	20·4	Fredrikshald	1879	J.Brevig	[illegible]	T.J.Heftye & Son	Christian
	658	N	**Schweigaard** *Anderson*	W Bk	Norweg'n	625	194·9	34·1	17·9	Christiania	1870	H.C.Hansen	[illegible]	C.Dahl	Christian
	659		**Sciahini Baker**	W Sr	Turkish	120	77·1	22·1	13·1	Dolcigno	1881	HusenHansere		A.G.Ciobla	[illegible]
55930	660		**Scicluna** *Beringe* (ex Nuovo Rubino)	W Bk	British	339	122·5	28·5	15·8	Fiume	1865	G.G.B.Bellis	[illegible]	M.Apap	Malta
3996	661		**Scintilla** *Elena*	W Sr	Italian	109	90·9	21·4	10·8	SestriPonente	1881	Ag.Briasco fu L	[illegible]	F.Elena	Genoa
69190	662	BV	**Scotia** *Liebke* -84	W Bk	British	691	153·9	33·9	17·9	La Have, N.S.	1874	L.Hewitt	[illegible]	Jas.H.Smeltzer	Bridgew
99881	663	✱	**Scotia** *Jenkins* -87	W Bk	British	321	129·4	26·2	15·3	Kingston	1873	Kinloch	[illegible]	J.Duncan	Banff
78256	664	✱	**Scotia** *Conner* -83	W Bkn	British	172	115·6	22·8	11·9	Troon	1880	TroonShipB.Co.	[illegible]	J.A.Beeching	Ayr
67216	665	✱	**Scotia** *John* 88-88	W Bn	British	171	102·4	23·7	12·9	Dartmouth	1872	Moore	[illegible]	J.& W.Stewart	Greenock
76577	666	✱	**Scotia** *Wood*	W Sr	British	130	97·0	22·3	11·0	Appledore	1878	Cook	[illegible]	J.Wood	Banff
54097	667		**Scotia**	W Sp	British	111	96·4	22·8	7·4	Quebec	1872			W.M.Macpherson	Quebec
115836	668	R	**Scotia** *Shearer*	W 3mSr	American	386	138·0	33·2	12·2	Belfast,Me.	1883	C.P.Carter&Co.	[illegible]		
	669	N	**Scotia** [illegible] *Pedersen* -81	W Bk	Norweg'n	413	117·1	29·4	17·1	Abo	1846			J.S.Lorentzen, jun.	Drammen
83201	670	V	**Scotland** *Munroe*	W Bk	British	1499	208·0	41·0	24·6	Hantsport, NS	1881	J.Davison	[illegible]	Geo.W.Churchill	Windsor
72169	671	DV	**Scots Bay** *Steele* 83-83	W Bk	British	993	183·0	37·0	22·0	Cornwallis,NS	1875	J.E.Steele	[illegible]	Stephen Sheffield	Windsor
52574	672		**Scotscraig** *Brown* 84-84	W Bg	British	240	106·5	25·4	14·1	Dundee	1866	DundeeShipCo.	[illegible]	G.Smee	Maldon
74003	673		**Scotsman** *Rockfort* 82-82	W Bn	British	189	102·2	24·0	12·9	P.E.Island	1876	Duncan	[illegible]	G.Howatson & Co.	London
48841	674		**Scotsman** *Liddle*	W Bn	British	175	100·0	22·5	12·9	Arbroath	1864	Arbroath S.Co.		G.Widger	Arbroath
	675	N	**Scott** *Olsen* -84	W Bk	Norweg'n	630	146·6	33·7	17·1	Arendal	1884	O.M.Olsen	[illegible]	Tho.B.Olsen	Christian
77111	676	✱	**Scottish Admiral** *Ross* [83-85]	I Bk	British	939	217·8	32·8	18·1	Sunderland	1878	W.Doxford &S.		McIlwraith, McEacharn & Co.	London
68948	677	✱	**Scottish Bard** *Mowat* [illegible]	I Bk	British	803	189·2	33·0	19·2	Glasgow	1876	Lond.&Glas.Co.		McIlwraith, McEacharn & Co.	London
68932	678	✱	**Scottish Chief** *Davies* 88-88	I Bk	British	658	175·3	30·3	17·6	Sunderland	1875	J.Blumer&Co.		Tomlinson,Hodgetts& C	Liverpool
87802	679	✱	**Scottish Dales** *Magee* 73-83	I S	British	2059	277·7	40·3	24·5	Southampton	1883	Oswald,Mordaunt&Co.		W.H.Ross & Co.	Liverpool
76495	680	✱	**Scottish Fairy** *Ellis* 82-83	I Bk	British	727	187·1	31·4	18·4	Sunderland	1877	Austin&Hunter		A.Rae	Liverpool
91265	681	✱	**Scottish Glens** *Whiteford* [illegible]-88	I S	British	2061	279·6	40·3	24·5	Southampton	1885	Oswald,Mordaunt&Co.		W.H.Ross & Co.	Liverpool
73678	682	✱	**Scottish Hero** *Fraser* 79-81	I Bk	British	869	196·6	33·0	19·1	Sunderland	1876	W.Doxford&Sns		McIlwraith, McEacharn & Co.	London
87854	683	✱	**Scottish Hills** *Steel* 80-83	I S	British	2058	277·7	40·2	24·4	Southampton	1883	Oswald,Mordaunt&Co.		W.H.Ross & Co.	Liverpool
87821	684	✱	**Scottish Isles** *Nicoll* -88	I S	British	1998	269·0	41·6	24·3	Liverpool	1883	T.Royden&Sons		W.H.Ross & Co.	Liverpool
73682	685	✱D	**Scottish Knight** *Fowler* [88-88]	I Bk	British	875	196·0	32·9	19·3	Sunderland	1876	W.Doxford&Sns		McIlwraith, McEacharn & Co.	London
77001	686	✱	**Scottish Lassie** *Singer* [82-84]	I Bk	British	852	201·2	32·0	18·0	Aberdeen	1877	A.Hall&Co.		McIlwraith, McEacharn & Co.	London
93891	687	✱	**ScottishLochs** *Bryde* -88	I S	British	2573	308·0	42·0	24·8	Southampton	1888	Oswald, Mordaunt&Co.		W.H.Ross & Co.	Liverpool
3307	688		**ScottishMaid** (WRECKED)	W Sr	British	136	97·8	21·9	12·4	Aberdeen	1839			J.Crombie	Aberdeen
76548	689	✱R	**Scottish Minstrel** *Christie* 78-83	I S	British	1511	243·0	38·8	23·1	Stockton	1877	Richardson, Duck&Co.		W.H.Ross & Co.	Liverpool
85860	690	✱	**Scottish Wizard** *Orkney* 83-83	I Bk	British	1140	226·6	36·0	20·0	Aberdeen	1881	A.Hall&Co.		McIlwraith, McEacharn & Co.	London
[illegible]	691		**Scout** *Thomas* 77-82	W Sr	British	122	86·9	25·0	11·6	Dartmouth	1855			E.J.Hurdon(perM.Gey)	London
115346	692	R	**Screamer** *Berry* -87	W Bn	American	503	137·9	31·0	17·3	Yarmouth, Me	1877	G.Loring	[illegible]	William Sinnett	Portland
23358	693		**Scud** *Given*	W Bn	American	328	129·0	27·6	16·6	Portsm'th,NH	1868			L.W.&P.Armstrong	New York
1141	694	IV	**Scutolo** *Cafiero*	W Bk	Italian	460	131·8	27·9	18·7	Alimuri	1876	S.Mauro	Y 5.83	A.Lauro	Castell
85737	695		**Scylla** *Himes*	W Sr	British	102	83·8	24·0	9·0	MahoneBy.NS	1883	T.Langille		James Eisenhauer	Lunenb
28855	696		**Sea**	W Sr	British	115	85·2	22·2	9·4	London	1857			W.C.Middleton	Falmouth
	697	N	**Sea** *Skadberg* -88	W Bk	Norweg'n	768	161·3	32·6	20·4	StStephen's NB	1862	J.& C.Short	[illegible]	C.Bockman	Egersund
90285	698	✱	**Sea Belle** *Ryder* 85-85	W Sk	British	78	75·6	20·5	10·4	Rye	1884	G.&T.Smith		Mrs.E.Blake	Grimsby
61888	699	R	**Sea Bird** *Delfosse* -88	W Bn	British	202	99·6	26·0	12·9	Cornwallis NS	1869		[illegible]	H.Kelly	Canning
32142	700		**Sea Bird** *Coleman*	W Bn	British	155	98·0	22·8	10·0	Tasmania	1865	P.Oldham		George F.Mason	Sydney

✱ = **Lloyd's Register**; D = Nederlandsche Vereeniging; G = Germanischer Lloyd; I = Registro Italiano; R = American [illegible]; Á = Veritas Austro-Ungarico; H = Veritas Ellenico; N = Norske Veritas; V = Bureau Veritas.

In the 4th Column, C = Composite (iron frame, wood planked); I = Iron; Stl = Steel; W = Wood.

The Dates of Metalling (Col. 11) are based on the records of the Books in which Vessels are respectively classed, or on the Reports of the Surveyors of Lloyd's Re[gister].

Further particulars of Vessels classed by Lloyd's Register will be found in Lloyd's Register Book and Lloyd's Yacht Register.

2	3	4 Nom. Matériaux et Gréement. Capitaine et dates d'entrée au service des Armateurs actuels et à bord du navire. Ancien Nom.			5 Pavillon	6 Tonnage	7 Dimensions Longueur	Largeur	Creux	8 Construit. Où.	9 Quand	10 Par qui.	11	12 Armateurs ou Directeurs-Gérants.	13 Port d'Armement.
701		Sea Bird	*Petersen*	WSr	British	116	91·0	25·3	9·4	Cornwall	1875	Trethowan	Y 2.89	B.de Pass	Cape Town
702		Sea Bird	*Hogan*	WSr	American	179	100·8	29·4	8·1	Tuckahoe,N.J.	1869			J.Steelman	Millville
703		Sea Breeze	*Simmons 86-*	WBk	American	307	123·3	29·0	16·4	N.Bedford,Mass	1853			J.McKenna	San Francisco
704	N	Sea Breeze	*Kettlesen 83-83*	WBk	Norweg'n	668	150·1	32·2	20·2	Sunderland	1856		Y 7.88	C.H.Johnsen	Arendal
705	V	Sea Flower	*Revert*	WBk	French	352	133·2	27·9	15·3	Paspebiac, [Canada	1873	Renouf		Revert	St.Servan
706	✻	Sea Foam	*Jones*	WBk	British	319	121·3	27·2	16·9	Sunderland	1868	W.Pickersgill	Y 1.88	J.R.Down & Co.	Swansea
707		Sea Foam	*Arey*	WSr	American	150	105·3	26·1	8·5	Bangor,Me.	1855			J.H.Crosby	Bangor,Me.
708		Sea Fox	*Jenkins*	WBk	American	166	93·4	22·0	12·1	N.Bedford,Mass	1851	S.Andrews		J.P.Knowles	N.Bedford,Mass
709	✻	Sea Gull		WSk	British	87	78·4	20·5	10·7	Hull	1887	W. & J.M'Cann		J.Munby	Hull
710	R	Sea Gull	*Thompson -88*	WBk	American	313	121·5	29·0	16·0	Fairhavn.Mass.	1853		Y 1.89	J.F.Brooks	Boston,Mass.
711		Sea Gull (ex Columbus)		WBn	Ecuador	257	107·6	21·3	12·7	Brake	1870	F.Oltmann's Wwe.	Y 7.82	Ecuadorian Lloyd	Guayaquil
712		Sea Hopper	*Hussein*	WBg	Turkish	300	..	..	..	Greece	1872			Trebizondlee Kutchuk Osman & Co.	Trebizonde
713		Sea Horse	*Rasmussen*	WBk	Danish	329	128·0	26·2	16·1	Greenock	1859	R.Steele	Y 3.81	H.Iversen	Svendborg
714	RV	Sea King	*Getchell*	WS	American	1492	210·6	39·4	24·0	Bowdoinham, [Me.	1877	G.H.Theobald	Y 7.86	G.H.Theobald	Bath,Me,
715		Sea Lark	*Green*	WSr	British	151	94·1	22·7	12·5	Brixham	1874	Dewdney	Y 6.86	H.Green	Brixham
716		Sea Nymph (ex Ivan)	*Brady*	WBk	British	327	132·2	26·3	14·6	Guernsey	1866	J.Sebire	Y 1.82	W.A.Seward	Sydney,N.S.W.
717		Sea Nymph		WBk	British	240	120·0	24·5	13·7	Weymouth	1858			Norman McLeod	Melbourne
718		Sea Queen (CONDEMNED)	*Thompson*	WBk	American	185	99·2	24·0	14·0	Dartm'th,Mass.	1851			J.McCullough	N.Bedford,Mass
719	✻	Sea Queen	*Biermann -87*	IBk	Norweg'n	846	180·7	30·0	20·7	Glasgow	1858	A.Stephen&Sns.		Actieselskabet "Sea Queen" (T.I. Andersen)	Mandal
720		Sea Ranger	*Holmes*	WBk	American	259	106·8	27·3	16·3	Rochester, [Mass.	1856			G.F.Bartlett	N.Bedford,Mass
721		Sea Swallow	*Maher*	WBk	British	332	120·5	27·3	16·8	Sunderland	1866	R.Thompson	Y 1.82	C.A.L.Dunn	Shanghai
722	G	Sea Waif (ex Zelda)	*Simonson 81-*	WBg	American	260	105·0	27·6	13·0	Nantes	1873	Dubigeon&Sns.	Y 7.82	L.Ford	San Francisco
723		Sea Witch	*Foster 87-87*	WSr	British	136	88·3	20·9	12·2	Blyth	1856			Starbuck & Backstraw	London
724	RV	Sea Witch	*Drew -77*	WS	American	1288	197·0	37·8	24·2	East Boston	1872	R.E.Jackson	Y 12.88	Est.of Rd.Baker, jr.	Boston,Mass.
725		Seabird		WSr	British	124	78·0	20·8	9·0	Quebec	1853	N.Germain		David Germain	Quebec
726		Seabird		WSr	British	104	80·8	22·8	8·5	Quebec	1875	G.Barabé		Olivier Perrault	Quebec
727	✻	Seafarer	*Jackson 88-88*	Stl Bk	British	2163	281·6	41·2	24·6	Liverpool	1888	W.H.Potter & Sons		W.H.Potter	Liverpool
728		Seaflower		WSr	British	113	81·2	25·1	10·1	Tracadie,N.S.	1868			Ben.A.Smith	Picton,N.S.
729	✻	Seaforth	*Griffith 83-83*	IBk	British	1189	218·0	34·8	23·6	Liverpool	1862	Jones,Quiggin [&Co.		Seaforth ShippingCo.Ld (Craig, Scott & Co.)	Greenock
730	N	Seagull	*Christiansen -88*	WBk	Norweg'n	391	122·1	29·5	17·3	Drammen	1866			N.Olsen	Arendal
731	N	Seagull	*Pedersen -87*	WS	Norweg'n	360	131·1	31·2	12·4	Söndeled,Sogn	1886	E.Knudsen [Lindstöl	Y 3.87	J.W.Prebensen	Risör
732	N	Seagull	*Föyen*	WBg	Norweg'n	244	105·6	27·3	12·9	Sarpsborg	1866			O.A.Föyen	Tönsberg
733		Searsville	*Messervey*	WSr	American	210	107·6	27·4	9·9	Dennis,Mass.	1854			W.A.Elwell	St.George,Me.
734	✻	Seatoller	*Lister 85-85*	IBk	British	558	166·6	27·1	17·8	Liverpool	1866	R.&J.Evans&Co		J.Gambles	Liverpool
735	R	Sebago	*Clark 83-83*	W3mSr	American	232	123·2	31·0	10·6	E.Deering	1883	G.Russell		J.S.Winslow & Co.	Portland
736		Sebastian		WBn	Greek	396	126·3	27·1	..	Venice	1878			N.Georgiadis	Skiathos
737		Sebastian		WBn	Spanish	145	83·0	24·3	10·2	Palamos	1869			S.Lloret	Palamos
738	V	Sebastian Bach	*Schneider*	WBk	German	812	149·2	33·1	18·7	Bremerhaven	1868	J.C.Tecklenborg	Y 12.86	F.Reck & Co.	Bremen
739		Sebastian Gumá		WBk	Spanish	450	134·5	32·3	16·7	Palma	1875			Guma,hermanos	Havana
740	V	Sebastiano	*Giorgiadi -87*	WBg	Greek	431	132·8	30·2	20·6	Chioggia	1876	G.Poli	Y 7.81	N.Giorgiadi	
741		Sebastiano (ex Gabriella, ex B. F. Williams)	*Questa -87*	WBk	Italian	386	124·0	28·7	15·0	Bath,Me.	1872		Y 3.80	B.Barrera	CivitaVecchia
742	IV	Sebastiano Dall' Orso	*Cremonini*	WBk	Italian	1074	193·5	35·8	23·9	Varazze	1882	B.Cerruti	Y 5.88	S.Dall'Orso & figli	Genoa
743		Sebastiano S. (ex Angelina Guerriero)	*Spampinato*	WBn	Italian	221	104·0	23·6	15·4	Varazze	1870		Y 2.82	S.Spampinato	Catania
744		Sebastopol		WBk	British	610	149·6	31·9	18·1	Quebec	1855			T.J.M'Grath	London
745		Secondo S. Giovanni (ex Carlotta Argentina)	*Carletti*	WBk	Italian	373	122·3	29·2	18·0	Savona	1867	F.Sirello	Y 12.81	F[lli.] Cignoni di L.	Porto Ferraio
746		Secondo TreFanciulli	*Bollo*	WBk	Italian	966	187·0	36·1	22·5	Savona	1873	E.Ciarlo		P.Zeneglio	Genoa
747	✻	Secret	*Jones*	WSr	British	93	86·0	19·6	10·7	Appledore	1863	Pickard		Mrs.E.Owen & Co.	Carnarvon
748	G	Secundus	*Lillia 88-88*	WSr	Russian	290	112·0	29·2	11·1	Sastmola	1879	J.Elg		O.Blom	Nystad
749	I	Sedici D'Agosto	*Bottino -88*	W3mSr	Italian	336	149·2	27·6	14·2	Savona	1881	F.Sirello	Y 1.86	C.Anselmi	Savona
750	V	Séduisante	*Agneray*	WSr	French	111	90·5	19·7	11·1	Dunkirk	1878	Hecquet		Hecquet van Rapenbusche	Dunkirk

✻=Lloyd's Register; D=Nederlandsche Vereeniging; G=Germanischer Lloyd; I=Registro Italiano; R=American Record; [A]=Veritas Austro-Ungarico; H=Veritas Ellenico; N=Norske Veritas; V=Bureau Veritas.

Dans la 4me Colonne, C=Composite (membrure en fer, bordé en bois); I=Fer; Stl=Acier; W=Bois.

Les Dates de doublage en métal (Col. 11) ressortent des livres dans lesquels les navires sont respectivement classés ou des rapports des Inspecteurs du Lloyd's Register.

Pour plus amples renseignements sur les navires classés par le Lloyd's Register, voir le Lloyd's Register Book et le Lloyd's Yacht Register.

Official No.	No. in Book	Register in which classed	Ship's Name, Master, and Dates of Appointment to present Owner's Service and to Ship. Ship's Late Name.	Material and Rig	Flag	Net Tonnage	Length	Breadth	Depth	Built Where	When	By whom	Dates of Metalling	Owners or Managers	Port of registry
62892	751	✻	Sedwell Jane *Tulloch*	W Bkn	British	188	115·5	20·0	12·5	Bideford	1868	RolleCanalCo.	Y 9.85	J.Darling,jun.	Ad...
	752	R	Seeja *Blomster*	W3mSr	Russian	363	128·7	30·6	15·0	Luckalaks	1883			J.Ahlroos-Pietila	Luckalaks
	753	G	Seelent (ex Nordstern) *Warker -78*	I Bn	German	142	99·7	22·5	11·8	Glasgow	1861	AStephen&Sons		D.A.Andersen	Heligoland
	754	G V	Seenymphe *Jute -88*	W Bg	German	293	123·2	28·1	14·2	Reiherstieg	1866	E.Dreyer		E.B.Crasemann	Hamburg
7070	755		Segesta *D'Amico -87* (ex Nuovo S. Francesco)	W Bn	Italian	350	110·8	29·1	18·1	Varazze	1868	B.Craviotto	Y	G.D'Ali	Trapani
23103	756	R	Seguin *Hunterman*	W Sr	American	157	91·8	25·8	9·8	Bath,Me.	1866		Y	Poitevant & Favre	Shieldsboro
	757		Segunda Cantinera	W Bn	Spanish	146	100·4	27·2	9·0	SanFeliu	1862			Sainz,Alemañy&Co.	Havana
	758		Segunda Mercedes	W Bn	Spanish	165	96·5	25·1	10·9	SanFeliu	1864			P.Carreras	Palamos
	759		Segunda Sebastiana (ex Davis)	W Sr		143	79·4	25·1	12·1	Lloret	1851				
	760		Segundet (ex Prisca) *Pagès*	W Bn	Spanish	132	86·0	22·8	10·0	Blanes	1851			Segundo Pagès	Barcelona
	761		Segundo Barcelo (ex Virgen de la Caridad del Cobre)	W Bg	Spanish	193	90·5	25·3	13·5					Miguel Roig	Santiago
	762		Segundo Emilio	W Bg	Spanish	120	72·8	21·9	9·1	Barcelona	1840			F.Sus & Co.	Mataro
	763	V	Segundo Romano *Macia*	W Bg	Spanish	191	93·8	25·6	12·2	Lloret	1869		C	P.Gallart y Soller	Barcelona
34230	764	I	Sei Amici *Gerrato -88*	W Bk	Italian	668	164·0	31·9	20·5	Lavagna	1871	A.Briasco fu G.B.	Y 12.87	C.Gandolfo fu G.B.	Genoa
12686	765		Sei Fratelli *Olivari* (BROKEN UP)	W Bk	Italian	554	136·8	31·2	20·6	Varazze	1870	G.Scah	Y	F[lli] Degregori	Genoa
	766	D N V	Seier *Beck*	W Bk	Norweg'n	476	137·3	31·3	16·2	Risör	1885	T.Taraldsen	Y	A. & F.Smith	Tvedestrand
	767	V	Seierskrandsen *Schaadt*	W Bn	Danish	177	102·3	25·9	11·8	Marstal	1883	J.O.Christensen		H.C.Christensens Enke	Sönderborg
	768		Seiffi Derya	W Bg	Turkish	112	..	..	..	Oonia	1870			Tchezondliee Yelkenji (Seyh Temal Effendi & Co.)	Trebizond
398	769		Seifoo Maru	W Sr	Japanese	218	101·1	22·6	13·1	Tokio	1880	KawasakiSBYd		Teishin Sho	Ohio
189	770		Seikai Maru	W Sr	Japanese	119	81·0	18·3	10·4	Osaka	1878	NagataSanjuro		Kato Hikobei	Owari
7	771		Seilani Bari	W Bg	Montngrn	121	71·2	19·7	10·9	Marcero	1870			Eten Razas & Co.	Dulcigno
	772	✻	Seine *Le Bras 86-86*	I Bk	French	735	184·5	30·1	19·0	Stockton	1870	M.Pearse&Co.		Ant.Dom.Bordes & fils	Bordeaux
76740	773	✻	Seiriol Wyn *Storm 85-87*	I Bk	British	1065	211·3	34·0	21·1	PortGlasgow	1877	Russell&Co		H.Pritchard	Glasgow
	774		Seis Irmaos	W Bn	Brazilian	393	..	..	..					B.B. de Souza	Pernambuco
822	775		Seisho Maru	W Bk	Japanese	259	121·8	22·9	13·1	Sonderburg	1865			Rikni Tayemon	Sakai
	776		Seitunie	W Bg	Turkish	120	77·1	22·1	13·1	Dulcigno	1870	Salih		Casem Isuffi & Co.	[illegible]
	777		Seivahnie	W Bg	Turkish	112	..	..	..	Roumania	1875			Vassilikoslee Anastas Yanaki	Vassiliko
27893	778		Sela *Ongham 80-81*	W Bn	British	198	109·7	24·7	12·0	P.E.Island	1859	Hayes		F.W.Fairbrass	Faversham
	779	N	Sela *Tellefsen* (ex Hope Harmony)	W Bk	Norweg'n	327	116·4	26·8	16·2	Sunderland	1863	Taylor		O.Tellefsen	Grimstad
	780		Sela Kowda *Larconow 86-* (ex Johannes)	W Sr	Russian	182	96·7	23·6	11·3	Fünfhausen	1866	J.F.Strenge		I.Klementjeff	Archangel
	781	N V	Seladon *Thorsen -88*	W S	Norweg'n	1066	189·4	38·8	21·8	Drammen	1877	Jürgensen& [Knudsen	Y 10.87	G.Gundersen	Stavanger
	782		Selamet Hhair *Hussein*	W Bg	Turkish	155	..	..	..	Italy	1885			Tijar Hajee Ali Effendi	Constantinople
	783		Selametti Hhair	W Bg	Turkish	155	..	..	..	England	1883			Iskenda Sakkiz Hajee Ali Effendi Zevjesi Kadine Hafiseh Hanoum	[illegible]
	784		Selametti Hhair	W Sr	Turkish	145	..	..	..	Aides	1884			Hoshatilee Emir zadeh Hajee Yacoob Bey & Sons	Heshab
63731	785	✻	Selene (ex Isabel Craven) *England 84-88*	I S	British	996	215·0	34·3	20·3	Glasgow	1869	C.Connell & Co.		J.D.Clink	Greenock
46684	786	✻	Selene	C SrYt	British	168	116·3	23·4	12·8	Greenock	1865	Steele&Co.	C 9.87	Thomas Henderson	Glasgow
29172	787		Selina *Mann 78-85*	W Sr	British	124	89·4	22·9	11·7	Rye	1863	Hood		W.Dusting	Penzance
78081	788	✻	Selina Mary *Hopper*	W Sk	British	48	59·0	20·2	7·1	Calstock	1878	Brooming		C.B.Hamley	Plymouth
729	789	V	Selina Stanford *Sturila*	W Bg	Italian	349	116·1	26·5	17·7	Cassano	1868	A.de Rosa	Y	G.Maresca	[illegible]
	790		Selinda *Österlund -72*	W Sr	Russian	264	..	..	..		1872			J.G.Österlund	Wärtsilä
89173	791	V	Selkirk *Crowe 81-86*	W S	British	1757	236·6	43·7	24·0	S.Maitland,NS	1886	W.P.Cameron	Y 11.86	Martin Dickie	Maitland
80411	792	✻	Selkirkshire *Embauder* (9585)	I Bk	British	1192	228·4	35·8	20·4	Dumbarton	1878	Birrell,Sten-[housed&C.		T.Law & Co.	Glasgow
56677	793	✻	Selma *Bach 67-78*	W Bk	British	349	120·9	27·5	17·0	Sunderland	1867	R.Thompson&S.	Y	D.F.C.Beck	Newcastle
	794	D N V	Selma *Melvig -87* (ex John ...)	W Bk	Norweg'n	406	123·8	27·9	17·3	Sunderland	1868	Lumsden	Y	J.Melvig	Fredrikstad
	795	R	Selma *Andersson 83-86*	W Sr	Russian	271	108·0	29·2	15·3	Westanfjerd	1875			O.F.Ölander	Kimito
	796		Selma *Holmstrom* (CONDEMNED)	W Sr	Russian	204	102·5	27·2	12·2	Abo	1866			E.F.Lifvendel	Pargas
	797		Selma *Mattson*	W Sr	Russian	112	79·3	25·0	9·3	Dänöby	1877	E.Malmqvist		M.A.Söderström	Geta
	798		Selma *Svensson*	W Bg	Swedish	297	138·4	28·6	15·2	Vifsta	1866			S.Jönsson	Arboga
	799		Selma *Johansson -78*	W Bg	Swedish	256	108·6	25·0	14·4	Hanström	1850			J.Berntsson	[illegible]
	800		Selma *Pettersson*	W3mSr	Swedish	177	93·3	23·0	11·9	Orust	1871	A.Samuelsson		J.A.Samuelsson	[illegible]

✻=Lloyd's Register; D=Nederlandsche Vereeniging; G=Germanischer Lloyd; I=Registro Italiano; R=American Record; A=Veritas Austro-Ungarico; H=Veritas Ellenico; N=Norske Veritas; V=Bureau Veritas.

In the 4th Column, C=Composite (iron frame, wood planked); I=Iron; Stl=Steel; W=Wood.

The Dates of Metalling (Col. 11) are based on the records of the Books in which Vessels are respectively classed, or on the Reports of the Surveyors of Lloyd's Register.

Further particulars of Vessels classed by Lloyd's Register will be found in Lloyd's Register Book and Lloyd's Yacht Register.

No.	Registre dans lequel le navire est classé	Nom. Matériaux et Gréement. Capitaine et dates d'entrée au service des Armateurs actuels et à bord du navire. Ancien Nom.	Pavillon.	Tonnage net.	Longueur.	Largeur.	Creux.	Construit. Où.	Année.	Par qui.	Doublage métal	Armateurs ou Directeurs-Gérants.	Port d'Armement.
801		Selma *Nostrom* W Sr	Swedish	101	74·8	22·9	9·1	Åbo	1870			E.Storm	Öregrund
802		Seloni *Hendrikson* -82 W Bn Sr	Russian	291	..	..	..		1876			P.Häggblom	Åland,Skarpnåt
803		Selva Canapaddy W Bg	British	109	84·0	20·4	10·5	Tinnavelvasel	1865			Ahamadolebbe Marikar (Hydrudelbe)	Colombo
804	DN	Semafor *Hansen* W Sr (ex General Gonsalez)	Norweg'n	224	107·9	27·0	11·4	Windsor,N.S.	1865			N.C.Hansen	Brevig
805	✻	Semantha *Dart* 82-88 Stl Bk	British	2211	296·7	43·2	23·8	Port Glasgow	1888	R.Hamilton&Co.		Ship "Semantha" Co. Ld. (J.R.Haws & Co.)	Liverpool
806	N	Semaria *Yggesath* W Bk (ex George B.B)	Norweg'n	601	148·3	32·8	18·9	Clare,N.S.	1863			John Asker	Christiania
807	V	Seminole *De Lano* W Bk	American	1439	196·5	41·6	25·0	Mystic,Conn.	1865		Y 1.86	A.M.Simpson	San Francisco
808		Sémiramis *Rio* W Sr	French	160	83·3	23·3	11·7	Means	1868	Olliveand		J.M.Rio	Nantes
809	DNV	Semiramis *Andreassen* 82-87 W Bk	Norweg'n	1018	193·9	37·3	22·0	Stavanger	1876	Strömstenens [Vaerft	Y 11.87	Ths.S.Falck	Stavanger
810	GV	Semmy Cohn *Maas* -88 W Bk	German	435	130·1	29·6	17·8	Rostock	1876	O.Ladewig	Y 6.88	E.Burchard	Rostock
811	V	Sempiternel *Cellier* -87 W Bk	French	278	107·6	24·9	13·8 / 13·7	Nantes	1871	Sevestre frères	Y 1.87	Sevestre frères	Nantes
812	I	Sempre Avanti *Tortarolo* -87 W Bn	Italian	143	..	..	..	Varazze	1886	G.Scala	Y 1.87	G.Ravecca fu G.	Savona
813		Sempre S. Giovanni W Bg (ex Giorgio Washington Capassi)	Italian	214	101·7	25·6	14·6	Sestri Ponente	1861	D.Bianchi		L.Cignoni	Porto Ferraio
814	✻	Senator *Pall* 82-82 I S	British	1695	256·2	41·1	23·6	Sunderland	1874	Mounsey&Fostr		T.& J.Harrison	Liverpool
815	V	Senator *M'Kenzie* 87-88 W S	British	1474	222·3	40·5	24·0	Maitland,N.S.	1878	J.Montieth	Y 9.86	Martin Dickie	Maitland,N.S.
816		Senator Grimes *Martin* W Sr	American	118	92·0	26·2	8·0	Perry,Maine	1866	Frost		J.G.Murchie	Calais,Maine
817	G	Senator Iken *Holzholz* -87 W Bk	German	1247	182·4	38·1	23·6	Vegesack	1860	J.Lange	Y 11.86	Siedenburg,Wendt & Co	Bremen
818		Senator Saulsbury *McKenzie* W Sr	American	102	86·6	23·3	9·0	Essex,Mass.	1887			B.Low	Gloucester, Mass
819	R	Senator Weber *Winck* W S (ex Wellfleet)	Swedish	1297	197·7	39·4	26·1	Boston,U.S.A.	1853		Y 2.86	P.Pettersson	Helsingborg
820		Senba Maru W Bg	Japanese	187	97·7	23·2	9·5	England	1871			Hayashi Tokutaro	Kagoshima
821	N	Sendemanden *Einarsen* W Bk	Norweg'n	418	133·4	31·2	16·2	Porsgrund	1870	C.Olsen	Y 7.87	Jörgensen & Jorkjend	Tvedestrand
822	DV	Senior *Van Renten* -85 I Bk	Dutch	1088	200·1	35·1	21·0	Katendrecht	1885	Rykee&Co.		J.Vroege	Alblasserdam
823	A	Senno *Balzano* -88 W Bk	British	625	151·8	32·3	20·5	Sestri	1874	S.Bruzzone	Y 8.83	S.Gauci	Gibraltar
824	GV	Señora Quintana *Martini* W Bg	German	332	120·0	26·0	11·0	Brake	1879	J.Oltmanns Wwe	Y 4.84	Seetzen Gebrüder	Bremen
825		Sensat *Maristany* -87 W Pol (ex Hughes)	Spanish	168	91·2	25·5	12·3	Blanes	1866		C 10.84	Gerardo Sensat	Barcelona
826		Sensation *Morris* W Bn	British	189	100·3	23·6	12·8	Rye	1865	Hoad		J.W.Bussey	Dover
827	DV	Senta *Kellenberg* -87 C S (ex Ctesiphon)	German	2069	239·8	44·9	23·8 / 30·8	Bolnes	1868	Pot Bros.	Y 1.88	J.Hintze	Hamburg
828	V	Senta *Tiemann* 72-86 I Bk	German	1037	200·1	34·1	20·0	Hamburg	1884	Blohm&Voss		P.Breckwoldt,sen.	Blankenese
829	RV	Sentinel *Morrel* -88 W Bkn	British	489	148·0	32·0	13·0	Beliveau's Cove [N.S.	1885	W.D.Lovitt	Y 10.83	W.D.Lovitt	Yarmouth,N.S.
830	V	Sépet *Hourdel* W Bn	French	122	93·5	20·3	10·6	St.Malo	1877	L.Gautier	Y 2.82	Saint Martin Legasse	Bayonne
831	✻	Sepia (ex Dalhousie) *Beckett* 82-86 I Bk	British	696	177·3	29·4	18·0	Hartlepool	1864	Denton Gray&C.		Taylor,Sons & Co.	London
832	DN	Septentrio *Corneliussen* -86 W Bk	Norweg'n	346	120·3	28·6	15·4	Risör	1869	G.Pedersen	Y 6.83	J.F.Juell	Risör
833		Septimus (BROKEN UP) W Bk	British	329	121·5	30·5	17·6					Salvatore Gauci	Malta
834	D	Sequel *Richards* 82-82 W Bkn	British	535	160·0	37·7	17·9	P.E.Island	1877	W.Richards	Y 6.86	William Richards (per T.P.Richards & Co.)	Ch'rlottetwn, PEI
835		Serafina *Jansson* W Bk	Russian	358	..	..	..	Kimito	1871			M.Mattsson	Wårdö
836		Seraph *Svensson* -85 W Sr	Swedish	134	88·9	21·6	10·5	Bandholm	1857	H.J.Möller		P.A.Olsson	Karlskrona
837	R	Seraphine *Clute* -88 W Sr	British	120	93·6	24·2	9·3	Bear River, NS.	1888	J.D.Vroom		J.Ditmars Vroom	Digby,N.S.
838		Séraphine (LOST) *Bonnet* W Bg	French	292	102·6	26·2	15·9	Agde	1868	Allemand	Y 4.85	H.Chedru	Fécamp
839	✻	Serapis *Tompsett* 87-87 I S	British	995	224·0	35·5	20·2	Greenock	1876	J.E.Scott		W.Lund	London
840	V	Sereia *Dos Reis* W Bk (ex Rose Osborne)	Portug'se	397	132·0	26·7	16·7	Hartlepool	1862	Denton	Y 10.87	J.T.Costa Basto	Oporto
841		Seren (see Siren) *Sjögren* W Sr	Russian	236	..	..	..		1873			D.F.Söderlund	Lumparland
842	✻	Serena (ex Elvie) *Fox* 84-84 I Bk	British	505	153·4	27·4	17·5	Birkenhead	1864	G.R.Clover&Co.		A.H.Fox	Plymouth
843		Serena S. Kendall W Sr	American	129	95·2	27·0	7·6	Brewer,Maine	1876	J.Oakes&Son		O.E.Kendall	Bangor,Maine
844		Serena Thayer *Wheeler* W Sr	American	196	124·8	30·8	9·8	Port Discovery, [Cal.	1872	Clark&Webster		S.H.Mott	Wilmington,Cal.
845		Serene *Legerman* W Bk	American	522	148·1	31·0	17·5	Baltimore	1867		Y 8.87	T.Corner	Baltimore
846	V	Seretha *Harvey* W Bn (ABANDONED)	British	144	90·5	24·0	10·5	Port Clyde, NS.	1882	T.Coffin	Y 10.85	Peter G.Tessier	St.Johns,N.F.L.
847		Serffee Bahri W Bmb	Turkish	100	..	..	..	Barten	1879			Muttab Bashizadeh Hajee Rifaat Effendi	Barten
848		Sergei Hermann W Sr	Russian	174	..	..	..		1878			W.Luostari	Sordavala
849		Sergipe *Silva* -87 W Sr (ex Hermann)	Brazilian	114	96·3	24·5	11·3	Papenburg	1858		Y 8.85	J.P.S.Moreira	Bahia
850	✻	Serica *Lockit* I Bk	British	913	201·8	32·8	19·8	Dumbarton	1875	A.M'Millan&Sn		Flinn,Main & Montgomery	Liverpool

✻ = Lloyd's Register; D = Nederlandsche Vereeniging; G = Germanischer Lloyd; I = Registro Italiano; R = American Record; A = Veritas Austro-Ungarico; H = Veritas Ellenico; N = Norske Veritas; V = Bureau Veritas.

Dans la 4me Colonne, C = Composite (membrure en fer, bordé en bois); I = Fer; Stl = Acier; W = Bois.

Les Dates du doublage en métal (Col. 11) ressortent des livres dans lesquels les navires sont respectivement classés ou des rapports des Inspecteurs du Lloyd's Register.

Pour plus amples renseignements sur les navires classés par le Lloyd's Register, voir le Lloyd's Register Book et le Lloyd's Yacht Register.

Official No.	No. in Book	Registry in which classed	Ship's Name, Master, and Dates of Appointment to present Owner's Service and to Ship; Ship's Late Name	Material and Rig	Flag	Net Tonnage	Length	Breadth	Depth	Built: Where	Built: When	Built: By whom	Date of Metalling	Owners or Managers	Port of Registry
	851	A	Serif Bahri *Zembulis*		Turkish	106	..	..	..	Port Said	1875			A. Zembulis	Constan
71935	852	V	Seringa *Waaler* -88 (ex Seringapatam)	W S	Norweg'n	1128	198·5	38·0	22·7	Quebec	1875	F.X. Marquis	[illegible]	Actieselskabet "Seringa"	Christia
73361	853	V	Serrano *Edgett* -86	W Bk	British	594	145·8	31·3	18·2	Harvey, N.B.	1876	G.S. Turner	12.87	James W. Edgett	St.John
77981	854	V	Servia *Smith*	W S	British	1369	196·3	40·1	23·7	Parrsboro', N.S	1878	A. Leoner	[illegible]	Thomas E. Kenny	Windsor
43973	855	RV	Servia *Gilmore*	W S	American	1773	234·1	41·1	26·7	Bath, Maine	1883	Houghton Bros.	[illegible]	Houghton Bros.	Bath
	856		Servolito *Belleza*	W Bn	Brazilian	182	..	..	..	P.E.Island	1869		[illegible]	J.B. Belleza	Bahia
	857		Servulo (ex Judith, ex Nova Orleans) *Silva*	W Bn	Portug'se	174	92·5	23·6	10·5	Villa Nova	1869	J.F. Lapa	[illegible]	S.A. da Silva	Lisbon
63562	858		Sesame (WRE ...) *Blampied*	W Bn	British	175	100·0	24·8	12·7	Sunderland	1870	Crown		White Cliffe Shipping Co. Ld	London
	859	N	Sestri *Restrup*	W Bk	Norweg'n	497	131·1	31·4	18·5	Sestri Ponente	1874	Traverso	[illegible]	S. Eyde	Arendal
23458	860		Setagawa	W S	American	103	87·7	22·6	8·5	Kennebunk, Me	1865			A.C. Gay	Rockland
	861		Sete d'Agosto *Jorge*	W Sr	Portug'se	133	80·0	20·3	10·5	Caminha	1871	A.J. Gonçalves		M.A. Torres Silva & D.J. Pereira	Caminha
115379	862	R	Seth M. Todd *Clark*	W Bn Sr	American	185	114·0	29·6	9·6	Calais, Maine	1875	Boardman Bros.		Boardman Bros.	Calais
115160	863	R	Seth W. Smith *Martin*	W Sr	American	188	114·0	29·6	9·6	Calais Maine	1872	Nickerson & Rideout		J. Murchie & Sons	Calais
	864		Setiem Ma Salud	Bn	Spanish	132	93·8	20·5	9·6	Tayabas	1887			Doña S. Borromeo	Manila
	865		Sette fratelli *Sakir* (ex St. Nicolo)	W Bn	Turkish	155	..	..	..	Galaxidi	1863	Cariati	Z 5.86	Emin Sakir & Co.	Dulcig
	866		Seura *Rosenberg* -1	W Sr	Russian	240	107·8	28·9	12·4	Laitakari	1877	J.A. Sandelin		G. Martiila	Luvia
116296	867		Seven Sisters *Lunewald* -88	W Sr	American	123	97·0	27·3	7·7	Benicia	1888	M. Turner		Iversen & Jansen	San Fr
23427	868	R	Seventy-six *Feel* 64-86	W Sr	American	187	98·4	27·0	9·8	Thomast'n, Me	1864	Buckman & Co.		Dunn & Elliot	Thomas
	869		Severena *Sundstrom*	W Bk	Russian	443	..	..	..	Finström	1863			J. Mattsson	Finstr
85448	870	V	Severn *Gormley* -87	W Sr	British	446	148·0	32·6	15·9	Newport, N.S.	1884	W.H. Mosher	[illegible]	Rufus Curry	Windsor
[illegible]	871	✻	Severn *Thomas* 79-84	W Bkn	British	397	113·6	29·1	15·9	P.E.Island	1884	John Yeo	12.84	C.C. Wyllie (T.P. Richards & Co.)	Ch'town
	872	✻	Severn	W Sk	British	87	78·2	21·1	10·6	[illegible]	1888	J. Garside		T. Hamling	Hull
	873		Severus *Kasbadt*	W Bk	German	446	117·4	29·5	17·9	Elbing	18^{47}_{78}	Mitzlaff		R.C. Gribel	Stettin
27737	874	R	Sevilla *Brokensha*	W Bk	British	598	156·2	28·6	19·0	Jersey	1859	Clarke	[illegible]	J.H. Clement	Glasgow
	875		Sevilla *Linares* -87 (ex Farazon)	W Bn	Spanish	225	105·9	23·7	13·7	Bilbao	1868			Juan Murga	Seville
366	876		Sfacteria	W Bk	Greek	417	121·3	24·9	11·7	England	1860			L. Marias	Syra
	877		Shabko	W Bg	Turkish	180	..	..	..	England	1884			Iskenderende Saklan Madjarush	Consta
	878		Shaboon Bahri	W Bg	Turkish	210	..	..	..	Gallipoli	1884			Kerestli [illegible] Ahmed Effendi ibn Mehemed [illegible]	Gallipo
30597	879		Shah Allum (ex Harriet LeJeune)	W S	British	498	123·5	27·8	21·2	Portsmouth, U.S.	1832			Hajee Sagoin H. Jaffer	Bombay
42854	880		Shah Jehan	W S	British	538	111·3	28·2	17·9					Cummoo Yacob	Bombay
	881		Shah Yunnus	W Bg	Turkish	165	..	..	..	Barten	1875			Dillee zadeh Halim	Barten
	882		Shah Yunnus *Halim*	W Bg	Turkish	165	..	..	..	Barten	1880			Aktasse [illegible] Escol [illegible] & Co.	Aktasp
	883		Shahin Bahri	W Bg	Turkish	250	..	..	..	Oonia	1866			[illegible] Cheriff [illegible] & Co.	Allabe
	884		Shahin Bahri	W Bg	Turkish	216	..	..	..	Evranieh	1883			Tiyar Yorghanji zadeh Hajee Mehmed Bey	Sinope
	885		Shahin Bahri	W Sr	Turkish	200	..	..	..	Oonia	1880			Haitah [illegible] Messou- [illegible] Suleiman Agha	Masso
	886		Shahin Bahri	W Sr	Turkish	200	..	..	..	Kara Sou	1878			Altun zadeh Ismael Pasha	Kara S
	887		Shahin Bahri	W Bg	Turkish	187	..	..	..	Barten	1880			Toglaji Hajee Mustapha Effendi	Barten
	888		Shahin Bahri	W Bg	Turkish	179	..	..	..	Alabnah	1880			Yorghanji zadeh Hajee Mehmed Effendi & Co.	Alabn
	889		Shahin Bahri	W Bg	Turkish	162	..	..	..	Evranyieh	1880			Manoli oglou Ismael & Co.	Evran
	890		Shahin Bahri	W Bg	Turkish	162	..	..	..	Barten	1878			Hattab [illegible] Nalbour [illegible] Effendi	Barten
	891		Shahin Bahri	W Bg	Turkish	156	..	..	..	Greece	1872			Rizelilee Terlnik oglou Rahi	Rizeh
	892		Shahin Bahri	W Bg	Turkish	137	..	..	..	Sakarieh	1871			Altun zadeh Ismael	Kara
	893		Shahin Bahri	W Bg	Turkish	112	..	..	..	Barten	1870			[illegible] zadeh Mus-	Barten
	894		Shahin Bahri	W Bg	Turkish	105	..	..	..	Kara Agatch	1873			[illegible] Kirkeekis [illegible] & Co.	Inebol
	895		Shahin Bahri	W Sr	Turkish	100	..	..	..	Terebolice	1879			[illegible] Hussein Effendi	Tereb
	896		Shahin Deriya	W Bg	Turkish	192	..	..	..		1865			Halil Agha & Co.	Consta
	897		Shahin Deriya	W Bg	Turkish	170	..	..	..	Monastir Skel	1884			Shereefis Khadich [illegible]	Monas
	898		Shahin Derya *Mustapha*	W Bg	Turkish	147	..	..	..	Greece	1873			Rizelilee Mollah oglou Mustapha	Rizeh
	899		Shahin Derya	W Bg	Turkish	137	..	..	..	Alabnah	1875			[illegible] zadeh [illegible] Pappas	Alabn
	900		Shahin Derya	W Bg	Turkish	106	..	..	..	Terebolice	1863			Anton oglou Andon Hajee [illegible] & Co.	Tereb

✻ = **Lloyd's Register**; D = Nederlandsche Vereeniging; G = Germanischer Lloyd; I = Registro Italiano; R = American R[ecord]; A = Veritas Austro-Ungarico; H = Veritas Ellenico; N = Norske Veritas; V = Bureau Ve[ritas]

In the 4th Column, C = Composite (iron frame, wood planked); I = Iron; Stl = Steel; W = Wood.

The Dates of Metalling (**Col. 11**) are based on the records of the Books in which Vessels are respectively classed, or on the Reports of the Surveyors of Lloyd's Re[gister]

Further particulars of Vessels classed by Lloyd's Register will be found in Lloyd's Register Book and Lloyd's Yacht Register.

1	3	4 Nom. Matériaux et Gréement. Capitaine et dates d'entrée au service des Armateurs actuels et à bord du navire. Ancien Nom.			5 Pavillon.	6 Tonnage net.	7 Dimensions. Longueur	Largeur	Creux	8 Construit. Où.	9 Quand.	10 Par qui.	11 Date du doublage en métal	12 Armateurs ou Directeurs-Gérants.	13 Port d'Armement.
901		Shahin i Bahri		W Bg	Turkish	187	..	..	..	Barten	1885			Toolaji Hajee Mustapha Effendi	Barten
902		Shahin i Bahri	Ismael	W Bg	Turkish	180	..	..	..	Abahnah	1875			Abahnahlee Dervisch oglou Ismael & Co.	Abahnah
903		Shahin i Bahri	Ismael	W Bg	Turkish	134	..	..	..	Abahnah	1883			Abahnahlee Ismael	Abahnah
904		Shahin i Bahri		W Bmb	Turkish	126	..	..	..					Kumurdji Ethem Agha zadeh Ismael Hakki & Co.	Barten
905		Shahin i Bahri	Hourshid	W Bg	Turkish	100	..	..	..	Kooroojeh Sh	1885			Hattab Djari Redwan oul. h Mehemed Effendi	Kooroojehsahe
906	✻	Shakespeare	Ewen 81-85	I Bk	British	767	191·8	34·8	19·2	Liverpool	1876	R.&J.Evans&Co		E.C.Friend & Co.	Liverpool
907	✻	Shakespeare		W Sk	British	84	77·5	20·4	10·7	Hull	1885	W.&J.M·Cann		R.Loram,jun.	Hull
908	✻V	Shakespeare (ex Clara Ilka)	Muller '86	I Bk	German	877	196·4	32·3	20·4	Newport,Mon.	1855	Bachelor		R.M.Sloman & Co.	Hamburg
909	N	Shakespeare	Meyer 80-86	W Bk	Norweg'n	632	148·8	31·0	19·9	Richmond,Me.	18 54/85	CharlesHeger	Y 11.85	M.Oppen	Laurvig
910	GV	Shakspere	Müller 81-84	W S	German	1183	193·8	39·0	22·3	Quincy,Mass.	1856	Thomas	Y 2.85	LouisF.Kalkmann & Co.	Bremen
911	✻D	Shalimar	Linklater 86-86	I Bk	British	985	206·3	33·6	20·0	PortGlasgow	1883	Russell&Co.		W.& J.Crawford	Greenock
912		Shamrock	Cole	W Bg	British	264	117·4	26·0	14·5	Shoreham	1862	Bailey		T.Gates & Co.	Shoreham
913		Shamrock	Wescott	W Bn	British	139	93·0	23·4	11·2	Carbonear,N.F.	1885	R.Horwood		John Rorke	St.John's,N.F.
914		Shamrock		W Sc Yt	American	118	111·8	25·5	11·3	Deptford	1872	Walker&Co.	C 1.85		
915	✻	Shandon	Mordaunt 86-86	I S	British	1397	245·9	37·8	21·3	PortGlasgow	1883	R.Duncan&Co.		Thom & Cameron	Glasgow
916		Shannon (ABANDONED)	Mitchell -87	W S	British	1292	217·0	37·7	22·8	London	1862	R.&H.Green	Y 12.85	J.C.Ellis	Sydney
917		Shannon	Sawyer -87	W Bn	American	374	119·8	29·7	11·9	Millbridge,Me.	1867			F.Sawyer	Machias,Me.
918		Shappiro		W Bg	Turkish	168	..	..	..	Greece	1885			Muttab Bashi zadeh Hajee Rifaat Effendi	Barten
919		Sharkieh		W Bg	Turkish	200	..	..	..	Chamjol	1876			Kaiseri Tujari Bedos Velideh Bessktek	Chamjol
920	✻	Sharpshooter (ex Madeleine)	Magnussen	I Bk	British	489	149·2	27·5	17·0	Sunderland	1860	T.R.Oswald		James C.Ellis	Sydney,N.S.W.
921	R	Shawmut	Small 66-84	W Bk	American	445	130·2	30·0	17·5	E.Boston	1866	W.Burkett	Y 10.88	J.S.Emery & Co.	Boston,Mass.
922		Shealtiel		W Bn	British	140	81·8	24·2	12·5	Sunderland	1847			W.W.Bird,jun.&Co.	Guernsey
923		Shearwater (LOST)	Depledge 86-86	W Sr	British	116	78·0	28·0	11·7	BurtonStather	1858	Wray		W.Cass	London
924		Sheemaga		I Barge	British	362	197·2	25·2	8·5	PortGlasgow	1880	R.Duncan & Co.		Irrawaddy Flotilla Co. Ld	Glasgow
925		Shefher	Tipho	W Bg	Turkish	182	..	..	..	Roumania	1884			Velhenji Hajee Ibrahim Effendi	Barten
926		Shefher		W Bg	Turkish	138	..	..	..	Barten	1883			Velhenji Hajee Ibrahim Effendi	Barten
927	N	Sheffield	Morch 69-81	W Bk	Norweg'n	620	139·6	32·2	19·0	Oromocto,N.B.	1863	Bryerson & Co.		L.A.Morch	Fredrikstad
928	✻D	Sheila	Pearse 81-81	I S	British	1198	228·7	36·8	21·0	Glasgow	1877	C.Connell&Co.		Sandbach,Tinne & Co.	Liverpool
929	V	Shelburne	Murphy -88	W S	British	1165	182·4	37·2	21·7	Shelburne,N.S.	1874	F.Pence	Y 8.85	Annie Willett	Shelburne,N.S.
930	✻	Shenir	Stirling 80-80	I Bk	British	1173	225·5	34·8	21·2	Glasgow	1876	A.Stephen&Sns		Smith & Service	Glasgow
931		Shepherdess	Wetheridge [86	W Bg	British	215	109·1	23·8	13·9	Salcombe	1865	Booker	Y 10.85	T.May	Salcombe
932		Shereff Bahri		W Bg	Turkish	171	..	..	..	Irmak Hagoosy	1885			Tijar Hajee Shefki Effendi	Irmak Hagoosy
933		Shereff Yildiz		W Bg	Turkish	302	..	..	..	Oonia	1873			Kiritlee Mahmood Mahmoud	Kirit
934	✻	Sherwood	Beakie -88	I S	British	1823	261·3	39·3	24·1	Whitehaven	1883	Whitehaven [S.B.Co.,Ld.		Fisher & Sprott	Liverpool
935	DV	Sherwood	Douglas	W Bk	British	998	179·0	36·7	21·5	Maitland,N.S.	1878	A.Putnam	Y 8.86	Alfred Putnam	Maitland,N.S.
936	RV	Shetland	Haskell 71-78	W Bk	American	612	143·0	33·3	17·6	Ellsworth,Me.	1878	C.W.Conners	Y 11.85	J.S.Emery & Co.	Boston,Mass.
937	✻	Shiantelle	Macdonald -87	W Cr	British	39	60·2	18·2	8·0	Fraserburgh	1887	J.Webster&Co.		J.B.H.Brown	Fraserburgh
938		Shichiho Maru		W Sr	Japanese	231	107·6	24·5	11·7	Osaka	1882	Tuda Shojiro		Hamanaka Hatesaburo	Osaka
939		Shillee		W Bg	Turkish	144	..	..	..	Shillee	1868			Shilleelee Ibrahim&Co	Shillee
940		Shillee		W Bmb	Turkish	130	..	..	..	Shillee	1868			Hairie Djari Hajee Achmet Effendi	Shillee
941		Shillee		W Bmb	Turkish	100	..	..	..	Shillee	1866			Kattirleelee Tujarleree Chrbacki & Co.	Katirlee
942		Shillee		W Bg	Turkish	100	..	..	..	Shillee	1867			Mehemet Ali Bey	Shillee
943		Shillee		W Bmb	Turkish	100	..	..	..	Shillee	1868			Hairie Djari Hajee Achmet Effendi	Shillee
944		Shinji Maru		W Sr	Japanese	560	139·4	30·8	16·6	Hiogo	1880	SuzukiKichizo		Seiko Kwaisha	Hiogo
945		Shinkoku Maru		W S	Japanese	144	77·2	22·1	10·5	Hiogo	1881	OmatsuJugoro		Hamanaka Hatesaburo	Osaka
946		Shinpo Maru		W Sr	Japanese	463	127·6	28·0	16·3	Hiogo	1882	WakinoHiojiro		Tatsui Teru	Hiogo
947		Shinyei Maru		W Sr	Japanese	126	69·7	21·6	11·0	Owari	1882	ItoKinpei		Sugiye Hanzaburo	Owari
948		Shinyei Maru		W Sr	Japanese	120	84·6	21·3	10·4	Ichikawa	1886	Ichikawa [Genkichi		Ichikawa Genkichi	Ise
949		Shioko Maru (ex Hayashi Maru)		W Sr	Japanese	200	106·4	24·0	11·2	Fushiki	1884	Ishido Shojiro		Hayashi Navgoro	Kagoshima
950	✻	Shirley	Quaintrell	W Sr	British	87	86·0	20·6	9·7	Falmouth	1877	Burt		E.D.Anderton	Falmouth

✻ Lloyd's Register; D = Nederlandsche Vereeniging; G = Germanischer Lloyd; I = Registro Italiano; R = American Record; … Veritas Austro-Ungarico; H = Veritas Ellenico; N = Norske Veritas; V = Bureau Veritas.

Dans la 4me Colonne, C = Composite (membrure en fer, bordé en bois); I = Fer; Stl = Acier; W = Bois.

… du doublage en métal (Col. 11) ressortent des livres dans lesquels les navires sont respectivement classés ou des rapports des Inspecteurs du Lloyd's Register.

Pour plus amples renseignements sur les navires classés par le Lloyd's Register, voir le Lloyd's Register Book et le Lloyd's Yacht Register.

Official No.	No. in Book	Registry in which classed	Ship's Name. Material and Rig. Master, and Dates of Appointment to present Owner's Service and to Ship. Ship's Late Name.	Rig	Flag	Net Tonnage	Length	Breadth	Depth	Built: Where	When	By whom	Date of Metalling	Owners or Managers	Port
28707	951		Shirley *Hayden -87*	W S	American	997	170·6	36·0	24·0	Medford, Mass.	1850	P. Curtis		G. E. Atkinson	[illegible]
85691	952		Shoay Doung	I Barge	British	791	194·0	25·5	8·2	Glasgow	1875	A. & J. Inglis		Irrawaddy Flotilla Co. Ld.	Glasgow
86142	953	✳	Shoaymyo *Dyer*	W Sk	British	86	78·0	20·3	10·9	[illegible]	1883	Beeching Bros.		G. Beeching & Co.	Hull
636	954		Shohei Maru	W Sr	Japanese	230	107·9	25·9	12·8	[illegible]				Hayashi Tokutaro	Kagoshima
632	955		Shotoku Maru	W Sr	Japanese	101	82·9	17·5	10·5	[illegible]	1881	Nagata Sanjuro		Masumura Gitaro	Yetehi[illegible]
51739	956	✳ D	Shun Lee *Morgan 83-87*	C Bk	British	650	158·0	31·5	19·4	London	1866	Walker	Y 7.87	J. Jenkins (R. Conaway & Co.)	London
143	957		Siam	W Bk	Siamese	226	104·0	26·0	12·0	[illegible]	1865			Chessua Kow	Bangkok
3	958		Siamese Crown	W S	Siamese	690	166·0	32·3	17·8	Bangkok	1862			Siamese Government	Bangkok
	959	N	Sibal *Hauger*	W Bk	Norweg'n	561	146·2	31·6	18·7	Arendal	1876		Y 5.88	T. Thommesen & Sön	Arendal
332	960		Sibilla *Maiorazzo*	W Bg	Italian	264	113·1	28·9	15·2	Gaeta	1870	P. Gallinaro	Y 5.84	N. Di Fonzo	Gaeta
	961	V	Sibirien *Kuelzo -87*	W Bk	German	367	132·1	28·3	13·8	Arnis	1880	F. Johansen	Y 1.86	H. A. Petersen	Arnis
	962		Siblas *Lecholc*	W Bk	French	322	126·3	25·6	13·0	La Seyne	1873	Caroul fils		A. Fabre & fils	Marseille
42138	963		Sibyl (ex Trujillo)	W Bg	British	152	88·5	24·4	9·3					William T. Frink	Turks Is.
	964	N	Sichem *Wroldsen 86-86*	W Bk	Norweg'n	523	141·7	31·7	17·7	Arendal	1880	T. A. Morland	Y 5.88	J. M. A. Marcussen	[illegible]
	965	V	Sicié *Erilland -87*	W Bn	French	119	86·0	18·4	10·1	St. Malo	1876		Y 4.87	Huet	St. Malo
340	966	V	Sicilia *Morando*	W Bn	Italian	147	..	..	..	Warazze	1884	B. Cerruti	Y [illegible]	Cerruti & Co.	Savona
77894	967		Siddartha *Rogers -88*	W Bk	British	463	146·5	32·5	12·9	Sackville, N.B.	1880	E. W. Ogden	Y 6.88	Josiah Wood	Sackville
58234	968	✳ D	Sidlaw *Lynn 88-88*	I Bk	British	499	156·0	26·7	17·2	Dundee	1867	Brown & Simpson		Dundee [illegible] Owners' Co.	Dundee
	969		Sidney *Knudsen -88*	W Bk	Norweg'n	822	170·8	34·5	19·9	Grimstad	1888			P. Pedersen	Grimstad
	970	V	Sidney *Lundh -87*	W Bk	Swedish	658	156·8	30·4	18·1	Gefle	1869	O. A. Brodin	Y 7.86	Engström & Luth	Gefle
116178	971		Sidney G. Hupper (WRECKED) *Watts -87*	W 4m Sr	American	926	189·9	38·0	18·4	Thomaston Me.	1887	G. Mehan		Samuel Watts	Thomaston
	972	N	Sidon *Jorgensen*	W Bk	Norweg'n	398	130·8	29·7	16·1	Arendal	1875		Y 9.86	Gunder Tellefsen	Grimstad
	973	V	Sidonie & Madeleine *Joubert*	W Bg	French	209	95·5	25·3	12·5	Nantes	1869	Tillé	Y 2.88	J. Joubert	Nantes
	974		Sie Mangoe (STRANDED) *Mohamad Habib*	W Sr	Dutch	128	75·5	20·6	8·0	[illegible]	1878	Alhiernie[illegible]		Hajee Mohamed Habib	Bawean
74162	975	✳	Sierra Blanca (ex Antiga) *Lyon -88*	I S	British	1518	249·4	38·0	23·3	Greenock	1875	Scott & Co.		Sierra Shipping Co. Ld.	Liverpool
78785	976	✳	Sierra Colonna *Bowles 86-86*	I S	British	1435	238·6	38·0	22·8	Stockton	1878	Richardson, Duck & Co.		Sierra Shipping Co. Ld.	Liverpool
76496	977	✳	Sierra Cordova (ex [illegible]) *McClure 80-83*	I S	British	1430	232·0	37·9	22·8	Liverpool	1877	T. Royden & Sons		Sierra Shipping Co. Ld.	Liverpool
81346	978	✳ D	Sierra Estrella *McMaster 82-83*	I S	British	1436	238·6	38·0	22·7	Stockton	1880	Richardson, Duck & Co.		Sierra Shipping Co. Ld.	Liverpool
87394	979	✳	Sierra Lucena (ex Integrity) *Murdock 77-81*	I S	British	1684	260·4	39·0	23·5	Greenock	1883	R. Steele & Co.		Sierra Shipping Co. Ld.	Liverpool
74461	980	✳	Sierra Madrona *Scott 81-81*	I S	British	1430	230·0	38·1	23·5	Stockton	1875	Richardson, Duck & Co.		Sierra Shipping Co. Ld.	Liverpool
87957	981	✳	Sierra Miranda *Maynard 80-83*	Stl S	British	1808	264·7	39·1	23·9	Port Glasgow	1884	J. Reid & Co.		Sierra Shipping Co. Ld.	Liverpool
74409	982	✳ D	Sierra Morena *Spencer 86-86*	I S	British	1385	231·2	37·2	23·5	[illegible]	1875	[illegible]		Sierra Shipping Co. Ld.	Liverpool
78711	983	✳ D	Sierra Nevada *McWhirr [illegible]*	I S	British	1474	23[illegible]	37·6	23·5	[illegible]	1877	[illegible]		Sierra Shipping Co. Ld.	Liverpool
86212	984	✳ D	Sierra Parima *[illegible] 79-83*	Stl S	British	1518	2[illegible]·8	38·2	23·1	[illegible]	1882	[illegible]		Sierra Shipping Co. Ld.	Liverpool
87911	985	✳	Sierra Pedrosa *McWhirr [illegible]*	Stl S	British	1620	249·0	38·1	23·6	[illegible]	1883	[illegible]		Sierra Shipping Co. Ld.	Liverpool
	986	D V	Sieverine *Binderhagen -70*	W Sr	German	126	85·3	20·4	9·5	Grossefehn	1870	A. J. Schapp		J. H. Waterborg	[illegible]
	987	N	Sif (ex Melina, ex A. V. Marzo) *Jensen 71-83*	W Bk	Norweg'n	862	171·7	34·3	20·8	Quebec	1861	[illegible]		H. A. Lohre	[illegible]
	988	N	Sif *Mar. Janese -82*	W Bk	Norweg'n	495	143·1	32·6	17·6	Fredrikshald	1873	J. Brevig	Y 8.87	R. M. Peterson	Moss
	989	N	Sif *Hansen 56-82*	W Bk	Norweg'n	317	105·8	27·6	16·3	Kragerö	1856	T. Pedersen		T. Jensen	Kragerö
	990	N	Sif *Olsen -87*	W Bn	Norweg'n	186	108·0	25·5	11·1	Bergen	1876		Y 3.88	J. E. Lehmkuhl	Bergen
	991	N	Sigdal *Aasvedt 77-82*	W Bk	Norweg'n	436	143·5	31·6	15·9	Drammen	1877		Y 12.86	Chr. Lingsom	Drammen
	992	V	Signal (ex [illegible]) *Svendsen 88-88*	W Bk	Norweg'n	967	182·8	34·7	22·2	Hamburg	1878	Amsinck		Hohlt & Isachsen	Stav[illegible]
	993	V	Signe *Hansen 82-82*	W Sr	Danish	111	88·2	20·1	9·9	Thurø	1882	J. Ph. Jorgensen		J. Jorgensen	Svend[illegible]
	994		Signe (STRANDED) (ex [illegible]) *[illegible]ntzen -88*	W Bk	Norweg'n	994	177·0	37·2	21·8	Carleton, N.B.	1861			H. M. Rummelhoff	Christiania
	995	R	Signe (ex Rangoon) *Boyter -87*	W Bk	Swedish	512	146·3	24·7	18·5	Liverpool	1859	J. Steel	[illegible]	E. Elfgren	Piteå
	996	G	Signe (ex Ernst Brockelmann) *Johansson -88*	W Bk	Swedish	344	114·4	2·52	16·1	Rostock	1853	A. Schön[illegible]		E. Andersson	[illegible]
	997	V	Signe *Petersson 86-85*	W Bn	Swedish	162	100·1	22·9	10·4	Oscarshamn	1876	P. Petersson		J. E. Hagström	[illegible]
75655	998	R V	Signet *Doty*	W Bk	British	574	146·3	31·1	18·3	Quebec	1877	J. E. Gingras	Y [illegible]	J. G. Ross	Quebec
	999	R	Sigrid (ex Proteus) *Gregersen*	W Bk	Norweg'n	1183	174·3	38·3	24·5	St. Martins N.B.	1863	J. H. Moran	Y 12.86	Nils Christensen	Arendal
	1000	N	Sigrid (ex [illegible]) *Larsen*	W Bk	Norweg'n	1003	195·0	35·9	22·9	Whitehaven	1865	T. & J. Brocklebank	Y 5.88	L. Hofgaard & Co.	Tønsberg

✳ = Lloyd's Register; D = Nederlandsche Vereeniging; G = Germanischer Lloyd; I = Registro Italiano; R = American [illegible];
A = Veritas Austro-Ungarico; H = Veritas Ellenico; N = Norske Veritas; V = Bureau Veritas.

In the 4th Column, C = Composite (iron frame, wood planked); I = Iron; Stl = Steel; W = Wood.

The Dates of Metalling (Col. 11) are based on the records of the Books in which Vessels are respectively classed, or on the Reports of the Surveyors of Lloyd's [illegible]

Further particulars of Vessels classed by Lloyd's Register will be found in Lloyd's Register Book and Lloyd's Yacht Register.

1 Numéro d'ordre	3 Registres	4 Nom. Matériaux et Gréement. Capitaine et dates d'entrée au service des Armateurs actuels et à bord du navire. Ancien Nom.	5 Pavillon.	6 Tonnage net.	7 Dimensions. Longueur	Largeur	Creux	8 Construit. Où.	9 Quand.	10 Par qui.	11 Date du doublage en métal	12 Armateurs ou Directeurs-Gérants.	13 Port d'Armement.
1001	NV	Sigrid (ex Minnie Camill?) Olsen --- WBk	Norweg'n	574	150·1	33·4	16·2	St.John,N.B.	1872	W.Hogg	5.86	S.Christensen	Drammen
1002		Sigrid Collin WBk	Norweg'n	319	128·0	28·1	13·1	America	1852			L.Hansen	Christiania
1003	N	Sigrid (ex Martha) Nielsen WBg	Norweg'n	261	104·8	26·8	13·9	Greifswald	1863	W.Sadelkow		O.Lydersen	Tvedestrand
1004		Sigrid Collin WBk	Norweg'n	253	111·5	26·6	12·6	Grimstad	1873	J.Tellefsen	Y 8.85	S.Hansen	Christiania
1005	R	Sigrid Henrikson -86 WBg	Swedish	258	115·1	25·1	12·5	Ljungnäs	1876	A.F.Bring		J.D.Björkegren	Cimbrishamn
1006		Sigrid Elisabeth WBn	Swedish	198	102·0	25·0	11·9	Halifax,N.S.	1863			L.P.Strömberg	Cimbrishamn
1007	N	Sigrlinn Christiansen WBg	Norweg'n	221	105·0	26·1	12·1	Bordeaux	18⁶⁹		Y 8.85	J.Bang	Grimstad
1008	DV	Sigurd Amundsen 86-86 WBn	Norweg'n	221	111·0	25·8	13·4	Stavanger	1871	M.Magnus	Y 8.85	A.Lothe	Haugesund
1009	N	Sigurd Skogland -83 WSr	Norweg'n	121	91·0	23·9	10·5	Stordöen,Nwy	1877	H.Madsen		E.H.Kongshavn	Haugesund
1010	N	Sigurd Jarl Olsen 82-83 WBk	Norweg'n	435	136·4	28·2	14·7	Hammelw'rdn	1868	Menke&Botter	Y 1.86	Ths.S.Falck	Stavanger
1011	V	Sigyn Elander 87-87 WBk	Swedish	356	139·8	27·0	12·4	Gothenburg	1887	J.E.Hübe	Y 9.87	G.D.Kennedy	Gothenburg
1012		Siivo Korström -86 WBk	Russian	324	129·7	28·8	13·1	Korpo	1875	A.Holmberg		A.B.Nyström	Åbo
1013	V	Sikeå Varf Carlsson 85-85 WBk	Swedish	230	121·5	22·6	10·9	Sikeå	1875	Oberg	Y 8.86	Fr.Jacobsen	Robertsfors,Westerbotten
1014	R	Silas Fish (WRECKED) Crook -87 WBk	American	667	154·0	33·0	19·6	Mystic,Conn.	1864	Maxon&Fish	Y 8.87	J.W.Covert	New York
1015		Silas H. Morse Pike 79-79 WSr	British	114	84·1	23·2	10·0	Pugwash,N.S.	1877	J.Nicholson		William Duff	St.Johns,N.F.L.
1016		Silas McLoon Merrill WSr	American	116	86·8	21·2	7·5	Rockland,Me.	1871			P.C.Merrill	Camden,Me.
1017	✻	Silberhorn Irish -87 I 4mBk	British	1853	267·5	40·2	24·0	PortGlasgow	1884	Russell&Co.		J.R.De Wolff & Son	Liverpool
1018	N	Silentia [illegible] -87 WBk	Norweg'n	407	110·7	30·6	18·1	Amsterdam	1815			H.Basberg	Christiania
1019	I	Silenzio Gardella -87 WBk	Italian	700	164·2	32·8	21·6	Varazze	1875	M.Baglietto	Y 1.86	F.Denegri	Genoa
1020	✻	Silhet (MISSING) Partridge 86-88 I Bk	British	1131	220·2	35·2	21·0	Glasgow	1876	BrclayCurle&C.		Foley & Co.	London
1021	V	Silicon Bartlett -87 WBk	American	298	138·5	33·0	8·8	Bath,Me.	1887	Kelly&Spear		McKay & Dix	New York
1022		Silistria McLean -83 WS	British	642	148·5	30·3	20·0	Miramichi	1854			R.P.Moser	Lancaster
1023		Sillistria WBk	Turkish	500	..	..	..	Greece	1877			Ethem Bey & Co.	Constantinople
1024	N	Silo Pedersen WBk	Norweg'n	377	119·0	28·7	16·3	Grimstad	1872	Ole Ribe	Y 8.86	Gunder Tellefsen	Grimstad
1025		Siloam Nicholls WBn	British	146	94·6	22·1	11·1	P.E.Island	1859	Coffin		J.Reeves	Faversham
1026		Siloam Corfe WBn	British	138	95·5	22·4	11·3	P.E.Island	1868	Davies		J.Oakes	Dundalk
1027	A	Siloe Cattarinich -88 WBg	Austrian	422	120·4	27·7	18·1	Lussinpiccolo	1865	G.S.Cattarinich		S.Q.Cosulich	Lussinpiccolo
1028	N	Silome Smith -84 WBk	Norweg'n	426	126·6	28·2	18·1	Sunderland	1860	W.Naizby		A.Wilhelmsen	Christiania
1029	✻D	Silurian CBk	Chilian	475	119·9	26·2	17·0	Topsham	1870	Holman		Barra,Fisher & Co.	Valparaiso
1030		Silva I. Louro WBn	Brazilian	140	..	..	..	Campos	18₈₁		Y 8.81	Cia.Cabotagem de S.João da Barra	S.João da Barra
1031		Silvanus WBk	Russian	315	117·0	28·0	14·0	Libau	1859	Bachhoff		Gebrüder Gutschmidt	Windau
1032		Silver Cloud Balle W3mSr	British	292	130·6	27·0	12·8	Sunderland	1874	R.Thompson		J.Orr	Auckland,N.Z.
1033	✻	Silver Cloud Davey -77 WSr	British	75	78·2	19·5	9·9	Jersey	1873	Picot		J.Cantell & Co.	Jersey
1034	✻	Silver Fir Wendott WK	British	92	80·7	21·3	10·7	Galmpton	1885	T.Mathews		A.Walton	Grimsby
1035		Silver Heels McMullen WSr	American	112	87·8	25·7	7·5	Tremont,Me.	1872			H.E.Newman	Tremont,Me.
1036	✻	Silver King Smith 86-86 WSk	British	87	76·5	21·0	10·5	Goole	1885	CottinghamBrs.		J.Rutter	Hull
1037	✻	Silver King WSk	British	85	78·5	20·0	10·5	Grimsby	1886	G.W.Outwin		G.W.Outwin	Grimsby
1038	✻	Silver Lining WK	British	96	82·0	21·0	10·5	Ponsharden	1885	E.Martin		W.R.Leyman	Hull
1039	✻	Silver Sea Rennie -87 WBkn	British	177	103·0	23·3	12·8	Brixham	1878	E.Barter	5.87	Wm.H.Mare	St.Johns,N.F.L.
1040	✻	Silver Spray Pearn 73- WSr	British	169	100·6	23·3	12·9	Fowey	1874	Slade		N.Hocken	Fowey
1041	✻	Silver Spray Foale 84- WSr	British	114	86·0	21·6	11·0	Dartmouth	1876	Moore		N.Trute	Salcombe
1042		Silver Spray (ex Cutty Sark) WSrYt	British	105	87·1	18·1	10·6	Cowes	1874	J.White		Leonard R. West	Cowes
1043		Silver Spray Lindsey WSr	American	173	105·0	27·7	8·9	Addison,Me.	1873	T.Levok		H.B.Nash	Machias,Me.
1044		Silver Spray Russell WSr	American	118	77·1	23·1	10·0	Damrsctta,Me.	1872			J.A.Creighton	Thomaston,Me.
1045		Silver Stream (HULK) WBk	British	323	111·7	26·7	16·5	Southwick	1860	Douglas	10.81	David Bower	Adelaide,S.A.
1046		Silver Stream Jones WSr	British	147	97·9	22·6	12·3	Brixham	1873	Upham		T.Jones	Guernsey
1047	✻	Silver Thread Croft WSk	British	79	73·9	20·3	10·7	Elmshorn	1886	D.W.Kremer		J.W.Croft	Grimsby
1048	✻	Silvercrag Peters 88-88 StlS	British	1897	264·0	39·0	23·6	Workington	1888	R.Williamson & Son		R.Williamson & Son	Workington
1049	✻	Silverdale Miller -88 IS	British	1877	263·9	39·0	23·7	Workington	1887	R.Williamson & Son		R.Williamson & Son	Workington
1050													

✻=Lloyd's Register; D=Nederlandsche Vereeniging; G=Germanischer Lloyd; I=Registro Italiano; R=American Record; A=Veritas Austro-Ungarico; H=Veritas Ellenico; N=Norske Veritas; V=Bureau Veritas.

Dans la 4me Colonne, C=Composite (membrure en fer, bordé en bois); I=Fer; Stl=Acier; W=Bois.

Les Dates du doublage en métal (Col. 11) ressortent des livres dans lesquels les navires sont respectivement classés ou des rapports des Inspecteurs du Lloyd's Register.

Pour plus amples renseignements sur les navires classés par le Lloyd's Register, voir le Lloyd's Register Book et le Lloyd's Yacht Register.

Official No.	No. in Book	Class	Ship's Name. Master, and Dates of Appointment. Material and Rig. Ship's Late Name.	Flag.	Net Tonnage.	Length.	Breadth.	Depth.	Built: Where.	When.	By whom.	Date of Metalling	Owners or Managers.	Port
60719	1051	✻	Silverhow *Dougall 73-73* I S	British	1221	226·0	36·1	22·1	Whitehaven	1875	WhitehvnSBCo		G.Nelson & Sons	Whiteh
63431	1052		Silverlands *Hodge* W Sr	British	116	87·4	20·9	11·2	Sunderland	1868	A.Simey&Co.		T.Hodge	Fowey
87794	1053	✻	Silvery Cloud *Little* W K	British	81	77·0	20·2	10·2	Galmpton	1884	A.Gibbs		J.Little	Grimsby
89373	1054	✻	Silvery Crest *Westerly-5-86* W K	British	88	79·4	20·8	10·3	Galmpton	1884	W.A.Gibbs		R.W.Roberts	Grimsby
74866	1055		Silvery Wave *Estie 4-84* W Bn	British	104	94·4	21·4	8·7	MyallLkNSW	1876	J.Kinnaird		James Campbell	Brisban
	1056	✻	Silvery Wave *Brinham* W Sk	British	87	78·0	20·9	10·5	Goole	1886	CottinghamBrs.		W.Brinham	Hull
950	1057	Λ	Silvestro *WBE KEle Belli* W Bg (ex Fortuna Tonissimo)	Italian	113	82·3	22·2	11·4	Leghorn	1855	Pichiotti.		Silvestro Belli	Leghor
65212	1058	✻	Silvia *Evans* W Sr	British	153	97·8	22·1	12·3	Appledore	1871	Pickard		H.Evans	Braun
76910	1059	✻	Silvia *Beavis 87-87* W Bn	British	113	94·3	21·3	10·6	Bridport	1875	Cox	[illegible]	C.T.Bowring & Co.	Liverp
493	1060	I	Silvia *Ruggiero* W Bk (ex Luigina Bruno)	Italian	388	127·3	29·7	18·8	Castellamare diStabia	1857		Y I.S.	L.Petriccone	Naples
679	1061	V	Silvia *Leboffe -61* W Bn	Italian	103	82·0	21·5	10·4	Gaeta	1866	Gallinaro		Flli L.& S.Burgio-Gianfrida	Catani
764	1062	I	Silvia B. *Colallo 87-88* W Bk (ex Maria Boggiano)	Italian	772	160·7	31·8	19·0	[illegible]	1873	[illegible] Durante	[illegible]	G.B.Beverino	Spezia
3418	1063	DIV	Silvio *Albino 85-86* W Bk	Italian	1082	189·9	35·7	24·3	Sestri Ponente	1885	C.Bozzano	[illegible]	A.Benvenuto	Genoa
	1064		Silyda *Jansson* W Sr	Russian	239	..	..	..					K.J.Jansson	Lemland
	1065	DNV	Simeon *Olsen -84* W Bk	Norweg'n	749	158·6	35·2	18·0	[illegible]	1884	T.A.Mörland	Y 4.88	C.Wroldsen	Tvedes
632	1066	Λ	Simeto *Landini* (ex Villa de Caracas, ex May Fly)	Italian	288	107·1	21·1	13·8	[illegible]	1867	Penrose	[illegible]	F.Landani	[illegible]
	1067	GV	Simon *Gallas -85* W Bk	German	587	122·3	32·0	19·2	[illegible]	1856	L.Grott		Aug.Wolff & Co.	[illegible]
	1068		Simon *Olsson -87* W Sr (ex S. Gasper Ehrenborg)	Swedish	106	79·6	21·0	9·3	[illegible]	1858			C.O.Carlsson	[illegible]
3404	1069		Simonetti B. *Simonetti* W Bn (ex Giuseppe)	Italian	202	92·8	25·2	15·7	Castellamare	1849		Z 85	P.Simonetti	[illegible]
724	1070	I	Simpatia *Sambuceti -87* W Bk (ex Angelo Padre, ex Deva)	Italian	685	164·3	32·1	20·4	Sestri Ponente	1871	A.Briasco fuG.B.	Y 12.85	G. Dall'Orso di Andrea & G. Dall'Orso di Michelangelo	Spezia
18	1071		Simpson *Cilento* W Bk (ex Vittorio) (PONTOON)	Italian	323	111·8	26·2	16·2	[illegible]	1862	A.Castellano	Y 8.86	A.Garginlo & Co.	Naples
72988	1072		SinChinSan *Fiben -88* CmLr	British	433	143·5	31·0	11·0	Shanghai	1888	Boyd & Co.		A.Fiben	Shangh
72780	1073		Sin Kolga *Jensen 4-88* W Bk (ex Kolga)	British	542	137·5	31·0	17·6	Lubeck	1880	J.Meyer	C 4.88	P.V.Grant	[illegible]
89604	1074		Sin Nam Oh W Tongkang	British	121	..	..	..	Singapore	1885			Tan Tye	Singap
	1075		Sincora W Sr	Brazilian	159	..	..	..					A.J.S.Bellens	[illegible]
70501	1076	✻	Sindbad *Owen 88-88* I Bk	British	615	171·0	29·2	17·5	Sunderland	1875	R.Thompson,jr.		Swansea Ship. Co.Ld.	Swanse
	1077	✻	Sindia *Mackenzie 73-88* StlS	British	3007	329·3	45·2	26·7	Belfast	1887	Harland&Wolff		T.&J.Brocklebank	Liverp
	1078	DV	Sindoro *Abel 12-88* W Bk (ex Theybecke)	Norweg'n	813	177·8	31·4	19·9	Nieuwediep	1874	W.H.Meursing	Y 10.88	Actieselskabet"Sindoro" (M.Malvad)	Fredrik
	1079		Sine (ex Constantia) *Andersen* W Sr	Danish	153	90·3	21·4	12·5	Kiel	1858	AntonConradi		H.J.Andersen	Fano
	1080		Sine *Pettersen* W3mSr (TO BE BROKEN UP)	Norweg'n	370	120·1	27·4	17·1	[illegible]	1857			L.Pettersen	[illegible]
	1081		Sing Tai *Douglas* W Bk	Chilian	293	119·6	24·1	15·0	Yarmouth	1872	J.B.Beeching	Y 1.88	R.Douglas	Valpara
77628	1082		Singapore W Sr	British	186	106·0	25·4	9·9	Kingston,Ont.	1878	W.Power		William Power	Kingst
	1083	✻	Singapore *Wiggers 88-88* W Bk (ex Dunbar Castle)	German	922	182·4	34·4	21·7	Sunderland	1864	J.Laing	Y 1.89	Gildemeister & Ries	Bremen
	1084		Sinhasinha W Sr	Brazilian	176	..	..	..					C.M.G. da Silva	Rio de J
	1085		Sinken *Solsberg 87-87* W Bk	Norweg'n	675	143·7	30·2	21·1	America	1847			L.Meldahl	Laurvig
	1086		Sinope W Bg	Turkish	390	..	..	..	Sinope	1869			Hairi tijari Ethem Effendi & Co.	Sinope
	1087		Sinope *Ibrahim* W Bg	Turkish	300	..	..	..	Sinope	1885			Rizehlizi Kiril oglou Ibrahim	Rizeh
	1088		Sinope W Bg	Turkish	390	..	..	..	Sinope	1876			Tefyk Effendi	Sinope
	1089		Sinope W Bg	Turkish	262	..	..	..	Sinope	1870			Keresteh tijar Kara Agah oglou [illegible] Effendi	Sinope
	1090		Sinope W Bg	Turkish	254	..	..	..	Sinope	1878			Keresteh tijari Isaak Gumbet	Constan
	1091		Sinope W Bg	Turkish	225	..	..	..	Sinope	1870			Keresteh tijar Kara Agah oglou [illegible] Effendi	Sinope
	1092		Sinope W Bg	Turkish	212	..	..	..	Sinope	1868			Stambolice Kerestiji Hajee Ali Effendi	Stambo
	1093	✻D	Sinquasi *Richards* W Bk	French	421	144·3	28·7	14·6	London	1875	Salisbury	Y 10.86	R.Moquet	Dieppe
115575	1094	RV	Sintram *Woodside* W S	American	1674	215·4	42·9	24·2	Freeport,Me.	1877	E.C.Soule	Y 1.84	E.C.Soule	Portland
89893	1095	✻	Sir Albert *George -88* W K	British	89	80·3	20·9	10·7	Brixham	1886	S.Dewdney&Ss.		J.Roach	Hull
78432	1096		Sir C. T. Van Straubenzee W Bkn	British	317	127·7	26·2	12·0	St.Catharine's	1875	L.Shickluns		John B.Miller	St.Cath
	1097	✻	Sir Frederick Roberts *Walker* W K	British	90	79·8	21·1	10·1	Galmpton	1886	W.A.Gibbs		J.Thorpe	Grimsby
87811	1098	✻	Sir Garnet *Carr 83-83* W Sr	British	88	79·3	21·3	10·0	Goole	1883	CottinghamBrs.		G.Kilner & Co.	Goole
13031	1099		Sir GeorgeF.Seymour *Hollis* W Bk	British	268	108·6	24·0	14·3	Bermuda	1853			Hy.J.Watlington	Bermuda
22148	1100		Sir Henry Havelock *Kent 7-84* W Bg	British	198	92·0	24·0	14·5	Whitby	1859	Hobkirk		G.Milburn	Whitby

✻ = Lloyd's Register; D = Nederlandsche Vereeniging; G = Germanischer Lloyd; I = Registro Italiano; R = American Record; Λ = Veritas Austro-Ungarico; H = Veritas Ellenico; N = Norske Veritas; V = Bureau Veritas.

In the 4th Column, C = Composite (iron frame, wood planked); I = Iron; Stl = Steel; W = Wood.

The Dates of Metalling (Col. 11) are based on the records of the Books in which Vessels are respectively classed, or on the Reports of the Surveyors of Lloyd's Register.

Further particulars of Vessels classed by Lloyd's Register will be found in Lloyd's Register Book and Lloyd's Yacht Register.

No.	Cl.	Nom. Capitaine et dates d'entrée au service des Armateurs actuels et à bord du navire. Ancien Nom.	Matériaux et Gréement	Pavillon	Tonnage net	Longueur	Largeur	Creux	Construit: Où	Quand	Par qui	Date du doublage en métal	Armateurs ou Directeurs-Gérants	Port d'Armement
1101	✻	Sir Henry Lawrence *Gunning 85-85*	I S	British	1221	211·4	35·9	22·9	Liverpool	1865	T.Royden & Sons		J.Herron & Co.	Liverpool
1102	✻	Sir Herbert Maxwell *Morris*	W3mSr	British	232	125·1	24·8	12·8	Ardrossan	1877	Barr & Shearer	Y 1.88	J.Goodwin	Ardrossan
1103	R	Sir Humphry Davy *Lerbert*	W Bk	British	294	123·3	24·5	15·4	Bristol	1864	Symons	Y 8.83	F.Gauntlett	Bristol
1104		Sir John Franklin	W Bkn	British	245	121·5	23·0	14·3	Jervis Bay, N.S.W.	1884	A.Coulon		Andrew Coulon	Sydney, N.S.W.
1105		Sir John Franklin *Mortensen*	W Bk	Danish	286	112·7	26·8	16·0	Hornestrand	1856	S.Josephsen		Jorgen Hoppe	Copenhagen
1106	DN	Sir John Lawrence *Larsen*	W S	Norweg'n	1090	188·5	36·0	22·9	Liverpool	1859	T.Royden	Y 1.87	O.L.Röd	Tönsberg
1107	✻	Sir Lancelot *Noel*	C S	British	886	197·6	33·7	21·0	Greenock	1865	R.Steele	Y 2.87	Visram Ebrahim & Co.	Bombay
1108	✻	Sir Robert Fernie	Stl 4m Bk	British	2564	313·6	42·0	24·5	Port Glasgow	1889	Russell & Co.		W.J.Fernie	Liverpool
1109		Sir Robert Hodgson *Somerville*	W Bn	British	168	108·5	24·6	12·0	P.E.Island	1869	Coffin		W.F.Hannan	Fowey
1110		Sir Robert Peel	W Bk	American	764	157·8	36·3	21·6	New York	1846	W.H.Webb		N.F.Larrabee	New York
1111		Sir Robert Peel	W Bk	Norweg'n	443	124·2	32·1	16·6	Sundsvall	1848	S.G.Jensen		E.Berentsen	Stavanger
1112		Sir Robert Sale (WRECKED) *Taylor 84-84*	W Bk	British	662	138·3	29·9	20·5	Moulmein	1843		Y 7.83	E.G.Williams	London
1113	✻	Sir Walter Raleigh *Mathis*	I S	British	1492	243·4	38·9	21·9	Glasgow	1876	J.& G.Thomson		D.Rose & Co.	Aberdeen
1114		Sir William Wallace *Mellon*	W Bk	British	968	195·6	34·2	21·1	Aberdeen	1866	J.Duthie Ss.& Co	Y 8.82	J.C.Ellis	Newcastle, N.S.W.
1115	✻	Siren *Shapland 81-81*	I S	British	1482	218·0	38·0	23·0	Glasgow	1881	Barclay Curle & C°		A.& J.H.Carmichael & C°	Greenock
1116		Siren *Sjögren*	W Sr	Russian	236	..	..	..		1877			D.F.Söderlund	Lumparland
1117	V	Sirena (ex Zanon) *Cosulich*	W Bk	Austrian	568	141·4	26·2	19·2	Venice	1877	G.Zanon	Y 2.87	V.Dollenz	Trieste
1118		Sirena	W Bk		755	164·3	33·8	21·5	Pra	1874	L.Durante	Y 8.84		
1119	I	Sirena *Calera -88*	W3mSr	Italian	319	121·3	25·1	16·7	Cassano	1877	A.Castellano	Y 3.83	Ciampa & Jaccarino	Castellamare
1120	N	Sirena *Grøtte*	W Sr	Norweg'n	253	108·5	26·9	12·6	Granville	18 58/77			W.M.Michaelsen	Moss
1121		Sirène *Vauhille*	W Sr	French	162	86·3	21·3	10·7	Dunkirk	1873	D.van Cauwenberghe		Van Cauwenberghe-Lemaire	Dunkirk
1122		Sirène *Le Rohellec*	W Sr	French	147	93·8	23·0	11·8	Dunkirk	1854	Lefebvre	Z 81	A.Laffetay	Caen
1123	GV	Sirene (ex W. Mitzlaff) *Arndt 86-*	W Bk	German	501	136·9	30·6	20·0	Elbing	18 57/65	Gebr. Mitzlaff		C.Arndt	Stettin
1124	G	Sirene *Busch 68-68*	W Bg	German	213	86·0	25·5	14·6	Barth	1840	Bohl		J.J.Wallis & Sohn	Barth
1125	N	Sirene (ex Gaetano & Maria, ex Battistina) *Salstad -88*	W Bk	Norweg'n	693	165·6	31·3	18·9	Sestri Ponente	1867	A.Briasco fu G.B.	Y 8.86	Winge & Co.	Christiania
1126		Sirene *Dorn*	W Sr	Russian	130	80·5	26·4	10·6	Angern	1876	K.Muzeneck		Dorn, Romik & Ohsoling	Riga
1127	✻	Sirenia (STRANDED) *McIntyre*	Stl S	British	1598	264·8	38·1	22·6	Glasgow	1885	C.Connell & Co.		R.M'Millan	Glasgow
1128		Siri *Olsson -88*	W Bg	Swedish	227	96·4	24·2	13·8	Skellefteå	1846			O.A.Ohlsson	Uddevalla
1129	V	Siri *Malmberg*	W Sr	Swedish	197	108·9	23·2	10·1	Söderhamn	1876	Wannberg	Y 9.86	J.F.Schöning	Söderhamn
1130		Sirian Star *Morison -88*	W Bk	British	567	147·9	32·4	18·5	S.Martins, N.B.	1865	N.B.Gardner	Y 11.87	"Sirian Star" Ship Co. (Smith, Wright & Co.)	Belfast
1131	I	Sirio (ex Girasello) *Cichero -88*	W Bk	Italian	598	144·3	30·1	20·0	Voltri	1872	D.Salamone	Y 1.87	Vittorio Bertolotto	Genoa
1132		Sirius (BROKEN UP) *Drake*	W Sw	British	200	82·0	25·0	16·0	Shields	1824			J.Drake	S.Shields
1133		Sirius *Sigsworth 82-82*	W Bn	British	122	89·1	22·4	9·6	P.E.Island	1875	D.Gordon		Daniel Gordon	Charlttetn, P.E.I.
1134	V	Sirius *Meyer*	W Sr	Danish	166	97·8	22·3	11·2	Svendborg	1886	P.Troensegaard		I.A.Meyer	Rudkjöbing
1135	✻	Sirius (ex River Boyne) *Moller 86-86*	I Bk	German	479	154·1	27·8	17·3	Glasgow	1867	Dobie & Co.		J.O.Williamson	Hamburg
1136		Sirius (ex Hafenmeister Lüdingsdorf) *Buselow 79-86*	W Bk	German	371	123·6	28·4	17·2	Rostock	1856	A.Schönemann		E.Burchard	Rostock
1137	V	Sirius *Metz 60-78*	W Bg	German	310	114·5	24·0	13·1	Elsfleth	1877	J.Jurgens	Y 8.81	E.tom Dieck	Elsfleth
1138	NV	Sirius *Fornebo 88-88*	W S	Norweg'n	847	171·0	36·1	19·3	Baltimore	1858		Y 4.86	Westergaard & Co.	Christiania
1139	N	Sirius *Olsen -85*	W Bk	Norweg'n	388	125·9	29·3	15·5	Brevig	1865	P.Knudsen	Y 3.84	Chr.Th.Boe	Arendal
1140	NV	Sirius *Helms 83-83*	W Bkn	Norweg'n	343	136·6	28·7	14·8	Bergen	1876	J.Trumpy & Søn	Y 11.87	Bergh & Helland	Bergen
1141	N	Sirius (ex Marie Auguste) *Rabe -85*	W Bk	Norweg'n	277	112·9	28·3	14·5	Luleå	1852			J.C.Rabe	Drammen
1142	V	Sirius *Grikwan*	W3mSr	Russian	252	113·3	26·8	13·0	Dondangen	1881	R.Tum		K.Sandfeld	Riga
1143	R	Sirius *Olsson -73*	W Sr	Swedish	187	95·8	22·9	12·9	Sölblön	1873			J.H.Gren	Fiskebäck-kil
1144		Siro *Roloff 88-88*	W Bk	Chilian	492	138·4	31·2	18·4	Cornigliano	1873			F.Peede	Valparaiso
1145		Sirocco *Riglan*	W S	British	1133	186·3	37·1	22·9	St.John, N.B.	1856			H.Milvain (R.S.Allen)	Newcastle
1146	R	Sirocco *Cahoon*	W Bn	British	176	94·7	25·3	10·5	Pt.Medway NS	1884	J.Hutt	Y 10.87	Walter Mitchell	Medway, N.S.
1147	N	Sirrah (STRANDED) *Larsen*	W Bk	Norweg'n	560	145·5	31·4	18·5	Arendal	1873	J.Smith	Y 8.81	T.Thommesen & Søn & Smith	Arendal
1148	V	Sisal *Nebendahl 87-87*	W3mSr	German	300	132·3	27·9	12·1	Hamburg	1878	C.C.A.Dreyer	Y 5.86	E.B.Crasemann	Hamburg
1149		Sisters *Hayden*	W Bn	British	114	91·6	22·7	10·8	Trinity B., N.F.	1873			John Munn & Co.	St.Johns, N.F.L.
1150		Sitjá (ex Dolores) *Dios*	W Sr	Spanish	133	84·3	24·7	10·0	Blanes	1858			Sitja & Coca	Barcelona

✻ = Lloyd's Register; D = Nederlandsche Vereeniging; G = Germanischer Lloyd; I = Registro Italiano; R = American Record; A = Veritas Austro-Ungarico; H = Veritas Ellenico; N = Norske Veritas; V = Bureau Veritas.

Dans la 4me Colonne, C = Composite (membrure en fer, bordé en bois); I = Fer; Stl = Acier; W = Bois.

Les Dates du doublage en métal (Col. 11) ressortent des livres dans lesquels les navires sont respectivement classés ou des rapports des Inspecteurs du Lloyd's Register.

Pour plus amples renseignements sur les navires classés par le Lloyd's Register, voir le Lloyd's Register Book et le Lloyd's Yacht Register.

Official No.	No. in Book	Registry in which classed	Ship's Name. Master, and Dates of Appointment to present Owner's Service and to Ship. Material and Rig. (Ship's Late Name)	Flag	Net Tonnage	Length	Breadth	Depth	Built: Where	Built: When	Built: By whom	Dates of Metalling	Owners or Managers	Port…
	1151	N	Sivah *Ellingsen -81* W3mSr	Norweg'n	296	124·1	29·0	12·4	Arendal	1878		V 9.87	J.E.Hanssen	Arend
	1152		Sivia *Nyholm 81-* WBg	Russian	269	107·4	30·5	13·2	Åbo	1873			J.G.Abrahamsson	Gustafs
	1153		Sivia WSr	Russian	197	100·4	27·7	11·1	Laitakari	1876	P.Sandström		N.Heimilä	Lavia
	1154		Sixtus (WREMER) *Olsson 81-81* WBg	Swedish	206	89·2	22·6	13·4	Rostock	1841			P.Olsson	Cimbrit
	1155		Sjofna WS	Norweg'n	493	..	..	..					L.Ellingsen	Christia
	1156	N	Sjofna *Olsen* W3mSr	Norweg'n	330	125·9	28·0	15·1	Arendal	1871	H.S.Hansen		S.Eyde	Arendal
	1157	N	Sjofroken *Lydersen* WBk	Norweg'n	274	121·7	27·5	12·0	Tvedestrand	1875			Stian Salvesen	Tvedestr
	1158	R	Sjofroken *Sundberg 84-84* WS	Swedish	105	75·4	20·3	10·0	Upperud	1874			G.J.Sundberg	Lysekil
	1159		Sjögudinnan (ex Hilleron) *Holmberg* WBk	Swedish	855	130·5	27·8	15·0	Finland	1868			G.D.Kennedy	Gothenb
	1160	N	Sjohvat *Eqvist* WSr	Norweg'n	235	105·0	27·5	12·0	Tvedestrand	1877	E.Gundersen	V 3.86	Hagbert Nielsen	Tvedestr
	1161	N	Sjomanden *Lunde* WBk	Norweg'n	368	118·1	27·6	16·8	Sunderland	1867	Wood	V 7.88	M.G.Monsen	Stavange
	1162	N	Sjufna *Nielsen* WBk	Norweg'n	316	115·3	29·2	15·1	Sandefjord	1874			H.A.Moe	Sandefjo
115670	1163		Skagit *Robinson 86-86* WBkn	American	482	156·2	38·3	12·3	P.Ludlow W.T.	1883	Doncaster		Puget Sound Commercial Co.	Pt.Townsend
	1164		Skandia *Andersson* W3mSr	Russian	272	110·0	27·0	12·0	Åbo	1873			N.Sittkoff	Mariehan
	1165		Skandia WSr	Russian	230	..	..	..		1868			M.Mattsson	Wårdö
	1166	G	Skandia *Backe* WBk	Swedish	366	123·6	27·9	14·5	Gefle	1867	A.Brodin	V 3.81	P.N.Horndahl	Nyhamn
	1167	V	Skandia *Tahlberg 86-88* WBk	Swedish	330	131·5	26·2	12·6	Hernösand	1874	Fr.Hägglund		H.Nordlander	Hernösa
	1168	NV	Skandinav *Johannesen* WBn	Norweg'n	180	104·2	25·9	11·8	Porsgrund	1867	Resch	V 5.86	Carl Konow	Bergen
	1169	G	Skandinav *Pahlsson -72* WBkn	Swedish	204	113·1	23·9	11·7	Oscarshamn	1870	C.Thorén		L.Pahlsson	Lerberge
	1170	N	Skandinavien *Heidsten [-83* WBg	Norweg'n	174	91·1	23·9	12·3	Sandefjord	1857	Hansen		L.Olsen	Laurvig
	1171		Skåne *Andersson* WBg	Swedish	182	97·4	22·7	11·5	Nyhamn	1863			J.P.Andersson	Nyhamn
	1172	N	Skarbo *Hansen -81* WBg	Norweg'n	176	94·8	23·6	11·8	Kragerö	1868	J.Andreassen		O.F.Jahnsen	Kragerö
	1173	NV	Skaregrom *Olsen* WBk	Norweg'n	705	166·7	34·4	19·1	Grimstad	1877	J.Baag	V 8.87	S.Eyde	Arendal
	1174		Skatan (ex Hero) *Bengtsson -84* WBk	Swedish	359	126·6	25·7	15·8	P.E.Island	1863			C.Henckel	Helsing
	1175		Skeldervik (ex Odense) *Pahlsson [70-73* W3mSr	Swedish	261	127·9	24·5	12·3	Odense	1866			B.M.Andersson	Jonstor
87388	1176	✳	Skelmorlie *Halliday 76-83* IS	British	1528	254·1	39·0	22·9	PortGlasgow	1883	R.Duncan&Co.		Skelmorlie Ship Co.Ld. (T. O. Hodder & Co)	Greenock
53983	1177		Sketty Belle *Foxworthy* WBn	British	184	102·8	22·6	12·7	Quebec	1866	Labbie		William D.Moore	Fremant
	1178	NV	Skibladner *Børresen 85-85* WBk	Norweg'n	381	122·6	29·0	15·7	Risör	1875	P.Gundersen	V 10.85	O.Caspersen	Risör
	1179	N	Skibladner *Olsen* WSr	Norweg'n	174	100·4	23·8	11·5	Oscarshamn	1875			Carl Dahl	Fredriks
	1180	N	Skien *Andreassen 86-86* WBk	Norweg'n	648	134·8	32·6	20·4	Vegesack	1847		Y 7.84	Nils P.Höyer	Skien
	1181	N	Skien *Pettersen 84-84* WBk	Norweg'n	476	110·5	32·8	19·2	Skien	1842		Y 1.87	F.Olsen	Dröbak
	1182	V	Skirner *Jepsen -84* W3mSr	Danish	232	113·4	25·1	12·9	Marstal	1884	H.J.&L.J.Bager		L.J.Bager	Marstal
	1183	V	Skirner *Nielsen 81-81* WSr	Danish	131	90·1	22·1	10·3	Troense	1881	R.V.Møller		C.R.Møller	Svendbo
	1184	N	Skirner *Gundersen 74-82* WBk	Norweg'n	405	124·2	32·1	16·3	Helsingfors	1848			S.A.Dedekam	Arendal
	1185	N	Skjald *Tønnesen -86* WBk	Norweg'n	460	136·2	29·8	17·1	Lillesand	1877	O.Tönnesen	Y 11.85	B.Birkeland	Lillesan
	1186		Skjelskör (ex Friedrich Moser ex Fiennmore) *Svane 88-88* WBk	Danish	428	116·9	27·8	17·0	Vegesack	1865	A.Cerruti		P.Mahler	Skjelskö
	1187	DV	Skjold *Friis* W3mSr	Danish	232	112·5	24·9	12·9	Marstal	1884	F.Hansen	Z 2.8	N.Petersen	Marstal
	1188		Skjold *Clausen* WSr	Danish	130	85·5	22·3	11·3	Thurø	1884	C.Bom		Chr.Bom	Svendbo
	1189	N	Skjold *Begge* WBk	Norweg'n	444	137·8	31·3	17·0	Tönsberg	1869	S.Josephsen	V 2.8	H.H.Christiansen	Tönsber
	1190	N	Skjold *Hansen -88* WBk	Norweg'n	381	120·7	29·0	17·1	Risör	1864	Johan Smith	V 1.87	L.Hansen Klodeborg	Arendal
	1191		Skjold (SE N Kaiser Laura) *Olsen 83-88* WBg	Norweg'n	283	105·3	26·4	16·1	Sunderland	1861	Sykes		S.Andersen	Drammen
	1192	N	Skjold (ex Peter U. Warwick) *Knudsen -88* WBk	Norweg'n	262	119·5	27·6	11·1	Baltimore	1858	Cooper & Slicer		H.Larsen	Kragerö
	1193	N	Skjold *Olsen 83-85* WBg	Norweg'n	227	91·8	26·5	14·1	Grimstad	1861	N.Olsen Gjömle		K.Torstensen	Holmest
	1194	NV	Skjold *Tobiassen -82* WSr	Norweg'n	209	108·6	25·7	13·0	Christiansand	1873	O.Olsen Hagen	Z 9.8	O.A.Olsen	Christia
	1195	N	Skjoldmöen (ex Mary) *Lund 84-84* WBk	Norweg'n	409	139·4	29·4	14·8	Halifax,N.S.	1869	M.C.Norris	Z 8.8	T.Ihlen	Holmest
	1196	NV	Skjoldmöen *Haagensen -87* WSr	Norweg'n	283	118·5	29·5	12·6	Soon	1879	W.Schimmerud	Z 12.8	H.A.Haagensen	Drammen
828	1197	✳	Sklad (CONDEMNED) ex Mary Wilson) *Sepich* CBk	Austrian	497	140·2	28·2	17·4	Sunderland	1866	R.Thompson	Y 1.8	Ursich & Co.	Buccari
115838	1198	R	Skobeleff *Tucker 78-82* WBk	American	590	151·5	33·5	12·2	E.Deering,Me.	1882	G.Russell		J.S.Winslow & Co.	Portland
	1199		Skobeleff (ex Czar) WS	Russian	1147	184·3	32·2	22·6	Quebec	1854			M.Sjölund	Lemland
	1200	N	Skudesnæs *Eriksen -84* WBk	Norweg'n	452	139·1	30·2	16·5	Skudesnæs	1876	H.C.Knudsen	V 7.8	K.Larsen	Skudesn

✳=**Lloyd's Register**; A=Veritas Austro-Ungarico; D=Nederlandsche Vereeniging; G=Germanischer Lloyd; H=Veritas Ellenico; I=Registro Italiano; N=Norske Veritas; R=American; V=Bureau Veritas

In the 4th Column, C=Composite (iron frame, wood planked); I=Iron; Stl=Steel; W=Wood.

The Dates of Metalling (**Col. 11**) are based on the records of the Books in which Vessels are respectively classed, or on the Reports of the Surveyors of Lloyd's Re

Further particulars of Vessels classed by Lloyd's Register will be found in Lloyd's Register Book and Lloyd's Yacht Register

No.	Cote	Nom	Capitaine et dates d'entrée au service des Armateurs actuels et à bord du navire. Ancien Nom.	Matériaux et Gréement	Pavillon	Tonnage net	Longueur	Largeur	Creux	Construit où	Quand	Par qui	11	Armateurs ou Directeurs-Gérants	Port d'Armement
1201		Skulda		W Bk	Norweg'n	423	128·8	30·7	16·6	[illegible]	1868	H.Nielsen		Steen & Ström	Christiania
1202	N	Skulda	*Jörgensen 87-87*	W Bk	Norweg'n	247	108·9	26·1	15·2	[illegible]	1858	P.Pedersen		J.Kildahl	Risör
1203	N	Skulda	*Kjolne*	W Sr	Norweg'n	144	85·0	24·3	10·0	[illegible]	1851			A.Andersen	Tonsberg
1204		Skylark		W Sr	American	322	118·2	30·2	12·7	[illegible], Mass.	1850			S.B.Hinckley	Boston,Mass
1205	A	Skylla	*Meuras*	W B	Greek	140	85·9	22·3	..	[illegible]	1870		[illegible]	G.Meuras	Zante
1206	DV	Slamat	*Schots 79-79*	C Bk	Dutch	918	184·0	34·8	19·7	Amsterdam	1876	W.H.Meursing	[illegible]	A.Hendrichs & Co.	Amsterdam
1207	❊	Slavia	*Cosmovich*	W Bk	Austrian	452	130·2	27·2	17·5	Trieste	1871	G.Tonello	[illegible]	A.Radonicich	Dobrato
1208	N	Sleipner	*Knudsen -87*	W Bk	Norweg'n	467	127·3	31·2	18·0	Tvedestrand	1870			Winge & Co.	Christiania
1209	N	Sleipner (CONDEMNED)	*Lauritzen -87*	W Bk	Norweg'n	289	99·6	27·7	16·5	[illegible]	1840			Chr.Hannevig	Christiania
1210	N	Sleipner	*Knudsen -74*	W Bg	Norweg'n	227	108·1	26·6	12·3	Arendal	1866	H.Stiansen		A.D.Fürst	Arendal
1211	N	Sleipner (WRECKED)	*Berg -83*	W 3m Sr	Norweg'n	135	97·6	23·3	9·8	Sandefjord	1868	Chr.Christensen		L.G.S.Larsen	Sandefjord
1212	N	Sleipner	*Pedersen -84*	W Sr	Norweg'n	127	86·3	23·1	10·3	Lillesand	1876	O.& P.Knudsen		O. & P.Knudsen	Lillesand
1213	G	Sleipner	*Andersson*	W Sr	Swedish	111	85·3	21·4	9·2	[illegible]	1870	P.J.Pettersson		S.Andersson	Lerberget
1214	V	Slesvig (ABANDONED)	*Jensen -88*	W Sr	Danish	100	81·2	19·8	9·8	Svendborg	1874	J.R.Andersen		N.Jensen	Svendborg
1215	❊DV	Slieve Bawn	*Crowley 77-77*	I S	British	1667	257·1	38·3	23·1	Belfast	1877	Harland& Wolff		W.P.Sinclair & Co.	Liverpool
1216	❊	Slieve Donard	*Campbell (71-83)*	Stl S	British	1788	267·0	39·2	23·6	Port Glasgow	1883	J.Reid&Co.		W.P.Sinclair & Co.	Liverpool
1217	❊V	Slieve Roe	*McMullan 76-82*	I S	British	1667	257·1	38·3	23·1	Belfast	1878	Harland & Wolff		W.P.Sinclair & Co.	Liverpool
1218		Sligo		W Sr	British	284	137·0	23·0	11·8	St.Catharine's	1874	L.Shickluna		P.Kerwin	St.Catharine's
1219	A	Sloboda	*Begna*	W Bk	Austrian	460	138·4	26·6	17·4	Fiume	1871	G.Brazzoduro	[illegible]	S.Clescovich	Trieste
1220		Sly Boots	*Tyrer*	W Bkn	British	178	101·0	23·5	13·0	Dartmouth	1868	Philip		W.S.Allport & Co.	Brixham
1221	N	Smaragd	*Klausen 83-83*	W Bg	Norweg'n	328	129·9	28·7	13·7	[illegible]	1870	Carl Haastad	[illegible]	Ths.S.Falck	Stavanger
1222	N	Smaragd	*Jorgensen*	W Sr	Norweg'n	246	112·5	26·8	12·1	Kragerö	1880	T.G.Litangen	[illegible]	A.Terkeldsen	Kragerö
1223	DN	Smaragd	*Grande -87*	W Sr	Norweg'n	109	90·1	23·3	9·7	Bergen	1876			Brödr.Jacobsen	Haugesund
1224	DV	Smeroe	*Schenk 83-83*	C Bk	Dutch	1000	183·0	36·1	21·5	Amsterdam	1878	W.H.Meursing	[illegible]	A.Hendrichs & Co.	Amsterdam
1225	GV	Smidt	*Gerdes*	I Bk	German	1673	201·1	37·4	31·2	Vegesack	1867	J.Lange		Siedenburg,Wendt&Co.	Bremen
1226		Smjeli	*Croce*	W Bn	Russian	153	..	..	..	[illegible]	1868	G.Ghioni		L.Fetter	Berdiansk
1227	V	Smolensk	*Rodenius*	W Bk	Russian	406	181·7	30·8	21·0	[illegible]	1856		[illegible]	J.S.Smolensk	Archangel
1228		Smut (LOST)	*Tonnesen -87*	W Sr	Norweg'n	185	98·5	25·8	10·4	Arendal	1884	S.Johnsen	[illegible]	A.Tönnesen	Lillesand
1229	❊	Smyrna (COLLIDED)	*Taylor 68-87*	I S	British	1305	232·3	38·5	22·2	Aberdeen	1876	Hood		G.Thompson & Co.	Aberdeen
1230	❊A	Smyrna (ex Sabique Bo)	*Manolondis*	W Bk	Turkish	472	114·0	30·7	19·8	Trieste	1866	Stabilimento Tecnico	[illegible]	G.Chrissafidis	Constantinople
1231	NV	Snæfrid	*Olsen 85-88*	W 3m Sr	Norweg'n	340	127·8	30·1	12·2	Risör	1885	E.Lindstöl	[illegible]	J.Olsen	Kragerö
1232	N	Snar (ex Eleni)	*Brynildsen*	W 3m Sr	Norweg'n	344	117·8	27·1	16·9	[illegible]	1866	G.Barker	[illegible]	Rederiselskabet "Snar" (R. Salvesen)	Fredrichshald
1233	N	Snar (ex [illegible])	*Johannsen -83*	W Sr	Norweg'n	204	111·0	25·3	11·8	Amsterdam	1866		[illegible]	And. & Chr.Ellefsen	Fredrikstad
1234		Sneezer		W Bn	British	129	91·6	21·9	11·6	Nova Scotia	1858			John Hennessey	St.Johns,N.F.L.
1235	N	Snefrid	*Ulstrup -87*	W Bk	Norweg'n	520	142·8	33·7	16·2	Grimstad	1887			J.Borresen	Grimstad
1236	N	Snefrid	*[illegible] -85*	W Bk	Norweg'n	410	128·0	30·5	16·1	[illegible]	1880	Brinch	[illegible]	C.J.Christensen	Arendal
1237	N	Snefrid	*Pedersen -87*	W Sr	Norweg'n	101	80·7	22·4	10·4	Stavanger	1872		[illegible]	S.& P.Pedersen	Flekkefjord
1238	DNV	Sneklokke	*Reinertsen*	W Bk	Norweg'n	604	149·6	31·6	18·5	Grimstad	1875	F.Smith Petersen	[illegible]	M.Smith Petersen	Grimstad
1239	❊	Snow Flake	*Stephens*	W Sr	British	96	88·2	21·8	9·7	Runcorn	1886	Brundritt		J.Foulkes,jun.	Runcorn
1240		Snow Flake	*Curtis*	W Sr	American	158	99·0	28·3	8·7	Newburg,N.Y.	1855			E.R.Kirk	New York
1241	DV	Snow Queen	*McDougall-87*	W Bk	British	984	183·0	37·1	21·5	Maitland,N.S.	1872	A.Roy	[illegible]	Adam Roy	Halifax,N.S.
1242	V	Snow & Burgess	*Brown*	W S	American	1578	228·5	41·5	24·7	Thomaston Me	1878	S.Watts	[illegible]	Snow & Burgess	New York
1243	❊	Snowdon	*Lewis 82-82*	I Bk	British	1065	207·3	34·0	21·0	Port Glasgow	1877	Russell&Co.		Iron Sailing Ship "Snowdon" Co., Ld. (R. & D.Jones)	Liverpool
1244	❊	Snowdrop	*Foster 75-84*	I Bk	British	609	170·7	30·8	17·0	Sunderland	1884	W.Pickersgill&Ss.		J.Tedford & Co.	Belfast
1245	❊	Snowdrop	*Butel*	W Bn	British	149	99·8	22·2	11·9	Jersey	1873	LeHuquet	[illegible]	D.Bisson,jun.	Jersey
1246	❊	Snowdrop	*Howard*	W Sk	British	95	84·6	20·8	10·6	Rostock	1885	E.Burchard&Co.		J.Calvert	Hull
1247		Soberano	*d'Almeida*	W Bg	Portug'se	230	103·0	23·0	12·5	[illegible]	18 54/68	J.R.Pestano		J.do Nascimento & [illegible]	Setubal
1248		Soberano	*Maristany -88*	W Bg	Spanish	317	104·9	29·6	15·5	Blanes	1850	Vegnet		Juan Maristany	Barcelona
1249	V	Soberano Tercero	*Isern*	W Bn	Spanish	337	112·8	29·0	15·3	Blanes	1871	Vieta	[illegible]	Pedro Maristany	Barcelona
1250	NV	Söblomsten	*Larsen -87*	W Bk	Norweg'n	696	153·7	32·6	18·7	Maitland,N.S.	1874	G.Smith&Son	[illegible]	J.D.Havstad	Grimstad

❊ = Lloyd's Register; D = Nederlandsche Vereeniging; G = Germanischer Lloyd; I = Registro Italiano; R = American Record;
A = Veritas Austro-Ungarico; H = Veritas Ellenico; N = Norske Veritas; V = Bureau Veritas.

Dans la 4me Colonne, C = Composite (membrure en fer, bordé en bois); I = Fer; Stl = Acier; W = Bois.

Les Dates du doublage en métal (Col. 11) ressortent des livres dans lesquels les navires sont respectivement classés ou des rapports des Inspecteurs du Lloyd's Register.

Pour plus amples renseignements sur les navires classés par le Lloyd's Register, voir le Lloyd's Register Book et le Lloyd's Yacht Register.

Official No.	No. in Book	Registry in which classed	Ship's Name. Material and Rig. Master, and Dates of Appointment to present Owner's Service and to Ship. Ship's Late Name.	Flag	Net Tonnage	Length	Breadth	Depth	Where	When	By whom	Date of Metalling	Owners or Managers	Port
	1251	N	Soblomsten *Johansen* W Bk	Norweg'n	318	99·4	28·4	16·7	Arendal	1867			Arent Larsen	Aren
54080	1252	✻	Sobraon *Elmslie 67-87* C S	British	2131	272·0	40·0	27·0	Aberdeen	1866	Hall	Y 3.88	Devitt & Moore	Londo
	1253		Social (ex Fernanda Nora) -87 W Sr	Brazilian	217	..	..	..	Brake	1880			J.B.Cassius Almeida	Rio Gr
	1254	V	Social (ex Loss) *Santos* W Bn	Portug'se	148	87·6	22·3	10·5	SanMartinho	1876	A.J.Gonzalves		Associação Industrial Mathiense	Fayal
	1255		Social *Pereira* W Sr	Portug'se	125	81·0	21·1	9·4	Villa do Conde	1887	J.M.de Araujo		J.F.Pereira	Aveiro
	1256		Social II. *Ribeiro* -87 W S (ex Novo Destino)	Brazilian	161	..	..	..	Weenel	1870		Y 10.86	J.B.Cassius Almeida	Pelota
	1257		Socorro W3mSr		195	94·5	21·9	11·5	Dundee	1868				
	1258		Socrates *Scoularikis* W Bg	Greek	258	..	..	..	Galaxidi	1872			G.Scoularikis	Cephal
	1259		Söderhamn *Moas* W Bk (ex Hoekswijk)	Dutch	719	144·6	34·4	21·4	Slikkerveer	1856	Smit	3.86	W.Pont	Edam
	1260		Soderhamn *Tellefsen* W Bk	Norweg'n	447	116·9	30·0	18·7	Söderhamn	18 54/61			Fr.Tellefsen	Arenda
90304	1261	V	Sodium *Manson* -88 W Bk	British	296	136·0	30·0	11·6	Eatonville,NS	1885	J. Parsons		James G.Ross & Co.	Quebec
	1262	V	Sofala *Bengtson* -86 W Bkn (ex Kathe ... Maria)	Swedish	352	136·1	25·6	13·8	Dordrecht	1865	Gebr.Pott	Y 7.86	C.Bruce	Pitkeå
65943	1263	✻	Sofia *Brakenshaw 81-86* I Bn (ex Jubilee)	British	143	100·6	21·7	11·2	Llanelly	1872	Nevill Bros.		Sailing Ship "Sofia" Co. Ld (R.F.Hickman)	Londo
	1264	A	Sofia *Jerojanni* W Sr (ex Sokol ... Emma)	Greek	400	121·0	26·9	17·5	Fiume	1875	G.B.Bachich	Y 10.87	G.Jerojanni	Syra
	1265	A	Sofia *Cosma* W Bg	Greek	355	..	..	..	Syra	1874		Y 3.88	D.Yourdis	Syra
	1266		Sofia *Marcantonis* W Bg	Greek	100	..	..	..	Genoa	1870			E.Marcantonis	Syra
	1267		Sofia *Durchman* W Bk	Russian	470	..	..	..	Kaskö	1876			N.N.Durchman	Brahes
	1268		Sofia W Sr	Russian	231	103·3	28·1	13·0	Ingå	1874	J.Kjellman		S.Rehnberg	Ingå
	1269		Sofia W Sr	Russian	171	..	..	..		1874			E.Seppala	Säkkijä
	1270		Sofia *Karlsson* W Sr	Russian	163	..	..	..		1873			J.E.Michelsson	Lumpa
	1271		Sofia *Roman 72-88* W Sr	Russian	135	80·8	28·5	10·2	Sideby	1872	F.W.Johansson		L.Olofsson & Co.	Wärd
	1272	V	Sofia *Zaragoza 86-86* W Bkn	Spanish	189	83·7	26·6	13·8	Lloret	1854		Y 3.88	Maria Jimenez	Barce
	1273	V	Sofia *Oman* -87 W Bk	Swedish	334	131·8	25·8	12·8	Hernösand	1873	Strandberg	Y 8.84	Hernösands Skeppsred. aktieb (Wilh. Åkerlund)	Hernö
	1274		Sofia *Persson 81-85* W Bk (STRANDED)	Swedish	243	122·0	22·1	11·6	Altona	1862	F.Dreyer		J.P.Jönsson	Lands
	1275		Sofia *Öfverberg* 75- W Bg	Swedish	219	108·2	24·0	11·4	Gefle	1865	L.F.Berggren		N.B.Öfverberg	Lerhan
	1276	V	Sofia *Larsson 59-71* W Bg	Swedish	219	104·6	24·3	12·4	Gefle	1871	O.A.Brodin		A.Jakobsson	Mollös
	1277		Sofia *Hagland* -83 W Sr	Swedish	134	80·2	25·3	10·0	Sastmola	1874	J.Elfy		J.E.Johansson	Stockh
	1278		Sofia *Levin* W3mSr	Swedish	117	86·2	21·2	9·4					A.B.Rosell	Köping
	1279		Sofia *Andersson 86-86* W Sr (ex Hansine)	Swedish	113	82·3	19·7	10·1	Stettin	1850	A.E.Nüske		J.E.Larsson	Lingsk
	1280		Sofia *Hagman* -88 W Sr	Swedish	106	75·1	20·2	9·9	Kälboda	1849	M.Olsen		F.O.Olsson	Bergq
36	1281	✻	Sofia B. (ex Falcon) *Orebich* W Bk	Austrian	794	191·4	32·2	20·0	Greenock	1859	R.Steele	Y 3.87	G.Brailli & Co.	Orebić
	1282		Sofia Fredrika *Lundström 57-83* W Sr	Swedish	116	75·4	21·6	10·2		1845			O.H.Nyman	Vesterv
	1283		Sofia Guzman *Rasmussen* W Bk (ex Matilda)	Chilian	596	154·8	32·0	17·7	P.E.Island	1874	Richards		M.G.Velasquez	Valpara
	1284	GR	Sofia Karolina *Hellgren* W Bg (ex Colleen Bawn)	Swedish	302	106·9	24·7	15·8	Sunderland	1861	Robinson	Y 12.86	J.E.Hellgren	Helsing
	1285	G	Sofie *Hansen 86-86* W Bn (ex Matthew F. Maury)	Danish	149	101·3	24·4	10·3	Skien	1858	L.Jørgensen		M.Rasmussen	Odense
	1286	N	Soflid *Jensen* -88 W Bk	Norweg'n	385	105·8	28·8	18·6	Risör	1865	Thor Jensen		J.Jörgensen	Risör
1	1287		Sofoula W Bg	Greek	251	98·4	25·0	..	Naples	1867			A.Vagliano	Corfu
	1288	N	Sognedalen *Johannesen* W Bk (ex Hinsen)	Norweg'n	625	158·9	33·7	18·3	Uleåborg	1864	J.Pettersson		T.Johannesen	Drøbak
	1289		Sohod Maskoor W Bk	Dutch	125	86·9	22·3	..	Java	1834			Hadji Mohamet Saleh Arsat	Banjar
807	1290		Sokko Maru W Bg	Japanese	151	95·0	22·4	10·4	Le	1880	Yoshikawa [Sadakichi		Fukuzawa Tatsuzo	Tokio
93741	1291	✻	Sokoto *Cross 87-87* I m Bk	British	2193	278·5	41·9	24·5	PortGlasgow	1887	Russell&Co.		Ship Sokoto Co. Ld (G.M.Steeves)	Liverp
	1292		Sol *Fernandes* W Bn	Brazilian	142	..	..	..	Campos	18 21		Y 4.86	F.S.Ferreira	Rio de
	1293	✻N	Sola (ex Glamorganshire) W Bk	Norweg'n	447	148·4	28·0	17·6	Neyland	1869	Gaddem	Y 11.87	S.Otto & Co.	Christ
	1294	N	Solafide *Clemensen* -88 W Bk	Norweg'n	345	115·1	29·7	15·7	Grimstad	1871	OleRibe	Y 4.88	H.O.Ribe	Grimst
	1295	V	Solamith *Holl 74-74* W Sr	Swedish	162	97·4	21·7	10·1	Ljungnäs	1875	S.Beich		M.Jonsson	Kirk
	1296		Solange *Lechapelain* W Sr	French	106	81·7	18·7	10·7	St.Vaast	1873	E.Costey		Legoulon Deneve	Granv
80272	1297	R	Solario *Brown* -87 W Bn	British	220	114·0	25·2	12·5	Mt.St'wrt PEI	1871	E.Collin	Y 12.88	W.H.Dean	Kingst
9.4	1298	I	Sole *Vervena* W Bk	Italian	411	132·8	27·6	18·3	Albauri	1872	F.S.Mauro	Y 12.87	Flli Vervena	Castell
	1299		Soledad (ex Carlotta) W Bn	Spanish	139	92·8	19·5	9·6	Mindoro	1876			E.José	Manila
	1300													

✻=Lloyd's Register; D=Nederlandsche Vereeniging; G=Germanischer Lloyd; I=Registro Italiano; R=American Re...
A=Veritas Austro-Ungarico; H=Veritas Ellenico; N=Norske Veritas; V=Bureau Veri...

In the 4th Column, C=Composite (iron frame, wood planked); I=Iron; Stl=Steel; W=Wood.

The Dates of Metalling (Col. 11) are based on the records of the Books in which Vessels are respectively classed, or on the Reports of the Surveyors of Lloyd's Re...

Further particulars of Vessels classed by Lloyd's Register will be found in Lloyd's Register Book and Lloyd's Yacht Register.

1 N°	2 Registre dans lequel le navire est coté	4 Nom. Capitaine et dates d'entrée au service des Armateurs actuels et à bord du navire. Ancien Nom.	Matériaux et Gréement	5 Pavillon.	6 Tonnage net	7 Dimensions. Longueur	Largeur	Creux	8 Construit. Où.	9 Quand.	10 Par qui.	11 Date du doublage en métal	12 Armateurs ou Directeurs-Gérants.	13 Port d'Armement.
1301		Soledad (ex Pr...)	W Bn	Spanish	137	90·5	21·2	10·2	Albay	1859			Tomas Osmena	Manila
1302		Soledad	W Pol	Spanish	109	86·6	21·0	9·8	Palma	1851			Doña M.T.Tomas	Majorca
1303		Soler *Lloret*	W Bn	Spanish	114	80·0	23·0	11·4	Barcelona	1848		Z 11.84	V.A.Planelles	Valencia
1304	N	Solertia *Antie -87*	W Bg	Norweg'n	257	104·4	28·9	12·9	Porsgrund	18 $\frac{}{80}$			Olans Antie	Tönsberg
1305	NV	Solglyt *Andersen -88*	W Bk	Norweg'n	843	176·5	34·6	20·0	Grimstad	1888	M.Svensen	V 6.88	O.R.Föreid	Grimstad
1306	NV	Solgran *Tollefsen 84-85*	W Bk	Norweg'n	366	111·1	29·6	12·7	Stavanger	1885	M.A.Magnus	V 12.88	K.S.Bertelsen	Stavanger
1307	GV	SoliDeoGloria *Abendroth (-85*	W Bk	German	426	125·1	28·8	17·3	Danzig	1863	Klawitter	Z 9.88	W.Lüdke & Co.	Stettin
1308	V	Solid *Engel -83* (ex Far Graf Attems)	I Sm Sr	German	340	120·1	23·9	15·0	Hamburg	1869	Reiherstieg (SchiffsWerfte		J.Breckwoldt	Hamburg
1309	N	Solid *Arentsen -87*	W Bk	Norweg'n	493	139·4	32·6	16·5	Tvedestrand	1879	G.Berntsen	V 9.88	Hagbert Nielsen	Tvedestrand
1310	N	Solid *Kongsaven 82-85*	W Sr	Norweg'n	114	92·6	22·9	9·9	Bergen	1878	J.Melkevig		E.H.Kongshavn	Haugesund
1311	DN	Solid *Selliken -88*	W Sr	Norweg'n	110	84·6	20·5	9·3	Hanover	1875			G.Kuhnle	Bergen
1312	V	Solid *Buzin 85-86*	W Bk	Swedish	513	138·1	28·2	17·6	Gefle	1881	P.Bergforss&Sn	V 9.87	P.Bergforss & Son	Gefle
1313	G	Solid (ex Alfred & Marie) *Person*	W Bg	Swedish	227	91·2	22·8	12·8	Germany	1848		V 8.83	J.C.Hulthén	Helsingborg
1314		Solid *Klein 74-74*	W Bk	Swedish	206	110·2	23·8	11·5	Malmö	1843			N.P.Klein	Skanör
1315		Solid *Sorensson*	W Sr	Swedish	137	81·4	21·5	11·5	Oroust	1866			A.Abrahamsson	Grundsund Bohle
1316	*V	Solide *Schumacher 72-84*	I Bk	German	825	188·2	32·0	18·5	Vegesack	1884	J.Lange		Gebrüder Hustede	Elsfleth
1317		Solidor *Jensen 84-86*	W Bk	British	240	117·5	23·9	13·0	St.Servan	1876			John Sharp	Shanghai
1318		Soliman Bakri	W Bg	Turkish	265	..	..	..					El Said Ali El Masri	Alexandria
1319	RV	Solitaire *Sewall 86-86*	W S	American	1462	213·7	40·1	24·1	Bath, Me.	1879	E.Sewall	V 5.86	A.Sewall & Co.	Bath, Me.
1320	I	Sollecito *Capaero -87*	W Bg	Italian	367	118·1	29·4	18·4	Alimuri	1868	M.Mauro	V 3.87	A.Paturzo	Castell.di Stab
1321	N	Solnær (ex Brothers) *Nohre 83-83*	W Sr	Norweg'n	237	116·5	26·7	12·2	Economy, N.S.	1871	T.F.Mackenzie	V 4.83	H.L.Petersen	Christiania
1322		Solo	W Bn Sr	Italian	463	139·4	28·7	19·9	SestriPonente	1863	D.Bianchi		V.Firpo	Genoa
1323		Solo	W Bk	Norweg'n	342	109·6	29·5	15·3	Luleå	1849			J.Thiis & Co.	Fredrikstad
1324	V	Solon *Jensen 74-74*	W Sr	Danish	131	83·0	21·8	10·8	Troense	1872	C.R.Møller		F.W.Valentin	Svendborg
1325	N	Solon *Pedersen -88*	W Bk	Norweg'n	524	128·2	32·2	18·9	Arendal	18$\frac{55}{78}$		V 9.88	E.Dedekam	Arendal
1326		Solon *Zachariassen*	W Bk	Norweg'n	518	128·8	28·6	20·2	Bath, Me.	1853			L.Rode	Tönsberg
1327	G	Solon (ex Castor) *Palm -88*	W Bg	Swedish	219	106·9	21·9	13·3	Lödöse, (Gothenburg	1864	G.A.Lundgren		A.J.Dannberg	Gothenburg
1328		Solon	W Bn	Turkish	211	..	..	..	Genoa	1854			Isak Mehemet	Constantinople
1329	N	Solvang *Nielsen*	W Sr	Norweg'n	154	96·0	23·4	12·0	Stavanger	1875		V 10.87	E.Berentsen	Stavanger
1330	N	Solveig *Gjertsen -86*	W Bk	Norweg'n	602	158·2	33·0	17·7	Porsgrund	1877		[illegible]	J.Jeremiassen	Porsgrund
1331	NV	Solveig *Larsen*	W Bn	Norweg'n	296	120·2	27·4	13·6	Christiansand	1878	M.Knudsen	V 9.88	H.Natvig	Christiansand
1332	NV	Sölyst *Johnsen*	W Bk	Norweg'n	341	123·3	28·6	15·0	Tvedestrand	1877	J.P.Pedersen	V 1.82	J.A.Pedersen	Tvedestrand
1333	V	Sömand *Nickelsen 88-88* (ex Noordzeekanaal)	W Bk	Norweg'n	730	158·2	29·7	20·2	Nieuwendam	1870	W. & A.H.Mearsing	V 6.85	N.M.Gram	Drammen
1334	DN	Somerset *Sorensen 86-86*	W Sr	Norweg'n	305	122·3	29·7	12·8	Oak Point, N.S	1869	J.E.Woodworth	V 3.84	H.L.Petersen	Christiania
1335		Sønderjylland *Johansen* (ex Hjalmar, ex Lavinia) *87-87*	W Sr	Danish	105	84·0	20·7	9·1	Oscarshamn	1857	R.Andersen		A.W.Skibsted's Enke	Copenhagen
1336	N	Sondre *Lie*	W Bk	Norweg'n	585	160·7	31·8	17·6	Sandefjord	1876	S.Olsen	[illegible]	Knud Bache	Sandefjord
1337	N	Söndre Norge *Larsen* (ex Ermina)	W Bk	Norweg'n	507	143·7	30·3	18·3	Shelburne, NS	1867	J.McAlpine		H.A.Olsen	Mandal
1338		Sonnabend *Past*	W Bk	German	342	122·1	27·3	15·0	Ueckermünde	1866	C.Radmann		A.R.Bieback	Stettin
1339	R	Sonntag *Haskell*	W Bk	American	954	171·0	36·3	21·0	Stockton, Me.	1870	A.P.Goodhue	V 11.86	F.Harrington	Boston, Mass.
1340	V	Sonoma *Griffiths 87-88*	W Bk	American	998	177·5	36·2	22·4	Chelsea, Mass.	1868	Pierce& Michael	V 8.82	G.F.Smith & Co.	SanFrancisco
1341	GV	Sonora *Spille 83-83* (ex Bode...)	W Bk	German	558	139·2	31·3	18·9	Rostock	1879	E.Burchard&Co	V 5.86	Oetling Gebrüder	Hamburg
1342		Soodol Hair (ex Telegraph)	W Bk	Dutch	287	118·1	26·2	..					Maha	Pontianak
1343		Sooeddin (ex Ingmo)	W Bk	Dutch	156	92·5	19·6	..					Said Alwi bin Abdulla Madeisie	Palembang
1344		Soon (SUNE) *Andersen -87*	W Bk	Norweg'n	448	140·2	33·3	15·8	Söon	1870	Schinderu	V 1.82	H.Jensen	Drammen
1345		Soon Hoat	W Bg	British	173	103·5	20·0	11·0	Bangkok				Koh Cheng Kee	Singapore
1346		Soopreamaneapuravy	W Bk	British	242	122·3	24·6	14·0	Valuetturrie	1878			Navana Chena Thana Sithengarenehitty Nattaha	Jaffna
1347		Sooroori Bahri	W Sr	Turkish	106	..	..	..	Rizeh	1868			TrebizondleeFeizi Haies Achmet & Co.	Trebizonde
1348		Sooroori Bahri	W Sr	Turkish	106	..	..	..	Rizeh	1885			Rizehlee Yunnus & Co.	Rizeh
1349		Sophia *Tezzel*	W Bg	British	168	86·0	22·9	13·5	Belfast	1841			D.T.Sharper, jun.	W.Hartlepool
1350		Sophia	W Bg	Brazilian	124	..	..	..					J.J.F.Sarmento	Rio Janeiro

*=Lloyd's Register; D=Nederlandsche Vereeniging; G=Germanischer Lloyd; I=Registro Italiano; R=American Record; A=Veritas Austro-Ungarico; H=Veritas Ellenico; N=Norske Veritas; V=Bureau Veritas.

Dans la 4me Colonne, C=Composite (membrure en fer, bordé en bois); I=Fer; Stl=Acier; W=Bois.

Les Dates du doublage en métal (Col. 11) ressortent des livres dans lesquels les navires sont respectivement classés ou des rapports des Inspecteurs du Lloyd's Register.

Pour plus amples renseignements sur les navires classés par le Lloyd's Register, voir le Lloyd's Register Book et le Lloyd's Yacht Register.

Official No.	No. in Book	Registry	Ship's Name. Master, and Dates of Appointment to present Owner's Service and to Ship. Ship's Late Name.	Material and Rig	Flag	Net Tonnage	Length	Breadth	Depth	Built: Where	Built: When	Built: By whom	Date of Metalling	Owners or Managers	Port
	1351		Sophia *Nehls 61-61*	W Bg	German	210	85·5	21·7	11·1	Barth	1841	Bohl		J.F.Nehls	Barth
81	1352		Sophia	W Bg	Greek	148	91·8	21·7	..	Syra	1856			N.Vlangalis	Piræus
	1353		Sophia *Langbein*	W 3mSr	Russian	265	101·9	22·0	12·0	Libau	1858			Demolaki	
	1354		Sophia *Michael 88-88* (ex [illegible])	W Bk	Turkish	418	131·0	31·0	17·5	Leghorn	1857	Mancini	[illegible]	G.Chrissafidis	Constan
	1355		Sophia	W Bg	Turkish	180	..	..	..	Greece	1880			Madame Sophia Teodori (Turkande)	Constan
23833	1356		Sophia Godfrey *McCarty*	W Sr	American	244	109·5	29·3	8·3	Dorchester N.J	1869			F.L.Godfrey	Bridge
48748	1357	✻	Sophia Joakim *Braddon*	I Bk	British	1007	204·8	32·8	21·0	Sunderland	1864	Pile,Hay&Co.		Hajee H.Esmile	Calcutta
68904	1358		Sophia R. Luhrs (WRECKED)	W Bk	British	661	144·7	34·0	18·2	Millbridge U.S	1874			Union S.S.Co. of New Zealand	Welling
	1359	V	Sophie *Petersen*	W Sr	Danish	142	96·2	20·9	10·5	Troense	1876	C.R.Møller		P.O.Petersen	Rørvig
	1360	V	Sophie *Mikkelsen -88* (ex Zampa)	W Sr	Danish	138	93·1	20·5	10·7	Troense	1871	C.J.Jensen		N.Jensen	Svendbg
	1361		Sophie *Raoult*	W Bk	French	682	141·0	32·1	21·3	Marseilles	1867	V.Jouvin		H.Bergasse & Co.	Marseille
	1362		Sophie *Boye*	W Bk	German	315	111·6	29·1	15·4	Hamburg	1850	J.C.Godeffroy & [Sohn		P.Kock	Heiligen
	1363	G	Sophie *Brinkmann 86-*	W Sr	German	127	78·5	22·9	10·2	Greifswald	1874	Aug.Spruth		E.W.Ohrloff Sohn	Rostock
	1364	DG	Sophie *Willms 75-*	W Sr	German	123	86·3	20·2	10·3	Veendam	1860	J.J.Pick		J.E.Willms	Timmel
	1365	GV	Sophie (ex Dexter)	W S	Norweg'n	1257	189·9	38·1	24·0	Quincy	1867	G.Thomas	[illegible]		
	1366	N	Sophie *Jansen 87-87*	W Bk	Norweg'n	533	135·2	33·7	18·2	Helsingfors	1857		[illegible]	[illegible] Selskabet "Sophie" (Hansen)	Christian
	1367	GV	Sophie (ex Ida) *Winkler -88*	W Bk	Norweg'n	371	124·0	26·9	15·1	Elsfleth	1873	Joh.Ahlers	[illegible]	J.L.Winkler	Dronthei
	1368	N	Sophie *Andersen -88*	W Bk	Norweg'n	349	124·5	27·4	15·7	P.E.Island	1869		[illegible]	H.U.Singdahlsen	Laurvig
	1369		Sophie *Popp 75-*	W Sr	Russian	118	81·9	24·5	10·3	Haynasch	1873	M.Weide		W.L.Sternberg	Haynasch
	1370	G	Sophie Helene	W Bk	Norweg'n	533	146·3	31·3	18·6	Apenrade	1873	W.Jacobsen	[illegible]	T.Bassöe	Christian
	1371	G	Sophie Josepha *Larsen* (ex Georgiana II., ex Triton) [-84	W Bk	Norweg'n	402	125·5	29·0	16·0	Dunkirk	1866	DeRyckefreres		L.B.Larsen	Flekkefjo
61266	1372		Sophie Rickmers	W Bk	British	865	160·2	34·0	20·8	Bremerhaven				Peter A.Rickmers	Singapor
	1373	DNV	Sophie Wilhelmine (ex Ben Lomond) *Boye -87*	W Bk	Norweg'n	976	185·3	34·6	22·1	Dumbarton	1862	A.McMillan& [Sons	[illegible]	P.B.Engelschiön	Christian
	1374	N	Sophie Wilhelmine *Fos 71-71*	W Sr	Norweg'n	111	81·6	20·0	10·2	Brevig	1855	O.Røe		T.Isaksen Eeg	Brevig
	1375		Sophie & Marie *Royter* (WRECKED) [-77	W Bk	Norweg'n	338	104·7	27·7	17·9	Arendal	1788			G.Gussow	Fredriks
	1376	N	Sophie & Nicoline *Michelsen -81*	W Bg	Norweg'n	244	108·1	26·4	14·0	Mandal	1870	C.Weyergang	[illegible]	Sv.Svensen	Mandal
77455	1377	✻	Sophocles *Smith 76-81*	I S	British	1120	223·4	34·7	21·7	Aberdeen	1879	Hood		G.Thompson & Co.	Aberdeen
	1378	V	Sørane *Hansen*	W Bg	Danish	229	108·0	23·5	12·9	Nordby,Fanø	1875	C.Graff	[illegible]	AktSelsk (J.N.Hansen)	Nordby
63170	1379	✻	Sorata *Ager 85-88*	I Bk	British	786	188·2	32·0	19·7	Liverpool	1869	Bowdler		J.B.Walmsley & Co.	Liverpool
85725	1380		Sorato	W Sr	British	360	121·0	31·8	12·8	Pictou,N.S.	1876			William Dewar	Lunenbu
	1381	G	Søren Peter *Kyhn 82-86* (ex Triumph, ex Louisa Adelaide, ex Julius)	W Bn	Danish	108	82·2	18·8	10·8	Rønne	18 53/74			H.P.Lund	Rønne
2899	1382		Sori (ex Vittorio S.) *Consigliere*	W Bk	Italian	408	123·6	26·1	18·1	Savona	1878	F.Sirello	[illegible]	P.Consigliere di G.B	Genoa
	1383	N	Soridderen *Clausen -87*	W Bk	Norweg'n	468	110·5	31·5	19·6	Arendal	18 [illegible]	Olsen		H.U.Singdahlsen	Laurvig
	1384	N	Soridderen *Pedersen*	W Bk	Norweg'n	384	131·1	28·6	15·4	Grimstad	1876	O.&L.Dömmesmoen	[illegible]	J.Börresen	Grimstad
	1385	N	Söridderen *Carlsen* [-88	W Bg	Norweg'n	243	104·1	26·2	13·8	Risör	1859	O&G.Halvorsen		Rederiskabet Söridderen (Z.[illegible])	Fredriks
	1386	GV	Sorkholm (ex Undine) *Roller -86*	W Sr	Russian	232	105·7	24·0	14·0	Rostock	1874	W.Zeltz	[illegible]	H.D.Schmidt	Pernau
	1387		Sorpresa *Vidal -87*	W Sr	Spanish	223	130·5	27·1	12·9	Barcelona	1855			J.P.Ordal	Barcelona
22826	1388		Sorrento	W S	American	1584	197·0	39·3	27·8	Kennebunk Me	1863			H.A.Wilson	New
	1389		Sosima Savatin *Gunin -86*	W Sr	Russian	113	68·0	18·0	10·0	Onega	1866	J.Gunin		J.Gunin	Onega
	1390	D	Soskummeren *Larsen*	W Bk	Norweg'n	290	92·0	26·9	16·7	Arendal	1839			H.U.Singdahlsen	Laurvig
	1391	NV	Söskummeren *[illegible] -88*	W Bg	Norweg'n	214	103·5	26·6	11·8	Risör	1870	P.Thorvaldsen	[illegible]	A.& F.Smith	Tvedest
182	1392		Sospir (CONDEMNED) *Nicolich*	W Bk	Austrian	409	114·1	26·9	17·0	Lussinpiccolo	1869	N.Martinolich	[illegible]	Olivetti & Comuzzi	Trieste
	1393		Sossana	W Bg	Greek	144	..	..	..	Scopelos				A.Pierros	Ithaca
	1394	N	Söstrene *Föyen*	W Bk	Norweg'n	490	134·2	30·7	17·8	Kragerö	1864	J.Andreasen	[illegible]	L.Föyen	Tönsberg
42	1395	H	Sotèr (ex Vassiliki) *Costelos*	W Bg	Greek	239	91·8	27·1	..	Syra	1869			D.E.Costelos	Piræus
	1396		Sotèr *Matakos*	W Bg	Greek	191	81·0	23·5	14·8	Scopelos	1869	P.Vareta		D.A.Matakos	Poros
45	1397	A	Sotèr *Tsipouras*	W Bg	Greek	182	90·8	23·0	13·9	Galaxidi	1880	J.Canatas		J.Tsipouras	Galaxidi
67288	1398	✻	Soteria (ABANDONED) *Jarvis*	W Bk	British	598	162·7	31·1	17·9	Falmouth	1875	Trethowan	[illegible]	Balkwill & Co.	Salcombe
87	1399	A	Sotirion Skafos *Coutzoukos*	W Bn	Greek	410	123·6	29·5	..	Syra	1878		[illegible]	C.A.Coutzoukos	Syra
	1400		Souakim Bachri	W Bg	Turkish	200	..	..	..	Caragatch	1868			Ismail Pacha	

✻ = Lloyd's Register; D = Nederlandsche Vereeniging; G = Germanischer Lloyd; I = Registro Italiano; R = [illegible]
A = Veritas Austro-Ungarico; H = Veritas Ellenico; N = Norske Veritas; V = [illegible]

In the 4th Column, C = Composite (iron frame, wood planked); I = Iron; Stl = Steel; W = Wood.

The Dates of Metalling (Col. 11) are based on the records of the Books in which Vessels are respectively classed, or on the Reports of the Surveyors of Lloyd's

Further particulars of Vessels classed by Lloyd's Register will be found in Lloyd's Register Book and Lloyd's Yacht Register

No.		Nom. Capitaine et dates d'entrée au service des Armateurs actuels et à bord du navire. Ancien Nom.	Matériaux et Gréement	Pavillon	Tonnage net	Dimensions: Longueur	Largeur	Creux	Construit: Où	Quand	Par qui	Doublage	Armateurs ou Directeurs Gérants	Port d'Armement
1401	✻	Soudan *Donald 85-85*	I S	British	1699	250·8	39·0	24·1	Stockton	1885	Richardson, Duck&Co.		British & Eastern Shipping Co.	Liverpool
1402	✻	Soudan *Rickers 85-85*	I S	British	1517	251·5	38·2	23·0	PortGlasgow	1885	Russell&Co.		Ship Soudan Co.,Ld. (G.M.Steeves)	Liverpool
1403	R	Soudan	W Sr	British	119	86·0	35·1	9·5	PortMulgrave, N.S.	1884	R.McVicar		ChristopherMcDonald	Hawkesbury
1404	✻	Soukar *Croker 73-79*	I S	British	1304	229·3	33·3	23·0	PortGlasgow	1864	Reid		Shaw,Savill & Albion Co. Ld.	Southampton
1405	H	Soupoute Bachri	W Bg	Turkish	210	..	..	..		1864			Emir Efendis	Constantinople
1406		Sourabaya	W S	British	103	68·0	29·0	6·0	Singapore	1876	Wee Ah Oh Smiths		Tan Tye	Singapore
1407	V	South American *Connolly -88*	W S	American	1694	227·5	41·6	25·2	Boston,Mass.	1876	Townsend	Y 10.87	H.Hastings	Boston,Mass.
1408	✻	South Australian *Adams 87-87*	C S	British	1040	201·0	36·0	20·0	Sunderland	1868	W.Pile & Co.	Y 1.89	Ship "South Australian" Co. Ld. (Woodside & Workman)	Belfast
1409		South Boston (NOT IN RAISTEN-T)	W Bn	British	197	99·3	22·2	12·1	N.Glasgow,NS	1853			William Elliott	Pictou,N.S.
1410		South Downs *Lilburn 88*	W Bn	British	178	94·1	22·6	13·0	Rye	1857			R.J.Head	Folkestone
1411	✻	South Glen *Rawlins 84*	I Bk	British	598	173·8	29·2	17·5	Glasgow	1874	Dobie&Co.		L.Hodgson	Liverpool
1412	R	South Shore *Jones 84-*	W 3mSr	American	345	126·0	30·8	14·6	Newburyport, Mass.	1873	G.E.Currier, jun	Y 9.85	C.A.Cousens	Boston
1413		Southampton	W Sr	British	319	134·5	26·1	11·5	GardenIsland	1861	H.Roney		Samuel Fraser	Kingston,Ont.
1414	✻	Southerfield *Brisco 81-83* (BURNT)	W Bk	British	707	177·5	33·5	17·9	Maryport	1881	R.Ritson&Co.	Y 1.83	Ritson & Co.	Maryport
1415	R	Southern Belle *Phillips -86*	W Bk	British	1120	185·0	36·9	23·5	Quebec	1862	Oliver	Y 1.83	Burgess & Co.	Bristol
1416	V	Southern Belle *Durkee -88*	W Bk	British	588	146·0	31·4	18·8	Digby,N.S.	1871	J.Mulcaha	Y 8.85	Abel C.Robbins	Yarmouth,N.S.
1417		Southern Belle *Most 78-78*	W Bk	British	336	123·3	29·2	13·5	St.Martin's,NB	1866		Y 7.83	John Bickers	Adelaide,S.A.
1418	✻	Southern Belle *Golden 85-85*	W Barge	British	80	85·0	20·6	7·0	Ipswich	1885	G.P.Gildea&Co.		Hawkins,Holman&Co.	London
1419		Southern Chief *Bailey -86*	W Bk	American	1218	186·9	38·3	23·5	Bath,Me.	1857	Trufant		F.Barnard & Co.	San Francisco
1420	RV	Southern Cross *Bailey 82-82*	W S	American	1087	176·8	37·5	23·0	East Boston	1868	CurtisSmith&C.	Y 1.86	E.Lawrence	New York
1421		Southern Home	W Sr	British	201	109·5	28·6	10·4	Cornwallis,NS	1873			Robert Gordon	Windsor,N.S.
1422	D	Southern Queen *Rickard -87*	W Bk		789	157·2	33·2	21·8	St.John,N.B.	1866	Ruddock	Y 7.86		
1423	✻	Southesk *Hill 87-87*	I Bk	British	1154	225·2	35·0	21·8	Dundee	1877	A.Stephen&Sns.		D.Bruce & Co.	Dundee
1424		Souto	W S	Brazilian	884	..	..	..					M.A.Cardoso	RioJaneiro
1425	V	Souvenir *Collis 85-85*	W Bk	British	482	142·0	29·6	17·2	Lillesand	1875	T.Aslagsen	Y 4.88	J.E.Cole	Bristol
1426		Souvenir *Bruno*	W Bn	Frenca	187	75·8	20·7	11·6	Nantes	1864	Boju&Bertrand		L.Jacquier	Nantes
1427	N	Souvenir *Sorensen*	W Bk	Norweg'n	504	127·9	31·6	17·9	Arendal	1881		Y 5.88	H.S.Sörensen	Arendal
1428	V	Souverain *Bourdase*	W Bk	French	665	151·5	29·8	18·7	Marseilles	1876	Jonvain	Y 4.86	L.Senglar	Marseilles
1429	XV	Souverain *Nielsen 84-87* (ex Charles H. Marshall)	W S	Norweg'n	1688	196·5	42·0	28·3	New York	1869	W.H.Webb	Y 5.87	Actieselskabet "Souverain" (A.Mele)	Stavanger
1430	V	Souveraine *Jonnquain -87*	W Sr	French	109	82·7	20·0	10·1	Dunkirk	1867	B.Deryeke	Y 7.84	Bonvarlet frères	Dunkirk
1431	✻	Sovereign *Putnam*	W S	British	1193	201·6	39·0	22·2	Londonderry, N.S.	1879	J.Geddes	Y 4.88	McLelan & Blackie	Halifax,N.S.
1432	✻	Sovereign *Cook 84-84*	W Bk	British	524	140·0	30·3	18·0	Sunderland	1876	R.Thompson	Y 4.82	W.Pickford	Newport
1433	R	Sovereign *Kearon 79-79*	W Bkn	British	332	155·0	25·0	14·0	PortGlasgow	1836	Wood		E.Kearon	Dublin
1434	D	Sovereign *Flett 77-77*	W Sr	British	109	89·9	21·9	10·5	Inverness	1877	Macgregor	Y 3.81	W.Corner	Inverness
1435	✻	Sovereign *Darragh -84*	W Sr	British	95	84·4	21·3	10·1	Peterhead	1880	Carnegie		Mrs.A.M.Howard	Runcorn
1436	✻	Sovereign	W Sk	British	86	80·0	20·0	10·7	Grimsby	1886	G.Outwin		A.J.Beeson	Grimsby
1437		Sovinto *Michelson*	W Bg	Russian	239	..	..	..	Sideby	1867			J.M.Michelson	Korpo
1438		Sovinto *Ridderstad* (WELLO)	W Sr	Russian	129	..	..	..	Wederlako	1872			A.Ridderstad	Rauma
1439	R	Sower *Dexon 81-83*	W Sr	British	143	88·0	27·2	9·2	ThreeSisters, N.S.	1883	D.R.Eaton		Scammell Bros.	Parrsboro',N.S.
1440	✻	Spark *Davies*	W Sr	British	197	112·4	24·0	12·6	Rye	1868	Head	Y 10.88	D.Davies	Liverpool
1441		Sparkling Foam *Mills 85-85*	W 3mBn	British	196	115·8	24·4	12·8	Sandwich	1868	Pain	Y 6.86	C.Hoult & Co.	Faversham
1442	✻	Sparkling Glance *Piggis -86*	W Sr	British	159	95·0	25·0	12·8	Howden	1878	Banks	Y 5.83	A.Meek	London
1443	✻	Sparkling Nellie *Rough*	W K	British	63	70·6	19·2	9·0	Bideford	1885	R.M.Restarick		T.& J.Lelen	Yarmouth
1444	R	Sparkling Water *Beddows*	W Bn	American	303	108·0	28·0	14·0	Stockton,Me.	1869	N.G.Nichborn	Y 12.86	N.W.Bridge	Boston
1445	✻	Sparkling Wave	W K	British	93	80·1	21·4	10·5	Galmpton	1885	T.Matthews		J.Haylock	Grimsby
1446		Sparrow *Sheehan*	W Sr	American	187	119·5	28·5	9·7	Oakland,Cal.	1869	Simpson		J.Vance	Eureka,Cal.
1447	G	Spartan *Pottle 85-85*	W S	American	1376	206·6	40·5	24·3	East Boston	1874	R.E.Jackson	Y 9.85	H.Cairns	SanFrancisco
1448	RV	Spartan *Colley*	W Bk	American	749	151·1	34·2	20·2	Bath,Me.	1876	Hawley&Palm'r	Y 4.87	A.Palmer	Bath,Me.
1449		Spartan *St. John*	W 3mSr	American	592	154·0	35·4	15·8	Bath,Me.	1880	Goss,Sawyer & Packard		L.W.&P.Armstrong	Newhaven,Con.
1450	R	Spartel *Hallowell*	W Sr	American	126	91·0	27·0	7·6	Pembroke,Me.	1868	Parrott		W.T.Hobart	Pembroke,Me.

✻=Lloyd's Register; D=Nederlandsche Vereeniging; G=Germanischer Lloyd; I=Registro Italiano; R=American Record; A=Veritas Austro-Ungarico; H=Veritas Ellenico; N=Norske Veritas; V=Bureau Veritas.

Dans la 4me Colonne, C=Composite (membrure en fer, bordé en bois); I=Fer; Stl=Acier; W=Bois.

Les Dates du doublage en métal (Col. 11) ressortent des livres dans lesquels les navires sont respectivement classés ou des rapports des Inspecteurs du Lloyd's Register.

Pour plus amples renseignements sur les navires classés par le Lloyd's Register, voir le Lloyd's Register Book et le Lloyd's Yacht Register.

Official No.	No. in Book	Registry in which classed	Ship's Name. Material and Rig. Master, and Dates of Appointment to present Owner's Service and to Ship. Ship's Late Name.	Flag.	Net Tonnage	Length	Breadth	Depth	Built: Where.	When	By whom.	Date of Metalling	Owners or Managers.	Port
85446	1451	✻	Specimen Methorth 81-87 WK	British	82	76·5	21·2	10·6	BurtonStather	1882	J.Wray&Son		J.Craven	Grimsb
10334	1452		Speck Dyer 86-88 WBg	British	154	99·2	23·1	10·8	Sackville,N.B.	1837			J.Dyer	Portsm
	1453	GV	Speculant Kumpehl 78-82 WBk	German	628	117·9	29·2	18·5	Hammel-{warde	1876	W.Hagens	[illegible]	Adolph Schiff	Elsfleth
53013	1454		Speculation (or Bk) WBk	British	371	112·0	32·0	14·0					Jas.J.Swaine	[illegible]
	1455	N	Speculation Tarraldsen 76-81 WBk	Norweg'n	283	108·0	27·0	14·1	Grimstad	1881	N.O.Gjömle	[illegible]	N.Thue Johnsen	Grimst
	1456		Speculation WBg	Norweg'n	173	80·2	24·3	14·2	Risör	1811			S.Wesmann	Christia
	1457	N	Speculationen Olsen -88 WBg	Norweg'n	162	77·1	24·0	13·5	Grimstad	18 29/73			E.Suhrke	Fredriks
49139	1458		Speed WBk	British	228	95·0	23·5	15·0	Isle of Man	1842			John Armstrong	Nassau
	1459	N	Speed Corneliusen WBk	Norweg'n	336	112·4	27·1	16·0	Wallace,N.S.	1856			O.Corneliusons Enke	Dramm
	1460	N	Speed Næss WBg	Norweg'n	251	116·2	26·6	12·7	Brevig	1875		[illegible]	T.Jacobsen Næss	Sandefj
	1461	N	Speed Tonnesen 79-84 WBk	Norweg'n	219	112·8	27·0	11·6	Kragerø	1868	Brandt		J.Bilstad	Egersund
13683	1462		Speedwell Nixon IBk	British	350	138·7	23·0	14·0	Cork	1856	G.Robinson&Co		James C.Ellis	Newcast
76578	1463	✻	Speedwell Noel 83-83 WBkn	British	234	118·2	24·8	12·9	Kingston	1878	Spence	[illegible]	J.Mill	Banff
72061	1464		Speedwell WSr	British	182	110·7	24·1	9·5	S.Marysburgh	1875	G.Dickson		Alexander Ure	Pictou
87280	1465		Speedwell Osten WK	British	117	91·6	22·6	8·8	Sittingbourne	1883	J.Taylor		J.Smith	Colches
92796	1466	✻	Speedwell Hill WK	British	91	80·0	21·2	10·3	Galmpton	1885	W.A.Gibbs		W.Hill	Grimsby
113051	1467	R	Speedwell Weed WSr	American	142	96·0	25·6	8·7	Frankfort,Me	1872			Cobb,Wight & Co.	Rocklan
27770	1468		Speedy Hardy WBg	British	274	101·6	24·1	15·0	Cork	1859	G.Robinson&C.		W.H.Smith & Sons,Ld.	Melbou
	1469		Spekulation Lyth 76-81 WBg	Swedish	185	82·0	21·9	13·0	Wismar	1834			F.Nyström	SliteGo
1519	1470	I	Speme Castello -87 WBk	Italian	454	135·8	30·6	19·2	Varazze	1876	S.Craviotto	[illegible]	G.B.Degregori	Genoa
	1471	N	Spener Aanansen -83 WBnSr	Norweg'n	271	118·8	27·9	11·9	Risör	1882	T.Tarraldsen	[illegible]	P.D.Prebensen	Risör
213	1472		Spera in Dio Farlon -87 WBn	Italian	145	94·2	23·8	13·4	Chioggia	1877	G.Poli	y.88	Eredi Poli Minei	Venice
	1473	✻	Speranza Dupont -88 WSr (or Heather Bell)	French	118	87·6	21·8	10·8	Ardrossan	1879	Barr		Beust & Fils	Granvill
472	1474	H	Speranza WBg	Greek	254	101·3	25·9	..	Syra	1869			M.Kolakis	Syra
	1475	N	Speranza Pedersen 82-86 WBk	Norweg'n	486	128·9	32·1	17·3	Tvedestrand	1870	J.P.Pedersen	[illegible]	Pedersen & Salvesen	Tvedes
	1476	N	Speranza Andersen -77 WBk	Norweg'n	446	127·2	29·9	16·6	Grimstad	1877	K.Andersen	Y.78	K.Andersen	Grimsta
	1477	N	Speranza Nielsen 83-83 WBk	Norweg'n	341	115·3	27·0	16·8	Sunderland	1852		[illegible]	T.M.Nielsen&Son	Laurvig
	1478	N	Speranza Pedersen WBg	Norweg'n	176	105·5	23·0	11·5	Brevig	1864			P.S.Pedersen	Farsund
1287	1479	I	Speranza C. Carletti -87 WBk (or Museum, or Enrica Morello)	Italian	506	140·0	32·2	20·4	Sestri Ponente	1871	L.Marcenaro	[illegible]	Frat.Cignoni	Leghorn
	1480		Sperchios WSp	Greek	312	131·2	..	13·6	Austria	1876			Theophilatos&Stathatos	Ithaca
3900	1481	I	Spero Razeto -88 WBk	Italian	389	117·4	30·2	19·2	Savona	1867	F.Sirello	[illegible]	G.B.Demarchi	Genoa
	1482		Spersiyah WBg	Turkish	187	..	..	..	Roumania	1880			Vassilikoslee Yanna sehdeh Tekfurd	Vassil
	1483		Spes WBk		726	152·8	34·1	18·8	Leghorn	1854				
	1484	N	Spes Gundersen -87 WBk	Norweg'n	427	128·7	29·8	16·3	Risör	1873	E.J.Giving	[illegible]	O.Caspersen	Risör
	1485	G	Spes Hallgren -85 WBn	Swedish	159	96·8	23·9	10·3	Westervik	1874	C.R.Fredell		G.Ideström	Wester
	1486		Spes Jansson 85-86 WBn	Swedish	139	90·0	22·1	9·9	Westervik	1846	P.Risholm		G.R.Ohman	Norrtel
	1487	N	Spes&Fides Andersen -88 WBg	Norweg'n	391	110·9	27·6	15·5	Christiania	1865	L.Mathiesen	[illegible]	M.Gjein	Tonsberg
	1488	V	Spica Block 87-87 IBk	German	915	185·7	32·5	19·9	Vegesack	1878	H.F.Ulrichs		W.A.Fritze & Co.	Bremen
	1489	N	Spica Museus -76 WBn	Norweg'n	184	95·8	21·2	12·0	Christiansund	1876	G.Wessell	[illegible]	T.O.Larsens Enke	Christi
64022	1490		Spinaway Garrick 85- WBkn	British	325	134·8	28·1	12·8	Sunderland	1875	R.Thompson,jnr	[illegible]	John Bateman	Freman
77380	1491	✻	Spinaway Richards 73-77 WSr	British	95	83·2	20·4	10·6	Fowey	1877	Stephens	[illegible]	A.Luke	Fowey
14946	1492		Spirit of the Age Johnson WBk	British	347	128·6	24·6	16·3	Barnstaple	1857			William O.Gilchrist	Sydney
63151	1493	✻	Spirit of the Dawn Judd 84-86 IBk	British	692	175·1	30·7	19·7	Sunderland	1869	Oswald&Co		J.Bell & Son	Liverpo
50283	1494	✻	Spirit of the Morning Windell IBk	British	599	168·6	29·2	19·3	Sunderland	1864	Oswald&Co		J.McFee	Liverpo
2637	1495	A	Spirito Raicich 86-86 WBk	Austrian	491	133·5	28·5	19·2	Volosca	1863	Fll Schiavoni	[illegible]	G.Minach	[illegible]
305	1496		Spirito Santo Fresco WBg	Italian	100	68·9	23·0	11·8	Pisa	1853	M.Tisi		B.Olivari	[illegible]
	1497		Spiro Elia WBg	Turkish	227	..	..	..	Casho	1872			Yorghi Elia	[illegible]
	1498		Spiro WBg	Turkish	168	..	..	..	Greece	1879			Mutter Rashi zadeh Bens ...	[illegible]
	1499		Spiros Falouris WBg	Greek	334	107·2	27·0	..	Scopelos	1867	Panagetti		D.Caliamida	[illegible]
40018	1500		Splendid WBk	British	358	109·5	27·2	18·3	Massachusetts	1835			William Elder	[illegible]

✻=Lloyd's Register; D=Nederlandsche Vereeniging; G=Germanischer Lloyd; I=Registro Italiano; R=American Re[cord]
A=Veritas Austro-Ungarico; H=Veritas Ellenico; N=Norske Veritas; V=Bureau Veri[tas]

In the 4th Column, C=Composite (iron frame, wood planked); I=Iron; Stl=Steel; W=Wood.

The Dates of Metalling (Col. 11) are based on the records of the Books in which Vessels are respectively classed, or on the Reports of the Surveyors of Lloyd's ...

Further particulars of Vessels classed by Lloyd's Register will be found in Lloyd's Register Book and Lloyd's Yacht Register.

1	2	4 Nom. Capitaine et dates d'entrée au service des Armateurs actuels et à bord du navire. Anciens Noms.	Matériaux et Gréement	5 Pavillon	6 Tonnage net	7 Dimensions. Longueur	Largeur	Creux	8 Construit. Où.	9 Quand.	10 Par qui.	11 Date du doublage en métal	12 Armateurs ou Directeurs-Gérants.	13 Port d'Armement.
1501	I	Splendido *Peragala* -88	W Bg	Italian	364	117·7	30·5	18·8	Varazze	1867	A. Boilo	V 1.88	Ant. Schiaffino	Genoa
1502		Spotless *Schauer* -88	W 3mSr	American	397	115·0	32·4	13·2	Baltimore	1875	Stevens & [Newman	V 1.87	S. B. Marts & Co.	Baltimore
1503		Spray *Martin* 85	W Sr	American	119	100·0	17·0	8·0	Hamburg, Pa.	1863			J. A. Anderson	Trenton, N.J.
1504		Spring *Piper* 86-86	W Sr	British	132	89·0	22·0	12·2	Kingsbridge	1867	W. Date	V 3.86	E. Rillston	Fowey
1505		Spring Bird	W Sr	Venezuela	177	99·0	26·0	10·8	Canning, N.S.	1868		V 3.83	Tinoco & Co.	Venezuela
1506		Spruce Bud *Arnold*	W Sr	British	119	75·2	19·1	11·0	Ilfracombe	1861	Cook		P. Farnill & G. Arnold	London
1507		Spruit	W Bn	Turkish	198	95·8	21·3	12·1	Pekela	1871	J. W. Kuiper	[illegible]		
1508		Spunkie (COLLISION)	W Sr	British	132	90·0	21·5	11·1	Alloa	1864			Henry Beattie	Sydney, N.S.W.
1509		Spyridon *Cancilogiannis*	W Bg	Greek	241	..	..	..	Galaxidi	1881			A. Bacogeorgios	Zante
1510	✻	Squirrel *Edwards*	W Sr	British	98	83·8	19·5	11·0	Bristol	1862	Peters		R. Edwards	Aberystwith
1511	V	Sreena P. *Scurich*	W Bk	Austrian	565	118·3	30·2	18·7	Fiume	1871	Flli Schiavoni	V 2.86	G. Gargurevich	Ragusa
1512		Sredstvo *Kisdich* -87 (ex James Duncan) (Mission)	W Bk	Austrian	699	182·4	32·8	17·9	P. E. Island	1870	Duncan	V 1.83	F. Czar di G.	Buccari
1513		Sree Pas Sair	W Sr	British	112	85·0	20·2	8·0	Singapore	1883	T. Hong		E. W. Streeter	London
1514		Sri Bintang Salamat *Daeng Pasundreh*	W Sr	Dutch	132	94·8	21·0	..					Daeng Pasundreh	Passir
1515		Sri Passir	W Sr	Dutch	156	92·8	21·0	..					Daeng Machmoed bin Abdul Karim	Passir
1516		Srie Kailie Sennanghati	W Sr	Dutch	104	87·3	21·5	8·4					Hadji Ali bin Samatah	Dongala, Celebes
1517		Staatsrath von Broock (WRECKED) *Zander* 72-82	W Bg	German	240	99·5	25·5	14·7	Rostock	1851	F. Dethloff		G. Kindler	Rostock
1518	N	Stabil *Lindtner* -88 (WRECKED)	W 3mSr	Norweg'n	250	116·4	27·1	13·1	Stavanger	1873	R. & T. Ras-[mussen	V 2.88	J. E. Gjemre	Stavanger
1519	R	Stacy Clark *Whitney*	W Bn	American	374	136·0	30·2	12·0	Waldoboro' Me	1879	E. O. Clark	V 8.86	E. O. Clark	New York
1520	D	Stad Steenwijk *Leijer* (ex Emma Laura) 85-83	W 3mSr	Dutch	294	117·4	27·6	13·0	Nova Scotia	1863			S. Tromp Meesters	Steenwijk
1521	V	Stadacona *Falker* -82	W Bk	British	1011	217·7	39·4	19·7	Parrsboro', NS	1885	G. E. Pettis	Y 10.85	Allen Haley	Parrsboro', N.S.
1522		Stadacona *Bergeson*	W Sr	British	132	85·1	24·0	10·4	Les Eboule-[ments	1875	M. Bergeron		Narcisse Blais	Quebec
1523		Stadsman *Mattson* (ex Statsman)	W Bk	Russian	345	120·0	26·6	16·0	Sunderland	1860	J. Robinson		M. Michelson	Lumparland
1524		Stadt *Leondreisen* (ex Weymouth)	W Bk	Norweg'n	690	152·3	33·7	18·9	Weymouth NS	1864	E. Everett		L. Pedersen	Farsund
1525		Stadt Frankfurt (BROKEN UP)	W Bk	French	507	123·6	27·9	19·7	Colberg	1859			Mathieu frères	Cherbourg
1526	DN	Stærk *Olsen* 80-82	W Sr	Norweg'n	152	90·3	23·1	10·7	Fünfhausen	1855			R. Olsen	Fredrikstad
1527	N	Stærkodder *Wright* -88	W Bg	Norweg'n	271	111·1	28·7	15·2	Sandefjord	1858	S. Andersen		W. Wright	Porsgrund
1528	RV	Staffa *Wry* -88	W Bk	British	921	167·0	34·0	22·2	Quebec	1861	Gingras	Y 10.86	G. Windram	Liverpool
1529		Stafford	W Bk	American	148	93·4	23·6	10·6	Kingston Mass	1848			J. & W. R. Wing	New Bedfrd Mass
1530	✻	Staffordshire *Meckison* 86-86	I Bk	British	1168	207·1	34·3	23·3	Liverpool	1862	Jones, [Quiggin & Co.		H. Fernie & Sons	Liverpool
1531		Stag	W Bk	British	258	124·8	23·5	15·0	Leith	1856			James Martin, jun.	Melbourne
1532		Stag *Burniston* 85-85	W Sr	British	209	111·4	24·2	12·9	Dartmouth	1871	Phillips	V 3.87	Whitstable Shipping Co. Ld.	Faversham
1533		Stagno *Antola* (ex Fabe...) (WRECKED)	W Bk	Italian	450	129·6	30·2	19·4	Lavagna	1871	D. Briasco	V 3.87	Giovanni Stagno	Genoa
1534	✻	Stagshaw *Crozier* 88-88	W Sw	British	229	98·3	25·5	16·0	Sunderland	1859	T. Stonehouse	Z 12.84	T. McVeagh, junr.	Belfast
1535	V	Stalwart *Cann* 78-85	W S	British	1545	216·0	40·0	22·8	Barton, N.S.	1885	J. Bingay	V 7.87	J. Bingay	Yarmouth, N.S.
1536		Stamatios Sevastopoulos *Vergotes*	W Sr	Greek	288	116·8	23·7	14·4	Syra	1875		V 6.83	G. Sevastopoulos	Syra
1537		Stambool	W Bg	Turkish	250	..	..	..	Sinope	1871			Mehemed Ali & Co.	Sinope
1538		Stambool	W Bg	Turkish	190	..	..	..	Barten	1870			Oon Cappanlee Tijar Veli Agha & Co.	Oon Cappan
1539		Stambool	W Bg	Turkish		..	..	..	Austria	1873			Tutunji Yordan & Co	Constantinople
1540	V	Stamboul *Cann* 76-81	W S	British	1248	193·5	38·1	22·3	Pt. Gilbert, N.S	1875	J. Bingay	V 4.86	Lovitt & Co.	Yarmouth, N.S.
1541		Stamboul	W Bk	American	247	106·2	25·0	14·5	New Bedford, [Mass.	1843			Wright, Browne & Co.	San Francisco
1542		Standard *Christie* 83-83	W Bn	British	197	101·4	23·6	12·7	Montrose	1868	Petrie	V 1.82	J. Mitchell	Montrose
1543	✻	Standard *Jeanes* 77-	W K	British	60	70·5	18·7	8·7	Jersey	1877	Picot		Mrs. E. A. Wright	Jersey
1544	RV	Standard *Perry*	W S	American	1493	212·0	40·2	24·3	Phippsburgh, [Me.	1878	C. V. Minott	Y 10.87	C. V. Minott	Bath, Maine
1545	R	Standard *Oram*	W 3mSr	American	331	125·0	32·0	10·0	Newcastle, Me.	1882	Haggett		A. Hall	Damariscotta, Me.
1546	V	Standard *Garlichs* 84-86 (ex Queenstown)	W S	German	1471	199·3	39·1	26·1	Richmond, Me.	1876	J. M. Hagar	V 5.88	D. Cordes & Co.	Bremen
1547	✻	Stanfield *Clark* 89-89	W Bk	British	570	153·1	30·7	18·8	Sunderland	1869	W. Adamson	V 8.88	J. Adamson	London
1548	V	Stanislas Emmanuel *Loundel*	W Bg	French	209	92·5	23·6	12·9	Nantes	1865	Boju & Clergeau	Z 8.85	L. Avril	Nantes
1549		Stanley *Macdonald* -86	W Bn	British	344	135·8	30·7	12·9	Parrsboro'	1882	J. Balconi		D. Ross	Auckland, N.Z.

[illegible] = Lloyd's Register; D = Nederlandsche Vereeniging; G = Germanischer Lloyd; I = Registro Italiano; R = American Record; [illegible] = Veritas Austro-Ungarico; H = Veritas Ellenico; N = Norske Veritas; V = Bureau Veritas.

Dans la 4me Colonne, C = Composite (membrure en fer, bordé en bois); I = Fer; Stl = Acier; W = Bois.

[illegible] Dates du doublage en métal (Col. 11) ressortent des livres dans lesquels les navires sont respectivement classés ou des rapports des Inspecteurs du Lloyd's Register.

Pour plus amples renseignements sur les navires classés par le Lloyd's Register, voir le Lloyd's Register Book et le Lloyd's Yacht Register.

Official No.	No. in Book	Registry in which classed	Ship's Name. Material and Rig. Master, and Dates of Appointment to present Owner's Service and to Ship. Ship's Late Name.	Flag	Net Tonnage	Length	Breadth	Depth	Where	When	By whom	Date of Metalling	Owners or Managers	Port of [Registry]
63404	1551		Stanley WSr	British	113	90·6	20·5	10·0	Auckland	1866			David Kirkwood	Auckland
66835	1552	✳	Stanley *Peck* [illegible] WSr	British	110	85·3	21·5	10·4	Bosham	1879	Apps		W.H.Barnaby	Yarmouth
	1553	DNV	Stanley *Rasmussen* -87 WBk	Norweg'n	663	161·0	34·1	19·0	Bergen	1880	Jens Gran	Y [illegible]	Rasmus F.Olsen	Bergen
	1554		Stanley (ex [illegible]) WBk	Norweg'n	541	159·5	31·8	18·0	Nantes	1856	G.Goun		J.Spörck	Fred[rikstad]
	1555	N	Stanley *Johnsen* -87 WBk	Norweg'n	481	138·4	30·5	16·9	Risör	1875	O.Schjeldal	Y [illegible]	J.Klöcker	Arendal
	1556	N	Stanley *Olsen* -87 WSnSr	Norweg'n	303	138·3	30·7	11·0	Porsgrund	1878	J.Jeremiassen		F.K.Flörenes	Porsgrund
	1557	N	Stanley *Thorbjörnsen* WSnSr	Norweg'n	300	123·3	28·2	13·2	Risör	1879	B.Tellefsen	Y [illegible]	C.H.Mollach	Risör
	1558		Stanley *Abrahamsen* -88 WGaliot	Norweg'n	121	95·9	24·1	10·2	Skonevig	1888			K.F.Langfelt	Christia[nia]
53486	1559	✳D	Stanley Sleath (ex Gaston) *Austin* [17-88] I Bk	British	763	169·1	33·0	20·2	Bordeaux	1865	Kichou		B.M.Hennessy	Swans[ea]
	1560	✳	Stanmore *Bruckenshaw* 86-88 I Bk	British	1768	268·6	38·2	23·3	Belfast	1886	Harland& Wolff		W.J.MyersSon&Co.	Liverp[ool]
55030	1561	✳	Star (ex Estrella) *Lloyd* WBk	British	530	149·7	29·9	19·0	Bridport	1866	Cox	Y [illegible]	J.Lloyd	Liverp[ool]
12704	1562		Star WSw	British	205	92·9	23·8	14·4	Sunderland	1851			T.W.Spencer	Lond[on]
97381	1563		Star I Sr	British	195	103·7	21·2	11·7	Liverpool	1885	R&J Evans & Co		Mersey Docks and Harbour Board	Liverp[ool]
22116	1564		Star WSr	American	130	86·0	25·0	9·0	Bristol,R.I.	1851			H.T.Carmen	Deer Is[le]
51300	1565		Star Queen *Abbott* WBk	British	264	117·4	25·3	14·1	Kingsbridge	1866			Barrington C. Wood	Freman[tle]
93152	1566	✳	Star of Austria *Willis* [11-86] Stl S	British	1693	261·7	38·7	23·1	Belfast	1886	Workman, Clark & Co		J.P.Corry & Co.	Belfast
63957	1567	✳D	Star of Bengal *Legg* I S	British	1797	262·8	40·2	23·5	Belfast	1874	Harland& Wolff		J.P.Corry & Co.	Belfast
44440	1568	✳	Star of China *M. Donald* 87-88 WBk	British	794	183·0	32·0	20·2	Aberdeen	1862	Hall	Y [illegible]	G.Milne & Co.	Aberdeen
47182	1569	✳	Star of Denmark *M. Williams* 87-87 I Bk	British	998	208·7	32·1	21·9	Belfast	1863	Harland& Wolff		J.P.Corry & Co.	Belfast
	1570		Star of Devon *Lovett* WBk	Hawaian	340	129·0	26·2	16·2	Plymouth	1864	Shilston	Y [illegible]	P.Buttler	Honolu[lu]
	1571	R	StarofEngland *Hendriksen* WS (ex E. Creighton) (CONDEMNED)	Norweg'n	1544	196·6	38·8	27·3	Thomaston, Me.	1860		Y [illegible]	H.Henriksen	Porsgrund
45123	1572	✳	Star of Erin *Smyth* 85-85 I Bk	British	949	200·0	32·0	21·9	Belfast	1862	Harland& Wolff		J.P.Corry & Co.	Belfast
78105	1573	✳	Star of France *Wilson* 84-82 I S	British	1569	258·0	38·0	22·8	Belfast	1877	Harland& Wolff		J.P.Corry & Co.	Belfast
63340	1574	✳	Star of Germany *Reed* 83-85 I S	British	1284	232·0	35·0	22·0	Belfast	1872	Harland& Wolff		J.P.Corry & Co.	Belfast
60313	1575	✳	Star ofGreece *Harrower* -87 I S (WRECKED)	British	1227	227·0	35·0	22·2	Belfast	1868	Harland& Wolff		J.P.Corry & Co.	Belfast
43925	1576	✳	Star of India *Mickelsen* WBk	Norweg'n	1040	190·1	34·2	22·1	Dundee	1861	Stephen	Y [illegible]	Vetteseiskab (H.C.A. Michelsen & Co.)	Sandefj[ord]
78102	1577	✳	Star of Italy *Cotter* 86-87 I S	British	1571	257·1	38·0	22·8	Belfast	1877	Harland& Wolff		J.P.Corry & Co.	Belfast
6867	1578		Star of Peace *Webster* WBk	British	1114	215·2	35·9	22·5	Aberdeen	1855	W.Hood&Co.	Y [illegible]	Burns,Philp & Co.	Sydney
60312	1579	✳	Star of Persia *Russell* 87-85 I S	British	1227	227·0	35·0	22·2	Belfast	1868	Harland& Wolff		J.P.Corry & Co.	Belfast
63958	1580	✳	Star of Russia *Legg* 76-87 I S	British	1892	275·5	40·2	24·2	Belfast	1874	Harland& Wolff		J.P.Corry & Co.	Belfast
74629	1581	✳	StaroftheEast *Harrow* -84 I Bk	British	734	183·3	30·7	18·5	Glasgow	1876	C.Connell&Co.		Sir R.W.Cameron	London
11207	1582		Star of the North *Carlton* 73-73 WBn	British	164	111·4	22·4	11·9	Bathurst, N.B.	1853			F.W.Fairbrass	Faversham
44912	1583		Star of the Ocean *Clothier* [illegible] WBn	British	176	100·1	22·6	13·0	Southampton	1863	Langley		F.W.Fairbrass	Faversham
34161	1584		Star of the Sea WBg	British	127	93·1	22·0	11·6	Heart'sCont'nt [N.F.L.	1858			James Goodfellow	St.John's
19157	1585		Star of the Sea *Ely* 84-84 WBn	British	103	84·9	21·3	11·0	St.John,N.B.	1856			F.A.Christie	Ipswich
58583	1586		Star of the West *Sanders* WBn	British	199	109·7	25·5	12·1	Guernsey	1869	Domaille		R.J.Dorey	Guernsey
51137	1587		Starbeam *Whiten* 8-86 WBg	British	228	99·7	26·3	15·5	Sunderland	1864	Pearson		J.P.Mountain	Roster
76501	1588	✳	Starlight *Roberts* WSr	British	82	80·8	20·8	9·1	Bridgwater	1877	Gough		F.Prince	Liverp[ool]
115280	1589		Starlight *Sears* WBn	American	243	116·0	29·0	11·1	E.Haven,Conn	1874			L.W.&P.Armstrong	New Y[ork]
27589	1590		Starling *Best* WSr	British	112	76·0	19·5	10·3	Barton	1859	Wray		J.W.Craven	Goole
58131	1591		Startled Fawn WK	British	118	100·0	23·2	8·0	Brightlingsea	1860			Mrs.A.Richmond	Harwich
	1592	V	Stassfurt (ex Edinburgh) *Vogelsang* -87 WS	German	1572	212·0	40·0	25·0	Quebec	1872	Gingras	[illegible]	Rob.S.Carr	Hamburg
115632	1593	V	State of Maine *Hilyard* -87 WS	American	1146	216·0	40·1	24·4	DamariscottaMe	1878	Haggett&Co	Y [illegible]	E.D.T.Nickels	[illegible]
481	1594		Stathopoulos WSr	Greek	118	86·9	21·4	..	Syra	1884			C.N.Papalas	Syra
	1595	N	Statsminister Selmer *Husted* WBk	Norweg'n	607	152·4	32·5	16·6	Arendal	1883	J.Halvorsen	[illegible]	Arendal's Söforsikrings-selskab	Arendal
	1596	N	Statsminister Stang *Fairbenberg* 85-88 WBk	Norweg'n	425	133·9	30·2	16·3	Kragerö	1873	T.G.Litangen	[illegible]	W.Hopp	Kragerö
	1597	N	StatsraadBroch *Jacobsen* WBk	Norweg'n	312	117·2	28·5	14·2	Svelvig	1869		Y [illegible]	Winge & Co.	Chris[tiania]
	1598	N	Staubo *Christophersen* WBk	Norweg'n	321	120·1	28·0	14·9	Tvedestrand	1873	J.P.Pedersen.	Y [illegible]	J.A.Pedersen	Tv[edestrand]
	1599	V	Staut *Olsen* WBk	Norweg'n	581	155·1	32·8	17·2	Sarpsborg	1876	Olsen	Y [illegible]	H.H.Pettersen	[illegible]
	1600	DNV	Stavanger *Berg* -89 WBk	Norweg'n	743	179·5	35·4	18·8	Bergen	1882	Brancker & [illegible]	[illegible]	K.S.Bertelsen	[illegible]

✳=Lloyd's Register; D=Nederlandsche Vereeniging; G=Germanischer Lloyd; I=Registro Italiano; K=American [Record]; A=Veritas Austro-Ungarico; H=Veritas Ellenico; N=Norske Veritas; V=Bureau Veritas.

In the 4th Column, C=Composite (iron frame, wood planked); I=Iron; Stl=Steel; W=Wood.

The Dates of Metalling (Col. 11) are based on the records of the Books in which Vessels are respectively classed, or on the Reports of the Surveyors of Lloyd's [Register].

Further particulars of Vessels classed by Lloyd's Register will be found in Lloyd's Register Book and Lloyd's Yacht Register.

		Nom. Matériaux et Gréement. Capitaine et dates d'entrée au service des Armateurs actuels et à bord du navire. Ancien Nom.	Pavillon.	Tonneaux net.	Dimensions. Longueur.	Largeur.	Creux.	Construit. Où.	Quand.	Par qui.	Date de doublage en métal	Armateurs ou Directeurs-Gérants.	Port d'Armement.
1604	DV	Stavanger [illegible] W Bk	Swedish	939	183·4	33·4	20·6	Sikea	1880	O.P. Åberg	[illegible]	Fr. Jacobsen	Robertsfors
1605		Stavros W Sr	Greek	203	95·4	24·6	..	Chios	1874			I. Stathopoulos	Syra
1606	✻	Steelfield *Robbins 83-85* I Bk	British	1251	232·0	35·1	21·6	Belfast	1876	Harland & Wolff		W.H.Ross & Co.	Liverpool
1607		Stefano *Mazini 87* W Bg	Italian	315	112·8	27·2	17·1	Chiavari	1870	L. Gotuzzo	[illegible]	F[lli] Sanguineti fu G.	Spezia
1608		Stefano Kovasebitz *Angeletto* W Bk	Greek	414	128·9	29·5	..	Fiume	1872		[illegible]	P. Lascaridis	Piræus
1609	I	Stefano Padre *Olivari* W 3m Sr	Italian	407	125·6	29·2	18·3	Savona	1872	G. Calamaro	[illegible]	P. Cavassa	Genoa
1610	I	Stefano Repetto W Bk	Italian	610	149·2	31·9	20·3	Alimari	1875	G.M. Mauro	[illegible]	G.B. Repetto fu A.	Genoa
1611	✻	Steinvora *McKenzie* W S	British	1107	187·5	37·5	22·4	New Glasgow, N.S.	1875	J.W. Carmichael	[illegible]	James W. Carmichael	Pictou, N.S.
1612	✻	Stella *Nisbet 76-84* W 3m Bn	British	221	121·7	25·8	13·0	Ayr	1872	Weir	[illegible]	J. & W. Stewart	Greenock
1613	R	Stella *Reich* W Bn	American	477	135·0	34·0	17·0	Setauket, N.Y.	1874	W. Bacon	[illegible]	E.M. Davis	New York
1614	V	Stella *Ancoignard* W Bg	French	168	86·9	22·0	11·5	Nantes	1873	Tilh	[illegible]	J.J. Ancoignard	Granville
1615	V	Stella *Leleitu* W Sr	French	129	90·9	24·3	10·0	Barrington N.S.	1877		[illegible]	Legasse & Vidart	Bordeaux
1616		Stella W Sr	French	115	86·0	19·3	10·9	Dunkirk	1852	Pyotte Bayaert		Cauzic	Vannes
1617	GV	Stella *Hagemann 87* W S	German	1193	187·0	37·1	24·1	Newburyport, Mass.	1861	E. Marson	[illegible]	Joh. Fr. Arens	Bremen
1618	V	Stella *Rosenkranz 82-84* W Bk	German	501	134·8	33·0	16·7	Wolfville, C.S.	1870	J.F. Harris	[illegible]	W. Maack	Rostock
1619	V	Stella *Wiese 86-86* W Bn	German	194	97·7	21·0	11·2	Hamm'lwarden	1877	K. Gross	[illegible]	K. Gross	Brake
1620	R	Stella *Pedersen* W Bk	Norweg'n	655	153·0	32·8	18·7	Dorchester, N.B.	1872	Palmer	[illegible]	Eyde & Hein	Arendal
1621	N	Stella *Jacobsen* W Bk	Norweg'n	406	138·9	29·7	15·1	Drammen	1869		[illegible]	L. Föyen	Tönsberg
1622	N	Stella *Bugge* W Bk	Norweg'n	336	123·4	29·4	15·7	Hernösand	1853			G.C. Hellesen's Enke	Drammen
1623	I	Stella B. *Mascarini 84* W Bk	Italian	861	164·0	24·6	23·0	Chiavari	1876	M. Tappani	[illegible]	G. Bertollo fu G.B.	Genoa
1624	R	Stella M. Kenyon *Fendleton 89* W 3m Sr	American	356	143·3	37·7	12·2	Belfast, Maine	1888	Brown & McDonald		J. Lennon	Belfast
1625	DV	Stella Maris *Greenwald* W Bn	Dutch	177	98·1	24·7	12·0	Oude Pekela	1875		[illegible]	B.B. Drenth	Oude Pekela
1626	V	Stella Maris *Steinmintz 84-84* W Bk	Swedish	569	145·0	30·4	17·2	Bygdeå	1869	O.P. Åberg	[illegible]	G.R. Lagergren	Stockholm
1627	I	Stella Mattutina *Lavagna* W Sr	Italian	304	124·3	26·7	16·0	Sestri	1876	A. Cardevaro	[illegible]	V. Lavagna	Savona
1628	I	Stella Sorrento *Gargiulo* W Bk	Italian	457	134·5	31·2	19·7	Sestri	1868	Tixi	[illegible]	F.S. Gargiulo fu A.	Castell. di Stab.
1629	V	Stella d'Italia T. *Tonietti 87* W Bk	Italian	251	120·7	24·1	14·4	Meta	1865	F.S. Mauro	[illegible]	G. & E. Tonietti fu L.	Porto Ferraio
1630	V	Stephan *Schirtle 73-88* W S	German	1267	199·1	37·4	24·2	Newburyport	1873	J. Currier	[illegible]	D. Cordes & Co.	Bremen
1631		Stephan W Bg	Turkish	237	..	..	..	Oonia	1871			ChebookjiOgluGarabet	Turkey
1632	GV	Stephanie (CONDEMNED) W Bk		302	127·5	26·3	13·5	Vegesack	1863	P. Sager	[illegible]		
1633		Stephen Bennett *Hodgdon 89* W 3m Sr	American	231	127·0	30·0	9·0	Rockport, Me.	1872		[illegible]	T.B. Hodgman	New York
1634	R	Stephen Bishop *Rivers* W Bn	American	281	108·5	28·1	16·0	N. Haven Conn	1867		[illegible]	C.C. Rivers	Baltimore, Md.
1635	V	Stephen D. Horton W S	British	1626	231·7	42·0	23·8	Parrsboro' NS.	1883	Spencers Isld. Co	[illegible]	Johnson Spicer	Parrsboro', N.S.
1636	RV	Stephen G. Hart W Bk	American	589	160·0	33·9	16·2	Waldob'ro' Me	1879	Welt & Flanders	[illegible]	S.G. Hart	St. George, Me.
1637		Stephen G. Hart W 3m Sr	American	243	119·0	29·0	10·0	Waldob'ro' Me	1871			S.G. Hart	St. George, Me.
1638	R	Stephen G. Loud *Matthews 88* W 3m Sr	American	475	144·0	34·0	12·0	Waldob'ro' Me	1882	W. Fisk	[illegible]	H.W. Loud	New York
1639		Stephen Hart W Sr	British	220	107·6	28·3	7·6	Greenport, L.I.	1859			J.M. Harris	Liverpool
1640		Stephen J. Fooks *Henderson* W 3m Sr	American	409	135·4	35·0	11·6	Solomon's I., Md	1874			T.C. Craumer	Baltimore
1641		Stephen J. Watts *Kelley* W Sr	American	100	84·0	25·8	7·3	Jonesport, Me.	1873			W.F. Mansfield	Machias, Me.
1642		Stephen Morgan *McLaffey* W 3m Sr	American	339	134·8	30·4	8·9	Noank, Conn.	1864			S.W. Walker	Boston, Mass.
1643		Stephen Morris *Dutch* W 3m Sr	American	245	112·4	29·8	10·0	Philadelphia	1866			J.H. Allen	Philadelphia
1644		Stephen & Sarah *Fo[illegible]* W Bn	British	178	96·5	22·0	13·4	Rye	1854	Hassell		W. Kemp	Faversham
1645	G	Stephenson *de Buhr* W Bg	German	175	98·2	21·5	12·1	Papenburg	1859	F.W. Meyer		Heinr. de Buhr	[illegible]
1646		Sterling *Brooks* W Sr	British	169	99·4	23·6	12·6	Brixham	1863	Dewdney		J. Johns	Fowey
1647		Sterling *Goodwin 81-84* W S	American	1663	208·4	42·7	26·1	Bath, Maine	1873	A. Sewall & Co.	[illegible]	J. Rosenfeld's Sons	San Francisco
1648	DNV	Sterling *Bue 77-77* W Bk	Norweg'n	783	161·9	32·8	20·0	Quebec	1876	P. Baldwin	[illegible]	Ths. S. Falck	Stavanger
1649		Sterna *Jensen 82-82* W Bg	Norweg'n	288	107·2	27·7	15·3	Arendal	1856			A. Arvesen	Arendal
1650	A	Stilicone *Franich 86* W Bg	Austrian	498	131·5	29·2	19·1	Fiume	1868	G. Brazzoduro	[illegible]	C. cav. Cosulich	Fiume
1651	DV	Still Water *Fritts* W Bk	British	1090	186·4	37·4	22·5	Portland, N.B.	1879	D. Lynch	[illegible]	Howard D. Troop	St. John, N.B.
1652	V	Stine *Madsen 86-86* W Sr	Danish	103	75·5	20·6	10·8	Bandholm	1853	H.J. Møller		C.P. Clemmensen	Svendborg

=Lloyd's Register; D=Nederlandsche Vereeniging; G=Germanischer Lloyd; I=Registro Italiano; R=American Record; =Veritas Austro-Ungarico; H=Veritas Ellenico; N=Norske Veritas; V=Bureau Veritas.

Dans la 4me Colonne, C=Composite (membrure en fer, bordé en bois); I=Fer; Stl=Acier; W=Bois.

Dates de doublage en métal (Col. 11) ressortent des livres dans lesquels les navires sont respectivement classés ou des rapports des Inspecteurs du Lloyd's Register.

Pour plus amples renseignements sur les navires classés par le Lloyd's Register, voir le Lloyd's Register Book et le Lloyd's Yacht Register.

Official No.	No. in Book	Registry in which classed	Ship's Name. Master, and Dates of Appointment to present Owner's Service and to Ship. Ship's Late Name.	Material and Rig	Flag	Net Tonnage	Length	Breadth	Depth	Where built	When	By whom	Date of Metalling	Owners or Managers	Port
76[illegible]	1651	✻	Stirlingshire *[illegible]*	I S	British	1221	230·8	35·8	20·5	Dumbarton	1877	Birrell, Stenhouse & Co.		T. Law & Co.	Glasg[illegible]
	1652	N	Stjernen *Borgen –88*	W Bk	Norweg'n	822	118·4	28·4	14·9	Drammen	1852			W. M. Michaelsen	[illegible]
	1653	N	Stjernen *Jensen 85-86*	W Sr	Norweg'n	188	101·9	25·6	11·7	Nova Scotia	1854			H. A. Lohre	[illegible]
	1654		Stjernen (WRECKED) *Berentsen*	W Sr	Norweg'n	121	[illegible]	22·8	10·5	Stavanger	1876	F. Imsland		J. Einarsen	[illegible]
87[illegible]	1655	✻	Stockbridge *Bolger 85-85*	I S	British	2348	284·0	41·2	24·6	Port Glasgow	1883	W Hamilton & Co.		Ship Stockbridge Co. Ld. ([illegible])	Liv[illegible]
[illegible]	1656	R	Stockton *Allen*	W Bn	American	394	126·6	27·6	17·1	Stockton, Me.	1863	Richborn	[illegible]	N. W. Bridge	[illegible]
45708	1657	✻ D	Stonehouse *[illegible] –86*	W Sr	British	96	92·5	20·5	10·2	Teignmouth	1863	Mansfield	[illegible]	S. Truscott	Fowey
	1658	N	Stord *Carlsen*	W Bg	Norweg'n	23[illegible]	113·4	25·8	15·[illegible]	[illegible]	1866	S. Bjerkelund		C. Knudsen	[illegible]
	1659	V	Storfursten *Engblom –88*	W Bk	Russian	58[illegible]	117·9	31·2	16·[illegible]	[illegible]	1879	E. P. Kjaldström	[illegible]	V. A. Reims	[illegible]
	1660	V	Storken *Høglund –88*	W Bk	Swedish	517	149·9	27·5	17·5	[illegible]	1880	F. O. Bergström	[illegible]	Rederi-Aktiebol Ornen (J. F. Strandberg)	[illegible]
[illegible]	1661		Storm *Skibb 85-*	W Sr	American	11[illegible]	99·6	17·6	8·4	Hamburg, Pa.	1864			J. A. Anderson	[illegible]
	1662		Storm Bird	W Bn	Chinese	16[illegible]	88·6	26·0	10·6	Spry Bay, N.S.	1875			A. Crocker	[illegible]
41[illegible]	1663		Storm Nymph *M'[illegible]*	W Bn	British	143	100·0	26·0	11·8	Liverpool, N.S.	1858			W. Hinde	Belfas[illegible]
72319	1664	V	Stormy Petrel *[illegible]*	W Bk	British	748	167·7	35·1	18·6	Portland, N.B.	1877	D. Lynch	11·86	Wm. Thomson & Co.	St. Joh[illegible]
56[illegible]	1665	✻	Stormy Petrel (ex Lady of the Lake) *Reed*	I Bk	British	497	158·0	26·1	17·2	Lancaster	1867	Lanc. Ship Co.		J. T. Read	Liv[illegible]
75187	1666	✻	Stracathro *Smith 86-86*	I Bk	British	1159	230·6	35·0	21·1	Dundee	1876	Gourlay Bros.		D. Bruce & Co.	Dund[illegible]
	1667	I	Stradella *Arcidiacono*	W Bn	Italian	117	94·2	22·3	9·6	Sestri	1873	[illegible] Fasco fu G. B.	[illegible]	G. Puglisi fu A.	Catan[illegible]
	1668	G	Stralsund *Oelberg 76-74*	W Bk	German	581	140·0	30·7	19·7	Stralsund	1874	O. Kirchhoff		C. Lobeck	Stra[illegible]
[illegible]	1669		Stranger	W Sr	American	11[illegible]	93·0	31·8	6·1	San Francisco	1869			J. Deming	San F[illegible]
	1670	G	Stranger (ex Helena) *Limberg –78*	W Bg	Russian	25[illegible]	113·0	26·6	13·9	Lokola	1872			Graf E. Ungern Sternb'rg	Dag[illegible]
	1671	D N	Stratford (ex Arctic) *Anke –87*	W S	Norweg'n	1120	176·2	38·8	23·0	Willmsbg N Y	1850		[illegible]	Et Actieselskab A. O. (Olsen)	Stav[illegible]
6[illegible]	1672	✻	Strathblane *Cuthbert 86-86*	I S	British	1364	235·4	37·4	22·5	Glasgow	1868	Barclay, Curle & Co.		J. & A. Allan	Glasg[illegible]
[illegible]	1673	✻	Strathearn *Willis –86*	I S	British	1705	255·3	40·4	24·0	Glasgow	1871	Barclay, Curle & Co.		J. & A. Allan	Glasg[illegible]
73391	1674	✻	Strathearn *Anderson 85-85*	I Bk	British	1101	220·6	34·7	21·5	Dundee	1875	Gourlay Bros.		W. Thomson & Sons	Dund[illegible]
6[illegible]	1675	✻	Strathearn *Tomlinson 85–8*	W Bk	British	247	117·3	25·1	12·8	Sunderland	1871	J. Crown	[illegible]	T. Harrison & Co.	Liv[illegible]
[illegible]	1676	✻	Strathendrick *[illegible] –86*	W Bn S[illegible]	British	18[illegible]	113·0	22·3	12·0	Troon	1882	Troon Sh. B. Co.	[illegible]	E. Bindon	Liv[illegible]
9[illegible]	1677	✻	Strathmuir *Fleming 75–85*	W Bk	British	1175	198·6	38·2	22·0	Maitland, N.S.	1885	A. Putnam	[illegible]	Alfred Putnam	Mait[illegible]
83277	1678	V	Strathome *Urquhart –83*	W Bk	British	1098	193·5	37·5	22·2	Maitland, N.S.	1882	A. Putnam	[illegible]	Alfred Putnam	Mait[illegible]
[illegible]	1679	✻	Strathspey *McKenzie 79-86*	Stl Bk	British	319	116·8	26·2	12·6	Grangemouth	1886	Grangemouth Dockyard Co.		Anderson, Webster & Co.	[illegible]
73127	1680	D V	Strathtay *Urquhart*	W Bk	British	1024	182·5	37·3	22·5	Maitland, N.S.	1881	A. Putnam	[illegible]	Alfred Putnam	Tr[illegible]
	1681	A	Stratigos Favieros *Zaimon*	W Bn	Greek	321	109·5	..	..	Scopelos	1874	K. Plakavis	[illegible]	P. Siskos	Scop[illegible]
	1682	N V	Strauss *Gunnufsen –84*	W Bk	Norweg'n	835	168·0	35·5	19·1	Arendal	1885	A. Aanonsen	[illegible]	Thv. B. Olsen	[illegible]
[illegible]	1683		Stree Amothalechmy	W Bg	British	124	103·0	21·2	11·1	Cochin	1876			Kathiryveli Ganderelle	Jaffn[illegible]
[illegible]	1684		Stree Vencatayswerloo	W Bk	British	481	115·7	28·3	7·0	Coringa	1868			Jovnidiram Jayagopaul Bros.	Madr[illegible]
	1685	N	Strilen *Olsvik*	W Sr	Norweg'n	13[illegible]	91·5	25·2	10·8	Bergen	1882			K. Kleppe Olsirk	Berge[illegible]
	1686		Stromstad *Hofgren*	W Bk	Swedish	210	108·1	24·8	12·1	Nova Scotia	1856			L. O. Asberg	Stromstad
[illegible]	1687	✻	Stronsa *Brooks 82–2*	I S	British	2009	269·0	41·6	24·3	Liverpool	1882	T. Royden & Sons		Sandbach, Tinne & Co.	Liv[illegible]
	1688		Strossmayer *Perlainovich*	W Bk	Russian	261	97·8	23·0	13·3	Baccari	1869	M. Matergian	[illegible]	G. Cambosetli	[illegible]
72[illegible]	1689	V	Struan *Scott*	W S	British	1473	211·8	39·3	24·2	St. John, N.B.	1877	J. Fraser & Son	[illegible]	A. Gilson & Co.	Liv[illegible]
76[illegible]	1690	✻	Stuart *Williams 88-88*	I Bk	British	881	202·5	34·2	19·1	Dundee	1877	A Stephen & Sons		Doward, Dickson & Co.	Liv[illegible]
[illegible]	1691	A H	Stylianos Chrysakis	W Bn	Greek	424	121·3	27·8	11·7	[illegible]	1875		[illegible]	J. Eliadis	Syra
	1692	V	Suaheli *Jacobs 84–1*	W Bk	German	36[illegible]	135·8	28·3	15·3	Apenrade	1872	N. Jacobsen		Hansing & Co.	Ham[illegible]
9[illegible]	1693	✻	Suakim *Seabrook 78-86*	I S	British	1618	259·0	38·2	23·0	Port Glasgow	1886	Russell & Co.	[illegible]	Ship Suakim Co. Ld (G. M. Steeves)	Liv[illegible]
[illegible]	1694		Subansiri	I Lr	British	547	220·0	32·0	9·5	Calcutta	1888	India Genl Steam Nav. Co. Ld.		Irrawaddy Flotilla Co. Ld	[illegible]
	1695		Subiminneh	W Sr	Turkish	136	..	..	..	Barten	1871			Bakkal Stephan Yaan[illegible]	[illegible]
	1696	N V	Subra *Knudsen –*	W Bk	Norweg'n	562	146·4	31·7	18·4	Arendal	1875		[illegible]	T. Thommessen & Son	Aren[illegible]
[illegible]	1697		Success *Benger*	W Bn	British	148	93·7	23·7	10·9	St. John, N.B.	1858	S. Smith		H. D. Dennis	[illegible]
[illegible]	1698	R	Success (LOST) *King*	W S	American	1080	184·0	37·5	21·5	Damariscotta, Me.	1857		[illegible]	I. E. Stafford	New [illegible]
[illegible]	1699		Success *Matthews*	W Sr	American	12[illegible]	92·3	23·9	7·2	Wilmington, Del.	1854			B. T. Thurlow	[illegible]
	1700	N	Success *[illegible]*	W Bk	Norweg'n	31[illegible]	118·9	28·7	15·7	Grimstad	1873	S. Johnsen	[illegible]	J. D. Havstad	Grim[illegible]

✻ = Lloyd's Register; D = Nederlandsche Vereeniging; G = Germanischer Lloyd; I = Registro Italiano; R = American [illegible]
A = Veritas Austro-Ungarico; H = Veritas Ellenico; N = Norske Veritas; V = Bureau [illegible]

In the 4th Column, C = Composite (iron frame, wood planked); I = Iron; Stl = Steel; W = Wood.

The Dates of Metalling (Col. 11) are based on the records of the Books in which Vessels are respectively classed, or on the Reports of the Surveyors of Lloyd's [illegible]

Further particulars of Vessels classed by Lloyd's Register will be found in Lloyd's Register Book and Lloyd's Yacht Register.

1	2	3 Nom. Matériaux et Gréement. Capitaine et dates d'entrée au service des Armateurs actuels et à bord du navire. Ancien Nom.	4 Pavillon.	5 Tonnage net.	6 Dimensions. Longueur.	7 Largeur.	Creux.	8 Construit. Où.	9 Quand.	10 Par qui.	11 Date de doublage	12 Armateurs ou Directeurs-Gérants.	13 Port d'Armement.
1701	N	Success *Larsen -87* W S	Norweg'n	305	121·1	28·7	11·9	Arendal	1886	S. Iterdie Vaer	Y.s.s.	T.S.Larsen	Lillesand
1702		Success *Christensen* W Bg	Norweg'n	217	86·7	26·4	14·8	Arendal	1848			Chr.Taraldsen	Arendal
1703	✻	Sudbourn *Dean 80-85* I S	British	1703	251·0	39·0	21·0	Stockton	1884	Richardson, Duck&Co.		C.W.Kellock & Co.	Liverpool
1704	AV	Sudore *Antunovich -88* W Sr	Austrian	286	97·4	25·4	13·6	Lussinpiccolo	1886	N.Martinolich	Y.s.s.	G.F.Tonietti fu G.	Lussinpiccolo
1705	R	Sue Williams *Pierce* W 3m Sr	American	63?	162·8	36·5	15·4	City Island NY	1883	D.Carll		E.L.Pierce	Richmond,Va.
1706		Suez *Bohme* W Bk	British	422	128·5	27·0	18·7	Nantes	1859			John Whitaker	Bombay
1707	A	Suez *Scarpa -88* W Bk	Italian	433	127·6	30·1	19·4	Savona	1866		Y.s.s.	G.Scarpa	Venice
1708	N	Suez *Larsen -82* W Bk	Norweg'n	348	123·5	30·2	14·8	Stavanger	1870	K.Nielsen	Y.s.s.	D.Davidson & H.Didriksen	Haugesund
1709	✻	Suffolk *Curd 87-87* W Bk	British	975	198·5	33·5	20·8	Northam	1857	Wigram	Y. 10.s.s.	A.E.Isaacks	Alleppey
1710	V	Suffren *Faunis 83-83* W Sr (ex Kuryet...)	French	133	97·1	20·0	11·5	Galmpton	1866	Calley		J.Le Pomellec	Binic
1711		Suleiman Bahri W Bg	Turkish	490	..	..	..	Oonia	1875			Iskenderiehlee Hjar Essaid Ali & Misirie	Iskenderich
1712	✻V	Sulitelma *McGregor 79-86* W Bk	British	964	187·9	36·3	21·0	N.Glasgow NS	1880	J.W.Carmichael	Y.s.s.	James W.Carmichael	Pictou,N.S.
1713	R	Sullivan *Dorick -88* W Bn	American	328	115·5	28·5	15·2	Gouldsboro Me	1866	G.W.Whitaker	Y.s.s.	J.S.Emery & Co.	Boston,Mass.
1714		Sullivan Sawin *Rich* W 3m Sr	American	643	147·9	33·9	16·6	Bath,Me.	1874	Jewell Bros.		E.H.Atwood	Boston,Mass.
1715	V	Sully *Chapel 85-85* W Bk	French	378	132·8	29·0	13·9	La Mailleraye	1869	E.Lefranc	Y.s.s.	H.Auger aîné	Havre
1716	DV	Sully (ex Mets Rostock...) W 3m Sr	French	348	128·6	27·8	13·0	Brake	1876	Lühmann&Wst.	Y. 11.s.s.	H.Monnier	Fécamp
1717	G	Sulo *Karlsson 79-* W Sr	Russian	246	113·4	28·9	11·9	Pyhäjmaa	1882	J.Palmros		D.Inppo	Pyhämaa
1718	V	Sultan *Armstrong* W S	British	1323	203·1	40·0	23·9	Kempt,N.S.	1879	D.Caldwell	Y.s.s.	George Armstrong	Windsor,N.S.
1719		Sultan *Newman 85-88* W Bn	British	226	104·0	25·5	13·0	Portsmouth	1868	White		J.T.Crampton	Portsmouth
1720		Sultan *Lewis* W Bn	British	121	76·2	20·7	12·5	P.E.Island	1856			G.Clyde,per J.M.Barkley	Belfast
1721		Sultan W Bg	Turkish	125	..	..	..	Terebollee	1876			Bigadoslee Yarrar Simon & Co.	Bigados
1722		Sultan Achmad W Sr (ex Nieuw Bevrin)	Dutch	191	95·5	22·0	13·0	Veendam	1866	K.J.Wilkens	Y.s.s.	C.W.R.van Renesse van Duyvenbode	Ternate
1723		Sultan Mahyed Deen W Bk (ex Spaniel)	British	374	125·4	27·6	16·9					Pana Ahmeed Jellalubdeen	Rangoon
1724		Sultan i Deriya *Hussein* W Bg	Turkish	242	..	..	..	Barten	1885			Rizehlee Chollak oglou Hussein & Co.	Rizeh
1725	V	Sultana *Mosher* W Bk	British	812	174·3	35·6	21·0	Kempt,N.S.	1877	D.Caldwell	Y.s.s.	Mark Curry	Windsor,N.S.
1726		Sultana *John -88* W Sr	British	138	80·0	21·3	13·0	Jersey	1840			Clement & Co.	Jersey
1727		Sultana W Bg	British	104	87·7	22·7	10·7	Trinity By.N.F.L.	1859			John Chapman	St.Johns,N.F.L.
1728	V	Sultana *de Almeida* W Bk (ex Aurora)	Portug'se	410	116·1	29·5	15·1	Villa Nova	18 74/85	C.M.da Silva Santos	Y.s.s.	F.A.de Lima & Co.	Oporto
1729	✻	Sultana Reina *Parry 87-87* W Bn	British	129	82·0	20·0	14·5	Yarmouth	1857			P.Jones & Co.	Carnarvon
1730		Sultanieh W Bg	Turkish	300	..	..	..	Dardanelles	1867			Bricklee Calliadi	Constantinople
1731		Sultanieh W Bg	Turkish	128	..	..	..	Dardanelles	1868			Focchialee Chakir Hassan oglou Ahmed & Co.	Focchia
1732		Sultany W Bk	British	351	117·3	29·2	15·0	Quilong	1880	J.Laurence		Choona Leena Kavenna Bebo...	Calcutta
1733		Sultão *Rangel* W Sr	Brazilian	128	..	..	..	S.João da Barra	18 86		Y.s.s.	Ca. Cabotagem de S.João da Barra & Campos	S.João da Barra
1734		Sulun Bahri W Sr	Turkish	125	..	..	..	Chattal Hagooz	1883			Hekakji oglou Hassan & Co.	Chattal Hagooz
1735	NV	Sumarlide *Wiig 76-84* W S	Norweg'n	943	179·1	35·9	20·8	Drammen	1877	Jörgensen& Knudsen	Y.s.s.	Ths.S.Falck	Stavanger
1736	✻	Sumatra *Rae 86-86* C Bk	British	740	171·4	32·2	18·7	Sunderland	1866	J.Laing	Y.s.s.	J.Willis & Son	London
1737		Sumatra *Nerrick 86-87* W S	American	1016	180·0	36·8	23·5	Chelsea,Mass.	1856	J.Taylor		J.Jensen	San Francisco
1738		Sunbeam *Clayburn* I Bk	British	443	141·5	27·1	16·8	Kirkcaldy	1857	J.Brown&Co.		J.Huddart	Melbourne
1739	✻	Sunbeam *Woolgar 78-79* W Bn	British	195	114·1	25·1	12·1	Bridport	1876	Cox	Y.s.s.	Edwin J.Duder	St.Johns,N.F.L.
1740	✻	Sunbeam *Potter* W 3m Sr	British	135	102·2	23·1	11·1	Runcorn	1883	Brundrit&Co.	Y.s.s.	J.Foulkes,junr.	Runcorn
1741		Sunbeam *Johnson* W Sr	British	113	93·2	20·3	11·8	Brixham	1868	Ferneaux		T.H.Colby	Bristol
1742	✻	Sunbeam *Trembath 83-84* W 3m Sr	British	123	99·8	23·0	11·6	Portreath	1878	Massey	Y.s.s.	J.P.Dunstan	Truro
1743	✻	Sunbeam *Sinclair 79-79* W Sr	British	94	80·7	20·6	9·8	Stromness	1878	Copland		G.Swanson	Wick
1744	✻	Sunbeam *Merch* W Barge	British	65	83·0	18·6	6·3	Strood	1886	G.H.Curel		J.D.Lee	Rochester
1745	V	Sunbeam *Morrel* W Bk	American	845	172·3	33·4	20·2	St.John,N.B.	1874	Rowan		J.B.Guimares	Philadelphia
1746		Sunbeam *Moulton* W Bk	American	255	106·4	27·3	15·2	Mattapoisett, Mass.	1856	J.Holmes		J.& W.R.Wing	N.Bedford,Mass.
1747	A	Sunce *ex Luigi. 83 Mercich* W Bk	Austrian	498	129·9	28·2	18·9	Fiume	1873	Flli. Schiavon	Y.s.s.	Teresa Ivetta	Trappano
1748		Sunderland Packet *Mellor 82-88* W Bkn	British	171	96·3	23·9	11·1	Sunderland	1856	Green		H.Smith	London
1749		Sundown W Sr	British	144	91·0	22·0	10·3	St.Anne de la Parade	1864			Charles Gariepy	Quebec
1750	V	Sundsvall *Olsson -88* W Bk	Swedish	315	128·6	25·7	12·1	Sundsvall	1871	J.V.Hägglund	Y.s.s.	J.Olsson	Hellevik

...Lloyd's Register ; D = Nederlandsche Vereeniging ; G = Germanischer Lloyd ; I = Registro Italiano ; R = American Record ; ... Austro-Ungarico ; H = Veritas Ellenico ; N = Norske Veritas ; V = Bureau Veritas.

Dans la 4me Colonne, C = Composite (membrure en fer, bordé en bois) ; I = Fer ; Stl = Acier ; W = Bois.

... de doublage en métal (Col. 11) ressortent des livres dans lesquels les navires sont respectivement classés ou des rapports des Inspecteurs du Lloyd's Register.

Pour plus amples renseignements sur les navires classés par le Lloyd's Register, voir le Lloyd's Register Book et le Lloyd's Yacht Register.

Official No.	No. in Book	Registry in which classed	Ship's Name. Material and Rig. / Ships Late Name	Master, and Dates of Appointment to present Owner's Service	Material and Rig	Flag	Net Tonnage	Length	Breadth	Depth	Built: Where	When	By whom	Date of Metalling	Owners or Managers	Port
	1751	D	Sundswall (ex Catharina Wilhelmina)	Ypenes	WBk	Dutch	941	157·0	37·0	22·2	Middelsburg	1842	Commercie Co.	[illegible]	W. Post	Edam
[illegible]	1752		Sungoo		StlLr	British	480	220·0	30·0	8·4	Dumbarton	1888	W. Denny & Bros		IrrawaddyFlotillaCo.Ld	Glasgow
[illegible]	1753		Sunlight	Pierce -89	WBn	American	355	125·1	29·2	15·3	Bath, Me.	1881	Goss, Sawyer & Packard		L. W. & P. Armstrong	New York
115206	1754	R	Sunlight	Coley -87	W3mSr	American	349	122·5	32·0	15·6	Fair Haven, Conn.	1873			G. J. Doe	New York
85579	1755	V	Sunshine	Simpson 88-88	WBn	British	349	131·6	30·0	13·1	MossGlen, N.B.	1888	G. Merritt	[illegible]	Herbert J. Olive	St. John
70526	1756	*	Sunshine	Jack 88-86	WSr	British	89	77·9	20·0	9·7	Perth	1877	Wallace		G. Sinclair	Wick
	1757		Sunshine (STRANDED)	Knudsen	WBk	Norweg'n	462	123·9	28·9	17·2	Pembroke, Me.	1864		[illegible]	A. Haneborg	Christiania
15	1758	A	Suocero	Mortola -88	WBk	Italian	424	126·6	30·6	18·9	Varazze	1866	G. Cerruti	[illegible]	L. Laurin	Genoa
	1759		Suomalainen	Rengström (-88)	WBg	Russian	397	121·8	29·2	15·1	Töfsala	1874	A. Lindström		G. Grönholm	Abo
	1760		Suometar	Salminen	W3mSr	Russian	259	107·9	26·9	13·2	Riga	1866	J. Lindell		Ylistyyna	Wiborg
	1761		Suometar	Seppä 75-	WSr	Russian	156	80·8	18·9	12·0	England	1849			A. Mäkelä	Wiborg
	1762		Suomi (ABANDONED)	Berg (ex C[illegible])	WS	Russian	938	165·0	35·0	22·7	Vegesack	1847	J. Lange		L. W. Wendelin	Christinestad
	1763		Suomi	Nordberg	WBk	Russian	418	..	..	..	Pyttis	18[illegible]			A. G. Hamberg	Lovisa
	1764		Suomi	Kurki 79-84	W3mSr	Russian	369	..	..	..		1885			V. Pikoff	Wiborg
	1765		Suomi	Lindroos	WSr	Russian	232	..	..	..	Westanfjerd	1873			E. J. Hollsten	Westanfjerd
	1766		Suomi	Holmberg 81-	WSr	Russian	220	102·8	26·3	12·6	Luvia Capell	1873	Sandström		B. Henriksson	Luvia
	1767		Suomi	Jansson	WSr	Russian	223	..	..	..	Luvia	1870			J. S. Jansson	Gustaf
	1768		Suomi		WSr	Russian	122	..	..	..		1868			A. J. Öhberg	Kumlinge
	1769		Suomi	Raggblom	WSr	Russian	115	..	..	..	Gustaf	1873			B. Gustafsson	Gustaf
	1770		Suosio		WSr	Russian	220	..	..	..		1877			J. Kämäri	Pyhäjoki
	1771		Suovu	Miettinen	WGlt	Russian	101	..	..	..	Kuopio	1875			Chr. Tavast	Kuopio
54755	1772	*	Superb	Evans 88-88	IS	British	1360	230·3	37·9	23·1	London	1866	R. & H. Green		Superb Ship Co. Ld. (Harry, Jones & Co.)	Liverpool
	1773	NV	Superb	Andreasen	WS	Norweg'n	759	166·1	35·3	19·8	Stavanger	1874	R. O. Hetem	[illegible]	E. Berentsen	Stavanger
	1774	N	Superb	Amundsen	WBk	Norweg'n	305	111·8	26·8	15·6	Kragerö	1859			O. T. Amundsen	Grimstad
9[illegible]	1775		Superbe	De Haz 67-81	WBn	Peruvian	544	140·0	31·1	18·9	Savona	1871			T. Shute	Callao
2403	1776	A	Superbo	Valentin -89	WBk	Austrian	481	130·5	29·5	17·9	Fiume	1865	Flli Schiavoni	[illegible]	N. Gelletich fu G.	Fiume
8593	1777		Superbo (ex Hosanna)	Mango	WSr	Italian	129	102·0	27·2	8·7	San Francisco	1875		[illegible]	S. Soracco	Genoa
48200	1778	R	Superior	Perkins	WS	British	1369	201·7	38·4	25·0	Quebec	1864	Valin	[illegible]	Trayes & Co.	Liverpool
74234	1779		Superior		WSp	British	112	92·0	22·8	7·2	Quebec	1875			Joseph Marchand	Montreal
	1780	N	Superior (ex Augusta)	Nygaard	WBk	Norweg'n	784	152·1	32·8	21·2	Bath, Me.	1857		[illegible]	J. Rods Sønner	Tönsberg
	1781		Superior	Jonasson 80-87	WBk	Swedish	516	137·4	29·4	18·5	Gefle	1848			J. C. Hulthén	Helsingborg
	1782	V	Superior	Larsson 78-86	WBk	Swedish	483	138·7	28·1	17·1	Gefle	1879	P. Bergforss & Son	[illegible]	P. Bergforss & Son	Gefle
	1783		Supreme	Smedsrig 87-87	WBk	Norweg'n	736	163·7	32·8	20·0	Quebec	1876	Baldwin	[illegible]	J. Thorsen	Haugesund
2870	1784	I	Supremo	Massone	WBk	Italian	549	140·1	30·7	21·0	Pra	1878	L. Durante	[illegible]	T. Massone fu G.	Genoa
	1785		Surat i Bahri		WBg	Turkish	225	..	..	..	Sinope	1869			Hairie Hijari Hadjee Ali & Co.	Sinope
22596	1786		Surge		WSr	American	111	99·6	17·6	8·0	Hamburg, Ct.	1864			J. A. Anderson	Trenton
	1787		Suro (ex John)		WBg	Russian	185	82·3	22·7	13·9	Rostock	1841			O. Carlson, junr.	Rauma
	1788		Surpreza (ex Maria IV.)	Borst -88	WBk	Portug'se	356	123·6	26·4	13·9	Villa Nova	1847	B. J. F. Azevedo		Mendes & Irmãos	Funchal
44942	1789		Surprise (ex [illegible])	Hornby 72-85	WBn	British	172	98·5	23·9	13·8	France	1835			G. F. Peek	Guernsey
55387	1790		Surprise		WSrYt	British	161	98·7	22·3	12·3	London	1851	R. & H. Green	[illegible]	Lady Vivian	Cowes
81520	1791		Surprise		WSr	British	125	81·7	22·9	11·1	Vannes	1857	Ridel		R. O. Sully	Bristol
89280	1792		Surprise		WSr	British	115	95·8	23·1	7·1	Bellinger River [N.S.W.	1884	J. Anderson		John Anderson	Sydney
[illegible]	1793		Surprise		WBg	British	113	89·1	23·0	9·7	France				Alphonse E. Nagoon	Cochin
62001	1794		Surprise	Lamb	WSr	British	100	81·5	20·1	10·1	Arbroath	1869	Arbroath Sh. Co.		J. & K. Murchison	Arbroath
	1795		Surprise	Curtis	WBk	Bolivian	772	158·1	31·1	15·0	La Bastide	1860			Heath, Gallup & Co.	Bolivia
[illegible]	1796	*	Surto (ex [illegible])	Mitsach	WBk	Austrian	465	152·1	28·5	16·0	Dundee	1870	Dundee S.B. Co.	[illegible]	M. Manco	Fiume
	1797		Surutoin	Ekblom 81-	WSr	Russian	222	104·6	25·8	12·8	Christinestad	1864	[illegible]		P. A. Petterson & Co.	Kerro
	1798	V	Survivor	Hicard	WBn	French	205	95·1	27·2	11·0	Cornwallis, NS	1880	E. Bigelow & Son	[illegible]	C. Landry	[illegible]
19452	1799		Susan	Finch	WBn	British	144	95·0	22·2	12·6	Rye	1857	[illegible]		T. W. Hutchings	Teignmouth
49[illegible]	1800		Susan	Ryan	WBn	British	134	96·7	22·8	10·8	Tatamagouche [N.S.	1864			Alfred Smith	St. John

*=Lloyd's Register; D=Nederlandsche Vereeniging; G=Germanischer Lloyd; I=Registro Italiano; R=American [illegible]
A=Veritas Austro-Ungarico; H=Veritas Ellenico; N=Norske Veritas; V=Bureau Veritas

In the 4th Column, C=Composite (iron frame, wood planked); I=Iron; Stl=Steel; W=Wood.

The Dates of Metalling (Col. 11) are based on the records of the Books in which Vessels are respectively classed, or on the Reports of the Surveyors of Lloyd's [illegible]

Further particulars of Vessels classed by Lloyd's Register will be found in Lloyd's Register Book and Lloyd's Yacht Register.

1 N° dans le livre	2 Registre dans lequel le navire est coté	4 Nom. Matériaux et Gréement. Capitaine et dates d'entrée au service des Armateurs actuels et à bord du navire. Ancien Nom		5 Pavillon	6 Tonnage net	7 Dimensions. Longueur	Largeur	Creux	8 Construit. Où.	9 Quand.	10 Par qui.	11 Date du doublage en métal	12 Armateurs ou Directeurs-Gérants.	13 Port d'Armement.
1801	R	Susan B. Ray *Burlingham -84*	W3mSc	American	374	130·0	32·5	14·0	Millville, N.J.	1874			E.C.Boice	Philadelphia
1802		Susan B Thompson (ex George Albert)	WBn	British	231	102·2	26·6	9·8	Thomast'n, Me				Charles Shepherd	Nassau, N.P.
1803		SusanBergen *Lewis*	WBn	American	239	104·0	26·0	13·0	P.JeffersonNY	1868			Miller, Bull & Co.	New York
1804	R	Susan H. Ritchie *Foring*	W3mSr	American	513	142·0	35·8	12·8	East Boston	1881	J.N.Grant	Y 10.85	C.R.Bartlett	Boston, Mass.
1805		Susan L. Campbell	WBk	Greek	727	151·5	33·0	9·8	WeymouthNS	1875	C.Campbell			
1806	R	Susan N. Pickering *Haskell*	W3mSr	American	319	135·5	32·0	10·9	Belfast, Me.	1882	Cotterall	Y 8.83	J.P.Ellicott & Co.	Deer Isle. Me.
1807	R	Susan P. Thurlow *Smith*	W3mSr	American	440	126·1	31·1	10·6	HarringtonMe	1872		Y 12.86	J.E.Brett	New York
1808		Susan Stetson *Frisbee*	WSr	American	134	89·2	26·0	8·2	Damariscotta, Me.	1867	Norris&Co.		D.F.Frisbie	Boston. Mass.
1809		SusanVittery *Pollard 83-86*	WSr	British	108	100·6	21·4	12·1	Dartmouth	1859	Kelly		W.J.M.Hawkey	Brixham
1810	I	Susanna *Gandarilella*	WBk	Italian	394	121·0	28·6	17·8	Milazzo	1880	G.B.Bonifacio	Y 1.86	A.Miceli fu A.	Messina
1811		Susanna *Olsson 73-83* (ex Susan Gray)	WBk	Swedish	226	100·7	22·2	13·4	P.E.Island	18^{57}_{73}			A.Svensson	Gothenburg
1812		Susannah Thrift *Thurlow 83-84*	WBg	British	199	102·9	25·0	11·3	Newcastle	1865	Fulton		M.Pearson	South Shields
1813	G	Susanne *Haskagen -88* (ex Anna ...)	WS	German	1433	203·0	40·0	24·7	Waldoboro'	1868	Reed, Welt&Co.	Y 1.87	E.C.Schramm & Co.	Bremen
1814	N	Susanne *Eoel 76-76* (ex ...)	WBk	Norweg'n	453	130·1	30·0	17·8	Sestri	1868		Y 5.85	Winge & Co.	Christiania
1815	N	Susanne *Olsen -87*	WBk	Norweg'n	451	124·5	31·8	17·1	Uleåborg	1853			C.A.Leschbrandt	Mandal
1816	DN	Susanne *Martinsen 76-76*	WSr	Norweg'n	284	123·2	29·6	10·9	America	1865			H.L.Petersen	Christiania
1817	N	Susanne *Eriksen 83-87*	WBn	Norweg'n	184	109·5	25·0	11·1	Bergen	1875	J.Haavesholm	Y 7.87	E.Schmidt, junr.	Bergen
1818	N	Susanne (Låst) *Solum*	WSr	Norweg'n	140	85·7	22·8	11·4	Maitland, N.S.	1857			L.Solum	Drammen
1819	R	Susie H. Davidson	W3mSr	American	519	149·0	35·0	14·1	Camden, N.J.	1883	Morris&Mathis		B.F.Haley	Bridgeton, N.J.
1820	✻D	Susie May *May*	WSr	British	124	94·2	22·3	11·6	Padstow	1876	Rawle		T.May	Padstow
1821	R	Susie P. Oliver *Curtis*	W3mSr	American	259	128·6	32·3	9·2	Bucksport, Me	1882	W.Beasley&Co.	Y 1.87	Morse & Co.	Bangor, Me.
1822	✻	Susie Spurgeon *Dixon*	WK	British	93	80·7	21·3	10·6	Galmpton	1885	T.Mathews		W.Klingsick	Grimsby
1823	✻D	Sussex *Malcolm 83-88*	IBk	British	1256	230·0	35·2	22·0	Glasgow	1866	Barclay, Curle &Co.		G.G.Macfarlane&C.	Glasgow
1824		Sussex Maid *Robinson 83-83*	WBg	British	171	90·2	20·6	13·7	Newhaven	1856			G.Robinson	Newhaven
1825	✻	Sutherlandshire *Beattie (76-81)*	IS	British	1549	258·9	38·2	22·9	Greenock	1882	Russell&Co.		T.Law & Co.	Glasgow
1826	✻	Sutlej *Pitt -1-88*	IS	British	1672	260·3	38·2	23·1	PortGlasgow	1888	Russell&Co.		Foley&Co.	London
1827	D	Suzanna *Woudsma -82* (ex ...)	WBg	Dutch	284	102·6	26·6	16·2	Sunderland	1853	B.Hodgson&Co.		A.Kranenborg	Oostwold
1828		Suzanne	WBg	British	252	110·0	23·2	14·0	Bordeaux	1865	Raymond freres & fils		John P.Ellis	Port Louis
1829		Suzanneh	WBg	Turkish	100	..	..	..	Greece	1874			Mahounahji Nicola&Co.	Constantinople
1830	✻	Suzerain *Williams 87-87*	WBk	British	393	130·0	28·4	17·5	Sunderland	1870	Robinson	Y 2.87	W.Price & Co.	Liverpool
1831		Sv. Nicolaja *Gheorghi*	WBn	Russian	228	..	..	..	Galaxidi	1872		Z 9.85	M.Barbati	Kertch
1832	V	Svafva *Höysted 74-81*	WBn	Danish	178	103·0	23·0	10·8	Svendborg	1876	C.C.Hansen& H.P.Hansen.jr		C.C.Hansen	Svendborg
1833		Svalan *Pålsson*	WBg	Swedish	210	103·6	23·4	12·3	Westervik	1851			M.Backelin	Fjellbacka
1834		Svalan *Jansson -88*	WBkn	Swedish	171	92·5	23·8	11·2	Vätö	1875			J.E.Jansson	Kulla, Roslagen
1835		Svalen *Christensen 79-79*	WSr	Danish	100	82·3	20·4	9·8	Marstal	1873	J.J.Bager		H.C.ChristensensEnke	Marstal
1836	N	Svalen *Haglund -86*	WBk	Norweg'n	560	153·4	31·1	17·7	Risör	1876		Y 10.86	H.B.Svensen	Tvedestrand
1837	N	Svalen *Jakobsen 88-*	WBk	Norweg'n	502	139·9	32·3	17·8	Drammen	1865	Flatter	Y 12.88	L.G.S.Larsen	Sandefjord
1838		Svan *Karlsson*	WSr	Russian	118	..	..	..		1872			Erik Sjöblom	Finström
1839		Svan *Jansson*	WGlt	Russian	110	..	..	..	Westanfjerd	1883			O.R.Jansson	Hiitis
1840		Svan *Pettersson 74-86*	WSr	Swedish	160	89·9	25·2	11·7	Norrtelje	1873			P.Jansson	Vätö, Roslagen
1841	R	Svan *Herm 79-80*	WSr	Swedish	132	91·5	21·0	10·9	Stavanger	18_{80}			G.J.Sundberg	Lysekil
1842		Svan *Pettersson*	WSr	Swedish	123	83·3	22·5	10·0	Axmár	18^{58}_{70}			P.Lundgren	Hamrånge, Gefle
1843	G	Svanen *Lyhre -88* (ex ...)	WBg	Danish	172	87·9	21·9	10·4	Stralsund	1862	J.P.Juhl		R.A.Holst	Svaneke
1844		Svanen *Thomsen 84-71*	WBn	Danish	114	74·5	18·4	10·5	Eckernförde	$\frac{1777}{1865}$		Y 2.83	H.A.Clausen	Copenhagen
1845	N	Svanen *Mikkelsen -84*	WBk	Norweg'n	452	132·6	31·3	16·6	Tvedestrand	1878	M.Eldrup	Y 10.87	Brödr.Thorvildsen	Tvedestrand
1846	DNV	Svanen *S...sen*	WBk	Norweg'n	321	117·5	27·4	15·0	Grimstad	1874	N.Osmundsen	Y 10.87	J.E.Einersen	Grimstad
1847	DN	Svanen *Danielsen*	WBg	Norweg'n	263	101·0	28·1	13·8	Tönsberg	1856			D.Danielsen	Tönsberg
1848	G	Svanen *Schlin 82-*	WSr	Swedish	158	87·2	21·0	11·9	Horsens	1856	M.F.Möller		G.Förselius	Stockholm
1849		Svanhvide *Wilson*	WBk	Norweg'n	472	124·3	32·1	18·7	Arendal	1853			E.A.Thoules Enke	Arendal
1850														

✻=Lloyd's Register; D=Nederlandsche Vereeniging; G=Germanischer Lloyd; I=Registro Italiano; R=American Record; A=Veritas Austro-Ungarico; H=Veritas Ellenico; N=Norske Veritas; V=Bureau Veritas.

Dans la 4me Colonne, C=Composite (membrure en fer, bordé en bois); I=Fer; Stl=Acier; W=Bois.

Les Dates du doublage en métal (Col. 11) ressortent des livres dans lesquels les navires sont respectivement classés ou des rapports des Inspecteurs du Lloyd's Register.

Pour plus amples renseignements sur les navires classés par le Lloyd's Register, voir le Lloyd's Register Book et le Lloyd's Yacht Register.

Official No.	No. in Book	Registry in which classed	Ship's Name. Master, and Dates of Appointment. Ship's Late Name.	Material and Rig	Flag	Net Tonnage	Length	Breadth	Depth	Where built	When	By whom	Date of Metalling	Owners or Managers	Port
	1851		Svante *Stenberg -70*	W Bg	Swedish	191	97·1	22·2	11·6	Stettin	1858			C. Molander	Visby
	1852	V	Svante *Eriksson -84*	W Sr	Swedish	178	109·5	22·6	9·8	Hernösand	1867	F. Hägglund		M. Lundqvist	Ostersund
	1853		Svartvik *Hermansson 86-77*	W Bg	Swedish	311	108·9	25·1	15·1	Svartvik	1839	A. Ahlberg		C. R. Holst	Nösund
	1854		Svartvik *Hermtsson -86*	W Bg	Swedish	178	92·8	23·0	11·1	Hernösand	1866			C. Bruce	Fiskebäck
	1855	V	Svava *Jensen 79-*	W Bn	Danish	184	107·9	24·0	10·8	Ripen	1864	P. Beck		C. Jensen	Stubbek
	1856		Svea *Egeriis 74-74* (White Kite)	W Bk	Norweg'n	535	161·1	35·4	16·2	Hernösand	1855			I. A. Berggrav	Fredriks
	1857	N	Svea *Ericson 74-88*	W Bk	Norweg'n	453	128·0	31·6	17·2	Bielle	1867	J. Steenersen	[illegible]	J. W. Prebensen, sen.	Risör
	1858	DN	Svea *Falch -87*	W Bg	Norweg'n	328	120·2	29·5	15·6	Drammen	1868	Ths. Lorentzen	V 7 87	T. & H. Lorentzen	Drammen
	1859	N	Svea *Olsen*	W Bg	Norweg'n	263	99·8	27·4	15·0	Kalix, Sweden	1840			N. Olsen	Fred
	1860	V	Svea *Thorkildsen 78-86*	W Bk	Norweg'n	772	167·0	34·0	20·9	Gefle	1872	J. A. Bang	V 3 87	Actieselskabet "Svea" (O. Bessen)	Christiania
	1861	V	Svea *Krook*	W Bk	Swedish	604	160·4	27·4	18·2	Sundsvall	1881	J. Sjölen	V 7 80	Sundsvalls Skeppsr.-Bolag (G. L. Wahlsberg)	Sundsvall
	1862	DV	Svea *Andersson*	W Bk	Swedish	603	144·6	29·6	18·4	Umeå	1874	F. Holmström	V 11 81	A. F. Scharin's Söner	Umeå
	1863	V	Svea *Eriksson -88*	W Bkn	Swedish	454	140·7	25·9	17·2	Gefle	1880	Bondhvfs Akt.-Bol.	V 5 86	O. S. Hedberg	Malmö
	1864	G	Svea *Bengtsson -88*	W Bg	Swedish	415	125·6	27·2	18·0	Trieste	1866	G. Spadon		Th. Olsson	Timmern
	1865	N	Svelvig *Johansen 85-86* (ex Regina) (ABANDONED)	W Bk	Norweg'n	344	117·7	26·2	16·8	Stockton	1852			O. Jörgensen	Drammen
	1866	DV	Svend *Andersen 77-77*	W Sr	Danish	121	90·4	20·6	10·8	Landeborg	1874	I. Bens		L. Bonnelycke	Svendborg
	1867	N	Svend Foyn *Berg 89-80*	W Bk	Norweg'n	281	114·9	28·4	13·0	Christiania	1865			J. A. Berg	Christiania
	1868	N	Sverdrup *Olsen*	W Bn	Norweg'n	147	96·1	23·4	11·3	Stavanger	1883	M. A. Magnus	V 3 87	T. Kleiberg	Stavanger
	1869		Sverre *Pedersen* (ex Maria Angela)	W Bk	Norweg'n	724	152·8	32·5	20·9	Swansea Mass.	1854			Akt. Selsk. "Sverre" (H. P. Petersen)	Fredrikstad
	1870	NV	Sverre *Dietriksen -87*	W Bk	Norweg'n	601	156·2	32·4	18·7	Christiansand	1880	M. Knudsen	V 1 87	H. Natvig	Christiansand
	1871	N	Sverre *Schøug 69-87* (ex Constance)	W Bk	Norweg'n	404	127·6	31·9	15·9	Åbo	1852		V 10 87	Johan Thorne	Moss
94	1872	V	Svisda Mimbelli *Tolich -88*	W Bk	Austrian	470	171·2	28·8	15·9	Trieste	1862	Tonello	V 7 87	G. Dobrillovich fu G.	Dubrovnik
	1873	V	Svitiod *Wahlgren 82-82*	W Bkn	Swedish	351	129·4	25·8	14·7	Bergqvara	1876	S. M. Bech	V 5 82	J. E. Gadelius	Gothenburg
60412	1874		Swallow *Rockett*	W Sr	British	100	83·0	20·2	10·6	Howden	1864	Banks		E. J. Foster	Goole
22240	1875		Swallow	W S	American	1177	201·7	38·8	23·0	Boston, Mass.	1854	J. Hall		M. E. Allen	New York
23000	1876		Swallow *McInnis 79-83*	W Bk	American	311	118·6	28·3	17·3	Fairhaven Mass.	1856			Aiken & Swift	N. Bedford
115010	1877		Swallow *Cummings*	W Sr	American	100	85·2	24·4	7·2	Perry, Me.	1870	Hinds		J. H. Ames	Machias
20195	1878		Swan *Baramech 83-83*	W Bk	British	140	85·0	23·1	12·1	Sunderland	1858			J. Saul	Lowestoft
	1879	✻	Swanmore *Brown 68-86*	I Bk	British	1766	268·6	38·2	23·3	Belfast	1886	Harland & Wolff		W. J. Myers Son & Co.	Liverpool
75408	1880	V	Swansea *Sanford -87*	W Bk	British	746	163·6	33·6	20·5	Hantsport, N.S.	1877	J. Davison	V 8 87	Geo. W. Churchill	Windsor
71091	1881	R	Swansea Castle *Evans*	I Bk	British	575	160·2	28·2	18·0	Dumbarton	1874	M'Kellar, M'Millan & Co.		Jacobs Brs. & Co.	London
	1882	V	Swantewit *Brodhering*	W Bg	German	269	105·9	27·0	14·4	Rostock	1867	O. Ludewig		J. C. Janentzky	Rostock
	1883		Swätoi Wassilly Wiliky *Sokerschkoff 75*	W Sr	Russian	120	84·0	20·6	11·0	Archangel	1875	Antenoff		W. Zaborschikoff	Archangel
54706	1884		Swift *Parkins 75-75*	W Sw	British	179	101·0	24·1	12·4	Shoreham	1867	May		E. Bannister	Grimsby
9023	1885		Swift	W Bk	British	158	84·6	21·9	13·0	Jersey	1826			William Kidston	Halifax N.S.
62193	1886		Swift *Coad*	W Sr	British	126	88·5	22·3	10·6	Portsmouth	1871	Beal		J. G. Coad	Fowey
	1887	N	Swift *Beer*	W Bk	Norweg'n	440	134·2	29·7	16·8	Arendal	1866	S. Johnsen	[illegible]	S. Eyde	Arendal
	1888	N	Swift *Knudsen -86*	W Bn	Norweg'n	234	111·4	25·5	12·2	Christiansand	1877	P. van de Lippe	V 1 87	H. Natvig	Christiansand
	1889		Swift *Christiansen*	W Bk	Norweg'n	168	98·0	25·3	9·8	Boston	1849			J. N. Jacobsen & Co.	Fredrikstad
	1890	R	Swiftsure (CONDEMNED)	W Bn		199	106·2	25·2	12·6	P. E. Island	1874	Dingwall			
90428	1891	✻	Swordfish *Evans 85-85*	W Sr	British	140	114·8	24·0	10·5	Summerside, P.E.I.	1885	J. W. Richards	[illegible]	H. Hoskin	Swansea
38811	1892		Swordfish	W Bn	British	111	92·7	20·9	9·4	McLeay River	1866			Edw. Taylor	Melbourne
88426	1893	V	Syanara *McIntosh 86-86*	W Sr	British	318	132·3	29·0	12·1	Portland, N.B.	1884	J. A. Likely	[illegible]	Joseph A. Likely	St John N.B.
78883	1894		Sybil *Fernandez*	W Sr	British	150	95·0	23·8	10·1	Aratapu, N.Z.	1879	J. Barbour	[illegible]	James Macfarlane	Auckland
63523	1895	✻	Sydenham *Richards*	I Bk	British	1063	210·1	34·1	21·0	Glasgow	1870	A. Stephen & Sons		Goldberg & Co.	Swansea
26367	1896		Sydney Griffiths	W Bk	British	333	111·1	25·1	17·6	Launceston, Tas.	1850			Government of Victoria	Melbourne
1192	1897		Sydney & Jane *Jones 80-80*	W Bn	British	178	97·2	24·4	12·5	Pwllheli	1841			R. Griffith	Carnarvon
	1898		Sydpolen *Andersson -86*	W Bg	Swedish	258	108·2	25·2	15·2	Stockholm	1851			L. B. Edberg	Lidköping
	1899	V	Sylfid *Törnell -85*	W Bn	Swedish	342	125·3	27·1	13·2	Sundsvall	1885	J. Sjölen	V 8 86	Sundsvalls Skeppsrederi Aktiebolag (G. L. Wahlberg)	Sundsvall
	1900	V	Sylfid *Ohlson 79-*	W 3m Sr	Swedish	208	105·3	23·5	12·7	Ystad	1879	J. Bang	[illegible]	Ystads Warfs-Aktiebolag (... Sundberg)	Ystad

✻ = Lloyd's Register; D = Nederlandsche Vereeniging; G = Germanischer Lloyd; I = Registro Italiano; R = American Record; A = Veritas Austro-Ungarico; H = Veritas Ellenico; N = Norske Veritas; V = Bureau Veritas.

In the 4th Column, C = Composite (iron frame, wood planked); I = Iron; Stl = Steel; W = Wood.

The Dates of Metalling (**Col. 11**) are based on the records of the Books in which Vessels are respectively classed, or on the Reports of the Surveyors of Lloyd's.

Further particulars of Vessels classed by Lloyd's Register will be found in Lloyd's Register Book and Lloyd's Yacht Register.

N°	Cote	Nom. Capitaine et dates d'entrée au service des Armateurs actuels et à bord du navire. Ancien Nom.	Matériaux et Gréement	Pavillon	Tonnage net	Longueur	Largeur	Creux	Construit: Où	Quand	Par qui	Doublage	Armateurs ou Directeurs-Gérants.	Port d'Armement.
1901	V	Sylfis [illegible]	W Bn	Greek	240	106·3	22·6	12·2	[illegible]	1877	J. Sjöden	[illegible]	H.G. Nicolakis	Syra
1902	✻	Sylhet (exx [illegible]) Evans	I Bk	British	498	155·0	26·7	17·2	Liverpool	1866	R.&J. Evans & Co.		Richards, Mills & Co.	Liverpool
1903		Sylph Nolley 51-68	W Bk	British	599	140·7	28·7	19·7	[illegible]	1842			T. Daniel & Sons	London
1904	✻	Sylph Owens 73-73	W Sr	British	108	78·1	21·2	12·6	[illegible]	1851			J. Owens	Beaumaris
1905	DN	Sylphide Andersen 87-88	W Bk	Norweg'n	392	120·2	29·5	16·7	Lillesand	1868	P. Knudsen	[illegible]	J. Westergaard	Arendal
1906	N	Sylphide Tellefsen	W Bn	Norweg'n	131	95·2	22·3	10·7	Copenhagen	1865	Popp & Michelsen		Johan Tröye	Bergen
1907	G	Sylphiden Kromann 81-88	W Sr	Danish	115	82·8	20·3	11·1	Naskov	1854	H.P. Ridersborg		Actieselskab (S.&R. Harbou)	Skjelskør
1908	NV	Sylphiden Fredriksen	W Bk	Norweg'n	417	137·5	27·8	15·5	Grimstad	1873	J. Terjesen	[illegible]	J. Börresen	Grimstad
1909	N	Sylphiden Nilsen -88	W Sr	Norweg'n	192	97·5	25·8	12·4	Fredrikstad	1857		[illegible]	N. Svendsens Enke	Laurvig
1910	N	Sylphiden Zernichow	W Sr	Norweg'n	154	95·0	24·2	10·8	Tönsberg	1862	H. Skoug		P. Zernichow	Tönsberg
1911	N	Sylphiden Olsen -84	W Sr	Norweg'n	131	94·7	23·6	11·2	Stavanger	1860			K. Finsdahl	Mandal
1912		Sylphis	W Sr	Greek	228	106·9	23·6	..					H.G. Nicolan	Syra
1913	V	Sylvan McDougall -88	W Bk	British	1046	186·6	35·9	22·0	Maitland, N.S.	1879	A. McDougall	Y 6.88	McDougall & Co.	Maitland, N.S.
1914		Sylvanus Hansen 79-86	W Bg	Norweg'n	176	89·2	21·2	12·7	Drammen	18 55/77	O. Olsen		H. Kierulf	Drammen
1915		Sylvester Neelon	W Sr	British	231	137·0	26·2	12·0	St. Catharines	1875	L. Shickluna		William B. Searth	St. Catharines
1916		Sylvester Whalen Whalen -87	W Sr	American	112	93·0	24·3	9·5	E. Boston, Mass.	1886	D.J. Lawler		Judith Whalen	Boston, Mass.
1917	NV	Sylvia Gabrielsen -85 (ex [illegible])	W Bk	Norweg'n	1141	191·5	37·5	22·8	Quebec	1878	W. Charland	[illegible]	Johan Bilstad	Egersund
1918	N	Sylvia Nielsen	W Bk	Norweg'n	436	125·7	29·8	17·1	Bordeaux	1848			C.A. Hansen	Fredriksstad
1919	DN	Sylvia Larsen 68-80	W Bk	Norweg'n	323	108·6	29·2	15·5	Carleby	18 /74			H.F. Gjessing	Drammen
1920		Sylvia Bengelsdorff 81-	W Sr	Russian	128	..	..	..	Wiborg	1865			S.A. Antman	Borgå
1921	D	Symbol Dew	W Bg	British	197	105·5	25·5	12·8	P.E. Island	1864	Hunt		W. Dew	South Shields
1922	AH	Symeon Tsipouras	W Bg	Greek	321	107·6	27·9	15·5	Galaxidi	1880	C. Papapetris		E. Tsipouras	Galaxidi
1923		Sympathia	W Bn	Brazilian	184	..	..	..					A.O. Santos	Rio Janeiro
1924	R	Syra Pattengall	W Bk	American	438	131·0	33·1	17·7	Pembroke, Me.	1870	Hobart & Pettingall	[illegible]	Jas. E. Brett	Calais, Me.
1925		Syren Howard -86 (ex Ryped)	W Bn	British	162	105·0	22·5	11·8	Tallemarken (Norway)	1867			Brazilian Coal Co. Ld	Swansea
1926		Syren Smith	W Bg	British	157	88·5	22·9	12·4	Rochester	1855	W. Langton		George F. Mason	Sydney, N.S.W.
1927		Syren (CONDEMNED) Mallett	W S	American	822	177·0	36·0	22·0	Medford, Mass.	1851			W.H. Besse	New Bdfrd. Mass.
1928	R	Syringa Given -88	W Bk	British	605	112·5	32·6	18·5	Indian Town, N.B.	1872	Rowan	[illegible]	Alfred Mills	St. John, N.B.
1929	N	Syttende Mai Tollefsen	W Bk	Norweg'n	213	100·9	24·9	12·1	Christiansand	1845			J.P. Tollefsen	Drammen
1930	N	17 de Mai Tomqvist 88-88	W Bk	Norweg'n	466	133·8	30·8	16·8	Grimstad	1883	N.O. Gjömle	[illegible]	H. Omundsen	Grimstad

ADDITIONAL SHIPS ARE INSERTED IN THE SUPPLEMENT AT THE END OF THE REGISTER.

✻ = Lloyd's Register; D = Nederlandsche Vereeniging; G = Germanischer Lloyd; I = Registro Italiano; R = American Record;
A = Veritas Austro-Ungarico; H = Veritas Ellenico; N = Norske Veritas; V = Bureau Veritas.

Dans la 4me Colonne, C = Composite (membrure en fer, bordé en bois); I = Fer; Stl = Acier; W = Bois.

Les Dates du doublage en métal (**Col. 11**) ressortent des livres dans lesquels les navires sont respectivement classés ou des rapports des Inspecteurs du Lloyd's Register.

Pour plus amples renseignements sur les navires classés par le Lloyd's Register, voir le Lloyd's Register Book et le Lloyd's Yacht Register.

Official No.	No. in Book	Registers in which classed	Ship's Name. Material and Rig. Master, and Dates of Appointment to present Owner's Service and to Ship. Ship's Late Name.	Flag	Tonnage	Dimensions: Length	Breadth	Depth	Built: Where	When	By whom	Date of Metalling	Owners or Managers	Port
[illegible]	1		T. A. Stuart *Kelley* WSr	American	170	105·0	27·0	8·8	Machias, Me.	1874			T. J. Kelley	[illegible]
[illegible]	2	[illegible]	T. A. Lambert *[illegible]* W4mSr (ALEX'S GONDRA)	American	[illegible]	[illegible]	[illegible]	[illegible]	[illegible]	1887	B. W. & H. F. Morse		B. W. & H. F. Morse	Bath, Me
[illegible]	3		T. Benedict *Linscott* WSr	American	[illegible]	[illegible]	[illegible]	[illegible]	[illegible]	1850			A. Linscott	Bath, Me
	4	V	T. C. Berg *Kriegel* WBk	German	[illegible]	[illegible]	[illegible]	17·[illegible]	[illegible]	1870	A. Witte	[illegible]	J. N. Rodbertus	Barth
[illegible]	5		T. C. Lee WBk	British	213	110·1	20·1	10·5	Quebec	1854	T. C. Lee		T. S. Brown	Quebec
[illegible]	6	✻	T. F. Oakes *Reed* IS	American	1896	255·0	40·6	23·5	Philadelphia	1883	American S B Co.		W. H. Starbuck	New [illegible]
[illegible]	7		T. G. V. *Blackmore* -88 WBg	British	184	109·5	22·4	12·7	Jersey	1859	Carey		G. Shepherd & Co.	[illegible]
[illegible]	8	R	T. H. A. Pitt *White* WBn	British	166	87·2	23·8	9·9	P. E. Island	1863		[illegible]	T. H. Pitt	[illegible]
[illegible]	9		T. Harris Kirk *Walker* W3mSr	American	[illegible]	[illegible]	[illegible]	9·0	Pt Jefferson, NY	1873			S. E. Birdsall, Exctr.	
[illegible]	10		T. Morris Perot *Randolph* W3mSr	American	[illegible]	[illegible]	33·8	11·6	Jamestown [illegible]	[illegible]	J. Matthis		S. E. Birdsall	New [illegible]
[illegible]	11		T. P. Dixon *Perry* WSr	American	[illegible]	[illegible]	[illegible]	[illegible]	City Island [illegible]	[illegible]			S. L. Keen	New York
[illegible]	12		T. R. Merritt WBk	British	[illegible]	[illegible]	[illegible]	12·2	Pt. Dalhousie [illegible]	1874	S. Andrews		J. W. King	St. [illegible]
[illegible]	13	RV	T. Remick *[illegible]* -8[illegible] WBn	American	[illegible]	[illegible]	29·5	15·2	D[illegible]	1871	G. C. Merry	[illegible]	J. H. Hilton	[illegible]
[illegible]	14	R	T. Towner *Bayton* WBn	American	157	[illegible]	25·6	10·4	Fair Haven Con / Me.	18[illegible]	W. A. Wright	[illegible]	B. B. Mathis	New [illegible]
[illegible]	15		T. W. Allan *Morang* WSr	American	[illegible]	[illegible]	26·0	7·5	[illegible]	1870			N. S. Allan	[illegible]
[illegible]	16		T. W. Dodd *Whitsall* WBn	British	[illegible]	[illegible]	[illegible]	12·[illegible]	[illegible]	1864	Cousins		F. W. Fairbrass	[illegible]
[illegible]	17	R	T. W. Dunn *McFarland* WSr	American	673	[illegible]	[illegible]	[illegible]	[illegible] Me	1884	Dunn & Elliot	[illegible]	Dunn & Elliot	[illegible]
[illegible]	18		T. W. Lucas *Kroger* WBg	American	[illegible]	121·9	[illegible]	10·7	[illegible]	1857			J. Madison	[illegible]
[illegible]	19		T. W. McKay *Roberts* 83-83 WSr	British	113	[illegible]	26·0	[illegible]	[illegible] N.S	1883	V. B. Roberts		T. W. McKay	[illegible]
[illegible]	20		Ta Hongkong WS	Siamese	636	157·2	31·6	17·[illegible]	[illegible]	1865			Chessua Poh Seng	Bangkok
	21		Tabatinga WBk	Brazilian	171	..	..	..					Barao de Maua	Amazon[illegible]
	22	DV	Tabitha *[illegible]* WBk (ex Lady Dufferin)	Norweg'n	997	182·0	36·3	21·[illegible]	[illegible] N. B.	1873	R. Titus	[illegible]	N. S. Andersen	Fredr[illegible]
	23	N	Tabor (ex [illegible]) *Bache* W Bg	Norweg'n	[illegible]	118·7	[illegible]	[illegible]	[illegible] Italy	1869	B. Barbieri	[illegible]	T. J. [illegible]	Tönsberg
	24	N	Tabor *Thorbjornsen* -87 W3mSr	Norweg'n	[illegible]	[illegible]	[illegible]	[illegible]	[illegible]	1884	T. Jensen	[illegible]	J. [illegible]	Risør
	25	N	Tabor *Orie* -78 W3mSr	Norweg'n	[illegible]	128·7	27·7	[illegible]	[illegible]	1878	K. Nielsen	[illegible]	N. W. Olsen	[illegible]
	26		Tabor (ex Argo) *Lefaitzky* W Bg	Russian	[illegible]	..	..	..	Spain	1866		[illegible]	M. Mendel	Poti
	27		Taborda (ex São Luiz, ex [illegible]) *Netto* WBn	Brazilian	132	..	..	..	England	1867		[illegible]	A. d'Oliveira Maia	Pernam[illegible]
[illegible]	28	A	Tacito (ex Vela) *Chersiunna* WBn	Austrian	588	132·2	31·2	19·8	Fiume	1875	G. Bachich	[illegible]	A. Gelletich	Lussin
[illegible]	29	I	Tacito (ex Giuseppe B.) *Marchese* -8[illegible] WBk	Italian	346	130·5	27·8	17·5	Savona	1874	L. Cabinnaro	[illegible]	F[illegible] Drago	Genoa
[illegible]	30	RV	Tacoma *Otis* -81 WS	American	1672	222·2	41·0	26·0	Bath, Me.	1881	Goss & Sawyer	[illegible]	C. Davenport	Bath, Me
[illegible]	31	✻	Tacora *Gordon* 84-88 I 3mSr	British	874	204·7	35·6	18·3	Port Glasgow	1888	J. Reid & Co.		Nicholson & M'Gill	Liverpool
	32		Tafalla (ex Annina) *Roig* W Bk	Spanish	1000	191·0	35·0	23·7	Sestri Ponente	1877	A. Briasco fu L.	[illegible]	Balcells & Subiran	Barcelona
	33		Taful (ex Lopes) (CONDEMNED) *Pereira* W Bk	Portug'se	367	112·8	25·6	16·[illegible]	[illegible]				F. Ricca	Lisbon
	34	NV	Tagal (ex John [illegible]) *Sörensen* 78-86 W Bk	Norweg'n	1023	183·8	37·6	22·0	Stavanger	1876	M. Magnus	[illegible]	T. M. Nielsen & Son	Laurvig
[illegible]	35	A	Taganrog (ex [illegible]) *Radonicich* W Bk	Austrian	513	138·7	27·8	18·7	Fiume	1859		[illegible]	P. Tripcovich	Meglin[illegible]
	36	V	Tage *Lorho* -88 W Bk	French	557	131·8	28·5	17·3	Bordeaux	1869		[illegible]	Lorho	Havre
[illegible]	37		Tagoon StlSr	British	215	160·2	22·1	7·3	Dumbarton	1886	W. Denny & Bros		Irrawaddy Flotilla C. Ld	Glasgow
[illegible]	38		Tagus (STRANDED) *Bust* W Bk	British	248	125·3	24·2	12·6	[illegible]	1866	Adamson		T. B. Gates	[illegible]
[illegible]	39		Tagus W Bg	British	[illegible]	91·3	22·9	15·0	Liverpool	1841			P. Bannerman	Aberdeen
	40	A	Tahir i Bahri *Hussein* W Bk	Turkish	397	..	..	..	[illegible]	1881			H. F. Bostandjoglu	Constant[illegible]
	41		Tahir i Bahri WS	Turkish	[illegible]	..	..	..	[illegible]	1884			Bindbley Bostanji zadeh Hadje Hassan & Co.	Rizeh
	42		Tahir i Bahri W Bg	Turkish	[illegible]	..	..	..	[illegible]	1872			Limanlee Dimitri [illegible]	Lemnos
	43		Tahir i Bahri WSr	Turkish	112	..	..	..	[illegible]	1873			Bartenlee Emin oglou [illegible]	Barten
[illegible]	44		Tahiti *Turner* 76-80 W Bn	American	[illegible]	121·[illegible]	[illegible]	12·[illegible]	[illegible]	1881	M. Turner		M. Turner	San Fr[illegible]
[illegible]	45		Tahmiroo *[illegible]* WSr	American	105	[illegible]	[illegible]	[illegible]	[illegible] Me	18[illegible]	Bucksport		R. C. Davis	Boston [illegible]
	46		Tahti *Idman* -76 W Bk	Russian	498	126·0	34·0	17·0	Nikolaistad	1857	Wasa Werfsbolag		J. Kirten	Nik[illegible]
	47		Tahti *Stenroos* 81- W3mSr	Russian	270	110·0	28·5	12·4	Nystad	1863	H. E. Ekdström		A. Forsström	Nystad
	48		Tahti *Lundström* -[illegible] W Bg	Russian	[illegible]	..	..	..		1862			F. F. Söderlund	Raum[illegible]
	49		Tahwa StlSr	British	[illegible]	[illegible]	[illegible]	7·[illegible]	Dumbarton	1886	W. Denny & Bros		Indian Government	
13[illegible]	50		Taihei Maru WSr	Japanese	[illegible]	122·[illegible]	24·[illegible]	13·2	Hiogo	1878	Kossakuikuo [illegible] Kioku		Tatsuma Kiyo	Hiogo

✻ = **Lloyd's Register;** A = Veritas Austro-Ungarico; D = Nederlandsche Vereeniging; G = Germanischer Lloyd; H = Veritas Ellenico; I = Registro Italiano; N = Norske Veritas; R = American [illegible]; V = Bureau Veritas.

In the 4th Column, C = Composite (iron frame, wood planked); I = Iron; Stl = Steel; W = Wood.

The Dates of Metalling (Col. 11) are based on the records of the Books in which Vessels are respectively classed, or on the Reports of the Surveyors of Lloyd's [illegible]

Further particulars of Vessels classed by Lloyd's Register will be found in Lloyd's Register Book and Lloyd's Yacht Register.

		Nom. Matériaux et Gréement. Capitaine et dates d'entrée au service des Armateurs actuels et à bord du navire. Ancien Nom.		Pavillon.	Tonnage net.	Dimensions. Longueur.	Largeur.	Creux.	Construit. Où.	Quand.	Par qui.	Date du doublage en métal	Armateurs ou Directeurs-Gérants.	Port d'Armement.
51	[illegible]	Taikun *Bertelsen 75-86*	W Bk	Danish	336	123·7	[illegible]	11·[illegible]	[illegible]	1868	A.P. [illegible]	V 7.88	Actieselskab (C. Andersen)	Vejle
52		Taimi *Forsström*	W Bg	Russian	322	115·0	[illegible]	17·[illegible]	[illegible]	1872	[illegible]		K.Forsström	Finby
53	V	Taiwan *Héroy -88*	W Bk	French	373	127·9	25·9	14·6	Sonderburg	1875	H. [illegible]	V 8.88	H.Angerainé	Havre
54		Taiyo Maru	W Sr	Japanese	368	135·0	26·2	15·7	Hiogo	1879	SuzukiKichizo		SeikoKwaisha	Hiogo
55		Taiyo Maru	W Bg	Japanese	138	97·7	20·5	9·8	Osaka	1880	NagataSanjiro		Abe Mototaro	Osaka
56		Taiyo Maru No. 2	W Bkn	Japanese	36[illegible]	1[illegible]	26·8	14·5	Osaka	1882	NagataSanjuro		Abe Mototaro	Osaka
57		Takasago Maru	W Bg	Japanese	2[illegible]	[illegible]	22·5	12·8	Tokio	1883	KawasakiShip-[BuildingYard		Moroto Seiroku	Hiogo
58		Takkerort *Weide 85-88*	W3mSr	Russian	2[illegible]	116·0	27·0	12·0	Gutsmanns-[bach	18[illegible]	[illegible]		A.Weide	Riga
59		Takouhi (ex [illegible]) *Apostol*	W Bn	Turkish	3[illegible]	1[illegible]·2	29·5	17·4	St.Martin'sNB	1878	J.M[illegible]	V 4.85	[illegible] zadeh Apik	Constantinople
60	✻	Talarvor *Lloyd*	W Sr	British	128	86·0	21·6	11·9	Portmadoc	1861	W[illegible]		C.Phillips	Fowey
61	✻	Talavera *Robson 76-82*	I 4mS	British	1730	266·2	40·0	23·6	Dumbarton	1882	[illegible] [Dennys&Co.		J.Hardie & Co.	Glasgow
62		Talavera	W Bk	British	378	127·5	25·4	17·2	Sunderland	1853			W.Lund	London
63		Talbot *Pett 84-84*	W Bg	British	187	107·6	19·7	13·9	Northam	1851			Newman,Hunt&Co.	London
64		Talca (ex Roderick Dhu)	W Bk	Chilian	342	119·4	21·9	..	Dumbarton	1853			F.Peede	Valparaiso
65		Talete (ex Nuovo Talete)	W Bkn	Italian	215	110·8	25·1	13·1	Savona	1887	E.Calamaro		D.Fascia	Civitavecchia
66		Talieboe Sooot	W Sr	British	144	66·0	22·0	10·9	Pontianak				D.Maugherit	Singapore
67		Talio *Jonsson*	W Sr	Swedish	127	84·3	20·4	10·6	Risör	1859	D.R.Andersson		N.R.Jonsson	Djurswik
68		Talip Djoedool Karim (ex [illegible])	W Bk	Dutch	197	9[illegible]·2	23·0	..	Java	1834		V 1.78	Said Hassan bin Abdul [illegible]	Sourabaya
69	✻	Talisman (ex Cursetjee Furdoonjee) *Pult 87-87*	I S	British	1159	213·8	35·2	22·0	Liverpool	1864	T.Vernon&Son		Williamson,Milligan&Co	Liverpool
70	V	Talisman *Baker*	W Bk	British	954	173·7	36·2	21·6	[illegible] N.S	1872	[illegible]		Albert Baker	Yarmouth. N.S.
71	R	Talisman *Delane*	W Bk	British	366	128·6	27·6	14·4	Baltimore,Md.	1860		V 9.87	H.B.Phillips	Trinidad
72	N	Talisman *Andersen -87*	W Bk	Norweg'n	446	111·3	30·8	16·5	Tönsberg	1869	A[illegible]	[illegible]	N.Bugge	Tönsberg
73	DN	Talisman *Pedersen*	W Bg	Norweg'n	241	11[illegible]·7	27·4	11·7	Grimstad	1865	[illegible]	[illegible]	W.Erichsen	Moss
74		Talma *Rousseau*	W Bg	French	13[illegible]	87·[illegible]	23·0	12·3	Nantes	1863	Boju&Bertrand		F.Rousseau	Nantes
75	N	Talma *Johansen -87*	W3mSr	Norweg'n	313	111·2	25·4	15·3	Boston	18[illegible]			H.J.Holker	Drammen
76	✻	Talookdar *Morley 85-88*	Stl S	British	2953	292·[illegible]	39·7	2[illegible]·5	Belfast	1885	Harland&Wolff		T. & J.Brocklebank	Liverpool
77	V	Tam O'Shanter *Harding (-87)*	W S	American	1522	213·8	[illegible]1·7	21·[illegible]	Freeport,Me.	1875	E.C.Soule	[illegible]	E.C.Soule	Portland,Me.
78		Tam O'Shanter *Petersen 75*	W Bkn	American	562	170·0	39·0	13·8	CoosBay,Oreg	1875			A.M.Simpson	San Francisco
79	V	Tamar E. Marshall *Parker*	W Bk	British	1270	200·5	38·3	22·7	BearRiver,NS.	1883	A.Marshall	V 1.88	A.Marshall	Digby,N.S.
80	✻	Tamaya *Rugby*	I Bk	British	579	162·0	26·5	18·0	Liverpool	1862	T.Vernon&Sons		Balfour,Williamson&Co	Liverpool
81	✻	Tamerlane *Beattie 88-88*	I Bk	British	768	196·4	32·1	18·6	Greenock	1861	Scott & Co.		Thomson,Dickie & Co.	Glasgow
82		Tamerlane *Briggs*	W Bk	American	354	108·9	27·3	19·2	Wiscasset,Me.	1824	Johnson		J.P.Knowles	New Bdford Mass.
83	N	Tamerlane *Gogstad 72-87*	W S	Norweg'n	915	175·0	38·0	22·0	Wiscasset,Me	18[illegible]		V 1.81	J.M.Bryde	Sandefjord
84		Tametomo Maru (ex Summer Moon)	W Bkn	Japanese	330	126·[illegible]	21·8	13·6	Plymouth	1875	Marshall	[illegible]	Nippon Yusen Kaisha	Tokio
85		Tamora *Stoconib*	W Bk	British	497	142·7	31·1	16·1	New Glasgow,[N.S.	1872	[illegible]	V [illegible]	A.Gibson & Co.	Liverpool
86	R	Tamos *Maule*	W3mSr	American	369	141·2	34·2	10·2	Bath,Me.	1881	Goss,Sawyer& [Packard	V [illegible]	J.Booth	New York
87		Tampa	Sr	American	162	..	..	..						Wilmington
88	V	Tampico *Delay 81-84*	Stl3mSr	French	292	127·3	23·9	11·2	Glasgow	1884	A.Stephen&Sns		Jencquel frères	Bordeaux
89		Tan Choon Seng	W Junk	British	151	95·0	24·0	9·0	[illegible]	1872			Chin Chow Seang	Penang
90	✻	Tana (ex [illegible]) *Hartvigsen 88*	I Bk	Norweg'n	890	202·3	32·1	18·8	[illegible]	1878	A.Stephen&Sons		T.Lassen	Arendal
91	N	Tana (ex Tolo) *Pedersen*	W Bg	Norweg'n	249	101·9	27·5	12·8	[illegible]	1861			Stian Salvesen	Tvedestrand
92	N	Tancred *Bjørløw 83-83*	W Bk	Norweg'n	456	128·8	30·6	17·2	Arendal	1872	[illegible]	V 1.87	C.Boe	Arendal
93		Tancredi (ex Fredd [illegible] Lamon) *Zennaro*	W3mSr	Italian	223	108·9	23·7	15·2	[illegible]	1867	F.Zanon	V 8.81	P.Ardisson	Venezia
94		Tang Seng Lee	W Junk	British	118	80·0	23·0	9·0	Penang	18..			Rahman Chitty	Penang
95		Tangen (-TEA DED) *Cyelstad*	W Bg	Norweg'n	163	82·6	24·7	12·2	Drammen	1853			H.Johnsen	Laurvig
96		Tangen *Nilsen 81-87*	W Sr	Norweg'n	151	93·3	23·5	10·7	Kragerö	1881	[illegible] Knoph	Y 12.83	B.Knoph	Kragerö
97		Tango Maru	W Sr	Japanese	164	107·4	19·2	10·9	Hiogo	1877	[illegible]		Kamada Seizo	Osaka
98		Tanisarie (ex [illegible]) *Van Elsbroek* (BROKEN UP)	W Bk	Dutch	928	161·4	[illegible]4·4	..	Salmien,N.S.	1864			A.G.van Elsbroek	Batavia
99	✻	Tanjore *Richard*	W Bk	British	933	172·4	35·1	21·1	Levis,Quebec	1877	W.Charland,jun	V 7.88	J.G.Ross & Co.	Quebec
100		Tankau Maru	W Sr	Japanese	114	93·3	24·2	8·4	Hakodate	1881	[illegible]		OkadaZenji	Tokio

✻=Lloyd's Register; D=Nederlandsche Vereeniging; G=Germanischer Lloyd; I=Registro Italiano; R=American Record; A=Veritas Austro-Ungarico; H=Veritas Ellenico; N=Norske Veritas; V=Bureau Veritas.

Dans la 4me Colonne, C=Composite (membrure en fer, bordé en bois); I=Fer; Stl=Acier; W=Bois.

Les Dates du doublage en métal (Col. 11) ressortent des livres dans lesquels les navires sont respectivement classés ou des rapports des Inspecteurs du Lloyd's Register.

Pour plus amples renseignements sur les navires classés par le Lloyd's Register, voir le Lloyd's Register Book et le Lloyd's Yacht Register.

Official No.	No. in Book	Registry in which classed	Ship's Name. Master, and Dates of Appointment to present Owner's Service and to Ship. Ship's Late Name.	Material and Rig	Flag	Net Tonnage	Length	Breadth	Depth	Built: Where	When	By whom	Date of Metalling	Owners or Managers	Port belonging to
40852	101		Tankerton Tower	WSr	British	110	95·0	20·5	9·1	Whitstable	1884	Goldfinch Bros.		A.J. Anderson	Faversh…
24451	102		Tanner *Williams*	WBg	American	277	129·0	29·7	11·7	New York	1855			S.H. Mott	Wilm…
	103		Tantah	WBg	Turkish	212	..	..	..					Abdul Magid Eff Khatab	
24098	104		Tantamount	WSr	American	190	99·0	28·0	9·0	South Thomaston, Me.	1855			M. Pendleton	Islesbo…
212	105		Tantivy *Mutschel 86-86*	W3mSr	Russian	316	112·5	27·1	13·5	St. Andrews, N.B.	1866	Grieg & Downing		M. Breede	Riga
82256	106		Tapaing (ex Steamer)	I Barge	British	771	126·7	30·0	6·3	Dumbarton	1877	W. Denny & Bros.		Irrawaddy Flotilla Co. Ld.	Glasgow
	107		Tapperheten *Andersson*	WSr	Swedish	155	91·5	26·1	9·7	Sundsvall	1864			A. Andersson	Hakn…
76749	108	✳	Taranaki *Gordon 84-84*	IS	British	2497	228·2	35·2	20·0	Port Glasgow	1877	R. Duncan & Co.		J. Leslie	Glasgow
45387	109		Tarapaca *Kennett -88*	WBk	British	495	150·8	28·2	17·9	Liverpool	1862	Cato		China Borneo Co. Ld.	Hong K…
	110	✳V	Tarapaca *Voisin 77-*	I 4mS	French	2497	305·0	44·6	23·3	[illegible]	1886	W.B. Thompson		Ant. Dom. Bordes & fils	Bordeau…
233	111	A	Tare *Stuch -84*	WBk	Austrian	682	148·3	32·1	18·6	[illegible]	1871	G. Tonelli		G. Stuch & Co.	Orebich
24627	112		Taria Topan *Trumbull*	WBk	American	600	145·6	31·2	18·6	Salem, Mass.	1870	J. Bertram		G.R. Emmerson	Salem…
24012	113	R	Tarifa *Sawyer*	WBn	American	536	129·3	31·6	17·2	Millbridge, Me.	1872	J.W. Sawyer		Simpson, Clapp & Co.	New York
	114		Tarifa *Ismai 87-87*	WBt	French	189	100·7	25·0	11·4	P.E. Island	1869	Douse		Jules Corbel	Courseu…
	115	N	Taritta *Knutsen -78*	WBg	Norweg'n	281	113·2	27·6	13·3	Tvedestrand	1875	T. Gundersen		A. Biering Olsen	Tvedes…
62446	116		Tarlair *Cruickshank 85-85*	WSr	British	85	77·0	20·8	10·0	Banff	1870	Watson		J.S. Wood	Peterhead
	117	V	Tarmow *Bossmo -87*	WBn	German	153	89·6	21·6	10·5	Rostock	1879	G.W. Tesdorpf		P. Kneppe	Weener
	118	V	Tarnan *Fröckberg 80-86*	W3mSr	Swedish	371	143·9	25·2	13·1	Gothenburg	1874	Fr. O. Bergström		Rederi Akt.-Bol.-Ownen (J.F. Strandberg)	Gothenb…
93366	119		Taroup	StlLr	British	215	160·0	22·0	7·3	Dumbarton	1887	W. Denny & Bros.		Irrawaddy Flotilla Co. Ld.	Glasgow
44230	120		Tarragona *Simon -79*	WBg	British	145	92·2	21·6	12·2	Whitehaven	1863	Kennedy		J. Vaughan	Whiteha…
24846	121		Tarry Not *Luker*	WSr	American	234	112·8	28·8	8·4	[illegible], Me.	1870	L.A. Knowles		H. Bothmann	New York
47270	122		Tartar *Rale*	WBg	British	179	100·6	22·9	12·8	Rye	1864	Hoad		F.C. Clothier	Faversham
86151	123	✳	Tartar *Mayorey 85-85*	WSk	British	89	79·4	21·2	10·4	Hull	1883	Hunt & Fowler		R. Roach	Hull
758	124	A	Tartaro *Deselj*	WBg	Austrian	300	102·6	25·9	16·1	Fiume	1856	G. Brazzoduro		C. cav. Cosulich	Fiume
	125		Tarujo *Faz*	WBg	Portug'se	270	101·0	24·9	14·4	[illegible]	1869	M. da Silva		V. Tarujo & filhos	Lisbon
85890	126	✳R	Tasma *Nicholson -83*	WBn	British	249	112·5	27·8	12·9	Mt. Stewart, P.E.I.	1884	E. Coffin		L.C. Owen	Charlotte…
93330	127	✳	Tasmania *Kohn 83-87*	I 4mS	British	2175	278·1	41·9	24·5	Port Glasgow	1886	Russell & Co.		P. Denniston & Co.	Glasgow
	128	✳	Tasmania *Schuth -79*	WBn	Argentine	160	90·5	22·1	11·5	[illegible]	1863	J.A. Hooites			Buenos A…
	129	DN	Tasmania *Hansen*	WBk	Norweg'n	367	117·1	31·1	16·0	Zaandam, Holland	1855			Winge & Co.	Christiania
35076	130	R	Tasmanian *Thompson 87-87*	WBk	British	1137	193·3	37·1	21·2	Quaco, N.B.	1856			W.T. Dixon & Sons	St. John…
20398	131		Tasso	WBk	British	261	122·0	24·6	14·2	Rye	1857	Hoad Bros.		The Mount Kemble Coal and Oil Co. Ld.	Sydney…
640	132	A	Tasso *Giudice -88*	WBg	Italian	249	101·7	26·7	17·1	[illegible]	1856	A. Castellano		F. Midolo	Catania
798	133		Tatsu Maru No. 1	WSr	Japanese	174	96·6	20·7	11·6	Kiel	1865			Tatsuma Miyo	Settsu
263	134		Tatsu Maru No. 2	WBk	Japanese	690	165·2	31·2	18·7					Seiko Kwaisha	Hiogo
	135	V	Taube *Aden 87-87*	WBn	German	218	95·1	24·0	10·7	[illegible]	1874	J.F. Nicolai		E.C. Aden	Grossefehn
	136		Taube (STRANDED) *Gerdes -88*	WSr	German	153	85·9	22·7	10·1	Leer	1875	B.D. Middendorf		C. Onnecken	Grossefehn
60830	137	✳	Taunton *Thomas*	CBk	British	688	181·5	32·0	17·7	Hartlepool	1868	Denton, Gray & Co.		G.H. Menger	Swansea
	138	✳	Taurus (ex Otter) *Brandt 87-87*	IBk	German	499	156·6	28·2	16·8	Sunderland	1864	Oswald & Co.		J.O. Williamson	Hamburg
	139	N	Taurus *Johnsen*	WBk	Norweg'n	504	145·9	29·8	17·9	Grimstad	1878	J.N. Ribe		J. Klocker	Arendal
	140		Taxiarchi	WSr	Turkish	124	..	..	..	[illegible]	1884			Sakvaire Yorghi Gasmalis & Co.	Scio
12	141		Taxiarchis	WBg	Greek	250	100·7	28·9	17·7	Syra	1870			D. Coucoudakis	Hydra
	142		Taxiarchis (ex Agios Anargyros)	WBg	Greek	249	..	..	..		1863			M.V. Zannes	Santorin
25	143	V	Taxiarchis *Lotsos*	WBn	Greek	225	93·8	27·2	15·4	Paros	1875	P. Bonis		N. Lotsos	Hydra
	144		Taxiarchis *Russo*	WBg	Greek	198	92·2	23·9	15·8	Syra	1868			S. Domaxa	Santorin
46	145		Taxiarchis	WBn	Greek	145	..	..	..	Syra	1865	P.M. Coumis		G. Korneggos	Zante
52	146		Taxiarchis	WBn	Greek	137	77·4	21·8	..	[illegible]	1878			J. Makesakakis	Syra
28	147		Taxiarchis	WBg	Greek	125	72·5	22·0	..	[illegible]	1870			A.A. Papaliodios	Syra
35	148	H	Taxiarchis *Platys*	WSr	Greek	126	110·2	22·5	12·1	Syra	1880			G. Platys	Santorin
	149	A	Taxiarchis (ex Kurtchas) *Valandassi*	W	Greek	110				Phaleron	1876			C. Valandassi	Syra
48	150		Taxiarchis (BALLISSON) *Canachi*	WSr	Greek	1[illegible]	68·9	20·0	..	Rhodes	1881			C.D. Cancartsis	Santorin

✳ = Lloyd's Register; D = Nederlandsche Vereeniging; G = Germanischer Lloyd; I = Registro Italiano; R = American [illegible]; A = Veritas Austro-Ungarico; H = Veritas Ellenico; N = Norske Veritas; V = Bureau Veri[tas]

In the 4th Column, C = Composite (iron frame, wood planked); I = Iron; Stl = Steel; W = Wood.

The Dates of Metalling (Col. 11) are based on the records of the Books in which Vessels are respectively classed, or on the Reports of the Surveyors of Lloyd's [illegible]

Further particulars of Vessels classed by Lloyd's Register will be found in Lloyd's Register Book and Lloyd's Yacht Register.

N° dans le livre	Société de classification	Nom. Capitaine et dates d'entrée au service des Armateurs actuels et à bord du navire. Ancien Nom	Matériaux et Grément	Pavillon	Tonnage net	Longueur	Largeur	Creux	Où	Quand	Par qui	Date du doublage en métal	Armateurs ou Directeurs-Gérants	Port d'Armement
1	2	3	4	5	6	7			8	9	10	11	12	13
151	✱	Tay *Kerr*	I S	British	1591	257·0	38·2	23·2	PortGlasgow	1885	Russell&Co.		P. M'Farlane	Port Glasgow
152		Tay	W	British	154	111·7	22·4	11·8	Quebec	1880	A. Miller		J. Wood	Ottawa, Ont.
153		Tay *McIntyre -87*	W Sr	British	125	93·7	27·7	7·1	BlackRiverNB	1887	J.&R. McLeod		P. McIntyre	St. John, N.B.
154	R	Taylor-Dickson *Lake*	W 3m Sr	American	570	146·4	36·2	15·2	CoopersPt.NJ	1880	Morris&Mathis	[illegible]	N. B. Lake	Camden, N.J.
155	R	Taylor & Mathis *Brewer*	W 3m Sr	American	266	119·5	29·5	9·1	CoopersPt.NJ	1860			J. F. Sheppard	Philadelphia
156	A	Tea *Naccari*	W Bn	Austrian	147	85·3	21·2	12·7	Lussinpiccolo	1882	N. Martinolich	[illegible]	A. Naccari	Lussinpiccolo
157	I	Teandro *Morasso -88* (ex [illegible])	W 3m Sr	Italian	248	114·1	25·2	13·0	Sestri	1878	A. Briasco fu L.	[illegible]	F[lli] Drago fu F.	Genova
158	✱	Teazer *Blake 73-73*	W Sk	British	50	62·0	19·3	8·4	Barnstaple	1873	Westacott		W. Gould & Co.	Padstow
159		Tefyk *Ibrahim Gerim*	W Bg	Turkish	180	..	..	..	Cumi	1863			Souleim Effendi	Constantinople
160	H	Tefyk Bachri *Hassan [illegible]*	W Bg	Turkish	240	..	..	..	Iraclia	1881			Housseinali Moust	Iraclia
161		Tefyk Barri	W Bg	Turkish	300	..	..	..	Heraclea	1877			Heracleaien Bulukji zade Mehemed Ali	Heraclea
162		Tefyk Barri	W Bg	Turkish	287	..	..	..	Heraclea	1882			Heracleaiee Mustapha [illegible]	Heraclea
163		Tefyk Barri	W Sr	Turkish	247	..	..	..	Heraclea	1879			Heracleaiee Sayn Beken [illegible]	Heraclea
164		Tefyk Huda	W Bg	Turkish	123	..	..	..	Greece	1878			[illegible] Hadji Ahmed [illegible]	Constantinople
165		Tefyk Ikhsan	W Bg	Turkish	247	..	..	..	Heraclea	1867			Heracleaiee Hadji Yussuf Ismail & Sons	Heraclea
166		Tefyk i Bahri	W Bg	Turkish	212	..	..	..	Allaplee	1868			Tekkeh Kebirlee Yasidji oglou Halil & Omer	Tekkeh Kebir
167		Teignmouth	W Sr	British	109	81·2	23·5	9·3	ShipHarborNS	1877			D. H. Pitts	Halifax, N.S.
168		Teil		French	141	..	..	..						Marseilles
169	V	Teixeira *Cunha* (ex [illegible] Norton)	W Bk	Portug'se	403	143·7	27·8	13·6	St. Martin N.B.	1874		[illegible]	A. T. Rodrigues	Lisbon
170		Teixeirinha *Lopes*	W Sr	Brazilian	178	..	..	..	Rio Janiero	1875		[illegible]	Cia Cabotagem de S. João da Barra	S. João da Barra
171		Tejo (ex Mercurio) *Machado*	W Bn	Brazilian	134	..	..	..	Macahe	1846		[illegible]	J. Machado	S. João da Barra
172		Tek Lee (ex Padang)	W Sr	British	122	70·0	20·6	10·0	Padang	1869			Pana Akhamud Jalla-ladin	Penang
173		Teka Asiah *Kontang*	W Bk	Dutch	415	134·5	26·4	..					Go Tjoen Bioen Achmad bin Abdul Djalil	Bandjarmassin
174	G	Tekla *Petersson* (ex Charlotte & Anne)	W Bk	Swedish	378	126·3	25·7	17·4	Cowes	1871		[illegible]	A. B. Elfversson	Gothenburg
175		Tekla Maria *Johansson -85*	W 3m Sr	Swedish	185	95·1	22·7	12·0	Oroust	1874	B. Samuelsson		C. R. Holst	Nösund, Bhln.
176	N	Telefon *Petersen -87* (ex [illegible])	W S	Norweg'n	1200	195·5	37·1	21·5	Annapolis, NS.	1877	Delap&Co.	[illegible]	Axel J. Myhre	Porsgrund
177	DNV	Telefon *Vinje*	W Bk	Norweg'n	754	167·9	34·3	19·8	Stavanger	1878	H. C. Knudsen	[illegible]	Brödr. Berner	Stavanger
178	N	Telefon *Bergreen 74-78*	W Bn	Norweg'n	220	114·2	26·4	12·2	Sandefjord	1878	AlbertGrön		P. A. Grön	Sandefjord
179		Telefon *Månson 67-81* (ex Julius Daniel)	W 3m Sr	Swedish	313	121·7	24·7	14·6	P. E. Island	1869			O. Nilsson	Helsingborg
180	N	Telegraf *Mauritzen -87*	W Bg	Norweg'n	265	90·1	27·5	15·3	Arendal	1844	V. Furst		O. Ledien	Arendal
181		Telegrafen *Nilsson -36*	W Sr	Swedish	106	71·8	21·2	10·7	Wenersborg	1857			O. Nilsson	Carlshamn
182		Telegrafen *Droth -86*	W Sr	Swedish	105	82·3	20·7	9·2	Oscarshamn	1874			G. & F. Johansson	Oscarshamn
183		Telegramma	W Bg	Brazilian	141	..	..	..					M. G. de Medeiros	Alcobaca
184		Telegraph *Thorndike*	W Sr	American	113	77·0	22·0	12·0	Boston, Mass.	18 47/77			G. R. Post	St. George, Me.
185		Telegraph *Vicuña* (ex Telegraphe)	W Bk	Chilian	452	132·8	29·0	..	Bordeaux	1858			Dubart Hermanos	Coronel
186	N	Telegraph *Jensen*	W 3m Sr	Norweg'n	361	118·2	29·0	16·3	Amsterdam	1852		Z 5.81	Carl Dahl	Fredrikshald
187	V	Télémaque *Miteraux*	W Bg	French	175	85·0	22·3	12·1	Nantes	1874	E. Clergeau	Z 8.86	Pitre Legal	Nantes
188	✱	Telephone *Clarke 80-80*	W Sr	British	100	90·0	22·1	11·1	Plymouth	1878	Banks		J. Westcott	Plymouth
189	✱	Telephone *Couch 82-82*	W Sk	British	44	59·7	19·1	8·5	Barnstaple	1878	Westacott		M. Guy	Padstow
190	N	Tell *Christensen -87*	W 3m Sr	Norweg'n	492	142·8	33·9	15·5	Arendal	1883	O. M. Olsen	[illegible]	B. A. Olsen	Risör
191	N	Tell *Olsen 85-85*	W Bn	Norweg'n	277	116·1	28·1	13·9	Christiansand	1870	H. Olsen	[illegible]	S. Otto & Co.	Christiansand
192		Tellash *Vassil*	W Bg	Turkish	250	..	..	..	Allaplee	1880			Midehlee Seralim oglou Chakal Vassil	Mideh
193		Tellassoporos	W Bg	Turkish	106	..	..	..	Greece	1877			Bulbuljehlee Stavri Yanni Tachaki	Bulbuljeh
194	✱	Tellus *Nielsen -88* (ex Zanoni)	W Bk	German	430	137·5	27·4	17·6	Ayr	1865	Fullerton	[illegible]	Wachsmuth & Krogmann	Hamburg
195	N	Tellus *Olsen -87*	W Bk	Norweg'n	358	113·0	28·1	16·6	Söderhamn	1839			Aktieselskabet "Tellus" [illegible]	Sarpsborg
196	R	Telos *Coney*	W Bn	American	371	127·0	30·0	16·0	Bangor, Me.	1883	J. H. Crosby	[illegible]	Crosby Bros.	Bangor, Me.
197	R	Telumah *Marshall*	W Sr	American	218	110·3	28·1	10·1	Damariscotta, Me.	1866			A. E. Fernald	Winterport, Me.
198		Tema *Lycke*	W Sr	Swedish	161	87·3	21·9	12·2	Bastholmen	1867			J. A. Johansson	Bastholmen
199	V	Temerario *Cunha*	W 3m Sr	Portug'se	278	130·2	24·9	10·8	Temerario	1883	J. F. da Lapa	[illegible]	M. J. d'Amorim	Oporto
200		Temerario *Freitas*	W Pol	Spanish	191	98·1	14·7	12·6	SanFelice	1856			Carlos Roses	Barcelona

✱=Lloyd's Register; A=Veritas Austro-Ungarico; D=Nederlandsche Vereeniging; G=Germanischer Lloyd; H=Veritas Ellenico; I=Registro Italiano; N=Norske Veritas; R=American Record; V=Bureau Veritas.

Dans la 4me Colonne, C=Composite (membrure en fer, bordé en bois); I=Fer; Stl=Acier; W=Bois.

Les Dates du doublage en métal (Col. 11) ressortent des livres dans lesquels les navires sont respectivement classés ou des rapports des Inspecteurs du Lloyd's Register.

Pour plus amples renseignements sur les navires classés par le Lloyd's Register, voir le Lloyd's Register Book et le Lloyd's Yacht Register.

Official No.	No. in Book	Registry in which Classed	Ship's Name, Master, Material and Rig. (Ship's Late Name)	Flag	Net Tonnage	Length	Breadth	Depth	Built Where	When	By whom	Date of Metalling	Owners or Managers	Port
[illegible]	201	A	Temi *Tomassich* W Bg	Austrian	413	123·6	27·2	18·2	Fiume	1864	F[lli] Schiavoni	[illegible]	F.Minach	Fiume
	202		Temo W Bk	Argentine	[illegible]	..	..	..	[illegible]	1864	[illegible]		A.Stefano	Buenos
[illegible]	203		Templar *Cardell* 81-87 W S	American	910	155·0	35·2	23·[illegible]	[illegible]	1858			P.B.Cornwall	San F.
[illegible]	204	V	Temple Bar *Creighton* -88 W Bk	British	808	175·7	35·0	20·9	St. Martins, N.B.	1878	W.Vaughan&Co	[illegible]	G.C.Stewart	Liverpool
	205		Temur Cappu W Schieppe	Turkish	1[illegible]	..	..	..	Hasskioi	1881			[illegible]	Haskioi
[illegible]	206		Ten Brothers W S	British	122	87·9	22·0	8·5	Rochester	1860			J.Smith	[illegible]
[illegible]	207	R	Tena A. Cotton *Cranmer* W 3mSt	American	358	[illegible]	[illegible]	[illegible]	Milton, Del.	1883	C.C.Davidson		H.W.Cranmer	Philadelphia
[illegible]	208	✻	Tenasserim *Lindsay* 83-87 I S	British	1413	23[illegible]	[illegible]	23·6	Belfast	1866	Harland&Wolff		T. & J.Brocklebank	Liverpool
	209	N	Tenax Propositi *Eilertsen* W Bk	Norweg'n	403	133·0	30·2	16·2	Kragerö	1868		Y 10.88	P.& C.Römer	Kragerö
[illegible]	210	✻D	Tenby Castle *Davies* I Bk	British	588	174·0	28·6	18·0	Liverpool	1868	R.& J.Evans&Co.		Richards, Mills & Co.	Liverpool
	211		Tenedos W Bmb	Turkish	13[illegible]	..	..	..	Tenedos	1869			Bostelia Adalre Has-[illegible]	Bogtcha
[illegible]	212	R	Teneriffe *Small* -88 W Bn	American	518	130·5	31·5	17·5	Millbridge, Me.	1871	J.W.Sawyer	[illegible]	Miner & Houghton	New York
	213	V	Tentadora *Dos Santos L.* W Bk (ex Lyndhurst)	Portug'se	394	131·8	25·4	10·3	Sunderland	1872			L. Marques de Nova	Oporto
	214	✻	Tentativa *Belliconi* W Bk (ex [illegible])	Portug'se	373	13[illegible]·7	2[illegible]	17·[illegible]	[illegible]	1865	Ritson	Y 3.86	L.T.Mendes	Lisbon
	215	V	Tentativa *Sarmiento* -88 W Bn	Portug'se	222	110·8	23·9	10·5	Fão	1879	M.DaiesS.Borda	Y 12.84	M.M.Marinhas	Oporto
[illegible]	216		Tenun Maru W S	Japanese	173	112·[illegible]	22·[illegible]	[illegible]·2	[illegible]	1880	S[illegible]Kinjiro		Kimura Yenkichi	Awomori
[illegible]	217	I	Teocle *Sbrila* -87 W Bk	Italian	746	142·3	30·8	21·0	Cassano	1875	F.De Rosa	[illegible]	F. & S.Maresca	Castellamare
[illegible]	218		Teodosia (LOST) *Aronetti* W Bk	Italian	465	131·8	30·6	19·4	Savona	1868	A.Calcagno	Y [illegible]	T Pignone	Genoa
	219		Terceirense W Bk	Portug'se	198	92·5	23·1	12·6	San M[illegible]	1857			F.Mariano	Terceira
	220	V	Tercer Barcelo W Bg		242	1[illegible]3·0	27·6	13·8	[illegible]	1866				
	221		Terebolie W Bg	Turkish	3[illegible]	..	..	..	Terebolie	1871			Ibrahim Halil Bey	Terebolie
	222		Terebolie W Bg	Turkish	284	..	..	..	Terebolie	1869			Anastas Andriaki	Terebolie
	223		Terebolie W Bg	Turkish	275	..	..	..	Terebolie	1868			Yelkon [illegible]zadeh Hadjee Mustapha & Co.	Terebolie
	224		Terebolie W Bg	Turkish	112	..	..	..	Terebolie	1870			Yorghanoglu Hussein Effendi	Constantinople
[illegible]	225	V	Teresa [illegible] *Ruts* W B[illegible]	Italian	492	133·8	29·6	19·7	Sorrento	1872	M.Patorzo	[illegible]	Ang.Castellano	Naples
[illegible]	226		Teresa (ex Kayabik) W S	Italian	135	97·6	21·7	12·1	[illegible]	1865	Calley		G.Figaro di F.	Genoa
[illegible]	227	I	Teresa *Lucignani* W S	Italian	114	89·2	22·6	10·[illegible]	[illegible]	1855		[illegible]	Stef.Gargava	Civita V.
	228		Teresa W Bg		567	138·7	33·1	18·1	[illegible]	1864	Viato			
	229	V	Teresa *Bofill* W Bk (ex Mary E. Chapman)	Spanish	296	120·9	26·2	13·6	Rockland, N.B.	1873	R.A.Chapman	[illegible]	J.R.Basterrechea	Cienfuegos
	230		Teresa (ex Narcisa) W Bn	Spanish	201	91·5	24·8	13·0	Nantes	1863			Luis Pereda & Co.	Havana
	231		Teresa W S	Spanish	121	78·4	24·5	10·7	M[illegible]	1856			J.Fonrodona y Lloveras	Majorca
	232		Teresa W Bn	Spanish	115	..	..	..	Santander	1858			Jimenez Sobrinos & Co.	Havana
[illegible]	233	I	Teresa Accame *Rosetto* W Bk	Italian	928	181·1	31·8	23·3	Sestri Ponente	1879	G.Calcagno	[illegible]	E. Accame & figli	Genoa
	234		Teresa B. (ex Elena) *Ascercio* W Bk	Chilian	802	170·6	32·5	21·6	Sestri	1873	A.Briasco		A.Baffico	Valparaiso
	235		Teresa Bianchi W Bk (ex Petronila R., ex Bordeaux)	Chilian	581	165·3	31·3	18·2	Bordeaux	1872	Rochet and Largeteau	[illegible]	G.Bianchi & Heirs of J. Bianchi	Valparaiso
[illegible]	236	I	Teresa C. *Kergren* -88 W Bk	Italian	401	123·0	29·2	17·7	Chiavari	1878	L.Gotuzzo	[illegible]	Gius.Copello fu L.	Spezia
[illegible]	237		Teresa C. *Mutti* -88 W Bg (ex E. Rossi)	Italian	2[illegible]	10[illegible]·6	26·6	15·8	Savona	1871			L. & E.Cignoni F[lli]	Leghorn
[illegible]	238		Teresa Carmela *Guida* W Bk	Italian	391	121·7	29·8	19·0	Fiume	1855		Y [illegible]	F[lli] Cordillo	Naples
	239		Teresa Cosulich Stl Bk	Austrian	452	14[illegible]	28·5	14·7	Lussinpiccolo	1888	N.Martinolich		C. di A.Cosulich & Co.	Lussinpiccolo
	240	V	Teresa Figueras *Juan* W Bk	Spanish	526	128·[illegible]	30·5	17·9	[illegible]	1867		[illegible]	M.Fabra	Barcelona
[illegible]	241	I	Teresa G. (ex [illegible]) *Galliano* W Bk	Italian	630	151·2	30·9	21·1	Sestri Ponente	1873	A.Briasco fu L.	[illegible]	G.Mortola & V.Galliano	Genoa
	242	✻	Teresa Garnham *Green* -88 Comp Bk (ex Fairlie)	Chilian	539	148·0	29·1	18·8	Glasgow	1865	Henderson & Co.	Y 3.87	J.A.Verdugo	Valparaiso
[illegible]	243		Teresa Gemma W Bn	Italian	10[illegible]	86·6	22·9	11·6	Spotorno	1877	F[lli] Cadenaccio		A.Felugo	Genoa
[illegible]	244	I	Teresa Lo Vico *Ferrari* W Bk	Italian	563	14[illegible]	2[illegible]	19·3	Sestri Ponente	1874	S.Debarbieri	[illegible]	Andrea Lo Vico	Palermo
[illegible]	245	I	Teresa Madre *T.Gazzolo* W Bk (ex Marini A.)	Italian	478	132·8	30·2	19·4	Sestri Ponente	1869	F[lli] Cadenaccio	[illegible]	F.Gazzolo fu A.	Genoa
[illegible]	246	A	Teresa Mignano *Mazzella* W Bg	Italian	188	88·6	25·1	15·8	Minori	1860		[illegible]	A.Mignano	Naples
[illegible]	247	D I	Teresa Olivari *Marchese* W Bk (ex Ave Maria)	Italian	[illegible]	173·8	34·1	22·1	Sestri Ponente	1870	S.Debarbieri	[illegible]	Fort.Olivari	Genoa
[illegible]	248	I	Teresa Picasso *Picasso* W Bk	Italian	461	135·1	31·0	19·6	Sestri Ponente	1868	M.Rossi	Y 12.85	Enrico Picasso	Genoa
[illegible]	249	A	Teresa Prima *Pascinto* W Bg	Italian	197	97·1	25·5	15·2	Procida	1858	A.Lubrano	[illegible]	L.Talano	Gaeta
[illegible]	250	✻A	Teresa Rocca *Terissano* W Bk	Italian	598	16[illegible]·7	33·1	18·9	Lavagna	1869	D.Briasco fu G.B.	Y 12.86	Rocca & Lavello	Genoa

✻ = Lloyd's Register; D = Nederlandsche Vereeniging; G = Germanischer Lloyd; I = Registro Italiano; R = American [illegible]; A = Veritas Austro-Ungarico; H = Veritas Ellenico; N = Norske Veritas; V = Bureau Veritas.

In the 4th Column, C = Composite (iron frame, wood planked); I = Iron; Stl = Steel; W = Wood.

The Dates of Metalling (Col. 11) are based on the records of the Books in which Vessels are respectively classed, or on the Reports of the Surveyors of Lloyd's R[egister].

Further particulars of Vessels classed by Lloyd's Register will be found in Lloyd's Register Book and Lloyd's Yacht Register.

1 No.	2 Registre	3-4 Nom. Capitaine et dates d'entrée au service des Armateurs actuels et à bord du navire. Ancien Nom.	Matériaux et Gréement	5 Pavillon	6 Tonnage net	7 Longueur	Largeur	Creux	8 Construit. Où.	9 Quand.	10 Par qui.	11 Date du doublage en métal	12 Armateurs ou Directeurs-Gérants.	13 Port d'Armement.
251		Teresa Tubino *Felugo* (ex Ulysse) (DIDLES)	W Bn	Italian	258	100·1	27·3	16·7	Savona	1856	F.Calaoaro		F[lli] Tubino fu G.	Genoa
252		Teresia (BURNT) *Saverman*	W Sr	Swedish	114	82·8	24·9	9·5	Hille	1874			P.Pettersson	Blidö.Rslgn.
253	IV	Teresina *Laborano*	W Bk	Italian	591	148·9	32·0	20·5	Castellamare	1877	G.Bonifacio	Y 3.88	A. & S.Mazzella	Naples
254	I	Teresina *Cuccace*	W Bg	Italian	456	132·5	27·2	19·0	Cassano	1880	Castellano	Y 11.81	F.S.Ciampa & Co.	Castellamare
255	V	Teresina *Emanuel* -87	W Bg	Italian	322	112·5	26·2	17·5	Sorrento	1865	A.Castellano	Y 10.88	S.Maresca	Castellamare di Stab.
256	A	Teresina *Vianello*	W Bn	Italian	241	117·1	26·6	14·9	Malta	1875	A.Busuttil	Y 6.83	J.Vianello-Moro&Nipoti	Venice
257		Teresina (ex Fiorence)	W Bn	Italian	174	85·3	23·9	15·9	Prà	1856	M.Tisi		S.Lubrano	Naples
258	DIV	Teresina Bruno *Scotto* (ex Bruno D.)	W Bk	Italian	747	167·6	32·8	22·6	Savona	1877	F.Sirello	Y 3.86	G.Bruno fu S.	Naples
259	I	Teresina P. *Palomba*	W Sr	Italian	101	83·0	20·5	9·8	Torre del Greco	1885	Trombarulo	Y 3.85	L.Palomba	Naples
260		Teresinella (ex Rosa) *Lauro*	W Bk	Italian	484	139·1	30·7	18·6	Cassano	1874	F.deRosa	Y 3.88	G.Esposito	Castellam.diStab
261		Teresita	W Bn	Spanish	244	..	..	..	England	1879			Miguel Moll	Valencia
262		Teresita	W Sr	Spanish	109	78·7	23·2	10·2	SanSebastian				Jaime Morant & Co.	Valencia
263	DN	Ternen *Larsen* -85	W Sr	Norweg'n	109	87·9	23·3	10·1	Bergen	1884	A.T.Digernes		J.Larsen	Haugesund
264	❄	Terpsichore *Warncke*	I S	German	1999	269·0	41·5	24·3	Liverpool	1883	T.Royden&Sons		B.Wencke Söhne	Hamburg
265	DN	Terpsichore *Nielsen* -88 (ex M. B. Lindsay)	W Bk	Norweg'n	1188	177·2	37·2	24·0	Warren,Me.	1856		Y 10.86	Ole Nielsen	Sandefjord
266		Terpsichore *Olsen* -87 (WRECKED)	W Sr	Norweg'n	248	107·5	28·9	12·4	Arendal	1866	S.Johnsen		S.Eyde	Arendal
267		Terrapin *Hannah*	W Sr	American	121	95·0	26·0	7·4	Calais,Me.	1870	Hinds		G.W.Hinds	Calais,Me.
268	N	Terra Nova *Andresen* -87	W Bg	Norweg'n	156	76·5	24·1	11·9	Bolstad,Swedn	1816			I.H.Buvig	Sarpsborg
269		Terrier *Carr* 85-85	W Bg	British	144	101·3	21·2	12·5	Dartmouth	1873	Philip		Newman,Hunt&Co.	London
270	DG	Terwisch *de Vries* -81 (ex Johanna Reinholdie)	W Bn	Dutch	184	99·6	24·1	12·2	Elshout	1858			G.Wirtjes	Nieuwe Schans
271	N	Terzo *Christiansen* 86-86	W Bk	Norweg'n	1025	181·9	37·2	22·1	Genoa	1865		Y 5.88	J.Ericksen	Skien
272		Tessalia	W Bg	Turkish	250	135·1	31·4	16·7	Austria	1882			Bulbulichioe Stavraki Favlaki & Bros.	Constantinople
273		Test *O'Neill* -84	W Bn	British	154	90·6	22·5	12·4	Portmadoc	1862	Williams	Y 2.82	T.Gill	Carnarvon
274		Tesver *Westerling*	W Sr	Russian	151	94·8	27·7	9·0	Björnö	1865	J.Jansson		B.F.Westerling	Raumo
275	N	Tetens *Marcussen* -87	W Bk	Norweg'n	483	135·1	31·4	16·7	Risör	1875	JacobHansen	Y 6.83	J.Kildahl	Risör
276		Tetuan *Brown* 89-89	W Bk	British	438	131·0	27·5	17·6	Sunderland	1860	W.Briggs&Sons	Y 86	P.V.Grant	Shanghai
277	GV	Teutonia *Spieske* 82-82	W Bk	German	632	150·9	31·6	18·5	Hammelwardn	1881	C.Luhring	Y 12.85	Gebr.Hustede	Elsfleth
278		Teviot *Newton*	W	British	134	101·7	21·2	11·7	Berwick	1848	A.B.Gowan&Sns		M.Fenwick	Scarborough
279	R	Tewksbury L. Sweat *Gooding*	W Bk	American	523	127·5	32·2	17·1	Yarmouth,Me.	1874	Hutchins&Stubbs	Y 8.85	T.L.Sweat	Portland,Me.
280		Texada *L'huillier* -88 (BROKEN UP)	W Bn	French	126	85·3	22·6	9·1	Canso,N.S.	1874		C 2.83	Folquet & fils	St. Malo
281		Texas *Nemesch* -87 (CONDEMNED)	W Bk	German	591	132·8	31·8	14·5	Vegesack	1867	J.Lange	Y 3.86	Von Riegen&Stindt	Bremerhaven
282	N	Th. Thoresen *Ness* 87	W Bk	Norweg'n	438	144·6	29·3	16·0	Porsgrund	1880	Stephensen	Y 11.85	O.Thoresen	Tonsberg
283	❄D	Thalassa *Stokje* 80-80	W3mSr	Dutch	238	128·6	22·6	11·9	Hoogezand	1872	J.A.Hooites	Y 8.88	J.A.Hooites	Hoogezand
284	V	Thalassa *Otto* 87-87	I Bk	German	647	160·0	29·0	18·0	Hamburg	1875	Reiherstieg Schiffs-Werfte		Wachsmuth & Kregmann	Hamburg
285	N	Thalassa *Jacobsen* -86 (ex Beck Lyon)	W Bk	Norweg'n	779	165·2	32·2	20·8	Quebec	1864	Dinning	Y 10.84	J.Janssen	Fredrikshald
286	❄	Thalatta *Pennicuik* 79-80	I S	British	1748	263·0	39·1	24·1	Liverpool	1881	W.H.Potter&Sn		McDiarmid,Greenshields & Co.	Liverpool
287	❄V	Thalia *Christiansen* 88-88	I Bk	German	1416	247·8	37·6	21·5	Rostock	1887	Rostock S.B.Co.		B.Wencke Söhne	Hamburg
288	V	Thalia (BURNT) *Stolp* 87-87	I S	German	1060	210·0	34·2	21·0	Flensburg	1878	Flensburg Schiffsb.Ges.		J.C.Pflugk	Hamburg
289	A	Thalia Theocharidou *Andreadi* (ex Salamina)	W Bk	Greek	296	106·9	27·5	..	Syra	1876		Y 86	Coliroi Fabbiati	Syra
290	D	Thames *Reece* -88	W Bk	British	413	132·6	27·3	17·0	Scarborough	1861	Tindalls	Y 1.84	C.F.B.Birchall	London
291	❄	Thames *Goldfinch* 88-88	W Bn	British	162	103·7	24·0	12·7	Plymouth	1875	Marshall		Whitstable Shipping Co.Ld.	Faversham
292		Thames	W	British	151	110·7	22·3	7·0	Quebec	1880	A.Miller		James Wood	Ottawa,Ont.
293		Thames	I K	British	134	80·8	21·2	10·3	London	1882	Halsey&Co.		C.Halsey	London
294		Thamine	I Flat	British	777	197·0	25·7	8·3	PortGlasgow	1878	RbtDuncan&Co		Irrawaddy Flotilla Co. Ld.	Glasgow
295		Thanaletchmy *Sylvestry*	W Bg	British	141	94·0	21·8	13·6	Moulmein	1856			Lana Ana Chena Chinnayachitty	Jaffna
296		Tharros	W Bg	Greek	107	..	..	..	Malta	1856			E.Bouriki	Zante
297		Thassos		Greek	179	121·3	19·7	9·8	Austria	1871			Theophilatos&Stathatos	Ithaca
298		Thatone	I Barge	British	358	196·4	25·2	8·4	PortGlasgow	1881	R.Duncan&Co.		Irrawaddy Flotilla Co. Ld.	Glasgow
299		Thaubo	I St	British	361	195·2	25·0	8·5	Port Glasgow	1880	R.Duncan&Co.		Irrawaddy Flotilla Co. Ld.	Glasgow
300		Thayebnayahy *Caloopulle*	W Dhoney	British	148	100·0	23·2	15·5	Valvetiturai	1888	K.K.Valoopulle		Therumamear Ramasamy	Jaffna

❄=Lloyd's Register; D=Nederlandsche Vereeniging; G=Germanischer Lloyd; I=Registro Italiano; R=American Record; A=Veritas Austro-Ungarico; H=Veritas Ellenico; N=Norske Veritas; V=Bureau Veritas.

Dans la 4me Colonne, C=Composite (membrure en fer, bordé en bois); I=Fer; Stl=Acier; W=Bois.

Les Dates du doublage en métal (Col. 11) ressortent des livres dans lesquels les navires sont respectivement classés ou des rapports des inspecteurs du Lloyd's Register.

Pour plus amples renseignements sur les navires classés par le Lloyd's Register, voir le Lloyd's Register Book et le Lloyd's Yacht Register.

Official No.	No. in Book.	Registers in which classed.	Ship's Name. Material and Rig. Master, and Dates of Appointment to present Owner's Service and to Ship. Ship's Late Name.		Flag.	Net Tonnage.	Dimensions. Length.	Breadth.	Depth.	Built. Where.	When.	By whom.	Date of Metalling.	Owners or Managers.	Port
[illegible]	301		Thaynaletchemy	W Bg	British	178	[illegible]	22·2	11·[illegible]	[illegible]	1867			Firumoneer Tampen-[illegible]	[illegible]
84045	302	⁕	The Bee *Jackson 87*	W Sk	British	24	[illegible]	[illegible]	[illegible]		1887	G. & T. Smith		J.C. Little	Grim[illegible]
[illegible]	303	⁕	The Bruce *Eastcott 88-88*	I S	British	1146	22[illegible]·7	35·6	22·3	Glasgow	1866	Aitken & Mansel		J. Nourse	London
58080	304	⁕	The Douglas *Hellt* ([illegible])	I S	British	1428	22[illegible]·2	37·8	25·[illegible]	Liverpool	1869	T. Royden & Sons		W. & H. Wright ([illegible] Trustee)	Liv[illegible]
70179	305	⁕	TheFrederick *Simpson*	I Bk	British	812	18[illegible]	32·0	19·[illegible]	[illegible]	1875	Osbourne {Graham & Co.		J.N. Ward & Co.	Flee[illegible]
	306	⁕	The Governor *Higginbottom*	W Sk	British	96	81·5	21·2	10·[illegible]	[illegible]	1885	Hunt & Fowler		Hunt, Fowler & Co.	Hull
88170	307	⁕	The Hahnemann *Rogers*	I S	British	1937	269·8	39·5	24·1	[illegible]	1882	Oswald, Mordaunt & Co.		Stuart & Douglas	[illegible]
3483	308	⁕	TheLordWarden *Ossoinack 88*	W S	Austrian	1337	21[illegible]	36·7	22·7	Sunderland	1862	Pile, Hay & Co.	v. 88	G. Ossoinak di G.	[illegible]
82549	309	⁕	The Macbain *Taylor 88*	W Bkn	British	285	121·5	25·1	14·[illegible]	[illegible]	1881	J. Roney	v. 88	Robertson Bros.	[illegible]
54770	310		The Saint *Dodd 83-86*	W Sr	British	107	87·2	21·3	10·7	[illegible]	1870	Greenlon		R. Clark	[illegible]
61508	311		The Star *Robbins 80*	W Sr	British	118	82·5	25·[illegible]	8·1	[illegible]	1871			I. Letteney	A[illegible]
47427	312	⁕	The Tweed *Moore 80-85* (CONDEMNED) (ex Punjaub)	W S	British	1646	25[illegible]	39·6	24·7	Bombay	1857		v. 88	J. Willis & Son	London
	313	V	Thea *Jensen 77-77*	W Bn	Danish	240	116·6	25·1	15·[illegible]	[illegible]	1877	Helsingborg {Skepps. Verk.	v. 87	Moses & Son & Melchior	Copen[illegible]
	314	N	Theckla (ex Tecla) *Brenne*	W Bk	Norweg'n	329	125·1	29·0	15·7	M[illegible]	1854			P. Christiansen	Sarp[illegible]
38	315		Thecla	W Bg	Greek	260	101·3	27·9	16·1	Syra	1872	A. Lovonbria		P. Kontoulis	Ang[illegible]
	316	N	Thecla *Dahl 86-86*	W Bg	Norweg'n	292	106·6	27·3	14·1	Flensborg	1844			C.M. Sundt	F[illegible]
74863	317	⁕	Theda *Pedwyth*	W BmSr	British	150	100·0	23·6	12·4	Pwllheli	1876	Prichard	v. 87	Mrs. M. Jones	Carn[illegible]
	318	V	Thedea *Behrens*	W Sr	German	107	82·7	20·9	9·1	Jemgum	1866	G.H. Ader		F.H. Behrens	Ho[illegible]
	319	NR	Theis Lundegaard (ex Diana) *Reinertsen*	W BmSr	Norweg'n	462	12[illegible]	30·2	18·4	Chiavari	1870	L. Gotuzzo		F.M. Reinertsen (S. Abrahamsen)	Farsund
	320	V	Thekla (ex Ethel Gem) *Hansen 88*	W S	Norweg'n	1479	211·0	38·7	24·0	St. John	1876	M'Fee	v. 88	G.C. Hansen	Tons[illegible]
	321	N	Thela *Mathiesen 86-86*	W Bg	Norweg'n	362	122·8	29·1	15·3	Arendal	1871		v. 88	M. Chr. Mathiesen	Arendal
	322		Themis (ex Polly) *Haslewille*	W Bkn	French	246	123·0	25·0	12·8	Sunderland	1878	W. Pickersgill		E. Follin	Fécamp
	323	V	Themis *Buschen 88-88*	W Bg	German	330	118·3	26·5	12·[illegible]	[illegible]	1874	W. Hagens	v. 88	D. Haye	Brake
6	324	A	Themis	W Bg	Greek	358	115·1	29·3	..	Syra	1876	A. Lividoco	v. 88	J.K. Iksas	Piræus
	325	N	Themis *Weiby*	W Bk	Norweg'n	527	139·7	30·6	17·5	Grimstad	1873	J. Tellefsen	v. 88	P. Pedersen	Grims[illegible]
	326	N	Themis *Sørensen*	W Bk	Norweg'n	423	124·7	31·2	16·7	Wasa	1840			C. Sørensen	Tonsb[illegible]
	327	N	Themis *Jacobsen 87*	W Bg	Norweg'n	211	99·3	23·5	13·6	W[illegible]	1850			H. Tambs	Fredr[illegible]
1195	328		Themis D (WRECKED)	W Bk	Italian	491	139·1	30·7	19·1	[illegible]	1871	E. Taxi	v. 88	G. D'Ottone	Leghorn
31	329	A	Themistocles	W Bg	Greek	303	107·2	26·2	15·9	Syra	1874	Epitanis	v. 86	S. Siscos	Scopelo
	330	A	Themistocles *Asimacis*	W Bg	Greek	267	..	..	..	Sunderland	1866		v. 86	G. Glicco	Syra
318	331		Themistocles	W Sr	Greek	135	78·1	22·5	..	Syra	1870	C. Cristali		S.C. Ragias	Syra
24546	332		Theobald *Collins 87*	W Bk	American	932	172·0	34·0	23·6	[illegible]	1861	Farran		Renton, Holmes & Co.	San F[illegible]
	333		Theodor (ex [illegible]) *Grube 86-88*	I Bk	German	418	119·2	26·6	16·[illegible]	[illegible]	1868	W. Pile & Co.		A. Schiff	Elsfleth
	334		Theodor *Sassen*	W Sr	German	116	83·3	18·7	9·7	Edewecht	1882	H. Kramer		Meinhard Sassen	Barssel
	335	N	Theodor *Andreassen 83-83*	W Bk	Norweg'n	293	127·7	27·8	12·8	[illegible]	1871	J.T. Tryde	v. 88	J.M.A. Marenssen	Lyngør
	336	N	Theodor (ex [illegible]) *Johannesen 78*	W Bg	Norweg'n	225	119·9	24·4	12·2	[illegible]	18[illegible]			J.M. Johannesen	Moss
408	337	G	Theodor *Kalnbehrsing*	W Sr	Russian	207	100·0	27·6	12·8	[illegible]	1884	A. Boschmann		T.A. Kalnbehrsing	Riga
	338		Theodor Behrend (STRANDED) *Lorenz 88*	W Bk	German	675	130·1	33·0	22·4	Danzig	1862	J.W. Klawitter		Schoenberg & Domansky	Danzig
	339	G V	Theodor Fischer (ex Loading Fish) *Van Essen*	W S	German	1944	212·0	42·2	21·3	Thomaston, Maine	1869	S. Watts & Co.	v. 88	Siedenburg, Wendt & C.	Bremen
	340	G V	Theodor Körner (ex Tabor) *Kohnenkamp 83-83*	W Bk	German	1091	177·6	3[illegible]	23·4	Boston, Mass.	1864	J. Taylor		Gildemeister & Ries	Bremen
	341	V	Theodor Rüger (ex Lightning) (COLLISION) *Meyer*	W S	German	1576	219·1	40·0	25·7	Portland (St. John, N.B.)	1872	T. Hilyard	v. 88	H. Wessels	Bremen
93025	342	⁕	Theodora *Grunill*	W Sk	British	84	76·[illegible]	20·4	10·[illegible]	[illegible]	1886	J. Kremer		F. Grunill	Grimsby
173	343	V	Theodora *Caliza 88*	W Bg	Greek	238	109·2	29·0	17·1	Syra	1879	P. Bungis	v. 88	Th. Caliza	Hydra
	344	D N	Theodora *Landmark 71-80*	W Sr	Norweg'n	170	102·1	23·9	11·4	Drammen	1871			P. Laz. Aass	Dram[illegible]
93631	345	⁕	Theodore *Banks*	W Sk	British	84	76·1	20·1	10·6	Elmshorn	1886	J. Kremer		A.E. Banks	Grimsby
	346	V	Théodore *Le Diabat 76*	W Bn	French	196	91·3	22·3	12·3	Nantes	1876	T. Dubigeon & fils	v. 88	T. Dubigeon & fils	Nantes
	347	⁕ V	Theodore (ex County [illegible]) *Nicolaisen 88-88 Arevic*	I S	German	699	182·5	30·0	18·9	Glasgow	1865	C. Connell & Co.		C.C. Koch	Brake
23112	348		Theodore Dean *Blake*	W Sr	American	332	133·0	31·4	11·4	[illegible] N.Y.	1864			J.B. Phillips	Fall R[illegible]
	349		Théodore Ducos (BURNT) *Despain 77*	W Bk	French	603	133·5	29·6	22·4	Bordeaux	1866	Chaigneau frères		J.H. Tandonnet & frères	Bordeaux
	350	⁕ V	Théodore Engels *De Vries 87*	I Bk	Belgian	498	163·[illegible]	27·9	16·4	S. Shields	1869	Readhead, Softley & Co.		T.C. Engels & Co.	Ant[illegible]

⁕ = Lloyd's Register; D = Nederlandsche Vereeniging; G = Germanischer Lloyd; I = Registro Italiano; R = American [illegible]
A = Veritas Austro-Ungarico; Π = Veritas Ellenico; N = Norske Veritas; V = Bureau Veritas

In the 4th Column, C = Composite (iron frame, wood planked); I = Iron; Stl = Steel; W = Wood.

The Dates of Metalling (Col. 11) are based on the records of the Books in which Vessels are respectively classed, or on the Reports of the Surveyors of Lloyd's [illegible]

* *Further particulars of Vessels classed by Lloyd's Register will be found in Lloyd's Register Book and Lloyd's Yacht Register.*

No.		Nom. Matériaux et Gréement. Capitaine et dates d'entrée au service des Armateurs actuels et à bord du navire. Ancien Nom.	Pavillon.	Tonnage net.	Dimensions. Longueur	Largeur	Creux.	Construit. Où.	Quand.	Par qui.	Date du doublage en métal	Armateurs ou Directeurs-Gérants.	Port d'Armement.
351	DV	Theodore H. Rand *Morris* W S	British	1198	191·0	38·8	22·6	Three Sisters, N.S.	1878		V 1.88	Scammell Bros.	Parrsboro', N.S.
352	A	Theodoros *Vassilaki 83* W Bk	Greek	545	..	..	..	Sestri	1859		V 1.88	D. Callinico	Ithaca
353		Theodoros W Bn	Greek	427	124·6	..	..	England	1869			P. Lemanas	Ithaca
354	A	Theodosios *Travos* W BnSr	Turkish	226	100·1	24·1	..	Chios	1880			M. Ligomalas & Co.	Chios
355	A	Theologos *Pieris* W Sr	Greek	102	77·8	21·0	..	Syra	1884			Emm. Pieris	Syra
356	✻	Theophane *Nixon 82-83* I S	British	1519	218·4	38·0	23·7	Liverpool	1868	R & J Evans & Co.		Australasian Shipping Co. Ld.	Liverpool
357		Théophile Marie *Fardel* W Sr	French	127	74·1	21·3	10·5	Nantes	1857	Chauvelon		A. Fardel	Nantes
358	A	Theotokos W Bk	Greek	479	135·5	30·8	..	Syra	1876		V 2.86	N. Stefanakis	Syra
359		Theresa *Glass* W BnSr	American	195	114·0	27·0	10·0	Noank, Conn.	1874			T. D. Miller	New York
360		Theresa Wolf *Robinson* W BnSr	American	292	126·2	33·7	10·0	May's Landing [N.J.	1874			E. B. Champion	Somer's Pt., N.J.
361	R	Therese *Wood -88* W S	American	986	163·0	35·2	24·0	Richmond, Me.	1854		Y 10.86	E. Dunham's Nephew & C.	New York
362	V	Therese *Brouwer -88* W Bk	Dutch	413	139·4	25·4	15·0	Altona	1863	E. Dreyer	V 3.81	S. van der Hei	Heveskes
363	V	Thérèse *Lebreton [illegible]* W Bg	French	152	103·0	21·0	11·3	Granville	1875	Le Carpentier	V 8.87	Riotteau & fils	Granville
364		Therese *Möller* W Bk	German	391	128·0	30·7	15·2	Apenrade	1865	J. Paulsen		H. W. Clausen	Eckernförde
365	G	Therese *Gart 79-81* W Bk	German	319	115·0	28·3	14·7	Greifswald	1866	J. W. Sadelkow		C. Graedener	Greifswald
366	GV	Therese *Bellmer 87-87* W Bg	German	318	118·4	26·8	13·0	Brake	1879	J. Nicolai	Y 12.83	Gerd Heye	Elsfleth
367	✻	Therese *Seetzen* W BnSr (ex Kourier ex Rosebuds)	German	158	112·8	22·2	11·5	Tayport	1869	Dickie	V 7.86	P. F. Seetzen	Elsfleth
368		Therese (WRECKED) *Kruse 77-* W Sr	German	151	74·5	23·2	12·0	Stralsund	1862	O. Kirchoff		F. Böttcher	Stralsund
369		Therese *Prohn 77-77* W Sr	German	108	72·9	21·2	11·6	Zingst	1857	H. Drossel		J. J. Wallis & Sohn	Barth
370	N	Therese (ex Indomitable) *Wang -88* W Bk	Norweg'n	635	180·7	29·4	19·4	Sunderland	1853			H. M. Rummelhoff	Christiania
371	V	Thérèse Victor *Philippe -87* W Bg	French	213	89·2	22·6	13·8	St. Malo	1875	Gautier	Z 2.88	A. Le Gualès de Mezaubran	Saint Brieuc
372	V	Thérèse & Nelly *Fleuriot* W Bk	French	356	120·0	25·3	15·7	Nantes	1874	J. Sevestre	V 8.88	André Fleuriot	Nantes
373	✻D	Thermopylæ *Jackson 84-88* C S	British	948	212·0	36·0	21·0	Aberdeen	1868	W. Hood & Co.	V 4.87	G. Thompson & Co.	Aberdeen
374	✻	Thermuthis *Grimwood -88* W K	British	91	80·7	21·0	10·8	Brixham	1888	J. W. & A. Upham		J. W. Hellyer	Hull
375		Thessalia W Sr	Greek	197	96·5	24·4	..	Syra	1880			I. Damlassis	Syra
376	✻	Thessalus *Bennett 64-74* I S	British	1782	269·0	41·2	23·7	Glasgow	1874	Barclay, Curle & Co.		A & J H Carmichael & Co.	Greenock
377	NR	Theta (LOST) *Olsen 85-88* W Bk	Norweg'n	518	154·0	29·4	18·7	Bideford	1865	Cox	V 8.85	Møller & Norbom	Christiania
378	V	Theta (WRECKED) *Hansen* W Bk (ex Hebe ex Glendalar)	Norweg'n	381	123·3	27·1	16·7	Sunderland	1862	J. Gill	V 1.88	Actieselskab (Chr. Hansen)	Christiania
379	G	Theta *Wenger -84* W BnSr	Russian	183	114·3	28·4	10·4	Harra	1884	L. Bootsmann		Count N. Stenbock	Harra
380	✻	Thetis *Shearer 80-85* Stl Bk	British	1295	218·5	35·4	21·6	Dundee	1885	A. Stephen & Sns.		A. Stephen & Sons	Dundee
381	✻	Thetis *Lennard* W Bkn	British	380	139·9	29·5	14·0	Charlottetown, P.E.I.	1888	J. Yeo	V 1.89	J. Yeo	Charlottetown, P.E.I.
382		Thetis (BROKEN UP) *Morris 83* W Cr	British	148	78·6	22·8	12·6	Yarmouth, N.S.	1883			C. Weatherburn	South Shields
383	V	Thetis *Frederiksen 85-85* W Bk	Danish	242	100·7	21·1	14·4	Rønne	1855	S. P. Bech		Jørgen Hoppe	Copenhagen
384	V	Thetis *Pill 76-76* W Bn	Danish	144	94·8	21·6	10·3	Troense	1877	C. J. Jensen		C. J. Jensen	Svendborg
385	V	Thetis (ex Anna Marie) *Kolb -88* W Bn	German	125	78·7	21·6	10·2	Rendsburg	18 57/74	H. Frahm		Zerssen & Co.	Rendsburg
386	G	Thetis (CONDEMNED) *Paschen 82* W Sr	German	111	81·0	20·5	10·2	Barth	1877	J. S. Kraeft		J. N. Rodbertus	Barth
387		Thetis *Corneliussen* W Bg	Norweg'n	190	83·1	25·5	11·1	Kragerö	1847			M. Corneliussen	Lillesand
388		Thetis W Bg	Spanish	170	113·1	19·3	13·0	Liverpool				José de Novas	Barcelona
389		Thettalomagnissia W Bg	Greek	210	..	..	..	Syra	1870			D. S. Calamidas	Poros
390	A	Thia Charis *Pavlou* W Bg	Greek	157	79·7	20·2	11·3	Spezzia	1867			N. J. Pavlou	Spezzia
391	N	Thinca *Weigar -88* W Bk	Norweg'n	438	137·0	31·6	15·6	St. John, N.B.	1863			M. Weyergang	Mandal
392	✻	Thiorva *Fraser 84-84* W S	British	1174	198·6	38·5	22·7	N. Glasgow N.S.	1876	J. W. Carmichael & Co.	Y 11.87	J. W. Carmichael	Pictou, N.S.
393	✻	Thirlmere *Brown 85-85* I S	British	1711	260·0	39·1	23·1	Whitehaven	1874	Whitehaven Ship B. Co.		Fisher & Sprott	Liverpool
394		Thirza *Pont 88-88* W Bn	British	200	104·1	23·8	12·7	P. E. Island	1865	Longworth		G. H. Gann	Whitstable
395		Thistle *Roberts -87* W BnSr	British	139	109·3	23·6	9·5	Eagleton NSW	1886	Muirhead Bros.		Neil Brodie	Sydney, N.S.W.
396		Thistle *Osten -87* W K	British	137	103·0	23·5	8·9	Sittingbourne	1887	J. Taylor		J. Smith	Colchester
397		Thistle *Mearns -79* W Sr	British	114	86·0	20·0	11·0	Bowling	1863	Scott & Macgill		R. Lewis, sen.	Greenock
398		Thistle W Sr	British	114	82·2	23·8	10·1	Chicoutimi	1887	E. J. Price		E. J. Price	Quebec
399		Thistle Stl Cr Yt	British	191	98·0	20·3	14·1	Glasgow	1887	D. & W. Henderson & Co.		J. Bell	Glasgow
400													

✻ = Lloyd's Register; D = Nederlandsche Vereeniging; G = Germanischer Lloyd; I = Registro Italiano; R = American Record; A = Veritas Austro-Ungarico; H = Veritas Ellenico; N = Norske Veritas; V = Bureau Veritas.

Dans la 4me Colonne, C = Composite (membrure en fer, bordé en bois); I = Fer; Stl = Acier; W = Bois.

Les Dates du doublage en métal (Col. 11) ressortent des livres dans lesquels les navires sont respectivement classés ou des rapports des Inspecteurs du Lloyd's Register.

Pour plus amples renseignements sur les navires classés par le Lloyd's Register, voir le Lloyd's Register Book et le Lloyd's Yacht Register.

Official No.	No. in Book	Registry in which classed	Ship's Name. Material and Rig. Master, and Dates of Appointment to present Owner's Service and to Ship. Ship's Late Name.	Flag	Net Tonnage	Length	Breadth	Depth	Built Where	When	By whom	Date of Metalling	Owners or Managers	Port
15552	401		Thomas *Rushmer 60-66* WSr	British	142	87·3	20·5	11·5	Barnstaple	1837			T.Stewart & Co.	Lon[illegible]
	402	V	Thomas *Bang* WBn (ex Johanne [illegible])	Danish	133	97·5	20·6	10·4	Aalborg	1867	N.P.Vang		J.Chr.Bang	[illegible]
	403	DN	Thomas *Amundsen* WBg	Norweg'n	305	107·5	28·9	14·6	Risör	1855			J.Amundsen	Fre[illegible]
14505	404	RV	Thomas A. Goddard WBk *Sears*	American	686	154·9	34·0	18·5	Duxbury,Mass	1874	MerrittBros.	[illegible]	Laforme & Frothingham	Bosto[illegible]
23098	405		Thomas Arbuthnot WS	British	621	131·0	26·5	19·1	Aberdeen	1841			F.Bryant	Bombay
29659	406		Thomas Aylan *Taylor* WSr	British	115	85·0	21·6	11·7	Fowey	1860	N.&J.Butson		T.H.Knight	Fowey
145252	407	R	Thomas B. Garland W3mSr *Kelley 79-*	American	303	126·6	32·0	11·4	Bath,Maine	1881	Goss,Sawyer [&Packard		B.F.Nealley	Pe[illegible]
45304	408	✳	Thomas Bell *Lee -4* I S	British	838	183·0	30·6	21·7	PortGlasgow	1862	J.Reid&Co.		J.Swanson	Liverp[illegible]
24079	409		Thomas Booz *Somers* W3mSr	American	295	116·9	29·8	8·7	Baltimore	1867	S.W.Tilton		J.C.Cottingham	Philad[illegible]
24174	410		Thomas Borden *Conary 85-86* WSr	American	199	105·0	28·0	8·5	Dartm'th,Mas	1860	D.Brown		R.P.Reynard	Fall [illegible]
82200	411	✳	Thomas Boustead *Jones 86-87* I Bkn	British	172	114·3	24·2	11·0	Berwick	1878	Gowan		W.Thomas	Liverpo[illegible]
24038	412		Thomas Brooks *Johnson* WBk	American	442	135·9	31·8	14·8	Setauket, N.Y.	1873	Hand&Son		F.Waydel	Machi[illegible]
68630	413	✳	Thomas C. Seed WBkn *Orange 88-88*	British	296	131·3	26·9	12·9	Sunderland	1873	Pickersgill	[illegible]	R.Gray	Fleetw[illegible]
74372	414		Thos. Clark Street WBk	British	338	138·4	25·6	11·5	St.Catharines, [Ont	1869	L.Shickluna		J.H.G.Haggarty	St.Cath[illegible]
24422	415		Thomas Clyde *Frazier -87* W3mSr	American	320	120·0	29·8	9·6	Milford,Del.	1866	W.A.[illegible]		J.W.Hall	Wilm[illegible]
18686	416		Thomas Connolly WBn (ex Cinque frères) *Davidson 85-86*	British	111	79·2	23·4	11·3	Brest	1837			W.Hinde	Maryp[illegible]
24867	417	RV	Thomas Dana *Gerrish 86* WS	American	1388	203·5	39·5	21·2	Newburyport, [Mass	1873	J.Currier,junr.	[illegible]	Thayer & Lincoln	Boston[illegible]
15223	418		Thomas Daniel *Dagns 81-82* WBk	British	292	124·0	24·0	15·1	Barnstaple	1856			Rider,Son & Andrew	Bristol
24585	419		Thomas G. Benton WSr *Anderson*	American	155	90·5	28·9	8·2	CapeMay,N.J.	1868			F.D.Averill	Ro[illegible]
145331	420	R	Thomas G. Smith W3mSr *Adams -86*	American	488	153·2	34·5	14·9	Wilmington, [Del	1883	Jacks'n&Sharpe [Co.		J.Swain	Camden
17497	421	✳	Thomas Hamlin *[illegible] -87* I Bk	British	688	161·0	30·1	20·0	Newcastle	1851	Coutts&Par-[kinson		R.W.Leyland	Liverp[illegible]
	422	✳	Thomas Henry WSk *Agerskow -88*	British	90	81·6	20·8	10·6	Rostock	1888	[illegible]Burchard&Co.		W.Agerskow	Hull
66953	423	V	Thomas Hilyard *Jones 82-82* WS	British	1500	209·7	39·5	21·4	St.John,N.B.	1874	HilyardBros.	[illegible]	D.V.Roberts	St.Joh[illegible]
145293	424	R	Thomas J. May W3mSr *Thewsing -88*	American	213	121·4	30·2	7·8	Milford,Del.	1882	Abbott		J.C.Davis	Philad[illegible]
71030	425	V	Thomas Keillor *Thomson* WBk	British	1096	184·0	37·1	21·8	DorchesterNB	1875	W.Hickman	[illegible]	W.Hickman	Dorch[illegible]
145245	426	R	Thomas L. James W3mSr *Lounsbury 83-83*	American	377	138·0	35·0	11·0	N.Hav'n,Conn	1883	H.H.Hanscom		N.Easterbrook,jun.	N.Hav[illegible]
24057	427		Thomas Lord *Brody* WS	American	1316	190·0	34·9	24·0	Bath,Maine	1864		[illegible]	J.H.Whitaker	Boston
145215	428	R	Thomas M. Reed *Work* WS (BURNT)	American	1988	227·4	42·1	28·0	Bath,Maine	1880	A.Sewall	[illegible]	A.Sewall & Co.	Bath [illegible]
24612	429		Thomas N. Stone W3mSr *Bryne 88-88*	American	375	126·0	30·8	14·8	Chelsea,Mass.	1867			R.R.Freeman	Boston
80230	430	✳	Thomas Owen *Williams* WSr	British	81	71·2	21·0	10·8	Portmadoc	1880	Williams		J.Williams	Carnarv[illegible]
24873	431		Thomas P. Ball *Ryder* W3mSr	American	408	133·0	32·1	15·7	PJefferson,NY	1871		[illegible]	Evans,Ball & Co.	New Yo[illegible]
75730	432	RV	Thomas Perry *Grant -87* WBk	British	1192	198·0	37·5	22·8	GreenCove NS	1879	J.W.Raymond	[illegible]	W.Law	Yarmo[illegible]
24337	433		Thomas Pope *Fisher* WBk	American	227	100·0	27·9	14·6	Mattapoisett, [Mass	1856	J.Holmes		J.N.Knowles	New Bed[illegible]
24248	434		ThomasPotter *Humphrey* WSr	American	114	80·4	25·7	7·8	NLondonConn	1852			W.Handy	NewBed[illegible]
230	435		Thomas R. Foster *Rugg* WBk (ex Hermann, ex Uncle Toby [illegible])	Hawaian	1122	195·0	35·0	26·0	Freeport,Me.	1866			J.Campbell	Honolu[illegible]
24982	436	R	Thomas R.Pillsbury W3mSr *Fisher*	American	501	140·7	32·1	12·6	ThomastonMe	1874	S.Watts	[illegible]	S.Watts	Thom[illegible]
	437	N	Ths.S.Falck *Jacobsen 83-83* WBk	Norweg'n	519	139·7	30·3	18·5	Stavanger	1875	M.A.Magnus	[illegible]	Holdt & Isacksen	Stava[illegible]
45667	438	✳	Thomas S. Stowe *Fraser* I Bk	British	686	191·3	28·1	17·1	Sunderland	1863	Oswald&Co.		Hargrove & Hellon	Liverpo[illegible]
24110	439		Thomas Sinnickson WSr	American	247	114·8	30·0	8·5	Milford,Del.	1866			A.Dickerson	[illegible]
63190	440	✳	Thomas Stephens I S *Gillespie*	British	1469	263·0	38·2	23·1	Liverpool	1869	W.H.Potter&Co.		T.Stephens & Sons	London
24293	441		Thomas Turull *Edwards* WBg	American	190	99·2	25·5	10·4	N.HavenConn	1861	E.Lane		L.W.& P.Armstrong	NewHa[illegible]
24043	442		Thomas van Gilder WSr *Van Gilder*	American	215	107·2	29·4	9·1	Tuckahoe,N.J.	1873			T.van Gilder	Somers[illegible]
24996	443		Thomas W.H.White W3mSr	American	203	111·6	29·6	9·2	Tuckerton,NJ	1867			S.H.Haines	New Y[illegible]
24552	444		Thomas W. Haven W3mSr *Hutchings*	American	299	126·6	32·7	10·2	Bath,Me.	1872	Goss&Sawyer		Rackett & Co.	Greenp[illegible]
145366	445		Thomas W. Holder W3mSr *Stevens*	American	220	121·0	28·0	9·7	Bucksport,Me	1875			Joseph Stevens	Malde[illegible]
145248	446	R	Thomas W. Hyde W3mSr *[illegible]*	American	489	117·9	33·2	12·0	Bath,Me.	1881	Adams&Hitch-[cock		Adams&Hitchcock	Bath,Me
24546	447		Thomas W. Rowland WBg	American	366	117·3	29·0	12·0	Brooklyn	1855			D.D.Baker	New Yo[illegible]
29530	448		Thomas Whitney WBk	British	230	112·8	26·4	13·3	Boston,U.S.	1860			J.Carnegie	Turks I[illegible]
46032	449		Thomas & Henry WBg	British	215	91·9	23·7	11·2	ManningRiver [N.S.W.	1858			J.Mill	Dunedi[illegible]
26258	450		Thomas & Mary WSw	British	264	103·1	25·8	15·8	Sunderland	1861	S.P.Austin&Son		R.L.Manners	Sunde[illegible]

✳ = Lloyd's Register; D = Nederlandsche Vereeniging; G = Germanischer Lloyd; I = Registro Italiano; R = American [illegible]
A = Veritas Austro-Ungarico; H = Veritas Ellenico; N = Norske Veritas; V = Bureau V[illegible]

In the 4th Column, C = Composite (iron frame, wood planked); I = Iron; Stl = Steel; W = Wood.

The Dates of Metalling (Col. 11) are based on the records of the Books in which Vessels are respectively classed, or on the Reports of the Surveyors of Lloyd's [illegible]

Further particulars of Vessels classed by Lloyd's Register will be found in Lloyd's Register Book and Lloyd's Yacht Register.

1	2	4 Nom. Matériaux et Gréement. Capitaine et dates d'entrée au service des Armateurs actuels et à bord du navire. Ancien Nom.	5 Pavillon.	6 Tonnage net.	7 Dimensions. Longueur.	Largeur.	Creux.	8 Construit. Où.	9 Quand.	10 Par qui.	11 Date du doublage en métal.	12 Armateurs ou Directeurs-Gérants.	13 Port d'Armement.
451	*	Thomasina MacLellan *Mann 83-83* I S	British	1788	262·6	40·7	23·8	Dumbarton	1873	A.McMillan&Sn.		W.T.MacLellan	Glasgow
452		Thongwa I Flat	British	777	197·0	25·7	8·3	PortGlasgow	1878	Robt. Duncan & Co.		Irrawaddy Flotilla Co. Ld.	Glasgow
453		Thoodaw I Barge	British	211	115·0	19·3	5·0	PortGlasgow	1879	Robt. Duncan & Co.		Irrawaddy Flotilla Co. Ld.	Glasgow
454		Thoon Kramon WS	Siamese	475	151·0	28·0	15·0	Bangkok	1866			Siamese Government	Bangkok
455		Thor *Strömberg -88* WBk (ex Meta, ex Uekalina)	Danish	460	128·2	29·7	18·2	Memel	1861	B.Pieper		C.P.Strömberg	Dragör
456		Thor *Kofoed 72-72* WBk	Danish	240	100·9	24·2	14·1	Rønne	1854	S.P.Bech		C.Thorsen	Rønne
457		Thor *Nielsen 72-72* W3mSr (ex Ethel)	Danish	218	125·8	23·8	10·3	Sundsvall	1862	J.Lindberg		J.Z.Levy	Horsens
458	V	Thor *Andersen* WBn	Danish	145	98·9	21·2	10·0	Aalborg	1868	S.L.Sørensen	Y 4.88	Chr.Simoni	Aalborg
459	G	Thor *Levinsen 80-80* WSr	Danish	121	84·5	22·7	10·6	Nykjøbing	1849	E.C.Benzon		H.P.Levinsen	Marstal
460	NV	Thor *Steinert 82-82* WS (ex Aldborough)	Norweg'n	1106	187·3	37·0	22·3	Riv.John,N.S.	1875	J.Kitchin	N 11.86	M.& H.Ingier	Christiania
461	N	Thor (WRECKED) *Gundersen* WBk	Norweg'n	345	107·8	28·6	16·9	Kaskö	1840			H.Gundersen	Sandefjord
462	N	Thor *Amundsen -87* WBk	Norweg'n	338	126·6	28·9	15·1	Kragerö	1871	J.Andreasen	Y 9.88	Chr.Ellefsen	Kragerö
463	N	Thor *Christensen 86-86* WSr	Norweg'n	150	99·6	23·3	11·1	Bergen	1870	P.Kaaven		Carl J.Christensen	Stavanger
464		Thor (ex Guldkasin) WBk	Swedish	341	113·5	25·1	16·1	Finland	1876			C.Westöö	Storungs,Gotlnd
465	V	Thor *Lundwall 81-81* WSr	Swedish	223	109·5	23·2	11·7	Oscarshamn	1874	C.Thoren	Y 6.88	J.E.Hagström	Oscarshamn
466	V	Thor *Andersson -88* WSr (ex Hemmin)	Swedish	157	85·6	21·0	11·7	Åmal	18 62/77			B.Bengtsson	Falkenberg
467		Thor *Backe -85* WSr	Swedish	145	99·1	22·2	9·8	Oscarshamn	1864	C.Thorén		P.Hansson	Råå
468	DN	Thor Hovland *Jensen* WBg	Norweg'n	214	100·0	26·4	13·2	Drammen	1853	Knudsen		C.Lingsom	Drammen
469	DN	Thor Wærner *Stern 83-86* WBk	Norweg'n	400	136·0	31·7	14·7	Uleåborg	1858			S.Larssen	Drammen
470	V	Thora *Jessen 67-67* WBn	Danish	123	84·7	19·7	9·6	Bandholm	1868	Wolters	Z 5.87	A.H.Schade	Nykjøbing,Jylld.
471	DNV	Thora *Nielsen -88* WBk	Norweg'n	646	156·2	33·5	18·2	Tvedestrand	1884	J.P.Pedersen	Y 5.84	J.A.Pedersen	Tvedestrand
472		Thora *Söderman* WSr	Swedish	103	75·1	22·7	9·7	Skeppsmyra	1872			P.Söderman	Tjockö,Rslgn.
473	DV	Thorbecke *Kuypers 85-85* WBk	Dutch	826	176·5	34·1	20·1	Nieuwendam	1875	A.H.Meursing	Y 3.87	A.H.Meursing	Amsterdam
474	DV	ThorbeckeII. *Steerenberg 83-83* WBk	Dutch	900	179·4	35·1	20·8	Nieuwendam	1877	A.H.Meursing	Y 2.87	A.H.Meursing	Amsterdam
475	DV	ThorbeckeIII. *Koster 85-85* WBk	Dutch	915	181·4	35·4	20·7	Nieuwendam	1878	A.H.Meursing	Y 4.87	A.H.Meursing	Amsterdam
476	DV	ThorbeckeIV. *DeJong 79-79* WBk	Dutch	917	182·4	35·4	20·9	Nieuwendam	1880	A.H.Meursing	Y 2.87	A.H.Meursing	Amsterdam
477	DV	Thorbecke V. *Karst 83-83* WBk	Dutch	793	174·5	35·4	20·4	Nieuwendam	1883	A.H.Meursing	Y 2.86	A.H.Meursing	Amsterdam
478	DV	Thorbecke VI. *Vander Mey 86-81* WBk	Dutch	911	194·5	35·4	20·1	Nieuwendam	1884	A.H.Meursing	Y 1.89	A.H.Meursing	Amsterdam
479	DV	Thorbecke VII. *Heijer 85-85* WBk	Dutch	929	194·5	35·8	20·4	Nieuwendam	1885	A.H.Meursing	Y 3.88	A.H.Meursing	Amsterdam
480	N	Thorfinn *Nilsen* WSr	Norweg'n	114	82·7	26·3	9·0	Tromsö(rebuilt)	18 85			O.Kavli	Tromsö
481	N	Thorgny *Andreasen -87* WBk	Norweg'n	448	137·6	30·1	16·6	Kragerö	1876	T.G.Litangen	Y 3.87	P.Rinde	Kragerö
482	*	Thorne *Hatchwell 78-78* I Bk	British	877	191·3	33·0	19·0	Sunderland	1878	R.Thompson,jun		J.Dowie & Co.	Liverpool
483		Thorney Island *Stoner 70-76* WBnFlat	British	185	102·0	24·7	11·4	Emsworth	1870	Foster		H.& J.Ash	Portsmouth
484		Thornhill *Simmons 64-80* WS	British	919	195·5	36·6	20·7	Quebec	1855			W.Simmons	Cork
485	*	Thornhill *Tozer 85-85* I Bkn	British	275	137·4	26·2	12·4	Hayle	1880	Harvey&Co.		T.C.Guthrie	Glasgow
486	*	Thornliebank *Henderson -88* I Bk	British	1405	224·5	37·6	21·8	Port Glasgow	1886	Russell & Co.		A.Weir & Co.	Glasgow
487	DN	Thorvald Heyerdahl *Christensen* WBk	Norweg'n	316	130·2	29·1	12·8	Bathurst,N.B.	1866			Winge & Co.	Christiania
488		Thorvaldsen *Schmidt 78-83* WBg (ex Janet, ex Summer...)	Danish	319	105·3	26·1	16·0	Grimstad	18 64/75	J.Tellefsen		P.P.Schmidt	Dragör
489		Thorvaldsen *Amundsen 54-72* WBk	Danish	275	117·4	25·7	12·7	Kolboda	1866	H.Olsen		Den.Kgl.Grønl.Handel	Copenhagen
490	R	Thrasher *Whelpley 86-86* WSr	British	123	87·0	26·1	8·4	MossGlen,NB	1881	G.Merritt		Scammell Brothers	St.John.N.B.
491		Three Brothers WBk (ex Kim Guan Bin)	British	367	135·0	27·0	17·0	Bangkok	1859			Chan Hye Kim	Singapore
492		Three Brothers *Peterson* WSr	American	134	92·4	29·6	7·0	Tuckahoe,N.J.	1872			E.D.Bailey	Somers Pt.,N.J.
493		Three Cheers *Francis -86* WSr	British	100	93·2	22·2	9·6	Omaha,N.Z.	1882	D.Matheson	Y 1.82	Mrs. C. Matheson & D. H. McKenzie	Auckland,N.Z.
494	*	Three Janes *Bennett 87-87* WSr	British	81	83·0	21·0	9·5	Conway	1864	Roberts		R.Mitchell & Co.	Swansea
495	*	Three Sisters *Owens -88* WSp	British	33	55·0	17·0	7·2	Ceibach	1882	E.Daniel		T.Thomas	Cardigan
496	R	Three Sisters *Simpson* W3mSr	American	288	124·0	30·6	11·1	Milford,Del.	1872			J.May & Sons	Philadelphia
497		Three Sisters *Zachrisson -86* WBg	Swedish	248	93·5	21·1	15·0	England				P.A.Lindberg	Gothenburg
498	V	Thrine *Brandt -75* WBn	Danish	155	94·5	23·5	11·3	Nakskov	1875	Ridersborg & Tischmann		I.F.Holmblad	Nakskov
499		Thrine *Jacobsson 88-88* WBk (ex Dagmar)	Swedish	454	147·0	27·9	16·2	Bjerno	1874	G.Isaksson		J.Barfoed	Stockholm
500	N	Thrudvang *Fredriksen -81* WBk	Norweg'n	354	120·7	29·0	15·7	Mandal	1861	S.Ericksen		Sv.Svenson	Mandal

=Lloyd's Register ; D=Nederlandsche Vereeniging ; G=Germanischer Lloyd ; I=Registro Italiano ; R=American Record ;
Veritas Austro-Ungarico ; H=Veritas Ellenico ; N=Norske Veritas ; V=Bureau Veritas.

Dans la 4me **Colonne**, C=Composite (membrure en fer, bordé en bois) ; I=Fer ; Stl=Acier ; W=Bois.

Les Dates du doublage en métal (**Col. 11**) ressortent des livres dans lesquels les navires sont respectivement classés ou des rapports des Inspecteurs du Lloyd's Register.

Pour plus amples renseignements sur les navires classés par le Lloyd's Register, voir le Lloyd's Register Book et le Lloyd's Yacht Register.

Official No.	No. in Book	Registry in which classed	Ship's Name. Master, and Dates of Appointment to present Owner's Service and to Ship. Ship's Late Name.	Material and Rig	Flag	Net Tonnage	Length	Breadth	Depth	Built: Where	Built: Date	Built: By whom	Date of Metalling	Owners or Managers	Port
43460	501		**Thrush**	W	British	281	[illegible]	[illegible]	9·[illegible]	[illegible]	1862			T. Hart	[illegible]
	502		**Thule**	W Bk	Norweg'n	371	117·9	29·5	16·5	[illegible]	1858	B. Tellefsen		M. Danielsen	Risör
	503	DN	**Thule** *Olsen*	W Bg	Norweg'n	264	98·1	28·2	15·6	[illegible]	1855	N.O. Gjömle		J. Munch	[illegible]
	504		**Thule** (LOST) *Anderson* -85	W Bg		261	107·9	25·5	13·8	Carlshamn	1853	Benghson			
47410	505	✳	**Thunderbolt** *Johnson*	I Bk	British	1193	210·0	35·[illegible]	23·[illegible]	[illegible]	1863	[illegible]		Burgess & Co.	London
	506		**Thure** *Sjoholm*	W Sr	Russian	236	..	..	..	W[illegible]	1876			C.M. Eriksson	Kimito
56002	507	✳	**Thurland Castle** *Chamberlain* 86-84	I Bk	British	1244	226·1	34·8	21·[illegible]	Belfast	1876	Harland & Wolff		Lancaster Shipowners Co. Ld.	Lancaster
	508	V	**Thurø** (ex Kastelsek) *Petersen* 80-80	W Sr	Danish	125	88·9	21·3	10·5	Karrebæk-[illegible]	1877	J.C. Sorensen		N.W. Knus	Svendb.
56034	509	✳	**Thurso** (ex Lucius) *Scoullar*	I Bk	British	490	146·5	27·2	18·2	Sunderland	1864	Pile, Hay & Co.		C.W. Turner	Newcastle
	510		**Thusnelda**	W Bn		161	103·1	22·3	10·[illegible]	[illegible]	1876	R.H. Møller	Y 1.82	W.H. Lean	
	511		**Thusnelde** *Schleebecker* [illegible]	W Bk	German	374	103·0	28·5	16·4	Memel	1837	[illegible]		J.H. Bulcke	Memel
56908	512	✳	**Thyatira** *Philip* 85-88	C S	British	962	204·0	34·0	21·3	Aberdeen	1867	W. Hood & Co.	Y 6.87	G. Thompson & Co.	Aberdeen
77674	513	✳	**Thyra** (WRECKED) *Thompson*	W Bn	British	250	116·4	25·9	12·9	P.E. Island	1878	[illegible]	Y 6.86	C.F. Orr	Melbourne
	514	G	**Thyra** (ex Sarah Wascoe) *Larsen*	W Bk	Danish	319	116·4	24·1	17·0	Sunderland	1856			J.P. Folkmann	Rosso
	515		**Thyra** (ex Johannes) *Rasmussen* 83-85	W Bg	Danish	162	96·5	21·9	11·[illegible]	[illegible]	1863	A.F.H. Conradi	Y [illegible]	Martin Jensen	Nyborg
	516	V	**Thyra** *Troensegaard* 79-79	W Sr	Danish	143	97·2	21·3	10·3	Troense	1879	C.J. Jensen		N. Nielsen	Svendb.
	517	N	**Thyra** *Hansen*	W Bk	Norweg'n	612	161·5	28·5	18·7	Quebec	1857			Sören Larsen	[illegible]
	518		**Thyra** (CONDEMNED) (ex Catharina Roberts) *Nilsson* 85-85	W Sr	Swedish	256	120·7	25·3	16·6	P.E. Island	1857			J.C. Hulthén	Helsingb.
	519		**Thyra** (ex Thorshavn) *Sagergren*	W Sr	Swedish	116	81·7	19·7	10·7	Norway				J. Brunn	Calmar
	520		**Ti Sodskende** (MISSING)	W Bk	Norweg'n	211	101·4	24·5	13·7	Havre	1854			A. Knoph	Porsgrund
1989	521	✳V	**Tiber** (ex Steamer) *Forbes* 87-87	I Bk	British	924	224·9	29·1	20·2	Port Glasgow	1851	J. Reid & Co.		Tiber Shipping Co. Ld. (E.K. Ellison)	Liverpool
75820	522	V	**Tiber** *Johnson* -88	W Bn	British	213	105·6	26·9	11·1	Newport, N.S.	1878	[illegible] Mosher		R.I. Hart	Halifax
	523	N	**Tichborne** *Asdahl*	W Bn	Norweg'n	196	106·6	24·6	12·7	P.E. Island	1853		Z 4.83	H. Cappelen's Enke	Skien
	524	V	**Tichytte** *Larroque* -89	I Bk	French	311	118·7	27·1	12·6	Bordeaux	1873	Chaigneau frères		H. Raband & Co.	Bordeaux
41664	525		**Ticino** *Bassel* 82-82	W Sr	British	115	90·6	23·0	11·0	[illegible]	1859			S. Hare	Bideford
24780	526		**Tidal Wave** *Wilson* -87	W Bk	American	573	161·5	37·2	13·7	[illegible]	1869			W.P. Sayward	[illegible]
	527		**Tidjaret Hhair**	W Bg	Turkish	450	..	..	..	[illegible]	1869			Hairie tijari Emin zadeh Hus[illegible] Sadi Effendi	Constant.
	528		**Tidjaretti Bahri**	W Sr	Turkish	192	..	..	..	Stephan Skell	1875			Stephanlee Kadi oglou [illegible]	[illegible]
	529		**Tidjaretti Bahri**	W Bg	Turkish	125	..	..	..	[illegible]	1871			[illegible]	[illegible]
	530		**TidjarettiBahri** *Mahmoud*	W Sr	Turkish	126	..	..	..	[illegible]	1876			[illegible] zadeh Mus[illegible] [illegible] Vasalji & Co.	Samsun
	531	N	**Tiger** *Jensen* -83	W Bg	Norweg'n	161	93·2	23·6	10·1	[illegible]				E. Jensen	Brevig
	532		**Tigre** *Carneiro*	W Sr	Brazilian	352	..	..	..	Villa do Conde	1875			B. Oliveira & Co.	Pernamb.
	533		**Tigre II.**	W Bk	Brazilian	216	..	..	..					V.L. de Mattos	Rio G.
	534		**Tijaret i Bahri**	W Bk	Turkish	209	..	..	..	[illegible]	1883			[illegible] Tijari Bahri [illegible] Mehemed Effendi	Constant.
	535		**Tijaret i Bahri**	W Sr	Turkish	260	..	..	..	[illegible]	1883			Ismeh Bey Keremessi Hajee [illegible]	Heraclea
	536		**Tijaret i Haliss** *Reshid*	W Bg	Turkish	125	..	..	..	Spet Skellessi	1884			Rizehlee Hajee Temmel & Sons	Rizeh
	537	✳	**Tijuca** *Chevalier* 81-88	I Bk	French	827	178·0	34·5	21·[illegible]	Nantes	1866	E. Gouin & Co.		Ant. Dom. Bordes & fils	Bordeaux
	538		**Tika Asiah** (ex Gemini)	W Bk	Dutch	414	137·8	27·5	17·0	Jersey	1864	Le Vesconte	Y [illegible]	Hajee Abdul Djalil	Banjoew.
74405	539	✳	**Tikoma** *Pugh* 85-85	W Bk	British	799	166·2	33·8	19·5	Richibucto, N.B.	1877	J. & T. Jardine	Y [illegible]	J. & T. Jardine	Chatham
222	540	A	**Tilde** *Colazio*	W Bk	Austrian	636	156·2	33·2	18·9	Fiume	1873	F[lli] Schiavoni	Y [illegible]	A.F. Poschich fu A.	Fiume
78974	541	✳	**Tilkhurst** *Hinton* -87	I S	British	1527	239·1	39·3	22·9	Dumbarton	1877	A. McMillan & S.		Tilkhurst Ship Co. Ld. (W.R. Price & Co.)	London
	542	D	**Tillid** *Stetson* -87	W Bk	American	438	132·0	30·2	16·6	[illegible]	1871	L. Grettres	Y [illegible]	H.H. Stetson	Boston
	543	N	**Tillid** *Nielsen* -88	W Bk	Norweg'n	283	133·2	29·2	12·1	Drammen	1878		Y [illegible]	L. Gundersen	Porsgrund
145118	544	RV	**Tillie Baker** *Carter*	W Bk	American	684	150·1	34·0	18·7	Harrington, Me.	1876	G. Emery	Y 6.88	J.E. Brett	New York
85793	545	R	**Tillie E.** *Warner* 84-84	W Sr	British	106	85·5	23·1	9·7	P.E. Island	1883	[illegible]		W. Wood	Faversham
145333	546	RV	**Tillie E. Starbuck** *Curtis*	I S	American	1931	257·0	42·7	23·0	Chester, Pa.	1883	J. Roach & Son		W.H. Starbuck	New York
145088	547		**Tillie Vanderherchen** *Robens*	W 3m Sr	American	439	130·9	31·3	11·7	Mauricetown, N.J.	1875			C. Bateman	Bri[illegible]
	548	G	**Tilly** *Svenson* -86	W 3m Sr	Swedish	149	89·6	23·0	9·9	Fillingenabin	1866	O. Nordstrom		A.A. Pihl	Kullen
90076	549	✳	**Timandra** *Andrews*	I S	British	1500	245·6	38·8	22·4	Port Glasgow	1885	R. Duncan & Co.		G.F. Smith	Glasgow
49137	550		**Timandra**	W Bn	British	165	89·5	24·4	10·0	Port Gilbert, N.S.	1865			R. Neill & Sons	Belfast

✳ = Lloyd's Register; D = Nederlandsche Vereeniging; G = Germanischer Lloyd; I = Registro Italiano; R = American Re[cord]; A = Veritas Austro-Ungarico; H = Veritas Ellenico; N = Norske Veritas; V = Bureau Veritas.

In the 4th Column, C = Composite (iron frame, wood planked); I = Iron; Stl = Steel; W = Wood.

The Dates of Metalling (Col. 11) are based on the records of the Books in which Vessels are respectively classed, or on the Reports of the Surveyors of Lloyd's Re[gister].

Further particulars of Vessels classed by Lloyd's Register will be found in Lloyd's Register Book and Lloyd's Yacht Register.

No.	Registre dans lequel le navire est coté	Nom	Capitaine et dates d'entrée au service des Armateurs actuels et à bord du navire	Matériaux et Gréement	Pavillon	Tonnage net	Longueur	Largeur	Creux	Construit: Où	Quand	Par qui	Date du doublage en métal	Armateurs ou Directeurs-Gérants	Port d'Armement
551	❊	Timaru	Fullarton 76-81	I S	British	1306	239·3	36·1	21·1	Greenock	1874	Scott&Co.		Shaw,Savill & Albion Co.Ld.	Glasgow
552		Timbre II.	de Telha	W Sr	Portug'se	131	80·1	20·7	9·8	Fão	1874	M.D.dosS.Borda		A.dos.S.Gomes	Oporto
553	❊	Times	Bullen 84-84	I Bk	British	582	180·4	29·8	19·1	Sunderland	1868	W.Doxford		G.N.Gardiner & Co.	Liverpool
554		Times (FOUNDERED)	Masterton	W Bkn	British	280	129·5	21·3	15·2	Guernsey	1854		Y 2.82	A.&F.Manuelle	Guernsey
555		Times		W Bn	British	152	92·2	23·8	9·4	Waldoboro,U.S.	1847			J.H.Oviatt	St.Andrew's N.B.
556		Timoteo I°	Torres	W Bn	Spanish	143	86·3	25·5	10·8	Pineda	1855			F.Carreras	Barcelona
557	R	Timothy Field	Stratton 187-88	W Bn	American	167	95·5	27·4	8·0	Breider,Me.	1861	J.Dunning	Y 8.86	J.S.Emery & Co.	Boston,Mass.
558	N	Tinten (ex Sophie Koningin der Nederlanden)	Olsen -87	W Bk	Norweg'n	639	143·4	35·7	20·0	Gravenshage	1856	H.de Hoog	Y 2.88	S.Rumi	Moss
559		Tinto	Swart -88	W Bk	Chilian	464	137·8	26·8	18·0	Liverpool	1852	J.Steel jr.		Gütschow & Piza	Valparaiso
560	❊	Tinto Hill	Hall 84-86	I 4m Bk	British	2067	283·1	40·5	24·6	PortGlasgow	1888	Russell & Co.		J.R.Dickson & Co.	Glasgow
561		Tipo (ex Ferrari Antonio)	Schiaffino	W Bk	Italian	460	127·3	31·2	19·6	SestriPonente	1869	Bruzzone	Y 10.82	Gius.Degregori	Genoa
562	R	Tirfing	Johansson	W Sr	Swedish	174	100·0	21·7	11·5	Oscarshamn	1875	P.Pettersson		A.V.Castegren	Oscarshamn
563		Tirzah	Brookbank	W Bg	British	239	103·0	25·0	14·5	Sunderland	1862	R.H.Potts&Brs.		J.W.Woods	North Shields
564		Tit Bit	Jenkins	W Bn	British	134	81·7	21·0	12·5	Bridport	1849			T.Jenkins	Cardigan
565	V	Titan	Allyn 84-84	W S	American	1176	180·4	37·5	23·5	KennebunkMe	1869	W.L. Thompson	Y 4.86	C.H.Allyn	Barnstable.Mass.
566	V	Titan	Witt -72	W Bk	German	353	120·0	28·0	16·6	Wismar	1867	M.Maren		J.C.Thormann	Wismar
567	❊	Titania	Dana	C S	British	879	200·0	36·0	21·0	Greenock	1866	R.Steele&Co.	Y 12.85	Hudson's Bay Co.	London
568		Titania	Brown 88-83	W Sr	British	185	103·0	22·8	12·8	Kingsbridge	1865	Date		J.Ennor	Salcombe
569	❊	Titania	Foot	W K	British	86	79·3	20·8	10·6	Brixham	1883	S.Dewdney&Ss.		C.Hellyer	Hull
570	N	Titania	Christoffersen -86	W Bk	Norweg'n	445	123·9	28·3	18·1	Newcastle	1852		Y 2.82	H.T.Hauff	Tönsberg
571	N	Titania (CONDEMNED)	Christophersen	W Bk	Norweg'n	322	113·4	28·2	15·2	Risör	1855			C.Christophersen	Christiansand
572	N	Titania	Terjesen -88	W Bk	Norweg'n	303	118·2	29·0	14·2	Porsgrund	1868	F.T.Flood	Y 1.88	A.Albrethsen	Tvedestrand
573	N	Titania	Hansen	W Bk	Norweg'n	245	121·6	26·9	11·2	Hamburg	1865	JHN.Wichhorst	Y 12.81	H.Schanning	Kragerö
574	V	Titania	Mutzneck	W 3m Sr	Russian	324	127·7	31·0	13·5	Kurland	1886	Muzeneck		Mutzneck	Riga
575	V	Titia	Poljer	W 3m Sr	German	160	98·4	22·5	11·6	Hoogezand	1867	J.A.Hoelles	Y 3.81	R.F.Aden	Grossefehn
576		Tito	Fasarone	W Bg	Italian	239	101·9	26·6	15·8	Gaeta	1871	L.Contreras	Y 3.84	Vinc.Nocella	Gaeta
577	A	Tito	Millini	W Bg	Italian	141	81·1	23·8	14·1	Pra	1851	Tixi	Y 1.80	S.Millini	Porto Ferraio
578		Tito		W Bg	Portug'se	209								A.J.Soares	Oporto
579	I	Tito Secondo	Cavassa	W Bk	Italian	341	115·1	29·2	19·8	Savona	1867	G.B.Guastavino	Y 10.87	Flli.Cavassa	Genoa
580		Tivola		W Bk	British	184	94·8	24·0	9·4	Bangor,U.S.	1847			T.O'Leary	Nassau.N.P.
581		Tivoli	Harding -88	W Bk	British	1111	178·0	37·0	22·7	St.John	1860	Ruddock	Y 10.85	L.Pugh	Bermuda
582	D	Tjaduwine (ex Catharina & Agnes)	Dijk -88	W 3m Sr	Dutch	221	110·5	22·5	12·7	Veendam	1867	H.&J.Wilkens		U.F.Zuiderveen	Oude Pekela
583	N	Tjalf (ex Eric)	Lassen 75-75	W Bn	Norweg'n	212	105·5	25·3	12·7	P.E.Island	1875	Peake	Y 2.84	P.A.Christensen	Langesund
584	N	Tjalf	Hopp	W Bg	Norweg'n	168	89·7	22·4	11·7	Kragerö	1855			O.H.Sollie	Horten
585		Tjalfe	Ipsen 80-84	W Bg	Danish	197	92·7	25·5	13·4	Svendborg	1853	P.Brandt		Den.Kgl.Grønl.Handel	Copenhagen
586	V	Tjalfe	Lillelund 74-74	W Sr	Danish	131	94·1	21·8	10·8	Assens	1874	HChristophersn		Odense Skibs.-Selskab (H.F.Hellwege)	Odense
587	G	Tjelvar (ex Fraulein)	Sundström	W Sr	Swedish	183	95·1	22·1	11·9	Oldenburg	1857			G.Wykström	Oscarshamn
588	DV	Tjeremai	Hoogenbroek 75-83	C Bk	Dutch	1013	188·3	36·7	21·1	Amsterdam	1883	J.F.Meursing	Y 2.85	A.Hendrichs & Co.	Amsterdam
589	N	Tjomö	Torgeln -88	W Bk	Norweg'n	565	146·0	31·9	17·8	Tönsberg	1872	Jensen	Y 7.86	O.L.Röd	Tönsberg
590		Tlalok		W Sr	Mexican	148	103·5	24·4	8·6	Reiherstieg	1877	C.Dreyer		Hidalgo Carrago & Co.	Mazatlan
591	N	To Brödre (ex ...)	Olsen -86	W Bk	Norweg'n	617	140·4	32·1	17·4	Nappan,N.S.	1865	Fullerton		R.Olsen	Fredriksstad
592		To Mellon		W Bg	Greek	203	88·6	24·3	14·9	Syra	1867				Scopelos
593		Tó Pepromenon	Nicola	W Bn	Greek	234	..	..	..	Syra	1877	L.Iseppo	Y 3.86	C.Coggia	Syra
594	N	To Venner	Andersen -87	W Bk	Norweg'n	308	108·7	28·3	14·7	Liban	1853		Y 2.87	J.Olsen	Farsund
595		Tobin		I Sr	British	171	99·2	21·3	..	Liverpool	1850	T.Vernon&Son		Mersey Docks & Harbour Board	Liverpool
596	DV	Tobique	Gagné -88	W S	British	1414	199·1	39·2	24·0	St.John,N.B.	1876	J.Stewart	Y 2.86	G.E.Berteaux	St,John.N.B.
597	❊	Tocopilla	Garland 88-88	I Bk	British	496	157·5	26·7	17·5	Glasgow	1865	A.Stephen&Sns		J.Jose & Co.	Falmouth
598		Todrakhan		W Schleppe	Turkish	175	..	..	..	Todrakhan	1869			Todrakhanlee Cantarjee zaduk Ali	Todrakhan
599	N	Tofteland (ex Queen)	Aanensen 88-88	W Bn	Norweg'n	316	109·9	27·9	16·4	Rockland,N.S.	1873	Smith & Mackelvie		L.Knudsen	Christiansand
600	DV	Toivo	Fergersen -88	W Bk	Norweg'n	944	169·9	37·1	22·2	Uleåborg	1871		Y 1.87	O.Eriksen&Gjersöe	Christiania

❊=Lloyd's Register; A=Veritas Austro-Ungarico; D=Nederlandsche Vereeniging; G=Germanischer Lloyd; H=Veritas Ellenico; I=Registro Italiano; N=Norske Veritas; R=American Record; V=Bureau Veritas.

Dans la 4me Colonne, C=Composite (membrure en fer, bordé en bois); I=Fer; Stl=Acier; W=Bois.

Les Dates du doublage en métal (Col. 11) ressortent des livres dans lesquels les navires sont respectivement classés ou des rapports des Inspecteurs du Lloyd's Register.

Pour plus amples renseignements sur les navires classés par le Lloyd's Register, voir le Lloyd's Register Book et le Lloyd's Yacht Register

Official No.	No. in Book	Registry in which classed	Ship's Name. Master, and Dates of Appointment. Ship's Late Name.	Material and Rig	Flag	Net Tonnage	Length	Breadth	Depth	Built: Where	When	By whom	Date of Metalling	Owners or Managers	Port
	601	N	Toivo *Jacobsen -81*	W Bk	Norweg'n	499	147·3	34·6	16·7	Ny Carleby	1860	G.Fredriksen		A.M.Thoresen	Fred
	602		Toivo *Andersson -77*	W Bk	Russian	399	122·7	31·1	15·2	Kimito	1865	M.T.Lindqvist		P.E.Grandell	Naga
	603		Toivo *Pihlman*	W Bk	Russian	375	..	..	..	Welkas	1875			B.H.Haukka	Welk
	604		Toivo *Lassma -81*	W Bn	Russian	218	97·6	26·2	12·9	Fredrikshamn	1875	Mekkelet		D.Brunila	Fred
	605		Toivo *Peusa*	W Bg	Russian	215	106·5	25·6	12·7	Björkö	1873	S.Koli		G.Maurunsen	Björk
	606		Toivo *Karlson*	W Bg	Russian	202	..	..	..	Wiborg	1873			A.Maatson	Naga
	607		Toivo *Penttilä*	W Sr	Russian	164	..	..	..		18[illegible]			B.W.Langerstam	Wibo
	608		Toivo *Holmros*	W Glt	Russian	161	..	..	..	Gustaf	1884			F.J.Holmros	Gustaf
	609		Toivo *Blomqvist*	W Sr	Russian	160	..	..	..					H.Ostmans Enke	Rau
	610		Toivo *Holmberg 84-84*	W Sr	Russian	146	85·6	27·2	10·9	Tölösala	1876	J.Palmros		A.Abrahamsson	Gustaf
	611		Toivo	W Sr	Russian	120	..	..	..		1876			M.Mannonen	Lav
	612		Toivo	W Sr	Russian	106	..	..	..		1877			A.Waittinen	Imp
	613		Toivotar *(ex Iog?) Wikström*	W Sr	Russian	219	101·3	27·1	12·9	Hackalo	1875	W.Gelius		E.V.Heerman	Nyst
643	614		Tokuho Maru	W Bg	Japanese	168	99·7	21·8	10·9	Shima	1881	FukuzawaSBYd		Morita Kinzayemon	Owa
145419	615		Tom Beattie	W Sr	American	133	93·7	28·3	7·4	Bath,Me.	1886	G.Hawley		J.Beattie	New
55380	616		Tom Henry *Clements 83-83*	W Sr	British	124	92·0	22·4	11·0	Fraserburgh	1867	Webster		J.Ennor	Kin
16675	617		Tom Roper *Taysome 83-83*	W Sr	British	100	80·0	20·5	11·0	Barrow	1857			J.Geldart	
24905	618		Tom Williams *Mills*	W 3m Sr	American	349	118·0	32·5	11·1	Fairhaven Con	1874	W.O.Nettleton		N.Edwards	New
35421	619		Tomas Terry	W Bk	British	404	125·7	30·1	12·6	St.Andrew,N.B	1862			O.B.Rideout	St.A
	620		Tomaso (BROKEN UP)	W Bk		468	137·8	30·5	20·3	Sestri Ponente	1868	F.Tixi	[illegible] 1881		
581	621	I	Tomaso *G. Carletti*	W Bg	Italian	174	108·2	23·8	13·3	Limite	1875	N.Pichiotti	[illegible]	F[lli] Carletti fu T.	Porto
	622	V	Tombola *Cholard 87-88*	W Bn	French	144	101·0	22·3	10·3	St.Malo	1884	Scoland	[illegible]	Anatole Lemoine	St.Mal
	623		Tomma *Lagerqvist*	W Kf	Swedish	100	75·1	17·5	9·6	Elsfleth				J.Uddenberg	Ore
3470	624		Tommaseo *Magnano* (ex Giovanni)	W Bk	Italian	645	152·5	33·1	20·7	Lerici	1869	Faggioni	[illegible]	F.Magnano	Gen
1115	625	I	Tommasino *Trapani*	W Bk	Italian	568	144·3	30·9	20·4	Equa	1876	G.Mauro	[illegible]	M.Cafiero & Co.	Cast
907	626		Tommasino *Spampinato* (ex La Fortuna, ex Nuovo Michelino, ex Bighellino)	W Bg	Italian	128	77·8	23·9	13·7	Fiume	1850			L.Trapani fu T.	Catan
618	627	H V	Tommasino (LOST) *Greece*	W Bg	Italian	127	78·7	20·7	10·8	Gaeta	1870	F.Contreras		N.A.Di Fonzo	Gaeta
310	628		Tommaso *Corbo -88*	W 3m Sr	Italian	263	131·5	25·6	14·6	Trieste	1868	S.Strudthoff		G.Puglisi	Catan
2302	629		Tommaso Padre (ex Il Vecchio Tommaso) *Gaglione*	W Sr	Italian	105	79·4	20·5	10·4	Torre d'lGreco	1881	Trambarulo		O.Gaglione	Naple
806	630		Tomoye Maru	W Sr	Japanese	137	84·3	22·2	10·2	Hakodate	1883			Tsuji Matsunojio	Hako
54578	631		Ton Mawr *Nichol 85-85*	W Sr	British	100	92·0	22·4	10·5	Padstow	1865	Willmett		J.Nichol	
51114	632	R V	Tongoy *Morris 85-85*	C Bk	British	478	153·6	27·4	18·4	Sunderland	1866	W.Pile&Co.	[illegible]	W.O.M'Cormick	Dublin
	633	G	Toni *Brodthering 87-87*	W Bk	German	468	129·9	30·3	17·8	Danzig	1858	J.W.Klawitter		E.Burchard	Rost
163	634	A D	Tonina II. *Cosulich -85*	W Bn	Austrian	274	101·7	25·0	16·4	Lussinpiccolo	1876	G.S.Cattarinich	[illegible]	S.G.Nicolich	Lus
1003	635		Tonino *Ciano*	W Bg	Italian	303	108·2	27·9	17·7	Savona	1860			R.Ciano fu C.	Leghor
3244	636	D I V	Tonio *Serra 85-86*	W Bk	Italian	1032	191·6	35·1	23·7	Voltri	1882	G.Fava	[illegible]	E.R.Zino	Gen
	637	G V	Tonjes *Warring 85-*	W Sr	German	124	88·3	20·6	9·2	Grossefehn	1883	J.J.Cassens&Shn	[illegible]	Tonjes J.Cassens & Sohn	Gro
44446	638		Tonkin *Neal*	W Bn	British	179	111·0	23·0	12·1	P.E.Island	1862	Hayden		T.N.Gudge	Pet
	639	D V	Tonna *Eckhoff 88-88*	W Tn	German	159	91·8	22·7	10·6	Papenburg	1884	A.Beckmann	[illegible]	Harm K.Buss	Ihren
77424	640	✳	Tony Krogmann *Evans*	W Bkn	British	251	119·0	26·0	14·7	Portmadoc	1878	Roberts	[illegible]	E.Roberts & Co.	Carn
2775	641		Toowoomba (ex Jane Henderson) *King*	I Bk	British	585	169·9	28·3	17·9	Greenock	1859	Scott&Co.		Eastern & S.African Teleg. Co. Ld.	
24683	642		Top Gallant *Wickberg -84*	W S	American	1229	187·5	38·5	24·0	East Boston	1863	P.Curtis		Renton,Holmes & Co.	
74010	643	✳	Topaz *Cully 85-85*	W Bn	British	198	103·7	21·6	12·9	P.E.Island	1876	Morrow	[illegible]	A.King	Liver
59778	644		Topaz *Grigg*	W Bn	British	196	105·1	21·6	12·7	P.E.Island	1870	McDonald		P.Furlong	Swan
3015	645		Topaz *Bogan*	W Sr	British	140	82·3	23·0	13·0	Greenock	1842	J.Scott&Sons		Mrs.E.Bogan (per F.E. Carter)	Flee
85367	646	R	Topaz *Bartling -85*	W Sr	British	104	85·5	24·5	9·1	Pt.Clyde,N.S.	1883	T.Coffin		W.McMillan	Hali
6529	647	✳	Topaz *Matthias*	W Sk	British	42	50·6	15·6	8·6	Portmadoc	1850	Evans		W.J.Canton	Milford
	648	D N V	Topdal *Anderson*	W Bk	Norweg'n	598	147·9	32·3	18·3	Grimstad	1877	S.Johnsen	[illegible]	H.Hansen	Lilles
6346	649	G	Topsy *Bowes*	W Sr	British	139	89·2	19·1	11·4	P.E.Island	1853			C.Thompson	Hull
48626	650		Torbay Lass *Down*	W Sr	British	131	90·6	21·2	12·1	Brixham	1864	Upham		Mrs.L.Roberts	Brix

✳ = Lloyd's Register; D = Nederlandsche Vereeniging; G = Germanischer Lloyd; I = Registro Italiano; R = American [...]
A = Veritas Austro-Ungarico; H = Veritas Ellenico; N = Norske Veritas; V = Bureau Ve[ritas]

In the 4th Column, C = Composite (iron frame, wood planked); I = Iron; Stl = Steel; W = Wood.

The Dates of Metalling (Col. 11) are based on the records of the Books in which Vessels are respectively classed, or on the Reports of the Surveyors of Lloyd's [...]

Further particulars of Vessels classed by Lloyd's Register will be found in Lloyd's Register Book and Lloyd's Yacht Register.

1 Numéro officiel	2 Nº dans le livre	3 Registre dans lequel le navire est coté	4 Nom. Matériaux et Gréement. Capitains et dates d'entrée au service des Armateurs actuels et à bord du navire. Ancien Nom.	5 Pavillon	6 Tonnage net	7 Dimensions. Longueur	Largeur	Creux	8 Construit. Où.	9 Quand	10 Par qui	11 Date du doublage en métal	12 Armateurs ou Directeurs Gérants.	13 Port d'Armement.
	651	V	Tordenskjold (ex Ravnekjaer) Schou... W Bk	Danish	362	112·7	25·8	17·6	Altona	18 47/80			M.Christiansen	Ronne
	652	V	Tordenskjold Eilertsen W Bk	Norweg'n	1225	172·4	41·2	26·5	Copenhagen	18 ..		y.88	Smith & Thommesen	Arendal
	653	N	Tordenskjold Jensen W Bk	Norweg'n	559	142·6	32·6	17·8	Tvedestrand	1877	O.Gundersen	y.88	Jörgensen & Jorkjend	Tvedestrand
	654		Tordenskjold Pettersen W Bk	Norweg'n	433	120·2	30·7	17·8	N.Bedford Ma	1827			Chr.Hannevig	Christiania
	655	N	Tordenskjold Fahlell [.... W Bg	Norweg'n	227	115·5	26·6	13·1	Drontheim	1860			A.Jenssen & Co.	Drontheim
	656		Tordo (ex Norton) Hall W Bk	Chilian	424	124·5	28·1	..	Altona	1866	J.J.Diecks		Sanchez & Cardenas	Valparaiso
	657		Torgny (ex Formosa) Hammarlund -88 W Bk	Swedish	334	127·6	24·0	13·0	Berks	1863	J.F.Nicolai		C.Carlsson	Torp. Bhln.
	658		Torpado Söderberg W Sr	Swedish	122	90·5	20·0	9·1	Bergqvarn	1877	Hammerlund		C.A.Svensson	Bergqvarn
	659	✳	Torrens Angel 67-75 C S	British	1276	222·1	38·1	21·5	Sunderland	1875	J.Laing	y.88	A.L.Elder & Co.	London
	660	V	Torrhamn Ohlsson -86 W 3m Sr	Swedish	193	98·4	23·0	12·0	Syltenäs	1874	J.Abrahamson		S.H.Korn	Gothenburg
	661		Torrid Zone W Bn	British	188	100·0	21·5	11·6	Shelburne,NS	1864			T.H.Pitt	Hamilton,Bermuda
	662	✳	Torridon Shepherd 85-85 I S	British	1502	246·0	38·1	22·0	Aberdeen	1885	A.Hall&Co.		A.Nicol & Co.	Aberdeen
	663		Torridon Heckman -87 W Sr	British	105	80·3	23·9	9·6	Pleasantville, [N.S.	1887	A.McKean		A.McKean	Lunenburg, NS.
	664		Totuuas W Sr	Russian	144	..	..	..		1865			K.Östmanne	Rammo
	665	✳	Touch Me Not Pryan W Sr	British	120	81·6	23·3	13·7	Fowey	1841			S.Truscott	Fowey
	666		Tour de Malakoff W Bg	French	129	81·4	21·3	12·3	Fécamp	1855	Brument	z.88	H.Lecharpentier	Granville
	667		Tourlianos (WRECKED) W Bg	Greek	246	92·8	26·6	..	Syra	1872			C.D.Cotzia	Syra
	668		Tourterelle W Sp	British	102	95·9	22·8	6·5	St.Thomas de [Pierreville.	1887	G.Mousran		N.Beauchemin	Montreal
	669		Tove Rasmussen W Bk	Danish	370	129·1	27·2	14·9	Molde	1855			C.L.J.Fischer	Nyborg
	670	V	Town of Liverpool (ex Raleigh) W Bg	British	236	110·0	26·6	16·0	Sunderland	1869	Wheatly		C.François fils	Port Louis
	671	✳	Towy Gower 87-87 I Bk	British	304	132·8	24·9	14·1	Llanelly	1866	Nevill Bros.		Llanelly Iron Ship Co. Ld	Llanelly
	672	N	Toxa Sørensen 80-80 W Bg	Norweg'n	181	96·8	25·1	11·9	Arendal	1875			N.B.Petersons Enke (P.A.Petersen)	Kragerö
	673	✳V	Toxteth Whiting 87-87 I S	British	2526	305·5	41·7	24·1	Southampton	1887	Oswald, Mor- [daunt & Co.		Sailing Ship "Toxteth" Co. Lim. (R. W. Leyland & Co.)	Liverpool
	674		Toyokuni Maru W Sr	Japanese	303	106·0	26·5	14·0	Tokio	1879	Hirano Sh. B. Yd.		Shiba Toranosuke	Hiogo
	675		Tprruuh Surring -85 W Sr	Russian	197	96·0	27·9	12·5	Haynasch	1882	J.Mikkelson		J.Mikkelsohn	Haynasch
	676	V	Trabant Scheel W Bk	German	380	116·2	29·8	16·8	Danzig	1863	F.Devrient		J.N.Rodbertus	Barth
	677		Trade Wind W Sr	British	181	100·5	21·1	9·0	Port Hope	1868	H.Hacken		W.Cann	Whitby
	678	✳	Trafalgar Auld 68-86 I 4m S	British	1696	271·5	39·3	23·4	Glasgow	1877	C.Connell&Co.		A.Brown	Glasgow
	679	✳	Trafalgar Moir 6--77 I S	British	1429	242·0	38·1	22·0	Greenock	1875	J.E.Scott		D.Rose & Co.	Aberdeen
	680		Trafalgar Johnsson 73-73 W Bk	Swedish	287	104·6	23·7	16·3	Sunderland	1862			H.Andersson	Nyhamn
	681	N	Trafik Johnsen 73-74 W Sr	Norweg'n	105	79·2	21·8	10·2	Kragerö	1874			P.&C.Römer	Kragerö
	682		Trafik W Sr	Russian	144	..	..	..		1851			K.W.Burmeister	Abo
	683		Trajano W Bn	Spanish	137	78·1	21·9	11·8	Arlay	1850			Doña L.Liquinco	Manila
	684	V	Tranquebar Cochery -82 W Bk	French	376	123·3	26·7	16·4	Nantes	1870	F.Gouin	v.88	P.Debry	St.Malo
	685	N	Transatlantic Tønnesen [80-81 W Bk	Norweg'n	598	152·5	29·4	17·9	Aberdeen	18 57/78	W.Hood & Co.	y.88	J.L.Ugland	Arendal
	686	DV	Transit Morton W Bk	British	856	166·1	35·2	20·0	Granville,N.S.	1876	J.E.Reed	v.88	J.Mills	Annapolis,N.S.
	687		Transit Jarman W Bn	British	165	78·8	19·9	14·0	Brixham	1848			J.Drew	Teignmouth
	688	✳	Transit Ellis W K	British	86	80·7	20·9	10·6	Brixham	1885	S.Dewdney&Ss.		Th.J.Boyd	Hull
	689	N	Transit Larsen W Bn	Norweg'n	132	91·7	21·6	11·0	Arendal	1858	Aanonsen		L.Larsen	Christiansand
	690		Transport (BROKEN UP) Kavanagh W Bg	British	307	111·5	27·1	14·8	Pt George,NS	1865	G.B.Reid		J.C.Ellis	Newcastle,NSW
	691	V	Travailleuse Claeyssen W Sr	French	111	83·0	20·0	11·3	Dunkirk	1877	Vanderiele		Maurois & Co.	Dunkirk
	692	✳D	Traveller McManus 87-87 I S	British	1429	228·1	38·2	24·0	Dumbarton	1879	Birrell, Sten- [house & Co.		J.R.Haws & Co.	Liverpool
	693		Traveller Simpson 68- W Sr	British	196	102·0	24·3	12·8	Peterhead	1868	Robinson		A.Simpson	Peterhead
	694	R	Traveller Frayne W Sr	American	164	95·0	28·0	8·0	Essex,Conn.	1869			W.Frayne	Middletwn, Conn.
	695		Traviata W Bk	Brazilian	536	..	..	..					J.R. de Souza	Rio Janeiro
	696	N	Tre Brodre Thorsen W Bg	Norweg'n	147	81·4	24·5	11·1	Tönsberg	1855			H.& M.Thorsen	Tönsberg
	697	A	Tre Fratelli Ballarin W Bn	Italian	201	93·5	24·8	14·2	Venice	1859	G.Grasso	y.88	F.Ballarin & Co.	Venice
	698		Tre Fratelli (ex Pietro Costa) Castellano W Bg	Italian	200	108·2	25·3	14·1	Sestri Ponente	187.	F[lli] Cichenaccio		G.Paredi fu D.	Genoa
	699	IV	Tre Fratelli Salvo (ex Matilde Maggiolo Rossando) W Bn Sr	Italian	347	125·3	28·1	15·7	Pra	1883	L.Durante	y.88	F[lli] Salvo	Trapani
	700													

✳=Lloyd's Register; D=Nederlandsche Vereeniging; G=Germanischer Lloyd; I=Registro Italiano; R=American Record; A=Veritas Austro-Ungarico; H=Veritas Ellenico; N=Norske Veritas; V=Bureau Veritas.

Dans la 4me Colonne, C=Composite (membrure en fer, bordé en bois); I=Fer; Stl=Acier; W=Bois.

Les Dates du doublage en métal (Col. 11) ressortent des livres dans lesquels les navires sont respectivement classés ou des rapports des Inspecteurs du Lloyd's Register.

Pour plus amples renseignements sur les navires classés par le Lloyd's Register, voir le Lloyd's Register Book et le Lloyd's Yacht Register.

Official No.	No. in Book	Registry	Ship's Name. Material and Rig. Master, and Dates of Appointment to present Owner's Service and to Ship. Ship's Late Name.	Flag	Net Tonnage	Length	Breadth	Depth	Where built	When	By whom	Date of Metalling	Owners or Managers	Port
[illegible]	701	A	Tre Fratelli Z *Zennaro* W Tri	Italian	100	81·0	23·1	10·2	Chioggia	1881	G. Poli	[illegible]	F.lli Zennaro	Venice
[illegible]	702		Tre Sorelle *Consiglieri* (BROKEN UP) W Bk	Italian	476	137·8	32·3	19·0	[illegible]	1867	F. Celesia	[illegible]	Crist. Dapueto	Genoa
[illegible]	703	I	TreSorelleM. *Masse* -84 (ex [illegible]) W Bk	Italian	408	128·6	30·4	19·5	Pra	1869	L. Durante	[illegible]	F. Masse	Genoa
	704	V	Tre Søstre *Hintze* 80-87 W Bn	Danish	160	100·0	22·2	10·9	Nexø	1870	J.H. Brandt	[illegible]	M. Christiansen	Rønne
	705	V	Tre Syskon *Blom* -88 (ex [illegible]) W Bk	Russian	677	163·0	35·0	18·0	Jacobstad	18 64/75	J. Lof	[illegible]	O.A. Fardig	Nystad
	706	N	Tre Venner *Hansen* -87 W Bg	Norweg'n	212	86·5	26·0	14·2	Grimstad	1847			H. Melsoms Born	[illegible]
[illegible]	707	D V	Treasurer *Downey* 77-84 W S	British	1386	225·8	40·5	23·6	Economy, N.S.	1884	D.P. Soley	[illegible]	D.P. Soley	[illegible]
	708		Trebizonde *Mossin* W Bg	Turkish	225	..	..	..	Austria	1872			Hajee Hashim zadeh Mustafa Effendi	Con[illegible]
	709		Trebizonde W Bg	Turkish	125	..	..	..	[illegible]	1866			Achmet & Co.	Trebi[illegible]
	710		Trebizonde W Bg	Turkish	125	..	..	..	Trebizonde	1870			Trebizondlee Hajee Hafis Ali Effendi	Trebi[illegible]
	711		Trebizondlee W Bg	Turkish	106	..	..	..	Trebizonde	1869			Hajee Dervisch & Co	Trebi[illegible]
[illegible]	712		Tregea *Crothers* -87 W Sr	British	133	88·0	25·0	11·0	[illegible]	1872	Massey		D.W. Bain & Co.	Hay[illegible]
	713	N	Trekroner *Hansen* 87-87 W Bg	Norweg'n	238	91·3	25·7	14·9	[illegible]	1844	T. Jensen		C. Hansen	Holm[illegible]
[illegible]	714	R	Tremont *Brophy* 87-87 W Bk	American	459	131·7	30·0	16·0	Kennebunk, M.	1868	D. Clark	[illegible]	J.S. Emery & Co.	Bos[illegible]
	715	N	Trenmor *Olsen* W Bg	Norweg'n	365	115·7	29·4	15·6	Grimstad	1871	O. Kittelsen	[illegible]	J. Jörgensen	Grimstad
	716		Trent *Borjesson* -87 W Bk	Swedish	364	117·2	26·5	16·1	Vestervik				J.C. Hulthen	Helsing[illegible]
[illegible]	717		Trenton *Stanley* W Sr	American	123	81·5	24·9	8·8	Eden, Maine	18 [illegible]			A.L. Cotton	Mach[illegible]
	718	N	Treport *Christiansen* W Bg	Norweg'n	297	106·8	27·5	16·5	Genoa	1862			N. Christensen Solberg	Horten
[illegible]	719		Tres Amigos W Sr	British	150	89·1	26·1	8·5					M.A.P. Colloman	Belize
	720		Tres Amigos W Sr	Brazilian	152	..	..	..					J.F.R. Pinto	Rio J[illegible]
	721	✻	Tres Carmencitas I Sr	Spanish	145	92·5	23·5	7·5	[illegible]	1884	A. Jack & Co.		M. Ramos & hermanos	S. Nicol[illegible]
	722		TresHermanos *Rosenmuth* (ex [illegible], ex Hambla) 87-87 W Bk	Chilian	163	84·0	29·6	11·5	Briand	1844			A. Sanguino	Callao
	723		Tres Hermanos *Spina* W Bn	Spanish	130	87·6	22·8	11·1	Almeria	1874			R. Bono Guarner	Alicante
	724		Tres Irmãos W Sr	Brazilian	712	..	..	..					E.J. de Freitas	Rio Jan[illegible]
	725		Tres Irmãos *De Mello* W Bn	Brazilian	118	..	..	..	St. João da Barra	18 43/73		[illegible]	Cia Cabotagem de S. João da Barra	S. João
	726		Treue *Nehls* 79- W Bg	German	258	106·5	26·8	11·1	Barth	1860	Braunschweig		J.C. Beug	Barth
[illegible]	727	✻	Trevellas *Radcliffe* ([illegible]) W 3m Sr	British	98	97·0	22·3	11·1	St. Agnes	1876	Hitchins	[illegible]	M.T. Hitchins	Hayle
[illegible]	728	✻ D	Trevelyan (MISSING) *Roberts* 88-88 I Bk	British	1042	203·6	33·4	21·2	Sunderland	1863	Pile, Hay & Co.		Shaw, Savill & Albion Co. Ld.	South[illegible]
[illegible]	729	A	Tri Sina *Vucic* W Bk	Austrian	839	159·0	34·5	21·8	PortoRe	1875	I. Archianin	[illegible]	V. Turkovich	Porto R[illegible]
	730		Triade *Nicola* W Bk	Greek	764	152·5	33·0	23·1	Lussinpiccolo	1872	N. Martinolich	[illegible]	Fillini Bros.	Syra
[illegible]	731	A	Triade Tarabochia *Tarabochia* W Bk	Austrian	829	144·3	32·8	23·5	Lussinpiccolo	1879	N. Martinolich	[illegible]	M. Ivancich	Lussin[illegible]
[illegible]	732		Trial W Bg	British	183	102·6	25·4	11·9	Antigonish, N.S.	1859			Chas. Kennedy	[illegible]
	733		Triandifil W Bg	Turkish	271	..	..	..	Abalmah	1872			Kerestehji Tujari Nicola	Abalm[illegible]
	734	V	Trias *Paap* -84 W Bn	Dutch	205	96·4	21·6	11·0	Varel	1884	A. Schoon	[illegible]	C.E. Duijts	Am[illegible]
	735	V	Tricana *Henriques* W Sr	Portug'se	184	95·8	22·5	10·9	[illegible]	1882	M. D. dos Santos	[illegible]	A. Henriques & Co.	[illegible]
	736	V	Trientje *de Buhr* -83 W Sr	German	103	82·3	20·4	9·2	Leer	1867	B.D. Middendorf		S. Wreesmann Wwe.	Leer
	737	D	Trijntje *Wortel* -88 (ex [illegible]) W 3m Sr	Dutch	300	115·8	24·5	15·6	Ilfracombe	1852	Dennis		R. Gosselaar	Farmsum
[illegible]	738	✻	Trim *Green* 81-81 W K Barge	British	110	95·3	22·1	8·3	Harwich	1881	J.H. Vaux		W. Christie & Co.	Harwich
	739	V	Trimpus *Salla* -87 (ex [illegible]) W Bk	Russian	640	143·0	33·2	20·1	Rönnebeck	1862	Gebr. Christoffers		Navigation Co. "Austra"	Riga
[illegible]	740		Trinacria *Mirabella* (ex [illegible]) W Bn	Italian	124	93·5	22·3	11·5	Catania	1842		[illegible]	S. Vasta	Catania
[illegible]	741	✻	Trinidad *Dyer* 83-83 I Bk	British	676	172·4	29·6	19·0	Sunderland	1866	Doxford		Shallcross & Higham	Liverpool
	742		Trinidad (ex [illegible]) W Sr	Spanish	142	86·4	22·9	11·9	Palma	1850			E. Furio	Alicante
	743		Trinidad W Sr	Spanish	142	86·9	23·6	10·8	Palma	1867	José Coll		G.A. Alzamora	Majorca
	744		Trinidad W Sr	Spanish	126	80·0	23·8	12·5	Palma	1850	José Coll		V. Terrasa	Majorca
	745		Trinita *Cabrera* (ex Fils [illegible]) W Bk	Spanish	289	101·7	25·9	15·3	Paimboeuf	1856			R. Gomez Alvarez	Havana
	746		Trinitario W Bn	Spanish	193	86·3	21·8	8·5	Samar	1857			Marcelino Cineo	Manila
[illegible]	747	A	Trino *Cippuri* -83 W Bk	Austrian	488	130·9	28·3	18·5	Lussinpiccolo	1866	G.S. Cattarinich	[illegible]	M.V. Premuda	Lussin[illegible]
[illegible]	748		Trio W Sr	British	101	71·0	21·0	..	Walsoken	1849			D.W. Bain & Co.	Penzance
[illegible]	749	✻	Trio *Young* 81-84 W Sr	British	78	77·6	19·6	9·2	Jersey	1876	Le Sueur		A. McMahon & Co.	Guernsey
	750	N	Trio *Printz* -87 (ex Garden Island) W Bk	Norweg'n	889	169·1	35·9	21·5	Garden Island (Ont.)	1877	Calvin & Breck	[illegible]	H.T. Hanff	Tönsberg

✻ = Lloyd's Register; D = Nederlandsche Vereeniging; G = Germanischer Lloyd; I = Registro Italiano; R = American [illegible]
A = Veritas Austro-Ungarico; H = Veritas Ellenico; N = Norske Veritas; V = Bureau V[illegible]

In the 4th Column, C = Composite (iron frame, wood planked); I = Iron; Stl = Steel; W = Wood.

The Dates of Metalling (Col. 11) are based on the records of the Books in which Vessels are respectively classed, or on the Reports of the Surveyors of Lloyd's [illegible]

Further particulars of Vessels classed by Lloyd's Register will be found in Lloyd's Register Book and Lloyd's Yacht Register.

N° dans le Livre	Cote	Nom. Matériaux et Gréement. Capitaine et dates d'entrée au service des Armateurs actuels et à bord du navire. Ancien Nom.	Pavillon	Tonnage net	Dimensions. Longueur	Largeur	Creux	Construit. Où.	Quand.	Par qui.	Date du doublage en métal	Armateurs ou Directeurs-Gérants.	Port d'Armement.
751	DN	Trio *Johansen* 85-88 W Bk	Norweg'n	365	115·8	28·7	16·1	Sandefjord	1854			Chr.Olsen	Christiania
752	N	Trio *Rolfsen* -88 W Bg	Norweg'n	320	130·3	27·9	14·1	Sundsvall	1870	J.A.Nordström		M.Rolfsen	Christiansand
753	N	Trio *Olsen* -86 W Bk	Norweg'n	239	96·7	26·4	14·1	Rodon	1861			D.Hultgreen	Drammen
754	N	Trio *Henriksen* W Bk	Norweg'n	220	114·7	23·9	14·6	Montrose	1852			H.Olsen	Tönsberg
755		Trio (LOST) *Olafsson* W Bk	Swedish	267	106·6	25·4	14·2	Westervik	1839			P.A.Johansson	Florön,Bhsln
756	R	Trio *Wikholm* -86 W 3m Sr	Swedish	211	109·2	23·7	12·2	Oscarshamn	1871	P.Pettersson		A.Castegren	Oscarshamn
757	A	Tris Adelphi *Colakis* W Bk	Greek	249	102·3	25·3	..	Sunderland	1866		Y 2.81	N.Colakis	Syra
758	A	Tris Adelphi W Bg	Greek	193	88·6	24·6	..	Syra	1861			Z.Patronis	Syra
759	A	Tris Adelphi *Manolakis* W Sr	Greek	169	86·9	23·3	..	Syra	1880		I 11.85	E.Manolakis	Syra
760		Tris Adelphi *Caroussakis* W Sr	Greek	102	..	..	..	Syra	1885			A.Gaitanas	Syra
761		Tris Adelphi *Barbati* W Bg	Russian	300	..	..	..	Castellamare	1869			S.A.Mussuri	Taganrog
762		Tris Adelphi W Bg	Russian	194	..	..	..	Syra	1866			G.Russetto	Taganrog
763		Tris Fili (ex Aspasia) W Bk	Greek	278	105·6	26·1	..	Galaxidi	1866			A.Nicolia & G.Mamuzi	Syra
764		Tris Ierarchai *Catranis* 85-85 W Bn	Greek	280	116·8	26·6	17·7	England	1865			G.Svoronos	Argostoli
765		Tris Ierarchai W Sr	Greek	277	110·8	26·6	..	Syra	1885			S.I.Svoronos	Syra
766	V	Tris Ierarchai *Carvatsis* W Sr	Greek	102	71·8	22·3	11·0	Syra	1885	Jonseph Bros.		A.Goitsias	Syra
767		Triton *Childs* W Bk	American	264	96·1	26·0	17·0	Fairhaven, [Mass.	18 15/57			J.& W.R.Wing	New Bedfd.Mass.
768	A	Triton *Scopinich* -87 W Bk	Austrian	545	127·3	29·3	22·0	Lussinpiccolo	1871	G.S.Cattarinich	Y 9.85	A.L.Cattarinich & Co.	Lussinpiccolo
769	✱	Triton *Reinicke* 80-81 I Bk	German	739	188·0	31·5	18·4	Sunderland	1878	Austin&Hunter		G.Bolte	Elsfleth
770	V	Triton *Witt* -71 W Bk	German	470	126·7	28·8	16·1	Rostock	1871	O.Ludewig	I 12.82	J.C.Thormann	Wismar
771	A	Triton *Mantucas* W Bn	Greek	326	114·1	27·9	..	Syra	1874	S.Koufoudakis	Y 3.83	L.Vrontisis	Andros
772	N	Triton *Sanne* W Bk	Norweg'n	383	137·4	30·2	15·5	Helle	1854			P.Anker	Fredrikshald
773	N	Triton *Öfstie* -73 W Bg	Norweg'n	222	117·0	26·4	12·1	Aalesund	1868	C.I.Haastad		G.Ofstie	Aalesund
774	N	Triton *Jorgensen* -88 W Bg	Norweg'n	219	99·7	27·2	12·7	Kragerö	1875		Y 12.87	T.Jensen	Kragerö
775		Triton *Nilsson* W Bg	Swedish	226	97·1	23·5	13·8	Carlshamn	1837			N.Nilsson	Kivik
776		Tritone (PONTOON) *Chiozza* W Bk	Italian	339	117·7	27·8	17·1	Loano	1858	G.Amico	Y 2.84	F[lli] Chiozza	Savona
777	✱	Tritonia W Sr Yt	British	107	98·3	20·4	12·1	Gosport	1881	Camper & [Nicholson	C 5.88	Peter Stubs	Portsmouth
778		Triumph *Taylor* W Bn	British	150	83·3	20·3	13·0	Ipswich	1837	Bayley		E.Dixon	Newcastle
779	✱	Triumph *Hocken* W K	British	90	79·9	21·8	9·7	Newquay	1881	T.&J.Clemens		S.Mayor	London
780	N	Triumph *Olsen* -87 W Bk	Norweg'n	552	137·9	31·8	18·7	Risör	1871	B.Tellefsen	5.88	M.Danielsen	Risör
781		Triumpho W Sr	Brazilian	238	..	..	..					A.L.de Carvalho	Resende
782		Triumpho W Sr	Brazilian	157	..	..	..					J.A.V.de Amorim	Rio de Janeiro
783	V	Triumpho *Arnellas, jun.* (ex Kage Heat, ex Villa de Morgao, ex Triumpho de Gaya) W Bk	Portug'se	427	127·9	26·9	15·7	Bare	1871			J.J.R.de Magalhaes	Oporto
784		Triunfo *Lesvilla* W Bk	Spanish	232	105·9	24·7	13·7	Palma, Santa [Cruz	1873	J.Rodriguez		Rodriguez & Co.	Palma
785	A	Troas *Pappales* W Bk	Greek	480	132·8	29·2	..	Syra	1877		Y 12.87	S.Calojannas	Syra
786		Trois Frères W Bk	French	249	109·5	27·6	14·7	Fécamp	1879	Brument & [Capon	C 11.75	BailleTorquebian & fils	Cette
787		Trois Sœurs W Bn		146	84·3	20·3	11·2	Granville	1870	Goubert			
788	V	Trojan *Mosher* W S	British	1595	223·0	43·0	23·8	Kempt,N.S.	1884	G.Armstrong	Y 1.85	G.Armstrong	Windsor,N.S.
789	A	Trojednica *Marochini* W Bk	Austrian	479	134·5	27·9	18·8	PortoRe	1873	I.Archianin	Y 5.85	M[a] Battagliarini ved.	Fiume
790	DN	Tromso *Simonsen* 85-85 W Bg	Norweg'n	186	100·5	22·7	11·9	Christiansund	1852			Grönvold Brug (Thorleif Bache)	Drammen
791	✱D	Trongate *Campbell* 84-84 I Bk	British	949	204·0	33·6	20·0	Glasgow	1878	Dobie&Co.		W.D.Denny	Glasgow
792	✱	Troop *Corning* 85-89 I S	British	1526	243·0	38·9	22·4	Dumbarton	1884	A.M[c]Millan&Sn		Troop & Son	Liverpool
793	V	Tropic Bird *Barn* 72- W Bkn	American	331	136·2	31·5	11·7	CoosBay,Oreg	1882	SimpsonBros.		A.Crawford & Co.	San Francisco
794		TropicBird (HULK) (ex Mary Webster, ex La Plata, ex Andalas) *Cobb* W Bn	American	172	100·8	23·2	12·2	Georgetown, [P.E.I.	1864			M[c]Collam Fishing & Trading Co.	San Francisco
795	N	Tros *Jacobsen* -87 W S	Norweg'n	687	161·5	33·0	18·4	Grimstad	1870	S.Johnsen	Y 6.85	Chr.Herlofson	Arendal
796	✱	Trossachs *Batchelor* 77-77 W Bk	British	553	148·6	29·4	18·2	Littlehampton	1877	Harvey	Y 1.86	J.& W.B.Harvey	Littlehampton
797	N	Trosvik *Jensen* 81-85 W Bk	Norweg'n	331	125·5	28·8	15·3	Brevig	1875			Interessentskab Trosvik (E.Larsen)	Brevig
798	V	Troubadour *Gahrs* 84-84 W 3m Sr	German	295	118·0	26·9	12·6	Hamburg	1884	J.H.Wichhorst	Y 11.86	J.H.N.Wichhorst	Blankenese
799		Trovador W Sr	Brazilian	193	..	..	..					J.J.F.Magalhaes	Rio de Janeiro
800		Trovador (ex India) W Bg	Portug'se	146	79·4	20·2	10·9	Lisbon	1853	E.N.Shirley		M.R.de Lima	Lisbon

✱ = Lloyd's Register; D = Nederlandsche Vereeniging; G = Germanischer Lloyd; I = Registro Italiano; R = American Record;
A = Veritas Austro-Ungarico; H = Veritas Ellenico; N = Norske Veritas; V = Bureau Veritas.

Dans la 4me Colonne, C = Composite (membrure en fer, bordé en bois); I = Fer; Stl = Acier; W = Bois.

Les Dates du doublage en métal (Col. 11) ressortent des livres dans lesquels les navires sont respectivement classés ou des rapports des Inspecteurs du Lloyd's Register.

Pour plus amples renseignements sur les navires classés par le Lloyd's Register, voir le Lloyd's Register Book et le Lloyd's Yacht Register.

1 Official No.	2 No. in Book	3 Registry in which classed	4 Ship's Name. Material and Rig. Master, and Dates of Appointment to present Owner's Service and to Ship. Ship's Late Name.	5 Flag.	6 Net Tonnage.	7 Dimensions. Length.	Breadth.	Depth.	8 Built. Where.	9 When.	10 By whom.	11 Date of Metalling	12 Owners or Managers.	Port belonging to.
[illegible]	801	✳	Trowbridge Thomson 81-86 I Bk	British	1287	221·4	36·8	22·3	Sunderland	1868	Oswald&Co.		S. Trowbridge Co. Ld. (Owners Duckam & Co.)	London
	802	N	Trudvang Jacobsen 84-84 W Bk	Norweg'n	374	114·8	30·1	16·5	Grimstad	1865	E.Jorgensen		O.Jorgensen	Grimstad
68845	803	✳	True Blue (ex Hatton) Davison 84-86 I Bk	British	449	152·5	26·0	15·8	Sunderland	1869	Oswald&Co.		W.R.Cave & Co.	Adelaide
[illegible]	804		True Blue W Bn	British	152	96·0	24·1	12·0	P.E.Island	1858			R.H.Prowse	St.John
51542	805	V	True Briton Torrance -88 W S	British	1365	204·0	38·0	24·2	Quebec	1865	Dinning		R.Hunter	Liverpool
5701	806		Truelove W Bk	British	285	96·8	26·9	17·0	Philadelphia	1764			J.S.Ward	London
69053	807	D	Truro Phillips 86-86 W Bk	British	895	170·4	35·2	20·8	Colchester N.S	1873	Sanderson		Martin Dickie	Halifax
	808		Trust Brau W Bn	German	506	138·0	31·1	18·0	Maitland,N.S.	1876	Frieze& Roy		E.Burchard & Co.	Rostock
70275	809	✳	Trust Quere -88 W Sr	British	79	81·8	17·5	8·8	Jersey	1880	G.Deslande s & [illegible]		J.Machon	Jersey
84274	810	✳	Trusty Apsey -88 W Bg	British	155	101·6	23·4	12·6	Bridport	1872	Cox		John Munn	St.John's
	811	DN	Truth Skjelbred -87 W Bk	Norweg'n	496	146·5	29·8	17·9	Sunderland	1874	J.Gardner		O.A.T.Skjelbred	Christiania
80293	812	✳	Truthseeker Garrett W Sr	British	99	90·0	21·8	10·1	Northwich	1879	Thompson		J.Heath & Co.	Runcorn
	813	N	Try Kreutz 71-71 W Bk	Norweg'n	581	118·9	32·3	17·8	Archangel	1862			M.Mulvad	Fredrik...
	814	NR	Try (ex Kirkland) Aanonsen W Bk	Norweg'n	472	135·4	27·4	18·1	Sunderland	1859	G.V.Peverall		T.Aanonsen	Arendal
55337	815		Try Again (LIGHTER) Ferrier W	British	227	104·8	24·0	13·8	Montrose	1867	Strachan		A.Marr	Aberdeen
58802	816		Try Again Clements 77-83 W Sr	British	139	95·0	22·1	12·9	Kippford	1868	Cuming		W.J.M.Hawkey	Dumfries
69876	817	V	Trygve O'Neil -88 W Bn	British	221	107·3	24·2	13·0	Stavanger	1872	K.Nielsen		J.W.Edgett	Turks Is.
	818	N	Trygve Grammes W Bk	Norweg'n	352	107·8	29·1	17·7	Amsterdam	1842			J.M.Pauss	Drammen
[illegible]	819		Tryphena (SUNK) Harris W Bn	British	178	100·3	24·5	12·4	P.E.Island	1865	Cousins		W.Milburn	North S...
[illegible]	820		Tsalay I Flat	British	777	197·0	25·7	8·3	Port Glasgow	1876	R.Duncan & Co.		Irrawaddy Flotilla Co., Ld.	Glasgow
[illegible]	821	AH	Tsamados Coulouras W Bg	Greek	189	90·5	25·1	..	Syra	1880			I.P.Eleutherion	Syra
[illegible]	822		Tsampo Stl Lr	British	484	220·0	30·0	8·3	Dumbarton	1887	W.Denny&Bros.		Irrawaddy Flotilla Co., Ld.	Glasgow
13	823		Tsuho Maru W Bg	Japanese	126	79·5	20·3	10·1	Toba,Shima	1879	FukuZawaTatsuzo		Fukuchi Zenkichi	Tokio
720	824		Tsuho Maru W Bg	Japanese	113	85·0	17·9	9·9	Osaka	1882	Murache [Tamekichi		Morita Kinzayemon	Owari
[illegible]	825		Tsunemoto Maru W Sr	Japanese	423	126·6	27·8	15·8	Tokio	1883	KawasakiSBYd		Hiroumi Nisaburo	Osaka
745	826		Tsurunoye Maru W Bk	Japanese	428	127·4	26·2	17·4					Kono Saiji	Nagasaki
182	827		Tsusai Maru W Bk	Japanese	259	116·1	24·8	12·8	Hamburg	1864			Hisazumi Kohei	Tsuda
64494	828	V	Tuck Sing Laura -88 W Bk	British	388	133·1	29·0	14·0	St.John,N.B.	1870	J.S.King		R.Matthew	St.John,N.B.
[illegible]	829		Tukeing I Barge	British	211	115·0	19·3	5·0	Port Glasgow	1879	R.Duncan & Co.		Irrawaddy Flotilla Co., Ld.	Glasgow
24594	830		Tula Allen W Bn	American	171	98·0	26·5	9·5	Newburyport, Mass	1868			F.W.Brightman	Philadelphia
33	831	RV	Tula Berdios -87 W Bk	Spanish	425	134·1	29·2	16·8	Ferrol	1882			L.Fernandez&J.Colomez	Ferrol
	832		Tulla W Bn		357	111·8	30·1	18·3	Varazze	1864	A.Craviotto			
	833		Tulli Antoine Tulli W Bn	French	110	74·5	20·0	10·2	Martigues	1866	A.Paul		A.Tulli	Marseille
65357	834	✳	Tullochgorum Crocker 83-84 W 3mSr	British	175	107·8	23·5	11·7	Peterhead	1873	Stephen		J.Crocker	Peterhead
55425	835		Tullochgorum Mitchell (84-84) W Bn	British	157	94·0	22·2	12·3	Garmouth	1867	Duncan		T.S.Honey	Plymouth
3465	836	I	Tunisi Laurent -88 W Bn	Italian	258	126·3	26·9	11·0	SestriPonente	1873	A.Briasco fu G.B.		Piombino,Marre & Co.	Genoa
	837		Tunnah Mustapha W Schleppe	Turkish	175	..	..	..	Tödrakhen	1869			[illegible]	Todrakhan
	838		Tunnah W Schleppe	Turkish	162	..	..	..	Rustchuck	1868			[illegible]	Todrakhan
	839		Tunnah W Sr	Turkish	115	..	..	..	Danube	1869			[illegible]	Rustchuk
60352	840	✳	Turakina (ex City of Perth) Power 77-83 I S	British	1189	232·5	35·4	22·2	Glasgow	1868	C.Connell&Co.		New Zealand Shipping Co.	London
83206	841	R	Turban Hill -88 W Sr	British	159	96·5	25·5	10·9	Kempt,N.S.	1881	G.Armstrong		E.E.Walker	Halifax,N.S.
	842	V	Ture Johansson -85 W Bn	Swedish	171	101·3	22·2	11·5	Oscarshamn	1880	C.Thoren		N.J.Carlsson	Misterhult
	843	DN	Turist Berentsen 86-86 W Bk	Norweg'n	307	112·3	28·1	11·7	Grimstad	1868	J.Tellefsen		Jacob Holst	Grimstad
	844		Turki (STRANDED) Sundell W Bk	Russian	492	..	..	..	Åbo	18 62/67			G.A.Lindblom	Åbo
87398	845	✳	Turkistan Scott 75-87 I S	British	1593	259·7	38·2	23·1	PortGlasgow	1883	Russell&Co.		W. & J.Crawford	Greenock
	846		Tursina (ex Sholomith) Ahmed Hussein Boonsom W S	Arabian	816	140·8	30·2	22·9	Rangoon	1852			Mahomed Cossem	Jeddah
83435	847	V	Tuskar Curry W S	British	1556	226·6	41·3	24·0	Newport,N.S.	1883	F.Curry		F.Curry	Windsor
	848		Tuspan Lindberg W Bg	Swedish	254	103·6	23·5	14·2	Brake	18 [illegible]/67			A.Nilsson	Gothenburg
	849		Tuticorin (ex Berryer) W Bk	British	369	122·0	28·3	16·0	Bordeaux	1871	Raymond, frères & [illegible]		J.M.B.RocheVictoria	Tuticorin
	850	V	Tuya Roig W Bk	Spanish	264	104·3	28·3	14·9	Arenys	1851			J.Roig Carran	Barcelona

✳ = Lloyd's Register; D = Nederlandsche Vereeniging; G = Germanischer Lloyd; I = Registro Italiano; R = American [Record]; A = Veritas Austro-Ungarico; H = Veritas Ellenico; N = Norske Veritas; V = Bureau Veritas.

In the 4th Column, C = Composite (iron frame, wood planked); I = Iron; Stl = Steel; W = Wood.

The Dates of Metalling (Col. 11) are based on the records of the Books in which Vessels are respectively classed, or on the Reports of the Surveyors of Lloyd's Register.

Further particulars of Vessels classed by Lloyd's Register will be found in Lloyd's Register Book and Lloyd's Yacht Register.

1	2	3 Nom. Capitaine et dates d'entrée au service des Armateurs actuels et à bord du navire. Ancien Nom.	4 Matériaux et Grément	5 Pavillon	6 Tonnage	7 Dimensions. Longueur	Largeur	Creux	8 Construit. Où.	9 Quand.	10 Par qui.	11 Date du doublage en métal	12 Armateurs ou Directeurs-Gérants.	13 Port d'Armement.
851	N	Tvende Brödre *Olsen -83*	W Bk	Norweg'n	322	97·7	27·6	18·5	Holmestrand	1797			J.C.Svendsen	Christiansand
852		Tvende Brodre	W Bg	Norweg'n	231	89·0	24·6	15·2	Krageró	1844			G.Abrahamsen	Krageró
853	N	Tvende Brödre *Holm [-87*	W Bg	Norweg'n	132	82·4	21·2	11·7	Strömstad	1810			S.&H.Holm	Drammen
854		Twantay	I Flat	British	777	197·0	25·7	8·3	Port Glasgow	1878	R.Duncan & Co.		IrrawaddyFlotillaCo.Ld.	Glasgow
855	DV	Twee Gebroeders *Hosscane*	W Bg	Dutch	233	108·6	25·1	12·1	Harlingen	1871	D.L.Alta	12.81	J.Alta L[zn.]	Harlingen
856	D	Twee Vrienden (ex Christina) *Warlington*	W Bk	Dutch	480	111·0	28·2	17·1	Elsfleth	1874	J.Jürgens	V 5.84	S.B.Mesritz	Sourabaya
857	D	Twee Zusters *Harding* (ex Wild Hunter) [-85 (ABANDONED)	W Bk	Dutch	377	129·9	30·5	12·7	Windsor,N.S.	1864	W.Fitch	V 80	J.S.Visser	Lemmer
858		Tweed (ex J. M. Sawyer)	W Bn	British	204	99·0	26·2	10·4	Damariscotta, [Me.	1854			W.Morrice	Nassau.N.P.
859		Tweed *Cooper*	W Sr	British	199	106·8	24·5	12·7	Newburgh	1865	Scrimgeour		A.Nicholls	Faversham
860	*	Tweed *Flak*	W Sr	British	100	92·0	21·5	10·1	Arbroath	1879	Roney	V 5.84	G.C.Ross	Glasgow
861	*	Tweedsdale *Buchanan [87-87*	I 4m Bk	British	1403	244·4	37·4	22·6	Glasgow	1877	Barclay,Curle & [Co.		J. & A.Roxburgh	Glasgow
862	*	Twelve Apostles *Williams 77-77*	W Sr	British	118	86·2	22·2	11·4	Pwllheli	1858			H.Hughes & Co.	Carnarvon
863	D	Twig *Millikin 84-84*	W Sr	British	110	90·7	21·7	10·6	Fareham	1875	Purkis		D.Macdonald	Liverpool
864		Twilight *Brown 80-80*	W Bk	British	756	166·4	32·3	20·8	Quebec	1866	Charland		Cooke Bros.	Londonderry
865	V	Twilight *Jones 86-88*	I Bk	British	631	167·3	30·0	20·0	Glasgow	1855	R.Napier & Sons		R.W.Leyland & Co.	Liverpool
866	R	Twilight *Mackenzie*	W Bn	British	144	92·0	24·4	10·7	Shelburne,N.S	1880	J.A.McGowan,jr	V 6.83	J.Mackenzie	Shelburne,N.S.
867		Twilight *Van Vleaf*	W 3m Sr	American	357	125·5	32·0	10·2	E.Haven,Conn	1874			L.W.&P.Armstrong	New Havn.Conn.
868		Twilight *Larsen 87-87*	W Sr	American	175	112·0	30·0	9·0	PortLudlow, [Wash.	1874			S.B.Peterson	San Francisco
869	*	Twins *Williams*	W Sr	British	86	71·5	19·7	9·2	Fowey	1837	G.Nickels		E.Williams	Falmouth
870		Two Boys (ex Minta Heros Bounn)	W 3m Sr	American	164	91·6	22·3	11·3	Anclam	1873	C.Hansen	V 7.84	O'Keep	Yap,Caroline Isl.
871		Two Brothers	W Sr	British	122	92·9	21·6	7·6	PortBurwell	1868	L.McDermott		G.Wright	Port Hope,Ont.
872	*	Two Brothers *Pritchard*	W Sr	British	76	60·7	17·7	10·0	Pwllheli	1851			D.Pritchard	Carnarvon
873		Two Brothers *McCartney [88-88*	W S	American	1343	198·0	38·0	24·9	FrmingdaleMe	1868	P.G.Bradstreet	V 1.82	S.Blair	San Francisco
874	R	Two Brothers *Goudy*	W Sr	American	114	84·4	24·0	7·0	Damariscotta, [Me.	1883	F.C.Clark		E.H.Goudy	Damariscotta,Me.
875		Two Friends *Ploy le 82-82*	W Bn	British	170	95·8	22·8	12·2	Guernsey	1850	Marquand		M.Tonzeau	Guernsey
876		Two Sisters	W Sr	British	115	94·8	21·8	9·7	Bosham	1882	T.Smart		Kate Smart	Portsmouth
877	*	Tyburnia *Devitt*	W Bk	British	948	185·2	34·2	22·1	Glasgow	1857	A.Stephen&S'ns	V 5.82	Burns,Philp & Co.	Sydney,N.S.W.
878	DV	Tycho Brahe *Krarup [86-88*	W Bn	Danish	175	105·9	22·3	11·0	Veile	1872	P.H.Illum		S.Krarup	Svendborg
879		Tyne *Lloyd -87*	W Bn	British	157	92·1	22·1	12·5	S.Shields	1867	Winlo	V 5.84	O.Lloyd	North Shields
880	N	Tyne *Knudsen*	W Sr	Norweg'n	181	92·3	24·2	11·7	St.John,N.B.	1858			C.Knudsen	Drammen
881	N	Typhon *Gundersen 85-85*	W Bg	Norweg'n	198	97·6	25·1	12·4	Tvedestrand	1864	A.Jensen		Jörgensen & Jorkjend	Tvedestrand
882	N	Tyr *Mathisen 83-*	W Sr	Norweg'n	135	86·4	22·3	10·5	Drammen	18 5/83			B.A.Ellingsen	Christiania
883		Tyr *Jeppsson -86*	W Bn	Swedish	153	88·6	22·5	10·4	Figeholm	1882	N.P.Andersson		G.Hammarberg	Figeholm
884		Tyri *Smebye 87-87*	W 3m Sr	Norweg'n	189	104·0	22·5	12·4	Porsgrund	1868			H.Smebye	Skien
885		Tyro	W Sr	Brazilian	161	..	..	..					D.M.B.Verany	Rio de Janeiro
886		Tyrone	W Bg	British	155	79·1	20·8	12·3	Granville,N.S	1848			Kappa & Co.	Liverpool
887		Tyrone *Heckman 86-*	W Sr	British	107	82·4	23·9	9·5	PleasntvlleNS	1883	A.McKeen		A.McKean	Lunenburg,N.S
888	*	Tythonus *Lambert 85-85*	I Bk	British	1152	201·0	36·0	22·9	W.Hartlepool	1862	Pile Spence&Co		Gillison&Chadwick	Liverpool
889	*	Tzinfara *Andersen* (ex M. Rene) (STRANDED)	W S	Norweg'n	867	180·2	32·3	22·5	Workington	1863	Fell	V 9.80 / 3.84	H.N.Gjersöe&O.Eriksen	Christiania

ADDITIONAL SHIPS ARE INSERTED IN THE SUPPLEMENT AT THE END OF THE REGISTER.

= Lloyd's Register; D = Nederlandsche Vereeniging; G = Germanischer Lloyd; I = Registro Italiano; R = American Record;
A = Veritas Austro-Ungarico; H = Veritas Ellenico; N = Norske Veritas; V = Bureau Veritas.

Dans la 4me Colonne, C = Composite (membrure en fer, bordé en bois); I = Fer; Stl = Acier; W = Bois.

Les Dates du doublage en métal (Col. 11) ressortent des livres dans lesquels les navires sont respectivement classés ou des rapports des Inspecteurs du Lloyd's Register.

Pour plus amples renseignements sur les navires classés par le Lloyd's Register, voir le Lloyd's Register Book et le Lloyd's Yacht Register.

Official No.	No. in Book	Registry in which classed	Ship's Name. Material and Rig. Master, and Dates of Appointment to present Owner's Service and to Ship. Ship's Late Name.	Flag	Net Tonnage	Length	Breadth	Depth	Where built	When	By whom	Date of Metalling	Owners or Managers	Port
40651	1	✻	U. Larsing *Thomas* W Sr	British	96	75·5	20·2	10·6	Newquay	1864	Davies		W.O.Morris	Carn
61924	2	R	Ubaldiena *Sponagle* W Bn	British	266	112·2	29·0	12·3	W.Dublin,N.S.	1871	R.B.Currie	[illegible]	J.S.Sponagle	Liver
	3	R	Uddevalla *Lycke* W Bkn (ex [illegible])	Swedish	260	111·5	24·7	12·2	Nova Scotia	1863		[illegible]	J.Berntsson	Kalland
	4	N	Udsire *Warsland* -87 W Bk	Norweg'n	417	126·8	28·5	16·8	Richibucto	1863		[illegible]	S.Bergesen	Stavan
68062	5	✻ D	Udston *Gill* 86-86 I S	British	1066	209·0	33·8	23·7	Glasgow	1874	Barclay,Curle & Co.		Watson Bros.	Glasgow
	6		Ufven W Sr	Swedish	105	97·1	20·6	7·3	Harlstad				A.F.Andersson	Torsby
	7		Uffo *Paterson* 72- W Bg (LOST) (ex [illegible])	Danish	176	88·5	22·6	13·9	Gaeta	1854			P.J.Pedersen	Svendbo
47152	8		Uganda *Skandelbery* 71-81 W Sr	British	129	89·5	20·8	11·7	Bideford	1864	Johnson		W.Dusting	Penzan
	9	V	Uggenzeems *Puhling* W 3m Sr	Russian	267	113·0	31·0	13·1	Uggenzeem	1886	C.Ohrmann		H.Puling	Riga
	10	V	Ugglan *Gudd* 77-85 W S (ex [illegible])	Swedish	895	178·4	31·8	21·3	Genoa	1873		[illegible]	Rederi Akt Bol Ornen (J.F.Lundberg)	Gothen
	11	N	Ugland *Engebrethsen* -77 W Bk	Norweg'n	477	122·2	31·9	18·2	Grimstad	1858	Jens Tellefsen		W.Engebrethsen	Lourtig
	12		Ukko *Kurki* W Bk	Russian	560	150·0	29·0	19·0	Sunderland	1866			W.& H.Martelin	Wiborg
	13		Ukko *Borqvist* W Bk	Russian	421	..	..	..	Lovisa	18^{58}_{75}			C.M.Lindroth	Borgå
	14		Ukko *Kystein* W Glt	Russian	113	..	..	..		1883			J.Lenkkeri	Björkö
58	15	AV	Ukraine *Mircovich* W Bk	Austrian	691	152·8	30·8	20·0	Bath,Me.	1865	Mallet,Meicher,	[illegible]	N.Mircovich fu A.	Lussin
74420	16	✻	Ulelia *Phillips* 77-87 W S	British	69	75·4	19·9	9·4	Truro	1877	Dyer		J.Estlick	Truro
	17	N	Uller *Lydersen* -88 W Bk (ex Amber Witch)	Norweg'n	517	139·1	30·0	18·1	Quebec	1861	T.Valin		O.Lydersen,junr.	Tvedes
	18	N	Uller *Olsen* -88 W 3m Sr	Norweg'n	256	118·2	25·9	12·6	Kragerö	1865		[illegible]	J.Dahll	Kragerö
70955	19	✻ D	Ullock *Brown* 84-84 I Bk	British	779	186·0	32·1	18·8	Aberdeen	1875	A.Hall & Co.		P.Iredale & Son	Liverpool
87399	20	✻	Ulrica *Scott* 70-84 I 4m S	British	1923	276·5	41·2	24·0	Glasgow	1884	Barclay,Curle & Co.		W.Lethem	Greenock
8580	21		Ulrica *Bell* 86-5 W Sw	British	288	99·0	26·5	16·8	Sunderland	1851	Austin & Mills		G.Marshall	North
25203	22		Ulrica R. Smith *Sawyer* W Sr	American	119	93·0	26·0	7·8	Jonesboro'Me.	1873			Albert Smith	Machias
183	23		Ulrika *Lacke* 72-77 W Bk	Russian	919	182·2	36·0	20·5	Kristinestad	18^{65}_{77}	Hannas		O.Wendelin	Kristines
	24		Ulrika *Zendeman* 72-76 W Bk	Russian	648	160·0	33·0	17·5	Kaskö	1871	A.Granskog		L.Olofsson	Wardö
	25		Ulrika (LIGHTER) *Andersson* W Bk	Swedish	381	119·4	26·4	16·1	Dalkarlså	1868	E.A.Löfgren		G.Andersson	Fjellbac
	26	V	Ulrika *Bulér* W Bk	Swedish	300	128·6	26·9	12·0	Sundsvall	1871	J.A.Nordström	[illegible]	C.Bulér	Stockholm
	27		Ulrika *Brandt* 77-77 W Bn	Swedish	142	83·3	22·0	11·0	Gefle	1852			A.Gudmundsson	Jonstorp
81974	28	✻	Ulster *Shekel* 75-84 W Bkn	British	290	128·7	27·2	12·9	Belfast	1879	M'Laine	[illegible]	C.T.Bowring & Co.	Liverpool
	29		Ultima *Rombolti* 81- W Bk	Russian	406	126·2	30·0	17·0	[illegible]	1875	M.Reponen		W.& H.Martelin	Wiborg
	30	V	Ulysse *Fauran* W Bk	French	342	109·5	23·9	13·0	[illegible]	1875	Dubigeon & fils	[illegible]	A.Fabre & fils	Marseille
	31		Ulysses *De Mel'o* W Bn	Brazilian	150	..	..	..	[illegible]	1867		[illegible]	Cia Cabotagem de S.José da Barra	St.John
	32	V	Uman *Stenbäck* 87-87 W Bk	Russian	404	133·5	25·9	16·0	[illegible]	1855		[illegible]	C.Stenbäck	Jacobst
2808	33	D I V	Umberto I. *Traverso* W Bk	Italian	766	170·6	33·5	21·3	[illegible]ri Ponente	1878	C.Lunano	[illegible]	A.Badino	Genoa
1186	34	I	Umberto I. *Canziano* -88 W Bk	Italian	509	117·3	27·9	18·1	[illegible]	1878	F.De Rosa	[illegible]	N.Jaccarino	Castellam
	35		Umeå *Pahlberg* W Bg	Swedish	234	110·3	21·0	12·0	Umeå	1864	Jonsson		L.Glas	Umeå
51005	36		Umgeni *Davies* W Bk	British	276	139·0	26·0	12·7	Liverpool	1864	Robinson	[illegible]	M.Davies	Liverpool
	37	N V	Umkomanzi *Bache* 78-84 W Bk	Norweg'n	318	135·7	27·3	12·8	Sunderland	1873	J.Crown	[illegible]	C.M.Bache	Drammen
60697	38	✻	Umvoti *Buck* 88-88 I Bk	British	442	158·9	28·7	15·9	Aberdeen	1869	Hall,Russell & C		J.T.Rennie & Son	Aberdeen
65918	39	✻	Umzinto *Bruckshaw* -88 I Bk	British	280	138·1	25·7	12·6	Sunderland	1871	W.Pile & Co.		Bullard,King & Co.	London
41481	40		Una W S	British	768	144·0	33·1	21·0	Tynem'th,N.B.	1846			T.G.White	Melbourne
25189	41		Una W Bk	American	793	158·0	32·0	20·0	Bath,Me.	1869			C.T.Weaver	Bath,Me.
25216	42		Una *Smith* W 3m Sr	American	197	118·2	31·5	9·8	Pt Blakely W.T.	1874			J.Knowland	San F
81	43	✻	Una *Claccevich* W Bk	Austrian	352	134·2	27·2	15·0	Littlehampton	1871	Harvey	[illegible]	Matovich & Claccevich	Dobrota
59205	44	R V	Unanima *Korff* W Bk	British	747	156·0	31·2	19·9	St Martins N.B.	1868	Bradshaw	[illegible]	G.Bedell	St.John
25246	45		Uncle John *Munson* W Bkn	American	299	138·0	33·4	10·4	Eureka,Cal.	1882	Murray		J.Vance	Eureka
56344	46		Uncle Ned *Nutt* W Bkn	British	189	118·7	23·5	12·8	Ipswich	1867	Robertson		H.Roach & Co.	Faversham
	47	N	Unda *Olsen* W Sr	Norweg'n	266	115·8	25·9	11·7	Arendal	1874	B.Sorensen	[illegible]	O.B.Sörensen	Arendal
33804	48		Undaunted *Rowe* W Bn	British	148	99·0	23·4	12·1	Tatamagouche N.S.	1861	Kent		W.F.Ball	Dartmouth
5461	49	✻	Undaunted *Keast* W Sr	British	98	73·5	18·1	11·4	Newport	1839	Young		F.Keast	Plymouth
	50													

✻ = Lloyd's Register; D = Nederlandsche Vereeniging; G = Germanischer Lloyd; I = Registro Italiano; R = American [Register]; Δ = Ver[itas] Austro-Ungarico; H = Veritas Ellenico; N = Norske Veritas; V = Bureau V[eritas]

In the 4th Column, C = Composite (iron frame, wood planked); I = Iron; Stl = Steel; W = Wood.

The Dates of [Meta]lling (Col. 11) are based on the records of the Books in which Vessels are respectively classed, or on the Reports of the Surveyors of Lloyd's [Register].

✻ *Further particulars of Vessels classed by Lloyd's Register will be found in Lloyd's Register Book and Lloyd's Yacht Register.*

No.	Classe	Nom. Capitaine et dates d'entrée au service des Armateurs actuels et à bord du navire. Ancien Nom.	Matériaux et Grèement	Pavillon	Tonnage net	Longueur	Largeur	Creux	Construit. Où	Quand	Par qui	Doublage	Armateurs ou Directeurs-Gérants.	Port d'Armement.
51	RV	**Undaunted** *Hamilton 78-86*	WS	American	1764	207·3	41·1	27·8	Bath,Me.	1869	E.&A.Sewall	[illegible]	A.Sewall & Co.	Bath,Me.
52		**Undine**	WSr	British	207	115·7	23·0	12·0	Gaspart Hauls	1869	Cowper & Nicholson		C.G.Millar	Melbourne
53		**Undine**	WSr	British	196	107·8	22·6	11·4	Hamilton,Ont	1868	D.P.Lavallee		T.Myles	Hamilton,Ont.
54	✻	**Undine** *Beale*	WBn	British	174	101·1	24·0	13·2	Fowey	1875	Butson	Y 3.86	T.Sherwill	Fowey
55	✻	**Undine** *Colcerich*	CBk	Austrian	818	183·7	35·1	19·6	Sunderland	1867	W.Pile&Co.	Y 10.86	M.Ivetta	Ragusa
56	GV	**Undine** *Kruse 67-73*	WBk	German	968	174·1	34·8	22·5	Bremerhaven	1877	J.C.Tecklenborg	[illegible]	F.Tecklenborg	Bremen
57	V	**Undine** *Bielenberg 83-85*	IS	German	760	169·0	30·5	19·7	Hamburg	1867	Reiherstieg [Schiffs Werfte		Wachsmuth & Krogmann	Hamburg
58	R	**Unexpected** *Brinkman [-86*	WSr	British	117	91·4	25·0	9·3	Cornwallis,NS	1883	J.E.Bigelin		Mrs.H.A.Benekman	Windsor,N.S.
59	N	**Ungdoms Venner** (ex Carrie Hart) *Andersen -81*	WBk	Norweg'n	534	130·5	31·3	18·0	MossGlen,N.S	1869	Wetmore & [Merritt	Y 3.86	Th.Northaus	Egersand
60	V	**União** (ex Santos) *Nielsen*	WBk	German	231	111·5	24·9	13·8	Arenzano	1871		Y 2.86	J.C.A.Zerrenner	Hamburg
61		**Unico** *Malmberg*	WBg	Swedish	314	112·2	25·1	14·6	Venice	1855			Johanson & Co.	Halmstad
62	✻	**Unicorn** *McDougall 81-81*	IBk	British	499	163·0	27·0	16·2	Stockton	1863	Richardson, [Duck&Co.		P.H.Dixon & Co.	Glasgow
63	✻	**Unicorn** *Davies -68*	WSr	British	119	72·2	19·9	12·1	Carnarvon	1840			E.Davies	Carnarvon
64		**Unicorn** *Dick 82-82*	WBn	British	110	87·6	25·0	10·7	Kempt,N.S.	1865	J.Starrat		A.F.MacFarlane	Irvine
65	✻	**Unie** *Franken 78-78*	WBmSr	Dutch	234	108·5	25·8	12·6	Pekela	1878	Kooerts&Kolk	Y 1.87	J.J.Koerts & Zoon	Oude Pekela
66		**Union** *Pettigrew*	WBk	British	263	125·8	28·2	12·9	Liverpool,N.S.	1864	McLeod		Miss M.A.Finnie	Ardrossan
67		**Union** *Briard 86-86*	WBg	British	193	111·5	23·0	13·0	Jersey	1865	Picot		G.Le Gros	Jersey
68	✻	**Union** *Vince, jun.*	WSr	British	100	84·7	19·0	10·0	Ipswich	1819	Bayley		A.Beaumont & Co.	Ipswich
69		**Union** *Reed*	WSr	American	115	82·0	25·2	8·7	St.John,N.B.	18 67/72			G.Reed	Ellsworth,Me.
70		**Union** *Cole*	WSr	American	110	87·5	24·7	8·4	E.Machias,Me.	1851			P.A.Cator	Machias.Me.
71	V	**Union** *Rasmussen -76*	WSr	Danish	153	96·2	21·4	11·1	Svendborg	1876	P.Troensegaard		P.Troensegaard	Svendborg
72	V	**Union** (ex Nakskov) *Harboe 80-80* (LOST)	WSr	Danish	133	90·4	21·4	10·6	Nakskov	1867	Ridersborg& [Trochmann		Akt. Selsk. (B. P. Christensen)	Fanø
73	V	**Union** *Lauritzen -72*	WSr	Danish	111	85·1	21·2	10·8	Nakskov	1872	Ridersborg& [Trochmann		H.P.Riddersborg	Marstal
74	✻	**Union** *Martin 72-85*	I 4mS	French	2120	289·8	43·0	23·8	Greenock	1882	Russell&Co.		Ant.Dom.Bordes & fils	Bordeaux
75	✻	**Union** (ex Idaho)	WBk	French	365	142·5	27·2	13·9	Sunderland	1875	J.Crown	Y 11.86	Larchevêque	St.Valery
76		**Union** *Clement*	WBg	French	140	78·7	21·0	13·5	St.Malo	1857	Hervé		Vve Marie neé Brehier	Binic
77	V	**Union** (ex Semiramis) *Eekhen 75-83*	WS	German	1152	182·3	36·5	23·5	Portsm'thNH.	1863	Toby&Littlefield	Y 10.87	Joh.Fr.Arens	Bremen
78	V	**Union** (ex Indian Chief) *Boge 87-87*	WBk	German	364	108·8	26·4	18·3	Sunderland	1847			C.F.Eschricht	Swinemünde
79	I	**Union** (ex Mia Madre) *Scriaffino*	WBk	Italian	737	162·1	33·1	20·6	Finale	1875	F.Ponzielo	Y 9.86	N.Cuneo	Genoa
80	N	**Union** *Olsen*	WS	Norweg'n	901	190·3	35·7	21·6	Bergen	1875	JensGran	Y 12.87	Rasmus F.Olsen	Bergen
81	N	**Union** (ex Eden) *Staalhaas*	WBk	Norweg'n	483	141·5	28·6	17·5	Quebec	1866	Baldwin	Y 2.87	G.Synnestvedt	Kragerö
82	V	**Union** *Petersen 74-81*	WBk	Norweg'n	467	135·8	28·6	16·6	Bremen	1868	D.Regge	Y 10.85	Stavanger Skibsrederiselskab (Holst & [illegible])	Stavanger
83	N	**Union** *Borge -84*	WBg	Norweg'n	198	92·0	26·6	12·3	Christiansund	1847			A.Schive	Tönsberg
84	G	**Union** *Wahlroos*	WBg	Russian	264	104·3	28·0	14·4	Sideby	1874			G.W.Wahlroos	Raumo
85		**Union** *Eriksson -83*	WBg	Russian	264	109·7	29·2	13·5	Åbo	1876	Lindqvist		J.Korsström	Nagu
86		**Unión** (ex Basturiano, ex Lavinia)	WBn	Spanish	248	105·6	23·5	13·1	Sunderland	1866	W.Doxford		V.Villanueva	Manila
87		**Union** *Andersson*	WSr	Swedish	112	75·1	21·1	10·2	Flensburg				N.P.A.Andersson	Högsholla,Calmar,ln.
88	N	**Union Grove** *Tobiassen [-88*	WBg	Norweg'n	151	88·5	23·8	11·3	Banff	1858	J.Smith		C.Tobiassen	Christiansand
89	✻	**Union Jack** *Croft*	WSk	British	75	74·6	20·6	10·6	Hamburg	1886	C.C.A.Dreyer		F.Croft	Grimsby
90		**Union des Chargeurs** *Lechapelain -87*	WBk	French	564	155·0	30·1	19·0	Honfleur	1865	Leviels	Y 2.87	C.Leclère	Granville
91		**Unione** (CONDEMNED) (ex Caterina Assunta)	W3mSr	Italian	631	154·8	32·5	20·8	Loano	1872	A.Cerruti	Y 12.84	P.A.Rosasco di P.	Genoa
92	I	**Unione** *Ansaldo -88*	WBk	Italian	544	140·7	31·8	20·5	Chiavari	1870	M.Tappani	Y 12.85	Raggio & Porcella	Spezia
93	V	**Unione** *Castellano*	WBk	Italian	418	131·3	29·9	18·2	Alimuri	1870	M.Matteo	Y 2.84	Salv.Cafiero & Co.	Castellamare
94		**Unione** (ex Antonio Lamato) *Riccardo -87*	WBk	Italian	327	115·1	26·5	17·7	Castellamare [di Stab.	1856		Y 5.86	Salv.Amedeo & Co.	Trapani
95	I	**Unione** *Vianello -88*	WBn	Italian	190	102·6	25·8	13·2	Chioggia	1885	F.Menetto	Y 6.85	N.Salvagno	Venice
96		**Unione** *Carletti*	WBn	Italian	130	77·4	23·6	13·3	Varazze	1847	D.Craviotto		Flli.Carletti fu T.	Porto Ferraio
97	I	**Unione P.** (ex St. [illegible], ex Rosalia) *Zino*	WBk	Italian	529	139·7	31·5	21·0	SestriPonente	1871	G.Casanova	Y 10.87	G.Zino fu G.	Genoa
98	I	**Unione S.** (ex Orsa Terza) *Villa -87*	WBk	Italian	431	136·1	30·2	17·2	Sestri Ponente	1878	A Briasco fu G.B	Y 9.88	Frat.Sanguineti fu Gius.	Spezia
99	AD	**Unione V.** (ex Tri Braco) *Ballarin*	W	Italian	358	..	..	..	Portoré			Y 2.81	G.Ballarin & Co.	Venice
100	A	**Unione & Perseveranza** (ex Caterina St.lla)	WBk	Italian	535	144·3	32·0	20·5	Voltri	1872	N.A.Canepa	Y 9.88	A.Pertusio fu G.	Savona

✻ = Lloyd's Register; D = Nederlandsche Vereeniging; G = Germanischer Lloyd; I = Registro Italiano; R = American Record;
A = Veritas Austro-Ungarico; H = Veritas Ellenico; N = Norske Veritas; V = Bureau Veritas.

Dans la 4me Colonne, C = Composite (membrure en fer, bordé en bois); I = Fer; Stl = Acier; W = Bois.

Les Dates de doublage en métal (Col. 11) ressortent des livres dans lesquels les navires sont respectivement classés ou des rapports des Inspecteurs du Lloyd's Register.

Pour plus amples renseignements sur les navires classés par le Lloyd's Register, voir le Lloyd's Register Book et le Lloyd's Yacht Register.

1 Official No.	2 No. in Book	3 Registry in which classed	4 Ship's Name. Material and Rig. Master, and Dates of Appointment to present Owner's Service and to Ship. Ship's Late Name.	5 Flag.	6 Net Tonnage	7 Dimensions: Length.	Breadth.	Depth.	8 Built: Where.	9 When	10 By whom.	11 Date of Metalling	12 Owners or Managers	Port
	101		Unionen *Ruttler -86* W Bn	Swedish	127	82·7	21·6	11·1	S...org	1856			P.N.Ruttler	Råå
91406	102	✳	Unique W Sk	British	84	77·1	21·0	10·9	Hull	1885	G.W.Brown		W.J.Robins	Hull
[illegible]	103	A	Unitas *Vianello* W Pol	Italian	209	96·1	23·8	15·8	Chioggia	1866	F[lli] Baldo	Y 1.85	F[lli] Vianello	Venice
3118	104	I	Uniti (ex Clementina S.) *Marino -87* W Bk	Italian	594	145·0	32·3	21·3	Varazze	1875	M.Baglietto	Y 11.86	F[lli] Ferrari	Genoa
3583	105		Unition *Stewart* W Sw	British	185	97·6	24·5	12·6	Sunderland	1849			G.F.Peek	Guernsey
72995	106	R	Unity *Saunders -88* W Bk	British	420	133·0	30·4	12·9	Richibucto,N.B.	1876	H.O'Leary	Y ...	R.Caie	Chatham
91429	107	✳	Unity *Matthew 85-86* I Sk	British	89	81·0	20·2	9·5	Hull	1885	Cook, Welton& [Gemmell		F. & T.Ross	Hull
86827	108	✳	Unity *Hands* W K Barge	British	79	85·4	20·6	6·8	Ipswich	1885	W.J.Curtis		G.Hayward	Ipswich
	109	N	Uno *Haavig 80-80* W S	Norweg'n	824	166·3	33·5	20·9	Frankfort, Me.	1855			Sömme & Wattne	Stavanger
322	110		Unohi Maru No. 2 W Sr	Japanese	145	89·2	19·7	11·2	Hiogo	1880	Masumoto [Matsujiro		Yokoyama Hikoichi	Osaka
8486	111		Unrestricted *Robinson* W Bn	British	166	206·0	22·2	13·3	Granville	1854			J.Tyrrell	Dundee
	112	RV	Unto *Hansen -88* W Bk	Norweg'n	616	144·1	34·5	18·5	Brahestad	18$^{66}_{84}$	E.Danielsson		O.Eriksen & Gjersoe	Christiania
	113	N	Urædd *Larsen* W Sr	Norweg'n	254	118·9	27·9	12·5	Fredrikstad	1880		Y 1.88	O.A.Foyen	Tönsberg
	114		Urania (ex Maria da Moraes) W Sr	Portug'se	167	..	..	..	Villa de Conde	1877		Y 12.83	F.de Campos Moraes	Lisbon
	115	G	Urania *Larsen -82* W Sr	Danish	165	90·7	21·8	12·6	Kiel	1854	A.F.H.Conradi		F.E.Larsen	Nyborg
	116	V	Urania *Hansen 75-75* W Sr	Danish	142	91·7	22·1	10·6	Troense	1875	R.W.Møller		F.W.Valentin	Svendborg
	117	✳	Urania *Gahde 87-87* I S	German	1091	219·4	34·4	21·5	Hamburg	1878	Reiherstieg [Schiffs Werfte		B.Wencke Söhne	Hamburg
	118	G	Urania (STRANDED) *Geyken 75-86* W Sr	German	141	89·2	23·0	10·1	Grossefehn	1878	T.Cassens& [Söhne		Harm F. Geyken	Grossefehn
36	119	A	Urania (ex Ioannis) *Foccas* W Bg	Greek	225	101·0	25·3	14·3	Malta	1870	F.C.German	Y 3.86	Andr.Phocas Cosmitato	Argostoli
	120	DN	Urania (ex Socotra) *Sorensen* W S	Norweg'n	1287	196·0	39·0	24·0	St.John, N.B.	1874	J.Stewart	Y 3.87	A.P.Andersen	Moss
	121	N	Urania *Johannesen* W Bk	Norweg'n	312	113·2	27·6	15·0	Brevig	1865	H.Nielsen	Y 2.82	G.Johannesen & Co.	Fredrikstad
	122	N	Urania *Olsen -88* W Bg	Norweg'n	264	92·8	26·8	15·4	Memel	1843			Chr.Swane	Christiania
	123		Urano (ex Deus te Ajude) W Bn	Brazilian	176	..	..	..	Oporto	1871			I.P.Lemes & Co.	Desterro
	124	N	Uranos *Harris 81-81* W Bk	Norweg'n	490	155·3	32·5	17·3	Skien	1870	N.Nielsen	Y 10.85	L.Föyen	Tönsberg
25248	125	R	Uranus *Peters 74-87* W 3m Sr	American	344	115·9	29·9	10·1	E.Deering, Me.	1882	G.Russell		J.S.Winslow & Co.	Portland
25683	126		Urbana *Tyler* W Sr	American	100	83·3	24·9	7·5	Westerly, R.I.	18$^{43}_{68}$			B.V.Tyler	Pt.Jefferson
	127		Urbana W Bg	Spanish	155	86·0	24·6	12·6	Barcelona	1853			M.Vaquer y Paleron	Majorca
2410	128	IR	Urbano (WRECKED) *Olivari* W Bk	Italian	698	158·0	30·3	20·2	Varazze	1873	A.Cerruti	Y 12.83	Prosp. Olivari	Genoa
	129	N	Urd (ex Thorne) *Övergard -3-83* W Bn	Norweg'n	199	105·5	24·8	11·5	Elmshorn	1869	J.H.Kremer	Y 9.87	I.A.Berggrav	Fredrikstad
	130	V	Urda *Nielsen 83-83* W Sr	Danish	135	92·2	22·7	10·8	Troense	1883	R.W.Møller		F.W.Valentin	Svendborg
	131	V	Urda *Olsen* W Sr	Danish	105	85·2	21·5	9·7	Korsør	1874	A.Petersen		P.Hermansen	Næstved
	132	N	Urda *Hurum -88* W Bk	Norweg'n	498	123·4	30·8	19·6	Boston, Mass.	1850			Miss J.Lorentzen	Drammen
	133		Urda *Ball* W Bk	Norweg'n	382	122·1	30·8	16·9	Drammen	1867	Ole Svensen		H.Aasaker	Drammen
	134		Urda (CONDEMNED) *Salvesen 87-75* W Bn	Norweg'n	167	105·4	25·0	11·4	Bergen	1875	J.Hauge	Y 8.8.	P.G.Pettersen	Bergen
	135	N	Urda *Torkilesen 83-83* W Sr	Norweg'n	161	93·6	24·2	10·9	Drammen	1869			Julius Solum	Drammen
	136	N	Urda *Olsen* W Sr	Norweg'n	124	90·1	22·7	9·3	Stavanger	1852	K.Knisen		S.Olsen	Christiania
	137	G	Urho *Fagerholm -87* W 3m Sr	Russian	273	116·9	26·5	13·5	Raumo	1870	J.Wigelius		J.E.Panelius' Enka	Raumo
84156	138		Uribes I Bk	British	250	110·8	24·1	13·8	Stockton	1868	M.Pearse&Co.		Adelaide Stm.Tug Co.	Adelaide
556	139	V	Urinj (ex Mandita) *Stipanovich* W Bk	Austrian	464	139·7	30·5	19·2	Varazze	1864	G.Scala	Y 10.8.	Mme.Stipanovich	Buccari
1782	140	A	Uros (ex Isola) *Adrianich* W Bk	Austrian	606	148·9	29·8	18·4	Lussinpiccolo	1874	G.S.Cattarinich	Y 8.8.	C.Andrianich fu A	Buccari
	141	G	Ursula *Morisse 84-86* I S	German	1506	217·0	37·0	23·0	Vegesack	1885	Bremen [Schiffsb.Ges.		D.H.Watjen & Co.	Bremen
	142		Ursula W Bn	Spanish	136	90·5	21·0	9·3	Mindoro	1885			P.Lardizabal	Manila
	143		Ursule W Lr	French	107	67·2	19·4	11·0	Abbeville	1865	Sonatte		Mdme Thébaud le Bark	Vannes
21451	144		Useful W Sw	British	244	100·6	25·5	15·5	Sunderland	1858	G.Bartram		J.Davison	Shields
76882	145	✳	Useful *Wright 79-79* W Sr	British	84	89·3	21·3	9·1	Barrow	1879	Ashburner		T.Ashburner & Co.	Barrow
	146		Usko *Petrell -87* W S	Russian	829	174·6	34·1	20·7	Uleåborg	1872	G.Hihlunen		G.&C.Bergbom	Uleåborg
	147		Usko *Ekholm 79-87* W Bk	Russian	471	132·5	29·7	17·0	Nystad	1875	H.Nordling		O.Blom	Nystad
	148		Usko *Ekroth* W Sr	Russian	191	..	..	..	Lavia	1874			B.Ekroth	Raumo
	149		Usko *Lundgren* W Sr	Russian	150	87·3	25·1	12·0	Pyhämaa	18$^{56}_{70}$	J.H.Justen		P.J.Sallmén	Raumo
	150		Uteh Azizler *Kondoidis Anastasis* W	Turkish	135	..	..	..	Syra	1884			Tifar Ahmed Effendi & sisters	Constantinople

✳ = Lloyd's Register; D = Nederlandsche Vereeniging; G = Germanischer Lloyd; I = Registro Italiano; R = American Re...
A = Veritas Austro-Ungarico; Π = Veritas Ellenico; N = Norske Veritas; V = Bureau Ver...

In the 4th Column, C = Composite (iron frame, wood planked); I = Iron; Stl = Steel; W = Wood.

The Dates of Metalling (Col. 11) are based on the records of the Books in which Vessels are respectively classed, or on the Reports of the Surveyors of Lloyd's Re...

Further particulars of Vessels classed by Lloyd's Register will be found in Lloyd's Register Book and Lloyd's Yacht Register.

No. dans le livre	Registre dans lequel le navire est coté	Nom. Materiaux et Gréement. Capitaine et dates d'entrée au service des Armateurs actuels et à bord du navire. Ancien Nom.	Pavillon.	Tonnage net.	Dimensions. Longueur	Largeur	Creux	Construit. Où.	Quand	Par qui.	Date du doublage en métal	Armateurs ou Directeurs-Gérants.	Port d'Armement.
151		Utch Kardash W Bg	Turkish	240	..	..	..	England	1866			Latin Millitlee Magazazi (Sinope)	Latin Millit
152		Utch Kardash W Sr	Turkish	187	..	..	..	Sinope	1883			Kerasti-tijari Isaak Kamyan	Sinope
153		Utility W Sr	British	217	107·2	20·9	10·6	Quebec	1854	Davison & Goudie		E. Berry	Quebec
154	N	Uto *Andreasen 69-76* W Sr	Norweg'n	144	92·7	23·4	10·9	Lillesand	1870	T. Langstrom	V 1.87	A. Tönnesen	Lillesand
155		Utopia *Johns 86* W Bgn	British	184	105·0	23·5	12·6	Rye	1860	Hood	V 7.86	Edwin Roberts	Aberystwith
156		Utopia *Carmichan* W Bn	British	116	91·3	24·9	10·4	St. George, N.B.	1856			J. Stewart	Belfast
157	✻	Uzziah *Henry* W Bk	British	219	125·5	24·6	12·9	Salcombe	1869	Vivian	V 9.84	T. J. Ferguson	Salcombe

ADDITIONAL SHIPS ARE INSERTED IN THE SUPPLEMENT AT THE END OF THE REGISTER.

✻ = Lloyd's Register; D = Nederlandsche Vereeniging; G = Germanischer Lloyd; I = Registro Italiano; R = American Record; A = Veritas Austro-Ungarico; H = Veritas Ellenico; N = Norske Veritas; V = Bureau Veritas.

Dans la 4me Colonne, C = Composite (membrure en fer, bordé en bois); I = Fer; Stl = Acier; W = Bois.

Les Dates du doublage en métal (Col. 11) ressortent des livres dans lesquels les navires sont respectivement classés ou des rapports des Inspecteurs du Lloyd's Register.

Pour plus amples renseignements sur les navires classés par le Lloyd's Register, voir le Lloyd's Register Book et le Lloyd's Yacht Register.

Official No.	No. in Book	Classed	Ship's Name. Material and Rig. Master, and Dates of Appointment to present Owner's Service and to Ship. Ship's Late Name.	Flag.	Net Tonnage	Length	Breadth	Depth	Built: Where.	When.	By whom.	Date of Metalling	Owners or Managers.	Port
	1	V	V. A. *Mourean -88* W Bg	French	211	[illegible]	24·9	12·8	Nantes	1869	T. Dubigeon		J. Vallée	Nantes
	2	V	V. G. T. *Nédélec* W Bg	French	198	87·6	23·0	12·6	Mesquer	1866	Mahé		L. Jacquier	Vannes
00669	3	R	V. T. H. *Marshall -88* W Sr	British	148	95·1	26·3	9·7	Bear River, N.S.	1888	A. Marshall		A. Marshall	Digby
	4		Va (ex Cristoforo Bonazzo) (WRECKED) *Ceppont* W Bn	Austrian	222	99·7	23·2	14·9	Fiume	1862	P. Burmello		Silvio Suttora	Trieste
	5	V	Va-donc *Bégo -87* W Bg	French	201	94·8	25·6	11·3	Nantes	1869	Clairgeau		F. Rousseau	Nantes
	6	N	Vaan *Lindtner -87* W 3m Sr	Norweg'n	375	127·7	29·4	15·4	Stavanger	1875	Strömsten & [V. ...]		G. C. Gabrielsen	Stavanger
	7	N	Vaar *Sokkberg 79-79* W Bk	Norweg'n	261	125·3	26·0	12·2	Brevig	1876	P. Knudsen		S. C. Larsen	Brevig
	8	V	Vaarbud (ex Kirsborn) (STRANDED) *Nelson 86-87* W Bn	Norweg'n	270	105·7	25·9	12·4	Elsfleth	1869	J. Jürgens		A. Meling	Stavanger
	9	N	Vaaren *Danielsen -86* W Bg	Norweg'n	227	99·4	27·0	12·8	Grimstad	1866	A. Frivold		J. Halvorsen	Grimstad
15922	10	✳	Vacuna (ex Olive Branch) *Eastwood 85-85* W K	British	46	74·6	16·4	9·2	Littlehampt'n	1855	Oliver		F. G. Cock	Bideford
	11	N	Væni (ex Frans Scharling) *Olsen* W S	Norweg'n	960	175·6	36·8	21·4	Gefle	1884	J. A. Banr		Jacob Somme	Stavanger
	12	N	Værge *Gundersen* W Bn	Norweg'n	181	88·9	26·5	12·0	Grimstad	1867	Jorgen Tellefsen		C. Hvistendahl	Kragerö
	13	DN	Væring *Hansen 85-85* W 3m Sr	Norweg'n	302	128·5	28·2	12·9	Stavanger	1876	A. Birkeland		Somme & Wattne	Stavanger
	14	N	Væringer *Aanansen* W Bg	Norweg'n	247	113·7	27·1	11·6	Grimstad	1865	M. Smith Petersen		F. Smith Petersen	Grimstad
	15		Våhlberg *Anderson* W Sr	Swedish	116	97·1	21·2	7·9	Vühlberg	1874			P. Olefsson	Motala
	16		Vahtohyidi W Bg	Turkish	233	..	..	..	Danube	1879			Nicola Franco	Constantinople
	17	V	Vaillante (WRECKED) *Hemminck* W Sr	French	140	92·2	21·6	12·0	Dunkirk	1859	Sagary		A. Bellais	Dunkirk
	18	V	Val de Saire *Letour -87* W Bk	French	295	111·5	23·0	15·9	Cherbourg	1865	Hamel		L. Baudouin	Dunkirk
	19	GV	Valborg *Albertsen 77-77* W Sr	Danish	133	90·7	20·7	10·3	Veile	1877	Gylding [& Lindtner		[illegible]	Veile
	20	N	Valborg (ex Early Morn) *Olsen* W Bk	Norweg'n	1025	182·2	36·0	23·0	Quebec	1864	M'Kay		[illegible] A. Amundsen	Sarpsborg
	21	V	Valborg (ex Johanna) *Haakonsen* W Bk	Norweg'n	758	159·1	31·2	21·3	Ronnebeck	1864	Piercke & Co.		C. Haakonsen	Tönsberg
	22		Valborg (CONDEMNED) W Bk	Norweg'n	363	108·6	28·3	17·3					M. Berentsen	Stavanger
	23		Valborg (ex Mohawk) W S	Norweg'n	399	125·0	30·0	15·0	St. John, N.B.	1866	D. J. MacLachlan		J. Jansen	Bergen
94743	24		Valdare *Farris* W Sr	British	100	83·0	27·0	7·2	Cambridge, N.B.	1888	J. S. Robinson		P. McIntyre	St. John
	25		Valdemar (WRECKED) *Jensen* W Bg	Norweg'n	167	76·4	23·7	13·6	Arendal	1807			G. Jensen	Grimstad
	26		Valdemar *Edlund -86* W Sr	Swedish	100	78·7	20·3	10·6	Kalleda	1850			C. G. Edlund	Vesta
	27		Valdemar (ex Status) *Eriksson -88* W 3m Sr	Swedish	195	103·0	21·4	12·7	[illegible]	1861			J. E. Eriksson	Rådmansö
	28	V	Valdemar Seir *Møller 83-83* W Sr	Danish	133	89·8	21·8	10·5	Faaborg	1883	R. Møller		Akt. Selsk "Valdemar Seir" (K. H. Mogensen)	Faaborg
84155	29	✳	Valdivia *Purley 86-89* Stl Bk	British	877	202·3	33·1	19·9	[illegible]	1881	J. Reid & Co.		Nicholson & McGill	Liverpool
	30	N	Vale *Olsen* W Bk	Norweg'n	536	118·7	31·6	16·8	Kragerö	1878	Alf Johnsen		J. Dahll	Kragerö
	31	N	Vale *Gundersen 73-78* W Bk	Norweg'n	418	126·0	29·3	16·5	[illegible]	1873	N. Gjertsen & [L. Olsen		Christen Boe	Arendal
89648	32	✳	Vale Royal *Wood* I Bk	British	355	113·1	26·1	13·0	Port Glasgow	1884	Murdoch [& Murray		G. Wood	London
63211	33	✳	Vale of Doon *Thomas 84-84* I Bk	British	669	176·5	29·5	18·5	Port Glasgow	1869	McCulloch, [Patterson & Co.		W. B. Jones	Liverpool
63257	34	✳	Vale of Nith *Sledge* I Bk	British	672	176·5	29·5	18·5	Port Glasgow	1870	McCulloch, [Patterson & Co.		Hayton & Simpson	Liverpool
	35		Valencia W Sr	Spanish	171	95·1	23·7	12·8	Valencia	1852			J. José Sister	Barcelona
	36		Valentina W S	Spanish	416	138·7	26·5	21·0	United States				Joaquin Gurri	Barcelona
	37	✳	Valentine (ex River Clyde) *Harang -88* I Bk	French	741	178·8	31·2	19·0	Port Glasgow	1864	L. Hill & Co.		Ant. Dom. Bordes & fils	Bordeaux
	38		Valentine (BROKEN UP) (ex Dacca) *Lebeau -88* W Bg	French	466	128·9	24·7	15·3 / 31·4	Shoreham	1868	Shuttleworth [& Co.		F. Foerster	Havre
	39		Valentine *Simonsen* W Bk	Norweg'n	289	126·3	27·9	12·5	St. John's NFL.	1864			S. Wiborg	Kragerö
	40	✳	Valentine & Hélène *Jouet 88-88* I Bk	French	640	176·9	29·4	17·9	Glasgow	1870	A. Stephen & Sons		A. C. Le Quellec	Bordeaux
90752	41		Valetta *W. Akerly -86* W Sr	British	108	81·0	27·0	7·4	Rothesay, N.B.	1885	H. Titus		V. S. White	St. John
	42	N	Valhalla *Johnsen -81* W Bk	Norweg'n	274	92·2	27·4	16·6	[illegible]	1836			A. Olsen	Christiania
	43	NV	Valhalla (ex Anna) *Larsen 84-86* W Bkn	Norweg'n	266	113·0	22·6	11·2	Vegesack	1869	F. Sager		L. L. Sunde	Farsund
48347	44	V	Valiant *Douglas* W S	British	1582	216·7	39·6	24·0	Damariscotta, [U.S.	1874	Thayer & Lincoln		Nauth Mull Dugar	Calcutta
	45	N	Valkyria *Wilhelmsen* W Bk	Norweg'n	690	158·5	32·7	18·6	New Glasgow, [N.S.	1867	Carmichael		C. G. Hansens Enke	Christiania
60653	46	✳	Valkyrien *Baikie 79-87* W Bk	British	498	147·8	30·0	18·1	Sunderland	1868	W. Adamson		J. Adamson	London
	47		Valkyrien *Nielsen* W Bn	Danish	161	94·3	22·6	11·3	Troense	1868	C. R. Møller			Svendborg
	48	N	Valkyrien *Andreassen* W Bk	Norweg'n	575	146·6	33·0	17·3	Bergen	1868	J. Trumpy & Son		C. H. Johnsen	Arendal
	49		Valkyrien (LOST) *Steen 77-82* W Bk	Norweg'n	280	100·7	27·7	15·6	Grimstad	1848	J. Erlandsen Vig		L. Pettersen	Drammen
	50	N	Valkyrien *Gundersen* W Bk	Norweg'n	252	118·6	26·7	17·2	Kragerö	1870	P. Christensen		J. Dahll	Kragerö

✳ = Lloyd's Register; D = Nederlandsche Vereeniging; G = Germanischer Lloyd; I = Registro Italiano; R = American [Record]
A = Veritas Austro-Ungarico; H = Veritas Ellenico; N = Norske Veritas; V = Bureau Veritas

In the 4th Column, C = Composite (iron frame, wood planked); I = Iron; Stl = Steel; W = Wood.

The Dates of Metalling (Col. 11) are based on the records of the Books in which Vessels are respectively classed, or on the Reports of the Surveyors of Lloyd's [Register].

Further particulars of Vessels classed by Lloyd's Register will be found in Lloyd's Register Book and Lloyd's Yacht Register.

Nos	Registre	Nom. Capitaine et dates d'entrée au service des Armateurs actuels et à bord du navire. Ancien Nom.	Matériaux et Grément	Pavillon	Tonnage net	Longueur	Largeur	Creux	Construit: Où	Quand	Par qui	Date du doublage en métal	Armateurs ou Directeurs-Gérants	Port d'Armement
51	D	Valkyrien *Bryde -87* (WRECKED)	W Bk	Norwegian	[illegible]	112·8	28·1	12·6	Laurvig	1874	L.Backer&Co.	[illegible]	T.M.Bryde	Sandefjord
52	N	Valkyrien (LOST) *Grönn*	W Bg	Norwegian	228	116·6	26·5	12·7	Arendal	1856	A.R.Andersen		J.O.Grönn	Skien
53	V	Valladares *Pinto*	W Bn	Portuguese	160	91·9	20·6	10·0	Caminha	1880	P.Netto	[illegible]	J.M.Valladares	Caminha
54	V	Valladares *Ponteira*	W S	Portuguese	157	90·2	22·6	9·8	Caminha	1882	J.M.Branco	[illegible]	J.M.Valladares	Caminha
55	✻	Vallejo *Potter 87-87*	I Bk	British	652	173·6	31·0	19·6	[illegible]	1871	R.Williamson&S.		J.Gambles	Workington
56		Valley Forge *Reed 86-87*	W S	American	1226	187·0	37·6	21·0	Pittston,Me.	1862			Middlemos & Boole	San Francisco
57	V	Vally *Wolter 81-87* (ex Emma)	W Bk	German	708	147·6	28·6	20·2	Bowdoinham, Me.	1866	J.Harward	[illegible]	J.J.Wallis & Sohn	Barth
58	V	Valoise *Le Valois*	W S	French	109	81·4	19·7	10·1	St.Vaast	1876	E.Levêque		E.T.Levêque	La Hogue
59	✻	Valona *Andrews 77-84*	W Bk	British	800	173·8	33·5	20·2	Kingston,N.B.	1884	J.&T.Jardine	[illegible]	J.&T.Jardine	Richibucto,N.B
60	✻	Valparaiso *Robinson*	I Bk	British	730	188·5	31·1	21·0	Liverpool	1866	T.Vernon&Son		Balfour,Williamson&Co.	Liverpool
61		Valparaiso	W Bk	Chilian	296	100·4	24·3	..	Whitehaven	1841			A.Sutherland	Valparaiso
62	✻	Valparaiso *Leonetti* (ex Sultan, ex Pt. of Montevideo, ex Workington)	I Bk	French	1239	233·3	36·1	21·0	PortGlasgow	1874	Macfayden&Co.		Ant.Dom.Bordes & fils	Bordeaux
63	V	Valparaiso *Ahrens -86*	W Bk	German	486	140·4	29·2	17·1	Brake	1870	L.F.Paulsen	[illegible]	Eduard Burchard	Rostock
64	NV	Valuta *Nathanielsen 66-83*	W Bk	Norweg'n	940	189·0	37·2	21·1	Stavanger	1878	Ths.Tonning	[illegible]	Ths.S.Falck	Stavanger
65		Valuta *Bjørnstad -87* (ex Ernst ... ABANDONED)	W Bk	Norweg'n	491	141·6	28·9	18·2	Sunderland	1866	Taylor		L.Larsen Sunde	Farsund
66	DGV	Van den Bergh *Rehberg 75-75*	W Bk	German	436	128·8	29·3	17·2	Rostock	1875	H.Rickmann	[illegible]	A.Burchard	Rostock
67	R	Van Name & King *Curtis*	W 3m Sr	American	699	160·5	36·5	16·6	Fairhaven	1886	J.Doyle		Van Name & King	Newhaven
68		Vanadis *Said Oesin Eidroes*	W Bk	Dutch	618	145·0	32·6	18·7	Sundsvall	1869	J.A.Nordström	[illegible]	Sech Ibrahim bin Mohamat Badjoeber	Sourabaya
69	DNV	Vanadis *Samme 73-84*	W Bk	Norweg'n	754	173·2	34·5	19·6	Haugesund	1877	J.Hauge	[illegible]	K.S.Bertelsen	Stavanger
70	N	Vanadis *Thommesen 82-86*	W Bk	Norweg'n	468	144·8	30·3	16·9	Kragerö	1875	L.Larsen	[illegible]	H.Larsen	Kragerö
71	N	Vanadis *Pettersen -75*	W Bk	Norweg'n	376	126·2	29·0	15·8	Arendal	1875		[illegible]	S.E.Ellingsen	Arendal
72	N	Vanadis *Christensen*	W Bk	Norweg'n	322	137·2	26·4	12·5	Chicago	1863			B.Broch	Drammen
73	DV	Vanadis *Roosgren 77-82*	W S	Russian	1019	190·5	36·1	22·9	Jacobstad	1874	K.Löfbjelm	[illegible]	P.Malm,junr.	Jacobstad
74	R	Vanadis *Sandin -87* (ex Anne Nilsson)	W Bk	Swedish	314	107·6	25·2	16·2	Sunderland	1865	Gray&Young	Y 6.83	C.Sundin	Råå
75	DRV	Vancouver *McNeill -86*	W S	British	1376	198·0	38·0	23·8	GrosCoqueNS	1876	G.H.Lovitt	[illegible]	G.H.Lovitt	Yarmouth,N.S.
76	DV	Vandalia *Chanan 83-83*	W S	British	1422	210·3	40·2	24·3	Harvey,N.B.	1883	G.S.Turner	[illegible]	G.F.Smith	St.John,N.B.
77	✻	Vanduara *Skinner 88-88*	I 4m S	British	2012	291·8	42·0	24·0	Glasgow	1882	C.Connell&Co.		J.D.Clink	Greenock
78	RV	Vanduara *Durkee -88*	W S	British	1868	201·2	38·3	24·0	GrosCoqueNS	1882	G.H.Lovitt	[illegible]	G.H.Lovitt	Yarmouth,N.S.
79		Vanessa (NOT NOW SEAGOING)	W Sw	British	176	88·8	24·5	14·0	Sunderland	1856			J.G.Hill	Southampton
80	N	Vang *Meling 83-85*	W Bn	Norweg'n	235	117·4	27·4	12·5	Bergen	1875	K.Skaalur	[illegible]	A.Meling	Stavanger
81		Vangelista	W Bg	Turkish	187	..	..	..	Sakarieh	1872			Terzi bashi Yorghaki Koriaspolo	Sakarieh
82		Vangelista	W Bg	Turkish	125	..	..	..	Trebizonde	1876			Bulbuljehlee Vangelli Stavako	Bulbuljeh
83		Vangelista	W Bg	Turkish	117	..	..	..	Greece	1874			Bulbuljehlee Dimitri Yorghi	Bulbuljeh
84		Vangelista	W Bg	Turkish	111	..	..	..	Roumania	1873			[illegible] Yakin [illegible] & Co.	Marmara
85		Vangelista	W Sr	Turkish	116	..	..	..	Skyathos	1876			[illegible] Vangeli Simoni & Co.	Skyathos
86		Vangelista	W Sr	Turkish	100	..	..	..	Lerios	1886			Strati Dimitri	Lerios
87		Vangelistra *Anastes*	W Bk	Turkish	250	..	..	..	Greece	1886			Kerestch tijari Christophor & Co.	Constantinople
88		Vangelistra	W Bg	Turkish	153	..	..	..	Greece	1882			Sakyzlee Costanti Zakkilsoglu & Co.	Scio
89		Vangelistra	W Bg	Turkish	137	..	..	..	Roumania	1883			[illegible]	Constantinople
90		Vangelistra	W Bg	Turkish	128	..	..	..	Crete	1880			[illegible] Zadeh Suleiman [illegible]	Constantinople
91		Vangelistra	W Bg	Turkish	115	..	..	..	Vassilikos	1883			[illegible] koslee Sawah [illegible] Yanni & Co.	Vassilikos
92	✻	Vanguard *Patterson 85-85*	I Bk	British	688	171·0	26·8	19·1	PortGlasgow	1852	J.Reid & Co.		J.Bommphrey & Co.	Liverpool
93		Vanguard (ex Dacotah, ex Florence)	W Bk	British	396	139·2	29·6	11·1	BayVerte,N.B.	18[illegible]			T.D.Findlay	Moulmein
94	✻	Vanguard *Evans 80-80*	W Bk	British	332	129·3	26·4	15·7	Montrose	1866	Birnie	[illegible]	W.Balfour	Arbroath
95		Vanguard	W S	British	133	101·5	23·0	8·6	Sittingbourne	1885	J.Taylor		J.Smith	Colchester
96	✻	Vanguard	W Sk	British	77	75·3	20·0	10·4	Elmshorn	1885	J.Kremer		J.G.Stoakes	Grimsby
97		Vanilla *Pomkey*	W Sr	British	102	80·5	23·2	9·2	LaHave,N.S.	1883	S.Leary		J.Romkey	Lunenburg,N.S.
98	RV	Vanloo *Gouley -88*	W S	British	1497	216·0	40·6	23·8	Digby,N.S.	1884	G.H.Lovitt	[illegible]	G.H.Lovitt	Yarmouth,N.S.
99		Vanora	W Bk	Chilian	729	173·6	32·5	19·5	Sunderland	1873	J.Crown	[illegible]	C.J.Holm	Valparaiso

…=Lloyd's Register; D=Nederlandsche Vereeniging; G=Germanischer Lloyd; I=Registro Italiano; R=American Record;
…=Veritas Austro-Ungarico; H=Veritas Ellenico; N=Norske Veritas; V=Bureau Veritas.

Dans la 4me Colonne, C=Composite (membrure en fer, bordé en bois); I=Fer; Stl=Acier; W=Bois.

…dates du doublage en métal (**Col. 11**) ressortent des livres dans lesquels les navires sont respectivement classés ou des rapports des Inspecteurs du Lloyd's Register.

Pour plus amples renseignements sur les navires classés par le Lloyd's Register, voir le Lloyd's Register Book et le Lloyd's Yacht Register.

Official No.	No. in Book	Registry in which classed	Ship's Name; Ship's Late Name	Master, and Dates of appointment to present Owner's Service and to Ship	Material and Rig	Flag	Net Tonnage	Length	Breadth	Depth	Where Built	When	By whom	Date of Metalling	Owners or Managers	Port
48731	101		Vanquisher (BROKEN UP)	Abley	WSw	British	219	99·0	25·0	15·[illegible]	Sunderland	1861	Green [Richardson		H. Vane	North
	102	N	Vanse (ex Dagmar)	Danielsen -7-87	WBk	Norweg'n	445	127·2	30·9	16·8	WilmngtnDel.	1867			I.L.Olsen Maberg	Farsund
	103		Vapna (BROKEN UP)		WBk	Norweg'n	287	102·9	25·4	16·0	Aberdeen	1844		Y 7.86	M.Hansen	Tönsberg
23034	104		Vapor	Woodhull	WSr	American	230	113·0	31·2	10·0	Madison,Conn.	1854			J.C.Chamberlain	Philadelphia
25557	105		Vapor		WSr	American	112	99·6	17·6	7·7	Hamburg,Pa.	1864			J.A.Anderson	Trenton
	106		Varde	Andersen	WSr	Danish	127	85·8	21·0	10·7	[illegible]	1866	[illegible]		B.K.Eskesen	Varde
	107		Varna	Chrissafidi	WBg	Turkish	110	..	..	..	[illegible]	1884			Madame Zoetcha Stargani	Vassiliko
50169	108	✳D	Varuna	Groot -87	I Bk	British	1271	237·0	33·2	22·7	London	1864	[illegible]		W.H.Ross & Co.	Liverpool
83483	109	R	Varuna	Wolf -82	WBn	British	201	93·6	25·2	11·9	[illegible],N.S.	1882	Anderson & [illegible]	[illegible]	S.T.R.Bill	Liverpool
161508	110	R	Varuna	Birdsall	W3mSr	American	463	144·2	34·0	13·5	Camden,N.J.	1882	S.W.Tilton		S.E.Birdsall	New Ha[illegible]
	111	V	Varuna	Dade 84-	WBk	German	486	144·2	29·2	18·5	Sunderland	1867	Briggs	Y 8.86	H.Bauer	Rostock
	112	NV	Varuna	Andersen -84	W3mSr	Norweg'n	257	114·3	28·1	12·4	Risör	1877	BerntTellefsen	Y 6.88	Holmens Interessentskab	Risör
	113	N	Vasa	Tallaksen -87	WBk	Norweg'n	530	137·8	31·5	17·9	Grimstad	1869	MSmithPetersen	Y 4.87	F.Smith Petersen	Grimstad
	114	V	Vasa	Olsson -86	WSr	Swedish	226	113·2	23·5	11·6	[illegible]	1886	G.Hammarberg	Y 5.8	G.Hammarberg	Figeholm
	115	N	Vasco de Gama	Olsen -85	WBk	Norweg'n	595	146·7	31·8	19·0	Risör	1875	T.Jensen	Y 1.87	P.D.Prehensen	Risör
	116	N	Vasco de Gama	Krogh [73-83	WBk	Norweg'n	484	135·3	30·5	17·1	Grimstad	1873	J.Bang	Y 6.85	Christen Boe	Arendal
	117	V	Vasco da Gama	Cocho	WBk	Portug'se	491	135·1	30·2	5·1	Villa de Conde	1865	M.G.Rodrigues	Y 11.86	Cia. Alliança Maritimo Portuense	Oporto
106	118	H	Vasilissa		WBn	Greek	389	122·7	31·2	..	Syra	1872			E.B.Caffos	Piraeus
2251	119	I	Vassallo (ex Alma)	Ferrello 76-84	WBk	Italian	583	151·2	32·1	20·7	Savona	1872	F.Sirello	Y 3.86	P.Vassallo fu A.	Genoa
103	120	A	Vassiliki		WSr	Greek	128	74·8	23·6	12·8	Chios	1872	D.Drapetis		G.Talotis	Syra
	121		Vassiliko		WSr	Turkish	125	..	..	..	Koyun Ada	1873			Koyun Adalee Nicola Lebush	Koyun [illegible]
	122	V	Vasso (ex Kate Howe)	Saliaris	WBk	Greek	549	143·0	28·2	18·2	Kingsport,N.S.	1875	J.E.Woodsworth	Y 2.88	G.Saliaris	Syra
	123	N	Vats	Tilland -87	WBn	Norweg'n	210	112·5	25·8	12·1	Stavanger	1877		Y 8.87	Elias Olsen	Florö
	124	V	Vauban (LOST)		WBk	French	819	169·9	28·9	23·6	Honfleur	1868	Leviels	Y 8.83	A.Postel & ses fils	Havre
	125		Vauquelin	Vauquelain -88	WBk	French	204	94·8	23·0	13·1	Dieppe	1854	P.Corne		A.Leroux	Saint [illegible]
21352	126		Vecta	Hawes 83-83	WBn	British	132	83·7	22·0	11·5	Dunkirk	1842			J.H.Vaux	Harwich
	127	V	Vedele	Friis	WBn	Danish	164	100·3	22·5	11·0	Veile	1867	P.H.Illum	Y 11.85	A.T.Friis	Veile
	128		Vedette (ex Nuova Buon Cassoneno)	Duquesa -88	WBn	French	111	74·8	21·6	11·1	Gaeta,Italy	1862	L.Contreras		A.Lemoine	Saint [illegible]
3515	129	I	Vedova R. (ex Antonia Gattorno)	Figaro -87	WBk	Italian	541	141·7	31·5	20·3	Sestri Ponente	1871	F[lli] Cadenaccio	Y 1.87	P.Razeto	Genoa
18730	130		Vedra	Sayers	WSw	British	272	101·6	26·0	16·0	Sunderland	1861	Austin		R.F.Ellerby	Sunderland
	131	D	Veendam	Bos -87	W3mSr	Dutch	175	98·8	22·4	12·2	Hoogezand	1865	H.Hoites		T.van Buuren	Delfzyl
25982	132		Vega	Theroe	W3mSr	American	233	117·0	32·0	10·0	Fairhaven,Cal.	1881			B.H.Madison	San F[illegible]
	133	V	Vega	Rasmussen 80-86	WBr	Danish	184	92·7	24·7	12·4	Marstal	1880	H.J.Bager		H.J.Bager	Marstal
	134	V	Vega	Olsen	WBk	Norweg'n	559	147·2	32·8	17·1	Drammen	1873	E.Flater	Y 3.86	H.W.Finne & Sönner	Drammen
	135	N	Vega	Larsen -87	WBk	Norweg'n	393	131·9	28·8	15·5	Kragerö	1879		Y 3.87	Jens Olsen	Kragerö
	136		Vega	Pedersen -88	WBg	Norweg'n	302	119·1	26·2	13·6					L.Pedersen	Fredrik[illegible]
	137	NV	Vega	Levorsen 74-85	W3mSr	Norweg'n	196	106·6	25·1	12·2	Tönsberg	1884	T.Hansen	Y 12.84	N.Bugge	Tönsberg
	138	V	Vega	Lunden	WBk	Russian	399	148·3	28·2	15·6	Raumo	1880	E.Cronberg	Y 1.88	J.W.Söderlund	Raumo
	139	G	Vega (ex Trudel)	Olsson 87-87	WBk	Swedish	398	132·2	27·3	15·9	Lübeck	1867	P.Beitmann	Y 11.83	L.V.Åkerson	Sölle
	140		Vega (ex Marion)	Wickman 80-83	WS	Swedish	291	123·3	23·4	12·8	P.E.Island	1870			J.D.Björkegren	Cimbrishamn
	141		Vega	Skoglund	W3mSr	Swedish	101	71·7	21·9	9·4	Hudikswall	1882			O.Skoglund	Hudiksvall
	142	NV	Vegar	Andersen	WBk	Norweg'n	372	123·6	27·7	15·5	Tvedestrand	1873	A.Andersen	Y 1.88	A.&F.Smith	Tvedestrand
	143		Veloidosa (ex David Ewan)	Lerch	WSr	Chilian	138	85·6	21·5	10·4	Peterhead	1862	Stephen		D.Fuentes	Talcahuano
80227	144	✳	Velin-heli	Williams 66-78	WSr	British	65	75·2	21·3	9·3	Port Dinorwic	1878	Jones		J.Dean	Carnarvon
3025	145	A	Velizid (ex Anastasia V[illegible])	Svich	WBk	Austrian	536	130·5	30·2	19·0	Buccari	1857	Matterglian	Y 10.84	E.Giusti fu V.	Fiume
	146	NV	Velkommen	Gundersen -85	WBk	Norweg'n	361	146·2	29·6	11·9	Stavanger	1883	M.A.Magnus	Y 8.86	K.S.Bertelsen	Stavanger
	147		Vellamo	Norberg -86	WSr	Swedish	131	86·0	20·6	10·5	Finland	1872			J.E.Jansson	Skeppsholm
	148	V	Velleda	Topin -87	W3mSr	French	252	116·8	24·1	13·7	Leith	1869			Vicomte de Chambray	Granville
250	149		Vellimiro	Cattugerovich	WBg	Austrian	343	107·6	27·3	18·3	Fiume	1830			M.Bosevich	Trieste
78356	150	✳	Vellore	Yeo 81-81	I S	British	1631	246·3	40·1	24·0	Stockton	1876	Richardson, Duck&Co.		Eyre,Evans & Co.	Liverpool

✳ = Lloyd's Register; D = Nederlandsche Vereeniging; G = Germanischer Lloyd; I = Registro Italiano; R = American Re[cord]; A = Veritas Austro-Ungarico; H = Veritas Ellenico; N = Norske Veritas; V = Bureau Veritas.

In the 4th Column, C = Composite (iron frame, wood planked); I = Iron; Stl = Steel; W = Wood.

The Dates of Metalling (Col. 11) are based on the records of the Books in which Vessels are respectively classed, or on the Reports of the Surveyors of Lloyd's Re[gister].

Further particulars of Vessels classed by Lloyd's Register will be found in Lloyd's Register Book and Lloyd's Yacht Register.

No.	Registres dans lesquels le navire est coté	Nom / Ancien Nom	Capitaine et dates d'entrée au service des Armateurs actuels et à bord du navire	Matériaux et Gréement	Pavillon	Tonnage net	Longueur	Largeur	Creux	Construit: Où	Quand	Par qui	Doublage	Armateurs ou Directeurs-Gérants	Port d'Armement
151	A	Veloce (ex General Faidherbe)	Premuda	W Bk	Austrian	313	100·4	25·6	18·1	Bilbao	1855	M.Cortabitarte	Y 1.87	G.C.Gerolomich	Lussinpiccolo
152		Véloce		W Bk	French	425	120·4	26·0	17·8	Nantes	1859	P.Guibert		E.Lavalley	Nantes
153	H I	Velocifero (ex Consiglieri)	Guerello	W Bk	Italian	980	180·7	35·1	23·5	Prà	1876	L.Durante	Y 11.85	D.Bozzo	Genoa
154	✻	Velocity	Martin	C Bk	British	491	142·0	28·5	17·5	Sunderland	1866	R.Thompson,jr.	Y 12.85	R.Martin	London
155		Velocity	Guthrie	W3mSr	British	177	100·6	23·3	12·9	Garmouth	1865	Kinloch	Y 7.84	W.J.Coveney	Leith
156		Velocity		W Bg	British	139	78·6	20·8	13·8	Shaldon	1827			James Mulcahy	Hobart Town
157		Velox (BROKEN UP)	Hjøme 75-75	W Bn	Danish	109	87·1	20·4	9·3	Porsgrund	18⁶₆			P.Berg	Nexø
158	N	Velox (ex Grev Berg) (CONDEMNED)	Christensen -83	W Bk	Norweg'n	623	169·1	34·2	17·7	Cleåborg	1859		Y 1.86	P.A.Storsen	Brevig
159	N	Velox	Olsen	W Bk	Norweg'n	390	124·0	29·6	16·2	Arendal	1868	H.Stiansen	Y 4.86	A.O.Olsen	Stavanger
160		Vencedor	Pinlero	W Bn	Spanish	133	88·3	22·5	11·8	Bilbao	1859		Y 9.83	F.A.Pineiro	San Sebastian
161	✻	Venedocian	Jones	W Sr	British	170	102·8	22·8	12·7	Portmadoc	1873	G.Th.Williams &Co.	Y 1.86	H.Jones & Co.	Carnarvon
162	D N	Venerata	Persen 76-81	W Bk	Norweg'n	626	158·9	33·4	18·4	Arendal	1880	Tellef Lassen	Y 1.88	Tellef Lassen	Arendal
163	I V	Venere (ex ... Marina)	Vianello	W Bk	Italian	434	135·1	28·9	18·5	Sestri Ponente	1873	S.Delarbieri	Y 1.86	A.A.Vianello	Venice
164		Venersborg	Jacobson -86	W3mSr		149	90·8	21·4	11·7	Wenersborg	1867				
165	I	Venezian (ex Duilio)	Defferari 87-88	W Bk	Italian	846	182·4	33·5	23·4	Varazze	1876		Y 3.83	G.B.Ottone	Genoa
166	V	Vénézolana	Calvé -83	W Bk	French	281	117·1	27·2	12·4	Bordeaux	1868	Raymond frères	Y 1.88	K.Morvant	Bordeaux
167	V	Venezuela	Perry 84-84	W Bk	British	984	182·4	35·8	20·8	St.Martins N.B.	1879	A.Parks, junr.	Y 5.87	T.Killam	Yarmouth.N.S.
168	V	Veni	Karlsson 84-86	W3mSr	Russian	312	121·6	30·0	13·7	Kimito	1884	G.A.Kronström	Z 1.83	G.A.Kronström	Kimito
169	R	Venice	Dugas	W Bn	British	149	92·0	25·6	10·2	Church Pnt, NS	1879	H.P.Boudreau		J.B.Belliveau	Weymouth,N.S.
170		Venio	Lindell -81	W Bk	Russian	404	130·0	31·0	18·0	Bjorneborg	18⁶⁰₆₉			M.Th.Lindqvist	Nagu
171		Venio	Grandell 77-80	W Sr	Russian	248	..	..	..	Kimito	1879			A.Gustafsson	Kimito
172		Venner (ex White Hall)	Steen	W Bk	Norweg'n	444	128·3	29·7	16·6	Parrsboro N.S.	1865	J.Durant		S.Steen & Co.	Laurvig
173	N	Venskabet	Kjernliff 71-82	W Bk	Norweg'n	418	124·0	29·7	16·9	Libau	1856		Y 5.87	M.Oppen	Laurvig
174	N	Venskabet	Nielsen -87	W Bk	Norweg'n	264	118·7	28·3	11·5	Arendal	1868	S.Johnsen		S.Eyde	Arendal
175	✻	Venti Settembre		I Sr	Italian	100	80·0	21·2	9·7	Savona	1884	F^lli Migliardi		F^lli Cavassa	Genoa
176	V	Ventilia (LOST)	Asmus -87	W3mSr	German	304	130·0	28·0	12·8	Hamburg	1875	C.Dreyer	Y 4.86	J.Breckwoldt	Blankenese
177	✻D	Ventura	Lillie 88-88	Stl S	British	1669	258·5	39·1	22·8	Dumbarton	1886	A.M'Millan&Sn		Scotia Shipping Co.	Glasgow
178	V	Ventura (ex Sewal)	Gibernau	W Bn	Spanish	207	96·8	27·0	13·5	Blanes	1866		Y 5.81	Masso, Primos & Co	Barcelona
179		Ventura		W Pol	Spanish	140	86·6	25·1	11·5	Palma	1840			Jaime Vidal	Majorca
180	✻	Venture	Reid 88-88	W Bkn	British	249	119·3	25·1	12·9	Garmouth	1878	Geddie,junr.	Y 12.86	J.Geddie	Banff
181		Venture		W Bn	British	167	91·2	25·0	10·1	Haton, N.S.	1871	J.B.North		Thomas S.Brown	Sydney,N.S.W.
182		Venture (FOUNDERED)	Hatfield	W Sr	British	119	80·0	25·1	9·9	Parrsboro' N.S.	1875			J.A.Hatfield	Parrsboro',N.S.
183		Venture		W Bg	British	117	79·1	19·1	10·8	Pennfield, N.B.	1847			G.Robertson	Sydney,N.S.W.
184		Venture		W Sr	British	116	101·1	22·2	6·6	Les Eboul mnt.	1855			T.Forsyth	Quebec
185	R	Venturer	Frelick	W Bn	British	338	145·0	29·4	11·8	Brooklyn, N.S.	1886	J.Gardner	Y 10.86	J.V.& J.H.Dexter	Liverpool,N.S.
186		Venturita		W Sr	Spanish	153	86·6	25·1	11·1	Blanes	1855			M.Comba	Barcelona
187	V	Venturosa (ex Princess Louise)	De Fonséca	W Bk	Portug'se	451	138·5	28·6	17·0	Sunderland	1871	R.Thompson	Y 10.85	J.Nogueira Pinto	Oporto
188		Venturosa (ex Praia da Funchal)		W Sr	Portug'se	115	74·8	19·3	10·5					M.J.Gonçalves	Lisbon
189	✻	Venus	Ramsey 88-88	W Bk	British	311	125·0	22·3	16·4	Bristol	1853		Y 1.88	J.Winship & Co.	Bristol
190	G	Venus	Accoparai 86-86	W Bk	British	301	102·2	27·2	15·9	Malta	1847		Y 7.83	P.Apap & figli	Malta
191		Venus	Bright 81-81	W Bn	British	211	113·0	24·0	12·9	Wexford	1863	Sparrow		F.W.Fairbrass	Faversham
192		Venus	Hodge 81-86	W Sr	British	159	84·4	20·0	13·7	Poole	1854			W.H.Shilston	Plymouth
193		Venus	Quinsloe -88	W Sr	British	134	86·7	22·8	10·5	Svendborg	1875	J.B.Andersen		L.C.Nairac	Port Louis, Mauritius
194	✻	Venus	Hughes	W Sr	British	107	83·0	22·1	11·5	Nevin	1880	Owen		H.Hughes	Carnarvon
195	V	Venus (ex Sophie)	Holm 78-87	I Bk	Danish	473	151·1	26·9	16·1	Hamburg	1877			Aktieselskab Ithederi (J. P. Holm)	Nordby,Fanø
196	V	Venus (ABANDONED)		W Bn	German	190	110·4	23·7	11·2	Middelfart	1870	H.Illum	Y 1.88	Reents & Lübben	Itzehoe
197	✻N	Venus (ex Devon)	Klaveness -88	W Bk	Norweg'n	1018	207·2	37·5	21·7	Sunderland	1864	J.Smurthwaite	Y 6.81	A.F.Klaveness	Sandefjord
198	N	Venus	Olsen -88	W Bg	Norweg'n	270	102·8	28·7	15·6	Rhodosto	1858			H.Andersen	Tönsberg
199	N	Venus	Henriksen 81-81	W Bg	Norweg'n	243	97·2	26·5	14·5	Arendal	1854			B.Henrichsen	Risör
200	N	Venus	Thorsen -88	W Bg	Norweg'n	195	99·1	2[illegible]·2	12·2	Kragerö	1872		Y 11.81	H.Björn, junr.	Kragerö

[A =] Lloyd's Register; D = Nederlandsche Vereeniging; G = Germanischer Lloyd; I = Registro Italiano; R = American Record;
[A =] Veritas Austro-Ungarico; H = Veritas Ellenico; N = Norske Veritas; V = Bureau Veritas.

Dans la 4me Colonne, C = Composite (membrure en fer, bordé en bois); I = Fer; Stl = Acier; W = Bois.

[Les dates] du doublage en métal (**Col. 11**) ressortent des livres dans lesquels les navires sont respectivement classés ou des rapports des Inspecteurs du Lloyd's Register.

Pour plus amples renseignements sur les navires classés par le Lloyd's Register, voir le Lloyd's Register Book et le Lloyd's Yacht Register.

Official No.	No. in Book.	Registry in which classed.	Ship's Name. Master, and Dates of Appointment to present Owner's Service and to Ship. Ship's late Name.	Material and Rig.	Flag.	Net Tonnage.	Length	Breadth	Depth	Where built	When	By whom	Date of Metalling	Owners or Managers.	Port
	201	V	Venus *Rosa* -88	WBk	Portug'se	305	102·6	26·6	14·3	RioMondego	18 47/74		Y [illegible]	M.J.G.Rosa	Lisbon
	202		Venus *Johansson*	WSr	Russian	235	..	..	..	Dragsfjerd	1871			K.J.Blomberg	Drag…
46[illegible]	203		Venus *Lohde*	WSr	Russian	187	92·6	26·2	12·6	Czunzeom	1885	J.Ohrmann		Leepa,Balloit & Rewsky	Riga
	204		Venus	WSr	Spanish	169	105·6	29·2	8·2	New York	1879			Antonio Bayona	Havana
	205		Venus *Jacobsen* -78	WSr	Swedish	174	96·1	22·2	11·7	Hernösand	18 56/87			J.Berntsson	Kal…
	206	V	Vera (ex Victoire) *Lübke* 74-74	WBk	German	364	115·1	28·6	15·3	Cherbourg	18 62/78	Hausel		O.Münsterberg	Rüg…
	207	N	Vera *Timm* 84-84	WBg	Norweg'n	175	111·0	22·8	11·0	Skien	1872			Westergaard & Co.	Christ…
	208	G	Vera *Andersson* -87 (ex A[illegible] ex Albert Tesen)	WBg	Swedish	307	110·2	25·3	15·1	Wolgast	1862	F.Schelle		J.P.Jönsson	[illegible]
	209	V	Vera *Norrman* 80-84	WBg	Swedish	300	119·7	26·0	14·2	Helsingborg	1884	S.Nordström	Y [illegible]	C.Sylvan	Helsing…
	210	N	Veranda *Olsen* 86-86	WBk	Norweg'n	388	128·5	29·8	16·0	Grimstad	1873	JacobBoe	Y 12.87	Christen Boe	Arend…
	211		Veranda *Stendal* -85	WBk	Norweg'n	362	119·4	29·9	15·4	Lillesand	1855	Knudsen		B.Birkeland	Lill…
	212		Veranda (LIGHTER)	WBg	Norweg'n	163	76·9	23·9	13·1	Kragerö	1843			L.Larsen	Krag…
	213	N	Veranda *Jorgesen* -88	WSr	Norweg'n	113	92·3	22·8	9·9	Stavanger	1855			P.Jorgensen	[illegible]
	214		Verdad *Losillas*	WBk	Spanish	511	128·9	30·7	16·0	Palma	1872	S.Avocena		Viuda & hijosdeJ.Yanes	Palma
	215	V	Verdande *Bergsen* -87	W3mBn	Danish	299	121·4	24·8	13·9	Nordby,Fanø	1884	S.Abrahamsen	Y [illegible]	Akt.Selsk.(C.H.Nielsen)	Nord…
	216	V	Verdandi *Hagberg* -8-86	WSr	Swedish	307	140·7	25·9	11·1	Siked	1884	P.O.Åberg	Y [illegible]	Fr.Jacobsen	Rob…
	217		Vere *Kastman* -81	WBk	Swedish	394	129·8	27·2	17·2	Sunderland	1857	W.Watson	Y [illegible]	E.A.Enhorning	Sund…
	218	N	Verena *Nielsen* -83	WBk	Norweg'n	418	129·8	27·5	17·5	Sunderland	1865	S.P.Austin&Son	Y [illegible]	P.D.Prebensen	Risör
1311	219	✳	Vergine Della Guardia (ex James Shepherd)	WBk	Italian	843	180·4	32·7	21·2	Sunderland	1865	Briggs	Y 10.85	Michele Lubrano	Castel…
72620	220		Veritas *McCully*	WBk	British	1000	183·0	37·0	21·0	LondonderryNS	1876	A.Fraser	Y [illegible]	J.G.Jordan	Maitla…
79513	221		Veritas *Crichton* -88	WBk	British	748	173·8	33·3	18·2	Helsingborg	1876	J.Nordstrom	Y 11.84	H.R.Reid	Melb…
29084	222		Veritas	WSr	British	142	95·2	21·2	11·3	Cowes	1861	John White		C.Ratsey	Cowes
73765	223		Veritas *Terris*	WSr	British	111	82·9	25·9	9·9	Brooklyn,N.S.	1877			Annie A.Terrio	Arichat
	224	N	Veritas (STRANDED) *Andersen* -83	WBk	Norweg'n	518	147·0	30·5	17·2	Gefle	1859			Nils Christensen	Arendal
	225	N	Veritas *Fredriksen* 75-77	WBk	Norweg'n	487	136·9	31·3	17·2	Christiansand	1874	M.Knudsen	Y 11.85	M.F.Stray	Christ…
80173	226	V	Veritas (ex Ideal) *Leite*	W3mSr	Portug'se	187	98·4	23·0	10·9	Villedo Conde	1871	C.M.de Araujo		L.Marques da Nova	Oporto
	227	V	Veritas (ex Sandade) *De Souza*	WBn	Portug'se	167	91·2	21·6	10·8	Vianna	1873	M.D.dosSantos [Borda	Y [illegible]	L.Marques da Nova	Oporto
	228	V	Veritas *Schantz* 75-80	WBk	Russian	673	165·0	34·0	19·7	Jacobstad	18 65/80	P.Malm,junr.	Y [illegible]	P.Malm,junr.	Jacobst…
	229		Veritas *Hagen* -73	WBk	Russian	555	150·0	36·0	17·5	Arholme	18 69/80			W.Rosenlew & Co.	Björ…
	230		Veritas *Langström*	W3mSr	Russian	247	..	..	..	Korpo	1867			J.Mattson	Korpo
	231	V	Veritas *Möller* -85	W3mSr	Swedish	287	119·1	24·8	13·3	Helsingborg	1871	J.Nordström	Y [illegible]	J.F.Göransson	Råå
	232	R	Veritas *Nilsson* -79	WBg	Swedish	260	104·0	24·5	13·8	King'sLynn	18 57/74		Y [illegible]	G.Nilsson	Helsing…
	233	A	Vero F. *Fiorillo*	WBk	British	553	143·3	31·4	20·5	Spotorno	1870	F.Cadenaccio	Y [illegible]	S.Gauci	Gibralt…
[illegible]	234		Verona (DISMANTLED) *Dix*	WBk	American	649	153·0	31·9	18·5	Quebec	1870	M'Kay	Y [illegible]	Alfred Mills	Quebec
7[illegible]	235	DRV	Veronica *Payson* -73	WBk	British	1137	186·4	37·8	22·1	IndianTwnNB	1879	J.Rowan	Y [illegible]		St.Joh…
	236	N	Veronica (ex Pekin) *Pettersen* -87	WBk	Norweg'n	631	142·8	32·1	18·8	Moss Glen,N.S.	1869	Wetmore & [Merritt	Y [illegible]	A.D.Westergaard	Arend…
	237	DG	Vertrouwen (ex Hendrika Huisman) *Nieboer* 76-77	WBn	Dutch	153	89·9	22·0	10·9	Hoogezand	1862	M.Kroeze		G.W.Pott	[illegible]
10652	238		Verulam *Farrell* 82-85	WBk	British	493	111·2	28·9	19·2	Sunderland	1858	S.P.Austin&Son	Y [illegible]	J.L. & W.A.Simpson & Elder, Smith & Co.	Adel…
	239	V	Verveine *Moniaudry*	WBk	French	449	140·0	28·5	14·4	La Seyne	1871	Curet	Y [illegible]	A.Huleau	Bord…
83482	240		Vesper *Mosher*	WSr	British	132	86·0	24·4	10·4	Brooklyn,N.S.	1881	H.Smith		Philip Bagnell	Live…
57091	241		Vesta *Keefe*	WBn	British	148	96·8	23·7	11·3	P.E.Island	1867			J.Munn & Co.	St.Jo…
161509	242		Vesta *Platt*	W3mSr	American	272	127·8	32·0	10·0	SanFrancisco	1882	M.Turner		A.Rowe	San F…
25644	243		Vesta *Ayer*	WSrYt	American	114	110·0	24·7	8·5	CityIsland,NY	1866			S.M.Mills	New …
	244	V	Vesta (ex Justine Helene ex Auguste Elise) *Brinck* 88-88	IBk	Danish	355	139·1	26·2	15·7	Sunderland	1871	W.Watson		O.Hirth	Kold…
	245	G	Vesta *Frick* 87-87	WBk	German	489	137·9	30·1	18·1	Memel	1866	B.Pieper		J.G.Gerlach	Memel
	246		Vesta (BROKEN UP)	WBg	German	257	101·8	27·1	14·3	Greifswald	1848	J.D.Garde		J.H.Reptz & Co.	Danz…
	247		Vesta *Møller* -83	WBg		128	93·7	21·9	11·0	MurrayHarb'r	1855				
	248	NV	Vesta (ex Nabob) *Jacobsen*	WBk	Norweg'n	695	147·9	32·9	20·2	PortsmouthNJ	1867	D.Marcy	Y [illegible]	A.J.Jacobsen	[illegible]
	249	NV	Vesta *Olsen*	WBk	Norweg'n	613	149·4	32·0	18·5	Medway,N.S.	1871	E.Dolliver	Y [illegible]	T.Thommesen & Søn	Arend…
	250	N	Vesta *Andersen* 74-74	WBk	Norweg'n	423	121·7	30·9	17·1	Drammen	1865	Jorgensen& [Knudsen		A.Andersen	Dram…

✳=**Lloyd's Register**; D=Nederlandsche Vereeniging; G=Germanischer Lloyd; I=Registro Italiano; R=American Rec…
A=Veritas Austro-Ungarico; H=Veritas Ellenico; N=Norske Veritas; V=Bureau Ve…

In the 4th Column, C=Composite (iron frame, wood planked); I=Iron; Stl=Steel; W=Wood.

The Dates of Metalling (Col. 11) are based on the records of the Books in which Vessels are respectively classed, or on the Reports of the Surveyors of Lloyd's…

Further particulars of Vessels classed by Lloyd's Register will be found in Lloyd's Register Book and Lloyd's Yacht Register.

1	3	4 Nom. Matériaux et Gréement. Capitaine et dates d'entrée au service des Armateurs actuels et à bord du navire. Ancien Nom.	5 Pavillon	6 Tonnage net	7 Dimensions: Longueur	Largeur	Creux	8 Construit: Où.	9 Quand	10 Par qui.	11 Date du doublage en métal	12 Armateurs ou Directeurs-Gérants.	13 Port d'Armement.
251	N	Vesta *Christiansen* WBk	Norweg'n	345	108·7	29·0	16·5	Arendal	1850		Y 3.83	E.Dedekam	Arendal
252	N	Vesta *Rode 86-86* WBg	Norweg'n	345	108·7	29·0	16·5	Arendal	1851			S.C.Larsen	Brevig
253	N	Vesta (WRECKED) *Halstensen* WBk	Norweg'n	304	119·9	26·7	14·7	Bremen	1854			H.Halstensen	Drammen
254	N	Vesta *Olsen* WBg	Norweg'n	246	92·4	27·1	15·2	Tvedestrand	1847	A.Smith		S.M.Salvesen	Tvedestrand
255	N	Vesta *Jensen* WBk	Norweg'n	215	102·6	25·4	11·5	Maine	1848			J.Lorange	Fredrikshald
256	N	Vesta *Iversen* WBg	Norweg'n	215	108·7	26·3	12·4	Porsgrund	1880	Hansen	Y 12.84	Baarud	Horten
257	N	Vesta (ex Polstjernan Swedish) *Olsen -74* WSr	Norweg'n	172	97·5	25·7	11·5	Archangel	1857			M.Stillesen	Drammen
258	V	Vesta (ex Wasa) *Kaidin* WBk	Russian	533	144·0	35·4	19·5	Wasa	1867	J.Lofhjelm	Y 8.84	Wasa Rederi Bolag (C.W.Jahnsson)	Wasa
259		Vesta *Jansén* WBk	Russian	309	..	..	..	Sideby	1865			J.Jansson	Korpo
260	DGV	Vesta *Ottmann 86-86* WSr	Russian	154	90·9	26·4	10·4	NeuSalis	1878	R.Graf		J.Jansohn	Riga
261		Vesta *Melara* WBg	Russian	105	..	..	..	Syra	1879	A.Livadara		C.G.Vanderjuhter	Taganrog
262	R	Vesta *Svensson -88* WBn	Swedish	226	103·4	26·2	11·9	Nystad	1884			J.D.Bjorkegren	Cimbrishamn
263		Vestalinden *Olsen* WBk	Norweg'n	459	135·5	29·0	17·5	Kragerö	1859			H.J.Olsen	Drammen
264	N	Vestalinden *Hansen 75-79* WBk	Norweg'n	401	120·9	30·2	17·7	Drammen	1865	Morberg		Petter Olsen	Dröbak
265		Vestalinden *Langfeldt [79-85* WBg	Norweg'n	299	99·5	28·1	15·8	Grimstad	1846	Bent Aanonsen		N.T.Langfeldt	Christiansand
266	N	Vestalinden *Egbertsen [-89* WSr	Norweg'n	134	87·9	22·5	11·0	Tönsberg	1857	Larsen		C.Zernichow	Tönsberg
267	N	Vesterlide *Ellertsen 79-78* WBk	Norweg'n	387	129·8	29·9	16·1	Drammen	1867	Jörgensen& [Knudsen	Y 1.88	Ths.S.Falck	Stavanger
268	V	Vesternorrland *Lindholm [-88* WBk	Swedish	318	129·6	25·9	12·2	Sundsvall	1873	J.A.Nordström	Y 6.88	N.A.Horney	Sundsvall
269	N	Vestfold *Gledsen -88* WBk	Norweg'n	501	154·4	32·0	16·0	Tönsberg	1868		Y 11.88	J.Harbitz	Tönsberg
270		Vesuvio WBn	Brazilian	163	..	..	..					J.P.de Araujo	Bahia
271	V	Vesuvio *Longobardo* WBk	Italian	539	142·4	31·4	19·0	TorredelGreco	1870	A.Castellano	Y 7.84	Poleze & d'Amato	Castellamare
272		Vesuvius (LOST) *Marshall* WBk	American	772	160·8	32·6	21·1	Richmond,Me.	1870	TJSouthard&Sn	Y 3.86	T.J.Southard & Son	Richmond,Me.
273	V	Vétéran *Martin -80* WBk	French	337	113·2	26·2	15·5	Nantes	1874	TDubigeon&fils	Cl 81	J.B.Demange	Nantes
274	V	Vettern *Andersson -86* WSr	Swedish	140	96·7	21·1	10·2	Sjötorp	1886	S.Groth		J.F.Olofson	Motala
275	V	Veturia *Caldas -87* W3mSr	Portug'se	160	90·9	23·6	9·5	Caminha	1882	A.A.Netto	Y 1.88	F.da S.Caldas & J.L.da Costa	Figueira
276		Vetus *Sharp* WBn	British	136	84·4	21·7	12·4	P.E.Island	1851			S.W.Coe	Liverpool
277		Viajante *Da Costa* WS	Portug'se	357	109·2	26·9	15·5	Damão	1850		Y 10.81	Cia Alliança Fabril	Lisbon
278		Viajante (ex La Flandre) *Santos Silva* WBn	Portug'se	191	93·8	21·6	12·1	Dunkirk	1876	A.Pyotte	Y 5.80	JoasdosSantosSilva	Lisbon
279	N	Viator *Björnsen 70-77* WBk	Norweg'n	418	134·2	30·7	16·8	Bergen	1862	Brunchorst& [Dekke		Brödr.Berner	Stavanger
280	R	Vibilia *Beveridge -88* WBk	British	553	146·0	32·0	15·8	ClydeRiver,NS	1870	J.D.Coffin	Y 10.88	G.J.Troop	Yarmouth,N.S.
281	N	Vibilia *Larsen 86-86* WSr	Norweg'n	189	87·0	25·5	13·4	Grimstad	1854	N.OlsenGjöiule		N.Thue Johnsen	Grimstad
282		Vicksburg *Higgins* WSr	American	145	95·1	26·5	8·9	Bath,Me.	1865			N.Hopkins	Bangor,Me.
283	✻	Vicomtesse (ex Nellie Harland) WSr	French	78	83·7	19·0	8·7	Birkenhead	1884	Harland&Co.	Y 4.85	A LeG de Mégoubran	St.Brieux
284	GR	Victor *Leighton* WBk	British	510	133·8	31·8	17·7	Arendal	1880	Larsen	Y 2.87	E.C.Adderley	Nassau,N.P.
285		Victor (ex Concordia) WBn	British	152	96·1	22·7	10·6					R.Farrell	Youghal
286		Victor WSp	British	115	104·0	22·7	7·0	Yamaska	1872	D.Richard		Thos.Foreman	Montreal
287	✻	Victor *Tell 86-86* WSr	British	111	90·0	22·4	10·6	Banff	1877	Watson		C.Taylor	Banff
288		Victor *Peters* WSr	American	158	105·6	27·2	9·2	Newburyport, [Mass.	1870			E.C.Peters	New York
289		Victor *Donovan* WBg	American	139	96·0	26·2	7·9	Addison,Me.	1870			C.L.Donovan	Machias,Me.
290	✻	Victor *Wilhjelm -87* WBk	Danish	403	133·9	27·9	16·7	Montrose	1874	Petrie	Y 12.86	P.A.Wilhjelm	Rudkjobing
291	G	Victor *Granz* WSr	German	177	92·6	23·8	13·0	Wolgast	1858	W.Lübke		A.Garnn	Greifswald
292	N	Victor *Jensen* WBk	Norweg'n	639	159·3	32·0	19·1	Varazze	1870		Y 11.87	A.Lothe	Haugesund
293	N	Victor *Thue -87* WBk	Norweg'n	562	138·9	29·8	19·1	N.Brunswick	1849			N.Bugge	Tönsberg
294		Victor (ex Figueirense, ex Camponeza) *Pinto -88* WSr	Portug'se	151	87·9	21·3	10·3	Villa do Conde	1873	J.M.d'Araujo	Y 12.81	A. da Silva Pinto	Lisbon
295		Victor Covacevich *G. Moccia* WBk	Russian	442	..	..	..	Fiume	1867		Y 7.81	G.Antoniades	Marianopolis
296		Victor Emmeline (SUNK) *Bertaud* WBg	French	106	73·2	20·0	10·8	Cherbourg	1855			B.Leclerc	Cherbourg
297	R	Victor Hugo (ex Mahone) WBn	French	152	96·8	24·7	10·1	MahoneBayNS	1884	J.H.Zwicker	Y 7.86	J.Clement père	St.Malo
298		Victor Manuel *Oliveira da Velha* WSr	Portug'se	115	72·5	21·3	9·8	VilladoConde	1863	F.A.Arreiro		J.J.daEncarnação	Oporto
299		Victor Puig *Harris* WSr	American	148	94·0	25·1	9·0	BoothBay,Me.	1875	J.McDougall		C.McKown	Booth Bay,Me.
300		Victor & Eugène *Eon 88-88* WBk	French	206	104·0	23·6	11·3	Granville	1865	Pétel		H.Lecharpentier	Granville

…Lloyd's Register; D=Nederlandsche Vereeniging; G=Germanischer Lloyd; I=Registro Italiano; R=American Record;
…Veritas Austro-Ungarico; H=Veritas Ellenico; N=Norske Veritas; V=Bureau Veritas.

Dans la 4me Colonne, C=Composite (membrure en fer, bordé en bois); I=Fer; Stl=Acier; W=Bois.

…dates du doublage en métal (Col. 11) ressortent des livres dans lesquels les navires sont respectivement classés ou des rapports des Inspecteurs du Lloyd's Register.

Pour plus amples renseignements sur les navires classés par le Lloyd's Register, voir le Lloyd's Register Book et le Lloyd's Yacht Register.

Official No.	No. in Book	Registry in which classed	Ship's Name. Ship's Late Name	Master, and Dates of Appointment to present Owner's Service and to Ship	Material and Rig	Flag	Net Tonnage	Length	Breadth	Depth	Built Where	Built When	Built By whom	Date of Metalling	Owners or Managers	Port of Registry
71199	301	✻	Victoria	Davies 71-86	W Bk	British	748	169·9	34·6	19·3	P.E.Island	1874	W.Richards	[illegible]	W.Richards (T.T.Richards & [illegible])	[illegible]
63243	302	✻	Victoria	Jones 84-[illegible]	I Bk	British	680	167·3	29·5	19·1	Sunderland	1867	W.Doxford		SwanseaShippingCo.Ld.	Swans
75813	303		Victoria	Shimmins	W Bk	British	365	128·6	27·6	15·4	Lubeck	1860			Shimmins & Rapp	Hobart
6827	304		Victoria	Christie 86-86	W Sr	British	164	102·2	23·5	13·1	Aberdeen	1849	Hood		A.Farquhar	Aberdee
4489	305		Victoria		W Sr	British	118	86·9	19·4	10·9	Dartmouth	1837			J.A.Row	Penzance
73903	306	✻	Victoria	Watt	W Sr	British	99	86·5	21·7	10·3	Banff	1876	Watson		H.Munro	Banff
25882	307		Victoria	Hammond -87	W Bn	American	260	120·2	28·0	12·0	Madison,Conn.	1874			H.Trowbridge's Sons	New Ha
	308		Victoria	Caldeira	W Bn	Brazilian	142	..	..	..	[illegible]	1859		[illegible]	V.Caldeira	[illegible]
	309		Victoria		W Sr	Brazilian	120	..	..	..					J. V. de Mattos	Aracaju
	310		Victoria (ex Madagascar)	Biagini	W Bk	[illegible]	404	119·1	26·9	..	[illegible]	18 58/72	Guibert		J.S.Vicuña	Valpa
	311		Victoria (ex Apollo)		W Bn		174	97·0	20·3	11·1	Geestemünde	1865				
	312	DG	Victoria	Wiersma -87	W Bg	Dutch	272	107·9	27·8	14·2	Greifswald	1866	J.W.Sadelkow		J.Wiersma	Delfzijl
	313	G	Victoria	Lehnert -82	W Bk	German	804	170·6	33·1	20·8	Newport	1862		[illegible]	D.Heinrichs	Bremen
	314	DG V	Victoria	Winter 84-82	W Bk	German	742	165·0	34·8	19·6	Hammelwarden	1884	J.F.Strenge&Sn	[illegible]	Gerh.Eilers	Brake
	315	✻	Victoria (ex La France, ex [illegible])	Frahm -87	I Bk	German	552	175·2	27·2	17·4	W.Hartlepool	1863	Pile,Spence&Co.		Wilkins	Hambu
	316	V	Victoria	Plessentien -68	W Bk	German	408	121·5	28·5	17·5	Damgarten	1867	H.Dierling	[illegible]	F.Crull	Rostock
	317	N	Victoria	Aangensen	W Bk	Norweg'n	855	175·4	34·5	19·6	Arendal	1881	Ole J. Herlofson	[illegible]	Ole J.Herlofson	Arendal
	318	N	Victoria (ex Venus)	Børresen	W Bk	Norweg'n	725	158·2	34·0	19·6	Avondale,N.S.	1869	N.Mosher	[illegible]	H.R.Astrup	Christia
	319	N	Victoria	Thoresen	W Bk	Norweg'n	402	110·4	28·4	19·4	Gefle	1827			H.P.Samuelsen	Drøbak
	320	N D	Victoria	Arnesen -85	W Bk	Norweg'n	381	128·0	29·5	15·7	Holmestrand	1868	G.Gulbrausen	[illegible]	J.Petersen	Holme
	321	V	Victoria (ex Viktoria)	Nessvig -86	W3mSr	Norweg'n	275	119·7	25·2	12·2	Calmar	1881	E.Vireen	[illegible]	S.Bergesen	Stavang
	322	N	Victoria	Magnusen -8[illegible]	W Sr	Norweg'n	245	118·3	27·1	12·2	Sandefjord	1881	AlbertGrim		P.A.Grön	Sande
	323		Victoria		W Bg	Norweg'n	148	80·9	25·0	12·1	Bergen	1840			Parelius & Lossius	Christ
	324		Victoria (ex Albatros)	Dos Reis	W Bk	Portug'se	426	159·1	27·2	15·0	Vegesack	1862	H.F.Ulrichs	[illegible]	V.A.d'Oliveira	Lisbon
	325	A	Victoria	Reis	W Bn	Portug'se	161	89·2	20·7	11·1	Norway	1875		[illegible]	Costa Pinto Filhos	Oporto
	326		Victoria		W Bg	Spanish	747	101·0	27·1	14·6	Palamos	1855			Eusebio Matas	Barce
	327	V	Victoria	Barbusano -88	W Bk	Spanish	510	149·9	30·6	15·6	SantaCruz de Tenerife	1879	A.Marquez	[illegible]	Juan Cumella	Canary
	328		Victoria (ex San José)		W Sr		139	86·9	21·3	9·8	Palma	1850				
96008	329	✻	Victoria Bay	Scoble 85-87	I Bk	British	1118	215·8	35·1	21·3	PortGlasgow	1885	Russell&Co.		Hatfield,Cameron&Co.	Glasgow
47516	330	✻	Victoria Nyanza	Eaton [illegible]	I Bk	British	1022	204·4	32·5	22·2	Belfast	1863	Harland&Wolff		Lowden,Edgar&Co.	Liverpool
84491	331	✻	Victoria Regina	Causey	I S	British	1942	269·9	39·6	24·1	Southampton	1881	Oswald,Mordaunt&Co.		J.Coupland	Liverpool
	332	✻D	Victorine	Lequerhic -88	I Bk	French	1197	233·5	36·2	21·3	[illegible]	1879	A.Stephen&Sns		Ant.Dom.Bordes & fils	Bordeaux
	333	✻	Victory	Gibbs 83-88	W Bg		194	99·7	24·5	12·9	P.E.Island	1870	Yeo	[illegible]		
64124	334		Victory		W Bn	British	142	91·2	24·2	9·1	Jervis B. NSW	1873	G.Dent,jun.		E.Carter	Sydney
89100	335		Victory	Russell -87	W Sr	British	131	84·4	25·0	9·6	HopeTown	1885	H.A.Fisher		W.H.Albury	Nassau
25542	336		Victory	Milliken	W Sr	American	107	88·8	25·3	8·5	Friendship,Me	1854			I.Blaisdell	Ells
	337	N	Victory	Emanuelsen	W S	Norweg'n	1403	182·1	42·4	26·8	Newburyport, Mass.	1857			J.A.Andersen	Mandal
	338	N	Vidar	Swane 87-87	W Bg	Norweg'n	348	107·4	29·1	16·6	Grimstad	1857	H.O.Ribe		A.Swane	Christi
	339	N	Vidar	Paulsen 81-81	W Bk	Norweg'n	278	121·9	27·0	13·6	Kragerö	1870	A.Johnsen	[illegible]	L.Larsen	Kragerö
	340	N	Vidar	Pedersen -80	W Bg	Norweg'n	223	97·0	26·5	12·6	Grimstad	1859	J.Tellefsen		A.Tönnesen	Lilles
	341	N	Vidar (W.R.B.K.E.D.)	Jorgensen	W3mSr	Norweg'n	218	109·1	25·6	12·1	Risör	1867	K.Nielsen		P.Pedersen	Risör
	342	N	Vidar	Jacobsen 79-83	W3mSr	Norweg'n	177	111·1	23·5	10·4	Horten	18 66/74	Norwegian Gvt.	[illegible]	T.Ihlen	Holme
25703	343		Vidette	Gillman -87	W Bk	American	586	155·0	34·6	13·8	Bath,Me.	1865			W.P.Sayward	Pt.T
161592	344	R	Vidette	Tunnell	W Bkn	American	540	141·6	33·3	16·0	Millbridge,Me	1884	J.W.Sawyer&Ss	[illegible]	S.L.Tunnel	Phila
	345		Vidfarle		W S	Norweg'n	268	..	..	..					I.Waxen	
	346	N	Vidfarne (ex Hengist)	Ness -86	W Bk	Norweg'n	505	142·4	30·7	17·0	Quebec	1865	W.Cotnam	[illegible]	Ole Nielsen	[illegible]
	347	N	Vidfarne	Sørensen 80-82	W Bn	Norweg'n	188	[illegible]	25·5	11·8	Brevig	1877	N.W.Coch	[illegible]	N.W.Coch	Brevig
198	348	A	Vidianos	Rambos -87	W Bk	Greek	823	110·2	27·7	..	Syra	1879	S.Cristali	[illegible]	T.E.Vamvas	Spezia
56293	349	✻	Vidonia	Wilson 85-87	W3mBn	British	196	114·6	23·2	12·6	[illegible]	1877	Cox	[illegible]	C.T.Bowring & Co.	Liver
	350	V	Vieira	De Fonseca	W3mSr	Brazilian	250	117·4	25·5	11·2	Italy	18 58/63		[illegible]	J.de D.Vieira	Rio Ja

✻ = Lloyd's Register; D = Nederlandsche Vereeniging; G = Germanischer Lloyd; I = Registro Italiano; R = American Re…; A = Veritas Austro-Ungarico; H = Veritas Ellenico; N = Norske Veritas; V = Bureau Ver…

In the 4th Column, C = Composite (iron frame, wood planked); I = Iron; Stl = Steel; W = Wood.

The Dates of Metalling (Col. 11) are based on the records of the Books in which Vessels are respectively classed, or on the Reports of the Surveyors of Lloyd's Re…

Further particulars of Vessels classed by Lloyd's Register will be found in Lloyd's Register Book and Lloyd's Yacht Register.

1	2 N° dans le livre.	3 Registre dans lequel le navire est coté.	4 Nom. Matériaux et Gréement. Capitaine et dates d'entrée au service des Armateurs actuels et à bord du navire. Ancien Nom.			5 Pavillon	6 Tonnage net.	7 Dimensions. Longueur	Largeur	Creux	8 Construit. Où.	9 Quand.	10 Par qui.	11 Date du doublage en métal	12 Armateurs ou Directeurs-Gérants.	13 Port d'Armement.
	351	V	Vier Bruder	Albrandt 77-77	W Bk	German	774	150·0	35·5	18·5	Rostock	1877	E.Burchard & Co	V [illegible]	E.Burchard & Co.	Rostock
	352	N	Vif (ex Ménagère)	Ellingsen 84-84	W Bg	Norweg'n	146	83·8	23·8	10·9	Granville	1857			A.Ellingsen & Son	Holmestrand
	353	V	Vifstavarf	Jansson	W Bk	Swedish	457	131·8	29·8	17·1	Vifstavarf	1870	A.Kjellin	[illegible]	A.B.Elfversson	Gothenburg
	354	A	Vigani (ex [illegible])	[illegible]	W Bk	Austrian	519	139·4	28·4	19·1	Fiume	1870	F[lli] Schiavoni	[illegible]	A.Covacevich	Orebich
	355		Vigia		W	Spanish	107	105·0	22·3	8·0	North America				Zulueta y Sobrino	Habana
	356	✱	Vigil	Chalk 86-87	I Bk	British	522	163·9	27·2	18·1	Liverpool	1862	T.Vernon&Son		Thomas, Watkins & Jenkins	Swansea
	357	✱	Vigilant	Williams -87	W Bkn	British	303	125·4	25·4	14·9	Kingston	1871	Geddie	Y 12.86	M.Jones	Banff
	358	RV	Vigilant	Gould 84-	W S	American	1800	224·0	40·0	28·0	Kennebunk Me	1877	N.L.Thompson	[illegible]	Edward Lawrence	Boston, Mass.
	359	V	Vigilant	Sandberg	W Sr	Swedish	235	110·2	26·4	11·3	Sundsvall	1869		V [illegible]	C.M.Åberg	Sundsvall
	360		Vigo (ex Reina)		W Bk	French	811	165·0	35·7	21·3	Gefle	1866	J.A.Bang		P.Poujade	Nantes
	361	N	Vigo	Ustri 82-82	W Bk	Norweg'n	505	136·0	30·5	18·2	Philadelphia	1845			J.Ustri	Laurvig
	362	N	Vigo	Ellefsen -83	W Bg	Norweg'n	173	91·7	24·8	12·0	Tvedestrand	1862	O.Gundersen		And. & Chr.Ellefsen	Fredriksstad
	363		Vigoureuse	Gens -87	W Sr	French	145	95·1	21·6	11·6	Dunkirk	1874	Vanderiele		A.Bellais	Dunkirk
	364	N	Viig	Lase -87	W Bk	Norweg'n	496	135·0	30·8	17·2	Grimstad	1874	S.Johnsen	V 3.86	O.A.Olsen	Arendal
	365		Vik	Mattson -84	W S	Swedish	265	101·9	27·5	13·4	Roslagen	1865			C.M.Wiklund	Norrsund, Ralgn
	366	N	Vikar	Magnesen 82-82	W Bk	Norweg'n	843	170·3	36·7	21·5	Haugesund	1875	J.Hauge	V 3.87	K.H.Rönnevig	Haugesund
	367	N	Vikedal	Tönnesen 85-85	W Sr	Norweg'n	118	86·4	22·8	11·1	Stavanger	1877			Th.Aarstad	Egersund
	368	N	Viken	Hansen 84-85	W3mSr	Norweg'n	337	171·8	29·9	13·1	Greenock	1845			Chr.Swane	Christiania
	369	N	Viken (WRECKED)	Treu	W Bg	Norweg'n	211	112·9	25·9	12·1	Sandefjord	1873			Carl Treu	Holmestrand
	370		Viking (ex Czarina)		W Sr Yt	British	151	99·3	22·8	11·8	Gosport	1853	Camper & Nicholson		John Brett, A.R.A.	Cowes
	371	R	Viking	Church	W4mSr	American	966	189·6	40·1	18·5	Bath, Me.	1888	New England S.B.Co.		A.J.Newbury	New York
	372		Viking		W Sr	American	140	108·0	30·0	8·2	Marshfield, Oreg.	1882			E.P.Nissen	San Francisco
	373	DN	Viking	Höye	W3mSr	Norweg'n	265	120·9	27·1	12·4	Stavanger	1875	M.Magnus	V [illegible]	S.Berners Sönner	Stavanger
	374	N	Viking	Olsen	W Bk	Norweg'n	259	108·4	27·7	12·6	Grimstad	1866	J.Tellefsen		J.Börresen	Grimstad
	375	N	Vikingen (ex Jason, ex Arnold Boulinger)	Hannestad 86-86	W Bk	Norweg'n	903	169·8	24·4	22·1	Vegesack	1852	J.Lange	V [illegible]	J.M.Bryde	Sandefjord
	376	N	Vikingen	Backe	W Bk	Norweg'n	328	118·3	28·3	15·4	Moss	1854			T.J.Bache	Tönsberg
	377	N	Vikingen	Halstensen	W Bk	Norweg'n	277	113·5	27·0	14·2	Sunderland	18[illegible] 71			C. Günther & Ad Halstensen	Drammen
	378	V	Vikingen	Arfvedsson [86-86]	W3mSr	Swedish	199	100·0	22·0	12·2	Wenersborg	1876	J.V.Plate		S.Berntsson	Hellevikstrand
	379		Viktor	Malmqvist 84-86	W Bk	Russian	478	148·0	31·0	17·0	Åbo	1863			V.A.Reims	Åbo
	380		Viktor		W Sr	Russian	182	..	..	..	Wederlacks	1874			D.Brunila	Kotka
	381		Viktor	Svensson -87	W Bk	Swedish	360	122·3	23·9	16·4	England	1855			A.Samuelsson	Carlshamn
	382		Viktor	Österlund	W Sr	Swedish	235	99·4	27·0	13·1	Vätö	1852			A.P.Olsson	Vaster Edsvik, Ralgn.
	383		Viktor	Fröberg -81	W Bg	Swedish	161	95·7	23·8	11·0	Libau	1847	J.Puhre		J.G.Fröberg	Norrteljie
	384		Viktoria		W Sr	Russian	174	..	..	..		1864			K.A.Lundell	Lamparland
	385		Viktoria	Backe	W Bk	Swedish	273	114·8	26·5	12·4	Finland	1868			J.P.Backe	Råå
	386	V	Viktoria (ex Wilhelm)	Wickman -85	W Bg	Swedish	268	110·9	24·8	12·5	Stettin	1864	E.Lieske		J.D.Björkegren	Cimbrishamn
	387		Viktoria	Mattsson -81	W Sr	Swedish	187	94·2	24·6	11·1	Wäddö	1880	Sundberg		J.Z.Wennerborg	Uddevalla
	388		Viktoria	Magnusson -86	W3mSr	Swedish	134	95·1	20·8	10·0	Carlstad	1852			C.J.Andersson	Wenersborg
	389		Viktoria (ex Regina Wilhelmina)	Smaugberg	W Sr	Swedish	103	73·4	23·9	8·9	Roslagen	1848			H.Cornelius	Gamleby
	390	A	Vila (ex Osmi Dubrovacki)	Stanese	W Bk	Austrian	699	160·1	30·2	19·5	Gravosa	1871	N.A.Bradicich	Y 10.84	S.Bielovucich	Ragusa
	391		Vilasar	Martinez 86-86	W Bn	Spanish	184	93·2	26·6	12·6	Mataro	1856	F.Sagarro		M.Barreras	Vigo
	392		Vilhelm		W Bg	Swedish	179	90·2	22·7	12·2	Germany	1854			P.A.Eriksson	Calmar
	393	DN	Vilhelm Birkedal (ex York)	Stangebye 74-83	W Bk	Norweg'n	895	172·4	33·0	21·7	Sunderland	1854	Austin & Mills	V 1.88	Edv.Stangebye	Fredriksstad
	394	V	Vilhelm Gynther	Fröberg 86-86	W Bk	Swedish	520	146·6	28·8	17·5	Hernösand	1864	I.O.Strandberg	V 6.87	H.Nordlander	Hernösand
	395		Vilhelmine	Sprensen 70-70	W Bg	Danish	194	111·9	23·4	11·8	Paskallavik	1870	N.J.Carlsson		E.B.Ohlsson	Kjöge
	396	V	Villa Adele	Tomé	W Bn	Italian	112	86·0	23·6	10·2	Viareggio	1873	A.Raffaelli	Z 6.88	Carlo Fontanini	Leghorn
	397		Villa Flor	Garcia	W Bn	Brazilian	139	..	..	..	Oporto	1868		Y 10.81	G.Assburg	Itajahy
	398		Villa de Comillas		W Bk	Spanish	505	136·8	28·2	18·7	St.John, N.B.	1852			Joaquin Benitez	Havana
	399	DV	Villa de Llanes (ex R. Von Baumgarten)	Osole	W Bg	Spanish	696	116·1	26·6	12·9	Geestemünde	1875	Schaw & Oltmanns	V 2.88	Roman Romano	Barcelona
	400															

[illegible]=Lloyd's Register; [illegible]=Veritas Austro-Ungarico; D=Nederlandsche Vereeniging; G=Germanischer Lloyd; H=Veritas Ellenico; I=Registro Italiano; N=Norske Veritas; R=American Record; V=Bureau Veritas.

Dans la 4me Colonne, C=Composite (membrure en fer, bordé en bois); I=Fer; Stl=Acier; W=Bois.

Les Dates du doublage en métal (Col. 11) ressortent des livres dans lesquels les navires sont respectivement classés ou des rapports des Inspecteurs du Lloyd's Register.

Pour plus amples renseignements sur les navires classés par le Lloyd's Register, voir le Lloyd's Register Book et le Lloyd's Yacht Register.

Official No.	No. in Book	Registry in which classed	Ship's Name. Material and Rig. Master, and Dates of Appointment to present Owner's Service and to Ship. Ship's Late Name.	Flag	Net Tonnage	Dimensions: Length	Breadth	Depth	Built: Where	When	By whom	Date of Metalling	Owners or Managers	Port
	401		Villa de Ribadavia WSr	Spanish	274	104·9	24·5	15·1	Albay	1860			Ramon Aeulle	[illegible]
	402		Villa del Pilar *Garibaldi* WSr	Argentine	156	106·6	27·6	16·1	SestriPonente	1882	C.Bozzano	[illegible]	L.Risso & Co.	[illegible]
	403		Villafranca WBn (*ex Jovem Sancho, ex São Lourenço*)		474	126·3	31·2	16·5	Rheaden	1861				
[illegible]	404		Village Belle WBn	British	196	106·9	24·8	11·4	Londond'y,NS	1861			W.Douglas	[illegible]
[illegible]	405	⁎	Village Belle *Webber 82-82* WSr	British	145	101·6	21·3	11·3	Portreath	1882	W.Davies		W.Davies & Co.	[illegible]
[illegible]	406		Village Belle *Arnold* WSr	British	102	85·4	18·7	9·8	Knottingley	1867	Cliffe		B.Arnold	[illegible]
[illegible]	407	⁎	Village Belle *Kitching -84* WSk	British	90	81·9	21·3	10·3	Hull	1885	Hunt&Fowler		R.Roach	[illegible]
[illegible]	408		Village Gem WSr	British	118	77·0	22·0	7·3	Mass.,U.S.	1853			J.Poppleton	[illegible]
[illegible]	409		Village Girl *Hunter* WBg	British	199	89·4	24·6	15·2	Douglas	1841	Williamson		W.Robinson	[illegible]
[illegible]	410	⁎	Village Maid *Lloyd* WSr	British	131	87·7	22·3	11·8	Inverness	1864	Stewart		R.Lloyd	[illegible]
[illegible]	411		VillagePride *Georgie 83-86* WBg	British	159	96·0	26·0	12·3	St.Mary's,N.S.	1859	McDaniel		G.Sinclair	[illegible]
[illegible]	412	⁎	Villalta *Kitchin 86-86* Stl Bk	British	866	199·5	32·7	20·5	PortGlasgow	1883	J.Reid&Co.		Nicholson & McGill	[illegible]
	413		Ville Besnard WBk (BROKEN UP)	French	384	117·4	23·3	11·4	St.Malo	1863	Parent		Sire fils	[illegible]
	414		Ville d' Ostende W3mSr	Belgian	244	111·5	26·2	13·1	Antwerp	1881		[illegible]	Belgian Government	
	415	V	Ville de Dieppe Stl&IBk (*Petroleum Vessel*) *Chaillou -88*	French	1279	217·2	36·4	21·0	Southampton	1888	Oswald, Mordaunt & Co.		L.Robbe,fils	[illegible]
	416	V	Ville de Marseille *Millot* WBk	French	533	131·2	30·2	17·8	Marseilles	1869	V.Jouvin	[illegible]	L.Senglar	[illegible]
	417		Ville des Couëts *Ledelin* WLr	French	119	77·8	20·0	11·5	Nantes	1858	Sevestre		Clore	[illegible]
	418	V	Ville du Temple WBk *Pasquier -87*	French	430	137·4	29·5	16·9	Hamburg	1863	C.Stülcken	[illegible]	J.Bachelier	[illegible]
25943	419	RV	Vilora H. Hopkins *Blood* WBk	American	933	167·8	36·1	21·6	Millbridge,Me.	1877	E.Young	[illegible]	Swan & Son	[illegible]
	420	⁎N	Vimeira *Olsen -87* WBk	Norweg'n	925	165·7	33·0	22·9	Sunderland	1851	J.Laing	[illegible]	N.Olsen	[illegible]
1163	421	I	Vincenzino *Schiano* WBg	Italian	219	101·0	25·4	14·6	Schiano	1865	Schiano	[illegible]	A.Schiano di Pepe	[illegible]
813	422	A	Vincenzino *Gargiulo* WBg (*ex Luigino La Monica*)	Italian	213	98·4	24·9	12·9	Equa	1870	A.Gargiulo	[illegible]	A. & V.Burgio	[illegible]
2780	423	I	Vincenzino *Delli Sante* WBn (*ex La Stella di Barletta*)	Italian	124	75·1	23·1	14·0	Barletta	1879			V.Albanese fu M.	[illegible]
3269	424	I	Vincenzo *Ansaldo* WBn (*ex Veneranda, ex Valle Argentina*)	Italian	145	91·9	23·5	11·4	Alassio	1876	I.Airaldi	[illegible]	V.Galetto&M.Moncapitta	[illegible]
310	425	I	Vincenzo Accame W3mSr *Rembado -88*	Italian	523	113·7	31·7	19·4	Loano	1880	A.Cerruti	[illegible]	P.Accame & figli	[illegible]
2029	426	I	Vincenzo Galatola WBk *Mazzola*	Italian	535	142·6	31·4	20·6	Castellamare	1875	A.Bonifacio	[illegible]	A.Galatola	[illegible]
1470	427	I	Vincenzo Mazzella WBg (*ex Carolus*) *Mazzella*	Italian	350	111·5	29·1	18·2	Procida	1869	N.Lubrano	[illegible]	F.Romeo di M.	[illegible]
500	428		Vincenzo Padre WBn *Catanzaro*	Italian	128	80·7	21·3	11·8	Gaeta	1859	L.Contreras		S.Campisi fu G.	[illegible]
55429	429		Vinco *Baxter* WSr	British	222	111·0	23·8	14·0	Garmouth	1867	Kinloch		W.Brander	Leith
54219	430		Vindex *Kearon 76-84* WBn	British	172	104·5	24·4	12·8	P.E.Island	1866	Jenkins		R.Kearon	Dublin
43881	431		Vine *George* WSr	British	153	92·5	22·0	12·0	Arbroath	1862	ArbroathSh.Co.		J.Little	Newport
	432	GV	Vineta *Köhler -88* WBk	German	665	153·2	32·5	18·7	Elsfleth	1877	J.Ahlers	[illegible]	Joh.D.Ahlers	Elsfleth
25898	433		Vineyard *Cummings -80* W3mSr	American	205	115·0	30·0	9·3	Gouldsboro', Me.	1874			D.J.Sawyer	[illegible]
	434	N	Vingolf *Aanonsen -86* W3mSr	Norweg'n	392	139·9	28·7	15·3	Arendal	1874	H.Kallevig	[illegible]	S.M.Klöcker	[illegible]
	435		Vingolf *Zachariassen* WBn	Norweg'n	140	95·1	24·5	10·6	Bergen	1873	O.Eie		R.G.Hagland	[illegible]
	436	N	Vingul (*ex Follum*) *Peltz* WBk	Norweg'n	345	112·0	27·5	16·8	Sunderland	1861	Rawson & Watson		Carl Dahl	[illegible]
2616	437	A	Vinka *Medanich* WBk (*ex R. M. B. Park*)	Austrian	407	126·6	29·0	18·0	Sunderland	1868	Sykes	[illegible]	Va.Medanich	[illegible]
	438	N	Vinland *Clementsen* WBk	Norweg'n	482	133·5	31·0	17·9	Arendal	1870		[illegible]	J.Hanssen	[illegible]
	439		Vint'oito de Outubro WSr	Brazilian	122	..	..	..					A.F.Passos	[illegible]
	440	DN	Vinteren *Jacobsen* WBk	Norweg'n	478	115·8	31·2	19·0	Grimstadt (re-built)	18[illegible]			T. & L.Bache	[illegible]
[illegible]	441	⁎	Viola *Price 77-77* I Bk	British	595	171·8	27·9	17·5	Liverpool	1868	T.Royden&Sons		C.T.Bowring & Co.	Liverpool
[illegible]	442	⁎	Viola *Jolliffe 77-78* WBkn	British	182	107·4	23·4	13·0	Bristol	1872	Peters	[illegible]	A.Goodridge & Sons	St.John's
[illegible]	443	⁎	Viola *Male 84-84* WSr	British	159	99·1	22·7	12·5	Appledore	1872	Pickard		J.H.Male	Padstow
[illegible]	444		Viola *Russell* WBn	British	122	84·0	22·3	10·9	Whitehaven	1868	Hugh Williamson		C.Bennett	[illegible]
[illegible]	445		Viola *Sealoger* WSr	British	109	80·9	24·2	9·5	LunenburgNS	1882	J.Young		H.A.N.Kaulback	[illegible]
	446		Viola WSr	Argentine	168	..	..	..	Lussinpiccolo	1879	A.Cattarinich	[illegible]	E.Anselmo	[illegible]
	447		Viola (*ex Bettie*) *Woods* WSr	Danish	100	81·0	20·3	9·5	BarringtonNS	1872	Doane	[illegible]	A.Lugo	[illegible]
[illegible]	448		Viola May *Fisher* WSr	American	164	103·0	27·8	8·8	Dennysville, Me	1874	Watt		W.T.Hobart	[illegible]
[illegible]	449	R	Viola Reppard W3mSr *Smith*	American	387	148·0	35·2	12·0	Camden,Me.	1882	Coombs&Day	[illegible]	M.P.Smith	[illegible]
[illegible]	450	R	Viola W. Burton W3mSr *Smith 86*	American	120	95·0	25·0	7·0	Camden,N.J.	1884	L.Burton		J.A.J.Sheets	[illegible]

⁎ = Lloyd's Register; D = Nederlandsche Vereeniging; G = Germanischer Lloyd; I = Registro Italiano; R = American [illegible]; A = Veritas Austro-Ungarico; H = Veritas Ellenico; N = Norske Veritas; V = Bureau Veritas.

In the 4th Column, C = Composite (iron frame, wood planked); I = Iron; Stl = Steel; W = Wood.

The Dates of Metalling (**Col. 11**) are based on the records of the Books in which Vessels are respectively classed, or on the Reports of the Surveyors of Lloyd's [illegible]

Further particulars of Vessels classed by Lloyd's Register will be found in Lloyd's Register Book and Lloyd's Yacht Register.

N° dans le livre	Registres dans lesquels le navire est coté	Nom. Matériaux et Gréement. Capitaine et dates d'entrée au service des Armateurs actuels et à bord du navire. Ancien Nom.	Pavillon	Tonnage net	Dimensions. Longueur	Largeur	Creux	Construit. Où.	Quand.	Par qui.	Date du doublage en métal	Armateurs ou Directeurs-Gérants.	Port d'Armement.
451	V	Violet *Harding* -89 W Bk	British	846	161·8	33·4	19·0	IndianTown, N.B.	1875	J.Rowan	Y 9.86	W.H.Harding	St.John,N.B.
452	*	Violet *West* 79-88 W Bkn	British	398	148·2	28·9	15·4	P.E.Island	1880	Richards	Y 11.88	Wylie & Gordon (T.P.Richards & Co.)	London
453	*	Violet *Shapland* 75-82 W Bkn	British	149	107·8	23·0	11·1	Sunderland	1874	J.&J.Gibbon	Y 12.86	C.T.Bennett	Bristol
454	N	Violet (WRECKED) *Hansen* -86 W Sr	Norweg'n	169	86·0	22·2	13·5	Aberdeen	1836			H.S.Hansen	Dröbak
455	A	Violetta (FOUNDERED) *Œconomidis* W Sr	Greek	101	72·1	21·2	..	Syra	1882			P.Œconomidis	Syra
456	V	Violette *Lahourcau* -87 W Bk	French	401	137·1	28·5	13·4	Pugwash,N.S.	1869	Wm.McLean	V 7.88	A.Hubeau	Bordeaux
457	V	Violette (ex Polli) *Lelieu* W Bn	French	201	115·1	21·2	12·6	Capodistria	1873	L.Polli	B 3.85	H.Chédru	Fécamp
458	V	Violette *Carcilan* W Sr	French	129	92·8	20·3	11·3	St.Malo	1884	Gautier		Vve.L.Buhot-Launay	Paimpol
459		Virago (ex NORMNER) *Sjöberg* 79-79 W Bg	Swedish	251	99·1	24·0	14·6	St.Malo	1852			A.R.Andersson	Calmar
460		Virgen de Begona W Bn	Spanish	104	65·3	20·1	10·5	Bilbao	1854			V.Artiñano	Bilbao
461		VirgendeMonserat (ex Centro de Exportacion, ex Sovereign) W Bk	Spanish	457	133·5	28·3	15·8	Kennebunkport, Me.	18[illegible]		Y 8.82	R.Tartabully Borras	Barcelona
462		Virgen de la Barca W Sr	Spanish	388	134·5	30·8	12·0	England				D.Romero	Corunna
463		Virgen del Camino W Bn	Spanish	212	86·9	25·4	11·8	Greece	1865			B.C.Ferrer	Majorca
464		Virgen del Carmen (ex Los Sans) *Mirondal* W Pol	Spanish	142	82·0	25·0	11·9	Palamos	1855		Y 11.83	B.López Carballés	Tarragona
465		Virgenes *Casals* W Bg	Spanish	219	95·1	27·2	13·4	Mataro	1855			Pedro Oliver	Barcelona
466	V	Virgilio C. (ex Corriere d'Italia) *Schezzini* W3mSr	Italian	263	108·9	27·8	16·0	SestriPonente	1862	A.Briasco fuG.B	V 7.88	G.Cignoni fu L.	Porto Ferraio
467	DRV	Virginia *Pettigrew* W Bk	American	750	151·2	33·5	19·5	Machias,Me.	1877	J.Shaw	Y 5.88	Miller & Houghton	Machias,Me.
468		Virginia W Sr	American	221	110·0	29·0	9·0	PJeffersonNY	1856		Y 3.82	P.Burgess	Barnstable,Mass.
469	H	Virginia *Filarelos* W Bk	Greek	476	..	..	..	Baccari	1856		Y 10.85	Zografos Nicolaou	
470		Virginia W Bg	Greek	159	76·4	24·6	..	Syra	1869			D.A.Georgalas	Syra
471	I	Virginia (ex Fisco) *Schiaffino* W3mSr	Italian	498	137·8	30·3	20·3	Chiavari	1867	M.Tappani	Y 11.88	G.A.Schiaffino	Genoa
472	I	Virginia (ex Venti Settembre, ex Benedetto) *Rebua* -88 W Bk	Italian	416	135·8	30·3	18·2	Lavagna	1873	G.Brignole	Y 8.85	R.Nottari	Leghorn
473	G	Virginia (ex Victor) *Poch* W Bg	Mexican	251	110·1	26·4	14·6	Risör	1871	T.Aamonsen	Y 8.83	O.Poch	Tonala
474		Virginia W Bg	Turkish	174	..	..	..	Syra	1869			A.Maliadis	Chios
475		Virginia Dare *Delaney* W Sr	American	105	93·1	26·6	7·0	TalbotCounty, Md.	1866			J.Hilles	Havre de Grace,Md.
476	RV	Virginia L. Stafford *Lawrence* -87 W Bk	British	483	136·5	30·5	12·5	Horton,N.S.	1870	J.B.North	Y 1.86	J.W.Parker	Windsor,N.S.
477	R	Virginia Lee Hickman (WRECKED) *Fisher* W3mSr	American	321	115·6	30·8	13·8	Chester,Pa.	1873			T.Winsmore	Philadelphia
478		Virginia Rulon *Rulon* W3mSr	American	280	130·0	32·7	9·7	Nyack,N.Y.	1874			Jarrard Wilson&Co	Belfast,Me.
479		Virginie *Bushford* 77-77 W Bn	British	104	86·1	23·0	10·2	Quebec	1848			J.M.Barkley	Belfast
480		Virginie W Sr	British	104	83·1	22·5	8·3	Lotbiniere	1855			F.Brunet	Quebec
481	V	Virginie *Hars* -88 W Sr	French	135	93·8	21·0	11·7	Dunkirk	1865			Ricquier-Debats	Dunkirk
482	A	Virgo *Gladulich* W Bk	Austrian	470	128·9	28·0	18·0	Lussinpiccolo	1868	N.Martinolich	Y 8.87	G.F.Gladulich	Lussinpiccolo
483	N	Virgo *Michelsen* 60-71 W Bk	Norweg'n	559	156·6	31·8	16·6	Moldö	1868		Y 7.85	H.A.C.Sunne	Arendal
484	V	Virgo (LOST) *Sandborg* 80-81 W Bk	Swedish	730	167·6	31·6	18·5	Sikeå	1881	O.P.Åberg	Y 1.86	Fr.Jacobsen	Robertsfors
485	I	Virgo Maria (ex La Maria) *Gazzolo*	Italian	269	111·5	24·1	14·8	Chiavari	1879	L.Gotuzzo	Y 11.88	J.E.Gazzolo	Genoa
486	I	Virtuosa Maria *Molmenmico* W Bn	Italian	100	..	..	..	Torred'lGreco	1874	G.Savastano		B.Miele	Naples
487	*	Viscount *Hay* 78-88 I S	British	1271	222·7	37·8	22·3	Liverpool	1873	T.Royden &Sns.		J.B.Walmsley & Co.	Liverpool
488		Visenta *Giannoni* W Bn	Italian	140	89·9	23·6	12·5		1863			G.Giannoni fu A.	Porto Ferraio
489		Vision W Bg	British	159	105·9	22·0	11·2	ClydeRiver, N.S.W.	1866			J.L.Campbell	Auckland,N.Z.
490		Visitor *McIlgorm* W Bn	British	183	105·5	24·7	12·8	P.E.Island	1866	Douse		M.McIlgorm	Liverpool
491		Visitor (MISSING) *Durrant* W Sr	British	128	77·7	19·7	12·1	Ipswich	1840	W.Bayley		D.F.Goddard	Ipswich
492		Visitor W Bn	British	114	69·0	18·8	11·8	Aberdeen	1829			J.Kelly	Dublin
493		Vistula *Lorie* 86-86 W Sr	British	125	88·0	22·1	11·2	Banff	1870	Geddie	Y 8.84	R.Morrison	Banff
494	G	Visurgis *Meyer* 63-77 I Bk	German	1089	211·2	34·4	20·5	Glasgow	1877	A.Stephen&Sns.		D.H.Wätjen & Co.	Bremen
495		Vitalis *Vesterlund* W Sr	Swedish	190	97·1	26·2	13·0	Jersö	1874			O.Persson	Jersö
496		Vitas W Bn	Spanish	190	102·3	22·8	10·8	Manila	1885	J.Rodriguez		Juan Rodriguez	Manila
497	AR	Vitez (ex Indiaman) *Scattich* -88 C Bk	Austrian	1221	231·2	34·1	22·8	London	1865	C.Langley	Y 1.8[illegible]	G.Gargurevich	Gravosa
498	I	Vito *Vassallo* W Bn	Italian	126	78·7	22·0	9·3	Castellam di Stab.	1886	V.Bonifacio	Y 2.86	E. Rizzo & F.A.V. Buccellato	PortEmpedocl
499	I	Vittoria *Bastile* -86 W Bk	Italian	410	126·6	30·0	19·0	Alimuri	1871	S.Mauro	Y 11.88	G.Demartino	Castell.di Stab.
500	A	Vittoria *Pidatella* -88 W Bn	Italian	151	96·1	22·3	13·4	Catania	1848		Y 7.8[illegible]	G.Pidatella	Catania

*=Lloyd's Register; D=Nederlandsche Vereeniging; G=Germanischer Lloyd; I=Registro Italiano; R=American Record; A=Veritas Austro-Ungarico; H=Veritas Ellenico; N=Norske Veritas; V=Bureau Veritas.

Dans la 4me Colonne, C=Composite (membrure en fer, bordé en bois); I=Fer; Stl=Acier; W=Bois.

Les dates du doublage en métal (Col. 11) ressortent des livres dans lesquels les navires sont respectivement classés ou des rapports des Inspecteurs du Lloyd's Register.

Pour plus amples renseignements sur les navires classés par le Lloyd's Register, voir le Lloyd's Register Book et le Lloyd's Yacht Register.

1 Official No.	No. in Book	2 Registry in which classed	3 Ship's Name, Material and Rig. Master, and Dates of Appointment to present Owner's Service and to Ship. Ship's Late Name.			5 Flag	6 Net Tonnage	7 Dimensions. Length	Breadth	Depth	8 Built. Where	9 When	10 By whom	11 Date of Metalling	12 Owners or Managers	Port
	501		Vittoria		W Bn		120	97·8	25·5	9·3	SestriPonente	1884	C.Bozzano			
1003	502	I	Vittoria	*Gardella*	W Bn	Italian	114	85·3	23·2	10·3	[illegible]	1867	[illegible]		Luigi Rozzo fu G. B.	Genoa
1654	503	I	Vittoria M.	*Demartino*	W Bk	Italian	492	137·4	30·7	19·9	[illegible]	1875	[illegible]		L. Demartino	Castellamare
329	504		Vittorina	*Duce*	W Bk	Italian	441	..	..	..	Loano	1864			G.Perluso fu A.	Savona
	505	N	Viva	*Christensen 74-81*	W Bk	Norweg'n	476	133·8	30·9	16·6	Grimstad	1881	N.O.Gjömle		G.Husland	Grimstad
	506		Viva (MISSING)	*Nielsen*	W Bk	Norweg'n	382	125·1	29·2	16·3	Arendal	1875	S T.Olsen		K.J.Nielsen	Risör
	507		Vivandiere		W Bkn	Chilian	198	107·9	23·3	12·7	Rye	1866	Hood		E.Eastman	Coquimbo
	508		Vivax		W Bk	Norweg'n	1100	191·1	36·6	21·6	[illegible]	1888			Thommesen & Smith	Arendal
90820	509	✻	Vivid	*Jones 86-86*	W Bkn	British	424	145·0	30·5	14·6	P.E.Island	1885	J.Yeo		T.P.Richards & Co.	Swansea
68493	510		Vivid		W Bk	British	238	128·2	27·2	11·0	Drammen	1870	Jorgensen & [Knudsen		H.C.Pigot	Melbourne
2850	511		Vivid		W Sw	British	230	88·8	23·4	16·0	Sunderland	1846			C.Marwood	Whitby
11736	512		Vivid		W Bn	British	162	91·1	22·9	13·3	Rye	1852			F.C.Clethier	Faversham
4525	513		Vivid	*Brown*	W Bn	British	153	78·6	20·3	13·6	Bideford	1849			J.Drew	Teignmouth
58806	514		Vixen	*Stephen*	W Sr	British	180	109·7	23·4	12·5	Peterhead	1870	Stephen		J.E.Prettyman	Peterhead
22742	515		Vixen		W Sr	British	106	78·7	20·0	10·6	Sculcoates	1833			R.Frend	Faversham
88120	516	✻	Vixen	*Berry 83.*	W Sk	British	87	78·7	21·1	10·4	Hull	1884	Hunt, Fowler [&Co.		R.Roach	Hull
16940	517		Vizcaya	*Smith -86*	W Bk	British	287	109·3	27·3	14·8	Bilbao				G. & J.Robinson	Arundel
3016	518	A	Vjeran (ex Matteo L.)	*Randich*	W Bg	Austrian	343	125·9	27·2	18·6	Fiume	1857	Schiavoni	Y 12.82	N.Randich	Costrena
2531	519	A	Vladimir (ex Ludovico Cerdan)	*Cergogna*	W Bg	Austrian	349	118·1	25·9	16·8	LesSables, [France	1864	Noleau	Y 4.83	F.Olivieri	Segna
	520		Voador (ex Apollo)	*Godinho*	W Bn	Brazilian	127	..	..	..	Pelotas	1885		Y 5.85	J.M.R.Brito	Rio de Janeiro
90376	521	✻	Vol au Vent	*Springall 87-87*	W Sk	British	89	77·1	21·4	10·8	BurtonStather	1884	J.Garside		W.T.Dunlin	Grimsby
78380	522	✻	Volador	*Davies 88-88*	W Bkn	British	196	117·0	22·5	11·8	Troon	1884	TroonShp.B.Co.	Y 2.87	T.Steele	Ayr
	523		Voladora (ex Bella Juana, ex Wild Pigeon)	*Serra -87*	W Bk	Spanish	719	167·0	35·8	19·6	P'rtsmouth,NH	1851		Y 9.85	Balcells y Subirana	Barcelona
57078	524		Volant	*Scallon*	W Bn	British	183	101·0	25·2	12·9	P.E.Island	1867	Welsh		M.Fitzmaurice	Limerick
70520	525	✻	Volant	*Sinclair 75-83*	W Sr	British	99	85·8	21·7	10·3	Banff	1875	Geddie		K.Sinclair	Wick
161515	526		Volant	*Christ 84-87*	W3mSr	American	164	122·8	29·0	8·9	Eureka,Cal.	1883	H.D.Bendixon		E.Christ	San Francisco
	527	V	Voldtofte	*Lehmann -87*	W Bn	Danish	136	96·0	21·0	10·4	Troense	1875	C.R.Møller	Y 5.87	Brødrene Hiort	Copenhagen
	528	✻	Volga	*Hopps 82-88*	I S	British	1620	257·5	38·2	23·2	PortGlasgow	1887	Russell&Co.		J.Nourse	London
	529	DV	Volharding (ex Geertje)		W Sr	Urug'yan	129	84·3	21·3	13·1	N.Pekela	1868	G.H.Holtenan			Montevideo
2567	530	DI	Volonta' di Dio	*Peragallo*	W Bk	Italian	884	180·7	34·0	23·4	SestriPonente	1874	G.Casanova	Y 8.86	G.Olivari	Genoa
3947	531		Volpe (ex Sophie) (MISSING)	*Pazzallacqua*	W Bk	Italian	608	141·0	32·4	21·3	Marseilles	1867	F.Jonvin	Y 7.85	F.Oneto	Genoa
	532		Volta		W Sr	Brazilian	193	..	..	..					J.J.O.Monteiro	Rio de Janeiro
2583	533	I	Volturno	*Maresca*	W Bk	Italian	571	137·8	29·8	20·2	Castellamare	1871	B.Bonifacio	Y 8.86	G.Molle	Naples
	534	✻D	Voluna	*Thorsen*	W Bk	Norweg'n	245	106·9	24·7	15·0	Southwick	1864	J.Hardie	Y 5.86 pt 7.83	Abr.Thorsen	Sandefjord
	535	N	Voluntas	*Dahl*	W3mSr	Norweg'n	343	113·9	29·0	15·8	Fiume	1863			L.O.Dahl	Tönsberg
60882	536		Volunteer	*Milne 86-86*	W Sr	British	147	99·7	23·8	11·0	Kingston	1873	Duncan	Y 5.85	W.V.Kelloe	Fowey
35900	537		Volunteer	*Strickland*	W Bn	British	125	90·7	23·0	11·3	St.Mary's,N.S	1855			W.Wallace	Dublin
46543	538		Volunteer (ex sch)	*Wilcocks*	W Bg	British	139	89·0	22·5	10·2					New Zealand S.N.Co.	Auckland
44249	539	✻	Volunteer	*Jones -86*	W Sr	British	116	81·1	21·7	11·5	Portmadoc	1863	Griffiths		Mrs.J.Jones	Carnarvon
28308	540	✻	Volunteer	*Jones*	W Sr	British	65	71·1	18·6	8·2	Aberystwith	1861	James		E.Jones	Aberystwith
161573	541		Volunteer	*Swan -87*	W4mSr	American	571	..	..	..	Gray's Har- [bour	1887	R.Hawkes		A.M.Simpson	San Francisco
	542		Volzy		W Bg	French	106	80·4	23·6	10·4	Leghorn	1853		Z 8.86	G.M.Lorho	Bayonne
	543	G	Von Berg	*Hartung 88-88*	W Bk	German	665	142·1	32·1	20·2	Motzen	1857	D.Oltmanns Wwe	Y 8.86	O.Thyen	Bremen
	544		Von der Lühe-Rohlstorf	*B.Mandt 88-88*	W Bk	German	352	112·2	27·8	14·6	Wismar	1858	J.Maren		P.Grampp	Rostock
	545		Von Buch-Wendorf	*Jess 79-8*	W Bg	German	249	101·2	26·2	13·5	Ribnitz	1856	J.Miebrodt		E.Burchard	Rostock
	546		Von Moltke	*Cox*	W Bk	Bolivian	687	163·0	33·1	18·4	L'ndnderryNS	1874				
	547	DGV	Von Roon	*Hollander 81-81*	W Bg	German	270	110·5	25·6	12·7	Brake	1871	F.Nicolai	Y 8.86	U.C.Koch	Brake
	548	DV	Vooruit	*van Heukelom*	W Bn	Dutch	155	96·9	22·2	11·6	Hoogezand	1866	M.Kroese		J.S.van Heukelom	Helvoirsluis
	549	✻	Voorwaarts	*Schuth 76-76*	W Bn	Dutch	184	104·7	22·5	11·8	Hoogezand	1876	J.A.Hooites	Y 8.86	J.A.Hooites	Hoogezand
	550		Voorwaarts		W Bg	Dutch	111	99·4	15·3	9·2	Pontianak	1879	Overleden		Song Pian Lan & Pin-[neel]	Pontianak

✻ = **Lloyd's Register**; D = Nederlandsche Vereeniging; G = Germanischer Lloyd; I = Registro Italiano; R = American Rec[ord]; A = Veritas Austro-Ungarico; H = Veritas Ellenico; N = Norske Veritas; V = Bureau Veritas.

In the 4th Column, C = Composite (iron frame, wood planked); I = Iron; Stl = Steel; W = Wood.

The Dates of Metalling (Col. 11) are based on the records of the Books in which Vessels are respectively classed, or on the Reports of the Surveyors of Lloyd's Re[gister].

Further particulars of Vessels classed by Lloyd's Register will be found in Lloyd's Register Book and Lloyd's Yacht Register.

No.	Reg.	4 Nom. Materiaux et Gréement. Capitaine et dates d'entrée au service des Armateurs actuels et à bord du navire. Ancien Nom.	5 Pavillon	6 Tonnage net.	7 Dimensions. Longueur.	Largeur.	Creux.	8 Construit. Où.	9 Quand.	10 Par qui.	11	12 Armateurs ou Directeurs-Gérants.	13 Port d'Armement.
351		Vorwarts *Dreyer* -88 W Bk	German	494	124·7	31·6	19·1	Memel	1856	B.Pieper		J.Storch	Danzig
352	V	Vorwarts *Oltmanns* 76-78 W Bk	German	362	119·5	24·2	14·0	Elsfleth	1870	G.Wempe	V 5.88	E.tom Dieck	Elsfleth
353		Vorwärts (ex Anna) *Woywode* W Sr	German	139	82·5	22·1	12·4	Damgarten	1858	H.Dierling		A.Woywode	Stralsund
354	*	Voyager *Geddes* 85-85 W Bkn	British	242	119·0	22·9	13·5	Kingston	1873	Kinloch	V 12.8	D.Reid	Banff
355	*	Voyageur (WRECKED) *Blackmore* [97-87 W3mSr	British	196	112·9	24·1	12·9	Brixham	1868	Upham		F.le Boulanger	Swansea
356	V	Voyageuse *Gautier* 85-85 W Sr	French	101	80·0	23·4	10·0	Meteghan,NS.	1885	M.L.Deveau		M.F.Cordon & fils	St.Malo
357		Vrachtzoeker (ex Willem Daniel) (BROKEN UP) *Bakker* [-88 W3mSr	Dutch	420	121·7	32·0	18·7	Amsterdam	1854	F.T.Groen	V 1.82	W.Velthuis	
358	D	Vrede (ex F. C. Clarke) *Visser* W Bk	Dutch	648	156·1	30·2	20·3	Jersey	1852			G.Smit Dzn.	Barsingerhorn
359	DV	Vrede *Stam* -84 W3mSr	Dutch	256	117·1	25·6	12·1	Brake	1884	F.F.Nicolai	V 5.84	A.Hoogendijk Jzn	Vlaardingen
360	G	Vredenhoven (ex Europe) *Strijker* [89-81 W Bk	Dutch	335	116·4	26·6	16·7	Sunderland	1863	G.Bartram		Van Kregten Wiemann & Co.	Winschoten
361	DV	Vriendschap (ex Maria Anna Catharina Elisabeth) *Rab* -87 W Bn	Dutch	199	94·2	17·1	10·3	Amsterdam	1866	Meijjes & Zoonen	Z 3.88	D.Visser	Amsterdam
362		Vrisiis *Pramatefti* W Bn	Greek	323	117·1	26·7	..	Syra	1878	A.Livadaro	V 12.82	I.Cosmas	Syra
363		Vulcan (ex Soumbayn) (MISSING) W S	British	573	125·5	32·8	20·6					Anglo - Egyptian Nav. Co., Ld.	London
364		Vulcan *Dreman* -84 I Sr	German	109	90·1	18·5	9·0	Paisley	1855	Blackwood & Gordon		A.Tobias	Bremen
365	V	Vulkan *Lenander* 82-82 W Bk	Swedish	373	125·6	27·6	13·1	Gefle	1879	O.A.Brodin	V 9.81	O.A.Brodin	Gefle

Additional Ships are inserted in the Supplement at the end of the Register.

A=Lloyd's Register; D=Nederlandsche Vereeniging; G=Germanischer Lloyd; I=Registro Italiano; R=American Record; A=Veritas Austro-Ungarico; H=Veritas Ellenico; N=Norske Veritas; V=Bureau Veritas.

Dans la 4me Colonne, C=Composite (membrure en fer, bordé en bois); I=Fer; Stl=Acier; W=Bois.

Dates du doublage en métal (Col. 11) ressortent des livres dans lesquels les navires sont respectivement classés ou des rapports des Inspecteurs du Lloyd's Register.

Pour plus amples renseignements sur les navires classés par le Lloyd's Register, voir le Lloyd's Register Book et le Lloyd's Yacht Register.

Official No.	No. in Book	Registry in which classed	Ship's Name. Material and Rig. Master, and Dates of Appointment to present Owner's Service and to Ship. Ship's Late Name.	Flag.	Net Tonnage	Length	Breadth	Depth	Built: Where.	When	By whom.	Date of Metalling	Owners or Managers.	Port
42395	1		W. A. Black WBn	British	140	90·4	23·4	11·0	[illegible] N.S.	1863			Geo.N.McLellan	Halifax
77815	2		W. A. Woodbury WBarge	British	150	107·0	22·3	6·9	Quebec	1879	J. Bolduc		Denis Murphy	Ottawa
[illegible]	3	✳	W.B.Godfrey (ex Mm) Dabel WBk	Hawaiian	623	174·5	29·8	19·3	Greenock	1861	Steele	Y 5.87	T.R.Foster&others	Hon...
	4	G	W. Brügmann & Sohn No. 3 (ex Victoria) Wessels WBmSr	German	236	113·8	29·3	10·5	Newcastle	1864			Bernh.Tholen	Pa...
	5	G	W. Brügmann & Sohn No. 2 (ex Agathe) Egberts WBg	German	187	108·9	22·6	11·6	Hoogezand	1859	J.A.Hoites		G.H.Egberts	Grootk...
	6	G	W. Brügmann & Sohn No. 1 (ex Adriana et Susanna) Faerber WBn	German	165	86·9	24·0	12·4	Vlaardingen	1850	S.vanGyn&Zoon		Joh.Eisenhart&Joh.Free	Pape...
	7	G	W. Brügmann & Sohn No. 5 (ex Mathilde) Tebbens WSr	German	149	92·8	22·9	11·2	Brake	1860	D.Denker		Wilh.Hilling	Paper...
80793	8	R	W. C. Warner Warner WBn (ABANDONED)	British	144	86·4	24·0	9·8	Plympton,NS.	1881	W.C.Warner		William C. Warner	Digby
46439	9		W. C. Wentworth WBk (ex John Bowes)	British	275	114·6	26·6	12·0	Belfast,U.S.	1855			James C. Ellis	Newcas...
77418	10	✳	W. D. Potts Williams WSr	British	97	83·0	22·0	10·4	Pwllheli	1878	Williams		R.Williams	Carn...
69214	11		W.E.Stowe Shawenwhite -87 WBn	British	160	88·0	24·0	10·7	LunenburgNS	1875	J.Young		James Eisenhauer	Lunenb...
80961	12	R	W. E. & W. L. Tuck Harding W3mSr	American	395	142·0	34·0	13·0	East Boston	1883	R.Crosbie&Son		H.Mays	Boston
80935	13	RV	W. F. Babcock Murphy (82-82) WS	American	2029	240·8	43·8	28·0	Bath, Me.	1882	A.Sewall&Co.	7.87	A.Sewall & Co.	Bath
81100	14		W.F.Jewett Sprague-87 W3mSr	American	452	155·7	37·7	12·0	Pt.Ludlow, [U.S.A.	1887	W.A.Boole		J.Knowland	San Fra...
75793	15		W. G. Perley WBarge	British	164	107·0	22·0	7·1	Quebec	1877	A.Miller		Denis Murphy	Ottawa
53480	16	R	W. G. Russell Robinson (86-88) WS	British	1248	200·9	37·4	22·9	Quebec	1865	Russell	Y 5.87	Ship "W.G.Russell" Co. Ld. (C.E.De Wolf & Co.)	Liverpool
	17	R	W. H. B. WBn	Urugu'y'n	187	105·0	24·6	12·8	P.E.Island	1871	McMillan	Y 6.85		
80465	18		W. H Card Wooster WSr	American	119	91·0	27·0	7·0	Franklin,Me.	1874			J.G Moseley	Boston
72193	19	V	W. H. Corsar Brown 85-82 WS	British	1410	220·3	40·4	23·4	Maitland,N.S.	1874	Brown&Crowe		W.T.Dixon & Sons	St.John
80803	20	V	W.H.Dimond Drew 81-87 WBkn	American	391	155·0	35·0	11·0	SanFrancisco	1881	M.Turner	Y 8.84	Oceanic S.S. Co.	San Fra...
80881	21		W. H. Fredson Nickerson 86-88 W4mSr	American	622	154·0	31·0	19·0					P.H.Crowell	Boston
	22	N	W. H. Green Evensdal -88 W3mSr (COLLISION)	Norweg'n	330	132·2	28·8	12·3	Tvedestrand	1885	T.A.Mørland	11.85	C.Wroldsen	Tvedes...
93226	23	✳	W. H. L. Lawrence 86-86 WK	British	100	81·2	21·0	11·1	Falmouth	1885	W.H.Lean		W.H.Lean	Falmouth
75795	24		W. H. Lewis WBarge	British	156	108·0	22·7	7·6	Ottawa	1878	A.Clarke		Margaret W.Burll	Ottawa
80856	25	RV	W. H. Lincoln Daly 82- WS	American	1727	223·4	40·5	26·4	N'wb'rypt.Ma	1881	J.Currier,jun.	Y 1.87	Thayer & Lincoln	Boston
36639	26		W. H. Merritt WSr	British	266	123·0	25·5	10·5	StC'th'rineOnt	1855			Thos. Nelson Smith	Quebec
80046	27		W. H. Meyer Whalman WBg	American	256	122·6	31·5	10·4	SanFrancisco	1869			A.Andersen	San F...
80768	28	R	W. H. Oler Crowel -80 W3mSr	American	659	150·9	35·0	17·4	Bath,Me.	1880	Blaisdell		C.&J.F.Baker	Dennis...
26869	29		W. H. Prescott WS	American	1667	198·5	41·0	20·1	Chelsea,Mass.	1856			W.Bramhall	Boston
68918	30	✳	W. H. Watson Smith (87-87) I Bk	British	514	160·0	28·5	16·7	Sunderland	1875	ShortBros.		W.Porter & Sons	Liverp...
78969	31	✳	W. J. C. Burt WSr	British	80	79·2	20·8	9·5	Truro	1880	Dyer		W.H.P.Martin	Truro
80805	32		W.J.Phillips (ex Dashaway) WBn	American	192	100·0	25·5	12·6	St.Peter'sBay. [P.E.I.	1868			E.C.Lyles	Mobile
88339	33	✳	W. J. Pirrie Luther 86-86 I 4mS	British	2516	308·2	42·8	25·0	Belfast	1883	Harland&Wolff		S.Lawther	Belfast
71274	34		W. J. Suffel WSr	British	238	120·0	26·0	9·8	PtBarwellOnt	1874	S.McDermand		Robert C.Smith	Port...
80151	35		W. L. Bradley Chase W3mSr	American	411	130·4	30·0	16·1	East Boston	1870	A.Gove		C.& J.F.Baker	Boston
66652	36	R	W. L. J. Rennard 85-88 WBkn	British	302	120·4	26·7	14·6	P.E.Island	1870	Ramsey	Y 3.86	D.Jenkins	Swansea
81023	37	R	W. L. White White W3mSr (ABANDONED)	American	626	157·0	35·0	17·0	Rockland,Me.	1884	A.F.Ames	Y 1.87	A.F.Ames	Rockland
58032	38	✳	W. M. J. Evans 85-85 WSr	British	97	79·0	21·0	10·3	BartonStather	1867	Wray		W.M.Jeffreys	Swansea
64562	39	V	W. N. H. Clements Chandler -83 WBn	British	429	121·3	30·0	17·2	Hopewell,N.B	1871	Pye	Y 1.86	Charles T.Smith	St.John
80792	40		W. R. Chester Thompson WSr	American	121	85·7	24·8	8·1	Machias,Me.	1880	Holway		W.C.Holway	Machias
80379	41	V	W. R. Grace Walcott WS	American	1799	218·1	42·1	28·0	Bath,Maine	1873	Chapman&Flint	Y 8.86	B.Flint	New York
72864	42		W. R. Taylor WSr	British	322	131·2	25·2	12·5	SouthMarys- [burgh,Ont.	1877	J.Dickson		S.H.Dunn	Picton
81093	43		W.S.Bowne Bluhm 87-87 W3mSr	American	391	144·1	34·4	12·6	PortBlakely	1885	Hall		G.F.Smith & Co.	San Fra...
80341	44		W. S. Jordan Crowell -86 W3mSr	American	355	136·5	32·6	14·5	Nwburypt.Mass	1873	G.E.Currier		C.&J.F.Baker	Mass
70304	45	✳	W. W. Lloyd Griffiths WBg	British	244	116·0	24·5	14·4	Portmadoc	1875	Roberts	Y 3.86	O.Lloyd	Carnarvon
	46		W. Wallace Ward Vogler 88-88 W4mSr	American	1182	193·0	40·0	18·3	NewHaven,Conn	1888	H.Sutton		H.Sutton	NewH...
71288	47		W. Y. Emery WSr	British	154	104·3	25·0	10·8	PtBarwellOnt	1867	Geo.Dakin		Edward Burge	Pt.B...
71374	48	GR	W. & H. Wotherspoon Chase -87 WSr	British	194	108·2	27·1	11·2	Parrsboro'N.S	1875	J.E.Pettis	Y 1.88	A.Shaw	Parrs...
	49		Wa Adannoor (ex Djamoor Hassan) WBk	Dutch	221	95·1	21·6	..					Said Alie bin Mohamad Alwi	Palemb...
	50													

✳ = Lloyd's Register; D = Nederlandsche Vereeniging; G = Germanischer Lloyd; I = Registro Italiano; R = American
A = Veritas Austro-Ungarico; H = Veritas Ellenico; N = Norske Veritas; V = Bureau Ver

In the 4th Column, C = Composite (iron frame, wood planked); I = Iron; Stl = Steel; W = Wood.

The Dates of Metalling (Col. 11) are based on the records of the Books in which Vessels are respectively classed, or on the Reports of the Surveyors of Lloyd's

Further particulars of Vessels classed by Lloyd's Register will be found in Lloyd's Register Book and Lloyd's Yacht Register.

1 No. du livre	2-3 Registre dans lequel le navire est coté	4 Nom. Matériaux et Gréement. Capitaine et dates d'entrée au service des Armateurs actuels et à bord du navire. Ancien Nom.	5 Pavillon.	6 Tonnage net.	7 Dimensions. Longueur	Largeur	Creux	8 Construit. Où.	9 Quand.	10 Par qui.	11 Date du doublage en métal	12 Armateurs ou Directeurs-Gérants.	13 Port d'Armement.
51		Waaja *Dahlqvist -88* W Bk	Russian	484	110·0	32·0	16·0	Åbo	1871	C.H.Stromberg		K.L.Lindberg	Nagu
52	NV	Waaland *Eriksen 72-77* W Bk	Norweg'n	584	150·0	33·3	18·8	Arendal	1877	Waaland	Y 7.81	Stian Eriksen	Arendal
53	R	Wacamaw *Squires* W3mSr	American	436	137·7	33·3	14·5	PtJeff'rs'nNY	1881	Bayles		A.E.Squires	New York
54	V	Wachusett *Oakland 86-86* W S	American	1519	214·7	40·2	26·4	Kenneb'nk,Me	1878	Crawf'd&Perkns	Y 11.86	G.C.Lord & Co.	Boston,Mass.
55		Wadiatool Rachman *Said Hamed Alkaf* W Bk	Dutch	265	..	..	..					Said Hamed bin Aloei Alkaf	Palembang
56		Waft *Saunders* W Sr	British	131	90·0	20·7	12·3	Brixham	1866	Dewdney	Y 5.86	W.G.Kendrick&Co.	Brixham
57		Wagrien *Vincent* W Sr	British	179	101·5	24·0	12·1	Heiligenhafen	1873	J.H.Tornoe	Y 12.81	Charles Vincent	Hong Kong
58	✻	Waimate *Canese 83-85* I S	British	1123	219·7	35·1	20·7	Sunderland	1874	J.Blumer&Co.		N.Zealand Shipping Co.	London
59	V	Waimea (ex Isabelle) *Pollinger 87-87* I S	British	848	194·3	31·7	19·0	Hamburg	1868	Godeffroy		N.Zealand Shipping Co.	London
60	✻	Wainfleet *Smith* W K	British	90	80·0	21·0	10·4	Galmpton	1886	W.A.Gibbs		Butt Bros.	Grimsby
61		Waino *Nylund* W3mSr	Russian	260	..	..	..		1872			A.W.Nylund	Nykyrko
62		Waino *Spitsin* W Sr	Russian	168	..	..	..	Raumo	1873			H.Östman's Enke	Raumo
63		Waino *Penttilä* W Sr	Russian	100	..	..	..		1883			W.Penttilä	Björkö
64	✻	Waipa *Forsdick 85-87* I S	British	1017	204·1	34·2	20·0	Newcastle	1875	Palmers' S&ICo.		N.Zealand Shipping Co.	London
65	✻	Wairoa *Bungard 85-87* I S	British	1015	204·1	34·2	20·0	Newcastle	1875	Palmers' S&ICo.		N.Zealand Shipping Co.	London
66	✻	Waitangi *Leeman 84-84* I S	British	1128	222·0	35·1	20·8	Sunderland	1874	J.Blumer&Co.		N.Zealand Shipping Co.	London
67	✻	Waitemata (ex Eyldo) *Ohlson -84* W Bkn	British	365	146·9	29·0	12·9	Sunderland	1875	R.Thompson	Y 3.84	David B. Cruickshank	Auckland
68		Wake *Ingersoll* W Sr	American	168	91·3	28·2	9·0	Tuckahoe,N.J.	1867			G.H.Squire & Co.	Philadelphia
69	RV	Wakefield *Crowell* W Bk	American	842	163·6	34·5	21·0	Newb'ryp't Ma	1873	Atkinson& [Fillmore	Y 3.87	J.Baker & Co.	Boston,Mass.
70		Walamo *Bask* W	Russian	179	..	..	..		1859			Höyrylai	Sordavala
71		Walcheren *Lindblad 78-78* W Bk	Dutch	374	115·1	28·9	18·1	Middelburg	1852	Commercie [Compagnie		Den Bouwmeester Borsius & van Lelju	Middelburg
72	R	Waldemar *Parker 72-72* W3mSr	American	269	117·0	29·0	14·5	Bucksport,Me.	1872	Genn&Hill	Y 12.88	J.S.Emery & Co.	Bucksport,Me.
73	GV	Waldemar *Wallis 88-88* W Bg	German	274	101·3	26·6	14·3	Demmin	1865	Hassen		C.A.Beug	Stralsund
74	✻	Walden Abbey *Williams [79-82* I S	British	1754	263·0	39·1	24·1	Liverpool	1881	W.H.Potter&Sn		J.Poole & Co.	Liverpool
75		Wales W Sp	British	152	110·0	21·7	10·0	Kingston,Ont.	1881	JamesBell		Muskoka Mill & Lumber Co.	Toronto
76	R	WalkerArmington,jr. W Bk	American	495	133·0	30·0	8·0	Rockland,Me.	1874	Goss&Sawyer		S.M.Bird	Rockland
77	✻R	Wallace (BROKEN UP) *Smith 78-82* W S	British	1583	223·6	41·5	23·5	Horton,N.S.	1882	J.B.North	Y 11.85	M.Shaw & Co.	Windsor.N.S.
78	R	Wallace (DISMANTLED) *McCormick 60-60* W Bk	American	613	142·0	30·4	20·5	C.Elizabth.Me	1865		Y 5.80	L.S.Davis	New York
79	RV	Wallace B. Flint *De Winter -87* W Bk	American	793	178·4	35·4	17·7	Bath,Maine	1885	J.McDonald	Y 6.88	B.Flint	New York
80	R	Wallace J. Boyd *Bates* W3mSr	American	602	144·4	38·4	17·2	Mauricet'n NJ	1879	J.W.Vannemann [& Bros.		A.Lake	Philadelphia
81	✻	Wallacetown *Wortley* I S	British	1626	252·1	38·7	23·5	Sunderland	1874	T.R.Oswald		Hargrove & Hellon	Liverpool
82	N	Walle (ex Eva Carroll) *Wichard* W Bk	Norweg'n	574	153·3	33·0	16·2	St.John,N.B.	1872	W.Hogg	Y 7.86	M.Pedersen	Mandal
83	V	Wallis & Sohn *Sprengberg -83* W Bk	German	469	125·3	29·2	17·3	Barth	1872	J.Holzerland	Y 5.86	J.J.Wallis & Sohn	Barth
84		Walpas (ex Wenn) *Karsten -86* W Bk	Russian	317	113·0	29·2	..	Nystad	1866			Otto Carlsson	Raumo
85		Walraven Melchers (ex Birmingham) (ABANDONED) *Wachter* W Bk	Dutch	1408	188·9	34·1	24·0	Kenneb'nk,Me	1860		Y 8.86	S.J.Melchers	Schiedam
86	R	Walrus *Musson* W Sr	British	160	97·0	26·0	10·0	JordanRiver,N.S.	1888	R.McDonald		R.W.Freeman	Shelburne
87	N	Walter *Næss 79-86* W Bk	Norweg'n	144	98·0	25·5	8·8	Wilmington	1846			F.A.Kierulf	Drammen
88	V	Walter D. Wallet *Healy -88* W S	British	1413	202·5	40·0	24·0	Tusket,N.S.	1875	S.H.Jeffery	Y 5.88	W.H.Ross & Co.	Liverpool
89		Walter F. Parker *Kelly* W3mSr	American	371	132·0	32·0	14·6	Belfast,Maine	1874			S.A.McCaulley	Greenport,N.Y.
90	✻	Walter H. Wilson *Sproul 87-87* I S	British	2461	308·1	42·7	24·9	Belfast	1882	Harland&Wolff		S.Lawther	Belfast
91	V	Walter Holly *Holder [88-88* W3mSr	British	273	128·4	30·5	10·5	Gardner's [Creek,N.B.	1888	F.E.Sayre		F.E.Sayre	St.John.N.B.
92		Walter J. Cummins W Bn	British	134	91·3	23·1	11·6	Miramichi	1865	Brown		T.O'Brien	Wigtown
93	R	Walter L. Plummer *Plummer* W3mSr	American	317	128·0	29·0	12·0	Addis'n,Maine	1876	J.Wilder	Y 4.88	V.C.Miller	Machias,Maine
94	R	Walter S. Massey *Phelan 85* W Bkn	American	548	142·8	35·0	17·0	Camden,N.J.	1883	Burtn&Vnnemn	Y 2.83	M.P.Summers	Philadelphia
95	V	Walter Siegfried *Lawrence 86-86* W Bk	British	394	135·2	27·9	15·8	Reihersteig	1871	ErnstDreyer		P.V.Grant	Shanghai
96		Walter Smith W Bn	American	413	128·6	29·0	15·5	Stockton,Me.	1868			E.W.White	New York
97	✻	Walter Ulric *Owen 85-85* W Sr	British	98	85·8	22·5	10·9	Nevin	1875	Owens		R.Evans & Co.	Carnarvon
98	V	Waltikka *Waldstedt 88-88* W S	Russian	971	174·2	33·1	21·3	Nystad	1873	H.Kjäldström	Y 1.86	G.Rivell	Nystad
99	✻	Wanderer (LOST) *Bogg* W Sr	British	256	121·3	25·0	13·9	Kingston	1874	Kinloch	Y 11.84	D.Reid	Leith
100		Wanderer *McKenzie* W Bg	British	152	83·8	23·0	13·0					Peynah Keyther Pulley	Penang

✻=Lloyd's Register; D=Nederlandsche Vereeniging; G=Germanischer Lloyd; I=Registro Italiano; R=American Record; ‡=Veritas Austro-Ungarico; H=Veritas Ellenico; N=Norske Veritas; V=Bureau Veritas.

Dans la 4me Colonne, C=Composite (membrure en fer, bordé en bois); I=Fer; Stl=Acier; W=Bois.

Les Dates du doublage en métal (**Col. 11**) ressortent des livres dans lesquels les navires sont respectivement classés ou des rapports des Inspecteurs du Lloyd's Register.

Pour plus amples renseignements sur les navires classés par le Lloyd's Register, voir le Lloyd's Register Book et le Lloyd's Yacht Register.

1 Official No.	2 No. in Book.	3 Registry in which classed.	4 Ship's Name. Material and Rig. Master, and Dates of Appointment to present Owner's Service and to Ship. Ship's Late Name.	5 Flag.	6 Net Tonnage	7 Dimensions. Length.	Breadth.	Depth.	8 Built. Where.	9 When.	10 By whom.	11 Date of Metalling.	12 Owners or Managers.	Port of Registry
[illegible]	101	R	Wanderer *Corrigan 88-88* W Bk	American	288	116·2	27·6	15·8	Mattapoisett, [Mass.	1878	J.B.Helmes		Pacific Steam Whaling Co.	San F…
	102	G	Wanderer *Boggs -88* W Bk (CONDEMNED)(ex Eenlinkeighe)	Liberian	599	140·3	31·4	19·3	Varazze	1870	A.Bollo		Robt.E.N.Boggs	Monr…
	103	N	Wanderer *Meyer -88* W Bk	Norweg'n	439	123·9	30·3	16·3	Grimstad	1869	JörgenTellefs'n	Y 7.86	H.Omundsen	Grim…
401	104		Wanderer *Ahl -86* W3mSr	Russian	266	109·0	29·5	13·4	Uggenzeem	1881	RetikTam		Puhing Bros.	Riga
73898	105	✳	Wandering Chief *Harris* W Bk	British	418	147·0	27·7	17·0	Kingston	1876	Kinloch	Y 4.86	W.Kinloch	Banff
80645	106	V	Wandering Jew *Nichols* W S	American	1650	219·2	40·0	29·0	Camden,Me.	1877	J.Pascal	Y 10.86	Carleton,Norwood&Co	…
80749	107	R	Wandering Jew *Wildey -88* W Bkn	American	634	142·0	34·0	18·0	GreenportN.Y	1880	Smith&Terry	Y 7.88	R.R.Baizley	Philad…
65303	108		Wandering Minstrel *Wright* W3mSr	British	362	140·0	27·5	14·5	Peterhead	1875	Stephen&Forb's	Y 5.88	Shark Fishery Co.Ld. of Hong Kong	Hong K…
85917	109	V	Wandrian *Woods -88* W3mSr	British	371	135·3	32·3	12·5	Parrsboro'N.S	1883	L.Hatfield	Y 12.84	Allen Haley	Parrs…
	110	G	Wangerland *Eecs 75-75* W Bg	German	216	95·3	24·6	12·9	[illegible]	1866	F.Nicolai		T.Knupper	Geeste…
73911	111	✳	Wanlock *Cooper 84-84* I Bk	British	745	189·7	31·3	19·1	Glasgow	1875	London&Glas.C		Bramwell & Gardiner	London
60409	112		Wansbeck *Harroway 86-86* W Bk	British	392	128·6	28·1	17·1	Sunderland	1868	Hodgson		R.L.Manners	North…
26949	113		Wapella *Bengger* W Sr	American	359	109·5	36·0	16·8	GreenPnt,N.Y	1864			Rackett & Bros.	New Yo…
26075	114		War Steed *Gookin 79-81* W Sr	American	114	93·6	24·6	8·4	P.Jefferson,N.Y	1854			W.S.Gookin	Saco …
57325	115	✳	Waratah *Krause* W Bk	British	202	123·0	25·0	11·0	HobartTown	1872	JohnLucas	Y 11.84	Alex.J.Soutar	Ade…
48850	116		Warblington *Taylor84-84* W Bg (ABANDONED)	British	202	105·7	24·3	12·9	P.E.Island	1863	M·Millan		G.Robinson	Newh…
63824	117		Warden Law *Graham* W Bk	British	460	138·0	29·3	17·7	Sunderland	1871	J.Gill		GreatGrimsbyIceCo.Ld.	Grimsby
80987	118	R	Warner Moore *Crockett* W3mSr	American	421	143·3	33·5	13·9	Newburyport, [Mass	1883	Atkinson & [Fillmore		A.F.Crockett & Co.	Richm…
	119		Warnow W Bg		187	85·9	21·7	13·5	Ribnitz	1862	J.H.Wilken			
1408	120		Warree *Lamb* W Bn	British	103	81·0	19·8	9·6	Liverpool	1837			J.Farrell	Dundalk
81039	121	R	Warren Adams *Colcord 84-* W3mSr	American	634	163·4	35·1	13·0	Bowdoinham, [Me.	1884	J.T.Rideout&C.	Y 7.88	J.B.Drake	Bath,…
80723	122	R	Warren B. Potter *Andrews 84-84* W3mSr	American	350	126·6	32·4	12·0	Bath,Maine	1879	Goss&Sawyer		E.Gibbs	New B…
80604	123		Warren J. Crosby *Carroll* W Sr	American	108	90·0	24·1	8·7	Essex,Mass.	1876	JamesBros.		F.Carroll	Glouc…
90638	124	✳	Warren W. *Buchanan* W Bn (ABANDONED)	British	268	113·1	29·2	11·5	MontaguePEI	1886	G.Wightman		G.Wightman	Charlott…
83163	125	RV	Warrior *Kitchin 84-84* W S	British	1687	221·0	40·5	24·2	RiverJohn,NS	1884	J.Kitchin	Y 5.87	Alex.C.Macdonald	Pictou,…
61583	126		Warrior *Jackson* W Sr	British	102	80·0	22·1	10·0	JordanRvr,NS	1875	JohnMarten		M.Robinson	Halifax
73189	127	V	Warsaw *Leblanc* W S	British	1347	203·6	39·0	23·6	Newport,N.S.	1876	N.Mosher	Y 12.86	Frederick Curry	Windsor
51848	128		Warwick *Mullan* W Bn	British	113	86·4	23·4	10·3	P.E.Island	1865	Cox		J.Kingsberry	Belfast
47486	129	✳	Warwick Castle *Hutton* I Bk	British	1156	209·4	34·1	22·3	Glasgow	1863	R.Napier&Sons		Bk.Warwick Castle Co. Ld. (Woodside & Workman)	…
45407	130	✳	Warwickshire *Mallinson 88-88* I Bk	British	647	179·9	29·0	18·5	Liverpool	1862	T.Vernon&Son		W.J.MyersSon&Co.	Liverpool
	131	V	Wasama *Kyntzen 87-87* W Bk	Russian	635	152·9	35·1	19·3	Brahestad	1870	E.Klubb		Johan Lang	Brahestad
92378	132		Wascano *Balser -88* W Sr	British	115	88·2	26·3	8·2	Coverdale,N.B.	1888	J.M.Steeves		J.M.Steeves	St.John
84165	133	✳	Wasdale *Brown 74-86* I S	British	1821	262·3	39·3	24·1	Whitehaven	1881	WhitehvnS.B.C		J.D.Newton & Co.	Liverpool
26474	134	R	Washington (BARGE) *North* W S	American	1814	201·5	44·0	28·9	Williamsburg, [N.Y.	1849	J.Williams		T. Dunham's Nephew & Co.	New York
80318	135		Washington *Jordan* W3mSr	American	295	127·0	30·7	11·2	Brooklyn	1872	C.&R.Poillon	Y 6.82	Tupper & Beattie	New York
1029	136	I	Washington *Cafiero* W Bk	Italiano	635	141·0	31·2	20·6	Castellamare	1874	A.Bonifacio	Y 1.87	M.Cacace	Castell…
	137		Wasili *Pitkänen* W Sr	Russian	168	..	..	..	[illegible]	1873			B.Lisitzin	Kexholm
50384	138		Watch *Thomas 86-86* W Bn	British	282	124·8	25·6	14·1	Guernsey	1869	Sebire	Y 3.88	John Grace & Co.	Guernsey
26845	139		Watchful *Hearn 84-84* W Sr	American	132	108·6	22·6	8·6	Clayton,N.Y.	1858			E.P.Burton	…
90111	140	✳	Watchman *Smith* I Bk	British	449	171·1	28·1	13·5	Belfast	1884	Workman,Clark [&Co.		Watchman Ship.Co.Ld. (Woodside & Workman)	Belfast
58131	141	✳	Water Bird *Williams* W Sr (ex James M…)	British	97	83·2	21·7	10·3	Fraserburgh	1868	Webster		W.H.Stott & Co.	Liverpool
50807	142		Water Lily *Tuggly* W Sr	British	128	90·1	21·4	11·4	Bideford	1867	Pickard		J.Helme	Lancaster
70521	143	✳	Water Lily *Bennett* W Sr	British	99	86·4	21·6	10·3	[illegible]	1876	Watson		Mrs.A.M.Howard	Liverpool
32376	144		Water Witch *McGrath* W Bk	British	236	99·5	22·6	17·4	Pembroke	1820	British Govt.		A.McGregor	Hobart
26852	145	R	Water Witch *Miller -88* W Bn	American	226	98·5	20·8	14·0	Baltimore	1864	Cooper&Slicer	Y 1.88	W.F.Greene & Son	Boston
101	146		Waterford *Rosa 76-83* W Bn	British	122	84·4	21·6	10·5	Bridport	1818			R.A.Ayles	Weym…
	147		Watergeus *Kuiper* W Bn (ex Willey)	Dutch	253	116·1	24·5	12·6	CardiganRiver [P.E.I.	1868	D.Stewart		Fontein&Tjallengii	Harling…
58165	148		Waterlily *Davies* W3mBn	British	158	95·5	22·1	11·9	Newburgh	1866	Scrimgeour		H.Lamb	Fowey
78631	149	✳	Waterloo *Johnston 87-87* I 4mS	British	1018	282·5	40·2	24·2	Glasgow	1878	C.Connell&Co.		A.Mackay & Co.	Glasgow
17041	150		Waterloo *W.Jury*	British	136	72·2	23·8	11·0	Naples	1836			C.Weatherburn	S.Shields

✳ = Lloyd's Register; D = Nederlandsche Vereeniging; G = Germanischer Lloyd; I = Registro Italiano; R = American Rec…
A = Veritas Austro-Ungarico; H = Veritas Ellenico; N = Norske Veritas; V = Bureau Veritas.

In the 4th Column, C = Composite (iron frame, wood planked); I = Iron; Stl = Steel; W = Wood.

The Dates of Metalling (**Col. 11**) are based on the records of the Books in which Vessels are respectively classed, or on the Reports of the Surveyors of Lloyd's Re…

Further particulars of Vessels classed by Lloyd's Register will be found in Lloyd's Register Book and Lloyd's Yacht Register.

1	2-3	4 Nom. Capitaine et dates d'entrée au service des Armateurs actuels et à bord du navire. Ancien Nom.	Matériaux et Gréement.	5 Pavillon.	6 Tonnage net.	7 Dimensions. Longueur.	Largeur.	Creux.	8 Construit. Où.	9 Quand.	10 Par qui.	11 Date du doublage en métal.	12 Armateurs ou Directeurs-Gérants.	13 Port d'Armement.
151		Waterloo *Gray*	WSr	American	111	84·0	20·6	8·0	Waldoboro'Me	1848			C.Comery	Waldoboro',Me.
152		Waterwitch	WSr	British	289	131·8	23·6	11·0	Montreal	1850	A.Cautin		Samuel Fraser	Quebec
153		Waterwitch	WBk	British	279	104·0	25·0	17·6	Calcutta				Francis B.Johnson	Hong Kong
154		Waterwitch *Gardener 83-84*	WBn	British	194	112·0	25·8	12·8	Poole	1871	Meadus		H. & J.Ash	Portsmouth
155		Waterwitch	WBn	British	165	115·5	25·0	9·4	Cape Hawk, N.S.W.	1873	R.P.Egan		C.J.Norris	Rockhampton
156	✻	Waterwitch	WSrYt	British	101	100·1	19·2	12·3	Gosport	1880	Camper& Nicholson	[illegible]	Lord Revelstoke	Portsmouth
157	V	Waterwitch *Nissen 88-88* (ex Jens Hintze)	WBn	Danish	187	109·5	22·1	11·6	Rønne	1873	G.N.Sandorpf	Y 1.80	M.Nissen	Copenhagen
158		Wau-bun *Welch 79-84*	WBn	American	491	135·0	30·0	16·9	Machias,Me.	1873		Y 10.83	W.C.Holway	Machias,Me.
159		Wave *Adams*	WSr	British	110	82·2	20·7	10·7	Montrose	1863	Birnie		D.M'Gregor	Arbroath
160		Wave	WSr	British	108	74·6	20·6	10·0	Maldon	1848			W.R.Smith	Maldon
161		Wave *Masher 87-87*	WBk	American	142	82·7	24·6	10·0	DuxburyMass.	1841			W.Lewis	NewBedfrd,Mass.
162		Wave Crest	WSr	British	182	110·1	22·5	9·7	Brockville,Ont	1867	D.Anderson		RobertC.Smith,jun.	Port Hope,Ont.
163		Wave Crest *Hewitt -83* (STRANDBID)	WSr	American	285	117·5	29·0	13·0	Wlmngtn,N.C	1865			S.S.Hewitt	New York
164	✻	Wave Queen *Kelly 72-86*	IBk	British	853	190·0	30·0	20·8	Glasgow	1861	A.Stephen&Sns		Park Bros.	London
165	✻	Wave Queen *Boys -88*	WBk	British	313	130·6	27·2	14·1	Littlehampton	1870	Harvey	Y 10.86	T.B.Gates	Shoreham
166	N	Wavelet *Eeeaten*	WBk	Norweg'n	239	118·3	28·4	11·0	Pembroke	1859			Winge & Co.	Christiania
167	✻D	Waverley *Dixon 88-88*	IBk	British	1113	215·4	35·1	21·2	PortGlasgow	1885	Russell&Co.		Russell & Pinkerton	Glasgow
168	✻	Wavertree *Trotts 88-88* (ex Southgate)	IS	British	2118	279·0	40·2	24·4	Southampton	1885	Oswald,Mordaunt&Co		Sailing Ship Wavertree Co.,Ld.(R.W.Leyland&Co.)	Liverpool
169		Wawbeek (HURST) *Balser*	WSr	British	100	76·4	24·8	8·2	HopewellCape N.B.	1883	C.Dowling		Joshua M.Steeves	St.John,N.B.
170	✻	Wayfarer *Currie 86-86*	IS	British	2040	278·0	40·1	24·2	Liverpool	1886	W.H.Potter&Ss		W.H.Potter & Sons	Liverpool
171	NV	Wayfarer *Gram 88-88*	WBk	Norweg'n	642	140·8	33·4	18·0	Digby,N.S.	1874		Y 3.87	Actieselskabet "Wayfarer"	Drammen
172	✻	Weathersfield *Sherris 77-77*	IBk	British	1047	214·5	34·7	21·0	Port Glasgow	1865	R.Duncan&Co.		Greenshields,Cowie&Co.	Liverpool
173	✻	Weaver Belle *Cave*	WSr	British	98	90·3	20·6	10·3	Appledore	1876	Cook	Y 2.81	J.Steer	St.Johns
174		Web Foot *Lewis*	WBkn	American	344	146·5	31·0	10·8	CoosBay,Oreg.	1869			A.M.Simpson	San Francisco
175		Webster Barnard *Marshall -82*	WSr	American	143	87·7	24·8	8·9	Bucksport,Me.	1868	Parker		J.S.Marshall	Bangor,Me.
176	V	Wega *Leopold 80-90*	IS	German	1115	206·7	34·4	20·3	Vegesack	1877	H.F.Ulrichs		W.A.Fritze & Co.	Bremen
177	G	Wega *Kuhl -85* (ex Francis Banfield)	WBk	German	344	120·3	26·6	17·1	Sunderland	18 56/79	R.Reay		Retzlaff & Schober	Stettin
178	V	Wega (ex Anne) *Sweers*	WBn	German	171	99·2	21·4	11·5	Nordby,Fanø	1874	N.Nielsen	Y 8.88	G.Sweers	Neermoor
179		Wejdenhjelm *Gidlöf -86*	WS	Swedish	556	150·9	28·7	17·8	Hernösand	1863	Strandberg		N.J.Nyberg	Bollsta
180	R	Welaka *Cottrell -83*	W3mSr	American	411	134·2	31·9	14·0	Belfast,Me.	1873			C.Y.Cottrell	Belfast,Me.
181		Welcome *Morrissey -85*	WSr	British	131	90·8	21·2	12·1	Brixham	1864	Upham		M.Morrissey	Guernsey
182		Welcome *Fry -82*	WSr	British	112	88·0	21·8	10·1	Auckland,N.Z.	1882	J.Bigelow&Son	[illegible]	J.Bigelow & Son	Auckland,N.Z.
183	✻	Welcome *Banks 85-*	W3mSr	British	100	94·0	22·4	9·6	Freckleton	1885	Rawstrone		Mrs.M.Ashcroft	Preston
184	✻	Welcome Home *Nichol -88*	WSr	British	100	85·0	21·5	10·6	Stornoway	1881	A.M.Mackenzie	Y 5.86	J.M'William	Banff
185	✻	Welcome Home *Robertson 85-88*	WSr	British	56	69·5	18·5	8·3	Peterhead	1885	Steph'n&Forbes	Y 12.85	J.Robertson	Peterhead
186		Welcome R. Beebe *Smith -87*	W3mSr	American	386	123·5	30·6	16·0	Belleville,N.J.	1860			W.Blair,jun.	New York
187	✻	Welfare	WSk	British	88	76·5	21·0	10·4	Grimsby	1885	SmithBros.		E.Bacon	Grimsby
188		Weljekset	WBk	Russian	500	118·9	29·2	16·4	Pargas	1875			M.E.Jansson	Nadendal
189	D	Weljekset *Jansson*	WBk	Russian	482	146·7	32·8	16·5	Åbo	1875	Sundström		A.Abrahamsson	Gustafssocken
190	R	Wellamo *Flinkenberg -88*	WBk	Russian	646	139·5	33·8	20·0	Brahestad	18 71/78	F.Pettersson		Z.Durchman	Brahestad
191		Wellamo *Andersson*	WSr	Russian	184	..	..	..	Luvia	1869			M.Michelsson	Sund
192		Wellgunde *McArthur -87* (ex Campon)	WBk	American	324	117·0	27·9	17·0	Sunderland	1865	Robinson		Mark Golinsky	Fernandina
193	✻	Wellington *Cowan 71-74* (ex Compton)	IS	British	1247	239·9	36·0	20·8	PortGlasgow	1874	R.Duncan&Co.		Shaw, Savill & Albion Co.,Ld.	Glasgow
194		Wellington *Ramsey*	WBg	British	232	101·7	25·1	13·8	P.E.Island	1856			C.Marwood	Whitby
195	✻	Welsh Belle *Davies*	WSr	British	103	90·5	21·4	10·8	Bideford	1875	Cox	Y 3.82	R.Mitchell & Co.	Swansea
196	✻	Welsh Girl *Davies*	WBn	British	173	100·8	24·1	12·5	Pill	1875	Francis	Y 7.87	W.Francis	Swansea
197		Welsh Girl *Wyatt*	WSr	British	123	108·0	23·5	11·3	Hayle	1863	Harvey&Co.		S.Morse	London
198	D	Weltevreden *Douwes -87*	WSr	Dutch	139	87·9	22·0	10·5	Sappemeer	1859	J.Berg		J.Douwes	Sappemeer
199		Wemyss Castle	CS		715	183·0	31·0	17·9	Glasgow	1866	C.Connell&Co.	Y 7.83		
200	V	Wemyss Castle *Albertsen -87*	WSr	Danish	106	83·5	21·0	9·8	Marstal	1875	F.Hansen		R.Albertsen	Marstal

✻=Lloyd's Register; D=Nederlandsche Vereeniging; G=Germanischer Lloyd; I=Registro Italiano; R=American Record; A=Veritas Austro-Ungarico; H=Veritas Ellenico; N=Norske Veritas; V=Bureau Veritas.

Dans la 4me Colonne, C=Composite (membrure en fer, bordé en bois); I=Fer; Stl=Acier; W=Bois.

Les Dates du doublage en métal (Col. 11) ressortent des livres dans lesquels les navires sont respectivement classés ou des rapports des Inspecteurs du Lloyd's Register.

Pour plus amples renseignements sur les navires classés par le Lloyd's Register, voir le Lloyd's Register Book et le Lloyd's Yacht Register.

1 Official No.	2 No. in Book	3 Registry in which classed	4 Ship's Name. Material and Rig. Master, and Dates of Appointment to present Owner's Service and to Ship. Ship's Late Name.	5 Flag	6 Net Tonnage	7 Dimensions: Length	Breadth	Depth	8 Built: Where	9 When	10 By whom	11 Date of Metalling	12 Owners or Managers	13 Port
	201	G	Wendeline Cristiane *Ihlen 78-78* WSr	German	120	83·6	19·6	10·5	Papenburg	1859	Venn&Baeren		EhmeClaassenAden	Gr
	202		Wendla *Mattson* W3mSr	Russian	341	..	..	..		1875			E.P.Eriksson	Lon
	203		Wendla *Lindblom* W3mSr	Russian	246	..	..	..		1875			P.Sittkoff	Hamo
89256	204	✳	Wendur *Corrance 88-88* I4mS	British	1982	292·7	42·0	23·8	Glasgow	1884	C.Connell&Co.		A.Mackay & Co.	Glasgow
87659	205	✳	Wendur StlYwlYt	British	115	102·0	17·9	14·3	Glasgow	1883	D&W.Henders'n		T.B.C.West	Glasgow
	206	R	Wenkul *Hitroff* W3mSr	Russian	189	100·0	25·5	11·9	Narva	1884	C.Hitroff		C.Hitroff	Narva
83374	207		Wenona *Davis -84* WBk	British	511	148·9	34·5	13·2	Canning,N.S.	1882	J.E.Bigelow		Donald Ross	
77878	208	V	Wenonah *McCann 81-86* WBk	British	767	162·4	32·9	20·5	Quebec	1879	Baldwin	V 6.88	B.H.McCorkell	Lon
	209	NV	Wenonah *Hirsch 77-77* (ex Lottie Jenkins) WBn	Norweg'n	292	112·1	27·7	12·9	St.John,N.B.	1875	T.Jenkins	V 10.87	A.Ervig	Christ
36600	210		Wentworth WSr	British	100	77·6	22·0	9·7	Newport,NS.	1859			Robert McKenzie	Wind
	211		Wera *Marksimoff* W	Russian	177	..	..	..		1885			Gromofsarvin	Sord
	212	N	Wergeland *Gregersen -87* WBg	Norweg'n	228	112·3	26·6	12·2	Porsgrund	1877	Holmens [&Skibsbygtri	V 9.88	F.Thorstensen	Fred
	213	✳D	Werklust *Poljewijk 88-88* (ex Ambulant) W3mSr	Dutch	211	118·1	22·3	11·4	Hoogezand	1875	J.A.Hooites	C 1.88	J.A.Hooites	Hoog
76311	214	✳	Wern *Jones 60-87* WBn	British	175	102·2	24·2	13·2	Portmadoc	1876	Roberts	V 1.88	M.Jones & Co.	
	215	GV	Werner *Hendorff 78-83* WBk	German	892	176·1	35·1	20·8	Elsfleth	1883	J.Ahlers	V 7.87	J.D.Ahlers	Elsfleth
	216	G	Werra *Horstmann 80-86* IBk	German	932	191·1	32·0	19·6	Glasgow	1876	A.Stephen&Sns.		D.H.Wätjen & Co.	Brem
	217	✳	Weser *Harde 79-83* (ex Summer Cloud) IBk	German	705	168·4	30·9	18·3	Hartlepool	1859	Pile,Spence&Co		E.F.C.Tobias	Brake
80135	218		Wesley Abbott *Whittaker* WSr	American	131	95·0	27·0	7·6	Ellsworth,Me.	1870	J.M.Grant		J.M.Higgins	Ells
24000	219	✳	West Australian *Thomas (76-76)* IBk	British	600	160·0	28·5	17·0	Hartlepool	1859	Pile,Spence&Co		Williamson,Milligan & Co.	Liverp
73435	220	✳	West Australian *Broom [-89* (ex G.B.S.) IBk	British	572	164·7	29·3	17·1	Sunderland	1876	J.L.Thompson		Ship "West Australian" Co. Ld. (C. Bethell & Co.)	London
74538	221	✳	West Glen *Urquhart 87-87* IBk	British	700	188·4	31·1	18·6	Glasgow	1876	Dobie&Co.		L.Hodgson	Liverp
86721	222	✳D	West Lothian *Falconer 83-88* IS	British	1814	279·5	40·2	23·6	Glasgow	1882	C.Connell&Co.		J.Boyd	Glasg
	223		West Nord West (DISMANTLED) ICr	German	103	81·2	39·5	9·6	Renfrew	1866	Henderson, [Coulborn&Co.		Hamburg-Amer.Packet Akt. Ges.	Hamb
70824	224	✳	West Riding *Thomas 78-* IBk	British	913	203·9	33·6	19·9	Glasgow	1875	Dobie&Co.		T.Thompson	Liverp
80215	225		West Side *Gaunon* WSr	American	148	103·1	28·8	7·3	Pt.Jeffrsn.NY	1871			W.Gaunnon	Glouc
74496	226	✳	West York *Irving 83-* IBk	British	689	179·5	29·9	19·1	Sunderland	1876	W.Doxford&Ss.		T.Thompson	Liverp
	227		Westa *Bonsss 79-* WBkn	Russian	187	97·6	23·4	12·9	Kuolemojarwi	1875	S.Koli		P.Halonen	Kuole
92451	228	V	Westaway *McDonald -87* WBn	British	276	111·0	29·0	12·4	Sturgeon,PEI	1886	R.Westaway	V 12.88	R.Westaway	Charl
56787	229	✳	Westbury *Poole 78-78* IBk	British	493	151·2	29·1	17·0	Sunderland	1867	W.Pile&Co.		T.B.Walker	London
	230	N	Westerbotten *Andreasen 87-87* WBk	Norweg'n	376	117·1	30·6	16·2	Umeå	1852			L.Hansen	Christ
68327	231	✳	Western Belle *Kerr* WBg	British	229	117·2	24·9	12·6	Bideford	1874	J.Johnson	V 1.87	Mackay & Co.	Leith
80577	232	RV	Western Belle *Batman* WBk	American	1084	183·3	38·1	22·5	Bath,Me.	1876	Goss&Sawyer	V 12.86	W.H.Besse	Portland
	233	V	Western Chief *Sander* WBk	German	744	172·9	32·2	19·6	Plymouth	1865	Hill	V 12.86	H.Vogemann	Hamb
80421	234		Western Home *Nelson* WSr	American	128	96·5	29·0	7·9	MainePrairie	1874			B.Steffens	San F
86027	235	✳	Western Lass *Williams -88* WSr	British	100	90·3	21·1	11·1	Plymouth	1885	W.H.Shilston	[illegible]	W.H.Shilston	Plymo
40800	236		Western Maid *Corkhill* WSr	British	137	91·0	22·0	12·2	Peel	1870	Graves	[illegible]	H.Graves & Co.	Peel I.M.
73384	237	✳	Western Monarch *Thomas 88-88* IS	British	1315	237·0	37·2	22·2	Barrow	1875	BarrowS.B.Co.		R.Thomas & Co.	London
45898	238		Western Ocean *Berwick* (ex Zaluesta) WBk	British	1231	182·5	38·2	23·9	Bath,Me.	1856			F. de Anastasio y Balbastena	South
84951	239		Western Star WBn	British	124	99·0	23·0	8·8	Riverton, N.Z.	1885	J.Beer		B.Corrigan	Sydney
60774	240	✳D	Westfa *Gore 86-86* WBn	British	142	103·0	22·5	11·4	Llanelly	1876	Rees	V 10.87	D.Paton	Llanelly
91188	241	✳	Westgate *Watkins 86-85* IS	British	1874	259·0	40·4	24·2	Liverpool	1885	R.&J.Evans		Gracie,Beazley&Co.	Liverp
80423	242	✳	Westland *Scotland 77-87* IS	British	1116	222·8	35·1	21·0	PortGlasgow	1878	R.Duncan&Co.		Shaw,Savill & AlbionCo. Ld.	Glasg
87868	243		Westmoreland CFlat	British	136	88·5	20·5	9·2	Northwich	1881	J.Thompson		Salt Union,Ld.	Liverp
	244	N	Westphalia *Kjelsen -88* WBk	Norweg'n	421	115·6	30·6	17·9	Bremen	1841			R.Jacobsen	Fred
55445	245	✳	Westward *Madge -88* WSr	British	87	82·7	20·2	9·6	Bowling	1867	Scott		W.G.Crinks	Plymouth
90315	246	✳	Westward Ho *Hastings 187-85* IBk	British	1219	225·5	34·9	20·9	Sunderland	1884	J.Laing		R.H.Gayner	Sun
29166	247		Westwood *Ruddle* (DISMANTLED) WBk	British	333	121·0	26·5	16·0	Gloucester	1861	Pickersgill	V 8.88	MedlockShippingCo.Ld.	Liverp
26503	248		Weybosset *Towne 86-88* W4mSr	American	598	167·0	31·0	16·0	Mystic,Conn.	1863 [72			S.R.Crowell	Boston
26730	249		Wheatland *Cochran -87* WBk	American	343	131·0	29·0	14·0	Baltimore,Md.	1856		V	G.Ropes	Boston
51332	250		Whim *Mathews 79-82* WSr	British	135	96·5	21·5	12·3	Brixham	1865	Dewdney		W.J.M.Hawkey	Brixham

✳ = Lloyd's Register; D = Nederlandsche Vereeniging; G = Germanischer Lloyd; I = Registro Italiano; R = American
A = Veritas Austro-Ungarico; H = Veritas Ellenico; N = Norske Veritas; V = Bureau Veritas

In the 4th Column, C = Composite (iron frame, wood planked); I = Iron; Stl = Steel; W = Wood.

The Dates of Metalling (**Col. 11**) are based on the records of the Books in which Vessels are respectively classed, or on the Reports of the Surveyors of Lloyd's

Further particulars of Vessels classed by Lloyd's Register will be found in Lloyd's Register Book and Lloyd's Yacht Register.

1 N° dans le livre.	2-3 Registres dans lesquels le navire est coté.	4 Nom. Matériaux et Gréement. Capitaine et dates d'entrée au service des Armateurs actuels et à bord du navire. Ancien Nom.	5 Pavillon	6 Tonnage net.	7 Dimensions. Longueur	Largeur	Creux	8 Construit. Où.	9 Quand.	10 Par qui.	11 Date du doublage en métal.	12 Armateurs ou Directeurs-Gérants.	13 Port d'Armement.
251		White Cloud *Jenkins* (ex Sétimana B. ...) *(81-84* WBk	American	557	118·6	32·8	16·0	East Boston	1874	A.Gove		N.B.Mansfield	Boston, Mass.
252		White Oak WSr	British	180	111·0	24·8	9·5	Oakville, Ont.	1867	D.Chisholm		Joseph Dix	Kingston, Ont.
253	NV	White Rose *Tofte 86-86* WS	Norweg'n	1513	212·0	39·9	24·1	Quebec	1874	Dunn	V 11.87	C.Zernichow	Christiania
254	R	White Sea *Lee* W3mSr	American	169	102·0	27·0	9·1	Waldoboro, Me.	1862			J.Grothgar	Galveston, Tex.
255		Wick Lassies *Lyon* WSr	British	118	86·7	22·4	11·3	Inverness	1865	Stewart		T.Brown	Wick
256		Widal *Sjölund* WBk	Russian	413	..	..	..	England				B.Sjölund	Finström
257	R	Wide Awake *Wingfield* WSr	American	112	86·5	26·0	8·6	Carleton, N.B.	1882	J.Olive		Snow & Pearsons	Rockland, Me.
258		Widgeon *Gair 77-88* WSw	British	229	101·3	25·9	15·1	Sunderland	1863	Gray&Young		H.Andrews	North Shields
259	V	Widja *Pehlsson* (ex P. ... Rickmers *(72-81)* WBk	Swedish	678	147·6	28·8	19·7	Bremerhaven	1868	R.C.Rickmers	V [illegible]	N.M.Pahlsson	Helsingborg
260	G	Wieland *Klocking 83-* WBk	German	603	144·7	31·4	17·6	Vegesack	1862	J.Lange	V [illegible]	E.Burchard	Rostock
261		Wifredo WBg	Spanish	191	..	27·6	13·5	Masnou	1849			J.Ferrer y Oliver	Barcelona
262		Wigwam (ABANDONED) WSr	American	118	86·2	25·5	7·5	Cherryfi'ld Me.	1871			W.M.Nash	Machias, Me.
263		Wik *Andersson -72* WSr	Swedish	172	88·9	22·1	12·0	Strömstad	1871			C.Andersson	Strömstad
264	✱	WildDaisy *Hitchins 86-88* WSr	British	100	91·0	19·9	10·1	Southampton	1888	J.Dible & Sons		J.Dible & Sons	Southampton
265		Wild Dayrell WBk		300	121·9	27·0	14·3	Shoreham	1856	Balley			
266		Wild Fire *Morrissey* WSr	American	103	84·9	23·5	8·8	Danversport, Mass.	1866			G.Steele	Gloucester, Mass.
267	✱	Wild Flower *Clatworthy* WK	British	89	80·1	21·3	10·6	Brixham	1885	S.Dewdney&Ss.		F.D.Barrett	Hull
268		Wild Goose WBarge	British	130	105·0	22·5	7·0	Quebec	1871			Jean B.Blanchard	Quebec
269		Wild Hunter *Maher* WSr	British	104	79·1	22·3	9·7	Parrsboro, N.S.	1864			E.Y.Houghton	Youghal
270		Wild Pigeon *Paine* WSr	American	368	128·6	30·8	15·6	Mystic, Conn.	1854			S.N.Staples	Fall River, Mass.
271	✱N	Wild Rose *Price -87* WBg	British	263	102·2	25·9	15·7	Sunderland	1863	Gardner	V 7.87	W.Jones	Carnarvon
272	✱	Wild Rose *Abbott 86-88* WSr	British	130	95·3	21·4	11·8	Plymouth	1888	C.Gent		C.Gent	Plymouth
273	✱	Wild Rose *Sharp* WSk	British	77	74·6	20·7	10·7	Hamburg	1886	C.C.A.Dreyer		E.Butler	Grimsby
274		Wild Wave *McArthur* WBk	British	237	119·1	27·1	12·8	Liverpool	1875	Croft	V 3.81	Facy & Fisher	Hobart
275		Wild Wave WBg	British	173	111·2	22·0	11·0	Guernsey	1854	H.Marquand		J.E.Ellis	Sydney, N.S.W.
276		Wild Wave *Hannah* WBn	British	160	100·2	22·6	11·4	P.E.Island	1861	Cox		J.H.Hocken	Fowey
277	DV	Wildina Christina (ex Pitbet) *Kramer -88* WSr	Dutch	135	85·0	22·1	10·4	Hoogezand	1860	J.A.Hooites	Z 8.88	G.Kramer	Groningen
278	V	Wildwood *Saunders 69-87* WS	British	1549	222·3	40·1	24·0	St.John, N.B.	1883	J.K.Dunlop	V 3.86	Wm.Thomson&Co.	St.John, N.B.
279	RV	Wildwood *Sawyer 76* WBk	American	1056	198·4	40·0	20·8	PortTownsend, Wash.	1871	Meigs&Gawley	V 1.86	C.L.Taylor & Co.	San Francisco
280		Wiley Smith WBn	British	129	82·9	23·2	9·7	RverPhilipNS	1866			Thomas Donnison	Pugwash, N.S.
281		Wilhelm (ex Little Willie, ex Luisa Canevaro, ex North America) *Ballestaer -83* WS	Chilian	1479	194·2	38·0	24·9	Boston, Mass.	1851			G.R.Gepp	Valparaiso
282	D	Wilhelm *Stenger -88* (ex Vilhelm, ex Wilhelm) WBg	Dutch	208	88·2	22·1	14·0	Rostock	1840	F.Dethloff		Joh.J.Stenger	Farmsum
283	✱DG	Wilhelm *Wijrdeman 85.* WSr	Dutch	114	81·0	21·4	9·8	Stettin	1877	A.Sieg	Z 12.86	J.J.Sijdzes	Delfzyl
284	GV	Wilhelm *Wilmsen 77-79* (ex North Star) WS	German	1315	201·4	38·4	23·9	Bath, Me.	1871	W.Moses&Son	V 7.87	Boyes & Ruyter	Bremen
285	N	Wilhelm *Haage* WBk	Norweg'n	724	162·0	35·1	18·8	Fredrikshald	1875	J.Brevig	V 1.88	H.A.Thorbjörnsen	Tönsberg
286		Wilhelm *Grönroos* WBg	Russian	321	..	..	..	Dragsfjerd	1874			G.A.Fredriksson	Dragsfjerd
287	D	Wilhelm *Wegemann 76-76* W3mSr	Russian	242	109·2	27·3	12·4	Kürlas	1876	J.Ormann		Y.C.Schneider&Co.	Libau
288		Wilhelm *Sjölander* WBk	Russian	206	97·1	20·0	12·7	Raumo	18$^{59}_{73}$			G.W.Wahlroos	Raumo
289	DGV	Wilhelm Anton *Hedergott 82-82* WBk	German	999	178·4	34·2	21·7	Bremerhaven	1877	C.Lange	V 10.87	W.A.Riedemann	Geestemünde
290		Wilhelm Foss *Olsen* (COLLISION) WBk	Norweg'n	365	119·1	29·4	16·2	Lillesand	1870	H.Jacobsen	V 3.86	P.Josephsen	Arendal
291	G	Wilhelm Linck *Utech* WS	German	668	129·8	33·5	22·4	Danzig	1877	J.W.Klawitter		C.Utech	Danzig
292	N	Wilhelm Ludvig *Sundbye* W3mSr	Norweg'n	381	123·4	26·8	16·6	Lübeck	1854		V 11.81	J.Sundbye	Fredriksstad
293	V	Wilhelm Maack *Freitag 77-78* WBk	German	421	131·2	29·0	17·3	Rostock	1878	J.Miller	V 1.88	W.Maack	Rostock
294		Wilhelm Ruthern WSr	Russian	112	83·0	24·4	9·8	Ruthern	1878	P.Kraus		M.Adamson	Riga
295	GV	Wilhelm & Joseph *Hille 69-* WBn	German	188	95·8	23·2	12·2	Papenburg	1866	F.W.Meyer		F.W.Meyer	Papenburg
296		Wilhelmina *Simmons -86* (ABANDONED) WBn	British	166	92·0	24·9	10·7	LunenburgNS	1878		V 9.86	William N.Zwicker	Lunenburg, N.S.
297	DGV	Wilhelmina *Tesman* (ex Fecunica) *-88* WBg	Dutch	318	116·0	26·5	14·2	Elsfleth	1866	F.Deetjen	V 11.85	N.Kranenborg	Groningen
298	✱D	Wilhelmina *Smit -87* (ex Suse Jacobs) WSr	Dutch	138	88·3	22·1	10·1	Martenshoek	1863	H.Nyhuis, jun.		J.J.Blokmeer	Delfzijl
299		Wilhelmina *Johansson 74-86* WBk	Russian	374	119·3	21·4	11·2	Kaskö	1874	A.Granskog		L.Olofsson	Wardö
300													

Lloyd's Register; D = Nederlandsche Vereeniging; G = Germanischer Lloyd; I = Registro Italiano; R = American Record; Veritas Austro-Ungarico; H = Veritas Ellenico; N = Norske Veritas; V = Bureau Veritas.

Dans la 4me Colonne, C = Composite (membrure en fer, bordé en bois); I = Fer; Stl = Acier; W = Bois.

Dates du doublage en métal (Col. 11) ressortent des livres dans lesquels les navires sont respectivement classés ou des rapports des Inspecteurs du Lloyd's Register.

Pour plus amples renseignements sur les navires classés par le Lloyd's Register, voir le Lloyd's Register Book et le Lloyd's Yacht Register.

Official No.	No. in Book.	Registry in which classed.	Ship's Name. Material and Rig. Master, and Dates of Appointment to present Owner's Service and to Ship. Ship's Late Name.	Flag.	Net Tonnage.	Dimensions. Length.	Breadth.	Depth.	Built. Where.	When.	By whom.	Date of Metalling.	Owners or Managers.	Port
1	2	3	4	5	6	7			8	9	10	11	12	
[illegible]	301		**Wilhelmina Agatha** W Bkn (STRANDED)	British	151	101·8	25·6	10·0	New York	1856			H. Craig & Co.	Belfas
[illegible]	302		**Wilhelmine** *Hewitt* W Bk	British	339	111·8	27·2	15·4	New York, U.S.	1856			Henry Thackray	Turks
	303		**Wilhelmine** *Gelves* W Bk	Chilian	309	105·9	27·2	16·4	Hobart	1848			J. McLean	Valpa
	304	G V	**Wilhelmine** *Thedens 80-83* I S	German	846	182·0	32·1	17·3	Bremen	1875	Act. Ges. Weser		Oetling Gebrüder	Hamb
	305	V	**Wilhelmine** *Schobi 82-80* W Bn	German	270	109·5	25·6	12·8	Motzen	1870	C. Oltmann	Y 12.88	W. Maack	Rostock
	306	V	**Wilhelmine** *Hasledt (87-87)* W3mSr	German	260	117·4	27·1	11·4	Hammelwarden	1883	C. Lühring	Y 1.88	U. C. Koch	Brake
	307	G	**Wilhelmine** *Pyhel 69-* W Sr	German	141	88·6	22·2	10·5	Papenburg	1862	H. B. Zurmeyer		L. Jongebloed Wwe.	Papen
	308	N	**Wilhelmine** *Olsen -88* W Bg	Norweg'n	232	102·0	28·8	13·5	Fredrikshavn	1848			T. Olsen	
	309		**Wilhelmine** *Kreusen -87* W Sr (ex Emilie) (WRECKED)	Norweg'n	179	81·5	24·0	13·4	Rostock				T. Torkildsen	Fevig
	310		**Wilhelmine** *Thorsson -88* W Bg	Swedish	256	97·7	24·3	15·4	Rostock	1856	W. Zeltz		L. H. Sjöström	Cimbri
	311	G V	**Wilhelmine Pust** *Beese 86-88* W Bk	German	441	123·8	30·0	17·9	Rostock	1873	J. Möller	Y 3.88	W. Maack	Rostock
	312		**Wilhelmine Waitz von Eschen** *Witt 87-87* W Bk	German	366	119·5	28·0	16·7	Rostock	1857	H. Rickmann		F. Schultze	Rostoc
80661	313	R V	**Will W. Case** *Johnson* W Bk	American	555	143·1	31·7	17·6	Rockland, Me.	1878	S. Starrat	Y 5.88	Arctic Packing Co.	San F
[illegible]	314	V	**Willard Mudgett** *Crocket* W Bk	American	832	160·6	34·0	21·0	Stockton, Me.	1874	Mudgett	Y 7.88	N. W. Bridge	New Y
26851	315		**Willard Saulsbury** *Pendleton* W Sr	American	161	105·0	27·0	8·0	Milton, Mass.	1855			S. F. Smith	Bango
	316	D	**Willem** *Meensma -87* W Bn (ex Vrouwe Hilden-rust en Balthus Pole)	Dutch	302	121·0	28·5	12·9	Ruthern	1873	C. Mühlberg		M. Meensma	Harlin
	317		**Willem** *Brahms* W Sr (CONDEMNED)	German	100	82·1	18·2	9·2	Emden	1867	H. Klattenhoff		H. O. Brahms	Bockum
91873	318	V	**Willem Barendts** *Dawit* W Sr	British	107	82·2	20·5	9·2	Amsterdam	1878			W. J. Adie	London
	319	✳ D V	**Willem Christiaan** *Jager 87-87* I Bk	Dutch	396	132·9	29·1	14·3	Middelburg	1887	Den Bouwmeester [Bosschaert & v.d. Loge]		Den Bouwmeester, Bosschaert & van der Loige	Middel
	320	✳ D	**Willem Eggerts** *Gerdk [86-8]* I Bk	Dutch	1276	230·0	36·0	21·0	Paisley	1885	H. M'Intyre & Co		N. Brantjes	Purme
84580	321		**Willemina** *Howes* W Bn	British	169	102·3	20·8	12·0					Bessey & Palmer	Yarmo
	322	V	**Willemina** *Middendorf [-86]* W Sr (ex Maria Sinnige)	German	155	86·9	22·3	11·4	Hoogezand	1865	J. U. vanderWerf		J. G. Ukena	Leer
	323	D V	**Willemina Cornelia** *de Jonge* W S (ex A. & W. C.)	Dutch	1269	188·0	38·1	25·1	Alblasserdam	1874	FransHarms [vonLindern]	Y 1.88	Von Lindern & Von Vissem	Rotte
196	324		**Willens** *Grüneisen -86* W Sr	Russian	134	73·2	23·2	10·4	Ekkendorf	1874			K. Grüneisen	Riga
	325	D	**William** (HULK) W Bk		713	162·0	33·2	20·0	Cornwallis, NS	1869	J. E. Woodworth	Y 6.88		
84314	326	✳	**William** *Foote 73-86* W Bg	British	198	114·5	24·2	12·9	Bridport	1872	Cox	Y 3.88	J. Munn & Co.	St. Joh
11372	327		**William** *Foote* W Bg	British	197	104·5	23·9	13·6	Jersey	1856			F. W. Picot	Jersey
6512	328		**William** *Mobbs* W Bn	British	138	85·2	20·5	12·0	Shaldon	1828			Bessey & Palmer	Yarmo
38581	329		**William** W Bn	British	123	83·2	22·1	11·6	P. E. Island	1855			P. Kehoe	Waterf
8868	330		**William** *Fogan 84-84* W Bn	British	116	77·0	19·0	12·8	P. E. Island	1846			A. Murdoch	Drogh
	331		**William** (ex Victory) *Larsen* W Bg	Danish	129	81·4	20·7	10·3	Wiken	1808	O. Pehrsson		P. Larsen	Copenh
	332		**William** *Meltz* W Bn	Danish	126	89·6	21·8	10·5	Nyborg	1868	P. Petersen		L. D. Meltz	Nyborg
	333	G	**William** *Barske 80-80* W Bk	German	246	104·6	27·3	13·0	Ueckermünde	1857	L. Wittenberg		G. Krasemann	Stettin
	334		**William** W Sr	Norweg'n	102	84·8	20·3	10·3	Stavanger	1873			Interessentskab "Spidsbergen"	Christ
	335		**William** *Heinberg 62-62* W Sr	Russian	120	88·0	22·0	10·0	Windau	1862	C. Fullhaas		W. Heinberg	Windau
	336		**William** *Bank* W Bk	Swedish	350	120·0	24·5	16·7	Sunderland	1869	R. Pace		N. M. Bank	Arilb
	337	R	**William** *Berndtsen* W Sr (ex William Forest)	Swedish	316	103·6	23·7	17·2	Sunderland	1856		Y 7.88	A. Berntsson	Ma
26045	338		**William A. Campbell** *Trask -88* W S	American	1494	206·0	41·3	23·9	ThomastonMe	1867	E. O'Brien		E. E. O'Brien	Thoma
26081	339		**William A Crocker** *Hutchins 83* W Sr (LOST)	American	187	113·0	29·0	8·0	Dighton, Mass.	1857			G. Thacker & Co.	Bosto
26095	340		**William A. Grozier** *Roberts* W Sr	American	111	89·0	23·9	10·9	KennebunkMe	1865			A. Marvel	Provi
80580	341	R	**William A. Marburg** *Pillsbury* W3mSr	American	678	163·4	33·7	19·0	Baltimore	1883	Woodall & Co.		B. A. Pillsbury	Baltim
26992	342		**William A. Morrell** *Randall* W Sr	American	104	84·6	27·0	7·2	ElizabthprtNJ	1868			E. E. Rackett	New Y
80252	343		**William A. Platinius** W Bk	American	615	144·0	30·0	18·0	Providence, RI	1865			W. M. Peck	New Y
70483	344	✳	**William Ashburner** *Charnley 72-76* W Sr	British	194	115·1	25·1	12·8	Barrow	1876	Ashburner	Y 1.88	T. Ashburner & Co.	Barrow
80446	345		**William B. Herrick** *Chase 86-* W3mSr	American	475	116·6	34·0	19·0	Nwbrypt Mas	1874	G. E. Currier		C. & J. F. Baker	Bosto
80708	346		**William B. Steelman** *English* W3mSr	American	419	130·6	32·7	17·7	Abscom, N.J.	1875			R. Chamberlain	Philad
80746	347	R	**William B. Wood** *Davidson 83* W3mSr	American	569	140·0	35·0	16·0	Camden, N.J.	1880	Morris & Mathis		J. B. Davidson	Bridge
	348	G	**William Bateman** *Knüther -* W Bk	German	504	129·0	30·7	19·3	Danzig	1873	J. W. Klawitter		Ch. Kriegel	Barth
81129	349	R	**William Baylies** *Montague 86-86* W Bk	American	309	118·5	28·0	16·5	Bath, Me.	1886	NewEng. S B Co.	Y 9.88	William Lewis	New B
80845	350	R	**William Beazley** *Farrow -88* W3mSr	American	227	113·7	29·3	9·9	Bucksport Me.	1881	W. Beazley	Y 11.88	J. Cassidy	Bango

✳ = **Lloyd's Register**; D = Nederlandsche Vereeniging; G = Germanischer Lloyd; I = Registro Italiano; R = American R…
A = Veritas Austro-Ungarico; H = Veritas Ellenico; N = Norske Veritas; V = Bureau Ver…

In the 4th Column, C = Composite (iron frame, wood planked); I = Iron; Stl = Steel; W = Wood.

The Dates of Metalling (**Col. 11**) are based on the records of the Books in which Vessels are respectively classed, or on the Reports of the Surveyors of Lloyd's R…

Further particulars of Vessels classed by Lloyd's Register will be found in Lloyd's Register Book and Lloyd's Yacht Register.

No.	Cote	Nom. Matériaux et Gréement. Capitaines et dates d'entrée au service des Armateurs actuels et à bord du navire. Ancien Nom.	Pavillon	Tonnage net	Longueur	Largeur	Creux	Construit: Où	Quand	Par qui	Date du doublage en métal	Armateurs ou Directeurs-Gérants	Port d'Armement
351		William Bell *Thompson* W Bn (WRECKED)	British	131	98·0	21·6	10·0	P.E.Island	1856			R.Dunn	Belfast
352		William Bement *Smith* W Sr	American	177	105·1	28·2	8·4	Madison, Conn	1855			N.Koplan	Greenport, N.Y.
353		William Boardman *Buell* W Sr	American	132	108·0	27·6	7·5	Essex, Conn.	1864	N.Starkey		S.J.Buell	Hartford, Conn.
354		William Butman *Sprout* W Sr	American	104	87·2	24·1	8·6	Searsport, Me.	1861	J.H.Lane		J.C.Nickels	Searsport, Me.
355	✻	William Butt *Baker* W Trawler	British	88	79·0	20·8	10·4	Galmpton	1884	W.A.Gibbs		W.Butt	Grimsby
356	R	William C. Bee W3mSr	American	332	119·7	29·0	14·3	Noank, Conn.	1870	R.Palmer		E.Rathbun	Stonington, Conn
357	R	William C. French W3mSr *Sherman*	American	388	142·4	33·7	11·9	Nwbyprt Mas	1883	Atknsn & Fllmre		J.Walter	Boston, Mass.
358	R	William C. Green W3mSr	American	343	130·0	32·6	11·3	Bath, Me.	1882	Deering & Donall		S.C.Hart	New Bedfrd, Mass
359	●	William Cobb *Chase* W3mSr	American	493	128·4	32·0	15·5	Dighton, Mass	1874			O.T.Chase	Dennis, Mass.
360	DV	William Cochran *Bernier* W Bk	British	1154	192·8	38·1	22·5	Dorchester N B	1879	W.Hickman	Y 3.88	William Hickman	Dorchester, N.B.
361		William Crow W Sr	British	109	73·6	20·5	12·0	[illegible]	1841			Bessey & Palmer	Yarmouth
362		William Cundall W Sr	British	244	113·0	25·2	13·9	P.E.Island	1863	Clark		H. & E.Crundall	Dover
363		William D. W Sr	British	144	87·5	26·1	10·3	Little Bras [d'Or, C.B.	1886	S.Deveaux		S.Deveaux	Sydney, C.B.
364		William D.Cargill *Fogg* W Sr	American	133	94·8	26·3	8·2	Brooklyn, N.Y.	1854			S.G.Haskell	Deer Isle, Me.
365		William D. Hilton W3mSr *Harner -87*	American	309	124·5	32·2	11·1	Bath, Me.	1871	Goss & Sawyer		J.H.Cady	Providence, R.I.
366		William D. Marvel W3mSr	American	432	139·1	33·7	14·0	Wilmingtn Del	1872			D.O'Keefe	Taunton, Mass.
367		William D. S. Hyer W Sr	British	100	71·0	24·1	6·0	Barnegat, U.S.	1862			George P. Bell	Nassau, N.P.
368	✻	William Davie *Bowen 82-85* C Bk	British	795	187·0	31·6	19·3	Glasgow	1866	A Stephen & Sons	[illegible]	J.C.&G.S.Richardson	Swansea
369	R	William Deming W Sr	American	170	104·0	27·6	8·1	Calais, Me.	1869	W.Robinson		H.B.Eaton	Calais, Me.
370	DV	William Douglas W S *McDougall*	British	1263	201·4	39·6	23·1	Maitland, N.S.	1875	A.Roy	[illegible]	William J.Stairs	Maitland, N.S.
371		William Douglas *McIndoe* W Sr	American	126	97·5	25·3	8·1	City Island NY	1874			S.McIndoe	New York
372	✻	William Duthie W Bk	British	954	183·0	34·5	21·0	Aberdeen	1862	Duthie	[illegible]	H.W.French	Aberdeen
373		William Dyer W Bn	British	186	97·9	23·6	13·3	Sunderland	1865	Lumsden		C.W.Lovell	Weymouth
374		William E. Barnes W Sr *Perkins*	American	116	81·0	24·0	7·0	Orland, Me.	1867			W.Barbour, junr.	Bucksport, Me.
375		William E. Clowes W3mSr *Penney -87*	American	542	130·0	33·0	13·0	Pt Jeffersn NY	1875			A.E.Penny	Pt.J'ff'rson, N.Y
376	R	William E. Downes W3mSr *Haskell -82*	American	716	164·0	37·3	18·0	Bath, Me.	1885	Geo.Hawley		E.H.Weaver	Newhaven, Con.
377	R	William E. Lee *Lynch* W3mSr	American	476	138·0	34·0	12·0	Dennisville NJ	1877	G.P.Wentzel		L.H.Babcock	Philadelphia
378	✻V	William Engels *Hansen [81-]* I Bk	Belgian	742	178·4	31·3	18·5	Newcastle	1876	C.Mitchell & Co.		T.C.Engels & Co.	Antwerp
379		William F.Burden *Adams* W Sr	American	123	95·7	26·6	7·5	Norwalk, Conn	1858			L.L.Adams	Providence, R.I.
380	R	William F. Collins W3mSr *Higgins*	American	289	128·2	30·3	9·1	Bath, Me.	1882	Goss & Sawyer		D.Higgins	Boston, Mass.
381	R	William F. Greene W3mSr (ex John S. Taylor) *Crockett*	American	254	116·0	28·5	12·0	Ellsworth, Me.	1879	J.H.Grant		Wm.F.Greene & Son	Boston, Mass.
382		William F. Phelps W Sr	American	224	112·8	30·0	9·0	Tuckerton, NJ	1863			A.R.Pharo	Tuckerton, N.J.
383		William F. Witzemann W3mSr *Johnson -87*	American	450	160·0	35·0	12·2	Fairhaven, Cal	1887	Benderson		J.Madison	San Francisco
384		William Fisher *Young* W Bk	British	300	104·4	26·3	18·1	Liverpool	1844		[illegible]	W.R.Kelly	Whitehaven
385		William Flint *Dodge* W Sr	American	238	108·3	28·4	9·9	Waldoboro Me	1863			J.H.Crosby	Bangor, Me.
386	✻	William Frederick *Jones* W Sr	British	91	78·0	19·8	10·5	Conway	1867	Roberts		W.Roberts & Co.	Beaumaris
387		William Frederick W3mSr *Burgess -87*	American	430	131·9	32·3	15·0	Belfast, Me.	1874		[illegible]	J.W.Frederick	Belfast, Me.
388		William G. Bartlett W3mSr (WRECKED) *Darby*	American	208	115·7	28·6	9·0	May's Ladg NJ	1858			C.W.Visel	New York
389	RV	William G. Davis W S *Morse -83*	American	1669	213·7	41·7	25·4	E.Deering, Me.	1877	G.Russell	[illegible]	J.H.Winchester & Co.	Portland, Me.
390	V	William G. Irwin W Bn *McCulloch*	American	348	135·0	32·8	13·0	San Francisco	1881	Turner & Rundll		Oceanic S.S.Co.	San Francisco
391		William G. Lewis W3mSr *Kelley*	American	398	140·0	33·0	15·0	Nwbrypt Mas	1874	G.E.Currier		Doane & Crowell	Boston, Mass.
392		William G. R. Mowry W Sr	American	151	99·4	27·3	8·3	Calais, Me.	1870	O.B.Rideout		B.L.Eaton	Calais, Me.
393	✻	William Geake *Hocken* W Bkn	British	277	127·5	25·8	13·8	Bideford	1876	Cox	[illegible]	W.Geake & Co.	Fowey
394	✻	William George W3mSr *Garrett*	British	140	97·7	24·0	11·9	Rhyl	1876	Jones		S.Vickers	Chester
395		William Gillum *Clark* W Sr	American	137	98·0	27·0	7·0	Middletwn Cn.	1864			J.Carroll	Hartford
396		William Gilmore *Abram* W Sr	British	109	86·2	21·7	10·7	Troon	1871	Troon Shp Bg Co		E.Porter	Fleetwood
397	V	William Gordon W Bk	British	732	163·7	33·8	19·1	Buctouche, NB	1876	James McNairn	[illegible]	W.& H.McLeod	Chatham, N.B.
398	✻	William Graham W Bk *Atkinson*	British	276	121·0	24·9	14·8	Bideford	1866	Cox	[illegible]	R.Atkinson	Liverpool
399	R	William H. Allison W3mSr	American	453	141·2	33·2	12·6	Camden, Me.	1883	H.Bean		W.Moore	Richmond, Va.
400	R	William H. Bailey W3mSr	American	405	146·6	34·6	15·1	City Island NY	1878	D.Carll		G.Bailey	New York

✻=Lloyd's Register; D=Nederlandsche Vereeniging; G=Germanischer Lloyd; I=Registro Italiano; R=American Record; [illegible]=Veritas Austro-Ungarico; H=Veritas Ellenico; N=Norske Veritas; V=Bureau Veritas.

Dans la 4me Colonne, C=Composite (membrure en fer, bordé en bois); I=Fer; Stl=Acier; W=Bois.

Les Dates du doublage en métal (Col. 11) ressortent des livres dans lesquels les navires sont respectivement classés ou des rapports des Inspecteurs du Lloyd's Register.

Pour plus amples renseignements sur les navires classés par le Lloyd's Register, voir le Lloyd's Register Book et le Lloyd's Yacht Register.

Official No.	No. in Book	Register in which classed	Ship's Name. Material and Rig. Master, and Dates of Appointment to present Owner's Service and to Ship. Ship's Late Name.	Flag.	Net Tonnage	Length	Breadth	Depth	Built: Where.	When.	By whom.	Date of Metalling	Owners or Managers.	Port of Registry
[illegible]	401	RV	William H. Connor W S Sullivan -86	American	1496	210·0	40·0	24·0	Searsport, Me.	1877	M. Packard	[illegible]	B. F. Pendleton	Sear
[illegible]	402	R	William H. Deitz W Bkn Hooper	American	463	140·0	32·2	16·1	Philadelphia	1875		[illegible]	W. H. Deitz	Philad
[illegible]	403	R	William H. Fredson W 4m Sr (ex Vestal (ex W. H. Fredson)	American	622	140·0	31·6	19·0	Jacobstadt Fin	1886		[illegible]	P. H. Crowell	Boston
[illegible]	404	R	William H. Hopkins W 3m Sr Barrett	American	308	134·0	32·6	10·0	Mystic, Conn.	1876	Greenman		M. Townsend	Somers
[illegible]	405		William H. Jones W 3m Sr Schultz	American	283	110·0	29·0	14·0	Wilmingtn Del	1867			S. L. Falkenberg	New Yo
[illegible]	406	R	William H. Jourdan W 3m Sr (LAST) Thrasher 85-86	American	473	139·4	34·2	15·5	Bath, Me.	1873			N. D. Thrasher	Fall Riv
[illegible]	407		William H. Keeney W 3m Sr Smith -88	American	298	126·0	32·9	11·0	P. Jefferson NY	1873			J. R. Beers	New Yo
[illegible]	408	R	William H. Kenzel W 3m Sr Kinch	American	439	129·0	32·9	10·5	Staten Isld. NY	1865			John van Buskirk	New Yo
[illegible]	409	RV	William H. Macy W S Amesbury -86	American	2092	254·9	43·2	28·	Rockport, Me.	1883	Carleton, Norwood & Co.	[illegible]	Carleton, Norwood & Co.	Camden
[illegible]	410		William H. Mitchell W Sr (WRECKED) Cole	American	198	95·2	26·5	11·2	Harrington Me	18[illegible]			W. Mitchell	Machias
[illegible]	411		William H. Rowe Morgan W Sr (FOUNDERED)	American	149	92·0	25·0	8·0	Greenport, NY	1857			E. T. Morgan	New
[illegible]	412		William H. Rutan Dodd W Sr	American	120	93·6	26·0	7·9	Staten Isld. NY	1851			A. L. Burton	Wilming
[illegible]	413	R	William H. Shubert W 3m Sr Remy -86	American	611	146·7	35·3	18·0	Camden, N.J.	1884	Morris & Mathis		L. R. King	Philad
[illegible]	414	RV	William H. Smith Merritt W S	American	1908	252·4	43·3	26·2	Bath, Me.	1883	Goss, Sawyer & Packard	[illegible]	F. H. Smith & Co.	New Yo
[illegible]	415	RV	William H. Starbuck W S Reed	American	1272	194·1	39·0	24·3	Bath, Me.	1882	Goss, Sawyer & Packard	[illegible]	G. L. Caldwell	New Yo
[illegible]	416	R	William H. Stuart W 3m Sr (WRECKED) Sparks	American	532	158·3	35·8	13·1	Richmond, Me.	1884	G. H. Ferren		G. H. Theobald	Richmo
[illegible]	417	RV	William Hales Buckley W Bk -88	American	833	164·8	35·0	19·7	Newburypt Mas	1877	Atknsn & Filmre	[illegible]	J. H. Conant & Co.	New Yo
[illegible]	418	R	William Hays Smith W 3m Sr -88	American	382	139·0	33·0	12·0	Camden, Me.	1882	H. M. Bean	[illegible]	William Smith	New Yo
[illegible]	419		William Henry W Bg	British	314	112·0	25·0	15·5	M[illegible]	1854			William Burnard	Nassau
[illegible]	420		William Henry W	British	185	111·8	22·9	8·4	Sorel, P.Q.	1885	P. Dauphinais		James Williamson	Montreal
[illegible]	421		William Henry W Bg	American	303	110·0	25·8	11·0	Surry, Me.	1854			T. A. R. Webster	New Yo
	422		William Hunter W Sr Knudsen 87-87	Norweg'n	385	125·3	30·0	10·1	Deep Riv. Conn	1859			Thommesen & Son	Arendal
81078	423	R	William J. Lermond W 4m Sr Hagger	American	843	181·8	36·0	18·5	Thomaston Me	1885	S. Watts & Co.	[illegible]	S. Watts	Thomas
80817	424	RV	William J. Rotch Gibbs W S	American	1503	218·2	42·1	24·2	Bath, Me.	1881	Goss, Sawyer & Packard	[illegible] 1887	W. H. Besse	New Bed
75912	425		William Jamieson W Sr	British	144	100·0	25·4	8·6	Mill Point, Ont.	1878	H. B. Rathburn & Son		D. Galbraith	Whitby
[illegible]	426		William Jenkins Lewis W Sr	British	106	83·0	24·0	9·5	Pt. Gilbert, NS	1872			George Dickson	Yarmouth
[illegible]	427	✻	William Jones Owen W Sr	British	79	77·4	19·8	10·3	Portmadoc	1870	Griffiths		M. Owen (J. Williams)	Carnarv
[illegible]	428	R	William Jones Colling W 3m Sr	American	296	129·2	30·0	10·6	E. Boston, Mas	1872	H. M. Bean	[illegible]	W. C. Haskins & Sons	Boston
71060	429	V	William K. Chapman W Bk (ex ELLISON) Smith	British	1077	184·7	38·0	21·0	Dorchester NB	1878	Wm. Hickman	[illegible]	William Hickman	Dorches
27438	430	✻	William Keith Jones 73-80 W Sr	British	99	79·5	20·2	10·7	Pt. Dinorwic	1859	Jones		R. Jones & Co.	Carnarv
80227	431		William L. Abbott W 3m Sr Lodlow	American	233	115·0	29·8	9·5	Dennisville NJ	1871			L. Matthews	Philad
[illegible]	432		William L. Beebe W 3m Sr Henderson	American	281	133·8	33·3	9·8	P. Ludlow W.T.	1875	Hall Bros.		J. J. Smith	San Fra
[illegible]	433		William L. Burroughs W 3m Sr Cook	American	487	138·6	30·0	19·5	Greenport, NY	1855			O. P. Lovell	New Yo
[illegible]	434		William L. Elkins W 3m Sr Fowler -88	American	229	121·4	31·0	8·4	Dennisville NJ	1873			H. Matthews	Philad
[illegible]	435	R	William L. Roberts W 3m Sr Lacy	American	725	160·8	35·3	16·3	Bath, Me.	1881	Goss & Sawyer		L. Williams	Fall Riv
[illegible]	436	R	William L. Walker W 3m Sr Crocker	American	562	150·0	35·6	15·6	East Boston	1882	R. Crosbie & Son		G. H. Rhodes	Fall Riv
[illegible]	437		William Le Lacheur W S Veitch	British	573	165·8	30·5	17·7	Guernsey	1864	J. Sebire	[illegible]	Andrew McIntyre	Melbour
	438	V	William Leavitt W Bk (WRECKED) Grunt 87-87	Norweg'n	1184	179·9	37·6	23·8	St. John, N.B.	1863		[illegible]	Actieselskabet "William Leavitt" (H. C. [illegible])	Laurvig
[illegible]	439	R	William M. Bird Reed W 4m Sr	American	768	167·2	36·6	19·0	Camden, N.J.	1884	S. W. Tilton		P. R. Reed	Somers
[illegible]	440		William M. Wilson Bragg W Sr	American	240	114·0	30·4	10·2	Camden, N.J.	18[illegible] 75			A. R. Pharo	Tucker
[illegible]	441	V	William McGilvery W S Dunbar	American	1206	192·0	38·2	24·0	Searsport, Me.	1870	W. McGilvery	[illegible]	J. C. Nickels & Co.	Searsp
[illegible]	442	✻	William Manson Kendred W Bk	British	366	117·7	27·1	15·7	Aberdeen	1872	J. Duthie S. & Co	[illegible]	James Ewan	Sydney
[illegible]	443		William Marshall W 3m Sr [illegible]	American	291	113·4	31·0	11·0	Milford, Del.	1874	Abbott, Hrpr & C		M. Cain	Philad
[illegible]	444	✻	William Martin Allenby W Sk	British	76	74·8	20·2	10·3	Rye	1883	J. C. Hoad		J. T. Stroud	Grimsby
[illegible]	445	✻	William Martyn W Sr Scammell	British	93	87·1	22·2	10·4	Newquay	1878	Clemens		J. Crocker	Padstow
80263	446		William Mason W 3m Sr Chase -87	American	442	141·6	33·5	14·9	Wilmingtn Del	1872			W. Chase, junr.	Fall Riv
[illegible]	447		William Mason Hardy W Sr	American	284	110·6	27·1	13·6	Castine, Me.	1857			Scammell Bros.	Portland
[illegible]	448		William Matheson W Sr McDonald -86	American	106	88·3	23·8	9·0	Essex, Mass.	1875			A. McKay	Provinc
[illegible]	449		William Mogee W Sr	American	114	98·4	17·6	8·1	Hamburg, Pa	1869			I. Ramsey	Philad
[illegible]	450	R	William O. Snow W 3m Sr	American	560	152·9	35·2	15·2	Bath, Me.	1881	[illegible]		G. H. Rhodes	Fall Riv

✻ = Lloyd's Register; D = Nederlandsche Vereeniging; G = Germanischer Lloyd; I = Registro Italiano; R = American Rec[ord]
A = Veritas Austro-Ungarico; H = Veritas Ellenico; N = Norske Veritas; V = Bureau Verit[as]

In the 4th Column, C = Composite (iron frame, wood planked); I = Iron; Stl = Steel; W = Wood.

The Dates of Metalling (Col. 11) are based on the records of the Books in which Vessels are respectively classed, or on the Reports of the Surveyors of Lloyd's Re[gister]

Further particulars of Vessels classed by Lloyd's Register will be found in Lloyd's Register Book and Lloyd's Yacht Register.

1	2 Registres dans lesquels le navire est classé	4 Nom. Matériaux et Gréement. Capitaine et dates d'entrée en service des Armateurs actuels et à bord du navire. Ancien Nom.	5 Pavillon.	6 Tonnage net.	7 Dimensions. Longueur.	Largeur.	Creux.	8 Construit. Où.	9 Quand.	10 Par qui.	11 Dates du doublage	12 Armateurs ou Directeurs-Gérants.	13 Port d'Armement.
451	*	William Owen W Bk	British	644	160·0	31·7	17·9	P.E.Island	1874	J. McDougall	[illegible]	J. Atherton	Liverpool
452	R	William P. Hood W3mSr	American	633	158·4	36·4	16·9	Bath, Me.	1880	Goss, Sawyer & Packard		J.F. & C.A. Davis	Fall River, Mass.
453		William Parker W3mBt	British	120	72·4	21·6	11·4	Wells	1840			A. Beaumont & Co.	Ipswich
454	*	William Phillips W Sr	British	73	78·0	19·9	9·5	Plymouth	1874	Banks	[illegible]	C.W. Spreckley & G.A. Brenakley	Fowey
455	RV	William Phillips W Bkn	American	592	145·0	34·0	16·9	N. Bedford Mass.	1874	Delano	[illegible]	W. Phillips & Son	New Bedfrd, Mass.
456		William Philpot W Sr	British	123	81·0	23·7	11·3	Hawkesbry NS	1871			Jeremiah Philpot	Hawkesbury, N.S.
457	R	William R. Drury W3mSr	American	346	130·8	32·8	11·5	East Boston	1872			J.H. Conant & Co.	Boston, Mass.
458		William R. Huston W3mSr	American	521	160·0	36·0	16·0	Pembroke, Me.	1874			S.M. Staples	Fall River, Mass.
459		William Renton W3mSr	American	425	145·4	36·4	11·9	P. Blakely Wsh	1882	Hall Bros.		J.J. Smith	San Francisco
460	R	William Rice Gregory W Sr	American	126	81·2	24·5	7·5	Bath, Me.	1869			E.A. Butler	Rockland, Me.
461	*	William Ritson Nelson W Bk	British	680	164·0	30·7	20·6	Maryport	1864	Ritson	[illegible]	Hayton & Simpson	Maryport
462		William S. Doughten W Sr	American	137	101·8	27·7	8·0	Bridgept. Con.	1861			E. Place	New York
463		William S. Hansell W Bk	British	307	111·8	27·5	10·8	Rhode Island	1852			William F. Ogilvie	Turks Island
464	*	William Shepherd W3mSr	British	133	91·0	22·8	11·3	Glasson	1870	Simpson		Porter's Shipping Co., Ld.	Fleetwood
465		William Slater Small W Sr	American	211	111·0	27·0	9·0	Thomaston Me.	1866			F.C. Pendleton	North Haven, Me.
466		William Smith W Sr	British	240	102·5	24·3	15·3	Brookhavn US	1853			Peter A. Spearwater	Nassau, N.P.
467	N	William Stephenson W Bk	Norweg'n	749	168·2	32·6	20·2	Belfast, Me.	1851	McWhite & Conner		J. Schaanning	Porsgrund
468		William Stonard W Sr	British	100	80·0	20·0	11·0	Ulverston	1858	Schollock		J. Helme	Fleetwood
469	R	William T. Donnell W3mSr	American	511	149·4	34·2	14·6	Bath, Me.	1883	Deering & Donnell		Deering & Donnell	Bath, Me.
470	R	William Tapscott W S	American	1593	193·0	41·4	20·0	Bath, Me.	1852	Trufant & Drummond	[illegible]	J.H. Whitaker	Boston, Mass.
471		William Thomas W Bn	British	172	105·8	23·7	12·4	P.E.Island	1865	McKay		P. Owens	Drogheda
472		William Todd Wood W Sr	American	129	96·0	27·3	8·4	Calais, Me.	1873	Rideout & Lord		Rideout & Lord	Calais, Me.
473	* D	William Turner Berry W Bk	British	431	114·6	27·2	17·4	Workington	1862	Fell	[illegible]	J.E. Ellis	Sydney, N.S.W.
474	G	William Volkens W S	Swedish	879	155·4	29·4	22·7	Newburyport	1850		[illegible]	T. Pettersson	Wiken
475	R	William W. Converse W3mSr	American	709	168·8	37·8	18·0	Newhaven, Con.	1886	H.H. Hanscom		H.H. Hanscom	Newhaven, Conn.
476	RV	William W. Crapo W Bk	American	1573	215·3	41·8	24·3	Bath, Me.	1880	Goss, Sawyer & Packard	[illegible]	W.H. Besse	New Bedfrd, Mass.
477		William Warren W Sr	American	112	79·5	24·5	9·5	Granville, N.S.	1870			W. Warren	Machias, Me.
478	R	William Wilcox W Bk		881	160·0	32·5	20·5	New Bedford, Mass.	1863	W. Wilcox	[illegible]	James G. Ross	Bridgetown
479	R	William Wiler Miller W3mSr	American	236	115·0	30·0	9·6	South Milford, Del.	1870	Deputy		H.D. May	Philadelphia
480	*	William Wilson W Bk	British	400	156·0	25·9	15·0	Bristol	1863	Hyde		W. McGowan	Whitehaven
481		William Wilson W3mSr	American	253	114·3	29·9	11·6	Rockland, Me.	1869			H.N. Johnson	Newhaven, Conn.
482		William Woodbury W S	American	1154	180·6	36·4	24·6	Portland, Me.	1856		[illegible]	P. Gustavus	Baltimore
483	R	William Wright Coal W Bk	British	724	157·0	32·5	21·8	St. John, N.B.	1856		[illegible]	L. Cook	Liverpool
484	*	William Wright Wright W K	British	90	80·0	20·9	10·4	Galmpton	1885	W.A. Gibbs		J. Wright	Grimsby
485		William & Alice Bayes W Sr	British	100	82·4	20·0	10·3	Goole	1862	Pearson		W. Whitaker	Goole
486	V	William & Anna W Bk	Swedish	528	111·7	29·4	17·3	Carlscrona	1856		[illegible]	Rederi-Aktieb Ornen (J.E.S. Sanberg)	Gothenburg
487		William & Annie Evans W Bn	British	187	105·0	23·9	12·9	P.E.Island	1868	McMillan		W. Milburn	Llanelly
488		William & Anthony W Bg	British	189	96·5	22·5	14·0	Charlestown	1862	Lake		T.H. Franks	Folkestone
489	R	William & Richard W Sr	American	253	108·9	29·0	10·0	St. George, Me.	18[illegible]		[illegible]	Wm. F. Greene & Son	Boston, Mass.
490		William & Sarah W3mSr (WRECKED)	British	140	93·4	21·7	12·2	Garmouth	1865	Duncan		T. Long	Drogheda
491	R	Williamine W3mSr	American	394	129·0	31·2	14·7	East Boston	1869	Campbell & Brooks	[illegible]	J.H. Conant & Co.	Boston, Mass.
492	R	Williams C. Wickham W3mSr	American	313	131·6	32·6	10·7	Milford, Del.	1880	T. Carlisle		Z.S. Steelman	Philadelphia
493	*	Willie Haffey W Bn	British	366	126·0	29·8	12·9	Sackville, N.B.	1874	Milner	[illegible]	William Milner & Co.	Sackville, N.B.
494		Willie W Bg	British	191	110·0	22·5	13·1	Kingsbridge	1860	Date	[illegible]	W.J. Lawson	Dartmouth
495		Willie W Bn	British	185	104·4	21·7	12·5	P.E.Island	1874	Burdett		Holden & Gann	Faversham
496	*	Willie George W Sr	British	83	80·9	19·9	10·4	Galmpton	1865	Gibbs	[illegible]	J.P. Dunstan	Dartmouth
497		Willie Gottzeit W Sr	German	275	123·6	25·3	11·5	Vegesack	1861	J. Lange		G.C.L. Blecher	Hamburg
498	V	Willie Larsson W Bn	Swedish	236	115·9	23·3	11·8	Oscarshamn	1888	C. Thorén		A.J. Thollander	Oscarshamn
499		Willie A. McKay W Sr	American	161	102·0	25·8	9·9	Essex, Mass.	1880			A. McKay	Provincetn, Mass.
500	R	Willie H. Child Giles W3mSr	American	595	175·5	36·4	13·1	Thomaston, Me.	1888	Dunn & Elliot		Dunn & Elliot	Thomaston, Me.

Lloyd's Register; D = Nederlandsche Vereeniging; G = Germanischer Lloyd; I = Registro Italiano; R = American Record; [illegible] Austro-Ungarico; H = Veritas Ellenico; N = Norske Veritas; V = Bureau Veritas.

Dans la 4me Colonne, C = Composite (membrure en fer, bordé en bois); I = Fer; Stl = Acier; W = Bois.

Les dates du doublage en métal (Col. 11) ressortent des livres dans lesquels les navires sont respectivement classés ou des rapports des Inspecteurs du Lloyd's Register.

Pour plus amples renseignements sur les navires classés par le Lloyd's Register, voir le Lloyd's Register Book et le Lloyd's Yacht Register.

1 Official No.	2 No. in Book.	3 Registry in which Classed.	4 Ship's Name. Material and Rig. Master, and Dates of Appointment to present Owner's Service and to Ship. Ship's Late Name.	5 Flag.	6 Net Tonnage.	7 Dimensions. Length.	Breadth.	Depth.	8 Built. Where.	9 When.	10 By whom.	11 Date of Metalling.	12 Owners or Managers.	Port…
80022	501	R	Willie H. Higgins W3mSr	American	563	118·0	35·1	14·3	Yarmouth, Me.	1882	G.Loring		Chas.F.Higgins	Bosto…
80840	502	R	Willie L. Newton W3mSr Coombs	American	366	142·0	32·4	12·0	Bangor, Me.	1881	Crosby Bros.		Crosby Bros.	Bang…
80230	503		Willie Luce WSr	American	126	92·5	24·1	8·4	Camden, Me.	1871			Carleton, Norwood & Co.	Camd…
85497	504		Willie McGowan WSr Goodrich -84	British	116	88·0	23·8	9·6	Shelburne, N.S.	1883	J.A. McGowan, [junr.		John A. McGowan, jr.	Shelb…
71462	505	✻	Willie McLaren Lowe WBk	British	536	145·4	31·6	17·0	P.E.Island	1874	J.LeBrocq	V 1.84	David Williams	New…
80781	506		Willie Martin WSr	American	144	95·3	28·3	7·8	Yarmouth, Me.	1866			J.E.Stafford	Portl…
80630	507	V	Willie Reed Yates WS	American	1380	210·0	39·6	24·0	WaldoboroMe	1877	A.R.Reed	V 8.87	Yates & Porterfield	New Y…
81087	508	RV	Willie Rosenfeld WS Dunphy	American	2353	266·5	44·8	27·8	Bath, Me.	1885	Sewall&Co.	V 8.87	A.Sewall & Co.	Ba…
	509		Willina D. Deveaux 85-86 WSr	British	144	87·5	26·1	10·3	Little Bras [D'Or, C.B.	1886	J.P. Deveaux		S.Deveaux	Sy…
33360	510	✻	Willing Clark WSr	British	100	81·3	19·7	10·3	Jersey	1868	Bellot	V 7.86	G.Noel	Jersey
80389	511	V	Willis S. Shepard W3mSr Reeves 71	American	452	141·5	32·4	15·1	Bath, Me.	1874	Deering & Donnell		T.L.Mayo	Bost…
26947	512		Willow Harp Norton WSr	American	164	95·5	26·7	8·0	PtJefferson, NY	1858			C.E.Norton	Pt.Jeff…
29841	513	✻D	Willowbank Wilson -88 I Bk (ex Airlie...)	British	811	190·0	32·0	20·0	Newcastle	1861	Wigham Richardson & Co.		A.Weir & Co.	Glasgo…
	514	V	Willy Rickmers WBk Bruhns 8…-8	German	802	157·6	36·5	21·4	Bremerhaven	1868	R.C.Rickmers	V 11.84	R.C.Rickmers	Brem…
80708	515	V	Wilna Slater 83-83 WS	American	1483	203·0	41·9	24·0	Freeport, Me.	1880	Briggs&Cushing	V 4.86	W.E.Mighell	Searsp…
9314	516		Wilson WBg	British	137	79·6	22·2	14·0	Whitehaven	1840	Kennedy		T.J.Troy	White…
80076	517	R	Wilson & Hunting W3mSr Potter	American	398	145·0	35·0	11·0	Alexandria Va	1883	Goss&Ward		G.W.Anderson	New…
60909	518	✻	Wiltshire Balfour 87-87 I S	British	1461	239·4	37·7	23·0	Glasgow	1869	Barclay, Curle [& Co.		G.Marshall & Sons	Londo…
	519	N	Wimburn Johansen 86-86 WBk	Norweg'n	489	138·6	30·0	16·7	N.Glasgow NS	1864	J.W.Carmichael		Act.Selsk. "Wimburn" (H.Larsen)	Kr…
	520	NR	Wimmera Pettersen -86 WBk	Norweg'n	984	205·7	35·9	19·6	Aberdeen	1866	Smith	V 5.88	J.Prebensen	Risör
	521		Winchoten II. WSr	Brazilian	137	..	..	..					D.Dwayer	Bahia
56087	522	✻	Windermere Bruce C Bk	British	482	144·9	28·3	17·6	Sunderland	1868	R.Thompson, jun	V 5.86	W.M.Deacon & Co.	New…
60396	523	✻	Windhover Planck 85-85 C Bk	British	847	201·1	34·0	19·8	Glasgow	1868	C.Connell&Co.	V 3.86	J.Kerr & Co.	Gree…
	524	N	Windsor Walloe -87 WBk	Norweg'n	301	124·3	28·8	13·0	Newport, N.S.	1866			A.N.Walloe	Chris…
70899	525	✻	Windsor Castle James I Bk	British	613	174·7	29·1	18·4	PortGlasgow	1875	R.Murray&Co.		Jacobs Bros. & Co.	Londo…
88163	526	✻	Windsor Castle WSk Maunder 85-85	British	78	75·6	21·2	10·4	Hull	1884	Hunt&Fowler		J.Edwards	Hull
85132	527	✻	WindsorPark Bentley 82-82 I S	British	1761	250·5	40·2	24·0	Liverpool	1882	T.Royden&Sons		"Windsor Park" Ship.Co. (John A. Hemden & Co.)	Londo…
24168	528	✻	Winefred Ellis 82-82 I S	British	1359	219·0	35·9	21·9	Chester	1855	G.Cram		Taylor, Sons & Co.	Leeds
80647	529		Wing & Wing WSr Moffatt	American	135	104·0	30·0	8·0	SanFrancisco	1881			A.M Simpson	San F…
79352	530	✻	Winifred Bennett WBn	British	192	119·2	24·1	10·8	Bideford	1879	Restarick	V 6.85	H.Jones	Live…
55301	531	✻	Winifred Jones -87 WBg	British	173	99·2	23·7	13·1	Portmadoc	1866	Williams	V 1.84	J.Owen & Co.	Carn…
48565	532	✻	Winifred Paterson 85-85 WSr	British	123	86·0	21·2	10·2	Lowestoft	1863	Metcalfe		W.Paterson	Aberd…
76270	533	✻	Winifred Noal -88 WK	British	44	62·0	17·4	8·0	Jersey	1883	J.F.Picot		P.Bailhache & Co.	Jerse…
80305	534		Winner Frye WSr	American	176	101·0	28·6	9·0	HarringtonMe	1871			E.Frye	Machi…
80430	535	R	Winnie Lawry McRitchie WSr	American	234	108·0	29·6	10·6	WaldoboroMe	1874		V 1.82	E.M.Dyer	Bosto…
48078	536		Winnifred Dutton WBk	British	329	127·0	28·0	11·5	Baltimore	1857	Abrahams		Joseph H.G.Millard	Yarm…
	537	D	Winnifred Larsen 88-88 WS	Norweg'n	1012	181·1	37·0	23·5	BeaverRiv NS	1868			A/S Rederiselskabet "Winnifred" (H.C. Magelssen)	Laurv…
83362	538	V	Winnipeg Hughes 78-82 WS	British	1303	205·5	37·6	23·1	Quebec	1882	E.Samson	V 3.83	W.H.Ross & Co.	Quebe…
	539	✻NR	Winona Larsson WBk	Norweg'n	683	155·1	31·5	21·5	Richibucto	1874	JardineBros.	V 2.85	Jacob C. Larsson	Arend…
80146	540		Winslow Morse WSr M'Donough	American	106	82·9	24·1	7·2	Bath, Me.	1870			Scott & Bridge	[illegible]
	541		Wiria Klenberg -86 W3mSr	Russian	300	138·0	27·3	12·6	Bengtsfors	1871	J.Wigelius		A.J.Bergroth	Abo
	542	G	Wismar (DISMANTLED) W3mSr (ex Meddergarden, ex Matcenson)	German	285	121·3	25·2	14·7	P.E.Island	1857			H.Kruse	Kiel
	543		Witch Pehrsson -88 WBk	Swedish	285	112·3	23·5	14·4	Perth	1866	PerthShipCo.	V 1882	L.V.Åkerson	Gefle
80267	544		Witch Hazel Arnold W3mSr	American	239	118·2	32·1	8·5	Bath, Me.	1872	Goss&Sawyer		A.W.Smith	Provi…
26154	545		Witch Queen WSr	American	116	85·3	22·7	8·5	[illegible]	1860			W.R.Taylor	New…
	546		Wittow (WRECKED) Lehmann WBg	German	205	90·4	26·0	14·0	Stralsund	1849	O.Kirchhoff		J.N.Rolfbertus	Barth
	547	N	Wladimir Pedersen -87 WBk	Norweg'n	503	139·9	31·7	17·1	Christiansand	1869	O.Olsen	V 5.84	O.C.Reinhardt	Christ…
116	548	G	Wladimir Latwing 86- W3mSr	Russian	257	118·8	27·4	12·7	Orenhof	1883	JurriSaal		P.Latwing	Oren…
487	549	V	Wladimir Adamson W3mSr	Russian	251	112·0	27·8	13·6	Pernigel	1886	A.Eschmann		J.Welling	Riga
	550		Wladimir Teigin 74-75 W3mSr (ex W…)	Russian	198	105·9	21·8	12·9	Yarmouth	1865	Geddie		F.Savin	Kem…

✻ = Lloyd's Register; D = Nederlandsche Vereeniging; G = Germanischer Lloyd; I = Registro Italiano; R = American …;
A = Veritas Austro-Ungarico; H = Veritas Ellenico; N = Norske Veritas; V = Bureau V…

In the 4th Column, C = Composite (iron frame, wood planked); I = Iron; Stl = Steel; W = Wood.

The Dates of Metalling (Col. 11) are based on the records of the Books in which Vessels are respectively classed, or on the Reports of the Surveyors of Lloyd's …

Further particulars of Vessels classed by Lloyd's Register will be found in Lloyd's Register Book and Lloyd's Yacht Register.

No.	Cote	Nom. Capitaine et dates d'entrée au service des Armateurs actuels et à bord du navire. Ancien Nom.	Matériaux et Gréement	Pavillon.	Tonnage net.	Longueur	Largeur	Creux	Construit Où.	Quand.	Par qui.	Date du doublage	Armateurs ou Directeurs-Gérants.	Port d'Armement.
551		Wodan *Hermansen (1885)*	W Bk	Danish	250	112·7	27·5	13·9	[illegible]	1860	Jul. Mendbrodt		Emil Kock	Marstal
552		Wodan *Jahnke*	W Bk	German	479	130·6	31·0	17·5	[illegible]	1865	F. Devrient	V 11.83	J.H.Rhetz & Co.	Danzig
553	H	Wodonga *Bullard* (ex Edgar Rossi)	W Bk	British	297	129·2	27·8	13·4	Harburg	1862	M.G. Ansorck		John Darling, junr.	Adelaide,S.A.
554		Woldemar *Gann 85-85*	W3mBn	British	323	125·5	28·8	13·7	Gutmanns-[Fisch	1871	G. Abrystea		A.Gann & Co.	Faversham
555	V	Woldemar *Behrtling -84*	W3mSr	Russian	293	120·0	28·1	13·8	Kalstsi	1884	J.Kapanen		B.Behrtling	Riga
556		Woldemar	W Sr	Russian	213	110·0	25·0	11·5	Reval	1869	J.Wickmann		J.Wickmann	Riga
557	D	Woldemar *Breede -88*	W Sr	Russian	211	100·0	26·6	12·9	[illegible]	1881	J.Jacobson	V 8.84	J.Andersohn	Riga
558	RV	Wolfe *Whalen -87*	W Bk	British	949	181·4	36·9	20·7	RiverJohn,NS	1881	Kitchin	V 11.88	David S.Crerar	Halifax,N.S.
559		Wolfgang *Freitmann 86-86*	W Bg	German	240	91·8	25·6	13·1	Rostock	1847	F.Dethloff		A.Burchard	Rostock
560		Wolgast *Jantzen -8-*	W Bg	German	196	96·7	26·0	13·8	Wolgast	1845	J.Ehrich	V 4.83	F.Neumann	Wolgast
561		Wollomai *Tulloch*	W Sr	British	143	107·6	23·6	9·5	JarraBank,SA	1876	R.Kennedy		William Tulloch	Adelaide,S.A.
562	*	Wolseley *Ligman* (Wh[illegible])	I Bk	British	341	110·0	27·0	13·5	Greenock	1876	J.E.Scott		Bullard, King & Co.	London
563	NE	Wolseley *Olsen -87* (ex Loch [illegible])	W Bk	Norweg'n	751	142·4	36·1	20·9	Middelburg	1865	Borsius & Van [der Leye	V 10.87	Niels Christensen	Arendal
564		Wonder *Sprakling*	W Bg	British	169	101·6	22·2	12·0	Cowes	1860	White		Mrs.C.Byerley	Portsmouth
565	R	Wonder *Foley*	W Sr	British	169	101·2	23·4	12·4	Padstow	1866	Tredwen		English's Coasting & Shipping Co.	Swansea
566		Woo Ling	W Sr	Chinese	221	120·3	28·7	10·0	Hankow	1879		V 2.81		
567	R	Wood Cock *Tibbetts*	W Bn	British	135	94·4	24·8	10·7	RolloBay,PEI	1871	Keefe		Estate of late M Williams	Halifax,N.S.
568	*D	Wood Hall *Thomas -84*	I Bk	British	721	179·0	30·5	19·3	PortGlasgow	1869	Blackwood & [Gordon		B.M.Hennessy	Swansea
569		Woodbine	W Bk	British	262	113·4	25·1	15·2	Banff	1857			J.E.Mitchell	Sydney,N.S.W.
570	*	Woodburn *Christie 88-88*	I S	British	1425	245·0	37·5	23·0	Glasgow	1872	Barclay,Curle [& Co.		R.Shankland & Co.	Greenock
571	R	Woodbury *Brown -88* (ex David Owen)	W Bn	American	364	112·7	27·6	17·2	Bath,Me.	18 66/81		V 3.84	J.W.Dresser	Castine,Me.
572	R	Woodbury M. Snow (ex Nel Sampter, ex Lunet)	W Sr	American	192	71·0	28·0	9·0		18 [illegible]/83		V 3.86	Snow & Pearson	Rockland,Me.
573	GV	Woodland *Lorange 83-*	W Sr	Norweg'n	456	135·7	29·7	11·8	DeepRiv.Conn	1862		V 2.86	Woodland's Akt.Selsk.	Christiania
574	*	Woodlark *Addison 88-88*	C Bk	British	869	182·4	32·4	19·2	Dundee	1870	A.Stephen&Sns	V 2.86	A.Stephen & Sons	Dundee
575		Woodlark	W Bg	British	259	90·5	25·4	18·0	Volman Am-[hirston,E.I.	1832			UnionS.S.Co.ofN.Z.	Dunedin
576		Woodside *Barber 86-86*	W Bg	British	176	92·8	23·9	12·9	Sunderland	1867	George Bartram	V 8.83	T.H.Franks	Folkestone
577	R	Woodside *Dunning 87-87*	W Bk	American	579	142·8	31·4	18·7	Cumberland, [Me.	1866	J.A.Woonside	V 3.85	R.Lewis & Co.	Portland,Me.
578	*	Woodville *Fea 85-85*	I Bk	British	714	191·2	30·8	18·6	Liverpool	1876	W.H.Potter&Co		"Woodville" Shipowners Co.,Ld.(J.S. Ward & Co.)	Liverpool
579		Woodville	W Bk	British	362	125·5	28·5	16·1	Arbroath	1856			William V.Brown	Sydney,N.S.W
580		Woodville	W Bg	British	131	75·0	22·8	13·0	Whitehaven	1839	Scott		J.Purfield	Whitehaven
581		Woodward Abrahams *Snow*	W3mSr	American	707	166·0	35·0	16·1	Waldoboro,Me	1881	A.Storer & Son		J.Baker & Co.	Boston,Mass.
582	*	Woollahra *Burneson*	I Bk	British	942	202·1	33·6	20·1	Sunderland	1875	Osbourne, [Graham & Co.		Cowlishaw Bros.	Sydney,N.S.W.
583	*	Woolton *Calder 85-86*	I S	British	2101	274·3	40·4	24·9	Southampton	1885	Oswald, Mor-[daunt & Co.		Sailing Ship "Woolton" Co.,Ld.(R.W.Leyland & Co.)	Liverpool
584	*	Woolton *Poole 75-85*	W3mSr	British	246	122·9	24·2	13·5	Southwick	1863	J.Hardie	V 8.8	J.Fisher & Sons	Barrow
585	*V	Woosung *Brandtner 88-88*	I S	German	745	174·5	31·4	19·3	Glasgow	1863	A.Stephen&Sns		P.Breckwoldt,sen.	Blankenese
586	V	Wopke *Tholen 84-84*	W Bn	German	128	91·0	21·3	11·2	Emden	1864	H.Klattenhoff		TönjesJ.Cassens & Sohn	Grossefehn
587	*	Workman *McLerkon 84-84* (STRANGFORD)	I Bk	British	371	156·5	27·7	12·8	Belfast	1884	Workman,Clark [& Co.		"Workman" Ship Co. Ld (Workman & Workman)	Belfast
588	*	Worrall *Sanders 86-95*	I Bk	British	483	166·3	25·3	16·9	Belfast	1862	Harland & Wolff		S.Wakeham & Sons	Liverpool
589	N	Woye *Andersen 75-75*	W Bk	Norweg'n	447	132·3	31·2	17·2	Arendal	1875		V 5.86	Axel Smith	Arendal
590	*	Wray Castle	Stl S	British		..	..	..	Workington	1889	R.Williamson & Son		"Wray Castle" Sailing Ship Co.,Ld.(J.Chambers & Co.)	Liverpool
591		Wrestler *Cook -87*	W Bkn	American	447	148·5	34·9	13·0	PortLudlow, [Wash.	1880	Hall Bros.		F.Haywood & Co.	SanFrancisco
592	G	Wrights *Svensson*	W Bg	Swedish	235	91·8	22·1	15·5	Newcastle	1849	Hall & Fulton		J.M.Åbrandt	Calmar
593	G	Wustrow *Borgwardt 84-84*	W Bg	German	301	106·2	26·8	15·9	Rostock	1865	W.Zeltz		A.Burchard	Rostock
594	R	Wyer G. Sargent *Sargent*	W3mSr	American	309	131·0	31·9	11·2	Sedgwick,Me.	1881	Billings	V 4.87	R.B.Sargent	Sedgwick,Me.
595	*	Wynnstay *McBryde 88-84*	I S	British	1601	258·8	38·2	23·1	PortGlasgow	1884	Russell & Co.		P.Macfarlane	Liverpool

ADDITIONAL SHIPS ARE INSERTED IN THE SUPPLEMENT AT THE END OF THE REGISTER.

*=Lloyd's Register; D=Nederlandsche Vereeniging; G=Germanischer Lloyd; I=Registro Italiano; R=American Record;
A=Veritas Austro-Ungarico; H=Veritas Ellenico; N=Norske Veritas; V=Bureau Veritas.

Dans la 4me Colonne, C=Composite (membrure en fer, bordé en bois); I=Fer; Stl=Acier; W=Bois.

Les Dates du doublage en métal (Col. 11) ressortent des livres dans lesquels les navires sont respectivement classés ou des rapports des Inspecteurs du Lloyd's Register.

Pour plus amples renseignements sur les navires classés par le Lloyd's Register, voir le Lloyd's Register Book et le Lloyd's Yacht Register.

Official No.	No. in Book	Registry in which classed	Ship's Name. Material and Rig. / Master, and Dates of Appointment to present Owner's Service and to Ship. / Ship's Late Name.		Flag	Net Tonnage	Dimensions. Length	Breadth	Depth	Built. Where.	When	By whom.	Date of Metalling	Owners or Managers.	Port of Registry
	1	V	X.	[illegible] W Bg	French	141	79·7	21·0	11·8	[illegible]	1852	[illegible]		Mme. E. Legal	Nantes
[illegible]	2		Xanthus	W Bg	British	200	108·8	26·2	12·7	P. E. Island	1866	[illegible]		W. O. McCormick	Dublin
[illegible]	3		Xebec	W S: Barge	American	220	161·0	22·0	8·0	Bordentown, N.J.	1873			J. A. Anderson	[illegible]
[illegible]	4	V	Xenia	Reynolds [illegible] W Bk	American	1135	184·9	36·1	22·6	Bath, Me.	1874	[illegible]	[illegible]	M. F. Pickering & Co.	Boston
	5		Xenia	W S:	Swedish	183	90·7	27·6	11·8	[illegible]	1876	B. F. [illegible]		Slite Warfsbolag	Slite
	6		Xenophon (CONDEMNED)	[illegible] W Bk	Norweg'n	[illegible]	152·5	34·2	17·1	Mystic	1880	[illegible]	[illegible]	Westergaard & Co.	Christiania

Additional Ships are inserted in the Supplement at the end of the Register.

✻ = Lloyd's Register; D = Nederlandsche Vereeniging; G = Germanischer Lloyd; I = Registro Italiano; R = American Rec[illegible]
A = Veritas Austro-Ungarico; H = Veritas Ellenico; N = Norske Veritas; V = Bureau Ver[illegible]

In the 4th Column, C = Composite (iron frame, wood planked); I = Iron; Stl = Steel; W = Wood.

The Dates of Metalling (**Col. 11**) are based on the records of the Books in which Vessels are respectively classed, or on the Reports of the Surveyors of Lloyd's [illegible]

Further particulars of Vessels classed by Lloyd's Register will be found in Lloyd's Register Book and Lloyd's Yacht Register.

		Nom.	Capitaine et dates d'entrée au service des Armateurs actuels et à bord du navire. Ancien Nom.	Matériaux et Gréement.	Pavillon.	Tonnage	Longueur	Largeur	Creux	Construit. Où.	Quand.	Par qui.	Date du doublage en métal	Armateurs ou Directeurs-Gérants.	Port d'Armement.
1		Yaffa	*Mahmoud*	W Bg	Turkish	125	..	..	..	Syra	1883			Trebizondlee Koolookir oglou Mahmoud & Co.	Trebizonde
2		Yale	*Simpson*	W 3m Sr	American	681	166·0	35·4	17·6	Belfast, Me.	1884	G.W.Cottrill		L. W. & P. Armstrong	N.Haven, Conn.
3	✻	Yallaroi	*Brown 74-85*	I S	British	1499	215·8	38·1	22·0	Aberdeen	1885	A.Hall&Co.		A.Nicol & Co.	Aberdeen
4		Yamato Maru		W Bg	Japanese	209	110·1	22·8	11·6					Odsu Yoyemon	Tokio
5		Yamoyden	*Walters -88*	W Bk	American	463	133·2	30·6	16·1	Baltimore	1868	Jones&Ashcroft	v 6.88	C.M.Stewart & Co.	Baltimore
6		Yampa	*Eldridge -81*	I Sr Yt	American	161	110·0	27·0	17·6	Wilm'ton Del.	1887	The Harlan and Hollingsworth Co.		C.W.Chapin	New York
7		Yandaboh		I Flat	British	777	197·0	25·7	8·3	Port Glasgow	1876	R.Duncan&Co.		Irrawaddy Flotilla Co. Ld.	Glasgow
8		Yandoon		I Barge	British	702	170·7	25·1	8·6	Port Glasgow	1874	R.Duncan&Co.		Irrawaddy Flotilla Co. Ld.	Glasgow
9		Yankee Blade	*Coombs*	W Sr	American	128	84·9	24·0	8·4	Bucksport, Me.	1849			J.H.Sprague	Bangor, Me.
10		Yankee Doodle	*Camp*	W Sr	American	204	138·0	32·5	11·8	Elizabethport, Me.	1853			E.R.Kirk	New York
11		Yantaba		I Lighter	British	362	197·2	25·2	8·2	PortGlasgow	1880	R.Duncan&Co.		Irrawaddy Flotilla Co. Ld.	Glasgow
12	✻	Yarkand	*Johnson*	I S	British	1311	222·2	37·2	22·4	Barrow	1877	BarrowS.B.C.Ld		E.Bates & Sons	Liverpool
13	N	Yarra Yarra (ex Gladiator)	*Pedersen (1882)*	W Bk	Norweg'n	668	138·1	33·0	20·8	New York	1836		v 8.86	J.L.Ugland	Arendal
14		Yau		Stl Lr	British	215	160·0	22·0	7·3	Dumbarton	1887	W.Denny & Brs.		Irrawaddy Flotilla Co. Ld.	Glasgow
15		Ydale	*Helmeksen*	W Bn	Norweg'n	168	102·6	25·5	11·2	Stavanger	1867	Nielsen	v 8.86	J.Johnsen	Stavanger
16		Ydra	*N. Xenos*	W Bg	Greek	292	..	..	..	Syra	1874			A.Xenos	Syra
17	V	Ydun	*Klingenberg 86-73*	W Bn	Danish	167	103·0	22·8	10·8	Svendborg	1873	C.C.H.Hansen& H.P.Hansen, jr		Actieselskab (J. Andersköw)	Svendborg
18		Ydun	*Larsen 74-74*	W Bn	Danish	142	94·5	22·3	11·3	Stege	1874	J.P.Johansen		Aktie-Selskab "Ydun" (R.Hertel)	Stege
19	DV	Ydun	*Andersen 79-88*	W Sr	Danish	100	82·7	20·0	9·8	Svendborg	1879	P.Troensegaard		A.K.Andersen	Marstal
20	N	Ydun	*Olsen*	W Bk	Norweg'n	582	164·5	33·0	18·7	Bergen	1871		v 7.88	J.E.Lehmkuhl	Bergen
21	N	Yduna	*Skadberg*	W 3m Sr	Norweg'n	239	122·8	25·1	11·1	Hernösand	1869			O.Skadberg	Egersund
22		Yeiraku Maru		I Sr	Japanese	499	200·0	24·2	16·2	Glasgow	1861	Wingate		Inouye Nihei	Osaka
23		Yeitai Maru		W Bg	Japanese	226	104·7	23·8	13·8	Lübeck	1867			HamanakaHachisaburo	Osaka
24	N	Yelva	*Jensen 83-83*	W Sr	Norweg'n	179	87·0	24·7	12·4	Cronstadt	1840			T. & H.Lorentzen	Drammen
25		Yeni Dunnia		W Bg	Turkish	108	..	..	..	Barten	1875			Bulbuljehlee Stati Karavangeli	Bulbuljeh
26		Yesashi Maru		W Sr	Japanese	139	88·5	20·5	9·6	Osaka	1880			Nishizawa Kizo	Yesashi
27		Yeshil Kanad		W Sr	Turkish	125	..	..	..	Terebolie	1879			Terebolielee tujar Hajee Emin Agha Vehbeh Mehmed & Co.	Terebolie
28		Yeshil Kanad	*Toossoon*	W Bg	Turkish	119	..	..	..	Terebolie	1885			Yelkendji Mustapha Effendi & Co.	Terebolie
29		Yeshil Kanad		W Sr	Turkish	100	..	..	..	Terebolie	1879			Tunnuslee Sali bin Said Billeh	Tunnus
30		Ygdrasil	*Johansen -88*	W Bk	Norweg'n	359	99·4	28·9	17·9	Christiania	1841			A.G.Samuelsen	Dröbak
31		Yhteys	*Starck 81-*	W Sr	Russian	221	98·3	28·5	12·5	Risby	1874	E.Norrgård	v 6.88	O.Nygård	Sastmola
32	DN	Ymer	*Andersen -85*	W Bk	Norweg'n	818	171·8	35·5	19·2	Arendal	1885	H.Stiansen	v 81	Chr.Th.Boe	Arendal
33	D	Ymuiden	*Wiersma 84-86*	W Bk	Dutch	666	152·8	32·4	19·9	Amsterdam	1864	P.Haverkamp		E.J.Bok & Zonen	Amsterdam
34		Ynez	*Andersson*	W Bg	Swedish	264	113·5	25·0	13·5	Wiborg	18 42/55			N.Olsson	Glimsås, Bohuslän
35		Yoeserin		W Bk	Dutch	278	111·5	21·8	..					Said Aloei bin Achmud Alsaf	Palembang
36		Yoeserin		W Bk	Dutch	209	101·7	24·0	..					Said Abdulrachman bin Alie Alsaf	Palembang
37		Yoeserin		W Bg	Dutch	133	69·9	20·7	..				v 8.87	Said Achmad bin Alie Alsaf	Palembang
38	✻	Yolande	*Watt -86*	W Bkn	British	395	110·6	31·0	12·8	Tatamagouche N.S.	1883	D.& A.Campbell		A.O.Knight	Auckland, N.Z.
39	A	Yonathan (ex Cavaliere Knud ex Consolidé)	*Andersen*	W Bk	Urugu'y'n	344	123·0	27·9	17·2	Sunderland	1869	G.Short	v 3.86	Williams & Co.	Montevideo
40		Yorghi Yuan	*Yanni*	W Bg	Turkish	222	..	..	..	Greece	1884			Christophor & Yorghi Vassilikano	Constantinople
41		Yorghios	*Anagosti*	W Bg	Turkish	226	..	..	..	Greece	1880			Limnilee Aristidi Mulrdlo & Co.	Lemnos
42		Yorinobu Maru		W Bkn	Japanese	487	131·9	28·8	16·2	Hiogo	1884	KawasakiShip Buildg. Yard	v 7.87	Nippon Yusen Kaisha	Tokio
43	BV	Yorktown	*Delap -85*	W S	American	1861	227·1	40·5	28·1	Richmond, Me.	1878	J.M.Hagar		P.M.Whitmore	Bath. Me.
44	✻	Yosemite	*Roberts 89-89*	I Bk	British	768	183·4	31·7	19·5	Glasgow	1868	Dobie&Co.		J.B.Walmsley & Co.	Liverpool
45		Yosemite	*Fullerton 83-87*	W S	American	1104	183·0	37·2	23·5	Portsmouth, N.H.	1868			S.Blair	San Francisco
46		Yosemite		W Sr	American	131	91·0	28·7	7·2	Damariscotta, Me.	1873			S.C.Evans	New York
47		Yoshiiye Maru		W Bk	Japanese	344	130·9	25·5	14·0	Hamburg	1876			Nippon Yusen Kaisha	Tokio
48	✻	Young Alfred	*Barrett*	W Bk	British	95	82·0	20·9	10·6	Rostock	1885	E.Burchard&Co.	3.88	T.Hamling	Hull
49	R	Young Eagle	*Anderson -88*	W Bk	British	532	140·0	31·2	18·0	Clifton, N.B.	1870	G.H.Flewelling		Wm.Thomson & Co.	St.John, N.B.
50		Young Greek		W Bk	British	424	132·0	27·9	17·0	Medford, U.S.	1855			Koh Mah Wah	Hong Kong

✻=Lloyd's Register; D=Nederlandsche Vereeniging; G=Germanischer Lloyd; I=Registro Italiano; R=American Record; A=Veritas Austro-Ungarico; H=Veritas Ellenico; N=Norske Veritas; V=Bureau Veritas.

Dans la 4me Colonne, C=Composite (membrure en fer, bordé en bois); I=Fer; Stl=Acier; W=Bois.

Les Dates du doublage en métal (Col. 11) ressortent des livres dans lesquels les navires sont respectivement classés ou des rapports des Inspecteurs du Lloyd's Register.

Pour plus amples renseignements sur les navires classés par le Lloyd's Register, voir le Lloyd's Register Book et le Lloyd's Yacht Register.

Official No.	No. in Book	Registry where classed	Ship's Name. Material and Rig. Master. Ship's Late Name.	Flag	Net Tonnage	Length	Breadth	Depth	Built Where	Built When	Built By whom	Date of Metalling	Owners or Managers	Port of Registry
177	51		Young Siam (BROKEN UP) W S	Siamese	701	160·0	36·0	16·6	Bangkok	1864			Chuessua Poh Hee	Bangkok
[illegible]	52	✻	Young Tom May W K	British	88	80·1	20·8	10·4	Brisham	1884	J.W.&A.Upham		T. May	Hull
74	53		Ypapanti W Bg	Greek	234	90·8	26·4	..	Spezzia	1865			P. Eustathopoulos	Piræus
[illegible]	54		Ypapanti (BRANY) W Sr	Greek	198	95·4	24·4	..	Syra	1877			N. Voulakis	Syra
	55	H	Ypiretria [illegible] W Bg		261	99·0	27·1	17·4	Scopelos	1867	J. Alexander			
	56		Ypome I Sr	Argentine	102	85·0	23·0	7·1	Paisley	1884	J.Fullerton&Co		LaPlatenseFlotillaCo.	Buenos Ay
[illegible]	57		Yreka Falkingham W Sr	American	114	87·7	25·5	7·3	Columbia Falls, Me.	1873			J. J. Faulkingham	Machias
	58		Yru Chulo Careaga W Bk (BROKEN UP) [illegible]	Spanish	296	116·4	23·6	14·5	Bilbao	1866	[illegible]		Mercader hijos & [illegible]	San Sebas
[illegible]	59		Ysidora Rionda Spec W Bn [illegible]	British	242	101·0	27·0	12·0	Richmond, N.S.	1866			D. E. Sands	Nassau, N
	60		Ystava Wilen W Sr	Russian	220	95·2	23·8	11·9	Pyhäjoki	1870	J. H. Sottka		E. Wanhoperä	Pyhäjoki
	61		Ystava Pukki W Sr	Russian	146	..	..	..		1875			J. Bukki	Sikajoki
	62	D	Ystavat Törnblom W Bk	Russian	654	155·8	34·1	19·8	Nystad	1869	H. Kjaldst Co.		F. W. Wahlberg	Nystad
	63		Ystavat Soderlund W Bg	Russian	232	..	..	..		1853			D. F. Soderlund	Lampp
	64	DN	Ystawa Seeuningsen W Bk	Norweg'n	667	155·5	33·6	17·1	Arendal	1877	J. Halvorsen	[illegible]	O. B. Sorensen	Arendal
[illegible]	65	R	Yuba McKenzie W Bk	British	879	167·9	35·1	20·0	St. John, N.B.	1873	C. McLennan	[illegible]	Dickson Anderson	Montreal
93706	66	V	Yucatan Date ✻ I Bm Bn [illegible]	British	319	133·4	24·1	14·2	Hamburg	1870	ReihersttegShipWrks		W. C. Jarvis	Liverpool
	67	V	Yvonne Durand W Bk	French	300	115·1	27·2	13·8	St. Malo	1866	Vve Dandin&fils	[illegible]	Cie de Transports Maritimes & Cabotage [illegible]	Nantes
	68		Yvonne Gobec W Bn	French	113	79·4	20·7	11·5	Dunkirk	1865	Beck		L. Verry	Binic
	69	V	Yvonne Marie Brohan I Bk	French	1024	205·7	33·5	20·4	Nantes	1876	Jollet&Babin		Victor Vincent	Nantes

Additional Ships are inserted in the Supplement at the end of the Register.

✻ = Lloyd's Register; D = Nederlandsche Vereeniging; G = Germanischer Lloyd; I = Registro Italiano; R = American Record; A = Veritas Austro-Ungarico; H = Veritas Ellenico; N = Norske Veritas; V = Bureau Veritas.

In the 4th Column, C = Composite (iron frame, wood planked); I = Iron; Stl = Steel; W = Wood.

The Dates of Metalling (Col. 11) are based on the records of the Books in which Vessels are respectively classed, or on the Reports of the Surveyors of Lloyd's [illegible]

Further particulars of Vessels classed by Lloyd's Register will be found in Lloyd's Register Book and Lloyd's Yacht Register.

N° dans ce livre	Registre dans lequel le navire est classé	Nom. Matériaux et Grèement. Capitaine et dates d'entrée au service des Armateurs actuels et à bord du navire. Ancien Nom.	Pavillon.	Tonnage net.	Longueur.	Largeur.	Creux.	Où.	Quand.	Par qui.	Date du doublage en métal	Armateurs ou Directeurs-Gérants.	Port d'Armement.
1	DV	Z. Ring M'Leod -88 WS	British	1371	198·9	39·4	24·4	St.John, N.B.	1872	Cruickshank	V [illegible]	Scammell Bros.	St.John, N.B.
2		Z. Stratton McFadden WSr	American	109	89·2	25·9	7·3	[illegible], N.J.	1849			G.S.Dow	Providence, R.I.
3	R	Zaccheus Sherman (Canada) W3mSr	American	729	171·5	28·5	17·2	Bath, Me.	1880	Goss, Sawyer & Packard		S.H.Walker	Fall River, Mass
4		Zacharias Lazopoulos -88 WBg	Greek	279	..	..	..	Scopelos	1871	M.Sarambalakis		D.Lazopoulos	Scopelos
5	V	Zacharie Lancelin -88 WBn	French	136	93·5	21·3	10·8	St.Malo	1883	F.Gautier	V 2.87	Vve.E.Thomazeau&Co.	St.Malo
6		Zachris Franzen WBk	Russian	512	..	..	..	Brahestad	1860			J.Lang	Brahestad
7	✻D	Zadok Schumacher 84-86 IBk	German	597	170·5	28·5	18·0	Liverpool	1868	Bowdler, Chaffer & Co.		Gebrüder Hustede	Elsfleth
8		Zaeta [illegible] WSr	Venezu'la	146	95·0	26·0	10·9	Canning, N.S.	1873			C.G.Pinedo	Maracaibo
9		Zaggodhopki WBg	Turkish	192	..	..	..	Greece	1883			Aktarpoleelee Yorghi Teodori Daudamicho	Aktarpoleo
10		Zaida D. Eddy W	British	150	110·7	22·6	7·0	Quebec	1882	A.Millar		J.B.Blanchard	Ottawa, Ont.
11	V	Zaima Rugberg -88 WBg	Swedish	335	125·0	23·8	15·5	Gothenburg	1883	J.E.Hube	V 9.86	G.D.Kennedy	Gothenburg
12		Zaire Guerra (ex Severino Candida I, ex Montenegro) WSr	Portug'se	128	83·7	21·6	9·2	Villa de Conde	1875	J.M.d'Aranjo		N.F.P.Abreu Guimerães	Oporto
13		Zaloon I Flat	British	777	197·0	25·7	8·3	Port Glasgow	1876	R.Duncan & Co.		Irrawaddy Flotilla Co., Ld.	Glasgow
14	✻	Zamora Richards 85-86 (MISSING) IBk	British	1180	222·6	34·3	20·5	Dumbarton	1876	Birrell, Stenhouse & Co.		Scotia Shipping Co., Ld.	Glasgow
15	R	Zamora Bird WSch	American	118	80·4	26·2	8·6	Quaco, N.B.	1880			S.L.Keene	New York
16		Zampa W3mSr	American	366	145·7	36·0	10·4	P.Maddisonwr	1887	Hughes		George W. Hinsdale and others	San Francisco
17		Zampa Lambert WSr	American	137	92·2	25·6	8·3	E.Machias, Me	1864			W.Pope & Sons	Machias, Me.
18		Zampa Larsen 82-82 WBg	Danish	198	102·3	23·6	12·3	Drammen	18 54/73			K.F.Marstrands Eftf.	Elsinore
19		Zampa Jörgensen -88 (ex Antilope) WBn	Danish	115	71·3	21·2	10·8	Kiel	1861	A.F.H.Conradi		J.M.Jörgensen	Marstal
20	N	Zampa Halvorsen -88 WBk	Norweg'n	430	126·4	31·0	16·9	Arendal	1874	J.Osmundsen	V 7.86	J.Lassen	Arendal
21	R	Zanoni French 84-85 WBn	British	245	117·2	26·1	12·9	P.E.Island	1872	Coffin		J.Woods	St.Johns, N.F.
22	NV	Zanrak Mathiesen WBk	Norweg'n	595	147·9	31·0	18·9	Arendal	1878		V 10.87	T.Thommesen & Smith	Arendal
23		Zant-Ara WBn	Greek	256	..	..	..					J.Marvilas	Santorin
24	V	Zante Lanco WBk	French	282	108·2	24·0	14·2	Nantes	1874	Tille	C 5.84	J.B.Demange	Nantes
25	✻	Zara Holgate 83-88 WBn	British	258	116·5	26·5	12·9	P.E.Island	1882	PeakeBros.&Co.	V 10.81	Gillies & Da Costa	Liverpool
26		Zaranak WS	Norweg'n	838	153·1	35·0	21·7	Philadelphia	1844			O.B.Sörensen	Arendal
27		Zarapito Serrano WBk	Chilian	465	124·6	29·3	..	Wexford	1851			E.H.Laugdon	Chiloe
28	NV	Zaritza Brustad WBn	Norweg'n	167	101·2	23·6	11·3	Christiansund	1875	O.Mœhle	V 3.83	H.F.Klingenberg	Drontheim
29	GV	Zaritza Dannberg (ex Vierge de Lisa) WS	Swedish	916	179·4	32·6	21·6	Portsmouth	1854		V 5.87	P.Leffler	Gothenburg
30	✻	Zayda Little 88-88 W3mSr	British	199	117·7	24·8	12·9	Bideford	1869	Johnson	V 11.88	Dennery Co., Ld.	London
31	✻	Zealandia Phillips 71-81 IS	British	1116	215·6	35·1	20·3	Glasgow	1869	C.Connell&Co.		Shaw, Savill & Albion Co., Ld.	Southampton
32	DV	Zebina Goudey Sanderson -88 WBk	British	1088	195·5	36·3	21·5	Annapolis, NS	1881	L.Delap	V 3.85	G.T.Sanderson	Yarmouth, N.S.
33		Zebra Rothenkiser WSr	British	142	91·0	24·2	11·1	LaHave, N.S.	1859		V 7.88	J.Bartlett, junr.	St.Johns, N.F.
34	✻	Zebra WSk	British	84	79·0	20·0	10·6	Grimsby	1886	G.Oatwin		J.Gidley	Grimsby
35	A	Zebra Sattora WSr	Austrian	210	98·4	23·6	13·6	Lussinpiccolo	1879	A.Catterinich	V 11.81	G.B.Stuparich	Lussinpiccolo
36		Zebrina Skinner 85-85 WSr	British	169	109·1	23·9	9·9	Whitstable	1873	Gann		Whitstable Shipping Co., Ld.	Faversham
37		Zebron WBn	British	100	73·1	20·1	9·9	NovaScotia	1847			I.Dart	Halifax, N.S.
38		Zedekiyeh Rahmet WBg	Turkish	232	..	..	..	Oonia	1877			OonialeeHajee Ahmed	Oonia
39	D	Zeemeeuw Koer -81 (ex Elizabeth, ex Mittwoch) W3mSr	Dutch	350	120·0	27·4	14·4	Danzig	1857	V.Deverient		J. & S.Wiarda	Harlingen
40	DV	Zeemeeuw Bos 82-82 WBn	Dutch	155	94·2	21·6	10·9	Harlingen	1855	D.&G.Alta		E.T.Bos	Delfzijl
41	✻D	Zeenymph de Groot 79-86 (ex Spirit of the [illegible]) WBk	Dutch	393	150·9	26·9	16·5	Belfast	1859	McLaine	V 3.88	C.van de Wetering	Rotterdam
42	✻I	Zeffiro Pellerano WBk	Italian	646	157·4	31·5	21·3	SestriPonente	1877	M.Bossi	V 5.83	G.Pellerano & G.Schiaffino	Genoa
43	I	Zefiro Russo WBk	Italian	584	146·6	28·5	19·5	Cassano	1875	M.Paturzo	V 6.85	M. & F.Russo	Castellamare
44	A	Zefs Tzissimos -88 WBg	Greek	227	..	..	..	Galaxidi	1877		V 7.83	I.G.Zissimos	Poros
45		Zehlima Oneto WBk	Italian	447	134·0	28·3	18·5	Sunderland	1860	RThompson, jun	V 5.84	L.Oneto & F. Degregori	Genoa
46		Zeila Hollowell -74 WSr	American	116	91·0	27·0	7·3	Pembroke, Me.	1874	W.T.Holbart&Co		B.Lincoln	Eastport, Me.
47	H	Zeilan Bachri WBg	Montng'n	121	..	..	..	Trebize	1870		[illegible] 2.86	Ismail Mehmet	Dulcigno
48		Zein (BROKEN UP) WBk	Dutch	168	86·0	22·1	..	Java	1832			Sech Oembarak bin Abdull. Badjerdana	Sourabaya
49		Zelee WSr	British	100	81·0	22·5	8·3	Grondinés	1862			A.T.Beaulieu	Quebec
50		Zelia WSr	British	126	83·3	21·9	10·2	St. Anne de la Pe.	1879	B.Auetil		E.J.Price	Quebec

[illegible] Lloyd's Register; D=Nederlandsche Vereeniging; G=Germanischer Lloyd; I=Registro Italiano; R=American Record; [illegible] Veritas Austro-Ungarico; H=Veritas Ellenico; N=Norske Veritas; V=Bureau Veritas.

Dans la 4me Colonne, C=Composite (membrure en fer, bordé en bois); I=Fer; Stl=Acier; W=Bois.

[illegible] Dates de doublage en métal (**Col. 11**) ressortent des livres dans lesquels les navires sont respectivement classés ou des rapports des Inspecteurs du Lloyd's Register.

Pour plus amples renseignements sur les navires classés par le Lloyd's Register, voir le Lloyd's Register Book et le Lloyd's Yacht Register.

1 Official No.	2 No. in Book	3 Registry in which classed	4 Ship's Name. Material and Rig. Master, and Dates of Appointment to present Owner's Service and to Ship. Ship's Late Name.	5 Flag.	6 Net Tonnage	7 Dimensions. Length	Breadth	Depth	8 Built. Where	9 When	10 By whom	11 Date of Metalling	12 Owners or Managers.	Port
47978	51		**Zelia** *Blackford 82-82* W Sr Yt	British	103	101·1	21·1	11·8	Lymington	1864	G. Inman	[illegible]	Martin Ridley Smith	South
	52	✳	**Zelia** *Grum* W 3msts	Columb'n	257	119·5	24·3	14·8	Barnstable	1862	Westacott	[illegible]	Cane Dugand Martinez & Co.	Rio
85797	53	✳R	**Zelica** *Murchison 86-86* W Bn	British	299	123·0	27·6	12·9	Mount Stewart (P.E.Island	1884	Peake Bros.& Co.	[illegible]	Peake Bros.& Co.	
638	54	I	**Zelinda** *Albore* W Bg	Italian	205	104·9	26·1	13·8	Sestri Ponente	1869	Fili Cadenaccio	[illegible]	D. Paino	
	55	G	**Zelmire** *Lima* W Bk (ex Rita Neves, ex Lodge)	Portug'se	865	165·0	34·5	21·4	Portsmouth, (N.H.	1854		[illegible]	W. Guimarães & Co.	Oporto
	56	N	**Zemach** *Riks -88* W Bk	Norweg'n	451	142·1	31·3	14·5	Grimstad	1888	O.H. Riks	[illegible]	J. Jörgensen	
91235	57	✳	**Zemindar** *Graham -88* Stl S	British	2053	292·6	39·7	23·5	Belfast	1885	Harland & Wolff		T. & J. Brocklebank	Liverp
55	58		**Zeni** W Bg	Greek	226	91·8	26·1	..	Piræus	1867			G. I. Psaroudis	Piræus
74343	59	V	**Zeno** *Le Blanc -81* W Bn	British	390	127·7	30·4	12·9	River John, NS	1876	J. Henry	[illegible]	T. E. Kenny	Picto
56111	60	✳	**Zenobia** *Davies 77-77* W Sr	British	178	105·7	24·2	12·1	Ipswich	1870	Lambert	[illegible]	E. Davies	Aberys
87080	61	✳	**Zenobia** *Redford* W K	British	84	77·5	20·5	10·5	Brixham	1884	S. Dewdney & Sns.		J. Guzzwell	Brixham
91902	62	✳	**Zenobia** *Douse* W K Barge	British	78	85·6	20·6	6·9	Ipswich	1886	W. Bayley & Sons		F.C.W. & A.G.T. Claydon	London
	63		**Zenobia** *Mangana -88* W Bk (ex Dolon Anna, ex Betti)	Greek	448	134·8	27·7	17·3	Fiume	1874	Schiavoni	[illegible]	J. D. Mangana	Piræus
	64	NV	**Zenobia** *Albrechtson 84-84* W Bk	Norweg'n	437	133·0	31·8	16·0	Laurvig	1874	N. Pedersen	[illegible]	M. Oppen	Laurvig
28058	65		**Zephaniah Steelman** *Bennett* W Sr	American	219	112·0	30·2	8·5	Absecom, N.J.	1868	Steelman		J. Rose	New Yo
82480	66		**Zephyr** W S	British	1312	189·1	37·3	25·7	E Boston, Mass	1854			Adelaide S.S.Co.	Adelai
51233	67		**Zephyr** *Ellis 83-83* W Sr	British	139	96·5	20·8	12·0	Jersey	1865	Bellot		J. P. Tocque	Jersey
38312	68		**Zephyr** W Bn	British	135	102·0	21·4	10·3	Sydney, NSW.	1866			H. L. Michee	Melbo
59727	69		**Zephyr** *Smyth* W Sr	British	117	85·4	23·4	10·8	P.E.Island	1868	Strong		R. Dunn	Belfas
28977	70		**Zephyr** *Smith 75-72* W Sr	British	112	81·0	21·6	10·7	Montrose	1861	Birnie		A. Mackie	Montros
	71		**Zephyr** *Clausen 85-85* W 3m Sr (WRECKED)	Danish	232	98·9	26·8	13·1	Marstal	1885	Brødr. Bager		L. J. Bager	Marstal
	72	V	**Zephyr** *Henriksen 85-85* W Sr	Danish	145	85·6	22·1	10·8	Kolbaek	1882	H. Olsen		W. Ronne	Ronne
	73		**Zephyr** *Christensen 78-82* W 3m Sr (CONDEMNED)	Norweg'n	236	112·5	27·6	12·0	Risör	1875	T. Gundersen	[illegible]	B. A. Olsen	Risor
	74	N	**Zephyr** *Halvorsen -88* W Sr	Norweg'n	152	93·9	24·3	11·0	Svelvig	1869	Helgesen		A. Olsen	Drammen
1	75	AH	**Zephyros** *Raptis* W Bg	Greek	330	110·8	..	17·4	Galaxidi	1878	C. Papetros	[illegible]	J. Raptis	Galaxi
454	76	V	**Zephyros** *Emrrios* W Bg	Greek	243	111·8	25·1	..	Syra	1885	Livadaros	[illegible]	P. Svoronos	Syra
	77	V	**Zephyrus** *Andersen 71-75* W Sr	Danish	125	91·5	20·6	9·9	Svendborg	1875	J. R. Andersen		R. P. Pedersen	Svendb
	78		**Zequinha** *da Silva* W 3m Sr (ex Mary Barbara)	Brazilian	220	111·4	23·9	12·3	Poole	1881	W. Meadus & Son	[illegible]	J. da S. Loyo, junr.	Pernam
1478	79	A	**Zerlino** *Monaro* W Bn	Italian	100	..	..	..	Chioggia	1884		[illegible]	D. Monaro	Genoa
477	80	A	**Zert** *Jukanovic -88* W Bk (ex ... d'Colon)	Austrian	355	120·3	24·6	16·0	St. Malo	1866	L. Mallard	[illegible]	Giacomo Gurcovich	Pola
76511	81	✳	**Zeta** *Harman 75-83* I Bk	British	640	185·2	29·7	17·1	Glasgow	1877	D. & W. Hend'rson		Swansea Merchant Ship-owners' Co. Ld.	Swans
	82		**Zeta** (ex Sophie) *Henriksson 76-* W Sr	Russian	101	75·0	18·6	8·6	N. Pekela	1860	G. Holtmanns		Graf N. von Steenbock	Reval
95071	83		**Zeus** *Walker -88* Stl Sr	British	219	115·2	28·0	9·2	Dumbarton	1888	W. Denny & Bros		La Platense Flotilla Co. Ld	Glasgo
	84	GV	**Zeus** *Gruber 86-86* W Bn	German	253	113·6	25·5	12·8	Brake	1880	D. Bogge	[illegible]	H. A. Hellmers	Brake
28	85		**Zeus** W Bg	Greek	239	..	..	..					E. Zisimos	Poros
	86	N	**Zeus** *Christophersen* W Bk	Norweg'n	445	125·5	30·8	17·4	Grimstad	1860	Ole Kittelsen	[illegible]	Gunder Olsen	Arend
	87	V	**Zeus** *Christophersen* W 3m Sr	Norweg'n	271	120·2	27·2	13·1	Stavanger	1875	K. Nielsen	[illegible]	E. Berentsen	Stavan
	88	V	**Zévallos** *Guidici -88* I Bk	French	264	120·7	22·5	..	Nantes	1868	Jollet & Babin		Servel & fils	Marseil
2579	89	A	**Zia Catterina** *Komanich -88* W Bk	Austrian	509	134·8	27·6	18·4	Fiume	1869	G. Brazzoduro	[illegible]	Catterina Chiachich	Fiume
	90	N	**Ziba** *Knudsen 81-81* W 3m Sr	Norweg'n	271	122·4	27·5	12·0	Arendal	1876	P. Kittelsen	[illegible]	O. P. Olsen	Arendal
56338	91	R	**Zimi** *Lloyd 83-83* W S	British	943	165·3	37·2	22·7	St. Martin's, NB	1864		[illegible]	Trayes & Co.	Cardiff
28890	92	R	**Zimri S. Wallingford** *Willetts 82-* W 3m Sr	American	296	129·7	30·3	9·7	Bath, Me.	1882	Goss & Sawyer		B. F. Nealley	Portsm
51280	93		**Zingara** *Le Brocq* W Bn	British	174	103·0	21·1	12·6	Jersey	1866	Le Vesconte	[illegible]	F. Le Sueur & Son	Jersey
	94	R	**Zingara** W Bn	French	150	94·0	25·3	10·8	Shelburne, NS	1882	W. S. Kelly	[illegible]	L. Cost	St. Malo
25936	95		**Zingarella** W Bk	American	272	117·0	28·1	10·5	Wood's Hole, (Mass.	1856			J. Stewart	Boston
28025	96		**Zingo** *Perkins* W Sr	American	101	80·3	24·1	6·3	Bath, Me.	1865			J. M. Seaver	Duxbu
58518	97	✳	**Zingra** *Couch* W Sr	British	99	91·3	20·8	11·7	Brixham	1866	Pearce & ...	[illegible]	W. C. Phillips	Fowey
	98		**Zinnet Deriyah** W Bg	Turkish	153	..	..	..	Cephalonia	1879			[illegible]	Tereb
940	99		**Zio** *Schiaffino* W Bk	Brazilian	330	113·8	27·9	17·6	Varazze	1864	...		Costa Pinto & filhos	Bahia
2688	100	I	**Zio Battista** *Olivari* W Bk	Italian	687	155·8	33·1	22·8	Savona	1875	F. Sirello	[illegible]	G. B. Olivari	Genoa

✳ = **Lloyd's Register**; D = Nederlandsche Vereeniging; G = Germanischer Lloyd; I = Registro Italiano; R = American ...
A = Veritas Austro-Ungarico; H = Veritas Ellenico; N = Norske Veritas; V = Bureau Veri...

In the 4th Column, C = Composite (iron frame, wood planked); I = Iron; Stl = Steel; W = Wood.

The Dates of Metalling (**Col. 11**) are based on the records of the Books in which Vessels are respectively classed, or on the Reports of the Surveyors of Lloyd's ...

Further particulars of Vessels classed by Lloyd's Register will be found in Lloyd's Register Book and Lloyd's Yacht Register.

Nos dans le livre	Registre dans lequel le navire est coté	Nom. Matériaux et Gréement. Capitaine et dates d'entrée au service des Armateurs actuels et à bord du navire. Ancien Nom.	Pavillon.	Tonnage net.	Dimensions. Longueur	Largeur	Creux	Construit. Où.	Quand.	Par qui.	Date des doublages	Armateurs ou Directeurs-Gérants.	Port d'Armement.
101	XV	Zippora *Olsen 84-84* WS	Norweg'n	890	182·5	35·7	21·1	Haugesund	1878	J.Hauge	Y 10.87	H.O.Sundför	Haugesund
102	N	Zippora *Nielsen* W3mSr	Norweg'n	334	125·1	29·7	13·7	Risör	1885	Tarald Olsen	Y 12.85	J.Prebensen	Risör
103	N	Zippora *Thommessen* WBg	Norweg'n	256	112·9	26·3	12·9	P.E.Island	1863			W.Wilhelmsen	Tönsberg
104		Zizi *Castanheira* WSr	Brazilian	130	..	..	..	England	1871		Y 2.84	J.P.S.Moreira	Bahia
105	IV	Zongli *Porcella* WBk (ex [illegible])	Italian	746	158·8	33·8	21·8	Sestri	1877	A.Briasco fu G.B.	Y 3.88	Raggio & Porcella	Spezia
106		Zodiac *Powell* WBk	British	316	117·2	25·0	16·0	Salcombe	1852	Hayden		J.P.Walker	Sydney.N.S.W.
107	*D	Zodiak *De Jonge -86* WSr	Dutch	142	87·9	22·1	11·7	Hoogezand	1875	J.A.Hooites	Y 7.86	J.A.Hooites	Hoogezand
108	*	Zoe *Skinner 88-88* IBk	British	742	187·9	30·8	18·8	Liverpool	1876	W.H.Potter&Co.		W.Nicol & Co.	Liverpool
109		Zoe *De la Haye -86* WBn	British	174	95·0	26·3	11·3	Oromocto.N.B.	1868			P.Colfer	Fleetwood
110		Zoe (ex Rudeta) (BROKEN UP) WBg	French	280	116·0	25·1	14·2	P.E.Island	18^{66}_{74}	Hayden		Vve Leroux	St. Malo
111	V	Zoila *McKay* WBk	British	932	173·0	35·0	21·4	C.St.Ignace	1873	Methot	Y 2.88	R.W.Morgan	London
112		Zoodochos Pigi WBg	Greek	247	92·1	26·9	..	Scopelos	1869	P.Barelas		I.F.Georgaris	Scopelos
113	A	Zoodochos Pigi *Zangaris* WSr	Greek	131	76·7	22·2	11·8	Scopelos	1883	N.Karaggiotzos		E.Parilari & Co.	Scopelos
114		Zoodochos Pigi WBg	Turkish	231	..	..	..	Syra	1869		Y 11.84	A.Cokinogenis	Samos
115	H	Zoodochos Pigi *Cardosotis* WBg	Turkish	217	..	..	..	Sinope	1868			G.T.Pantelemos	Vasiliko
116	N	Zorida *Berentsen* WBk	Norweg'n	543	145·7	29·4	17·0	Vegesack	1867	P.Sager	Y 5.88	M.Berentsen	Stavanger
117		Zoroya *Miller* WBk	British	383	130·5	27·3	17·0	Sunderland	1864	W.Briggs&Sons	Y 10.83	M.P.Blundell	Melbourne
118		Zorritos *Casanu 87-87* WSr (ex [illegible])	Peruvian	110	89·5	21·1	8·5	UnitedStates				J.D.Respaldiza	Callao
119	*	Zouave *Jones -74* WSr	British	127	93·2	17·9	10·7	Salcombe	1855		Y 1.85	T.H.Jones & Co.	Aberystwith
120	V	Zouave *Soper 76-85* WS (ABANDONED)	American	1151	182·5	37·4	23·8	Richmond,Me.	1862	Farrin	Y 8.86	M.F.Pickering&Co.	Boston
121		Zouave WBk	French	220	92·5	25·0	13·1	Vannes	1857				Narbonne
122	*	Zuleika *Nicol -86* IS	British	1092	211·5	35·2	21·1	Glasgow	1875	Aitken&Mansel		T.Law & Co.	Leith
123		Zuleika *Mulholland* WBn	British	212	111·7	25·4	12·4	P.E.Island	1866	Holman		W.Mulholland	Liverpool
124	N	Zuleima *Höyer -87* WBk	Norweg'n	469	139·3	31·0	16·6	Risör	1869	B.Tellefsen	Y 9.84	Actieselskabet Zuleima (J. C. Møller)	Christiania
125		Zulette *Kelly* WBk	British	763	164·0	33·3	20·1	St.John	1864	Milledge		J.Eskdale	North Shields
126		Zulette Kenyon *Buckingham* WSr	American	141	107·0	26·0	7·0	Hartford,Conn	1869			W.S.Williams	Hartford Conn.
127		Zulheig WSr	British	117	81·0	17·0	10·0	Calcutta				Ebrahim I. Myat	Moulmein
128		Zulma *Reymer -88* WBk	British	326	124·7	26·0	16·3	Yarmouth,N.S	1863			M.T.Chisholm	Turk's Island
129	V	Zulu *Corbett -87* WBk	British	888	176·1	35·7	20·3	Maitland,N.S.	1879	A.Putnam		A.Putnam	Maitland.N.S.
130	*	Zulu *Rutland 86-88* WBn	British	280	119·8	26·2	14·4	S.Shields	1872	Wildo	Y 2.88	Matthews & Luff	London
131		Zuurdijk *Olthof -82* W3mSr (ex ONDERNEMING)	Dutch	123	91·2	26·0	9·9	Sappemeer	1876	J.Berg Izn	Y 12.82	C.Doornbos	Zuurdijk
132	V	Zwei Carl *Jacobson* WSr	Russian	259	105·4	29·0	13·2	Kurland	1886	M.Sepp		P.Enke	Riga

Additional Ships are inserted in the Supplement at the end of the Register.

Lloyd's Register; D = Nederlandsche Vereeniging; G = Germanischer Lloyd; I = Registro Italiano; R = American Record;
Veritas Austro-Ungarico; H = Veritas Ellenico; N = Norske Veritas; V = Bureau Veritas.

Dans la 4me Colonne, C = Composite (membrure en fer, bordé en bois); I = Fer; Stl = Acier; W = Bois.

Dates de doublage en métal (Col. 11) ressortent des livres dans lesquels les navires sont respectivement classés ou des rapports des Inspecteurs du Lloyd's Register.

Pour plus amples renseignements sur les navires classés par le Lloyd's Register, voir le Lloyd's Register Book et le Lloyd's Yacht Register.

SUPPLEMENT TO REGISTER OF SAILING VESSELS.

SUPPLÉMENT DU REGISTRE DES NAVIRES À VOILES.

SAILING VESSELS.

No.	Class	Ship's Name. Master, and Dates of Appointment to present Owner's Service and to Ship. Ship's Late Name.	Material and Rig.	Flag.	Net Tonnage.	Dimensions. Length.	Breadth.	Depth.	Built. Where.	When.	By whom.	Date of Metalling	Owners or Managers.	Port of Registry.
1		Agios Markos		Greek	193	..	..	..	Syra	1879			C. Margaronis	Syra
2	H	Agios Mercurios *Comnenos*	W Bg	Turkish	225	..	..	..	Syra	1881			Hadji Commines	Castelor
3	H	AgiosSpyridon *Lazarou*	W Bk	Greek	273	..	..	..	Fiume	18 73 / 82			J. Lazarou	Spezzia
4	H	Agios Spyridon *Sotirakis*	W Bg	Greek	177	..	..	..	Galaxidi	1869			E.N. Sotirakis	Galaxidi
1		Bintang Antisfarie *Mohamet Osman*	W Sr	Dutch		..	..	..					PangeranBvendaHara	Banjermassin
1		Chyoan Maru	W Sr	Japanese	123	86·3	19·6	9·8	IdaKawajuno	1878	HiranoS.B.Yard		SukuSeizaemon	Osaka
1		Deus Comtigo	W Sr	Portug'se	110	..	..	..	SanMartinbo	1888			J. Rodrigues Ascenso (Filhos & Co.)	SanMartinbo
1		Evangelistria	W Sr	Greek	161	83·7	22·3	..	Skiathos	1888	C. Mitylinaios		C.&J. Raftakis	Skiathos
2		Evangelistria	W Sr	Greek	126	71·8	23·1	..	Syra	1888			G. Skyrianos	Syra
1	DV	Georgina *Stanley* (ex Cambridge)	W S	Chilian	1135	189·8	38·2	22·5	Metegiam.NS	1874	D.Smette	Y 10.83	W.S. Stanley & Co.	Valparaiso
1		Hairool Bario (ex Cambridge) *Moh Mosrah*	W Bg	Dutch	265	..	..	..	Siam	1865		Y 12.87	SechMohametbinOemar Arsan	Cheribon
1		Kate	W Bg	Greek	247	105·3	26·1	..	Syra	1888			A. Karpitsis	Syra
2		Krete	W Sr	Greek	214	97·8	25·7	..	Syra	1888			D. Vafiadakis	Syra
1	✱	Lamorna *Crane* -89 (ex Saint Marnock)	I Bk	British	1258	216·8	36·0	22·7	Liverpool	1867	T. Royden & Sons		Trinder, Anderson & Co.	London
2	G	Latwis *Mahlberg*	W Gallias	Russian	100	65·8	24·1	10·3	Mitau	1888	Saring		E. Mahlberg	Riga
1		Maimoon *Said Osman Barkia*		Dutch	177	..	..	..				V 1.88	SechAchmat bin Osman Barkia	Banjermassin
2		Masschoor Balgair *Sech Mohamat Tagie*	W Bg	Dutch	111	..	..	..					SaidAgel bin Aydroesbin Agel bin Salim	Sourabaya
1		Napoleon	W Sp	British	130	104·4	22·8	7·4	St. Thomas de Pierreville PQ	1888	G. Manseau		N. Meyette	Montreal
2	✱	Nellie *Roberts* 77-	W K Barge	British	70	85·3	19·1	6·8	Littlehampton	1882	J.& W.B. Harvey		J.& W.B. Harvey	Littlehampton
3		Noer Hassim *Sech Abdulla Basilla*		Dutch		..	..	..					SaidMohamat bin Alie Innesihae	Palembang
4		Novo Razoulo	W Sr	Portug'se	120	..	..	..					J.d'Oliveira da Vettro	Aveiro
1		Rosalina	W Sr	Portug'se	168	74·1	20·0	8·2	VilladoConde	1885	S.M.d'Araujo		J.deSouzaFaria	Oporto
1		Siak *Djapar*	W Sr	Dutch	216	..	..	..	Sourabaya	1872			SechSaleh bin Salim bin Abdoe Albatrie	Sourabaya
2		Suiniho Maru	W Bk	Japanese	299	130·5	24·1	13·3	America				Inouye Nihei	Osaka
3	A	Spirito (ex Spiro)	W Bn	Turkish	198	95·8	24·3	12·1	Pekela	1871	J.W. Kniper	Y 10.81	T. Mavronicola	Nicosia

Lloyd's Register; D = Nederlandsche Vereeniging; G = Germanischer Lloyd; I = Registro Italiano; R = American Record;
Veritas Austro-Ungarico; H = Veritas Ellenico; N = Norske Veritas; V = Bureau Veritas.

In the 4th Column, C = Composite (iron frame, wood planked); I = Iron; Stl = Steel; W = Wood.

Dates of Metalling (**Col. 11**) are based on the records of the Books in which Vessels are respectively classed, or on the Reports of the Surveyors of Lloyd's Register.

Further particulars of Vessels classed by Lloyd's Register will be found in Lloyd's Register Book and Lloyd's Yacht Register.

STEAMERS.

NAVIRES À VAPEUR.

2	3 Ship's Name. Material, Rig, &c. Master, and Dates of Appointment to present Owner's Service and to Ship. Ship's Late Name. No. of Decks, &c.	4 Flag. Port of Registry.	5 Tons. Net Gross	6 Dimensions. Length	Breadth	Depth	7 Built. Where. By Whom.	8 When	9 No. of Bulkheads. Particulars of Water Ballast.	10 Engines. Particulars. Maker's Name.	11 Owners or Managers.
	A. C.Cheney W ScTug *Helgea*	American New York	154	98·0	26·1	10·5	Newburgh Baldwin&Brown	1877			Cheney's Towing Line
	A. C. Freese	American	175 200	119·0	27·5	4·5	San Francisco	1886		40 HP.	A.C.Freese
V	A. E. Nordenskjold C ScSr *Kiseł -84* 1 *Dk*	Russian St.Petrsbg.	176 289	140·0	21·3	12·2	Malmö KockumsMek.Vks.	1879		2Cy.19"&35"–20" 60 HP. KockumsVks.Malmö	A. M. Sibiriakoff
	A. N. Hansen I ScSmSr *Thidemann 80-882 Dks&Spr Dk*	Danish Copenhagen	948 1505	260·1	32·1	17·2 23·9	Copenhagen Burmeister&Wain	1877	MT 585 FPT 37/ APT 37	1Cy.24"&48"–32" 150 HP. Burmeister&Wain,Cpnhgn.	Det Forenede Dmpskbs Selskab
	A. Schiaffino I ScSr *ex Nidelven) Dominici -88* 1 *Dk*	French Algiers	190 302	131·3	21·5	11·5	Drontheim Nidelvens Mek.Vrk.	1872	MT 56	2Cy.18"&36"–24" 45 HP Nidelvens Mek. Verks, Drontheim	Schiaffino frères.
✻	A. Strong I ScSr *Seamer* 1 *Dk 2 B*	British Cardiff	770 1198	240·0	30·7	19·6 *Q 39 f. B 50 f. F 35 f.*	Sunderland W.F.Pile&Co.	1871	5 Blds. Dbottom 382	2Cy.28½"&59"–36" 115 HP. R.&W.Hawthorn,N'wcastle	W. & R. H. Strong
	Aage W ScSmSr *Rom 81-87* 1 *Dk*	Danish Nyborg	247 393	119·6	25·5	13·0 *H—*	Odense H.L.Hansen	1873		2Cy.20"&36"–21" 50 HP Schweffel&Howaldt,Kiel	W.Krause
	Aalborg I PadSp *Steen 67-76* 1 *Dk*	Danish Aalborg	108 169	158·4	17·5	7·7 *Len. 1885.*	Copenhagen Burmeister&Wain	1876		1Cy.16"&32"–32" 60 HP. Burmeister&Wain,Cpnhgn.	Det Forenede Dmpskbs Selskab
	Aarhuus I ScSr *Fisgersen 81-88* 1 *Dk*	Danish Copenhagen	233 382	163·3	22·0	12·9 *P 94 f.*	Copenhagen Burmeister&Wain	$18\frac{65}{78}$		1Cy.18"&38"–20" 100 HP. Burmeister&Wain,Cpnhgn.	Det Forenede Dmpskbs Selskab
	Aawasaksa I ScSr *Tornberg 71-85* 1 *Dk*	Russian Tornea	179	..	..	..	Gothenburg	1871		80 HP.	Tornea Ångbåts Aktie Bolag
✻	Abana I ScBn *Bemert* 2 *Dks 3 B*	British Liverpool	1966 2998	325·2	40·0	26·5 *P 22 f. B 85 f. F 39 f.*	Stockton M.Pearse&Co.	1883	5 Blds. DBot 448	2Cy.40"&76"–54" 380 HP. Blair&Co.Ld.,Stockton	J. M. Wood
✻	Abana I ScSr *Torston 87-84* 1 *Dk 2 B*	British Sunderland	460 719	204·8	27·8	15·2 *Q 30 f. F 17 f. F 16 f.*	Sunderland J.Laing	1871	4 Blds. W Bos f	2Cy.25"&48"–33" 90 HP. J.Dickinson,Sunderland	J.Westoll
✻D	Abbotsford I ScBg *Figgans 88-93* 1 *Dk & Spr Dk 3 B*	British Leith	656 1039	227·5	29·1	15·7 23·1	Port Glasgow Blackwd.&Gordon	1870	5 Blds.	2Cy.26¾"&53¾"–36" 100 HP Blackwood&Gordon,Pt.Gls.	G.Gibson & Co.
V	Abd el Kader I ScBn *Beranrdoni -84 2 Dks & Spr Dk*	French Marseilles	992 1816	310·2	33·5	16·7 *H 175 f. P 46 f. 24·3*	Glasgow John Elder&Co.	1880		2Cy.42"&80"–48" 150 HP. John Elder&Co.,Glasgow	Compagnie Générale Transatlantique
✻	Abden I ScSr *McKendrick 78-84* 1 *Dk*	British Kirkcaldy	188 370	165·0	24·2	12·0	Dundee Gourlay Bros.&Co.	1884	4 Blds. DB 38 FPT 19	2Cy.21"&42"–30" 90 HP. Gourlay Bros.,Dundee	London & Kirkcaldy Shipping Co.
	Abeille No. 8 I TwnScTug *Howard 79* 1 *Dk*	French Havre	74 160	118·1	19·0	11·7	Port Glasgow Cunliffe&Dunlop	1879		1Cy.17"&30"–24" 180 HP. Cunliffe&Dunlop,Pt.Glsgw.	Soc. de Remorquage Les Abeilles
	Abeille No. 9 Stl TwnScTug *Delamer 78* 1 *Dk*	French Havre	75 161	123·0	19·0	12·1	Port Glasgow Cunliffe&Dunlop	1881	FPT 12 APT 8	1Cy.17"&32"–24" 180 HP. Cunliffe&Dunlop,Pt.Glsgw.	Soc. de Remorquage Les Abeilles
✻	Abeona Stl ScSr *Lewis* 1 *Dk 2 B*	British W.Hartlepl.	1405 2152	275·3	37·3	19·8 *P 30 f. B 108 f. F 30 f. Q 76 f.*	W. Hartlepool Gray&Co.	1886	4 Blds. DBot&f 531	3C.21"35"&57"–39" 180 HP. Central Marine E.Co.,W.Hpl.	Rickinson Son & Co.
✻	Abeona I ScSr *Temple -80-86* 1 *Dk 2 B*	British Newcastle	575 895	200·1	30·0	15·5 *Q 70 f. F 24 f. B 73 f.*	Sunderland Bartram,Haswell&Co.	1881	3 Blds. W Bos f APT	2Cy.27"&51"–33" 99 HP. J.Dickinson,Sunderland	J.Graham & Co.
✻	Aberdare I ScSr *Knowles* 1 *Dk 2 B*	British Cardiff	877 1348	240·0	33·2	16·7 *Q 92 f. B 80 f. F 26 f.*	Newcastle Palmers'Co.,Ld.	1882	4 Blds. APT 54	2Cy.30"&57"–39" 99 HP. Palmers'Co.,Ld.,Newcastle	"Aberdare" S.S.Co.,Ld. (Morel Bros. & Co.)
✻	Aberdeen I ScBk *Taylor 85-87 2 Dks & Spr Dk*	British Aberdeen	2371 3616	362·5	44·4	23·5 *B 43 f. 31·2*	Glasgow R.Napier&Sons	1881	5 Blds. APT 196	3Cy.30"45"&70"–54" 400 HP R.Napier&Sons,Glasgow	G.Thompson & Co.
✻	Aberdeen I ScSr *Turner 58-88* 1 *Dk 2 B*	British Aberdeen	685 1077	230·1	30·3	17·0 *P 121 f. F 31 f.*	Aberdeen Hall,Russell&Co.	1871	5 Blds.	3Cy.17"28½"&51" 123 HP. Tpl.'87,Hall,Russell&Co.,Abdn	Adam & Co.
✻	Aberdeenshire I ScSr *Cross* 1 *Dk*	British Hull	228 365	171·2	20·6	11·5 *P & B 83 f. F 16 f.*	Renfrew W.Simons&Co.	1855	5 Blds.	2Cy.17"&34"–26" 50 HP. Comp.79,Earles'Co.,Ld.,Hull	W. Rawson & J. Robinson
✻	Aberdour I ScSr *Wilson 79-87* 1 *Dk 2 B*	British Aberdeen	971 1519	250·1	34·2	17·7 *P 142 f. F 34 f.*	Aberdeen Hall,Russell&Co.	1883	5 Blds. DBot&f 200	2Cy.30"&59"–36" 150 HP. Hall,Russell&Co.,Aberdeen	Adam & Co.
✻	Aberforth Stl ScSr *Sinclair 86-88* 1 *Dk*	British Grng'mouth	423 694	190·3	28·6	13·2 *Q 57 f. B 46 f. F 46 f.*	Port Glasgow David J.Dunlop&Co.	1884	4 Blds. FPT 26 APT 12 MT 294	2Cy.24"&46"–36" 90 HP David J.Dunlop&Co.,PtGlsgw	A.& A. Y. Mackay
✻	Abergeldie I ScBkn *Murray 75-84* 1 *Dk & Spr Dk 3 B*	British Aberdeen	1878 2863	315·0	40·1	20·3 *F — 28·3*	Sunderland Sunderld.S.B.C.,Ld.	1882	5 Blds.	2Cy.40"&75"–48" 300 HP. N.E.Marine E.Co.,Ld.,Sdrlnd	Adam S.S.Co.,Ld.
✻	Abergrange I ScSr *Bell -88* 1 *Dk 2 B*	British Grng'mouth	381 602	191·0	25·6	14·2 *Q 100 f. F 32 f.*	Port Glasgow Cunliffe&Dunlop	1874	4 Blds. MT 206	2Cy.33"&44"–30" 85 HP. W.King&Co.,Glasgow	A.& A. Y. Mackay
✻	Aberlady Bay I ScSr *McGregor 82-83* 2 *Dks 3 B*	British N.Shields	1188 1845	267·2	35·3	23·8 *P 28 f. B 50 f. F 36 f.*	Newcastle A.Leslie&Co.	1883	5 Blds. APT 54	2Cy.32"&62"–39" 200 HP. N.E.Marine E.Co.,Ld.,Nwc.	W. J. Affleck & Co.
✻	Abermaed I ScSr *Phillips 86-86* 1 *Dk 2 B*	British Swansea	710 1111	225·7	31·2	17·0 *Q 113 f. B 34 f. F 29 f.*	Newcastle Palmers'Co.,Ld.	1883	4 Blds. APT 60	2Cy.28"&52"–33" 98 HP. Palmers'Co.,Ld.,Newcastle	"Abermaed" S.S.Co., Ld.(Letricheux & David)
✻	Abertawe I ScSr *Phillips 84-86* 1 *Dk 2 B*	British Swansea	917 1411	240·5	33·2	18·1 *Q 95 f. B 46 f. F 28 f.*	Newcastle Palmers'Co.,Ld.	1884	4 Blds.	2Cy.30"&57"–39" 150 HP. Palmers'Co.,Ld.,Newcastle	"Abertawe" S.S.Co., Ld.(Letricheux & David)

...loyd's Register; D = Nederlandsche Vereeniging; G = Germanischer Lloyd; I = Registro Italiano; R = American Record; ...itas Austro-Ungarico; H = Veritas Ellenico; N = Norske Veritas; V = Bureau Veritas.

...the 2nd Column, B.T., with figures following, signifies Date of report of bottom sighting by Board of Trade Surveyors ▬ condition of PASSENGER CERTIFICATE.

In the 3rd Column, C = Composite (iron frame, wood planked); I = Iron; Stl = Steel; W = Wood.

Further particulars of Vessels classed by Lloyd's Register will be found in Lloyd's Register Book and Lloyd's Yacht Register.

No. in book. / Official No.	Registry in which classed	Ship's Name. Material, Rig, &c. / Master, and Date of Appointment to present Owner's service and in Ship / Ship's Late Name. No. of Decks, &c.	Flag. / Port of Registry.	Tons Net / Gross	Dimensions. Length	Breadth	Depth	Built. Where. / By Whom.	When.	No. of Bulkheads. / Particulars of Water Ballast	Engines. Particulars. Maker's Name.	Owners or Managers
31 [illegible]	✱	Abertay [illegible] Logan 82-88 [illegible]	British [illegible]	635 [illegible]	221·0	33·1	12·[illegible]	Renfrew [illegible]	1888	[illegible] Blhds. [illegible]	3Cy.17",27"&44"–30" 98HP. W.Simons&Co.,Renfrew	A. & A.Y. Ma[illegible]
32 87333	✱	Abington [illegible] Wilson 82-83 2 Dks [illegible]	British [illegible]	1[illegible] [illegible]	2[illegible]	37·4	22·[illegible]	Sunderland Doxford&Sons	1882	[illegible] Blhds. [illegible]	2Cy.3[illegible]"&[illegible]"–45" 250 HP. W.Doxford&Sons,Sundrlnd	"Abington"S[illegible] (J. A. [illegible])
33		Åbo Norring 1 [illegible] 2 Dks [illegible]	Russian Åbo	283 [illegible]	160·0	26·0	12·2	Åbo W.Crichton & Co.	1873	[illegible] Blhds.	2Cy.20"&40"–24" 98 HP. W.Crichton&Co.,Åbo	Åbo LansK[illegible] Aktiebolag[illegible]
34 86106	✱ D	Aboraca 1 [illegible] Temple 1 Dk 2 [illegible]	British Newcastle	1387 2129	275·0	38·6	20·6	Sunderland Short Bros.	1883	[illegible] Blhds. [illegible]	2Cy.36"&67"–36" 180 HP. T.Richardson&Sons,Hrtlpl.	J.Graham & C[illegible]
35 [illegible]	✱	Abraham Sutton 1 ScSmSr Sutton 83-83 1 Dk [illegible]	British [illegible]	174 [illegible]	145·0	22·2	10·8	Dundee Pearce Bros.	1883	[illegible] Blhds. [illegible]	2Cy.20"&38"–27" 60 HP. Pearce Bros.,Dundee	G.A. Sutton
36	✱ D	Abril [illegible] Lezama 83-88 1 Dk 2 [illegible]	Spanish [illegible]	846 1314	240·0	34·8	15·4	Port Glasgow Cunliffe&Dunlop	1881	[illegible] Blhds. [illegible]	2Cy.30"&55"–36" 130 HP. Cunliffe&Dunlop,Pt.Glsgw.	Cia Bilbaina de [illegible]
37 86102	✱	Abrota 1 [illegible] Robson 86-86 1 Dk 2 [illegible]	British Newcastle	780 1204	228·0	32·7	16·5	Sunderland Short Bros.	1882	[illegible] Blhds. [illegible]	2Cy.29"&56"–36" 120 HP. N.E.Marine E.Co.Ld.,Sndrlnd	J.Graham & [illegible]
38	✱ V	Absalon 1 [illegible] Pedersen 83-83 2 Dks 3 [illegible]	Danish Copenhagen	903 [illegible]	245·7	32·3	16·9 24·2	Copenhagen Burmeister&Wain	1872	[illegible] Blhds. [illegible]	1Cy.22"&44"–30" 120 HP. Burmeister&Wain,Cpnhgn.	Dampsk Sels[illegible] mark"
39 65566	✱	Abydos 1 [illegible] Smith 2 Dks 3 [illegible]	British [illegible]	1069 [illegible]	260·0	31·4	21·8	Glasgow London&Glas.Co.	1871	[illegible] Blhds. [illegible]	2Cy.31"&54"–33" 150 HP. London&GlasgowCo.Ld.,Gls.	Raeburn & [illegible]
40	✱	Abydos [illegible] Siegmund 80-88 2 Dks 3 [illegible]	German Hamburg	1495 2[illegible]	295·0	36·2	23·[illegible]	Newcastle W[illegible]	1888	6 Blhds. [illegible]	3Cy.24½",38"&63"–42" 310 HP. Wigham Richardson & Co. [illegible]	Deutsche Da[illegible] Gesell. [illegible]
41 63703		Abyssinia 1 [illegible] Lee 87-87	British [illegible]	2346 3651	363·5	42·4	34·0	Glasgow J.&G.Thomson	1870		1Cy.31"&72"–48" 500 HP. Comp[illegible] J.Jones&Sons,Lpl.	GuionSteam[illegible]
42 77226	✱	Acacia 1 [illegible] Smith 86-86 1 Dk	British London	270 428	165·3	22·7	1·25	Port Glasgow W.Hamilton&Co.	1879	3 Blhds. [illegible]	2Cy.22"&40"–30" 80 HP. Dunsmuir&Jackson,Glsgw.	Arnati & Ha[illegible]
43 56164		Acadia 1 [illegible] Russell 80-88 2 Dks	British Glasgow	697 1082	264·8	26·4	20·2	Glasgow A.Stephen&Sons	1866	1 Blhds.	2Cy.29½"&58"–36" 200 HP. D.&W.Henderson,Glasgow	Henderson B[illegible]
44 77697		Acadia 1 [illegible] 2 Dks	British [illegible]	509 [illegible]	176·6	25·6	10·4	Hamilton,Ont. J.Malcomson	1867		1Cy.32"–32" 87 HP. Beckett&Co.,Hamilton,Ont.	J.W. Sutherl[illegible]
45 61157		Acadian 1 [illegible] 1 Dk 2 [illegible]	British Montreal	596 931	205·1	28·8	15·8	Sunderland J.Dennison	1872	1 Blhds. [illegible]	2Cy.26"&49"–30" 100 HP. T.Richardson&Sons,Hrtlpl.	H. & A. Alla[illegible]
46 105236	R V	Acapulco 1 [illegible] Pitts 78-87 2 Dks & Spr Dk	American New York	1759 2572	300·0	43·0	20·7 28·6	Wilmington Del Harlan&Hollingsworth Co.	1873		2Cy.50"&88"–42" 400 HP. J.Elder&Co.,Glasgow	Pacific Mail S[illegible]
47		Acara [illegible] 1 Dk	Brazilian Pará	334 429	177·0	27·5	8·2	Newcastle A.Leslie&Co.	1884		2Cy.30"&54"–51" 120 HP. Black,Hawthorn&Co.,Nwc.	Cia.de Nav.[illegible] Amazonas
48 [illegible]	✱	Acaster 1 [illegible] Phillips 88-88 1 Dk 2 [illegible]	British W. Hartlepl.	941 1450	251·0	34·6	17·5	Sunderland J.Priestman&Co.	1882	1 Blhds. [illegible]	2Cy.31"&58"–36" 140 HP. N.E.Marine E.Co.Ld.,Sdrlnd	J.Wood & Co.
49 86294	✱	Accomac 1 ScBg Windsor 86-86 2 Dks 3 [illegible]	British Liverpool	1629 2500	299·5	39·0	25·0	Stockton M.Pearse&Co.	1883	[illegible] Blhds. [illegible]	2Cy.38"&71"–45" 300 HP. Blair&Co.,Ld.,Stockton	Tapscott S.S[illegible]
50 81486	✱	Accretive 1 [illegible] 1 Dk & Web Frames	British Sunderland	754 1170	225·4	33·0	16·5	Sunderland S.P.Austin&Son	1880	[illegible] Blhds. [illegible]	2Cy.28"&56"–36" 99 HP. G.Clark,Sunderland	T.Stockdal[illegible]
51 [illegible]		Achille 1 ScSr Zelinka 57-87 2 Dks	Austrian Trieste	1249 1964	307·0	33·0	26·0	Dumbarton W.Denny&Bros.	1874		2Cy.41"&70"–42" 250 HP. Denny&Co.,Dumbarton	Lloyd Austr[illegible] rico
52	V	Achille 1 ScBg Naudin 1 Dk 2 [illegible]	French Dieppe	446 754	202·1	27·9	16·5	Renfrew Lobnitz&Co.	1883	[illegible]	2Cy.27"&50"–36" 120 HP. Lobnitz&Co.,Renfrew	Grandchamp[illegible]
53 94285	✱ BT8.88	Achille Adam SOT–ScSr Harrold 87-87 1 Dk	British London	236 434	190·0	27·0	11·8	London Samuda Bros. Ld.	1887	[illegible] Blhds.	1Cy.15"&35"–19" 119 HP. Maudslay Sons & Field,Lndn	South Easte[illegible] way Co.
54 [illegible]	BT12.88	Achilles 1ScBk Anderson 77-77	British Liverpool	1529 2359	309·5	38·8	20·6 28·1	Greenock	1866		2Cy.28"&56"–52" 300 HP. Greenock Foundry Co.Grnck	Ocean Steam[illegible]
55 51181	✱	Achilles 1 ScSr Nairn [illegible] 1 Dk	British Sunderland	913 1433	274·8	32·2	18·5	Sunderland J.Laing	1865	[illegible] Blhds.	2Cy.31"&58"–39" 130 HP. N.E.Marine E.Co.Ld.,Sdrlnd	Tyzack & Br[illegible]
56 70471	✱	Achilles 1 ScSmSr Billing 83-83 1 Dk	British Barrow	308 486	182·0	24·1	12·1	Barrow Barrow S.B.Co.	1874	3 Blhds. [illegible]	2Cy.20"&37½"–32" 70 HP. Barrow S.B.Co.,Barrow	J. & W.W.[illegible]
57 [illegible]	✱	Acme StlScSr Bentley –87 1 Dk 2 H	British London	1398 2148	275·3	37·9	19·0	Newcastle Palmers Co.Ld.	1887	[illegible] Blhds. [illegible]	3Cy.22½",36½"&60"–39" 200 HP. Palmers Co.Ld.,Newc.	Acme S.S.C[illegible] (Newman & [illegible])
58 [illegible]		Acme W ScSr	British Liverpool	103 151	85·2	21·1	9·2	Winsford G.Deakin Ld.	1883		2Cy.10"–12" 22 HP. Gordon,Manchester	G.Deakin,Ld
59 63990	✱ BT7.88	Aconcagua 1 ScBk Hamilton 3 Dks & Awn Dk	British Liverpool	2625 4104	404·8	41·4	35·3	Glasgow J.Elder&Co.	1872	[illegible] Blhds.	2Cy.61"&112"–48" 500 HP. J.Elder&Co.,Glasgow	Pacific S.Nav[illegible]
60	✱	Aconcagua StlScSr 2 Dks & Shade Dk	Chilian Valparaiso	1400 2300	340·0	40·2	[illegible]	Port Glasgow J.Reid&Co.	1889	5 Blhds. [illegible]	[illegible] J.&J.Thomson,Glasgow	Compania Sud [illegible]

✱ = **Lloyd's Register**; D = Nederlandsche Vereeniging; G = Germanischer Lloyd; I = Registro Italiano; R = American [illegible]
A = Veritas Austro-Ungarico; H = Veritas Ellenico; N = Norske Veritas; V = Bureau Ve[illegible]

In the 2nd Column, B.T., with figures following, signifies Date of report of bottom sighting by Board of Trade Surveyors as condition of PASSENGER CER[illegible]

In the 3rd Column, C = Composite (iron frame, wood planked); I = Iron; Stl = Steel; W = Wood.

Further particulars of Vessels classed by Lloyd's Register will be found in Lloyd's Register Book and Lloyd's Yacht Register.

Classe	Nom, Matériaux, Gréement, etc. Capitaine et dates d'entrée au service des Armateurs actuels et à bord du navire. Ancien Nom, nombre de ponts, etc.		Pavillon. Port d'Armement	Tonnage. Net / Total	Dimensions	Construit. Où. Par qui.	Quand	Nombre des cloisons. Détails sur le lest d'eau.	Machines. Détails. Nom du Constructeur.	Armateurs ou Directeurs-Gérants.
✻	**Açôr** *Satterley*	I ScSr 1 *Dk & Spr Dk 3 B*	Port'g'se Ponta Delgada, Azores	774 1211	240·0|30·7|15·4 22·7	Hull Earle's S. & E. Co.	1877	4 Blkds. [illegible]	2Cy.32" & 64"–36" 150 HP. Earle's S. & E. Co., Hull	Empreza Insulana de Navegação
	Acre	I PdSr 1 *Dk, Spt Hur Dk*	Brazilian Para	200	130·0|24·0|7·0	Wilmingtn Del. Pusey, Jones & Co.	1873		2Cy.18"–72" 100 HP. Pusey, Jones & Co., Wilmngtn.	Amazon S. Nav. Co.
	Actif *Auffret* –86	I ScSr 1 *Dk 2 B*	French Caen	565 883	205·0|31·6|14·2 *Q 67 f. B 48 f. F 22 f.*	Aberdeen A. Hall & Co.	1883	5 Blkds. [illegible]	2Cy.27" & 50"–33" 110 HP. Blaikie Bros., Aberdeen	J. Lamy
	Actif	I PadSr		80 146	.. .. ..	La Seyne	1857		125 HP.	
✻	**Activ** *Rersbech* 79-81	I ScSr 1 *Dk & Awn Dk*	Danish Randers	249 416	174·8|24·3|11·8	Dumbarton R. Chambers jun.	1880	5 Blkds. MT 64 t	2Cy.20" & 38"–27" 65 HP. M. Paul & Co., Dumbarton	DampskSelsk "Activ" (C. Hassager)
D X	**Activ** *Ellefsen* –87	I ScSr 1 *Dk*	Norweg'n Bergen	267 427	151·6|24·9|11·6 *P 81 f. F 29 f.*	Glasgow Aitken & Mansel	1872	4 Blkds. [illegible]	2Cy.19" & 33½"–24" 50 HP. Kincaid & Co., Greenock	P. G. Halvorsen
	Activ ex Johan	W ScSr 1 *Dk*	Swedish Vestra	112 133	102·3|20·0|11·5 *H 5 t.*	Christinehamn. A. Tr[illegible]	18 57/72		2Cy. 12 HP. Christinehamns Mek. Verks.	Johan Olsson
V	**Activa** *Johannsen* 87-87	I ScSr 1 *Dk*	German Flensburg	342 539	169·9|25·6|13·2 *Q 48 f. B 47 f. F 20 f.*	Flensburg Flensburg Schiffsties.	1883	WB	2Cy.21" & 38"–27" 60 HP. Flensburg Schiff. Ges., Flensburg.	H. A. Petersen
	Active *Robertson*	W ScBk	British Dundee	237 348	117·0|28·6|18·1	Peterhead Robertson	1852		2Cy.20" & 30"–21" 40 HP.	R. Kinnes
	Active ex Sailing Vessel	W ScK 1 *Dk*	British Melbourne	79 117	100·7|18·2|9·0	Como, N.S.W. H. Stephens	1885		2Cy.12" & 24"–30" 24 HP. Anderson, Campbell & Ss. Melb.	D. Munro & Co., Ld.
✻	**Activity** *Ralston* 85-86	I ScSr 1 *Dk 2 B*	British Cardiff	758 1178	225·5|31·8|16·2 *Q 82 f. F 23 f.*	Sunderland J. Laing	1879	4 Blkds. [illegible] APT FPT	2Cy.29½" & 55"–36" 140 HP. J. Dickinson, Sunderland	L. & H. Guéret
✻ 32.4.88	**Actor** *James* 81-86	Stl ScBn 1 *Dk & Spr Dk*	British Liverpool	1035 1600	250·0|34·0|22·9 *B 22 f. F 23 f.*	Middlesbro' R. Dixon & Co.	1885	4 Blkds.	3C 20"–33" & 54"–36" 140 HP. Blair & Co., Ld., Stockton	Charente S.S. Co., Ld.
✻	**Acuba** *Graham* 85-88	I ScSr 1 *Dk 2 B*	British Sunderland	1159 1793	260·0|36·2|20·2 *Q 83 f. P 78 f. B 132 f. F 38 f.*	Sunderland Short Bros.	1883	6 Blkds. [illegible] APT	2Cy.33" & 62"–42" 160 HP. J. Dickinson, Sunderland	T. Stockdale
	Acuna ex Mahoma	I Sc	French Oran	132 175	132·5|18·1|9·8	Greenock J. Scott & Sons	1854			F. Gonzalvez
✻	**Ada** *Gibb*	I ScSr 1 *Dk*	British London	555 876	220·0|30·0|13·9 *Q 91 f. B 40 f. F 21 f.*	Sunderland R. Thompson & Sons	1881	4 Blkds. [illegible] FPT 17t	2Cy.26" & 48"–36" 99 HP. G. Clark, Sunderland	Dansey & Robinson
✻	**Ada** ex Allan Macdonald *Swift* 84-84	I ScSr 1 *Dk*	British Liverpool	59 104	108·7|19·6|8·9 *Q 36 f. F 12 f.*	Campbeltown Campbeltown S.B. Co.	1880	3 Blkds. FPT 12t	2Cy.13" & 23½"–18" 25 HP. Ross & Duncan, Glasgow	S.S. "Ada" Co., Ld. (J. J. Mack & Sons)
✻	**Ada** *Taylor*	I ScTug 1 *Dk*	British London	13 47	70·2|13·9|7·0	London J. & W. Dudgeon	1875	3 Blkds.	2Cy.15¼" & 27⅞"–14" 30 HP. J. & W. Dudgeon, London	Phillipps & Graves
	Adam *Friberg*	I ScSr 1 *Dk*	Swedish Ljusne	199 285	115·4|21·7|13·5 *B 31 f. H 5 t.*	Ljusne Ljusne Mek. Verks.	1883		40 HP.	Ljusne-Voxna Akt. Bol. (W. H. Kempe)
✻	**Adam Smith** *Kinnear*	I ScSr 1 *Dk*	British Dundee	156 292	150·3|22·1|12·0 *Q 49 f.*	Port Glasgow H. Murray & Co.	1876	4 Blkds. DB 31 t APT FPT	2Cy.21" & 37"–21" 60 HP. Kincaid, Donald & Co., Grnck.	W. Wighton
✻	**Adara** *Hool* 86-88	I ScSr 1 *Dk 2 B*	British Shoreham	832 1303	236·2|31·5|17·7 *Q 86 f. F 29 f.*	Sunderland Mounsey & Foster	1877	4 Blkds. [illegible]	2Cy.29" & 55"–36" 120 HP. J. Dickinson, Sunderland	R. H. Penney
✻ D	**Addington** *Tapper* 82-87	I ScSr 1 *Dk 2 B*	British London	1091 1695	260·0|36·1|18·7 *Q 84 f. P 26 f. B 59 f. F 27 f.*	Newcastle Edwards Sns & Craig	1883	4 Blkds. [illegible]	2Cy.32" & 60"–39" 170 HP. Blair & Co., Ld., Stockton	T. Nelson & Son
	Adeh	I PadSr 1 *Dk*	Sarawak Sarawak	189 300	149·8|25·1|8·5 *P 25 t. F 8 t. H 33 t.*	Port Glasgow Blackwd. & Gordon	1884		2Cy.20" & 37"–48" 75 HP. Blackwood & Gordon, Pt. Gls.	Raja Brooke
✻ V	**Adela** *Torque*	I Sc3mSr 2 *Dks*	British London	381 617	200·4|23·2|17·8	Newcastle A. Leslie & Co.	1870	3 Blkds.	2Cy.22" & 44"–30" 75 HP. J. Taylor & Co., Birkenhead	A. J. Bird
	Adela *Foster*	I Sc3mSr 2 *Dks*	British Dublin	346 655	199·8|32·2|14·5 *P 23 f. F 16 f.*	Port Glasgow H. Murray & Co.	1878	5 Blkds. FPT	2Cy.32" & 59"–39" 150 HP. Kincaid, Donald & Co., Grnck.	R. Tedcastle & Co.
B.T.	**Adela** *McTaggart* 79-79	I PadSk	British Glasgow	33 206	207·7|19·2|7·4	Greenock Caird & Co.	1877		1Cy.49"–54" 90 HP. Blackwood & Gordon, Pt. Gls.	A. Campbell
	Adela	I	Spanish Sagua	382 622	156·5|28·4|13·1	United States	1878		130 HP.	Cuetara & Co.
✻	**Adelaide** *Lockyer*	I ScSr 2 *Dks & Awn Dk*	British Adelaide	917 1711	280·0|33·2|17·6 *H 34 t.* 25·4	Glasgow D. & W. Henderson & Co.	1883	4 Blkds. [illegible]	2Cy.35½" & 69"–48" 267 HP. D. & W. Henderson & Co., Gls.	Adelaide Steamship Co., Ld.
B.T.	**Adelaide** *Sherlock* 81-82	Stl PadSr 2 *Dks*	British Harwich	553 969	254·2|32·3|13·0 *P 22 t. F 16 t.*	Barrow Barrow S.B. Co.	1880	3 Blkds.	2Cy.45" & 87"–72" 300 HP. Barrow S.B. Co., Barrow	Great Eastern Railway Co.
	Adelaide	I PadBkn 1 *Dk*	British Adelaide	139 255	141·2|22·4|10·4 *H 8 t.*	Aberdeen Hall, Russell & Co.	1877		3Cy.26" 32" & 32"–54" 99 HP. Hall, Russell & Co., Aberdeen	Adelaide Steam Tug Co., Ld.
	Adelaide *Gerard* –88	W ScSr 1 *Dk*	British Port Darwin S.A.	75 124	107·0|18·3|7·5 *F, &c., 9 t.*	Hongkong G. Fenwick & Co.	1888		2Cy.12" & 24"–14" 25 HP. G. Fenwick & Co., Hongkong	H. C. Edwards

[illegible] = Lloyd's Register; D = Nederlandsche Vereeniging; G = Germanischer Lloyd; I = Registro Italiano; R = American Record;
[illegible] = Veritas Austro-Ungarico; H = Veritas Ellenico; N = Norske Veritas; V = Bureau Veritas.

Dans la 2me Colonne, B.T., avec les indications qui suivent, signifie la date du rapport de visite de la Carène par les experts du Board of Trade pour obtenir le certificat relatif au TRANSPORT DES PASSAGERS.

Dans la 3me Colonne, C = Composite (membrure en fer, bordé en bois); I = Fer; Stl = Acier; W = Bois.

Pour plus amples renseignements sur les navires classés par le Lloyd's Register, voir le Lloyd's Register Book et le Lloyd's Yacht Register.

No. in Book. Official No.	Registry Society & where classed.	Ship's Name. Material, Rig, &c. Master, and Dates of Appointment to present Owner's Service and to Ship. Ship's Late Name. No. of Decks, &c.	Flag. Port of Registry.	Tons. Net. Gross.	Dimensions. Length	Breadth	Depth	Built. Where. By Whom.	When	No. of Bulkheads. Particulars of Water Ballast.	Engines. Particulars. Maker's Name.	Owners or Ma[nagers]
91	✻	Adelaide Lavarello StlScSr Olivari 89 1 Dk & Awn Dk 3 B	Italian Genoa	1228 1887	300·0 P 36 f.	35·5 F 24 f.	17·1 25·7	Newcastle Wigham Richardson & Co.	1889	5 Bhds.	3Cy.27½",43½"&71"–48" [150 HP.] Wigham Richardson & Co. Newcastle	Fratelli Lavar[ello] Gio Batta
92	G	Adele I ScSr Kreutzfeldt 82-82 1 Dk	German Kiel	152 199	96·1	21·1	12·4	Kiel Geo. Howaldt	1881		2Cy.13½"& 21"–16" 25 HP. Geo. Howaldt, Kiel	Sartori & Ber[ger]
93 [illegible]	✻	Adelphi Couppa I ScSr Canducis [illegible] 1 Dk 2 B	Greek Piraeus	940 1386	245·0 Q 84 f.	33·0 B 58 f.	18·0 F 32 f.	Sunderland W. Pickersgill & Son	1880	4 Bhds. DB 880	2Cy.29½"&55"–36" 120 HP. G. Clark, Sunderland	R. Couppa
94 65	✻	Adelphi Vagliano I Sc Conremendas 83-83 1 Dk 2 B	Greek Cephalonia	1087 1677	258·5 Q 94 f. P 24 f.	36·0 B 50 f.	18·6 F 32 f.	Sunderland Bartram, Haswell & Co.	1883	5 Bhds. DB [illegible] FP [illegible] AP [illegible]	2Cy.33"& 62"–42" 170 HP. J. Dickinson, Sunderland	Vagliano Bro[s.]
95 [illegible]	BT 1.88	Adirondack StlScSr Summ [illegible]	British Liverpool	1415 2167	290·4 F 28 f.	37·0 B 54 f.	21·5	Glasgow [illegible]	1888		3C.25",41"&67"–42" 275 HP. J. & G. Thomson, Glasgow	J. Aitken
96 81292	✻	Adirondack I ScSr Green 81-81 2 Dks 3 B	British Liverpool	1302 1999	285·0 S 66 f.	35·1 P 26 f.	24·2 [illegible]	Newcastle [illegible]	1879	[illegible] Bhds. DB 960	2Cy.34"& 64"–42" 225 HP. Wallsend Slipway Co., New.	"Adirondack" Ld. (W. Tapsc[ott])
97	DG	Adler I ScSr Müller 84-88 2 Dks & Spr Dk	German Bremen	926 1336	240·0	33·1	17·2 24·5	Flensburg [illegible]	1884	[illegible]	2Cy.37"& 68"–40" 250 HP. Flensburger Schiffs-Ges. Flensb.	Norddeutscher
98	V	Adler I ScSr Lemcke 88-88 1 Dk	German Stettin	258 328	149·2 Q 72 f.	20·7	10·2	Stettin Act. Ges. "Vulcan"	1882	[illegible]	2C.17½"&34½"–19"? 50 HP. Act. Ges. "Vulcan," Stettin	R. C. Gribel
99	G	Adler ex Elbe Rohlff 77-83 1 Dk	German Kiel	167 231	159·1	19·2 H 23 f.	9·9	Greenock J. Scott & Sons.	1857		2Cy.26"&43"–42" 100 HP. G. Howaldt, Kiel	Sartori & Ber[ger]
100 86896		Admiral W PadSr 2 Dks	British Quebec	408 632	176·0	29·0 H 472 f.	17·6	Wilmington Del.	1880		1Cy.49"–96" 106 HP.	Robert H. Mc[…]
101 47774	BT 1.88	Admiral I ScSmSr Scorie 73-88	British Newcastle	383 698	207·5 Q 67 f.	26·0 F 30 f.	14·7	Newcastle Palmer Bros. & Co.	1864		2Cy.27"& 52"–33" 120 HP. Palmers Co., Ld., Newcastle	Tyne Steam S[hip] Co., Ld.
102 93757		Admiral W ScSr 1 Dk	British Sydney	82 121	102·0	20·6 H 10 f.	9·2	North Shore, [N.S.W.] W. Dunn	1883		2Cy.14½"& 25"–18" 50 HP. Plenty & Sons, Newbury, Eng.	W. Waterhou[se]
103 53461		Admiral I PadTug	British Glasgow	25 149	115·9	19·7	10·7	Newcastle T. Hepple & Co.	1866		1Cy.40"–60" 60 HP. Hepple & Co., Newcastle	D. Bevan
104 89824	BT 3.88	Admiral I PadSr 1 Dk	British Glasgow	20 111	100·7	19·3 B 10 f.	9·3	North Shields W. Hepple & Co.	1886		1Cy.36½"–54" 50 HP. W. Hepple & Co., N. Shields	J. Steel & Son[s]
105	V	Admiral (WRECKED) I PadSmSr Essmann 64- 1 Dk	Russian Riga	396 542	195·0	31·5 P 86 f.	11·4	Glasgow J. R. Napier	1858		1Cy.46"& 76"–50" 240 HP. Randolph, Elder & Co., Glasgw.	Riga S.S. Co.
106 74376		Admiral D. D. Porter W ScSr 1 Dk & Spr Dk	British Sarnia, Ont.	137 241	113·6	21·2	9·6	Wilmington Del.	1864		1Cy.30"–26" 150 HP. Davidson Doran & Co Kngstn Ont	Sarnia Navig[ation] & Wrecking Co.
107 82750	✻	Admiral Rooke I ScBr Graham 1 Dk 2 B	British Gibraltar	881 1369	240·2 Q 75 f.	34·2 B 72 f.	17·0 F 29 f.	Middlesbro' R. Dixon & Co.	1882	4 Bhds. DB aft 280	2Cy.31"& 58"–36" 130 HP. Blair & Co., Ld., Stockton	W. H. Smith
108	✻	Adolf Deppe IScSr Kersting 78-88 1 Dk 2 B	Belgian Antwerp	553 883	208·4 P 30 f.	28·7 F 26 f.	14·2	Aberdeen J. Duthie Sons & Co.	1888	4 Bhds. DB aft 108	2Cy.25"&48"–33" 96 HP. Hall, Russell & Co., Aberdeen	A. Deppe
109	✻	Adolph Meyer Stl ScSr Kinman 1 Dk 2 B	Swedish Gothenburg	594 785	190·9 Q 65 f. B 47 f.	26·5 F 17 f.	16·0 H 13 f.	Gothenburg Motala Co.	1880	5 Bhds. MT 130 APT 19	2C.23⅝"&42½"–29½" 110 HP. Motala Co., Gothenburg	Förnvade Ång[…] Bol. Svenska Lloyd
110	V	Adolph Woermann StlScSr Meinertz 83-85 2 Dks	German Hamburg	1287 1887	270·0 P 40 f.	35·1 B 69 f.	21·9 F 34 f.	Hamburg Blohm & Voss	1886	CDB [illegible] APT [illegible]	3Cy.20",31½"&50"–42" 200 [HP.] Blohm & Voss, Hmbgh.	Afrikanische D[ampfschiffs] A.G. (Woerman[n])
111 87314		Adonis I PadSr 1 Dk	British Brisbane	116 186	109·8	21·6 H 18 f.	7·4	Brisbane Dath, Henderson & Co.	1883		2Cy.16"–42" 55 HP. Robey & Co., Lincoln	Robert Dath
112	✻	Adour I ScSr Laine 82-82 1 Dk 2 B	French La Rochelle	789 1221	238·5 Q 93 f. P 27 f.	32·6 B 16 f.	16·4 F 28 f.	Newcastle Palmers SB & I Co Ld	1882	4 Bhds. DB aft 170 APT 17	2Cy.30"& 57"–36" 140 HP. Palmers S.B. & I. Co. Ld. Nwc.	D'Orbigny & F[…] fils
113		Adour No. 2	French Bayonne	198	..	..	..				150 HP.	
114 85192	✻	Adowa I ScSr Paton 87-87 2 Dks & Awn Dk 4 B	British London	1347 2087	266·0 P 46 f. B 40 f.	36·2 F 34 f.	22·4 29·5	Sunderland Strand Slipway Co.	1882	5 Bhds. DB aft 298	3C.25",38" 63"–39" 150 HP. Trpl. Ex. by [illegible] Marine E. Co. Ld. Sunderland	H. Edlmann & [Co.]
115 50628	✻	Adria I ScSr Hedder 73-75 1 Dk 2 B	British London	522 844	219·1 P 65 f.	29·5 F 33 f.	17·3	Middlesbro' Backhouse & Dixon	1864	5 Bhds.	2Cy.26"& 52"–33" 170 HP. Palmers Co. Ld. Newcastle	J. Fenwick & S[on]
116 1641	✻A	Adria Stl ScSr Merlato 88 1 Dk 2 B	Austrian Fiume	707 1039	242·7 Q 86 f.	32·5 F 31 f.	16·8	Port Glasgow Blackwd. & Gordon	1880	5 Bhds. DB aft [illegible]	2Cy.30"& 59"–36" 120 HP. Blackwood & Gordon, Pt. Gls.	"Adria" Hun[garian] Sea Nav. Co. Ld.
117 54	V	Adria ex Sempione ex Mexico, ex City of Bristol. IScSr Romanelli 86-87 ex Etna 3 Dks	Italian Genoa	1393 2399	344·5 Q 111 f. F 53 f.	37·1 H 167 f.	27·0 71 f.	Greenock Caird & Co.	1855		2Cy.48½"& 84½"–42" 342 HP. Laird Bros., Birkenhead	Navigazione G[enerale] Italiana
118	✻N	Adria I ScSr Rasmussen 69-82 1 Dk & Awn Dk	Norweg'n Bergen	519 687	175·5	26·3	12·8 19·6	Rostock Rostock S.B. & Eng. Co.	1882	5 Bhds. DB aft 102	2Cy.22"&41½"–27½ 70 HP. Rostock S.B. & E. Co., Rostock	J. L. Mowinck[el]
119 [illegible]	✻	Adriana StlScSrYt 2 Dks	British Glasgow	98 228	157·5	21·0	11·5	Glasgow A. & J. Inglis	1888	5 Bhds. DB 38	3Cy.11",18"&30"–24" [60 H] A. & J. Inglis, Glasgow	H. Bubb
120 91381		Adriana I ScK	British London	73 108	113·0	16·2 Q 64 f.	6·8	Utrecht A. F. Smulders	1887		2Cy.14½"&21"–17" 30 HP. A. F. Smulders, Utrecht	Watkins & C[o.]

✻ = **Lloyd's Register**; D = Nederlandsche Vereeniging; G = Germanischer Lloyd; I = Registro Italiano; R = American [Record]; A = Veritas Austro-Ungarico; H = Veritas Ellenico; N = Norske Veritas; V = Bureau Ve[ritas]

In the 2nd Column, B.T., with figures following, signifies Date of report of bottom sighting by Board of Trade Surveyors as condition of PASSENGER CERTIF[ICATE]

In the 3rd Column, C = Composite (iron frame, wood planked); I = Iron; Stl = Steel; W = Wood.

Further particulars of Vessels classed by Lloyd's Register will be found in Lloyd's Register Book and Lloyd's Yacht Register.

Cote	Noms, Matériaux, Gréement, etc. Capitaine et dates d'entrée au service des Armateurs actuels et à bord du navire. Ancien Nom, nombre de ponts, etc.	Pavillon. Port d'Armement	Tonnage. Net. Total	Dimensions. Longueur, Largeur, Creux (entrepont ou sur le Pont)	Construit. Où. Par qui.	Quand	Nombre des cloisons. Détails sur le lest d'eau.	Machines. Détails. Nom du Constructeur.	Armateurs ou Directeurs-Gérants.
V	**Adriana** W ScSr *Rizzomdo* 1 *Dk*	Austrian Capodistria	79 146	116·6 | 16·9 | 10·3	Capodistria L. Poli	1885		2Cy.16½" & 33"–18" 40 HP. Bartlett&Co.	Luigi Poli
	Adriatic I ScImBk *Lawson* -87	British Liverpool	2458 3888	437·2 | 40·9 | 31·0 *H 133 f.*	Belfast Harland&Wolff	1871	7 Blhds.	1Cy.41" & 78"–60" 600 HP. Maudslay Sns.&Field,Lond.	OceanicSteamNav.Co. Ld.
I	**Adriatico** I ScSr *Marcus* 70-87 2 *Dks & Awn Dk*	Italian Genoa	742 1154	236·8 | 28·4 | 16·5 *Lea. 70* | 23·5	Dundee Gourlay Bros.&Co.	1862		2Cy.27" & 45"–44" 113 HP. Pernam&Duncvern	Navigazione Generale Italiana
I	**Adriatico** I ScSr *Bagello*	Italian Genoa	142 216	.. | .. | ..	SestriPonente G.Ansaldo&Co.	1888		2Cy.14" & 27"–18" 61 HP. G.Ansaldo&Co.,Genoa	C.Queirolo
✻	**Advance** I ScSr *Garrick* 1 *Dk 2 B*	British Cardiff	881 1294	244·6 | 34·1 | 15·7 *Q 80 f. F 27 f.*	PortGlasgow R.Duncan&Co.	1880	5 Blhds. DB350	2Cy.29" & 58"–36" 150 HP. Rankin&Blackmore,Grnck.	"Advance" S.S.Co. (M.Thompson&Co.)
	Advance I ScCr *Rogers* -83 1 *Dk*	British Brisbane	12 181	126·0 | 21·0 | 13·0	W'mstw'n Vic PortPhillipE.,S.Dk. Co.	1884		2Cy.22" & 40"–30" 80 HP. W.King&Co.,Glasgow	Government of Queensland
RV	**Advance** I ScBn *Lord* 80-86 1 *Dk & Spr Dk 3 B*	American New York	1902 2605	295·0 | 38·4 | 14·5 | 23·6	Chester, Pa J.Roach&Son	1883		2Cy.30" & 60"–54" J.Roach&Son,New York	United States & Brazil Mail S.S.Co.
✻	**Advent** I ScSr *Fuller* 65-76 1 *Dk 2 B*	British Sunderland	564 876	213·5 | 29·4 | 15·9 *Q 120 f. F 17 f.*	Sunderland Short Bros.	1876	4 Blhds. DBaft 193t FPT 10t APT 14	2Cy.27" & 50"–33" 98 HP. J.Dickinson,Sunderland	J.Westoll
✻	**Adventure** I ScSr *Biron* 82-87 1 *Dk 2 B*	British Whitby	1327 2050	280·0 | 38·8 | 18·8 *Q 108 f. B 68 f. F 29 f.*	SouthShields J.Readhead&Co.	1886	4 Blhds. DBcf 160t	2Cy.34" & 64"–42" 170 HP. J.Readhead&Co.,S.Shields	R.Harrowing
✻	**Adventure** I ScK *Ferrand* 1 *Dk*	British Hull	50 122	89·8 | 20·7 | 10·7	Hull Cook,Welton&Gemmell	1886	3 Blhds.	2Cy.15" & 29"–21" 35 HP. Bailey&Leetham,Hull	Humber Steam Trawling Co.
✻	**Ægean** (COLLISION) I ScSr *Thomas* 81-82 1 *Dk & Spr Dk 3 B*	British Leith	750 1157	231·2 | 31·2 | 15·0 | 21·8	Glasgow Lon.&Glasgow Co.	1870	5 Blhds.	2Cy.31" & 54"–33" 150 HP. London&GlasgowCo.,Glas.	F.F.Reid
	Ægir I PaiSr *Brastrup* 83-88	Danish Korsør	311 533	200·5 | 23·5 | 11·5 *P 176 f. H 31 f.*	Copenhagen Burmeister&Wain	1883		1Cy.26" & 50"–48" 200 HP. Burmeister&Wain,Copngn.	De Danske Statsbaner
✻	**Æolus** I ScSr *Gonzales* 1 *Dk & Sh'd Dk*	Spanish Manila	303 497	166·8 | 25·0 | 13·6 *P 30 f. F 28 f.*	Leith Ramage&Ferguson	1879	4 Blhds.	2Cy.20" & 37"–23" 75 HP. Barclay,Curle&Co.,Glasgow	Macleod & Co.
✻	**Afghan** I ScSr *Roy* 2 *Dks 3 B*	British Glasgow	1439 2262	312·0 | 34·1 | 25·6 *F 27 f.*	Glasgow Aitken&Mansel	1877	6 Blhds.	2Cy.40" & 72"–42" 280 HP. J.&J.Thomson,Glasgow	Mogul S.S.Co.,Ld.
	Africa I Sc3mSr *Johnston* 80-88 3 *Dks*	British Glasgow	1318 2032	315·2 | 33·2 | 25·6	Dumbarton W.Denny&Bros.	1874	5 Blhds.	2Cy.41" & 70"–42" 250 HP. Denny&Co.,Dumbarton	British IndiaStm.Nav. Co.,Ld.
✻	**Africa** I Sc3mSr *Carre* 86-87 2 *Dks*	British London	1099 1717	295·3 | 34·3 | 23·2 *F 53 f.*	Birkenhead Laird Bros.	1871	7 Blhds.	2Cy.27" & 54"–36" 150 HP. Harland&Wolff,Belfast.	African Steam Ship Co.
	Africa W Sr *Rinier* -87 1 *Dk*	British Owen Sound	328 452	148·0 | 26·0 | 12·5 *H 120 f.*	Kingston, Ont. W.Power&Co.	1873		1Cy.36"–34" 35 HP. Davidson&Doran,Kingstn,On.	A.R.Christie
✻	**Africa** I ScSr *Cotterell* 1 *Dk*	British Hull	114 199	130·7 | 21·0 | 10·8	Hull Earles' Co.	1880	3 Blhds.	2Cy.21" & 40"–27" 60 HP. Earles' Co.,Ld.,Hull	Hull Steam Fishing & Ice Co.,Ld.
	Africa W ScSr	British Liverpool	86 127	85·6 | 20·5 | 8·8	Winsford T.Hulse	1876		2Cy.12"–18" 26 HP. J.Bolend&Sons,Winsford	R.Verdin
✻	**Africa** I ScSr *Bunsey* 1 *Dk*	British N.Shields	2 122	100·0 | 19·8 | 10·1	Newcastle Sir W. G. Armstrong, Mitchell & Co.	1883	4 Blhds.	2Cy.21" & 38"–24" 80 HP. Ross&Duncan,Glasgow	C.Dyble
DG	**Africa** I ScSr *Barber* 72-84 1 *Dk & Spr Dk*	German Bremen	1379 2131	289·7 | 37·7 | 25·8	Bremen Act.Ges.Weser	1883	DB.	2Cy.36½" & 66"–42" Act.Ges.Weser,Bremen	D.H.Wätjen & Co.
V	**Africa** I ScSr *Schoning* 1 *Dk*	German Lübeck	273 352	150·0 | 22·3 | 13·0 *Q 30 f. B 36 f. F 22 f.*	Hamburg Reiherstieg Schiffs Wrft	1871		2Cy.17" & 28"–24" 40 HP. Reiherstg.Schfs.Wrte.,Ham.	Lübeck-Wyburger Dampfschifffahrts-Ges.
I	**Africa** I ScBg *Garazzo* 88-88 1 *Dk 3 B & Awn Dk*	Italian Genoa	685 1112	240·1 | 27·4 | 17·2 | 25·3	Newcastle Palmer&Co.	1865		2Cy.29" & 53"–36" 225 HP. G.Ansaldo&Co.,Genoa	NavigazioneGenerale Italiana
✻	**Africa** I ScSr *Costa* 3 *Dks*	Portug'se Lisbon	902 1463	271·0 | 33·2 | 22·5 *P 68 f. F 40 f.*	Dumbarton W.Denny&Bros.	1875	5 Blhds.	2Cy.41" & 70"–42" 256 HP. Denny&Co.,Dumbarton	Portuguese Government
	Africa (ex Saratoga) I ScSr 2 *Dks*	Russian St.Petersbg.	2283	298·0 | 38·6 | 23·5	Chester, Pa. J.Roach&Son	1877		2Cy.34" & 60"–16" John Roach&Sns,ChesterPa	Russian Government
	Africa I ScSr 1 *Dk*	Spanish Barcelona	429 632	182·0 | 26·2 | 15·6 *P 44 f. F 22 f.*	Renfrew W.Simons&Co.	1869	6 Blhds.	2Cy.28" & 50"–33" 90 HP. Simons&Co.,Renfrew	Sociedad de Navegacion Industria
✻	**African** Stl ScSr *Smyth* 85-88 1 *Dk & Awn Dk*	British Southampton	866 1372	244·2 | 33·3 | 16·3	Middlesbro' RayltonDixon&Co.	1886	4 Blhds.	3Cy.21" 34½" & 55½"–39" T.Richardson&Sons,Hartlp.	Union Steam Ship Co.
✻DV	**Afrikaan** I ScSr *Vernes* 82-82 2 *Dks & Spr Dk*	Dutch Rotterdam	842 1187	232·0 | 31·2 | 22·0	Middlesbro' RayltonDixon&Co.	1881	5 Blhds.	2Cy.27" & 50½"–33" 100 HP. Blair&Co.,Ld.,Stockton	Nieuwe Afrikaansche Handels Vennootschap
V	**Afrique** I ScSr *Fauran* -87 2 *Dks & Spr Dk*	French Marseilles	682 1165	260·1 | 29·8 | 15·2 *H 80 f.* | 22·7	Greenock Scott&Co.	1872		2Cy.33½" & 60"–36" 250 HP. GreenockFoundryCo.,Grnk.	Compagnie Générale Transatlantique
✻	**Afton** I Sc3mSr *Mathieson* 85-85 1 *Dk*	British Glasgow	119 266	140·4 | 23·2 | 10·5 *Q 50 f. F 25 f. B 14 f.*	Paisley AbercornS.B.Co.	1877	4 Blhds. WB APT FPT	2Cy.20" & 36"–24" 60 HP. D.Rowan,Glasgow	D.Rowan

✻=Lloyd's Register; D=Nederlandsche Vereeniging; G=Germanischer Lloyd; I=Registro Italiano; R=American Record
=Veritas Austro-Ungarico; H=Veritas Ellenico; N=Norske Veritas; V=Bureau Veritas.

Dans la 2me Colonne, B.T., avec les indications qui suivent, signifie la date du rapport de visite de la Carène par les experts du Board of Trade pour obtenir le certificat relatif au TRANSPORT DES PASSAGERS.

Dans la 3me Colonne, C=Composite (membrure en fer, bordé en bois); I=Fer; Stl=Acier; W=Bois.

Pour plus amples renseignements sur les navires classés par le Lloyd's Register, voir le Lloyd's Register Book et le Lloyd's Yacht Register.

No. in Book. / Official No.	Registry in which classed.	Ship's Name. Material, Rig, &c. / Master, and Dates of Appointment to present Owner's Service and to Ship / Ship's Late Name. No. of Decks, &c.	Flag. / Port of Registry.	Tons. Net. / Gross	Dimensions. Length	Breadth	Depth	Built. Where. / By Whom.	When	No. of Bulkheads. / Particulars of Water Ballast.	Engines. Particulars. / Maker's Name.	Owners or Ma
151 54924	BT3.88	Agamemnon I ScBkn *Wilding 73-75*	British Liverpool	1523 2347	309·3	38·8	20·6 28·4	Greenock Scott&Co.	1865		2Cy.26"&55"-52" 300 HP. GreenockFoundryCo.,Grnk.	Ocean Steam S
152 65873		Agamemnon I PadSp 1 *Dk*	British London	6 129	106·3	18·4	10·0	S.Shields Readhd.,Softly&Co.	1872		2Cy.28"-18" 55 HP. Readhead,Softly&Co.S.Shds.	James Watki
153 78366	✳	Agate I ScSr *M'Nicol 85-85* 1 *Dk*	British Glasgow	68 187	120·6 *Q 39 f.*	20·1	9·6 *F 20 f.*	Glasgow T.B.Seath&Co.	1878	3 Bhds. FPT15	2Cy.16"&28"-22" 35 HP. W.King&Co.,Glasgow	W.Robertson
154	V	Agathe I ScSr *Kondruhn 84-86* 1 *Dk*	German Memel	334 535	168·6 *Q 50 f.*	24·9 *B 32 f.*	14·9 *F 21 f.*	Flensburg Flensbg.SchiffbGes	1880	WB(f)45 AFT71	2Cy.20"&34½"-24" 50 HP. Flensbg.Schiffs.&Ges.Flens.	Memeler Dam A.G. (R.Sch
155	N	Agder I ScSr *Hanson 84* 1 *Dk*	Norweg'n Christiania	228 324	135·9 *P 32 f.*	23·2 *B 37 f.*	11·1 *F 18 f.*	Christiania AkersMekVerkstel	1884	DB37t AFT	2Cy.16½"&29"-18½" 35 HP. AkersMekVerksted,Cia.	Östlandske (Rolf Andvord)
156 92959	✳	Age Stl ScSr *Tindal 88* 1 *Dk*&W*b Frms*	British Melbourne	1493 2284	289·8 *P 31 f. Q 86 f.*	38·1 *B 118 f.*	19·2 *F 34 f.*	W.Hartlepool E.Withy&Co.	1888	5 Bhds. DB780t AFT44	3C21½"&35"&59"-39" 200 HP. T.Richardson&SonsHartpl.	W.H.Smith&S
157 86974	✳	Agenoria I ScSr *Sargent 85-86* 1 *Dk*2 *B*	British W. Hartlepl.	880 1360	244·0 *Q 92 f.*	35·0 *B 61 f.*	15·9 *F 27 f.*	W.Hartlepool Irvine&Co.	1883	4 Bhds. DB315t	2Cy.30"&57"-33" 130 HP. T.Richardson&Sons,Htlepl.	RickinsonSon
158	G	Agersøsund StlScSr *Bayer 75-85* 2 *Dks*	Danish Skjelskør	63 122	119·3 *Len. 85*	18·6	5·7 11·5	Kiel G.Howaldt	1884	[illegible]	2Cy.16½"&28½"-15½" 45 HP. HowaldtBros.,Kiel	Dampski[illegible] for skjelsk[illegible] (R. Harbar)
159		Aghami I PadSr 1 *Dk*	Turkish Suez	209	125·0	22·0	9·0	Birkenhead LairdBros.	1864		2Cy.38"-54" 100 HP. LairdBros.,Birkenhead	KhediviehMa
160 20430	BT12.88	Agia Sofia I ScimBk 3 *Dks*	British Liverpool	1695 2586	338·4	35·5	29·2	Glasgow J.&G.Thomson	1857	6 Bhds.	1Cy.26"&56"-42" 300 HP. G.Forrester&Co.,Liverpool	Papayanni &
161 44932		Agincourt (ex Talbot) I Pad	British Liverpool	31 161	103·7	19·4	11·2	S.Shields W.Marshall	1861		2Cy.30"-48" 70 HP. MarshallBros.,SouthShields	W.R.Chishol
162		Ägir I ScSr *Aminoff 77* 1 *Dk*	Swedish Gothenburg	340 438	144·0 *B 37 f.*	23·6 *H 3 f.*	15·4	Gothenburg GöteborgsMkVkstad	1874		2Cy.14"&30"-35" 52 HP. GöteborgsMekVerks.,Gotbg.	Förnyade Å Bol. Göta
163		Ägir C ScSr *Lundgren* 1 *Dk*	Swedish Edsvalla	165 223	102·3	20·4 *B 18 f.*	13·0	Spangen	1878		30 HP.	Nors Jernbr (T.Holländer)
164 3		Aglaia I ScBk *Mettel 63-86* 2 *Dks*	Austrian Trieste	1258 1983	282·6	35·2	25·4	Trieste A'streUng'ricoLl'yd	1879		2Cy.35¾"&61⅜"-48" 172 HP. ArsenaleLloyd,Trieste	Lloyd Austr rico
165	✳V	Aglaia StlSc *Christiansen 88* 2 *Dks*3 *B*	German Hamburg	1666 2302	309·8 *B 84 f.*	38·9 *P 62 f.*	22·8 *F 36 f.*	Flensburg FlensburgS.B.Co.	1888	6 Bhds. CDB45t	3C24"36½"&65"-45" 350 HP. FlensburgS.B.C.,Flensburg	Deutsche D Rhederei
166	✳N	Agnar I ScSr *Andersen 77-85* 1 *Dk*2 *B*	Norweg'n Stavanger	305 432	151·2 *Q 48 f.*	24·0 *B 44 f.*	12·2 *F 19 f.*	Stavanger StavangerStöberi&Dk	1884	4 Bhds. DB50t AFT13	2Cy.17"&32"-21" 45 HP. J.Wigham,Sunderland	T.S.Falck
167 87132		Agnes I TwnScSr	British London	135 322	139·0	27·6	11·9	Renfrew W.Simons&Co.	1883		1Cy.14"&27"-42" 60 HP. W.Simons&Co.,Renfrew	G.Furness
168 74912		Agnes W Pad 1 *Dk*	British Sydney	85 110	82·0	16·4 *H 55 f.*	5·8	Moama,NSW G.B.Air	1877		1Cy.9½"-18" 18 HP. Marshall&Sons,Gainsbo.Eng.	Th.H.Freem
169	✳	Agnes I ScSr *Barber* 1 *Dk*	German Danzig	532 840	200·6 *Q 83 f.*	29·0 *B 43 f.*	16·7 *F 17 f.*	Middlesbro' Backhouse&Dixon	1870	4 Bhds. DB(a)y	2Cy.25½"&47½"-33" 99 HP. Blair&Co.,Ld.,Stockton	T.Rodenacker
170		Agnes StlScSr *Lithman* 1 *Dk*2 *B*	Swedish Gothenburg	371 492	150·5 *B 43 f.*	23·9 *H 13 f.*	15·6	Norrköping MotalaCo.	1885	DB(a)64t FPT15 AP.f29t	2Cy.19"&34"-20½" 60 HP. MotalaCo.,Norrköping	Angfartygs A "Stella" (T
171 82806	✳	Agnes Otto I ScSr *Maxwell 87* 1 *Dk*2 *B*	British London	846 1319	249·9 *Q 95 f.*	32·6 *B 50 f.*	17·5 *F 26 f.*	S.Shields J.Readhead&Co.	1881	4 Bhds. DBa&f305t	2Cy.28"&53"-33" 115 HP. J.Readhead&Co.,S.Shields	G.Otto
172		Agnhammar C ScSr *Johansson 83-86* 1 *Dk*	Swedish Agnhammar	130 186	100·7 *Q 55 f.*	20·5 *P 20*	8·6 *H 4 f.*	Oroust J.H.Sebbe	1883		2Cy.10½"&22½"-15" 25 HP. P.Larsson,Torskog	A.Elieson
173	V	Aguadillana I ScSr *Gitray* 1 *Dk*&*Spr Dk*2 *B*	Spanish SntgodeCuba	835 1250	227·4	31·3	14·1 21·6	Geestemünde J.C.Tecklenburg	1885	WB	2Cy.26¾&50½"-31½" 125 HP. Mas.Fab.Bucka,Magdeburg.	Antillana Lin (G.A.Mendez &
174 93352	✳	Aguan Stl ScSr *Adair 86-86* 2 *Dks*&*Awn Dk*	British Glasgow	1013 1618	235·8	35·0	18·5 26·9	PortGlasgow R.Duncan&Co.	1886	4 Bhds.	3C22"35½"&57"-33½" 180 [H] Muir&Houston,Glasgw	Honduras & C AmericanS.S.C
175 84090	✳	Ahdeek I ScSr *Hutton 82-82* 1 *Dk*2 *B*	British Sunderland	970 1498	245·6 *Q 58 f. B 56 f.*	34·5 *P 28 f.*	18·5 *F 29 f.*	Sunderland ShortBros.	1881	6 Bhds. DBa&f223t	2Cy.32"&61"-36" 140 HP. N.E.MarineE.Co.Ld.Sndrld.	G.Butchart
176	✳	Ahter (ex Olga) I PadTg *Giuseppe* 1 *Dk*	Turkish Constant'pl	25 73	93·0	15·0	7·4	Pesth	1883	4 Bhds.	1Cy.24"-60" 25 HP.	Osman Bey
177 95035		Aid I&StlSc *Smith 88*	British Glasgow	99 223	112·3	26·0 *Q 46*	11·5	Renfrew W.Simons&Co.	1888		2Cy.15"&30"-21" 40 HP. W.Simons&Co.,Renfrew	W.Hill
178 24691		Aid W PadSr	British London	16 112	89·2	18·0	9·9	London	1855		50 HP.	Board of Trade
179 88		Aida W PadSc *Flori* 1 *Dk*	Austrian Trieste	74 135	115·2	16·6	14·2	Capo d'Istria L.Petti	1875		2Cy.21"&36½"-30¼" 50 HP. Bartlett&Co.	Cesare & Dan
180		Aida I ScSr 1 *Dk*	Turkish Alexandria	253 723	213·0	26·2	15·4	Southampton Day,Summers&Co.	1882		2Cy.27"&52"-36" 130 HP. Day,Summers&Co.,Southptn	The Khedive of

✳ = **Lloyd's Register**; D = Nederlandsche Vereeniging; G = Germanischer Lloyd; I = Registro Italiano; R = American Re
A = Veritas Austro-Ungarico; H = Veritas Ellenico; N = Norske Veritas; V = Bureau Ver

In the 2nd Column, B.T., with figures following, signifies Date of report of bottom sighting by Board of Trade Surveyors as condition of PASSENGER CERTI

In the 3rd Column, C = Composite (iron frame, wood planked); I = Iron; Stl = Steel; W = Wood.

Further particulars of Vessels classed by Lloyd's Register will be found in Lloyd's Register Book and Lloyd's Yacht Register.

No.	Cote	Nom, Matériaux, Gréement, etc. Capitaine et dates d'entrée au service des Armateurs actuels et à bord du navire. Ancien Nom, nombre de ponts, etc.	Pavillon. Port d'Armement	Tonnage Net / Total	Dimensions (Longueur / Largeur / Creux) Constructions sur le Pont	Construit. Où. Par qui.	Quand.	Nombre des cloisons. Détails sur le lest d'eau.	Machines. Détails. Nom du Constructeur.	Armateurs ou Directeurs-Gérants.
	✻D	**Aidar** (ex ...) I S... *Karebaum 86-89 1 Dk & Sp Dk 3 B*	British Liverpool	1583 2355	300·0 | 37·2 | 19·5 *B 90 f.* 273	Newcastle Wigham Richardson & Co.	1883	5 Blhds. DB ... FPT ... APT ...	2Cy.35" & 70"–48" 300 HP. Wigham Richrdsn.&C.,Nwc.	G. Tweedy & Co.
		Aigle I T-Sc *Lerey 83-83 Spar Dk*	French Saïgon	120 180	98·4 | 15·4 | 6·6		1872		2Cy.11" & 12"–12"	Compagnie des Messageries Fluviales de Cochinchine
	✻	**Ailsa** I ScSr *Borrie 88 2 Dks 3 B*	British Liverpool	1259 1963	290·6 | 34·2 | 23·9 *F 26 f.*	Glasgow Aitken & Mansel	1877	5 Blhds. FPT ... APT ...	2Cy.34" & 64"–42" 200 HP. J.&J.Thomson,Glasgow	Atlas Steam Ship Co.
		Ailsa Craig StlSc	British Glasgow	 345	350·0 | 44·5 | 25·7	Dundee W.B.Thompson & Co. Ld.	1889		3Cy.23" 39" & 63"–54" 250 HP W.B.Thompson&Co.,Ld.,Dun	Clyde Shipping Co.
185		**Airali** I Sc	French Algiers	 141	86·9 | 15·6 | 7·6	Inverkeithing	1869		20 HP.	J. Gemearo
186	✻	**Aire** Stl ScSr *Woodhead 87-87 1 Dk*	British Goole	324 625	212·5 | 28·8 | 13·8 *P 46 f. F 23 f.*	Newcastle W.Dobson & Co.	1886	5 Blhds. FPT ... APT ...	2Cy.27" & 54"–33" 130 HP. N.E.Marine Co.,Ld.,Nwcastl	Goole S.S. Co., Ld.
187	✻	**Airlie** I ScBg *Ellis 78-88 2 Dks 3 B*	British London	1492 2387	314·0 | 38·3 | 23·6 *P 47 f. B 55 f. F 44 f.*	Sunderland W.Doxford & Sons	1884	6 Blhds. ...	2Cy.40" & 73"–48" 340 HP. Hawks,Crawshay&Co.,Nwc.	Eastern & Australian S.S. Co.
188	V	**Ajaccio** I ScSr *Marini 84 2 Dks & Spr Dk*	French Marseilles	682 1166	261·1 | 29·8 | 15·2 *B 60 f.* 227	Greenock Scott & Co.	1872		2Cy.33½" & 66"–36" 250 HP. Greenock Foundry Co.,Grnk.	Compagnie Générale Transatlantique
		Ajax I ScBk *Riley 83-87*	British Liverpool	1525 2355	309·6 | 38·8 | 20·6 28·4	Greenock Scott & Co.	1866		2Cy.28" & 56"–52" 300 HP. Greenock Foundry Co.,Grnk.	Ocean Steam Ship Co.
		Ajax I ScSr	British Liverpool	133 162	108·0 | 23·1 | 8·9	Glasgow Tod & McGregor	1867		2Cy.14½"–18" 30 HP. Tod&McGregor,Glasgow	Inman & International S.S. Co., Ld.
191		**Ajax** I ScSr *1 Dk*	British Rangoon	117 228	120·0 | 24·0 | 12·0 *P 14 f.*	Rangoon	1874		2Cy. 90 HP.	C. E. L. Dawson
192		**Ajax** W ScS *Holt 88 2 Dks*	American San Francisco	534 838	212·0 | 33·2 | 15·0	San Francisco Boole & Bratou	1888		3Cy.18", 32" & 48"–34" 300 HP. Hinckley,Spiers&Hayes, San Francisco	Newport Coal Co.
193	V	**Ajax** I ScSr *Christiansen 83-84 1 Dk*	Danish Copenhagen	176 292	119·7 | 23·4 | 11·8 *P 18 f. B 28 f.*	Thorskog P.Larsson & Prvl.ppe	1884		2Cy.15" & 26"–18" 45 HP. Bolinder Mk. Werks,Stkhlm.	J. Christiansen
194	DV	**Ajax** I ScSr *Zeemer 66-77 1 Dk*	German Bremen	658 882	206·3 | 32·1 | 18·4 *Q 11 f. B 35 f. F 29 f.*	Sunderland ...	18..		4Cy.15½" & 39½"–23½" 95 HP. F.Schichau,Elbing	J. H. Meyer
		Ajikawa Maru I ScSr	Japanese Osaka	329 530	160·2 | 22·2 | 19·1	Kobe ... E. Yard	1885		56 HP.	Osaka Shosen Kaisha
	✻	**Ajudante** (ex David) StlScSr *Harris 87-87 1 Dk Spt Shd Dk*	Brazilian Para	145 214	129·5 | 24·9 | 7·7	Greenock Russell & Co.	1887	5 Blhds. FPT ...	2Cy.16" & 32"–20" 43 HP. Kincaid&Co.,Greenock	Amazon Steam Navigation Co.
197	✻	**Akaba** I ScBn *Mandeno 2 Dks 3 B*	British Liverpool	2473 3751	350·0 | 42·0 | 28·5 *P 32 f. B 90 f. F 40 f.*	Stockton Pearse & Co.	1885	6 Blhds. DB ...	3C.27½" 45" 74"–48" 450 HP. Blair&Co.,Ld.,Stockton	J. M. Wood
198	✻	**Akassa** Stl ScSr *Wharton 86-87 2 Dks*	British London	883 1389	295·5 | 30·3 | 18·9 *B 11 f. F 46 f. H 96 f.*	Belfast Harland & Wolff	1881	5 Blhds.	2Cy.27" & 52"–36" 120 HP. Harland&Wolff,Belfast	African Steam Ship Co.
199	✻	**Akbar** (ex Nederlandsche Oranje) I ScSr *Collin 86 2 Dks 3 B*	British Bombay	1441 2244	300·4 | 37·0 | 25·6 *P 39 f. B 30 f. F 30 f.*	Amsterdam Konin Fabrick	1883	4 Blhds. DB ...	2Cy.38" & 70"–36" 210 HP. Kon.Fab.v.Stm.Werk,Amstrdm.	Hajee Cassum Joosub
200		**Akers Kanal** Stl Sc *Gillix 1 Dk*	Swedish Stockholm	71 107	88·3 | 16·5 | 6·6 *Q 31 f. B 30 f. F 9 f. H 41 f.*	Motala Motala Mek. Verk. ...	1884		2Cy. 35 HP. Motala Co.,Motala	J. Lundin
201		**Akitsu Maru** W S	Japanese Osaka	120 193	123·3 | 16·7 | 13·3	Hiogo I.Yeijiro	1881		34 HP.	Osaka Shosen Kaisha
		Akola I ScSr *1 Dk & Shd Dk 3 B*	Zanzibar Zanzibar	315 578	201·1 | 25·1 | 19·1 *P 36 f. F 42 f.*	Greenock Caird & Co.	1875	4 Blhds.	2Cy.25" & 50"–30" 120 HP. Caird&Co.,Greenock	Sultan of Zanzibar
	R	**Al-ki** W ScSr *Bennett 88 1 Dk 2 B*	American S. Francisco	572 943	200·1 | 38·2 | 16·2	Bath, Me. Goss Marine Works	1884	...	2Cy.22" & 44"–36" 250 HP. Goss Marine Works	Pacific Coast S.S. Co.
	✻	**Alacrity** Stl ScSr *Robertson 85-85 1 Dk*	British Bristol	755 1182	225·2 | 32·3 | 15·2 *Q 93 f. B 42 f. F 26 f.*	Milford Castle Stl & I Wks. C.	1883	4 Blhds. DB ... FPT ... APT ...	2Cy.28" & 52"–36" 99 HP. North E. Mar. Eng. Co.,Nwc.	M. Thompson & Co.
	✻	**Alagoas** StlScSr *Pessoa 88 2 Dks & Shd Dk*	Brazilian Rio Janeiro	1254 1989	276·0 | 38·2 | 20·1 *F 36 f.*	Newcastle R. & W. Hawthorn, Leslie & Co., Ld.	1888	5 Blhds. DB ... APT ...	3C.26" 41" & 67"–45" 250 HP R.&W.Hawthorn,Leslie&Co., Ld., Newcastle	Brazilian Steam Navigation Co.
		Alameda I ScBn *Morse 83-83 2 Dks & Awn Dk 3 B*	American S. Francisco	1939 3158	314·0 | 41·0 | 17·3 *B 222 f.* 26·0	Philadelphia W.Cramp & Sons	1883		3Cy.43" & (2) 61"–51" W.Cramp&Sons,Philadelph.	Oceanic S.S. Co.
	R	**Alamo** I ScSr *Risk 78-85 2 Dks*	American New York	2237 2943	329·0 | 40·5 | 21·5	Chester, Pa. J.Roach & Sons	1883		2Cy.40" & 78"–54" J.Roach&Sons,Chester,Pa.	C. H. Mallory & Co.
		Alarm I TwnScSr *1 Dk*	British Liverpool	94 225	118·8 | 21·2 | 11·2	Liverpool W.H.Potter & Sons	1885		4Cy.13" & 26"–24" 56 HP. J.Jones&Sons,Liverpool	Mersey Docks & Harbour Board
	✻	**Alaska** I ScIrnBkn *Murray 80-82 4 Dks*	British London	3554 6932	500·0 | 50·0 | 38·0 *B 328 f.*	Glasgow J.Elder & Co.	1881	9 Blhds.	3C.68" & (2) 100"–72" 1800 HP. J.Elder&Co.,Glasgow	Guion Steam Ship C. Ld.
	✻	**Alaska** I ScSr *Robertson 1 Dk 2 B*	British Cardiff	1349 2065	275·0 | 38·6 | 20·7 *P 70 f. B 151 f. F 39 f.*	Sunderland Short Bros.	1883	5 Blhds. WB ...	2Cy.33" & 62"–42" 180 HP. Carr&Co.,Ld.,Sunderland	M. Thompson & Co.

✻ = Lloyd's Register; D = Nederlandsche Vereeniging; G = Germanischer Lloyd; I = Registro Italiano; R = American Record; A = Veritas Austro-Ungarico; H = Veritas Ellenico; N = Norske Veritas; V = Bureau Veritas.

Dans la 2me Colonne, B.T., avec les indications qui suivent, signifie la date du rapport de visite de la Carène par les experts du Board of Trade pour obtenir le certificat relatif au TRANSPORT DES PASSAGERS.

Dans la 3me Colonne, C = Composite (membrure en fer, bordé en bois); I = Fer; Stl = Acier; W = Bois.

Pour plus amples renseignements sur les navires classés par le Lloyd's Register, voir le Lloyd's Register Book et le Lloyd's Yacht Register.

No. in Book. / Official No.	Classed in which Register	Ship's Name. Master, and Dates of Appointment to present Owner's Service and to Ship. Ship's Late Name.	Material, Rig, &c. No. of Decks, &c.	Flag. / Port of Registry.	Tons. Net / Gross	Dimensions. Length	Breadth	Depth	Built. Where. / By Whom.	When	No. of Bulkheads. / Particulars of Water Ballast.	Engines. Particulars. / Maker's Name.	Owners or Managers
211 10002		**Alaskan** *Martin*	I & Stl PadSr *2 Dks*	American Portland,Ore.	1259 1718	275·5	40·0	13·4	Chester, Pa. J.Roach&Son	1883		1Cy.72"–134" J.Roach&Sons,Chester,Pa.	Oregon Railway Navigation Co.
212 86532	✻	**Alassio** *Pearson*	I ScSr *1 Dk*	British Newport	812 1261	221·8	34·3	15·9	[illegible]	1883	4 Blhds. DB[illegible] AP[illegible]	2Cy.28"& 52"–36" 99 HP. North E.Mar.Eng.Co.,New.	R.W.Jones & Co.
213	✻	**Alava** *Albeniga 86-88*	I ScBg *2 Dks 3 B*	Spanish Bilbao	1448 2244	320·5	35·0	24·7	Sunderland W.Doxford&Sons	1879	5 Blhds.	2Cy.40"& 70"–48" 250 HP. W.Doxford&Sons,Sunderland	Larrinaga & Co.
214		**Alava**	I ScSr *2 Dks & pl Shd Dk*	Spanish Havana	340 500	191·1	27·2	11·0	Port Glasgow Cunliffe&Dunlop	1875		2Cy.26"& 46"–33" 90 HP. Cunliffe&Dunlop,Pt.Glas.	Viuda de J. de [illegible]
215	✻ I	**Alba** (ex Gallo) *Serra 87 (CONDEMNED)*	I ScSmSr *1 Dk 2 B*	Arg'ntine BuenosAyres	499 792	254·2	28·2	16·5	Greenock J.Scott&Co.	1860	5 Blhds.	2Cy.26"& 51"–30" 160 HP. [illegible]	E. Piaggio & Y[illegible] carich
216 85360	✻	**Albania** *Simmons 83-86*	I ScSr *1 Dk 2 B*	British W. Hartlep'l	1212 1877	263·8	35·8	26·0	W. Hartlepool W.Gray&Co.	1883	4 Blhds. DB[illegible]	2Cy.32"& 62"–46" 160 HP. Black,Hawthorn&Co.,Gatshd	W.Gray
217	✻	**Albania** *Raftopoulo 83-88*	Stl ScSr *2 Dks*	Greek Piræus	881 1481	264·0	34·0	19·8	Dumbarton A.McMillan&Son	1887	5 Blhds. FPT[illegible] APT[illegible]	3Cy.25"38"& 62"–42" 230 HP D.Rowan&Son,Glas.	Panhellenic S. Navigation.
218 19715	✻	**Albanian** *Higgins 85-85*	I Sc3mSr *2 Dks*	British Hull	869 1350	286·3	30·2	19·8	Port Glasgow R.Reid&Co.	1855	5 Blhds.	2Cy.27"& 58"–36" 118 HP. Comp.'76,Holmes&Co.,Hull	W. Brown, Atkinson & Co.
219 87360	✻ D	**Albano** *Murrey*	I ScSr *1 Dk & Spr Dk 3 B*	British Sunderland	1522 2330	287·7	38·5	19·6 27·1	Sunderland J.Laing	1883	4 Blhds. DB[illegible]	2Cy.33"& 63"–42" 164 HP. G.Clark,Sunderland	Neptune S.N.Co.
220 29316	BT3.88	**Albano** (ex Albion) *Johnson*	I Sc3mSr *1 Dk 2 B*	British Hull	691 1069	245·2	29·0	17·0	Hull M.Samuelson&Co.	1861	4 Blhds.	2Cy.29"& 55"–26" 150 HP. Comp.'80,Earles Co.,Hull	T. Wilson, Sons & Co.
221 87695	✻	**Albany** *Gough 77-84*	I ScSr *2 Dks*	British Liverpool	1629 2432	300·7	39·2	26·4	Birkenhead Laird Bros.	1883	6 Blhds. [illegible]	2Cy.35"& 64"–42" 260 HP. Laird Brothers,Birkenhead	S.S."Albany" Co. (D.MacIver & Co.)
222 84947	✻	**Albany** *Parker 84-84*	I ScSr *2 Dks 3 B*	British Leith	1489 2276	310·9	38·1	25·3	Glasgow A.Stephen&Sons.	1882	4 Blhds. DB[illegible]	2Cy.38"& 68"–48" 300 HP. A.Stephen&Sons,Glasgow	J.Warrack & Co.
223 86642	✻	**Albany** *Godfrey 83-83*	I ScSr *1 Dk 2 B*	British Hull	933 1456	243·0	34·0	17·8	Whitby T.Turnbull&Son	1883	4 Blhds. DB[illegible]	2Cy.31"& 58"–36" 130 HP. Blair&Co.,Ld.,Stockton	T. D. Woodhead
224 44863		**Albany** (ex Claud Hamilton)	I ScSr *1 Dk*	British Adelaide	460 878	231·3	28·1	16·5	Newcastle [illegible]	1862	4 Blhds.	2Cy.27"& 54"–36" 100 HP. C.&W.Henderson,Glasgow	Adelaide S.S.Co.
225 94263	✻	**Albatros** *Elliott 87*	I Sc3mSr *1 Dk*	British London	202 341	111·7	23·1	10·9	Penarth [illegible]	1887	3 Blhds. FPT[illegible]	2Cy.17½"& 34"–24" 55 HP. Penarth S.B.Co.,Ld.,Penarth	T. A. Walker.
226 89641	✻ BT3.88	**Albatross** *Randell*	Stl ScSr *1 Dk 2 B*	British London	937 1436	250·0	37·2	17·4	Newcastle Palmers'Co.,Ld.	1884	4 Blhds. CDB[illegible]	2Cy.34"& 66"–45" 184 HP. Palmers'Co.,Ld.,Newcastle	General Steam N.
227 59222	✻	**Albatross** *Campbell 76-82*	I ScSr *1 Dk*	British Carlisle	162 366	170·0	22·4	12·2	Leith Ramage&Ferguson	1878	4 Blhds.	2Cy.23"& 44"–30" 85 HP. Hawthorns&Co.,Leith	R. Young
228 87860	✻	**Albatross**	I ScK *1 Dk*	British Grimsby	141 244	133·3	21·7	10·9	Hull Earles'Co.,Ld.	1884	4 Blhds.	2Cy.21"& 40"–24" 68 HP. Earles'Co.,Ld.,Hull	Gt. Grimsby Ice
229 82958		**Albatross** *Sown*	I ScSr	British Scarboro'	84 123	101·6	19·0	10·3	London W.Walker&Co.	1882	3 Blhds.	2Cy.18"& 34"–28" 50 HP. W.Lochhead&Co.,Glasgow	Middleton & Co.
230 78600	BT2.89	**Albatross**	I PadSp	British Glasgow	42 139	111·0	19·1	7·7	S.Shields J.Softley&Co.	1878		1Cy.40½"–54" 60 HP. Hepple&Co.,SouthShields	J.Steel & Son
231 74651		**Albatross** *MacMeckan 74-77*	I PadBg *1 Dk*	British Melbourne	31 191	131·4	21·1	10·7	Glasgow Barclay,Curle&Co.	1875	3 Blhds.	2Cy.34"–46" 80 HP. Barclay,Curle&Co.,Glasgow	James Paterson
232 50678	✻	**Albatross** *Fowens 85-85*	W ScK	British Berwick	23 34	72·3	17·2	8·0	Granton Hawthorns&Co.	1884	3 Blhds.	2Cy.8"& 16"–12" 11 HP. Hawthorns&Co.,Leith	T. F. Robertson
233	✻ G	**Albatross** *Ihle 77-87*	I ScSr *2 Dks & Awn Dk*	German Bremen	816 1040	225·0	28·1	17·0 24·0	Stockton M.Pearse&Co.	1872	4 Blhds. DB[illegible]	2Cy.28"& 56½"–36" 120 HP. Blair&Co.,Stockton	Norddeutscher L.
234 16663	✻	**Albert** *Spray 82-85*	I ScSr *1 Dk 2 B*	British Penzance	334 560	178·5	27·8	15·1	Hull M.Samuelson	1856	4 Blhds.	2Cy.27"& 50"–30" 98 HP. Humphrys&Pearson,Ld.,Hull	G. Bazeley & Sons
235 42985	✻ BT1.89	**Albert** (ex Viscount St. Arlington) *Hendry 67-67*	I ScSr *1 Dk 2 B*	British Hull	174 407	199·3	25·7	12·6	Hull C.&W.Earle	1856	4 Blhds.	2Cy.27"& 53"–33" 98 HP. Earles'Co.,Ld.,Hull	G. Lawson
236		**Albert** *Vannorbeeke 87-87*	I Sc3mSr *1 Dk 2 B*	French Dunkirk	287 431	180·4	25·3	12·6	Newcastle C.Mitchell&Co.	1855		2Cy.24"& 42"–34" 70 HP. H.Normand,Havre	Cie. Générale des Bateaux à vap. à hél.
237	✻ N	**Albert** *Tellefsen 84-88*	I ScSr *1 Dk*	Norweg'n Bergen	545 741	185·3	30·4	13·8	Oscarshamn	1884	4 Blhds. APT[illegible] MT[illegible]	2Cy.26"& 49"–30" 95 HP. Oscarshamns M.Vks.	Bergh & Helland
238		**Albert**	W ScBk *2 Dks*	Norweg'n Tönsberg	349 545	143·2	32·9	18·5	Bremerhaven F.W.Wencke	1867		95 HP.	F. V. Hansen
239 45021	BT6.88	**Albert Edward**	I PadSr	British London	198 371	205·3	24·0	11·5	London Samuda Bros.	1862		2Cy.52"–54" 220 HP. J.Penn&Sons,Greenwich	South Eastern R. Co.
240 76914	BT3.88	**Albert Edward**	I PadCr	British Portsmouth	163 269	169·4	20·5	9·0	Southampton Oswald,Mord'nt&C.	1878		2Cy.26"& 50"–51" 120 HP. Oswald,Mordaunt&C.Sthptn	L.&S.W.R.& S.C.R.Cos.

✻ = **Lloyd's Register**; D = Nederlandsche Vereeniging; G = Germanischer Lloyd; I = Registro Italiano; R = American [illegible]
A = Veritas Austro-Ungarico; H = Veritas Ellenico; N = Norske Veritas; V = Bureau Ve[illegible]

In the 2nd Column, B.T., with figures following, signifies Date of report of bottom sighting by Board of Trade Surveyors as condition of PASSENGER CERT[illegible]

In the 3rd Column, C = Composite (iron frame, wood planked); I = Iron; Stl = Steel; W = Wood.

Further particulars of Vessels classed by Lloyd's Register will be found in Lloyd's Register Book and Lloyd's Yacht Register.

Cote	Nom, Matériaux, Gréement, etc. Capitaine et dates d'entrée au service des Armateurs actuels et à bord du navire. Ancien Nom, nombre de ponts, etc.	Pavillon Port d'Armement	Tonnage Net Total	Dimensions Longueur	Largeur	Creux	Construit Où Par qui	Quand	Nombre des cloisons Détails sur le lest d'eau	Machines Détails, Nom du Constructeur	Armateurs ou Directeurs-Gérants
	Albert Edward I PdlSp Kyles (BROKEN UP)	British London	168 252	219·4	20·2	8·4	Greenock Caird&Co.	1865		2Cy.44"–45" 110 HP. Caird&Co.,Greenock	River Thames Stmbt Co.,Ld.
✻	Albert Edward I Tw.ScTug Eyers 1 Dk	British S'thampton	100 160	120·0	20·1	9·6	Northam Day,Summers&Co.	1886	1 Bhds.	1Cy.17"& 32"–24" 100 HP. Day,Summers&Co.,Southam	Sthmptn., I.of W.,& S. of Eng. St. Pkt. Co., Ld.
	Albert Edward I ScSr Rydell 80-83 1 Dk & 1 on Dk 3 H	Swedish Gothenburg	740 1011	234·2	28·9	17·2 24·3	Newcastle H 46 f. Schlesinger, Davis &Co.	1872	WB APT	2Cy.30"& 60"–30" 170 HP. Thompson,Boyd&Co.,Nwcl.	Angf.Aktie Bolaget Thule (O.Melin)
	Albert Ehrensvard I ScSr Klith (ex Alex. v. Humboldt) 1 Dk	Swedish Gothenburg	146 284	122·3 Q 49 f.	21·4 F 32 f.	9·5 H 15 f.	Nyköping NyköpingMek.Vks.	1859		70 HP.	Göteborg& Strömstads Angf.Aktb (O. Lindberg)
	Albert Victor Stl Pad Haywood 86-87	British London	359 782	250·0	29·2 H 51 f.	14·3	London Samuda Bros.	1880		2Cy.72"–60" 350 HP. J.Penn&Sons,Greenwich	South Eastern Railway Co.
	Albert Victor I Pad	British London	67 106	129·0	17·5 Q 8 f.	6·6	London Westwood	1879		2Cy.25"–30" 36 HP. Lon.SteamboatCo.,Ld.,Lon.	Managers of Metropolitan Asylum District
✻	Alberta StlScSr Murray 88-88 2 Dks 3 B	British Liverpool	2214 3398	340·0 P 32 f.	42·2 B 78 f.	27·3 F 38 f.	Newcastle R. & W. Hawthorn, Leslie & Co.	1888	7 Bhds. DB&f1061 FPT 108 APT250	3C.27"43"&70"–45" 280 HP. R.&W.Hawthorn,Leslie&Co.	G.L.Tapscott
	Alberta Stl ScSr Anderson 83-84 2 Dks	British Montreal	1155 1779	263·5	38·2	23·3	Glasgow C.Connell&Co.	1883		2Cy.35"& 70"–48" 300 HP. D.Rowan,Glasgow	Canadian Pacific Railway Co.
✻	Albertina I ScSr Parsons 84-88 1 Dk	British Newcastle	453 768	185·0 Q 35 f.	30·3 B 18 f.	13·0 F 22 f.	Sunderland D.Baxter&Co.	1882	5 Bhds. DB&f 134 FPT 28 APT 08	3C.14¼"20½"40"–33" 90 HP. Douglas&Grant,Kirkcaldy	Fisher, Renwick & Co.
	Albertito (ex Fortyninth) I ScSr 1 Dk	Spanish Bilbao	105 168	118·3	18·2	10·4	Middlesbro' Backhouse&Dixon	1870	3 Bhds. WB f	2Cy.18"–18" 35 HP. D.Joy&Co.,Middlesbro'	R.Garcia & Co.
DV	Albertus I ScSr Hega 2 Dks	German Königsberg	731 912	200·1	28·0	22·0	Elbing F.Schichau	1882	WB 210 f	2Cy.22"& 44"–25½" 90 HP. F.Schichau,Elbing	Markus Cohn & Sohn
✻B	Albicore I ScSr Driver 83-85 1 Dk	British Glasgow	277 467	186·3	24·6 Q 77 f.	13·3	Port Glasgow H.Murray&Co.	1869	1 Bhds.	2Cy.19"& 34"–36" 75 HP. Lees,Anderson&Co.,Glasgw.	J.Rankine & Sons
V	Albingia I ScSr Lauderer 86-86 2 Dks	German Hamburg	1201 1848	286·0	34·4 F 32 f.	24·0 H 13 f.	Glasgow Dobie&Co.	1881	WB a 120	2Cy.31"& 62"–42" 180 HP. J.Howden&Co.,Glasgow	Hamburg-Amerik. Packetf.-Aktien-Gesellsch.
	Albion I ScBg 1 Dk & Spr Dk	British Sydney	591 806	218·3	27·2 P 55 f.	15·4	Greenock Scott&Co.	1863		2Cy.44"–30" 180 HP. GreenockFoundryCo.,Grnk.	James C.Ellis
	Albion W Pad 1 Dk	British Quebec	107 171	123·0	22·0	9·7	Point Levis	1863		1Cy. 45 HP.	St.Lawrence Steam Nav.Co.
	Albion W Sc Hulse 1 Dk	British Liverpool	103 151	89·0	21·0	10·6	Winsford G.Deakin,Ld.	1887		3Cy10"13½"&23"–14" 34 HP. G.Deakin,Ld.,Winsford	G.Deakin,Ld.
✻	Albion W ScSrYt	British Portsmouth	88 120	120·3	18·2	11·6	Gosport Camper & Nicholson	1887		3Cy.12"17"&30"–20" 56 HP. W.B.Thompson&Co.,Dundee	P.Thellusson
✻	Albion I TwnSc Beverley 1 Dk	British Hull	45 85	81·3	18·2	7·2	Beverley Cochrane&Co.	1884	1 Bhds.	2Cy.18"–18" 30 HP. Amos&Smith,Hull	T.F.Wood
✻	Albireo I ScSr Jupp 75-85 1 Dk 2 B	British Shoreham	847 1325	237·0 Q 66 f.	31·6 B 62 f.	18·3 F 30 f.	Sunderland Mounsey&Foster	1878	5 Bhds. DB 64 f 357 f	2Cy.29"& 55"–36" 120 HP. J.Dickinson,Sunderland	R.H.Penney
✻D	Alblasserdam I ScSr Polder -86 1 Dk 2 B	Dutch Rotterdam	747 1035	231·2 Q 9 f.	31·5 B 46 f.	14·7 F 23 f.	Rotterdam Rykee&Co.	1883	4 Bhds. DB a f 210	2Cy.27"& 50"–36" 100 HP. Maatsch.deMaas,Delfshaven	Vrœge & de Wys
✻	Albula I ScSr Johnson 84-84 2 Dks 3 B	British S.Shields	1221 1885	284·5	34·2	24·4	Newcastle C.S.Swan&Co.	1877	5 Bhds. DB a f 223	2Cy.32"& 62"–36" 180 HP. Black,Hawthrn&C°,Gateshd.	Cay,Hall & Co.
	Alcantara I Sc	Brazilian Maranham	220	125·0	19·0	8·1	Birkenhead Laird Bros.	1872		70 HP.	"Maranham" S.S.Co.
	Alcatraz W ScSr Johnson -87	American S.Francisco	194 256	130·2	31·0	10·8	San Francisco C.G.White	1887		2Cy.12"&24"–20" 150 HP. Hinckley,Spiers&Hayes, San Francisco	L.E.White
	Alcazar W ScSr Gerdau -87 1 Dk	American S.Francisco	202 264	132·0	31·5	10·3	San Francisco A.Hay	1887		2Cy.12"&24"–20" 150 HP. Hinckley,Spiers&Hayes, San Francisco	L.E.White
✻	Alcester I ScBn Bruce 76-83 1 Dk 2 B	British London	1047 1615	260·0 P 26 f.	36·0 B 60 f.	18·8 F 60 f.	Stockton Richrdsn,Duck&Co.	1883	4 Bhds. C DB 388	2Cy.32"& 59"–33" 180 HP. T.Richardson&Sons,Htlepl.	Christie & Co.
A	Alcibiades I Sc Carcameni	Russian Taganrog	255 322	..	..	..	Trieste Stabilimento Tecnico	1883		50 HP.	G.N.Serebrakoff
✻	Alcides Stl ScSm Bella 69-86 2 Dks & Spr Dk	British Glasgow	2244 3421	340·0	47·1 B 100 f.	30·0 F 36 f.	Glasgow Napier, Shanks & Bell	1886	4 Bhds. DB f 150 FPT 38	3C.29"43"&68"–54" 400 HP. J.&J.Thomson,Glasgow	Donaldson Bros.
	Alcira I	Spanish Valencia	455 745	191·6	26·9	16·1	Belfort	1864		110 HP.	Antonio Devesa
✻	Aldborough StlScSr O'Neil 88-88 1 Dk H'h Frms	British London	1518 2314	289·8 P 39 f. Q 86 f.	38·1 B 118 f.	19·4 F 30 f.	Port Glasgow Russell&Co.	1888	5 Bhds. C DB 383 APT 20	3Cy.22½"35½"&58½"–39" 220 HP. J.Howden&Co., Glasgow	Aldborough S.S. Co., Ld. (F.Woods)
N	Alden I ScSr Tücken -88 1 Dk	Norweg'n Bergen	317 433	154·8 F 18 f.	23·9 B 21 f. P 67 f.	12·4 H 22 f.	Bergen Laxevaags Maskin & Jernskibsbyggeri	1888		3C.15½"25"–40"&27" 80 HP. Hall,Russell&Co.,Aberdeen	Nordre Bergenhus Amt Dampskibsselskab (Manager H.Lund)

Lloyd's Register; D = Nederlandsche Vereeniging; G = Germanischer Lloyd; I = Registro Italiano; R = American Record;
A = Veritas Austro-Ungarico; H = Veritas Ellenico; N = Norske Veritas; V = Bureau Veritas.

Dans la 2me **Colonne**, B.T., avec les indications qui suivent, signifie la date du rapport de visite de la Carène par les experts du Board of Trade pour obtenir le certificat relatif au TRANSPORT DES PASSAGERS.

Dans la 3me Colonne, C = Composite (membrure en fer, bordé en bois); I = Fer; Stl = Acier; W = Bois.

Pour plus amples renseignements sur les navires classés par le Lloyd's Register, voir le Lloyd's Register Book et le Lloyd's Yacht Register.

No. in Book / Official No.	Registry in which classed	Ship's Name. Material, Rig, &c. / Master, and Dates of Appointment to present Owner's Service and to Ship / Ship's Late Name. No. of Decks, &c.	Flag. / Port of Registry.	Tons. Net. / Gross.	Dimensions. Length / Breadth / Depth. Deck Erections.	Built. Where. / By Whom.	When.	No. of Bulkheads. / Particulars of Water Ballast.	Engines. Particulars. / Maker's Name.	Owners or M...
271 72346	BT5.88	**Alderney** I ScSr	British S'thampton	395 642	175·0|25·2|19·5	Newcastle Wigham Richardson	1875		2Cy.26½" & 42"–27" 90 HP. J.Shaw&Co.,Newcastle	London & S.W. Railway Co.
272 56970		**Aldinga** I ScSr 1 Dk	British Sydney	259 447	202·6|24·6|13·4 H 25 f. Q 41 f.	Greenock Scott&Co.	1860		2Cy.44"–48" 140 HP. GreenockFoundryCo.,Grnk.	Ebenezer Vicke...
273	✻	**Alejandro** Stl&I ScSr *Garcia* 1 Dk&Awn Dk	Mexican Mazatlan	620 979	198·8|32·1|12·1 19·4	Sunderland Sunderland S.B.Co.	1886 Ld.	4 Blds. APT FPT DB	2Cy.29"&56"–36" 140 HP. N.E.MarineEng.Co.,Ld.Sld.	J.Redo
274 84043	✻	**Alene** I ScSr *Seibert 80-84* 2 Dks	British Liverpool	1369 2104	291·5|36·3|23·8 B 93 f. F 27 f.	Glasgow Aitken&Mansel	1880	4 Blds.	2Cy.36"& 68"–48" 300 HP. J.&J.Thomson,Glasgow	Atlas Steamsh...
275 30370	BT7.88	**Aleppo** I ScBg *Ferguson 82-86*	British Glasgow	1386 2143	292·5|38·2|26·2 F 74 f.	Glasgow J.&G.Thomson	2865		2Cy.38"& 68"–48" 212 HP. J.Jack&Co.,Liverpool	Cunard Steam... Co.Ld.
276 352		**Alert** I ScSr *Cherry 36-86*	British Hull	209 329	176·0|21·4|10·8 P 38 f.	Hull	1855		60 HP.	W.&C.L.Ring...
277 67797		**Alert** (ex Jasper) W Sc3mSr	British Dublin	159 277	146·8|25·2|12·9 B 20 f.	London	1862		2Cy.30"–18" 80 HP. Laird,Birkenhead	Commissioners Irish Lights
278 76188		**Alert** I ScBk *Schull 84-85* 1 Dk	British Melbourne	116 243	169·0|19·6|9·8 Q 40 f.	PortGlasgow R.Duncan&Co.	1877	6 Blds. FPT APT	2Cy.24"& 46"–26" 90 HP. Rankin&Blackmore,Grnck.	Huddart,Park...
279 68808		**Alert** I ScSr	British London	98 144	104·0|20·0|10·7	Glasgow	1869		2Cy. 35 HP.	Corporation of Trinity House
280 28880	BT11.88	**Alert** I PadSr 1 Dk2 B	British Liverpool	87 172	117·8|21·0|10·7	Birkenhead LairdBros.	1865		2Cy.35"–48" 70 HP. LairdBros.,Birkenhead	Mersey Docks & Harbour Board
281 90871		**Alert** (ex George J. Stevens) W Sp *Gonbey -86* 1 Dk	British S.JohnsNFL	63 105	98·3|19·1|7·4 H 24 f.	Maine,U.S.	1880		1Cy.18"–18" 35 HP. Kendall&RobertsCam.Pt.Mass	Newfoundland... Co.,Ld.
282 86450	✻ BT4.86	**Alert** I ScSr *Lihou 56-85* 1 Dk&Shade Dk	British Guernsey	13 122	110·0|19·0|10·1	Paisley J.Fullerton&Co.	1885	4 Blds.	2Cy.20"& 38"–24" 60 HP. W.Kemp,Glasgow	Guernsey Stm. & Trading Co.Ld.
283	✻ N	**Alert** I ScSr *Kanitz 87-87* 1 Dk	Norweg'n Bergen	559 783	194·4|28·9|13·8 Q 57 f. B 60 f. F 24 f.	Bergen BergensMek.Verks	1884	5 Blds. DB&ATrs	2Cy.25"& 47"–30" 85 HP. BergensMek.Verks,Bergen	Bergh & Helle...
284	✻	**Alesia** I ScBg *Vallat 81-84* 2 Dks3 B	French Marseilles	2131 2851	328·0|40·4|27·1 P 27 f. B 70 f. F 42 f.	Liverpool T.Royden&Sons	1882	5 Blds. DB475t	2Cy.40"& 76"–54" 300 HP. G.Forrester&Co.,Liverpool	Compagnie F... de Nav. à Vap.
285 4	✻	**Alessandro** I Sc *Sturlese* 1 Dk	Italian Genoa	234 385	145·0|23·1|12·0 Q 70 f. B 14 f. F 21 f.	Dundee PearceBros.	1882	4 Blds. FPT35t APT25t	2Cy.26"& 35"–24" 60 HP. PearceBros.,Dundee	G.Sturlese
286 33	I	**Alessandro Volta** I ScSr *Desideri 85-85* 1 Dk2 B	Italian Genoa	384 609	203·4|25·4|13·7	Glasgow Randolph,Elder&Co.	1862		2Cy28½"&48½"–45"169HP. McNab,Greenock	Navigazione G... Italiana
287 58704	✻	**Alethea** I ScSr *Donovan 85-85* 1 Dk2 B	British Hartlepool	798 1244	240·0|31·9|17·6 Q 29 f. F 28 f.	Hartlepool Withy,Alx'der&Co.	1872	4 Blds. DB	2Cy.29½"& 59"–36"130HP. Blair&Co.Ld.,Stockton	G.Horsley & So...
288 72071		**Alexander** W PadTug 1 Dk	British Victoria,BC	192 332	170·0|27·2|12·5 H 16 f.	Pt. Essington,B.C J.McAllister	1876		2Cy.38"–60" 160 HP. Fawcett,Preston&Co.,L'pool	Robert Dunsmu...
289		**Alexander** I Sc3mSr 1 Dk	Russian Helsingfors	311 431	169·0|23·2|12·0	Hull C.&W.Earle	1858		80 HP.	O.Eklund
290		**Alexander** I Pad	Russian Odessa	113 160	.. .. ..	London	1860		60 HP.	Bezasnoff,M... & Co.
291	V	**Alexander** I Sc3mSr *Härstrom -87* 1 Dk	Swedish Gothenburg	352 411	149·6|23·4|15·4 B 25 f. H 3 f.	Gothenburg GötbrgsMek.Verks	1882		2Cy.16"&28"–19½" 45 HP. Götbrgs.Mek.VerksGthnbg	J.F.Olsson
292 105505		**Alexander Duncan** W ScSr *Lyons* 2 Dks	American S. Francisco	260 372	140·5|25·3|12·8	SanFrancisco	1876		2Cy.12"& 20"–20" 75 HP. RisdonIrnWks S.Francisco	Pacific Coast S.S...
293	V	**Alexander Keiller** I ScSr *Anderson 84* 1 Dk	Swedish Gothenburg	341 434	145·3|23·2|15·4 B 20 f. H 3 f.	Gothenburg GötbrgsMek.Verks	1874		2Cy.13" & 30"–20" 52HP. GotborgsMek.VerksGothenbg	Fornyade Angf... Aktie-Bolag ...
294 65102	✻ BT4.88	**Alexander Pirie** I ScSr *Linklater 62-78* 1 Dk	British Aberdeen	250 460	180·0|26·0|14·0 B 49 f. Q 47 f. F 30 f.	Aberdeen A.Hall&Co	1873	4 Blds.	2Cy.38"& 48½"–24" 90 HP. Gourlay Bros.&Co.,Dundee	Aberdeen,New... Hull Steam Na...
295	✻ V	**Alexander II.** I Sc *Brose* 1 Dk&Awn Dk3 B	German Hamburg	528 729	205·6|25·2|13·5 Lrn. 69 19·7	W.Hartlepool Pile,Spence&Co.	1860	3 Blds.	4Cy.18½"&40½"–33" 100HP. [illegible]	O.Schmidt
296	V	**Alexander II.** I PadSr *Lindholm 67-*	Russian Riga	424 518	199·1|24·9|11·7 P 63 f.	London C.Langley	1858		2Cy.33½"&48"–57"152 HP. Compounded	Riga S.S.Co.
297		**Alexander II.** W ScBk *Sandmann* (ex Alexander) 2 Dks	Russian Vladivostck	295 331	160·4|21·0|13·5	New York W.A.Webb	18⁵⁵⁄₇₉		2Cy.18"& 34"–21" 90 HP. RisdonIrnWks S.Francisco	Hutchinson,Ko... Philippeus & Co.
298 45957		**Alexandra** I ScSr 2 Dks	British Sydney	496 681	243·8|27·1|14·3 P 68 f. F 17 f.	PortGlasgow Blackwood&Gordon	1863		4Cy.24"& 42"–30" 150 HP. AustralasianS.N.Co.	Australasian St... Nav.Co.Ld.
299 45563	BT3.88	**Alexandra** I PadSr *Beaumont 55-63* 2 Dks	British Dublin	266 703	226·6|28·3|16·0 P 93 f.	Birkenhead LairdBros.	1863	5 Blds.	2Cy.56"–69" 320 HP. LairdBros.,Birkenhead	London & N.We... Railway Co.
300										

✻ = **Lloyd's Register;** D = Nederlandsche Vereeniging; G = Germanischer Lloyd; I = Registro Italiano; R = American Re...
A = Veritas Austro-Ungarico; H = Veritas Ellenico; N = Norske Veritas; V = Bureau Ve...

In the 2nd Column, B.T., with figures following, signifies Date of report of bottom sighting by Board of Trade Surveyors as condition of PASSENGER CERTIF...

In the 3rd Column, C = Composite (iron frame, wood planked); I = Iron; Stl = Steel; W = Wood.

Further particulars of Vessels classed by Lloyd's Register will be found in Lloyd's Register Book and Lloyd's Yacht Register.

Cote	Nom, Matériaux, Gréement, etc. Capitaine et dates d'entrée au service des Armateurs actuels et à bord du navire. Ancien Nom, nombre de ponts, etc.	Pavillon. Port d'Armement	Tonnage. Net. Total	Longueur	Largeur	Creux	Construit. Où. Par qui.	Quand	Nombre des cloisons. Détails sur le lest d'eau.	Machines. Détails. Nom du Constructeur.	Armateurs ou Directeurs-Gérants.
	Alexandra I PadSc	British London	203 746	211·8	23·9	11·5	London J.D'A.Samuda	1864		2Cy.58"–53" 220 HP. J.Penn&Sons,Greenwich	South Eastern Railway Co.
	Alexandra I PadSc	British London	149 278	230·5	22·2	8·5	PortGlasgow Kirkpatrick,M'Intyre&Co.	1865		2Cy.44"–66" 140 HP. Smith&Co.,Greenock	River Thames Steamboat Co.,Ld.
	Alexandra I PadSr Bryning -80 1 Dk	British Calcutta	125 363	157·1	25·1	14·6	Birkenhead LairdBros.	1864		2Cy.52"–72" 200 HP. LairdBros.,Birkenhead	George J.Scott
	Alexandra I PadSr	British Liverpool	107 327	204·5	23·7	11·0	Greenock Caird&Co.	1863		2Cy.52"& 53"–57" 200 HP. Caird&Co.,Greenock	R.&D.Jones
	Alexandra I PadSp	British Portsmouth	97 235	171·0	20·2½	8·5	Greenock Scott&Co.	1879		2Cy.25"& 50"–54" 120 HP. Scott&Co.,Greenock	L. & S. W. Rail. Co. & L.B.& S.C.Railway Co.
	Alexandra StlPadSp Adams -88	British Plymouth	80 127	125·7	17·3 P 37 f.	5·5	Preston W.Allsup&Sons	1888		2Cy.20"&38"–36" 64 HP. W.Allsup&Sons,Preston	Devon and Cornwall Tamar SteamPacketCo.,Ld
	Alexandra I ScCr	British Southamptn	62 121	110·0	19·1	8·9	Northam Day, Summers&Co.	1876		2Cy.20"& 40"–21" 60 HP. Day,Summers&CoSouthptn	Sthmptn.,I.of W.,& S. of Eng.RoyalMailS.P.Co.
	Alexandra I Pad	British London	37 158	100·0	20·6½	10·7	London SamudaBros.	1877		2Cy.32½"–54" 80 HP. Stewart&Co.,London	London & St.Katharine Docks Co.
✻	Alexandra IScTug Blackburn 85-86 1 Dk	British Hull	11 63	72·6	16·6	8·5	Hull Earle'sCo.	1885	1 Blds.	3Cy.11½"17"&30"–21"45HP Earle'sCo.,Ld.,Hull	Hull & Barnsley Railway & Dock Co.
✻	Alexandra I ScSk Johnson 88-88 1 Dk	British Liverpool	1 142	92·0 B 7 f.	20·1	11·1	Ayr S.M'Knight&Co.	1888	4 Blds. AP f 25t	2Cy.21"&42"–27" 75 HP. Muir&Houston,Glasgow	Alexandra TowingCo., Ld.
V	Alexandra I Sc3mSc Stowahre 66-81 1 Dk	German Stettin	373 496	196·8 Len. 67.	24·3 B 21 f.	15·1	Stettin Früchtenicht&Brock	1856		2Cy.20½"&41"–23¾"80 HP. Stettiner MasAnsVulcanStettin	Rud.Christ.Gribel
	Alexandra No. 2 I PadSp Richardson -86 1 Dk	British Liverpool	82 114	142·0	17·2½	6·7	Glasgow	1866		2Cy.30"–15" 50 HP. J.Howden&Co.,Glasgow	TheRock Ferry Co.,Ld.
✻	Alexandre Bixio I Sc3mSr Christophle 82-88 2 Dks3 B	French Havre	1473 2189	298·2 P 28 f.	35·8 F 36 f.	26·7	Stockton M.Pearse&Co.	1880	4 Blds. DBaf 328t	2Cy.36½"&68"–42"250 HP. Blair&Co.,Ld.,Stockton	Compagnie Générale Transatlantique
	Alexandria I ScBk Laird 88-88 2 Dks&AwaDk	British Glasgow	1247 2017	300·5 F 33 f.	33·2	22·6 29·4	PortGlasgow R.Duncan&Co.	1870	5 Blds.	2Cy.38"& 68"–42" 424 HP. FinniestonWks.Co.,Glasgw	Henderson Bros.
N	Alf W ScSc Kiddelsen 84-89 1 Dk2 B	Norweg'n Stavanger	253 337	134·1 Q 32 f.	24·8 B 43 f.	12·6	Stavanger StrömstenensVerft	1880		2Cy.15¼"&29¾"–24½"35HP AkersMek.Vrks.Christiana.	T.S.Falck
	Alfa (ex T-r) I ScSr Carlsson 1 Dk	Swedish Rödesund	108 161	97·8 Q 31 f.	20·3 F 16 f.	9·5 H 16 f.	Amål	1857		2Cy. 32HP	P.A.Palmqvist
✻	Alfonso I ScBg Hutchinson 1 Dk2 B	British Liverpool	856 1325	246·5 P 263 f.	31·9 F 27 f.	17·4	Southampton Oswald,Mordaunt&Co.	1878	5 Blds. WBaf	2Cy.30"& 60"–39" 150 HP. Oswald,Mordaunt&CoSthn.	Strong,Reid & Page
✻	Alfonso XII StlSc4m Choquert -88 2 Dks&AwaDk4 B	Spanish Barcelona	3268 5063	408·5 F 62 f.	47·8 B 98 f.	24·9 33·2	Newcastle Wigham Richardson&Co.	1888	6 Blds. CDB 512t	3C.31"55"&90"–60"700HP WighamRichrdsn.&C.,New.	Cia.Trasatlantica
	Alfonso XIII. StlSc1mSr	Spanish Barcelona	 4400	400·0	48·0	32·0	Dumbarton W.Denny&Bros.	1888		3Cy.31"57"&94"–60" W.Denny&Co.,Dumbarton	Cia.Trasatlantica
	Alford I ScBk 2 Dks	British London	638 771	216·0 Q 20 f.	28·2	16·9	Newcastle PalmerBros&Co.	1863	4 Blds. WB	2Cy.38"–30" 90 HP. R.Stephenson&Co.N'castle.	General Steam Nav. Co.
G	Alfred (ex Teutonia) I ScSr Blanck 87-87 (or Rudskvall) 1 Dk	German Königsberg	133 182	114·1	22·8	10·2	Stockholm O.Felander&Hammar	1857		2Cy.14¼"&27¼"–15¾"35HP	Storrer & Scott
	Alfredo I Sc	Brazilian Para	91 146	110·0	22·1½	8·2	Liverpool R.&J.Evans&Co.	1883		2Cy.15"& 30"–20" 38 HP. D.Rollo&Sons,Liverpool	J.L.da Silva e Souza
	Alfredo Capellini W PadSc 1 Dk	Italian Palermo	70 145	117·7	17·4½	9·7	SestriPonente E.Starlese	1860		2Cy.17½"&29¾"–36"40 HP. Stephenson,London	Navigazione Generale Italiana
✻	Algarve I Sc G. da C.Esteveus -86 1 Dk	Portug'se Lisbon	546 846	200·0 Q 103 f.	30·1 B 20 f.	14·0 F 21 f.	Sunderland J.Priestman&Co.	1885	4 Blds. DBaf 46t	2Cy.24"& 48"–36" 96 HP. T.Clark&Co.,Newcastle	A.Centeno
	Alger I Sc3m 2 Dks	French Marseilles	463 799	230·9 P 54 f.	27·9 F 31 f.	17·5	LaSeyne Forges&C.delaMédit	1865		2Cy.28½"&57"–28¾"150 HP. Forges&C.de la Medit, Marsl	Cie.de Nav.Mixte.
	Algerian I ScSr Davies 82-87 2 Dks	British Liverpool	1160 1771	312·0 F 28 f.	31·2 B 54 f.	23·9	Liverpool Bowdl'rChaffer&Co.	1876	5 Blds.	2Cy.24"& 64"–36" 200 HP. J.Jones&Sons,Liverpool	S.S."Algerian"Co.,Ld. (F.Leyland & Co.)
	Algérie I ScSr C.Besque 76-88 2 Dks	French Marseilles	589 880	236·2 P 62 f.	28·1 F 32 f.	17·8	Greenock Caird&Co.	1856		2Cy.26¼"&49¾"–30"180HP Fraissinet&Co.,Marseilles	Fraissinet & Co.
F	Algérien (ex Anda) I ScSr Angles 1 Dk2 B	French Marseilles	713 1075	231·3	29·2 B 43 f.	20·7	Inverkeithing J.Scott&Sons	1875	6 Blds.	2Cy.30"& 60"–36" 150 HP. J.Scott&SonsInverkeithing	Caillol & St.Pierre
	Algerine W ScSr (ex H.M.S. Algerine) 1 Dk	British Hong Kong	161 230	125·7	22·2 H 12 f.	8·6	London Pitcher	1857		1Cy.18"–18½" 80 HP. MaudslaySons&Field,Lndn.	George Mitchell
R	Algiers ISeSr Percy 2 Dks	American N.Orleans,La	1305 2287	288·0	38·4	28·5	WilmingtnDel Harlan&Hollingsworth	1876	3 Blds. WB	1Cy.50"-60" 500 HP. Harlan&Hollingsworth,WlmDel	Morgan R.R.&S.S.Co.

✻=Lloyd's Register; D=Nederlandsche Vereeniging; G=Germanischer Lloyd; I=Registro Italiano; R=American Record;
A=Veritas Austro-Ungarico; H=Veritas Ellenico; N=Norske Veritas; V=Bureau Veritas.

Dans la 2me Colonne, B.T., avec les indications qui suivent, signifie la date du rapport de visite de la Carène par les experts du Board of Trade pour obtenir le certificat relatif au TRANSPORT DES PASSAGERS.

Dans la 3me Colonne, C=Composite (membrure en fer, bordé en bois); I=Fer; Stl=Acier; W=Bois.

Pour plus amples renseignements sur les navires classés par le Lloyd's Register, voir le Lloyd's Register Book et le Lloyd's Yacht Register.

No. in Book / Official No.	Registry in which classed	Ship's Name. Material, Rig, &c. / Master, and Dates of Appointment to present Owner's Service and to Ship / Ship's Late Name. No. of Decks, &c.	Flag. / Port of Registry.	Tons. Net. / Gross	Length	Breadth	Depth	Built. Where. / By Whom.	When	No. of Bulkheads. / Particulars of Water Ballast.	Engines. Particulars. / Maker's Name.	Owners
331 73048	❋	Algitha (MISSING) I ScSr Colville 2 Dks 3 B	British North Shlds.	1219 1872	280·0 F 36 f.	34·9 F 40 f.	24·1	Newcastle A.Leslie&Co.	1877	5 Blhds. [illegible]	2Cy.33"& 65"–39" 200 HP. Thompson&Co.,Newcastle	H.E.P.Ad[illegible]
332 91241	❋	Algoma StlSc3mSr Murray -83 2 Dks 3 B	British Liverpool	1892 2914	320·0 F 38 f.	42·0 B 14 f.	27·0	[illegible] [illegible]	1885	[illegible]	3Cy.26"42&66"–42" 300 HP. T.Richardson&Sons,HtlepI.	Alleghany S.S.[illegible]
333		Algoma Stl ScSr	British Montreal	1153 1773	262·1	38·2	23·2	Glasgow [illegible]	1883		2Cy.35"& 70"–48" 300 HP. D.Rowan,Glasgow	Canadian Pacific [illegible]way Co.
334 95051	❋	Algonquin StlSc3mSr Moore 11-88 1 Dk Sp Dk & Wh Fms	British Glasgow	1172 1806	245·0	40·1 F 26 f.	18·9 20 f.	Glasgow [illegible]	1888	5 Blhds. [illegible]	3Cy.21"33"&54–36" 150 HP. Dunsmuir&Jackson,Glasgow	T. Marks & Co.
335		Algorta WPadTug (ex Reina del Mar)	Spanish Bilbao	105	89·9	18·6	9·3	NorthShields	1870		40 HP.	Fernando C[illegible]
336 80160		Alguada Stl TwnSr Rockell	British Glasgow	188 302	165·2	27·0	9·3	Dumbarton W.Denny&Bros.	1886		4Cy.21½"& 38"–24" 42 HP. Denny&Co.,Dumbarton	Irrawaddy Fl[illegible] Co.Ld.
337		Alhajuela I ScSr Hagen (ex Thomas A. Whitelaw)	C'sta R'ca P'ntaArenas	287	143·9	25·9	12·7	SanFrancisco Riley&Hayes	1878		2Cy.15"& 28"–20" 40 HP. [illegible]	Luis Batres
338 51464		Alhambra (ex Ada Wilson) I ScSr Fraser (BROKEN UP) 3 Dks	British Quebec	722 [illegible]	219·0	28·0	15·9 22·8	Preston Preston I.S.B.Co.	1865	5 Blhds. DB / St	2Cy.42"–30" 120 HP. J.Jones&Sons,Liverpool	Quebec & Gulf [illegible] S.S.Co.Ld.
339 24758		Alhambra (FOUNDERED) I ScBk 1 Dk	British Sydney	497 768	209·8 B 92 f.	27·1 F 135 f.	16·3	London SamudaBros.	1855		2Cy. 140 HP.	Aaron Wheeler
340		Alhassannee I ScBkn Hale 1 Dk & Awn Dk 3 B	Morocco	1001 1478	240·8	34·0	17·9 24·7	Middlesbro' R.Cr[illegible]	1882	[illegible] Blhds. [illegible] FPT[illegible]	2Cy.31"& 58"–36" 130 HP. Blair&Co.,Ld.,Stockton	Emperor of M[illegible]
341		Ali Saib Pasha I Sc3mSr (ex Somerset) Omer Bey -88 3 Dks	Turkish Cnstntinple.	1240 1923	284·4	35·8 B 90 f.	25·0	Stockton Richardson,Duck&Co.	1875	5 Blhds.	2Cy.40"& 75"–45" 250 HP. Blair&Co.,Ld.,Stockton	Idarei Mass[illegible] (Ottoman [illegible])
342	❋	Alianza I ScTug Esketh 1 Dk	British Preston	2 26	52·4	12·7	6·3	Preston R.Smith	1884	3 Blhds.	2Cy.11"& 20"–15" 17 HP. R.Smith,Preston	Cory Bros. & Co.
343 31830		Alice (HULK) I PadSr	British Southamptn	294 [illegible]	231·6	26·2 Q 43 f.	13·4	Greenock Caird&Co.	1857		2Cy.60"–72" 250 HP. Caird&Co.,Greenock	London & S.W.[illegible] way Co.
344 63828		Alice I ScSr	British Glasgow	193 [illegible]	187·1	25·3	8·8	Glasgow A.&J.Inglis	1871		2Cy.35"–18" 90 HP. A.&J.Inglis,Glasgow	G.Turnbull & [illegible]
345 87498	❋	Alice I ScBarge Fuller 1 Dk	British [illegible]	97 [illegible]	115·4 Q 33 f.	17·7 F 14 f.	8·8	Beverley H.J.J.Scarr	1882	3 Blhds.	1Cy.26"–18" 20 HP. Good&Menzies,Hull	J.Bland & Co.
346 91073	❋	Alice W ScSr Sanderson 1 Dk	British Leith	21 46	71·0	17·5	7·1	Leith MarrBros.	1885		2Cy.8"& 15"–12" 9 HP. J.Cran&Co.,Leith	P.Fairley
347		Alice W Machado	Brazilian Rio d J aneiro	232	..	..	..	RiodeJaneiro PontadAreaS.B.Co	18[illegible]			C'ia de Nav& [illegible] pirito Santo& [illegible]
348	V	Alice [illegible] I Sc3mSr Brisson 1 Dk 2 B	French St. Nazaire	587 [illegible]	226·3	30·8 F 58 f.	17·5	Glasgow A.Stephen&Sons	1872	[illegible]	2Cy.26"& 45"–30" 100 HP. D.Rowan,Glasgow	Compagnie G[illegible] Transatlantique
349	V	Alice I ScSr Pettersson -88 1 Dk	Swedish [illegible]	263 [illegible]	130·8 Q 52 f.	22·5 B 34 f.	12·7 H 11 f.	Oscarshamn MotalaCo.	1887		3Cy. MotalaCo.	C.Nilsson
350	❋	Alice Depeaux StlScSr Thomas -87 1 Dk 2 B	French Rouen	571* 916	220·5 Q 80 f.	32·0 B 58 f.	14·9 F 25 f.	Newcastle Palmers'Co.,Ld.	1887	5 Blhds. [illegible]	3Cy.19"31½"& 51"–36" 110 HP. PalmersCo.Ld.,Newcstle	F.Depeaux & [illegible]
351 105264		Alice Garratt	American S.Francisco	431 [illegible]	150·0	38·9	6·3	Stockton,Cal.	1873		90 HP.	W.Martin
352	V	Alice Krohn I ScSr Pretau 68-84 1 Dk	German Lübeck	395 [illegible]	176·8 Q 35 f.	25·5 B 51 f.	15·0 F 20 f.	Lübeck H.Koch	1884	[illegible]	2Cy.22"& 36"–26" 80 HP. D.Hennigrs&Co.,Berlin	Neue Dampfer [illegible] (Ges.[illegible])
353 65426	❋	Alice Otto I ScSr Holm 1 Dk 2 B	British North Shlds.	817 1267	250·8 P 31 f.	32·2 B 79 f.	19·0 F 32 f.	Sunderland Oswald&Co.	1871	[illegible] Blhds. [illegible]	2Cy.28"& 55"–39" 130 HP. Oswald&Co.,Sunderland	G.Otto
354 81515	❋	Alicia I Sc3mSr Jones 87-87 1 Dk 2 B	British W.Hartlepl.	923 142[illegible]	245·8 P 46 f.	33·1 F 31 f.	16·6	Hartlepool E.Withy&Co.	1880	4 Blhds. [illegible]	2Cy.30"& 55"–36" 120 HP. Blair&Co.Ld.,Stockton	R. Ropner & Co.
355	❋V	Alicia I Sc3mSr Aldamiz -88 2 Dks 3 B	Spanish Bilbao	1839 [illegible]	345·5	37·7 B 96 f.	25·[illegible] F 36 f.	Glasgow Lon.&Gls.Eng.&I.S.B.Co.	1883	5 Blhds. WB[illegible]	2Cy.38"& 71"–48" 300 HP. Lon.&Gls.Eng.&ISBCo.,Gls.	Linea de Vap[illegible]
356 868		Alida W P[illegible]	American Albany,N.Y.	51[illegible]	272·0	29·0	10·0	New York	1847			Cornell Steam[illegible]
357 86771	❋	Aline Stl ScSr Forth 71-84 1 Dk 2 B	British Whitby	1143 1[illegible]74	258·2 Q 74 f.	37·0 I 24 f. B 54 f.	16·7 F 28	Sunderland Doxford&Sons	1883	5 Blhds. [illegible]	2Cy.33"& 62"–42" 164 HP. W. Doxford&Sons,Sunderld.	T.Marwood & [illegible]
358 81641	❋	Aline I Sc3mSrYt	British London	156 [illegible]	165·7	22·1	11·5	Leith Ramage&Ferguson	1880	4 Blhds.	2Cy.19"& 38"–26" 65 HP. WalkerHenderson&Co.,Gls.	Col.D.Gamb[illegible]
359		Aline I Sc	Sarawak	101 17[illegible]	142·5	21·2	9·0	PortGlasgow Blackwd.&Gordon	1875		2Cy.28"–18" 40 HP. Blackwood&Gordon,P.Gls.	Rajah Broo[illegible]
360 106115	R	Alleghany I ScSr Hollett 3 Dks	American Baltimore	1433 [illegible]	250·0	38·5	27·0	Philadelphia Cramp&Sons	1881		2Cy.34"& 60"–45" 500 HP. Cramp&Sons,Philadelphia	Merchants' & [illegible] Transportation [illegible]

❋ = Lloyd's Register; D = Nederlandsche Vereeniging; G = Germanischer Lloyd; I = Registro Italiano; R = American [illegible]
A = Veritas Austro-Ungarico; H = Veritas Ellenico; N = Norske Veritas; V = Bureau Veri[illegible]

In the 2nd Column, B.T., with figures following, signifies Date of report of bottom sighting by Board of Trade Surveyors as condition of PASSENGER [illegible]

In the 3rd Column, C = Composite (iron frame, wood planked); I = Iron; Stl = Steel; W = Wood.

Further particulars of Vessels classed by Lloyd's Register will be found in Lloyd's Register Book and Lloyd's Yacht Register.

Cote	Nom, Matériaux, Gréement, etc. — Capitaine et dates d'entrée au service des Armateurs actuels et à bord du navire. — Ancien Nom, nombre de ponts, etc.		Pavillon. — Port d'Armement	Tonnage Net. — Total	Longueur	Largeur	Creux	Construit. Où. — Par qui.	Quand.	Nombre des cloisons. — Détails sur le lest d'eau.	Machines. — Détails. Nom du Constructeur.	Armateurs ou Directeurs-Gérants.
	Allegra (ex Melrose)	I Sc3mSr 1 *Dk*	British Malta	307 377	169·8	23·0	9·0 *P 16 f. H 12 f.*	Newcastle	1860	1 Blkd.	1Cy.26"–30" 45 HP. R.&W.Hawthorn,Newcastle	G. F[lli] Agius.
V	Allegro *Mingaten 83-88*	I ScSr 1 *Dk2 H*	Swedish Stockholm	845 1072	217·8	28·6	17·7 *P 161 f. F 23 f. H 13 f.*	Stockholm BergsundsMk.Vks.	1874		1Cy.18"&42½"–27" 130 HP. BergsundsMek.Verk.Stkhlm	Stockholms Angf. Rederibol. (L.Peyron)
V	Allemannia *Schröder 81-87*	I ScSr 2 *Dks*	German Hamburg	1375 1841	285·4	34·4	23·8 *P 32 f. B 35 f. Q 48 f.*	Glasgow Dobie&Co.	1881	DBal261	2Cy.34"&62"–42" 195 HP. J.Howden&Co.,Glasgow	Hamburg-Amerik. Packett.-Aktien-Ges.
✻	Allendale *Thompson 86-88*	I ScSr 1 *Dk2 B*	British Sunderland	551 857	211·8	28·5	16·1 *Q 69 f. H 44 f. B 86 f.*	Sunderland J.Laing	1875	3 Blkds. DBaf	2Cy.25"&48"–33" 98 HP. T.Clark&Co.,Newcastle	P.H.Laing
✻D	Allende *Pyres*	I ScSr 1 *Dk2 B*	British Cardiff	1069 1644	258·1	36·2	17·3 *Q 91 f. B 60 f. F 31 f.*	Newcastle Palmers Co.,Ld.	1883	1 Blkds. CDB361t APT21t	2Cy.31"&60"–39" 99 HP. Palmers Co.,Ld.,Newcastle	Forest S.S.Co.,Ld.
R	Allentown (FOUNDERED) *Osborne 83-87*	I ScSr 1 *Dk2 B*	American Philadelphia	910 1283	250·6	37·0	20·2 *P 29 f.*	Philadelphia W.Cramp&Sons	1874	4 Blkds. WBaf&b	1Cy.42"–42" 160 HP. W.Cramp&Sons	Philadelphia & Reading R.R.Co.
V	Aller *Christoffers 79-86*	I&St. Sc1mSr	German Bremen	2779 5381	438·0	48·1	34·6 *P 50 f. B 140 f. F 72 f.*	Glasgow Fairfield,S.B.&Co.	1885		3C.44"70"&108"–72" 1300 HP. FairfieldE.&S.B.C.Gls.	Norddeutscher Lloyd
✻	Allerwash (ex Ariadne) *Leafster*	I Sc3mSr 1 *Dk*	British Newcastle	284 389	167·5	25·8	11·1 *Q 23 f.*	Millbrook Kerman&Co.	1861	4 Blkds.	2Cy.18"&36"–24" 50 HP. Gilbert&Cooper,Hull	W.Benson
RV	Aliança *Bears 83-86*	I ScBn 2 *Dks & Awn Dk*	American New York	2205 2985	310·0	42·0	24·2	Chester, Pa. DlwreRiverISB&E.Wks	1886		2Cy.38"&74"–54" DelawareRiv.I.S.B.&Eng.W	U.S. & Brazil Mail S.S. Co.
	Alliance	I PadSr 1 *Dk*	British Nassau, N.P.	342 407	174·7	29·8	7·2	Glasgow Tod&McGregor	1857		2Cy. 90 HP. Tod&McGregor,Glasgow	L.G.Watson
✻	Alliance *Kavanagh 85-86*	I ScSr 1 *Dk*	British Dublin	198 316	144·5	22·6	11·6 *Q 54 f. F 17 f.*	PortGlasgow H.Murray&Co.	1874	4 Blkds.	2Cy.18½"&32"–22" 47 HP. Kemp&Hume,Glasgow	J.Weatherill
	Alliance	I PadSr	British Southampton	189 400	175·5	23·7	11·6	London C.J.Mare&Co.	1855		2Cy.34"&57"–54" 116 HP. Day,Summers&Co.,Sthptn.	London & S.W. Railway Co.
R	Alliance (ex Emma Hume) *McKenna*	WSc3mSr 1 *Dk2 B*	American SanFrancisco	141 253	128·0	26·6	10·0	Astoria, Oreg. J.Stephens	1884			J.McKenna
N	Alliance *Holmre 72-80*	I ScSr 1 *Dk*	Norweg'n Bergen	431 583	175·5	24·5	13·1 *P 105 f. F 27 f.*	Newcastle C.Mitchell&Co.	1880	4 Blkds. DBa&cb72ft	2Cy.20"&40"–28" 75 HP. T.Clark&Co.,Newcastle	J.L.Mowinckel & Sön
✻	Allie *Hansen 85-88*	I ScBn 1 *Dk2 B*	British Hartlepool	1134 1760	257·4	34·6	19·4 *Q 62 f. B 59 f. P 28 f. F 30 f.*	W. Hartlepool E.Withy&Co.	1882	4 Blkds. DBa&f318t	2Cy.32"&60"–39" 150 HP. Black,Hawthrn&Co.Gatshd	C.Nielsen & Son
	Alligator *Horner 74-82*	I ScSr	British Glasgow	409 932	250·2	30·0	14·7 *P 154 f. F 38 f.*	Glasgow Barclay,Curle&Co.	1881		2Cy.38½"&69"–48" 280 HP. Barclay,Curle&Co.,Glasg'w	G.&J.Burns
✻	Allonby *Gardener*	I ScSr 1 *Dk2 B*	British Sunderland	838 1291	240·0	34·1	15·4 *Q 94 f. B 140 f. F 32 f.*	Sunderland Kish,Boolds&Co.	1883	WBa&f APT	2Cy.31"&57"–39" 140 HP. Carr&Co.,Ld.,Sunderland	L.&H.Guéret
	Allowrie	Stl TwnScSr 1 *Dk & Awn Dk*	British Sydney	294 594	179·4	24·8	8·0 16·0	Glasgow NapierShanks&Bell	1880	6 Blkds. 3DT125t	2Cy.25"&48"–27" 98 HP. Rankin&Blackmore,Grnck.	Illawarra Steam Nav. Co.
GV	Alma *Stillewasser 83-84*	I ScSr 1 *Dk*	German Danzig	470 746	191·9	27·0	16·5 *P 47 f. B 40 f.*	Danzig J.W.Klawitter	1884	DT f 190t	2Cy.24½"&46"–27½" 90 HP. Aron&Gollnow,Stettin	F.G.Reinhold
	Alma	I ScSr 1 *Dk*	Russian Odessa	400 498	210·0	29·0	10·4	London J.&W.Dudgeon	1872		1Cy.15½"&27¼"–14" 60 HP. J.&W.Dudgeon,London	Russian Steam Nav. & Trading Co.
	Alma Munro	W ScSr 2 *Dks*	British PStanleyOn.	581 891	176·4	26·1	11·4 *H 166 f.*	Pt.Dalhousie Andrews&Sons	18[73/84]		1Cy.30"–36" 175 HP. G.N.Oille,St.Catharine'sOnt	Elgin Transportation Co.
	Almagrera	I	Spanish Cartagena	68 103	76·8	18·2	8·8	England	1874		15 HP.	L.Figuera y Silvela
✻D	Almandine *Collings 81-85*	I ScBn 1 *Dk2 B*	British Hull	1236 1897	260·3	36·0	21·0 *P 26 f. B 64 f. F 30 f.*	Stockton M.Pearse&Co.	1883	4 Blkds. DBa&f331t	2Cy.32"&60"–39" 170 HP. Blair&Co.,Ld.,Stockton	W.Tulley & Co.
✻	Almora *Hey 78-85*	I Sc3mSr 3 *Dks*	British Glasgow	1719 2613	350·4	36·4	28·1	Dumbarton W.Denny&Bros.	1873	5 Blkds.	2Cy.47"&82"–42" 316 HP. Denny&Co.,Dumbarton	British India Associated Steamers, Ld.
✻	Almsford *Dobinson 88-88*	I ScSr 1 *Dk2 L*	British W.Hartlep'l	919 1440	240·0	36·2	15·4 *Q 69 f. P 23 f. B 52 f. F 28 f.*	Sunderland R.Thompson&Sons	1883	1 Blkds. DBa f17t FPT50t	2Cy.28"&54"–36" 99 HP. Black,Hawthrn&Co.Gatshd	J.Wood & Co.
✻	Alne Holme *Markham 78-78*	I ScSr 1 *Dk2 L*	British Maryport	658 1036	220·3	31·0	16·9 *Q 123 f. F 29 f.*	Sunderland J.L.Thompson&Sn	1876	4 Blkds. DBa 37ft f 61ft	2Cy.27"&51"–33" 98 HP. J.Dickinson,Sunderland	Hine Bros.
	Alnö *Fogelqvist*	I Sc 1 *Dk*	Swedish Sundsvall	67 100	86·3	15·3	8·1 *Q 32 f. B 20 f. F 17 f. H 4 f.*	Stockholm W.LindbergsMk.Vks.	1874		2Cy10½"&22¼"–15⅝" 25 HP. W.LindbergsMk.Vks,Stkholm	Alnösunds Akt.Bol.
✻	Alonso *Neilsen 85-85*	I ScSr 1 *Dk2 B*	British Hull	845 1315	240·0	32·1	19·5 *P 200 f. F 12 f.*	Stockton Richardson,Duck&Co.	1869	5 Blkds. WBa&f	2Cy.29"&56"–33" 98 HP. Cooper&Co.,Hull	E.Leetham
	Aloungpyah *Rimmer*	I PadSp	British Glasgow	385 569	245·0	26·2	8·0	Port Glasgow R.Duncan&Co.	1872		2Cy.31"&51"–48" 125 HP. Rankin&Blackmore,Grnck.	Irrawaddy Flotilla Co. Ld.
	Alpha *Crowell 88-78*	I ScBk	British Glasgow	514 653	221·8	27·6	14·9 *P 38 f.*	Glasgow Barclay,Curle&Co.	1863		2Cy. 112 HP.	W.A.Black

✻=Lloyd's Register; D=Nederlandsche Vereeniging; G=Germanischer Lloyd; I=Registro Italiano; R=American Record;
A=Veritas Austro-Ungarico; H=Veritas Ellenico; N=Norske Veritas; V=Bureau Veritas.

Dans la 2me Colonne, B.T., avec les indications qui suivent, signifie la date du rapport de visite de la Carène par les experts du Board of Trade pour obtenir le certificat relatif au TRANSPORT DES PASSAGERS.

Dans la 3me Colonne, C=Composite (membrure en fer, bordé en bois); I=Fer; Stl=Acier; W=Bois.

Pour plus amples renseignements sur les navires classés par le Lloyd's Register, voir le Lloyd's Register Book et le Lloyd's Yacht Register.

1 No. in Book Official No.	2 Registry in which classed	3 Ship's Name. Material, Rig, &c. Master, and Dates of Appointment to present Owner's Service and to Ship. Ship's Late Name. No. of Decks, &c.	4 Flag. Port of Registry.	5 Tons Net Gross	6 Dimensions Length, Breadth, Depth	7 Built. Where. By Whom.	8 When	9 No. of Bulkheads. Particulars of Water Ballast.	10 Engines. Particulars. Maker's Name.	11 Owners or Managers
391 71413	R	**Alpha** W Sc	British Yarm'th N.S.	212 397	129·6 | 22·6 | 17·1	Prince Edw. Is. H. Ramsay	1874		2Cy.18"&36"–24" 100 HP. Burrell Johnson&Co Yar.NS.	Yarmouth Stm. Co.Ld.
392	DGV	**Alpha** I Sc3mSr *Bendfeldt* 84	German Lübeck	300 300	159·7 | 21·3 | 12·1 P 48 f. F 11 f.	Newcastle W.B.Hornby	1869		2Cy.19"&38"–26" 90 HP. Rostock Sch. & M. Co. Rostock	Akt.-Ges. St. P. (Joh.Siegm.Olsen)
393	DN	**Alpha** I ScSr *Nielsen* 2 *Dks*	Norweg'n Christiania	458 607	180·8 | 25·0 | 16·7	Dumbarton A.McMillan&Son	1872	4 Blkds. AU 120	2Cy.22"&38"–27" 70 HP. M.Paul&Co. Dumbarton	Ostlandske Ll. (Rolf Andvord)
394		**Alpha** W ScSr *Jansen* 1 *Dk* 2 *B*	Norweg'n Bergen	216	119·3 | 25·9 | 13·3 Q 30 ft. B 24 f.	Bergen Trampe & [illegible]	1872		2Cy.15"&30"–18" 32 HP. Bergens Mek Verks., Bergen	J. Amelu
395		**Alpha** ex Gestaldy I ScSr	Russian St.Peterb'g	66 108	86·2 | 16·1 | 9·1	Oscarshamn J.Fletcher	1875		2Cy. 20 HP. Oscarshamn Mek. Verks	P. Morch
396		**Alphée** I Sc 3 *Dks*	French Marseilles	1215 1831	331·8 | 33·1 | 25·4 P 67 f. H 51 f. F 33 f.	La Ciotat Messageries Marit.	1861		2Cy.68"–46½" 400 HP. Messageries Mar., La Ciotat	Messageries Maritimes
397	✻	**Alphonse Conseil** I ScSr ex Port of Alicante, Himese 85, 1 *Dk* & *Spr Dk* 3 *B*	French Bordeaux	1090 1683	241·0 | 34·3 | 17·1 24·8	Newcastle Palmers Co., Ld.	1884	4 Blkds. DB [illegible]	2Cy.30"&62"–42" 158 HP. Palmer's Co., Ld., Newcastle	Les fils de T.Ge
398		**Alps** (ex Alpes, ex Alps, *McKnight* 84-87 ex Mandingo) 3 *Dks* I ScSr	British Kingston, Jam	1117 1725	295·9 | 32·4 | 19·3 27·0	Glasgow Randolph, Elder&C	1865		2Cy.32"&60"–56" 180 HP. J.&J.Thomson, Glasgow	Atlas S.S.Co.Ld
399		**Alsace** I Sc *Marroc* 80-80 2 *Dks*	French Marseilles	697 1047	212·6 | 29·0 | 24·3	La Seyne Forges&Ch.dela Medit	1866		2Cy.33"–24½" 100 HP. Forges&Ch.dela Medit. Marseille	Société Générale Transports Ma.
400 78817		**Alsatia** I ScSr *Brown* 81-86 3 *Dks*	British Glasgow	1793 2773	356·7 | 36·2 | 29·4	Glasgow D.&W.Henderson&Co.	1876	5 Blkds. WB	3C.24"37"&58"–48" 240 HP. D&W Henderson&Co.Glasgw	Henderson Bro.
401 [illegible]		**Alsatian** I ScSr *Christian* 83-83 2 *Dks*	British Liverpool	1133 1732	312·0 | 31·2 | 23·9 F 72 f.	Liverpool F.Leyland&Co.	1877	5 Blkds.	2Cy.23"&60"–39" 220 HP. Fawcett, Preston&Co., Lvpl	S.S. "Alsatian" C. (F.Leyland & Co.)
402	✻	**Alsen** Stl Sc *Kühr* 82-88 1 *Dk* & *B'b Fms*	German Kiel	572 742	195·0 | 30·0 | 13·8 B 44 f. F 29 f.	Kiel G.Howaldt	1888 6mo	5 Blkds. [illegible]	3Cy14½"23½"&39"–23½" G.Howaldt, Kiel. 100 HP	Lange Gebrüder
403		**Alta Paraguay** I Pad *Fernandez*	Arg'ntine B'enos Ayres	140 260	135·0 | 26·0 | 5·0	Buenos Ayres Marshall&Co.	1870		92 HP.	Gigliani & Co.
404		**Altata** I ScTug 1 *Dk*	Mexican Mazatlan	79 116	101·5 | 16·4 | 9·2	London W.Walker&Co.	1881	4 Blkds.	2Cy.14"&28"–18" 35 HP. J.Watt&Co., Birmingham	Sinalva & Durango Railroad Co.
405		**Alten** I ScSr (ex Dannebrog) 2 *Dks*	Norweg'n Arendal	92 140	100·9 | 16·3 | 12·2	Christiania	1870		32 HP.	O.B.Sörensen
406 77439	✻D	**Altmore** I ScSr *Stephen* 82-84 2 *Dks* 3 *B*	British Aberdeen	1352 2073	291·5 | 35·2 | 24·8 B 58 f. F 24 f.	Sunderland R.Thompson, jr.	1877	5 Blkds. DB aft 122 FPT 38 APT 38	3Cy.32"36"&66"–36" 200 HP. Tyne [illegible] N. E. Marine E. Co. Ld. Sunderland	Adam S.S.Co.
407 84361	✻D	**Altnacraig** I Sc3mSr *Buyers* 76-82 1 *Dk* & *Spr Dk* 3 *B*	British Aberdeen	1872 2858	315·0 | 40·1 | 28·3	Sunderland Sunderland S.B.C., Ld.	1882	6 Blkds. CDB [illegible]	2Cy.37"& 72"–48" 300 HP. Black, Hawthorn&Co. Gatesd	Adam S.S.Co.
408 67821	✻	**Altona** I ScSr *Eccles* 1 *Dk*	British Leith	385 670	202·1 | 28·2 | 13·0 Q 39 f. F 29 f. B 16 f.	Dundee W.B.Thompson	1877	5 Blkds. DB [illegible]	2Cy.28"&52"–33" 98 HP. W.B.Thompson, Dundee	Yorkshire Coal Co., Ld.
409 77462	✻D	**Altonower** I Sc3mSr *Barnet* 82-84 2 *Dks* & *Spr Dk*	British Aberdeen	1611 2476	300·0 | 37·0 | 21·8 28·9	Newcastle A.Leslie&Co.	1880	6 Blkds. [illegible] FPT 22 APT 12	[illegible] & [illegible]–42" 220 HP. [illegible]	Adam S.S.Co.
410 77447	✻	**Altyre** I ScSr *Watson* 86-86 1 *Dk* 2 *B*	British Aberdeen	670 1041	231·5 | 28·5 | 15·0 Q 76 f. B 80 f. F 27 f.	Sunderland Mounsey&Foster	1878	5 Blkds. DB aft 154 [illegible]	2Cy.37"& 17"–30" 98 HP. N.E.Marine E.Co.Ld. Sunderld	Adam & Co.
411 81703	✻D	**Aludra** I ScSr *Rule* 80-87 1 *Dk* 2 *B*	British Shoreham	966 1503	249·5 | 33·5 | 18·4 P 28 f. Q 34 f. F 34 f. B 71 f.	Sunderland Mounsey&Foster	1879	4 Blkds. [illegible]	2Cy.31"& 57"–36" 140 HP. J.Dickinson, Sunderland	R.H.Penney
412 106448	✻	**Alva** Stl ScBkn *Morrison* 2 *Dks*	American New York	601 1151	256·0 | 32·3 | 18·5	Wlmingtn. Del Harlan&Hollingsworth	1886	6 Blkds.	3C.32"&(2)45"–42" [illegible] H Harlan&Hollingsworth [illegible]	W.K.Vanderbilt
413 77449	✻	**Alvah** I ScSr *Young* 78-84 2 *Dks* 3 *B*	British Aberdeen	1511 2332	300·0 | 36·0 | 25·2 P 33 f. B 52 f. F 35 f.	Sunderland Mounsey&Foster	1878	5 Blkds. DB aft 290 FPT 12 APT 28	[illegible] 47 & [illegible]–42 [illegible] HP. N.E.Marine E.Co.Ld. Sunderld	Adam S.S.Co.Ld
414	✻	**Alvarado** I Sc3mSr *Fernandez* 1 *Dk* & *Awn Dk*	Spanish Barcelona	772 1017	210·3 | 28·9 | 21·1	Leith S.&H.Morton&Co.	1883	4 Blkds.	2Cy.23"& 46"–30" 85 HP. S.&H.Morton&Co., Leith	J.Roca & Co.
415 63790		**Alvena** ex Glengyle I ScBg *Mackay* 86-86 3 *Dks*	British Liverpool	1142 1741	274·2 | 32·9 | 24·4 P 76 f.	Glasgow London&Glas Co.	1870	6 Blkds.	3Cy.21½"35½"&58½"–39" Blair&Co., Ld. Stktn 200 HP	Atlas S.S.Co.Ld
416 81509	✻	**Alverton** I ScSr *Harris* 87-87 1 *Dk* 2 *B*	British London	898 1387	240·0 | 33·0 | 16·5 P&B 144 f. F 31 f.	Hartlepool E.Withy&Co.	1880	4 Blkds. CDB aft	2Cy.29½"&55"–36" 130 HP. [illegible] Sunderland	Jackson Bros.&
417 92059	✻	**Alvina** I ScSrYt	British S'hampton	132 194	138·8 | 20·2 | 11·7	Southampton Oswald, Mordaunt & Co.	1887	4 Blkds.	3Cy.12"18"&30"–21" 56 HP. Oswald, Mordaunt&Co. S'hmptn.	R.Clifford Smith
418 81296	✻	**Alvo** I ScSr *Williams* 75-82 2 *Dks* 3 *B*	British Liverpool	1304 2009	291·9 | 34·4 | 24·0 B 94 f. F 26 f.	Glasgow Aitken&Mansel	1879	5 Blkds. FPT APT	2Cy.36"& 68"–42" 240 HP. J.&J.Thomson, Glasgow	Atlas S.S.Co.Ld
419	G	**Alwine** I ScSr *Saunders* 87-87 1 *Dk* & *Awn Dk*	German Kiel	400 611	162·4 | 22·7 | 17·8	Kiel G.Howaldt	1881		2Cy.20"& 36"–21" 50 HP. G.Howaldt, Kiel	M.Jebsen
420										

✻ = **Lloyd's Register**; D = Nederlandsche Vereeniging; G = Germanischer Lloyd; I = Registro Italiano; R = American Rec.; A = Veritas Austro-Ungarico; H = Veritas Ellenico; N = Norske Veritas; V = Bureau Veritas.

In the 2nd Column, B.T., with figures following, signifies Date of report of bottom sighting by Board of Trade Surveyors as condition of PASSENGER CERTIF.

In the 3rd Column, C = Composite (iron frame, wood planked); I = Iron; Stl = Steel; W = Wood.

Further particulars of Vessels classed by Lloyd's Register will be found in Lloyd's Register Book and Lloyd's Yacht Register.

1-2	3 Nom, Matériaux, Gréement, etc. / Capitaine et dates d'entrée au service des Armateurs actuels et à bord du navire. / Ancien Nom, nombre de ponts, etc.	4 Pavillon. / Port d'Armement	5 Tonnage. Net. / Total	6 Dimensions. Longueur	Largeur	Creux	7 Construit. Où. / Par qui.	8 Quand.	9 Nombre des cloisons. / Détails sur le lest d'eau.	10 Machines. / Détails. Nom du Constructeur.	11 Armateurs ou Directeurs-Gérants.
DV	Alwine Seyd 1 [illegible] *[illegible]* 1 *Dk*	German [illegible]	558 740	181·1 *B 51 f.*	27·1 *F 22 f.*	18·0	Vegesack J.Lange	1884	WB[illegible]	2Cy.17¾"&33½"-27½" 75 IP. J.Lange,Vegesack	J.D.Bischoff
	Amadeo 1 SeSr *Rice*	British Liverpool	258 411	151·8 *B 41 f.*	24·0 *F 18 f.*	11·9	Liverpool Liv.ForgeCo.,Ld.	1884		2Cy.18"&36"-30" 54 IP. J.Jones&Sons,Liverpool	Argentine Steam Lighter Co.,Ld.
*	Amadeo StlSeSTug *Lewis 88-89* 1 *Dk*	Brazilian Pernambuco	15 70	81·3	16·0	9·5	Glasgow Scott&Co.	1888	3 Blds.	3Cy.9¾"16"&25"-18" 35 IP. Alley&M'Lellan,Glasgow	Livramento & Co.
	Amador *Brown 74-*	American	757 988	199·0	39·0	10·6	SanFrancisco	1869		300 IP.	E.H.Miller, Jun.
	Amal CSeSp *Larsson* 1 *Dk*	Swedish Åmål	127 165	100·4 *Q 28 f.*	19·4 *F 24 f.*	10·1 *H 10 f.*	Åmal	1876		25 IP.	P.Larsson
*	Amalfi 1 Se3mSr *Krafft -88* 2 *Dks3B*	German Hamburg	1792 2345	300·5 *P 28 f.*	36·1 *F 36 f.*	25·2	Stockton M.Pearse&Co.	1881	4 Blds. DB[illegible]	2Cy.36½"&68"-42" 200 IP. Blair&Co.,Ld.,Stockton	R.M.Sloman & Co.
V	Amalia 1 Se3mSr *Rathke 72-78* 1 *Dk*	German Stettin	154 192	121·7 *Q 9 f.*	18·9 *Len 78.*	9·4	Kiel NddtscheSchiffA.G.	18$^{71}_{78}$		2Cy.18"&24"-18" 30 IP. Aren&Gottnow,Stettin	R.C.Gribel
A	Amalia 1 Sc *Angelini*	Russian Taganrog	289 375	..	..	..	Trieste Stabilimento Tecnico	1883		50 IP.	Azoff Steam Nav.Co.
	Amalia	Spanish Barcelona	242 382	160·1	12·5	10·9	England	1860			P.Pi
	Amalia 1 Sr 1 *Dk*	Spanish Alicante	167 267	139·7	22·9	11·3	Glasgow	1856		80 IP.	Sra.Viuda & hijos de Carratala
	Amalia CSeSp *Kvick -87* 1 *Dk*	Swedish Gothenburg	128 187	101·8 *B 18 f.*	21·0 *H 6 f.*	10·5	Sjötorp S.Groth	1886		2Cy.12"&21"-16" 25 IP. Eriksbergs,M.V.Gothenburg	C.F.Sandberg
*	Amalie (ex Llanelli) 1 SeTrawler *Reuter 88-88* 1 *Dk*	German Geestemünde	55 83	90·0	20·1	10·6	Hull Cook,Welton&Gemmell	1886	3 Blds.	2Cy.14"&26"-20" 30 IP. C.D.Holmes&Co.,Hull	W. Bade & H. Eissfeld
V	Amalienborg 1 SeSr *Van Deurs 84-86* 1 *Dk*	Danish Copenhagen	903 1415	247·0 *P 150 f.*	33·2 *F 50 f.*	15·5	Copenhagen Burmeister&Wain	1884	WB 480 APT 130	2Cy.30"&52"-33" 120 IP. Burmeister&Wain,Copnhag	Dampsk-Selsk. "Dannebrog" (CKHansen)
*	Amana 1 SeBn *Leadbeater* 2 *Dks3B*	British Liverpool	2049 3122	325·5 *P 21 f.*	40·0 *B 82 f.*	27·5 *F 29 f.*	Stockton M.Pearse&Co.	1885	5 Blds. DB[illegible] 470	3Cy.24"40"&65"-48" 400 IP. Blair&Co.,Ld.,Stockton	J.M.Wood
	Amand-Dumeau I&StlPadSr *Masqueles*	French Bordeaux	128 262	181·1	20·0	7·6	Arles HenriSatre	1885		4Cy.21"&45"-47" 200 IP. HenriSatre,Lyons	Dufilhot & Co.
*	Amanda 1 SeSr *Kler* 1 *Dk2B*	British Hartlepool	811 1261	237·6 *Q 76 f.*	31·6 *B 57 f.*	17·2 *F 32 f.*	W.Hartlepool W.Gray&Co.	1877	4 Blds. DB[illegible]	2Cy.28"&53"-33" 110 IP. T.Richardson&Sons,Hartpl	O.Trechmann
*	Amaranth StlSeSr *Saunders -88* 2 *Dks*	British Glasgow	1753 2677	310·0 *P 29 f.*	40·1 *B 60 f.*	24·8 *F 37 f.*	Glasgow A.Stephen&Sons	1888	5 Blds. CDB 517t	3Cy.23"37"&59"-42" A.Stephen&Sons, Glasgow [240 IP.	J.B.Murray
*	Amarapoora 1 Se3mSr *Carrin 84-84* 3 *Dks*	British Glasgow	1619 2464	350·2	36·3	27·0	Greenock Scott&Co.	1874	5 Blds.	2Cy.45"&78"-45" 300 IP. GreenockFoundryCo.,Grnk.	British & Burmese Steam Nav.Co.,Ld.
*	Amaryllis 1 Se3mSr *Sanders 88-88* 2 *Dks3B*	British NthShields	1109 1714	264·4	34·0	24·3	Newcastle ColeBros.	1875	5 Blds. DB[illegible]	2Cy.34"&58"-33" 150 IP. N.E.MarineE.Co.,Ld.,Sudld.	J.Robinson & Sons
*	Amazonas 1 SeSr *Dickinson 80-80* 1 *Dk2B*	British Sunderland	673 1057	229·2	29·5	20·8	Sunderland A.Simey&Co.	1874	4 Blds. WB[illegible]	2Cy.27"&53"-33" 110 IP. Hawks,Crawshy&SonsGtshd	J.Laing
	Amazone 1 Se 3 *Dks*	French Marseilles	2080 3120	370·9 *H 104 f.*	39·4 *F 50 f.*	30·6	LaCiotat MessageriesMarit.	1869		3C.43¼"&60¼"-13¼" 500 IP. Messgries.Mr.LaCiotat.	Messageries Maritimes
*	Amazonense 1 SeBg *Kempthorne 85-88* (ex Windsor)	British Liverpool	1081 1692	279·4 *P 68 f.*	35·2 *ShltrDk 184 f.*	20·8	Newcastle Wgham.Richardson & Co.	1869	3 Blds. WB[illegible]	2Cy.44"&78"-36" 300 IP. Cmp.81,C.D.Hlms&Co.,Hull	R.Singlehurst & Co.
	Amazonia 1 TwnSeSr	Brazilian Para	237 570	193·0	32·2	9·2	Chester,Pa. J.Roach&Son	1883		2Cy.15"&28"-27" J.Roach&Son,Chester,Pa.	S.G.de Oliveira & Co.
*	Ambassador StlSeSr *Aikman 71-88* 2 *Dks3B*	British Newcastle	1641 2511	300·0 *P 30 f.*	38·7 *B 60 f.*	23·6 *F 58 f.*	Newcastle TyneI.S.B.Co.,Ld.	1888	6 Blds. CDB 508 APT 321	3Cy.24"38"&62"-42" R.&W.Hawthorn, Leslie&Co.,Ld.,Nwe. [220 IP.	Hall,Brothers
*	Amber StlSeSr *Greey 85-88* 2 *Dks,&AwnDk*	British London	556 1034	217·5	31·1	14·3 21·5	Glasgow R.Napier&Sons	1888	1 Blds. DB[illegible]	3Cy.21"33"&46" 250 IP. R.Napier&Sons,Glasgow	Eastern TelegraphCo. Ld.
*D	Ambient 1 SeSr *Errington 87-88* 1 *Dk2B*	British Sunderland	668 1033	216·8 *Q 74 f.*	30·1 *B 69 f.*	16·5 *F 27 f.*	Sunderland ShortBros.	1879	4 Blds. WB[illegible] FPT APT	2Cy.27"&51"-33" 98 IP. J.Dickinson,Sunderland	J.Westoll
	Amboina 1 SeSr *Van Gorcum -87* 2 *Dks*	Dutch Batavia	487 917	225·9	28·0 *F 28 f.*	15·1	PortGlasgow Blackwd.&Gordon	1874		2Cy.32"&54"-35" 155 IP. Blackwood&Gordon,Pt.Gls.	NederlandschIndische Stoomvaart Maatschappij
*	Ambriz (ex Asiatic) 1 Se3mSr *Westray 85-88* 2 *Dks3B*	British London	1373 2130	324·6	35·2 *H 18 f.*	25·6	Liverpool T.Royden&Sons	1871	8 Blds.	2Cy.37"&75"-36" 233 IP. LairdBros.,Birkenhead	AfricanSteamShipCo.
	Ambrose 1 SeBg *Bisson 81-82*	British Liverpool	756 1168	242·0	29·6	24·7	Newcastle A.Leslie&Co.	1869		2Cy.21"&48"-40" 90 IP. Fawcett,Preston&Co.,Livpl	BoothStm.ShipCo.,Ld
A	Ambrosios N. 1 Se	Russian Taganrog	258 325	..	..	..	Trieste Stabilimento Tecnico	1885		10 IP.	D.A.Negroponte

[illegible]=Lloyd's Register; D=Nederlandsche Vereeniging; G=Germanischer Lloyd; I=Registro Italiano; R=American Record;
[illegible]=Veritas Austro-Ungarico; H=Veritas Ellenico; N=Norske Veritas; V=Bureau Veritas.

Dans la 2me **Colonne**, B.T., avec les indications qui suivent, signifie la date du rapport de visite de la Carène par les experts du Board of Trade pour obtenir le certificat relatif au TRANSPORT DES PASSAGERS.

Dans la 3me Colonne, C=Composite (membrure en fer, bordé en bois); I=Fer; Stl=Acier; W=Bois.

Pour plus amples renseignements sur les navires classés par le Lloyd's Register, voir le Lloyd's Register Book et le Lloyd's Yacht Register.

No. in Book. / Official No.	Registry & where Classed.	Ship's Name. Material, Rig, &c. / Master, and Dates of Appointment to present Owner's Service and to Ship / Ship's Late Name. No. of Decks, &c.	Flag. / Port of Registry.	Tons. Gross / Net	Length	Breadth	Depth	Built. Where. / By Whom.	When.	No. of Bulkheads. / Particulars of Water Ballast.	Engines. Particulars. / Maker's Name.	Owners or M[anagers]
451 67524	*	Amcott I ScSr *Hildmon* 81 1 *Dk2 B*	British W. Hartlepl.	655 1019	225·0 *Q* 84*f.*	29·8 *B* 31*f.*	16·9 *F* 28*f.*	Hartlepool Withy, Alexander& Co.	1872	4 Bhds. [illegible]	2Cy. 27" & 51"–33" 99 HP. Blair&Co., Ld., Stockton	H. G. Harper &
452 86382		Amelia W PadSp 2 *Dks*	British Victoria, B.C.	222 431	150·0	26·0 *H* 98 *f.*	8·5	San Francisco	1863		1Cy. 36"–72" 50 HP. J. Kellwell, Jersey City, U.S.A.	John H. Turn
453 61		Amelia (ex Ligaria P., ex Nuovo Porto Maurizio) I ScSr *Vicari* 1 *Dk*	Italian Palermo	102 174	122·3	19·7	10·2	Pt. Glasgow R. Murray&Co.	1868		2Cy. 15¼" & 26¼"–23¼" 49 HP. New Engines 1877	P. Pastorino &
454		Amelia (ex Oseian, ex Deva) I Sc3mSrYt	Port'g'se Lisbon	69 111	111·4	17·1	10·0	Preston W. Allsup&Sons	1878		2Cy. 14" & 25"–21" 25 HP. J. H. Wilson & Co. Liverpool	Crown Prince Portugal
455		Amélie	French [illegible]	126								
456 48703	BT3.88	America (ex Cheshire) I Pad	British	291 532	150·0	30·0	11·0	Liverpool	1863		4Cy. 33"–42" 160 HP. H. M. Lawrence&Co., Liverpl.	Liverpool S[...] Co., Ld.
457 82480	*	America I ScSr *Salmon* 1 *Dk*	British Hull	114 198	130·4	21·0	10·8	Hull Earle's Co.	1880	3 Bhds.	2Cy. 21" & 40"–27" 60 HP. Earle's Co., Ld., Hull	Hull Stm Fi[...] Co. Ld.
458 87881		America W Sc	British Liverpool	80 117	82·9	20·5	8·9	Winsford T. H. [illegible]	1877		2Cy. 14"–18" 30 HP. J. Verdin&Sons, Winsford	R. Verdin
459 885		America W Pad	American New York	404 560	153·5	33·0	11·7	Brooklyn	1862			M. Bunker
460		America I PadSr *Franco* 88 1 *Dk*	Brazilian Rio de Janeiro	677 884	250·0	30·6 *P* 150 *f.*	16·2	Newcastle Wigham Richardson	1874		2Cy. 40" & 74"–78" 200 HP. Thompson, Boyd&Co., Newc.	Cia. de Nav. P[...]
461	G	America I ScBk *Heintze* 88-88 1 *Dks*	German Bremen	1794 2752	328·8	40·0	33·1	Greenock Caird&Co.	1863		2Cy. 56" & 96"–48" 700 HP. Day, Summers&Co. S'thamptn.	Norddeutsche
462		America W&I	Spanish Barcelona	439 778	206·7	28·4	18·8		1855		250 HP.	Sociedad de [...] cioné Indus[...]
463 29036	BT12.88	American I ScBn *Anson*	British Liverpool	1369 1839	274·9	34·0	22·3 29·7	Greenock Scott&Co.	1865		1Cy. 21" & 48"–46" 350 HP. G. Forrester&Co., Liverpool	West India & [...] S.S.Co., Ld.
464 2[...]	I	Amerigo Vespucci I ScSr *Renzi* 88-88 2 *Dks*	Italian Palermo	373 592	200·7	25·3	13·6	Port Glasgow Blackwd.&Gordon	1862		2Cy. 28½" & 48"–45" 180 HP. P. Dannovaro Genoa	Navigazione [...] Italiana
465	* V	Amérique *Dardignac* -86 I ScBk (ex Imperatrice Eugenie, ex Atlantique) 3 *Dks. & 1 awn Dk*	French Havre	2744 4538	393·7 *Len.* 73	43·9 *F* 33 *f.*	29·0 38·4	St. Nazaire Scott&Co.	1865	9 Bhds.	1Cy. 41" & 75"–51" 900 HP. Maudslay Sons, & Field, London.	Compagnie G[...] Transatlantique
466	*	Amérique I ScSr *Labie* 73-86 3 *Dks*	French Marseilles	1351 2055	308·7	36·1 *F* 32 *f.*	26·7	Dumbarton A. M'Millan&Son	1879	4 Bhds.	2Cy. 38" & 68"–45" 260 HP. Denny&Co., Dumbarton	Fraissinet & C[...]
467 7904[...]	*	Amethyst I Sc3mSr *Caiese* 85-85 1 *Dk2 B*	British Liverpool	872 1357	240·2 *P* 154 *f.*	32·0 *F* 31 *f.*	16·7	Stockton M. Pearse&Co.	1878	4 Bhds. DB[illegible]	2Cy. 28½" & 59"–38" 110 HP. Blair&Co., Ld., Stockton	"Amethyst" S[...] (T. R. Williams)
468 76891	*	Amethyst I ScSr *Middleton* 1 *Dk2 B*	British Dundee	601 934	218·0 *Q* 66 *f.*	29·0 *B* 53 *f.*	15·9 *F* 26 *f.*	Dundee Gourlay Bros.&Co.	1879	5 Bhds. DB[illegible] FPT APT	2Cy. 28" & 48"–30" 98 HP. Gourlay Bros.&Co. Dun.	Dundee Gem [...] Co. Ld.
469 87839	*	Amethyst I Sc3mSr *Donaldson* 88-88 1 *Dk*	British Glasgow	290 533	185·1 *Q* 89 *f.*	27·1 *B* 9 *f.*	11·5 *F* 26 *f.*	Paisley J. Fullerton&Co.	1883	4 Bhds. CDB 110 *ft.* 68 FPT 31 APT 5	2Cy. 23" & 45"–33" 85 HP. W. King&Co., Glasgow	W. Robertson
470 74451		Amethyst I Sc3mSrYt *Harvey* 2 *Dks*	British Cowes	129 201	148·0	22·2	11·0	Southampton Day&Summers	1877	4 Bhds.	2Cy. 20" & 40"–24" 60 HP. Day&Summers, S'thampton.	Capt. Thomas [...]
471 93331		Amherst Stl Pad *Rodgers*	British Glasgow	149 287	160·5	24·0	7·8	Dumbarton W. Denny&Bros.	1886		2Cy. 25" & 48"–42" 96 HP. Denny&Co., Dumbarton	Irrawaddy Fl[...] Co. Ld.
472 9906[...]	*	Amherst I ScSr *Macey* 86-88 1 *Dk & part Shd Dk*	British Singapore	108 257	130·1	25·1 *B* 7 *f.*	10·0 *F* 18 *f.*	Port Glasgow Blackw'd&Gordon	1886	4 Bhds. FPT [illegible]	2Cy. 18" & 35"–20" 52 HP. Bow&McLachlan, Paisley	Wee Boon T[...]
473	* N	Amicitia I ScSr *Aslaksen* 87-87 1 *Dk & 1 awn Dk*	Norweg'n Bergen	529 711	173·8	27·1	14·5 19·5	Bergen Martens, Olsen&Co.	1883	5 Bhds. DB[illegible]	2Cy. 23" & 41"–28" 75 HP. Martens, Olsen&Co., Bergen	Bergh & Hel[...]
474	V	Amigo I ScSr *Bundewadt* 84-87 2 *Dks & Spr Dk*	German Apenrade	771 1186	220·1 *Q* 27 *f.*	32·0 *B* 18 *f.*	15·4 21·4	Hamburg Blohm&Voss	1884	DB [illegible]	2Cy. 28" & 48"–36" 120 HP. Blohm&Voss, Hamburg	M. Jebsen
475 10510[...]		Amite W Sc *Morris*	American New Orleans	172 238	127·0	28·0	6·9	Madisonvile La	1872		180 HP.	Ch. R. Post
476 87370	*	Amity I ScSr *Hazleton* 86-86 1 *Dk2 B*	British Sunderland	923 1430	240·0 *Q* 96 *f.*	36·3 *B* 53 *f.*	15·7 *F* 24 *f.*	Sunderland J. Laing	1883	4 Bhds. DB[illegible]	2Cy. 29½" & 55"–36" 130 HP. G. Clark, Sunderland	"Amity" S.S.C[...] (G. Hair&Co.)
477 81554	* D	Amoor I ScSr *Bennett* 88-88 2 *Dks 3 B*	British London	1288 1987	285·0	35·2 *P* 30 *f.*	24·4	Newcastle C. S. Swan&Co.	1879	5 Bhds. DB[illegible]	2Cy. 34" & 62"–46" 200 HP. Black, Hawthorn&Co., Gtsd.	Mercantile S.S[...]
478 10526[...]	*	Amos C. Barstow W ScSr *Corwin* 71-71	American Providence, R.I.	204 268	140·0	23·0	9·1	Brooklyn, N.Y.	1873		120 HP.	Providence and [...] [...]ington Steamboat
479		Amoy I ScSr *Koehler* 88 2 *Dks 3 B*	German Hamburg	814 1250	249·0	31·6	22·0	Newcastle C. Mitchell&Co.	1873	4 Bhds.	2Cy. 31" & 61"–36" 160 HP. T. Clark&Co., Newcastle	Chinesische [...] fahrt Gesells[...]
480 52906		Amphion I ScSr 1 *Dk*	British Newry	143 301	151·4 *P* 14 *f.*	21·1 *F* 8 *f.*	12·3	Inverkeithing J. Scott&Sons	1867	9 Bhds.	2Cy. 20" & 39"–27" 66 HP. A.&J. Inglis, Glasgow	Dundalk&Ne[...] Packet Co. Ld.

*=Lloyd's Register; D=Nederlandsche Vereeniging; G=Germanischer Lloyd; I=Registro Italiano; R=American [...]
A=Veritas Austro-Ungarico; H=Veritas Ellenico; N=Norske Veritas; V=Bureau Veri[tas]

In the 2nd Column, B.T., with figures following, signifies Date of report of bottom sighting by Board of Trade Surveyors as condition of PASSENGER [...]

In the 3rd Column, C=Composite (iron frame, wood planked); I=Iron; Stl=Steel; W=Wood.

Further particulars of Vessels classed by Lloyd's Register will be found in Lloyd's Register Book and Lloyd's Yacht Register.

No.	Cote	Nom, Matériaux, Gréement, etc. Capitaine et dates d'entrée au service des Armateurs actuels et à bord du navire. Ancien Nom, nombre de ponts, etc.		Pavillon. Port d'Armement	Tonnage. Net. Total	Dimensions. Longueur	Largeur	Creux	Construit. Où. Par qui.	Quand.	Nombre des cloisons. Détails sur le lest d'eau.	Machines. Détails. Nom du Constructeur.	Armateurs ou Directeurs-Gérants.
481	*	Amphitrite Brockenbury 88-87	StlSeSr 1 Dk & W b f s	British W.Hartlepool	1503 2295	289·8 F 31 f.	38·1 F31 f. Q 85 f.	19·0 S 134 f.	W. Hartlepool E.Withy&Co.	1887	5 Blds. CDB138t APT35t	3C 21¼" 35" & 58"–39" 200 HP. Cntrl.MrneEng.Co.W.Hpol	Rickinson, Son & Co.
482	* D	Amphitrite Lemesich 68-87	Stl S Sr 2 Dks & Sp Dk 4 B	Austrian Trieste	2485 3631	380·0	42·5	24·0 31·8	Trieste LloydAustro-Ungarico	1884	6 Blds.	2Cy.50" & 90"–57" 450 HP. ArsenaleLloyd,Trieste	Lloyd Austro-Ungarico
483	D V	Amstelstroom Porleuliet 83-85	I SeSr 2 Dks	Dutch Amsterdam	479 787	222·7 P...	28·4 F 56 f.	14·7 H 40 f.	Fyenoord Nederl.Stoombt.Maats	1885		2Cy.25" & 45¼"–33" 118 HP. Nederl.StoombtMaats.Rott	Hollandsche Stoombt. Maatschappij
484	*	Amsterdam Turnbull 79-82	I SeSr 1 Dk 2 B	British Glasgow	252 549	196·5 P 33 f.	27·1 B 60 f.	14·4 F 52 f.	Dumbarton Birrell,Stenhouse& Co.	1878	5 Blds.	2Cy.22" & 40"–42" 115 HP. M.Paul&Co.,Dumbarton	J. Rankine & Son
485	* D V	Amsterdam (ex British Crown) Bolier	I Se 4 m Sr 3 Dks	Dutch Rotterdam	2650 3664	110·3 P 48 f.	39·0 F 78 f.	28·9	Belfast Harland&Wolff	1879	7 Blds.	4Cy.28" & 60"–54" 375 HP. G.Forrester&Co.,Liverpool	Nederland Amerik.Stoom.Maats.
486	D V	Amsterdam Hemmes ..-87	I Se 3 m Sr 1 Dk	Dutch Rotterdam	190 267	190·0	21·0	12·0	Rostock Tischbein	1853		2Cy.20"–20" 40 HP. Tischbein	P. A. Van Es & Co.
487	* D	Amsterdam Leveria -86	Stl SeSr 1 Dk 2 B	Swedish Stockholm	750 933	199·7 P 115 f.	29·1 F 17 f.	16·6 H 16 f.	Gothenberg Motala Co.	1880	4 Blds. MT29t APT29t	2Cy 22¼" & 42¼"–29¼" 120 HP. MotalaCo.,Gothenburg	Rederi Akt. Bolag Svea
488	*	Amulet Lundgreen 81-83	I SeSr 2 Dks	Dutch Rotterdam	627 1013	246·6 P 61 f.	28·8 B 29 f.	16·0 F 51 f.	Glasgow Aitken&Mansel	1876	5 Blds. MT27ft 16m	2Cy.30" & 60"–36" 150 HP. J.&J.Thomson,Glasgow	A. J. F. Burger
489	*	Amy Nichols 88-88	StlSeSr 1 Dk & Spar Dk & W b f s Trm	British London	1522 2336	286·0 F 18 f.	39·2 F 8 f.	20·2 23·2	Sunderland W.Pickersgill&Sns	1888	6 Blds. CBD100t APT30t	3Cy.23½" 39" & 64"–42" 230 HP. Black,Hawthorn&Co., Newcastle	R. Gordon & Co.
490	*	Amy Lumsden 75-84	I SeSr 1 Dk 2 B	British London	516 808	218·3	28·7 Q 33 f.	17·0	Sunderland W.Doxford&Sons	1870	3 Blds. W b & f	2Cy.27" & 47"–30" 90 HP. N.E.MarineE.Co.,Ld.,Sndld.	Fenwick & Co.
491	*	Amy Pond 87-87	Stl Se 3 m Sr Yt 2 Dks	British Glasgow	447 657	221·3	28·1	17·8	Glasgow D.&W.Henderson& Co.	1885	6 Blds.	2Cy.29" & 58"–36" 140 HP. D.&W.Henderson,Glasgow	N. B. Stewart
492	*	Amy Skarzel -86	I Pad 1 Dk	British London	46 72	100·0	16·2	7·0	London R.&H.Green	1886	3 Blds.	1Cy.21½–28" 32 HP. J.Penn&Son,London	Sproston Dock & Foundry Co.
493	* D	Amy Dora Thompson 83-83	I Se 3 m Sr 2 Dks 3 B	British N. Shields	1107 1708	265·0	34·0	24·4	Newcastle ColeBros.	1875	5 Blds. DBaft f36	2Cy.31" & 58"–33" 150 HP. N.E.MarineE.Co.,Ld.,Sndld.	J. Robinson & Sons
494		Amyen	Stl Stern Wheel	British Glasgow	148 173	130·0	28·0	5·6	Dumbarton W.Denny&Bros.	1888		2Cy 15" & 26"–54" 45 HP. Denny&Co.,Dumbarton	Irrawaddy Flotilla Co., Ld.
495		Anadolu Mustafa 82-84	W Pad	Turkish Constntnple	100	102·1	14·0	10·4					Idarei Massousieh (Ottoman Steam Nav.Co.)
496		Anadyr	I Se 3 Dks	French Marseilles	2358 3547	395·8	39·7 H 117 f.	30·8 F 51 f.	LaCiotat Messageries Maritimes	1873		3C 43¼" & (2) 80¼"–43¼" 600 HP. MessgriesMr.,LaCiotat	Messageries Maritimes
497		Anajas	I	Brazilian Para	175	..	..	..	Wilmington	1869		2Cy.20"–72" The Pusey&JonesCo.,Wilm.,Del.	C. Ludmann & Co.
498	*	Ananda Windows 86-86	I SeSr 2 Dks	British Greenock	192 353	149·9	22·3	14·9	Greenock Caird&Co.	1871	5 Blds.	2Cy.20" & 33"–30" 50 HP. Caird&Co.,Greenock	Shearer Bros.
499		Ananda Jacks	Stl T-Se Barge	British Glasgow	140 206	140·5	25·0	7·7	PortGlasgow R.Duncan&Co.	1883		2Cy.18" & 31½"–21" 25 HP. Denny&Co.,Dumbarton	Irrawaddy Flotilla Co., Ld.
500		Anatoli	I Se	Greek Syra	200	150·5	24·8	..	Scotland	1877			S. Chrysakis
501	V	Anatolie Bouchet	Stl & I SeBn 2 Dks 3 B	French Marseilles	1497 2309	308·4 P 55 f.	35·8 B 73 f.	23·8 F 55 f.	Havre Forg's&CiedeaMédit	1884	DB a & f 200t	2Cy.39" & 73"–42" 375 HP. Forges&C.delaMédit.Havre	Cie de Nav. Marocaine & Arménienne
502		Anatolie Ahmed 83-	W Pad	Turkish Constntnple	122	135·0	20·0	10·2	WestCowes J.White	1857		60 HP. Maudslay,Sons&Field,Lond.	Chirket-i-Hairie (Bosphorus Steam Nav.Co.)
503		Anchises Lapage 81-85	I SeSr	British Liverpool	1304 2021	314·4	35·2	26·0	Greenock Scott&Co.	1875		2Cy.26½" & 56"–52" 250 HP. GreenockFndry.Co.,Grnck.	Ocean Steam Ship Co.
504		Anchoria Brown 69-87	I SeBk 2 Dks & Awn Dk 4 B	British Barrow	2713 4157	408·0	40·1	33·8	Barrow BarrowS.B.C.	1874	6 Blds.	3Cy.36" 59" & 96–48" 617 HP. Tripled '87, D. & W. Henderson & Co., Glasgow	Barrow Steam Ship C.
505	*	Ancient Briton Pawley 86-86	I SeSr 1 Dk 2 B	British Cardiff	692 1092	225·0 Q 113 f.	31·2 B 44 f.	16·0 F 28 f.	Newcastle TyneI.S.B.Co.,Ld.	1882	4 Blds. DBa & f 215t APT8t	2Cy.28" & 52"–33" 98 HP. R.&W.Hawthorn,Newcstle.	Gibbs & Lee
506		Ancon Wallace	W PadBn 3 Dks	American S. Francisco	1208 1541	226·0	49·0	17·6	San Francisco PacificMailS.S.Co.	18 67/73	Y L 86	1Cy.50½"–120" 102 HP. MorganIronWks.,NewYork	Pacific Coast Steam-Ship Co.
507		Ancona Webber 81-87	I Se 3 m Sr 3 Dks	British Greenock	1878 3142	380·9 P 112 f.	38·3 F 48 f.	26·0	Greenock Caird&Co.	1879		2Cy.54" & 94"–60" 600 HP. Caird&Co.,Greenock	Peninsular & Oriental Steam Nav.Co.
508	*	Ancona Parker 72-88	Stl Se 3 m Sr 2 Dks	British Leith	773 1207	248·9 P 23 f.	33·2 B 40 f.	17·0 F 28 f.	Leith Ramage&Ferguson	1888	6 Blds. MT34t FPT30t	3C.18" 30" & 48"–36" 120 HP. Ramage & Ferguson, Leith	J. Currie & Co.
509	I	Ancona (ex Salerno) Cicone 77-80	I SeSr 2 Dks	Italian Palermo	427 678	204·0	26·2	14·5	PortGlasgow Blackwood&Gordn.	1861		2Cy.29" & 54"–30" 180 HP. Comp. '79, Blackwood&Gordon, P. Glasgow	Navigazione Generale Italiana
510	*	Andalusia Kinmond	I SeSr 1 Dk 2 B	British Newcastle	855 1354	250·0 Q 116 f. P 24 f.	36·6 B 53 f.	13·4 F 30 f.	Sunderland StrandSlipwayCo.	1881	4 Blds. CDB52t	2Cy.32" & 60"–39" 150 HP. G.Clark,Sunderland	H. Scholefield & Son

* = Lloyd's Register; D = Nederlandsche Vereeniging; G = Germanischer Lloyd; I = Registro Italiano; R = American Record;
= Veritas Austro-Ungarico; H = Veritas Ellenico; N = Norske Veritas; V = Bureau Veritas.

Dans la 2me Colonne, B.T., avec les indications qui suivent, signifie la date du rapport de visite de la Carène par les experts du Board of Trade pour obtenir le certificat relatif au TRANSPORT DES PASSAGERS.

Dans la 3me Colonne, C = Composite (membrure en fer, bordé en bois); I = Fer; Stl = Acier; W = Bois.

Pour plus amples renseignements sur les navires classés par le Lloyd's Register, voir le Lloyd's Register Book et le Lloyd's Yacht Register.

No. in Book. Official No.	Register in which classed.	Ship's Name. Material, Rig, &c. Master, and Dates of Appointment to present Owner's Service and to Ship. Ship's Late Name. No. of Decks, &c.	Flag. Port of Registry.	Tons. Net. Gross.	Dimensions. Length	Breadth	Depth	Built. Where. By Whom.	When.	No. of Bulkheads. Particulars of Water Ballast.	Engines. Particulars, Maker's Name.	Owners or Managers
511 47628	✻	**Andalusia** I Sc3mSr *Witten* 1 Dk	British Scarborough	160 289	147·7 Q31f.	19·8 F15f.	11·7 Len 79	Maryhill J.&R.Swan	1863	3 Bhds.	2Cy.19"& 31"–30" 60 HP. Hutson&Corbett,Glasgow	Scarborough & Shipping Co.Ld
512 76538		**Andalusian** I ScSr *Gould* 87-87 2 Dks	British Liverpool	1136 1733	312·0	31·3 F27f.	23·9	Liverpool F.Leyland&Co.	1877	6 Bhds.	2Cy.33" &60"–33" 150 HP. Fawcett,Preston&Co.Lvpl	"Andalusian" S. Ld.(F.Leyland &
513 [illegible]	[illegible]	**Andean** I ScBg *Highton* -82	British Liverpool	1375 2117	334·9	35·1	25·3	Liverpool T.Royden&Sons	1872		1Cy.26"& 50"–42" 130 HP. G.Forrester&Co.,Liverpool	West India & P. S.S.Co.,Ld.
514 54626		**Andes** (ex Ingrel) I ScSr *Gissing* 87-87 3 Dks	British Kngstn Jam	1104 1711	294·0	32·5 F23f.	19·3 271	Glasgow Randolph,Elder&Co	1866		2Cy.32"& 62"–36" 220 HP. J.Taylor&Co.,Birkenhead	Atlas S.S.Co.,L
515		**Andira** I Pad 1 Dk Spt Hur Dk	Brazilian Para	240 463	150·0	26·0	7·0	Wlmngtn. Del Pusey,Jones&Co	1871		2Cy.20"–72" 80 HP. The Pusey & Jones Co.,Wilm Del	Amazon S.Nav. Co.
516		**Andrade**	Spanish Gijon	89 173	99·7	19·1	8·5	North Shields	1884		30 HP.	J.M.Canosa
517 -48	✻D	**Andrea Vagliano** I Sc *Patrikios* 84-84 1 Dk2 B	Greek Argostoli	1045 1622	257·5 Q38f.	35·0 F[illegible]	18·7 F29f	Sunderland Bartram,Haswell&Co.	1880	1 Bhds. DBaf 168 FPT28 APT22	2Cy.32"& 60"–39" 160 HP. J.Dickinson,Sunderland	Vagliano Bros
518 [illegible]		**Andrew Bain** I PadSp 1 Dk	British N.Shields	25 111	98·0	18·5	9·0	NorthShields Hepple&Co.	1885		1Cy.31"–52" 40 HP. Hepple&Co.,NorthShields	John Bain
519	✻	**Andriana Stathatos** I ScSr *Zanetos* 1 Dk2 B	Greek Ithaca	1068 1644	259·3 Q[illegible]	36·0 B[illegible]	17·8 F[illegible]	SouthShields [illegible]	1884	3 Bhds. DBaA 308 APT36	2Cy.32"& 60"–36" 160 HP. J.Readhead&Co.,S.Shields	Stathatos Bros
520	✻	**Andriana** Stl ScSr *Embiricos* -88 1 Dk2 B	Greek Andros	1174 1823	274·8 Q23f.	37·2 B78f.	18·8 F29f.	Liverpool J.Jones&Sons	1889	6 Bhds. CBDB58 MT[illegible] FPT 46 APT38	3C.21"35"&58"–42" 160 HP. J.Jones & Sons, Liverpool	A.Embiricos
521 85865	✻	**Anerley** I ScSr *Sherbonne* 2 Dks3 B	British London	1256 1942	270·0 P30f.	37·2 B64f.	24·2 F30f.	Sunderland J.Blumer&Co.	1881	7 Bhds. DBaf&A 370 FPT38 APT25	2Cy.35"& 68"–48" 230 HP. T.Clark&Co.,Newcastle	Watts, Ward &
522		**Angantyr** (ex Ellida, ex Angantyr, ex Ida, ex Transport) W ScSr *Nilsander* -87 1 Dk	Swedish Agnlnamm'r	96 121	97·1 Q6f.	20·2 B17f.	8·7	Norrköping	1858		25 HP.	A.Eliason
523 82773	V	**Angelica** (ex Jose, ex Alma) I ScSr *Bailey* 85-85 (Gavar)	British London	752 1180	250·7	31·8 Len. 69	22·1	Greenock Caird&Co.	1857		1Cy.22"& 52½–36" 140 HP. J.Jack&Co.,Liverpool	Powley, Thos.
524	V	**Angeln** I ScSr *Stuke* 82-87 1 Dk	German Kiel	336 581	176·5	24·9 B44f.	15·7	Kiel G.Howaldt	1881	DB[illegible]	2Cy.23"& 37½"–27" 70 HP. G.Howaldt,Kiel	Lange Gebrüder
525 68721	BT11.88	**Angelo** I Sc3mSr *Bowse* 2 Dks	British Hull	993 1536	258·8 P38f.	33·6 B73f.	18·2 F32f.	Hull H'mphrys&Pears'n	1874	5 Bhds.	2Cy.37"& 72"–42" 300 HP. Humphrys & Pearson,Hull	T.Wilson,Sons
526		**Ångermanland** I ScSr *Hultkvist* 1 Dk&Awn Dk	Swedish Hernösand	522 647	172·2 B27f.	24·4	13·0 20·3	Stockholm Bergsunds mek Ver.k-s	1872		2Cy.26"&47"–27" 100 HP. W.LindbergsMek.V.Stkhlm	Hernösands Ång. Bolag. [illegible]
527 87027	✻D	**Angers** I ScSr *Pinkham* 82-86 1 Dk&Spr Dk3 B	British London	2077 3185	320·6 B63f.	40·0 F32f.	28·1	Hull D.P.Garbutt	1882	5 Bhds. DB[illegible]	2Cy.40"& 72"–48" 400 HP. C.D.Holmes&Co.,Hull	Angier Line,Ld
528 89520	✻	**Angerton** I ScBg *Morris* 83-87 1 Dk&Spr Dk3 B	British London	1823 2794	320·0	38·0	27·4	Newcastle W.Dobson&Co.	1884	6 Bhds. DBaf&A 568 FPT42 APT17	2Cy.38"& 74"–48" 375 HP. Wllsend Slpwy Co,Newcsl.	W.Milburn & Co
529 59054		**Anglesea** W Pad	British Quebec	97 153	129·0	30·0	9·0	Quebec	1870			L.Nadeau
530 93312	BT12.88	**Anglesey** StlScSr 2 Dks	British Dublin	46 877	301·5	33·1	13·4	Belfast Harland&Wolff	1888		6C19"29½"& 18"–33" 280 HP. Harland& Wolff,Belfast	London & N. Western Railway
531 96023	BT11.88	**Anglia** StlScSr *Mitchell* 84-88	British	2120 [illegible]	340·0 P[illegible]f.	43·0 B9[illegible]f.	27·8 F33f.	Glasgow D.&W.Henderson&Co	1888		3C.24"40"&66"–48" 300 HP. D.&W.Henderson,Glasgow	Henderson Bros
532 45816	✻ BT5.88	**Anglia** I ScSr *Smith* 63-83 2 Dks	British Leith	520 827	253·4 P51f.	28·0 F46f.	16·3	Dundee Gourlay Bros.	1863	4 Bhds.	2Cy.33"& 52"–30" 130 HP. Gourlay Bros.&Co.,Dundee	G.Gibson & Co.
533 54745		**Anglia** I Pad *Porter* 77-82	British London	77 [illegible]	144·1	22·2	12·4	London ThamesIronWorks	1866		2Cy.38"–40" 140 HP. J.Stewart,London	W.Watkins
534 [illegible]		**Anglia** I PadSk *Heads* 1 Dk	British N.Shields	9 120	100·6	18·7	9·7	South Shields [illegible]	1885		1Cy.35"–52" 40 HP. J.P.Rennoldson&Sons,S.Shds	Mrs. Isabella H.
535		**Anglia** I Pad	Turkish Cnstntnple	1759 2049	365·0	40·2	28·7	Hull M.Samuelson&Co.	1861		1000 HP.	Ottoman Govt
536 [illegible]	✻	**Anglian** I ScBg *Morton* 81-86 2 Dks3 B	British Southamptn	1413 2246	314·0	35·8 B49f.	26·4 F46f.	Glasgow Aitken& Mansel	1873	7 Bhds. WT58 FPT	3C.26"42"&68"–42" 270 HP. T.Richardson&Sons,Hrtlpl.	Union Steam S.
537		**Anglo-Dane** I ScSr *Lund* 72-84 1 Dk2 B	Danish Copenhagen	506 888	209·5 Q63f.	28·4 B51f.	14·8 F35f.	Newcastle A.Leslie&Co.	1866	4 Bhds. MT[illegible]	2Cy.19"& 38"–36" 90 HP. Burmeister & Wain,Copenhagen	Det Forenede Selskab,C.P.A.
538	✻	**Angola** I ScBg *Rosa* 81-81 1 Dk&Spr Dk3 B	Portug'se Lisbon	1263 1966	290·0	35·0	25·6	Hull Earle's S.&E.Co.	1881	6 Bhds. DB[illegible] MT[illegible]	2Cy.40"& 72"–45" 220 HP. Earle's Shipb.Eng.Co.,Hull	Empreza Nacional Nav.a Vap.
539	✻	**Ani** StlFwnScSr 1 Dk	Russian Astracan	224 374	140·0 P21f.	24·0 B23f.	11·3 F26f.	Sunderland Bodds,Sharer&Co.	1886	6 Bhds.	3Cy.8¾"15"&25"–18" 70 HP. N.E.MarineEng.Co.Ld.Shi.	J. Pitoeff
540												

✻ = **Lloyd's Register**; D = Nederlandsche Vereeniging; G = Germanischer Lloyd; I = Registro Italiano; R = American [Record]; A = Veritas Austro-Ungarico; H = Veritas Ellenico; N = Norske Veritas; V = Bureau Veritas

In the 2nd Column, B.T., with figures following, signifies Date of report of bottom sighting by Board of Trade Surveyors as condition of PASSENGER [Certificate].

In the 3rd Column, C=Composite (iron frame, wood planked); I=Iron; Stl=Steel; W=Wood.

Further particulars of Vessels classed by Lloyd's Register will be found in Lloyd's Register Book and Lloyd's Yacht Register.

No.	Registres dans lesquels le navire est coté	Nom, Matériaux, Gréement, etc. Capitaine et dates d'entrée au service des Armateurs actuels et à bord du navire. Ancien Nom, nombre de ponts, etc.		Pavillon. Port d'Armement	Tonnage. Net. Total	Dimensions. Longueur | Largeur | Creux. emplacements sur le Pont	Construit. Où. Par qui.	Quand	Nombre des cloisons. Détails sur le lest d'eau.	Machines. Détails. Nom du Constructeur.	Armateurs ou Directeurs-Gérants.
		Anita	W	Spanish	390 549	154·5|23·5|10·4	Philadelphia	1862		85 HP.	Juan Fernandez
		Anita (ex Aurora)	I SeSp 1 Dk	Spanish	93 125	102·3|21·0|9·1	WestHartlpl.			2Cy. 30 HP.	Antonio Millan
	✻	Anjer Head Pinkstain 85-89	I SeBn 2 Dk&3 B	British London	1300 2013	276·0|36·3|24·4 P 38 f. F 32 f.	Middlesbro' R.Dixon&Co.	1881	5 Blhds. DB	2Cy.36 & 67"–36" 240 HP. T.Richardson&Sons,Hrtlpl.	Augier Line, Ld.
544		Anjou	I Sc 2 Dks	French Marseilles	416 625	196·8|21·7|17·1	LaSeyne Forges&Ch.delaMedit.	1868		2Cy.22" & 44"–31½" 100HP. Fraissinet&Co.,Marseilles	Société Générale de Transports Maritimes
545	✻	Ann Brown	I SeSr 1 Dk	British London	188 301	150·8|21·7|11·6 Q 48 f.	Glasgow A.&J.Inglis	1876	1 Blhds. FPT	2Cy.18" & 31"–21" 50 HP. A.&J.Inglis,Glasgow	J. Burnett & Sons
546	✻	Ann Webster Stenbeck -87	I SeSr 1 Dk 2 B	British London	498 771	200·6|28·5|15·9 Q 47 f.	Hartlepool Withy, Alexander& Co.	1870	4 Blhds. DB	2Cy.25" & 48"–36" 96 HP. T.Richardson&Sons,Hrtlpl.	Lambert Bros.
547		Anna Gibson 81-81	W SeSp 1 Dk	British Kirkcaldy	65 111	88·4|22·1|8·6	Finland	1860		1Cy.12"–14" 25 HP. W.King&Co.,Glasgow	W.Jarvis
548	G	Anna (ex Gratitude) Mogell 86-86	C SeSr 1 Dk	Danish Copenhagen	174 252	125·4|24·6|11·1	Malmö J.Bonnesen	1871		2Cy.14" & 32"–24" 40 HP. KockumsMek.Works,Malmö	Dampskibsselskabet "Anna" (F.W.Kiørboe)
549	✻	Anna	StlTwnSc 1 Dk	French Senegal	25 68	80·1|13·0|8·2	Birkenhead Cochran&Co.	1887	3 Blhds.	1Cy.10" & 20"–14" 35 HP. Cochran&Co.,Birkenhead	L. Deseemet
550	N	Anna (ex Sedan) Mossens 73-87	Stl&I SeSr 1 Dk	Norweg'n Christ'ns'nd	296 491	143·3|24·3|12·8 Q 45 f. B 35 f. F 19 f. H 46 f.	Thorskog P.Larsson	1886	4 Blhds. MT70t APT8t	2C.16½" & 31¼"–20½ 55HP. J.&C.Belinders,Stockholm	Nicolay H. Knudtzon
551		Anna Pettersson	W SeSr 1 Dk	Swedish Svanvik	168 230	101·7|21·2|12·8 Erections 36 f.	Svanvik	1882		2Cy.13" & 21"–16" 25 HP. Eriksberg-Mek. Verks,Gothbg.	C.J.Pettersson
552		Anna (ex Vulkan)	C SeSp 1 Dk	Swedish Gothenburg	165 240	101·1|19·7|12·7 Q 16 f. H 19 f.	Dalsland Spångens Varf	1871		2Cy. 28 HP.	F.T.Appelberg
553		Anna Clara	I SeSr 1 Dk	Brazilian RioJaneiro	129 216	133·9|21·0|9·0 P 32 f. F 22 f.	PortGlasgow Blackwd.&Gordon	1875	4 Blhds.	2Cy.20" & 35"–24" 60 HP. Blackwood&Gordon,Pt.Gls.	A.M.dos Santos
554	V	Anna Woermann Toggenbrock 85-87	Stl SeSr 1 Dk.&SprDk	German Hamburg	818 1110	222·1|31·2|14·5 B 14 f. | 20·5	Hamburg ReiherstiegSchiffsWerfte	1884	DB	2Cy.27" & 50"–33" 140 HP. ReiherstiegSchiffsWerfte,Hbrg	Afrikanische Dampfsch-Actien-Ges. (Woermann Linie)
555	XV	Anna & Caroline Nielsen	W SeSr 1 Dk.& Awn Dk	Norweg'n Bodö	329 431	132·2|25·4|10·9 B 58 f.	Malmö J.Bonnesen	1875		1Cy.12" & 24¾"–17¾" 45HP. Burmeister&Wain,Copenhagen	J.C.Koch
556		Anna & Maria (ex Sylph)	I Sc 1 Dk.& Awn Dk	Spanish Havana	704 1081	227·5|29·6|13·9 | 21·1	Sunderland Austin	1876	1 Blhds.	2Cy.30" & 52"–33" 140 HP. Humphrys&Pearson,Hull	R.Herrara
557		Annam	StlTwnSc	French	103 191	120·0|23·0|7·3	Glasgow FairfieldS.B.&E.Co. Ld.	1886		1Cy.11"–12" 60 HP. FairfieldS.B.&E.Co.,Ld.,Gls.	
558	✻D	Annandale Deuchars 86-88	I SeSr 1 Dk 2 B	British London	1041 1607	261·5|33·4|18·9 Q 96 f. B 68 f. F 40 f.	Hartlepool E.Withy&Co.	1878	4 Blhds. DB 437t	2Cy.32" & 59"–33" 140 HP. T.Richardson&Sons,Hrtlpl.	Steel, Young & Co.
559	✻D	Anne Thomas Clegg 87-87	I SeSr 1 Dk 2 B	British Cardiff	909 1419	249·5|35·3|17·8 B 48 f. F 26 f.	Newcastle Palmers'Co.,Ld.	1883	4 Blhds. DB 333t APT 45t	2Cy.30" & 58"–36" 99 HP. Palmers'Co.,Ld.,Newcastle	"Anne Thomas" S.S. Co.Ld (E.Thos.Radcliffe&Co.)
560		Annei Maru	W SeSr	Japanese Osaka	334 451	174·6|22·8|15·3	Nagasaki KosakuBunKioku	1880		49 HP.	Osaka Shosen Kaisha
561		Annette (ex Primus) Johansson	I SeSp 1 Dk	Swedish LillaEdet	82 123	96·9|20·5|10·5 Q 7 f. H 9 f.	Motala	1876		50 HP.	O.A.Johansson
562	✻	Annie Ormston 86-86	I SeSr 2 Dks 3 B	British Whitby	1205 1872	273·5|34·7|24·5 P 26 f. F 37 f.	Whitby T.Turnbull&Son	1879	5 Blhds. DB 108 ft	2Cy.33" & 62"–39" 170 HP. Blair&Co.Ld.,Stockton	J.Gray & Co.
563	✻	Annie Bird 87-87	I SeSr 1 Dk 2 B	British Whitby	806 1242	240·0|32·0|17·5 Q 50 f. B 49 f. F 32 f.	SouthShields J.Readhead&Co.	1879	4 Blhds. DB 32 ft	2C.27½" & 52"–30" 110 HP. J.Readhead&Co.,S.Shields	R.Harrowing
564	✻	Annie Tointon	I SeSr 1 Dk 2 B	British London	785 1215	230·4|31·8|17·5 Q 85 f. B 45 f. F 27 f.	Sunderland J.L.Thompson&Ss.	1877	5 Blhds. DB 36 ft FPT APT	2Cy.28" & 53"–33" 110 HP. J.Dickinson,Sunderland	J.Taylor
565	✻	Annie (now named ...) Robertson 88-88	Stl SeSr 1 Dk 2 B	Turkish	274 443	188·2|24·0|12·8 Q 25 f.	Hull M.Samuelson&Co.	1864	1 Blhds.	2Cy.29¾" & 54¾"–24" 98 HP. M.Samuelson&Co.,Hull	
566	✻	Annie Barrett 81-84	I SeSr 1 Dk	British Liverpool	192 411	160·0|23·2|11·0 Q 56 f. B 31 f. F 17 f.	Barrow BarrowS.B.Co.Ld.	1883	4 Blhds. DB FPT 37t	2Cy.22" & 44"–27" 80 HP. BarrowS.B.Co.Ld.,Barrow.	S.S. "Annie" Co., Ld. (J.J.Mack & Son)
567	✻	Annie Messling 83-	I SeSr 1 Dk 2 B	German Danzig	576 844	200·7|28·8|15·7 Q 77 f. B 43 f. F 26 f.	Newcastle Swan&Hunter	1880	3 Blhds. MT 12t FPT 20t APT 38t	2Cy.26" & 49"–30" 99 HP Black,Hawthorn&Co.,Newc.	Th. Rodenacker
568	✻	Annie Ainslie	I SeSr 2 Dks 3 B	British London	847 1326	248·6|30·4|23·2	Stockton M.Pearse&Co.	1870	4 Blhds. DB 70 ft	2Cy.26" & 50"–30" 98 HP. N.E.MarineEng.Co.,Ld.,Sld.	Nelson, Donkin & Co.
569		Anselm Macartney 80-84	I SeBn	British Liverpool	998 1562	271·0|33·8|22·9 P 55 f.	Newcastle A.Leslie&Co.	1882		2Cy.28" & 55"–48" 140 HP. R.Stephenson&Co.,Nwcstle.	The Booth Steamship Co., Ld.
570		Anselmo (ex Paz) Navallega -75	I SeSr 1 Dk 2 B	Spanish Gijon	495 785	200·9|28·1|15·5 P 109 f. F 39 f.	Sunderland R.Thompson,jun.	1875	4 Blhds. DB 150t	2Cy.27" & 51"–33" 110 HP. Black,Hawthrn&Co.,Gtshd.	O.de Olavarria & Co.

✻ = Lloyd's Register; D = Nederlandsche Vereeniging; G = Germanischer Lloyd; I = Registro Italiano; R = American Record;
A = Veritas Austro-Ungarico; H = Veritas Ellenico; N = Norske Veritas; V = Bureau Veritas.

Dans la 2me Colonne, B.T., avec les indications qui suivent, signifie la date du rapport de visite de la Carène par les experts du Board of Trade pour obtenir le certificat relatif au TRANSPORT DES PASSAGERS.

Dans la 3me Colonne, C = Composite (membrure en fer, bordé en bois); I = Fer; Stl = Acier; W = Bois.

Pour plus amples renseignements sur les navires classés par le Lloyd's Register, voir le Lloyd's Register Book et le Lloyd's Yacht Register.

No. in Book / Official No.	Registry in which classed	Ship's Name. Material, Rig, &c. / Master, and Dates of Appointment to present Owner's Service and to Ship / Ship's Late Name. No. of Decks, &c.	Flag / Port of Registry	Tons Net / Gross	Dimensions Length	Breadth	Depth	Built Where / By Whom	When	No. of Bulkheads / Particulars of Water Ballast	Engines Particulars / Maker's Name	Owners or Managers
571	DV	**Ansgar** I ScSr *Larsen 81-84* 2 Dks	Danish [illegible]	750 1141	242·4	33·3	15·9 22·7	Copenhagen Burmeister & Wain	1879	APT[illegible]	1Cy.21" & 42"–30" 120 HP Burmeister & Wain, Copenhagen	Dampskibsselsk. "Danmark" (W.)
572 95235	✻	**Anshin Maru** I ScSr *Gollichan -88* 1 Dk	British [illegible]	191 [illegible]	145·0	22·1	10·9 B 21 f. F 17 f.	Dundee Pearce Bros.	1888	4 Bhds.	2Cy.20" & 38"–27" 60 HP Pearce Bros., Dundee	H. E. Reynell &
573 869		**Ansonia** W Pad	American New York	440 632	190·1	28·0	8·7	Brooklyn	1848			E. J. Brockett
574 87716	✻	**Ant** I ScSr *Nash 83-83* 1 Dk 2 B	British [illegible]	723 1133	225·0	31·5	14·9 [illegible] P 37 f. B 88 f. F 31 f.	W. Hartlepool W. Gray & Co.	1883	4 Bhds. DB[illegible] APT[illegible]	2Cy.27" & 50½"–33" 99 HP Blair & Co., Ld., Stockton	C. & G. Jamieson
575 65934	✻	**Ant** I ScSr *Owen 83-83* 1 Dk	British Liverpool	97 188	110·3	20·5	10·3 Q 15 f. F 8 f.	Port Glasgow J. Reid & Co.	1872	4 Bhds.	1Cy.8" & 15"–18" 52 HP Comp. '78 S. Baker, Liverpool	"Ant" S.S.Co. (R. & D. Jones)
576	✻	**Anta** Stl ScTug *Bennett 87-87* 1 Dk	Brazilian Para	1 43	64·8	14·1	7·1	Liverpool W. H. Potter & Sons	1887	4 Bhds.	3Cy.10½" 16" & 27" 16" 37 HP D. Rollo & Sons, Liverpool	R. Singlehurst &
577 63034	✻	**Antelope** (with [illegible]) I ScSr *Thompson* 2 Dks	British W. Hartlepl	509 721	201·6	28·8	15·8 Q 75 [illegible]	Hartlepool Withy, Alexander & Co.	1869	4 Bhds. DB[illegible] APT[illegible]	2Cy.25½" & 48"–33" 96 HP Blair & Co., Ld., Stockton	R. Irvine, jun. p'r Jackson Bros. & C.
578 1212		**Antelope** W Pad 3 Dks	American S. Francisco	448 581	206·6	28·0	8·9	New York J. Simonsen	1847		1Cy.48"–120" 200 HP F. F. Secor, New York	J. M. Donahue
579 65910	BT 8-88	**Antenor** I ScSr *Grier 88-86*	British Liverpool	1376 2113	322·0	33·7	25·6	Newcastle A. Leslie & Co.	1872		2C.28½" & 61"–45" 250 HP R. Stephenson & Co., N'castle	Ocean Steam
580 85073	✻	**Antilles** I ScSr *Cary 85-85* 1 Dk	British London	857 1341	270·3	33·6	17·1 P 44 f. F 31 f. B 51 f.	Glasgow Napier, Shanks & Bell	1881	4 Bhds. CDB[illegible]	2Cy.34" & 58"–36" 160 HP J. Howden & Co., Glasgow	"Antilles" S.S.Co. (Scrutton, Sons & C.)
581	G	**Anton** I ScSr *Arreboe 84-84* 2 Dks	German Kiel	396 565	163·0	22·7	17·8	Kiel G. Howaldt	1882	WB	2Cy.20" & 36"–21" 60 HP G. Howaldt, Kiel	Sartori & Berger
582 45985	✻	**Antona** I Sc3mSr *McDougall* 2 Dks apt Sh'lt Dk	British Glasgow	336 517	190·1	26·8	14·0 P 22 f.	Glasgow Barclay, Curle & Co.	1863	4 Bhds.	2Cy.21" & 47"–30" 90 HP Comp. '80 J. & T. Young, Ayr	J. Clark
583	G	**Antonie** I ScSr *Galster 76-83* 1 Dk	German Kiel	134 181	96·5	20·9	12·2	Kiel G. Howaldt	1879		2Cy.13¾" & 21"–16" 25 HP Schweffel & Howaldt	Sartori & Berger
584 77017	✻	**Antonio** I ScBg *Crawford -87* 2 Dks 3 B	British London	1214 1858	280·0	34·0	23·9 P 29 f. B 24 f. F 31 f.	Sunderland W. Doxford & Sons	1877	5 Bhds. [illegible]	2Cy.33" & 62"–39" 160 HP G. Clark, Sunderland	Steam Nav. Co. N.
585	✻	**Antonio Lopez** Stl ScBk *Dominguez 75-84* 3 Dks	Spanish Cadiz	1979 3460	384·0	42·2	27·4 P 101 f. B 48 f. F 58 f.	Dumbarton W. Denny & Bros.	1881	6 Bhds. CDB[illegible]	2C.57" & 102"–60" 750 HP Denny & Co., Dumbarton	Cia. Trasatlantica
586	✻	**Antonio Muñoz** I ScSr *Osoro* 1 Dk & Awn Dk	Spanish Manila	527 814	204·5	26·4	10·0 17·0	Glasgow R. Napier & Sons	1882	4 Bhds. CDB[illegible]	2Cy.26" & 45"–36" 100 HP R. Napier & Sons, Glasgow	Muñoz Hermanos
587 2	✻D	**Antonios Stathatos** I ScSr *Cardvios* 1 Dk 2 B	Greek Ithaca	976 1516	249·0	36·0	17·5 Q 95 f. B 32 f. F 26 f.	South Shields J. Readhead & Co.	1883	4 Bhds. DB[illegible]	2Cy.29" & 54"–36" 130 HP J. Readhead & Co., S. Shields	Stathatos Bros.
588 93173	✻	**Antrim** Stl ScSr *Brown 86-86* 1 Dk	British Belfast	171 478	174·0	26·2	12·5 F 27 f. Q & B 101 f.	Belfast McIlwaine, Lewis & Co.	1888	5 Bhds. DB[illegible] APT[illegible]	3Cy.17" 27" & 44"–30" 98 HP McIlwaine, Lewis & Co., Belfast	Antrim Iron Ore
589	✻	**Antwerpen** I & Stl ScSr *Caroc 77-87* 1 Dk 2 B	Danish Copenhagen	1112 1726	264·0	35·3	18·3 [illegible] B 70 f. P 30 f.	Newcastle C. S. Swan & Hunter	1887	4 Bhds. DB[illegible]	3Cy.20" 33" & 54"–36" 130 HP Wallsend Slipway & Eng. Co., Ld., Newcastle	Det Forenede [illegible] [illegible]
590 70008	BT 1-88	**Aorangi** Stl ScBk *Sutcliff 86-86* 2 Dks & Spr Dk	British London	2655 4163	389·0	46·0	23·7 P 35 f. B 121 f. F 57 f. 31·9	Glasgow J. Elder & Co.	1883	8 Bhds.	2Cy.46" & 88"–57" 600 HP J. Elder & Co., Glasgow	New Zealand S. Co., Ld.
591		**Apache** *Ferres -74*	American	620 773	170·0	42·2	8·0	Oakland, Cal.	1880		566 HP	E. H. Miller, jun.
592	V	**Apenrade** (ex Montevideo) I ScBg *Beers -88*	German Hamburg	1476 2238	310·0	36·1	27·3 Q B & F 131 f.	Newcastle C. Mitchell & Co.	1873	4 Bhds.	2Cy.40" & 72"–42" 250 HP Thompson, Boyd & Co., Newcastle	M. Jebsen.
593 97400	✻	**Apex** Stl ScSr *Smith -88* 1 Dk 2 B	British London	1380 2119	276·0	37·8	18·8 Q 75 f. B 117 f. F 29 f.	Newcastle Palmers' S.B.Co., Ld.	1888	4 Bhds. CDB[illegible]	3Cy.22" 35" & 58"–42" 200 HP Palmers' S.B.Co., Newcastle	Apex S.S.Co., Ld. (Newman & [illegible])
594 136		**Apis** I ScSr *Tellner 71-87* 2 Dks	Austrian Trieste	808 1235	248·0	31·0	21·5	Dumbarton W. Denny & Bros.	1869		2Cy.31" & 56"–36" 150 HP Denny & Co., Dumbarton	Lloyd Austro-Ungarico
595 93122		**Apollo** Stl ScSr *Douglas -87* 2 Dks 3 B	British Hull	2066 3181	330·4	41·2	29·0 B 70 f.	Hull Earles' S.B.Co., Ld.	1887	5 Bhds. DB[illegible]	3Cy.25" 40" & 66"–48" 300 HP Earles' S.B.Co., Ld., Hull	T. Wilson, Sons &
596 70468	✻	**Apollo** I Sc3mSr *Wards 87-87* 1 Dk	British Barrow	308 469	182·0	24·1	12·4 Q 47 f. F 24 f. B 28 f.	Barrow Barrow S.B.Co.	1874	3 Bhds. FPT	2Cy.20" & 37½"–32" 70 HP Barrow S.B.Co., Ld., Barrow	J. Little & Co.
597 190		**Apollo** I ScSr *Radimiri 61-87* 2 Dks	Austrian Trieste	1106 1718	270·0	35·0	24·5	Dumbarton W. Denny & Bros.	1866		1Cy.29½" & 51"–36" 160 HP Denny & Co., Dumbarton	Lloyd Austro-Ungarico
598	DGV	**Apollo** I ScSr *Schrage* 1 Dk & Spr Dk	German Bremen	517 643	168·9	26·3	11·3 18·4	Stettin Möller & Holberg	1883	DB[illegible]	2Cy.18" & 30"–21" 50 HP Möller & Holberg, Stettin	Dampfschifffahrts Ges. "Neptun"
599	✻	**Apollo** (ex Shotton) I ScSr *Kaylenstjernen* 1 Dk 2 B	Swedish Gothenburg	588 839	200·0	28·1	15·9 Q 75 f. B 45 f. F 25 f.	Hartlepool Withy, Alexander & C.	1870	3 Bhds. DB[illegible]	2Cy.25" & 48"–36" 96 HP T. Richardson & Sns, Hartlpl.	J. A. Waller
600		**Apolo** I Sc *Mazella*	Arg'ntine B'enos Ayres	115 203	126·0	21·0	7·6	Glasgow Aitken & Mansel	1869		60 HP Aitken & Mansel, Glasgow	S. Ribes & Co.

✻ = **Lloyd's Register**; D = Nederlandsche Vereeniging; G = Germanischer Lloyd; I = Registro Italiano; R = American [illegible]
A = Veritas Austro-Ungarico; H = Veritas Ellenico; N = Norske Veritas; V = Bureau Veritas

In the 2nd Column, B.T., with figures following, signifies Date of report of bottom sighting by Board of Trade Surveyors as condition of PASSENGER CERTI[FICATE]

In the 3rd Column, C = Composite (iron frame, wood planked); I = Iron; Stl = Steel; W = Wood.

Further particulars of Vessels classed by Lloyd's Register will be found in Lloyd's Register Book and Lloyd's Yacht Register.

Registres	Nom, Matériaux, Gréement, etc. Capitaine et dates d'entrée en service des Armateurs actuels et à bord du navire. Ancien Nom, nombre de ponts, etc.	Pavillon. Port d'Armement	Tonnage. Net. Total.	Dimensions. Long.	Larg.	Creux.	Construit. Où. Par qui.	Quand.	Nombre des cloisons. Détails sur le lest d'eau.	Machines. Détails. Nom du Constructeur.	Armateurs ou Directeurs-Gérants.
	Apolo Stl PadSr *Pacostovich* 1 *Dk & Awn Dk*	British Glasgow	693 1253	255·6 *P 167 t. F 112 t.*	30·1	12·7	Dumbarton W.Denny&Bros.	1883	4 Blhds.	2Cy.44" & 77"–72" 262 HP. W.Denny&Co.,Dumbarton	La Platense Flotilla Co.,Ld.
	Apostol I	Spanish Seville	119 207	150·9	22·8	13·2	England	1859		70 HP.	Portella White
	Aquila I PadSr *Paisley 82-82*	British Weymouth	90 270	180·1 *Q 20 t.*	21·0	10·9	Renfrew J.Henderson&Sons	1854		2Cy.42"–42" 110 HP. M'Nabb&Clarke,Greenock	Weymouth & Channel Islands S. Packet Co.,Ld.
	Aquila Imperiale I ScSr *Germani 71-87* (BROKEN UP) 2 *Dks*	Austrian Trieste	756 1194	266·0	31·5	21·0	Glasgow R.Napier&Sons	1856		2Cy.31" & 59½"–36" 140 HP. R.Napier&Sons,Glasgow	Lloyd Austro-Ungarico
*	**Aquitaine** I ScSr *Renaudeau 84-86* 1 *Dk 2 B*	French La Rochelle	1121 1619	260·1 *P 29 t. B 52 f. F 26 f.*	38·1	17·2	Sunderland J.Blumer&Co.	1883	6 Blhds. CDB156	2Cy.31" & 60"–39" 160 HP. T.Clark&Co.,Newcastle	Delmas frères
*	**Arab** I ScSr *Jones 81-88* 2 *Dks & Spr Dk*	British Southampton	2627 3145	350·0	40·2	23·6 31·0	Glasgow J.&G.Thomson	1879	6 Blhds. DBaft 230	2Cy.50" & 86"–54" 500 HP. J.&G.Thomson,Glasgow	Union Steam Ship Co.
	Arabia I ScSr *Laird 65-84* 2 *Dks & W h Frms*	British Glasgow	2315 3544	378·2 *P . . . f. B 85 f. F 45 f.*	40·2	28·6	Glasgow D.&W.Henderson& Co.	1883	8 Blhds.	2Cy.38" & 66"–60" 480 HP. D.&W.Henderson&Co.,Gls.	Henderson Bros.
	Arabia I ScBg *Simpson 87-88*	British Glasgow	675 1080	239·2 *P 119 t. F 83 t.*	30·2	19·0	Dumbarton W.Denny&Sons	1863		2Cy.33½" & 56"–36" 140 HP. Comp. '70 Denny&C.,Dmbrtn	British India Steam Navigation Co., Ld.
	Arabia W Sc	British Liverpool	106 156	90·0	20·9	10·4	Northwich H.Gibson	1883		2Cy.9"–16" 30 HP. E.Bates,Northwich	R.Verdin
I	**Arabia** I Sc3mSr *Raffo 74-86* 2 *Dks & Awn Dk*	Italian Genoa	890 1351	250·3	31·4	25·3	Newcastle Palmer'sCo.	1870	5 Blhds. M.S. MT38 ft	2Cy.30" & 60"–33" 262 HP. Palmers'Co.,Newcastle	Navigazione Generale Italiana
	Arabian I Sc3mSr *Richards 83-86* 2 *Dks*	British Liverpool	1355 2074	335·0	34·2	24·1	Belfast Harland&Wolff	1862	6 Blhds.	4Cy.24" & 30"–48" 220 HP. G.Forrester&Co.,Liverpool	S.S. "Arabian" Co.,Ld. (F. Leyland & Co.)
	Arabic Stl Sc4mBk *Smith ·88* 3 *Dks*	British Liverpool	2788 4368	430·2	42·2	31·5	Belfast Harland&Wolff	1881	7 Blhds.	4Cy.32" & 71"–60" 550 HP. J.Jack&Co.,Liverpool	"Oceanic" S. Nav. Co., Ld.
	Aracaju I ScSr	Brazilian RiodeJaneiro	160 350	156·0	22·0	10·0	Dumbarton	1866			Cia.de Nav.à Vap.de Bahia
*	**Arago** I ScBn *Sutherland* 2 *Dks 3 B*	British Liverpool	1061 1656	282·0	33·2	24·4	Newcastle A.Leslie&Co.	1876	5 Blhds.	2Cy.26" & 58"–42" 170 HP. R.Stephenson&Co.,Nwcstle.	Eills & Co.
V	**Arago** (COLLIER) Stl ScSr *Holt ·85* 1 *Dk 2 B*	American S. Francisco	620 828	200·0 *P 24 f. B 57 t. B 38 f.*	30·0	16·0	San Francisco UnionIronWorks	1885	DB 292 FPT 29t APT 19t	2Cy.24" & 48"–38" 120 HP. UnionIronWorks	R.B.Williams
	Araguahy Stl PadSr 1 *Dk*	Brazilian Para	371	177·0	27·5	8·2	Hebburn A.Leslie&Co.	1884		2Cy.30" & 54"–51" 120 HP. Black,Hawthorn&Co.,Gtsd.	Amazon Steam Nav. Co.
	Aranas W ScSr *Lundgren* 1 *Dk*	Swedish Arnäs, Mstd.	140 186	100·0 *Erections 17 t.*	18·9	10·6	Sjötorp	1874		20 HP.	D.Danielsson
	Aranci Stl TwnScSr *Brock*	British London	85 157	118·0 *B 9 t. Q 22 t.*	18·8	7·3	London W.Walker&Co.	1882		4Cy.11" & 21"–14" 40 HP. M.Pratt,Huddersfield	B.Piercy
*	**Aranmore** I ScSr *Tait 81-85* 2 *Dks Spt Shd Dk*	British Glasgow	395 813	220·2 *F 31 f.*	29·7	15·7	Renfrew W.Simons&Co.	1881	5 Blhds. CDB176 H 176 FPT 62	2Cy.26" & 54"–48" 180 HP. W.Simons&Co.,Renfrew	Clyde Shipping Co.
R	**Aransas** I TwnScSr *Smith* 2 *Dks*	American N.Orleans La.	679 1157	240·0	35·6	18·6	Wlmngtn Del. Harlan & Hellingswor thCo.	1877	WB	2Cy.30"–30" 350 HP. The Harlan & Hollingsworth Co., Wlm.,Del.	Morgan R.R.& S.S.Co.
	Arapixy I PadSr	Brazilian Para	250	140·0	23·0	7·6	Wilmington The Pusey & Jones Co.	1877		2Cy.18"–72" The Pusey&Jones Co., Wlm.Del.	Cia. de Marajó & Tocantins
*	**Arara** I ScBn *Mackinlay 85-87* 2 *Dks 3 B*	British Liverpool	2602 3981	370·7 *P 24 f. B 98 f. F 40 f.*	42·3	29·0	Stockton M.Pearse&Co.	1886	6 Blhds. DBaft 595	3C.28" 46" & 75"–54" 495 HP. Blair&Co.,Ld.,Stockton	J.M.Wood
	Ararat I ScBg	British Liverpool	1305 2016	323·4	33·4	26·0	Liverpool Liv.Sh.Bldg.Co.,Ld.	1871		2Cy.26" & 56"–36" 200 HP. G.Forrester&Co.,Liverpool	Papayanni & Co.
	Araruama (ex Barros & Ferreira) I *da Cunha*	Brazilian RiodeJaneiro	144	..	..	..	Glasgow	1875			Dr.C.A.de M.Jordao
	Araucania I ScBg 3 *Dks*	British Liverpool	1728 2771	354·8 *P 118 t.*	41·0	18·0 26·0	Glasgow J.Elder&Co.	1869	6 Blhds.	2Cy.60" & 96"–45" 500 HP. J.Elder&Co.,Glasgow	Pacific Steam Nav.Co.
	Arauco I Sc3mSr 1 *Dk 2 B*	British Liverpool	586 801	200·0 *P 60 t. F 26 t.*	29·2	15·5	Dundee GourlayBros.	1879	3 Blhds. WB	2Cy.24" & 48"–36" 105 HP. Gourlay,Bros.,&Co.,Dundee	Pacific Steam Nav.Co.
*	**Arawa** Stl Sc4mSr *Stuart 84-84* 3 *Dks*	British Glasgow	3268 5026	439·6 *P 64 f. B 188 f. F 48 f.*	46·3	28·9	Dumbarton W.Denny&Bros.	1884	8 Blhds. CDB 826	4C.37" 61" & 71" (2)–60" 800 HP Denny&Co. Dumbarton	Shaw, Savill & Albion Co.,Ld.
	Arawata I ScBg 3 *Dks*	British Sydney	623 1098	245·4 *H 13 t.*	30·1	14·2 21·9	Glasgow T.Wingate&Co.	1875		2Cy.40" & 80"–40" 300 HP. T.Wingate&Co.,Glasgow	Union Steamship Co.of NewZealand, Ld.
*	**Arbib Brothers** Stl ScSr *Marshall 89-89* 2 *Dks 3 B*	British London	1739 2653	320·0 *B 63 f. P 35 f. F 34 f.*	40·0	24·3	Newcastle C.S.Swan&Hunter	1889	5 Blhds. CDB 176	3C.24" 39" & 64"–42" 230 HP T.Richardson&Sons,Hartpl.	E.& E.Arbib
	Arboga (ex Oskar) C ScSp *Svensson 87-87* 1 *Dk*	Swedish Arboga	127 176	96·1 *Q 23 f. B 20 f. P 25 f. H 5 f.*	20·5	9·9	Stockholm LängholmsVarft.	$18\frac{58}{80}$		2Cy.15½" & 29"–24" 40 HP. Bergsunds Mek.Verks.,Stkholm	Arboga Rederi Aktie Bolag (L.W.Lundberg)

*=Lloyd's Register; D=Nederlandsche Vereeniging; G=Germanischer Lloyd; I=Registro Italiano; R=American Record;
V=Veritas Austro-Ungarico; H=Veritas Ellenico; N=Norske Veritas; V=Bureau Veritas.

Dans la 2me Colonne, B.T., avec les indications qui suivent, signifie la date du rapport de visite de la Carène par les experts du Board of Trade pour obtenir le certificat relatif au TRANSPORT DES PASSAGERS.

Dans la 3me Colonne, C=Composite (membrure en fer, bordé en bois); I=Fer; Stl=Acier; W=Bois.

Pour plus amples renseignements sur les navires classés par le Lloyd's Register, voir le Lloyd's Register Book et le Lloyd's Yacht Register.

No. in Book / Official No.	Registry in which classed	Ship's Name. Material, Rig, &c. / Master, and Dates of Appointment to present Owner's Service and to Ship / Ship's Late Name. No. of Decks, &c.	Flag / Port of Registry	Tons Net / Gross	Dimensions Length / Deck Erections	Breadth	Depth	Built Where / By Whom	When	No. of Bulkheads / Particulars of Water Ballast	Engines Particulars / Maker's Name	Owners or Managers
631		**Arboga II.** (ex ...) I ScSp *Törnqvist* ... 1 *Dk&Shltr Dk*	Swedish Stockholm	156 213	99·8 B 7 *f.*	20·9	10·0	Motala MotalaCo.	18 58/71		2Cy.14"&29"–21" 45 HP. MotalaCo. Motala	Motala Ström Aktiebol.
632 87...	✻	**Arbutus** I ScSr *Bovey* ... 1 *Dk2 B*	British Liverpool	929 1434	244·0 ... B 53 *f.* F 28 *f.*	33·8	17·3	Sunderland StrandSlipwayCo.	1884	4Bhds. DB ...	2Cy.30"& 56"–36" 120 HP. J.W.&F.Wilson,Sunderland	"Arbutus" S.S. ... (T.Stephens & ...)
633 1895		**Arbutus** I ScSr	British Dundee	185 336	179·9	22·6	12·3	Newcastle T.Toward	1854		2Cy.34½"–34" 70 HP. Hawthorn&Co.,Newcastle	R.Taylor
634 931..	✻ BT 5-88	**Arcadia** Stl ScIm Bkn *Andrews* 84-88 2 *Dks&Spr Dk4 B*	British Belfast	3175 6188	468·4 P 91 *f.* F 47 *f.*	52·1	26·8 846	Belfast Harland&Wolff	1887	8 Bhds. DB ...	3Cy.40½"63½"& 102"–72" Harland&Wolff, Belfast [1000 HP.]	Peninsular & ... Steam Nav.Co.
635 175..	BT 8-88	**Arcadia** I ScBg	British Liverpool	1221 1901	303·8	31·7	27·4	Glasgow J.&G.Thomson	1855		1Cy.24"& 48"–36" 230 HP. Thompson&Co.,Newcastle	Pajayanni & ...
636 105013		**Arcata** W ScImSr *Mountfort* ... 86 1 *Dk&Spr Dk*	American S. Francisco	416 560	180·0 Len. 80	26·2	8·6 166	San Francisco DickeyBros.	1876		2Cy.14"–26" 60 HP. DickeyBros.	Newport Coal ...
637 65390	✻	**Arch Druid** I ScSr *Leinster* 86-84 1 *Dk2 B*	British N. Shields	812 1275	240·5 Q 46 *f.* P 26 *f.*	30·9	19·6	Sunderland W.Pile&Co.	1871	5 Bhds. DB ...	2Cy.27"& 56"–36" 120 HP. R.&W.Hawthorn,Nwcastle	V.T.Thompson
638		**Archangelesk** (ex J. A. Hartley) ISeSr (COLLISION) 1 *Dk*	Russian Archangel	378 711	187·0 Q— P— B51 *f.* F—	27·2	13·4	Sunderland StrandSlipwayCo.	1884	4 Bhds. DB ... FPT 31 APT 24	2Cy.26"&50"–36" 160 HP. N.E.Marine E.Co.,Sunderld.	Archangel M... Steam Nav.Co.
639 87..	✻	**Archer** I ScSr *Molley* 1 *Dk*	British London	440 691	200·2 P 44 *f.* F 23 *f.*	29·0	13·2	Glasgow Lon.&Glas.S.B.Co.	1882	4 Bhds.	2Cy.27"& 48"–36" 95 HP. London&GlasgowCo.,Glgw	Australasian ... Steam Naviga...
640	V	**Archiduc Rodolphe** Stl&I ScSr *Callour* 84-85	Belgian Antwerp	814 1013	228·6 Q 70 *f.* B 60 *f.* F 32 *f.*	27·7	16·7	Hoboken Soc. Cockerill à Seraing	1883	W Bt ... MT ...	2Cy.27"&45"–33" 90 HP. Soc.Cockerill à Seraing	Soc.Cockerill
641 28..	I	**Archimede** I ScBk *Pirandello* 72-87 3 *Dks*	Italian Palermo	1849 2839	350·1 P 76 *f.* B 25 *f.* F 48 *f.*	40·0	26·1	Glasgow A.Stephen&Sons	1881	5 Bhds. CDB 457	2Cy.48"& 90"–48" 762 HP. A.Stephen&Sons,Glasgow	Navigazione G... Italiana
642 7088.	✻	**Archimedes** I ScBg *Carnan* 88-88 2 *Dks&Spr Dk*	British Liverpool	985 1561	270·7	32·1	19·1 I 245	Aberdeen Hall,Russell&Co.	1874	6 Bhds.	2Cy.26"& 58"–30" 150 HP. Hall,Russell &Co.,Aberdeen	Liverpool, B... RiverPlate S...
643		**Archimedes** I Sc3mSr *Markwardt* 81-85 (ex Sigmund Robinow) 1 *Dk*	German Stettin	240 319	166·0	23·6	11·1	Greenock Caird&Co.	1856		2C.19½"&39½"–25½" 60 HP. Aron&Gollnow	Neue Dampfer... pagnie, Actien...
644 84044	BT 1-88	**Architect** I ScSr *Bond* 85-88	British Liverpool	1186 1859	302·0 P 59 *f.* F 46 *f.*	34·5	23·1	PortGlasgow Cunliffe&Dunlop	1880	5 Bhds. W Bt ... MT FPT APT	2Cy.34"& 68"–42" 200 HP. Cunliffe&Dunlop,P.Glsgw.	Charente S.S. ...
645 76		**Arciduchessa Carlotta** I PadSr *Serra* 54-87 2 *Dks*	Austrian Trieste	515 842	230·0	28·1	21·5	London Mare&Co.	1857		2Cy.35"& 61"–48" 400 HP. J.Watt&Co.,London	Lloyd Austro- rico
646	V	**Arcona** I Sc3mSr *Stoebase* 80-80 1 *Dk*	German Stettin	374 489	213·2 Len74 P 30 *f.* B 28 *f.* F 30 *f.*	21·7	12·6	Stettin Actien Ges. "Vulcan"	1863		2Cy.19"& 42"–24" 65 HP. BottcherMaschin&Schiffsb.	Neue Dampfer... pagnie Actien...
647	V	**Arcona** (ex Dalarne) I ScSr *Lange* (not named Smaland) 1 *Dk*	German Greifswald	111 169	100·0 B 35 *f.*	20·9	9·0	Christiania AkersMek.Verks	1868		2Cy.19"& 31"–24" 60 HP. AkersMek.Verks.	A.Spruth
648 9469.	✻	**Arctic** I ScK *Bryant* ... 88	British Grimsby	93 154	100·6 Q 16 *f.* F 18 *f.*	20·0	11·6	Hull Earles'SB&E Co. Ld.	1888	3 Bhds.	3C.12" 18"&30"–18" 45 HP. Earles'S.B.&E.Co. Ld.,Hull	Grimsby Steam... ing Co. Ld.
649	GN	**Arctic** (ex San Francisco, ex Sailing Vessel) W ScBn *Gjertsen* ... 85 1 *Dk2 B*	Norweg'n Christiania	175 254	113·1 Q 43 *f.*	25·9	12·4	Vegesack H.F.Ulrichs	1868		2Cy.11"& 21"–14" 25 HP. GebrüderHowaldt,Kiel	Bottlenose A... "Arctic" (R...)
650 7859.	✻	**Ardanach** I ScSr *Glore* 88-88 1 *Dk2 B*	British Glasgow	535 843	210·0 B 37 *f.* Q 79 *f.* F 36 *f.*	29·4	14·9	PortGlasgow H.Murray&Co.	1878	4 Bhds. DB ... FPT APT	2Cy.27"& 50"–33" 98 HP. W.King&Co.,Glasgow	Ardan S.S.Co. (Clark & Service)
651	✻	**Ardanaz** (ex Blyth) I ScSr *Ocampa* 84-84 2 *Dks*	Spanish Santander	479 751	200·5	28·2	17·7	Sunderland J.Blumer&Co.	1874	4 Bhds. ...	2Cy.27"& 50"–30" 95 HP. Fowler&McCollen,Hull	Luis Calisalve
652 823..	✻	**Ardanbhan** I Sc3mSr *Edmondson* 86-87 2 *Dks*	British Glasgow	718 1132	235·5 B 50 *f.* Q 84 *f.* F 34 *f.*	31·5	17·1	PortGlasgow H.Murray&Co.	1880	4 Bhds. DB ... FPT ...	2Cy.28"& 50"–36" 98 HP. W.King&Co.,Glasgow	Ardan S.S.Co. (Clark & Service)
653 85..	✻	**Ardancorrach** Stl Sc3mSr *Bean* 86-88 1 *Dk2 B*	British Glasgow	948 1488	244·5 Q 83 *f.* B 59 *f.* F 36 *f.*	34·2	17·2	PortGlasgow Blackwd.&Gordon	1882	4 Bhds. DB ... FPT ... APT ...	2Cy.30"& 54"–36" 125 HP. Blackwood&Gordon,P.Glw.	Ardan S.S.Co. (Clark & Service)
654 82288	✻D	**Ardandhu** I Sc3mSr *Darey* 86-88 1 *Dk2 B*	British Glasgow	757 1190	235·9 B 50 *f.* Q 84 *f.* F 34 *f.*	31·1	17·0	PortGlasgow H.Murray&Co.	1879	5 Bhds. DB ... FPT ... APT 30	2Cy.28"& 50"–36" 98 HP. W.King&Co.,Glasgow	Ardan S.S.Co. (Clark & Service)
655 85831	✻	**Ardangorm** I Sc3mSr *Cameron* 86-88 1 *Dk2 B*	British Glasgow	1043 1612	260·2 Q 115 *f.* B 52 *f.* F 33 *f.*	35·3	17·9	Leith Ramage&Ferguson	1881	6 Bhds. CDB ... FPT ... APT 18	2Cy.32"& 60"–39" 180 HP. Walker,Henderson&C.,Glw	Ardan S.S.Co. (Clark & Service)
656 83737	✻	**Ardanrigh** I Sc3mSr *Anderson* 86-86 1 *Dk2 B*	British Glasgow	864 1356	244·3 Q 89 *f.* B 50 *f.* F 34 *f.*	34·2	17·3	Glasgow Napier, Shanks & Bell	1882	4 Bhds. DB ... FPT 45 APT ...	2Cy.28"& 50"–36" 120 HP. W.King&Co.,Glasgow	Ardan S.S.Co. (Clark & Service)
657		**Ardantiene** I Sc 1 *Dk2 B*	Japanese Japan	411 649	190·7 Q 69 *f.* F 29 *f.*	25·7	14·2	PortGlasgow H.Murray&Co.	1878	4 Bhds. ...	2Cy.24"& 44"–36" 90 HP. W.King&Co.,Glasgow	J.Walter
658 91087	✻D	**Ardbeg** I ScSr *Greenfield* 83-87 1 *Dk&Spr Dk3 B*	British London	1467 2255	280·0 F 17 *f.*	38·7	25·3	Sunderland SndrlandS.B.Co.Ld	1887	5 Bhds. DB ...	2Cy.34"& 64"–42" 180 HP. J.Dickinson,Sunderland	R.Gordon & Co.
659 6394.	✻	**Ardclinis** I ScSr *Parry* 78-87 1 *Dk*	British Aberdeen	81 161	129·2	19·1	9·0	Glasgow Doble&Co.	1870	3 Bhds.	2Cy.16"& 26"–22" 32 HP. W.King&Co.,Glasgow	Mrs. J. M. H... (J. & W. Hen...)
660 78904		**Arden** (ex Hossea) I TwnScSr 1 *Dk*	British Liverpool	332 544	193·4 Q 45 *f.* F 23 *f.*	26·2	12·1	Glasgow A.Stephen&Sons	1878	4 Bhds.	1Cy.17"& 28"–27" 90 HP. A.Stephen&Sons,Glasgow	AtlasSteamship...

✻ = **Lloyd's Register**; D = Nederlandsche Vereeniging; G = Germanischer Lloyd; I = Registro Italiano; R = American ...; A = Veritas Austro-Ungarico; H = Veritas Ellenico; N = Norske Veritas; V = Bureau Ver...

In the 2nd Column, B.T., with figures following, signifies Date of report of bottom sighting by Board of Trade Surveyors as condition of PASSENGER CERTI...

In the 3rd Column, C = Composite (iron frame, wood planked); I = Iron; Stl = Steel; W = Wood.

Further particulars of Vessels classed by Lloyd's Register will be found in Lloyd's Register Book and Lloyd's Yacht Register.

No.	Registre	Nom. Matériaux, Gréement, etc. — Capitaine et dates d'entrée au service des Armateurs actuels et à bord du navire. — Ancien Nom, nombre de ponts, etc.	Pavillon. — Port d'Armement	Tonnage Net. — Total	Longueur	Largeur	Creux	Construit. Où — Par qui.	Quand.	Nombre des cloisons. — Détails sur le lest d'eau.	Machines. — Détails. Nom du Constructeur.	Armateurs ou Directeurs-Gérants.
	✻	**Ardilla** I ScSr *Recken 75-82* 1 *Dk*	Chilian Valparaiso	46 94	83·1	17·2	8·4	Preston W. Allsup & Sons	1882	5 Blhds. BT2/ft/29	2Cy.14"& 28"–21" 40 HP. W. Allsup&Sons, Preston	Cia. Sud-Americana de Vapores
	✻	**Ardle** Stl ScSr *Leighton 84-88* 1 *Dk2 B*	British Dundee	794 1245	215·5	34·2	16·1	Dundee Gourlay, Bros.&Co. *Q86 f. B65 f. F30 f.*	1888	4 Blhds. DB28ft FPT22	3C.20"31"&50"–36" 150 HP. Gourlay, Bros.&Co., Dundee	North Sea Steam Shipping Co., Ld.
	✻	**Ardlethen** I ScSr *Murray 83-83* 1 *Dk & Spr Dk3 B*	British Aberdeen	1024 1579	249·0	34·1	23·3	Sunderland Sunderland S. B. Co. *F 23 f.*	1883	5 Blhds. DB63ft/83ft/25ft	2Cy.30"& 58"–36" 160 HP. Hawks, Crawshay&Sns, Gtshead.	Adam & Co.
	✻	**Ardmellie** I ScSr *Russell 79-87* 1 *Dk2 B*	British Aberdeen	973 1518	251·9	31·3	17·9	Aberdeen Hall, Russell&Co. *P 112 f. F 34 f.*	1884	5 Blhds. DB54ft/80ft/25ft	2Cy.30"& 59"–36" 180 HP. Hall, Russell&Co., Aberdeen	Adam & Co.
	✻D	**Ardmore** I ScSr *Scoggans 82-87* 1 *Dk2 B*	British Maryport	597 928	215·8	30·0	15·2	Sunderland Bartram, Haswell & Co. *P & B 132 f. F 24 f.*	1872	4 Blhds. DB65ft/30ft	2Cy.25"& 48"–30" 90 HP. Fowler&McCollin, Hull	Hine Bros.
	✻	**Ardnamult** I ScSr *Goggin* 1 *Dk*	British London	725 1123	226·0	33·1	15·0	North Shields T.&W. Smith *Q87 f. B50 f. F26 f.*	1883	4 Blhds. DBa7ft/65ft/26ft FPT29 APT7	2Cy.26"& 52"–33" 99 HP. WilsonShipCo., Ld., Ncastle.	C. Morley
	✻	**Ardoe** I ScSr *Wilson 82-86* 1 *Dk2 B*	British Aberdeen	890 1388	213·4	32·7	17·3	Aberdeen Hall, Russell&Co. *P 140 f. F 34 f.*	1880	5 Blhds. DBa50ft/82ft/25ft	2Cy.30"& 56"–36" 135 HP. Hall, Russell&Co., Aberdeen	Adam & Co.
	✻V	**Arecibo** (ex Scindia) I ScSr *Mendialdua* 2 *Dks3 B*	Spanish Bilbao	1323 2021	285·0	36·0	23·9	Newcastle C.S.Swan& Hunter *P 28 f. B 52 f. F 31 f.*	1881	6 Blhds. DBa&cft124 MT888	2Cy.35"& 68"–45" 250 HP. WilsonSlipwayCo., Ld., Nwc.	D. de Eizaga
	✻	**Arecuna** Stl ScSr *Strickland 84-84* 1 *Dk*	British Glasgow	1106 1720	275·5	36·1	21·8	Glasgow J.&G. Thomson *P44 f. B56 f. F30 f.*	1884	4 Blhds. APT25	2Cy.32"& 63"–45" 176 HP. J.&G. Thomson, Glasgow	Caw, Prentice, Clapperton & Co.
570	N	**Arendal** Stl ScSr *Bastrup* 1 *Dk & Awn Dk*	Norweg'n Arendal	401 518	159·0	27·0	17·7	Gothenburg MotalaCo.	1882		2C.28⅔"& 43"–23⅔" 110 HP. MotalaCo., Motala	Arendals Dampskibs-Selskab
571		**Arequipa** (HULK) I PadSr	British Liverpool	662 1065	231·8	35·2	14·7	Glasgow J. Elder&Co. *P 168 f. F 48 f.*	1870		4Cy.38"& 76"–51" 300 HP. J. Elder&Co., Glasgow	Pacific Steam Nav. Co.
572	✻	**Arethusa** I Sc3mSr *Thompson 77-84* 2 *Dks3 B*	British London	1181 1841	282·8	33·3	24·2	Newcastle Palmers'Co. *B 40 f. F 34 f.*	1875	5 Blhds. DB95ft	2Cy.31"& 61"–36" 150 HP. Palmers'Co., Newcastle	Hall Bros.
573		**Aréthuse** I Sc 2 *Dks*	French Marseilles	685 1184	263·7	33·5	21·8	London SamudaBros. *B 105 f. F 42 f.*	1864		2Cy.30"& 59"–39½" 250 HP. MessageriesMar., LaCiotat	Messageries Maritimes
574	V	**Argentina** I ScBg *Reteike 86-86* 2 *Dks3 B*	German Hamburg	1552 2114	299·5	36·1	27·1	Newcastle C. Mitchell&Co. *F 38 f. B 226 f. Q29 f.*	1873	5 Blhds.	2Cy.40"& 72"–42" 250 HP. Thompson, Boyd&Co., Newc.	Hamburg-Südamerik. Dampfschiff-Gesells.
575	✻D	**Argentino** I ScSr *Rand 89-89* 2 *Dks3 B*	British London	918 1432	248·0	31·9	23·8	Sunderland W. Pile&Co.	1872	5 Blhds. DBa80ft/100	2Cy.32¼"& 58¼"–33" 160 HP. N.E. MarineEng.Co., Sndrld.	Avis Steamships, Ld.
576	✻	**Argo** I Sc3mSr *Sowell 68-86* 2 *Dks Spt Awn Dk*	British Bristol	516 928	230·7	30·3	16·2 23·5	Glasgow Lond.&Glasg.Co.	1871	5 Blhds.	2Cy.38"& 68"–36" 210 HP. London&GlasgowCo., Glasg.	Bristol Stm. Nav. Co., Ld.
577		**Argo** I ScSr *Hedgecock*	British Hull	489 758	210·5	27·2	16·2	Hull C.&W. Earle *Q 75 f.*	1860	5 Blhds.	2Cy.31"& 58"–30" 120 HP. C.&W. Earle, Hull	T. Wilsons, Sons & Co.
578		**Argo** I ScSrYt 2 *Dks*	British Liverpool	203 360	141·0	23·4	12·7	Greenock Scott&Co. *Q20 f. B12 f. F17 f.*	1885		2Cy.16"& 32"–28" 200 HP. Scott&Co., Greenock	Alfred Holt
579		**Argo** I ScSr *Cocacich 72-87* 2 *Dks*	Austrian Trieste	793 1250	243·6	31·5	21·8	Trieste Austro-Ungarico Lloyd	1877		2Cy.31"& 56"–36" 180 HP. ArsenaleLloyd, Trieste	Lloyd Austro-Ungarico
580	✻N	**Argo** I ScSr *Kehnle 81-83* 1 *Dk2 B*	Norweg'n Bergen	930 1333	234·9	34·1	16·8	Bergen Martens, Olsen&Co. *Q84 f. B56 f. F25 f.*	1883	5 Blhds. DBaft 25ft APT16	2Cy.29½"& 57"–33" 130 HP. Martens, Olsen&Co., Bergen	S. M. Kuhnle
581		**Argolis** (ex Vine) I ScSr *Cauzoundaki 83-83* 1 *Dk2 B*	Greek Piraeus	247 488	183·0	25·5	13·3	Glasgow D.&W. Henderson *P 51 f. F 29 f.*	1878		2Cy.27"& 49"–36" 234 HP. D.&W. Henderson, Glasgow	Panhellenic Steam Ship Navigation
582		**Argonarvis** I PadSr 1 *Dk*	Russian Odessa	271 430	206·0	25·0	13·1	London Samuda	1859		2Cy.48"–48" 160 HP. J. Penn&Son, London	Russian Steam Nav. & Trading Co.
583		**Argonauta** (ex Cuba) I ScSr	Spanish Cienfuegos	601 938	211·0	34·0	21·6	Philadelphia Neafie&Levy	1879		2Cy.30"& 50"–36" 200 HP. Neafie&Levy, Philadelphia	Menendez & Co.
584		**Argos** (BROKEN UP) I ScSr 1 *Dk*	British Liverpool	178 241	165·2	20·1	9·7	Newcastle A. Leslie&Co.	1868	4 Blhds.	2Cy.27"–18" 50 HP. R.&W. Hawthorn, Newcstle.	N. Rundell
585	✻	**Argosy** I ScBn *Atchison 87-87* 2 *Dks3 B*	British Newcastle	1259 1855	275·0	35·5	24·0	Stockton M. Pearse&Co. *F 34 f.*	1879	4 Blhds. DBa85ft/52ft	2Cy.33"& 62"–39" 150 HP. Blair&Co., Ld., Stockton	J. Temperley & Co.
586		**Argosy** W Sc *Atherton*	British Liverpool	69 101	82·0	19·7	8·4	Northwich	1872		2Cy.10½–14" 24 HP. W.P. Goulton, Manchester	Salt Union, Ld.
587	✻	**Argus** Stl ScSr *Arthur 88-88* 2 *Dks3 B*	British Melbourne	1822 2792	325·0	40·4	24·0	Newcastle Palmers'Co., Ld. *B 82 f. P34 f. F41 f.*	1889	5 Blhds. DBa&c6227	3C.26"43"&70"–45" 400 HP. Palmers'Co., Ld., Newcastle	A. Currie & Co.
588	✻	**Argus** I ScSr *Cowey 68-88* 1 *Dk2 B*	British Barrow	746 1156	240·0	34·2	14·1	Barrow BarrowS.B.Co., Ld. *Q92 f. B52 f. F28 f.*	1883	4 Blhds. CDB388 FPT11 APT7	2Cy.31"& 58"–36" 150 HP. Westray, Copeland&Co, Brw	T. Kish
589		**Argus** I PadSr	British London	208 331	172·5	23·9	13·3	London M. Wigram&Sons	1856		2Cy.48"–48" 160 HP. J. Penn&Sons, Greenwich	Corporation of the Trinity House
590												

✻ = **Lloyd's Register**; D = Nederlandsche Vereeniging; G = Germanischer Lloyd; I = Registro Italiano; R = American Record; A = Veritas Austro-Ungarico; H = Veritas Ellenico; N = Norske Veritas; V = Bureau Veritas.

Dans la 2me Colonne, B.T., avec les indications qui suivent, signifie la date du rapport de visite de la Carène par les experts du Board of Trade pour obtenir le certificat relatif au TRANSPORT DES PASSAGERS.

Dans la 3me Colonne, C = Composite (membrure en fer, bordé en bois); I = Fer; Stl = Acier; W = Bois.

Pour plus amples renseignements sur les navires classés par le Lloyd's Register, voir le Lloyd's Register Book et le Lloyd's Yacht Register.

No. in Book / Official No.	Registry in which classed	Ship's Name. Material, Rig, &c. / Master, and Dates of Appointment to present Owner's Service and to Ship. / Ship's Late Name. No. of Decks, &c.		Flag. / Port of Registry.	Tons. Net / Gross	Dimensions. Length / Breadth / Depth. Deck Erections.			Built. Where / By Whom.	When	No. of Bulkheads. / Particulars of Water Ballast.	Engines. Particulars. / Maker's Name.	Owners or Managers
691 87644	✻	Argus *Welsh 85-86*	I ScSr 1 *Dk*	British Glasgow	79 1[illegible]	115·0	21·1	9·0 F 18 f.	Paisley J.Fullerton&Co.	1883	3 Blds.	2Cy.17"& 30"–22" 40 HP. W.King&Co.,Glasgow	W.Robertson
692 [illegible]	✻	Argyle *Walters 74-85*	I ScSr 1 *Dk*	British Hull	707 11[illegible]	211·6	30·9	15·8 F 29 f.	Hull Gilbert&Cooper	1872	[illegible]	2Cy.27"& 53"–36" 98 HP. Gilbert&Cooper,Hull	W.S.Bailey
693 [illegible]		Argyle	W ScSr 1 *Dk*	British Auckland	129 1[illegible]	114·3	20·0	7·3 B 25 f.	Auckland McQuarrie& McCall	1876		1Cy.30"& 16"–20" 45 HP. Fraser&Tinne,Auckland	Northern S.S.
694 [illegible]	BT4.88	Argyle *McEwan 81-81*	I P[illegible]	British Glasgow	23 1[illegible]	177·3	17·5	7·6	Glasgow Barclay,Curle&Co.	1866		1Cy.47"–42" 75 HP. Hutson&Corbett,Glasgow	J.Gillies&A.
695 [illegible] ✻	BT3.88	Argyll *Mitchell 86-86*	I ScSr 1 *Dk*	British [illegible]	45 [illegible]	139·9	23·1	9·1	Port Glasgow [illegible]	1886	[illegible]	2Cy.22"& 44"–30" 80 HP. Muir&Houston,Glasgow	Argyll S.S.Co. (J.Little & Co.)
696 17674		Ariadne (ex Dom Pedro) *McCallum 83-83*	I ScSr 1 *Dk*	British Barrow	152 2[illegible]	158·0	20·1	12·1	Greenock [illegible]	1856		2Cy.20"& 35"–21" 55 HP. Westfield,Copeland&C.,Barrow	J.Little & Co.
697	✻D	Ariadne (ex Zulu) *Sharp 82-86*	I ScSr 1 *Dk2 B*	Dutch Rotterdam	1008 1467	214·0	32·0	18·5 Q 88 f.	Stockton M.[illegible]	1879	[illegible]	2Cy.30"& 55"–36" 120 HP. Blair&Co.,Ld.,Stockton	Rotterdamsche
698 84121		Arica	Stl ScSr 2 *Dks*	British Liverpool	1310 1771	300·0	36·2	19·3 F 23 f.	Birkenhead [illegible]	1881	[illegible] Blds.	2Cy.34"& 60"–42" 220 HP. LairdBros.,Birkenhead	Pacific Steam
699 91432	✻D	Ariel *Brown 88-88*	Stl Sc3mSr 1 *Dk2 B*	British Hull	1452 2226	300·0	42·2	19·9 Q 82 f. B115 f. F 62 f.	Hull Earles Co.,Ld.	1885	[illegible] Blds.	3C.23"35"&60"–57"300HP. Earles Co.,Ld.,Hull	E.Leetham
700 83078	✻	Aries	I Sc3mSrYt	British Barrow	107 187	145·0	20·1	10·3	Barrow Barrow S.B.Co.	1880	[illegible] Blds.	2Cy.16"& 32"–24" 50 HP. Barrow S.B.Co.,Barrow	W.C.Clark
701 [illegible]	✻	Aries *Bruce 88-88*	I ScTr 1 *Dk*	British Grimsby	67 118	95·4	20·3	10·2	Grimsby T.Charlton	1881	[illegible] Blds.	2Cy.16½"& 31"–20" 35 HP. T.Charlton,Grimsby	Grimsby & N. Steam Trawling
702 293		Aries *Briggs 81-87*	I Sc3mSr 2 *Dks*	American Boston,Mass.	559 882	197·0	27·6	15·9 22·0	Glasgow	1862		1Cy.40"–48" 332 HP. Pusey&JonesCo.,Wilm.,Del.	Boston & Phil. S.S.Co.
703 106247		Aries	I Sc	American New York	57 115	89·5	20·1	9·0	Philadelphia American Shipbuilding Co.	1884		2Cy.17"&30"–24 Neafie&Levy,Philadelphia	American Shipping Co.
704 [illegible]	I	Arietta Zigomala (ex Betty, ex Sea Gull)	I ScSr 2 *Dks*	Greek Syra	653 1010	216·5	29·2	16·8 P 125 f. F 26 f.	Sunderland Hill&Mounsey	1873	[illegible] Blds.	2Cy.27"& 50"–30" 99 HP. G.Clark,Sunderland	J.Costalas
705	DV	Arion *Hinrichs -87*	I ScSr 1 *Dk*	German Bremen	202 271	117·4	20·6	11·7 F11 f. Lca76. P27 f.	Bremen Act.Ges."Weser"	1874		2Cy.12"& 21½"–19" Janssen&Schmilinsky,Hmbrg.	Dampfschiffs Ges." Neptun"
706 [illegible]	✻	Ariosto *Mollard 75-87*	StlScSr 1 *Dk,SprDk3 B*	British Glasgow	2342 2992	320·3	40·2	19·4 P12 f. B70 f. 27·3	Greenock Russell&Co.	1887	5 Blds.	3Cy.23"30"&64"–42" 250 HP. J.Howden&Co.,Glasgw.	R.McAndrew &
707	R	Aripuana	I PadSr	Brazilian Pará	371	160·0	30·0	8·0	Wilmington Pusey&JonesCo.	1884		2Cy.24"&72" Pusey&JonesCo.,Wlm.,Del.	Cia de Nav. do Amazonas
708 85956	✻	Arisaig *Currie 87-87*	Stl ScSr 1 *Dk2 B*	British Glasgow	842 1328	240·3	33·1	16·7 Q 83 f. B75 f. F31 f.	Port Glasgow Blackwood&Gordn	1882	[illegible] Blds.	2Cy.28"& 54"–36" 120HP. Blackwood&Gordon,P.Glas.	J.Gardiner &
709 81271	✻ BT10.88	Arizona *Brooks 86-82*	I Sc4mSr 4 *Dks*	British London	2657 5164	450·2	45·4	35·7 B 31 f. F 74 f.	Glasgow J.Elder&Co.	1879	8 Blds.	3C.62"&90"(2)–66"1200HP. J.Elder&Co.,Glasgow	Guion Steam S. Ld.
710		Arizona Maru	W PadBg	Japanese	2793	362·0	46·0	30·0	Greenpnt, L.I.	1865			
711	✻	Arla (ex Cebino) *Fris 88-88*	I ScSr 1 *Dk2 B*	Swedish Stockholm	789 1021	214·8	29·8	16·8 P121 f. F23 f. H23 f.	Sunderland J.L.Thompson&Ss	1871	4 Blds.	2Cy.27"&50"–30" 98HP. G.Clark,Sunderland	Rederibolaget (A.O.Fredholm)
712		Arla *Röyen*	C ScSr 1 *Dk*	Swedish Gefle	185 258	119·7	21·2	10·7 Q39 f. B29 f. F24 f. H40.	Oscarshamn	1870		50 HP.	G.R.Thalin
713		Arlindo (ex Calderon) *Cundill -87*	I ScSr 2 *Dks*	Brazilian Rio de Janeiro	660 1054	249·2	29·3	16·9 H 87 f.	Newcastle A.Leslie&Co.	1871	1 Blds.	2Cy.27"&52"–36" 170HP. R.Stephenson&Co.,N'castle	J.N.de Vincenzi
714 [illegible]	BT4.88	Armagh (ex Talbot) *Morrison 73-74*	I PadSr	British [illegible]	229 631	263·5	27·0	12·0 Q91 f. B80 f. F32 f.	Renfrew J.Henderson&Sons	1860		2Cy.48"–60" 260 HP. Henderson&Co.,Renfrew	J.Little & Co.
715 74758	✻	Armathwaite *Evans*	I ScSr 1 *Dk2 B*	British Whitehaven	958 1487	255·0	34·2	20·5 P26 f. B48 f. F29 f.	Whitehaven W'havenS.B.Co.Ld.	1881	4 Blds.	2Cy.30"& 60"–36" 150 HP. J.Jones&Sons,Liverpool	S.S." Armathwaite" Co.,Ld.(W.Dan[illegible])
716 84314	BT7.88	Armenia *Read 76-88*	I ScSr 2 *Dks3 B*	British Glasgow	2218 3396	364·3	38·0	28·6 P21 f. B238 f. F21 f.	Glasgow D&W Hendrsn&Co	1881	[illegible] Blds.	2Cy.38"& 66"–60" 500 HP. D.&W.Henderson&Co.Glas.	Henderson Bros.
717 [illegible]		Armenia *Brien*	W Sc1m 1 *Dk*	British Toronto	403 643	172·0	25·0	11·6 F 17 f.	Chatham,Ont. D.Chisholm&J.Simpson	1873		1Cy.34"–36" 38HP. T.Wilson&Son,Dundas	J.C.Hume.
718 71200		Armenia	W Sc 1 *Dk*	British Picton,Ont.	85 110	100·0	18·0	7·0	Picton,Ont. D.McMurchy	1876		1Cy.16"–18" 50 HP. J.Brokenshire,Kingstn,Ont.	S.S.Mutton
719	V	Arménie *Duchesne*	I ScBn 2 *Dks3 B*	French Marseilles	1395 2128	296·5	35·8	23·7 P24 f. B91 f. F32 f.	LaSeyne Forges&ChtsdelaMedt	1879	W.B.[illegible]	2Cy.37½"&94½"–39¼"350HP. Forges&Chts deMedit.,Mars.	Cie.de Nav. Armenienne & Mars.
720	✻	Armin *Vogel -84*	Stl ScSr 1 *Dk&WhFrms*	German Hamburg	810 1285	227·6	33·5	21·6 Q79 f. B48 f. F28 f.	Vegesack BremerSchiffwerft	1884	4 Blds.	2Cy.25½"&46"–33½"120HP. GuteHoffnungsHtt.	Rudolf Rettich

✻ = Lloyd's Register; D = Nederlandsche Vereeniging; G = Germanischer Lloyd; I = Registro Italiano; R = American R[illegible]
A = Veritas Austro-Ungarico; H = Veritas Ellenico; N = Norske Veritas; V = Bureau Veritas

In the 2nd Column, B.T., with figures following, signifies Date of report of bottom sighting by Board of Trade Surveyors as condition of PASSENGER C[illegible]

In the 3rd Column, C = Composite (iron frame, wood planked); I = Iron; Stl = Steel; W = Wood.

Further particulars of Vessels classed by Lloyd's Register will be found in Lloyd's Register Book and Lloyd's Yacht Register.

Class.	Nom, Matériaux, Gréement, etc. — Capitaine et dates d'entrée au service des Armateurs actuels et à bord du navire. — Ancien Nom, nombre de ponts, etc.		Pavillon. — Port d'Armement	Tonnage Net — Total	Dimensions. Long.	Larg.	Creux	Construit. Où. — Par qui.	Quand.	Nombre des cloisons. — Détails sur le test d'eau.	Machines. — Détails. Nom du Constructeur.	Armateurs ou Directeurs-Gérants.
*	Armine *Southward* -82	W ScTg 1 *Dk*	British Falmouth	13 [illegible]	103·6	17·1	9·6	Falmouth C.Burt	1883		2C. 20½"&38½"–22" 85 HP. Sara&Burgess,Penryn	J.H.Rusden
V	Armorique *Lemilliant* -87	I ScSr	French Brest	258 [illegible]	154·2	21·3	12·4	Bordeaux *P 24 f. B 15 f. F 12 f.* Chantiers & Ateliers de l'Océan	1871	W Bs&[illegible]	2Cy.18"&29½"–19¾" 40 HP. Chantiers&A.de l'Oc.,Bord	Chevillotte Frères
	Arnao *Corrales*	I ScSr	Belgian	90 132	109·9	18·1	9·0	Port Glasgow Cunliffe&Dunlop	1873		2Cy.17"& 30"–20" 40 HP. Cunliffe&Dunlop,P.Glasgw.	Soc. Royale Asturiene des Mines
*	Arno *Westlake* 85-85	I Sc3mSr 1 *Dk* 2 *B*	British Sunderland	691 1083	232·1	30·1	17·1	Sunderland *P 211 f. F 29 f.* J.Laing	1871	4 Blds. W B[illegible]	2Cy.27"& 58"–30" 98 HP. Thompson,Boyd &Co.,Nwel.	J.Westoll
G	Arnold *Krall* 89-89	I ScSr 2 *Dks*	German Stettin	919 1150	222·7	31·3	21·7	Kiel *P 5 f. H 30 f.* G.Howaldt	1884		2Cy.25¼"&43¼"–30" 110 HP G.Howaldt,Kiel	"Arnold" Dampfchiffs Ges. (Rud.Christ.Gribel)
	Aron	C Sc3mSr	Norweg'n Bergen	181 240	101·7	21·9	10·1	Wennersborg *Q 30 f. B 54 f. F 15 f.*	1869		80 HP.	Döscher & Co.
	Aros *Ekeberg*	I ScSr 1 *Dk*	Swedish Vesteras	229 296	136·8	20·6	9·2	Oscarshamn *P 92 f. F 22 f. H 11 f.* Oscarshamns Mek. Vrks.	1872		80 HP. Oscarshamn Mek.Vrks.Oscrhm.	Vesteras Ångf-Aktie-Bolag.
	Aros *Steer*	I Pad 1 *Dk*	Swedish Luleå	70 121	144·6	14·5	8·2	Norrköping *Erections 15 f.*	1852		60 HP.	Bergman, Hummel & Co.
*	Arracan *Dunbar* 82-84	S[illegible] ScSr 3 *Dks*	British Glasgow	1856 2880	346·3	39·9	25·2	Port Glasgow *P 41 f. B 73 f. F 45 f.* R.Duncan&Co.	1882	8 Blds. CDB [illegible]	2Cy.38"& 68"–51" 290 HP. Denny&Co.,Dumbarton	British & Burmese Steam Nav.Co.,Ld.
*	Arran *Brown*	I ScSr 1 *Dk*	British Liverpool	283 462	170·3	25·1	10·9	Port Glasgow *P 25 f. F 24 f.* Murdoch& Murray	1878	4 Blds. FET APT	2Cy.19"& 34"–36" 75 HP. Lees,Anderson&Co.,Glasgw	Atlas S.S.Co.,Ld.
	Arran (ex Dunoon Castle) *McLachlan* 85-85	I PadSk	British Glasgow	30 171	191·7	18·2	7·5	Glasgow T.Wingate&Co.	1867		1Cy.50"–39" 85 HP. T.Wingate&Co.,Glasgow	A.Campbell
*	Arranmore *May* 88-88	I ScSr 1 *Dk* 2 *B*	British Liverpool	896 1394	253·1	33·0	18·0	Sunderland *B 42 f. F 32 f. Q 102 f.* W.Doxford&Sons	1879	4 Blds. DB [illegible] MT	2Cy.27"& 56"–39" 150 HP. W.Doxford&Sons,Sunderld.	
*	Arratoon Apcar *Olifant*	I Sc3mSr 2 *Dks* 3 *B*	British London	1392 2153	323·2	35·5	25·7	Newcastle C.Mitchell&Co.	1873	4 Blds.	2Cy.38"& 68"–45" 200 HP. Maudsly Sons&Field,Londn.	S.A.Apcar
	Arroio de Pelotas		Brazilian Porto Alegre	115	..	..	..					M.Bromberg
*	Arrow *Coates* 85-85	I ScSr 1 *Dk* 2 *B*	British Cardiff	1201 1837	250·4	36·2	20·4	Chepstow *Q 89 f. B 72 f. F 54 f.* E.Finch&Co.,Ld.	1884	4 Blds. WB [illegible] FET APT	2Cy.30"& 60"–36" 150 HP. Westray,Copeland&Co.Brw.	Cardiff & West of England S.S.Co.,Ld.
	Arrow	I ScSr 1 *Dk*	British Bombay	130 192	120·4	20·2	10·6	Glasgow *H 8 f.* T.B.Seath&Co.	1869		2Cy.18"–24" 38 HP. D.Rowan,Glasgow	Joomabhoy Lalljee
	Arrow	Stl TwnScTug	British	100	102·5	21·0	8·8	Dundee W.B.Thompson&Co. Ld.	1888		3Cy.9"13"&24"–14" W.B.Thompson&Co.,Dundee	Shropshire Union Canal & Rail Co.
*	Artemis *Lee* 85-87	I Sc3mSr 1 *Dk* 2 *B*	British Hull	408 659	179·7	27·8	14·6	Newcastle *P 113 f. F 29 f.* Marshall Bros.	1865	3 Blds. FET APT	2Cy.25"& 48"–27" 90 HP. Comp.82,Cooper&Co.,Hull	Good, Fledman & Duncan
*D	Arthur *Kirby* 87-87	I ScSr 1 *Dk* 2 *B*	British Whitby	817 1260	240·0	32·6	17·5	SouthShields *Q 90 f. B 43 f. F 26 f.* J.Readhead&Co.	1879	4 Blds. DB [illegible]	2Cy.27½"&52"–33" 110 HP. J.Readhead&Co.,S.Shields	H.Baxter & Co.
*	Arthur *Dryden* 84-84	I ScSr 1 *Dk* 2 *B*	British N. Shields	778 1221	242·0	31·2	17·3	SouthShields *P 124 f. F 37 f.* Readhd. Softly&Co.	1872	5 Blds. DB [illegible]	2Cy.27"& 54"–33" 110 HP. Thompson,Boyd&Co.,Newc.	Charlton, McAllum & Co.
	Arthur	I PadSr	British Glasgow	165 207	150·5	22·1	7·5	Port Glasgow Cunliffe&Dunlop	1871		1Cy.38"–48" 60 HP. Cunliffe& Dunlop,Pt.Glasgw	G.Turnbull & Co.
	Arthur		Brazilian Pará		..	..	..					L.H.Campos
V	Arthur *Tolle* 75-77 (ex Marquis of Lorne)	I ScSr 1 *Dk*	German Stettin	123 168	120·0	18·9	9·2	SouthShields *Q 27 f. Len. 75.* J.T.Eltringham	1870		2C."13½"&26¾"–18½" 35 HP. Möller&Holberg,Stettin	R.C.Gribel
	Artois *Bresq* 74-87	I Sc 2 *Dks*	French Marseilles	697 1016	212·6	29·0	24·3	LaSeyne Forges&C.de la Médit	1866		2Cy.33"–24½" 100 HP. Forges&C.dela Médit.,Mars.	Société Générale de Transports Maritimes
DV	Artushof *Hoppe* 82-83	I ScSr 1 *Dk*	German Danzig	520 728	192·5	26·8	16·2	Glasgow *B 52 f. F 31 f. Q 57 f.* J.G.Lawrie	1873	W B 130t	2Cy.28"& 50"–33" 110 HP. Blackwood&Gordon,Pt.Glas	Danziger Schiffahrts-Aktien Ges.(Alex.Gibsone)
	Arvika II. *Ledgren* 84	C ScSp 1 *Dk*	Swedish Arvika	120 179	98·4	21·2	9·7	Thorskog *Erections 31 f.* P.Larsson	1883		2Cy. 30 HP.	F.G.Olsen
*D	Ary Scheffer *Lambach* 61	I ScSr 1 *Dk* 2 *B*	Dutch Rotterdam	364 528	170·0	24·2	13·9	Rotterdam *Q 84 f. B 21 f. F 11 f.* Bykee&Co.	1883	4 Blds. DB [illegible]	2Cy.22"& 40"–30" 70 HP. Maats.de Schelde,Vlissingen	C.Balguerie & Zoon
G	Arzach *Bairowitch* -83	I ScSr	Russian Kolomna	240	130·0	25·0	10·0	Kolomna Kolomna Engnrg Co.	1883		2Cy.17½"& 31"–18½"	Kolomna Engineering Co.
*	Asagao *Tirant* 89-89	Stl ScSr 1 *Dk* 2 *B* & *Wch Frms*	Japanese Tokio	1568 2408	308·0	39·0	23·3	Sunderland *P 45 f. B 83 f. F 35 f.* J.Laing	1889	5 Blds. CDB [illegible] MT [illegible] MT [illegible]	3C.23"38"&62"–42" 220 HP. G.Clark,Sunderland	Iawasaki Hisaya

*=Lloyd's Register; D=Nederlandsche Vereeniging; G=Germanischer Lloyd; I=Registro Italiano; R=American Record;
[illegible]=Veritas Austro-Ungarico; H=Veritas Ellenico; N=Norske Veritas; V=Bureau Veritas.

Dans la 2me Colonne, B.T., avec les indications qui suivent, signifie la date du rapport de visite de la Carène par les experts du Board of Trade pour obtenir le certificat relatif au TRANSPORT DES PASSAGERS.

Dans la 3me Colonne, C=Composite (membrure en fer, bordé en bois); I=Fer; Stl=Acier; W=Bois.

Pour plus amples renseignements sur les navires classés par le Lloyd's Register, voir le Lloyd's Register Book et le Lloyd's Yacht Register.

No. in Book. / Official No.	Registry in which classed.	Ship's Name. Material, Rig, &c. / Master, and Dates of Appointment to present Owner's Service and to Ship. / Ship's Late Name. No. of Decks, &c.	Flag. / Port of Registry.	Tons. Net / Gross	Length	Breadth	Depth	Built. Where. / By Whom.	When.	No. of Bulkheads. / Particulars of Water Ballast	Engines. Particulars. / Maker's Name.	Owners
751 830		Asahi Maru 1 ScSr	Japanese	387	161·4	22·5	18·9	Kobe [illegible]	1884		2Cy.20" & 36"–22" 56 H.P.	Osaka Shosen
752 9375[illegible]	*	Asama StlScSr Nevell [illegible] 2 Dks3 B	British Liverpool	2460 3118	470·0	42·3	27·4	Stockton [illegible]	1886	[illegible] Blds. [illegible]	3C.28"46"&75"–51" 495 H.P. Blair&Co.,Ld.,Stockton	J.M.Wood
753	V	Ascania StlScSr Metzenthin [illegible]-87 2 Dks3 B	German Hamburg	1480 2080	290·0	36·2	22·6	Newcastle [illegible]	1887		[illegible]	Hamburg [illegible]
754	GV	Ascania 1 ScSr Schultz -84 1 Dk	German Stettin	255 [illegible]	131·2	21·0	10·7	Stettin [illegible]	1884	[illegible]	[illegible] H.P. Aron&Gollnow,Stettin	A.Stemann[illegible]
755 84049		Ascanius 1 ScCr	British Liverpool	74 187	74·0	18·1	7·5	Birkenhead [illegible]	1880		1Cy.15"–16" 25 H.P. Cochran&Co.,Birkenhead	Ocean Steam[illegible]
756 72358	*	Ascupart 1 ScSr Weaver 86-86 2 Dks3 B	British Southampton	858 [illegible]	250·0	32·2	23·0	Stockton M.Pearse&Co.	1875	4 Blds. [illegible]	3C.18"29"&56"–40" 120 H.P. Tripd. '88	G.T.H[illegible]
757 81548	*	Ashbrooke 1 ScSr Pollenger 1 Dk2 B	British London	954 1478	245·4	33·5	18·2	Sunderland Mounsey&Foster	1879	4 Blds. [illegible]	2Cy.31" & 57"–36" 150 H.P. J.Dickinson,Sunderland	T.G.Greenwell
758 [illegible]	*	Ashburne 1 ScSr Lambert -88 2 Dks3 B	British Sunderland	1613 2469	297·5	38·0	25·0	Sunderland ShortBros.	1879	4 Blds. [illegible]	2Cy.36" & 68"–44" 300 H.P. J.Dickinson,Sunderland	J.S.Barwick&[illegible]
759 86748	*D	Ashdale 1 Sc3mSr Templeton 1 Dk	British Glasgow	131 291	145·0	22·6	9·9	Grangemouth Dobson&Charles	1882	3 Blds. [illegible]	2Cy.18"& 36"–28" 60 H.P. W.B.Lochhead&Co.,Glasgw	Clydesdale S.S.[illegible]
760 89592	*	Ashdell 1 ScSr Main 1 Dk2 B	British London	1135 1741	258·0	36·0	18·5	Sunderland S.ofEnglandS.B.Co.	1884	5 Blds. [illegible]	2Cy.33" & 62"–42" 200 H.P. J.Dickinson,Sunderland	Britannic Ste[illegible]ping Co.,Ld
761 84986	*D	Ashdene 1 ScSr Thompson 84-84 1 Dk2 B	British Sunderland	734 1178	230·1	33·0	16·3	Sunderland Osbourne,Graham&Co.	1881	4 Blds. [illegible]	2Cy.28" & 53"–36" 99 H.P. G.Clark,Sunderland	Bulman & D[illegible]
762 67540	*	Ashfield 1 ScSr Cornwall 80-87 2 Dks	British Stockton	994 1[illegible]	250·0	33·0	23·5	Stockton Richardsn,Duck&Co.	1878	5 Blds. [illegible] FPT APT	2Cy.31½" & 58"–36" 150 H.P. Blair&Co.,Ld.,Stockton	F.Binnington
763 84898	*	Ashford 1 Sc Andrews 82-82 1 Dk2 B	British Sunderland	1278 1964	276·0	37·5	19·6	Sunderland J.L.Thompson&Sns	1881	6 Blds. [illegible]	2Cy.34" & 64"–42" 180 H.P. J.Dickinson,Sunderland	Second Steam[illegible] Co.,Ld.,[illegible]
764	*	Ashington (ex Harry S. Edwards) 1 Sc3mSr Ziadel 1 Dk2 B	German Hamburg	809 1[illegible]	255·5	33·0	18·8	Sunderland Hill,Mounsey&Co.	1872	5 Blds. [illegible]	2Cy.31" & 57"–45" 180 H.P. R.Steele&Co.,Greenock	Sietossen & Co[illegible]
765 787		Ashland W ScSr Tunnell 2 Dks	American Philadelphia	487 762	196·5	30·2	18·8	Philadelphia	1863		1Cy.42"–36" 166 H.P. Neafie&Levy,Philadelphia	W.P.Clyde & [illegible]
766 86171	*	Ashleigh Brook 1 ScSr Fanquier 84-84 (ex Abana) 2 Dks3 B	British Liverpool	1875 2863	325·0	38·0	26·5	Stockton M.Pearse&Co.	1882	4 Blds. [illegible]	2Cy.40" & 75"–48" 300 H.P. Blair&Co.,Ld.,Stockton	"Ashleigh Br[illegible]" Co.,Ld.
767 82261		Ashley Eden 1 PadSp Ramage	British Glasgow	396 548	247·2	28·0	7·5	PortGlasgow R.Duncan&Co.	1873		2Cy.34" & 60"–54" 156 H.P. Rankin&Blackmore,Grnck.	IrrawaddyFl[illegible] Ld.
768 85145	*D	Ashton 1 ScBn McPhee [illegible] 1 Dk2 B	British London	1069 1[illegible]	257·4	34·6	17·8	Hartlepool E.Withy&Co.	1882	4 Blds. [illegible]	2Cy.32" & 62"–40" 150 H.P. Black,Hawthorn&Co.,Gtsd.	W.&E.S.L[illegible]
769 80365	* BT11.88	Ashton 1 ScSr Miller 83-87 1 Dk2 B	British Grimsby	650 1067	238·6	32·3	14·1	W. Hartlepool E.Withy&Co.	1884	5 Blds. [illegible]	3Cy.22"37"58"–36" 170 H.P. T.Richardson&Sons,Hrtlpl	Manchester S[illegible] Lincolnshire [illegible]
770 85113	*	Ashurst (FOUNDERED) 1 ScSr Rawson 1 Dk	British Dundee	93 1[illegible]	94·0	17·4	9·7	London Edwards&Symes	1881	3 Blds. WB FPT APT	2Cy.11" & 18"–22" 30 H.P. Hamilton,Woods&Co.,Mlm.	B.L.Nairn.
771 89081	BT5.88	Asia 1 ScSr Russell 80-88	British Glasgow	2323 3[illegible]	378·2	40·2	28·6	Glasgow D.&W.Henderson&Co.	1884	8 Blds.	2Cy.33" & 66"–40" 480 H.P. D.&W.Henderson&Co.,Glas.	Henderson Br[illegible]
772 52586		Asia 1 ScBg Stewart 84-88	British Glasgow	1139 1766	274·8	34·2	25·2	Dumbarton W.Denny&Bros.	1865		2Cy.44" & 78"–36" 200 H.P. Denny&Co.,Dumbarton	British India[illegible] Nav.Co.,Ld.
773 82474	*	Asia 1 ScSc Carlisle 1 Dk	British Hull	121 1[illegible]	131·0	20·0	10·8	Hull Earles'Co.	1880	3 Blds.	2Cy.21" & 40"–27" 60 H.P. Earles Co.,Ld.,Hull	Hull Steam F[illegible] Ice Co.,Ld.
774	*DV	Asia 1 ScSr Dipeup 77-77 1 Dk&Spr Dk3 B	Danish Copenhagen	995 1582	239·4	32·7	16·8 24·0	Kiel Nordd.Schiffb.A.G.	1873	5 Blds. [illegible]	2Cy.33" & 52"–36" 125 H.P. Schweffel&Co.,Kiel	Dampskibssel[illegible] "Kjobenhavn"
775	DG	Asia 1 ScSr Horstmann 11-83 1 Dk&Spr Dk	German Bremen	986 1[illegible]	268·4	34·4	24·9	Bremen Act. Ges. "Weser"	1882	[illegible]	2Cy.34½" & 59"–42" 198 H.P. Act.Ges. "Weser," Bremen	D.H.Wätjen&C[illegible]
776 12	*	Asia (ex Principe, ex Rio Negro, Calvert, ex Eugenio, ex Sein) 1 Sc3mSr 1 Dk&Spr Dk3 B	Italian Bari	1044 1[illegible]	278·7	30·0	17·5 24·4	Sestri Ponente F.Odero	1874	5 Blds. [illegible]	2Cy.27" & 47½"–27½" 140 H.P. F.Odero,Genoa	L.Scorcia & F[illegible]
777 39	I	Asia 1 Sc3mSr De Micheli 85-87 3 Dks&Awn Dk	Italian Genoa	897 1384	268·0	31·0	24·2	Newcastle Wighm Richardson &Co.	1870	5 Blds.	2Cy.28¾" & 54"–36" 268 H.P. Comp.R.&W.Hawthrn.Nwc	Navigazione G[illegible] Italiana
778 68812	*	Asiatic 1 ScSr Leigh 78-78 2 Dks3 B	British Southampton	1330 2[illegible]	299·9	34·2	23·3	Whitehaven WhitehvnS.B.Co.	1872	4 Blds.	2Cy.40" & 76"–36" 274 H.P. J.Jack,Rollo&Co.,Liverpool	H.Martini
779 85250	*	Asiatic Prince Stl ScSr Barge [illegible] 1 Dk W.[illegible]	British Newcastle	1416 2[illegible]	292·0	39·1	18·9	Sunderland ShortBros.	1888	6 Blds. [illegible]	3C.22"36"&59"–42" 200 H.P. J.Dickinson,Sunderland	J.Knott
780	N	Ask W S[illegible] Stoltz 87-87 1 Dk	Norweg'n Bergen	305 1[illegible]	150·7	24·1	14·4	Bergen Brunchorst&Dekke	1867		3Cy.11"18"29"–24" 45 I.P. Akers Mek.Vks.Christiania	J.C.Giertsen

*=Lloyd's Register; D=Nederlandsche Vereeniging; G=Germanischer Lloyd; I=Registro Italiano; R=American [illegible]
A=Veritas Austro-Ungarico; H=Veritas Ellenico; N=Norske Veritas; V=Bureau Ve[illegible]

In the 2nd Column, B.T., with figures following, signifies Date of report of bottom sighting by Board of Trade Surveyors as condition of PASSENGER CERT[illegible]

In the 3rd Column, C=Composite (iron frame, wood planked); I=Iron; Stl=Steel; W=Wood.

Further particulars of Vessels classed by Lloyd's Register will be found in Lloyd's Register Book and Lloyd's Yacht Register

	Nom, Matériaux, Grèement, etc. Capitaine et dates d'entrée service des Armateurs actuels et à bord du navire. Ancien Nom, nombre de ponts, etc.	Pavillon Port d'Armement	Tonnage Net. Total	Dimensions Longueur, Largeur, Creux Constructions sur le Pont	Construit Où. Par qui.	Quand.	Nombre des cloisons. Détails sur le lest d'eau.	Machines. Détails. Nom du Constructeur.	Armateurs ou Directeurs-Gérants.
	Askur C ScSr *Nordahl* -75 1 *Dk*	Swedish Stockholm	242 333	139·1 | 22·8 | 11·5 Q17 f. B26f. F24 f. H61	Oscarshamn SvenOlsson	1872		2Cy.19"& 33"–18½" 45 HP. BergsundsMek.Verkst.Sklm.	Nordström & Thulin
	Askurland ex-Soliel C ScSr *Andersson* -87 1 *Dk*	Swedish Strœmstad	121 163	100·2 | 21·2 | 9·2 Q41f. B26f. F12f. H21	Thorskog P.Larsson	1886			V.Vallin
✳	**Aslan** ex-Mirkezi Ticaret H I ScBn *Hussein Bey* -88	Turkish Constntinple	1491 2510	350·4 | 36·8 | 27·5	Glasgow Barclay,Curle&Co.	1872	5 Blds.	2Cy.50"&86"–48" 450 HP. Barclay,Curle&Co.,Glasg'w	Idarei Massousieh
✳	**Aspatria** I ScBq *Teale* 76-83 2 *Dks*	British W.Hartlepl.	1894 2907	325·0 | 40·0 | 26·5 P32 f. B30f. F37 f.	Stockton M.Pearse&Co.	1883	5 Blds. DB. 1/381	2Cy.42"& 77"–48" 250 HP. Blair&Co.Ld.,Stockton	Herskind & Woods
✳	**Assagai** I ScSrYt *Cabell* 1 *Dk*	British Salcombe	48 71	91·0 | 14·2 | 8·3	Southampton Day,Summers & Co.	1876		2Cy.10"20"–18" 20HP. Day,Summers&Co.Southamptn	R.P.Crozier
	Assaish W Pad *Frango* 61-	Turkish Cnstntinple.	140	160·0 | 21·0 | 10·0	London MoneyWigram	1865		90 HP. Maudslay,Sons&Field,Lnd.	Chirket-i-Hairie (Bosphorus Steam Nav.Co.)
	Assam ex-Fath-i-Mouchell Molke I ScBg *Bourne* 87-87	British Greenock	1611 3060	350·0 | 39·4 | 30·8	Greenock Caird&Co.	1873		2Cy.50"& 86"–54" 500 HP. Caird&Co.,Greenock	Peninsular & Oriental Steam Nav.Co.
	Assecuradeur I ScSp *Neyer* 84-88 1 *Dk*	German Bremen	66 172	110·5 | 21·1 | 12·5	Hamburg Reiherstieg Schiffs-Werfte	1871		2Cy.28"& 48"–30" 120 HP. Reiherstg.Schffs.Wrfte.,Hmg	Bugsirgesellschaft Union
I	**Assiria** I ScBg *Mastellotti* 84-88 1 *Dk&Spr Dk3 B*	Italian Palermo	1055 1670	276·5 | 30·1 | 25·2	Sestri Ponente N.Odero	1874	5 Blds. DB132f	2Cy.35"& 66"–30" 325 HP. N.Odero,SestriPon.	Navigazione Generale Italiana
✳	**Assistance** I ScTug *Lihou* -85 1 *Dk*	British Guernsey	20 91	96·7 | 17·3 | 8·7 B 87 f.	Paisley J.McArthur&Co.	1886	4 Blds.	2Cy.18"& 35"–20" 52 HP. Bow,McLachlan&Co.,Paisly	Guernsey Steam Towing & Trading Co.,Ld.
✳	**Assyria** I ScSr *Spittal* 75-83	British Glasgow	1249 2023	300·5 | 33·2 | 22·6 P105 f. F32 f. | 29·4	Port Glasgow R.Duncan&Co.	1871	5 Blds.	2Cy.38"& 68"–42" 424 HP. FinniestonCo.,Glasgow	Henderson Bros.
✳	**Assyria** I ScSr *Whitehead* 88-88 2 *Dks3 B*	British Glasgow	970 1495	269·8 | 30·6 | 24·2	Renfrew W.Simons&Co.	1872	5 Blds.	2Cy.35½"& 60"–39" 200 HP. W.Simons&Co.,Renfrew	British India Steam Nav.Co.,Ld.
✳ BT. 88	**Assyrian** ex-Assyrian Monarch Stl ScInSr *Dodds* -87 2 *Dks&Shlt Dk4 B*	British Glasgow	2608 3970	360·0 | 42·7 | 25·1 | 32·2	Hull Earle's Co.	1880	6 Blds. CDB300f 600t	2Cy.46"& 87"–57" 500 HP. Earle's Co.Ld.,Hull	J.& A. Allan
✳	**Aster** I ScSrYt	British Greenock	137 201	135·9 | 20·1 | 12·2	Greenock R.Steele&Co.	1883	4 Blds.	2Cy.18"& 36"–24" 55 HP. R.Steele&Co.,Greenock	Sir Wm.Collins
	Aston I ScSr *Davies* 84-84 1 *Dk*	British Chester	78 132	100·0 | 20·0 | 9·5	Newcastle Schlsngr.Davis	1867	3 Blds.	2Cy.18"–15" 25 HP. Thompson,Boyd&Co.,Newc.	Coppack, Carter & Co.
✳	**Aston Hall** I ScSr *Hill* 2 *Dks3 B*	British Liverpool	2333 3568	360·0 | 42·6 | 28·0 P49 f. B76 f. F50 f.	Dundee Gourlay Bros.&Co.	1882	5 Blds.	2Cy.40"& 76"–54" 400 HP. Gourlay Bros.&Co.,Dundee	Sun Shipping Co.,Ld.
	Astrachan Stl Sc3mSr 1 *Dk*		566 716	212·1 | 29·2 | 10·9	Stettin VulcanCo.	1880		1Cy.17½"&31½"–19¾" 40HP VulcanCo.,Stettin	
✳	**Astræa** Stl ScSr *Nicholas* 87-87 1 *Dk&Wb Frms*	British W.Hartlepl.	1390 2140	275·2 | 37·3 | 20·0 Q76f. P29f. B108f. F27f.	W.Hartlepool W.Gray&Co.	1886	4 Blds. DBf/485t	3C.21"35"&57"–39" 220HP. CentralMarineE.Co.W.Hpl.	Rickinson, Son & Co.
DV	**Astrea** (STRANDED) I ScSr *Potkoff* 77-87 2 *Dks*	Dutch Amsterdam	381 553	186·0 | 26·0 | 15·6	Stockton Richardson,Duck& Co.	1867		2Cy.23"& 43"–30" 90 HP. A.&J.Inglis,Glasgow	K'ninklijkeNederlndsche. Stoomboot Maatschappij
A	**Astrea** I Sc	Russian Taganrog	305 384	.. .. ..	Trieste Stblmento.Tecnico	1883		50 HP.	J.E.Komneno
✳	**Astrion** I ScSr *Tilker* 88-88 1 *Dk2 B*	British Hull	1046 1612	261·0 | 35·1 | 18·1 Q94 f. B62 f. F28 f.	Newcastle C.S.Swan&Hunter	1883	5 Blds. DB. f/298 AP.T.24	2Cy.32"& 62"–39" 170 HP. WallsendSlipwayCo.,Nwc.	W.Tulley & Co.
✳	**Astronom** ex-Kennett I Sc3mSr *Michelsen* 85-86 1 *Dk2 B*	German Hamburg	542 700	201·4 | 28·4 | 16·5 Q30f. B66f. F32f. H30f	Hull M.Samuelson&Co.	1863	4 Blds.	2Cy.26"& 48"–30" 90 HP. Humphreys&Pearson,Hull	H.J.Perlbach & Co.
	Astronomer I ScSr *Wallace* 71-85 2 *Dks3 B*	British Liverpool	2029 3075	365·2 | 40·0 | 26·6 P25 f. B 112 f. F42 f.	Glasgow Aitken&Mansel	1883	6 Blds.	2Cy.38"& 75"–54" 375 HP. J.&J.Thomson,Glasgow	Charente S.S.Co.,Ld.
✳	**Asturiano** ex-Nellie Martini I Sc *Guerricaechevarria* (LOST) 2 *Dks3 B*	Spanish Bilbao	1187 1804	275·0 | 34·2 | 24·8 F 34 f.	Middlesbro' R.Dixon&Co.	1875	4 Blds. DBa/455t	3Cy.21½"34" & 55"–36" Tripl.'87 D.Rollo&Sns.,L'pl	M.M.de Arrotegui
	Asturias I ScSrTg 1 *Dk*	Mexican Frontera	90 113	100·0 | 18·0 | 7·6 H 36 f.	Wlmngtn.Del Pusey&JonesCo.	1881		2Cy.12"& 21"–18" 25 HP. Pusey&JonesC.WlmgtnDel	Bulnes Herm.
	Asturias I Sc	Spanish Gijon	469 628	196·9 | 26·3 | 12·9 B45 f. Q33 f. F27 f.	Dundee Gourlay Bros.	1869	4 Blds.	2Cy.25"& 44"–30" 90 HP. Gourlay Bros.&Co.,Dundee.	Meliton,Gonzalez&Co.
	Atago Maru I Pad	Japanese Yokohama	1033 1640	309·9 | 36·1 | 22·3 Q 107 f.	London Samuda Bros.	1860		2Cy.72"–84" 400 HP. J.Penn&Sons,London	Nippon Yusen Kaisha
✳	**Atalanta** I ScSr *Granger* 1 *Dk2 B*	British W.Hartlepl.	743 1164	235·2 | 31·7 | 15·1 Q84 f. B58 f. F31 f.	W.Hartlepool W.Gray&Co.	1883	4 Blds. DBa/225t	2Cy.27"& 50"–33" 99 HP. T.Richardson&Sons,Hartpl.	J.S.Allison & Co.
	Atalanta I ScSp	British Bristol	90 113	118·2 | 19·7 | 7·5	Bristol G.K.Stothert&Co.	1884		1Cy.22"–18" 16 HP. G.K.Stothert&Co.,Bristol	Severn&CanalCarrying,Shippng&Stm.TowingCo.
BT. 89	**Atalanta** I Pad 1 *Dk*	British Grimsby	60 108	135·0 | 15·2 | 7·7	Gainsborough H.Smith&Son	1851		2Cy. 40 HP. J.Penn&Son,Greenwich	Gainsborough United Steam Packet Co.,Ld.

✳=Lloyd's Register; D=Nederlandsche Vereeniging; G=Germanischer Lloyd; I=Registro Italiano; R=American Record;
A=Veritas Austro-Ungarico; H=Veritas Ellenico; N=Norske Veritas; V=Bureau Veritas.

Dans la 2me Colonne, B.T., avec les indications qui suivent, signifie la date du rapport de visite de la Carène par les experts du Board of Trade pour obtenir le certificat relatif au TRANSPORT DES PASSAGERS.

Dans la 3me Colonne, C=Composite (membrure en fer, bordé en bois); I=Fer; Stl=Acier; W=Bois.

Pour plus amples renseignements sur les navires classés par le Lloyd's Register, voir le Lloyd's Register Book et le Lloyd's Yacht Register.

No. in Book. / Official No.	Class	Ship's Name. Master, and Dates of Appointment to present Owner's Service and to Ship / Ship's Late Name.	Material, Rig, &c. / No. of Decks, &c.	Flag. / Port of Registry.	Tons. Net / Gross	Length	Breadth	Depth	Built. Where. / By Whom.	When	No. of Bulkheads. / Particulars of Water Ballast.	Engines. Particulars. / Maker's Name.	Owners
811 106173		**Atalanta** *Shackford*	I ScYt	American New York	264 368	235·0	26·5	15·3	Philadelphia W.Cramp&Sons	1888		2Cy.30"& 60"–30" 1000 IP. WmCramp&Sns,Philadelph.	Jay Gould
812	V	**Atalanta**	I ScBg 1 *Dk2 H*	Japanese Tokio	788 [illegible]	229·6	30·6	16·9	W.Hartlepool Denton,Gray&Co.	1872	4 Bhds.	2Cy.27"& 50"–33" 110 IP. T.Richardson&Sons,Hartpl.	Cement Actie... seischaft
813	V	**Atalanta** *Ekstrom 81-81*	I ScSr 1 *Dk2 B*	Swedish Stockholm	767 [illegible]	209·6	28·5	17·2 *P 122 f. F 23 f. H 131 f.*	Stockholm BergsundM.Works	1872		2C 18⅜"& 41⅞"–26" 120 IP. BergsundM.Works,Stockholm	Stockholms A... Rederibol...
814 87898		**Ataran** *Tinniswood*	Stl TwnScBarge	British Glasgow	140 266	140·5	25·0	7·7	PortGlasgow R.Duncan&Co.	1883		4Cy.18"& 8½"–21" 25 IP. Denny&Co.,Dumbarton	Irrawaddy F... Co.Ld.
815 84185	✳	**Athabasca** *Davies*	I ScBn 2 *Dks3 H*	British Liverpool	1645 2147	299·6	38·6	26·9 *F 30 f.*	Stockton M.Pearse&Co.	1881	4 Bhds.	2Cy.38"& 71"–45" 300 IP. Blair&Co.Ld.,Stockton	Tapscott S.S.C...
816 85794		**Athabaska** *Foote 84-84*	Stl ScSr 2 *Dks*	British [illegible]	1148 1773	262·8	38·2	23·3	Glasgow Aitken&Mansel	1883		2Cy.35"& 70"–48" 300 IP. D.Rowan,Glasgow	Canadian P... Railway Co.
817 513		**Athan** *Stathatos*	I Sc	Greek Syra	1170	264·1	35·1	..	England	1888			Stathatos Br...
818 65436	✳	**Athelstane** *Foreman 82-82*	I ScSr 1 *Dk2 H*	British North Shields	709 1122	225·5	30·7	17·4 *P 112 f. F 37 f.*	Newcastle A.Leslie&Co.	1872	5 Bhds.	2Cy.26"& 52"–30" 99 IP. Thompson,Boyd&Co.,Ncl.	H.E.P.Adam...
819 82125	✳	**Athenian** *Bainbridge 77-87*	I Sc3mSr 3 *Dks*	British Southampton	2428 3782	365·0	45·8	29·0 *B 102 f. F 42*	Glasgow Aitken&Mansel	1881	8 Bhds.	3C 36"58"&94"–54" 600 IP. Tripl.Ex.,T.Richrdsn.&Sons,Hpl	Union Steam...
820 70973		**Athenian** *Prior 86-87*	I ScSr	British Liverpool	1030 1579	301·6	31·2	23·3 *F 30 f.*	Liverpool T.Royden&Sons	1875		2Cy.22½"&60"–36" 150 IP. J.Jack,Rollo&Co.,Liverpool	S.S."Athenian" (F.Leyland &...)
821 77089	✳	**Athens** *Robertson 82-82*	I Sc3mSr 2 *Dks3 B*	British London	1494 2343	315·9	35·7	26·1 *P 44 f. B 68 f. F 45 f.*	Newcastle C.Mitchell&Co.	1877	5 Bhds.	3Cy.21" 33½" & 55"–45" Tripled '88, Wallsend Slipway Co.,Wallsend	Nelson, Donk...
822 120		**Athens** (ex Elizabeth Martin) *Satiari 87-88*	I Sc3mSr 2 *Dks3 B*	Greek Piraeus	809 1246	250·6	30·2	21·5	Glasgow R.Napier&Sons	1872	6 Bhds.	2Cy.30"& 54"–36" 130 IP. R.Napier&Sons,Glasgow	Panhellenic S... Ship Naviga...
823		**Athinai**	I Pad	Greek Syra	510	..	..	..	Greece	1837			Hellenic S.S.C...
824		**Athlet** *Timm 78-84*	I ScCr 1 *Dk*	German Hamburg	29 110	85·3	18·7	11·9	Hamburg Janssen&Schmilinsky	1880		2Cy.18"& 30"–25" 75 IP. Janssen&Schmilinsky	Vereinigte Bu... Dampfschiff Co...
825 53394	BT 1.88	**Athole** *Brown 86-86*	I PadSp	British Glasgow	41 [illegible]	192·1	18·5	7·8	Glasgow Barclay,Curle&Co.	1866		1Cy.48"–48" 80 IP. Barclay,Curle&Co.,Glasgow	A.M'Lean
826 81322	✳	**Athos** *Law 74-80*	I ScSr 2 *Dks*	British Liverpool	1262 1944	290·2	35·2	23·2 *B 85 f. F 34 f.*	Glasgow C.Connell&Co.	1879	5 Bhds. FPT APT	2Cy.36"&68"–48" 275 IP. J.&J.Thomson,Glasgow	Atlas Steam S...
827		**Atjeh** (ex Glenfarne) *Bouwbijs*	I ScBg 2 *Dks*	Dutch Batavia	625 1312	272·8	33·1	24·2	Glasgow A.Stephen&Sons	1873		2Cy.34"& 60"–36" 160 IP. J.&J. Thomson,Glasgow	Nederland.-Ind... Stoom.-Maats...
828 18781	✳	**Atlantic** (now Kaplan) *Musserich*	I ScBg	Turkish Constantinople	993 1420	254·6	30·4	24·1	Hull Earle&Co.	1857	4 Bhds	2Cy.30"& 53"–33" 150 IP. T.Brassey&Co.,Birkenhead	Idarei Massous... (Ottoman State...)
829 85491		**Atlantic**	W Sc 2 *Dks*	British Colingwd,Ont.	442 683	147·0	30·0	11·0 *P 176 f.*	OwenSound J.Simpson	1883		1Cy.36"–34" 75 IP. G.N.Oliver,St.Catharine'sOnt	Great Northern Transit Co.
830	✳	**Atlantique** (ex Alsace) *Jonas 81-86*	I ScBn 2 *Dks3 B*	French Havre	1917 2434	320·0	36·2	25·5 *P 28 f. F 38 f.*	Glasgow C.Connell&Co.	1881	4 Bhds.	2Cy.36"& 68"–42" 300 IP. J.&J.Thomson,Glasgow	Cie Maritime du Pacifique
831	V	**Atlantique** *Marisen 82-82*	I Sc3mSr 1 *Dk2 B*	French La Rochelle	605 911	209·9	30·2	17·8 *P 83 f. F 35 f.*	Nantes Jollet&Babin	1879	[illegible]	3C.19¾"&28¾"44"–27½" 100 IP. Jollet&Babin,Nantes	D'Orbigny A... fils
832 89818	✳	**Atlantis** *Wandless 83-84*	Stl ScSr 2 *Dks*	British London	950 1476	280·0	33·1	19·2 *P 41 f. B 52 f. F 34 f.*	Glasgow J.&G.Thomson	1884	4 Bhds. APT 311	2Cy.30"& 56"–36" 131 IP. J.&G.Thomson,Glasgow	"Atlantis" S.S... (Scrutton Sons &...)
833 28477	BT 10.88	**Atlas** *Dawson 86-88*	I ScBk	British Glasgow	1553 2393	339·0	36·5	25·7 *F 53 f.*	Glasgow J.&G.Thomson	1860		2Cy.44"& 78"–42" 300 IP. J.&G.Thomson,Glasgow	Cunard Steam... Co.,Ld.
834 69282		**Atlas** (ex Antillas, ex Atlas) *Tobin 87-87*	I ScSr	British Liverpool	902 1187	250·3	29·0	17·7 *Q 140 f. F 35 f.*	Dumbarton M'Kellar,M'Millan & Co.	1873		2Cy.30"& 60"–36" 150 IP. Rait&Lindsay,Glasgow	Atlas S.S.Co.,Ld.
835 91280		**Atlas** *Manley 85-85*	I TwnSc	British Liverpool	217 457	131·0	48·3	9·0	Newcastle SirW.G.Armstrong Mitchell&Co.Ld.	1884		4Cy.[illegible] 43 IP. Ross&Duncan,Glasgow	Mersey Docks & Harbour Board
836 1134		**Atlas**	W Pad	American [illegible]	[illegible]	243·5	58·0	10·6	Hoboken,N.J.	1852			J.S.Buckalew
837		**Atlas** *Kruse 82*		German Hamburg	209	115·0	20·6	13·9	Hamburg	1886			Vereinigte-Bu... Dampfschiffahrt...
838	V	**Atlas** *Andersson 88*	C ScSr 1 *Dk2 B*	Swedish Stockholm	378 599	157·8	24·7	15·9 *B 45 f. H 41*	Westervik C.R.Pridell	1873		2Cy.17"& 33"–23⅝" 60 IP. BergsundM.Vrks,Stockhlm.	A.Broström
839 95512	✳ BT 8.88	**Atrato** *Dickenson 78-88*	Stl Sc3mSr 3 *Dks&Sp Dk*	British London	2865 4140	431·2	50·0	25·4 33·4	Glasgow R.Napier&Sons	1888	9 Bhds.	3Cy.42"66"&92"–66" 1000 IP. R.Napier&Sons,Gla.	Royal Mail Steam Packet Co.
840 27334	✳	**Atrato** *Gordon 79-83*	I ScSr 1 *Dk&Spr Dk3 B*	British Sunderland	2056 3122	329·8	42·4	28·6 *B 7 f.*	Sunderland J.Laing	1883	9 Bhds.	2Cy.35"& 74"–45" 350 IP. G.Clark,Sunderland	The Pinkney &... S.S.Co.,Ld.

✳ = **Lloyd's Register**; D = Nederlandsche Vereeniging; G = Germanischer Lloyd; I = Registro Italiano; R = American ...; A = Veritas Austro-Ungarico; H = Veritas Ellenico; N = Norske Veritas; V = Bureau Veritas.

In the 2nd Column, B.T. with figures following, signifies Date of report of bottom sighting by Board of Trade Surveyors as condition of PASSENGER CERT...

In the 3rd Column, C = Composite (iron frame, wood planked); I = Iron; Stl = Steel; W = Wood.

Further particulars of Vessels classed by Lloyd's Register will be found in Lloyd's Register Book and Lloyd's Yacht Register.

Registre dans Lloyd's	Nom. Capitaine et dates d'entrée au service des Armateurs actuels et à bord du navire. Ancien Nom, nombre de ponts, etc.	Matériaux, Gréement, etc.	Pavillon. Port d'Armement	Tonnage. Net. Total	Dimensions. Long. Larg. Creux	Construit. Où. Par qui	Quand.	Nombre des cloisons. Détails sur le lest d'eau.	Machines. Détails. Nom du Constructeur.	Armateurs ou Directeurs-Gérants.
	Atrek	I ScSr	Russian St Petersburg	247 322	142·5 | 23·5 | 10·5	Rostock RostockShipB.Co.	1880		4Cy.12¼" & 26½"–17½"	Steam Ship Co. "Lebed."
	Attalo *Guy 86-88*	I ScSr 1 *Dk2 B*	French Saïgon	215 363	173·7 | 21·9 | 13·3 *Q 30 f.*	Glasgow T.Wingate&Co.	1869	1 Blhds.	2Cy.21½"&36½"–30" 85 HP. T.Wingate&Co.,Glasgow	Messageries Fluviales de Cochinchine
	Auby No. 1	I	French Brest	83 103	.. .. ..	St.Denis	1888			Société Générale des Forges et Chantiers
*	Auckland *Smith 87-88*	I ScSr 1 *Dk2 B*	British Newcastle	542 850	209·5 | 31·1 | 14·3 *Q 100 f. B 25 f. F 24 f.*	Dundee PearceBros.	1883	1 Blhds. DBcf/232 FPT22t APT8t	2Cy.25"&49"–33" 99 HP. PearceBros.,Dundee	Hannay,Boyd & Co.
*	Auckland *Wilkin -87*	Stl ScSr 1 *Dk2 B*	British Hull	91 210	132·4 | 22·0 | 10·8	Hull Earle'sS.B&EngCo.Ld.	1887	5Blhds.	2Cy.18"&36"–21" 45 HP. Earle'sCo.,Ld.,Hull	Humber Conservancy Commissioners
*	Auckland Castle *Rowlands*	I ScSr 1 *Dk*	British Newport	714 1130	225·0 | 32·1 | 16·0 *Q 56 f. B 48 f. F 26 f.*	Sunderland S.P.Austin&Son	1883	1 Blhds. DBcf/301t APT30t	2Cy.28"&53"–36" 99 HP. G.Clark,Sunderland	Mordey,Carney & Co.
*	Aude *ex Life Guard* *Puginier 77-78*	I ScSr 1 *Dk*	French Marseilles	106 184	125·5 | 20·2 | 9·0 *Q 41 f. F 18 f.*	Glasgow A.&J.Inglis	1879	3 Blhds. FPT	2Cy.16¼"&28"–22" 40 HP. A.&J.Inglis,Glasgow	Fraissinet & Co.
G	August *Iwffs 75-82*	I ScSr 1 *Dk*	German Kiel	366 458	162·4 | 22·6 | 14·9	Kiel G.Howaldt	1881	WB	2Cy.20"&36"–28" 60 HP. SchweffelHowaldt,Kiel	Sartori & Berger
V	August *Bolle 81-84*	I ScSr 1 *Dk*	German Bremen	342 536	149·9 | 25·3 | 17·3 *B 41 f. F 19 f.*	Vegesack J.Lange	1883	PT80t APT6	2Cy.22"&42"–22" 50 HP. J.Lange,Vegesack	J.D.Bischoff
*N	August *Isberg 81-81*	I ScSr 2 *Dks3 B*	Norweg'n Bergen	1026 1599	260·7 | 33·5 | 23·7 *B 19 f. F 30 f.*	Middlesbro' R.Dixon&Co.	1875	1 Blhds. MT40t	2Cy.33"&61"–33" 160 HP. T.Richardson&Sons,Hartpl.	Jacob Christensen
*	Augusta *Peterson 87-87*	I Sc3mSr 1 *Dk2 B*	British Newcastle	568 902	221·0 | 28·0 | 17·2 *Q 56 f. F 30 f.*	Newcastle Palmers'Co.	1872	4 Blhds. DB 24t	2Cy.24"&48"–30" 90 HP. Palmers'Co.,Ld.,Newcastle	R.Gordon & Co.
*	Augusta *Nash 86-86*	I ScTg 1 *Dk*	British Newport	12 41	60·5 | 13·5 | 8·5	Newport UskS.B.Co.,Ld.	1884	1 Blhds.	2Cy.13"&24"–12" 30 HP. Forges&C.delaMedit.,Hav. Made '81	Mordey,Carney & Co.
V	Augusta *(ex Astrid, ex Sedan)* *Madsen 88-88*	I ScSr 1 *Dk*	Danish Thisted	75 132	93·1 | 18·7 | 9·5 *P 12 f. B 26 f. F 13 f.*	Kiel Norddl.Schiffsb.A.G.	1871	FPT10t	2Cy.13"&21"–18" 25 HP. Schweffel&Howaldt,Kiel	M.Lund
G	Augusta *Woortman 86-*	Stl Pad	German Leer	75 119	124·0 | 16·5 | 7·4	Papenburg J.L.Meyer	1886		2Cy.19"&31"–33"	Leer Dampfschiffs-Gesellschaft
N	Augusta *Johannesen 77-83*	W ScSr 1 *Dk2 B*	Norweg'n Bergen	211 289	128·1 | 23·3 | 11·6 *Q 37 f. B 41 f.*	Bergen S.K.Flage	1882		2Cy.17"&32"–22" 40 HP. LaxevaagsVerk.,Bergen	Albert Gran
G	Auguste *Arp 71-83*	I ScSr 1 *Dk*	German Kiel	285 452	161·7 | 22·4 | 14·8	Kiel G.Howaldt	1879	WB	2Cy.20"&36"–21" 60 HP. Schweffel&Howaldt	Sartori & Berger
*	Auguste Conseil *Rolier (ex Lady Tredegar)*	I Sc 1 *Dk2 B*	French Bordeaux	483 732	198·2 | 28·4 | 15·1 *Q 67 f.*	Middlesbro' W.Harkess	1874	1 Blhds. DBakf	2Cy.27"&50"–30" 99 HP. G.Clark,Sunderland	Les Fils de Th.Conseil
G	Auguste Victoria *Petersen 66-81 (ex Topaze, ex Ville de Bruges)*	I PadSr 2 *Dks*	German Kiel	173 242	167·6 | 20·8 | 10·2	Antwerp Soc.J.Cockerill	1847		3Cy.(2)17¾"&(1)22"–42" 100 HP. G.Howaldt,Kiel	Sartori & Berger
	Augustine *Harwood 82-84*	I ScSr	British Liverpool	814 1106	213·5 | 29·6 | 25·1	Liverpool Hart&Sinnott	1865		2Cy.21"&48"–40" 94 HP. J.Taylor&Co.,Birkenhead	Booth Steam Ship Co.,Ld.
	Augusto	S	American Stmfd.Conn.	68 136	.. .. ..	Philadelphia	1884			W.de Puy
	Augusto		Brazilian Para	203	.. .. ..				70 HP.	J.Ginny
*	Augusto	Stl ScSr *Sp Dk*	Spanish Bilbao	1161 1810	275·0 | 36·0 | 13·1 *F 26 f.* ; 21·2	Sunderland W.Pickersgill&Sons	1889		3Cy.21"35"&57½"–39" 180 HP. Black,Hawthorn&Co., Gateshead	Compania Bilbaina de Navegacion
N	Aukathor *Brown 83-85*	W ScSr 1 *Dk2 B*	Norweg'n Bergen	344 462	147·7 | 24·7 | 14·4 *Q 49 f. B 34 f. F 17 f.*	Bergen S.K.Flage	1885		2Cy.16"&31"–25" 40 HP. BergensMekVerks.,Bergen	P.G.Pettersen
	Auld Reekie *Primrose 70-82*	I Pad	British Granton	103 163	141·5 | 19·1 | 9·1	London	1847		2Cy.34"–33" 68 HP. Miller&Ravenhill,London	North British Railway Co.
*	Aunis *ex Poitou* *Govin 86-86*	I ScSr 1 *Dk2 B*	French La Rochelle	693 1040	235·3 | 32·4 | 14·7 *Q 84 f. B 42 f. F 28 f.*	Sunderland A.Simey&Co.	1877	5 Blhds. DBakf	2Cy.28"&52"–36" 110 HP. G.Clark&Co.,Sunderland	Delmas Frères
	Aura J. P.	Sr	Russian Åbo	154 408	192·7 | 23·0 | 12·6	Nyköping Mek.Verkstad	1858		150 HP.	Nya Ångf.Aktie-bolaget
	Aurania *Haines 67-83*	Stl ScBk	British Liverpool	4030 7269	470·0 | 57·2 | 37·2 *F 102 f.*	Glasgow J.&G.Thomson	1883		3C68"&(2)91"–72" 1500 HP. J.&G.Thomson,Glasgow	Cunard Steam Ship Co.,Ld.
*	Auretta *Read 83-83*	I ScSr 2 *Dks3 B*	British London	1729 2634	301·0 | 38·2 | 26·6 *P 29 f. B 62 f. F 36 f.*	Newcastle Palmers'Co.	1883	7 Blhds. DBfaf/457t FPT80t APT38t MT40t	2Cy.36"&68"–45" 280 HP. Palmers'Co.,Ld.,Newcastle	Oriental S.S.Co.,Ld.
*	Auric *Nelson 84-84*	I ScSr 1 *Dk*	British Belfast	222 424	173·5 | 25·2 | 12·2 *Q 21 f. B 41 f. P 30 f.*	Belfast WorkmanClark&Co.	1884	5 Blhds. FPT30t APT20t	2Cy.21"&42"–30" 74 HP. V.Coates&Co.,Ld.,Belfast	H.J.Scott & Co.
*	Auricula *Ross 87-87*	I ScSr 1 *Dk2 B*	British Sunderland	899 1384	240·0 | 33·2 | 16·2 *P 146 f. F 28 f.*	Sunderland J.Blumer&Co.	1887	4 Blhds. DB5f/233t	2Cy.28"&53"–36" 120 HP. J.W.&F.Wilson&Co.,Sunderland	J.Blumer & Co.

* = Lloyd's Register; D = Nederlandsche Vereeniging; G = Germanischer Lloyd; I = Registro Italiano; R = American Record;
A = Veritas Austro-Ungarico; H = Veritas Ellenico; N = Norske Veritas; V = Bureau Veritas.

Dans la 2me Colonne, B.T., avec les indications qui suivent, signifie la date du rapport de visite de la Carène par les experts du Board of Trade pour obtenir le certificat relatif au TRANSPORT DES PASSAGERS.

Dans la 3me Colonne, C = Composite (membrure en fer, bordé en bois); I = Fer; Stl = Acier; W = Bois.

Pour plus amples renseignements sur les navires classés par le Lloyd's Register, voir le Lloyd's Register Book et le Lloyd's Yacht Register.

No. in Book. — Official No.	Registry in which classed.	Ship's Name. Material, Rig, &c. — Master, and Dates of Appointment to present Owner's Service and to Ship — Ship's Late Name. No. of Decks, &c.	Flag. — Port of Registry.	Tons. Net. — Gross.	Dimensions. Length	Breadth	Depth	Built. Where. — By Whom.	When	No. of Bulkheads. — Particulars of Water Ballast.	Engines. — Particulars. Maker's Name.	Owners or M...
871 75184	✻	**Aurora** I ScSr *Archer*	British Dundee	823 1268	244·1 *P 241 f.*	31·4 *B 56 f.*	17·8	Dundee Gourlay Bros.	1876	5 Blhds.	2Cy.31"& 54"–36" 140 HP. Gourlay Bros&Co.,Dundee	J.& R.Mudie
872 75186		**Aurora** W ScBk *Fairweather 81-82*	British Dundee	386 580	165·2	30·6 *Q 32 f.*	18·9	Dundee Stephen	1876		2Cy.25"& 48"–30" 98 HP. Cunliffe&Dunlop,PGlasgow	A.Stephen & ...
873 93317		**Aurora** Stl PadlSp *J. Dumbreck* 3 Dks	British Glasgow	267 400	175·2 *Q 36 f.*	27·1 *F 33 f.*	7·6	Dumbarton W.Denny&Bros.	1886		2Cy.21"& 37"–42" 60 HP. Denny&Co.,Dumbarton	La Platense F... Co.
874 106369		**Aurora**	American	329 406	145·0	30·0	7·5	San Francisco	1885		165 HP.	A.Nelson
875 128		**Aurora** I Sc *Gelcich 58-87* 2 Dks	Austrian Trieste	1175 1824	274·5	34·0	18·7	Dumbarton W.Denny&Bros.	1869		1Cy.29¾"&57"–39"400 HP. Denny&Co.,Dumbarton	Lloyd Austr... rico
876		**Aurora** I Sc3mSr *Baltzersen 67-67* 1 Dk	Danish Copenhagen	118 198	144·8 *Q 14 f.*	20·3 *B 8 f.*	9·9	Renfrew J.Henderson&Sons	18⁵⁸⁄₇₅		1Cy.15"& 30"–20" 60 HP. Lindholmens Wks.,Gothenburg	Det Forenede ... kibs Selskab ...
877		**Aurora** (THE BROKEN UP) StlPadlSr *(ex Stad Middelburg)*	Dutch Amsterdam	779 1498	274·1	33·1	23·8	Liverpool Jones Quiggin&Co.	1865		2Cy.64"–78" 300 HP. J.Jack,Rollo&Co.,Liverpool	Stoomvaart M... schappij Zee...
878 115	✻	**Aurora** I Sc3mSr *Viča* 2 Dks	Italian Genoa	450 714	210·8	28·3 *Q 52 f.*	17·0	Newcastle Palmer&Co.	1863	4 Blhds. [illegible]	2Cy.26"& 51"–36" 160 HP. Palmers'Co.,Newcastle	E.Zenoglio
879	✻N	**Aurora** I ScSr *Engelsen* 1 Dk & Spr Dk 3 B	Norweg'n Bergen	883 1203	230·4	32·8	16·2 22·6	Bergen Bergens Mek.Verks	1874	5 Blhds. [illegible]	2Cy.30"& 60"–34" 140 HP. Bergens Mek.Verks,Bergen	Brødrene Kal...
880	G	**Aurora** I ScSr *Olsson —88* 1 Dk	Swedish Gothenburg	193 263	111·9 *B 27 f.*	21·2 *H 4 f.*	12·7	Gothenburg Götheborgs Mek.Verkst	1881		2Cy.12"& 26"–17" 40 HP.	Angf. Aktie-... "Aurora" (T...
881 84334	✻D	**Austerlitz** I ScSr *Frazier 80-87* 1 Dk 2 B	British Glasgow	1076 1653	260·0 *Q 52 f.*	35·3 *B 66 f.*	19·4 *F 31 f.*	Aberdeen Hall,Russell&Co.	1881	5 Blhds. [illegible]	2Cy.32"& 61"–39" 165 HP. Hall,Russell&Co.,Aberdeen	A.C.Gow & Co.
882 44655	✻	**Austin Friars** I Sc3mSr *Brooker 87-87* 2 Dks	British London	839 1314	250·0	30·3	23·3	Sunderland [illegible]	1868	4 Blhds.	2Cy.27"& 50"–30" 98 HP. T.Richardson&Sons,Hartpl.	Galbraith, Pe... & Co.
883 85923	✻ BT12,88	**Austral** StlSc1mSr *Ruthven 81-87* 4 Dks	British Glasgow	3214 5524	456·0	48·2	33·9	Glasgow J.Elder&Co.	1881	[illegible]	3C 62"&(2)86"–60"1000 HP. J.Elder&Co.,Glasgow	Orient Steam... Co.,Ld.
884 88659	✻ BT1,89	**Australasian** I ScBk *Simpson 84-84* 2 Dks & Spr Dk	British Aberdeen	2343 3630	361·6 *B 34 f.*	44·2 *B 167 f.*	21·3 29·1	Glasgow R.Napier&Sons	1884	6 Blhds. DB 2404 A.P.T.80	3Cy.32"46"& 70"–54" 400 HP. R.Napier&Sons,Glasgow	G.Thompson ...
885 63614		**Australia** I ScBk	British Greenock	2085 3047	381·9	41·7	33·0	Greenock [illegible]	1870		2Cy.62"& 96"–54" 600 HP. Caird&Co.,Greenock	Peninsular & O... Steam Nav. ...
886 63750	BT7,89	**Australia** I ScS *McRitchie 74-84* 2 Dks & Awn Dk	British Glasgow	1455 2252	324·6	35·2	22·5 20·3	Port Glasgow R.Duncan&Co.	1870	5 Blhds.	2Cy.41¼"& 73"–42"170 HP. Comp.77, D.& W.Henderson&Co.,Glasgow	Henderson Br...
887 89920	✻	**Australia** Stl TwnScSr *Crawthorn* 1 Dk & Awn Dk	British Dunedin	260 459	150·4	26·1	8·9 15·6	Dumbarton Burrell&Sons	1884	4 Blhds. FPT 28 APT 18	4Cy.16"& 30"–22" 77 HP. M.Paul&Co.,Dumbarton	Union S.S.Co... Zealand
888 82510	✻	**Australia** I ScK *Calvert* 1 Dk	British Hull	146 238	135·2 *Q 18 f.*	22·6 *F 23 f.*	11·0	Hull Earles Co.Ld.	1882	4 Blhds.	2Cy.21"& 40"–27" 60 HP. Earles Co.,Ld.,Hull	Hull Steam Fi... Ice Co.,Ld.
889 3827		**Australia** W ScSr	British Liverpool	87 128	85·5	20·9	8·8	Northwich C.Gibson	1877		2Cy.14"& 24"–12" 28 HP. [illegible] Friendenstal&Co.	J.Verdin
890 83685		**Australia** I ScSr *Glanville 84-87*	British London	18 128	90·4	20·6	11·3	Kinderdyk L.Smit&Zoon	1882		2Cy.22"–22" 70 HP. Appleby Bros.,E.Greenwich	W.Watkins
891	✻V	**Australia** I ScSr *Bauer 84-88* 2 Dks 3 B	German Hamburg	1603 2119	297·9 *P 25 f.*	37·1 *B 30 f.*	22·9 *F 30 f.*	Newcastle C.Mitchell&Co.	1881	4 Blhds. [illegible]	2Cy.31½"& 64"–42"250 HP. Blair&Co.,Stockton	Hamburg-Am... Packetfahrt Actien...
892	DG	**Australia** I ScSr *Basson 69-84* 1 Dk & Spr Dk	German Bremen	1619 2134	289·7	37·7	25·8	Bremen Act.Ges."Weser"	1884	4 Blhds.	2Cy.36½"& 66"–42" 300 HP. Act.Ges."Weser,"Bremen	D.H.Wätjen ...
893	✻	**Australia** I Sc1mSr *Houdlette 81-86* 2 Dks & Spr Dk	Hawaiian Honolulu	1715 2737	376·9	37·4	18·7 26·3	Glasgow J.Elder&Co.	1875	5 Blhds.	3C 43"&62"(2)–51"500 HP. J.Elder&Co.,Glasgow	Oceanic S.S.Co.
894 58896	BT5,88	**Australian** I ScBg *Peter 64-73*	British Liverpool	1699 2499	324·5	39·8 *P 52 f.*	26·4	W.Hartlepool Pile,Spence&Co.	1867		1Cy.26"& 58"–48" 350 HP. Forrester&Co.,Liverpool	West India & P... Steam Ship Co.
895 78092		**Australian** I ScSr 1 Dk	British Sydney	232 363	160·3 *Q 61 f.*	22·5 *B 30 f.*	10·5 *F 21 f.*	Dundee Gourlay Bros.&Co.	1879	5 Blhds. [illegible]	2Cy.21"& 36"–24" 70 HP. Gourlay Bros.&Co.,Dundee	John See
896 75129	✻	**Australind** I ScSr *Abrahamson 83-88* 1 Dk & Sd Dk 3 B	British Fremantl W.A.	554 1010	224·8 *P 90 f.*	32·0 *B 30 f.*	14·6 *F 42 f.*	Port Glasgow Blackwd.&Gordon	1886	4 Blhds. DB 4 ... APT 98 FPT 20	3C 19"30"&50"–36"125 HP. Blackwood&Gordon,P.Glas.	West Australia... Nav. Co., Ld.
897 87883		**Austria** W Sc	British [illegible]	91 140	86·6	20·7	9·7	Northwich C.Gibson	1880		2Cy.9"–12" 18 HP. Bates,Northwich	R.Verdin
898 6		**Austria** I ScBn *Cattarinich 69-87* 2 Dks	Austrian Trieste	1097 1697	276·0	35·0	24·6	Trieste Austro Ungarico Lloyd	1865		1Cy.29½"& 54"–36" 400 HP. Denny&Co.,Dumbarton	Lloyd Austro-U... rico
899 56190	BT10,88	**Austrian** I ScBk *Bentley —81*	British Glasgow	1681 2682	319·0	38·5	32·5	Glasgow Barclay,Curle&Co.	1867		3Cy.23"39"64"–42"220 HP. Tpl 88,J.Howden&C.,Glgw	J.& A.Allan
900 81327		**Author** I ScSr *Everett 81-87*	British Liverpool	838 1339	250·8 *P 35 f.*	32·1 *B 18 f.*	21·2 *F 35 f.*	Newcastle C.Mitchell&Co.	1880		2Cy.28"& 56"–39" 170 HP. Wallsend Slipway Co.,Nwcl.	Charente S.S. ...

✻ = **Lloyd's Register**; D = Nederlandsche Vereeniging; G = Germanischer Lloyd; I = Registro Italiano; R = American Re...
A = Veritas Austro-Ungarico; H = Veritas Ellenico; N = Norske Veritas; V = Bureau Veri...

In the 2nd Column, B.T., with figures following, signifies Date of report of bottom sighting by Board of Trade Surveyors as condition of PASSENGER CERTIF...

In the 3rd Column, C = Composite (iron frame, wood planked); I = Iron; Stl = Steel; W = Wood

Further particulars of Vessels classed by Lloyd's Register will be found in Lloyd's Register Book and Lloyd's Yacht Register.

	1	2. Nom, Matériaux, Gréement, etc. Capitaine et dates d'entrée au service des Armateurs actuels et à bord du navire. Nom, nombre de ponts, etc.		4. Pavillon. Port d'Armement	5. Tonnage. Net. Total	6. Dimensions. Long.	Larg.	Creux	7. Construit. Où. Par qui.	8. Quand.	9. Nombre des cloisons. Détails sur le lest d'eau.	10. Machines. Détails. Nom du Constructeur.	11. Armateurs ou Directeurs-Gérants.
901	*D	**Autocrat** *Brownleg*	I ScSr 1 *Dk & Spr Dk 3 B*	British Newcastle	996 1539	262·0	33·1	16·8 23·7	Sunderland W.Doxford&Sons	1875	4 Blhds. DBa91t MT302t	2Cy.29½"&55"-36" 170 HP. G.Clark,Sunderland	W.Forster
902	*	**Auvergne** ex ... *Langrais 77-86*	I ScSr 1 *Dk & Spr Dk 3 B*	French Marseilles	984 1472	254·9	32·0	17·2 23·8	Dundee W.B.Thompson	1879	4 Blhds. DBa ... FPT ...	2Cy.35"& 70"-39" 250 HP. W.B.Thompson,Dundee	Société Générale de Transports Maritimes
903		**Auxilia** ex Ríanes *Bilbao 84-4*	I Pad 1 *Dk*	Spanish Bilbao	62 193	115·1	20·2	10·4	London	1875		2Cy.32½"-61¾" 60 HP.	Remolcadores Bilbainos
904		**Ava**	Stl PadLe	British Glasgow	250 674	250·0	35·1	9·8	Dumbarton W.Denny&Bros.	1888		2Cy.37½"65"-72" 375 HP. Denny&Co.,Dumbarton	Irrawaddy Flotilla Co., Ld.
905		**Ava**	I ScSr 1 *Dk*	British Moulmein	207 278	141·4 P 45 f.	23·3	11·9	Dumbarton W.Denny&Bros.	1862		2Cy. 60 HP. Denny&Co.,Dumbarton	Bernhard Fischer
906		**Ava**	I Sc 3 *Dks*	French Marseilles	2077 3129	371·4 R 102 f. F 53 f.	39·4	30·5	LaCiotat Messageries Maritimes	1870		3C(13¼"&60¼"(2)-43¼" 500 HP. Mssgrs.Mar.,LaCiotat	Messageries Maritimes
907		**Avagyee** *Collins 87-87*	I ScBg	British Glasgow	247 427	196·4 P 54 f. F 21 f.	23·8	12·2	Dumbarton W.Denny&Bros.	1869		2Cy.34"-27" 76 HP. Denny&Co.,Dumbarton	British India Steam Nav.Co.,Ld.
908	*	**Avalon** *Towill -88*	StlScSr 1 *Dk & Wb Frms*	British London	1354 2083	280·0 Q 96 f. B 120 f. F 27 f.	37·5	19·2	Newcastle C.S.Swan&Hunter	1888	5 Blhds DB ...	3C.22"36"&59"-39" 170 HP. Blair&Co.,Ld.,Stockton	English & American S.S.Co.,Ld.
909		**Avalon**	I PadSr	British Harwich	495 677	239·8	27·0	13·5	London J.&W.Dudgeon	1865		2Cy.56"-54" 220 HP. VictoriaG.Dk.Co.,London	Earles' Co.,Ld.
910	*D	**Avance** *Tüchsen 81-84*	I ScSr 1 *Dk*	German Flensburg	263 344	139·1 B 26 f.	23·0	11·5	Kiel G.Howaldt	1884	3 Blhds. DBc&64t MT60t	2Cy 15¾"&27⅝"-15¾" 40 HP. G.Howaldt,Kiel	H.A.Petersen
911		**Avanti** *Rasmussen 82-86*	I ScSr 1 *Dk & Hur Dk*	Danish Frederikshn	140 230	134·1 Len. 19.	19·8	8·7 15·3	Gothenburg GothnbrgMek.Wrks	1875		2Cy.16"& 35"-12½" 60 HP. GothenbgM.Wrks,Gothnbg	Det Forenede Dampskibs Selskab(C.P.A.Koch)
912	*	**Avena** ex Glenisla *Tegner 85*	I ScSr 1 *Dk*	Swedish Uddevalla	522 747	200·0 P 103 f. F 30 f.	28·3	15·0	Sunderland Davison&Stokoe	1873	4 Blhds. DBc34t FPT ... MT ...	95 HP. OuseburnE.WksCo.,N'castle	Rederi-Aktie-Bol. "Avena," (W.Thorburns Söner)
913		**Avenir** ex Olmo, ex Two Roses) *Thomas 85-87*	I ScSr 1 *Dk*	French St.Brieuc	50 111	83·3 B 9 f.	18·0	8·5	Hull C.Norfolk&Co.	1882		2Cy.12"&22"-28" 30 HP. Blair&Co.,Stockton	Dupuis Robial
914	*	**Aviemore** *Hodge 83-88*	I ScSr 1 *Dk & Awn Dk*	British Dundee	617 954	204·5	29·9	13·8 20·6	Dundee W.B.Thompson	1882	5 Blhds. DBaf&ab125t APT18t	2Cy.24"& 48"-30" 90 HP. W.B.Thompson,Dundee	Albany Ship Co.,Ld.
915		**Aviken** *Norden*	W ScSr 1 *Dk*	Swedish Avike bruk	101 129	96·1 Q 29 f. F 25 f.	21·0	9·6				10 HP.	M.Bellander
916		**Aviles** *Bandaja*	I ScSr 1 *Dk*	Spanish Barcelona	425 567	173·8 P 38 f. F 27 f.	26·4	12·8	PortGlasgow Cunliffe&Dunlop	1873		2Cy.23"&40"-28" 75 HP. Cunliffe&Dunlop,Pt.Glsgw.	José Viñas
917	*	**Avilesino** *S.Lanza*	I Sc 1 *Dk*	Spanish Aviles	249 366	139·7 Q 73 f. B 10 f. F 12 f.	22·6	11·0	Newport Us&S.B.Co.,Ld.	1886	4 Blhds. FPT19t APT25t	2Cy.20"& 38"-24" 65 HP. Westray,Copeland&Co.,Brw	A.Guturrez
918	*	**Avis** *Turner 86-86*	I ScSr 1 *Dk 2 B*	British W.Hartlepl.	732 1150	231·4 Q 92 f. B 51 f. F 60 f.	31·4	14·9	W.Hartlepool W.Gray&Co.	1883	4 Blhds. DBaf 221t	2Cy.27"& 50½"-33" 99 HP. Blair&Co.,Ld.,Stockton	Hardy, Wilson & Co.
919	*	**Avlona** *Dawick 81-88*	I ScSr 2 *Dks 3 B*	British Dundee	1194 1833	284·1 P 38 f. B 62 f. F 35 f.	34·5	24·1	Dundee Gourlay Bros.&Co.	1880	5 Blhds. DBaf&a ... FPT23t	2Cy.31"& 64"-39" 190 HP. GourlayBros.&Co.,Dundee	W.Thomson & Sons
920	*	**Avoca** *Westcott 79-83*	I ScSr 1 *Dk 2 B*	British London	995 1539	250·0 Q 92 f. B 52 f. F 28 f.	35·2	18·0	Sunderland Strnd.Slpwy.Co.	1883	5 Blhds. DBaf 245t	2Cy.29"& 56"-36" 120 HP. N.E.MarineE.C.Ld.Sunderld	Westcott & Laurance
921	BT1.89	**Avoca** *Archbold 85-86*	I ScSr	British Dublin	509 844	228·0 P 36 f. F 35 f. B 12 f.	31·0	16·0	Cork G.Robinson&Co.	1865		2Cy.45"-36" 150 HP. GreenockFndryCo.,Greenk.	British & Irish Steam Packet Co.,Ld.
922	*D	**Avocet** *Stubbs 74-85*	StlSc3mSr 2 *Dks*	British Cork	583 1097	260·0 P 105 f. F 44 f. B 27 f.	34·3	15·6	Dundee W.B.Thomson	1885	4 Blhds. DBaf 136t	2Cy.31"& 62"-48" 187 HP. W.B.Thompson,Dundee	Cork S.S.Co.,Ld.
923	*	**Avochie** *Smith 80-84*	I ScSr 1 *Dk & Spr Dk 3 B*	British Aberdeen	1034 1591	249·0	34·0	16·9 23·9	Newcastle W.Dobson&Co.	1884	5 Blhds. DBa ... FPT27t	2Cy.31"& 58"-36" 200 HP. N.E.MarineE.C.Ld.Sunderld	Adam & Co.
924	BT4.89	**Avon** *Milner*	I ScSr 1 *Dk & Awn Dk 3 B*	British Sunderland	1417 2225	283·3	35·2	29·3	Sunderland J.Laing	1880	5 Blhds. DBaf	2Cy.37"&68"-42" 200 HP. N.E.MarineE.C.Ld.Sunderld	Royal Mail Steam Packet Co.
925	* BT8.89	**Avon** *Kennedy 81-86*	I ScSr 2 *Dks*	British Glasgow	372 749	219·9 P 125 f. F 33 f.	29·1	15·3	PortGlasgow Cunliffe&Dunlop	1877	6 Blhds.	2Cy.21"33"53"-36" 140 HP. Tripled '88, D.J.Dunlop&Co., Port Glasgow	W.Sloan
926		**Avon** *Sutherland 88-88*	I Sc3mSr 2 *Dks*	British Grangemth.	277 572	192·3 Q 22 f. B 43 f. Q 18 f. F 17 f.	28·0	14·9	Glasgow Barclay,Curle&Co.	1865	4 Blhds.	2Cy.31"& 54"-32" 120 HP. Barclay,Curle&Co.,Glasgow	Avon S.S. Co.,Ld. (T.Jack,jun.)
927		**Avon**	I PadSr 1 *Dk*	British Melbourne	72 115	100·3 F 3 f. B 9 f. H 3 f.	18·2	6·8	Melbourne H.J.McGee	1867		2Cy.18"-30 30 HP. T.Wingate,Glasgow.	The Bairnsdale Steam Nav.Co.,Ld.
928	*	**Awarua** *Cunningham*	I PadTg 1 *Dk*	British Invrcgl.N.Z.	100 159	109·8	20·3	10·0	London R.H.Green&Co.	1884	3 Blhds.	2Cy.28"-48" 32 HP. J.Stewart&Co.,London	Bluff Harbour Board
929		**Awhina** *Campbell*	W ScCr 1 *Dk*	British Auckland	5 136	100·0 H 18 f.	21·2	10·4	Auckland H.McQuarrie	1884		2Cy.18½"& 33"-24" 50 HP. D.McLaughlin&Co.,Paisley	Tug Boat Co.,Ld.
930	N	**Axel** *Johnsen 57-72*	W ScSr 1 *Dk 2 B*	Norweg'n Bergen	192 272	129·6 Q 35 ft. B 31 ft.	23·0	11·7	Bergen	1873		2Cy.15½"&30"-17¾ 30 HP. LaksevaagsVerks,Bergen	J.E.Lehmkuhl

*=Lloyd's **Register**; D=Nederlandsche Vereeniging; G=Germanischer Lloyd; I=Registro Italiano; R=American Record; A=Veritas Austro-Ungarico; H=Veritas Ellenico; N=Norske Veritas; V=Bureau Veritas.

Dans la 2me Colonne, B.T., avec les indications qui suivent, signifie la date du rapport de visite de la Carène par les experts du Board of Trade pour obtenir le certificat relatif au TRANSPORT DES PASSAGERS.

Dans la 3me Colonne, C=Composite (membrure en fer, bordé en bois); I=Fer; Stl=Acier; W=Bois.

Pour plus amples renseignements sur les navires classés par le Lloyd's Register, voir le Lloyd's Register Book et le Lloyd's Yacht Register.

1 No. in Book Official No.	2 Registry in which classed	3 Ship's Name. Material, Rig, &c. Master, and Dates of Appointment to present Owner's Service and to Ship. Ship's Late Name. No. of Decks, &c.	4 Flag. Port of Registry.	5 Tons Net. Gross.	6 Dimensions. Length. Breadth. Depth. Deck Erections.	7 Built. Where. By Whom.	8 When	9 No. of Bulkheads. Particulars of Water Ballast.	10 Engines. Particulars. Maker's Name.	11 Owners or Ma[nagers]
931	G	**Axel** C ScSmSc *Andersson 84-84* 1 Dk	Swedish Gothenburg	189 294	121·0|21·7|12·1 *B25f. H1c.*	Gothenburg J.A.Svensson	1884		2Cy.15½"& 27"-18" 45 HP. J&C.G.Bolinders M.W.Stkhlm	Angf.Aktie B[olaget] "Axel" (A.Bu…)
932	V	**Axelhuus** I ScSr *Gall 84-84* 2 Dks	Danish Copenhagen	404 636	183·1|27·6|11·7 *P80f. B32f. 31f. lst.*	Renfrew Lobnitz&Co.	1884	WB100t FPT APT	2Cy.23"& 44"-33" 90 HP. Lobnitz&Co.,Renfrew	Det forenede S[…] skibsselskab (C…)
933 60215		**Ayacucho** I ScSmSr *(ex New Granada)*	British Liverpool	1208 1916	311·8|38·2|21·1 *F55f.*	Glasgow T.Wingate&Co.	1873	6 Bhds. WB f	2Cy.51½"&84½"-42" 400 HP. Wingate&Co.,Glasgow	Pacific Steam […]
934 82767	✳	**Aycliffe** I ScSr *Hogg* 1 Dk 2 B	British Middlesbro'	1039 1611	250·0|36·0|17·5 *Q80f. B70f. F29f.*	Middlesbro' R.Dixon&Co.	1883	4 Bhds. DB…	2Cy.30"& 55"-36" 120 HP. Blair&Co. Ld.,Stockton	Aycliffe S.S.Co. (Hogg & He…)
935	✳	**Aymore** Stl ScSr *Brown* 1 Dk	Brazilian RiodeJaniro	389 605	200·5|28·2|12·1 *P67f. B44f. F30f.*	Newcastle Sir W.G.Armstrong Mitchell&Co.	1883	5 Bhds. APT…	2Cy.25"& 50"-27" 120 HP. Wallsend Slipway C.N'castle	Cia.de Nav.Pa[…]
936 78631		**Ayr** I PadSp *Boyd 86-86*	British Ayr	13 111	96·2|18·6|9·7	South Shields J.T.Eltringham	1878		2Cy.26½"-18" 30 HP. J.P.Rennoldson,S.Shields	Trustees of Ayr [Har]bour
937 76783	✳	**Ayrshire** I ScSmSr *Jenkins* 2 Dks 3 B	British Glasgow	871 1354	260·3|32·3|22·5	Port Glasgow W.Hamilton&Co.	1877	5 Bhds. …	2Cy.30"& 55"-36" 140 HP. Walker,Henderson&Co.,Gls	Turnbull,M[…]
938 88861	✳	**Ayrshire** I ScSr *McPhail 88-88* 1 Dk	British Belfast	500 774	200·6|29·2|13·1 *Q93f. B20f. F25f.*	Dundee Pearce Bros.	1882	4 Bhds. DB… FPT… APT…	2Cy.23"& 45"-30" 90 HP. Pearce Bros.,Dundee.	G.A.Essayan
939 79634	✳	**Azalea** I ScSmSr *Petrie 81-85* 2 Dks 3 B	British N.Shields	1188 1828	280·5|34·0|24·6	Newcastle C.Mitchell&Co.	1877	5 Bhds. DB…	2Cy.32"& 61"-36" 190 HP. N.E.MarineE.Co.,Ld.,Sndld.	J.Robinson & […]
940 91895	✳	**Azalea** I ScSr *Fife* 1 Dk	British …	314 503	162·0|23·6|13·1 *Q87f. F16f. B42f.*	Sunderland Sundrlnd S.B.Co.Ld	1886	4 Bhds. …	2Cy.20"&40"-27" 80 HP. N.E.MarineE.Co.Ld.,Sndld.	C.F.Leach
941 78846	BT…	**Azalea** I ScSr	British Glasgow	222 583	217·6|30·2|15·1 *P46f. B74f. F46f.*	Glasgow A.&J.Inglis	1878		2Cy.36"& 67"-42" 200 HP. A.&J.Inglis,Glasgow	Glasgow, Dubl[in &] LondonderryS[…]
942		**Azimet** *Goujon 52-*	Turkish Constantinople	90	140·0|18·0|9·6	London Maudslay,Sons&Field	1870		60 HP. Maudslay,Sons&Field	Chirket-i-Hair[iye] (BosphorusStea[m…])
943		**Azoff** I ScSr 1 Dk 2 B	Russian Odessa	1057 1411	256·0|33·0|21·2	Newcastle A.Leslie&Co.	1867	…18	4Cy.21"& 41"-30" 120 HP. Maudslay,Sons&Co.,London	Russian Steam […] Trading Co.

Additional Ships are inserted in the Supplement at the end of the Register.

✳ = **Lloyd's Register**; D = Nederlandsche Vereeniging; G = Germanischer Lloyd; I = Registro Italiano; R = American R[ecord];
A = Veritas Austro-Ungarico; H = Veritas Ellenico; N = Norske Veritas; V = Bureau Ver[itas].

In the 2nd Column, B.T., with figures following, signifies Date of report of bottom sighting by Board of Trade Surveyors as condition of Passenger Cert[ificate].

In the 3rd Column, C = Composite (iron frame, wood planked); I = Iron; Stl = Steel; W = Wood.

Further particulars of Vessels classed by Lloyd's Register will be found in Lloyd's Register Book and Lloyd's Yacht Register

1	2 Registres où le navire est classé	3 Nom, Matériaux, Gréement, etc. — Capitaine et dates d'entrée au service des Armateurs actuels et à bord du navire — Ancien Nom, nombre de ponts, etc.	4 Pavillon. — Port d'Armement	5 Tonnage. Net — Total	6 Dimensions. Longueur	Largeur	Creux	7 Construit. Où. — Par qui.	8 Quand.	9 Nombre des cloisons. — Détails sur le test d'eau.	10 Machines. — Détails. Nom du Constructeur.	11 Armateurs ou Directeurs-Gérants.
1	✻	**B.D.No.1** Stl ScHopperDredger *Nicholas* -86 1 *Dk2 B*	British Bristol	415 1049	214·3	43·0	15·1	Renfrew W.Simons&Co.	1886	6 Blhds. FPT24t APT436t	6C.17"27"&43"–30" 250 HP. W.Simons&Co.,Renfrew	Corporation of the City of Bristol
2	✻	**B. Granger** I ScSr *Mills* 86-86 1 *Dk2 B*	British Whitby	943 1474	243·0	34·0	17·9	Whitby T.Turnbull&Son *Q63f. B59f. P25f. F33f.*	1882	4 Blhds. DBo&f26ft	2Cy.31"& 58"–36" 130 HP. Blair&Co.,Ld.,Stockton	Robinson Bros.
3	✻A	**B. Kemeny** I Sc2m *Knirsch* 1 *Dk2 B*	Austrian Fiume	767 1129	245·4	35·1	16·9	Dumbarton Burrell&Son *Q 77 f. B 62 f. F 30 f.*	1881	4 Blhds. DBaf&c63tt	2Cy.30"&55½"–36" 150 HP. M.Paul&Co.,Dumbarton	"Adria" Hungarian Sea Navigation Co.,Ld.
4		**B. W. Morse** W Sc	American Bath,Me.	216 432	147·0	30·8	16·3	Bath, Me. R.W.&H.F.Morse	1887		2Cy.31"&58"–36"	B.W.&H.F.Morse
5		**Babushka** I Pad 1 *Dk*	Russian Odessa	120	130·0	22·5	7·1	Sebastopol Rus.S.S.&TradgCo.	1871		2Cy.18"–14" 60 HP. Russian S.S.&Trading Co.	Russian Steam Nav. & Trading Co.
6		**Bacolod** I Pad Sr	Spanish Manila	113 187	119·1	22·4	6·7	Glasgow A.&J.Inglis *B 20 f.*	1874		1Cy.38"–45" 45 HP. A.&J.Inglis,Glasgow	Inchausti & Co.
7	✻	**Badsworth** I ScSr *Fargher* 1 *Dk&Spr Dk3 B*	British Liverpool	1273 1943	263·8	36·1	24·2	Southampton Oswald,Mordaunt&Co. *P 20 f. F 28 f.*	1883	5 Blhds. DBeb125t MT330t	2Cy.30"& 60"–42" 170 HP. Oswald,Mordaunt&Co.,S'hampton	Badsworth S.S. Co.,Ld. (W.Wheatley & Co.)
8	✻	**Baghdad** I ScBg *Frokaerk* 85-85 2 *Dks3 B*	British Glasgow	809 1271	251·5	31·5	23·6	Renfrew W.Simons&Co.	1871	5 Blhds.	2Cy.34"& 60–36" 180 HP. W.Simons&Co.,Renfrew	British India Steam Navigation Co.,Ld.
9	✻	**Baghdadi** I ScSr *Scott* -87 1 *Dk&Spr Dk3 B*	British London	1104 1704	255·0	35·1	17·0 21·5	Newcastle W.Richardson&Co. *B 48 f.*	1885	4 Blhds. DBo&c6t FPT28t APT57t	3C.21½"33"&55"–39" 200 HP. W.Richardson&Co.,Newc.	Persian Gulf Steamship Co.,Ld.
10	I	**Bagnara** ex Mimosa I ScBg *Anatra* 86-88 1 *Dk2 B*	Italian Palermo	553 867	218·5	29·2	17·2	Liverpool R.&J.Evans&Co. *Q 74 f. F 30 f.*	1872		2Cy.25"& 50"–36" 185 HP. DeWinton&Co.,Carnarvon	Navagazione Generale Italiana
11	V	**Bahia** I &Stl ScBg *Schatterow* 86-88 2 *Dks*	German Hamburg	1587 2172	307·4	36·1	25·0	Newcastle SirW.G.Armstrong, Mitchell&Co. *B 73 f. P 52 f.*	1886		3Cy.24"39"& 64"–42" 350 HP. WallsendSlipCo.	Hmburg-Südamerik Dampfschiff Gesells.
12		**Bahia Honda** I & W	Spanish Havana	268 450	156·1	27·2	10·5	Philadelphia Neafie&Levy	1862		1Cy.34"–30" 100 HP. Neafie&Levy,Philadelphia	Cosme de Toca
13	DV	**Bahrenfeld** I ScSr *Hinrichs* -85 2 *Dks&Spr Dk*	German Neumühlen	946 1468	233·6	31·6	16.8 21·4	Stettin Act.Gesc Vulcan" *F28f. B52f. Q28f.*	1882	DT222t	2Cy.29½"&54"–36" 130 HP. Act.Gese Vulcan,"Stettin	Gläfeke & Hennings
14		**Bahri-Jeddid** I Sc3mSr (ex Olenda, ex Florence) *Mustafa Galib Bey* -88 3 *Dks*	Turkish Constantinple	895 1316	249·7	32·8	22·3	London SamudaBros. *P 52 f. F 40 f.*	1873	4 Blhds.	2Cy.26"& 54"–36" 150 HP. Humphrys,Tennant&C.,Lon	Idarei Massousich (Ottoman Steam Nav.Co.)
15	✻	**Baidar** I ScSr *Bolton* -88 1 *Dk2 B*	British Liverpool	778 1003	232·0	28·0	15·6	Hull Humphrys&Co. *P 133 f. F 30 f.*	1871	5 Blhds.	2Cy.26"& 46"–30" 95 HP. Humphrys&Co.,Hull	S.Hough
16	✻	**Baikal** (ex Ebenholm) I ScSr *Lemasechiffsky* 1 *Dk&Awn Dk*	Russian Vladivostok	713 1109	227·0	29·5	15·9 22·0	Sunderland JonesLaing	1879	4 Blhds. DBo55ft250	2Cy.27"& 50"–36" 99 HP. GeorgeClark,Sunderland	M.G.Sheveleff
17	✻	**Baines Hawkins** I ScSr *Irvine* 88-88 1 *Dk*	British Newcastle	464 723	189·6	30·1	13·6	Blyth Hodgson&Soulsby *Q104 f. B 18 f. F 21 f.*	1881	4 Blhds. DBo&f150 APT19t	2Cy.24"& 48"–33" 95 HP. Hawks,Crawshay&Co.,Gateshead	Fisher, Renwick & Co.
18		**Bais** W Sc *Mil* 86-86	Spanish Iloilo	93 137	109·9	..	..	Hong Kong	1884			J.Montenegro
19		**Baklanoff** Stl Pad	Russian Rostoff	127 176	..	..	..	Neupest H.Schoenichen	1881		55 HP.	Societa di Nav. Tanais
20	✻	**Bakuin** I ScSr3m *Beyer* 87-87 2 *Dks*	British London	1093 1669	260·4	36·0	17·5	W. Hartlepool W.Gray&Co. *P161 f. F38 f.*	1886	7 Blhds. DB&c4t FPT39t APT7t	3C.21"35"&57"–39" 200 HP. CentralMar.Eng.Co.,W.Hpl.	A.Suart
21	✻D	**Bala** I ScSr *McMullen* 85-87 1 *Dk2 B*	British Cardiff	1288 2011	270·5	35·7	20·1	W. Hartlepool W.Gray&Co. *P28f. Q12f. B103f. F31f.*	1884	4 Blhds. DBo&f365t	2Cy.34"& 64"–39" 175 HP. T.Richardson&Sons,Hrtlpl.	Bala S.S.Co.,Ld. (E.Thomas Radcliffe & Co.)
22	V	**Balæna** W ScBk *Smith* 82-84 2 *Dks3 B*	American SnFrancisco	390 524	149·5	32·0	17·0	San Francisco DickieBros.	1883		3Cy.12"20"32"–24" 245 HP. RisdonIronWks,SanFranco.	Pacific Steam Whaling Co.
23		**Balbirnie** I Pad *Mackie* 73-82	British Granton	357 533	199·6	40·4	8·9	Leith H.Morton&Co.	1861		1Cy.44"–42" 200 HP. H.Morton&Co.,Leith	North British Railway Co.
24	✻D	**Balbus** I ScSr *Leask* 84-86 1 *Dk2 B*	British Barrow	746 1156	240·0	34·2	14·1	Barrow BarrowS.B.C.,Lim. *Q 92 f. B 52 f. F 28 f.*	1883	4 Blhds. CDB300t FPT11t APT7t	2Cy.31"& 58"–36" 150 HP. Westray,Copeland&Co.,Barrow	T.Kish
25	✻D	**Balcarres Brook** I ScSr *Burgess* 86-86 1 *Dk&Spr Dk3 B*	British Liverpool	1306 2001	264·8	36·1	24·2	Southampton Oswald,Mord't&Co. *P 20 f. F 28 f.*	1883	5 Blhds. DBeb125t MT330t	2Cy.30"& 60"–42" 170 HP. Oswald,Mordaunt&Co.,S'hampton	"Balcarres Brook" S.S. Co.,Ld. (W.G.Killick & Co.)
26	✻N	**Balder** Stl ScSr *Molta* 82-82 1 *Dk2 B*	Norweg'n Bergen	305 485	159·4	24·3	13.8	Gothenburg MotalaCo. *Q 52 f. B 35 f. F 18 f.*	1882	4 Blhds. MT130t APT FPT	2Cy.19½"&35½"–23½" 60 HP. MotalaCo.,Gothenburg	J.C.Giertsen
27	N	**Balder** W ScSr *Gundersen* 67-81 1 *Dk*	Norweg'n Egersund	119 187	100·1	21·3	12·1	Sjötorp (Rebuilt) *B 28 f.*	1870		2Cy.15"& 30"–18" 30 HP. BergensMek.Verks,Bergen	O.Puntervold
28	V	**Balder** I ScSr *Skoldebrand* 76-83 1 *Dk&Awn Dk*	Swedish Gefle	473 595	152·0	25·3	14·3 21·5	Norrköping MotalaCo. *H 4 t.*	1872		2Cy.17"&36"–50" 60 HP. A.Tham,Norrköping	Ångf.Rederibolaget Norden (P.J.Hagerstrand)
29	G	**Balder** C Sc *Olsson* 1 *Dk*	Swedish Gothenburg	259 322	126·6	22·1	14·0	Lödöse *B 28 f. H 4 t.*	1872		40 HP.	Ångf.Aktie Bolaget Arctic (J.F.Olsson)
30	✻	**Baldomero Iglesias** I ScSr (ex Ebro, ex Bakalai) *Gaudana* -88 2 *Dks3 B*	Spanish Barcelona	1025 1567	273·0	34·1	25·8	Glasgow RandolphElder&Co.	1866	5 Blhds.	2Cy.35"& 70"–48" 300 HP. G.Forrester&Co.,Liverpool	Cia. Trasatlantica

✻ = Lloyd's Register; D = Nederlandsche Vereeniging; G = Germanischer Lloyd; I = Registro Italiano; R = American Record;
A = Veritas Austro-Ungarico; H = Veritas Ellenico; N = Norske Veritas; V = Bureau Veritas.

Dans la 2me Colonne, B.T., avec les indications qui suivent, signifie la date du rapport de visite de la Carène par les experts du Board of Trade pour obtenir le certificat relatif au TRANSPORT DES PASSAGERS.

Dans la 3me Colonne, C = Composite (membrure en fer, bordé en bois); I = Fer; Stl = Acier; W = Bois.

Pour plus amples renseignements sur les navires classés par le Lloyd's Register, voir le Lloyd's Register Book et le Lloyd's Yacht Register.

1 No. in Book / Official No.	2 Registry in which classed	3 Ship's Name. Material, Rig, &c. / Master, and Dates of Appointment to present Owner's Service and to Ship / Ship's Late Name. No. of Decks, &c.	4 Flag / Port of Registry.	5 Tons. Net / Gross	6 Dimensions. Length	Breadth	Depth	7 Built. Where. / By Whom.	8 When	9 No. of Bulkheads. / Particulars of Water Ballast.	10 Engines. Particulars. Maker's Name.	11 Owners or Managers
31		Baldur I ScSr Olsen 77-81 1 Dk	Danish Frederikshavn	282 466	168·7	23·5	11·6 P 193 f. F 24 f.	Copenhagen Burmeister&Wain	1876	[illegible]	1Cy.16½"&31"–20" 80 HP. Burmeister&Wain,Cpnhgn.	Det Forenede [illegible] skibs Selskab [illegible]
32 77450	✻	Balgownie I ScSr McDonald 73-86 1 Dk2 B	British Aberdeen	726 1134	235·1	30·1	17·0 P 126 f. F 34 f.	Aberdeen A.Hall&Co.	1880	5 Blkds. DB [illegible]	2Cy.28"&52"–34" 120 HP. SmithBros.,Glasgow	Grampian S.S.Co.
33	✻	Balkan I Sc Michel 67-83 1 Dk&Spr Dk3 B	French Marseilles	1192 1733	260·0	37·1	23·0 P 24 f. B 28 f. F 29 f.	Dumbarton A.McMillan&Son	1882	6 Blkds.	2Cy.36½"&63"–48" 250 HP. Denny&Co.,Dumbarton	Fraissinet & Co.
34 81638	BT1.87	Ballaarat Stl ScBg Ashdown 80-87 2 Dks&Spr Dk4 B	British Greenock	2663 4748	420·2	43·0	26·4 34·5	Greenock Caird&Co.	1882	8 Blkds.	2Cy.56"&100"–66" 900 HP. Caird&Co.,Greenock	Peninsular & [illegible] Steam Nav. Co.
35 77463	✻	Ballochbuie I ScSr Webster 1 Dk2 B	British Aberdeen	408 650	190·3	31·2	13·6 Q 35 f. B 40 f. F 22 f.	Aberdeen Hall,Russell&Co.	1880	5 Blkds. [illegible]	2Cy.28½"&46"–30" 90 HP. Hall,Russell&Co.,Aberdeen	J.&A.Davidson
36 84280	✻ BT3.88	Ballycotton I ScSr Davies 78-82 2 Dks	British Glasgow	450 883	225·0	30·3	15·7 P 127 f. F 54 f.	Renfrew W.Simons&Co.	1880	[illegible] Blkds. [illegible]	2Cy.35"&51"–54" 200 HP. W.Simons&Co.,Renfrew	Clyde Shipping [illegible]
37 89591	✻	Balmain Stl ScSr Hammer 1 Dk	British Melbourne	377 604	186·0	27·1	10·8 P 96 f. F 29 f.	Glasgow Dobie&Co.	1883	[illegible] Blkds. [illegible]	2Cy.23"&45"–30" 85 HP. J.Jack&Co.,Liverpool	W.H.Smith&[illegible]
38 83775		Balmain W Pad 1 Dk	British Sydney	112 177	120·0	20·0	7·3 B 34 f.	Balmain,NSW S.Briggs	1883		2Cy.15½"&28"–48" 45 HP. Mort's Dk.&Eng.Co.,Balmain	Balmain Steam Ferry Co.,Ld.
39 3178	BT6.88	Balmoral (ex Lady Brisbane) I PolSk McNicol 81-87	British Glasgow	18 127	136·7	18·2	7·9 Q 4 f. 21·1	Paisley Barr&McNab	1842		1Cy.47"–50" 70 HP. Barr&McNab,Paisley	W. Buchanan
40 76038	✻	Balmoral Castle I ScSr Cringle (ex San Augustin) 2 Dks3 B	British London	1988 3050	344·8	39·4	29·1 B 92 f. F 52 f.	Glasgow R.Napier&Sons	1876	[illegible] Blkds.	3Cy.28" 47½" & 75½"–48" Triple [illegible] Rowan& [illegible] Glasgow [383 HP.	Balmoral Castle Co.,Ld.(J.Kilgour)
41 78696	✻ D	Balmuir I ScSr Spence 75-80 2 Dks	British Dundee	996 1531	258·1	32·0	20·2 P 36 f. B 57 f. F 30 f.	Dundee Gourlay Bros.	1879	7 Blkds. DB [illegible] FPT APT	2Cy.32"&56"–36" 150 HP. GourlayBros.&Co.,Dundee	J. Machan
42 93690	✻	Balniel I ScSr M. Cowell 85-86 1 Dk	British Liverpool	217 450	174·0	26·1	12·6 Q 95 f. F 27 f.	Belfast M'Ilwaine,Lewis&Co.	1886	5 Blkds. [illegible]	2Cy.22"&44"–30" 80 HP. MacIlwaine,Lewis&Co.,Ld.	The Wigan Coal Iron Co.
43 92346		Baltic (ex Frances Smith) W PadSr Robertson -88 2 Dks	British Cllngwd.Ont	834 1324	181·8	27·9	12·0	OwenSndOnt M.Simpson	1867		1Cy.40"–144" C.Micklem,Chippawa,Ont.	Great Northern Transit Co.
44 94096	✻	Baltic I ScK Bryant 88-88 1 Dk	British Grimsby	93 154	100·6	20·1	11·6 Q 10 f. F 18 f.	Hull Earle's S.B.&E.Co.,Ld.	1889	3 Blkds.	3Cy.12"18"&30"–18" 45 HP. Earle's S.B.&E.Co.,Ld.,Hull	Grimsby Steam [illegible] ing Co.,Ld.
45	N	Baltic C ScSr Ellingsen 83-83 1 Dk2 B	Norweg'n Moss	272 363	133·0	23·6	14·2 B 37 ft.	Westervik R.Triddel	1875		2Cy.13¾"&28⅝"–16¾" 40 HP. Bergsunds Verks,Stockholm	H.A.Reinert
46	V	Baltic (ex Kurland ex Riga Lübeck) I PadSr Taube 1 Dk	Russian Riga	278 362	195·0	26·0	11·5 Len 31. P 64 f.	Greenock Caird&Co.	1852		1Cy.23"&45"–48" 125 HP. Cmp.Lange&Skuge,Riga	A.Storongenko
47	G	Baltic C ScSr Rothlin 72-82 1 Dk	Swedish Carlskrona	216 293	127·9	21·8	11·1 [illegible]	Carlshamn O.R. Andersson	1872		2Cy.13"&30"–23" 40 HP.	C.G.O.Winter
48 91142	✻	Baltimore I ScSr Trenery 81-84 2 Dks3 B	British Liverpool	2459 3730	350·5	43·1	28·7 P 32 f. B 96 f. F 38 f.	Newcastle M'Intyre&Co.,Ld.	1884	6 Blkds. [illegible] FPT APT	2Cy.43"&84"–54" 360 HP. Blair&Co.Ld.,Stockton	S.S."Baltimore" (W.Johnston [illegible])
49 3329	R	Baltimore W ScSr Nickle -73	American Baltimore	1013 1414	223·0	35·7	22·2	Baltimore W.E.Woodall&Co.	1885	[illegible]	2Cy.30"&57"–36" C.Reeder&Sons,Baltimore	Baltimore,Ches[illegible] & Richmond St[illegible]
50	GV	Baltimore I ScBg Reimkasten 86-88 2 Dks&Spr Dk	German Bremen	1676 2321	297·5	39·4	23·0 Q 32 f. 30·8	Greenock Caird&Co.	1868		2Cy.42"&77"–48" 300 HP. Caird&Co.,Greenock	Norddeutscher [illegible]
51 92930	✻	Baltimore City Stl ScSr Benn 81-88 2 Dks	British W.Hartlep'l	1534 2334	290·0	39·1	22·8 P 28 f. B [illegible] f. F 34 f.	Glasgow A.Stephen&Sons	1888	5 Blkds. [illegible]	3C.21"33"&52"–42" 210 HP. A.Stephen&Sons,Glasgow	C.Furness
52	V	Baltzar von Platen I ScSr Swensson 69 1 Dk	Swedish Helsingborg	548 732	186·3	27·9	11·6 Q63f. B32f. F23f. H20f.	Newcastle Cole Bros.	1872		2Cy.22"&42"–33" 75 HP. Ouseburn Eng.Wrks.Gtshd	C.Henckel
53		Baltzar von Platen I ScSp Hanson 85-85 1 Dks&Awn Dk	Swedish Stockholm	167 225	97·4	21·6	8·3 H 6 f.	Motala MotalaCo.	1871		2Cy.18"&32"–26" 45 HP. Bergsd.Mek.Verks,Stkhlm.	Motala Ströms [illegible] atkiebolag [illegible]
54		Ban Hok (ex Siok Tjia) (now named Baron Van Tuyll) I ScSr Mendt -82 1 Dk	Dutch Batavia	173 282	152·3	21·3	13·3 [illegible]	Amsterdam Kon.Fab.V.Stoom.	1880	4 Blkds.	2Cy.18"&32"–28" 45 HP. Kon.Fab.V.Stoom,Amsterdam	Billiton Maats[illegible]
55 60704	✻ BT8.88	Ban Righ I ScSr Tarriff 69-73 2 Dks	British Aberdeen	490 940	241·0	30·2	16·6 P 72 f. F 22 f.	Glasgow J.Elder&Co.	1870	5 Blkds.	2Cy.46"&80"–39" 275 HP. J.Elder&Co.,Glasgow	AberdeenStmN[illegible]
56 93020	✻	Ban Seng Guan I ScSr Geary 84-86 2 Dks	British Singapore	501 801	225·8	30·0	15·5 P 26 f. B 92 f. F 27 f.	Leith Ramage&Ferguson	1886	4 Blkds.	2Cy.28"&54"–36" 125 HP. Ramage&Ferguson,Leith	Tan Yong Se[illegible]
57 89012	✻	Ban Whatt Hin I ScSr Vincent 1 Dk	British Singapore	195 361	169·6	25·0	10·4 F 24 f.	PortGlasgow Blackwo'd&Gordon	1885	4 Blkds.	2Cy.22"&44"–30" 85 HP. Blackwood&Gordon,Pt.Gls	Wee Boon Tek
58 82045	✻	Ban Whatt Soon I ScSr Grant 1 Dk	British Singapore	199 336	170·7	25·1	9·6 F 18 f. H 37 f.	PortGlasgow Blackwo'd&Gordon	1881	4 Blkds.	2Cy.22"&44"–24" 80 HP. Blackwood&Gordon,Pt.Gls.	Tan Yong Se[illegible] & others
59 63298		Ban Yong Seng I ScSr Hole 2 Dks	British Singapore	367 540	180·4	25·4	14·0 P 47 f. F 25 f.	Paisley J.Fullerton&Co.	1872	4 Blkds.	2Cy.22"&40"–30" 75 HP. Walker,Henderson&Co.,Gls.	Wee Boon Teck
60	✻	Banan Stl ScSr Hansen 88-89 1 Dk2 B	Norweg'n Christiania	632 1007	216·1	30·2	15·7 B 51 f. Q 72 f. F 30 f.	Sunderland W.Doxford&Sons	1889	4 Blkds. [illegible]	3Cy.16½"27½"&44"–30" W.Doxford&Sons,Sunderland [120 HP.	Sondefjeld Nord[illegible] Dampskibs Selsk[illegible]

✻ = **Lloyd's Register**; D = Nederlandsche Vereeniging; G = Germanischer Lloyd; I = Registro Italiano; R = American Rec[ord]
A = Veritas Austro-Ungarico; H = Veritas Ellenico; N = Norske Veritas; V = Bureau Veritas

In the 2nd Column, B.T., with figures following, signifies Date of report of bottom sighting by Board of Trade Surveyors as condition of PASSENGER CERTIFIC[ATE]

In the 3rd Column, C = Composite (iron frame, wood planked); I = Iron; Stl = Steel; W = Wood.

Further particulars of Vessels classed by Lloyd's Register will be found in Lloyd's Register Book and Lloyd's Yacht Register.

N°	Classe	Nom, Matériaux, Gréement, etc. / Capitaine et dates d'entrée au service des Armateurs actuels et à bord du navire. / Ancien Nom, nombre de ponts, etc.	Pavillon. / Port d'Armement	Tonnage. Net. / Total	Dimensions. Longueur	Largeur	Creux	Construit. Où. / Par qui.	Quand	Nombre des cloisons. / Détails sur le lest d'eau.	Machines. / Détails. / Nom du Constructeur.	Armateurs ou Directeurs-Gérants.
61		**Banca** (ex Eikenholmer) I SeSr *Dunlop* 87-87 1 *Dk*	 Sandaker	235 363	160·4 Q 45 *f.*	21·3	11·8	Newcastle W.E.Boutland	1873	4 Blhds. DBa31*ft*	2Cy.20"& 36"–24" 56 HP. Marshall,Osborne&Co.S.Sds	A.S.Johnston & Co.
62	*	**Banchory** I SeSr *Kerr* 87-88 1 *Dk2 B*	British Aberdeen	372 576	174·8 Q 52 *f.*	26·1 B 35 *f.*	14·5	Aberdeen Hall,Russell&Co.	1877	5 Blhds. DBa74	2Cy.23"& 40"–27" 70 HP. Hall,Russell&Co.,Aberdeen	Grampian S.S. Co., Ld.
63	*	**Bancoora** Stl SeBmSr *Brown* 86-85 2 *Dks*	British Glasgow	2170 2880	322·8 P 44 *f.*	40·3 B 25 *f.*	25·9 F 42 *f.*	Dumbarton W.Denny&Bros.	1880	6 Blhds. CDB267t	2Cy.36"& 68"–51" 290 HP. Denny&Co.,Dumbarton	British India Steam Nav.Co.,Ld.
64		**Bangalore** I SeBn *Baldwin* -86	British Bombay	1310 2342	318·4 P 180 *f.*	38·4 F 92 *f.*	26·4	Dumbarton W.Denny&Bros.	1867		2Cy.57"& 95"–45" 500 HP. Denny&Co.,Dumbarton	Hajee Cassum Joosub
65		**Bangkok** Stl TwnScSr *Meyer*	British Glasgow	197 307	140·0	28·0	9·5	Glasgow J.Elder&Co.	1885		4Cy.15"& 30"–18" 100 HP. J.Elder&Co.,Glasgow	Scottish Oriental Stm. Ship Co.,Ld.
66		**Bangkok** I PadSr 1 *Dk*	British Singapore	32 149	115·3	20·0	10·0	North Shields J.Softley&Co.	1877		1Cy.30½"–54" 80 HP. Hepple&Co.,N.Shields	Tanjong Pagar Dock Co.,Ld.
67		**Bangor Castle** I PadSk (ex Palmerston)	British Belfast	71 240	191·3	22·1	6·9	Glasgow T.Wingate&Co.	1866		2Cy.36"–42" 100 HP. T.Wingate&Co.,Glasgow	Belfast, Bangor & Larne SteamboatCo.Ld.
68	*	**Banjermassin** I SeSr *Lamlester* 87-87 1 *Dk & Shd Dk*	Dutch Banjermassin	289 428	174·0	23·7	12·2	Newcastle Wghm.Richardson &Co.	1886	4 Blhds.	3C14½"22½"&38"–30"80HP WighamRichrdsn&Co.Nwc.	Kho Soen Tjio & Ang Eng Tjio & Others
69	N	**Bankchef Henriksen** I SeSr *Benzenius* 86-86 1 *Dk & Awn Dk*	Norweg'n Bergen	529 709	173·9	27·2	13·3 19·2	Bergen Martens,Olsen&Co.	1883	5 Blhds. DBa&f 138t APT12t	2Cy.23"&41"–28" 75 HP. Martens,Olsen&Co.,Bergen	P.G.Halvorsen
70	*	**Banks Peninsula** StlTwScSr 1 *Dk*						Grangemouth Grangemouth Dockyard Co.	1889		4Cy.13½"&27"–18" Hutson&Corbett,Glasgow	
71		**Banner** I SeSr *McMillan* 1 *Dk*	British Newpt.Mon.	100 228	130·0 Q 47 *f.*	21·1 F 17 *f.*	9·9	Paisley J.Fullerton&Co.	1881	4 Blhds. FPT29t	2Cy.18"& 34"–28" 50 HP. ScottBros.,Glasgow	H.Burton
72		**Banshee** I PadSr *Taylor* 57	British Dublin	246 1109	310·0 F 197 *f.*	34·1	14·3	Birkenhead LairdBros.	1884		2Cy.80"–84" 420 HP. LairdBros.,Birkenhead	L.& N.W.Railway Co.
73	*DV	**Bantam** Stl SeSr *Scholten* 2 *Dks 3 B*	Dutch Amsterdam	1484 2132	290·2 P 54 *f.*	37·2 B 82 *f.*	24·2 F 30 *f.*	Glasgow A.&J.Inglis	1884	5 Blhds.	2Cy.36"& 70"–45" 250 HP. A.&J.Inglis,Glasgow	Maatschappy Holland
74		**Barambio** I Sc *Piñole* 1 *Dk2 B*	Spanish Gijon	483 754	195·8 Q 66 *f.*	29·0 B 44 *f.*	15·5 F 23 *f.*	Sunderland Austin&Hunter	1874	5 Blhds. DBa&f 80t	2Cy.27"& 47"–30" 90 HP. N.E.MarineEng.Co.,Sndlnd.	Meliton Gonzales y Ca
75		**Barão de Sao Diogo** I SeSr 1 *Dk*	Brazilian RioJaneiro	261 471	180·6 P 55 *f.*	23·2 F 34 *f.*	12·7	Newcastle C.Mitchell&Co.	1879	4 Blhds. DBa41*f*	2Cy.28"& 50"–33" 100 HP. Easton&Anderson,London	Cia de Estrada de ferro de Macahé & Campos.
76	V	**Barawa** I SeSr 1 *Dk*	Zanzibar	234 330	141·0 P 17 *f.*	22·1 B 19 *f.*	12·0 F 25 *f.*	Flensburg Flensbrg. Schiffsb. Ges.	1885	4 Blhds.	2Cy.18"& 33"–21" 45 HP. FlensburgSchiffsb Ges,Flensbrg	Sultan of Zanzibar
77		**Barbara Boscowitz** W SeSr 2 *Dks*	British Victoria,B.C	239 269	119·0	22·0	16·0	Victoria J.M.C.Smith	1883		2Cy.9½"& 18"–18" 12 HP. T.Gowan,Victoria,B.C.	T.H.Cooper
78	*	**Barbaras** W SeK *Gibson* 84-88 1 *Dk*	British Kirkcaldy	57 93	85·0	20·5	9·0	Anstruther W.Jarvis	1888		4Cy.7½"&11"–12" 20HP. W.King&Co.,Glasgow	W.Jarvis
79	*	**Barcelona** I SeSr *Boyle* 88-88 2 *Dks 3 B*	British Dundee	1180 1862	284·2 P 28 *f.*	34·2 F 32 *f.*	21·3	Dundee Gourlay Bros.	1878	DBa&f 278t FPT22t APT16t	2Cy.34"&64"–39" 190 HP. GourlayBros.&Co.,Dundee	W.Thomson & Sons
80	*V	**Barcelona** I SeSr *Dau* -88 2 *Dks*	German Hamburg	883 1446	265·7 B 28 *f.*	33·9 F 23 *f.*	19·9	Glasgow A.Stephen&Sons	1879	4 Blhds.	2Cy.30"& 57"–36" 150 HP. A.Stephen&Sons,Glasgow	R.M.Sloman & Co.
81	*	**Barcoo** Stl SeSr *Hampden* 1 *Dk & Sp Dk & Wb Frms*	British London	745 1505	250·1 B 16 *f.*	30·1 F 41 *f.*	15·5 23·2	Dumbarton W.Denny&Bros.	1885	4 Blhds. CDB210t	3Cy25"41"&68"–48"308HP Denny&Co.,Dumbarton	Australasian United Steam Nav.Co.,Ld.
82	*	**Barden Tower** (ex Gettysburg) I SeSr *Wilson* 1 *Dk & Spr Dk 3 B*	British Liverpool	1484 2273	290·5 B 70 *f.*	37·2	27·5	Newcastle Wghm.Richardson &Co.	1881	5 Blhds. DBa&f 258t	2Cy.35"& 70"–48" 300HP. W.Richardson&Co.,Newc.	G.Bell & Co.
83	A	**Bari** (ex Liburnia) I Sc *Alfonso* 86-88	Italian Bari	167 326	149·0	22·0	12·6	Fiume Schlick	1873		45 HP.	Soc.Nav.Puglia
84	*	**Barion** I SeBn *Andriola* 86-86 2 *Dks*	Italian Bari	567 882	223·5 P 39 *f.*	28·1 B 28 *f.*	18·2 F 29 *f.*	Newcastle C.Mitchell&Co.	1880	4 Blhds. DBa 55t	2Cy.27"& 54"–32" 184 HP. Wallsend SlipwayCo.,Newc.	Soc.Nav.Puglia
85	*	**Barmore** I SeSrYt *McKinnon* 82-85 1 *Dk*	British Glasgow	49 120	110·3 Q 46 *f.*	18·0 F 19 *f.*	9·0	Paisley J.Fullerton&Co.	1879	4 Blhds.	2Cy.18"& 32"–22" 45 HP. W.King&Co.,Glasgow	R.Thom
86	*	**Barmston** Stl SeSr *Brown* 84-88 1 *Dk & Wb Frms*	British Sunderland	868 1374	250·0 Q 68 *f.*	35·5 B 18 *f.* F 31 *f.*	16·0 P 27 *f.*	Sunderland S.P.Austin&Son	1888	4 Blhds. CDB34t FPT32t APT6t	3Cy.20"32½"&53"–36" 150HP. J.Clark,Sunderlnd	Marquess of Londonderry (S.J.Ditchfield,Manager)
87	*	**Barnesmore** I SeSr *Parry* 85-85 1 *Dk2 B*	British Liverpool	981 1518	259·5 F 40 *f.*	33·9 B 120 *f.*	17·9 Q 87 *f.*	Liverpool T.Royden&Sons	1880	4 Blhds. WBa&f MT FPT APT	2Cy.28"& 56"–39" 150 HP. J.Jack&Sons,Liverpool	S.S. "Barnesmore" Co. Ld. (W. Johnston & Co.)
88	*	**Barnsley** I SeSr *Bee* 88-88 1 *Dk2 B*	British Grimsby	374 603	185·2 P 45 *f.*	27·0 F 32 *f.*	15·0	Glasgow J.Elder&Co.	1876	4 Blhds.	2Cy.28"& 50"–30" 90 HP. J.Elder&Co.,Glasgow	Manchester, Sheffield & Lincolnshire Raily. Co.
89	*	**Baron Ardrossan** I SeSr *Russell* 87-87 1 *Dk2 B*	British Ardrossan	917 1451	243·5 Q 87 *f.*	34·0 B 121 *f.*	17·9 F 33 *f.*	Whitby T.Turnbull&Son	1881	4 Blhds. DBa&f 205t	2Cy.31"& 58"–36" 130 HP. Blair&Co.,Ld.,Stockton	H.Hogarth
90	*	**Baron Belhaven** Stl SeSr *Wiseman* 2 *Dks*	British Ardrossan	1511 2356	300·5 P 32 *f.*	40·0 B 70 *f.*	21·9 F 42 *f.*	Port Glasgow R.Duncan&Co.,	1887	5 Blhds. CDB367t APT50t	3Cy.22"35½"&57"–39" 250 HP. Duncan,Stewart&Co., Glasgow.	H.Hogarth

*=Lloyd's Register; D=Nederlandsche Vereeniging; G=Germanischer Lloyd; I=Registro Italiano; R=American Record;
A=Veritas Austro-Ungarico; H=Veritas Ellenico; N=Norske Veritas; V=Bureau Veritas.

Dans la 2me Colonne, B.T., avec les indications qui suivent, signifie la date du rapport de visite de la Carène par les experts du Board of Trade pour obtenir le certificat relatif au TRANSPORT DES PASSAGERS.

Dans la 3me Colonne, C=Composite (membrure en fer, bordé en bois); I=Fer; Stl=Acier; W=Bois.

Pour plus amples renseignements sur les navires classés par le Lloyd's Register, voir le Lloyd's Register Book et le Lloyd's Yacht Register.

No. in Book / Official No.	Registry in which classed	Ship's Name. Material, Rig, &c. / Master, and Dates of Appointment to present Owner's Service and to Ship / Ship's Late Name. No. of Decks, &c.	Flag / Port of Registry.	Tons Net / Gross	Length	Breadth	Depth	Built Where / By Whom	When	No. of Bulkheads / Particulars of Water Ballast	Engines. Particulars. / Maker's Name.	Owners or Managers
91 86882	*	**Baron Clyde** I ScSr *Wiseman 83-87* 1 *Dk2 B*	British Ardrossan	1155 1696	260·1 *Q 98 f.*	37·1 *B 58 f.*	18·5 *F 29 f.*	Whitby T.Turnbull&Son	1883	4 Blhds. DB87&73TB	2Cy.33"& 62"–39" 160 HP. Blair&Co.,Ld.Stockton	H.Hogarth
92 92870	*	**Baron Douglas** Stl ScSr *Shields 1288 2 Dks,& Wb F'rms*	British Ardrossan	1725 2730	302·3 *P 47 f.*	40·2 *B 68 f.*	23·5 *F 27 f.*	Port Glasgow [illegible]	1888	5 Blhds. CDB[illegible] APT[illegible]	3Cy.22",36"& 57"–42" [illegible] 200 HP.	H.Hogarth
93 29793		**Baron Hambro'** I Sc3mSr *Hale 85-88* 1 *Dk2 B*	British W.Hartlep'l	367 [illegible]	209·0 *Q 30 f.*	25·2	14·5	London C.Langley	1861	3 Blhds.	2Cy.20½"& 40"–26" 70 HP. C.&W.Earle,Hull	E.Jenkins & Co.
94 94382	* V BT3.88	**Baron Osy** I PadSr *Grant* 1 *Dk2 B*	British London	573 963	237·9 *P 71 f.*	30·2 *F 44 f.*	14·9	Newcastle C.Mitchell&Co.	1875	4 Blhds.	2Cy.42"& 78"–72" 250 HP. Thompson,Boyd&Co.,Newc.	Arnati & H[illegible]
95 187	A	**Baron de Pino** I ScSr *Andreis* (ex Salvo) 1 *Dk*	Austrian Zara	147 233	140·1	17·0	11·0	Trieste StabilimentTecnico	1883		2Cy.18½"& 32"–21" 50 HP. Stab[ilime]ntoTecnico,Trieste	F[lli] Rismondo
96		**Baron van Tuijll** I ScSr *Mault -88* (ex Ban Hok, ex Sing Tjin) 1 *Dk*	Dutch Batavia	173 282	132·3 *P 20 f.*	21·3	13·3	Amsterdam Kon.Fab.v.Stoom	1880	4 Blhds.	2Cy.18"&32"–28" 45 HP. Kon.Fab.v.Stoom,Amstrdm	Billiton Maatsch.
97 44	I	**Barone Ricasoli** I ScSr *Bonomolo -88* 1 *Dk*	Italian Palermo	96 164	127·6	17·1	9·3	Glasgow J.G.Lawrie	1865		2Cy.14"& 27½"–20½" 53 HP. Comp.Lawrie,Glasgow	Navigazione Gen. Italiana
98 64795	*	**Barrabool** I ScBk *Tindall* 1 *Dk2 B*	British Melbourne	588 942	224·8 *Q 84 f.*	30·5 *B 44 f.*	15·2 *F 36 f.*	Newcastle Palmers'S.B.C.	1874	5 Blhds.	2Cy.30½"&57½"–36" 140 HP. Comp.'81,Sydney	W.H.Smith & S[ons]
99 87184	*	**Barracouta** Stl ScSr *Hubbard 86-86* 2 *Dks&pt Shd Dk*	British London	1082 1650	275·0 *P 74 f.*	35·1 *B 58 f.*	19·9 *F 30 f.*	Glasgow J.&G.Thomson	1883	4 Blhds. APT35t	2Cy.32"& 63"–45" 176 HP. J.&G.Thomson,Glasgow	S.S. "Barracouta" Ld.(Leagerat [illegible])
100 67382		**Barrington** I ScSr *Taplin 75-77* 1 *Dk*	British Newcastle	236 374	159·2 *Q 70 f.*	21·3 *F —*	12·1	Newcastle Coulson,Cook&Co.	1873	4 Blhds. DBa31t	2Cy.19"&36"–24" 50 HP. Thompson,Boyd&Co.,Newc	Humble & Th[illegible]
101 81157	*	**Barrowmore** I ScBn *Inch 80-87* 2 *Dks3 B*	British Liverpool	2454 3715	350·0 *P 33 f.*	42·7 *B 94 f.*	28·8 *F 68 f.*	Stockton Richardson,Duck&Co.	1884	6 Blhds. DBa[illegible] [illegible] FPT[illegible] APT[illegible]	2Cy.43"& 84"–54" 400 HP. Blair&Co.,Ld.,Stockton	S.S. "Barrowmore" Ld.(W.Johnston [illegible])
102	*	**Basic** I ScSr 1 *Dk*	Spanish Gijon	106 169	101·0 *Q 33 f.*	19·1 *B 8 f.*	8·6 *F 12 f.*	NorthShields T.&W.Smith	1884	3 Blhds. FPT14	2Cy.14"& 27"–20" 30 HP. Baird&Barnsley,N.Shields	M.Canosa
103 65649	BT4.88	**Basil** (ex Mirfield) I ScSr *Thompson 80-81* 2 *Dks3 B*	British Liverpool	748 1185	221·4	30·4	17·6 244	Greenock Scott&Co.	1871	4 Blhds. WBaMT	2Cy.27"&55"–39" 130 HP. GreenockFoundryCo.,Grnk.	Booth Steam S[hip Co.] Ld.
104 [illegible]		**Bassano** (ex Altona) I Sc3mSr *Rea* 2 *Dks3 B*	British Hull	1187 1822	280·0 *Q 110 f.*	35·0 *F 30 f.*	24·5	Middlesbro' Backhouse&Dixon	1872	4 Blhds.	3C.21",35"&67"–36" 260 HP. Cmp.'93C.D.Holmes&C.,Hull	T.Wilson, Sons [&Co.]
105 52629		**Bassein** I PadSr 2 *Dks*	British Calcutta	208 340	205·0	21·2	10·3	Glasgow Barclay,Curle&Co.	1865	6 Blhds.	2Cy.25"& 40"–60" 110 HP. Barclay,Curle&Co.,Glasgow	India General S[team] Navigation Co.Ld.
106	N	**Bastanta** (ex Sailing Vessel) W ScSr *Falck 65-85* 1 *Dk*	Norweg'n Bergen	104 150	95·6 *P 72 t.*	23·5 *Q 22 t.*	11·4	Stavanger	1872		1Cy.11"–12" 12 HP. BergensMekaniskeVerksted	Bergh & Helland
107	V	**Bastia** I ScSr *Bargilliat -88* 2 *Dks,&Spr Dk*	French Marseilles	682 1166	260·5 *H 80 t.*	29·9	15·2 22·7	Greenock Scott&Co.	1872		2Cy.33"& 66"–36" 250 HP. GreenockFoundryCo.,Grnk.	Compagnie Gén[érale] Transatlantique
108 65911		**Batanga** I Sc3mSr 1 *Dk*	British Liverpool	113 201	130·3 *Q 18 t.*	20·1 *F 4 t.*	8·8	Liverpool T.Royden&Sons	1871	3 Blhds.	2Cy.15"& 30"–18" 40 HP. Fawcett,Preston&Co.,Lvpl	T.W.& E.H.Co[ok]
109		**Batangas** Stl Sc (ex Battambang) 1 *Dk*	Spanish Manila	116 185	118·1	20·0	8·7	La Seyne F.&C.de la Medit.	1880		2C.18⅛"&35⅛"–17¾" 70 HP. F.&C.delaMedit.,LaSeyne	F.L.Roxas
110 63751		**Batavia** I ScBg *Wotton 87-87*	British London	1662 2549	327·4 *P 201 t.*	39·3 *F 45 t.*	26·5	Dumbarton W.Denny&Bros.	1870		4C.23",26"&58"–42" 250 HP. J.Elder&Co.,Glasgow	Guion Steam S[hip Co.] Ld.
111	* D	**Batavia** I ScBk *de Gruiter -87* 3 *Dks*	Dutch Rotterdam	1625 2223	292·3	36·9	26·3	Flushing Kninkl.My.deSchelde	1883	5 Blhds. DBa410t	2Cy.38"& 70"–36" 240 HP. Kon.Maats.deSchelde,Flushing	Rotterdamsche L[loyd]
112		**Batavia** PadSr	Dutch Sourabaya	83 106	..	..	..				18 HP.	Hoijnck & Co.
113	D	**Batavier** I ScSr *Sauer 74-83* 2 *Dks*	Dutch Rotterdam	502 678	226·3 *P 44 f.*	29·9 *Q 19 f.*	13·6 *B 37 f.*	Rotterdam Nederl.Stoomb.Maatsch	1872		2Cy.25"& 45"–30" 115 HP. Nederl.Stmbt.Maats.,Rtdm.	Nederlandsche S[toom-] boot Maatschappij
114		**Batavier** I ScSr *Nagtigaal* 1 *Dk*	Dutch Macassar	234 340	140·3 *B 21 f.*	22·0 *F 21 f.*	12·0	Slikkerveer P.Smit,jr.	1882	4 Blhds. DBa[illegible]	2Cy.15¾"&30¾"–21" 40 HP. P.Smit,jr.,Slikkerveer	Macassarsche S[toom-] vaart Vereeniging
115 87125	* D	**Batley** I ScSr *Green* 1 *Dk2 B*	British London	550 865	214·6 *Q 98 f.*	31·1 *B 18 f.*	14·7 *F 26 f.*	Campbeltown Campb'town S.B.C.	1883	4 Blhds. DB[illegible] FPT[illegible]	2Cy.26"& 50"–33" 99 HP. Kincaid&Co.,Greenock	J.T.& J.Taylor
116 79548		**Batman** I ScSr 1 *Dk*	British Melbourne	169 388	153·6	26·1	12·2	Renfrew W.Simons&Co.	1883		1Cy.22"& 40"–27" 75 HP. W.Simons&Co.,Renfrew	Melbourne Harb[our] Trust Commission[ers]
117		**Batoum** I Sc *Michailitziano*	Russian Rostoff	403	..	..	..	Rostoff R.Tulsental	1885		70 HP.	J.Ferraponte Mas[illegible]
118	V	**Battambang** Stl ScSr *Noury -88* 1 *Dk&Awn Dk*	French Saigon	317 688	191·7 *B 31 f.*	28·0	21·0 14·1	Nantes A.Dubigeon	1888		3Cy.18",29"&47"–30" J.Voruz ainé,Nantes 150 HP.	Cie.des Messageries Fluviales de Cochinchine
119 90085	*	**Battle Isle** I Sc *McFarlane 84-85* 1 *Dk*	British Glasgow	44 188	125·0 *Q 52 f.*	20·1 *B 5 f.*	9·5 *F 19 f.*	Glasgow Scott&Co.	1885	3 Blhds. FPT15t	2Cy.21"& 40"–27" 68 HP. W.King&Co.,Glasgow	W.& R.Hay
120		**Battler** I ScSr *O'Niel*	American Baltimore	139	116·0	22·0	11·0	Philadelphia Neafie&Levy	1881		2Cy.20"& 34"–26" Neafie&Levy,Philadelphia	G.W.Beck

***=Lloyd's Register;** D=Nederlandsche Vereeniging; G=Germanischer Lloyd; I=Registro Italiano; R=American Rec[ord];
A=Veritas Austro-Ungarico; H=Veritas Ellenico; N=Norske Veritas; V=Bureau Veritas.

In the 2nd Column, B.T., with figures following, signifies Date of report of bottom sighting by Board of Trade Surveyors as condition of PASSENGER CERTIFICA[TE]

In the 3rd Column, C=Composite (iron frame, wood planked); I=Iron; Stl=Steel; W=Wood.

Further particulars of Vessels classed by Lloyd's Register will be found in Lloyd's Register Book and Lloyd's Yacht Register.

1	2	3 Nom, Materiaux, Gréement, etc. Capitaine et dates d'entrée au service des Armateurs actuels et à bord du navire. Ancien Nom, nombre de ponts, etc.	4 Pavillon. Port d'Armement	5 Tonnage. Net. Total	6 Dimensions. Longueur / Largeur / Creux Constructions sur le Pont	7 Construit. Où. Par qui	8 Quand.	9 Nombre des cloisons. Détails sur le test d'eau.	10 Machines. Détails. Nom du Constructeur.	11 Armateurs ou Directeurs-Gérants.
[illegible]		**Banan** W Sc	Spanish Manila	114 210	110·0 | 21·8 | 8·5	Manila	1885		60 HP.	Inchausti & Co.
[illegible]	V	**Baumwall** I ScSr *Holm 87-88* 1 *Dk & Spr Dk*	German Hamburg	1051 1629	257·8 | 33·8 | 17·7 *B* [illegible] *f.* 1 21 9	Flensburg FlensburgerSchiffsb.Ges	1881	DB f 1351 WBa 130 APT101	2Cy.32½"&60"–36" 160 HP. FlensburgerSchiffsb.Ges.	Dampfschiff-Rhederei "Hansa"
[illegible]	V	**Bavaria** I ScSr *Fielsch 88-88* 2 *Dks*	German Hamburg	1243 1748	273·9 | 33·8 | 23·5 *P 48 f. F 32 f.*	Flensburg FlensburgerSchiffsb.Ges	1879		2Cy.34"& 62"–42" 180 HP. FlensburgerSchiffsb.Ges.	Hmbg.AmerikPacketf Aktien-Gesellschaft
124		**Bavarian** I Sc4mSr *Leask 81-87* 3 *Dks*	British Liverpool	2027 3111	100·0 | 37·2 | 28·5	Belfast Harland&Wolff	1869	7 Bhds.	2Cy.32"& 66"–54" 300 HP J.Jack&Co.,Liverpool	S.S."Bavarian"Co.,Ld. (F.Leyland& Co.)
125	*	**Bawean** Stl ScSr *Van Borssen* 1 *Dk2 B*	Dutch Batavia	649 923	210·0 | 30·5 | 13·8 *P 180 f. F 32 f.*	Sunderland J.Laing	1886		3C.19"31"&51"–33" 150HP. G.Clark,Sunderland	Nederland.-Indische Stoomvaarts Maats.
126		**Bay City** W	American S.Francisco	933 1283	230·0 | 36·8 | 13·6	San Francisco	1878		800 HP.	C.S.Neal
127	*	**Bay Fisher** I ScSr *Bannister 81-87* 1 *Dk*	British Barrow	177 352	162·5 | 23·9 | 11·9 *B 35 f. F 29 f. Q 21 f.*	Belfast M'Ilwaine, Lewis & Co.	1884	4 Bhds. FPT211 APT271	2Cy.19"& 36"–30" 60 HP. M'Ilwaine,Lewis&Co.,Blfst.	"Bay Fisher"SS.Co.Ld. (J.Fisher &Sons)
128		**Bay Queen** W Pad	American Providec.RI	557 680	184·3 | 29·7 | 9·6	Brooklyn	1865		300 HP.	B.Buffum
129		**Bayazid** W Pad *Osman 55-72*	Turkish Constntinple.	64	120·0 | 17·0 | 8·1	West Cowes J.White	1860		40 HP. JohnPenn&Sons	Chirket-i-Hairie (BosphorusSteamNav.Co.)
130	G	**Bayern** StlScBg *Mergell 83-88* 3 *Dks*	German Bremen	2880 4577	388·4 | 45·0 | 32·6 *P79f. B161f. F56f.*	Stettin Act.Ges."Vulcan"	1886		3Cy.36"57"&91"–59" Act.Ges."Vulcan"	Norddeutscher Lloyd
131		**Bayfield** (ex Edsall, ex General U.S. Grant) W Sc 1 *Dk*	British Toronto	94 150	110·0 | 18·5 | 9·3	Buffalo,U.S. W.Crosthwait	1864		1Cy.28"–30" 75 HP. D.Bell,Buffalo,U.S.	William J.Murray
132	*	**Bayley** Stl ScSr *Child -86* 2 *Dks3 B*	British London	1696 2607	329·5 | 40·0 | 24·5 *B80f. P38f. F40f.*	Newcastle C.S.Swan&Hunter	1886	5 Bhds. [illegible] FPT114 APT114	3Cy.26"42"&69"–48" Wallsend Slipway & Eng. Co. Ld. [400 HP.	C.C.Barton
133		**Bayonés** (ex Ugarte No. 1.) I ScSr 1 *Dk*	Spanish Bilbao	120 177	118·2 | 20·6 | 9·8 *Q 20 f. F 19 f.*	Dundee W.B.Thompson	1876	4 Bhds.	2Cy.17"&34"–20" 45 HP. W.B.Thompson,Dundee	P.Pedrósa
134	*	**Bayswater** I ScSr *Davidson 85-85* 2 *Dks3 B*	British London	1039 1623	251·4 | 35·1 | 21·5 *P26f. B50f. F30f.*	Sunderland J.Blumer&Co.	1880	4 Bhds. CDB281	2Cy.30"&58"–39" 160 HP. T.Clark&Co.,Newcastle	Watts,Ward & Co.
135	*	**Bazalgette** StlTwnSc *Vireash* 1 *Dk*	British London	613 990	230·2 | 38·2 | 13·8	Barrow BarrowS.B.Co.,Ld.	1887	5 Bhds.	3Cy.15"22"&23"–24" BarrowS.B.Co.Ld. [120HP.	Metropolitan Board of Works
136	*	**Bazan** I Sc3mSr *Gamecho 74-84* 1 *Dk2 B*	Spanish Barcelona	496 805	194·5 | 29·6 | 15·5 *P66f. B9f. F27 f.*	SouthShields J.Readhead&Co.	1880	4 Bhds.	2Cy.25"&46"–30" 90 HP. J.Readhead&Co.,S.Shields	J. Roca & Co.
137		**Beacon** I PadSr	British London	181 262	146·9 | 22·2 | 12·5	London M.Wigram&Sons	1862		2Cy.44½"&49½"–36" 120HP J.Penn&Sons,Greenwich	Corporation of Trinity House
138	*	**Beaconsfield** I ScSr *Cotton 77-77* 2 *Dks3 B*	British S.Shields	1118 1736	270·0 | 34·0 | 24·5 *Q 40 f. F 38 f.*	Newcastle TyneI.S.B.Co.	1877	4 Bhds. DBa&b 1961	2Cy.32"&61"–36" 160 HP. N.E.Marine Eng.Co.Ld.,Sld.	G.Cleugh & Co.
139		**Beagle** WSc3mSr 1 *Dk*	British Brisbane	156 229	147·1 | 21·0 | 8·8	Reid R.NSW W.Yabsley	1877		2Cy.21"–15" 35 HP.	JohnM.Brydon
140	*	**Beamish** I ScSr *Watson 79-83* 1 *Dk2 B*	British London	846 1318	240·0 | 35·2 | 16·0 *Q90f. B50f. F30f.*	SouthShields J.Readhead&Co.	1883	4 Bhds. DBa&f 260t APT14t	2Cy.30"& 57"–36" 135 HP. J.Readhead&Co.,S.Shields	J.Fenwick & Son
141	*	**Bear** I ScSr *Howrie* 1 *Dk2 B*	British Middlesbro'	370 593	174·4 | 25·6 | 14·3 *Q 66 f. F 20 f.*	Middlesbro' R.Dixon&Co.	1877	4 Bhds. DBa&f 149t	2Cy.21"& 45"–26" 80 HP. T.Richardson&Sons,Htlpl.	J.Watson & Co.
142		**Bear** I ScSr *Macchie 74-85*	British Glasgow	317 658	220·6 | 28·1 | 14·1 *P 64 t. F 50 t.*	Glasgow J.&G.Thomson	1870		2Cy.33½"&58½"–36" 150HP J.&G.Thomson,Glasgow	G.&J.Burns
143	*V	**Béarn** I ScBg *de Ferry* 3 *Dks*	French Marseilles	2616 4134	398·9 | 40·2 | 31·6 *B 91 f. F 73 f.*	Barrow BarrowS.B.Co.,Ld.	1881	8 Bhds. DBa&f 598t	2C.49¼"&90½"–47¼" 500HP Barrow S.B.Co.,Ld., Barrow	Société Générale de Transports Maritimes
144	*	**Beatrice** StlScSr *Grate -88* 1 *Dk2 B*	British London	1473 2241	290·0 | 39·0 | 18·7 *F33f. Q98f. B138f.*	Sunderland Bartram,Haswell&Co.	1888	6 Bhds. CDB102t APT21t	3Cy.22"36½"&60"–42 J.Dickinson,Sld. [200 HP.	C.Howard & Son
145		**Beatrice** I Pad	British London	38 159	100·0 | 20·6 | 10·7	London SamudaBros.	1881		2Cy.32½"–54" 80 HP. J.Stewart&Son,London	London& StKatharine DockCo.
146		**Beautiful Star** I ScSr 1 *Dk*	British Dunedin	146 177	119·2 | 17·2 | 10·6 *H 21 t.*	Newcastle Wghm.Richardson &Co.	1862	4 Bhds.	2Cy.18"–13" 30 HP.	Union S.S.Co.of New Zealand
147		**Beaver** (STRANDED) W Pad 1 *Dk*	British Victoria, BC	109 187	101·4 | 20·0 | 11·0	London Green,Wigram & Green	1835		70 HP.	E.C.Baker
148		**Beaver** I ScSr *Le Maistre* 1 *Dk*	British Quebec	100 147	109·5 | 19·1 | 8·9 *Q 12 t. F 13 t.*	Glasgow Dobie&Co.	1872	4 Bhds.	25 HP.	A.Frazer & Co.
149	*	**Beaver** Stl T-ScSrTug *Mitchell -86* 1 *Dk*	British Brisbane	73 222	135·2 | 21·6 | 10·6 *B 55 f.*	Leith Ramage&Ferguson	1886	4 Bhds.	4Cy.17"& 34"–18" 96 HP. Ramage& Ferguson,Leith	Webster & Co.
150		**Bebington** I Double TwnSc *Hobby 57-*	British Liverpool	140 435	130·0 | 45·2 | 12·4	Preston Allsup&Sons	1880		4Cy.19"& 34"–24" 98 HP. W.Allsup&Sons,Preston	MunicipalCorporation of Birkenhead

* = Lloyd's Register; D = Nederlandsche Vereeniging; G = Germanischer Lloyd; I = Registro Italiano; R = American Record;
‡ = Veritas Austro-Ungarico; H = Veritas Ellenico; N = Norske Veritas; V = Bureau Veritas.

Dans la 2me Colonne, B.T., avec les indications qui suivent, signifie la date du rapport de visite de la Carène par les experts du Board of Trade pour obtenir le certificat relatif au TRANSPORT DES PASSAGERS.

Dans la 3me Colonne, C = Composite (membrure en fer, bordé en bois); I = For; Stl = Acier; W = Bois.

Pour plus amples renseignements sur les navires classés par le Lloyd's Register, voir le Lloyd's Register Book et le Lloyd's Yacht Register.

No. in Book. / Official No.	Registry in which classed.	Ship's Name. Material, Rig, &c. / Master, and Dates of Appointment to present Owner's Service and to Ship / Ship's Late Name. No. of Decks, &c.	Flag. / Port of Registry.	Tons Gross / Net	Length	Breadth	Depth	Built Where. / By Whom.	When	No. of Bulkheads. / Particulars of Water Ballast.	Engines. / Particulars. Maker's Name.	Owners or Managers
151 82782	✻ D	**Beckton** I ScSr *Gibson 84-87* 1 *Dk2 B*	British London	855 1477	218·0 *P 20 f.*	35·3 *B 26 f.*	17·6 *F 26 f.*	Sunderland J.L.Thompson&Sons	1880	6 Blkds. DBle[illegible]	2Cy.33"& 63"–39" 160 HP. N.E.Marine Eng.Co.Ld.,Shd.	R.Gordon & Co.
152 77352	✻ D	**Bedale** I ScSr *Mackwood 82-82* 1 *Dk2 B*	British Scarbro'	822 1213	242·0 *Qs8.*	32·6 *B60 f.*	17·6 *F26 f.* 19·9	SouthShields J.Readhead&Co.	1879	4 Blkds. DBa[illegible]	2Cy.27½"&52"–33" 110HP. J.Readhead&Co.,S.Shields	P.Hick,jun.
153 93337	✻	**Bede** I ScSr *Buck 86-86* 1 *Dk.&Spar Dk3 B*	British London	2134 3210	322·0 *B 80 f.*	42·1 *F 21 f.*	19·9 27·1	Sunderland Short Bros.	1886	5 Blkds. DB[illegible]	3C.25"40"&66"–45" 160HP. J.Dickinson,Sunderland	J.S.Barwick
154 78229	✻	**Bedford** I ScSr *Aitkenhead 80-88* 2 *Dks3 B*	British N.Shields	1201 1840	270·0 *P25 f.*	35·0 *B 98 f.*	24·1 *F 32 f.*	Newcastle Tyne I.ShipB.Co.	1879	5 Blkds. DB[illegible]	2Cy.32"& 63"–39" 180 HP. N.E.MarineEng.Co.,Ld.,Shd.	Cay, Hall & Co.
155 89170	✻	**Bedlington** I ScSr *Owen 84-87* 1 *Dk2 B*	British W.Hartlepl.	889 1379	240·1 *Q102f.*	34·2 *B[illegible]*	16·3 *H26f. F31f.*	Barrow BarrowS.B.Co.,Ld.	1884	4 Blkds. WB	2Cy.30"& 60"–36" 150 HP. BarrowS.B.Co.,Ld.,Barrow	Lilly, Wilson & Co.
156 79937	✻	**Bedlormie** I Sc3mSr *Sealey* 1 *Dk*	British Glasgow	243 432	155·0 *P 58 f.*	23·6 *F 23 f.*	12·4	Middlesbro' R.Dixon&Co.	1878	3 Blkds. DB[illegible]	2Cy.23½"& 42"–30" 70 HP. Blair&Co.,Ld.,Stockton	P. M'Morland
157 81270	✻	**Bedouin** I ScSr *Ridson 80-87* 2 *Dks3 B*	British Liverpool	1277 1967	286·0 *P 27 f.*	35·0 *F 37 f.*	21·2	Newcastle C.Mitchell&Co.	1879	5 Blkds. DB[illegible]	2Cy.31"& 61"–42" 225 HP. WallsendS.Co.Ld.Wallsend	BedouinStm.Nav. Ld.
158 63288		**Bee** (WRECKED) I ScSr *Jones 80-88*	British Liverpool	88 1[illegible]	126·0 *Q 12 f.*	18·6 *F 3 f.*	8·0	Pt.Glasgow H.Murray&Co.	1870		2Cy.12"& 24"–18" 25 HP. McNab&Co.,Greenock	"Bee"S.S.Co.,Ld. (R. & D.Jones)
159 85007	✻ D	**Beechville** I ScSr *Watson 78-83* 1 *Dk2 B*	British Sunderland	1128 1747	257·1 *Q 98 f.*	35·7 *B 67 f.*	18·8 *F 33 f.*	W. Hartlepool W.Gray&Co.	1882	4 Blkds. DB[illegible]	2Cy.32"& 60"–39" 150 HP. Blair&Co.,Ld.,Stockton	R.Shadforth & Co.
160 93392		**Beeloo** Stl Pad	British Glasgow	650 1033	310·0	40·1	11·7	Dumbarton W.Denny&Bros.	1886		2Cy.37½"&65"–72"375HP. Denny&Co.,Dumbarton	Irrawaddy Flotilla Co.,Ld.
161 87048	✻	**Beeswing** Stl ScSr *McIntyre* 1 *Dk2 B*	British Melbourne	716 1119	230·0 *Q 77 f.*	33·2 *B 61 f.*	15·5 *F 32 f.*	Grangem'th Dobson&Charles	1882	5 Blkds. DB[illegible]	2Cy.27"& 53"–36" 120 HP. Hutson&Corbett,Glasgow	Melbourne Coal Shipping & Eng.Co.,Ld.
162 95262	✻ D	**Beethoven** Stl ScSr *Sawyer 80-88* 1 *Dk.& Web Frms*	British Sunderland	1407 2216	290·6 *P 30 f.*	38·6 *B 12 f. F 23 f.*	19·5 *Q 80 f.*	Sunderland J.L.Thompson&Sons	1888	6 Blkds. CDB[illegible]	3C.22"37"&60"–39" 200HP. J.Dickinson,Sunderland	JennesonTaylor
163 89763	✻	**Bega** Stl TwnScSr *Edwards* 1 *Dk.&Awn Dk*	British Sydney	305 507	189·4	24·6	8·0 16.0	Greenock Russell&Co.	1883	4 Blkds. [illegible]	2Cy.25"& 48"–27" 98 HP. Rankin&Blackmore,Grnk.	IllawarraStm.N.
164		**Beharyeh** W Pad *Osmanh 53-*	Turkish Cnstntinple.	140	160·0	21·0	10·0	Poplar M.Wigram	1865		90 HP. Maudslay,Sons&Field	Chirket-i-Hairie (BosphorusSteam)
165 49619	✻	**Behera** (ex Fanny David) I ScSr *McDonald 85-87* 2 *Dks.&Spr Dk*	British Glasgow	829 1384	239·5	32·7 *B 33 f.*	17·0 24·9	Newcastle MarshallBros.	1864	4 Blkds.	2Cy.30"& 54"–36" 127 HP. Cmp.'79, M.Paul&C.,Dmbtn	Maclay & McIntyre
166		**Behera** I ScBn *Spr Dk*	Turkish Alexandria	1080	274·0	34·6	22·0	London SamudaBros.	1865		2Cy.38"& 70"–48" 300 HP. JamesJack&Co.,Liverpool	KhedivichMail S.
167 63816	✻ D	**Beignon** I ScBn *Read* 1 *Dk2 B*	British Cardiff	868 1332	250·0 *Q 79 f.*	35·0 *F 29 f.*	15·2 *B 70 f.*	Stockton Richardson,Duck&Co.	1880	4 Blkds. CDB[illegible]	2Cy.30"& 55"–36" 99 HP. Blair&Co.,Ld.,Stockton	"Beignon"S.S.Co. (MorelBros.&Co.)
168 82278	✻	**Belair** I ScSr *Dunlop 84-84* 1 *Dk*	British Glasgow	908 1419	243·8 *Q 51 f.*	32·7 *F 11 f. B 56 f.*	16·6 *F 33 f.*	Glasgow J.&G.Thomson	1879	5 Blkds. DB[illegible]	2Cy.29"& 52"–36" 150 HP. J.&G.Thomson,Glasgow	Caw, Prentice, Clapperton & Co.
169		**Belbek** (ex Dowerin) I ScSr 2 *Dks*	Russian Odessa	410 582	195·0	27·0	11·5	Linz Mayer	1871		2Cy.20"& 31"–26" 60 HP. Escher,Wyss&Co.,Zurich	Russian Steam Nav. & Trading Co.
170	✻	**Bele** Stl ScSr *Ericsson 82-83* 1 *Dk.&Spr Dk3 B*	Swedish Gothenburg	1044 1339	231·9 *Erections 32 f.*	30·6	16·2 23·8	Gothenburg MotalaCo.	1879	5 Blkds. [illegible]	2Cy.31½"&54½"–35"190HP. MotalaCo.,Gothenburg	Angf. Aktb. Thule
171 88894	✻ BT1.88	**Belfast** I ScSr *Cleminger 87-87* 2 *Dks*	British Dublin	147 425	192·3	24·2 *F 29 f.*	13·5	W. Hartlepool Irvine&Co.	1884	5 Blkds.	2Cy.28½"&53"–33" 110HP. Blair&Co.,Ld.,Stockton	City of Dublin S. Packet Co.
172	✻ V	**Belgenland** I Sc *Heynon 84-84* 3 *Dks4 B*	Belgian Antwerp	2364 3692	402·9	40·2	30·6	Barrow BarrowS.B.Co.	1878	7 Blkds.	2Cy.52"&90"–54" 600HP. BarrowS.B.Co.,Barrow	Soc. Anon.deNav. Belge-Americaine
173 91208		**Belgic** Stl & I Sc4mSr *Walker* 3 *Dks*	British Liverpool	2695 4212	420·3	42·4	29·6	Belfast Harland&Wolff	1885	7 Blkds.	2Cy.43"& 86"–51" 500 HP. Harland& Wolff,Belfast	OceanicSteamNav. Ld.
174		**Belgique** I Pad	Belgian Antwerp	269	182·1	21·3	6·8	Antwerp Soc.J.Cockerill	1863			Belgian Government
175		**Belgrano** I Sc3mSr *Lequeux 82-86* 2 *Dks*	French Havre	1647 2121	321·5 *P 75 f.*	33·1 *B 22 f.*	26·2 *F 55 f.*	Havre Forges & Cdc l'Ocean	1872		2Cy.34"&59"–47¼" 150HP. B.Normand,Havre	Chargeurs Réunis
176	V	**Belgrano** Stl ScSr *Riedel 84-88* 2 *Dks*	German Hamburg	1849 2578	321·0 *P[illegible]*	39·8 *B[illegible]*	25·8 *F 54 f.*	Hamburg [illegible]	1888		3Cy.25,"40½"&65"–42" 380IP. Reiherstieg Schiffswerfte Hamburg	Hamb.-Sud.-Amerik. Dampfs.-Ges.
177 83011	BT9.89	**Belgravia** I ScBk *Harris 73-88* 2*DksSpr Dk& WebFrms*	British Glasgow	3275 4977	400·0 *P[illegible]*	44·8 *B[illegible]*	33·0 *F[illegible]*	Glasgow D.&W. Henderson	1881	7 Blkds. WB FPT APT	2Cy.57"&103"–48" 495HP. D.&W.Henderson&C.,Glw.	Henderson Bros.
178 0383		**Belize** (ex Elbon) I ScSr	British London	632 1058	231·1	28·2	22·0	Glasgow J.&G.Thomson	1871		2Cy.33½"&58½"–42" 160HP. J.&G.Thomson,Glasgow	Royal Mail Steam Co.
179 96062	✻	**Bellarden** Stl ScSr *Inscidem 88-88* 2 *Dks3 B*	British Glasgow	1770 2745	310·0 *B 62 f.*	39·2 *P 32 f.*	23·8 *F 40 f.*	Glasgow D.&W.Henderson&Co.	1888	5 Blkds. [illegible]	3Cy.22",36"&59"–42" 250HP. D.&W.Henderson&Co.,Glasgow	BellBros.&M.Lelland
180 87691	✻	**Bellarena** I ScSr *Maclean 83-88* 1 *Dk2 B*	British Glasgow	1008 1588	260·0 *Q 127 f.*	36·6 *F 35 f.*	19·4	Dumbarton H.Murray&Co.	1883	4 Blkds. DB[illegible]	2Cy.30"& 60"–42" 160 HP. Hutson&Corbett,Glasgow	"Bellarena"S.S. Co. Ld. (BellBros.&M.Lelland)

✻ = **Lloyd's Register**; D = Nederlandsche Vereeniging; G = Germanischer Lloyd; I = Registro Italiano; R = American Bureau; A = Veritas Austro-Ungarico; H = Veritas Ellenico; N = Norske Veritas; V = Bureau Veritas.

In the 2nd Column, B.T., with figures following, signifies Date of report of bottom sighting by Board of Trade Surveyors as condition of PASSENGER CERTIFICATE.

In the 3rd Column, C = Composite (iron frame, wood planked); I = Iron; Stl = Steel; W = Wood.

Further particulars of Vessels classed by Lloyd's Register will be found in Lloyd's Register Book and Lloyd's Yacht Register

		Nom, Matériaux, Gréement, etc. / Capitaine et dates d'entrée au service des Armateurs actuels et à bord du navire. / Anciens Noms, nombre de ponts, etc.	Pavillon. / Port d'Armement	Tonnage Net / Total	Dimensions: Longueur	Largeur	Creux	Construit: Où. / Par qui.	Quand.	Nombre des cloisons. / Détails sur le lest d'eau.	Machines. / Détails. / Nom du Constructeur.	Armateurs ou Directeurs-Gérants.
	✻	**Bellaura** StlScSr *Meppell 87-87* 2 *Dks3 B*	British Glasgow	1769 2714	310·0 *P32 f.*	39·2 *B62 f.*	23·8 *F40 f.*	Glasgow D.&W.Henderson&Co.	1887	6 Bhds. DBo&cd489t FPT86t	3C.22"36"&59"–42"250HP D.&W.Henderson&Co.,Glsg	"Bellaura" S.S.Co.,Ld. (BellBros.& McLelland)
	✻1	**Belleairn** ISeBu *Marlin 88-88* 1 *Dk2 B*	British W.Hartlpool	870 1364	244·6 *Q91 f.*	33·3 *B61 f.*	15·8 *F27 f.*	W. Hartlepool Irvine&Co.	1883	4 Bhds. CDB315t	2Cy.30"& 57"–33" 130 HP. T.Richardson&Sons,W.Hpl.	Jackson Bros.& Cory
	✻	**Belle** ISeTg *Collins 84-84* 1 *Dk*	British Plymouth	7 49	70·3	11·0	8·1	Chepstow E.Finch&Co.,Ld.	1884	1 Bhds.	2Cy.15"& 28"–20" 40 HP. Newall&Co.,Bristol	Barbican Stm.Tow-ingCo.,Ld.
		Belle Isle IPadTg *Jamais* 1 *Dk*	French St.Nazaire	36 178	120·0	20·3 *B 68 f.*	10·2	Nantes Gouin&Co.	1863		2Cy.33"–37" 90 HP. Verus,Nantes	Compagnie Générale Transatlantique
		Belle Wilson W ScBge 1 *Dk*	British Picton,Ont.	119 185	103·3	24·1 *F 6 t.*	10·1	Picton,Ont. W.Redmond	1881		1Cy.16"–18" 100 HP. Davidson,Doran&C.Ktn.Ot.	Blind River Nav. Co. Ld.
	✻	**Belle of Dunkerque** ISeSr 1 *Dk*	British London	323 527	180·0 *B45 f.*	25·1 *P18 f.*	12·5 *F22 f.*	Sunderland ShortBros.	1887	4Bhds.	3Cy.15"25"&40"–27" 80HP J.Dickinson,Sunderland	Weatherley,Mead & Hussey
	✻	**Bellena** StlScSr *Blacklock 85-87* 2 *Dks3 B*	British Glasgow	1770 2715	310·0 *P32 f.*	39·2 *B62 f.*	23·8 *F40 f.*	Glasgow D.&W.Henderson&Co.	1887	6 Bhds. CDB480tFPT80t	3C.22"36"&59"–42"250HP. D.&W.Hendrsn&Co.,Glsgw.	"Bellena" S.S.Co.,Ld. (BellBrothers & M'Lelland)
188	✻	**Bellenden** StlScSr *Fraser 87-87* 2 *Dks3 B*	British Glasgow	1770 2715	310·0 *P32 f.*	39·2 *B62 f.*	23·8 *F40 f.*	Glasgow D.&W.Henderson&Co.	1887	6 Bhds. DBo&EB480t FPT80t	3C'22"36"&59"–42"250HP. D.&W.Hendrsn&Co.,Glsgw	"Bellenden" S.S.Co., Ld.,(BellBrothers & M'Lelland)
189	D	**Bellerophon** IScSr *Guthrie 84-86*	British Liverpool	1397 2155	320·0	34·3	26·0	Greenock Scott&Co.	1880		2Cy.27"&56"–52" Scott&Co.,Greenock	Ocean Steam Ship Co.
	✻	**Bellinger** IScSr *Reid* 1 *Dk*	British Melbourne	134 225	125·0 *B7f.*	22·1 *F15 f.*	8·7 *Q64f.*	Paisley J.McArthur&Co.	1884	4 Bhds. FPT9t APT15t	2Cy.17"& 32"–21" 45 HP. W.Kemp,Glasgow	Belfast& KoroitSteam Nav.Co.,Ld.
191	✻	**Bellingham** IScSmSr *McGregor -83* 2 *Dks3 B*	British N.Shields	1066 1652	260·0	34·2 *P40 f.*	23·3 *F 36 f.*	Newcastle A.Leslie&Co.	1879	5 Bhds. DB[illegible] APT[illegible] MT210t	2Cy.32"& 61"–36" 180 HP. N.E.MarineEng.Co.Ld.,Sld.	J.Coull
192	✻	**Bellini** IScSr *Robertson* 1 *Dk2 B*	British London	1134 1730	269·6 *B60f.*	36·0 *Q64f.*	18·2 *P32f. F28f.*	Sunderland ShortBros.	1880	6 Bhds. CDB[illegible] APT[illegible]	2Cy.33"& 62"–42" 180 HP. J.Dickinson,Sunderland	J.Taylor
193	✻	**Bellmore** IScSr *White 88-88* 1 *Dk2 B*	British Glasgow	717 1121	235·4 *Q 116 f.*	32·8	16·3 *F 31 f.*	Paisley H.M'Intyre&Co.	1882	5 Bhds. DBo&f250t	2Cy.26"& 52"–42" 99 HP. Muir&Houston,Glasgow	"Bellmore"S.S.Co.,Ld. (BellBrothers&M'Lelland)
194	✻	**Bellona** StlScSr *Dunlop 85-87* 2 *Dks3 B*	German Hamburg	1721 2647	330·0 *P43 f.*	38·2 *B70f.*	22·5 *F[illegible]*	Newcastle [illegible]&Hunter	1887	5 Bhds. DBA[illegible] FPT[illegible] APT[illegible]	3C21"37½"&64"–42"200HP R.&W.Hawthorn,Leslie&Co., Newcastle	DeutscheDampfschiffs Rhederei
195		**Bellona** IScBg 1 *Dk2 B*	Japanese Tokio	790 1055	230·9 *P& B. 138 f.*	30·1 *F28 f.*	17·0	W. Hartlepool Denton,Gray&Co.	1872	4 Bhds. WB[illegible]	2Cy.27"& 50"–33" 110 HP. T.Richardson&Sons,Hrtlpl.	Cement Actien Gesell-schaft
196	✻	**Bellucia** Stl ScSr *Aagaard 87-88* 2 *Dks2 B*	British Glasgow	1770 2715	310·6 *B62 f.*	39·2 *P31 f.*	23·8 *F40 f.*	Glasgow D&W.Henderson&Co.	1888	6 Bhds. DBo&[illegible] FPT80t	3C22",36"&59"–42" 250HP D.&W.Henderson&Co.,Glw	Bell Bros.& M'Lelland
197		**Bellver** IScBg 1 *Dk&SprDk3 B*	Spanish Palma,Mjrc.	857 1261	247·8	30·6	15·2 22·9	Glasgow J.&G.Thomson	1879	5 Bhds. MT20t	2Cy.34"& 60"–39" 160 HP. J.&G.Thomson,Glasgow	José Astier
198	✻	**Belmont** IScSr *Stagg 64-89* 1 *Dk2 B*	British Sunderland	577 777	210·3 *Q 23 t.*	28·7 *B29 t.*	16·7	Sunderland J.Laing	1865	1 Bhds. DB[illegible]	2Cy.24"& 46"–27" 80 HP. [illegible] N. E. Marine Eng.Co., Ld., Sunderland	Fenwick & Co.
199		**Belos** StlScSr *Wright 84-87* 1 *Dk*	Swedish Stockholm	185 362	146·6	24·8 *Erections 11 t.*	13·4	Gothenburg MotalaCo.	1885	FPT37t APT16t	2Cy.26"& 48"–30" 160 HP. MotalaCo.,Gothenburg	Bergn.och Dykeribol. "Neptun" (E.Liljewalch)
200		**Beluga** (ex Mary & Helen) W ScBk *Brooks 86-89* 2 *Dks*	American S.Francisco	409 508	144·8	31·3	16·7	Bath, Me. Goss&Sawyer	1882	VT.82	1Cy.22"–30" DelamaterIronWorks,Phil.	Pacific S.Whaling Co.
201	R	**Belvedere** WScBk *Sherman 84-87* 2 *Dk3 B*	American N.BedfordM	440 440	140·6	31·3	17·2	Bath,Me Goss,Sawyer&Packard	1880		1Cy.22"–26" 95 HP. C.H.Delamater&Co.N.York	W.Lewis
202		**Bembridge** IPad *Jones 85-85 (ex Princess of Wales)*	British Portsmouth	73 104	137·4	16·1 *Q 10 t.*	6·8	London London Steamboat Co.	1878		2Cy.24½"–30" 40 HP. Kincaid,Donald&Co.,Grnk.	F.M.Coldwells
203	✻	**Bempton** IScSr *Turner 82-85* 1 *Dk2 B*	British London	1307 2013	272·7 *P26f.Q72f.*	36·1 *B116 f.*	20·0 *F33f.*	W.Hartlepool E.Withy&Co.	1885	4 Bhds. DB[illegible] FPT36t APT66t	2Cy.34"& 63"–36" 170 HP. T.Richardson&Sons,Hartpl.	W.& E.S.Lamplough
204	✻	**Ben Avon** (WRECKED) IScSr *Seorgie 85-86* 1 *Dk*	British Aberdeen	402 632	188·9 *Q 48 f.*	26·2 *B 42 f.*	14·4	Aberdeen Hall,Russell&Co.	1875	5 Bhds. DB[illegible]	3Cy.14"24"&40"–30" 80HP Tpld.'86.HallRussell&C.,Abrdn	Grampian S.S.Co.,Ld.
205	✻D	**Ben Clune** Stl ScSr *Brock 77-88* 1 *Dk&WLFrms*	British N. Shields	1334 2051	273·5 *Q111 f.*	38·2 *F30 f.*	19·3 *B106 f.*	Sunderland J.Laing	1888	DB[illegible]233t	3C.20"33"&54"–36" 80HP. G.Clark,Sunderland	J.Morrison & Son
206	✻	**Ben Ledi** IScSr *Cornish 83-88* 1 *Dk2 B*	British N.Shields	1279 1986	270·0 *Q98 f.*	36·4 *B 56 f.*	21·0 *F34 f.*	Sunderland J.Laing	1884	4 Bhds. DBa&f[illegible]	2Cy.32"& 60"–39" 160 HP. G.Clark,Sunderland	J.Morrison & Son
207		**Ben Ledi** IPadK *Wake*	British Aberdeen	10 114	90·0	18·6	10·2	Aberdeen Hall,Russell&Co.	1878		2Cy.25"–48" 50 HP Hall,Russell& Co.,Aberdeen	R.L.Cook
208	✻	**Ben Lomond** IScSr *Nichols 86-86* 1 *Dk2 B*	British N. Shields	754 1171	234·9	31·7 *Q 33 t.*	19·5	Sunderland J.Laing	1871	4 Bhds. DB[illegible]	2Cy.27"& 53"–33" 98 HP. Thompson,Boyd&Co.,Nwc.	J.Morrison & Son
209	✻	**Ben Macdui** IScSr *Wood 81-87*	British Aberdeen	378 592	184·8 *Q 55 f.*	26·2 *B 38 f.*	14·4	Aberdeen Hall,Russell&Co.	1874	5 Bhds. DB[illegible]	2Cy.24"&40"–30" 72HP Hall,Russell& Co.,Aberdeen	Grampian S.S.Co.,Ld.
210												

✻=Lloyd's Register; D=Nederlandsche Vereeniging; G=Germanischer Lloyd; I=Registro Italiano; R=American Record;
[illegible]=Veritas Austro-Ungarico; H=Veritas Ellenico; N=Norske Veritas; V=Bureau Veritas.

Dans la 2me Colonne, B.T., avec les indications qui suivent, signifie la date du rapport de visite de la Carène par les experts du Board of Trade pour obtenir le certificat relatif au TRANSPORT DES PASSAGERS.

Dans la 3me Colonne, C=Composite (membrure en fer, bordé en bois); I=Fer; Stl=Acier; W=Bois.

Pour plus amples renseignements sur les navires classés par le Lloyd's Register, voir le Lloyd's Register Book et le Lloyd's Yacht Register.

No. in Book. — Official No.	Registry in which classed.	Ship's Name. Master, and Dates of Appointment to present Owner's Service and to Ship. Ship's Late Name.	Material, Rig, &c. No. of Decks, &c.	Flag. — Port of Registry.	Tons. Gross. — Net.	Length	Breadth	Depth	Built. Where. — By Whom.	When	No. of Bulkheads. Particulars of Water Ballast.	Engines. Particulars. Maker's Name.	Owners
211 26139	*	Ben More Wright 77-88	I ScSr 2 Dks & Spr Dk	British N. Shields	882 1362	228·0	32·7	18·0 23·1	Sunderland J. Laing	1860	5 Blhds.	2Cy. 27"&53"–33 98 HP. Gilbert & Cooper, Hull	J. Morrison &
212 67289	BT5.89	Ben My Chree McQueen	I PadSr	British Douglas, I.M.	523 1140	310·0	31·5	13·0	Barrow [illegible]	1875		2Cy. 65"–90" 120 HP. Barrow S.B. Co., Ld., Barrow	Isle of Man S.
213 65441	*	Ben Nevis Oram 81-88	I ScSr 1 Dk 2 B	British N. Shields	639 1003	232·7	30·2	17·7	Sunderland J. Laing	1872	4 Blhds.	2Cy. 27"&53"–33" 98 HP. Thompson, Boyd & Co., New.	J. Morrison &
214 89654	*	Ben Nevis McAlpine 85-85	I ScSr 1 Dk	British [illegible]	73 208	121·0 Q 13 f.	20·1 B 9 f.	10·3 F 13 f.	Paisley [illegible]	1884	3 Blhds. DB [illegible] FPT [illegible]	2Cy. 18"&36"–24" 55 HP. Hanna, Donald & Co., Paisley	D.P. Macdon
215 70997	*	Ben Venue Hoy 79-86	I ScSr 2 Dks 3 B	British N. Shields	989 1515	257·4	34·5 F 19 f.	23·9	Sunderland J. Laing	1877	4 Blhds. DB [illegible]	2Cy. 32"&60"–39" 150 HP. Thompson & Co., Newcastle	J. Morrison &
216 76541	* D	Ben Voirlich Dyer 84-88	I ScSr 2 Dks 3 B	British N. Shields	984 1506	257·2 B 8 f.	34·1 F 18 f.	23·8	Sunderland J. Laing	1877	4 Blhds. DB [illegible]	2Cy. 32"&60"–36" 150 HP. N.E. Marine E. Co., Ld., Sndld.	J. Morrison &
217 77464	*	Benabourd Keith 84-88	I ScSr 1 Dk	British Aberdeen	430 680	190·3 Q 55 f.	31·2 B 42 f.	13·0 F 22 f.	Aberdeen Hall, Russell & Co.	1881	5 Blhds.	2Cy. 26"&49"–30" 99 HP. Hall, Russell & Co., Aberdeen	Grampian S.S.
218 89174	*	Benacre Keith	I ScSr 1 Dk 2 B	British Liverpool	1030 1598	246·3 P 375 t.	34·2	17·0 F 38 t.	Barrow Caird & Purdie	1882	3 Blhds. DTa [illegible] FPT APT	2Cy. 30"&55"–39" 160 HP. W. Kemp, Glasgow	S.S. "Benacre" (J. Hoult)
219 82011	* D	Benalder Thomson 87-87	I ScSr 2 Dks 3 B	British Leith	1331 2054	309·5 P 32 f.	35·0 B 36 f.	23·6 F 36 f.	Glasgow A. Stephen & Sons	1880	5 Blhds. [illegible]	2Cy. 38"&68"–45" 240 HP. A. Stephen & Sons, Glasgow	W. Thomson &
220 60707	* D	Benalder (ex Thomas Adam) Davidson 71-87	I ScBg 1 Dk	British Aberdeen	562 886	201·4 P 51 f.	29·2 B 8 f.	16·9 F 28 f.	Aberdeen Hall, Russell & Co.	1870	5 Blhds. DBa [illegible]	2Cy. 26"&47"–33" 95 HP. Hall, Russell & Co., Aberdeen	J. & A. Davids
221 81237	V	Benalla (ex Acorianna, ex Benalla) Edwards 88-88	I ScSr 1 Dk 2 B	British W. Hartlepool	786 1229	222·5 F 43 f.	33·2 B 103 f.	15·8 Q 115 f.	Liverpool W.H. Potter & Son	1880	BT APT FPT	2Cy. 23"&46"–33" 95 HP. J. Jones & Sons, Liverpool	J. Wood & Co.
222 77446	* D	Benamain French 84-88	I ScSr 1 Dk	British London	247 381	150·2 Q 45 f.	23·2 B 26 f.	13·0 F 16 f.	Aberdeen Hall, Russell & Co.	1878	5 Blhds. [illegible]	2Cy. 19"&36"–22" 43 HP. Hall, Russell & Co., Aberdeen	J. & C. Harris
223 78895	*	Benan Aspley 86-86	I Sc 1 Dk 2 B	British Liverpool	812 1273	240·2	33·2	16·6	Whitehaven Whitehaven S.B. Co.	1878	3 Blhds. DBa [illegible] FPT [illegible] APT [illegible]	2Cy. 25"&50"–36" 110 HP. J. Jones & Sons, Liverpool	S.S. "Benan" (J. Hoult)
224 70781	*	Benarty Le Boutillier 81-81	I ScBg 2 Dks 3 B	British Leith	1111 1716	291·8	33·2 F 22 f.	24·6	Glasgow Barclay, Curle & Co.	1876	5 Blhds.	2Cy. 36"&68"–39" 190 HP. Barclay, Curle & Co., Glasgow	W. Thomson &
225 81328	* D	Benayo (WRECKED) Gulliver 85-85	I ScSr 1 Dk 2 B	British Cardiff	784 1229	222·5 Q 115 t.	33·2 B 103 t. H 15 t.	15·8 F 42 t.	Liverpool W.H. Potter & Son	1880	3 Blhds. DTa [illegible] FPT APT	2Cy. 23"&46"–33" 95 HP. J. Jones & Sons, Liverpool	Cardiff & W. S.S. Co., Ld.
226 29921	* D	Benbow Rose	I Sc3mSr 1 Dk 2 B	British London	582 894	224·8 Q 31 f.	29·1	17·2 F 29 f.	Newcastle Tyne I. S.B. Co.	1865	4 Blhds. DBa [illegible]	2Cy. 31"&54"–30" 125 HP. Gourlay Bros. & Co., Dundee	Gen. Stm. Nav.
227 29549	*	Benbrack (ex Bellona) May 87-87	I Sc3mSr 2 Dks	British Liverpool	1251 1944	300·3 P 156 t.	34·2 F 66 t.	24·3 H 81 t.	Glasgow Smith & Rodger	1862	7 Blhds. DTa [illegible]	2Cy. 30"&57"–42" 220 HP. J. Jones & Sons, Liverpool	S.S. "Benbrack" Ld. (J. Hoult)
228 93749	*	Bencroy Brock 78-87 1 Dk & Spr Dk & Web Frms	Stl ScSr	British Liverpool	1667 2517	288·7 P 12 f. B 66 f.	38·3 F 31 f.	17·2 25·4	Liverpool J. Jones & Sons	1887	5 Blhds. CDB [illegible] APT [illegible]	3C. 21"35"&58"–42" 160 HP. J. Jones & Sons, Liverpool	S.S. "Bencroy" (J. Hoult)
229 47475		Bendigo (ex Milly)	I PadSc 1 Dk	British Nassau, N.P.	178 258	162·1	20·1	10·9	Preston J. Meckin	1863		2Cy. 90 HP. Fawcett, Preston & Co., Lvpl.	Henry Adde
230 84048	* D	Benefactor Murray 87-87	I ScSr 1 Dk 2 B	British Liverpool	647 1034	203·0 Q 94 f. F 45 f.	32·4 H 16 f.	14·3 B 97 f.	Liverpool W.H. Potter & Son	1880	3 Blhds. DTa [illegible] FPT APT	2Cy. 23"&46"–33" 95 HP. J. Jones & Sons, Liverpool	S.S. "Benef Ld. (J. Hoult)
231 2722	R	Benefactor Tribon	I ScSr 2 Dks	American New York	638 843	167·6	36·2	18·4	Chester, Pa. Reaney, Son & Archbold	1870		1Cy. 34"–28" 125 HP. Reaney, Son & Archbold	New York & ton S.S. Co.
232 84996	* D	Beneficent Mait 78-82	I ScSr 1 Dk 2 B	British Sunderland	1336 2024	275·0 Q 90 f. H 160 f.	37·1 F 69 f. H 48 f.	20·6 F 30 f.	Sunderland Short Bros.	1881	4 Blhds. WBa & f	2Cy. 36"&67"–46" 180 HP. T. Richardson & Sons, Hrtlpl.	J. Westoll
233 87857	*	Benevolent Walker 86-87	I ScSr 1 Dk 2 B	British Liverpool	790 1244	221·6 Q 112 t.	33·3 B 124 t.	15·8 F 38 t.	Whitehaven Whitehaven S.B. Co. L.	1883	3 Blhds. DTa [illegible] FPT APT	2Cy. 23"&46"–42" 130 HP. J. Jones & Sons, Liverpool	S.S. "Benevolent" Ld. (J. Hoult)
234 87424	* BT11.88	Bengal Barratt 76-88	Stl Sc3mSr 3 Dks	British Greenock	2377 4344	400·3 F 79 f.	45·3 P 261 f.	29·2	Greenock Caird & Co.	1885	8 Blhds.	3C. 35"56"&89"–66" 800 HP. [illegible], Greenock	Peninsular & Steam Nav. Co.
235 93826		Bengal Holland	I ScFlat	British Liverpool	117 173	90·4	20·8	10·6	Winsford J. Verdin & Sons	1885		[illegible]"–12" 20 HP. W.E. Bates, Northwich	J. Verdin
236 37		Bengala (ex Livorno) Marchini 83-85 (ex Moses)	I ScBg 2 Dks	Italian Palermo	1039 1567	243·1	31·2 H 78 f.	24·3	Sunderland Iliff, Mounsey & Co.	1871	4 Blhds. WBa	2Cy. 29"&53"–35" 254 HP. [illegible] by [illegible]	Navigazione Italiana
237 93704	*	Bengar Appleton 80-87	StlScSr 1 Dk 2 B	British Liverpool	946 1461	236·0 Q 87 f.	33·7 H 59 f.	16·7 F 26 f.	Stockton Craig, Taylor & Co.	1887	4 Blhds. CDB [illegible] FPT [illegible]	[illegible] 18"29"&48"–[illegible] 120 HP. Carr & Co., Ld., Sunderland	S.S. Bengar C (J. Hoult)
238 73749		Bengkalis	W ScSr 1 Dk	British Singapore	90 133	111·3	17·0	8·2	Hong Kong Tam Amow	1882		2Cy. 12"&22"–14" 20 HP. Fenwick & Morrison, Hg-Kg.	Wee Boon T
239 79150	*	Bengloe Farquhar 85-86	I ScSr 2 Dks 3 B	British Leith	1198 1854	301·5 B 16 f.	33·2 F 41 f.	25·1	Glasgow Barclay, Curle & Co.	1878	5 Blhds.	2Cy. 36"&63"–45" 230 HP. Barclay, Curle & Co., Glasgow	W. Thomson &
240 90101	*	Bengore Head Smith 79-89	I ScSr 1 Dk & Spr Dk 3 B	British Belfast	1607 2459	324·0 B 68 f.	37·2	17·7 24·9	Glasgow A. & J. Inglis	1884	5 Blhds. CDB [illegible] APT [illegible]	2Cy. 34"&68"–45" 275 HP. A. & J. Inglis, Glasgow	Ulster S.S. Co.

***** = **Lloyd's Register**; D = Nederlandsche Vereeniging; G = Germanischer Lloyd; I = Registro Italiano; R = American [Record]; A = Veritas Austro-Ungarico; H = Veritas Ellenico; N = Norske Veritas; V = Bureau Ve[ritas].

In the 2nd Column, B.T., with figures following, signifies Date of report of bottom sighting by Board of Trade Surveyors as condition of PASSENGER CERTI[FICATE].

In the 3rd Column, C = Composite (iron frame, wood planked); I = Iron; Stl = Steel; W = Wood.

Further particulars of Vessels classed by Lloyd's Register will be found in Lloyd's Register Book and Lloyd's Yacht Register.

	Nom, Matériaux, Gréement, etc. Capitaine et dates d'entrée au service des Armateurs actuels et à bord du navire. Ancien Nom, nombre de ponts, etc.	Pavillon. Port d'Armement	Tonnage. Net Total	Dimensions. Longueur	Largeur	Creux	Construit. Où. Par qui.	Quand.	Nombre des cloisons. Détails sur le lest d'eau.	Machines. Détails. Nom du Constructeur.	Armateurs ou Directeurs-Gérants.
*	**Benguela** I ScSr *Jolly* -87 2 *Dks*	British Glasgow	1157 1790	302·6 *P 101 f.*	33·9 *F 52 f.*	23·9 *B 87 f.*	Glasgow J.Elder&Co.	1874	5 Blds.	3Cy.21"31½"&35"–36" Todd'87 Fawcett Preston&Co.,Lpl. [250 HP	British&Afrien.Steam Nav.Co.
*	**Benguella** (ex ...) I ScBg *Furtado*	Portug'se Pon. Delgada St Michael	990 1308	245·0	31·7	24·0	Southampton Day&Summers	1862		2Cy.27"&48"–33" 200 HP. Comp.'76 Barclay&Co. ...	Empreza Insulana de Navegação
*	**Benholm** Stl ScSr *Doyle* 83-87 1 *Dk2 B*	British Liverpool	944 1438	236·0 *Q 87 f.*	33·6 *B 60 f.*	16·7 *F 26 f.*	Stockton Craig,Taylor&Co.	1887	4 Blds. CDB 326t. FPT 60t APT 18t	3C.18"29"&48"–36" 120 HP. Carr&Co.,Ld.,Sunderland	S.S."Benholm"Co.,Ld (J.Hoult)
*	**Benicarlo** I ScBn (ex Piraller, ex City of Baltimore)	Spanish Valencia	1509 2388	330·7	39·0	18·2 (26·0	Glasgow Tod&McGregor	1855		2Cy.44"&82"–45" 300 HP. J.Jack,Rollo&Co.,Liverpool	Joaquin Ripallès
*	**Benin** Stl ScSr *Addison* 84-84 2 *Dks*	British London	1437 2223	330·7 *F 55 t.*	36·7 *F 48 t.*	21·7 *B 111 t.*	Belfast Harland&Wolff	1884	5 Blds.	2Cy.34"&68"–42" 200 HP. Harland&Wolff,Belfast	African Steam Ship Co.
*	**Benita** Stl ScSr *Santanbari* -87 2 *Dks*	Spanish Bilbao	1048 1610	261·0 *B 68 f.*	34·3 *F 38 f.*	20·7	Southampton Oswald,Mordaunt&Co	1887	4 Blds. DB(a)& FA 32·40t.	3C.21½"33"&55"–39" 150 HP Oswald,Mordaunt&Co. Sthmptn	Linea de Vapores Serra
*	**Benito** I ScSr *Parkinson* 84-87 1 *Dk*	British Glasgow	424 712	240·9 *P 28 f.*	28·2 *F 23 f.*	13·5	PortGlasgow D.J.Dunlop&Co.	1884	5 Blds.	2Cy.24"&46"–33" 90 HP. D.J.Dunlop&Co.,P.Glasgow	British&Afrien.Steam Nav.Co.
R	**Benito Estenger** I ScSr	Spanish Snt'go de Cb.	219 439	146·3	25·1	11·9	Camden,N.J. J.H.Dialogne	1882		2Cy.20"&40"–28" 250 HP. J.H.Dialogne,Camden	Estenger, Mesa & Gallego
	Benjamin Atwood I Sc (ex Brandon)	Turkish Const'nti'pl	495 763	216·1	27·3	17·1	Glasgow Barclay,Curle&Co.	1854		164 HP.	B.J.Giustiniani
*	**BenjaminWhitworth** I ScSr *Matthew* 88-88 1 *Dk2 B*	British Middlesbro'	654 1016	216·2 *B 44 f.*	31·2 *Q 75 f.*	16·5 *F 32 f.*	Newcastle Palmers'Co.	1871	4 Blds. DB(a)f &66	2Cy.25"&50"–33" 99 HP. Palmers'Co.,Newcastle	Wilson & Symonds
*	**Benlarig** I ScSr *Freeman* 81-87 2 *Dks3 B*	British Leith	1482 2265	318·4 *P 36 f.*	38·1 *B 20 f.*	25·2 *F 36 f.*	Glasgow A.Stephen&Sons	1881	5 Blds.	2Cy.38"&68"–48" 300 HP. A.Stephen&Sons,Glasgow	W.Thomson & Co.
*	**Benlawers** Stl ScSr *Webster* 68-86 2 *Dks3 B*	British Leith	1513 2314	319·7 *P 37 f.*	38·0 *B 84 f.*	25·3 *F 36 f.*	Glasgow Barclay,Curle&Co.,Ld.	1886	6 Blds. DB(a)&180t	3C.24"40"&61"–45" 350 HP Barclay,Curle&Co.,Ld.,Glsg.	W.Thomson & Co.
*	**Benledi** Stl ScSr *Clark* 73-88 2 *Dks3 B*	British Leith	1454 2238	320·0 *P 35 f.*	38·0 *F 38 f.*	25·1 *B 84 f.*	Glasgow Barclay,Curle&Co.	1888	6 Blds. DB(a)&&165t	3Cy.24",38"&63"–45" Barclay,Curle&Co., Glasgow [350 HP	W.Thomson & Co.
*	**Benmore** Stl ScSr *Watson* -88 1 *Dk & Wb Frms*	British Aberdeen	974 1521	255·2 *P 30 f. Q 54 f.*	34·7 *B 54 f.*	16·9 *F 60 f.*	Newcastle W.Dobson&Co.	1888	4 Blds. CDB 230t	3Cy.20",33"&55"–36" N.E.Marine Co.,Ld., Newcastle [150 HP	Grampian S.S. Co.,Ld.
	Benmore I ScSr *Carrell* 80-87 2 *Dks & Awn Dk*	British Singapore	361 569	160·0 *P 36 f.*	24·0 *F 24 f.*	14·1 (21·2	PortGlasgow Blackwood&Gordn.	1872	3 Blds.	2Cy.22"&40"–27" 80 HP. Blackwood&Gordon,Pt.Gls.	Kim Seng & Co.
	Benmore I PadSk *Gillies* 77-87	British Glasgow	32 235	201·2	19·1 *Q 37 t.*	7·3	Glasgow T.B.Seath&Co.	1876		1Cy.50"–56" 85 HP. W.King&Co.,Glasgow	W.Buchanan
*	**Bentinck** (FOUNDERED) I ScSmSr *Richardson* 86-86 1 *Dk2 B*	British Whitehaven	555 901	232·7 *P 66 f.*	29·2 *F 34 f.*	16·1 *B 15 f.*	Whitehaven WhitehavenS.B.Co.	1873	2 Blds. DB(f 186t	3Cy.15"20"&46"–24" 99 HP Trip'86, J.Jones&Sns,Lvrpl.	C.Jones
*	**Benton** I ScSr *Rankin* 85-85 1 *Dk2 B*	British London	587 911	229·0 *Q 24 f.*	28·7 *B 70 t.*	16·2 *F 26 f.*	Sunderland W.Doxford&Sons	1872	4 Blds. DB(a)&f 237t	2Cy.26"&48"–30" 90 HP. Hawks,Crawshay&Ss,Gtshd.	Chapman & Miller
*	**Benvenue** I ScSr *Thomson* 83-85 2 *Dks3 B*	British Leith	1497 2286	318·5 *F 36 f.*	38·1 *B 26 f.*	25·2 *F 35 f.*	Glasgow A.Stephen&Sons	1883	4 Blds. DB(a)&180t	2Cy.38"&68"–48" 320 HP. A.Stephen&Sons,Glasgow	W.Thomson & Co.
*	**Benwell** I ScSr *Grant* -95 1 *Dk2 B*	British Newcastle	450 695	201·6 *Q 28 t.*	28·3 *B 10 t.*	15·7	Sunderland J.Laing	1871	4 Blds. WB(a)&f	2Cy.38"–30" 90 HP. G.Clark,Sunderland	AdamBros. & Co.
*	**Beny** I ScSr *George* 1 *Dk2 B*	British London	776 1222	220·7 *Q 103 t.*	33·3 *B 128 t.*	15·8 *F 44 t.*	Whitehaven WhitehavenS.B.Co.,Ld.	1882	3 Blds. DT(a)&f APT FPT	2Cy.23"&46"–36" 88 HP. J.Jones&Sons,Liverpool	Pierce,Becker & Hardi
*	**Berbice** I ScSr *Campbell* 83-97	British Liverpool	787 1246	221·6 *Q 112 t.*	33·3 *B 123 t.*	15·8 *F 39 t.*	Whitehaven WhitehavenS.B.Co.,Ld.	1883	3 Blds. DT(a)&f FPT APT	2Cy.23"&46"–42" 130 HP. J.Jones&Sons,Liverpool	S.S."Berbice"Co., Ld.(J.Hoult)
	Berbice I PadSr 1 *Dk2 B*		178 282	181·5	21·6 *Q 24 t.*	9·9	Newcastle Palmers'Co.	1859	4 Blds.	2Cy.30"&52"–48" 120 HP. R.&W.Hawthorn,N'weastle	F.R.Olivares
*	**Berenice** I ScBk *Perini* 6?-87 2 *Dk3 B*	Austrian Trieste	1707 2623	340·0	37·5	24·9	Trieste Stabilimento Tecnico	1882	5 Blds.	2Cy.39"& 70"–48" 400 HP. StabilimentoTecnico,Trste.	Lloyd Austro-Ungarico
W	**Berenice** I ScBg *Wilmink* 87-92 1 *Dk2 B*	Dutch Amsterdam	549 775	195·0	29·0	17·2	Stockton Richardson,Duck& Co.	1864		2Cy.26"& 48"–36" 120 HP. Richrdsn,Duck&Co.,Stcktn	Koninklyke NederlandschStoomboot Maats.
*	**Beresford** Stl ScSr *Vickers* 86-88 1 *Dk & Web Frms*	British W. Hartlep'l	1379 2108	285·3 *Q 86 f. P 28 f.*	36·5 *B 104 f.*	18·5 *F 29 f.*	W.Hartlepool E.Withy&Co.	1885	5 Blds. CDB 38t	3Cy21"35"&57"–36" 170 HP T.Richardson&Sons,Htlpol.	G.Horsley & Son
*	**Bergamo** I ScSr *Pearson* 1 *Dk & Web Frms*	British Newport	778 1208	225·2 *Q 83 f.*	32·2 *B 46 f.*	16·3 *F 22 f.*	Sunderland W.Pickersgill&Son	1883	4 Blds. DB(a)&f 270t APT	2Cy.28"& 52"–36" 116 HP. N.MarineE.C.Ld.,Gateshead	R.W.Jones & Co.
XV	**Bergen** I ScSr *Kjorland* -88 1 *Dk & Awn Dk3 B*	Norweg'n Bergen	1065 1366	232·0	30·6	25·0	Copenhagen Burmeister&Wain	1871	MT 35 ft	4Cy.21½"& 44"–30" 120 HP. Burmeister&Wain,Cpnhgn.	Jacob Christensen
N	**Bergen** I ScSr *Nieuwejaar* -83 1 *Dk2 B*	Norweg'n Bergen	284 397	159·4 *Q 71 f.*	21·2 *B 24 f.*	12·0 *F 12 f.*	Newcastle Coulson,Cooke&Co.	1873	4 Blds. DB(a)31ft	2Cy.19"& 36"–24" 70 HP. Thompson,Boyd&Co.,Nwc.	J.C.Gjertsen

Lloyd's Register; D = Nederlandsche Vereeniging; G = Germanischer Lloyd; I = Registro Italiano; R = American Record;
Veritas Austro-Ungarico; H = Veritas Ellenico; N = Norske Veritas; V = Bureau Veritas.

Dans la 2me **Colonne,** B.T., avec les indications qui suivent, signifie la date du rapport de visite de la Carène par les experts du Board of Trade pour obtenir le certificat relatif au TRANSPORT DES PASSAGERS.

Dans la 3me Colonne, C = Composite (membrure en fer, bordé en bois); I = Fer; Stl = Acier; W = Bois.

Pour plus amples renseignements sur les navires classés par le Lloyd's Register, voir le Lloyd's Register Book et le Lloyd's Yacht Register.

No. in Book. / Official No.	Registry in which classed.	Ship's Name. Master, and Dates of Appointment to present Owner's Service and to Ship / Ship's Late Name	Material, Rig, &c. / No. of Decks, &c.	Flag. / Port of Registry.	Tons. Net. / Gross.	Length	Breadth	Depth	Built. Where. / By Whom.	When	No. of Bulkheads / Particulars of Water Ballast.	Engines. Particulars. / Maker's Name.	Owners or Managers
271	V	**Bergenhuus** *Sirenzen 84-88*	I ScSr 1 Dk1 B	Danish Copenhagen	322 444	156·7	22·2	14·0	Copenhagen Burmeister&Wain	1870	[illegible] Blkds. [illegible]	4Cy.15"& 30"–20" 55 HP. Burmeister&Wain,Cpnhgn.	Det Forened[illegible] [illegible]
						P 49. L. M. F 22. H 71							
272	✻N	**Bergenseren** *Mortensen 84-84*	I ScSr 1 Dk2 B	Norweg'n Bergen	386 615	179·5	27·3	14·2	Flensburg FlensburgS.B.Co.	1884	4 Blkds. MT AFT	2Cy.22½"& 42"–27" 70 HP. FlensburgS.B.Co,Flensburg	Et Interess[illegible] (Bergenske [illegible])
						Q 52 f. B 49 f. F 25 f.							
273	GV	**Berger Wilhelm** *(ex Wilhelm [illegible]) Hein 86*	I Sr 1 Dk2 B	German Hamburg	362 500	213·0	26·2	13·8	Amsterdam P.V.Vlissingen	1856		2Cy.21"& 34"–23" [illegible]Howaldt,Kiel	Nordischer [illegible] Gesellschaft
274	DNV	**Bergliot** *Jacobsen 84-84*	W ScSr 1 Dk2 B	Norweg'n Stavanger	373 504	145·4	26·5	17·3	Stavanger K.Knochsen	1881		2Cy.18"& 32½"–24" 45 HP. Akers M. Verks Christiania	Act. Selsk.[illegible] (Sømme [illegible])
						Q 44 ft. B 40 ft.							
275	V	**Bergsund** *Aminoff 77-77*	I ScSr 1 Dk2 B	Swedish Stockholm	666 888	201·7	28·9	16·3	Stockholm BergsundsMk.Vk.	1869		2Cy.24"& 43"–25" 75 HP. BergsundsMekVerksStkhm	Stockholms[illegible] deribol. [illegible]
						P 111 f. F 22 f. H 94 Len. 82							
276 2995	R	**Berks** *Adams 86-86*	I ScSr 1 Dk2 B	American Philadelphia	377 505	189·0	29·0	14·4	Chester, Pa. J.Roach&Sons	1874	[illegible]	2Cy.20½"&34"–30" 150 HP. Phildlpha&Rdng R.R.Co.Pa.	Philadelphia[illegible] ing R.R.Co.
277 3175	R	**Berkshire** *Howes*	I ScSr 3 Dks	American Baltimore	1433 2014	250·0	38·5	27·0	Philadelphia Cramp&Sons	1881		2Cy.34"& 60"–45" 500 HP. Cramp&Sons,Philadelphia	Merchant [illegible] Transportation
						F 38 t.							
278 67827	✻	**Berlin** *Danby*	I ScSr 1 Dk	British Goole	397 678	202·3	28·0	13·0	Dundee W.B.Thompson	1879	[illegible] Blkds. [illegible]	2Cy.30"& 54"–33" 130 HP. W.B.Thompson,Dundee	Yorkshire C[illegible] ShipCo.Ld.
						Q 99 f. B 18 f. F 25 f.							
279 47014	BT1.89	**Berlin** *White 83-83*	I ScSr	British Leith	380 628	216·6	27·1	14·4	Glasgow Barclay, Curle&Co.	1863		2Cy.28"& 47"–33" 90 HP. J.Penn&Sons,Greenwich	Leith, Hull[illegible] Stm.Pkt.Co.
						F 27 t.							
280	GV	**Berlin** *Von Cöllen 83-83*	I ScBg 2 Dks & Awn Dk	German Bremen	1637 2333	297·2	39·0	23·0 30·5	Greenock Caird&Co.	1868		2Cy.42"& 77"–42" 300 HP. Act.Gesells.Weser,Bremen	Norddeutsch[illegible]
281	V	**Berlin** *Hoppe 82-86*	I ScSr 1 Dk	German Stettin	597 844	217·5	28·3	15·6	Stettin Act. Gesell. "Vulcan"	1881	[illegible]	2Cy.25"&47"–29½" 100 HP. Act.Gesells"Vulcan"Stettn	Neue Dampf[illegible] pagnie Akt.G.
						P 105 f. F 32 f.							
282 68019	✻	**Bermuda** *Fraser 76-84*	I ScSr 1 Dk & Spr Dk3 B	British Sunderland	823 1284	233·0	29·2	16·7 22·9	Sunderland W.Doxford&Son	1874	[illegible] Blkds. [illegible]	3C.18"29"& 48"–33" 160 HP. W.Doxford&Sons,Sndrland	Quebec S.S.C[illegible]
283		**Bermudez** (COLLISION) (ex Maracaibo, ex Gresford)	I ScSr	Venezuln	174 256	142·5	24·6	11·0	PortGlasgow H.Murray&Co.	1873		2Cy.17"& 30"–22" 60 HP. Kemp&Hume,Glasgow	
284 89772	✻	**Bernard** *Jones*	I ScSr 1 Dk2 B	British Cardiff	1109 1726	260·0	37·1	18·5	Whitby T.Turnbull&Son	1884	[illegible]	2Cy.32"& 66"–39" 154 HP. Blair&Co.,Ld.,Stockton	Turnbull Bro[illegible]
						Q 96 f. B 68 f. F 31 f.							
285 63318	BT6.88	**Bernard** *Churchward 85-85*	I ScSr	British Liverpool	577 915	237·2	26·2	19·9	Liverpool T.Royden&Sons	1870		2Cy.21"& 48"–40" 90 HP. Fawcett,Preston&Co.,Liv.	Booth Stms[illegible]
286 84050	BT5.88	**Bernard Hall** *Alexander 80-84*	I ScBg	British Liverpool	1740 2678	338·0	38·0	26·0	Sunderland J.Laing	1880		2Cy.29"& 59"–42" 320 HP. G.Forrester&Co.,Liverpool	West India [illegible] Steamship C[illegible]
						B 93 t. F 30 t.							
287	V	**Bernhard** *Schrader 82-85*	I ScSr 1 Dk	German Kiel	386 519	150·0	24·8	17·4	Vegesack J.Lange	1884	[illegible]	2Cy.18"& 33"–28" 50 HP. J.Lange&Sons,Vegesack	Sartori & [illegible]
						B 41 f. F 19 f.							
288 86107	✻	**Bernicia** *Jones 82-83*	I ScSr 2 Dks3 B	British Newcastle	1702 2619	300·8	39·3	25·2	Newcastle Palmers'Co.,Ld.	1883	[illegible] Blkds. [illegible]	2Cy.38"& 71"–48" 300 HP. Palmers'Co.,Ld.,N'wcastle	"Bernicia" [illegible] (J. E. [illegible])
						P 25 f. B 76 f. F 34 f.							
289 87829	✻	**Bernina** *James*	I ScBn 2 Dks	British Liverpool	955 1483	256·2	33·1	20·6	Liverpool R.&J.Evans&Co.	1883	[illegible] Blkds. [illegible]	2Cy.28"& 50"–36" 120 HP. DeWinton&Co.,Carnarvon	Richards, [illegible]
						P 59 t. B 91 t. F 31 t.							
290 53117		**Berrington** *Scarfe 82-84*	I ScSr 1 Dk2 B	British London	508 794	210·3	28·1	16·7	Newcastle Palmers'Co.,Ld.	1865	[illegible] Blkds. [illegible]	2Cy.25"& 48"–30" 99 HP. Cmp.'76,Palmers'Co.,Nwc.	J.Fenwick [illegible]
						Q 24 f.							
291	✻	**Berry** *Débrien 79-88*	I ScSr 1 Dk & Spr Dk3 B	French Marseilles	1045 1625	255·2	34·1	16·7 24·5	Glasgow W.B.Thompson	1884	[illegible] Blkds. [illegible]	2Cy.36"& 72"–42" 250 HP. W.B.Thompson,Dundee	Société Gé[illegible] Transports [illegible]
						F 25 f.							
292 68297	✻	**Bertha** *Bailes 81-87*	I ScSr 1 Dk2 B	British Middlesbro'	401 636	195·0	27·0	14·9	Middlesbro' Backhouse&Dixon	1873	[illegible] Blkds. [illegible]	2Cy.24½"& 48½"–33" 90 HP. Blair&Co.,Ld.,Stockton	Fox Brother[illegible]
						Q 72 f. F 27 f.							
293 3397		**Bertha** *Andersen 88*	W Sc3mSr 1 Dk	American S.Francisco	270 389	138·0	32·0	15·0	Benicia M.Turner	1888		2Cy.16"&32"–24" 125 HP. Wm.Deacon&Co.,SFrncisco	Karluk Pack[illegible]
294	GV	**Berthilde** *Hoffmann 88*	I ScSalvageSr 1 Dk	German Hamburg	285 514	170·7	27·3	14·7	Rostock RostockS.B.Co.	1886	[illegible]	2Cy.31¼"&59"–35¾ 106 HP. RostockS.B.Co.,Rostock	Nordischer [illegible] Verein
						Q 17 f. B 93 f. H 51.							
295 81059	✻D	**Bertie** *Ormandy*	I ScBn 1 Dk2 B	British Hartlepool	1109 1716	257·5	34·5	18·0	Stockton M.Pearse&Co.	1881	[illegible] Blkds. [illegible]	2Cy.32"& 60"–39" 140 HP. Blair&Co.,Ld.,Stockton	O.Trechmann
						Q 68 f. F 27 f. F 30 f.							
296 72487	✻	**Berwick** (ex Richard Trevithick) *Farnes 82-84*	I ScSr 1 Dk	British London	262 421	160·4	24·5	13·2	PortGlasgow Cunliffe&Dunlop	1874	[illegible] Blkds.	2Cy.21"& 38"–24" 50 HP. Cunliffe&Dunlop,P.Glasgw.	J.T.Harris[illegible]
						Q 74 f. F 23 f.							
297 82465	✻	**Beryl** *Winship 79-83*	I ScSr 1 Dk2 B	British Hull	868 1340	240·4	32·0	16·7	Stockton M.Pearse&Co.	1880	[illegible] Blkds. [illegible]	2Cy.28½"&53"–33" 140 HP. Blair&Co.,Ld.,Stockton	W.Tulley & C[illegible]
						F 164 f. F 31 f.							
298 82746	✻	**Beryl** *Campbell 82-82*	I ScSr 1 Dk2 B	British Dundee	774 1190	234·0	32·2	15·3	Dundee W.B.Thompson	1881	5 Blkds. CDB[illegible] AFT[illegible]	2Cy.27"& 54"–36" 120 HP. W.B.Thompson,Dundee	Dundee Gen[illegible] Co.,Ld.
						F 136 f. F 37 f.							
299		**Beshiktash** *Mustapha 81-81*	W Pad	Turkish Constantinople	62	120·0	18·6	9·3	West Cowes J.White	1852		60 HP. Maudslay,Sons&Field,Lond	Chirket-i-H[illegible] (Bosphorus [illegible])
300		**Beshiktash** (ex Washington) *Luca 73-88*	W Pad	Turkish Constantinople		..	..	..					Idarei Mas[illegible] (Ottoman [illegible])

✻ = Lloyd's Register; D = Nederlandsche Vereeniging; G = Germanischer Lloyd; I = Registro Italiano; R = American [illegible]
A = Veritas Austro-Ungarico; H = Veritas Ellenico; N = Norske Veritas; V = Bureau Ve[illegible]

In the 2nd Column, B.T., with figures following, signifies Date of report of bottom sighting by Board of Trade Surveyors as condition of PASSENGER CE[illegible]

In the 3rd Column, C = Composite (iron frame, wood planked); I = Iron; Stl = Steel; W = Wood.

Further particulars of Vessels classed by Lloyd's Register will be found in Lloyd's Register Book and Lloyd's Yacht Register.

Registre	Nom, Matériaux, Grèement, etc. / Capitaine et dates d'entrée au service des Armateurs actuels et à bord du navire. / Ancien Nom, nombre de ponts, etc.	Pavillon. / Port d'Armement	Tonnage. Net. / Total	Dimensions (Longueur, Largeur, Creux)	Construit. Où. / Par qui.	Quand.	Nombre des cloisons. / Détails sur le lest d'eau.	Machines. Détails. / Nom du Constructeur.	Armateurs ou Directeurs-Gérants.
	Besós I Sc	Spanish Alicante	277 408	172·9|23·0|12·8	England	1862		60 HP.	Hijos de G.Carratala
✻	Bessarabia I ScSr Martin 80-84 2 Dk 3 B	British Liverpool	1058 1644	261·0|33·6|23·4 P 24 f. F 36 f.	W. Hartlepool W.Gray&Co.	1877	4 Blds. [illegible]	2Cy.32"&60"–39" 170 HP. Blair&Co.,Ld.,Stockton	Taylor,Cameron & Co.
	Bessarabia I ScSr 1 Dk	Russian Odessa	322 423	219·0|27·0|10·4	Norrköping Motala Werks	1861		2Cy.26"&41"–24" 60 HP. RussianS.S.&Nav.Co.	Russian Steam Nav. & Trading Co.
	Bessbrook I ScSr	British Newry	161 [illegible]	182·0|27·2|14·0	Glasgow A.&J.Inglis	1877		2Cy.32"&58"–36" 140 HP. A.&J.Inglis,Glasgow	Dundalk& NewryStm. Pkt.Co.,Ld.
	Bessell I ScSr Wilson 72-[illegible]	British Liverpool	1151 1841	306·9|33·7|24·4	Newcastle A.Leslie&Co.	1878		2Cy.26"&58"–45" 180 HP. R.Stephenson&Co.,Newcstl	Liverp'l, Brazil & Riv. Plate Steam Nav.Co.,Ld.
	Bessemer I ScK Long 82-82 1 Dk	British Stockton	183 286	140·0|20·0|12·1 Q 24 f.	Stockton Richardson,Duck& Co.	1872	3 Blds. WB	2Cy.18½"&34"–24" 50 HP. C.D.Holmes&Co.,Hull	J.M.Lennard & Son
	Bessie I ScBn Hill 78-82	British Hayle	250 287	131·9|25·9|13·1	Hayle Harvey&Co.	1865		26 HP. Harvey&Co.	Harvey & Co.
✻	Bessie Morris I ScBn Gillies 1 Dk 2 B	British Liverpool	1119 1729	280·2|34·0|21·9 P 30 f. F 32 f.	PortGlasgow W.Hamilton&Co.	1880	5 Blds. [illegible] APT 22t	2Cy.32"&62"–42" 200 HP. Dunsmuir&Jackson,Glasg.	S.S."Bessie Morris" Co.,Ld.(E.C.Thin)
	Bestante (ex Sailing Vessel) W ScSr Fiick [illegible] 1 Dk	Norweg'n Bergen	104 150	95·6|23·5|11·4 P 72 f. Q 22 f.	Stavanger	1872		12 HP.	Bergh & Helland
✻	Beta I ScSr Butler 1 Dk 2 B	British Hull	856 1317	259·5|31·8|18·6 P 197 t. F 24 t.	Hull Humphreys&Pearson	1871	5 Blds.	2Cy30"&56"–33" 130 HP. Humphreys&Co.,Hull	G.R.Sanderson & Co.
	Beta I Sc3mSr Smith 88-88 2 Dks & Spr Dk	British Glasgow	677 1087	235·1|28·5|14·7 |22·1	Glasgow Aitken&Mansel	1873	4 Blds.	2Cy33½"&58½"–36" 160 HP. J.&G.Thomson,Glasgow	W.A.Black
✻	Beta I ScSk Griffiths	British Liverpool	225 395	151·0|25·2|12·3	Renfrew W.Simons&Co.	1884	4 Blds. WB	2Cy.22"&42"–27" 75 HP. W.Simons&Co.,Renfrew	Corporation of Liverpool
	Betsy I ScSr 1 Dk	British Penang	113 167	119·0|17·5|10·5 B 25 t.	Rotterdam J.& K.Smit	1875		2Cy.14"&24"–20" 36 HP. Diepveen,Lels&Smit,Kndjk	Ong Beng Tek
✻	Betty I ScTrawler Stallbohm 87 1 Dk	German Rostock	72 119	101·7|20·6|10·5 F 18 f.	Rostock RostockS.B.&E.Co.	1887	3 Blds.	2Cy.17¾"&33½" 19⅝" 60 HP. RostockS.B.&Co.,Ros.	F.A.Pust
	Betty C ScSr Jacobsen 82-82 1 Dk	Norweg'n Bergen	193 262	123·3|22·6|10·6 Q 39 f. B 25 f. F 12 f.	Oscarshamn S.Olsen	1870		2Cy.11"&23½"–20" 50 HP.	Jacob R.Olsen
V	Betty I ScSr Sjögren 87 1 Dk	Swedish Stockholm	283 379	135·5|22·0|13·8 B 31 f. H 4 t.	Oscarshamn OscarshamnMkVrks	1882	WB	2Cy.15½"&32"–22" 60 HP. OscarshamnMek.Verks., Oscarshamn	F.G.Althaintz
	Betty I ScSr Ericsson 86 1 Dk	Swedish Tunrå	128 155	113·0|16·8|9·5 B 24 f. Lw 86	Sjöberg G.F.Cawallin	1883		2Cy.11"&19"–18" 25 HP. G.F.Cawallin,Sjöberg	G.F.Cawallin
✻	Betty Sauber I ScSr Korner 1 Dk 2 B	German Hamburg	657 1015	218·8|31·0|18·4 Q 98 f. F 28 f.	Hamburg Rbst.Schiffs-Wrfte	1876	4 Blds. [illegible]	2Cy.27"&50"–33" 99 HP. ReiherstiegS.W.,Hamburg	Sauber Gebrüder
✻	Bewick I Sc3mSr Ellegaard 85-85 2 Dks	British Newcastle	945 1463	253·3|32·4|24·0	Sunderland W.Doxford&Sons	1872	5 Blds. [illegible]	2Cy.31"&58"–33" 130 HP. N.E.MarineCo.Ld.,Sundrld.	Eeles & Dryden
	Beylerbey (HCLK) W Pad Hussein 65-78	Turkish Constantpl.	62	120·0|18·6|9·3	WestCowes J.White	1852		60 HP. Maudslay,Sons&Field,Lond	Chirket-i-Hairie (Bosphorus Stm.Nav.Co.)
	Bezerra de Menezes I Sc De Menezes 1 Dk	Brazilian RioJaneiro	283 485	184·0|22·6|12·5	London Lewis&Stockwell	1874	5 Blds.	2Cy.28"&50"–33" 100 HP. Ravenhill,Easton&Co.Lond	Cia de Estrada de Ferro de Macahé & Campos
	Bheema I ScSr	British Bombay	166 357	180·6|22·1|10·8 Q 20 t. F 8 t.	Glasgow T.Wingate&Co.	1878		2Cy.20"&40"–30" 75 HP. T.Wingate&Co.,Glasgow	Shepherd & Co.
✻	Bhundara Stl Sc3mSr Jacobs 85-88 2 Dks	British Glasgow	2190 2889	319·9|40·3|25·9 P 44 f. B 25 f. F 42 f.	Dumbarton W.Denny&Bros.	1880	5 Blds. [illegible]	2Cy.38"&68"–51" 290 HP. Denny&Co.,Dumbarton	British India Steam Nav.Co.,Ld.
	Biafra I ScBk Bates 84-88	British London	1133 1487	312·0|32·0|19·5 P 133 t. F 35 t.	Glasgow RandolphElder&Co	1868		2Cy.34"&57"–36" 150 HP. LairdBros.,Birkenhead	African Steamship Co.
V	Bianca I ScSr Witt 88 1 Dk & Spr Dk 3 B	German Hamburg	1174 1745	258·5|34·8|17·1	Flensburg FlensburgS.B.Co.	1886	[illegible]	3C.22"34"&58"–42" 250 HP. FlensburgS.B.Co.,Flensbrg.	Hamburg Pacific Dampfs.Linie
	Bickerstaffe I PadSr Seed	British Fleetwood	107 197	155·2|21·6|8·6	Birkenhead LairdBros.	1879		2Cy.38"–48" 95 HP. LairdBros.,Birkenhead	J.Bickerstaffe
✻D	Bidar (ex Castlecraig) I ScSr Cunningham 86-86 1 Dk & Spr Dk	British Liverpool	1891 2876	330·0|38·3|27·7 B 80 f.	Newcastle Wghm,Richardson &Co.	1884	6 Blds. DB&c [illegible] [illegible]	2Cy.38"&75"–48" 235 HP. WighamRichrdsn&Co,Nwc	G.Tweedy & Co.
D	Biddick I ScSr Roark 85-85 1 Dk 2 B	British Sunderland	561 874	211·0|29·0|16·9 Q 22 t. B 86 t.	Sunderland J.Laing	1864	4 Blds. [illegible]	2Cy.26"&47"–36" 95 HP. G.Clark,Sunderland	H.T.Morton & Co.
	Biela I ScBg Graham 82-84	British Liverpool	1400 2182	316·0|34·9|29·4	Newcastle A.Leslie&Co.	1870		2Cy.34"&68"–48" 250 HP. J.Jones&Sons,Liverpool	Liverp'l, Brazil & Riv. Plate Steam Nav.Co.,Ld.
V	Biene I Sc3mSr Jantzen 88-88 1 Dk	German Danzig	167 235	141·0|17·6|10·1	Vegesack J.Lange	1874	WB	2Cy.13"&22½"–15" 45 HP. J.Lange,Vegesack	F.G.Reinhold

[illegible] = Lloyd's Register; D = Nederlandsche Vereeniging; G = Germanischer Lloyd; I = Registro Italiano; R = American Record;
[illegible] = Veritas Austro-Ungarico; H = Veritas Ellenico; N = Norske Veritas; V = Bureau Veritas.

Dans la 2me Colonne, B.T., avec les indications qui suivent, signifie la date du rapport de visite de la Carène par les experts du Board of Trade pour obtenir le certificat relatif au TRANSPORT DES PASSAGERS.

Dans la 3me Colonne, C = Composite (membrure en fer, bordé en bois); I = Fer; Stl = Acier; W = Bois.

Pour plus amples renseignements sur les navires classés par le Lloyd's Register, voir le Lloyd's Register Book et le Lloyd's Yacht Register.

No.	Registry	Ship's Name. Material, Rig, &c. / Master, and Dates of Appointment to present Owner's Service and to Ship / Ship's Late Name. No. of Decks, &c.	Flag. / Port of Registry.	Tons. Net. / Gross.	Length	Breadth	Depth	Built. Where. / By Whom.	When.	No. of Bulkheads / Particulars of Water Ballast.	Engines. / Particulars / Maker's Name.	Owners or
					Dimensions. Deck Erections.							
331	V	**Bifrost** I ScSr *VonBelow 78-89* 1 *Dk & Awn Dk3 B*	Swedish Gothenburg	1092 1332	231·9	30·8 *B 20 f.*	17·7 246	Gothenburg MotalaCo.	1874		2Cy.27½"&55"–36" 150 HP. MotalaCo.,Gothenburg	Ångf.Aktie "Svithiod"
332 89168	✱	**Bilbao** I ScSr *Donald 82-83* 1 *Dk2 B*	British Cardiff	783 1211	230·3 *Q 88 f.*	32·0 *B 50 f.*	15·6 *F 24 f.*	Sunderland R. Thompson&Sons	1883	4 Bhds.	2Cy.28"& 54"–36" 99 HP. Black,Hwthrn&Co.,Gatshd.	"Bilbao" S.S. (E.C.Bre…)
333		**Bilbao** I Sc3mSr *Tejeiro 84-86* 2 *Dks*	Spanish Barcelona	370 597	165·6 *Q 65 f.*	28·2 *F 24 f.*	14·0	Greenock Caird&Co.	1856		2Cy.35"& 50"–24" 75 HP. LairdBros.,Birkenhead	Aniceto La…
334		**Bilbao No. 2** I ScBarge	Spanish Manila	76 120	92·0	19·0	8·2				2Cy.14"–16" 25 HP.	H.MacDon…
335 81614		**Billiton** I ScSr 1 *Dk*	British Singapore	195 336	159·9 *F 19 f.*	25·1 *B 33 f.*	9·1	PortGlasgow Blackwd.&Gordon	1880		2Cy.22"& 42"–24" 80 HP. Blackwood&Gordon,P.Glas	Tan Ben Gu…
336		**Billiton** (ex ? Goens) I ScS. *Meyer 79-73* 2 *Dks*	Dutch Batavia	173 424	137·8	23·6	18·9	Amsterdam MaatschappydeAtlas	1874		2Cy.20"& 36"–21" 75 HP. Kkl.MaatsdeSchelde,Vlisgn	"Billiton"… chappy
337 84527	✱ D	**Billow** I Sc *Douglass 86-86* 1 *Dk2 B*	British London	1065 1650	257·1 *Q 92 f.*	34·4 *B 68 f.*	18·1 *F 33 f.*	Hartlepool R.Withy&Co.	1881	4 Bhds.	2Cy.34"& 68"–36" 160 HP. T.Richardson&Sons,Hartpl	Jackson Bro…
338	✱	**Biobio** StlTwinScSr *Wakely 73-88* 1 *Dk & Shltr Dk*	Chilian Valparaiso	443 713	170·2	32·1	9·9 10/8	Glasgow R.Napier&Sons	1888	5 Bhds. FPT28 APT29	3C.11"17"&28"–21" 100 HP. R.Napier&Sons,Glasgow	Cia Sud Am… de Vapore…
339 87394	✱	**Birch** I ScSr *Clift 84-83* 1 *Dk2 B*	British Sunderland	1094 1674	260·0 *Q 8 f. P 22 f.*	35·3 *B 54 f.*	18·7 *F 28 f.*	Sunderland StrandSlipwayCo.	1882	4 Bhds.	2Cy.32"& 60"–39" 140 HP. J.W.&F.Wilson,Sunderlnd.	Thompson &…
340	✱	**Birchfield** StlScSr *Ferguson 84-88* 1 *Dk & Wh Frms*	British Liverpool	1480 2327	290·0 *P 28 f.*	38·7 *Q 80 f. F 33 f.*	20·9 *B 122 f.*	Sunderland L.Thompson&Sons	1888	6 Bhds. APT71	3C.23"38"&62"–39" 220 HP. J.Dickinson,Sunderland	J. Brown
341		**Birger** C ScSr *Johansson* (ex Birger Jarl) 1 *Dk*	Swedish Skälda	130 177	98·8	20·5 *H 15 f.*	12·6	Motala	1855		30 HP.	J.Andersson
342 65052	BT11.88	**Birkenhead** I Pad *Brett 73-*	British Liverpool	285 458	148·8	30·0	11·7	Liverpool R.&J.Evans&Co.	1872		4Cy.32"–48" 140 HP. J.Jones&Sons,Liverpool	The Municip… Birkenhead
343 77452	✱	**Birkhall** I ScSr *Profit 87-87* 1 *Dk & Spr Dk3 B*	British Aberdeen	937 1447	250·6	32·3	17·5 246	Aberdeen Hall,Russell&Co.	1878	5 Bhds.	3C17"31"&57"–36" 140 HP. Hall,Russell&Co.,Aberdn.	Grampian S.S…
344 79337	✱	**Birksgate** I ScSr *Fleetwood* 1 *Dk & Spr Dk*	British Sydney	916 1438	255·0	33·0	21·7	Glasgow D.&W.Henderson	1882	5 Bhds.	2Cy.30½"&63"–48" 163 HP. D.&W.Henderson&Co.,Gls.	Australasian Steam Nav.Co…
345 68376	✱	**Birling** I ScSr *Arrough* 1 *Dk*	British London	195 360	141·1 *P 70 f.*	22·1 *F 15 f.*	11·5	SouthShields J.Kitringham	1872	4 Bhds.	2Cy.18"& 34"–24" 45 HP. W.&J.Almond,N.Shields	W.WilsonS…
346 58	I	**Birmania** I ScBn *De Marchi 85-87* 2 *Dks3 B*	Italian Genoa	1562 2364	290·0 *P 90 f. 181*	36·1 *F 47 f.*	26·6 *H 55 f.*	Leghorn F.lli Orlando	1882		2Cy.35"& 70"–36" 420 HP. Palmer&Co.,Newcastle	Navigazione Italiana
347 104	✱ I	**Bisagno** I ScSr *Cognasso 85-87* 2 *Dks3 B*	Italian Genoa	1474 2290	301·3	37·2	23·3	Dumbarton Barclay&Son	1884	5 Bhds.	2Cy.35"& 67"–42" 200 HP. D.Stewart&Co.,Glasgow	Navigazione Italiana
348	N	**Biscaya** (ex Oregrund) W Sc3mSr *Ellingsen 82-84* 1 *Dk2 B*	Norweg'n Moss	536 683	176·5 *P 53 f.*	30·8 *B 38 f.*	15·8 *F 25 f.*	Gefle A.Brodin	1873		2Cy.17"& 34"–19½" 40 HP. Oscarshamn'sMek.Verksted	H.A.Reinert
349	✱	**Biscaye** I ScSr *Haulst 84-84* 1 *Dk2 B*	French La Rochelle	1106 1626	260·1 *P 29 f.*	38·0 *B 52 f.*	17·2 *F 26 f.*	Sunderland J.Blumer&Co.	1882	6 Bhds.	2Cy.31"& 60"–39" 150 HP. T.Clark&Co.,Newcastle	Delmas frère…
350 95502	✱	**Bishopsgate** Stl ScSr *Willis 88-88* 1 *Dk & Wh Frms*	British London	1464 2242	290·0 *P 29 f. Q 80 f.*	38·7 *B 122 f.*	20·9 *F 33 f.*	Sunderland J.L. Thompson & Sons	1888	6 Bhds. APT85	3C.22"37"&60"–39" 230 HP. J.Dickinson,Sunderland	Bishopsgate S… Co.Ld. (H.W…
351	✱	**Bissau** (ex Lottie Kershaw) I ScSr *Queiroz 81-83* 1 *Dk*	Portug'se Lisbon	285 458	159·0 *Q 73 f.*	24·0 *B 16 f.*	12·2 *F 23 f.*	London Forrest&Son	1882	4 Bhds.	2Cy.20"& 38"–30" 62 HP. R.H.Pearson&Co.,Glasgow	Empreza Nac… Nav.áVap.
352 77084	✱ BT1.88	**Bittern** I Sc3mSr *Jarrett* 2 *Dks*	British London	519 948	230·0 *F 44 f.*	29·0 *F 48 f.*	17·0	Stockton M.Pearse&Co.	1878	5 Bhds.	2Cy.35"& 65"–42" 165 HP. Blair&Co.,Ld.,Stockton	General Stea…
353 80214	✱	**Bittern** I ScSr *Cotter 80-80* 1 *Dk*	British Cork	424 657	180·3 *Q 88 f.*	28·1 *B 13 f.*	13·3 *F 43 f.*	Paisley H.McIntyre&Co.	1880	4 Bhds. APT36	2Cy.24"& 48"–33" 96 HP. Hutson&Corbett,Glasgow	Cork S.S.Co.…
354 65762	D	**Bivouac** I Sc3mSr *Hanley 79-83* 2 *Dks3 B*	British Bristol	706 1122	216·4	29·3	21·6	Glasgow Aitken& Mansel	1871	6 Bhds. DT	2Cy.26"& 48"–36" 99 HP. Hawthorns&Co.,Leith	BristolStm.N…
355	G	**Bjelbo** C ScSr *Arvidsson 85* 1 *Dk*	Swedish Motala	145 210	100·7	21·1 *B 17 f.*	11·7 *H 34.*	Sjötorp S.Groth	1884		2Cy.13"& 24"–17" 50 HP.	Motala Rede… bolag (C.N…
356	N	**Björgvin** I ScSr *Scheitlie 86-86* 1 *Dk2 B*	Norweg'n Drontheim	283 390	138·4 *P 75 f.*	22·7 *F 25 f.*	13·2	Bergen BergensMek.Verks.	1870		2Cy.22"& 44"–22" 60 HP. BergensMek.VerksBergen	Det Nordenf… Dampskibs…
357	N	**Bjórn** (ex Jemtland, ex Brage) W SrSr *Wettergreen88-89* 1 *Dk2 B*	Norweg'n Moss	387 522	164·8 *F 19 f. B 53 f.*	27·4 *Q 35 f.*	14·1 *H 9 f.*	Sundswall W.Hägglund	1872		2Cy17½"&37"–27½" 60 HP. KockumsMek.Verks.,Malmö	Rederiselskab… "Bjorn" (H.A…
358	V	**Bjórn** I ScSr *Larsson 72-79* 1 *Dk2 B*	Swedish Gothenburg	763 1005	218·5 *Q 88 f. B 46 f.*	25·6 *F 25 f.*	17·0 *H 10 f.*	Sunderland W.Watson	1872		2Cy.28"& 52"–30" 100 HP. Thompson,Boyd&Co.,Nwc.	Angf.Aktie-B… "Thule"(O.…
359		**Björneborg** (ex Auxiliary) W ScBk *Ekblad*	Russian Åbo	328 404	152·8	29·2	14·4	Björneborg J.Bandorff	1873			J.G.Wikestr…
360 30937		**Black Boy** I Sc3mSr 1 *Dk2 B*	British Sunderland	343 504	166·2	26·7 *Q 36 f.*	15·7	Newcastle PalmerBrothers	1854	4 Bhds.	2Cy.23"& 42"–26" 50 HP. N.E.Mar.Eng.Co.,Sunderland	J.Wood

✱ = **Lloyd's Register**; D = Nederlandsche Vereeniging; G = Germanischer Lloyd; I = Registro Italiano; R = American…
A = Veritas Austro-Ungarico; H = Veritas Ellenico; N = Norske Veritas; V = Bureau Ve…

In the 2nd Column, B.T., with figures following, signifies Date of report of bottom sighting by Board of Trade Surveyors as condition of PASSENGER CERT…

In the 3rd Column, C = Composite (iron frame, wood planked); I = Iron; Stl = Steel; W = Wood.

Further particulars of Vessels classed by Lloyd's Register will be found in Lloyd's Register Book and Lloyd's Yacht Register.

	Nom, Matériaux, Gréement, etc. / Capitaine et dates d'entrée au service des Armateurs actuels et à bord du navire / Ancien Nom, nombre de ponts, etc.	Pavillon. / Port d'Armement	Tonnage Net / Total	Dimensions Longueur	Largeur	Creux	Construit. Où. / Par qui.	Quand.	Nombre des cloisons. / Détails sur le lest d'eau.	Machines. Détails. / Nom du Constructeur.	Armateurs ou Directeurs-Gérants.
	Black Diamond I ScSc	British	408 684	191·8	28·2	16·4	Sunderland J.Laing	1855		2Cy. 70 HP.	Fenwick & Co.
	Black Diamond W Sc3mSc *Postag*	British Belfast	109 274	148·2 Q 22 f.	21·6	11·6	Troon PortlandS.B.Co.	1864		2Cy.30"–22" 74 HP. V.&D.Coates,Belfast	J.Howden
✻	Black Head I ScSr *Hughes 81-89* 1 *Dk*2 *B*	British Belfast	720 1145	249·7 Q 90 f. B 71 f. F 33 f.	31·2	15·5	Belfast Harland&Wolff	1881	4 Bhds. CDB	2Cy.28"& 58"–38" 140 HP. J.Howden&Co.,Glasgow	Ulster S.S.Co.,Ld.
✻	Black Pearl Stl ScSrYt *Band 85-88* 1 *Dk*	British Southampton	145 240	144·0	23·1	13·1	Greenock Caird&Co.,E.&E.Co.	1885	1 Bhds.	2Cy.19"& 36"–24" 55 HP. Rankin&Blackmore,Grnck.	Earl of Pembroke
✻	Black Prince (ex York Minster) I ScSr *Milburn -87* 1 *Dk*2 *B*	British N.Shields	960 1480	250·7 Q 36 f. B 52 f. F 30 f.	35·3	17·3	Newcastle Palmers'Co.,Ld	1883	4 Bhds. CDB330t APT70t	2Cy.30"& 58"–36" 142 HP. Palmers'Co.,Ld.,N'wcastle	J.Knott
	Black Prince I PadSp	British Liverpool	55 212	139·0	20·7	11·9	Glasgow Dobie&Co.	1867		2Cy.36"–60" 90 HP. Rankin&Blackmore,Grnck.	E.Griffiths
	Black Prince W Pad *Johnston 88-88*	British S.Shields	14 164	97·8	18·9	9·8	NorthShields Stobbs	1874		1Cy.35½–50" 42 HP. T.R.Scott&Co.,N.Shields	High&Shallcross
✻D	Black Sea I ScSr *Morrow 82-87* 1 *Dk*2	British Glasgow	582 922	226·2 P 128 f. F 28 f.	30·7	16·8	Sunderland J.Thompson	1871	4 Bhds. DBa65f f29t	2Cy.29½"&55"–33" 120 HP. G.Clark,Sunderland	J.&P.Hutchison
✻	Black Swan I ScBn *Lawton* 1 *Dk*2 *B*	British Newcastle	412 650	199·1 Q 55 f. F 16 f.	27·3	15·6	NorthShields T.&W.Smith	1864	4 Bhds.	4Cy.19"& 34"–26" 80 HP. Chip.'79,Thompson&Boyd,Nwe.	T.&W.Smith
✻D	Black Watch I ScSr *Lowe 76-86* (FOUNDERED) 2 *Dks*3 *B*	British Cardiff	943 1468	250·0 P 29 f.	33·0	22·9	Middlesbro' R.Dixon&Co.	1874	4 Bhds. DBo&b140t MT270t	2Cy.30"& 57"–33" 130 HP. T.Richardson&Sons,Htlpol.	Christie & Co.
	Blackcock I ScSr *Griffiths 76-86*	British Liverpool	26 228	146·2 F 14 f.	21·8	10·8	Birkenhead LairdBros.	1886		3C19"28"&47½"–30" 200 HP LairdBros.,Birkenhead	LiverpoolScrewTwng. Lighterage Co.
✻D	Blackhalls (STRANDED) I ScSr *Strachan 86-86* 1 *Dk*2 *B*	British W.Hartlpool	724 1142	235·0 B 58 f. Q 92 f. F 31 f.	31·6	15·0	W.Hartlepool W.Gray&Co.	1880	4 Bhds. BDa&f215t	2Cy.27"& 50"–33" 99 HP. T.Richardson&Sons,Htlpol.	R.Ropner & Co.
	Blackstone W ScSr *Taylor* 3 *Dks*	American Baltimore	864 1147	179·0	30·6	19·7	Mystic,Conn.	1863		1Cy.48"–28" 125 HP. CorlissSteamEngineCo.	Merchants & Miners Transportation Co.
✻	Blackwater I ScSr *Morgan 75-83* 1 *Dk*	British Dublin	342 538	176·0 Q 82 f. B 17 f. F 22 f.	24·6	12·9	Belfast Workman,Clark&Co.	1883	4 Bhds. DBa50t FPT75t	2Cy.22"& 44"–33" 80 HP. W.King&Co.,Glasgow	S.S.M'Cormick
✻D	Blaenavon I ScSr *Caslall* 1 *Dk*2 *B*	British Cardiff	842 1226	240·9 Q 89 f. B 50 f. F 28 f.	33·2	16·5	Newcastle Palmers'Co.	1881	4 Bhds. DBa&f231t APT15t	2Cy.28"& 54"–36" 99 HP. Palmers'Co.,Ld.,Newcastle	"Blaenavon" S.S.C.Ld. (Morel Bros. & Co.)
✻D	Blagdon I ScSr *Bunn 65-81* 2 *Dks*	British Newcastle	1339 2062	285·0 P 30 f. B 55 f. F 29 f.	36·0	24·0	Blyth Hodgson&Soulsby	1881	6 Bhds. DBa&b270t	2Cy.35"& 68"–45" 200 HP. R.&W.Hawthorn,Newcstle.	R.Bell & Co.
✻	Blakemoor StlSc	British S.Shields	1768	257·4 Q 76f. F 29f. B 88f. F 33f.	35·8	20·9	S. Shields J.Readhead&Sons	1889		3Cy.20½" ½"&55"–36" J.Readhead&Sons, S.Shields [170 HP.	W.Runciman
✻	Blanche I ScSr *Cruickshanks 88-88*	British W.Hartlepl	147 246	146·1 Q 30 f.	20·0	12·0	Glasgow A.&J.Inglis	1863	4 Bhds.	2Cy.22½"–22" 40 HP. A.&J.Inglis,Glasgow	C.Furness
	Blanche I Sc3mSr	British Londnderry	138 235	120·0	20·0	12·3	Newport UsksideI.Co.	1872		2Cy.18"–18" 40 HP. UsksideI.Co.,Newport	H.Lane
	Blanche Stl PadSp	British W.Hartlpool	12 112	95·6	19·2	9·8	SouthShields J.T.Eltringham	1883		2Cy.28"–50" 75 HP. J.P.Renaoldson&Son,S.Shld.	North Eastern Rail. Co.
V	Blanche I ScBn *Mignot -87* 2 *Dks*	French Havre	560 837	220·0 P 52 f. F 34 f.	30·2	16·2	Renfrew Henderson,Coulborn&Co	1869	6 Bhds. MT300t	2Cy.30"& 54"–36" 120 HP. Soc.desCnstr.Navales,Hvre.	Worms,Josse & Co.
	Blanche I Sc3mSr *Moré* 1 *Dk*2 *B*	French Rouen	317 505	164·3 P 22 f.	27·6	15·7	Hull M.Samuelson	1857	WB	2Cy.36¾"&55"–26½" 80 HP. Lobnitz,Coulborn&Co.Ren.	A.Grandchamp fils & Co.
	Blanche Henderson (ex Chase) (BARGE) W Sc3mSr *Peterson*	American Philadelphia	289 505	146·5	27·5	18·0	Norwich,Conn	18 64/83			D.K.Joselyn
✻	Blanchland I ScSr *Manson* 1 *Dk*2 *B*	British Newcastle	797 1238	230·5 Q 75 f. B 68 f. F 25 f.	34·1	16·2	Sunderland Osbourne,Graham&Co.	1883	4 Bhds. DBo&f250t FPT35t APT13t	2Cy.28"& 53"–36" 120 HP. Carr&Co.,Ld.,Sunderland	Hannay,Boyd & Co.
✻	Blandusia W ScSrYt *Brown -78*	British Cowes	45 107	108·0	17·5	10·7	Plymouth W.S.Kelly	1878		2Cy.15¼"&26¼"–21" 30 HP. WilloughbyBros.,Plymouth	Camper & Nicholson
	Blankenese (ex Albert Victor) I PadCr *Faje 62-76* 1 *Dk*	German Hamburg	232 398	230·9	24·3	8·8	Glasgow Aitken&Mansel	1866		2Cy.17¾"–28½" 200 HP. Aitken&Mansel,Glasgow	Hamburg-Amerik Packetf.-Aktien-Ges.
	Blekinge I ScSr *Ohlsson* 1 *Dk*	Swedish Ronneby	91 127	94·2 Q25f. B41f. F30f. H4f.	17·2	7·0	Oscarshamn OscarshamnMk.Vks.	1881		2Cy.13"&27"–18" 25 HP. Fletscher,Oscarshamn	P.Siegbahn
✻	Blencowe I ScSr *Gray 87-87* 1 *Dk*	British Newcastle	304 519	165·0 Q 85 f. B 17 f. P 21 f.	26·1	13·0	SouthShields J.T.Eltringham	1881	4 Bhds. DBa&f110t	2Cy.22"& 40"–30" 70 HP. J.P.Rennoldson,S.Shields	Fisher,Renwick & Co.
	Blenda I TwnScSr *Forsberg 72-75* 1 *Dk*	Swedish Warberg	178 242	135·8 Q50f. B25f. F37f. H4f.	20·1	9·6	Gothenburg Gothenburg Mek.Verks.	1875		4Cy.12¼"&26½"–18" 80 HP. Gothenburg Mek.Verkst.	Hallands Ångbåts-Aktiebolag
✻V	Bléville Stl Sc *Soutie* 2 *Dks*3 *B*	French Havre	1645 2518	300·0 P 48 f. B 64 f. F 36 f.	40·0	24·5	Glasgow A.Stephen&Sons	1886	5 Bhds. DBeb162tMT535t APT43t	3C.21"33"&52"–42" 210 HP. A.Stephen&Sons,Glasgow	E.Corblet

Lloyd's Register; D = Nederlandsche Vereeniging; G = Germanischer Lloyd; I = Registro Italiano; R = American Record;
Veritas Austro-Ungarico; H = Veritas Ellenico; N = Norske Veritas; V = Bureau Veritas.

Dans la 2me **Colonne**, B.T., avec les indications qui suivent, signifie la date du rapport de visite de la Carène par les experts du Board of Trade pour obtenir le certificat relatif au TRANSPORT DES PASSAGERS.

Dans la 3me **Colonne**, C = Composite (membrure en fer, bordé en bois); I = Fer; Stl = Acier; W = Bois.

Pour plus amples renseignements sur les navires classés par le Lloyd's Register, voir le Lloyd's Register Book et le Lloyd's Yacht Register.

No. in Book / Official No.	Registry in which classed	Ship's Name. Material, Rig, &c. / Master, and Dates of Appointment to present Owner's Service and to Ship / Ship's Late Name. No. of Decks, &c.	Flag / Port of Registry	Tons. Net / Gross	Dimensions: Length	Breadth	Depth	Built: Where / By Whom	When	No. of Bulkheads / Particulars of Water Ballast	Engines: Particulars / Maker's Name	Owners
391		**Blidah** I ScSmSr *Lombardiere 76-88* 2 *Dks*	French Marseilles	312 485	183·4 *B 50 f.*	28·0 *F 18 f.*	15·5	Greenock Caird&Co.	1856		2Cy.21½"&42"–27" 100 HP. Fraissinet&Co.,Marseilles	Fraissinet &…
392		**Blido** (ex Gogs…) I Sc *Österman 82-86* 1 *Dk*	Swedish Stockholm	82 166	104·0 *Q29f. B37f.*	15·9 *F11f.*	6·8 *B31.*	Nyköping Nyköping,M.V.	1866		1Cy.14"–18" 20 HP. Bergsunds Mk. Vks, Stkhlm.	Ljusterö Ång… (C.Petterson)
393 56878	✳	**Blonde** I ScSr *Evans 85-87* 1 *Dk*	British Whitehaven	175 282	147·5	22·1 *Q 34 f.*	11·5	Port Glasgow Blackwood&Gordon	1867	4 Bhds.	2Cy.24"–24" 40 HP. Blackwood&Gordon,Pt.Gls.	G.Nelson&…
394		**Blonde** I ScSr *Brettschneider -81* 1 *Dk2 B*	German Danzig	462 585	177·8	28·7	14·9	London [illegible]	1863		1Cy.16"& 36"–27" 40 HP. C.Langley,London	Th.Rodenack…
395 76839		**Blosse Lynch** I Pad	British London	384 [illegible]	220·4	29·1	6·9	London J.&G. Rennie	1876		1Cy.18½"&85"–54" 100 HP. J.&G.Rennie,London	Euphrates &… Stm.Nav.Co…
396 89882		**Blücher** I PadSp	British Nth.Shields	28 [illegible]	95·8	18·4	8·9	North Shields Hepple&Co.	1885		2Cy.34½"–52" 40 HP. Hepple&Co.,North Shields	W.Stephenson…
397 62312	✳	**Blue Cross** (WRECKED) I ScBk *Robertson* 1 *Dk2 B*	British Dundee	694 1075	219·5 *P 21 f.*	30·4 *F 119*	18·2	North Shields T.&W. Smith	1869	4 Bhds. DBe… FPT26 APT…	2Cy.27"& 52"–36" 160 HP. Hawks,Crawshay&Sn.Gtshd	R.Taylor
398 87481	✳D	**Blue Jacket** I ScSr *Webb 85-85* 2 *Dks3 B*	British Cardiff	1440 2285	282·7 *P 42 f.*	36·6 *B 85 f.*	23·9 *F 26 f.*	Sunderland [illegible]	1883	7 Bhds. DBa&Cast FPT46 APT31	2Cy.35"& 60"–42" 200 HP. J.Dickinson,Sunderland	"Blue Jacket"… Ltd. (G.Muller)
399 78402	✳	**Blythville** I ScSr *Ling 87-87* 1 *Dk2 B*	British W.Hartlepool	897 1384	218·9 [illegible]	32·1 [illegible]	18·4 [illegible]	W.Hartlepool W.Gray&Co.	1877	4 Bhds. DBa…	2Cy.29"& 56"–33" 120 HP. N.E.Marine E.Co.Ld.,Sndrld.	W.Gray
400 63638	✳	**Blythwoode** I ScSr *Gibson* 1 *Dk2 B*	British London	770 1244	230·0 *P 223 f.*	32·0 *F 28 f.*	17·2	W.Hartlepool Denton,Gray&Co.	1870	4 Bhds. [illegible]	2Cy.28"& 56"–36" 120 HP. R.&W.Hawthorn,Newcastle	Watts,W…
401 85937		**Boadicea** Stl ScScYt	British Glasgow	166 [illegible]	170·0	24·0	12·3	Glasgow J.Elder&Co.	1882		2Cy.23"& 46"–30" 115 HP. J.Elder&Co.,Glasgow	Mrs.Church…
402		**Bocognano** I ScSr	French Marseilles	490 829	217·0	29·0	17·5	Rouen Claparede&Co.	1884		2C23½"&59½"–35½" 150 HP. Claparede&Co.,Rouen	F.Morelli&C…
403 1071		**Bohcho Maru** W ScSr	Japanese Hiogo	238 381	143·4	20·4	18·5	Kobe [illegible]	1888		65 HP.	Kyuyei Sha…
404	V	**Bohemia** I ScSr *Kopff 80-88* 2 *Dks&Spr Dk*	German Hamburg	2216 3410	350·5 *B 84 f.*	40·7	23·0 30·0	Glasgow A.&J.Inglis	1881	[illegible]	2Cy.40"& 77"–54" 360 HP. A.&J.Inglis,Glasgow	Hamburg-A… Packett. A…
405	G	**Bohus** C ScSr *Olsson* 1 *Dk*	Swedish Gothenburg	139 [illegible]	96·5 *B 23 f.*	20·6	10·5 *H 4 f.*	Gäddvik E.Jansson	1883		2Cy.13"& 16"–18" 30 HP.	A.Bratt
406 80155	✳	**Boileau** I ScSr *Crosby 75-84* 1 *Dk2 B*	British Cardiff	1231 1885	258·0 *B 56 f.*	36·4 *F 31 f.*	21·7	Newcastle Palmers'Co.	1884	4 Bhds. DBa… 331 FPT61 APT30	2Cy.33"& 62"–42" 180 HP. Palmers'Co.Ld.,Newcastle	"Boileau" S… (H.Guéret)
407 68397		**Bokhara** I Sc3mSr *Bason 84-88*	British Greenock	1711 2932	361·5	39·0	29·0	Greenock Caird&Co.	1873		2Cy.59½"&86"–54" 500 HP. Caird&Co.,Greenock	Peninsular &… Steam Nav.C…
408 76494		**Boko** I PadTugSr *Arscott* 1 *Dk*	British Brisbane	70 203	125·0	21·1	10·4	Newcastle A.Leslie&Co.	1877	3 Bhds.	2Cy.37"–54" 99 HP. Thompson&Co.,Newcastle	Frederick…
409	✳	**Bolama** I ScSr *Rodrigues 81-83* 1 *Dk*	Portug'se Lisbon	337 530	180·0 *P 32 f.*	26·0 *B 23 f.*	13·6 *F 28 f.*	Hull Earles'S.&E.Co.	1883	4 Bhds. WBay…	2Cy.24"& 45"–33" 90 HP. Earles'S.&E.Co.,Hull	Empreza Na… Nav. à Vap…
410 89590	✳	**Bolderaa** I ScSr *Hannah 85-85* 1 *Dk2 B*	British London	1061 1638	244·0 *Q100f.H25f.*	36·3 *B50f.*	18·1 *F20f.*	Sunderland J.Laing	1884	4 Bhds. WBay	2Cy.29½"&55"–36" 140 HP. G.Clark,Sunderland	C.M.Norwo…
411	✳	**Bolinao** I ScSr *Eiguren* 1 *Dk*	Spanish Manila	120 247	134·0	24·1 *F 17 f.*	7·9	Paisley Abercorn S.B.Co.	1884	4 Bhds.	2Cy.18"& 34"–24" 50 HP. Hanna,Donald&Wilsn.Pais	F.L.Roxas
412 3314		**Bolivar** I TScSr	American New York	117 234	127·1	23·2	9·6	Philadelphia Neafie & Levy	1885		1Cy.15"&26"–22" Neafie & Levy,Philadelphia	Panama R.R…
413	A	**Bolivar** (ex Saidi) I ScSr *Gerolimich -88* 1 *Dk*	Austrian Trieste	444 [illegible]	190·0	27·4	18·1	Hull Samuelson&Co.	1863		2Cy.31"–30" 100 HP. Samuelson & Co.,Hull	G.Dr.Martin…
414 69081		**Bolivia** I ScBk *Baxter 84-88* 2 *Dks,& Awn Dk1 B*	British Barrow	2626 4050	400·0	40·0	25·1 33·0	Port Glasgow R.Duncan&Co.	1873	7 Bhds.	2C.59"& 107"–48" 1120 HP. D.&W.Henderson&Co.Glsw	Barrow Ste… Ld.
415 69349		**Bolivia** I Sc3mSr 2 *Dks,& Shade Dk*	British Liverpool	1215 1925	311·8	38·2 *F 54 f.*	21·1	Glasgow T.Wingate&Co.	1874	6 Bhds.	2Cy.41½"&84½"–42" 160 HP. T.Wingate&Co.,Glasgow	Pacific Ste…
416	✳	**Bombay** Stl ScSr 2 *Dks*	British Greenock	 3250	350·0 *P 17 f.*	42·3 *B 104 f.*	26·2 *F 46 f.*	Greenock Caird & Co.Ld.	1889	7 Bhds. [illegible]	3Cy.24",40"&62"–54" Caird & Co.,Greenock	Peninsular… Steam Nav.…
417 86076	✳D	**Bombay** I Sc3mSr *Willis* 2 *Dks3 B*	British London	2044 3158	337·0 *P 40 f.*	41·4 *B 82 f.*	26·5 *F 46 f.*	Newcastle C.Mitchell&Co.	1882	6 Bhds. CDB108 FPT30 APT30	2Cy.40"& 74"–54" 350 HP. N.E.Marine E.Co.Ld.,Sndrld	Nelson,Do…
418 63084	✳	**Bon-Accord** (WRECKED) I ScSr *Talbot* 1 *Dk*	British Aberdeen	297 [illegible]	169·9	26·2 *Q 56 f.*	14·2	Aberdeen Hall,Russell&Co.	1871	4 Bhds. DB 68	2Cy.22"& 37"–30" 60 HP. Hall,Russell&Co.,Aberdeen	J.& A.Davi…
419 87966	✳	**Bonavista** I ScSr *Anderson* 1 *Dk2 B*	British Montreal	837 1306	240·4 *P 52 f.*	33·5 *B 53 f.*	18·3 *F 26 f.*	Newcastle Wigham Richardson&Co.	1884	4 Bhds. DBa… FPT61 APT…	2Cy.31"& 62"–42" 160 HP. W.Richardson&Co.,Newcstl.	Black Diam… Co.,Ld.
420 3182	RV	**Bonita** W ScSr *Hall* 2 *Dks3 B*	American San Francisco	356 [illegible]	175·0	27·0	8·9 16·0	San Francisco Dickie Bros.	1881	[illegible]	2Cy.18"& 34"–27" 50 HP. Risdon Iron Wks.,S.Francisco	Pacific Coa…

✳ = **Lloyd's Register**; D = Nederlandsche Vereeniging; G = Germanischer Lloyd; I = Registro Italiano; R = Ameri…; A = Veritas Austro-Ungarico; H = Veritas Ellenico; N = Norske Veritas; V = Bureau …

In the 2nd Column, B.T., with figures following, signifies Date of report of bottom sighting by Board of Trade Surveyors as condition of PASSENGER CERT…

In the 3rd Column, C = Composite (iron frame, wood planked); I = Iron; Stl = Steel; W = Wood.

Further particulars of Vessels classed by Lloyd's Register will be found in Lloyd's Register Book and Lloyd's Yacht Register.

Classe	Nom, Matériaux, Gréement, etc. / Capitaine et dates d'entrée au service des Armateurs actuels et à bord du navire / Ancien Nom, nombre de ponts, etc.		Pavillon. / Port d'Armement	Tonnage Net / Total	Dimensions Longueur	Largeur	Creux	Construit. Où. / Par qui.	Quand.	Nombre des cloisons. / Détails sur le lest d'eau.	Machines. Détails. / Nom du Constructeur.	Armateurs ou Directeurs-Gérants.
✻	Bonito *Thompson 84-88*	I ScTrawler *1 Dk*	British Leith	23	90·0	20·1	9·0	Dundee W.B.Thompson	1885	3 Blhds.	3C.10½"15"&27"–16" 38 HP. W.B.Thompson, Dundee	Mackay & Co.
	Bonnie Doon *Nicholas -86*	I PadSk	British Bristol	38	218·0	20·0	7·5	Glasgow T.B.Seath&Co.	1876		1Cy.50"–72" 96 HP. A.Campbell&Son, Glasgow	G.Nourse
✻	Bonnie Kate *Needham 86-87*	I ScSr *1 Dk 2 B*	British Hull	539	208·1	28·2	15·9 *P 118 f. F 15 f.*	Sunderland Oswald&Co.	1870	1 Blhds. MT60 APT20	4Cy.19"&34"–30" 80 HP. Comp. 75 C.D.Holmes&Co., Hull	E.J.Sutton & Co.
	Bonnie Princess	I PadSr	British Liverpool	92	210·0	26·2	9·3	Glasgow T.B.Seath&Co.	1882		2Cy.49"–72" 160 HP. A.Campbell&Son, Glasgow	Liv.Lland'dno&Welsh Coast Steamboat Co., Ld.
✻	Bonnington *Reine 85-88*	Stl Sc3mSr *2 Dks 3 B*	British Glasgow	1385 2125	289·2	38·2	23·1 *B 72 f. F 39 f.*	PortGlasgow R.Duncan&Co.	1883	5 Blhds. OB375t	2Cy.36"&68"–45" 200 HP. Dunsmuir&Jackson, Glasw.	S.S."Bonnington"Co., Ld. (Raeburn & Verel)
	Bonny *Walsh -88*	I ScSr	British Glasgow	798 1277	261·2	30·1	23·0	Glasgow Randolph,Elder&C.	1869		2Cy.38"&68"–33" 250 HP. Randolph,Elder&Co., Glasw	British & African Steam Nav.Co.
	Booldana *Windebank 86-88*	Stl Sc3mSr *2 Dks*	British Glasgow	2134 2873	332·2	40·2	25·5 *P 48 f. B 26 f. F 41 f.*	Dumbarton W.Denny&Bros.	1880	6 Blhds. CDB62t	2Cy.38"&68"–51" 300 HP. Denny&Co., Dumbarton	British India Steam Nav.Co., Ld.
	Boomerang	I ScSr *1 Dk*	British Nw'cleNSW	446 653	211·9	21·8	19·0	Glasgow T.Wingate&Co.	1854	5 Blhds.	2Cy.40"–30" 106 HP. T.Wingate&Co., Glasgow	Newcastle S.S.Co., Ld.
	Booth	W Pad *1 Dk*	British Ottawa	194 235	120·0	37·0	8·0 *H 88 t.*	Nor. Bay, Ont. P.Lamorthe	1885		1Cy.26½–81" 120 HP. J.&R.Wier, Montreal	John R.Booth
✻	Bordeaux *Giles*	I ScSr *1 Dk*	British London	353 560	170·0	26·9	12·8 *Q 50 f. B 44 f. F 21 f.*	Leith Ramage&Ferguson	1884	5 Blhds. DB 62t FPT21t	2Cy.22"&41"–30" 72 HP. Ramage&Ferguson, Leith	C.F.Leach
	Bordeaux	I PadSr	British Newhaven	307 483	214·0	23·7	10·8 *Q 34 f.*	London C.Langley	1865		2Cy.52"–57" 200 HP. C.Langley, London	London Brighton & South Coast Railway Co.
DV	Bordeaux *Gandillon 87-87*	I ScImSr *2 Dks & Spr Dk*	French Havre	2662 3417	379·6	40·0	21·7 *P 37 f. 29·6*	Havre Forges&C. de la Medit.	1882		2Cy.41"&71" 43½" 400 HP. Forges&C. de la Medit, Hvre	Cie Commerciale de Transps. à vap. Français
V	Bordeaux *Nielsen -87*	Stl ScSr *2 Dks*	Swedish Gothenburg	736 843	195·8	27·2	21·3 *H 8 f.*	Gothenburg MotalaCo.	1881	MT 250t APT 35t	2Cy.23"&44"–30" 110 HP. MotalaCo., Gothenburg	FörnyadeÅngf.Akt. BolSvenskaLloyd(A.Meyer)
✻	Border Chieftain *Rowell 77-85*	I ScBg *2 Dks 3 B*	British Newcastle	990 1519	261·0	31·4	23·8 *B 8 f. F 34 f.*	Sunderland W.Pile&Co.	1873	5 Blhds. DB ef 20t	2Cy.30"&60"–36" 160 HP. W.Pile&Co., Sunderland	Clapham S.S.Co., Ld.
✻	Borderer *Bailey 79-86*	I ScImSr *1 Dk & Spr Dk 3 B*	British Liverpool	3135 4748	400·0	44·2	23·7 *P 20 f. F 6 f. B 101 f. 31·3*	Barrow BarrowCo., Ld.	1884	5 Blhds. DB af & ef 518t MT 112t	2Cy.50"&90"–60" 354 HP. Barrow S.B.Co., Ld., Barrow	Atlantic & Eastern S.S. Co., Ld.
	Bore *Reskow -86*	I TwnScSr *1 Dk*	Swedish Stockholm	435 563	175·2	26·0	11·5 *P 111 f. F 28 f. H 21 t.*	Norrköping MotalaCo.	1853		4Cy.16"&31"–21" 160 HP. W.Lindbergs Mk Vks., Stkhm	Rederi Aktie-Bolaget "Svea" (K.Ahlborg)
✻	Borghese *ex Galatea* *Hettie 83-85*	I ScSr *3 Dks*	British Glasgow	1331 2044	289·6	35·4	25·9	Sunderland W.Pile&Co.	1870	5 Blhds. efDB	2Cy.36½"&64"–42" 200 HP. J.Howden&Co., Glasgow	Raeburn & Vérel
	Borgholm *Rosenström*	I ScSr *1 Dk*	Swedish Oscarshamn	112 143	95·5	17·6	7·3 *Q 25 f. B 42 f. F 25 f. H 3 t.*	Oscarshamn Oscarshamn Mek. Vrks	1877		30 HP.	Oscarshamns Rederi Akt.Bol. (P.Deurell)
V	Borinquen *Arrotegui -88*	I ScSr *2 Dks & Awn Dk*	Spanish Bilbao	804 1262	248·5	30·1	16·9 *24·2*	Liverpool R.&J.Evans	1878		2Cy.26"&45"–42" 120 HP. J.Jack&Co., Liverpool	D.T.de Eizaga
✻I	Bormida *Denegri 88-88*	I ScSr *2 Dks 3 B*	Italian Genoa	1474 2290	299·7	37·2	23·6 *P 44 f. B 82 f. F 40 f. H 102 t.*	Dumbarton Barrell&Son	1884	5 Blhds. [illegible]	2Cy.35"&67"–42" 220 HP. D.Stewart&Co., Glasgow	Navigazione Generale Italiana
	Borneo *Edwards*	I ScSr *1 Dk*	British Singapore	349 562	184·3	26·0	13·9 *Q 85 f. F 26 f.*	Newcastle Campbell, Mackintosh & Bowstead	1881	5 Blhds.	2Cy.25½"&48"–55" 100 HP. T.Clark&Co., Newcastle	Oh Swee Keat
✻	Borneo *Willems*	Stl ScSr *2 Dks 3 B*	Dutch Batavia	1494 2140	290·2	37·2	24·2 *P 54 f. B 82 f. F 30 f.*	Glasgow A.&J.Inglis	1884	5 Blhds.	2Cy.36"&70"–45" 250 HP. A.&J.Inglis, Glasgow	Nederlandsch-indische Stoomvaart-Maatschappy
DGV	Bornholm *Jensen -88*	I ScSr *1 Dk & Spr Dk*	Danish Svaneke	541 860	208·6	26·3	13·4 *20·5*	Kiel G.Howaldt	1883	W B 70t	2Cy.25"&46"–27" 85 HP. HowaldtBros., Kiel	Dampsk.-Selskabet "Bornholm" (D.Torm)
	Borodino *Dennison*	Stl ScSr *1 Dk 2 B*	British Hull	991 1284	239·8	33·0	16·9 *Q 85 f. H 37 f. F 36 f.*	Hull Earles'Co.	1880	4 Blhds. DB	2Cy.21"&53"–36" 98 HP. Earles'ShipBldg.Co., Hull	T.Wilson, Sons & Co.
V	Borussia *Magin 86-88*	I ScSr *2 Dks*	German Hamburg	1306 1776	271·0	33·8	24·2 *P 46 f. B 32 f.*	Hamburg ReiherstiegSchiffsWerfte	1878		2Cy.34"&62"–42" 180 HP. ReiherstiegSchiffsWerfte	Hamburg-Amerik. Packetf.-Akt.Ges.
	Borwick Rails	I PadSp	British Liverpool	48 136	102·6	19·1	9·6	Liverpool LiverpoolS.B.Co.	1871		2Cy.36"–30" 70 HP. G.Forrester&Co., Liverpool	W.J.Barratt
I	Bosforo *ex Winlow Hall* *Rustica 79-88*	I Sc3mSr *2 Dks 3 B*	Italian Palermo	1703 2531	360·5	37·1	26·4 *B 35 f. F 31 f.*	Glasgow C.Connell&Co.	1878	5 Blhds.	2Cy.44"&78"–54" 400 HP. J.&J.Thomson, Glasgow	Navigazione Generale Italiana
✻D	Boskenna Bay *Legg 83-83*	I ScSr *1 Dk & Spr Dk 3 B*	British Penzance	1500 2300	290·6	37·2	26·8	Newcastle Schlesinger, Davis &Co.	1882	5 Blhds. DB af 228t APT 32t	2Cy.37"&70"–45" 280 HP. R.&W.Hawthorn, Newcastle	"Boskenna Bay" S.S. Co., Ld. (F.Banfield & Sons)
✻	Bosphorus (COLLISION) *Smith 75-83*	I ScSr *2 Dks 3 B*	British Newcastle	1630 2490	295·0	37·2	25·2 *P 31 f. B 62 f. F 38 f.*	Newcastle W.Dobson&Co.	1883	6 Blhds. DB af & ef 331t FPT 69t APT 38t	2Cy.36"&68"–45" 212 HP. R.&W.Hawthorn, Newcastle	Hall Bros.
	Bosphorus	I PadSp	British Sth.Shields	16 139	112·2	19·2	9·7	SouthShields J.T.Eltringham	1884		1Cy.36¼"–54" 70 HP. J.P.Rennoldson&Ss, S.Slds	J.Wilson

✻=Lloyd's Register; D=Nederlandsche Vereeniging; G=Germanischer Lloyd; I=Registro Italiano; R=American Record;
A=Veritas Austro-Ungarico; H=Veritas Ellenico; N=Norske Veritas; V=Bureau Veritas.

Dans la 2me Colonne, B.T., avec les indications qui suivent, signifie la date du rapport de visite de la Carène par les experts du Board of Trade pour obtenir le certificat relatif au TRANSPORT DES PASSAGERS.

Dans la 3me Colonne, C=Composite (membrure en fer, bordé en bois); I=Fer; Stl=Acier; W=Bois.

Pour plus amples renseignements sur les navires classés par le Lloyd's Register, voir le Lloyd's Register Book et le Lloyd's Yacht Register.

No. in Book. / Official No.	Registry	Ship's Name. Material, Rig, &c. / Master, and Dates of Appointment to present Owners' Service and to Ship. / Ship's Late Name. No. of Decks, &c.	Flag. / Port of Registry.	Tons.	Dimensions. Length	Breadth	Depth	Built. Where. / By Whom.	When	No. of Bulkheads. / Particulars of Water Ballast.	Engines. / Particulars, Maker's Name.	Owners or Managers
451 54032		**Boston** I Sc3mSr *Bristow*	British London	547 [illegible]	226·0	27·7	17·1	Newcastle [illegible]	1866		[illegible] 24" & 44"–30" 80 HP. [illegible] 72, Palmers' Co., Newctl.	Commercial S.S. Ld.
452 84538	✻	**Boston City** I ScSr *Smith* 81-88 2 *Dks* 3 *B*	British W. Hartlpool	1864 2333	211·8	[illegible]	22·9	W. Hartlepool [illegible]	1884	[illegible]	2Cy. 38" & 71"–[illegible]" 200 HP. Blair & Co. Ld. Stockton	C. Furness
453 [illegible]	✻ BT [illegible]	**Bostonian** StlScSr [illegible] *Trent* 85-88 3 *Dks*	British Liverpool	[illegible] [illegible]	421·8	43·2	30·9	Belfast Harland & Wolff	1888	7 Blds. [illegible]	3C. 29" 45" & 71"–60" 500 HP. Harland & Wolff, Belfast	S.S. Bostonian (F. Leyland & Co.)
454 70235	✻ D	**Bothal** I ScSr *Smith* 85-86 2 *Dks* 3 *B*	British Newcastle	[illegible] 1908	280·0	34·3	24·0	Newcastle [illegible]	1878	5 Blds. DB aft [illegible]	2Cy. 32" & 62"–42" 180 HP. R. & W. Hawthorn, Newcastle.	R. Bell & Co.
455 68094	BT 2.86	**Bothnia** I ScBk *Button* 83-86	British Glasgow	2923 4535	422·3	42·2	34·5	Glasgow J. & G. Thomson	1874		2Cy. 60" & 104"–54" 600 HP. J. & G. Thomson, Glasgow	Cunard Steam Co. Ld.
456 47832	✻	**Bothwell Castle** I ScBo *Ted* 78-82 2 *Dks* 3 *B*	British London	1653 2542	329·1	38·3	26·0	Middlesbro' [illegible]	1881	4 Blds. [illegible]	2Cy. 40" & 75"–42" 300 HP. T. Richardson & Sons, Hartlpl	Bothwell Castle Owners, Ld. [illegible]
457		**Botnia** (ex W. T. Marstoll) WScSr *Nilsgsen* -88 (ex Karl Abraham Axel) 1 *Dk* 2 *B*	Norweg'n Moss	493 843	183·0	30·6	16·0	Sundsvall [illegible]	1873		2C. 30½" & 44"–24" 80 HP. W. Lindberg's Mk. Vrks. Sthlm	Rederiselskabet nis (H. A. Rehder)
458		**Bottenhafvet** WSc3mSr *Westerin* -87 (ex Sanna Bryner) 1 *Dk*	Swedish Södermhmn	240 285	119·7	29·1	11·3	Sunderland	18[illegible]		45 HP. Bergsunds Mk. Vk. Stckhlm	A. F. Hillman
459	✻	**Boucau** I ScSr *Pinaud* 85-88 1 *Dk* 2 *B*	French La Rochelle	789 1216	240·6	33·2	16·1	Newcastle Palmers' Ship. & Iron Co.	1882	[illegible]	2Cy. 30" & 57"–36" 140 HP. Palmers' S. B. & I. Co. Ld. Nwcl	D'Orbigny & F. fils
460		**Boug** ISc Bk		667 1000	211·2	25·3	21·0	Newcastle J. R. Palmer	1857		144 HP.	
461 79835	BT 2.88	**Boulogne** I TwnScSr	British London	222 407	182·0	25·0	12·0	Hull Earles' Co.	1878		1Cy. 20" & 35"–21" 110 HP. Earles' Co. Ld. Hull	South Eastern Co.
462		**Bourgogne** I Sc *Iperti* 76-87 3 *Dks*	French Marseilles	1480 1902	318·2	31·8	22·6	Sunderland James Laing	1865		2Cy. 36¼" & 73¾"–36" 200 HP. R. Napier & Sons, Glasgow	Société Générale Transports Mar.
463 93107	✻	**Bournemouth** I ScTrawler *Chant* 82-87 1 *Dk*	British Hull	62 135	97·0	20·6	11·0	Hull Cook, Welton & Gemmell	1887	3 Blds.	2Cy. 16" & 30"–22" 40 HP. C. D. Holmes & Co. Hull.	C. Hellyer
464 87119		**Boussa** Stl Stern Wheel *Hyson*	British London	395 478	170·0	32·0	5·8	Erith Shuttleworth & Chapman	1883		1Cy. 13" & 24"–30" 60 HP. Shuttleworth & Chapman, Erith	National African Ld.
465 67514	✻ D	**Bowesfield** I ScBo *Lightfoot* 87-87 1 *Dk* 2 *B*	British Stockton	995 1524	243·0	34·7	18·6	Stockton Richardson, Duck & Co.	1880	3 Blds. DB [illegible]	2Cy. 31" & 58"–36" 156 HP. Blair & Co. Ld., Stockton	F. Binnington
466 89247	✻	**Bowra** W ScSr 1 *Dk*	British Melbourne	140 [illegible]	123·0	23·6	8·2	Terrigal, N.S.W. T. Davis	1883		3Cy. 23" 14" & 9"–14" 40 HP. A. Shanks & Son, Arbroath	J. Mackintosh
467 85047	✻ D	**Boyne** I ScSr *Ryan* 87-88 2 *Dks* 3 *B*	British London	1306 [illegible]	285·0	35·5	24·4	W. Hartlepool W. Gray & Co.	1881	5 Blds. DB [illegible]	2Cy. 35" & 65"–39" 200 HP. T. Richardson & Sons, Hartlpl	Mercantile S.S.
468	✻	**Brabo** I ScSr *Pardage* 1 *Dk.* Spt [illegible]	Belgian Antwerp	842 [illegible]	224·7	33·7	17·5	Dumbarton H. Murray & Co.	1883	5 Blds. DB [illegible]	2Cy. 27" & 54"–36" 120 HP. Hutson & Corbett, Glasgow	Walford & Co.
469 83909	✻ D	**Bracadaile** I ScSr *Nelmes* 81-87 3 *Dks*	British Newcastle	1380 2141	292·0	37·2	23·6	Newcastle Palmers' Co.	1881	7 Blds. [illegible]	2Cy. 36" & 68"–45" 270 HP. Palmers' Co. Ld., Newcastle	J. Temperley & Co.
470 63743	✻	**Brackley** (ex Anne Smith) I ScSr *Murray* 81-88 1 *Dk* 2 *B*	British Liverpool	542 [illegible]	209·8	28·8	14·0	Newcastle W. Cleland	1872	5 Blds. WB [illegible]	2Cy. 26" & 50"–33" 98 HP. Ouseburn Eng. Wks., Newcstl	Ocean Queen S.S. Ld. (Wake & Sons)
471 28182		**Brackley** I PadSr	British Liverpool	46 112	116·0	17·6	7·6	Liverpool T. Vernon & Sons	1859		2Cy. 29"–36" 50 HP. R. Daglish, Warrington	Manchester Ship Co.
472 54805	✻ BT 7.88	**Bradford** I ScSr *Howell* 88-88 1 *Dk* 2 *B*	British Grimsby	393 [illegible]	205·2	24·4	12·0	Glasgow [illegible]	1865	4 Blds.	2Cy. 27" & 50"–30" 97 HP. Cmp. '79, Earles' Co., Hull	Manchester, Sheffield & Lincolnshire Ry.
473 63389	✻	**Bradley** (ex Jones Brothers) I Sc3mSr *Fletcher* 86-88 1 *Dk* 2 *B*	British Liverpool	482 [illegible]	189·6	28·2	15·3	Middlesbro' W. [illegible]	1871	4 Blds. [illegible]	3Cy. 13" 19" & 35"–24" 90 HP. N. E. Marine Co. Ld. Sundrld.	R. Tiernan
474 56947	✻	**Bradley** (COLLISION) I Sc3mSr *Lawson* 75-88 1 *Dk* 2 *B*	British Newcastle	374 [illegible]	189·7	26·7	15·1	North Shields [illegible]	1866	5 Blds. [illegible]	2Cy. 22" & 42"–30" 70 HP. W. Richardson & Co. Newcstle	T. & W. Smith
475 7044[illegible]	✻	**Braemar** I ScSr *Grubb* 81-82 1 *Dk* 2 *B*	British Aberdeen	462 [illegible]	200·5	27·1	15·4	Aberdeen Hall, Russell & Co.	1875	5 Blds. DB [illegible]	3Cy. 13" 24½" & 45"–30" 90 HP. Tripled 1888	Grampian S.S.
476 63324	✻ V	**Braganza** I ScBg *Hendry* 75 2 *Dks*	British Liverpool	755 [illegible]	258·2	29·7	20·0	Birkenhead Laird Bros.	1870	4 Blds.	2Cy. 36" & 56"–36" 180 HP. Laird Bros., Birkenhead	Liverp. & M[illegible] S.S. Co. Ld.
477		**Brage** I ScSr *Hansen* 71-87 1 *Dk*	Danish Kolding	123 217	143·6	19·8	10·4	Dundee Gourlay Bros. & Co.	18[illegible]		1Cy. 15" & 30"–20" 60 HP. Burmeister & Wain, Cphgn.	Det Forenede Selskab (C. P. A. [illegible])
478	N	**Brage** Stl ScSr *Meidell* 83-83 1 *Dk* 2 *B*	Norweg'n Bergen	356 [illegible]	161·2	24·5	13·8	Malmö Kockums Verksted	1883	[illegible]	2Cy. 20" & 35"–24" 65 HP. Kockums Verksted, Malmö	Bergh & Helland
479		**Brahestad** I ScSr 1 *Dk*	Russian Baku	161 [illegible]	125·3	19·5	9·6	London J. & W. Dudgeon	1874		4Cy. 12" & 22½"–13" 40 HP. J. & W. Dudgeon, London	
480	✻	**Braila** I ScSr *Imace* 65-87 1 *Dk.* & *Spr Dk* 3 *B*	French Marseilles	1175 [illegible]	259·2	37·0	22·7	Dumbarton A. McMillan & Son	1881	5 Blds.	2Cy. 36" & 63"–48" 300 HP. D. Rowan, Glasgow	Fraissinet & Co.

✻ = Lloyd's Register; D = Nederlandsche Vereeniging; G = Germanischer Lloyd; I = Registro Italiano; R = American [...]
A = Veritas Austro-Ungarico; H = Veritas Ellenico; N = Norske Veritas; V = Bureau Ver[itas]

In the 2nd Column, B.T., with figures following, signifies Date of report of bottom sighting by Board of Trade Surveyors as condition of PASSENGER CERTIF[ICATE]

In the 3rd Column, C = Composite (iron frame, wood planked); I = Iron; Stl = Steel; W = Wood.

Further particulars of Vessels classed by Lloyd's Register will be found in Lloyd's Register Book and Lloyd's Yacht Register.

No.	Class.	Nom, Matériaux, Gréement, etc. / Capitaine et dates d'entrée au service des Armateurs actuels et à bord du navire / Ancien Nom, nombre de ponts, etc.		Pavillon. / Port d'Armement	Tonnage Net / Total	Dimensions Longueur	Largeur	Creux	Construit. Où. / Par qui.	Quand.	Nombre des cloisons. / Détails sur le lest d'eau.	Machines. / Détails. / Nom du Constructeur.	Armateurs ou Directeurs-Gérants.
481	V	Brake *Loeb* 86-86	I ScSr 3 *Dks*	German Oldenburg	371 519	167·3 *B* 24 *f.*	26·0	15·4	Kiel Act.Ges."Germania"	1885	DT113 APT9	2Cy.26"&57½"–27½" 80 IP. Act.Ges."Germania," Kiel	Oldenburg-Portugiesische Dampfs. Rhederei
482		Brama *Lasig* 81-83	2m	Russian St. Petersburg	714				Gotheuburg MotalaCo.	1881		120 IP.	Naphta Productions Gesellschaft (Gebr.Noble)
483	✱	Bramble *Hutchinson* 88	StlScSr 1 *Dk*2 *B*	British Glasgow	962 1508	260·0 *Q*99 *ft.* *B*72 *ft.* *F*34 *ft.*	37·1	16·9	Campbelt'wn Campbeltown Shipbuilding Co.	1888	4 Blds. DB238 APT30	3C.18"29"&46"–36" 170IP Hutson&Corbett,Glasgow	P.H.Dixon & Co.
484	✱D	Brampton *Kydd* 88-88	Stl ScSr 1 *Dk*2 *B*	British Newcastle	1377 2105	275·5 *P*28*f* *Q*76*f* *B*110*f* *F*34*f*	37·2	19·9	S.Shields J.Readhead&Co.	1887	4 Blds. DB97/330	3C.21"35"&57"–39"200IP. J.Readhead&Co.,S.Shields	Chapman & Miller
485		Brandon *Begg* 83-83	I ScSr 1 *Dk*	British Limerick	141 307	154·0 *B* 30 *f.* *F* 37 *f.*	21·4	12·5	Newcastle W.Richardson&Co.	1875	1 Blds. DB82	2Cy.32"–30" 80 IP. J.Shaw&Co.,Newcastle	A.Russell
486		Brandywine 1 *Dk*.*SaloonDk*&*HurDk*	I Sc	American Philadelphia	215 407	175·0	25·0	9·2	WilmingtnDel Hrln&.Hollingswr	1885		2Cy.24"& 42"–24" Harlan&Hollingswth,C.Wil	I.Shields Wilson
487	✱	Brankelow *Mandeno* 87	I ScBn 2 *Dks*3 *B*	British Liverpool	1661 2531	300·0 *B*46 *f.* *P*26 *f.* *F*35 *f.*	38·0	25·5	Stockton M.Pearse&Co.	1882	4 Blds. DB243B MT298	2Cy.38"& 71"–45" 300 IP. Blair&Co.,Ld.,Stockton	"Brankelow" S.S.Co., Ld.(W.Wheatley & Co.)
488	✱	Branksome Hall *Barnes* 88	Stl ScSr 2 *Dks*3 *B*	British Liverpool	2576 3950	380·0 *P*43 *f.* *B* 130 *f.* *F*47 *f.*	45·3	27·1	Newcastle PalmersCo.,Ld.	1888	6 Blds DB951 APT37	3C.29"47"&76"–51"550IP Palmers Co.,Ld.,Newcastle	Sun Shipping Co., Ld.
489		Brat	I Pad 1 *Dk*	Russian Odessa	190 291	170·0	25·0	8·0	Linz Mayer	1882		2Cy.24"& 36"–42" 70 IP. Escher,Wyss&Co.,Zurich	Russian Steam Nav.& Trading Co.
490	N	Bratsberg *Nielsen*	I ScSr 1 *Dk*&*Spr Dk*	Norweg'n Porsgrund	1100 1423	232·8 *B* 59 *f.*	33·3	22·0	Rostock RostockS.B.&Eng.Co.	1884	DB189 APT FPT	2C.27¾"&51⅛"31½"100 IP. RostockS.B.&Eng.Co.,Rost.	J.C.& G.Knudsen
491	✱N	Bratten (ex Dakota) *Ueberweiss* 87-88	I ScSr 1 *Dk*2 *B*	Norweg'n Bergen	467 751	189·1 *P*19*f.* *B*10*f.* *F*30*f.*	28·2	15·9	Newcastle Schlesinger Davis&Co	1878	3 Blds. DB80 FPT APT	2Cy.24"& 45"–33 90 IP. T.Clark&Co.,Newcastle	J.L.Mowinckel & Sön
492	G	Braunschweig *Boedeker* 86-87	I ScBg	German Bremen	2150 3079	351·6	39·0	31·7	Greenock R.Steele&Co.	1873		2Cy.49"& 86"–48" 491 IP. R.Steele&Co.,Greenock	Norddeutscher Lloyd
493		Bravo *Avery*	I ScBg 1 *Dk*2 *B*	British Hull	698 1074	240·1 *Q* 24 *f.* *F*—	29·2	15·9	Glasgow Lond.&GlasgowCo.	1865	5 Blds.	2Cy.25½"&47½"–30" 95IP. Amos&Smith,Hull	T.Wilson,Sons & Co.
494		Brazil	W ScFlat	British Liverpool	111 142	87·7	21·1	10·0	Witton H.Gibson	1881		2Cy.10"–11" 16 IP. E.Bates,Northwich	R.Verdin
495		Brazil *Cottrell*	I ScSr	American Philad'lphia	123	105·0	20·0	11·0	Camden,N.J. J.H.Dialogue	1878		2Cy.18"& 30"–22" IP. J.H.Dialogue,Camden,N.J.	Mackey Scott & Co.
496	✱	Brazilian (ex Brazilian) *Chambers* 80-80	I ScBg 2 *Dks*3 *B*	British Newcastle	886 1360	240·0 *B* 30 *f.*	32·4	24·2	Newcastle C.Mitchell&Co.	1869	4 Blds. DB67 [illegible] DB [illegible] 27	2Cy.29"& 58"–33" 130 IP. Thompson&Co.,Newcastle	Clapham S.S.Co.,Ld.
497	R	Breakwater *Walker* 79-88	I ScSr 2 *Dks*	American New York	794 1066	221·0	30·0	19·8	Chester, Pa. J.Roach&Son	1880		1Cy.38"–48" J.Roach&Son,Chester,Pa.	Old Dominion S.S.Co.
498	✱	Breconshire (ex Yomida) *Waring* 86-85	I ScSr 2 *Dks*3 *B*	British London	1648 2544	299·7 *P*&*B* 178 *f.* *F* 30 *f.*	37·2	24·3	Sunderland SunderlandS.B.Co.,Ld.	1883	5 Blds. DB67&[illegible]	2Cy.38½"&68"–42" 250 IP. Blair&Co.,Ld.,Stockton	Jenkins & Co.
499	✱	Breeze *Allen* 86-86	I ScSr 1 *Dk*2 *B*	British W.Hartlep'l	628 987	220·0 *Q* 85 *f.* *B* 84 *f.* *F* 27 *f.*	29·0	16·3	Hartlepool E.Withy&Co.	1874	1 Blds. DB68 [illegible] 17	2Cy.27"& 50"–33" 99 IP. T.Richardson&Sons,Hartpl	T.Appleby & Co.
500		Breeze *Bennett*	Stl Pad	British London	144 373	201·4	25·2	11·6	London M.Wigram&Sons	1863		2Cy.56"–45" 160 IP. Ravenhill&Co.,London	London Chatham & Dover Rail.Co.
501	DGV	Brema *Ott* 87-87	I ScBg 2 *Dks*	German Bremen	1342 1843	261·1 *Q*80*f.* *B*117*f.* *F*24*f.*	33·8	25·8	Vegesack H.F.Ulrichs	1882	DB28	2Cy.33"& 60"–36" 165 IP. O.Henniges,Berlin	Gildemeister & Ries
502	V	Bremen *Claussen* 84-88	I ScSr	German Oldenburg	812 1060	197·1 *B* 57 *f.*	29·7	20·7	Lubeck H.Koch	1886	DB108 APT WB [illegible]	3C.15"26"&43"–27" 100IP. H.Paucksch,Landsberg	Oldenburgh-Portugiesische Dampf. Rhederei
503	N	Bremen *Lorentzen*	I ScSr 1 *Dk*	Norweg'n Christiania	244 348	139·9 *B* 10 *f.*	22·1	13·7	Blyth	1881	DB37 APT FPT	2Cy.20"& 36"–24" 60 IP. R.H.Pearson&Co.,Glasgow	Winge & Co.
504	V	Bremeña *Mendoza* 88	I ScSr 1 *Dk*&*Spr Dk*3 *B*	Spanish SantdeCuba	898 1255	224·7	29·8	16·1 23·4	Geestemünde J.C.Tecklenborg	1884	WB	2Cy.27"& 50"–31" 125 IP. Magdeburg-BuckauMasch	Antillana Linie (G.A.Münder&Koch)
505		Bremerhaven	I Pad	German Bremen	60 117	164·0	15·5	7·6	Vegesack J.Lange	1846		2Cy.33"& 45" IP.	Norddeutscher Lloyd
506	✱	Brenda *Merrifield* 83-84	I ScBn 1 *Dk*2 *B*	British W.Hartlepl	1053 1641	257·5 *Q*88*f.* *B*68*f.* *P*24*f.* *F*31*f.*	34·6	17·9	Hartlepool E.Withy&Co.	1881	1 Blds. DB365 APT51	2Cy.33"& 61"–33" 150 IP. T.Richardson&Sons,Hartpl	Herskind & Woods
507		Brenda *Parinoli*	I Sc2mSr	British Liverpool	258 411	154·8 *F* 18 *f.*	24·0	11·9	Liverpool LiverpoolForgeCo.	1884		2Cy.18"& 36"–30" 54 IP. J.Jones&Sons,Liverpool	Argentine Steam Lighter Co.,Ld.
508	✱	Brenner *Cotton* 88-88	StlScSr 1 *Dk*2 *B*	British Sth. Shields	1185 1827	265·0 *P*28*f.* *Q*80*f.* *B*120*f.* *F*28*f.*	37·1	17·4	Sunderland W.Doxford&Sons	1888	4 Blds. DB391 APT 90	3C.20"33"&54"–36" 160IP W.Doxford&Sons,Sunderld.	Cay, Hall & Co.
509	✱	Breslau *Tannox* 83-83	Stl ScSmSr 2 *Dks*	British Leith	829 1305	254·5 *P*32*f.* *B*50*f.* *F*135*f.*	33·2	16·8	Greenock R.Steele&Co.	1882	6 Blds. FPT20 DT390	2Cy.38"& 60"–42" 175 IP. R.Steele&Co.,Greenock	J.Currie & Co.
510	V	Breslau *Holm* 73-82	I ScSr 2 *Dks*	German Stettin	753 1195	231·6	29·4	22·1	Grabow Möller&Holberg	1882	DT390	2C.27½"&47"–29½" 105 IP. Möller&Holberg,Grabow	Neue Dampfer-Compagnie Akt.-Ges.

✱ = Lloyd's Register; D = Nederlandsche Vereeniging; G = Germanischer Lloyd; I = Registro Italiano; R = American Record;
I = Veritas Austro-Ungarico; H = Veritas Ellenico; N = Norske Veritas; V = Bureau Veritas.

Dans la 2me Colonne, B.T., avec les indications qui suivent, signifie la date du rapport de visite de la Carène par les experts du Board of Trade pour obtenir le certificat relatif au TRANSPORT DES PASSAGERS.

Dans la 3me Colonne, C = Composite (membrure en fer, bordé en bois); I = Fer; Stl = Acier; W = Bois.

Pour plus amples renseignements sur les navires classés par le Lloyd's Register, voir le Lloyd's Register Book et le Lloyd's Yacht Register.

No. in Book. / Offl. No.	Registry in which classed.	Ship's Name. Master, and Dates of Appointment to present Owner's Service and to Ship. / Ship's Late Name.	Material, Rig, &c. / No. of Decks, &c.	Flag. / Port of Registry.	Tons. Net. / Gross.	Length	Breadth	Depth	Built. Where. / By Whom.	When.	No. of Bulkheads. / Particulars of Water Ballast	Engines. Particulars. / Maker's Name.	Owners or Ma[illegible]
511	V	**Brestois** *Rolurier*	I Sc3mSr	French Brest	224 376	152·8	22·0	12·6 Q 42 f. F 23 f.	Dumbarton McKellar&Co.	1873	[illegible]	2Cy.30"&33"-21" 50 HP W.King&Co.,Glasgow	Chevilotte frères
512 84293		**Bretagne**	W ScBk	British London	432 836	214·6	32·8	16·7 Q 20 f.	Baltimore U.S.A. Malster & Reany	1881		2Cy.28"&50"-33" 100 HP Malster&Reany,Baltimore	Sir W.M.Call[illegible]
513		**Bretagne** *Escarras* 82-84	I Sc 2 Dks	French Marseilles	1397 2095	288·7	39·0	25·6	La Seyne Forges&Cdeds de la Medit.	1877		2Cy.37½"&69½"-35½" 380 HP Forges&Cdeds de la Medit.,Mars	Société Générale Transports Mar[illegible]
514 [illegible]		**Bretagne** *Merdrignac*	I ScSr 2 Dks	French Saint Malo	67 117	87·9	21·6	6·1	Havre Soc. Anon. de Const. Navales, Havre	1881		1Cy.21"&40"-30" 30 HP Soc. Anon. de Const. Navales, H[illegible]	A.R.Dubdah
515 94353	*	**Breton** *Briand* 86	I ScSr 1 Dk2 B	British London	862 1330	243·0	32·2	18·6 Q 94 f. F 33 f.	Newcastle Schlesinger,Davis&Co.	1878	4 Blkds. DB[illegible]	2Cy.29"&55"-36" 130 HP Hawks,Crawshay&Ss,Nwl.	R.Gordon & Co.
516	*	**Breton** *Le Marchand* 81-84	I ScSr 1 Dk	French Havre	96 164	110·3	19·0	8·5 Q 35 f. F 17 f.	Leith S.&H.Morton&Co.	1884	3 Blkds.	2Cy.13"&25"-18" 30 HP S.&H.Morton&Co.,Leith	R.Nicole
517	G	**Breton** *Thomas* 83-83	I ScSr 1 Dk	French Nantes	76 137	110·1	17·1	9·5 Gen. 87	London Westwood, Baillie & Co.	1873		2Cy.16"&22"-16"	P. & A.Legal &[illegible]
518 83917	* D	**Bretwalda** *Howsager*	I ScSr 1 Dk & Awn Dk3 B	British Newcastle	1902 2907	320·0	39·2	20·5 27·6	Newcastle Palmers'Co.	1881	7 Blkds. [illegible]	2Cy.38"&71"-48" 300 HP Palmers'Co.,Ld.,Newcastle	Hall Bros.
519 33292	*	**Bride** *Graham* 67-67	I ScSr 1 Dk2 B	British Hull	796 1227	252·6	30·1	15·9 P 182 f. F 30 f.	Hull Earle&Co.	1866	4 Blkds.	2Cy.26"&50"-30" 105 HP Comp. 84, Earle's Co.,Ld.,Hull	W. Brown, Atki[illegible] Co.
520 39737		**Bride** *Urmo* 78-82	I Sc3mSr	British Hayle	199 292	163·7	23·1	12·1	Hayle Harvey&Co.	1863		2Cy.35"-24" 100 HP Harvey&Co.,Hayle	Harvey & Co.
521 91865		**Bridesmaid**	I ScK	British London	84 139	103·6	20·2	9·6	London Hewett&Co.	1885		2Cy.12"&24"-18 24 HP J.Stewart&Son,London	Hewett & Co.,L[illegible]
522 2166		**Bridgeport**	W Pad 1 Dk2 B	American Bridgpt.Con.	931 1063	232·0	30·5	11·0	New York	1857		291 HP	Bridgeport Stea[illegible] Co.
523 10571		**Bridgewater**	I Pad	British Liverpool	56 126	109·2	20·1	8·1	Liverpool T.Vernon&Son	1857		2Cy.34½"-30" 70 HP Sanderson&Warrington	Manchester Shi[illegible] Co.
524 86704	BT10.88	**Brier**	I ScSr	British Glasgow	198 676	209·8	30·2	15·3 P 71 f. F 28 f.	Glasgow D.&W.Henderson	1882		2Cy.31"&60"-42" 177 HP D.&W.Henderson&Co,Glasg	Glas., Dublin & [illegible]donderry Stm.P[illegible]
525 68364	*	**Brighton** *Gibb*	I ScSr 1 Dk2 B	British London	561 900	214·4	30·3	15·3 Q 29 f. F 24 f.	Sunderland J.Blumer&Co.	1872	4 Blkds. DBaft 350	2Cy.27"&50"-30" 98 HP Fowler&McCollin,Hull	Commercial S.S. Ld.
526 82030	BT6.88	**Brighton** *Hale* 81-88	Stl PadSr 1 Dk	British Newhaven	316 531	221·3	27·7	10·6 Q 95 f.	Glasgow J.Elder&Co.	1878	4 Blkds.	2Cy.48"&83"-60" 300 HP J.Elder&Co.,Glasgow	London Brighton South Coast Ra[illegible]
527 83792		**Brighton**	I & Stl PadSr 1 Dk	British Sydney	127 417	220·2	23·0	10·7	Glasgow T.B.Seath&Co.	1883		2Cy.31"&60"-60" 160 HP Campbell&Son,Glasgow	Port Jackson S.S.
528 78056	BT3.88	**Brighton** *Lacer*	W Pad	British Shoreham	34 100	105·4	17·5	9·0	North Shields Johnston	1878		1Cy.29½"-48" 28 HP T.R.Scott&Co.,N.Shields	J.Lee
529 47981	*	**Brilliant** *Berry*	I Sc3mSrYt 1 Dk2 B	British Southamptn	250 331	192·3	21·1	14·7	Southampton C.A.Day&Co.	1863	7 Blkds.	[illegible] 150 HP [illegible]	Sir B.Samuelson M.P.
530 76388		**Brilliant Star**	I PadSp	British Liverpool	25 205	158·4	23·2	12·4	South Shields J.Readhead&Co.	1876		2Cy.40"-66" 180 HP J.Readhead&Co.,S.Shields	Strong & Sons
531 81478	BT6.88	**Brindisi** *Hassell* 81-88	I ScBk	British Sunderland	2074 3474	360·2	40·0	23·0 F 42 f. 31·0	Sunderland W.Doxford&Sons	1880		2Cy.48"&84"-54" 550 HP W.Doxford&Sons,Sunderld.	Peninsular & O[illegible] Steam Nav. C[illegible]
532	* D	**Brinio** *Hoogewerff* 87	I ScSr 1 Dk2 B	Dutch Rotterdam	438 661	184·1	27·0	14·1 Q 62 f. B 44 f. F 25 f.	Port Glasgow Murdoch&Murray	1882	4 Blkds. [illegible]	2Cy.19"&34"-36" 80 HP M.Paul&Co.,Dumbarton	D.Burger & Zo[illegible]
533 32616		**Brisbane**	I PadSr 1 Dk	British Sydney	65 103	136·6	18·6	6·3	Glasgow Randolph,Elder&Co.	1856		2Cy.18"-30" 40 HP J.G.Thompson,Glasgow	Illawarra S.Na[illegible]
534		**Brisbane** *Mulamy*	I Tug	British Brisbane	100	86·0	16·0	9·6	Clyde Inglis	1863			Gov'ment of [illegible]land
535 86081	*	**Briscoe** *Webster* 83-87	I ScSr 2 Dks3 B	British Newcastle	1442 2220	292·5	38·3	23·7 P 29 f. B 81 f. F 29 f.	Newcastle Palmers'Co.	1882	6 Blkds. [illegible]	2Cy.36"&68"-45" 250 HP Palmers'Co.,Ld.,Newcastle	J.E.Bowser, O[illegible] & Co.
536 2168		**Bristol** *Simmons*	W Pad 2 Dks3 B	American Fall R.,Mass	2064 2962	362·0	48·0	16·5 Breadth over all 83 f.	New York W.H.Webb	1865		1Cy.110"-144" 2800 HP Morgan Iron Works,N.York	Old Colony Stea[illegible] Co.
537 93186	BT9.88	**Britannia** *Hector* 85-88	Stl Sc4mBq 2 Dks & Spr Dk4 B	British Greenock	3145 6257	465·8	52·0	18·7 P 44 f. F 94 f. 34·1	Greenock Caird&Co.	1887	8 Blkds. [illegible]	3C.41"63"&100"-72" 1000 [illegible] Caird&Co.,Greenock [B]	Peninsular & O[illegible] Steam Nav.Co.
538 69291	BT12.88	**Britannia** *Hussey*	I ScBk 3 Dks & Awn Dk	British Liverpool	2474 4105	411·4	43·8	33·8 H 53 f.	Birkenhead Laird Bros.	1873	7 Blkds.	2Cy.60"&103"-54" 650 HP Laird Bros.,Birkenhead	Pacific Steam N[illegible]
539 91193	BT [illegible]	**Britannia**	I Sc1mBk	British Liverpool	2041 3129	353·0	39·3	28·1	Liverpool T.Royden&Sons	1885		2Cy.29"&55"-42" 280 HP G.Forrester&Co.,Liverpool	Papayanni & Co.
540 2258	BT7.88	**Britannia** *Blight* 80-82	I Sc3mSr 2 Dks3 B	British Glasgow	1991 3069	350·0	38·5	28·7 H 156 f.	Glasgow D.&W.Henderson	1879	5 Blkds.	2Cy.38"&68"-60" 160 HP D.&W.Henderson&Co.,Glw	Henderson Bro[illegible]

*** = Lloyd's Register;** D = Nederlandsche Vereeniging; G = Germanischer Lloyd; I = Registro Italiano; R = American [illegible]
A = Veritas Austro-Ungarico; H = Veritas Ellenico; N = Norske Veritas; V = Bureau Ve[illegible]

In the 2nd Column, B.T., with figures following, signifies Date of report of bottom sighting by Board of Trade Surveyors as condition of PASSENGER [illegible]

In the 3rd Column, C = Composite (iron frame, wood planked); I = Iron; Stl = Steel; W = Wood.

Further particulars of Vessels classed by Lloyd's Register will be found in Lloyd's Register Book and Lloyd's Yacht Register.

No.		Nom, Matériaux, Gréement, etc. Capitaine et dates d'entrée au service des Armateurs actuels et à bord du navire. Ancien Nom, nombre de ponts, etc.		Pavillon. Port d'Armement	Tonnage. Net Total	Dimensions. Longueur	Largeur	Creux Constructions sur le Pont	Construit. Où. Par qui.	Quand.	Nombre de Cloisons. Détails sur le bord d'eau.	Machines. Détails. Nom du Constructeur.	Armateurs ou Directeurs-Gérants.
541	✻	Britannia *Weymouth 85-85*	I ScSr *1 Dk2 B*	British	1100 [illegible]	278·5	36·1	18·5 *Q38 f. B58 f. F28 f. F30 f.*	Sunderland Short Bros.	1882	4 Blds.	2Cy.33"& 62"–42" 150 HP. J.Dickinson,Sunderland	United Kingdom S.S. Co.,Ld.
542	✻	Britannia *Kennedy 85-*	Stl TwnScSr *1 Dk & Spr Dk3 B*	British	859	247·2	31·3	17·4 24·8	Birkenhead Laird Bros.	1885	5 Blds.	4Cy."26"& 45"–30"180 HP. Laird Bros.,Birkenhead	Telegraph Constructn. and Maint.Co.,Ld.
543	✻	Britannia *Crisp 88-88*	I ScSr *1 Dk2 B*	British	632	212·0	30·1	15·9 *Q77 f. B40 f. F24 f.*	Sunderland Short Bros.	1877	4 Blds.	2Cy.27"& 50"–33" 98 HP. J.Dickinson,Sunderland	Taylor & Sanderson
544	✻	Britannia *Packham*	I ScSr *2 Dks*	British Dundee	490	251·8	29·5	16·7 *P40 f. B75 f. F42 f.*	Dundee Gourlay Bros.	1877	6 Blds.	2Cy.45"& 73"–36" 250 HP. Gourlay Bros.&Co.,Dundee	Dundee, Perth & London Shipping Co.
545	✻	Britannia *Robinson 11-85*	Stl ScSr *2 Dks*	British Leith	359	210·1	27·1	14·5	Leith S.&H.Morton&Co.	1885	6 Blds.	2Cy.30"& 52"–36" 120 HP. S.&H.Morton&Co.,Leith	J.Currie & Co.
546		Britannia	W Pad *1 Dk*	British Auckland	108 197	119·0	21·5	5·5	Auckland C.Bailey	1885		2Cy.16"–42" 45 HP.	Devonport Steam Ferry Co.
547	✻	Britannia *Pattison*	I ScTrawler *1 Dk*	British Hull	51 122	89·8	20·7	10·9	Hull Cook,Welton&Gemmell	1886	5 Blds.	2Cy.15"& 29"–21" 35 HP. C.D.Holmes&Co.,Hull	Humber Steam Trawling Co.
548		Britannia *Lyon -88*	Stl TwnScSr *1 Dk*	British	36 123	120·1	19·0	7·0	Grangemouth Grngmth Dk Yd Co.	1888		4Cy.12"&21"–18" 50 HP. Hawthorns&Co.,Leith	J.Brown
549	✻	Britannia *Martin*	I ScTug *1 Dk*	British London	6 76	80·0	17·1	9·6	Hayle Harvey&Co.,Ld.	1887	4 Blds.	2Cy.17"&32"–21" 90 HP. Harvey&Co.,Hayle	W.A.Walker
550	✻	Britannia *Parazol 83-84*	I Sc *2 Dks3 B*	French Marseilles	1846 2190	328·0	40·4	23·4	Liverpool T.Royden&Sons	1881	5 Blds.	2Cy.40"& 73"–48" 375 HP. G.Forrester&Co.,Liverpool	Compagnie Française de Nav. à Vap.
551		Britannic *Parsell*	I Sc4mBk *3 Dks*	British Liverpool	3152 5004	455·0	45·2	26·2 33·7	Belfast Harland&Wolff	1874	8 Blds.	4Cy.48"& 83"–60" 760 HP. Maudslay Sons&Field,Lndn.	Oceanic Steam Nav. Co.,Ld.
552	✻	Britannic *Main 88-88*	Stl ScSr *1 Dk & Wb Frms*	British Sunderland	1507 2302	290·0	39·0	19·4	Sunderland Bartram,Haswell&Co.	1888	6 Blds.	3Cy.22",36½"&60"–42" 200 HP. J.Dickinson,Sunderland	Britannic S.S.Co.,Ld. (Greenwell&Parsons)
553	✻	British King *Kelly 81-81*	Stl Sc4mBk *3 Dks*	British Liverpool	2289 3559	410·3	39·0	28·9 *P50 f. F56 f.*	Belfast Harland&Wolff	1881	6 Blds.	4Cy.28"& 60"–54" 400 HP. J.Jack&Co.,Liverpool	British Ship Owners' Co.,Ld.
554		British King *(ex Flying Hurricane)*	I ScTg *1 Dk*	British Liverpool	28 154	112·5	20·2	12·0	Paisley Fullerton&Co.	1877	4 Blds.	2Cy.28"&48"–30" 100 HP. J.Howden&Co.,Glasgow	H.J.Ward
555	✻	British Prince *Nowell 81-81*	Stl Sc4mBkn *3 Dks*	British Liverpool	2548 3974	420·1	42·2	29·0 *P55 f. F88 f.*	Belfast Harland&Wolff	1882	7 Blds.	2Cy.43"& 86"–60" 450 HP. Harland&Wolff,Belfast	British Ship Owners' Co.,Ld.
556	✻	British Prince *Innes 88-88*	StlScSr *1 Dk & Wb Frms*	British Newcastle	1439 2217	292·0	39·1	18·9 *P32 f.*	Sunderland Short Bros.	1888	6 Blds.	3C.22"36"&59"–42" 200 HP. G.Clark,Sunderland	Prince Steam Shipping Co.,Ld.(J.Knott,Manager)
557	✻	British Prince *Anderson*	W ScK *1 Dk*	British Nth. Shields	8 46	70·1	17·3	7·1	Leith Marr Bros.	1885		2Cy.11"& 22"–16" 20 HP. Baird&Barnsley,N.Shields	W.H.Storey
558	✻	British Princess *Forth 82-82*	Stl Sc4mBk *3 Dks*	British Liverpool	2558 3995	420·1	42·2	29·0 *B 150 f.*	Belfast Harland&Wolff	1882	9 Blds.	2Cy.43"& 86"–60" 450 HP. Harland&Wolff,Belfast	British Ship Owners' Co.,Ld.
559	✻	British Queen *Wills 83-83*	Stl Sc4mBk *3 Dks*	British Liverpool	2277 3558	410·3	39·0	28·9 *P50 f. F56 f.*	Belfast Harland&Wolff	1880	6 Blds.	4Cy.28"& 60"–54" 400 HP. J.Jack&Co.,Liverpool	British Ship Owners' Co.,Ld.
560		British Queen *Harlow 88-88*	I Sc3mSr	British Glasgow	483 772	195·0	29·0	18·0	Dumbarton W.Denny&Bros.	1849		2Cy.29"& 54"–36" 125 HP. J.Jack&Co.,Liverpool	Cunard Steam Ship Co.,Ld.
561	✻	British Queen *Cooper 15-84*	I ScSr *1 Dk2 B*	British W.Hartlepl	440 687	230·2	28·5	15·9	Hartlepool T.H.Pile&Co.	1865	5 Blds.	2Cy.29"& 55"–33" 120 HP. T.Richardson&Sons,Hrtlpl.	W. Hartlepool Steam Nav.Co.
562		Briton *Thomas 73-86*	I PadSr	British Bristol	277 179	207·1	24·1	11·7 *P70 f.*	Glasgow Tod& McGregor	1862		2Cy.48"–60" 150 HP. Tod&McGregor,Glasgow	Bristol Steam Nav.Co.
563		Briton	I Sc	British Falmouth	50 139	79·0	26·8	8·5	Cardiff	1861		2 60 HP.	R.H.Michell
564	✻	Briton *Sledden 86-86*	I ScTug *1 Dk*	British Falmouth	19 66	84·8	15·7	9·1	Falmouth Cox&Co.	1880	4 Blds.	2Cy.16"& 30"–21" 46 HP. Cox&Co.,Falmouth	Falmouth Fishery Co. Ld.
565	✻	Briton *Tester 84-*	I ScTug *1 Dk*	British Cardiff	2 64	76·1	16·1	9·2	Hayle Harvey&Co.,Ld.	1884	4 Blds.	2Cy.15"& 28"–20" 55 HP. Harvey&Co.Ld.,Hayle	E.Handcock & Co.
566	✻	Brittany *Evans 89-89*	I ScSr *1 Dk2 B*	British Cardiff	554 876	205·0	28·6	17·2 *Q92 f. F27 f.*	Newcastle Schlesinger,Davis &Co.	1871	4 Blds.	2Cy.26"& 52"–30" 95 HP. Cmp. 1881, R. & W. Hawthorn. Newcastle	"Brittany" S.S.Co.,Ld. (W.J.Tromner)
567		Brittany *Mortimer*	I PadSr	British Southamptn	283 625	236·0	25·3	13·0 *Q41 f. F31 f.*	London J.Ash&Co.	1864		2Cy.40"& 68"–60" 230 HP. Day,Summers&Co.,S'thptn.	London & S.W.Ry.Co
568		Brittany *Sharp 86-86*	Stl PadSr *2 Dks*	British Newhaven	239 579	231·0	27·7	10·6 *Q142 f.*	Glasgow J.Elder&Co.	1882	4 Blds.	2Cy.46"& 83"–60" 350 HP. J.Elder&Co.,Glasgow	London, Brighton & SouthCoast RailwayCo.
569	✻	Brixham *Debrix 85-85*	I ScSr *1 Dk2 B*	British Brixham	396 626	183·0	27·1	11·2 *P26 f. B41 f. F22 f.*	Sunderland Boolds,Sharer&Co.	1885	4 Blds.	2Cy.24"& 48"–33" 96 HP. J.Dickinson,Sunderland	Brixham S.S.Co.,Ld. (F.W.Baddeley & Sons)

✻=Lloyd's Register; D=Nederlandsche Vereeniging; G=Germanischer Lloyd; I=Registro Italiano; R=American Record. [illegible]=Veritas Austro-Ungarico; H=Veritas Ellenico; N=Norske Veritas; V=Bureau Veritas.

Dans la 2me Colonne, B.T., avec les indications qui suivent, signifie la date du rapport de visite de la Carène par les experts du Board of Trade pour obtenir le certificat relatif au TRANSPORT DES PASSAGERS.

Dans la 3me Colonne, C=Composite (membrure en fer, bordé en bois); I=Fer; Stl=Acier; W=Bois.

Pour plus amples renseignements sur les navires classés par le Lloyd's Register, voir le Lloyd's Register Book et le Lloyd's Yacht Register.

No. in Book / Official No.	Class	Ship's Name; Master, and Dates of Appointment; Ship's Late Name	Material, Rig, &c.; No. of Decks, &c.	Flag; Port of Registry	Tons Net; Gross	Dimensions (Length, Breadth, Depth)	Built Where; By Whom	When	No. of Bulkheads; Particulars of Water Ballast	Engines: Particulars; Maker's Name	Owners or Managers
571 93[0]31	BT3.88	Brock	I TwnScSp	British Fleetwood	9 122	95·2 | 19·1 | 9·0	Glasgow T.B.Seath&Co.	1883		2Cy.13"&31"–24" 50 HP. Rankin&Blackmore,Greenock	Lane, & Yorks. [R]ail. & N.W.Ry.C[o.]
572 76914	BT3.88	Brodick Castle *Edward 87-87*	I PadlSr	British Poole	81 283	207·6 | 21·7 | 7·5 *Q 19 f.*	Paisley H.M'Intyre&Co.	1878		2Cy.38½"–66" 96 HP. W. King&Co.,Glasgow	Bournemouth, &c. [illegible]
573	D V	Bromo *Boon 88*	StlScBn *3 Dks*	Dutch Rotterdam	1810 2307	320·0 | 36·9 | .. *P 96 f. B 76 f. F 32 f.*	Flushing Koninkl. Maats. de Schelde	1888	[5] Blhds. DB[illegible]	1C23"35"45"63"–42"300 HP. Koninklijke Maatschappij de Schelde,Flushing	Rotterdamsche [Lloyd]
574		Bromo (WEIR KEM) *Poljet*	I ScSr *2 Dks & Shd Dk*	Dutch Batavia	779 1213	256·8 | 31·7 | 17·2 *P 80 f. F 78 f.*	Greenock Caird&Co.	1875	4 Blhds.	2Cy.34"&60"–39" 180 HP. Caird&Co.,Greenock	Nederlandsch-Indische Stoomvaart-Maats.
575 62477	✻	Bromsgrove *Philip*	I ScSr *1 Dk 2 H*	British Southamptn	428 658	191·1 | 28·6 | 16·5 *Q 27 f.*	Sunderland Oswald&Co.	1868	4 Blhds. DB a&f 130t	2Cy.25"&50"–20" 90 HP. Comp. 1845. Day, Summers & Co., Southampton	J.G. Hill
576 82093	✻	Brookfield *Moffat 83-88*	I ScSr *2 Dks 3 H*	British Newcastle	1570 2405	317·0 | 38·0 | 24·7 *P 69 f. F 24 f.*	Newcastle C.S.Swan&Hunter	1882	5 Blhds. CDB[illegible]	2Cy.38"&71"–48" 275 HP. Blair&Co.,Ld.,Stockton	J. Temperley & Co.
577 845[illegible]	✻	Brooklands *Moore 86-86*	I ScBn *1 Dk 2 H*	British W.Hartlepl	689 1096	234·1 | 31·6 | 14·9 *Q 84 f. B 58 f. F 28 f.*	W. Hartlepool W.Gray&Co.	1882	4 Blhds. DB[illegible]	2Cy.27"&50½"–33" 99 HP. Blair&Co.,Ld.,Stockton	Hardy, Wilson & [Co.]
578 78407	✻	Brooklyn City *Fitt 85-85*	I ScBn *2 Dks 3 H*	British Bristol	1122 1720	260·0 | 34·5 | 22·5 *P 26 f. F 33 f.*	Stockton Richardson,Duck&Co.	1881	5 Blhds. DB[illegible] MT[illegible]	2Cy.33"&61"–33" 180 HP. T.Richardson&Sons,Hrtlpl.	C. Hill & Sons
579 8390[illegible]	✻ D	Broomhaugh *Weightman 15-87*	I ScSr *2 Dks 3 H*	British Newcastle	1362 2095	286·1 | 36·1 | 24·5 *P 28 f. B 67 f. F 29 f.*	Newcastle C.S.Swan&Hunter	1881	[5] Blhds. DB a f &[illegible]	2Cy.34½"&64"–42" 249 HP. R.&W.Hawthorn,Newcastle	Elliott, Lowrey & Dunford
580 79987	✻	Broomhill *Meikin 81-87*	I ScSr *1 Dk*	British N.Shields	378 609	174·7 | 29·0 | 12·8 *Q 86 f. B 7 f. F 21 f.*	Glasgow T.Wingate&Co.	1878	4 Blhds. DB[illegible]	2Cy.25"&50"–33" 95 HP. Wingate&Co.,Glasgow	H. Andrews
581 70214		Brothock (FOUNDERED) *Haubner 85-85*	W ScSr	British Newcastle	70 116	93·5 | 20·3 | 9·0	Newcastle Lindsay	1874		2Cy.14"–18" 36 HP. Ouseburn Eng.W.Co.,N'wcle	T. Robson
582 93158	✻	Broughshane *Brown 86-86*	I ScSr *1 Dk*	British Belfast	132 325	152·2 | 23·1 | 11·8 *Q 21 f. B 10 f. F 27 f.*	Belfast Workman,Clark & Co.	1886	5 Blhds. DT[illegible] APT[illegible]	2Cy.19"&38"–30" 70 HP. J.&J.Thompson,Glasgow	Antrim Iron [Ore Co.] Ld.
583 8[illegible]910	✻	Bruce *Combe 85-85*	I ScDy *1 Dk*	British Granton	43 103	92·8 | 18·0 | 9·4 *F 12 f.*	Leith Hawthorns&Co.	1883	4 Blhds. FPT[illegible]	2Cy.15"&27"–21" 32 HP. Hawthorns&Co.,Leith	General Steam [F.] Co.,Ld.
584 90340	✻	Bruges *Eggers 87-87*	I ScSr *1 Dk*	British Sunderland	397 617	170·3 | 29·1 | 11·7 *Q 84 f. B 16 f. F 17 f.*	Sunderland R.Thompson&Sons	1884	4 Blhds. DB[illegible] FPT[illegible] APT[illegible]	2Cy.22"&42"–30" 80 HP. J.W.&P.Wilson,Sunderland	R. Thompson (V.T. Thompson[illegible])
585 80857	✻	Brunel *Henderson 83-88*	StlScSr *2 Dks 3 B*	British N.Shields	1308 2003	286·7 | 37·0 | 22·1 *P 25 f. B 56 f. F 27 f.*	Newcastle TyneI.S.B.Co.,Ld.	1888	5 Blhds. CDB[illegible] APT[illegible]	3C22"36½"&60"–39"200 HP. N.E.MarineEng.Co.,New.	J. Wait
586	✻ D	Brunette *Trapp 83-87*	I ScSr *1 Dk 2 H*	German Danzig	518 792	201·7 | 28·6 | 15·7 *P & B 113 f. F 22 f.*	Newcastle C.S.Swan&Hunter	1881	4 Blhds. FPT[illegible] APT[illegible] MT[illegible]	2Cy.26"&49"–36" 99 HP. Black,Hawthorn&Co.,Gtshd	Th. Rodenacker
587 8440[illegible]	✻	Brunner *Highman 88-88*	StlScSr *1 Dk*	British [illegible]	333 540	172·7 | 26·1 | 13·1 *P 28 f. F 24 f. B 44 f.*	Paisley J.M'Arthur&Co.	1888	4 Blhds. DB[illegible]	3Cy.17"27"&43"–30" 95 HP. Bow & M'Lachlan,Paisley	Union Steam [S. Co.] of New Zealand
588 80360		Bruno	W ScSr *1 Dk*	British Toronto	300 475	136·0 | 25·5 | 12·1 *H 153 f.*	Montreal B.Leighson	1863		2Cy.26½"–30" 60 HP. Barclay,Montreal	James Mathe[ws]
589 84092	✻	Brunswick *Mawer 88-88*	I ScSr *1 Dk 2 B*	British W.Hartlepl	904 1389	246·0 | 33·0 | 18·0 *Q 84 f. B 58 f. F 32 f.*	Sunderland W.Pickersgill&Sns	1881	4 Blhds. DB a&f [illegible]	2C.29½"&55"–36" 130 HP. G.Clark,Sunderland	Guthe, Murdoch
590 63316	✻ V	Brunswick *Roach 65-87*	I ScBg *2 Dks*	British Liverpool	757 1171	258·2 | 29·7 | 20·0 *Len. 74.*	Birkenhead LairdBros.	1870	4 Blhds.	2Cy.30"&56"–36" 180 HP. LairdBros.,Birkenhead	Liverpool & M[ississippi] Steam Ship Co.,Ld.
591	✻	Brutus *Blumberg*	I ScSr *1 Dk 2 B*	German Kiel	460 630	190·1 | 27·0 | 15·8 *F 20 f.*	Glasgow A.Stephen&Sons	1877	3 Blhds. MT[illegible]	2Cy.26"&45"–30" 90 HP. A.Stephen&Sons,Glasgow	Andersen Geb[r.]
592		Bryderen *Christensen 84-85*	I&Stl ScSr *1 Dk*	Danish Copenhagen	112 417	138·1 | 32·2 | 14·1 *Q 43 f. F 18 f. H 21 f.*	Malmö Kockums Mek.Werks.	1884	APT[illegible] FPT[illegible]	2Cy.35"&64½"–35" 185 HP. Kockums Mek. Werks,Malmö	Det Forenede [Dampsk]ibs Selskab (C[openhagen])
593 80356	✻	Bryn Glás *Bethell 85-85*	I ScSr *1 Dk & Spr Dk 3 B*	British Newport	1504 2308	296·0 | 37·9 | 20·1 *F 17 f. 27 f.*	Sunderland J.Laing	1883	4 Blhds. DB a & f 250t	2Cy.37"&68"–42" 280 HP. N.E.Marine Eng.Co.,Ld.,Shl.	"Bryn Glás" S.S. [Co.] (Jones Bros.&Co.)
594 91871	✻	Buccaneer	Stl ScSr *1 Dk & Awn Dk*	British London	460 785	190·0 | 28·2 | 12·8 *20 0*	Newcastle W.Richardson&Co.	1885	4 Blhds. DB[illegible] APT[illegible]	2Cy.30"&61"–39" 180 HP. W.Richardson&Co.,N'wcstle	India Rubber, Gutta Percha & Telegraph W[orks]
595 84005	✻	Buccleuch *Jorgensen 86-86*	I ScDy *1 Dk*	British Granton	33 103	92·3 | 18·0 | 9·4 *F 12 f.*	Leith Hawthorns&Co.	1883	4 Blhds. FPT[illegible]	2Cy.15"&27"–21" 32 HP. Hawthorns&Co.,Leith	General Steam [F.] Co.,Ld.
596 45322		Buccleugh *Joy 72*	W ScSr	Spanish [illegible]	109 184	134·0 | 21·4 | 9·8 *Q 44 f. B 19 f. F 19 f.*	Newcastle Adamson	1863		2Cy.24"–24" 40 HP. R.&W.Hawthorn,N'wcastle	J.J. Wallis & Co.
597 70232	✻ D	Bucentaur *Steele 73-87*	I ScSmSr *2 Dks 3 H*	British Newcastle	1176 1836	282·3 | 33·3 | 24·2 *B 40 f. F 32 f.*	Newcastle PalmersCo.	1875	5 Blhds. DB[illegible]	2Cy.30½"&61"–36" 150 HP. PalmersCo.,Newcastle	Hall Bros.
598 83919	✻	Bucephalus *Service*	I ScSr *2 Dks*	British Melbourne	1192 1818	276·0 | 36·2 | 21·9 *F 28 f. F 30 f.*	Newcastle PalmersCo.,Ld.	1884	4 Blhds. DB[illegible]	2Cy.39"&61"–42" 174 HP. PalmersCo.,Ld.,Newcastle	Archibald Cu[rrie]
599 25111		Buda *Cameron 87-87*	I ScSr	British Leith	523 769	225·1 | 29·6 | 15·4 *P 41 f.*	Glasgow Barclay,Curle&Co.	1865		2Cy.27"&47"–33" 98 HP. R.Napier&Sons,Glasgow	Leith, Hull & [Hamburg] Steam Packet Co.
600 87739	✻	Budapest *Thomson 87-88*	I ScSr *1 Dk & Spr Dk 3 B*	British Glasgow	1091 1678	240·0 | 34·1 | 17·5	Dumbarton Burrell&Son	1883	[5] Blhds. CDB[illegible]	2Cy.30"&55"–36" 150 HP. D.Stewart&Co.,Glasgow	Burrell & Son

✻ = Lloyd's Register; D = Nederlandsche Vereeniging; G = Germanischer Lloyd; I = Registro Italiano; R = American [Record]; A = Veritas Austro-Ungarico; H = Veritas Ellenico; N = Norske Veritas; V = Bureau V[eritas].

In the 2nd Column, B.T., with figures following, signifies Date of report of bottom sighting by Board of Trade Surveyors as condition of PASSENGER C[ERTIFICATE].

In the 3rd Column, C=Composite (iron frame, wood planked); I=Iron; Stl=Steel; W=Wood.

Further particulars of Vessels classed by Lloyd's Register will be found in Lloyd's Register Book and Lloyd's Yacht Register.

Registre	Nom, Matériaux, Gréement, etc. Capitaine et dates d'entrée au service des Armateurs actuels et à bord du navire	Matériaux, Gréement, nombre de ponts, etc.	Pavillon / Port d'Armement	Tonnage Net / Total	Longueur	Largeur	Creux	Construit: Où / Par qui	Quand	Nombre des cloisons / Détails sur le lest d'eau	Machines: Détails / Nom du Constructeur	Armateurs ou Directeurs-Gérants
V	Budapest *Sukovich -85*	I ScSr *1 Dk*	Austrian Fiume	60 119	118·4	15·7	9·7	Porto Rase Selnavon	1883		2Cy.17"& 28"–18" 35 HP. M.Bartlett	L.Schwartz&G.Prister
	Budda *Baermann 83-85*	2m	Russian St Petersburg	430				Norrköping Motala Co.	1879		100 HP.	Naphta Productions Gesellschaft (Gebr.Nobel)
✱	Buena Ventura *Berthagoi*	I S·Bg *1 Dk2 B*	Spanish Bilbao	1155 1699	278·6 *P150f.*	31·1 *B61f.*	23·7 *F39f.*	Sunderland Oswald & Co.	1871	4 Blds. FPT	2Cy.36"& 61"–48" 200 HP. J.Jack.Rollo&Co.,Liverpool	Larrinaga & Co.
	Buenaventura	I	Spanish Barcelona	168 280	160·1	21·9	11·9	England	1871		70 HP.	Magin Soler Caprera &Co.
	Buenaventura	I	Spanish Barcelona	115	115·1	15·5	9·1	England	1860		60 HP.	Ministeriode Fomento
V	Buenos Aires *Lutz 83-86*	I ScBg *2 Dks3 B*	German Hamburg	1564 2438	331·1 *P72f.*	36·1 *F42f.*	25·3	Sunderland JamesLaing	1872	5 Blds.	2Cy.42 & 79"–42" 300 HP. Humphrey,Tennant&Co.,Lond.	Hamburg-Südamerik. Dampfschiff.-Gesells.
✱	Buenos Aires *Tenera 86-87*	Stl Sc3mSr *3 Dks*	Spanish Barcelona	3367 5195	440·6 *P 47 F 81*	48·2 *B364f.*	22·4 30·5	Dumbarton W.Denny&Bros.	1887	6 Blds.	4Cy32"46½"64½"& 92"–60" 627 HP Denny&Co.,Dumbarton	Compania Trasatlantica
	Buenos Ayrean *Scott -87*	Stl ScBk	British Glasgow	2560 4005	385·2	42·2	23·3 30·8	Dumbarton W.Denny&Bros.	1879	5 Blds.	2Cy.51"& 88"–54" 500 HP. Denny&Co.,Dumbarton	J.&A.Allan
	Buffalo *Nabel*	Stl Sc4mS *2 Dks3 B*	British Hull	2895 4427	385·0 *P&B241f.*	45·3 *F50f.*	27·8	Newcastle PalmersCo.	1885	7 Blds. CDB321/26 at	3Cy33"54"&86"–60" 150 HP PalmersCo.Ld.,Newcastle	T.Wilson, Sons & Co.
✱	Buffalo *Wright*	I PadTug *1 Dk*	British EastLondon	34 107	101·6	17·8	8·9	NorthShields Hepple&Co.	1883	3 Blds.	2Cy.24½"–46" 60 HP. Hepple&Co.,NorthShields	Kaffrarian Stm.Landing.Shipping&ForwardingCo
	Buffon *Lyons 72-83*	I ScS	British Liverpool	1504 2322	302·0 *P168f.*	37·3	24·5	Newcastle A.Leslie&Co.	1883		2Cy.31"& 66"–48" 250 HP. R.Stephenson&Co.,Nwestle.	Liverpool, Brazil & RiverPlateS.Nav.Co.,Ld.
✱	Bulgaria *Kombout*	Stl ScSr	Russian Odessa	389 572	190·0 *P48f.*	26·1 *B41f.*	12·6 *F30f.*	Newcastle Wghm.Richardson &Co.	1886	4 Blds.	3C14½"22½"&38"–30" 80 HP Wigham,Rchrdsn&Co.,Nwe.	Black Sea & Danube Nav.Co.
	Bulgarian *Leask -88*	I Sc4mBk *3 Dks*	British Liverpool	2031 3118	400·0	37·2	20·5 28·1	Belfast Harland&Wolff	1870	7 Blds.	2Cy.32"& 66"–54" 300 HP. J.Jack&Co.,Liverpool	S. S. "Bulgarian" Co., Ld. (F.Leyland & Co.)
	Bulimba *Clarke 81-83*	Stl ScSr *3 Dks*	British Glasgow	1607 2510	315·8 *P64f.*	38·2 *B76f.*	17·3 *F37f.* 25·1	Glasgow A.&J.Inglis	1882	5 Blds.	2Cy.40"& 73"–51" 300 HP. A.&J.Inglis,Glasgow	British India Associated Steamers,Ld.
✱	Bull *Gair 88-88*	I Sc *1 Dk2 B*	British Middlesbro'	370 593	174·8 *Q66f.*	25·6 *B39f.*	14·3 *F20f.*	Middlesbro' R.Dixon&Co.	1877	4 Blds. DBa429/40/DT	2Cy.24"& 45"–26" 80 HP. T.Richardson&Sons,Hrtlpl.	J.Watson & Co.
	Bull Dog *Storer*	I ScK *1 Dk*	British Bridgwater	92 144	99·6	20·1	7·1	Newcastle W.Richardson&Co.	1866	2 Blds.	2Cy.12"–15" 40 HP. Richardson&Co.,Newcastle	J.W.Sully
	Bull Dog	Stl TwnSc	British Bristol	62 105	87·9	20·0	8·7	Chepstow E.Finch&Co.,Ld.	1884		4Cy.12 & 24"–14" 60 HP. E.Finch&Co.,Ld.,Chepstow	Bristol Corporation
✱	Bulldog *Rook*	I ScTug *1 Dk*	British London	3 72	77·0	16·5 *B42f.*	9·3	London LondonDryDk.Co.	1884	4 Blds.	2Cy.15¼"& 29"–20" 40 HP Winshs&Dunsmore,Milwal	London Steam Tug Co.,Ld.
	Bungaree	I Pad *1 Dk*	British Newc.,NSW.	85 163	121·5	22·5	9·4	Glasgow Barclay,Curle&Co.	1861		2Cy.44"–54" 150 HP A.&J.Inglis,Glasgow	J.& A.Brown
✱	Buninyong *Joy*	Stl ScBg *1 Dk2 B*	British Melbourne	1289 2070	279·9 *P147f.*	38·1 *F36f.*	20·2	Barrow Caird,Purdie&Co.	1883	5 Blds. DB&462581 MT323	2Cy.36"& 70"–48" 300 HP. J.Jack&Co.,Liverpool	W.H.Smith & Sons,Ld.
	Bunya-Bunya	W Pad *2 Dk*	British Sydney	127 202	121·0	21·0	7·0	Pyrmont,NSW DavidDrake	1885		2Cy.22"& 48"–35" Chapman&Co.,Sydney	North Shore Steam Ferry Co.
	Bunyip	I ScSr *1 Dk*	British Melbourne	152 317	160·5	28·2	9·8	Renfrew W.Simons&Co.	1879		2Cy.22"& 38"–27" 65 HP W.Simons&Co.,Renfrew	Melbourne Harbour Trust Commissioners
V	Bur *Petterson 78-84*	I ScSr *1 Dk*	Swedish Stockholm	648 867	202·1 *B40f.*	27·9 *H84*	18·5	Gefle O.A.Brodin	1882	WB	2Cy.21"& 42"–27" 80 HP. MotalaCo.,Motala	Rederi-Bolag."Union" (G.O.Wallenberg.)
V	Burg *Johannsen 81-81*	I Sc3mSr *2 Dks*	German Lübeck	159 216	128·6 *B42f.*	16·5 *Lon. 81*	7·7	Hamburg Blohm&Voss	1879		2Cy.14"& 24"–18" 27 HP Blohm&Voss,Hamburg	Lübeck- Königsberger Dampfschiff.-Ges.
✱DV	Burgemeester den Tex *Bruijne 74-81 (cs Insulinde)*	I Sc3mSr *2 Dks,&SprDk*	Dutch Amsterdam	2036 3074	350·0 *F41f.*	39·1	21·2 29·2	Glasgow J.Elder&Co.	1882	6 Blds.	2Cy.42"& 80"–48" 400 HP. J.Elder&Co.,Glasgow	Stoomvaart Maatschappij "Nederland"
✱D	Burgemeester van Vollenhoven *Scholtens -85*	I Sc *1 Dk*	Dutch Rotterdam	128 215	119·8 *B28f.*	19·4 *F15f.*	8·7	Slikkerveer P.Smit,jun.	1884	4 Blds.	2Cy 15½"& 28"–16" 30 HP. P.Smit,jr.,Slikkerveer	Duinkerksche Stoomboot Reederij
✱	Burgundia *Dulac 82-82*	I Sc *2 Dks3 L*	French Marseilles	2169 2858	328·0 *P27f.*	40·3 *B70f.*	27·1 *F42f.*	Liverpool T.Royden&Sons	1882	5 Blds. CHB475	2Cy.40"& 76"–54" 375 HP. G.Forrester&Co.,Liverpool	Compagnie Française de NapàVap.
✱	Burlington *Gibson 81-88*	I ScSr *1 Dk2 B*	British W.Hartlepl.	534 839	202·7 *Q60f.*	29·1 *B61f.*	16·3 *F26f.*	Sunderland W.Doxford&Sons	1872	4 Blds. DBa&f213	2Cy.25"& 47½"–30" 90 HP. T.Richardson&Sons,Hrtlpl.	J.Lohden & Co.
	Burma *Butterwick*	Stl Pad	British Glasgow	345 476	249·9	30·1	8·8	PortGlasgow R.Duncan&Co.	1879		2Cy.31"& 59"–54" 155 HP. Rankin&Blackmore,Greenk	IrrawaddyFlotilla Co., Ld.
	Burmah	W S·Flat	British Liverpool	86 127	86·9	20·8	8·8	Northwich C.Gilson	1875		2Cy.8"–13" 24 HP S.Bates Northwich	R.Verdin

✱=Lloyd's Register; D=Nederlandsche Vereeniging; G=Germanischer Lloyd; I=Registro Italiano; R=American Record;
1=Veritas Austro-Ungarico; H=Veritas Ellenico; N=Norske Veritas; V=Bureau Veritas.

Dans la 2me Colonne, B.T., avec les indications qui suivent, signifie la date du rapport de visite de la Carène par les experts du Board of Trade pour obtenir le certificat relatif au TRANSPORT DES PASSAGERS.

Dans la 3me Colonne, C=Composite (membrure en fer, bordé en bois); I=Fer; Stl=Acier; W=Bois.

Pour plus amples renseignements sur les navires classés par le Lloyd's Register, voir le Lloyd's Register Book et le Lloyd's Yacht Register.

No. in Book. / Official No.	Registry in which classed.	Ship's Name. Material, Rig, &c. / Master, and Dates of Appointment to present Owner's Service and to Ship. / Ship's Late Name. No. of Decks, &c.	Flag. / Port of Registry.	Tons. Net / Gross	Length	Breadth	Depth	Built. Where / By Whom.	When	No. of Bulkheads. / Particulars of Water Ballast.	Engines. Particulars. / Maker's Name.	Owners or M...
631 73560		**Burmah** ex Ben Lomond 1 PadK *Morris 83-87*	British London	26 180	124·7	21·2	11·0	Aberdeen Hall, Russell & Co.	1876		2Cy.33"&54" 180 HP. Hall, Russell & Co., Aberdeen	W. Watkins
632 87685	✱ B.T. 1888	**Burnley** Stl ScSr *Hanslip 85-88* *Dk & Wb Frames*	British Glasgow	987 1556	275·0 P 34 f.	33·2 B 62 f.	19·8 F 30 f.	Glasgow J. & G. Thomson	1883	4 Blds. [illegible]	2Cy.32"&63"-45" 176 HP. J. & G. Thomson, Glasgow	"Burnley" S.S. [illegible]
63? 91467	✱	**Burrumbeet** Stl ScSr *Bell 86-87* 1 Dk2 B	British Melbourne	1561 2420	300·0 P 151 f.	40·1 F 48 f.	19·7	Newcastle C. S. Swan & Hunter	1885	8 Blds. DB [illegible] FPT [illegible] APT [illegible] MT [illegible]	3Cy.25"38"&60"-42" 290 HP. Wallsend Slipway Co., Newc.	Huddart, Park...
634 68775	✱	**Burry** I ScSr *Rees 87-87* 1 Dk	British Liverpool	152 326	167·0 Q 30 f.	23·8 B 44 f.	11·9 F 32 f.	Preston Allsup & Sons	1879	4 Blds. [illegible]	2Cy.20"&36"-27" 70 HP. W. Allsup & Sons, Preston	South Wales A... Steam Ship Co. L...
635 82706	✱	**Burslem** I ScSr *Dingle 88-88* 1 Dk	British London	225 405	149·2 Q 48 f.	23·5 B 37 f.	11·3 F 24 f.	Port Glasgow Russell & Co.	1883	4 Blds. FPT 18 APT [illegible]	2Cy.18"&36"-24" 55 HP. Alley & Maclellan, Glasgow	T. A. Walker
636 83780	✱	**Burwah** Stl ScSr *Smith* 2 Dks	British Sydney	568 982	220·4	30·1 B 38 f.	19·6	Kinghorn J. Key & Sons	1883	4 Blds. DB [illegible] FPT [illegible]	2Cy.31"&63"-42" 280 HP. J. Key & Sons, Kirkcaldy	W. H. Smith & S...
637 48881		**Busheer** I ScBg *Simpson 83-88* 2 Dks	British Glasgow	539 792	216·0	29·1	16·7	Glasgow Aitken & Mansel	1863	4 Blds.	2Cy.42"-30" 130 HP. J. Aitken & Co., Glasgow	British India S... Nav. Co. Ld.
638 85001	✱	**Bussorah** I ScSr *Legsdad 87-88* 1 Dk & Awn Dk3 B	British London	1292 1986	268·2	35·2	20·5 28·4	Sunderland Sunderland S.B. Co. Ld.	1881	5 Blds. DB [illegible] f 586	2Cy.32"&61"-39" 200 HP. N. E. Marine Eng. Co. Ld. [illegible]	Persian Gulf S... Ld.
639 51356		**Busy Bee** I Sc3mSr *Williamson 78-88* 1 Dk2 B	British Newcastle	597 930	231·8 Q 49 f.	28·2 F 27 f.	17·4	Newcastle Palmer Bros. & Co.	1865	3 Blds. WB	2Cy.28"&50"-30" 150 HP. R. & W. Hawthorn, N'castle	Tyne Steam Sh... Co. Ld.
640 76102	✱	**Bute** I ScSr *M'Neil 79-81* 1 Dk	British Greenock	39 143	95·1 Q 39 f.	19·0 F 12 f.	9·2	Bowling [illegible]	1879	3 Blds. FPT 28	2Cy.16"&28"-22" 35 HP. W. King & Co., Glasgow	J. Rodger
641 78358	✱ D	**Buteshire** I Sc3mSr *Coull 87* 2 Dks3 B	British Glasgow	872 1355	260·3	32·3 B 29 f.	22·7	Port Glasgow W. Hamilton & Co.	1877	5 Blds. DB 240 ft	2Cy.29"&55"-36" 140 HP. Walker, Henderson & Co., Gls.	Turnbull, Mar...
642	✱	**Butuan** I Sc *[illegible] 88* 1 Dk & Awn Dk	Spanish Manila	328 502	160·0	21·1	11·0 17·8	Dundee Gourlay Bros.	1873	4 Blds.	2Cy.27"&48"-24" 90 HP. Gourlay Bros. & Co., Dundee	Macleod & Co.
643		**Buyukdereh** W Pad *Hassan 79-*	Turkish Constantinople	100	140·0	19·6	9·2	West Cowes J. White	1860		60 HP. Maudslay, Sons & Field	Chirket-i-Hair... (Bosphorus Stea...)
644 34787	✱	**Bwllfa** I Sc3mSr *Winrop* 1 Dk2 B	British London	516 812	209·0	28·1 Q 62 f.	17·5	Newcastle Palmers Co.	1865	4 Blds. DB 166 ft	2Cy.27"&50½"-30" 90 HP. Palmers Co., Newcastle	Locket & Judk...
645 78617		**Byculla** I Sc3mSr *Kendall 81-88* 1 Dk2 B	British Glasgow	1072 1464	264·1	32·2	21·7	Dumbarton W. Denny & Bros.	1878	4 Blds. DB [illegible] APT 12 [illegible]	2Cy.27"&54"-36" 122 HP. Denny & Co., Dumbarton	British India S... Nav. Co. Ld.
646	N	**Bygdo** I ScSr *Olsen 86-86* 1 Dk & Spr Dk3 B	Norweg'n Christiania	966 1216	226·3	31·3	15·3 21·9	Malmö Kockums Werkstad	1882	4 Blds. MT 230 DT 458	2Cy.28"&48"-30" 100 HP. Kockums Werkstad, Malmö	Kristiania Da... Selskab [illegible]
647		**Byglia** (ex Skonelly) I Pad *Ahrnlem 88* 1 Dk & Shlt r Dk	Swedish Gefle	63 100	88·7	16·9 H 36 f.	4·9	Hudikswall	1885			O. A. Brodin
648		**Bylgia** (ex Sjötorp) C ScSp *Pärsson 86-86* 1 Dk	Swedish Linköping	127 181	104·0 P 46 f.	20·6 H 1 f.	11·4	Sjötorp	1878		2Cy. 25 HP.	Angf. Aktiebol... (Alsen & Son)
649	✱	**Byron** Stl ScSr *Cook 89-89* 1 Dk Spd Awn Dk & Wb Frm	British London	1319 2057	275·0 P 29 f. Q 72 H 1 f. [illegible]	38·2	20·6	Sunderland J. L. Thompson & Sons	1889	6 Blds. DB [illegible] APT [illegible]	3Cy.22",37"&60"-39" 250 HP. J. Dickinson, Sunderland	Glover Bros.
650		**Byzantion** (WRECKED) I Sc	Greek Syra	351 558	194·5	25·4	15·5	England	1860		1Cy.24"&42"-3[illegible] 170 HP.	Hellenic S.S. ...

ADDITIONAL SHIPS ARE INSERTED IN THE SUPPLEMENT AT THE END OF THE REGISTER.

✱ = Lloyd's Register; D = Nederlandsche Vereeniging; G = Germanischer Lloyd; I = Registro Italiano; R = American ...; A = Veritas Austro-Ungarico; H = Veritas Ellenico; N = Norske Veritas; V = Bureau V...

In the 2nd Column, B.T., with figures following, signifies Date of report of bottom sighting by Board of Trade Surveyors as condition of PASSENGER CERTI...

In the 3rd Column, C = Composite (iron frame, wood planked); I = Iron; Stl = Steel; W = Wood.

Further particulars of Vessels classed by Lloyd's Register will be found in Lloyd's Register Book and Lloyd's Yacht Register.

No.	Registre	Nom, Matériaux, Gréement, etc. / Capitaine et dates d'entrée au service des Armateurs actuels et à bord du navire. / Ancien Nom, nombre de ponts, etc.		Pavillon. / Port d'Armement	Tonnage Net / Total	Dimensions (Longueur | Largeur | Creux)	Construit. Où. / Par qui.	Quand.	Nombre des cloisons. / Détails sur le lest d'eau.	Machines. Détails. / Nom du Constructeur.	Armateurs ou Directeurs-Gérants.
1	*G	C. A. Bade Wilken 88	Stl ScSr 1 Dk	German Bremen	378 592	170·0|26·2|12·5 P 100 f. F 28 f.	Sunderland Sunderland S.B.Co. Ld.	1883	4 Blkds. DB & FPT & APT	3Cy.14",23"&39"–27" 85 HP. N.E.Marine E.Co.Ld.Sundld.	Dampfs. Gesellschaft "Neptun"
2	GV	C. R. Bishop Davis 84 (ex Kaipo)	W ScSr 1 Dk	Hawaiian Honolulu	183 389	114·0|26·0|9·5 Q 21 f. F 13 f. B 34 f.	Pt. Ludlow, [W.T. Hall Bros.	1880		2Cy.12" & 22"–18" 25 HP. Hinckley, Spires & Hays	Inter. Island Stm. Nav. Co.
3		C.W.Chamberlain ex N. Pratt	W Sr 1 Dk	British Windsor, Ont.	243 385	127·0|26·6|9·7	Adelaide H. Jenkings	1881		1Cy.24"–24" 120 HP. Clougharge Wks., Clevld, Org	Simon Wiglo
4		C. W. Eborall	I TwnScSr 1 Dk	British London	236 428	190·2|25·1|12·0 Q 36 f. F 33 f.	Hull Earle's Co. Ld.	1882	5 Blkds.	4Cy.20" & 39"–21" 110 HP. Earle's Co. Ld. Hull	South Eastern Railway Co.
5		Cabenda	I Sc3mSc 1 Dk	British Liverpool	113 204	130·3|20·1|8·8 Q 18 f. F 4 f.	Liverpool T. Royden & Sons	1871	3 Blkds.	2Cy.15" & 30"–18" 40 HP. Fawcett, Preston & Co. L'pool	T. W. & E. H. Cookson
6	*	Cabo Creux (ex Lindens) Elorgoyen 86-88	I ScSr 1 Dk 2 B	Spanish Seville	1032 1517	254·0|32·8|18·8 Q 82 f. B 70 f. F 32 f.	Newcastle Schlesngr, Davis & C	1880	4 Blkds. DB & FPT & APT	2Cy.30" & 58"–36" 150 HP. R. & W. Hawthorn, Newcstle.	Ybarra & Co.
7	*	Cabo Machichaco de la Torre 81-86 (ex Benisa)	I ScSr 2 Dks	Spanish Seville	1148 1689	260·0|33·7|19·4 Q & B 158 f. F 35 f.	Newcastle Schlesngr, Davis & C	1882	4 Blkds. DB & FPT & APT	2Cy.32" & 62"–42" 200 HP. R. & W. Hawthorn, Newcstle.	Ybarra & Co.
8	*	Cabo Ortegal Fano 85-85 (ex Cypriano)	I ScSr 1 Dk 2 B	Spanish Seville	1055 1551	254·0|32·8|18·9 Q 82 f. B 70 f. F 32 f.	Newcastle Schlesngr, Davis & C	1880	4 Blkds. DB & FPT & APT	2Cy.30" & 58"–36" 150 HP. R. & W. Hawthorn, Newcstle	Ybarra & Co.
9	*	Cabo Palos Rolanda 85-87	I ScK 1 Dk 2 B	Spanish Seville	1004 1562	250·0|35·1|17·0 P & B 148 f. F 65 f.	Sunderland J. L. Thompson & Sns	1885	5 Blkds. CDB	3Cy.19",31½"&51"–36" 130 HP. Blair & Co. Ld. Stockton	Ybarra & Co.
10	*	Cabo Penas Valtis 82-87	Stl ScSr 2 Dks & Awn Dk	Spanish Seville	1116 1742	250·0|36·0|16·6 24·1	Sunderland J. L. Thompson & Sons	1887	6 Blkds. DB	3C.19",31½"&51"–36" 130 HP. Blair & Co. Ld. Stockton	Ybarra & Co.
11	*	Cabo Quejo de Zarraga 83-87	Stl ScSr 2 Dks & Awn Dk	Spanish Seville	1118 1747	250·0|36·0|16·6 24·1	Sunderland J. L. Thompson & Sons	1887	6 Blkds. DB	3C.19",31½"&51"–36" 130 HP. Blair & Co. Ld. Stockton	Ybarra & Co.
12	*	Cabo San Antonio Gorostiaga 88-88	Stl ScSr 2 Dks & Awn Dk	Spanish Seville	1124 1756	250·0|36·0|16·7 24·5	Sunderland J. L. Thompson & Sons	1888	5 Blkds. DB 273	3C.19",31½"&51"–36" 130 HP. Blair & Co. Ld. Stockton	Ybarra & Co.
13	*	Cabo Santa Maria Berenguer 85-85	I ScSk 1 Dk	Spanish Seville	105 155	100·0|19·1|8·5 Q 77 f.	Middlesbro' R. Dixon & Co.	1885	3 Blkds. FPT	2Cy.14" & 25"–18" 25 HP. J. P. Rennoldson & Ss. S. Shlds	Ybarra & Co.
14	*	Cabo Torres (ex Phoebe, Gertrude) Mauecos 87-87	I ScSr 1 Dk	Spanish Seville	157 244	120·2|22·2|10·5 Q 36 f. F 15 f.	Newport Usk S.B. Co. Ld.	1882	3 Blkds. FPT	2Cy.15" & 27"–21" 45 HP. Rose, Downes & Thomson, Hl.	Ybarra & Co.
15	*	Cabo Trafalgar Leizaudi 86-88	I ScSr 1 Dk 2 B	Spanish Seville	992 1542	250·4|35·1|16·7 P 150 f. F 25 f.	Newcastle C. S. Swan & Hunter	1885	5 Blkds. DB	3Cy.19",31½"51"–36" 130 HP. Blair & Co. Ld. Stockton	Ybarra & Co.
16	*	Cabo Verde Tito 81-85	I ScBg 1 Dk & Spr Dk 3 B	Portug'se Lisbon	1460 2261	310·0|37·1|26·0	Hull Earle's Co. Ld.	1883	5 Blkds. DB & MT	2Cy.40" & 74"–48" 250 HP. Earle's Co. Ld. Hull	Empreza Nacional de Nav. à Vap.
17		Cabot	I PadSp 1 Dk	British St Johns, NFL	21 128	110·5|19·7|10·2	N. Shields C. W. Dodgin	1872		1Cy.38½–52" 50 HP. Hickles Bros., N. Shields	Newfoundland Screw Steam Tug Co.
18		Cabral Ratcliffe 88	Stl ScSc	British London	467 753	231·1|33·1|12·8 B 96 f. P 16 f. F 19 f.	Greenock Scott & Co.	1888		3Cy.16",26"&43"–31" [98 H Scott & Co. Green'ck	Megaw & Norton.
19		Cabral	I TwnScSc 1 Dk	Brazilian Maranhão	305 445	185·0|26·0|12·0 P 51 f. B 62 f. F 44 f.	Birkenhead Laird Bros.	1884		4Cy.19" & 35"–24" 120 HP. Laird Bros. Birkenhead	Cia de Nav. à Vapor do Maranhão
20		Caçador	ScSc	Portug'se Lisbon	114	65·6|16·4|9·8	Havre Forges & C. de l'Océan			70 HP.	F. G. Burnay
21		Cachapoal Chalers 81-87	I Sc3mSr 2 Dks & Shd Dk	Chilian Valparaiso	1449 2160	334·6|40·2|20·8	Birkenhead Laird Bros.	1881	5 Blkds. DB 282	2Cy.45" & 80"–54" 400 HP. Laird Bros., Birkenhead	Cia Sud-Americana de Vapores
22	V	Cachar Escarras 83-85	I & Stl ScBg 2 Dks 3 B	French Marseilles	2231 3645	344·5|41·0|29·1 P 36 f. F 44 f.	La Seyne Forges & C. de la Medit	1884	DT A 300	2Cy.39½" & 73"–42" 450 HP. Forges & C. de la Medit., Mars.	Cie Nationale de Navigation
23		Cacheiorano	I Pad 1 Dk	Brazilian Bahia	102 175		Newcastle W'gham, Richdsn & C	1874		2Cy.32"–46" 80 HP. W'gham, Richardson & C. Nwc.	Comp. de Nav. à Vapor de Bahia
24	V	Cachemire Aubert 86-87	I & Stl ScBg 2 Dks 3 B	French Marseilles	2198 3369	344·5|41·0|29·1 P 36 f. F 44 f.	La Seyne Forges & C. de la Medit	1884	DT A 125	2Cy.39½" & 73"–42" 450 HP. Forges & C. de la Medit., Mars.	Cie Nationale de Navigation
25	*	Cacouna McDonald	I ScSc 1 Dk 2 B	British Montreal	931 1451	250·0|35·1|16·4 Q 8 f. P 22 f. B 82 f. F 26 f.	Newcastle W'gham, Richardson & Co.	1880	4 Blkds. DB & APT	2Cy.33" & 58"–36" 112 HP. W'gham, Richardson & C. Nwc.	Black Diamond S.S. Co. of Montreal, Ld.
26		Cadell	C Pad 1 Dk	British Adelaide	92 171	122·3|20·3|8·0 B 11 f.	Goolwa, S.A. W. Gordon	1876		2Cy.18" & 13"–36" 45 HP. Blackwood & Gordon, Pt. Gls.	Alexander A. Scott
27	*	Cadiz (ex Lacydon) Drummond 73-75	I ScSr 2 Dks & Spr Dk	British London	901 1400	255·0|32·2|17·6 21·8	Sunderland W. Pile & Co.	1869	5 Blkds. DB	2Cy.31½" & 61"–42" 160 HP. Blair & Co. Ld. Stockton	J. Hall, jun., & Co.
28	*	Cadiz (ex Independence, ex Wallace) Garca 80-86	I Sc3mSr 2 Dks	Spanish Bilbao	1731 2682	331·6|38·2|25·5 B 142 f. P 207 f. F 48 f.	Sunderland Oswald & Co.	1873	5 Blkds.	3Cy.24½",39½&64"–48" [300 HP J. & J. Thomson, Gw.	Larrinaga & Co.
29		Cadiz	I ScSp 1 Dk	Spanish Seville	77 107	85·6|18·1|7·7	Glasgow	1859		2Cy. 20 HP.	A. Millan
30	*D	Cadoxton Stacey	I ScSr 2 Dks 3 B	British Cardiff	1140 1744	281·0|33·8|23·1 P 21 f. B 28 f. F 32 f.	Newcastle Schlesngr, Davis & C	1880	5 Blkds. DB & FPT & APT	2Cy.32" & 62"–40" 200 HP. R. & W. Hawthorn, Newcstle	M. Cope

*** = Lloyd's Register;** D = Nederlandsche Vereeniging; G = Germanischer Lloyd; I = Registro Italiano; R = American Record; A = Veritas Austro-Ungarico; H = Veritas Ellenico; N = Norske Veritas; V = Bureau Veritas.

Dans la 2me Colonne, B.T., avec les indications qui suivent, signifie la date du rapport de visite de la Carène par les experts du Board of Trade pour obtenir le certificat relatif au TRANSPORT DES PASSAGERS.

Dans la 3me Colonne, C = Composite (membrure en fer, bordé en bois); I = Fer; Stl = Acier; W = Bois.

Pour plus amples renseignements sur les navires classés par le Lloyd's Register, voir le Lloyd's Register Book et le Lloyd's Yacht Register.

No. in Book. / Official No.	Registry in which classed.	Ship's Name. Material, Rig, &c. / Master, and Dates of Appointment to present Owner's Service and to Ship / Ship's Late Name. No. of Decks, &c.	Flag. / Port of Registry.	Tons. Net. / Gross.	Dimensions. Length, Breadth, Depth / Deck Erections.	Built. Where. / By Whom.	When.	No. of Bulkheads. / Particulars of Water Ballast.	Engines. Particulars. / Maker's Name.	Owners or Managers.
31 70037	✱	**Caduceus** I ScSchSr *Larkin 73-83* *2 Dks 3 B*	British London	1202 1860	282·0|35·5|24·2 *B 46 f.*	Newcastle Palmers Co.	1874	5 Blkds. DBaft & ab231	2Cy.30½"&61"-36"-150 HP. Palmers Co.,Ld.,Newcastle	Hall Bros.
32 72661	✱D	**Cædmon** I ScSr *Lowe 85-87* *1 Dk 2 B*	British Whitby	771 1225	236·2|32·1|17·1 *B 55 f. F 33 f.*	Whitby T.Turnbull&Son	1877	4 Blkds. DBaft	2Cy.28½"&55"-33" 130 HP. Blair&Co.,Ld.,Stockton	T. Turnbull & S
33 52444	✱	**Caerau** (ex Hardwood, ex Rebeck Reborn) I ScSr *Berry 84-85 -M1884 NO* *1 Dk 2 B*	British Newport	361 581	189·5|27·1|14·2	Sunderland	1865		2Cy.34½"-28" 70 HP. G.Clark,Sunderland	Stephens, Ma Ross
34 68146	✱	**Caerleon** I ScSr *Lefevre* *1 Dk 2 B*	British Cardiff	522 841	205·3|28·5|17·4 *Q 34 f. F 27 f.*	Newcastle Schlesinger,Davis&Co.	1873	W Bak	2Cy.25"&50"-30" 90 HP. Simpson&Co.,London	M. Cope
35 84266	✱	**Caerloch** I ScSr *Rynan 82-82* *1 Dk*	British Aberdeen	422 694	180·5|30·2|13·0 *Q 93 f. B 18 f. F 21 f.*	Sunderland J.Priestman&Co	1882	4 Blkds. DBaft	2Cy.28"&46"-33" 88 HP. WelshRichardson&Co.,Newc	A. A. S. Marshall
36		**Cæsar** StlScSr *1 Dk & Wb Frms*	German Stettin		*B&P 181 f. Q 98 f.*	Newcastle Swan&Hunter	1888		3Cy.21",35"&57"-39" Wallsend Slipway & Eng.Co. Newcastle	A. Brodt
37 68147	✱	**Cæsarea** (STRANORD) I ScSr *George* *1 Dk 2 B*	British Cardiff	646 1010	210·3|30·2|18·0 *Q 78 f. F 36 f. B 54 f.*	Newcastle Schlesinger,Davis&Co.	1873	4 Blkds. W Bak	2Cy.26"&50"-33" 98 HP. Hawks,Crawshay&Co.,Gts.	Hacquoil Bros.
38 131	✱IV	**Caffaro** (ex Prinz Alexander) I ScSr *Vitali* *1 Dk & Spr Dk 3 B*	Italian Genoa	1927 2927	311·3|40·3|28·2 *B 61 f.*	Kiel N.DeutscheWerft	1882	7 Blkds.	2Cy.38½"&66"-43½" 300HP. Markische Maschinen	G. B. Repetto
39 85011	✱D	**Caffila** I S S *Poun l* *2 Dks 3 B*	British Sunderland	1430 2179	285·5|37·2|24·4 *P 33 f. B 70 f. F 35 f.*	Sunderland Bartram,Haswell&Co	1882	6 Blkds. DBaft & ab	2Cy.36"&68"-45" 220 HP. R.&W.Hawthorn,Newcastle	G. J. Hay & Co.
40		**Caïd** (ex Empress) I ScSchSr *1 Dk 2 B*	French Marseilles	728 1087	267·0|29·5|16·4 *P 55 f. F 16 f.*	Newcastle Mitchell&Co.	1867		2Cy.28½"&57"-30" 150HP. Cope,Fraissinet&Co.,Mar	Cie.de Navig.
41 91489	✱	**Cairndhu** I ScSr *Anderson 85-8* *1 Dk & Wb Frms*	British Newcastle	895 1368	231·0|34·6|17·2 *R 129, B 29 f. Q 31, F 27 f.*	Newcastle H.S.Edwards&Son	1885	4 Blkds. W Bak	3C.18½"30"48"-36" 120HP. Blair&Co.,Ld.,Stockton	Cairns & Young
42 87175	✱	**Cairngorm** I ScSr *Bunn 83-84* *2 Dks & Wb Frms*	British London	1166 1790	275·7|36·0|22·7 *P 30 f. B 30 f. F 35 f.*	Dumbarton A.M'Millan&S.	1883	4 Blkds. DBaft	2Cy.31"&60"-42" 250 HP. D.Rowan,Glasgow	S.S. "Cairngorm" Co.
43 87257	✱	**Cairngorm** I ScSr *Watson 86-83* *1 Dk 2 B*	British Leith	730 1125	240·6|32·9|15·6 *Q 63 f. B 52 f. F 30 f.*	Glasgow Dobie&Co.	1883	4 Blkds. MT 283 AFT 30	2Cy.28"&54"-36" 98 HP. W.Kemp,Glasgow	R. & D. Cairns
44 88733	✱	**Cairngowan** I ScSr *Baker* *1 Dk & Wb Frms*	British Newcastle	827 1286	225·4|34·5|16·2 *Q 76 f. B 57 f. F 23 f.*	Sunderland S.P.Austin&Son	1883	5 Blkds. DBaft FPT AFT	2Cy.28½"&53"-33" 99 HP. Blair&Co.,Ld.,Stockton	Cairns & Young
45 95203	✱	**Cairnryan** StlScSr *Humbly 88-88* *1 Dk & Wb Frm*	British Leith	1060 1662	260·0|36·3|16·5 *P 31, Q 64 f. B 110 f. F 27 f.*	Newcastle Palmers Co.Ld.	1888	4 Blkds. DBaft FPT AFT	3Cy.18½",31"&51"-33" 150HP. Palmers Co.,Ld.,Newcastle	R. & D. Cairns
46 87476	✱D	**Cairo** I ScSr *Collins 86-87* *1 Dk & Spr Dk 3 B*	British Cardiff	1469 2268	295·0|37·4|20·0 *B 42 f.*	Sunderland J.Laing	1883	4 Blkds. DBaft	2Cy.35"&66"-42" 230 HP. G.Clark,Sunderland	Tellefsen, Wills
47 86638	✱	**Cairo** I ScBk *Sample 86-86* *1 Dk 2 B*	British Whitby	1110 1728	260·4|37·1|18·5 *Q 104 f. F 32 f. B 58 f.*	Whitby T.Turnbull&Son	1882	4 Blkds. DBaft	2Cy.32"&60"-39" 150 HP. Blair&Co.,Ld.,Stockton	T. Turnbull & S
48 82542	✱	**Cairo** Stl ScSr *1 Dk 2 B*	British Hull	1112 1711	270·8|34·1|18·3 *B 186 f. F 38 f. Q 93 f.*	Hull Earles Co.Ld.	1882	4 Blkds.	2Cy.25"&50"-45" 140 HP. Earles Co.,Ld.,Hull	T. Wilson, Sons
49 82967		**Cakobau** I PadSr *1 Dk*	British Sydney	112 178	111·3|19·3|7·6 *B 26 f.*	Berry Bay, W.Ford [N.S.W.]	1882		2Cy.17"-48" 40 HP. F.A.Young,London	Edward W. Kno
50 87679	✱ BT4.88	**Calabar** Stl ScSr *Keene 74-86* *2 Dks*	British Glasgow	1127 1758	300·3|36·2|20·1 *P 113 f. F 43 f. B 37 f.*	Glasgow J.Elder&Co.	1883	7 Blkds.	2Cy.33"&63"-42" 250 HP. J.Elder&Co.,Glasgow	The British & Steam Nav.Co.
51 19492	BT12.88	**Calabria** I ScBk *(ex Australasian)*	British London	2031 3321	338·5|42·1|36·7 *P 184 f. F 81 f.*	Glasgow J.&G.Thomson	1857		2Cy.37½"&67½"-48" 220HP. J.&G.Thomson,Glasgow	Telegraph Cons. Maintenance Co.
52 86372	✱	**Calabria** I ScSchSr *Rees* *1 Dk 2 B*	British Dundee	977 1526	259·6|36·6|16·6 *Q 8 f. B 54 f. F 28 f.*	Dundee W.B.Thompson	1883	4 Blkds. DB 337 FPT AFT	2Cy.28"&56"-42" 140 HP. W.B.Thompson,Dundee	D. Scott & Son
53 76		**Calabria** I ScSchSr *Valle 75-88 (ex Wellesley)* *1 Dk 2 B*	Italian Palermo	781 1201	231·9|30·3|19·0 *Q 212 f. F 21 f.*	Newcastle Palmers Co.	1870	4 Blkds.	2Cy.25"&50"-34" 185 HP. Palmer&Co.,Newcastle	Navigazione Gen. Italiana
54 68126		**Calais** I S S	British Dover	128 209	161·2|22·2|12·2	London J.&W.Dudgeon	1874		2Cy.21½"&15"-21" 80 HP. J.&W.Dudgeon,London	London Chatham & Dover R.Co.
55		**Calais-Douvres** I TwnPadK *1 Dk 2 B*		559 1921	302·0|61·0|13·9	Newcastle A.Leslie&Co.	1877	6 Blkds.	1Cy.63"-72" 600 HP. Black,Hawthorn&Co.,Newc	London Chatham & Dover R.Co.
56 83985	✱	**Calanas** I S S *Henderson 87-87* *1 Dk 2 B*	British Newcastle	1025 1588	265·2|35·5|16·8 *Q 72, P 24 f. B 68 f. F 29*	Newcastle Palmers Co.	1882	4 Blkds. DB 240 AFT 37	2Cy.30"&57"-39" 140 HP. Palmers Co.,Ld.,Newcastle	Scott Bros.
57 93861		**Calcutta** W Sc	British Liverpool	78 111	81·8|19·5|8·3	Northwich J.Gibson	1868		2Cy.12"-14" 24 HP. E.Bates,Northwich	R. Verdin
58 87978		**Calder** I ScSr *Hughes* *1 Dk & Iron Dk*	British Bombay	406 636	171·6|25·1|13·7	Dumbarton J.&R.Swan	187[illegible]	4 Blkds.	2Cy.25"&44"-33" 85 HP. Blackwood&Gordon,P.Glas	Bombay & Persia Nav.Co.Ld.
59 91306	✱	**Calder** StlScSr *King 65-7*	British	323 631	215·5|28·8|13·8	Newcastle W.Dobson&Co.	1887		2Cy.27"&51"-33" 130 HP. [illegible] Newcastle	Goole Steam S.
60 84338	✱	**Calder** I ScSr *Jones 87-7* *1 Dk*	British Whitehaven	94 200	110·0|18·0|9·8 *Q 21 f.*	London Edwards&Symes	1862		2Cy.12½"&22"-18" 30 HP. [illegible] Burrell&Son,Thetford	G. Nelson & Sons

✱ = **Lloyd's Register**; D = Nederlandsche Vereeniging; G = Germanischer Lloyd; I = Registro Italiano; R = American Re[gister]
A = Veritas Austro-Ungarico; H = Veritas Ellenico; N = Norske Veritas; V = Bureau Ver[itas]

In the 2nd Column, B.T., with figures following, signifies Date of report of bottom sighting by Board of Trade Surveyors as condition of PASSENGER CERT[IFICATE]

In the 3rd Column, C = Composite (iron frame, wood planked); I = Iron; Stl = Steel; W = Wood.

Further particulars of Vessels classed by Lloyd's Register will be found in Lloyd's Register Book and Lloyd's Yacht Register.

Registre	Nom, Matériaux, Gréement, etc. Capitaines et dates d'entrée au service des Armateurs actuels et à bord du navire. Ancien Nom, nombre de ponts, etc		Pavillon. Port l'Armement	Tonnage. Net. Total	Dimensions. Longueur	Largeur	Creux	Construit. Où. Par qui.	Quand	Nombre des cloisons. Détails sur le lest d'eau.	Machines. Détails. Nom du Constructeur.	Armateurs ou Directeurs-Gérants.
	Calderon *Urbina 86-86*	I Sc3mSr *2 Dks*	Spanish Seville	392 668	183·1	25·0	15·2	Birkenhead Bowdler, Chaffer & Co.	1869		2Cy.22" & 41"–24" 70 HP. J.Taylor&Co.,Birkenhead	M.Saenz & Co.
	Caledonia *Hasenstein 80-88*	I ScBk *2 Dks & Awn Dk*	British Glasgow	1396 2151	310·6	33·3	21·1 28·5	Glasgow Tod&McGregor	1863	5 Blds.	2Cy.38" & 68"–42" 421 HP. FinniestonCo.,Glasgow	Henderson Bros.
*D	Caledonia *Honeywarne 85-85*	I ScSr *1 Dk 2 B*	British Cardiff	1256 1978	270·5	38·0	18·6 *Q 24 f. B 62 f. F 30 f.*	Southampton Oswald, Mordaunt & Co.	1883	4 Blds. DB 636 f 540t	2Cy.32" & 60"–42" 155 HP. Oswald, Mordaunt&Co.,Sou.	United Kingdom S.S. Co., Ld.
*	Caledonian *McCallum 70-78*	I Sc3mSr *2 Dks & pt Awn Dk*	British Ardrossan	263 521	201·0	26·1	12·9 *P 74 f.*	Glasgow J.Elder&Co.	1874	5 Blds.	2Cy.30½" & 55"–36" 150 HP. J.Elder&Co.,Glasgow	Ardrossan Shipping Co.
	Caledonian	I Sc	British Grangemouth	122 249	133·6	24·6	12·2	Renfrew W.Simons&Co.	1875		2Cy.18" & 30"–24" 60 HP. W.Simons&Co.,Renfrew	Caledonian Railway Co.
F	Calédonien	I ScBk *3 Dks*	French Marseilles	2682 4181	413·9	41·3	32·8	LaCiotat MessageriesMaritimes	1882		3C.43¼" 65(2)"–48¼" 640 HP. MessageriesMar.,LaCiotat	Messageries Maritimes
	Caliban	I Sc3mSrYt	French Marseilles	135	111·7	16·3	10·5	London Edwards&Symes	1881	3 Blds.	2Cy.14" & 25"–16" 30 HP. A.Wilson&Co.,London	Georges Borelli.
	California *Garvie 77-87*	I ScSr *2 Dks & Awn Dk*	British Glasgow	2214 3410	361·5	40·5	24·2 31·6	Glasgow A.Stephen&Sons	1872	6 Blds.	2Cy.33" & 66"–60" 480 HP. D.&W.Henderson&Co.,Glas	Henderson Bros.
V	California *Woelder 88-88*	I ScSr *2 Dks & Spr Dk*	German Hamburg	2163 2690	300·5	38·7	22·5 30·7 *B 62 f.*	Newcastle SirW.G.Armstrong, Mitchell&Co.,Ld.	1883	WB 672t	2Cy.38" & 74"–48" 350 HP. WallsendSlipwayCo.,Newc.	Hamburg Amerika- nische Packetfahrt Act. [illegible]
	Californian *Pickthall 85-86*	I ScSr	British Liverpool	1287 1832	274·9	34·0	29·7	Greenock Scott&Co.	1865		4Cy.21" & 48"–46" 200 HP. G.Forrester&Co.,Liverpool	West India & Pacific S.S.Co.,Ld.
	Calipso (ex Conte di Bismark Vers.) *Crisagono 72-87*	I ScBg *2 Dks*	Austrian Trieste	1092 1742	299·6	32·5	21·1 *P 93 f. F 50 f.*	Stockton RichardsonDuck&Co.	1867		2Cy.38" & 68"–33" 200 HP. MillwallIronWorks,London	Lloyd Austro-Ungarico
*	Calliope *Hosking 87-87*	I ScBg *1 Dk 2 B*	British Hull	907 1409	260·0	32·0	17·1 *P 135 f. F 28 f.*	Hull Humphrys&Co.	1871	WB FPT APT	2Cy.30" & 56"–33" 140 HP. Humphrys&Co.,Hull	J.Wild & Co.
*D	Callisto *Nathusius 84-88*	StlScSr *1 Dk 2 B*	Dutch Rotterdam	1272 1721	250·0	36·0	17·1 *Q 82 f. B 102 f. F 23 f.*	Sunderland W.Doxford&Sons	1888	4 Blds. CDB 264t APT 16t	3Cy.18", 29" & 48"–33" 130 HP. W.Doxford&Sons, Sunderland	Hudig & Veder
*D	Calon *Eguibel 86*	I ScSr *1 Dk 2 B*	Spanish Bilbao	1009 1567	250·0	35·3	17·9 *Q 61 f. B 52 f. F 27 f.*	Sunderland StrandSlipwayCo.	1882	4 Blds. DB 315 f 296t	2Cy.31" & 57"–36" 130 HP. G.Clark,Sunderland	J.Duranona
*	Caloric *Ransom 88-88*	Stl ScSr *1 Dk*	British Glasgow	1138 1747	260·0	36·8	16·6 *Q 92 f. B 102 f. F 21 f.*	Glasgow A.Stephen&Sons	1887	4 Blds. CDB 336t APT 28t	3C.18", 29" & 46"–39" 170 HP. A.Stephen&Sons,Glasgow	E.Kemp (Maclay&MacIntyre,Mgrs.)
*	Caloric *Davidson 79-86*	Stl ScSr *2 Dks*	British Belfast	301 893	243·9	31·9	15·3 *P 42 f. B 92 f. F 37 f.*	Belfast Harland&Wolff	1885	4 Blds. APT	2Cy.35" & 70"–45" 200 HP. Harland&Wolff,Belfast	Belfast Steam Ship Co.,Ld.
V	Calvados (ex Adriana) *Lenormand*	I Sc	French Havre	45 107	108·3	15·9	6·8	Marseilles AteliersduRhône	1885		2Cy.11" & 20"–20" 25 HP. A.F.Smulders,Utrecht	Lenormand
*	Calvilla *Webster 85-86*	I ScSr *1 Dk 2 B*	British Scarbrough	913 1411	242·8	34·2	18·3 *Q 57 f. B 46 f. F 20 f.*	Sunderland J.L.Thompson&Son	1879	5 Blds. DB 262 ft 161t FPT APT	2Cy.31" & 57"–36" 130 HP. J.Dickinson,Sunderland	T.E.Hick
	Calypso *Edwards*	I ScBg *1 Dk 2 B*	British Hull	835 1283	251·4	32·1	20·5 *H 68 f. B 68 f. F 28 f.*	Hull C.&W.Earle	1865	4 Blds.	4Cy.20" & 40"–30" 150 HP. Cp.74,Humphrys&Co.,Hull	T.Wilson,Sons & Co.
*	Calypso	I Sr *1 Dk*	British Liverpool	328 582	190·0	30·6	11·6 *P 31 f. F 26 f.*	Greenock Scott&Co.	1889		2Cy.18½" & 38"–30" Scott&Co.,Greenock	A.Holt
	Calypso *Reed 88*	I Sc3mSr *1 Dk 2 B*	British Bristol	300 563	207·1	25·3	14·3 *P 36 f. F 22 f.*	Glasgow Barclay, Curle&Co.	1865	4 Blds.	2Cy.27" & 47"–32" 97 HP. Barclay,Curle&Co.,Glasgow	Bristol Stm.Nav.Co., Ld.
*	Cam *Reed 83-83*	I ScSr *1 Dk*	British Lynn	93 175	120·4	20·1	9·6 *Q 35 f. F 14 f.*	Paisley J.McArthur&Co.	1883	3 Blds. FPT 13t	2Cy.15" & 27"–21" 30 HP. J.Gilmour&Co.,Glasgow	East Coast S.S.Co.,Ld.
	Camara	I ScSr *1 Dk 2 B*	Spanish Seville	562 826	210·7	30·0	15·9 *P 43 f. F 50 f.*	Sunderland A.Simey&Co.	1875	4 Blds. DB 100 f & 33 f.	2Cy.27" & 52"–33" 110 HP. J.Dickinson,Sunderland	Vinnesa & Co.
*	Camargo *Prance 81-83*	I ScSr *1 Dk*	British Newport	588 919	208·5	30·1	15·2 *Q 82 f. B 42 f. F 24 f.*	Sunderland W.Pickersgill&Son	1880	4 Blds. DB 155 f 294t FPT 16t APT 8t	2Cy.26" & 47"–36" 90 HP. G.Clark,Sunderland	R.W.Jones & Co.
	Camargue		French Bordeaux		..	..	..					Buhan père et fils, J.E. & A.Tesseire
F	Cambodge	I Sc *3 Dks*	French Marseilles	1639 2574	318·9	38·7	31·4 *P 77 f.*	LaSeyne Forges&CdelaMdt.	1861		2Cy.76½"–52½" 500 HP. Forges&Cde la Médit.,Mars.	Messageries Maritimes
*D	Cambodia *Waldgnose 79-82*	I ScSr *1 Dk & Spr Dk 3 B*	British Liverpool	1969 3066	327·3	40·1	29·1 *B 66 f.*	Sunderland R.Thompson&Son	1882	5 Blds. DB 88 f 236t FPT	2Cy.40" & 74"–48" 350 HP. G.Clark,Sunderland	S.S.Cambodia Co.,Ld. (Crow, Rudolf & Co.)
V	Cambrai (ex Louise) *Dubiblet 83-83*	I ScSr *1 Dk 2 B*	French Dunkirk	666 1066	219·1	30·2	18·4 *Q 114 f. F 34 f.*	Newcastle SchlesingerDavis&Co.	1875	WB	2Cy.28" & 53"–35½" 120 HP. T.Clark&Co.,Newcastle	Cie.Générale des Bateaux à Vap.du Nord
*	Cambria	StlScSr	British Whitby		..	..	.. *Q 74 f. P 22 f. pt Awning Dk 181 f*	Sunderland J.L.Thompson&Sons	1889		3C.21", 35" & 57"–39" 180 HP. J.Dickinson,Sunderland	International Line S.S.Co.,Ld.
	Cambria *(ex Champion)*	WPaSr *1 Dk*	British Owen Sound	487 715	131·2	23·3	10·7	Levis, Canada G.T.Davie	1877		1Cy.40"–81 80 HP. Owen Sound Dry Dock and S.B.Co.,Ld.	Owen Sound Dry Dock and S.B.Co.,Ld.

* = Lloyd's Register; D = Nederlandsche Vereeniging; G = Germanischer Lloyd; I = Registro Italiano; R = American Record; = Veritas Austro-Ungarico; H = Veritas Ellenico; N = Norske Veritas; V = Bureau Veritas.

Dans la 2me Colonne, B.T., avec les indications qui suivent, signifie la date du rapport de visite de la Carène par les experts du Board of Trade pour obtenir le certificat relatif au TRANSPORT DES PASSAGERS.

Dans la 3me Colonne, C = Composite (membrure en fer, bordé en bois); I = Fer; Stl = Acier; W = Bois.

Pour plus amples renseignements sur les navires classés par le Lloyd's Register, voir le Lloyd's Register Book et le Lloyd's Yacht Register.

No. in Book / Official No.	Class	Ship's Name, Material, Rig, &c. / Master, and Dates of Appointment / Ship's Late Name. No. of Decks, &c.	Flag / Port of Registry	Tons Net / Gross	Dimensions (Length, Breadth, Depth) / Deck Erections	Built: Where / By Whom	When	No. of Bulkheads / Particulars of Water Ballast	Engines: Particulars / Maker's Name	Owners
91 60784	✳ BT7.88	Cambria 1 Sc3mSr *Packham* *3 Dks*	British [illegible]	460 909	260·0 | 28·2 | 16·1 P 30 f. F 42 f.	Dundee Gourlay Bros.	1868	5 Blds.	3C 21"37" & 50"–36" 200 HP. Gourlay Bros.&Co., Dundee	Dundee, Pent[land] don S.S.Co.
92 76840	BT6.88	Cambria 1 PadSr *Fortunato* *1 Dk*	British Liverpool	74 176	118·7 | 20·1 | 9·9	Preston T.Redhead&Son	1879	4 Blds.	2Cy.33½"–45" 70 HP. R.Smith, Preston	T. Redhead & [S]…
93 63823		Cambria 1 PadSr *Anson 72-77*	British London	55 209	138·8 | 20·0 | 11·5	London ThamesE.&I.S.B.Co.Ld.	1870		2Cy.34"–54" 100 HP. J.Stewart, London	W. Watkins
94 53354	✳	Cambria 1 [illegible] *Brown 85-89* *1 Dk*	British Llanelly	47 134	114·6 | 17·1 | 8·8	Llanelly [illegible]	1866	4 Blds.	2Cy.16"&32"–16" 40 HP. Comp.'85, R.Nevill, Llanelly	D. Evans
95 84276	✳	Cambrian 1 ScSr *English 85-88* *1 Dk*	British [illegible]	228 380	155·0 | 22·9 | 12·0 Q 75 f. B 16 f. F 28 f.	Paisley J.Fullerton&Co.	1880	4 Blds. FPT [illegible] APT [illegible]	2Cy.20"&36"–24" 50 HP. W.King&Co., Glasgow	J.&W.W.C[…]
96 72728	BT3.88	Cambridge S[t]lTwinSc[r] *Nickerson 63-88*	British [illegible]	553 1190	280·5 | 31·0 | 15·2	Hull Earles'S.B.&Eng.Co.Ld.	1886		4Cy.30"& 57"–36" 275 HP. EarlesCo.Ld., Hull	Great Eastern [Ry.] Co.
97 79512	✳	Camden 1 ScSr *James* *2 Dks3 B*	British London	970 1492	252·0 | 33·0 | 22·8 B 50 f. F 39 f.	Newcastle [illegible]	1878	4 Blds. DBob[illegible]	2Cy.30"& 58"–36" 150 HP. WallsendSlipwayCo., Newc.	Watts, Ward & [Co.]
98 63644	✳	Camel 1 ScSr *Bannister* *1 Dk*	British Belfast	226 356	170·5 | 21·9 | 12·2 P 18 f.	Belfast Harland&Wolff	1870	4 Blds.	2Cy.18"& 38½"–24" 70 HP. Greenock FoundryCo., Grnk	Matthews & [Co.]
99 86751	✳	Camel 1 ScSr *Lee* *1 Dk*	British Newport	182 293	135·8 | 22·2 | 10·0 Q 57 f. B 15 f. F 16 f.	Newport Mordey,Carney&Co.Ld	1883	4 Blds. FPT [illegible] APT [illegible]	2Cy.18"& 33"–24" 50 HP. Plenty&Son, Newbury	Ponsonby & Co.
100 78208	✳	Camellia 1 ScSr *Pinney 87-87* *1 Dk2 B*	British N. Shields	880 1363	258·0 | 32·0 | 21·2 B 17 f.	Newcastle TyneI.S.B.Co.Ld.	1878	5 Blds. DBott [illegible]	2Cy.28"& 54"–33" 130 HP. N.E.MarineEng.Co.Ld.Sld.	J. Robinson & [Co.]
101 70780	✳	Camelot 1 ScSr *Daily 69-78* *2 Dks3 B*	British Leith	1049 1608	259·8 | 32·6 | 23·5 F 21 f.	Glasgow Aitken&Mansel	1878	5 Blds. DBottom FPT [illegible] APT [illegible]	2Cy.31"& 63"–36" 180 HP. J.&J.Thomson, Glasgow	J. Laing & Co. (F.F.Reid,Manager)
102 93078	✳	Camenes 1 ScK *Chapman* *1 Dk*	British Hull	79 153	98·8 | 20·1 | 11·0 Q 16 f. F 20 f.	Hull Head&Riley	1886	3 Blds.	2Cy.17"& 34"–21" 45 HP. E.Wales, Hull	W. R. Leyman & [Co.]
103 84031	✳	Cameo 1 ScSr *Rich 82-83* *2 Dks3 B*	British [illegible]	1485 2280	300·0 | 39·1 | 23·7 P 32 f. B 28 f. F 34 f.	Glasgow A.Stephen&Sons	1881	4 Blds. DBott [illegible]	2Cy.35"& 65"–42" 200 HP. A.Stephen&Sons, Glasgow	Tellefsen, W[…]
104 75329	BT5.88	Cameo 1 Sc3mSr *Langlands* *2 Dks*	British Hull	833 1272	245·0 | 31·0 | 17·3 P 155 f. F 28 f.	Hull Earles Co.	1876	6 Blds.	2Cy.33½"&64½"–42" 170 HP. EarlesCo., Hull	T. Wilson, Son[s] …
105 87082	✳	Cameo (ex [illegible]) 1 Sc3mSr *McLean 86-86* *1 Dk*	British Glasgow	148 338	160·2 | 23·2 | 10·9 Q 49 f. B 7 f. F 24 f.	Paisley J.Fullerton&Co.	1863	4 Blds. FPT [illegible] APT [illegible]	2Cy.19"&36"–27" 60 HP. J.&T.Young, Ayr	W. Robertson
106 65106	✳	Camerata 1 ScSr *Cook -88* *1 Dk2 B*	British Aberdeen	622 982	210·5 | 29·5 | 15·9 P 34 f. B 47 f. F 31 f.	S. Shields J.Readhead&Co.	1874	4 Blds. DBa&f [illegible]	2Cy.26"& 49"–33" 98 HP. J.Readhead&Co., S.Shields	Adam & Co.
107 68115	✳ BT8.88	Cameroon 1 ScBg *Turnbull 86-87* *2 Dks*	British Glasgow	1185 1862	302·5 | 33·8 | 24·0 P 106 f. H 28 f. F 53 f.	Glasgow J.Elder&Co.	1874	6 Blds.	2Cy.46"& 71"–36" 280 HP. J.Elder&Co., Glasgow	British & Afric[an] Nav.Co.Ld.
108		Cameta 1 P[illegible] *1 Dk.&HurDk*	Brazilian Para	340 478	181·0 | 28·0 | 9·5	Birkenhead Laird Bros.	1881		2Cy.32"& 53"–54" 121 HP. LairdBros., Birkenhead	Amazon Stm.[…]
109		Camiguin 1	Spanish Manila	99 169	130·2 | 22·2 | 6·6	England	1873		2Cy.11"& 26"–18" 25 HP. Thompson&Co., Newcastle	P. E. Martin[…]
110 54730	✳	Camilla 1 Sc3mSr *George 86-88* *1 Dk2 B*	British London	722 956	239·7 | 28·1 | 17·0 Q 70 f. F 37 f.	Newcastle Palmers'Co.	1866	4 Blds. DBott [illegible]	2Cy.26"& 51"–36" 160 HP. Comp.'74, VictoriaDk Eng Co. [illegible]	Collier & Ge[…] Co.,Ld.
111		Camille (BROKEN UP)	French Nantes	87 152						P. & A. Légal[…]
112 86938	✳	Camma 1 ScSr *Paulsen 88-88* *1 Dk2 B*	British W.Hartlepl.	524 840	199·5 | 30·0 | 11·0 Q 119 f. F 20 f.	Stockton M.Pearse&Co.	1882	4 Blds. DBa f [illegible] FPT [illegible] APT [illegible]	2Cy.24"& 45½"–30" 90 HP. Blair&Co.Ld., Stockton	Herskind & W[…]
113 84285	✳	Camorta 1 ScSr *Trewin 87-86* *3 Dks*	British Glasgow	1355 2087	285·2 | 35·2 | 24·1 P 42 f. B 44 f. F 38 f.	Glasgow A.&J.Inglis	1880	5 Blds. DBa f &c. [illegible]	2Cy.34"& 60"–45" 200 HP. A.&J.Inglis, Glasgow	E. S. Dawes & [Mac] kenzie
114 51646	R	Campana 1 ScSr	British London	876 1288	240·8 | 35·3 | 20·7	Glasgow Aitken&Mansel	1873		2Cy.26"& 52"–33" 225 HP. D.Rowan, Glasgow	A. M. Smith
115	✳	Campeador 1 Sc3mSr *Albisuri* *1 Dk.&SprDk*	Spanish Barcelona	976 1258	228·0 | 32·3 | 13·9 E 8 f. 217	Sunderland J.Laing	1884	4 Blds. APT [illegible]	2Cy.26"& 52"–36" 95 HP. J.Dickinson, Sunderland	J. Roca & Co.
116	✳	Campechano Stl TwinSc *De Garry* *1 Dk.&ShdDk*	Mexican Campeche	187 293	130·1 | 25·2 | 9·3 F 20 f.	Dumbarton MurrayBros.	1883	1 Blds.	4Cy.15"& 24"–18" 50 HP. Alley&Maclellan, Glasgow	Leandro, Reg[…]
117 24	I	Campidoglio 1 ScSr *Parinello 65-86* *2 Dks*	Italian Palermo	341 691	195·8 | 26·6 | 15·7 Q 54 f.	Greenock Scott&Co.	1862		2Cy.28"& 55"–30" 150 HP. FonderiaOretea, Palermo	Navigazione [Generale] Italiana
118	V	Campinas Stl&I ScSr *Barrelet 86-88* *2 Dks3 B*	German Hamburg	1634 2265	309·0 | 36·3 | 25·8 P 68 f. B 70 f. F 47 f.	Hamburg Reiherstieg Schiffs Wrft.	1886	WB [illegible]	3C 24"38"&62"–42" 300 HP. ReiherstiegSchiffsWrft, Hmbrg	Hamburg-S[üd]… Dampfschiff[…]
119 47060	BT[illegible]	Canada (ex Pennsylvania) 1 ScBkn *Robinson 79-88* *4 Dks*	British Liverpool	2709 [illegible]	391·6 | 41·2 | 35·2	Newcastle PalmerBros.&Co.	1863		2Cy.50"& 85"–48" 400 HP. LairdBros., Birkenhead	National Stea[m…] Co.,Ld.
120 [illegible]	✳	Canada 1 Sc *Archer* *1 Dk*	British Hull	103 191	117·2 | 22·2 | 11·2 Q 20 f. F 20 f.	Beverley Cochrane&Co.	1886	3 Blds.	3C 13½"19"&28"–24" 150 HP. C.D.Holmes&Co., Hull	Hull Steam F[ishing &] Ice Co.,Ld.

✳ = Lloyd's Register; D = Nederlandsche Vereeniging; G = Germanischer Lloyd; I = Registro Italiano; R = American [Record]; A = Veritas Austro-Ungarico; H = Veritas Ellenico; N = Norske Veritas; V = Bureau Ve[ritas].

In the 2nd Column, B.T., with figures following, signifies Date of report of bottom sighting by Board of Trade Surveyors as condition of PASSENGER CERT[IFICATE].

In the 3rd Column, C = Composite (iron frame, wood planked); I = Iron; Stl = Steel; W = Wood.

Further particulars of Vessels classed by Lloyd's Register will be found in Lloyd's Register Book and Lloyd's Yacht Register.

Registre	Nom, Matériaux, Gréement, etc. Capitaine et dates d'entrée au service des Armateurs actuels et à bord du navire. Ancien Nom, nombre de ponts, etc.	Pavillon. Port d'Armement	Tonnage Net. Total.	Longueur	Largeur	Creux	Construit. Où. Par qui.	Quand	Nombre des cloisons. Détails sur le lest d'eau.	Machines. Détails. Nom du Constructeur.	Armateurs ou Directeurs-Gérants.
	Canada W ScSr	British Liverpool	76 112	81·8	19·5	8·9	Northwich H.Gibson	1869		2Cy.12"–14" 24 HP. F.Butts,Northwich	R.Verdin
✻V	**Canada** (ex Panama) I ScBk Godet [illegible] 3 Dks & Awn Dk	French Havre	2262 4202	354·3 Len. 76.	43·9	30·7 38·1	St.Nazaire J.Scott&Co.	1865	9 Blhds.	1Cy.44 & 75"–51" 900 HP. Maudslay,Sons&Field,Lon.	Compagnie Générale Transatlantique
	Canada Atlantic Transfer W Pad 1 Dk	British Ottawa	395 619	171·0	70·0	8·6	Pittsburg,U.S. J.Dunbar	1884		2Cy.22½"–96" 270 HP. J.Rees,Pittsburg,U.S.	Canada Atlantic Railway Co.
	Canadian I Sc3mSr McKenzie 3 Dks	British Glasgow	1872 2911	349·9	35·6	25·1	Liverpool T.Royden&Sons	1872	6 Blhds.	2Cy.44"&82"–45" 280 HP. J.Jack,Rollo&Co.,Liverpool	J.& A.Allan
	Canara I ScSr Williams 75-88 3 Dks	British Glasgow	1229 1903	298·0	33·3	25·7	Dumbarton W.Denny&Bros.	1874	5 Blhds.	2Cy.36½"&63"–42"206 HP. Denny&Co.,Dumbarton	British India Steam Nav.Co.
✻	**Cancer** I ScDy Yarmouth 88-88 1 Dk	British Grimsby	74 126	95·0 Q 18 f.	20·1	10·4	Grimsby T.Charlton	1883	3 Blhds.	2Cy.16¾"&31¾"–20"35 HP. T.Charlton,Grimsby	Grimsby & North Sea Stm.Trawling Co.,Ld.
✻	**Candia** I ScSr Work 88 1 Dk2 B	British Newport	880 1366	248·5 Q 89 f. B 54 f. F 30 f.	34·2	17·0	Newcastle W.Dobson&Co.	1888	4 Blhds. DB[illegible]	3C.17"29"&48"–36" 125 HP. Black,Hawthorn&Co.,Gts.	Stephens, Mawson & Goss
I	**Candia** (ex Calabro) I ScSr De Simone 86-87 2 Dks	Italian Palermo	573 1042	218·0	27·9	20·1	Newcastle Palmer Bros.&Co.	1862	3 Blhds.	3C.24"38"&62"–36"168 HP. R. & W. Hawthorn, Leslie & Co.	Navigazione Generale Italiana
✻	**Canning** (ex J. Golconda) Stl ScSr Bocalan 3 Dks	British	1357 2112	285·0 P 41 f. B 48 f. F 30 f.	36·2	23·7	Glasgow A.&J.Inglis	1882	5 Blhds. CDB[illegible]	2Cy.34"& 60"–45" 200 HP. A.&J.Inglis,Glasgow	Indian Government
	Canning Stl ScSr Booth 73-84	British Liverpool	403 645	213·2 Q 5 f.	32·1	12·6	Greenock Scott&Co.	1883		2Cy.23"& 50"–30" 100 HP. Scott&Co.,Greenock	Liverpool,Brazil & River Plate S.N.Co.,Ld.
✻	**Canonbury** (LOST) I ScSr Hobert 2 Dks3 B	British London	1080 1676	258·0 P 26 f. B 52 f. F 30 f.	35·0	21·3	Sunderland J.Blumer&Co.	1880	4 Blhds. CDB[illegible]	2Cy.32"& 60"–39" 150 HP. G.Clark,Sunderland	Watts,Ward & Co.
	Canonicus W Pad	American Fall Riv.Mas	365 541	178·3	28·1	9·2	New York	1849			J.Borden
	Canopus I Sc1mBk Lockett 87 2 Dks3 B	British Liverpool	1818 2802	384·8 F 33 f.	37·0	27·5	Hull C.&W.Earle	1870	5 Blhds.	2Cy.32"& 74"–54" 217 HP. Hawthorn&Co.,Newcastle	Moss Steam Ship Co., Ld.
✻	**Canosa** (ex Harold) I ScSr 1 Dk	Spanish Gijon	85 170	110·0 Q 39 f. F 13 f.	18·1	9·9	Paisley H.McIntyre&Co.	1877	3 Blhds. FPT	2Cy.15"& 28"–24" 40 HP. Muir&Houston,Glasgow	Viuda y Hijos de J. M.Canosa
	Cantabria I ScBarge 1 Dk	Spanish Corunna	184 219	124·3	19·2	10·0	LaSeyne	1854		1Cy.35½"–26" 50 HP. LaSeyne	R.L.Perez & Co.
✻	**Canton** Stl ScSr 2 Dks	British Greenock	3250	350·0 P 17 f. B 104 f. F 46 f.	42·3	26·2	Greenock Caird&Co.,Ld.	1889	7 Blhds. CBD[illegible]	3Cy.24",140"&62"–54" Caird&Co.,Greenock	Peninsular & Oriental Steam Navigation Co.
✻	**Canton** (ex Alnwick Castle) I ScSr Parke 2 Dks3 B	British London	1176 1826	280·0 Len. 78	33·9	25·1	Newcastle C.Mitchell&Co.	1869	5 Blhds.	2Cy.30"& 60"–30" 150 HP. Thompson,Boyd&Co.Newc.	Corfield & Robson
✻	**Canton** I ScSr Bremner 85-85 1 Dk & Spr Dk3 B	British HongKong	1110 1737	285·0 H 85 f.	34·3	16·3 23·6	Newcastle W.glam,Richardson&Co.	1880	5 Blhds. DB[illegible]	2Cy.35"& 68"–45" 250 HP. W.glam,Richardson&C.,Nwe	Indo-China Steam Navigation Co.
	Canton Stl PadLr	British Glasgow	250 674	250·0	35·1	9·8	Dumbarton W.Denny&Bros.	1886		4C.23"35"46"&67"–72"250 HP. Denny&Co.,D'umbarton	Irrawaddy Flotilla Co.,Ld.
V	**Canton** I ScBg Gadail 88-88 2 Dks & Spr Dk	French Marseilles	2937 3721	279·6 B 28 f. F 54 f.	40·0	22·2 29·6	LaSeyne Forges&C.dela Medit	1882		2Cy.39½"& 73"–42"400 HP. Forges&C.de la Medit Marsls	Cie. Nationale de Navigation
	Cantonnais I ScSr Prugna 85-86 2 Dks	French Saigon	102 178	135·7 H 39 f.	19·9	7·2	Greenock Scott&Co.	1881		2Cy.15"& 27"–15" 40 HP. Scott&Co.,Greenock	Messageries Fluviales de Cochin Chine
✻	**Canute** I ScSr Mitchell 1 Dk2 B	British Newcastle	708 1110	225·5 P 119 f. F 27 f.	30·6	17·4	Newcastle A.Leslie&Co.	1871	5 Blhds. DB[illegible]	2Cy.26"& 52"–30" 99 HP. R.&W.Hawthorn, Newcastle	M.S.Carr
	Cap Nor W ScBk 1 Dk	Norweg'n Drammen	226 346	136·0 Q 33 f.	29·2	15·7	Drammen J.Jorgensen	1871		45 HP. Nylands Verks,Christiania	Soren S.Svendsen
	Cape Charles I Pad	American Wlmngtn Dl	649 941	252·5	36·1	13·0	Wlmngtn.Del The Harlan & Hollingsworth Co.	1885	4 Blhds.	1Cy.54"–131. 115 HP. The Harlan & Hollingsworth Co., Wlm. Del.	New York, Philadelphia & Norfolk Railway Co.
	Cape Clear I ScSr 1 Dk	British Singapore	99 168	118·5	22·1	10·2	Southampton Day&Summers	1876	4 Blhds.	2Cy.20"& 40"–24" 60 HP. Day&Summers,S'thampton	Borneo Co.,Ld.
	Capella I ScSr Lee 70-82	British Liverpool	2189 3359	367·0 P 322 f. F 36 f.	39·1	19·4	Newcastle A.Leslie&Co.	1882		4Cy.28"& 60"–54" 450 HP. R.Stephenson&Co.,N'castle	Star Nav.Co.,Ld.
	Capella I ScBg Kniesberg 1 Dk2 B	German Hamburg	556 818	224·5 P 83 B 46 f. F 24 f. H 30 f.	29·5	16·9	Newcastle C.Mitchell&Co.	1865	4 Blhds. DB[illegible] EB[illegible]	2Cy.29"& 52"–30" 120 HP. [illegible] Reiherstieg Schiff & Masch Fab.Hamburg	A.Kirsten
✻	**Capella** Stl & I ScSr 1 Dk & Awn Dk3 B	Norweg'n Bergen	640 887	191·9	28·8	14·7 20·3	Bergen Martens,Olsen&Co.	1885	4 Blhds. CDB[illegible]	2Cy.32"&59"–36" 140 HP. Martens,Olsen&Co.,Bergen	Det Bergenske Dampskibsselskab
V	**Capella** Stl ScSr Boqvist 1 Dk & Spr Dk	Russian Helsingfors	787 1050	195·5	29·8	12·8	Lübeck H.Koch	1888	DB[illegible] MT [illegible] APT [illegible]	3C.18"28"&44"–27½"125 HP. H.Taucksch,Landsberg W.	Finska Angfartygs Actie-Bolaget

[illegible]=Lloyd's Register; D=Nederlandsche Vereeniging; G=Germanischer Lloyd; I=Registro Italiano; R=American Record;
[illegible]=Veritas Austro-Ungarico; H=Veritas Ellenico; N=Norske Veritas; V=Bureau Veritas.

Dans la 2me Colonne, B.T., avec les indications qui suivent, signifie la date du rapport de visite de la Carène par les experts du Board of Trade pour obtenir le certificat relatif au TRANSPORT DES PASSAGERS.

Dans la 3me Colonne, C=Composite (membrure en fer, bordé en bois); I=Fer; Stl=Acier; W=Bois.

Pour plus amples renseignements sur les navires classés par le Lloyd's Register, voir le Lloyd's Register Book et le Lloyd's Yacht Register.

No. in Register / Official No.	Registry in which classed	Ship's Name / Master / Ship's late Name	Material, Rig, &c. / No. of Decks, &c.	Flag / Port of Registry	Tons Net / Gross	Length	Breadth	Depth	Built: Where / By Whom	When	No. of Bulkheads / Particulars of Water Ballast	Engines: Particulars / Maker's Name	Owners
151	V	Capella *Brydé*	W ScBk 1 *Dk*	Swedish Gothenburg	262 372	137·4	24·5	16·3 Q 37 *f.*	Bremerhaven F.W.Wencke	1872		2Cy 20¾" & 35"–20½" 60 HP Z.U.Burse,Magdeburg	Aktie-Bolag (C.D.Pedersen)
152 87600		Capercailzie	I ScScY	British Glasgow	214 365	192·2	24·1	13·0	Glasgow Barclay,Curle&Co.	1883		2Cy 23" & 42"–30" 76 HP Barclay,Curle&Co.,Glasgow	John Burns
153		Capitan *Piaggio*	I Pad	Arg'ntine BuenosAyres	188 304	160·0	19·0	9·0	Glasgow Rowan&Co.	1869		100 HP Rowan&Co.,Glasgow	E.Piaggio & C
154 5181		Capitol	W Pad	American S.Francisco	1522 1989	277·3	17·0	10·9	SanFrancisco	1866		900 HP	E.H.Miller,jun
155		Cappy		French Nantes	90 164	..	..	..	Nantes	1887			
156 22	I	Caprera (ex Express) *Dagnino* 85-87	I ScS 2 *Dks*	Italian [illegible]	324 584	193·2	24·9	15·1	Glasgow J.&G.Thomson	1856	5 Bhds.	2Cy 25" & 50"–33" 200 HP Palmer&Co.,Newcastle	Navigazione Generale Italiana
157	*	Capri *Bohn* 87-87	I ScS 1 *Dk* 2 *B*	German [illegible]	763 1181	231·0 Q 120 *f.*	35·0 B 19 *f.*	14·7 F 26 *f.*	Sunderland J.Blumer&Co.	1881	3 Bhds. [illegible]	2Cy 28" & 54"–33" 110 HP N.E.MarineCo.,Sunderland	R.M.Sloman &
158		Capri (ex Faustin, ex Winifold)	I ScS	Italian Naples	122 191	..	..	..	Fiume	1873		35 HP	Societa Napoletana Navigazione, C
159 58783	*	Captain Cook *Childs*	I ScSr 1 *Dk*	British Dublin	153 247	141·0 Q 22 *f.*	22·2 F 21 *f.*	10·4 B 22 *f.*	Newcastle W.B.Hornby	1870	4 Bhds. DB 35 *ft*	2Cy 20" & 35"–22" 50 HP T.Richardson&Sons,Hartpl.	M.Murphy,jun
160 89000	*	Captain M'Clure *Dalton*	I Sc 1 *Dk & Web Frames*	British Dublin	304 503	169·9 Q 72 *f.*	23·6 B 23 *f.*	14·2 F 21 *f.*	PortGlasgow Murdoch&Murray	1885	4 Bhds. DB [illegible] FPT 15t APT 8	2Cy 24" & 45"–36" 90 HP J.&T.Young,Ayr	M.Murphy,jun
161 67806	*	Captain Parry *Smith*	I ScSr 1 *Dk* 2 *B*	British Dublin	318 477	170·5 Q 21 *f.*	23·7 F 22 *f.*	14·6	Stockton Richardson,Duck&Co	1877	4 Bhds. DB 35 *ft* [illegible]	2Cy 24" & 45"–26" 80 HP T.Richardson&Sons,Hartpl.	M.Murphy,jun
162		Capua	Stl ScSr	German Hamburg		285·0	37·0	23·0	Glasgow A.Stephen&Son	1889	DB *t*	3Cy 21", 33" & 54"–42" A.Stephen&Sons,Glasgow	R.M.Sloman &
163 87236	*	Capulet *Ellis* 85-85	I Sc 2 *Dks* 3 *B*	British Liverpool	1461 2246	300·0 P 24 *f.*	37·0 B 79 *f.*	23·1 F 89 *f.*	Middlesbro' R.Dixon&Co.	1884	5 Bhds. CDB 450	2Cy 36" & 67"–45" 300 HP T.Richardson&Sons,Hartpl.	"Capulet" S.S. Co. (C.T.Bowring &
164 70382		Capulet *Halfpenny*	I ScSr 1 *Dk*	British London	202 338	151·1	22·1 Q 30 *f.*	12·1	S.Shields Softley&Co.	1874	4 Bhds. DB 35 *ft*	2Cy 18½" & 36"–30" 35 HP J.P.Rennoldson,S.Shields	General Steam
165		Caqueta *Sermento*	I PadCr *Main,Saloon & Awn Dks*	Brazilian Para	 216	118·0	26·0	5·0	Wlmngtn.Del Pusey Jones&Co	1879	4 Bhds.	2Cy 14"–60" PuseyJones&Co.Wilmingtn	Cincinato F & Co.
166 65451	*	Caradoc *Coglin* 87-87	I ScSr 1 *Dk & Spr Dk* 3 *B*	British N.Shields	880 1368	240·4	30·8	17·6 214	Newcastle A.Leslie&Co.	1872	5 Bhds. DB [illegible]	2Cy 36" & 54"–33" 120 HP Thompson,Boyd&Co.,Newc.	H.E.P.Adams
167 87186		Caraibe (WRECKED) *Midlin*	I Sc3mSr	British London	431 699	212·9 P 104 *f.*	27·8 F 28 *f.*	16·1	Dumbarton A.Denny	1856		2Cy 40"–42" 99 HP A.Denny&Co.,Dumbarton	D.H.Morgan
168		Carangola *Gonçalves*	W	Brazilian JoaodaBrra.	284 486	..	..	..	P.d'ArêaCorte	1885	Y6.85		Cia. Cabotagem Joao da Barra
169		Caravellas		Brazilian Bahia	320	..	..	..				100 HP	Cia. de Nava Bahia
170	V	Caravelle *Lemonnier*	I ScBkn	French Bordeaux	355 729	200·0	29·0	17·5	London SamudaBros.	1855		2Cy 33" & 57"–28" 200 HP Forges&Chantiers,Havre	F.Dupuis-Re
171 83255	*	Carbis Bay *Tregarthen* 81-84	I ScSr 1 *Dk* 2 *B*	British Penzance	1002 1538	265·3 B 54 *f.*	36·2 F 28 *f.*	19·4	Newcastle Schlesngr,Davis&Co	1882	5 Bhds. [illegible]	2Cy 32" & 62"–42" 200 HP R.&W.Hawthorn,Nwcastle	"Carbis Bay" S.S. Co. Ld.(F.Bazeley
172 26865		Carbon *Atchinson* 85-85	I Sc3mSr	British Newcastle	459 714	194·7	27·6 P 67 *f.*	19·1	Liverpool J.Clayton	1855		2Cy 38"–39" 90 HP R.&W.Hawthorn,Nwcastle	W.& G.Hedley
173 87133	*	Cardiganshire *Clark* 87-87	I ScBk 2 *Dks* 3 *B*	British London	1623 2488	317·3 P 28 *f.*	38·1 B 31 *f.*	24·5 F 33 *f.*	Newcastle Swan&Hunter	1883	5 Bhds. DB 602	2Cy 38" & 70"–45" 275 HP R.&W.Hawthorn,Nwcastle	Jenkins & Co
174		Cargador	I TwnSc3mSr	Spanish Cuba	112 204	100·1	27·7	8·4	PortGlasgow Cunliffe&Dunlop	1873	4 Bhds.	4Cy 12" & 18"–15" 40 HP R.Steele&Co.,Greenock	Cuba Sugar T
175 63965	*	Cargan *Ross* 85-85	I ScSr 1 *Dk*	British Belfast	96 167	123·1 Q 14 *f.*	19·2 B 38 *f.*	9·3	Paisley Fullerton&Co.	1873	3 Bhds. FPT 15t	2Cy 15" & 27½"–22" 32 HP J.&T.Young,Ayr	Aberdeen,Leith ray Firth Steam Sh
176 84642	*	Carib *Sharpe* 88-88	I ScBn 1 *Dk* 2 *B*	British Stockton	912 1437	245·3 Q 28 *f.* D 38 *f.*	34·0 T 36 *f.*	16·4 F 30 *f.*	Stockton Richdsn,Duck&Co	1882	4 Bhds. DB 323t	2Cy 39" & 55"–33" 99 HP T.Richardson&Sons,Hartpl.	Anderson,A & Co
177 54911	BT 1/88	Caribbean *Daniel* 84-84	I ScBn	British Liverpool	1186 1853	276·0	34·4	29·6	W.Hartlepool Pile,Spence&Co.	1866		4Cy 21" & 48"–46" 350 HP G.Forrester&Co.,Liverpool	West India & Steam Ship Co.
178	*	Caridad	I ScSc 2 *Dks*	Venezul'n	413 832	233·3 F 37 *f.*	31·0 F 69 *f.*	14·8	Dumbarton McMillan&Sons	1883	4 Bhds.	2Cy 38" & 72"–48" 260 HP D.Rowan,Glasgow	Venezuelan
179 0	I	Cariddi *Cassovich* 86-87	I ScBn 2 *Dks*	Italian Palermo	773 1228	269·0 P & F 127 *f.*	30·2 Len. 78	18·4	Greenock MacNab&Co.	1866		2Cy 21" & 52"–36" 338 HP MacNab&Co.,Greenock	Navigazione Generale Italiana
180 85931	* D	Carina *Blackburn*	I ScBn 1 *Dk* 2 *B*	British London	1089 1687	257·5 Q 68 *f.*	34·5 B 64 *f.*	18·1 F 30 *f.*	Stockton M.Pearse&Co.	1882	4 Bhds. [illegible]	2Cy 32" & 60"–39" 160 HP Blair&Co.Ld.,Stockton	H.T.Cleake,jun

***=Lloyd's Register;** D=Nederlandsche Vereeniging; G=Germanischer Lloyd; I=Registro Italiano; R=American ...
A=Veritas Austro-Ungarica; H=Veritas Ellenico; N=Norske Veritas; V=Bureau Ver...

In the 2nd Column, B.T., with figures following, signifies Date of report of bottom sighting by Board of Trade Surveyors as condition of PASSENGER CERTIF...

In the 3rd Column, C=Composite (iron frame, wood planked); I=Iron; Stl=Steel; W=Wood.

Further particulars of Vessels classed by Lloyd's Register will be found in Lloyd's Register Book and Lloyd's Yacht Register

	Nom, Matériaux, Gréement, etc. Capitaine et dates d'entrée au service des Armateurs actuels et à bord du navire. Ancien Nom, nombre de ponts, etc.	Pavillon. Port d'Armement	Tonnage Net / Total	Dimensions: Longueur (constructions et Ponts)	Largeur	Creux	Construit. Où. Par qui.	Quand	Nombre des cloisons. Détails sur le test d'eau.	Machines. Détails. Nom du Constructeur.	Armateurs ou Directeurs-Gérants.
✻D	Carisbrook 1 ScBr Kibblewhite 81-81 1 Dk2 B	British Whitby	1108 1724	256·8 Q100f. B63f. F23f.	34·6	19·4	W. Hartlepool W.Gray&Co.	1881	4 Blkds. DBaf581t	2Cy.32"&60"–39" 150 HP. Blair&Co.Ld.,Stockton	J. Foster
✻	Carisbrooke 1 ScSr Grant 1 Dk & Awn Dk	British HongKong	973 1514	250·2	32·0	17·7	Sunderland Davison&Stokoe	1873	4 Blkds.	2Cy.29½"&57"–30 140 HP. OuseburnEng.W.C.N'castle	E.C.Ray
	Carisbrooke 1 Pad Skorl	British Southamptn	95 198	165·7	20·1	8·0	Glasgow	1876		2Cy.23"&46"–42" 70 HP. Barclay,Curle &Co.,Glasgw	Southampton, I. of W., & S. of Eng.S.Pkt.Co.,Ld.
G	Carl 1 ScSr Neumann 70-83 1 Dk	German Kiel	152 196	96·5	21·0	12·2	Kiel	1883		2Cy.18½"&21"–16" 25 HP. Gebr.Howaldt,Kiel	Sartori & Berger
N	Carl W ScSr 1 Dk	Norweg'n Stavanger	163 227	116·5 Q34f. B39f.	24·7	10·2	Drammen Jürgensen&Knudsen	1871		2Cy.15"&29"–24½" 35 HP. Stavanger Stoberi & Dok	M.Gabriel Monsen
✻	Carl Hirschberg StlScSr Mohlese 88-88 1 Dk W&b Frames	German Hamburg	828 1310	232·0 Q53f. B93f. P28f. F30f.	35·1	15·0	Sunderland S.P.Austin&Son	1888	4 Blkds.	3C.19"31"&51"–33" 130 HP. NEMarineEngCo.Ld.,Sld.	C.Hirschberg
✻N	Carl Konow 1 ScSr Sahes 84-84 1 Dk2 B	Norweg'n Bergen	379 540	169·7 Q41f. B51f. F31f.	25·2	13·0	Bergen BergensMekVerks	1883	5 Blkds. APT15t MTest FPT13t	2Cy.20½"&39"–27" 60 HP. BergensMekVerksBergen	Rasmus F.Olsen
✻	Carl Rahtkens 1 ScSr Spiegelberg 85-85 1 Dk2 B	British Middlesbro	1126 1730	259·0 Q105f. B64f. F28f.	37·0	17·8	S.Shields J.Readhead&Co.	1885	4 Blkds. DBaf405t	2Cy.30"&60"–36" 150 HP. J.Readhead&Co.,S.Shields	F.Rahtkens & Co.
V	Carl Woermann 1 ScSr Türck 88-88 2 Dks	German Hamburg	1467 1941	291·0 F39ft. B31t.	36·4	21·9	Hamburg ReiherstiegSchiffW erfte	1881		2Cy.34"&62"–42" 190 HP. Reiherstg.SchiffsWfteHbg.	Afrikanische Dampfschiffs Act.Gesell. (Woermann Linie)
✻	Carlisle 1 ScSr Ratt 2 Dks3 B	British London	1397 2141	286·3 P30f. B60f. F33f.	37·2	24·2	Newcastle Palmers'Co.	1882	5 Blkds. DBaf&cb270t APT30t	2Cy.36"&68"–45" 220 HP. Palmers'Co.,Ld.,Newcastle	S.S. "Carlisle" Co.,Ld. (BlundellBros.&Co.)
✻	Carlo 1 ScSr Gern 85-85 1 Dk2 B	British Cardiff	786 1233	239·5 Q73f. B38f. F27f.	34·0	15·9	Glasgow A.Stephen&Sons	1880	4 Blkds.	2Cy.28"&52"–36" 130 HP. A.Stephen&Sons,Glasgow	Tellefsen, Wills & Co.
✻I	Carlo R. 1 ScSr Testori -86 2 Dks3 B	Italian Genoa	1759 2674	310·5 P28f. B62f. F32f.	39·1	25·1	Newcastle SirW.G.Armstrong, Mitchell&Co.,Ld.	1886	5 Blkds. APT30t	3C.24"39"&64–42" 300 HP. WallsendSlpwyCo.,N'castle.	Ditta C.Raggio
V	Carlo Risso 1 Sc Molinari 1 Dk	Italian Genoa	136 215	126·9 P15f. B12f. F23f.	23·0	8·8	Sampierd'rena FlliRoncallo	1885		4Cy.11"&21"–18" 68 HP. FlliRoncalloSampierdarena	G.Risso
DGV	Carlos 1 ScSr Plath 83-84 2 Dks & Spr Dk	German Danzig	763 961	201·1 B6f.	28·2	15·0 24·5	Kiel F.A.Egells&Co.	1881	DT790t DT25t	2C2½"&12½"–27½" 90 HP. F.A.Egells&Co.,Berlin	F.G.Reinhold
	Carlos 1	Spanish Seville	81 106	..	..	..	Chepstow	1880		30 HP.	Don Carlos Cañal
	Carlos Pacheco W ScSr Nelson 1 Dk	American S.Francisco	104 172	100·0	23·3	11·4	San Francisco A.Hay	1886		2Cy.11"&22"–18" Hinckley,Spiers&Hays	T.P.H.Whitelaw
	Carlos Roberto 1 ScSr (ex Don Mariano, ex Alberto) 1 Dk2 B	Chilian Valparaiso	464 644	191·2 Q68f. B51f. F28f.	26·1	18·0	Middlesbro' Backhouse&Dixon	1872	4 Blkds.	2Cy.24"&45–30" 81 HP. Black,Hawthr'n&Co.Newcl.	Cia.Esplotadora de Lota y Coronel
N	Carlotta 1 ScSr Herzberg 88-88 1 Dk	Norweg'n Bergen	475 647	190·3 Q100f. B14f. F23f.	27·4	13·6	Sunderland J.Blumer&Co.	1873	4 Blkds. DBaf118t	2Cy.25"&44"–30" 86 HP. Fowler&McCollin,Hull	P.G.Halvorsen
✻D	Carlton 1 ScSr Storey 87-88 1 Dk Spl Awn Dk3 B	British Newcastle	1322 2011	250·0 Q95f. F39f.	37·8	19·0	S.Shields J.Readhead&Co.	1885	4 Blkds. DBaf457t FPT80t APT18t MT10t	2Cy.32"&60"–36" 160 HP. J.Readhead&Co.,S.Shields	Chapman & Miller
✻	Carmarthenshire StlSc3mSr Dreyer 87-87 2 Dks3 B	British London	1776 2726	329·5 P49f. B80f. F45f.	40·0	25·1	Newcastle C.S.Swan&Hunter	1887	7 Blkds. DT/5136 DT&356t	3C.27"43"&70"–45" 500 HP. R.&W.Hawthorn,Leslie&C.Nwc	Jenkins & Co.
✻	Carmen 1 ScSrYt	British Southamptn	67 112	116·3	20·1	11·5	PortGlasgow J.Reid&Co.	1884	3 Blkds.	2Cy.14"&28"–21" 35 HP. Walker,Henderson&Co.Gls.	Sir Richard Williams Bulkeley, Bart.
✻	Carn Brea 1 ScSr Jenkin 70- 1 Dk2 B	British Falmouth	1167 1805	265·0 Q70f. P28f. B62f. F25f.	36·0	19·0	Newcastle TyneIronS.B.Co.Ld.	1882	5 Blkds. DBaf&cb441t FPT13t APT	2Cy.33"&63"–39" 180 HP. N.E.MarineE.Co.Ld.,Sundld	E.Handcock & Co.
✻D	Carn Marth 1 ScSr Harvell 82- 1 Dk2 B	British Falmouth	1174 1818	265·0 Q70f. P28f. B62f. F26f.	36·1	19·0	Newcastle TyneIronS.B.Co.Ld.	1884	5 Blkds. DBaf&cb440t FPT13t APT12t	2Cy.32"&62"–42" 162 HP. N.E.MarineE.Co.,Ld.,Nwc.	E.Handcock & Co.
✻	Carnarvon 1 ScTg Ebert 1 Dk	British Southamptn	47 104	90·1 F10f.	17·1	8·0	Northam Day,Summers&Co.	1884	1 Blkds. FPT1t	2Cy.16"&30"–18" 40 HP. Day,Summers&CoSouthptn	Union S.S.Co.
	Carnsew 1 Sc3mSr Clemo -88	British Penzance	212 340	141·6 Q33f. B9f.	23·1	12·0	Hayle Harvey&Co.,Ld.	1888		2Cy.19"&31½"–21" 51 HP. Harvey&Co.,Ld.,Hayle	Harvey & Co., Ld.
✻	Carolina 1 ScSr Aldamiz 2 Dks3 B	Spanish Bilbao	1355 2099	331·9	34·1	25·0	Glasgow Lon.&Gls.I.S.B.&EngCo.	1876	5 Blkds.	2Cy.38"&71"–48" 300 HP. Lon.&GlsI.S.B.&EngCo.Gls	Linea de Vapores Serra
	Carolina 1 ScSr 1 Dk	Spanish Carthagena	145 234	135·0 Q48f.	20·0	9·5	Hartlepool E.Withy&Co.	1879	1 Blkds. DT/	2Cy.20"&35"–21" 50 HP. T.Richardson&Sons,Hartpl.	Sociedad Anon. Mtma. y Mercantil de la Costa de Levante
✻	Carolina P. (ex Liguria, ex Heathercliff) 1 ScSr Schiaffino 1 Dk	Italian Genoa	322 509	162·7 Q59f. B40f. F24f.	25·7	13·1	PortGlasgow W.Hamilton&Co.	1883	4 Blkds. DBaf&f115t	2Cy.21½"&43"–29½" 150 HP J.Gilmour&Co.,Glasgow	G.Pastorino fu P.
✻	Caroline 1 ScSr Miller 85-87 2 Dks	British Grangemth.	308 618	192·8 P27f. B48f. F25f.	28·1	16·1	Liverpool T.Royden&Sons	1878	5 Blkds.	2Cy.29"&58"–36" 150 HP. LairdBros.,Birkenhead	The Carron Co.
	Caroline 1 Sc3mSr Berry 1 Dk	British London	251 478	153·8 F27f.	28·2	15·1	London J.ScottRussell&Co.	1853	3 Blkds.	2Cy.39"–31" 70 HP. T.Hosking,London	W.H.Sollas

Lloyd's Register; D = Nederlandsche Vereeniging; G = Germanischer Lloyd; I = Registro Italiano; R = American Record;
Veritas Austro-Ungarico; H = Veritas Ellenico; N = Norske Veritas; V = Bureau Veritas.

Dans la 2me Colonne, B.T., avec les indications qui suivent, signifie la date du rapport de visite de la Carène par les experts du Board of Trade pour obtenir le certificat relatif au TRANSPORT DES PASSAGERS.

Dans la 3me Colonne, C = Composite (membrure en fer, bordé en bois); I = Fer; Stl = Acier; W = Bois.

Pour plus amples renseignements sur les navires classés par le Lloyd's Register, voir le Lloyd's Register Book et le Lloyd's Yacht Register.

No. in Book. Official No.	Registry in which classed	Ship's Name. Material, Rig, &c. Master, and Dates of Appointment to present Owner's Service and to Ship Ship's Late Name. No. of Decks, &c.	Flag Port of Registry.	Tons. Net. Gross.	Dimensions (Length, Breadth, Depth) Deck Erections.	Built. Where By Whom.	When	No. of Bulkheads Particulars of Water Ballast	Engines. Particulars. Maker's Name.	Owners
211 71503		Caroline I ScSr 1 Dk	British Singapore	141 228	135·0|24·0|9·0	Nantes			2Cy.15"&18"–36" 36 HP. Brevettes,Nantes	Ebrahim I.B...
212	✻R	Caroline Miller I ScSr Miller (ex Dawn) 1 Dk	American [illegible]	416 681	185·4|29·1|12·9 Q22f. B26f. F19f.	Grangemouth Dobson&Charles	1883	4 Blkds. [illegible]	2Cy.25"&48"–30" 90 HP. Ross&Duncan,Govan	Georgetown &... tern Railw...
213 81630	✻	Caroline Robert de Massy StlScSr Barclay –88 1 Dk	British Stockton	1130 1742	254·3|37·0|17·9 [illegible] F27f.	Middlesbro' R.Craggs&Sons	1888	4 Blkds. [illegible]	[illegible] [140 HP	R.Craggs & ...
214 125106		Carondelet W ScSr Connor –88 3 Dks	American New York	1044 1598	241·0|35·5|21·0	MysticR.Conn Mallow	1873		1Cy. 300 HP. [illegible] Delamater&Co.N.York	C.H.Mallory &...
215		Carpio I Sc3mSr Aspe 84 86–86	Spanish [illegible]	445 655	185·3|27·0|16·0	Liverpool Bowdler,Chaffer&Co	1870		2Cy.22"&44"–24" 70 HP. J.Taylor&Co.,Birkenhead	M.Saenz & Co...
216	✻D	Carranza I Sc Laguardia 83–88 1 Dk2 B	Spanish Bilbao	1034 1583	255·0|36·5|18·0 Q91f. B64f. F28f.	Sunderland J.Priestman&Co.	1883	4 Blkds. [illegible]	2Cy.32"&60"–36" 140 HP. J.W.&P.Wilson&Co.,Sndld	Sociedad M... Vizcaya
217 88005	✻	Carrie StlScSr Holgaard 73–88 1 Dk&WhFrms	British W.Hartlepl.	1163 1789	260·0|36·7|18·2 F27f. Q107f. B[illegible]f.	W. Hartlepool W.Gray&Co	1888	4 Blkds. [illegible]	3Cy.19",30½"&51"–36" C. [illegible] Marine Eng.Co. W. Hartlepool [150HP	C. Nielsen & ...
218 22020		Carrier I Pad Walker	British [illegible]	131 243	124·1|24·7|8·5	Greenock Scott&Co.	1858		2Cy.42"–36" 120 HP. GreenockFoundryCo.,Grnk	London Bri... Coast Railw...
219 4071		Carroll (ex Trenton) W ScBk Bratsen 3 Dks	American [illegible], Mass	822 1372	206·7|35·3|20·3	New York	1857		2Cy.44"–36" 497 HP. D.Esler&Co.,Brooklyn	H. Loring
220 83156	✻	Carron StlScSr Bain 74–87 2 Dks	British Grangemouth	300 636	195·0|29·0|15·7 P21f. B61f. F25f.	Liverpool T.Royden&Sons	1883	[illegible]	2Cy.31"&60"–36" 170 HP. J.Jack&Co.,Liverpool	The Carron Co...
221 28670		Carron I Sc3mSr McDonald 2 Dks	British [illegible]	214 [illegible]	181·0|25·7|14·5 F14f.	Glasgow Barclay,Curle&Co.	1860	[illegible]	2Cy.42"–36" 85 HP. Barclay,Curle&Co.,Glasgow	P. Barr
222 84304	✻	Carron Park I ScSr Pull 80–82 1 Dk	British Glasgow	257 406	155·0|23·1|11·9 Q73f. F21f. B21f.	Campbeltown Campbeltown S.B.Co	1881	[illegible]	2Cy.18"&36"–24" 54 HP. Kincaid,Donald&Co.,Grnk.	Cathcart S.S... (J.&J.Denholm)
223 81574	✻	Cartagena I Sc3mSr Searle 85–86 2 Dks3 B	British London	1533 [illegible]	301·0|37·2|25·5 P29f. B25f. F32f.	Sunderland Bartram,Haswell&Co	1879	[illegible]	2Cy.36"&68"–42" 250 HP. J.Dickinson,Sunderland	W. McMurray (Marley&...)
224 63553	✻	Cartago Nova I Sc3mSr Brown 80–84 2 Dks3 B	British London	1293 [illegible]	265·6|33·2|24·3	Newcastle Palmers'Co.	1870	[illegible]	2Cy.30"&54"–33" 120 HP. Palmers'Co.,Newcastle	Hall Bros.
225 93242	✻	Cartdale I ScSr M. Miller 1 Dk	British Glasgow	18 115	110·2|18·0|9·5 Q51f. F19f.	Paisley J.Fullerton&Co.	1886	[illegible]	2Cy.18"&36"–24" 51 HP. W.King&Co.,Glasgow	McKinney & ...
226 81825	BT11.88	Carthage I Sc4mBk Steed 71–87 3 Dks4 B	British Greenock	2454 [illegible]	430·1|44·4|33·5 P49f. F68f.	Greenock Caird&Co.	1881	8 Blkds.	1Cy.44"&82"–60" 1000 HP. Caird&Co.,Greenock	Peninsular&O... Steam Nav. Co.
227 85080	BT1.88	Carthaginian StlSc3mSr M'Nicol 3 Dks	British Glasgow	2755 [illegible]	386·4|45·2|21·6 B118f. F52f. 29·1	Glasgow Govan Shipbg.Co.	1884	[illegible]	2Cy.51"&88"–54" 520 HP. J.&J.Thomson,Glasgow	J.&A. Allan
228 84086	✻	Cascapedia I ScSr Fraser 81–81 1 Dk&Spr Dk3 B	British Liverpool	1924 [illegible]	322·5|40·0|21·8 B23f. F18f. 29·7	Sunderland Sld.S.B.Co.Ld.	1881	4 Blkds. [illegible]	2Cy.39"&73"–45" 300 HP. N.E.Marine [illegible] Ld.Sundld.	S.S."Cascapedia"... (Crow,Rudolf&...)
229 86301	✻	Casino I Sc3mSr Boyd 1 Dk	British [illegible]	274 [illegible]	160·4|24·1|10·2 Q61f. B40f. F35f.	Dundee Gourlay Bros&Co.	1882	[illegible]	2Cy.21"&36"–24" 65 HP. Gourlay Bros&Co.,Dundee	Belfast & K... Nav. Co. Ld.
230 78830		Casma StlSc3mSr 2 Dks&Shade Dk	British Liverpool	358 592	180·4|30·1|14·8 H32f.	Birkenhead Laird Bros.	1878	4 Blkds.	2Cy.25"&48"–36" 95 HP. Laird Bros.,Birkenhead	Pacific Steam...
231 126518		Casper W ScSr Anfindsen 1 Dk	American S.Francisco	234 400	132·5|33·0|11·0	San Francisco Hansen&Fraser	1888		[illegible].12"&20"–24" 125 HP. [illegible] Foundry San Francisco	Casper Lumb...
232 63758	BT1.88	Caspian I ScBk McDougall –87	British Glasgow	1718 2728	349·6|38·0|31·8 P136f. F84f.	Glasgow Lond.&Glasg.Co.	1870	[illegible] Blkds.	2Cy.46"&86"–48" 300 HP. Laird Bros.,Birkenhead	J. & A. Allan
233	✻	Cass StlScSr Pybus –88 2 Dks	Chinese Formosa	705 [illegible]	250·5|34·2|18·0 F46f.	Newcastle R.&W.Hawthorn, Leslie&Co.Ld	1888	[illegible] Blkds.	3C.19"30"&50"–33 300 HP. R.&W.Hawthorn,Leslie&Co., Ld.,Nwc.	Formosa T... poration
234	V	Cassandra I ScSr Thomsen 86–88 2 Dks3 B	German Hamburg	1097 1474	248·6|31·3|16·5 24·0	Glasgow A.Stephen&Sons	1873		2Cy.32"&51"–36" 140 HP. D.Rowan,Glasgow	Deutsche D... schiffs-Rhed...
235 80360	✻	Cassia I ScSr Kendall 84–85 1 Dk2 B	British Newport	749 1163	236·2|32·3|15·9 Q82f. B51f. F30f.	Port Glasgow Murdoch&Murray	1883	4 Blkds. [illegible]	2Cy.29"&57"–39" 99 HP. Kincaid&Co.,Greenock	Stephens, Ma... Goss
236	✻D	Cassius I ScSr Hecht 2 Dks	German Hamburg	1606 2449	300·0|37·2|25·3 P39f. B68f. F30f.	Gaarden NorddeutscheWerft	1881	5 Blkds. [illegible]	2Cy.38½"&66"–42" 210 HP. Märkische M.C.,Berlin	C. Andersen
237 87457	✻	Castanos I ScSr Le Feuvre –88 1 Dk2 B	British Cardiff	1113 1709	258·9|36·0|18·0 Q87f. B64f. F26f.	Southampton Oswald,Mordaunt&Co.	1882	4 Blkds. [illegible]	2Cy.30"&60"–42" 170 HP. Oswald,Mordaunt&Co.,Sou.	"Castanos" S... (Morel Bros.&Co)
238	V	Castellano (ex Honduras, ex Legislator) I ScBg Antonis 86–88 2 Dks	Spanish Bilbao	1375 2126	310·5|34·1|24·4 P112f. F45f.	London Maudslay,Sons&Field	1872	[illegible] Blkds.	2Cy.42"&72"–48" 280 HP. Maudslay,Sons&Field,London	M.M.de Arr...
239		Castellano I ScSr Rengoechea –83 1 Dk	Spanish Manila	178 287	139·8|23·2|9·6 Q47f. F31f.	Dumbarton R.Chambers jun.	1880	[illegible] Blkds.	2Cy.20"&36"–27" 60 HP. M.Paul&Co.,Dumbarton	J. Reyes
240										

✻ = **Lloyd's Register**; D = Nederlandsche Vereeniging; G = Germanischer Lloyd; I = Registro Italiano; R = American...
A = Veritas Austro-Ungarico; H = Veritas Ellenico; N = Norske Veritas; V = Bureau V...

In the 2nd Column, B.T., with figures following, signifies Date of report of bottom sighting by Board of Trade Surveyors as condition of PASSENGER ...

In the 3rd Column, C = Composite (iron frame, wood planked); I = Iron; Stl = Steel; W = Wood.

Further particulars of Vessels classed by Lloyd's Register will be found in Lloyd's Register Book and Lloyd's Yacht Register.

	Nom, Matériaux, Gréement, etc. — Capitaine et dates d'entrée au service des Armateurs actuels et à bord du navire. — Ancien Nom, nombre de ponts, etc.	Pavillon. — Port d'Armement	Tonnage. Net. — Total	Dimensions. Longueur	Largeur	Creux	Construit. Où. — Par qui.	Quand.	Nombre des cloisons. — Détails sur le test d'eau.	Machines. Détails. Nom du Constructeur.	Armateurs ou Directeurs-Gérants.
✻	Casterton StlScSr Holbeck 81-87 1 Dk & W. & Frms	British W. Hartlepool	1540 2360	289·6 Q62f. P29f. B129f. F35f	39·1	19·0	W. Hartlepool Irvine&Co.	1887	5 Bhds. CDB30t	3Cy.21½"35½"58½"–39" 200 [HP. Blair&Co.,Ld.,Stk.	R. Irvine & Co.
	Castilian 1 ScSr Mills 87-87	British Liverpool	398 635	240·0 Q 23 f. F 13 f.	24·0	15·5	Belfast Harland&Wolff	1862		2Cy.24"&46"–36" 89 HP. J.Jones&Sons, Liverpool	S.S. "Castilian" Co., Ld. (F. Leyland & Co.)
	Castilla ex Peruvian 1 ScBk ex Westminster, ex Southerner 3 Dks Bengal	Spanish Barcelona	1713 2245	281·1	38·1	20·2 F 29 f	Stockton M.Pearse&Co.	1863		2Cy.60"–33" 500 HP. Fossick&Hackworth,Stcktn	Cia. Catalana de Vapores Trasatlantica
✻	Castle Eden 1 ScSr Bay 86-88 1 Dk & Spr Dk 3 B	British Liverpool	1555 2354	300·2	37·2	26·8	Newcastle Wghm.Richardson &Co.	1882	6 Bhds. DBa&c4270t FPT38t APT22t	2Cy.35"&70"–48" 300 HP. Wghm.Richardson&C.,Nwe	"Castle Eden" S.S.Co., Ld.(J.Little & Co.)
✻	Castle Rising 1 ScSr Dunbar 1 Dk 2 B	British King'sLynn	797 1233	230·2 Q93f.	32·0 B59f.	15·4 F78f.	Stockton R.Craggs&Sons	1884	4 Bhds. DBaf271t APT23t	2Cy.26"&48"–36" 99 HP. Westgrth,English&C.Mdlbro	"Castle Rising" S.S. Co., Ld.(W.F.Beaumont.)
✻	Castledale 1 ScSr Weatherhead 1 Dk & Spr Dk 3 B 86-88	British Liverpool	1532 2358	300·4	37·2	19·2 F 26·6	Newcastle Wghm.Richardson &Co.	1882	6 Bhds. DB189t FPT28t APT22t	2Cy.38"&72"–48" 330 HP. Wghm.Richardson&C.,Nwe	"Castledale" S.S.Co. Ld.(J. Little & Co.)
✻	Castlegate 1 ScSr Morgan 2 Dks 3 B	British Liverpool	1146 1783	265·0 P34f.	36·1 B68f.	23·0 F33f.	Blyth Hodgson&Soulsby	1882	5 Bhds. DBa&c4323t	2Cy.33"&62"–39" 200 HP. Blair&Co.,Ld.,Stockton	"Castlegate" S.S. Co., Ld.(R.H.Wesencraft&Co.)
✻	Castlehill 1 ScSr Cumming 86-88 2 Dks 3 B	British Liverpool	1152 1783	265·0 P34f.	36·1 B68f.	23·0 F29f.	Blyth Hodgson&Soulsby	1882	5 Bhds. DBa&c4323t	2Cy.32"&62"–42" 200 HP. Black Hawthorn&Co.,Gtshd.	"Castlehill" S.S.Co., Ld. (R.H.Wesencraft&Co.)
✻D	Castlemaine 1 ScSr Nunn 77-89 1 Dk 2 B	British London	729 1111	230·5 Q 80 f.	31·2 B 34 f.	17·0	Newcastle Palmers'Co.	1877	1 Bhds. DBaf236t	2Cy.27"&51"–33" 98 HP. Palmers'Co.,Ld.,Newcastle	J. & E. Wilson
	Castor 1 Pad Simnett 2 Dks	British Glasgow	231 282	123·7	21·0	7·3	Nantes Ateliers&C.de la Loire	1883		2C.16"&27½"–30¾" 100 HP. Ateliers&C.dela Loire,N'tes	La Platense Flotilla Co., Ld.
DV	Castor 1 Sc3mSr Visser 77-80 3 Dks & Spr Dk	Dutch Amsterdam	1073 1520	253·8 F 40 f.	33·5 H 25 f.	25·5	Glasgow A.&J.Inglis	1870		2Cy.37"&64"–36" 180 HP. A.&J.Inglis,Glasgow	Kon. Nederlandsche Stoomb-Maats.
V	Castor ex Rügenwalde 1 ScSr Torkildsen -87 1 Dk	Norweg'n Bergen	338 524	169·4 Q 50 f.	25·0 B 47 f.	14·2 F 22 f.	Sunderland Short Bros.	1875	1 Bhds. Daf 8t	2Cy.22"&41"–28" 90 HP. T.Richardson&Sons,Htlepl.	Det Bergenske Damp- skibs-Selskab
V	Castor 1 ScSr Mattsson 1 Dk	Swedish Trelleborg	441 568	166·3 Q 53 f.	24·8 B 38 f.	15·0 H 5 f.	Gothenburg MotalaCo.	1874	MT140t	2Cy.15½"&33"–27" 70 HP. MotalaCo.,Gothenburg	Trelleborgs Nya Ångf. Akt.Bol.(F.Malmros)
	Castore 1 ScSr Moharsich 81-86 2 Dks	Austrian Trieste	1235 1861	272·4	35·0	25·4	Trieste Austro-UngaricoLloyd	1874		2Cy.41"&70"–36" 200 HP. ArsenaleLloyd,Trieste	Lloyd Austro-Unga- rico
	Castro ex Jose Fani 1 TwnScSr Uribea 83-83 (ex Principe de Espana)	Spanish Bilbao	95 173	120·6	17·9	8·9	Newcastle Jameson	1865		2Cy.15"–18" 42 HP. Jameson,Newcastle	Pedro Acha
✻	Catalan ex Zante 1 ScSr Guerricaechevarria 87-89 2 Dks 3 B	Spanish Bilbao	1022 1574	272·6	33·1 B 137 f.	22·9	Liverpool Bowdler,Chaffer & Co.	1875	3 Bhds.	2Cy.32"&64"–36" 200 HP. J.Jones&Sons,Liverpool	MM. de Arrotegui
	Catalonia 1 ScBk Atkins 82-88	British Liverpool	3093 4841	429·6	43·0 F 7 f.	33·8	Glasgow J.&G.Thomson	1881		2Cy.51"&88"–69" 600 HP. J.&G.Thomson,Glasgow	Cunard Steam Ship Co., Ld.
✻	Cataluña 1 ScBn Jauregnizar 3 Dks	Spanish Cadiz	1989 3488	384·2 P101f.	42·2 B48f.	27·5 F62f.	Dumbarton W.Denny&Bros.	1883	6 Bhds. CDB378t	2Cy.57"&102"–60" 750 HP. Denny&Co.,Dumbarton	Cia. Trasatlantica
✻	Cataluña 1 ScSr 2 Dks	Spanish Palma	678 1071	221·6 P36f.	31·1 B52f.	18·5 F29f.	Blyth Hodgson&Soulsby	1880	4 Bhds. FPT58t APT53t	2Cy.29½"&55"–36" 150 HP. BlackHawthorn&C.Newcsl.	La Harinera Mal- lorquina
✻	Catania 1 ScSr Franck 85-84 2 Dks 3 B	German Hamburg	1429 2198	315·1 P40f.	35·7 B37f.	22·6 F41f.	Glasgow A.Stephen&Sons	1881	4 Bhds. CDB152t	2Cy.35"&65"–42" 200 HP. A.Stephen&Sons,Glasgow	R. M. Sloman & Co.
✻	Catch Me ScK Wathne 88-88 1 Dk	Danish Seydisfjord	18 54	69·6	17·1	8·6	Glasgow R.Moir	1886		2Cy.12"&24"–15 25 HP. A.Hutcheson,Glasgow	O. Wathne
V	Catharina I & StlSc Deroo	Belgian Ostend	60 120	93·5	20·0	9·9	Ombret Jabon	1885		2Cy.16"&30"–20" 36 HP. C.Beer,Jemeppe	Soc. Anon. des Pêcheries Ostendaises
	Catharine Whit- W Sc3mSr ing Crowell 88-88 2 Dks 3 B	American Philadelphia	565 881	186·0	31·2	20·5	Malden, N.Y.	1864		1Cy.40"–36" 350 HP. G.Starbuck, Troy, N.Y.	Boston & Philadelphia S.S.Co.
	Cathay 1 Sc3mSr Nautes 83-89	British Glasgow	1873 2980	360·8	39·3	29·4	Dumbarton W.Denny&Bros.	1872		2Cy.50"&86"–54" 500 HP. Denny&Co.,Dumbarton	Peninsular & Oriental Steam Nav.Co.
✻	Cathay 1 ScSr Turner 88-88 1 Dk 2 B	British W.Hartlepool	1149 1773	257·4 Q66f. B66f. P38f. F62f.	34·7	19·7	W. Hartlepool W.Gray&Co.	1883	4 Bhds. DBaf333t	2Cy.32"&60"–39" 150 HP. Blair&Co.,Ld.,Stockton	Gladstone & Carnforth
✻	Catherina II. 1 ScSr Bottleman 82-84 2 Dks 3 B	Russian St.Petersb'g	811 1253	242·5	30·4 H 32 f.	22·5	Sunderland Davison&Stokoe	1874	5 Bhds. Wtlie	2Cy.29½"&56"–36" 130 HP. OuseburnEng.Wks.,Newcsl.	St. Petersburger DampfschifffahrtGesellschaft
✻	Catherine 1 Sc Challis 84-86 1 Dk	British Grimsby	60 113	89·1	19·5 Q 16 f.	10·3	Grimsby T.Charlton	1885	3 Bhds.	2Cy.14"&26"–18" 30 HP. T.Charlton,Grimsby	T. Charlton
V	Cathrine ex Sverige W ScSr Steen 87-87 1 Dk	Danish Copenhagen	261 366	138·7 Q 45 f.	24·9 B 40 f.	12·7	Kälboda H.Olsson	1873		2Cy.14½"&31½"–23½" 40 HP KockumsMek.WerkstadAct., Malmö	Dampskibs Selskabet "Cathrine" (E.Knox)
	Cato 1 ScSr King 1 Dk	British Hull	701 1076	245·4 B 165f.	29·5 F 13f.	13·2 Q 31f.	Glasgow Lond.&GlasgowCo.	1867	6 Bhds.	2Cy.27"&48"–26" 98 HP. Cmp.'76,Amos&Smith,Hull	T. Wilson, Sons & Co.
	Cato 1 Pad	British London	80 128	115·2	18·0	7·7	Liverpool P.Cato	1849		2Cy.30"–36" 56 HP. Fawcett,Preston&Co.,Livpl	London, Tilbury & Southend Rail.Co.

Lloyd's Register; D = Nederlandsche Vereeniging; G = Germanischer Lloyd; I = Registro Italiano; R = American Record;
Veritas Austro-Ungarico; H = Veritas Ellenico; N = Norske Veritas; V = Bureau Veritas.

Dans la 2me Colonne, B.T., avec les indications qui suivent, signifie la date du rapport de visite de la Carène par les experts du Board of Trade pour obtenir le certificat relatif au TRANSPORT DES PASSAGERS.

Dans la 3me Colonne, C = Composite (membrure en fer, bordé en bois); I = Fer; Stl = Acier; W = Bois.

Pour plus amples renseignements sur les navires classés par le Lloyd's Register, voir le Lloyd's Register Book et le Lloyd's Yacht Register.

No. in Book. / Official No.	Registry in which classed.	Ship's Name. Material, Rig, &c. / Master, and Dates of Appointment to present Owner's Service and to Ship. / Ship's Late Names. No. of Decks, &c.	Flag. / Port of Registry.	Tons. Net. / Gross.	Dimensions. Length	Breadth	Depth	Built. Where. / By Whom.	When.	No. of Bulkheads. / Particulars of Water Ballast.	Engines. Particulars. / Maker's Name.	Owners or Managers
271 82376	*	Catterthun I ScSr *Darke 85-88* *2 Dks 3 B*	British London	1406 2179	302·3	36·5	23·7	Sunderland W. Doxford & Sons	1881	7 Bhds.	2Cy.40" & 70"–48" 250 HP W.Doxford&Sons,Sunderld	East.& Austr[illegible] Ld.
272		Caucase Stl ScSr *Mir -88* *1 Dk & Spr Dk*	Belgian Antwerp	1321 1844	250·0	35·1	24·3 *F 20 f.*	Newcastle Sir W. G. Armstrong Mitchell & Co.	1888	DB[illegible]	3Cy.18½" 30" & 48"–33" 150 HP Wallsend Slipway & Eng.Co.,Ld Newcastle-on-Tyne	L.Lambert
273 76231	* D	Cavalier I ScSr *Borwick 83-78* *2 Dks 3 B*	British Newcastle	1197 1833	279·1	34·7	24·0	Sunderland W. Doxford & Sons	1878	[illegible] Bhds.	2Cy.27" & 58"–45" 170 HP R.Stephenson&Co.,Newcastle	W.Forster
274 87638	* BT 1.88	Cavalier I ScSr *McTavish 83-88* *2 Dks & pt Shd Dk*	British Glasgow	183 360	151·1	25·1	11·8	Glasgow [illegible]	1883	[illegible] Bhds.	2Cy.24" & 48"–48" 110 HP [illegible] Caldwell,Glasgow	D.McBrayne
275 67590		Cavalier I ScSr Yt (*ex Margherita*)	British [illegible]	183 312	187·3	23·7	12·3	Glasgow [illegible]	1873		2Cy.22" & 44"–27" 100 HP [illegible] Campbell&Son,Glasgow	Sir M.R.S.[illegible]
276 67794	BT 3.88	Cavan I PadSr *Taylor*	British Dublin	392 803	250·2	27·2	14·8 *F 87 f.*	Birkenhead Laird Bros.	1876		2Cy.46" & 73"–84" 300 HP Laird Bros. Birkenhead	City of Dubl[illegible] Pkt.Co.
277 89964	*	Cavendish I ScSr *Andrew 83-88* *1 Dk 2 B*	British Leith	1079 1656	248·3	34·8	20·8	Newcastle Schlesinger Davis&Co.	1883	[illegible] Bhds.	2Cy.29" & 56"–42" 150 HP N.E.Marine E.Co.Ld.Sndrld	R.& D.Cairn[illegible]
278 84129		Cavour I ScSr *Shurlock 81-81*	British Liverpool	396 632	209·0	32·2	12·6	Greenock Scott&Co.	1881		2Cy.29" & 50"–30" 100 HP Scott&Co.,Greenock	Livrpl.,Braz[illegible] PlateStmNav
279 87919	*	Caxton (*ex Test*) I ScBn *Pratt 86-88* *2 Dks 3 B*	British Liverpool	1794 2784	310·6	38·8	26·7 *F 36 f.*	Southampton Oswald Mordaunt&Co.	1883	[illegible] Bhds.	2Cy.37" & 74"–48" 300 HP Oswald,Mordaunt&Co.,Sthmtn	Livrpl.,Braz PlateStmNav
280	V	Cayor Stl ScSr *Baffin -88* *1 Dk & Awn Dk*	French Bordeaux	967 1493	255·0	36·2	21·6	Dumbarton A.McMillan&Son	1888	[illegible]	3Cy.20" 30" & 52"–36" HP A.McMillan&Son,Dunbrtn.	Devès & G.[illegible]
281	V	Ceara I ScSr *Goltsche 86-87* *2 Dks & Spr Dk*	German Hamburg	1471 1890	284·7	36·4	16·0 24·2	Newcastle C.Mitchell&Co.	1882		2Cy.34" & 66"–42" 200 HP Wallsend Slipway Co.,Nwcst	Hamburg-S[illegible] Dampfschiff[illegible]
282 63181	*	Cearense I ScBg *Heath 86-88* *3 Dks*	British Liverpool	890 1381	269·4	30·5	25·1	Liverpool T.Royden&Sons	1869	6 Bhds.	2Cy.32" & 64"–30" 136 HP Fawcett,Preston&Co.,Livrpl	R.Singlehurs[illegible]
283	*	Cebah Stl TwnScSr *MacKay* *1 Dk*	Russian Astracan	291 473	148·0	27·1	12·2 *F 26 f.*	Sunderland Bodds,Sharer&Co.	1886	6 Bhds.	3C.10½" 16½" & 27"–18" 80 HP N.E.Marine Co.,Ld.,Sndrlnd.	J.Pitoeff
284		Cebu (*ex Julieta*) I ScSr *1 Dk & Awn Dk*	Spanish Manila	310 498	159·9	25·3	16·1	Glasgow Dodds&Co.	1880	[illegible] Bhds.	2Cy.20" & 37"–24" 60 HP W.Kemp,Glasgow	Spanish Gov[illegible]
285 45919	*	Cecile I ScSr *Stewart* *2 Dks*	British Liverpool	457 563	207·4	24·2	15·7	Hull C.&W.Earle	1863	[illegible] Bhds.	2Cy.21" & 46"–26" 80 HP Earles' Co.,Ld.,Hull	Moss S.S.Co.,[illegible]
286 38180		Cecile I Sc3mSrYt	British Shoreham	91 185	139·0	20·8	11·3	Hull Earles' S.B.Co.	1872		2Cy.21" & 36"–21" 45 HP J.Penn&Sons,Greenwich	Capt.P.B.Ha[illegible]
287 86424	*	Cecily I ScDy *Nichols 79-86* *1 Dk*	British Grimsby	64 95	81·9	19·4	10·2	Grimsby T.Charlton	1882	3 Bhds.	2Cy.10" & 18"–18" 18 HP T.Charlton,Grimsby	T.Charlton
288 80430	BT 1.89	Cedar I ScSr	British Glasgow	218 693	217·7	30·4	15·4 *F 46 f.*	Glasgow D.&W.Henderson	1878		2Cy.36" & 68"–42" 197 HP D.&W.Henderson,Glasgow	Glasgow,Dub[illegible] Londonderry[illegible]
289 93336	*	Cedar Branch Stl ScSr *Ritson 81-86* *1 Dk & Spr Dk 3 B*	British Sunderland	1645 2516	290·8	39·0	20·2 27·5	Sunderland Bartram,Haswell & Co.	1886	6 Bhds. DB[illegible]	3C.23½" 38" & 62"–42" 300 HP Dickinson,Sunderland	Nautilus S.S[illegible]
290	*	Celebes Stl ScSr *Wilkens* *2 Dks 3 B*	Dutch Amsterdam	1649 3000	285·0	37·2	23·9	Glasgow A.&J.Inglis	1883	5 Bhds.	2Cy.36" & 70"–45" 250 HP A.&J.Inglis,Glasgow	Stoom.Maat[illegible]
291 2774.		Celerity I Sc3mSr *2 Dks*	British Bombay	142 331	140·5	25·8	14·4	Dumbarton W.Denny&Bros.	1859		2Cy.21" & 38"–30" 150 HP W.Denny&Bros.,Dumbartn.	Joomabhoyh[illegible]
292 79086		Celerity I ScSr *Walker 81-84* *1 Dk*	British London	114 193	129·5	21·0	10·5	Middlesbro' R.Dixon&Co.	1878	[illegible] Bhds.	2Cy.18" & 36"–24" 50 HP R.&W.Hawthorn,Newcastle	Hewett & Co
293 68453		Celestial I ScSr *White* *1 Dk & Awn Dk*	British Singapore	641 1013	205·0	27·7	14·6 21·8	Northfleet Blake&Co.	1873	[illegible] Bhds.	2Cy.27" & 50"–36" 120 HP W.[illegible]&Co.,Sunderland	Tan Bengw[illegible]
294 126224		Celia W ScSr *Dallness*	American S.Francisco	115 174	115·0	29·0	8·6	Benicia,Cal. M Turner	1884		2Cy.10" & 20"–18" 40 HP Hinckley,Spiers&Hayes	G.H.Collins
295	V	Celia I ScSr *Köpcke 85-88* *1 Dk & Spr Dk*	German Hamburg	831 1175	221·4	32·4	15·0 21·0	Hamburg Blohm&Voss	1883	[illegible]	2Cy.28" & 48"–36" 135 HP Blohm&Voss,Hamburg	Hamburg Pa[illegible] Dampfs. Lin[illegible]
296 25147	*	Celt (*ex Gothenburg*) I Sc3mSr *Barclay*	British Liverpool	363 804	179·4	25·4	17·2	London C.Langley	1855	4 Bhds.	2Cy.26" & 45"–30" 90 HP D.Rowan,Glasgow	J.Meek
297 65079	BT 8.88	Celtic I Sc4mBk *Davison -88* *3 Dks*	British Liverpool	2439 3867	437·2	40·9	23·4 31·0	Belfast Harland&Wolff	1872	7 Bhds.	1Cy.41" & 78"–60" 600 HP G.Forrester&Co.,Liverpool	Oceanic Stm[illegible] Ld.
298 71451		Celtic W Sc *2 Dks*	British Hamiltn,Ont	441 698	134·0	21·0	14·7	Hamiltn.Ont.	1874		1Cy.34"–34" 237 HP T.Wilson&Co.Dundas,Ont.	A.D.Mackay
299 88352	*	Celynen I ScSr *Jones 85-88* *1 Dk 2 B*	British Newport	1034 1595	250·0	36·1	17·2 *F 20 f.*	Sunderland W.Doxford&Sons	1882	6 Bhds.	2Cy.31" & 56"–36" 150 HP W.Doxford&Sons,Sndrlnd	T.Beynon&[illegible]
300	V	Cenisio (*ex Centro America*) I SBk *Carbone* (*ex Clementina*) *Spr Dk*		900	229·0	35·1	22·0	Newcastle [illegible]	1867	[illegible] Bhds.	2Cy.[illegible]"–29" [illegible] HP Palmers Co.,Newcastl	

*=Lloyd's Register; D=Nederlandsche Vereeniging; G=Germanischer Lloyd; I=Registro Italiano; R=American [illegible]
A=Veritas Austro-Ungarico; H=Veritas Ellenico; N=Norske Veritas; V=Bureau Ve[illegible]

In the 2nd Column, B.T., with figures following, signifies Date of report of bottom sighting by Board of Trade Surveyors as condition of PASSENGER CERTI[illegible]

In the 3rd Column, C=Composite (iron frame, wood planked); I=Iron; Stl=Steel; W=Wood.

Further particulars of Vessels classed by Lloyd's Register will be found in Lloyd's Register Book and Lloyd's Yacht Register.

1	2	3	4	5	6			7	8	9	10	11
Caractère dans le Registre	Nom, Matériaux, Gréement, etc. Capitaine et dates d'entrée au service des Armateurs actuels et à bord du navire. Ancien Nom, nombre de ponts, etc.		Pavillon. Port d'Armement	Tonnage. Net Total	Dimensions. Longueur	Largeur	Creux	Construit. Où. Par qui.	Quand	Nombre des cloisons. Détails sur le test d'eau.	Machines. Détails. Nom du Constructeur.	Armateurs ou Directeurs-Gérants.
					Constructions sur le Pont							
✱	Centauro *Scarrano*	Stl TwSc 1 Dk	Arg'ntine [illegible]	268 [illegible]	180·1 *P 44 f.*	30·2 *B 46 f.*	14·0 *F 38 f.*	PortGlasgow [illegible]	1884	4 Blds. [illegible]	1Cy.18 & 35–24 90 HP. [illegible]&Co.PtGls.	Casares & Dauber
R	Centipede *Hansen 86-86*	I ScSr 1 Dk2 B	American [illegible]	319 [illegible]	170·0	28·9	12·6	Chester,Pa. ReaneySn.& Archbld	1869	4 Blds. W B a f & d	1Cy.34"–28" 109 HP. ReaneySn&Archbld,Chst.Pa	Philadelphia & ReadingR.R.Co.
	Cephalonia *Walker*	I ScBk	British Liverpool	3490 5517	430·6	46·5 *F 79 f.*	34·5	Birkenhead LairdBros.	1882		2Cy.52" & 93"–66" 700 HP. LairdBros.,Birkenhead	Cunard Steam Ship Co.,Ld.
✱D	Cephalonia	I Sc 1 Dk2 B	Greek [illegible]	929 1284	240·2 *Q 97 f.*	31·2 *B 30 f.*	17·5 *F 25 f.*	Middlesbro' R.Dixon&Co.	1878	4 Blds. DB 211, [illegible]	2Cy.29" & 50"–33" 115 HP. R.&W.Hawthorn,Newcstle.	G.Stathopulo
R	Cepheus *Nichols*	I Pad	American New York	578 882	213·6	32·4	11·5	Chester,Pa. J.Roach&Son	1881		1Cy.53"–132" W.Fletcher,NewYork	Iron Steamboat Co.
✱	Cerdic *Anderson*	I ScBk 1 Dk2 B	British Newcastle	1313 2042	275·3 *Q 102 f.*	36·0 *B 68 f.*	20·0 *F 71 f.*	Stockton M.Pearse&Co.	1884	5 Blds. [illegible]	2Cy.31½" & 64"–42" 175 HP. Blair&Co.,Ld.,Stockton	Durham S.S.Co.
	Ceres (ex Sailing vessel)	W ScFlat	British Liverpool	92 135	87·1	21·2	9·3	Winsford G.Deakin,Ld.	1880		2Cy.11"–12" 23 HP. G.Deakin,Ld.,Winsford	G.Deakin,Ld.
	Ceres	I ScSr Yt	British Liverpool	87 175	143·0	21·6	10·5	Glasgow Tod&McGregor	1859		2Cy.17" & 34"–20" 50 HP. A.Campbell&Son,Glasgow	Duke of St.Albans
	Ceres *Ivancich 67-87*	I ScSc 2 Dks	Austrian Trieste	1180 1824	274·5	34·0	26·0	Dumbarton W.Denny&Bros.	1867		4Cy.24½" & 54"–36" 100 HP. Denny&Co.,Dumbarton	Lloyd Austro-Ungarico
	Ceres *Ripper*	W	Brazilian RioJaneiro	182				RioJaneiro (Rebuilt)	18[illegible] 70	Y9.86		J.G.de Souza
DV	Ceres (ex Phoebe) *Schaepel 79-81*	I ScSr 1 Dk & Awn Dk	Dutch Amsterdam	678 1045	221·4 *H 50 f.*	29·7	13·0 20·3	Paisley AbercornS.B.Co.	1876	4 Blds. [illegible]	2Cy.33" & 56"–33" 130 HP. J.&F.Young,Ayr	Kon.Nederlandsche Stoomb.-Maats.
V	Ceres *Heyden*	I ScSr 1 Dk & Spr Dk	German Bremen	510 622	189·9	25·0 *B 35 f.*	19·0	Bremen Act.-Ges."Weser"	1874	DT/7[illegible]	2Cy.17" & 29½"–22½" 60 HP. Act.-Ges."Weser"	Dampfschifffahrts-Ges."Neptun"
V	Ceres *Dröger*	I ScSr 1 Dk	German Elbing	107 149	126·6 *P 20 f.*	17·2 *B 19 f.*	8·7	Elbing F.Schichau	1866		2Cy.9" & 19"–16½" 40 HP. F.Schichau,Elbing	F.Schichau
✱N	Ceres (ex Swaledale) *Erjenth*	I ScSr 2 Dks3 B	Norweg'n Bergen	1142 1627	257·2	32·8	23·6	Sunderland ShortBros.	1875	5 Blds. [illegible] APT[illegible]	2Cy.33" & 60"–33" 160 HP. Hawks,Crawshay&Co.Gtsh.	P.Jebsen
✱	Ceres *Mattsson*	I ScSr 2 Dks & Awn Dk	Swedish Malmö	581 874	227·0 *B 32 f.*	30·2 *H 13 f.*	14·3 21·8	Malmö KockumsMek.Verks.	1882	4 Blds. [illegible] APT[illegible]	2Cy.35" & 63"–39" 200 HP. KockumsMek.Verks.Malmö.	SydsvenskaÅngb.Akt.Bolag.
	Ceres *Svensson 87-88*	I & Stl Sc 1 Dk & Shltr Dk	Swedish Stockholm	190 257	96·4 *Erections 41 f.*	21·5	8·2	Motala MotalaCo.	1885		2Cy.16½" & 30"–20" 50 HP. MotalaCo.,Motala	Motala Ströms Ångf.Akt.Bol. (E.B.Thorssell)
✱D	Cerigo *Ward 84-85*	I ScSr 1 Dk2 B	British Liverpool	1000 1547	259·5 *Qtr. Dk.*	34·0 *B 147 f.*	18·0 *F 27 f.*	Liverpool T.Royden&Sons	1881	4 Blds. [illegible] DTF APT	2Cy.28" & 58"–39" 140 HP. J.Jack&Co.,Liverpool	R.Nicholson & Sons
	Cervantes	I ScSr	Brazilian RioJaneiro	698 1131	251·3	29·6	17·4	Newcastle A.Leslie&Co.	1874		2Cy.28" & 52"–36" 170 HP. Thompson,Boyd&Co.,Nwcsl	
	Cervantes (ex Acor) *Mercena*	I ScSr 1 Dk	Spanish Barcelona	352 418	192·0	21·3	12·0	Glasgow T.Wingate&Co.	1853		2Cy.16" & 32"–24" 60 HP. J.Jack&Co.,Liverpool	J.R.Juarez Llanos
✱D	Cervin *Hughson 79-81*	I ScSr 2 Dks3 B	British Sth Shields	1242 1914	287·4	34·2	24·7	Newcastle C.S.Swan&Co.	1877	5 Blds. [illegible]	2Cy.32" & 62"–36" 180 HP. Black,Hawthorn&Co,Gtshd	Cay,Hall & Co.
✱	Cestria	Stl ScSr Yt	British Liverpool	102 174	136·6	20·0	12·4	Birkenhead LairdBros.	1886	5 Blds.	3C.13" 19½" & 33"–24" 70 HP. LairdBros.,Birkenhead	J.Johnson Houghton
✱	Ceto *Webb 84-84*	Stl ScSr Yt	British Cowes	62 124	102·6	15·0	13·0	London Forrestt & Son	1884	4 Blds	2Cy.17" & 32"–24" 44 HP. Westray,Copeland&Co,Brw	Sir Ed.C.Guinness,Bt.
✱	Ceto	Stl TwnScSr Yt 2 B	British Cowes	5 138	125·5	18·0	9·8	Glasgow Napier, Shanks & Bell	1888	4 Blds.	3C.10" 16" & 27"–18" 70 HP. Dunsmuir&Jackson,Glsgw.	Sir Ed.C.Guinness,Bt.
R	Cetus *Vangilder*	I Pad	American New York	581 848	212·0	32·0	11·6	Philadelphia W.Cramp&Sons	1881		1Cy.53"–132" W.Cramp&Sons,Philadelphia	Iron Steamboat Co.
	Ceylon *Lambton 85-81*	I ScBk	British London	1277 2104	306·1 *P 105 f.*	40·9	26·1	London SamudaBros.	1858		2Cy.37" & 71"–45" 300 HP. G.Clark,Sunderland	M.Drury-Lavin
	Ceylon	W ScSr 1 Dk	British London	138 203	136·5	23·6	13·4	Gosport Camper&Nicholson	1872		2Cy.25"–15" 45 HP. Maudslay,Sns.&Field,Lond.	C.C.Trevor
✱N	Ceylon *Böe -88*	Stl ScSr 1 Dk & Wh Frme	Norweg'n Bergen	1704 2191	289·3 *Q 66 f.*	38·0 *B 112 f.*	19·6 *F 31 f.*	Sunderland J.Blumer&Co.	1888	6 Blds. [illegible]	3Cy.21½" 36" & 59"–39" N.E.Marine Eng. Co. Ld. Sunderland 200 HP.	Harloff & Böe
	Cezarevitch	I Sc 2 Dks3 B	Russian Odessa	1584 2083	286·8 *Q 16 f.*	36·1 *F 35 f.*	26·7	Newcastle A.Leslie&Co.	1869	5 Blds.	2Cy.52"–42" 170 HP. Maudslay,Sons&Field,Lond	Russian SteamNav.& Trading Co.
	Cezarevna	I Sc 2 Dks & Spr Dk	Russian Odessa	980 1200	254·3	32·0	17·6 25·1	London SamudaBros.	1872	5 Blds.	2Cy.55"–39" 180 HP. SamudaBros.,London	Russian SteamNav.& Trading Co.
✱D	Chadwick *Kemp*	I ScSr 1 Dk2 B	British London	980 1512	250·0 *Q 34 f.*	35·0 *B 50 f.*	18·0 *F 29 f.*	Newcastle C.S.Swan&Hunter	1882	5 Blds. [illegible]	2Cy.32" & 62"–36" 150 HP. WallsendSlipwayCo,Newcsl	"Chadwick" S.S.Co., Ld. (R.B.Avery)

✱ = Lloyd's Register; D = Nederlandsche Vereeniging; G = Germanischer Lloyd; I = Registro Italiano; R = American Record; [illegible] = Veritas Austro-Ungarico; H = Veritas Ellenico; N = Norske Veritas; V = Bureau Veritas.

Dans la 2me Colonne, B.T., avec les indications qui suivent, signifie la date du rapport de visite de la Carène par les experts du Board of Trade pour obtenir le certificat relatif au TRANSPORT DES PASSAGERS.

Dans la 3me Colonne, C = Composite (membrure en fer, bordé en bois); I = Fer; Stl = Acier; W = Bois.

Pour plus amples renseignements sur les navires classés par le Lloyd's Register, voir le Lloyd's Register Book et le Lloyd's Yacht Register.

No. in Book. / Official No.	Registry in which classed.	Ship's Name. Material, Rig, &c. / Master, and Dates of Appointment to present Owner's Service and to Ship / Ship's Late Name. No. of Decks, &c.	Flag. / Port of Registry.	Tons. Net. / Gross.	Dimensions. Length	Breadth	Depth	Built. Where. / By Whom.	When.	No. of Bulkheads. / Particulars of Water Ballast.	Engines. Particulars. / Maker's Name.	Owners or Ma[nagers]
331 [illegible]		Chala Stl ScSmSc 2 Dks Shade Dk	British Liverpool	358 [illegible]	180·4	30·1 B 38 f.	14·8	Birkenhead Laird Bros.	1878		2Cy.25"& 48"–36" 95 HP Laird Bros.,Birkenhead	Pacific Stea[m]
332 42819		Challenge I Pad Sewers 1 Dk	British Calcutta	257 168	161·0	25·2	15·4	Dumbarton	1865		2Cy. 200 HP A.Denny,Dumbarton	Clive Steam T[ug] (Gladstone, W[...])
333 93656	✻	Challenge I ScY 1 Dk	British London	14 137	100·0	20·0	11·4	S.Shields J.Readhead&Co.	1884	4 Bhds.	2Cy.22"&42"–24" 95 HP J.Readhead&Co.,S.Shields	T.W.Elliott
334 92331		Challenger W ScSt 1 Dk	British Quebec	73 108	88·8	20·2	9·7	Levis,Quebec W.T.Davie	1886		2Cy.19"&34"–18" 86 HP Carrier,Laine&Co.,Levis	The Mnstr. of [...]
335 [illegible]		Challenger I PadSt	British London	28 142	113·5	19·2	9·6	S.Shields J.Readhead&Co.	1873		2Cy.20½"–54" 98 HP J.Readhead&Co.,S.Shields	W.Sandford
336 84995		Challenger I PadSk Richardson 85	British Sunderland	23 104	100·7	18·0	9·0	S.Shields J.T.Eltringham	1881		1Cy.36"–56" 50 HP Hepple&Co.,SouthShields	I.Hodgson
337 8412	✻ BT 6.88	Challenger I ScTug Pengelly	British Swansea	22 77	85·2	16·5	9·1	Falmouth Cox&Co.	1885	5 Bhds.	2Cy.16"& 30"–21" 40 HP Cox&Co.,Falmouth	P.Rousey
338 45432	BT5.88	Challenger I PadSp (ex Rattler)	British Liverpool	15 240	139·2	24·3	12·0	Liverpool	1862		2Cy.36"–60" 90 HP J.Jack&Co.,Liverpool	Liverpl.Stm.T[ug Co.] Ld.
339 125768	R	Chalmette I Sc3mSr Adams -88 3 Dks	American N.Orl'ns,La.	1931 2983	320·0	42·0	31·0	Philadelphia W.Cramp&Sons	1879	6 Bhds. W Bo	2Cy.35"&70"–45" 1800 HP W.Cramp&Sons,Philadelphia	Morgan's L.[...] & S.S.Co.
340 95436	✻	Chamois Stl ScSt Rowlands 81-88 1 Dk 2 B	British London	1223 1879	265·0 P 26 f.	37·0 B 120 f. Q 66 f.	17·4 F 28 f.	Sunderland W.Doxford&Sons	1888	1 Bhds. CDB 291 APT 66	3C.24"33"&54"–36" 150 HP W.Doxford&Sons,Sunderl'd	Jackson Bros
341 64095		Champion W Sc 1 Dk	British Quebec	90 185	129·0	22·0	12·3	Levis,Canada	1871		2Cy. 200 HP	Levis Mari[time &] Industrial Co.
342 76036		Champion I PadSp Sherwood 74-85	British Grimsby	54 124	102·6	19·3	10·0	S.Shields J.T.Eltringham	1876		2Cy.27"–50" 80 HP J.T.Eltringham,Sth.Shields	T.& S.Oates
343 86867		Champion (ex Flying Javelin) I PadSk	British Glasgow	19 132	108·4	18·6	9·7	South Shields J.T.Eltringham	1882		1Cy.36"–54" 80 HP J.P.Rennoldson&Sn,S.Shlds	A.M'Kinno[n]
314	V	Champion I Sc Van Audenaarde	Belgian Antwerp	27 110	101·0	20·3	10·7	London LondonE.&S.B.Co.	1870		2Cy.15"& 26"–16" 40 HP LondonE.&S.B.Co.,London	Soc.Anon.d'[...] d'Armateurs [...] et le sauvetage
345		Champion I Pad Mequin	French Calais	65 138	107·9	19·6 B 19 f.	10·0	SouthShields J.Readhead&Co.	1875		2Cy.28½"–51" 90 HP J.Readhead&Co.,SthShields	Soc. de Rem[orquage] au Port [...]
346 95031		Chamroen Stl TwnScSp Sextorph 1 Dk	British Glasgow	257 418	160·4	28·1 Q 38 f.	10·6	Leith Ramage & Ferguson	1887		1Cy.17"& 31"–18" 96 HP Ramage&Ferguson,Leith	Scottish Ori[ental] Co.Ld.
347 82753	✻	Chancellor I ScBn Lynas 81-84 2 Dks 3 B	British Middlesbro'	1377 2116	270·0 B 151 f.	37·2 P 26 f.	24·4 F 37 f.	Middlesbro' R.Dixon&Co.	1881	5 Bhds. DB 392	2Cy.35"& 65"–36" 170 HP T.Richardson&Sons,Hartpl.	"Chancellor" Ld.(E.Harris[...])
348 69232	✻DV	Chancellor I ScBg Whayman 75-87 2 Dks	British Sunderland	1329 2052	310·4 P 116 f.	34·2 F 44 f.	23·9 B 30 f.	Sunderland J.Laing	1872	6 Bhds.	2Cy.38"& 70"–42" 200 HP Thompson,Boyd&Co.,Nwcstl	J.Laing (J.W.Squance [...])
349 82208	BT4.88	Chancellor Stl Pad	British Glasgow	23 233	199·7	21·1	8·2	Dumbarton R.Chambers,jun.	1880		2Cy.36"–60" 140 HP M.Paul&Co.,Dumbarton	M.T.Clark
350 76759	✻	Chanda I ScSr Gavin 76-85 3 Dks	British Glasgow	1305 2022	316·0	33·2	18·2 | 25.8	Dumbarton W.Denny&Bros.	1877	5 Bhds.	2Cy.36½"&63"–45" 221 HP Denny&Co.,Dumbarton	British India [...] Co.Ld.
351	DV	Chandernagor I ScBg Aubrault 88-88 2 Dks & Spr Dk	French Marseilles	1975 3075	341·2 B 74 f.	40·3 F 34 f.	21·2 | 28.7	Dumbarton W.Denny&Bros.	1882		2Cy.41"& 72"–51" 324 HP Denny&Co.,Dumbarton	Cie.Nationale [de] Navigation
352 91952		Changsha Stl ScSr Williams 76-86	British London	1463 2269	315·2 P 46 f.	38·2 B 102 f.	23·8 F 43 f.	Greenock Scott&Co.	1886		3Cy.25"40"&62"–48" Scott&Co.,Greenock	China Nav.C[o.]
353 3621		Chanticleer I ScSr Dew 88-88	British W.Hartlep'l	421 539	186·8	26·1 P 35 f.	16·1	Dumbarton Tulloch& Denny	1853		80 HP	C.Furness
354 86235		Chanticleer W ScFlat	British Liverpool	77 113	88·0	19·1	8·7	Winsford Ann Deakin	1868		2Cy.12"–16" 21 HP J.Jack,Liverpool	G.Deakin,[...]
355	✻	Chanzy Stl ScSr Mériel -88 1 Dk	French Caen	302 646	189·7 B 14 f.	30·0 F 26 f.	13·9 Q 92 f.	Sunderland S'nderl'dS.B.Co.Ld.	1888	4 Bhds. CDB 108 AFT 78 [...]	3Cy.16"26"&43"–30" 90 HP N.E.MarineEn.Ld.,Sundrl'd	P.Allainguill[aume &] Co.
356		Chapman I TwnScSt Pattisson 86-77 1 Dk	Swedish Stockholm	438 574	196·8 P 102 f.	25·8 B 10 f.	12·0	Norrköping Motala Co.	1859		1Cy.19"&37"–21" 120 HP W.Lindberg's V.& Vks.Stockholm	Angf.Akt.-B[...] Sverige (J.[...])
357		Charente Stl ScBn 3 Dks	French Marseilles	2800 3800	371·4	43·0	34·4	Havre Forges & Chantiers de la Medit.		AFT 60 CDB 1391	3C.28"44"&70"–47" 525 HP Forges&Chantiers,Havre	Messageries M[aritimes]
358		Charkieh I ScBn Spr Dk	Turkish Alexandria	1160	274·0	34·6	24·0	London ThamesIronWorks	1865		2Cy.38"& 70"–48" 300 HP James Jack&Co.,Liverpool	Khedivieh M[ail] Co.
359		Charkow (ex Mdewa) I Sc3mSc Fruger 84-88 2 Dks	Danish Copenhagen	509 832	201·0 Q 20 f.	26·0 B 48 f.	15·3	Newcastle PalmerBros.&Co.	1857		2Cy.26"& 52"–36" 100 HP Palmer Bros.& Co.,Newcstle	Det Forened[e Damp-] skibs Selskab
360												

✻ = Lloyd's Register; D = Nederlandsche Vereeniging; G = Germanischer Lloyd; I = Registro Italiano; R = American [Record]; A = Veritas Austro-Ungarico; H = Veritas Ellenico; N = Norske Veritas; V = Bureau Ve[ritas]

In the 2nd Column, B.T., with figures following, signifies Date of report of bottom sighting by Board of Trade Surveyors as condition of PASSENGER CE[RTIFICATE].

In the 3rd Column, C = Composite (iron frame, wood planked); I = Iron; Stl = Steel; W = Wood.

Further particulars of Vessels classed by Lloyd's Register will be found in Lloyd's Register Book and Lloyd's Yacht Register.

Registre	Nom, Matériaux, Gréement, etc. Capitaine et dates d'entrée au service des Armateurs actuels et à bord du navire. Ancien Nom, nombre de ponts, etc.	Pavillon. Port d'Armement	Tonnage. Net. Total.	Longueur	Largeur	Creux	Construit. Où. Par qui.	Quand.	Nombre des cloisons. Détails sur le lest d'eau.	Machines. Détails. Nom du Constructeur.	Armateurs ou Directeurs-Gérants.
✻	**Charlaw** 1 ScSr *Hollingshead 83-87* 1 *Dk2 B*	British London	587 901	212·7 *Q 63 f.*	29·0 *F 15 f.*	16·5 *B 61 f.*	Sunderland R.Thompson,jun.	1871	3 Blds.	2Cy.25"&48"–30" 90 HP. Black,Hawthorn&Co.Gtshd	R.Gordon & Co.
V	**Charles** Stl ScSr *Lelièvre 88* 1 *Dk*	French Brest	541 834	209·9	20·8 *B 61 f.*	15·7	Sunderland [illegible]	1886		3C.17½"25"&47"–33" 120 HP. N.E.Marine Eng.Co.Sunderland	Chevilotte frères
	Charles Dickens 1 PadlSp	British Stockton	17 121	103·2	18·3	10·1	S.Shields Readhead,Softley&Co.	1872		2Cy.28"–48" 55 HP. Readhead,Softley&C.,S.Shls	Tees Conservancy Commissioners
	Charles Edward 1 TwnScSr 1 *Dk*	British Nelson,N.Z.	140 201	143·0	19·2 *H 30 f.*	8·5	Dumbarton A.Denny	1864	3 Blds.	4Cy.20"&24"–14" 48 HP. A.Brown,Nelson,N.Z.	Joseph H.Cock
R	**Charles F. Mayer** 1 ScSr 1 *Dk2 B*	American Wilmington,Del	885 1167	236·5	37·0	18·0	Wlmngtn, Del Harlan&Hollingwrth	1884		2Cy.30"&56"–36" 800 HP. Hrlan&Hingwth,Wlmtn,Dl	Consolidation Coal Co.
V	**Charles Godard** 1 ScBk *Metronas 78* 1 *Dk*	French St Nazaire	610 914	209·9 *P 37 f.*	29·8 *F 31 f.*	17·5	Newcastle Schlesinger,Davis&Co	1876	APT FPT DB92ft	2Cy.27"&51"–33" 98 HP. Black,Hawthorn&Co.Nwcsl	Charles Godard
✻	**Charles Howard** 1 ScSr *Davis 85-86* 1 *Dk&Spr Dk3 B*	British Sunderland	849 1301	249·0	30·1	16·9 23·4	Sunderland W.Pile&Co.	1866	5 Blds.	2Cy27⅛"&49⅛"–40" 120 HP. Kincaid,Donald&Co.Greenk	A.Suart
✻	**Charles Mitchell** 1 Sc *Chatham 86-86* 1 *Dk2 B*	British Cardiff	491 776	205·4	28·3 *F 40 f.*	16·6	Newcastle C.Mitchell&Co.	1865	4 Blds. DBaft[illegible]	2Cy.26"&47"–36" 95 HP. R.&W.Hawthorn,Newcstle.	Ogmore S.S.Co.,Ld.
✻	**Charles Morand** 1 ScSr *Marshall 83-87* 1 *Dk2 B*	British Liverpool	455 761	200·0 *Q 73 f.*	27·3 *B 48 f.*	15·3 *F 21 f.*	Glasgow C.Connell&Co.	1884	4 Blds. DBcb150t	2Cy.24"&48"–36" 99 HP. Hutson&Corbett,Glasgow	"CharlesMorand" S.S. Co.,Ld.(S.Spriggs)
V	**CharlesQuint** [illegible] 1ScBk 2 *Dks&Spr Dk*	French Marseilles	1034 1587	310·7 *B 100 f.*	33·5 *F 48 f.*	16·7 24·3	Glasgow A.&J.Inglis	1880		2Cy.42"&80"–48" 450 HP. A.&J.Inglis,Glasgow	Compagnie Générale Transatlantique
✻	**CharlesW. Anderson** 1 Sc3mSr 1 *Dk&Spr Dk3 B*	British N.th Shields	938 1452	253·0	33·2	16·1 22·7	Newcastle Wghm,Richardson &Co.	1872	5 Blds. DBaft162ft	2Cy.31⅛"&60"–36" 140 HP. Humphreys&Tennant,Lon.	J.E.Bowser Ormston & Co.
	Charleston W PadlBg *Berry (ex Rhode Island, ex Eagle)*	American New York	750 1227	223·0	37·2	28·8	New York	1860			I.W.Quintard
✻	**Charlestown** 1 PadlSr *Dickson 85-85* 1 *Dk*	British Gravesend	56 88	104·0	16·1	7·7	London R.&H.Green	1880	3 Blds.	2Cy.24½"–28" 30 HP. J.Penn&Sons,Greenwich	Sproston, Dock & FoundryCo.
✻	**Charlois** Stl Sc3mSr *[illegible] 87-88* 2 *Dks&Wb Frms*	British Liverpool	1978 3003	310·3 *B 28 f.*	39·2 *F 94 f.*	25·2 *F 34 f.* *Shade Dk 58 f.*	Greenock Russell&Co.	1888	15 Blds. FPT147t APT196t	3C.22"36"&58"–42" 250 HP. D.Stewart&Co.,Glasgow	S.S. "Charlois"Co.,Ld. (Hermann,Sturzberg&Co.,N.Y.)
	Charlottenberg W ScSp *Olson 78-78* 1 *Dk*	Swedish Amal	126 173	100·4	20·6 *H 49 f.*	10·4	Näs	1873		24 HP.	P.A.Pettersson
✻	**Charlton** Stl ScSr *Turner 71-88* 1 *Dk&Wb Frms*	British London	1232 1925	277·0 *P[illegible]f Q[illegible]f*	37·1 *B 118 f.*	19·5 *F 34 f.*	Sunderland J.L.Thompson&Sns	1888	6 Blds. DBaft[illegible] APT[illegible]	3Cy.21½",35"&48"–39" 180 HP. J.Dickinson,Sunderland	W.&E.S.Lamplough
	Charlton (ex Charles Kellogg) W ScTg 1 *Dk*	British Windsor,On	178 261	120·0	19·2 *H 18 f.*	8·2	Chicago T.Miller	1862		2Cy.24"–22" 77 HP. H.Warrington,Chicago	International Wrecking & Trans.Co.
✻	**Charrington** 1 ScSr *Dawson 86-88* 1 *Dk2 B*	British Sunderland	1188 1823	269·0 *Q 72 f. P 28 f.*	36·0 *B 48 f.*	20·6 *F 32 f.*	Sunderland J.Laing	1881	5 Blds. DBaft[illegible]	2Cy.32"&60"–39" 160 HP. G.Clark,Sunderland	Culliford & Clark
✻	**Chaseley** Stl ScSr *Cass 88-88* 1 *Dk*	British Liverpool	238 433	160·0 *Q 51 f.*	23·0 *B 12 f.*	11·3 *F 25 f.*	Port Glasgow Murdoch&Murray	1888	4 Blds.	2Cy.20"&40"–30" 65 HP. Dunsmuir&Jackson,Glsgw.	The Chubut Co.,Ld.
	Chateangay 1 Padl 2 *Dks*	American Burlington,Vt		195·0	30·0	10·6	Wilmingtn Del The Harlan&Hollingsworth Co.	1887		1Cy.44"–120" W.&A.Fletcher,NewYork	Champlin Transportation Co.
✻	**Château Lafite** 1 ScSr *Ollivier* 1 *Dk&Spr Dk3 B*	French Bordeaux	2253 3162	366·2 *B 23 f.*	41·1 *F 40 f.*	22·0 30·9	Southampton Oswald,Mordaunt&Co	1881	6 Blds.	2Cy.45"&82"–50" 450 HP. Oswald,Mordaunt&Co.,Son.	Cie.Bordelaise de Nav. à Vapeur
✻D	**Château Léoville** 1 Sc3mSr *La Chapelain* 1 *Dk&Spr Dk3 B*	British Halifax,N.S.	2167 3254	365·5 *B 26 f.*	41·0 *F 48 f.*	21·5 29·5	Sunderland Sunderland S.B.Co.,Ld.	1881	7 Blds.	2Cy.44"&83"–48" 350 HP. N.E.Marine E.Co.Ld.,Sndrld.	R.M.Hudson & Son
V	**Château Margaux** 1 Sc3mSr *Sénésine* 2 *Dks&Spr Dk*	French Bordeaux	3109 4035	386·5 *P 39 f.*	41·0 *B 150 f.*	22·3 *F 62 f.* 31·8	Bordeaux Chantiers&A.delaGironde	1883		2C.49¼"&80¾"–51¼" 600 HP. Schneider&Co.,Creuzot	Cie.Bordelaise de Nav. à Vapeur
V	**Château Yquem** 1 Sc3mSr *Le Chapelain 88* 2 *Dks&Spr Dk*	French Bordeaux	3109 4035	386·5 *P 34 f.*	41·0 *B 127 f.*	22·8 *F 35 f.* 31·0	Bordeaux Chantiers&A.delaGironde	1883		2C.49¼"&80¾"–51¼" 600 HP. Schneider&Co.,Creuzot	Cie.Bordelaise de Nav. à Vapeur
	Chatham Stl ScSr *Ohls 86-88*	British Liverpool	401 647	213·2	32·1 *Q 5 f.*	12·6	Greenock Scott&Co.	1883		2Cy.23"&50"–30" 100 HP. Scott&Co.,Greenock	Livrpl., Brazil & River Plate Stm.Nav.Co.,Ld.
	Chatham 1 ScSr	British Dover	109 278	160·7	22·2	12·9	London J.&W.Dudgeon	1873		2Cy.24¼"&45"–24" 80 HP. J.&W.Dudgeon,London	Lon.Chatham& Dover RailwayCo.
R	**Chatham** 1 ScSr *Hallett* 3 *Dks&Awn Dk*	American Baltimore	1909 2729	265·4	40·0 *F 90 f.*	15·6 30·2	Philadelphia AmericanS.B.Co.	1884		2Cy.40"&72"–48" 350 HP. Neafie&Levy,Philadelphia	Merchants & Miners TransportationCo.
✻	**Chatsworth** 1 ScSr *Johnson 85-85* 1 *Dk2 B*	British Cardiff	620 966	229·2 *Q 83 f.*	29·0 *B 47 f.*	17·0 *F 29 f.*	S.Shields J.Softley&Co.	1872	4 Blds. DBaft[illegible]69ft	2Cy.26"&49"–30" 95 HP. T.Richardson&Sons,Hartpl	Christie & Co.
R	**Chattahoochee** 1 ScSr *Daggett 81-87* 3 *Dks*	American Savannah,Ga.	1888 2676	280·0 *H —*	40·7	24·1	Chester, Pa. J.Roach&Son	1882		2Cy.38"&74"–54" 360 HP. J.Roach&Son	Ocean S.S.Co.
	Chaucer 1 ScSr *Askew 78-88* 2 *Dks3 B*	British Liverpool	1871 2846	315·0 *P 38 f.*	39·3 *F 44 f.*	27·0 *B 72 f.*	Newcastle R.&W.Hawthorn, Leslie&Co.,Ld.	1886	DBa190t, cb210t	3C.26"41"&70"–48" 350 HP. R.&W.Hwthne&Co.,Ld.,Nc	Livrpl., Brazil & River Plate Stm.Nav.Co.,Ld.

✻=Lloyd's Register; D=Nederlandsche Vereeniging; G=Germanischer Lloyd; I=Registro Italiano; R=American Record;
[illegible]=Veritas Austro-Ungarico; H=Veritas Ellenico; N=Norske Veritas; V=Bureau Veritas.

Dans la 2me Colonne, B.T., avec les indications qui suivent, signifie la date du rapport de visite de la Carène par les experts du Board of Trade pour obtenir le certificat relatif au TRANSPORT DES PASSAGERS.

Dans la 3me Colonne, C=Composite (membrure en fer, bordé en bois); I=Fer; Stl=Acier; W=Bois.

Pour plus amples renseignements sur les navires classés par le Lloyd's Register, voir le Lloyd's Register Book et le Lloyd's Yacht Register.

No. in Book / Official No.	Registry in which classed	Ship's Name. Material, Rig, &c. / Master, and Dates of Appointment to present Owner's Service and to Ship / Ship's Late Name. No. of Decks, &c.	Flag. / Port of Registry.	Tons. Net / Gross	Length	Breadth	Depth	Built. Where. / By Whom.	When	No. of Bulkheads / Particulars of Water Ballast	Engines. Particulars / Maker's Name.	Owners or Managers
391	✻	Chazalie W ScSmSrYt	Danish Copenhagen	239 [illegible]	157·1	28·2	15·4	Gosport Camper&Nicholson	1875	[illegible]	2Cy.20"&16"-24" 60 H Day,Summers&Co.,Sthmtn	H. Excellency Folbe.
392 63394	✻	Cheang Chew (ex Saint Louis) I ScBg Dunlop -88 1 Dk & Spar Dk	British Singapore	1213 1862	301·3	35·0	17·4 24·8	Birkenhead J.R.Glover&Co.	1870	4 Bhds.	2Cy.32"&61"-36" 180 HP. Cp '83,C.D.Holmes&Co.,Hll.	Khoo Tiong P
393 16980	✻	Cheang Hock Kian I ScBg Wale 75-83 (ex United Service) 2 Dks & Awn Dk	British Singapore	956 1438	263·4 H 24 f.	31·3	17·9 24·6	Sunderland J.Laing	1857	4 Bhds.	2Cy.31½"&57"-36" 150HP. A.&J.Inglis,Glasgow	Khoo Tiong P
394 62292	✻	Cheang Hye-Teng I ScBg Scott -88 (ex Atholl) 2 Dks 3 B	British Singapore	923 1430	241·7	32·3	24·0	Glasgow A.Stephen&Sons	1870	5 Bhds.	2Cy.30"&60"-39" 170HP. J.Howden&Co.,Glasgow	Khoo Tiong P
395 73626		Chefoo I ScSr Rohde	British London	684 1087	233·6	27·3	22·1	Greenock Scott&Co.	1876		2Cy.23"&46"-36" 98 HP Scott&Co.,Greenock	China Nav.Co.
396 87351	✻	Chelona I ScSr White 1 Dk 2 B	British Sunderland	1124 1731	258·0 Q66 f. P28 f.	36·1 B66 f.	18·6 F28 f.	Sunderland Bartram,Haswell&Co.	1883	4 Bhds. DBa 728 ft	2Cy.33"&61"-33" 160 HP. T.Richardson&Sons,Hrtlpl.	G.J.Hay & Co.
397 83641	✻	Chelsea I ScSr Howsted 85-85 1 Dk & Web Frames	British London	754 1171	230·0	33·2 B44 f.	14·9 F24 f.	Newcastle Palmers'Co.,Ld.	1884	4 Bhds [illegible]	2Cy.28"&54"-36" 120 HP. Palmers'Co.,Ld.,Newcastle	River Steam Co. Co.,Ld.
398 91880	✻	Chelydra Stl ScSr Peace 84-85 2 Dks 3 B	British London	1574 2467	301·0 B86 f.	40·0 F40 f.	25·3 P37 f.	Sunderland J.L.Thompson&Sons	1885	6 Bhds. CDB 357 ft	3C.24"39"&64"-42" 300HP. T.Richardson&Sons,Hrtlpl.	Angier Bros. (Managers)
399 94324	✻	Cheniston Stl ScSr Sheldrake 71-87 1 Dk & Web Frms	British London	1304 1987	278·0 Q121 f.	37·2 B64 f.	19·2 F31 f.	Sunderland J.Laing	1887	4 Bhds. CDB [illegible]	3C.22"36½"&60"-42" 200HP. G.Clark,Sunderland	Harris & Dix
400 63327		Chepstow I PadSk Roberts 71-84 1 Dk	British Bristol	62 188	140·6	20·1 B49 f.	8·2	Renfrew Hndrsn,Clbrn&Co.	1874		4Cy.28"-36" 100 HP Henderson&Co.,Renfrew	Great Western
401 71694	BT 2.88	Cherbourg I ScBk King 88-87	British Glasgow	1038 1614	251·2	32·1	26·5	Glasgow J.&G.Thomson	1874		2C.33½"&58½"-42" 170HP. J.&G.Thomson,Glasgow	Cunard Stm. Ld.
402 66826		Cherbourg I ScSr	British Southmptn	214 373	165·6 Q 64 f.	23·1 F 42 f.	11·9	Dumbarton J.&R.Swan	1873		2Cy.22"&38"-27" 65 HP. Smith Bros.&Co.,Glasgow	London & S.W.
403		Cheribon (ex Ooryla) I PadSr Cramer 2 Dks	Dutch Batavia	419 343	231·8	26·1	12·4	Greenock Scott&Co.	1875		2Cy.42"&80"-60" 400 HP. Scott&Co.,Greenock	Nederlandsch Stoomv.-Maats
404	V	Chéribon I ScBg Duchesne 88-88 2 Dks & Spr Dk	French Marseilles	1976 3075	311·2 B 74 f.	40·3 F 34 f.	21·2 28·7	Dumbarton W.Denny&Bros.	1882		2Cy.41 & 72"-51" 324 HP. Denny&Co.,Dumbarton	Cie. National Navigation
405 123391	R	Cherokee I ScSr Vogel 2 Dks & Shelter Dk	American New York	1967 2557	264·0	43·0	18·8	Philadelphia W.Cramp&Ss.S.&E.B.Co.	1886		3Cy.21"34"&56"-36" W.Cramp&Sons,S.&E.B.Co., Philadelphia	Wm.P.Clyde
406 62395	✻	Chesapeake I ScSr Watson 84-84 1 Dk.Spr Dk 3 B	British London	933 1476	240·0	32·7	21·7	Dundee Gourlay Bros.	1872	6 Bhds. WB [illegible]	2Cy.29"&48"-42" 130 HP. Gourlay Bros.&Co.,Dundee	Mercantile S.S.
407 93765	✻D	Chester Stl ScSmSr Wohlmuth 87-88 2 Dks & Wb Frms	British Liverpool	1872 2831	311·7 P 88 f.	39·2 B 20 f.	25·0 F 31 f.	Greenock Russell&Co.	1887	13 Bhds FT [illegible] APT [illegible]	3C.22"36"&58"-42" 250HP. D.Stewart&Co.,Glasgow	S.S. "Chester" (Hermann,Sturken)
408 65712	✻	Chester I ScSr Groth 86-82 1 Dk 2 B	British W.Hartlepl	722 1128	228·3 Q 105 f.	30·9 B 72 f.	16·5 F 24 f.	Hartlepool Withy,Alxndr&Co.	1871	4 Bhds. [illegible]	2Cy.28"&53"-33" 110 HP. T.Richardson&Sons,Hrtlpl.	Pyman,Bell &
409 [illegible]157	✻ BT 6.88	Chester I ScSr Green 85-86 1 Dk 2 B	British Grimsby	652 1010	238·6 P 23 f.	32·2 B 76 f.	14·1 F 47 f.	W. Hartlepool E.Withy&Co.	1884	5 Bhds.	3C.22"37"&58"-36" 170HP. T.Richardson&Sons,Hrtlpl.	Manchester, Sh & Lincoln Railw
410 [illegible]	BT 1.88	Chevalier I PadSk McMillan 84-84	British Glasgow	100 202	211·0	22·2	9·3	Glasgow J.&G.Thomson	1866		2Cy.42"-54" 150 HP. J.&G.Thomson,Glasgow	D.Mac Brayne
411 67350	✻	Chevington I ScSr Warman 85-86 1 Dk	British Newcastle	185 305	140·5 Q 70 f.	21·6 B 7 f.	11·5 F 19 f.	S.Shields J.Eltringham	1872	1 Bhds.	2Cy.18"&34"-30" 50 HP. W.&J.Almond,N.Shields	J.T.Harrison
412	N	Chevy Chase I ScSr Haraldsen -88 1 Dk	Norweg'n Tönsberg	263 301	167·6 B 29 f.	24·1 H 8 f.	12·7	Dumbarton W.Denny&Bros.	1848		2Cy.19"&34⅞"-26" 80 HP. Thompson,Boyd&Co.,Nwc.	Dampskibs-Sels "Chevy Chase" (H.A
413		Chi Yuen I ScSmSr Nall 84-84 2 Dks 3 B	Chinese Shanghai	1211 1879	285·0 B 48 f.	34·1	18·3 21·2	Renfrew Lobnitz,Coulborn&Co.	1881	6 Bhds. APT 100 MT 360	2Cy.34"&64"-48" 300 HP. Lobnitz&Co.,Renfrew	China Merchants Nav.Co.
414		Chibine (ex Octavia) I ScSr	Turkish Alexandria	678	236·0	29·0 P	17·0	London J.Ash&Co.	1865		2Cy.33"&56"-33" 160 HP. J.Stewart,London	Khedivieh M Co.
415 88657		Chicago (ex Lincoln City) I ScSr Jones 2 Dks 3 B	British Hull	1750 2729	301·0 P 199 f.	40·0 F 75 f.	22·6	W. Hartlepool W.Gray&Co.	1884	5 Bhds. FPT 37 ft	2Cy.38"&70"-39" 300 HP. T.Richardson&Sons,Hrtlpl.	T.Wilson,Sons
416 95894	✻	Chicklade Stl ScSr Mouttit -88 1 Dk & Web Frames	British W.Hartlepl	1554 2374	299·0 [illegible]	38·5 [illegible]	20·7 [illegible]	W. Hartlepool E.Withy&Co.	1889	5 Bhds. CDB 388 APT 38	3C.22"37"&61"-39" 200HP. T.Richardson&Sons,Hrtlpl.	G.Horsley & S
417 71178		Chieftain W PadSr	British Kingston Ont	197 335	136·8	24·1	9·4	Garden Island	1874		100HP.	Dexter D.Calv
418 [illegible]	✻	Chigwell I ScSr Masters 83-83 1 Dk 2 B	British London	1192 1821	258·8 [illegible]	36·2 [illegible]	18·7 [illegible]	Sunderland Bartram,Haswell&Co.	1883	4 Bhds. DBa [illegible] FPT 21 APT 20	2Cy.33"&61"-33" 200 HP. T.Richardson&Sons,Hrtlpl.	A.Stuart
419 [illegible]		Chili W Sr	British Liverpool	73 108	82·0	19·5	8·8	Northwich [illegible]	1869		2Cy.12"-14" 24 HP. H.Gibson,Northwich	R.Verdin
420 95651	✻D	Chilian (ex Ryhal Halli) I ScSmSr Curtis 88-88 2 Dks 3 B	British [illegible]	1415 2154	330·5	34·4	24·3	Glasgow [illegible]	1871	6 Bhds.	3C.22"32"&58"-42" 160HP [illegible] Harland&Wolff,Belfast	Carlisle & Co. (London)

✻ = Lloyd's Register; D = Nederlandsche Vereeniging; G = Germanischer Lloyd; I = Registro Italiano; R = American Re[cord]
A = Veritas Austro-Ungarico; H = Veritas Ellenico; N = Norske Veritas; V = Bureau Veritas

In the 2nd Column, B.T. with figures following, signifies Date of report of bottom sighting by Board of Trade Surveyors as condition of PASSENGER CERT[IFICATE]

In the 3rd Column, C = Composite (iron frame, wood planked); I = Iron; Stl = Steel; W = Wood.

Further particulars of Vessels classed by Lloyd's Register will be found in Lloyd's Register Book and Lloyd's Yacht Register.

Registre	Nom, Matériaux, Gréement, etc. — Capitaine et dates d'entrée au service des Armateurs actuels et à bord du navire. — Ancien Nom, nombre de ponts, etc.	Pavillon. — Port d'Armement	Ton. — Net	Dimensions.			Construit. Où. — Par qui.	Quand	Nombre des cloisons. — Détails sur le lest d'eau.	Machines. — Détails. Nom du Constructeur.	Armateurs ou Directeurs-Gérants.
*	Chilka 1 ScSr Cumming 80-86 2 Dks	British Glasgow	1450 1945	285·0	35·2	23·9	Dumbarton W. Denny & Bros.	1878	5 Blkds. DB235t	2Cy.34" & 68"–42" 185 HP. Denny & Co., Dumbarton	British India Steam Nav. Co., Ld.
	Chiloe 1 ScSmSr 2 Dk & 3 B	British Liverpool	1316 2300	321·0	37·2	24·7 B 185 f.	Glasgow J. Elder & Co.	1881	5 Blkds. DB	2Cy.42" & 80"–48" 400 HP. J. Elder & Co., Glasgow	Pacific Steam Nav. Co.
	Chiltern 1 ScBg Kendall	British London	832 1372	262·0	31·3	22·6	Dumbarton W. Denny & Bros.	1866		2Cy.48" & 72"–30" 200 HP. J. Watt & Co., London	Eastern Telegraph Co., Ld.
*	Chilton (STRANDED) 1 ScBn Freeman 85-87 1 Dk 2 B	British W. Hartlepl	717 1118	228·0	30·2	16·8 Q 82 f. F 30 f. B 58 f.	W. Hartlepool W. Gray & Co.	1879	4 Blkds. DB 467 ft A 73 ft	2Cy.27" & 50"–33" 90 HP. T. Richardson & Sons, Hrtlpl.	T. Appleby & Co.
*	Chimborazo 1 ScBk 3 Dks & Awn Dk	British Liverpool	2443 3847	384·0	41·3	35·3 H 81 f.	Glasgow J. Elder & Co.	1871	6 Blkds.	2Cy.60" & 104"–48" 550 HP. J. Elder & Co., Glasgow	Orient Steam Nav. Co., Ld.
	Chin tung (ex Sean Tonzi) 1 ScSr Wiesor 77-77 2 Dks	Chinese Shanghai	835 1330	216·2	39·2	16·4	Glasgow A. & J. Inglis	1870		2Cy.24" & 46½"–36" 180 HP. A. & J. Inglis, Glasgow	China Merchants' Steam Nav. Co.
*	China 1 ScSr Billing 88-86 2 Dks & W. b Frms	British Liverpool	1394 2151	285·3	36·0	24·4 P 90 f. B 62 f. F 30 f.	W. Hartlepool E. Withy & Co.	1883	5 Blkds. DB 50 ft 398t	2Cy.36" & 67"–36" 200 HP. T. Richardson & Sons, Hrtlpl.	City of Liverpool Stm. Nav. Co., Ld.
*	China 1 ScSr Pinligen 74-85 2 Dks	British Hull	1121 1720	290·3	35·3	20·9 P 138 f.	Newcastle Wghm. Richardson & Co.	1869	4 Blkds. DB 467 ft F 35 ft	2C.36" & 64½"–36" 185 HP. Comp.'77, Bailey & Co., Hull	W. S. Bailey
	China Stl PaddLr	British Glasgow	892 1224	310·0	46·0	10·7	Dumbarton W. Denny & Bros.	1888		2Cy.37½" & 65"–72" 375 HP. Denny & Co., Dumbarton	Irrawaddy Flotilla Co., Ld.
*	China 1 ScSr Ingledng 76-86 1 Dk	British Bodo	244 449	169·6	25·7	12·2 Q 77 f. F 19 f. B 14 f.	South Shields J. Readhead & Co.	1877	5 Blkds. DB 30 ft F 52 ft	2Cy.22" & 40"–27" 80 HP. J. Readhead & Co., S. Shields	J. Bennett
V	China 1 ScSr Haye 82-85 1 Dk & Spr Dk	German Flensburg	1275 1890	256·5	34·7	17·7 B 72 f. 23·2	Flensburg Flensb. Schiffsb. Ges.	1884	CDB 332t	2Cy.32½" & 60"–36" 160 HP. Flensb. Schiffsb. Ges., Flensburg	Flensburger Schiffsbau Ges.
V	China 1 ScSr Pilsow –87 1 Dk	German Hamburg	648 854	206·0	28·0	18·7	Hamburg Reiherst. Schiffs Wrft	1865		2Cy.22" & 38"–33" 70 HP. Reiherstieg Schiffs Werft, Hmbg	F. H. Hölmke
*	Chindwara 1 ScSr Smith 88-88 2 Dks	British Glasgow	1487 1983	285·0	35·2	23·9	Dumbarton W. Denny & Bros.	1879	5 Blkds. CDB 235t	2Cy.34" & 60"–42" 185 HP. Denny & Co., Dumbarton	British India Steam Nav. Co., Ld.
*	Chindwin 1 ScK Trawler Dyer –87 1 Dk	British Hull	59 127	100·0	20·1	10·3	Hull Cook, Welton & Gemm ell	1887	3 Blkds.	2Cy.17" & 33"–22" 45 HP. C. D. Holmes & Co., Hull	G. Beeching & Co.
	Chingtu Stl ScSr Arthur 85-87	British London	1459 2265	315·3	38·2	23·8 P 83 f. B 125 f. F 32 f.	Greenock Scott & Co.	1886		3C.25" 40" & 62"–48" 400 HP. Scott & Co., Greenock	China Nav. Co., Ld.
*	Ching-Wo 1 ScBn Macheugh 85-85 2 Dk 3 B	British London	1556 2400	320·0	38·0	25·7 P 24 f. B 60 f. F 41 f.	Middlesbro' R. Dixon & Co.	1885	6 Blkds. DB & 230t	3C.26" 42" & 66"–42" 400 HP. T. Richardson & Sons, Hrtlpl.	China Shippers Mutual Steam Nav. Co. Ld.
	Chinzei Maru W ScSr	Japanese Osaka	285 460	171·0	21·4	16·8	Osaka K. Genjiro	1882		50 HP.	Osaka Shosen Kaisha
	Chios (ex Portland) 1 ScSr Kaumenas 79- 2 Dks	Greek Syra	537 923	230·3	30·8	15·6 P 56 f. F 44 f.	Greenock R. Steele & Co.	1876		2Cy.31" & 57"–45" 150 HP. R. Steele & Co., Greenock	Hellenic S.S. Co.
*	Chios (ex Leonid) 1 ScSr Spyridess 81-81	Turkish Constntinple.	194 411	191·0	25·2	13·0	Port Glasgow Blackwd. & Gordon	1881	5 Blkds. DB FPT	2Cy.25" & 48"–36" 100 HP. Blackwood & Gordon, P. Glsg.	P. M. Courtgi & Co.
*	Chipchase 1 ScSmSr Reed 1 Dk 2 B	British Newcastle	438 689	213·0	27·1	15·4	Newcastle Wghm. Richardson & Co.	1864	3 Blkds. DB ft 80 ft FPT APT	2Cy.26" & 48"–33" 99 HP. Dunstan Eng. Wks. Co., Nwc.	W. & G. Hedley
	Chiri Maru (ex Chiloshi)	Japanese Tokio	772	246·0	34·6	21·0	Glasgow	1867		200 HP.	Nippon Yusen Kaisha
* D	Chirton 1 Sc Downs 73-87 1 Dk & Web Frames	British N. Shields	724 1115	225·0	33·1	14·9 Q 99 f. B 50 f. F 26 f.	North Shields T. & W. Smith	1882	4 Blkds. DB ft & f 273t FPT 28t APT 7t	2Cy.26" & 52"–33" 99 HP. Wallsend Slipway Co., Nwc.	Eskdale, Dodds & Co.
*	Chiswick 1 ScSr Gowing 81-87 1 Dk 2 B	British London	796 1261	239·5	32·1	17·6 H 53 f. P 155 f. F 35 f.	Newcastle C. Mitchell & Co.	1871	4 Blkds. W Ba & f	2Cy.30" & 59"–36" 130 HP. T. Clark & Co., Newcastle	Watts, Ward & Co.
	Chitose Maru (ex Tsura) 1 ScSr Tooji 1 Dk & Spr Dk	Japanese Tokio	326 456	160·4	23·1	11·0 18·0	Glasgow A. Stephen & Sons	1869		2Cy.19" & 36"–24" 56 HP. J. Howden & Co., Glasgow	Nippon Yusen Kaisha
*	Chittagong 1 ScBn McDougall 76-85 2 Dks 3 B	British Glasgow	1241 1912	276·4	36·2	23·0 P 35 f. F 38 f.	Middlesbro' R. Dixon & Co.	1881	5 Blkds. CDB 233t	2Cy.36½" & 68"–42" 180 HP. Blair & Co., Ld., Stockton	P. Henderson & Co.
	Chloe 1 ScSr 1 Dk	Russian Taganrog	177 240	165·2	20·1	9·8 Len. 72.	Newcastle A. Leslie & Co.	1868	4 Blkds.	2Cy.27"–18" 50 HP. R. & W. Hawthorn, N'weastle	Hoyland Bros.
*	Chollerton 1 ScSr Free 1 Dk & Spr Dk 3 B	British London	1734 2650	317·2	37·1	27·2	Newcastle A. Leslie & Co.	1883	5 Blkds. DB ft & 363t FPT 30t APT 23t	2Cy.36" & 70"–48" 300 HP. Wgham Richards'n & Co., Nwe	W. Milburn & Co.
*	Cholmley 1 ScBn Hunter 72-88 1 Dk 2 B	British Whitby	894 1402	240·0	34·0	17·9 B 110 f. Q 87 f. F 33 f.	Whitby T. Turnbull & Son	1880	4 Blkds. DB ft 227t	2Cy.30" & 57"–33" 140 HP. T. Richardson & Sons, Hrtlpl.	T. Turnbull & Son
	Choper 1 TwnSc		861	210·0	29·0	10·5	London J. & W. Dudgeon	1872		2Cy.15½" & 27½"–14" 60 HP. J. & W. Dudgeon, London	
V	Chow-Chow-Foo 1 ScSr Wendt 85-85 1 Dk & Spar Dk	German Hamburg	796 1027	202·1	27·6	14·6 B 47 f. Q 12 f. 21·6	Lübeck Henry Koch	1885	DT	2Cy.24" & 43"–30" 100 HP. A. Borsig, Berlin	Dampfschiffs-Ges. "Swatow"

…=Lloyd's Register; D=Nederlandsche Vereeniging; G=Germanischer Lloyd; I=Registro Italiano; R=American Record; …=Veritas Austro-Ungarico; H=Veritas Ellenico; N=Norske Veritas; V=Bureau Veritas.

Dans la 2me Colonne, B.T., avec les indications qui suivent, signifie la date du rapport de visite de la Carène par les experts du Board of Trade pour obtenir le certificat relatif au TRANSPORT DES PASSAGERS.

Dans la 3me Colonne, C=Composite (membrure en fer, bordé en bois); I=Fer; Stl=Acier; W=Bois.

Pour plus amples renseignements sur les navires classés par le Lloyd's Register, voir le Lloyd's Register Book et le Lloyd's Yacht Register.

No. in Book. / Official No.	Registry in which classed.	Ship's Name. Material, Rig, &c. / Master, and Dates of Appointment to present Owner's Service and to ship / Ship's Late Name. No. of Decks, &c.	Flag / Port of Registry.	Tons Gross / Net	Dimensions (Length / Breadth / Depth) / Deck Erections.	Built. Where. / By Whom.	When	No. of Bulkheads. / Particulars of Water Ballast.	Engines. / Particulars. Maker's Name.	Owners or Managers
451 [illegible]	V BT2.88	**Chow Fa** StlScrSr *Phillips* -88	British Glasgow	1655 1318	270·5|37·0|21·8 *E 18f. F 40f.*	Glasgow Fairfield Shipbldg & Eng. Co.	1888		3Cy.22",36"&57"–42" Fairfield Shipbldg & Eng. Co. Ld. 250 HP.	Scottish Orient… Co.,Ld.
452 71507		**Chow Phya** I Sc3mSr 1 *Dk*	British Penang	379 507	178·6|25·7|12·7 *P 40f. H 37f.*	W. Hartlepool	1858		2Cy.34"–22" 70 HP. T. Richardson&Co.,Hartlepl	Kheo Tiong P…
453 94372	* BT1.88	**Choy Sang** Stl ScSr *Halberoie* 83-88 1 *Dk Spr Dk&H*	British London	1194 1880	260·4|35·0|18·1 *B 61f. F 33f.* 21·8	Aberdeen Hall,Russell&Co.	1888	5 Bhds. Dbl 200.	3Cy.21",33"&54"–39" Hall,Russell&Co., Aberdeen 200 HP.	Indo-China S… Co.,Ld.
454	V	**Christian** I ScSr *Plähn* 83-83 1 *Dk*	German Tönning	645 854	202·0|27·6|15·6 *P 30f. B 44f. F 26f.*	Lübeck Henry Koch	1883	ST 75tn	2Cy.28½"&48½–29"100 HP. G.Henniges&Co.	Zerssen & Co.
455	*	**Christian Johnsen** I ScSr (ex Maria) *Benstad* 88-88 1 *Dk 2 H*	Norweg'n Christiansund.	344 519	170·8|24·7|14·4 *Q 54f. B 38f. F 34f.*	Middlesbro' R.Dixon&Co.	1874	4 Bhds. DBaf/10t.	2Cy.22"&44"–29" 80 HP. Christie,Gutch&Co.,N.Shields	Et. Interess… (J.Loennechen)
456		**Christian IX.** I Sc3mSr *Mollerup* 72-76 1 *Dk & Hur Dk*	Danish Copenhagen	787 1236	232·4|30·3|16·5 *P & F 291f.* 23·3	Copenhagen Burmeister&Wain	1874	6 Bhds. APT 82 / MT 127	2Cy.21"&44"–30" 120 HP. Burmeister&Wain,Copngn.	Det Forenede … skibs Selskab
457		**Christiania** I PadSr *Hansen* 85-88 1 *Dk & Hur Dk*	Danish Copenhagen	480 837	219·3|25·1|11·3 *H 42f.* 18·2	Copenhagen Burmeister&Wain	1875		2Cy.33"&60"–54" 260 HP. Burmeister&Wain,Copngn.	Det Forenede… skibs Selskab
458	DN	**Christiania** I Sc3mSr *Christensen* -86 1 *Dk & Awn Dk*	Norweg'n Christiania	374 460	139·8|24·8|17·9	Christiania Akers Mek. Verks.	1871		2C.22½"&40½"–24½"70 HP. Akers Mek Verks,Christiania	Winge & Co.
459	* D	**Christiansborg** I ScSr *Schultz* 83-83 1 *Dk 2 H*	Danish Copenhagen	842 1326	237·1|33·2|15·1 *P 148f. F 28f.*	Copenhagen Burmeister&Wain	1883	4 Bhds. DBaf&c. 236t / FPT 31 / APT 29	2Cy.30"&52"–33" 120 HP. Burmeister & Wain,Copenhagn	Dampskibs-S… "Dannebrog"
460	V	**Christianssund** I ScSr *Holm* 88-88 1 *Dk & Awn Dk*	Danish Copenhagen	356 581	185·2|25·6|13·1 *P 135f. B 53f. F 33f.* 20·2	Renfrew Lobnitz&Co.	1881	WB	2Cy.24"&44"–30" 85 HP. Lobnitz&Co.,Renfrew	Det Forenede … skibsselskab
461		**Christiansund** I ScSr 1 *Dk*	Norweg'n Christiansund	64 102	107·5|18·2|8·4 *Q 34f. B 22f.*	Christiania Sinding	1865		2Cy.11½"&22"–14" 25 HP. S.Wiedemann,Drontheim	N.H.Knudtzon
462	G	**Christina** (ex Kristina) I ScSr *Smith* 87-87 1 *Dk*	Danish Copenhagen	149 215	104·1|21·2|11·9	Gothenburg Götebg's Mek Wkst.	1877		2Cy.12"&26"–18" 40 HP. Göteborg's Mek. Werkst., Gothenburg	J.Smith
463 86714	*	**Christine** I ScSrYt	British Glasgow	95 140	126·0|17·2|9·9	Port Glasgow Cunliffe&Dunlop	1881	3 Bhds.	2Cy.17"&30"–24" 40 HP. Cunliffe&Dunlop,P.Glasgow	E.Salvin Bowl…
464 45723		**Christopher Thomas** I PdSp	British Bristol	41 168	132·0|20·2|8·1	Renfrew	1864		100 HP.	Great Western… Co.
465 82518	*	**Chrysolite** I ScBn *Hallsey* -86 2 *Dks 3 H*	British Hull	1045 1619	250·0|34·0|22·0 *P 26f. F 33f.*	Stockton M.Pearse&Co.	1881	4 Bhds. Dbl/73t	2Cy.31"&58"–36" 160 HP. Blair&Co.,Ld.,Stockton	W.Tulley & Co.
466 4870[illegible]		**Chrystenah** W Pad	American New York	418 571	196·5|30·2|9·3	Nyack,N.Y.	1866			J.Blauvett
467	*	**Chuen Tiao** StlTwnScrSr *Wyles* -88 1 *Dk*	Chinese	161 378	155·0|25·1|12·4 *P 20f. B 16f. F 29f.*	Newcastle SirW.G.Armstrong, Mitchell&Co.,Ld.	1888	5 Bhds.	3Cy.12½",20"&33"–21" Wallsend Slipway Co., Newcastle 150 HP.	Chinese Imperial Maritime Customs
468 85103		**Chung King** Stl TwnScSr *Hutchison* 76-86	British London	801 1258	241·3|32·4|20·1	Greenock Scott&Co.	1882		3Cy.25"&50"–30" 240 HP. Scott&Co.,Greenock	China Nav.Co.,Ld.
469 80431		**Chupra** I ScSr *Anderson* 87-88 2 *Dks*	British Glasgow	1432 1944	285·0|35·2|23·9	Dumbarton W.Denny&Bros.	1878	5 Bhds. DB 318t	2Cy.34"&60"–42" 180 HP. Denny&Co.,Dumbarton	British India S… Nav.Co.,Ld.
470 88204		**Churchill** I TwnScTug *Wellington* 72-84 1 *Dk*	British Natal	42 184	115·0|22·1|11·0 *B 23f. H 40f.*	Aberdeen Hall,Russell&Co.	1883	4 Bhds. FPT 11	4Cy.19"&34"–24" 100 HP. Hall,Russell&Co.,Aberdeen	Natal Harbour Commissioners
471 79248		**Churruca** I Sc	British N.Shields	107 158	104·6|23·4|8·6	Sunderland Osbourne, Graham &Co.	1881		2Cy.18½"&25"–27" 30 HP. Hawks,Crawshay&Co.,Gtsd.	J.Forrest
472	*	**Churruca** *M.B.* [illegible] -86 I ScSr 1 *Dk Spr Dk & Web Frames*	Spanish Barcelona	839 1263	227·0|32·2|21·7	South Shields J.Readhead&Co.	1886	4 Bhds. DB 152t	3C.18",28"&48"–30" 90 HP. J.Readhead&Co.,Sh.Shields	J.Roca & Co.
473	*	**Churruca** I ScSr *Ojinaga* 82-82 2 *Dks*	Spanish Bilbao	419 616	190·0|26·2|16·0	Newcastle A.Leslie&Co.	1878	4 Bhds. DBaf/52t FPT	2Cy.26"&52"–30" 150 HP. R.&W.Hawthorn,N'wcastle	Larrinaga & Co.
474 87414	* BT5.88	**Chusan** Stl Sc3mSr *Fraser* 73-88 3 *Dks*	British Greenock	2509 1348	400·3|45·2|28·9 *P 255f. F 78f.*	Greenock Caird&Co.	1884	8 Bhds.	2Cy.52"&96"–66" 800 HP. Caird&Co.,Greenock	Peninsular & … Steam Nav.Co.
475 71890	BT6.88	**Chyebassa** I Sc3mSr *Wilson* 81-85 3 *Dks*	British Glasgow	1714 2644	350·2|36·4|28·0	Dumbarton W.Denny&Bros.	1874	5 Bhds.	2Cy.47"&82"–42" 346 HP. Denny&Co.,Dumbarton	British India … ciated Steamers
476		**Cibao** (ex Cleveland) I ScSp 1 *Dk*	Haytian St. Domingo	103 167	107·3|20·3|8·4	South Shields J.Eltringham	1875	4 Bhds.	2Cy.16"–18" 25 HP. Pattison&Atkinson,N'wcstle	St. Domingo S.S…
477 92732		**Cibola** StlPad *McCorquodale* -88 2 *Dks*	British Toronto	521 861	252·0|28·4|10·7	Deseronto W.C.White	1888		2Cy.47"&85"–66" 314 HP. Rankin&Blackmore,Grnk	Niagara Nav.Co.
478 79906	*	**Cicero** I ScSr *George* 83-83 1 *Dk 2 H*	British Liverpool	1030 1588	301·5|32·4|21·8 *H 24f. P 11f. F 40f.*	Barrow Barrow S.B.Co.Ld.	1878	[illegible] Bhds.	2Cy.29"&59"–48" 180 HP. Barrow S.B.Co.,Ld.,Barrow	W.H.Dixon
479 80558	*	**Cid** I Sc3mSr *Searle* 79-83 2 *Dks 3 H*	British Newcastle	976 1518	245·0|33·1|23·0 *P 25f. F 28f.*	Stockton M.Pearse&Co.	1879	4 Bhds. DB[illegible]	2Cy.31"&58"–36" 130 HP. Blair&Co.,Ld.,Stockton	Durham S.S.Co.
480	*	**Cidade da Praia** I ScSr (ex Selinda) 1 *Dk 2 H*	Prtug'ese Lisbon	416 645	180·0|28·1|15·3 *P 25f. H 44f.*	Glasgow D.&W.Henderson&Co.	1883	4 Bhds.	2Cy.18"&37"–30" 90 HP. D.&W.Henderson&Co.,Gls.	Empreza Nacional de Navegação

*** = Lloyd's Register;** D = Nederlandsche Vereeniging; G = Germanischer Lloyd; I = Registro Italiano; R = American Re…
A = Veritas Austro-Ungarico; H = Veritas Ellenico; N = Norske Veritas; V = Bureau Veritas.

In the 2nd Column, B.T., with figures following, signifies Date of report of bottom sighting by Board of Trade Surveyors as condition of PASSENGER CERTIFI…

In the 3rd Column, C = Composite (iron frame, wood planked); I = Iron; Stl = Steel; W = Wood.

Further particulars of Vessels classed by Lloyd's Register will be found in Lloyd's Register Book and Lloyd's Yacht Register.

	Nom, Matériaux, Gréement, etc. / Capitaine et dates d'entrée au service des Armateurs actuels et à bord du navire. / Ancien Nom, nombre de ponts, etc.	Pavillon. / Port d'Armement.	Tonnage. Net. / Total	Dimensions. Longueur / Largeur / Creux / Constructions sur le Pont	Construit. Où. / Par qui.	Quand.	Nombre des cloisons. / Détails sur le lest d'eau.	Machines. / Détails. Nom du Constructeur.	Armateurs ou Directeurs-Gérants.
*	Cienfuegos ISeSr *Crofton -87 2 Dks & Spar Dk3 B*	American New York	1630 [illegible]	292·0|39·8|[illegible] *P 49 f. [illegible]*	Chester, Pa. J. Roach & Son	1883		2Cy. 38" & 71"–54" J. Roach & Son, Chester, Pa.	New York & Cuba Mail S.S. Co.
	Cienfuegos I	Spanish Cienfuegos	293 [illegible]	226·7|31·1|10·9	Wilmington	1864		120 HP.	Campos & Co.
	Ciño WPad *1 Dk*	Italian Naples	73 119	116·4|18·0|..	Procida A. Lubrand	1864		2Cy. 26"–36" 40 HP.	Soc. Anon. Procida Ischia
	Cifuentes ISeSr *Menendez -74 2 Dks*	Spanish Gijon	436 673	191·7|28·7|15·8 *P 50 f. F 31 f.*	Middlesbro' Backhouse & Dixon	1870	[illegible]	2Cy. 28" & 51"–36" 115 HP. T. Richardson & Sns, Hrtlepl.	O. de Olavarria & Co.
* D	Cilurnum ISeSr *Wise 74-81 2 Dks & 3 B*	British Newcastle	1427 2199	295·8|36·2|24·6 *P 28 f. B 67 f. F 29 f.*	Newcastle C.S. Swan & Hunter	1881	5 Bhds. [illegible]	2Cy. 34½" & 64"–42" 240 HP. R. & W. Hawthorn, N'wcastle	Elliott, Lowrey & Dunford
*	Cincora ISeSr *Lloyd -84 2 Dks*	British Liverpool	688 1081	210·0|29·2|20·1 *B 21 f.*	Liverpool R. & J. Evans	1874	5 Bhds.	2Cy. 28" & 50"–30" 110 HP. De Winton & Co., Carnarvon	Richards, Mills & Co.
	Cinderella ISeScYt	British Cowes	78 133	131·4|19·5|9·2	Rutherglen T.B. Seath & Co.	1874		2Cy. 17" & 34"–20" 68 HP. A. Campbell & Son, Glasgow	Joseph H. Atkey
*	Cintra ISeBg *Banks 1 Dk3 B*	British Sydney	1175 1979	270·3|37·2|22·7 *P 62 f. B 62 f. F 62 f.*	Glasgow Napier, Shanks & Bell	1882	1 Bhds. MT 260t	2Cy. 41" & 78"–54" 400 HP. J. & J. Thomson, Glasgow	Australasian United Steam Nav. Co., Ld.
	Cintra ISeSr *Pixley 63-75*	British Liverpool	300 482	178·8|24·2|14·8 *Q 41 f.*	Paisley Blackwood & Gordon	1854		2Cy. 21" & 40"–26" 80 HP. J. Jack & Co., Liverpool	Coteswoth, Lyne & Co.
	Cintra	German							Hamburg Sud Amerik Dampfs. Gesell.
*	Cipero ISeSr *Dockard 86-89 1 Dk*	British Glasgow	908 1419	243·8|32·7|16·6 *B 56 f. Q 92 f. P 31 f. F 22 f.*	Glasgow J. & G. Thomson	1879	5 Bhds. [illegible]	2Cy. 29" & 52"–36" 150 HP. J. & G. Thomson, Glasgow	Caw, Prentice, Clapperton & Co.
I	Cipro (ex Cavalier, ex Emma) ISeSr *Quartino 85-86 2 Dks*	Italian Genoa	592 900	203·4|28·8|16·5 *P 130 f. F 31 f. B 41 f.*	Sunderland R. Thompson	1870	4 Bhds. [illegible]	2Cy. 29" & 53"–36" 212 HP. N. Odero & Ferro, Sestri, Pon.	Navigazione Generale Italiana
	Circassia ISeBk *Campbell 65-83 2 Dks & Spr Dk 4 B*	British Barrow	2770 4272	399·7|42·0|25·1 33·2	Barrow Barrow S.B. Co.	1878	7 Bhds.	2Cy. 59" & 107"–48" 600 HP. Barrow S.B. Co., Ld., Barrow	Barrow Steam Ship Co., Ld.
*	Circassia ISeSr *Colcord 83-85 1 Dk2 B*	British W. Hartlepl	889 1381	241·0|33·3|18·2 *B 59 f. Q 84 f. F 31 f.*	W. Hartlepool W. Gray & Co.	1878	4 Bhds. [illegible]	2Cy. 29" & 55"–33" 120 HP. T. Richardson & Sons, Hrtlpl.	W. Gray
	Circassian ISeBk *Barrett -87*	British Glasgow	2356 3724	415·5|40·3|18·5 33·5	Greenock R. Steele	1872		2Cy. 55" & 92"–48" 500 HP. Laird Bros., Birkenhead	J. & A. Allan
*	Circe (ex Glamis Castle) ISeBn *Crighton 85-87 2 Dks 3 B*	British Glasgow	1559 2354	331·9|35·7|26·4 *F 39 f. B 90 f. P 14 f.*	Glasgow Aitken & Mansel	1874	5 Bhds.	3Cy. 25" 41" & 68–48" 300 HP. Barclay, Curle, & Co., Ld., Gls.	Donaldson Brothers
* D	Ciscar ISeSr *Mantarola 80-87 1 Dk2 B*	Spanish Barcelona	492 799	194·3|29·5|15·3 *B 9 f. P 56 f. F 27 f.*	South Shields J. Readhead & Co.	1880	4 Bhds.	2Cy. 23" & 46"–30" 90 HP. J. Readhead & Co., S. Shields	J. Roca & Co.
	Cisne IT-Sc *Maggio*	Argentine Buenos Ayres	138 221	162·0|21·0|9·0	Glasgow D. Rowan & Co.	1869		2Cy. 18" & 36"–20" 55 HP. D. Rowan, Glasgow	La Platence Flotilla Co., Ld.
	Citie of the Tribes IPadSp	British Galway	48 117	99·5|18·8|9·2	South Shields J. Eltringham	1872		2Cy. 25"–48" 50 HP. J.P. Rennoldson, S. Shields	Galway Bay Steam Boat Co., Ld.
	Citizen IPadSk	British	74 130	160·0|17·7|7·1	Port Glasgow Blackw'd & Gordon	1861		2Cy. 36"–36" 80 HP. Blackwood & Gordon, P. Gls.	Citizens' River Steamers Co., Ld.
V	Cito ISeSr *Nodop -88 1 Dk*	German Blankenese	387 588	168·9|25·6|13·3 *Q 48 f. B 47 f. F 20 f.*	Flensburg Flnsbrger Schiffsb. Ges.	1884	DT 113 AFT 71	2Cy. 21" & 38"–27" 65 HP. Flensburger Schiffsb. Ges.	H.M. Struve
N	Cito (ex Sailing Vessel) WSeSr *Kronsen -87 1 Dk2 B*	Norweg'n Sandefjord	123 182	101·4|24·2|11·6 *Q 6 f. B 19 f.*	Stavanger	1873		2Cy. 12½" & 18"–14" 15 HP. Frederikstads Mek. Verks.	Actieselskabet Oceana (Chr. Christensen)
V	Cito (ex Riga Lübeck) ISeSr *Reanenberg*	Russian Windau	194 253	151·9|20·7|9·3 *Len. 76. P 30 f. F 30 f.*	Kiel Nord-Schiffsb. act. Ges.	1870		1Cy. 13" & 26"–27" 50 HP. Schweffel & Howaldt, Kiel	W.G. Reincke
*	Citta di Bari (ex [illegible]) ISeSr *A. Penzo 1 Dk2 B*	Italian Bari	681 1108	230·0|30·2|16·7 *P 249 f. F 38 f.*	Stockton M. Pearse & Co.	1877	4 Bhds. *WBaf*	2Cy. 25½" & 51½"–33" 99 HP. Blair & Co., Ld., Stockton	Aicardi, Heynes & Co.
*	Citta di Barletta ISeSr *Giunta (ex Celeste) 1 Dk2 B*	Italian Bari	408 645	167·3|23·8|16·3 *B 78 f. Q 76 f. F 27 f.*	Middlesbro' R. Craggs & Sons	1875	4 Bhds. [illegible]	2Cy. 21" & 45½"–30" 80 HP. Blair & Co., Ld., Stockton	Aicardi, Heynes & Co.
*	Citta di Messina ISeBnSr *(ex Glenlyon ex [illegible]) Dunn 1 Dk2 B*	British London	959 1490	261·3|33·7|20·7 *P 37 f. F 36 f.*	Newcastle Schlesinger, Davis & Co.	1879	4 Bhds. [illegible] FPT APT	2Cy. 32" & 62"–40" 180 HP. Hawks, Crawshay & Co., Gatshd.	W.F. Becker
*	Citta di Milano ISeSr *1 Dk & Awn Dk 3 B*	Italian Spezia	723 1247	240·0|32·2|16·9 23·7	Sunderland R. Thompson & Sons	1886	5 Bhds. [illegible] DTA 103t	2Cy. 35" & 68"–42" 180 HP. G. Clark, Sunderland	Pirelli & Co.
* I	Citta di Roma Stl & ISeSr *Tiscorina -88 2 Dks 3 B*	Italian Genoa	1308 2032	285·0|36·9|24·0 *B 56 f. P 28 f. F 32 f.*	Newcastle C.S. Swan & Hunter	1888	5 Bhds. [illegible] MT 300t	3C. 22" 35" & 58"–39" 200 HP. Wallsend Slipwy. & E. Co., Ld., Newcastle	B. Mazzini (G.B. Gastaldi & Co., Managers.)
	Citta di Ventimiglia ISeSr *(ex Ville de Marseille, ex Stromboli) Lappoli -88*	Italian P'rto M'rizio	343 544	198·8|26·4|4·5 *F 16 f.*	Greenock Max Clark	1854	[illegible]	2Cy. 26" & 49"–30" 98 HP. Freycinet & Co., Marseilles	P. Viale
*	City of Aberdeen ISeSr *Marchant 67-73 2 Dks*	British Aberdeen	481 852	240·8|30·1|16·6 *P 80 f. F 50 f. B 60 f.*	Port Glasgow Cunliffe & Dunlop	1873	4 Bhds.	2Cy. 46" & 80"–39" 300 HP. Cunliffe & Dunlop, P. Glasgw	Aberdeen Steam Nav. Co.

Lloyd's Register; D = Nederlandsche Vereeniging; G = Germanischer Lloyd; I = Registro Italiano; R = American Record; Veritas Austro-Ungarico; H = Veritas Ellenico; N = Norske Veritas; V = Bureau Veritas.

Dans la 2me **Colonne**, B.T., avec les indications qui suivent, signifie la date du rapport de visite de la Carène par les experts du Board of Trade pour obtenir le certificat relatif au TRANSPORT DES PASSAGERS.

Dans la 3me Colonne, C = Composite (membrure en fer, bordé en bois); I = Fer; Stl = Acier; W = Bois.

Pour plus amples renseignements sur les navires classés par le Lloyd's Register, voir le Lloyd's Register Book et le Lloyd's Yacht Register.

No. in Reg. / Official No.	Registry in which classed	Ship's Name. Material, Rig, &c. / Master, and dates of Appointment to present Owner's Service and to Ship. / Ship's Late Name. No. of Decks, &c.	Flag. / Port of Registry.	Tons Gross / Net	Dimensions: Length / Breadth / Depth / Deck Erections	Built Where. / By Whom.	When	No. of Bulkheads / Particulars of Water Ballast.	Engines. Particulars. Maker's Name.	Owners or Managers
511 48251		City of Adelaide I ScSr *2 Dks & Spr Dk*	British Sydney	773 [illegible]	252·8 | 28·3 | 24·4	Glasgow J.&G.Thomson	1864		2Cy.54½"–28" 200 HP J.&G.Thomson,Glasgow	Australasian [illegible] SteamNav.Co.[illegible]
512 81475	✻ BT 5.88	City of Agra I ScBk *Gordon 83–88* *3 Dks*	British Glasgow	2133 [illegible]	385·4 | 38·7 | 28·6 *B 30 f. F [illegible]*	Glasgow C.Connell&Co.	1879	5 Blds.	2Cy.50"& 86"–54" 380 HP J.&J.Thomson,Glasgow	G.Smith & Sons
513 125780	RV	City of Alexandria I ScBk *Reynolds* *3 Dks*	American New York	1702 2180	307·0 | 38·5 | 26·3	Chester, Pa. J.Roach&Son	1879		2C.42"& 78"–54" 2000 HP J.Roach&Son	New York & Mail S.S.Co.
514 77381	✻D	City of Amsterdam I ScSr *Hoare 87–88* *2 Dks*	British Dublin	524 841	220·5 | 30·2 | 15·6 *P 33 f. F 29 f.*	Stockton Richardson,Duck& Co.	1877	5 Blds. [illegible]	2Cy.28"& 52"–33" 120 HP T.Richardson&Sons,Hrtlpl	Palgrave, M[illegible] Co.
515 125404	R	City of Atlanta I Sc3mSr *Hansen –88 2 Dks,Spt Shd Dk*	American New York	1155 1621	242·0 | 40·0 | 20·3 *F 96 f.*	Brooklyn J.English&Sons	1875		1Cy.48"–60" 120 HP QuintardIronWorks,N.Y.	NewYork&[illegible] S.S.Co.
516 125980	RV	City of Augusta I ScSr *Catherine 80–87 2 Dks,Spt Hur Dk*	American Savannah	1929 2861	323·0 | 40·0 | 17·0 *B 118 f. F 33 f. 259*	Chester, Pa. J.Roach&Son	1880		2Cy.42"& 82"–54" J.Roach&Son,Chester,Pa.	Ocean S.S.Co.
517 71094		City of Belleville W Sc *2 Dks*	British Bllville, Ont	69 101	89·7 | 15·4 | 7·0	St. Catharines	1878		1Cy. 60 HP	H.D.Purkis
518 70915	BT 11.88	City of Berlin I ScBk *Land 77–84*	British Liverpool	3302 5526	488·6 | 44·2 | 34·9 *F 67 f.*	Greenock Caird&Co.	1874		3C.41"63"&101"–66" 1000 [H LairdBros,Birkenhead	Inman&Inter[illegible] S.S.Co.Ld.
519	R	City of Birmingham IScSr *Rorg* *3 Dks*	American Boston,Mass	2153 2807	300·0 | 42·4 | 26·9	Chester, Pa. Delaware River Iron Shipbldng & Eng.Co.	1888		3Cy.24",38"&63"–45"	New Eng. & S[illegible] S.S.Co.
520 90098	✻ BT 10.88	City of Bombay I ScBk *Anderson 64–85* *3 Dks*	British Glasgow	2938 4492	404·0 | 48·0 | 29·2 *B 150 ft. F 44 ft.*	Belfast Workman, Clark & Co.	1885	7 Blds.	2Cy.50"& 96"–60" 650 HP J.&J.Thomson,Glasgow	G.Smith & Sons
521 5270		City of Boston W Pad	American New London, Conn	1457 1577	301·0 | 40·0 | 12·3	Brooklyn	1861		1400 HP	O.L.Johnson
522 125370	R	City of Bridgeport W Sc *Wilcox 86 1 Dk,Slight Dk*	American Bridgeport	274 518	148·8 | 28·4 | 10·0	P.Jeffersn.L.I. Bayles&Son	1886		2Cy.20"& 34"–26"	Bridgeport S[illegible] Co.
523 31459	✻	City of Bristol I ScSr *Walsh 76–83* *1 Dk 2 B*	British Dublin	799 1252	235·1 | 32·6 | 15·7 *P & B 139 f. F 29 f.*	Hartlepool E.Withy&Co.	1882	4 Blds. [illegible]	2Cy.31"& 58"–33" 150 HP T.Richardson&Sons,Hrtlpl	Palgrave, M[illegible] Co.
524 82793	✻	City of Brussels I ScSr *Rolf 76–84* *1 Dk*	British London	167 272	135·1 | 22·5 | 10·8 *Q 28 f. B 12 f. F 20 f.*	Dumbarton R.Chambers,jun.	1880	3 Blds.	2Cy.20"& 36"–27" 67 HP Kincaid,Donald&Co,Grnck	D.C.Thomas [illegible]
525 45019	✻	City of Cadiz I Sc3mSr *Kehoe 76–84 (ex Corcyra)* *1 Dk 2 B*	British Dublin	474 749	204·4 | 28·5 | 17·3 *P & F*	Stockton Richardson Duck& Co.	1862	5 Blds.	2Cy.26"& 49"–30" 98 HP T.Richardson&Sons,Hrtlpl	Palgrave, Mur[illegible] Co.
526 85852	✻ BT 5.88	City of Calcutta I Sc3mSr *Thoms 78–84* *3 Dks*	British Glasgow	2505 3836	400·0 | 42·1 | 30·1 *F 34 f.*	Glasgow C.Connell&Co.	1881	6 Blds.	2Cy.50"& 96"–60" 600 HP J.&J.Thomson,Glasgow	G.Smith & Sons
527 86720	✻ BT 5.88	City of Cambridge I Sc4mBk *Marr 75–85* *3 Dks*	British Glasgow	2473 3788	400·0 | 42·1 | 29·6 *B 156 f. F 31 f.*	Belfast Workman, Clark & Co.	1882	7 Blds.	2Cy.50"& 96"–60" 600 HP J.&J.Thomson,Glasgow	G.Smith & Sons
528 71704	✻ BT 10.88	City of Canterbury I ScBk *Gray 83–85* *3 Dks*	British Glasgow	2100 3212	379·7 | 38·2 | 29·2 *F 28 f.*	Glasgow Barclay,Curle&Co.	1875	5 Blds.	2Cy.50"& 86"–48" 440 HP Barclay,Curle&Co.,Gl'sgow	G.Smith & Sons
529 68068	✻ BT 1.88	City of Carthage I Sc3mSr *Griere 76–85* *2 Dks 3 B*	British Glasgow	1717 2651	360·3 | 36·8 | 26·4	Glasgow C.Connell&Co.	1873	5 Blds.	2Cy.46"& 84"–45" 350 HP J.&J.Thomson,Glasgow	G.Smith & Sons
530 92734		City of Chatham WSc2m *Comerlet –88* *3 Dks*	British Toronto	232 341	125·6 | 28·5 | 9·0	Toronto Polson Iron Works	1888		2Cy.18"&31"–24" 150 HP PolsonI.Works,Ld.,Toronto	Chatham Nav.[illegible]
531 69272	BT 8.88	City of Chester I ScBkn *Bond 88–88*	British Liverpool	2944 4770	444·6 | 44·2 | 34·6	Greenock Caird&Co.	1873		2Cy.68"&120"–66" 850 HP Caird&Co,Greenock	Inman&Inter[illegible] S.S.Co.Ld.
532 125473		City of Chester I ScBn *Wallace –83 COLLISION 2 Dks 3 B*	American Prtlnd,Oreg	785 1106	206·6 | 33·6 | 16·0	Chester, Pa. J.Roach&Son	1874		2Cy.21"& 44"–45" 100 HP J.Roach&Sons,Chester,Pa.	Oregon Ry.& S[illegible] Nav.Co.
533 87871	✻ BT 9.88	City of Chicago I Sc4mSr *Lewis 83–88* *4 Dks*	British Liverpool	3383 5202	430·6 | 45·0 | 33·6 *B 188 f.*	Glasgow C.Connell&Co.	1883	7 Blds.	3C.56"80"&80"–60" 900 HP J.&J.Thomson,Glasgow	Inman&Inter[illegible] S.S.Co.Ld.
534 125878	R	City of Columbia W ScSr *McKee –83* *2 Dks 3 B*	American New York	1285 1878	266·0 | 38·7 | 21·5	Greenport N.Y. J.English&Son	1880		2C.36"& 60"–54" 1000 HP QuintardIronWorks,N.Y.	NewYork&Ch[illegible] S.S.Co.
535 60574	✻	City of Cork (ex Marquis of Lorne) I ScSr *Hungerford 79–83* *1 Dk 2 B*	British Dublin	659 1019	229·5 | 30·0 | 17·2 *P 115 f. F 34 f.*	Sunderland Oswald&Co.	1870	4 Blds. W.Blldy	2Cy.28"& 52"–33" 120 HP T.Richardson&Sons,Hrtlpl	Palgrave, Mur[illegible] Co.
536 52560	BT 4.88	City of Cork I PadSp	British Cork	94 146	150·0 | 17·1 | 6·9 *Q 49 f.*	Cork H.Robinson&Co.	1866		2Cy.35"–36" 75 HP J.Penn&Sons,Greenwich	Citizens' Rive[illegible] Steamers Co.Ld.
537 126076	R	City of Dallas W Sc3mSr *Reed 83–84* *2 Dks 3 B*	American N.Orleans,L	479 815	185·8 | 35·0 | 18·0 *B 74 f.*	Portland,Con. C.Mellony	1872		1Cy.36"–42" 260 HP C.H.Delamater&Co.N.Y.	New Orleans &[illegible] Royal [illegible]
538 54048	✻	City of Dortmund I ScBg *Cuslit 84–87 (ex Lord Byron)* *1 Dk 2 B*	British Dublin	515 758	200·5 | 29·4 | 17·4 *P 35 f. F 36 f. B 28 f.*	Stockton Richardson,Duck& Co.	1865	5 Blds.	2Cy.26"& 49"–36" 98 HP T.Richardson&Sons,Hrtlp	Palgrave,Murp[illegible]
539 [illegible]	✻D	City of Dublin Stl ScSr *McNeil –88* *2 Dks*	British Glasgow	2150 3267	361·7 | 42·7 | 26·4 *B 92 f. F 46 f.*	Belfast Workman Clark&Co.	1888	7 Blds. [illegible] Twin	3C25½",43"&67½"–48" 350 [B J.&J.Thomson,Glasg	G.Smith & Sons
540 73890	✻ BT 7.88	City of Edinburgh I ScBk *Barholm 80–82* *3 Dks*	British Glasgow	2102 3230	381·5 | 38·5 | 29·0 *F 28 f.*	Glasgow C.Connell&Co.	1876	5 Blds.	3C.38"54"&73"–48" 500 HP Tripled'M,J.&J.Thomson,Glasg	G.Smith & Sons

✻ = **Lloyd's Register**; D = Nederlandsche Vereeniging; G = Germanischer Lloyd; I = Registro Italiano; R = American Rec[ord]; A = Veritas Austro-Ungarico; H = Veritas Ellenico; N = Norske Veritas; V = Bureau Verita[s]

In the 2nd Column, B.T., with figures following, signifies Date of report of bottom sighting by Board of Trade Surveyors as condition of PASSENGER CERTIF[ICATE]

In the 3rd Column, C = Composite (iron frame, wood planked); I = Iron; Stl = Steel; W = Wood.

Further particulars of Vessels classed by Lloyd's Register will be found in Lloyd's Register Book and Lloyd's Yacht Register.

No.		Nom, Matériaux, Gréement, etc. Capitaine et dates d'entrée au service des Armateurs actuels et à bord du navire. Ancien Nom, nombre de ponts, etc.	Pavillon. Port d'Armement	Tonnage. Net. Total	Dimensions. Longueur	Largeur	Creux	Construit. Où. Par qui.	Quand.	Nombre des cloisons. Détails sur le lest d'eau.	Machines. Détails. Nom du Constructeur.	Armateurs ou Directeurs-Gérants.
	✻	City of Exeter 1 Sc3mSr Law 87-88 (ex DUNDEE?) 1 Dk2 B	British Exeter	788 1034	228·2 P 134 f.	29·2 B 12 f.	17·1 F 28 f.	Stockton M.Pearse&Co.	1870	4 Blds.	2Cy.28"&53"–33" 110 HP. Blair&Co.,Ld.,Stockton	J.Holman & Sons
		City of Fitchburg W ScSr Sprague 2 Dks3 B	American FallRvr,Mas	466 822	188·0	33·0	13·5	Williamsbrg,NY C.H.Delemater&Co	1874		1Cy.36"–36" C.H.Delemater&Co.,N.Yk.	J.S.Brayton
	✻	City of Ghent 1 ScSr Bryan 81-82 1 Dk	British London	119 199	135·9	20· P 20 f.	9·7	Grimsby T.Charlton	1871	1 Blds.	2Cy.14"&28"–20" 40 HP. Windsr&Dunsmuir,London	D.C.Thomas & Son
	R	City of Gloucester W ScSr	American Boston,Mass	396 561	149·0	28·0	22·0	Brooklyn J.D.Leary	1883		2Cy.21"&36"–30"	H.M.Whitney
545		City of Grafton 1 PadSr Greer 1 Dk & Awn Dk	British Sydney	555 825	207·4	26·9	11·9 193	Glasgow A.Stephen&Sons	1876	4 Blds.	2Cy.33"&56"–60" 150 HP. A.Stephen&Sons,Glasgow	Clarence & Richmond Rivers Steam Nav.Co.
546	✻D	City of Hamburg 1 ScSr McCallaghan 88-88 2 Dks	British Dublin	742 1174	231·0 P 134 f.	32·6	14·5 F 20 f.	Sunderland J.Laing	1881	1 Blds.	2Cy.31"&58"–33" 150 HP. T.Richardson&Sons,Hrtlpl	Palgrave, Murphy & Co.
		City of Jacksonville 1 Sc Shaw	American Wilmington	396 460	170·0	32·6	7·9	Wilmington,Del Harlan&Hollingsworth	1882		2Cy.30"–72" TheHarlan&HollingsworthCo.,Wilm.,Del.	De Bary Mer.Line
	✻	City of Khios 1 ScBk McPherson 80-84 3 Dks	British Glasgow	2121 3246	381·5	38·4 B 29 f.	29·2 F 30 f.	Glasgow Barclay,Curle&Co.	1878	6 Blds.	2Cy.50"&86"–54" 500 HP. Barclay,Curle&Co.,Glasgw.	G.Smith & Sons
		City of Lawrence 1 Pad	American New London	1351 1878	243·0	62·5	11·9	Wilmington,Del TheHarlan&HollingsworthCo.,Wilm.,Del	1867		1Cy.65"–132" 1000 HP. TheHarlan&HollingsworthCo.,Wilm.,Del.	O.L.Johnson
550	✻	City of Lincoln 1 ScSr Fry -87 (ex Selis, ex Massachusetts, ex Manhattan) 3 Dks	British Liverpool	2104 3182	335·0 P 184 f.	42·5	27·7 F 49 f.	Newcastle Palmers'Co.,Ld.	1866	6 Blds.	3Cy.26,42&68"–42"259HP W.Doxford&Sons,Sld.	W.H.Ross & Co.
551	✻	City of Lisbon 1 ScSr Kelly 85-86 (ex Anglian)	British Dublin	389 633	204·4 P 46 f.	26·3	15·9 F 48 f.	London C.Langley	1864	4 Blds.	2Cy.26"&49"–30" 98 HP. T.Richardson&Sons,Hrtlpl.	Palgrave,Murphy & Co.
552		City of Liverpool W Sc Stubbs	British Liverpool	100 147	88·8	20·8	10·2	Winsford G.Deakin,Ld.	1885		2Cy.13½"&22"–16" 32 HP. E.Foden,Sandbach	Salt Union,Ld.
553	✻	City of London 1 ScBk Black 83-82 3 Dks	British Glasgow	2097 3212	381·6	38·8 F 28 f.	29·0	Glasgow C.Connell&Co.	1876	5 Blds.	2Cy.52"&89"–48" 450 HP. J.&J.Thomson,Glasgow	G.Smith & Sons
554	✻	City of London 1 ScSr Chalmers 76-85 1 Dk2 B	British Aberdeen	477 844	241·9 P 70 f.	30·5	16·7 F 70 f.	Glasgow J.Elder&Co.	1871	5 Blds.	2Cy.46"&80"–39" 275 HP. J.Elder&Co.,Glasgow	AberdeenStm.Nav.Co.
555		City of London 1 PadSp	British Liverpool	 143	114·5	19·1	10·2	Poplar	1863		2Cy. 140 HP.	R.Strong
556	R	City of Macon 1 ScSr Kelly -87 2 Dks & Awn Dk	American S'vannah,Ga	1071 2093	257·0	38·7	24·8	Chester,Pa. J.Roach&Son	1877		2Cy.38"&68"–54" 400 HP. J.Roach&Son,Chester,Pa.	Ocean S.S.Co.
	✻D	City of Malaga 1 ScSr Gough 81-87 (ex Borthwick) 1 Dk2 B	British Dublin	650 1026	220·0 P 29 f.	30·2	16·9 F 24 f.	Sunderland J.L.Thompson	1872	4 Blds.	2Cy.27"&50"–30" 98 HP. G.Clark,Sunderland	Palgrave,Murphy & Co.
558	✻D	CityofManchester 1 Sc3mSr Brophy 83-88 2 Dks3 B	British Whitby	2029 3120	337·0 P 40 f.	41·4 B 82 f.	26·4 F 47 f.	Sunderland Sundrld.S.B.Co.,Ld.	1883	5 Blds.	2Cy.40"&74"–54" 350 HP. N.E.MarineE.Co.,Ld.,Sndld.	J.Gray & Co.
559		City of Melbourne 1 ScBk Thompson 1 Dk	British Sydney	580 838	250·4	28·2 P 59 f.	16·6	Glasgow J.&G.Thomson	1862		2Cy.40"&70"–33" 200 HP. J.Inglis,Glasgow	Australasian United Steam Nav.Co.,Ld.
560		City of Monticello 1 PadSc McKee (ex City of Norfolk) 2 Dks, Spt Hur Dk	American New York	478 893	232·0	32·0	10·9	Wilmington,Del Harlan&Hollingsworth	1866		1Cy.50"–120" 220 HP. Hrln&Hllngwrth,Wlmgtn.	Florida S.S.Co.
561		City of Mount Clemens W ScSr 1 Dk	British Mt.Clmns,Ont	69 102	95·0	21·8	6·0	Mt.Clmns,C.S.	1880		1Cy.14½"–16" 85 HP.	R.T.Riddle
		City of New Bedford W ScSr Fish 2 Dks3 B	American FallRvr,Mas	433 822	180·0	32·6	13·0	Williamsbrg,NY C.H.Delemater&Co.	1874		1Cy.36"–36" C.H.Delemater&Co.,N.Yk.	Old Colony Steamboat Co.
	✻	CityofNewYork StlTSc3mSr Watkins 87-88 4 Dks & Prom Dk 5 B	British Liverpool	5930 10499	527·6 P 75 f.	63·2 F 109 f.	22·0 39·2	Glasgow J.&G.Thomson	1888	15 Blds.	6C.45"71"&113"–60" 2000 HP. J.&G.Thomson,Glsgw	Inman& International S.S.Co.,Ld.
	RV	City of New York 1 ScBk Searle 72-84 4 Dks	American New York	1964 3020	339·0	40·2	20·5 28·9	Chester,Pa. J.Roach&Son	1875		2Cy.39"&68"–42" 700 HP. J.Roach&Son,Chester,Pa.	Pacific Mail S.S.Co.
		City of New York W Pad	American New London, Conn.	1468 1592	301·0	40·0	12·3	Brooklyn	1861			O.L.Johnson
	✻D	City of Newcastle 1 ScSr Townsend 86-86 2 Dks3 B	British Newcastle	1298 1985	285·5 P 28 f.	36·1 B 52 f.	24·1 F 32 f.	Newcastle C.S.Swan&Hunter	1882	6 Blds.	2Cy.36"&68"–42" 200 HP. R.&W.Hawthorn,Nwcastle	Tyne&BlythS.S.Owning Co.,Ld.
		City of Newport W Pad	American Prvdnce,R.I	469 562	177·5	28·5	8·8	New York	1863		250 HP.	B.Buffum
		City of Norwich W Pad	American New London, Conn.	499 997	200·7	36·0	13·2	New York	1862		700 HP.	O.L.Johnson
	✻	City of Oporto 1 ScSr Scott 80-80 (ex Marcasite) 1 Dk2 B	British Dublin	560 874	215·7 Q 20 f.	30·1	15·4 F 23 f.	Sunderland J.Blumer&Co.	1872	1 Blds.	4Cy.18"&37"–36" 95 HP. T.Bates&Co.,SowrbyBridge	Palgrave Murphy&Co.
	✻	City of Oxford 1 ScBk Miller 75-82 3 Dks	British Glasgow	2573 3959	400·0 B 62 f.	43·2	29·8 F 34 f.	Glasgow Barclay,Curle&Co.	1882	6 Blds.	2Cy.52"&96"–60" 660 HP. Barclay,Curle&Co.,Glasgw.	G.Smith & Sons

✻=Lloyd's Register; D=Nederlandsche Vereeniging; G=Germanischer Lloyd; I=Registro Italiano; R=American Record;
=Veritas Austro-Ungarico; H=Veritas Ellenico; N=Norske Veritas; V=Bureau Veritas.

Dans la 2me Colonne, B.T., avec les indications qui suivent, signifie la date du rapport de visite de la Carène par les experts du Board of Trade pour obtenir le certificat relatif au TRANSPORT DES PASSAGERS.

Dans la 3me Colonne, C=Composite (membrure en fer, bordé en bois); I=Fer; Stl=Acier; W=Bois.

Pour plus amples renseignements sur les navires classés par le Lloyd's Register, voir le Lloyd's Register Book et le Lloyd's Yacht Register.

Reg. No. / Official No.	Registry in which classed	Ship's Name. Material, Rig, &c. / Master and Dates of Appointment to present Owner's Service and to Ship / Ship's Late Name. No. of Decks, &c.	Flag. / Port of Registry	Tons Net. / Gross	Dimensions (Length, Breadth, Depth) / Deck Erections	Built. Where. / By Whom	When	No. of Bulkheads / Particulars of Water Ballast	Engines. Particulars. / Maker's Name	Owners or Managers
511 49291		City of Adelaide I ScSc 2 Dks & Spr Dk	British Sydney	779 1212	252·8 28·3 24·4	Glasgow J.&G.Thomson	1864		2Cy.54½"–38" 200 HP J.&G.Thomson, Glasgow	Australasian [illegible] Steam Nav.Co.Ld
512 84476	✻ BT1.88	City of Agra I ScBk Gordon 69–0 3 Dks	British Glasgow	2133 3374	385·4 38·7 28·6 B 30 f. F 35 f.	Glasgow C.Connell&Co.	1879	5 Blds.	2Cy.50"& 86"–54" 300 HP J.&J.Thomson, Glasgow	G.Smith & Sons
513 [illegible]780	RV	City of Alexandria I ScBn Reynolds 3 Dks	American New York	1702 2480	307·0 38·5 26·3	Chester, Pa. J.Roach&Son	1879		2C.42"& 78"–54" 2000 HP J.Roach&Son	New York & [illegible] Mail S.S.Co.
514 77388	✻D	City of Amsterdam I ScSc Hoare 87-81 2 Dks	British Dublin	524 841	220·5 30·2 15·6 P 33 f. F 29 f.	Stockton Richardson, Duck & Co.	1877	5 Blds. [illegible]	2Cy.38"& 52"–33" 120 HP T.Richardson&Sons, Hrtlpl	Palgrave, Murphy & Co.
515 125464	R	City of Atlanta I Sc3mSr Hansen -88 2 Dks Spt Shd Dk	American New York	1155 1621	242·0 40·0 20·3 F 26 f.	Brooklyn J.English&Sons	1875		1Cy.48"–60" 120 HP Quintard Iron Works, N.Y.	New York & Char[illegible] S.S.Co.
516 125858	RV	City of Augusta I ScSc Catherine 80-87 2 Dks Spt Hur Dk	American Savannah	1929 2861	323·0 40·0 17·0 B 118 f. F 43 f. 25·9	Chester, Pa. J.Roach&Son	1880		2Cy.42"& 82"–54" J.Roach&Son, Chester, Pa.	Ocean S.S.Co.
517 71094		City of Belleville W Sc 2 Dks	British Blville, Ont	69 101	89·7 15·4 7·0	St. Catharines	1878		1Cy. 60 HP	L.D.Purkis
518 70213	BT11.88	City of Berlin I ScBn Land 77-84	British Liverpool	3302 5526	488·6 44·2 34·9 F 97 f.	Greenock Caird&Co.	1874		3C.41"65"&101"–66" 1000 HP Laird Bros., Birkenhead	Inman & Internat. S.S.Co.,Ld.
519	R	City of Birmingham I ScSr Bury 3 Dks	American Boston, Mass	2153 2867	300·0 42·4 26·9	Chester, Pa. Delaware River Iron Shipbldng & Eng.Co.	1888		3Cy.24".38"&63"–45"	New Eng. & Sav. S.S.Co.
520 90098	✻ BT10.88	City of Bombay I ScBk Anderson 84-85 3 Dks	British Glasgow	2938 4492	404·0 48·0 29·2 B 150 ft. F 44 ft.	Belfast Workman, Clark & Co.	1885	7 Blds.	2Cy.50"& 96"–60" 650 HP J.&J.Thomson, Glasgow	G.Smith & Sons
521 5270		City of Boston W Pad	American New London, Conn.	1457 1577	301·0 40·0 12·3	Brooklyn	1861		1400 HP	O.L.Johnson
522 126376	R	City of Bridgeport W Sc Wilcox 86 1 Dk Slight Dk	American Bridgeport	274 518	148·8 28·4 10·0	P.Jeffersn.L.I. Bayles&Son	1886		2Cy.20"& 34"–26"	Bridgeport Steamboat Co.
523 81429	✻	City of Bristol I ScSc Walsh 76-83 1 Dk 2 B	British Dublin	799 1252	235·1 32·6 15·7 P & B 138 f. F 29 f.	Hartlepool E.Withy&Co.	1882	4 Blds. [illegible]	2Cy.31"& 58"–33" 150 HP T.Richardson&Sons, Hrtlpl	Palgrave, Murphy & Co.
524 82793	✻	City of Brussels I ScSr Rolf 78-84 1 Dk	British London	167 272	135·1 22·5 10·8 Q 28 f. B 12 f. F 20 f.	Dumbarton R.Chambers, jun.	1880	3 Blds.	2Cy.20"& 36"–27" 67 HP Kincaid, Donald&Co., Grnck	D.C.Thomas & [illegible]
525 45019	✻	City of Cadiz I Sc3mSr Kehoe 76-84 (ex Corcyra) 1 Dk 2 B	British Dublin	474 749	204·4 28·5 17·3 P & F	Stockton Richardson, Duck & Co.	1862	5 Blds.	2Cy.26"& 45"–30" 90 HP T.Richardson&Sons, Hrtlpl	Palgrave, Murphy & Co.
526 85862	✻ BT5.88	City of Calcutta I Sc3mSr Thoms 78-84 3 Dks	British Glasgow	2505 3836	400·0 42·1 30·1 F 34 f.	Glasgow C.Connell&Co.	1881	6 Blds.	2Cy.50"& 96"–60" 600 HP J.&J.Thomson, Glasgow	G.Smith & Sons
527 84729	✻ BT5.88	City of Cambridge I Sc4mBk Marr 75-85 3 Dks	British Glasgow	2473 3788	400·0 42·1 29·6 B 156 f. F 34 f.	Belfast Workman, Clark & Co.	1882	7 Blds.	2Cy.50"& 96"–60" 600 HP J.&J.Thomson, Glasgow	G.Smith & Sons
528 71704	✻ BT10.87	City of Canterbury I ScBk Gray 83-85 3 Dks	British Glasgow	2100 3212	379·7 38·2 29·2 F 28 f.	Glasgow Barclay, Curle&Co.	1875	7 Blds.	2Cy.50"& 86"–48" 440 HP Barclay, Curle&Co., Glasgow	G.Smith & Sons
529 63068	✻ BT1.88	City of Carthage I Sc3mSc Grieve 76-85 2 Dks 3 B	British Glasgow	1717 2651	360·3 36·8 26·4	Glasgow C.Connell&Co.	1873	5 Blds.	2Cy.46"& 84"–45" 350 HP J.&J.Thomson, Glasgow	G.Smith & Sons
530 92734		City of Chatham W Sc2m Cameted -88 3 Dks	British Toronto	232 341	125·6 28·5 9·0	Toronto Polson Iron Works	1888		2Cy.18"&31"–24" 150 HP Polson I. Works, Ld., Toronto	Chatham Nav. [illegible]
531 63272	BT8.88	City of Chester I ScBkn Band 88-88	British Liverpool	2944 4770	444·6 44·2 34·6	Greenock Caird&Co.	1873		2Cy.68"&120"–66" 850 HP Caird&Co., Greenock	Inman & Internat. S.S.Co.,Ld.
532 125473		City of Chester I ScBn Wallace -88 COLLISION 2 Dks 3 B	American Prtlnd, Oreg	785 1196	206·6 33·6 16·0	Chester, Pa. J.Roach&Son	1874		2Cy.24"& 44"–45" 400 HP J.Roach&Sons, Chester, Pa.	Oregon Ry. & St. Nav.Co.
533 [illegible]871	✻ BT9.88	City of Chicago I Sc4mSr Lewis 83-88 4 Dks	British Liverpool	3383 5202	430·6 45·0 33·6 B 188 f.	Glasgow C.Connell&Co.	1883	7 Blds.	3C.56"80"&80"–60" 900 HP J.&J.Thomson, Glasgow	Inman & Internat. S.S.Co.,Ld.
534 125879	R	City of Columbia W ScSr McKee -87 2 Dks 3 B	American New York	1285 1878	266·0 38·7 21·5	Greenport N Y J.English&Son	1880		2C.36"& 60"–54" 1000 HP Quintard Iron Works, N.Y.	New York & Char[illegible] S.S.Co.
535 60574	✻	City of Cork (80' Mangrove & Lemon) I ScSc Hungerford 79-83 1 Dk 2 B	British Dublin	659 1019	229·5 30·0 17·2 P 115 f. F 34 f.	Sunderland Oswald&Co.	1870	4 Blds. [illegible]	2Cy.28"& 52"–33" 120 HP T.Richardson&Sons, Hrtlpl	Palgrave, Murphy & Co.
536 32560	BT1.88	City of Cork I PadSp	British Cork	94 140	150·9 17·1 6·9 Q 19 f.	Cork R.Robinson&Co.	1866		2Cy.35"–36" 75 HP J.Penn&Sons, Greenwich	Citizens' River Steamers Co.,Ld.
537 125670	R	City of Dallas W Sc3mSr Rea.1 83-84 2 Dks 3 B	American N.Orleans, L	479 915	185·8 35·0 18·0 B 74 f.	Portland, Con. C.Mallory	1872		1Cy.36"–42" 260 HP C.H.Delamater&Co., N.Y.	New Orleans & Royal Mail Steamship [illegible]
538 64840	✻	City of Dortmund I ScBg Codd 84-87 (ex Lord Byron) 1 Dk 2 B	British Dublin	515 758	200·5 29·4 17·1 P 35 f. B 36 f. B 28 f.	Stockton Richardson, Duck & Co.	1865	5 Blds.	2Cy.26"& 49"–30" 98 HP T.Richardson&Sons, Hrtlp	Palgrave, Murphy
539 [illegible]945	✻D	City of Dublin Stl ScSr McNeil -88 2 Dks	British Glasgow	2150 3267	361·7 42·7 26·4 B 92 f. F 38 f.	Belfast Workman, Clark&Co.	1888	7 Blds. [illegible]	3C.25½".43"&67½"–48" 350 HP J.&J.Thomson, Glasg.	G.Smith & Sons
540 73886	✻ BT7.88	City of Edinburgh I ScBk Barham 80-82 3 Dks	British Glasgow	2102 3238	381·5 38·5 29·0 F 28 f.	Glasgow C.Connell&Co.	1876	5 Blds.	3C.30"44"&73"–48" 300 HP Tripled 86, J.&J.Thomson, Glasg.	G.Smith & Sons

✻ = Lloyd's Register; D = Nederlandsche Vereeniging; G = Germanischer Lloyd; I = Registro Italiano; R = American Re[illegible]
A = Veritas Austro-Ungarico; H = Veritas Ellenico; N = Norske Veritas; V = Bureau Ve[illegible]

In the 2nd Column, B.T. with figures following, signifies Date of report of bottom sighting by Board of Trade Surveyors as condition of PASSENGER CERTI[illegible]

In the 3rd Column, C = Composite (iron frame, wood planked); I = Iron; Stl = Steel; W = Wood.

Further particulars of Vessels classed by Lloyd's Register will be found in Lloyd's Register Book and Lloyd's Yacht Register.

N°	Registre	Nom, Matériaux, Gréement, etc. — Capitaine et dates d'entrée en service des Armateurs actuels et à bord du navire — Ancien Nom, nombre de ponts, etc.	Pavillon. — Port d'Armement	Tonnage Net — Total	Dimensions (Construction sur le Pont)	Construit. Où. Par qui.	Quand.	Nombre des cloisons. Détails sur le lest d'eau.	Machines. Détails. Nom du Constructeur.	Armateurs ou Directeurs-Gérants.
	✻	**City of Exeter** I Sc3mSr *Lyon 87-88* (FOUNDERED) 1 *Dk2 B*	British Exeter	788 1054	228·2|29·2|17·1 P 114 f. B 12 f. F 28 f.	Stockton M.Pearse&Co.	1870	4 Blkds. [illegible]	2Cy.28"& 53"–33" 110 HP. Blair&Co.,Ld.,Stockton	J.Holman & Sons
		City of Fitchburg W ScSr *Springer* 2 *Dks3 B*	American Fall Rvr,Mas	466 822	188·0|33·0|13·5	Williamsbrg,NY C.H.Delemater&Co	1874		1Cy.36"–36" C.H.Delemater&Co.,N.Yk.	J.S.Brayton
	✻	**City of Ghent** I ScSr *Bryan 82-82* 1 *Dk*	British London	119 199	135·9|20·4| 9·7 F 20 f.	Grimsby T.Charlton	1871	4 Blkds. [illegible]	2Cy.14"& 28"–20" 40 HP. Windas&Dunsmuir,London	D.C.Thomas & Son
544	R	**City of Gloucester** W ScSr	American [illegible]	396 561	149·6|28·0|22·0	Brooklyn J.D.Leary	1883		2Cy.21"&36"–30"	H.M.Whitney
545		**City of Grafton** I PadSr *Creer* 1 *Dk & Awn Dk*	British Sydney	555 825	207·4|26·9|11·9 193	Glasgow A.Stephen&Sons	1876	4 Blkds.	2Cy.33"& 56"–60" 150 HP. A.Stephen&Sons,Glasgow	Clarence & Richmond Rivers Steam Nav.Co.
546	✻D	**City of Hamburg** I ScSr *McCallaghan* [illegible] 2 *Dks*	British Dublin	742 1174	231·0|32·6|14·5 P 134 f. F 26 f.	Sunderland J.Laing	1881	4 Blkds. [illegible]	2Cy.31"& 58"–33" 150 HP. T.Richardson&Sons,Hrtlpl.	Palgrave, Murphy & Co.
547		**City of Jacksonville** I Sc *Shaw*	American Wilmington	396 460	170·0|32·6| 7·9	Wilmington,Del Harlan&Hollingsworth	1882		2Cy.30"–72" The Harlan&HollingsworthCo., Wilm.,Del.	De Bary Mer.Line
548	✻	**City of Khios** I ScBk *McPherson 80-84* 3 *Dks*	British Glasgow	2121 3246	381·5|38·4|29·2 B 29 f. F 30 f.	Glasgow Barclay,Curle&Co.	1878	6 Blkds.	2Cy.50"& 86"–54" 500 HP. Barclay,Curle&Co.,Glasgw.	G.Smith & Sons
549		**City of Lawrence** I Pad	American New London	1351 1678	243·0|62·5|11·9	Wilmington,Del The Harlan&HollingsworthCo.,Wilm.,Del.	1867		1Cy.65"–132" 1000 HP. The Harlan&HollingsworthCo., Wilm.,Del.	O.L.Johnson
	✻	**City of Lincoln** I ScSr *Fry* [illegible] (ex Sofia, ex Massachusetts, ex Manhattan) 3 *Dks*	British Liverpool	2104 3182	335·0|42·5|27·7 P 184 f. F 43 f.	Newcastle Palmers Co.,Ld.	1866	6 Blkds.	3Cy.26,42&68"–42" 259 HP. W.Doxford&Sons,Sld.	W.H.Ross & Co.
551	✻	**City of Lisbon** I ScSr *Kelly 85-86* (ex Anglian)	British Dublin	389 633	204·4|26·3|15·9 P 46 f. F 38 f.	London C.Lungley	1864	4 Blkds.	2Cy.26"& 49"–30" 98 HP. T.Richardson&Sons,Hrtlpl.	Palgrave,Murphy & Co.
552		**City of Liverpool** W Sc *Stubbs*	British Liverpool	100 147	88·8|20·8|10·2	Winsford G.Deakin,Ld.	1885		2Cy.13½"& 22"–16" 32 HP. E.Foden,Sandbach	Salt Union,Ld.
553	✻	**City of London** I ScBk *Black 82-82* 3 *Dks*	British Glasgow	2097 3212	381·6|38·8|29·0 F 28 f.	Glasgow C.Connell&Co.	1876	5 Blkds.	2Cy.52"& 89"–48" 450 HP. J.&J.Thomson,Glasgow	G.Smith & Sons
554	✻	**City of London** I ScSr *Chalmers 76-85* 1 *Dk2 B*	British Aberdeen	477 944	241·9|30·5|16·7 P 70 f. F 50 f.	Glasgow J.Elder&Co.	1871	5 Blkds.	2Cy.46"& 80"–39" 275 HP. J.Elder&Co.,Glasgow	AberdeenStm.Nav.Co.
555		**City of London** I PadSp	British Liverpool	 143	114·5|19·1|10·2	Poplar	1863		2Cy. 140 HP.	R.Strong
556	R	**City of Macon** I ScSr *Kelly* [illegible] 2 *Dks & Awn Dk*	American S'vannah,Ga	1071 2093	257·0|38·7|24·8	Chester,Pa. J.Roach&Son	1877		2Cy.38"& 68"–54" 400 HP. J.Roach&Son,Chester,Pa.	Ocean S.S.Co.
	✻D	**City of Malaga** I ScSr *Gough 87-87* (ex Berwick) 1 *Dk2 B*	British Dublin	650 1026	220·0|30·2|16·9 P 207 f. F 24 f.	Sunderland J.L.Thompson	1872	4 Blkds. [illegible]	2Cy.27"& 50"–30" 98 HP. S.Clark,Sunderland	Palgrave,Murphy & Co.
	✻D	**CityofManchester** I Sc3mSr *Brophy 83-88* 2 *Dks3 B*	British Whitby	2029 3126	337·0|41·4|26·4 P 40 f. B 82 f. F 47 f.	Sunderland Sundrld.S.B.Co.,Ld.	1883	5 Blkds. [illegible]	2Cy.40"& 74"–54" 350 HP. N.E.MarineE.Co.,Ld.,Sndld.	J.Gray & Co.
559		**City of Melbourne** I ScBk *Thompson* 1 *Dk*	British Sydney	580 838	250·4|28·2|16·6 P 59 f.	Glasgow J.&G.Thomson	1862		2Cy.40"& 70"–33" 200 HP. J.Inglis,Glasgow	Australasian United Steam Nav.Co.,Ld.
560		**City of Monticello** I PadSr *McKee* (ex City of Norfolk) 2 *Dks & Spt Hur Dk*	American New York	478 893	232·0|32·0|10·9	Wilmington,Del Harlan&Hollingsworth	1866		1Cy.50"–120" 220 HP. Hrln&Hllngwrth,Wlmgtn.	Florida S.S.Co.
561		**City of Mount Clemens** W ScSr 1 *Dk*	British St.Thomas,Ont	69 102	95·0|21·8| 6·0	Mt.Clmns,U.S.	1880		1Cy.14¼"–16" 85 HP.	R.T.Riddle
		City of New Bedford W ScSr *Fish* 2 *Dks3 B*	American Fall Rvr,Mas	433 822	180·0|32·6|13·0	Williamsbrg,NY C.H.Delemater&Co	1874		1Cy.36"–36" C.H.Delemater&Co.,N.Yk.	Old Colony Steamboat Co.
	✻	**CityofNewYork** StlTSc3mSr *Watkins 87-88* 4 *Dks & Prom Dk5 B*	British Liverpool	5930 10499	527·6|63·2|22·0 P 75 f. F 109 f. 39·2	Glasgow J.&G.Thomson	1888	15 Blkds. [illegible]	6C.45"71"&113"–60" 2000 HP. J.&G.Thomson,Glsgw	Inman&International S.S.Co.,Ld.
	RV	**City of New York** I ScBk *Searle 72-84* 4 *Dks*	American New York	1964 3020	339·0|40·2|20·5 289	Chester,Pa. J.Roach&Son	1875		2Cy.39"& 68"–42" 700 HP. J.Roach&Son,Chester,Pa.	Pacific Mail S.S.Co.
		City of New York W Pad	American New London,Conn.	1468 1592	301·0|40·0|12·3	Brooklyn	1861			O.L.Johnson
	✻D	**City of Newcastle** I ScSr *Townsend 86-86* 2 *Dks3 B*	British Newcastle	1298 1985	285·5|36·1|24·1 P 28 f. B 52 f. F 32 f.	Newcastle C.S.Swan&Hunter	1882	6 Blkds. [illegible]	2Cy.36"& 68"–42" 200 HP. R.&W.Hawthorn,Nwcastle	Tyne&BlythS.S.Owning Co.,Ld.
		City of Newport W Pad	American Prvdnce,R.I	469 562	177·5|28·5| 8·8	New York	1863		250 HP.	B.Buffurn
		City of Norwich W Pad	American New London,Conn.	499 997	200·7|36·0|13·2	New York	1862		700 HP.	O.L.Johnson
	✻	**City of Oporto** I ScSr *Scott 80-80* (ex Marseilles) 1 *Dk2 B*	British Dublin	560 874	215·7|30·1|15·4 Q 29 f. F 23 f.	Sunderland J.Blumer&Co.	1872	4 Blkds. [illegible]	4Cy.18"& 37"–30" 95 HP. T.Bates&Co.,SowrbyBridge	Palgrave Murphy&Co.
	✻	**City of Oxford** I ScBk *Miller 75-82* 3 *Dks*	British Glasgow	2573 3959	400·0|43·2|29·8 B 62 f. F 34 f.	Glasgow Barclay,Curle&Co.	1882	6 Blkds.	2Cy.52"& 96"–60" 660 HP. Barclay,Curle&Co.,Glasgw.	G.Smith & Sons

✻ = Lloyd's Register; D = Nederlandsche Vereeniging; G = Germanischer Lloyd; I = Registro Italiano; R = American Record; [illegible] = Veritas Austro-Ungarico; H = Veritas Ellenico; N = Norske Veritas; V = Bureau Veritas.

Dans la 2me Colonne, B.T., avec les indications qui suivent, signifie la date du rapport de visite de la Carène par les experts du Board of Trade pour obtenir le certificat relatif au TRANSPORT DES PASSAGERS.

Dans la 3me Colonne, C = Composite (membrure en fer, bordé en bois); I = Fer; Stl = Acier; W = Bois.

Pour plus amples renseignements sur les navires classés par le Lloyd's Register, voir le Lloyd's Register Book et le Lloyd's Yacht Register.

No. in Book. / Official No.	Societies in which classed	Ship's Name. Material, Rig, &c. / Master, and Dates of Appointment to present Owner's Service and to Ship / Ship's Late Name. No. of Decks, &c.	Flag. / Port of Registry.	Tons. Net. / Gross.	Dimensions. Length	Breadth	Depth	Built. Where. / By Whom.	When	No. of Bulkheads. / Particulars of Water Ballast.	Engines. Particulars. / Maker's Name.	Owners or M…
571 125204	RV	**City of Panama** I ScSr *White 76-87* *2 Dks 3 B*	American New York	1046 1490	248·0	36·1	20·2	Chester, Pa. J. Roach & Son	1873		2Cy. 30" & 58"–54" 216 HP. J. Roach & Son, Chester, Pa.	Pacific Mail S.S.
572 125084	RV	**City of Para** I ScBk *Shackford 77-87* *3 Dks*	American New York	2505 3532	345·0	38·6	28·3	Chester, Pa. J. Roach & Son	1878		2Cy. 42½" & 76"–60" 700 HP. J. Roach & Son, Chester, Pa.	Pacific Mail S.S.
573	*	**City of Paris** StlTwnSc3mSr	British Liverpool	5930 10499	527·6 *P 78 f.*	63·2 *F 1…*	22·0 39·2	Glasgow J.&G. Thomson	1889		6Cy. 45", 71" & 113"–60" J. & G. Thomson, Glasgow [2… HP.	Inman & Intern. S.S. Co., Ld.
574 125285	RV	**City of Peking** I Sc4mS *Seabury 75-88* *4 Dks*	American New York	3130 5080	423·0	48·0	27·8	Chester, Pa. J. Roach & Son	1874		1Cy. 51" & 88"–54" 1000 HP. J. Roach & Son, Chester, Pa.	Pacific Mail S.S.
575 125979	RV	**City of Puebla** I ScBa *Debney* *3 Dks*	American S. Francisco	1713 2624	319·0	38·6	26·3	Philadelphia W. Cramp & Sons	1881		2Cy. 43" & 86"–60" W. Cramp & Sons	J. N. Knowles
576 68266	BT 11.88	**City of Richmond** I ScBkn *Redford 84-86*	British Liverpool	2957 4780	440·8	43·5 *F 99 f.*	34·1	Glasgow Tod & MacGregor	1873		2Cy. 68" & 120"–60" 700 HP. Tod & MacGregor, Glasgow	Inman & Intern. S.S. Co., Ld.
577 125850		**City of Richmond** W Pad *Kemble*	American Hartford	704 1001	240·0	34·6	11·0	Philadelphia Birely, Hillman & Streeker	1880		1Cy. 54"–144" 806 HP.	Hartford Steam. Co.
578 5020		**City of Richmond** W Pad	American P'rtland, Me	601 940	227·5	30·6	10·0	Athens, N.Y.	1865		350 HP.	Portland, Bang. S.B. Co.
579 126346	RV	**City of Rio de Janeiro** I ScBkn *Ward 88-88* *3 Dks*	American New York	2275 3548	344·0	38·0	28·9	Chester, Pa. J. Roach & Son	1878		2Cy. 42½" & 76"–60" 700 HP. J. Roach & Son, Chester, Pa.	Pacific Mail S.S.
580 63906	*	**City of Rochester** I ScSr *Hardy 17-79* *1 Dk 2 B*	British Sunderland	907 1395	240·0 *Q 61 f.*	33·0 *B 44 f.*	18·9 *F 27 f.*	Sunderland Osbourne, Graham & Co.	1879	4 Bhds. DB & f'rac AP T 52t FP T 40t MT 96t	2Cy. 31" & 58"–36" 140 HP. N. E. Marine E. Co., Ld., Sudld	W. A. Watson
581 84148	BT 3.88	**City of Rome** I Sc4mSr *Young 68-87* *4 Dks*	British Barrow	3453 8144	560·2 *F 106 f.*	52·3 *B 84 f.*	37·0	Barrow Barrow S.B. Co., Ld.	1881	10 Bhds. DB a 376t	6Cy. 43" & 86"–72" 1500 HP. Barrow S.B., Ld., Barrow	Barrow Steam Co., Ld.
582 81447	*	**City of Rotterdam** I ScSr *Stack 81-86* *2 Dks*	British Dublin	303 656	210·0 *Q 77 f.*	29·2 *B 46 f.*	12·1 *F 25 f.*	Glasgow Dobie & Co.	1883	4 Bhds. DB 194	2Cy. 28" & 52"–36" 120 HP. Hutson & Corbett, Glasgow	Palgrave Murph…
583 125048	R	**City of San Antonio** I Sc3mSr *Wilder 83-83* *2 Dks 3 B*	American New York	1204 1605	228·0	36·6	20·3	Chester, Pa. J. Roach & Son	1872		1Cy. 48"–48" 250 HP. J. Roach & Son, Chester, Pa.	C. H. Mallory & C…
584 125629	R	**City of Savannah** I ScSr *Smith 81-87* *3 Dks*	American S'vannah, Ga	1359 2029	254·0	38·7 *H 327 f.*	24·8	Chester, Pa. J. Roach & Son	1877		2Cy. 38" & 68"–54" 350 HP. J. Roach & Son, Chester, Pa.	Ocean S.S. Co.
585 22571		**City of Springfield** W Pad	American H'rtf'rd, Con	1047 1417	286·0	26·0	9·6	Brooklyn	1867			E. H. William…
586 125470	RV	**City of Sydney** I ScBk *Pride 85-85* *4 Dks*	American New York	1966 3017	339·0	40·2	20·5	Chester, Pa. J. Roach & Son	1875		2Cy. 51" & 88"–60" 700 HP. J. Roach & Son, Chester, Pa.	Pacific Mail S.S.
587 126241	V	**City of Topeka** I ScSr *Rogers 85-85* *2 Dks 3 B*	American Boston, Mass	747 1057	198·0 *F 17 f.*	35·2 *H 25 f.*	18·0	Chester, Pa. J. Roach & Son	1884		2Cy. 34" & 44"–45" 800 HP.	Atchison, Topeka Santa Fe Railroad
588 88747	*	**City of Truro** I ScSr *Fulcher 81-86* *2 Dks 3 B*	British Newcastle	1340 2051	281·0 *P 20 f.*	37·3 *B 58 f.*	23·1 *F 29 f.*	Newcastle Palmers' Co., Ld.	1883	4 Bhds. DB a & f 300t	2Cy. 33" & 64"–42" 200 HP. Palmers' Co., Ld., Newcastle	Tyne & Blyth Sailing Co., Ld.
589 86129	*	**City of Truro** I ScSr *Chellew* *1 Dk 2 B*	British Falmouth	1141 1767	257·0 *Q 66 f. P 29 f.*	34·8 *B 66 f.*	19·7 *F 34 f.*	W. Hartlepool W. Gray & Co.	1883	4 Bhds. DB a & f 333t	2Cy. 32" & 60"–36" 150 HP. T. Richardson & Sons, H'tlepl.	Cornwall S.S. Co.
590 71726	* BT 11.88	**City of Venice** I ScBk *Moffat 78-80* *3 Dks*	British Glasgow	2201 3372	379·5 *B 132 f.*	38·2 *F 28 f.*	29·2	Glasgow Barclay, Curle & Co.	1875	5 Bhds.	4Cy. 30" 40" 52" & 70"–48" Quadrupled '87. J. Howden & Co., Glasgow	G. Smith & Sons
591 70736	*	**City of Verviers** I ScSr *Berry 79-82* *1 Dk*	British London	169 267	135·3	22·2 *Q 63 f.*	10·2	South Shields J. Readhead & Co.	1875	4 Bhds.	2Cy. 14" & 28"–20" 40 HP. J. Readhead & Co., S. Shields	D. C. Thomas
592 126025	RV	**City of Washington** I ScBn *Reynolds -87* *3 Dks*	American New York	1650 2618	300·0	38·0	27·0	Chester, Pa. J. Roach & Son	1877		2Cy. 40" & 74"–72" 650 HP. J. Roach & Son, Chester, Pa.	New York & … Mail S.S. Co.
593 78022		**City of Westminster** I ScK *(ex Strenus, ex Sailing Vessel)*	British Hull	58 107	85·6	20·3 *Len. 86.*	9·6	Hull Baltic Eng. Works	1878		2Cy. 14" & 26"–20" 30 HP. C. D. Holmes & Co., Hull	G. Bowman
594 77481		**City of Winchester** I ScTrawl *(ex Sailing Vessel, ex Cyclops)* *1 Dk*	British Hull	52 110	85·3	20·3 *Len. 86*	10·1	Hull Gilbert	1877		2Cy. 14" & 26"–20" 30 HP. C. D. Holmes & Co., Hull	G. Bowman
595 125941		**City of Worcester** I Pad	American New London, Conn.	1922 2490	328·0	46·0	14·5	Wlmngtn, Del The Harlan & Hollingsworth Co.	1881		1Cy. 90"–144" The Harlan & Hollingsworth Co., Wilm. Del.	O. L. Johnson
596	V	**Ciudad Condal** I ScBg *Emeterio 82-87 (ex Washington)* *3 Dks*	Spanish Barcelona	1751 2576	306·8 *B 68 f. F 60 f.*	38·2 *H 234 f.*	21·3 28·3	Sunderland Oswald & Co.	1872	3 Bhds.	2Cy. 51" & 86"–42" 350 HP. Oswald & Co., Sunderland	Cia. Trasatlantic…
597	*	**Ciudad de Cadiz** I ScBk *Chaquert 82-84* *3 Dks & Awn Dk 4 B*	Spanish Barcelona	1844 3084	363·6	38·2	24·2 31·7	Renfrew Lobnitz, Coulborn & Co.	1878	4 Bhds.	2Cy. 57" & 103"–54" 600 HP. Lobnitz, Coulborn & Co., Ren.	Cia. Trasatlantic…
598	*	**Ciudad de Ciudadela** I Sc *(ex Elon)* *Amingual 88-88* *1 Dk*	Spanish Ciudadela	140 249	140·0 *B 36 f.*	21·1 *F 21 f.*	11·5	Glasgow T. B. Seath & Co.	1884	4 Bhds. FPT…	2Cy. 18" & 36"–25" 65 HP. A. Campbell & Sons, Glasgow	G. Amingual
599	*	**Ciudad de Santander** I Sc3mSr *Cinciano 65-83* *3 Dks*	Spanish Cadiz	1985 3488	384·2 *P 101 f.*	42·2 *B 48 f.*	27·5 *F 62 f.*	Dumbarton W. Denny & Bros.	1883	6 Bhds. CDB 333t	2Cy. 57" & 102"–60" 600 HP. Denny & Co., Dumbarton	Cia. Trasatlantic…
600												

*= Lloyd's Register; D = Nederlandsche Vereeniging; G = Germanischer Lloyd; I = Registro Italiano; R = American Re… A = Veritas Austro-Ungarico; H = Veritas Ellenico; N = Norske Veritas; V = Bureau Veri…

In the 2nd Column, B.T., with figures following, signifies Date of report of bottom sighting by Board of Trade Surveyors as condition of PASSENGER CERTIF…

In the 3rd Column, C = Composite (iron frame, wood planked); I = Iron; Stl = Steel; W = Wood.

Further particulars of Vessels classed by Lloyd's Register will be found in Lloyd's Register Book and Lloyd's Yacht Register.

1	2	3 Nom, Matériaux, Gréement, etc. Capitaine et dates d'entrée au service des Armateurs actuels et à bord du navire. Ancien Nom, nombre de ponts, etc.	4 Pavillon. Port d'Armement	5 Tonnage Net. Total	6 Dimensions (Longueur, Largeur, Creux) Constructions sur le Pont	7 Construit. Où. Par qui.	8 Quand	9 Nombre des cloisons. Détails sur le lest d'eau.	10 Machines. Détails. Nom du Constructeur.	11 Armateurs ou Directeurs-Gérants.
601		Civility W ScSmSr 1 Dk	British Sydney	169 249	132·0 | 25·5 | 10·7 B 38 f.	Brisbane Rock Davis	1872		2Cy.15"–24" 40 HP. Vale&Lacy,Sydney	George Hill, jun.
602		Clacton StlPadSp Jeffs '88	British London	55 241	189·5 | 22·2 | 8·3	Kinghorn J.Scott&Co.	1888		2Cy.28"&57"–57" 200 HP. J.Scott&Co.,Kirkcaldy	London, Woolwich & Clactonon-Sea Steamboat Co. Ld.
603	✻	Clan Alpine I ScSr Elliott 2 Dks3 B	British Glasgow	1349 2080	305·6 | 34·8 | 24·2 P 13 f. F 36 f.	Glasgow A.Stephen&Sons	1878	5 Blds. MT a176t f 19t	2Cy.34" &63"–42" 250 HP. A.Stephen&Sons,Glasgow	Cayzer, Irvine & Co.
604	✻	Clan Buchanan I Sc Millar 82-88 2 Dks & W'ch Frames	British Glasgow	1913 2933	330·3 | 40·1 | 26·3 P 64 f. B 78 f. F 38 f.	Dumbarton A.M'Millan&Sons	1882	6 Blds. CDB30t	2Cy.39"& 75"–48" 400 HP. D.Rowan,Glasgow	Cayzer, Irvine & Co.
605	✻D	Clan Cameron I ScSr Wilson 2 Dks	British Glasgow	1585 2433	324·5 | 38·1 | 23·8 P 12 f. B 74 f. F 28 f.	Glasgow A.Stephen&Sons	1882	6 Blds. CDB88t	2Cy.38"& 68"–45" 300 HP. A.Stephen&Sons,Glasgow	Cayzer, Irvine & Co.
606	✻	Clan Drummond I ScSr Harrison 2 Dks & W'ch Frames	British Glasgow	1904 2922	330·3 | 40·1 | 26·3 P 64 f. B 78 f. F 38 f.	Dumbarton A.M'Millan&Son	1882	6 Blds. DB30t	2Cy.39"& 75"–48" 400 HP. D.Rowan,Glasgow	Cayzer, Irvine & Co.
607	✻	Clan Forbes I ScSr Taitt 2 Dks & W'ch Frames	British Glasgow	1591 2441	324·5 | 38·1 | 23·8 P 12 f. B 74 f. F 38 f.	Glasgow A.Stephen&Sons	1882	6 Blds. CDB88t	2Cy.38"& 68"–45" 300 HP. A.Stephen&Sons,Glasgow	Cayzer, Irvine & Co.
608	✻	Clan Fraser I ScBg Roberts 83-84 2 Dks3 B	British Glasgow	1349 2083	305·6 | 34·8 | 24·2 P 13 f. F 36 f.	Glasgow A.Stephen&Sons	1878	5 Blds. MT a176t f 19t	2Cy.34" & 63"–42" 250 HP. A.Stephen&Sons,Glasgow	Cayzer, Irvine & Co.
609	✻	Clan Gordon I ScBg Bartram 86-86 2 Dks3 B	British Glasgow	1343 2062	305·6 | 34·8 | 24·2 P 13 f. B 38 f. F 36 f.	Glasgow A.Stephen&Sons	1879	5 Blds. MT a176t f 19t	3Cy.22",34"&59"–42" 200 HP Tripld. '88, J.Howden&Co., Glasgow	Cayzer, Irvine & Co.
610	✻	Clan Graham I ScSr Broadfoot [88-87] 2 Dks & W'ch Frames	British Glasgow	1911 2926	330·5 | 40·2 | 26·3 P 62 f. B 80 f. F 38 f.	Glasgow Napier, Shanks & Bell	1882	6 Blds. CDB512t	2Cy.39"& 75"–48" 400 HP. D.Rowan,Glasgow	Cayzer, Irvine & Co.
611	✻	Clan Grant I Sc3mSr Darley 82-88 2 Dks3 B	British Glasgow	2306 3545	350·5 | 41·1 | 26·6 P 240 f. F 48 f.	Glasgow A.Stephen&Sons	1883	4 Blds.	2Cy.44"& 80"–48" 400 HP. A.Stephen&Sons,Glasgow	Cayzer, Irvine & Co.
612	✻	Clan Lamont I ScSr Alexander 87-87 2 Dks3 B	British Glasgow	1354 2091	305·6 | 34·8 | 24·2 P 13 f. F 36 f.	Glasgow A.Stephen&Sons	1879	4 Blds. MT 127 f 10t	2Cy.34" & 63"–42" 250 HP. A.Stephen&Sons,Glasgow	Cayzer, Irvine & Co.
613	✻	Clan Macarthur Stl ScSr McMahon 82-87 3 Dks	British Glasgow	2635 3984	381·2 | 43·3 | 27·9 B 144 f. F 48 f.	Greenock Scott&Co.	1883	6 Blds. CDB490t	2Cy.48"& 86"–60" 600 HP. Scott&Co.,Greenock	Cayzer, Irvine & Co.
614	✻	Clan Macdonald I ScBg Rule 79-86 3 Dks	British Glasgow	1716 2612	335·5 | 39·3 | 26·9 B 74 f. F 30 f.	Greenock Scott&Co.	1882	7 Blds. DB a138t MT 31 f 52t	2Cy.40" &78"–48" 400 HP. Scott&Co.,Greenock	Cayzer, Irvine & Co.
615	✻	Clan Macgregor I ScSr Palmer 80-86 2 Dks & W'ch Frames	British Glasgow	1927 2956	329·6 | 40·3 | 26·4 P 64 f. B 78 f. F 38 f.	Leith Ramage&Ferguson	1882	6 Blds. CDB536t	2Cy.39"& 75"–48" 400 HP. D.Rowan,Glasgow	Cayzer, Irvine & Co.
616	✻	Clan Macintosh Stl ScSr McAllister 82-86 3 Dks	British Glasgow	2636 3985	381·2 | 43·3 | 27·9 B 144 f. F 48 f.	Greenock Scott&Co.	1883	6 Blds. CDB490t	2Cy.48"& 86"–60" 600 HP. Scott&Co.,Greenock	Cayzer, Irvine & Co.
617	✻	Clan Mackay I ScSr Jones 82-88 2 Dks3 B	British Glasgow	1423 2171	285·7 | 37·2 | 24·4 P 33 f. B 70 f. F 35 f.	Sunderland Bartram, Haswell & Co.	1882	6 Blds. DB279t	2Cy.36"& 68"–45" 300 HP. R.&W.Hawthorn,N'weastle	Cayzer, Irvine & Co.
618	✻	Clan Mackenzie I ScSr Beer 2 Dks & W'ch Frames	British Glasgow	1930 2954	329·6 | 40·2 | 26·3 P 64 f. F 38 f.	Leith Ramage&Ferguson	1882	6 Blds. CDB536t	2Cy.39"& 75"–48" 400 HP. D.Rowan,Glasgow	Cayzer, Irvine & Co.
619	✻	Clan Maclean I ScBk Roberts 83-89 (ex City of Cambridge) 3 Dks	British Glasgow	1496 2336	326·5 | 36·1 | 27·7	Glasgow Barclay,Curle&Co.	1870	5 Blds.	2Cy.39"& 68"–42" 200 HP. Barclay,Curle&Co.,Glasgow	Cayzer, Irvine & Co.
620	✻	Clan Macpherson Stl ScSr Gourlay 82-88 3 Dks	British Glasgow	2587 3921	380·7 | 43·2 | 28·3 B 136 f. F 40 f.	Glasgow Napier, Shanks & Bell	1883	6 Blds. CDB524t	2Cy.48"& 86"–60" 600 HP. D.Rowan,Glasgow	Cayzer, Irvine & Co.
621	✻	Clan Matheson Stl ScSr Watson 83-83 3 Dks	British Glasgow	2581 3917	380·7 | 43·2 | 28·3 B 138 f. F 40 f.	Glasgow Napier, Shanks & Bell	1883	6 Blds. CDB524t	2Cy.48" &86"–60" 600 HP. D.Rowan,Glasgow	Cayzer, Irvine & Co.
622	✻	Clan Monroe I ScSr Jones 2 Dks3 B	British Glasgow	1437 2197	285·5 | 37·2 | 24·4 P 33 f. B 70 f. F 35 f.	Sunderland Bartram, Haswell & Co.	1881	6 Blds. DB a & cb275t	2Cy.36"& 68"–45" 300 HP. R.&W.Hawthorn,N'weastle	Cayzer, Irvine & Co.
623	✻	Clan Murray I ScSr Norfolk 83-84 (ex Muriel) 2 Dks	British Glasgow	1373 2108	290·0 | 36·2 | 23·2 P 28 f. B 68 f. F 28 f.	Newcastle TyneI.S.B.Co.	1881	5 Blds. CDB40t FPT33t APT13t	2Cy.38"& 71"–45" 250 HP. N.E.MarineE.Co.,Ld.Sund'ld	Cayzer, Irvine & Co.
624	✻	Clan Ranald I ScSr Miller 82-84 2 Dks3 B	British Glasgow	1339 2068	305·0 | 34·8 | 24·4 P 14 f. B 56 f.	Dumbarton A.M'Millan&Son	1878	5 Blds. MT f170t a168t	2Cy.36"& 66"–42" 280 HP. D.Rowan,Glasgow	Cayzer, Irvine & Co.
625	✻	Clan Sinclair I ScSr Kewley 82-88 2 Dks & W'ch Frames	British Glasgow	1912 2933	330·5 | 40·2 | 26·3 P 62 f. B 80 f. F 38 f.	Glasgow Napier, Shanks & Bell	1882	6 Blds. CDB512t	2Cy.39"& 75"–48" 400 HP. D.Rowan,Glasgow	Cayzer, Irvine & Co.
626	✻	Clan Stuart I ScBg Whyte 86-87 2 Dks3 B	British Glasgow	1355 2091	305·4 | 34·8 | 24·4 P 13 f. F 36 f.	Dumbarton A.McMillan&Son	1879	5 Blds. MT a207t f 17t	2Cy.36"& 66"–42" 280 HP. D.Rowan,Glasgow	Cayzer, Irvine & Co.
627	✻	Clandeboye StlScSr Ball 83-84 1 Dk & W'b F'ms	British London	1143 2136	275·0 | 37·3 | 19·1 Q 78 f. P 2 f. B 110 f. F 29 f.	Stockton Richardson,Duck&Co.	1888	4 Blds. CDB521t	3C22"36"&59"–39" 170 HP. Blair&Co.Ld.,Stockton	Clandeboye S.S.Co., Ld. (Carlisle&Co.)
628		Clandeboye StlPadSk Hinsdale '87	British	5 401	225·0 | 24·3 | 8·6	Belfast Workman,Clark & Co. Ld	1887		2Cy.44"–66" 230 HP. Hutson&Corbett,Glasgow	Belfast, Bangor and Larne Steamboat Co., Ld.
629	✻	Clansman Stl Sc3mSr Farquhar '85 1 Dk	British Auckland	336 591	191·7 | 26·1 | 10·8 F 20 f. P & B 138 f.	PortGlasgow Blackwd.&Gordon	1884	4 Blds. FPT36t APT13t	2Cy.26"& 48"–36" 99 HP. Blackwood&Gordon,Pt.Gls.	Northern S.S.Co., Ld.
630		Clansman I ScSr McKinnon 88-88	British Glasgow	308 600	211·3 | 27·3 | 13·4 P 112 f.	Glasgow J.&G.Thomson	1870		2Cy.33½"&58½"–36" 150 HP. J.&G.Thomson,Glasgow	D. MacBrayne

✻=Lloyd's Register; D=Nederlandsche Vereeniging; G=Germanischer Lloyd; I=Registro Italiano; R=American Record; ✻=Veritas Austro-Ungarico; H=Veritas Ellenico; N=Norske Veritas; V=Bureau Veritas.

Dans la 2me Colonne, B.T., avec les indications qui suivent, signifie la date du rapport de visite de la Carène par les experts du Board of Trade pour obtenir le certificat relatif au TRANSPORT DES PASSAGERS.

Dans la 3me Colonne, C=Composite (membrure en fer, bordé en bois); I=Fer; Stl=Acier; W=Bois.

Pour plus amples renseignements sur les navires classés par le Lloyd's Register, voir le Lloyd's Register Book et le Lloyd's Yacht Register.

1 No. Official No.	2 Registry in which classed	3 Ship's Name, Material, Rig, &c. Master, and Dates of Appointment to present Owner's Service and to Ship. Ship's Late Name. No. of Decks, &c.	4 Flag. Port of Registry.	5 Tons. Net. Gross	6 Dimensions. Length, Breadth, Depth	7 Built. Where. By Whom.	8 When	9 No. of Bulkheads. Particulars of Water Ballast.	10 Engines. Particulars. Maker's Name.	11 Owners
631		**Clapet No. XI.** Stl TScBrgs 1 Dk		182 [illegible]	133·5|25·0|11·5	Nwc.-on Tyne Palmers'Co.Ld.	1887		4Cy.14"& 28"–24" [illegible] HP. Palmers'Co.,Ld.,Newcastle	Panama Canal
632		**Clapet No. XII.** Stl TScBrgs 1 Dk		182 [illegible]	133·5|25·0|11·5	Nwc.-on-Tyne Palmers Co.Ld.	1887		4Cy.14"& 28"–21" [illegible] HP. Palmers'Co.,Ld.,Newcastle	Panama Canal
633 70[illegible]	✳ D	**Clapeyron** [illegible] I ScSr [illegible] 2 Dks3 B	British [illegible]	1163 18[illegible]	269·5|34·5|24·1 F 27 f.	Sunderland Short Brothers	1877	4 Bhds. [illegible] APT FPT	2Cy.33"& 62"–39" 180 HP. J.Dickinson,Sunderland	D.H.Morgan
634 56[illegible]	✳	**Clapton** [illegible] I ScBrgs Bartlett 80-80 1 Dk2 B	British London	509 796	201·3|28·1|17·4 Q54 f. F 25 f.	Newcastle [illegible]	1867	4 Bhds. [illegible]	4Cy.19"& 38"–30 98 HP. Comp.Gen.,L.S.Coll.Co.Lon.	J.&C.Harrison
635 91[illegible]	✳	**Clara** Stl ScSr Cox 8[illegible] 1 Dk&Spr Dk3 B	British [illegible]	1165 1803	270·0|35·0|17·0 F 15 f. B 40 f. 24·7	Newcastle Wigham Richardson&Co.	1886	4 Bhds. [illegible]	3C.22"34"&57"–39" 250 HP. Wghm.Richardsn&Co.,Nwc	G.Tweedy & Co
636 50[illegible]		**Clara** I ScSr Kirwan 87-87 1 Dk2 B	British Glasgow	160 257	165·4|20·9|12·0 Q 55 f.	Glasgow [illegible]	1864	4 Bhds.	2Cy.18"& 32"–22" 50 HP. Walker,Henderson&Co.Gls	J.&P.Hutchison
637	G	**Clara** I ScSr Christensen 85-85 2 Dks	German [illegible]	684 [illegible]	203·1|26·0|20·2 F 11 f.	Kiel Gebr.Howaldt	1882	[illegible]	2Cy.24½"& 40"–27" 80 HP. G.Howaldt,Kiel	M.Jebsen
638	V	**Clara** I ScSr Marx	German Stettin	201 299	135·5|20·8|11·1 P 28 f. Q 12 f. B 40 f. F 14 f.	Grabow Möller&Holberg	1881	[illegible]	2C.17¾ & 30¾"–17¾" 15 HP. Möller&Holberg	C.J.F.Braunlich
639		**Clara** I	Spanish Havana	488 [illegible]	190·6|28·4|12·6	England			100 HP.	Ramon de Herrera
640 87[illegible]	✳ D	**Clare** I ScSr Pollexfen 1 Dk&Spr Dk3 B	British Liverpool	1333 2034	270·4|38·4|24·1	Sunderland J.Blumer&Co.	1883	5 Bhds. [illegible] FPT 28 APT 30	2Cy.35"& 68"–48" 200 HP. T.Clark&Co.,Newcastle	A.L.Jones (Elder,Dempster&Co.)
641 83[illegible]	✳	**Claremont** I ScSr Adie 86-88 1 Dk	British Newcastle	432 677	180·7|29·1|13·3 Q 95 f. B 18 f. F 22 f.	NorthShields T.&W.Smith	1881	4 Bhds. DBaft 130 FPT 20 APT 20	3Cy.14½"20½"& 46" 90 HP. Douglas&Grant,Kirkcaldy	Fisher,Renwick
642 54610		**Clarence** (ex Lightning) I PadSp Bell 84-84	British London	13 165	97·3|18·4|9·3	Newcastle	1866		2Cy.27"–46" 60 HP. Hepple&Co.,Low Walker	T.P.White
643 5823		**Claribel** (ex Avila) I ScSr Clintsbel 87-87 1 Dk2 B	British Liverpool	883 1134	250·0|28·2|21·1 B 26 f.	Liverpool T.Royden&Sons	1872	4 Bhds.	2Cy.29"& 56"–30" 135 HP. J.Jack,Rollo&Co.,Liverpool	Atlas S.S.Co.,Ld
644		**Claris** I	Spanish Barcelona	1494 23[illegible]	326·7|35·1|23·9	England	1871		270 HP.	Cinde Transportes Maritimos
645	✳	**Clarissa Radcliffe** StlScSr	British Cardiff	1622 2487	296·0|40·2|20·6 Q 82 f. B 119 f. P 31 f. F 33 f.	Newcastle Palmers'Co.,Ld.	1889		3Cy.23½ 38 & 62"–42" Palmers' Co., Ld., Newcastle 224 HP.	E.Thomas,Radcliffe & Co.
646 74431	BT 5.88	**Claud Hamilton** I PadSr Cross 76-84	British Harwich	565 982	251·6|30·2|13·7 P 153 f. F 115 f.	Glasgow J.Elder&Co.	1875		2Cy.54"& 85"–63" 350 HP. J.Elder&Co.,Glasgow	Great Eastern R
647 81480	✳ D	**Claudius** I ScSr Gaylard 77-81 1 Dk2 B	British London	1110 1767	260·0|36·0|18·5 Q 64 f. B 58 f. P 23 f. F 26 f.	Sunderland J.L.Thompson&Sons	1880	5 Bhds. [illegible] APT [illegible] FPT [illegible]	2Cy.33"& 63"–39" 180 HP. N.E.Marine E.Co.,Ld.,Sndld	T.Kish
648 74491	BT 10.88	**Claughton** I Pad Hunt 55-76	British Liverpool	150 306	141·0|31·2|12·9	Glasgow D. & W. Henderson &Co.	1875		4Cy.30"–48" 120 HP. D.&W.Henderson&Co.Gls.	Municipal Corp. of Birkenhead
649 77224		**Clavis** (ex Walrus) I TwnScSk 1 Dk	British Port Louis	20 110	113·8|20·0|9·4	PortGlasgow R.[illegible]&Co.	1877		2Cy.18"& 34"–24" 50 HP. Rankin&Blackmore,Grk.	Government of Mauritius
650 87102	✳ BT 6.88	**Claymore** I ScSr Craig 83-87 2 Dks3 B	British London	1658 2554	314·5|38·7|24·8 P 42 f. B 58 f. F 39 f.	Glasgow D. & W. Henderson &Co.	1883	5 Bhds. [illegible]	2Cy.29"& 58"–60" 420 HP. D.&W.Henderson&Co.Gls.	Claymore S.S.Co. (J.Potter)
651 86644	✳	**Claymore** I ScSr Storm 1 Dk2 B	British Whitby	1064 1649	259·7|36·0|18·6 Q 100 f. B 58 f. F 26 f.	Sunderland J.L.Thompson&Sons	1883	5 Bhds. [illegible]	2Cy.32"& 60"–39" 150 HP. J.Dickinson,Sunderland	J.H.Barry
652 83852	BT 3.88	**Claymore** I ScSr McNeill 84-86	British Glasgow	424 726	227·0|29·6|14·5 P 131 f.	Glasgow J.&G.Thomson	1881		2Cy.32"& 61"–45" 200 HP. J.&G.Thomson,Glasgow	D.MacBrayne
653 62613	✳	**Cleadon** I ScSr Mohrke 64-87 1 Dk2 B	British Sunderland	489 760	196·8|29·0|16·7 Q 22 f. B 27 f.	Sunderland W.Watson	1871	4 Bhds. W.Ball f.	2Cy.38"–30" 90 HP. G.Clark,Sunderland	H.T.Morton
654 62[illegible]	✳	**Cleanthes** I ScSr Sayles 87-87 1 Dk2 B	British Sunderland	569 878	210·5|29·5|16·5 P 48 f. F 16 f.	Sunderland R.Thompson,junr.	1871	4 Bhds. [illegible]	2Cy. 26"& 47"–30" 90 HP. G.Clark,Sunderland	Culliford & Cl
655 87467	✳ D	**Cleddy** I ScSr Seaward 74-83 2 Dks3 B	British Cardiff	1398 2180	282·5|36·6|23·8 P 32 f. B 58 f. F 26 f.	Sunderland J.L.Thompson&Sons	1883	6 Bhds. [illegible] APT [illegible]	2Cy.35"& 66"–42" 230 HP. J.Dickinson,Sunderland	J.Marychurch
656 89462	✳ D	**Clematis** I ScSr Owen 84 1 Dk2 B	British W.H'pool	790 1234	230·2|32·0|15·4 Q 99 f. B 63 f. F 28 f.	Middlesbro' R.Craggs&Sons	1884	4 Bhds. [illegible]	2Cy.26"& 48"–36" 99 HP. Westgarth,English&Co.Mid	Lilly,Wilson &
657 70479	BT 8.88	**Clement** (ex Barbary) I ScSr Lancaster 84-88 1 Dk2 B	British Liverpool	785 1227	230·4|31·1|19·6 Q 58 f. P 239 f. F 34 f.	Birkenhead LairdBros.	1877	5 Bhds.	2Cy.28½"& 50"–33" 130 HP. LairdBros.,Birkenhead	Booth Steam S Ld.
658 128440		**Cleone** W ScSr Thompson 87-87 1 Dk	American S.Francisco	141 197	119·0|31·2|9·5	S.Francisco [illegible]	1887		2Cy.12"& 24"–20" 75 HP. [illegible] San Francisco	Pollard & Dod
659 125[illegible]		**Cleopatra** W Sc3mSr Colby 3 Dks	American Portland,Me	523 1045	184·1|34·8|23·6	FairHaven,Conn.	1865		600 HP.	T.C.Hersey
660 89719	✳ D	**Cleveland** Stl ScSr Jacob 87-87 1 Dk2 B	British Middlesbro'	1401 2173	275·1|37·3|20·0 Q 26 f. P 29 f. B 103 f. F 27 f.	W.Hartlepool W.Gray&Co.	1886	4 Bhds. [illegible]	3C.21"35"&57"–39" 200 HP. Central Marine Eng.Co.Hrtp.	W.Taylor

✳ = Lloyd's Register; D = Nederlandsche Vereeniging; G = Germanischer Lloyd; I = Registro Italiano; R = American R[...]
A = Veritas Austro-Ungarico; H = Veritas Ellenico; N = Norske Veritas; V = Bureau Veritas

In the 2nd Column, B.T., with figures following, signifies Date of report of bottom sighting by Board of Trade Surveyors as condition of PASSENGER CERTIF[...]

In the 3rd Column, C = Composite (iron frame, wood planked); I = Iron; Stl = Steel; W = Wood.

Further particulars of Vessels classed by Lloyd's Register will be found in Lloyd's Register Book and Lloyd's Yacht Register.

1	3 Nom, Matériaux, Gréement, etc. Capitaine et dates d'entrée au service des Armateurs actuels et à bord du navire. Ancien Nom, nombre de ponts, etc.		4 Pavillon. Port d'Armement	5 Tonnage. Net. Total.	6 Dimensions. Longueur | Largeur | Creux. Constructions sur le Pont	7 Construit. Où. Par qui.	8 Quand.	9 Nombre des cloisons. Détails sur le lest d'eau.	10 Machines. Détails. Nom du Constructeur.	11 Armateurs ou Directeurs-Gérants.
*	Cleveland *Rest 82-82*	I ScSr *1 Dk2 B*	British [illegible]	762 1218	255·5 | 29·5 | 16·4 *P 112 f. F 32 f.*	Newcastle Cole Brothers	1872	4 Bhds. [illegible] DB	2Cy.27" & 49½"–33" 98 HP. N.E.Marine E.C.,Ld.,Sundld.	Nelson, Donkin & Co.
	Cleveland	I PadSr	British Middlesbro'	54 177	120·0 | 20·6 | 10·6	South Shields J.T.Eltringham	1881		2Cy.32"–54" 95 HP. J.P.Rennoldson,S.Shields	North Eastern Rail. Co.
*	Clievedon *Gibb -86*	I ScSr *1 Dk2 B*	British [illegible]	1115 1723	258·0 | 36·0 | 18·2 *Q 71 f. P 27 f. B 54 f. F 32 f.*	Sunderland J.Blumer&Co.	1883	4 Bhds. DBcl&ct685	2Cy.32" & 60"–39" 140 HP. N.Marine E.C.Ld.,Gateshd.	M.Angel
	Clifton	I PadSr	British Preston	14 138	135·6 | 18·6 | 8·2	Preston W.Allsop	1871		2Cy.26"–48" 60 HP. W.Allsup,Preston	Blackpool Pier Co.
*	Clifton Grove *Morgans 87-87*	I ScSr *1 Dk*	British Bristol	148 278	134·7 | 22·6 | 11·6 *Q 89 f. F 17 f.*	Bristol G.K.Stothert&Co.	1883	3 Bhds.	2Cy.18½" & 32"–18" 50 HP. G.K.Stothert&Co.,Bristol	W.Butler & Co.
	Clinton	W Sc *1 Dk*	British St.Catherines	284 364	138·0 | 23·8 | 12·0 *P 49 f.*	St.Catherines M.Simpson	1874		1Cy.	J. Matthews
	Clinton	I Pad	American New Orleans,La.	718 1187	220·0 | 34·5 | 18·2	Wilmington,Del The Harlan & Hollingsworth Co.	1862		1Cy.50"–132" The Harlan & Hollingsworth Co., Wilm.,Del.	J.B.Richardson
*	Clintonia *Bulman 77-81*	I ScSr *2 Dks3 B*	British N.Shields	1287 1970	287·0 | 37·0 | 21·8 *P 28 f. B 28 f. F 22 f.*	Newcastle Tyne I.S.B.Co.,Ld.	1881	6 Bhds. CDB[illegible] FPT[illegible] APT[illegible]	2Cy.33" & 63"–39" 180 HP. N.E.Marine Eng.Co.,Ld.,Sld.	J.Robinson & Sons
D	Clio *Woolcock*	I ScSr *2 Dks*	British Bristol	470 760	230·3 | 28·5 | 16·0 *P 11 f. F 26 f.*	Kinghorn J.Key	1873	5 Bhds. DBe78t FPT30t	2Cy.28" & 54"–33" 120 HP. J.Key,Kirkcaldy	Bristol Steam Nav.Co. Ld.
	Clio *Mariani 72-87*	I ScSr *2 Dks*	Austrian Trieste	865 1360	255·6 | 31·5 | 21·6	Trieste Austro-Ungarico Lloyd.	1880		2Cy.31" & 56"–36" 126 HP. Arsenale Lloyd,Trieste	Lloyd Austro-Ungarico
*	Clio *Bjorkqvist 84-70*	I ScSr *1 Dk2 B*	Russian Nikolaistad	557 862	204·9 | 29·9 | 15·0 *Q 56 f. B 63 f. F 29 f.*	Port Glasgow Murdoch & Murray	1884	4 Bhds. MT15t	2Cy.24" & 46"–30" 100 HP. Dunsmuir&Jackson,Glasg.	A.Hedman
*	Clitus *Currie*	Stl ScSr *2 Dks3 B*	British Melbourne	1588 2485	300·0 | 38·2 | 23·5 *P 34 f. B 58 f. F 34 f.*	Newcastle Palmers'Co.Ld.	1886	5 Bhds. DBc&c216t	3C23½"37" & 61"–42" 200 HP Palmers'Co.,Ld.,Newcastle	A.Currie & Co.
*	Clive *Ware*	I TwnScTg *1 Dk*	British Greenock	91 372	192·0 | 30·2 | 14·0	Port Glasgow R.Duncan&Co.	1884	5 Bhds. APT27t	4Cy.27" & 52"–33" 230 HP. Rankin&Blackmore,Grnck.	"Clive" Steam Tug Co.
*	Cloch	I ScSr *1 Dk2 B*	British Glasgow	388 726	216·0 | 28·7 | 15·0 *P 35 f. B 88 f. F 23 f.*	Glasgow Dobie&Co.	1883	5 Bhds. DBo30t FPT5t	2Cy.25" & 46"–42" 105 HP. J.&J.Thomson,Glasgow	Clyde Shipping Co.
*	Cloncurry *Deason 87-87*	I ScSr *2 Dks3 B*	British London	1639 2579	300·0 | 40·0 | 24·4 *P 28 f. B 74 f. F 32 f.*	Sunderland W.Doxford&Sons	1884	6 Bhds. DBc&c285t MT386t	2Cy.35" & 70"–48" 260 HP. W.Doxford&Sons,Sunderld.	M'Ilwraith M'Eacharn & Co.
*	Clunie *Brown -89*	I ScK *1 Dk*	British Aberdeen	98 207	122·4 | 20·3 | 10·0 *F 16 f.*	Aberdeen J.Duthie,Sons&Co.	1889	1 Bhds. FPT[illegible]	2Cy.15" & 30"–20" 40 HP. Hall,Russell&Co.,Aberdeen	W.T.Moffatt
	Clutha *Marshall -82*	I Sc3mSr *2 Dks*	British Middlesbro'	394 534	191·5 | 27·5 | 15·0 *F 18 f.*	Glasgow Barclay,Curle&Co.	1864	4 Bhds.	2Cy.44"–42" 140 HP. Barclay,Curle&Co.,Glasgow	E.W.Harkess
*	Clutha *Russell*	I Sc3mSr *2 Dks*	British Glasgow	337 515	190·3 | 26·9 | 14·0 *P 22 f.*	Glasgow Barclay,Curle&Co.	1864	4 Bhds.	2Cy.25" & 47"–30" 95 HP. Comp.'73,Muir&Houstn,Gls	J.Clark
	Clyde *Moule 83-88*	Stl Sc3mSr *2 Dks & Spr Dk*	British Glasgow	2198 4089	390·0 | 42·1 | 24·5 *B 136 f. 32·1*	Dumbarton W.Denny&Bros.	1881	7 Bhds.	2Cy.58" & 100"–63" 800 HP. Denny&Co.,Dumbarton	Peninsular & Oriental Steam Nav.Co.
*	Clyde *Williams*	I ScSr *1 Dk*	British Liverpool	172 295	133·2 | 20·7 | 11·5 *Q 44 f. F 11 f.*	Whitehaven Whitehaven SBC.,L	1880	4 Bhds.	2Cy.17" & 34"–30" 70 HP. J.Jones&Sons,Liverpool	A.Cook
	Clyde *Duncan*	I ScSk	British P.Glasgow	152 337	159·3 | 30·0 | 9·4	Renfrew W.Simons&Co.	1882		2Cy.23" & 24"–30" 80 HP. W.Simons&Co.,Renfrew	The Trustees of the Clyde Lighthouses
	Clyde *Plummer*	I PadTg *1 Dk*	British Scarbro'	56 164	116·0 | 20·2 | 10·3	Newcastle T.&W.Toward	1882	3 Bhds.	1Cy.38"–58" 60 HP. G.P.Hepple,North Shields	"Britannia" S.S.Co., Ld.
	Clyde	I Pad	British Glasgow	14 124	103·0 | 20·1 | 8·4	Glasgow A.&J.Inglis	1851		2Cy.34"–48" 80 HP. A.&J.Inglis,Glasgow	The Trustees of the Clyde Navigation
	Clyde *Clark 87-88*	I ScBkn *2 Dks3 B*	American New York	1406 2017	263·0 | 37·0 | 20·2 *Len. 576*	Philadelphia Cramp&Sons	1870		1Cy.43"–42" 480 HP. Cramp&Sons,Philadelphia	Pacific Mail S.S.Co.
*	Clydesdale *Johnston 86-88*	Stl ScSr *1 Dk2 B*	British Glasgow	608 972	214·4 | 31·2 | 13·6 *Q 51 f. B 66 f. F 31 f.*	Port Glasgow Blackwd.&Gordon	1881	5 Bhds. CDB[illegible] FPT[illegible]	2Cy.26" & 49"–36" 100 HP. Blackwood&Gordon,Pt.Gls.	R.Mackill & Co.
	Clydesdale *McCallum 83-83*	I ScSr	British Glasgow	238 447	196·7 | 24·1 | 13·5 *Q 23 f.*	Glasgow J.&G.Thomson	1862		2Cy.40"–30" 120 HP. J.&G.Thomson,Glasgow	D.MacBrayne
*	Clymene *Elliot 83-83*	I ScBn *2 Dks3 B*	British London	974 1509	245·0 | 33·0 | 22·0 *P 25 f. F 31 f.*	Stockton M.Pearse&Co.	1879	4 Bhds. DBc[illegible] FT[illegible]	2Cy.32" & 60"–39" 140 HP. Blair&Co.,Ld.,Stockton	H.T.Cloake,junr.
*	Clytha *Willmott 81-87*	I ScSr *1 Dk2 B*	British Newport	519 819	220·8 | 30·8 | 13·3 *B 44 f. F 21 f.*	Newcastle Tyne I.S.B.Co.,Ld.	1881	4 Bhds. [illegible] APT[illegible]	2Cy.28" & 53"–33" 98 HP. J.Dickinson,Sunderland	Stephens, Mawson & Goss
*	Coanwood *Purvis 88-88*	I ScSr *1 Dk2 B*	British N.Shields	710 1105	226·8 | 30·0 | 16·6 *P 136 f. B 10 f. F 23 f. Q 56 f.*	Sunderland Osbourne,Graham&Co.	1872	3 Bhds. WB[illegible]	2Cy.26¼" & 48"–30" 95 HP. M.Samuelson,Hull	T.J.Eltringham
*	Coanza *Haynes 83-86*	Stl ScSr *1 Dk2 B*	British Glasgow	979 1578	290·4 | 36·2 | 18·0 *B 63 f. P 77 f. F 37 f.*	Glasgow J.Elder&Co.	1880	9 Bhds.	2Cy.34" & 62"–36" 175 HP. J.Elder&Co.,Glasgow	British & African Stm. Nav.Co.,Ld.

[illegible] = Lloyd's Register; D = Nederlandsche Vereeniging; G = Germanischer Lloyd; I = Registro Italiano; R = American Record;
[illegible] = Veritas Austro-Ungarico; H = Veritas Ellenico; N = Norske Veritas; V = Bureau Veritas.

Dans la 2me Colonne, B.T., avec les indications qui suivent, signifie la date du rapport de visite de la Carène par les experts du Board of Trade pour obtenir le certificat relatif au TRANSPORT DES PASSAGERS.

Dans la 3me Colonne, C = Composite (membrure en fer, bordé en bois); I = Fer; Stl = Acier; W = Bois.

Pour plus amples renseignements sur les navires classés par le Lloyd's Register, voir le Lloyd's Register Book et le Lloyd's Yacht Register.

No. in Book / Official No.	Registry	Ship's Name. Material, Rig, &c. / Master, and Dates of Appointment to present Owner's Service and to Ship / Ship's Late Name. No. of Decks, &c.	Flag. / Port of Registry.	Tons. Net. / Gross	Length	Breadth	Depth	Built. Where. / By Whom.	When	No. of Bulkheads. / Particulars of Water Ballast.	Engines. / Particulars. Maker's Name.	Owners or Managers
691 86071	✻	Coban I ScSr Fraser 84-84 1 Dk	British Montreal	689 1063	230·0 F35f.	33·1 F21f.	16·0	Sunderland W.Doxford&Sons	1882	4 Bhds. Cfl 1535t	2Cy.31"&56"–36" 130 HP. W.Doxford&Sons,Sunderld.	"Black Diamon Co. of Montreal
692 53915		Cobden I ScSr Battram 76-76 1 Dk	British Middlesbro'	618 [illegible]	240·3	29·0 F 205 t.	16·3	Stockton Richardson,Duck & Co.	1866	5 Bhds. WB[illegible]	2Cy.48"–30" 110 HP. Fossick,Blair&Co.,Stockton	J. W. Pease
693 88777	✻	Coble Dene I PadTug Wilkinson 73-83 1 Dk	British Newcastle	23 125	90·0	19·6	9·5	NorthShields Hepple&Co.	1883	3 Bhds.	2Cy.30"–48" 60 HP. Hepple&Co.,NorthShields	Tyne Improvement Commissioners
694 87245	✻ BT1.88	Coblenz Stl Sc3mSr Baird 81-84 2 Dks	British Leith	825 1307	254·5 P32f.	33·2 B50f.	16·8 F135f.	Greenock R.Steele&Co.	1883	6 Bhds. FPT[illegible] MT[illegible]	2Cy.38"&50"–42" 175 HP. R.Steele&Co.,Greenock	J. Currie & Co.
695 67995		Coconada I ScSr Blake 88-88 2 Dks	British Glasgow	547 862	230·4 P85f.	28·6 F46f.	16·4	Glasgow A.&J.Inglis	1872	4 Bhds.	2Cy.28"&50"–36" 120 HP. A.&J.Inglis,Glasgow	British India S. Co. Ld.
696 87351	✻	Cogent I ScSr Wilson 73-88 1 Dk2 B	British Sunderland	1394 2140	275·0 Q90f.	38·6 B31f. F11f.	20·6 B172f. F44f.	Sunderland ShortBros.	1883	5 Bhds. WB[illegible]	2Cy.36"&67"–36" 186 HP. T.Richardson&Sons,H'pool	J. Westoll
697 29152		Cognac I Sc3mSr Aiken 68-73 2 Dks	British Liverpool	303 495	170·0	25·1	16·4	Liverpool T.Vernon&Sons	1860	3 Bhds.	2Cy.17"&32"–27" 50 HP. Lees,Anderson&Co.,Glasgow	Charente S.S.Co
698 93246	✻	Coila I ScSr Macdonald 84-84 1 Dk	British Glasgow	177 341	150·0 Q53f.	23·0 B33f.	11·2 F19f.	Ayr M'Knight,M'Credie &Co.	1884	4 Bhds. [illegible]	2Cy.18"&36"–28" 55 HP. J.Gilmour&Co.,Glasgow	T. M'Laren & C
699 78007		Colaba I Sc3mSr Holt 86-88 1 Dk2 B	British Glasgow	1071 1652	264·6	32·2	21·6	Dumbarton W.Denny&Bros.	1878	4 Bhds. Dbot[illegible] APT12[illegible]	2Cy.37"&54"–36" 122 HP. Denny&Co.,Dumbarton	British India S. Nav. Co. Ld.
700 89460	✻	Colac I ScSr Marchant 1 Dk2 B	British Adelaide	958 1,[illegible]	245·2 Q34f.	34·2 B57f.	17·2 F31f.	Hartlepool E.Withy&Co.	1884	4 Bhds. Dbcel[illegible]	2Cy.30"&57"–38" 140 HP. T.Richardson&Sons,H'pool.	Adelaide S.S.Co
701	V	Colberg I ScSr Hildebrandt 82-83 1 Dk	German Colberg	155 213	125·3 Q39f.	19·5 B39f.	10·2 F22f.	Stettin Stettiner Mas. Akt Ges. "Vulcan"	1881	Dbot[illegible]	2Cy.15"&27½"–19½" 35 HP. Stettiner Masch Akt.-Ges. "Vulcan," Stettin	Colberger. Dampf Akt. Ges. (E.Beu
702		Colchester Stl ScwSr	British Harwich	907 1160	280·8 B 133f.	31·0 P58f.	15·2 F47f.	Hull Earles Co.,Ld.	1888		1Cy.30"&57"–36" 270 HP. Earles Co.,Ld.,Hull	Great Eastern R
703 67511	✻	Coleridge I ScBn Symons 78 1 Dk2 B	British Exeter	714 1120	215·0 B57f.	32·0 Q71f.	16·5 F31f.	Stockton M.Pearse&Co.	1879	4 Bhds. Dbf[illegible]	2Cy.28"&51½"–33" 98 HP. Blair&Co.,Ld.,Stockton	J. Wild & Co.
704 125153	RV	Colima I ScBg Caverly 65-87 2 Dks&Spr Dk	American New York	2144 2906	299·7	40·0	20·3 29·1	Chester, Pa. J.Roach&Son	1873		2Cy.49¾"&88¾"–42" 450 HP. J.Roach&Son,Chester,Pa.	Pacific Mail S.S
705 68022	✻	Colina I ScBg Jennings 86-87 3 Dks	British Glasgow	1297 2094	319·3	34·8	25·6	Glasgow Barclay,Curle&Co.	1872	5 Bhds.	2Cy.44"&78"–46" 265 HP. Barclay,Curle&Co.,Glasgow	Donaldson Bros
706 81502	✻	Colina I ScBn Bell 82-85 1 Dk2 B	British W.H'pool	1068 1658	259·0 Q105f.	34·1 B60f.	19·6 F30f.	W. Hartlepool Irvine&Co.	1879	4 Bhds. DBaf[illegible]	2Cy.32"&60"–39" 140 HP. Blair&Co.,Ld.,Stockton	R. Irvine & Co.
707 32278		Collaroy (now a Sailing Vessel) I PadSr 1 Dk	British Sydney	264 419	180·9 Q 35 f.	23·3 H 27 f.	11·1	Liverpool Fawcett&Preston	1859		2Cy.42"–36" 140 HP. Fawcett&Preston,Liverp'l.	A. Burns
708 25060	BT6.88	Colleen Bawn I PadSr Brannagan	British Drogheda	297 715	221·4 Q—	28·1 F	15·3	Glasgow Randolph,Elder&C.	1862	4 Bhds.	2Cy.45"&81"–72" 303 HP. J.Jack&Co.,Liverpool	Drogheda S.Pa
709 15700		Collier I Sc3mSr Reid 1 Dk	British Swansea	126 206	127·7	20·2	10·5	PortGlasgow J.Reid&Co.	1849	4 Bhds.	2Cy.20"–21" 40 HP. Caird&Co.,Greenock	W. Pockett
710 70257	✻	Collingwood I ScSr Luen 85-85 1 Dk2 B	British Newcastle	531 880	215·0 Q74f.	29·4 F23f.	15·4	Newcastle ColeBros.	1876	4 Bhds. DBa[illegible]	[illegible] 98 HP. [illegible]	P. Rowe & Sons
711 89208	✻D	Collivaud Stl ScSr H. Brown 86 1 Dk2 B	British Cardiff	950 1447	240·9 Q86f.	33·1 B50f.	18·0 F26f.	Cardiff Bute S.E. & Dry Dk. Co.	1886	4 Bhds. Dbt[illegible] APT25t	[illegible] HP. [illegible] Ld., Newcastle	"Collivaud" S.S [illegible]
712	V	Cöln I ScSr Bening 88-88 2 Dks&Spr Dk	German Stettin	844 1107	223·7	30·3	15·1 20·3	Stettin Stettiner Mas. Akt Ges. "Vulcan"	1884	Dbot[illegible]	2Cy.26¾"&49"–31½" 110 HP. Stettiner Masch.Akt.-Ges. "Vulcan," Stettin	Neue Dampfer-Com- pagnie Akt. Ges
713 25171	BT5.88	Cologne I PadSr	British London	379 521	190·3	25·6	13·7	London Westwood,Baillie& Co.	1858		[illegible] 46"–42" 136 HP. [illegible] SteamNav.Co.,London	General Stm. N
714	FV	Colomba I ScSr Charret 88 2 Dks&Spr Dk	French St. Nazaire	537 843	199·4	26·8 B 38 f.	23·1	St. Nazaire Scott&Co.	1865		2Cy.23½"&50¼"–30" 120 HP. Scott&Co.,Greenock	Compagnie Gé Transatlantique
715 60242		Colombia I Sc3mSr 1 Dk2 B	British Liverpool	1137 1823	308·0 H 295 f.	38·1 F 51 f.	20·4	Birkenhead LairdBros.	1873		2Cy.51¼"&84½"–42" 375 HP. LairdBros.,Birkenhead	Pacific Stm. N
716 50828		Colombia W PadSr (ex Quaker City) 2 Dks	British Kingst'n,Ja.	758 1137	224·4	33·6	20·5	Philadelphia			1Cy. 1000 HP.	William McE
717	F	Colombie (ex Floride) I ScBk Banjou 88	French St. Nazaire	1696 2694	278·8	37·7	36·1	Greenock Caird&Co.	1862		2Cy.53½"&93"–42" 500 HP. Comp.SocdesConstr.Nav.Hav	Compagnie Gé Transatlantique
718		Colombo I PadSr 1 Dk	Brazilian Maranhão	266 423	168·4 B63f.	25·5 F31f.	9·9 P48f.	Birkenhead LairdBros.	1880		2Cy.32"&59"–54" 120 HP. LairdBros.,Birkenhead	Cia. de Nav. a V Maranhão
719	V	Colombo I ScBg Marrantoni 84-86 2 Dks&Spr Dk	French Marseilles	2948 3733	379·6 B 97 f.	39·7 F 34 f.	24·9 29·5	La Seyne Forges&Ch. de la Méditt.	1882		2Cy.39½"&72½"–42½" 400 HP. Forges&Ch.delaMédit.Mars'lls	Cie. Nationale Navigation
720 125000	RV	Colon I ScBg Henderson 86-87 3 Dks	American New York	1843 2686	280·0	40·0	28·6	Chester, Pa. J.Roach&Son	1873		2Cy.50"&88"–42" 400 HP. J.Roach&Son,Chester,Pa.	Pacific Mail S

✻ = Lloyd's Register; D = Nederlandsche Vereeniging; G = Germanischer Lloyd; I = Registro Italiano; R = American R
A = Veritas Austro-Ungarico; H = Veritas Ellenico; N = Norske Veritas; V = Bureau Veri

In the 2nd Column, B.T., with figures following, signifies Date of report of bottom sighting by Board of Trade Surveyors as condition of PASSENGER CERTIF

In the 3rd Column, C = Composite (iron frame, wood planked); I = Iron; Stl = Steel; W = Wood.

Further particulars of Vessels classed by Lloyd's Register will be found in Lloyd's Register Book and Lloyd's Yacht Register.

Classe	Nom, Matériaux, Gréement, etc. — Capitaine et dates d'entrée au service des Armateurs actuels et à bord du navire. — Ancien Nom, nombre de ponts, etc.		Pavillon — Port d'Armement	Tonnage Net — Total	Dimensions			Construit — Où — Par qui	Quand	Nombre des Chaudières — Détails sur le lest d'eau	Machines — Détails — Nom du Constructeur	Armateurs ou Directeurs-Gérants
	Colón *Eberle 83-84*	I [illegible] *2 Dks*	Spanish	476	191·0	26·9	16·7	Liverpool	1871		2Cy.23"&46"–21" 80 HP. J.Taylor&Co.,Sunderland	J.Roca & Co.
V	Colon *Sichet*	I S	Uruguay'n	271	172·9	26·9	9·8	Elbing	1884		1Cy.14"&27"–16" 70 HP. F.Schichau,Elbing	Galeano & Soto
✻	Colonel Smith *Edwards*	I S *1 Dk*	British	114 271	134·8	22·4	11·9 *B 24 f. F 12 f.*	London R.&H.Green&Co.	1884	3 Bass.	2Cy.23"&44"–28" 82 HP. A.W.Roberts'n&Co,London	Gt. Northern. S.S. FishingCo.,Ld.
V	Colonia *Rieke 83-84*	Stl [illegible] *2 Dks*	German	1479	290·0	35·2	22·9 *P 63 f. F 32 f. B 27 f.*	Newcastle SirW.G.Armstrong, Mitchell&Co.,Ld.	1887		[illegible] 250 HP. Newcastle	Ham.Amer.Packetf. Act. Ges.
	Colorado *2 Dks*	Stl S	British	2787	370·0	44·7	28·4 *P 230 f. F 39 f.*	Hull Earle'sCo.,Ld.	1887	8 Blds.	3C.31"50"&82"–57" 450 HP. Earle'sCo.,Ld.,Hull	T.Wilson Sons & Co.
R	Colorado *Evans 76-88*	I S *2 Dks*	American New York	2022	306·0	39·3	21·6	Chester, Pa. J.Roach&Son	1879		2Cy.38"&68"–54" 1200 HP. J.Roach&Son,Chester,Pa.	C.H.Mallory & Co.
✻	Colstrup *Le Gresley*	I S *1 Dk*	British	319	159·4	21·1	11·2 *Q 90 f. B 16 f. F 22 f.*	Newcastle C.S.Swan&Co.	1874	4 Blds.	2Cy.21"&38"–30" 60 HP. Ouseburn Eng.Wks,Newcastle	T.E.Brigham (W.Swanston&Sons,Managers)
✻	Columba *Rodd 85-85*	I ScSr	British	645 1031	221·4	29·5	16·8 *P 219 f. F 29 f.*	Sunderland J.Laing	1872	4 Blds.	2Cy.26"&49"–33" 98 HP. T.Clark&Co.,Newcastle	Avis Steamships Co.
	Columba *Campbell 6-81*	StlPadSk	British	126 543	301·4	27·1	9·4	Glasgow J.&G.Thomson	1878		2Cy.53"–66" 220 HP. J.&G.Thomson,Glasgow	D.MacBrayne
	Columba *ex L'Espiègle*	I ScSrYt	British Glasgow	90 174	143·1	21·5	9·5	Glasgow J.&G.Thomson	1869		2Cy.18"&36"–24" 60 HP. Lees,Anderson&Co.,Glasgow	HenryMacIver
	Columbia *Garrick 84-88* *2 Dks&AwnDk*	I ScBuS	British Glasgow	1322 2030	283·3	33·6	22·2 29·5	Glasgow A.Stephen&Sons	1866	6 Blds.	2Cy.33"&60½"–48" 380 HP. D.&W.Henderson,Glasgow	Henderson Bros.
	Columbia *Scott 81-84*	Stl ScK	British London	42 244	121·6	22·0	12·5	Delfshaven DeMaasCo.,Ld.	1884		2Cy.23"&47"–24" 120 HP. DeMaasCo.,Ld.,Delfshaven	W.Watkins
✻	Columbia *Edwards -88*	I ScK *1 Dk*	British Hull	121 251	137·2	21·5	11·0 *Q 51 f. F 22 f.*	Stockton M.Pearse&Co.	1886	3 Blds.	3C.15"24½"&40"–24" 70 HP. Blair&Co.,Ld.,Stockton	Great NorthernSteamshipFishingCo.,Ld.
RV	Columbia *Bolles 66-80*	I ScBg *2 Dks 3 B*	American Portland, Oreg.	1746 2722	310·0	38·4	23·2	Chester, Pa. J.Roach&Son	1880		2Cy.42½"&82"–54" 3111 HP. J.Roach&Son,Chester,Pa.	OregonRailway&Nav. Co.
	Columbia	W Pad *1 Dk*	American New York	1098 1465	260·6	39·0	12·0	Green Point, NY	1877			J.E.Hoffmire
	Columbus *Stone -86*	I PadSp *1 Dk*	British Calcutta	97 179	128·5	20·4	10·0	Birkenhead LairdBros.	1862		2Cy. 90 HP. LairdBros.,Birkenhead	Inder Chunder Sing & N.K.Ghose
	Columbus	I PadSp	British Liverpool	32 191	134·2	20·4	10·6	Birkenhead LairdBros.	1865		2Cy.38"–54" 100 HP. LairdBros.,Birkenhead	W.H.Dodd
R	Comal *Rick 79-86* *2 Dks&promenadeDk*	I ScSr	American New York	2251 2934	320·0	42·8	21·5	Chester, Pa. J.Roach&Son	1885		2Cy.38"&74"–54" J.Roach&Son,DelawareR.	C.H.Mallory & Co.
✻	Comeragh *Minards 85-88* *2 Dks&Spt.AwnDk*	I Sc3mSr	British Waterford	467 818	230·1	30·2	16·0 *P 16 f.*	Glasgow London&Glas.Co.	1879	5 Blds.	2Cy.30"&56"–45" 200 HP. London&Glas.Co.,Glasgow	Waterford S.S.Co.
	Comercio *Guiliani*	I Sc	Arg'ntine Buenos Ayres	192 295	160·0	21·9	11·2	Glasgow A.&J.Inglis	1870		50 HP. Cmp.'85,G.Harley,Mntvido.	Dgo.Guiliani
	Comet *Nichols*	I PadSp	British Sunderland	24 132	106·7	19·4	9·8	Cardiff J.&M.Gunn&Co.	1876		2Cy.28"–18" 56 HP. J.&M.Gunn&Co.,Cardiff	R.M.Hudson & Son
	Comet	I PadSp	German Bremen	66 107	112·5	18·1	6·9	Newcastle	1858		2Cy.28½"–45¾" 73 HP.	Norddeutscher Lloyd
✻	Cometa *Ogg -88* *1 Dk&AwnDk*	StlTwinScS	British Glasgow	718 1151	220·0	32·1	12·5 19·7	Dumbarton A.M'Millan&Son	1888	1 Blds.	6C.12"19"&31"–21" 90 HP. D.Rowan&Sons,Glasgow	Hill,Gomes & Co.
	Comillas *ex Infanta Isabel*	I ScBk	Spanish Barcelona	1383 1982	281·4	38·5	26·0 *Q 13 f. B 101 f. F 48 f.*	Dumbarton W.Denny&Bros.	1863		2Cy.66"–42" 300 HP. Denny&Co.,Dumbarton	Cia.Trasatlantica
V	Commandant Franchetti *Perthuis -81*	I ScSr *1 Dk 2 B*	French Bordeaux	619 993	219·8	30·5	17·2 *Q 67 f. B 35 f. F 26 f.*	Sunderland Oswald&Co.	1869		2Cy.30"&60"–36" 160 HP. Lobnitz&Co.,Renfrew	Worms,Josse & Co.
	Commentry *Babault 82-83*	I ScSr *1 Dk 2 B*	French La Rochelle	954 1501	251·3	37·2	17·3 *Q 28 f. B 38 f. F 35 f.*	Sunderland Osbourne,Graham & Co.	1882	1 Blds.	2Cy.32"&60"–40" 150 HP. Hawks,Crawshay&Co,Nwc.	D'Orbigny & Faustin fils
	Commerce *Collings 82-82*	W ScSr	British Guernsey	30 128	105·4	20·6	9·6	Guernsey Ogier	1874		2Cy.18"&29"–16" 40 HP. Cmp.'75,J.Stewart,London	Plymouth, Channel Islands and BrittanyS.S.Co.,Ld.
V	Commerce *Hertz 82-83*	I ScSr *1 Dk*	Swedish Gothenburg	369 591	149·6	23·8	15·5 *B 35 f.*	Gothenburg EriksbergsMek.Verks.	1882		2Cy.16"&33"–19½" 50 HP. EriksbergsMek.Verks.	Ångf.Aktie-Bolag. "Commerce" (T.Ahrenberg)
✻	Commercial *Weaver*	I ScK *1 Dk*	British N.Shields	64 113	100·7	18·3	8·6	SouthShields J.Softley&Co.	1874	3 Blds.	2Cy.11½"&23"–18" 25 HP. W.&J.Almond,N.Shields	Burt, Boulton & Haywood
V	Commercial *Kühr 82-83*	I ScSr *1 Dk*	German Kiel	407 592	166·0	24·6	13·0 *Q 98 f. F 29 f.*	Kiel Act.Ges. "Germania"	1883		2Cy.17¾"&32¾"–24" 60 HP. Act.Gesell."Germania,"Berlin	Paulsen & Ivers

Lloyd's Register; D = Nederlandsche Vereeniging; G = Germanischer Lloyd; I = Registro Italiano; R = American Record;
Veritas Austro-Ungarico; H = Veritas Ellenico; N = Norske Veritas; V = Bureau Veritas.

Dans la 2me Colonne, B.T., avec les indications qui suivent, signifie la date du rapport de visite de la Carène par les experts du Board of Trade pour obtenir le certificat relatif au TRANSPORT DES PASSAGERS.

Dans la 3me Colonne, C = Composite (membrure en fer, bordé en bois); I = Fer; Stl = Acier; W = Bois.

Pour plus amples renseignements sur les navires classés par le Lloyd's Register, voir le Lloyd's Register Book et le Lloyd's Yacht Register.

No.	Class	Ship's Name, Material, Rig, &c. / Master, and Dates of Appointment to present Owner's Service and to Ship / Ship's Late Name. No. of Decks, &c.	Flag. / Port of Registry.	Tons. Net. / Gross	Dimensions (Length, Breadth, Depth) / Deck Erections	Built. Where. / By Whom.	When	No. of Bulkheads / Particulars of Water Ballast.	Engines. Particulars. / Maker's Name.	Owners or [illegible]
751	V	Commercial ex [illegible] W ScSr Mietsch 1 Dk	German Stettin	122 154	107·2 / 20·7 / 9·8 Len. 74	NorthShields J.Wilson	1872		2Cy.15"& 27"–16" 55 HP [illegible], Stettin	E.Hambus[illegible]
752	G	Commerzienrath Fowler I ScSr Kutziel 84-86 1 Dk	German Memel	721 881	217·2 / 28·2 / 14·5	Kiel G.Howaldt	1884	[illegible]	2Cy.23"&50"–27"	Memel'er [illegible] Akt.-Ges.
753		Commilla I ScSr Jones 83-84 2 Dks & Shltr Dk	British Glasgow	550 [illegible]	230·4 / 28·3 / 16·3 P53 f. F53 f. B[illegible] 23·7	Glasgow A.&J.Inglis	1875	4 Blkds.	2Cy.32"& 54"–36" 110 HP A.&J.Inglis, Glasgow	British India [illegible] Co., Ld.
754		Commissioner W Pad Tuer 70-83	British Newcastle	34 102	90·6 / 19·5 / 9·5	Newcastle T.Hepple	1862		1Cy.36½"–52" 45 HP T.Hepple, Newcastle	Tyne Improv[illegible] Commissioners
755	✻	Commodore I Sc3mSr Elgar 77-86 1 Dk 2 B	British Newcastle	565 882	219·7 / 29·9 / 16·8 P31 f. B42 f. F24 f.	Newcastle Wigham Richardson&Co	1870	4 Blkds. [illegible]	2Cy.25"& 48"–30" 96 HP R.&W.Hawthorn, N'castle	Pyman, Bell & [illegible]
756 71986		Commodore I PadTug Hesellon 1 Dk	British Sydney	84 187	130·9 / 20·7 / 10·3	SouthShields J.T.Eltringham	1877	4 Blkds.	2Cy.34"–54" 95 HP J.P.Rennoldson, S.Shields	Port Jackson [illegible] Ld.
757 15450	BT2.88	Commodore I PadSp Grant (ex Great Emperor)	British Liverpool	30 225	139·7 / 23·2 / 11·3	Newcastle Marshall Bros.	1863		2Cy.39"–57" 110 HP Marshall Bros., Newcastle	W. & T. Jolli[illegible]
758 1954		Commodore W PadTug McCrorey 83-87	British Belfast	10 114	107·8 / 19·7 / 9·8	Newcastle T.Hepple	1864		1Cy.38½"–56" 60 HP T.Hepple, Newcastle	Union Steam T[illegible] Co., Ld.
759 88761	✻	Commonwealth I ScSr Faulkner 84-84 1 Dk 2 B	British Newcastle	603 958	215·0 / 31·7 / 14·3 Q107 f. B26 f. F31 f.	Dundee Pearce Bros.	1883	4 Blkds. DB[illegible] FPT[illegible] APT[illegible]	2Cy.26"& 50"–33" 99 HP Pearce Bros., Dundee	"Commonwe[illegible] Co. Ld. (W[illegible])
760 [illegible]	BT5.88	Como I ScBg 1 Dk 2 B	British Hull	979 1492	272·6 / 32·0 / 17·7 P117 f. F38 f.	Hull Earle's Co.	1871	4 Blkds.	2Cy.32"& 60"–33" 200 HP Earle's Co., Hull	T.Wilson, Son[illegible]
761 91298	✻	Como COLLISION StlScSr Ghiraldi 1 Dk	British Liverpool	297 477	150·3 / 30·1 / 11·9 B33 f. F21 f.	Barrow BarrowShipBdg.Co.	1885	4 Blkds. APT[illegible]	2Cy.18"& 36"–30" 67 HP Barrow Shipbuilding Co.	Argentine S[illegible] er Co., Ld.
762 [illegible]	✻	Comorin I ScSr Hamilton 84-84 2 Dks	British Glasgow	307 534	210·3 / 25·0 / 14·2 P& B140 f. F32 f.	Renfrew W.Simons&Co.	1863	4 Blkds.	2C25$\frac{9}{16}$"&48$\frac{3}{16}$"–27" 93 HP Cmp.'78, Hamilton&Co., Gls	J. & P. Hutch[illegible]
763	DV	Comorin I ScBg Ponet 83-88 2 Dks & Spr Dk	French Marseilles	2949 3742	379·2 / 40·0 / 21·9 B97 f. F34 f. P28 f. 29·6	LaSeyne Forges & C. de la Medit.	1882		2C.39½"&72¼"–42½" 400 HP Forges&C.dela Medit M'rslls	Cie. National[illegible] Navigation
764		Compta I ScSr Lammers 3 Dks	Dutch Batavia	1290 2135	285·2 / 35·2 / 24·1 P42 f. B44 f. F38 f.	Glasgow A.&J.Inglis	1881	6 Blkds. DB[illegible]	2Cy.34"& 60"–45" 200 HP A.&J.Inglis, Glasgow	Nederlandsch[illegible] Stoomv.-Maats
765		Comte Baciocchi I ScSr 2 Dks	French Marseilles	331 517	173·8 / 29·5 / 14·6 Q31 f.	Greenock Scott&Co.	1860		2Cy.42"–30" 120 HP Greenock Foundry Co., Grnk	F.Morelli & Co.
766		Comte d'Eu I PadSr	Arg'ntine B.Ayres	823	228·7 / 31·6 / 19·8	Newcastle Stephenson	1864		260 HP	C.de Mot & Co.
767		Comte d'Eu I PadSr	Brazilian RiodeJaneiro	596	227·1 / 35·0 / 9·7	London	1864		220 HP	V.J.Frietsch[illegible]
768		Comte de Flandre I Pad	Belgian	419	200·0 / 24·8 / 7·3	Antwerp	1870			Belgian Gove[illegible]
769	V	Comte de Hainaut I ScSr Knudtsen 85-88 2 Dks	Belgian Antwerp	525 738	261·7 / 28·2 / 16·8 Q111 f. F13 f.	Amsterdam Konin Fabriek	18$^{60}_{80}$		2Cy.26"& 50"–36" 100 HP Gebr.Howaldt, Kiel	A.Deppe
770		Comtesse de Flandre I Pad	Belgian	419	200·0 / 24·8 / 7·3	Antwerp	1870			Belgian Gove[illegible]
771		Concepcion I TwnSc3mSr (ex Maryland, ex Coquette)	Spanish Havanna	 531	227·0 / 24·0 / 14·0	Renfrew Henderson, Coulborn&Co.	1863			D.F.de Castro
772 87616	✻	Concetta W ScLaunch 1 Dk	British Malta	3 54	70·4 / 15·3 / 8·0	Malta P.Grech	1884		2Cy.12½"& 25"–16" 25 HP A.Wilson&Co., London	Soc. di Pontoni Lanzie
773	V	Concha Stl&I ScSr Krüger -82 1 Dk 2 B	German Hamburg	630 895	213·2 / 27·3 / 16·7 Q117 f. F33 f.	Hoboken Société Cockerill	1877	DB[illegible] APT[illegible]	2Cy.27"& 45"–33" 90 HP Société Cockerill, Seraing	O.Schmidt
774 84342	✻	Concordia I Sc3mSr McLean 81-86 2 Dks 3 B	British Glasgow	1617 2544	319·6 / 41·0 / 25·1 B94 f. F31 f.	Glasgow Barclay, Curle&Co.	1881	6 Blkds. DB[illegible]	2Cy.36"& 68"–54" 296 HP Barclay, Curle&Co., Glasgow	Donaldson Bro[illegible]
775		Concordia I PadCh Ritter 1 Dk	German Stade	76 125	138·0 / 17·8 / 7·5 260	Rostock A.Tischbein	1870		3Cy.21"& 31"–42" 60 HP A.Tischbein, Rostock	Stade Altländer schifff. & Rhed[illegible]
776	N	Concordia I ScSr Andersen 1 Dk 2 B	Norweg'n Christiania	400 597	182·2 / 24·9 / 15·2 Q65 f. B44 f. F30 f.	Sunderland [illegible]&Co.	1863	4 Blkds.	2Cy.24"& 47"–30" 90 HP Thompson, Boyd&Co., Newc.	Gerardo Mon[illegible]
777		Concordia Berrisso	Urug'yan Montevideo	140 210	115·0 / 20·0 / 7·4	Montevideo Harley&Co.	1875		55 HP Harley&Co., Montevideo	Sarda, Coll & Co.
778		Condé (ex Conde d'Eu) I ScSr 1 Dk & Awn Dk	French Bordeaux	896 1168	239·5 / 34·1 / 12·6 19·5	PortGlasgow W.Hamilton&Co.	1874	4 Blkds.	2Cy.28"& 54"–33" 150 HP J.&T.Young, Ayr	Manrel & H.P[illegible]
779	✻	Conde Wifredo StlScBg Gorordo 88-89 2 Dks & Spar Dk	Spanish Cadiz	2335 3555	363·0 / 44·1 / 20·1 B38 f. P[illegible] f. F30 f. (27·6)	Sunderland J.L.Thompson&Sons	1889	7 Blkds. [illegible] APT[illegible]	3Cy.31",51"&85"–54" J.Dickinson, Sunderland 1500 HP	Pinillos, Saenz & [illegible]
780		Conde d'Eu	Brazilian Corumba	622	.. .. ..				250 HP	C.ia Nacional de Vaps.

✻ = **Lloyd's Register;** D = Nederlandsche Vereeniging; G = Germanischer Lloyd; I = Registro Italiano; R = American Re[illegible]
A = Veritas Austro-Ungarico; H = Veritas Ellenico; N = Norske Veritas; V = Bureau Veri[illegible]

In the 2nd Column, B.T., with figures following, signifies Date of report of bottom sighting by Board of Trade Surveyors as condition of PASSENGER CERTIF[illegible]

In the 3rd Column, C = Composite (iron frame, wood planked); I = Iron; Stl = Steel; W = Wood.

Further particulars of Vessels classed by Lloyd's Register will be found in Lloyd's Register Book and Lloyd's Yacht Register.

1	2 Nom, Materiaux, Gréement, etc. Capitaine et dates d'entrée au service des Armateurs actuels et à bord du navire. Ancien Nom, nombre de ponts, etc.	4 Pavillon. Port l'Armement	5 Tonnage Net. Total	6 Dimensions. Longueur	Largeur	Creux	7 Construit. Où. Par qui.	8 Quand	9 Nombre des cloisons. Détails sur le lest d'eau.	10 Machines. Détails, Nom du Constructeur.	11 Armateurs ou Directeurs-Gérants.
	Conde-d'Eu I TwnSc 1 Dk & Hur Dk	Brazilian Para	303 443	181·0	28·0	9·5	Port Glasgow Blackwd.&Gordon	1883		1Cy.21"& 38"-27" 125 HP. Blackwood&Gordon,Pt.Gls.	Amazon Stm.Nav.Co.
	Conde de Vilana I ScBk Tines [illegible]	Spanish Barcelona	1577 2468	338·7	36·3	25·6 F 75 f. Lon 71.	Glasgow R.Napier&Sons	1869		2Cy.41"&78" -42" 270 HP. N.E.M., J.&G.Thomson, Glsgw.	Conde de Vilana
✻	Condor I ScSr Ferris 1 Dk 2 H	British London	434 682	189·6	26·5	14·5 Q 54 f. F 23 f. B 46 f.	Dundee Gourlay Bros.	1875	5 Blkds. [illegible]	2Cy.21"& 40"-30" 80 HP. Gourlay Bros.,Dundee	General Steam Nav.Co.
✻	Condor I ScTug Verdun 81-85 1 Dk	British London	8 123	99·7	19·1	11·2	Leith S.&H.Morton&Co.	1884	3 Blkds.	2Cy.20½"& 40"-21" 70 HP. A.Shanks&Son,Arbroath	Todd & Son
✻	Condor Scott 1 Dk	French Havre	 28	60·7	13·1	6·1	North Shields Hepple&Co.	1885		2Cy.12"-14" 30 HP. W.Hepple&Co.,N.Shields	Soc. Havraise de Remorquage
V	Condor I ScSr Wolters 82-88 2 Dks & Spar Dk	German Bremen	1020 1230	221·7	30·1	16·1 24·1	Geestemünde J.C.Tecklenborg	1883	DB/[illegible] DB.[illegible]	2C.26¾"&50½"-30½" 120 HP. F.Schichau,Elbing	Norddeutscher Lloyd
	Condor I ScTug Maciel	Arg'ntine Buenos Ayres	60 110	..	..	..	Buenos Ayres Bularacca&Sons	1885		26 HP. Wilson&Co.,London	T.Sacconi
	Conductor PadTug	Portug'se Lisbon	100	116·1	19·3	10·5	South Shields	1880		98 HP.	F.G.Burnay
✻	Confiance I TwnScTug (ex Navigator) 1 Dk	British London	 141	107·1	19·4	10·3	Northwich E.Foden	1884	3 Blkds.	1Cy.16"&30"-19" 85 HP. E.Foden,Sandbach	The Admiralty
✻	Confidence I ScSr Shaw 86-86 1 Dk	British Runcorn	99 282	120·5	22·1	10·9 Q 38 f. B 14 f. F 14 f.	Northwich E.Foden	1885	3 Blkds. [illegible]	2Cy.16"& 30"-22" 65 HP. W.E.Bates,Northwich	Confidence Steam Ship Co.(E.Finlay)
	Confidence W Pad	British Workington	37 103	98·8	18·4 B 35 f.	9·1	Middlesbro' F.J.Leach	1862		2Cy.25"-46" 50 HP. J.P.Rennoldson,S.Shields	R.A.Robinson
✻	Congella I ScSr Cringle 1 Dk & Spr Dk 3 B	British London	1033 1585	255·0	34·6	22·9	Sunderland W.Doxford&Sons	1881	6 Blkds. [illegible]	3C.21"31"&56"-42" 200 HP. G.Clark,Sunderland	Bullard, King & Co.
✻D	Congo I ScSr Williams 80-83 1 Dk & Awn Dk 3 B	British Cardiff	1423 2178	289·5	36·0	20·0 27·5	Sunderland J.Laing	1881	4 Blkds. DB[illegible]	2Cy.33"& 62"-42" 160 HP. G.Clark,Sunderland	Tellefsen, Wills & Co.
✻	Congo Stl ScSr Rutherford 83-88 2 Dks	British Glasgow	1078 1683	300·7	36·2	19·6 H 97 f. P 70 f. F 42 f.	Port Glasgow Cunliffe&Dunlop	1882	8 Blkds.	2Cy.33"& 63"-36" 175 HP. Cunliffe&Dunlop,Pt.Glasg.	British & African Steam Nav.Co.,Ld.
V	Congo Stl&I ScSr Nielsen 83-88 1 Dk & Spr Dk	Belgian Antwerp	1325 1764	267·5	32·9	21·2 16·5	Hoboken Société J.Cockerill	1888	DB/[illegible]	3Cy.26⅝"45¼"& 52½"-33¾" Soc. J.Cockerill [180 HP.	Société Cockerill
	Congo I ScBk 3 Dks	French Marseilles	2444 3666	391·6	39·7	32·8	La Ciotat Messageries Maritimes	1878		3C(60½(2)43½"-43¼" 600 HP. Messageries Mar.,La Ciotat.	Messageries Maritimes
V	Congo I ScSr Vendorf 76-87	French Dunkirk	824 1235	261·2	30·2	23·0	Glasgow John Elder	1869		2Cy.38"& 68"-33" 200 HP. John Elder,Glasgow	Cie Gen. des Bateaux à Vap à hél du Nord
✻	Congress I ScBk Woodhill 79-80 1 Dk 2 B	British Liverpool	781 1222	250·4	32·0	15·5 Q 96 f. B 57 f. F 28 f.	Middlesbro' R.Dixon&Co.	1878	4 Blkds. DBaft/[illegible]	2Cy.30"& 55"-36" 130 HP. Blair&Co.,Ld.,Stockton	New Biscay S.S.Co.,Ld.
✻D	Coningsby I ScSr Galley 87-87 1 Dk 2 B	British W.Hartlepl.	1137 1757	258·1	31·6	19·5 Q 68 f. P 28 f. F 30 f.	W.Hartlepool W.Gray&Co.	1881	4 Blkds. DBaft/33ft	2Cy.33"& 61"-33" 150 HP. T.Richardson&Sons,Htlpl.	Gladstone & Cornforth
✻	Coniston Fell Stl ScSr Fisher 85-86 1 Dk	British Liverpool	159 323	144·6	22·0	10·0 P 23 f. B 35 f. F 21 f.	Portsmouth H.Tipping&Co.	1882	4 Blkds.	2Cy.20"& 38"-27" 65 HP. D.Rollo&Sons,Liverpool	"Coniston Fell" S.S.Co. Ld. (G.Nelson&Sons)
	Connaught I PadSt	British Dublin	511 1627	338·0	35·2	18·8	Birkenhead J.Laird,Sons&Co.	1860		2Cy.98"-78" 700 HP. Ravenhill&Co.,London	City of Dublin Steam Packet Co.
	Conoho I Sc Walker	American Plymouth N.C.	367	176·6	26·6	9·0	Philadelphia Neafie&Levy	1881		2Cy.20"& 36"-28" Neafie&Levy,Philadelphia	Roanoke, Norfolk & Baltimore S.Co.
	Conqueror W PadSp 1 Dk	British St.John,N.B.	57 146	104·3	18·6	10·5	St.John,N.B.	1866		2Cy. 100 HP.	Intercolonial Coal Mining Co.
	Conqueror (TO BE BROKEN UP) I PadSp ex Friburg, ex Lady Londesborough	British Liverpool	30 102	97·7	17·9	9·3	Shields	1857		2Cy. 60 HP.	W.H.Dodd
	Conqueror I PadSp 1 Dk	British Quebec	24 233	136·6	21·7	11·8	Renfrew Henderson,Coulborn&Co	1871		2Cy.36"-66" 150 HP. Hndson.,Coulborn&C.,Rfw.	James Walsh
	Conqueror I PadSk 1 Dk	British Quebec	15 199	136·5	20·2	11·6	Renfrew W.Simons&Co.	1865		2Cy. 120 HP. W.Simons&Co.,Renfrew	James G.Ross & Co.
	Conqueror I PadTug Short 1 Dk	British W.Hartlepl.	14 144	107·0	20·4	10·2	W.Hartlepool Pile,Spence&Co.	1861	3 Blkds.	2Cy.30"-54" 60 HP. W.Huntley	North Eastern Ry.Co.
	Conqueror I TwnScSk	British Glasgow	9 184	122·0	20·9	11·0	Glasgow Napier,Shanks&Bell	1879		2Cy.21"& 40"-26" 98 HP. Rankin&Blackmore,G'nck.	Clyde Shipping Co.
	Conqueror I PadTug Hall 1 Dk	British London	8 167	123·5	20·1	10·1	North Shields Hepple&Co.	1884		2Cy.32"-58" 140 HP. Hepple&Co.,North Shields	W.Sandford
✻	Conquistador I ScSr Uñes 88-88 2 Dks	Spanish Valencia	933 1372	240·0	33·0 B 58 f.	19·1 F 29 f.	Dumbarton A.M'Millan&Son	1888	5 Blkds. DB[illegible]	3Cy.18"27"&45"-36" D.Rowan&Son,Glsgw. [103 HP.	Perez Hermanos & Mayens

✻=Lloyd's Register; D=Nederlandsche Vereeniging; G=Germanischer Lloyd; I=Registro Italiano; R=American Record; [illegible]=Veritas Austro-Ungarico; H=Veritas Ellenico; N=Norske Veritas; V=Bureau Veritas.

Dans la 2me Colonne, B.T., avec les indications qui suivent, signifie la date du rapport de visite de la Carène par les experts du Board of Trade pour obtenir le certificat relatif au TRANSPORT DES PASSAGERS.

Dans la 3me Colonne, C=Composite (membrure en fer, bordé en bois); I=Fer; Stl=Acier; W=Bois.

Pour plus amples renseignements sur les navires classés par le Lloyd's Register, voir le Lloyd's Register Book et le Lloyd's Yacht Register.

No. in Book / Official No.	Registry in which classed	Ship's Name. Master, and Dates of Appointment to present Owner's Service and to Ship. Ship's Late Name.	Material, Rig, &c. No. of Decks, &c.	Flag. Port of Registry.	Tons. Net. Gross.	Length	Breadth	Depth	Built. Where. By Whom.	When	No. of Bulkheads. Particulars of Water Ballast	Engines. Particulars. Maker's Name.	Owners
811		Conquy *Trivier*	Stl&I ScSr	French [illegible]	80 [illegible]	131·9	21·6	5·2	A..les	1884		1Cy.11"&18"-11" 30 IP. H.Satre,Lyons	A.Conquy aîné …
812	✻DV	Conrad *Adriani 80-84*	I Sc3mSr 2 Dks & Spr Dk	Dutch Amsterdam	[illegible]	352·0 H 137 f.	39·5	21·8 20·5	Glasgow J.Elder&Co.	1871	[illegible] Blkds.	2Cy.50"&86"-42" 300 IP. J.Elder&Co.,Glasgow	Stoomvaart-M… "Nederland"
813	✻	Conscript *Manuel -88*	CompScSr 1 Dk & Spr Dk 3 B	British Glasgow	440 [illegible]	180·3	30·0	21·3	Dumbarton A.M'Millan&Son	1888		3Cy.17"27"&44"-36" 98 IP. D.Rowan&Son,Glasgow	A.J.Harvey
814	✻V	Conseil (ex Helmdal) *Le Riche*	I ScBn 2 Dks 3 B	French Bordeaux	1323 [illegible]	286·0 B 16 f. F 26 f.	35·1	25·0	Glasgow Dobie&Co.	1882	5 Blkds. [illegible]	2Cy.35"&68"-42" 200 IP. WalkerHenderson&Co.,Gls.	Les Fils de T.C…
815	✻	Consent *Marshall 81-87*	I ScSr 1 Dk 2 B	British Sunderland	923 [illegible]	251·0 Q60 f. P28 f. B86 f. F29 f.	34·1	18·2	Sunderland Bartram,Haswell&Co.	1879	3 Blkds. [illegible] APT	2Cy.32"&59"-33" 110 IP. T.Richardson&Sons,H'lepl.	J.Westoll
816	✻	Consett *Kibble 71-83*	I ScSr 1 Dk 2 B	British [illegible]	522 [illegible]	211·7 B 27 f. Q 28 f.	29·8	16·1	Sunderland Short Bros.	1871	4 Blkds. DB…	2Cy.27"&53"-33" 98 IP. [illegible]W.Co.,Newcastle	H.T.Morton &…
817	✻	Constance *Pickering 82-2*	I ScSr 1 Dk 2 B	British W.Hartlepool	685 [illegible]	230·0 Q 50 f. B 51 f. F 27 f.	32·2	13·8	W.Hartlepool R.Irvine&Co.	1882	4 Blkds. DB 338	2Cy.37"&50"-33" 99 IP. T.Richardson&Sons,H'lepl.	G.Pyman & Co.
818	D	Constance (STRANDED) *Woodcock*	I ScSr 1 Dk 2 B	British Bristol	563 [illegible]	208·3 P 37 f. F 27 f.	30·3	17·1	Glasgow A.&J.Inglis	1871	4 Blkds. [illegible]	2Cy.27"&46"-36" 98 IP. A.&J.Inglis,Glasgow	Bristol Steam… Co.,Ld.
819	V	Constantia *Blanchard*	Stl PadScTug	Roum'nin Constantia	108 [illegible]	118·1 B 10 f.	19·0	10·5	Nantes Ateliers&C. de la Loire	1886		2Cy.32"&50"-47" 150 IP. Ateliers&C.de la Loire,Paris	Roumanian Gov…ment
820	V	Constantin *Tenger 83-87*	I ScSr 2 Dks & Awn Dk	Danish Copenhagen	504 [illegible]	228·3	30·3	15·6 22·4	Malmö Kockums Mek. W'ks	1880	4 Blkds. [illegible]	2Cy.35"&61"-30" 200 IP. Kockums Mek.Verks.Malmö	Det Forenede D… Selskab (C.P.A.…)
821	✻	Constantin *DeFloris 86-87 (ex Haco)*	I Sc3mSr 1 Dk	French Havre	392 [illegible]	194·2 Q 68 f. B 24 f. F 24 f.	23·1	14·8	Greenock Scott&Co.	1870	5 Blkds.	3Cy.12½"20½"&31"-27" 65 IP. G.Clark,Sunderland	Cie. Havraise P… laire de Nav. A…
822		Constantin (STRANDED) *Jacobsen 85-88*	I ScSr 2 Dks	Russian Helsingfors	468 [illegible]	207·4	29·1	12·5	Newcastle Palmers'S.&I.Co.	1866	4 Blkds.	2Cy.44"-54" 120 IP. Palmers'S.&I.Co.,Jarrow	Sydfinska Ang… Aktiebolaget
823	V	Constantin *Ling 82-87*	I ScSr 1 Dk	Russian Arensburg	283 364	159·5 P 125 f. F 21 f.	21·7	11·0	Elbing F.Schichau	1875		2C.18¾"&41½"-21½" 80 IP. F.Schichau,Elbing	"Osilia" S.S.C…
824 51388	✻	Constantine *Porter*	I Sc3mSr 1 Dk 2 B	British London	513 [illegible]	204·9 Q 42 f. B 7 f. F 30 f.	28·3	16·6	Newcastle C.Mitchell&Co.	1866	4 Blkds. DB…	2Cy.27"&48"-30" 90 IP. R.&W.Hawthorn,Newcastle	J.Walton
825	✻D	Constantinos *Ristis -88*	I ScSr 1 Dk 2 B	Greek Andros	1109 [illegible]	260·0 Q 88 f. B 74 f. F 33 f.	36·2	18·0	Newcastle M'Intyre&Co.	1884	4 Blkds. [illegible]	2Cy.32"&60"-39" 140 IP. Blair&Co.,Ld.,Stockton	G.M.Embiricos
826 6035		Constitution (BROKEN UP)	I PadSp	British London	32 [illegible]	149·4	23·2	11·0	Newcastle	1853		2Cy. 120 IP.	W.P.Bullivant
827	I	Conte-Menabrea *Gallino -88*	I ScSr 1 Dk	Italian Genoa	97 [illegible]	123·6 Q 13 f. F 9 f.	17·1	9·2	Glasgow J.G.Lawrie	1865	4 Blkds.	2Cy.14"&28"-24" 62 IP. Ferrando&Danovaro,Genoa	Navigazione Ge… Italiana
828 81810	✻	Contest *Midgley 86-86*	I ScSr 1 Dk	British London	335 [illegible]	166·0 Q 53 f. B 37 f. F 21 f.	26·2	12·3	Sunderland J.Laing	1879	4 Blkds. DB…	2Cy.20"&40"-30" 60 IP. G.Clark,Sunderland	W.France,Ld.
829 87178	✻	Contest *Thompson*	I ScSr 1 Dk	British London	10 [illegible]	81·0	16·6	9·4	Hull E.Wales	1883	3 Blkds.	2Cy.17"&34"-21" 50 IP. E.Wales,Hull	T.W.Elliott
830 4631		Continental	W Pad	American New Haven, Con.	947 [illegible]	286·0	54·0	11·5	Brooklyn	1860			New Haven St… Co.
831 [illegible]	✻	Conway Castle	I ScBq 3 Dks	British London	1899 [illegible]	349·0 F 24 f.	39·3	28·9	Glasgow R.Napier&Sons	1877	5 Blkds. [illegible]	2Cy.45"&85"-48" 370 IP. R.Napier&Sons,Glasgow	D.Currie & Co.
832 [illegible]	✻ BT7.88	Coogee (ex Lancashire Witch)	Stl ScSr	British Melbourne	286 [illegible]	225·0 B 80 f. F [illegible]	30·2	13·5	Sunderland J.L.Thompson&Sons	1887		3C.23"37"&60"-36" 250 IP. J.Dickinson,Sunderland	Huddart, Park…
833 [illegible]	✻	Coomonderry *Seward*	Stl ScSr 1 Dk	British Sydney	58 147	110·2 Q 25 f. B 12 f. F 17 f.	20·1	7·6	Glasgow T.B.Seath&Co.	1886	4 Blkds.	2Cy.15"&30"-18" 37 IP. Kincaid&Co.,Greenock	J.Hay
834 [illegible]		Coonanbara	I PadSr 2 Dks	British Newcastle, N.S.W.	567 [illegible]	223·2	27·2	12·7	London Thames Iron Works	1862		2Cy.59"-57" 240 IP. J.Stewart,London	Newcastle S.S…
835 45037		Coorong	I ScSr 1 Dk	British Sydney	304 [illegible]	171·1 P 20 f. H 92 f.	22·4	12·2	Glasgow J.G.Lawrie	1862		2Cy.32"-21" 70 IP. Blackwood&Gordon,Pt.Gls.	John See
836 [illegible]		Coos Bay *Leland*	W ScSr 2 Dks 3 B	American San Francisco	194 [illegible]	140·0 H —	26·0	7·3 18·0	Marshfield,Oreg. H.R.Reed	1884		2Cy.16"&30"-20" 300 IP. Fulton Iron Works,S.Francisco	Pacific Coast S… Ship Co.
837 [illegible]	✻	Coot *Booth 79-86*	Stl ScSr 2 Dks	British Cork	1275 [illegible]	270·8 P29 f. Q80 f. B109 f. F31 f.	37·0	18·5	W.Hartlepool W.Gray&Co.	1885	4 Blkds. [illegible]	3C.19½"32½"&53"-36" 160 IP. Central Mar.Eng.Co.,H'tlpl.	Cork Steam Sh… Ld.
838 [illegible]	✻ BT7.88	Copeland (STRANDED) *Simpson 18..-88*	I ScSr 2 Dks Spt Awn Dk	British Leith	439 [illegible]	225·3 P 62 f. F 40 f.	29·3	15·5	Glasgow London&Glasgow Co.	1874	5 Blkds.	2Cy.29"&53"-36" 150 IP. London&Glasgow Co.,Glsgw.	R. & D.Sliman…
839		Copernic	I Sc 3 Dks	French Marseilles	1651 [illegible]	258·2 Q 73 f.	33·8	27·0	Newcastle Andrew Leslie&Co.	1861		2Cy.48"-50" 200 IP. R.Morrison&Co.,Newcastle	Messageries Ma…
840 [illegible]	✻	Copernicus *Taylor 73-88*	I ScBn 2 Dks 3 B	British Liverpool	2136 [illegible]	331·4 P 30 f. B 12 f. F 44 f.	42·1	26·3	Southampton Oswald,Mordaunt&Co.	1888	7 Blkds. [illegible]	3Cy.26½"41"&70"-48" [illegible] IP. [illegible]	Liverpool, Bra…

✻ = Lloyd's Register; D = Nederlandsche Vereeniging; G = Germanischer Lloyd; I = Registro Italiano; R = American R…
A = Veritas Austro-Ungarico; H = Veritas Ellenico; N = Norske Veritas; V = Bureau Ve…

In the 2nd Column, B.T. with figures following, signifies Date of report of bottom sighting by Board of Trade Surveyors as condition of PASSENGER CERTI…

In the 3rd Column, C = Composite (iron frame, wood planked); I = Iron; Stl = Steel; W = Wood.

Further particulars of Vessels classed by Lloyd's Register will be found in Lloyd's Register Book and Lloyd's Yacht Register.

Registres		Nom, Matériaux, Gréement, etc. Capitaine et date d'entrée au service des Armateurs actuels et à bord du navire. Ancien Nom, nombre de ponts, etc.	Pavillon. Port d'Armement	Tonnage Net / Total	Long.	Larg.	Creux	Construit Où. Par qui.	Quand	Nombre de cloisons. Détails sur le lest d'eau.	Machines. Détails. Nom du Constructeur.	Armateurs ou Directeurs-Gérants.
		Copiapo I ScSr Martin 83-? 2 Dks & Awn Dk	Chilian Valparaiso	888 1337	258·0	34·0	20·6	Liverpool Bowdler,Chaffer&Co.	1870		2Cy.38"& 71"–33" 180 HP. Fawcett,Preston&Co.,Lvpl.	Cia.Sud-Americana de Vapores
		Coptic Stl ScInBk 3 Dks	British Liverpool	2857 4448	430·2	42·2	24·4	Belfast Harland&Wolff	1881	7 Blhds.	1Cy.32"& 71"–60" 530 HP. J.Jack&Co.,Liverpool	Oceanic Steam Nav. Co.Ld.
					H 329 f.		H 5					
		Coquimbo I Sc3mSr 2 Dks & Shade Dk	British Liverpool	1075 1821	290·5	38·3	22·9	Glasgow J.Elder&Co.	1870	6 Blhds.	2Cy.46"& 80"–39" 300 HP. J.Elder&Co.,Glasgow	Pacific Steam Nav. Co.
					H 158 f.							
	✻D	**Cora Maria** I ScSr Ward 86-86 1 Dk2 B	British Cardiff	835 1290	235·0	31·2	18·3	Newcastle Schlsngr. Davis &Co.	1879	1 Blhds. DB&53f/60f	2Cy.28"& 53"–36" 110 HP. T.Clark&Co.,Newcastle	R.Gordon & Co.
					? f.	F 30 f.	B 54 f.					
215		**Coraki** I TwnScSr Webster 1 Dk	British Sydney	198 326	159·6	26·1	8·5	Dundee W.B.Thompson	1879	1 Blhds.	1Cy.15"& 27"–16" 65 HP. W.B.Thompson,Dundee	Clarence & Richmond Rivers Steam Nav.Co.
					P 40 f.	B 40 f.	F 15 f.					
		Coral Queen I ScSr Hepwood 83-82 (ex Antelope)	British W.Hrtlepool	468 738	236·9	26·4	16·6	Liverpool	1846		2Cy.36"–24" 90 HP. G.Forrester&Sons,Liverp'l	W.Hartlepool Steam Nav. Co.
					Q 34 f.							
847	V	**Cordelia** I S Sr Schwerns 86-86 1 Dk & Spr Dk3 B	German Hamburg	1174 1745	260·0	35·0	17·0	Flensburg FlensburgS.B.Co.	1886	DB500t	3C22½"33"&58"–42"250HP. FlensburgS.B.Co.,Flensbrg.	Hamburg Pacific Dampfs.Linie
					H 86 f.		24·5					
848	V	**Cordoba** Stl ScBn Baret 83-? 1 Dk & Spar Dk	French Havre	2207 2908	320·0	40·2	27·2	Stockton M.Pearse&Co.	1887	DBal67f/228t	3C.24"40"&66"–45" 250HP. Blair&Co.,Ld.,Stockton	Chargeurs Réunis
					F 44 f.	H 44 f.						
849		**Cordouan** I ScSr 3 Dks	French Marseilles	2635 3641	328·1	45·3	26·2	LaCiotat MessageriesMaritimes	1884		2Cy.36½"&65"–43¼"350HP. MessageriesMar.,LaCiotat	Messageries Maritimes
850	✻D	**Cordova** I S Sr Mathews ?-84 1 Dk2 B	British Sunderland	1111 1730	275·4	37·5	18·8	Newcastle TyneI.S.B.C.,Ld.	1880	1 Blhds. CDB?t FPT?t APT?t	2Cy.36"& 68"–42" 180 HP. J.Dickinson,Sunderland	J.Tully
					P 19 f.	F 17 f.						
851		**Corea** I ScSr Gratt 1 Dk & Awn Dk	British London	382 603	170·2	25·1	18·5	PortGlasgow Russell&Co.	1875	4 Blhds. DB?t	2Cy.22"& 40"–24" 80 HP. Kincaid,Donald&Co.,Gnck.	Australasian United SteamNav.Co.,Ld.
852		**Corean** I Sc3mSr M'Dougall	British Glasgow	2259 3488	360·1	41·0	23·0	Sunderland W.Doxford&Sons	1881	6 Blhds.	2Cy.48"& 84"–54" 400 HP. W.Doxford&Sons,Sndrland	J.& A.Allan
					B 78 f.	F 46 f.	Q 31 f.					
853	✻	**Coredo** I ScSr 1 Dk	Dutch Ternate	28 41	60·0	12·0	6·8	Glasgow W.S.Cumming	1884	3 Blhds.	2Cy.8"& 15"–12" 10 HP. Fisher&Co.,Paisley	C.W.R.van Renesse VanDreijvenbode
854	V	**Corfitz Beck-Friis** I ScSr Khyatt 83-84 1 Dk	Swedish Trelleborg	352 451	146·3	23·8	15·4	Gothenburg MotalaCo.	1870		2Cy.15"& 30"–22" 45 HP. S.Almqvist,Gothenburg	Trelleborgs Ångf.Akt. Bolag.(F.Malmros)
					B 34 f.	H 5 f.						
855	✻	**Corine** Pollet ?-87 I Sc3mSr (ex Bristol, ex Catalan, ex Colthia) 1 Dk	French Nantes	154 232	134·8	19·5	11·0	Glasgow J.&R.Swan	1862	1 Blhds.	3Cy.12"17"& 29"–22" 60HP. Jollet&Babin,Nantes	Louis Flornoy & fils.
					Q 11 f.	B 12 f.						
		Corinna I ScSr White 2 Dks	British LauncestnTas	820 1279	235·6	32·5	21·0	Glasgow D.&W.Henderson&Co.	1882	5 Blhds. DB?f	2Cy.27"& 54"–42" 141 HP. D.&W.Henderson,Glasgow	Tasmanian Steam Nav.Co.
					H 35 f.							
857	✻	**Corisande** I ScSr Fisher 80-87 1 Dk2 B	British Middlesbro'	768 1022	218·5	30·1	17·1	Sunderland W.Pile&Co.	1870	5 Blhds. WB?f	1Cy.20"& 40"–33" 98 HP. Cmp.75,G.Clark,Sundrland	"Corisande" S.S.Co.Ld. (Jervelund & Clephan)
					Q? f.	H? f.	F? f.					
858		**Cormoran** Debats 81	French Arcachon	89	..	..	..					H.S.Johnston & Co. (Pêcheriesdel'Océan)
859	✻D	**Cormorant** I ScSr Taylor, jun. 1 Dk2 B	British London	709 927	195·0	30·1	15·0	London W.Walker&Co.	1882	5 Blhds. DB?t FPT?t	2Cy.27"& 48"–30" 95 HP. GourlayBros.&Co.,Dundee	GeneralSteam Nav.Co.
					P 119 f.	F 29 f.						
860		**Cormorant** I Sc3mSr	British Cork	555 1035	259·9	36·2	17·5	Cork E.Pike	1853		2Cy.38"& 70"–36" 200 HP. J.Jack,Rollo&Co.,Liverpool	City of Cork Steam Packet Co.,Ld.
					P 74 f.							
861	✻	**Cormorant** I ScK 1 Dk	British Grimsby	141 244	133·4	21·7	11·0	Hull Earles'Co.,Ld.	1884	1 Blhds.	2Cy.21"& 40"–24" 68 HP. Earles'Co.,Ld.,Hull	Great Grimsby Ice Co.,Ld.
					Q 45 f.	F 25 f.						
862	✻	**Cormorant** W ScK Borthwick 85-85 1 Dk	British Berwick	22 55	71·2	17·3	8·3	Berwick Lee&Wight	1885		2Cy.8½"& 16"–12" 10 HP. Hawthorns&Co.,Leith	T.F.Robertson Carr
863	✻	**Cornelia** I ScSr Stonehouse 82-82 1 Dk2 B	British W.Hartlepl.	604 942	220·3	29·2	16·3	W.Hartlepool Denton,Gray&Co.	1872	4 Blhds. WB af	2Cy.27"& 50"–33" 99 HP. T.Richardson&Sons,Htlepl.	Herskind & Woods
					Q 68 f.	H 87 f.						
864		**Cornelia** I ScSr Reid 1 Dk	British Penang	194 323	143·5	22·1	11·5	Montrose Roy&Mitchell	1878	4 Blhds. FPT	2Cy.20"& 36"–24" 60 HP. W.B.Thompson,Dundee	Lim Leng Cheak
					Q 35 f.	B 12 f.	F 20 f.					
865		**Cornelia** I Sc3mSrYt	British Sunderland	190 341	187·4	25·1	13·6	Newcastle Palmers'S.B.Co.	1880		2Cy.26"& 52"–33" 120 HP. Palmers'S.B.C.Ld.,Newcstle	Wm.Mackinnon
866	✻	**Cornucopia** I ScSr Smith 83-84 1 Dk & Spr Dk3 B	British Sunderland	1052 1643	251·0	34·2	24·0	Sunderland Osbourne,Gr'ham&Co.	1882	5 Blhds DB&?t FPT?t APT?t	2Cy.31"& 57"–36" 150 HP. G.Clark,Sunderland	Argosy S.S.Co.,Ld. (T.Potts&Co.)
	✻	**Coromandel** Stl Sc3mSr Reeves 75-85 3 Dks	British Greenock	2383 4339	400·3	45·3	29·2	Greenock Caird&Co.	1885	8 Blhds.	3C.35"56"&89"–66"800HP. Caird&Co.,Greenock	Peninsular & Oriental Steam Nav.Co.
					P 260 f.	F 79 f.						
	GV	**Corona** I ScBa Wittenberg 84-84 2 Dks3 B	German Elsfleth	1396 1883	262·4	33·8	25·5	Vegesack H.F.Ulrichs	1882	CDB?t	2Cy.33"& 60"–36" 170 HP. O.Hennigges&Co.,Berlin	Dampfschiffs Rhederei "Columbus"
					Q 68 f.	B 72 f.	F 36 f.					
	✻D	**Coronet** I Sc3mSr Beere 3 Dks	British London	1122 1736	274·8	33·3	24·2	Newcastle Palmers'Co.,Ld.	1873	5 Blhds. DB?t	2Cy.30"& 60"–33" 140 HP. Palmers'Co.,Ld.,Newcastle	Hall Brothers
	✻	**Coronilla** I ScSr Urwin 83-83 1 Dk2 B	British Nth. Shields	875 1361	258·0	32·0	21·1	Newcastle TyneI.S.B.Co.,Ld.	1878	5 Blhds. BDB&?t FPT	2Cy.28"& 54"–39" 130 HP. N.E.MarineEng.Co.,Ld.,Shd.	J.Robinson & Sons
					B 17 f.							

✻=Lloyd's Register; D=Nederlandsche Vereeniging; G=Germanischer Lloyd; I=Registro Italiano; R=American Record; A=Veritas Austro-Ungarico; H=Veritas Ellenico; N=Norske Veritas; V=Bureau Veritas.

Dans la 2me Colonne, B.T., avec les indications qui suivent, signifie la date du rapport de visite de la Carène par les experts du Board of Trade pour obtenir le certificat relatif au TRANSPORT DES PASSAGERS.

Dans la 3me Colonne, C=Composite (membrure en fer, bordé en bois); I=Fer; Stl=Acier; W=Bois.

Pour plus amples renseignements sur les navires classés par le Lloyd's Register, voir le Lloyd's Register Book et le Lloyd's Yacht Register.

No. in Book / Official No.	Registry in which classed	Ship's Name. Material, Rig, &c. / Master, and Dates of Appointment to present Owner's Service and to Ship / Ship's Late Name. No. of Decks, &c.	Flag / Port of Registry	Tons Net / Gross	Length	Breadth	Depth	Built Where / By Whom	When	No. of Bulkheads / Particulars of Water Ballast	Engines Particulars, Maker's Name	Owners or Ma[nagers]
871 88917	✳	Corra Linn I ScSr *Clements 81-87* 1 *Dk*	British Glasgow	527 833	208·0	29·2	14·1	Belfast Workman, Clark & Co.	1884	4 Blhds. DBallast FPT APT	2Cy. 24" & 48"–36" 96 HP. W. Kemp, Glasgow	J. & A. Wyllie
872	V	Corral I ScSr *Steinfall*	Chilian Valdivia	334 493	168·9	24·7	12·8	Kiel Akt.Ges."Germania"	1884	WB	2Cy. 24" & 47"–28" 115 HP. Akt.Ges."Germania," Berlin	Asociacion de [illegible]dores de Valdivia
873		Correo de Alicante I ScSr *Salinas -87 (ex La Vendee)*	Spanish Alicante	170 243	154·8	20·8	11·7	Pt. Glasgow H. Murray & Co.	1872		2Cy. 28" & 38"–24" 50 HP.	Vincente Salinas
874		Correo de Cartagena I ScSr *(ex Lafarge)*	Spanish Carthagena	272	177·1	23·5	13·4	Newcastle	1856		80 HP.	G. Salinas & Co.
875		Correo de Cette I	Spanish Barcelona	153 274	128·9	19·7	11·7	Belfast	1866		60 HP.	M. C. Ferrer
876	V	Corrientes I ScSr *Tosling 87-88* 1 *Dk & Spr Dk*	German Hamburg	1455 1938	285·1	35·8	15·8 23·9	Newcastle C. Mitchell & Co.	1881		2Cy. 34" & 66"–42" 220 HP. Wallsend Slipway Co., Nwc.	Hamburg-Südam[erikanische] Dampfschiff-Ges[ellschaft]
877 118	✳	Corriere di Genova I TwnScSr *M. Speich* 1 *Dk*	Italian Genoa	119 201	125·5	24·1	7·0	North Shields T. & W. Smith	1886	4 Blhds.	1Cy. 14" & 25"–18" 50 HP. Wallsend Slipway Co., Nwc.	A. Centurini
878 143		Corriere di Livorno I T ScSr 1 *Dk*	Italian Genoa	127 226	125·0	23·0	6·0	North Shields T. & W. Smith	1887		2Cy. 13" & 24"–18" 50 HP. Wallsend Slipway & Eng. Co., Ld., Newcastle	A. Centurini
879 109	✳	Corriere di Roma I Twn Sc 1 *Dk*	Italian Genoa	154 182	125·0	23·0	7·0	N. Shields T. & W. Smith	1885		1Cy. 14" & 25"–18" HP 50. Wallsend Slpwy Co., Ld., Nwe.	A. Centurini
880 84934	✳	Corrwg I ScSr *Gowin 86-88* 1 *Dk*	British Cardiff	599 931	220·0	31·2	12·5	Sunderland W. Pickersgill & Co.	1881	4 Blhds. DBallast	2Cy. 28" & 52"–33" 99 HP. Palmers Co., Newcastle	W. Perch & Co.
881 [illegible]	✳ D	Corsair I ScSr *Gibson* 1 *Dk*	British Bristol	247 400	160·8	23·2	12·6	Sunderland J. Blumer & Co.	1876	4 Blhds. DB	2Cy. 20" & 38"–24" 65 HP. G. Clark, Sunderland	London & Rotte[rdam] S.S. Co. Ld.
882 125840		Corsair I ScSrYt *McKee*	American New York	124 247	185·0	23·8	14·0	Philadelphia W. Cramp & Sons	1880		2Cy. 24" & 44"–24" 150 HP. W. Cramp & Sons, Philadelphia	J. Pierpont Mor[gan]
883 14		Corsica (ex Liguria) I ScSr *Profeville -89* 1 *Dk* 2 *B*	Italian Genoa	140 222	144·3	18·4	10·8	London J. & W. Dudgeon	1866	3 Blhds.	2Cy. 19" & 28"–21" 65 HP. P. Ferro & Co., Genoa	Navigazione Ge[nerale] Italiana
884 93367	✳	Corsican Stl ScSr *Macfarlane 66-87* 1 *Dk*	British Glasgow	148 338	150·3	22·2	10·7	Belfast Workman, Clark & Co.	1887	5 Blhds. DBaft APT	3Cy 15" 24" & 40"–30" 95 HP. Hutson & Corbett, Glasgow	J. & J. Macfarlane
885 74518	✳	Corso I ScBg *Thomas 78-87* 1 *Dk* 2 *B*	British Liverpool	693 1106	246·5	31·3	17·7	Liverpool Bowdler, Chaffer & Co.	1876	5 Blhds. DBaft	2Cy. 30" & 60"–30" 150 HP. Fawcett & Co., Liverpool	Strong, Reid & Pa[ge]
886	✳	Cortés I ScSr *Rabia* 1 *Dk & Spr Dk*	Spanish Barcelona	978 1261	228·0	32·4	13·9 21·6	Sunderland J. Laing	1884	4 Blhds. APT	2Cy. 26" & 52"–36" 95 HP. J. Dickinson, Sunderland	J. Roca & Co.
887		Coruña I ScBr *(ex Marques de Nunez)* 3 *Dks*	Spanish Barcelona	1253 1843	297·5	35·7	28·8	Sunderland W. Pile & Co.	1872		2Cy. 48" & 81"–42" 300 HP. N.E. Marine Eng. Co., Sndld.	Cia. Trasatlantica
888 126443		Cosmopolis W ScSr 1 *Dk*	American S. Francisco	267 340	154·0	32·0	10·0	San Francisco G. Boole	1887		2Cy. 16" & 30"–20" 200 HP. Hinkley, Spiers & Hayes, San Francisco	Higgins & Collins
889	G	Cosmopolit I ScSr *Holm -85* 2 *Dks*	German Kiel	671 865	204·0	25·8	20·2	Kiel G. Howaldt	1882	WB	2Cy. 25" & 46"–27" 85 HP. Howaldt Bros., Kiel	Rhederei "Cosmo[polit]"
890 72120	✳	Cosmopolitan I ScSr *Lothian 77-83* 1 *Dk & Spr Dk* 3 *B*	British Whitby	1024 1592	274·0	33·7	17·5 24·4	Whitby T. Turnbull & Son	1875	4 Blhds. DBaft	3Cy. 19½" 32" & 52"–39" 150 HP. [illegible]	T. Turnbull & So[n]
891 62985		Cosmopolitan W PadTug *Leath*	British S. Shields	70 170	118·7	20·1	11·2	S. Shields Readhead, Softley & Co.	1869		2Cy. 30"–54" 60 HP. Readhd, Softley & Co., S. Shlds	J. R. Sibbald
892 82382		Cosmos Stl PadSr *Baker*	British Glasgow	734 1003	249·6	29·1	11·2	Glasgow A. & J. Inglis	1879	4 Blhds.	2Cy. 44" & 77"–72" 250 HP. A. & J. Inglis, Glasgow	La Platense Fl[otilla] Co., Ld.
893 27507		Cossack I ScBmSr *Mason*	British Hull	683 868	218·7	29·4	17·0	Hull C. & W. Earle	1859		4Cy. 20" & 36"–30" 98 HP. Comp. 75, Bailey & Leetham, Hull	W. S. Bailey
894 5098		Costa Rica I ScSr *Whiteberry* 3 *Dks*	American New York	826 1457	227·0	35·0	21·4 28·0	Wilmington, Del. Harlan & Hollingworth	1868		2Cy. 48"–66" 180 HP. Harlan & Hollingwth, Wlmgtn	Oregon Railway & Navigation Co.
895	✳	Costa Rica I ScBmSr *Salamon (ex Bristol, ex City of Valparaiso)* 2 *Dks* 3 *B*	Hawaiian Honolulu	1274 1983	278·0	38·3	24·0	Stockton M. Pearse & Co.	1875	5 Blhds. DB APT	2Cy. 40" & 66"–42" 250 HP. Hawks, Crawshay & Co., Gtshd	S. F. Alexander
896 91230	✳ BT1.88	Costa Rican Stl ScSr *Watson 64-85* 3 *Dks*	British Liverpool	2133 3251	360·6	41·2	26·8	Belfast Harland & Wolff	1885	7 Blhds.	2Cy. 40" & 80"–51" 266 HP. Harland & Wolff, Belfast	West India & [Pacific] S.S. Co. Ld.
897	✳	Cotentin I Sc 1 *Dk*	French Port Bail	67 132	90·0	18·1	8·4	London R. & H. Green	1884	4 Blhds.	2Cy. 15" & 27"–18" A. Wilson & Co., London	De La Roche
898 63265	BT7.88	Cotopaxi I ScBk 3 *Dks & Awn Dk*	British Liverpool	2577 4045	402·2	42·8	34·3	Glasgow J. Elder & Co.	1873	6 Blhds.	2Cy. 61" & 112"–48" 600 HP. J. Elder & Co., Glasgow	Pacific Steam Na[vigation Co.]
899	✳	Cotopaxi (ex Chaihuin) I Sc 1 *Dk*	Ecuadran Guayaquil	114 186	124·4	22·1	9·8	Port Glasgow D. J. Dunlop & Co.	1884	4 Blhds. FPT	2Cy. 18" & 35"–24" 50 HP. D. J. Dunlop & Co., P. Glasgow	Gov'rnm'nt of Ec[uador]
900												

✳ = Lloyd's Register; D = Nederlandsche Vereeniging; G = Germanischer Lloyd; I = Registro Italiano; R = American [Record]; A = Veritas Austro-Ungarico; H = Veritas Ellenico; N = Norske Veritas; V = Bureau Veritas

In the 2nd Column, B.T., with figures following, signifies Date of report of bottom sighting by Board of Trade Surveyors as condition of PASSENGER CERTIFICATE.

In the 3rd Column, C = Composite (iron frame, wood planked); I = Iron; Stl = Steel; W = Wood.

Further particulars of Vessels classed by Lloyd's Register will be found in Lloyd's Register Book and Lloyd's Yacht Register.

No.	Registre	Nom, Matériaux, Gréement, etc. Capitaine et dates d'entrée au service des Armateurs actuels et à bord du navire. Ancien Nom, nombre de ponts, etc.	Pavillon. Port d'Armement	Tonnage Net Total	Dimensions Longueur, Largeur, Creux	Construit. Où. Par qui.	Quand.	Nombre des cloisons. Détails sur le test d'eau.	Machines. Détails. Nom du Constructeur.	Armateurs ou Directeurs-Gérants.
901		Counsellor I Sc3mSr *Lane 75-84*	British Liverpool	1411 2161	311·0 31·1 25·6	Glasgow Aitken&Mansel	1877		2Cy.35"& 70"–48" 280 HP. J.&J.Thomson,Glasgow	Charente S.S.Co.,Ld.
902	✱D	Countess I ScSr *Waters 78-87* 1 *Dk2 B*	British Sunderland	1430 2187	275·0 40·4 22·4 *B 103f. Q 48f.*	Sunderland Short Bros.	1885	5 Bhds. CDB	2Cy.33"&62"–42" 180 HP. Clark&Co.,Ld.,Sunderland	Taylor & Sanderson
903	✱	Countess Evelyne I ScSr *Thomas 78-82* 1 *Dk2 B*	British Cardiff	864 1354	245·6 34·1 15·5 *Q 79f. B 85f. F 39f.*	Barrow Caird&Purdie	1882	4 Bhds. CDB8274 AFT284	2Cy.30"& 55"–36" 140 HP. Westray,Copeland&Co.,Brw	"Countess Evelyne" Co.,Ld. (J.Ware)
904	✱	Countess of Aberdeen I ScSr *M'Bain* 1 *Dk2 B*	British Aberdeen	291 580	200·1 26·8 13·9 *P56f. B 48f. F 37f.*	Barrow Barrow S.B.Co.,Ld.	1878	4 Bhds.	2Cy.33"& 66"–33" 180 HP. BarrowS.B.C.,Ld.,Barrow	Aberdn.,Newc.& Hull Steam Nav.Co.,Ld.
905	✱	Countess of Bantry I ScSr *Shanahan 84-84* 1 *Dk*	British Skibbereen	44 86	91·6 17·0 9·0	Belfast Workman,Clark&Co.	1884	4 Bhds.	2Cy.14"& 27"–20" 30 HP. W.Kemp,Glasgow	Bantry Bay S.S.Co.,Ld.
906		Countess of Derby	British Sierra Leone	86 136	125·0 18·1 8·2 *B 19f. F 30f. H 25f.*	NorthShields Hepple&Co.	1884	3 Bhds.	2Cy.27"–36" 48 HP. Hepple&Co.,NorthShields	Crown Agent for the Colonies
907	✱	Countess of Dublin I ScSr *Hill 86-88* 2 *Dks*	British Dublin	445 770	220·4 28·2 15·0 *H 50f. P 87f. F 22f.*	Dublin Walpole,Webb&Co.	1869	4 Bhds.	2Cy.41"–42" 150 HP. Thompsn,Boyd&Co.,Newc.	British & Irish Steam Packet Co.,Ld.
908		Countess of Erne I PaddSr *Carlan 73* 2 *Dks*	British Dublin	370 823	241·4 29·0 14·3 *H 32f. P 76f. F 39f.*	Dublin Walpole,Webb&Co.	1868	4 Bhds.	2Cy.64"–60" 300 HP. Fawcett,Preston&Co.,Livpl	London & N.W.Railway Co.
909	✱	Countess of Lisburne I ScSr *Jones* 1 *Dk*	British Aberystwith	57 170	110·0 19·9 8·2 *B8f. Q39f. F14f.*	Grangemouth Grangemouth Dockyard Co.	1886	4 Bhds. FPT186	2Cy.18"& 34"–24" 50 HP. Ross&Duncan,Glasgow	Aberystwith & Aberdovey Stm.PacketCo.,Ld.
910		Countess of Lathom StlScFerry	British London	230 338	120·0 40·0 10·1	London Steward&Latham	1888		4Cy.18"&31"–24" 92 HP. ApplebyBros.,London	F.N.Latham
911		Countess of Zetland StlScFerry	British London	230 338	120·0 40·0 10·1	London Steward&Latham	1888		4Cy.18"&34"–24" 92 HP. ApplebyBros.,London	F.N.Latham
912	✱D	County of Durham I ScSr *Ross -87* 1 *Dk*	British Newcastle	505 787	195·0 30·2 14·1 *Q 101f. B 20f. F 24f.*	Sunderland J.Priestman&Co.	1883	4 Bhds. DBA/ 1191	2Cy.24"& 48"–36" 90 HP. Black,Hawthorn&Co,Gtshd	E.Jobson
913	✱	County of Lancaster I ScSr *Phillips* 1 *Dk*	British Liverpool	130 214	120·0 20·0 9·8 *Q 44f. F 16f.*	Penarth PenarthS.B.Co.	1884	3 Bhds.	2Cy.13"& 24"–18" 40 HP. PenarthSh.Bg.Co.,Ld.,Pnth.	S.S."County of Lanc." Co.,Ld.(T.Duckworth&Co.)
914	✱	County of Salop I ScSr *Holmes* 2 *Dks3 B*	British Liverpool	1385 2164	285·1 36·1 24·2 *P 25f. B 60f. F 32f.*	Barrow BarrowS.B.Co.,Ld.	1884	7 Bhds. DBof 3301	2Cy.34"& 64"–42" 175 HP. BarrowS.B.C.,Ld.,Barrow	Taylor,Abrahams&Co.
915	✱	County of York I ScSr *Wylie* 2 *Dks.& WebFrms*	British Liverpool	1478 2282	285·2 38·3 24·2 *P 39f. B 61f. F 33f.*	Barrow BarrowS.B.Co.,Ld.	1884	6 Bhds. DB[illegible] FPT601 AFT301	4Cy28" 28½" 40"&57"–42" 200 HP BarrowS.B.Co.,Ld., Barrow	Taylor,Abrahams&Co.
916	✱	Courage I ScK *Hunt* 1 *Dk*	British London	109 266	123·8 21·0 11·6	Middlesbro' R.Dixon&Co.	1885	4 Bhds.	2Cy.19"& 36"–24" 50 HP. Blair&Co.,Ld.,Stockton	Great Yarmouth Steam Carrying Co.,Ld.
917		Courier StlScCr *Bennell -87* 1 *Dk*	British Melbourne	223 728	220·8 30·0 12·6 *Q 140f. F 37f.*	Newcastle C.S.Swan&Hunter	1887		3C30"46"&73"–36" 800 HP. R.&W.Hawthorn,Leslie&Co., Newcastle	Huddart,Parker & Co.
918		Courier I ScSr *Whates* 1 *Dk*	British Guernsey	49 140	129·8 19·2 9·7	Northam Day,Summers&Co.	1883	4 Bhds.	2Cy.20"& 40"–24" 65 HP. Day,Summers&Co,Sthmptn	T.N.Barbenson & Co.
919	N	Courier StlScSr *Torresen 83-83* 1 *Dk2 B*	Norwegian Fredrikstad	443 587	161·8 26·6 15·3 *Q 49f. B 36f. F 14f.*	Gothenburg MotalaCo.	1883	4 Bhds. APT41 MT1701 FPT41	2Cy19½"&35½"–23⅝" 80 HP. MotalaCo.,Gothenburg	Lisleby Fabrik (S.Wiese & Co.)
920	✱	Courland I Sc3mSr *Le Sueur 88-89* 2 *Dks3 B*	British London	760 1248	250·4 30·2 21·5	Glasgow R.Napier&Sons	1872	6 Bhds. MT 91	2Cy.30"& 54"–36" 130 HP. R.Napier&Sons,Glasgow	D.Currie & Co.
921	✱	Courlis I Sc2mTrawler 1 *Dk*	French Arcachon	44 113	101·5 18·0 9·4 *B 8f. F 9f.*	Leith Hawthorns&Co.	1884	4 Bhds.	2Cy.15"& 27"–21" 33 HP. Hawthorns&Co.,Leith	H.S.Johnston & Co.
922	V	Courrières I ScSr *Leleu 80-80* 1 *Dk*	French Dunkirk	268 463	173·5 23·3 13·8 *P 31f. F 28f.*	Glasgow T.Wingate&Co.	1871	FPT 90	2Cy.24"& 44"–30" 90 HP. T.Wingate&Co.,Glasgow	Cie.Générale des Bateaux à vap.A hél.du Nord.
923	✱	Cousins Arbib I ScBn *Tearcig 82-84* 2 *Dks3 B*	British London	1647 2147	300·2 35·2 24·0 *P36f. B83f. F34f. Len88*	Middlesbro' R.Dixon&Co.	1882	5 Bhds. DBof 2534	2Cy.35"& 65"–36" 170 HP. T.Richardson&SonsHartpl.	E.& E.Arbib
924		Covadonga I Sc 2 *Dks*	Spanish Gijon	545 801	190·0 28·2 14·5 *P 116f. F 36f.*	Dundee Gourlay Bros.&Co.	1870	5 Bhds.	2Cy.31"& 19"–42" 140 HP. GourlayBros.&Co.,Dundee	Meliton,Gonzalez & Cia.
925	✱	Coventina I ScSr *Arnott* 2 *Dks3 B*	British Newcastle	1062 1645	255·7 34·2 22·3 *P 21f. F 27f.*	Newcastle Palmers'Co.,Ld.	1880	4 Bhds. CDB27[illegible]	2Cy.32"& 60"–42" 200 HP. Palmers'Co.,Ld.,Newcastle	G.Cockburn
926	✱	Coventry I ScSr *Scott* 1 *Dk.& WebFrames*	British W.Hartlepl.	1100 1715	257·6 34·6 19·4 *Q 88f. B 78f. F 33f.*	W.Hartlepool E.Withy&Co.	1883	4 Bhds. DBof 3072	2Cy.33"& 64"–33" 150 HP. T.Richardson&SonsHartpl.	Sivewright,Bacon & Co.
927	✱	Cragside Stl ScSr *Smith -87* 1 *Dk.& WebFrms*	British London	1278 1974	284·0 38·0 18·2 *P25f. Q82f. B41f. F27f.*	Sunderland J.L.Thompson&Sons	1887	5 Bhds. CDB[illegible] APT[illegible]	3C21½"35"&58"–39" 180 HP J.Dickinson,Sunderland	T.Briggs
928	✱	Cragside I ScSr *Morris -87* 1 *Dk*	British Newcastle	75 129	92·0 20·1 8·3 *Q 30f. B 77f. F 13f.*	Newcastle Wood,Skinner&Co.	1887	3 Bhds. FPT 151	2Cy.12"& 22"–16" 30 HP. J.Abbot&Co.,Ld.,Newcastle	R.Mason
929	✱	Craigendoran I ScSr *Allan 87-87* 1 *Dk2 B*	British Glasgow	943 1471	250·0 34·2 17·7 *Q 80f. B 72f. F 26f.*	Leith Ramage&Ferguson	1883	4 Bhds. FPT[illegible] MT2001	2Cy.30"& 56"–39" 150 HP. Ramage&Ferguson,Leith	Maclay & McIntyre
930	✱	Craighill StlScSr *Stewart 86-86* 1 *Dk2 B*	British Glasgow	709 1116	240·5 33·2 14·0 *Q 82f. B 52f. F 30f.*	Grangemouth Dobson&Charles	1882	4 Bhds. CDB[illegible] APT16[illegible]	2Cy.30"& 51"–36" 150 HP. Hutson&Corbett,Glasgow	R.Mackill & Co.

✱=Lloyd's Register; D=Nederlandsche Vereeniging; G=Germanischer Lloyd; I=Registro Italiano; R=American Record;
A=Veritas Austro-Ungarico; H=Veritas Ellenico; N=Norske Veritas; V=Bureau Veritas.

Dans la 2me Colonne, B.T., avec les indications qui suivent, signifie la date du rapport de visite de la Carène par les experts du Board of Trade pour obtenir le certificat relatif au TRANSPORT DES PASSAGERS.

Dans la 3me Colonne, C=Composite (membrure en fer, bordé en bois); I=Fer; Stl=Acier; W=Bois.

Pour plus amples renseignements sur les navires classés par le Lloyd's Register, voir le Lloyd's Register Book et le Lloyd's Yacht Register.

No.	Class	Ship's Name. Material, Rig, &c. Master, and Dates of Appointment Ship's Late Name. No. of Decks, &c.	Flag. Port of Registry.	Tons. Net. Gross	Dimensions. Length, Breadth, Depth Deck Erections.	Built. Where. By Whom.	When	No. of Bulkheads. Particulars of Water Ballast.	Engines. Particulars. Maker's Name.	Owners
931 [illegible]	✻	Craiglands I ScSr *Fortune 85-88* 1 *Dk2 B*	British W.Hartlepl.	886 1373	245·2|35·|15·9 [illegible] F63 *f.*	W. Hartlepool [illegible]	1883	4 Bhds. [illegible]	2Cy.30"& 57"–33" 138 HP. T.Richardson&SonsHartpl	Hardy,Wilson
932 81368	✻ D	Craigmore I ScSr *Baxter 82-86* 1 *Dk2 B*	British Liverpool	987 1538	260·0|34·2|17·9 *B131 f. Q107 f. F33 f.*	Barrow Barrow S.B.Co.	1880	4 Bhds. [illegible]	2Cy.32½"&60"–36" 160 HP. Barrow S.B.Co.,Barrow	S.S. "Craig… Ld. (W.Johnston…)
933 58805	✻	Craignair I ScSr *Hughes 84-83* 1 *Dk*	British Liverpool	59 105	87·8|18·5|7·2 *Q 33 f. F 11 f.*	Grangemouth Dobson&Charles	1881	3 Bhds. [illegible]	2Cy.13"& 23½"–18" 35 HP. R.B.Pearson&Co.,Glasgow	"Craignair" S… Ld. (R.& D.Jones)
934 83715	✻	Craigton I ScSr *McGowan 87-87* 2 *Dks3 B*	British Glasgow	1325 2063	275·0|37·2|22·4 *P41 f. B50 f. F40 f.*	Leith Ramage&Ferguson	1882	[illegible] FPT… MT…	2Cy.32"& 61"–42" 200 HP. Blaikie Bros.,Aberdeen	Glasgow S.S.Co. (P.H.Dixon & Co.)
935 67841	✻	Cramlington I Sc3mSr *Grieve* 1 *Dk2 B*	British Newcastle	576 893	214·5|28·5|15·7 *Q 61 f. F 18 f.*	N.Shields T.&W.Smith	1872	5 Bhds. [illegible] APT…	2Cy.26½"& 55"–33" 99 HP. M.Samuelson,Hull	W. Mason & Co.
936 80551	✻ D	Crane I ScSr *Robinson 81-81* 2 *Dks3 B*	British Newcastle	1209 1860	275·5|35·0|21·0 *F 27 f. B 66 f.*	Sunderland J.L.Thompson&Sns	1879	6 Bhds. DB… FPT APT	2Cy.34"& 54"–42" 180 HP. J.Dickinson,Sunderland	Crane Shipping… (Dent & Co.)
937 95450	✻	Cranford StlScS. *Robinson -88* 1 *Dk& Wb Frms*	British London	1498 2283	293·0|38·2|18·7 *P29f.Q80f.B122f.F28f.*	W. Hartlepool W.Gray&Co.	1888	5 Bhds. CDB…30t. APT21t	3Cy.23"36½"&62"–39" Central Marine Eng. Co., W. Hartlepool 220 HP.	Pyman Bros.
938	G	Cranz I Sc *Hasiann 85* 1 *Dk*	German Memel	102 142	145·3|15·5|7·5	Kiel [illegible]	1885		2Cy.20"& 34"–30"	Mem.-Cranz Da…
939 88658	✻	Crathie I ScSr *Stephen 84-84* 1 *Dk*	British Aberdeen	281 475	160·4|24·2|13·0 *Q54 f. B42 f. F20 f.*	Kinghorn J.Key&Sons	1883	4 Bhds. [illegible]	2Cy.21"& 42"–27" 73 HP. J.Key&Sons,Kirkcaldy	W.T.Moffatt
940 92923	✻	Crathorne StlScSr *M'Culloch 84-88* 1 *Dk& Wb Frms*	British W.Hartlepl.	1811 2752	300·0|40·7|21·0 *P32f.B112f.F35f.Q82f.*	W. Hartlepool W.Gray&Co.	1888	5 Bhds. [illegible]	3Cy.23"37½"&61½"–39" Blair&Co.,Ld.,Stk. 250 HP.	R.Ropner & Co.
941 83094	✻	Creaden I ScSr *Kavanagh 83-88* 1 *Dk*	British Waterford	292 504	165·5|25·3|12·7 *Q34 f. B17 f. F20 f.*	S.Shields J.T.Eltringham	1883	4 Bhds. [illegible] FPT…	2Cy.22"& 40"–30" 70 HP. J.P.Rennoldson,S.Shields	Waterford Ste… Co.,Ld.
942	✻ V	Cremon (ex Bahia) I ScBg *Schwaner 77-87* 2 *Dks3 B*	German Hamburg	1567 2067	302·1|35·4|25·1 *B 74 f. F 36 f.*	Sunderland J.Laing	1871	5 Bhds. [illegible]	3Cy.20"31½"50"–42" 200 HP. Blohm&Voss,Hamburg	Dampschiff Re… "Hansa"
943 87252	✻ D	Cremona StlSc3mSr *Grierson 82-85* 2 *Dks*	British Leith	776 1225	248·2|33·2|17·3 *P24 f. B40 f. F33 f.*	Greenock R.Steele&Co.	1883	5 Bhds. FPT… APT…	2Cy.30"& 52"–36" 120 HP. R.Steele&Co.,Greenock	J.Currie & Co.
944 83804	✻	Creole I ScSr *O'Leary 83-83* 1 *Dk& Spr Dk3 B*	British Newcastle	1070 1639	248·0|35·0|15·5 22·5	Sunderland J.L.Thompson&Sns	1881	4 Bhds. CDB298t APT15t	2Cy.32"& 69"–39" 160 HP. G.Clark,Sunderland	Creole S… (Dent & Co.)
945 89784	✻ D	Crescent I & StlScSr *Jefferson 85-87* 1 *Dk& Wb Frms*	British Whitby	1340 2064	284·0|38·1|19·3 *P29f.Q82f.B118f.F28f.*	Sunderland J.L.Thompson&Sns	1887	6 Bhds. DB… 336t APT…	3Cy.21"35"& 58"–39" 180 HP. J.Dickinson,Sunderland	J.H. Barry
946 5427		Crescent City (ex Massachusetts) I ScBn *Doyle 88-88* 3 *Dks*	American New York	1462 2003	265·4|34·0|16·5	S.Boston, Mass H.Loring	1860		1Cy.62"–44" 140 HP.	Pacific Mail S.S.
947 89921	✻	Cressida Stl ScSrYt *Macintosh -84*	British Glasgow	59 150	129·4|18·3|11·7	Glasgow A.&J.Inglis	1884	5 Bhds.	2Cy.16"& 28"–22" 35 HP. A.&J.Inglis,Glasgow	Earl of Morton
948 95247	✻	Crest StlScSr *Grout 76-88* 1 *Dk& Wb Frms*	British Newcastle	1326 2064	284·0|38·0|19·4 *Q74f.B126f.F26f.P29f.*	Sunderland J.L.Thompson&Sns	1888	6 Bhds. DBc/290t APT22t	3Cy.21½"35"&58"–39" J.Dickinson,Sunderland 180 HP.	Crest Shipping… (Dent & Co.)
949 68700	✻	Crete I ScSr *James 76-84* 1 *Dk2 B*	British Newcastle	1121 1725	260·0|36·0|18·5 *Q72f.B58f.P23f.F26f.*	Sunderland J.L.Thompson&Sns	1883	6 Bhds. DBc/267t FPT… APT…	2Cy.32"& 62"–39" 140 HP. WallsendSlipCo.,Ld.,Nwctl.	CobanShipping… (Dent & Co.)
950		Crete I ScSr *Goudy 81-86* 1 *Dk*	Greek Spezzia	302 492	186·3|26·0|13·7 *B 29 f. F 24 f.*	Paisley H.M'Intyre&Co.	1881	4 Bhds.	2Cy.24"& 48"–33" 96 HP. Hutson&Corbett,Glasgow	D.P.Goudis
951	✻	Crete (ex Lizard) I ScSr *Dimetri* 2 *Dks*	Turkish Constanple.	194 411	191·0|25·2|13·0	Port Glasgow Blackwood&Gordn	1881	5 Bhds. MT APT FPT	2Cy.25"& 48"–36" 100 HP. Blackwood&Gordon,Pt.Gls.	P.M.Courtgi & …
952 78410	✻	Crimdon I ScSr *Johnson 87-88* 1 *Dk2 B*	British W.Hartlepl.	1060 1656	261·8|34·7|19·3 *R62 f. Q 99 f. F 35 f.*	W. Hartlepool W.Gray&Co.	1878	4 Bhds. [illegible]	2Cy.34"& 59"–33" 140 HP. T.Richardson&Sons Hartpl	R.Ropner & Co.
953	✻	Cristóbal Colón I Sc *Torrens 85-86 (ex Rossmore)* 2 *Dks3 B*	Spanish Barcelona	1747 2677	330·0|36·5|26·0 *P27 f. B 59 f. F 23 f.*	Sunderland W.Doxford&Sons	1880	5 Bhds. DB… FPT… APT… MT…	3Cy.24½"34"&68"–45" [illegible] W.Doxford&Sons,Sld. HP.	E.Pi & Co.
954		Cristóbal Colón I PadSr *(ex Arno)* 2 *Dks& Awn Dk*	Spanish Porto Rico	710 1143	254·9|32·6|18·2 *Q78 f. B28 f. F32 f.*	Greenock	1865		2Cy.40"& 72"–69" 250 HP. Day,Summers&Co.S'thamptn	Visande & Co.
955 7	I	Cristoforo Colombo I PadSr *Bolero 85-86* 2 *Dks*	Italian Genoa	294 508	216·8|25·4|12·3 *Q71 t. F29 t. H67 t.*	Renfrew W.Simons&Co.	1863		2Cy.36"& 30"–66" 212 HP. Simons&Co.,Renfrew	Navigazione… Italiana
956 83929	✻	Critic I ScSr *Davies 81-87* 1 *Dk& Spr Dk3 B*	British Newcastle	1701 2601	330·0|39·0|18·7 26·2	Sunderland J.L.Thompson&Sns	1881	7 Bhds. CDB457t APT…	2Cy.40"& 76"–48" 300 HP. WallsendSlipCo.Ld.Nwcstl.	Arrow Ship Co.… (Dent & Co.)
957		Croatia Stl ScSr 2 *Dks3 B*	German Hamburg	1486 2052	290·0|36·0|25·7	Hamburg Blohm& Voss	1888		3Cy.22",35"&58"–39⅜" Blohm & Voss, Hamburg 300 HP.	Hamburg Süd- Amerikan. Pack…
958 91174	BT1-88	Crocus StlDoubleTwnSc *Potter 65-85* (FERRY) 1 *Dk*	British Liverpool	35 308	130·9|35·1|10·6	Preston W.Allsup&Sons	1884	17 Bhds.	1Cy.18"& 37"–24" 113 HP. W.Allsup&Sons,Preston	Wallasey Local B…
959 85198	✻ D	Crœsus I ScSr *Kish 81-88* 1 *Dk2 B*	British London	763 1182	240·0|34·1|14·1 *Q90 f. B48 f. F26 f.*	Sunderland Kish,Boolds&Co.	1882	4 Bhds. CB… APT…	2Cv.28"& 54"–36" 130 HP. Sarr&Co.Ld.,Sunderland	T.Kish
960 88745	✻	Croma I Sc3mSr *Lord 71-87* 1 *Dk& Spr Dk3 B*	British Newcastle	2084 3170	330·3|43·7|27·3 *B 76 f. F 31 f.*	Sunderland J.L.Thompson&Sns	1883	6 Bhds. DB… APT…	2Cy.40"& 76"–48" 300 HP. WallsendSlipCo.Ld.Nwcstl.	Croma Ship Co. (Dent & Co.)

✻ = **Lloyd's Register**; D = Nederlandsche Vereeniging; G = Germanischer Lloyd; I = Registro Italiano; R = American Re…
A = Veritas Austro-Ungarico; H = Veritas Ellenico; N = Norske Veritas; V = Bureau Veri…

In the 2nd Column, B.T., with figures following, signifies Date of report of bottom sighting by Board of Trade Surveyors as condition of PASSENGER CERTIF…

In the 3rd Column, C = Composite (iron frame, wood planked); I = Iron; Stl = Steel; W = Wood.

Further particulars of Vessels classed by Lloyd's Register will be found in Lloyd's Register Book and Lloyd's Yacht Register.

		Nom, Matériaux, Gréement, etc. Capitaine et dates d'entrée au service des Armateurs actuels et à bord du navire. Ancien Nom, nombre de ponts, etc.	Pavillon. Port d'Armement	Tonnage Net. Total	Dimensions (Longueur | Largeur | Creux) Constructeurs sur le Pont	Construit. Où. Par qui.	Quand	Nombre des cloisons. Détails sur le lest d'eau.	Machines. Détails. Nom du Constructeur.	Armateurs ou Directeurs-Gérants.
961	✻	Crosshill — I ScSr Sharp 87-? — 1 Dk 2 B	British Glasgow	617 [illegible]	215·5 | 30·1 | 15·3 Q [illegible] B 58 f. F [illegible]	Leith [illegible]	1883	4 Blkds. DB 42t FPT 58t	2Cy.26"& 50"–36" 98 HP. Ramage&Ferguson,Leith	Macbeth & Gray
962	✻	Crown Prince — I ScSr Sawyer — 1 Dk 2 B	British N. Shields	1070 1655	258·3 | 36·2 | 17·3 Q 94 f. B [illegible] F 34 f.	Newcastle Palmers'Co.,Ld.	1883	4 Blkds. [illegible] APT [illegible]	2Cy.31"& 60"–39" 152 HP. Palmers'Co.,Ld.,Newcastle	J.Knott
963	✻	Crown of Arragon — I ScSr Ferguson 83-3 — 2 Dks 3 B	British Glasgow	1486 2254	300·5 | 37·3 | 26·1 B 63 f. F 35 f.	Greenock Scott&Co.	1883	7 Blkds. DT [illegible] DB [illegible]	2Cy.24"& 49"–54" 275 HP. Scott&Co,Greenock	Caw,Prentice,Clapperton & Co.
964		Croydon (ex Garnet) — I ScSr Adrian	British London	198 357	180·6 | 22·2 | 10·8 Q 26 f. F 8 f.	Glasgow T.Wingate&Co.	1878		2Cy.20"&40"–30" 75 HP. T.Wingate&Co.,Glasgow	Australasian United Steam Nav.Co.,Ld.
965		Cruizer — I Pad	British Liverpool	48 164	156·4 | 25·2 | 12·8	Liverpool	1861		2Cy.47"–60" 180 HP. J.W.Hoby&Co.,Renfrew	Liverpool Steam Tug Co.
966	✻	Crusader — I ScSr Syston — 2 Dks	British Glasgow	647 1023	221·4 | 30·1 | 17·6 P 118 f. F 30 f.	Greenock Scott&Co.	1871	5 Blkds. DB 67t MT 140t	2Cy.25"&50½"–30" 120 HP. GreenockFoundryCo.,Grnk.	A.G.Kidston & Co.
967	✻	Crystal — I ScSr Stannard 82-85 — 2 Dks & Awn Dk	British Newcastle	1708 2613	330·0 | 39·0 | 18·9 [illegible] 26·0	Sunderland J.L.Thompson&Sons	1881	7 Blkds. CDB 196t APT 21t	2Cy.40"& 76"–48" 300 HP. WallsendSlipwayCo.,Nwc.	Arrow Ship Co.,Ld. (Bent & Co.)
968	✻	Cuba — I ScSr Ayers 83-88 — 2 Dks 3 B	British Liverpool	1372 2119	285·6 | 36·1 | 24·2 P 28 f. B 61 f. F 32 f.	W.Hartlepool W.Gray&Co.	1883	5 Blkds. DB aft 3.9t	2Cy.36½"&68"–42" 200 HP. Blair&Co.,Ld.,Stockton	City of Liverpool Steam Nav.Co.,Ld.
969	BT [illegible]	Cuban — I ScBk Lawson 85-85	British Liverpool	845 1330	248·7 | 31·7 | 26·5	Newcastle A.Leslie&Co.	1865		1Cy.16"& 36"–42" 120 HP. G.Forrester&Co.,Liverpool	West India & Pacific S.S.Co.,Ld.
970		Cueva	Brazilian Riodaneiro	141	.. | .. | ..					J.J.de França
971	✻	Cufic — StlSc1mSr Smith 88-88 — 2 Dks 3 B	British Liverpool	3055 4639	430·7 | 45·2 | 30·0 P 48 f. B 89 f. F 41 f.	Belfast Harland&Wolff	1888	7 Blkds. CDB 94t	3Cy.27"41½"&74"–60" 520 HP. Harland&Wolff, Belfast	Oceanic Steam Nav. Co.,Ld.
972	✻	Cuhona — I Sc3mSrYt Douglas 82-83	British Liverpool	234 344	163·5 | 26·1 | 14·1	Hull Earles'Co.,Ld.	1882	4 Blkds.	2Cy.23"& 44"–24" 70 HP. Earles'Co.Ld.,Hull	Sir Andrew B.Walker
973		Cuirassier — StlScSr	British Boston	73 108	95·4 | 19·6 | 8·1	Middlesbro' Richardson,Duck&Co.	1860		2Cy.9½"–14" 25 HP. Carmichael&Co.,Dundee	Boston & Hull S.S.Co.,Ld.
974		Culna — I ScSr West 80-86 — 2 Dks	British Glasgow	1469 1984	285·0 | 35·2 | 23·9	Dumbarton W.Denny&Bros	1878	5 Blkds. CDB 328t	2Cy.31"& 60"–42" 185 HP. Denny&Co.,Dumbarton	British India Steam Nav.Co.,Ld.
975	✻ BT [illegible]	Cumberland — I Sc3mSr Bell 83-84 — 2 Dks & Spt Awn Dk	British Leith	552 865	249·5 | 28·0 | 15·5 P 38 f. F 33 f.	Glasgow Barclay,Curle&Co.	1866	5 Blkds.	2Cy.29"& 49"–36" 108 HP. Barclay,Curle&Co.,Glasgow	Leith,Hull& Hamburg Steam Packet Co.
976		Cumberland — W Pad Thompson	American Eastport,Me	1188 1606	252·2 | 37·0 | 14·8	Bath, Me. NewEnglndS.B.Co.	1885			International S.S.Co.
977	✻ BT [illegible]	Cumbrae — I ScSr Robertson 67-83 — 2 Dks	British Glasgow	448 916	230·2 | 31·9 | 15·6 P 46 f. F 39 f.	Glasgow Dobie&Co.	1882	5 Blkds. DB 89t FPT 73t	2Cy.28"& 57"–51" 190 HP. Lees,Anderson&Co.,Glasgw	Clyde Shipping Co.
978		Cumbrae — I ScBarge Campbell	British Pt.Glasgow	178 389	153·8 | 26·0 | 12·2	Renfrew W.Simons&Co.	1882		2Cy.19"& 36"–24" 55 HP. W.Simons&Co.,Renfrew	The Trustees of the Clyde Lighthouses
979	BT [illegible]	Cumbrae (ex Victory, ex Marquis of Lorne) — I PadSk	British Glasgow	19 126	176·7 | 17·6 | 6·8	Glasgow Barclay,Curle&Co.	1863		1Cy.54"–42" 75 HP. Barclay,Curle&Co.,Glasgow	J.Rodger
980	✻	Cumbrian — I ScSr Watson 76-85 — 1 Dk & Web Frms	British Whitehaven	782 1221	240·0 | 33·2 | 16·1 Q 88 f. B 40 f. F 28 f.	Whitehaven WhitehavenS.B.Co.	1882	4 Blkds. DB aft 330t APT 84t	2Cy.27"& 52"–36" 99 HP. Westray&Copeland,Barrow	Cumbrian S.S.Co.,Ld. (G.Nelson & Sons)
981		Cunga — I Pad	Portug'se Loanda	140	160·1 | 25·9 | 6·9	Glasgow London&Glsgw.Co.	1869		1Cy.27" 40 HP.	Nova Cia.de Quanza
982	✻	Curfew — StlScSr Sangster -88 — 1 Dk & Spr Dk 3 B	British Dundee	1261 1949	275·0 | 37·2 | 22·8 B 64 f. F 36 f. 15.8	Dundee W.B.Thompson&Co.,Ld.	1888	5 Blkds. CDB 283t FPT 67t	3C.21"33"&54"–42" 180 HP. W.B.Thompson&Co.,Ld.,Dundee	J.& R.Mudie
983	V	Curityba — StlScBk Barch 81-88 — 2 Dks	German Hamburg	1757 2430	309·0 | 40·0 | 25·0 P 86 f. B 76 f. F 48 f.	Hamburg ReiherstiegWerft	1887		3C24"39"&62"–42" 325 HP. ReiherstiegWerft	Hamburg-Südamerik Dampfschiff. Gesells.
984	✻	Curlew — I ScSr Jones — 1 Dk 2 B	British London	465 630	188·5 | 26·5 | 14·7 Q 53 f. B 47 f.	Dundee Gourlay Bros.	1875	5 Blkds. [illegible]	2Cy.24"& 40"–30" 80 HP. GourlayBros.,Dundee	General Steam Nav.Co.
985		Curlew — C ScSr Ash 81-88 — 2 Dks	British S.John'sNF.	293 473	156·8 | 23·9 | 19·2	Liverpool W.H.Potter&Son	1877		2Cy.21"& 40"–30" 80 HP. J.Taylor&Co.,Birkenhead	C.T.Bowring & Co.
986		Curlew — I Sc3mSc Hajoo — 1 Dk	British Bombay	157 305	145·3 | 25·1 | 11·4	Glasgow J.Elder&Co.	1876		2Cy.20"& 36"–21" 60 HP. J.Elder&Co.,Glasgow	Bombay Port Trust
987		Curlew — Stl PadSr Man -81 — 1 Dk	British Calcutta	154 109	224·3 | 24·0 | 11·4	Liverpool Jones,Quiggin&Co	1865		2Cy.47"–66" 160 HP. J.Watt&Co.,Birmingham	India General Steam Nav.Co.,Ld.
988	✻	Curlew — I Sc3mSc Candlis 84-88 — 1 Dk	British Whitehaven	104 303	141·7 | 23·3 | 10·8 Q 74 f. F 24 f.	Campbeltown CampbltownS.B.Co	1885	3 Blkds. FPT 23t APT 10t	3Cy13"–19"&36"–24" 60 HP. Walker,Henderson&Co.Gls	J.S.Simpson & H. Howie
989	V	Curonia — I ScSr Judeke 73-83 — 1 Dk	German Stettin	399 536	163·0 | 24·4 | 12·7 P 94 f. F 27 f.	Stettin Act.-Ges."Vulcan"	1883	DT/80t	2Cy21½"&39½"–27½" 70 HP. Act.Ges."Vulcan,"Stettin	"Kurland" Dampfschifffahrts-Akt-Ges.
990		Currajong — I TwnSc3mSr (ex Marquis of Lorne, ex Victory, ex The [illegible]) — 1 Dk	British Sydney	410 603	222·2 | 30·1 | 15·7 Q 63 f. P 13 f.	Glasgow T.Wingate&Co.	1875		1Cy.26"& 52"–20" 250 HP. T.Wingate&Co.,Glasgow	Australasian Steam Nav.Co.

✻ = Lloyd's Register; D = Nederlandsche Vereeniging; G = Germanischer Lloyd; I = Registro Italiano; R = American Record;
A = Veritas Austro-Ungarico; H = Veritas Ellenico; N = Norske Veritas; V = Bureau Veritas.

Dans la 2me Colonne, B.T., avec les indications qui suivent, signifie la date du rapport de visite de la Carène par les experts du Board of Trade pour obtenir le certificat relatif au TRANSPORT DES PASSAGERS.

Dans la 3me Colonne, C = Composite (membrure en fer, bordé en bois); I = Fer; Stl = Acier; W = Bois.

Pour plus amples renseignements sur les navires classés par le Lloyd's Register, voir le Lloyd's Register Book et le Lloyd's Yacht Register.

No. in Book. / Official No.	Registry in which classed	Ship's Name. Master, and Dates of Appointment to present Owner's Service and to Ship. Ship's Late Name.	Material, Rig, &c. No. of Decks, &c.	Flag. / Port of Registry.	Tons. Net. / Gross	Length	Breadth	Depth	Built. Where. / By Whom.	When	No. of Bulkheads. Particulars of Water Ballast.	Engines. Particulars, Maker's Name.	Owners or Managers
991		Cururipe		Brazilian Cururipe	193	..	..	..					J.M.A.Filho
992 81280	✳	Cushie Doo	I ScbmSrYt 2 Dks	British Glasgow	159 249	145·0	24·1	12·8	Glasgow D.&W.Henderson& Co.	1881	4 Bhds.	2Cy.19"& 38"–24" 60 HP. D.&W.Henderson&Co.,Gla.	W.H.S.Osmond
993 88178	✳	Cutch Crowder	I ScSr 1 Dk	British Bombay	123 324	180·0	23·2	11·7	Hull J.Bremner&Co.	1884	4 Bhds. DB&Aft FPT&AFT	2Cy.25"& 48"–30" 98 HP. J.Bremner&Co.,Hull	Hajeebhoy Al- hoy&Joomabhoy
						Q 36 f.	P 37 f.	F 39 f.					
994 87002		Cuvier Blair 74-83	I ScSr	British Liverpool	1437 2229	302·0	37·2	24·6	Newcastle A.Leslie&Co.	1883		2Cy.34"& 66"–48" 230 HP. R.Stephenson&Co.,N'castle	Liv.Brazil & R. Steam Nav.Co.L
						P 177 f.	F 29 f.						
995 67836	✳	Cuxhaven Calvert	I ScSr 1 Dk	British Goole	398 743	220·0	32·0	18·7	Dundee W.B.Thompson	1882	5 Bhds. DBaft 3'2010	2Cy.30"& 60"–33" 150 HP. W.B.Thompson,Dundee	Yorkshire Coal Steam Ship Co.
						P 40 f.	B 72 f.	F 32 f.					
996		Cuxhaven ex Mary & Ellen Röhrs	I PadSr 1 Dk	German Hamburg	256 377	221·7	22·4	10·3	Greenock Caird&Co.	1864		150 HP.	Unterelbe'sche Eisen- bahn Ges.
997		Cuyaba	I Sc	Brazilian Rio Janeiro	391 585	186·2	25·5	15·3	Greenock Robertson&Co.	1867		2Cy.40"–48" 110 HP. DavidRowan,Glasgow	A.J.dos Santos
998	✳	Cuyuni Phillips 78-87	I TwnScSc 1 Dk	British London	41 78	80·0	16·0	8·8	London R.&H.Green	1883	3 Bhds.	1Cy.16"& 26"–18" 50 HP. J.Penn&Son,London	Spriston Docks Foundry Co.
999 65801	✳ BT9.88	Cuzco Nixon 84-88	I ScBk 3 Dks & Awn Dk	British Liverpool	2519 3918	384·2	41·4	35·3	Glasgow J.Elder&Co.	1871	6 Bhds.	3Cy.35"57"&90"–54" 650 HP. Fairfield&Co.,Ld.,Glasgow	Orient Steam N. Co.,Ld.
							H 31 f.						
1000 82863	✳	Cyanus Henderson	Stl ScSr 1 Dk 2 R	British London	1061 1635	265·0	34·4	16·6	Hartlepool E.Withy&Co.	1880	4 Bhds. CB[illegible]	2Cy.33"& 61"–33" 150 HP. T.Richardson&Sons, Hrtpl.	Steel, Young & Co.
						P 165 f.	F 30 f.						
1001 81491	✳	Cybele Tucker 79-93	I ScBn 1 Dk 2 R	British London	819 1287	285·0	32·5	16·5	Stockton M.Pearse&Co.	1883	4 Bhds. DBof 256t	2Cy.28½"&53"–33" 110 HP. Blair&Co.Ld.,Stockton	H.T.Clarke,jun.
						Q 80 f.	B 61 f.	F 31 f.					
1002	✳	Cycle	StlScSr	British Liverpool					Sunderland J.L.Thompson&Sons	1889		3Cy.21½",35"&58"–39" J.Dickinson,Sndrld [180 HP.	R.Nicholson & Co.
1003 78005	✳	Cyclone Dawson 88-88	I ScSr 1 Dk 2 R	British Glasgow	632 997	231·3	32·2	18·2	Glasgow R.Napier&Sons.	1878	5 Bhds. DB[illegible]	2Cy.28"& 48"–39" 110 HP. R.Napier&Sons,Glasgow	D.Currie & Co.
						B 44 f.	F 23 f.						
1004	V	Cyclop Martens 88-88	I Sc3mSr 1 Dk	German Hamburg	758 938	230·6	29·1	17·0	Hamburg Reihst.Schiffs Wft	1873	DB[illegible]	2Cy.28"& 56"–33" 120 HP. ReiherstiegSchiffsWft.Hbg	Hamburg-Amerikanische Packetf.-Akt.-Ges.
						Q 34 f.	H 25 f.	F 32 f.					
1005		Cyclop	I PadSpTug	German Bremen	75 129	131·5	16·6	7·6	Bremen Act.-Ges."Weser"	1867		2Cy.32"–22" 41 HP. Act.Ges."Weser," Bremen	Norddeutscher L[loyd]
1006 81342	D BT1.88	Cyclops Nish 82-86	I ScSr	British Liverpool	1403 2167	321·4	34·3	26·0	Greenock Scott&Co.	1880		2Cy.24½"&56"–52" 300 HP. GreenockFoundryCo.,Grnk.	Ocean Steam Ship Co.
1007 70415	✳D	Cydonia Winspear 88-88	I Sc3mSr 2 Dks 3 R	British N.Shields	1093 1693	265·0	34·0	24·4	Newcastle ColeBros.	1876	5 Bhds. DB[illegible]	2Cy.31"& 58"–33" 150 HP. N.E. Marine Eng. Co. Ld., Sunderland.	J.Robinson & Sons
1008 81784	✳D	Cyfarthfa Newton	I ScSr 1 Dk 2 R	British Cardiff	750 1167	241·0	33·2	16·6	Newcastle Palmers'Co.,Ld.	1880	4 Bhds. CDB[illegible]	2Cy.28"& 54"–36" 99 HP. Palmers'Co.Ld.,Newcastle	"Cyfarthfa" S.S. Co. Ld. (Morel Bros.)
						Q 127 f.	F 27 f.	B 10 f.					
1009		Cygne Bulloin 52-	I PadSr 1 Dk	French Havre	78 143	138·1	19·0	9·3	Havre A.Normand	1854		2Cy.37½"–41½" 80 HP. Mazeline frères,Havre	Cie. Anonyme des [illegible]
1010 60420		Cygnet ex Sweden Church -87	I Sc3mSr 2 Dks & Awn Dk 3 R	British Liverpool	768 1200	236·6	30·5	16·3	Glasgow Barclay,Curle&Co.	1869	5 Bhds. WB	2Cy.31"& 54"–36" 129 HP. J.Jack&Co.,Liverpool	G.Nelson & Sons
								23½					
1011 87183	✳	Cygnet Witt	Stl ScSr 2 Dks	British London	759 1227	250·4	37·1	15·9	Dundee Gourlay Bros.&Co.	1883	6 Bhds. CDB[illegible]	2Cy.34"& 64"–42" 210 HP. Gourlay Bros.&Co.,Dundee	General Steam Nav.
						P 50 f.	B 80 f.	F 73 f.					
1012 91893	✳	Cygnet Poole	Stl ScSc	British Hobart Twn	66 124	120·0	17·3	8·7	Abingdon Davis&Clow	1885	5 Bhds.	2Cy.16"& 32"–18" 43 HP. Bow&McLachlan,Paisley	W.J.Watchorn
1013 32959	✳	Cygnet Nightingale	I ScSr 1 Dk	British Scarboro'	45 102	94·7	19·0	10·0	S.Shields J.T.Eltringham	1882	4 Bhds.	2Cy.17"& 32"–22" 40 HP. J.P.Rennoldson&Sn.S.Shlds.	M.J.Scobie
1014 17807	BT1.88	Cygnus Fisher 59-73	I PadSr	British Weymouth	85 250	182·0	21·4	9·7	Renfrew J.Henderson& Sons	1854		2Cy.42½"–42" 120 HP. M'Nabb&Clark,Greenock	Weymouth & Channel Islands Stm. Pkt. Co.
						Q 18 f.							
1015 125896	R	Cygnus	I Pad	American New York	857	212·6	31·4	11·3	Chester,Pa. J.Roach&Son	1881		1Cy.58"–132" W.&A.Fletcher,New York	Iron Steamboat Co.
1016 70470	✳	Cymbeline Hadaway	I ScBn 1 Dk 2 R	British Hull	899 1387	248·6	32·2	18·5	Stockton Richardson,Duck& Co.	1879	5 Bhds. DBaft [illegible]	2Cy.29"& 55"–38" 130 HP. T.Richardson&Sons, Hrtpl.	H. Briggs,Sons & Co.
						Q 79 f.	B 56 f.	F 29 f.					
1017 89173	✳	Cymmrodorion Evans	I ScSr Dk 2 R	British Cardiff	1108 1723	260·0	36·3	19·4	Newcastle Schlesinger,Davis &Co.	1884	4 Bhds. [illegible] FPT AFT	2Cy.35"& 65"–39" 160 HP. Black,Hawthorn&Co.,Gtshd.	"Cymmrodorion" S.S. Co.,Ld.(Jones & Co.)
						Poo.Q 103t.B 119t.F 34t							
1018 86226		Cynosure (ex Sailing Vessel)	WScFlat	British Liverpool	81 119	84·2	20·5	8·8	Winsford AnnDeakin	1877		2Cy.11"–17" 22 HP. Deakin&Co.,Winsford	G.Deakin,Ld.
1019 82317	✳	Cynthia Taylor 77-83	I Sc3mSr 2 Dks 3 R	British Glasgow	1410 2152	329·7	36·8	25·1	Glasgow D.&W.Henderson& Co.	1880	7 Bhds. CDB[illegible]	2Cy.36"& 68"–54" 400 HP. D.&W.Henderson&Co.,Gla.	Donaldson Bros.
						B 96 f.	F 35 f.						
1020 63891	✳	Cyphrenes Kelly 86-86	I ScSr 2 Dks 3 R	British Liverpool	1309 1894	300·0	34·1	25·5	Dundee A.Stephen&Sons	1872	5 Bhds. WB [illegible]	2Cy.40"& 71"–39" 250 HP. J.Elder&Co.,Glasgow	C.W.Kellock

✳ = Lloyd's Register; D = Nederlandsche Vereeniging; G = Germanischer Lloyd; I = Registro Italiano; R = American Record; A = Veritas Austro-Ungarico; H = Veritas Ellenico; N = Norske Veritas; V = Bureau Veritas.

In the 2nd Column, B.T., with figures following, signifies Date of report of bottom sighting by Board of Trade Surveyors as condition of PASSENGER CERTIFICATE.

In the 3rd Column, C = Composite (iron frame, wood planked); I = Iron; Stl = Steel; W = Wood.

Further particulars of Vessels classed by Lloyd's Register will be found in Lloyd's Register Book and Lloyd's Yacht Register.

	Nom, Materiaux, Gréement, etc. Capitaine et dates d'entrée au service des Armateurs actuels et à bord du navire. Ancien Nom, nombre de ponts, etc.	Pavillon. Port d'Armement	Tonnage. Net. Total	Dimensions. Longueur	Largeur	Creux	Construit. Où. Par qui.	Année.	Nombre des cloisons. Détails sur le lest d'eau.	Machines. Détails. Nom du Constructeur.	Armateurs ou Directeurs-Gérants.
✻	Cyprus I ScSr 2 Dk&3 B	British Sunderland	1205 1859	274·5 P80 f.	36·0 B63 f.	23·8 F34 f.	Sunderland Short Bros.	1878	5 Blhds. DB&c 171 FPT88 APT29	2Cy.35"& 62"–42" 180 HP. J.Clark,Sunderland	Taylor & Sanderson
✻ D	Cyprus I ScSr 1 Dk2 B	British Newcastle	832 1291	245·0 P28 f.	33·2 F27 f.	19·6	Newcastle Wigham Richardson & Co.	1878	5 Blhds. DB&c 221	2Cy.30"& 56"–36" 160 HP. J.Simon&Co.,Newcastle	Elliott, Lowrey & Dunford
✻	Cyrene Stl ScSr Bodley 71-88 1 Dk,&Spr Dk3 B	British Sunderland	1966 3032	302·0 B35 f.	42·0 F33 f.	27·4 19·4	Sunderland Short Bros.	1888	5 Blhds. DB&c 520 FPT70	3Cy.23½"38"&62"–42" [250 HP G.Clark,Sundrld.	Lumsden, Byers & Co.
	Cyrenian I ScSr 2 Dks	British Liverpool	940 1437	292·2	36·1	22·2	Liverpool Bowdler, Chaffer & Co.	1874	5 Blhds.	2Cy.32"& 61"–36" 170 HP. J.Jack,Rollo&Co.,Liverpool	S.S. "Cyrenian" Co., Ld. (F. Leyland & Co.)
	Cyril Stl ScSr Taylor W.&D Dk 1 Dk2 B	British Cardiff	1545 2363	289·2 Q76 f. P28 f.	38·3 B124 f.	20·7 F32 f.	Whitby T. Turnbull & Son	1888	6 Blhds DB&c 1132	3Cy.22"36"&59"–39" 175 HP. Blair&Co.,Ld.,Stockton	Turnbull Bros.
	Cyril (ex Parisian) I ScBn Johnson 83-83 1 Dk2 B	British Liverpool	751 1190	235·0 Q82 f.	31·3 B55 f.	17·1 F31 f.	Glasgow Lond.&Glasgow Co.	1882	4 Blhds. DB&c 1165 MT 158	2Cy.27"& 51"–36" 130 HP. Lon.&GlasgowCo.,Glasgow	Booth Steam Ship Co., Ld.
✻	Cyrus Stl ScSr Conlin 87-88 1 Dk2 B	British Grangem'th	721 1115	240·5 Q82 f.	33·1 B62 f.	14·0 F30 f.	Grangemouth Dobson & Charles	1883	4 Blhds. DB&c 280 APT 166	2Cy.30"& 54"–36" 130 HP. Hutson&Corbett,Glasgow	T. Kish
✻	Czar I ScSr Potter 82-82 2 Dks	British Leith	863 1347	243·0 P89 f.	32·5 B58 f.	17·8 F30 f.	Sunderland J. Laing	1879	4 Blhds.	2Cy.29½"&56"–36" 140 HP. N.E.MarineE.Co.Ld.,Sndld.	W. Thomson & Co.
✻	Czar (WRECKED) I SchnSr Sewell 82-86 1 Dk2 B	British Hull	324 452	199·3	26·5 Q 23 f.	12·2	Newcastle C. Mitchell & Co.	1857	4 Blhds.	2Cy.24"& 46"–26" 80 HP. Cmp'76 C.D.Holmes&Co.Hull	W. Rawson & J. Robinson
	Czar Stl ScBk 3 Dks	Russian Odessa	1491 2242	319·0 P59 f.	37·0 B94 f.	23·5 F55 f.	Newcastle Sir W. G. Armstrong, Mitchell & Co., Ld.	1883	8 Blhds. C 1380	2Cy.40"& 78"–48" 350 HP. WallsendSlipway&Eng.Co.	Russian Steam Nav. & Trading Co.
✻	Czarina (ex Sultan) W SchnSrYt Poland 78-78 2 Dks	British Portsmouth	225 332	152·8	29·3	15·2	Gosport Camper & Nicholson	1877	C12.86	2Cy.20"& 40"–24" 60 HP. Day,Summers&Co.,Sthpton	Albert Brassey

ADDITIONAL SHIPS ARE INSERTED IN THE SUPPLEMENT AT THE END OF THE REGISTER.

✻=Lloyd's Register; D=Nederlandsche Vereeniging; G=Germanischer Lloyd; I=Registro Italiano; R=American Record;
=Veritas Austro-Ungarico; H=Veritas Ellenico; N=Norske Veritas; V=Bureau Veritas.

Dans la 2me Colonne, B.T., avec les indications qui suivent, signifie la date du rapport de visite de la Carène par les experts du Board of Trade pour obtenir le certificat relatif au TRANSPORT DES PASSAGERS.

Dans la 3me Colonne, C=Composite (membrure en fer, bordé en bois); I=Fer; Stl=Acier; W=Bois.

Pour plus amples renseignements sur les navires classés par le Lloyd's Register, voir le Lloyd's Register Book et le Lloyd's Yacht Register.

No. in Book / Official No.	Society in which classed	Ship's Name. Master, and Dates of Appointment to present Owner's Service and to Ship. Ship's Late Name.	Material, Rig, &c. No. of Decks, &c.	Flag. / Port of Registry.	Tons. Net. / Gross.	Length	Breadth	Depth	Built Where. / By Whom.	When.	No. of Bulkheads. / Particulars of Water Ballast.	Engines. Particulars. Maker's Name.	Owners or Managers
1 83209		D. D. Calvin	W ScSp 1 Dk	British Kingston, Ont	483 756	166·0	32·0	15·1	Garden Is. Ont H. Roney	1883		2Cy. 27" & 50"–36" 300 HP. [illegible]	The Calvin [illegible]
2 87[illegible]	R	D. J. Foley *Rand -87*	W ScSr 2 Dks	American Chicago	348 341	169·0	30·0	16·0	Baltimore Beecham	1872		132 HP.	J. P. Godman
3 74091		D. P. Ingraham	W Sc 1 Dk	British S. John's, N.B.	69 113	102·0	21·4	8·1	Philadelphia	1864		1Cy. 30"–24" 75 HP. Neafie & Levy, Philadelphia	J. Green
4 71101		D. R. Van Allen	W Sc 1 Dk	British Chatham, Ont	216 318	136·0	26·0	10·0	Chatham, Ont Chisholm & Simpson	1871		1Cy. 87 HP. Hyslop & Ronald, Chatm., On.	G. Wright
5		D. V. Liljehook *Nylén 83-83*	C ScSp 1 Dk	Swedish Säffle, Wermld	126 168	100·7	20·8	11·3	Spångens Vrf.	1875		20 HP.	G. Silvén
6 84138	*	Dabulamanzi *Flint 88-88*	I ScSr 1 Dk & Spr Dk 3 B	British Aberdeen	980 1537	260·6	35·1	21·2	Aberdeen Hall, Russell & Co.	1882	4 Blds. [illegible]	3Cy. 21" 33" & 56"–42" 200 HP. Trpl. [illegible] Hall, Russell & Co., Abn	J. T. Rennie & [illegible]
7 85[illegible]	BT 3-89	Dacca *Stone 79-87*	Stl ScBg 3 Dks	British Glasgow	2545 3909	390·0	43·6	27·6	Glasgow A. & J. Inglis	1881	8 Blds.	2Cy. 48" & 87"–54" 500 HP. A. & J. Inglis, Glasgow	British India [illegible] ated Steam [illegible]
8 56307	*	Dacia *Barker 85-88*	I ScBk 1 Dk & Spr Dk 3 B	British London	1473 1858	283·0	34·7	17·9	Sunderland J. Laing	1867	5 Blds.	1Cy. 23" & 45"–36" 170 HP. Comp. '75, G. Clark, Sunderl.	India Rubber [illegible] percha & Teleg. Wo[illegible]
9 68301	*	Dafila *Neilsen*	I ScS 1 Dk 2 B	British Hong Kong	536 840	221·5	26·2	18·4	Sunderland J. Laing	1873	4 Blds. W[illegible]	2Cy. 27" & 50"–30" 93 HP. J. Key, Kirkcaldy	Kaw Hong T[illegible]
10 93350		Daga	Stl PadLc	British Glasgow	149 237	160·0	21·0	7·7	Dumbarton W. Denny & Bros.	1887		2Cy. 24" & 48"–42" 90 HP. Denny & Co., Dumbarton	Irrawaddy Fl[illegible] Co.
11		Dagestan *Brilkin 79-*	I ScSr	Russian S. Petersburg	210 278	143·0	21·3	10·6	Rostock Rost. Schiff- & Masch. Bau Ges.	1879		1Cy. 15½" & 26¾"–18" 75 HP. Rostocker Schiffs & Masch. Bau Ges.	Steamship Co.[illegible]
12 47614		Dagmar	I Pad	British Liverpool	19 183	138·0	21·0	8·4	Birkenhead Laird Bros.	1863		2Cy. 38"–48" 95 HP. Laird Bros., Birkenhead	Manchester S[illegible] Co.
13		Dagmar (ex Thebes) *Hansen 72-86*	I ScSr 1 Dk & Awn Dk	Danish Copenhagen	776 1223	225·5	30·4	17·1 23·9	W. Hartlepool Denton, Gray & Co.	1872	4 Blds. [illegible]	2Cy. 28" & 56½"–36" 120 HP. Blair & Co., Ld., Stockton	Det Forenede [illegible] skibs-Selskab [illegible]
14	N	Dagmar *Iversen 75-75*	W ScSt 1 Dk	Norweg'n Bergen	190 274	121·4	22·2	13·2	Sjötorp	1860		2Cy. 15" & 29½"–18" 30 HP. Laxevaags Verkst., Bergen	N. J. Michele[illegible]
15		Dagmar *Zöllner 85-*	I PadSr 1 Dk	Russian Pernau	385 504	188·5	25·0	11·8	Norrköping Motala Co.	1867		2Cy. 52"–48"	R. Barlehn & Co.
16	V	Dagmar *Rosen*	I ScS	Russian Windau	176 245	100·0	19·0	11·0	Elbing J. Schichau	1867		2Cy. 23½"–18½" 60 HP. F. Schichau, Elbing	W. G. Reincke
17		Dagmar	I Pad 3m Sr	Russian Taganrog	99 157	130·0	21·1	7·8	Newcastle Wigham Richardson	1869		2Cy. 30"–42" 60 HP. Black, Hawthorn & Co., Nwc.	J. Draper
18	G	Dagmar (WRECKED) *Orbom 87-87*	C ScSr 1 Dk	Swedish Gothenburg	261 368	139·8	21·5	13·4	Gamla Lödöse G. Pehrsson	1882		2Cy. 16½" & 30"–21" 50 HP.	G. E. Falck
19		Dagmar (ex Eskilstuna III, ex Boslagen) *Johansson -87*	I ScSr 1 Dk	Swedish Stockholm	163 211	112·5	18·7	7·9	Oscarshamn Oscarshamn M. V.	1883		2Cy. 45 HP. Oscarshamn M. V., Oscarshamn	L. L. Liberg
20 68965	*	Dago *Corner 85-87*	I ScSr 2 Dks 3 B	British Sunderland	1239 1914	283·0	33·8	24·8	Sunderland J. Laing	1877	4 Blds. [illegible]	2Cy. 32" & 60"–39" 180 HP. G. Clark, Sunderland	The Pinkney & [illegible] S.S. Co., Ld.
21 734		Daiki Maru	W ScSr	Japanese [illegible]	145 268	123·4	15·1	..	Yokohama Whitfield	1882		41 HP.	Hanyei Sha
22 544		Dairio Maru	W ScS	Japanese Osaka	181 292	139·0	18·6	14·7	Hiogo Kosaka Sen Kioku	1881		2Cy. 17" & 30"–24" 39 HP. James Long	Osaka Shosen [illegible]
23 33797	*	Daisy *Cromar*	I ScS 1 Dk	British Dundee	418 665	180·8	27·9	14·5	Whitby T. Turnbull & Son	1873	4 Blds.	2Cy. 23" & 46"–30" 80 HP. Blair & Co., Ld., Stockton	T. Muir, Son & [illegible]
24 89997	* BT 7-88	Daisy *Saultry*	I ScS 1 Dk	British Glasgow	85 216	130·0	21·1	9·3	Paisley H. McIntyre & Co.	1884	4 Blds.	2Cy. 17" & 34"–24" 50 HP. Hutson & Corbett, Glasgow	Glasgow, Dubl[illegible] Londonderry S[illegible]
25 81303	BT 1-89	Daisy *Maxwell 66-85*	I Pad	British Liverpool	82 285	150·2	25·3	10·3	Glasgow T. B. Seath & Co.	1879		2Cy. 36"–66" 99 HP. D. Rowan, Glasgow	Wallasey Loca[illegible]
26 321		Daiyu Maru	W ScSr	Japanese Okinawa	319 578	180·2	28·6	20·8	Glasgow T. Wingate & Co.	1869		71 HP.	Okinawa Kai[illegible]
27		Dakahlieh	I ScBn Spr Dk	Turkish Alexandria	1108	270·0	34·6	24·0	London Wigram & Sons	1865		2Cy. 38" & 70"–48" 300 HP. James Jack & Co., Liverpool	Khedevich Ma[illegible] Co.
28		Dalaröström *Alexandersson*	I Sc 1 Dk	Swedish Stockholm	67 102	106·3	18·0	8·0	Stockholm Bergsunds Mek. Verks	1876		2Cy. 13½" & 23½" 15½ & 19½" Bergsunds Mek. V., Stockholm. 30 H[illegible]	Anglaits Bol. Na[illegible] Dalarö (C.R[illegible]
29 76190	*	Dalbeattie *Hall 82-88*	I ScBn 1 Dk 2 B	British Middlesbro'	896 1393	239·9	33·2	17·6	Middlesbro' R. Dixon & Co.	1879	4 Blds. [illegible]	2Cy. 29" & 55"–33" 120 HP. T. Richardson & Sons, Hrtlpl.	"Dalbeattie" [illegible] Ld. (T. Harris & Co.[illegible]
30 65881	* D	Dale *Scoular 85-85*	I ScS 1 Dk 2 B	British Leith	645 998	221·5	30·1	17·0	Sunderland W. Pile & Co.	1871	5 Blds. [illegible]	2Cy. 27" & 50"–30" 100 HP. R. W. Hawthorn, Newcastle	London & Edi[illegible] Shipping Co., Ld.

***=Lloyd's Register;** D=Nederlandsche Vereeniging; G=Germanischer Lloyd; I=Registro Italiano; R=American [illegible]
A=Veritas Austro-Ungarico; H=Veritas Ellenico; N=Norske Veritas; V=Bureau Veritas

In the 2nd Column, B.T., with figures following, signifies Date of report of bottom sighting by Board of Trade Surveyors as condition of PASSENGER CERTIFIC[illegible]

In the 3rd Column, C=Composite (iron frame, wood planked); I=Iron; Stl=Steel; W=Wood.

Further particulars of Vessels classed by Lloyd's Register will be found in Lloyd's Register Book and Lloyd's Yacht Register.

No.	Class.	Nom, Matériaux, Gréement, etc. — Capitaine et dates d'entrée au service des Armateurs actuels et à bord du navire. — Ancien Nom, nombre de ponts, etc.	Pavillon. — Port d'Armement	Tonnage Net. — Total	Dimensions Long.	Larg.	Creux	Construit. Où — Par qui	Date	Nombre des cloisons. — Détails sur le lest d'eau.	Machines. — Détails. Nom du Constructeur.	Armateurs ou Directeurs-Gérants.
31		Dalhousie StlScSr	British	 1824	239·5	36·0	27·3	Greenock Caird&Co.	1886		3C.27½"42"&66"–45" 300IP. Caird&Co.,Greenock	Government of India
32	※	Dalhousie I ScSr Ross 87-87 1 Dk	British Dundee	25 [illegible]	90·0	20·1	9·1	Dundee W.B.Thompson	1886	3 Blkds.	3C.10½"15"&27"–16" 38IP. W.B.Thompson,Dundee	W.B.Thompson
33	※	Dalmazia I ScSr Mihanovich 1 Dk	Arg'ntine Buenos Ayres	238 [illegible]	141·3 Q 48 f.	29·1 B 28 f.	9·7 F 19 f.	Dumbarton Murray Bros.	1884	3 Blkds. FPT38t	2Cy.20"&40"–27" 70 IP. Muir&Houston,Glasgow	N.Mihanovich & Co.
34	※V	Dalmazia I ScSr Bark	Austrian Fiume	178 277	125·0 P 35 f.	21·2 Q 23 f.	10·4 F 7 f.	Sunderland J.Knox&Co.	1886	4 Blkds. FPT35t APT12t	2Cy.16"&30"–20" 40 IP. J.Clark,Sunderland	M.Sverljuga & Co.
35	※	Dalriada I ScSr McKeegan 82-85 1 Dk	British Glasgow	128 242	110·0 Q 15 f.	23·0 F 21 f.	10·5	Paisley J.Fullerton&Co.	1881	3 Blkds. FPT20t APT8t	2Cy.20"&36"–21" 56 IP. W.King&Co.,Glasgow	H.H.Smiley
36		Dalton I ScSr Russell 79-81	British Liverpool	1332 2030	315·4	34·9	24·9	Newcastle A.Leslie&Co.	1881		2Cy.26"&60"–45" 200 IP. R.Stephenson&Co.,Nwcstle	Liverpool, Brazil & River Pl.Stm.Nav.Co.,Ld.
37	※D	Dalton I ScSr Steel 79-79 1 Dk2 B	British Newcastle	840 1325	235·0 Q 80 f.	32·5 F 34 f.	17·9	Stockton M.Pearse&Co.	1877	4 Blkds. DBa63ft f80t	2Cy.30"&55½"–36" 120IP. Blair&Co.Ld.,Stockton	Pyman, Bell & Co.
38		Damanhour I ScBkSr (ex Bena)	Turkish Alexandria	627 [illegible]	225·0	30·0 P —	17·0	Glasgow Smith&Rodger	1857		2Cy.26"&52"–36" 120 IP. Day,Summers&Co,Sthamptn	Khedevieh Mail S.S. Co.
39	※	Damara I ScSr Dixon 87-87 1 Dk & Spr Dk 3 B	British W.Hartlep'l	1145 1779	275·0 B 84 f.	35·0	15·4 23·0	Glasgow A.Stephen&Sons	1885	6 Blkds. DBa411ft MT79t	2Cy.33"&62"–42" 200 IP. A.Stephen&Sons,Glasgow	C.Furness
40	※	Damascus Stl ScBtnSr Douglas 87-87 2 Dks & Spr Dk & W b Fms	British Aberdeen	2327 3609	362·0 B 98 f.	44·3 F 45 f.	23·6 81·6	Glasgow R.Napier&Sons	1887	9 Blkds. DBa270t APT60t	3C.28"42"&70"–54" 400IP. R.Napier&Sons,Glasgow	G.Thompson & Co.
41	V	Dan I ScBtnSr Christensen 82-87 1 Dk	Danish Copenhagen	986 1566	244·1 P 16 f.	34·1 F 25 f.	17·1	Renfrew Lobnitz&Co.	1882	WT58t APT30t	2Cy.32"&60"–36" 160 IP. Lobnitz&Co.,Renfrew	Dampskibs-Selskabet "Carl" (L.H.Carl)
42	V	Dana I ScSr Johansen 1 Dk & Spr Dk	Danish Elsinore	648 1021	217·5 F 16 f.	30·5	15·0 21·7	Malmö Kockums Mek.Wrk.	1875	5 Blkds. MT230t	2Cy.25"&50"–30" 100 IP. Kockums Mek.Wrks.,Malmö	Det Helsingørske Dampskibs-Selsk.(H.C.Nyholm)
43	※N	Dana I ScSr Michelsen 83-83 1 Dk2 B	Norweg'n Bergen	318 526	170·0 Q 46 f.	24·2 L 50 f.	12·9 F 28 f.	Bergen Bergens Mek Verks	1882	5 Blkds. APT12t MT94t	2Cy.20½"&39"–24" 55 IP. Bergens Mek Verks,Bergen	M.A.Schjelderup
44		Dana I ScSr Paulsen 36-83 1 Dk	Swedish Stockholm	318 405	162·4 P 94 f.	23·2 B 35 f.	10·4 H 40 f.	Stockholm W.Lindbergs Mk.Vks.	1858		2Cy.20½"&41"–24" 80 IP. W.Lindbergs Mek.Vks.,Stkhlm	Ångf.Bolaget Södra Sverige(J.G.Setterwall)
45		Danae I ScSr Brelich 72-87 2 Dks	Austrian Trieste	861 1356	249·6	31·5	21·8	Trieste Lloyd Austro-Ungarico	1879		2Cy.31"&56"–36" 126 IP. Arsenale Lloyd,Trieste	Lloyd Austro-Unga- rico
46	※	Danakil (ex Rose ex Amy) I ScSrYt	French Marseilles	71 137	144·4	19·3	9·5	Renfrew Henderson,Coulborn & Co.	1874	4 Blkds.	2Cy.21"&36"–24" 60 IP. Henderson,Coulborn&Co., Renfrew	Comte Pierre Pastré
47		Dandie Dinmont I PadSp King '88	British Glasgow	70 278	197·2	22·1	6·9	Glasgow A.&J.Inglis	1866		2Cy.40"–60" 110 IP. A.&J.Inglis,Glasgow	Southsea, Ventnor, Sandown, & Shanklin Steamboat Co.,Ld.
48		Dandy I PadSp	British Scarb'ro'	34 126	105·9	19·1	9·9	Newcastle Marshall Bros.	1863		Cy.40"–48" 60 IP. Rennoldson&Co.,Shields	J.W.Woodall
49	※	Danehill I ScSr Sheriff 75-87 1 Dk	British Newcastle	721 1124	220·5 Q&B 126 f.	35·1 F 57 f.	13·7	Newcastle Campbell, Mackintosh & Bowstead	1884	4 Blkds. DBa&rf410t FPT11t APT30t	2Cy.26"&52"–30" 125 IP. Wilsns Slipway Co.,Ld.,Nwc.	Danehill S.S.Co. (H.Andrews)
50	※	Dania I ScSr Johnson 83-83 1 Dk2 B	British W.Hartlepl.	607 958	220·0	29·0 Q 65 f.	16·3	W.Hartlepool Denton,Gray&Co.	1871	4 Blkds. WBa&f	2Cy.26½"&50"–33" 99 IP. Blair&Co.,Ld.,Stockton	Herskind & Woods
51	※	Daniel StlScSr Tulluck 88-88 1 W B Dk	British Belfast	1388 2125	280·0 P 29 f. Q 70 f.	38·2 B 126 f.	18·4 F 28 f.	Sunderland W.Doxford&Sons	1888	4 Blkds. CDB538t APT90t	3C.21"35"&57"–39" 180IP. W.Doxford&Sons,Sndrlnd.	O. Jaffe (Carlisle&Co.,Managers)
52		Daniel Webster W Pad	American Boston,Mass.	214 429	131·0	27·0	10·0	Boston, Mass.	1853			Mayor of Boston
53	※	Danish Monarch I ScSr Payne 87-87 1 Dk2 B	British London	867 1358	250·8 Q 124 f.	33·2 F 30 f.	18·5	Newcastle Wghm.Richardson &Co.	1878	4 Blkds. DBa 327 MT230t APT	2Cy.30"&56"–36" 150 IP. J.Shaw&Co.,Newcastle	J.Wood & Co.
54	※	Danish Prince I ScSr Jones 84-84 1 Dk2 B	British Newcastle	1044 1618	259·0 Q 99 f.	36·1 B 64 f.	17·5 F 28 f.	South Shields J.Readhead&Co.	1884	4 Blkds. DBa&f316t	2Cy.32"&60"–36" 151 IP. J.Readhead&Co.,S.Shields	Prince S.S.Co., Ld. (J.Knott, Manager)
55	※	Dankali I Sr 1 Dk	French Havre	49 72	77·1	16·0	8·5	Swansea Bainbridge&Co.	1884	5 Blkds.	2Cy.15"&30"–20" 40 IP. Bainbridge&Bckwth,Swans	J.Mesnier
56	※V	Danmark (ex Jan Breydel) I ScBkn Knudsen 88-88 3 Dks & Awn Dk	Danish Copenhagen	2262 3507	340·0	40·0	26·0 33·3	Newcastle C.Mitchell&Co.	1880	5 Blkds. DB558/117t MT30/125t	2Cy.44"&84"–48" 400 IP. T.Clark&Co.,Newcastle	Dampskibs-Selskabet Thingvalla
57	※	Danmark I ScBtnSr Holm 83-84 1 Dk	Danish Copenhagen	514 826	203·9 Q 55 f.	28·4 F 31 f.	14·8 B 72 f.	Renfrew Henderson, Coulborn &Co.	1867	5 Blkds. MT130t	4Cy.21"&42"–36" 100 IP. Hndrsn,Coulborn&Co,Rnfw	Dampskibs-Selskabet "Danmark"
58	V	Danmark I ScSr Zachau 87-87 1 Dk	Swedish Gothenburg	368 490	149·6 B 35 f.	23·8 H 4 f.	15·5	Gothenburg Eriksbergs Mek.Verkst	1883		2Cy.16"&31"–19" 50 IP. Eriksbergs Mek.Verkst.	A.Meyer
59	※	Danos (FOUNDERED) I ScSr Moulton 83-85 1 Dk2 B	British London	1390 2003	275·0 Q 76 f. P 28 f.	37·5 B 88 f.	19·5 F 29 f.	Sunderland J.L.Thompson&Sons	1882	6 Blkds. DBa&f 270t FPT36t APT75t	2Cy.34"&64"–42" 200 IP. J.Dickinson,Sunderland	R.Gordon & Co.
60	※	Dante I ScSr Mollard 74-77 2 Dks 3 B	British London	1302 1993	276·0 P 41 f.	34·2 F 34 f.	22·6	South Shields J.Readhead&Co.	1877	4 Blkds. DBa32ft f36ft	2Cy.29"&58"–36" 140 IP. J.Readhead&Co.,S.Shields	R.McAndrew & Co.

※ Lloyd's Register; D = Nederlandsche Vereeniging; G = Germanischer Lloyd; I = Registro Italiano; R = American Record; ※ = Veritas Austro-Ungarico; H = Veritas Ellenico; N = Norske Veritas; V = Bureau Veritas.

Dans la 2me Colonne, B.T., avec les indications qui suivent, signifie la date du rapport de visite de la Carène par les experts du Board of Trade pour obtenir le certificat relatif au TRANSPORT DES PASSAGERS.

Dans la 3me Colonne, C = Composite (membrure en fer, bordé en bois); I = Fer; Stl = Acier; W = Bois.

Pour plus amples renseignements sur les navires classés par le Lloyd's Register, voir le Lloyd's Register Book et le Lloyd's Yacht Register.

No. in Book / Official No.	Registry in which classed	Ship's Name, Material, Rig, &c. / Master, and Dates of Appointment to present Owner's Service and to Ship / Ship's late Name. No. of Decks, &c.	Flag / Port of Registry	Tons Net / Gross	Dimensions Length	Breadth	Depth	Built Where / By Whom	When	No. of Bulkheads / Particulars of Water Ballast	Engines Particulars, Maker's Name	Owners or Managers
61 54753		Danube (BROKEN UP) I ScBn *Hepworthy* *2 Dks & Awn Dk*	British Southampton	1462 2038	332·0	34·4	21·4	London Millwall Iron Wks.	1866		2Cy.36" & 72"–30" 300 HP. Day & Co., Southampton	I. Cohen
62 62279		Danube I ScBg *Hill* 85-88	British Glasgow	561 887	215·6	27·7	20·7	Glasgow John Elder	1869		2Cy.26" & 50"–36" 160 HP. J. Elder & Co., Glasgow	Scottish Oriental Co., Ld.
63 91238	✻ BT 7-88	Danube Stl ScStg *1 Dk*	British London	7 23	80·0	18·1	9·7	Port Glasgow D. J. Dunlop & Co.	1886	1 Bhds.	3C.15", 23" & 38"–24" 70 HP. D. J. Dunlop & Co., P. Glasgow	London & Tilbury Lighterage Co., Ld.
64 28		Danubio I ScSr *Bellen* 73-87 *2 Dks*	Austrian Trieste	547 849	198·0	30·5	18·2	Stettin "Vulcan" Works	1866		1C.28½" & 52"–28½" 150 HP. "Vulcan" Works, Stettin	Lloyd Austro-Ungarico
65 137088	R	Danville W ScSr *Billup* 81-86	American Baltimore	826 1234	210·0	36·0	21·6	Baltimore Woodall & Co.	1882	Y [illegible]	2Cy.30" & 50"–36" [illegible] Reeder & Sons, Baltimore	Baltimore, Chesapeake & Richmond S.S. Co.
66	V	Danzig (SUNK) I ScSr *Uckerbandt* 83-83 *1 Dk*	German Danzig	467 741	196·9	27·6	15·7	Stettin Act. Ges. "Vulcan"	1880	W B	2Cy.25" & 43½"–27½" 80 HP. Act. Ges. "Vulcan," Stettin	Danziger Schiffahrts Akt.-Ges. (A.G.)
67	G	Danzig Stl ScBg *Krahoffer* 86-86 *2 Dks*	German Bremen	1078 1815	260·0 *P 58 f.*	35·5 *B 95 f.*	24·5 *F 39 f.*	Stettin Act. Ges. "Vulcan"	1886	CDB 220 FPT 180 APT 131	3Cy.25½", 39" & 63"–49" Act. Ges. "Vulcan"	Norddeutscher Lloyd
68		Daoiz (ex Dante) I Sc3mSr *1 Dk & Spr Dk*	Spanish Barcelona	713 931	210·2	28·2	20·8	Liverpool Bowdler, Chaffer & Co.	1873		2Cy.23" & 46"–24" 85 HP. J. Taylor & Co., Birkenhead	J. Roca & Co.
69 72		Daphne I ScBk *Radonicich* 6-81 *2 Dks*	Austrian Trieste	1264 1927	289·5	35·2	25·4	Trieste Lloyd Austro-Ungarico	1881		2Cy.35¼" & 61½"–48" 172 HP. Arsenale Lloyd, Trieste	Lloyd Austro-Ungarico
70	✻V	Daphne Stl Sc *Voss* 84-87 *2 Dks 3 B*	German Hamburg	1395 1966	289·5 *P 52 f.*	36·6 *B 72 f.*	21·9 *F 44 f.*	Flensburg Flensburger S.B. Co.	1887	7 Bhds. CDB 320	3C.22½", 35½" & 54"–42" 320 HP. Flensburger S.G.-Flabrs	Deutsche Dampfschiffs Rhederei
71 93671	BT 7-88	Dardanus I ScSr *Purdy* 80-86	British Liverpool	1536 2374	321·0	36·5	26·0	Newcastle R. & W. Hwthn. Lslie & Co.	1886		2Cy.27" & 58"–60" 400 HP. R. Stephenson & Co., Newcstl.	Ocean Steam S.S. Co.
72 73128	✻	Darent I ScSr *Page* 7-87 *1 Dk 2 B*	British Whitby	636 1008	225·0 *Q 83 f.*	30·1 *B 48 f.*	17·0 *F 28 f.*	Whitby T. Turnbull & Son	1875	4 Bhds. DB 93 f. 80	2Cy.27" & 50½"–33" 99 HP. Blair & Co., Ld., Stockton	T. Turnbull & Son
73 87164	✻	Darenth I Sc3mSr *Davies* 83-84 *1 Dk*	British London	566 893	215·7 *Q 57 f.*	33·2 *F 27 f.*	13·9	Newcastle Palmers Co., Ld.	1883	4 Bhds. DB 291t FP 130	2Cy.27" & 51"–33" 99 HP. Palmers Co., Ld., Newcastle	J. Fenwick & Son
74 93821	BT 10-88 ✻	Darien Stl ScSr *Watson* 62-88 *3 Dks*	British Liverpool	2216 3362	370·0 *P 42 f.*	41·2 *B 126 f.*	27·0 *F 47 f.*	Glasgow Barclay, Curle & Co., Ld.	1888	7 Bhds. DB [illegible] FPT 148 APT 61	3Cy.28", 44" & 71"–51" Barclay, Curle & Co., Ld. 387 HP.	West India & Pacific Steamship Co. Ld.
75 82896	✻	Darien (MISSING) I Sc3mSr *McKie* 83-87 *3 Dks*	British London	1780 2743	316·6 *P 40 f.*	39·3 *B 72 f.*	26·7 *F 48 f.*	Newcastle Palmers Co., Ld.	1881	5 Bhds. DB [illegible] MT [illegible] APT [illegible]	2Cy.38" & 70"–48" 300 HP. Palmers Co., Ld., Newcastle	Nelson, Donkin & Co.
76 73743		Daring I PadSr *Nicholson*	British London	2 135	106·9	19·1	10·0	South Shields J. Readhead & Co.	1876		2Cy.28"–51" 70 HP. J. Readhead & Co., S. Shields	R. Redhead
77 27210		Darro I ScSr *Richardson* 82-87	British Barrow	145 305	156·2	21·2 *Q 42 f.*	12·3	Glasgow W. Simons & Co.	1859		2Cy.20" & 38"–24" 55 HP. Hall, Russell & Co., Aberdeen	J. Mawson
78 94205	✻	Dartmeet Stl ScSr *Adams* -88 *1 Dk*	British Dartmouth	327 653	183·7 *Q 66 f.*	28·2 *B 37 f.*	11·7 *F 27 f.*	Hayle Harvey & Co., Ld.	1888	5 Bhds. CDB 160 FPT 30 APT 21	3Cy.18", 29" & 48"–30" Harvey & Co., Ld., Hayle 115 HP.	Whiteway & Ball
79 86615	✻	Dartmoor I ScSr *Jaggs* 80-82 *1 Dk*	British Ipswich	154 249	135·2 *Q 32 f.*	20·6 *F 16 f.*	10·0	London Edwards & Symes	1882	3 Bhds.	2Cy.17" & 32"–22" 40 HP. M. Pratt, Huddersfield	Ipswich S.S. Co. (R. Packard, jun. & Co.)
80 81147	✻D	Dartmore I ScBn *Watson* 85-86 *1 Dk 2 B*	British Liverpool	1039 1609	260·0 *B 160 f., B 21 f., Q 31 f.*	36·0	18·2 *F 39 f.*	Stockton Richardson, Duck & Co.	1881	4 Bhds. W B 64 DB TF FPT APT	2Cy.32" & 60"–39" 160 HP. Blair & Co., Ld., Stockton	S.S. "Dartmore" Co. (W. Johnston & Co.)
81 90880		Dartmouth W Pad *1 Dk*	British Yarmouth, N.S.	196 311	136·0	28·4	10·7	Yarmouth N.S. Burrell Johnson Iron Co. Ld.	1888		2Cy.22" & 42"–60" 75 HP. Burrell Johnson Iron Co., Ld.	Halifax & Dartmouth Steam Ferry Co.
82		Darvin 2m *Berne* 79-82	Russian St. Petersburg	673	..	..	..	Gothenburg Motala Co.	1882		120 HP.	Naphta Production Gesellschaft
83 89056	✻D	Darwin I ScSr *Williams* 87-87 *2 Dks 3 B*	British London	1638 2506	317·0 *P 32 f.*	37·5 *B 71 f.*	23·8 *F 35 f.*	Newcastle Swan & Hunter	1884	5 Bhds. DB [illegible] APT [illegible]	2Cy.37" & 70"–45" 250 HP. R. & W. Hawthorn, Newcastle	"Darwin" S.S. Co. (A. Holland & Co.)
84 2		Dauno I ScSr *Roca* 87-88 *1 Dk*	Italian Bari	260 467	164·0	21·6 *F 23 f.*	12·8	Paisley McIntyre & Co.	1878	4 Bhds. DB 60 f	2Cy.20" & 38"–30" 73 HP. Muir & Houston, Glasgow	Soc. Nav. Puglia
85 89408	✻	Dauntless I ScSr *Wilkes* *1 Dk 2 B*	British W. Hartlepl.	1124 1740	258·0 *Q [illegible] f.*	35·8 *B 67 f.*	19·0 *F 35 f.*	W. Hartlepool W. Gray & Co.	1884	4 Bhds. DB [illegible]	2Cy.32" & 60"–39" 150 HP. Black, Hawthorn & Co., Gtesd.	G. B. Harland & Co.
86 86086	✻D	Dauntless I ScSr *Sparling* 80-88 *1 Dk 2 B*	British Newcastle	1098 1685	269·0 *Q 104 f.*	36·6 *B 56 f.*	17·5 *F 26 f.*	North Shields T. & W. Smith	1882	4 Bhds. DB [illegible] FPT [illegible] APT [illegible]	2Cy.32" & 60"–40" 200 HP. Hawks, Crawshay & Co., Gtsd.	Hall Bros.
87 90610		Dauntless W PadTug *1 Dk*	British Ottawa	216 324	149·6	12·0	7·9	Pembroke, Ont. P. Lamothe	1885		1Cy.32"–96" 80 HP. J. Brush, Montreal	Upper Ottawa Towing Co., Ld.
88 89904	BT 1-89	Dauntless I PadSp	British Waterford	15 136	108·5	18·9	9·8	South Shields J. T. Eltringham	1881		2Cy.28"–53" 80 HP. J. P. Rennoldson & Sons, S. Shields	Waterford S.S. Co.
89 29664	✻	Dauntless I ScSr *Paulon* *1 Dk*	British Hayle	14 120	107·0	19·1	10·2	Hayle Harvey & Co.	1884	6 Bhds.	2Cy.21" & 39"–21" 120 HP. Harvey & Co., Ld., Hayle	Wilson, Sons & Co.
93 89630	✻	Dauntless Stl ScTug *Bulbery* 85-85 *1 Dk*	British London	7 37	56·0	13·1	7·2	London Watson & Co.	1884	1 Bhds.	2Cy.12" & 22"–18" 30 HP. Deittiter, London	Ross & Deering

✻ = Lloyd's Register; D = Nederlandsche Vereeniging; G = Germanischer Lloyd; I = Registro Italiano; R = American Record; A = Veritas Austro-Ungarico; H = Veritas Ellenico; N = Norske Veritas; V = Bureau Veritas.

In the 2nd Column, B.T., with figures following, signifies Date of report of bottom sighting by Board of Trade Surveyors as condition of PASSENGER CERTIFICATE.

In the 3rd Column, C = Composite (iron frame, wood planked); I = Iron; Stl = Steel; W = Wood.

Further particulars of Vessels classed by Lloyd's Register will be found in Lloyd's Register Book and Lloyd's Yacht Register.

Marques	Nom, Matériaux, Gréement, etc. Capitaine et dates d'entrée au service des Armateurs actuels et à bord du navire. Ancien Nom, nombre de ponts, etc.		Pavillon. Port d'Armement	Tonnage. Net / Total	Dimensions. Long.	Larg.	Creux	Construit. Où. Par qui.	Quand.	Nombre des cloisons. Détails sur le lest d'eau.	Machines. Details. Nom du Constructeur.	Armateurs ou Directeurs-Gérants.
	Dauphiné Marchisi 85-88	I Sc 2 Dks	French Marseilles	685 1029	238·1	28·9	21·3	La Seyne Forges&Ch de la Médit.	1866		2Cy.33"–24½ 100 HP. Forges&C.de la Medit.,Mrslls	Société Générale de Transports Maritimes
✻ Diss.…	Davaar Kerr 85-85	Stl ScSr 1 Dk	British Campbeltn.	97 516	217·8	27·0	12·9 Q 56 f. B 63 f. F 48 f.	Glasgow Lond.&Glas.Co.Ld.	1885	1 Blds. DTa… DT f 47t	2Cy.29"&58"–42" 155 HP. London&GlsgowCo.Ld.,Gls.	Campbeltown & Glasgow StmPktJntStockCo.
	David G. Thomson	W Sc 1 Dk	British Montreal	75 182	103·3	17·3	9·8 H 47 f.	Kingston, Ont. R.Robertson	1883		2Cy.17"–24" 200 HP. J.Weir,Montreal	Montreal Transportation Co.
✻	Dawdon Robson 77-80	I ScSr 1 Dk 2 B	British Sunderland	891 1361	240·4	34·0	18·0 Q 94 f. B 48 f. F 26 f.	Sunderland Short Bros.	1878	5 Blds. DBaf 208t APT 46t	2Cy.29½"&55"–36" 130 HP. G.Clark,Sunderland	R.Thorman
	Dawn Windale	I ScSr 1 Dk & Awn Dk	British Prtlnd,Vict	328 522	154·6	24·0	12·5 H 12 f. (18·4)	Sunderland W.Pile&Co.	1876	1 Blds. …t	2Cy.20"&40"–27" 75 HP. Rankin&Blackmore,Grnck	Portland & Belfast Steam Nav.Co.,Ld.
✻	Daylesford McInarn…	I ScSr 1 Dk 2 B	British Glasgow	913 1419	240·7	34·1	17·0 Q 66 f P 23 f B 52 f F 28 f	Sunderland R.Thompson&Sons	1882	1 Blds. DB… APT 60	2Cy.30"&58"–39" 140 HP. T.Clark&Co.,Newcastle	Burrell & Son
✻	De Bay Hall	I ScSr 2 Dks 3 B	British Belfast	1083 1684	250·5	35·6	21·5 P 28 f. B 62 f. F 32 f.	Newcastle Palmers'Co.,Ld.	1881	5 Blds. CDB 295t	2Cy.30"&55"–42" 160 HP. Palmers'Co.,Ld.,Newcastle	"De Bay" S.S. Co., Ld. (Woodside & Workman.)
V	De Ruyter ex Tromp Arfsten 88-88	I ScBk 2 Dks 3 B	Belgian Antwerp	1618 2304	316·9	35·4	27·0 H 83 f.	Glasgow A.Stephen&Sons	1873		3C.21"38"&66"–42" 250 HP. Tripled Wall. Slip & E.C., Ld. Newcastle.	T.C.Engels & Co.
✻D	Deak Innes 88-88	I ScSr 1 Dk & Awn Dk 3 B	British Glasgow	992 1549	240·0	34·1	22·6	Dumbarton Burrell&Son	1883	5 Blds. CDB 44?t	2Cy.26"&51"–36" 110 HP. M.Paul&Co.,Dumbarton	Burrell & Son
✻	Dean (now named Akashi) Peacock -88	Stl ScSr 1 Dk 2 B	Japanese Tokio	856 1340	244·0	34·2	17·0 P 28 f. F 86 f. Q 109 f. B 69 f.	Dundee Gourlay Bros.&Co.	1887	5 Blds. DBaf 187t FPT 29t APT 9t MT 150t	3C.20"31"&50"–36" 150 HP. Gourlay Bros.&Co.,Dundee	Iwasaki Hisaya
R	Decatur H. Miller Howes	I ScSr 4 Dks	American Baltimore	1674 2296	258·0	38·0	16·0 (30·0)	Wilmingtn Del Harlan & Hollingsworth	1879		4Cy.24"&54"–48" 300 HP. Harlan&Hollingsworth	Merchants' & Miners' Transportation Co.
	Deccan Case 82-86	I ScBk	British Glasgow	2022 3429	368·3	42·5	30·3 P 104 f. F 54 f.	Dumbarton W.Denny&Bros.	1868		2Cy.56"&97"–54" 550 HP. R.&W.Hawthorn,Newcastle	Peninsular & Oriental Steam Nav.Co.
	Decempedes (ex Sailing Vessel)	Stl ScFlat	British Liverpool	141 169	90·0	21·1	10·5	Winsford H.E.Falk	1879		2Cy.10½"–14" 7 HP. W.E.Bates,Northwich	Salt Union, Ld.
V	Decima Breitung -88	I ScSr 1 Dk & Spr Dk	German Flensburg	1151 1497	245·0	32·5	17·7 (24·7)	Flensburg Flensburg Schiffsb. Ges.	1878	WB	2Cy.30"&57"–33" 140 HP. FlensburgS.B.Co.Flensburg	Flensburger Dampfs. Ges. von 1869 (C.N.Hansen)
	Dee (ex State of Minnesota, ex Minnesota)	I ScBk	British Glasgow	1196 1864	300·5	35·3	24·6	Glasgow T.Wingate&Co.	1872		3Cy.24"39½"&64"–39" 250 HP. Earles Co.,Ld. Hull	Royal Mail Steam Packet Co.
✻	Dee Smith 84-91	I ScSr 1 Dk	British Aberdeen	305 472	160·3	23·7	12·3 Q 43 f. B 45 f. F 19 f.	Aberdeen Hall,Russell&Co.	1883	4 Blds. DBaf 36f 60t	2Cy.20"&39"–27" 70 HP. Hall,Russell&Co.,Aberdeen	G.&W.Davidson
✻D	Deepdale Sharp 83-83	I ScSr 1 Dk & Awn Dk 3 B	British Sunderland	1715 2683	310·8	38·1	29·4	Sunderland J.Laing	1880	5 Blds. DBaf 243t	2Cy.37"&71"–45" 280 HP. G.Clark,Sunderland	P.H.Laing
✻	Deer Hill Jenkins 86-86	I ScBn 1 Dk 2 B	British W.Hartlepl	1118 1733	259·6	34·7	19·5 Q 100 f. B 67 f. F 35 f.	W.Hartlepool W.Gray&Co.	1882	4 Blds. DBaf 331t	2Cy.32"&60"–39" 150 HP. Blair&Co.,Ld.,Stockton	J.H.Murrell & Co.
✻	Deerhound Atkinson 86-86	I ScBn 1 Dk 2 B	British Hartlepool	1151 1783	257·9	35·7	19·0 Q 68 f P 28 f B 83 f F 32 f	W.Hartlepool W.Gray&Co.	1882	4 Blds. DBaf 316t	2Cy.32"&60"–39" 150 HP. Blair&Co.,Ld.,Stockton	W.Gray
✻	Deerhound Smith	I ScSr 1 Dk	British London	271 443	159·6	24·0	12·2 Q 73 f. B 15 f. F 23 f.	London Forrestt&Son	1882	3 Blds. FPT 16t APT 22t	2Cy.20"&38"–30" 70 HP. R.H.Pearson&Co.,Glasgow	Walker, Howard & Co.
✻	Deeside Spence 85-85	I ScSr 1 Dk	British Aberdeen	362 569	171·5	26·8	12·9 Q 52 f. B 44 f. F 20 f.	Aberdeen A.Hall&Co.	1883	5 Blds. DBaf 116t	2Cy.24"&42"–30" 75 HP. Blaikie Bros.,Aberdeen	"Deeside" S.S.Co.,Ld. (Mennie & Brown)
	Defiance Hand -88	I PadBg 1 Dk	British Calcutta	223 408	150·6	25·2	14·6	Dumbarton W.Denny&Bros.	1863		2Cy.54"–54" 250 HP. Denny&Co.,Dumbarton	Clive Steam Tug Co. (Gladstone,Wyllie & Co.)
	Defiance	I PadSr 1 Dk	British Moulmein	114 199	129·1	21·3	10·6	Liverpool J.Laird Sons&Co.	1862		2Cy. 90 HP. J.Laird,Sons&Co.,Birkenhd.	Burmah Steam Tug Co.,Ld.
	Defiance	I Pad	British London	27 114	91·0	19·0	9·0	Preston W.Allsup&Sons	1883		2Cy.30"–52" 60 HP. W.Allsup&Sons,Preston	E.&W.India Dock Co.
	Defiance	I PadSk	British Greenock	24 115	101·0	18·4	9·0	North Shields Hepple&Co.	1884		1Cy.36"–54" 80 HP. Hepple&Co.,North Shields	W.Park
 8T…	Defiance (ex Charley)	I PadSp	British Hull	10 106	87·8	17·0	10·5	Neath Abbey	1862		2Cy 28"–48" 60 HP.	T.Wilson, Sons & Co.
	Degrave Degrave	W ScDy 1 Dk	British Granton	46 113	104·0	20·2	9·4	Granton D.Allan&Co.	1885		2Cy.14"&28"–21" 35 HP. Hutson&Corbett,Glasgow	G.Hutson
	Del Norte Waleig -88	W ScSr 1 Dk	American S.Francisco	227 304	139·0	32·0	10·7	San Francisco G.Bodie	1888		2Cy.14"&28"–24" 125 HP. Fulton Iron W'rks,S.Frncsco	J.K.C.Hobbs
 8T…	Delambre Barton 87-88	I ScSr 2 Dks 3 B	British Liverpool	988 1308	233·7	31·2	24·4	Port Glasgow W.Hamilton&Co.	1873	3 Blds.	2Cy.28"&54"–33" 125 HP. J.&T.Young,Ayr	Liverpl.,Brazil&River Plate Steam Nav.Co.,Ld.
R	Delaware Tribou -87	W ScBn 2 Dks 3 B	American Philadelphia	1279 1846	251·8	37·0	17·8	Philadelphia Birely,Hill&Streaker	1880		2Cy.30"&50"–36" 600 HP.	W.P.Clyde & Co.

…Lloyd's Register; D = Nederlandsche Vereeniging; G = Germanischer Lloyd; I = Registro Italiano; R = American Record;
…= Veritas Austro-Ungarico; H = Veritas Ellenico; N = Norske Veritas; V = Bureau Veritas.

Dans la 2me Colonne, B.T., avec les indications qui suivent, signifie la date du rapport de visite de la Carène par les experts du Board of Trade pour obtenir le certificat relatif au TRANSPORT DES PASSAGERS.

Dans la 3me Colonne, C = Composite (membrure en fer, bordé en bois); I = Fer; Stl = Acier; W = Bois.

Pour plus amples renseignements sur les navires classés par le Lloyd's Register, voir le Lloyd's Register Book et le Lloyd's Yacht Register.

No. in Book / Official No.	Class	Ship's Name. Material, Rig, &c. / Master, and Dates of Appointment / Ship's Late Name. No. of Decks, &c.	Flag / Port of Registry	Tons Net / Gross	Length	Breadth	Depth	Built Where / By Whom	When	No. of Bulkheads / Particulars of Water Ballast	Engines. Particulars / Maker's Name	Owners
121	G V	**Delbruck** I ScSmSr *Hamer* [illegible] 1 *Dk*	German Bremen	288 371	145·6	22·6	13·9	Bremen Act.Ges."Weser"	1873		2Cy.17"&29½"-22" 40 HP Act.Ges."Weser," Bremen	J.D.Bischoff
122 [illegible]	✻	**Delcomyn** I ScSr *Ekins* 87-87 2 *Dks*3 *B*	British London	1184 1817	280·0	35·3	22·6	Newcastle Wigham Richardson &Co.	1880	5 Blhds. [illegible]	2Cy.31"&62"-48" 200 HP WighamRichardson&Co.,Newc	W.Lund
123		**Delfin** C ScSr *Andersson* 76-76 1 *Dk*	Swedish Norrkoping	108 163	97·8	20·8	9·1	Tosskog P.Larsson	1884		2Cy.11"&22"-18" 20 HP Motala&Co.	Angluitsaktie.. Skärgårdan..
124		**Delfin** (ex Morena) I ScSp *Rodehn* 88-88 1 *Dk*	Swedish Landskrona	80 114	81·7	17·8	8·0	Stockholm W.Lindbergs M.V.	1884		2Cy.11½"&19"-16" 25 HP W.Lindbergs M.V. Stkhlm.	A.C.Adelsk..
125		**Delfino** I ScSr *Vasiolio* 73-87 1 *Dk*	Austrian Trieste	500 752	198·0	27·0	18·7	Trieste Lloyd Austro-Ungarico	1874		2Cy.26"&36"-33" 60 HP Arsenale del Lloyd, Trieste	Lloyd Austr.. rico
126		**Delhi** I ScBg *Thomson* 87-87	British Glasgow	1293 2009	313·3	38·0	26·3	London M.Wigram&Sons	1864		2Cy.35"&60"-30" 200 HP Hutson&Corbett,Glasgow	Raeburn & V..
127 60226	✻	**Delos** I ScSr *Smith* 1 *Dk*2 *B*	British Hull	981 1501	258·5	31·1	19·7	Stockton Richardson Duck&Co.	1870	[illegible] Blhds.	2Cy.31"&60"-33" 140 HP C.D.Holmes&Co.,Hull	H.Briggs, So..
128 54506	✻ D	**Delta** (ex Michele Vaccaro) I Sc3mSr *Wood* 2 *Dks*	British Hull	788 1249	243·5	29·6	21·9	W.Hartlepool T.H.Pile&Co.	1865	[illegible]	2Cy.26"&50"-33" 99 HP R.&W.Hawthorn,Newcastle	London&Sou.. SteamShipCo..
129 1816	R	**Delta** I ScBk *Crowell* 75-	British Glasgow	552 831	208·6	27·0	15·8	Glasgow Barclay,Curle&Co.	1854		2Cy.20"&56"-24" 55 HP W.King&Co.,Glasgow	G.E.Franckly..
130 93083		**Delta** I TwnScSp	British Glasgow	182 280	135·6	28·1	7·9	Liverpool J.Jones&Sons	1886		2Cy.14"-30" 40 HP J.Jones&Sons,Liverpool	La Platense.. Co.,Ld.
131		**Delta** I ScBn 3 *Dks*	French Marseilles	802 1204	212·2	32·3	25·7	W.Hartlepool Pile,Spence&Co.	1862		2Cy.45¾"-27" 120 HP Liverpool	Messageries M..
132 89969	✻	**Denbighshire** I ScSr *Rickard* 85-87 2 *Dk*3 *B*	British London	1663 2538	317·0	38·0	24·5	Newcastle C.S.Swan&Hunter	1885	7 Blhds. [illegible] FPT228 APT23	2Cy.38"&72"-48" 300 HP R.&W.Hawthorn,Newcastle	Jenkins & Co.
133	✻ D	**Denderah** (ex Padang) I ScSr *Bruohr* 74-86 3 *Dks*	German Hamburg	1472 2285	300·0	37·0	25·7	Middlesbro' R.Dixon&Co.	1883	5 Blhds. [illegible]	2Cy.38"&70"-42" 250 HP R.&W.Hawthorn,Newcastle	Dampfschiff.. "Kosmos"
134 76234	✻	**Denham** I ScSr *Jewels* 86-86 1 *Dk*2 *B*	British Newcastle	561 869	210·0	30·2	16·0	Newcastle W.Richardson&Co.	1878	4 Blhds. DBcel [illegible] FPT APT	2Cy.27"&50"-33" 98 HP J.Shaw&Co.,Newcastle	J.H.Reah
135 54888	✻	**Denia** I ScSr *Jones* 87-87 1 *Dk*2 *B*	British Sunderland	469 733	190·6	27·2	15·8	Middlesbro' Candlish,Fox&Co.	1866	3 Blhds. [illegible]	2Cy.24"&44"-30" 85 HP Hall,Russell&Co.,Aberdeen	Fisher,Renwi..
136 50156	✻	**Denia** I ScK *Pybus* 1 *Dk*	British Middlesbro'	159 247	141·0	20·2	11·8	Greenock Scott&Co.	1866	3 Blhds.	2Cy.21⅜"&34½"-18" 40 HP Gourlay Bros.&Co.,Dundee	J.M.Lennard..
137 54843		**Denmark** I ScBkn *Rigby* 84-86 (ex Chilian) 1 *Dks*	British Liverpool	2326 3724	312·9	42·2	36·0	Stockton M.Pearse&Co.	1865		2Cy.28"&50¼"-42" 350 HP Corp.'83,J.Jones&Sons,Livi	National Stea.. Co.,Ld.
138 1085		**Denshiu Maru No.2** I TScSr	Japanese Osaka	205 331	128·8	19·7	15·9	Osaka Shinkawa Masajiro	1888		45 HP	Amasaki Isa.. Kaisha
139		**Der Kaiser** I PadSr	German Stettin	122 189	186·6	18·2	8·1		1865		2Cy. 80 HP	C.J.F.Braenli..
140		**Der Preusse** I ScSr *Bethmann* 82-83 1 *Dk*	German Stettin	221 290	144·0	19·8	11·5	London J.Scott Russell&Co.	18[illegible]		2Cy.20½"&35½"-25" 60 HP StettinMas.Act.Ges."Vulcan"	Neue Dampfer.. pagnie Akt.-Ge..
141		**Derjava** W Pad3mSrYt	Russian S.Petersburg		332·8	42·2	26·6	St. Petersburg Dimitrieff	1871		2Cy.96"&165"-96" 720 HP [illegible] Wks.,St.Petersburg	H.I.M.the E.. of Russia
142		**Dérocheuse** I ScSr 1 *Dk*	Egyptian Port Said	352 693	180·2	40·2	11·0	Renfrew Lobnitz&Co.	1887		4Cy.(2)14"&(2)28"-24" Lobnitz&Co.,Renfrew	Compagnie U.. du Canal Mari..
143 68658	✻ D	**Deronda** I ScSr *Andrew* 82-86 2 *Dks*3 *B*	British Sunderland	1090 1683	260·5	34·0	23·8	Sunderland J.L.Thompson&Sons	1877	4 Blhds. [illegible]	2Cy.33"&62"-39" 180 HP G.Clark,Sunderland	Culliford & ..
144 81693		**Derwent** I Sc3mSr *Hansler* 2 *Dks*3 *B*	British Sunderland	1558 2402	329·1	36·0	25·5	Sunderland R.Thompson,jun.	1879	5 Blhds. [illegible] FPT [illegible]	2Cy.42"&73"-48" 330 HP N.E.Marine E.Co.,Ld.,S.Shields	Royal Mail St.. Co.
145 66680	✻ D	**Derwent** I ScBg *Orchard* 87-87 1 *Dk*&*Spr Dk*3 *B*	British London	986 1539	239·3	32·7	17·3 217	Hull [illegible]	1873	5 Blhds. [illegible]	2Cy.30"&56"-36" 130 HP Cp.'83,C.D.Holmes&Co.,Hull	Mercantile S.S..
146 91316	✻ BT 8-88	**Derwent** StlScSr 1 *Dk*	British Goole	345 719	230·0	30·7	14·7	Newcastle W.Dobson&Co.	1888	[illegible] Blhds. [illegible]	3Cy.21"33½"&54"-33" [illegible] [1664]	Goole Steam .. Co.
147 68177	✻	**Derwent** I Sc3mSr *Hutton* 77-87 2 *Dks*	British Grangemouth	307 508	193·7	28·0	15·9	Glasgow Barclay,Curle&Co.	1874	[illegible] Blhds.	2Cy.31"&54"-36" 120 HP Barclay,Curle&Co.,Glasgow	Carron Co.
148 50005		**Derwent** I ScBk *Bristow* 1 *Dk*	British Sydney	351 478	196·0	25·3	13·3	Glasgow A.&J.Inglis	1864	[illegible] Blhds.	2Cy.38"-30" 120 HP A.&J.Inglis,Glasgow	W.H.Smith & .. Ld.
149 79124	✻	**Derwent** Stl&I Sc3mSr *Robinson* 1 *Dk*	British Workington	111 237	144·2	23·0	10·4	Workington R.Williamson&Son	1883	3 Blhds. [illegible]	2Cy.18"&34"-21" 49 HP J.T.Young&Co.,Ayr	Derwent S.S.. (Gibson & Gra..
150 82861	✻ D	**Derwentdale** I ScSr *Nicholson* 1 *Dk*2 *B*	British Scarboro'	1034 1603	259·5	36·0	17·5	South Shields J.Readhead&Co.	1883	4 Blhds. [illegible]	2Cy.30"&58"-36" 145 HP J.Readhead&Co.,S.Shields	P.Hick,jun.

✻ = **Lloyd's Register**; D = Nederlandsche Vereeniging; G = Germanischer Lloyd; I = Registro Italiano; R = American ..
A = Veritas Austro-Ungarico; H = Veritas Ellenico; N = Norske Veritas; V = Bureau V..

In the 2nd Column, B.T., with figures following, signifies Date of report of bottom sighting by Board of Trade Surveyors as condition of PASSENGER C..

In the 3rd Column, C = Composite (iron frame, wood planked); I = Iron; Stl = Steel; W = Wood.

Further particulars of Vessels classed by Lloyd's Register will be found in Lloyd's Register Book and Lloyd's Yacht Register.

Cote	Nom, Matériaux, Gréement, etc. Capitaine et dates d'entrée au service des Armateurs actuels et à bord du navire. Ancien Nom, nombre de ponts, etc.	Pavillon. Port d'Armement	Tonnage Net. Total.	Dimensions. Long. × Larg. × Creux constructions sur le Pont	Construit. Où. Par qui.	Quand.	Nombre des cloisons. Détails sur le lest d'eau.	Machines. Détails. Nom du Constructeur.	Armateurs ou Directeurs-Gérants.
✻	**Derwentwater** I ScSr *Prger 79-89* 2 *Dks3B*	British Newcastle	805 1253	230·0|32·2|21·7 *F 23 f.*	Newcastle C.S.Swan&Co.	1878	4 Bhds. DBaft	2Cy.29"&55"–36" 140 HP. Black,Hawthorn&Co,Gtshd	H.Strachan & Co.
✻V	**Desi** (exMarie,exPaul(Phoenix)) I ScSr *Klemke 88-88*	German Dantzig	149 252	125·5|21·0|10·5 *Q 15 f. B 6 f. F 13 f.*	Newcastle T.&W.Toward&Co.	1884	4 Bhds. FPT	2Cy.16"&30"–20" 40 HP. Blair&Co.,Ld.,Stockton	F.G.Reinhold
	Desierto (ex CorreraSemanero) I	Spanish Bilbao	288 438	168·6|24·6|13·1	London	1863			D.Blas Alzaga
	Desierto (ex Carlos Alzaa) I ScSr *Urgilles 88-88* 1 *Dk*	Spanish Bilbao	200	178·0|21·5|13·1	Hayle, Eng.			2Cy.20"&40"–28" 45 HP.	J.Martinez de las Rivas
	Désirade I ScBk (exCityofBrussels) *Corno -85* 2 *Dks*	French Marseilles	931 1440	268·0|32·7|20·6 *P 73 f. F 43 f.*	Glasgow Randolph,Elder&C.	1868		2Cy.32½"&71"–36" 250HP. Randolph,Elder&Co.,Glsgw	Compagnie Générale Transatlantique
✻	**Despatch** (COLLISION) I Sc3mSr *Mosey* 1 *Dk2B*	British London	463 734	200·5|28·1|16·9 *Q 55 f.*	Newcastle PalmerBros.&Co.	1863	3 Bhds. DBaft	2Cy.25"&48"–30" 90 HP. Cp.75,Vict.Dk.E.Wrks,Lond	H.James (J.Fenwick & Son, Managers)
	Despatch (FOUNDERED) I Sc3mSr *Munro* 1 *Dk2B*	British Glasgow	340 541	217·1|25·1|14·6 *P 61 f. F 27 f.*	Shields T.W.Smith&Co.	1855	4 Bhds.	2Cy.25"&47"–30" 90 HP. Muir&Houston,Glasgow	J.Clark
	Despatch I ScSr *Webb 87-87* 1 *Dk*	British Melbourne	114 237	151·0|20·1|10·3 *Q 39 f.*	Glasgow T.B.Seath&Co.	1869		2Cy.18"&36"–22" 90 HP. Rankin&Blackmore,Grnck.	Huddart Parker & Co.
BT	**Despatch** I PadSp	British Liverpool	78 210	129·0|20·2|10·0	Liverpool Jones	1856		2Cy.34"–48" 80 HP. Fawcett,Preston&Co.,Livpl	Lvrpool.Stm.Tug Co.
	Despatch I Pad *Porter*	British Lowestoft	36 130	99·1|18·5| ..	London J.Stewart	1875		2Cy.30"–42" 60 HP. J.Stewart,London	Great Eastern Rail.Co.
✻	**Despina G. Michalino** Stl ScSr *Senios -89* 1 *Dk&WbFcs*	Greece Syra	1141 1763	260·6|36·6|16·5 *P 29 f. B 110 f. Q 70 f. F 30 f.*	W.Hartlepool Irvine&Co.	1888	4 Bhds. CDB167t	3C.20"33"&54"–36" 140HP. Blair&Co.,Ld.,Stockton	A.Michalinos
	Dessalines (ex Ethel) W ScBn	American Philadelphia	306 570	167·0|31·0|17·5	Camden,N.J. B.G.Hillman&Co.	1883		2Cy.24"&42"–36" Neafie&Levy,Philadelphia	Haytian Government
R	**Dessoug** I Sc3mSr *Smith 81* 3 *Dks*	American Savannah	892 1367	233·0|32·3|23·6 *B 10 f.*	W.Hartlepool Denton,Gray&Co.	1864		2Cy.28"&57"–48" 160 HP. W.Cramp&Son,S.&E.B.Co.	Ocean Steam Ship Co.
V	**Desterro** I ScBn *Dreyer 82-87* 2 *Dks*	German Hamburg	1515 2011	286·7|37·1|24·6 *Q 77 f. B 76 f. F 32 f.*	Hamburg Blohm&Voss	1884		2Cy.34"&59"–42" 225 HP. Blohm&Voss,Hamburg	Hamburg Südamerik Dampfsch.-Gesells.
BT	**Deucalion** I ScSr *Asquith 86-86*	British Liverpool	1374 2111	322·0|33·7|25·6	Newcastle A.Leslie&Co.	1872		2Cy.28½"&61"–45" 250HP. R.Stephenson&Co.,Newcstl	Ocean Steam Ship Co.
DV	**Deuteros** I ScSr *Iversen 87-87* 1 *Dk&SprDk*	German Flensburg	1251 1630	257·8|33·8|17·8 *P 24 f. B 10 f.* | 24·8	Flensburg Flensburger Schiffsb. Ges.	1881	WB	2Cy.32½"&60"–36" 160HP. FlensburgSchiffsb.-Gesell.	Flensburger Dampfs. Gesell.von1869(C.N.Hansen)
DGV	**Deutscher Kaiser** I Sc3mSr *Gethn 87-87* (ex Shepperton) 1 *Dk&SprDk*	German Rostock	813 1283	253·0|30·3|23·5 *Len. 75*	Sunderland J.Laing	18 68/75	4 Bhds. WB	2Cy.29"&57"–36" 110 HP. OuseburnEng.Wks.,Newcsl.	J.A.Bredt
G	**Deutschland** I ScSr *Steffen 77-82* 1 *Dk*	German Lübeck	541 692	194·8|26·7|15·1	Stettin Möller&Holberg	1882	WB	2Cy.23"&46"–27" 100 HP. Möller&Holberg,Stettin	RigaLübeckerDampfschiffahrts Gesellschaft
V	**Deutschland** (LOST) I ScSr *Hildebrand -87* 2 *Dks*	German Hamburg	294 372	138·1|22·9|12·0 *Q 35 f. F 19 f.*	Stettin Möller&Holberg	1883	WB	2C.18"&31"–20½" 50HP. Möller&Holberg,Stettin	H.C.E.Meyer
V	**Devawongse** Stl ScSr *Loff 88-87*	British Glasgow	1057 1643	270·5|37·0|21·8 *B 18 f. F 30 f.*	Glasgow FairfieldS.B.&E.Co.,Ld.	1888		3C.22"36"&57"–42" 250HP. FairfieldS.B.&Eng.Co.,Ld.,Glsw	Scottish Oriental S.S. Co.,Ld.
	Development I ScFlat	British Liverpool	100 147	88·5|19·8|7·8 *Len. Q 26 f.*	Winsford [illegible]	18 65/87		2Cy.9½"–12" 6HP. W.E.Bates,Leftwich	Salt Union,Ld.
	Devon I ScSr *Tait* 1 *Dk*	British Launceston, Tas.	130 191	123·7|19·1|9·0 *P 40 f.*	Paisley J.Fullerton&Co.	1876	4 Bhds.	2Cy.19"&38"–24" 50 HP. W.King&Co.,Glasgow	Launceston & North West Coast S.Nav.Co.Ld.
✻	**Devonhurst** I ScSr *Houthoff* 2 *Dks3B*	Dutch Batavia	1235 1559	280·0|33·3|24·6 *P 56 f. F 41 f.*	Newcastle C.Mitchell&Co.	1874	4 Bhds. DBaft 526 f	2Cy.35"&63"–36" 200 HP. N.E.MarineCo.,Ld.,Sundrld.	Nederland-Indische Stoomv.Maats.
BT3.88	**Devonia** I ScBk *Craig 75-88* 2 *Dks&SprDk4B*	British Barrow	2772 4270	400·3|42·0|25·1 *H 149 f.* | 33·2	Barrow BarrowS.B.Co.	1877	6 Bhds.	2Cy.59"&107"–48" 600HP. BarrowS.B.Co.,Ld.,Barrow	BarrowSteamShipCo.,Ld.
✻	**Devonia** I ScSr *Borey 86-86* 1 *Dk2B*	British London	908 1387	242·3|33·2|20·1 *P 29 f. B 14 f. F 33 f.*	Newcastle Palmers'Co.	1879	5 Bhds. DBa f & cb134 f FPT APT	2Cy.31"&58"–36" 140 HP. Palmers'Co.,Ld.,Newcastle	Oriental S.S.Co.,Ld.
	Devonia I PadSk	British Leith	14 112	100·0|18·5|9·5	SouthShields J.T.Eltringham	1885		1Cy.35"–51" 40 HP. Baird&Barnsley,N.Shields	J.M.Clarke
✻	**Devonshire** I ScBn *Purvis 83-83* 2 *Dks3B*	British London	1513 2317	316·6|36·4|25·6 *P 33 f. F 27 f.*	Stockton Richardson,Duck&Co.	1878	5 Bhds. DBa f & cb423t	2Cy.38"&71"–45" 250 HP. Blair&Co.,Ld.,Stockton	Stm.Nav.Co."Nord"
✻	**Devonshire** I Sc3mSr *Bell* 1 *Dk*	British London	431 678	186·3|30·0|14·3 *Q 38 f.*	Sunderland J.Laing	1875	3 Bhds. WB f	2Cy.24"&46"–30" 90 HP. J.Dickinson,Sunderland	Milnes&Chaytor
✻D	**Dewdrop** I ScBn *Packer 86-86* 1 *Dk2B*	British W.Hartlep'l	718 1144	230·2|32·1|13·8 *Q 60 f. B 120 f. F 25 f.*	W.Hartlepool Irvine&Co.	1882	1 Bhds. FDB329t	2Cy.27"&50"–33" 99 HP. T.Richardson&Sons,Hrtlpl.	J.Wood & Co.

[illegible] = Lloyd's Register; D = Nederlandsche Vereeniging; G = Germanischer Lloyd; I = Registro Italiano; R = American Record; [illegible] = Veritas Austro-Ungarico; H = Veritas Ellenico; N = Norske Veritas; V = Bureau Veritas.

Dans la 2me Colonne, B.T., avec les indications qui suivent, signifie la date du rapport de visite de la Carène par les experts du Board of Trade pour obtenir le certificat relatif au TRANSPORT DES PASSAGERS.

Dans la 3me Colonne, C = Composite (membrure en fer, bordé en bois); I = Fer; Stl = Acier; W = Bois.

Pour plus amples renseignements sur les navires classés par le Lloyd's Register, voir le Lloyd's Register Book et le Lloyd's Yacht Register.

No. in Book / Official No.	Registry	Ship's Name / Master, and Dates of Appointment to present Owner's Service and to Ship / Ship's Late Name	Material, Rig, &c. / No. of Decks, &c.	Flag / Port of Registry	Tons Net / Gross	Length	Breadth	Depth	Built Where / By Whom	When	No. of Bulkheads / Particulars of Water Ballast	Engines Particulars, Maker's Name	Owners or Managers
181	✻	Dewdrop *Rand 85-88*	W Sc *1 Fc*	British Dundee	14 35	71·6	19·1	7·8	Anstruther W. Jarvis	1883		2Cy.15½&27"–16" 32 HP W.B.Thompson,Dundee	W. H. Burn
182	✻	Dewsland *Hunter 84-84*	I ScSr *1 Dk 2 H*	British Cardiff	1310 2086	275·0 *Q 6f. B 114f. F 25f.*	37·6	19·6 *Q 28f.*	Sunderland J.L.Thompson&Sons	1883	Blkds.	2Cy.34"&64"–42" 175 HP Richards,Sunderland	J. Marychurch
183		Dhu Heartach *Dew 83-87*	I PadSp	British W.Hartlep'l	34 148	125·5	18·0	9·8	Newcastle Wigham Richardson&Co.	1868		2Cy.36"–48" 75 HP Hawthorn&Co.,Newcastle	C. Furness
184	✻ D	Diadem *Parker 87-87*	I ScSr *2 Dks 3 H*	British London	1177 1842	282·8	33·3 *F 30f.*	24·2	Newcastle Palmers Co.	1874	5 Blkds.	2Cy.30"&60"–33" 140 HP Palmers Co.,Newcastle	Hall Bros.
185	✻	Diamante *McCaslin 85-85*	I ScSr *1 Dk & Awn Dk 3 H*	British HongKong	514 809	200·6 *B 27f.*	28·2	13·7 20·3	Aberdeen Hall,Russell&Co.	1878	4 Blkds.	2Cy.30"&52"–33" 120 HP Hall,Russell&Co.,Aberdeen	China & Manila Co.,Ld.
186	V	Diamantino *Meyer –88*	Stl PadSr *1 Dk*	Brazilian Rio de Jan'ro	540 760	220·1	29·5	13·2	Stettin "Vulcan" Co.	1888		1C 14"&32·5"&51·5"–39·1" 120 HP "Vulcan" Co.,Stettin	Companhia de Navegação
187		Diamantinos		Brazilian	150	..	..	..					A. J. Malheiros
188 67998		Diamond (ex Jeddah) *Geary*	I ScBg *2 Dks & Spr Dk*	British Penang	1030 1578	280·0	33·2	16·0 23·4	Dumbarton W.Denny&Bros.	1872	5 Blkds.	2Cy.35"&60"–42" 200 HP Hall,Russell&Co.,Aberdeen	Chong Moh
189 82744	✻	Diamond *Barnetson 85-85*	I ScSr *1 Dk & Awn Dk 3 H*	British Dundee	729 1126	218·0	30·2	15·7 21·4	Dundee W.B.Thompson	1881	4 Blkds. cellular	2Cy.25"&50"–42" 98 HP W.B.Thompson,Dundee	Dundee Gem Co.,Ld.
190 93273	✻	Diamond *Walker*	I Sc *1 Dk*	British Glasgow	186 444	175·0 *Q 88f. B 10f. F 23f.*	25·1	10·3	Glasgow Scott&Co.	1886	3 Blkds.	2Cy.22"&42"–30" 80 HP Ross&Duncan,Glasgow	W. Robertson
191 82348		Diana *Delon*	Stl Pad	British Glasgow	372 519	165·3 *P 93f. F 58f.*	26·0	8·7	Dumbarton W.Denny&Bros.	1880		2Cy.31½"&55"–40" 150 HP Denny&Co.,Dumbarton	La Platense Co.,Ld.
192 72370	BT 1-88	Diana *Merrills*	I ScSr *2 Dks*	British Southampton	313 723	232·6 *Q 132f. F 53f.*	28·2	14·1	Glasgow Aitken&Mansel	1876		2Cy.37"&68"–39" 210 HP J.&J.Thompson,Glasgow	London & S.W. Railway Co.
193 60416		Diana *Riches*	C Sc3mSr	British Liverpool	99 179	115·4	21·5	12·4	Glasgow Stephen&Sons	1869		2Cy.13½"&26"–18" 30 HP J.Howden,Glasgow	G. Rodrigues
194 47		Diana *Wulnschnig 70-87*	I ScSr *2 Dks*	Austrian Trieste	1106 1740	270·0	35·0	24·5	Dumbarton W.Denny&Bros.	1866		4Cy.29½"&54"–36" 400 HP Denny&Co.,Dumbarton	Lloyd Austro-Ungarico
195	V	Diana *Hammje 85-*	I ScSr *1 Dk*	German Bremen	283 372	145·7	22·8	13·8	Bremen Act.Ges."Weser"	1872		2Cy.17"&29½"–22½" 160 HP Act.Ges."Weser," Bremen	Dampschiffs. "Neptun"
196	V	Diana *Lehmann 83-86*	I ScSr *1 Dk*	German Flensburg	241 397	138·4	24·4	13·7	Oscarshamn Oscarshamn's Mek.Verkst.	1883		2Cy.17"&34"–23" 50 HP Oscarshamn's Mek.Verkst.	F. M. Bruhn
197	G	Diana *Ehlers 88-88*	I ScSp *1 Dk*	German Bremen	41 130	99·7	21·6	11·7	Newcastle Wigham Richardson&Co.	1868		2Cy.33"–34" Wigham Richardson&Co.,Newc.	Bugsirgesellschaft "Union"
198	✻ N	Diana *Sars*	I ScSr *1 Dk*	Norweg'n Bergen	277 382	153·7 *Q 16f. B 21f.*	21·6	12·0	Newcastle Coulson,Cook&Co.	1873	4 Blkds.	2Cy.18½"&36"–21" 51 HP Pattinson&Atkinson,Newcastle	Johan Trüse
199		Diana	W ScBk *1 Dk*	Norweg'n Sandefjord	224 341	135·5	29·0	15·8	Drammen J. Jorgensen	1871		1Cy.28"–21" 45 HP Nylands M.Verks Christiania	H. S. Iversen & ...cussen
200		Diana (ex Hudiksvall) *Ekhom 83*	I ScSr *1 Dk*	Swedish Gothenburg	285 370	149·3	23·5	11·4	Stockholm Lindbergs Mek. Verk.	18[illegible]		2Cy.20"&40"–24" 70 HP Bergsund M.V.,Stockholm	C. A. Moller
201 90064	BT 5-88	Diana Vernon *McNeill 85-85*	I PadSk	British Glasgow	45 193	180·5	18·1	7·1	Glasgow Barclay,Curle&Co.	1885		1Cy.43"–60" 103 HP Barclay,Curle&Co.,Glasgow	North British Packet Co.
202		Dib-el-Bahr (ex Aquarius)	I PadSr	Turkish Alexandria	220	135·0	25·0	12·0	Glasgow R.Napier&Sons,Glasgow	1865		2Cy.50"–72" 250 HP R.Napier&Sons,Glasgow	Khedivieh M.
203 88841	G	Dicky *Hochreuter –85*	I ScSr *2 Dks*	British HongKong	144 226	96·6	21·0	15·0	Kiel G.Howaldt	1883		2Cy.21"&36"–18" 36 HP G.Howaldt,Kiel	John M. Bryde
204 45106		Dido	I Sc1mSr *1 Dk 2 H*	British Hull	911 1409	278·0	30·4	21·0	Hull C.&W.Earle	1862	7 Blkds.	4Cy.23"&40"–36" 130 HP Comp.'79,C.&W.Earle,Hull	T. Wilson, Sons
205 30301	✻	Dido (ex Anemone) *Bray 76-85*	Stl ScSr *1 Dk*	British Bristol	405 646	185·0 *Q 92f. B 20f. F 19f.*	29·2	12·9	Grangemouth Dobson&Charles	1884	Blkds.	[illegible]	Bristol Stm. N. Co.,Ld.
206 478		Dido *Tomsich 69-87*	I ScSr *1 Dk*	Austrian Trieste	773 1224	247·0	31·0	17·0	Sunderland W.Pile&Co.	1873		2Cy.28"&[illegible] 120 HP F.Black&Co.,Sunderland	Lloyd Austro-Ungarico
207		Dido (ex South Western)	I ScSr *1 Dk*	Brazilian Brazil	127 209	138·7	20·1 *Q 43f.*	10·0	Paisley Blackwood&Gordon	1858	5 Blkds.	2Cy.14½"&32"–20" 50 HP Blair&Co.,Ld.,Stockton	
208	N	Dido *Westergaard*	W Sc *1 Dk 2 H*	Norweg'n	244 344	132·9	25·0 *B 38f.*	14·1	Stavanger	1879		2Cy.14"&28"–21" 35 HP Nylands Verkst.,Christiania	Stromstenens
209		Die Dievenow *Volkmann*	I PadSr	German	68 104	119·1	16·2	7·8	Stettin Früchtenicht&Brock	1852			J. F. Braunlich
210	V	Die Erndte *Fischer 83-*	I ScSr *1 Dk*	German Stettin	109 151	132·8	18·7	8·7	Stettin Act.Ges."Vulcan"	1864		2Cy.21½"&43½"–16½" 40 HP Act.Ges."Vulcan,"Stettin	E. Haabuss

✻ = **Lloyd's Register;** D = Nederlandsche Vereeniging; G = Germanischer Lloyd; I = Registro Italiano; R = American [Record];
A = Veritas Austro-Ungarico; H = Veritas Ellenico; N = Norske Veritas; V = Bureau V[eritas].

In the 2nd Column, B.T., with figures following, signifies Date of report of bottom sighting by Board of Trade Surveyors as condition of PASSENGER C[ertificate].

In the 3rd Column, C = Composite (iron frame, wood planked); I = Iron; Stl = Steel; W = Wood.

Further particulars of Vessels classed by Lloyd's Register will be found in Lloyd's Register Book and Lloyd's Yacht Register.

No.	Cl.	Nom, Materiaux, Grèement, etc. — Capitaine et dates d'entrées au service des Armateurs actuels et à bord du navire. — Ancien Nom, nombre de ponts, etc.	Pavillon. — Port d'Armement	Tonnage Net. — Total	Dimensions Longueur	Largeur	Creux	Construit Où. — Par qui.	Quand.	Nombre des cloisons. — Détails sur le lest d'eau.	Machines. — Détails. Nom du Constructeur.	Armateurs ou Directeurs-Gérants.
		Dieppe I ScSr *Phelipot 67-87*	French Dieppe	165 327	168·4	21·1	11·9	Havre Forges&Chantiers	1875		2C.24¼" & 39½"–23¾" 120 HP. Forges&Chantiers,Havre	London, Brighton & South Coast Railway Co.
		Digby Grand I PadSp	British Swansea	22 103	93·0	18·9	8·7	S.Shields J.T.Eltringham	1871		1Cy.33½"–51" 40 HP. J.Reynoldson,SouthShields	W.Fender
	V	Dijmphna (ex Lidköping) W ScSmSr *Tonnesen 81-88* 2 *Dks*	Danish Copenhagen	138 187	100·9	20·6	12·2	Sjötorp *P 13 f. B 18 f.*	18..		2Cy.9¼" & 20¼"–15½" 24 HP.	Carl C.Trane
		Dikili I Sc 1 *Dk*	Turkish Smyrna	85 161	120·8	18·0	9·0	Preston *Q 32 f. B 12 f. F 32 f.* W.Allsup&Sons	1885		2Cy.17 & 34"–20" 60 HP. W.Allsup&Sons,Preston	Essayou Shahum & Co.
	✱	Diligent Stl ScSr *Cook 87-88* 1 *Dk & Wb Frms*	British Sunderland	1413 2184	292·3	39·1	18·9	Sunderland Short Bros.	1888	6 Blkds. cellular	3C.21½" 35" & 58"–39" 200 HP. J.Dickinson,Sunderland	J.Westoll
	✱	Dilsberg Stl ScSr *Black 81-86* 2 *Dks 3 B*	British Glasgow	967 1514	265·2	34·1	22·3	PortGlasgow *F 36 f.* Murdoch&Murray	1882	4 Blkds. DB	2Cy.30" & 58"–45" 160 HP. Kincaid,Donald&Co.,Grnck	Raeburn & Vérel
	✱	Dilston Castle I ScSmSr *Baxter 71-86* 2 *Dks 3 B*	British London	1054 1639	265·8	32·7	24·0	Newcastle Palmers'Co.	1871	5 Blkds. DB	2Cy.30" & 60"–33" 150 HP. Palmers'Co.,Newcastle	Hall Bros.
	✱	Dimitri Stl ScSr *Kaloagin -87* 1 *Dk & Awn Dk*	Russian Odessa	680 1000	210·0	30·2	20·8	Newcastle W.Dobson&Co.	1887	5 Blkds. DB	3C.17" 28" & 47"–30" 100 HP. N.E.MarineE.Co.,Ld.,Nwcstl	Russian Co. for Sea, River and Land Insurance and Conveyance of Goods.
	✱	Dingadee Stl ScSmSr *Osborne* 1 *Dk*	British Sydney	393 640	180·4	28·1	13·6	Cork *Q 100 f. B 7 f. F 30 f.* Royal Victoria Dks. Co.	1883	3 Blkds.	2Cy.22" & 44"–33" 80 HP. Kincaid&Co.,Greenock	Australasian United Steam Nav. Co.,Ld.
	✱	Dinnington I ScSr *Coates 87-87* 1 *Dk*	British Goole	234 371	159·0	21·4	12·2	Newcastle *Q 70 f. F —* Coulson,Cooke&Co.	1873	4 Blkds. DB	2Cy.18½" & 36"–24" 50 HP. Pattison&Atkinson,Newc.	Yorkshire Coal & S.S. Co.,Ld.
221	✱	Diolibah I ScSr *Lebatut 84-84* 1 *Dk & Spr Dk*	French Marseilles	1264 1584	263·7	35·8	21·6	Dumbarton *B 32 f. H 69 f.* A.McMillan&Son	1880	3 Blkds. DB	Tripled by Soc. Forges et Chantiers de la Med. La Seyne.	Compagnie Française de Nav. à Vap.
222		Diomed I ScSr *Bigley 86-86*	British Liverpool	1471 2254	326·8	34·8	28·1	Newcastle A.Leslie&Co.	1868		2Cy.28" & 58"–45" 170 HP. R.Stephenson&Co.,Newcstl	Ocean Steam Ship Co.
223	✱	Diomedea I ScSr *Way* 1 *Dk 2 B*	British London	791 1240	233·6	32·1	21·2	Sunderland *F 31 f. Q 26 f.* J.Laing	1875	4 Blkds. WB	2Cy.30" & 57"–36" 150 HP. J.Dickinson,Sunderland	"Avis" Steamships,Ld.
224		Dione I ScSr *Bond 19-79* 1 *Dk*	British Stockton	493 830	191·1	29·0	16·8	Stockton *Q 29 f. B 43 f.* Richardson Duck&Co.	1868	4 Blkds. DB	2Cy.28½" & 53"–33" 120 HP. Blair&Co.,Ld.,Stockton	Tees Union Shipping Co.
225		Discoverer I ScSmSr *Hughes 77-79*	British Liverpool	1411 2162	311·3	34·0	25·6	Glasgow *F 24 f.* Aitken&Mansel	1877		2Cy.35" & 70"–48" 280 HP. J.&J.Thomson,Glasgow	Charente S.S.Co.,Ld.
226	✱D	Discovery StlScSr *Woodall 88-88* 1 *Dk & Wb Frms*	British Whitby	1224 1916	284·0	38·1	18·2	Sunderland J.L.Thompson&Sons	1888	6 Blkds. cellular	3Cy.21" 35" & 58"–36" 180 HP. T.Richardson&Sons, Hartlepool	Rowland & Marwood
227		Dispatch I Sc *Pryor 77-84* 1 *Dk*	British London	114 193	129·5	21·0	10·5	Middlesbro' R.Dixon&Co.	1873	4 Blkds.	2Cy.18" & 36"–24" 50 HP. R.&W.Hawthorn,Newcstle	Hewett & Co.,Ld.
228		Dispatch I ScSk	British Glasgow	96 167	119·0	21·1	10·6	PortGlasgow R.Duncan&Co.	1870		2Cy.24"–24" 70 HP. FinniestonCo.,Glasgow	Henderson Bros.
229	✱	Dithmarschen I ScSr *Hansen 84-84* 2 *Dks*	German Tönning	614 960	235·0	28·1	15·0	London *P 39 f. F 37 f.* C.&H.Green	1872	1 Blkds.	2C.29½" & 49½"–35½" 150 HP. Geb.Howaldt Bros.,Kiel	Tönninger Dampfschifffahrts Ges.
230		Ditmarsia II. I ScSr *Jochimsen -83*	German Kappeln	88 138	..	..	..		1873		60 HP.	Kappeln-Kieler Dampfschiffsrhederei
231		Djadja (ex Samara) I Pad 1 *Dk*	Russian Odessa	65 124	100·0	18·0	8·7	London Rennie&Co.	1862		2Cy.33"–48" 60 HP. Rennie&Co.,London	Russian Steam Nav. & Trading Co.
232		Djanik I Sc *Mustafa Bey -84*	Turkish Constantinple	668 1048	221·6	28·3	21·5	Middlesbro' Backhouse&Dixon	1871		2Cy.28" & 53"–33" 120 HP. T.Richardson&Sons,Hrtlpl.	Idarei-Massousieh (Ottoman Steam Nav.Co.)
233		Djedushka I Pad	Russian Sebastopol	 215	116·0	22·2	7·8	Sebastopol RussischeDampfschiffs &Handelsges.	1871		2Cy.30"–41" 60 HP. Russ.Dampfs&HndlsGes,Sbl	Russian Steam Nav. & Trading Co.
234		Djemnah I ScBk 3 *Dks*	French Marseilles	2371 3717	394·6	39·7	32·8	La Ciotat Messageries Maritimes	1875		3C.60¼"(2)43¼"–13½" 600 HP. MessageriesMar.,LaCiotat	Messageries Maritimes
235		Dnepr I ScSr 1 *Dk*	Russian Odessa	377 484	212·3	29·0	9·4	London Samuda	1872		2Cy.23" & 41"–24" 60 HP. Penn&Sons,Greenwich	Russian Steam Nav. & Trading Co.
236	✱	Dobhran I ScSrYt *Ellis 64-84* 2 *Dks*	British London	180 319	193·0	22·0	13·0	PortGlasgow Cunliffe&Dunlop	1876	1 Blkds.	2Cy.26" & 46"–27" 100 HP. Cunliffe&Dunlop,Pt.Glsgw.	T.V.Smith
237	G	Dr. Friedrich Witte IScSr *Ahrens -87* 1 *Dk & pt Awn Dk*	German Rostock	167 238	123·0	22·0	8·5	Rostock RostockS.&E.Co.	1887	FPT	2Cy.13" & 26¾"–18½" 160 HP. RostockS.&E.Co.,Rostock	Nordische Dampfschifffahrts Act.Ges.(F.Petersen)
238		Dodder I PadSr	British Dublin	88 181	118·1	17·0	9·4	Dublin F.Barrington	1863		2Cy.34"–42" 80 HP. F.Barrington,Dublin	London & N.W.Railway Co.
239		Dodo I ScSmSr	British Cork	754 1170	254·9	36·0	16·2	Cork *P 56 f.* E.Pike	1855		2Cy.36" & 66"–36" 180 HP. J.Jack&Co.,Liverpool	City of Cork Steam Packet Co.,Ld.
240	✱	Dodo I ScBg *Griffiths -88* 1 *Dk*	British Glasgow	326 531	195·8	25·2	12·3	PortGlasgow *P 28 f. F —* Cunliffe&Dunlop	1879	1 Blkds. FPT	2Cy.24" & 42"–27" 80 HP. Cunliffe&Dunlop,Pt.Glsgw	British & African Stm. Nav.Co.,Ld.

✱=Lloyd's Register; D=Nederlandsche Vereeniging; G=Germanischer Lloyd; I=Registro Italiano; R=American Record;
[illegible]=Veritas Austro-Ungarico; H=Veritas Ellenico; N=Norske Veritas; V=Bureau Veritas.

Dans la 2me Colonne, B.T., avec les indications qui suivent, signifie la date du rapport de visite de la Carène par les experts du Board of Trade pour obtenir le certificat relatif au TRANSPORT DES PASSAGERS.

Dans la 3me Colonne, C=Composite (membrure en fer, bordé en bois); I=Fer; Stl=Acier; W=Bois.

Pour plus amples renseignements sur les navires classés par le Lloyd's Register, voir le Lloyd's Register Book et le Lloyd's Yacht Register.

No. in Book / Official No.	Class	Ship's Name. Material, Rig, &c. / Master, and Dates of Appointment to present Owner's Service and to Ship. / Ship's Late Name. No. of Decks, &c.	Flag / Port of Registry	Tons Gross / Net	Dimensions (Length | Breadth | Depth)	Built Where / By Whom	When	No. of Bulkheads / Particulars of Water Ballast	Engines / Particulars / Maker's Name	Owners or Managers
241 82055	✻D	Dodona I SeS *Major 76-85* 1 *Dk 2 B*	British [illegible]	993 [illegible]	245·6|35·0|18·4	Sunderland [illegible]	1882	[illegible] Blkds.	2Cy.32"& 60"–39" 140 HP. J.Dickinson,Sunderland	T. E. Hick
242 93143	✻	Dogger Bank I SeY *Cargill -88* 1 *Dk*	British [illegible]	55 [illegible]	95·3|20·9|11·0	Hull CookW.Bend&Gemmell	1888	3 Blkds.	2Cy.17"&33"–22" 45 HP. C.D.Holmes&Co.,Hull	G. Bowman
243 86409	✻D	Dolcoath I SeS *Jago 82-82* 1 *Dk 2 B*	British [illegible]	1171 1790	266·3|36·5|19·2 *Q 74 f. B 34 f. D 56 f. F 26 f.*	Sunderland [illegible]	1882	[illegible] Blkds.	2Cy.33"& 65"–39" 150 HP. N.E.Marine Eng.Co.,Ld.,Sld.	"Dolcoath" S.S. [illegible]
244		Doli I Sc *Zuanich*	Arg'ntine [illegible]	102 108	98·9|16·9|8·7	Dumbarton [illegible]	1883		40 HP.	N. Milanovich
245 87092	✻	Dollie I SeS *Edwards 84-84* 1 *Dk*	British Liverpool	201 403	160·0|23·2|11·0 *Q 56 f. B 38 f. F 17 f.*	Barrow BarrowS.B.Co.,Ld.	1884	4 Blkds. [illegible]	2Cy.22"& 44"–27" 80 HP. BarrowS.B.C.,Ld.,Barrow	S.S. "Dollie" [illegible] (J.J.Mack & Sons)
246		Dolma-Bagtchi I Sc *Mehemed Bey -88*	Turkish [illegible]	834	217·0|28·0|17·8	Glasgow A.&J.Inglis	1856		2Cy.42"–36" M'Nabb&Co.,Greenock	Idarei-Massousie (Ottoman [illegible])
247 95797		Dolphin W SeSchGal *Molyneux*	British Liverpool	97 143	89·7|20·9|10·0	Northwich I.Woodcock	1887		1Cy.8"&15¼"–19¾" 40 HP. J.H.Wilson&Co.,Liverpool	Salt Union, Ld.
248 79336		Dolphin I SeSr 1 *Dk*	British Adelaide	63 147	100·4|17·6|9·4 *Q 10 f.*	Adelaide A.E.B.Brown	1882		2Cy.18"& 33"–24" 49 HP. Plenty&Sons,Newbury,Eng.	D. Symon
249 91025	✻	Dolphin I SeSr *Lake 88-88* 1 *Dk*	British Swansea	38 115	100·0|20·1|10·0	Dundee W.B.Thompson	1885	3 Blkds.	2Cy.17"& 34"–24" 45 HP. W.B.Thompson,Dundee	E. Benson Bros.
250 93483	✻	Dolphin (WRECKED) I SeSr *Normandale -87* 1 *Dk*	British Scarboro'	26 110	95·2|19·1|10·2	S.Shields J.Eltringham	1888	3 Blkds.	2Cy.17"& 32"–22" 40 HP. J.P.Rennoldson&Sns.S.Shlds	G. A. Smith
251		Dom Augusto Pad	Portug'se Lisbon	174	.. .. ..				100 HP.	Cia.de Sul & [illegible]
252	V	Dom Pedro I SeBg *Segond 78-83* 2 *Dks & Spr Dk*	French Havre	1957 2939	329·8|39·4|31·3 *B 77 f. F 41 f.*	Havre Forges&Ch.de la Medit.	1878		2C.35½"&71"–39⅜" 325 HP. Forges&Cie de la Medit.,Hvre	Chargeurs Réunis
253		Dom Pedro II. I Pad 1 *Dk & Hur Dk*	Brazilian Para	250 397	170·0|26·0|9·5	Port Glasgow Blackwd.&Gordon	1883		2Cy.33"–18" 90 HP. Blackwood&Gordon,Pt.Gls.	Amazon Stm. Nav.
254 50	I	Domenico Balduino I SeBn *Odero 50 82 (exCh'na)* 2 *Dks & Spr Dk*	Italian Genoa	3044 4530	400·3|41·6|31·2 *B 102 f. F 52 f. H 274 f.*	Newcastle Palmer&Co.	1882	[illegible] Blkds. [illegible]	2Cy.54"& 96"–60" 1112 HP. Palmer&Co.,Newcastle	Navigazione [illegible] Italiana
255 86317	✻	Domingo (ABANDONED) I SeSr *Wilson 85-85* 1 *Dk & Wb Frms*	British Sunderland	899 1388	249·5|35·4|18·1 *B 55 f.*	Sunderland J.Laing	1884	3 Blkds. WB [illegible]	2Cy.29½"&55"–36" 129 HP. G.Clark,Sunderland	The Pinkney & [illegible] S.S.Co.,Ld.
256 69342	✻ BT 1887	Dominion I SeBk *Cross -88* 3 *Dks & Awn Dk*	British Liverpool	2032 3176	335·0|38·4|25·7	Dumbarton A.M'Millan&Son	1873	5 Blkds.	2Cy.47"& 84"–45" 300 HP. J.Jack,Rollo&Co.,Liverpool	Mississippi & Dominion S.S.Co.,Ld.
257 81913		Dominion (ex Linda) W SeSr 3 *Dks*	British Yarmth,N.S.	410 584	150·0|26·5|22·3	Connecticut	1864		1Cy. 24 HP. Barrell,Jobson&Co.,Yar.,NS	Yarmouth [illegible] Co.,Ld.
258 90707		Dominion W SeLin 2 *Dks*	British S.Cath'ne's,Ont	304 478	135·0|25·8|11·1	St.Catherine's L.Shickluna	1868		1Cy.26¼"–28" 26 HP. G.N.Ollie,St.Catharines	S. Neelon
259 77469	✻ BT 1888	Domino I Sc3mSr *Tholander* 2 *Dks*	British Hull	610 937	225·4|28·2|14·7	Newcastle Schlesinger,Davis&Co.	1877	4 Blkds.	2Cy.30"& 58"–33" 120 HP. Amos&Smith,Hull	T. Wilson, Sons & Co.
260 95047	✻	Domira Stl ScS *Wilson 88-88* 1 *Dk*	British Glasgow	1174 1791	260·0|36·8|16·6 *Q 60 f. B 140 f. F 32 f.*	Glasgow A.Stephen&Sons	1888	4 Blkds. [illegible] APT 90	3C.18"29"&46"–39" 170 HP. A.Stephen&Sons,Glasgow	S.S. "Domira" [illegible] (Maclay & M'Intyre)
261 63227		Don (ex Corcovado) I ScBkn 3 *Dks & Awn Dk*	British London	2406 3805	387·5|43·1|33·9 *B 33 f.*	Birkenhead LairdBros.	1872	7 Blkds.	2Cy.60"&103"–54" 600 HP. LairdBros,Birkenhead	Royal Mail S. Pkt.Co.
262		Don Isabel I Pad	Brazilian Para		180·0|27·0|9·0	Port Glasgow Blackwd.&Gordon	1883		Blackwood&Gordon,Pt.Gls.	Amazon Stm. Nav.
263	✻	Don Juan (ex Kronshtadt) I SeSr 1 *Dk & Shd Dk*	Spanish Manila	482 780	217·0|27·8|13·6 20·6	London Lewis&Stockwell	1872	4 Blkds.	2Cy.30"& 60"–30" 160 HP. T.Wingate&Co.,Glasgow	F. L. Roxas
264	DV	Donar I ScS *Kuhn 82-82* 1 *Dk & Spr Dk*	German Bremen	1041 1611	263·4|34·4|17·5 21·9	Bremen Act.Ges.Weser	1882		2Cy.34"& 59"–42" 160 HP. Act.Ges.Weser,Bremen	Köper Decker [illegible]
265	V	Donata I ScS *Aguirre* 1 *Dk 2 B*	Spanish [illegible]	524 843	210·0|28·8|16·0 *Q 80 f. B 44 f. F 3 f.*	Glasgow Lon&Gls.Eng&I.S.Co.	1880	4 Blkds. DB [illegible] FT	2Cy.25"& 46"–36" 100 HP. Lon.&Gls.Eng.I.S.B.Co.,Gld	Linea de Vapores [illegible]
266 55040		Donati I ScBg *Austin 83-88*	British Liverpool	887 1392	257·6|31·1|26·9 *F 18 f.*	Newcastle A.Leslie&Co.	1866		2Cy.22"& 50"–42" 94 HP. Fawcett,Preston&Co.,Livpl	Livrp'l, Brazil & River Plate Stm.Nav.
267		Donau I ScBg *Toper 87-88* 2 *Dks & Spr Dk*	German Bremen	1771 2896	347·8|40·0|33·5	Greenock Caird&Co.	1868		2Cy.60"&100"–54" 600 HP. Caird&Co.,Greenock	Norddeutscher Lloyd
268	✻N	Donau I ScS *Hansen 82-88* 1 *Dk 2 B*	Norweg'n Bergen	1032 1452	240·1|35·1|16·8 *P 140 f. F 24 f.*	Sunderland R.Thompson&Sons	1883	4 Blkds. [illegible] FPT 36	2Cy.31"& 58"–36" 130 HP. J.Dickinson,Sunderland	M. A. Schjøtt [illegible]
269 75559		Doncaster I PadSp	British Hull	165 216	160·5|18·7|8·1	Hull M.Samuelson&Co.	1856		2Cy.26"–36" 80 HP. LairdBros.,Birkenhead	Manchestr, Sheffield & Lincolnshire Ry.
270 52949	BT [illegible]	Donegal (ex Buffalo) I PadS *Harrison 15-84*	British [illegible]	289 676	241·1|26·3|13·4 *Q [illegible] B 55 f. F 37 f.*	Greenock Caird&Co.	1865		3C 45"&(2) 58"–60" & 72 [illegible] 220 HP. [illegible]	J. Little & Co.

✻ = **Lloyd's Register**; D = Nederlandsche Vereeniging; G = Germanischer Lloyd; I = Registro Italiano; R = American Record;
A = Veritas Austro-Ungarico; H = Veritas Ellenico; N = Norske Veritas; V = Bureau Veritas.

In the 2nd Column, B.T., with figures following, signifies Date of report of bottom sighting by Board of Trade Surveyors as condition of PASSENGER CERTIFICATE.

In the 3rd Column, C = Composite (iron frame, wood planked); I = Iron; Stl = Steel; W = Wood.

Further particulars of Vessels classed by Lloyd's Register will be found in Lloyd's Register Book and Lloyd's Yacht Register.

Numéros dans le livre	Cote	Nom, Matériaux, Gréement, etc. Capitaine et dates d'entrée au service des Armateurs actuels et à bord du navire. Ancien Nom, nombre de ponts, etc.		Pavillon. Port d'Armement	Tonnage Net / Total	Longueur	Largeur	Creux	Construit. Où. Par qui.	Quand	Nombre des cloisons. Détails sur le lest d'eau.	Machines. Détails. Nom du Constructeur.	Armateurs ou Directeurs-Gérants.
	F	Donnai	I ScSr 3 Dks	French Marseilles	1445 2469	303·4	38·1	32·8	La Seyne Forges&Cᵉ de la Médit.	1862		2Cy.76½"–52½" 500 HP. Forges&Cᵉ de la Médit.Mrslls	Messageries Maritimes
	*	Doon (ex Dean) Nicholson -84	I ScSmSr 1 Dk2 B	British Glasgow	92	133·0 Q 50 f.	21·0 B 18 f.	11·5	Chester RhodeeS.B.Co.	1857	4 Bhds.	2Cy.17"& 33"–20" 50 HP. Comp. 78 D.Rowan,Glasgw.	D.Rowan
		Doowoon Matthews	I PadSp	British Glasgow	425 688	248·7	30·3	8·5	Dumbarton W.Denny&Bros.	1878		2Cy.34"&59"–54" 125 HP. W.Denny&Bros.,Dumbrtn.	Irrawaddy Flotilla Co. Ld.
274	*	Dora Mills 83-87	St ScSr 2 Dks3 B	British Whitby	1558 2376	301·3 P 29 f.	38·1 B 70 f.	24·1 F 35 f.	Whitby T.Turnbull&Son	1887	5 Bhds. DBfo f 220t	3C22½"36½,&60"–39" 180 HP. Blair&Co.,Ld.,Stocktn	T.Turnbull & Son
275	*	Dora Readhead	St ScSr 1 Dk & W B Frames	British W. Hartlepool	1133 1735	260·3 P 28 f.	36·5 B 116 f.	16·4 F 30 f. 470 f.	W. Hartlepool R.Irvine&Co.	1888	4 Bhds. CDB 521	3C.20"33"&54"–36 140 HP. Blair&Co.,Ld.,Stockton	O.Trechmann
276	* D	Dora Davies 73-82	I ScSr 1 Dk	British Sunderland	408 645	186·1 Q 37 f.	28·4 B 33 f.	14·8 F 24 f.	Sunderland J.Blumer&Co.	1873	4 Bhds. DBa&f 150t	2Cy.25"& 44"–30" 86 HP. Fowler&McCollin,Hull	Earl of Durham
277		Dora Hogue 1880	W ScBn 1 Dk2 B	American San Francisco	136 198	112·0	27·2	13·2	San Francisco M.Turner	1880		2Cy.11"&26"–18" 40 HP. W.McAffee,San Francisco	Alaska Commercial Co.
278	*	Dora Ewing Stafford 84-87	I ScSr 1 Dk	British Cardiff	229 379	144·7 Q 42 f.	25·1 F 16 f.	11·4	Newport Uaks,B.Co.,Ld.	1884	3 Bhds. FPT28 APT25	2Cy.20"& 35"–24" 55 HP. Westray,Copeland&Co.Brs.	R.Taylor
279	R	Dorcas Estabrooks -88	W ScSr 1 Dk	British Sackville, N.B.	120 215	110·1	24·1	9·9	Sackville, N.B. J.Purdy	1887		2Cy.21½"&24"–24" 80 HP. Gourlay Bros.&Co.,Dundee	J.Wood
280	*	Dordogne King 82-88	Stl ScSr 1 Dk & Web Frames	British Cardiff	546 850	200·0 P 114 f.	29·3 F 34 f.	15·2	Sunderland Osbourne, Graham & Co.	1885	4 Bhds. DBa&t FPT22t	2Cy.26"& 47"–36" 96 HP. G.Clark,Sunderland	Cardiff S.S.Co. (Hooper,Campbell&Co.)
281		Dordogne	Stl ScBn 3 Dks	French Marseilles	2800 3800	371·4	43·0	34·4	Havre Forges & Chantiers de la Médit.		APT 90 CDBe&A 128t	3C.28"44"&70"–47" 525 HP. Forges & Chantiers de la Médit., Havre	Messageries Maritimes
282		Dorian McFarlane 87-88	I ScBk 2 Dks	British Glasgow	667 1039	237·5 Q 90 f.	30·2 F 36 f.	19·9	Port Glasgow R.Duncan&Co.	1868	5 Bhds.	4Cy.20"& 35"–36" 207 HP. Comp. 78 D. & W.Henderson & Co.,Glasgow	Henderson Bros.
283		Doric Jennings	Stl Sc4mS 3 Dks	British Liverpool	3071 4784	440·9 B 46 f.	44·2 H 30 f.	31·5	Belfast Harland&Wolff	1883	7 Bhds.	4Cy.32"&71"–60" 550 HP. Harland&Wolff,Belfast	Oceanic Steam Nav. Co.,Ld.
284	*	Doris York 88-88	St 18ScSr 1 Dk & Spar Dk3 B	British Glasgow	1156 1785	259·8	37·2	22·4 15·5	P. Glasgow R.Duncan&Co.	1888	4 Bhds. CDB 381	3Cy.18"29½"&47"–39" 120 HP. Muir&Houston,Gls.	S.S. "Doris" Co.,Ld. (Aitken & Walker)
285		Doris Davis	W Sc	American Providence	947 1300	219·0	35·0	16·0	New York	1864		800 HP.	Central Vermont Railway Co.
286	G V	Doris Bahes 87-87	I ScSr 1 Dk & Awn Dk	German Apenrade	817 1025	219·1	28·5	14·6 20·9	Kiel Geb.Howaldt	18 82/84	W B	2Cy.23"&43½"–35½" 90 HP. Geb.Howaldt,Kiel	M.Jebsen
287	* V	Dorset Daniel 82-87	I Sc3mSr 2 Dks & Awn Dk	British Bristol	1716 2638	331·0	39·0	19·0 26·4	Sunderland J.L.Thompson & Son	1881	7 Bhds. CDB 440 APT 20 ft	2Cy.40"& 76"–48" 300 HP. G.Clark,Sunderland	T.B.Williams (W.Lamplough&Co.,Mgrs.)
288	*	Dorunda Sayers 76-87	I Sc3mSr 2 Dks3 B	British Glasgow	2033 3136	368·5	37·3	28·5	Dumbarton W.Denny&Bros.	1875	6 Bhds.	2Cy.47"& 82"–48" 400 HP. Denny&Co.,Dumbarton	British India Associated Steamers,Ld.
289		Dotshka	I Pad 2 Dks	Russian Odessa	108 201	144·0	21·0	8·2	Linz Mayer	1879		2Cy.24"& 36"–44" 70 HP. Escher,Wyss&Co.,Zurich	Russian Steam Nav. & Trading Co.
290	* D	Dotterel Dunn	I Sc3mSr 2 Dks	British Cork	910 1401	264·8	32·1	21·8	Liverpool W.H.Potter&Son	1878	6 Bhds. DBa&t FPT on MT 290	3C.20"32"&52"–36" 150 HP. Tripled '88, D. Rollo & Sons, Liverpl.	Cork S.S.Co.,Ld.
291	*	Dotterel Chalmers -87	I ScSrYt	British Leith	112 165	119·6 Q 17.	20·1	11·7	Leith Ramage&Ferguson	1887	4 Bhds.	3C.9½"15"&24½"–18" 40 HP. Ramage&Ferguson,Leith	C.Combe
292	*	Douglas (WRECKED) Franks 85-87	I ScSr 1 Dk2 B	British W.Hrtlpool	648 1022	225·3 Q 84 f.	30·1 F 29 f.	16·7	Whitby T.Turnbull&Son	1873	4 Bhds. DBa&f	2Cy.25½"&51½"–33" 99 HP. Blair&Co.,Ld.,Stockton	Pyman, Bell & Co.
293		Douglas Swanson 71-87	I ScSr 1 Dk2 B	British Sunderland	481 768	205·0 Q 30 f.	29·0 F 20 f.	17·2 B 7 f.	Shields Marshall Brothers	1861	4 Bhds. DB 86 ft	2Cy.26"& 48"–30" 90 HP. G.F.Pattison&Co.,Nwcastle	H.T.Morton & Co.
294		Douglas	I PadSr	British Douglas	249 610	227·4 Q 38 f.	26·2	14·2	Greenock Caird&Co.	1864		2Cy.58"–60" 250 HP. Caird&Co.,Greenock	Isle of Man Stm.Pkt. Co.
295	*	Douglas Jarvie 86-88	I ScSr 1 Dk	British Granton	48 101	97·2 Q 38 f.	17·2 F 11 f.	7·9	Port Glasgow Murdoch&Murray	1876	4 Bhds.	2Cy.15"&26"–21" 30 HP. Muir,Houston&Co.,Glasgw	General Stm.Fishing Co.
296	*	Douglas Brown -88	Stl PadSp 1 Dk	British Preston	26 132	120·8	20·2	6·5	Preston W.Allsup&Sons	1888	4 Bhds.	2Cy.30"–48" 70 HP. W.Allsup & Sons,Preston	Corporation of Preston
297	*	Douro Branfoot 86-86	I ScSr 1 Dk2 B	British Sunderland	1060 1644	263·0 B 101 f.	33·7 Q 34 f.	20·7 F 30 f.	Sunderland J.Laing	1879	3 Bhds. DB 114 ft	2Cy.31"& 57"–36" 150 HP. G.Clark,Sunderland	Culliford & Clark
298	*	Douro Thomas 81-84	I ScSr 1 Dk & Awn Dk	British Bristol	277 430	171·3	20·1	11·1 17·4	Bristol G.K.Stothert&Co.	1879	4 Bhds.	2Cy.20"& 38"–24" 65 HP. G.K.Stothert&Co.,Bros.	G.K.Stothert & Co.
299		Douro (ex Blair Athole) Dazilra	I Sc	Portug'se Lisbon	69 105	85·9 Q 6 f.	18·7	8·2	Paisley Hannah,Donald & Wilson	1875		2Cy.10½"& 17"–12" 15 HP. Hannah,Donald&Wilsn,Paisley	Les fils de Th.Conseil
300		Dous de Julho	I PadSr	Brazilian Bahia	261	135·0	20·0	8·5	Stockton Richardson,Duck & Co.	1862		2Cy.29"–36" 50 HP. Ritson&Hewitson,Leeds	Cia.de Nav. à Vap. de Bahia

*=Lloyd's Register; D=Nederlandsche Vereeniging; G=Germanischer Lloyd; I=Registro Italiano; R=American Record;
A=Veritas Austro-Ungarico; H=Veritas Ellenico; N=Norske Veritas; V=Bureau Veritas.

Dans la 2me Colonne, B.T., avec les indications qui suivent, signifie la date du rapport de visite de la Carène par les experts du Board of Trade pour obtenir le certificat relatif au TRANSPORT DES PASSAGERS.

Dans la 3me Colonne, C=Composite (membrure en fer, bordé en bois); I=Fer; Stl=Acier; W=Bois.

Pour plus amples renseignements sur les navires classés par le Lloyd's Register, voir le Lloyd's Register Book et le Lloyd's Yacht Register.

No. in Book. / Official No.	Registry in which classed.	Ship's Name, Material, Rig, &c. / Master, and Dates of Appointment to present Owner's Service and to Ship / Ship's Late Name. No. of Decks, &c.	Flag / Port of Registry	Tons Net / Gross	Dimensions (Length / Breadth / Depth)	Built Where / By Whom	When	No. of Bulkheads / Particulars of Water Ballast	Engines Particulars / Maker's Name	Owners or Managers
301 83817	✳	**Dowlais** I ScSr *Hodge* 1 *Dk2 B*	British Cardiff	763 1174	241·6 | 33·2 | 14·5	Newcastle [illegible]	1880	4 Blkds. [illegible]	2Cy.28"&54"–36" 99 HP. Palmers'Co.,Ld.,Newcastle	"Dowlais" S.S. [illegible] (Morel Bros.&Co.)
302	✳ DGV	**Drachenfels** I ScSr *Groot* [illegible] 2 *Dks3 B*	German Bremen	1459 2251	298·8 | 36·7 | 23·6 [illegible] P28 f. F30 f.	Sunderland SunderlandS.B.Co.	1882	7 Blkds. [illegible]	2Cy.36"&68"–42" 220 HP. Black,Hawthorn&Co.,Newc	Deutsche Dampf-[illegible] schiffahrts Ges.
303 [illegible]		**Draco** Stl ScSr *Douglass* 1 *Dk2 B*	British Hull	1113 1713	276·2 | 34·1 | 18·4 Q 93 f. B 184 f. F 38 f.	Hull Earles'Co.,Ld.	1882	4 Blkds. [illegible]	3C.21"32"&56"–36" 140 HP. Earles'Co.,Ld.,Hull	T. Wilson,Sons [illegible]
304 [illegible]	✳	**Dracona** Stl ScSr *Tait* 81-88 2 *Dks3 B*	British Dundee	1245 1963	284·0 | 35·7 | 24·1 P 28 f. B 62 f. F 34 f.	Dundee GourlayBros.&Co.	1884	5 Blkds. [illegible]	2Cy.34"&64"–42" 210 HP. Gourlay,Bros&Co.,Dundee	W. Thomson & [illegible]
305 91[illegible]	✳	**Dragoman** Stl ScSr *Rose* 80-85 1 *Dk&SprDk3 B*	British Liverpool	1818 2783	310·0 | 39·2 | 19·4 B 71 f. F 37 f. 26.9	Newcastle SirW.G.Armstrong, Mitchell&Co.	1885	6 Blkds. [illegible] FPT 35 APT 18	3C.24"39"&64"–42" 300 HP. Wallsend Slipway Co.,Ld.,Nwc	Bedouin Stm. [illegible]
306 43081		**Dragon** (BROKEN UP) W PadSp 1 *Dk*	British Pictou,N.S.	59 130	112·2 | 19·6 | 10·0	Miramichi,NB.	1863		2Cy. 80 HP.	General Mining [illegible] of London
307 84018	✳ 87-88	**Dragon** Stl ScTug 1 *Dk*	British London	8 40	66·9 | 13·5 | 7·0	Bristol Newall&Co.	1886	4 Blkds.	2Cy.13"&26"–18" 40 HP. Newall&Co.,Bristol	C. E. Gaselee
308 87186	✳	**Dragonfly** I ScSr *Howling* –83 1 *Dk2 B*	British London	1134 1761	257·2 | 34·9 | 19·6 Q 66 f. P 28 f. B 66 f. F 31 f.	W. Hartlepool W.Gray&Co.	1883	4 Blkds. [illegible]	2Cy.32"&60"–39" 150 HP. Black,Hawthorn&Co.,Gts.	Galbraith,Pembroke Co.
309 49641	84-86	**Dragoon** (WRECKED) I Sc3mSr *Walker* 70-76	British Newcastle	583 924	230·0 | 28·0 | 17·4 Q 37 f.	Newcastle PalmerBros&Co.	1864		2Cy.27½"&54"–30" 120 HP. R.&W.Hawthorn,Newcastle.	Tyne S.S.Co.,[illegible]
310	V	**Dragut** I ScSr *Sylvestre* 1 *Dk&SprB*	French Marseilles	312 533	182·4 | 25·6 | 9·3 16.6	Greenock Scott&Co.	1880		2Cy.25"&50"–30" 80 HP. GreenockFoundryCo.,Grk.	Compagnie Générale Transatlantique
311	DN	**Drammen** I ScSr *Lorentzen* 1 *Dk&AwnDk*	Norweg'n Christiania	486 695	160·5 | 26·2 | 20·1	Bergen BergensMek.Verks	1875		2Cy.25"&45"–24" 75 HP. BergensMek.Verks,Bergen	Winge & Co.
312 [illegible]		**Dreadnought** I PadSp *Holland* 86-86	British London	41 145	112·6 | 19·3 | 10·3	London T.A.Young	1867		2Cy.30"–54" 80 HP. T.A.Young,London	A. Tolhurst
313	✳D	**Drenthe** I Sc3mSr *Le Clercq* 86-86 2 *Dks3 B*	Dutch Rotterdam	1738 2288	295·9 | 36·4 | 26·1 P 86 f. B 68 f. F 46 f.	Newcastle C.Mitchell&Co.	1875	4 Blkds. [illegible]	3C.20½"38"&70"–42" 250 HP. N.E.MarineCo.,Ld.,Sundld.	Rotterdamsche [illegible]
314 17	I	**Drepano** I ScBn *Dadero* 86-87 2 *Dks3 B*	Italian Venice	1026 1558	278·8 | 32·8 | 23·6 B 60 f.	Sunderland W.Watson	1873	4 Blkds. [illegible]	2Cy.36"&68"–36" 365 HP. GeorgeClark,Sunderland	Navigazione Generale Italiana
315 52907		**Dresden** I ScSr *McVicar* 87-87	British Leith	489 772	225·0 | 29·7 | 15·5 P 42 f.	Glasgow Barclay,Curle&Co.	1865		2Cy.27"&47"–33" 117 HP. R.Napier&Sons,Glasgow.	Leith,Hull,& [illegible] Stm.Pkt.Co.
316 67846	✳	**Dresden** Stl ScSr *Kilwood* 74-78 1 *Dk*	British Goole	425 809	230·0 | 32·0 | 13·6 P 51 f. B 74 f. F 25 f.	Dundee W.B.Thompson	1884	5 Blkds. [illegible]	2Cy.30"&60"–33" 150 HP. W.B.Thompson,Dundee	Yorkshire Coal [illegible] Co.,Ld.
317	✳ [illegible]	**Dresden** Stl ScSr *Bruns* 81-89 3 *Dks*	German Bremen	2850 4500	390·5 | 46·7 | 33·0 P 70 f. B 131 f. F 76 f.	Glasgow FairfieldS.&E.Co.Ld.	1889		3Cy.30",50"&80"–54" FairfieldS.&E.Co.,Ld.,Glsg.	Norddeutscher [illegible]
318 [illegible]	✳	**Drever** Stl ScSr *Mutton* –88 1 *Dk&WhPoop*	British London	559 887	220·0 | 32·4 | 14·7 Q 88 f. B 43 f. F 24 f.	Newcastle Palmers'Co.,Ld.	1888	4 Blkds. [illegible]	3C.20",33"&54"–36" 163 HP. [illegible]	J. Fenwick & [illegible]
319 82547	✳	**Drewton** (ex Port de Rouen I ScSr *Dudding* 84-84 ex Java) 1 *Dk2 B*	British Hull	901 1394	250·0 | 32·0 | 21·0 Q 139 f. B 28 f. F 25 f.	Middlesbro' Backhouse&Dixon	1873	[illegible]	2Cy.29"&55"–32" 98 HP. N.E.MarineE.Co.,Ld.,Sn'dld.	Massey & Sawyer
320 84644	✳D	**Driffield** I ScBn *Goundry* 86-92 2 *Dks*	British Stockton	1179 1803	250·0 | 36·2 | 23·5 P 52 f. F 44 f.	Stockton Richardson,Duck&Co.	1882	4 Blkds. [illegible]	2Cy.31"&58"–36" 150 HP. Blair&Co.,Ld.,Stockton	F. Binnington
321 84310	BT.88	**Dromedary** I ScSr *Davies* 57-81	British Glasgow	397 822	250·2 | 30·0 | 14·7 P 186 f. F 31 f.	Glasgow Barclay,Curle&Co.	1881		2Cy.38½"&69"–48" 280 HP. Barclay,Curle&Co.,Glasgw	G.&J. Burns
322 [illegible]	✳	**Dromore** (ex Dawn) I ScSr *Dudding* 84-84 1 *Dk2 B*	British Hull	838 1110	246·6 | 30·0 | 16·4 F 230 f. F 27 f.	Hull Humphrys&Pearson	1872	4 Blkds. [illegible]	2Cy.27"&52"–33" 130 HP. Humphrys&Pearson,Hull	Massey & Sawyer
323	N	**Dronning Gyda** W ScSr *Travik* 82- 1 *Dk*	Norweg'n Christiansund	133 188	101·6 | 22·5 | 10·4 Q 8 f.	Sweden	1871		2Cy.16½"&25"–17½" 35 HP. KristinehamnsMek.Verks.	N. H. Knudtzon
324		**Dronning Lovisa** I ScSr *Thomsen* 86-88 (ex Drotning Lovisa) 1 *Dk*	Danish C'penhagen	384 626	200·8 | 23·6 | 12·4 P 72 f. B 57 f. F 30 f. 119.4	Norrköping MotalaCo.	1881	[illegible]	2Cy.34"&58"–32" 150 HP. MotalaCo.,Motala	Det Forenede Damp- skibs Selskab [illegible]
325	N	**Dronning Sophie** W ScSr *Stephensen* 1 *Dk2 B*	Norweg'n Stavanger	348 478	163·9 | 24·7 | 16·1 B 51 f.	Bergen J.Gran	1873		2Cy.19½"&37"–24" 55 HP. BergensMek.Verks,Bergen	J. Berner,junr.
326	✳N	**Dronning Sophie** Stl ScSr *Lindtner* 83-83 1 *Dk2 B*	Norweg'n Stavanger	270 363	132·7 | 23·4 | 12·6 Q 43 f. B 32 f. F 14 f.	Norrköping MotalaCo.	1878	4 Blkds.	2Cy.13½"&28½"–20½" 45 HP. MotalaCo.,Norrköping	F. Meling
327	N	**Dronningen** I PadSr *Rüse* 58-88 1 *Dk*	Danish Copenhagen	260 412	181·0 | 22·2 | 10·5 P 31 f. B 71 f.	Middlesbro' R.Dixon&Co.	1876	4 Blkds. [illegible]	2Cy.29"&55"–48" 130 HP. T.Richardson&Sons,Hrtlpl.	Det Forenede Damp- skibs Selskab [illegible]
328		**Drottning Sofia** I ScSr *Bengtsson* 75-75 2 *Dks&ShltrDk*	Swedish Stockholm	491 600	165·2 | 24·2 | 12·9 H 22 f. 20.2	Stockholm BergsundsMek.Verkst.	1875		2Cy.24"&48"–27" 100 HP. BergsundsMek.Vks,Stockholm	Ångf. Akt. Bol. [illegible] Sverige (J.G.[illegible])
329 [illegible]		**Drudge** Stl ScSt *Robinson* –87	British Newcastle	278 438	125·0 | 35·2 | 12·5	Newcastle SirW.G.Armstrong, Mitchell&Co. [illegible]	1887		3C.12"19½"&33"–30" 60 HP. WallsendSlipwayCo.Ld.,Newcastle	W. D. Cruddas
330	V	**Druentia** I ScSr *Jaubert* 83-86 2 *Dks*	French Marseilles	794 1183	234·9 | 32·5 | 21·2 B 52 f.	Rouen Claparede&Co.	1883	[illegible]	2Cy.33"&57"–41¼" 195 HP. Claparede&Co.,Rouen	Compagnie Fraissinet de Nav. à Vap.

✳ = **Lloyd's Register**; D = Nederlandsche Vereeniging; G = Germanischer Lloyd; I = Registro Italiano; R = American [illegible];
A = Veritas Austro-Ungarico; H = Veritas Ellenico; N = Norske Veritas; V = Bureau Veritas.

In the 2nd Column, B.T., with figures following, signifies Date of report of bottom sighting by Board of Trade Surveyors as condition of PASSENGER CERTIFICATE.

In the 3rd Column, C = Composite (iron frame, wood planked); I = Iron; Stl = Steel; W = Wood.

Further particulars of Vessels classed by Lloyd's Register will be found in Lloyd's Register Book and Lloyd's Yacht Register.

No.	Reg.	Nom, Matériaux, Gréement, etc. / Capitaine et dates d'entrée au service des Armateurs actuels et à bord du navire. / Ancien Nom, nombre de ponts, etc.		Pavillon. / Port d'Armement	Tonnage Net / Brut	Dimensions Long.	Larg.	Creux	Construit. Où. / Par qui.	Quand.	Nombre des Cloisons. / Détails sur le lest d'eau.	Machines. Détails. / Nom du Constructeur.	Armateurs ou Directeurs-Gérants.
		Druid *Bakey 86-88*	I ScSr *1 Dk 2 B*	British	446 703	191·4 *Q 30 f.*	28·7 *F 28 f.*	16·4	Middlesbro' Candlish,Fox&Co.	1856	4 Bhds.	1Cy.18"& 36"–30" 90 HP. W.Pile&Co.,Sunderland	J. Beng Smith
		Druid	I PadSr *1 Dk*	British	166 239	160·0 *B 18 f.*	21·6	9·6	Glasgow Tod&McGregor	1856		2Cy.44"–52" 170 HP. Tod&McGregor,Glasgow	Government of Canada
		Druid	I PadSr *1 Dk*	British	137 [illegible]	116·0 *Q 10 f.*	20·1	9·1	Glasgow	1856		2Cy. 135 HP.	George Chambers
	✻	Drumburlie *Sutherland 88-88*	Stl ScSr *2 Dks & 1 B*	British Liverpool	1608 2457	305·0 *P 28 f.*	39·7 *B 72 f.*	24·2 *F 31 f.*	Sunderland R.Thompson&Sons	1888	6 Bhds. CDB Aft FPT 67t APT 8t	3Cy.23½"38"&62"–42" 220 HP. [illegible]	S.S.DrumburlieCo.,Ld. (Gillison&Chadwick)
	✻	Drumfell *Chubb 76-88*	Stl ScSr *2 Dks & 1 B*	British Liverpool	1588 2500	305·0 *P 28 f.*	39·7 *B 72 f.*	24·2 *F 34 f.*	Sunderland R.Thompson&Sons	1888	5 Bhds. [illegible] FPT 67t APT 8t	3Cy.23½"38"&62"–42" 220 HP. North Eastern Marine Eng.Co.,Sunderland	S.S.Drumfell Co. (Gillison & Chadwick)
336	✻	Drumhendry *Harris 84-87*	I ScSr	British Glasgow	77 165	111·8 *P 45 f.*	19·8 *B 35 f.*	9·0 *F 16 f.*	Paisley AbercornS.B.Co.	1875	3 Bhds.	2Cy.16"& 25"–18" 27 HP. Fisher&Co.,Paisley	Nobel's ExplosivesCo., Ld.
337	✻	Drummond Castle *Winchester 75-85*	I ScBg *3 Dks*	British London	2352 3663	365·0 *B 96 f.*	43·5 *F 48 f.*	31·3	Glasgow J.Elder&Co.	1881	8 Bhds. DB [illegible] WT [illegible]	3C.32"58"&88"–57" 500 HP. [illegible]	D.Currie & Co.
338		Dryden *Graham 79-85*	I ScSr	British Liverpool	1848 2812	310·6 *P 79 f.*	38·8 *F 35 f.*	27·0	Newcastle [illegible]	1885		2Cy.35"& 70"–48" 300 HP. R.&W.Hawthorn,N'wcastle	Liv.,Brazil& Riv.Plate Stm.Nav.Co.,Ld.
339	✻	Duart Castle *Harris 84-87 (ex Adjutant)*	I ScBg *3 Dks*	British London	1180 1839	301·5 *F 22 f.*	33·2	25·1	Glasgow Barclay,Curle&Co.	1878	5 Bhds.	2Cy.36"& 63"–45" 230 HP. Barclay,Curle&Co.,Glasgw	D.Currie & Co.
340		Dublin (COLLISION) *Fenn*	I ScSmSr *1 Dk 2 B*	British Dublin	298 593	175·6 *F 18 f.*	28·0 *Q 10 f.*	13·9	Dublin Walpole,Webb&Co	1866	3 Bhds.	2Cy.26"& 45"–30" 90 HP. V.Coates&Co.,Ld.,Belfast	R.Tedcastle&Co
341	✻	Dubrovnik *Padich -86*	I ScSr *1 Dk*	Austrian Ragusa	139 220	131·2 *P 31 f.*	20·2 *B 6 f.*	9·5 *F 24 f.*	Dumbarton A.McMillan&Son	1879	4 Bhds. FPT 10t	2Cy.19"& 35"–24" 50 HP. Hutson&Corbett,Glasgow	A.T.Opuich&Co.
342	DV	Duburg *Bertelsen 85-86*	I ScSr *1 Dk & Spr Dk*	German Flensburg	921 1433	235·5 *P 20 f.*	32·8	16·5 23·9	Flensburg Flensbrgr.Schiffsb.Ges	1883	WB	2Cy.31"& 59"–33" 150 HP. FlensburgerSchiffsb.Ges.	DampfschiffahrtsGes. "Globus" (H.A.Petersen)
343	✻I	Duca di Galliera *Ricera -87 (ex Cosmos)*	Stl Sc *3 Dks & Spr Dk*	Italian Genoa	2841 4304	400·0 *F 46 f.*	44·0	23·1 31·6	Glasgow R.Napier&Sons	1883	5 Bhds. DB 4&L 222 MT 192	3C.40"64"&92"–60" 1000 HP. R.Napier&Sons,Glasgow	LaVeloce Nav. Italiana
344	✻	Duchess *Hammond 85-87*	I ScSr *1 Dk*	British Sunderland	1251 1926	284·5 *B 40 f. P 36 f.*	37·2 *Q 80 f.*	20·3 *F 46 f.*	Sunderland Short Brothers	1883	6 Bhds. WBa & f	2Cy.33"& 62"–42" 165 HP. Carr&Co.,Ld.,Sunderland	Taylor & Sanderson
345	✻	Duchess *Willis*	I ScSr *1 Dk 2 B*	British W. Hartlepl.	920 1430	245·5 *Q 94 f.*	34·0 *B 57 f.*	17·3 *F 31 f.*	W.Hartlepool E.Withy&Co.	1883	4 Bhds. DBaf 380	2Cy.30"& 57"–33" 130 HP. T.Richardson&Sons,Hrtlpl.	G.B.Harland & Co.
346	✻	Duchess *Watson 86-86*	Stl PadSk *1 Dk*	British N'rthShields	2 104	96·0	18·1	9·0	SouthShields J.Lawson & Co.	1886	3 Bhds.	1Cy.32½"–50" 42 HP. R.B.Findley&Ss.,N.Shields	J.T.Davison
347		Duchess of Con-naught	Stl PadSr *1 Dk*	British Portsmouth	149 342	190·6	26·1	8·8	Glasgow Aitken&Mansel	1884	6 Bhds.	2Cy.32"& 55"–60" 200 HP. J.&J.Thomson,Glasgow	L.& S.W.& L.B.& S.C. Ry.Cos.
348		Duchess of Edin-burgh	Stl PadSr *1 Dk*	British Portsmouth	149 342	190·6	26·1	8·8	Glasgow Aitken&Mansel	1884	6 Bhds.	2Cy.32"& 55"–60" 200 HP. J.&J.Thomson,Glasgow	L.& S.W.& L.B.& S.C. Ry.Cos.
349	✻	Duchess of Marl-borough *Farrier 81-84*	I ScSmSr *1 Dk*	British London	223 402	172·0 *Q 46 f.*	23·0 *F 30 f.*	12·5	Barrow BarrowS.B.Co.,Ld.	1874	3 Bhds.	2Cy.22"& 38"–32" 80 HP. BarrowS.B.Co.,Ld.,Barrow	West India & Panama Tel.Co.,Ld.
350		Duchess of Suther-land *Hobbins 79*	I TScSr *2 Dks*	British Dublin	111 775	251·0 *P 111 f.*	30·1	14·4	Newcastle A.Leslie&Co.	1869	4 Bhds.	3C.16½"26"&41"–30" 260 HP. LairdBros.,Birkn'd	London & N.W.Rail- way Co.
351	✻	DuchessadiGenova *Mataratini -88 (ex Mexico)*	StlScBk *2 Dks & Sp Dk*	Italian Genoa	2793 4304	400·0 *F 46 f.*	44·0	23·1 31·6	Glasgow R.Napier&Sons	1884	5 Bhds. DB a & E 222 MT 190t	3C.40"64"&92"–60" 1000 HP. R.Napier&Sons,Glasgow	LaVeloce Nav.Italiana
352		Duckenfield *Hardy*	I ScSr *1 Dk*	British NwcstlNSW	251 368	161·2 *Q 29 f.*	24·0 *P 23 f.*	12·0	London J.&W.Dudgeon	1875	4 Bhds.	2Cy.21"& 40"–27" 60 HP. J.&W.Dudgeon,London	J.& A.Brown
353	✻	Duddon *Evans 87-87*	I ScSr *1 Dk*	British Whitehaven	144 326	144·7 *Q 79 f.*	23·1 *B 10 f.*	10·6 *F 19 f.*	Preston R.Smith	1882	3 Bhds. DBaf 50t	2Cy.17"& 34"–24" 60 HP. R.Smith,Preston	G.Nelson & Sons
354		Duddon	I PadSp	British Liverpool	44 106	94·4	18·1	9·3	Belfast Harland&Wolff	1866		2Cy.27½"–46" 50 HP. G.K.Stothert&Co.,Bristol	W.J.Barratt
355		Dudley *Mountain*	I ScSr *1 Dk 2 B*	British S.Shields	527 825	219·0 *Q 56 f.*	28·2 *B 30 f.*	16·0	NorthShields T.&W.Smith	1865	3 Bhds. DBaf 36t FPT 6t APT 3t	2Cy.25"& 48"–30" 95 HP. Comp.'79. R. & W. Hawthorn, Newcastle	W.Runciman
356	GV	Duesternbrook *Oefjeng 82-82*	I ScSr *1 Dk*	German Kiel	366 581	176·1 *B 40 f.*	25·1	15·7	Kiel G.Howaldt	1882		2Cy.23"& 37½"–27" 70 HP. G.Howaldt,Kiel	Lange Gebrüder
357		Dufferin	Stl Pad	British Glasgow	650 1033	310·0	40·1	11·7	Dumbarton W.Denny&Bros.	1886		2Cy.37½"& 65"–72" 375 HP. Denny&Co.,Dumbarton	Irrawaddy Flotilla Co.,Ld.
358		Dugong	I ScCr *1 Dk*	British Brisbane	177 260	155·3	23·1	10·3	Maryborough J.Walker&Co.	1878		2Cy.18"&33"–20" 47 HP. J.Walker&Co,Maryborough	GovernmentofQueens- land
359		Duguay Trouin *Henon 84-84*	W Pad *1 Dk*	French St.Malo	45 145	95·8 *B 20 f.*	16·2	6·9	St.Malo F.Gautier	1886		2Cy.17½"&28½"–35½" 35 HP. Normend,Havre	Fichet & Co.
360	✻	Duitschland *Bronwer 82-87*	Stl PadSr *1 Dk & Awn Dk*	Dutch Flushing	870 1653	286·5	35·3	22·9	Glasgow FairfieldS.B.&E.C.Ld	1887	4 Bhds.	2Cy.60"& 104"–81" 600 HP. FairfieldS.B.&Eng.Co.,Ld.	Stoomvaart Maats- chappij "Zeeland"

✻=Lloyd's Register; D=Nederlandsche Vereeniging; G=Germanischer Lloyd; I=Registro Italiano; R=American Record;
A=Veritas Austro-Ungarico; H=Veritas Ellenico; N=Norske Veritas; V=Bureau Veritas.

Dans la 2me Colonne, B.T., avec les indications qui suivent, signifie la date du rapport de visite de la Carène par les experts du Board of Trade pour obtenir le certificat relatif au TRANSPORT DES PASSAGERS.

Dans la 3me Colonne, C=Composite (membrure en fer, bordé en bois); I=Fer; Stl=Acier; W=Bois.

Pour plus amples renseignements sur les navires classés par le Lloyd's Register, voir le Lloyd's Register Book et le Lloyd's Yacht Register.

No. in Book. / Official No.	Registry in which classed	Ship's Name. Material, Rig, &c. / Master, and Dates of Appointment to present Owner's Service and to Ship / Ship's Late Name. No. of Decks, &c.	Flag. / Port of Registry.	Tons. Net. / Gross	Dimensions. Length / Breadth / Depth	Built. Where. / By Whom.	When	No. of Bulkheads. / Particulars of Water Ballast.	Engines. Particulars. / Maker's Name.	Owners or Managers
361 68413	BT12.88	**Duke of Argyll** I Sc4mBk *Prentice* 86-86 3 *Dks*	British Barrow	2037 3115	379·8|38·3|28·5	PortGlasgow R.Duncan&Co.	1873	7 Blkds.	2Cy.46"&78"–48" 400 HP. D.Rowan&Co.,Glasgow	Eastern S.S.Co.
362 67767	BT1.88	**Duke of Argyll** I PadSr	British Dublin	386 749	250·0|29·2|15·1 *F* 28 *f.*	PortGlasgow R.Duncan&Co.	1873		2Cy.42"&75"–84" 300 HP. Sup.'81,A.&J.Inglis,Glasgw.	Dublin & Glasgow Sailing & Steam Packet Co.
363 68236	D	**Duke of Buccleuch** I Sc4mBk *Langlois* 86-86 3 *Dks*	British Barrow	2024 3106	379·8|38·3|28·5	Barrow Barrow S.B.Co.,Ld.	1873	8 Blkds.	2Cy.54"&90"–48" 500 HP. BarrowS.B.Co.,Ld.,Barrow	Eastern S.S.Co.
364 76892	D	**Duke of Buckingham** I Sc4mBk *Burnside* 86-86 3 *Dks*	British Barrow	2020 3128	384·0|38·3|28·5 *P* 82 *f.*	Barrow BarrowS.B.Co.,Ld.	1880	7 Blkds. WB·4	2Cy.42"&80"–48" 400 HP. BarrowS.B.Co.,Ld.,Barrow	Eastern S.S.Co.
365 73746	BT1.86	**Duke of Connaught** I PadSr *Crayther* 86-88 2 *Dks*	British Fleetwood	497 1052	267·5|30·2|15·7 *P* 138 *f.*	Barrow BarrowS.B.Co.,Ld.	1875	4 Blkds.	1Cy.34"&64"–72" 400 HP. BarrowS.B.Co.,Ld.,Barrow	Lan.& Yorks.& L.&N.W.Ry.Co.
366 40565		**Duke of Connaught** I Pad (ex Ardencaple) (BROKEN UP)	British London	71 127	159·1|16·1|6·4 *Q* 23 *f.*	PortGlasgow R.Duncan&Co.	1866		2Cy.38"–36" 50 HP. Rankin&Blackmore,Greek.	River Thames Steamboat Co.,Ld.
367 92237	✳	**Duke of Cornwall** StlScSr *Richards* 1 *Dk & Web Frames*	British Truro	1178 1815	260·5|36·6|18·2 *P*29*f.* *B*89*f.* *F*29*f.*	W. Hartlepool W.Gray&Co.	1888	4 Blkds.	3Cy.19"30½"&51"–36" 150 HP CentralMarineE.Co.W.Hpl	Duke of Cornwall S.S. Co.,Ld.(J.& B.B.)
368 68222		**Duke of Devonshire** I Sc4mBk *Long* 82-88 3 *Dks*	British Barrow	2024 3100	379·8|38·3|28·5	Barrow BarrowS.B.Co.,Ld.	1873		2Cy.46"&78"–48" 400 HP. BarrowS.B.Co.,Ld.,Barrow	Eastern S.S.Co.
369 68707		**Duke of Edinburgh** I ScSrYt 1 *Dk*	British Hull	72 106	114·1|19·7|8·2 *Len.* '81	Hull Humphrys&Co.	1873	3 Blkds.	2Cy.15"&30"–15" 35 HP. Humphrys&Co.,Ld.,Hull	Corporation of Trinity House
370 63444	BT5.88	**Duke of Edinburgh** I Pad (ex Craigrownie)	British London	68 123	175·0|17·1|6·8	PortGlasgow R.Duncan&Co.	1870		2Cy.28"–44" 70 HP. Rankin&Blackmore,Greek.	Victoria Steamboat Association,Ld.
371 58436	BT2.88	**Duke of Leinster** I PadSr *McClure*	British Dublin	299 721	250·0|29·2|15·0 *F* 26 *f.*	PortGlasgow R.Duncan&Co.	1870		2Cy.42"&72"–84" 529 HP. A.&J.Inglis,Glasgow	Dublin & Glasgow Sailing & Steam Packet Co.
372 68220	BT10.88	**Duke of Sutherland** I Sc4mBk *Cox* 74-87 3 *Dks*	British Barrow	2031 3116	379·8|38·3|28·5	PortGlasgow R.Duncan&Co.	1873	7 Blkds.	2Cy.46"&78"–48" 400 HP. D.Rowan&Co.,Glasgow	Eastern S.S.Co.,Ld.
373 58402		**Duke of Sutherland** I PadSr *Lewis* (TO BE BROKEN UP) 2 *Dks*	British Dublin	335 853	244·6|30·1|14·3 *F* 99 *f.*	Newcastle A.Leslie&Co.	1868	4 Blkds.	2Cy.63"–66" 270 HP. R.Stephenson&Co.,Nwcstl.	J.J.King
374 83942	BT5.88	**Duke of Westminster** Stl Sc4mSr *Reynolds* 88-88 3 *Dks*	British Barrow	2427 3726	400·0|40·3|28·7 *H* 300 *f.*	Barrow BarrowS.B.Co.,Ld.	1882	6 Blkds.	2Cy.43"&86"–54" 450 HP. BarrowS.B.Co.,Ld.,Barrow	Eastern S.S.Co.,Ld.
375	V	**Duna** I ScSr *Junker* 76- 1 *Dk*	Russian Riga	447 556	181·2|26·3|17·6	Glasgow Randolph, Elder&Co.	1868		2Cy.33"&40"–21" 80 HP. Randolph,Elder&Co.,Glasg.	Riga S.S.Co.
376		**Dunaj** I ScSr 1 *Dk*	Russian Odessa	410 546	215·0|29·0|9·7	London SamudaBros.	1871		2Cy.23"&41"–24" 60 HP. Penn&Sons,London	Russian Steam Nav. & Trading Co.
377 70319	BT2.88	**Dunara Castle** I ScSr *McEwan*	British Prt.Glasgow	203 123	180·4|24·5|12·5 *P* 70 *f.*	PortGlasgow Blackwood &Gordn	1875		2Cy.30"&52"–30" 130 HP. Comp.'85 Blackwood&Gordon PortGlasgow	M.Orme
378 70251	✳	**Dunbar** (ex Poet of the Rental) I Sc *Wigzell* 2 *Dks*&3 *H*	British Cardiff	1172 1802	280·2|34·0|24·0 *B* 31 *f.* *F* 16 *f.*	Sunderland W.Doxford&Sons	1876	5 Blkds. DB&70 MT&181	2Cy.26"&58"–42" 170 HP. R.Stephenson&Co.,Nwcstl.	J.& M.Gunn
379 87163	✳	**Dunbar Castle** I ScBg *Rigby* 78-87 3 *Dks*	British London	1741 2682	335·0|38·2|27·2 *B* 86 *f.* *F* 40 *f.*	Glasgow Barclay, Curle&Co.	1883	6 Blkds. MT200 T&80	2Cy.36"&68"–48" 270 HP. Barclay,Curle&Co.,Glasgw	D.Currie & Co.
380 80866	BT2.88	**Dunbrody** I Sc3mSr *Burns* 86-86 2 *Dks*	British Waterford	267 856	245·0|32·1|14·7 *P* 77 *f.* *B* 54 *f.* *F* 52 *f.*	Glasgow London&Glasgw.Co.Ld.	1886	4 Blkds. APT&20 DB&36 FPT&18	3C.24½"39"&62"–42" 250 HP London&GlasgowC.,Ld,Gls.	Waterford S.S.Co.
381 93476	BT9.88	**Duncan** Stl T–ScSr *Smith* –88 1 *Dk*	British Dundee	41 221	120·0|24·1|10·6	Dundee PearceBros.	1888	4 Blkds.	3Cy.12"18"&30"–20" 90 HP Pearce,Brothers,Dundee	Dundee Harbour Trustees
382 89997	BT5.88	**Dundee** StlScSr *Speedy* 2 *Dks*	British Dundee	649 1267	270·0|34·1|15·7 *P* 58 *f.* *B* 66 *f.* *F* 33 *f.*	Dundee Gourlay Bros.&Co.	1885	5 Blkds.	3C.26"48"&70"–42" 300 HP GourlayBros.&Co.,Dundee	Dundee,Perth & London Shipping Co.
383 72530	BT10.88	**Dundee** I Pad *Edward*	British Dundee	160 284	149·3|27·1|7·1	Renfrew W.Simons&Co.	1875		2Cy.28"&42"–42" 80 HP W.Simons&Co.,Renfrew	Trustees of the Harbour of Dundee
384 86541	✳	**Dundela** I & Stl ScSr *Long* 1 *Dk*	British Belfast	566 876	224·5|28·2|14·0 *B* 124 *f.* *H* 41 *f.*	Belfast Harland&Wolff	1883	2 Blkds.	2Cy.26"&50"–36" 98 HP. Harland&Wolff,Belfast	Harland & Wolff (Carlisle & Co.,Managers)
385 91492	✳	**Dundonald** (ex Edendale) I ScSr *Hadden* 87-87 1 *Dk*	British Glasgow	319 510	165·8|26·1|12·4 *Q & B* 105 *f.* *F.* 19 *f.*	Blyth BlythS.B.Co.,Ld.	1885	4 Blkds.	2Cy.22"&42"–27" 80 HP. M.Pratt,Huddersfield	J.& W.W.C.Smith
386 89400	✳	**Dunedin** I ScSc *Cumming* 83-83 2 *Dks*&3 *H*	British London	1418 2198	286·0|37·3|22·9 *B* 64 *f.* *F* 26 *f.*	Sunderland W.Doxford&Sons	1883	4 Blkds.	2Cy.34"&68"–45" 250 HP. W.Doxford&Sons,Sundrld.	J.& M.Gunn
387 87295	✳	**Dunedin** I ScSr *Campbell* 84-84 1 *Dk*2 *H*	British Leith	858 1326	240·6|34·1|17·3 *Q* 84 *f.* *B* 57 *f.* *F* 28 *f.*	Glasgow A.Stephen&Sons	1884	4 Blkds. APT&38 MT&86	2Cy.28"&52"–36" 140 HP. A.Stephen&Sons,Glasgow	"Dunedin" S.S.Co. (Henderson & McIntosh)
388 68952	✳	**Dunelm** I ScSr *Laming* 79-79 1 *Dk*2 *H*	British Sunderland	513 797	209·5|29·7|15·8 *P* 16 *f.* *F* 22 *f.* *B* 33 *f.*	Sunderland J.Blumer&Co.	1877	4 Blkds. DB&f 194 APT&8	2Cy.27"&51"–33" 99 HP. J.Dickinson,Sunderland	W.A.Watson
389 86335	✳	**Dungonnell** I ScSr *Cooper* 85-87 1 *Dk*	British Belfast	129 273	158·0|21·7|11·1 *Q* 21 *f.* *B* 30 *f.* *F* 21 *f.*	Belfast McIlwaine & Lewis Co.,Ld.	1883	5 Blkds. FPT&8 APT&4	2Cy.19"&32"–27" 50 HP. McIlwaine&Lewis,Belfast	Antrim Iron Ore Co.
390 89978	✳	**Dunholme** I ScSr *Wilkinson* 1 *Dk*2 *H*	British W.Hrtlpool	1142 1762	256·2|34·3|19·9 *Q*50*f.B*66*f.P*28*f.F*33*f.*	W.Hartlepool W.Gray&Co.	1883	4 Blkds. DB&f 528	2Cy.32"&66"–39" 150 HP. Blair&Co.,Ld.,Stockton	Britain S.S.Co.

✳ = **Lloyd's Register**; D = Nederlandsche Vereeniging; G = Germanischer Lloyd; I = Registro Italiano; R = American Record;
A = Veritas Austro-Ungarico; H = Veritas Ellenico; N = Norske Veritas; V = Bureau Veritas.

In the 2nd Column, B.T., with figures following, signifies Date of report of bottom sighting by Board of Trade Surveyors on condition of PASSENGER CERTIFICATE.

In the 3rd Column, C = Composite (iron frame, wood planked); I = Iron; Stl = Steel; W = Wood.

Further particulars of Vessels classed by Lloyd's Register will be found in Lloyd's Register Book and Lloyd's Yacht Register.

No.		Nom, Matériaux, Gréement, etc. — Capitaine et dates d'entrée au service des Armateurs actuels et à bord du navire. — Ancien Nom, nombre de ponts, etc.		Pavillon. — Port d'Armement	Tonnage. Net. — Total	Dimensions. Longueur | Largeur | Creux. (constructions sur le Pont)	Construit. Où. — Par qui.	Quand.	Nombre des cloisons. — Détails sur le lest d'eau.	Machines. — Détails. Nom du Constructeur.	Armateurs ou Directeurs-Gérants.
	*	Dunkeld *Broadfoot 7-80*	I ScSr *2 Dks*	British London	742 1158	240·8|32·3|18·8 *B 57 f. F 21 f.*	Glasgow R.Napier&Sons	1878	7 Bhds. Tc53ft	2Cy.32"& 55"–36" 150 HP. R.Napier&Sons,Glasgow	D.Currie & Co.
	*	Dunluce *Wolgery 88-88*	I & Stl ScSr *1 Dk*	British Belfast	568 877	224·5|28·2|14·0 *B 32 f.*	Belfast Harland&Wolff	1883	3 Bhds. CDB	2Cy.26"& 50"–36" 98 HP. Harland&Wolff,Belfast	Harland & Wolff (Carlisle & Co.,Managers)
	*	Dunmail *Bennett*	I ScSr *1 Dk & Web Frames*	British Newcastle	592 952	216·0|31·3|14·1 *Q 49 f. B 55 f. P 36 f. S 24 f.*	Sunderland S.P.Austin&Son	1883	4 Bhds. CDB52ft FPT30t APT32t	2Cy.28½"&53"–33"120 HP. Blair&Co.,Ld.,Stockton	Sharp & Co.
	*	Dunmore *Campbell 81-88*	Stl ScSr *1 Dk 2 B*	British Leith	820 1282	245·0|34·1|15·7 *Q 85 f. B 78 f. F 32 f.*	Ayr S.M.Knight&Co.	1888	4 Bhds. CDB540t APT17t	3C.18"30"&48–36" 117 HP. Hutson&Corbett,Glasgow	Henderson & M'Intosh
	*	Dunrobin *Peebles 85-88*	I Sc3mSr *1 Dk*	British Glasgow	152 340	160·0|23·1|11·0 *Q 55 f. F 22 f.*	Bowling Scott&Co.	1885	3 Bhds. FPT49t APT10t	2Cy.21"& 40"–30" 70 HP. Muir&Houston,Glasgow	W.E.Davies
		Dunrobin *White*	I PadSp	British Cardiff	19 109	95·4|18·2|9·6	Cardiff J.&W.Gunn	1876		1Cy.33½"–52" 37 HP. J.&W.Gunn&Co.,Cardiff	R.H.Strong & Co.
397	*	Dunrobin Castle *Duncan 72-88*	I ScBu *3 Dks*	British London	1778 2783	342·3|38·3|28·3	Glasgow R.Napier&Sons	1875	5 Bhds.	2Cy.47"& 75"–48" 300 HP. R.Napier&Sons,Glasgow	D.Currie & Co.
	*	Dunstanborough *Eason*	I Sc3mSr *1 Dk 2 B*	British N.Shields	108 1028	216·0|30·5|17·5 *Q 51 f. F 31 f.*	NorthShields T.&W.Smith	1871	4 Bhds. DBa&f	2Cy.27"& 50"–27" 98 HP. R.&W.Hawthorn,Newcastle	W.H.A.Johnson
399	*	Dunvegan *English 87-87*	I Sc3mSr *1 Dk*	British Glasgow	108 265	140·0|22·2|10·5 *Q 45 f. B 7 f. F 19 f.*	Bowling Scott&Co.	1883	4 Bhds. FPT3t	2Cy.18"&34"–28" 50 HP. J.Gilmour&Co.,Glasgow	J.& W.W.C.Smith
		Dunvegan Castle *Richardson 81-81*	I ScSr	British Barrow	177 254	151·5|20·6|11·0 *Q 21 f.*	PortGlasgow Blackwood&Gordn	1868		2Cy32"–30" 60 HP. Blackwood& Gordon,P.Gls.	J.Mawson
	F	Dupleix	I Sc *3 Dks*	French Marseilles	816 1410	292·0|32·1|24·2 *F 45 f. H 30 f.*	LaCiotat Messageries Maritimes	1862		3C.36¾"&51¼"(2)-31½"310 HP. MsagriesMar.,LaCiotat	Messageries Maritimes
402	V	Dupuy de Lôme *Destruille 85-87*	I Sc4mSr *2 Dks & Spr Dk*	French Havre	2662 3447	379·2|40·0|21·7 *F 37 f. B 39 f. P 296*	Havre Forges&C.delaMédit.	1882		2Cy.39½"&70½"–43"375 HP. Forges&C.delaMedit.Havre	Cie.Commerciale de Transports à vap.Français
403	*	Duquesa de Vista-hermosa	Stl ScSr	Spanish Bilbao	1526 2356	290·0|40·0|20·2 *B 108 f. P 35 f. F 64 f.*	Newcastle Palmers'Co.,Ld.	1889		3Cy.22"35"&58"–42" Palmers'Co.,Ld.,Newcastle	C. de Murrieta & Co.
404		Durance *Eharcal 68-83*	I Sc3mSr *2 Dks*	French Marseilles	319 481	183·4|28·0|15·5 *B 28 f. F 18 f.*	Greenock Caird&Co.	1855		2C.24½"&42"–26¼" 100 HP. Fraissinet&Co.,Marseilles	Fraissinet & Co.
405	*	Durban *Mann 87-88*	I ScSr *2 Dks 3 B*	British Southampton	1563 2809	360·0|38·5|28·5 *B 116 f. F 41 f.*	Sunderland J.Laing	1877	5 Bhds.	2Cy.56 & 97"–54" 550 HP. R.&W.Hawthorn,Newcastle	Union Steam Ship Co. Ld.
406	*	Durham (ex Smyrna) *Dobson 85-85*	I Sc3mSr *2 Dks*	British Hull	1444 1813	299·3|31·2|24·5 *P 25 f. B 27 f.*	Newcastle A.Leslie&Co.	1864	4 Bhds.	2Cy.28"& 54"–42" 120 HP. Bailey&Leetham,Hull	W.S.Bailey
407		Durham *Young 54-66*	I Pad	British N.Shields	37 141	94·1|26·4|8·2	Newcastle	1866		2Cy.25½"–45" 44 HP. T.Hepple&Co.,Newcastle	Tyne Improvement Commissioners
408	*	Durham City *Galbraith 86-86*	I ScBu *2 Dks & Spr Dk*	British W.Htlpool.	2037 3092	314·0|43·0|27·4 *B 81 f. F 37 f.*	W.Hartlepool W.Gray&Co.	1882	4 Bhds.	2Cy.45"& 83"–48" 300 HP. Blair&Co.,Ld.,Stockton	C.Furness
409		Duro (ex Hijos de Tomabo) *Vigil 81*	I ScS	Spanish Gijon	311 476	179·4|26·9|15·0	Glasgow	1874		2Cy.19"& 41"–28" 72 HP.	O.de Olavarria & Co.
410		Dux *Willborg 84-87*	Stl ScSr *1 Dk 2 B*	Swedish Stockholm	769 1011	208·1|29·3|17·0 *P 126 f. F 28 f.*	Norrköping MotalaCo.	1879	4 Bhds.	2Cy.24⅝"&48⅛"–33"130 HP. MotalaCo.,Gothenburg	Stockholms Angf.Red. Bolag.(L.Peyron)
411		Dwina *Tune 86-86*	I ScSr *1 Dk 2 B*	British London	817 978	233·2|28·9|14·9 *P 184 f. Q 9 f.*	Hull C.&W.Earle	1862		2Cy.34"–30" 80 HP. C.&W.Earle,Hull	C.M.Norwood & Co.
412	*	Dwina *Cockburn 85-87*	I Sc3mSr *1 Dk 2 B*	British Leith	464 729	241·6|27·7|15·0 *P 32 f.*	Glasgow Barclay,Curle&Co.	1861	5 Bhds. MT19t	2Cy.24"& 42"–30" 70 HP. J.Penn&Sons,London	Leith,Hull & Hamb. Stm.Pkt.Co.
413	G	Dwina *Bollnr -83*	I ScSr	Russian St.Petersburg	64 121	117·9|18·1|8·6	Rostock N.H.Witte	1870		1Cy.15"26"113½" N.H.Witte,Rostock	Peter Berg
414	*	Dynamic *Downes 74-86*	I ScSr *2 Dks*	British Belfast	324 884	243·3|31·9|15·3 *P 35 f. F 20 f.*	Belfast Harland&Wolff	1883	5 Bhds. Tanks 36ft	2Cy.35"& 70"–45" 200 HP. Harland&Wolff,Belfast	Belfast S.S.Co.,Ld.
415		Dynamo *Hewitt*	Stl Sc3mSr *1 Dk 2 B*	British Hull	319 501	175·7|25·0|13·8 *Q 28 f. B 24 f. F 16 f.*	Hull Earle's Co.,Ld.	1884	5 Bhds. DB/65t APT15t	3Cy17"27"&48"–27"110 HP. Earle's Co.,Ld.,Hull	T.Wilson,Sons & Co.
416	*	Dyo Fili (ex Sheffield) *Kakarouka*	I ScSr *2 Dks*	Greek Syra	706 945	240·8|29·0|16·7 *Q 33 f. F 28 f.*	W.Hartlepool Denton,Gray&Co.	1870	4 Bhds. pt DB	2Cy.25"& 45"–36" 96 HP. T.Richardson&Sons,Hpl.	N.Calamaris

ADDITIONAL SHIPS ARE INSERTED IN THE SUPPLEMENT AT THE END OF THE REGISTER.

*=Lloyd's Register; D=Nederlandsche Vereeniging; G=Germanischer Lloyd; I=Registro Italiano; R=American Record; A=Veritas Austro-Ungarico; H=Veritas Ellenico; N=Norske Veritas; V=Bureau Veritas.

Dans la 2me Colonne, B.T. avec les indications qui suivent, signifie la date du rapport de visite de la Carène par les experts du Board of Trade pour obtenir le certificat relatif au TRANSPORT DES PASSAGERS.

Dans la 3me Colonne, C=Composite (membrure en fer, bordé en bois); I=Fer; Stl=Acier; W=Bois.

Pour plus amples renseignements sur les navires classés par le Lloyd's Register, voir le Lloyd's Register Book et le Lloyd's Yacht Register.

No. in Book / Official No.	Registry in which Classed	Ship's Name. Material, Rig, &c. / Master, and Dates of Appointment to present Owner's Service and to Ship. / Ship's Late Name. No. of Decks, &c.	Flag. / Port of Registry.	Tons. Net. / Gross.	Length	Breadth	Depth	Built. Where. / By Whom.	When	No. of Bulkheads. / Particulars of Water Ballast.	Engines. Particulars. / Maker's Name.	Owners or Ma[nagers]
1 58783	✳	**E. S. Jobson** I ScSr *Shadforth 85-88* 1 *Dk2 B*	British Hartlepool	719 1113	230·8 *Q 88 f.*	30·1 *B 52 f.*	16·7 *F 30 f.*	Hartlepool Withy, Alexander&Co.	1871	4 Blds. [illegible]	2Cy.28"&[illegible]" 110 HP T.Richardson&Sons,Hrtlp'l	G. Horsley & S[ons]
2 135958		**E. V. McCaulley** I	American Philadelphia	136	105·0	20·6	11·6	Camden, N.J. J.H.Dialogue	1887		3Cy.13"21"&35"-24" J.H.Dialogue,Camden	James McCau[lley]
3 77230	✳	**Eagle** I ScSr *Samuels 88-88* 1 *Dk2 B*	British Glasgow	773 1214	229·7 *P 131 f.*	30·0	17·8 *F 33 f.*	Port Glasgow R.Duncan&Co.	1879	4 Blds. [illegible]	2Cy.27"&50"-36" 99 HP Rankin&Blackmore,Gr'nck	P. H. Dixon & [Co.]
4 59[illegible]		**Eagle** W ScBk *Jackman 69-5* 1 *Dk*	British St. Johns, NF	344 506	156·1	28·7 *F 26 f.*	18·1	Dundee A.Stephen&Sons	1871		2Cy. 70 HP	C. T. Bowring & [Co.]
5 23427	BT5.88	**Eagle** (TO BE BROKEN UP) I PadSr	British London	177 332	200·0	24·1	11·3	Northfleet Pitcher&Co.	1853		2Cy.49"-48" 130 HP Gen.Steam.Nav.Co.,Deptford	Gen. Stm. Nav. C[o.]
6 41073		**Eagle** W Sc3mSr 1 *Dk*	British Sydney	147 227	151·0 *Q 1 [illegible]*	19·1	10·0	Pyrmont, N. S. W.	1849		2Cy. 80 HP	William Su[illegible]
7 87543		**Eagle** Stl Padl[illegible] 1 *Dk*	British Auckland, NZ	138 219	135·7	21·3 *H 70 f.*	6·6	Auckland [illegible]	1886		2Cy.18"&36"-42" 70 HP McLachlan&Co.,Paisley	The Devonpo[rt] Ferry Co. Ld.
8 18031	BT5.88	**Eagle** I PadSk *McGregor 84-84*	British Glasgow	47 208	219·5	20·5	7·8	Glasgow [illegible]	1864		2Cy.50"-56" 85 HP W.King&Co.,Glasgow	W. Buchanan
9 85095	✳	**Eagle** Stl ScSr *Wilson 86-87* 1 *Dk*	British Melbourne	25 208	125·3	22·1	12·5	Dundee [illegible]	1885	4 Blds.	2Cy.23½"&47"-27" 95 HP A.Shanks&Son,Arbroath	Huddart, Park[er]
10 80128	✳ BT6.88	**Eagle** I ScTug *Kitt* 1 *Dk*	British Falmouth	18 102	100·2	17·1	10·0	Hayle Harvey&Co.	1883	4 Blds.	2Cy.16½"&31½"-21" 80 HP Harvey&Co.,Hayle	J. H. Deeble
11 89540	✳	**Eagle** I ScTug *Bruce 85-85* 1 *Dk*	British London	6 95	82·4	17·5	9·7	Delfshaven Maats. de Maas	1884	3 Blds.	2Cy.19"&35"-22" 70 HP Maats.de Maas,Delfshaven	Todd & Son
12 85116	✳D	**Ealing** I ScBk *Raeburn* 2 *Dks3 B*	British London	1345 2063	285·0 *P 31 f.*	36·0 *B 62 f.*	24·0 *F 38 f.*	Blyth Hodgson&Soulsby	1882	5 Blds. [illegible]	2Cy.35"&68"-45" 250 HP R.&W.Hawthorn,N'wcastle	Watts, Ward & [Co.]
13 51302	BT5.88	**Earl Percy** (COLLISION) I Sc3mSr *Clay 54-72*	British Newcastle	590 928	236·7 *P 25 f.*	28·2 *Q 27 f.*	16·2	Newcastle Palmer Bros.&Co.	1865		2Cy.28"&56"-36" 130 HP R.&W.Hawthorn,N'wcastle	Tyne S.S. Co. L[d.]
14 70020	BT2.88	**Earl Spencer** I Pad *Clay 80*	British Dublin	374 855	253·7 *Q 76 f.*	29·1 *B 61 f.*	14·7 *F 72 f.*	Birkenhead Laird Bros.	1874		2Cy.64"-72" 850 HP Laird Bros.,Birkenhead	London & N.W[.R.] Co.
15 94527	✳ BT11.88	**Earl of Aberdeen** StlScSr *McBain -89* 1 *Dk2 B*	British Aberdeen	326 734	210·0 *B 46 f.*	30·2 *P 55 f.*	14·8 *F 35 f.*	Aberdeen Hall, Russell&Co.	1889	4 Blds.	3Cy.21"33"&51"-36" 205 HP Hall,Russell&Co., Aberdeen	Aberdeen, New[castle &] Hull Stm. Nav. [Co.]
16 89209	BT3.88	**Earl of Bute** I PadSk *Cate 86-86* 1 *Dk*	British Cardiff	14 132	108·0	18·6	9·7	South Shields J.T.Eltringham	1886		1Cy.36½"-54" 80 HP J.P.Rennoldson&Sn.,S.Shlds.	Martin & Mar[illegible]
17 87485	✳D	**Earl of Chester** I ScSr *Richards 79-83* 1 *Dk2 B*	British Cardiff	940 1449	246·5 *Q 92 f.*	35·1 *B 50 f.*	16·3 *F 27 f.*	Sunderland J.Blumer&Co.	1883	4 Blds. [illegible]	2Cy.29½"&56"-39" 125 HP T.Clark&Co.,Newcastle	"Earl of Che[ster"] Co., Ld.([illegible])
18 84478	✳	**Earl of Dumfries** I ScSr *Arthur 82-82* 1 *Dk& WebFrms*	British Cardiff	971 1509	244·6 *Q 86 f.*	36·1 *B 48 f.*	17·3 *F 30 f.*	Sunderland S.P.Austin&Son	1882	4 Blds. [illegible]	2Cy.31"&57"-36" 99 HP J.Dickinson,Sunderland	Martin & Mar[illegible]
19 95157	BT5.88	**Earl of Dunraven** I PadSp *West 88-88*	British Cardiff	48 174	122·5	20·3	10·7	South Shields J.T.Eltringham	1888		1Cy.41"-60" 98 HP J.P.Rennoldson&Ss,S.Shlds.	Martin & Mar[illegible]
20 2837	✳	**Earl of Durham** I ScSr *Henderson 88-88* 1 *Dk2 B*	British Sunderland	403 593	189·6	28·2 *Q 24 f.*	16·2	Newcastle C.Mitchell&Co.	1854	4 Blds. [illegible]	2Cy.36"-26" 70 HP G.Clark,Sunderland	Earl of Durha[m]
21 20774		**Earl of Ellesmere** I Pad	British Liverpool	52 114	109·9	20·0	8·0	Preston Ogle&Co.	1857		2Cy.31"-42" 60 HP Ogle&Co.,Preston	Manchester Sh[ip] Co.
22 22810	BT5.88	**Earl of Erne** I PadSr	British Dundalk	277 602	219·4	31·5 *Q 26 f.*	14·7	Glasgow R.Napier&Sons	1855		2Cy.37"&70"-72" 300 HP A.&J.Inglis,Glasgow	Dundalk & Ne[wry] Packet Co. Ld.
23 89205	BT5.88	**Earl of Jersey** I PadSk *James 60-86*	British	14 133	108·0	18·5	9·8	South Shields J.T.Eltringham	1886		1Cy.36½"-54" 80 HP J.P.Rennoldson&SnsS.Shlds.	Martin & Ma[illegible]
24 13861		**Earl of Malmesbury** W Pad	British Plymouth	59 108	107·4	17·2	9·7	Cowes	1825		2Cy. 50 HP	Countess of M[illegible] combe
25 15873		**Earl of Mar and Kellie** (ex Sailing Vessel) W ScBk *Marks*	British Peterhead	278 430	144·3	25·8 *F 22 f.*	16·6	Alloa	1856		2Cy.20"&40"-24" 70 HP W.B.Thompson,Dundee	J. Duthie
26 70359	BT10.88	**Earl of Powis** I TwnScSk 1 *Dk*	British Chester	116	105·6	21·3	7·6	Port Glasgow R.Duncan&Co.	1882	4 Blds.	2Cy.18"&34"-24" 50 HP Rankin&Blackmore,Gr'nck	Shropshire Uni[on] Railway & Can[al]
27 83800	✳D	**Earl of Rosebery** I ScSr *Gregory 82-89* 1 *Dk2 B*	British Cardiff	725 1125	239·5 *Q 111 f.*	33·4 *F 28 f.*	15·1 *B 27 f.*	Glasgow A.Stephen&Sons	1880	4 Blds. [illegible]	2Cy.28"&52"-36" 99 HP A.Stephen&Sons,Glasgow	Martin & Mar[illegible]
28 73758	BT6.88	**Earl of Ulster** I PadSr *Cook 80-87* 2 *Dks*	British Fleetwood	467 1084	275·0 *P 145 f.*	30·2 *F 21 f.*	14·4	Barrow Barrow S.B.Co.	1878	4 Blds.	4Cy.31"&64"-72" 350 HP Barrow S.B.Co.,Barrow	Lan. & Yorks. & N.W. Railway [Co.]
29 68187	✳ BT1.89	**Earl of Zetland** I ScSr *Nicolson 72-77* 1 *Dk*	British Lerwick	162 254	144·8 *P 53 f.*	20·2 *B 25 f.*	8·9 *F 28 f.*	Paisley J.Fullerton&Co.	1877	4 Blds.	2Cy.18½"&33"-21" 50 HP J.Howden&Co.,Glasgow	Shetland Isles S[tm.] Nav. Co. Ld.
30 89049		**Earn** I ScSr	British Glasgow	74 109	78·2	22·0	8·0	Port Glasgow Blackwd.&Gordon	1884		2Cy.11"-16" 32 HP D.Gordon,Port Glasgow	W. Arrol & Co.

✳ = **Lloyd's Register**; D = Nederlandsche Vereeniging; G = Germanischer Lloyd; I = Registro Italiano; R = American R[ecord];
A = Veritas Austro-Ungarico; H = Veritas Ellenico; N = Norske Veritas; V = Bureau Veri[tas]

In the 2nd Column, B.T., with figures following, signifies Date of report of bottom sighting by Board of Trade Surveyors as condition of PASSENGER [CERTIFICATE.]

In the 3rd Column, C = Composite (iron frame, wood planked); I = Iron; Stl = Steel; W = Wood.

Further particulars of Vessels classed by Lloyd's Register will be found in Lloyd's Register Book and Lloyd's Yacht Register.

No.	Registre	Nom, Matériaux, Gréement, etc. — Capitaine et dates d'entrée au service des Armateurs actuels et à bord du navire. — Ancien Nom, nombre de ponts, etc.		Pavillon. — Port d'Armement	Tonnage Net — Total	Dimensions Long.	Larg.	Creux	Construit. Où. — Par qui.	Quand	Nombre des cloisons. — Détails sur le lest d'eau.	Machines. Détails. — Nom du Constructeur.	Armateurs ou Directeurs-Gérants.
31	✱	Earnest *Sansom*	W ScK 1 *Dk*	British Leith	23 30	62·2	17·5	6·8	Leith [illegible]	1884		2Cy.8"& 15"–12" 40 IP. J.Cran&Co.,Leith	A.Cook & Co.
32	✱	Earnholm *Black 81-88*	I ScSr *2 Dks & pt Awn Dk*	British Aberdeen	175 307	170·0 *Pdps.Awn D 98f. F 16f.*	21·3	13·0 *T 36*	Glasgow A.&J.Inglis	1874	5 Blds. *DB aft & 35t*	3C.17"27"& 48"–30"110IP. Tryd&C°87 HallRussell&Co.Abn	Aberdeen,Leith & Moray Firth Steam Shipping Co., Ld.
33	✱	Earnmoor *Gray 87-87*	Stl ScSr *2 Dks 3 B*	British Newcastle	1320 2010	280·0 *B 62 f.*	36·2 *F 29 f.*	23·0 *P 42 f.*	Newcastle Palmer'sCo.,Ld.	1887		3C.22"35"&58"–42"250IP. Palmer'sCo.,Ld.,Newcastle	"Earnmoor" S.S.Co., Ld.(A.Earnshaw)
34	✱	Earnwell *Mumford 87-87*	Stl ScSr *2 Dks 3 B*	British Newcastle	1323 2013	280·5 *B 62 f.*	36·2 *F 29 f.*	23·0 *P 43 f.*	Newcastle Palmer'sCo.,Ld.	1886	5 Blds. *DB aft 482t*	3C.22"35"&58"–42"250IP. Palmer'sCo.,Ld.,Newcastle	"Earnwell" S.S.Co., Ld.(A.Earnshaw)
35	✱D	Earsdon *Arthur 84-84*	I ScSr *1 Dk 2 B*	British NrthShields	825 1268	226·0 *Q 88 f.*	33·2 *B 49 f.*	17·6 *F 26 f.*	NorthShields T.&W.Smith	1883	4 Blds. *DB aft 273t*	2Cy.26"& 52"–33" 112 IP. WallsendSlipCo.,Ld.,Newc.	Eskdale,Dodds & Co.
36	✱	Easby *Prideaux 85*	I ScSr *1 Dk 2 B*	British Middlesbro'	989 1489	250·7 *Q 185 f.*	32·0 *H 26 f.*	21·4 *P 25 f.*	Stockton Richardson,Duck&Co.	1872	5 Blds. *DB aft*	2Cy.28"& 56"–36" 140 IP. C.D.Holmes&Co.,Hull	James Paterson & Co.
37	✱	Easdale *M'Phedron*	I ScSr *1 Dk*	British Glasgow	45 102	80·7 *Q 37 f.*	18·5 *B 5 f.*	8·2 *F 12 f.*	Port Glasgow Russell&Co.	1876	3 Blds.	2Cy.12"& 21"–21" 20 IP. Walker,Henderson&Co.,Gls	J.M'Phedron
38	✱	Easington *Morgan 83-85*	I ScSr *1 Dk 2 B*	British W.Hartlepl.	1164 1801	256·5 *Q 68 f. P 38 f.*	34·6 *F 31 f.*	19·7 *B 150 f.*	W.Hartlepool W.Gray&Co.	1880	4 Blds. *DB aft 335t*	2Cy.32"& 60"–39" 150 IP. Blair&Co.,Ld.,Stockton	J.Wood & Co.
39	✱D	East Anglia *Dowell 85*	I ScSr *2 Dks 3 B*	British Newcastle	1450 2230	290·9 *P 63 f.*	37·3 *B 53 f. F 25 f.*	24·7 *H 24 f.*	Newcastle Schlesinger, Davis	1882 &Co.	4 Blds. *DB aft*	2Cy.35"& 65"–42" 250 IP. T.Clark&Co.,Newcastle	International Line S.S. Co., Ld.
40	✱	East Anglian *Beacher 76-82*	I ScSr *1 Dk*	British K.Lynn	149 304	140·2 *Q 49 f.*	21·8 *B 26 f.*	11·1 *F 32 f.*	Paisley J.Fullerton&Co.	1882	4 Blds. *FPT [illegible] APT [illegible]*	2Cy.18"& 34"–28" 50 IP. J.&T.Young,Ayr	East Coast S.S. Co., Ld.
41	✱	Eastbourne *Watson 85-85*	I ScSr *2 Dks 3 B*	British Hull	1463 2210	300·2 *P 26 f.*	36·0 *B 52 f.*	25·3 *F 28 f.*	Sunderland R.Thompson,jun.	1878	5 Blds. *DB aft 22t FPT 163t*	2Cy.36"& 68"–42" 200 IP. G.Clark,Sunderland	Cammell,Woolf & Haigh
42		Eastern Oregon *(ex City of Palatka)*	I T ScSr *1 Dk & Hur Dk*	American Yaquina,Oreg	578 883	200·0	31·2	12·0	Chester,Pa. J.Roach&Son	1883		4Cy.18"& 34"–36" 100IP. J.Roach&Son,Chester,Pa.	Oregon Development Co.
43	✱	Eastern Prince *Knudsen 87-87*	Stl ScSr *1 Dk & Wch Frms*	British Newcastle	1408 2147	292·6 *Q 76 f. B 110 f.*	39·0 *P 33 f.*	18·9 *F 28 f.*	Sunderland Short Bros.	1887	4 Blds. *CDB 301t*	3C.22"35"&59"–42"200IP. G.Clark,Sunderland	Prince Steam S.Co.,Ld. (J.Knott,Manager)
44		Eastern Shore	I Pad	American Wilmington	655 792	176·0	35·0	9·5	WilmingtnDel Harlan&Hollingswrth Co.	1883		1Cy.38"–108" The Harlan & Hollingsworth Co., Wilm., Del.	Eastern Shore Steamboat Co.
45	✱	Eastern Star *Tate 84*	I Sc *1 Dk 2 B*	British W.Hartlepl.	775 1215	230·2 *Q 86 f.*	32·6 *B 57 f.*	15·4 *F 24 f.*	Sunderland R.Thompson&Son	1883	4 Blds. *DB aft 71t APT 18t*	2Cy.27"& 51"–33" 99 IP. J.Dickinson,Sunderland	V.T.Thompson & Co.
46		Eastham Fairy	I Pad	British Liverpool	80 115	125·0	19·2	7·7	Chester	1861		2Cy.30½"–33" 60 IP. H.M.Lawrence,Liverpool	T.W.Thompson
47	✱	Eastward *Archer 87-87*	I ScK *1 Dk*	British Hull	144 235	135·4 *Q 19 f.*	22·0	11·0 *F 23 f.*	Hull Earles'Co.,Ld.	1882	4 Blds.	2Cy.21"& 40"–27" 60 IP. Earles'Co.,Ld.,Hull	Gt.Nrthn.S.S.Fishing Co.,Ld.
48	✱	Eastwood *Wilkinson 85-85*	I ScSr *1 Dk 2 B*	British Hull	934 1447	256·8 *Q 83 f.*	34·0 *B 136 f.*	17·0 *F 37 f.*	Hull Earles'Co.,Ld.	1885	4 Blds. *WB aft & ch FPT APT*	3C.19½"30"&52"–33"130IP. Earles'Co.,Ld.,Hull	G.R.Sanderson & Co.
49		Ebba Munck *Tengström 88-*	Stl ScSr *1 Dk & Awn Dk*	Russian Nystad	250	130·0	22·0	10·0	Abo W.Crichton&Co.	1888	*APT 12t FPT 10t*	3C.10"17½"&26"–21" 80IP. W.Crichton&Co.,Åbo	Lonnas Suomi S.S.Co.
50	✱D	Ebbwvale *Richardson*	I ScSr *1 Dk 2 B*	British Cardiff	880 1350	240·0 *Q 86 f.*	33·4 *B 50 f.*	16·5 *F 29 f.*	Newcastle Palmers'Co.,Ld.	1881	4 Blds. *CDB 258t APT 13t*	2Cy.30"& 57"–36" 99 IP. Palmers'Co.,Ld.,Newcastle	"Ebbwvale"S.S.Co. (Morel Bros.& Co.)
51	DG	Eberstein *Eggerding 83-88*	I ScSr *1 Dk*	German Bremen	754 1031	229·3 *B 50 f.*	30·2 *F 25 f.*	16·5	Geestemünde G.C.Tecklenborg	1883	*WB*	2Cy.26"& 48"–31½"110IP. O.Henniges,Berlin	DeutscheDampfschifffahrts Ges."Hansa"
52	✱D	Ebor *Lennard*	I ScSr *1 Dk 2 B*	British Middlesbro'	441 701	201·7	29·0	17·0	Hartlepool Denton,Gray&Co.	1868	4 Blds. *DB aft & aft f 15t*	2Cy.24"& 45"–30" 90 IP. G.Clark,Sunderland	J.M.Lennard & Sons
53		Ebre	I Sc *3 Dks*	French Marseilles	1165 1758	314·3 *P 72 f.*	32·8 *F 41 f.*	24·1 *H 27 f.*	LaCiotat MessageriesMarit.	1868		3Cy.51¼"–31½" 280 IP. MessageriesMar.,LaCiotat	Messageries Maritimes
54	✱	Ebro *Skipper 82-84*	I ScSr *1 Dk 2 B*	British Sunderland	716 1115	224·5 *Q 84 f.*	32·0 *F 23 f.*	16·3 *B 42 f.*	Sunderland J.Laing	1879	4 Blds. *DB aft & aft f 49t*	2Cy.27"& 50"–36" 120 IP. G.Clark,Sunderland	The Pinkney & Sns S.S. Co.,Ld.
55		Ecaterina	I Sc	Russian Rostoff	247 316	..	..	..	Trieste StabilimentoTecnico	1880		40 IP.	Anisenco & Johnson
56		Ecaterini Coupa *(ex Llandough)*	I ScSr *1 Dk 2 B*	Greek Argostoli	883 1387	240·5 *Q 122 f.*	32·7 *F 34 f.*	20·4	Newcastle Schlesinger,Davis & Co.	1878	5 Blds. *DB a & f FPT APT*	2Cy.29"& 56"–36" 130 IP. T.Clark&Co.,Newcastle	A.Vagliano
57	✱	Echo *(ex Loda)*	I ScSr *1 Dk*	British Sunderland	27 58	69·0	15·0	6·9	Sunderland SunderlandS.B.Co.,Ld	1886	3 Blds. *FPT 2t*	2Cy.10"& 20"–12" 30IP. Fisher&Co.,Paisley	The Admiralty
58	✱D	Echo *Oosteen 88*	I ScSr *1 Dk 2 B*	Dutch Rotterdam	621 985	213·9 *Q 81 f.*	30·6 *B 42 f.*	15·4 *F 27 f.*	Slikkerveer P.Smit,jun.	1883	4 Blds. *DB aft 150t*	2Cy.27"& 47"–30" 100 IP. P.Smit,jun.,Slikkerveer	Hudig & Veder
59		Echo *Agostini 83-84*	I Sc3mSr *1 Dk*	French Marseilles	155 233	170·9 *P 16 f.*	19·8 *F 50 f.*	10·4	Newcastle A.Leslie&Co.	1868		2Cy.23½"& 37½"–18"50IP. Fraissinet&Co.,Marseilles	Fraissinet & Co.
60	✱	Echuca *Pankhurst 18-88*	Stl Sc3mSr *2 Dks 3 B*	British London	1881 2907	324·8 *P 37 f.*	40·2 *B 90 f.*	25·1 *F 42 f.*	Middlesbro' R.Dixon&Co.	1889	6 Blds. *CDB 401t*	3Cy.27½"43½"&71"–48" 400 IP. T.Richardson&Sons, Hartlepool	W.Lund

✱ = Lloyd's Register; D = Nederlandsche Vereeniging; G = Germanischer Lloyd; I = Registro Italiano; R = American Record;
A = Veritas Austro-Ungarico; H = Veritas Ellenico; N = Norske Veritas; V = Bureau Veritas.

Dans la 2me Colonne, B.T., avec les indications qui suivent, signifie la date du rapport de visite de la Carène par les experts du Board of Trade pour obtenir le certificat relatif au TRANSPORT DES PASSAGERS.

Dans la 3me Colonne, C = Composite (membrure en fer, bordé en bois); I = Fer; Stl = Acier; W = Bois.

Pour plus amples renseignements sur les navires classés par le Lloyd's Register, voir le Lloyd's Register Book et le Lloyd's Yacht Register.

No. in Book. Official No.	Registry in which classed	Ship's Name. Material, Rig, &c. Master, and Dates of Appointment: present Owner's Service and to Ship Ship's Late Name. No. of Decks, &c.	Flag. Port of Registry.	Tons Gross Net	Length	Breadth	Depth	Built. Where. By Whom.	When.	No. of Bulkheads. Particulars of Water Ballast.	Engines. Particulars. Maker's Name.	Owners or Managers
61 47884		**Eclair** (ex [illegible]) I PadSr 1 *Dk*	British London	135 237	179·8	20·2	8·5	PortGlasgow Kirkpatrick&Co.	1885	5 Blhds.	2Cy.40"–42" 120 HP W.Simons&Co.,Renfrew	River Thames [boat] Co.,Ld.
62	✻	**Eclair** W ScLaunch 1 *Dk*	French Lasseyne	 31	66·1	10·0	6·3	London J.Mutch	1885	 YM&81	2Cy.10½"& 18"–12" Vulcan Eng. Wrks.,Stockton	M.Michel
63 78490	✻	**Eclipse** I ScSr Wood 84-84 1 *Dk3 B*	British W.Hartlepl	1014 1583	257·8	34·2	19·5	Stockton M.Pearse&Co.	1877	4 Blhds. BB118*ft*	2Cy.32"& 60"–39" 140 HP Blair&Co.,Ld.,Stockton	T.Appleby & Co.
64 55345		**Eclipse** W ScS *Gray 49-67*	British Peterhead	296 435	141·5 *F* 36 *f.*	29·4	16·3	Aberdeen	1867		2Cy.23"& 40"–27" 72 HP Hall,Russell&Co.,Aberdeen	D.Gray
65 87249	✻	**Ecossaise** Stl ScSr *Dundas 83-83* 1 *Dk*	British Leith	536 848	210·5 *Q* 70 *f.*	30·2 *B* 57 *f.*	13·3 *F* 23 *f.*	Leith Ramage&Ferguson	1883	4 Blhds. [illegible]	2Cy.27"& 50"–30" 98 HP Ramage&Ferguson,Leith	Stevenson & [illegible]
66 84127		**Ecuador** Stl Sc3mSr 2 *Dks & Shade Dk*	British Liverpool	1310 1768	300·0 *F* 24 *f.*	36·2	19·3	Birkenhead LairdBros.	1881	5 Blhds. DB	2Cy.34"& 60"–42" 190 HP LairdBros.,Birkenhead	Pacific Stm.Nav.
67	V D	**Edam** Stl ScBg *Bakker* -88 2 *Dks & Spr Dk*	Dutch Rotterdam	2267 3130	328·1 *B* 19 *f.*	41·3 *H* 135 *f.*	29·5	Rotterdam Ned.Stoomb.Maats.	1883		2Cy43"& 76"–48" 300 HP Ned.Stoomb.Maats.Rttrdm.	Ned.American Stoomv.Maats.
68 87129	✻ D	**Edda** I ScSr *Muir 88-88* 1 *Dk*	British Grangemouth	629 997	216·0 *Q* 48 *f.* *B* 48 *f.*	31·0 *P* 26 *f.*	13·9 *F* 28 *f.*	Sunderland S.P.Austin&Son	1881	4 Blhds. DB215t	2Cy.27"& 50"–36" 98 HP J.Dickinson,Sunderland	J.T.Salvesen & [illegible]
69 84357	✻ D	**Eddlethorpe** I ScBn *Houlfleur 79-83* 1 *Dk2 B*	British Hartlepool	1119 1735	259·6 *Q* 68 *f.* *B* 68 *f.*	34·7 *P* 29 *f.*	19·5 *F* 33 *f.*	W.Hartlepool W.Gray&Co.	1882	4 Blhds. DB[illegible]	2Cy.32"& 60"–39" 150 HP Black,Hawthorn&Co.,Gatsh'd	J.Coverdale & [illegible]
70 91942	✻	**Eddystone** I ScSr *Thompson* 1 *Dk2 B*	British London	1313 2040	275·0 *Q* 103 *f.*	36·0 *B* 68 *f.*	20·0 *F* 72 *f.*	Stockton M.Pearse&Co.	1886	4 Blhds. DB[illegible]	3C.21"35"&57"–39" 200 HP Blair&Co.,Ld.,Stockton	Farrar,Groves & [illegible]
71 93263	BT 2.88 ✻	**Eddystone** Stl ScSr *Boyd 76-87* 2 *Dks*	British Glasgow	451 1003	233·3 *P* 142 *f.*	33·2 *F* 54 *f.*	15·7	Dundee W.B.Thompson	1886	5 Blhds. FPT[illegible] DB[illegible]	3C.23"37"&58"–48" 250 HP W.B.Thompson,Dundee	Clyde Shipping [illegible]
72 67019		**Eden** I ScSr *Armstrong* 2 *Dks & Spr Dk*	British London	1374 2145	309·5 *B* 77 *f.*	36·2	16·7 [illegible] 24·6	Barrow BarrowS.B.Co.,Ld.	1882	8 Blhds.	2Cy.38"& 74"–54" 350 HP BarrowS.B.Co.,Ld.,Barrow	Royal Mail S.P. [illegible]
73 81503	✻	**Eden** I ScBn *Levitt 85-87* 1 *Dk2 B*	British W.Hartlep'l.	904 1403	246·0 *Q* 97 *f.*	34·2 *B* 57 *f.*	17·2 *F* 32 *f.*	W.Hartlepool W.Gray&Co.	1879	4 Blhds. DB[illegible]	2Cy.29"& 55"–33" 120 HP T.Richardson&Sons,H'rtlpl	R.Ropner & Co.
74 87841	✻	**Eden Hall** I ScSr *Hughes* 2 *Dks3 B*	British Liverpool	2273 3507	360·0 *P* 49 *f.*	42·6 *B* 76 *f.*	28·2 *F* 50 *f.*	Dundee Gourlay Bros.&Co.	1883	5 Blhds.	2Cy.40"& 76"–51" 180 HP Gourlay Bros.&Co.,Dundee	The Eden S.S. [illegible] (R.Alexander&Co.)
75 95267	✻	**Edendale** Stl ScSr *Humphrey 77-88* 1 *Dk2 B & Wb Fms*	British Sunderland	1566 2405	308·0 *B* 82 *f.*	39·0 *P* 43 *f.*	23·4 *F* 42 *f.*	Sunderland J.Laing	1888	5 Blhds. CDB411t	3C.23"38"&62"–42" 250 HP G.Clark,Sunderland	P.H.Laing
76 81294	✻ D	**Edgar** I ScBn *Hodgson 87-87* 2 *Dks3 B*	British Whitby	922 1445	243·0 *P* 25 *f.*	33·0 *B* 21 *f.*	22·0 *F* 30 *f.*	Whitby T.Turnbull&Son	1879	1 Blhds. DB[illegible]	2Cy.31"& 58"–36" 130 HP Blair&Co.,Ld.,Stockton	T.Turnbull & Son
77 62370	✻	**Edgar** I ScSr *Matthews 83-83* 1 *Dk2 B*	British Newcastle	667 1070	225·0 *P* 123 *f.*	30·1 *F* 36 *f.*	17·3	Newcastle A.Leslie&Co.	1870	5 Blhds. [illegible]	2Cy.26"& 52½"–30" 99 HP Thompson,Boyd&Co.,Nwstle	H.E.P.Adamson
78 10286	✻	**Edina** I ScSr *Robertson 87-87* 2 *Dks*	British Leith	232 403	191·0	23·7	12·9	Glasgow Aitken&Mansel	1865	3 Blhds.	2Cy.23"& 40"–27" 70 HP Hawthorns&Co.,Leith	Leith, Hull & [illegible] Stm.Pkt.Co.
79 11136		**Edina** I ScSr *Matthews 83-83* 2 *Dks*	British Melbourne	223 380	171·0 *B* 54 *f.*	23·6 *F* 31 *f.*	12·7	Glasgow Barclay,Curle&Co.	1854		2Cy.30"& 22"–26" 85 HP D.Rollo&Sons,Liverpool	W.H.Smith & Sons
80		**Edinburgh** I ScBk (ex Amsterdam) 2 *Dks & Spr Dk*	Italian	1737 2520	300·5 *F* 93 *f.*	39·8	25·0	Glasgow Tod&McGregor	1855	4 Blhds.	2Cy.44"& 78"–42" 250 HP LairdBros.,Birkenhead	Italian Government
81 86774	✻	**Edinburgh** I ScSr *Milne 88-88* 1 *Dk2 B*	British Newcastle	1072 1653	248·0 *Q* 88 *f.* *H* 18 *f.*	34·5 *B* 80 *f.*	20·6 *F* 30 *f.*	Stockton M.Pearse&Co.	1884	4 Blhds. WB[illegible]	2Cy.31"& 58"–36" 144 HP Blair&Co.,Ld.,Stockton	R.H.Wesencraft
82 80491	BT 1.88	**Edinburgh Castle** I PadSk	British Glasgow	28 234	205·3	19·9	7·6	PortGlasgow R.Duncan&Co.	1879		1Cy.50"–66" 83 HP Rankin&Blackmore,Grnck	M.T.Clark
83 91098		**Edinburgh Castle** Stl PadSp *Cragie*	British Leith	23 158	160·0	20·1	7·0	Kinghorn J.Scott&Co.	1886		1Cy.46"–60" 85 HP J.Scott&Co.,Kirkcaldy	Galloway Saloon Packet Co.
84 50309	BT 5.88	**Edith** I Sc3mSr	British Hull	225 341	218·5 *F* 47 *f.*	25·2	12·4	Renfrew Henderson,Coulborn &Co.	1864		4Cy.21"& 40"–27" 150 HP Earle's Co.,Ld.,Hull	G.Lawson
85 63963	BT 12.86	**Edith** I PadSr *Owen -73* 2 *Dks*	British Dublin	275 758	250·6 *F* 109 *f.*	30·1	14·4	Newcastle A.Leslie&Co.	1870	4 Blhds.	2Cy.63"–72" 300 HP R.Stephenson&Co.,N'wcstle	London & N.W. [illegible] Co.
86 87781	✻	**Edith** I ScSr *Woods 86-86* 1 *Dk*	British Grimsby	259 405	149·2 *Q* 48 *f.*	23·5 *B* 37 *f.*	11·3 *F* 24 *f.*	PortGlasgow Russell&Co.	1883	1 Blhds. FPT[illegible] APT[illegible]	2Cy.18"& 36"–24" 55 HP Alley&Maclellan,Glasgow	W.France,Ld.
87 94391	✻	**Edith** I TwnScSr *Popham -88*	British London	97 183	110·0 *Q* 14 *f.*	23·1	7·3	Barrow Barrow Co.,Ld.	1888		2Cy.10"&20"–18" 35 HP BarrowCo.,Ld.,Barrow	T.A.Walker
88 79982	✻	**Edith** W ScK *Watson* 1 *Dk*	British G.Yarmouth	53 89	82·0	20·2	8·7	Berwick Gowan	1878		2Cy.12"–16" 30 HP Wrdbaugh&Blnr.,N'wcstle	J.T.B.Kell
89 91091	✻	**Edith** W ScK *J. Strachan* 1 *Dk*	British Leith	20 49	69·0	17·4	7·6	Leith McKenzie&Co.	1886		2Cy.10"& 20"–1[illegible] 18 HP J.Cran&Co.,Leith	Haworth & [illegible]
90 73724	✻	**Edith Godden** I ScSr *Bennett* 1 *Dk & Awn Dk*	British London	350 553	170·8	25·2	11·3 18·0	Glasgow A.&J.Inglis	1876	4 Blhds. FPT[illegible]	2Cy.27½"& 50"–30" 110 HP A.&J.Inglis,Glasgow	J.E.Kerr

✻ = **Lloyd's Register**; D = Nederlandsche Vereeniging; G = Germanischer Lloyd; I = Registro Italiano; R = American [illegible]; A = Veritas Austro-Ungarico; H = Veritas Ellenico; N = Norske Veritas; V = Bureau Veritas

In the 2nd Column, B.T., with figures following, signifies Date of report of bottom sighting by Board of Trade Surveyors as condition of PASSENGER CERTIFICATE

In the 3rd Column, C = Composite (iron frame, wood planked); I = Iron; Stl = Steel; W = Wood.

Further particulars of Vessels classed by Lloyd's Register will be found in Lloyd's Register Book and Lloyd's Yacht Register.

Registres		Nom, Matériaux, Gréement, etc. Capitaine et dates d'entrée au service des Armateurs actuels et à bord du navire. Ancien Nom, nombre de ponts, etc.	Pavillon. Port d'Armement	Tonnage Net / Total	Longueur	Largeur	Creux	Construit. Où. Par qui.	Quand.	Nombre des cloisons. Détails sur le lest d'eau.	Machines. Détails. Nom du Constructeur.	Armateurs ou Directeurs-Gérants.
	✻	Edith Hough I ScSr Fokes 84-84 (ex Zakynthos) 1 Dk2 B	British Liverpool	493 758	219·6	26·2	16·9 P 41 f.	Liverpool Bowdler, Chaffer & Co.	1869	5 Blkds. WB DTF APT FPT	2Cy.25"& 50"–36" 104 HP. D.Rollo&Sons,Liverpool	S.Hough
92	✻	Editor (Chandler 80-85) St ScSr 1 Dk Spr Dk & Web Frames	British Liverpool	1039 1618	250·0	33·9	15·1 B 21 f. F 35 f. 23·0	Stockton M.Pearse&Co.	1885	4 Blkds.	3C.20"33" 54"–36" 116 HP. Blair&Co.,Ld.,Stockton	Charente S.S.Co.
93	✻	Edmondsley I ScSr Gibson 85-86 1 Dk & Web Frames	British Sunderland	980 1526	245·2	35·6	17·5 Q 89 f. F [illegible] B 42 f.	Sunderland J.L.Thompson & Sons	1884	5 Blkds.	2Cy.31"& 57"–36" 140 HP. J.Dickinson,Sunderland	Tyzack & Branfoot
94		Edsvalla I ScSr Kristiansson 1 Dk	Swedish [illegible]	135 [illegible]	101·3	21·9	11·0 P 16 f. F 13 f. M 7 f.	Gothenburg	1866		30 HP.	F.Wall
95	✻G	Eduard (ex Jubilee) I ScSr Dahms 81-86 1 Dk	German [illegible]	642 987	220·0	31·1	13·6 B 44 f. Q 77 f. F 29 f.	Middlesbro' W.Harkess	1877	5 Blkds. DA[illegible]	2Cy.25½"& 48"–33" 90 HP. Blair&Co.,Ld.,Stockton	R.C.Gribel
96	✻	Eduardo I ScBk Larrauri 2 Dks & B	Spanish [illegible]	1503 2[illegible]	325·3	34·7	25·3	Hull Earles'Co.,Ld.	1878	5 Blkds. DB[illegible]	2Cy.36"& 72"–48" 250 HP. Earles'Co.,Ld.,Hull	Cia.de Nav.La Flecha
97	✻	Edward Eccles I Sc3mSr Wilkinson 1 Dk2 B	British Newcastle	504 [illegible]	210·9	28·5	15·3 B 64 f. Q 69 f. F 25 f.	Sunderland Short Bros.	1872	4 Blkds. DB[illegible]	2Cy.27"& 50"–30" 99 HP. R.&W.Hawthorn,N'wcastle	E.Eccles
98		Edward Finch I PadSr ex Lord Derby	British Swansea	27 157	122·8	19·5	10·0	S.Shields J.T.Eltringham	1877		2Cy.30½"–54" 99 HP. Hepple&Co.,S.Shields	P.C.Rousey
99	✻	Edward Williams I ScSr Robb 88-99 1 Dk2 B	British W.Hartlep'l	455 712	191·0	28·8	16·2 Q 69 f.	W.Hartlepool Denton,Gray&Co.	1871	4 Blkds. DB[illegible]	2Cy24½"&47½"–30" 96 HP. T.Richardson&Sons,H'rtlpl	Maclay & McIntyre
100	✻	Edwin I ScSr Stacey 72-79 2 Dks & Shd Dk3 B	British N.Shields	1204 [illegible]	285·2	34·7	24·2 P 40 f. F 40 f. S B 54 f.	Newcastle A.Leslie&Co.	1878	5 Blkds. BD[illegible]	2Cy.33"& 66"–42" 200 HP. R.&W.Hawthorn,N'wcastle	H.E.P.Adamson
101		Edwin Forrest I Pad	American Philadelphia	503 607	196·5	27·8	7·6	Chester,Pa. Reaney,Son&Archbld	1865		1Cy.32"–144" 109 HP. Reaney,Sn&Archbld,Chester,Pa	J.Cone
102	✻	Eeta I ScSr [illegible] 84-86 1 Dk & Web Frames	British London	765 1191	235·0	32·3	14·6 P 134 f. F 34 f.	Newcastle W'gham Richardson &Co.	1883	4 Blkds. CDB227t FPT34t APT[illegible]	2Cy.27"& 54"–36" 150 HP. W'gham Richardson&Co.,N'wcstl	Demerara & Berbice S.S.Co.,Ld.
103	✻	Effective I ScSr Crosby 84-87 2 Dks3 B	British Sunderland	1264 1936	274·4	35·0	24·2 P 28 f. B 49 f. F 30 f.	Sunderland Short Bros.	1878	5 Blkds. DB[illegible]/170t APT20t	2Cy.33"& 62"–39" 180 HP. G.Clark,Sunderland	Anderson,Horan & Co.
104	✻	Efficient I ScSr Harris 78-82 2 Dks3 B	British Sunderland	1058 1644	260·0	34·0	24·0 B 48 f. F 27 f.	Sunderland Short Bros.	1877	5 Blkds. DB[illegible]/230t	2Cy.33"& 62"–39" 160 HP. G.Clark,Sunderland	Anderson,Horan & Co.
105	✻	Effort W ScK Mackay 86-88 1 Dk	British Leith	21 33	61·6	17·4	6·4	Leith A.G.Gifford&Co.	1884		2Cy.8"& 15"–12" 9 HP. J.Cran&Co.,Leith	A.Cook & Co.
106		Eftychia Lambrinidi I Sc	Greek Ithaca	299	..	..	..					Lambrinidis
107	✻I	Egadi (ex Galatea, Marmion 75-80 ex Surprise) I Sc3mSr 3 Dks	Italian Palermo	1239 1950	278·2	34·1	24·1 [illegible]	Renfrew Henderson Colbourn &Co.	1872	5 Blkds.	2Cy.30"& 67½"–48" 620 HP. Hndrsn,Clbrn.&Co.Rnfrw.	Navigazione Generale Italiana
108		Egeon I Sc	Greek Syra	502	..	..	..	England	1856			Hellenic S.S.Co.
109	✻	Egglestone Abbey Stl ScSr Pyrgyn 83-88 1 Dk & Web Frms	British Cardiff	1567 2401	290·0	38·2	20·3 [illegible]	W.Hartlepool W.Gray&Co.	1888	5 Blkds. DB[illegible]/152t	3C.22"35"&59"–39" 200 HP. Cent.MarineEng.Co.,West Hartlepool	Pyman,Watson & Co.
110		Egida I ScSr Mulle	Austrian Trieste	51 100	111·9	16·2	8·39	Kiel	1879		30 HP.	Dr.T.Ricchetti fu E.
111		Egitto I Sc3mSr Tarabotto 82-86 2 Dks	Italian Genoa	733 1112	232·6	30·0	19·3 Q 50 f. F 24 f. M 87 f.	Newcastle Palmers'Co.	1869	4 Blkds. DB/112t	2Cy.29"& 53"–30" 190 HP. Cmp. N.Odero & Ferro,Sestri Ponente	Navigazione Generale Italiana
112	✻	Eglantine I ScSr Grimwade 79-87 1 Dk2 B	British Nrth.Shields	874 1357	258·0	32·0	21·1 B 17 f.	Newcastle TyneI.S.B.Co.	1878	5 Blkds. DB[illegible]	2Cy.28"& 54"–33" 130 HP. N.E.MarineEng.C.,Ld.,Sndl.	J.Robinson & Sons
113		Egmont I Sc3mSr Selmes 1 Dk & Spr Dk	British Sydney	419 670	200·0	24·8	18·3	Renfrew Henderson Colbourn &Co	1864	4 Blkds.	2Cy.26"& 50"–24" 100 HP. J.&W.Dudgeon,London	Australasian United Steam Nav.Co.,Ld.
114		Egret I Sc3mSr Croft 88-88 1 Dk2 B	British W.Hartlepl.	555 816	211·5	28·1	17·4 Q 49 f.	Newcastle Palmers'Co.	1865		4Cy.19"& 38"–30" 90 HP. Cmp.'73,Palmers'Co.,N'wcst.	J.Wood & Co.
115	✻D	Egret Stl ScSr Colton 1 Dk2 B	British London	464 723	190·6	27·7	14·4 Q 70 f. B 40 f. F 26 f.	Newcastle Sir W.G.Armstrong, Mitchell&Co.,Ld.	1883	5 Blkds. [illegible] FPT[illegible] APT[illegible]	2Cy.25"& 50"–27" 98 HP. Wllsd SlipCo.,Ld.,N'wcastle	General Stm.Nav.Co.
116		Egypt I Sc4mBk Sumner 70-86	British Liverpool	2959 4670	440·3	44·3	36·5	Liverpool Liverpool S.B.Co.	1871		2Cy.60"& 106"–54" 600 HP. J.Jack,Rollo&Co.,Liverpool	National S.S.Co.,Ld.
117		Egyptian I Sc3mSr Shepherd 78-86	British Liverpool	1342 2012	335·0	34·2	24·0 F 36 f.	Belfast Harland&Wolff	1861		4Cy.24"& 50"–48" 220 HP. G.Forrester&Co.,Liverpool	S.S."Egyptian"Co.,Ld (F.Leyland & Co.)
118		EgyptianMonarch I Sc4mBk Bristow 88 3 Dks & Shlt Dk	British London	2552 3916	360·0	43·1	24·8 32·3	Dumbarton A.McMillan&Son	1880	8 Blkds. [illegible]	2Cy.46"& 87"–57" 500 HP. D.Rowan,Glasgow	T.Wilson,Sons & Co.
119		Egyptien I ScBk	French Marseilles	401	185·0	26·2	17·3	LaSeyne	1853		120 HP.	Durand
120	✻DV	Ehrenfels I ScSr Bauern 8-88 1 Dk & Spr Dk3 B	German Bremen	1522 2345	303·6	36·2	16·5 DB[illegible] P 31 f. 26·6	Newcastle Wigham Richardson &Co.	1882	6 Blkds. DB[illegible]/228t FPT30t APT20t	2C.25"38"&72"–48" 300 HP. Triple[illegible],Wigham Richardson&Co.,Newcastle	Deutsche Dampfschiffahrts Ges."Hansa"

✻ = Lloyd's Register; D = Nederlandsche Vereeniging; G = Germanischer Lloyd; I = Registro Italiano; R = American Record;
[illegible] = Veritas Austro-Ungarico; H = Veritas Ellenico; N = Norske Veritas; V = Bureau Veritas.

Dans la 2me Colonne, B.T., avec les indications qui suivent, signifie la date du rapport de visite de la Carène par les experts du Board of Trade pour obtenir le certificat relatif au TRANSPORT DES PASSAGERS.

Dans la 3me Colonne, C = Composite (membrure en fer, bordé en bois); I = Fer; Stl = Acier; W = Bois.

Pour plus amples renseignements sur les navires classés par le Lloyd's Register, voir le Lloyd's Register Book et le Lloyd's Yacht Register.

No. in Book / Official No.	Registry in which Classed	Ship's Name. Material, Rig, &c. / Master and Dates of Appointment to present Owner's Service and to Ship / Ship's Late Name. No. of Decks, &c.	Flag / Port of Registry	Tons Net / Gross	Length	Breadth	Depth	Built: Where / By Whom	When	No. of Bulkheads / Particulars of Water Ballast	Engines: Particulars, Maker's Name	Owners or Managers
121 [illegible]	✻	**Eident** I ScSr *Kieffer 75-87* *1 Dk 2 B*	British Sunderland	1064 1624	260·5 *Q 100 f.*	34·0 *B 62 f.*	18·7 *F 30 f.*	Sunderland Short Bros.	1880	[illegible]	2Cy.33" & 61"–33" 150 HP. T. Richardson & Sons, H'rtlpl	J. Westoll
122 54700		**Eider** (BROKEN UP) I Pad *Edwards* *2 Dks*	British London	456 723	235·3	28·2	14·6	Dundee Gourlay Bros.	1866	4 Blkds.	2Cy.58"–54" 220 HP. Gourlay Bros. & Co., Dundee	General Stm. N[illegible]
123 8409	✻	**Eider** I ScSr *Jarre* *1 Dk*	British Newcastle	139 227	124·3	20·1	11·6	London C. Langley	1856	4 Blkds.	2Cy.15" & 30½"–20" 38 HP. Bl[illegible]&Co., Ld., Stockton	Blaydon & Lo[illegible] Co., Ld.
124	V	**Eider** I ScInSr *Baur 80-87* *3 Dks*	German Bremen	2952 4719	429·8 *P 28 f.*	46·9 *B 138 f.*	34·5 *F 47 f.*	Glasgow J. Elder & Co.	1884		3C 62" & (2) 86"–60" 1000 HP. J. Elder & Co., Glasgow	Norddeutscher [illegible]
125		**Eider** (ex Caster) I ScSr *Hobeck 85* *1 Dk*	German Stettin	92 153	95·0	17·5	10·0	London Edwards & Symes	1881		1Cy.18"–22" 30 HP. Hamilton & Woods, Manchst.	Stenzel & Co.
126		**Eider** I Pad	Haytian Hayti	1018 1568	296·0	32·1	23·8	Greenock [illegible]aird & Co.	1864		2Cy.42" & 72"–78" 310 HP. [illegible] & Co., Southampton	General Jean B[illegible]
127	N	**Einar** I ScSr *Hansen* *1 Dk 2 B*	Norweg'n Bergen	148 217	94·5	22·8 *B 26 f.*	11·9	Motala W. Tham	1872		2Cy.15" & 30"–22" 45 HP. Motala Co., Motala	J. A. Kruger
128		**Ekholmen** I Sc *Lindberg* *1 Dk*	Swedish Stockholm	103 138	104·6 *Q 32 f.*	17·6 *B 36 f.*	8·2 *H 8 f.*	Stockholm Bergsunds Mek. Verk. Akt. Bol.	1876		2C 13½" 23¾" 15½"–19½" 32 [HP.] Bergsunds Mek. Vk. Stockholm	Angbåtsbol[illegible] holm-Fittja [illegible]
129		**Ekolsund** I Sc *Nyman 80-83* *1 Dk*	Swedish Stockholm	119 155	110·2 *Q 26 f.*	17·7 *B 44 f.*	8·2 *F 13 f. H 10 f.*	Stockholm Bergsunds Mek. Verks Akt. Bol.	1883		35 HP. Bergsunds Mek. Vk. Stockholm	Angf. Bol. S[illegible] lander (K. A. Wa[illegible])
130 [illegible]	✻	**El Callao** I ScSr *Breen 86-88* *1 Dk & Awn Dk*	British Glasgow	642 1012	210·2	31·2 *B 35 f.*	12·0 18·4	Leith Ramage & Ferguson	1885	4 Blkds. [illegible]	2Cy.26" & 50"–36" 106 HP. Ramage & Ferguson, Leith	Walker, Donald
131	V	**El-Caratal** I ScSr *Jasseau*	Venezuln. LaGuayra	140 271	146·9	20·3 *B 98 f.*	9·8	Rotterdam Maatschappij de Maas	1886		2Cy.19" & 35"–22" 75 HP. Maatschappij de Maas Rtdm	Société Anony[illegible] de la Mo[illegible]
132 [illegible]		**El Cid** W ScSr *2 Dks 3 B*	American Norwich, Ct.	 312	135·3	32·7	15·0	E. Haddam Ct.	1863		175 HP.	Peter Charles
133	✻ D	**El-Djezzer** (ex [illegible]) I ScSr *Plessis -88* *1 Dk 2 B*	French Algiers	417 661	187·4	28·2	16·9	London Millwall S. B. Co.	1866	4 Blkds.	1Cy.17" & 36"–27" 80 HP. Comp. by Earles Co., Hull	La Société Ma[illegible] Algérienne
134 86738	✻	**El Dorado** I ScSr *McGorera* *1 Dk 2 B*	British Newcastle	975 1518	259·9 *Q 72 f.*	38·0 *B 48 f.*	14·4 *P 28 f. F 26 f.*	Sunderland J. L. Thompson & Sons	1883	4 Blkds. [illegible]	2Cy.32" & 60"–39" 140 HP. [illegible], Sunderland	H. Scholefield & [illegible]
135 70049		**El Dorado** (ex Doona Isabel) I ScSr *Payne 86-87* *1 Dk & Awn Dk*	British London	892 1182	240·5	32·9	12·8 20·0	Port Glasgow Cunliffe & Dunlop	1874	4 Blkds.	2Cy.31" & 54"–33" 130 HP. Cunliffe & Dunlop, Pt. Glasgw.	Indo-China St[illegible] Nav. Co., Ld.
136 [illegible]	✻	**El Dorado** I Sc *Sampson 85-85* *1 Dk & Web Frms*	British London	833 1298	240·2 *P 30 f.*	34·0 *B 54 f.*	17·5 *F 28 f.*	North Shields T. & W. Smith	1882	4 Blkds. [illegible]	2Cy.39" & 66"–36" 150 HP. R. & W. Hawthorn, Newcstle	"El Dorado" S.S. [illegible] (Scrutton, Son[illegible])
137 [illegible]	R	**El Dorado** I ScIn *Byrne* *2 Dks & Awn Dk*	American New York	2562 3531	331·8	42·8	22·6	Philadelphia W. Cramp & Sons	1884	[illegible]	2Cy.38" & 76"–54" W. Cramp & Son, Philadelphia	Southern Deve[illegible] ment Co.
138 [illegible]	R	**El Monte** I ScIn *Hawthorne* *3 Dks*	American New York	2562 3531	330·0	42·6	32·6	Philadelphia W. Cramp & Sons	1886	[illegible]	2Cy.38" & 76"–54" 300 HP. W. Cramp & Sons, Philadlpha.	Southern Develop ment Co.
139		**El Oriental**	Urug'yan MonteVideo	810	212·0	28·0	10·2	Glasgow Randolph, Elder & Co.	1864			H. Formosa
140 [illegible]	R	**El Paso** I ScIn *Quick* *2 Dks & Awn Dk*	American New York	2562 3531	351·0	42·8	32·6	Philadelphia W. Cramp & Sons	1884	4 Blkds. [illegible]	2Cy.38" & 76"–54" W. Cramp & Sons, Philadlpha.	Southern Devel[illegible] ment Co.
141		**El Progresso** I PadSr	Venezuln Maracaibo	225	150·0	25·0	7·0	Wilmington Pusey & Jones Co.	1878		2Cy.18"–72" Pusey & Jones Co. Wilmingtn	H. L. Boulton
142		**El Siglo** I Pad	Spanish Bilbao	48 121	95·0	18·5	9·5	South Shields Lawson, Ellingham	1883		50 HP. H. P. Hepple, N. Shields	F. Carranza
143 81913		**Elagh Hall** I ScSr *Glackin 78-83* *1 Dk*	British Londonderry	126 237	125·0 *Q 44 f.*	21·2 *F 12 f.*	10·0	Ayr McKnight & McCredie	1883	[illegible]	2Cy.18" & 36"–24" 55 HP. Lees, Anderson & Co., Glasgow	J. Christy
144 63381	✻	**Elaine** (ex Yeres) I ScIn Sr *Kolesar 79-79* *1 Dk*	British Liverpool	343 544	195·5 *P 41 f.*	23·6 *F 20 f.*	15·3	Belfast Harland & Wolff	1869	[illegible]	2Cy.22" & 42"–28" 80 HP. V. & D. Coates, Belfast	J. Ellis
145 55434	BT1.88	**Elaine** I Pad *McPherson 85-85* *1 Dk*	British Glasgow	39 118	175·0	17·1	6·6	Port Glasgow R. Duncan & Co.	1867		2Cy.38"–44" 70 HP. Rankin & Blackmore, Grnk.	W. Buchanan
146 [illegible]		**Elamang** I TwnSc Bg *Butcher* *2 Dks*	British Sydney	490 946	229·9	29·2 *B 34 f.*	18·8	Port Glasgow Blackwood & Gordon	1876	[illegible]	1Cy.27" & 47"–30" 200 HP. Blackwood & Gordon, Pt. Gls.	Australasian U[illegible] Steam Nav. Co., Ld.
147 66		**Elba** I ScSr *Cipozza 83-88*	Italian Genoa	105 207	132·8	18·1	11·0	Newcastle Wigham Richardson & Co.	1862		2Cy.21" & 40"–24" 76 HP. P. Ferro & Co., Genoa	Navigazione Gen[illegible] Italiana
148 63545	BT12.88	**Elbe** I ScBq	British London	1732 3033	350·0	40·2	33·3	Glasgow J. Elder & Co.	1870		3C 33" 33" & 88"–48" 390 HP. Triple, '87, [illegible] & Co., Southampton	Royal Mail Stea[illegible] Packet Co.
149 87074	✻	**Elbe** I ScSr *Johnson 80-82* *2 Dks*	British London	492 774	210·0 *P 29 f.*	29·1 *B 42 f.*	15·9 *F 29 f.*	Glasgow Dobie & Co.	1883	4 Blkds. [illegible]	2Cy.26" & 49"–36" 98 HP. W. Kemp, Glasgow	Coverley & West[illegible]
150	G V	**Elbe** I ScIn *Sander 72-88* *3 Dks & Sp. Dk*	German Bremen	2810 4510	418·2	44·9 *B 90 f.*	35·4	Glasgow J. Elder & Co.	1881		3C 60" & (2) 85"–60" 1000 HP. J. Elder & Co., Glasgow	Norddeutscher [illegible]

✻ = Lloyd's Register; D = Nederlandsche Vereeniging; G = Germanischer Lloyd; I = Registro Italiano; R = American R[illegible]
A = Veritas Austro-Ungarico; H = Veritas Ellenico; N = Norske Veritas; V = Bureau Veritas

In the 2nd Column, B.T., with figures following, signifies Date of report of bottom sighting by Board of Trade Surveyors as condition of PASSENGER CERTIFICATE

In the 3rd Column, C = Composite (iron frame, wood planked); I = Iron; Stl = Steel; W = Wood.

Further particulars of Vessels classed by Lloyd's Register will be found in Lloyd's Register Book and Lloyd's Yacht Register.

Classe	Nom, Matériaux, Grèement, etc. Capitaine et dates d'entrée au service des Armateurs actuels et à bord du navire. Ancien Nom, nombre de ponts, etc.	Pavillon. Port d'Armement	Tonnage. Net. Total	Dimensions. Longueur | Largeur | Creux. Constructions sur le Pont	Construit. Où. Par qui.	Quand.	Nombre des cloisons. Détails sur le lest d'eau.	Machines. Détails. Nom du Constructeur.	Armateurs ou Directeurs-Gérants
*	Elbe I ScSr Hoy -88 1 Dk W b Frms	German Hamburg	755 1191	225·0|33·1|16·3 Q 78 f. B 45 f. F 27 f.	Sunderland S.P.Austin&Son	1884	5 Bhds. DBaft78t	2Cy.28"&53"–36" 115 HP.	Wm.Zoder
V	Elbe Stl ScSr Elfers -87	German Lübeck	519 658	180·1|26·0|15·2 P 110 f. F 23 f.	Lübeck HenryKoch	1886	W Ball FPT30t APT20t	2Cy.23"&39½"–25½" 80 HP. Act. Akten-Gesellsch. Vormals "Freund Berlin"	Hanseatische Dampf-schifffahrt Gesellschaft
	Elbe I PadCr Kleinwort 1 Dk	German Stade	86 138	143·0|20·0|7·5 B 30 f.	Hamburg Blohm&Voss	1879		2Cy.25"&44"–39" 70 HP. Blohm&Voss,Hamburg	Std.-Altländer Dampf-schiff. & Rhederei-Ges.
V	Elbruz StlSc3mSr Morris -88 1 Dk & Spr Dk	British London	2101 2113	309·5|40·4|28·2	Newcastle Sir W.G.Armstrong, Mitchell&Co.	1888		3Cy.23"37"&60"–39" 250 HP Wallsend Slipway&Eng. Co. Newcastle	Elbruz S.S.Co.,Ld. (Lamb&Macandrew)
*	Elcano I ScSr Iquirre 79-85 2 Dks	Spanish Bilbao	518 812	212·0|30·2|17·8 F 40 f.	Dumbarton A.McMillan&Sons	1880	4 Bhds.	2Cy.29"&56"–36" 150 HP. D.Rowan,Glasgow	Larrinaga & Co.
*	Elderslie I ScBk Cuthbert 2 Dks 3 B	British Glasgow	1801 2764	300·0|40·0|23·9 P 184 f. F 28 f.	Newcastle Palmers'Co.,Ld.	1884	5 Bhds. DB157t	2Cy.38"&72"–48" 300 HP. Palmers'Co.,Ld.,Newcastle	"Elderslie" S.S.Co.,Ld. (Turnbull,Martin&Co.)
	Eldorado I ScSr Rosch 2 Dks & Awn Dk	British Hull	828 1382	249·6|33·2|15·7	Hull Earle's Co.,Ld.	1886	5 Bhds. DB60t FPT10t APT55t	3C.28"43"&70"–39" 250 HP. Earle's Co.,Ld.,Hull	T.Wilson Sons & Co.
*	Eleanor Stl ScSr Clark -88 1 Dk & W b Frms	British S.Shields	1228 1914	270·0|36·5|19·7 Q 39 f. B 113 f. P 32 f. F 31 f.	Newcastle H.S.Edwards&Sons	1888	5 Bhds. CBB171t MT136t	3Cy.21"34½"&55"–36" J.P.Rennoldson&Sons,S.Shields 160 HP.	J.Ridley,Son & Tully
	Eleanor Stl PadSr Thomas 82	British Dublin	337 834	254·2|30·0|14·2 P 116 f.	Birkenhead LairdBros.	1881		1Cy.36"&60"–78" 350 HP. LairdBros.,Birkenhead	London & N.W.Railway Co.
	Eleanor I PadSr 1 Dk	British Adelaide	99 167	130·3|20·2|10·6	Newcastle	1864		2Cy. 80 HP.	Adelaide Steam Tug Co.,Ld.
	Eleanora W ScSr Bragg 3 Dks	American Portland,Me	582 988	186·4|35·0|16·6	Greenpoint, J.English&Son [N.Y.	1874		1Cy.42"–60" 300 HP. C.H.Delamater&Co.,N.York	Maine S.S.Co.
*	Eleanore I ScSr Ferguson 63-83 1 Dk	British Glasgow	59 185	120·2|21·2|9·8 Q 50 f. F 19 f.	Paisley J.McArthur&Co.	1883	1 Bhds. FPT15t	2Cy.20"&36"–24" 56 HP. W.King&Co.,Glasgow	M'Kinney & Rafferty
*	Electra Stl ScSr Smith 79-87 2 Dks & Awn Dk	British London	656 1249	237·4|32·1|16·6 23·8	Glasgow R.Napier&Sons	1885	1 Bhds. DTa12t DB&c 1681	2Cy.32"&64"–42" 200 HP. R.Napier&Sons,Glasgow	Eastern Telegraph Co.
*	Electra I ScSr Davies 1 Dk 2 B	British Liverpool	305 495	176·6|25·4|14·8	SouthShields J.Readhead&Co.	1876	4 Bhds. DBa48ft f 37ft	2Cy.20"&39"–30" 65 HP. J.Readhead&Co.,S.Shields	J.H.Lewis
*	Electra Stl T ScSr Watson 87-87 1 Dk	British Greenock	210 393	160·5|27·0|10·9	Port Glasgow Russell&Co.	1887	4 Bhds. DT/22t DTa1M FPT10t APT33t	2Cy.23"&42"–30" 76 HP. Rankin&Blackmore,Grnck	Clarence & Richmond Rivers S.N.Co.
*	Electra I ScTrawler Owners 86-86 1 Dk	British Hull	48 102	90·0|20·4|10·6	Hull Cook,Welton&Gemmell	1886	3 Bhds.	2Cy.14"&26"–20" 35 HP. C.D.Holmes&Co.,Hull	F.& T.Ross
	Electra W Sc Mott	American Prdnce.,R.I.	1091 1567	232·0|40·3|15·3	New York	1864		890 HP.	D.S.Babcock
	Electra Stl ScSr	American New York	191 304	174·0|23·0|12·0	Wilmington Harlan&Hollingsworth			2Cy.26"&44"–26" 200 HP Harlan & Hollingsworth, Wilmington	E.T.Gerry
DV	Electra I ScSr Möller 85-88 2 Dks	German Hamburg	1162 1624	260·8|31·9|24·1 F 45 f. B 31 f. P 51 f.	Glasgow Dobie&Co.	1879	W B	2Cy.35"&58"–36" 160 HP. J.Howden&Co.,Glasgow	Deutsche Dampf-schiffs-Rhederei
	Electric I ScSr McEachnie 88-88	British Belfast	295 638	210·1|27·1|14·4 P 61 f.	Glasgow Tod&McGregor	1863		3C 19½"31"&49"–36" 150 HP A.&J.Inglis,Glasgow	J.Hay,Sons & Co.
	Electric I PadBg Heath 1 Dk	British Calcutta	201 395	162·3|26·0|14·2	Newcastle Wgham Richardson &Co.	1864	1 Bhds.	2Cy.52"–72" 180 HP. Thompson,Boyd&Co.,Newc.	Clive Steam Tug Co. (Gladstone,Wyllie & Co.)
*	Electrician Stl ScSr Reynolds 67-87 1 Dk & Spr Dk 3 B	British Liverpool	1862 2847	325·0|40·2|20·2 B 68 f. F 41 f. 28·0	Middlesbro' R.Dixon&Co.	1887	6 Bhds.	3C.24"40"&66"–45" 375 HP Blair&Co.,Ld.,Stockton	Charente S.S.Co.,Ld.
	Electro Stl Sc3mSr Dryden 1 Dk 2 B	British Hull	305 510	179·2|25·0|13·8 Q 28 f. B 24 f. F 16 f.	Hull Earle's Co.,Ld.	1884	5 Bhds. DB... FPT...	3Cy18"27"&48"–27" 112 HP Earle's Co.,Ld.,Hull	T.Wilson,Sons & Co.
	Elektra I ScBk Hesat 63-87 3 Dks	Austrian Trieste	2095 3123	375·0|39·0|28·0	Trieste Lloyd Austro-Ungarico	1884		2Cy.45"&80"–52" 450 HP. ArsenaleLloyd,Trieste	Lloyd Austro-Ungarico
*	Elemore I ScSr Golding 83-84 1 Dk	British Newcastle	613 963	216·0|31·0|13·8 Q 56 f. P 26 f. B 42 f. F 22	Sunderland S.P.Austin&Son	1881	5 Bhds. FPT... APT...	2Cy.28"&54"–36" 99 HP. N.E.Marine Eng.Co.,Ld.,Sld.	Sharp & Co.
	Elena I TwnScSp Pratt	British Glasgow	197 289	135·6|28·1|7·9 Q 23 f.	Liverpool J.Jones&Sons	1886		2Cy.15"–30" 40 HP. J.Jones&Sons,Liverpool	La Platense Flotilla Co.,Ld.
V	Elena I ScSr Aguirre	Spanish Bilbao	417 591	179·1|27·2|15·6	Stockton Richardson,Duck&Co.	1862		2Cy.23"&43"–24" 90 HP. ... Blackwood & Wood, Liverpool	Linea de Vapores Serra
*	Elena Oms Stl ScSr Bilbao -88 1 Dkpt Shade Dk	Spanish Sagua la Grande	384 617	200·0|31·0|12·3 P 44 f. F 27 f.	Sunderland Sunderland S.B.Co.	1888	4 Bhds.	3Cy.18"29"&48"–33" 140 HP. N.E.Marine E.Co.,Ld., Sunderland,	Demetrio Cuetara & Co.
*I	Elettrico Stl ScSr Compagnie 88-88 2 Dks & Awn Dk	Italian Palermo	572 1246	259·8|33·0|14·9 22·8	Glasgow A.Stephen&Sons	1887	4 Bhds.	3Cy.30"48"&77"–48" 500 HP A.Stephen&SonsGlasg.	Nav.Generale Italiana

*=Lloyd's Register; D=Nederlandsche Vereeniging; G=Germanischer Lloyd; I=Registro Italiano; R=American Record; ‡=Veritas Austro-Ungarico; H=Veritas Ellenico; N=Norske Veritas; V=Bureau Veritas.

Dans la 2me Colonne, B.T., avec les indications qui suivent, signifie la date du rapport de visite de la Carène par les experts du Board of Trade pour obtenir le certificat relatif au TRANSPORT DES PASSAGERS.

Dans la 3me Colonne, C=Composite (membrure en fer, bordé en bois); I=Fer; Stl=Acier; W=Bois.

Pour plus amples renseignements sur les navires classés par le Lloyd's Register, voir le Lloyd's Register Book et le Lloyd's Yacht Register.

No. in Book. / Official No.	Registry in which classed.	Ship's Name. Material, Rig, &c. / Master, and Dates of Appointment to present Owner's Service and to Ship / Ship's Late Name. No. of Decks, &c.	Flag. / Port of Registry.	Tons. Net. / Gross.	Dimensions. Length	Breadth	Depth	Built. Where. / By Whom.	When.	No. of Bulkheads. / Particulars of Water Ballast.	Engines. Particulars. Maker's Name.	Owners or Managers
181 69290	✻	Elf I ScSr *Hawkins 1 Dk2 B*	British Leith	789 1334	262·7	32·4	16·2 *P 35 f. F 37 f.*	Sunderland W.Doxford&Sons	1871	5 Blds. WT 281	2Cy.30"&56"–36" 120 HP. Gilbert&Cooper,Hull	Stevenson & [illegible]
182		Elfangen (ex Tordalla ex Runan) 1 Sc *Sköldberg -88 1 Dk*	Swedish Gothenburg	80 108	93·5	16·1	7·5	Stockholm	1865		2Cy.12"&25¼"–19½" 80 HP BergsundsM.V.Stockholm	A.G.Friländer
183 81850	✻	Elfie I ScSr *Robinson 82-84 1 Dk2 B*	British Hartlepool	800 1277	258·0	32·1	17·0 *P 33 f.*	W.Hartlepool W.Gray&Co.	1880	4 Blds. DB 284	2Cy.28½"&58"–33" 110 HP. Blair&Co.,Ld.,Stockton	C.Nielsen & [illegible]
184 96133	✻	Elfrida St ScSr *Rasmussen -89 1 Dk & W'h Frame*	British Newcastle	1446 2201	290·0	38·0	20·1 *Q 93 f. B 114 f. P 27 f. F 33 f.*	Sunderland Osbourne,Graham&Co.	1889	6 Blds. DB	3Cy.22"36½"&60"–42" 200 HP. N.E.Marine E.Co.,Ld.,Sunderland	Pyman,Bell & [illegible]
185 84290	✻	Elfrida I ScS *McNeil 83-85 1 Dk*	British	310 498	165·0	24·6	12·9 *Q 77 f. F 40 f. B 17 f.*	Paisley Campbell&Co.	1880	4 Blds. DB FPT	2Cy.22"&40"–30" 70 HP. Fleming&Ferguson,Paisley	Colvils, Lowd[illegible]
186 84944	✻	Elfrida I ScSr Yt	British	66 109	118·1	17·2	9·1	Leith Ramage&Ferguson	1882	4 Blds.	2Cy.18"&27"–22" 30 HP. Ramage&Ferguson,Leith	J.McDonald
187	N	Elgen W ScSr *Lindtner 82-82 1 Dk2 B*	Norweg'n	386 594	156·8	28·2	17·2 *B 37 f.*	Gothenburg MotalaCo.,Gothenburg	1870		2Cy.17¾"&34½–22½" 66 HP. MotalaCo.,Gothenburg	T.Meling
188 85590	✻	Elginshire I ScSr *Wallace 2 Dks*	British	1353	285·5	36·0	22·7 *B 55 f. F 33 f.*	Hartlepool E.Withy&Co.	1881	4 Blds.	2Cy.34½"&64"–42" 200 HP. Blair&Co.,Ld.,Stockton	Turnbull,M[illegible]
189 93289	✻	Elgiva I ScS *Lewes 88-89 1 Dk*	British	429 668	180·6	26·2	13·7 *Q 50 f. B 39 f. F 28 f.*	Belfast Workman,Clark&Co.	1886	4 Blds. DB FPT	3C.15"24"&40"–33" 95 HP. Hutson&Corbett,Glasgow	Macgregor & [illegible]
190		Elias	Brazilian	397	..	..	..					E.E.P.de Am[illegible]
191		Elias Lönnroth Sc *Lagercrantz -87*	Russian	146	..	..	..	Wiborg	1877		57 HP.	A.A.B.Saima
192		Elias Sehlstedt I ScSr *Fehlman -87 (ex Emil Posse) 1 Dk*	Swedish Hernösand	190	119·4	19·5	10·1 *Q 33 f. B 38 f. P 38 f. H 13 f.*	Oscarshamn OscarshamnsMek.Verks	1882		2Cy.15"&33" 60 HP. Oscarshamn Mek.Verks	Ångf. Aktieb[illegible] land(E.R.Lu[illegible])
193	N	Elida W ScSr *Fagerli -88 1 Dk2 B*	Norweg'n	147	116·5	25·2	11·8 *Q 33 f.*	Sandefjord Chr.Christensen	1883		30 HP.	Chr.Christens[illegible]
194 92865	✻	Elingamite St ScSr *Rull 85-87 1 Dk & W'h Frame*	British	1675	310·5	40·8	19·7 *P 159 f. F 57 f.*	Newcastle CS Swan&Hunter	1887	7 Blds. DB FPT APT	3C.25"41"&67"–42" 250 HP. Wallsend Slipway Co.,Ld.,Newcastle	Huddart,Park[illegible]
195 11	V	Elisa Anna I ScSr *Muzio (ex Cordova) 1 Dk & Awn Dk*	Italian Leghorn	827	246·0	30·2	17·0 236	Waterford Malcomson Bros.	1864		2Cy.30"&56"–36" 220 HP. J.Jones&Sons,Liverpool	S.S.Coen fu D.
196	V	Elisabeth I ScS *Frandsen 86-86 1 Dk*	Danish	277 459	114·8	23·9	14·6 *B 41 f. F 10 f.*	Gothenburg Eriksbergs Mk.Werks.	1884	APT WB	2Cy.16"&31"–19" 50 HP. Eriksbergs Mk.Werks.,Goth.	S.A.Philipsen
197 9	✻	Elisabetta I ScSr *Scotto (ex Adeline Schulz) 1 Dk*	Italian Leghorn	245 390	162·0	23·2	13·2 *Q 74 f. B 12 f. F 22 f.*	Paisley H.McIntyre&Co.	1880	4 Blds. FPT 70t	2Cy.21"&40"–30" 71 HP. Fleming&Ferguson,Paisley	Eredi F.Raffo
198		Elispondos I S	Greek	102	..	..	..					T.Stathatos
199 65601	✻	Eliza I Pd *Olsen 87-87 1 Dk*	British Demerara	36 69	100·0	16·1	6·9	London Green	1871	3 Blds.	2Cy.24½"–33" 37 HP. J.Stewart,London	Sproston Dock & Foundry Co.
200 63564	✻	Elizabeth (ex Nineveh) I ScSr *Turnbull 1 Dk2 B*	British Hull	679 1041	225·5	30·7	16·9 *P 194 f. F 22 f.*	Sunderland Oswald&Co.	1869	4 Blds. WB	1Cy.20"&39"–33" 128 HP. Bailey&Leetham,Hull	W.S.Bailey
201 27789		Elizabeth (ex Paquete Correo, ex General Miramon) W ScSr *1 Dk*	British	170 250	131·5	25·1	9·0 *P 50 f.*	Glasgow	1859		2Cy. 50 HP.	John H.Ham[illegible]
202 8281		Elizabeth W P	American New York	829	215·0	35·3	12·7	Brooklyn	1867		1Cy.50"–144" Fletcher,Harrison&Co.,New York	S.Knox
203 82894	D ✻	Elizabeth Allen I ScS *Allen 84-84 1 Dk2 B*	British	1049 1614	257·2	35·0	18·7 *Q 58 f. B 62 f. F 34 f. P 29 f.*	Sunderland Bartram,Haswell & Co.	1881	5 Blds. DB FPT APT	2Cy.32"&60"–39" 160 HP. J.Dickinson,Sunderland	J.Allen
204 85101	✻	Elk I ScS *Taylor 88-88 1 Dk2 B*	British	945 1476	246·5	34·1	17·4 *Q 76 f. B 76 f. F 29 f.*	Sunderland Osbourne,Grah'm&Co.	1881	4 Blds. DB	2Cy.30"&55"–36" 120 HP. Blair&Co.,Ld.,Stockton	Jackson Bros. & [illegible]
205 85554	✻	Ella I S S *Pines 84-86 1 Dk2 B*	British	1375 2153	270·0	39·3	19·0 *Q 65 f. B 70 f. P 28 f. F 39 f.*	Sunderland N.ofEng.S.B.Co.,Ld.	1884	5 Blds. DB FPT APT	2Cy.34"&65"–42" 190 HP. N.E.Marine Eng.Co.,Ld.,Shi.	R.Gordon & Co.
206 62294	✻	Ella I ScSrS *Philpott 1 Dk & 1 [illegible] Dk*	British	443 693	197·5	25·3	11·5 18·2	Inverkeithing J.Scott&Sons	1870	6 Blds. APT	2Cy.20"&36"–27" 75 HP. J.Scott&Sons,Inverkeithing	London & R[illegible] Steamship Co.,Ld.
207 82420	BT	Ella I ScS *2 Dks*	British	355 797	235·5	29·1	14·2 *Q 126 f. F 60 f.*	Glasgow Aitken&Mansel	1881	5 Blds.	2Cy.37"&69"–39" 220 HP. J.&J.Thomson,Glasgow	London & S.W[illegible] way Co.
208 76129	✻	Ella I ScS *Steele 84-84 1 Dk*	British Middlesbro'	290 456	160·2	24·3	13·0 *Q 58 f. B 37 f. F 17 f.*	Stockton Richardson,Duck&Co.	1878	4 Blds. DB	2Cy.20"&37½"–24" 55 HP. Blair&Co.,Ld.,Stockton	J.M.Lennard & [illegible]
209	✻	Ella I ScSr *Bartels 1 Dk2 B*	German Danzig	678 1058	220·0	31·1	16·5 *P 123 f. F 25 f.*	Glasgow A.Stephen&Sons	1883	4 Blds. FPT APT WT	2Cy.28"&52"–36" 140 HP. A.Stephen&Sons,Glasgow	Th.Rodenacker
210 21173	✻	Ella Constance I ScSmSr *Wake 1 Dk2 B.S pt Awn Dk*	British Sunderland	420 653	192·0	26·3	15·5 *Q 25 f. B 76 f. F 13 f.*	Stockton Richardson,Duck&Co.	1858	4 Blds.	2Cy.22"&49"–36" 80 HP. Greenock Foundry Co.,Grnk.	W.Jacks & Co.

✻ = Lloyd's Register; D = Nederlandsche Vereeniging; G = Germanischer Lloyd; I = Registro Italiano; R = American [illegible]; A = Veritas Austro-Ungarico; H = Veritas Ellenico; N = Norske Veritas; V = Bureau Veritas

In the 2nd Column, B.T., with figures following, signifies Date of report of bottom sighting by Board of Trade Surveyors as condition of PASSENGER [illegible]

In the 3rd Column, C = Composite (iron frame, wood planked); I = Iron; Stl = Steel; W = Wood.

Further particulars of Vessels classed by Lloyd's Register will be found in Lloyd's Register Book and Lloyd's Yacht Register.

No.	Cl.	Nom, Matériaux, Gréement, etc. Capitaine et dates d'entrée au service des Armateurs actuels et à bord du navire Ancien Nom, nombre de ponts, etc.		Pavillon Port d'Armement	Tonnage Net Total	Dimensions: Longueur	Largeur	Creux	Construit: Où Par qui	Quand	Nombre des cloisons Détails sur le lest d'eau	Machines Détails Nom du Constructeur	Armateurs ou Directeurs-Gérants
		Ella Ross (ex Ghost)	I Pad 1 Dk & Spr Dk	British Toronto	190 [illegible]	99·2	27·8	6·1	Montreal W. Webster	1873		1Cy. 22"–48" 85 HP. Davidson, Doran & Co., Kgstn.	W. Garrett
	*	Ella Sayer Turnbull 80-84	I ScSr 1 Dk 2 B	British Newcastle	1096 1692	258·0	37·0	16·7	Sunderland Doxford & Sons	1883	1 Bhds. DB 330t APT 70t	2Cy. 33" & 62"–42" 161 HP. W. Doxford & Sons, Sundrlnd.	Fisher, Renwick & Co.
		Ella Warley	W PadSr 3 Dks	British Nassau, N.P.	576 1848	212·6	33·5	24·0	Baltimore	1848		1Cy. 500 HP.	Edwin C. Adderley
	V	Ella Woermann Dittmer 82-85	I ScSr 2 Dks	German Hamburg	1025 1694	271·3	31·8	21·2	Hamburg Blohm & Voss	1883		2Cy. 30" & 52"–42" 160 HP. Blohm & Voss, Hamburg	Africanische Dampfschiffs-Act. Ges. (Woermann Linie)
215		Ellan Vannin (ex Mona's Isle)	I TwnScSr	British Douglas	97 342	198·6	22·2	10·7	Glasgow Tod & McGregor	1860		1Cy. 18" & 34"–24" 100 HP. Westray, Copeland & Co. Brw	Isle of Man Steam Packet Co.
216	*	Ellida	W ScSrYt	British Cowes	81 119	107·0	20·6	12·2	Cowes Batsey & Co.	1875		2Cy. 10" & 20"–18" 20 HP. Day, Summers, & Co., Sthptn.	A. W. Fulcher
217	DV	Ellida Adamsen 83-83	I ScSr	German Flensburg	340 537	168·9	25·5	14·1	Gothenburg Eriksberg Mek. Verkst.	1883	WB	2Cy. 20" & 39"–23" 75 HP. Eriksberg Mek. Verks. Gtbg.	F. M. Bruhn
218		Ellida	W ScSr 1 Dk	Norweg'n Bergen	146 225	101·7	24·1	11·9	Wenersborg	1870		30 HP.	Statens Havnevæsen
219		Ellida	I PadSr 1 Dk	Swedish Stockholm	140 220	179·1	17·7	9·6	Norrköping Motala Co.	1862		2Cy. 12" & 25½"–19" 100 HP.	A. T. Edlund
220		Ellie Knight	W ScSr	American Key West	207 289	155·0	23·0	16·6	Philadelphia Birely, Hillman & Co.	1862			J. H. Menge
221	*	Ellington Glook 83-83	I ScSr 1 Dk	British Newcastle	467 724	184·5	30·4	13·0	Sunderland D. Baxter & Co.	1882	5 Bhds. DBaft 130t FPT 25t APT 30t	2Cy. 24" & 48"–33" 95 HP. Baird & Barnsley, N. Shields	Humble & Thompson
222		Elliot & Jeffery	I ScSp	British Cardiff	10 138	105·2	19·8	10·7	Cardiff Elliot & Jeffery	1884		2Cy. 19" & 35"–20" 70 HP. Elliot & Jeffery, Cardiff	Elliot & Jeffery
223		Ellora Johnston 86-88	I ScSc 2 Dks	British Glasgow	1471 1970	285·0	35·2	23·9	Glasgow R. Napier & Sons	1879	5 Bhds. DBaft & 250 ft	2Cy. 34" & 60"–42" 185 HP. R. Napier & Sons, Glasgow	British India Steam Nav. Co., Ld.
224		Elm	Stl ScSr	British Glasgow	155 189	182·2	27·7	12·8	Glasgow A. & J. Inglis	1884		2Cy. 26" & 50"–36" 130 HP. A. & J. Inglis, Glasgow	Glasgow, Dublin & Londonderry Stm. P. Co.
225		Elm City	W Pad	American New Haven, Conn.	897 1049	282·6	36·7	11·3	New York	1855			R. Peck
226	*D	Elma (ex Parsee) Campbell 84-87	I ScBg 1 Dk 2 B	British Hull	992 1536	250·0	34·2	19·6	Newcastle Schlesinger, Davis & Co.	1872	4 Bhds. W Bay	2Cy. 30" & 60"–36" 150 HP. Jackson & Watkins, London	W. S. Bailey
227	*	Elmfield Ching 87-87	I ScBn 2 Dks	British Stockton	1180 1805	250·0	36·2	23·5	Stockton Richardson, Duck & Co.	1883	3 Bhds. WBaft	2Cy. 31" & 58"–36" 150 HP. Blair & Co., Ld., Stockton	F. Binnington
228	*	Elmina Cawthorne 75-87	Stl ScSr 2 Dks	British London	1143 1764	300·5	36·2	19·8	Belfast Harland & Wolff	1885	7 Bhds.	2Cy. 32" & 64"–42" 200 HP. Harland & Wolff, Belfast	African Steamship Co.
229	V	Eloby Ludwig 85-85	I & Stl ScSr	German Stettin	118 168	114·8	21·3	8·9	Stettin Act. Ges. "Vulcan"	1884	WB	2Cy. 12" & 22"–15" 25 HP. Act. Ges. "Vulcan," Stettin	Africanische Dampfschiffs Actien Gesellschaft. (Woermann Linie)
230	*D	Elphinstone Marshall 87-87	I ScSr 2 Dks 3 B	British Newcastle	1146 1776	280·5	33·5	24·9	Newcastle Palmers Co.	1874	5 Bhds. DBaft & 260t	2Cy. 30" & 60"–33" 140 HP. Palmers Co., Ld., Newcastle	Macintyre Bros. & Co.
231	*	Elpis Burnett 86-88	I ScSr 2 Dks 3 B	British W. Hartlepl.	1302 2010	286·0	35·9	23·9	W. Hartlepool W. Gray & Co.	1878	4 Bhds. DBaft & 420t	2Cy. 35 & 64½"–36" 180 HP. T. Richardson & Sons, Hrtpl.	R. Ropner & Co.
232		Elpis (ex Truthful) V. Hissarios 79-	I ScSr 1 Dk 2 B	Greek Syra	606 956	240·2	30·0	16·1	Barrow Barrow S. B. Co.	1877	4 Bhds. DBa 79t FPT	2Cy. 33" & 66"–36" 160 HP. Barrow S. B. Co., Barrow	Hellenic Steam Ship Co.
233	*	Elpitha (ex Pessagno) Melrena	I Sc3mSr	British Glasgow	462 733	218·8	28·0	16·6	Glasgow C. Connell & Co.	1862	4 Bhds.	2Cy. 28" & 52"–30" 120 HP. Cmp. 73, J. & T. Young, Ayr	J. Clark
234	*	Elsa Barron 88-88	I ScSr 1 Dk 2 B	British Glasgow	525 850	215·0	31·1	14·7	Campbeltown Campbeltwn. S. B. C.	1882	4 Bhds. DBa 110t FPT 30t	2Cy. 26" & 50"–33" 99 HP. Kincaid & Co., Greenock	S.S. "Elsa" Co., Ld. (R. Monteith & Co).
235	G	Elsa Wiese 84-87	I ScSr 1 Dk	German Kiel	577 749	196·5	24·7	18·6	Kiel G. Howaldt	1883	WB	2Cy. 23" & 37½"–25" 70 HP. Howaldt Gebr., Kiel	F. Lange
236	V	Elsa Petersen 88-88	Stl ScSr 1 Dk	German Flensburg	571 786	112·0	30·5	15·7	Flensburg Flensburger Schiff Gesell.	1888	DB 164t	3Cy. 14" 25" & 40½"–27" Flensb. Schiff. Ges., Flensb.	H. Schmidt
237	G	Else Jebsen, Jr. 87	Stl ScSr	German Hamburg	747 975	205·0	30·0	20·9	Hamburg Blohm & Voss	1888	5 Bhds. W B 248t	3C 15½" 24½" 39½"–33" 90 HP Blohm & Voss, Hamburg	Mich. Jebsen
238	*	Elsie White 77-88	I Sc 1 Dk & Awn Dk 3 B	British Whitby	1543 2374	285·0	37·6	21·2 28·4	Sunderland J. L. Thompson & Sn	1881	7 Bhds. DBaf 293t FPT 28t APT 21t	2Cy. 36" & 68"–42" 220 HP. J. Dickinson, Sunderland	J. Gray & Co.
239	*	Elspeth (ex Sarinella, ex Elspeth)	I ScSrYt	British S'thampton	79 131	130·0	17·6	9·3	Glasgow J. & G. Thomson	1877	4 Bhds.	2Cy. 17" & 32"–21" 45 HP. J. & G. Thomson, Glasgow	Hon. H. C. Denison
240	*	Elstow Gray	I ScSr 1 Dk 2 B	British Sunderland	1140 1766	258·0	36·1	18·5	Sunderland N. of England S. B. C.	1883	6 Bhds. DBa 290t FPT 9t APT 36t	2Cy. 33" & 62"–42" 200 HP. J. W. & F. Wilson & Co., Sndld.	Britannic Steam Shipping Co., Ld.

* = Lloyd's Register; D = Nederlandsche Vereeniging; G = Germanischer Lloyd; I = Registro Italiano; R = American Record;
= Veritas Austro-Ungarico; H = Veritas Ellenico; N = Norske Veritas; V = Bureau Veritas.

Dans la 2me Colonne, B.T., avec les indications qui suivent, signifie la date du rapport de visite de la Carène par les experts du Board of Trade pour obtenir le certificat relatif au TRANSPORT DES PASSAGERS.

Dans la 3me Colonne, C = Composite (membrure en fer, bordé en bois); I = Fer; Stl = Acier; W = Bois.

Pour plus amples renseignements sur les navires classés par le Lloyd's Register, voir le Lloyd's Register Book et le Lloyd's Yacht Register.

No. in Book. Official No.	Registry in which classed.	Ship's Name. Material, Rig, &c. Master, and Dates of Appointment to present Owner's Service and to Ship. Ship's Late Name. No. of Decks, &c.	Flag. Port of Registry.	Tons. Net. Gross.	Dimensions. Length. Breadth. Depth. Deck Erections.	Built. Where. By Whom.	When.	No. of Bulkheads Particulars of Water Ballast.	Engines. Particulars. Maker's Name.	Owners or Managers.
241	*	**Elsy** I ScK *Hargrave 86*	British	65	86·4 19·3 8·5	Hull R. Wales	1883	3 Blkds.	2Cy.11"& 22"-15" 20 HP E. Wales, Hull	J. E. Crisp
242	*	**Elton** St ScS *Campbell 86-88* 1 Dk & Web Frames	British W. Hartl'pool	1725 259[illegible]	[illegible] 38·6 22·0	W. Hartlepool W. Gray & Co.	1888	[illegible]	[illegible] 37½"&61½"-39" 250[HP] [illegible] Ld. Stockton	R. Ropner & Co.
243	*	**Elvaston** St ScS *Richardson 79-87* 1 Dk & Web Frames	British London	1367	278·0 37·2 19·2 B 96 f. F 31 f.	Sunderland J. Laing	1887	4 Blkds.	[illegible] & 60"-42" 300 HP [illegible] Sunderland	Harris & Dixon
244	DGV	**Elve** I ScS *Laribij 17-86* 2 Dks	Dutch Rotterdam	320	161·4 24·0 11·5 Q 98 f. B 6 f.	Kiel [illegible]	1885		2Cy.18"& 34"-24" 60 HP G. Howaldt, Kiel	P. A. van Es & Co.
245	V	**Elvira** I ScS *De Aspuru*	Spanish [illegible]	677	231·5 30·3 16·6	Glasgow Tod & Macgregor	1871		2Cy.28"& 48"-36" 140 HP Tod & Macgregor, Glasgow	Linea de Vapores Serra
246	*	**Ely Rise** I ScS *Jones 83-85* 1 Dk 2 B	British [illegible]	784 1212	225·0 31·5 17·8	Sunderland Short Bros.	1877	5 Blkds. DB [illegible] FT [illegible] AP T [illegible]	2Cy.27"& 50"-36" 110 HP J. Clark, Sunderland	Ely Rise S.S. Co. (Hull) Matthews
247 68053	BT 12.88	**Elysia** I ScB *Woodhouse 88-88* 2 Dks & Awn Dk	British Glasgow	1745 2714	351·0 35·0 29·7	Glasgow J. Elder & Co.	1873	4 Blkds.	2Cy.33"& 60"-48" 436 HP D. & W. Henderson, Glasgow	Henderson Bros.
248	V	**Em. Z. Svitzer** I ScS *Mogensen 86-87 (SALVAGE)* 1 Dk	Danish Copenhagen	129 [illegible]	142·0 25·3 12·5 B 10 f.	Copenhagen Burmeister & Wain	1885	MT AP T [illegible] FT [illegible]	2Cy.30"& 52"-33" 120 HP Burmeister & Wain, Cp'nhgn	Em. Z. Svitzer's [illegible]
249		**Emanuel** C ScS *Mattsson* 1 Dk	Swedish [illegible]	128 154	99·1 20·5 12·5					E. Mattsson
250 81123	*	**Embla** I ScS *Irons 81-88* 1 Dk & Web Frames	British Grangemouth	742 1176	226·9 33·0 14·8 P 27 f. Q 48 f. B 83 f. F 38 f.	Leith S. & H. Morton & Co.	1882	5 Blkds. C DB 29 ft. FT 38 AP T [illegible]	2Cy.28½"& 53"-33" 99 HP Blair & Co., Ld. Stockton	J. T. Salvesen & Co.
251	N	**Embla** W ScS *Olsen* 1 Dk 2 B	Norweg'n Bergen	288 395	142·9 24·1 14·3 Q 47 f. B 30 f. F 19 f.	Bergen Bonnevie & Dekke	1883		2Cy.17"& 32"-22" 40 HP Martens, Olsen & Co., Bergen	Johan C. Giertsen
252 84521	*D	**Embleton** I ScBk *Simmonds 82-85* 1 Dk 2 B	British W. Hartlepool	1134 1762	256·2 34·7 19·7 Q 68 f. P 38 f. F 27 f.	W. Hartlepool W. Gray & Co.	1881	4 Blkds. DB af 35 ft.	2Cy.32"& 60"-39" 150 HP Blair & Co., Ld. Stockton	J. Wood & Co.
253 91310	*	**Emden** I ScSc *Ayre -87* 1 Dk	British Goole	357 501	220·0 32·1 13·9 F 32 f. F 40 f. B 80 f.	Hull Earles Co., Ld.	1887	5 Blkds. DB af & cases	3C 21½"31"& 52"-36" 150 HP Earles Co., Ld. Hull	Yorkshire Coal S.S. Co., Ld.
254 62349	*	**Emerald** I ScS *Scott 64-75* 1 Dk 2 B	British Sunderland	616 [illegible]	217·8 29·5 16·5 Q 83 f.	Sunderland [illegible]	1872	5 Blkds. [illegible]	2Cy.27"& 50"-36" 98 HP J. Clark, Sunderland	E. J. Weatherley
255 78676	*	**Emerald** I ScS *Jones 87-87* 1 Dk 2 B	British Dundee	400 618	189·0 26·0 14·5 B 43 f. Q 58 f. F 13 f.	Dundee Gourlay Bros.	1877	3 Blkds. [illegible]	2Cy.24"& 46"-30" 75 HP Gourlay Bros. & Co., Dundee	Dundee Gem Line Co., Ld.
256	*	**Emerald** W ScK 1 Dk	British Nth. Shields	3 13	70·2 17·2 7·2	Leith Marr Bros.	1886		2Cy.11"& 22"-16" 20 HP Baird & Barnsley, N. Shields	Baird & Barnsley
257 [illegible]	BT 5.88	**Emerald Isle** I PadSr	British Dundalk	260 725	235·5 28·2 15·1 Q 36 f.	Glasgow J. & G. Thomson	1862		2Cy 63½"-66" 320 HP J. & G. Thomson, Glasgow	Dundalk & Newry Steam Packet Co.
258		**Emeraude**	Belgian	192	169·0 20·0 6·0	England	1858			Belgian Government
259		**Emile** (ex Fleur[illegible]) I ScSr 1 Dk	French Gravelines	128 211	130·0 19·4 11·7 P 23 f.	London Lewis & Stockwell	1867	4 Blkds.	2Cy.26"-18" 50 HP Humphrys, Hodgson & Co. Ld.	E. Buisine & Co.
260		**Emile Eloise** I ScSr 1 Dk	French Dellys, Algeria	158 [illegible]	124·3 20·2 9·0	Montrose Black & Noble	1877	3 Blkds.	2Cy.18"& 30"-22" 45 HP Kemp & Hume, Glasgow	Franceschi
261		**Emiliana** (ex Goiana[illegible]) I Sc *Santos -87*	Brazilian [illegible]	132 [illegible]	119·5 20·1 9·3 Q 8 f. F 6 f.	Paisley J. Fullerton & Co.	1871	4 Blkds.	2Cy.16"& 26"-22" 40 HP W. King & Co. Glasgow	M. Monteiro & Co.
262	*	**Emiliano** I ScBg *Beagoa 85-87* 2 Dks Spl Awn Dk	Spanish Bilbao	1427 [illegible]	306·0 34·9 24·0 B 81 f. F 32 f.	Sunderland Oswald & Co.	1871	6 Blkds.	2Cy.40"& 70"-48" 280 HP J. Jack & Co., Liverpool	Larrinaga & Co.
263 91842	*	**Emilie** I ScS *Cox 87-87* 1 Dk & Web Frames	British London	1099 1693	260·0 36·1 17·9 P 23 f. Q 21 f. B 80 f. F 31 f.	W. Hartlepool E. Withy & Co.	1885	4 Blkds. DB [illegible]	3Cy 21"34"& 56"-36" 180 HP Blair & Co., Ld. Stockton	Burdick & Cook
264 [illegible]		**Emilie** W PadSp (ex Wm. Seabrook) 2 Dks	British Nassau, N.P.	202 352	136·7 24·1 16·7	Charleston, US	1839		1Cy. 125 HP	G. A. Stewart
265	V	**Emilie** I ScSh *Balzan 84-84* 1 Dk 2 B	French Dunkirk	493 [illegible]	214·2 28·2 16·2 P 106 f. F 28 f.	S. Denis, Seine Claparède & Co.	1877	DB [illegible]	2Cy. 25½"& 43½"-29½" 125 HP Claparède & C[illegible], St. Denis, Seine	G. Beck
266 78865	*D	**Emily** I ScSt *Gosling 83-83* 1 Dk 2 B	British Whitby	777 1202	240·0 31·0 17·4 Q 89 f. P 39 f. B 19 f.	S. Shields J. Readhead & Co.	1878	4 Blkds. DB [illegible]	2Cy.26½"& 51"-36" 110 HP J. Readhead & Co., S. Shields	R. Harrowing
267 [illegible]		**Emily** I ScSc *French* 1 Dk	British Liverpool	96 [illegible]	116·5 16·2 8·4 Q 33 f.	Liverpool C. McConnochie & Co.	1868		2Cy.14"& 25"-16" 27 HP C. McConnochie & Co., Liverpl.	J. Burnett & Son
268 135939		**Emily** W ScS *Roberts -87* 1 Dk	American San Francisco	209 [illegible]	131·5 32·2 10·5	San Francisco C. G. White	1887		2Cy.16"& 30"-20" 175 HP Hinckley, Spiers & Hayes, S. Francisco	Fort Bragg Redwood Co.
269		**Emir** I ScS 2 Dks & Spr Dk	French Marseilles	889 [illegible]	246·0 31·7 23·6	Newcastle Wigham Richardson & Co.	1882	DB [illegible]	2Cy.34"& 61"-41½" 250 HP Wigham Richardson & Co., Newc.	Cie. de Nav. Mixte
270		**Emla** I ScS *Wetterlund* 1 Dk	Swedish Gothenburg	432 [illegible]	161·4 25·1 15·5 Q 9 f. B 18 f. P 36 f. W 25 f.	Gothenburg A. Brodin	1872		2Cy.18"& 35"-24" 80 HP [illegible] M. Verkstäder	Angf. Aktiebolaget Nordsjön[illegible]

*=Lloyd's Register; D=Nederlandsche Vereeniging; G=Germanischer Lloyd; I=Registro Italiano; R=American Record; A=Veritas Austro-Ungarico; H=Veritas Ellenico; N=Norske Veritas; V=Bureau Veritas.

In the 2nd Column, B.T., with figures following, signifies Date of report of bottom sighting by Board of Trade Surveyors as condition of PASSENGER CERTIFICATE.

In the 3rd Column, C=Composite (iron frame, wood planked); I=Iron; Stl=Steel; W=Wood.

Further particulars of Vessels classed by Lloyd's Register will be found in Lloyd's Register Book and Lloyd's Yacht Register.

No.	Cl.	Nom, Materiaux, Gréement, etc. Capitaine et dates d'entrée au service des Armateurs actuels et à bord du navire. Ancien Nom, nombre de ponts, etc	Pavillon. Port d'Armement	Tonnage Net Total	Dimensions Longueur	Largeur	Creux	Construit. Où. Par qui.	Quand.	Nombre des cloisons. Détails sur le lest d'eau.	Machines. Détails. Nom du Constructeur.	Armateurs ou Directeurs-Gérants.
	V	**Emma** (LGSTD) I ScSr *Lambert* -88 1 *Dk2 B*	British Sunderland	529 825	204·1	27·3	16·5 *P 109 f.*	Sunderland J.Blumer&Co.	1872	*T.44 f.aa*	3Cy.26½ & 43½"-37" 80 HP. Soc.desCostr.Navales,Hvre.	J.Laing
	V	**Emma** I ScBr *Rosrogers* -88 2 *Dks*	French Havre	508 789	208·3	30·2	17·4 *P 30 f. B 8 f. F 40 f.*	Hartlepool Denton,Gray&Co.	1867	1 Blkds.	2C.36½"& 55½"-27" 120 HP. Comp.73,Nillus,Havre	Worms,Josse & Co.
	G	**Emma** I ScS *Kreitsfeldt, Jr.* 84-83 2 *Dks*	German Kiel	671 868	201·0	26·9	19·3	Kiel G.Howaldt	1883	W.B.	2Cy.25"& 40"-27" 85 HP. Howaldt Bros.	Sartori & Berger
		Emma St ScS 1 *Dk*	German	505 693	180·0	26·5	13·9 *Q 60 f. B 49 f. F 21 f.*	Rostock RostockS.B.& E.Co.	1888	*DB* [illegible] *FPT* [illegible]	3Cy.12½".19⅛"&33½"-23⅝" Rostocker [illegible] Co., Rostock 350 HP.	Zerssen & Co.
	GDV	**Emma** I ScS *Wunderlich* 83-84 1 *Dk*	German Danzig	493 683	182·4	26·0	17·4	Danzig J.W.Klawitter	1883	W.B.	2Cy.23"& 44"-25" 90 HP. F.Schichau,Elbing	F.G.Reinhold
	✻D	**Emma Lawson** I ScS *Hills* 90-88 1 *Dk2 B*	British Whitby	636 1008	225·0	30·0	17·1 *Q 84 f. F 26 f. B 48 f.*	Whitby T.Turnbull&Son	1875	4 Blkds. *DBaf* [illegible]	2Cy.27"& 50½"-33" 99 HP. Blair&Co.,Ld.,Stockton	T.Turnbull & Son
	✻	**Emma Sauber** I ScSr *Farre* -88 1 *Dk2 B*	German Hamburg	855 1307	255·2	32·8	18·9 *B 48 f. F 29 f.*	Hamburg Reiherstieg Schiffs Wft.	1883	7 Blkds. *DB* [illegible]	2Cy.30"& 56"-36" 130 HP. ReiherstiegSchiffsWfte,Hbg	Sauber Gebrüder
		Emmanuel I ScS *Harforth* 86-89 1 *Dk2 B*	British Middlesbro'	254 401	145·0	23·0	12·1 *B 49 f. Q 34 f. F 17 f.*	Stockton M.Pearse&Co.	1877	3 Blkds. W.B.	2Cy.18½"& 34"-24" 45 HP. Blair&Co.,Ld.,Stockton	E.Harris & Co.
	✻	**Emmanuel Scicluna** I ScSr *De Marin* 1 *Dk2 B*	British Malta	622 971	200·6	31·0	15·3 *Q 63 f. F 50 f. F 25 f.*	Barrow Caird,Purdie&Co.	1884	4 Blkds. *DBaf* [illegible]	2Cy.27"& 52"-36" 114 HP. Westray,Copeland&Co,Brw	Marquis J.Scicluna
		Emperor I ScS *Leighton* 1 *Dk2 B*	British Hull	987 1503	251·2	40·8	19·3 *P 142 f. F 42 f.*	Glasgow R.Napier&Sons	1849	4 Blkds.	2Cy.35"& 66"-36" 170 HP. Gilbert&Cooper,Hull	Earles'S.B.& E.Co.,Ld.
		Emperor W ScSr 1 *Dk*	British Sydney	74 108	82·5	23·0	7·3	Blackwall RockDavis [N.S.W.	1884		2Cy.10"-15" 20 HP. Plenty&Sons,Newbury,Eng	Francis Buckle
		Emperor I ScTug 1 *Dk*	British London	56 106	90·1	18·0	10·5	Hayle Harvey&Co.	1880	4 Blkds.	2Cy.21"& 38"-21" 65 HP. Harvey&Co.,Hayle	Wilson,Sons & Co.,Ld.
		Emperor I ScSp *Caddy* 1 *Dk*	Arg'ntine Bue'osAyres	35 98	85·8	18·1	10·0	Preston Allsup&Sons	1882	4 Blkds.	2Cy.20"& 36"-24" 60 HP. W.Allsup&Sons,Preston	Fierro Carril del Sud
	✻ BT.88	**Emperor** I ScTug *Thomas* 77-83 1 *Dk*	British Falmouth	16 114	100·2	19·2	10·2 *Len.* 86	Hayle Harvey&Co.	1883	4 Blkds.	2Cy.21"& 38"-21" 80 HP. Harvey&Co.,Hayle	P.Thomas
	✻	**Empire** I ScS *Davies* 2 *Dks3 B*	British Liverpool	1375 2130	280·0	35·9	24·8 *P 287, B 80 f, F 34 f, Q 102 f.*	Liverpool R.&J.Evans&Co.	1881	5 Blkds. *DB* [illegible]	2Cy.33"& 66"-42" 200 HP. D.Rollo&Sons,Liverpool	S.S. "Empire" Co.,Ld. (W.Thomas & Co.)
		Empire W ScSr *Butler* 86-87 2 *Dks3 B*	American S.Francisco	526 732	170·8	32·9	20·0 *F 46 f. H 124 f.*	P.MadisonWa O.Reed	1873		1Cy.34"-34" 225 HP. W.Deacon	R.D.Chandler
		Empire State W Pad *Phillips* -76 3 *Dks*	American Taunton,Ma.	1579 1692	312·0	41·0 *Breadth over all 71 f.*	13·0 31·6	New York Sneden&Co.	18^{48}_{54}		1Cy.76"-144" 1522 HP. MorganIronWks.,NewYork	J.M.Phillips
288	✻	**Empress** I ScSr *Rippon* 86-86 1 *Dk2 B*	British Sunderland	1431 2183	291·0	40·2	20·2 *P 75 f, B 152 f, F 31 f, Q 102 f.*	Sunderland Short Bros.	1882	6 Blkds. *DBaf*	2Cy.34"& 64"-42" 200 HP. J.Dickinson,Sunderland	Taylor & Sanderson
289	✻	**Empress** ex Kolan I ScS *Docker* 1 *Dk & Awn Dk3 B*	British Dundee	775 1205	230·1	30·2	17·8 25·4	Middlesbro' Candlish,Fox&Co.	1864	1 Blkds. *DBaf*	2Cy.28"& 55"-36" 130 HP. W.B.Thompson,Dundee	W.B.Thompson
290		**Empress** W Pad 1 *Dk*	British Montreal	372 678	185·3	27·6 *H 270 f.*	8·1 27·1	Montreal W.C.White	1886		1Cy.38"-120" 152 HP. G.Brush,Montreal	Ottawa River Nav.Co.
291	BT.88	**Empress** Stl PadSr 2 *Dks*	British London	151 1059	324·6	34·9	13·5	Glasgow FairfieldS.&E.Co.,Ld.	1887		2Cy.58"& 101"-72" 900 HP. FairfieldS.&E.Co.,Ld,Glsgw	London,Chatham and Dover Railway Co.
292	BT.88	**Empress** I PadSr *Bowering* 74-84	British Weymouth	88 164	160·1	18·1	8·3	London Samuda Bros.	1879		2Cy.30"-33" 52 HP. J.Penn&Sons,Greenwich	Cosens & Co.,Ld.
293	BT.88	**Empress** I PadSp	British Plymouth	54 101	115·0	16·2	4·9	Preston W.Allsup&Sons	1880		2Cy.16"& 30"-30" 40 HP. W.Allsup&Sons,Preston	Devon & Cornwall Tamar Stm.Pkt.Co.,Ld.
294	V	**Empress of India** I TwnScBkn ex Coromandel, ex La Pampa 2 *Dks & Spr Dk*	British Glasgow	1370 2184	346·5	34·2 *H 58 f.*	17·5	London J.&W.Dudgeon	1870	5 Blkds.	3Cy.17"27& 45"-30" 275 HP. Blaikie Bros.,Aberdeen	Macfarlane,Strang & Co.,Ld.
295		**Empress of India** W Pad 3 *Dks*	British Toronto	336 570	170·0	26·0	15·5	MillPoint,Ont H.R.Rathbone&Sn.	1876		1Cy.33½"-96" 75 HP. Rathbone&S.MillPoint,Ont.	Arthur W.Hepburn
296	BT.88	**Empress of India** I PadSp *Mathews*	British Cardiff	17 160	117·4	19·7	10·3	Cardiff J.&M.Gunn	1877		2Cy.28½"-48" 70 HP. J.&M.Gunn,Cardiff	W.& R.H.Strong
297		**Empusa** I ScSr *Wister*	British Liverpool	732 1153	230·0	30·2	17·0 *P 134 f. F 32 f.*	Stockton M.Pearse&Co.	1878		2Cy.30"& 55"-36" 120 HP. Blair&Co.,Ld.,Stockton	Mersey S.S.Co.,Ld.
298	V	**Ems** I ScImSr *Jungst* 77-88 2 *Dks & Spr Dk*	German Bremen	2893 5192	429·8	47·0	34·8 *P 51 f. B 136 f. F 82 f.*	Glasgow J.Elder&Co.	1884		3C 62"& 12 86"-60" 1000 HP. J.Elder&Co.,Glasgow	Norddeutscher Lloyd
299		**Emu** I ScS *Creer* 1 *Dk & Awn Dk*	British Adelaide	362 616	179·4	25·1	11·7 18·7	PortGlasgow Blackwood&Gordn	1876	1 Blkds.	2Cy.21"& 46"-33" 95 HP. Blackwood&Gordon,PtGls.	Adelaide S.S.Co.,Ld.
300		**Emu** (WEKERD) W ScSr 1 *Dk*	British Sydney	131 165	120·8	22·3	7·8 *Q 9 f. H* [illegible]	Shortland,N.Z. R.Stone	1874		2Cy.10"& 20"-10" 25 HP. Fraser&Tinne,Auckland	W.T.Yeager

✻=Lloyd's Register; D=Nederlandsche Vereeniging; G=Germanischer Lloyd; I=Registro Italiano; R=American Record;
A=Veritas Austro-Ungarico; H=Veritas Ellenico; N=Norske Veritas; V=Bureau Veritas.

Dans la 2me Colonne, B.T., avec les indications qui suivent, signifie la date du rapport de visite de la Carène par les experts du Board of Trade pour obtenir le certificat relatif au TRANSPORT DES PASSAGERS.

Dans la 3me Colonne, C=Composite (membrure en fer, bordé en bois); I=Fer; Stl=Acier; W=Bois.

Pour plus amples renseignements sur les navires classés par le Lloyd's Register, voir le Lloyd's Register Book et le Lloyd's Yacht Register.

No. in Book / Official No.	Registry	Ship's Name. Master, and Dates of Appointment to present Owner's Service and to Ship. Ship's Late Name.	Material, Rig, &c. No. of Decks, &c.	Flag. / Port of Registry.	Tons Net. / Gross.	Dimensions: Length	Breadth	Depth	Built: Where / By Whom.	When.	No. of Bulkheads. Particulars of Water Ballast.	Engines. Particulars. Maker's Name.	Owners or Managers
301		Emu	I PadK 1 Dk	British Melbourne	42 [illegible]	82·6	21·7	5·8	Melbourne D.&R.Buchanan	1886		2Cy.15"&35"-36" 30 HP D.&R.Buchanan,S.Melbrne	D. Munro & Co.
302	✻	Emulator *Marshall*	I Sc 1 Dk	British Runcorn	164	95·0	18·0	10·0	Carrickfergus P.Rodgers	1885	[illegible] Blkds.	2Cy.18"&34"-21" 50 HP	Weston Point Towing Co.
303		Enak *Timmermann 76-84*	I S Cr 1 Dk	German Hamburg	37 118	90·9	20·0	11·3	Hamburg Janssen & Schmilinsky	1874		2Cy.20"&40"-21" 75 HP Janssen&Schmilinsky,Hbrg	Vereinigte Bugsir-Dampfschiff-Ges.
304 [illegible]	✻	Endeavour *North 77-88*	Stl ScSr 1 Dk&Spr Dk3 B	British Newcastle	1640 2516	290·0	40·2	18·0	Newcastle R.Stephenson&Co.,Ld	1888	7 Blkds. [illegible]	3Cy.22½"37"&61"-42" 250 H Blair&Co.,Ld.Stockton	M'Intyre Bros.
305 [illegible]	✻	Endeavour	I ScK 1 Dk	British London	108 203	123·5	21·0	11·6	Middlesbro' R.Dixon&Co.	1884	[illegible] Blkds.	2Cy.19"&36"-24" 55 HP Blair&Co.,Ld,Stockton	Gt. Yarmouth Carrying Co.,Ld.
306 [illegible]	✻	Endeavour *Mackay 85-88*	W ScK 1 Dk	British Leith	19 27	62·5	17·1	6·7	Leith A.G.Gifford&Co.	1884		2Cy.8"&15"-12" 9 HP J.Cran&Co.,Leith	W. M'Glashan
307 [illegible]	✻	Endymion *Sellerley 62-86*	I ScSr 2 Dks3 B	British London	1495 2283	291·3	37·3	24·5	Newcastle Palmers'Co.	1880	[illegible] Blkds.	2Cy.36"&68"-45" 250 HP Palmers'Co.,Ld.Newcastle	Hall Bros.
308 [illegible]	✻	Energia *Pope*	I ScSr 2 Dks3 B	British London	2084 3177	337·0	41·2	26·5	Newcastle TyneI.S.B.Co.,Ld.	1883	6 Blkds.	2Cy.40"&74"-54" 350 HP N.E.MarineEng.Co.,Ld,Sld	S.S. "Energia" (Newson, Bros.)
309	V	Energie	I ScCr	Belgian Antwerp	50 115	80·0	17·4	11·6	Hamburg Janssen & Schmilinsky	1872		2Cy.16½"& 33-"18" Janssen&Schmilinsky,Hmb.	Soc. Anon. de Remorquage à hélice
310		Energie *Poolmann*	Stl Sc3mSr 2 Dks3 B	German Geestemünde	2102 2714	310·0	40·1	29·5	Newcastle Sir W.G. Armstrong, Mitchell & Co.	1888		3Cy.23"37"&60"-39" Wallsend,Slipway&Eng.Co.,Ld. Newcastle	W. A. Riedemann
311	✻	Energie *Arp 80-87*	Stl T Sc 1 Dk	German Cologne	403 649	200·0	30·1	11·1	Newcastle Sir W.G.Armstrong, Mitchell&Co.	1887	5 Blkds.	3Cy.10½"17"&28"-20" 90 H Wallsend Slipway Co.,Ld,Newcastle	Badische Schleppschifffahrts-Ges.
312 89557	✻	Energy *Hunt*	I ScK 1 Dk	British London	108 203	123·5	21·0	11·6	Middlesbro' R.Dixon&Co.	1884	4 Blkds.	2Cy.19"&36"-24" 55 HP Blair&Co.,Ld.,Stockton	Gt. Yarmouth S. Carrying Co.,Ld.
313 92882	✻ D	Enfield *Vasey 82-85*	I ScSr 1 Dk2 B	British W. Hartlepl.	1405 2158	279·7	37·4	20·0	W. Hartlepool W.Gray&Co.	1885	5 Blkds.	3C.21"35"&57"-39" 200 HP Central Mar.E.Co.W.Hartpl.	Pyman Bros.
314 81607	✻ D	Engadine *Morris 82-82*	I ScSr 1 Dk2 B	British London	803 1258	240·5	33·2	18·0	Newcastle Palmers'Co.	1879	4 Blkds.	2Cy.28"&54"-36" 130 HP Palmers'Co.,Ld.,Newcastle	Wilson & Co.
315	V	Engeland *Thys*	I PadSpTug	Belgian Antwerp	48 136	108·2	19·0	9·7	South Shields J.Readhead&Co.	1876		2Cy.28"-51" J.Readhead&Co.,S.Shields	Soc. Anon. d'Armement [illegible]
316	✻	Engeland *Staase 75-*	Stl PadSr 2 Dks&Awn Dk2 B	Dutch Flushing	829 1648	286·5	35·3	22·9	Glasgow Fairfield S.&E.Co.,Ld.	1887		2Cy.60"& 104"-84" FairfieldShip.&Eng.Co.,Ld.	Stoomvaart Maatschappij "Zeeland"
317 85175		Engineer *Jeffrey 65-82*	I Sc3mSr 2 Dks3 B	British Liverpool	1683 2591	350·2	38·0	25·3	Glasgow Aitken&Mansel	1882	6 Blkds.	2Cy.38"&75"-54" 375 HP J.&J.Thomson,Glasgow	Charente S.S. Co.
318 94286	✻	Engineer *Jerry 84-88*	Stl Sc3mSr 1 Dk	British London	466 723	195·2	28·2	12·6	Sunderland J.Laing	1888	3 Blkds.	3C.18"28"&45"-36" 120 HP J.Clark,Sunderland	Engineer S.S. Co.
319 28886	B.T. 2-88	England *Hesley 87-87*	I ScBkn 1 Dks	British Liverpool	3022 4898	437·9	42·5	35·0	Newcastle Palmers'Co.,Ld.	1865		2Cy.55"&90"-48" 420 HP G.Forrester&Co.,Liverpool	National Steam Ship Co.,Ld.
320 72540	✻	England *Holt*	I ScSr 1 Dk2 B	British Cardiff	681 1068	220·5	30·2	18·5	Newcastle Schlesinger,Davis&Co	1877	4 Blkds.	2Cy.28"&53"-30" 120 HP T.Clark&Co.,Newcastle	Short & Dunn
321 [illegible]		England *Henderson*	W PadTug	British S.Shields	12 115	100·6	19·5	9·9	South Shields Rennoldson	1876		1Cy.37"-51" 60 HP J.Rennoldson&S.Shields	C. Dykle
322		Engso (ex Knngsor, ex Nya Engso) *Solerstrom 82-86*	I Sr 1 Dk	Swedish Strengnas	122 187	107·6	17·3	7·4	Stockholm Bergsunds M.V.	1875		2Cy.15¼"23¾"&15¼"-19¾" Bergsunds M.V. Stockholm 30 HP	Strengnäs Ebbs Bolag
323		Enighed *Hansen*	W ScSr 1 Dk	Danish Skjelskør	51 110	99·6	19·1	9·8	Odense H.L.Hansen	1874		2Cy.16"&32"-18" 30 HP Schweffel&Howaldt,Kiel	Dampsk.-Selsk. Skjelskør og Omegn
324		Enkoping I. *Pettersson*	I ScSp 1 Dk	Swedish Enköping	130 183	109·5	18·4	9·6	Stockholm W.Lindbergs M.V.	1885		2Cy.17"&30"-21" 45 HP W.Lindbergs M.Vrks,Stckhlm	A. Östlund
325		Enkoping II. (ex Nya Enköping) *Larsson*	I Sc 1 Dk	Swedish Enköping	117 134	109·2	17·2	7·5	Stockholm W.Lindbergs M.V.	1871		2Cy.11"&24"-21" 30 HP W.Lindbergs M.Vrks,Stckhlm	A. Östlund
326	I	Enna *Rocco 78-83*	I ScBkn 2 Dks&Spr Dk	Italian Palermo	918 1775	276·5	32·5	25·1	Leghorn Flli Orlando	1874		2Cy.41"&76"-42" 549 HP Thompson,Boyd&Co.,Nwcl	Navigazione Generale Italiana
327	✻	Ennerdale *Pitmer 83-87*	I ScSr 1 Dk&Spr Dk3 B	British Newcastle	798 1250	233·3	30·6	17·5	Sunderland Short Bros.	1872	5 Blkds.	3Cy.17"27"&52"-33" 110 HP	Milburn Bros.
328	✻	Ennismore *Bee*	I ScSr 1 Dk2 B	British Liverpool	1002 1560	260·0	34·2	17·9	Barrow Barrow S.B.Co.,Ld.	1880	4 Blkds.	2Cy.30"&60"-36" 160 HP Barrow S.B.Co.,Ld.,Barrow	G. R. Sanderson
329		Enrico	I ScTug 1 Dk	Italian Leghorn	43 [illegible]	90·5	16·8	10·7	Genoa E.Cravero&Co.	1883	3 Blkds.	2Cy.15"&2[illegible]"-24" 60 HP E.Cravero&Co.	E. Campes
330	✻	Enrique *Alcrrsaturi*	I Sc 2 Dks3 B	Spanish Bilbao	1465 2258	325·3	34·7	25·3	Hull Earles Co.,Ld.	1878	5 Blkds.	2Cy.36"&72"-48" 250 HP Earles Co.,Ld.,Hull	Cia. de Nav. "La Flecha"

✻ = Lloyd's Register; D = Nederlandsche Vereeniging; G = Germanischer Lloyd; I = Registro Italiano; E = American [illegible];
A = Veritas Austro-Ungarico; H = Veritas Ellenico; N = Norske Veritas; V = Bureau Veritas.

In the 2nd Column, B.T., with figures following, signifies Date of report of bottom sighting by Board of Trade Surveyors as condition of PASSENGER CERTIFICATE.

In the 3rd Column, C = Composite (iron frame, wood planked); I = Iron; Stl = Steel; W = Wood.

Further particulars of Vessels classed by Lloyd's Register will be found in Lloyd's Register Book and Lloyd's Yacht Register.

No.		Nom, Matériaux, Gréement, etc. Capitaine et dates d'entrée au service des Armateurs actuels et à bord du navire. Ancien Nom, nombre de ponts, etc.		Pavillon. Port d'Armement	Tonnage Net / Total	Dimensions (Longueur, Largeur, Creux)	Construit. Où. Par qui.	Quand	Nombre des cloisons. Détails sur le lest d'eau.	Machines. Détails. Nom du Constructeur.	Armateurs ou Directeurs-Gérants.
	✻	**Enrique Barozzo** *Irvine 88-88*	Stl ScSr 1 *Dk*	Urug'yan Montevideo	313 580	199·5|32·1|10·1 *P 77 f. F 30 f.*	Grangemouth GrangemouthDkyd.Co	1889		3Cy.15"21"&40"–30" 80 H Hutson&Corbett,Glasgow	Lanfranco & Ramo
	V	**Ensenada I.** *Wink*	I ScSr	Dutch Haarlem	31 106	85·3|18·0|10·8	Haarlem Th.Figée	1886		2Cy.11"& 21"–20" Th.Figée,Haarlem	Th.Figée
	V	**Ensenada II.**	I ScSr	Dutch Haarlem	31 100	85·3|18·0|10·8	Haarlem Th.Figée	1886		2Cy.11"& 21"–20" Th.Figée,Haarlem	Th.Figée
334	✻	**Ensign** *Wilson 87-87*	I Sc3mSr 1 *Dk*	British Glasgow	146 318	115·0|23·1|10·6 *Q 16 f. B 28 f. F 22 f.*	Glasgow T.B.Seath&Co.	1882	1 Blhds. FPT30t	2Cy.18"&34"–28" 50 HP. J.&T.Young,Ayr	W.Robertson
335	I	**Entella** *Bruno 77-77*	I ScBk 2 *Dks* 3 *B*	Italian Palermo	1479 2258	299·2|37·2|23·6 *P 30 f. B 54 f. F 43 f.*	PortGlasgow Blackwood&Gordn	1883	1 Blhds. CDB370t	2Cy.35"& 67"–42" 200 HP Blackwood&Gordon,Pt.Gls.	Navigazione Generale Italiana
336		**Enterprise**	W ScCr 1 *Dk* & *Spr Dk*	British S.Catherines	587 915	177·2|28·1|12·2	St.Catherines. Alex.Muir [Ont.	1864		1Cy.50"–54" 200 HP. A.Gartshore,Dundas,Ont.	David R.Dobbie
337		**Enterprise**	I PadSr	British Dundalk	258 696	221·6|27·8|15·0 *Q 35 f.*	Newcastle T.&W.Smith	1856		2Cy.44"& 76"–78" 300 HP. A.&J.Inglis,Glasgow	Dundalk & Newry Steam Packet Co.,Ld.
338	✻	**Enterprise** *Lepage 81-87*	I TwnScK 1 *Dk*	British Yarmouth	127 206	120·4|21·1|9·7 *Q 43 f. F 21 f.*	Hull Earles'Co.,Ld.	1884	1 Blhds. FPT28t	1Cy.17"&32"–18" 68 HP. Earles'Co.,Ld.,Hull	J.Campbell
339		**Enterprise**	I PadSk	British London	8 157	118·1|19·7|10·2	North.Shields Hepple&Co.	1885		2Cy.31½"–58" 120 HP. Hepple&Co.,N.Shields	Elizabeth Martin
340	✻	**Enterprize** *Storm 83-88*	Stl ScSr 1 *Dk* & *Web Frms*	British Whitby	1274 1981	281·0|38·1|18·2 *Q 74 f. B 126 f. P 29 f. F 36 f.*	Sunderland J.L.Thompson&Sons	1888	6 Blhds. CDB379t APT28t	3Cy.21"35"&58"–36" T.Richardson&Sons, Hartlepool [180 H	Rowland & Marwood
341		**Enterprize**	W Sc 1 *Dk*	British Quebec	90 133	90·0|24·5|9·0	Montreal	1863		1Cy. 30 HP.	Harbour Commissioners of Quebec
342	V	**Entre Rios** *Clequer 83-87*	Stl ScBk 2 *Dks* 3 *B*	French Havre	2135 2853	322·5|40·0|27·1 *P 93 f. F 45 f.*	St.Nazaire Ateliers&Chantiers de la Loire	1887	DBF138t DBA170t	3Cy.25"43"55"–31"350HP. Ateliers&C.delaLoire	Chargeurs Réunis
343	✻	**Envoy** *Taylor 79-87*	I ScSr 1 *Dk* 2 *B*	British Hull	881 1353	253·8|31·9|18·0 *P 136 f. F 42 f.*	Hull Gilbert&Cooper	1872	5 Blhds.	2Cy.31⅝"&58"–36"130 HP. Gilbert&Cooper,Hull	W.S.Bailey
344		**Eol** (ex Drottning Sophie) *Borg 87-87*	I ScSr 1 *Dk*	Swedish Stockholm	90 149	124·9|16·6|10·9 *B 17 f.*	Christiania NylandsMek.Verkst.	1876		2Cy.18"&31"–24" 60 HP. NylandsMk.Verkst.,Christiania	Bergn.O.Dykeribed Neptun(E.Lilewalch)
345	✻	**Eolo** *M'Lean*	Stl PadSr 1 *Dk* & *Awn Dk*	British Glasgow	969 1749	299·0|35·1|12·8	Dumbarton W.Denny&Bros.	1886	4 Blhds.	3Cy.44"& 87"–72" 300 HP. Denny&Co.,Dumbarton	La Platense Flotilla Co.,Ld.
346		**Eolus** *Sederquist 75-85*	Stl TwnScSr 1 *Dk* & *Awn Dk*	Swedish Stockholm	625 857	204·7|28·6|13·2 *B 74 f.*	Gothenburg MotalaCo.	1884		4Cy.26"& 48"–27" 240 HP. MotalaCo.,Gothenburg	Angf.Aktie Bol.Södra Sverige
347	✻	**Eos** *Nyberg*	Stl ScSr 1 *Dk*	Russian Helsingfors	214 293	116·3|21·2|11·6 *Q 30 f. F 17 f. B 25 f. H 6 f.*	Thorskog P.Larsson	1886	4 Blhds. FPT6t APT5t	2Cy.13⅜"&24⅜"–17½"85HP P.Larsson,Thorskog	Nya Angfartygs Aktiebolaget
348		**Eos**	C ScSr 1 *Dk*	Swedish Lidköping	142 186	101·0|18·1|12·5 *B 24 f. B 3 f.*	Wenersborg G.Bäährnhjelm	1873		2Cy.10¾"&24¼"–20"30 HP. MotalaCo.,Gothenburg	E.Pettersson
349		**Eothen**	I Sc3mSrYt	British Leith	123 191	152·0|22·2|12·2	London J.Ash	1864		2Cy.20"& 40"–24" 70 HP. J.Stewart,London	Earl de la Warr
350	✻	**Epidauro** *Marinovich 88-88*	Stl ScSr 1 *Dk*	Austrian Trieste	194 342	148·0|22·8|10·9 *P 35 f. B 32 f. F 30 f.*	Kinghorn J.Scott&Co.	1888	4 Blhds. FPT29 APT14	3Cy.13"19"&34"–27"70HP. J.Scott&Co.,Kirkcaldy	Consorzio Vapore Epidauro
351		**Epirus** (ex Estremadura) *Concomilaki 83-87*	I Sc	Greek Piraeus	460 615	213·0|24·3|13·1 *P 43 f. F 50 f.*	Newcastle A.Leslie&Co.	1862	1 Blhds. DB46t	2Cy.23"&44"–36" 85 HP. N.E.MarineCo.,Ld.,Sndeld	PanhellenicSteamShip Nav.
352	✻	**Eppleton** *Kind 86-86*	I ScSr 1 *Dk* & *Web Frames*	British Newcastle	562 902	216·2|31·0|13·6 *B 45 f. Q 66 f. F 28 f.*	Sunderland S.P.Austin&Son	1882	1 Blhds. CDB29t APT30t	2Cy.28"& 54"–33" 99 HP. J.Dickinson,Sunderland	Sharp & Co.
353		**Eptanisos** *Bosa 76-*	I ScSr	Greek Syra	332 631	185·0|26·5|15·5	W.Hartlepool T.H.Pile&Co.	1860		1Cy.24"&45"–30" 125 HP.	HellenicSteamShipCo.
354		**Eptanisos**	I Pad 1 *Dk*	Greek Cephalonia	36 121	105·8|18·9|9·5	S.Shields J.Eltringham	1879	3 Blhds.	1Cy.37⅜"–54" 60 HP. J.P.Rennoldson,S.Shields	C.A.Foca
355		**Equateur**	I ScBk 3 *Dks*	French Marseilles	2497 3725	394·6|39·7|32·8	LaCiotat Messageries Marit.	1875		3C60¼"(2)13¼"–43¼"600HP MessageriesMar.LaCiotat	Messageries Maritimes
356		**Equator** (ex Hor. Buckingham) *Nelson -87*	W Sc3mSr 2 *Dks*	American Philadelphia	628 959	187·6|34·0|18·0	Mystic,Conn.	1863		125 HP.	W.P.Clyde & Co.
357	✻	**Equity** *Talley 86-88*	I ScrSr 1 *Dk*	British Goole	346 761	220·7|32·4|13·7 *B 71 f. P 43 f. F 30 f.*	Hull Earles'S.B.&E.Co.,Ld.	1888	5 Blhds. CDB128t FPT APT8t	3Cy.24"39"&64"–33" Earles'S.B.&E.Co., Ld.,Hull [240 HP	Co-operative Wholesale Society,Ld.
358	✻	**Era** *Bannatyne -88*	Stl ScSr 1 *Dk* & *Web Fms*	British Melbourne	1550 2379	299·2|38·5|20·7 *P 35 f. B 122 f. F 34 f. Q 38 f.*	W. Hartlepool E.Withy&Co.	1888	5 Blhds. CDB330t APT39	3Cy.23"36½"&62"–39" CentralMarineE.Co. West Hartlepool [250 HP	W.H.Smith&Sons,Ld.
359	✻	**Era** *Emerson -88* 1 *Dk* & *Spr Dk* & *Web Frms*	Stl Sc3mSr	British London	1195 1851	271·0|37·2|15·2 *B 36 f. | 22·5*	Newcastle Palmers'Co.,Ld.	1887	1 Blhds. FPT18t APT53t	3Cy.21"34"&57"–39" 180 Palmers'Co.,Ld.,Newcle [HP	"Era" S.S.Co.,Ld.
360											

✻ = Lloyd's Register; D = Nederlandsche Vereeniging; G = Germanischer Lloyd; I = Registro Italiano; R = American Record; A = Veritas Austro-Ungarico; H = Veritas Ellenico; N = Norske Veritas; V = Bureau Veritas.

Dans la 2me Colonne, B.T., avec les indications qui suivent, signifie la date du rapport de visite de la Carène par les experts du Board of Trade pour obtenir le certificat relatif au TRANSPORT DES PASSAGERS.

Dans la 3me Colonne, C = Composite (membrure en fer, bordé en bois); I = Fer; Stl = Acier; W = Bois.

Pour plus amples renseignements sur les navires classés par le Lloyd's Register, voir le Lloyd's Register Book et le Lloyd's Yacht Register.

No. in Book / Official No.	Registry in which classed	Ship's Name. Material, Rig, &c. / Master, and Dates of Appointment to present Owner's Service and to Ship / Ship's Late Name. No. of Decks, &c.	Flag. / Port of Registry.	Tons. Net. / Gross	Length	Breadth	Depth	Built. Where. / By Whom.	When	No. of Bulkheads / Particulars of Water Ballast.	Engines. Particulars. / Maker's Name.	Owners
361 20545		**Era** I ScSr Shotwell	British London	351	201·5	24·1	15·6	Newcastle PalmerBros.	1861		2Cy.34"–30" 90 HP. PalmerBros.,Newcastle	General Steam
362	✳D	**Erasmus** I ScSr Meyer –83 I Dk2 B	Dutch Rotterdam	891 1263	238·2	32·9	16·5	Rotterdam J.Smit,Czn	1884	4 Blkds.	2Cy.28"&52"–33" 120 HP. Kon.MaatsdeSchelde,Vlgn.	Bouke & Visser
363 73052	✳	**Erasmus Wilson** I Sc3mSr Bond 76-76 I Dk2 B	British London	497 805	199·6	30·1	18·2	Hartlepool E.Withy&Co.	1876	4 Blkds.	2Cy.27"&50½"–33" 99 HP. Blair&Co.,Ld.,Stockton	S.Clarke & Co.
364 60217	BT1.89	**Erato** I ScSr Elsom 2 Dks	British Hull	996 1515	252·4	34·8	17·5	Hull Humphrys&Co.	1870	5 Blkds.	2Cy.32"&60"–33" 150 HP. Humphrys&Co.,Hull	T.Wilson,Sons&
365 85848	✳	**Erato** I ScBn Simmonds 88-88 I Dk2 B	British W.Hartlepl.	1137 1764	259·1	34·7	19·5	W.Hartlepool W.Gray&Co.	1883	4 Blkds.	2Cy.32"&60"–39" 150 HP. Blair&Co.,Ld.,Stockton	Gladstone & C.
366	✳	**Erato** (ex Baron ...) I Sc3mSr Grundin –88 I Dk2 B	Swedish Gothenberg	700 987	217·0	27·8	16·8	Sunderland W.Watson	1872	4 Blkds.	2Cy.27"&50"–30" 90 HP. N.E.MarineE.Co.,Ld.,Sunderld	K.O.F.Dalman
367		**Eridan** I Sc 2 Dks	French Marseilles	1169 1756	320·5	32·3	24·0	LaCiotat MessageriesMarit.	1866		2Cy.62½"–46½ 280 HP. MessageriesMar.,LaCiotat	Messageries Ma.
368 53045		**Erik** W ScS Shaw –88	British London	412 583	157·8	29·5	18·5	Dundee	1865		2Cy.23"&44"–24" 80 HP. Gourlay Bros.&Co.,Dundee	Hudson's Bay C.
369	✳D	**Erik** I Sc3mSr Wrisberg 84-88 I Dk2 B	Danish Copenhagen	846 1343	236·7	32·1	16·9	Renfrew Hndrsn,Colbrn&Co.	1872	5 Blkds.	2Cy.27½"&60"–36" 150 HP. Hndrsn,Coulbrn,&Co,Rnfw	Dampskibet Rederi H.H.Cas.
370		**Erik Berentsen** I ScSr I Dk & AwnDk	Norweg'n Stavanger	329 438	139·1	24·0	17·5	Christiania NylundsVerksted	1881		2Cy.18"&32"–24" 45 HP. NylandsVerkst,Christiania	E.Berentsen
371 63896		**Erimus** I ScSr Harrison I Dk	British Middlesbro'	124 203	137·8	20·2	9·6	Middlesbro' Backhouse&Dixon	1870	3 Blkds.	2Cy.13½"&30½"–20" 35 HP. Blair&Co.,Ld.,Stockton	W.Hanson
372 89724	✳	**Erimus** I&Stl Pad Allan –88	British Middlesbro'	86 183	95·2	30·4	6·1	Middlesbro' R.Dixon&Co.	1888		2Cy.26"–42" 45 HP. Westgarth English&Co.,Middbr.	Corporation of Middlesbrough
373 30274	BT5.88	**Erin** I ScBkn Tyrer 81-88 4 Dks	British Liverpool	2848 4577	418·8	41·2	35·3	Newcastle PalmerBros.	1864	6 Blkds.	2Cy.55"&90"–48" 420 HP. J.Penn&Son,London	National Steam Co.,Ld.
374 43985	✳	**Erin** (ex Andalusia) I ScSr Hughes (FOUNDERED)	British Glasgow	278 443	171·5	25·1	15·1	Glasgow J.G.Lawrie	1861	4 Blkds.	2Cy.22"&44"–27" 80 HP. Blackwood&Gordon,P.Glas.	Western Steam Co.,Ld.
375 49718	BT5.88	**Erin** I PadSp	British Belfast	88 210	205·2	21·0	7·7	PortGlasgow Kirkpatrick	1864		2Cy.40"–54" 110 HP. J.Howden&Co.,Glasgow	Belfast, Bangor & LarneSteamboat
376 73550	✳	**Erith** I ScS Scott 77-77 I Dk2 B	British London	409 655	187·7	28·2	15·1	Sunderland J.Blumer&Co.	1875	4 Blkds.	2Cy.27"&47"–30" 90 HP. N.E.MarineE.Co.,Ld.,Sndrld	W.A.Watson
377 52820	✳	**Erl King** I ScBg Priske 84-84 2 Dks & Spr Dk	British Glasgow	1418 2193	305·6	34·1	21·5 / 28·3	Glasgow A.&J.Inglis	1865	5 Blkds.	2Cy.36"&70"–48" 250 HP. D.Rollo&Sons,Liverpool	W.Ross & Co.
378		**Erna** I ScSr Ødbergsen 85-85 I Dk Spt Hur Dk	Danish Nexö	97 188	124·5	19·1	9·4 / 15·7	Malmö KockumsMek.Vrks	1877		2Cy.23"&41"–20½" 70 HP. KockumsMek.MalmöVrks.	Østbornholmsk Dampsk.Selsk.
379	DNV	**Erna** I ScSr Nielsen –88 I Dk	Norweg'n Moss	363 488	165·1	24·7	12·6	Kiel Act.Ges."Germania"	1884		2Cy.17¼"&30¼"–24" 50 HP. Act.Ges."Germania,"Berlin	R.M.Petersen
380	V	**Erna Woermann** Stl ScSr Busch 83-87 I Dk & Spr Dk	German Hamburg	826 1119	222·4	31·3	14·5 / 22·0	Hamburg ReiherstiegSchiffs Wfte.	1884		2Cy.27"&50"–33" 140 HP. Reiherstieg S.B.Co.	Afrikanische Dampfschiffs-Act.-Ges. (Woermann Linie)
381 90402	✳	**Erne** I ScSrYt	British Southamptn	83 122	119·7	17·0	10·3	Southampton Day&Summers	1884	5 Blkds.	2Cy.16"&30"–18" 37 HP. Day&Summers,South'pton	Henry Evans
382 63613	✳	**Ernest** I ScSr Warren	British London	464 726	200·0	28·6	17·0	Sunderland W.Doxford&Sons	1870	4 Blkds.	2Cy.26"&47"–33" 90 HP. G.Clark,Sunderland	D.W.Stobart
383 63	✳D	**Ernesta Foscolo** I ScBg (ex Paris Lambrinidis, ex Olga) I Dk2 B	Greek Argostoli	1044 1421	251·2	32·2	17·2	Sunderland J.Laing	1870	5 Blkds.	2Cy.31"&58"–30" 130 HP. Thompson,Boyd&Co.,Nwc.	Foscolo & Vals
384	V	**Ernestine** (ex Aurelia, ex Express) I ScSr Hémine –87 I Dk2 B	French Nantes	486 756	188·6	26·4	16·4	Sunderland W.Pile&Co.	1870		2Cy.23"&39½"–28" 100 HP. T.Clark&Co.,Newcastle	L. Flornoy & fils
385	V	**Ernestine** I Sc3mSr Godard 79-87 I Dk2 B	French Dieppe	391 589	184·0	27·5	16·3	Hull C.&W.Earle	1865	5 Blkds.	2Cy.37"–30" 90 HP. C.&W.Earle,Hull	A.Grandchamp & Co.
386	✳	**Ernst** (ex Wolf) I ScSr Neumann –88 I Dk	German Kiel	393 628	170·4	25·3	13·2	Flushing RoyalS.&E.W.Co.	1884		2Cy.22"&41"–30" 90 HP.	Sartori & Berger
387		**Ernst** C ScSr Deurell I Dk	Swedish Christianstd	218 276	111·8	19·8	9·5	Oscarshamn	1872		35 HP.	C.J.F.Ljunggren
388		**Ernst Gunther** I ScSr Jensen –83 I Dk	German Flensburg	89 133	110·0	27·6	8·0	Flensburg FlensburgSchiffsb.Ges.	1885		2Cy.18½"&34"–21" 55 HP. FlensburgSchiffsbauGes.,Flensbg	Flensburg-Ekensund Dampschiffahrt
389 88139	✳D	**Eros** I ScBn Hicken I Dk2 B	British Hull	1203 1842	260·0	36·0	20·0	Stockton Richardson,Duck&Co.	1883	4 Blkds.	2Cy.32"&60"–39" 160 HP. Blair&Co.,Ld.,Stockton	H.Briggs,Sons&
390	RV	**Eros** Stl ScSrYt Briand	French Nice	380 633	219·6	26·8	15·4	Erith	1885		2Cy.32"&60"–36" 155 HP. Day,Summers&Co.,Southampton	Baron A.de Roth.

✳ = Lloyd's Register; D = Nederlandsche Vereeniging; G = Germanischer Lloyd; I = Registro Italiano; R = American Record; A = Veritas Austro-Ungarico; H = Veritas Ellenico; N = Norske Veritas; V = Bureau Veritas.

In the 2nd Column, B.T., with figures following, signifies Date of report of bottom sighting by Board of Trade Surveyors as condition of PASSENGER CERTIFICATE.

In the 3rd Column, C = Composite (iron frame, wood planked); I = Iron; Stl = Steel; W = Wood.

Further particulars of Vessels classed by Lloyd's Register will be found in Lloyd's Register Book and Lloyd's Yacht Register.

No.	B.T.		Nom, Matériaux, Gréement, etc. — Capitaine et dates d'entrée au service des Armateurs actuels et à bord du navire. — Ancien Nom, nombre de ponts, etc.		Pavillon. — Port d'Armement	Tonnage Net / Total	Dimensions	Construit. Où. — Par qui.	Quand	Nombre des cloisons. — Détails sur le lest d'eau.	Machines. — Détails. Nom du Constructeur.	Armateurs ou Directeurs-Gérants.
391			**Erymanthe**	I Sc 3 Dks	French Marseilles	1234 1511	314·8 | 32·8 | 25·2 P 69 f. F 32 f. H 43 f.	LaCiotat Messageries Marit.	1862		3Cy.36¼" & 51¼"(2)-31½" 316 HP Messagers.Mar.,LaCiotat	Messageries Maritimes
392			**Esaias Tegnér** *Petterson -88*	I ScSp 1 Dk	Swedish Linköping	84 134	96·1 | 20·5 | 9·8 Q 37 f. F 16 f. H 15 f.	Motala Motala Co.	1851		40 HP	Linköpings Rederi AktieBolag(C.Thorngrén)
393			**Esbern Snare** *Fischer 82-84*	I ScSr 2 Dks & Hur Dk	Danish Copenhagen	242 405	168·3 | 23·3 | 12·0 P 36 f. F 23 f. H 13 f.	Copenhagen Burmeister & Wain	1875	APT97 MT20	1Cy.16¼" & 34"-20" 80 HP Burmeister & Wain,C'phagn	Det Forenede Dampskibs Selskab (C.P.A.Koch)
394			**Esbjerg** *Kolster 85-87*	I ScSr 1 Dk & Hu Dk	Danish Copenhagen	290 478	175·9 | 27·1 | 12·8 P 37 f. F 22 f. B 29 f. 19·5	PortGlasgow Cunliffe & Dunlop	1872	DB 74	2Cy.23¾" & 40¼"-28" 75 HP Cunliffe & Dunlop,P'Glasgow	Det Forenede Dampskibs Selskab (C.P.A.Koch)
395		✻	**Escalona** *Cummings 85-88*	Stl ScSr 2 Dks & B	British Dundee	1234 1880	281·0 | 35·7 | 23·9 P 28 f. B 62 f. F 34 f.	Dundee Gourlay Bros.&Co.	1884	5 Blds. DBaf & aft21 FPT22 APT19	2Cy.34" & 64"-42" 240 HP Gourlay Bros.&Co.,Dundee	W.Thomson & Sons
396		✻	**Escurial** *Langwill 85-85*	I ScSr 2 Dks 3 B	British Glasgow	762 1185	230·2 | 30·2 | 22·0 B 16 f. F 33 f.	Glasgow A.Stephen & Sons	1879	4 Blds. DBass	2Cy.28" & 52"-36" 130 HP A.Stephen & Sons,Glasgow	Raeburn & Vérel
397		✻	**Eshcol** *Crowza 79-79*	I ScSr 2 Dks 3 B	British N.Shields	973 1509	251·7 | 34·0 | 23·6 B 56 f. F 28 f.	Sunderland J.Laing	1875	4 Blds. DBass	2Cy.32" & 60"-36" 150 HP G.Clark,Sunderland	J.F.Middleton
398			**Esk** *Buckler*	I Sc3mSc 2 Dks & Spr Dk	British Glasgow	1374 2145	309·5 | 36·2 | 16·7 B 77 f. 21·6	Barrow Barrow S.B.Co.,Ld.	1882	8 Blds.	2Cy.38" & 74"-54" 350 HP Barrow S.B.Co.,Ld.,Barrow	Royal Mail Steam Packet Co.
399			**Esk** *Burgher -85*	I ScSr 1 Dk 2 B	British S.Shields	447 706	201·5 | 29·0 | 16·9 Q 21 f.	W.Hartlepool Denton,Gray&Co.	1869	4 Blds. DBaf	2Cy.24" & 45"-30" 90 HP G.Clark,Sunderland	T.S.Blues
400			**Esk**	I Sc	British Whitby	96 224	112·0 | 26·1 | 11·4	Renfrew W.Simons&Co.	1887		2Cy.15" & 28"-21" 35 HP W.Simons&Co.,Renfrew	Whitby Port and Harbour Trustees
401		✻	**Esk Holme** *Halliday -89*	I ScSr 1 Dk 2 B	British Maryport	595 925	206·2 | 30·5 | 16·0 Q 73 f. F 25 f. B 40 f.	Sunderland J.L.Thompson&Sons	1877	4 Blds. DBa51 ft f 6 ft APT	2Cy.26½" & 49"-33" 99 HP J.Dickinson,Sunderland	Hine Bros.
402		✻	**Eskdale** (STRANDED) *Backe 86-88*	I ScSr 1 Dk 2 B	British Hartlep'l	737 1133	229·4 | 30·8 | 17·6 Q 85 f. F 32 f. B 50 f.	SouthShields J.Readhead&Co.	1873	4 Blds. DBaf 139 ft	2Cy.26½" & 50"-33" 99 HP J.Readhead&Co.,S.Shields	Maclean, Doughty & Co.
403			**Eskilstuna I.** *Tjulin -86*	I Sc 1 Dk	Swedish Eskilstuna	164 211	101·7 | 20·6 | 8·5 P 71 f. F 12 f. H 21 f.	Stockholm W.Lindberg's M.V.	1886		2Cy.14" & 26"-18" 35 HP W.Lindberg's MV Stockhlm.	"Eskilstuna" Ångf. Bol.
404			**Eskilstuna II.** *Olsson 74-86*	I ScSp 1 Dk	Swedish Eskilstuna	110 140	102·2 | 18·7 | 7·0 H 33 f. F 5 f. Q 52 f. H 15 f.	Motala Motala Co.	1868		1Cy.16"-22" 30 HP Motala Co.,Motala	"Eskilstuna" Ångf. Bol.
405		✻	**Eslington** *Parton -86*	I ScSr 1 Dk 2 B	British S.Shields	673 1054	224·0 | 30·2 | 18·4 Q 117 f. F 29 f. B 12 f.	Newcastle Schlesinger,Davis&Co	1876	4 Blds. DB 139 f	2Cy.28" & 53"-36" 120 HP T.Clark&Co.,Newcastle	W.Runciman & Co.
406		✻	**Esme** (WRECKED) *Winter*	I ScBn 1 Dk 2 B	British London	1061 1657	255·0 | 35·0 | 18·0 Q 86 f. B 70 f. F 33 f.	Stockton M.Pearse&Co.	1882	4 Blds. DBaf 270 ft	2Cy.32" & 60"-39" 140 HP Blair&Co.,Ld.,Stockton	Collings & Co.
407		✻	**España** (ex Edinburgh Castle) *Gordon 85-87*	I ScBg 3 Dks	Spanish Cadiz	2006 2678	335·3 | 37·7 | 28·2	Glasgow R.Napier&Sons	1872	5 Blds.	2Cy.44" & 72"-42" 270 HP R.Napier&Sons,Glasgow	C^ia^ Trasatlantica
408			**España** (ex Congress)	I ScBk 3 Dks	Spanish Barcelona	1333 1983	279·8 | 38·7 | 24·7 Q 77 f. B 65 f. F 76 f.	Antwerp Soc.J.Cockerill	1856		2Cy.50" & 82"-48" 400 HP D.Rowan,Glasgow	C^ia^ Trasatlantica
409		✻	**España**	I Sc 1 Dk & Shd Dk	Spanish Manila	595 875	200·0 | 29·2 | 19·9	Barrow Caird&Purdie	1881	4 Blds. DBa 38 f	2Cy.30" & 52"-36" 150 HP Westray,Copeland &C.,B'rw	F.L.Roxas
410		✻	**Esparto** *Harris 82-83*	I ScSr 1 Dk 2 B	British Leith	803 1245	241·2 | 33·1 | 15·8 B 61 f. Q 95 f. F 30 f.	Leith S.&H.Morton&Co.	1880	4 Blds. CDB 200	2Cy.29" & 56"-40" 145 HP S.&H.Morton&Co.,Leith	London & Edinboro' Shipping Co.
411		✻	**Esperança** *Reeis -87*	Stl T.ScSr 1 Dk & Shd Dk	Brazilian Para	340 500	182·0 | 28·1 | 9·6 F 28 f.	Birkenhead LairdBros.	1887	7 Blds.	6Cy,13" 19½" & 33"-24" 150 HP LairdBros.,Birkenhead	Amazon S.Nav.Co.,Ld.
412	BT1.88	✻	**Esperanza** (ex Tornado) *Turney -88*	I ScSr 1 Dk 2 B	British Hull	375 729	220·0 | 28·6 | 15·6 P 122 f. F 40 f.	Glasgow J.G.Lawrie	1871	5 Blds.	2Cy.28" & 52"-36" 120 HP Blackwood&Gordn,Pt.Glas.	W.S.Bailey
413		✻	**Esperanza** (ex Queen)	I ScSr	Spanish Almeria	178 284	159·9 | 22·1 | 13·1	Glasgow R.Napier&Sons	1854		2Cy.29" & 36"-22½" 58 HP W.B.Thompson,Dundee	Compania Hispano-Algerina
414			**Esperanza** (ex Diana)	W	Spanish Sagua	52 103	81·4 | 23·2 | 6·6	Cuba	1877			Larrondo & Co.
415			**Espero** *Bogenovich 69-87*	I ScSr 2 Dks	Austrian Trieste	1181 1826	274·5 | 34·0 | 26·0	Dumbarton W.Denny&Bros.	1869		4Cy.29¾" & 57"-39" 400 HP Denny&Co.,Dumbarton	Lloyd Austro-Ungarico
416			**Espirito Santo**	I ScBg 2 Dks & Shd Dk	Brazilian RioJaneiro	1181 1737	251·1 | 38·2 | 12·6 F 49 f. 20·1	Glasgow J.&G.Thomson	1875		2Cy.46" & 80"-42" 350 HP J.&G.Thomson,Glasgow	C^ia^ Braziliana de Nav. à Vapor
417			**Esquimaux** *Milne 83-83*	W ScS	British Dundee	466 533	157·3 | 29·5 | 19·3	Dundee Stephen	1865		2Cy. 70 HP	D.Bruce & Co.
418	BT1.89		**Essequibo** *Powles* (ex Bladworth)	I ScSr 3 Dks	British Sunderland	1329 1831	275·4 | 34·6 | 25·5	Stockton Richardson,Duck & Co.	1871		2Cy.35" & 66"-36" 170 HP C.D.Holmes&Co.,Hull	Royal Mail Steam Packet Co.
419		✻	**Essex** *Burgess*	I ScBn 2 Dks 3 B	British Middlesbro'	1675 2504	300·0 | 40·3 | 25·2 P 44 f. B 83 f. F 44 f.	Middlesbro' R.Dixon&Co.	1882	5 Blds.	2Cy.48" & 78"-42" 250 HP T.Richardson&Sons,H'rtlpl.	M.Wigram & Sons, Ld.
420		✻	**Essex** *Kershaw 75-77*	I ScSr 1 Dk 2 B	British Hull	884 1181	238·0 | 34·7 | 17·0 P 124 f. F 25 f.	Hull Humphreys&Pearson	1869		2Cy.36" & 83" 98 HP Comp. '79. Humphreys & Pearson,Hull	W.S.Bailey

✻ = **Lloyd's Register**; D = Nederlandsche Vereeniging; G = Germanischer Lloyd; I = Registro Italiano; R = American Record;
A = Veritas Austro-Ungarico; H = Veritas Ellenico; N = Norske Veritas; V = Bureau Veritas.

Dans la 2me Colonne, B.T., avec les indications qui suivent, signifie la date du rapport de visite de la Carène par les experts du Board of Trade pour obtenir le certificat relatif au TRANSPORT DES PASSAGERS.

Dans la 3me Colonne, C = Composite (membrure en fer, bordé en bois); I = Fer; Stl = Acier; W = Bois.

Pour plus amples renseignements sur les navires classés par le Lloyd's Register, voir le Lloyd's Register Book et le Lloyd's Yacht Register.

No. in Book / Official No.	Registry	Ship's Name. Material, Rig, &c. / Master, and Dates of Appointment to present Owner's Service and to Ship / Ship's Late Name. No. of Decks, &c.	Port of Registry	Tons Net / Gross	Length	Breadth	Depth	Built. Where. / By Whom.	When	No. of Bulkheads / Particulars of Water Ballast.	Engines. Particulars. / Maker's Name.	Owners or Managers
421	✻	**Ester** Stl TwinScSr *Sels* 1 *Dk*	Chilian Talcahuano	[illegible]	112·0	21·5	7·6 *B 31 f. Q 36 f. F 14 f.*	Sunderland Bodie, Shorer & Co.	1886	4 Blkds. FPT36t APT23t	2Cy.11"& 22"–18" 40 HP. J.Dickinson,Sunderland	J.Bunster
422 [illegible]		**Esther** I ScSr *Lewis*	British [illegible]	208 [illegible]	145·5	25·7	10·1 *Q 16 f.*	Paris A.Cochot	1864		2Cy.22½"–30½" 50 HP. A.Cochot,Paris	A.J.Hutchins
423		**Estonia** (DTRNT) W ScSr *Zorn* (ex Germania) 1 *Dk*	Russian Reval	325 [illegible]	115·3	25·0	12·4 *P 50 f. B 75 f.*	Lübeck J.Steffen & Sohn	1867		2Cy.15"–12" 30 HP. Schweffel & Howaldt,Kiel	A.G.Senglbusch
424 933[illegible]	✻	**Estrella** I ScSr *Wright* 87-87 1 *Dk & Shade Dk*	British London	325 [illegible]	175·0	25·0	10·4 *P 32 f. F 30 f.*	Port Glasgow Blackwood & Gordon	1887	4 Blkds.	2Cy.22"& 44"–30" 85 HP. Blackwood & Gordin,PGlsgw	Hill,Gomes & Co.
425 9602[illegible]	✻	**Ethel** Stl ScSr *Foshaw* 88-88 1 *Dk & Wh Frms*	British Glasgow	551 [illegible]	250·0	35·1	16·7 *Q 181 f. B 70 f. F 31 f.*	Grangemouth Grangemouth Dkyd.Co.	1888	4 Blkds. FDB330t APT31t	3C.18"30"& 48"–36" 120 HP. Hutson & Corbett,Glasgow	W.Christie & Co.
426 65248	✻	**Ethel** I ScSr *Coysh* 71-86 1 *Dk 2 B*	British Hull	669 1036	227·2	29·3	16·4 *P 121 f. F 25 f.*	Sunderland R.Thompson, jun.	1871	4 Blkds. [illegible]	2Cy.26"& 47"–33" 95 HP. G.Clark,Sunderland	F.B.Grotrian
427 76[illegible]	✻	**Ethel** Stl ScSr *Hebblack* 87-87 1 *Dk 2 B*	British Newcastle	587 [illegible]	210·6	30·1	16·5 *P 27 f. B 42 f. F 27 f.*	Newcastle C.Mitchell & Co.	1878	4 Blkds. [illegible]	2Cy.26"& 52"–30" 100 HP. WallsendSlipwayCo.Nwcstl	Clapham S.S.Co.
428 81[illegible]	✻ BT.88	**Ethel** (ex Cloch) I ScSr *Carmichael* 80-88 1 *Dk*	British Glasgow	162 [illegible]	150·1	21·7	10·7 *B 28 f. F 26 f.*	Belfast Workman, Clark & Co.	1880	3 Blkds.	2Cy.19"& 38"–30" 60 HP. J.Rowan & Sons,Ld.,Belfast	D.MacBrayne
429 67[illegible]	✻	**Ethel** I ScSr *Whittington* 84-84 1 *Dk*	British Stockton	107 [illegible]	120·2	19·9	8·4	Middlesbro' R.Craggs & Sons	1878	3 Blkds.	2Cy.15"& 28"–18" 30 HP. Pattison & Atkinson,N'castle.	Cleveland S.S.Co.
430 [illegible]		**Ethel** I ScSp	British [illegible]	58 [illegible]	83·2	18·7	9·3	VauxhallSom. J.Payne	1870		1Cy.18"–16" 25 HP. J.Payne,London	H.Burton
431 8[illegible]	✻	**Ethel Gwendoline** I ScSr *Head* 87-87 1 *Dk*	British Cardiff	155 249	120·1	22·0	10·5 *Q 36 f. F 14 f.*	Newport Cox & S.B.Co.	1883	3 Blkds. FPT[illegible]	2Cy.16"& 30"–22" 50 HP. Sheryn & Lawrie,Newport	R.Taylor
432 8140[illegible]	✻	**Ethel Horatia** I ScSr *Wetherstone* (WRECKED) 1 *Dk 2 B*	British [illegible]	741 11[illegible]	227·0	32·7	16·5 *Q 76 f. B 42 f. F 26 f.*	Sunderland Osbourne, Graham & Co.	1881	4 Blkds. [illegible] FPT APT	2Cy.28"& 54"–36" 99 HP. Hawks,Crawshay & C.,G'tshd	J.F.Marshall
433 74971		**Ethel Jackson** CPad	British Melbourne	90 132	115·2	22·0	8·4	Echuca,Vict. R.Wilson	1876		2Cy.18"–42" Roberts & Sons,Sandhurst	D.Munro
434 93[illegible]	✻	**Ethelbald** I ScSr *Smith* 82-86 1 *Dk*	British Glasgow	423 [illegible]	180·6	26·1	13·7 *Q 60 f. B 38 f. F 28 f.*	Belfast Wrkmn,Clark & Co.	1886	4 Blkds. DB[illegible] FPT[illegible]	3C.15"24"& 40"–33" 96 HP. Hutson & Corbett,Glasgow	MacgregoryS.S.Co. Ld.
435 84301	✻	**Ethelbert** I ScSr *Cargill* 86-87 1 *Dk*	British Glasgow	328 [illegible]	170·0	24·6	12·8 *Q 80 f. B 18 f. F 38 f.*	Belfast Workman, Clark & Co.	1881	3 Blkds. DB[illegible] FPT[illegible]	2Cy.22"& 40"–30" 75 HP. Muir & Houston,Glasgow	Colvils,Lowden & [illegible]
436 8978[illegible]	✻	**Ethelburga** Stl ScSr *Willis* -87 1 *Dk 2 B*	British Whitby	1446 2228	290·0	38·5	18·7 *[illegible] B 118 f. F 33 f.*	S.Shields J.Readhead & Co.	1877	6 Blkds. DB[illegible]	3C,23"37½"&61½"–39" 250 HP. J.Readhead & Co.SShlds	Dillon,Harrowi[illegible]
437 [illegible]	✻	**Ethelred** I ScSr *Inch* 82-83 1 *Dk 2 B*	British [illegible]	708 1121	225·5	30·6	17·4 *P 119 f. F 37 f.*	Newcastle A.Leslie & Co.	1872	5 Blkds. DB[illegible]	2Cy.26"& 52"–30" 99 HP. Thompson,Boyd & Co.,Nwc.	H.E.P.Adamson
438 [illegible]	BT.7.88	**Ethiopia** I ScBk *Wilson* [illegible] 2 *Dks & 1 Spar Dk 4 B*	British Glasgow	2604 4005	402·0	40·2	24·8 33·0	Glasgow A.Stephen & Sons	1873	7 Blkds.	2Cy.38"& 76"–60" 720 HP. D.&W.Henderson & Co.Glas.	BarrowSteamShip Co. Ld.
439 [illegible]		**Ethiopia** I ScSr *Paterson* 87-87 3 *Dks*	British Glasgow	1314 20[illegible]	315·6	33·3	25·5	Dumbarton W.Denny & Bros.	1874	5 Blkds.	2Cy.41"& 70"–42" 250 HP. Denny & Co.,Dumbarton	British India Steam Nav.Co.,Ld.
440 47[illegible]		**Etna** (ex St. Thomas) I ScBk *Haydon* 85-85	British Liverpool	826 12[illegible]	226·0	31·2	24·7	Sunderland Oswald & Co.	1863		1Cy.20"& 15"–30" 150 HP. G.Forrester & Co.,Liverpool	A.Laming & Co.
441 [illegible]	✻	**Etna** I ScSr *Dreaper* 85-85 1 *Dk 2 B*	British W.Hartlepool	702 1119	235·0	31·1	17·5 *B 16 f. Q 93 f. F 31 f.*	Stockton M.Pearse & Co.	1874	4 Blkds. [illegible]	3C.17"28½"& 46½"–33" 110 HP. [illegible] Blair & Co.,Ld.Stn	Herskind & Wo[illegible]
442	G	**Etna** (ex A. C. de Freitas) I ScSr [illegible] 2 *Dks & Spar Dk 4 B*	Danish Copenhagen	676 1067	222·0	28·6	14·0 *B 29 f. B 21 f.* 22·7	Kiel G.Howaldt	1883	MT[illegible] [illegible]	2Cy.28"& 43"–35" 160 HP. Howaldt Bros.,Kiel	Det Forenede [illegible] skibs Selskab
443	D V	**Etna** I ScSr *De Jonge* 80-81 1 *Dk*	Dutch Amsterdam	434 [illegible]	194·5	23·0	14·8 *B 96 f.*	Greenock Scott & Co.	1871		2Cy.29"& 41"–36" 90 HP. GreenockFoundryCo.	Koninklijke Nederlandsche Stoomv. [illegible]
444	✻ V	**Etna** I ScSr *Pape* 87-87 2 *Dks*	German Hamburg	1478 [illegible]	301·5	37·4	23·8 *P 40 f. B 68 f. F 32 f.*	Kiel Norddeutsche Werft	1882	6 Blkds. DB[illegible]	2Cy.38"& 66"–43" 240 HP. Markische Schiffsb.Co.,Berlin	B.Wencke Söhne
445 [illegible]	I	**Etna** I ScBk *Pernigotti* -88 2 *Dks*	Italian [illegible]	346 [illegible]	198·4	26·6	15·7	Greenock Scott & Co.	1862		1Cy.13"(2)26"& 53"–30" Scott & Co.,Greenock (altered 1883)	Navigazione Generale Italiana
446 93718	✻ BT.1.89	**Etolia** Stl & I ScSr *Gill* 83-87 2 *Dks*	British Liverpool	2108 3211	315·6	40·9	26·7 *P 33 f. B 92 f. F 43 f.*	Belfast Harland & Wolff	1887	6 Blkds. CBD[illegible] APT[illegible]	3C24½"37"&61"–48" 320 HP. Harland & Wolff, Belfast	City of Liverpool S.S. Co.
447 94088	✻	**Eton** I ScK *Raynor* -88 1 *Dk*	British [illegible]	91 1[illegible]	101·6	20·4	10·6 *F 21 f. Q 17 f.*	Hull Earle's Co.,Ld.	1888	3 Blkds.	3C.10½"17"&30"–18" 44 HP. Earle's Co.,Ld.,Hull	H.Smethurst,jun.
448 91186	BT.5.88	**Etruria** Stl ScSr *Cook* 53-85	British Liverpool	3331 [illegible]	501·6	57·2	38·2 *P 56 f. B 200 f. F 39 f.*	Glasgow J.Elder & Co.	1884		3C.71"&(2)105"–72" 2500 HP. J.Elder & Co.,Glasgow	CunardSteamShip Co. Ld.
449 [illegible]	✻ D	**Etta** I ScSr *Pillmore* 1 *Dk 2 B*	British W.Hartlepl.	698 1093	234·2	31·7	14·8 *Q 92 f. B 50 f. F 60 f.*	W. Hartlepool W.Gray & Co.	1882	4 Blkds. [illegible]	2Cy.27"& 50½"–33" 99 HP. Blair & Co.,Ld.Stockton	T.Appleby & Co.
450 [illegible]		**Ettore** I ScSr *Lazarich* 54-87 2 *Dks*	Austrian Trieste	1241 [illegible]	307·0	33·0	26·0	Dumbarton W.Denny & Bros	1874		2Cy.41"& 70"–42" 250 HP. Denny & Co.,Dumbarton	Lloyd Austro-Ungarico

✻ = **Lloyd's Register**; D = Nederlandsche Vereeniging; G = Germanischer Lloyd; I = Registro Italiano; R = American [illegible]
A = Veritas Austro-Ungarico; H = Veritas Ellenico; N = Norske Veritas; V = Bureau Veritas.

In the 2nd Column, B.T., with figures following, signifies Date of report of bottom sighting by Board of Trade Surveyors as condition of PASSENGER CERTIFICATE.

In the 3rd Column, C = Composite (iron frame, wood planked); I = Iron; Stl = Steel; W = Wood.

Further particulars of Vessels classed by Lloyd's Register will be found in Lloyd's Register Book and Lloyd's Yacht Register.

	Nom, Matériaux, Gréement, etc. Capitaine et dates d'entrée au service des Armateurs actuels et à bord du navire. Ancien Nom, nombre de ponts, etc.		Pavillon. Port d'Armement	Ton. Net Total	Long.	Larg.	Creux	Construit. Où. Par qui.	Quand.	Nombre des cloisons. Détails sur le lest d'eau.	Machines. Détails. Nom du Constructeur.	Armateurs ou Directeurs-Gérants.
	Euclid Phelan	I ScBg 1 Dk & Spr Dk3 B	British Liverpool	1001 1560	270·3	32·7	19·1 24·5	Aberdeen Hall, Russell & Co.	1877	5 Blkds.	2Cy. 26" & 58"–39" 150 HP. Hall, Russell & Co., Aberdeen	Liverpl., Brazil & River Plate Steam Nav. Co., Ld.
	Euclid Knott 82-2	I ScSr 1 Dk 2 B	British	433 688	196·4	29·1	16·6	Sunderland Oswald & Co.	1866	4 Blkds.	2Cy. 38"–29" 90 HP. B. Clark, Sunderland	E. J. Weatherley
✻	**Eugalia** Bewerell 88-88	Stl ScSr 1 Dk	Belgian Antwerp	348 551	175·0	26·6	12·3	Middlesbro' R. Dixon & Co.	1887	4 Blkds.	3Cy. 15" 25" & 40"–30" 80 HP. Blair & Co., Ld., Stockton	M. Matthys
V	**Eugène** (ex Bluebell) Lewis 79-88	I Sc3mSr 1 Dk 2 B	British Liverpool	498 800	196·9	28·1	16·5	Newcastle Schlesinger, Davis & Co.	1872	3 Blkds.	2Cy. 26" & 48"–30" 99 HP. Hawks, Crawshay & Co., Nwc	"Eugene" S.S. Co., Ld. (G. Nelson & Sons)
V	**Eugene Krohn** Jochen 83-	I ScSr 2 Dks & Spr Dk	German Rostock	465 722	187·5	27·0	13·0	Lübeck H. Koch	1883		2Cy. 21" & 48"–30" 90 HP.	F. W. Fischer
V	**Eugène Pereire** Lota -88	Stl ScSr 2 Dks & 1 Spr Dk	French Marseilles	1156 2038	334·6	35·1	23·9 16·4	St. Nazaire Compagnie Générale Transatlantique	1888		3Cy. 31.5" 49" & 79"–49" Cie Générale Transatlantique, St. Nazaire 850 HP.	Cie Générale Transatlantique
	Eugenie (ex Anna J. Lyman)	W PadSr 1 Dk	British Nassau, N.P.	92 176	119·2	22·1	7·8	Keyport, U.S.	1862		1Cy. 45 HP.	R. W. Simpson
✻	**Eugenie** Müller 78-88	Stl ScSr 1 Dk 2 B	Norweg'n Bergen	1507 2300	295·0	38·2	20·1	Middlesbro' R. Dixon & Co.	1888	5 Blkds.	3C. 22" 35" & 59"–39" 190 HP. T. Richardson & Sons, Hartpl.	Actieselskabet "Eugenie" (Mohr & Son)
✻	**Euphrates** Edwards 86-86	I ScSr 2 Dks 3 B	British London	1300 2025	287·0	34·6	24·7	Newcastle Tyne I. Ship B. Co.	1877	5 Blkds.	2Cy. 38" & 64"–42" 200 HP. N. E. Marine E. Co., Ld., Smfld.	W. H. Tindall & Co.
	Euphrates Brown 87-87	I ScBg 2 Dks	British Glasgow	508 803	239·0	28·3	16·4	Glasgow A. & J. Inglis	1863	5 Blkds.	2Cy. 28" & 50"–36" 120 HP. A. & J. Inglis, Glasgow	British India Steam Nav. Co., Ld.
✻	**Euphrates** Bodder	I ScSr 1 Dk	British Hull	29 102	87·0	20·2	10·4	Dundee W. B. Thompson	1886	3 Blkds.	2Cy. 15" & 27"–16" 35 HP. W. B. Thompson, Dundee	H. Rippon
✻	**Eureka** Hole 87-87	I ScSr 1 Dk 2 B	British Cardiff	932 1439	244·0	33·8	17·2	Sunderland Strand Slipway Co.	1882	4 Blkds.	2Cy. 29" & 56"–36" 130 HP. N. E. Marine E. Co., Ld., Sndld.	"Eureka" S.S. Co., Ld. (Osborn & Wallis)
R	**Eureka** (COLLISION) Reed 87	I ScIm 2 Dks & Awn Dk	American New York	2562 3531	351·0	42·8	32·6	Philadelphia W. Cramp & Sons	1884	4 Blkds.	2Cy. 58" & 76"–54" W. Cramp & Sons, Philadelphia	Southern Development Co.
	Eureka (ex California) Hall	W ScSr 2 Dks 3 B	American S. Francisco	514 689	168·0	28·5	15·5	Mystic R. Con.	1864		2Cy. 21" & 42"–34" 88 HP. Risdon I. Wks., S. Francisco	Pacific Coast S.S. Co.
	Eureka	I ScTug	Italian Genoa	47 114	..	..	..	Sestri Ponente	1885		50 HP.	Belleno Benedetto
	Eurimbla Nelson	Stl ScSr 1 Dk & Spr Dk	British Sydney	598 1055	214·2	31·2	19·2	Kinghorn J. Key & Sons	1884		2Cy. 30" & 60"–39" 180 HP. J. Key & Sons, Kirkcaldy	Australasian United Steam Nav. Co., Ld.
✻	**Euripides** Herd 77-83	I ScSr 2 Dks	British Liverpool	1153 1782	250·7	35·1	21·8	Barrow Chas. J. Purdie & Co.	1883	4 Blkds.	2Cy. 31" & 60"–42" 160 HP. W. Kemp, Glasgow	S.S. "Euripides" Co., Ld. (Layborn & Legge)
✻	**Europa** (ex Hamlet) Jones 86-86	I ScBkn 2 Dks 3 B	British Liverpool	1493 2308	300·0	36·6	25·5	Stockton M. Pearse & Co.	1881	4 Blkds.	2C. 36½" & 68"–42" 280 HP. Blair & Co., Ld., Stockton	City of Liverpool Stm. Nav. Co., Ld.
DG	**Europa** Willnebeu 74-82	I ScBn 2 Dks 1 Har Dk	German Bremen	1542 2388	287·7	36·1	24·7	Vegesack H. F. Ulrichs	1881		3Cy. 22" 35" & 60" 190 HP. J. C. Tecklenborg, Geestemunde	D. H. Wätjen & Co.
✻V	**Europa** Stiefel 86-86	I ScSr 2 Dks	German Hamburg	1002 1532	260·1	33·8	22·2	Newcastle C. Mitchell & Co.	1879	3 Blkds.	2Cy. 32" & 60"–39" 150 HP. Blair & Co., Stockton	Edwd. Carr
G	**Europa** Stegelmann 80-82	I ScSr	German Lübeck	503 655	175·6	25·3	18·4	Kiel	1883		2Cy. 23" & 37½"–27" 100 HP. Howaldt Bros., Kiel	Akt.-Ges. "Europa" (Schröder & Pape)
V	**Europa** Burlando -88	I Sc4mSr 3 Dks & Spr Dk	Italian Genoa	1264 2202	313·0	35·4	27·3	Newcastle Wigham Richardson & Co.	1873		2Cy. 52" & 91½"–41½" 780 HP. Wigham, Richardson & Co., Nwctle.	La Veloce Navigazione Italiana
✻	**Europa** (ex Camilla, ex Granada -87 ex Maranhense) Grandia	I Sc3mSr 3 Dks	Italian Bari	873 1334	269·3	30·5	16·9 25·1	Liverpool T. Royden & Sons	1869	6 Blkds.	2Cy. 32" & 64"–30" 136 HP. Fawcett, Preston & Co., Lvrpl	L. Scorcia & Sons
	Europa	I	Spanish Barcelona	380 733	219·4	28·3	18·3				260 HP.	Sociedad de Navé. Industria
✻	**Europe** Richardson	I ScSr 1 Dk	British Hull	120 198	130·8	21·0	10·9	Hull Earle's Co.	1880	3 Blkds.	2Cy. 21" & 40"–27" 60 HP. Earle's Co., Ld., Hull	Hull Steam Fishing & Ice Co., Ld.
V	**Europe** (ex Tamar) Portal 71-88	I ScBn 2 Dks & Awn Dk A B	French Marseilles	1298 2226	309·0	35·4	24·1 32·5	Glasgow Dobie & Co.	1873	4 Blkds.	2C. 47½" & 94½"–48" 350 HP. Howden & Co., Glasgow	Nouvelle Cie Marseillaise de Navigation à Vapeur (Fraissinet & Co.)
✻	**European** (ex William Penn) Johnson 82-89	I ScBn 2 Dks & Awn Dk	British Liverpool	1736 2639	326·8	36·3	21·5 28·6	Waterford Malcomson Bros.	1866	9 Blkds.	2Cy. 34" & 68"–46" 250 HP. Oswald, Mordaunt & Co., Sou.	W. Ross & Co.
D	**European** Anderson 85-86	I Sc3mSr	British Hull	331 523	196·0	25·3	14·3 18·7	Glasgow Smith & Rodger	1850		2Cy. 23" & 44"–27" 75 HP. C. D. Holmes & Co., Hull	W. & C. L. Ringrose
✻V	**Euskaro** de Telleria 88-88	I ScSr 1 Dk & Spr Dk 3 B	Spanish Bilbao	1613 2472	290·0	38·6	26·3	Sunderland R. Thompson & Sons	1886	5 Blkds.	2Cy. 38" & 72"–48" 250 HP. J. Dickinson, Sunderland	M. M. de Arrotegui
✻D	**Eustace** Hosking 83-87	I ScSr 1 Dk 2 B	British W. Hartlepl.	1076 1665	258·6	34·1	19·9	Whitby T. Turnbull & Son	1878	5 Blkds.	2Cy. 32" & 60"–39" 140 HP. Blair & Co., Ld., Stockton	G. Pyman & Co.

✻ = Lloyd's Register; D = Nederlandsche Vereeniging; G = Germanischer Lloyd; I = Registro Italiano; R = American Record; A = Veritas Austro-Ungarico; H = Veritas Ellenico; N = Norske Veritas; V = Bureau Veritas.

Dans la 2me Colonne, B.T., avec les indications qui suivent, signifie la date du rapport de visite de la Carène par les experts du Board of Trade pour obtenir le certificat relatif au TRANSPORT DES PASSAGERS.

Dans la 3me Colonne, C = Composite (membrure en fer, bordé en bois); I = Fer; Stl = Acier; W = Bois.

Pour plus amples renseignements sur les navires classés par le Lloyd's Register, voir le Lloyd's Register Book et le Lloyd's Yacht Register.

No. in Book / Official No.	Registry in which classed	Ship's Name. Master, and Dates of Appointment to present Owner's Service and to Ship. Ship's Late Name.	Material, Rig, &c. No. of Decks, &c.	Flag. Port of Registry.	Tons Net. Gross.	Length	Breadth	Depth	Built Where. By Whom.	When	No. of Bulkheads. Particulars of Water Ballast.	Engines. Particulars. Maker's Name.	Owners
481 [illegible]		**Eutaw** *Hopkins*	W Pad	American Charleston,S.C.	504 [illegible]	156·0	29·0	7·0	Charleston, S.C. S.J.Pregnall&Bro.	1882		180 HP	J. T. Foster
482	*	**Euterpe** *Frenzio 64-86*	Stl ScSr 2 Dks3 B	Austrian Trieste	1444 [illegible]	315·0	38·1	24·3	Sunderland J.L.Thompson&Sons	1888	[illegible] Blds. [illegible]	3Cy.26"42"&69"-48" 160 HP T.Richardson&Sons,Hrtlpl.	Lloyd Austriaco
483 87498	*	**Euterpe** *Tonkin*	I ScSr 1 Dk2 B	British Cardiff	990 1522	260·1	36·0	16·4	Glasgow A.Stephen&Sons	1883	[illegible] Blds. [illegible]	2Cy.32"&60"-39" 180 HP A.Stephen&Sons,Glasgow	"Euterpe" S. [illegible] (Osborne & [illegible])
484	*	**Euterpe** (ex [illegible]) *Gratfos 85-86*	I Pad Tug 1 Dk	Greek Ithaca	45 158	110·8	20·3	10·6	SouthShields J.T.Eltringham	1877		1Cy.41½"-56" 80 HP J.P.Rennoldson,S.Shields	[illegible]
485		**Euxène** *Arnaud 72-83*	I ScSr 2 Dks	French Marseilles	704 1051	249·3	30·0	20·6	Greenock [illegible]&Co.	1864		2C.30½"&64"-33" 130 HP Fraissinet&Co.,Marseilles	Fraissinet & [illegible]
486 65500	*	**Euxine** *Reid 81-87*	I ScSr 2 Dks&Spr Dk	British London	978 1531	244·8	32·1	17·9 23·6	Newcastle C.Mitchell&Co.	1871	1 Blds. [illegible]	2Cy.26"&56"-33" 180 HP N.E.MarineEngCo.,Ld.,Sld.	Nelson, Donk[illegible]
487 93816	*	**Euxine** *Brock 88*	Stl ScSr 1 Dk	British Liverpool	84 166	120·0	20·0	8·5	Seacombe J. F. Waddington & Co.	1888	3 Blds.	2Cy.13"&26"-18" 28 HP J.F.Waddington&Co.Seacmb	C.G.Lambros
488 74441		**Evangeline** (ex Bessie)	I ScSrYt	British [illegible]	68 100	94·7	17·6	10·6	Cowes J.White	1876		2Cy.16"&30"-14" 20 HP Belliss&Co.,Birmingham	Laurence Ca[illegible]
489	V	**Evelina** (ex Vettore) *Sullora*	I ScSr	Austrian Trieste	154 258	166·7	20·0	11·7	London SamudaBrothers	1853		2Cy.32"-30" 100 HP J.Watt&Co.	B & A [illegible]
490 91494	*	**Eveline** *Rasmussen 82-85*	Stl ScSr 1 Dk2 B	British Newcastle	844 1312	245·0	34·0	16·0	Newcastle C.S.Swan&Hunter	1885	1 Blds. [illegible]	3Cy.18½"31"50"-36" 130 HP WallsendSlipwayCo.,Ld.	[illegible]
491 92924	*	**Evelyn** *Morris 73-88*	Stl ScSr 1 Dk&Web Frms	British W.Hartlepl	1529 2346	290·0	38·2	20·3	W. Hartlepool W.Gray&Co.	1888	[illegible] Blds. [illegible]	3Cy.22½" 36½" & 60"-39" 200 HP Blair[illegible]	T. Appleby & C[illegible]
492 91083	*	**Evelyn** *Martin*	W ScK 1 Dk	British Leith	20 47	67·3	17·3	7·5	Leith J.M'Kenzie&Co.	1885		2Cy.8½"&16"-12" 11 HP J.Cran&Co.,Leith	S.H.Payne
493		**Evénement**	I ScSr 2 Dks	French Marseilles	221 334	171·6	23·6	12·9	Greenock Scott&Co.	1862		2Cy.32"-30" 120 HP GreenockFoundryCo.Grnk	F. Morelli & C[illegible]
494 84535	*	**Everest** *Douglass 82-82*	I ScBn 1 Dk2 B	British W.Hartlepl.	1078 1676	261·6	34·6	19·5	W. Hartlepool W.Gray&Co.	1881	1 Blds. [illegible]	2Cy.32"&60"-39" 150 HP Blair&Co.Ld.,Stockton	Murrell & Y[illegible]
495 84677	* D	**Everilda** *Story*	I ScBn 1 Dk2 B	British Cardiff	894 1401	243·0	34·0	17·9	Whitby T.Turnbull&Son	1882	1 Blds. [illegible]	2Cy.30"&57"-33" 130 HP T.Richardson&Sons,Hrtlepl	Turnbull Bro[illegible]
496 64346		**Examiner**	W ScBk 1 Dk	British Sydney	181 266	139·6	29·7	9·8	Richmond River N.S.W.	1870		1Cy. 35 HP	John McK[illegible]
497	N	**Excellencen** (ex Excellencen Toll) *Thommessen*	I ScSr 2 Dks	Norweg'n Christnsand.	296 414	159·2	23·1	16·5	Norrköping MotalaCo.	1859		2Cy.29"&56"-22" 110 HP MotalaCo.,Gothenburg	J. W. Prebe[illegible]
498 81454	*	**Excellent** (COLLISION) *Taylor 79-84*	I ScSr 1 Dk2 B	British Sunderland	1083 1659	260·0	36·0	18·5	Sunderland ShortBros.	1880	6 Blds. [illegible]	2Cy.33"&62"-42" 180 HP J.Clark,Sunderland	Anderson, H[illegible]
499 95662	*	**Excelsior** *Cornforth 88-88*	Stl ScSr 1 Dk2 B	British Whitby	1202 1861	259·0	37·0	18·1	Whitby T.Turnbull&Son	1888	[illegible] Blds. [illegible]	3Cy.20"33"&54"-36" 140 HP Blair&Co.Ld.Stockton	H. Baxter & C[illegible]
500 60560	*	**Excelsior** *Deakin*	I ScBg 2 Dks3 B	British Cardiff	917 1417	239·7	34·0	17·0 23·9	Sunderland W.Pile&Co.	1870	[illegible] Blds. [illegible]	2Cy.28"&57"-36" 150 HP T.Clark&Co.,Newcastle	W. & R. H. S[illegible]
501 86320		**Excelsior** *Patrick 82-87*	Stl ScCr 1 Dk	British Melbourne	172 340	186·6	21·1	10·7	Southampton Day,Summers&Co.	1882	[illegible] Blds. [illegible]	2Cy.28"&54"-30" 160 HP Day,Summers&Co.,Sthptn	Huddart, P[illegible]
502 76525		**Excelsior**	W ScF[illegible]	British Liverpool	96 122	82·0	19·5	10·0	Winsford W.Cross	1864		2Cy.9½"-12" 10 HP W.Gardiner,Manchester	Salt Union,L[illegible]
503 90004	*	**Excelsior** *McKinlay 86-86*	I ScSr 1 Dk	British Glasgow	83 218	130·0	21·0	9·6	Paisley J.Fullerton&Co.	1885	[illegible] Blds. [illegible]	2Cy.19"&36"-24" 55 HP Ross&Duncan,Glasgow	H. W. F. Nic[illegible]
504 58734		**Excelsior**	I PadSr	British N.Shields	20 105	90·5	17·2	9·4	London	1866		2Cy.26"-42" 45 HP T.A.Young,London	F. Warren
505 135620	R	**Excelsior** *Higgins*	I ScIn Sr 3 Dks	American N.Orleans	2407 3264	330·0	42·6	32·6	WilmingtnDel Harlan&Hollingswth Co.	1882	[illegible] Blds. [illegible]	2Cy.38"&75"-54" Harlan&HollingsworthCo.,Wilmington,Del.	Morgan R.R[illegible]
506 135463		**Excelsior**	W Pad 3 Dks	American Georgetown,Del.	1016 1326	232·0	37·0	10·1	WilmingtnDel Harlan&Hollingswth Co.	1880		2Cy.40"-120" 1050 HP Harlan& HollingsworthCo.,Wilmington,Del.	M. Robinson
507 87965	*	**Exchange** *Farrell 84-86*	I ScSr 1 Dk	British Liverpool	98 275	135·0	22·3	9·9	Amlwch W.Thomas	1884	3 Blds.	2Cy.18"&34"-24" 50 HP DeWinton&Co.,Carnarvon	W. Thomas & C[illegible]
508 94368	*	**Exe** *Clark 78-87*	Stl ScSr 1 Dk&Web Frms	British London	1368 2180	275·0	38·2	20·7	Sunderland J.L.Thompson&Sons	1887	[illegible] Blds. [illegible]	3Cy.21½"35½"&58½"-39" 250 HP Blair&Co.Ld.Stockton	Mercantile S[illegible]
509 91984	*	**Exeter City** *Weiss 74-87*	Stl ScSr 2 Dks3 B	British Bristol	1434 2198	285·0	38·0	23·9	Blyth BlythS.B.Co.Ld.	1887	6 Blds. [illegible]	3Cy.22½"37"&61"-42" 260 HP Blair&Co.Ld.Stockton	C. Hill & Son[illegible]
510 87927	*	**Exmoor** (ex [illegible]) (LOST) *Osborne 85-88*	I ScSr 1 Dk	British Ipswich	234 381	150·8	22·2	12·0	Preston W.Allsup&Sons	1883	3 Blds. [illegible]	2Cy.20"&36"-24" 60 HP W.Allsup&Sons,Preston	[illegible] & H. W. [illegible]

* = Lloyd's Register; D = Nederlandsche Vereeniging; G = Germanischer Lloyd; I = Registro Italiano; R = American [illegible]
A = Veritas Austro-Ungarico; H = Veritas Ellenico; N = Norske Veritas; V = Bureau V[illegible]

In the 2nd Column, B.T., with figures following, signifies Date of report of bottom sighting by Board of Trade Surveyors as condition of PASSENGER [illegible]

In the 3rd Column, C = Composite (iron frame, wood planked); I = Iron; Stl = Steel; W = Wood.

Further particulars of Vessels classed by Lloyd's Register will be found in Lloyd's Register Book and Lloyd's Yacht Register.

Registre et classe	Nom, Materiaux, Greement, etc. Capitaine et dates d'entrée au service des Armateurs actuels et à bord du navire. Ancien Nom, nombre de ponts, etc.	Pavillon. Port d'Armement	Tonnage Net. Total	Longueur	Largeur	Creux	Construit. Où. Par qui.	Quand.	Nombre des cloisons. Détails sur le lest d'eau.	Machines. Détails. Nom du Constructeur.	Armateurs ou Directeurs-Gérants.
	Exmouth (ex Wasa) I ScBrig Pearson 75-?? 1 Dk & 1 ?? Dk	British Liverpool	713 1108	261·1 Iss. 73	26·2	13·4 20·7	Glasgow J.&G.Thomson	1866		1Cy.21"&12"–33" 280 HP. D.&W.Henderson, Glasgow (NE&B78)	G.Nelson & Sons
	Experiment I Sc	British London	126 211	113·0	26·1	11·1	Glasgow Aitken&Mansel	1875		2Cy.15"&30"–16" 40 HP. Bait&Lindsay,Glasgow	Millwall Dock Co.
	Experiment I Sc	British Liverpool	70 103	86·8	17·9 Q 24.	7·0	Chester H.E.Falk	1864		2Cy.10"–14" 20 HP. H.E.Falk,Winsford	Salt Union,Ld.
	Explorateur I ScSr (ex Gemina)	Zanzibar	138 212	138·6	19·8	10·6	London G.H.Harrington	1868		2Cy.26"–20" 50 HP. J.Stewart,London	Sultan of Zanzibar
※	**Explorer** I ScSr Brown 82-?? 2 Dk&3 B	British Liverpool	1304 2011	300·3 F 32 f. B 45 f.	34·7	25·2	Liverpool R.&J.Evans&Co.	1873	5 Blds.	3Cy.20"33"&54"–36" 225 HP Blair&Co.Ld.Stockton	Charente S.S.Co.,Ld.
	Expres I ScSr Thomsen 79-81 2 Dks & 1 ?? Dk	Danish Aalborg	396 672	205·2 P18f. F29f. H21f.	27·5	13·6 20·1	Flensburg Flensburger Schiffsbau Gesellschaft	1881	APT81 FPT18t MT13&13t DB 31t	2Cy.32½"&60"–36" 160 HP. FlensburgSchiffsbauGesels.	Det Forenede Damp-skibs Selskab (C.P.A.Koch)
	Express I ScSr Willis 79-81	British Hayle	195 248	146·5	22·7 Q 7 f.	12·2	London Ditchburn&Mare	1847		20 HP.	Harvey & Co.
	Express I TwnSc	British Glasgow	169 299	150·0	25·1	10·7	Glasgow D.&W.Hudson& Co.	1880		4Cy.16"&32"–24" 170 HP. D.&W.Henderson&Co.,Gls.	Henderson Bros.
	Express I ScSr Callin 1 Dk 2 B	British Dublin	145 398	191·4 P 30 f.	23·8 F 33 f.	13·2	W. Hartlepool Irvine&Co.	1874	5 Blds.	2Cy.27"&54"–33" 110 HP. Blair&Co.,Ld.,Stockton	City of Dublin Steam Packet Co.
※	**Express** I ScSr Robertson 77-?? 1 Dk	British Kirkwall	119 228	138·1 Q 28 f.	19·1 B 44 f.	10·2	SouthShields Readhead Softley & Co.	1869	4 Blds. DB ?? 47t	2Cy.18"&34½"–22" 50 HP. Comp. '75, Readhead, Softley & Co., South Shields	G.Robertson
	Express Stl ScSp	British London	71 142	158·0	20·1	8·2	London SamudaBros.	1883		4Cy.7"&12"–9"&14¾"& 25½"–11" 95 HP A.M.Perkins&Sons,London	W.Finlay&L.Perkins
※	**Express** W Pad Brown 1 Dk	British Salcombe	32 115	102·6	19·0 Q 24 f.	9·3	Kingsbridge W.Date&Son	1885		2Cy.28½"–36" 68 HP. Willoughby Bros.,Plymouth	Beer & Co.
	Express I ScSr Grenroth	Russian Hangö	218 298	137·6	22·5	11·9	Oscarshamn	1877		100 HP.	C.Korsman
	Express (ex Sherboat) W Pad Gutierrez 88-88 1 Dk	Spanish Bilbao	118 239	152·5	21·9	11·6	WestCowes White	1870		2Cy.45"–40" 120 HP Humphreys,Tennant&Co., London	Compania Cantabrica de Navegacion
V	**Eyriaud des Vergnes** I ScSr Stronde 83-83 1 Dk & Spr Dk 3 B	French Dunkirk	938 1458	250·3 B 45 f.	33·1	15·1 23·1	S.Shields J.Readhead&Co.	1884	DB 91t MT225t	2Cy.32"&61"–36" 160 HP. J.Readhead&Co.,S.Shields	Cie. Gén. des Bateaux à Vap. du Nord

ADDITIONAL SHIPS ARE INSERTED IN THE SUPPLEMENT AT THE END OF THE REGISTER.

※=Lloyd's Register; D=Nederlandsche Vereeniging; G=Germanischer Lloyd; I=Registro Italiano; R=American Record; A=Veritas Austro-Ungarico; H=Veritas Ellenico; N=Norsko Veritas; V=Bureau Veritas.

Dans la 2me Colonne, B.T., avec les indications qui suivent, signifie la date du rapport de visite de la Carène par les experts du Board of Trade pour obtenir le certificat relatif au TRANSPORT DES PASSAGERS.

Dans la 3me Colonne, C=Composite (membrure en fer, bordé en bois); I=Fer; Stl=Acier; W=Bois.

Pour plus amples renseignements sur les navires classés par le Lloyd's Register, voir le Lloyd's Register Book et le Lloyd's Yacht Register.

No. / Official No.	Registry	Ship's Name, Master, Late Name	Material, Rig, &c. / No. of Decks	Flag / Port of Registry	Tons Net / Gross	Length	Breadth	Depth	Built: Where / By Whom	When	No. of Bulkheads / Water Ballast	Engines: Particulars, Makers' Name	Owners
1 [illegible]	✻	F. T. Barry *Rennie 82-82*	I ScSr 2 *Dks*	British London	545 [illegible]	220·0	31·2	15·1	Middlesbro' Backhouse & Dixon	1872	4 Blkds.	2Cy.27" & 50"–30" 90 HP. T. Brassey & Co., Birkenhead	Harris & Dixon
2 84156		Fabian *Hamblin 76-81*	I ScSr 2 *Dks*	British Liverpool	1474 [illegible]	342·0	34·1	23·5	Southampton Oswald, Mordaunt & Co.	1881	5 Blkds.	1Cy.24" & 50"–48" 250 HP. Oswald, Mordaunt & Co., Son.	S.S. "Fabian" Co. (F. Leyland & Co.)
3	✻ D N	Fædrelandet *Falsen 88-88*	I ScSr 1 *Dk2 B*	Norweg'n Bergen	1223 1611	254·4	37·2	17·0	Bergen Martens, Olsen & Co.	1884	5 Blkds.	2Cy.34" & 63"–39" 160 HP. Martens, Olsen & Co., Bergen	Bergh & Helland
4	✻ N	Færder *Egeness -86*	Stl ScSr 1 *Dk*	Norweg'n Christiania	415 [illegible]	161·4	26·0	12·7	Lübeck Henry Koch	1885	4 Blkds.	2Cy.18½" & 36"–21½" 50 HP. H. Paucksch, Landsberg	Actieselskabet "Færder" (Jens Meinich)
5	✻	Faidherbe (ex Chiswell) *Mayairal*	I ScSr 1 *Dk & Shd Dk*	French Bordeaux	987 [illegible]	240·0	35·0	15·1	Sunderland R. Thompson & Sons	1882	4 Blkds.	2Cy.31" & 58"–36" 150 HP. J. Dickinson, Sunderland	J. E. Buhan, père & fils & A. Teisseire
6 78140	✻	Fair Head *Brennan 82-88*	I ScSr 1 *Dk2 B*	British Belfast	734 1173	232·3	31·2	17·0	Belfast Harland & Wolff	1879	5 Blkds.	2Cy.27" & 51"–38" 120 HP. D. Rowan, Glasgow	Ulster S.S. Co.
7 84347	✻	Fairfield *Millington 88-88*	I ScBn 1 *Dk & Web Frame*	British Stockton	1157 1775	260·0	38·0	16·3	Stockton Richardson, Duck & Co.	1884	4 Blkds.	2Cy.30" & 57"–36" 130 HP. Blair & Co., Ld., Stockton	F. Binnington
8 [illegible]	✻ D	Fairfield *Jones 88-88*	I ScSr 1 *Dk2 B*	British Liverpool	1001 1568	250·8	34·3	20·4	Newcastle Palmers Co., Ld.	1881	4 Blkds.	2Cy.30" & 60"–36" 160 HP. Palmers Co., Ld., Newcastle	J. Brown
9 74326		Fairlight	I PadSr 1 *Dk*	British Sydney	177 313	171·4	22·2	10·1	Glasgow T. Wingate & Co.	1878		2Cy.26" & 50"–60" 130 HP. T. Wingate & Co., Glasgow	Port Jackson S.S. Co., Ld.
10 87606	✻	Fairport *Thomson*	I ScSr 1 *Dk*	British Glasgow	232 353	150·2	22·2	11·8	Dundee W. B. Thompson	1877	5 Blkds.	2Cy.20" & 36"–24" 60 HP. W. B. Thompson, Dundee	McFarlane S.S. Co., Ld.
11 89580	✻	Fairway *Gould 87-89*	I ScSr 1 *Dk*	British London	374 615	172·0	26·2	13·4	Dundee Pearce Bros. & Co.	1884	4 Blkds.	2Cy.20" & 40"–30" 66 HP. Pearce Bros., Dundee	"Fairway" S.S. Co. (J. Forsey & Sons)
12 60123	✻ BT3.89	Fairy *Buckley 83-84*	I ScSr 1 *Dk2 B*	British Hull	407 633	195·2	26·4	15·3	Hull C. & W. Earle	1868	4 Blkds.	2Cy.26" & 48"–26" 90 HP. Comp.'78, Earles Co., Ld., Hull	W. S. Bailey
13 29982	BT3.89	Fairy Queen	I Pad	British Liverpool	106 149	135·0	20·1	7·7	Belfast Harland & Wolff	1865		2Cy.35"–42" 70 HP. Blackwood & Gordon, Pt. Gls.	T. W. Thompson
14 65891	✻ BT3.88	Faithful *Smith 82-85*	I ScSr 1 *Dk2 B*	British Liverpool	555 869	230·1	28·7	16·1	Glasgow Dobie & Co.	1871	5 Blkds.	2Cy.20" & 40"–33" 135 HP. Rankin & Blackmore, Greenock	J. Ellis
15 83089	✻	Faithlie *Fairweather 87-87*	I ScSr 1 *Dk*	British Fraserburgh	321 506	165·1	24·2	12·5	Aberdeen A. Hall & Co.	1884	5 Blkds.	2Cy.22" & 40"–27" 70 HP. Hall, Russell & Co., Aberdeen	Fraserburgh & S.S. Co., Ld.
16 86169		Falaba	I PadSp	British Liverpool	81 128	121·1	23·1	5·0	Liverpool R. & J. Evans & Co.	1882		2Cy.14" & 28"–36" 35 HP. Fawcett, Preston & Co., Livpl.	T. W. & E. H. C.
17	V	Falabah *Gaudy -88*	I ScBn 1 *Dk & Spr Dk3 B*	British Shanghai	1066 1585	261·1	36·1	13·2	Rouen Claparede & Co.	1883		2Cy.35½" & 59½"–42" 250 HP. Claparede & C., S. Denis, Seine	J. Walter
18 87358	✻ D	Falcon *Childs 83-83*	I ScSr 1 *Dk2 B*	British Sunderland	781 1210	240·0	34·2	14·0	Sunderland Kish, Boolds & Co.	1883	3 Blkds.	2Cy.31" & 57"–36" 130 HP. G. Clark, Sunderland	"Falcon" S.S. Co. (J. M. Smith, ma.)
19 469		Falcon *Kerr 88-88*	I Sc3mSr	British Glasgow	376 613	211·0	28·0	15·9	Cork E. Pike	1854		2Cy.26" & 48"–30" 160 HP. Cp.'83, City of Cork S.P. Co.	D. MacBrayne
20 73616	✻ BT3.88	Falcon *Watts*	I ScSr 2 *Dks*	British London	352 651	200·0	27·2	14·0	Newcastle C. Mitchell & Co.	1876	4 Blkds.	2Cy.30" & 56"–33" 140 HP. Thompson & Co., Newcastle	General Steam Nav. Co.
21 75964		Falcon (ex Greenland) *Knee 88-88*	W ScBk 1 *Dk*	British St. Johns, NF.	311 458	160·1	26·6	17·2	Bremerhaven F. W. Wencke	1872		1Cy. 80 HP.	C. T. Bowring & Co.
22 5330		Falcon *Fowler*	I ScSr 1 *Dk2 B*	British Hull	263 442	174·2	24·1	14·1	Dumbarton W. Denny & Bros.	1854	4 Blkds.	2Cy.28" & 41"–27" 75 HP. C. D. Holmes, Hull	T. Wilson, Sons & Co.
23 89631	✻	Falcon *Moore 85-87*	I ScTug 1 *Dk*	British Melbourne	25 154	96·9	19·3	11·2	Leith S. & H. Morton & Co.	1884	3 Blkds.	2Cy.20½" & 40"–24" 90 HP. A. Shanks & Son, Arbroath	Huddart, Parker & Co.
24	✻	Falcon	W Sc 1 *Dk*	Spanish S'Sebastian	29 91	87·0	18·4	10·1	Leith A. G. Gifford & Co.	1882		2Cy.14" & 28"–20" 35 HP. J. Cran & Co., Leith	Mercader & Sa
25 84092		Falernian *Lawrenson 82-86*	I ScSr 2 *Dks*	British Liverpool	1453 2208	343·4	34·5	23·5	Southampton Oswald, Mordaunt & Co.	1880	5 Blkds.	1Cy.24" & 50"–48" 250 HP. Oswald, Mordaunt & Co., Son.	S.S. "Falernian" Co. Ld. (F. Leyland & Co.)
26		Falke *Schütze 84-87*	I ScSr 1 *Dk1 B*	German Bremen	570 728	201·4	28·0	17·1	Bremen J. H. Wätjen & Co.	1866		2Cy.30" & 56"–36" 213 HP. Humphrys & Pearson, Hull	Norddeutscher Lloyd
27		Falken (ex Eskilstuna) *Andersson 74-76*	I ScSr 1 *Dk*	Swedish Halmstad	167 230	131·2	20·1	8·9	Stockholm Bergsund Mek. Verks.	18[illegible]		2Cy.19" & 38½"–20" 60 HP. Kockums Mek. Verk., Malmo	Hallands Ang. Bolag.
28	D G V	Falkenburg *Weber 87-88*	I ScSr 1 *Dk & Spr Dk3 B*	German Bremen	988 1565	230·9	34·4	17·4	Elbing F. Schichau	1884		3Cy.20½", 32½" & 50"–33" F. Schichau, Elbing [175 HP.	Deutsche Dampfschiff-fahrts-Gesellschaft Hansa
29 86753	✻	Falloden *Gibb -87*	I ScSr 1 *Dk2 B*	British Newcastle	1085 1685	260·0	36·2	18·1	North Shields T. & W. Smith	1883	3 Blkds.	2Cy.32" & 62"–40" 200 HP. R. & W. Hawthorn, Newc'stle	J. Knott
30													

✻ = Lloyd's Register; D = Nederlandsche Vereeniging; G = Germanischer Lloyd; I = Registro Italiano; R = American Record; A = Veritas Austro-Ungarico; H = Veritas Ellenico; N = Norske Veritas; V = Bureau Veritas.

In the 2nd Column, B.T., with figures following, signifies Date of report of bottom sighting by Board of Trade Surveyors as condition of PASSENGER CERTIFICATE.

In the 3rd Column, C = Composite (iron frame, wood planked); I = Iron; Stl = Steel; W = Wood.

Further particulars of Vessels classed by Lloyd's Register will be found in Lloyd's Register Book and Lloyd's Yacht Register.

Class.	Nom, Matériaux, Gréement, etc. Capitaine et dates d'entrée au service des Armateurs actuels et à bord du navire. Ancien Nom, nombre de ponts, etc.	Pavillon. Port d'Armement	Tonnage. Net. Total	Dimensions. Longueur	Largeur	Creux	Construit. Où. Par qui.	Quand.	Nombre des cloisons. Détails sur le lest d'eau.	Machines. Détails. Nom du Constructeur.	Armateurs ou Directeurs-Gérants.
✻	Falls of Inversnaid Stl ScSr Hardistone 1 Dk & Wd k Frms	British Glasgow	1758 2669	312·6	40·2	19·5	Greenock Russell & Co.	1888	5 Blhds. DB.... FPT... APT271	4Cy.18"26"36&52"–39" 280 IP Ross & Blackmore, Greenock	S.S. "Falls of Inversnaid" Co. Ld. (Wright & Breakenridge)
	Falmouth W Pad	American Eastport, Me	850 1156	232·1	36·4	14·2	New York	1872		1000 IP.	G. Hayes
✻	Falshaw Stl ScSr Bennett 88-88 1 Dk 2 B	British Whitby	1580 2432	288·0	38·3	21·0	Whitby T. Turnbull & Son	1888	6 Blhds. DB...	3C.22"36"&59"–39" 175 IP Blair & Co. Ld. Stockton	H. Baxter & Co.
	Falster 1 S S ... 1 Dk	Danish ...	93 139	137·1	18·6	9·0	Copenhagen Burmeister & Wain	1866		1Cy.14"& 28"–18" 50 IP. Burmeister & Wain, Cp'hagn	Det Forenede Dmpskbs Selskab. (C.P.A. Koch)
	Fame I TwnScSr ... 1 Dk	British ...	140 226	133·2	22·3	11·8	South Shields T.D. Marshall	1857		1Cy.15"&30"–18" 74 IP. Hong Kong & Whampoa Dock Co. Hong Kong	Hong Kong & Whampoa Dock Co., Ld.
✻ I	Fanfulla (ex Clan ... Tower) Stl ScSr Carbone ·87 2 Dk & B	Italian Genoa	1772 2745	300·0	40·2	24·5	S. Shields J. Readhead & Co.	1887	5 Blhds. DB...	3C.22"36"& 60"–39" 180 IP J. Readhead & Co., S. Shields	Schiaffino & Solari
	Fanita I S S Nelson	American Phil'd'lphia	289 432	161·0	26·0	8·8	Wilmingtn Del The Pusey & Jones Co.	1868		1Cy.28"–28" The Pusey & Jones Co. Wlm. Del.	W.P. Clyde & Co.
✻ D	Fannie I ScSr Nielsen 82-83 1 Dk 2 B	British Hartlepool	1132 1754	256·5	35·6	18·9	W. Hartlepool W. Gray & Co.	1882	4 Blhds. DB... f 316	2Cy.32"& 60"–39" 150 IP. Blair & Co., Ld. Stockton	C. Nielsen & Son
	Fannie I PadSr	British South'mpt'n	278 616	231·5	26·2	13·3	Greenock Caird & Co.	1859		2Cy.60"–66" 250 IP. Caird & Co., Greenock	Lon. & S.W. Rlwy. Co.
✻	Fannie Bertha I ScSr Bolls 80-83 1 Dk	British Newcastle	479 741	189·6	30·1	13·7	Blyth Hodgson & Soulsby	1882	4 Blhds. DB...	2Cy.24"& 48"–33" 95 IP. Hawks, Crawshay & Ss., Nwe.	Fisher, Renwick & Co.
	Fannie Woodall W Sc 1 Dk	American Baltimore	147 167	112·0	21·5	7·0	Wilmington Enoch Moore	1878		1Cy.18"–18" 280 IP. Pusey, Jones & Co., Wlmngtn.	Andrew Woodall
✻	Fanny I ScSr Holmes 1 Dk	British London	192 308	151·0	21·6	11·8	Glasgow A. & J. Inglis	1878	4 Blhds. FPT30 APT30	2Cy.18"& 34"–24" 50 IP. A. & J. Inglis, Glasgow	J. Burnett & Sons
V	Fanny I ScSr Schultz 83-83 1 Dk	Swedish Gothenburg	350 470	149·9	23·8	15·1	Gothenburg Gothenburg Mek. Verkst.	1883		2Cy.19"& 31"–20" 60 IP. Gothenburg Mek. Verkst.	Ångf. Aktie-Bolag. Marine (T. Ahrenberg.)
	Fanny & Jenny I PadSr (ex Scotia, ex General Banks) 2 Dks	British Nassau, N.P.	297 518	202·2	28·4	13·7		1847		2Cy. 150 IP.	Augustus J. Adderley
	Fanrik Stål I PadSp Wellholm 83 1 Dk	Swedish Sollefteå	75 110	118·7	17·1	7·7	Stockholm Bergsunds Mek. Vks	1866		2C.29¾"&29¼"–29¾" 50 IP. Bergsunds M.V., Stockholm	L. Sundelins
	Fantee Stl TwnScSr Highton 1 Dk	British Liverpool	98 167	120·2	18·0	8·3	Glasgow A. Stephen & Sons	1879	4 Blhds.	1Cy.13"&20"–21" 50 IP. A. Stephen & Sons, Glasgow	"Ocean" S.S. Co.
	Far East (ex Augusta) I Sc		113 163	99·7	20·0	7·5	Port Glasgow Blackwd. & Gordon	1870		2Cy.14"–20" 30 IP. Blackwood & Gordon, P.Glsg.	
✻	Faraday I T-Sc3mSr Le Fanu 82-82 2 Dks 4 B	British London	3123 4917	360·4	52·3	34·7	Newcastle C. Mitchell & Co.	1874	3 Blhds. DB...	4Cy.39"& 68"–48" 500 IP. T. Clark & Co., Newcastle	Siemens Bros. & Co., Ld.
	Faraday I ScSr Proud 85-85 1 Dk 2 B	British Newcastle	556 830	214·5	28·5	15·2	North Shields T. & W. Smith	1873	4 Blhds. DB...	2Cy.26"& 50"–33" 90 IP. Ouseburn Eng. Wrks., Nwcst.	G. Reid & Co.
	Fardig I ScSp Flinkholdt 1 Dk	Swedish Gothenburg	53 108	86·3	18·4	10·9	Motala	1871		50 IP.	Ångf. Aktie-Bolag. "Herkules" (J.E. Wennerlund)
N	Farmanden I ScSr Sneby 74-74 1 Dk	Norweg'n Drammen	80 112	101·5	20·6	7·6	Motala Motala Co.	1865		1Cy.11½"–25" 20 IP. Motala Co., Motala	Drammens Kystdampskibs-Selskab.
✻ D	Farnley Hall I ScSr Smith 1 Dk	British W. Hartlep'l	606 946	213·0	29·2	16·1	W. Hartlepool R. Irvine & Co.	1871	4 Blhds. DB...	2Cy.27"& 50"–33" 98 IP. T. Richardson & S'ns, Hrtlp'l	R. Irvine & Co.
I	Faro (ex Head Quarters) I Sc3mSr Pipilo 88-88 1 Dk 2 B	Italian Palermo	618 963	210·2	29·2	16·6	Glasgow Aitken & Mansel	1868	4 Blhds. FPT...	2Cy.26½"& 46"–36" 174 IP. James Aitken & Co., Glasgow	Navigazione Generale Italiana
	Farralhor W ScSr Bongfield ·88 1 Dk	American San Francisco	286 368	158·5	33·8	10·3	San Francisco A. Hay	1888		3C.14"22"&36"–24" 175 IP Fulton I. Wrks. San Francisco	Dolbeer & Carson
	Fashion (WRECKED) W Pad	American Trenton, N.J.	 604	..	..	..	Philadelphia	1847			J. S. Buckalew
✻	Fastnet Stl ScSr Barley ·87 1 Dk & Wd k Frms	British London	1498 2296	290·0	38·0	19·9	Stockton M. Pearse & Co.	1887	6 Blhds. DB...	3C.21½"35½"&58½"–39" 200 IP Blair & Co., Ld., Stockton	Farrar, Groves & Co.
✻	Fastnet Stl ScSr McMenemy 87-87 1 Dk	British Glasgow	155 338	115·4	23·5	11·3	Glasgow T. Wingate & Co.	1878	4 Blhds. DB...	2Cy.20"& 40"–30" 75 IP. T. Wingate & Co., Glasgow	Clyde Shipping Co.
	Fatfield I ScSr Metcalf 88-88 1 Dk 2 B	British London	417 658	185·2	28·7	16·6	Sunderland Oswald & Co.	1865	4 Blhds. DB...	2Cy.36"–26" 80 IP. G. Clark, Sunderland	London & South Wales S.S. Co.
	Fatshan Stl T Sc Goggin 82-87	British Hong Kong	1425 2260	280·0	51·0	10·2	Leith Ramage & Ferguson	1887		3Cy.17"27"&42"–24" 225 IP Ramage & Ferguson, Leith	Hong Kong, Canton & Macao Steamboat Co., Ld. and China Nav. Co., Ld.
	Fauvette I S S Yt 1 Dk	French Cherbourg	97	124·6	19·7	13·1	Havre B. Normand	1869		2C.20"& 31½"–23½" 50 IP. Nillus & Normand, Havre	E.A. Pérignon

✻ = Lloyd's Register; D = Nederlandsche Vereeniging; G = Germanischer Lloyd; I = Registro Italiano; R = American Record; A = Veritas Austro-Ungarico; H = Veritas Ellenico; N = Norske Veritas; V = Bureau Veritas.

Dans la 2me Colonne, B.T., avec les indications qui suivent, signifie la date du rapport de visite de la Carène par les experts du Board of Trade pour obtenir le certificat relatif au TRANSPORT DES PASSAGERS.

Dans la 3me Colonne, C = Composite (membrure en fer, bordé en bois); I = Fer; Stl = Acier; W = Bois.

Pour plus amples renseignements sur les navires classés par le Lloyd's Register, voir le Lloyd's Register Book et le Lloyd's Yacht Register.

No. in Book / Official No.	Registry in which classed	Ship's Name. Material, Rig, &c. / Master, and Dates of Appointment to present Owner's Service and to Ship / Ship's Late Name. No. of Decks, &c.	Flag. / Port of Registry.	Tons Net / Gross	Length	Breadth	Depth	Built Where / By Whom	When	No. of Bulkheads / Particulars of Water Ballast	Engines / Particulars. Maker's Name.	Owners
61 84136		Favonian 1 ScBg Walker 85-86 2 Dks	British Liverpool	1476 [illegible]	342·5	34·5	23·5 F 31 f.	Southampton Oswald, Mordaunt & Co	1881	[illegible] Bhds	3Cy.24"&39"-48" 250 HP. Oswald, Mordaunt&Co., Son.	S.S. "Favonian" (F. Leyland&Co.
62 79549		Fawkner 1 ScSr 1 Dk	British Melbourne	169 [illegible]	153·6	26·1	12·2	Renfrew [illegible]	1883		2Cy.22"&40"-27" 75 HP. W. Simons&Co., Renfrew	Melbourne Harb. Trust Commis.
63 87106	✻	Fawn 1 ScSr Nutt 88-89 1 Dk2 H	British London	746 1175	230·0	32·2	15·3	Beverley Vulcan I.W.Co., Ld.	1883	1 Bhds. [illegible]	2Cy.27"&50"-36" 99 HP. Vulcan I.W.Co., Ld., Hull	Jackson Bros.
64		Fayoum 1 PadBo	Turkish Alexandria	1070	274·0	34·0	21·6	London [illegible]	18[illegible]		2Cy.57"&97"-72" 300 HP. J. Penn&Sons, Greenwich	Khedivieh Mail Co.
65	✻D	Febrero (ex Cornucopia) 1 Sc3mSr (Nost. name) Alcorta 1 Dk2 H Madariaga 79-89	Spanish Bilbao	871 [illegible]	245·0	34·5	15·6 Q 81 f. B 65 f.	Barrow [illegible]	1881	4 Bhds. [illegible]	2Cy.30"&55"-36" 110 HP. Westray, Copeland&Co., B'rrow	R. de la Lota
66 82677	✻	Fedele Primavesi 1 ScSr Stevens 86-86 2 Dks3 H	British London	1067 [illegible]	250·4	33·8	23·6 P 20 f. B 111 f.	Newcastle Palmers Co. Ld.	1881	4 Bhds. [illegible]	2Cy.31"&55"-36" 150 HP. Palmers Co., Ld., Newcastle	Oriental S.S.Co.
67 91965	✻	Federation Stl Sc Pinkham 79-86 2 Dks3 H	British London	1578 2472	300·0	40·1	25·1 P 37 f. B 86 f.	Sunderland [illegible]	1886	4 Bhds. [illegible]	3C.24"39"&64"-42" 300 HP. J. Richardson&Sons, Htpl.	Angier Bros. (Managers)
68 91307	✻	Federation Stl ScSr Sherrwood 82-86 1 Dk	British [illegible]	327 [illegible]	220·0	30·7	12·4 B 65 f. P 35 f.	Sunderland [illegible]	1886	4 Bhds. [illegible]	3C.19"31"&51"-33" 130 HP. N.E. Marine Eng. Co., Ld., Shl.	Co-operative Whole- sale Society
69	✻	Federico (ex Tropic) 1 Sc3mSr Garteiz 2 Dks3 H	Spanish Bilbao	1423 [illegible]	324·6	35·2	17·9	Liverpool [illegible]	1871	5 Bhds.	2Cy.40"&80"-36" 250 HP. Laird Bros., Birkenhead	Cia. de Naveg. Florh"
70	✻	Fee Cheu Stl ScSr 2 Dks	Chinese	643 1031	220·0	32·1	20·0 F 26 f.	Sunderland [illegible]	1887	1 Bhds.	3Cy.19"31½"&52"-36" 150 HP. W. Doxford&Sons, Sunderld	
71	✻	Fehmarn Stl Sc Heppe 86-88 1 Dk & Web Frms	German Kiel	572 744	195·1	30·0	13·8 B 44 f. F 29 f.	Kiel G. Howaldt	1888	4 Bhds. [illegible]	3Cy.14¼", 23¾"&39"-23⅝" 100 HP. Gebrueder Howaldt, Kiel	Lange Gebr.
72 85110	✻	Fei Lung 1 ScSr Crawford 2 Dks	British London	752 1180	240·0	33·0	15·0 Q 67 f. P 23 f. B 44 f. F 21 f.	Sunderland J. Blumer&Co.	1882	4 Bhds. [illegible]	2Cy.29½"&56"-39" 130 HP. T. Clark&Co., Newcastle	H. J. Such
73 72804	V	Fei Ma (ex Watergeus) 1 ScSr Kleppel 88	British Shanghai	685 [illegible]	231·8	26·8	13·9 Q 82 f. B 68 f. F 35 f. Len'81	Antwerp Soc. Anonyme des Chantiers & c.	1873		2Cy.25"&48"-33" 97 HP. J. Redhead&Co., S. Shields	H. J. Such
74 93455	✻	Felbridge Stl ScSr Rathen 88 1 Dk & Web Frms	British London	1336 [illegible]	280·0	37·5	19·3 P 24 f. Q 90 f. B 112 f. F 32 f.	Newcastle C.S. Swan&Hunter	1888	4 Bhds. [illegible]	3Cy.22",36"&59"-39" 250 HP. [illegible] & K. Co., Newcastle	"Felbridge" S.S. Co. (W. R. Price&Co.
75	✻DG	Felicia 1 ScSr Wonck 76-87 1 Dk2 H	German [illegible]	823 1127	236·1	31·0	17·8 Q 111 f. F 29 f.	Glasgow A. Stephen&Sons	1878	4 Bhds. [illegible]	2Cy.27"&50"-33" 110 HP. A. Stephen&Sons, Glasgow	Sartori & Berger
76		Felix C ScSr Carlon 1 Dk	Swedish Gothenburg	117 [illegible]	98·4	20·9	10·0 B 18 f.	Thorskog P. Larsson	1885		2Cy.8½"&17"-14" 42 HP. P. Larsson, Thorskog	P. Larsson
77		Feliz 1 ScTug Catoni	Arg'ntine Buen's Ayres	70 101	84·0	18·0	6·0	Greenock Kincaid&Co.	1875		22 HP.	José Catoni
78		Felix Melliza W	Spanish [illegible]	100 [illegible]	121·7	18·0	7·7	Donzol	1886		30 HP.	Salas M. Mel.
79		Feliz Naposta 1 Sc Catoni	Arg'ntine Buen's Ayres	127 240	113·0	20·3	6·8	Greenock Kincaid&Co.	1864		27 HP.	José Catoni
80 63645	✻	Fenella 1 Sc3mSr Hutten 6-86 1 Dk2 H	British London	655 1018	246·6	29·1	17·5 Q 88 f. F 31 f.	Newcastle Palmers Co. Ld.	1870	5 Bhds. [illegible]	2Cy.30"&52"-33" 130 HP. Palmers Co., Ld., Newcastle	J. Fenwick & Son
81 76303	BT10.88	Fenella 1 TwinScSr Gibson 2 Dks	British Douglas, I.M.	189 [illegible]	200·0	26·0	13·0 P 85 f. B 49 f. F 30 f.	Barrow Barrow S.B.Co., Ld.	1881	7 Bhds.	4Cy.23"&42"-21" 154 HP. Barrow S.B.Co., Ld., Barrow	Isle of Man S.P. Co.
82		Fenix 1 ScSr Abonuza	Mexican Frontera	128	98·0	20·0	8·6	Camden, N.J. J. H. Dialogue	1880	4 Bhds.	2Cy.15"&26"-20" J. H. Dialogue, Camden, N.J.	Romano Her.
83	R	Fenix 1 Sc3mSr Olsson 86-86 1 Dk	Swedish Christinehamn	177 244	126·9	16·7	9·7 B 33 f.	Gothenburg	1885	4 Bhds. [illegible]	40 HP.	Anglatsaktie- "Ferm" (A. [illegible]
84		Fennia (SUNK) 1 ScSr Favoria 84-76 1 Dk	Russian Nikolaistad	251 406	165·0	23·2	13·5 P 31 f. F 31 f.	Port Glasgow W. Hamilton&Co.	1874	4 Bhds.	2Cy.19"&34"-24" 50 HP. J.&T. Young, Ayr	A. Hedman
85 73783		Fenton 1 Sc3mSr Herring 1 Dk	British London	496 783	200·8	30·9	14·1 Q 94 f. F 28 f.	Sunderland Austin&Hunter	1876	4 Bhds. [illegible]	2Cy.26"&48"-33" 95 HP. G. Clark, Sunderland	Milnes & Cha.
86	G	Ferdinand 1 ScSr Lege 75-83 1 Dk & Hur Dk	German Kiel	427 512	162·0	23·7	17·0	Kiel G. Howaldt	1883	[illegible]	2Cy.17¼"&35⅛"-23⅝" 60 HP. Howaldt Bros., Kiel	Sartori & Berger
87	✻V	Ferdinand de Lesseps 1 ScBk Geffroy 86 (ex Saint Laurent) 3 Dks	French Havre	1978 2863	350·0	38·3	27·8 F 50 f.	Glasgow A.&J. Inglis	1875	5 Bhds.	2Cy.50"&88"-48" 450 HP. A.&J. Inglis, Glasgow	Compagnie Générale Transatlantique
88	✻V	Ferdinand Vandertaelen 1 ScBg Gombeer 82-82 2 Dks3 H	Belgian Antwerp	1237 1818	271·0	34·2	24·5 P 46 f. F 41 f.	South Shields J. Softley&Co.	1874	4 Bhds. [illegible]	2Cy.32"&60"-36" 150 HP. [illegible] Co., S. Shields	J. P. Best & Co.
89 72		Ferdinando Massimiliano 1 PadSr 2 Dks	Austrian Trieste	516 [illegible]	230·0	28·1	21·5	London Mare&Co.	1856		2Cy.55"&61"-48" 300 HP. J. Watt&Co., Birmingham	Lloyd Austriaco
90 83973	✻	Fergusons (ex Calliope Nicolapulo) 1 ScSr Young 1 Dk2 H	British Middlesbro'	973 1501	250·0	34·5	18·8 Q 83 f. P 25 f. B 82 f. F 28 f.	Sunderland Bertram, Haswell&Co.	1880	4 Bhds. [illegible]	2Cy.30"&57"-36" 140 HP. J. Dickinson, Sunderland	J. M. Lennard

✻ = Lloyd's Register; D = Nederlandsche Vereeniging; G = Germanischer Lloyd; I = Registro Italiano; R = American [illegible]
A = Veritas Austro-Ungarico; H = Veritas Ellenico; N = Norske Veritas; V = Bureau Ve[ritas]

In the 2nd Column, B.T., with figures following, signifies Date of report of bottom sighting by Board of Trade Surveyors as condition of PASSENGER CERTI[FICATE]

In the 3rd Column, C = Composite (iron frame, wood planked); I = Iron; Stl = Steel; W = Wood.

Further particulars of Vessels classed by Lloyd's Register will be found in Lloyd's Register Book and Lloyd's Yacht Register.

Registres	Nom, Matériaux, Gréement, etc. Capitaine et dates d'entrée au service des Armateurs actuels et à bord du navire. Ancien Nom, nombre de ponts, etc.		Pavillon. Port l'Armement	Tonnage. Net. Total	Dimensions Longueur	Largeur	Creux	Construit. Où. Par qui.	Quand	Nombre des cloisons. Détails sur le lest d'eau.	Machines. Détails. Nom du Constructeur.	Armateurs ou Directeurs-Gérants.
G	Ferm Andersson 83-84	C ScSmSr 1 Dk	Swedish Christinehamn	213 268	130·2	21·4 B 21 f.	12·1	Thorskog P. Larsson	1882		2C. 11¼" & 18½"–15½" 40 HP.	Angbatsaktiebolaget "Ferm" (A. Erostrom)
	Fern (ex Gaisl)	I ScSr 1 Dk	British Glasgow	137 260	147·4 P 10 f.	21·2	11·0 F 12 f.	Port Glasgow McCulloch, Paters'n & Co.	1871		2Cy. 19" & 38"–28" 60 HP. Rankin & Blackm're, Gr'n'ck	Glasgow Dublin & Londonderry S.P.Co.
✱ D	Fern Holme (STRANDED) Ritchie 71-84	I ScSr 2 Dks 3 B	British Maryport	1715 2610	310·6 P 40 f.	40·2 B 72 f.	24·5 F 28 f.	Sunderland J. L. Thompson & Sons	1883	6 Blds. DB[illegible] APT[illegible]	2Cy. 38" & 72"–48" 325 HP. J. Dickinson, Sunderland	Hine Bros.
	Fernandez Sanz	I ScSr 1 Dk	Spanish Santander	109 172	133·1	19·1	9·5	Port Glasgow Blackwd. & Gordon	1875	4 Blds.	2Cy. 14" & 25"–18" 30 HP. Blackwood & Gordon, Pt. Gls.	A. F. Sanz & Co.
G	Fernanda (ex Aron) Brandt	I ScSr 1 Dk	Norweg'n Haugesund	159 234	101·0	22·4 B 31 f.	12·5 H 3 f.	Thorskog P. Larsson	1882	B 31 ft	2C. 12½" & 25½"–22½" 35 HP. P. Larsson, Thorskog	Brödrene Jacobsen
✱	Fernbrook James 79-84	I ScSr 1 Dk 2 B	British Whitby	1099 1702	257·5 Q 9 f.	35·5 B 67 f.	19·0 F 32 f.	W. Hartlepool W. Gray & Co.	1883	4 Blds. DB[illegible] FPT[illegible] APT[illegible]	2Cy. 32" & 60"–39" 150 HP. Blair & Co., Ld., Stockton	J. Foster
✱ D	Ferncliffe Pawley	I ScSr 1 Dk 2 B	British Newport	959 1484	251·3	37·8 B 52 f.	17·2 F 33 f.	Newcastle Palmers' Co., Ld.	1883	4 Blds. DB[illegible] APT[illegible]	2Cy. 31" & 60"–39" 160 HP. Palmers' Co., Ld., Newcastle	"Ferncliffe" S.S.Co., Ld. (D. H. Morgan & Co.)
✱ D	Ferndale Lambert 83-86	I ScSr 1 Dk 2 B	British Sunderland	604 937	212·2 Q 70 f.	30·5 B 31 f.	16·0 F 24 f.	Sunderland J. Laing	1877	4 Blds. DB[illegible]	2Cy. 27" & 50"–33" 95 HP. J. Dickinson, Sunderland	P. H. Laing
✱ D	Fernlands Sellers 84-86	Stl Sc 1 Dk 2 B	British W. Hartlep'l	1355 2087	276·1 Q 76 f. P 29 f.	37·2 B 108 f.	19·9 F 27 f.	W. Hartlepool W. Gray & Co.	1885	4 Blds. DB[illegible]	3C. 21" 35" & 57"–39" 170 HP. Central Mar. Eng. Co., Hrtlpl.	Hardy, Wilson & Co.
	Fernmount	W ScSr 1 Dk	British Sydney	183 269	134·0	25·6 H 20 f.	8·4	Scot's Creek, N.S.W. W. M'Culloch	1884		1Cy. 9½" & 16½"–12" 40 HP. Perdrian & West, Blmn NSW	John See
DV	Fero (WRECKED) Sørensen -84	I ScSr	German Flensburg	754 1025	213·5	29·9	15·0 22·1	Elsinor Helsingörs, B. Co.	1883	WB	2Cy. 24" & 45½"–30" 90 HP. Helsingörs, B. Co., Elsinor	H. Schuldt
✱	Ferrando Betts 82-88	Stl ScSr 1 Dk 2 B	British Newcastle	1333 2061	285·5 P 29 f. F 72 f.	39·2 Q 76 f.	18·9 F 35 f.	Newcastle Palmers Co., Ld.	1888	5 Blds. [illegible] APT[illegible]	3Cy. 22" 35" & 58"–42" 200 HP. Palmers Co., Ld., Newcastle	Scott Bros.
✱ D	Ferrara Parker 73-81	Stl ScSmSr 2 Dks	British Leith	735 1162	248·3 P 21 f.	33·2	17·0 F 29 f.	Greenock R. Steele & Co.	1880	6 Blds. MT[illegible]	2Cy. 30" & 52"–36" 120 HP. R. Steele & Co., Greenock	J. Currie & Co.
	Ferret	I ScSr 1 Dk	British Adelaide	246 445	170·9	23·2	12·7	Glasgow J. & G. Thomson	1871	4 Blds.	2Cy. 23" & 40"–33" 90 HP. J. & G. Thomson, Glasgow	Adelaide S.S.Co., Ld.
✱	Ferric Holmes 83-84	I ScSr 1 Dk	British Belfast	133 307	154·0 Q 21 f.	22·6 B 33 f.	11·5 F 28 f.	Belfast M'Ilwaine, Lewis & Co.	1883	5 Blds. FPT[illegible] APT[illegible]	2Cy. 19" & 36"–30" 60 HP. M'Ilwaine, Lewis & C., Belfst.	H. J. Scott & Co.
	Fez Forward	I ScSr 2 Dks	British Liverpool	560 860	217·7	27·6 F 33 f.	16·4	Glasgow Aitken & Mansel	1878		2Cy. 27" & 47"–36" 99 HP. J. & J. Thomson, Glasgow	Mersey S.S.Co., Ld.
✱	Fiado Jones	I ScSr 1 Dk 2 B	British Sydney	640 985	210·3 B 40 f.	30·1 Q 82 f.	16·4 F 24 f.	Sunderland R. Thompson, jun.	1878	4 Blds. DB[illegible] FPT[illegible] APT[illegible]	2Cy. 28" & 51"–33" 110 HP. J. Dickinson, Sunderland	W. H. Smith & Sons, Ld.
V	Fidelio Brock	I ScSr 1 Dk & Spr Dk 4 B	German Hamburg	853 1160	221·1	32·4	15·4 21·0	Hamburg Blohm & Voss	1884		2Cy. 26" & 46"–33" 115 HP. Blohm & Voss, Hamburg	Dampfschiff. Gesellschaft "Swatow"
✱	Fides Petersen 87-87	I ScSr 1 Dk	German Flensburg	262 341	139·4	23·0 B 26 f.	11·4	Kiel Gebrüder Howaldt	1883	4 Blds. DB[illegible] MT[illegible]	2Cy. 15¼" & 27½"–15¾" 40 HP. Gebrüder Howaldt, Kiel	J. L. Lassen
	Fido Marshall	I ScSr 1 Dk 2 B	British Hull	610 954	217·8 F 34 f.	29·0 B 50 f.	16·8	Hull C. & W. Earle	1867		2Cy. 24" & 48"–30" 95 HP. C. & W. Earle, Hull	T. Wilson, Sons & Co.
DV	Fiducia Siewert 88-88	I ScSr 1 Dk	German Flensburg	337 528	167·9 Q 50 f.	24·7 B 12 f.	11·0 F 21 f.	Flensburg Flensbrg. Schiffsb. Ges.	1880	WB[illegible] MT[illegible]	2Cy. 20" & 34½"–24" 50 HP. Flnsbg. Schiffsb. Ges. Flnsbg	J. Jost
✱	Fieramosca Milella 77-78 (ex Sophie Johnson)	I ScSr 1 Dk	Italian Bari	227 600	170·0 Q 54 f.	27·1 B 44 f.	13·8 F 27 f.	Sunderland Davison & Stokoe	1873	4 Blds. W[illegible]	2Cy. 22" & 42"–33" 71 HP. [illegible] Eng. Wrks., N'westl.	Soc. Nav. Puglia
	Fiery Cross Black	W PadTug	British S. Shields	 105	93·7	19·1	9·8	North Shields Stobbs	1873		1Cy. 35"–50" 45 HP. T. R. Scott & Co., N. Shields	C. Dyble
✱	Fifeshire Miller 84-87	Stl ScSg 2 Dks & W'th Frms	British Glasgow	2425 3720	345·0 P 206 f.	47·6	24·1 F 48 f.	Newcastle C. S. Swan & Hunter	1887	6 Blds. CDB[illegible]	3Cy. 27" 41" & 71"–48" 400 HP. Blair & Co., Ld., Stockton	Elderslie S.S.Co., Ld. (Turnbull, Martin & Co.)
	Fifeshire Landsman 85-85	I PadSk	British Dundee	153 243	134·4	30·1	6·6	Glasgow R. Napier & Sons	1858		2Cy. 36"–42" 80 HP. R. Napier, Glasgow	Trustees of the Harbour of Dundee
✱	Fijian Mowatt	Stl ScSr 1 Dk 2 B	British London	633 988	215·0 Q 74 f.	30·2 B 57 f.	15·2 F 23 f.	Newcastle Palmers' Co., Ld.	1886	4 Blds. DB[illegible] FPT[illegible] APT[illegible]	3C. 18½" 31" & 49"–36" 130 HP. Palmers' Co., Ld., Newcastle	G. W. Nicoll & Co.
✱	Filey Laurie	I ScSr 1 Dk 2 B	British Newcastle	773 1221	230·4 H 11 f.	32·2 P 140 f.	17·6 F 37 f.	Newcastle Schlesinger, Davis & Co.	1870	4 Blds. DB[illegible]	2Cy. 29" & 55"–30" 120 HP. Th'mps'n, Boyd & Co., N'westl	H. E. Moss & Co.
	Filgate	I Pad	British Montreal	154 263	158·0	25·4	7·8	Montreal W. White	1879		2Cy. 32"–36" 100 HP. Bartley & Co., Montreal	S. Filgate
N	Fin Gundersen	I ScSr 1 Dk	Norweg'n Skien	150 210	115·7 P 74 f.	20·8	7·2 F 8 f.	Gothenburg	1862	4 Blds.	2C. 14¼" & 28¾"–24⅛" 35 HP. Verkst. Nylands, Christiania	H. Gundersen
RV	Finance Baker -88	I ScBg 1 Dk & Shd Dk 3 B	American New York	1920 2603	300·0	38·4	23·6	Chester, Pa. J. Roach & Son	1883	4 Blds.	2Cy. 36" & 66"–54" J. Roach & Son, Chester, Pa.	United States & Brazil Mail S.S. Co.

[illegible]=Lloyd's Register; D=Nederlandsche Vereeniging; G=Germanischer Lloyd; I=Registro Italiano; R=American Record; [illegible]=Veritas Austro-Ungarico; H=Veritas Ellenico; N=Norske Veritas; V=Bureau Veritas.

Dans la 2me Colonne, B.T., avec les indications qui suivent, signifie la date du rapport de visite de la Carène par les experts du Board of Trade pour obtenir le certificat relatif au TRANSPORT DES PASSAGERS.

Dans la 3me Colonne, C=Composite (membrure en fer, bordé en bois); I=Fer; Stl=Acier; W=Bois.

Pour plus amples renseignements sur les navires classés par le Lloyd's Register, voir le Lloyd's Register Book et le Lloyd's Yacht Register.

1 No. in Book. / Official No.	2 Registry	3 Ship's Name. Material, Rig, &c. / Master, and Dates of Appointment to present Owner's Service and to Ship / Ship's Late Name. No. of Decks, &c.	4 Flag. / Port of Registry.	5 Tons. Net. / Gross.	6 Dimensions. Length	Breadth	Depth	7 Built. Where. / By Whom.	8 When.	9 No. of Bulkheads / Particulars of Water Ballast.	10 Engines. Particulars. / Maker's Name.	11 Owners
121 92329		Finchale I ScSr Meltenton 83-84 1 Dk2 B	British Sunderland	549 714	196·0 Q 22 f. F 17 f.	29·0	16·7	Sunderland J.Laing	1869	4 Blds. [illegible]	1Cy.19"&38"-28" 90 HP N.E.MarineE Co.,Ld.,Sndld.	H.T.Morton[illegible]
122 44960		Fingal I Sc3mSr Hardwick 83-88 1 Dk2 B	British Hull	598 948	237·4 P 225 f.	28·1	16·0	Leith S.&H.Morton&Co.	1862	5 Blds.	2Cy.28"& 50"-30" 95 HP Cmp. '69 Tennant&Co.,Leith	F.Hagerup [illegible]
123 63844		Fingal (ex Lancashire Witch) I ScSrYt Macdonald -80	British Glasgow	66 103	110·5	17·9	8·9	Birkenhead LairdBros.	1872		2Cy.18"& 32"-16" 40 HP LairdBros.,Birkenhead	T.H.T.Potts
124 76335	BT5.86	Fingal I ScSk Campbell 83-85	British Glasgow	54 124	115·3 Q 27 f.	18·1	7·3	Glasgow A.&J.Inglis	1877		2Cy.15"&26"-20" 30 HP Muir&Caldwell,Glasgow	D.MacBrayne
125	✳	Finistère (ex Scots) I ScSr Haupais 81-82	French Havre	252 453	179·5 P 35 f. F 17 f.	24·1	14·3	Dundee W.B.Thompson	1875	4 Blds.	2Cy.27"& 50"-33" 99 HP W.B.Thompson,Dundee	Cie.Anon.des [illegible] bots du Finis[illegible]
126 88108	✳D	Finland I ScSr Eshelby 86-86 1 Dk2 B	British Hull	1242 1904	271·1 P 169 f. F 29 f.	35·1	18·6	Hull Earles'Co.,Ld.	1884	5 Blds. [illegible]	3C.20"32"&56"-36" 152 HP Earles'Co.,Ld.,Hull	G.R.Sanderson
127		Finland I ScSr Fogelholm -81 2 Dks	Russian Åbo	357	175·0 P 125 f.	26·0	12·8	Åbo W.Crichton&Co.	1876	3 Blds.	2Cy.20"&40"-24" 120 HP W.Crichton&Co.,Åbo.	Åbo Läns Ka[illegible] Aktiebolag
128	G	Finland I ScSr Strömberg 86-86 3 Dks	Russian Nikolaistad	208 284	147·6 P 24 f.	20·2	12·0	Hull Brownlw.,Lumsden&Co.	1855		3Cy.16"&28½"-23½" 60 HP LübeckerMasch.Akt.-Ges.	Alfred Hedm[illegible]
129		Finmarken I Sc3m 2 Dks	Norweg'n Bergen	428 565	171·6	24·4	19·9	Hull Sampson&Co.	1864	4 Blds.	2Cy.25"& 48"-24" 80 HP HumberIronWorks	Det Bergensk[illegible] Kibsselskab
130 82795	✳	Finsbury I ScSr Graystone 2 Dks3 B	British London	1230 1909	270·0 P 30 f. B 50 f. F 36 f.	35·1	24·5	Newcastle Swan&Hunter	1880	5 Blds. [illegible]	2Cy.32"&64"-42" 200 HP T.Clark&Co.,Newcastle	Watts,Ward[illegible]
131 89289	✳	Fiona Stl TwnScBn Curphey 1 Dk2 B	British Sydney	414 617	211·1 B 64 f. F 40 f.	30·2	15·4	PortGlasgow J.Reid&Co.	1883	5 Blds. [illegible]	4Cy.24"&46"-30" 178 HP Rankin&Blackmore,Greenock	Colonial Sugar ing Co.
132 87420	✳	Fiona W Sc3mSrYt	British Portsmouth	164 241	147·2	22·4	13·9	Gosport Camper&Nicholson	1882	[illegible]	2Cy.20"&40"-24" 65 HP Day,Summers&Co.,Shmptn.	The Governm[illegible] Newfoundland
133	✳I	Firdene I ScSr Reeves 1 Dk2 B	British London	921 1415	250·0 Q 28 f. P 28 f. B 58 f. F 26 f.	35·3	16·4	Newcastle Wghm.Richardson&Co.	1883	4 Blds. [illegible]	2Cy.30"&58"-36" 98 HP Wghm.Richardson&C.,Nwc.	T.Stephens[illegible]
134 [illegible]		Fire Fay Stl ScSrYt	British Glasgow	113 246	178·0	22·0	11·5	Renfrew Lobnitz&Co.	1886	8 Blds.	3C.15"21"&42"-24" 99 HP Lobnitz&Co.,Renfrew	James Stevens
135		Fire Fly I Pad (ex Tristam Shandy, ex Boxer)	American Philadelphia.	636	217·7	23·0	17·5	Greenock	1863		100 HP	J.Middleton
136 63785	✳	Fire King I Sc3mSr Campbell 78-78 1 Dk2 B	British Glasgow	243 438	170·3 Q 33 f. F 11 f.	24·3	14·1	Renfrew Hndrsn.Coulborn&Co.	1870	4 Blds.	2Cy.24"&50"-33" 90 HP Hndrsn.Coulborn&Co.,Rnf.	H.Lamont&[illegible]
137 15052		Fire King I PadSp	British Scarboro'	34 121	105·6	18·8	10·4	SouthShields T.D.Marshall	1856		2Cy.30"-32" 60 HP T.D.Marshall,SouthShields	W.McBean
138 49805	✳	Fire Queen I ScSr Whiteman 88-88 1 Dk& Spr Dk3 B	British W.Hartlepl.	914 1208	249·5 22·0	29·6	14·1	Hartlepool Pile,Spence&Co.	1864	4 Blds.	2Cy.25"&50"-36" 101 HP J.Jones&Co.,Liverpool	C.Furness
139 80409	✳	Fire Queen I ScSr Beaton 88-88 1 Dk2 B	British Glasgow	297 474	179·7 P 27 f. F 29 f.	25·0	14·2	Greenock J.E.Scott	1879	3 Blds.	2Cy.28"&48"-33" 110 HP J.Howden&Co.,Glasgow	H.Lamont&[illegible]
140		Fire Queen (ex Carlas) I ScSrYt 1 Dk	British Leith	135 268	157·6	23·6	12·6	Leith Ramage&Ferguson	1881	3 Blds.	2Cy.24½"& 44"-30" 95 HP WalkerHenderson&Co.,Gls.	Admiralty Te[illegible] HMS "Duke of [illegible]
141 17590		Firefly W ScSrYt (Auxiliary)	British Cowes	129 164	107·9	21·3	11·3	Cowes J.S.White	1857		2Cy.14"&25"-14½" 27 HP	T.R.B.L.M.G[illegible] wright
142 93744	BT3.88	Firefly I T ScSp Ashworth -87 1 Dk	British Liverpool	44 117	133·5	18·2	8·5	Seacombe J.P.Waddington&Co.	1887	4 Blds.	1Cy.12½"& 25"-18" 57 HP D.Rollo&Sons,Liverpool	R.A.Macfie
143 45		Firenze I ScBa Cutrono 84-86 2 Dks	Italian Palermo	225 468	198·4 18·3	24·5	11·9	LaSeyne Forges&C.dela Med it.	1862		1Cy.14"&34"-24" 140 HP Forges&Chantiers,LaSeyne	Navigazione [illegible] Italiana
144 93402	✳	Fishtoft I ScTrawler Jordan 88-88 1 Dk	British Boston	52 105	85·0 F 17 f.	20·0	9·9	Hull Earle'sCo.,Ld.	1886	3 Blds.	2Cy.12"& 22"-20" 25 HP Earle'sCo.,Ld.,Hull	Boston Deep [illegible] ing & Ice Co.[illegible]
145	N	Fiskeren I ScSr Nielsen -85 1 Dk& Awn Dk3 B	Norweg'n Bergen	389 499	153·7 18·3	25·3	11·9	Stavanger StavangerStöberi&Dok	1884	4 Blds. [illegible]	2Cy.21"&40"-26" 60 HP N.E.MarineE.Co.,Ld.,N'c'stle	Vesteraalensk[illegible] skibsselskab
146 71663	✳	Fitzclarence I Sc3mSr Tunie 88-88 1 Dk2 B	British Glasgow	584 918	231·5 Q 127 f. F 27 f.	28·1	16·1	Sunderland T.R.Oswald	1874	4 Blds. [illegible]	2Cy.27½"&51½"-33" 90 HP T.R.Oswald,Sunderland	Burrell & So[illegible]
147 66048	✳	Fitzgerald I ScSr Ruggier 87-87 1 Dk	British Glasgow	248 398	165·6 Q 84 f. P 42 f.	23·1	11·9	Dumbarton J.&R.Swan	1873	4 Blds.	2Cy.22"&88"-27" 65 HP W.King&Co.,Glasgow	C.P.Ascink
148 83733		Fitzpatrick I Sc3mSr Skinner 2 Dks& Awn Dk	British Penang	597 912	225·7 P 111 f. B 46 f. F 17 f.	29·2	17·3	PortGlasgow Blackwood&Gordn	1869	7 Blds.	2Cy.27½"&47½"-36" 120 HP Blackwood&Gordon,Pt.Gls.	Lee Pee Chu[illegible]
149 [illegible]	✳D	Fitzroy I Sc3mSr Studholm 87-87 2 Dks	British N.Shields	1065 1628	260·7 P 47 f. B 24 f. F 31 f.	33·2	23·6	Sunderland Bowman,Haswold&Co.	1877	3 Blds. [illegible]	2Cy.32"&60"-36" 150 HP J.Dickinson,Sunderland	Wilkie & Tu[illegible]
150 83641		Fitzroy Stl ScSr Reaburn 2 Dks	British Sydney	504 870	209·5 B 40 f. F 25 f.	28·9	18·6	Kirkcaldy J.Key&Sons	1883	5 Blds. [illegible]	2Cy.30"&60"-39" 180 HP J.Key&Sons,Kirkcaldy	Australasian [illegible] Steam Nav.Co.

✳ = Lloyd's Register; D = Nederlandsche Vereeniging; G = Germanischer Lloyd; I = Registro Italiano; R = American [illegible]
A = Veritas Austro-Ungarico; H = Veritas Ellenico; N = Norske Veritas; V = Bureau Ve[illegible]

In the 2nd Column, B.T., with figures following, signifies Date of report of bottom sighting by Board of Trade Surveyors as condition of PASSENGER CERTI[illegible]

In the 3rd Column, C = Composite (iron frame, wood planked); I = Iron; Stl = Steel; W = Wood.

Further particulars of Vessels classed by Lloyd's Register will be found in Lloyd's Register Book and Lloyd's Yacht Register.

1 Nº	2 Registre	3 Nom, Matériaux, Gréement, etc. — Capitaine et dates d'entrée au service des Armateurs actuels et à bord du navire. — Ancien Nom, nombre de ponts, etc.		4 Pavillon. — Port l'Armement	5 Tonnage Net — Total	6 Dimensions. Longueur	Largeur	Creux	7 Construit. Où. — Par qui.	8 Quand.	9 Nombre des cloisons. — Détails sur le lest d'eau.	10 Machines. — Détails. Nom du Constructeur.	11 Armateurs ou Directeurs-Gérants.
151	*	Fitzwilliam *Paxman 89-89*	I Sc3mSr 1 *Dk*	British Middlesbro'	189 307	160·6 *Q 59 f.*	19·7	11·8	Glasgow J.&R.Swan	1866	3 Bhds.	2Cy.18"&32"–21" 45 IP. Kincaid,Donald&Co.,Grnk.	J.E.Swan & Brs.,Ld.
152		Fiume *Mrak 73-87*	I ScSr 1 *Dk*	Austrian Trieste	420 684	211·2	29·2	19·2	Newcastle Miller,Prav&Co.	1853		2Cy.20"&34"–30" 160 IP. Miller,Prav&Co.,Newcastle	Lloyd Austro-Ungarico
153		Fjalir	I ScSr 1 *Dk*	Norweg'n Bergen	176 243	131·7 *Q 44 f. F 38 f.*	20·9	11·1	Dumbarton A.Denny	1858	3 Bhds.	2Cy.17½"&35"–24½" 45 IP. Cmp.'74,NylandsVerks,Cia	Nordre Bergenhuus Amtskommune
154	*	Flachat (ex Akaba) *Bechoiffe -88*	I Sc3mSr 2 *Dks3 B*	French Havre	1495 2277	300·0 *P 28 f. F 36 f.*	26·0	25·5	Stockton M.Pearse&Co.	1880	1 Bhds. DBa& f.338t	2Cy.36½"&68"–42" 250 IP. Blair&Co.,Ld.,Stockton	Compagnie Générale Transatlantique
155	*	Flamboro *Russell 82-84*	Stl ScSr 1 *Dk & Spr Dk 3 B*	British Newcastle	1259 1933	265·0	39·9	22·4	Newcastle Palmers'Co.,Ld.	1885	4 Bhds. CDB138t APT32t	3Cy.22"35"&58"–42" 200 IP. Palmers'Co.,Ld.,Newcastle	"Flamboro" S.S.Co. (J.E.Bowser,Ormston & Co.)
156		Flamborough *Fraser 82-84*	I ScSr 1 *Dk 2 B*	British Bermuda	639 993	198·7 *B 32 t.*	28·4	23·6	Sunderland Oswald&Co.	1867	4 Bhds. DBa MT76t APT32t	2Cy.28"&54"–33" 130 IP. N.E.MarineE.Co.,Sundrlnd.	Quebec S.S.Co.
157	*	Flambro *Reinold 83-87*	Stl ScSr 1 *Dk & Web Frms*	British W.Hart'pool	1449 2224	284·2 *P 27 f. B 104 f. F 27 f. Q 90 f.*	37·8	19·5	W.Hartlepool W.Gray&Co.	1887	5 Bhds. CDB498t	3C.22"35"&59"–39" 200 IP. Central Marine E.Co.,W.Hartlepl	W. Hartlepool S. Nav. Co.,Ld.
158		Flamingo (ex Leopard) *Rea*	I ScSr 1 *Dk 2 B*	British Hull	550 849	223·1 *Q 68 f.*	28·9	16·5	Hull Brownlow,Pearson &Co.	1858	5 Bhds.	2Cy.26"&47"–36" 99 IP. Earles'Co.,Ld.,Hull	T.Wilson,Sons & Co.
159	*	Flamingo	I Sc 1 *Dk*	British Grimsby	134 255	137·9 *Q 16 f.*	21·5	11·2	Hull Earles'Co.,Ld.	1885	1 Bhds.	3C.15½"24"&40"–24" 80 IP. Earles'Co.,Ld.,Hull	Gt.GrimsbyIce Co.,Ld.
160		Flaminian *Alexander 86-87*	I ScBg 2 *Dks*	British Liverpool	1391 2131	333·0 *F 35 t.*	33·3	23·6	Newcastle Palmers'Co.,Ld.	1880	7 Bhds.	2Cy.24"&50"–48" 250 IP. Palmers'Co.,Ld.,Newcastle	S.S. "Flaminian" Co., Ld. (F.Leyland & Co.)
161		Flandria	Stl ScSr 2 *Dks 3 L*	German Hamburg	1533 2038	290·0 *P 46 f. B 46 f. B 26 f.*	36·0	25·7	Hamburg Reiherstieg Schiffswerfte	1888		3Cy.22½",36"&58"–42" 300 IP. Reiherstieg Schiffwerft, Hamburg	Hamburg-Amerikanische Paketf.Ges.
162		Flavian *Fry 81-84*	I ScBg 2 *Dks*	British Liverpool	1399 2139	333·0 *F 38 t.*	33·3	23·6	Newcastle Palmers'Co.,Ld.	1880	7 Bhds.	2Cy.24"&50"–48" 250 IP. Palmers'Co.,Ld.,Newcastle	S.S. "Flavian" Co.,Ld. (F.Leyland & Co.)
163	I	Flavio Gioja *Sapelli -88*	I PadSr 2 *Dks*	Italian Palermo	300 517	213·5 *Erections 94 t.*	23·9	11·8	Renfrew W.Simons&Co.	1863		2Cy.29"&46"–45" 190 IP. W.Simons&Co.,Renfrew	Navigazione Generale Italiana
164	*	Flaxman *Royce 81-88*	I ScSr 2 *Dks 3 B*	British Liverpool	1418 2175	299·3 *B 48 f.*	36·1	23·9	Southampton Oswald,Mordaunt&Co.	1882	5 Bhds.	2Cy.32"&64"–42" 200 IP. Oswald,Mordaunt&Co.,Sthtn.	Livrpl., Brazil & River Plate S.N.Co.
165		Fleetwood (ex Ibrahimieh)	I TwnSc	British Fleetwood	45 122	82·0	24·4	8·4	Renfrew W.Simons&Co.	1872		2Cy.18"&30"–24" 60 IP. W.Simons&Co.,Renfrew	Lan.& Yorks.& Lon.& N.W.Railway Co.
166	NV	Flid (ex Thorskog) *Helland*	W ScSr 1 *Dk 2 B*	Norweg'n Stavanger	372 501	157·3 *Q 40 f. B 36 f. F 20 f.*	27·1	16·7	Gothenburg P.Larsson	1871		2Cy.13½"&30½"–20½" 52 IP. GöteborgsMek.Verks.,Gbrg.	R.G.Ramsland
167		Flinders *Drysdale*	I ScSr 2 *Dks*	British Hobart	499 948	227·1 *B 68 t.*	28·3	20·1	Glasgow A.&J.Inglis	1878	5 Bhds. DBa& f.93t	2Cy.37"&66"–42" 200 IP. A.&J.Inglis,Glasgow	Tasmanian Steam Nav.Co.
168		Flinders *Crocker*	I ScSr 1 *Dk*	British Melbourne	278 521	195·7 *P 47 f. B 45 f. F 22 f.*	25·0	12·1	Sunderland J.Laing	1874	5 Bhds.	2Cy.26"&50"–30" 90 IP. G.Clark,Sunderland	Adelaide S.S.Co.,Ld.
169	V	Flink *Martinolich*	I ScSr	Austrian Lussinpiccolo	70 107	99·4	14·6	9·2	Lussinpiccolo N.Martinolich	1886		2Cy.13"&23"–15" 25 IP. Bartlett&Greenham,Trieste	N.Martinolich
170	*	Flintshire *Dwyer 87-88*	Stl Sc3mSr 2 *Dks 3 B*	British London	1871 2879	344·2 *P 50 f. B 72 f. F 47 f.*	40·1	26·5	Glasgow Lon.&Gls.S.&E.Co,Ld	1888	6 Bhds. DT,28ft. 240t	3C.33"48"&75"–51" 650 IP. London&Glas.S.&E.Co.,Ld.	Jenkins & Co.
171	*	Flintshire *Habekost 84-88*	I ScSr 2 *Dks 3 B*	British Singapore	1017 1563	270·7	32·8	23·8	Glasgow London&GlasgowS Co.	1872	6 Bhds.	2Cy.31"&54½"–33" 150 IP. London&Glas.Co.,Ld.,Glas.	Lim Tiang Hee
172		Flora *Bennison*	I ScSr 2 *Dks*	British Hobart	838 1273	250·8 *P 280 t. F 34 t.*	32·2	16·6	Hull Earles'Co.,Ld.	1882	3 Bhds. DBa FPT	2Cy.34"&64"–39" 180 IP. Bailey&Leetham,Hull	TasmanianStm Nav.C.
173	G	Flora (ex Dinnington) *Ahrens*	I ScSr 1 *Dk*	German Lubeck	235 330	140·4	19·4	11·4	NorthShields C.W.Dodgin&Co.	$18\frac{72}{82}$	WB	2Cy.18"&36"–28" 60 IP.	Libau-Lübecker Dampfschiffsf.-Ges.
174	*	Flora *Juravloff*	I ScSr 1 *Dk*	Russian Odessa	492 723	189·5 *Q 62 f. B 46 f. F 23 f.*	29·2	11·6	Newcastle Wghm.Richardson &Co.	1885	4 Bhds. DBa& f.20t FPT2t APT1t	2Cy.25"&48"–33" 97 IP. Wghm.Richardson&C.,Nwc.	L.Brodsky
175		Flora L. *Miletich 68-87*	I ScSr 2 *Dks*	Austrian Trieste	909 1375	255·4	31·7	22·4	Trieste Austro-Ungarico Lloyd	1871		2Cy.34½"&59½"–36" 180 IP. ArsenaleLloyd,Trieste	Lloyd Austro-Ungarico
176	*	Florence *Whiteway 87-87*	I ScSr 1 *Dk & Spr Dk 3 B*	British London	1576 2401	265·3 *P 28 f. B 52 f. F 25 f.*	38·5	20·9 28·1	Sunderland J.L.Thompson&Sn.	1880	DBa51ft f.32ft FPT APT	2Cy.37"&68"–42" 200 IP. N.E.MarineE.Co,Ld.,Sndlnd	R.Gordon & Co.
177	* D	Florence *Sanley 85-87*	I ScBn 2 *Dks 3 B*	British London	1430 2213	281·2 *P 32 f. F 35 f.*	38·0	23·8	Whitby T.Turnbull&Son	1881	5 Bhds. DBa& f.352t	2Cy.36½"&68"–42" 180 IP. Blair&Co.,Ld.,Stockton	Pyman Bros.
178	*	Florence *Creaghe 87-87*	I ScSr 1 *Dk 2 B*	British London	431 694	219·3 *P 87 f. B 37 f. F 35 f.*	25·5	14·2	Glasgow R.Napier&Sons	1865	4 Bhds.	2Cy.24"&42"–30" 70 IP. J.Penn&Sons,London	D.Currie & Co.
179	*	Florence *Cameron*	I ScSr 1 *Dk*	British Liverpool	115 268	149·8 *P 17 f. B 28 f. F 23 f.*	22·2	10·7	Glasgow T.B.Seath&Co.	1882	4 Bhds. FPT9t APT10t	2Cy.20"&36"–24" 56 IP. W.King&Co.,Glasgow	S.S. "Florence" Co.,Ld. (J.J.Mack & Sons)
180	*	Florence *Ecclesfield 86-86*	I ScSr 1 *Dk*	British Stockton	123 207	121·8 *Q 20 f.*	21·2	8·7	Middlesbro' R.Craggs&Sons	1881	4 Bhds.	2Cy.16"&30"–20" 35 IP. Blair&Co.,Ld.,Stockton	Cleveland S.S.Co.,Ld.

*=Lloyd's Register; D=Nederlandsche Vereeniging; G=Germanischer Lloyd; I=Registro Italiano; R=American Record;
A=Veritas Austro-Ungarico; H=Veritas Ellenico; N=Norske Veritas; V=Bureau Veritas.

Dans la 2me Colonne, B.T., avec les indications qui suivent, signifie la date du rapport de visite de la Carène par les experts du Board of Trade pour obtenir le certificat relatif au TRANSPORT DES PASSAGERS.

Dans la 3me Colonne, C=Composite (membrure en fer, bordé en bois); I=Fer; Stl=Acier; W=Bois.

Pour plus amples renseignements sur les navires classés par le Lloyd's Register, voir le Lloyd's Register Book et le Lloyd's Yacht Register.

No. in Book / Official No.	Registry in which classed	Ship's Name. Material, Rig, &c. / Master, and Dates of Appointment to present Owner's Service and to Ship. / Ship's Late Name. No. of Decks, &c.	Flag / Port of Registry	Tons Net / Gross	Length	Breadth	Depth	Built Where / By Whom	When	No. of Bulkheads / Particulars of Water Ballast	Engines / Particulars. Maker's Name.	Owners or Ma…
181		Florence W ScSr 1 Dk	British Quebec	90 1..	91·2	22·8	9·5	Quebec H.Dinning	1883		1Cy.11"-11" 25 HP. F.Drolet,St.Rocks,Quebec	Henry Brown
182		Florence W Sc 1 Dk	British Quebec	77 113	91·0 B 30 t.	19·8	9·0	Levis,Quebec Maritime&IndustrialCo	1885		2Cy.18"& 36"-24" 54 HP. Maritime&IndlCo.,Levis	Henry Jewell
183		Florence I Pad Sp	British Sunderland	42 137	113·8	19·2	9·9	SouthShields J.T.Eltringham	1879		1Cy.37"-54" 85 HP. J.P.Rennoldson,S.Shields	W.W.Douglas
184	✳	Florence I PadTug *Maunsey 71-80* 1 Dk	British Maryport	26 137	94·6	18·6	9·5	Harrington R.Williamson &Son	1879	2 Blhds.	2Cy.31"-48" 70 HP. J.P.Almond,N.Shields	Maryport S.S
185		Florence Annie I Pad	British Sydney	137 1.7	77·5	25·6	6·2	Echuca,Va. S.Thomas	1882		2Cy.12"-30" 28 HP. Scott&Young,Melbourne	John H. Brow
186		Florence Maud I ScCr 1 Dk	British Melbourne	71 105	96·5	18·8	9·0	Melbourne RobisonBros.&Co.	1876		2Cy.15"-16" 35 HP. RobisonBros.&Co.,Mlbrne.	The Colonial Refining Co.
187	✳	Florence Nightingale I ScSr *Gould 85-86* 1 Dk2 B	British Sunderland	629 958	213·0 Q 73 f.	30·5 B 30 f.	16·3 F 33 f.	Sunderland Osbourne,Graham &Co.	1878	1 Blhds. DBd&OH FPT50 APT150	2Cy.27"&50"-36" 99 HP. J.Clark,Sunderland	J.F.Marshall
188	✳	Florence Richards I ScSr *Gaskall 86-86* 1 Dk2 B	British Liverpool	668 1038	221·5 P 122 f.	30·1	16·8 F 27 f.	Sunderland J.L.Thompson	1873	1 Blhds. DB… f.	3C.18½",27"&47"-33" 112 HP. Tpld 86,J.Jack&Co.,Lvpl.	R.Nicholson &
189 [illegible]	✳ BT…	Florida I Sc3mSr *Beard 75-86* 2 Dks3 B	British London	2044 …138	336·6 P 40 f.	41·2 B 82 f.	26·5 F 46 f.	Newcastle C.Mitchell&Co.	1882	6 Blhds. … FPT… APT…	2Cy.46"& 74"-54" 350 HP. N.E.MarineE.Co.,Ld.,Sudld.	Nelson,Donk
190	✳	Florida Stl Sc 1 Dk	British London	11 28	56·5	11·4	6·6	Leith J.Cran&Co.	1887	3 Blhds.	2Cy.10"& 20"-15" 17 HP. J.Cran&Co.,Leith	T.A.Walker
191 [illegible]	✳ BT9.88	Floridian I Sc3mSr *Sundry -88* 3 Dks	British Liverpool	2136 3257	360·7 P56 f	41·2 B118 f	26·9 F48 f H287 f	Belfast Harland&Wolff	1884	7 Blhds.	2Cy.40"& 80"-51" 266 HP. Harland&Wolff,Belfast	West India & Steam Nav.Co
192	✳	Flos I ScSr *Pindar* 1 Dk2 B	British Hull	1100 1414	247·7 P 222 t.	32·8	18·9 F 33 t.	Stockton Richardson,Duck&Co.	1870	5 Blhds. DBd	2Cy.27"& 54"-33" 110 HP. C.D.Holmes&Co.,Hull	H.Briggs,Son
193	✳	Flowergate I ScBn *Nicholson 82-85* 2 Dks3 B	British Whitby	1340 2053	286·0 P 26 f.	36·0 B 68 f.	24·1 F 35 f.	Whitby T.Turnbull&Son	1884	5 Blhds. DB…	2Cy.34½"& 61"-42" 160 HP. Blair&Co.Ld.,Stockton	T.Turnbull &
194		Flushing (ex Ingersoll) W ScSr 1 Dk	British St.JohnN.B.	175 257	115·9 H 102 t.	24·0	8·2	Athens,N.Y.	1882		1Cy.20"-22" 43 HP. McEntee&Dillon,Rondout,N.Y.	E.Gaskill
195	✳	Flygarn (ex Tweeddale) I ScSk 1 Dk	Swedish Gtnbrg.	11 46	60·2	16·0	8·4	Leith Hawthorns&Co.	1886	1 Blhds.	2Cy.8"&16"-12" 10 HP. Hawthorns&Co.,Leith	A.F.Ouchter
196		Flying Arrow I PadSp	British Glasgow	19 131	108·0	18·5	9·8	SouthShields J.T.Eltringham	1882		1Cy.36"-54" 80 HP. J.P.Rennoldson&S.,S.Shlds.	Clyde Shippi
197 71620		Flying Arrow I PadSp *T. Anderson*	British N.Shields	8 123	108·6	18·2	9·3	SouthShields J.T.Eltringham	1875		1Cy.36"-54" 60 HP. J.P.Rennoldson&S.,S.Shlds.	Mrs.D.Ander
198		Flying Bat I PadSp	British Glasgow	16 130	108·3	18·9	9·7	SouthShields J.T.Eltringham	1884		1Cy.36"-54" 80 HP. J.P.Rennoldson&S.,S.Shlds.	Clyde Shippi
199		Flying Cloud I Pad *Harvey*	British Swansea	22 101	96·5	18·1	9·2	Newcastle T.Hepple&Co.	1871		1Cy.34½"-54" 40 HP. T.Hepple&Co.,Newcastle	Mrs.E.J.Chi
200		FlyingCormorant StlT-ScSr *Lindsay -88*	British Glasgow	5 209	135·2	21·0	12·0	Port Glasgow R.Duncan&Co.	1888		2Cy.25"& 48"-30" 98 HP. Rankin&Blackmore,Grnck.	Clyde Shippi
201		Flying Dart I PadSk	British Glasgow	9 111	100·0	18·1	9·4	SouthShields J.T.Eltringham	1882		1Cy.34"-54" 70 HP. J.P.Rennoldson&S.,S.Shlds.	Clyde Shippi
202		Flying Dragon Stl TwnScSrTug	British Glasgow	5 260	134·9	24·0	11·9	PortGlasgow R.Duncan&Co.	1886		2Cy.25"& 48"-30" 98 HP. Rankin&Blackmore,Grnck.	Clyde Shippi
203 58313		Flying Dutchman I Pad	British Sunderland	20 1.3	117·7	19·3	10·4	Newcastle T.D.Marshall	1867		90 HP.	W.W.Dougl
204 82332		Flying Dutchman I PadSp	British Glasgow	7 112	100·0	18·1	9·4	SouthShields J.T.Eltringham	1880		1Cy.34"-54" 70 HP. J.P.Rennoldson&S.,S.Shlds.	Clyde Shippi
205 86210		Flying Eagle I PadSk (ex Mount Etna) 2 Dks	British Glasgow	34 282	139·6	24·9	11·4	Dumbarton R.Chambers	1880	3 Blhds.	2Cy.25"& 48"-60" 98 HP. Rankine&Blackmore,Grnck	Clyde Shippi
206 70802	BT7.88	Flying Falcon I PadSp *Williamson (ex Lord Bangor)*	British Glasgow	30 1.1	122·2	19·7	10·3	S.Shields J.T.Eltringham	1878		2Cy.30½"-54" 95 HP. Hepple&Co.,S.Shields	Clyde Shippi
207 83014		Flying Fish I PadSp *Standalough*	British Sunderland	59 151	116·0	19·6	10·3	SouthShields J.T.Eltringham	1882		1Cy.38"-53" 75 HP. J.P.Rennoldson&S.,S.Shlds.	F.Downs
208 93248	BT…88	Flying Fish I PadSk	British Glasgow	40 177	122·6	20·5	10·7	SouthShields J.T.Eltringham	1886		1Cy.41"-60" 98 HP. J.P.Rennoldson&S.,S.Shlds.	Clyde Shippi
209		Flying Foam I ScSk	British Glasgow	8 …	100·2	18·1	9·2	PortGlasgow Murdoch&Murray	1877		2Cy.20"& 40"-24" 70 HP. Rait&Lindsay,Glasgow	Clyde Shippi
210		Flying Fox I Pad	British Glasgow	47 173	122·5	20·3	10·9	SouthShields J.T.Eltringham	1885		1Cy.41"-… 98 HP. J.T.Eltringham,S.Shields	G.J.Kidston

✳ = Lloyd's Register; D = Nederlandsche Vereeniging; G = Germanischer Lloyd; I = Registro Italiano; R = American …
A = Veritas Austro-Ungarico; H = Veritas Ellenico; N = Norske Veritas; V = Bureau V…

In the 2nd Column, B.T., with figures following, signifies Date of report of bottom sighting by Board of Trade Surveyors as condition of PASSENGER CERTI…

In the 3rd Column, C = Composite (iron frame, wood planked); I = Iron; Stl = Steel; W = Wood.

Further particulars of Vessels classed by Lloyd's Register will be found in Lloyd's Register Book and Lloyd's Yacht Register.

	Nom, Matériaux, Gréement, etc. — Capitaine et dates d'entrée au service des Armateurs actuels et à bord du navire. — Ancien Nom, nombre de ponts, etc.	Pavillon — Port d'Armement	Tonnage Net — Total	Longueur	Largeur	Creux	Construit Où — Par qui	Quand	Nombre des cloisons. — Détails sur le lest d'eau.	Machines. — Détails. Nom du Constructeur.	Armateurs ou Directeurs-Gérants.
	Flying Huntress 1 PadSp	British Glasgow	12 164	122·5	20·2	10·6	South Shields J.T.Eltringham	1878		1Cy.41"–60" 98 HP. J.P.Rennoldson&S.,S.Shlds.	Clyde Shipping Co.
	Flying Huntsman 1 PadSk Hall	British Limerick	39 172	122·2	20·1	10·7	South Shields J.T.Eltringham	1881		1Cy.41"–60" 98 HP. J.P.Rennoldson&S.,S.Shlds.	J.Spaight
	Flying Irishman 1 PadSk 1 Dk	British Glasgow	19 152	116·0	19·6	10·3	South Shields J.T.Eltringham	1885		1Cy.38"–56" 90 HP. J.P.Rennoldson&S.,S.Shlds	Clyde Shipping Co.
	Flying Kestrel 1 TWnScSk Roberts 87-87 (ex Narwhal)	British Liverpool	 135	114·6	20·8	10·0	Port Glasgow R.Duncan&Co.	1878		2Cy.21"&10"–26" 70 HP. Rankin&Blackmore,Grnck	Alexandra Towing Co. Ld.
	Flying Owl 1 PadSk	British Glasgow	13 129	108·7	18·6	9·7	South Shields J.T.Eltringham	1883		1Cy.36"–54" 80 HP. J.P.Rennoldson&S.,S.Shlds	Clyde Shipping Co.
	Flying Scotchman 1 PadSp	British Milford	8 111	98·8	18·2	9·5	South Shields J.T.Eltringham	1879		1Cy.34½"–54" 70 HP. J.P.Rennoldson&S.,S.Shlds.	Pembrokeshire Steam Trawling Co.,Ld.
	Flying Scotsman 1 PadSk	British Glasgow	14 131	108·0	18·6	9·9	South Shields J.T.Eltringham	1886		1Cy.36"–54" 80 HP. J.P.Rennoldson&S.,S.Shlds	Clyde Shipping Co.
	Flying Serpent Stl TwnScSr 1 Dk	British Glasgow	5 260	134·9	24·0	11·9	Port Glasgow R.Duncan&Co.	1886		2Cy.25"&48"–30" 98 HP. Rankin&Blackmore,Grnck.	Clyde Shipping Co.
	Flying Spear 1 PadSk	British Glasgow	5 111	100·3	18·1	9·4	South Shields J.T.Eltringham	1883		1Cy.34"–54" 70 HP. J.P.Rennoldson&S.,S.Shlds.	Clyde Shipping Co.
	Flying Sportsman 1 PadSp	British Glasgow	38 174	122·0	20·2	10·7	South Shields J.T.Eltringham	1882		1Cy.41"–60" 98 HP. J.P.Rennoldson&S.,S.Shlds	Clyde Shipping Co.
	Flying Sprite 1 PadSp	British Scarboro'	6 103	98·2	17·9	9·4	South Shields J.T.Eltringham	1869		65 HP.	Frank Ness
	Flying Squadron 1 PadSk (or Fergus Macivor)	British Glasgow	14 139	109·4	20·0	10·1	London R.&H.Green	1868		2Cy.31"–54" 70 HP. J.Stewart,London	Clyde Shipping Co.
	Flying Squall 1 PadSp	British Scarboro'	17 116	108·2	18·1	9·2	South Shields J.T.Eltringham	1871		1Cy.36"–54" 80 HP. J.P.Rennoldson,S.Shields	R.Middleton
	Flying Squirrel 1 PadSk 1 Dk	British Glasgow	14 131	108·5	18·5	9·7	South Shields J.T.Eltringham	1886		1Cy.36"–54" 80 HP. J.P.Rennoldson&S.,S.Shlds.	Clyde Shipping Co.
	Flying Swallow 1 PadSp	British Glasgow	15 132	108·0	18·6 B87.	9·8	South Shields J.T.Eltringham	1887		1Cy.36"–54" 80 HP. J.P.Rennoldson&S.,S.Shlds.	Clyde Shipping Co.
	Flying Sylph 1 PadSp	British Scarboro'	8 116	108·2	18·0	9·2	South Shields J.T.Eltringham	1869		1Cy.36"–54" 80 HP. J.P.Rennoldson&S.,S.Shlds.	H.L.Woodger
	Foam Stl PadSr 2 Dks	French Calais	234 512	230·6	26·5 B24f.	14·5	London Samuda Bros.	1862		2Cy.62"–54" 240 HP. Ravenhill,Salkeld&Co.London	London,Chatham & Dover Railway Co.
	Fohr Stl Sc 1 Dk Dethlefsen 88-89	German Kiel	572 712	195·1 B41f.	30·0 F32f.	13·8	Kiel G.Howaldt	1888	1 Blkds. CD1270	3Cy.14⅛"23⅝"&36⅝"–23⅝" G.Howaldt,Kiel 100 HP.	Gebruder Lange
*	Fokien 1 ScSr Lewis 88 1 Dk & Awn Dk	British Hong Kong	509 814	200·3 B22f.	28·6	13·0 19·9	Aberdeen Hall,Russell&Co.	1880	1 Blkds. [illegible]	2Cy.30"&52"–38" 120 HP. Hall Russell&Co.,Aberdeen	Douglas S.S.Co.,Ld.
*N	Foldin 1 ScSr Dahl 1 Dk 2 H	Norweg'n Christiania	508 679	183·8 Q26f.	29·3 B44f.	14·0 F23f.	Elsinore Helsingörs I. S. B. & E.C.	1884	1 Blkds. [illegible]	2Cy.23½"&44"–27" 85 HP. Helsingörs I. S. B. & E.C. Elsore.	Östlandske Lloyd (Rolf Andvord)
	Folgefonden 1 ScSr 1 Dk	Norweg'n Bergen	141 197	120·5 P45f.	19·1 F24f.	9·5	Newcastle C.Mitchell&Co.	1875		2Cy.17"&32"–21" 45 HP. Thompson&Co.,Newcastle	Hardanger Söndhordlandske Dampskibsselskab
	Folkestone 1 TwnScSr Hall 86-86	British London	219 398	190·0	25·0 Q37f.	11·7	Hull Earle's Co.,Ld.	1878		1Cy.20"&39"–24" 110 HP. Earle's Co.,Ld.,Hull	South Eastern Railway Co.
	Fomento (ex Glengarnock) 1 ScSr 1 Dk	French Bordeaux	113 205	129·4 H15f.	20·1 Q49f.	9·9	Paisley J.Fullerton&Co.	1870	3 Blkds. [illegible]	2Cy.25"–24" 45 HP. W.King&Co.,Glasgow	J.Aubian
	Foochow Stl PadL	British Glasgow	526 731	261·0	35·0	9·9	Glasgow Alley&M'Lellan	1884		1Cy.27"&16"–54" 220 HP. Alley&M'Lellan,Glasgow	Irrawaddy Flotilla Co., Ld.
	Fook Ching W ScSr 1 Dk Christian 88	British Hong Kong	77 126	98·5	17·0	8·6	Hong Kong Tsin Ug	1888		2Cy.12"&24"–16" 24 HP. Kwong Hep Leung,Hong Kong	Kaw Hong Take
	Fooksang 1 ScSr Sawer 85-88 2 Dks & Shd Dk 4 B	British London	991 1557	250·0 B48f.	35·2 F29f.	22·3	Aberdeen Hall,Russell&Co.	1881	1 Blkds. [illegible]	2Cy.33"64"–39" 175 HP. Hall,Russell&Co.,Aberdeen	Indian-China Steam Nav.Co.,Ld.
*	Footah 1 ScTug 1 Dk Robertson	British Sierra Leone	21 36	65·2 Q30f.	14·0	5·9	Liverpool J.Jones&Sons	1884	3 Blkds.	2Cy.12"&22"–15" 20 HP. J.Jones&Sons,Liverpool	Fisher & Randall
	Forcados 1 ScBg Goudie 88	British Glasgow	327 456	185·6 P22f.	25·1 F18f.	11·4	Port Glasgow Cunliffe&Dunlop	1874		2Cy.23"&40"–24" 80 HP. Cunliffe&Dunlop,Pt.Glasg.	British & African Steam Nav.Co.
G	Forelle 1 PadSp 1 Dk Hogemann	German Bremen	180 293	201·1	20·2	9·0	Vegesack H.F.Ulrichs	1881		2Cy.27"&44"–18" H.F.Ulrichs,Vegesack	Norddeutscher Lloyd
*D	Foreningen (ex Elgiva) 1 ScSr 1 Dk Paulsen 73-82	Danish Copenhagen	287 469	161·6 E15f.	23·7 Q70f.	12·1 F28f.	Port Glasgow Murdoch&Murray	1878	1 Blkds. [illegible]	2Cy.20"&36"–27" 60 HP. Dunsmuir&Jackson,Glasg.	Dampsk. Sels. "Foreningen" (H.J.Rønne)

A=Lloyd's Register; D=Nederlandsche Vereeniging; G=Germanischer Lloyd; I=Registro Italiano; R=American Record;
A=Veritas Austro-Ungarico; H=Veritas Ellenico; N=Norske Veritas; V=Bureau Veritas.

Dans la 2me Colonne, B.T., avec les indications qui suivent, signifie la date du rapport de visite de la Carène par les experts du Board of Trade pour obtenir le certificat relatif au TRANSPORT DES PASSAGERS.

Dans la 3me Colonne, C=Composite (membrure en fer, bordé en bois); I=Fer; Stl=Acier; W=Bois.

Pour plus amples renseignements sur les navires classés par le Lloyd's Register, voir le Lloyd's Register Book et le Lloyd's Yacht Register.

No. / Official No.	Class	Ship's Name. Material, Rig, &c. Master, and Dates of Appointment to present Owner's service and to Ship. Ship's Late Name. No. of Decks, &c.	Flag. Port of Registry.	Tons. Net. Gross.	Length	Breadth	Depth	Built, Where. By Whom.	When.	No. of Bulkheads. Particulars of Water Ballast.	Engines. Particulars. Maker's Name.	Owners or M[anagers]
241		**Forerunner** I ScSp *Jewell -83* 1 *Dk*	British Port Natal	54 150	100·0	21·0	9·4	London Westwood&Baillie	1873		2Cy.34"–48" 80 HP. Westwood&Baillie, London	Natal Govern[ment]
242 9513		**Forest City** W Pad *Donovan -83* 2 *Dks*	American Portland, Me	1072 1134	234·7	33·0	15·0	Greenpnt. N Y J. Englis	1854			Portland Steam[ship] Co.
243 47631	✻D	**Forest Queen** I ScBk *Small 83-88* 1 *Dk2 B*	British Hull	279 441	171·4	24·7	13·2	Renfrew Henderson, Coulbor[n] & Co.	1863	4 Bhds.	2Cy.27"&46"–28" 90 HP. Gilbert&Cooper, Hull	W. Rawson & [S]on
244 20436	BT 2-88	**Forfarshire** Stl Pad	British Dundee	72 115	120·5	20·1	6·6	Dundee Gourlay Bros.&Co.	1861		2Cy.22"–36" 40 HP. Gourlay Bros.&Co., Dundee	Trustees of the [Har]bour of Dundee
245		**Foria** (ex Cyrene) I ScSr *Briand 88-88* 2 *Dks*	French Marseilles	722 1085	230·1	31·2	20·8 F 25 f.	Liverpool Bowdler, Chaffer & Co.	1875	5 Bhds. DB[illegible]	2Cy.20½"&40"–30" 110 HP. Smith Bros.&Co., Glasgow	Compagnie Fr[ançaise] de Nav. à Vap.
246 86840	✻	**Formosa** I ScSr *Hall* 1 *DkSprDk & WebFrames*	British Hong Kong	674 1097	221·0 B 41 f.	32·2	12·4 19·4	Newcastle Wghm. Richardson & Co.	1885	4 Bhds. DB[illegible]	3C.20½"32"&53"–36" 200 HP. Wghm. Richardson&C., Nwc.	Douglas S.S.Co.
247 31232		**Formosa** I ScSr 2 *Dks*	British Hong Kong	458 837	203·5 P 59 f. F 44 f.	25·4	16·5	Glasgow Smith&Rodger	1852		2Cy.48"–42" 154 HP. Smith&Rodger, Glasgow	China Nav. Co.
248 73822		**Formoso** I ScBg *Patterson 88*	British Glasgow	328 461	185·4 P 43 f. F 13 f.	25·2	11·3	Port Glasgow Cunliffe&Dunlop	1876		2Cy.23"&40"–24" 75 HP. Cunliffe&Dunlop, Pt. Glasg.	British & Afri[can] Nav. Co., Ld.
249		**Forshaga** I ScSp *Svartling -87* 1 *Dk*	Swedish Trollhättan	153 204	102·0 P 32 f. H 4 f.	21·4	10·3	Gothenburg Eriksbergs Mek. Verk.	1887			J. Olsson
250		**Forsög** W ScSr 1 *Dk*	Norweg'n Christiansund	154 222	128·2 Q 26 f. B 30 f. F 14 f.	22·8	10·1	Gothenburg	1853		50 HP.	C. Johnsen
251 95090	✻	**Fort William** Stl ScSr *Pearson 86-88* 1 *Dk & Spar Dk*	British Glasgow	1179 1897	271·2 F 28 f. B 42 f.	37·1	16·0 23·1	Belfast Workman, Clark & Co., Ld.	1888	4 Bhds. CDB[illegible]	3C.19"32"&52"–42 136 HP. W. King&Co., Glasgow	Clark & Servi[ce]
252 83853	✻	**Fortescue** I ScSr *Ballaine* 1 *Dk2 B*	British Exeter	754 1184	225·0 Q 76 f. B 57 f. F 31 f.	33·2	15·5	Port Glasgow Murdoch&Murray	1884	4 Bhds. DB[illegible]	2Cy.27"&52"–33" 115 HP. Muir&Houston, Glasgow	"Fortescue" S.S. Co. (J. Holman & [Sons])
253 81139		**Forth** I TwnSc	British Grangem'th	363 601	163·3	34·2	13·7	Renfrew W. Simons&Co.	1883		4Cy.18"&33"–21" 94 HP. W. Simons&Co., Renfrew	Caledonian R[ailway]
254 28679		**Forth** I Sc3mSr *Longstaff -83* 1 *Dk2 B*	British Middlesbro'	328 513	188·1	27·1	14·8	Glasgow Barclay, Curle&Co.	1862	6 Bhds.	2Cy.43"–36" 130 HP. Barclay, Curle&Co., Glasgow	T. Simmons
255 90902	✻ BT 2-88	**Forth** Stl ScSr *Carne 87-87* 2 *Dks*	British Grangem'th	263 827	230·0 B 105 f. P 50 f. F 31 f.	31·2	14·3	Glasgow A.&J. Inglis	1886	5 Bhds. DB[illegible]	3Cy26"42"&65"–42" 250 HP. A.&J. Inglis, Glasgow	Carron Co.
256 45008	✻	**Fortuna** I Sc3mSr *Berry* 2 *Dks*	British Glasgow	432 697	206·5 Q 67 f.	26·1	17·2	Newcastle Palmer Bros.	1862	4 Bhds.	2Cy.26½"&48"–30" 90 HP. D. Rowan, Glasgow	J. Clark
257	V	**Fortuna** I ScSr *Petersen 80-83* 1 *Dk*	German Flensburg	320 504	167·9 Q 49 f. B 42 f. F 21 f.	25·1	13·3	Flensburg Flensburg Schiffsb. Ges.	1881		2Cy.20"&34½"–24" 50 HP. Flensbg Schiffsb. Ges., Fensbg	F. M. Bruhn
258		**Fortuna** I PadSp *Greizer 85-* 1 *Dk*	German Danzig	144 194	144·0	22·6	7·3	Danzig Danziger Schiffsw.	1882			Dampfschiffs[…] "[illegible]"
259	N	**Fortuna** W Sc3mSr *Larsen* 1 *Dk2 B*	Norweg'n Sandefjord	105 165	100·5 Q 27 f. F 26 f.	22·0	11·6	Sandefjord	1885		2Cy.12½"&22½"–15½" 20 HP. Akers Mek. V., Christiania	Chr. Christens[en]
260	G	**Fortuna** C ScSr *Magnusson 83-85* 1 *Dk*	Swedish Christinehamn	242 332	183·5 B 23 f. H 9 f.	22·1	12·8	Gothenburg J. A. Svensson	1883		2Cy.16"&29"–19½" 30 HP.	Ångbåtsaktieb[olaget] "Fern" (Manager […])
261		**Fortuna** W Sc *Howard*	Urug'ayn Montevideo	206 374	170·4	22·0	11·9	Boston Jones&Co.	1879		90 HP.	J. Jones
262	✻I	**Fortunata R.** (ex Bell Rock) I ScSr *Contardi* 1 *Dk & Spr Dk3 B*	Italian Genoa	2317 3532	340·5 B 75 f. F 34 f.	42·1	27·8	Hull D. P. Garbutt	1883	7 Bhds. DB[illegible]	2Cy.44"&84"–54" 813 HP. D. P. Garbutt, Hull	Ditta C. Raggi
263 89464	✻D	**Fortunatus** I ScSr *Davies 86-88* 1 *Dk2 B*	British W. Hartlep'l	883 1371	245·0 Q 92 f. B 61 f. F 31 f.	33·0	16·0	W. Hartlepool Irvine&Co.	1884	5 Bhds. CDB[illegible]	2Cy.30"&57"–35" 130 HP. T. Richardson&Sons, Hrtpl.	R. Irvine & Co.
264 81570	✻	**Foscolia** I ScSr *Le Templier -83* 2 *Dks3 B*	British London	1025 1573	252·0 B 9 f. F 18 f.	34·0	22·7	Newcastle C. Mitchell&Co.	1879	5 Bhds. DB[illegible]	2Cy.30"&58"–39" 160 HP. Wallsend S. & E. Co. Ld., Nwcl.	Watts, Ward & [Co.]
265 84997	✻D	**Foscolino** I ScSr *Thirkell 87-87* 1 *Dk2 B*	British Sunderland	716 1124	221·9 Q 75 f. B 69 f. F 29 f.	31·1	15·8	Sunderland W. Pickersgill&Sons	1881	4 Bhds. [illegible]	2Cy.27"&50"–36" 99 HP. G. Clark, Sunderland	Wear S.S. Co., L[d.]
266	✻	**Foulah** I ScBk 1 *Dk & Spr Dk3 B*	French Marseilles	1156 1492	242·0	34·2	15·0 22·0	Sunderland J. Laing	1878	4 Bhds. [illegible]	2Cy.30"&57"–36" 140 HP. G. Clark, Sunderland	Compagnie du Sénégal
267		**Foulazi Osmani** I ScBkn (ex Giuseppe Garibaldi, ex G. Lanza, Djevad Bey, ex Corinthian, ex Damascus)	Turkish Constantpl.	1159 1307	280·5 Len. '70.	31·2	22·1	Dumbarton W. Denny&Bros.	1856		2Cy.38½"&66"–36" 165 HP. Greenock Foundry Co., Grnk.	Idarei Massous[é] (Ottoman S[team]…)
268 81770	✻D	**Fountains Abbey** I ScSr *Smith 86-88* 1 *Dk2 B*	British Cardiff	893 1395	243·0 Q 86 f. B 64 f. F 32 f.	33·2	18·0	Newcastle Palmers Co., Ld.	1879	3 Bhds. [illegible]	2Cy.30"&57"–36" 130 HP. Palmers Co., Ld., Newcastle	Pyman, Wats[on & Co.]
269	V	**Fournel** I ScSr *Dumayne -88* 2 *Dks3 B*	French Havre	1339 2068	300·2 H 42 f. F 32 f.	36·0	24·9	Glasgow A.&J. Inglis	1880	[illegible]	2Cy.38½"&68"–41½" 220 HP. A.&J. Inglis, Glasgow	Compagnie Gé[nérale] Transatlantique
270												

✻ = Lloyd's Register; D = Nederlandsche Vereeniging; G = Germanischer Lloyd; I = Registro Italiano; R = American [Lloyd's];
A = Veritas Austro-Ungarico; H = Veritas Ellenico; N = Norske Veritas; V = Bureau Veri[tas].

In the 2nd Column, B.T., with figures following, signifies Date of report of bottom sighting by Board of Trade Surveyors as condition of PASSENGER CE[RTIFICATE].

In the 3rd Column, C = Composite (iron frame, wood planked); I = Iron; Stl = Steel; W = Wood.

Further particulars of Vessels classed by Lloyd's Register will be found in Lloyd's Register Book and Lloyd's Yacht Register.

No.	Cote	Nom, Matériaux, Gréement, etc. Capitaine et dates d'entrée au service des Armateurs actuels et à bord du navire. Ancien Nom, nombre de ponts, etc.		Pavillon. Port d'Armement	Tonnage Net. Total	Dimensions. Longueur, Largeur, Creux	Construit. Où. Par qui.	Quand	Nombre des cloisons. Détails sur le lest d'eau.	Machines. Détails. Nom du Constructeur.	Armateurs ou Directeurs-Gérants.
	✻	Fox *Anderson*	I ScSp 1 *Dk*	British Hull	94 133	108·7|20·2|8·8	Middlesbro' Backhouse&Dixon	1869	3 Blhds.	2Cy.15½"-18" 25 HP. Pe..ys&Co. Middlesbro'	Boston & Hull Steam Ship Co.,Ld.
	V	Fox *Olsen 74-74*	W ScSmSr 1 *Dk*	Danish Copenhagen	110 186	120·5|24·3|11·5 *Q 10 f.*	Aberdeen Alex.Hall&Co.	1855		2Cy.14"&24"-12" 25 HP. Burmeister&Wain, Copenhagen	Kryolith-Mine og Handels-Selskabet(J.P.Suhr&Søn)
	✻	Foxhall *Dalby*	Stl ScSr	British London	538 843	190·0|28·2|12·8 *B 66 f.* 20·0	Newcastle Wghm. Richardson &Co.	1885	4 Blhds. DB&FPK APT156	2Cy.30"&61"-39" 200 HP. Wghm.Richardson&C.Nwc.	Phipps & Co.
274	✻	Foyle (ex Miranda) *Corson 85-88*	Stl ScSr 2 *Dks* 3 *B*	British London	1584 2454	300·0|40·2|25·3 *P 32 f. B 68 f. F 30 f.*	Sunderland J.L.Thompson&Sons	1887	6 Blhds. CDB1031 APT86	3Cy24"39"&64"-42"300HP T.Richardson&Sns,Hrtlepl	Mercantile S.S.Co.,Ld
275	✻	Foyle (ex Ellen Kaun Maru)	Stl ScSr 1 *Dk*	Japanese	200 370	160·4|23·0|12·0 *P 24 f. F 22 f.*	Londonderry C.J.Bigger	1887	4 Blhds. FPT27t APT17t	3Cy.15 22"&38"-27"90HP Dunsmuir&Jackson,Glsgw	
276	✻	Foyle *Meier*	I ScSmSr 1 *Dk*	British Dublin	199 320	160·0|21·1|12·1 *Q 22 f. B 17 f. F 26 f.*	Newcastle Cole Bros.	1872	5 Blhds. MT32t	2Cy.28"-24" 56 HP. Mrshall,Osborn&Co,S.Shlds	M.Murphy,jun.
277	N	Fram *Gullormsen*	I ScSr 1 *Dk*	Norweg'n Fredrikstad	389 528	159·1|25·9|12·8 *Q 41 f. B 43 f. F 23 f.*	Lübeck Henry Koch	1884	4 Blhds. DB80t APT FPT	2Cy17¾"&30¾"-21⅝"70HP. IronFndry.&Eng.Co.,Berlin	Brödrene Thorvildsen
278		Framnæs	I ScSr 1 *Dk*	Norweg'n Bergen	185 253	131·9|21·0|11·1 *Q 43 f. F 37 f.*	Dumbarton A.Denny	1858	3 Blhds.	2Cy.18"&25"-24" 45 HP. A.Denny,Dumbarton	Nordre Bergenhuus. Amtskommune
279	✻N	Framnes *Thorbjørnsen -88*	Stl ScSr 1 *Dk* 2 *B*	Norweg'n Christiania	1146 1497	262·6|35·0|16·6 *B 58 f. Q 97 f. F 32 f.*	Newcastle Wood,Skinner&Co.	1888	4 Blhds. CDB320t APT30t	3Cy.18½",30"&49"-33" N.E.Marine E.Co.,Ld., Newcastle [110HP	Actieselskabet "Framnes"(Fearnley&Eger
280	✻	Frampton *Shates 87-87*	I ScTrawler 1 *Dk*	British Hull	52 105	85·0|20·0|9·9 *F 11 f.*	Hull Earle'sCo.,Ld.	1886	3 Blhds.	2Cy.12"&22"-20" 25 HP. Earle'sCo.,Ld.,Hull	Boston Deep Sea Fishing & Ice Co.,Ld.
281		France *Hadley 80-85*	I ScBk 4 *Dks*	British Liverpool	2713 4281	385·6|42·4|28·7 36·1	Liverpool T.Royden&Sons	1867	2 Blhds.	2Cy.30"&63"-45" 400 HP. J.Jones&Sons,Liverpool	National Steam Ship Co.,Ld.
282		France	W ScFlat	British Liverpool	86 120	86·5|20·7|9·0	Northwich C.Gibson	1878		2Cy.9½"-12" 24 HP. E.Bates,Northwich	R.Verdin
283	✻V	France *Collier -84*	I ScBk 3 *Dks & Awn Dk*	French St.Nazaire	2744 4538	395·2|43·3|28·7 36·5	St.Nazaire Scott&Co.	1865	10 Blhds.	4Cy.41"&75"-51" 850 HP. Maudslay Ss.&Field,London	Compagnie Générale Transatlantique
284		France	I PadSc 2 *Dks*	French Calais	164 365	206·0|24·0|12·5 *B 24 f.*	London J.Ash&Co.	1864		2Cy.52½"-48" 180 HP. J.Penn&Son,Greenwich	London, Chatham & Dover Railway Co.
285		Frances *Thompson 87-87*	I PadSr 2 *Dks*	American Prvdnce.R.I	718 988	225·0|32·0|9·0	WilmingtnDel Harlan & Hollingsworth Co.	1864		1Cy.50"-132" 800 HP. The Harlan&HollingsworthCo., Wilm.,Del.	Providence&Stonington S.S.Co.
286	✻	Francesca	C ScSrYt	British Cowes	151 194	135·0|23·0|11·5	Hull Earle's S.B.Co.	1874	C.50	2Cy.17"&28"-18" 25 HP. J.Penn&Son,Greenwich	Earl of Caithness
287		Franche Comté *Allègre 80-86*	I 4mSc 2 *Dks*	French Marseilles	682 1026	238·1|28·9|21·3	LaSeyne Forges & C. de la Medit.	1866		2Cy.33"-24½" 100 HP. Forges&C.de la MeditMarsls	Société Générale de Transports Maritimes
288	V	Francia *Karnopp 86-88*	I ScSr 2 *Dks*	German Hamburg	1590 2138	290·0|36·0|23·8 *P 80 f. B 51 f. F 40 f.*	Hamburg Reiherstieg Schiffswrft.	1886	CDB378t	3Cy22"33"&58"-42"250HP ReiherstiegSchiffswrft.,Hm.	Hamb-Amer.Packetf. Act.Ges.
289		Francis Cadell	I PadSp	British Brisbane	105 140	120·0|19·9|6·5	Pt.Adelaide	1866		2Cy. 60 HP.	Reginald Bright
290		Francis Garnier *Marchem 82-87*	I ScSr 2 *Dks*	French Saigon	102 308	135·7|19·9|7·2 *H 39 f.*	Greenock Scott&Co.	1881		2Cy.15"&27"-30" 60 HP. GreenockFoundryCo.,Grnk.	Messageries Fluviales de Cochinchine
291	✻V	Francisca *Arribi -88*	I ScSmSr 2 *Dks* 3 *B*	Spanish Bilbao	1839 2854	315·2|37·1|25·8 *B 96 f. F 37 f.*	Glasgow Lon.&Glsgw.Eng.&I.S.B.Co.	1883	5 Blhds. WB...	2Cy.38"&71"-48" 300 HP. Lon.&Gls.Eng.&I.S.B.C.Gls.	Linea de Vapores Serra
292	✻	Francisco Reyes	I Sc 1 *Dk & Awn Dk*	Spanish Manila	480 703	194·7|27·3|17·0	Glasgow A.&J.Inglis	1880	4 Blhds. CDB112t	2Cy.27"&50"-30" 110 HP. A.&J.Inglis,Glasgow	F.L.Roxas
293	N	Franco (ex Vine) *Waage*	I ScSmSr 1 *Dk*	Norweg'n Stavanger	457 581	185·8|25·4|13·9 *P 77 f. F 32 f.*	Hull C.&W.Earle	1866	3 Blhds. APT	1Cy.18"&30"-26" 80 HP. Cmp.'76,BalticE.Wks.,Hull	H.Madsen
294	✻	Franco Belge *Kegn 87-87*	I Sc 1 *Dk*	Belgian Ostend	62 109	90·0|19·3|10·4 *Q 16 f.*	Grimsby T.Charlton	1884	3 Blhds.	2Cy.15"&29"-21" 35 HP. C.D.Holmes&Co.,Hull	Harbeaux Fileants
295		François Marie	I ScSr 2 *Dks*	French Marseilles	177 239	144·4|20·4|11·8 *Q 14 f.*	Greenock Scott&Co.	1861		2Cy.30"-26" 89 HP. GreenockFoundryCo.,Grnk.	F.Morelli & Co.
296		Francois I. *Manfort 73-86*	I PadSr 1 *Dk*	French Havre	90 159	149·9|15·7|9·6	Havre A.Normand	1866		2C.31½"&50½"-33½"110HP B.Normand,Havre	Cie.Anonyme des Paquebots à Vapeur entre le Havre, Caen, Honfleur et les Ports de la Normandie
297	✻	Francoli *Ydoyaga -88*	I Sc 1 *Dk* 2 *B*	Spanish Barcelona	786 1210	236·5|31·2|18·4 *P 98 f. F 44 f.*	Sunderland J.Laing	1865	6 Blhds.	2Cy.27"&52"-36" 180 HP. T.Richardson&Sons,Hartpl	P.M.Tintoré & Co.
298		Franconia *Bennett 81*	W ScSmSr 2 *Dks*	American Portland,Me	445 673	179·0|30·0|17·5	KennebunkMe	1863		2Cy.30"-13" 250 HP. AtlanticWorks,Boston	Maine S.S.Co.
299		Frank Buckland *Sly*	I PadSk	British Sunderland	43 154	116·0|19·6|10·4	SouthShields J.T.Eltringham	1884		1Cy.38"-56" 75 HP. J.P.Rennoldson&Ss.S.Shlds.	J.J.Usher
300	✻	Frankfort *Newlove -88*	Stl ScSr 1 *Dk* 2 *B*	British Goole	626 983	240·0|33·0|15·3 *P 23 f. B 35 f. F 26 f.*	Hull Earles'Co.,Ld.	1888	5 Blhds. DB...	3Cy.20½",31"&54"-33" Earles'Co.,Ld.,Hull [150HP	Yorkshire Coal & S.S. Co.,Ld.

✻=Lloyd's Register; D=Nederlandsche Vereeniging; G=Germanischer Lloyd; I=Registro Italiano; R=American Record; A=Veritas Austro-Ungarico; H=Veritas Ellenico; N=Norske Veritas; V=Bureau Veritas.

Dans la 2me Colonne, B.T., avec les indications qui suivent, signifie la date du rapport de visite de la Carène par les experts du Board of Trade pour obtenir le certificat relatif au TRANSPORT DES PASSAGERS.

Dans la 3me Colonne, C=Composite (membrure en fer, bordé en bois); I=Fer; Stl=Acier; W=Bois.

Pour plus amples renseignements sur les navires classés par le Lloyd's Register, voir le Lloyd's Register Book et le Lloyd's Yacht Register.

No. in Book. / Official No.	Registry in which classed	Ship's Name. Material, Rig, &c. / Master, and Dates of Appointment to present Owner's Service and to Ship / Ship's Late Name. No. of Decks, &c.	Flag / Port of Registry.	Tons Net / Gross	Dimensions: Length, Breadth, Depth / Deck Erections	Built: Where. / By Whom.	When	No. of Bulkheads / Particulars of Water Ballast.	Engines. Particulars. Maker's Name.	Owners or Managers
301		Frankfurt I ScBg Steenken 76-86 2 Dks & Hur Dk 1 B	German Bremen	1979 2582	311·6 | 39·4 | 31·2	Greenock Caird & Co.	1869		2Cy.37" & 68"–48" 300 HP. Caird & Co., Greenock	Norddeutscher [illegible]
302 62542	✻	Frankland I ScSc Bragg 80-82 1 Dk	British Sunderland	541 706	196·8 | 28·9 | 16·8 Q 22 f.	Sunderland J. Laing	1869	4 Bhds. [illegible]	2Cy.26" & 48"–30" 90 HP. Comp. 76, J. Dickinson, Sndld	H. T. Morton [illegible]
303 78238		Franklin I ScSc Brown 1 Dk & Awn Dk	British Melbourne	395 720	200·0 | 26·3 | 11·7 B 26 f. 19·4	Glasgow D. & W. Henderson	1880	5 Bhds.	2Cy.28½" & 57"–36" 284 HP. D. & W. Henderson, Glasgow	Adelaide S.S. [illegible]
304	N	Franklin (ex Sailing Vessel) W ScSmSr Kopstad (87 Brazelles) 1 Dk 1 B	Norweg'n Sandefjord	100 148	99·9 | 22·4 | 10·3 P 18 f.	Arendal	1872		2Cy 10¼" & 18½"–10½" 12 HP. Akers Mek. Verks, Christiania	Actieselskabet [illegible] (Chr. Christensen)
305	G	Franz I ScSc Hein 76-83 2 Dks	German Kiel	657 850	203·7 | 26·1 | 20·3	Kiel G. Howaldt	1881		2Cy.25" & 40"–27" 85 HP. Howaldt Bros., Kiel	Sartori & Berger
306	G	Franziska I ScSc Kratzfeldt 81-82	German Kiel	556 860	201·0 | 26·0 | 20·2	Kiel G. Howaldt	1882		2Cy.25" & 39½"–27" 85 HP. G. Howaldt, Kiel	Sartori & Berger
307	✻	Fratelli Prinzi I ScSc Tomaselli 87 (ex Queensferry) 1 Dk 2 B	Italian Catania	621 1111	225·0 | 32·5 | 16·0 [illegible] B 42 f. F 28 f.	Sunderland R. Thompson & Sons	1881	[illegible]	2Cy.28" & 54"–36" 125 HP. Black, Hawthorn & Co., Gteshd.	Frat. Prinzi
308 93788	✻	Freda Stl ScSc Curran 88-88 1 Dk	British Liverpool	310 498	158·0 | 30·2 | 11·9 B 39 f. F 19 f.	Barrow Naval Construction & Armaments Co., Ld.	1888	4 Bhds. FPT 91 APT 23	2Cy.18" & 36"–30" 54 HP. Naval Construction & Armaments Co., Ld., Barrow	Argentine S. [illegible] Lighter Co., Ld.
309	✻	Freda I ScSc Schmidt 1 Dk 2 B	German Danzig	710 1004	215·0 | 29·6 | 16·2 P 120 f. F 21 f.	Newcastle Swan & Hunter	1883	4 Bhds. FPT 38 APT [illegible]	2Cy.28½" & 55"–33" 110 HP. Blair & Co., Stockton	Th. Rodenacker
310	✻V	Frédéric Franck I ScSc Séraing 77-87 2 Dks	French Havre	559 846	231·2 | 30·6 | 16·6 P 31 f. F 33 f.	Renfrew Lobnitz, Coulborn & Co.	1875	3 Bhds. MT 208 APT 10	2Cy.32" & 60"–42" 170 HP. Lobnitz, Coulborn & Co., Rfw.	Worms, Josse & [illegible]
311	V	Frédéric Morel I ScSc Binquet 84-84 1 Dk & Spr Dk 3 B	French Dunkirk	970 1468	250·3 | 33·1 | 15·0 B 44 f. 23·1	South Shields J. Readhead & Co.	1883	[illegible]	2Cy.32" & 61"–36" 160 HP. J. Readhead & Co., S. Shields	Cie. Générale de Bateaux à vapeur
312		Frederic de Bary I Pad Shaw 87	American Wilmingt(n)D	323 395	161·6 | 24·2 | 8·0	Wilmington Del Pusey & Jones Co.	1881	4 Bhds.	1Cy.30"–72" Pusey & Jones Co., Wilmington Del	F. de Bary & Co.
313		Fredericia I Pad Nielsen 82-82 1 Dk	Danish Fredericia	132 304	138·9 | 25·4 | 10·8	Elbing F. Schichau	1877		2Cy.36"–45" 90 HP. F. Schichau, Elbing	De Danske S. [illegible] (N. Holst)
314 53017	✻D	Frederick Snowdon I ScSc Watson 1 Dk 2 B	British Newcastle	477 784	200·0 | 29·2 | 15·9 Q 32 f.	Middlesbro' Backhouse & Co.	1866	2 Bhds. [illegible]	2Cy.26" & 48"–33" 98 HP. Dunston Eng. Wks. Co., New.	W. & G. Hedley
315		Frederik I ScSmSr Harboe 67-72 1 Dk 2 B	Danish Copenhagen	702 1113	232·1 | 30·0 | 16·5 P 150 f. F 33 f.	Copenhagen Burmeister & Wain	1872	4 Bhds. [illegible]	4Cy.24" & 44"–30" 120 HP. Burmeister & Wain, Cpnhgn.	Det Forenede [illegible] Selskab.
316	✻D	Frederiksborg I ScSc Olsen 86-86 1 Dk 2 B	Danish Copenhagen	841 1328	237·1 | 33·2 | 15·1 P 144 f. F 28 f.	Copenhagen Burmeister & Wain	1883	4 Bhds. [illegible]	2Cy.30" & 52"–33" 120 HP. Burmeister & Wain, Cpnhgn.	Dampsk.-Selsk. "Dannebrog"
317		Frederikshavn I ScSalvageSr Pii 77-79 1 Dk	Danish Copenhagen	105 214	133·9 | 21·7 | 12·0 B 8 f.	Copenhagen Burmeister & Wain	1875		4Cy.16½" & 34"–20" 80 HP. Burmeister & Wain, Cpnhgn.	Em. Z. Svitzer's [illegible] nings Entreprise
318		Fredrik C ScSc Jönsson 74-83 1 Dk	Swedish Gefle	232 318	129·6 | 21·0 | 13·5 B 29 f. H 7 f.	Eckerna Ahlberg	1872		45 HP. Gothenburg Mek. Verkst.	L. V. Åkerson
319		Fredrikstad I ScSc 1 Dk	Norweg'n Fredrikstad	91 137	104·9 | 16·4 | 8·6 Q 11 f. B 26 f.	Bergen	1867		30 HP.	Et Interessentskab (Th. Cappelen)
320 68299	✻	Free Lance I ScSc Gibson 85-85 1 Dk 2 B	British Middlesbro'	756 1180	262·6 | 30·3 | 17·6 Q 66 f. B 81 f. F 13 f.	Middlesbro' R. Dixon & Co.	1873	5 Bhds. [illegible]	2Cy.28" & 53"–33" 99 HP. T. Richardson & Sons, Hartpl.	Christie & Co.
321	V	Freia StlPadSc Wahlen 2 Dks	German Hamburg	421 682	235·9 | 26·6 | 13·4 B 184 f.	Hamburg Blohm & Voss	1885		2Cy.40" & 67"–72" 400 HP. Blohm & Voss, Hamburg	Blohm & Voss
322 93401	✻	Freiston I ScTrawler Harrod 88-88 1 Dk	British Boston	52 105	85·0 | 20·0 | 9·9 F 17 f.	Hull Earle's Co., Ld.	1886	3 Bhds.	2Cy.12" & 22"–20" 25 HP. Earle's Co. Ld., Hull	Boston Deep Sea Fishing & Ice Co., Ld.
323	G	Frej C ScSmSr Söderström 82-86 1 Dk	Swedish Vänersvik vid Oscarshamn	148 188	99·1 | 20·3 | 10·1 B 29 f. H 29	Oscarshamn	1876		20 HP.	C. J. Schylander
324	✻DN	Freja (Hamburg, ex Bampton) I ScBg Hauge 88-88 2 Dks 3 B	Norweg'n Bergen	1344 1848	282·5 | 33·4 | 24·0 [illegible]	Sunderland [illegible]	1873	[illegible]	2Cy.35" & 67"–36" 200 HP. T. Richardson & Sons, Hartpl.	Chr. Michelsen [illegible]
325		Freja C ScSc Ramström 84-86	Russian Reval	100 156	101·3 | 21·7 | 10·5	Motala Motala Co.	1859		2 45 HP. Motala Co., Motala	Russisch-B[illegible] [illegible]
326	✻	Freja (ex Rievaulx Abbey, ex Ithaca) I ScSc Ryeggren 1 Dk 2 B	Swedish Stockholm	852 1124	220·3 | 30·1 | 16·8 P 129 f. F 30 f. H 13 f.	Sunderland R. Thompson, jun.	1873	4 Bhds.	2Cy.27" & 51"–33" 98 HP. J. Dickinson, Sunderland	Rederi Aktie[bolaget] "Svea"
327	G	Freja C ScSc Lonnquist 86-86 1 Dk	Swedish Christineh'n	167 213	102·6 | 20·9 | 12·6 P 21 f.	Christinehamn E. J. Ericsson	1878		2Cy 11½" & 27½"–16¼" 35 HP.	Ångbåtsaktie[bolaget] "Ferm"
328	V	Frejr I ScSc Lund 85-86 2 Dks & Awn Dk	Danish Randers	419 676	165·6 | 26·6 | 13·3 B 13 f. 19·7	Renfrew Lobnitz & Co.	1883	[illegible]	2Cy.24" & 44"–30" 85 HP. Lobnitz & Co., Renfrew	Randers Damp[skibs] Selskab af 18[illegible]
329 93470	✻	Fremona Stl ScSc Anderson 72-87 2 Dks & W. h Frame	British Dundee	1840 2837	328·0 | 42·2 | 25·9 P 32 f. B 80 f. F 78 f.	Dundee Gourlay Bros. & Co.	1887	[illegible]	3C.26" 41" & 66"–48" 360 HP. Gourlay Bros. & Co., Dundee	W. Thomson & [illegible]
330	N	Frey W ScSc Clausen 85-86 1 Dk 2 L	Norweg'n Bergen	329 460	164·1 | 25·8 | 15·5 B 40 f.	Bergen Brunchorst & Dekke	1873		2Cy.19½" & 37"–21" 55 HP. Bergens Mek. Verkst, Bergen	Actieselskabet (Berner, Bach [illegible])

✻ = Lloyd's Register; D = Nederlandsche Vereeniging; G = Germanischer Lloyd; I = Registro Italiano; R = American [illegible]
A = Veritas Austro-Ungarico; H = Veritas Ellenico; N = Norske Veritas; V = Bureau V[eritas]

In the 2nd Column, B.T., with figures following, signifies Date of report of bottom sighting by Board of Trade Surveyors as condition of PASSENGER CERT[IFICATE].

In the 3rd Column, C = Composite (iron frame, wood planked); I = Iron; Stl = Steel; W = Wood.

Further particulars of Vessels classed by Lloyd's Register will be found in Lloyd's Register Book and Lloyd's Yacht Register.

N°	Registre dans lequel le navire est classé	Nom, Matériaux, Gréement, etc. Capitaine et dates d'entrée au service des Armateurs actuels et à bord du navire. Ancien Nom, nombre de ponts, etc.		Pavillon. Port d'Armement	Tonnage. Net. Total	Dimensions. Longueur, Largeur, Creux. Constructions sur le Pont	Construit. Où. Par qui.	Quand.	Nombre des cloisons. Détails sur le lest d'eau.	Machines. Détails. Nom du Constructeur.	Armateurs ou Directeurs-Gérants.
331	N	Frey *Ofte* -88	W ScSr 1 *Dk*	Norweg'n Aalesund	207 291	131·1|25·3|11·2 Q 31 *f.* B 42 *f.*	Malmö J.Bonnesen	1873		1Cy.12"&26"–22" 45 HP. Burmeister&Wain,Cpnhgn.	C.E.Rönneberg & Sönner
332		Freya *Arelsen*	I PadSr 1 *Dk*	Danish Kjöge	205 372	202·9|21·4|10·5	Copenhagen Baumgarten & Burmeister	1862		2Cy.50"–48" 160 HP. Burmeister&Wain,Cpnhgn.	De Danske Statsbaner
333		Freyr	I PadSr 1 *Dk*	Danish Fredericia	113 189	151·9|17·9|10·2	London J.Maudslay	1853		2Cy.40½"–42¼" 100 HP. J.Maudslay,London	De Danske Statsbaner
334	✱D	Friary *Evans*	I ScSr 1 *Dk*&*Spr Dk* 3 *B*	British Newport	1506 2387	295·7|37·9|20·0 27·5	Sunderland J.Laing	1882	6 Blds. DB&a / APT	2Cy.37"&68"–42" 280 HP. N.E.MarineE.Co.Ld.,Sundld	Jones Bros.& Co.
335		Fridolf	C ScSr 1 *Dk*	Swedish Thorskog	124 165	101·3|20·3|10·6 H 11 *f.*	Thorskog P.Larsson	1884			P.Larsson
336		Fridolf *ex Helfrid* *Andersson* -86	C ScSr 1 *Dk*	Swedish Gothenburg	122 174	101·9|20·6|11·0 P 14 *f.* B 4 *f.*	Norrköping	1872		25 HP.	A.Liljequist
337	NV	Fridtjof *Refsdal* -88	I ScSr 1 *Dk*&*Awn Dk*	Norweg'n Bergen	725 951	190·4|29·3|22·6 B 20 *f.*	Bergen BergensMekVerkst	1872	5 Blds.	2Cy.29"&54"–28" 120 HP. BergensMek.Verkst,Bergen	P.G.Halvorsen
338		Friede *ex Marseille, ex Baron Bohlen, ex Bergen, ex* *Bankow/Kayser* -88	I ScSr 1 *Dk*&*Awn Dk*	German Elsfleth	353 556	175·2|23·1|17·5	Liverpool LairdBros.	1864	4 Blds.	3Cy.18"&(2)28"–30" 70 HP. A.&J.Inglis,Glasgow	Joh.Aug.Visser
339	✱D	Friedrich Krupp *Gehrke* 80-89	I ScSr 1 *Dk* 2 *B*	German Essen	919 1240	240·1|32·5|19·6 B F 34 *f.*	Stockton M.Pearse&Co.	1874	4 Blds. DB	2Cy.28"&56½"–36" 120 HP. Blair&Co.,Stockton	Friedr.Krupp
340	G	Friese *ex Sabine, ex* *John* -86 *Ville d'Ostende)*	I PadSr 1 *Dk*	German Kiel	147 219	167·6|21·3|10·6	Antwerp J.Cockerill	1849		3Cy.(2)22"&48"–42½"	Friesische Dampfsch. Ges.(J.Mestorff)
341	G	Frigg *Andersson* 84-84	C ScSr 1 *Dk*	Swedish Christ'hamn	157 212	102·6|20·9|12·7 F 19 *f.*	Sjötorp S.Groth	1882		2Cy.10¼"&25"–17½" 28 HP.	A.Broström
342	✱V	Frigga *Nagel* 76-87	Stl Sc 2 *Dks* 3 *B*	German Hamburg	1400 1970	289·5|36·6|21·9 P 52 *f.* B 72 *f.* F 44 *f.*	Flensburg FlensburgS.B.Co.	1887	7 Blds. DB	3Cy.22½"40"&64"–42" 320 HP. FlensburgS.B.Co.,Flensburg	Deutsche Dampfs. Rhederei
343	V	Frigga *Ericsson* -87	I ScSr 1 *Dk*	Swedish Gothenburg	275 385	135·5|23·4|13·9 B 55 *f.* H 4 *f.*	Oscarshamn Oscarshamns, M.V.	1878		2Cy.16"&32"–22" 50 HP. R.Martin,Oscarshamn	J.F.Olson
344	V	Frisia *Willycombe* -88	I ScSr 2 *Dks*&*Spr Dk*	British London	2083 3268	349·5|40·2|17·7	Greenock Caird&Co.	1872		2Cy.62"&104"–54" 500 HP. Caird&Co.,Greenock	T.B.Williams (W.Lamplough&Co.,Mngers)
345	V	Frithiof *Rhodin* 77-80	I ScSr 1 *Dk*&*Awn Dk*	Swedish Gothenburg	604 766	193·2|23·9|14·7 H 14 *f.* Awn 74	Newcastle Schlesinger,Davis&Co.	$18\frac{70}{80}$		2Cy.22½"&13⅝"–28¼" 60 HP. T.Clark&Co.,Newcastle	Ångf.Aktiebolag. "Thule" (O.Melin)
346		Frithiof	W ScSr 1 *Dk*	Swedish Tösse	89 111	94·2|18·0|8·2 H 14 *f.*					E.G.Vassberg
347	N	Frithjof *Hansen*	W ScBk 1 *Dk* 2 *B*	Norweg'n Tönsberg	164 255	121·4|25·9|12·9 Q 10 *f.* F 25 *f.*	Tönsberg O.Ellefsen	1884		2Cy.18½"&33"–24¾" 50 HP. NylandsVerkst.,Christiania	Red'riselsk."Frithjof" (A.Ellefsen)
348	✱	Frogmore *Fulton*	I ScSr 1 *Dk* 2 *B*	British Liverpool	1000 1547	259·5|33·9|17·9 Q 87 *f.* B 148 *f.* F 39 *f.*	Liverpool T.Royden&Sons	1880	4 Blds. DB&MT APT	2Cy.28"&56"–39" 140 HP. J.Jack&Co.,Liverpool	S. S. "Frogmore" Co., Ld. (W.Johnston & Co.)
349		Frontera	I ScSr	Mexican Frontera	90 113	100·0|18·0|7·6 H 30 *f.*	WilmingtnDel Pusey&JonesCo.	1878		2Cy.12"&21"–18" 25 HP. Pusey&JonesCo.Wilmingtn	Bulnes Hermanos
350		Frost	I ScSr	British London	142 209	130·6|21·1|10·6	Stockton M.Pearse&Co.	1866		2Cy.28¼"–24" 50 HP. Fossick&Hackworth,Stktn.	Hewett & Co.,Ld.
351	V	Frostburg *Mills*	I ScSr 2 *Dks*	American Baltimore	564 927	204·0|33·0|16·0	P Richmond Pa AmericanS.B.Co.	1884	4 Blds.	2Cy.26"&48"–36" 100 HP. Neafie&Levy,Philadelphia	Consolidation Coal Co.
352	✱D	Frutera *Symons* 72-84	I ScSr 1 *Dk*& *Web Frames*	British Liverpool	638 1001	216·4|31·1|15·5 Q 75 *f.* B 64 *f.* F 45 *f.*	Paisley H.M'Intyre&Co.	1883	4 Blds. DB&a / 303t FPT 30t APT 18t	2Cy.28"&54"–36" 124 HP. J.Howden&Co.,Glasgow	"Frutera"S.S.Co.,Ld. (Goodyear & Co.)
353		Fuh Le *M'Caslin*	I Sc 1 *Dk*	British Shanghai	38 101	95·2|18·1|9·1	Hartlepool R.Irvine&Co.	1870	3 Blds.	2Cy.21"–18" 60 HP. D.Joy&Co.,Middlesbro'	ShanghaiTugBoat Co. Ld.
354		Fuh Wo *Davis* 82-87	I ScSr	British London	764 992	204·0|26·1|17·8	Shanghai Boyd&Co.	1879		2Cy.16"&36"–36" 52 HP. Boyd&Co.,Shanghai	Indo-China Steam Nav.Co.,Ld.
355	GV	Fulda *Bingk* 72-85	I ScImSr 3 *Dks*&1 *B*	German Bremen	2864 5124	429·8|45·9|27·0 P 51 *f.* B 138 *f.* F 72 *f.* 35·8	Glasgow J.Elder&Co.	1882		3Cy.62"&86"–60" 1000 HP. J.Elder&Co.,Glasgow	Norddeutscher Lloyd
356		Fulda	I PadSp	German Bremen	68 158	150·0|17·4|7·9	Bremen C.J.Wätjen&Co.	1858		2Cy.27"–48"	Norddeutscher Lloyd
357	✱	Fulford *Tomlinson* 56-88	Stl ScSr 1 *Dk*& *Wb Frms*	British Middlesbro'	1165 1786	257·0|36·0|18·1 P 24 *f.* B 112 *f.* Q 65 *f.* F 30 *f.*	Middlesbro' R.Dixon&Co.	1888	4 Blds. DB&a / 477t	3Cy.20",33"&54"–36" 150 HP. Blair&Co.,Ld.,Stk.	J.M.Lennard & Sons
358	✱	Fulmar *O'Brien* -88	Stl Sc3mSr 2 *Dks*	British Cork	524 1063	250·0|32·2|15·4 P 109 *f.* B 28 *f.* F 46 *f.*	Dundee W.B.Thompson&Co. Ld.	1888	5 Blds. CDB	3Cy.21",34"&55"–42" 175 HP. W.B.Thompson&Co., Ld.,Dundee	Cork S.S.Co.,Ld.
359	✱	Fulmar *Forbes* 87-	W ScK 1 *Dk*	British Granton	20 48	67·5|17·6|7·5	Leith J.M'Kenzie&Co.	1885		2Cy.8"&15"–12" 10 HP. J.Cran&Co.,Leith	J.L.Cunliffe
360	✱	Funchal	I ScBn 1 *Dk*&*Spr Dk* 3 *B*	Portug'se Ponta Delgada St. Michaels	1134 1743	280·0|34·6|14·1 B 12 *f.* F 26 *f.* 22·6	Sunderland J.Laing	1884	5 Blds. CDB	2Cy.34½"&64"–42" 180 HP. Blair&Co.,Ld.,Stockton	Empreza Insulana de Navegação

✱=**Lloyd's Register**; D=Nederlandsche Vereeniging; G=Germanischer Lloyd; I=Registro Italiano; R=American Record; A=Veritas Austro-Ungarico; H=Veritas Ellenico; N=Norske Veritas; V=Bureau Veritas.

Dans la 2me Colonne, B.T., avec les indications qui suivent, signifie la date du rapport de visite de la Carène par les experts du Board of Trade pour obtenir le certificat relatif au TRANSPORT DES PASSAGERS.

Dans la 3me Colonne, C=Composite (membrure en fer, bordé en bois); I=Fer; Stl=Acier; W=Bois.

Pour plus amples renseignements sur les navires classés par le Lloyd's Register, voir le Lloyd's Register Book et le Lloyd's Yacht Register.

No. in Book. / Official No.	Registry in which classed.	Ship's Name. Material, Rig, &c. / Master, and Dates of Appointment to present Owner's Service and to Ship. / Ship's Late Name. No. of Decks, &c.	Flag. / Port of Registry.	Tons. Net / Gross	Dimensions. Length	Breadth	Depth	Built. Where. / By Whom.	When	No. of Bulkheads. / Particulars of Water Ballast.	Engines. / Particulars. Maker's Name.	Owners of Vessel
361		Fung shun I ScSr *Tisdall 74-?* *2 Dks*	Chinese Shanghai	863 1388	225·0	34·0	21·0	Dumbarton W.Denny&Bros.	1875		2Cy.37½"&65"-36" 187 HP. Denny,Dumbarton	China Merchants Steam Nav.Co.
362 78[illegible]	BT[illegible]	Furnessia I ScBr *Hedderwick 87-83* *3 Dks&I [illegible]*	British Barrow	3613 5495	445·1	44·8	34·5	Barrow BarrowS.B.Co.,Ld.	1880	[illegible] Bhds.	2Cy.43"&100"-[illegible] [illegible]HP. BarrowS.B.Co.,Ld.,Barrow	Henderson Bros.
363		Fu shun Stl ScSr *Wood 74-?* *3 Dks*	Chinese Shanghai	1504 2362	280·0 *B76f.*	40·0 *F59f.*	17·9 25·4	Glasgow W.B.Thompson	1883		2Cy.36"&72"-48" 2[illegible] HP. W.B.Thompson,Dundee	China Merchants Steam Nav.Co.
364		Fu Yew (ex Paris) I ScBunS *Graham 87-88* *2 Dks&pt Awn Dk*	Chinese Shanghai	936 1251	262·5 *Len. 1874*	28·4	15·2 22·5	Glasgow Barclay,Curle&Co.	1865	[illegible] Bhds.	2Cy.32¼"&57¼"-33" 1[illegible] HP. LairdBros.,Birkenhead	China Merchants Steam Nav.Co.
365 1047	✻	Fushiki Maru I ScSr *Periam 87-87* *1 Dk2 H*	Japanese Tokio	1119 1919	275·0 *P29f.*	37·5 *B60f.*	20·1 *F30f.*	Newcastle TyneS.B.Co.,Ld.	1885	[illegible] Bhds. [illegible]	2Cy.35"&68"-45" 1[illegible] HP. [illegible] Hawthorn,Leslie&Co.,Ld.,Newcastle	Nippon Yusen [illegible]
366 45[illegible]	✻	Fusi Yama I ScSr *Cain* *2 Dks3 H*	British London	632 994	214·3 *B48f.*	29·3 *F36f.*	20·5	Stockton M.Pearse&Co.	1862	[illegible] Bhds.	[illegible] 27"&[illegible] [illegible] HP. M.Samuelson,Hull	E.L.Evan-Thomas
367 9[illegible]	BT[illegible]	Fusilier I&StlPadSk *Baxter 78-88*	British Glasgow	21 251	202·0	21·6	8·1	Paisley M.Arthur&Co.	1888		2Cy.49"-90" 1[illegible] HP. Hutson&Corbett,Glasgow	D.MacBrayne
368 1011	✻	Fuyo (ex Prinz Heinrich) I ScSr *Trewitt* *1 Dk2 H*	Japanese Nagasaki	790 1274	238·5 *Q92f.*	35·0 *B44f.*	18·0 *F27f.*	Sunderland ShortBros.	1878	[illegible] Bhds. [illegible]	2Cy.31"&57"-36" 1[illegible] HP. J.Dickinson,Sunderland	Iwasaki Kiuya
369	V	Fyen I ScBunSr *Petersen 81-83* *2 Dks&SprDk*	Danish Copenhagen	900 1399	232·9 *H18f.*	32·5	17·8 21·6	Kiel Nordd.Schiffsb.A.G.	1874	WB330t	2Cy.32"&52"-36" 125 HP. Schweffel&Co.,Kiel	Dampskibs-Selsk. "Kjobenhavn" (C.B. [illegible])
370		Fyen I ScSr *Feidemann 67-87* *1 Dk*	Danish Korsør	65 166	129·8	19·0	11·8	Copenhagen Burmeister&Wain	1867		1Cy.15"&30"-20" 60 HP. Burmeister&Wain,Cpnhgn.	De Danske S[illegible] (N.Holst)
371 73764	BT[illegible]	Fylde I TwnScSk	British Fleetwood	59 216	122·4	22·5	10·8	PortGlasgow J.Reid&Co.	1881		2Cy.25"&48"-27" 98 HP. Rankin&Blackmore,Grnk	Lancashire & Yorkshire & London & N.W. [illegible]
372 86[illegible]	✻	Fylingdale I ScSr *Donovan* *1 Dk2 H*	British Whitby	1072 1657	259·7 *Q100f.*	36·0 *B58f.*	18·6 *F26f.*	Sunderland J.L.Thompson&Sns	1883	[illegible] Bhds. [illegible]	2Cy.32"&59"-33" 150 HP. T.Richardson&Sons,Hartpl	J.Rowland
373		Fylla I ScSr *Olsen 78-81* *1 Dk*	Danish Nykjøbing, Jylland	123 212	133·9 *Q38f.*	21·2 *B8f.*	9·8	Copenhagen Baumgarten & Burmeister	1863		1Cy.15"&30"-20" [illegible] HP. Burmeister&Wain,Cpnhgn	Det Forenede [illegible] skibs Selskab (C.[illegible])
374		Fylla I ScSr *Christopfursen 73-83* *1 Dk*	German Sonderburg	111 157	114·5 *Q42f.*	19·4 *B27f.*	9·7 *F32f.*	Flensburg FlnsbrgrSchiffsb.Gs	1881		2Cy.18"&31"-21" [illegible] HP. FlensburgerSchiffsb.Ges.	Sonderburger Dampfschifffahrts Akt.Ges.
375		Fyris I ScSp *Karlin 60-76* *1 Dk*	Swedish Upsala	119 152	104·6 *Q32f. B38f.*	17·8 *F18f.*	8·0 *H6f.*	Stockholm BergsundMekVrks.	1876		2Cy.17½"&31½"-19½ 10 HP. BergsundMekVerks,Stkhlm	G.F.Stadenberg
376		Fyris I TwnScSr *Lundbäck* *1 Dk*	Swedish Hernösand	102 101	95·3 *Q29f. B40f.*	18·2 *F12f.*	8·6 *H7f.*	Stockholm W.LindbergMekVrks.	18[illegible]		1Cy.12½"&28"-16" [illegible] HP. W.LindbergsMekVerks,Stk.	N.J.Nyberg

Additional Ships are inserted in the Supplement at the end of the Register.

✻ = Lloyd's Register; A = Veritas Austro-Ungarico; D = Nederlandsche Vereeniging; G = Germanischer Lloyd; H = Veritas Ellenico; I = Registro Italiano; N = Norske Veritas; R = American Record; V = Bureau Veritas.

In the 2nd Column, B.T., with figures following, signifies Date of report of bottom sighting by Board of Trade Surveyors as condition of PASSENGER CERTIFICATE.

In the 3rd Column, C = Composite (iron frame, wood planked); I = Iron; Stl = Steel; W = Wood.

Further particulars of Vessels classed by Lloyd's Register will be found in Lloyd's Register Book and Lloyd's Yacht Register.

Registre dans lequel le navire est coté		Nom, Matériaux, Gréement, etc. / Capitaine et dates d'entrée au service des Armateurs actuels et à bord du navire. / Ancien Nom, nombre de ponts, etc.		Pavillon. / Port d'Armement	Tonnage. Net / Total	Dimensions. Longueur	Largeur	Creux	Construit. Où. / Par qui.	Quand	Nombre des cloisons. / Détails sur le lest d'eau.	Machines. Détails. / Nom du Constructeur.	Armateurs ou Directeurs-Gérants.
1		G. B. Patten	C Pad 1 Dk	British Ottawa	172 [illegible]	139·0	36·0	7·2	Pembr'ke, Ont A.Miller&W.C.White	1887		1Cy.25"-60" 21 HP. J.&R.Weir, Montreal, P.Q.	Upper Ottawa Towing Co.,Ld.
2	✱D	G. E. Wood *Hunter 87-87*	I ScSr 1 Dk2 B	British Cardiff	678 1081	221·1	30·5	16·9	Sunderland *P 125 f. F 38 f.* J.L.Thompson	1873	4 Blds. DBcf 87ft f35ft	2Cy.27"&51"-33" 99 HP. J.Dickinson,Sunderland	J.Marychurch & Co.
3	✱	G. M. B. *Dunlop*	Stl Sc3mSr 1 Dk2 B	British Middlesbro'	350 [illegible]	175·4	25·7	14·1	Middlesbro' *P 61 f. F 26 f.* R.Dixon&Co.	1878	3 Blds. DB f148t	2Cy.25"&48"-30" 90 HP. T.Richardson&Sons,Hartpl.	J.Watson & Co.
4	✱	G. N. Wilkinson *Hodgson 71-87*	I S Sr 1 Dk2 B	British Cardiff	476 719	201·8	28·3	16·3	W.Hartlepool *Q 22 f. F 4 f.* Denton,Gray&Co.	1869	4 Blds. DBa60ft f35ft	2Cy.26"&47"-36" 90 HP. G.Clark,Sunderland	J.G.Hill (Harrison Bros. & Moore)
5	✱	G. W. Jones *Lapsley 87-87*	I S Sr 1 Dk2 B	British London	627 983	220·3	31·6	13·8	Sunderland *Q 108 f. B 20 f. F 24 f.* J.Blumer&Co.	1883	4 Blds. DBaf70t FPT23t APT15t	2Cy.27"&50"-33" 95 HP. N.E.MarineE.Co.,Ld.,Nwc.	R.M.Phillips
6	✱N	Gaa Paa *Iverson*	Stl ScSr 1 Dk2 B	Norweg'n Moss	370 509	162·9	24·5	13·7	Gothenburg *Q 56 f. B 37 f. F 15 f.* MotalaCo.	1883	4 Blds. APT28t FPT16t	2Cy13½"&35½"-23⅝"70HP. MotalaCo.,Gothenburg	R.M.Peterson
7	✱D	Gabalva *Coleman -88*	Stl ScSr 1 Dk2 B	British Cardiff	1091 1662	259·5	36·0	17·7	Newcastle *Q 129 f. B 29 f. F 28 f.* Wood,Skinner&Co.	1888	4 Blds. DB f176t	3C20",33"&54"-36"150HP N.E.MarineE.Co.,Ld.,Nwc.	J.H.Wilson
8	✱	Gabo *Payne*	Stl ScSr 1 Dk2 B	British Melbourne	1246 2060	279·7	38·4	20·0	Kinghorn *P&B 140 f. F 43 f.* J.Key&Sons	1883	5 Blds. DBa&cf628t MT32t	2Cy.36"&70"-48" 300 HP. J.Key&Sons,Kirkcaldy	W.H.Smith & Sons,Ld.
9	✱	Gaboon *Hartje -88*	I ScSr 2 Dks	British Glasgow	1126 1788	302·3	34·2	23·6	Glasgow *P 106 f. F 50 f. H 42 f.* J.Elder&Co.	1878	6 Blds.	2Cy.40"&71"-36" 280 HP. J.Elder&Co.,Glasgow	British & African Steam Nav.Co.
10	V	Gabrielle *Corbière*	Stl & I ScSr Yt 1 Dk	French Havre	101 202	131·2	19·7	11·1	Antwerp Soc.JohnCockerill	1884		2Cy.20"&32"-24" 90 HP. Soc.JohnCockerill,Seraing	F.Sieber
11	✱D	Gaditano (ex Triumph) *Goicoechea 85-88*	I ScSr 2 Dks3 B	Spanish Bilbao	1797 2749	340·0	37·3	27·2	Middlesbro' R.Dixon&Co.	1880	5 Blds. DBa&cf E&1398t	2Cy.48"&80"-51" 400 HP. Blair&Co.,Ld.,Stockton	M.M.de Arrotegui
12		Gael	I PadSr	British Milford	107 403	211·0	23·2	10·6	Greenock *Q 27 f.* Robertson&Co.	1867		2Cy.45"-63" 150 HP. Rankin&Blackmore,Grnk.	Great Western Railway Co.,Ld.
13		Gaelic *Pearne*	Stl 4mScSr 3 Dks	British Liverpool	2691 4206	420·3	42·4	29·6	Belfast Harland&Wolff	1885	7 Blds.	2Cy.43"&86"-60" 500 HP. Harland&Wolff,Belfast	Oceanic Steam Nav. Co.,Ld.
14	✱	Gairloch *Danvers 87-88*	Stl ScSr 1 Dk Web Frms	British Glasgow	1422 2178	282·0	37·0	19·2	Glasgow *P 24 f. Q 86 f. B 112 f. F 31 f.* A.Stephen & Sons	1887	5 Blds.	3C21",33"&54"-42"220HP. A.Stephen&Sons,Glasgow	J.Gardiner & Co.
15	✱	Gairloch *McArthur 82*	Stl TwnSc3mSr 1 Dk	British Auckland	211 373	164·9	23·0	8·7	PortGlasgow *Q 66 f. B 37 f. F 18 f.* Blackwd.&Gordon	1884	3 Blds.	1Cy.17"&31"-21" 85 HP Blackwood&Gordon,Pt.Gls.	Northern S.S.Co.,Ld.
16	V	Gaiser *Schacht*	I ScSr 1 Dk	German Hamburg	197 277	158·8	24·2	11·2	Hamburg *P 37 f. F 21 f.* RhrstgSchiffs Wfte	1878		2Cy.16"&28"-21" 50 HP. ReiherstiegSchiffsWerfte	G.L.Gaiser
17		Gaiyret *Simoli 69-*	I Pad	Turkish Cnstntinple.	137	160·0	20·0	9·6	London R.&H.Green	1872		80 HP. J.Penn&Sons,London	Chirket-i-Hairie (Bosphorus Steam Nav.Co.)
18		Galata *Yanni 85-85*	W Pad	Turkish Cnstntinple.	64	120·0	17·6	8·1	WestCowes J.White	1860		40 HP. J.Penn&Sons,London	Chirket-i-Hairie (Bosphorus Steam Nav.Co.)
19	✱	Galatea *Lumley -87*	I ScSr 1 Dk	British London	348 559	170·3	27·0	12·8	Sunderland *Q 63 f. B 47 f. F 25 f.* SunderldS.B.Co.Ld.	1887	4 Blds.	3C16",26"&43"-30" 95HP. NEMarineECoLd.Sundrlnd.	C.F.Leach
20		Galatea (ex Irene)	I PadSr	British London	292 507	219·5	26·2	13·8	Greenock Caird&Co.	1868		2Cy.52"-66" 200 HP. Caird&Co.,Greenock	Corporation of Trinity House
21		Galatea *Mitis 73-87*	I ScSr 2 Dks	Austrian Trieste	886 1352	255·4	31·7	23·0	Trieste AustroUng'oLloyd	1871		2Cy.33¾"&64"-36"180 HP. ArsenaleLloyd,Trieste	Lloyd Austro-Ungarico
22	✱	Galatz *Protheroe*	I Sc3mSr 1 Dk2 B	British N.Shields	571 884	212·5	30·2	17·2	Newcastle *H 28 f. Q 34 f.* Schlsngr.Davis&Co	1870	5 Blds. WBcf	2Cy.27"&51"-33" 99 HP. T.Clark&Co.,Newcastle	C.E.Stallybrass
23	✱	Galatz *Maigre 63-86*	I ScBn 1 Dk&Spr Dk3 B	French Marseilles	1175 1727	259·2	37·1	23·0	Dumbarton *P 25 f. B 28 f. F 26 f.* A.McMillan&Son	1881	5 Blds.	2Cy.36"&63"-48" 300 HP. D.Rowan,Glasgow	Fraissinet & Co.
24	✱D	Galea (ex Godrevy) *Goyoaga 86-86*	I ScSr 1 Dk2 B	Spanish Bilbao	714 1106	221·0	32·0	16·1	Sunderland *Q 84 f. B 46 f. F 28 f.* J.Laing	1879	4 Blds. DBa55f&f84f	2Cy.27"&50"-36" 100 HP. G.Clark,Sunderland	Soc.Maritima de Vizcaya
25	✱	Galgo *Fiancireda*	StlScSrTug 1 Dk	Portug'se Oporto	23 75	84·2	17·1	8·6	Irvine *B 5 f.* D.Macgill & Co.	1887	4 Blds.	2Cy.14½"&28"-22" 25HP. Ross & Duncan,Glasgow.	CiaM'ritima Portuense do Oporto Reboques e Salvadego
26	✱	Galgorm Castle *Darcy 80-88*	I ScSr 1 Dk	British Belfast	75 183	111·0	19·4	9·0	Belfast *Q 43 f. F 20 f.* Harland&Wolff	1879	3 Blds.	2Cy.14"&26"-24" 45 HP. J.Rowan&Sons,Ld.,Belfast	C.Marshall
27		Galicia *Park*	I ScBk 3 Dks&Awn Dk	British Liverpool	2319 3834	383·4	43·0	33·6	Glasgow *H 34 f.* R.Napier&Sons	1873	6 Blds.	2Cy.60"&104"-54"600 HP. R.Napier&Sons,Glasgow	Pacific Steam Nav.Co.
28	✱	Galicia *Robinson 79-79*	I ScBg 1 Dk2 B	British London	499 794	210·0	28·3	16·9	Hartlepool *P 39 f. F 31 f.* Denton,Gray&Co.	1866	4 Blds. DBa84ft f18ft	2Cy.27"&50½"-33" 98 HP. Blair&Co.,Ld.,Stockton	J.Hall,jun.,& Co.
29		Galicia *Lopez*	I ScSr 1 Dk	Spanish Bilbao	112 208	135·1	19·2	11·9	Newcastle *Q 31 f. F 10 f.* T.&W.Smith	1853		2Cy17¾"&33½"-21¼"40HP. Chantiers&A.deBacal.,Brdx.	José Pedros & Co.
30	V	Galilée *Jore 84-84*	I ScSr 1 Dk	French La Rochelle	397 508	182·4	26·6	13·6	Havre *P 22 f. F 22 f.* Forges&C.delaMedit	1874		2C21⅛"&37½"-27½" 80 HP. Forges&C.delaMedit,Havre	Delmas frères

✱=Lloyd's Register; D=Nederlandsche Vereeniging; G=Germanischer Lloyd; I=Registro Italiano; R=American Record; A=Veritas Austro-Ungarico; H=Veritas Ellenico; N=Norske Veritas; V=Bureau Veritas.

Dans la 2me Colonne, B.T., avec les indications qui suivent, signifie la date du rapport de visite de la Carène par les experts du Board of Trade pour obtenir le certificat relatif au TRANSPORT DES PASSAGERS.

Dans la 3me Colonne, C=Composite (membrure en fer, bordé en bois); I=Fer; Stl=Acier; W=Bois.

Pour plus amples renseignements sur les navires classés par le Lloyd's Register, voir le Lloyd's Register Book et le Lloyd's Yacht Register.

No. in Book. / Official No.	Registry	Ship's Name. Material, Rig, &c. / Master, and Dates of Appointment to present Owner's Service and to Ship. / Ship's Late Name. No. of Decks, &c.	Flag / Port of Registry	Tons Net. / Gross	Dimensions. Length, Breadth, Depth. Deck Erections.	Built. Where / By Whom.	When	No. of Bulkheads. / Particulars of Water Ballast.	Engines. / Particulars. Maker's Name.	Owners or Managers
31 82605		Galileo I ScBg Moger 2 Dks3 B	British Hull	2018 3071	350·7|41·2|27·1 P 86 f. F 41 f. B 184 f.	Hull Earles Co., Ld.	1881	4 Blkds. C 78	2Cy.34" & 68"–63" 350 HP. Earles Co., Ld., Hull	C. H. Wilson
32	BT1·88	Galileo I ScBg Farley 86-81	Belgian Antwerp	1446 2263	343·2|35·1|27·7	Newcastle A. Leslie & Co.	1873		2Cy.30" & 70"–48" 280 HP. Maudslay Ss. & Field, London	Soc. Royale Belge Sud-Américaine
33 41		Galileo Galilei I PadSc Capezzi 85-85 2 Dks	Italian Palermo	342 588	215·5|24·8|12·8	Renfrew Hudson Coulborn & Co.	1862		2Cy.36" & 60"–62" 330 HP. Henderson Coulborn & C., Rfw	Navigazione Gen. Italiana
34	V	Galindo I ScSr Confournis 85-85 1 Dk2 B	French Dunkirk	578 1051	235·5|31·5|15·8 P 28 f. B 49 f. F 30 f.	Havre Forges & C. de la Medit	1881	Fe & 150 DB 70	2C 27½" & 47¼"–33½" 141 HP. Forges & C. de la Medit, Havre	Soc. Anonyme & Demain & Anis
35	✳	Gallego I Sc Arribalzaga 86-87 2 Dks3 B	Spanish Bilbao	1716 2690	329·5|36·0|25·5 P 46 f. B 63 f. F 37 f.	Sunderland R. Thompson, jun.	1880	6 Blkds. DB & f. FP T 75	2Cy.42" & 73"–48" 350 HP. D. Clarke Sunderland	M. M. de Arrotegui
36 65851	✳	Galley of Lorne I ScSr Pomeroy 3 Dks	British London	1380 2147	321·8|36·4|26·4 B 16 f.	Glasgow R. Napier & Sons	1871	5 Blkds.	2C.39½" & 68"–42" 300 HP. R. Napier & Sons, Glasgow	Shaw, Bushby & Co.
37 78837	BT7·88	Gallia I ScBk Murphy 67-84	British Liverpool	2898 4809	430·1|44·6|34·4	Glasgow J. & G. Thomson	1879		3C.64" & (2)80"–60" 700 HP. J. & G. Thomson, Glasgow	Cunard Steam S. Co., Ld.
38		Gallia (ex Loch-na-gar) I ScSr Coste 86-88 1 Dk2 B	French Marseilles	701 1057	231·9|30·5|16·7 P 124 f. F 36 f.	Aberdeen J. Duthie & Son	1872	WB	2Cy.27" & 54"–36" 110 HP. Black, Hawthorn & Co., Gts.	Compagnie Fse. de Nav. à Vap.
39 79654	✳	Gallina I Sc3mSr Harper 78-81 2 Dks3 B	British London	1210 1879	270·0|31·5|23·8 P 30 f. B 52 f. F 28 f.	Sunderland Mounsey & Foster	1878	6 Blkds. DB a f. t ... 310	2C.34½" & 64"–42" 180 HP. Blair & Co., Ld., Stockton	"Avis" Steamship
40 63389	✳	Galtee I ScSr Sinnott 83-83 1 Dk	British Waterford	299 508	165·0|25·4|12·9 Q 84 f. B 17 f. F 20 f.	South Shields J. T. Eltringham	1881	4 Blkds. DB a f. FP T 20	2Cy.22" & 40"–30" 70 HP. J. P. Rennoldson, S. Shields	Waterford S.S. Co.
41 49727		Galvanic I ScSr Wallace 82-8	British Belfast	240 670	235·5|27·1|14·5 P 66 f.	Greenock Scott & Co.	1867		3Cy.18½" 32½" & 18¼"–33" MacIlwaine, Lewis & Co., Belfast 130 HP.	W. A. Grainger
42 85018	✳D	Galveston I ScSr Chater 82-82 1 Dk & Spr Dk3 B	British Sunderland	1301 2002	277·0|37·6|23·2	Newcastle Campbell, Mackintosh & Bowstead	1882	6 Blkds. DB a f. 772	2Cy.32" & 60"–45" 180 HP. T. Clark & Co., Newcastle	Culliford & Clark
43	✳N	Gambetta I ScSr Hensen 83-83 1 Dk2 B	Norweg'n Bergen	404 563	170·0|25·3|13·0 P 32 f. F 31 f.	Bergen Bergens Mek Verkst	1888	5 Blkds. AP T 15 MT 60 FP T 34	2Cy.20½" & 39"–27" 60 HP. Bergens Mek. Verks, Bergen	S. M. Kuhnle
44 89090	✳	Gambier (ex Geraut) I ScSr Brown 2 Dks & Spr Dk	British Melbourne	1030 1578	280·1|31·9|16·8 1 239	Dumbarton J. & R. Swan	1874	4 Blkds.	2Cy.32" & 60"–39" 155 HP. Kincaid, Donald & Co., Grnck	W. H. Smith & Sons
45 61355		Game Cock Stl TwnScSr Sumner 84-84	British Liverpool	59 370	156·1|25·1|13·0	Birkenhead T. Brassey & Co.	1880		4Cy.26" & 45"–30" 200 HP. T. Brassey & Co., Birkenhead	W. B. Hill
46		Gamleby (ex Per Ennes, ex Norra Viken) I ScSr Ugarph 79-79 1 Dk	Swedish Westervik	139 167	107·3|18·1|6·9 Q 26 f. B 16 f. F 30 f. H 10 f.	Oscarshamn Gustaf Tillberg	1872		2Cy.12½" & 23"–16" 30 HP. G. Tillberg, Oscarshamn	C. Carlbom
47 65291	✳	Gamma (STRANDED) I ScSr Robson 73-83 2 Dks3 B	British Hull	1453 1868	309·8|32·9|24·0 Q 58 f. B 18 f. F 44 f.	Hull Humphrys & Pearson	1872	5 Blkds.	2Cy.30" & 62"–33" 170 HP. Comp.'81, Earles Co., Ld., Hull	Smith, Hill & Co.
48		Gange (ex Black Swan) (BROKEN UP) I Sc 2 Dks	French Marseilles	839 1440	256·5|36·4|27·0 H 67 f.	Greenock J. Scott & Sons	1853		2C.36½" & 59"–39½" 250 HP. Messageries Mar., La Ciotat.	Messageries Mar.
49	✳DN	Ganger Rolf I ScSr Møller 84-84 1 Dk & Spr Dk3 B	Norweg'n Stavanger	1194 1545	238·6|35·2|22·2	Bergen Martens, Olsen & Co.	1884	5 Blkds. DB a f. 311	2Cy.30" & 57"–33" 130 HP. Martens, Olsen & Co., Bergen	Dampskibsselskabet "Ganger Rolf" (Hoddas)
50		Ganger Rolf I ScSr 1 Dk	Norweg'n Christiansund	73 105	104·9|16·7|7·8 Q 29 f. B 16 f.	Bergen Bergens Mek Verkst	1876		2Cy.15" & 29"–19" 35 HP. Bergens Mek Verks, Bergen	N. R. Parelius
51 82078	BT3·88	Ganges Stl Sc3mSr Ablerton 75-88 4 Dks	British Barrow	2111 4168	390·0|42·2|32·8 B 130 f.	Barrow Barrow S. B. Co., Ld.	1881	7 Blkds.	2C.58" & 100"–63" 800 HP. Barrow S. B. Co., Ld., Barrow	Peninsular & Oriental Steam Nav. Co.
52 82738	✳D	Ganges I ScSr Roberts 84-84 2 Dks3 B	British Dundee	1490 2265	310·2|36·3|23·6 P 40 f. B 95 f. F 36 f.	Dundee Gourlay, Bros. & Co.	1880	5 Blkds. CB 88 Sa FP T 38	2Cy.34" & 64"–48" 220 HP. Gourlay Bros. & Co., Dundee	D. Bruce & Co.
53 60141	✳	Ganges I ScSr D'Ath 85-5 2 Dks3 B	British London	1242 1903	267·5|33·4|25·6	Glasgow Lond. & Glasgow Co.	1868	5 Blkds. DB a f.	2Cy.33" & 62"–42" 200 HP. W. Doxford & Sons, Sundrlnd	Mercantile Steam Co. Ld.
54 81589	BT6·88	Gannet I Sc3mSr Taylor 2 Dks	British London	804 1248	245·3|33·0|17·7 P 61 f. F 32 f.	Stockton M. Pearse & Co.	1879	4 Blkds. DB a f. 178	2Cy.33" & 62"–39" 150 HP. Blair & Co., Ld., Stockton	General Steam Nav.
55 90353	✳	Gannet I ScK 1 Dk	British Grimsby	140 244	133·7|21·7|11·0 Q 45 f. F 23 f.	Hull Earles Co., Ld.	1884	4 Blkds.	2Cy.21" & 40"–24" 68 HP. Earles Co., Ld., Hull	Great Grimsby ... Ld.
56 88924	✳	Gannet I PadTugSr Freebody 1 Dk	British Melbourne	59 246	134·1|22·1|11·0 B 42 f.	Barrow Barrow S. B. Co., Ld.	1884	3 Blkds.	4Cy.20" & 35"–48" 108 HP. Barrow S. B. Co., Ld., Barrow	Melbourne Harbour Trust Commissioners
57 84909	V	Gannet W ScSr Sanderson	British Granton	34 102	96·0|18·2|9·8	Granton D. Allan & Co.	1883		2Cy.14½" & 26"–20" 33 HP. J. Cran & Co., Leith	J. L. Cunliffe
58 81299		Ganymede I ScSk	British Liverpool	236 405	166·5|28·0|10·8 P 33 f. F 29 f.	Greenock Scott & Co.	1879		1Cy.21"–34" 80 HP. Scott & Co., Greenock	Ocean Steam S. Co.
59	✳	Garcia de Vinuesa I ScSr 2 Dks & Spr Dk	Spanish Seville	732 1077	212·0|30·2|15·4 22·7	Newcastle A. Leslie & Co.	1883	5 Blkds. DB a f. FP T 20	2Cy.28" & 55"–33" 120 HP. Black, Hawthorn & Co., Gts.	Vinuesa & Co.
60	N	Gardar W ScSr Hansken 82-87 1 Dk2 B	Norweg'n Stavanger	312 441	154·1|26·0|14·1 B 39 f.	Stavanger K. Knudsen	1883		2Cy.18" & 32"–24" 45 HP. Aker Mek Verks, Christiania	Holdt & Isachsen

✳ = Lloyd's Register; D = Nederlandsche Vereeniging; G = Germanischer Lloyd; I = Registro Italiano; R = American Record;
A = Veritas Austro-Ungarico; H = Veritas Ellenico; N = Norske Veritas; V = Bureau Veritas.

In the 2nd Column, B.T., with figures following, signifies Date of report of bottom sighting by Board of Trade Surveyors as condition of PASSENGER CERTIFICATE.

In the 3rd Column, C = Composite (iron frame, wood planked); I = Iron; Stl = Steel; W = Wood.

Further particulars of Vessels classed by Lloyd's Register will be found in Lloyd's Register Book and Lloyd's Yacht Register.

No.	Cl.	Nom, Matériaux, Gréement, etc. / Capitaine et dates d'entrée au service des Armateurs actuels et à bord du navire. / Ancien Nom, nombre de ponts, etc.	Pavillon. / Port d'Armement	Tonnage Net / Total	Dimensions. (Longueur, Largeur, Creux) / Constructions sur le Pont	Construit Où. / Par qui.	Année	Nombre des cloisons. / Détails sur le lest d'eau.	Machines. / Détails. / Nom du Constructeur.	Armateurs ou Directeurs-Gérants.
61	✱	**Gardenia** I S.Sr *Rutherford 85-87* *2 Dks 3 B*	British N.Shields	1276 1997	280·0|35·0|24·7 *Q 49 f. B 26 f.*	Newcastle TyneIronS.B.Co.	1879	6 Blds.	2Cy.33"&63"–39" 180 IP. N.E.MarineE.Co.,Ld.,Smlld.	J.Robinson & Sons
62	✱	**Gardenia** Stl ScSr *Keir* *1 Dk*	British Glasgow	127 445	172·0|27·1|12·6 *P 26 f. B 60 f. F 48 f.*	Glasgow D.&W. Henderson &	1885	5 Blds.	2Cy.23"&46"–36" 120 IP. A.Stephen&Sons,Glasgow	Glasgow, Dublin, and LondonderryStm.Pkt.Co.
63	✱	**Gardépée** I ScSr *Miller* *1 Dk 2 B*	British Cardiff	1126 1733	258·0|36·2|18·3 *Q 34 f. B 66 f. F 33 f.*	Sunderland W.Pickersgill&Sons	1882	5 Blds.	2Cy.32"&60"–39" 140 IP. G.Clark,Sunderland	"Gardépée" S.S.Co.,Ld. (Morel Bros.&Co.)
64		**Gareloch** I PadSk *McTaggart 83-85*	British Glasgow	55 178	180·0|18·2|6·8 *Q 25 f.*	Port Glasgow H.Murray&Co.	1872		2Cy.35"–51" 85 IP. D.Rowan,Glasgow	North British Steam Packet Co. (R.Young)
65	I	**Gargano** (ex ...) I Sc *Gambardella 85-86* *1 Dk & Awn Dk*	Italian Bari	590 787	172·0|28·2|12·2 19·2	Dundee PearceBros.	1884	4 Blds.	2Cy.20"&40"–30" 80 IP. PearceBros.,Dundee	Soc.Nav.Puglia
66	✱	**Garigliano** StlSc3mSr *Gallurd ...-87 (ex Engineer)* *1 Dk*	Italian Naples	449 661	187·2|28·2|12·5 *F 24 f. Q 80 f.*	Middlesbro' R.Dixon & Co.	1887	4 Blds.	3C18"–28"&43"–36" 120 IP G.Clark,Sunderland.	Italian Government
67		**Garnet** W Pad *Gillespie* *2 Dks*	British Cornw'll Ont	160 213	121·6|35·0|8·6	Cornwall, Ont. O.Gillespie	1884		1Cy.30"–72" 55 IP. Brush&Co.,Montreal	O.Gillespie
68	✱	**Garonne** I ScBk *White 86-87* *3 Dks & Awn Dk*	British Liverpool	2468 3876	382·1|41·4|35·7 *B 31 f.*	Glasgow R.Napier&Sons	1871	6 Blds.	2C.60"&104"–48" 550 IP. R.Napier&Sons,Glasgow	Orient Steam Nav.Co., Ld.
69	✱	**Garonne** *Garvin 85-85* I ScSr *1 Dk & Spr Dk & Web Frames*	British Cardiff	684 1052	210·5|29·6|15·2 22·1	Newcastle C.S.Swan&Hunter	1884	4 Blds.	2Cy.26"&47"–36" 96 IP. G.Clark,Sunderland	Cardiff S.S.Co.,Ld. (Hooper,Campbell&Co.)
70	N	**Garonne** I ScSr *Brekke -82* *1 Dk*	Norweg'n Christiania	450 661	181·9|28·2|13·8 *Q 60 f. B 42 f. F 24 f.*	Malmö KockumsMekVerks	1883		2Cy.21"&42"–26" 80 IP. KockumsMek.Verks,Malmö	Fearnley & Eger
71		**Garrick** I Sc2mSr *Garvin 77-85*	British Liverpool	1651 2527	321·0|37·8|25·1 *P 88 f. F 25 f.*	Newcastle A.Leslie&Co.	1885		2Cy.34"&68"–48" 300 IP. R.&W.Hawthorn,N'weastle	Liverpl., Brazil & River Plate Steam Nav.Co.,Ld.
72		**Garrison** I Sc3mSr *Burrell 84-87*	British Cardiff	668 852	238·3|28·6|16·7 *P 60 f. F 26 f.*	Glasgow Aitken&Mansel	1867		2Cy.38"–30" 95 IP. J.Aitken&Co.,Glasgow	"Garrison" S.S.Co.,Ld. (Pewley,Thomas & Co.)
73	✱	**Garron Tower** I ScSr *Horlock 78-87* *1 Dk*	British Sunderland	403 650	180·0|28·2|13·6 *Q 23 f. B 38 f.*	W. Hartlepool E.Withy&Co.	1876	4 Blds.	2C.25½"&47½"–36" 90 IP. T.Richardson&Sons,Hrtlpl.	Marquis of Londonderry, (S.J.Ditchfield,Mngr.)
74	✱	**Garry** I ScSr *Batchelor* *1 Dk 2 B*	British Dundee	628 973	217·9|29·5|15·5 *P 120 f. F 29 f.*	Dundee W.B.Thompson	1880	5 Blds.	2Cy.27"&50"–33" 98 IP. W.B.Thompson,Dundee	North Sea S.S.Co.,Ld.
75	✱ D	**Garth** I ScSr *Storm 81-85* *1 Dk 2 B*	British Cardiff	941 1446	240·0|33·1|18·2 *Q 91 f. B 50 f. F 28 f.*	Sunderland W.Pickersgill&Sons	1884	4 Blds.	2C.29½"&55"–36" 130 IP. G.Clark,Sunderland	GarthS.S.Co. (Hurley Matthews&Co.)
76	✱	**Garth Castle** I ScBg *Jeffries 79-87* *2 Dks 3 B*	British London	2381 3705	365·0|43·5|31·3 *B 96 f. F 48 f.*	Glasgow J.Elder&Co.	1880	8 Blds.	3C.33"55"&88"–57" 500 IP Tripl.'83 T.Richardson & Sons, Hartlepool	D.Currie & Co.
77	✱	**Gassendi** I ScSr *Smith 85-87* *1 Dk & Awn Dk 3 B*	British Liverpool	801 1249	238·1|30·3|23·9	Aberdeen Hall,Russell&Co.	1872	5 Blds. WB	2Cy.29¾"&53½"–36" 98 IP. Hall,Russell&Co.,Aberdeen	G.Willison
78		**Gaston** I Sc	American Baltimore	464 847	212·0|31·5|19·0	WilmingtnDel Harlan&HollingsworthCo.	1881	WB	2Cy.26"&44"–36" 540 IP. Harlan&HollingsworthCo.	Baltimore Steam Packet Co.
79	R	**Gate City** I ScSr *Hedge* *3 Dks*	American Boston	1112 1997	254·0|38·7|24·1 *B 293 f.*	Chester, Pa. J.Roach&Son	1878		2Cy.38"&68"–54" 207 IP. J.Roach&Son,Chester,Pa.	Boston & Savannah S.S.Co.
80	✱	**Gathorne** (ex ...) I ScSr *Griffiths 86-88* *1 Dk 2 B*	British Newcastle	765 1195	243·9|29·1|17·5 *P 246 f. F 27 f. B 5 f.*	Cardiff C.Hill&Sons	1873	4 Blds.	2Cy.25"&52"–30" 99 IP. GreenockFndryCo.,Grnock.	Fisher, Renwick & Co.
81		**Gatineau** W Pad *1 Dk*	British Quebec	79 175	118·2|23·4|6·8	Quebec	1873		55 IP.	Comp. de Nav. de Boucherville
82	V	**Gaulois** (ex ...) I ScSr *Garnier* *1 Dk*	French Brest	183 350	134·5|22·6|11·6 *Q 23 f. B 5 f. F 11 f.*	Hull Humphrys&Pearson	1874		2C.18"&29½"–20½" 45 IP. Humphreys&Pearson,Hull	Chevillotte frères
83	GV	**Gauss** I Sc Sc *Carstens 86-86* *1 Dk*	German Bremen	324 416	148·3|22·5|13·7	Stettin Möller&Holberg	1882	WB	2Cy.17¼"&30"–20" 45 IP. Möller&Holberg,Stettin	Dampfschifff.-Ges. "Neptun"
84		**Gauthiod** I ScSr *Rydell* *1 Dk & Shltr Dk*	Swedish Stockholm	562 730	191·0|24·0|11·9 *P 123 f. F 19 f. B 75 f.* 19·1	London J.&W.Dudgeon	1873		3C.18½",30"&50"–30" 160 IP W.LindbergsM.V.Stkm.	Rederi-Aktie Bolaget "Svea" (K.Ahlborg)
85	✱	**Gaviota** I ScSrYt *Céspedes 85-*	Spanish Malaga	64 121	124·5|16·7|7·8	Paisley J.Fullerton&Co.	1882	4 Blds.	2Cy.14"&28"–22" 32 IP. T.Reid&Sons,Paisley	Cia Arrendataria Tabacos
86	✱	**Gaw Quan Sia** I ScBg *Morris 86-88* *(ex Valencia, ex Walmer Castle)* *3 Dks*	British W.Hartlep'l	1607 2459	325·8|36·4|27·9	Glasgow Barclay,Curle&Co.	1872	5 Blds.	3Cy.21½",35"&58"–39" 190 IP Cntl.M.E.Co.,W.Hpl.	T.Cuthbertson
87	✱	**Gazelle** I ScSr *Rees 76-84* *1 Dk 2 B*	British Llanelly	298 407	179·8|28·5|15·3 *Q 36 f. B 49 f. F 24 f.*	Hartlepool Withy,Alexander & Co.	1869	4 Blds.	2C.24¼"&47½"–30" 80 IP. T.Richardson&Sons,Hrtlpl.	Samuel Bros.
88		**Gazelle** Stl PadSr *Gillard 83-* *1 Dk*	French Havre	79 170	154·2|20·3|9·6	Havre A.Normand	1884		2C.29½"&47¼"–38" 130 IP. A.Normand,Havre	Cie. Anonyme des Paquebots à Vapeur entre Le Havre, Caen, Honfleur et les Ports de la Normandie
89	✱	**Gédéon** (MISSING) I ScSr *1 Dk*	French Rochefort	323 507	172·1|25·6|13·5 *F 32 f. B 14 f. F 19 f.*	Paisley H.McIntyre&Co.	1882	4 Blds.	2Cy.20"&38"–36" 70 IP. R.H.Pearson&Co.,Glasgow	G.Coudert & fils
90	✱	**Gédéon Coudert** I ScSr *Colomdeau, (ex Bengo, ex Hayti, ex Ruminates)* *2 Dks*	French Bordeaux	1009 1379	260·0|30·3|20·0 *Q 253 f. B 98 f. F 12 f. len 76*	Glasgow Smith&Rodger	1854	7 Blds.	2Cy.30"&58"–28" 98 IP. Cp 82, Bailey&Leetham,Hull	G.Coudert & fils

✱ = **Lloyd's Register**; D = Nederlandsche Vereeniging; G = Germanischer Lloyd; I = Registro Italiano; R = American Record; A = Veritas Austro-Ungarico; H = Veritas Ellenico; N = Norske Veritas; V = Bureau Veritas.

Dans la 2me Colonne, B.T., avec les indications qui suivent, signifie la date du rapport de visite de la Carène par les experts du Board of Trade pour obtenir le certificat relatif au TRANSPORT DES PASSAGERS.

Dans la 3me Colonne, C = Composite (membrure en fer, bordé en bois); I = Fer; Stl = Acier; W = Bois.

Pour plus amples renseignements sur les navires classés par le Lloyd's Register, voir le Lloyd's Register Book et le Lloyd's Yacht Register.

No.	Class	Ship's Name. Material, Rig, &c. / Master / Ship's Late Name. No. of Decks, &c.	Flag / Port of Registry	Tons Net / Gross	Length	Breadth	Depth	Built Where / By Whom	When	No. of Bulkheads / Particulars of Water Ballast	Engines Particulars / Maker's Name	Owners or Managers
91		Geelong I ScBg Whale 75-84	British London	1139 1835	275·0	34·3	25·0	Dumbarton W.Denny&Bros.	1866		2Cy.41"&70"-36" 228 HP. Denny&Co.,Dumbarton	P.& O.Steam Na[illegible]
92	✻	Geelong (WRECKED) I Sc3mSr O'Callaghan (ex Thomas Powell) 1 Dk2 B	British Melbourne	239 431	162·3	24·8	14·6	Bristol G.K.Stothert&Co.	1856	1 Bhds.	[illegible]&16"-30" 30 HP. Stothert&Co.,Bristol	W.H.Smith&S[illegible]
93	V	Geestemunde I PalBkn Müller -87	German [illegible]	147 187	150·9	21·6	7·2	Vegesack J.Lange	1887		2Cy.15"&28½"-27½" 35 HP. J.Lange,Vegesack	Ch.Kulken
94		Gefion I PadSr Krogh 75-75 1 Dk	Danish Copenhagen	150 277	164·4	23·2	10·2	Copenhagen Burmeister&Wain	1874		2Cy.22"&41"-42" 100 HP. Burmeister&Wain,Copnhgn	Det F[illegible] Selskab. (C.F.A.K.)
95	V	Gefle I ScSr Rickm -87 1 Dk	German Hamburg	415 634	181·4	27·1	12·3	Elsinore Helsingør Schiffs og Maschinenbau Ges.	1888	[illegible]	2Cy.25"&47"-30" 130 HP. [illegible]	H.M.Gehrckens
96		Gefle I ScSr Brandt 1 Dk&AwnDk	Swedish Gefle	398 495	164·0	24·4	11·5	Stockholm Bergsund's Mek. Verks	1871		2Cy.24"&46"-27" 100 HP. Bergsund'sMek.Verks,Stklm	R.G.Kronberg
97	N	Geiranger I StlScSr Korsvig 74-85 1 Dk	Norweg'n Aalesund	107 157	110·0	18·3	7·6	Bergen Martens,Olsen&Co.	1885	APT FPT	2Cy.15½"&29"-18" 30 HP. Martens,Olsen&Co.,Bergen	Söndmore Damp[illegible] Selskab
98		Geiser (COLLISION) I Sc3mSr Møller 86-86 2 Dks&Spr Dk	Danish Copenhagen	1818 2831	324·4	39·3	22·0 29·7	Copenhagen Burmeister&Wain	1881	FPT [illegible] APT [illegible]	2Cy.36"&72"-42" 360 HP. Burmeister&Wain,Copnhgn	Dampskibs-Selsk[illegible] "Thingvalla"
99	✻D	Gelderland I ScBg v.d. Sprong v. Lee 2 Dks&B	Dutch Rotterdam	1600 2177	298·8	36·0	25·1	Middlesbro' R.Dixon&Co.	1878	4 Bhds.	2Cy.38"&70"-36" 200 HP. T.Richardson&Sons,Hrtlpl.	Rotterdamsche [illegible]
100 87462	✻D	Gelert I ScSr Winter 86-86 1 Dk2 B	British London	1108 1717	250·8	36·2	20·1	Newcastle Palmers'Co.,Ld.	1883	5 Bhds. [illegible]	2Cy.30"&55"-42" 160 HP. Palmers'Co.,Ld.,Newcastle	OrientalS.S.Co
101	V	Gellert I ScSr Kahlewein 76-83 2 Dks&Spr Dk	German Hamburg	2350 3533	374·0	40·0	32·2	Glasgow A.Stephen&Sons	1874		2Cy.60"&106"-54" 800 HP. A.Stephen&Sons,Glasgow	Hamburg-Ame[illegible] Packetf.-Akt.-Ges.
102 95428	✻	Gellivara StlTwnScSr Thompson -88 1 Dk&Spr Dk&WbFas	British London	1675 2608	332·5	45·0	14·1 21·6	Newcastle C.S.Swan&Hunter	1888	6 Bhds. [illegible]	3Cy.18"28½"&46"-33" 300 HP Wallsend Slipway & Eng. Co.,Ld.,Newcastle	Anglo Scandina[illegible] Steamship Co.,L[illegible]
103 92867	✻	Gem StlScSr Nicol 72-87 1 Dk2 B	British Newcastle	1361 2098	280·0	38·8	18·3	South Shields J.Readhead & Co.	1887	6 Bhds. [illegible]	3C.21½"34"&59"-39" 200 HP J.Readhead&Co.S.Shls.	Clapham S.S.C[illegible]
104 9339	✻	Gem ISc3mSr McNicol 88-88 1 Dk	British Glasgow	185 432	175·2	25·2	10·2	Paisley J.Fullerton & Co.	1887	3 Bhds. [illegible]	3C.15½"24"&40½"-33" 80 HP. W.King & Co.Glasgow	W.Robertson
105 53986		Gem C Pad 1 Dk	British Melbourne	92 145	129·4	17·8	8·4	Melbourne	1868		2Cy. 60 HP.	Arthur Devlin
106 77451	✻	Gem I ScSr Henderson 82-82 1 Dk	British Aberdeen	88 180	120·2	19·0	9·0	Paisley J.Fullerton&Co.	1878	1 Bhds. FPT [illegible]	2Cy.17"&31"-22" 45 HP. Lees,Anderson&Co.,Glasgw.	Mitchell & Rae
107 72756		Gem I Sc 1 Dk	British Shanghai	33 120	80·0	16·5	8·7	Shanghai Boyd&Co.	1880		3Cy.18"&36"-16" 35 HP. Boyd&Co.,Shanghai	George Mitchell
108 459		Gembu Maru W ScSr Watanabe	Japanese Tokio	434 700	194·6	28·0	16·0	Brooklyn,N.Y.	1872		2Cy.34"&42"-34" 78 HP.	Nippon Yusen K[illegible]
109 95277	✻	Gemini I ScSr Snowdon 88-88 1 Dk2 B	British Sunderland	644 990	210·0	31·0	15·5	Sunderland J.Priestman & Co.	1888	1 Bhds. [illegible]	3Cy.17",28"&46"-30" 100 HP N.E.Marine Eng.Co., Sunderland	J.G.Hill
110 87753	✻	Gemini I ScK Trawler Barber 84-88 1 Dk	British Grimsby	73 128	95·5	20·1	10·5	Hull Earles'Co.,Ld.	1883	3 Bhds.	2Cy.17"&32"-18" 35 HP. Earles'Co.,Ld.,Hull	Grimsby & North[illegible] Steam Trawling [illegible]
111	✻	Gemma Stl ScSr Hansen 84-88 1 Dk	German Flensburg	613 832	199·9	30·0	14·0	Flensburg Flensb. Schiffb.Ges.	1888	4 Bhds.	3Cy.15",24"&48"-30" 110 HP Flensburg Schiffb.Ges. Flensburg	Holm & Molzen
112	D	Gemma (ex Claremont) I ScSr Pagels -87 1 Dk2 B	German Hamburg	523 813	213·9	29·2	17·5	Newcastle C.Mitchell&Co.	1868	4 Bhds. [illegible]	2Cy.29"&5[illegible]"-27" 100 HP. Thompson,Boyd&Co.	A.Kirsten
113	✻	Generaal Pel I ScSr Van Der Ham 2 Dks&Shd Dk	Dutch Batavia	886 1391	255·9	31·7	17·3	Dumbarton W.Denny&Bros.	1876	4 Bhds.	2Cy.34"&[illegible]"-36" 172 HP. Denny&Co.,Dumbarton	NederlandschInd[illegible] Stoom. Maats.
114		General Bolivar I Sc Long	Arg'ntine Buen'sAyres	500 870	192·0	24·0	11·1	Glasgow J.&G.Thomson	1864		90 HP. Bolton&Watt	Wells & Barb[illegible]
115	✻	Général Chanzy I ScSr Bernard 81-85 1 Dk2 B	French Nantes	949 1475	250·8	35·3	16·2	Newcastle Palmers'Co.,Ld.	1883	4 Bhds. [illegible]	2Cy.31"&58"-36" 110 HP. Palmers'Co.,Ld.,New.	Pergeline fils & C[illegible]
116 82786	✻	General Eliott I ScBu Davidson 1 Dk2 B	British Gibraltar	880 1388	240·0	34·2	17·0	Middlesbro' R.Dixon&Co.	1882	4 Bhds. [illegible]	2Cy.31"&58"-36" 130 HP. Blair&Co.,Ld.,Stockton	Smith, Imossi & C[illegible]
117 90071	✻	General Gordon Stl ScSr Allardyce 85-88 1 Dk2 B	British Glasgow	807 1255	240·4	34·1	16·0	Glasgow A.Stephen&Sons	1885	4 Bhds. [illegible]	3C.15"24"&40"-36" 140 HP. A.Stephen&Sons,Glasgow	S.S.Gordon Co.,L[illegible] (Maclay & McIntyre)
118 89907	✻ BT2.88	General Gordon Stl Sc Boland 2 Dks	British [illegible]	233 502	230·0	31·2	15·3	Glasgow A.&J.Inglis	1885	5 Bhds.	3C.26"41"&60"-42" 250 HP. A.&J.Inglis,Glasgow	Dublin & Glasgo[illegible] Steam Packet Co.
119 62496	BT5.88	General Havelock I ScSr Douglas 74-74 1 Dk2 B	British Sunderland	439 673	190·0	26·3	14·4	Sunderland J.Laing	1868	5 Bhds. WB	2Cy.26"&53"-36" 99 HP. G.Clark,Sunderland	R.M.Hudson & S[illegible]
120		General Kotzebue I PadSr 2 Dks	Russian Odessa	501 709	223·0	27·0	15·1	Newcastle C.Mitchell&Co.	1866		2Cy.55"-60" 180 HP. Penn&Sons,London	Russian Steam & Trading Co.

✻ = Lloyd's Register; D = Nederlandsche Vereeniging; G = Germanischer Lloyd; I = Registro Italiano; R = American Rec[ord]; A = Veritas Austro-Ungarico; H = Veritas Ellenico; N = Norske Veritas; V = Bureau Veritas.

In the 2nd Column, B.T., with figures following, signifies Date of report of bottom sighting by Board of Trade Surveyors as condition of PASSENGER CERTIFICATE.

In the 3rd Column, C = Composite (iron frame, wood planked); I = Iron; Stl = Steel; W = Wood.

Further particulars of Vessels classed by Lloyd's Register will be found in Lloyd's Register Book and Lloyd's Yacht Register.

No.	Cote	Nom, Matériaux, Gréement, etc. / Capitaine et dates d'entrée au service des Armateurs actuels et à bord du navire. / Ancien Nom, nombre de ponts, etc.		Pavillon. / Port d'Armement	Tonnage. Net. / Total	Dimensions. Longueur	Largeur	Creux	Construit. Où. / Par qui.	Quand.	Nombre des cloisons. / Détails sur le lest d'eau.	Machines. Détails. / Nom du Constructeur.	Armateurs ou Directeurs-Gérants.
	✻	General Roberts *McNair 85-85*	I ScSr 1 *Dk2 B*	British Cardiff	936 1450	218·9 *P 96 f.*	35·2 *B 44 f.*	16·7 *F 27 f.*	Newcastle PalmersCo.,Ld.	1881	5 Blds. DB53a	2Cy.30"& 58"–36" 160 HP. PalmersCo.,Ld.,Newcastle	SS."General Roberts" Co.,Ld.(Barnes,Guthrie&Co.)
	✻	General Roberts *Russell*	I ScK 1 *Dk*	British Hull	62 142	100·0	20·5 *Q 44 f.*	11·0	Hull Cook, Welton & Gemmell	1888	4 Blds.	2Cy.17"&33"–22" 45 HP. C.D.Holmes&Co.,Hull	Pickering & Haldane
	G	General Werder *Schackmann 86–82 Dks Hur Dk S B*	I ScBr	German Bremen	1820 3020	352·7	39·4	31·1	Greenock Caird&Co.	1874		2Cy.50"& 86"–54" Caird&Co.,Greenock	Norddeutscher Lloyd
124	R	General Whitney *Bearse*	I ScSr 2 *Dks*	American New York	1196 1849	225·0	40·0	18·0	WilmingtnDel TheHarlan & Hollingsworth Co.	1873		2Cy.36"–60" TheHarlan&HollingsworthCo. Wilm.,Del.	H.F.Dimock
125	✻	Geneva *Cromarty 85-85*	I ScSr 2 *Dks*	British Leith	441 708	198·2 *P 34 f.*	28·3 *F 18 f.*	15·5	PortGlasgow Blackwood&Gordon	1878	5 Blds.	2Cy.27"&47"–33" 98 HP. Blackwood&Gordon,Pt.Gls.	J.Currie & Co.
126		Genil *Lloyd*	I ScSr 1 *Dk*	British Sydney	180 309	151·8 *P 49 t.*	23·1 *F 5 t.*	11·3	Glasgow	1854		2Cy.30"–30" 60 HP.	J.W.Cliff
127		Genkai Maru	W PadSr	Japanese Tokio	1740 2493	261·5	36·8	20·0	New York W.H.Webb	1863		1Cy.81"–144" 300 HP.	Nippon Yusen Kaisha
128	✻	Genoese *Carwell 86-87*	I ScSr 1 *Dk2 B*	British Liverpool	719 1114	232·0 *P 143 f.*	28·9 *F 34 f.*	17·2	Southampton Oswald,Mordaunt&Co.	1878	5 Blds.	2Cy.28"& 56"–36" 150 HP. Oswld.Mordnt.&C.,Sthmptn	Atlantic & Eastern S.S.Co.
129		Genova *Robinson*	I Sc3mSr	British Middlesbro'	276 391	176·0 *Q 110 f.*	22·1 *F 12 f.*	13·5	Glasgow C.Connell&Co.	1863		2Cy.20½"&41½"–26" 65 HP. Blair&Co.,Ld.,Stockton	C.E.Muller
130		Genoves *Vicente*	I ScTug	Arg'ntine Buen'sAyres	78 110	87·0	18·0	6·6	Greenock Kincaid&Co.	1874		25 HP. Kincaid&Co.,Greenock	E.Piaggio & Co.
131	✻	Genua *ex Balcarres* *Bache -88*	I ScSr 1 *Dk2 B*	German Hamburg	1128 1385	250·3 *P 125 f.*	32·0 *F 36 f.*	20·2	Barrow BarrowS.B.Co.	1877	5 Blds. FPT 69 APT 71 MT 364	2Cy.30"& 60"–36" 150 HP. BarrowS.B.Co.,Barrow	R.M.Sloman & Co.
132		Genyo Maru	I ScSr	Japanese Kobe	297 480	156·0	23·9	16·4	Scotland	1871		66 HP.	Kyogei Sha
133	V	Georg *Nielsen 82-83 2 Dks & 1 Awn Dk*	I ScSr	Danish Copenhagen	491 787	209·1 *B 11 H F 21 H*	27·2 *B 97*	14·7 21·5	Kiel Norddt.Schifsb.A.G.	1875	WF 2·04 FPT 84	2Cy.30"& 56"–36" 130 HP. Schweffel&Co.,Kiel	Det Forenede Dampskibs Selskab.(C.P.A.Koch)
134	V	Georg *Reins 81-83*	I ScSr 1 *Dk & B*	German Bremen	558 733	180·1 *Q 52 f.*	26·9 *B 32 f.*	18·2 *F 24 f.*	Vegesack H.F.Ulrichs	1883	WB	2C.21"& 37½"–25½" 70 HP. Hannoversche Masch. Fabrik, Hanover	J.D.Bischoff
135		Georg Stage	I ScS 1 *Dk*	Danish Copenhagen	125 206	103·0 *B 3, 73 t.*	25·4 *F 2, 64 t.*	13·2	Copenhagen Burmeister&Wain	1882		1Cy.10"& 20"–15" 25 HP. Burmeister&Wain,Copenhagen	Stiftelsen "Georg Stages Minde"
136		George Appold *Loveland (WRECKED)*	W ScSr 3 *Dks*	American Baltimore	845 1456	206·0	37·5 *F 25 t.*	22·0	Philadelphia J.Englis&Co.	1864		1Cy.55"& 41" 130 HP. Neafie&Levy,Philadelphia	Merchants' & Miners' Transportation Co.
137		George Clarkson	Stl ScSr	British Whitby	1162 1806	258·8 *P 28 f. B 112 f.*	37·6 *F 32 f.*	18·1 *Q 89 f.*	Whitby T.Turnbull&Son	1889		3C.26",33"&54"–36" 140 HP. Blair&Co.,Ld.,Stockton.	H.Baxter & Co.
138		George Deakin	W ScFlat	British Liverpool	71 105	81·9	19·1	8·6	Winsford Ann Deakin	1866		2Cy.12"–14" 24 HP. G.Deakin,Winsford	A.Houghton
139	DV	George Dittmann *Bruntz 87-87*	I ScSr 1 *Dk & B*	German Hamburg	355 471	159·7 *Q 50 f.*	23·9 *B 35 f.*	13·5 *F 22*	Flensburg FlensburgSchiffsb.Ges.	1882	WB	2Cy.20"& 34½"–24" 50 HP. FlensburgerSchiffsb.-Ges.	L.F.Mathies & Co.
140	✻D	George Fisher *Weatherill -87*	I ScSr 1 *Dk2 B*	British Cardiff	821 1275	236·1 *B 51 f.*	31·2 *Q 73 f.*	18·5 *F 29 f.*	Stockton Richardson, Duck&Co.	1877	5 Blds. DB a & f 249 t	2Cy.28"& 53"–33" 120 HP. T.Richardson&Sons,Hrtlpl.	CapperAlexander&Co.
141	✻	George Heaton *Richards 79-86*	I ScSr 2 *Dks 3 B*	British N.Shields	1428 2215	284·6 *P 27 f.*	36·9 *B 55 f.*	21·4 *F 29 f.*	Newcastle TyneI.S.B.Co.,Ld.	1882	5 Blds. DB a & f 293 t	2Cy.35"& 68"–45" 250 HP. R.&W.Hawthorn,N'wcastle	"George Heaton" S.S. Co.,Ld.(Stephens & Mawson)
142		George Leary	W Pad	American Georgetown	621 810	237·0	33·0	10·6	Brooklyn C.Leary	1864		1Cy.50"–144" 275 HP. J.Murphy&Co.,NewYork	M.Robinson
143	✻D	George Locket *Martin 85-85*	I ScSr 1 *Dk2 B*	British London	668 869	221·0 *P 41 t.*	30·0 *F 23 t.*	16·1	Middlesbro' Backhouse & Dixon	1870	5 Blds. W B a & f	2Cy.27"& 48"–36" 90 HP. T.Richardson&Sons,Hrtlpl.	Harris & Dixon
144		George Peabody *Davies 80-88*	I Pad	British Llanelly	22 115	100·4	18·2	9·9	SouthShields Readhead,Softley&Co.	1867		2Cy.27"–18" 55 HP. Readhead,Softley&C.S.Shl.	W.Bowen,jun.
145		George Shattuck *Farquhar*	W ScSr 2 *Dks*	British Halifax,N.S.	231 362	129·2	23·8	17·8	Boston	1862		1Cy. 75 HP.	G.W.Churchill
146		George Stanton	I TwnScSp 1 *Dk*	British Chester	40 124	100·0	21·5	7·9	Birkenhead LairdBros.	1880		1Cy.15"& 26"–15" 70 HP. LairdBros.,Birkenhead	Shropshire Union Railway & Canal Co.
147		George T. Hope	W	American Clevel'd,Ohio	1275 1539	263·0	39·1	19·0	West BayCity, Mich.	1883			T.Fitzpatrick
148	R	George W. Clyde *Holmes*	I ScBkn 3 *Dks*	American New York	1470 1865	256·0	35·0	19·6	Philadelphia W.Cramp&Sons	1872		2Cy.24"&38"–36" 230 HP. W.Cramp&Sons,Phil.	W.P.Clyde & Co.
149	RV	George W. Elder *Angerstein 2 Dks & Hur Dk*	I ScBn	American Portland,Or.	1255 1710	242·0	38·0	20·0	Chester, Pa. J.Roach&Son	1874		2Cy.30"& 56"–54" 161 HP. J.Roach&Son,Chester,Pa.	Oregon Railway & Nav. Co.
150		George 1st	W PadTug 1 *Dk*	Greek Cephalonia	33 125	108·6	18·0	10·1	NorthShields Woodhave	1879		1Cy.37"–54" 60 HP. Hepple&Co.,S.Shields	C.Compethecra

✻ = Lloyd's Register; D = Nederlandsche Vereeniging; G = Germanischer Lloyd; I = Registro Italiano; R = American Record; A = Veritas Austro-Ungarico; H = Veritas Ellenico; N = Norske Veritas; V = Bureau Veritas.

Dans la 2me Colonne, B.T., avec les indications qui suivent, signifie la date du rapport de visite de la Carène par les experts du Board of Trade pour obtenir le certificat relatif au TRANSPORT DES PASSAGERS.

Dans la 3me Colonne, C = Composite (membrure en fer, bordé en bois); I = Fer; Stl = Acier; W = Bois.

Pour plus amples renseignements sur les navires classés par le Lloyd's Register, voir le Lloyd's Register Book et le Lloyd's Yacht Register.

No. in Book. / Official No.	Registry in which classed.	Ship's Name. Material, Rig, &c. / Master, and Dates of Appointment to present Owner's Service and to Ship. / Ship's Late Name. No. of Decks, &c.	Flag. / Port of Registry.	Tons. Gross. / Net.	Dimensions. Length, Breadth, Depth.	Built. Where. / By Whom.	When.	No. of Bulkheads. / Particulars of Water Ballast.	Engines. Particulars. / Maker's Name.	Owners or Managers.
151	✻ V	Georges & Henri 1 ScSp *Borny* ·87 (ex Black Sea...) 1 *Dk*	French Philippeville	67 123	105·0 17·6 8·2 B 5 f. F 12 f.	Paisley R. McIntyre & Co.	1879	[illegible] Blkds.	2Cy. 13 & 23½"–18" 25 HP Muir & Houston, Glasgow	H. Tessier
152 [illegible]	✻	Georgetown 1 Pad *Goodridge* 87–81 1 *Dk*	British Georgetown, De.	125 297	189·6 21·1 8·8	Newcastle W. Richardson & Co.	1873	1 Blkds.	2Cy. 28 & 50"–45" 150 HP J. Shaw & Co., Newcastle	Spreston Dock & Foundry Co.
153	V	Georgette 1 ScBkn *Masson* 1 *Dk* 2 *H*	French Dieppe	406 612	177·8 26·0 14·7 P 104 f. F 22 f.	Newcastle C. Mitchell & Co.	1876	DB [illegible]	2Cy. 27 & 52"–36" 100 HP T. Clark & Co., Newcastle	A. Grandchamp & Co.
154 89472	✻	Georgia 1 ScSc *Green* 76–85 1 *Dk* 2 *H*	British W. Hartlepl.	1168 1822	261·5 36·8 18·2 Q 65 f. P 29 f. B 106 f. F 28 f.	W. Hartlepool W. Gray & Co.	1885	1 Blkds. DB [illegible]	2Cy. 32 & 60"–39" 150 HP Black, Hawthorn & Co., Gts.	W. Gray
155		Georgia 1 Sc *Robinson* 3 *Dks*	American Baltimore	1188 1749	280·0 40·0 17·6	Wilmington Del The Harlan & Hollingsworth Co.	1887		2Cy. 34 & 64"–42" The Harlan & Hollingsworth Co., Wilmington	Baltimore Stm. P.
156 63058	✻	Georgian 1 ScSc *Nixon* 84–85 2 *Dks*	British Liverpool	712 1124	239·5 30·0 16·9 P 215 f. F 27 f.	Sunderland J. Laing	1872	1 Blkds. W B [illegible]	2Cy. 27 & 52"–36" 115 HP G. Clark, Sunderland	R. McAndrew & Co.
157 80504		Georgian W ScSp 1 *Dk* & *Spr Dk*	British Toronto	227 377	130·0 21·8 11·5 B 59 f.	Georgian Bay J. Potter	1864		1Cy. 24"–24" 75 HP Gartley & Co., Montreal	Canadian Pacific Railway Co.
158		Georgios Stl & 1 Sc (ex Baronne Rudich)	Greek Syra	72 123	.. .. ..	Trieste Stabilimento Technico	1878		30 HP	J. B. Paterson & Co.
159	V	Georgios I. Stl ScSc *Bohde* ·88 2 *Dks*	Danish Copenhagen	1067 1617	260·8 33·8 16·2 P 167 f. F 64 f.	Copenhagen Burmeister & Wain	1888	[illegible]	3Cy. 18, 30 & 48"–36" 150 HP Burmeister & Wain, Coph'gn	Det Forenede Dampskibsselskab
160 112		Georgy 1 PadSp 1 *Dk*	Russian Archangel	143	137·0 21·0 10·0	Hamburg Godefroy	1860		2Cy. 85 HP J. & G. Thomson, Glasgow	Gellermann
161		Gequia	Brazilian Pernambuco	223	.. .. ..				80 HP	Cia. de Nav. de Pernambuco
162 93791	✻	Gerda Stl ScSc *Gunther* ·88 1 *Dk*	British Liverpool	310 498	158·0 30·2 11·9 B 39 f. F 19 f.	Barrow [illegible]	1888	1 Blkds. [illegible]	2Cy. 18 & [illegible] 54 HP [illegible]	Argentine Steam Lighter Co. Ld.
163	V	Gerda (WRECKED) 1 ScSc *Ericksen* ·83 1 *Dk* & *H*	German Flensburg	340 537	169·6 25·6 13·2 Q 48 f. B 47 f. F 20 f.	Flensburg Flensburger Schiff Ges.	1883	WB	2Cy. 22½ & 38"–27" 60 HP Flensburger Schiffsb.-Ges., Flensburg	H. Schuldt
164 [illegible]	✻	Gerent Stl ScSc *Eggleton* 65–88 1 *Dk* 2 *H*	British Sunderland	1467 2283	290·0 39·0 21·8 Q 32 f. F 31 f. B 74 f.	Sunderland Short Bros.	1888	1 Blkds. [illegible]	3Cy. 21½, 35 & 58"–39" J. Dickinson, Sndld. 200 HP	J. Westoll
165		Gerente 1 Pad 1 *Dk*	Brazilian Rio de Janiro	324 375	170·0 25·5 12·0	London M. Wigram & Sons	1862	1 Blkds.	2Cy. 47½"–42" 140 HP Renhill, Salkeld & Co., London	Cia. da Estrada de ferro de Macahé
166	V	Gergovia 1 ScBn *Bourge* 87–87 2 *Dks* 3 *B*	French Marseilles	1376 2024	280·1 36·7 24·6 P 33 f. B 43 f. F 24 f.	Greenock Scott & Co.	1883	DB [illegible]	2Cy. 36 & 68"–45" 375 HP Greenock Foundry Co., Grnk.	Compagnie Française de Nav. à Vap.
167 76834	✻ BT 5.88	German 1 ScSc *Reynolds* 88–88 3 *Dks*	British Southampton	1822 3008	350·6 39·4 29·6	Dumbarton W. Denny & Bros.	1877	5 Blkds. [illegible]	2Cy. 50 & 86"–54" 300 HP Denny & Co., Dumbarton	Union Steam Ship Co. Ld.
168 62850	✻	German Emperor 1 Sc3mSc *Bend* 87–88 1 *Dk* 2 *H*	British London	587 909	210·3 29·5 16·5 Q 73 f. B 40 f.	Sunderland Short Bros.	1872	1 Blkds. W B [illegible]	2Cy. 27 & 50"–33" 98 HP Gourlay Bros., Dundee	J. S. Barwick & Co.
169 67537	✻ BT 5.88	German Empire 1 ScSc *Nash* 65–88 1 *Dk* 2 *H*	British W. Hartlepl.	454 705	209·5 27·8 15·8 P 53 f. F 39 f.	W. Hartlepool Denton, Gray & Co.	1872	1 Blkds. W B [illegible]	2Cy. 29 & 56"–33" 120 HP N.E. Marine E. Co. Ld. Sndld.	West Hartlepool Nav. Co.
170 88708	✻ D	Germania 1 ScBn *Thomas* ·85 1 *Dk* & *Spr Dk* 3 *H*	British Newcastle	1948 2970	300·0 40·1 28·5 B 70 f. F [illegible]	Stockton M. Pearse & Co.	1883	5 Blkds. DB [illegible] AP T[illegible]	2Cy. 38 & 71"–48" 246 HP Blair & Co. Ld., Stockton	International Line Co.
171	✻	Germania (ex Sylph) 1 ScSc *Wolters* 84–86 1 *Dk* 2 *H*	German Hamburg	439 681	213·2 27·7 16·0 [illegible]	Sunderland J. Laing	1856	1 Blkds.	2Cy. 27 & 52"–33" 140 HP Humphries & Pearson, Hull	H. J. Perlbach & Co.
172		Germania (ex Hughes) 1 ScSc *Morin* ·88 1 *Dk*	Swedish Söderhamn	317 120	150·7 24·1 11·4	Nyköping Nyköping M.V.	1864		2C. 14½ & 33½"–24" 50 HP Nyköping M.V., Nyköping	Ångf. Aktieb. "Germania" (A. Bros.)
173 70932	BT 5.88	Germanic 1 Sc4mBk *Irving* ·88 3 *Dks*	British Liverpool	3150 5008	455·0 45·2 33·7	Belfast Harland & Wolff	1874	8 Blkds.	1Cy. 48 & 83"–60" 760 HP Maudslay Sons & Field, Lond.	Oceanic Stm. Nav. Co. Ld.
174 93475	✻	Gerona Stl ScS. *Sangster* 82–88 2 *Dks* & *W h Frms*	British Dundee	1958 3056	345·0 43·2 26·4 F 99 f. P 37 f. B 160 f.	Dundee Gourlay Bros. & Co.	1888	5 Blkds. [illegible] FPT 34 APT 24	3Cy. 29, 44 & 71"–48" Gourlay Bros. & Co., Dundee 120 HP	W. Thomson & Co.
175	V	Gertrud Woermann StlScSc *Jensen* ·85–88 2 *Dks*	German Hamburg	1359 1818	270·0 35·0 23·5 P 72 f. B 76 f. F 33 f.	Stettin Act. Ges. "Vulcan"	1885	[illegible]	3Cy. 19, 30 & 49"–35" 180 HP Act. Ges. "Vulcan" Stet.	Afrikan. Damp. Act. Ges. (Woermann)
176 70130	✻	Gertrude 1 ScBn *Thoren* 1 *Dk* 2 *H*	British Hartlepool	869 1347	241·0 32·2 18·4 B 43 f. Q 86 f. F 32 f.	W. Hartlepool W. Gray & Co.	1879	1 Blkds. [illegible]	2Cy. 29 & 55"–33" 120 HP T. Richardson & Sons, Hrtlpl.	O. Trechmann
177 60303	✻	Gertrude 1 ScK *Robinson* 85–88 1 *Dk*	British Middlesboro	139 223	130·0 20·5 11·3	Stockton M. Pearse & Co.	1873	3 Blkds. W B [illegible]	2Cy. 16½ & 33"–20" 40 HP Blair & Co. Ld., Stockton	J. M. Lennard & Sons
178 82921	✻	Gervase 1 ScSc *Hodge* 82–84 1 *Dk* 2 *H*	British [illegible]	417 660	185·5 25·5 14·0 P 53 f. B 35 f. F 23 f.	Sunderland Osbourne, Graham & Co.	1880	1 Blkds. [illegible] APT [illegible]	2Cy. 23½ & 42"–30" 80 HP G. Clark, Sunderland	G. Bazeley & Sons
179		Gestrikland (ex [illegible]) 1 ScSc *Granath* ·86 1 *Dk*	Swedish Gefle	275 348	140·4 22·0 10·5 P 80 f. F 38 f. B 39 f.	Norrköping Motala Co.	18 68/75		2Cy. 18½ & 37½"–21" 70 HP W. Lindberg's Mek. V. Stkhlm	Ångf. Bolaget Tr[illegible]
180	V	Gevalia 1 ScSc *Akerson* 68–72 1 *Dk*	Swedish Gefle	393 595	154·8 23·6 15·2 Q 40 f. B 37 f. [illegible]	Gothenburg Motala Co.	1872		2Cy. 17 & 34"–24" 60 HP Motala Co., Motala	Rederi Aktieb. Svea (L. V. Akerson)

✻ = Lloyd's Register; D = Nederlandsche Vereeniging; G = Germanischer Lloyd; I = Registro Italiano; R = American Record;
A = Veritas Austro-Ungarico; H = Veritas Ellenico; N = Norske Veritas. V = Bureau Veritas.

In the 2nd Column, B.T., with figures following, signifies Date of report of bottom sighting by Board of Trade Surveyors as condition of PASSENGER CERTIFICATE.

In the 3rd Column, C = Composite (iron frame, wood planked); I = Iron; Stl = Steel; W = Wood.

Further particulars of Vessels classed by Lloyd's Register will be found in Lloyd's Register Book and Lloyd's Yacht Register.

Classe	Nom, Materiaux, Gréement, etc. Capitaine et dates d'entrée au service des Armateurs actuels et à bord du navire. Ancien Nom, nombre de ponts, etc.	Pavillon. Port l'Armement	Tonnage Net. Total	Dimensions Longueur	Largeur	Creux	Construit. Où. Par qui.	Quand.	Nombre des cloisons. Détails sur le lest d'eau.	Machines. Détails. Nom du Constructeur.	Armateurs ou Directeurs-Gérants.
	Geysir W ScSr 1 Dk 2 B	Norweg'n Tönsberg	243 375	133·3	30·2	15·6	Tönsberg	1870		1Cy.24¼"–24¾" 45 HP. Nylands Verkst., Christiania	Aktie-Selskabet "Geysir" (G. Knap)
✻	Ghazee I ScSr Scotland 2 Dks 3 B	British Rochester	1764 2712	340·0 P 37 f. B 40 f. F 42 f.	38·2	26·3	Glasgow Aitken & Mansel	1883	6 Blds.	2Cy.43" & 75"–51" 350 HP. J. & J. Thomson, Glasgow	Mogul S.S. Co., Ld.
	Giafferieh I ScSr 1 Dk & Spar Dk	Egyptian Alexandria	600	230·0	30·0	16·3	London J. Ash & Co.	1862		2Cy.52"–33" 160 HP. T. Stewart, London	H.H. The Khedive of Egypt
✻	Gian Paolo (wreck'd) I Sc Rebosi 1 Dk	Italian Spezia	288 461	152·6 Q 55 f. B 40 f. F 20 f.	23·5	12·4	Sunderland Sunderland S.B. Co. Ld.	1885	1 Blds. DB 45 FPT 43	2Cy.20" & 38"–27" 74 HP. N.E. Marine Co., Ld., Sundrld.	Giacopini Bros.
✻	Giang Ann Stl ScSr Wooldridge -88 pt Shd Dk 1 Dk 2 B	British Singapore	557 873	230·0 P 56 f. F 38 f.	31·1	15·2	Newcastle Wigham Richardson & Co.	1888	4 Blds.	3Cy.17½", 27½" & 46"–33" 130 HP. Wigham Richardson & Co., Newcastle	Tan Beng Wan
	Gianrinaldo Carli W ScSr Palay -88 1 Dk	Austrian Capodistria	54 104	110·1	15·9	9·5	Capo d'Istria L. Polli	1883		2Cy.16½" & 28½"–18" 35 HP. Bartleth & Co.	Societa cittadina di Nav. di Capodistria
	Gião	Brazilian Para	118	..	..	..				35 HP.	E.J.C. Brito
I	Giava I ScBg Vernengo 83-85 2 Dks 3 B	Italian Genoa	1818 2713	333·9 B 72 f. F 42 f.	36·2	26·7	Newcastle A. Leslie & Co.	1881	5 Blds. DB 80/680	2Cy.35" & 70"–48" 417 HP. R. & W. Hawthorn, Newcstle	Navigazione Generale Italiana
✻	Gibel Tarik Stl TwnScSr Charlton 1 Dk	British Gibraltar	123 228	135·0 F 20 f.	22·1	10·0	Newcastle Sir W.G. Armstrong, Mitchell & Co., Ld.	1884	3 Blds. DB 80	1Cy.17" & 32"–20" 106 HP. Wallsend Slipwy Co., Ld. Nwe	M.H. Bland & Co.
✻	Gibraltar (ex Ariadne) I ScSr Harvey 76-78 Spr Dk	British London	915 1412	254·0	32·2	17·3 24·9	Sunderland W. Pile & Co.	1869	5 Blds. pDB	2Cy.32" & 60"–39" 140 HP. Blair & Co. Ld., Stockton	J. Hall, jun., & Co.
	Gijon I ScSr 2 Dks	Spanish Gijon	499 790	198·7 P & B 121 f. F 36 f.	28·0	15·1	Dundee Gourlay Bros. & Co.	1875	5 Blds.	2Cy.31" & 51"–30" 120 HP. Gourlay Bros. & Co., Dundee	Meliton, Gonzalez & Co.
	Gijon I	Spanish Gijon	130 155	129·9	19·5	10·4	Liverpool	1868		30 HP.	B. Barbachano
✻	Gilsland I ScSr Robinson 2 Dks 3 B	British London	1058 1639	265·0	34·0	24·5	Newcastle Cole Bros.	1874	4 Blds. W.B. & A.B.	2Cy.33" & 65"–36" 200 HP. N.E. Marine E. Co., Ld., Sndrld	W. Milburn & Co.
V	Ginevra I ScSr Kahl 79-	Russian Riga	160 221	129·9 B 23 f.	19·3	10·9	Stockton M. Pearse & Co.	1873	WB	2Cy.15" & 30"–21" 35 HP. Blair & Co., Stockton	A.G. Sengbusch & Co. Soc. di Nav. Italiana.
✻ I	Giovanni Batta Lavarello Stl ScSr Jonas -88 1 Dk & Awn Dk 3 B	Italian Genoa	1222 1862	270·0	34·3	17·7	Newcastle Wighm. Richardson & Co.	1886	5 Blds.	3C.24", 37" & 62"–42" 300 HP. Wigham Richrdsn. & C. Nwe.	Frat. Lavarello fu Gio. Batta
	Giovanni da Procida I Sc 1 Dk	Italian Naples	85 135	115·1	20·5	10·7	England	1858		2Cy.22"–15" 34 HP.	Soc. Anon. Procida Ischia
	Gipsey Queen I Pad 1 Dk	British Liverpool	106 149	135·0	20·1	7·7	Belfast Harland & Wolff	1865		1Cy.35"–42" 70 HP. Blackwood & Gordon, Pt. Gls.	T.W. Thompson
✻	Gipsy I STrawler Bluth 86-86 1 Dk	British Aberdeen	68 161	126·5	19·3	10·1	Aberdeen J. Duthie, Sons & Co.	1883	4 Blds.	2Cy.19" & 38"–27" 62 HP. Hall, Russell & Co., Aberdeen	W. Leslie & Co.
	Gipsy W ScSr Plummer 2 Dks	American S. Francisco	215 294	102·0	28·3	9·1	San Francisco	1868		2Cy.12" & 20"–20" 21 HP. Risdon Iron Wks, S. Francsco	Pacific Coast S.S. Co.
	Gipsy King I PadSr (ex Princess of Wales)	British Liverpool	27 104	98·8	17·1	9·1	Bristol G.K. Stothert & Co.	1868		2Cy.28"–44" 70 HP. G.K. Stothert & Co., Bristol	H.J. Ward
	Gipsy Queen I ScSr Wilkie 86-86	British W. Hartlepl.	413 667	207·1 P 54 f. B 65 f. F 60 f.	26·3	15·7	Stockton M. Pearse & Co.	1856		2Cy.26" & 50"–30" 100 HP. T. Richardson & Sons, Hrtlpl.	W. Hartlepool Steam Nav. Co.
✻	Giralda I ScSr Englefield 75-87 1 Dk 2 B	British Newcastle	732 1148	225·0 Q 75 f. B 52 f. F 41 f.	33·0	16·1	Sunderland Osbourne, Graham & Co.	1887	1 Blds. DB 60/240	2Cy.28" & 53"–36" 120 HP. G. Clark, Sunderland	Bilton, Williams & Co.
✻	Giralda I Sc3mSrYt (ex Fair Geraldine)	British Shoreham	130 191	148·0	21·1	11·1	Leith Ramage & Ferguson	1880	1 Blds.	2Cy.18" & 36"–22" 55 HP. Walker, Hudrsn & Co., Glsgw	Capt. Pryce B. Hamilton
✻	Girdleness I ScSr Bannerman 84-84 1 Dk	British Aberdeen	367 578	171·7 Q 52 f. B 44 f. F 20 f.	26·7	13·1	Aberdeen A. Hall & Co.	1883	5 Blds. DB 80 & FPT 100	2Cy.22" & 44"–30" 80 HP. Hall, Russell & Co., Aberdeen	J. Cook & Son
	Girgenti Stl ScSr Sparr Dk	German Hamburg		250·0	32·2	22·9	Sunderland J. Blumer & Co.	1889		3Cy.19", 31½" & 51½"–36" N.E. Mar. Eng. Co., Nweastle.	R.M. Sloman & Co.
✻	Gironde I ScSr Boudeon 37-87 1 Dk 2 B	British Cardiff	441 701	200·9 P & B 114 f. F 42 f.	29·2	15·1	Newcastle C.S. Swan & Hunter	1881	1 Blds. DB 63/312 AT 27/8	2Cy.26" & 47"–36" 99 HP. G. Clark, Sunderland	Cardiff S.S. Co., Ld. (Hooper, Campbell & Co., Managers)
DV	Gironde I ScSr de Boer 82-84 2 Dks	Dutch Rotterdam	450 667	188·3	25·5	14·2	Port Glasgow J. Reid & Co.	1852		2Cy.20" & 40"–26" 60 HP. N.E. Marine E. Co., Ld., Sndrld	Smith & Co.
	Gironde I ScSr 3 Dks	French Marseilles	2052 3206	388·3 F 38 f. B 264 f.	40·0	28·5	La Ciotat Messageries Maritimes	1869		3Cy.60½–43¼ 500 HP. Messageries Mar. La Ciotat.	Messageries Maritimes
✻	Gitana (ex Inmaculada Concepcion) I Sc3mSr Ray 2 Dks	British Glasgow	219 177	187·2 P 35 f. F 16 f.	27·0	15·0	London Seaward & Co.	1859	4 Blds.	2Cy.25" & 50"–24" 96 HP. Cp. 83 C.D. Holmes & C. Hull	J. Clark
✻	Gitana I Sc3mSrYt	British Leith	205 314	157·8	25·9	14·1	Leith Ramage & Ferguson	1882	5 Blds.	2Cy.23" & 43"–24" 85 HP. M. Paul & Co., Dumbarton	Henry Jameson

✻=Lloyd's Register; D=Nederlandsche Vereeniging; G=Germanischer Lloyd; I=Registro Italiano; R=American Record; A=Veritas Austro-Ungarico; H=Veritas Ellenico; N=Norske Veritas; V=Bureau Veritas.

Dans la 2me Colonne, B.T., avec les indications qui suivent, signifie la date du rapport de visite de la Carène par les experts du Board of Trade pour obtenir le certificat relatif au TRANSPORT DES PASSAGERS.

Dans la 3me Colonne, C=Composite (membrure en fer, bordé en bois); I=Fer; Stl=Acier; W=Bois.

Pour plus amples renseignements sur les navires classés par le Lloyd's Register, voir le Lloyd's Register Book et le Lloyd's Yacht Register.

No. in Book / Official No.	Registry in which classed	Ship's Name. Material, Rig, &c. / Master, and Dates of Appointment to present Owner's Service and to Ship / Ship's Late Name. No. of Decks, &c.	Flag. / Port of Registry.	Tons. Net. / Gross.	Length	Breadth	Depth	Built. Where. / By Whom.	When	No. of Bulkheads. / Particulars of Water Ballast.	Engines. / Particulars. Maker's Name.	Owners or Managers
211 82481		Gitano Stl SeSr Wood 1 Dk 2 [illegible]	British [illegible]	972 1243	243·1	33·1	16·9	Hull Earle'sCo.,Ld.	1880	4 Blkds.	2Cy.24½"&50"–36" 98 HP. Earle'sCo.,Ld.,Hull	T.W[illegible],Sons
212 134		Giulietta A. (ex Madonna del Carmine) ISeSr Pinello 1 Dk	Italian Genoa	108 [illegible]	130·0	20·1	11·2	Newcastle C.S.Swan & Hunter	1883	1 Blkd [illegible]	2Cy.16"&30"–22" 64HP. Ross & Duncan,Glasgow	S.Anselmo
213 128	✻ I	Giulio Mazzino I SeSr Gorasgio 1 Dk & pt. Awn Dk 3 R	Italian Genoa	1238 [illegible]	257·4	36·1	18·8	SouthShields J.Readhead&Co.	1884	4 Blkds. [illegible]	2Cy.38"&61"–36" 160 HP. J.Readhead&Co.,S.Shields	B.Mazzino (G.B.Gastaldi [illegible])
214 90538		Glacial W Sc	British Montreal	74 [illegible]	103·2	24·0	7·8	Sorel,P.Q. J.Metton	1885		4C15"(2)18"36".18" 70HP J.A.Pontbreand,Sorel,P.Q.	C.Bourgeois
215 29018	✻	Gladiator I Sc3mSr Kerr 87-87 1 Dk 2 H	British Glasgow	422 [illegible]	192·8	27·8	15·7	Stockton M.Pearse&Co.	1860	4 Blkds	2Cy.23"&42"–30" 80 HP. Dunsmuir&Jackson,Glasgw.	D.McBrayne
216		Gladiator I SeSrYt	British London	111 164	119·6	20·1	11·7	Leith Ramage&Ferguson	1886	4 Blkds.	3C.9¼"15"&24½"–18" 40HP. Ramage&Ferguson,Leith	Sir J.W.McK[illegible]
217 70861	BT5.88	Gladiator I PadSp	British Liverpool	37 [illegible]	144·7	20·7	11·8	Birkenhead T.Brassey&Co.	1874		2Cy.35"–60" 99 HP. T.Brassey&Co.,Birkenhead	E.Griffiths
218 79237	✻	Gladiolus I SeSr Wright 84-88 2 Dks 3 H	British N.Shields	1258 [illegible]	280·6	36·0	22·6	Newcastle [illegible]	1880	5 Blkds [illegible]	2Cy.33"&63"–35" 180 HP. N.E.MarineE.Co.Ld.,Sndrld.	J.Robinson &[illegible]
219 73787	BT1.88	Gladstone I PadSp (ex Flying Meteor)	British [illegible]	14 138	118·0	19·8	10·4	S.Shields J.T.Eltringham	1875		1Cy.40"–51" 95 HP. J.P.Rennoldson,S.Shields	J.M'Kean
220 74413	✻	Gladwyn I Sc3mSrYt (ex Franziska)	British [illegible]	97 143	132·0	18·2	10·8	Southampton Day, Summers	1876	3 Blkds	2Cy.16"&32"–21" 45 HP. Day, Summers,S'hmpton.	F.A.Gerwyn J[illegible]
221 58794	✻ D	Gladys I SeSr M'Gregor 2 Dks 3 H	British [illegible]	1022 1661	258·2	33·0	24·1	W. Hartlepool Denton,Gray&Co.	1872	4 Blkds. [illegible]	2Cy.32"&60"–33" 140 HP. N.E.MarineE.Co.Ld.,Sndrld.	T.Marwood &[illegible]
222 82948		Gladys I SeSrYt McGregor 85-85	British [illegible]	97 143	128·1	17·5	9·1	PortGlasgow J.Reid&Co.	1883	4 Blkds.	2Cy.14"&28"–21" 33 HP. Walker,Henderson&Co.,Gls.	Mrs.R.E.Stew[illegible]
223 89651	✻	Glamorganshire I Sc3mSr Davies 83-86 2 Dks 3 H	British London	1843 [illegible]	340·2	40·2	26·5	Glasgow London&GlasgwCo	1884	6 Blkds. [illegible]	2Cy.43"&78"–54" 450 HP. London&GlasgowCo.	Jenkins & Co.
224	V	Glaneuse Stl SeSr Gadebert -88 1 Dk	French Brest	83 158	95·0	20·1	7·9	Aberdeen A.Hall&Co.	1888		2Cy.15"&30"–20" 40 HP. A.Hall&Co.,Aberdeen	Chevillotte Fr[illegible]
225 92013	✻ BT1.88	Glanmire I SeSr Browne -87 2 Dks	British Cork	466 [illegible]	242·2	33·2	15·3	Dundee W.B.Thompson&Co.Ld.	1888	[illegible] Blkds. [illegible]	3C.23"38"&62"–48" 300HP. W.B.Thompson&Co.Ld.Dun	City of Cork[illegible] Co.,Ld.
226 70236	✻	Glannibanta I Sc3mSr Owens 86-86 1 Dk 2 H	British Newcastle	535 853	200·6	30·5	15·5	Sunderland J.L.Thompson	1875	4 Blkds [illegible]	2Cy.27"&52"–33" 99 HP. J.Dickinson,Sunderland	C.F.H.Leslie (Fenwick & Co.[illegible])
227 85241	✻	Glanrheidol I SeSr Francis 84-84 1 Dk 2 H	British Aberystwith	638 1003	212·0	30·1	15·5	Sunderland W.Doxford&Sons	1883	4 Blkds. [illegible]	2Cy.25"&48"–33" 97 HP. W.Doxford&Sons,Sunderld.	"Glanrheidol"[illegible] Ld.(J.Mathias)
228 85262	✻	Glanwern I SeSr Davies 69-82 1 Dk 2 H	British Aberystwith	641 [illegible]	212·0	30·1	15·5	Sunderland W.Doxford&Sons	1882	4 Blkds [illegible]	2Cy.25"&48"–33" 90 HP. W.Doxford&Sons,Sunderld.	Aberystwith [illegible]
229 83898		Glanworth I SeSr Johnstone 1 Dk & Awn Dk	British [illegible]	558 [illegible]	211·2	29·1	12·1	Kirkcaldy J.Key&Sons	1882	4 Blkds [illegible]	2Cy.30"&60"–39" 160 HP. J.Key&Sons,Kirkcaldy	Australasian [illegible] Stm.Nav.Co.Ld.
230 83299	✻	Glanystwyth Stl SeSr Davies -88 1 Dk & WebFrms	British Aberystwith	1167 1821	255·0	37·0	18·8	Newcastle Palmers Co.,Ld.	1888	4 Blkds. [illegible]	3C.20"33"&54"–36" 160HP. Palmers'Co.,Ld.,Newcastle	"Glanystwith"[illegible] Ld.(T.B.Jen[illegible])
231 63430	✻	Glasgow I SeSr Scobbie 82-82 1 Dk 2 H	British Glasgow	289 471	170·0	25·0	14·4	Middlesbro' Backhouse&Dixon	1869	4 Blkds. [illegible]	2Cy.24"&42"–30" 85 HP. Lees,Anderson&Co.,Glasgw.	J.Rankine & S[illegible]
232 22935		Glasgow (ex Garland) I SeSr Patterson 79	British Sligo	191 [illegible]	187·9	23·4	12·2	Glasgow A.Stephen&Sons	1855		2Cy.25"&50"–25" 110 HP. Muir&Houston,Glasgow	Sligo Stm.Nav. Ld.
233 84374	✻	Glassalt I SeSr Joiner 86-86 1 Dk 2 H	British Aberdeen	417 [illegible]	183·1	26·3	14·1	Aberdeen J.Duthie,Sons&Co.	1883	5 Blkds.	2Cy.23"&44"–30" 80 HP. Hall,Russell&Co.,Aberdeen	Aberdeen & C[illegible] S.S.Co.,Ld.
234 63897	BT5.88	Glaucus I SeSr Hannah 85-85	British Liverpool	1382 2118	322·0	33·7	25·6	Newcastle A.Leslie&Co.	1871		2Cy.28½"&61"–45" 250HP. R.Stephenson,Newcastle	Ocean Steam[illegible]
235 [illegible]	✻	Glaucus I SeSr Sanderson 81-86 1 Dk 2 H	British [illegible]	884 1363	238·5	34·2	18·0	Sunderland J.Laing	1878	4 Blkds. [illegible]	2Cy.31"&57"–36" 150 HP. G.Clark,Sunderland	James Paters[illegible]
236 18504	R	Glaucus W Sc Colman -87 2 Dks	American Boston,Mass	1206 1809	245·0	40·0	17·6	New York	18[illegible]			Metropolitan S[illegible]
237 81591	✻	Gleadowe I SeSr Mehegan 84-87 2 Dks 3 H	British [illegible]	1381 [illegible]	285·6	37·2	24·4	Newcastle Palmers'Co.,Ld.	1879	[illegible]	2Cy.36"&68"–45" 260 HP. Palmers'Co.,Ld.,Newcastle	Hunting & P[illegible]
238 [illegible]	✻	Gledholt I SeSr Whitburn 82-83 1 Dk 2 H	British W.Hartlepl.	1084 1688	250·0	36·1	17·5	W.Hartlepool E.Withy&Co.	1883	4 Blkds. [illegible]	2Cy.32"&59"–33" 99 HP. T.Richardson&Sons,Hrtlpl.	R.Ropner & Co.
239 93377	✻	Glen Caladh Tower Stl SeSr Hamlyn 88-88 2 Dks 3 H	British London	2401 [illegible]	340·0	42·0	26·8	Newcastle R.Stephenson&Co.Ld.	1888	6 Blkds. [illegible]	3Cy.25½"41"&67"–42" 400 HP R.Stephenson&Co.,Ld.,Newcastle	F. Stumore & C[illegible]
240												

✻ = Lloyd's Register; D = Nederlandsche Vereeniging; G = Germanischer Lloyd; I = Registro Italiano; R = American [illegible]
A = Veritas Austro-Ungarico; H = Veritas Ellenico; N = Norske Veritas; V = Bureau Ver[illegible]

In the 2nd Column, B.T., with figures following, signifies Date of report of bottom sighting by Board of Trade Surveyors as condition of PASSENGER CER[illegible]

In the 3rd Column, C = Composite (iron frame, wood planked); I = Iron; Stl = Steel; W = Wood.

Further particulars of Vessels classed by Lloyd's Register will be found in Lloyd's Register Book and Lloyd's Yacht Register.

		Nom, Matériaux, Gréement, etc. — Capitaines et dates d'entrée au service des Armateurs actuels et à bord du navire. — Ancien Nom, nombre de ponts, etc.		Pavillon. — Port d'Armement	Tonnage. Net. — Brut.	Dimensions. Longueur	Largeur	Creux	Construit. Où. — Par qui.	Quand	Nombre des cloisons. — Détails sur le lest d'eau.	Machines. — Détails. Nom du Constructeur.	Armateurs ou Directeurs-Gérants.
	✻D	Glen Dochart *Purvis*	I ScSr 1 *Dk2 B*	British	832	236·0 Q141f.	32·2 B12f.	18·1 F28f.	Newcastle Schlesinger,Davis&Co.	1882	3 Blds. WB.. FPT APT	2Cy.28"&53"–36" 120 HP. Hawks,Crawshay&Sns,Gts.	H.Baxter & Co.
	✻	Glen Gelder *Munro* 84-83	I ScSr 1 *Dk2 B*	British Aberdeen	498	195·2 B6f.	31·2 B40f.	13·5 F23f.	Aberdeen A.Hall&Co.	1881	5 Blds. DBa..	2Cy.26"&49"–30" 99 HP. Hall,Russell&Co.,Aberdeen	Aberdeen Glen Line S.S.Co.,Ld.
	✻	Glen Holme (ex Margaret Banks) *Morwick* 77-82	I ScSr 1 *Dk2 B*	British Maryport	532	213·6 Q 80f.	29·2 F 22f.	16·2	W. Hartlepool Denton,Gray&Co.	1870	1 Blds. DBa..	2Cy.27"&50"–30" 96 HP. T.Richardson&Sons,Hrtlpl.	Hine Bros.
		Glen Rosa	I PdlSk	British London	99	206·1	20·11 Q 29f.	7·5	Greenock Caird&Co.	1877		1Cy.50"–72" 120 HP. Caird&Co.,Greenock	Victoria Steamboat Association,Ld.
	✻	Glen Tanar *Evans* 81-84	I ScBn 1 *Dk&Spr Dk3 B*	British Aberdeen	1028	250·7	33·1	17·7 24·8	Aberdeen A.Hall&Co.	1884	5 Blds. DBa..	2Cy.30"&57"–36" 150 HP. Hall,Russell&Co.,Aberdeen	Aberdeen Glen Line S.S.Co.,Ld.
	✻	Glen Tilt *Allan* 86-86	I ScSr 1 *Dk*	British Aberdeen	303 477	165·6 Q49f.	23·6 B43f.	12·4 F21f.	Aberdeen A.Hall&Co.	1882	5 Blds. DBa..	2Cy.20"&38"–24" 62 HP. Hall,Russell&Co.,Aberdeen	Aberdeen Glen Line S.S.Co.,Ld.
	✻	Glenarm *Campbell*	Stl ScSr 1 *Dk*	British Glasgow	105 247	130·3 Q14f.	22·0 B9f.	9·8 F24f.	Paisley H.M·Intyre&Co.	1881	3 Blds. FPT.. APT..	2Cy.20"&38"–30" 61 HP. R.H.Pearson&Co.,Glasgow	Eglinton Chemical Co.,Ld.
	✻	Glenartney *Mackintosh* 83-85	I Sc3mSr 2 *Dks3 B*	British Glasgow	1400 2143	331·2	35·3	24·7	Glasgow London&Glsgw.Co.	1873	5 Blds.	2Cy.44"&79"–45" 320 HP. London&GlasgowCo.,Glasg.	M·Gregor,Gow & Co.
	✻	Glenavon *Jacobs*	I Sc3mSr 2 *Dks3 B*	British Glasgow	1936 2985	360·4 P42f.	43·3 B70f.	24·5 F50f.	Glasgow London&Glsgw.Co.	1881	5 Blds. CDB..	2Cy.48"&88"–54" 530 HP. London&GlasgowCo.,Glasg.	M·Gregor,Gow & Co.
	✻	Glenavon *Davis* 86-86	I Sc3mSr 1 *Dk*	British Newcastle	545 864	213·5 Q 70f.	28·6 F 21f.	15·1	N.Shields T.&W.Smith	1872	4 Blds. DB..	2Cy.25"&48"–30" 90 HP. Hawthorn,Hodgson&Co.,Lon	Elliott, Lowrey & Dunford
	✻	Glenbervie *Wallace* 83-88	I ScSr 2 *Dks3 B*	British Leith	1013 1572	250·0 P..	33·7 F38f.	23·0 B16f.	Newcastle Palmers'Co.,Ld.	1878	4 Blds. DB..	2Cy.30"&55"–36" 140 HP. Palmers'Co.,Ld.,Newcastle	Lindsay,Gracie & Co.
	✻	Glencairn *Crosby* 82-82	I ScSr 1 *Dk*	British Middlesbro'	176 314	156·0	22·0	13·5	Middlesbro' Backhouse&Dixon	1868	3 Blds. DB..	2Cy.18½"&34"–24" 45 HP. Blair&Co.,Ld.,Stockton	J.M.Lennard & Sons
253	✻	Glencoe (MISSING) *Cooke* 82-86	I Sc3mSr 2 *Dks3 B*	British Glasgow	1901 2913	387·2 P50f.	38·2 B62f.	26·5 F56f.	Glasgow London&Glsgw.Co.	1878	5 Blds.	2Cy.48"&88"–60" 550 HP. London&GlasgowCo.,Glasg.	M·Gregor,Gow & Co.
254	✻	Glencoe *Philip* 85-85	I Sc3mSr 1 *Dk*	British Middlesbro'	190 310	144·7	22·0 Q 20f.	13·2	Middlesbro' Backhouse&Dixon	1867	3 Blds. DB..	2Cy.19"&36"–24" 50 HP. Blair&Co.,Ld.,Stockton	J.E.Swan & Bros., Ld.
255		Glencoe ex Mary Jane *Buchanan* 84-88	I PdlSk	British Glasgow	122 226	165·4	20·2	9·4	Glasgow Tod&McGregor	1846		1Cy 56"–54" 120 HP. Tod&McGregor,Glasgow	D.MacBrayne
256	✻	Glendale *Willis* 87-87	I Sc3mSr 1 *Dk2 B*	British W.Hartlep'l	714 1118	230·3 Q20f.	30·6 B53f.	17·3 F23f.	W. Hartlepool R.Irvine&Co.	1874	4 Blds. DBa..	2Cy.28"&53"–33" 110 HP. T.Richardson&Sons,Hrtlpl.	G.B.Harland & Co.
257	✻	Glendale *Beckman* 82-88	Stl ScSr 1 *Dk&WebFrms*	British Sunderland	641 989	211·0 F27f.	30·5 Q84f.	14·6 B86f.	Sunderland J.Laing	1886	4 Blds. CDB..	3C.17½"29"&47"–33"99HP. G.Clark,Sunderland	P.H.Laing
258	✻	Glendevon *Peterson* 87-87	I ScSr 2 *Dks3 B*	British Leith	1147 1760	260·8 P32f.	35·3 B..	23·3 F30f.	Newcastle Palmers'Co.,Ld.	1880	3 Blds. DB..	2Cy.32"&60"–42" 190 HP. Palmers'Co.,Ld.,Newcastle	Lindsay,Gracie & Co.
259	✻	Glendower *Hodge* 85-85	I ScSr 1 *Dk&Awn Dk3 B*	British Leith	1650 2496	291·8 E..	37·3	20·7 28·0	Newcastle Palmers'Co.,Ld.	1882	5 Blds. CDB.. APT..	2Cy.36"&68"–45" 260 HP. Palmers'Co.,Ld.,Newcastle	"Tyne Glendower" S.S.Co.,Ld.(Lindsay,Gracie&Co.)
260	✻	Glendower *Cornish* 83-83	I ScSr 1 *Dk&WebFrames*	British Cardiff	794 1236	225·4 Q72f.	32·3 B..	16·1 F29f.	Sunderland W.Pickersgill&Sns.	1883	1 Blds. DBa.. APT..	2Cy.28"&54"–33" 99 HP. N.E.MarineEng.Co.,Ld.,Sbl.	"Glendower" S.S.Co., Ld.(Hurley,Matthews&Co.)
261	✻D	Gleneagles *Park* 80-85	I Sc3mSr 2 *Dks3 B*	British Glasgow	1838 2798	367·1 P47f.	37·1 F48f.	26·4	Glasgow London&Glsgw,Co.	1877	5 Blds.	2Cy.50"&88"–48" 330 HP. London&GlasgowCo.,Glasg.	M·Gregor,Gow & Co.
262	✻	Gleneagles *Stoto* 86-88	I ScSr 1 *Dk*	British Malta	86 207	135·9	23·2	11·3	Aberdeen Hall,Russell&Co.	1884	1 Blds. MPT..	2Cy.22"&41"–27" 72 HP. Hall,Russell&Co.,Aberdeen	Gustaf Gollcher
263	✻	Glenearn *Brass* 87-87	I Sc3mSr 2 *Dks3 B*	British Glasgow	1410 2151	330·1	35·3 F42f.	24·8	Glasgow London&Glsgw.Co.	1873	6 Blds.	2Cy.44"&79"–45" 320 HP. London&GlasgowCo.,Glasg.	M·Gregor,Gow & Co.
264	✻	Glenelg *Morgan* 86-88	Stl ScSr 2 *Dks3 B*	British Glasgow	1617 2470	300·0 B 66f.	40·0 P28f.	24·1 F 36f.	Glasgow C.Connell&Co.	1888	6 Blds. CDB.. APT..	3Cy23"37"&59"–42"300HP. Duncan,Stewart&Co.Glasg.	J.Gardiner & Co.
265		Glenelg *Johnson* 84	I Sc3mSr 1 *Dk*	British Auckland	156 288	134·8 Q 53f.	22·2 B 62f.	9·1	PortGlasgow Blackwd. &Gordon	1878	4 Blds. APT FPT	2Cy.22"&40"–27" 75 HP. Blackwood&Gordon,Pt.Gls.	Northern S.S.Co.,Ld.
266		Glenelg *Brismage*	I TwnSc3mSr 1 *Dk*	British Sydney	76 210	135·8	21·1	11·3	Glasgow Aitken&Mansel	1875		1Cy.15"&30"–21" 80 HP. Gait&Lindsay,Glasgow	Port Jackson S.S.Co., Ld.
267	✻	Glenfalloch *Cormack* 86-86	I Sc3mSr 2 *Dks3 B*	British Glasgow	1419 2157	330·0	34·3 F41f.	24·8	Glasgow Lond.&Glas.Co.,Ld.	1873	6 Blds.	2Cy.40"&70"–45" 275 HP. London&GlasgowCo.,Glasg.	M·Gregor,Gow & Co.
268	✻D	Glenfield *Newdick* 87-87	I ScSr 1 *Dk&WebFrms*	British Stockton	1403 2127	277·0 P26f.Q72f.	37·4 B113f.	18·4 F29f.	Stockton Richrdsn,Duck&Co.	1887	4 Blds. CDB..	3C21"35"&57"–39"170HP. Blair & Co.Lim.Stockton	F.Binnington & Co.
269	✻	Glenfinlas *Ferguson*	I Sc3mSr 2 *Dks3 B*	British Glasgow	1409 2154	330·4	35·3 F42f.	24·8	Glasgow Lond.&Glas.Co.,Ld.	1874	6 Blds.	2Cy.44"&79"–45" 320 HP. London&GlasgowCo.,Glasg.	M·Gregor,Gow & Co.
270	✻	Glenfruin *Norman* 80-83	I Sc3mSr 2 *Dks3 B*	British Glasgow	1936 2985	360·4 P49f.	43·3 B70f.	24·5 F59f.	Glasgow Lond.&Glas.Co.,Ld.	1880	5 Blds. CDB..	2Cy.48"&88"–54" 530 HP. London&GlasgowCo.,Glasg.	M·Gregor,Gow & Co.

✻ = Lloyd's Register; D = Nederlandsche Vereeniging; G = Germanischer Lloyd; I = Registro Italiano; R = American Record; ⁂ = Veritas Austro-Ungarico; H = Veritas Ellenico; N = Norske Veritas; V = Bureau Veritas.

Dans la 2me Colonne, B.T., avec les indications qui suivent, signifie la date du rapport de visite de la Carène par les experts du Board of Trade pour obtenir le certificat relatif au TRANSPORT DES PASSAGERS.

Dans la 3me Colonne, C = Composite (membrure en fer, bordé en bois); I = Fer; Stl = Acier; W = Bois.

Pour plus amples renseignements sur les navires classés par le Lloyd's Register, voir le Lloyd's Register Book et le Lloyd's Yacht Register.

No.	Register	Ship's Name. Material, Rig, &c. Master, and Dates of Appointment to present Owner's Service and to Ship. Ship's Late Name. No. of Decks, &c.	Flag. Port of Registry	Tons Net / Gross	Dimensions Length / Breadth / Depth	Built Where / By Whom	When	No. of Bulkheads. Particulars of Water Ballast.	Engines. Particulars. Maker's Name.	Owners or Managers
271	*	Glengarnock I ScSmSr Scott 1 Dk	British Glasgow	127 202	118·6 | 22·3 | 11·1 P 18 f. F 28 f.	Glasgow T.B.Seath&Co.	1874	5 Blkds.	2Cy.21"&35"–24" 50 HP. A.Campbell&Sons,Glasgow	H.Burden,jun.,&
272	*	Glengarry I ScSmSr Taylor 75-83 2 Dks&3 B	British Glasgow	1956 3034	369·2 | 43·3 | 24·5 P 72 f. B 93 f. F 54 f.	Glasgow Lond&GlasCo.Ld.	1883	5 Blkds.	2Cy.48"&88"–57" 330 HP. London&GlasgowCo.,Glasg.	M'Gregor,Gow &
273		Glengarry (ex Argyle) W ScSr Murray 1 Dk	British Montreal	305 495	170·9 | 26·0 | 11·2 H 107 f.	St.Cthrn'sOnt. L.Shickluna	18[illegible]		1Cy.36"–30" 300 HP. [illegible],St.Catherine,Ont	Montreal Transportation Co.
274		Glengarry (ex Edinburgh) I PaSSk Cameron 87-87 (Scotia)	British Glasgow	62 124	148·5 | 16·3 | 7·2 Q 8 f.	Glasgow Smith&Rodger	1844		1Cy.36"–42" 45 HP. Smith&Rodger,Glasgow	D.MacBrayne
275	*	Glengoil I ScSr Holman 78-83 1 Dk & Awn Dk 3 B	British Leith	1934 2959	320·0 | 40·0 | 21·8 276	Sunderland J.L.Thompson&Sons	1882	6 Blkds.	2Cy.38"&72"–48" 300 HP. J.Dickinson,Sunderland	"Glengoil" S.S.Co. (Lindsay,Gracie&Co.)
276	*	Glengyle Stl ScSmSr Garson 2 Dks&3 B	British Glasgow	2244 3495	370·0 | 45·1 | 26·6 P 76 f. B 69 f. F 53 f.	Glasgow Lond.&GlasgowCo.Ld.	1886	5 Blkds.	3C.33"53"&87"–54" 550 HP. Lon.&GlasgowCo.,Ld.,Glas.	M'Gregor,Gow &
277	*	Gleniffer I ScSmSr Strdidmann 2 Dks&3 B	British London	1412 2165	291·0 | 34·7 | 24·6 P 34 f. F 41 f.	Sunderland Mounsey&Foster	1877	5 Blkds.	2Cy.26"&58"–45" 250 HP. EarlesCo.,Hull	G.C.Stewart
278	*	Glenisla I ScSr Ronald 83-83 2 Dks	British Dundee	859 1318	250·1 | 32·7 | 20·3 P 33 f. F 34 f. B 64 f.	Dundee W.B.Thompson	1878	5 Blkds.	2Cy.33"&54"–38" 140 HP. W.B.Thompson,Dundee	J.Mitchell
279	* D	Glenlivet I ScBa Steeves 86-86 1 Dk 2 B	British Leith	1026 1589	265·5 | 36·3 | 16·5 Q 74 f. P 30 f. B 56 f. F 29 f.	Newcastle PalmersCo.,Ld.	1883	4 Blkds.	2Cy.32"&60"–39" 160 HP. PalmersCo.,Ld.,Newcastle	"Glenlivet" S.S.Co. (Lindsay,Gracie & Co.)
280	*	Glenlivet I ScSr Brenes 1 Dk	British Glasgow	76 155	120·0 | 20·1 | 9·1	Middlesbro' Backhouse&Dixon	1870	3 Blkds.	2Cy.19"&31"–20" 42 HP. Cup[illegible]D.Joy&Co.Middlsbro	Eglington Chemical Co.,Ld.
281	*	Glenlyon I ScSmSr Sumner 83-85 2 Dks&3 B	British Glasgow	1410 2160	329·0 | 34·3 | 24·8	Glasgow London&GlasgwCo.	1872	5 Blkds.	2Cy.40"&70"–45" 275 HP. London&GlasgowCo.,Glasg.	M'Gregor,Gow &
282	*	Glenmanna I ScSmSr Halsman -88 1 Dk	British Newcastle	473 738	209·4 | 28·2 | 14·5 P 30 f. F 32 f.	N.Shields Smith	1871	4 Blkds.	2Cy.25"&48"–30" 98 HP. R.&W.Hawthorn,Newcastle	W.J.Tillett
283	*	Glenmavis I ScSr Wyse 85-85 2 Dks&3 B	British Leith	1385 2114	286·4 | 37·2 | 23·1 P 28 f. B 69 f. F 32 f.	Newcastle PalmersCo.	1881	5 Blkds.	2Cy.36"&68"–45" 260 HP. PalmersCo.,Ld.,Newcastle	"Glenmavis" S.S.Co. Ld.(Lindsay,Gracie&Co.)
284	*	Glenmore Stl ScSr Cave 82-85 1 Dk 2 B	British Liverpool	1216 1863	265·0 | 37·1 | 18·7 Q 84 f. B 78 f. F 33 f.	Kinghorn J.Key&Sons	1882	5 Blkds.	2Cy.30"&60"–39" 160 HP. J.Key&Sons,Kirkcaldy	S.S."Glenmore" Co. (W.Johnston&Co.)
285	* D	Glenmore I ScSr Budd 1 Dk	British Middlesbro	240 393	160·1 | 23·2 | 13·7 Q 16 f. B 9 f.	Middlesbro' Backhouse&Dixon	1869	4 Blkds.	2Cy.21"&39½"–26" 70 HP. Blair&Co.,Ld.,Stockton	J.M.Lennard &
286	*	Glenmorven I ScSr Wright 82-89 2 Dks&3 B	British Leith	1826 2777	320·5 | 40·3 | 25·0 P 41 f. B 84 f. F 34 f.	Newcastle PalmersCo.,Ld.	1882	6 Blkds.	2Cy.38"&72"–48" 300 HP. PalmersCo.,Ld.,Newcastle	"Glenmorven" S.S.Co. Ld.(Lindsay,Gracie&Co.)
287	*	Glenochil I ScSr Hay 80-83 1 Dk & Awn Dk 3 B	British Leith	1581 2424	286·9 | 37·3 | 27·4	Sunderland J.L.Thompson&Sons	1882	7 Blkds.	2Cy.36"&68"–45" 260 HP. J.Dickinson,Sunderland	"Glenochil" S.S.Co. Ld.(Lindsay,Gracie&Co.)
288	*	Glenogle I ScBk Duke 77-87 2 Dks&3 B	British Glasgow	2000 3749	420·5 | 45·1 | 25·3 P 60 f. B 144 f. F 59 f.	Glasgow London&GlasgwCo.	1882	6 Blkds.	3C.59"&(2)88"–60" 700 HP. London&GlasgowCo.,Glasg.	M'Gregor,Gow &
289	*	Glenorchy Hodge 81-83 I ScSmSr ex Quang See Ravensbourne, ex William III 2 Dks&Spr Dk	British Glasgow	1822 2827	321·3 | 39·6 | 24·2 294	Glasgow J.Elder&Co.	1872	6 Blkds.	2Cy.50"&86"–42" 400 HP. J.Elder&Co.,Glasgow	M'Gregor,Gow &
290	*	Glenrath I ScSr Anderson 87-87 2 Dks&3 B	British Leith	1080 1663	255·0 | 34·6 | 23·6 Q 40 f. F 31 f. B 11 f.	Newcastle PalmersCo.,Ld.	1879	4 Blkds.	2Cy.30"&60"–39" 160 HP. PalmersCo.,Ld.,Newcastle	Lindsay,Gracie &
291	*	Glenroy I ScSmSr Webster 2 Dks&3 B	British Glasgow	1411 2150	331·2 | 34·4 | 24·4 F 41 f.	Glasgow London&GlasgwCo.	1871	5 Blkds.	2Cy.40"&70"–42" 250 HP. London&GlasgowCo.,Glasg.	M'Gregor,Gow &
292	*	Glenshiel Stl ScSmSr Donaldson -87 2 Dks&3 B	British Glasgow	2240 3455	370·0 | 45·1 | 26·6 P 76 f. B 69 f. F 53 f.	Glasgow Lond&GlasgowCo.Ld.	1887	5 Blkds.	3C.34"53"&87"–54" 550 HP. London&GlasCo.,Ld.,Glasg.	M'Gregor,Gow &
293	*	Gloamin Stl ScSr Sangster 1 Dk & Spr Dk 3 B	British Dundee	999 1541	250·6 | 34·6 | 22·7 B 57 f. F 36 f.	Aberdeen Hall,Russell&Co.	1886	5 Blkds.	3C.19"30"&50"–42" 170 HP. Hall,Russell&Co.,Aberdeen	J.&R.Mudie
294	N	Glommen I ScSr Isaksen 1 Dk	Norweg'n Fredrikstad	77 122	99·5 | 17·7 | 7·7 B 41 f.	Christiania NylandsVerksted	1868		2Cy.13½"&24"–18½" 20 HP. FredrikstadM.Vrks.,Frs.	Glommens Dampskibs-Selskab
295		Gloria W	Spanish Cienfuegos	1350 1867	232·6 | 31·1 | 12·7	New York	1874		100 HP.	Menendez & Co.
296	*	Glow Worm W ScSrYt 2 Dks	British Southampton	118 182	119·8 | 23·4 | 12·8	Northam Day,Summers&Co.	1876		2Cy.16"&32"–24" 45 HP. Day,Summers&Co.,Northam	C.H.Fauvel
297	V	Gluckauf I&Stl ScSr Szymansky 2 Dks	German Geestemünde	1503 2307	300·5 | 37·2 | 23·2 P & B 163 f.	Newcastle W.G.Armstrong Mitchell&Co.	1886		3C.22"35"&58"–36" 200 HP. WallsendSlipwayCo.,Newc.	W.A.Riedemann
298	V	Glucksburg I ScSr Schultz 83-88 1 Dk & Spr Dk & B	German Flensburg	916 1434	235·9 | 32·8 | 16·5 238	Flensburg FlnsbrgrSchiffs.Ges	1882		2Cy.31"&59"–33" 150 HP. Flnsbg.Schiffs.Ges.,Flnsbg.	Dampfschiff. "Globus"
299		Goa I ScSr MacLachlan 85-88 3 Dks	British Glasgow	1232 1906	297·8 | 33·3 | 25·7	Dumbarton W.Denny&Bros.	1874	5 Blkds.	2Cy.36½"&63"–42" 250 HP. Denny&Co.,Dumbarton	British India S.N.Co.,Ld.
300		Goalpara Stl ScSr Russell 75-[illegible] 3 Dks	British [illegible]	1355 2114	285·0 | 36·2 | 23·7 P 41 f. B 88 f. F 34 f.	Glasgow A.&J.Inglis	1883	5 Blkds.	2Cy.34"&60"–45" 300 HP. A.&J.Inglis,Glasgow	British India S.N.Co.,Ld.

* = Lloyd's Register; D = Nederlandsche Vereeniging; G = Germanischer Lloyd; I = Registro Italiano; R = American R[illegible]
A = Veritas Austro-Ungarico; H = Veritas Ellenico; N = Norske Veritas; V = Bureau Veri[tas]

In the 2nd Column, B.T., with figures following, signifies Date of report of bottom sighting by Board of Trade Surveyors as condition of PASSENGER CERTIF[ICATE]

In the 3rd Column, C = Composite (iron frame, wood planked); I = Iron; Stl = Steel; W = Wood.

Further particulars of Vessels classed by Lloyd's Register will be found in Lloyd's Register Book and Lloyd's Yacht Register.

1–2	3 Nom, Matériaux, Gréement, etc. Capitaine et dates d'entrée au service des Armateurs actuels et à bord du navire. Ancien Nom, nombre de ponts, etc.	4 Pavillon. Port d'Armement	5 Tonnage. Net. Total.	6 Dimensions. Longueur. Largeur. Creux. (enregistrées sur le Port)	7 Construit. Où. Par qui.	8 Quand.	9 Nombre des cloisons. Détails sur le lest d'eau.	10 Machines. Détails. Nom du Constructeur.	11 Armateurs ou Directeurs-Gérants.
✻	Godalming 1 SeSh Shiels 82-82 1 Dk2 H	British London	834 1292	240·0|34·0|17·0 Q38 P23 B46 F29	Sunderland Bulmer&Co.	1881	4 Blhds. DBaft 299	2Cy.30" & 58"–39" 130 HP T.Clark&Co.,Newcastle	"Godalming" S.S.Co., Ld. (Scrutton,Sons&Co.)
	Godavari Stl TwScSr	British Glasgow	14 182	159·9|25·6|7·5	PortGlasgow R.Duncan&Co.	1884		1Cy.11" & 28"–21" 65 HP Dunsmuir&Jackson,Glasgw	Shepherd & Co.
	Godavery 1 Sc 2 Dks	French Marseille	816 1409	292·6|32·0|23·9 P31 f. B44 f.	Bordeaux Arman	1863		3C 36¼" & 51¼"(2)–31¼" 310HP.	Messageries Maritimes
✻	Godiva 1 SeSt Smith 1 Dk2 H	British London	887 1386	240·5|34·1|17·5 Q38 f. B33 f. F28 f.	Middlesbro' R.Craggs&Sons	1882	4 Blhds. DBaft 300	2Cy.28" & 54"–36" 115 HP Westgarth English&Co. Midd	Demerara & Berbice S.S.Co.
✻	Godmunding 1 SeSr 1 Dk&H Frms	British London	850 1316	242·0|33·10|16·5 Q87 B64 F20	Blyth BlythS.B.Co.	1888	1 Blhds. DB 239 AP	3Cy.18" 29" & 48"–33" N. E. Marine E. Co. Newcastle 150 HP.	T.B.Williams (W.Lamplough&Co.Managers)
✻	Godrevy 1 SeSr 1 Dk2 L	British London	1066 1628	255·7|35·0|18·4 F28 f.	Sunderland J.Priestman&Co.	1884	1 Blhds. DBaft 300	2Cy.38" & 62"–42" 160 HP J.W.&F.Wilson&Co.,Sndlrd.	T.Briggs.
✻	Goeland 1 ScK 1 Dk	French Bordeaux	56 115	95·8|18·1|8·6 Q 38 f. B f. F 15 f.	Amlwch W.Thomas&Son	1884	3 Blhds.	2Cy.14" & 27"–20" 35 HP DeWinton&Co.,Carnarvon	E.Balguerie
✻	Gogo 1 SeSmSt Landells 75-75 1 Dk	British London	395 648	193·6|24·5|13·4 P66 f. B30 f. F31 f.	Greenock Scott&Co.	1872	1 Blhds. DB 128	2Cy.21" & 43"–36" 90 HP GreenockFoundryCo.,Grnk	Mattheson & Co.
✻	Golconda Stl SeIm Leslie 75-88 2 Dks&Spar Dk	British Glasgow	3361 6037	422·0|48·1|32·5 P BA F 24	Sunderland W.Doxford&Sons	1887	7 Blhds. DB M Tank	3C 34" 56" & 90"–60" 550 HP W.Doxford&Sons,Sndrlnd.	British India Steam Nav.Co.,Ld.
✻D	Goldcliffe 1 SeSr West 1 Dk2 L	British Newport	832 1316	240·5|33·2|16·4 B30 f. F28 f.	Newcastle Palmers Co.,Ld.	1881	4 Blhds. DB AP	2Cy.28" & 54"–36" 99 HP Palmers Co.Ld.,Newcastle	"Goldcliffe" S.S.Co. Ld. (D.H.Morgan & Co.)
✻	Golden Eagle C SeScYt	British Cowes	145 213	134·0|21·3|13·3	Cowes J.S.White	1877		2Cy.19" & 32"–22" 40 HP G.E.Belliss,Birmingham	Lady E.H.Scott
	Golden Fleece C ScSYt	British	115 169	119·1|22·5|12·9	E.Cowes J.S.White	1880		2Cy.13½" & 22"–13" 20 HP G.E.Belliss,Birmingham	B.D.Harrison
✻D	Golden Grove 1 SeSr Thompson 85-85 1 Dk2 H	British Whitby	932 1455	253·2|33·8|19·1	Whitby T.Turnbull&Son	1877	5 Blhds. DBaft 270	2Cy.34" & 58"–36" 130 HP Blair&Co.Ld.,Stockton	T.Turnbull & Son
✻	Golden Horn 1 SeBg Leish 83-84 2 Dks3 H	British W.Hartlepl.	1027 1585	274·8|31·7|23·6	Hartlepool Denton,Gray&Co.	1865	4 Blhds. DB	2Cy.29½" & 59"–36" 130 HP Blair&Co.Ld.,Stockton	W.Gray
R	Goldsboro' W ScSmSt Jones	American Philadelph'a	486 681	188·6|28·8|8·1	Philadelphia Birely Hillman & Streaker	1882		2Cy.20" & 34"–26" 300 HP Neafie&Levy,Philadelphia	Clyde S.S.Co.
✻	Goldsbro' Stl SeSr 1 Dk&W&B Frms	British W.Hartlepl.	1458 2247	282·3|37·3|19·5 B F	W.Hartlepool W.Gray & Co.	1887	5 Blhds. DB	3Cy. 22" 35" & 59"–39" 200 HP Central Marine Eng.Co. W.Hartlepool	W.Hartlepool Steam Nav.Co.
	Goldsworthy 1 SeSk	British Liverpool	41 106	117·0|13·5|8·8	Liverpool T.Turton	1881		2Cy.14"–21" 58 HP J.Taylor&Co.,Birkenhead	G.D.Peters
	Goliah W ScSt 1 Dk	British Halifax	100 147	88·4|18·4|14·7	Philadelphia	1863		2Cy.18" & 34"–26" 200 HP A.Moir&Sons,Halifax,N.S.	William Lawson
	Gomes IV. 1 Sc 1 Dk	Portug'se Lisbon	248 420	155·6|26·1|10·5 Q 60 f. F 21 f.	Grangemouth Dobson&Charles	1882	1 Blhds. DB AP FP	2Cy.21" & 40"–30" 70 HP W.Kemp,Glasgaw	A.Gomes
✻	Gomes V. 1 SeSr (ex Tornado) 1 Dk	Portug'se Lisbon	447 736	184·3|30·1|13·3 P22 Q35 B41 F22	Dumbarton Burrell&Son	1883	1 Blhds. DB FP AP	2Cy.23" & 45"–33" 85 HP Hibb&Hogg,Glasgow	A.Gomes
	Goncalves Martio 1 ScS	Brazilian Bahia	374 184	200·0|26·0|11·6	Stockton Richardson,Duck&Co.	1858			Cia.de Navegação a Vap.de Bahia
✻	Gondola 1 SeSr Tregarthen 82-84 1 Dk2 H	British Cardiff	939 1449	255·1|35·0|17·1 Q98 f. B51 f. F27 f.	Sunderland W.Pickersgill&Sons	1882	1 Blhds. DBaft 216 AP	2Cy.32" & 60"–39" 99 HP J.Clark,Sunderland	"Gondola" S.S.Co.,Ld. (W.Lamplough&Co.)
	Gondolier 1 PadSk Carmichael 84-87	British Glasgow	80 173	148·2|20·2|7·8	Glasgow J.&G.Thomson	1866		2Cy.33"–39" 80 HP J.&G.Thomson,Glasgow	D.MacBrayne
✻	Good Hope 1 SeBg Kerr 82-83 3 Dks	British Sunderland	1027 1588	268·6|32·1|25·0	Sunderland J.Laing	1868	1 Blhds.	2Cy.32" & 60"–33" 150 HP N.E.MarineE.Co.Ld.,Sndld.	J.Laing
	Goole No. 1 1 PadSp	British Goole	20 110	96·6|18·4|9·4	S.Shields J.Softley&Co.	1874		2Cy.25½"–46" 40 HP J.Hepple&Co.,SouthShields	Goole & Hull Steam Towing Co.,Ld.
✻	Goole No. 5 1 SeTug 1 Dk	British Goole	1 69	75·2|16·2|8·9	Goole T.Scott	1884	1 Blhds.	2Cy.17" & 31"–21" 50 HP T.Scott,Goole	Goole & Hull Steam Towing Co.,Ld.
✻	Goole No. 6 1 SeTug Mummery 1 Dk	British Goole	 77	80·0|16·0|8·9	Goole T.Scott	1885	1 Blhds.	2Cy.17" & 31"–21" 50 HP T.Scott,Goole	Goole & Hull Steam Towing Co.,Ld.
✻	Goole No. 7 1 ScSpTug Mummery 1 Dk	British Goole	10 138	100·0|19·7|11·6	S.Shields J.T.Eltringham	1883	1 Blhds.	3Cy.14" 22" & 36"–21" 77 HP J.P.Rennoldson&Son, S.Shields	Goole & Hull Steam Towing Co.,Ld.
	Goolwa 1 Pad 1 Dk	British Port Adelaide	116 191	130·0|21·1|10·3	Newcastle Palmer Bros.&Co.	1864		2Cy. 90 HP. J.Thompson,Newcastle	J.& A.Brown
✻	Goorkha Stl SeSmSr Henderson 13-85 2 Dks&Spr Dk	British Glasgow	2660 4194	390·0|42·2|21·6 B132 f. F28 f. 29·6	Dumbarton W.Denny&Bros.	1882	5 Blhds.	2Cy.50" & 86"–54" 500 HP Denny&Co.,Dumbarton	British India Steam Nav.Co.,Ld.

✻=Lloyd's Register; D=Nederlandsche Vereeniging; G=Germanischer Lloyd; I=Registro Italiano; R=American Record; ✻=Veritas Austro-Ungarico; H=Veritas Ellenico; N=Norske Veritas; V=Bureau Veritas.

Dans la 2me Colonne, B.T., avec les indications qui suivent, signifie la date du rapport de visite de la Carène par les experts du Board of Trade pour obtenir le certificat relatif au TRANSPORT DES PASSAGERS.

Dans la 3me Colonne, C=Composite (membrure en fer, bordé en bois); I=Fer; Stl=Acier; W=Bois.

Pour plus amples renseignements sur les navires classés par le Lloyd's Register, voir le Lloyd's Register Book et le Lloyd's Yacht Register.

1 No. in Book	2 Registry by which Classed	3 Ship's Name. Material, Rig, &c. / Master, and Dates of Appointment to present Owner's Service and to Ship / Ship's Late Name. No. of Decks, &c.	4 Flag. / Port of Registry.	5 Tons. Net. / Gross	6 Dimensions Length, Breadth, Depth / Deck Erections	7 Built. Where. / By Whom.	8 When	9 No. of Bulkheads / Particulars of Water Ballast	10 Engines. Particulars. / Maker's Name.	11 Owners or M[anagers]
331	✱	Gordon Castle I ScBg Clark 87-87 3 Dks	British Glasgow	1320 2031	307·7|34·3|25·4	Glasgow J.&G.Thomson	1871	5 Blkds.	2Cy.36"&67"–45" 250 HP J.&G.Thomson,Glasgow	N.McLean & Co.
332	✱	Gordonia I ScSr Cook 84-85 1 Dk & Spr Dk 3 B	British London	1555 2343	275·0|38·5|20·6	Sunderland J.L.Thompson&Sons	1881	5 Blkds. [illegible]	2Cy.37"&68"–45" 200 HP Black,Hawthorn&Co.Gtshd	R.Gordon & Co.
333 | 72	I	Gorgona (ex Caeglio) I ScSr Burgio -88 1 Dk	Italian Palermo	109 183	130·2|17·0|9·9	Greenock MacNab&Co.	1863		2Cy.14"&28"–24" 50 HP P.Ferro&Co.,Genoa	Navigazione G[enerale] Italiana
334	BT1.88	Gorilla I ScSr Rankin 85-81	British Glasgow	406 929	250·2|30·0|14·7 P 157 f. F 33 f.	Glasgow Barclay,Curle&Co.	1881		2Cy.36½"&69"–48" 285 HP Barclay,Curle&Co.,Glasgow	G.& J.Burns
335	✱	Gorji I ScSr Miller -84 1 Dk & Spr Dk 3 B	British London	1111 1716	255·5|35·1|24·5 B 48 f.	Newcastle Wigham Richardson &Co.	1883	4 Blkds. [illegible]	2Cy.31"&62"–42" 180 HP WighamRichardson&Co.,N.S.	H.Jones
336		Gota I ScSr Hartzell 1 Dk	Swedish Norrköping	181 237	105·3|22·7|8·6 P 58 f. F 33 f. B 106	Norrköping MotalaCo.	1877		2Cy.15½"&31"–22" 45 HP MotalaCo.,Gothenburg	Ångf. Aktie Bo[lag] "Norden" (G.F.A.)
337		Gota (ex Forshaga) C ScSp Gustafsson -87 1 Dk	Swedish Gothenburg	120 170	100·7|20·6|9·0 P 21 f. F 14 f. B 21	Spången	1877		2" 25 HP	H.Edgren
338		Gota Elf I ScSp 1 Dk	Swedish Lilla Edet	83 119	93·4|17·7|9·0 B 15 f.	Gothenburg	1885			J.F.Bodén
339		Gota Kanal (ex Neptun) C ScSp Nikander -88 1 Dk	Swedish Steneby s.n.	115 168	100·0|20·4|11·8 P 12 f. B 11 f.	Norrköping Hägglund	1872		25 HP MotalaCo.	A.Nilsson
340		Gota Kanal I. I ScSp Andersson 86-88 1 Dk	Swedish Stockholm	112 161	97·3|21·8|8·1 P 33 f. B 31.	Gothenburg MotalaCo.	1887		2Cy.11"&22"–16" 30 HP	Motala Ström[s] Aktiebol. (E.R.)
341		Gota Kanal II. Stl & I Sc2m Danielsen 88-88 1 Dk	Swedish Stockholm	107 165	97·3|21·8|8·1 B 34 f. F 8 f.	Gothenburg MotalaCo.	1888	[illegible]	2Cy.11"&22"–16" 30 HP MotalaCo.,Gothenburg	Motala Ström[s] Aktiebol. (E.R.)
342		Gota Kanal III. Stl & I Sc2m Hansson 88-88 1 Dk	Swedish Stockholm	107 165	97·3|21·8|8·1	Gothenburg MotalaCo.	1888	[illegible]	2Cy.11"&22"–16" 30 HP MotalaCo.,Gothenburg	Motala Ström[s] Aktiebol. (E.R.)
343		Gota Kanal IV. C Sc2m Strandberg 88-88 1 Dk	Swedish Stockholm	108 163	97·7|20·8|9·0	Thorskog P.Larsson	1884		2Cy.11"&22"–18" 30 HP MotalaCo.,Gothenburg	Motala Ström[s] Aktiebol (E.R.)
344		Goteborg C ScSr Nissen 72-89 1 Dk 2 B	Swedish Gothenburg	445 584	167·3|26·3|16·4 B 37 f. H 31.	Torskog J.Berg	1873		2Cy.16"&31"–28" 60 HP MotalaCo.,Gothenburg	Ångf.Aktie Bo[lag] "Triton" (A.M.)
345 | 91315	✱ BT5.89	Gotha Stl ScSr Newlove -88 1 Dk	British Goole	257 626	220·5|29·1|13·7 P 40 f. B 77 f. F 33 f.	Hull Earles Co.Ld.	1888	5 Blkds. [illegible]	3Cy.20½"31"&54"–33" 150 HP EarlesCo.Ld.,Hull	Yorkshire Coal Steamship Co.L[d.]
346	G	Gothe C ScSr Ternström 84-84 1 Dk	Swedish Oscarshamn	341 440	147·9|23·6|12·3 Q 11 f. B 62 f. P 22 f. H 5 f.	Oscarshamn S.Olsson	1874		2Cy.18"&36"–21" 60 HP	Ångf.Aktie F[..] "Fama" (P.L.)
347 | 89465	✱	Gothenburg City I ScSr Harrison 86-84 2 Dks & Web Frames	British W.Hartlepl.	1658 2529	301·5|38·0|23·8 P 36 f. B 76 f. F 28 f.	W.Hartlepool E.Withy&Co.	1884	5 Blkds. CDB[illegible]	2Cy.40"&73"–42" 300 HP T.Richardson&Sons,Hartpl.	C.Furness
348	V	Gothia (ex Jacobus) I ScBn Franck 88-88 2 Dks 3 B	German Hamburg	1562 2433	314·5|34·1|25·1 P 53 f. B 71 f. F 36 f.	Middlesbro' R.Dixon&Co.	1884	5 Blkds. [illegible]	3C.22"37"&58"–36" 300 HP T.Richardson&Sons,Hartpl.	Hmbg.Ameri[k.] Packetfahrt.Act.
349 | 62298	✱	Gothland I Sc3mSr Hutton 72-75 1 Dk & Spr Dk	British Liverpool	952 1482	251·6|32·0|16·7 23·9	Glasgow J.&G.Thomson	1871	4 Blkds.	2Cy.33½"&58½"–36" 150 HP J.&G.Thomson,Glasgow	D.Currie & Co.
350		Gotland I ScSr Smittenberg 76-86 1 Dk	Swedish Visby	224 282	120·7|23·8|10·5 Q 40 f. B 35 f. F 28 f. H 9 f.	Stockholm Bergsunds Mek.Verks.	1868		2Cy.19½"&39"–23½" 60 HP Bergsunds Mek.Verks.,Stklm	Ångf.Akt.Bolag "Gotland" (A.L.)
351 | 56	I ✱	Gottardo I ScBk Caffiero 82-88 3 Dks	Italian Palermo	1839 2847	350·0|40·0|26·1 P 68 f. B 32 f. F 53 f.	Glasgow A.Stephen&Sons	1883	6 Blkds. [illegible]	2Cy.48"&90"–48" 370 HP A.Stephen&Sons,Glasgow	Navigazione Ge[nerale] Italiana
352	G	Gottorp I ScSr Lütschen -83 1 Dk	German Schleswig	213 384	139·1|23·1|12·9	Kiel G.Howaldt	1883		2Cy.15¾"&27½"–15¾" Gebr.Howaldt,Kiel	Th.Reimers
353		Gouverneur Generaal Loudon I Sc de Bruyn 2 Dks & Shd Dk	Dutch Batavia	860 1298	256·8|31·7|17·4 P [illegible] F 78 f.	Greenock Caird&Co.	1875	4 Blkds.	2Cy.34"&60"–39" 180 HP Caird&Co.,Greenock	Ned.Indische S[..] Maats.
354		Gouverneur Generaal Mijer I Sc van Woortman 1 Dk & Awn Dk 3 B	Dutch Batavia	442 960	227·1|29·7|16·1 22·8	Glasgow [illegible]	1871	4 Blkds.	2Cy.30"&59"–36" 135 HP J.Howden&Co.,Glasgow	Ned.Indische S[..] Maats.
355	✱ D V	Gouverneur Generaal's Jacob Stl ScSr Swart -88	Dutch Batavia	1569 2176	287·2|36·2|23·6 P 58 f. B 73 f. F 37 f.	Greenock Caird&Co.	1883	5 Blkds.	2Cy.36"&70"–45" 260 HP Caird&Co.,Greenock	Ned.Indische S[..] Maats.
356	✱	Gouverneur Generaal van Lansberge (ex Puttialla) I ScSr Delf 2 Dks & Shd Dk	Dutch Batavia	510 824	230·7|28·2|16·2 P 51 f. F 42 f.	Greenock Caird&Co.	1875	7 Blkds.	2Cy.32"&51"–36" 120 HP Caird&Co.,Greenock	Ned.Indische S[..] Maats.
357 | 81352	✱	Goval I ScSr Stephen 85-85 1 Dk 2 B	British Aberdeen	272 427	160·0|23·2|12·3 Q 48 f. B 40 f. F 19 f.	Aberdeen Hall,Russell&Co.	1881	4 Blkds. [illegible]	2Cy.20"&38"–24" 60 HP Hall,Russell&Co.,Aberdeen	W.Leslie & Co.
358 | 81696	BT5.88	Governor I Sc3mSr Valiant 59-81	British Liverpool	1665 [illegible]	347·0|37·0|25·5 F 30 f.	Glasgow Aitken&Mansel	1881		2Cy.38"&75"–54" 300 HP J.&G.Thomson,Glasgow	Charente S.S.C[o.]
359 | 61300		Governor Blackall I ScSr McAulay -87 2 Dks	British Sydney	331 487	194·4|23·8|15·6	Balmain,NSW Mort&Co.	1871		2Cy.25"&50"–28" 90 HP Mort's Dk. & Eng. Co. Ld. Balmain, N.S.W.	North Illawar[ra] MiningCo.
360		Governor Maclean Stl TwScSr 1 Dk	British [illegible]	128 [illegible]	135·0|23·0|9·0 Q 24 f. F 28 f.	Barrow BarrowS.B.& E.Co.	1886		4Cy.16"&32"–20" 80 HP BarrowS.B.&E.Co.,Barrow	The Crown Ag[ents] for the Colonies

✱ = Lloyd's Register; D = Nederlandsche Vereeniging; G = Germanischer Lloyd; I = Registro Italiano; R = American R[ecord];
A = Veritas Austro-Ungarico; H = Veritas Ellenico; N = Norske Veritas; V = Bureau Veri[tas].

In the 2nd Column, B.T., with figures following, signifies Date of report of bottom sighting by Board of Trade Surveyors as condition of PASSENGER CERTIF[ICATE].

In the 3rd Column, C = Composite (iron frame, wood planked); I = Iron; Stl = Steel; W = Wood.

Further particulars of Vessels classed by Lloyd's Register will be found in Lloyd's Register Book and Lloyd's Yacht Register.

No.	Registre	Nom, Matériaux, Gréement, etc. Capitaine et dates d'entrée au service des Armateurs actuels et à bord du navire. Ancien Nom, nombre de ponts, etc.		Pavillon. Port d'Armement	Tonnage Net. Total.	Dimensions. Longueur	Largeur	Creux	Construit. Où. Par qui.	Quand.	Nombre des cloisons. Détails sur le lest d'eau.	Machines. Détails. Nom du Constructeur.	Armateurs ou Directeurs-Gérants.
361	✻	Govino *Brothers 83-87*	I ScSr *1 Dk&Spr Dk3 B*	British Sunderland	1454 2220	287·8	37·1	27·3 *F 22 f.*	Sunderland J.Laing	1882	7 Blds. DBaf&f 2151	2Cy.33"&62"–42" 250 IP. G.Clark,Sunderland	The Pinkney & Sons S.S.Co.,Ld.
362		Gower	I Pad	British Liverpool	10 143	119·8	20·5	8·2	Liverpool	1868		2 80 IP.	Manchester Ship CanalCo.
363	✻	Goya *Egaña 76-85*	I ScSr *1 Dk&Awn Dk*	Spanish Seville	824 1056	210·5	29·8	13·9 21·1	Dundee W.B.Thompson	1882	1 Blds. DB	2Cy.23&46"–30" 90 IP. W.B.Thompson,Dundee	Miguel Saenz & Co.
364		Goytacaz	I Sc	Brazilian Rio de Janeiro	275 476	187·8	23·1	12·7 *P 73 f. F 28 f.*	Newcastle C.Mitchell&Co	1874	4 Blds.	2Cy.28"&50"–33" 100 IP. Ravenhill,Easton&Co.,Lon.	Companhia da Estrada de ferro de Macahe & Campos
365		Gozo *Dawe*	I ScSr *1 Dk2 B*	British Hull	657 1008	217·8	29·0	16·8 *Q 85 f. F 14 f. B 72 f.*	Hull C.&W.Earle	1868		2Cy.25"&50"–26" 95 IP. C.&W.Earle,Hull	T.Wilson,Sons & Co.
366		Graaf van Bylandt *Blomberg*	I ScSr *2 Dks&Shd Dk*	Dutch Batavia	794 1321	256·2	31·6	17·3	Greenock Caird&Co.	1876	4 Blds.	2Cy.34"&60"–30" 180 IP. Caird&Co.,Greenock	Ned.Indische Stoom. Maats.
367	✻	Grace *Crombie*	W ScK *1 Dk*	British Leith	22 43	63·2	17·3	7·8	Granton Hawthorns&Co.	1884		2Cy.8"&16"–12" 10 IP. Hawthorns&Co.,Leith	R.Irvin
368	✻	Grace *Carter*	I PadTug *1 Dk*	British Workington	17 167	110·3	22·1	9·6	Whitehaven WhitehnS.B.Co.Ld.	1884	3 Blds.	2Cy.35"–60" 80 IP. Dunsmuir& Jackson,Glsgw.	Rt.Hon.J.Lowther
369	✻	Grace Darling	Stl ScSrYt *2 Dks*	British Portsmouth	108 171	143·0	19·5	10·3	Paisley Fleming&Fergusn.	1887	5 Blds.	1C.10"14"20"28"–20" 70IP. Fleming & Ferguson,Psly.	J.C.Evans
370	✻V	Gracia *Cirarda -88*	Stl ScSr *2 Dks3 B*	Spanish Bilbao	2017 3076	345·5	40·2	24·6 *B 89 f. P 39 f. F 40 f.*	Newcastle Wigham Richardson & Co.	1888	6 Blds. CDB&ADB3MT452 DB032t	3Cy.26"42"&68"–48" 350 IP. Wigham,Richardson &Co.,Newcastle	Linea de Vapores Serra
371	✻	Gracie *Brown 74-79*	I ScSr *1 Dk2 B*	British London	863 1346	261·0	32·1	17·0 *B 12 f. Q 142 f. F 31 f.*	Hartlepool E.Withy&Co.	1879	4 Blds. DBaf&f	2Cy.36½"&68"–42" 180 IP. Blair&Co.,Ld.,Stockton	S.Clarke & Co.
372	G	Graf Bismarck *Warnkes 85-88*	I ScBg *2 Dks Hur Dk & B*	German Bremen	1761 2406	318·5	39·4	31·2	Greenock Caird&Co.	1871		2Cy.34"&66¾"–48" Caird&Co.,Greenock	Norddeutscher Lloyd
373		Graf Totleben	I PadSr *1 Dk*	Russian Odessa	171 311	215·5	25·0	13·3	London Samuda	1880		2Cy.35"&62½"–51" 130 IP. MaudslaySons&Co.,London	Russian Steam Nav. & Trading Co.
374	✻	Grafton *Soffen*	I ScSr *1 Dk&Spr Dk3 B*	British London	1158 1802	260·0	36·3	23·3	Sunderland R.Thompson&Sons	1881	6 Blds. CDB624 FPT47	2Cy.31"&61"–42" 180 IP. G.Clark,Sunderland	Grafton S.S.Co.,Ld. (E.W.Morgan & Co.)
375		Grafton *Nordstrom -87*	I TwnScSr *1 Dk*	British Dunedin	297 548	173·0	27·3	13·6 *Lon. 84 B 51 f. B 18*	Birkenhead LairdBros.	18 $^{64}_{77}$		1Cy.18"&34"–20" 123 IP. Robertson&Co.,Wellington	Union S.S.Co.of New Zealand,Ld.
376		Grampus (ex NorthWestern) *Cochrane 82-82*	I ScSr	British Glasgow	272 654	211·2	28·3	14·6 *P 58 f. B 53 f. F 46 f.*	Glasgow A.&J.Inglis	1877		2Cy.36"&64"–36" 180 IP. A.&J.Inglis,Glasgow	G.&J.Burns
377		Grampus (ex Richard Rush) *Dexter 86-86*		American S.Francisco	191								Pacific Steam Whaling Co.
378	✻	Granada (WRECKED) *Nesbitt 84-85*	I ScSr *1 Dk2 B*	British Glasgow	609 958	215·6	30·1	15·1 *Q 76 f. B 58 f. F 30 f.*	Leith Ramage&Ferguson	1883	4 Blds. DB&f 251 FPT	2Cy.26"&50"–36" 98 IP. Ramage&Ferguson,Leith	H.Lamont & Co.
379	RV	Granada *Austin 71-88*	I ScBg *3 Dks*	American New York	1798 2572	280·0	40·0	28·6	WilmingtnDel Harlu&Hollngswth	1873		2Cy.50"&86"–42" 400 IP. J.Reach&Son,Chester,Pa.	Panama R.R.Co.
380		Grand Duc Constantin	I ScSr *2 Dks*		679 987	232·5	29·8	22·5	Marseille Forges&Chantiers	1857		2Cy.50"–36" 160 IP. J.Penn&Sons,London	
381		Grand Duc Michael	I PadSr *1 Dk2 B*	Russian Odessa	489 626	210·0	28·2	13·9 *P & F 99 f.*	Newcastle C.Mitchell&Co.	1865	4 Blds.	2Cy.52"–54" 185 IP. J.Penn&Son,London	Russian Steam Nav.& Trading Co.
382		Grand Duchess Olga	StlScSr *2 Dks&Awn Dk*	Russian Odessa	499 1488	266·6	34·2	15·8 23·3	Newcastle A.Leslie&Co.	1880	4 Blds.	2Cy.38"&70"–48" 300 IP. MaudslaySons&Field,Lond.	Russian Steam Nav.& Trading Co.
383		Grand Duke Alexis	I ScSr *2 Dks*	Russian Archangel	302 444	151·0	24·3	16·0	Newcastle C.Mitchell&Co.	1871		2Cy.20"&38"–24" 60 IP. Thompson,Boyd&Co.,Nwc.	White Sea Steam Nav.Co.
384	✻	Grande Rivière *Williams 87-87*	Stl ScSr *1 Dk*	Haytian Port au Prince	146 310	156·0	22·1	9·5 *B & C 91 f. F 26 f.*	Grangemouth Grngmth DkydCo.	1887	1 Bhds.	2Cy.18"&34"–24" 50IP. Dunsmuir & Jackson,Glasg	General B.Rivière
385	✻	Grandholm *Mason 84-85*	I ScSr *1 Dk2 B*	British Aberdeen	904 1408	243·2	32·8	17·3 *P 141 f. F 33 f.*	Aberdeen Hall,Russell&Co.	1884	5 Blds. DBaf&f 220t	2Cy.30"&57"–36" 150 IP. Hall,Russell&Co.,Aberdeen	S.S."Grandholm"Co., Ld.(W.Leslie)
386	✻	Grangemouth *Gentle 81-85*	I ScSr *2 Dks*	British Glasgow	242 541	189·4	27·2	14·4 *P 33 f. B 60 f. F 52 f.*	Dumbarton A.McMillan&Son	1877	5 Blds. DBaf&f	2Cy.24⅜"&42"–42" 125IP. M.Paul&Co.,Dumbarton.	J.Rankine & Sons
387	✻	Grangense *Good 81-87*	Stl ScSr *1 Dk*	British Liverpool	223 420	180·0	26·2	10·4 *P 11 f. B 17 f. F 26 f.*	Barrow BarrowS.B.Co.,Ld.	1887	4 Blds. DB&f 35t FPT 28t	3C.17"26½"&42"–24" 91IP. BarrowS.B.Co.,Ld.,Barrow	R.Singlehurst & Co.
388	G	Granit (ex Svea) *Melan -88*	W ScSr *1 Dk*	Russian Helsingfors	159 215	101·1	21·0	12·2 *P 24 f. B 31.*	Sjötorp S.Groth	1886		2C.10½"&25½"–17½" 30IP.	Actiebola. "Granit"
389		Granite City *Newton*	W PadTug *1 Dk*	British Aberdeen	21 100	95·5	18·7	10·4	N.Shields Wlhave&Johnson	1882		1Cy.34¼"–52" 40 IP. R.B.Findlay&Sons,N.Shlds.	J.N.Newton
390	✻D	Granton *Taylor -87*	I Sc3mSr *2 Dks*	British London	601 931	235·4	30·1	16·8 *P 160 f. F 33 f.*	Dundee GourlayBros.	1866	1 Blds.	2Cy.27"&52"–33" 120 IP. GourlayBros.&Co.,Dundee	GeneralSteamNav.Co.

✻ = **Lloyd's Register**; D = Nederlandsche Vereeniging; G = Germanischer Lloyd; I = Registro Italiano; R = American Record; A = Veritas Austro-Ungarico; H = Veritas Ellenico; N = Norske Veritas; V = Bureau Veritas.

Dans la 2me Colonne, B.T., avec les indications qui suivent, signifie la date du rapport de visite de la Carène par les experts du Board of Trade pour obtenir le certificat relatif au TRANSPORT DES PASSAGERS.

Dans la 3me Colonne, C = Composite (membrure en fer, bordé en bois); I = Fer; Stl = Acier; W = Bois.

Pour plus amples renseignements sur les navires classés par le Lloyd's Register, voir le Lloyd's Register Book et le Lloyd's Yacht Register.

No.	Society classed	Ship's Name. Material, Rig, &c. / Master, and Dates of Appointment to present Owner's Service and to Ship / Ship's Late Name. No. of Decks, &c.		Flag. / Port of Registry.	Tons.	Dimensions.	Built. Where. / By Whom.	When.	No. of Bulkheads. / Particulars of Water Ballast.	Engines. Particulars. Maker's Name.	Owners or Managers.
391	✻ D	Grantully / Rogie [illegible]	I ScSr / 1 Dk2 B	British / W.Hartlepool	[illegible]	[illegible]	S.Shields / [illegible]	1881	[illegible]	[illegible] / J.Readhead&Co.,S.Shields	Maclean,Dougla[illegible]
392	✻ BT [illegible]	Grantully Castle / Young 77-82	I ScBg / 3 Dks	British / London	[illegible]	359·6 / 43·8 / 29·4	Glasgow / [illegible]	1879	[illegible] Blkds.	3C 31"51"&82"-57" 550 HP. / [illegible]	D.Currie & Co.
393	✻	Granville / Ferrani 83-88	I ScBn / 1 Dk2 B	British / Middlesbro	5[illegible]	210·0 / 31·2 / 15·9 / Q[illegible] B73 F28	Middlesbro' / R.Dixon&Co.	1880	[illegible] Blkds.	2Cy.33"&58"-36" 130 HP. / Blair&Co.Ld.Stockton	S.S.GranvilleC[illegible] / (F.Woods)
394		Granville / Reader 83-83	I PadTug / 1 Dk	British / Dover	[illegible] / 157	101·0 / 19·1 / 10·4	London / Westwood&[illegible]	1876	[illegible] Blkds.	2Cy.36"-54" 70 HP. / J.Stewart,London	Dover Harbour [illegible]
395	✻	Grao / Cano [illegible]-85	I ScSr / 1 Dk&Spr Dk	Spanish / Valencia	9[illegible]	240·0 / 32·6 / 15·4 / 23·6	S.Shields / J.Readhead&Co.	1885	[illegible] Blkds.	2Cy.30"&56"-36" 130 HP. / J.Readhead&Co.,S.Shields	Cia.Valenciana [illegible] Navegacion
396	✻	Graphic / Bamman[illegible] 80-85	I ScSr / 1 Dk2 B	British / Hull	661 / 1036	228·9 / 29·6 / 16·8 / P130 F33	Aberdeen / A.Hall&Co.	1872	[illegible] Blkds.	2Cy.35"&50"-36" 98 HP. / [illegible] Earle's Co.,Hull	Mussey & Sawy[illegible]
397	✻	Grappler / Farrier	I ScSr / 2 Dks	British / London	4[illegible]	208·5 / 29·3 / 16·2 / Q62 B24 F29	Sunderland / J.Laing	1880	[illegible] Blkds.	2Cy.26"&50"-36" 100 HP. / R.&W.Hawthorn,Newcastle	West India & [illegible] Tel.Co.Ld.
398	V	Grasbrook / Schmidt 87-87	I ScSr / 2 Dks	German / Hamburg	1274 / 1839	290·0 / 36·6 / 25·8 / B42 P42	Hamburg / [illegible]	1882		2Cy.34"&62"-42" 130 HP. / [illegible]	Dampfschiff[illegible] Hansa
399	✻	Gratitude / Seaborne 87-87	I ScSr / 1 Dk&Spr Dk3 B	British / London	12[illegible]	260·2 / 36·6 / 17·6 / 23·3	Sunderland / [illegible]	188[illegible]	[illegible]	2Cy.34"&64"-42" 180 HP. / G.Clark,Sunderland	"Gratitude" S.S. Ld.(E.W.Morgan)[illegible]
400	✻ D	Gravina M.Barandika 75-85 / 1 Dk&Spr Dk&Web Frames	I ScSr	Spanish / Barcelona	8[illegible]	227·0 / 32·2 / 14·0 / 21·7	South Shields / [illegible]	188[illegible]	[illegible]	3C 18"29"&48"-30" 99 HP. / J.Readhead&Co.,S.Shields	J.Roca & Co.
401	✻	Gravina / Mendiola 80-83	I ScSr / 3 Dks	Spanish / Bilbao	4[illegible] / 2[illegible]	190·0 / 26·1 / 16·0	Newcastle / A.Leslie&Co.	1878	[illegible]	2Cy.26"&52"-36" 150 HP. / R.&W.Hawthorn,Newcastle	Larrinaga & Co.
402	✻	Grazia (ex Santander) / S.opuich	I ScSr / 1 Dk	Austrian / Lussinpiccolo	445	201·0 / 28·2 / 13·6 / Q37 B47 F26	Sunderland / Mounsey&Foster	1873	[illegible] Blkds.	2Cy.23"&40"-32" 70 HP. / [illegible]	G.L.Premuda
403	BT5.88	Great Britain	I PadSk	British / Liverpool	64	154·1 / 23·7 / 12·7	Renfrew / W.Simons&Co.	1876		2Cy.38"-48" 130 HP. / W.Simons&Co.,Renfrew	W.& T.Jolliffe
404		Great Britain / Browe	W PadTug	British / S.Shields	100	99·7 / 18·2 / 9·9	S.Shields / Wind	1874		2Cy.35"-30" 60 HP. / J.Readhead&Co.,S.Shields	J.Winlo & Co.
405		Great Conquest	W Pad	British / Liverpool	116	98·0 / 18·4 / 10·4	N.Shields	1857		2Cy. 60 HP.	W.R.Chisholm
406		Great Eastern / Compton 86-88 (BROKEN UP)	I Sc&Pad6mBk	British / London	13344 / 18915	679·6 / 82·8 / 48·2	London / J.Scott Russell&Co.	1858		4Cy.84"-48" 1000 HP. / 4Cy.74"-174" 1000 HP. / Watt&Co.,Birmingham	H.Bath & Sons
407		Great Emperor	I PadSp	British / Liverpool	51	143·7 / 23·1 / 11·8	Newcastle	1864		2Cy.40"-40" 110 HP. / G.Marshall&Co.,S.Shields	W.& T.Jolliffe
408	✻	Great Northern / Dillon 87-87	I ScSr / 2 Dks3 B	British / London	754 / 1348	240·0 / 32·3 / 23·9	W.Hartlepool / Denton,Gray&Co.	1870	[illegible] Blkds.	2Cy.31½"&53"-30" 100 HP. / T.Richardson&Sons,Hartlpl.	Eastern & S.[illegible] Tel.Co.L[illegible]
409		Great Western	W Pad / 1 Dk	British / Windsor,Ont.	662	220·0 / 40·2 / 13·6	Windsor,Ont. / H.H.Jenkins	1866		2Cy.42"-96" [illegible] HP. / [illegible]	Grand Trunk R[illegible] of Canada
410	BT5.88	Great Western	I PadSr	British / Milford	207 / 466	220·4 / 25·2 / 12·4	Renfrew / W.Simons&Co.	1867		2Cy.48"-45" 190 HP. / W.Simons&Co.,Renfrew	Great Western [illegible] way Co.,Ld.
411	BT5.88	Great Western	I PadSk	British / Liverpool	59	154·2 / 23·7 / 12·7	Renfrew / W.Simons&Co.	187[illegible]		2Cy.38"-66" 130 HP. / W.Simons&Co.,Renfrew	W.& T.Jolliffe
412	✻	Greata / Taylor [illegible]	I Sc3mSr / 2 Dks	British / Glasgow	288	192·3 / 28·0 / 15·1 / B40 F26	Glasgow / Barclay,Curle&Co.	1867	[illegible] Blkds.	2Cy.31"&58"-36" 120 HP. / Barclay,Curle&Co.,Glasgw.	Napier Shipp[illegible] Ld.
413	✻ D	Grebe / Taylor -87	Stl ScSr / 1 Dk2 B	British / London	513 / 844	210·0 / 30·1 / 15·0 / F31	Dundee / Gourlay Bros.& Co.	1887	[illegible] Blkds.	3C 19"29"&48"-33" 120 HP. / Gourlay,Bros.& Co.,Dundee	General Steam [illegible]
414	BT3.89	Grecian	I Sc3mSr	British / Glasgow	2375	360·[illegible] / 40·1 / 22·8	Sunderland / W.Doxford&Sons	1879	[illegible]	2Cy.48"&84"-54" 400 HP. / W.Doxford&Sons,Sunderld.	J.& A.Allan
415 / 45170	BT [illegible]	Greece (ex Virginia) / Jeffrey 83-85	I ScSr / 1 Dks	British / Liverpool	2712 / 4310	390·7 / 41·3 / 35·3	Newcastle / Palmer Bros.&Co.	186[illegible]		2Cy.60"&85"-48" 400 HP. / Laird Bros.,Birkenhead	National Steam Co.,Ld.
416 / 65383		Greenland / Curtis 75-81	W ScBk / 2 Dks	British / St.Johns N.F.	259 / 442	151·4 / 27·3 / 16·4 / Q37 F24	Aberdeen / A.Hall&Co.	1872		2Cy.24½"-42½" 75 HP. / D.Brown,Glasgow	J.Munn & Co.
417 / 60417		Greenwood / Bell 83-83	I Sc3mSr / 1 Dk2 B	British / [illegible]	727	232·3 / 30·3 / 17·6 / F38	Newcastle / Palmers'Co.	186[illegible]	[illegible]	2Cy.25"&60"-33" 99 HP. / Palmers'Co.,Jarrow	E.Harris & Co.
418 / 85021		Greenwood	W / 1 Dk	American / [illegible]	144	111·0 / 31·0 / 9·6	St.Francisco / C.G.White	188[illegible]		2Cy.11"&22"-18" 75 HP. / Hinkley,Spiers&Hayes	L.E.White
419	✻	Greetlands / Delisle 78-80	I ScSr	British / Quebec	733 / 1094	230·0 / 33·2 / 16·1 / B59 F27	Sunderland / D.Baxter&Co.	188[illegible]	[illegible]	2Cy.28"&52"-36" 116 HP. / [illegible]	J.G.Ross
420											

✻ = Lloyd's Register; D = Nederlandsche Vereeniging; G = Germanischer Lloyd; I = Registro Italiano, R = American [illegible]
A = Veritas Austro-Ungarico; H = Veritas Ellenico; N = Norske Veritas, V = Bureau Veri[illegible]

In the 2nd Column, B.T., with figures following, signifies Date of report of bottom sighting by Board of Trade Surveyors as condition of PASSENGER CERTIF[illegible]

In the 3rd Column, C = Composite (iron frame, wood planked); I = Iron; Stl = Steel; W = Wood.

Further particulars of Vessels classed by Lloyd's Register will be found in Lloyd's Register Book and Lloyd's Yacht Register.

	Registre dans lequel le navire est classé	Nom, Matériaux, Gréement, etc. Capitaine et dates d'entrée au service des Armateurs actuels et à bord du navire. Ancien Nom, nombre de ponts, etc.		Pavillon. Port d'Armement	Tonnage Net. Total.	Dimensions: Longueur	Largeur	Creux	Construit. Où. Par qui.	Quand.	Nombre des cloisons. Détails sur le lest d'eau.	Machines. Détails. Nom du Constructeur.	Armateurs ou Directeurs-Gérants.
421		Grenadier *Clark 19-86*	I ScSr	British Newcastle	539 821	236·6 *P39f.*	28·2 *Q71f.*	16·8 *F5f.*	Newcastle Palmer Bros.&Co.	1865		2Cy.28"&51"-36" 120 HP. R.&W.Hawthorn,N'weastle	Tyne Steam Shipping Co.,Ld.
422		Grenadier *McCallum 85-85*	Stl PadSk	British Glasgow	99	222·0	23·1	9·3	Glasgow J.&G.Thomson	1885		2Cy.30"&58"- 150 HP. J.&G.Thomson,Glasgow	D.MacBrayne
423	N	Grenmar *Meier 82-83*	I ScSr 1 Dk	Norweg'n Bravig	336 470	152·8 *Q45f.*	25·4 *B35f.*	12·7 *F20f.*	Stockholm W.Lindberg's Warfs & Works, Alderholmar	1883	APT MT15f	2Cy.20½"&35½"-23½" 55 HP. W.Lindberg,Stockholm	S.C.Larsen
424	*	Grenville *Tucker -88*	Stl ScSr 1 Dk,& Wb Frames	British London	1003 1565	250·0 *Q31f.P25f.*	35·0 *B100f.*	17·1 *F31f.*	W. Hartlepool E.Withy&Co.	1888	4 Blhds. CDB300 APT180	3C.18½"30"&48"-36" 110 HP. Blair&Co.,Ld.,Stockton	J.Holman & Sons
425	*	Greta *Poulton 83*	I ScSr 1 Dk	British W.Hartlepl.	398 635	180·4 *Q56f.*	28·4 *B44f.*	14·3	S.Shields J.Softley&Co.	1873	4 Blhds. orDB	3Cy.23"&46"-30" 80 HP. Blair&Co.,Ld.,Stockton	H.G.Harper & Co.
426		Gretzia	I ScSr 1 Dk	Russian Baku	318 468	175·0	27·7	11·4	Newcastle C.Mitchell&Co.	1881		2Cy.14"&25"-18" 30 HP. WallsendS.&E.Co.,N'castle	C.Theophylactos
427	*	Greyhound	I ScSr 1 Dk	Japanese	227 372	154·3 *Q74f.*	22·7 *F24f.*	10·9	PortGlasgow R.Duncan&Co.	1880	4 Blhds. DB[illegible] FPT[illegible]	2Cy.19"&34"-24" 60 HP. Smith Bros.,Glasgow	
428		Greyhound *Donaldson -88*	W ScCr 1 Dk,&Spar Dk	British Toronto	219 337	130·0	25·2	9·0	HamiltonOnt. M.Simpson	1888		2Cy.14"&24"-20" 40 HP. F.G.Beckett&Co.,Hamilton	Harriet A.Simpson
429	*	Greystoke *Sherwood 82-85*	I ScSr 1 Dk,& Web Frames	British W.Hartlepl.	1377 2119	275·9 *P25f.Q82f.*	37·3 *B106f.*	19·0 *F30f.*	W.Hartlepool E.Withy&Co.	1885	4 Blhds. CDB552	3C.21"35"&56"-36" 160 HP. T.Richardson&Sons,Hartpl	R.Ropner & Co.
430		Griffin *Tuttle*	I ScSr	British Southampton	145 272	155·0 *F52f.*	20·1 *F43f.*	11·8	PortGlasgow L.Hill&Co.	1858		2Cy.22"&44"-26" 70 HP. Comm'76 Day Summers&Co., Southampton	London & S.Western Railway Co.
431		Griffin	I ScSrYt 2 Dks	British London	75 111	126·1	16·7	9·3	Greenock Scott&Co.	1879	3 Blhds.	2Cy.12½"&24"-21" 40 HP. Greenock Foundry Co.,Grnk.	C.E.Dashwood
432		Grimaneza	I Sc 1 Dk	Peruvian Callao	78 140	109·8 *Q22f.*	18·8 *F—*	8·6	Paisley J.Fullerton&Co.	1873	3 Blhds.	2Cy.13½"&23"-18" 25 HP. W.King&Co.,Glasgow	M.Cotés
433		Grimsby *Jowres*	I Sc3mSr 1 Dk2 B	British Grimsby	456 713	208·9 *P77f.*	28·5 *B11f.*	16·2	Hull M.Samuelson&Co.	1857	4 Blhds.	2Cy.37"&26" 95 HP. C.&W.Earle,Hull	G.R.Nesbitt
434		Grimsby	StlPad	British Grimsby		180·0 *AwnDk100f.*	25·0	7·3	Hull Earles S.B.&E.Co.,Ld.	1888		2Cy.35"&60"-48" Earles'S.B.&E.Co.,Ld.,Hull	Manchester,Sheffield& Lincolnshire Railway Co.
435	*	Grimsel *Clark 87-87*	I ScSr 1 Dk2 B	British S. Shields	902 1398	251·0 *Q26f.*	34·2 *F25f.*	17·5 *B50f.*	S.Shields J.Readhead&Co.	1880	4 Blhds. DBa&f260	2Cy.29"&55"-36" 130 HP. J.Readhead&Co.,S.Shields	Cay,Hall & Co.
436		Grinkle *Verrill 76-76*	I Sc3mSr	British Newcastle	187 293	140·5	25·1	12·0	Newcastle Palmers'Co.	1867		2Cy.27"-24" 48 HP. Palmers'Co.,Newcastle	Palmers'Shipbuilding & Iron Co.,Ld.
437	*	Grip Fast *Rogers 83-85*	I ScSr 1 Dk2 B	British Aberdeen	544 860	200·1 *F104f.*	28·3 *F26f.*	15·0	Aberdeen Hall,Russell&Co.	1883	5 Blhds. DBa&f138	2Cy.25"&47"-30" 95 HP. Hall,Russell&Co.,Aberdeen	W.Leslie & Co.
438		Gripsholm *Ekstrom*	I Sc 1 Dk	Swedish Stockholm	122 157	108·2 *Q25f.*	17·6 *B42f.*	7·9 *H8f.*	Stockholm W.Lindbergs Mek.Vks.	1878		2Cy.16"&31"-21" 45 HP. W.Lindbergs Mek.Verks,Stklm.	Ångf.Bol."Gripsholm"(A.G.Förselius)
439		Grodno *Forth*	Stl ScSr 1 Dk2 B	British Hull	1116 1695	269·9 *Q94f.*	34·1 *B185f.*	18·4 *F29f.*	Hull Earles Co.,Ld.	1882	DB	2Cy.25"&50"-45" 130 HP. Earles'Co.,Ld.,Hull	T.Wilson,Sons & Co.
440	G	Grossherzog, Friedrich Franz *Schüning*	Stl PadSr	German Rostock	164 263	170·1	20·2	11·4	Elbing F.Schichau	1885		2Cy30"&55"-43¼" 160 HP.	Norddeutscher Lloyd
441		Guadalquiver	Stl ScSr 3 Dks	French Marseilles	1723 2612	334·5	37·7	28·7	Havre Forges&Chantiers		7 Blhds. DB	3Cy.25½"37"&55"-43" Forges&Chantiers,Havre	Messageries Maritimes
442		Guadeloupe *Intruch -88* (ex Empereur, ex Imperator)	I ScBk 2 Dks	French St.Nazaire	1133 1709	249·6	34·4 *F90f.*	24·2	Liverpool J.Laird	1854		2Cy.46"&80"-39" 400 HP. J.Elder&Co.,Glasgow	Compagnie Générale Transatlantique
443		Guadiana *Lemoine -88*	Stl ScBn 3 Dks	French Marseilles	1723 2560	334·6	37·7 *B52f.*	29·4	Havre Forges & Chantiers de la Medit.	1888	CDB[illegible]	3C.25"37"&55"-43" 400 HP. Forges & Chantiers de la Medit., Havre	Messageries Maritimes
444		Guahy	I Pad	Brazilian Bahia	149 336	165·0 *P52f.*	22·0 *B54f.*	9·6 *F29f.*	Greenock Caird&Co.	1885		2Cy.26"&52"-54" 170 HP. Caird&Co.,Greenock	Comp.de Nav.á Vapor de Bahia
445		Guama	I PadSr	Brazilian Para	120	120·0	22·0	5·6	Wilmington The Pusey&Jones Co.	1861		2Cy.16"-60" The Pusey&Jones Co.,Wilm.,Del.	C.Ludmann & Co.
446		Guaniguanico (ex Domingo Hernandes)	W	Spanish Havana	148 207	238·5	24·9	8·4	Cuba	1874		200 HP.	P.Saez
447		Guarani *Colozzi*	I Sc	Arg'ntine Bn'os Ayres	141 225	136·0	22·0	13·9	Belfast Harland& Wolff	1870		16 HP.	La Platense Flotilla Co.,Ld.
448		Guarapiranga		Brazilian Maranhão	140	..	..	..				45 HP.	J.M.F.Vasconcellos & Co.
449		Guayacan	I ScSr 1 Dk2 B	Chilian Guayacan	879 1186	240·0 *Q131f.*	32·0 *B26f.*	20·6 *F36f.*	Stockton N.Pearse&Co.	1872	4 Blhds. WBa&f	2Cy.30"&60"-36" 150 HP. Jackson&Watkins,London	Sociedad Chilena de Fundiciones
450													

*=Lloyd's Register; D=Nederlandsche Vereeniging; G=Germanischer Lloyd; I=Registro Italiano; R=American Record;
Å=Veritas Austro-Ungarico: H=Veritas Ellenico; N=Norske Veritas; V=Bureau Veritas.

Dans la 2me Colonne, B.T., avec les indications qui suivent, signifie la date du rapport de visite de la Carène par les experts du Board of Trade pour obtenir le certificat relatif au TRANSPORT DES PASSAGERS.

Dans la 3me Colonne, C=Composite (membrure en fer, bordé en bois); I=Fer; Stl=Acier; W=Bois.

Pour plus amples renseignements sur les navires classés par le Lloyd's Register, voir le Lloyd's Register Book et le Lloyd's Yacht Register.

No. in Book. / Official No.	Registry in which classed.	Ship's Name. Material, Rig, &c. / Master, and Dates of Appointment to present Owner's Service and to Ship / Ship's Late Name. No. of Decks, &c.	Flag. / Port of Registry.	Tons. Net. / Gross.	Dimensions. Length | Breadth | Depth / Deck Erections.	Built. Where. / By Whom.	When	No. of Bulkheads / Particulars of Water Ballast.	Engines. Particulars. / Maker's Name.	Owners or M[illegible]
451 68831	BT11.86	Guernsey I ScSr Pay	British Southampton	256 521	195·5|26·0|13·4 Q [illegible] F [illegible]	London J.&W.Dudgeon	1874		2Cy.35"&61"–32" 130 HP J.&W.Dudgeon,London	London [illegible]
452 81587	✻	Guiana I PadSr George 85-88 1 Dk	British Georgetown, Dem	152 241	175·6|19·1|7·5 B 30 f.	London R.&H.Green	1879	3 Blds.	2Cy.28"&47"–42" 70 HP J.Penn&Sons,London	[illegible] Foundry Co.
453	✻	Guido Stl ScSr Lachiondo 2 Dks	Spanish Bilbao	2065 3133	360·4|41·2|26·4	Belfast Harland&Wolff	1883	5 Blds. W [illegible] APT FPT	2Cy.49"&86"–51" 400 HP Harland&Wolff,Belfast	Cia. de Nav. La[illegible]
454		Guikson (BTGS) W Pad Pietro 84-85	Turkish Const'nople	51	120·0|18·0|8·5	West Cowes J.White	1852		60 HP Maudslay,Sons&Field	Chirket-i-H[illegible] [illegible]
455 86073	✻D	Guildford I ScSr Stabell 88-88 2 Dks3 B	British Newcastle	1435 [illegible]	298·6|37·2|23·7 P 26 f. B 58 f. F 34 f.	Newcastle Palmers Co.,Ld.	1882	5 Blds. [illegible] APT [illegible]	2Cy.36"&68"–45" 270 HP Palmers Co.,Ld.,Newcastle	J.Temperley &C[illegible]
456 89827	✻	Guillemot W ScK Fawens 85-87 1 Dk	British S.Shields	10 53	71·5|17·0|8·2	Leith J.M'Kenzie&Co.	1887		2Cy.10"&20"–11" 17 HP Hawthorns&Co.,Leith	T.F.Robertson
457		Guillermo (COLLISION) I ScSr 2 Dks3 B	Spanish Bilbao	1314 2044	314·3|34·5|24·2 P 36 f. B 52 f. F 40 f.	Liverpool T.Royden&Sons	1879	5 Blds. DB [illegible] APT	2Cy.35"&70"–48" 250 HP J.Key&Sons,Kirkcaldy	Cia.de Nav. La[illegible]
458 62061	BT5.88	Guinevere I Pad McIntyre 85-85	British Glasgow	43 109	200·3|19·1|6·8	Port Glasgow R.Duncan&Co.	1869		2Cy.36"–54" 86 HP Rankin&Blackmore,Grnck.	W.Buchanan
459		Guipuzcoa I ScBk (ex Principe Alfonso) 3 Dks	Spanish Barcelona	1377 1927	280·8|38·5|26·0 B 104 f. F 53 f.	Dumbarton W.Denny&Bros	1863		2Cy.56"–42" 400 HP Denny&Co.,Dumbarton	Cia.Trasatlantic[illegible]
460		Guipuzcoa (ex [illegible]) I ScSp Be Arana 86-86 1 Dk	Spanish Bilbao	88 106	99·7|20·5|8·7	Hull [illegible]	1885		1Cy.10"&15"–16" 35 HP Cochrane&Co.,Birkenhead	J.B.de Longa
461 10523		GulfStream (ex Queen, ex Victory, ex Carolina) I ScSmSr Kimble 3 Dks	American New York	709 908	215·0|32·0|13·6	New York	1861		1Cy.42"–42" 280 HP John Baird	New York & Wi[illegible] ton S.S.Co.
462 83103	✻	Gulf of Aden Stl ScSr Hudson 87-88 2 Dks3 B	British Greenock	1572 2470	300·6|40·2|23·9 P 83 f. B 76 f. F [illegible]	Middlesbro' R.Dixon&Co.	1887	6 Blds. DB [illegible] FPT [illegible] APT [illegible]	3C.26"42"&69"–48" 450 HP Blair&Co.,Ld.,Stockton	Greenock S.S.[illegible]
463 87408	✻	Gulf of Akaba I ScBn Mauro 88-88 1 Dk2 B	British Greenock	1273 1975	275·0|37·8|19·6 Q 74 f. [illegible]	Middlesbro' R.Dixon&Co.	1883	5 Blds. DB [illegible] FPT [illegible] APT [illegible]	2Cy.34½"&64"–42" 177 HP Blair&Co.,Ld.,Stockton	Greenock S.S.[illegible]
464 93294	✻ BT8.88	Gulf of Corcovado Stl ScSr Warden 86-88 2 Dks3 B	British Greenock	1482 2361	310·0|40·3|22·6 P 82 f. F 82 f. B [illegible]	Newcastle R.&W.Hawthorn, Leslie&Co.,Ld.	1888	6 Blds. DB [illegible] FPT [illegible]	3Cy.26"42"&69"–48" Blair&Co.,Ld., Stockton [300 HP	Greenock S.S.C[illegible]
465 93269	✻ BT11.88	Gulf of Guayaquil Stl ScSr Friend 85-88 (LOST) 2 Dks3 B	British Greenock	1478 2365	310·0|40·3|22·6 P 82 f. B 8[illegible] F [illegible]	Newcastle R.&W.Hawthorn, Leslie&Co.,Ld.	1888	6 Blds. DB [illegible] FPT [illegible]	3Cy.26"42"&69"–48" Blair&Co.,Ld., Stockton [300 HP	Greenock S.S.C[illegible]
466 93266	✻ BT9.88	Gulf of Guinea Stl ScSr Allan 75-88 2 Dks3 B	British Greenock	1541 2438	320·6|40·3|22·6 P 83 f. B 88 f. F [illegible]	Newcastle R.&W.Hawthorn, Leslie&Co.,Ld.	1888	6 Blds. DB [illegible] FPT [illegible]	3Cy.26"42"&69"–48" Blair&Co.,Ld., Stockton [300 HP	Greenock S.S.C[illegible]
467	✻	Gulf of Martaban Stl ScSr Swan 80-88	British Greenock	2800	310·0|40·3|22·6 P 82 f. B 89 f. F 83 f.	Newcastle R.&W.Hawthorn, Leslie&Co.,Ld.	1889		3Cy.26"42"&69"–48" Blair&Co.,Ld., Stockton [300 HP	Greenock S.S.C[illegible]
468 87387	✻ BT9.88	Gulf of Mexico I ScBn Pepper 88-88 1 Dk & Spr Dk3 B	British Greenock	2059 3172	330·0|42·3|26·0 B 96 f. F 28 f.	Middlesbro' R.Dixon&Co.	1883	5 Blds. DB [illegible] FPT [illegible] APT [illegible]	2Cy.43"&80"–51" 450 HP Blair&Co.,Ld.,Stockton	Greenock S.S.[illegible]
469 87402	✻	Gulf of Papua I ScBn Fox 88-88 1 Dk2 B	British Greenock	1266 1971	275·3|37·5|19·1 Q 74 f. P 37 f. B 62 f. F 72 f.	Middlesbro' R.Dixon&Co.	1883	5 Blds. DB [illegible] FPT [illegible] APT [illegible]	2Cy.34½"&64"–42" 176 HP Blair&Co.,Ld.,Stockton	Greenock S.S.C[illegible]
470 81532	✻	Gulf of St.Vincent I Sc Kennedy 88-88 1 Dk & Spr Dk3 B	British Greenock	1570 2421	289·7|38·2|17·7 B 70 f. 25·9	W. Hartlepool W.Gray&Co.	1881	5 Blds. DB [illegible] FPT [illegible] APT [illegible]	2Cy.38"&71"–45" 300 HP Blair&Co.,Ld.,Stockton	Greenock S.S.C[illegible]
471 81700	✻D	Gulf of Suez I ScBn Cook 88-88 2 Dks3 B	British Greenock	976 1535	260·0|34·0|21·7 P 16 f. F 35 f.	W. Hartlepool W.Gray&Co.	1880	4 Blds. DB [illegible]	3C19½"32"&51"–39" 170 HP Triple? Blair&Co.,Ld.,Stkn	Greenock S.S.C[illegible]
472 93197	✻ BT10.88	Gulf of Trinidad Stl ScSr Sanderson 84-87 2 Dks3 B	British Greenock	1495 2362	300·0|40·2|23·9 P 82 f. B 76 f. F 90 f.	Middlesbro' R.Dixon&Co.	1887	6 Blds. [illegible] [illegible] FPT [illegible] APT [illegible]	3Cy.26"42"&69"–48" 450 HP Blair&Co.,Ld.,Stockton	Greenock S.S.C[illegible]
473 87369	✻ BT7.88	Gulf of Venice I ScBn Lagertwood 84-88 1 Dk & Spr Dk3 B	British Greenock	1984 3022	331·1|42·3|25·7 B 30 f. F 40 f.	W. Hartlepool W.Gray&Co.	1883	6 Blds. DB [illegible] FPT [illegible] APT [illegible]	2Cy.43"&80"–54" 450 HP Blair&Co.,Ld.,Stockton	Greenock S.S.C[illegible]
474 77671	✻	Gulnare C ScBn McLeod 79-86 1 Dk	British Charlotteto.	120 247	137·7|21·8|11·1 Q 71 f. B 15 f. F 14 f.	Glasgow J.Connell&Co.	1878	[illegible]	2Cy.23"&38"–27" 70 HP J.&J.Thomson,Glasgow	Alexander M'Le[illegible]
475 51415		Gunga I ScBn Calder 2 Dks	British Sydney	798 1257	257·6|32·2|20·1 P 61 f. F 54 f.	Hull Earle&Co.	1864	3 Blds. DB aft [illegible]	1Cy.26"&47"–30" 220 HP Comp.'73,Humphreys,Hull	Australasian [illegible] Stm.Nav.Co.,Ld.
476		Gurupy	Brazilian Maranhão	410	.. .. ..					Comp.de Mara[illegible]
477 85214		Gussie W Pad Staples	American N.Orleans La	576 998	221·6|33·8|16·8	Wilmington Del Harlan&Hollingsworth Co.	1872		1Cy.44"–132" 387 HP The Harlan Hollingsworth Co., Wilm.,Del.	J.B.Richardson
478	G	Gustaf C ScSr Johansen -88 1 Dk	Norweg'n Christiania	164 220	101·7|22·2|11·6 B 32 f.	Gothenburg G.Pehrsson	1884		2Cy.12"&23"–15" 25 HP	O.Jorgenson
479		Gustaf II. Adolf I ScSr Dahlström 1 Dk	Swedish Sundsvall	312 411	157·4|23·0|13·4 Q 52 f. B 38 f. P [illegible]	Gothenburg	1854		2Cy.20½"&41"–21" 80 HP W.Lindbergs Mk. Vrks.,Stckhlm	A.P.Sundqvist
480										

✻ = Lloyd's Register; D = Nederlandsche Vereeniging; G = Germanischer Lloyd; I = Registro Italiano; R = American R[illegible]
A = Veritas Austro-Ungarico; H = Veritas Ellenico; N = Norske Veritas; V = Bureau Verita[illegible]

In the 2nd Column, B.T., with figures following, signifies Date of report of bottom sighting by Board of Trade Surveyors as condition of PASSENGER CERTIFI[illegible]

In the 3rd Column, C=Composite (iron frame, wood planked); I=Iron; Stl=Steel; W=Wood.

Further particulars of Vessels classed by Lloyd's Register will be found in Lloyd's Register Book and Lloyd's Yacht Register.

1	2	3 Nom, Matériaux, Grément, etc. Capitaine et dates d'entrée au service des Armateurs actuels et à bord du navire. Ancien Nom, nombre de ponts, etc.		4 Pavillon. Port d'Armement.	5 Tonnage. Net. Total	6 Dimensions. Longueur	Largeur	Creux	7 Construit. Où. Par qui.	8 Quand.	9 Nombre des cloisons. Détails sur le lest d'eau.	10 Machines. Détails. Nom du Constructeur.	11 Armateurs ou Directeurs-Gérants.
	V	Gustaf Tillberg *Agelsen*	I ScSr *1 Dk 2 B*	Swedish Trelleborgs	618 760	187·3	26·1	18·6 *B 82 f. H 8 t.*	Oscarshamn OscarshamnsMk.Verks	1874		2Cy.20"&35"-24" 60 HP. KockumsMek.Verks,Malmö	Trelleborgs Nya Ångf. Akt.Bol.(T.Malmros)
		Gustaf Vasa *Bergström 85-85*	I ScSr *1 Dk*	Russian Nikolaistad	244 [illegible]	172·2	24·1	12·5	Gothenburg MotalaCo.	1866		80 HP.	A.Hedman
		Gustafsberg II. *Ackland*	I Sc *1 Dk*	Swedish Gustafsberg	63 106	83·3	18·2	7·5 *Q 28 f. B 25 f. H 2 t.*	Motala MotalaCo.	1878		30 HP. MotalaCo.,Motala	Gustafsbergs Fabriks-Intressenter (V.Odelberg)
		Gustafsberg IV.	I Sc *1 Dk*	Swedish Gustafsberg	79 118	99·7	18·0	8·5 *B 25 f. H 8 t.*	Stockholm BergsundsMek.Verks.	1884		2Cy14½"&25½"-19½" 35 HP. BergsundsMk.Vrks.Stkhlm	Gustafsbergs Fabriks-Intressenter (V.Odelberg)
		Gustafsberg V.	I Sc *1 Dk*	Swedish Gustafsberg	88 127	102·1	18·1	8·4 *B 33 f.*	Stockholm BergsundsMekWerk	1888		2Cy.18"&32"-20" 45 HP. BergsundsMekWrk.Stklm	Gustafsbergs Fabriks-Intressenter (V.Odelberg)
486	✱	Gustav Bitter *Hemmings 81-86*	I ScSr *1 Dk*	British Newcastle	314 538	169·5	27·3	12·4 *Q 80 f. B 46 f. F 22 f.*	Newcastle C.S.Swan&Co.	1878	4 Bhds. DBcdo ft f 34 ft	2Cy.21½"&41"-30" 80 HP. T.Clark&Co.,Newcastle	Fisher,Renwick & Co.
487	GN	Gustav Nermann *Olsen -79*	C ScSr *1 Dk*	Norweg'n Stavanger	232 307	132·9	23·3	11·7 *B 35 f.*	Oscarshamn S.Olssen	1872		2Cy.11½"&23"-16" 30 HP.	L.Larsen
488	V	Gut Heil *Hebner 87-87*	Stl ScSr *2 Dks 3 B*	German Geestemünde	2096 2700	310·0	40·2	28·2	Newcastle Sir W. G. Armstrong, Mitchell & Co.	1888	DB/10ft MT600t FPT100t	3Cy.23"-37"&60"-39" 220 HP. WllsndSlip.&E.Co.,Ncl.	W.A.Riedemann
489		Gutenfels	Stl ScSr	German Bremen	1816 2766	309·0	40·0	26·7	Sunderland SunderlandS.B.Co.	1889		3Cy. WallsendSlip.Co.,Ld.,Nwcl.	Deutsche Damps Gesellshaft Hansa.
490	✱	Guthrie *Shannon*	I ScBg *2 Dks 3 B*	British London	1494 2338	314·0	38·3	23·6 *P 46 f. B 56 f. F 46 f.*	Sunderland Doxford&Sons	1884	6 Bhds. CDB161t	2Cy.40"&73"-48" 340 HP. Hawks,Crawshay&Sns.,Gts.	W.Mactaggart
491		Guttenberg *Buhle*	I PadCr *1 Dk*	German Stade	92 146	140·4	20·0	7·5 *B 31 f.*	Rostock Act.Gesell.Hansa	1875		3Cy.23½"&(2)32½"-42" HP. A.Tischbein,Rostock	Stade-Altländer Dampfschifff.Rhed.-Ges.
492		Gutzlaff	I Sc *1 Dk*	British Shanghai	18 135	105·5	19·4	9·2	Shanghai Boyd&Co.	1886		2Cy.20"&38"-27" 65 HP. D.Rowan,Glasgow	P.&O.Steam Nav.Co.
493	✱	Guy Colin *Leach -88*	Stl ScSr *1 Dk & Wb Frms*	British London	1463 2234	290·0	38·4	19·1 *P29f. Q121f. F112f.*	Middlesbro' R.Dixon&Co.	1888	5 Bhds. CDB442t	3Cy.22",35"&59"-39" 250 HP. T.Richardson&Sns,Hpl	English & American Shipping Co., Ld.
494	✱	Guy Mannering *Hall -87 (ex Maddaboni)*	I ScIrnSr *2 Dks 3 B*	British Bristol	1829 2842	380·0	35·8	27·0 *F 72 f.*	Newcastle A.Leslie&Co.	1873	5 Bhds. WB f	2Cy.30"&70"-48" 240 HP. Maudslay&Co.,London	T.B.Williams (W.Lamplough&Co.,Mgrs.)
495		Guy Mannering *Bowie 82-82 (ex Shells)*	I PadSt	British Glasgow	40 256	205·5	20·0	7·7 *Q 30 f.*	Greenock Caird&Co.	1877		1Cy.50"-72" 120 HP. Caird&Co.,Greenock	North British S.P.Co. (R.Young)
496	R	Guyandotte *Kelley 72-82*	I ScSr *3 Dks*	American New York	1651 2351	267·0	40·0	24·4	Chester, Pa. J.Roach&Son	1882		2Cy.38"&74"-54" J.Roach&Sons,Chester,Pa.	Old Dominion S.S.Co.
497	✱	Guyenne *Senan 82-82*	I ScSr *1 Dk 2 B*	French LaRochelle	1121 1592	260·1	38·1	17·2 *P 29 f. B 52 f. F 26 f.*	Sunderland J.Blumer&Co.	1882	6 Bhds. CDB65t	2Cy.31"&60"-39" 160 HP. T.Clark&Co.,Newcastle	Delmas frères
498	✱	Guyers *Watson 87-87*	I ScSr *1 Dk*	British Newcastle	337 531	180·5	23·7	13·0 *Q 89 f. F 20 f. B 29 f.*	Newcastle Coulson,Cook&Co.	1873	4 Bhds. DBa39ft f 40 ft	2Cy.21"&38"-30" 60 HP. Ouseburn.Eng.W.Co,Nwcstle	P.Brown
499	✱D	Gwalia *Mead 82-82*	I ScSr *1 Dk 2 B*	British Cardiff	876 1354	240·0	33·0	16·7 *Q 81 f. B 62 f. F 26 f.*	Sunderland W.Pickersgill&Sns	1881	4 Bhds. [illegible] APT [illegible]	2Cy.29"&56"-36" 120 HP. N.E.MarineEng.Co.,Ld.,Sld.	"Gwalia" S.S.Co.,Ld. (Hurley,Matthews & Co.)
500		Gwalior *Symons 73-88*	I ScSr	British Glasgow	1602 2733	351·6	38·3	27·7 *F 48 f.*	Dumbarton W.Denny&Bros.	1873		2Cy.19"&86"-54" 500 HP. Denny&Co.,Dumbarton	Peninsular & Oriental Steam Nav.Co.
501	✱	Gwendoline *Milburn 84-84*	I ScBn *1 Dk 2 B*	British Cardiff	1110 1729	260·2	37·1	18·5 *Q 56 f. B 68 f. F 31 f.*	Whitby T.Turnbull&Son	1883	4 Bhds. DBc&f 251t	2Cy.32"&60"-39" 150 HP. Blair&Co.,Ld.,Stockton	Turnbull Bros.
502	✱D	Gwendoline *Pigeon 75-82*	I ScSr *1 Dk 2 B*	British London	932 1467	260·4	35·2	18·7 *P 27 f. B 50 f. F 32 f.*	Newcastle PalmersCo.Ld.	1882	4 Bhds. DBcy&c&a132t APT15t	2Cy.31"&59"-42" 180 HP. PalmersCo.,Ld.,Newcastle	Wilson & Co.
503	✱	Gwendoline *Leader 85-85*	I ScSr *1 Dk*	British Middlesbro'	328 525	165·7	24·8	14·3 *B 31 f. Q 64 f. F 20 f.*	Middlesbro' R.Dixon&Co.	1876	4 Bhds. DBc&f 116t	2Cy.23½"&42"-30" 70 HP. Blair&Co.,Ld.,Stockton	J.E.Swan & Bros.,Ld.
504	✱	Gwenllian Thomas *Jones 87-87*	I ScSr *1 Dk 2 B*	British Swansea	688 1032	225·4	31·2	17·0 *Q113f. B44f. F28f.*	Newcastle PalmersCo.Ld.	1882	4 Bhds. DBa&f220t APT40t	2Cy.28"&52"-33" 99 HP. PalmersCo.,Ld.,Newcastle	"GwenllianThos."S.S. Co(EvanThomas,Radcliffe&Co)
505	✱N	Gwent *Reimers 88-88*	I ScSr *1 Dk*	Norweg'n Bergen	447 652	190·0	28·0	13·9 *Q 27 f. F 25 f.*	Hartlepool E.Withy&Co.	1878	4 Bhds. DBa30 ft f 32 ft &c b31 ft APT	2Cy.25½"&48"-30" 90 HP. Blair&Co.,Ld.,Stockton	J.Christensen
506	✱D	Gwentland *Kendrick*	I ScSr *1 Dk 2 B*	British Cardiff	1012 1553	250·9	35·2	17·5 *Q96 f. B 56 f. F 25 f.*	SouthShields J.Readhead&Co.	1881	5 Bhds. DBc&f395t APT17t	2Cy.29"&55"-39" 120 HP. J.Readhead&Co.,S.Shields	"Gwentland"S.S.Co., Ld. (Morel Bros.&Co.)
507	✱	Gwydir *Hobler*	Stl ScSr *1 Dk*	British Sydney	323 518	186·5	28·1	12·1 *B38f. Q30f. F20f.*	SouthShields J.Readhead&Co.	1886	4 Bhds. DBc f 123t APT73t	2Cy.21"&48"-30" 90 HP. J.Readhead&Co.,S.Shields	Hunter River Steam Nav.Co.
508	✱	Gwynfaen *Owens*	I ScIrnSr *1 Dk*	British Liverpool	99 255	139·8	20·3	9·9 *Q 76 f. B 7 f. F 28 f.*	Liverpool J.Jones&Sons	1887	3 Bhds. FPT33t APT7t	2Cy.18"&36"-24" 54 HP. J.Jones&Sons,Liverpool	Kneeshaw, Lupton & Co.
509		Gylfe *Prahl 74-74*	I PadSr *1 Dk*	Danish Copenhagen	149 275	164·4	23·2	10·2 *P 33 t. H 57 t.*	Copenhagen Burmeister & Wain	1874		4Cy.22"&44"-42" 100 HP. Burmeister&Wain, Cpnhgn	Det Forenede Dmpskbs Selskab.(C.P.A.Koch)
510	V	Gylfe *Gillbdon*	I ScSr *1 Dk*	Swedish Gothenburg	210 302	120·4	21·7	13·2 *B 31 f. H 8 t.*	Gothenburg GöteborgsMek.V.	1884		2Cy.16"&28"-20" 45 HP. GöteborgsMk.Werkst.,G'burgh	Ångf.Aktieb.Bol. "Gylfe" (T.Ahrenberg)

✱ = Lloyd's Register; D = Nederlandsche Vereeniging; G = Germanischer Lloyd; I = Registro Italiano; R = American Record;
A = Veritas Austro-Ungarico; H = Veritas Ellenico; N = Norske Veritas; V = Bureau Veritas.

Dans la 2me Colonne, B.T., avec les indications qui suivent, signifie la date du rapport de visite de la Carène par les experts du Board of Trade pour obtenir le certificat relatif au TRANSPORT DES PASSAGERS.

Dans la 3me Colonne, C = Composite (membrure en fer, bordé en bois); I = Fer; Stl = Acier; W = Bois.

Pour plus amples renseignements sur les navires classés par le Lloyd's Register, voir le Lloyd's Register Book et le Lloyd's Yacht Register.

No. in Book / Official No.	Registry in which Classed	Ship's Name. Material, Rig, &c. / Master, and Dates of Appointment to present Owner's Service and to Ship. / Ship's Late Name. No. of Decks, &c.	Flag. / Port of Registry.	Tons Net. / Gross.	Dimensions Length / Breadth / Depth / Deck Erections	Built Where. / By Whom.	When	No. of Bulkheads. / Particulars of Water Ballast.	Engines. / Particulars. Maker's Name.	Owners or Managers
511 / 87189		Gympie StlScSr / Michael 1 Dk	British / London	126 / 220	120·5 / 22·1 / 8·0 / Q 43, B 22, F 23	Campbeltown / Campbltwn S.B.Co.	1883	4 Blkds. / [illegible]	2Cy.18" & 33"-30" 45 HP. / Dunsmuir & Jackson, Glsgw	
512 / 98376		Gyne StlTScLr	British / Glasgow	181 / 287	165·3 / 27·2 / 9·2	Dumbarton / W. Denny & Bros.	1887		2Cy.21½"&38"-24" 42 HP / Denny&Co., Dumbarton	Irrawaddy F[illegible] Ld.
513	V	Gyptis IScSr / Ranch 69-87 2 Dks	French / Marseilles	702 / 1032	255·9 / 30·0 / 21·8 / P 40, B 15, F 36	Greenock / Caird & Co.	1865		2C 34½"& 56"-33" 130 HP / Fraissinet&Co., Marseilles	Fraissinet & [illegible]

Additional Ships are inserted in the Supplement at the end of the Register.

✱ = Lloyd's Register; D = Nederlandsche Vereeniging; G = Germanischer Lloyd; I = Registro Italiano; R = American [illegible]
A = Veritas Austro-Ungarico; H = Veritas Ellenico; N = Norske Veritas; V = Bureau Veritas

In the 2nd Column, B.T., with figures following, signifies Date of report of bottom sighting by Board of Trade Surveyors as condition of Passenger Ce[illegible]

In the 3rd Column, C = Composite (iron frame, wood planked); I = Iron; Stl = Steel; W = Wood.

Further particulars of Vessels classed by Lloyd's Register will be found in Lloyd's Register Book and Lloyd's Yacht Register.

	Classe	Nom, Matériaux, Arrimage, etc. Capitaine et dates d'entrée au service des Armateurs actuels à bord du navire. Ancien Nom, nombre de ponts, etc.		Pavillon. Port d'immatricule.	Net. Brut.	Dimensions.	Construit. Où. Par qui.		Nombre des cloisons. Détails sur le lest d'eau.	Machines. Détails. Nom du Constructeur.	Armateurs ou Directeurs-Gérants.
1	✻N	H. A. Hartmann Westberg	I ScSr 1 Dk 2 B	Norweg'n Flensburg	383	155·9 / 25·2 / 15·9 Q 47 f. B 45 f. F 23 f.	Flensburg FlensburgerS.B.C.	1882	4 Blhds. APT 97 MT 150	2Cy. $22\frac{1}{2}$" & 42"–27" 80 HP. Flensburger Schiffsb. Ges., Flnsb.	Et Interessentskab (H.C.Clausen)
2	V	H. C. Ørsted Ørsted	I ScSr 1 Dk	Danish Copenhagen	451 719	183·4 / 28·1 / 15·5 P 91 f. F 38 f.	Copenhagen Burmeister & Wain	18 72/82	APT 90 MT 370	1Cy. 21" & 41"–30" 120 HP. Burmeister & Wain, Cpnhgn.	"Store Nordiske Telegraf-Selskab" (E.Suenson)
3	R	H. F. Dimock Berry –87	I ScSr 2 Dks & Spr Dk	American Boston Mass.	1795 2029	271·8 / 41·6 / 19·2 29·3	Philadelphia W.Cramp&Sons	1884	5 Blhds.	2Cy. 36" & 68"–54" W.Cramp&Sons, Philadelphia	Metropolitan S.S.Co.
4	✻DV	H. I. Pallisen Hansen 77-77	I ScSr 1 Dk & Spr Dk 3 B	Danish Copenhagen	714 1133	221·2 / 30·3 / 14·9 21·8	Copenhagen Burmeister & Wain	1877	5 Blhds. MT 270 APT 38 FPT 36	1Cy. 21" & 42"–27" 120 HP. Burmeister & Wain, Cpnhgn.	Dampsk.-Selsk. "H.I. Pallisen" (H.J.Rønne)
5		H. P. Prior Toft 87-78	I PadSr 1 Dk	Danish Aalborg	308 518	201·9 / 23·5 / 14·2 B 29 f. F 18 f. Z 25 f.	Copenhagen Burmeister & Wain	1878		4Cy. 26" & 58"–18" 200 HP. Burmeister & Wain, Cpnhgn.	Det Forenede Dmpskbs Selskab. (C.P.A.Koch)
6		H. S. Pickands	W Sc	American Grand Haven, Mich.	506 [illegible]	181·4 / 32·5 / 13·1	Grand Haven, [Mich.	1884			M.E.Kirby
7		H. T. Baya	W Pad	American New York	411 670	205·0 / 32·0 / 9·2	Philadelphia Birely, Hillman & Streaker	1882		1Cy. 52"–120" Neafie&Levy, Philadelphia	G.Poggenburg
8	V	H. von Witt Wegner –80	I Sc 1 Dk 2 B	German Rostock	559 879	206·0 / 27·7 / 15·7 P 76 f. F 33 f.	Port Glasgow Cunliffe&Dunlop	1875	WB	2Cy. 26" & 46"–30" 100 HP. Cunliffe&Dunlop, Pt.Glsgw.	Neue Dampfschiffs-Akt.-Ges. (M.Petersen)
9		Haakon Adalstein Rienan 77-82	I ScSr 2 Dks & Awn Dk	Norweg'n Drontheim	514 868	180·6 / 26·4 / 16·7 21·2	Newcastle Mitchell&Co.	1873	4 Blhds.	2Cy. 27" & 50"–30" 100 HP. N.E.MarineEng.Co., Ld., Sld.	Nordenfjeldske Dampskibs-Selskab.
10		Haakon Jarl Berg 88-81	Stl ScSr 2 Dks & Awn Dk	Norweg'n Drontheim	602 789	181·2 / 27·4 / 21·7	Gothenburg MotalaCo.	1879	DB 83	2C. $29\frac{1}{2}$" & $51\frac{1}{2}$"–$29\frac{1}{2}$" 160 HP. MotalaCo., Gothenburg	Nordenfjeldske Dampskibs-Selskab.
11		Haardraade Vastberg	W ScBk 2 Dks	Norweg'n Christiania	245 414	144·3 / 30·0 / 16·1 Q 35 f.	Christiania Brinchs Vertt.	1871		45 HP. Nylands Verks., Christiania	Et Interessentskab (Carl Hansen)
12	V	Habana (ex Ernst Moritz Arndt) Criado 83-87	I ScBg 3 Dks	Spanish Barcelona	1766 2537	317·0 / 37·2 / 28·5 B 70 f. F 44 f.	Sunderland Oswald&Co.	1872	3 Blhds.	2Cy. 51" & 86"–42" 350 HP. T.R.Oswald&Co., Sunderlnd	C[ie] Trasatlantica
13	G	Habsburg Thomann 86-89	I ScBg 2 Dks & Hur Dk	German Bremen	1916 3094	351·1 / 39·1 / 32·5	Hull Earles'S.&E.Co.	1875		2Cy. 52" & 96"–48" 370 HP. Earles'S.&E.Co., Hull	Norddeutscher Lloyd
14		Hae-an (ex Shinnkiang) Lancaster 78-88	I Sc 2 Dks	Chinese Shanghai	869 1378	244·0 / 33·0 / 18·2	Glasgow A.&J.Inglis	1873		2Cy. 36" & 62"–39" 150 HP. Laird Bros., Birkenhead	China Merchants' Steam Nav.Co.
15		Hae-shin (ex Chihli) Wells 77-88	I ScSr 2 Dks & Spr Dk	Chinese Shanghai	763 1115	212·6 / 33·5 / 18·8	Glasgow A.&J.Inglis	1871		2Cy. 31" & 60"–36" 150 HP. A.&J.Inglis, Glasgow	China Merchants' Steam Nav.Co.
16		Hae-ting Patterson 75-84 (ex Paou-ting)	Stl & I ScSr	Chinese Shanghai	1099 1713	253·0 / 33·0 / 20·9	Glasgow J.Elder&Co.	1873		2Cy. 32" & 62"–33" 150 HP. J.Elder, Glasgow	China Merchants' Steam Nav.Co.
17	XV	Hægholmen Korseth 88-88	Stl & I ScSr 1 Dk & Flush Dk	Norweg'n Christiania	286 385	137·2 / 23·1 / 12·9 B 14 f. F 14 f.	Christiania P. von der Lippe	1885		2Cy. 13" & 23"–18" 25 HP. AkersM.V., Christiania	Dampskibs-Selskabet "Hægholmen"
18	V	Hafis Wiezfeldt –88	Stl ScSmSr 1 Dk 2 B	German Harburg	1091 1689	270·0 / 36·0 / 18·7 P 35 f. F 30 f. B 12 f.	Newcastle W.s.Hawthorn Leslie & Co., Ld.	1886		3C. 20" 32" & 52"–36" 160 HP. R.&W.Hawthorn, Leslie&C., Nwc.	Renck & Hessenmüller
19		Hafnia Petersen 79-88	Stl ScSr 2 Dks	Danish Copenhagen	1039 1683	257·5 / 35·2 / 16·7 P 154 f. F 24 f.	Elsinore Helsingørs Jernskibs & Maskin Byggeri	1888		3Cy. $19\frac{1}{2}$" $31\frac{1}{2}$" & $51\frac{1}{2}$"–36" [160 HP. Helsingørs Jernskibs & Mak. Bg.	Dampskibs-Selskabet "Kjöbenhavn" (C.Hecksher)
20		Hai-phong	Stl Sc 2 Dks	French Marseilles	874 1493	278·5 / 33·1 / 22·4 H 133 f.	LaCiotat Messageries Mritms	1885		2C. $31\frac{1}{2}$" & $56\frac{1}{4}$"–$43\frac{1}{4}$" 250 HP. Messageries Mar. LaCiotat	Messageries Maritimes
21		Haidar Pacha	I PadSr	British London	257 342	151·6 / 22·0 / 9·6	London Samuda Bros.	1882		2Cy. 37"–36" 80 HP. Maudslay Sons & Field, Lond.	C.Morrison
22	✻	Hailoong Beach –88	Stl ScSr 1 Dk Spr Dk & Wb Frms	British HongKong	783 1253	230·0 / 33·2 / 13·2 B 35 f. F 46 f. 20·2	Leith Ramage&Fergus'n	1888	4 Blhds. DB 80 FPT 37 APT 20	3Cy. $22\frac{1}{2}$" 37" & 59"–39" [180 HP. Ramage & Ferguson, Leith	Douglas S.S.Co., Ld.
23	✻	Hailoong Hall	I ScSr 1 Dk & Awn Dk	Japanese Hiogo	277 446	155·0 / 24·2 / 10·2 16·5	Aberdeen Hall, Russell&Co.	1871	4 Blhds.	2Cy. 22" & 37"–30" 60 HP. Hall, Russell&Co., Aberdeen	H.E.Reynell & Co.
24	✻	Haiphong Harris	I ScSr 1 Dk & Spr Dk 3 B	British HongKong	1122 1743	255·3 / 35·2 / 24·4 B 46 f.	Newcastle W.Richardson & Co.	1885	4 Blhds. DB 175 FPT 38 APT 17	2Cy. 31" & 62"–42" 170 HP. W.Richardson&Co., Nwcstle	Douglas S.S.Co., Ld.
25	✻	Haitan Ashton –87	Stl ScBn 2 Dks 3 B	British HongKong	1183 1856	275·0 / 35·0 / 24·0 B 36 f. F 56 f.	Middlesbro' R.Dixon&Co.	1887	4 Blhds. CDB 227 FPT 51 APT 18	3C. $25\frac{1}{2}$" 42" & 67"–42" 350 HP. T.Richardson&Sons, Hrtlpl.	Douglas S.S.Co., Ld.
26	✻	Hajeen Newey 83-85	Stl ScSr 1 Dk & Spr Dk 3 B	British Liverpool	1821 2799	310·5 / 39·0 / 19·4 B 71 f. F 27 f. 26·9	Newcastle Sir W.G.Armstrong Mitchell&Co.	1885	6 Blhds. CDB 325 FPT 51 APT 12	3Cy 24" 39" & 64"–42" 325 HP. W.Rstd.Slipway Co., Nwestle.	Bedouin Stm. Nav.Co.
27		Hakodate Maru	W ScSr	Japanese Tokio	295 476	172·8 / 21·4 / 13·4	Yokosuka	1875		2Cy. 20" & 36"–22" 56 HP.	Sumihara Kakuhei
28	✻N	Hakon Adelsten Clausen –87	I ScSr 1 Dk & Spr Dk 3 B	Norweg'n Bergen	1049 1464	250·6 / 32·8 / 17·5 B 27 f. 24·1	Middlesbro' Backhouse, Dixon & Co.	1873	4 Blhds. MT 398	2Cy. 29" & 55"–33" 150 HP. N.E.MarineEng.Co., Ld., Sld.	P.Jebsen
29	✻D	Halcyon Smith 88-88	I ScSmSr 1 Dk 2 B	British Hartlepool	1021 1580	264·6 / 33·4 / 19·0 B 62 f. Q 107 f. F 27 f.	Middlesbro' R.Dixon&Co.	1877	4 Blhds. DB at Sch 212 f	2Cy. 31" & 58"–33" 140 HP. T.Richardson&Sons, Hrtlpl.	G.Horsley & Son
30											

✻ = Lloyd's Register; D = Nederlandsche Vereeniging; G = Germanischer Lloyd; I = Registro Italiano; R = American Record; A = Veritas Austro-Ungarico; H = Veritas Ellenico; N = Norske Veritas; V = Bureau Veritas.

Dans la 2me Colonne, B.T., avec les indications qui suivent, signifie la date du rapport de visite de la Carène par les experts du Board of Trade pour obtenir le certificat relatif au TRANSPORT DES PASSAGERS.

Dans la 3me Colonne, C = Composite (membrure en fer, bordé en bois); I = Fer; Stl = Acier; W = Bois.

Pour plus amples renseignements sur les navires classés par le Lloyd's Register, voir le Lloyd's Register Book et le Lloyd's Yacht Register.

No. in Book. Official No.	Registry in which classed	Ship's Name. Material, Rig, &c. Master, and Dates of Appointment to present Owner's Service and to Ship Ship's Late Name. No. of Decks, &c.	Flag. Port of Registry.	Tons. Nett. Gross.	Dimensions Length, Breadth, Depth Feet. Tenths.	Built. Where. By Whom.	When	No. of Bulkheads. Particulars of Water Ballast.	Engines. Particulars. Maker's Name.	Owners
31 94[illegible]	BT 4.88	Halcyon Stl Pad 1 Dk	British London	323 553	209·2 | 26·2 | [illegible] Q 7 f.	[illegible] J. Scott&Co.	1887		2Cy.33"& 61"–60" 250 HP. J.Scott&Co.,Kinghorn	General Ste[illegible]
32 95[illegible]	✻ BT [illegible]	Halifax Stl ScSr Hill -88	British Glasgow	958 1738	250·3 | 35·0 | 21·5	Glasgow Lon.&Glas.Co.,Ld.	1888		3C.30"48"&75"–48" 350 HP. Lon.&Glas.Co.,Ld.,Glasgow	J.E.Chipman
33 72840	✻	Hallamshire I ScSr Reenkon 87-87 1 Dk2 B	British W.Hartlepl.	876 1358	243·2 | 32·3 | 18·9 Q 34 f. B 61 f. F 27 f.	Hartlepool E.Withy&Co.	1876	4 Blkds. [illegible]	2Cy.30"& 57"–33" 130 H T.Richardson&Sons,Hrtlpl.	Steel,Young&[illegible]
34		Halland Stl TwnScSr Petersson 77-86 1 Dk & Awn Dk	Swedish Halmstad	297 430	158·8 | 23·5 | 10·7 H 21 f. 18·5	Gothenburg MotalaCo.	1884		4Cy.20"& 35"–25" 120 HP. MotalaCo.,Motala	Hallands Ang[illegible]tiebolag.
35 54885		Hallett I ScSr Keddell 75-86	British [illegible]	166 262	157·8 | 21·3 | 10·5	Stockton M.Pearse&Co.	1866		2Cy.28"–24" 50 HP. Fossick&Hackworth,Stktn.	Hewett & Co.
36 29968		Halley I ScBg Holt 84-87	British Liverpool	1028 1621	267·8 | 32·1 | 21·3 P 157 f. 28·2	Newcastle A.Leslie&Co.	1865		2Cy.35"& 56"–42" 120 HP. Fawcett,Preston&Co.,Lvpl.	Liverp'l,B[illegible] Plate Steam [illegible]
37 58[illegible]	✻	Halling (ex Emma Trautmann) I ScSr Nicholson -88 1 Dk2 B	British London	515 803	194·9 | 28·6 | 15·9 Q 60 f. B 44 f. F 25 f.	Hartlepool Withy,Alexander&	1871	4 Blkds. DBof 170	2Cy.25"&48"–36" 96 HP. T.Richardson&Sons,Hartpl.	Lambert Bro[illegible]
38		Halmstad I TwnScSr Hullen 73-84 1 Dk & Awn Dk	Swedish Halmstad	282 404	160·1 | 23·6 | 10·6 H 15 f.	Stockholm BergsundsMek.Ve[illegible]	1882		1Cy.19"& 33"–21" 120 HP. [illegible]gsundsMek.Verkst., Stockholm	HallandsAng[illegible]tiebolag.
39		Hamarona I Pad	French Marseilles	95 151	130·0 | 23·0 | 5·8	Liverpool Liverpool ForgeCo.	1882		42 HP. G.Forrester&Co.,Liverpool	Cie. de Séné[illegible] Côte Occident[illegible]
40	V	Hamburg I ScSr Albert 88-88 2 Dks	German Hamburg	1284 1644	270·3 | 33·8 | 23·6 P 36 f. H 77 f.	Newcastle C.Mitchell&Co.	1878		2Cy.34"& 62"–42" 200 HP. WigandSlipwayCo.,Nwcstle.	Hamburg-Süd[illegible] Dampfschiff[illegible]
41	✻	Hamburg I ScSr Krabbe 73-79 1 Dk2 B	German Hamburg	545 778	207·0 | 27·8 | 15·2 P 100 f. F 28 f.	Hull Humphreys&Pearson	1872	5 Blkds. [illegible]	[illegible]7"& 52"–32" 136 HP. [illegible] & Pearson,Hull	H.J.Perlbach[illegible]
42	✻ V	Hamburg I Sc3mSr Wolters 78-83 1 Dk	German Hamburg	430 574	189·6 | 25·6 | 15·0 Q 54 f. F 16 f.	Hamburg Reiherstieg Schiffs Werfte	1871	4 Blkds.	2Cy.21"& 38"–26" 80 HP. Reiherstieg[illegible] Hamburg	R.M.Sloman[illegible]
43	✻	Hamburg Stl ScSr Wetterström 75-86 1 Dk2 B	Swedish Gothenburg	433 757	199·0 | 27·1 | 13·0 P & B 132 f. F 48 f.	Sunderland W.Doxford&Sons	1888	5 Blkds. [illegible]	[illegible]	Svenska Llo[illegible]
44 114[illegible]		Hamilton W Pad	American New York	391 585	153·2 | 33·0 | 11·9	Brooklyn	1862			M.Bunker
45	V	Hammonia Stl Sc3mSr Hebich 68-87 4 Dks	German Hamburg	2563 3969	373·7 | 45·3 | 32·0	Glasgow J.&G.Thomson	1882		[illegible]	Hamburg-A[illegible] Packett.Akt.G[illegible]
46 828[illegible]	✻	Hampshire I Sc3mSr Kerrnish 2 Dks	British London	1700 2597	316·1 | 38·0 | 25·1 P 34 f. B 70 f. F 42 f.	Sunderland W.Doxford&Sons	1880	4 Blkds. [illegible]	2Cy.38"& 70"–48" 300 HP. W.Doxford&Sons,Sunderld.	G.Marshall&[illegible]
47 5129[illegible]	✻	Hampshire I ScSr Harland 87-87 1 Dk2 B	British Southampton	425 625	188·7 | 28·8 | 16·3 Q & F 24 f.	Sunderland Oswald&Co.	1866	3 Blkds. [illegible]	2Cy.24"&44"–30" 85 HP. W.B.Thompson,Dundee	J.L.Hill
48 943[illegible]	✻	Hampstead Stl ScSr Jones -87 1 Dk & Web Frames	British London	1448 2245	282·3 | 37·6 | 19·5 P 28 f. Q 88 f. B 112 f. F 27 f.	W.Hartlepool W.Gray&Co.	1887	5 Blkds. [illegible]	3C 22"36"& 59"–39" 225 HP. Blair&Co.,Ld.,Stockton	Britain S.S.C[illegible]
49 9437[illegible]	✻	Hampton Stl ScSr Page 86-88 1 Dk & Web Frames	British London	1383 2128	279·0 | 38·2 | 19·0 Q [illegible] P 26 f. B 114 f. F 26 f.	Sunderland J.Blumer&Co.	1888	5 Blkds. [illegible]	3C 21½"36"&59"–39" 200 HP. N.E.MarineCo.,Ld.,Sundld.	Commercial[illegible] Co., Ld.
50 77220	BT 5.88	Handa (ex Ana Castle) I ScSr Martin 87-87	British Glasgow	62 140	84·2 | 20·1 | 9·5 F 8 f.	Port Glasgow Blackwood&Gordon	1878		2Cy.16"& 28"–21" 35 HP. Blackwd&Gordn,P.Glasgw	D.M'Brayne
51 85142	✻ D	Handel I ScSr Matthews 1 Dk & Spr Dk3 B	British London	1382 2123	265·0 | 37·2 | 27·7	Sunderland Bartram,Haswell& Co.	1882	5 Blkds. [illegible]	2Cy.34"& 64"–42" 200 HP. J.Dickinson,Sunderland	J.Taylor
52 84135	BT 10.88	Handel I ScSr Braithwaite 81-82	British Liverpool	1294 1988	290·0 | 36·2 | 24·5	Newcastle A.Leslie&Co.	1881		2Cy.26"& 60"–45" 200 HP. R.Stephenson&[illegible]	Liverp'l,B[illegible] Plate Steam [illegible]
53 91878		Hangchow Stl ScSr Glasson -87	British [illegible]	999 1572	259·7 | 33·1 | 23·2 B 62 f. F 31 f.	Greenock Scott&Co.	1885		2Cy.29"& 58"–42" 150 HP. Scott&Co.,Greenock	China Nav.Co[illegible]
54		Hango I ScSr Andstén 1 Dk	Russian [illegible]	175	125·0 | 25·0 | 11·0	Motala MotalaMek.Werkst.	1863			F.Hultman
55 [illegible]516	✻ BT 11.88	Hankow I Sc3mSr Laws 2 Dks3 B	British London	2332 [illegible]	389·0 | 42·1 | 28·8 P 86 f. F 52 f.	Newcastle C.Mitchell&Co.	1873	4 Blkds.	2Cy.52"& 96"–51" 300 HP. R.&W.Hawthorn,Newcastle	W.Milburn&C[illegible]
56 68[illegible]		Hankow I Pad Sr	British London	2252 [illegible]	308·5 | 42·3 | 15·2	Glasgow A.&J.Inglis	1874		1Cy.72"–108" 400 HP. A.&J.Inglis,Glasgow	China Nav.Co[illegible]
57 84139	✻	Hannah Beckett W Sc Flat Atherton 1 Dk	British Liverpool	33 [illegible]	77·0 | 19·6 | 8·5	Northwich Bracegirdle	1881		2Cy.15"–16" 36 HP. W.E.Bates,Northwich	W.E.Davies
58	G V	Hannover I ScBg Gathemann 86-86 2 Dks Hur Dk & B	German Bremen	1933 [illegible]	311·0 | 39·4 | 31·5	Greenock Caird&Co.	1869		2Cy.38½"&73"–48" 360 HP. Caird&Co.,Greenock	Norddeutsche[illegible]
59 73416		Hanoi W ScSr 1 Dk	British Singapore	97 1[illegible]	115·3 | 17·2 | 8·0 F 31 f.	Hong Kong Amoo	1881		2Cy.15"& 30"–15" 44 HP. HongKong&WhampoaDk. Co.,Hong Kong	Khoo Tiong P[illegible]
60		Hans Jost Stl ScSr Ziesmer -88		1330		Flushing Vlissingsche Scheepsbouw Gesellschaft				J. Jost

✻ = Lloyd's Register; D = Nederlandsche Vereeniging; G = Germanischer Lloyd; I = Registro Italiano; R = American [illegible]
A = Veritas Austro-Ungarico; H = Veritas Ellenico; N = Norske Veritas; V = Bureau Ve[illegible]

In the 2nd Column, B.T., with figures following, signifies Date of report of bottom sighting by Board of Trade Surveyors as condition of PASSENGER [illegible]

In the 3rd Column, C = Composite (iron frame, wood planked); I = Iron; Stl = Steel; W = Wood.

Further particulars of Vessels classed by Lloyd's Register will be found in Lloyd's Register Book and Lloyd's Yacht Register.

Registre	Nom, Matériaux, Gréement, etc. / Capitaine et dates d'entrée au service des Armateurs actuels et à bord du navire / Ancien Nom, nombre de ponts, etc.	Pavillon. / Port d'Armement	Tonnage. Net. / Total	Dimensions. Longueur	Largeur	Creux	Construit. Où. / Par qui.	Quand.	Nombre des cloisons. / Détails sur le test d'eau.	Machines. Détails. / Nom du Constructeur.	Armateurs ou Directeurs-Gérants.
V	Hans Krohn 1 ScSr *Keller* 81- 1 *Dk & Spr Dk*	German Rostock	785 1238	232·6	31·1	14·0 21·1	Stettin Act.Ges. "Vulcan"	1882		2Cy.27½ & 51 - 31½ 126 HP. Act.Ges. "Vulcan", Stettin	Dampfer Gesellschaft von 1881 (M.Petersen)
DV	Hans & Kurt Stl & I ScSmSr *Brackmann* -88 1 *Dk*	German Hamburg	1528 2153	270·0	39·0 *P* 157 *f.* *F* 33 *f.*	21·2	Newcastle [illegible]	1887		3Cy.20½ 34 & 56 - 36 170 HP. [illegible] Ld., Newcastle	G.J.H.Siemers & Co.
❋	Hansa 1 ScSr *Kruger* 75-84 1 *Dk* 2 *B*	German Hamburg	553 736	200·4	29·5 *P & F* 71 *f.*	16·8	Stockton Richardson,Duck & Co.	1865	5 Blhds.	2Cy.42½"–27" 120 HP. T.Richardson & Sons, Hrtlpl.	H.J.Perlbach & Co.
	Hansa 1 ScSr *Lemmel* 82-82	German Hamburg	342 528	156·8	32·5	15·8	Glasgow J.&G.Thomson	1881		2Cy.24"& 44"–36" 100 HP. J.&G.Thomson, Glasgow	Hamburg-Amerik. Packetf. Akt.Ges.
G	Hansa 1 ScSr *Piersdorff* -80 1 *Dk* 2 *B*	German Lübeck	295 391	159·1	21·6	12·1	Stettin Act.Ges. "Vulcan"	1864		2Cy.19½ & 35½–24" 55 HP. RostockS.B.& E.Co., Rostock	Liban-Lübecker Dampfschiffahrts-Ges.
❋	Harald 1 ScSmSr *Bonde* 75-76 1 *Dk & Spr Dk* 3 *B*	Danish Copenhagen	972 1532	257·7	32·1	17·4	Copenhagen Burmeister & Wain	1875	6 Blhds. MT88 AT [illegible]	1Cy.24"& 48"–32" 150 HP. Burmeister & Wain, Cpnhgn.	Dampskibs-Selskab. "Carl"
❋N	Harald Haarfager 1 ScSr *Olsen* -89 2 *Dks* 3 *B*	Norweg'n Bergen	1531 2070	294·3	36·0 *F* 33 *f.*	26·4	Middlesbro' Backhouse, Dixon & Co.	1872	5 Blhds. [illegible]	2Cy.41"& 73"–39" 220 HP. N.E.Marine Eng.Co., Ld., Sld.	Et Interessentkab (Harloff & Boe, Managers)
N	Harald Haarfager 1 ScSr *Heyerdahl* 86-86 1 *Dk & Awn Dk*	Norweg'n Christiania	339 441	161·6	22·6	17·5	Drontheim Throndhjems M.Verkst	1871	DB [illegible]	1Cy.17"& 32"–22½" 80 HP. Throndhjems Mek.Verkst.	Christiania Kystdampskibs-Sel.(Jens Meinich & Co)
	Harald Haarfager W ScBk *Mæn* 86-86 2 *Dks*	Norweg'n Christiania	317 419	144·2	30·3 *P* 34 *f.*	16·2	Christiania M.C.Hansen	1872		2Cy.20"& 40"–24½" 60 HP. Nylands Mek.Verk, Chrst'nia	Act.Selsk. "Harald Haarfager"
❋	Harbinger 1 ScSr *Beer* 87-87 1 *Dk* 2 *B*	British Sunderland	1206 1800	271·0	36·2	20·7	Sunderland Sunderland S.B.Co., Ld.	1881	7 Blhds. DB [illegible]	2Cy.33"& 63"–39" 170 HP. N.E.Marine E.Co., Ld., Sndld.	"Harbinger" S.S.Co., Ld. (J.L.Browne)
	Hardangeren 1 ScSr 1 *Dk*	Norweg'n Bergen	171 230	144·4	19·7 *Q* 48 *f.* *F* 43 *f.*	9·6	Bergen Bergens Mek Verkst	1869		1Cy.15"& 28"–18" 50 HP. Bergens Mek.Verks, Bergen	Hardanger Söndhordlandske Dampskibsselskab
❋	Hardwick 1 ScSr *Williams* 88-88 1 *Dk* 2 *B*	British W.Hartlepl.	715 1122	236·0	31·6 *Q* 90 *f.* *F* 30 *f.*	13·7	Hartlepool E.Withy & Co.	1880	4 Blhds. [illegible]	2Cy.27"& 50"–33" 99 HP. T.Richardson & Sons, Hartpl.	R.Ropner & Co.
	Hare Stl ScSr *Neill* -86 2 *Dks*	British Glasgow	312 771	216·0	29·0 63 *f.* *B* 57 *f.* *F* 41 *f.*	14·7	Glasgow Barclay, Curle & Co., Ld.	1886		3C.24"38"& 58"–42" 216 HP. Barclay, Curle & Co., Ld., Glasg.	G.& J.Burns
❋	Harefield 1 ScSr *Harrison* 1 *Dk* 2 *B*	British W.Hartlepl.	589 924	213·5	29·1 *Q* 90 *f.* *F* 27 *f.*	16·1	W.Hartlepool [illegible]	1871	4 Blhds. DB [illegible]	2Cy.25½"& 51⅞"–33" 98 HP. Blair & Co., Ld., Stockton	W.Runciman
❋	Harima Maru 1 ScSr *Teeri* 1 *Dk & Awn Dk*	Japanese Tokio	447 721	172·5	26·2	13·0 20·0	Paisley H.McIntyre & Co.	1884	5 Blhds. FPT45t APT35t MT [illegible]	2Cy.22"& 41"–30" 72 HP. Fleming & Ferguson, Paisley	Nippon Yusen Kaisha
	Harlan 1 PadSr *Brown* 2 *Dks*	American New Orleans	707 1163	219·9	34·1	16·0	Wilmingtn, Del. Harlan & Hollingsworth Co	1866		1Cy.50"–132" 500 HP. The Harlan & Hollingsworth Co., Wilm, Del.	J.B.Richardson
❋	Harlaw 1 ScSr *Farquhar* 88-88 1 *Dk*	British Windsor, NS	267 451	165·1	24·5 *P* 38 *f.* *F* 31 *f.*	11·8	PortGlasgow Cunliffe & Dunlop	1881	4 Blhds. [illegible]	2Cy.23"& 42"–30" 71 HP. Cunliffe & Dunlop, PGlasgow	W.A.Black
❋	Harley 1 ScSmSr *Creamer* 1 *Dk*	British Liverpool	281 455	168·8	24·3 *P* 41 *f.* *B* 38 *f.* *F* 21 *f.*	11·6	PortGlasgow Russell & Co.	1880	4 Blhds. FPT55t	2Cy.24"& 44"–30" 85 HP. Walker, Henderson & Co, Gls.	G.C.Stewart
❋D	Harlsey 1 ScBu *Marquest* 82-83 1 *Dk* 2 *B*	British W.Hartlepl.	1078 1677	260·6	34·8 *B* 91 *f.* *Q* 100 *f.* *F* 37 *f.*	19·4	W.Hartlepool W.Gray & Co.	1879	4 Blhds. DB [illegible]	2Cy.33"& 61"–33" 150 HP. T.Richardson & Sons, Hartpl.	R.Ropner & Co.
❋V	Harmonie Stl 1 ScSr *Krutzfeldt* (ex Industrie 71) 1 *Dk*	German Cologne	200 335	149·6	24·9	10·6	Flushing Kon.Maats. "De Schelde"	1886	5 Blhds.	3Cy.16"16"& 26"–16" 70 HP. Koninklyke Maatschappy de Schelde, Flushing	Badische Schrauben Dampfschiffahrts Ges.
❋	Harold 1 ScSr *Bergwitz* 88-88 1 *Dk* 2 *B*	British Hull	807 1257	238·0	31·2 *P* 123 *f.* *F* 28 *f.*	16·9	Stockton Richrdsn, Duck & Co.	1877	5 Blhds. DB [illegible]	2Cy.27"& 51"–35" 99 HP. C.Holmes & Co., Hull	H.Briggs, Sons & Co.
❋	Harold 1 ScSr *Bedlington* 1 *Dk* 2 *B*	British Newcastle	682 1073	225·5	30·6 *P* 119 *f.* *F* 37 *f.*	17·4	Newcastle A.Leslie & Co.	1871	5 Blhds. DB [illegible]	2Cy.26"& 52"–30" 90 HP. R.& W.Hawthorn, Newcastl.	Charlton, McAllum & Co.
❋	Harold 1 ScSr *Hutton* 88-88	British Glasgow	536 833	190·3	27·1	21·4	Belfast Workman, Clark & Co., Ld.	1887	4 Blhds. DB [illegible] FPT [illegible]	3Cy17"27"& 44"–36" 98 HP. Hutson & Corbett, Glasgow	S.S.Harold Co., Ld. (Colvils, Lowden & Co.)
❋	Harold (ex Hampstead) 1 ScSr *Skantze* 86-86 1 *Dk* 2 *B*	Swedish Stockholm	1314 1688	257·5	34·2 *Q* [illegible] *B* 62 *f.* *F* 32 *f.* *H* 9 *t.*	19·5	Stockton M.Pearse & Co.	1878	4 Blhds. DB [illegible]	2Cy.32"& 60"–39" 140 HP. Blair & Co., Ld., Stockton	E.Liljewalch
❋	Harperly Stl ScSr *Freeman* -88 1 *Dk* 2 *B*	British London	1451 2233	290·0	39·1 [illegible] *B* 118 *f.* *F* 33 *f.*	20·4	South Shields J.Readhead & Co.	1888	6 Blhds. [illegible]	3Cy.22"36"& 60"–39" 250 HP. J.Readhead & Co., South Shields	W.Wright
❋	Harraton 1 ScSr *Burton* 77-77 1 *Dk* 2 *B*	British Sunderland	461 678	192·0	28·9 *Q & F* —	16·7	Sunderland J.Laing	1867	4 Blhds. DB [illegible]	2Cy.18"& 36"–26" 80 HP. G.Clark, Sunderland	H.T.Morton & Co.
	Harriet 1 ScSmSr *Robinson* 1 *Dk* 2 *B*	British Middlesbro'	381 504	199·4	25·1	14·7	N.Shields T.& W.Smith	1857	4 Blhds.	2Cy.34"–26" 70 HP. Hawks, Crawshay & Co, Gtshd.	Hogg & Henderson
	Harriet (ex Glenmore) 1 PadSc *Petersen* 89-89 1 *Dk*	Danish Kastrup	59 108	102·2	16·8	8·8	Stockton	1856		2Cy.24½"–34" 65 HP. Thompson & Woods, Nwcstl.	Oscar Petersen's Bjergnings Enterprise
❋	Harrington 1 ScSr *Kerrison* 87-87 1 *Dk* 2 *B*	British W.Hartlepl.	690 1072	230·0	31·2 *P* 123 *f.* *F* 33 *f.*	17·6	PortGlasgow W.Hamilton & Co.	1872	4 Blhds. DB [illegible]	2Cy.28"& 50"–33" 125 HP. Walker, Henderson & Co, Gls.	Hardy, Wilson & Co.
	Harrisburg 1 ScSmSr *Atherton* 73-87 1 *Dk* 2 *B*	American Philadelphia	910 1283	250·0	37·0 *F & B* —	20·6	Philadelphia Cramp & Sons	1874	4 Blhds. WB [illegible]	1Cy.42"–42" 166 HP. Cramp & Sons, Philadelphia	Philadelphia & Reading R.R.Co.

❋ = Lloyd's Register; D = Nederlandsche Vereeniging; G = Germanischer Lloyd; I = Registro Italiano; R = American Record;
A = Veritas Austro-Ungarico; H = Veritas Ellenico; N = Norske Veritas; V = Bureau Veritas.

Dans la 2me Colonne, B.T., avec les indications qui suivent, signifie la date du rapport de visite de la Carène par les experts du Board of Trade pour obtenir le certificat relatif au TRANSPORT DES PASSAGERS.

Dans la 3me Colonne, C = Composite (membrure en fer, bordé en bois); I = Fer; Stl = Acier; W = Bois.

Pour plus amples renseignements sur les navires classés par le Lloyd's Register, voir le Lloyd's Register Book et le Lloyd's Yacht Register.

No. in Book / Official No.	Registry in which classed	Ship's Name. Master, and Dates of Appointment to present Owner's Service and to Ship. Ship's Late Name.	Material, Rig, &c. No. of Decks, &c.	Flag / Port of Registry	Tons Net / Gross	Length	Breadth	Depth	Built Where / By Whom	When	No. of Bulkheads. Particulars of Water Ballast.	Engines. Particulars. Maker's Name.	Owners & Managers
91 / 87174	✱	Harrogate / Cutler 86-86	I S[illegible] / 1 Dk[illegible]	British	1[illegible]	27[illegible]	[illegible]	[illegible]	Sunderland / Short Bros.	1883	[illegible] Blkds.	2Cy.36"&67"-36" 2[illegible] HP / T.Richardson&Sons,Hartpl.	"Harrogate"[illegible]
92 / [illegible]	✱	Harrogate / Hoyland [illegible]	I S[illegible] / 1 Dk[illegible]	British	1[illegible]	247·5	35[illegible]	1[illegible]·7	W.Hartlepool / [illegible] Withy&Co.	1883	[illegible] Blkds.	2Cy.33"&62"-39" 180 HP / Blair&Co.Ld.Stockton	Gladstone&C[illegible]
93 / 94[illegible]	✱D	Harrow / Kerlan 18-88	[illegible] S[illegible] / 1 Dk & W.b.Frms	British	1702	300·0	38·6	2[illegible]	W.Hartlepool / W.Gray&Co.	1888	[illegible] Blkds.	3Cy.23"36½"&62"-39" [illegible] 220HP / Central Marine Eng.Co. W.Htpl.	Galbraith,P[illegible] & Co.
94	V	Harry / Svensson 84-84	I ScSr / 1 Dk	Swedish / Gothenburg	378	149·9	23·2	15·7	Gothenburg / Göteborgs Mek.Wek.	1884		2Cy.18"&3[illegible]"-2[illegible]" [illegible] / Göteborgs Mek.Werk.	Angf.Aktie-[illegible] "Vidar" [illegible]
95 / 94623		Harry Bohe	W S / 2 Dks	British / Ottawa	184	108·0	25·8	7·2	Ottawa, Ont. / P.H.Waters	1886		1Cy.14"-14" 40 HP / [illegible]	H.A.Harris
96 / 29114		Harry Clasper	I Pad	British / Liverpool	49	115·6	16·7	7·5	Newcastle	1861		2Cy.24"-36" 40 HP / Hawks,Crawshay&Co.Gtshd.	S.Davies
97 / 70129	✱	Hart / [illegible] 84-86	I ScSr / 1 Dk 2 [illegible]	British	889	247·0	32·1	18·[illegible]	W.Hartlepool	1878	[illegible] Blkds.	2Cy.29"&5[illegible]"-33" 120 [illegible] / T.Richardson&Sons,Hartpl.	J.Wood & Co.
98 / 87836	✱	Hart Fell / Jones 87-87	Stl ScSr / 1 Dk	British / Liverpool	139	139·3	21·6	[illegible]	Port Glasgow	1883	[illegible]	2Cy.1[illegible]"&31"-27" 50 HP / Dunsmuir&Jackson,Glasgow	G.Nelson & S[illegible]
99 / 86415	✱D	Hartburn / Bateics	I ScB[illegible] / 1 Dk 2 [illegible]	British / W.Hartlepl.	1245	275·5	35·7	19·[illegible]	W.Hartlepool	1882	[illegible] Blkds.	2Cy.34"&6[illegible]"-42" 200 HP / R.&W.Hawthorn,Newcastle	R.Ropner & C[illegible]
100 / 79983	✱	Harter / Townsend	I S[illegible] / 2 Dks 3 [illegible]	British	1126	280·5	34·8	24·[illegible]	[illegible]	1879	[illegible] Blkds.	2Cy.40"&72"-45" 300 HP / [illegible] Holmes&Co.,Hull	Shaw,Bushby[illegible]
101 / 89471	✱D	Hartington / Hall 81-81	I S[illegible] / 2 Dks 3 [illegible]	British / W.Hartlepl.	1554	281·0	37·2	23·[illegible]	Newcastle	1883	[illegible] Blkds.	2Cy.33"&61"-42" 200 HP / Palmers'Co.Ld.,Newcastle	Hudson Shipp[illegible] Ld.
102 / 81507	✱	Hartlepools (wel. kelm.) / Evans 84-85	I ScBo / 1 Dk 2 [illegible]	British / W.Hartlepl.	1131 / 1734	258·5	31·6	19·7	W.Hartlepool	1880	[illegible] Blkds.	2Cy.33"&61"-33" 150 HP / T.Richardson&Sons,Hartpl.	Hudson Shipp[illegible] Ld.
103 / 67351	✱	Harton / [illegible]	I S[illegible] / 1 Dk	British	326	171·0	28·1	13·7	Sunderland	1872	[illegible] Blkds.	[illegible] / [illegible] Sndld.	Harton Coal C[illegible]
104 / [illegible]	✱	Hartside / Dryden 85-85	I S[illegible] / 1 Dk	British	440	181·9	29·2	13·[illegible]	Sunderland	1883	[illegible] Blkds.	2Cy.25"&48"-36" 95 HP / R.&W.Hawthorn,Newcastl.	Charlton,M[illegible] & Co.
105 / 81504	✱	Hartville / Petersen 85-88	I ScBo / 1 Dk 2 [illegible]	British	1141	258·0	31·6	19·7	W.Hartlepool / W.Gray&Co.	1879	[illegible] Blkds.	2Cy.33"&61"-33" 150 HP / T.Richardson&Sons,Hrtpl.	Maclean,Dou[illegible]
106 / 64524	✱	Harvest / Walker 86-86	I S[illegible] / 1 Dk 2 [illegible]	British	853	215·0	33·1	16·0	W.Hartlepool	1881	[illegible] Blkds.	2Cy.30"&57"-33" 130 HP / T.Richardson&Sons,Hrtpl.	English & Co.
107 / 28331		Harvest Queen / (ex Petro Bey) Hautrods 78-78	I ScSmSr	British / Sunderland	344	205·3	26·2	13·[illegible]	W.Hartlepool / Pile,Spence&Co.	1860		2Cy.27"&46"-30" 80 HP / J.Dickinson,Sunderland	Marquis of L[illegible] derry
108 / 50653		Harwich / Chilver 85-87	I TwnScSr / 2 Dks	British / Harwich	390	215·0	27·0	17·[illegible]	London / Simpson&Co.	1864	[illegible]	1Cy.22"&42"-27" 150 HP / Earle's Co.Ld.,Hull	Great Eastern [illegible] way Co.
109 / 74437		Harwich / Keeble 76-77	W Pad	British / Harwich	37	100·6	18·9	10·4	Harwich / Vaux	1877		1Cy.27"-16" [illegible] HP	J.H.Vaux
110 / 84520	✱	Hasland / Berry 85-85	I ScBo / 1 Dk 2 [illegible]	British / W.Hartlepl.	1134	258·6	31·6	19·5	W.Hartlepool / W.Gray&Co.	1881	[illegible] Blkds.	2Cy.33"&61"-33" 150 HP / T.Richardson&Sons,Hrtpl.	Maclean,Dou[illegible]
111		Hassan Pasha / (ex Cornwall) Hassan Bey	I S[illegible] / 3 Dks	Turkish	1295	280·0	35·1	24·5	Stockton / Richardson,Duck&Co.	1873	[illegible] Blkds.	2Cy.36"&72"-42" 200 HP / Blair&Co.,Ld.,Stockton	Idarei-Massou[illegible] (Ottoman [illegible])
112	GV	Hassia / Plasse 77-83	I ScB[illegible] / 2 Dks	German / Bremen	1405	269·0	35·8	25·7	Vegesack / H.F.Ulrichs	1883		2Cy.33"&60"-36" 200 HP / Hannover Mas.Act.Ges.Han.	Gildemeister & [illegible]
113 / 50182	✱	Hastings (Collision) / Whitnop -79	I Sc3mSr / 1 Dk 2 [illegible]	British / Newcastle	473	199·3	28·2	16·0	North Shields / T.&W.Smith	1864	[illegible] Blkds.	2Cy.38"-30" 96 HP / Palmers'Co.,Newcastle	J.O.Scott
114 / 53322		Hasty	I Pad / 1 Dk	British / Bombay	64	87·0	16·6	9·0	Glasgow	1867		[illegible] 27"-40" 40 HP / R.Napier&Sons,Glasgow	British India [illegible] Nav.Co.Ld.
115 / 43620	✱	Haswell / Holley 83-85	I Sc3mSr / 1 Dk 2 [illegible]	British / London	454	196·0	29·0	16·8	Sunderland / J.Laing	1861	[illegible] Blkds.	3C.16½"26"&43"-36" 99 HP / Wallsend Slipway Co.,N'castle	J.Fenwick & S[illegible]
116 / 86937	✱	Hatfield / Surtees	I ScB[illegible] / 1 Dk 2 [illegible]	British	1194	264·2	36·0	19·6	W.Hartlepool / W.Gray&Co.	1882	[illegible] Blkds.	2Cy.33"&6[illegible]"-39" 180 HP / Blair&Co.,Ld.,Stockton	The Hudson S[illegible] Co.Ld.
117 / 78420	✱D	Hathersage / Dennis 85-85	I ScB[illegible] / 1 Dk 2 [illegible]	British	1059	258·0	31·8	19·3	W.Hartlepool / W.Gray&Co.	1878	[illegible] Blkds.	2Cy.33"&61"-33" 150 HP / T.Richardson&Sons,Hartpl.	J.Merryweather
118	R	Hattie Fullerton	[illegible]	Brazilian / Para	155	100·0	18·6	[illegible]	Wilmington / The Pusey&Jones Co.	1884		2Cy.15"&33"-28" / The Pusey&Jones Co. Wilmington	Sears & Co.
119		Haukelid	I S[illegible] / 1 Dk	Norweg'n / Stavanger	69	107·8	17·5	8·1	Bergen / Bergens Mek.Verkst.	1869		2Cy.18"&31½"-2[illegible]" [illegible] HP / Bergens Mek.Verkst.	Stavangerske [illegible]
120 / 84479	✱	Hauroto / Kennedy	Stl S[illegible] / 2 Dks 2 [illegible]	British / Dunedin	1276	281·8	36·3	22·6	Dumbarton / W.Denny&Bros.	1882	[illegible]	2Cy.[illegible]"&[illegible]"-45" [illegible] HP / Denny&Co.,Dumbarton	Union S.S.Co.of [illegible] Zealand,Ld.

✱ = Lloyd's Register; D = Nederlandsche Vereeniging; G = Germanischer Lloyd; I = Registro Italiano; R = American R[illegible]; A = Veritas Austro-Ungarico; H = Veritas Ellenico; N = Norske Veritas; V = Bureau Verit[illegible]

In the 2nd Column, B.T., with figures following, signifies Date of report of bottom sighting by Board of Trade Surveyors as condition of PASSENGER C[illegible]

In the 3rd Column, C=Composite (iron frame, wood planked); I=Iron; Stl=Steel; W=Wood.

Further particulars of Vessels classed by Lloyd's Register will be found in Lloyd's Register Book and Lloyd's Yacht Register.

1	2	3 Nom, Matériaux, Gréement, etc. Capitaine et dates d'entrée au service des Armateurs actuels et à bord du navire. Ancien Nom, nombre de ponts, etc.		4 Pavillon. Port d'Armement	5 Tonnage. Net Total	6 Dimensions. Longueur	Largeur	Creux	7 Construit. Où. Par qui.	8 Quand	9 Nombre des cloisons. Détails sur le test d'eau.	10 Machines. Détails. Nom du Constructeur.	11 Armateurs ou Directeurs-Gérants.
		Havana	1 ScSr	Peruvian	1400	240·0	35·0	..	Philadelphia Neafie&Levy	1863		1Cy.60"–60" Neafie&Levy,Philadelphia	Peruvian Government
	❊	Haverstoe Stanburn –87	Stl ScSr 1 Dk2 B	British Grimsby	1372 2105	275·0	39·5	18·6	Sunderland W.Doxford&Sons	1887	4 Bhds. CDB38ft	3C.21"35"& 57"–39" 160 HP. W.Doxford&Sons,Sundrld	Bennetts & Co.
	❊	Haverton Rowe –87	1 ScSr 1 Dk&Spr Dk3 B	British London	1645 2531	300·0	37·3	27·5	Newcastle TyneI.S.B.Co.,Ld.	1883	5 Bhds. DBaf&c63ft FPT33t APT28t	2Cy.37"& 76"–45" 300 HP. R.& W.Hawthorn,Newcastle.	W.Milburn & Co.
		Havilah	1 Sc3mSr 2 Dks	British Sydney	166 301	151·0	22·0	13·4	Newcastle	1853		2Cy. 55 HP.	Mount Kembla Coal & Oil Co.
	V	Havre Laporte 82-82	1 Sc1mSr 2 Dks&Spr Dk	French Havre	2663 3417	378·2	40·0	21·7	Havre Forges&Cde laMédit	1882		1Cy.41"&71"–43½" 400 HP. Forges&C.de laMédit.Havre	Cie.Commerciale de Transports à Vap.Française
	❊D	Hawarden Wilson 83-83	1 ScBg 1 Dk2 B	British W. Hartlepl.	1167 1816	263·5	35·8	19·5	W.Hartlepool W.Gray&Co.	1882	4 Bhds. DBaf&c337t	2Cy.34"& 63"–36" 180 HP. T.Richardson&Sons,Hartpl.	The Hudson Shipping Co.,Ld.
	❊	Hawarden Castle Harrison 76-88	1 ScBg 3 Dks	British London	2669 4134	380·6	48·2	31·4	Glasgow J.Elder&Co.	1883	7 Bhds. MT290t DT152t	2Cy.50"& 90"–60" 600 HP. J.Elder&Co.,Glasgow	D.Currie & Co.
	❊	Hawea (WRECKED) Heasby	1 ScSr 2 Dks	British Dunedin	462 721	215·7	27·3	14·1	Dumbarton W.Denny&Bros.	1875	4 Bhds.	2Cy.35"& 66"–36" 160 HP. Denny&Co.,Dumbarton	Union S.S.Co.of New Zealand,Ld.
	❊	Hawk Whittle	1 ScSr 2 Dks	British London	348 648	200·2	27·4	14·1	Dundee Gourlay Bros.	1876	4 Bhds. DB63t APT11t	2Cy.34"& 56"–30" 140 HP. Gourlay Bros.&Co.,Dundee	General Steam Nav.Co.
	❊	Hawk (ex Lady Alma)	1 ScSrYt	British London	146 289	167·0	23·7	12·8	Leith Ramage&Ferguson	1884	4 Bhds.	2Cy.20"&38"–24" 75 HP. Ramage&Ferguson,Leith	The Admiralty
	❊	Hawk	Stl ScTug 1 Dk	British London	10 51	70·4	15·0	8·1	Penarth PenarthS.B.&c.Co.	1888	4 Bhds.	2Cy.13½"&29"–18" 50 HP. J.Stewart & Son,London	T.A.Walker
	❊	Hawk	W ScSr 1 Dk	Spanish SanSebastian	24 83	87·2	18·1	9·5	Leith A.G.Gifford&Co.	1882		2Cy.13"&24"–18" 25 HP. Walker,Henderson&Co,Gls.	Mercader & hijos
		Hawkesbury Shakeshaft –86	W ScSr	British Sydney	80 117	92·0	20·0	7·4	Botany,N.S.W. G.Dent,Sen.	1886		2Cy.11"& 22"–17" 32 HP. J.Pollock,London	The Hawkesbury Stm. Nav.Co.
	❊	Hawkhurst Robertson 88-88	Stl ScSr 1 Dk&WebFrms	British London	1538 2340	297·2	40·1	19·0	Stockton Richardson,Duck&Co.	1888	5 Bhds. CDB55ft	3Cy.21½"35"&59"–39" T. Richardson & Sons, Hartlepool [200 HP.	Hawkhurst Steamship Co.,Ld. (Carlisle&Co.)
	❊D	Hawthorn Hodge 88-88	1 ScSr 1 Dk2 B	British Grangemth.	751 1164	224·6	32·0	16·2	Sunderland W.Doxford&Sons	1879	4 Bhds.	2Cy.28"& 54"–36" 125 HP. W.Doxford&Sons,Sundlnd.	Crawford & Co.
	❊	Hawthorns Allen 83-83	1 Sc3mSr 1 Dk2 B	British London	473 752	199·0	28·0	17·4	Newcastle Palmer Bros.&Co	1861	4 Bhds.	2Cy.25"& 48"–30" 95 HP. Cmp.'73,Vict'riaDck.C.,Lon.	J.Fenwick & Son
	❊	Hay Green Blacklaw 83-83	1 ScSr 1 Dk2 B	British W. Hartlepl.	1209 1869	264·2	35·6	19·9	W.Hartlepool W.Gray&Co.	1883	4 Bhds. DBaf&c355t	2Cy.33"& 62"–39" 170 HP. Blair&Co.,Ld.,Stockton	J.H.Murrell & Co.
	❊	Hayburn (ex Shap Fell) Fisher 86-88	1 ScSr 1 Dk	British Liverpool	149 351	155·7	24·1	10·8	Campbeltown Campbltwn.S.B.Co.	1882	4 Bhds. FPT20t APT20t	2Cy.20"& 38"–27" 65 HP. Dunsmuir&Jackson,Glasg.	G.Nelson & Sons
	❊	Hayle Gordon 68-68	1 ScSr 1 Dk2 B	British Aberdeen	237 373	147·1	24·7	13·8	Hayle Harvey&Co.	1867	4 Bhds. DBaft 44ft	2Cy.20"&37"–24" 60 HP. Hall,Russell&Co.,Aberdeen	Adam & Co.
		Haytian Gawrd –83	1 ScBg	British Liverpool	1490 2336	347·5	35·1	25·7	Sunderland J.Laing	1875		1Cy.28"& 56"–42" 300 HP. G.Forrester&Co.,Liverpool	West India & Pacific Steam Ship Co.,Ld.
	RV	Haytien Republic Compton	W ScBu 2 Dks	American Boston	780 1089	191·0	36·0	29·0	Bath,Me. NewEnglandS.B.Co.	1885		2Cy.22"& 44"–36" Goss MarineIronWorks,BathMe.	
	❊	Heather Bell	1 Sc3mSrYt 2 Dks	British Glasgow	140 205	154·0	20·0	13·1	Glasgow T.B.Seath&Co.	1881	5 Bhds.	2Cy.17"& 34"–26" 60 HP. A.Campbell&Sons,Glasgow	J.E.Chandler
		Heather Bell	1 PaiSk 1 Dk	British Portsmouth	126 271	207·7	21·0	8·8	PortGlasgow Blackwood&Gordn.	1871		2Cy.40"–66" 150 HP. Blackwood&Gordon,Pt.Gls.	London & S.W.& L.B.& S.C.Railway Cos.
		Heather Bell Pennant 82-85	1 Pad	British Liverpool	91 205	159·8	21·2	9·0	Liverpool T.Vernon&Sons	1865		2Cy.36"–42" 80 HP. Jones&Sons,Liverpool	Wallasey Local Board
		Heather Belle	W PadSp 1 Dk	British Charlottetown (P.E.I.)	189 282	120·0	34·0	7·8	Mt.Stewart JamesWhite	1883		1Cy.30"–60" 75 HP. Todd&McGregor,Glasgow	Inland Steam Nav.Co.
	❊	Heathfield M'Fee 87-87	Stl ScSr 1 Dk&WebFrms	British London	1398 2140	283·3	37·0	19·1	W.Hartlepool E.Withy&Co.	1887	5 Bhds. CDB129t APT36t	3C.21½"35"&58"39" 190 HP. CentralMarineE.Co.,W.Hartlepl.	S.S."Heathfield" Co., Ld. (F.Woods)
	❊D	Heathmore Jago 87-88	Stl ScSr 1 Dk2 B	British Liverpool	1218 1864	265·2	37·1	18·7	Kinghorn J.Key&Sons	1883	6 Bhds. DBaf&c197t FPT53t APT25t MT300t	2Cy.30"& 60"–39" 160 HP. J.Key&Sons,Kirkcaldy	S.S."Heathmore"Co., Ld.(W.Johnston & Co.)
		Heathpool Tebbutt 85-85	1 ScSr 1 Dk&WebFrames	British Sunderland	620 974	210·0	32·3	14·6	Sunderland SundrldS.B.Co.,Ld.	1885	5 Bhds. DBaf&c	2Cy.28"& 54"–33" 110 HP. N.E.MarineCo.,Ld.,Sundrld	H.T.Morton & Co.
		Hebe	1 ScSp	British Liverpool	346 545	190·0	30·6	11·5	Greenock Scott&Co.	1885		2Cy.21"&38"–30" 90 HP. Scott&Co.,Greenock	Ocean Steam Ship Co.
		Hebe	W ScSrYt	British Cowes	117 173	126·6	23·6	12·8	Cowes J.S.White	1867		40 HP.	S.G.Smith

❊=Lloyd's Register; D=Nederlandsche Vereeniging; G=Germanischer Lloyd; I=Registro Italiano; R=American Record;
⚓=Veritas Austro-Ungarico; H=Veritas Ellenico; N=Norske Veritas; V=Bureau Veritas.

Dans la 2me Colonne, B.T., avec les indications qui suivent, signifie la date du rapport de visite de la Carène par les experts du Board of Trade pour obtenir le certificat relatif au TRANSPORT DES PASSAGERS.

Dans la 3me Colonne, C=Composite (membrure en fer, bordé en bois); I=Fer; Stl=Acier; W=Bois.

Pour plus amples renseignements sur les navires classés par le Lloyd's Register, voir le Lloyd's Register Book et le Lloyd's Yacht Register.

No. / Official No.	Registry in which classed	Ship's Name. Material, Rig, &c. / Master, and Dates of Appointment to present Owner's Service and to Ship. / Ship's Late Name. No. of Decks, &c.	Flag / Port of Registry.	Tons Net / Gross	Length	Breadth	Depth	Built Where / By Whom	When	No. of Bulkheads / Particulars of Water Ballast	Engines / Particulars / Maker's Name.	Owners
151 [illegible]		Hebe W ScSrYt	British Liverpool	165 [illegible]	145·6	21·7	9·7	Glasgow J.&G.Thomson	1889		2Cy.20"&[illegible]"–24" 70 HP J.&G.Thomson,Glasgow	Mrs [illegible]
152 86534	✻	Hebe [illegible] Green 86-86 I ScSr 1 Dk	British Liverpool	68 [illegible]	101·5	18·1	9·2	Belfast M'Ilwaine&Lewis	1882	3 Blhds. [illegible]	2Cy.14"& 26"–21" 30 HP M'Ilwaine&Lewis,Belfast	W.Rowland
153 117		Hebe Milis 73-87 I ScSr 1 Dk	Austrian Trieste	155 246	148·0	18·7	11·5	Trieste Austro-Ungaric Lloyd	1877		2Cy.20"&37½"–24" 50 HP ArsenaleLloyd,Trieste	Lloyd Austro[illegible]rico
154	✻	Hebe [illegible] Strubberg -88 [illegible] I PadSr 1 Dk	Danish Copenhagen	124 227	166·5	20·2	8·7 Q 8f. H 19f.	Delfshaven Maatschpy.de Maas	1884	4 Blhds.	2Cy.25"&44"–36" 70 HP Maats.deMaas,Delfshaven	Det Forened[illegible]skibsselskab
155		Hebe Nystrom 88-88 Stl ScSr 2 Dks	Russian Abo	170	120·0	19·6	9·3 P 85f. F 20f.	Abo W.Crichton&Co.	1888	4 Blhds. [illegible]	3C.12"17½"&26"–21" 80 HP W.Crichton&Co.,Abo	Abo LänsKu[illegible] Aktiebolag
156 [illegible]	✻ BT12.88	Hebridean Campbell I ScSr 2 Dks	British Glasgow	127 314	139·8	23·6	12·0 [illegible] B43f. F 36f.	Glasgow T.B.Seath&Co.	1881	4 Blhds. AFT[illegible]	2Cy.22"&40"–30" 70 HP W.King&Co.,Glasgow	J.M'Callum &[illegible]
157 91238		Hecate I ScSr	British Liverpool	607 868	237·4	34·2	15·4 P 39f. F 38f.	Liverpool W.H.Potter&Sons	1885		2Cy.23"& 46"–36" 88 HP Fawcett,Preston&Co.,Lvrpl	Ocean Steam[illegible]
158	G	Hecht Wurfmann Stl PadSp 1 Dk	German Bremen	158 243	180·0	20·0	9·0	Bremen Act.-Ges."Weser"	1885		2Cy.33½"&57½"–37½"	Norddeutsche[illegible]
159		Hecla I Sc 3 Dks	British	1879 3348	392·8	39·0	28·7 F 76f.	Belfast Harland&Wolff	1878	7 Blhds.	4Cy.28"& 60"–54" 300 HP Forrester&Co.,Liverpool	The Admiralty
160 67530	✻	Hecla Ternquist 83-86 I ScSr 1 Dk2 H	British W.Hartlepl.	749 1185	235·8	31·0	16·9 Q 80f. F 30f.	W.Hartlepool W.Gray&Co.	1874	4 Blhds. [illegible]	3C17"28"&46½"–33" 110 HP Trpld'87 Blair&Co.Ld.Stcktn	Heskind & [illegible]
161 78729	BT8.88	Hector Ball 84-84 I ScSr 2 Dks3 H	British Liverpool	1590 2040	310·0	34·2	25·3	Newcastle A.Leslie&Co.	1877		2Cy.37"& 58"–45" 250 HP R.Stephenson&Co.,Nwcstle	Ocean Steam S[illegible]
162 44550	✻D	Hector Geddes 85-89 I Sc3mSr 3 Dks	British Sunderland	1280 1981	284·4	34·8	27·8	Sunderland J.Laing	1863	4 Blhds. AFT29 MT39[illegible]	2Cy.32"& 62"–42" 200 HP J.Dickinson,Sunderland	W.H.Loverid[illegible]
163 59082		Hector White 79-79 W ScBk 1 Dk	British S.John'sNFL.	291 427	151·2	27·1	16·6 F 22f.	Dundee A.Stephen&Sons	1870		2Cy. 70 HP	AvalonSteam[illegible] Ld.
164 81977		Hector W PadSp	British So'thShields	15 101	94·3	18·3	9·9	SouthShields J.R.Maxwell	1882		1Cy.34"–51" 38 HP Baird&Barnsley,N.Shields	Miss G.White
165		Hector I PadTug	Spanish Bilbao	35 168	95·0	18·6	9·6	Cardiff	1871		30 HP	F.Carranza
166 86152		Hecuba I ScSr	British Liverpool	590 918	233·4	32·6	15·8 F 29f.	Greenock Scott&Co.	1882		2Cy.23"& 46"–36" 98 HP Scott&Co.,Greenock	Ocean Steam S[illegible]
167 90306	✻	Hedworth Todd 79-84 I ScSr 1 Dk&Wch Frames	British Sunderland	726 1149	225·0	33·5	16·2 Q123f. H46f. F26f.	Sunderland S.P.Austin&Son	1884	5 Blhds. W[illegible]	2Cy.28½"& 53"–33" 120 HP Blair&Co.,Ld.,Stockton	H.T.Morton &[illegible]
168 684		Heian Maru W ScSr	Japanese [illegible]	299 482	168·5	22·8	17·2	Hiogo SuzukiSeijiro	1881		45 HP	Osaka Shosen [illegible]
169		Heimdal Mortensen 73-85 I ScSr 1 Dk	Danish Rønne	136 231	137·6	22·0	9·7 Q45f. B11f. F22f. H13f.	Copenhagen Burmeister&Wain	1873		1Cy.17"& 34"–20" 80 HP Burmeister&Wain,Cpnhgn.	Dampsk.Selsk[illegible] Bornholm af18[illegible]
170	DN	Heimdal Krohn I ScSr 1 Dk2 H	Norweg'n Bergen	431 571	169·9	25·2	12·9 P 92f. F 31f.	Bergen BergensMekVerkst	1882	5 Blhds. AFT13 MT68 FFT11	[illegible] 80 HP [illegible]	M.J.Schjeld[illegible]
171	N	Heimdal Mohn W Sc3mSr 1 Dk2 H	Norweg'n Kragerö	281 388	151·6	26·3	15·3 P 38f.	Kragerö	1872		2Cy.14"& 33½"–24½" 45 HP NylandsVerkst,Christiania	Johan Dahl
172		Heimdal Herig -88 Stl ScSr 1 Dk	Norweg'n Christiansand	69 112	83·4	16·1	10·4 B 91f.	Christiansand	1888			Et Interessen[illegible]
173 216		Heion Maru W ScSr	Japanese [illegible]	122 197	116·0	16·2	13·7	Hiogo KosakuBunKichi	1879	[illegible]	16 HP	Osaka Shosen [illegible]
174	V	Hekla Thomsen 81-85 I Sc3mSr 3 Dks	Danish Copenhagen	2113 3258	333·2	41·6	22·0 P35f. B63f. F46f. 29·9	Greenock Scott&Co.	1884	WT[illegible] DB[illegible] FPT	2Cy.46"& 84"–51" 450 HP Scott&Co.,Greenock	Dampskibs[illegible] "Thingvalla"[illegible]
175		Hekla W ScBk 1 Dk2 H	Norweg'n Tönsberg	239 375	139·6	28·9	15·6 Q 33f.	Drammen	1872		1Cy.24"–24½" 35 HP NylandsVerks,Christiania	N.Bugge
176 50715		Helen W Pad 1 Dk	British Quebec	85 135	120·0	19·5	8·0	ThreeRivers	1864		2Cy. 60 HP	St.Lawrence [illegible] Nav.Co.
177 28035		Helen I Pad	British Liverpool	10 154	140·0	20·0	8·3	Liverpool W.C.Miller	1860		2Cy.38"–42" 100 HP Fawcett,Preston&Co.,Livpl	Manchester [illegible] Co.
178 81468	✻D	Helen Newton Rawson -88 I ScSr 1 Dk2 H	British Hull	828 1288	240·0	32·0	17·0 P 149f. F 28f.	Sunderland R.Thompson,jun.	1880	4 Blhds. [illegible]	2Cy.28"& 54"–33" 115 HP Blair&Co.,Ld.,Stockton	Newton Bros [illegible]
179 86303	✻	Helen Nicoll Chudleigh I ScSr 1 Dk	British Sydney	246 384	157·0	22·0	10·3 Q 63f. B 31f. F 16f.	Dundee Gourlay,Bros.&Co.	1882	4 Blhds. [illegible]	2Cy.21"& 36"–24" 65 HP GourlayBros.&Co.,Dundee	John See
180 87193	✻	Helen Otto Oliver 77-[illegible] I ScSr 1 Dk2 H	British London	976 1518	249·0	36·0	17·6 Q 26f. B 52f. F 26f.	SouthShields J.Readhead&Co.	1883	4 Blhds. [illegible]	2Cy.29"& 54"–36" 99 HP J.Readhead&Co.,S.Shields	G.Otto

✻ = Lloyd's Register; D = Nederlandsche Vereeniging; G = Germanischer Lloyd; I = Registro Italiano; R = American [illegible]
A = Veritas Austro-Ungarico; H = Veritas Ellenico; N = Norske Veritas; V = Bureau Ve[illegible]

In the 2nd Column, B.T., with figures following, signifies Date of report of bottom sighting by Board of Trade Surveyors [illegible] condition of PASSENGER CE[illegible]

In the 3rd Column, C=Composite (iron frame, wood planked); I=Iron; Stl=Steel; W=Wood.

Further particulars of Vessels classed by Lloyd's Register will be found in Lloyd's Register Book and Lloyd's Yacht Register.

1	2	3 Nom, Matériaux, Gréement, etc. Capitaine et dates d'entrée au service des Armateurs actuels et à bord du navire. Ancien Nom, nombre de ponts, etc.	4 Pavillon. Port d'Armement	5 Tonnage Net. Total	6 Dimensions. Longueur	Largeur	Creux	7 Construit. Où. Par qui.	8 Quand	9 Nombre des cloisons. Détails sur le lest d'eau.	10 Machines. Détails. Nom du Constructeur.	11 Armateurs ou Directeurs-Gérants.
181	✻	Helena I ScSr Gill 88-88 1 Dk2 B	British	803 1243	240·0 Q 97 f.	32·1 F 26 f.	17·6 B 49 f.	South Shields J.Readhead&Co.	1878	4 Blhds. DB 96 f.	2Cy.27½"&52"–33" 110 HP. J.Readhead&Co.,S.Shields	R.Harrowing
182	V	Helene St ScSr Petersen 88-88 1 Dk	German Flensburg	573 786	200·0 F 26 f.	30·5 B 44 f.	15·7 Q 52 f.	Flensburg Flensburger Schiffsbau-Gesellschaft	1888	DB f 120	3C.14"25"&40½"–27" 90 HP. Flensb.Schiffs.Gesell.Flensburg	H.Schmidt
183	G	Helene I ScSr Lehmkuhl 85-85 1 Dk	German Kiel	151 198	96·1	21·0	12·4	Kiel G.Howaldt	1881		2Cy.15"& 21"–14" 25 HP. Howaldt Bros.,Kiel	Sartori & Berger
184	✻	Helene Rickmers St ScSr 1 Dk Sp Dk,& Wb Frms	German Bremerhaven	2000 3100	318·0 P 12 f.	42·7 B 80 f.	26·3 F 32 f.	Greenock Russell&Co.	1889	6 Blhds. DB	3Cy.24"39"&64"–45" Dunsmuir&Jackson,Glasgow	R.C.Rickmers
185	✻V	Helene Sauber Stl Sc Lemmel -88 1 Dk (Sh) 2 B	German Hamburg	731 988	209·4 Q 102 f.	31·1 F 23 f.	14·3	Flensburg Flensburg S.B.Co.	1888	4 Blhds. CDB	3C.19½"28"&51"–33" 160 HP. Flensburg S.B.Co.Flensburg	Sauber Gebruder
186	N	Helga W ScSr Sörensen -88 1 Dk	Norweg'n Tromsö	115 160	97·2	22·2 P 23 f.	10·3	Carlstad	1875		2Cy.16¼"&25"–17½" 30 HP. Christinehamns Mek. Vrkst.	M.W.Holmboe & Sön
187		Helga I ScSr Kullman -86 1 Dk	Swedish Thurö	129 153	112·8 B 25 f.	17·7 Len. '86.	9·5	Sjöberg G.F.Cawallin	1883		2Cy.11"& 19"–18" 25 HP. G.F.Cawallin,Sjöberg	G.F.Cawallin
188	✻	Helge I Sc3mSr Pii -86 1 Dk	Danish Copenhagen	529 849	205·4 Q 58 f.	28·1 F 31 f.	15·2 P 77 f.	Renfrew Hudson,Coulborn&Co.	1869	5 Blhds. AT 219	4C.21⅞"&42⅞"–36" 110 HP. Hudson,Coulborn&Co.,Rfw.	Dampskibs-Selskab. "Danmark"
189	✻	Heliades Carr 84-86 I ScBn 1 Dk & Spr Dk 2 B & Wb Frms	British Liverpool	1922 2965	320·0	40·2 B 68 f.	19·4 27·3	Stockton Richardson,Duck & Co.	1885	5 Blhds. DB 202, AT 111, MT 102	3C.24"40"&64"–45" 350 HP. T.Richardson&Ss.W.Hartpl.	R.P.Houston
190	✻	Helios I Sc3mSr Johnson -87 2 Dks 3 B	British Hull	1249 1911	280·0	35·1	26·1	Stockton Richardson,Duck & Co.	1873	5 Blhds. DT 101	2Cy.32"& 62"–36" 160 HP. C.D.Holmes&Co.,Hull	H.Briggs,Sons & Co.
191		Helios I ScBk Martinolich 57-87 2 Dks	Austrian Trieste	1289 1959	297·5	35·2	25·4	Trieste Lloyd Austro-Ungarico	1881		2Cy.35¼"&61¾"–18" 172 HP. Arsenale Lloyd,Trieste	Lloyd Austro-Ungarico
192	G	Helix I ScSr Hannemann 82-82 1 Dk	German Lübeck	366 499	181·7	25·1	12·1	Glasgow Smith&Rodger	1857		2C.21½"&41¾"–26¾" 300 HP. Rostock S.B.&E.Co.,Rostock	Wm.Minlos
193	A	Hellas (ex President) I Sc • Monarchidi 1 Dk H2	Greek Syra	447 690	206·8	28·2 P & F 32 f.	16·0	Newcastle C.Mitchell	1869	5 Blhds.	2Cy.20"& 38"–30" 90 HP. Thompson,Boyd&Co.,Nwc.	P.Sugdury
194		Hellas I ScSr Caloyeros 80-83 1 Dk	Greek Piræus	209 326	162·1 P 31 f.	22·4 B 19 f.	11·5 F 16 f.	Paisley H.M'Intyre&Co.	1880		2Cy.22"& 40"–30" 75 HP. Hutson&Corbett,Glasgow	D.P.Goudis
195		Hellekis (ex Emil, ex Östera) C ScSr 1 Dk	Swedish Hellekis	80 125	96·1 Q 7 f.	20·8 H 17 f.	9·1	Motala	1854		2Cy. 24 HP.	F.Sköldebrand
196	✻	Hellenes I ScSr Austin -88 1 Dk & Wb Frms	British Liverpool	1294 1984	276·2 Q 96 f.	40·3 B 60 f.	18·7 F 32 f.	Sunderland W.Pickersgill&Sons	1884	4 Blhds. AT 291, MT 399, DT 321	2Cy.33"& 62"–42" 164 HP. G.Clark,Sunderland	S.S."Hellenes"Co.,Ld. (R.P.Houston & Co.)
197		Hellespontos W Sc 1 Dk	Greek Piræus	102	116·8	17·1	8·7		1865			Theophilatos & Stathatos
198	✻D	Helmsley I ScBn Tindle 81-83 1 Dk 2 B	British W.Hartlepl.	1151 1797	265·1 Q 68 f. P 28 f.	36·0 B 68 f.	19·7 F 63 f.	W. Hartlepool W.Gray&Co.	1883	4 Blhds. DB	2Cy.33"& 62"–39" 180 HP. Blair&Co.,Ld.,Stockton	Hudson Shipping Co., Ld.
199	✻D	Helmstedt I ScSr Strachan 86-88 1 Dk 2 B	British W.Hartlepl.	974 1530	267·0 Q 100 f.	33·3 F 30 f.	18·9	Hartlepool R.Withy&Co.	1877	4 Blhds. DB f 312	2Cy.31"& 59"–33" 110 HP. T.Richardson&Sons Hartpl.	R.Ropner & Co.
200		Helsingborg Stl ScSr Bergendorff 80-83 1 Dk	Danish Copenhagen	103 179	144·2 P 59 f. B 81.	20·9 F 17 f.	9·5 Len.—	Gothenburg Götaborgs Mek. Vk.	1884		1Cy.14"& 28"–18" 70 HP. Burmeister&Wain,Cpnhgn.	Det Forenede Dmpskbs Selskab,(C.P.A.Koch)
201		Helsingfors I ScSr Nyberg 1 Dk	Russian Helsingfors	165 246	141·0	23·0	10·3	Wiborg Wiborg Mek. Verkst	1870		2Cy.17"& 31"–22" 70 HP. Wiborg Mek. Verkst. Wiborg	Helsingfors Ångfartys Aktiebolag,(J.Nyman)
202	V	Helsingør I ScSr Becker 1 Dk & Spr Dk	Danish Elsinore	686 1050	218·8	29·9	15·0 22·1	Elsinore Jrnsk. & Maskabyg	1883	DB	2Cy.26"& 50"–33" 110 HP. Jernsk& Maskbygg Elsinore	Helsingørske Dampskibs-Selskab,(H.C.Nyholm)
203	V	Helsingør I ScSalvageSr Heckscher 81-81 1 Dk	Danish Copenhagen	111 213	133·4	21·4 B 5 f.	12·1	Copenhagen Burmeister & Wain	1868		1Cy.16½"& 34"–20" 80 HP. Burmeister&Wain,Cpnhgn.	Em.Z.Svitzer's Bjergnings Enterprise(H.P.Lyngbye)
204	DT 189	Helvetia I ScBkn Cochrane 79-86 4 Dks	British Liverpool	2855 1588	419·2	41·2	35·6	Newcastle Palmer Bros.	1864	4 Blhds.	2Cy.55"& 90"–48" 420 HP. P.Penn&Sons,London	National Steam Ship Co.,Ld.
205	✻	Helvetia Stl ScSr Landtrer 89-89 2 Dks 3 B	German Hamburg	1796 2763	316·0 P 28 f.	40·2 B 58 f.	24·4 F 32 f.	Newcastle Tyne I.S.B.Co.,Ld.	1889	6 Blhds. CDB	3C.24"40"&64"–45" 300 HP. Wallsend Slip.Co.,Newcstle.	Hamburg-Amerik. Paketf.Act.Ges.
206	V	Hengest I ScS Olsen 63-76 2 Dks & Spr Dk	Danish Aarhuus	415 723	207·4 P 34 f.	27·4 F 37 f.	15·8 23·2	Renfrew Lobnitz,Coulborn&Co.	1876	WB	2Cy.22"& 54"–42" 135 HP. Lobnitz,Coulborn&Co.Rfrw.	Jydsk-Engelske Dampskibs-Selskb.(E.Bergmann)
207	✻D	Henley Stl ScSr Valder 87-87 1 Dk & Wb Frms	British London	1446 2213	282·0 P 29 f. B 112 f.	37·6 F 27 f.	19·6 Q 77 f.	W. Hartlepool W.Gray&Co.	1887	5 Blhds. DB 308	3C.22"36"&59"–39" 225 HP. Blair&Co.,Ld.,Stockton	Britain S.S.Co.
208	V	Henri IV. I ScBk Paussett 1 Dk & Spr Dk 3 B	French Havre	1189 1605	247·7 B 13 f.	36·1 F 15 f.	18·9 25·7	La Seyne Forges&Ch.dela Médit.	1871		2Cy31½"&55"–35½" 150 HP. Forges&C.dela Médit,Mrslls.	Bossiero frères & Co.
209	✻	Henrietta I ScBn Coulson 1 Dk 2 B	British London	925 1447	242·5 Q 88 f.	34·1 B 56 f.	18·1 F 33 f.	Whitby T.Turnbull&Son	1882	4 Blhds. DB f 264	2Cy.30"& 55"–36" 120 HP. Blair&Co.,Ld.,Stockton	T.Marwood & Sons
210		Henrietta I Pad 1 Dk	Dutch Rotterdam	21 118	93·1	19·0	8·6	Middlesbro' R.Dixon&Co.	1882	3 Blhds.	2Cy.24"–36" 45 HP. Kincaird,Donald&Co.,Grnk.	Samarang Stoomboot Veer

✻ = Lloyd's Register; D = Nederlandsche Vereeniging; G = Germanischer Lloyd; I = Registro Italiano; R = American Record; A = Veritas Austro-Ungarico; H = Veritas Ellenico; N = Norske Veritas; V = Bureau Veritas.

Dans la 2me Colonne, B.T., avec les indications qui suivent, signifie la date du rapport de visite de la Carène par les experts du Board of Trade pour obtenir le certificat relatif au TRANSPORT DES PASSAGERS.

Dans la 3me Colonne, C = Composite (membrure en fer, bordé en bois); I = Fer; Stl = Acier; W = Bois.

Pour plus amples renseignements sur les navires classés par le Lloyd's Register, voir le Lloyd's Register Book et le Lloyd's Yacht Register.

No. / Official No.	Class	Ship's Name. Material, Rig, &c. / Master, and Dates of Appointment / Ship's Late Name. No. of Decks, &c.	Flag. / Port of Registry.	Tons. Gross / Net	Length	Breadth	Depth	Built. Where. / By Whom.	When	No. of Bulkheads. / Particulars of Water Ballast.	Engines. Particulars. Maker's Name.	Owners or Managers
211	G	**Henriette** [illegible] I ScSr *Engelhardt* [illegible]	German Lübeck	319 [illegible]	165·5	22·2	10·5	Rostock Act.Gesell.Hansa	1857		2Cy.17"&30"–23" 230 HP.	Wm. Minks
212 [illegible]	✱	**Henriette H.** I ScSr *Voss* 1 Dk&SprDk3 B	British Newcastle	1468 [illegible]	280·0	38·8	26·7	Sunderland Laing	1886	4 Blkds. DB f [illegible]	3C.21"33"&56"–42" 250HP. J.Clark Sunderland	M. J. de Ham
213	V	**Henriette Schlusser** I ScSr *Plath* –79 2 Dks	German Rostock	526 781	191·2	26·6	20·1	Kiel G.Howaldt	1879		2Cy.25"&40"–27" 80 HP. G.Howaldt,Kiel	F. Petersen
214	N	**Henrik Wergeland** I ScSr *Jensen* 86-86 1 Dk&AwnDk	Norweg'n Drontheim	377 498	150·5	21·5	19·1	Kiel GeorgHowaldt	1883	DB86 ft APT	2Cy.21½"&85½"–23½" 70 HP. [illegible] Howaldt,Kiel	Nordenfjeldsk Skibsselskab
215 [illegible]	✱	**Henry** (ex [illegible]) I Sc3mSr *Naile* 88-88 1 Dk2 B	British Liverpool	653 1013	221·2	29·8	20·0	Liverpool Potter&Hodgkinson	1873	3 Blkds. MT [illegible]	3C.15"23½"&39"–33" 90 HP. Westgarth,English&Co. Middlesbro	Curwen Bros.
216 69910		**Henry Allen** I TwnScSr *Washbourne* 1 Dk	British [illegible]	138 225	127·5	21·5	10·3	Bristol J.Payne	1874	4 Blkds.	2Cy.19"–18" 40 HP. J.Payne,Bristol	S. Morris
217 77219	✱	**Henry Anning** I ScSr *Partridge* 83-83 2 Dks3 B	British Cardiff	1075 1650	264·0	33·2	22·1	Newcastle Palmers Co.	1878	4 Blkds. DB [illegible]	2Cy.30"&59"–36" 140 HP. Palmers'Co.,Ld.,Newcastle	Anning Bros.
218 [illegible]	✱	**Henry Bolckow** I ScSr *Hughes* 1 Dk2 B	British Bombay	639 1006	214·0	30·6	16·9	Middlesbro' Backhouse&Dixon	1869	4 Blkds. DB [illegible]	2Cy.26"&47"–36" 120 HP. G.Clark,Sunderland	Bombay & P[illegible] Nav.Co.,Ld.
219 62756	✱	**Henry Brand** I ScSr *Sealey* 79-84 1 Dk2 B	British Southamptn	445 899	200·8	29·0	16·1	Hartlepool Denton,Gray&Co.	1868	2 Blkds.	2Cy.26"&47"–36" 98 HP. G.Clark,Sunderland	J. L. Hill
220 70220	✱	**Henry Fisher** I ScSr *Bain* [illegible] 1 Dk	British Newcastle	333 355	166·0	25·7	12·7	Newcastle C.S.Swan&Co.	1878	4 Blkds. [illegible]	2Cy.21½"&40"–30" 80 HP. T.Clark&Co.,Newcastle	Fisher, Renw[illegible]
221 [illegible]	✱	**Henry Morton** I Sc3mSr *Swan* 68-84 1 Dk2 B	British London	634 893	229·8	32·2	17·5	Newcastle Palmer Bros.&Co.	1860	4 Blkds.	2Cy.26"&51"–30" 99 HP. [illegible]	H. James (J. Fenwick & Son [illegible])
222		**Henry Wright** W Pad	British Shields	42 113	93·8	18·0	9·7	SouthShields Woodhouse	1852		50 HP.	H. G. Graham
223 87692	✱	**Henry Wright** C ScSr *Harnes* 85-86 1 Dk	British London	18 [illegible]	85·7	16·0	8·5	London R.&H.Green	1883	3 Blkds.	2Cy.16"&30½"–21" 40 HP. A.Wilson&Co.,London	Church Missi[illegible] Society
224 82289		**Henzada** I ScSr *Avern* 66-84 2 Dks3 B	British Glasgow	1341 2078	284·6	35·2	24·1	Glasgow A.&J.Inglis	1879	5 Blkds.	2Cy.34"&60"–45" 200 HP. A.&J.Inglis,Glasgow	British India S[illegible] Nav.Co.,Ld.
225 65620	✱	**Heptarchy** I Sc3mSr *Goldsmith* 1 Dk2 B	British London	502 786	201·2	26·5	17·2	London Samuda Bros.	1871	4 Blkds. FPT DBFT [illegible]	2Cy.26"&45"–26" 90 HP. J.Jack,Rollo&Co.,Liverpool	J. Wright
226 [illegible]	BT 5.85	**Her Majesty** I PadSk	British Southamptn	131 285	185·2	20·1	8·3	Glasgow Barclay,Curle&Co	1885		2Cy.25"&47"–54" 140 HP. Barclay,Curle&Co.,Glasgow	Sthmptn., Isl[illegible]
227 93115	✱	**Her Majesty** I ScK *Chalis* –87 1 Dk	British Hull	55 142	100·2	20·1	10·9	Hull Head&Riley	1887	3 Blkds.	2Cy.17"&34"–21" 45 HP. [illegible] Wates,Hull	W. Wolfe
228	✱	**Heraclée** (ex Himalaya) I ScSr (new moved Pars) 1 Dk2 B	Turkish Cnstntinple	515 808	200·8	28·5	15·9	Hartlepool Withy,Alexander&Co.	1871	4 Blkds. DB [illegible] APT [illegible]	2Cy.25"&48"–36" 96 HP. T.Richardson&Sons,Hrtlpl.	A. Caramanian& hin oglou & Co.
229		**Heraclee** W Pad *Hadji Mohammed*	Turkish Cnstntinple.	75	166·5	20·7	10·0	PortGlasgow W.Hamilton&Co.	1869		2Cy. 80 HP. W.Hamilton&Co.,Pt.Glasg.	Idarei-Masso[illegible] (Ottoman [illegible])
230 93680	✱	**Heraclides** I ScSr *Dovers* 82-86 1 DkSprDk&Wbfrms	British Liverpool	1898 2947	320·0	40·4	19·2	Sunderland Bonds,Sharer&Co.	1886	6 Blkds. [illegible] APT [illegible] MT [illegible]	3C.27"43"&69"–45" 450 HP. T.Richardson&Sons,Hrtlpl.	R. P. Houston &
231 [illegible]	✱	**Herald** Stl ScSr *Jones* 86-86 1 Dk	British Auckland	370 573	172·0	28·2	13·3	Dundee Pearce Bros.	1884	4 Blkds. DB [illegible] FPT [illegible]	2Cy.23"&45"–30" 85 HP. Pearce Bros.,Dundee	C. B. Stone
232 [illegible]	BT [illegible]	**Herald** I PadSr *Bigerstaff* 86-86	British [illegible]	252 446	221·6	22·0	10·4	Greenock Caird&Co.	1866		2Cy.44"–57" 150 HP. Caird&Co.,Greenock	J. Little & Co.
233 84157		**Herald of Peace** I ScFlat	British Liverpool	71 111	84·6	19·9	8·2	Northwich J.Thompson&Son	1877		2Cy.13"–14" 27 HP. Watt Bros.,Liverpool	Salt Union,Ld
234		**Herault** (ex Isla Christina) Stl ScSr *Vis* 78-88 1 Dk&SparDk	French Marseilles	418 600	207·0	23·9	16·9	Sunderland	1864	APT [illegible]	2Cy.22"&42" 80 HP. Pile&Co.,Sunderland	Fraissinet & Co
235 87320	✱	**Herbert** I ScSr *Lewis* 1 Dk	British Launcstn.Tas	139 233	110·4	25·1	7·2	Newcastle McIntyre&Co.	1884	4 Blkds. FPT [illegible] APT [illegible]	2Cy.18"&34"–24" 60 HP. R.H.Pearson&Co.,Glasgow	J. Williams
236 84146	✱	**Herbert** I ScSr *Iddon* 1 Dk	British Liverpool	90 185	115·4	19·2	9·3	Liverpool R.&J.Evans&Co.	1881	3 Blkds. FPT [illegible]	2Cy.14"&28"–20" 35 HP. De Winton&Co.,Carnarvon	S.S. "Herbert" (J. J. Mack & [illegible])
237 [illegible]	✱	**Herbert** W ScTug *Cock* 84-84 1 Dk	British Plymouth	10 45	65·4	15·5	7·6	Plymouth D.Banks&Co.	1884		2Cy.13"&26"–18" 30 HP. S.Sara,Penryn	J. Westcott
238		**Hercule**	French Calais	71 150	109·9	19·2	10·5	SouthShields	1877			Soc. Anonyme d[illegible] [illegible]
239 51540		**Hercules** W Pad 1 Dk	British Quebec	126 200	119·0	22·0	9·9	Quebec	1865		1Cy 60 HP.	Francois Samso[illegible]
240 [illegible]		**Hercules** W ScSr *Christopher* 1 Dk	British Shields N.T.L.	90 216	113·1	22·7	11·3	SouthShields J.P.Rennoldson	1873		2Cy.20"&39"–30" 60 HP. J.P.Rennoldson,S.Shields	D. Condon

✱ = **Lloyd's Register**; D = Nederlandsche Vereeniging; G = Germanischer Lloyd; I = Registro Italiano; R = American [illegible]
A = Veritas Austro-Ungarico; H = Veritas Ellenico; N = Norske Veritas; V = Bureau Veri[tas]

In the 2nd Column, B.T., with figures following, signifies Date of report of bottom sighting by Board of Trade Surveyors as condition of PASSENGER C[ERTIFICATE].

In the 3rd Column, C = Composite (iron frame, wood planked); I = Iron; Stl = Steel; W = Wood.

Further particulars of Vessels classed by Lloyd's Register will be found in Lloyd's Register Book and Lloyd's Yacht Register.

No.	Cote	Nom, Matériaux, Gréement, etc. / Capitaine et dates d'entrée au service des Armateurs actuels et à bord du navire. / Ancien Nom, nombre de ponts, etc.	Pavillon / Port d'Armement	Tonnage Brut / Net	Dimensions (Longueur / Largeur / Creux)	Construit Où / Par qui	Quand	Nombre des cloisons / Détails sur le lest d'eau	Machines / Détails / Nom du Constructeur	Armateurs ou Directeurs-Gérants
		Hercules I ScSr	British Liverpool	81 [illegible]	120·5 | 20·2 | 10·6	Glasgow T.B.Seath&Co.	1870		2Cy.18"–30" 21 HP. D.Rowan,Glasgow	W. Williams
		Hercules I PadSp	British Hull	23 113	105·3 | 19·2 | 10·6	SouthShields J.T.Eltringham	1877		2Cy.34"–60" 80 HP. J.P.Rennoldson,S.Shields	Dock Co. of Kingston-upon-Hull
		Hercules I PadSp	British Liverpool	20 143	112·6 | 19·0 | 10·2	Birkenhead Laird&Co.	1861		2Cy.33"–48" 70 HP. Laird&Co.,Birkenhead	W.H. Dodd
		Hercules I TwnScTug	British Glasgow	3 107	125·0 | 21·0 | 11·2	Glasgow [illegible]	1885		2Cy.23"&42"–30" 98 HP. Rankin&Blackmore,Grnck.	Clyde Shipping Co.
		Hercules I ScSmSr *Badeff* 88-89 1 Dk 2 B	American Philad'lphia	584 764	200·7 | 37·0 | 13·5 F— B—	Chester, Pa. Reaney,Son&Archbold	1870		1Cy.40"–30" 150 HP. Reaney,Son&Archbold,Chs.Pa.	Philadelphia & Reading R.R.Co.
		Hercules W Sc *Lasini*	Arg'ntine Buen'sAyres	80 110	85·0 | 15·5 | 9·2	BuenosAyres Parody	1884		40 HP. Fontanas,B.Ayres	E. Fülle & Co.
247	✻D	**Hercules** I PadTug 1 Dk	Dutch Amsterdam	97 [illegible] F— B—	142·2 | 22·1 | 12·1	Newcastle C.Mitchell&Co.	1877	5 Blkds.	2Cy.40"–66" 150 HP. Thompson&Co.,Newcastle	A.D. Zurmuhlen
248	✻D	**Hercules** I ScSr *Gelderandt* 80-88 1 Dk 2 B	German Danzig	742 1155	212·3 | 34·2 | 16·6 Q 70 f. B 50 f. F 19 f.	Whitehaven WhitehavenSBCo.,Ld	1881	4 Blkds. [illegible] FPT 30 APT 20	2Cy.27 & 54"–36" 150 HP. J.Jones&Son,Liverpool	Danziger Schiff. Act. Gesellschaft (A.Gibsone)
249	V	**Hercules** I ScSr *Bauer* -88	German Bremen	454 701	173·8 | 25·9 | 13·9 F 20 f.	Lübeck H.Koch	1888	DB 63 FPT 31		Dampfsch. Ges. "Neptun"
250	V	**Hercules** Stl ScYt *Harmsen* -88	Russian Riga	55 135	97·7 | 21·3 | 9·9	Motala Motala Mekaniska Werkstad	1884		2Cy.24 & 41"–23½" 90 HP.	Börsen Committee
251		**Hercules** I *Parial* -85	Spanish Coruñna	46 185	155·1 | 19·0 | 8·5	England			110 HP.	Lopez, Perez & Co.
252		**Hercules** I PadSp	Spanish Santander	24 104	91·9 | 18·2 | 9·5	Cardiff J.Batchelor	1871	4 Blkds. WB [illegible]	1Cy.34½"–54" 50 HP. R.Marshall,S.Shields	Hijos de Doriga
253	✻	**Hereward** I ScSpTug *Richardson* -88 1 Dk	British London	12 108	98·2 | 18·9 | 10·0	Hull Cook, Welton & Gemmell	1888	4 Blkds.	3Cy.12½" 19½" & 31½"–22" C.D.Holmes&Co.,Hull [60 HP.]	Peninsular & Oriental Steam Navigation Co.
254		**Herga** I ScSr *Henderson* 1 Dk	British Sydney	135 221	125·2 | 24·2 | 10·6 Q 17 f. F 28 f.	Port Glasgow Cunliffe&Dunlop	1878	3 Blkds. FPT	2Cy.20" & 35"–24" 55 HP. Cunliffe&Dunlop,Pt.Glasgw.	Alexander Stuart
255	G	**Heridia** (ex Nicaragua) I ScSr *Price* -82	Costa Rica PortLimon	138 241	140·1 | 22·0 | 8·5	Montrose Black&Noble	1877		2Cy.26" & 34"–24" 58 HP. Kincaid,Donald&Co.,Grnck.	M.C. Keith
256	G	**Herkules** I PadTug *Behrens* -87 1 Dk	German Bremen	121 244	137·5 | 23·5 | 11·0 B 36 f.	Geestemünde J.C.Tecklenborg	1887	FPT 9 APT 21	1Cy.22½" & 44"–45" 120 HP. J.C.Tecklenborg,Geestemde.	Norddeutscher Lloyd
257	R	**Herman Winter** I ScSr *Bartlett* -87 2 Dks & Spr Dk	American New York	1786 2626	271·8 | 41·6 | 19·2 29·5	Philadelphia W.Cramp&Sons,S.&E.Co.	1886		2Cy.38" & 68"–54" 350 HP. W.Cramp&Sons,S.&E.B.Co., Philadelphia	Metropolitan S.S.Co.
258	✻V	**Hermann** I Sc *Meyer* 83-88 1 Dk & Awn Dk 3 B	Belgian Antwerp	1879 2879	322·2 | 40·0 | 22·6 29·5	Sunderland [illegible]	1881	4 Blkds. DB [illegible]	2Cy.39" & 74"–45" 220 HP. G.Clark,Sunderland	Steinmann & Ludwig
259	G	**Hermann** I ScBg *Kohlmann* 89-88 3 Dks & Spr Dk	German Bremen	1677 2713	336·6 | 40·0 | 33·1	Greenock Caird&Co.	1865		2Cy.56" & 96"–48" Day,Summers&Co.,Sthptn	Norddeutscher Lloyd
260		**Hermes** I ScSr *Bojer* 82-87 1 Dk	Swedish Stockholm	164 317	142·7 | 24·0 | 12·7 H 12 f.	Norrköping MotalaCo.	1877	Yes	2Cy.30½" & 58"–24" 150 HP. MotalaCo.,Norrköping	Berg. och Dyk Bol. "Neptun" (E.Liljewalch)
261		**Hermes** I ScSp *Hjärne* 77-80 1 Dk	Swedish Christinehamn	141 212	102·3 | 21·1 | 12·1 B 19 f. H 7 f.	Stockholm Bergsund Mek. Verkst.	1867		2Cy.12" & 14"–20" 28 HP. BergsundMek.Verkst,Stckh'lm	Angf. Akt. Bolag. "Hermes" (C.J.Andersson&C°)
262		**Herminia** I ScSr *Aranalegui* -86 1 Dk	Spanish Bilbao	184 266	141·4 | 19·1 | 12·3 B 40 f. F 18 f.	Sunderland A.Simey&Co.	1873		2Cy.18" & 30"–20" 45 HP. SmithBros.&Co.,Glasgow	Aldecoa & Co.
263	✻	**Hermione** (for Sailing, ex Bessemer) I ScK *Howard* -88 1 Dk	British Hull	55 120	96·7 | 20·1 | 10·0 Len.88	Hull Cook, Welton & Gemmell	1885	4 Blkds.	2Cy.16" & 30"–22" 40 HP. C.D.Holmes&Co.,Hull	C. Hellyer
264		**Hermod** I PadSr *Poulsen* 87-87 1 Dk	Danish Korsör	149 271	181·9 | 20·1 | 10·4 B 26 f.	Copenhagen Baumgarten & Burmeister	1864		2Cy.42¾"–47½" 120 HP. Burmeister&Wain,Copenhagen	De Danske Statsbaner (N.Holst)
265	✻	**Hermod** Stl ScSr *Hofschlund* 74-78 1 Dk 2 B	Swedish Gothenburg	919 1167	235·5 | 32·8 | 17·0 Q 77 f. B 51 f. H 57 f. F 32 f.	Gothenburg MotalaCo.	1878	4 Blkds. DB [illegible]	2Cy.27½" & 55"–36" 150 HP. MotalaCo.,Gothenburg	Ångf. Aktie. Bolag Svithiod
266		**Hermod** (ex Triton) W ScSp *Gottbrandt* -87 1 Dk	Swedish Mariestad	103 149	97·1 | 20·7 | 10·1 Q 5 f. B 30 f.	Sjötorp Andersson	1871		1Cy.20"–14" 15 HP. GöteborgM.V.,Gothenburg	S.J. Nilsson
267		**Hermoupolis** (Castillo 79-) I ScSr (ex Wicklow, ex Oalsten) pt Awn Dk	Greek Syra	493 780	221·0 | 29·2 | 15·1 P 199 f. F 26 f. 22·2	PortGlasgow Blackwd.&Gordon	1860		2Cy.29¼ & 57"–36" 150 HP. R.Steele&Co.,Greenock	Hellenic Steam Ship Co.
268	✻	**Hernán Cortés** (ex Stuart King) I Sc *Orts* -87 1 Dk & Spr Dk 3 B	Spanish Barcelona	2294 3537	340·3 | 42·1 | 27·5 B 75 f. F 34 f.	Hull D.P.Garbutt	1883	7 Blkds. DB & [illegible] 475 t	2Cy.44" & 84"–54" 400 HP. D.P.Garbutt,Hull	E. Pi & Co.
269		**Hernani** I ScSr *Jaque* 1 Dk	Spanish Santander	263 387	160·4 | 24·3 | 11·8 P 71 f. F 18 f.	Port Glasgow Cunliffe&Dunlop	1876		2Cy.23" & 40"–27" 70 HP. Cunliffe&Dunlop,Pt.Glsgw.	Jose Pedros & Co.
270										

✻ = Lloyd's Register; D = Nederlandsche Vereeniging; G = Germanischer Lloyd; I = Registro Italiano; R = American Record;
A = Veritas Austro-Ungarico; H = Veritas Ellenico; N = Norske Veritas; V = Bureau Veritas.

Dans la 2me Colonne, B.T., avec les indications qui suivent, signifie la date du rapport de visite de la Carène par les experts du Board of Trade pour obtenir le certificat relatif au TRANSPORT DES PASSAGERS.

Dans la 3me Colonne, C = Composite (membrure en fer, bordé en bois); I = Fer; Stl = Acier; W = Bois.

Pour plus amples renseignements sur les navires classés par le Lloyd's Register, voir le Lloyd's Register Book et le Lloyd's Yacht Register

No.	Society in which classed	Ship's Name, Master, and Dates of Appointment to present Owner's Service and to Ship; Ship's Late Name	Material, Rig, &c.; No. of Decks, &c.	Flag; Port of Registry	Tons Net; Gross	Length	Breadth	Depth	Built Where; By Whom	When	No. of Bulkheads; Particulars of Water Ballast	Engines, Particulars, Maker's Name	Owners
271		Herno (ex Malden)	I Pad 1 Dk	Swedish Hernösand	59 [illegible]	133·2	15·2 B 15 f.	7·2	England	18[illegible]		60 HP.	E. & V. Becker
272		Hernosand II. Pettersson	I ScSr 1 Dk	Swedish Hernösand	357 487	162·6 Q41f. B58f.	23·2 F9f.	12·5 H94f.	Stockholm Bergsund Mek. Verkst	1865		2Cy.24"&42"–26 100 HP. W. Lindbergs Mek. Verkst. Stkhlm.	Hernösands Ång. Aktie Bolag
273 33285	BT9.88	Hero Thorlander	I ScSr 2 Dks	British Hull	671 1083	229·3	30·6	20·8	Hull C.&W. Earle	1866	[illegible]	4Cy.27"&47"–30" 180 HP. Comp. 73, C.D. Holmes, Hull	T. Wilson, Sons
274 29825		Hero	I ScBk 1 Dk	British Dunedin	640 [illegible]	224·3 P75f.	29·2 B134f.	15·7 F107f.	Hull C.&W. Earle	1861	5 Blhds.	2Cy.46⅜"–26" 150 HP. C.&W. Earle, Hull	J.E. Mitchell
275 51308		Hero Wood 88–88	I Sc3mSr	British Middlesbro'	370 [illegible]	186·0	24·8	14·5	Newcastle [illegible]	1866		2Cy.23"&41"–24" 70 HP. Black, Hawthorn & Co., Gtshd.	E. Harris & Co.
276 42725		Hero	W Pad 1 Dk	British Quebec	131 [illegible]	140·0	20·4	10·0	Point Levis	1860		1Cy. 75 HP.	St. Lawrence Nav. Co.
277 21381		Hero	I PadSr	British [illegible]	44 157	181·0	19·1	7·1	Glasgow T. Wingate & Co.	1858		1Cy.48"–42" 80 HP. T. Wingate & Co., Glasgow	W. Buchanan
278 29799		Hero	W Pad	British [illegible]	11 105	96·8	17·5	10·7	London	1861		2Cy. 60 HP.	W. & T. Jolliffe
279	✳	Hero	I ScSk 1 Dk	British London	5 38	59·8	14·1	7·4	Hull E. Wales	1888	4 Blhds.	2Cy.11"&22"–18" 30 HP. E. Wales, Hull	J. Constant
280	DV	Hero Klofkorn –87	I ScSr 1 Dk	German Bremen	436 563	161·4	25·1	17·0	Lübeck H. Koch	1883		2Cy.18"&31"–22" 50 HP. O. Hendrichs & Co., Berlin	Dampfschifffahrts Ges. "Neptun"
281	V	Hero Pettersson	I ScSr 1 Dk 2 B	Swedish Gothenburg	249 346	128·2 Q 40f.	22·4 B 31f.	13·0 H [illegible]	Stockholm W. Lindbergs M.V.	1884	[illegible]	2Cy.16"&31"–21" 40 HP. W. Lindbergs Mk. V. Stkhlm.	Ångf. Akt. Bol. (O. Lindberg)
282 26215		Heron (TO BE BROKEN UP) Arnold	I ScSr	British London	380 630	210·2	27·1	14·1	Dumbarton W. Denny & Bros.	1860		2Cy.45"–33" 150 HP. Tulloch & Denny, Dumbartn.	General Steam
283 67077		Heron Crawlie –87	I TwnSc 1 Dk	British [illegible]	56 141	120·1	20·0	8·1	Glasgow Lon. & Glasg. E. & I.S. B.Co.	1872		2Cy.24"–21 60 HP. Lon. & Gls. E. & I.S. B.Co., Gls.	J. Watts
284 93073	✳	Heron Foster	I ScTrawler 1 Dk	British Hull	35 106	90·0	20·0	10·5	Hull Cook, Welton & Gemmell	1886	3 Blhds.	2Cy.15"&30"–20" 40 HP. Muir & Houston, Glasgow	W.H. Putt & Co.
285		Heron De Gruchy	I ScSr 1 Dk	Belgian Antwerp	36 114	106·5 Q 46f.	18·3 F 14f.	8·9	Bowling Scott & Co.	1881	4 Blhds. FPTH.	2Cy.18"&32"–22" 45 HP. R. Harvey & Co., Glasgow	Assocn. Interna. du Congo
286		Héron Dalue –84		French [illegible]		..	..	..					Johnston & Co.
287 94280	✳	Herongate Reynolds 88–88	I ScSr 1 Dk	British London	322 525	180·0 P18f.	25·1 B 48f.	12·5 F22f.	Sunderland Short Bros.	1887	4 Blhds.	3Cy15"25"& 40"–27" 80 HP. J. Dickinson, Sunderland	Weatherley, Mason & Hussey
288	✳	Herrera Morales 78–88	I Sc 1 Dk 2 B	Spanish Seville	670 1041	210·0 P 110f.	30·3 F31f.	17·2	Berwick Gowan & Wilson	1878	4 Blhds. [illegible]	2Cy.26"&52"–30" 90 HP. W.B. Thompson, Dundee	Miguel Saenz
289 81277	BT12.88	Herschel	I ScSr	British Liverpool	1186 1886	311·0	34·4	24·4	Newcastle A. Leslie & Co.	1879		2Cy.26"&58"–45" 180 HP. R. Stephenson & Co., N'castle.	Liverp'l, Braz. & Plate Steam N.
290 89901	✳D	Herschel Lascelles 79–81	I ScSr 1 Dk 2 B	British N. Shields	1152 1783	260·0 Q 95f.	36·0 B 74f.	18·6 F 29f.	Sunderland Bartram, Haswell & Co.	1884	6 Blhds. DBok f 335t	2Cy.33"& 61"–33" 160 HP. T. Richardson & Sons, Hrtlpl.	Wilkie & Turn
291		Hertha Medanich	I ScSrYt	Austrian Trieste	165 291	157·7	20·0	13·2	Kiel Otto Schlick	1877		100 HP.	Prince of Liechtenstein
292		Hertha Jensen 77–84	I ScSr 1 Dk	German Sonderberg	84 124	106·3	18·8	9·7	Gothenburg Eriksberg Mek. Werkstads Aktiebolag	1877		50 HP.	Sonderburger Dampfschifffahrts-Ges.
293	✳N	Hertha Nygaard	I ScSr 1 Dk 2 B	Norweg'n Bergen	322 531	169·9 Q 41f.	24·3 B 51f.	12·9 F 31f.	Bergen Bergens Mek. Verkst.	1880	4 Blhds. MT[illegible] APT	2Cy.20½"&39"–24" 55 HP. Bergens Mek. Verks. Bergen.	M.J. Schjelderup
294	N	Hertha Petersen –88	W ScBk 1 Dk 2 B	Norweg'n Sandefjord	164 254	117·8	25·3 Q 35f.	13·1	Sandefjord Chr. Christensen	1884		2Cy.16"&29"–18½" 35 HP. Nylands Verkst., Christiania	C. Christensen
295 89225	✳	Hesketh Rowe	Stl TwnScSr 1 Dk	British Sydney	393 640	180·4 Q 100f.	28·1 B 7f.	13·6 F 30f.	Cork Royal Victoria Dk. Co.	1883	3 Blhds. [illegible]	2Cy.22"&44"–33" 80 HP. Kincaid & Co., Greenock	Australasian U. Steam Nav. Co. Ld.
296 72651	✳	Hesleden Payne 88–88	I ScSr 1 Dk 2 B	British W. Hartlepl	957 1491	259·1 B 60f.	33·0 Q 96f.	19·2 F 35f.	W. Hartlepool W. Gray & Co.	1876	4 Blhds. DBok f Seb 176t	2Cy.30"& 55"–36" 120 HP. Blair & Co., Ld. Stockton	R. Ropner & Co.
297 86089	✳	Hesleyside Russell 86–86 (ex Turgeniet)	I ScSr 1 Dk	British Newcastle	440 688	184·8 Q 62f.	29·2 B 46f.	13·2 F 23f.	Sunderland D. Baxter & Co.	1882	4 Blhds. DB [illegible] FPT [illegible] APT [illegible]	2Cy.25"&48"–30" 98 HP. R.&W. Hawthorn, N'castle	Charlton, McAllum & Co.
298 81958	✳	Hesper Wilson 79–83	I ScSr 1 Dk 2 B	British Hartlepool	1070 1653	257·8 B 162f.	34·7 Q 92f.	17·8 P 31f.	Hartlepool E. Withy & Co.	1881	4 Blhds. [illegible]	2Cy.32"&60"–39" 150 HP. Blair & Co., Ld., Stockton	G. Horsley & Son
299 82[illegible]	BT1.89	Hesperia Linquister 73–83	I ScSr 1 Dk & Spar Dk 3 B	British Barrow	1982 3037	340·5	37·7 P 31f.	29·5	Barrow Barrow S.B. Co.	1882	7 Blhds. [illegible]	2Cy.53"&86"–60" 182 HP. Barrow S.B. Co., Barrow	Barrow Steam S. Co., Ld.
300	V	Hesperia Madsen 76–88	I ScSr 2 Dks	German Hamburg	1136 1627	262·1 P41f.	32·1 F58f.	23·8 H50f.	Kiel Norddeutsche Schiffb.	1876 Wrft.		2Cy.33"&56"–36" 160 HP. Schweffel & Howaldt, Kiel	Deutsche Dampfschiffs-Rhederei

✳ = Lloyd's Register; D = Nederlandsche Vereeniging; G = Germanischer Lloyd; I = Registro Italiano; R = American Re[illegible]
Δ = Veritas Austro-Ungarico; H = Veritas Ellenico; N = Norske Veritas; V = Bureau Veritas

In the 2nd Column, B.T., with figures following, signifies Date of report of bottom sighting by Board of Trade Surveyors as condition of PASSENGER CERTIF[illegible]

In the 3rd Column, C = Composite (iron frame, wood planked); I = Iron; Stl = Steel; W = Wood.

Further particulars of Vessels classed by Lloyd's Register will be found in Lloyd's Register Book and Lloyd's Yacht Register.

No.	Cl.	Nom, Matériaux, Gréement, etc. — Capitaine et dates d'entrée au service des Armateurs actuels et à bord du navire. — Ancien Nom, nombre de ponts, etc.		Pavillon. — Port d'Armement	Tonnage. Net. — Total	Dimensions. Longueur — Largeur — Creux — Constructions sur le Pont	Construit. Où. — Par qui.	Quand	Nombre des cloisons. — Détails sur le lest d'eau.	Machines. — Détails. Nom du Constructeur.	Armateurs ou Directeurs-Gérants.
301	✻	Hesperides *Bill 84-86*	I ScSr *2 Dks3 B*	British Liverpool	1553 2404	286·5|38·3|24·3 *P 36 f. B 66 f. F 38 f.*	Liverpool G.&J.Evans&Co.	1884	5 Blhds. APT 154 MT 333&375	2Cy.35"&65"–48" 182 HP. G.Forrester&Co.,Liverpool	S.S."Hesperides" Co., Ld. (R.P.Houston & Co.)
302	✻	Hessle *Wilson -89*	I ScSr *1 Dk2 B*	British W.Hartlepl.	1259 1925	270·0|37·2|18·1 *B 106 f. F 31 f. Q 103 f.*	Stockton Craig,Taylor&Co.	1889	4 Blhds. DBcf 867t FPT 90 APT 21t	3C.20"33"&54"–36" 180 HP. W.Allan(?)English&Co., Middlesbro'	Hudson Shipping Co., Ld.
303	✻	Hetty *Harrison*	I ScSr *1 Dk2 B*	British Middlesbro'	334 532	165·7|24·7|13·8 *B 46 f. Q 59 f. F 21 f.*	Middlesbro' R.Dixon&Co.	1875	4 Blhds. DBaft / 60t	2Cy.22"&44"–26" 65 HP. Blair&Co.,Ld.,Stockton	J.F.Wilson (Paterson, Fox & Co., Mngrs.)
304		Hevelius *Carroll 66-81*	I ScBg	Belgian Antwerp	1682 2011	349·8|37·0|27·8	Newcastle A.Leslie&Co.	1874		2Cy.30"&76"–54" 300 HP. R.Stephenson&Co.,Nwcstle.	Soc. Royale Belge Sud-Américaine
305	V	Hever *Belkefoten 84-84*	I ScSr *1 Dk*	German Husum	388 539	169·2|25·6|13·3 *Q 48 f. B 47 f. F 20 f.*	Flensburg FlensburgerSchiffsb. Ges.	1884		2Cy.22½"& 38"–27" 60 HP. Flensburg.Schiffsb.Ges.,Flnsbg.	Husumer-Rhederi Ges. (Lor.Hinrichsen)
306		Hewett *Suell 85-86*	I ScK	British London	115 206	130·2|21·2|10·8	Stockton M.Pease&Co.	1877		2Cy.28½"–24" 50 HP. Blair&Co.,Ld.,Stockton	Hewett & Co., Ld.
307	✻	Hexham *Adams 81-81*	I Sc3mSr *2 Dks,& Shd Dk3 B*	British N.Shields	1115 1715	260·9|34·2|23·3 *Shd Dk 48 f. P 40 f. F 36 f.*	Newcastle A.Leslie&Co.	1880	5 Blhds. DBcf&f 288t APT 36t	2Cy.32"&61"–36" 180 HP. N.E.MarineEng.Co.,Ld.,Sld.	J.Coull
308	✻	Hiawatha *(ex Nora)*	I Sc3mSrYt *2 Dks*	British Southamptn	328 529	176·8|26·1|13·4	Newcastle C.Mitchell&Co.	1879	6 Blhds.	2Cy.25"&45"–27" 90 HP. R.&W.Hawthorn,Newcastle	D.H.Macfarlane
309		Hiawatha	W ScSr *1 Dk,& Spr Dk*	British Windsor,NS.	148 230	132·3|21·2| 8·0 15·0	Hantsport NS J.Davison	1882		2Cy.28"&16"–20" 225 HP. Burrell,Johnson&C.,Yar.NS	George W.Churchill
310		Hiawatha	W Sc *1 Dk*	British Sarnia,Ont.	111 163	92·7|20·0| 7·6 *H 43 f.*	Dresden, Ont. J.A.Jenking	1874		1Cy.20"–24" 150 HP. Farrar&Trefts,Buffalo,U.S.	Sarnia Transportation Co.,Ld.
311		Hibellee *Mehemmed Ali 83-87*	W Pad	Turkish Constantnple	75 436	116·5|20·7|10·0	PortGlasgow W.Hamilton&Co.	1869		2Cy. 80 HP. W.Hamilton&Co.,Pt.Glasg.	Idarei-Massousieh (Ottoman Steam Nav.Co.)
312		Hibernia *Buchan 77-84*	I ScK	British London	42 214	121·0|22·1|12·5	Delfshaven Maatschappy de Maas	1884		2Cy.23"& 47"–24" 120 HP. DeMaasCo.,Ld.,Delfshaven	W.Watkins
313		Hibernian *Brown*	I ScBk *4 Dks*	British Glasgow	1873 2997	351·2|37·7|31·9	Dumbarton W.Denny&Bros.	1861		2Cy.38"& 75"–54" 380 HP. J.&J.Thomson,Glasgow	J.& A.Allan
314	✻	Hibernian *Brown 82-82*	I ScSr *1 Dk2 B*	British Middlesbro'	204 334	145·3|22·7|11·8 *Q 54 f. F 26 f.*	Port Glasgow H.Murray&Co.	1875	4 Blhds. DBaft /	2Cy.18½"& 32"–22" 47 HP. Kemp&Hume,Glasgow	Craig & Peterson
315		Hidalgo *Jenkins*	I ScBg *1 Dk2 B*	British Hull	1220 1580	272·6|32·0|17·8 *P 174 f. B 42 f. F 32 f.*	Hull Earles'Co.	1872	5 Blhds.	3C.22"32"&60"–33" 140 HP. Alt.'82,Amos&Smith,Hull	T.Wilson,Sons & Co.
316	✻	Hideyoshi Maru *(ex Orduna)*	I ScSr *1 Dk2 B*	Japanese Nagasaki	466 672	190·1|28·8|15·0 *Q 26 f. F 23 f.*	Sunderland W.F.Pile&Co.	1877	4 Blhds. DB 122t	2Cy.23½"& 44"–30" 65 HP. Pattison&Atkinson,Sndlnd.	Mitsui Yonosuke
317	✻	Highfield *Simpson*	I ScBn *2 Dks3 B*	British Liverpool	1665 2539	300·0|39·6|25·5 *P 28 f. B 68 f. F 38 f.*	Stockton Richardson,Duck& Co.	1883	5 Blhds. DBc&f 414t MT 306t	2Cy.38"& 71"–45" 300 HP. Blair&Co.,Ld.,Stockton	"Highfield" S.S.Co. (W.Wheatley & Co.)
318	✻D	Highgate *Thompson 85-85*	I ScBn *1 Dk2 B*	British London	890 1395	243·0|34·0|17·9 *Q 63 f. P 26 f. B 54 f. F 34 f.*	Whitby T.Turnbull&Son	1882	4 Blhds. DBc&f 212t	2Cy.30"& 57"–38" 130 HP. T.Richardson&Sons,Hrtlpl.	Turnbull,Scott & Co.
319		Highland Maid *Mosher -87*	W Sc *1 Dk*	British Winnipeg	73 106	82·0|16·8| 5·1	Fort Francis, Ont.	1887		1Cy.10"–12" 8 HP. Wilson&Son,Dundas,Ont.	R.Mosher
320	✻D	Highland Prince *Harvey -87*	I Sc3mSr *1 Dk2 B*	British Newcastle	978 1516	240·0|36·1|16·9 *P 28 f. B 64 f. Q 68 f. F 25 f.*	Sunderland ShortBros.	1883	6 Blhds. WBaf	2Cy.31"& 58"–36" 141 HP. N.E.MarineEng.Co.,Ld.,Sld.	Prince S.S.Co.,Ld. (J.Knott,Manager)
321	✻	Highlands *McGregor 88-88*	I ScBn *1 Dk2 B*	British W.Hartlepl.	795 1240	230·0|33·1|16·2 *Q 90 f. B 57 f. F 24 f.*	Aberdeen A.Hall&Co.	1884	4 Blhds. DBa&f 320t	2Cy.28"& 54"–36" 123 HP. BlaikieBros.,Aberdeen	Hardy,Wilson & Co.
322	✻	Higo Maru *Jones*	Stl ScSr *1 Dk,& Awn Dk3 B*	Japanese Tokio	871 1405	235·0|33·1|22·7	Glasgow London & Glasgow Co.Ld.	1884	5 Blhds. DBc& f 250t	2Cy.30"& 56"–39" 150 HP. London&GlasgowCo.,Ld.	Nippon Yusen Kaisha
323	✻	Hilda *Greig 82-83*	I ScBn *1 Dk2 B*	British W.Hartlepl.	1110 1720	258·5|34·0|19·7 *Q 72 f. P 24 f. F 30 f.*	W. Hartlepool W.Gray&Co.	1880	4 Blhds. DBa&f 335t	2Cy.33"& 61"–33" 150 HP. T.Richardson&Sons,Hrtlpl.	Herskind & Woods
324		Hilda *Merrells*	I ScSr *2 Dks*	British Southamptn	368 790	235·5|29·1|14·2 *Q 127 f. F 60 f.*	Glasgow Aitken&Mansel	1882	5 Blhds.	2Cy.37"& 69"–39" 220 HP. J.&J.Thomson,Glasgow	London & S.W.Ry.Co.
325	✻	Hilda *M'Ewen*	I Sc3mSr *2 Dks*	British Glasgow	351 663	212·8|25·1|16·1 *P 25 f. B 35 f. F 24 f.*	Paisley J.Fullerton&Co.	1884	5 Blhds.	2Cy.24"& 43"–30" 95 HP. J.S.Hume&Co.,Glasgow	J.Clark
326		Hilda *(ex Eugenie)* *Fishenden*	I PadSr	British London	190 428	225·0|24·2|11·9	Hull M.Samuelson&Co.	1862		2Cy.58"–54" 220 HP. M.Samuelson&Co.,Hull	GeneralSteamNav.Co.
327		Hilda *Wallcock*	I ScSr *1 Dk*	British Sydney	135 222	125·2|21·2|10·6 *Q 17 f. F 18 f.*	PortGlasgow Cunliffe&Dunlop	1878	3 Blhds. FPT	2Cy.20"& 35"–24" 55 HP. Cunliffe&Dunlop,Pt.Glsgw.	Alexander Stuart
328	✻	Hildegarde *Stevens 83-88*	I ScSr *2 Dks3 B*	British London	1268 1940	262·4|36·2|23·6 *P 26 f. B 16 f. F 33 f.*	Newcastle Palmers'Co.,Ld.	1882	5 Blhds. DBc f & a 635t FPT 149t APT 48t	2Cy.32"& 60"–42" 180 HP. Palmers'Co.,Ld.,Newcastle	Oriental S.S.Co.,Ld.
329	G	Hilding *Sjögren -77*	C ScSr *1 Dk*	Swedish Calmar	289 388	145·0|24·6|11·7 *Q 18 f. B 35 f. F 16 f. H 41.*	Oscarshamn S.Olsson	1873		2Cy.18"& 36"–18" 60 HP. G.Tillberg,Oscarshamn	A.Söderström
330		Hilding *Plombolt*	C ScSp *1 Dk*	Swedish Tessa	110 143	98·6|21·6| 8·5 *Q 4 f. B 19 f. H 21.*		1884			J.F.Kallin

✻=Lloyd's Register; D=Nederlandsche Vereeniging; G=Germanischer Lloyd; I=Registro Italiano; R=American Record; A=Veritas Austro-Ungarico; H=Veritas Ellenico; N=Norske Veritas; V=Bureau Veritas.

Dans la 2me Colonne, B.T., avec les indications qui suivent, signifie la date du rapport de visite de la Carène par les experts du Board of Trade pour obtenir le certificat relatif au TRANSPORT DES PASSAGERS.

Dans la 3me Colonne, C=Composite (membrure en fer, bordé en bois); I=Fer; Stl=Acier; W=Bois.

Pour plus amples renseignements sur les navires classés par le Lloyd's Register, voir le Lloyd's Register Book et le Lloyd's Yacht Register.

No. in Book. Official No.	Registry in which classed.	Ship's Name. Material, Rig, &c. Master, and Dates of Appointment to present Owner's Service and to Ship. Ship's Late Name. No. of Decks, &c.	Flag. Port of Registry.	Tons. Net. Gross.	Dimensions. Length.	Breadth.	Depth.	Built. Where. By Whom.	When.	No. of Bulkheads. Particulars of Water Ballast.	Engines. Particulars. Maker's Name.	Owners or Managers.
331		Hillersjo I ScSr Pettersson 83-81 1 Dk	Swedish Stockholm	123 165	98'8 B 48 f.	18'0 F 30 f.	8'3 H 30 f.	Gefle [illegible]	1884		2Cy.18¼"&27½"–17½ 45 HP. [illegible]	Angf. Bol. Sv[illegible] landet (G.O.W[illegible]
332	N	Hillevaag (ex Advance) W ScSr Hansen 83-83 1 Dk	Norweg'n Stavanger	182 340	125'7	2[illegible]'8	10'3 F 13 f.	Gothenburg J. Kjaulo	1873		2Cy.11¼"&23"–16½" 35 HP. GothenburgsMek.Verksted	R.G. Ramsland
333 60021		Himalaya I ScBg Bagill 86-86	British Glasgow	910 1427	263'8 P 70 f.	292'6	36'0 F 32 f.	W. Hartlepool Pile, Spence&Co.	1867	[illegible] Bhds.	2Cy.37"&69"–42" 193 HP. Comp.'76, Denny&[illegible]	British India [illegible] Nav. Co. Ld.
334	✻	Hindoo Stl ScSr	British Hull		367'1 B 126 f.	42'8 P 21 f.	31'2 F 80 f.	Newcastle [illegible]	1889		3C.27"43"&70"–48" 400 HP. R.Stephenson&Co.,Ld,Nwc.	T. Wilson, Son[illegible]
335 95232	✻	Hindoo Stl ScSr Poole 87 1 Dk&Shade Dk	British Newcastle	25 [illegible]	205'[illegible]		15'7 12'1	Newcastle Wigham Richardson&Co.	1887	[illegible] Bhds.	3C.21"32"&53"36"–300 HP. Wigham Richardson&Co. Newcastle	P. Richardson
336	✻	Hindoustan I ScSr Lat 81-81 1 Dk&Awn Dk 3 B	French Marseilles	2157 2953	340'7	40'2 B 44 f.	27'9	Newcastle Palmers Co. Ld.	1881	7 Bhds. [illegible]	2Cy.38"&71"–48" 300 HP. Palmers Co. Ld. Newcastle	Cie. Nationale [illegible] Navigation
337 62016		Hinemoa I Sc3mSr Watson 2 Dks	British Wellington	282 542	207'4	25'2	15'2	Greenock Scott&Co.	1876	[illegible] Bhds.	2Cy.33"&58"–36" 150 HP. Greenock Foundry Co., Grnk.	Government of [illegible] Zealand
338 491		Hiogo Maru (ex [illegible]) I Sc3mSr Nye 2 Dks	Japanese Tokio	940 1517	265'8	32'9 F 32 f.	21'1	Sunderland R. Thompson, jun.	1874	[illegible] Bhds.	2Cy.36"&68"–42" 197 HP. G. Clark, Sunderland	Nippon Yusen [illegible]
339		Hipparchus I ScBk Lindegan 85-88	Belgian Antwerp	1486 1863	290'7	34'5	28'2	Newcastle A. Leslie&Co.	1867		2Cy.28"&58"–45" 170 HP. R. Stephenson, Newcastle	Soc. Royale Be[illegible] Sud-Américaine
340	V	Hirondelle I ScSr Vassela 1 Dk	French Dieppe	199 337	172'5	21'3 Q 59 f.	12'6	Hull C.&W. Earle	1857		2Cy.21½"&38"–27" 60 HP. F. Charlton, Grimsby	A. Grandchamp [illegible] Co.
341		Hirondelle I PadSr Leconte 76- 1 Dk	French Havre	94 157	145'0	19'0	9'6	Havre A. Normand	1875		2C.29½"&47½"–38" 180 HP. A. Normand, Havre	Cie. Anonyme des [illegible] Vapeur entre Le Havre [illegible] [illegible]
342 2		Hiroshima Maru W PadSr Hardis	Japanese Tokio	1398 2452	268'8	41'6	22'7	New York	1853		1Cy.81¾"–144" 231 HP.	Nippon Yusen [illegible]
343 82316	BT 1.88	Hispania I ScSr Boothby 82-83 2 Dks 3 B	British Glasgow	2213 3389	364'3 H 151 f.	38'0 P 29 f.	28'6 F 16 f.	Glasgow D.&W. Henderson	1880	7 Bhds.	2Cy.38"&66"–60" 300 HP. D.&W. Henderson&Co., Gla.	Henderson Bros.
344 63780	✻	Hispania I Sc3mSr Short 84-85 1 Dk	British Liverpool	263 420	177'0 Q 60 f.	22'8 F 15 f.	14'0	Greenock Scott&Co.	1870	[illegible] Bhds.	2Cy.47"&38"–24" 60 HP. Greenock Foundry Co., Grnk.	J. Ellis
345	✻ D	Hispania Stl ScSr Heidemann 83-83 1 Dk 2 B	Dutch Rotterdam	940 1327	250'0 P 28 f.	33'6 B 90 f.	18'6 F 34 f.	Vegesack H.F. Ulrichs	1883	[illegible] Bhds. [illegible]	2Cy.30¾"&56"–39¼" 150 HP. Gute Hoffnungs Hütte, Ruhr	W. H. Müller & Co.
346	V	Hispania I ScSr Kuester 80-85 2 Dks	German Stettin	546 867	198'8	27'4	20'8	Stettin Act. Ges. "Vulcan"	1883		2Cy.25"&45"–29½" 95 HP. Act. Ges. "Vulcan," Stettin	Pommersche [illegible] schiffs-Ges. (F. [illegible])
347	V	Hispania I & Stl ScSr Claze 1 Dk&Spr Dk	Swedish Gothenburg	814 1104	210'2	30'1 B 33 f.	21'6	Newcastle Tyne Iron S.B. Co.	1885	[illegible]	3Cy.20"31"58"–36" 150 HP. Wigham Richardson&C. Nwc.	Förnyade Ång. [illegible] Bol. Svenska Lloyd
348 63237	BT 1.88	Historian I ScBk Buckle 85-88 2 Dks	British Liverpool	1202 1830	323'5	33'4	24'0	Belfast Harland&Wolff	1870	6 Bhds.	2Cy.38"&70"–42" 200 HP. J. Jack&Co., Liverpool	Charente S.S. Co.
349	G	Hjalmar Stl ScSr Lund 81-86 1 Dk	Danish Høsse	99 169	125'7 P 37 f.	21'7 F 33 f.	9'5 H 46 f. 16'1	Gothenburg Motala Co.	1881		2Cy.15¾"&28"–17¾" 60 HP. Motala Co.	Dampsk. Selsk. [illegible] Bruhlinaf 1869 [illegible]
350 89128	✻	Ho Kwéi I ScSr Bush 88-88 1 Dk	British Penang	209 376	169'6 B 26 f.	25'0 F 24 f.	10'4	Port Glasgow Blackwood&Gordon	1885	[illegible] Bhds.	2Cy.22"&44"–30" 85 HP. Blackwood&Gordon	Lee Phee Yeow [illegible] Yu Kay
351 76987	BT [illegible]	Hoboken I Pad Sargent	British London	131 413	222'1	22'9	10'5	Glasgow R. Napier&Sons	1873		2Cy.40"&68"–72" 200 HP. R. Napier&Sons, Glasgow	General Stm. N[illegible]
352 90546		Hochelaga W Pad 1 Dk	British Montreal	163 419	178'9	26'3	9'2	Sorel, Quebec H. Beauchemin	1886		1Cy.36"–108" 85 HP. Slackle&Strewson, Mntrl.	Richelieu & Onta[illegible] Nav. Co.
353	DGV	Hochfeld I Sc3mSr Düsing -88 1 Dk	German Stettin	651 855	215'5 Q 78 f.	29'8 F 27 f.	15'5	Stettin Act. Ges. "Vulcan"	1878		2C.29½"&52–31½" 100 HP. Act. Ges. "Vulcan," Stettin	F. Krupp
354		Hodeidah (ex Leda) I Sc3mSr	Turkish Alexandria	626	221'6 F—	30'0	16'9	Glasgow Smith&Rodger	1857		2Cy.26"&52"–36" 120 HP. Day, Summers&Co. S'thampton	Khedivieh Mail S[illegible] Co.
355 87849	✻	Hodgson I TwnScSp Houghton 1 Dk	British Liverpool	8 110	90'0	18'2	8'9	Liverpool R.&J. Evans&Co.	1883	[illegible] Bhds. [illegible]	2Cy.15"&30"–18" 90 HP. J. Jack&Co., Liverpool	Mersey Docks & [illegible] bour Board
356 86216		Hogarth I ScSr Eddes 79-85	British Liverpool	1341 2057	295'4	36'2	24'5	Newcastle A. Leslie&Co.	1882		2Cy.26"&60"–45" 200 HP. R. Stephenson&Co. Nwcstle.	Liverpool, Brazil [illegible] River Plate Stm Nav. [illegible]
357	G	Hohenstaufen I ScBg Kepler 83-85 2 Dks Hur Dk&B	German Bremen	2107 3098	353'0	39'4	32'5	Hull Earles Co.	1874		2Cy.52"&95"–48" 300 HP. Earles Co., Hull	Norddeutscher L[illegible]
358	G	Hohenzollern I ScBg Meier 81-83 2 Dks Hur Dk&B	German Bremen	2107 3098	353'0	39'4	32'5	Hull Earles Co.	1873		2Cy.52"&95"–48" 300 HP. Earles Co., Hull	Norddeutscher L[illegible]
359 83800		Hoihow I ScSr Gyles 84-84	British London	896 1411	248'9	31'3	23'2	Greenock Scott&Co.	1880		2Cy.28"&56"–42" 160 HP. Scott&Co., Greenock	China Nav. Co. Ld.
360	✻	Hok Canton (ex Egeria) C ScSr 2 Dks	Dutch Batavia	305 507	176'6 F 28 f.	26'0	14'0	Glasgow J. Elder&Co.	1875	[illegible]	2Cy.23"&40"–24" 75 HP. J. Elder&Co., Glasgow	Thio Tiaun S[illegible] Boen Sin

✻ = **Lloyd's Register**; D = Nederlandsche Vereeniging; G = Germanischer Lloyd; I = Registro Italiano; R = American Rec[ord]
A = Veritas Austro-Ungarico; H = Veritas Ellenico; N = Norske Veritas; V = Bureau Veritas

In the 2nd Column, B.T., with figures following, signifies Date of report of bottom sighting by Board of Trade Surveyors as condition of PASSENGER CERTIFICAT[E]

In the 3rd Column, C=Composite (iron frame, wood planked); I=Iron; Stl=Steel; W=Wood.

Further particulars of Vessels classed by Lloyd's Register will be found in Lloyd's Register Book and Lloyd's Yacht Register.

		Nom, Matériaux, Gréement, etc. Capitaine et dates d'entrée au service des Armateurs actuels et à bord du navire. Ancien Nom, nombre de ponts, etc.		Pavillon. Port d'Armement	Tonnage Net. Total	Dimensions Long.	Larg.	Creux	Construit Où. Par qui.	Quand.	Nombre des cloisons. Détails sur le lest d'eau.	Machines. Détails. Nom du Constructeur.	Armateurs ou Directeurs-Gérants.
		Hokkaido Maru	I ScSr	Japanese Hakodate	398 643	192·5	26·3	15·7	Stkton-on-Tees M.Pearse&Co.	1857		88 HP.	Hakodate Kisen Kaisha
	✻	Hokuriku Maru (ex Beauchamp)	I ScBn DEκ Aw n Dk	Japanese Yokohama	381 615	185·0	26·2	12·3 18·5	Aberdeen Hall,Russell&Co.	1874	1 Bhds.	2Cy.26½"&48"-30" 102HP. Hall,Russell&Co.,Aberdeen	Hiroumi Nisaburo
		Hokuyo Maru	W ScSr 2 Dks	Japanese Niigata	140 198	119·0	15·4	8·0 B 11 f.	Hiogo J.B.Hannah	1881		2Cy.17"&30"-21" 40 HP. J.Lang	Hokuyo Kaisha
		Holbein *Glassfoole 86-89*	I ScSr	British Liverpool	1338 2053	295·7	36·2	24·5 P 141 f.	Newcastle A.Leslie&Co.	1882		2Cy.26"&60"-45" 200 HP. R.Stephenson&Co.,Nwcstle.	Liverp'l,Brazil&River Plate Steam Nav.Co.,Ld.
	✻D	Holderness *Searle 83-83*	I ScSr 1 Dk 2 B	British Hull	983 1519	255·1	31·6	17·7 Q 86 f. B 60 f. F 31 f.	Hull Earles Co.,Ld.	1882	5 Bhds.	2Cy.30"&55"-36" 130 HP. EarlesCo.,Ld.,Hull	"Holderness" S.S.Co., Ld. (G.R.Sanderson & Co.)
		Holland (ex Louisiana) *Reed 85-86*	I ScBkn 4 Dks	British Liverpool	2419 3847	395·1	40·8	33·2	Newcastle PalmerBros.&Co.	1862	5 Bhds.	2Cy.46"&86"-48" 300 HP. J.Jack&Co.,Liverpool	National Steam Ship Co.,Ld.
357	✻	Holland *Ainsworth 87-88*	I ScTrawler 1 Dk	British Boston	72 145	100·4	20·2	10·3 Q 16 f. P 21 f.	Hull Earles Co.,Ld.	1885	3 Bhds.	3C.11½"17"&30"-21" 45 HP. Earles Co.,Ld.,Hull	Boston Deep Sea Fishing & Ice Co.,Ld.
358	D	Holland *Smidt 76-76*	I ScSr 2 Dks	Dutch Rotterdam	521 726	216·5	28·6	13·8 P 33 f. B 40 f. H 30 f.	Rotterdam Ned.Stoomb.Maats.	1873		2Cy.35"&45"-27" 100 HP. Ned.Stoom.Maats.,Rotrdm.	Nederlandsche Stoomb Maats.
359	DV	Hollander *Van Emmerik 76-81*	I ScSr 2 Dks	Dutch Rotterdam	556 804	201·1	28·0	15·7 Q 115 f. B 7 f. F 18 f. H 24 f.	Rotterdam Ned.Stoomb.Maats.	1884		2Cy.23"&43"-30" 100 HP. Ned.Stoom.Maats.,Rotrdm.	Smith & Co.
360	✻D	Hollandia *De Jong 82-82*	Stl ScSr 1 Dk 2 B	Dutch Rotterdam	1033 1504	249·3	33·8	18·7 H 21 f.	Rotterdam Ned.Stoomb.Maats.	1882	4 Bhds.	2Cy.30½"&56"-39" 150 HP. Ned.Stoom.Maats.,Rotrdm.	W.H.Müller & Co.
361		Holly *Aitken*	I ScSr	British Glasgow	172 368	170·4	23·2	13·4 Q 36 f. B 54 f. F 40 f. Q 25 f.	Glasgow D.&W.Henderson	1875		2Cy.22"&44"-30" 151 HP. T.Wingate&Co.,Glasgow	Glasgow Dublin & Londonderry Steam Packet Co.
362		Holmdel	W Sc	American New York	353 501	161·4	28·7	8·4	Keyport, N.J.	1866			J.Cornell
373	✻	Holme Force *Evans 86-87*	Stl & I Sc3mSr 1 Dk	British Whitehaven	86 266	144·0	23·0	10·2 Q 46 f. B 7 f. F 19 f.	Workington R.Williamson&Son	1887	3 Bhds.	3C 12½" 20"&33"-24" 55 HP. Dunsmuir&Jackson,Glasgw	W.S.Kennaugh & Co.
374	✻	Holmrook *Baines 80-80*	I ScSr 1 Dk	British Newcastle	254 404	150·0	22·3	13·1 F 11 f. Q 76 f. B 14 f.	Newcastle C.S.Swan&Co.	1875	3 Bhds.	2Cy.18½"&36"-24" 50 HP. Pattison&Atkinson,Nwcstl.	J.G.Charlton & Co.
375	V	Holsatia *Droescher 85-86*	I ScSr 2 Dks	German Hamburg	1398 1867	280·8	34·1	23·8	Hamburg ReiherstiegSchiffs. Wrft.	1880		2Cy.34"&62"-42" 180 HP. ReiherstiegSchiffsWrft.,Hmbg.	Hamburg-Amerik. Packetf.Akt.Ges.
376	G	Holsatia *Baner -88*	I ScSr 2 Dks	German Kiel	211 300	151·2	20·0	14·4	Kiel NorddeutscheAct.Ges.	1868		2Cy.17"&36"-36" 60 HP. Schweffel&Howaldt,Kiel	Sartori & Berger
377	GV	Holstein *Haeseker 63-82*	I ScSr 2 Dks	German Hamburg	985 1540	259·1	32·1	23·5	Kiel G.Howaldt	1882	WB	2Cy.27½"&52"-43" 160 HP. HowaldtBros.,Kiel	Th.Wille
378	GV	Holtenau *Ivens 88-88*	I ScSr 1 Dk	German Kiel	422 578	176·5	25·2	15·7 B 40 f.	Kiel G.Howaldt	1882	WB	2Cy.23"&37½"-27" 70 HP. HowaldtBros.,Kiel	Lange Gebrüder
379	✻	Homer *Thompson*	I ScSr 1 Dk 2 B	British London	892 1384	250·0	34·2	17·4 Q 95 f. B 50 f. F 25 f.	SouthShields J.Readhead&Co.	1881	4 Bhds.	2Cy.30"&57"-36" 145 HP. J.Readhead&Co.,S.Shields	Dick & Page
380		Homer Ramsdell	I Sc	American NewYork	611 841	200·0	32·1	11·9	NewburghNY	1887			J.A.P.Ramsdell
381		Honam *Lefaucon 77-87*	Stl PadSr 1 Dk & Awn Dk	British HongKong	1378 2364	270·0	38·1	13·1 21·9	Glasgow A.&J.Inglis	1882		2Cy.40"&72"-120" 500 HP. A.&J.Inglis,Glasgow	Hong Kong, Canton & Macao S.B.Co.,Ld.
382	✻	Hondo *Williams -88*	Stl ScSr 2 Dks Awn Dk 1 B	British Glasgow	1013 1619	235·8	35·0	18·5 26·9	PortGlasgow R.Duncan&Co.	1887	4 Bhds.	3Cy.22"35½"&57"-39" 180 HP. Muir&Houston,Glasg.	Honduras & Central American S.S.Co.,Ld.
383		Honer	I ScSr 1 Dk	Russian Odessa	399 498	210·0	29·0	10·5	London J.&W.Dudgeon	1872		4Cy.15¼"&27½"-14" 60 HP. J.&W.Dudgeon,London	Russian Steam Nav. & Trading Co.
384		Honfleur *Hartfield 75-75*	I Sc3mSr 1 Dk	British London	230 306	150·4	20·9	11·7 Q 43 f. F 18 f.	Dundee GourlayBros.&Co.	1875		2Cy.18"&34"-18" 50 HP. GourlayBros.&Co.,Dundee	London Brighton & S. Coast Railway Co.
385		Honfleur *Merrells*	I ScSr	British Southamptn	107 398	176·4	24·1	12·3 Q 91 f. F 25 f. B 11 f.	Glasgow Aitken&Mansel	1873		3Cy.19½"31"&51"-33" 125 HP. J.&G.Thomson,Glsgw.	Lon.& S.W.Rail.Co.
386	✻	Hong Ann *Moore 87-87*	I ScSr 1 Dk & Awn Dk	British Singapore	299 440	150·0	23·1	10·5 17·1	Newcastle WighamRichardson&Co.	1879	4 Bhds.	2Cy.22"&40"-30" 75 HP. Wghm.Richardson&C.Nwc.	Wee Boon Tek
387	✻	Hong Kong	Stl Sr 2 Dks	British Greenock	 3250	350·0	42·3	26·2 P 17 f. B 164 f. F 46 f.	Greenock Caird&Co.,Ld.	1889	7 Bhds.	3Cy.24"40"&62"-54" Caird&Co.,Ld.,Greenock	Peninsular & Oriental Steam Navigation Co.
388	✻D	Hoopoe (ex Norway ex Beaumont (COLLIS (US)) *Hannagan*	I Sc3mSr 2 Dks & Awn Dk	British Cork	823 1297	236·6	31·6	16·8 24·0	Glasgow Barclay,Curle&Co.	1868	5 Bhds.	3C.21"33"&53"-30" 150 HP. D.Rollo&Sons,Liverpool	Cork S.S.Co.,Ld.
389	✻	Hope (ex Donna Antonia) *Leighton 80-83*	I ScBk 1 Dk 2 B	British Hull	1219 1536	302·5	30·2	18·7 P 186 f. F 27 f.	PortGlasgow L.Hill&Co.	1860	6 Bhds.	2Cy.27"&53"-36" 98 HP. Gilbert&Cooper,Hull	W.S.Bailey
390		Hope *Gray 63-72*	W ScS	British Peterhead	307 452	149·5	28·1	17·0 Q 82 f.	Aberdeen Hall	1873		2Cy.23"&44"-24" 70 HP. GourlayBros.&Co.,Dundee	R.J.Kidd

✻=Lloyd's Register; D=Nederlandsche Vereeniging; G=Germanischer Lloyd; I=Registro Italiano; R=American Record; Á=Veritas Austro-Ungarico; H=Veritas Ellenico; N=Norske Veritas; V=Bureau Veritas.

Dans la 2me Colonne, B.T., avec les indications qui suivent, signifie la date du rapport de visite de la Carène par les experts du Board of Trade pour obtenir le certificat relatif au TRANSPORT DES PASSAGERS.

Dans la 3me Colonne, C=Composite (membrure en fer, bordé en bois); I=Fer; Stl=Acier; W=Bois.

Pour plus amples renseignements sur les navires classés par le Lloyd's Register, voir le Lloyd's Register Book et le Lloyd's Yacht Register.

No. in Book. / Official No.	Registry in which classed	*Ship's Name. Material, Rig, &c. / Master, and Dates of Appointment to present Owner's Service and to Ship / Ship's Late Name. No. of Decks, &c.	Flag. / Port of Registry	Tons. Net. / Gross.	Dimensions. Length / Breadth / Depth / Deck Erections	Built. Where. / By Whom.	When	No. of Bulkheads / Particulars of Water Ballast	Engines. Particulars. / Maker's Name.	Owners or M[…]
391 77502		Hope W Pad 1 Dk	British Montreal	162 [illegible]	134·0|23·0| 7·9 H 20 f. F 122 f.	Sorel F.Gerard	1878		1Cy.32"–96" 50 HP. A.Cantin,Montreal	Sincennes McN[…] ton Line
392 71232		Hope W Sc 1 Dk	British [illegible]	116 170	95·0|25·0| 8·2	Detroit,U.S. J.G.Campbell	1870		2Cy.16"& 18"–18" 150 HP. Kerr Bros.,Walkerville,Ont.	Detroit & Win[…] Ferry Co.
393 52640	✻	Hope I ScSp Simpleton 1 Dk	British Sligo	40 [illegible]	85·5|16·5| 8·6 Q 5 f. P 5 f. B 16 f. F 14 f.	Glasgow T.B.Seath&Co.	1869	4 Bhds.	2Cy.12"& 24"–21" 25 HP. A.Campbell&Son,Glasgow	Sligo Harbour [C…] missioners
394 94283	✻	Hopetoun Stl ScSr Freeman -87 1 Dk2 B	British London	1365 2115	275·2|37·8|19·0 B 108 f. P 35 f. Q f. Fcg	Newcastle Palmers Co.,Ld.	1887	4 Bhds. DB 1166	3Cy.22"35"&57"–39" 200 HP. Palmers Co.,Ld.,Newcastle	W. Wright
395 78418	✻	Horace I Sc Dalziel 82-87 1 Dk2 B	British W.Hartlepl.	1060 1634	259·6|34·5|19·4 B 60 f. Q 95 f. F 30 f.	SouthShields J.Readhead&Co.	1878	4 Bhds. DB al 1/0 f 101 0	2Cy.31"& 59"–36" 145 HP. J. Readhead&Co.,S.Shields	Maclean, Doug[…] Co.
396		Horai Maru (ex Migoto) I Sc 1 Dk & Awn Dk	Japanese Yokohama	407 663	184·6|25·0|14·0 21·5	Renfrew W.Simons&Co.	1870	4 Bhds.	2Cy.33"–30" 90 HP. W.Simons&Co.,Renfrew	Nippon Yusen [K…]
397 89823	✻	Horatia Stl Pad Thornhill 85-87 1 Dk	British Georgetown, Dem.	93 211	175·0|19·5| 7·5	Newcastle C.S.Swan&Hunter	1886	4 Bhds.	2Cy.28"& 47"–42" 99 HP. Black,Hawthorn&C.,G'tshd.	Sproston Dock & Foundry Co.
398 67261	✻	Horatio (WRECKED) I ScSr Gorley 87-88 1 Dk	British Maryport	262 418	156·2|23·7|13·9 F —	Newcastle [illegible]	1873	4 Bhds. [illegible]	2Cy.18"& 36"–24" 50 HP. Bowden Bros.,Newcastle	Hine Bros.
399		Hordaland I ScSr 1 Dk	Norweg'n Bergen	99 143	111·4|18·6| 9·5 Q 32 f. B 58 f. F 3[…]	Malmö [illegible] Mek Verks	1879		2C.19½"&34½"–19½" 50 HP.	Hardanger Sön[d…] landske Damp[…]
400 72658	✻ D	Horden I ScSr Geddes 86-88 1 Dk2 B	British W.Hartlepl.	1034 1616	263·0|33·0|20·0 B 62 f. Q 100 f. F 36 f.	W. Hartlepool W.Gray&Co.	1877	4 Bhds. [illegible]	2Cy.31"&58½"–33" 140 HP. T.Richardson&Sons,Hartpl.	R.Ropner & Co.
401 90103	✻ D	Horn Head I ScSr Thompson 80-84 1 Dk,&Spr Dk3 B	British Belfast	1559 2379	321·8|37·3|25·0 B 72 f.	Belfast Harland&Wolff	1884	5 Bhds. [illegible]	2Cy.34"& 68"–45" 275 HP. Harland&Wolff,Belfast	Ulster S.S.Co.,L[d.]
402 89463	✻ D	Horneburg I ScSr Clark 82-84 1 Dk	British W.Hartlepl.	687 1078	230·5|32·1|13·2 Q 90 f. B 55 f. F 26 f.	Sunderland J.Blumer&Co.	1884	4 Bhds. [illegible]	2Cy.27"& 50"–33" 110 HP. N.E.MarineE.Co.,Ld.,N'wcstl.	J. Lohden & Co.
403		Hornelen I ScSr Hatteberg -87 1 Dk	Norweg'n Bergen	191 266	130·4|20·8|11·2 Q 47 f. B 37 f. H 35 f.	Newcastle C.Mitchell&Co.	1866		2Cy.17½"& 34"–24" 45 HP. Nylands Verks,Christiania	Nordre Bergenh[…] Amtskommune
404	✻	Hornet I TwnScTug Ramsey 63-86 1 Dk	British London	58 86	80·5|18·1| 8·3	Southampton Day,Summers&Co.	1886	4 Bhds.	1Cy.16"& 30"–18" 80 HP. Day,Summers&C.,S'thaptn.	E. & W.I.Dock C[o.]
405 76547		Horrox I ScBn Rendle 83-85	British Liverpool	1091 1714	281·0|33·8|23·9	Southampton T.R.Oswald	1877		2Cy.30"& 60"–42" 180 HP. T.R.Oswald,Southampton	Liverpool, Bra[zil &] River Plate S.N.[…]
406	V	Horsa I ScSr Beuse 75-81 2 Dks&Spr Dk	Danish Aarhus	367 630	212·2|28·0|14·7 P 39 f. F 32 f. 21·1	Renfrew Lobnitz&Co.	1881	[illegible]	2Cy.32"& 60"–36" 160 HP. Lobnitz&Co.,Renfrew	Jydsk Engelske[…] skibs-Selskab(C[…])
407		Horsens I ScSr Hahn 75--1 1 Dk	Danish Svendborg	131 227	153·8|20·6| 9·3 Q 41 f. B 8 f. F 19 f.	Copenhagen Burmeister&Wain	18[…] […]82		1Cy.15"& 30"–26" 60 HP. Burmeister&Wain,C'pnhagn	Det Forenede D[amp-] skibsSelskab(C[…])
408 70123	✻	Horsley I ScSr Bruce 86-88 1 Dk2 B	British Hartlepool	871 1352	243·2|32·8|18·0 B 142 f. Q 84 f. F 27 f.	Hartlepool E.Withy&Co.	1876	4 Bhds. [illegible]	2Cy.29"&54½"–33" 120 HP. T.Richardson&Sons,Hartpl.	G. Horsley & So[n]
409 85179	✻ D	Horton I ScSr Anderson 1 Dk2 B	British London	1187 1827	260·0|36·1|20·2 Q 20 f. P 75 f. B [illegible] B 132 f. F 32 f.	Sunderland Short Bros.	1882	5 Bhds. [illegible]	2Cy.33"& 62"–42" 180 HP. J.Dickinson,Sunderland	J.S.Barwick
410 62304	✻	Hotspur I Sc3mSr Korn 1 Dk2 B	British S.Shields	714 1081	221·2|30·5|18·2 P 106 f. F 21 f.	N.Shields T.&W.Smith	1868	4 Bhds. DB[…] /261t FPT 19t	2Cy.28"& 52"–30" 98 HP. Comp. 72.	"Hotspur" S.S.C[o.] (G.P.Cuttiss)
411 74545	BT1.88	Hotspur I PadSk	British Liverpool	32 224	143·2|20·6|11·8	PortGlasgow W.Hamilton&Co.	1876		2Cy.36"–66" 99 HP. D.Rowan&Co.,Glasgow	E.Griffiths
412 65423		Hotspur I PadSp	British Glasgow	8 163	97·4|18·1| 9·2	Newcastle T.Hepple&Co.	1872		4Cy.35"–54" 45 HP. T.Hepple&Co.,Newcastle	J.Steel & Son
413 85002	✻	Howards I ScSr Earl 1 Dk2 B	British Sunderland	722 1133	225·0|32·5|16·0 Q 66 f. F 28 f. B 46 f.	Sunderland R.Thompson&Sons	1882	4 Bhds. DB[…] /112t FPT 28t APT 29	2Cy.28"& 56"–36" 99 HP. Black,Hawthorn&Co.,G'tshd.	J.F.Marshall
414 76456	✻	Howick I ScSr Ellis 84-84 2 Dk3 B	British N.Shields	1010 1543	255·0|33·3|23·0 F 33 f.	Newcastle A.Leslie&Co.	1878	5 Bhds. DB[…] /280t	2Cy.31"& 58"–33" 150 HP. N.E.MarineE.Co.,Ld.,S'ndrld	H.E.P.Adamson
415 76		Hozui Maru I Bk	Japanese Hiogo	478 774	189·4|27·5|15·6	England	1858		96 HP.	Shimadzu Tada[…]
416		Huacho I Sc3mSr	Ecuador	249 329	149·4|25·0|14·3	Liverpool T.Royden&Sons	1870		2Cy.20"& 36"–18" 50 HP. J.Jack&Co.,Liverpool	Ecuadorian Gov[ern-] ment
417		Huanay (BF1.85) I PadSr Roberts 85-86 1 Dk	Chilian Valparaiso	323 113	165·0|29·3|10·7	Newcastle Palmer Bros.&Co.	1864	4 Bhds.	2Cy.30"–48" 100 HP.	Cia.Sud.-Americ[ana] de Vapores
418 91912	✻ BT3.88	Hubbuck Stl Sc3mSr Brodie 86-88 2 Dks3 B	British London	1776 2718	325·0|40·0|25·9 P 37 f. B 95 f. F 40 f.	Sunderland J.L.Thompson&Sons	1886	6 Bhds. [illegible] FPT[…]	3Cy.27"43"68"–47" 400 HP. T.Richardson&Sons,Hrtlpl.	W.Lund
419 67097	✻ BT1.88	Huddersfield I Sc3mSr Jowers 82-88 1 Dk2 B	British Grimsby	681 1082	231·0|30·2|16·4 P 163 f. F 27 f.	Glasgow J.Elder&Co.	1872	4 Bhds.	2Cy.30½"&57"–36" 120 HP. J.Elder&Co.,Glasgow	Manchester,Sh[effield] & Lincolnshire [R…]
420		Hudiksvall I TwnScSr Sjurrman (ex Oruskoldsvik) 1 Dk	Swedish Hudiksvall	381 480	157·4|22·0|11·7 P 100 f. F 16 f. H 32 f.	Norrköping MotalaCo.	1849		4Cy.11"& 24"–21" 100 HP. W.[illegible]	E.Frisk

✻ = Lloyd's Register; D = Nederlandsche Vereeniging; G = Germanischer Lloyd; I = Registro Italiano; R = American Rec[ord]; A = Veritas Austro-Ungarico; H = Veritas Ellenico; N = Norske Veritas; V = Bureau Verita[s]

In the 2nd Column, B.T., with figures following, signifies Date of report of bottom sighting by Board of Trade Surveyors as condition of PASSENGER CERTIFICATE.

In the 3rd Column, C = Composite (iron frame, wood planked); I = Iron; Stl = Steel; W = Wood.

Further particulars of Vessels classed by Lloyd's Register will be found in Lloyd's Register Book and Lloyd's Yacht Register.

No.	Cote	Nom, Matériaux, Gréement, etc. Capitaine et dates d'entrée au service des Armateurs actuels et à bord du navire. Ancien Nom, nombre de ponts, etc.		Pavillon. Port d'Armement	Tonnage Net Total	Dimensions. Longueur	Largeur	Creux Constructions sur le Pont	Construit. Où. Par qui.	Quand.	Nombre des cloisons. Détails sur le lest d'eau	Machines. Détails, Nom du Constructeur.	Armateurs ou Directeurs-Gérants.
	✻	Hudson *Woodless 84-85*	I ScSr 1 *Dk2 B*	British W.Hartlepl	1105 1720	260·6	34·8	19·4 *B 61 f. Q 100 f. F 37 f.*	W.Hartlepool W.Gray&Co.	1878	4 Bhds. DBa84 ft f91 ft	2Cy.33"& 61"–39" 150 HP. [illegible]	Hudson S.Co.,Ld.
	R	Hudson *Pressman 74-78*	I ScSr 2 *Dks3 B*	American NewYork	1310 1873	287·6	34·0	23·0	WilmingtnDel ThePusey&JonesCo.	1874		1Cy.48"–72 ThePusey&JonesCo.,Wilm.Del.	Cromwell S.S.Line
		Hugh Bell *Allen 76-84*	I Pad	British Middlesbro'	90 176	94·8	30·3	6·1	N.Shields Hepple&Co.	1884		2Cy.24½"–42" 40 HP. Hepple&Co.,NorthShields	The Corporation of Middlesbro'
	✻D	Hugh Sleigh *McNair 87-87*	I ScSr 1 *Dk2 B*	British London	906 1411	242·0	33·2	20·1 *P 18 f. B 44 f. F 27 f.*	Newcastle Palmers'Co.,Ld.	1879	4 Bhds. DBa76 ft f60 ft &a34 ft	2Cy.31"& 58"–36" 140 HP. Palmers'Co.,Ld.,Newcastle	Oriental S.S.Co.Ld.
		Hugh Taylor *Wilson 73-86*	I ScSr 1 *Dk2 B*	British Newcastle	615 978	224·5	29·2	17·6 *P 119 f. F 28 f.*	Newcastle C.Mitchell&Co.	1868	4 Bhds. DBa&f230t	2Cy.27"& 52"–36" 110 HP. R.&W.Hawthorn,Newcastle.	J.C.Straker (Adam,Bros.&Co.)
	✻	Hughenden (ex ...) *Bews 83-85*	I ScBn 1 *Dk2 B*	British W.Hartlepl.	1153 1802	264·3	35·6	19·6 *Q 68 f. P 29 f. F 33 f.*	W.Hartlepool W.Gray&Co.	1881	4 Bhds. DBa&f237t	2Cy.33"& 62"–39" 180 HP. Blair&Co.,Ld.,Stockton	Hudson Shipping Co., Ld.
427	✻N	Hugin *Lohe 83-83*	I ScSr 1 *Dk*	Norweg'n Bergen	305 489	162·7	24·2	12·5 *Q 44 f. B 40 f. F 23 f.*	Hull VulcanIronWksCo.	1883	4 Bhds. DBa&f29t APT7t	2Cy.21"& 40"–27" 65 HP. VulcanIronWorksCo.,Hull	Jacob R.Olsen
428	V	Hugin *Beckman*	I ScSr 1 *Dk*	Swedish Gothenburg	578 751	191·6	25·2	18·7 *B 39 f. H 8 t.*	Norrköping MotalaCo.	1871		2Cy.24"& 48"–48" 90 HP. MotalaCo.,Motala	Ångf.Aktiebol. Svithiod (A.Carlson)
429		Hugo (ex Gaelic) *Hugica*	I ScBk 3 *Dks*	Spanish Bilbao	1713 2652	370·0	36·3	27·9	Belfast Harland&Wolff	1872	5 Bhds.	2Cy.47"& 84"–45" 300 HP. J.Jack,Rollo&Co.,Liverpool	Cie de Nav.La Flecha
430		Hugo *Patterson -87*	I ScSp 1 *Dk*	Swedish Gothenburg	128 190	102·7	20·5	10·7 *P 24 f. H 3 t.*	Gothenburg GöteborgMek.Verkst.	1887		30 HP.	S.Aman
431		Huia	I ScSr 1 *Dk*	British Wellington	90 133	113·2	16·0	8·7 *H 35 t.*	Auckland R.H.Yeoman	1878		2Cy.13"& 23"–15" 25 HP. Masefield&Co.,Auckland	Charles Seagar
432	V	Hulda *Fischer 86-88*	I ScSr 1 *Dk*	German Flensburg	248 319	139·7	22·0	11·2	Stettin Müller&Holberg	1883	WB	2Cy.17"& 30½"–20½" 40 HP. Möller&Holberg,Stettin	F.M.Bruhn
433	✻	Humacao (LOST)	I Sc 2 *Dks3 B*	Spanish Bilbao	1650 2525	321·0	37·0	25·0 *P 36 f. B 72 f. F 38 f.*	Glasgow C.Connell&Co.	1883	6 Bhds. MT43t APT27t DT300t	2Cy.36"& 68"–42" 240 HP. J.&J.Thomson,Glasgow	D.de Eizaga
434		Humayta *Natividade*	I Pad 1 *Dk*	Brazilian Rio Grande do Sul	187 297	171·5	21·6	7·8 *P 49 f.*	PortGlasgow Murdoch&Murray	1878	3 Bhds.	2Cy.27"& 48"–54" 106 HP. D.Rowan,Glasgow	Companhia Nacional de Nav.a Vap.
435	✻	Humber *Couche 87-88*	Stl ScSr 1 *Dk & Wb Frms*	British London	1417 2198	281·4	38·1	20·5 *P 29 f. B 116 f. F 33 f. Q 78 f.*	W.Hartlepool W.Gray&Co.	1888	6 Bhds. DBa341t APT27t	3Cy.22"35"&59"–39" [250 HP. CentralMarineECo., WestHartlepool	Mercantile Steam Ship Co.,Ld.
436	✻	Humber	Stl ScSr	British Glasgow	470 900	230·0	32·1	14·3 *P 136 f. F 35 f.*	P.Glasgow D.J.Dunlop&Co.	1889		3Cy.21½"34"&57"–42" [170 HP. D.J.Dunlop&Co., P.Glasgow	W.Sloan & Co.
437		Humber	I ScSr 1 *Dk2 B*	British Hull	384 596	211·5	26·2	14·2 *Q 36 f. B 88 f.*	Dumbarton A.Denny	1854	4 Bhds.	2Cy.17"& 36"–36" 90 HP. Earles'ShipBldg.Co.,Hull	T.Wilson,Sons & Co.
438	✻D	Humbert *Transon 82-89*	I ScSr 2 *Dks3 B*	British London	1241 1913	270·0	35·1	24·5 *P 32 f. F 34 f.*	Newcastle C.S.Swan&Co.	1878	5 Bhds. DBa&f310t	2Cy.32"& 62"–40" 180 HP. R.&W.Hawthorn,Newcastle	Dixon,Robson & Co.
439		Humboldt *Grimes 86-86*	I ScBg	British Liverpool	1027 1022	268·2	32·1	21·3 *P 156 f. 28·2*	Newcastle A.Leslie&Co.	1866		2Cy.25"& 56"–42" 120 HP. Fawcett,Preston&Co.,Livpl	Liverpool,Brazil & River PlateStmNav.Co.Ld.
440		Humboldt *Manson 75-86*	W ScSr 1 *Dk*	American Eureka,Cal.	294 389	150·5	25·8	9·6	Eureka.Cal. A.Cookson	1875		2Cy.20"& 36"–24" 200 HP. Hinkly,Spiers&Hayes,S.Frcisco	W.J.Sweasey
441	✻	Hungaria *Stephen*	I Sc2mSr 1 *Dk2 B*	British Newcastle	1066 1656	247·5	34·7	20·7 *Q 101 f. H 18 t B 110 f. F 35 f.*	Newcastle H.S.Edwards,Sons& Craig	1884	3 Bhds. WBa&f	2Cy.29"& 56"–42" 150 HP. NorthernMar.Eng.Co.Gtshd	Adriatic S.S.Co.,Ld.
442		Hungaria *Egger 63-87*	I ScBg 2 *Dks*	Austrian Trieste	1273 1976	286·6	35·0	26·3	Trieste LloydAustro-Ungarico	1869		4Cy.29¾"&59"–39" 400 HP. StabilimatoTecnico,Trieste	Lloyd Austro-Ungarico
443	V	Hungaria *Leithäuser 82-86*	Stl ScSr 2 *Dks*	German Hamburg	1513 2036	288·0	36·4	15·8 *P 44 f. B 71 f. F 38 f. 22·7*	Hamburg ReiherstiegSchiffs Wfte.	1884		2Cy.35"& 66"–42" 250 HP. ReiherstiegSchiffsWerfte	Hamburg-Amerik. PacketAkt.Ges.
444	✻	Hungarian *Langlois 86-86*	I ScSr 1 *Dk2 B*	British Glasgow	984 1552	260·7	35·2	18·7 *Q 53 f. F 37 f.*	PortGlasgow Blackwood&Gordn.	1879	5 Bhds. DBa f298t	2Cy.34"& 59"–36" 160 HP. Blackwood&Gordon,P.Glsw	Burrell & Son
445		Hunsdon *Arden*	I PadSr 1 *Dk*	British Calcutta	275 488	166·5	25·2	15·5 *Q 29 f.*	Dumbarton A.Denny	1865		2Cy.54"–54" 200 HP. Denny&Co.,Dumbarton	Clive Steam Tug Co. (Gladstone,Wyllie & Co.)
446	✻D	Hunstanton *Brown 84-86*	I Sc3mSr 1 *Dk2 B*	British King'sLynn	606 942	212·2	31·0	14·7 *P 61 f. F 22 f.*	Middlesbro' R.Craggs&Sons	1882	3 Bhds. DBa f&B146t FPT13t APT20t	2Cy.26"& 48"–36" 98 HP. Wstgrth,English&Mdlbro.	King's Lynn S.S.Co. (W.F.Beaumont,Manager)
447		Hunter	I PadSr 1 *Dk*	British Sydney	203 310	181·5	20·5	11·5	Greenock Scott&Co.	1854		2Cy. 120 HP.	P.Hunter
448	✻	Huntingdon *Braustrom*	I ScSr 2 *Dks3 B*	British London	1464 2224	300·0	36·3	24·7 *B 68 f. F 40 f.*	Newcastle A.Leslie&Co.	1881	5 Bhds. DBa f&a335t	2Cy.36"& 68"–42" 276 HP. R.&W.Hawthorn,Newcstle.	W.Milburn & Co.
449	✻D	Huntsman *Whyte*	I ScSr 2 *Dks3 B*	British N.Shields	1348 2050	281·0	37·2	23·0 *P 26 f. B 58 f. F 29 f.*	Newcastle Palmers'Co.,Ld.	1883	4 Bhds. DBa&f300t	2Cy.33"& 64"–42" 200 HP. Palmers'Co.,Ld.,Newcastle	W.J.Jobling
450		Huron	I TwnSc 1 *Dk*	British Sarnia,Ont.	638 1052	238·5	53·9	12·8	PntEdwd.Ont JohnSmith	1875		2Cy.30"–30" 300 HP. T.Wilson&Co.,Dundas,Ont.	Grand Trunk Rail.Co. of Canada

✻ = Lloyd's Register; D = Nederlandsche Vereeniging; G = Germanischer Lloyd; I = Registro Italiano; R = American Record;
A = Veritas Austro-Ungarico; H = Veritas Ellenico; N = Norske Veritas; V = Bureau Veritas.

Dans la 2me Colonne, B.T., avec les indications qui suivent, signifie la date du rapport de visite de la Carène par les experts du Board of Trade pour obtenir le certificat relatif au TRANSPORT DES PASSAGERS.

Dans la 3me Colonne, C = Composite (membrure en fer, bordé en bois); I = Fer; Stl = Acier; W = Bois.

Pour plus amples renseignements sur les navires classés par le Lloyd's Register, voir le Lloyd's Register Book et le Lloyd's Yacht Register.

No. in Book / Official No.	Registry in which Classed	Ship's Name. Material, Rig, &c. / Master, and Dates of Appointment / Ship's Late Name. No. of Decks, &c.	Flag / Port of Registry	Tons Net / Gross	Length	Breadth	Depth	Built Where / By Whom	When	No. of Bulkheads / Particulars of Water Ballast	Engines Particulars / Maker's Name	Owners
451	V	**Hurricane** I ScSr *Leander* 1 Dk &2 [illegible]	Russian Riga	616 959	231·2	32·2	18·0	Glasgow R. Napier&Sons	1878	5 Blkds. [illegible]	2Cy.33"&48"–39" 110 HP. [illegible]	Riga Steam [illegible]
452 [illegible]	❋	**Hurworth** Stl ScSr *Peterson* 86-88 1 Dk & Web Frms	British W. Hartlepl.	1602 2437	[illegible]	[illegible]	20·3	W. Hartlepool W. Gray&Co.	1888	5 Blkds. [illegible]	3Cy.21½"35½"&58½"–39" Blair&Co.,Ld.,Stk [illegible] HP.	R. Ropner & Co
453 [illegible]		**Hutchinson** I TwnSc *Baker*	American New Orleans	910 1435	252·0	35·0	18·0	W'mingtn Del Harlan&Hollingsworth	1870		2Cy.[illegible] [illegible] HP. [illegible],Philadelphia	Morgan L.&T. [illegible] S.S.Co.
454 [illegible]		**Huzara** Stl ScSr *Campbell* 79-86 2 Dks&3 B	British Glasgow	1336 2078	284·6 P39f.	35·2 B27f.	24·1 F38f.	Glasgow A.&J.Inglis	1880	5 Blkds. [illegible]	2Cy.34"&60"–45" 300 HP. A.&J.Inglis,Glasgow	British India [illegible] Nav.Co.Ld.
455		**Hveen** I ScSr *Svendsen* 85-85 1 Dk	Danish Copenhagen	118 201	142·8	23·0	8·9	Gothenburg Motala Co.	1882		2C.[illegible] 160 HP. Motala Co.,Gothenburg	Det Forenede [illegible] -skibs-Selskab [illegible]
456	V	**Hvidbjørnen** W ScBk *Jensen* -87 1 Dk	Danish Copenhagen	241 387	136·8	24·9	15·3	Elsinore [illegible]	1887		2Cy.19"&35"–21" 50 HP. Helsingørs Jernskibs-og Maskinbyggeri,Elsinore	Den Kongelige [illegible] landske Handel [illegible]
457 [illegible]		**Hyæna** W ScSr *Hill* 74-79	British Liverpool	74 158	108·3	22·0		London	1856		2Cy.16"–12" 60 HP. [illegible],Liverpool	Liverpool Salv [illegible] Association
458 [illegible]	BT 88	**Hydaspes** I Sc3mSr *Thompson* 80-88	British Glasgow	1899 2898	361·5	39·4	29·4	Dumbarton W. Denny&Bros.	1872		2Cy.50"&[illegible]"–54" 300 HP. Denny&Co.,Dumbarton	Peninsular & [illegible] Stm.Nav.Co.
459 [illegible]	BT 88	**Hydra** I ScSr *White* -86	British Liverpool	619 990	237·3 [illegible]	34·2 P27f.	15·1 F40f.	Liverpool W.H.Potter&Sons	1888		3C.17"29"&46"–30" 120 HP. D.Rollo&Sons,Liverpool	Ocean Steam S [illegible]
460 [illegible]	❋	**Hydra** I ScSr *Wadsworth* 80-86 1 Dk	British Goole	240 385	154·7	23·8 F —	13·8	Newcastle E. Lindsay	1873	4 Blkds. [illegible]	2Cy.18½"&36"–21" 50 HP. Pattison&Atkinson,Nwcstl.	J. Bennett
461		**Hydra** I ScSr *Charopulos* -86 1 Dk	Greek Syra	218 347	148·3	21·9	12·5	Port Glasgow Blackwood&Gordn.	1856		4Cy.15"&32"–21" 70 HP.	Hellenic Steam [illegible] Co.
462 [illegible]		**Hye Leong** I ScSr *Mappett* 1 Dk	British Singapore	296 492	180·6 F24f.	25·0 H61f.	12·2	Port Glasgow Blackwood&Gordn.	1882	4 Blkds.	2Cy.28"&50"–30" 110 HP. Blackwood&Gordon,P.Glsw	Tan Beng Gam
463 [illegible]	❋	**Hypatia** I ScBg *White* 80-88 2 Dks & Web Frames	British W. Hartlepl.	1299 2011	285·0 B78f.	36·0 F32f.	22·7	W. Hartlepool E. Withy&Co.	1882	4 Blkds. [illegible]	2Cy.38"&70"–36" 200 HP. T.Richardson&Sons,Hrtlpl.	Herskind & W [illegible]
464		**Hyperion** [illegible] I ScSr *Friberg* 85-86 1 Dk & Shltr Dk	Swedish Stockholm	472 685	165·0	24·5 H27f.	12·3	Norrköping Motala Co.	1884		2C.[illegible]"&48"–26" 120 HP. Motala Co.,Motala	Ångfartygs Bo [illegible] Sverige [illegible]
465	V	**Hypolite Worms** I Sc3mSr *Lemonnier* 2 Dks	French Havre	717 1055	238·0 P42f.	31·6 B81f.	16·8 F39f.	Renfrew Lobnitz&Co.	1882	4 Blkds. [illegible]	2Cy.32"&62"–42" 180 HP. Lobnitz&Co.,Renfrew	Worms, Josse & [illegible]

Additional Ships are inserted in the Supplement at the end of the Register.

❋ = Lloyd's Register; D = Nederlandsche Vereeniging; G = Germanischer Lloyd; I = Registro Italiano; R = American [illegible]
A = Veritas Austro-Ungarico; H = Veritas Ellenico; N = Norske Veritas; V = Bureau Veritas

In the 2nd Column, B.T., with figures following, signifies Date of report of bottom sighting by Board of Trade Surveyors as condition of Passenger Certificate.

In the 3rd Column, C = Composite (iron frame, wood planked); I = Iron; Stl = Steel; W = Wood.

Further particulars of Vessels classed by Lloyd's Register will be found in Lloyd's Register Book and Lloyd's Yacht Register.

No.	Registre	Nom, Matériaux, Gréement, etc. / Capitaine et dates d'entrée au service des Armateurs actuels et à bord du navire. / Ancien Nom, nombre de ponts, etc.		Pavillon. / Port d'Armement	Tonnage. Net / Total	Longueur	Largeur	Creux	Construit. Où. / Par qui.	Quand	Nombre des cloisons. / Détails sur le test d'eau.	Machines. Détails. / Nom du Constructeur.	Armateurs ou Directeurs-Gérants.
1		I. C. Clark	W Sc 2 Dks	British Sarnia	93 115	85·0	20·0	7·0	Newport, U.S.	1864		1Cy.20"–21" 70 HP. W.Moses,Buffalo,U.S.	William B. Clark
2		I. L. Murphy	W Sc 2 Dks	British Ottawa	109 173	101·5	22·0	8·6	SandPnt., Ont A.Miller	1888		1Cy.20"–22" 13 HP. L.&R. Weir,Montreal	I.L.Murphy
3	※	Iago	Stl ScSr	British Hull					Middlesbro' R.Dixon&Co.	1889		3Cy.23"37"&61"–42" G.Clark,Sunderland	T.Wilson, Sons & Co.
4	※	Ianthe *Johnston*	I ScSr 2 Dks	British Glasgow	223 423	177·0 P 25 f.	25·3 F 32 f.	13·4	Glasgow A.Stephen&Sons	1883	4 Blhds.	2Cy.26"&35"–24" 65 HP. PearceBros.,Dundee	J.Bell
5	※	Ianthe *Evans 76-76*	I ScSr 1 Dk	British Aberystwith	83 139	115·0 P 21 f.	18·0 F 16 f.	9·5	PortGlasgow H.Murray&Co.	1873	3 Blhds.	2Cy.12½"&22"–20" 22 HP. Kincaid,Donald&Co.,Grnck.	E.Williams
6	※ I	Iapigia *Belliaanti 85-87*	I ScSr 1 Dk 2 H	Italian Bari	789 1251	229·0 P 136 f.	30·1 F 48 f.	16·8	Sunderland Sndrlnd.S.B.Co.,Ld.	1887	4 Blhds. DBaf 185	3C.18"29"&50"–36" 173 HP. N.E.MarineCo.,Ld.,Sundrld.	Soc. Anon di Nav. à Vap. Puglia
7	※	Ibaizabal (ex Ybarra No. 2) *Gacrastacka 83-87*	I ScSr 1 Dk 2 H	Spanish Seville	466 739	198·0 P 52 f.	28·1 F 3 f.	16·4	Middlesbro' Backhouse&Dixon	1871	4 Blhds. DBa&f	2Cy.26"&52"–36" 115 HP. C.A.Day&Co.,Southampton	Ybarra & Co.
8		Iberia *Sagoes*	I ScBk 3 Dks & Awn Dk	British Liverpool	2954 4661	433·5	45·0	35·1	Glasgow J.Elder&Co.	1873	7 Blhds.	3Cy.56"&(2)78"–60 750 HP. J.Elder&Co.,Glasgow	Pacific Steam Nav. Co.
9	※	Iberia *Corning 83-83*	I ScHmSr 1 Dk & Spr Dk	British Ardrossan	675 1072	230·0	30·1	21·6	PortGlasgow Black'w'd&Gordon	1871	6 Blhds.	2Cy.28"&48"–36" 120 HP. Blackwood&Gordn,P.Glsgw	H.Hogarth
10	※ D	Iberia *Hyker 84-84*	Stl ScSr 1 Dk 2 H	Dutch Rotterdam	963 1260	250·8 B 60 f.	32·2 Lon. 37	15·5 F 28 f.	Kinderdyk L.Smitt & Zoon	1884	4 Blhds. DBa&f&c	2Cy.28"&52"–33" 120 HP. Kon.Maats.deSchelde,Flushing	W.H.Müller & Co.
11		Iberia (COLLISION)	I Sc 2 Dks	French Marseilles	923 1331	254·6 B 22 f.	36·0 F 28 f.	19·5	Leith S.&H.Morton&Co.	1881	6 Blhds. DB&H	2Cy.34"&61"–42" 180 HP. S.&H.Morton&Co.,Leith	Cyprien Fabre & Co.
12	※	Ibex *Osborn 84-84*	I ScSr 1 Dk 2 H	British London	834 1300	245·5 Q 82 f.	34·0 F 32 f.	15·4	Hartlepool E.Withy&Co.	1880	4 Blhds. DB&H	2Cy.31"&58"–36" 130 HP. Blair&Co.,Ld.,Stockton	Jackson Bros. & Cory
13		Ibiapapa	W	Brazilian	163	..	..	..					J.C.Freire & Irmão
14	※ D	Ibis *Jennings 88*	Stl ScSr 2 Dks	British Cork	528 1069	250·0 P 103 f.	32·2 B 28 f.	15·1 F 46 f.	Dundee W.B.Thompson&Co., Ld.	1888	5 Blhds. DB&H	3C.21"36"&58"–42" 175 HP. W.B.Thompson&Co.,Ld.,Dundee	CorkSteamShipCo., Ld.
15		Ibis	I Sc	British Newport	97 169	110·2 Q 8 f.	20·1	9·0	Bristol J.Payne	1881		2Cy.14"–14" 30 HP. J.Payne,Bristol	W.E.Davies
16		Ibis	I ScSrYt	British Glasgow	65 111	106·0	19·9	11·0	Glasgow Tod&McGregor	1866		2Cy.23"–24" 30 HP. Tod&McGregor,Glasgow	J.Ware
17	※	Ibo (ex Conde d'Eu ex Powhatan) *Tavares 68-88*	I ScBn 2 Dks	Portug'se	1060 1533	265·4 F 30 f.	32·1	22·1	Liverpool T.Roydon&Sons	1878	4 Blhds. BD&f	2Cy.32"&61"–39" 170 HP. Fawcett,Preston&Co.L'pool	Mala Real Portugueza
18		Ica	I PadSr 1 Dk & Hur Dk	Brazilian Para	343 481	181·0	28·0	9·5	Birkenhead LairdBros.	1881		2Cy.32"&52"–54" 124 HP. LairdBros.,Birkenhead	Amazon Stm.Nav.Co.
19		Icamyaba	I Pad 1 Dk	Brazilian Para	200	160·0 P 48 f.	24·0	11·0	Birkenhead LairdBros.	1862		2Cy.39"–48" 100 HP. LairdBros.,Birkenhead	Amazon Stm.Nav.Co.
20		Iceberg	I ScCr 1 Dk	British Brisbane	206 214	127·5	25·0	9·2	Brisbane J.W.Sutton&Co.	1882		2Cy.11"–22" 25 HP. A.W.Wilson,London	W.H.Smith&Sons, Ld.
21		Iceland *Pike 74-82*	W ScBk 1 Dk	British S.Johns,NFL	287 423	150·5 F 22 f.	27·3	16·4	Dundee A.Stephen&Sons	1872		2Cy.22"–24" 70 HP. PearceBros.,Dundee	John Munn & Co.
22	※	Ich Dien *Sparling*	I ScTug 1 Dk	British Aberdeen	24 87	94·8	17·1	8·7	PortGlasgow Cunliffe&Dunlop	1877	3 Blhds.	2Cy.17"&36"–20" 40 HP. Cunliffe&Dunlop,Pt.Glsgw.	Aberdeen Steam Nav. Co.
23		Ichang	I PadSr	British London	1049 1782	242·6	36·1	20·8	Glasgow A.&J.Inglis	1873		1Cy.62"–132" 250 HP. A.&J.Inglis,Glasgow	China Nav.Co., Ld.
24		Ichi Maru	W ScSr	Japanese Osaka	106 170	113·5	19·8	12·5	Hiogo YagiMasajiro	1881		15 HP.	Osaka Shosen Kaisha
25	※	Ichnusa (ex Aleha) *Ponza di San Martino*	I ScSr	Italian Genoa	452 722	179·6	27·7	13·5	Rotterdam Nes.Stoom-Maat.	1884	4 Blhds. DBa&f FPTest APT	2Cy.23"&43"–30" 80 HP. NederlandscheStoomMaats.,Rot.	G.Ponza San Martino
26	※	Ida *Carter 82-88*	I ScBn 1 Dk 2 H	British Newcastle	799 1255	239·9 Q 92 f.	34·0 B 52 f.	16·1 F 32 f.	Aberdeen A.Hall&Co.	1883	4 Blhds. DBa&f&c	2Cy.30"&55"–36" 120 HP. Blair&Co.,Ld.,Stockton	Pyman, Bell & Co.
27		Ida (ex Mortje Flore)	I ScSr	British London	171 221	137·7 Q 6 f.	20·2	..	Stockton Richardson,Duck& Co.	1858		2Cy.25"–24" 55 HP. J.Jones&Co.,Liverpool	London, Brighton & South Coast Railway Co.
28		Ida	W Sc 1 Dk & Spr Dk	British Kngstn,Ont.	153 247	95·6 H 59 f.	20·6	6·2	Deser'nto, Ont RathbunCo.	1884		1Cy.12"–14" 75 HP. Black Bros.,Brockville,Ont.	John E. Arnold
29		Ida	I PadSp	British Waterford	87 172	148·6	19·1	8·7	Waterford MalcomsonBros.	1867		2Cy.33"–39" 75 HP. J.Jack&Co.,Liverpool	Waterford Steam Ship Co., Ld.
30		Ida *Linse 85-*	I ScSr 1 Dk	German Danzig	268 426	168·0	24·1	13·8	Danzig J.W.Klawitter	1859		1Cy.13½"&33"–24" 76 HP. F.Schichau,Elbing	T.Rodenacker

※ = Lloyd's Register; D = Nederlandsche Vereeniging; G = Germanischer Lloyd; I = Registro Italiano; R = American Record;
A = Veritas Austro-Ungarico; H = Veritas Ellenico; N = Norske Veritas; V = Bureau Veritas.

Dans la 2me Colonne, B.T., avec les indications qui suivent, signifie la date du rapport de visite de la Carène par les experts du Board of Trade pour obtenir le certificat relatif au TRANSPORT DES PASSAGERS.

Dans la 3me Colonne, C = Composite (membrure en fer, bordé en bois); I = Fer; Stl = Acier; W = Bois.

Pour plus amples renseignements sur les navires classés par le Lloyd's Register, voir le Lloyd's Register Book et le Lloyd's Yacht Register.

No. in Book / Official No.	Registry in which classed	Ship's Name, Material, Rig, &c. / Master, and Dates of Appointment to present Owner's Service and to Ship / Ship's Late Name. No. of Decks, &c.	Flag / Port of Registry	Tons Net / Gross	Dimensions Length	Breadth	Depth	Built Where / By Whom	When	No. of Bulkheads / Particulars of Water Ballast	Engines Particulars / Maker's Name	Owners or M[anagers]
31		Ida C Sc[illegible] 1 Dk	Swedish	111 157	97·8	21·4	8·9 Erections 1º,42 f.	Torskog P.Larsson	1882		25 HP.	P. Larsson
32 12045		Idaho W ScBk Hunter 3 Dks	American S.Francisco	800 1077	198·1	31·2	16·9 23·0	Bath, Me.	1866		2Cy.21"& 40"–34" 160 HP. RisdonIronWorks,Frisco	Pacific Coast
33 12296		Idaho W Sc	American New York	355 496	153·5	33·0	11·8	Brooklyn	1864			J. Affleck
34 93790	✳	Idar Stl ScSc MacDougall 2 Dks3 B	British Liverpool	2595 3950	400·0 P 45 f.	45·2 B 78 f.	28·2 F 10 f.	Belfast Harland&Wolff	1888	6 Bhds. APT	3Cy.29"45"&74"–54" 450 HP Harland&Wolff	"Idar" S.S.C[o.] (E.Bates&Sons)
35 68873	✳	Idlewild I Sc S[r] Lloyd 1 Dk2 B	British Sunderland	510 802	212·0 Q 32 f.	28·4 F 14 f.	15·9 B 39 f.	Sunderland Iliff&Mounsey	1872	3 Bhds.	2Cy.26"&47"–30" 90 HP. G.Clark,Sunderland	London & S[…] S.S.Co.
36	V	Iduna I ScSr 2 Dks	Japanese Osaka	302 478	144·0 P 80 f.	24·1 F 21 f.	12·8	Flensburg Flensburger,Schiffs	1879		2Cy.31"&50"–27" 60 HP. Flnsbrgr.Schiffs.Ges.Flnsbg.	Osaka Shosen
37 [illegible]	✳	Idzumo Maru I ScSr Kobayashi 1 Dk & A[wning] Dk	Japanese	446 720	175·6	25·5	19·8	Paisley H.M'Intyre&Co.	1884	Bhds. FPT, APT	2Cy.22"&41"–30" 73 HP. Fleming&Ferguson,Paisley	Nippon Yusen
38		Iequitaia I PadSr	Brazilian Bahia	250	143·6	17·0	6·6	Dumbarton	1851			Cia. de Nav. [illegible] Bahia
39		Ifigenia I PadSr 2 Dks	Russian Odessa	191 319	160·0	25·0	11·0	Birkenhead JohnLaird	1857		2Cy.47½"–54" 140 HP. Day&Co.,Newcastle	Russian Steam Trading Co.
40		Igor (ex Regesis E.) I ScSr 1 Dk	Russian Odessa	373 531	171·1	25·7	13·1	Paisley H.M'Intyre&Co.	1879		2Cy.20"&38"–30" 80 HP. Muir&Houston,Glasgow	Russian Steam Trading Co.
41	A	Ika I ScSr Schneditz –85 1 Dk	Austrian Fiume	57 110	123·0	14·0	6·0	Trieste Stabilimento Tecni[co]	1885		2Cy.15"&25"–18" 30 HP.	M.Svenljuga & […]
42 813		Iki Maru (now Kane Maru) W Sr	Japanese Tokio	131 228	118·6	19·1	11·5	Nagasaki KosakuBunKioku	1883		80 HP.	Nippon Yusen
43		Iksan I Pad Kadri 82-85	Turkish Constntnple	90	..	..	..		1879			Idarei-Massos[…] (Ottoman Steam […])
44 85017	✳	Ilios I ScSr Potts 1 Dk & Spr Dk3 B	British Sunderland	1297 2020	274·7	37·0	27·1	Sunderland ShortBros.	1882	5 Bhds. APT	2Cy.33"&62"–42" 180 HP. G.Clark,Sunderland	Lumsdon, By[…]
45		Ilissus (BROKEN UP) I Sc 2 Dks		778 1173	296·5 P 74 f.	29·6 F 29 f.	20·0 B 24 f.	LaCiotat Messageries Mariti[mes]	1865		2Cy.62½"–46½" 200 HP. MessageriesMar.,LaCiotat	
46 89340	✳	Illaroo Stl ScSr Kempton 1 Dk2 B	British Sydney	318 515	165·0 Q 21 f.	26·2 B 15 f.	14·5 F 25 f.	Newcastle Wghm. Richardson & [Co.]	1884	5 Bhds.	2Cy.25"&44"–33" 82 HP. Wghm.Richardson&C.Nwc.	Bulli Coal M[ining]
47 79202		Illawarra I PadSr Poole 2 Dks	British Sydney	313 522	190·0	23·7	16·5	Newcastle Wghm. Richardson & Co.	1878		2Cy.32"&58"–60" 190 HP. R.&W.Hawthorn,Newcastle	Illawarra Steam Co.
48 100082		Illinois I ScBn Spencer –89 3 Dks & Spr Dk	American Phil'delphia	1936 3102	343·0	43·0	24·4 31·9	Philadelphia Cramp&Sons	1873		2Cy.48"&90"–48" 425 HP. Cramp&Sons,Philadelphia	American S.S.
49 65932		Ilo I ScSr 2 Dks & Shade Dk	British Liverpool	1129 1794	289·7	38·2 F 59 f.	21·1	Glasgow J.Elder&Co.	1872	5 Bhds.	2Cy.46"&80"–39" 300 HP. J.Elder&Co.,Glasgow	Pacific Steam N[av.]
50 75036		Iluka I PadCr 1 Dk	British Sydney	81 128	110·9	18·3	7·2	BluesBayNSW W.M.Ford	1879		2Cy.18"–46" 30 HP. D.&W.Henderson&Co.,Glas.	Edward W.K[…]
51	V	Imatra (ex Malaren) I ScSr Karstedt –88 1 Dk2 B	German Lübeck	279 383	163·3 Q 27 f.	21·8 B 26 f.	12·7 B 5 f.	Glasgow Dobie&Co.	1872	3 Bhds.	2Cy.20"&35"–30" 60 HP. Kincaid,Donald &Co.,Greenock	Lübeck-Wi[…] Dampfsch.-Gesell
52		Imbitiba I Sc 1 Dk	Brazilian RiodJaneiro	283 485	184·0 P 35 f.	22·6 F 36 f.	12·5	London Lewis&Stockwell	1874	5 Bhds.	2Cy.28"&50"–36" 100 HP. Ravenhill,Eastons&Co.,Lon.	Companhia da […] de ferro […]
53 68868	✳	Imbro I ScSmSr Redman 87-87 1 Dk2 B	British Sunderland	780 1222	243·4 Q & B 138 f.	30·4 F 33 f.	17·9	Sunderland Iliff,Mounsey&Co.	1873	4 Bhds.	2Cy.28"&52"–33" 130 HP. G.Clark,Sunderland	The Pinkney & […] S.S.Co.,Ld.
54 73354	✳	Imbros I ScBg Pashley 2 Dks3 B	British Hull	1276 1958	280·5	36·1 B 24 f.	25·8	Stockton Richardson,Duck & Co.	1876	5 Bhds.	2Cy.32"&62"–39" 170 HP. Blair&Co.,Ld.,Stockton	H. Briggs,Sons & […]
55	NV	Imbs (ex Hjalmar) W ScSr Ramsland 1 Dk	Norweg'n Stavanger	229 315	131·5	25·6 B 32 f.	12·5	Gothenburg A.Samuelsson	1871		2Cy.15½"&30½"–21½" 45 HP. MotalaCo.,Gothenburg	R.G. Ramsland
56 17		Imera (ex Marmion) I ScBn Jannelle 84-88 2 Dks3 B	Italian Venice	822 1258	225·7	31·0	24·6	Sunderland W.Pile&Co.	1870	5 Bhds.	2Cy.26"&55"–30" 198 HP. N.E.MarineEng.Co.,Sndlnd.	Navigazione […] Italiana
57	✳	Imperator Stl Sc3mSr Bechtinger 64-87 2 Dks & Spr Dk	Austrian Trieste	2441 4070	390·0	45·0	25·5 32·1	Trieste LloydAustro-Unga[rico]	1886	Bhds.	1C36"62"(2)72"–60" 780 HP. LloydAustroUngarico,Trieste	Lloyd Austro-U[nga]rico
58		Imperator Alexander II. I Sc[Sr] 2 Dks	Russian Odessa	784 1118	225·6	32·0	19·0	Birkenhead JohnLaird	1858		2Cy.36"&62"–36" 160 HP. J.Penn&Sons,London	Russian Steam Trading Co.
59		Imperatrix Stl Sc3mSr 2 Dks & Spr Dk	Austrian Trieste	4076	390·0	45·0	24·7	Trieste LloydAustro-Ungarico	1888			Lloyd Austro-U[nga]rico
60		Imperatriz Theresa I Pad 1 Dk & Hur D[k]	Brazilian Para	341 481	180·0	30·0	9·5	Birkenhead LairdBros.	1883		2Cy.32"&52"–54" 131 HP. LairdBros.,Birkenhead	Amazon Stm. N[av.]

✳ = Lloyd's Register; D = Nederlandsche Vereeniging; G = Germanischer Lloyd; I = Registro Italiano, R = American R[ecord]; A = Veritas Austro-Ungarico; H = Veritas Ellenico; N = Norske Veritas; V = Bureau Veri[tas]

In the 2nd Column, B.T., with figures following, signifies Date of report of bottom sighting by Board of Trade Surveyors as condition of PASSENGER C[ertificate].

In the 3rd Column, C = Composite (iron frame, wood planked); I = Iron; Stl = Steel; W = Wood.

Further particulars of Vessels classed by Lloyd's Register will be found in Lloyd's Register Book and Lloyd's Yacht Register.

		Nom, Matériaux, Gréement, etc. Capitaine et dates d'entrée au service des Armateurs actuels et à bord du navire. Ancien Nom, nombre de ponts, etc.	Pavillon. Port d'Armement	Tonnage. Net Total.	Dimensions. Longueur | Largeur | Creux	Construit. Où. Par qui.	Quand.	Nombre des cloisons. Détails sur le lest d'eau.	Machines. Détails. Nom du Constructeur.	Armateurs ou Directeurs-Gérants.
		Imperial W ScSr William -86 2 Dks	British Toronto	125 189	109·0 | 22·0 | 7·0 H 4 f.	Toronto W.E.Redway	1886		1Cy.16"–18" 45 HP. J.Doty Engine Co., Toronto	Muskota & Nipissing Nav.Co.
	✻	Imperial Stl Sc3mSr Shade Dk	Chilian Valparaiso		.. .. ..	Birkenhead Laird Bros.	1889		3Cy.31"49"&75"–54" Laird Bros., Birkenhead	Compania Sud Americana de Vapores en el Pacifico
		Imperial I ScSr 1 Dk	Chilian Valparaiso	348 531	176·8 | 26·9 | 11·8 Q 70 f. F 29 f.	Port Glasgow J.Reid&Co.	1881	4 Bhds. FPT 10t APT 30t	2Cy.23"&43"–33" 85 HP. Kincaid, Donald&Co., Grnck.	Associacion de Armadores
		Improvement I Sc 	British Liverpool	105 154	88·7 | 20·1 | 7·9 B 30 f.	Winsford	1865		2Cy.10¾"&14" 25 HP. (H.E.Falk, Winsford)	Salt Union, Ld.
	✻	Inanda StScSr Stuart 83-88 1 Dk & Spr Dk3 B	British Aberdeen	1128 1758	270·4 | 35·2 | 23·0 B 38 f.	Aberdeen Hall, Russell&Co.	1888	5 Bhds. DBcb&f 189t MT 133t	3Cy.21"34"&56"–42" 220 HP. Hall, Russell&Co. Abn.	J.T.Rennie & Son
		Inca I Sc [illegible]	Argentine Buenos Ayres	114 192	103·0 | 16·0 | 9·2	Buenos Ayres Sherman	1870		42 HP. Fada&Pella, Buenos Ayres	J.G.Paez
		Inca	Brazilian Para	 111	.. .. ..					Amazon Stm. Nav. Co.
	✻	Incemore I ScBr Hargreaves 88-88 1 Dk2 B	British Liverpool	1118 1721	260·0 | 36·0 | 18·0 Q 103f. B 158f. F 31f. H 21f	Stockton Richardson, Duck & Co.	1883	4 Bhds. DBcb MT APT FPT	2Cy.32"&60"–39" 170 HP. Blair&Co., Ld., Stockton	S.S. "Incemore," Ld. (W.Johnston & Co.)
	✻D	Inchborva I ScSr [illegible] 2 Dks3 B	British Liverpool	1487 2301	285·2 | 38·1 | 26·2 B 58 f. F 32 f.	Newcastle C.Mitchell&Co.	1881	5 Bhds. DBcf&cb 312t FPT 126t MT 262t	2Cy.35"&68"–45" 300 HP. R.&W.Hawthorn, Newcastle	Hamilton, Fraser & Co.
70	✻D	Inchgarvie I ScSr Moss 2 Dks3 B	British Liverpool	1728 2672	310·5 | 39·0 | 25·2 P 26 f. B 58 f. F 33 f.	Newcastle Sir W. G. Armstrong, Mitchell & Co.	1882	5 Bhds. CDB 421t APT 54t	2Cy.37"&72"–48" 325 HP. R.&W.Hawthorn, Newcastle	Hamilton, Fraser & Co.
71	✻	Inchlonga (ex Madagascar) I ScSr Wacklock 2 Dks & Spr Dk	British Liverpool	1963 3025	320·0 | 40·2 | 28·0 B 75 f.	Sunderland J.Blumer&Co.	1882	7 Bhds. DBcb&f 365t	2Cy.40"&72"–54" 400 HP. C.D.Holmes&Co., Hull	Hamilton, Fraser & Co.
72	✻	Inchmaree I ScSr Major 80-87 2 Dks & Shade Dk3 B	British Liverpool	1287 1975	285·8 | 36·0 | 23·0 P 30 f. F 35 f. Shd. Dk. 64 f.	Newcastle C.Mitchell&Co.	1880	5 Bhds. CDB 350t	2Cy.34"&66"–42" 235 HP. Wallsend Slip Co., Ld., Nwc.	Hamilton, Fraser & Co.
73	✻D	Inchmornish I ScSr Ashby 2 Dks & Shade Dk3 B	British Liverpool	1293 1978	286·0 | 35·1 | 24·2 P 26 f. F 36 f.	Newcastle C.Mitchell&Co.	1879	5 Bhds. DBcb&f 356t	2Cy.34"&64"–42" 225 HP. Wallsend Slip Co., Ld., Nwc.	Hamilton, Fraser & Co.
74	✻D	Inchrhona I ScSr McDonald 2 Dks3 B	British Liverpool	1479 2266	285·0 | 38·1 | 26·3 B 58 f. F 32 f.	Newcastle C.Mitchell&Co.	1881	5 Bhds. DBcb&f 317t FPT 126t MT 262t	2Cy.35"&68"–45" 300 HP. R.&W.Hawthorn, Newcastle	Hamilton, Fraser & Co.
	✻D	Inchulva I ScSr Hough 87-87 2 Dks5 B	British Liverpool	1453 2229	285·0 | 38·1 | 26·3 B 58 f. F 32 f.	Newcastle C.Mitchell&Co.	1881	5 Bhds. DBcb 131t FPT 29t APT 13t MT 263t DT 427t	2Cy.35"&68"–42" 275 HP. Wallsend Slip Co., Ld., Nwc.	Hamilton, Fraser & Co.
76		Indalen I Pad 1 Dk & Shade Dk	Swedish Sundsvall	109 118	91·6 | 24·3 | 5·0 H 37 f.	Stockholm W.Lindberg's M.V.	1887		2Cy.13½"&21"–54" 40 HP. W.Lindberg's M.V., Stockholm	F.Bünsow
77		Independant I PadCr Combe -88	French Arles	115 287	179·1 | 22·9 | 9·2	Creusot Schneider&Co.	1853	APT 29t FPT 29t	1Cy.59"–79" 75 HP. Schneider&Co., Creusot	Cie Generale de Navigation
	DG	Independent I ScSr Shaefer -88 1 Dk & Spr Dk	German Kiel	1003 1318	232·6 | 31·1 | 16·5 24·0	Kiel G.Howaldt	1883	WB 263t	2Cy.29½"&49½"–35½ 140 HP. Howaldt Bros., Kiel	Akt. Ges. Rhederei "Independent"
79		India Stl Sc3mSr Day 63-82 2 Dks & Spr Dk	British Glasgow	2634 4065	390·0 | 42·2 | 21·6 B 132 f. F 28 f. 29·6	Dumbarton W.Denny&Bros.	1881	6 Bhds.	2Cy.50"&86"–54" 500 HP. Denny&Co., Dumbarton	British Ind. Stm. Nav. Co., Ld.
		India I ScBg Crockhart 87-88 2 Dks & Awn Dk	British Glasgow	1592 2477	311·6 | 36·7 | 23·6 30·9	Renfrew W.Simons&Co.	1868	5 Bhds.	2Cy.40"&72"–42" 469 HP. Henderson&Co., Glasgow	Henderson Brothers
81	✻	India I ScSr Powell 77-80 1 Dk	British Goole	212 308	160·0 | 23·8 | 12·5	South Shields J.Readhead&Co.	1876	5 Bhds. DBcb 30 [illegible]	2Cy.20"&38"–26" 70 HP. J.Readhead&Co., S.Shields	J.Bennett
82		India I PadSp Reader 68-76	British London	42 218	138·9 | 20·6 | 11·5	London Westwood&Baillie	1876		2Cy.30"–54" 90 HP. J.Stewart, London	W.Watkins
83	✻V	India I ScSr Nelsen 85-86 2 Dks	German Hamburg	1013 1551	260·8 | 33·6 | 21·9 P 25 f. B 48 f. F 31 f.	Newcastle C.Mitchell&Co.	1881	4 Bhds. DBcb&cb 103t	2Cy.32"&60"–39" 150 HP. Blair&Co., Stockton	Edward Carr
84	I	India I Sc3mSr Tarabotto -88 1 Dk & Awn Dk3 B	Italian Genoa	872 1331	257·2 | 31·0 | 16·8 24·8	Newcastle C.Mitchell&Co.	1870	5 Bhds. MT 34t	2Cy.33"&60"–33" 320 HP. Thompson, Boyd&Co., Nwc.	Navigazione Generale Italiana
85	✻	Indian Prince I ScSr Lawson 81-86 1 Dk2 B	British Newcastle	1233 1912	275·0 | 38·1 | 19·9 Q 28 f. B 60 f. F 32 f.	Sunderland Short Bros.	1886	4 Bhds. CDB 332t APT 79t	3C 21½"35"&58"–39" 180 HP. J.Dickinson, Sunderland	Prince Stm. Shipping Co., Ld.
86	V	Indiana I ScBn Boggs -87 3 Dks & Spr Dk	American Phil'delphia	1936 3102	357·6 | 43·0 | 24·4 31·9	Philadelphia Cramp&Sons	1873		2Cy.48"&90"–48" 425 HP. Cramp&Sons, Philadelphia	American S.S.Co.
87	I	Indipendente I ScBk Ionda 77-86 3 Dks	Italian Palermo	1814 2837	350·0 | 40·0 | 26·1 P 68 f. B 32 f. F 33 f.	Glasgow A.Stephen&Sons	1883	5 Bhds. CDB 187t	2Cy.48"&90"–48" 736 HP. A.Stephen&Sons, Glasgow	Navigazione Generale Italiana
88	✻	Indra Stl ScSr Leslie 88-88 2 Dks3 B	British Liverpool	2333 3582	361·8 | 44·3 | 27·1 P 24 f. B 84 f. F 41 f.	Liverpool T.Royden&Sons	1888	6 Bhds. CDB 711t MT 398t	3C 27"44"&71"–48" 350 HP. Fawcett, Preston&Co, Livpl.	"Indra" Steamship Co. Ld. (Macvicar, Marshall&Co.)
89	✻	Indra I TwnScTug Reid 2 Dks	British London	97 719	212·0 | 30·2 | 15·3	Dundee W.B.Thompson	1885	4 Bhds. DB [illegible] APT [illegible]	4Cy.30"&60"–33" 256 HP. W.B.Thompson, Dundee	The Admiralty
	✻	Indrani Stl ScSr Burkill -88 2 Dks3 B	British Liverpool	2337 3581	361·8 | 44·3 | 27·0 P 24 f. B 84 f. F 41 f.	Liverpool T.Royden&Sons	1888	6 Bhds. CDB 711t MT 398t	3C 27"44"&71"–48" 350 HP. Fawcett, Preston&Co, Livpl.	"Indrani" S.S. Co., Ld. (Macvicar, Marshall&Co.)

✻=Lloyd's Register; D=Nederlandsche Vereeniging; G=Germanischer Lloyd; I=Registro Italiano; R=American Record;
A=Veritas Austro-Ungarico; H=Veritas Ellenico; N=Norske Veritas; V=Bureau Veritas.

Dans la 2me Colonne, B.T., avec les indications qui suivent, signifie la date du rapport de visite de la Carène par les experts du Board of Trade pour obtenir le certificat relatif au TRANSPORT DES PASSAGERS.

Dans la 3me Colonne, C=Composite (membrure en fer, bordé en bois); I=Fer; Stl=Acier; W=Bois.

Pour plus amples renseignements sur les navires classés par le Lloyd's Register, voir le Lloyd's Register Book et le Lloyd's Yacht Register.

No. / Official No.	Class	Ship's Name, Material, Rig, &c. / Master / Ship's Late Name, No. of Decks, &c.	Flag / Port of Registry	Tons Gross / Net	Dimensions (Length, Breadth, Depth) / Erections	Built: Where / By Whom	When	No. of Bulkheads / Water Ballast	Engines: Particulars / Maker's Name	Owners
91		**Indus** [illegible] (ex Binorais) 1 Sc 3 Dks	French Marseilles	1074 1800	264·4|36·1|27·0 H 78 f.	Port Glasgow J. Reid & Co.	1853		2Cy.36½"&[illegible]-35⅜"250 HP Messageries Marit., La Ciotat	Messageries …
92 [illegible]	✻	**Industria** Jones -87 1 ScK 1 Dk	British Hull	62 133	95·0|20·7|10·7 F 16 f.	Hull [illegible]	1867	[illegible] Blkds.	3Cy.15"&[illegible]-24" 55 HP C.D. Holmes & Co., Hull	F. & T. Ross
93	✻D	**Industrie** Nielsen 85- Stl TwnSc 1 Dk	German Coblenz	409 649	[illegible]|28·2|11·4	Kinderdijk [illegible]	1885	[illegible] Blkds. [illegible]	1Cy.14½"&27"-18" 52 HP Dieperen, Lele & Smit, Kdk.	Badische Sch… Dampfsch. Ges.
94 93125	✻	**Industry** Harberry -87 1 ScSr 1 Dk 2 H	British Hull	1189 1830	259·5|35·6|18·3 Q 26 f. P 29 f. B 102 f. F 26 f.	Sunderland Strand Slipway Co.	1887	[illegible] Blkds. [illegible]	3C 19"31"&52"-36" 140 HP N.E. Marine E. Co., Sunderland	W. H. Cocker…
95 [illegible]	✻	**Industry** Shore 1 ScK 1 Dk	British London	108 203	123·5|21·0|11·6	Middlesbro' R. Dixon & Co.	1884	[illegible] Blkds.	2Cy.19"&[illegible]-21" 55 HP Blair & Co., Ld., Stockton	Gt. Yarmouth … Carrying Co. …
96 [illegible]	✻	**Inflexible** Shaw 68-82 1 ScSr 1 Dk & Spr Dk 3 B	British Sunderland	1489 2300	281·0|38·2|27·4	Sunderland Short Bros.	1882	[illegible] Blkds. [illegible]	2Cy.36"&68"-42" 200 HP J. Dickinson, Sunderland	Anderson, H…
97 87186		**Inflexible** Stl ScSr 1 Dk	British [illegible]	119 175	128·3|21·0|8·4 H 15 f.	London Edwards & Symes	1883		2Cy.18"&33"-22" 95 HP J. Hume & Co., Glasgow	James Halste…
98 [illegible]		**Inflexible** Appleton -87 W [illegible]	British [illegible]	86 137	87·6|20·7|9·0	Winsford Ano. Deakin	1878		2Cy.12"-14" 24 HP G. Deakin, Winsford	Wharton … River Salt Wor…
99 73455	V	**Ingeborg** (ex Nouvelle Bretagne) W ScSr 2 Dks	British Hong Kong	437 626	166·0|28·1|18·6 Hs 34 f.	Gothenburg Motala Co.	1873		2Cy.20½"&41"-26½" 70 HP Motala Co., Gothenburg	H. Hunter
100		**Ingeborg** Jespersen 72-82 1 Pad 1 Dk	Danish Fredericia	136 342	165·4|21·9|11·3 H 23 f.	Elbing F. Schichau	1883		2Cy.32"&62"-45⅜" 160 HP F. Schichau, Elbing	De Danske S… (N. Holst)
101	N	**Ingeborg** Hauge -88 W ScSr 1 Dk	Norweg'n Haugesund	158 211	100·9|22·[illegible]|11·4 [illegible]	Skarnberg	1870		2Cy.10¾"22½"-16⅞" 20 HP Gothenburg Mek. Verks., Gothenburg	Nils Hauge
102		**Ingeborg** Johansson -87 C ScSp 1 Dk	Swedish Gothenburg	109 150	99·3|20·8|8·7 Q 56 f. P 29 f.	Södra Garn. (Sweden)	1887		2Cy.9"&18"-12" 12 HP Wilhelmsbergs Mek. Werk. Akt., Gothenburg	J. A. Waller
103	✻D	**Ingerid** Leffers 1 ScSr 1 Dk 2 [illegible]	Dutch Rotterdam	369 588	169·4|26·1|13·7 Q 15 f. B 51 f. F 25 f.	Port Glasgow R. Murray & Co.	1882	[illegible] Blkds. [illegible]	2Cy.19"&34"-36" [illegible] HP Paul & Co., Dumbarton	D. Burger & So…
104	V	**Ingo** Jesselsen 80- 1 ScSr 1 Dk & Spr Dk	German [illegible]	672 865	201·4|28·6|19·2	Hamburg Blohm & Voss	1882	W B	2Cy.24"&42"-33" [illegible] HP Blohm & Voss, Hamburg	J. P. Massmann
105 83821	✻D	**Ingoldsby** Smith 1 ScSr 1 Dk 2 B	British Cardiff	810 1254	235·5|32·2|17·8 Q 38 f. B 50 f. F 25 f.	Newcastle Palmers' Co., Ld.	1881	[illegible] Blkds. [illegible]	2Cy.28"&54"-36" [illegible] HP Palmers' Co., Ld., Newcastle	Lee, Finch & C…
106	V	**Ingraban** Rager -87 2 Dks & Spr Dk & B 1 ScSr	German Heiligenhafen	894 1284	236·2|32·4|15·4 22·1	Hamburg Blohm & Voss	1884	W B	2Cy.28"&48"-30" 175 HP Blohm & Voss, Hamburg	J. P. Massmann
107 87091	✻	**Ingram** Gillings 80-86 1 ScSr 1 Dk 2 B	British London	1122 1748	257·4|34·6|19·4 Q 38 f. P 24 f. B 71 f. F 31 f.	W. Hartlepool E. Withy & Co.	1882	[illegible] Blkds. [illegible]	2Cy.33"&61"-33" 150 HP T. Richardson & Sons, Hrtlpl.	Steel, Young &…
108		**Ingul** 1 ScSr 1 Dk	Russian Odessa	377 484	212·3|29·0|9·4	London Samuda	1872		2Cy.23"&41"-24" [illegible] HP J. Penn & Sons, London	Russian Steam… Trading Co.
109 93155	✻	**Inishowen Head** McCalmont 76-86 Stl ScSr 2 Dks 3 B	British Belfast	1988 3050	311·6|40·2|26·4 B 88 f. F 40 f.	Belfast Harland & Wolff	1886	[illegible] Blkds. [illegible]	3C 24½"37"&61"-48" 320 HP Harland & Wolff, Belfast	Ulster S.S. Co.
110 [illegible]	✻ BT 3 -8	**Inishtrahull** Hetherington 77-85 Stl ScSr 2 Dks	British Glasgow	368 912	231·1|33·1|15·1 P & B 142 f. F 38 f.	Glasgow D. & W. Henderson & Co.	1885	[illegible] Blkds. [illegible]	3Cy 25"37"&58"-18" 240 HP D. & W. Henderson & Co., Glsgw.	Clyde Shipping…
111 [illegible]	I	**Iniziativa** Canzoneri [illegible] 1 ScSr 2 Dks 3 H	Italian Palermo	1293 2032	[illegible]|37·2|23·1 [illegible]	Glasgow A. Stephen & Sons	1881	[illegible] Blkds. [illegible]	2Cy.35"&65"-42" 231 HP A. Stephen & Sons, Glasgow	Navigazione Ge… Italiana
112 [illegible]	V	**Iniziativa** Zino -86 1 ScSr	Italian Savona	292 430	149·0|22·9|12·5 Q 53 f. B 24 f. F 19 f.	Sunderland J. Knox & Co.	1886	[illegible]	2Cy.21"&38"-26" 60 HP G. Clark, Sunderland	Zino Fratelli
113 [illegible]		**Inkrstate** 1 1 Dk	American New York	103	90·0|19·6|10·5	Camden, N.J. J. H. Dialogue			2Cy.15"&30"-22" J. H. Dialogue, Camden	New York Cen… R. R. R. Co.
114		**Inland** Holm [illegible] 1 ScSp 1 Dk	Norweg'n Skien	107 138	94·0|17·9|6·6 B 57 f.	Gothenburg	1876		20 HP	Et Interessent… ([illegible])
115		**Inmaculada Concepcion** Dejoie 1	Spanish Carthagena	61 110	100·0|15·6|9·0	Antwerp Soc. Cockerill	1879		2Cy.11"&23"-11" 75 HP Soc. J. Cockerill, Antwerp	L. Figueras y S…
116 81912	BT [illegible]	**Innishowen** 1 Pad	British Londonderry	53 120	120·3|20·2|6·7	Glasgow Napier, Shanks & Bell	1881		1Cy.31"-42" 21 HP Ross & Duncan, Glasgow	B. M'Crea & J… land
117	✻D	**Ino** Hemrink 84-84 1 ScSr 1 Dk 2 H	Dutch Rotterdam	725 996	214·5|30·2|15·2 F 23 f. B 104 f.	Slikkerveer [illegible]	1884	[illegible] Blkds.	2Cy.27"&47"-30" 106 HP F. Smit, jun., Slikkerveer	Hudig & Veder
118	N	**Ino** Eeede W ScSr 1 Dk 2 B	Norweg'n Stavanger	308 437	145·2|25·8|15·9 Q 30 f. B 40 f.	Stavanger	1883		2Cy.18"&35"-24" [illegible] HP [illegible], Bergen	Strömstenens …
119		**Insulaire** (ex Prince Napoleon) 1 ScSr 2 Dks	British London	352 692	193·9|25·6|18·2	Greenock Scott & Co.	1861		1Cy.20½"&[illegible]-36" 250 HP Comp. 77, Cia Valery [?]	J. MacFee
120		**Interior** Bergeta 1 Sc	Arg'ntine Buen's Ayres	210 380	171·0|24·0|11·3	Glasgow J. & G. Thomson	1867		55 HP	Torrado & Mo…

✻ = Lloyd's Register; D = Nederlandsche Vereeniging; G = Germanischer Lloyd; I = Registro Italiano; R = American [illegible]
A = Veritas Austro-Ungarico; H = Veritas Ellenico; N = Norske Veritas; V = Bureau V[illegible]

In the 2nd Column, B.T., with figures following, signifies Date of report of bottom sighting by Board of Trade Surveyors as condition of PASSENGER CERT[illegible]

In the 3rd Column, C = Composite (iron frame, wood planked); I = Iron; Stl = Steel; W = Wood.

Further particulars of Vessels classed by Lloyd's Register will be found in Lloyd's Register Book and Lloyd's Yacht Register

No.	Cote	Nom, Matériaux, Gréement, etc. / Capitaine et dates d'entrée au service des Armateurs actuels et à bord du navire. / Ancien Nom, nombre de ponts, etc.		Pavillon. / Port d'Armement	Tonnage Net / Brut	Dimensions. Longueur	Largeur	Creux	Construit. Où. / Par qui.	Quand.	Nombre des cloisons. / Détails sur le lest d'eau.	Machines. / Détails. / Nom du Constructeur.	Armateurs ou Directeurs-Gérants.
	*	International *Rogers* -83	I ScSr 1 *Dk2 B*	British Newcastle	1056 1616	248·6	33·8	20·9	Newcastle Schlesinger, Davis & Co.	1883	4 Bhds. DB 60 f. F 51 f.	2Cy.29" & 56"–42" 150 HP. T. Clark & Co., Newcastle	W.H.Cockerline
	*	International	I ScSr 2 *Dks*	British London	1004 1581	240·0	30·1	17·4	Stockton Richardson, Duck & Co.	1870	5 Bhds. W Ba f. & el	2Cy.28" & 54½"–33" 110 HP. Blair & Co., Ld., Stockton	India Rub., Gut. Per. & Telegraph Works Co., Ld.
		International (ex W.B.Pringle)	W Sc 1 *Dk*	British Osaping	174 236	120·0	19·0	7·6	Michigan, U.S. T. Arnold	1870		2Cy.22" & 40"–27" 300 HP. Bartley & Gilbert, Montreal	International Wreckg. & Trans. Co.
	* D	Intrepid *Hodge* 86-87	I ScSr 1 *Dk2 B*	British Sunderland	960 1470	251·3	31·0	18·5	Sunderland Short Bros.	1879	5 Bhds. DB ... & ... F P T ... A P T 10	2Cy.31" & 57"–36" 140 HP. G. Clark, Sunderland	Anderson, Horan & Co.
		Intrepido *Millich* -87	W ScSr 1 *Dk*	Austrian Trieste	74 132	118·1	16·0	11·1	Trieste A. Schiaroni	1876		2Cy.16⅝" & 28⅞"–18" 35 HP. Bartlett & Co., Trieste	A.Cesare & Co.
	*	Inva (ex Argo)	W ScSrYt	Russian St. Petersbg.	104 153	117·8	22·1	12·3	Gosport Camper & Nicholson	1878	...	2Cy.16" & 30"–18" 35 HP. Humphrys, Tennant & Co., Lndn.	Count S.Stroganoff
	BT1.88	Inventor *Jones* 74-87	I Sc3mSr	British Liverpool	1441 2222	325·8 F 28 f.	34·0	25·8	Glasgow Aitken & Mansel	1878		2Cy.35" & 70"–48" 280 HP. J. & J. Thomson, Glasgow	Charente S.S.Co., Ld.
	BT1.88	Inveraray Castle *McTavish* 76-76	I PadSr	British Glasgow	129 230	172·9	20·5	9·3	Glasgow Tod & McGregor	1839		1Cy.50"–48" 90 HP. Tod & McGregor, Glasgow	D.MacBrayne
	*	Invercargill *Sundstrom*	Stl ScSr 1 *Dk*	British Dunedin	123 224	128·7 Q 55 f. B 19 f. F 25 f.	20·1	8·7	Dunedin Kincaid, McQueen & Co.	1885	4 Bhds.	2Cy.13" & 33"–21" 50 HP. Kincaid, McQueen & Co., Dunedin	K.Ramsay
	*	Inverleith *McKay* 87-87	I ScBu 1 *Dk2 B*	British Glasgow	743 1152	225·0 Q 42 f. P 2 ... B 80 f. F 28 f.	32·8	14·6	Stockton Richardson, Duck & Co.	1883	4 Bhds. C D B 3 ...	2Cy.28½" & 53"–33" 99 HP. Blair & Co., Ld., Stockton	J.Mair
131	* D	Invermay *Mitchell* 87-87	I ScBu 1 *Dk2 B*	British Liverpool	952 1439	240·0 P 28 f. Q 54 f. B 82 f. F 30 f.	34·0	17·8	Whitby T. Turnbull & Son	1884	4 Bhds. DB ... f 268	2Cy.32" & 60"–39" 154 HP. Blair & Co., Ld., Stockton	S.S. "Invermay" Co., Ld. (T.Turnbull & Son)
132	*	Invertay *Leighton* 85-85	Stl ScSr 1 *Dk & Spr Dk 3 B*	British Dundee	827 1289	234·0	31·6	22·7	Dundee Gourlay Bros. & Co.	1883	5 Bhds. DB ... & c 131 F P T ...	2Cy.28" & 54"–36" 140 HP. Gourlay Bros. & Co., Dundee	C.Barrie
133		Investigator *Gilchrist*	I Sc3mSr 1 *Dk*	British Adelaide	345 584	210·3 Q 66 f. F 40 f.	28·3	12·5	Glasgow D. & W. Henderson	1882	7 Bhds. F P T 20 t. A P T 27 t.	2Cy.22" & 44"–36" 80 HP. D. & W. Henderson, Glasgow	Adelaide S.S.Co., Ld.
		Invicta *Bass*	Stl PadSr	French Calais	462 1252	312·3 Q 9 f.	33·6	16·0	London Thames I. W. Co., Ld.	1882		1Cy.80"–78" 700 HP. Maudslay, Sons & Field, Lndn.	London, Chatham & Dover Railway Co.
	*	Iolani *Russell* 84-88	I ScSr 2 *Dks 3 B*	British Glasgow	981 1543	264·4 B 16 f. F 25 f.	34·1	22·4	Port Glasgow Murdoch & Murray	1881	4 Bhds. DB ... A P T ...	2Cy.32" & 60"–39" 157 HP. Kincaid, Donald & Co., Grnk.	Raeburn & Verel
	*	Iolanthe	I Sc3mSrYt	American	261 412	189·3	27·1	15·0	Leith Ramage & Ferguson	1881	6 Bhds.	2Cy.27" & 50"–30" 110 HP. M. Paul & Co., Dumbarton	A.M.Singer
	*	Iolo Morganwg *Jenkins* 87-87	I ScSr 1 *Dk2 B*	British Cardiff	830 1292	241·3 P 21 f. B 45 f. F 26 f.	33·2	18·0	Newcastle Palmers Co., Ld.	1882	4 Bhds. DB ... f & c ... A P T ...	2Cy.28" & 54"–36" 99 HP. Palmers Co., Ld., Newcastle	"IoloMrgnwg." S.S.Co. Ld. (E.Thomas, Radcliffe & Co.)
	* BT1.88	Iona *Hosking* 61-87	Stl ScSr 2 *Dks*	British Leith	636 1180	260·7 P 66 f. B 62 f. F 70 f.	32·1	18·2	Glasgow J. & G. Thomson	1883	5 Bhds. DB ...	2Cy.38" & 75"–48" 300 HP. J. & G. Thomson, Glasgow	London & Edinburgh Shipping Co.
139		Iona *Amodio* -81	W Sc 1 *Dk*	British Auckland	159 220	140·0 Q 29 f. Hs 53 f.	19·2	8·6	Auckland	1875		2Cy. 65 HP.	Northern S.S.Co., Ld.
140	BT1.88	Iona *McArthur* 87-87	I PadSk	British Glasgow	113 393	255·5	25·6	9·0	Glasgow J. & G. Thomson	1864		2Cy.50½"–51" 180 HP. J. & G. Thomson, Glasgow	D.MacBrayne
141	*	Iona *Jackson* -88	I ScK 1 *Dk*	British Hull	67 145	100·8 Q 20 f.	20·5	11·0	Beverley Cochrane, Cooper & Schofield	1888	3 Bhds.	2Cy.17" & 33"–22" 45 HP. C.D. Holmes & Co., Hull	Pickering & Haldane
142		Iona (ex Ben Nevis) *Harris* 86-87	I PadK	British London	27 180	121·7	21·2	11·0	Aberdeen Hall, Russell & Co.	1876		2Cy.33"–51" 100 HP. Hall, Russell & Co., Aberdeen	W.Watkins
143	*	Ionia *Kissé* 83-87	Stl ScSr 2 *Dks*	Greek Pireus	881 1181	264·0 P 26 f. B 74 f. F 39 f.	34·0	19·8	Dumbarton A. McMillan & Son	1887	5 Bhds. F P T ... A P T ...	3C.25", 38" & 62"–42" 230 HP. D. Rowan & Son, Glasgow	Panhellenic Steam Ship Navigation.
144	BT1.89	Ionic	Stl Sc4mSr 3 *Dks*	British Liverpool	3070 4753	439·9	44·2	31·5	Belfast Harland & Wolff	1883	7 Bhds.	1Cy.32" & 71"–60" 500 HP. Harland & Wolff, Belfast	Oceanic Steam Nav. Co., Ld.
145	*	Ionio (ex St. Osvald) *Indero* 88-88	I ScSr 2 *Dks*	Italian Palermo	650 1019	239·8 Q 20 f. F 31 f.	28·0	19·8	Newcastle Palmer & Co.	1865	1 Bhds. DB ...	2Cy.27" & 50"–33" 170 HP. N.E. Marine Eng. Co., Sndlnd.	Navigazione Generale Italiana
146		Ionion *Vatis*	W Pad	Greek Syra	183 291	164·3	20·9	10·5	England	1842		2Cy.43"–42" 110 HP.	HellenicSteamShipCo.
147	I	Iosto (ex Iona) *Orengo* 88-88	I ScSr	Italian Palermo	623 1027	239·6 P 53 f. B 53 f. F 64 f.	30·1	17·6	Glasgow J. & G. Thomson	1866		2Cy.39" & 70"–36" 220 HP. Comp 75 J. Key, Kirkcaldy	Navigazione Generale Italiana
148	* V BT1.88	Iowa *Owens* 87-87	I Sc4mSr 3 *Dks & Awn Dk*	British Liverpool	2804 4329	378·2 H 28 f.	43·9	35·0	Liverpool R. & J. Evans & Co.	1879	1 Bhd.	2Cy.48" & 84"–60" 400 HP. J. Jack & Co., Liverpool	"Iowa" S.S.Co., Ld. (G.Warren & Co.)
149	V	Iphigenia *Vollmer* 86-86	I ScSr 2 *Dks*	German Hamburg	1059 1510	251·9 B 76 f. F 62 f.	34·1	21·0	Stettin Act. Ges. "Vulcan"	1883		2Cy.33½" & 63"–43" 220 HP. Act. Ges. "Vulcan," Stettin	Deutsche Dampfschiffs-Rhederei
150		Ipojuca	I Sc 1 *Dk*	Brazilian Pernamb'co	445 558	188·7 P 105 f.	28·0	11·0	Preston Preston I. S. Co.	1866	3 Bhds.	2Cy.43" & 48" 120 HP. J. Watt & Co.	Cia. de Nav. de Pernambuco

*=Lloyd's Register; D=Nederlandsche Vereeniging; G=Germanischer Lloyd; I=Registro Italiano; R=American Record; A=Veritas Austro-Ungarico; H=Veritas Ellenico; N=Norske Veritas; V=Bureau Veritas.

Dans la 2me Colonne, B.T., avec les indications qui suivent, signifie la date du rapport de visite de la Carène par les experts du Board of Trade pour obtenir le certificat relatif au TRANSPORT DES PASSAGERS.

Dans la 3me Colonne, C=Composite (membrure en fer, bordé en bois); I=Fer; Stl=Acier; W=Bois.

Pour plus amples renseignements sur les navires classés par le Lloyd's Register, voir le Lloyd's Register Book et le Lloyd's Yacht Register.

No. in Book. / Official No.	Registry in which Classed.	Ship's Name. Material, Rig, &c. / Master, and Dates of Appointment to present Owner's Service and to Ship. / Ship's Late Name. No. of Decks, &c.	Flag. / Port of Registry.	Tons. Net / Gross	Dimensions: Length	Breadth	Depth	Built. Where. / By Whom.	When.	No. of Bulkheads / Particulars of Water Ballast.	Engines. Particulars. / Maker's Name.	Owners or Managers.
151 [illegible]	BT [illegible]	**Ipswich** I TwnScSr *Henderson 76-83*	British [illegible]	556 1067	260·2 *B 58 f.*	31·3	15·0	Hull Earles Co.,Ld.	1883		2Cy.30"&57"-36" 270 HP. Earles Co.,Ld.,Hull	Great Eastern Railway Co.
152 [illegible]	✻	**Iran** Stl ScSr *Roland* 2 Dks3 B	British Liverpool	2317 3531	380·3 *P 41 f. B 76 f. F 47 f.*	43·0	27·4	Belfast Harland&Wolff	1886	6 Bhds. CDB33ft APT12t	3Cy.25"43"&71"-51" 400 HP. Harland&Wolff,Belfast	"Iran" S.S. Co. (E.Bates&Sons)
153 [illegible]		**Iran** W ScLaunch	British Bombay	109 [illegible]	101·0 *H 9 t.*	19·4	6·0	Bombay Pestonjee Nusserwanjee	1882		2Cy.9"&16"-15" 26 HP. Alison Miller,London	The Arabian S. Launch Co.,Ld.
154		**Iraouaddy** I Sc 3 Dks	French Marseilles	2347 3533	398·2 *F 53 f. H 119 t.*	39·7	30·8	LaCiotat MessageriesMarit.	1872		2Cy.48½"&60½"-48½" 600 HP. MessageriesMar.,LaCiotat	Messageries M[illegible]
155 [illegible]	BT [illegible]	**Ireland** StlPadScSr	British Dublin	533 1952	366·3 *P 283 t.*	38·3	19·2	Birkenhead LairdBros.	1885		2Cy.102"-102" 846 HP. LairdBros.,Birkenhead	City of Dublin Packet Co.
156 [illegible]		**Irene** [illegible] StPadSr [illegible] 1 Dk	British Nassau, N.P.	315 378	209·8 *H 37 t.*	20·3	11·3	Liverpool	1863	5 Bhds.	2Cy. 250 HP.	Charles M.Bum[illegible]
157 [illegible]	✻	**Irene** Stl TwnScSr *Liddicoat -83* 2 Dks	British [illegible]	110 897	301·2 *P 109 f. B 78 f. F 28 f.*	33·2	13·5	Belfast Harland&Wolff	1885	5 Bhds.	1Cy.28"&56"-36" 280 HP. Harland&Wolff,Belfast	London & N.W. Railway Co.
158 99811		**Irene** W ScFlat	British Liverpool	74 109	84·3	19·8	8·6	Northwich J.Cowley	1886		2Cy.12"&20"-16" 18 HP. W.Allen,Manchester	W.Hickson
159	DV	**Irene** [illegible] I Sc3mSr *Mulder 78-84* 2 Dks	Dutch Amsterdam	866 1184	252·0 *Len. 70*	30·0	18·0	Dumbarton W.Denny&Bros.	1863		2Cy.31"&56"-36" 150 HP. Denny&Co.,Dumbarton	Kon.Neder[illegible] Maats.
160	G	**Irene** I ScSr [illegible] -87 1 Dk	Swedish Gothenburg	191 252	108·6 *B 26 f. H 4 t.*	21·2	12·3	Gothenburg	1879		2Cy.11½"&25"-17½" 40 HP.	Ångf.Akt.B[illegible] (Th.Ahrenberg)
161 84281	✻	**Irene Morris** I ScSr *Halliburton* 1 Dk2 B	British Liverpool	929 1440	249·9 *P 26 f. F 29 f. B 46 f.*	33·4	21·7	Whitehaven WhitehavenS.B.Co.	1879	4 Bhds. [illegible]	2Cy.30"&60"-36" 160 HP. [illegible] Jones&Sons,Liverpool	S.S."Irene M[illegible]" Co.,Ld.([illegible])
162 [illegible]	✻	**Iris** I Sc3mSr *Arnold* 2 Dks	British London	760 1168	235·0 *Q 68 f. F 65 f.*	30·0	17·9	London Lewis&Stockwell	1872	4 Bhds. [illegible]	2Cy.46"&72"-30" 200 HP. J.Watt&Co.,London	GeneralSteam[illegible]
163 [illegible]		**Iris** I ScSr *Casa 73-87* 2 Dks	Austrian Trieste	742 1128	221·8	31·5	21·6	Trieste Lloyd Austro-Ungarico	1870		4Cy27½"&49½"-36" 200 HP. ArsenaleLloyd,Trieste	Lloyd Austro-Ungarico
164		**Iris** (ex Amelia Eugenia) I Pad *Filino 85-* 1 Dk	Greek Syra	400 635	201·4	28·1	15·2	England	1861		2Cy.48"-45" 175 HP.	HellenicSteam[illegible]
165	V	**Iris** I ScSr *Wallenius 82-82*	Russian [illegible]	360 453	164·3 *Q 41 f. B 49 f. F 21 f.*	24·3	12·9	Lübeck R.Koch	1884	[illegible]	2Cy.18"&31"-22" 50 HP. O.Henniges&Co.,Berlin	A.Hedman
166	V	**Iris** I ScSr *Eliasson* 1 Dk	Swedish Gothenburg	268 363	134·5 *B 41 f. H 5 t.*	21·7	13·4	Gothenburg GöteborgsMekVerks	1884		2Cy.16"&28"-20" 50 HP. GöteborgsMek.Vrks.Gthbg.	Ångf.Akt.Bol. (T.Ahrenberg)
167		**Iris** I Pad *Margetich*	Urug'yan MonteVideo	98 145	102·0	16·6	6·0	Salto,B.O. Hardy&Co.	1865		31 HP. Hardy&Co.,Salto	S.Ribes & Co.
168 [illegible]	✻	**Iron Acton** I ScBn *Wilcox 85-85* 1 Dk2 B	British Bristol	733 1148	225·2 *B 67 f. Q 82 f. F 22 f.*	32·7	16·0	Middlesbro' R.Dixon&Co.	1881	4 Bhds. [illegible]	2Cy.27"&56"-33" 99 HP. T.Richardson&Sons,Hartpl.	R.& W.King
169 [illegible]	BT 7.88	**Iron Duke** I PadSr	British Dublin	361 740	222·0	28·2	17·2	Liverpool T.Wilson	1844		2Cy.63"-84" 320 HP. Fawcett,Preston&Co.,Lvrpl.	W.Watson
170 [illegible]		**Iron Duke** I ScSk *Allen 87-90*	British [illegible]	95 140	108·3	19·2	8·4	Glasgow T.Wingate&Co.	1857		2Cy.17"-18" 38 HP. Avonside Eng.Wks.,Bristol	A.J.Smith
171 [illegible]	BT12.88	**Iron Duke** [illegible] I PadSp	British Liverpool	40 184	122·2	20·4	10·8	S.Shields MarshallBros.	1864		2Cy.34"-48" 100 HP. MarshallBros.,S.Shields	W.& T.Jolliffe
172 [illegible]	BT2.88	**Iron King** I PadSp	British Liverpool	42 169	124·6	19·4	10·8	St.Antony's T.Potts	1854		2Cy.36"-48" 86 HP. Marshall&Co.,S.Shields	H.J.Ward
173 [illegible]	BT [illegible]	**Iron King** (WRECKED) I PadSp	British N.Shields	17 112	100·8	18·2	9·6	S.Shields J.T.Eltringham	1880		1Cy.36½"-51" 45 HP. T.R.Scott&Co.,N.Shields	F.Warren
174 [illegible]	✻D	**Iron Prince** I ScSr *McKenna* 1 Dk2 B	British N.Shields	851 1328	245·4 *Q 84 f. B 63 f. F 27 f.*	34·1	17·3	Newcastle C.S.Swan&Hunter	1883	4 Bhds. [illegible] FPT [illegible] APT [illegible]	2Cy.29"&56"-36" 132 HP. N.E.MarineE.Co.,Ld.,Nwc.	J.Knott
175 [illegible]	R	**Iroquois** Stl ScSr *Kemble* 3 Dks	American NewYork	2236 2944	..	..	..	Philadelphia W.Cramp&Sons	1888		3Cy.23",36"&60"-36"	W.P.Clyde &Co.
176 [illegible]	✻	**Irrawaddy** I Sc3mSr *Hannah 84-84* 3 Dks	British Glasgow	1697 2622	350·8	36·3	27·1	Dumbarton W.Denny&Bros.	1873	5 Bhds.	4C.21½"32"43"&62½"-42" 202 HP. Qdd.'[illegible] Denny&Co., Dumbarton	British & Burmese Steam Nav.Co.[illegible]
177 [illegible]		**Irrawaddy** I PadSp *Wells*	British Glasgow	396 516	247·2	28·0	7·5	PortGlasgow R.Duncan&Co.	1873		2Cy.31"&59"-54" 162 HP. Rankin&Blackmore,Grnck.	Irrawaddy Fl[illegible] Co.,Ld.
178 [illegible]	✻	**Irrawaddy** I StTrawler *Dyer* 1 Dk	British Hull	55 128	100·0	20·0	10·3	Hull Cook,Welton & Gemmell	1885	3 Bhds.	2Cy.17"&33"-22" 45 HP. C.D.Holmes&Co.,Hull	G.Beeching
179 [illegible]		**Irresistible** Stl ScSr 1 Dk	British [illegible]	93 136	109·0	21·0	9·0	London Edwards&Symes	1883		2Cy.18"&33"-24" 90 HP. Plenty&Sons,Newbury,Eng.	Port Jackson S[illegible] Ld.
180												

✻ = **Lloyd's Register**; D = Nederlandsche Vereeniging; G = Germanischer Lloyd; I = Registro Italiano; R = American [illegible]
A = Veritas Austro-Ungarico; H = Veritas Ellenico; N = Norske Veritas; V = Bureau V[illegible]

In the 2nd Column, B.T., with figures following, signifies Date of report of bottom sighting by Board of Trade Surveyors as condition of PASSENGER CERTI[illegible]

In the 3rd Column, C = Composite (iron frame, wood planked); I = Iron; Stl = Steel; W = Wood.

Further particulars of Vessels classed by Lloyd's Register will be found in Lloyd's Register Book and Lloyd's Yacht Register.

		Nom, Matériaux, Gréement, etc. Capitaine et dates d'entrée au service des Armateurs actuels et à bord du navire. Ancien Nom, nombre de ponts, etc.	Pavillon. Port d'Armement	Tonnage. Net. Total	Dimensions. Longueur, Largeur, Creux. Constructions sur le Pont	Construit. Où. Par qui.	Quand.	Nombre des cloisons. Détails sur le lest d'eau.	Machines. Détails. Nom du Constructeur.	Armateurs ou Directeurs-Gérants.
	✻	Irthington I ScSc Kidd 2 Dk&3 B	British London	1290 1961	270·0|36·1|24·7 B 68 f. F 34 f.	Sunderland Osbourne, Graham & Co.	1880	5 Bhds. [illegible]	2Cy.32" & 62"–40" 200 HP. Black, Hawthorn & Co., Gts.	W. Milburn & Co.
	✻	Irwin I ScSr Humphrey 83-83 1 Dk & W'd Frames	British Newport	611 954	220·0|31·6|12·9 Q 75 f. B 52 f. F 20 f.	Sunderland Kish, Boolds & Co.	1883	4 Bhds. [illegible]	2Cy.27" & 53"–36" 99 HP. Carr & Co., Ld., Sunderland	"Irwin" S.S. Co., Ld. (T. Pugsley & Co.)
	✻	Isa I ScSrYt	British Newcastle	119 175	143·0|18·9|10·6 Len. 84	Montrose Black & Noble	1877	3 Bhds.	3Cy.10" 16" & 28"–24" 50 HP. Douglas & Grant, Kirkcaldy	Edward C. Healey
	✻	Isaac Pennock I ScSc Harris 86-86 1 Dk 2 B	British S. Shields	581 912	209·8|28·8|17·0 Q 74 f. B 46 f. F 30 f.	Whitby T. Turnbull & Son	1871	4 Bhds. [illegible]	2Cy.24½" & 48½"–33" 90 HP. Blair & Co., Ld., Stockton	G. W. Allen & Co.
	V	Isaac Péreire I ScBn Pheridony 87 2 Dks & Spr Dk	French Marseilles	991 1816	310·0|33·5|16·7 F 46 f. B 175 f. 24·3	Glasgow John Elder & Co.	1880		2Cy.42" & 80"–48" 300 HP. John Elder & Co., Glasgow	Compagnie Générale Transatlantique
	✻	Isabel I ScSc Harrison 85-88 1 Dk 2 B	British Whitby	820 1260	240·6|32·6|17·5 [illegible] B 49 f. F 26 f.	S. Shields J. Readhead & Co.	1879	4 Bhds. [illegible]	2Cy.27½" & 52"–33" 110 HP. J. Readhead & Co., S. Shields	R. Harrowing
		Isabel W Pad 1 Dk	British Victoria, BC	147 233	142·4|22·6|9·1	Victoria, B.C.	1866		1Cy. 80 HP.	Henry J. Macdonald
		Isabel II. W	Spanish Manila	80 120	100·7|20·1|5·9	Manila	1867		25 HP.	R. Pozas
		Isabella Stl PadSc Gordon 85	British Dublin	338 831	254·2|30·0|14·2 Q 76 f. B 50 f. F 82 f.	Birkenhead Laird Bros.	1877		2Cy.64"–72" 350 HP. Laird Bros., Birkenhead	London & N.W. Railway Co.
		Isabella W ScSc	British Singapore	84 100	116·0|16·6|7·0	Singapore	1881		30 HP.	Jacob Jens
	✻	Isabella Stl ScSr M'Intosh	British Ardrossan	30 103	95·0|18·6|7·6 Q 38 f. F [illegible]	Greenock [illegible] & Eng. Co.	1886	1 Bhds. [illegible]	2Cy.15" & 29"–21" 35 HP. J. & T. Young, Ayr.	J. & M. Murray
	✻V	Isabelle I ScSmSr Varlet 1 Dk 2 B	French Havre	375 583	184·0|27·5|15·7 P 28 f.	Hull C. & W. Earle	1866	5 Bhds.	2Cy.33½" & 51"–26" 90 HP. Cie. 73, Forges & C. de la Méd. Marsl.	Worms, Josse & Co.
	✻	Isbergues I ScSc Testard 77-83 1 Dk 2 B	French La Rochelle	943 1476	250·0|37·5|17·2 Q 29 f. B 31 f. F 26 f.	Newcastle Palmers' Co., Ld.	1882	4 Bhds. [illegible]	2Cy.31" & 60"–39" 150 HP. Palmers' Co., Ld., Newcastle	D'Orbigny & Faustin fils
		Isbjörnen W ScBk Hansen 2 Dks	Norweg'n Tönsberg	208 303	111·8|31·8|15·3	Tönsberg	1863		40 HP.	Svend Föyn
		Isbrytaren I ScSp Andersson 1 Dk	Swedish Gothenburg	171 405	128·6|33·8|17·1 H 59 f.	Gothenburg Motala Co.	1882		2Cy.33" & 60"–36" 220 HP. Motala Co., Gothenburg	Gothenburg Harbour Com. (G. v. Krusenstjerna)
	✻	Ise Maru I ScSr Nieri 1 Dk & Awn Dk 3 B	Japanese Tokio	771 1244	226·0|32·0|21·0	Paisley H. McIntyre & Co.	1883	4 Bhds. FPT 21 MT 130	2Cy.28" & 52"–36" 116 HP. Hutson & Corbett, Glasgow	Nippon Yusen Kaisha
197	A	Isea I ScSc Buttoraz 88 1 Dk	Austrian Trieste	178 255	140·0|19·1|10·6	Trieste Stabilimento Tecnico	1881		2Cy.18" & 30"–21" 70 HP. Stabilimento Tecnico, Trieste	Flli. Rismondo
	✻	Iser StlScSr Rargoyne 82-88 1 Dk & W'b Fms	British London	1415 2177	275·0|38·0|20·7 Q 72 P 229 B 114 F 32	Newcastle C. S. Swan & Hunter	1888	6 Bhds. [illegible]	3Cy.21½" 35½" & 58½"–39" 250 HP. Blair & Co., Ld., Stk.	Mercantile Steam Ship Co., Ld.
199		Isère I ScSmSr Christian 76-88 2 Dks	French Marseilles	318 479	172·9|28·2|14·3 B 28 f. F 11 f.	Greenock Caird & Co.	1856		2Cy.24½" & 42"–27" 100 HP. Fraissinet & Co., Marseilles	Fraissinet & Co.
200	N	Ishavet (ex Sailing Vessel) W ScBg Nelsen 85 1 Dk 2 B	Norweg'n Tönsberg	183 253	111·1|27·8|14·8 Q 6 f. P 25 f.	Christiania	1859		2Cy.9" & 18"–15" 15 HP. Nylands Verkst., Christiania	Alf Monsen
201		Ishigaki Maru No. 1 I ScSr	Japanese Kobe	1108 1787	259·0|34·6|16·9	Dumbarton W. Denny & Bros.	1866		220 HP.	Ishizaki Hikobaro
		Isidora de Cousino I Sc (ex Isidora) 1 Dk 2 B	Chilian Valparaiso	377 597	199·0|26·9|15·2 Q 68 f. F 28 f.	Middlesbro' Backhouse & Dixon	1873	1 Bhds. [illegible]	2Cy.25" & 48"–30" 90 HP. T. Richardson & Sons, Hartpl.	Cia. Esplotadora de Lota & Coronel
203	✻	Isis I ScBg Slobart 80-85 2 Dks 3 B	British Newcastle	1158 1789	302·5|33·0|25·6 B 81 f.	Hull C. & W. Earle	1862	7 Bhds.	2Cy.36" & 72"–36" 250 HP. C. & W. Earle, Hull	Clapham Steam Ship Co., Ld.
204	✻	Isla Stl ScSc Robertson 1 Dk & Spr Dk 3 B	British Dundee	723 1120	218·0|29·7|22·6	Dundee Gourlay Bros. & Co.	1883	5 Bhds. [illegible]	2Cy.26" & 48"–33" 98 HP. Gourlay Bros. & Co., Dundee	North Sea Steam Shipping Co., Ld.
206		Isla I ScSk	British Glasgow	74 109	78·4|22·1|8·1	Port Glasgow Blackwd. & Gordon	1885		2Cy.11"–16" 22 HP. D. Gordon, Port Glasgow	W. Arrol & Co.
		Isla de Cebu (ex [illegible]) I ScBn Portuondo 83-85 [illegible] 3 Dks	Spanish Barcelona	2120 3242	380·3|38·1|29·1 F 34 f.	Dumbarton W. Denny & Bros.	1878	5 Bhds.	2Cy.50" & 86"–54" 500 HP. Denny & Co., Dumbarton	Cia. Trasatlantica
207	✻	Isla de Luzón I ScBg Gerardo 84-87 2 Dks 3 B	Spanish Barcelona	2773 4252	381·9|44·2|32·6 B 132 f. F 42 f.	Southampton Oswald, Mordaunt & Co.	1882	7 Bhds.	2Cy.52" & 94"–60" 800 HP. Oswald, Mordaunt & Co., Sou.	Cia. Trasatlantica
208	✻	Isla de Mindanao I ScBn (ex Lisnaree) Galiano 84-86 2 Dks & Awn Dk 4 B	Spanish Barcelona	2773 4195	376·5|42·3|33·7 B 13 f.	Barrow [illegible]	1881	6 Bhds. [illegible]	2Cy.43" & 86"–54" 450 HP. Barrow S.B. Co., Barrow	Cia. Trasatlantica
209	✻	Isla de Panay I ScBg Orozty 2 Dks 3 B	Spanish Barcelona	2332 3550	362·4|43·2|29·2 B 86 f. F 35 f.	Greenock Scott & Co.	1882	8 Bhds. [illegible]	2Cy.44" & 82"–54" 500 HP. Greenock Foundry Co., Grnk.	Cia. Trasatlantica
210		Isla de Pinos W (ex Jose R. Rodriguez)	Spanish Habana	105 171	93·2|22·8|8·6	Philadelphia Neafie & Levy	1883		1Cy.18"–18" Neafie & Levy, Philadelphia	P. Garmendia

✻ = Lloyd's Register; D = Nederlandsche Vereeniging; G = Germanischer Lloyd; I = Registro Italiano; R = American Record; A = Veritas Austro-Ungarico; H = Veritas Ellenico; N = Norske Veritas; V = Bureau Veritas.

Dans la 2me Colonne, B.T., avec les indications qui suivent, signifie la date du rapport de visite de la Carène par les experts du Board of Trade pour obtenir le certificat relatif au TRANSPORT DES PASSAGERS.

Dans la 3me Colonne, C = Composite (membrure en fer, bordé en bois); I = Fer; Stl = Acier; W = Bois.

Pour plus amples renseignements sur les navires classés par le Lloyd's Register, voir le Lloyd's Register Book et le Lloyd's Yacht Register.

1 No. in Book / Official No.	2 Registry in which classed	3 Ship's Name. Material, Rig, &c. / Master, and Dates of Appointment to present Owner's Service and to Ship / Ship's Late Name. No. of Decks, &c.	4 Flag. / Port of Registry.	5 Tons Net / Gross	6 Dimensions. Length / Breadth / Depth (Deck Erections)	7 Built. Where. / By Whom.	8 When.	9 No. of Bulkheads / Particulars of Water Ballast.	10 Engines. Particulars. Maker's Name.	11 Owners
211		Isla de Puerto Rico I Sc		1104 1638	272·0 | 36·8 | 21·9	Liverpool John Laird	1855		2Cy.60"–42"	
212	V	Island I Sc3mSr Skjødt 83-84 2 Dks & Spr Dk	Danish Copenhagen	1819 2844	324·4 | 39·1 | 22·0 29·7	Copenhagen Burmeister&Wain	1882	[illegible]	4Cy.36"&72"–42" 400 H. Burmeister&Wain,Cpnhgn	Dampskibs-S. "Thingvalla" (D.)
213 12141		Island Home W Pad Baggett 74-85	American NewBedford	330 481	184·0 | 29·6 | 11·0	Green Pnt NY	1855		487 HP.	E.T.Pierce
214 85693		Islander Stl TwnScSr Robertson -86	British Glasgow	478 1495	240·0 | 42·0 | 14·0	Glasgow [illegible]	1888		3Cy.20"31"&52"–36" Dunsmuir&Jackson, Glasgow 289 HP	G.W.Roberts
215 69253		Islay I PadSr	British Liverpool	1109 1538	271·0 | 35·5 | 22·5	Glasgow J.Elder&Co.	1873		4Cy.38"&76"–51" 300 HP. J.Elder&Co.Glasgow	Pacific Steam
216 80327	BT5.88	Islay I PadSr McNeill 6 -68	British Glasgow	168 305	206·6 | 23·4 | 10·7 Q 26 f.	Glasgow Barclay,Curle&Co.	1867		2Cy.46"–60" 145 HP. Barclay,Curle&Co.,Glasgow	D. MacBrayne
217 93808	✳	Isle of Anglesey Stl S ScSr Williams 88-88 1 Dk 2 B	British Liverpool	783 1228	250·0 | 33·5 | 16·5 [illegible]	South Shields [illegible]	1888	[illegible]	3Cy.18"29"&48"–30" J.Readhead&Co., S.Shields 114 HP	Isle of Angles. Co.,Ld. (J.T.W.)
218 88109	✳	Isle of Cyprus I ScSr Pettersson 88-88 1 Dk 2 B	British Newcastle	744 1149	225·2 | 33·5 | 15·1 [illegible]	Sunderland [illegible]	1883	[illegible]	2Cy.26"&53"–33" 92 HP. WallsendSlipC.,Ld.,N'castle	Dixon, Robson
219 88736	✳	Isle of Dursey I ScSr Pelton 86-86 1 Dk	British Newcastle	620 963	210·0 | 31·3 | 11·1 Q74 f. B59 f. F26 f.	Sunderland J.Priestman&Co.	1883	[illegible]	3C.15g"22"&44"–33" 95 HP. WallsendSlipC.,Ld.,N'castle	Dixon, Robson
220 88742	✳D	Isle of Elba I ScBn Williamson 88-88 1 Dk & W h Frms	British Newcastle	629 976	210·9 | 32·4 | 14·5 Q78 f. B49 f. F25 f.	North Shields T.&W.Smith	1883	[illegible]	2Cy.25"&50"–33" 90 HP. WallsendSlipC.,Ld.,N'castle	Dixon, Robson
221 88769	✳D	Isle of France I ScSr Sands 85-86 1 Dk 2 B	British Newcastle	1048 1609	260·0 | 35·2 | 17·0 Q90 f. B62 f. F28 f.	Newcastle TyneI.S.B.Co.,Ld.	1883	[illegible]	2Cy.29"&56"–36" 132 HP. N.E.MarineEng.Co.,Ld.,Sld.	Dixon, Robson
222 91490	✳D	Isle of Georgia I ScSr Hodge 83-85 1 Dk 2 B	British Newcastle	870 1351	235·0 | 34·1 | 17·0 Q87 f. B55 f. F27 f.	SouthShields J.Readhead&Co.	1885	[illegible]	2Cy.28"&55"–36" 130 HP. Readhead&Co.,S.Shields	Dixon, Robson
223 91469	✳D	Isle of Hastings Stl ScSr Sutherland 84-85 1 Dk 2 B	British Newcastle	1078 1616	250·0 | 35·2 | 18·7 P29 f. Q60 f. B60 f. F32 f.	Newcastle R.S.Edwards&Sons	1885	[illegible]	3C.20"33"&53"–36" 140 HP. Black,Hawthorn&Co.,Gtshd	Dixon, Robson
224		Isly I ScBn Massères 1 Dk & Spr Dk 3 B	French Marseilles	951 1398	246·0 | 31·7 | 23·6 B 60 f.	Newcastle WighamRichdsn.&Co.	1883	[illegible]	2Cy.34"&64"–42" 250 HP. WighamRchrdson&Co.New	Cie.de Nav.Mi.
225		Ismail (ex Theodore R., ex Harald Hoper) I ScSr	Russian [illegible]	246 393	146·5 | 22·2 | 12·0	Sunderland OsborneGraham&Co.	1880		2Cy.20"&40"–30" 60 HP. Geo.Clark,Sunderland	P.Gagarin & C.
226 95308	✳ BT12.88	Ismailia Stl Sc3mSr Inch 71-88 2 Dks 3 L	British [illegible]	1940 3008	337·0 | 41·2 | 26·5 B63 f. F45 f. F46 f.	Newcastle Palmers'Co.,Ld.	1888	[illegible]	3C.25"41"&66"–45" 350 HP. Palmers'SB.Co.,Ld.,Newcastle	Nelson, Donkin
227		Ismid W P[illegible] Mustafa 88-84-8	Turkish Cnstntinple.	138	171·8 | 30·1 | 10·3	Constntinople. [illegible] Dockyard	1886		2Cy. 109 HP.	Idarei Massous. (Ottoman Steam N.)
228	V	Isolde I ScSr Boss -84 1 Dk	German Hamburg	174 229	95·5 | 23·0 | 12·3	Elbing F.Schichau	1883		2Cy.12½"&21½"–14" 140 HP. F.Schichau,Elbing	Amsinck & Ha.
229	✳	Istinié (ex Anole) Stl ScSr Hassein Bey -88 1 Dk 2 B	Turkish Cnstntinple.	274 443	188·2 | 24·0 | 12·8 Q 25 f.	Hull M.Samuelson&Co.	1864	4 Bhds.	2Cy.29½"&54½"–24" 98 HP. M.Samuelson&Co.,Hull	Idarei Massous. (Ottoman Steam N.)
230 55088		Istrian I Sc1mSr Ball -88 3 Dks	British Liverpool	1955 2963	390·0 | 37·2 | 29·3	Belfast Harland&Wolff	1867		2Cy.32"&63"–54" 300 HP. G.Forrester&Co.,Liverpool	S.S."Istrian"C. (F.Leyland & Co)
231		Itajahy I Sc 1 Dk	Brazilian Rio Janeiro	433 702	216·5 | 28·2 | 13·2 P 100 f. F 23 f.	PortGlasgow H.Murray&Co.	1871	5 Bhds.	4Cy.26"&45"–30" 180 HP. D.Rowan,Glasgow	A.J.dos Santos
232 68011		Italia I ScSr Craig 74-84 1 Dk & Awn Dk 3 B	British Glasgow	1442 2236	306·0 | 31·1 | 21·9 B 21 f.	PortGlasgow R.Duncan&Co.	1872	6 Bhds.	2Cy.38"&68"–42" 350 HP. Finnieston Co.,Glasgow	Henderson Bros.
233 69445	✳	Italia I Sc3mSr Waterman 1 Dk	British Middlesbro'	239 460	175·1 | 22·8 | 13·9 Q 60 f.	Greenock Scott&Co.	1869	5 Bhds.	2Cy.17¼"&35¼"–24" 51 HP. GreenockFoundryCo.,Grnk.	J.M.Lennard &
234	V	Italia I ScSr Pust 75-82 1 Dk & Hur Dk	German Stettin	666 874	199·8 | 27·4 | 13·7 20·8	Stettin Act.Ges."Vulcan"	1881	[illegible]	2Cy.24½"&43½"–27½" 90 HP. Act.Ges."Vulcan,"Stettin	Pommersche schiffs-Ges. (F.Ivers)
235	I	Italia I ScSr Pavelo 73-78 2 Dks	Italian [illegible]	307 569	183·0 | 25·6 | 16·7 P 9 f. H 38 f.	Chester Graham&Co.	1855		2Cy.23"&43"–30" 155 HP. G.Ansaldo&Co.,Sampirdarena	Navigazione Gen. Italiana
236		Italia (ex Goodly) I ScSr Villani 1 Dk	Italian Civitav'chia	122 197	116·8 | 20·5 | 8·9 Q36 f. B11 f. F12 f.	Paisley H.McIntyre&Co.	1880	[illegible]	2Cy.15"&28"–25" 50 HP. [illegible]	P.Fumaroli
237	✳	Italia I ScSr Daren 88-88 1 Dk Sp Dk & W h Frms	Norweg'n Bergen	710 922	195·0 | 28·0 | 20·6 B 29 f.	Newcastle W.Dobson&Co.	1888	[illegible]	3Cy.15"23"&40"–30" 93 HP. N.E.MarineE.Co.Ld.,Newcastle	J.L.Mowinckel
238	✳	Italia I ScSr da Silva 88-88 2 Dks	Portug'se Oporto	404 583	175·0 | 25·1 | 14·7 Q62 f. B33 f. F29 f.	London Black&Drury	1883	[illegible]	2Cy.21"&42"–30" 80 HP. GourlayBros.&Co.,Dundee	Cia.Portugueza vegação a Vapor
239	✳	Italica I ScSr Guerricaechevarria 85-87 1 Dk 2 B	Spanish Seville	750 1156	220·0 | 31·3 | 16·6 P 127 f. F 27 f.	Sunderland J.L.Thompson&Sons	1884	[illegible]	2Cy.37"&39½"–33" 100 HP. Blair&Co.,Ld.,Stockton	Ybarra & Co.
240	✳	Italie (ex Italy) Stl T ScSr Sylvester 84-87 1 Dk	French Dieppe	220 487	190·6 | 26·1 | 14·0 B 11 f. F 36 f.	Glasgow J.Elder&Co.	1885	[illegible]	3C.12 22"&(1)32"–21" [illegible] J.Elder&Co.,Glasgow [illegible] HP	London,Brighton South Coast Rail

✳ = Lloyd's Register; D = Nederlandsche Vereeniging; G = Germanischer Lloyd; I = Registro Italiano; R = American [illegible]
A = Veritas Austro-Ungarico; H = Veritas Ellenico; N = Norske Veritas; V = Bureau Veritas

In the 2nd Column, B.T., with figures following, signifies Date of report of bottom sighting by Board of Trade Surveyors as condition of PASSENGER CERTIFICATE

In the 3rd Column, C = Composite (iron frame, wood planked); I = Iron; Stl = Steel; W = Wood.

Further particulars of Vessels classed by Lloyd's Register will be found in Lloyd's Register Book and Lloyd's Yacht Register.

No.		Nom, Matériaux, Grément, etc. / Capitaine et dates d'entrée au service des Armateurs actuels et à bord du navire. / Ancien Nom, nombre de ponts, etc.		Pavillon. / Port d'Armement	Tonnage. Net. / Total	Dimensions. Longueur	Largeur	Creux	Construit. Où. / Par qui.	Quand.	Nombre des cloisons. / Détails sur le lest d'eau.	Machines. / Détails. Nom du Constructeur.	Armateurs ou Directeurs-Gérants.
		Italy *Pearce 76-88*	1 ScBkn	British Liverpool	2624 4169	389·0	42·3	28·7	Glasgow J.Elder&Co.	1870		2Cy.60"&96"–48" 600 HP. J.Elder&Co.,Glasgow	National Steam Ship Co.,Ld.
		Itamaraty	Stl PolSr	Brazilian RiodeJaneiro	253 401	200·5	30·4	7·5	Glasgow J.&G.Thomson	1885		2Cy.47"–48" 160 HP. J.&G.Thomson,Glasgow	Megaw & Norton
						B 101t.							
		Itata *Moore 85-87*	1 Sc *2 Dks & 1 trn Dk*	Chilian Valparaiso	1201 1706	290·7	39·9	19·7 27·2	Liverpool R.&J.Evans&Co.	1873	4 Blds. pDB	2Cy.46"&86"–42" 340 HP. Fawcett,Preston&Co.Lvrpl.	Cia.Sud.Americana de Vapores
		Itchen *Weyer*	1 ScK	British Southampton	148 193	106·1	22·3	10·8	Northam Day,Summers&Co.	1872	3 Blds. FPT28	2Cy.10"&20"–18" 20 HP. Day,Summers&Co.Southptn	Osborn & Wallis
		Ithaka	1 Sc *1 Dk*	Greek Syra	703	210·6	28·5	..	England	1873			P.Vaglianos
	✻	Itamba	1 ScSr *1 Dk*	British Liverpool	155 273	130·0 *P* 83 *f.*	25·2	9·3 *F* 23 *f.*	Barrow BarrowS.B.Co.,Ld.	1885		2Cy.20"&40"–24" 67 HP. BarrowCo.,Ld.,Barrow	T.W.& E.H.Cookson
	✻	Ituna *Rowely -79*	1 ScSr *1 Dk 2 B*	British London	442 684	181·2 *Q* 60 *f.*	28·3 *B* 26 *f.*	14·3 *F* 17 *f.*	Kinghorn J.Key	1877	3 Blds. DBaft 130	2Cy.23"&40"–36" 80 HP. Tayler Bros.,Birkenhead	G.R.Carter
	✻	Ituna *Telfer 88-88*	Stl ScSr *1 Dk*	British Grangemouth	244 408	160·0 *Q* 51 *f.*	25·2 *B* 38 *f.*	12·0 *F* 21 *f.*	Grangemouth Dobson&Charles	1884	4 Blds. FPT38 APT28	2Cy.20"&38"–30" 70 HP. Hutson&Corbett,Glasgow	T.Cowan
249	✻	Ituna	Stl ScSr Yt *2 Dks*	British Glasgow	98 172	137·6	19·7	11·3	Glasgow A.&J.Inglis	1886	4 Blds.	3C.10"16"&26"–22" 70 HP. A.&J.Inglis,Glasgow	R.H.Alexander
250	✻	Ituni *Young*	1 ScSr *1 Dk & W. L Frames*	British London	779 1226	237·0 *P* 130 *f.*	32·2	15·5 *F* 24 *f.*	Newcastle W.Richardson&Co.	1884	4 Blds. FPT30 APT14	2Cy.27"&54"–36" 120 HP. W.Richardson&Co.,N'castle	Demerara & Berbice S.S.Co.,Ld.
251	✻	Ivanhoe *Mock -83*	1 ScSr *1 Dk 2 B*	British Newcastle	624 974	210·5 *Q* 78 *f.*	31·8 *B* 45 *f.*	14·3 *F* 29 *f.*	Sunderland D.Baxter&Co.	1883	4 Blds. DBaft 185 APT38	2Cy.26"&47"–36" 96 HP. G.Clark,Sunderland	G.C.Ward
252	✻	Ivanhoe *Gray 85-86*	1 ScSr *1 Dk 2 B*	British Glasgow	599 942	215·4 *Q* 70 *f.*	29·9 *B* 47 *f.*	15·0 *F* 31 *f.*	PortGlasgow Murdoch&Murray	1881	4 Blds. DBaft 236 FPT24 APT18	2Cy.28"&52"–36" 98 HP. W.Kemp,Glasgow	Maclay & M'Intyre
253		Ivanhoe *Williamson 79-80*	1 PdlSk	British Glasgow	 262	225·3	22·2	8·3	Glasgow D.&W.Hndrsn&Co.	1880		2Cy.43"–36" 123 HP. D.&W.Henderson&Co.,Gls.	Frith of Clyde Steam Packet Co.,Ld.
254	✻	Ivanhoe *McFagan 83-84*	1 ScSr *1 Dk*	British Glasgow	58 172	120·2 *Q* 49 *f.*	19·1	9·1 *F* 17 *f.*	Paisley H.M'Intyre&Co.	1880	3 Blds. FPT24	2Cy.20"&36"–24" 60 HP. W.King&Co.,Glasgow	W.M'Lachlan & Co.
255	✻	Ivy *Campbell -88*	Stl ScSr *1 Dk*	British Glasgow	135 531	190·0 *P* 55 *f.*	28·2 *B* 66 *f.*	13·3 *F* 41 *f.*	Londonderry C.J.Bigger	1888	5 Blds.	3C.17½"29"&47"–36" 112 HP. Ross Anderson&Co.Glsgw	Glasgow, Dublin, & Londonderry Steam Packet Co.
256	✻	Ivy Holme *Wilson 88-88*	1 ScSr *1 Dk*	British Maryport	86 237	120·0 *Q* 41 *f.*	22·1 *B* 11 *f.*	9·5 *F* 14 *f.*	Sunderland J.L.Thompson&Sns	1883	3 Blds. FPT20	2Cy.16"&30"–21" 60 HP. J.W.&F.Wilson&Co.,Sndld.	Hine Bros.
257	G V	Iwalani *Freeman*	W ScSr	Hawaian Honolulu	283 588	146·0 *Q* 23 *f.*	27·9 *B* 10 *f.*	11·5 *F* 30 *f.*	SanFrancisco DickieBros.	1881		2Cy.18"&34"–28" 56 HP. RisdonIronWks.S.Francisco	Inter Island Steam Navigation Co.
258	✻	Ixia *Churnside 84-86*	1 ScSr *2 Dks 3 B*	British NorthShields	1284 1970	287·0 *P* 28 *f.*	37·0 *B* 28 *f.*	21·9 *F* 22 *f.*	Newcastle TyneI.S.B.Co.,Ld.	1881	6 Blds. CDB 303 FPT56 APT 61	2Cy.33"&63"–39" 180 HP. N.E.MarineEng.Co.,Ld.,Shl.	J.Robinson & Sons
259	✻	Ixia *Reed 81-81*	1 ScSr *1 Dk*	British Bristol	132 227	122·0	20·0 *Q* 36 *f.*	10·6	Bristol G.Stothert&Co.	1871	4 Blds.	2Cy.16"&32"–18" 40 HP. G.Stothert&Co.,Bristol	Turner, Edwards & Co.

ADDITIONAL SHIPS ARE INSERTED IN THE SUPPLEMENT AT THE END OF THE REGISTER.

✻ = Lloyd's Register; D = Nederlandsche Vereeniging; G = Germanischer Lloyd; I = Registro Italiano; R = American Record; A = Veritas Austro-Ungarico; H = Veritas Ellenico; N = Norske Veritas; V = Bureau Veritas.

Dans la 2me Colonne, B.T., avec les indications qui suivent, signifie la date du rapport de visite de la Carène par les experts du Board of Trade pour obtenir le certificat relatif au TRANSPORT DES PASSAGERS.

Dans la 3me Colonne, C = Composite (membrure en fer, bordé en bois); I = Fer; Stl = Acier; W = Bois.

Pour plus amples renseignements sur les navires classés par le Lloyd's Register, voir le Lloyd's Register Book et le Lloyd's Yacht Register.

No.	Register	Ship's Name. Material, Rig, &c. Master, and Dates of Appointment to present Owner's Service and to Ship. Ship's Late Name. No. of Decks, &c.	Flag. Port of Registry.	Tons.	Length	Breadth	Depth	Built. Where. By Whom.	When	No. of Bulkheads. Particulars of Water Ballast.	Engines. Particulars, Maker's Name.	Owners
1	V	J. A. Gripenstedt CScSr Carlsson -87 1 Dk	Swedish Gothenburg	337 451	117·6	23·8	15·4	Westervik C.R. Fridd	1872		2Cy.17"&33"-21" 60 HP Bergsunds Mek. Verkst. Stk.	G. E. Falck
2	✳	J. C. Stevenson I PadTg Legg 62-83 1 Dk	British Newcastle	94 183	115·2	19·7	10·0	North Shields Hepple & Co.	1883	4 Blhds.	2Cy.20½"-48" 60 HP Hepple & Co. North Shields	Tyne Impr. Commissioners
3		J. H. Jones WSc Chambers -88 1 Dk	British Goderich Ont.	142 208	107·0	21·4	9·5	Goderich, Ont. W. Marlton	1888		1Cy.18"-20" 100 HP [illegible]	J. Clark
4	V	J. H. Niemann I ScSr Greizmann -82 1 Dk	German Bremen	676 995	213·5	29·9	17·0	Flensburg Flensburger Schiffb. Ges.	1878	WB	2Cy.27½"&50"-33" 130 HP Flensburger Schiffsb. Ges.	Dampfschiff Ges. "Neptun"
5		J. H. Outhwaite Sc3mSr	American Cleveland, Oh.	1099 1304	224·0	37·4	18·0	Clevelnd, Ohio	1886			W. J. Riche[illegible]
6		J. L. Runeberg I ScSr Sandberg 1 Dk	Swedish Sundsvall	368 458	159·4	24·1	12·5	Stockholm [illegible]	1876		2Cy.19½"&39"-23½ 90 HP Bergsunds Mek. Verkst.	Sundsvall-[illegible]
7	✳DV	J. M. Lennard I ScSr Horne 86-86 1 Dk2 B	British Middlesbro'	285 491	165·0	24·2	14·5	Middlesbro' R. Dixon & Co.	1875	5 Blhds.	2Cy.22"&41"-26" 65 HP Blair & Co., Ld. Stockton	J. M. Lennard
8	✳	J. M. Lockwood I ScBn Walley 83-86 1 Dk2 B	British W. Hartlepl.	1146 1772	257·[illegible]	[illegible]	19·0	W. Hartlepool W. Gray & Co.	1882	4 Blhds.	2Cy.32"&60"-39" 150 HP Black, Hawthorn & Co., Gtshd.	J. H. Murrell & [illegible]
9	✳	J. M. Smith I ScSr Stephenson 73-87 1 Dk2 B	British Sunderland	1331 2055	288·0	38·7	19·3	Sunderland Short Bros.	1885	4 Blhds.	3C 21"35"&57"-36" 180 HP T. Richardson & Sons, Hartpl	J. Westoll
10		J. M. Strachan I Sc3mSr Scott -84	British London	495 789	199·8	28·1	17·4	Newcastle Palmer Bros. & Co.	1865		2Cy.35"&48"-30" [illegible] J. Stewart & Son, Blackwall	S. Clarke & Co.
11	DV	J. N. Madvig I ScSr Matthiessen 73-84 1 Dk & Spr Dk	Danish Copenhagen	1334 1779	272·6	35·3	16·5	Elsinore Jernsk. og Maskin Bygg.	1884	[illegible]	2Cy.32"&60"-39" 175 HP Jernsk. og Maskin Bygg. [illegible]	Dampskibs-S. "Danmark" [illegible]
12		J. P. Dillberg I TwnScSr Tornerhjelm 71-86 1 Dk & Awn Dk	Swedish Halmstad	296 411	159·4	23·6	10·5	Gothenburg Motala Co.	1876		1Cy.16"&42"-22" 120 HP Motala Co., Gothenburg	Hallands Ång. Aktiebolag
13		J. P. Rennoldson I PadSp	British S. Shields	23 137	110·5	19·1	9·8	South Shields J. T. Eltringham	1883		1Cy.36½"-54" 70 HP J. P. Rennoldson & Sons, S. Shlds	J. Wilson
14	V	J. P. Taylor I Sc3mSr Arkley -86	British Newcastle	164 270	132·0	20·6	11·0	Newcastle E. Lindsay	1873		2Cy.13½"&17"-20" 45 HP Pattison & Atkinson, N'castle	W. J. Taylor & [illegible]
15		J. R. Booth W Sc 1 Dk	British Ottawa, Ont.	71 132	96·0	18·8	8·4	Hull A. Millar	1882		1Cy.22"-20" 130 HP Stewart & Fleck, Ottawa	Denis Murph[illegible]
16	✳	J. R. Hinde I Sc3mSr Scott 80-85	British London	467 746	198·7	28·0	17·4	Newcastle Palmer Bros. & Co.	1864	3 Blhds.	3C 16½"26"&42"-30" 99 HP Palmers Co., Ld., Newcastle	J. Fenwick & S[illegible]
17		J. W. Garrett I Pad 1 Dk	American Baltimore	1276 1343	350·0	41·0	13·3	Wilmingtn Del The Harlan & Hollingsworth Co.	1887		2Cy.56"-108" The Harlan & Hollingsworth Co., Wilmington	Baltimore & [illegible] Co.
18		J W Steinhoff W Sc 1 Dk	British Chatham, Ont	209 312	123·0	21·0	8·0	Wallceburg, Ont J. A. Jenking	1874		1Cy.18"-18" 120 HP R. Parkinson, Chatham, Ont.	James W. Ste[illegible]
19	✳	J. W. Taylor Stl ScSr Cook 77-87 1 Dk & Spr Dk3 B	British Sunderland	1816 2785	316·5	40·0	19·2	Sunderland Short Bros.	1887	5 Blhds.	3C 23½"38"&62"-42" 220 HP J. Dickinson, Sunderland	Taylor & Sand[illegible]
20	✳	J. Y. Short Stl ScSr Milburn 71-88 1 Dk2 B	British Sunderland	1447 2217	290·0	39·0	21·9	Sunderland Short Bros.	1887	4 Blhds.	3C 21½"35"&58"-39" 200 HP J. Dickinson, Sunderland	J. Westoll
21	✳	Jacinth I ScSr Skinner 69-87 1 Dk	British Swansea	187 306	139·6	21·0	10·9	Port Glasgow Murdoch & Murray	1878	4 Blhds.	2Cy.18"&30"-24" 45 HP Muir & Houston, Glasgow	H. Hoskin
22	✳	Jacinth I Sc3mSr Walker -88 1 Dk	British Glasgow	178 453	175·2	26·5	10·5	Paisley J. Fullerton & Co.	1888	5 Blhds.	3Cy.15½"24"&40"-30" [illegible] HP W. King & Co. Glas.	W. Robertson
23		Jackal I PadSp	British Liverpool	38 166	127·8	20·2	9·8	Glasgow J. & G. Thomson	1853		2Cy.36"-36" 86 HP J. & G. Thomson, Glasgow	Cunard Steam[illegible] Co., Ld.
24		Jaco	Brazilian Para	354	..	..	..					Amazon Stm. N[illegible]
25	✳DN	Jacob Christensen I ScSr Bjerck 81-81 2 Dks3 B	Norweg'n Bergen	1144 1773	260·8	34·6	24·0	Middlesbro' Raylton Dixon & Co.	1881	5 Blhds.	2Cy.34"&63"-36" 180 HP T. Richardson & Sons, Hartpl	Jacob Christe[illegible]
26	DG	Jacoff Prosoroff I ScSr Sandt -86 1 Dk	German Düsseldorff	420 925	217·2	28·7	16·5	Kiel [illegible] Howaldt	1884		2Cy.25"&39½"-27"	W. H. Muller & [illegible]
27		Jacques Cartier W Pad 1 Dk	British Quebec	113 212	117·6	23·0	7·4	Levis [illegible] Paradis	1888		1Cy.30"-72" 130 HP	T. Paradis
28	✳	Jacuhype I TwnSc 1 Dk	Brazilian Pernambuco	383 631	200·0	28·1	10·7	Port Glasgow Murdoch & Murray	1883	4 Blhds.	[illegible] 120 HP D. Rowan, Glasgow	Companhia de N[illegible] Pernambuco
29		Jaguaribe I Pad 1 Dk2 B	Brazilian Pernambuco	251 399	186·0	24·2	12·0	Port Glasgow J. Reid & Co.	1860	4 Blhds.	2Cy.30"&53"-42" 120 HP D. Rowan, Glasgow	Companhia de N[illegible] Pernambuco
30		Jakob	Russian Taganrog	289 376	..	..	..	Trieste Stabilimento Tecnico	1883		50 HP	Soc. di Nav. a V[illegible] Azoff

✳ = Lloyd's Register; D = Nederlandsche Vereeniging; G = Germanischer Lloyd; I = Registro Italiano; R = American [illegible]; A = Veritas Austro-Ungarico; H = Veritas Ellenico; N = Norske Veritas; V = Bureau V[illegible]

In the 2nd Column, B.T., with figures following, signifies Date of report of bottom sighting by Board of Trade Surveyors as condition of PASSENGER C[illegible]

In the 3rd Column, C = Composite (iron frame, wood planked); I = Iron; Stl = Steel; W = Wood.

Further particulars of Vessels classed by Lloyd's Register will be found in Lloyd's Register Book and Lloyd's Yacht Register.

		Nom, Matériaux, Gréement, etc. Capitaine et dates d'entrée au service des Armateurs actuels et à bord du navire. Ancien Nom, nombre de ponts, etc.		Pavillon. Port d'Armement	Tonnage. Net. Total	Dimensions. Longueur	Largeur	Creux	Construit. Où. Par qui.	Quand.	Nombre des Cloisons. Détails sur le lest d'eau.	Machines. Détails. Nom du Constructeur.	Armateurs ou Directeurs-Gérants.
31		Jakobstad *Wallin 77-77*	Sr	Russian Jakobstad	384	..	..	..	Stockholm	1876		180 HP.	P. Malm, jun.
32		Jamaican *Catford 82-86*	I ScBk	British Liverpool	1297 2016	310·8	36·0	26·0	Glasgow Aitken&Mansel	1871		2Cy.45½"& 80"–42" 350 HP. J.&J.Thomson,Glasgow	West India & Pacific Steam Ship Co., Ld.
33		James	W Pad 1 Dk	British Quebec	31 104	105·2	20·1	7·7	St. Romauld	1872		1Cy. 50 HP.	J. Delisle
34		James A. Walker *Gaskin 87*	W ScBa 1 Dk	British Kingston, Ont	98 184	105·6	19·8	9·6	Kingston, Ont J. Gaskin	1887		2Cy.18"& 32"–26" 150 HP. CanadianLoco.,&Eng.Co., Kingston, Ont.	Montreal Transportation Co.
35	✻	James Barrass *Sims 83-87*	I Sc3mSr 2 Dks	British London	1032 1601	258·4	34·5	22·7	NorthShields T.&W.Smith	1874	4 Bhds. DBa72ft 181t	2Cy.30"& 57½"–36" 130 HP. T.Clark&Co.,Newcastle	P. Brown
36	✻D	James Cameron *Bures 78-87 (exPrinzAlbrecht)*	I ScSr 1 Dk2 B	British Sunderland	1177 1807	260·0	38·2	17·9	Sunderland Short Bros.	1882	6 Bhds. CDB	2Cy.33"& 62"–42" 150 HP. G.Clark,Sunderland	J. Westoll
37		James Comrie	W Sc	British Adelaide	73 102	93·8	17·5	8·0	Shoalhaven, [N.S.W.	1877		2Cy. 42 HP.	Yorke's Peninsula S.S.Co.Ld.
38	✻	James Drake *Bullock 86-86*	I ScSr 2 Dks3 B	British N.Shields	875 1351	241·3	32·6	23·1	Sunderland Bartram,Haswell&Co.	1874	4 Bhds. WBaf	1Cy.16"22½"32"& 45"–36" [130 HP. Tripled, 1888	Wilkie & Turnbull
39	✻	James Groves *Sowden 87-87*	I ScSr 1 Dk2 B	British W.Hartlepl.	818 1273	236·2	31·4	17·8	W. Hartlepool W.Gray&Co.	1877	4 Bhds. DBa&f214	2Cy.27"& 50"–33" 99 HP. T.Richardson&Sons,Hrtlpl.	Maclean, Doughty & Co.
40	✻	James Hall *McDonald 86-88*	I ScSr 1 Dk	British Aberdeen	184 346	174·0	21·0	13·3	Aberdeen Hall,Russell&Co.	1870	4 Bhds.	2Cy.23½"& 42"–30" 65 HP. Hall,Russell&Co.,Aberdeen	Aberdeen, Leith & Moray Firth Steam Shipping Co.
41		James Hogg *Carzon*	I ScSr 1 Dk2 B	British Middlesbro'	317 505	165·0	24·6	13·7	Middlesbro' R.Dixon&Co.	1874	4 Bhds. DBa&ft 58 ft	2Cy.21½"& 43"–30" 70 HP. Blair&Co.,Ld.,Stockton	James Hogg S.S.Co.Ld (Hogg&Henderson)
42	V	James J. Dickson *Sölan 84*	I ScSr 1 Dk	Swedish Gothenburg	659 845	195·2	27·5	15·0	Hull Earles'ShipB.Co.	1872		2Cy.26"& 48"–27" 90 HP. Earles'ShipBuildingCo.,Hull	Förnyade Angf. Akt. "Göta" (H.Sternhagen)
43	✻D	James Joicey *Harrison 85-86*	I Sc3mSr 1 Dk2 B	British London	459 728	200·6	28·1	16·8	Newcastle PalmerBros.&Co.	1863	4 Bhds.	3C.16½"26"& 43"–36" 99 HP. Palmers'Co.,Ld.,Newcastle	J. Fenwick & Son
44		James M. Donahue *White*	W Pad 1 Dk	American S.Francisco	592 731	219·0	32·1	9·7	San Francisco P. Donahue	1875		1Cy.48"–132" 290 HP. QuintardIronWks,Nw.York	J. M. Donahue
45		James Mackee *Freeman (ex Iris)*	W ScSr	Hawaiian Honolulu	105 301	110·9	25·6	9·2	Port Ludlow Hall Bros.	1879		2Cy.11"& 20"–18" 18 HP. Hinckley,Spiers&Hayes,San Francisco	Inter Island S.N.Co.
46	✻	James Malam *Whyte 87-87*	I ScSr 1 Dk2 B	British W.Hartlepl.	659 1042	225·0	29·7	16·7	Hartlepool Withy,Alexander&Co.	1873	4 Bhds. DBa&f 183t	2Cy.25½"& 51½"–33" 98 HP. Blair&Co.,Ld.,Stockton	Maclean, Doughty & Co.
47	✻	James Mutter *McPhail 85-85*	I ScSr 1 Dk	British Glasgow	41 157	100·0	19·1	8·8	Glasgow Scott&Co.	1884	4 Bhds. FPT7t	2Cy.16"& 28"–22" 35 HP. W.King&Co.,Glasgow	W. & J. Mutter
48		James Perry	I Sc3mSr 1 Dk	Urug'yan Monte Video	106 179	106·5	19·1	9·1	Chepstow Finch&Co.	1881	3 Bhds.	2Cy.13"& 24"–18" 26 HP. AvonsideEng.Co.,Bristol	Herrera & Co.
49	✻	James Searle *Buckingham*	C ScTug 1 Dk	British Pt.Elizabeth	29 56	71·0	16·0	8·5	London R.&H.Green	1881	Y3.53	2Cy.15"& 27"–18" 35 HP. A.Wilson&Co.,London	Union Boating Co., Ld.
50	✻D	James Shaw *Bitton 73-73*	I Sc3mSr 1 Dk2 B	British Stockton	428 559	200·1	26·2	13·9	Stockton M.Pearse&Co.	1865	4 Bhds.	2Cy.22"& 44"–26" 70 HP. Cmp.'72,Blair&C.,Ld.,Stktn	Tees & Thames Screw S.S.Co.,Ld.
51	✻	James Southern *Henslie*	I Sc3mSr 1 Dk2 B	British London	548 847	218·2	28·3	15·9	North Shields T.&W.Smith	1865	4 Bhds. DB+29t FPT20t APT3t	2Cy.38"–30" 90 HP. R.&W.Hawthorn,N'weastle	J. & C. Harrison
52	✻D	James Turpie *Smith 81-81*	I ScSr 2 Dks3 B	British N.Shields	1140 1767	270·8	34·7	23·0	Newcastle A.Leslie&Co.	1881	5 Bhds. DBa&f 230t APT30t	2Cy.32"& 61"–36" 180 HP. N.E.MarineEng.Co.,Ld.,Sld.	J. Coull
53	✻	James Watt *Petrie 83-83*	I Sc 2 Dks	British Greenock	1024 1598	260·5	36·1	20·9	Port Glasgow R.Duncan&Co.	1881	5 Bhds. CDB326t	2Cy.31"& 62"–39" 160 HP. Rankin&Blackmore,Grnck.	"James Watt" S.S. Co.,Ld.(Leitch & Muir)
54	✻	James Westoll *Dudley 81-84*	I ScSr 1 Dk2 B	British Sunderland	1274 1969	268·0	39·3	21·6	Sunderland Short Bros.	1884	6 Bhds. CDB	2Cy.36"& 67"–36" 192 HP. T.Richardson&Sons,Hrtlpl.	J. Westoll
55	✻	James Whishaw *Laverick 26-82*	I ScBa 1 Dk2 B	British Hull	977 1522	250·0	34·0	19·5	Stockton M.Pearse&Co.	1881	4 Bhds. DBa&f 268t	2Cy.30"& 55"–36" 150 HP. Blair&Co.,Ld.,Stockton	H. Baxter & Co.
56	✻D	Jane *Acteson*	I ScBa 1 Dk2 B	British Whitby	847 1337	240·2	34·0	17·7	Whitby T.Turnbull&Son	1879	4 Bhds. DBa&f ft f 91 ft	2Cy.30"& 57"–33" 140 HP. T.Richardson&Sons,Hrtlpl.	T. Turnbull & Son
57	✻	Jane Bacon *Rowles 78*	I Sc3mSr 1 Dk2 B	British Liverpool	438 674	191·0	26·7	15·4	Liverpool Bowdler, Chaffer & Co.	1865	4 Bhds. WB	2Cy.20"& 40"–30" 95 HP. D.Rollo&Sons,Liverpool	J. Wright
58	✻	Jane Clark *Taylor 88-88*	I ScSr 1 Dk	British Glasgow	532 839	208·0	29·1	14·1	Belfast Workmn,Clark&Co	1883	4 Bhds. DBa120t FPT63t APT28t	2Cy.24"& 48"–36" 96 HP. V.Coates&Co.,Ld.,Belfast	Clark & Service
59	✻	Jane Cory *Thompson 87-88*	I ScSr 1 Dk2 B	British W.Hartlepl.	519 808	201·7	28·3	15·9	Hartlepool Withy,Alexander&Co.	1870	4 Bhds. DBa&f 156t	2Cy.38"–30" 96 HP. G.Clark,Sunderland	R. Irvine, jun. (JacksonBros.&Cory,Mgrs
60		Jane Jolliffe	I ScSr	British Liverpool	14 235	133·0	22·6	12·9	South Shields J.Readhead&Co.	1888		3C.16"26"& 43"–30" 200 HP J.Readhead&Co.,S.Shields	W. & T. Jolliffe

✻=Lloyd's Register; D=Nederlandsche Vereeniging; G=Germanischer Lloyd; I=Registro Italiano; R=American Record;
A=Veritas Austro-Ungarico; H=Veritas Ellenico; N=Norske Veritas; V=Bureau Veritas.

Dans la 2me Colonne, B.T., avec les indications qui suivent, signifie la date du rapport de visite de la Carène par les experts du Board of Trade pour obtenir le certificat relatif au TRANSPORT DES PASSAGERS.

Dans la 3me Colonne, C=Composite (membrure en fer, bordé en bois); I=Fer; Stl=Acier; W=Bois.

Pour plus amples renseignements sur les navires classés par le Lloyd's Register, voir le Lloyd's Register Book et le Lloyd's Yacht Register.

No. in Book. Official No.	Registry	Ship's Name. Material, Rig, &c. Master, and Dates of Appointment to present Owner's Service and to Ship. Ship's Late Name. No. of Decks, &c.	Flag. Port of Registry.	Tons. Net. Gross.	Length	Breadth	Depth	Built. Where. By Whom.	When	No. of Bulkheads. Particulars of Water Ballast.	Engines. Particulars. Maker's Name.	Owners or [illegible]
61	*	Jane Kelsall I ScSr *Welch* 87-87 *2 Dks3 B*	British N.Shields	1269 1958	275·0 *P 29 f.*	35·6 *B 82 f.*	23·0 *F 23 f.*	SouthShields J.Readhead&Co.	1881	[illegible]	2Cy.35" & 63"–36" 150 HP. J.Readhead&Co.,S.Shields	W.D.C.Balls[illegible]
62	*	Janet Nicoll I ScSr *Hutton* *1 Dk2 B*	British Sydney	496 779	184·0 *Q 57 f.*	29·2 *B 51 f.*	13·8 *F 23 f.*	Newcastle [illegible]	1884	[illegible]	2Cy.20" & 42"–30" 90 HP. [illegible]	G.W.Nicoll
63	*	Janus Stl ScSr *Jackson* 85-88 *1 Dk2 B*	British Grangemouth	722 1117	240·5 *Q 8 f.*	33·2 *B 39 f.*	14·1 *F 30 f.*	Grangemouth Dobson&Charles	1883	[illegible]	[illegible]	T.Kish
64	*	Japan I ScBg *Gardner* 59-76 *2 Dks3 B*	British London	1865 2440	336·0	37·4	27·2	Newcastle [illegible]	1872	[illegible]	2C.36½" & 68"–48" 250 HP. Maudslay,Sons&Field,Lndn	Apcar & Co.
65	*D	Japanese I ScSr *Spedding* 83-87 *2 Dks3 B*	British London	1204 1838	271·0	34·2 *F 32 f.*	24·6	Newcastle Schlesinger,Davis & Co.	1878	[illegible] Bhds.	2Cy.32" & 61"–36" 170 HP. Hawks,Crawshay&Sn.Gtshd	Angier Line [illegible]
66		Japara I ScSr *Pot* *2 Dks & Shade Dk*	Dutch Batavia	727 1248	256·2 *P 31 f.*	31·6 *F 73 f.*	17·3	Greenock Caird&Co.	1876	[illegible] Bhds.	2Cy.34" & 60"–39" 150 HP. Caird&Co.,Greenock	Nederl.-Indis[illegible] Stoomv.-Maats.
67		Japura	Brazilian [illegible]	602	..	..	..					Comp.de Mana[illegible]
68	V	Jarl I ScSr *Davidsen* 70-85 *1 Dk Spt Awn Dk*	Danish Basle	115 233	131·7 *P 43 f. B 91 F 16 f.*	20·8	9·6 16·1	Elsinore Jernsk.&Mask.Byggeri	1885		2Cy.25" & 47"–27" 110 HP. Jern.sk.&Mask.Byg.,Elsinore	Østbornholmsk [illegible]
69		Jarlsberg I ScSr *Andersen* -87 *1 Dk*	Norweg'n Tonsberg	161 226	119·8 *P 71 f.*	19·3 *F 50 f.*	8·2	Christiania Akers Mek.Verkst.	1879		3Cy.–43" 50 HP. Akers Mek.Verkst, Christiania	Tönsbergs & [illegible] Dampsk[illegible]
70 81 [illegible]	BT 11.88	Jason I ScSr *Milligan* 87-87	British Liverpool	1412 2181	320·9	34·6	26·1	Newcastle A.Leslie&Co.	1880		2Cy.27" & 50"–52" 250 HP. R.Stephenson&Co.,Newcastle	Ocean Steam [illegible] Ld.
71 70882	*	Jason I Sc3mSrYt (*ex* Aies) *2 Dks*	British Barrow	196 288	161·7	23·7	13·3	Barrow Barrow S.B.Co.	1877	[illegible] Bhds.	2Cy.23" & 40"–24" 96 HP. Barrow S.B.Co.,Ld.,Barrow	F.Bibby
72		Jason (ex Belle of Dunkerque) I ScSr *Steckina* *1 Dk*	Austrian Trieste	257 411	169·0 *P 35 f.*	25·2 *F 26 f.*	12·2	Port Glasgow Blackwood&Gordon	1876	[illegible] Bhds.	2Cy.22" & 40"–27" 75 HP. Blackwd&Gordon,Pt.Glsg.	Società di Nav. [illegible] Danubio
73	DV	Jason I Sc3mSr *Kuyper* 80-88 *2 Dks*	Dutch Amsterdam	1228 1653	252·0 *Len.*	32·4 *71*	27·0	Glasgow A.&J.Inglis	1866		2Cy.36" & 62"–36" 150 HP. Konink.Fab.,Amsterdam	Kon.Nederl. [illegible] Maats.
74	N	Jason W ScBk *Jacobsen* *1 Dk2 B*	Norweg'n Sandefjord	358 495	147·0	30·6 *Q 37 f.*	17·1	Sandefjord	1881		2Cy.20⅝" & 37"–21⅝" 60 HP. Akers Mek.Verkst. Christiania	A.J.Freberg
75 [illegible]	*D	Jasper Stl ScSr *M'Millan* 82-83 *1 Dk & Spr Dk3 B*	British Dundee	811 1256	235·0	31·7	22·4	Dundee W.B.Thompson	1883	[illegible] Bhds. [illegible]	2Cy.25" & 50"–42" 99 HP. W.B.Thompson,Dundee	Dundee Gem Li[illegible] S.S.Co.,Ld.
76 [illegible]	*	Jasper (FOUNDERED) I Sc3mSr *Wilson* 87-87 *1 Dk*	British Glasgow	118 287	144·8 *Q 43 f.*	23·1 *F 23 f.*	10·7	Glasgow T.B.Seath&Co.	1880	1 Bhds. [illegible]	2Cy.20" & 36"–24" 60 HP. W.King&Co.,Glasgow	W.Robertson
77	*	Játiva I ScSr *Tonda* *1 Dk & Spr Dk3 B*	Spanish Valencia	793 1293	222·0	31·0	15·1 22·4	SouthShields J.Readhead&Co.	1880	[illegible] Bhds. [illegible]	2Cy.28" & 55"–36" 110 HP. J.Readhead&Co.,S.Shields	Cia.Valenciana [illegible] Nav.
78 [illegible]		Java I ScSr *Magenis* 78-88 *2 Dks3 B*	British Glasgow	943 1465	270·2	30·7	24·0	Glasgow A.&J.Inglis	1872	[illegible] Bhds.	2Cy.35½" & 62"–39" 180 HP. A.&J.Inglis,Glasgow	British India [illegible] Nav.Co.,Ld.
79		Javary	Brazilian Para	438	..	..	..					Amazon Stm. [illegible]
80	V	Jean Baptiste Say I ScSr *Crequel* *1 Dk*	French Nantes	197 354	161·7 *P 51 f.*	21·3 *F 26 f.*	12·3	Nantes Jollet&Babin	1876	[illegible]	3C.15¾" 22" 34½"–25½ 75 IP. Jollet&Babin,Nantes	Cie. Nantaise d[illegible] Navigation à vap[illegible]
81	V	Jean Bart I ScBn *Barbian* 84-84 *1 Dk & Spr Dk3 B*	French Dunkirk	456 683	187·6 *B 30 f.*	25·3	13·7 19·6	SouthShields Readhead,Softly & Co.	1870	[illegible]	2Cy.28"–16½" 80 HP. Readhead,Softly & C.,S.Shlds	Cie. Générale [illegible] Bateaux à vap. [illegible]
82		Jean Bart I PadTug *Passerant* 84-85 (ex Enterprise)	French [illegible]	68 144	115·1	19·3	9·8	London Charles Lungley	1859		2Cy.23" & 46"–30" 60 HP. J.Stewart,London	Société Havrai[illegible] Remorquage
83		Jean Dupuis I ScSr *Reullet* -85 *2 Dks*	French Saigon	102 176	135·7	19·9 *B 39 f.*	7·2	Greenock Scott&Co.	1881		2Cy.15" & 27"–15" 40 HP. Scott&Co.,Greenock	Messageries Flu[illegible] de Cochinchine
84		Jean Mathieu I ScSr	French [illegible]	255 412	183·7	23·3	13·3	Dunkirk	1851		120 HP.	F.Morelli & Co.
85 81080	*	Jeanie I ScBn *M'Dowall* 84-84 *1 Dk2 B*	British Hartlepool	1172 1816	275·0 *Q 76 f. P 39 f.*	35·6 *B 64 f.*	18·1 *F 30 f.*	Hartlepool E.Withy&Co.	1882	[illegible] Bhds. [illegible]	2Cy.35" & 65"–36" 180 HP. T.Richardson&Sons,Hrtlpl	J.Coverdale & S[illegible]
86 76389	RV	Jeanie W AuxSc3mSr *Humphrey* -88 *2 Dks*	American S.Francisco	860 1072	186·3	37·9	18·8	Bath, Me. Goss,Sawyer&Packard	1883	[illegible]	2Cy.20" & 38"–30" 350 IP. Neafie&Levy,Philadelphia	Pacific Steam [illegible] Co.
87 [illegible]	BT 5.88	Jeanie Deans I&Stl PadSr *McKinlay* 86-84	British Glasgow	69 271	210·0 *Q 70 f.*	20·1 *B 68 f.*	7·6	Glasgow Barclay,Curle&Co.	1884		1Cy.50"–72" 166 HP. Barclay,Curle&Co.,Glasgow	North British [illegible] Pkt.Co.[illegible]
88 82017		Jeanie Hope I ScDy *Wigg* -85 (ex Annie Hope)	British [illegible]	65 104	83·4 *Q 8 f.*	17·6	8·5	Leith Hawthorns&Co.	1881		2Cy.10" & 20"–15" 17 HP. Hawthorns&Co.,Leith	J.Crisp & Sons
89	V	Jeanne d'Arc Stl&I Sc3mSr *Boutemps* 86-88 (ex Harald Aguus) *1 Dk2 B*	French Dunkirk	428 624	194·8 *P 102 f.*	25·9 *F 19 f.*	15·2	Nyköping Nyköpings Mek. [illegible]	1866	[illegible]	2Cy.22" & 42"–27" 70 HP. R.&W.Hawthorn,Newcastle	Bossut-Plichon
90 93780	*	Jeannette Stl ScSrYt	British Liverpool	57 114	113·2	17·6	10·0	Birkenhead Cochrane & Co.	1888	[illegible] Bhds.	3C.11" 17" & 28"–20" 40 IP. Cochrane & Co.,Birkenhead	F.Platt

* = Lloyd's Register; D = Nederlandsche Vereeniging; G = Germanischer Lloyd; I = Registro Italiano; R = American Re[illegible]
A = Veritas Austro-Ungarico; H = Veritas Ellenico; N = Norske Veritas; V = Bureau Verita[illegible]

In the 2nd Column, B.T., with figures following, signifies Date of report of bottom sighting by Board of Trade Surveyors as condition of PASSENGER CERTIFICA[illegible]

In the 3rd Column, C = Composite (iron frame, wood planked); I = Iron; Stl = Steel; W = Wood.

Further particulars of Vessels classed by Lloyd's Register will be found in Lloyd's Register Book and Lloyd's Yacht Register.

		Nom, Matériaux, Gréement, etc. Capitaine et dates d'entrée au service des Armateurs actuels et à bord du navire. Ancien Nom, nombre de ponts, etc.		Pavillon. Port d'Armement	Tonnage. Net Total	Dimensions.	Construit. Où. Par qui.	Quand	Nombre des cloisons. Détails sur le lest d'eau.	Machines. Détails. Nom du Constructeur.	Armateurs ou Directeurs-Gérants.
		Jeannine *Hamon 86-88*	I ScTwSr 1 *Dk*	French Saïgon	318 431	.. .. ..	Liverpool Cato, Miller & Co.	1854		2Cy.35"& 60"–56" 70 HP. [illegible] Miller & Co.	Messageries Fluviales de Cochinchine
	V	Jehanne *Leroux*	I ScTwSr 1 *Dk* 2 *B*	French Dieppe	318 502	170·0|26·0|14·7	Newcastle C. Mitchell & Co.	1869	6 Blhds.	2Cy.30"–36" 80 HP. [illegible] & Co., Newcastle	A. Grandchamp, fils & Co.
	N	Jelo *Larsen 87*	I ScSr 1 *Dk*	Norweg'n Moss	244 336	135·5|23·0|12·1 *Q 37 f. B 24 f. F 14 f.*	Hartlepool Irvine & Co.	1880	APT FPT	2Cy.14"& 28"–21" 40 HP. W. B. Thompson, Dundee	R. M. Peterson
	✻	Jennie *(ex Princess Louise)*	I ScSrYt	British Grimsby	68 107	103·6|17·0|6·6	London Lewis & Stockwell	1873	3 Blhds.	2Cy.15"& 28"–17" 30 HP. Comp.'81, Earle's Co., Hull	Patrick J. M. Power
		Jenny *Millich 88*	I ScSr	Austrian Trieste	58 115	120·3|16·3|8·2	Trieste Stabilimento Tecnico	1888			G. de Ceconi
	✻	Jenny *Anderson*	I ScSr 1 *Dk* 2 *B*	German Danzig	544 776	201·0|28·0|15·4 *Q 71 f. B 43 f. F 24 f.*	Sunderland Austin & Hunter	1878	1 Blhds. APT [illegible] MT [illegible]	2Cy.25"& 48"–36" 90 HP. Black, Hawthorn & Co., Nwc.	T. Rodenacker
	✻	Jenny Otto *Parsons 85-87*	I ScSr 2 *Dks* 3 *B*	British London	942 1451	250·5|33·2|23·0 *F 29 f.*	Newcastle Wghm. Richardson & Co.	1877	1 Blhds. DB [illegible]	3C 20"32"& 57"–36" 160 HP. Tripl.'87, Central Mar. Eng. Co., W. Hartlepool	G. Otto
	✻ D	Jeranos *Frost*	I ScBg 2 *Dks* 3 *B*	British Hull	1288 1979	280·5|36·0|26·2	Stockton Richardson, Duck & Co.	1878	5 Blhds. WB DTF	2Cy.35½"& 66"–42" 190 HP. Blair & Co., Ld., Stockton	H. Briggs, Sons & Co.
	N V	Jernbarden *Rasmussen*	I ScSr 1 *Dk* 2 *B*	Norweg'n Stavanger	830 1038	204·1|31·2|16·8 *P 106 f. F 23 f.*	Gothenburg Göteborgs Mek. Verkst.	1882	MT [illegible] APT	2Cy.27"& 57"–36" 110 HP. Henderson, Coulborn & Co., Rfw.	Brödr. Berner
100		Jerome *Crimp 81-84*	I ScBg	British Liverpool	796 1090	213·8|29·7|25·3	Liverpool Hart & Sinnott	1866		2Cy.21"& 48"–40" 91 HP. J. Taylor, Birkenhead	Booth Steam Ship Co., Ld.
101	✻ D	Jersey *Gowan*	I ScSr 1 *Dk* 2 *B*	British Cardiff	1056 1627	260·0|36·2|17·0 *Q 94 f. B 60 f. F 31 f.*	Newcastle Palmers' Co., Ld.	1883	1 Blhds. DB [illegible] APT 20	2Cy.31"& 60"–39" 99 HP. Palmers' Co., Ld., Newcastle	"Jersey" S.S. Co., Ld. (Morel Bros. & Co.)
102	✻	Jersey City *Gosier 85-88*	I ScBg 2 *Dks* 3 *B*	British Bristol	1261 1936	270·4|36·0|23·5 *P 26 f. F 23 f.*	Stockton Richardson, Duck & Co.	1882	5 Blhds. DB [illegible]	2Cy.36"& 67"–36" 200 HP. T. Richardson & Sons, Hrtlpl.	C. Hill & Sons
103	✻	Jesmond *Hall*	I ScSr 2 *Dks* 3 *B*	British London	973 1495	252·0|33·0|22·5	Newcastle C. Mitchell & Co.	1878	1 Blhds. DB a 71 ft & cb 40 ft	2Cy.30"& 58"–36" 150 HP. Wllsnd Slipway Co., Nwcstle.	Watts, Ward & Co.
104	✻	Jesmond	I ScYt 1 *Dk*	British Newcastle	97 178	110·0|20·1|9·0 *Q 35 f. B 8 f. F 16 f.*	Newcastle Wood, Skinner & Co.	1888	3 Blhds. FPT 23	2Cy.15"& 28"–20" 50 HP. N. E. Marine E. Co., Ld., Nwc.	L. S. Carr & Co.
105	R	Jesse H. Freeman *Matson 87*	W AuxScSmSr 2 *Dks*	American S. Francisco	360 576	146·2|30·1|17·4	Bath, Me. Goss & Sawyer	1883	Y 8.54	1Cy.22"–36" 75 HP. C. H. Delamater & Co., NwYk.	J. N. Knowles
106	✻	Jessica *(ex Hamburg)* *Teegler 88*	I Sc Sr 1 *Dk* 2 *L*	German Hamburg	368 579	191·4|26·7|14·4 *Q 28 f. F 29 f.*	Glasgow London & Glasgow Co.	1878	4 Blhds. BT 108 t	2Cy.24½"& 46"–30" 95 HP. London & Glasgow Co., Gls.	A. Kirsten
107	✻	Jessie	I ScSrYt 1 *Dk*	British Leith	102 150	136·3|17·0|10·1 *Len. 83*	Glasgow Barclay, Curle & Co.	1879	4 Blhds.	2Cy.16"& 30"–21" 40 HP. Barclay, Curle & Co., Glasgow	James Currie
108	✻	Jessie Darling *Windlund*	I ScSr 1 *Dk*	British Melbourne	158 289	125·3|23·2|10·4 *Q 43 f. B 8 f. F 18 f.*	Port Glasgow Murdoch & Murray	1884	3 Blhds. FPT [illegible] APT [illegible]	2Cy.20"& 40"–27" 66 HP. Kincaid & Co., Greenock	John Darling & Son
109		Jessie May	I Pad	British London	58 149	79·0|27·0|8·0	London Edwards & Symes	1876		2Cy.23"–27" 30 HP. Maudslay, Sons & Field, Lon.	Robins & Co., Ld.
110	✻	Jessmore *Shaw 78-88*	Stl & I ScBq 1 *Dk* & *W* [illegible] *Frames*	British Liverpool	1573 2465	294·0|40·3|18·6 *Q 94 f. B 140 f. F 35 f.*	Southampton Oswald, Mordaunt & Co.	1888	7 Blhds. DB [illegible] MT 236 FPT [illegible] APT 100	3Cy.21½"36"& 61"–39" 230 HP. Oswald, Mordaunt & Co., Southampton	"Jessmore," S.S., Ld. (W. Johnston & Co.)
111		Jewel *Johnston 88*	W ScSr 1 *Dk*	American S. Francisco	202 265	134·0|33·0|10·3 *P—*	San Francisco A. Hay	1888		2Cy.12"& 24"–22" 100 HP. Miners Foundry, S. Frncsco.	Cooper Lumber Co.
112		Jiddah *(ex Scotland)* *Said Effendi*	I ScBg 1 *Dk* & *Spr Dk* 3 *B*	Turkish Cstntinple.	1190 1836	277·8|38·4|26·2 *F 22 f.*	Antwerp Soc. Cockerill	1856		2Cy.36"& 66"–42" 250 HP. Hawks, Crawshay & Co., Gts.	Idarei-Massousich (Ottoman Steam Nav. Co.)
113	✻	Jin-Ho *J. Vignod*	I ScSr 1 *Dk*	British Penang	93 203	119·7|22·1|9·0 *P 16 f. B 13 f. F 22 f.*	Glasgow Blackwd. & Gordon	1886	1 Blhds.	2Cy.17"& 34"–21" 45 HP. Blackwood & Gordon, P. Glsg.	Ong Beng Tek
114		Jisalmi	I ScSr 1 *Dk*	Russian Kuopio	121 176	96·0|21·6|10·7	Wiborg	1876		40 HP.	"Jisalmi" S.S. Co.
115		Jiuju Maru	I ScSr	Japanese Osaka	239 386	161·9|21·5|12·6	Newcastle W. E. Boutland	1873		58 HP.	Inouye Nihei
116	✻	João Alfredo *Jackson 82-88*	Stl TwnScSr 1 *Dk* & *Shd Dk*	Brazilian Para	378 556	200·6|34·3|9·7 *F 28 f.*	P. Glasgow Russell & Co.	1888	7 Blhds. APT [illegible]	6Cy.13"20"& 32"–24" 150 HP. Rankin & Blackmore, Greenock	The Amazon Steam Navigation Co., Ld.
117		João Augusto	I Pad 1 *Dk* & *Spd Hur Dk*	Brazilian Para	250	150·0|26·0|7·0	WilmingtnDel Pusey, Jones & Co.	1871		2Cy.20"–72" 78 HP. Pusey Jones & Co., Wlmngtn, Del	Amazon Steam Nav. Co.
118		Joaquin Ancona *Byrne (ex Molo)*	Stl TwnScSr 1 *Dk* & *Awn Dk* 3 *B*	Mexican Progreso	824 1024	201·6|32·1|12·2 *P 29 f. B 129 f. F 22 f. 19·5*	Paisley H. McIntyre & Co.	1881	1 Blhds. FPT 91 APT 100	4Cy.17"& 34"–30" 99 HP. Hutson & Corbett, Glasgow	N. Ancona
119	N	Johan Sverdrup *Monsen 81-85*	I ScSr 1 *Dk*	Norweg'n Bergen	494 703	190·3|27·2|13·2 *Q 75 f. B 48 f. F 22 f.*	W. Hartlepool Irvine & Co.	1880	1 Blhds. DB a & f 102 t	2Cy.24"& 46"–30" 80 HP. J. Dickinson, Sunderland	P. G. Halvorsen
120		Johan Tillberg *Winblad*	I ScSr 1 *Dk*	Swedish Stockholm	287 397	135·1|23·6|13·9 *B 32 f. F 17 f. H 3 f.*	Oscarshamn	1878		50 HP.	Rederi-Bol. Östersjön (F. Sköör)

✻ = Lloyd's Register; D = Nederlandsche Vereeniging; G = Germanischer Lloyd; I = Registro Italiano; R = American Record; A = Veritas Austro-Ungarico; H = Veritas Ellenico; N = Norske Veritas; V = Bureau Veritas.

Dans la 2me Colonne, B.T., avec les indications qui suivent, signifie la date du rapport de visite de la Carène par les experts du Board of Trade pour obtenir le certificat relatif au TRANSPORT DES PASSAGERS.

Dans la 3me Colonne, C = Composite (membrure en fer, bordé en bois); I = Fer; Stl = Acier; W = Bois.

Pour plus amples renseignements sur les navires classés par le Lloyd's Register, voir le Lloyd's Register Book et le Lloyd's Yacht Register.

No. in Book / Official No.	Class	Ship's Name. Master, and Dates of Appointment. Ship's Late Name.	Material, Rig, &c. No. of Decks, &c.	Flag. / Port of Registry.	Tons Net / Gross	Length	Breadth	Depth	Built Where / By Whom	When	No. of Bulkheads / Particulars of Water Ballast	Engines. Particulars. Maker's Name.	Owners
121	V	Johann *Krüger 76-86*	Stl ScSr *1 Dk & Spr Dk & B*	German Hamburg	1002 1333	[illegible]	[illegible]	13·2	Flensburg Flensburger Schiffsb. Ges.	1884		2Cy 28½" & 52"–30" 120 HP Flensburger Schiffsb. Ges. Flensbg	Witt & Busch
122	G	Johann *Hargt 75-88*	I ScSr *2 Dks*	German Kiel	428 [illegible]	162·4	23·7	18·1	Kiel G. Howaldt	1883		2C 17½" & 33¾"–23½" 60 HP Howaldt Bros., Kiel	Sartori & Berger
123	V	Johanna *Aoustin -87*	I ScBn *1 Dk*	French Nantes	236 348	164·0	22·3	13·1	Nantes Dubigeon & Fils	1875	WB	2Cy 17½" & 32½"–21" 30 HP Pattison & Atkinson, N'castl.	A. Riom
124	DNV	Johanna *Eriksen*	Stl & I ScSr *1 Dk*	Norweg'n Christiania	312 438	141·5	24·7	13·0	Christiania Akers Mek. Verkst. *Q 38 f. B 33 f. F 21 f.*	1885	MFCH APTR	2Cy 20½" & 36½"–24" 58 HP Shanks & Son, Arbroath	Winge & Co.
125	✻ N	Johannes Bruun *Wessenberg -88*	I ScSr *1 Dk*	Norweg'n Bergen	485 677	180·0	28·2	12·4	Bergen Martens, Olsen & Co. *Q 86 f. B 50 f. F 21 f.*	1883	1 Bhds. CBDBH APTR	2Cy 23" & 41"–28" 75 HP Martens, Olsen & Co., Bergen	P. G. Halvorsen
126		John	I [illegible]	British [illegible]	104 132	90·4	18·4	10·1	Neath Abbey	1849		1Cy 16"–16" 24 HP E. Roach & Co., Bristol	A. J. Smith
127 71[illegible]		John	W Pad *1 Dk*	British Montreal	99 204	107·5	21·6	8·0	Sorel	1872		1Cy 32"–96" 45 HP	Sincennes McNaughton Line
128	✻	John Adamson *Gibbs 82-87*	I Sc3mSr *2 Dks & B*	British Sunderland	1091 1693	266·5	35·1	24·5	Sunderland Short Bros. *F 30 f.*	1872	5 Bhds. WB & f	2Cy 33" & 61"–33" 150 HP T. Richardson & Sons, Hrtlpl.	J. Westoll
129	V	John B. Maas *Vauhoof -81*	I ScTug	Belgian Antwerp	72 145	99·0	19·0	10·9	Liege Michel	1880		2Cy 20" & 39"–22" H. de Ville Chatel & Co., Brus.	Soc. Anon. d'Assur. d'Armateurs [illegible]
130		John Batey *George 85-88*	I PadSp	British [illegible]	16 108	96·3	18·4	9·0	North Shields Hepple & Co.	1884		1Cy 34½"–52" 46 HP H. B. Findlay & Son, N. Shields	J. Batey & Son
131 68[illegible]	[illegible]	John Beaumont *Mill 78-78*	I ScSp	British [illegible]	53 166	125·4	21·3	7·1	Kinghorn J. Key & Sons	1876		2Cy 17" & 32"–30" 45 HP Lees, Anderson & Co., Glasgw	North British [illegible] Co.
132 29[illegible]		John Bowes *Cain*	I Sc[illegible]	British [illegible]	270 437	150·0	25·7	15·6	Newcastle Palmer Bros. & Co.	1852		2Cy 16" & 36"–26" 60 HP M. Paul & Co., Dumbarton	Barnett Bros.
133 13[illegible]		John Brooks *Mosscrop -81*	W Pad	American Portland, Me	917 1012	239·8	31·4	10·8	New York	1859		1000 HP	Portland Steam Packet Co.
134		John Bull *Holland*	W Pad	British London	35 115	96·0	18·2	9·7	London Watkins	1849		2Cy 20"–42" 66 HP J. Stewart, London	J. Jackson
135	V	John Bull *Huynen*	I Sc	Belgian Antwerp	94 187	100·0	21·0	9·0	Antwerp D. Dekker & Co.	1880		2Cy 24" & 41"–21" 70 HP J. Cenatto & Co., Liege	Soc. Anon de T. Maritimes Anglo[illegible]
136 84[illegible]	✻	John Burbery *Crofts*	I Sc3mSr *1 Dk*	British Liverpool	98 284	145·0	22·1	9·3	Campbeltown Campbeltown S. B. Co. *Q 90 f. B 13 f. F 20 f.*	1881	1 Bhds. FPTH	2Cy 19" & 36"–27" 55 HP J. & T. Young, Ayr	S.S. "John Burbery" Co., Ld. (J. J. Mas[illegible])
137 63031	✻	John Byng *Turnbull*	I ScSr *1 Dk 2 L*	British London	487 775	204·0	28·1	16·4	Newcastle C. Mitchell & Co. *P & F 49 f.*	1869	1 Bhds. DBaf	2Cy 27" & 56"–30" 90 HP T. Richardson & Sons, Hrtlpl.	Miss A. M. Clark (W. F. Conner, Man.)
138 67[illegible]	✻	John Dixon *Welch 87-87*	I ScSr *1 Dk 2 L*	British Newcastle	977 1522	250·1	33·2	19·4	Newcastle Schlesinger, Davis & Co. *P 23 f. F 36 f.*	1872	5 Bhds. WB & f	2Cy 30" & 58"–36" 130 HP C. D. Holmes & Co., Hull	H. Bucknall & S[illegible]
139 63318	BT 6.88	John Elder	I ScBg *3 Dks & Awn Dk*	British Liverpool	2431 4160	406·4	41·5	35·2	Glasgow J. Elder & Co.	1870		2Cy 60" & 104"–48" 550 HP J. Elder & Co., Glasgow	Pacific Steam N[illegible]
140		John Eriksson *Jansson*	C ScSp *1 Dk*	Swedish Grenna	103 153	98·8	20·6	8·7	Torskog P. Larsson *P 10 f. H 6 f.*	1878		2Cy 13" & 27"–18" 30 HP Motala Co., Gothenburg	C. L. Berggren
141 94[illegible]		John Fraser	W Pad *2 Dks*	British Ottawa	100 118	102·5	28·8	6·0	Nipissing, Ont L. & R. Weir	1888		1Cy 18"–48" 80 HP L. & R. Weir, Montreal	A. Fraser
142 13177	R	John Gibson *Young*	W Sc3mSr *2 Dks*	American Georgetown, Columbia	333 445	147·0	26·0	17·0	Philadelphia	1864		300 HP	Inland & Seaboard Coasting Co.
143 89[illegible]	✻	John Grafton *Kelsey -83*	I ScSr *1 Dk*	British London	367 592	158·0	28·8	13·8	Newcastle Palmers' Co., Ld. *Q 43 f. B 13 f. F 18 f.*	1883	1 Bhds. DB [illegible] FPTH	2Cy 28" & 45"–30" 80 HP Palmers' Co., Ld., Newcastle	S. Clarke & Co.
144 [illegible]2387		John Haggart *Moffat -88*	W Sc *2 Dks*	British Kingston, Ont.	59 112	78·3	17·5	6·0	Perth, Ont. R. Robinson	1887		1Cy 11"–12" 65 HP [illegible] & Co., Kingston, Ont.	T. Moffat
145 63[illegible]	✻	John Johnasson *Whyle*	I ScSr *1 Dk 2 B*	British London	348 534	179·3	27·0	15·0	Newcastle C. Mitchell & Co. *Q 31 f. B 39 f. F 8 f.*	1870	1 Bhds. DBaf	2Cy 26"–50" 80 HP N. E. Marine Eng. Co., Ld., Shd.	R. Gordon & Co.
146 7[illegible]		John King	Pad	American New York	821 1057	190·8	36·0	13·6	Phildlphia, Pa.	1886			A. R. Mordaway
147 47[illegible]	✻	John M'Intyre *Johnson 73-80*	I Sc3mSr *1 Dk 2 B*	British London	630 991	226·0	32·0	16·9	Newcastle Palmer Bros. & Co. *Q 53 f.*	1863	3 Bhds. [illegible] FPTH	2Cy 26" & 51"–36" 90 HP Comp. [illegible] Victoria Docks Eng. Works, London	J. Fenwick & So[illegible]
148 81[illegible]	✻ D	John Marychurch *Pollick 88-88*	I ScSr *1 Dk 2 B*	British Cardiff	973 1511	260·3	36·0	15·6	Sunderland J. L. Thompson & Sons *Q [illegible] P 42 f. B 50 f. F 26 f.*	1880	4 Bhds. [illegible] APTR	2Cy 31" & 57"–36" 130 HP J. Dickinson, Sunderland	J. Marychurch & Co.
149 84[illegible]	✻ D	John Morrison *James 83-88*	I ScSr *1 Dk 2 B*	British N. Shields	1286 1968	270·0	36·1	21·0	Sunderland J. Laing *Q 98 f. B 53 f. F 34 f.*	1883	4 Bhds. DB [illegible]	2Cy 32" & 60"–39" 151 HP N. E. Marine Eng. Co., Ld., Shd.	J. Morrison & Son
150 70[illegible]	✻	John O. Scott *Barnsley 86-86*	I Sc3mSr *1 Dk*	British Newcastle	563 834	216·8	28·7	15·4	North Shields T. & W. Smith *Q 71 f. F 23 f. B 42 f.*	1874	1 Bhds. [illegible]	2Cy 27" & 55"–30" 115 HP R. & W. Hawthorn, N'wcastle	J. O. Scott

✻ = **Lloyd's Register;** D = Nederlandsche Vereeniging; G = Germanischer Lloyd; I = Registro Italiano; R = American Record;
A = Veritas Austro-Ungarico; H = Veritas Ellenico; N = Norske Veritas; V = Bureau Veritas.

In the 2nd Column, B.T., with figures following, signifies Date of report of bottom sighting by Board of Trade Surveyors as condition of PASSENGER CERTIFICATE.

In the 3rd Column, C = Composite (iron frame, wood planked); I = Iron; Stl = Steel; W = Wood.

Further particulars of Vessels classed by Lloyd's Register will be found in Lloyd's Register Book and Lloyd's Yacht Register.

No.	Cote	Nom, Matériaux, Gréement, etc. Capitaine et dates d'entrée au service des Armateurs actuels et à bord du navire. Ancien Nom, nombre de ponts, etc.		Pavillon. Port d'Armement	Tonnage Net. Total	Dimensions. Longueur	Largeur	Creux	Construit. Où. Par qui.	Quand	Nombre des cloisons. Détails sur le lest d'eau.	Machines. Détails. Nom du Constructeur.	Armateurs ou Directeurs-Gérants.
151		John Ormston *Wedgler 81-89*	I ScSr	British Newcastle	561 934	230·0	29·1	16·8 *P 126 f. F 34 f.*	Newcastle C.Mitchell&Co.	1873		2Cy.32"& 60"-36" 160 HP. R.&W.Hawthorn,N'weastle	Tyne Steam Shipping Co.,Ld.
152	✻V	John P. Best *Donnelly 81-84 (ex Sarine)*	I ScSr 2 *Dks* 3 *B*	Belgian Antwerp	1126 1656	277·5	31·5	25·1	Antwerp J.Cockerill	1875	5 Blkds.	2Cy.39"& 63"-36" 183 HP. J.Cockerill,Antwerp	J.P.Best & Co.
153	✻	John Paterson *Nelson*	W PadTug 1 *Dk*	British Pt.Elizabeth	12 30	92·6	18·6	9·7	NorthShields Hepple&Co.	1882		2Cy.24½"-46" 36 HP. Hepple&Co.,N.Shields	Government of Cape of Good Hope
154	✻	John Pender *Perkins 82-86*	I ScSr 1 *Dk* 2 *B*	British London	706 1213	246·3	29·2	20·1 *F 31 f.*	Inverkeithing J.Scott&Sons	1875	5 Blkds. MT 28 f T 18 f Tails f	2Cy.27½"&48½"-36" 98 HP. J.Scott&Sons,Inverkeithing	Eastern Telegraph Co.,Ld.
155	✻	John Ray *Talbot 82-82*	I ScSr 1 *Dk*	British London	302 503	105·9	36·0	14·0 *B 38 f. F 17 f. Q 28 f.*	Dundee W.B.Thompson	1881	4 Blkds. DBot&c f 479	2Cy.24"& 48"-30" 90 HP. W.B.Thompson,Dundee	M.A.Ray & Sons
156	✻D	John Readhead *Rahtkens 77-84*	I ScSr 1 *Dk* 2 *B*	British Middlesbro'	1103 1703	257·0	34·7	19·6 *Q 104 f. B 60 f. F 34 f.*	SouthShields J.Readhead&Co.	1883	4 Blkds. DBot&c f 904	2Cy.33"& 61"-36" 160 HP. J.Readhead&Co.,S.Shields	F. Rahtkens & Co.
157		John Romer *Frenth*	W Pad	American New York	223 409	173·1	26·6	8·0	Keyport, N.J. B.C.Terry	1863			G.R.Eager
158		John Schöning	I ScSr 1 *Dk* & *Awn Dk* 3 *B*	Norweg'n Bergen	490 662	173·8	24·9	13·5	Greenock Caird&Co.	1875	5 Blkds. T f 58	2Cy.25"& 50"-30" 110 HP. Caird&Co.,Greenock	Det Bergenske Dampskibsselskab.
159	✻D	John Stevenson *Smailes 81-84*	I ScBn 1 *Dk* 2 *B*	British Whitby	899 1408	243·0	34·0	17·9 *Q 63 f P 24 f B 54 f F 31 f*	Whitby T.Turnbull&Son	1884	4 Blkds. DBot&c f 437	2Cy.31"&58"-36" 144 HP. Blair&Co.,Ld.,Stockton	T.Smailes
160		John Stirling *Elder 60-79*	I PadSp	British Granton	168 427	190·3	27·0	10·5	Kinghorn J.Key&Sons	1876		2Cy.54"-51" 250 HP. J.Key&Sons,Kinghorn	North British Railway Co.
161	✻	John Strachan *Goodsir*	I ScLighter 1 *Dk*	British Kirkcaldy	44 73	66·0	17·9	8·1	Glasgow W.Swan&Co.	1885	3 Blkds.	1Cy.15"-15" 22 HP. Clarkson&Becket,Glasgow	Kirkcaldy, Leith & GlasgowSteamPacketCo.
162	✻	John Wells *Redford 74-87*	I ScSr 1 *Dk*	British Goole	338 503	210·8	29·2	13·8 *Q 29 f. F 28 f.*	Sunderland Davison&Stokoe	1873	4 Blkds. DBot&c f 103	2Cy.27"& 51"-30" 98 HP. Black,Hawthorn&C.,Gtshd.	Goole S.S,Co.,Ld.
163	✻	John Williamson *Taylor 81-87*	I ScSr 1 *Dk* 2 *B*	British S.Shields	806 1272	244·8	32·0	17·6 *P 172 f. F 29 f.*	Newcastle C.Mitchell&Co.	1870	4 Blkds. DBot&c f	2Cy.27"& 54"-36" 120 HP. R.&W.Hawthorn,N'weastle	Cay, Hall & Co.
164		John Young	W Pad 1 *Dk*	British Montreal	103 164	125·0	22·0	8·1	Montreal	1875		2Cy.30"-96" 500 HP.	Sincennes McNaughton Line
165	✻	John & Albert *Knudtzen 7-87*	I Sc 1 *Dk* & *Wh Frms*	P'tuguese Oporto	1062 1063	250·0	36·2	17·7 *P 26 f. B 68 f. Q 52 f. T 37 f.*	Sunderland S.P.Austin&Son	1887	6 Blkds. DBot 3 ... APT ...	3C 19"31½"&51"-36" 130 HP. Blair&Co.,Ld.,Stockton	J.H.Andresen
166		John o' Groat *(ex Peter John o'Groat)*	I ScSr 1 *Dk*	British London	186 338	174·8	24·8	12·9 *Q 37 f. B 50 f. F 38 f.*	Dundee Gourlay Bros.	1877	5 Blkds.	2Cy.28"&52"-33" 116 HP. Gourlay Bros.&Co.,Dundee	T.W.Tamplin
167	R	Johns Hopkins *Hallett*	Stl Sc3mSr 3 *Dks*	American Baltimore	945 1471	243·3	38·0	16·7 *F 70 f.* | 24·2	Wilmington Del Harlan&Hollingsworth	1873		1Cy.60"-44" 200 HP. Harlan&Hollingsworth Wilmington D.	Merchant & Miners' Transportation Co.
168	✻A	Jokai *Fellenich -88*	I ScSr 1 *Dk* 2 *B*	Austrian Fiume	780 1148	245·6	34·0	15·4 *Q 82 f. B 59 f. F 29 f.*	Barrow Caird&Purdie	1882	4 Blkds. CD 25 ft	2Cy.30"&55½"-36" 150 HP. Westray&Copeland,Barrow	"Adria" Hungarian Sea Nav.Co.,Ld.
169		Jolantha *(ex Iolanthe)* *Kruuse 85-85*	I ScSr 1 *Dk*	Danish Copenhagen	489 773	188·6	28·8	13·8	Blyth Blyth Eng. & Shipbuilding Co. Ld.	1884	4 Blkds. DBot&c f 513 APT 24	2Cy.25½"&48"-33" 98 HP. Black Hawthorn&Co.,Newc.	Det Forenede Dampskibsselskab (C.P.A.Koch)
170		Jonas Ahlströmer *Edgren*	C ScSp 1 *Dk*	Swedish Gothenburg	146 263	101·3	20·2	12·9 *B 23 f. J 2 f.*	Åmål J.Olsson	1869		30 HP. MotalaCo.,Gothenburg	J.A.Waller
171		Jonas Lie	I ScSr 1 *Dk* & *Awn Dk* 3 *B*	Norweg'n Bergen	477 644	166·5	24·8	13·3 | 20·2	Bergen Bergens Mek.Verkst.	1875	5 Blkds.	2Cy.25"& 45"-24" 75 HP. BergensMek.Verkst.,Bergen	Det Bergenske Dampskibsselskab
172		Jonio	I ScSr 1 *Dk*	Austrian Trieste	424 648	211·2	29·2	18·6	Newcastle Miller,Prow&Co.	1853		2Cy.20"& 34"-30" 160 HP. Miller,Prow&Co.,Newcastle	Lloyd Austro-Ungarico
173		Jönköping *Jonsson*	C ScSp 1 *Dk*	Swedish Grenna	107 162	99·6	21·0	8·6 *Q ... B 22 f. F ... H 2 f.*	Gamla Lödöse P.Larssen	1885		2Cy.13"& 27½"-20" 24 HP. MotalaCo.,Gothenburg	C.L.Berggren
174	✻	Jordan	Stl ScSr	British London		..	..	.. *P 29 f. Q 72 f. & Awng Dk 171 f.*	Sunderland J.L.Thompson&Sns	1889		3Cy.21½"35"&59"-39" T.Richardson&Sons, Hartlepool 200 HP.	Mercantile S.S.Co., Ld.
175		José Baró	I ScBg 2 *Dks*	Spanish Barcelona	980 1543	272·0	32·0	21·1	Sunderland Oswald&Co.	1872	6 Blkds.	2Cy.35"& 68"-36" 170 HP. Oswald&Co.,Sunderland	L.Soler & Co.
176		José Garcia *(ex Villaclara)*	W	Spanish Cienfuegos	1195 1532	219·1	36·7	11·3	New York	1867		366 HP.	Menendez & Co.
177		José Ramon *Muniz -81 (ex Romero Ortiz)*	I ScSr	Spanish Gijon	323 493	174·5	25·6	15·0	England	1855		1Cy.45"& 30"-27" 75 HP.	O. de Olavarria & Co.
178	R	Josefita *Ygor -87*	I ScSr 2 *Dks*	Spanish Cienfuegos	1042	211·0	32·6	20·4	Philadelphia Neafie&Levy	1887	4 Blkds.	2Cy.30"& 50"-36" Neafie&Levy,Philadelphia	Menendez & Co.
179	✻	Joseph *Lawson -84*	I ScSr 1 *Dk* 2 *B*	British Dundee	569 880	205·4	28·9	16·0 *Q 75 f. F 25 f.*	W.Hartlepool Denton,Gray&Co.	1872	4 Blkds. DBot&c f 786	2Cy.24½"&47½"-30" 90 HP. T.Richardson&Sons,Hrtlpl.	Bell & Sime
180	✻	Joseph Arbib *Hill -82*	I ScBa 2 *Dks* 3 *B*	British London	1642 2139	300·3	35·2	24·0 *P 44 f. F 36 f. Len. '86*	Middlesbro' R.Dixon&Co.	1882	5 Blkds. DBot&c f ...	2Cy.35"& 65"-36" 170 HP. T.Richardson&Sons,Hrtlpl.	E.& E.Arbib

✻ = **Lloyd's Register;** D = Nederlandsche Vereeniging; G = Germanischer Lloyd; I = Registro Italiano; R = American Record;
A = Veritas Austro-Ungarico; H = Veritas Ellenico; N = Norske Veritas; V = Bureau Veritas.

Dans la 2me Colonne, B.T., avec les indications qui suivent, signifie la date du rapport de visite de la Carène par les experts du Board of Trade pour obtenir le certificat relatif au TRANSPORT DES PASSAGERS.

Dans la 3me Colonne, C = Composite (membrure en fer, bordé en bois); I = Fer; Stl = Acier; W = Bois.

Pour plus amples renseignements sur les navires classés par le Lloyd's Register, voir le Lloyd's Register Book et le Lloyd's Yacht Register.

No. in Book Official No.	Class	Ship's Name. Material, Rig, &c. Master, and Dates of Appointment to present Owner's Service and to Ship Ship's Late Name. No. of Decks, &c.	Flag Port of Registry.	Tons Net Gross.	Length	Breadth	Depth	Built Where, By Whom.	When	No. of Bulkheads Particulars of Water Ballast.	Engines. Particulars. Maker's Name.	Owners
181 76197	R	**Joseph Bartram** W ScSrTug Anselin	American New York	59 117	112·0	21·5	10·0	WilmingtnDel Jackson&SharpCo	1881			Bartram Bros.
182	✻	**Joseph John** Stl ScSr	British Belfast	1390 2125	280·0	38·2	18·4	Sunderland [illegible]	1889		3Cy.21"35"&57"–39" W.Doxford&Sons,Ltd. [180H	Jaffe Bros.& J. [illegible]
183 82421	✻ R	**Joseph Oteri, jr.** I ScSr Kreyegard (ex Anglian) 1 Dk2 B	British Glasgow	291 471	170·2	24·1	15·7	PortGlasgow Murdoch&Murray	1878	4 Blds.	2Cy.26"& 42"–30" 85 HP. W.King&Co.,Glasgow	S.Holmes
184 81548	✻	**Joseph Rickett** I ScSr Cockerton 76-79 1 Dk	British London	449 717	186·1	28·0	13·9	Kinghorn J.Key&Sons	1879	4 Blds. [illegible]	2Cy.24"& 46"–36" 90 HP. J.Key&Sons,Kirkcaldy	Gamman,Son&[illegible]
185 47913	✻	**Joseph Somes** I ScSr Cuthbert 83-83 1 Dk2 B	British Hull	654 1030	238·0	28·6	15·9	Hull C.&W.Earle	1863	4 Blds.	2Cy.26"& 50"–30" 90 HP. Comp. [illegible] C.D.Holmes,Hull	Brown,Atkinson[illegible]
186		**Josephine** (WRECKED) I ScS	French Havre	92 135	102·9	17·8	8·9	Paris Bourong&Co.	1879		2Cy.20"& 36"–20" 70 HP. Bourong&Co.,Paris	Hersent
187 70243	✻	**Joshua Nicholson** I ScSr Reguart 86-88 2 Dks3 B	British Newcastle	1196 1839	270·0	35·0	24·5	Newcastle TyneI.S.B.Co.,Ld	1880	5 Blds. [illegible]	2Cy.32"& 62"–40" 160 HP. R.&W.Hawthorn,Newcastle,	"JoshuaNich[illegible] Co.,Ld.(Stephens[illegible]
188 49804	✻	**Juan** (ex Galilea) I ScBn Westcott 2 Dks	British London	1019 1571	270·6	33·1	23·1	Newcastle A.Leslie&Co.	1864	5 Blds.	2Cy.37"& 66"–42" 200 HP. Jack,Rollo&Co.,Liverpool	Westcott & La[illegible]
189	✻	**Juan Cunningham** I ScSr Asqueta 76-87 1 Dk,& Awn Dk	Spanish Seville	739 1160	221·4	32·2	21·4	Glasgow W.B.Thompson	1883	4 Blds.	2Cy.26"& 52"–30" 110 HP. W.B.Thompson,Dundee	Miguel Saenz & C[illegible]
190	R	**Juan Mir** I ScSr Monar 2 Dks	Spanish Matanzas	423	164·0	28·0	13·1	Chester,Pa. John Roach&Son	1878		2Cy.18"& 34"–30" J.Roach&Son,Chester,Pa.	Juan Mir
191	V	**Juan Ramos** I ScBk Vidal (ex Chrysolite)	Spanish Carthagena	478 782	189·7	28·4	17·2	Middlesbro' Candlish,Fox&Co.	1865		2Cy.23"& 46"–30" 160 HP. J.Jones&Sons,Liverpool	J.J.Casanova
192	V	**Juana** (ex San Jacinto) I ScBg Gastanoga 2 Dks	Spanish Bilbao	757 1179	233·3	32·0	19·6	Dumbarton A.McMillan&Sons	1872	4 Blds. [illegible]	2Cy.32"& 57"–33" 160 HP. J.&J.Thomson,Glasgow	Linea de Vapores Serra
193 89624	✻	**Juanita** I ScSr Bird 1 Dk	British London	361 572	172·0	29·1	11·5	Paisley H.McIntyre&Co.	1884	7 Blds. [illegible]	2Cy.20"& 38"–28" 80 HP. W.G.Lochhead,Glasgow	J.T.North
194 66110		**Jubilant** (ex Gulas) W ScSp 1 Dk	British Newcastle	61 111	97·7	19·8	9·6	Hayle Harvey&Co.	1869		2Cy.18"&33½"–20" 30 HP. J.Abbot&Co.,Gateshead	J.&D.Morris
195 93105	✻	**Jubilee** Stl ScSr Cottell -87 1 Dk	British [illegible]	56 232	130·0	21·0	10·8	Dundee W.B.Thompson&CoLd	1887	3 Blds. [illegible]	2Cy.20"& 40"–27" 66HP. W.B.Thompson &Co.,Ld.Dundee	Great Northern FishingCo.,Ld
196 94294		**Jubilee** Stl ScSr Winer -87	British London	47 141	123·5	20·0	9·1	London SamudaBros.,Ld.	1887		2Cy.20"& 37"–21" 70 HP. T.A.Young&Son,London	S.Eastern Railw[illegible]
197	✻	**Jules Chagot** I ScSr Babin 1 Dk2 B	French Nantes	821 1270	230·0	32·2	17·4	Newcastle Palmers'S.B.&I.Co.,Ld	1880	4 Blds. [illegible]	2Cy.30"& 57"–36" 130 HP. Palmers'S.B.&I.Co.,Ld,New	Pergeline fils & C[illegible]
198 13570		**Julia** W Pad	American S.Francisco	407 503	170·0	32·0	8·8	SanFrancisco	1864		200 HP.	E.H.Miller,jun.
199	✻ D	**Julia** I ScSr Jansen 83- 1 Dk2 B	German Hamburg	676 1043	224·7	32·0	14·3	Sunderland S.P.Austin&Son	1880	5 Blds. [illegible]	2Cy.28"& 58"–36" 120 HP. J.Clark,Sunderland	A.C.de Freitas & [illegible]
200 76723		**Julia H. Ray** (LOST) W ScSr Jenks -88 1 Dk	American SnFrancisco	219 296	136·5	32·2	10·5	San Francisco C.G.White	1888		2Cy.14"&28"–24" 125 HP. Savage,Jones&Co.,SnFrancisco	W.S.Ray
201 73808		**Julia Percy** I ScSr 1 Dk	British Melbourne	335 589	219·0	24·2	13·0	Glasgow T.Wingate&Co.	1876		2Cy.33"& 60"–36" 140 HP. T.Wingate&Co.,Glasgow	Western Steam Co.,Ld.
202 73674	✻	**Julia Wiener** I ScSr Wight 87-87 1 Dk2 B	British London	628 976	220·0	29·3	17·6	Newcastle C.Mitchell&Co.	1876	4 Blds. [illegible]	3C.17"28"&45"–33" 110 HP. Black,Hawthorne&Co.,Gtsd	R.Gordon & Co.
203		**Julian** (ex Rosa) I ScSr Guinea -78	Spanish [illegible]	491 710	184·3	27·9	16·0	NorthShields	1849		2Cy.25"& 48"–30" 80 HP.	O.de Olavarria & [illegible]
204		**Julian de Zulueta** I ScSr	Spanish Havana	267 392	180·1	27·2	10·7	PortGlasgow Cunliffe&Dunlop	1873		2Cy.26"& 46"–27" 100 HP. Cunliffe&Dunlop,Pt.Glsgw	J.de Zulueta
205		**Julie** I Sc	Russian [illegible]	336 430	..	..	..	Trieste StabilimentoTecnico	1883		50 HP.	G.Rosotti
206	V	**Juliette Marie** W ScK 1 Dk	British Granton	39 100	96·3	20·0	9·3	Granton D.Allan	1884		2Cy.14"& 28"–21" 35 HP. Walker,Henderson&Co.Glsw	D.Allan & Co.
207	✻ D	**Julio** Stl ScSr Erausquin83-88 1 Dk,S,SpDk3 B	Spanish Bilbao	1161 1800	275·5	36·1	13·8 21·0	Sunderland W.Doxford&Sons	1888	6 Blds. [illegible]	3Cy.20",33"&54"–36" 160 W.Doxford&Sons,Ltd. [HP	Cia Bilbaina de N[illegible]
208		**Julio** (ex Segundo ex Arequipa) I ScSr Batlle 85-85 1 Dk	Spanish Vigo	352 516	209·3	22·8	16·4	Glasgow Tod&McGregor	1872		2Cy.25"& 48"–33" 90 HP. Tod&McGregor	J.Barreras y C[illegible]
209	V	**Julius Cæsar** (ex Stag) I ScSr Hasim 1 Dk2 B	Swedish Gothenburg	721 932	218·5	29·3	16·2	Hartlepool Denton,Gray&Co.	1870	4 Blds. [illegible]	2Cy.33"–33" 96 HP. T.Richardson&Sons,Hrtlpl.	Fornyade Ang[illegible] Bol.SvenskaLloyd
210 83291	✻ BT12.88	**Jumna** StlSc3mSr Burkitt 78-86 3 Dks	British Glasgow	3377 5197	410·5	48·2	22·4	Dumbarton W.Denny&Bros.	1886	6 Blds.	4C.30"42"60"&84"–60"532 Denny&Co.,Dumbarton [HP	British India As[illegible] Steamers, Ld.

✻ = **Lloyd's Register**; D = Nederlandsche Vereeniging; G = Germanischer Lloyd; I = Registro Italiano; R = American R[illegible]
A = Veritas Austro-Ungarico; H = Veritas Ellenico; N = Norske Veritas; V = Bureau Veritas

In the 2nd Column, B.T., with figures following, signifies Date of report of bottom sighting by Board of Trade Surveyors as condition of PASSENGER CERTIFICATE

In the 3rd Column, C = Composite (iron frame, wood planked); I = Iron; Stl = Steel; W = Wood.

Further particulars of Vessels classed by Lloyd's Register will be found in Lloyd's Register Book and Lloyd's Yacht Register.

	2	3 Nom, Materiaux, Grèement, etc. Capitaine et dates d'entrée au service des Armateurs actuels et à bord du navire. Ancien Nom, nombre de ponts, etc.	4 Pavillon. Port d'Armement	5 Tonnage. Net. Total	6 Dimensions. Longueur. Largeur. Creux. Constructions sur le Pont	7 Construit. Où. Par qui.	8 Quand.	9 Nombre des cloisons. Détails sur le lest d'eau.	10 Machines. Details. Nom du Constructeur.	11 Armateurs ou Directeurs-Gérants.
		Juma — I PadSp 1 Dk	British Calcutta	119 181	143·3|20·6|8·2	Calcutta			2C. 34" 24" 60 HP. Maudslay,Sons&Field,Lond.	Baboo H.M.Roy
		Juniata — I ScS 3 Dks	American	834 1320	206·5|31·8|16·6 H 105 f. 30·5	Boston H.Loring	18[illegible]		1Cy.62" 44" 362 HP. H.Loring,Boston	Ocean S.S.Co.
	✱D	Junio — Stl ScS Aguirre 88-88 1 Dk&Spr Dk3 B	Spanish Bilbao	1394 2140	275·0|39·5|16·7 23·7	Sunderland W.Doxford&Sons	1888	4 Blkds. [illegible]	3C.21"35"& 57"-39" 180 HP. W.Doxford&Sons,Sunderld	Compania Bilbaina de Navegacion
	✱	Juno — Stl ScSr Skinner 87-88 1 Dk2 B&W 6 Frms	British W.Hartlepl	1567 2403	300·0|38·5|20·6 Q 30 f. P 29 f. B 123 f. F 34 f.	W.Hartlepool E.Withy & Co.	1888	5 Blkds. [illegible]	3Cy.22"37"&61"-39" 226 HP. T.Richardson&Sns,Hpl	T.Robinson
215		Juno — I Sc3mSr Walker 70-88 1 Dk2 B	British Newcastle	794 1311	250·2|32·2|16·5 P 152 f. F 37 f.	Hull Earles'Co.,Ld.	1882	4 Blkds. FPT[illegible] Tons	2Cy.35"&69"-39" 200 HP. Earles'Co.,Ld.,Hull	Tyne Steam Shipping Co.,Ld.
216		Juno — I PadSr Stat -68	British Bristol	488 880	261·5|29·6|15·1 P 114 f. F 38 f. 22·3	Glasgow London&GlasgwCo	1868		2Cy.40"&66"-72" 350 HP. Comp[illegible],Lon.&Glas.Co.,Glasgow	Bristol Stm.Nav.Co., Ld.
217		Juno (ex Lady [illegible]) I Sc3mSr Yt	British Portsmouth	143 223	164·5|20·9|12·1	Newcastle Palmers'Co.,Ld.	1868		2Cy.17"&32"-24" 50 HP. Day,Summers&Co.,Shmptn	Frank Willan
218		Juno — W ScIm	British Windsr,Ont.	130 210	114·3|26·8|8·3	Wallaceburg W.J.McConnell	1885		2Cy.15"&24½"-30" 175 HP. KerrBros.,Walkerville	W.Allen
219		Juno — I ScSr Tremul 56-87 2 Dks	Austrian Trieste	1110 1700	264·0|34·0|26·0	Dumbarton W.Denny&Bros.	1864		2Cy.41"&70"-36" 400 HP. Denny&Co.,Dumbarton	Lloyd Austro-Ungarico
220	DV	Juno (ex [illegible]) I ScSr Visser 81-87 2 Dks3 B	Dutch Amsterdam	776 1066	218·4|28·3|22·0	Paisley AbercornS.B.Co.	1873		2Cy.34"&60"-36" 158 HP. Hanna,Donald&Co.,Paisley	Kon.Nederl.Stoomb. Maats.
221	✱D	Juno (ex [illegible]) I Sc3mSr Lundgren -88 1 Dk2 B	Swedish Gothenburg	820 1033	224·9|28·7|17·2 B 75 f. Q 83 f. F 23 f.	South Shields Readhead, Softley & Co.	1871	3 Blkds. DBu & f	2Cy.25½"&47½"-33" 95 HP. ReadheadSoftley&Co.,S.Shields	J.A.Waller
222		Juno — I & StlScSp Johansson 81-91 1 Dk&A wn Dk	Swedish Stockholm	127 190	97·3|21·5|9·2 H 10 f.	Motala MotalaCo.	1874		2Cy.17"&35"-22" 60 HP. MotalaCo.,Motala	Motala Ströms Angf. Aktiebol. (E.B.Thorsell)
223		Junon — I ScBn Paravagna 69-87 2 Dks	French Marseille	750 1125	251·9|29·5|21·2 P 45 f. F 30 f.	Hull Samuelson[illegible]	1861		2C.38"&84½"-26½" 200 HP. Fraissinet&Co.,Marseilles	Fraissinet & Co.
224		Junqueira	Brazilian Theresina	150	.. .. ..					Comp.Rio Parnahyba
225		Jupiter — I PadSp White	British Grangemth.	19 102	96·1|17·9|9·2	SouthShields J.T.Eltringham	1876		1Cy.33½"-50" 40 HP. Hepple&Co.,S.Shields	J.S.Mackay
226		Jupiter — I ScSr Lese 69-87 2 Dks	Austrian Trieste	1103 1750	264·0|34·0|26·0	Dumbarton W.Denny&Bros.	1864		1Cy.29½"&54"-36" 400 HP. Denny&Co.,Dumbarton	Lloyd Austro-Ungarico
227	DV	Jupiter — I ScSr Mulder 85-88 2 Dks	Dutch Amsterdam	685 931	220·0|30·6|15·6 B 48 f. Q 72 f. F 25 f.	Amsterdam Konin Fabriek	1880	4 Blkds. DBc) 174	2Cy.27"&50"-36" 100 HP. Kon.Fbk.vanStoom,Amstm.	Kon.Nederl.Stoomb. Maats.
228	V	Jupiter — I ScSr Köhler -88 1 Dk	German Bremen	310 393	144·6|22·4|14·0	Bremen Act.Ges."Weser"	1873		2C.17½"&29½"-22½" 145 HP. Act.Ges."Weser,"Bremen	Dampfschifffahrts-Ges."Neptun"
229		Jupiter — I ScSr 2 Dks	Norweg'n Bergen	488 648	180·5|26·2|20·1	Greenock Caird&Co.	1856	5 Blkds.	2Cy.26"&54"-28" 114 HP. Caird&Co.,Greenock	Det Bergenske Dampskibs-Selskab.
230		Jupiter — C ScSr Andersson -73 1 Dk	Swedish Gothenburg	124 172	99·7|21·0|9·1 P 22 f. E 3 f.	Gothenburg C.Bergström	1883		2Cy.13½"&27"-18" 35 HP. Erksbg.Mek.Verks.,Gthnbg	Ångf.Aktie Bol. "Jupiter"(J.D.Grönvall)
231		Jupiter — I Pad Julia	Urug'yan Montevideo	530 964	232·0|26·0|10·0	Glasgow J.&G.Thomson	1866		2Cy.54"-60" 185 HP.	La Platense Flotilla Co., Ld.
232		Jurua — I Pad 1 Dk,Spt Hur Dk	Brazilian Para	240	150·0|26·0|7·0	WilmingtnDel Pusey,Jones&Co.	1874		2Cy.26"-72" 73 HP. Pusey,Jones&Co.,Wilmngtn	Amazon Stm.Nav Co.
233		Juruty (ex Theotonio) I Sc 1 Dk,Spt Hur Dk	Brazilian Para	90	120·0|16·0|9·0		1871		2Cy.21"&36"-28" 60 HP.	Amazon Stm.Nav.Co.
234		Justitia — I Sc3mSr 1 Dk2 B	Ven'zul'n	535 831	201·3|29·1|17·2 P 42 f. F 34 f.	Stockton Richardson, Duck & Co.	1862	5 Blkds.	2Cy.25"&50"-33" 160 HP. Palmers'Co.,Ld.,Newcastle	
235	V	Jyden — I ScSr Petersen 83-86 1 Dk2 B	Danish Randers	227 378	156·9|24·2|13·4 B 49 f.	Rostock RostockSchiffsb.-Ges.	1880	WB[illegible]	2Cy.19"&35½"-24" 55 HP. RostockSchiffsb.-Ges.	Randers Dampskibs-Selskab af 1866 (A.Andersen)
236	✱DV	Jylland — I Sc3mSr Stark 88-88 1 Dk&Spr Dk	Danish Copenhagen	911 1430	233·7|32·3|18·0 H 18 f. 24·6	Kiel Norddt.-Schiffsb.A.G	1873	5 Blkds.	2Cy.32"&52"-36" 120 HP. Schweffel&Howaldt,Kiel	Dampskibs-Selskab. "Kjøbenhavn"

ADDITIONAL SHIPS ARE INSERTED IN THE SUPPLEMENT AT THE END OF THE REGISTER.

✱=Lloyd's Register; D=Nederlandsche Vereeniging; G=Germanischer Lloyd; I=Registro Italiano; R=American Record; A=Veritas Austro-Ungarico; H=Veritas Ellenico; N=Norske Veritas; V=Bureau Veritas.

Dans la 2me Colonne, B.T., avec les indications qui suivent, signifie la date du rapport de visite de la Carène par les experts du Board of Trade pour obtenir le certificat relatif au TRANSPORT DES PASSAGERS.

Dans la 3me Colonne, C=Composite (membrure en fer, bordé en bois); I=Fer; Stl=Acier; W=Bois.

Pour plus amples renseignements sur les navires classés par le Lloyd's Register, voir le Lloyd's Register Book et le Lloyd's Yacht Register.

No. in Series / Official No.	Class	Ship's Name. Material, Rig, &c. / Master, and Dates of Appointment to present Owner's Service and to Ship. / Ship's Late Name. No. of Decks, &c.	Flag. / Port of Registry.	Tons. Gross / Net	Length	Breadth	Depth	Built. Where / By Whom.	When	No. of Bulkheads. / Particulars of Water Ballast.	Engines. Particulars. / Maker's Name.	Owners
1	✱	**K. A. Niemann** I ScSr *Dapuy 76-84* 1 Dk	Dutch Rotterdam	[illegible]	142·0	22·7	10·2	Rotterdam J.Smit,Czn.	1884	4 Blds.	2Cy.19"&35"-27" 50 HP. Maats.deMaas,Delfshaven	Nieuwe Afrik. Handels-Venn.
2	V	**Kaala** W Sc *Teipp -87*	Hawaiian Honolulu	120	100·0	22·0	8·0	S.Francisco DickieBros.	1887		2Cy.12"&24"-15" 80 HP. UnionIronWrksS.Francisco	J.A.Cummings
3		**Kabatash** W Pad *Mustapha 78-79*	Turkish Cnstntinple.	95	140·0	20·6	9·6	WestCowes J.White	1866		1Cy. 80 HP. Maudslay,Sons&Field,Lond	Chirket-i-Hairie (Bosphorus steam.)
4		**Kachaloff** I Sc	Russian Archangel	127	110·0	20·2	10·5	Newcastle C.Mitchell&Co.	1871		2Cy.17"&31"-20" 40 HP. Pattison&Atkinson,Newcstl	White Sea S.N.
5		**Kadikioi** (DISMANTLED) W Pad	Turkish Cnstntinple.	75	161·2	20·0	9·8	PortGlasgow W.Hamilton&Co.	1869		2Cy. 120 HP. W.Hamilton&Co.,PtGlasg.	Idarei Massousie (Ottoman Steam.)
6		**Kadrinyeh** (ex Lyn) I PadSc *Nichola 68-87*	Turkish Cnstntinple.	150	190·4	20·2	8·2	PortGlasgow Blackwd.&Gordon	1878		2Cy.39"-66" 120 HP. Blackwood&Gordon,P.Glas.	Idarei Massousie (Ottoman Steam.)
7 57790	✱	**Kaffir** I Sc *Reimer* 1 Dk	British Glasgow	189	136·4	22·2	10·0	Newport Mosley,Carney&Co.Ld	1883	4 Blds. FPT APT	2Cy.19"&31"-24" 50 HP. SmithBros.,Glasgow	J.Napier
8 49017	✱ BT 1.88	**Kaffraria** I ScSr *Sarran 80-84* 1 Dk 2 B	British Hull	675 1032	237·8	29·0	16·2	Sunderland J.Laing	1864	Blds. WB	1Cy.21"&38"-24" HP. Humber Iron Wks.Hull	W.S.Bailey
9		**Kah-Byoo** Stl Pad *Smith*	British Glasgow	326	170·0	30·1	7·1	Dumbarton W.Denny&Bros.	1882		2Cy.24"&42"-72" 75 HP. Denny&Co.,Dumbarton	Irrawaddy Flotilla Ld.
10 37116	V	**Kahu** Stl&I Sc *Braddick -86* 1 Dk	British Napier	99 182	115·0	19·1	8·8	Newcastle H.S.Edwards&Sons	1886		2Cy.17"&32"-24" 40 HP. N.E.MarineEng.Co.,Nwc.	G.E.G.Richardson
11	✱	**Kai Pan** Stl TwnScSr *Heel -88* 1 Dk	Chinese	110	135·0	23·1	11·4	Newcastle SirW.G.Armstrong, Mitchell&Co.,Ld.	1888	5 Blds.	3C12"19½"&32"-20" 130 HP. WallsendSlipwayCo.Ld.Nwc.	Chinese Imperial Maritime Customs
12 93436	BT 6.88	**Kaifong** Stl TwnScSr *Dodd -88*	British London	998 1643	256·3	36·3	21·6	Greenock Scott&Co.	1888		6Cy14"23"&37"-30" 250 HP. Scott&Co.,Greenock	China Nav.Co.,Ld.
13 89027	✱ BT 11.88	**Kaikoura** Stl ScBk *Crutchley 83-84* 2 Dks,&Spr Dk	British London	2885 1174	430·0	46·0	24·0 32·2	Glasgow J.Elder&Co.	1884	7 Blds. FPT	2Cy.46"&86"-57" 600 HP. J.Elder&Co.,Glasgow	New Zealand Shipping Co.,Ld.
14	✱	**Kaimiloa** W Sc3mSr *Holman86-8 (exExplorer, exFireBrick)*	Hawaiian Honolulu	171 291	127·7	25·2	11·2	Blyth UnionCo-opSociety	1872		1Cy.24"-30" 40 HP. W.Hepple&Co.,S.Shields	HawaiianGovernment
15 816		**Kairiu Maru** W ScSr	Japanese Owari	277 447	162·1	21·2	18·1	Osaka OsakaIronWorks	1883		45 HP.	Hanyeisha
16 83175	✱D	**Kairos** I ScSr *Liddle* 1 Dk,&Spr Dk 3 B	British London	1912 2926	313·0	40·1	29·1	Sunderland J.L.Thompson&Sons	1882	6 Blds. FPT APT	2Cy.38"&69"-45" 270 HP. Black,Hawthorn&Co.,Gtshd	R.Gordon & Co.
17 76182	BT 7.88	**Kaisar-i-Hind** I ScBk *Briscoe 77-88* 2 Dks,&Spr Dk	British Greenock	2386 4008	400·7	42·3	24·3 31·8	Greenock Caird&Co.	1878	8 Blds.	2Cy.60"&104"-60" 700 HP. Caird&Co.,Greenock	Peninsular & Oriental Steam Nav.Co.
18 81513	✱ BT 1.88	**Kaiser** I ScSr *Beal 68-84* 1 Dk 2 B	British W.Hartlepl.	509 805	220·0	28·6	16·2	W.Hartlepool W.Gray&Co.	1880	4 Blds.	2Cy.30"&57"-33" 130 HP. T.Richardson&Sons,Hrtlpl.	W.Hartlepool S.S. Co.
19	G	**Kaiser Wilhelm** Stl PadSr *Zeissig 86-86* 1 Dk	German Rostock	203 390	191·4	21·7	10·8	Stettin Act.-Ges."Vulcan"	1885		2Cy.32½"&61"-49½" 280 HP. Act.-Ges."Vulcan,"Stettin	Deutsch-Nordischer Lloyd
20		**KaiserWilhelmII** Stl ScdmSr *Stürmer 84-89* 4 Dks,&Har Dk	German Bremen		449·6	51·0	35·6	Stettin StettinerMaschAktGes."Vulcan"	1889		3Cy.41"67"&106"-43" ActiengesellschaftVulcan,Stettin	Norddeutscher Lloyd
21		**Kaiserieh** (ex Princess Wa-[illegible], ex Pera) I ScBg *Giovanni 58-87* 2 Dks 3 B	Turkish Cnstntinple.	993 1303	244·3	32·1	24·9	Sunderland J.Laing	1868	4 Blds.	2Cy.28"&42"-30" 150 HP. Thompson,Boyd&Co.,Newc.	Idarei Massousie (Ottoman Steam Nav.)
22 81898	✱	**Kaisow** Stl Sc *Castle 85-88* 2 Dks 3 B	British London	1934 2659	350·0	40·9	25·8	Sunderland J.L.Thompson&Sons	1885	7 Blds.	3C.30"47"&75"-48" 600 HP. T.Richardson&Sons,Hrtlpl.	China Shippers Mutual S.N.Co.
23		**Kaitai** I Sc 1 Dk	Chinese Tientsin	121	90·0	19·0	10·3	Shanghai Boyd&Co.	1884		2Cy.16"&30"-21" 52 HP. A.&J.Inglis,Glasgow	TakuTug&Lighter
24 92611		**Kakabeka** W Sc *Marin -87* 2 Dks	British PortArthur	75 113	68·0	18·0	7·0	Toronto,Ont. M.Simpson	1885		1Cy.13"-14" 5 HP. J.Doty,Toronto,Ont.	T.Marks
25		**Kalahome** I Sc 1 Dk 2 B	Siamese Bangkok	176 278	144·6	22·1	13·6	Liverpool W.H.Potter&Co.	1867	4 Blds.	2Cy.30"-21" 60 HP. G.Forrester&Co.,Liverpool	Siamese Government
26		**Kaleva** Sc *Bögglund*	Russian	117				Wiborg	18?			P.Parviainen
27		**Kalinadi** I Sc	British	50 201	144·7	26·3	8·9	PortGlasgow R.Duncan&Co.	1884		2Cy.18"&31"-27" 49 HP. Dunsmuir&Jackson,Glsgw.	Shepherd & Co.
28 89946		**Kalmarsund No. 4** I ScSp *Nilsen* 1 Dk	Swedish Calmar	89 120	91·5	16·6	6·7	Motala MotalaCo.	1876		30 HP.	Kalmarsunds Ång. Aktie Bolag
29		**Kalmarsund No. 5** I ScSp *Knudsen* 1 Dk	Swedish Calmar	85 120	91·5	16·7	6·7	Motala MotalaCo.	1876		25 HP.	Kalmarsunds Ång. Aktie Bolag
30		**Kalmarsund No. 6** I ScSr *Pettersson -88* 1 Dk	Swedish Calmar	116 172	92·9	21·6	7·2	Motala Motala&Co.	1888			Kalmarsunds Ång. Aktie Bolag

✱ = Lloyd's Register; D = Nederlandsche Vereeniging; G = Germanischer Lloyd; I = Registro Italiano; R = American Record; A = Veritas Austro-Ungarico; H = Veritas Ellenico; N = Norske Veritas; V = Bureau Veritas.

In the 2nd Column, B.T., with figures following, signifies Date of report of bottom sighting by Board of Trade Surveyors as condition of PASSENGER CERTIFICATE.

In the 3rd Column, C = Composite (iron frame, wood planked); I = Iron; Stl = Steel; W = Wood.

Further particulars of Vessels classed by Lloyd's Register will be found in Lloyd's Register Book and Lloyd's Yacht Register.

1	2	3 Nom, Matériaux, Grément, etc. Capitaine et dates d'entrée au service des Armateurs actuels et à bord du navire. Ancien Nom, nombre de ponts, etc.	4 Pavillon. Port d'Armement	5 Tonnage. Net. Total	6 Dimensions. Long. Larg. Creux. Constructions sur le Pont	7 Construit. Où. Par qui.	8 Quand	9 Nombre des cloisons. Détails sur le lest d'eau.	10 Machines. Détails. Nom du Constructeur.	11 Armateurs ou Directeurs-Gérants.
		Kalmius (ex Morehod) I ScSr 2 Dks	Russian Odessa	439 569	204·5|27·0|13·7	Rostock Abendrod	1872		2Cy.19½"&38½"–21" 60 HP. Borzack	Russian Steam Nav. & Trading Co.
	✻	Kama (ex Lady ... Kama, ex ...) I ScSr Taylor 86-88 1 Dk 2 B	British Middlesbro	582 865	209·7|30·2|15·0 B 38 f. Q 75 f. F 30 f.	Leith Ramage&Ferguson	1881	4 Bhds.	2Cy.25"&48"–30" 95 HP. Lees,Anderson&Co.,Glasgw	E. Harris & Co.
	V	Kambyses I ScSr Prehn 86-88 2 Dks & B	German Hamburg	1294 1824	270·6|35·4|22·5 P 44 f. B 80 f. F 32 f.	Hamburg Reiherstieg Schiffs	1884	W.B.	2Cy.31"&62"–42" 180 HP. Reiherstg.Schffs.Wrfte.Hmbg.	Deutsche Dmpfschifff. Ges. "Kosmos"
		Kameruka Stl TwinScSr Rolick 1 Dk & Awn Dk	British Sydney	300 515	179·6|24·5|8·0 | 15·6	Greenock Russell&Co.	1880	4 Bhds.	1C 18"&36"–24" 98 HP. Rankin&Blackmore,Grnck.	Illawarra Stm. Nav. Co.
		Kamloops W Pad Sr	British N.Wstmnstr	268 426	126·0|27·4|5·6	Kamloops B.C. A.Watson	1885		1Cy.16"–54" 805 HP. Allison Iron Wrks.Victra.B.C.	John A. Mara
	✻	Kampokus I ScSr Johns 83-83 1 Dk	British Cork	279 484	160·1|25·1|13·5 Q 82 f. F 19 f.	Glasgow D.&W.Henderson	1882	4 Bhds. DB ... FPT ...	2Cy.17½"&36"–27" 53 HP. D.&W.Henderson&Co.Glsw	A. Slorach & Son
	✻	Kanahooka I ScSr Ball 1 Dk	British Sydney	239 386	150·0|24·2|13·2 Q 12 f. B 38 f. F 17 f.	Kinderdyk J.&K.Smit	1883	5 Bhds. DB ...	2Cy.22"&40"–30" 70 HP. Maats.de Maas,Delfshaven	F. Osborne
	V	Kanal Stl ScSr Jensen 86 1 Dk & Awn Dk	German Flensburg	206 264	94·2|23·3|10·0	Flensburg Flnsbg.Schffsb.Ges	1886	2 Bhds. WB ...	2Cy.16"&31"–18" 40 HP. Flensburg Ship B.Co.	Flensburg Ekensundr. Sønderbrgr. Dampf.Ges.
	R	Kanawha I ScSr Sears 87 2 Dks	American Nwpt.N.Vas	431 536	172·2|39·7|16·2	Bath, Me. W.Rogers	1881		1Cy.36"–30" 300 HP. C.H.Delamater&Co.,N.York	Chesapeake & Ohio Railway Co.
		Kangaroo I ScSr Phillips 88 3 Dks	British London	1099 1582	257·4|36·2|27·0 F 86 f.	Port Glasgow L.Hill&Co.	1853		2Cy.36"&60"–39" 160 HP. Laird Bros.,Birkenhead	Scott & Osborne
41	✻	Kangra I ScSr Ingram 85-86 2 Dks	British Glasgow	1469 1958	284·7|35·2|23·8 P 14 f. B 32 f. F 37 f.	Dumbarton W.Denny&Bros.	1880	5 Bhds. DB ...	2Cy.34"&60"–42" 180 HP. Denny&Co.,Dumbarton	British India Stm. Nav. Co., Ld.
		Kanieri Stl ScSr 1 Dk	British Dunedin	115 203	105·0|21·1|8·7 Q 36 f. B 34 f. F 13 f.	Dumbarton W.Denny&Bros.	1886		2Cy.15"&26"–15" 20 HP. Denny&Co.,Dumbarton	Union Steamship Co. of New Zealand, Ld.
		Kano Stl PadSr	British London	232 421	192·2|29·1|9·0 Q 12 f.	Port Glasgow David J.Dunlop&Co	1882		1C 19"&30"–48"&58" 110 HP David J.Dunlop&Co.PtGlas.	National African Co., Ld.
	✻V	Kansas I ScIn Sr Gleig 84-85 3 Dks & Awn Dk	British Liverpool	3455 5276	436·5|43·8|20·2 | 35·6	Glasgow C.Connell&Co.	1882	7 Bhds. MT 590 t.	2Cy.50"&94"–60" 600 HP. J.&J.Thomson,Glasgow	"Kansas" S.S.Co., Ld. (G.Warren & Co.)
	✻	Kant I ScSr Johannsen 81- 1 Dk	German Lübeck	228 263	130·0|20·1|10·2 Q 33 f. B 51 f. F 17 f.	Sunderland D.Baxter&Co.	1881	3 Bhds. DB ... FPT ...	2Cy.16"&32"–20" 45 HP. R.Smith,Preston	Lübeck-Königsberger Dampfschifff.-Ges.
		Kaplan (ex Atlantic) I ScBg Tahir Bey 88	Turkish Constantnple	993 1528	254·6|30·4|24·1	Hull Earle&Co.	1857	4 Bhds.	2Cy.30"&53"–33" 150 HP. T.Brassey&Co.,Birkenhead	Idarei Massousieh (Ottoman Stm. Nav. Co.)
	✻	Kapurthala Stl ScSr Turner 86-87 2 Dks & Shade Dk	British Glasgow	672 1122	240·2|34·2|16·3 P 44 f. B 73 f. F 53 f.	Dumbarton W.Denny&Bros.	1887	1 Bhds. DB ...	3C 20"33"&54"–42" 151 HP Denny&Co.,Dumbarton	British India Stm. Nav. Co., Ld.
	✻	Karagola Stl ScSr Gimblett 79-88 2 Dks & Shelter Dk	British Glasgow	593 1168	240·3|34·2|18·1 P 54 f. B 61 f. F 55 f.	Glasgow A.&J.Inglis	1887	5 Bhds. DB ...	3Cy 24"38"&57"–42" 280 HP A.&J.Inglis,Glasgow	British India Stm. Nav. Co., Ld.
		Karamania I ScSr Young 77-87 1 Dk & Spr Dk 3 B	British Barrow	2045 3148	340·0|39·2|29·6 F 44 f. B 88 f.	Glasgow D.&W.Henderson&Co.	1882	7 Bhds. WB ...	2Cy.33"&66"–60" 480 HP. D.&W.Henderson&Co.Glsw.	Barrow Steam Ship Co., Ld.
	✻	Karang I ScSr Kendler 1 Dk & Shd Dk	Dutch Batavia	209 326	161·0|23·0|10·8 P 28 f. F 39 f.	Greenock Caird&Co.	1875	6 Bhds.	2Cy.22"&44"–27" 85 HP. Caird&Co.,Greenock	Nederl.-Indische Stoomv.-Maats.
	✻	Karaweera I ScSr Rosen 82-82 1 Dk 2 B	British Adelaide	929 1477	250·0|35·1|17·6 P 30 f. B 62 f. F 33 f.	Blyth Hodgson&Soulsby	1882	5 Bhds. DB ...	2Cy.32"&60"–39" 170 HP. Blair&Co.,Ld.,Stockton	Harrold Bros.
		Karenee Stl PadLc	British Glasgow	250 674	260·0|35·1|9·8	Dumbarton W.Denny&Bros.	1888		2Cy.37½"&65"–72" 375 HP. Denny&Co.,Dumbarton	Irrawaddy Flotilla Co., Ld.
53		Karl I Sc	Russian Rostoff	247 320	.. .. ..	Trieste Stabilimento Tecnico Triestino	1882		40 HP.	O. Johnson & Co.
54	V	Karl Fredrik Wærn I Sc 3m Sr Söron 83 1 Dk	Swedish Gothenburg	305 382	144·0|22·2|15·4 B 8 f.	Gothenburg Göteborgs M.V.	1870		2Cy.13"&28"–20" 45 HP. J.Keiller,Gothenburg	Förnyade Ångf. Akt. Bol. "Göta" (H.Sternhagen)
55	✻	Karl O. Kjellberg Stl ScSr Möbius 83 1 Dk	Swedish Gothenburg	478 651	170·0|26·2|15·5 Q 40 f. B 45 f. F 21 f.	Gothenburg Göteborgs Mek. Verkst.	1883	5 Bhds. A P T ... M T ...	2Cy.24"&40"–30" 120 HP. Göteborgs Mek. Verkst.	Förnyade Ångf. Akt. Bol. "Göta"
56		Karl v. Linné Stl ScSr Lübeck 1 D	Swedish Stockholm	290 392	145·3|23·4|10·3 P 98 f. F 35 f. B 18 t.	Norrköping Motala Co.	1884		2Cy.21"&43"–24" 80 HP. Motala Co.	A. Larson
57	✻	Karl XV. I Sc 3m Sr Söderhjelm 85-85 1 Dk & Awn Dk 3 B	Swedish Gothenburg	894 1070	214·3|27·4|15·0 B 8 t. | 21·9	Newcastle T.&W.Smith	1865	1 Bhds.	2Cy.23"&47"–29½" 130 HP. Gotebrgs.Mek.Vrkst.Gtnbg	Ångf. Aktie. Bolag. "Thule" (O.Melin)
58		Karl XV. I ScSr Hedberg 1 Dk & Awn Dk	Swedish Nyland	415 520	162·0|24·0|11·5 B 35 t.	Stockholm Bergsunds Mek. Verk.	1872		2Cy.26"&47"–27" 100 HP. W.Lindbergs Mek. Vrks Stcklm.	Södra Ångermanlands Ångf. Akt-Bol. (J.P.Dahlberg)
59		Karl XV. C PadSr Blanck 1 Dk	Swedish Östersund	69 110	117·1|15·1|7·5 Q 30 f. B 10 t.	Mattmar	1873		40 HP.	Storsjö Ångb. Aktiebol.
60	G	Karlshamn C ScSr Fernstrom 75-83 1 Dk	Swedish Karlskrona	214 292	126·3|21·7|10·9 Q 40 f. B 35 f. F 12 f. H 4 t.	Carlshamn D.A.Andersson	1871		2Cy.16"&29"–17" 40 HP.	C.G.O. Winter

✻ = Lloyd's Register; D = Nederlandsche Vereeniging; G = Germanischer Lloyd; I = Registro Italiano; R = American Record; A = Veritas Austro-Ungarico; H = Veritas Ellenico; N = Norske Veritas; V = Bureau Veritas.

Dans la 2me Colonne, B.T., avec les indications qui suivent, signifie la date du rapport de visite de la Carène par les experts du Board of Trade pour obtenir le certificat relatif au TRANSPORT DES PASSAGERS.

Dans la 3me Colonne, C = Composite (membrure en fer, bordé en bois); I = Fer; Stl = Acier; W = Bois.

Pour plus amples renseignements sur les navires classés par le Lloyd's Register, voir le Lloyd's Register Book et le Lloyd's Yacht Register.

No. in Book / Official No.	Class	Ship's Name, Material, Rig, &c. / Master, and Dates of Appointment to present Owner's service and to Ship / Ship's Late Name, No. of Decks, &c.	Flag / Port of Registry	Tons Net / Gross	Dimensions (Length / Breadth / Depth)	Built: Where / By Whom	When	No. of Bulkheads / Particulars of Water Ballast	Engines: Particulars / Maker's Name	Owners
61		**Karlstad** I ScSr *Laurin* -87 1 *Dk*	Swedish Karlstad	165 221	104·3 | 21·6 | 10·4 *B* 34 *f.* *F* 12 *f.* *H* 21	Thorskog P. Larsson	1887		2Cy. 13" & 24"–17½" 35 HP.	Svensson & [illegible]
62 1142?		**Karluk** W ScBn *Andersen* 80-86 2 *Dks*	American S. Francisco	221 2?4	125·6 | 27·0 | 14·2	Benicia, Cal. M. Turner	1884		2Cy. 11" & 20"–18" 45 HP. W. MacAfee, San Francisco	"Karluk" Pa[illegible]
63		**Karmsund** I ScSr *Storksen* 1 *Dk*	Norweg'n Haugesund	102 152	106·1 | 18·8 | 8·0 *Q* 31 *f.* *B* 2? *f.* *F* 2? *f.* Len 3?	Drontheim Trondhjems Mek. V[illegible]	1874		2Cy. 16" & 28"–21½" 40 HP. Trond. Mek. Verkst., Dronth'm	Haugesunds [illegible] skibs Selskab
64	✱	**Karnak** Stl ScSr *Saggau* ?-88 2 *Dks* 3 *B*	German Hamburg	1727 2530	290·0 | 40·2 | 24·5 *B* 92 *f.* *F* 35 *f.*	Newcastle Tyne I. S. B. Co., Ld.	1887	7 Bhds. [illegible]	3C. 24" 37" & 62"–42" 280 HP. Wigham Richardson & Co., Newc'st	Deutsche Damp[illegible] farts Gesellsch.
65	✱	**Karoon** Stl ScSr	British W. Hartl'p'l		.. .. ..	W. Hartlepool W. Gray & Co.	1889		3Cy. 21" 35" & 57"–39" Blair & Co., Ld., Stockton	T. Appleby & Co.
66		**Kartal** W [illegible] *Khurshid* 86-87	Turkish [illegible]	50	129·3 | 22·0 | 10·3		1866		100 HP. C. A. Day & Co., Southampton	Idarei Massousa[illegible] (Ottoman Stm. Nav.)
67		**Karteria** I Sc *Koccinos* 75-	Greek [illegible]	317 [illegible]	184·3 | 24·9 | 15·0	England	1859		1Cy. 24" & 45"–30" 120 HP.	Hellenic Steam[illegible] Co.
68 83751		**Karuah** W ScSr *Tulloch*	British [illegible] N.S.W.	82 124	105·8 | 20·8 | 7·7 *Q* 8 *f.*	Berry's Bay. W. Dunn N.S.W.	1882		2Cy. 14" & 26"–18" 40 HP. Plenty & Sons, Newbury	G. W. Nicoll
69		**Kasan** Stl ScSr *Thomsen* 87-88 2 *Dks Spd Shd Dk*	Danish Copenhagen	716 1132	242·1 | 30·2 | 14·9 *P* 159 *f.* *F* 32 *f.* 21·7	Copenhagen Burmeister & Wain	1883	[illegible] FPT [illegible] APT [illegible]	1Cy. 24" & 48"–39" 150 HP. Burmeister & Wain, Cpnhgn.	Det Forenede [illegible] skibs Selskab [illegible]
70 95496	V	**Kasbek** StlSc 3mSr *Bevan* -88	British London	2098 2707	310·0 | 40·4 | 28·3	Newcastle Sir W. G. Armstrong Mitchell & Co.	1888		3C. 28" 37" & 60"–39" 250 HP [illegible]	Kasbek S.S. Co. L[illegible] (Lane & Macandrew)
71 68549		**Kashgar** I ScBg *Jephson* ?-88 2 *Dks* 3 *B*	British Sunderland	1515 2821	362·5 | 36·8 | 27·4 *F* 29 *f.*	Sunderland J. Laing	1873	6 Bhds.	2Cy. 36" & 66"–51" 450 HP. R. & W. Hawthorn, Newcastle.	Peninsular & [illegible] Steam Nav. Co.
72 85100		**Kasi** I PadSr *Young* -87	British Calcutta	198	145·7 | 23·1 | 7·5	Port Glasgow Cunliffe & Dunlop	1882		2Cy. 21" & 39"–57" 140 HP. Cunliffe & Dunlop, P. Glasgw.	Indian General Nav. Co. Ld.
73		**Kassed Kerim** I Sc3mSr (ex [illegible])	Turkish Alexandria	230 359	160·2 | 21·2 | 12·1	Newcastle W. E. Boutland	1873		2Cy. 20" & 36"–24" 56 HP. Marshall, Osbrn & Co., S. Shlds.	S. Bey Hullabo
74		**Kastrup** (ex [illegible]) I ScSalvageSp *Larsen* 80-? 1 *Dk*	Danish Kastrup	35 104	91·2 | 18·4 | 10·6	Malmö Kockums Mek. Werkst.	1874		2Cy. 18" & 30"–22" 35 HP. Kockums Mek. Werk. Malmö	Oscar Petersen's Bjergnings [illegible]
75 14028		**Katahdin** W Pad *Pierce* -86 2 *Dks*	American Boston	987 1235	241·4 | 35·4 | 11·6	New York	1863		1?0 HP.	Boston & Bangor Steamship Co.
76 81208	✱D	**Kate** I ScBn *Burkie* 82-82 2 *Dks* 3 *B*	British Whitby	1251 1934	273·5 | 34·5 | 24·5 [illegible]	Whitby T. Turnbull & Son	1879	6 Bhds. [illegible]	2Cy. 33" & 62"–39" 170 HP. Blair & Co., Ld., Stockton	J. Gray & Co.
77 43258		**Kate** W PadSp 2 *Dks*	British Nassau, N.P.	355 608	163·4 | 33·2 | 18·9	New York	1852		1Cy. 320 HP.	Augustus J. Ad[illegible]
78 48678		**Kate** I PadSr	British Brisbane	104 147	145·4 | 21·2 | 6·3	Hebburn	1864		[illegible] HP.	Queensland Gove[illegible] ment.
79 95152	✱D	**Kate B. Jones** Stl ScSr *Thomas* -87 1 *Dk* 2 *B*	British Cardiff	1285 1983	270·0 | 37·1 | 20·5 *Q* 62 *f.* *P* 25 *f.* *B* 105 *f.* *F* 27 *f.*	Newcastle Schlesinger Davis & Co.	1887	6 Bhds. CDB 288 APT 85	3C. 21" 35" & 58"–39" 1?0 HP. N. E. Marine [illegible], Newcastle	Kate B. Jones S.S. Ld. (Jones & Thom[illegible])
80 73687	✱	**Kate Fawcett** I ScSr *Young* 85-85 2 *Dks* 3 *B*	British London	887 1383	245·5 | 31·8 | 22·3	Hartlepool E. Withy & Co.	1876	4 Bhds. [illegible]	2Cy. 30" & 57"–33" 130 HP. T. Richardson & Sons, Hrtlpl.	T. G. Greenwell
81 77457	✱	**Kate Forster** ex Ballogie I ScSr *Hazard* 78-86 1 *Dk* 2 *B*	British Newcastle	643 1004	220·4 | 29·7 | 16·7 *P* 112 *f.* *F* 23 *f.*	Aberdeen Hall, Russell & Co.	1879	5 Bhds. [illegible]	2Cy. 26½" & 48"–30" 100 HP. Hall, Russell & Co., Aberdeen	Fisher, Renwick[illegible]
82 80162	✱D	**Kate Thomas** I ScSr *Thomas* 80-88 1 *Dk* 2 *B*	British Cardiff	1010 1558	260·0 | 36·3 | 18·0 *P* 22 *f.* *B* 46 *f.* *F* 26 *f.*	Newcastle Palmers Co., Ld.	1883	4 Bhds. [illegible] APT 7?	2Cy. 31" & 60"–39" 152 HP. Palmers Co., Ld., Newcastle	"Kate Thomas" S[illegible] Ld. (X. Thomas [illegible])
83 95013		**Katha** (LOST) Lr (Stern Wheel)	British Glasgow	131 154	120·0 | 25·3 | 5·7	Rangoon Irrawaddy Flotilla Co., Ld.	1887		2Cy. 13" & 24"–30" 33 HP. Irrawaddy Flotilla Co., Ld.	Irrawaddy Flotil[illegible] Ld.
84 93249	✱	**Kathleen** Stl ScSr *Storm* 78-88 1 *Dk* 2 *B*	British Newcastle	1086 1700	258·6 | 37·0 | 16·9 *Q* 62 *f.* *P* 33 *f.* *B* 62 *f.* *F* 30 *f.*	Sunderland Osbourne Graham & Co.	1888	4 Bhds. [illegible]	3Cy. 20" 33" & 54"–36" 1?0 HP G. Clark, Sund'rlnd.	Pyman, Bell & Co.
85 81521	✱D	**Kathleen** I ScBn *Mackay* 87-87 1 *Dk* 2 *B*	British W. Hartlep'l	1042 1607	257·1 | 34·6 | 17·9 *B* 60 *f.* *Q* 70 *f.* *P* 29 *f.* *F* 34 *f.*	W. Hartlepool R. Irvine & Co.	1880	4 Bhds. [illegible]	2Cy. 32" & 60"–39" 140 HP. Blair & Co., Ld., Stockton	R. Irvine & Co.
86 93380		**Kathleen** C Sc *Depencier* -88 3 *Dks*	British Kingston [illegible]	212 336	105·7 | 26·9 | 7·8 22·9	Kingston, Ont. M. Clayton	1888		3C. 10" 16" & 27"–15" 37 HP. Canadian Locomotive Co., Kingston, Ontario	W. Bajus
87 63157		**Kathleen** I Sc3mSrYt	British [illegible]	146 214	148·8 | 22·1 | 11·6	Birkenhead Laird Bros.	1869		2Cy. 27"–21" 60 HP. Laird Bros., Birkenhead	Earl Fitzwilliam
88 93183	✱	**Kathleen** I ScSr *Walters* 83-88 1 *Dk*	British Belfast	112 336	155·7 | 23·2 | 11·8 *Q* 30 *f.* *B* 40 *f.* *F* 26 *f.*	Belfast Workman Clark & Co.	1887	5 Bhds. APT 38 DT ? 38	2Cy. 19" & 38"–30" 70 HP. J. & J. Thomson, Glasgow	J. Milligen
89 93301	✱ BT 12 88	**Kathleen** Stl ScSrTug *Allen* -88 1 *Dk*	British [illegible]	190	121·0 | 19·5 | 12·3	Falmouth Cox & Co.	1888	5 Bhds.	3Cy. 13½" 20" & 42"–28" 95 HP Cox & Co., Falmouth	C. O'Brien
90 78125	✱ BT ? 88	**Kathleen Mavourneen** Stl PadSr 2 *Dks*	British Drogheda	409 928	260·5 | 31·1 | 15·4 *P* 48 *f.* *F* 33 *f.*	Seacombe A. Jack & Co.	1885	6 Bhds.	2Cy. 45" & 82"–72" 400 HP. A. Jack & Co., Liverpool	Drogheda S[illegible] Co.

✱ = Lloyd's Register; D = Nederlandsche Vereeniging; G = Germanischer Lloyd; I = Registro Italiano; R = American [illegible]
A = Veritas Austro-Ungarico; H = Veritas Ellenico; N = Norske Veritas; V = Bureau Veritas

In the 2nd Column, B.T., with figures following, signifies Date of report of bottom sighting by Board of Trade Surveyors as condition of PASSENGER CERTIFICATE

In the 3rd Column, C = Composite (iron frame, wood planked); I = Iron; Stl = Steel; W = Wood.

Further particulars of Vessels classed by Lloyd's Register will be found in Lloyd's Register Book and Lloyd's Yacht Register.

	Nom, Matériaux, Gréement, etc. Capitaine et dates d'entrée au service des Armateurs actuels ou à bord du navire. Ancien Nom, nombre de ponts, etc.		Pavillon. Port d'Armement.	Tonnage Net Total	Dimensions Longueur	Largeur	Creux	Construit. Où. Par qui.	Quand	Nombre des cloisons. Détails sur le lest d'eau.	Machines. Détails. Nom du Constructeur.	Armateurs ou Directeurs-Gérants.
	Katie *Roberts 85-85*	I ScCr	British London	73 107	76·2	18·8	9·5	Belfast Harland&Wolff	1875		2Cy.11"–17 20 HP. Harland&Wolff,Belfast	Price's Patent Candle Co.,Ld.
V	Katie *Randall 88-88*	I ScSr *2 Dks*	British W.Hartlepl.	2117 2796	320·8	40·5	23·2 *P 38 f. F 43 f.*	Glasgow A.Stephen&Sons	1880	DB500t	2Cy.38"&68"–45" 240 HP. A.Stephen&Sons,Glasgow	Christopher Furness
	Katoomba *Hatcher*	I TwnScBg *2 Dks*	British Sydney	489 1006	249·8	29·3	19·0 *H 41 f.*	PortGlasgow Blackwd.&Gordon	1878	4 Blhds. DB	4Cy.27"&47"–30" 200 HP. Blackwood&Gordon,Pt.Gls.	Australasian United Steam Nav.Co.
✳	Katrena	I ScSrYt	British Glasgow	137 268	167·6	22·6	12·3 *B 40 f.*	Leith Ramage&Ferguson	1885	4 Blhds.	2Cy.22"&40"–35" 70 HP. Ramage&Ferguson,Leith	John Anderson
	Katsha	I ScSr *1 Dk*	Russian Sebastopol	412 523	215·0	30·0	11·0	Sebastopol RussischeDampfschiff u. Handels-Ges.	1872		2Cy.28"&44"–24" 60 HP. J.Penn&Sons,London	RussianSteam Nav. & Trading Co.
	Kattegat *Bquren 79-84*	I ScSalvageSr *1 Dk*	Danish Copenhagen	133 275	142·7	25·5	12·3 *H 44 f.*	Copenhagen Burmeister&Wain	1873	APT & MT FPT	4Cy.21"&44"–24" 120 HP. Burmeister&Wain,Cpnhgn.	Em.Z.Svitzer's Bjergnings Enterprise(H.P.Lyngbye)
V	Kattegatt *Berndund 88-88*	I ScSr *1 Dk2 B*	Swedish Stockholm	320 420	144·3	22·5	15·3 *B 29 f. H 7 f.*	Gothenburg Göteborgs Mek. Verkst.	1872		2Cy.13"&28"–20" 45 HP. GöteborgsMek.Verkst.Gbg.	P.V.N.Slöör
✳ D	Katy *Paton 87-87*	I ScSr *1 Dk2 B*	British London	1418 2158	272·0	38·8	19·9 *Q 76 f. P 28 f. B 64 f. F 29 f.*	Sunderland J.L.Thompson&Sn	1883	6 Blhds. DB FPT APT	2Cy.34"&64"–42" 175 HP. J.Dickinson,Sunderland	R.Gordon & Co.
✳	Kaveri	I ScSr	British Glasgow		..	..	..	Whitehaven WhitehavenS.B.Co.,Ld.	1889		3Cy.16"26"&40"–30" Dunsmuir&Jackson,Glasgow	J.A.Shepherd
✳	Kawatiri *Metcalf*	I ScSr *1 Dk*	British Dunedin	286 453	170·5	26·1	11·2 *Q 83 f. F 21 f.*	Paisley H.McIntyre&Co.	1882	4 Blhds. FPT APT	2Cy.20"&40"–30" 70 HP. Fleming&Ferguson,Paisley	Union S. S. Co. of New Zealand,Ld.
	Kedar *Fenwick 85-88*	I ScBg	British Glasgow	1215 1876	275·8	36·2	25·7	Dumbarton W.Denny&Bros.	1860		2Cy.39"&68"–45" 212 HP. J.Jack,Rollo&Co.,Liverpool	Cunard Steam Ship Co.,Ld.
✳	Kedikler (ex Myriofton, ex Don)	I ScSr *1 Dk*	Turkish Constntnple	217 408	148·0	22·7	11·7 *Q 71 ft. F 16 ft.*	Aberdeen J.Duthie,Sons&Co.	1883	4 Blhds. DB	2Cy.19"&38"–27" 62 HP. Hall,Russell&Co.,Aberdeen	Idarei Massousich
A	Kefalonia (ex Success)	I Sc	Russian Odessa	473 618	..	..	..	Kinderdyk L.Smit&Zoon	1874		60 HP.	Flli.Teofani
V	Kehrwieder (ex Vandalia) *Benohr 82-88*	I ScSr *3 Dks & Spar Dk*	German Hamburg	2258 2815	330·7	39·0	23·0 30·9	Greenock Caird&Co.	1871		2Cy.36"&60"–48" 300 HP. Blohm&Voss,Hamburg	Dampfschiff-Rhederei Hansa
	Keicho Maru	W PadSr	Japanese Osaka	106 178	119·6	17·6	1·8	Hiogo KosakuBunKioku	1879		51 HP.	Osaka Shosen Kaisha
	Kelloe *Spence*	I ScSr *1 Dk2 B*	British London	340 501	164·7	28·3	14·9	Sunderland J.Laing	1866	4 Blhds. DB	2Cy.22½"&42"–36" 70 HP. G.Clark,Sunderland	J.Young,Ehlers & Co.
	Kelso *Deasc*	I Sc3mSr *1 Dk2 B*	British Hull	839 1050	220·7	31·1	15·6 *P 18 f. F 30 f.*	Hull C.&W.Earle	1869	4 Blhds.	2Cy.28"&56"–30" 95 HP. C.&W.Earle,Hull	T.Wilson,Sons & Co.
	Kemb	I PadSr *1 Dk*	Russian Archangel	239 325	160·5	24·2	10·0 *Q 38 f.*	Newcastle A.Leslie&Co.	1876		2Cy.24"&44"–54" 100 HP. R.&W.Hawthorn,Gatesh'd	Archangel Steam Nav. Co.
	Kembla	I PadSr *1 Dk*	British Nwcl.,NSW.	283 449	209·2	22·6	11·5 *B 20 f.*	PortGlasgow JohnReid&Co.	1860		2Cy.48"–52" 165 HP. McNab&Co.,Greenock	Newcastle S.S.Co.,Ld.
	Kenalee *Hussein 78-83*	I Pad	Turkish Cnstntinple.	153	180·0	27·0	10·3	London J.&W.Dudgeon	1874		2Cy.37" 120 HP.	Idarei Massousich (Ottoman Steam Nav.Co.)
	Kendat	Stl SternWheel *1 Dk*	British Glasgow	102 102	120·2	24·6	4·1	London Yarrow&Co.	1886		2Cy.13"–45" 60 HP. Yarrow&Co.,London	Irrawaddy Flotilla Co. Ld.,Glasgow
✳	Kenilworth *Graham*	I Sc3mSr *1 Dk*	British Glasgow	97 274	145·0	23·1	10·5 *Q 53 f. F 21 f.*	Bowling Scott&Co.	1885	4 Blhds. FPT APT	2Cy.21½"&42"–30" 80 HP. Muir&Houston,Glasgow	W.McLachlan & Co.
	Kenju Maru	W ScSr *2 Dks*	Japanese Totomi	210 339	129·5	19·7	17·5 *B 19 f.*	Hiogo KosakuBunKioku	1880		2Cy.17½"&32"–24" 41 HP. J.Lang	Koyekisha
✳	Kenley (COLLISION) *Udall 78-78*	I ScSr *1 Dk*	British London	421 667	180·1	30·1	13·5 *B 6 f. Q 86 f. F 22 f.*	Sunderland J.Blumer&Co.	1877	5 Blhds. DB FPT	2Cy.26"&47"–36" 90 HP. G.Clark,Sunderland	Lambert Bros.
✳	Kenmore *Prior*	I ScSr *1 Dk2 B*	British Dundee	495 769	202·6	27·1	14·6 *Q 96 f. B 51 f. F 20 f.*	Dundee W.B.Thompson	1880	5 Blhds. DB APT FPT	2Cy.24"&48"–30" 90 HP. W.B.Thompson,Dundee	Albany Shipping Co., Ld.
	Kennedy *Twineaux*	I TwnScSr *1 Dk*	British Nelson,N.Z.	147 189	138·4	19·4	8·9	Sydney AustralianS.N.Co.	1865		4Cy.12"&22"–14" 43 HP. Frazer&Tinne,Auckland	J.H.Cock
✳	Kennett (WRECKED) *Pearce 85-87*	I ScSr *2 Dks & Web Frms*	British London	1156 1795	259·4	34·7	22·7 *P 27 f. B 64 f. F 31 f.*	W.Hartlepool E.Withy&Co.	1883	4 Blhds. DB	2Cy.36"&67"–36" 160 HP. T.Richardson&Sons,Hrtlpl.	Steel,Young & Co.
✳	Kennett	Stl ScSr	British London					W.Hartlepool E.Withy&Co.	1889		3Cy.20½"33"&56"–36" T.Richardson&Sns,Hartlepool	Steel,Young & Co.
✳	Kent *Johnson 87-87*	I ScSr *2 Dk3 B*	British Glasgow	1620 2484	320·4	39·1	25·8 *B 58 f. F 36 f.*	Glasgow Lond.&Glas.S.B.C.	1883	4 Blhds.	2Cy.38"&68"–48" 300 HP. London&GlasgowCo.,Glasg.	M.Wigram & Sons,Ld.
✳	Kent *Patterson 85-88*	I Sc3mSr *1 Dk*	British London	666 1045	226·0	32·0	14·0 *Q 56 f. P 23 f. B 42 f. F 19 f.*	Sunderland S.P.Austin&Son	1881	5 Blhds. CB APT	2Cy.29"&56"–36" 99 HP. N.E.MarineEng.Co.,Ld.,Sd.	Lambert Bros.

= Lloyd's Register ; D = Nederlandsche Vereeniging ; G = Germanischer Lloyd ; I = Registro Italiano ; R = American Record ;
= Veritas Austro-Ungarico ; H = Veritas Ellenico ; N = Norske Veritas ; V = Bureau Veritas.

Dans la 2me Colonne, B.T., avec les indications qui suivent, signifie la date du rapport de visite de la Carène par les experts du Board of Trade pour obtenir le certificat relatif au TRANSPORT DES PASSAGERS.

Dans la 3me Colonne, C = Composite (membrure en fer, bordé en bois) ; I = Fer ; Stl = Acier ; W = Bois.

Pour plus amples renseignements sur les navires classés par le Lloyd's Register, voir le Lloyd's Register Book et le Lloyd's Yacht Register.

No. in Book / Official No.	Register	Ship's Name. Material, Rig, &c. / Master, and Dates of Appointment to present Owner's Service and to Ship / Ship's Late Name. No. of Decks, &c.	Flag. / Port of Registry.	Tons Net / Gross	Dimensions Length	Breadth	Depth	Built Where. / By Whom.	When	No. of Bulkheads / Particulars of Water Ballast.	Engines. Particulars. / Maker's Name.	Owners
121 95089	✻	**Kentigern** Stl ScSr *Campbell* [illegible] *2 Dks*	British Glasgow	1614 2403	300·0 *P32f.*	40·1 *B60f.*	23·1 *F34f.*	Glasgow A.Stephens&Co.	1888	[illegible] Blkds. [illegible]	3C.21"33"&54"-42"220HP A.Stephens&Sons,Glasgow.	R. M'Millan
122 76960	✻D	**Kepler** I ScBmSr *Flett 81-85* *2 Dks3 B*	British N.Shields	1205 1939	276·1 *B18f.*	35·2 *F22f.*	24·3	Sunderland Barton,Haswell&Co.	1878	[illegible] Blkds. [illegible]	2Cy.35"&65"-39" 170 HP J.Dickinson,Sunderland	Wilkie & Tur[illegible]
123		**Kepler** I ScSr *Tanner 76-87* *3 Dks*	Belgian Antwerp	1759 2258	320·5	32·9	30·5	Newcastle A.Leslie&Co.	1863	[illegible] Blkds.	2Cy.27"&56"-42" 180 HP R.Stephenson,Newcastle	Soc. Royale [illegible] Americaine
124	V	**Kepler** W ScSr *Jansen* *1 Dk2 B*	Swedish Gothenburg	540 703	188·7 *Q 94.*	26·9 *B28f.*	16·1 *H19f.*	Lödöse J.F.Olsson	1873		2Cy.23"&13"-19" 90 HP MotalaCo.,Motala	J.F.Olsson
125 84274	✻	**Kerbela** I ScSr *Stuart 82-87* *2 Dks*	British Glasgow	1458 2261	284·7 *P11f.*	35·2 *B52f.*	23·8 *F37f.*	Dumbarton W.Denny&Bros.	1880	[illegible] Blkds. [illegible]	2Cy.31"&60"-42" 180 HP Denny&Co.,Dumbarton	British India [illegible] Nav.Co.,Ld.
126 82812	✻D	**Keroula** I ScBn *Young* *1 Dk2 B*	British London	1002 1547	260·4 *Q79f.*	35·5 *B32f.*	17·1 *F27f.*	Sunderland S.P.Austin&Son	1880	[illegible] Blkds. [illegible]	2Cy.32"&60"-39" 150 HP [illegible]Clark,Sunderland	Avis Steamsh[illegible]
127 93233	✻	**Kesteven** I ScTrawler *Beadle 86-87* *1 Dk*	British Boston	72 144	100·2 *Q16f.*	20·2 *F21f.*	10·3	Hull EarlesCo.,Ld.	1885	[illegible] Blkds.	3Cy.11½"17"&26"-21" 45HP Earles Co.,Ld.,Hull	Boston Deep [illegible] ing & Ice Co.,Ld.
128 77085	✻ BT12.88	**Kestrel** I ScBmSr *Cotton* *2 Dks*	British London	586 960	230·1 *P60f.*	29·2 *B62f.*	17·2 *F29f.*	Dundee Gourlay Bros.	1878	[illegible] Blkds. [illegible]	2Cy.34"&56"-36" 155 HP Gourlay Bros.&Co.,Dundee	General Stm.N[illegible]
129		**Kestrel** I ScSr *1 Dk*	Spanish S.Sebastian	49 103	95·0	18·1	9·2	Leith S.&H.Morton&Co.	1884	[illegible] Blkds.	2Cy.15"&25"-21" 40 HP Canal Basin Foundry,[illegible]	Mercader & [illegible]
130 84107	✻	**Khalif** I ScSr *Nash 85-85* *2 Dks3 B*	British Liverpool	1475 2258	285·2 *B46f.*	38·1 *F21f.*	26·3	Newcastle C.Mitchell&Co.	1881	[illegible] Blkds. [illegible]	2Cy.35"&68"-45" 270 HP WallsendSlipwayCo.,Newc	Bedouin Stea[illegible] Co.,Ld.
131 81628		**Khalifah** I PadSk *Clements 79-86*	British London	391 470	210·1	33·3	5·4	London J.&G.Rennie	1879		[illegible]Cy.18½"&35"-24" 160 HP J.&G.Rennie,London	Euphrates & [illegible] Steam Nav.Co.
132 71890		**Khandalla** I ScSr *Alexander 87-88* *3 Dks*	British Glasgow	1318 2040	316·0	33·3	25·7	Dumbarton W.Denny&Bros.	1874	[illegible] Blkds.	2Cy.36½"&[illegible]"-45" 221 HP Denny&Co.,Dumbarton	British India [illegible] Nav.Co.,Ld.
133 63565	BT5.88	**Khedive** I ScBk *Leggin 80-85*	British Greenock	2132 3303	377·6	42·4	33·0	Greenock Caird&Co.	1871		2Cy.55"&95"-66" 600 HP Caird&Co.,Greenock	Peninsular & [illegible] Steam Nav.Co.
134 68491		**Khiva** (ex Antwerpen) I ScBg *Crew 87-88*	British Sunderland	1419 2609	361·4	36·8	27·4	Sunderland J.Laing	1873		2Cy.50"&86"-51" 430 HP R.&W.Hawthorn,Newcastle	Peninsular & [illegible] Steam Nav.Co.
135 32276		**Kiama** I PadSr *1 Dk*	British Sydney	111 175	154·5	20·4 *Q17f.*	7·8	Glasgow J.G.Lawrie	1854		2Cy.30"-36" 60 HP	Illawarra S[illegible] Co.
136 88019		**Kian Ann** W Sc	British Singapore	101 150	112·2	17·1	8·1	Singapore Riley,Hargreaves&Co.	1886		2Cy.15"&30"-18" 45 HP Ross,McLachlan&Co.,Paisly	Wee Boon T[illegible]
137		**Kiang foo** (ex Nankin) C Pad *Bassett 77-84* *1 Dk,&2 Hur Dks*	Chinese Shanghai	1468 2330	295·5	50·0	11·5	Shanghai ShanghaiS.Nav.Co.	1874		1Cy.80"-132" 360 HP M.Cargell,Shanghai	China Merc[illegible] Nav.Co.
138		**Kiang kwan** I Pad *Knights 76-88* *1 Dk,&2 Hur Dks*	Chinese Shanghai	1030 1647	230·1	40·2	12·8	Glasgow A.&J.Inglis	1876		2Cy.38"&69½"-84" 250 HP A.&J.Inglis,Glasgow	China Merch[illegible] Nav.Co.
139		**Kiang pian** (ex Nautilus, ex Tychow) W Pad *Graham 76-86* *1 Dk,&2 Hur Dks*	Chinese Shanghai	942 1493	245·2	..	12·7	New York	1866		1Cy.62"-144" 250 HP T.Harrison,New York	China Merc[illegible] Nav.Co.
140		**Kiang ping** (ex Japan, ex Soochow, ex Suan Kiang) I Pad *Holmes 80-80* *1 Dk,&Hur Dk*	Chinese Shanghai	392 580	177·2	21·0	9·4	Glasgow Hedderwick&Co.	1865		2Cy.36"-42" 80 HP	China Merc[illegible] Nav.Co.
141		**Kiang teen** (ex Moning) I Pad *Pratt 77-83* *1 Dk,&2 Hur Dks*	Chinese Shanghai	1368 2012	273·2	..	13·9	Glasgow A.&J.Inglis	1879		1Cy.66½"-144" 250 HP A.&J.Inglis,Glasgow	China Merc[illegible] Nav.Co.
142		**Kiang tung** (ex Honan) I Pad *Yankowsky 78-80* *1 Dk,&2 Hur Dks*	Chinese Shanghai	340 566	218·0	31·2	9·0	Shanghai ShanghaiS.Nav.Co.	1871		1Cy.50"-96" 200 HP New York	China Merc[illegible] Nav.Co.
143		**Kiang yu** Stl Pad *Yerse 76-88* *1 Dk,&2 Hur Dks*	Chinese Shanghai	2270 3093	300·0	42·0	13·4 29·9	Glasgow A.&J.Inglis	1862		2Cy.40"&75"-86" 300 HP A.&J.Inglis,Glasgow	China Merc[illegible] Nav.Co.
144		**Kiang yung** I Pad *Perkins 81-88* *1 Dk,&2 Hur Dks*	Chinese Shanghai	1511 2418	250·1	40·2	12·7	Glasgow A.&J.Inglis	1876		2Cy.38"&69½"-84" 250 HP A.&J.Inglis,Glasgow	China Merc[illegible] Nav.Co.
145	G	**Kiel** I ScSr *Thomsen* *2 Dks*	German Kiel	851 1104	217·8	30·3	19·8	Kiel G.Howaldt	1886	[illegible]	2Cy.26"&43"-30"	KielerDampfer[illegible] Act.Ges.
146	D	**Kiew** Stl ScSr *Rasmussen 76-88* *2 Dks,Spt Shd Dk*	Danish Copenhagen	714 1132	241·4 *P15f.*	30·2 *B29f.*	14·9 21·5	Copenhagen Burmeister&Wain	1881	[illegible]	1Cy.24"&48"-32" 150 HP Burmeister&Wain	Det Forenede [illegible] Selskab(C.[illegible])
147 3017	✻	**Kii Maru** I ScSr *Kilgour* *1 Dk,& Awn Dk3 B*	Japanese Tokio	868 1399	235·0	33·1	22·7	Glasgow Lond.&GlasgowCo.	1884	5 Blkds. [illegible]	2Cy.30"&56"-39" 135 HP London&GlasgowCo.,Glas.	Nippon Yusen [illegible]
148 633		**Kikaku Maru** W ScSr	Japanese Osaka	134 217	132·1	17·1	13·7	Kobe Kirby&Co.	1881		15 HP	Osaka Shosen [illegible]
149	V	**Kilauea Hou** W ScSr *Cameron 76-86*	Hawaiian Honolulu	271	119·0	24·6	10·0	Honolulu Tibbets&Sorensen	1878		2Cy.12"&20"-20" 18 HP RisdonIronWks.,S.Francisco	Wilder's S.S.C[illegible]
150 90807	✻	**Kildare** (ABANDONED) I ScSr *Johnson 75-84* *1 Dk,&Spr Dk3 B*	British Sunderland	2277 3463	334·0 *B50f.*	42·4	20·6 25·0	Sunderland J.Laing	1883	[illegible] Blkds. [illegible]	2Cy.41"&78"-54" 400 HP J.Dickinson,Sunderland	R.M.Hudson [illegible]

✻ = Lloyd's Register; D = Nederlandsche Vereeniging; G = Germanischer Lloyd; I = Registro Italiano; R = American [illegible]
A = Veritas Austro-Ungarico; H = Veritas Ellenico; N = Norske Veritas; V = Bureau Verit[illegible]

In the 2nd Column, B.T., with figures following, signifies Date of report of bottom sighting by Board of Trade Surveyors as condition of PASSENGER C[illegible]

In the 3rd Column, C = Composite (iron frame, wood planked); I = Iron; Stl = Steel; W = Wood.

Further particulars of Vessels classed by Lloyd's Register will be found in Lloyd's Register Book and Lloyd's Yacht Register

	Reg.	Nom, Matériaux, Gréement, etc. Capitaine et dates d'entrée au service des Armateurs actuels et à bord du navire. Ancien Nom, nombre de ponts, etc.	Pavillon. Port l'Armement	Tonnage Net. Total	Dimensions Longueur	Largeur	Creux	Construit. Où. Par qui.	Quand	Nombre des cloisons. Détails sur le lest d'eau.	Machines. Détails. Nom du Constructeur.	Armateurs ou Directeurs-Gérants.
		Kildare I PadSr	British Dublin	488 842	262·5	27·0	14·7 P73 f. F25 f.	Birkenhead LairdBros.	1867		2Cy.64"–72" 300 HP. LairdBros.,Birkenhead	City of Dublin Steam Packet Co.
	*	Kilkeel ex Celtic I ScSr *Mackinney 81-82* 1 *Dk*	British Newry	84 192	120·0	20·0	9·4 Q 42 f. F 18 f.	Paisley J.Fullerton&Co.	1882	4 Blds. FPT3t	2Cy.17"&30"–22" 40 HP. W.King&Co.,Glasgow	Newry&KilkeelS.S.Co Ld.
		Killingworth I ScSmSr	British East London	367 578	170·4	27·9	16·6	Newcastle C.Mitchell&Co.	1855		2Cy.34½"–26" 70 HP. T.Richardson&Sons,Hrtlpl.	J.Wood
	*	Kilwa I ScSr *Robertson -88* 2 *Dks3 B*	British Glasgow	993 1552	280·2	31·6	23·7	Glasgow A.&J.Inglis	1878	6 Blds.	2Cy.37"&66"–42" 200 HP. A.&J.Inglis,Glasgow	British India Steam Nav.Co.,Ld.
	V	Kilwa I ScSr 1 *Dk*	Zanzibar	216 350	140·0	22·0	12·0 P17 f. B19 f. F25 f.	Flensburg Flnsburg Schiffsb. Ges.	1885		2Cy.18"&36"–21" 45 HP. Flnsbrg.Schiffsb.Ges.,Flns.	Sultan of Zanzibar
	*	Kimberley I ScSmSr *Higgins 77-86* 1 *Dk&Spr Dk1 B*	British Pt.Glasgow	2464 3740	350·2	41·3	33·6 B 69 f.	PortGlasgow W.Hamilton&Co.	1884	7 Blds. DT327t DT(296t	2Cy.44"&82"–54" 400 HP. D.Stewart&Co.,Glasgow	Wm.Hamilton & Co.
	*	Kimberley I PadTg *Godfrey* 1 *Dk*	British East London	24 47	90·6	18·3	8·1 Q 22 f.	NorthShields Hepple&Co.	1881	4 Blds.	2Cy.24½"–46" 60 HP. Hepple&Co.,SouthShields	East London Landing & Shipping Co.
	V	Kinau I ScSr *Lorenzen 84-85*	Hawaiian Honolulu	669 993	201·0	32·8	15·5	Philadelphia W.Cramp&Sons	1883		2Cy.24"&44"–36" 85 HP. W.Cramp&Sons,Phildlphia.	Wilder's S.S.Co.
	*	King Arthur I ScBn *Baffin* 2 *Dks*	British Bombay	1047 1418	263·8	30·7	22·2 B 67 f.	Inverkeithing J.Scott&Sons	1874	1 Blds. APT	2Cy.30"&55"–36" 130 HP. J.Scott&Sons,Invrkeithng.	Bombay & Persia Steam Nav.Co.,Ld.
		King Ermyn I ScSr *Hughes 84-85* 1 *Dk2 B*	British Southamptn.	697 1110	226·0	30·7	16·9 P 139 f. P 28 f.	Stockton M.Pearse&Co.	1872	1 Blds. DB&F19t APT12t	2Cy.26"&52"–36" 99 HP. Day,Summers&Co.,Shmptn.	G.T.Harper & Co.,Ld.
		King Fisher I Pad	British Liverpool	56 264	143·1	23·7	12·0	SouthShields	1868		120 HP.	Liverpool Steam Tug Co.,Ld.
	*	King Ja Ja I ScSmSr *Anderson* 1 *Dk*	British Liverpool	99 198	125·3	19·8	9·3 Q 24 f. P 14 f. F 20 f.	Glasgow J.&R.Swan	1870	3 Blds.	2Cy.17"&28"–21" 30 HP. CanalBasinFoundryCo.,Gls	Liverpool, Carnarvon & MenaiStraitsS.S.Co.,Ld
		King Orry I PadSr	British Douglas	493 1221	290·2	29·4	14·7 P 169 B 219 Len.88	PortGlasgow R.Duncan&Co.	1871		2Cy.50"&92"–78" 365 HP. Westray,Copeland&Co.,Barrow	Isle of Man St.Pkt.Co. Ld.
	V	King-Tofa I ScSr *Petersen -88* 1 *Dk*	German Hamburg	211 288	118·1	25·2	10·3	Hamburg Janssen&Schmilinsky	1882		2Cy.16"&28"–21" 45 HP. Janssen&Schmilinsky,Hbg.	G.L.Gaiser
	*	Kingdom I ScSr *Roberts* 2 *Dks3 B*	British Liverpool	1414 2176	280·0	36·6	24·5 P 24 f. B 56 f. F 32 f.	Sunderland W.Doxford&Sons	1882	6 Blds. DB&c(?)	2Cy.33"&66"–42" 200 HP. W.Doxford&Sons,Sndrlnd.	S.S."Kingdom"Co.,Ld. (W.Thomas & Co.)
	*	Kingfisher I ScSr *Rinner* 1 *Dk*	British Sunderland	39 139	103·0	19·3	10·1 F 14 f. B 4 f.	Sunderland J.Knox&Co.,Ld.	1885	1 Blds. FPT	2Cy.20"&40"–21" 60 HP. W.Doxford&Sons,Sndrlnd.	J.Willcock
	*	Kinghorn I ScSr *Roberts 73-76* 2 *Dks*	British Leith	374 671	215·5	27·3	15·1 P 45 f. B 51 f. F 35 f.	Kinghorn J.Key	1865	5 Blds.	2Cy.29½"&56"–36" 130 HP. Bait&Lindsay,Glasgow	G.Gibson & Co.
	*	Kingmoor I ScSmSr *Lowe 84-85* 1 *Dk2 B*	British Newcastle	541 851	213·6	28·7	15·2 Q 80 f. F 19 f.	NorthShields T.&W.Smith	1872	1 Blds. DB&c/135t FPT19t APT18t	2Cy.24"&48"–36" 99 HP. R.Stephenson&Co.,Nowctle.	E.Eccles
	*	King's Cross I ScSr *Mills -86* 2 *Dks3 B*	British London	1407 2155	284·2	36·3	24·1 P 26 f. B 68 f. F 30 f.	Newcastle PalmersCo.,Ld.	1882	5 Blds. DB&c APT18t	2Cy.35"&64"–42" 250 HP. PalmersCo.,Ld.,Newcastle	J.A.Lee
	*	King's Lynn I ScSr *Jebbing 86-86* 1 *Dk2 B*	British King'sLynn	380 589	180·6	25·5	13·9 Q 55 f. F 12 f.	Middlesbro' R.Craggs&Son	1880	1 Blds. DB&c	2Cy.23½"&44"–30" 75 HP. Blair&Co.,Ld.,Stockton	W.F.Beaumont
		Kingsbridge Packet I ScSr	British Salcombe	28 110	102·0	19·5	9·0 Q 5 f.	Hayle Harvey&Co.	1879		2Cy.16"&30"–21" 36 HP. Harvey&Co.,Hayle	W.H.Prowse
	*	Kingscote I ScSr *Lambert -83* 1 *Dk2 B*	British Newcastle	666 1035	220·2	32·1	15·5 Q 75 f. B 48 f. F 26 f.	Sunderland R.Thompson&Sons	1883	1 Blds. DB&c/238 FPT38 APT33	2Cy.28"&54"–36" 120 HP. R.&W.Hawthorn,Newc'stle	E.Eccles
173	*D	Kingsdale I ScSr *Fair 79-84* 1 *Dk&Spr Dk3 B*	British Newcastle	1511 2316	292·1	38·3	27·3 F 48 f.	Sunderland Short Bros.	1883	6 Blds. DB&c/771t	2Cy.36"&68"–42" 180 HP. G.Clark,Sunderland	Milburn Brothers
174	*	Kingsley I ScSr *Adams 81-1* 1 *Dk*	British Newcastle	435 696	181·2	30·2	13·6 P 29 f. B 44 f. F 23 f.	NorthShields T.&W.Smith	1881	1 Blds. DB&c FPT13t APT27t	2Cy.25"&48"–30" 90 HP. R.&W.Hawthorn,Newc'stle	E.Eccles
175		Kingston I Pad	American New York	814 1065	193·5	36·0	13·3	Newburg,N.Y Ward,Stanton&Co.	1883			E.W.Rossiter
176	*	Kingswear Stl ScSr *Campbell 85-88* 1 *Dk*	British Glasgow	202 356	155·6	25·1	10·8 Q 73 f. B 14 f. F 21 f.	Campbeltown Campbelt'wnS.B.C.	1883	1 Blds. FPT20t APT15t	2Cy.20"&38"–27" 70 HP. Kincaid&Co.,Ld.,Greenock	Redfern,Alexander & Co.
177		Kinloch I Pad	British Granton	334 585	216·0	36·2	8·6	Glasgow A.&J.Inglis	1865		1Cy.42"–48" 260 HP. A.&J.Inglis,Glasgow	North British Railway Co.
178		Kinloch I ScSr *Muir 76-78* 2 *Dks*	British Campbeltn.	125 407	205·0	21·1	12·7 Q 23 f. F 25 f.	Glasgow A.&J.Inglis	1878	4 Blds. WB FPT APT	2Cy.29"&54"–42" 185 HP. A.&J.Inglis,Glasgow	Campbltwn.& Glasgw. Stm.Pckt.Joint Stock Co.
179		Kinrio Maru W T-ScSr 2 *Dks*	Japanese Osaka	329 531	172·9	22·6	18·3 B 22 f.	Hiogo KawasakiI.S.B.Co.	1884		2Cy.21½"&37"–27" 60 HP. J.Lang	Osaka Shosen Kaisha
180	*	Kinsembo I ScSr *Munro -88* 2 *Dks*	British Glasgow	1186 1868	301·5	34·5	23·4 B 31 f. P 112 f. F 57 f.	PortGlasgow Cunliffe&Dunlop	1876	5 Blds.	2Cy.40"&71"–36" 280 HP. Cunliffe&Dunlop,Pt.Glasg.	British & African Stm. Nav.Co.,Ld.

* = **Lloyd's Register**; D = Nederlandsche Vereeniging; G = Germanischer Lloyd; I = Registro Italiano; R = American Record;
A = Veritas Austro-Ungarico; H = Veritas Ellenico; N = Norske Veritas; V = Bureau Veritas.

Dans la 2me Colonne, B.T., avec les indications qui suivent, signifie la date du rapport de visite de la Carène par les experts du Board of Trade pour obtenir le certificat relatif au TRANSPORT DES PASSAGERS.

Dans la 3me Colonne, C = Composite (membrure en fer, bordé en bois); I = Fer; Stl = Acier; W = Bois.

Pour plus amples renseignements sur les navires classés par le Lloyd's Register, voir le Lloyd's Register Book et le Lloyd's Yacht Register.

No. in Book / Official No.	Class	Ship's Name / Master, and Dates of Appointment to present Owner's Service and to Ship / Ship's Late Name	Material, Rig, &c. / No. of Decks, &c.	Flag / Port of Registry	Tons Net / Gross	Dimensions	Built: Where / By Whom	When	No. of Bulkheads / Particulars of Water Ballast	Engines: Particulars / Maker's Name	Owners
181 62751	BT88	Kintyre *Kerr 78-78*	I ScSr	British Campbeltn.	88 204	184·7|22·9|11·5 *Q 9 f.*	Greenock Robertson&Co.	1868		2Cy.26"&48"-30" 90 HP. Kincaid,Donald&Co.,Grck.	Campbeltwn.&[illegible] Stm.Pckt.Join[illegible]
182 278		Kiorio Maru *Nakashima*	W ScSr	Japanese Hakodate	123 229	132·1|21·1|9·4	Brooklyn,N.Y Poillon	1872		1Cy.30"-30" 100 HP.	Nippon Yusen[illegible]
183 86268	✳	Kirby Hall *Petrie*	I ScSr *2 Dks&3 B*	British Liverpool	1759 2691	329·9|40·0|25·9 *P 28 f. B 66 f. F 38 f.*	Glasgow Lond.&GlasgowCo.	1882	4 Blds.	2Cy.36"&70"-48" 300 HP. London&Glas.Co.,Ld.,Glas.	"Kirby" S.S.[illegible] (R.Alexander &[illegible]
184 67834	✳	Kirkheaton *Moore 86-84*	I ScSr *1 Dk*	British Goole	314 485	170·6|26·9|11·7 *Q 83 f. B 9 f. F 19 f.*	Blyth Hodgson&Soulsby	1881	[illegible]	2Cy.21"&40"-30" 65 HP. J.W.&F.Wilson,Sunderland	W.France,Ld.
185 95100	✳	Kirklands *Wallace -88*	Stl ScSr *1 Dk&Spar Dk*	British Glasgow	1168 1801	271·2|37·1|16·0 *B 8 f. F 28 f. 23·4*	Belfast Workman,Clark&Co.	1888	4 Blds. [illegible] APT 10	3C.19"32"&52"-42" 190 HP. W.King&Co.,Glasgow	Kirklands S.S.[illegible] (J.B.Cuthb[illegible]
186 51497		Kirkless	I Sc3mSr *1 Dk*	British Liverpool	246 387	150·3|24·0|12·6 *P 31 f.*	Liverpool Bowdlr.Chaffr&Co.	1865		2Cy.28½"-22" 60 HP. Jones&Co.,Liverpool	S.Morris
187 63682		Kirkstall *Mallett 87-88*	I ScSr *1 Dk*	British Mdddlesbro'	298 470	173·3|24·7|14·3 *P 32 f.*	Northfleet Allibon&Co.	1870	5 Blds.	4Cy.12"&34"-24" 70 HP. Allibon,Noyes&Co.,N'fleet	J.Tollitt (E.Harris&Co.[illegible]
188 86355	✳	Kirtle *Ernst 85-8[illegible]*	I ScSr *1 Dk2 B*	British W.Hartlpl.	942 1462	245·2|34·2|17·2 *Q 94 f. B 56 f. F 31 f.*	W.Hartlepool E.Withy&Co.	1883	4 Blds. DB[illegible]	2Cy.36"&67"-36" 130 HP. T.Richardson&Sons,Htlpl.	G.Steel & Co.
189 93404	✳	Kirton *Ainscough 87-87*	I Sc *1 Dk*	British Boston	57 123	92·3|21·0|10·9 *F 17 f.*	Beverley Cochrane&Co.	1886	3 Blds.	2Cy.15"&29"-21" 35 HP. C.D.Holmes&Co.,Hull	Boston Deep[illegible] ing & Ice Co.,Ld.
190 93760	✳	Kisanga *Thompson -88*	Stl ScSr *2 Dks*	British Liverpool	928 1443	265·8|34·0|18·4 *P 30 f. B 70 f. F 38 f.*	Port Glasgow J.Reid&Co.	1887	5 Blds.	3C.25"39"&62"-36" 200 HP. D.Rollo&Sons,Liverpool	Hatton & Cook[illegible]
191	✳	Kisogawa *Ouwerkerk W[illegible] -87*	I ScBarge *1 Dk*	Japanese	210 321	137·8|28·3|9·4	Kinderdyk J.&K.Smit	1887	[illegible]	2Cy.18"&31"-21" 50 HP. [illegible]	Japanese Gove[illegible]
192 95078	✳ BT 8-88	Kistna *Coope 84-88*	Stl ScSr *2 Dks&Shltr Dk*	British Glasgow	481 1115	240·2|33·2|16·5 *P 46 f. B 74 f. F 51 f.*	Troon AilsaShipbldg.Co.	1888	4 Blds.	3C.25"40"&65"-42" 256 HP. [illegible]	British India [illegible] Nav.Co.,Ld.
193 73665		Kite (ex Norwegian) *Knee 75-88*	W ScBk *1 Dk*	British S.Johns,N.F.	190 280	117·6|26·4|14·1 *P & F 19 f.*	Brake,Germ'y	1873		1C[illegible] 50 HP.	C.T.Bowring[illegible]
194 96904	✳	Kittie *Finn 77-88*	Stl ScSr *1 Dk&WhFrms*	British Hartlepool	1446 2212	284·2|37·6|19·5 *Q 86 f. P 27 f. B 108 f. F 28 f.*	W.Hartlepool W.Gray&Co.	1888	[illegible] Blds. [illegible]	3Cy.21½",35"&58"-39" Central Marine Eng. Co.,W.Hartlepool [190 HP.	J.Coverdale [illegible]
195	✳	Kittiwake *Anderson 87-87*	W ScK *1 Dk*	British N.Shields	10 53	72·6|17·0|8·2	Leith J.McKenzie&Co.	1887		2Cy.10"&19-14" 20 HP. Hawthorns&Co.,Leith	T.F.Roberts[illegible]
196		Kitty	I ScTg *1 Dk*	Spanish Cadiz	7 33	60·0|12·6|6·6	Newcastle T.&W.Toward	1882	3 Blds.	2Cy.12"&22"-16" 30 HP. [illegible]Abbott&Co.,Newcastle	Spanish Gove[illegible]
197 76942		Kiu Kiang *Clarke 84-88*	W Pad *1 Dk*	British HongKong	418 706	241·6|36·4|15·8	New York	1864		1Cy.58" 300 HP.	Hong Kong,C[illegible] Macao Steamb[illegible]
198 915		Kiukwa Maru	W ScSr	Japanese Nagasaki	167 309	136·3|17·0|13·3	WakatsuChik	1885		42 HP.	Fukagawa Ki[illegible]
199 73448		Kiung Chow *McIver -88*	W ScSr *1 Dk*	British HongKong	289 425	155·0|22·5|10·5 *H 90 f.*	HongKong W.B.Spratt&Co.	1877		2Cy.16"&29"-18" 40 HP. Kwok,Acheong&Co.,H.Kong	Hong Kong,C[illegible] Macao Steamb[illegible]
200	✳	Kiushiu Maru *Perim -87*	I ScSr *2 Dks3 B*	Japanese Tokio	1199 1856	267·0|35·3|23·9 *P 28 f. B 90 f. F 38 f.*	Newcastle R.&W.Hawthorn,Leslie&Co.,Ld.	1887	5 Blds. [illegible]	2C.[illegible]"&62"-39" 140 HP. N.E.MarineCo.,Ld.,N'castle	Nippon Yusen[illegible]
201	✳	Kiuyo Maru *M'Carthy 87-87*	I ScSr *1 Dk*	Japanese Nippon	322 515	164·3|24·6|12·3 *Q 51 f. B 34 f. F 20 f.*	Sunderland Sunderland S.B.Co.Ld.	1887	4 Blds. [illegible]	2Cy.21"&42"-27" 80 HP. T.Clark&Co.,Newcastle	Kuchinda Ya[illegible]
202 68740		Kiwi *Campbell*	I Sc3mSr *1 Dk*	British Wellington	133 195	115·1|19·1|8·8 *Q 37 f. F 19 f.*	Hull Humphreys&Pearson,Ld.	1874	3 Blds.	2Cy.16"&30"-15" 30 HP. Humphreys&Prsn.,Ld.,Hull	W.H.Levin & C[illegible]
203		Kjøbenhavn *Brock 70-70*	I PaiSr *1 Dk*	Danish Aarhus	378 602	234·5|23·3|12·5 *Q 73 f. B 55 f. P 27 f. H 64 f.*	Renfrew Henderson,Coulborn&Co.	18 63/80		1Cy.27"&57"-60" 200 HP. Hndrsn.,Coulbrn.&Co.,Rnf.	Det Forenede [illegible] skibs Selskab [illegible]
204	V	Kléber *Thuillier -88*	I ScSr *2 Dks&Spr Dk*	French Marseilles	1038 1850	309·3|33·5|16·7 *F 48 f. B 179 f. 24·3*	Greenock Caird&Co.	1880		2Cy.42"&80"-48" 300 HP. Caird&Co.,Greenock	Compagnie Gé[illegible] Transatlantiq[illegible]
205		Klintehamn *Walferona -86*	I ScSr *1 Dk*	Swedish Visby	204 271	114·5|20·9|9·8 *P 79 f. F 38 f. H 7 f.*	Gefle O.A.Brodin	1875		2Cy.13½"&28½"-20" [illegible] HP. O.A.Brodin,Gefle	Angf.Akt.Bol.[illegible] land"(A.Ekman[illegible]
206 86085	✳	Klyde *Lane*	I ScSr *1 Dk2 B*	British London	1022 1573	256·0|35·0|17·9 *Q 96 f. B 62 f. F 28 f.*	Newcastle C.S.Swan&Hunter	1882	[illegible] Blds. [illegible]	2Cy.32"&62"-36" 150 HP. WallsendSlipwayCo.,Newc.	"Klyde" S.S.C[illegible] Glasgow,Ld. [illegible]
207 88746	✳	Knarwater *Johnson*	I ScSr *1 Dk2 B*	British Newcastle	1107 1705	260·3|35·4|18·2 *Q 70 f. P 29 f. B 54 f. F 30 f.*	Newcastle W.Armstrong Mitchell&Co.,Ld.	1883	[illegible] Blds. [illegible] FPT [illegible] APT 30	2Cy.32"&60"-39" 160 HP. Blair&Co.,Ld.,Stockton	W.Dickinson &[illegible]
208 14280	R	Knickerbocker *Kemble 71-74*	I ScSr *2 Dks3 B*	American New York	1140 1642	256·0|33·6|22·6	Wilmington,Del The Pusey & Jones Co.	1873		1Cy.44"&72" The Pusey&Jones Co.,Wilm.Del	Cromwell S.S.[illegible]
209 93239	✳	Knight Companion *Shiell 81-88*	Stl ScSr *2 Dks3 B*	British Liverpool	2543 3913	400·0|45·3|27·7 *P 37 f. B 95 f. F 46 f.*	Newcastle Palmers Co.,Ld.	1888	7 Blds. [illegible] APT 30	3C.29"47"&76"-51" 450 HP. Palmers Co.,Ld.,Newcastle	Kt.Companion S[illegible] Ld.(Greenshield[illegible]
210 91275	✳	Knight Errant *Matthews 87-87*	Stl ScSr *2 Dks3 B*	British Liverpool	2241 3380	330·0|43·5|29·0 *P 37 f. [illegible]*	Newcastle Palmers Co.,Ld.	1885	5 Blds. [illegible]	3C.27"44"&71"-48" 350 HP. Palmers Co.,Ld.,Newcastle	Knight S.S.Co.[illegible]

✳ = Lloyd's Register; D = Nederlandsche Vereeniging; G = Germanischer Lloyd; I = Registro Italiano; R = American S[illegible]; A = Veritas Austro-Ungarico; H = Veritas Ellenico; N = Norske Veritas; V = Bureau Verit[illegible]

In the 2nd Column, B.T., with figures following, signifies Date of report of bottom sighting by Board of Trade Surveyors as condition of PASSENGER CERTIFIC[illegible]

In the 3rd Column, C = Composite (iron frame, wood planked); I = Iron; Stl = Steel; W = Wood.

Further particulars of Vessels classed by Lloyd's Register will be found in Lloyd's Register Book and Lloyd's Yacht Register.

No.	Registre dans lequel le navire est classé	Nom, Matériaux, Gréement, etc. Capitaine et dates d'entrée au service des Armateurs actuels et à bord du navire. Ancien Nom, nombre de ponts, etc.	Pavillon. / Port d'Armement	Tonnage Brut / Net	Longueur	Largeur	Creux	Construit. Où. Par qui.	Quand.	Nombre des cloisons. Détails sur le lest d'eau.	Machines. Détails. Nom du Constructeur.	Armateurs ou Directeurs-Gérants.
		Knight Errant 1 Pad (ex Reliance)	British Liverpool	41 148	126·8	19·7	10·1	Hull M.Samuelson&Co.	1862		2Cy.36"–51" 80 HP. M.Samuelson&Co.,Hull	G.R.&R.Percival
	*	Knight Templar 1 ScSr Wackett (ABANDONED) 2 Dks 3 B	British Hull	1034 1567	262·0	32·2	23·9	Sunderland Austin&Hunter	1874	5 Blhds. DBaft 56t	2Cy.31½"&63"–39" 160 HP. Blair&Co.,Ld.,Stockton	H.Samman
		Knight Templar 1 PadSp (BROKEN UP)	British Liverpool	32 206	129·4	21·0	11·1	Liverpool W.C.Miller	1864		2Cy.38"–54" 160 HP. Fawcett,Preston&Co.,Lvpl.	Original Mersey Steam Tug Co.
		Knight of Malta 1 PadSp (ex Speedwell)	British Preston	12 182	121·0	18·3	8·7	Liverpool	1861		2Cy.31"–42" 64 HP. Fawcett, Preston&Co.,Lvpl.	D.W.Allsup
	*	Knight of St. George StlScSr Thomas 85-85 2 Dks 3 B	British Liverpool	1934 2935	329·7	39·2	28·0	Liverpool T.Royden&Sons P 40 f. F 40 f.	1883	5 Blhds.	2Cy.38"&71"–48" 300 HP. G. Forrester&Co.,Liverpool	Kt.of St.G.S.S.Co.,Ld. (Greenshields,Cowie & Co.)
		Knight of St. George 1 TwnScSr	British Liverpool	129	108·9	17·5	10·4	Birkenhead CanadaWorksCo.	1885		1Cy.14"28"–20" 59 HP. CanadaWorksCo.Birkenhd.	Knight of St. George Tug Co.,Ld.(J.Prendiville&Co.)
	*	Knight of St. John Stl ScSr Harris 88-88 2 Dks 3 B	British Liverpool	2248 3397	330·0	43·5	29·0	Newcastle Palmers'Co.,Ld. P 37 f. B 71 f. F 40 f.	1885	5 Blhds. DB 24 ft MT 132t Pa 33t	3Cy27"44"&71"–48" 350 HP. Palmers'Co.,Ld.,Newcastle	Knight S.S.Co.,Ld. (Greenshields,Cowie & Co.)
		Knight of St. John Stl TwnScSr	British Liverpool	36 275	147·2	22·6	12·3	Birkenhead W.A.Stevens	1884		1Cy.18"&36"–24" 99 HP. W.A.Stevens,Birkenhead	Kt.of St.John Tug Co.,Ld.(J.Prendiville & Co.)
		Knight of St. Patrick Stl TwnScSr	British Liverpool	2 271	147·2	22·6	12·1	Grangemouth Grangemth Dockyd Co.	1885		4Cy.18"&36"–24" 98 HP. Dunsmuir&Jackson,Glasgw	Kt.of St.Patrick Steam Tug Co.(J.Prendiville&Co.)
		Knight of the Cross 1 PadSp (ex Royal Arch) Duncan 88-88	British Middlesbro'	79 169	121·2	19·8	10·8	Newcastle Marshall Bros.	1862		2Cy.38"–48" 70 HP. Marshall Bros.,South Shields	C.C.Duncan
	*	Knud Jörgensen 87-87 1 Sc3mSr 1 Dk 2 B Spt A wn Dk	Danish Copenhagen	751 1190	231·2	30·2	16·1	Renfrew Hudson,Coulborn & Co. P 109 f. F 34 f.	1871	5 Blhds. MT 32t APT 8t	2Cy.30"&60"–36" 150 HP. Henderson Coulborn&C.Rfw	Dampsk."Knud's" Rederi (L.H.Carl)
		Knuthenborg 1 ScSr Thorpe 84-84 1 Dk & Shd Dk	Danish Bandholm	284 501	178·9	26·3	13·3	Malmö KockumsMekWerks P 39 f. B 61 f. F 21 f.	1880	MT 38t	2Cy.28"&48"–30" 120 HP. KockumsMek.Werks.	Det Forenede Dmpskbs Selskab. (C.P.A.Koch)
	*	Knutsford Stl ScSr Harrison 88-88 1 Dk & Wb Frms	British London	1350 2073	279·5	37·5	19·2	Newcastle C.S.Swan&Hunter P 24 f. Q 80 f. F 112 f. F 25 f.	1888	4 Blhds. CPB 110t	3C.22"36"&60"–39" 200 HP. Wallsend Slipway& Eng.Co.,Ld.,Newcastle	Knutsford S.S.Co.,Ld. (Carlisle&Co.)
224	*	Kobe Maru StlSc2mSr Boswell 88-88 2 Dks & Spr Dk	Japanese Tokio	1645 2901	325·0	41·2	18·9	Glasgow London&Glasgow S.&E. Co.,Ld.	1888	6 Blhds. FPT 42t APT 120t	3Cy.33½"54"&87"–54" 387 HP. London&Glasgow S.&E. Co.,Ld.,Glasgow	Nippon Yusen Kaisha
		Kobtshik 1 ScSr 1 Dk	Russian Sebastopol	299	198·0	29·0	9·0	Sebastopol Russische Dampfschiff und Handels Ges.	1870		1Cy.48"–15" 40 HP. Russische Dampfschiffs und Handels-Ges.,Sebastopol	Russian Steam Nav. & Trading Co.
226		Kodor 1 Sc3mSr		150	210·0	29·0	11·7	London Thames Iron Works	1872		60 HP.	
227	*	Kohinur Stl ScSr Bullock 88-88 2 Dks	British Liverpool	1935 2967	350·5	40·3	25·2	Belfast Harland&Wolff P 198 f.	1884	7 Blhds. W 65ft F&T APT	2Cy.36"&72"–48" 350 HP. Harland&Wolff,Belfast	Asiatic Steam Nav.Co.,Ld.
228		Kokkola 1 ScSr Akerberg 79- 1 Dk	Russian Gamla Karleby	159 228	125·7	19·9	10·0	South Shields R.Thomson	1874		30 HP.	Ångbåt Aktie Bolag.
229		Koldinghuus 1 PadSr Staal 82-88 3 Dks	Danish Copenhagen	823 1309	276·3	30·1	14·0	Renfrew Lobnitz&Co. P 95 f. B 39 f. 21·0 F 24 f. B 39 f.	1883	DT 123t DTa 160t FPT 9t APT 4t	4Cy.32"&53"–90" 100 HP. Lobnitz&Co.,Renfrew	Det Forenede Dmpskbs Selskab. (C.P.A.Koch)
230	*	Kolga 1 ScSr Appelgren 1 Dk & A wn Dk	Swedish Trelleborg	517 720	180·1	26·5	13·0	Sunderland Bartram,Haswell & Co. B 23 f. H 108. 19·7	1873	4 Blhds.	2Cy.23"&46"–33" 86 HP. Ouseburn E.W.Co.Newcastle	Ångf.Akt.Bol."Ägir" (F.Malmros)
231		Kolivan Stl ScTug 1 Dk	Russian Reval	60	86·6	15·0	8·9	Åbo W.Crichton&Co.	1887		2Cy.11"&21"–14" 40 HP. W.Crichton&Co.,Åbo	Russian Government
232	G	Koln 1 ScBg Meier 78-88 2 Dks	German Bremen	1736 2555	311·0	39·4	31·6	Greenock Caird&Co.	1870		2Cy.48½"–48" Caird&Co.,Greenock	Norddeutscher Lloyd
233		Komei Maru WPadSr	Japanese Hakodate	116 209	150·0	20·0	6·1	Yokosuka Zosen Kioku	1868		40 HP.	Miyagaki Saburo
234	DV	Komet 1 ScSr Bucholtz 1 Dk	German Königsberg	369 585	179·7	25·1	14·2	Elbing F.Schichau	1881	WB	2Cy.18¾"&37"–23¾" 75 HP. F.Schichau,Elbing	Königsberg.Dampfer-Compagnie Akt.-Ges.
235	N	Kong Alf 1 ScSr Ammesen 1 Dk 2 B	Norweg'n Christiania	581 805	194·5	30·2	15·3	Christiania Akers Mek.Verkst. Q 59 f. B 47 f. F 23 f.	1884	DB 220t APT MT 82t	2Cy.27"&49"–29" 106 HP. Akers Mek.Verkst.Christna.	Söndenfjelds-norske Dampskibsselskab.
236		Kong Beng Stl ScSr Phillips 87-87	British Glasgow	862 1359	240·2	34·1	22·8	Glasgow J.Elder&Co.	1883		2Cy.28"&56"–36" 175 HP. J.Elder&Co.,Glasgow	Scottish Oriental Steam Ship Co.,Ld.
237	N	Kong Bjorn 1 ScSr Dahl 2 Dks	Norweg'n Christiania	682 959	205·6	29·4	19·0	Malmö KockumsMekVerks Q 44 f. B 48 f. F 23 f.	1878	5 Blhds.	2Cy.27"&52"–29" 130 HP. KockumsMekVerks,Malmö	Söndenfjelds-norske Dampskibsselskab.
238		Kong Carl 1 ScSr Rasmussen -88 2 Dks	Norweg'n Bergen	498 653	170·2	25·2	21·0	Newcastle C.Mitchell&Co. P 146 f. F 23 f. H 20 f.	1868	4 Blhds.	2Cy.25"&46"–26" 81 HP. Thompson,Boyd&Co.,Nwc.	Det Bergenske Dampskibsselskab.
239	N	Kong Dag Stl&I ScSr Hille -87 1 Dk & A wn Dk	Norweg'n Christiania	697 981	207·0	29·7	21·5	Gothenburg MotalaCo.	1883	APT DB 33 ft	2Cy.33"&60"–36" 170 HP. MotalaCo.,Gothenburg	Söndenfjelds-norske Dampskibsselskab.
240		Kong Eystein 1 ScSr Aadahl 82-84 1 Dk & Spr Dk	Norweg'n Drontheim	233 394	125·1	21·5	16·2	Drontheim ThrondhjemsMkVrks	1870		1Cy.13"&24"–18" 45 HP. ThrondhjemsMek,Dm	Nordenfjeldske Dampskibsselskab.

*=Lloyd's Register; D=Nederlandsche Vereeniging; G=Germanischer Lloyd; I=Registro Italiano; R=American Record; A=Veritas Austro-Ungarico; H=Veritas Ellenico; N=Norske Veritas; V=Bureau Veritas.

Dans la 2me Colonne, B.T., avec les indications qui suivent, signifie la date du rapport de visite de la Carène par les experts du Board of Trade pour obtenir le certificat relatif au TRANSPORT DES PASSAGERS.

Dans la 3me Colonne, C=Composite (membrure en fer, bordé en bois); I=Fer; Stl=Acier; W=Bois.

Pour plus amples renseignements sur les navires classés par le Lloyd's Register, voir le Lloyd's Register Book et le Lloyd's Yacht Register.

No. in Book. / Official No.	Register	Ship's Name. Master, and Dates of Appointment to present Owner's Service and to Ship. Ship's Late Name	Material, Rig, &c. No. of Decks, &c.	Flag. / Port of Registry.	Tons. Net. / Gross.	Length	Breadth	Depth	Built. Where. / By Whom.	When	No. of Bulkheads. / Particulars of Water Ballast.	Engines. Particulars. / Maker's Name.	Owners
241	✻	Kong Frode *Kjerulf* 88-88	StlScSr *1 Dk 2 B*	Norweg'n Christiania	692 871	215·5 *Q72f.*	30·1 *B81f.*	15·1 *F30f.*	Sunderland W.Doxford&Sons	1889	1 Blds. [illegible]	3Cy.16½",27½"&44"-30" 99 HP. W. Doxford&Sons, Sunderland	Det Söndenfj[…] Norske Damp[…]
242	N	Kong Haakon *Michaelsen*	I ScSr *1 Dk & Awn Dk 3 B*	Norweg'n Stavanger	433 632	176·3	26·1	19·9	Christiania Nylands Verkst.	1878		2C.28½"&49½"-26½" 110HP. Nylands Verkst. Christiania	Stavanger Dampskibsselsk.
243		Kong Halfdan (ex Sirius) *Gierløner* 75-82	I ScSr *1 Dk & Awn Dk 3 B*	Norweg'n Drontheim	453 572	171·1	24·9	18·8	Stockholm [illegible]	1874		2C.22½"&51"-31½" 120 HP. Lindberg, Stockholm	Nordenfjeldske Dampskibsselsk.
244	N	Kong Magnus *Kjerulf* 87	I ScSr *1 Dk 2 B*	Norweg'n Christiania	421 598	189·6 *P98f.*	26·3	14·0 *F33f.*	Pt.Glasgow J.Reid&Co.	1872	1 Blds. MT390	2Cy.24"&42"-30" 85 HP. Walker, Henderson&Co., Gl.	Söndenfjelds-n[…] Dampskibsselsk.
245	N	Kong Olaf *Nissensen* 78-83	I ScSr *1 Dk & Awn Dk*	Norweg'n Stavanger	332 409	152·3	24·0	17·7	Stavanger Stavanger Stibsbyg. Dok.	1883	APT	2C.25½"&46½"-26½" 90 HP. Nylands Verksted, Christian.	Stavanger Dampskibsselsk.
246	N	Kong Oscar *Hellitsen* 85-85	W ScSr *1 Dk 2 B*	Norweg'n Stavanger	214 304	119·9	24·8 *B 26f.*	14·3	Stavanger	1873		2Cy.15½"&[illegible] HP. Nylands Verksted, Christiania	T. Meling
247		Kong Oscar	I ScSr *1 Dk*	Norweg'n Steenkjær	77 117	122·7	17·1	6·1	Christiania	1864		25 HP.	Stenkjærs Damp[…] selskabet
248	N	Kong Ring *Becker* 87	I ScSr *1 Dk & Spr Dk 3 B*	Norweg'n Christiania	861 1147	224·0	30·0 *B 27f.*	22·4	Malmö Kockums Verksted	1882	MT357 A[…]	2Cy.31"&54"-29" 150 HP. Kockums Verksted, Malmö	Söndenfjelds-n[…] Dampskibsselsk.
249	N	Kong Sigurd (ex Vlad) *Amundsen*	I ScSr *1 Dk & Spr Dk*	Norweg'n Christiania	505 672	190·2	26·6	18·0	Glasgow Randolph, Elder&Co.	1867		2Cy.25"&42"-21" 92 HP. Randolph, Elder&Co., Glasg.	Söndenfjelds-[…] Dampskibsselsk.
250	N	Kong Sverre *Pallesen* 87	I ScSr *1 Dk & Awn Dk*	Norweg'n Stavanger	479 602	176·2 *P58f.*	24·3 *B35f.*	19·5 *F34f.*	Christiania Nylands Verksted	1867		2Cy.26½"&48"-26½" 85 HP. Stavanger[illegible]	Stavanger Dampskibsselsk.
251 1053		Kongo Maru	I ScSr	Japanese Osaka	282 455	145·7	23·8	10·9	Flensburg	1879		60 HP.	Kaneda Tsune[…]
252		Kongsee	I ScSr *2 Dks*	Dutch Batavia	696 1072	248·0 *P30f.*	31·2 *B61f.*	17·9 *F44f.*	Newcastle C.Mitchell&Co.	1878	1 Blds. DB[illegible]	2Cy.33"&66"-42" 200 HP. R.&W.Hawthorn, Newcastle	Nederl.-Indisch Stoomv.-Maats.
253	G	Konig Christian *Jantzen* 86-86	Stl PadSr *1 Dk*	German Rostock	200 391	191·4	21·7 *B 50f.*	10·8 *H 40f.*	Stettin Act.-Ges. "Vulcan"	1885		2Cy.32½"&61"-40½" 280HP. Act.-Ges."Vulcan,"Stettin	Deutsch Nordis[…] Lloyd
254	DV	Konigsberg *Schwabe* 81-82	I ScSr *1 Dk*	German Stettin	580 915	217·5 *P104f.*	28·8	15·5 *F32f.*	Stettin Möller&Holberg	1881		2C.25½"&47"-27½" 100 HP. Möller&Holberg, Stettin	Neue Dampfer- Compagnie Akt.
255		Koning Willem III.	C ScSr *2 Dks*	Dutch Batavia	528 1055	231·0	31·5	7·6	Glasgow Randolph, Elder&Co.	1864		2Cy.53"-36" 200 HP. Rudolph, Elder&Co., Glsgw.	Nederl.-Indisch Stoomv.-Maats.
256	✻DV	Koningin Emma *Soomer* 83-87	I ScBk *3 Dks*	Dutch Amsterdam	1810 2531	316·5 *P 108f.*	36·8	25·5 *F 27f.*	Stockton Richardson, Duck&Co.	1879	5 Blds.	2Cy.44"&81"-45" 350 HP. Blair&Co., Stockton	Stoomv.-Maats. Nederland
257		Koningin Sophia	C ScSr *2 Dks*	Dutch Batavia	436 1055	231·0	31·5	7·6	Glasgow Randolph, Elder&Co.	1865		2Cy.53"-36" 200 HP. Rudolph, Elder&Co., Glsgw.	Nederl.-Indisch Stoomv.-Maats.
258 348		Konohana Maru	W ScSr	Japanese Osaka	275 341	138·0	20·1	18·1	Osaka W.Gaster	1879		21 HP.	Osaka Shosen […]
259 79530	✻	Konoowarra *Tozer*	StlScSr *1 Dk 2 B*	British Melbourne	809 1273	246·2 *P42f.*	33·5 *B66f.*	16·1 *F 26f.*	Greenock Russell&Co.	1880	1 Blds. DB[illegible]	2Cy.29"&56"-45" 120 HP. Kincaid, Donald&Co., Grnk.	W. H. Smith & S[…]
260		Konstantin	Stl ScSr *1 Dk*	Russian Baku		148·3 *Q21f.*	27·8 *B42f.*	.. *F26f.*	Norrkoping Motala Co.	1886	FPT[illegible]	2Cy.19½"&35"-23½" 70 HP. Motala Co., Norrköping	G. Tomateff
261		Konung Oskar *Åberg* 73-83	I ScSr *2 Dks & Shltr Dk*	Swedish Stockholm	484 696	165·6	24·1 *B 24f.*	13·1	Stockholm Bergsunds Mek. Vks.	1875		2Cy.24"&48"-27" 100 HP. Bergsunds Mek. Vrkst. Stkhlm	Ångf. Aktie Bo[…] "Södra Sverige"
262		Koonya	W ScSr *1 Dk*	British Hobart	80 119	107·2	19·4	9·6	Hobart W. Bayes	1887		2Cy.13"&26"-18" 30 HP. Ross & Duncan, Glasgow	A. E. Risby
263		Koping (ex Hjo, ex Hans Brask) *Blomquist*	I ScSp *1 Dk*	Swedish Koping	126 174	96·1 *Q26f.*	19·5 *B39f.*	10·2 *F15f. H6f.*	Motala			40 HP.	A. Bergendahl
264		Koran *Carlstedt* 82-83	2m	Russian St. Petersburg	692	..	..	..	Stockholm Bergsunds Mek. Werks.	1882		120 HP.	Naphta Produc[…] Gesellschaft […]
265 89013		Koranui *Oliver*	I ScSr *1 Dk*	British Dunedin	301 488	167·1	25·1 *F 15f.*	14·0	Paisley H.McIntyre&Co.	1883	1 Blds. DB[illegible]	2Cy.22½"&42"-30" 75 HP. Muir&Houston, Glasgow	Union S.S.Co.of[…] Zealand, Ld.
266		Korniloff	I ScBk *2 Dks 3 B*	Russian Odessa	1592 2079	286·8 *Q 48f.*	36·0	26·6 *F 51f.*	Newcastle A.Leslie&Co.	1869	5 Blds.	2Cy.33"&61"-42" 170 HP. Russian Steam Nav.&Trading Co.	Russian Steam N[…] & Trading Co.
267	V	Korrigan (ex St. Joseph) *Gabriel*	I Sc *1 Dk*	French St. Nazaire	101 201	147·7	19·5	10·5	Nantes Jollet&Babin	1878	5 Blds.	2C.30½"&30½"-25½" 120HP. Jollet&Babin, Nantes	Cie. de Boléo, Par[…]
268		Korshun	I ScSr *1 Dk*	Russian Odessa	280 376	190·5	29·0	8·2	Newcastle [illegible]	1868		2Cy.18"-15" 40 HP. Thompson, Boyd&Co. N'castle	Russian Steam […] & Trading Co.
269		Korsør *Dyrehavge* 86-87	I & Stl PadFerry *1 Dk*	Danish Korsør	435 671	251·2	34·1	14·7	Malmö Kockums Mk. Wrks.	1883		1Cy.33½"&63½"-54" 360HP. Kockums Mek. Wrks., Malmö	De Danske S[…]
270	✻	Kostroma *Ivanosky* 88	Stl ScwBk *1 Dk & Sp Dk 3 B*	Russian Odessa	2291 3513	375·0 *P24f.*	42·2 *B82f.*	20·9 *F53f.*	Newcastle R.&W. Hawthorn, Leslie&Co.	1888	5 Blds. DB[illegible] APT[illegible]	3C31"46"&78"-54" 600HP. Wigham, Richardson&Co. N'cse.	Russian Volun[…] Fleet Association

✻ = Lloyd's Register; D = Nederlandsche Vereeniging; G = Germanischer Lloyd; I = Registro Italiano; R = American [Record];
A = Veritas Austro-Ungarico; H = Veritas Ellenico; N = Norske Veritas; V = Bureau Veritas.

In the 2nd Column, B.T., with figures following, signifies Date of report of bottom sighting by Board of Trade Surveyors as condition of PASSENGER CERTI[FICATE].

In the 3rd Column, C = Composite (iron frame, wood planked); I = Iron; Stl = Steel; W = Wood.

Further particulars of Vessels classed by Lloyd's Register will be found in Lloyd's Register Book and Lloyd's Yacht Register.

	Registre	Nom, Matériaux, Gréement, etc. — Capitaine et dates d'entrée en service des Armateurs actuels et à bord du navire. — Ancien Nom, nombre de ponts, etc.	Pavillon. — Port d'Armement	Tonnage Net — Total	Dimensions (Longueur, Largeur, Creux)	Construit. Où. — Par qui.	Quand	Nombre des cloisons. — Détails sur le lest d'eau.	Machines. — Détails. Nom du Constructeur.	Armateurs ou Directeurs-Gérants.
271		**Kosuge Maru** W ScSr	Japanese Fushiki	928 1496	239·5|34·3|23·2	Nagasaki KosakuBunKioku	1883		2Cy.35"&61" 175 HP.	Yetchin Kisen Kaisha
272	*	**Kotka** I ScSr *Helgeson* 73-84 *1 Dk2 B*	British Hull	1099 1695	260·0|36·2|18·5 Q84f. P26f. B32f. F26f.	Newcastle H.S.Edwards&Son	1884	4 Blhds. DBaft335t APT28t	2Cy.32"& 60"–39" 140 HP. N.MarineEng.Co.Ld.,Gtshd.	W.S.Bailey
273		**Kotofuri Maru** W ScSr	Japanese Osaka	233 375	144.2|21.2|15.1	Osaka Mayck'waTamesit	1887		20 HP.	Ňagoshi
274	*	**Kottingham** (ex Virginia) I ScSr *Henderson* (Scaliotti) *1 Dk2 B*	British London	821 1290	254·0|32·0|17·8 P36f. Q96f. B34f.	South Shields Readhead,Softley&Co.	1872	4 Blhds. DBaft336t	2Cy.32"& 60"–36" 150 HP. Black,Hawthorn&C.,Gtshd.	Newton Bros.& Co.
275		**Koun Maru** W Sr	Japanese Osaka	263 357	169·3|24·4| 8·1	HongKong	1863		120 HP.	Osaka Shosen Kaisha
276		**Kovno** Stl ScSr *Neill* *1 Dk2 B*	British Hull	1119 1700	270·5|34·1|18·4 Q95f. B181f. F30f.	Hull Earles'Co.Ld.	1882	4 Blhds. CDB	2Cy.25"& 50"–45" 130 HP. Earles'Co.,Ld.,Hull	T.Wilson,Sons & Co.
277	*	**Kow Shing** I ScSr *Young* 82-87 *1 Dk&2 AwnDks*	British London	1355 2134	250·0|39·2|21·0 28·0	Barrow BarrowS.B.Co.	1883	5 Blhds. MT90t FPT72t APT36t	2Cy.38"& 76"–45" 244 HP. BarrowS.B.Co.,Ld.,Barrow	Indo-China Steam Nav.Co.,Ld.
278	V	**Kressmann** I ScSr *Kroll* 89-90 *1 Dk*	German Stettin	415 532	161·7|24·7|12·8 P124f. F29f.	Elbing F.Schichau	1875		2Cy.19½"& 42"–22" 75 HP. F.Schichau,Elbing	R.C.Gribel
279		**Kretshet** I ScSr *1 Dk*		289 379	190·5|29·0| 8·2	Newcastle A.Leslie&Co.	1868		4Cy.18"–15" 40 HP. Thompson,Boyd&Co.	
280		**Krishna** I ScSr	British Bombay	167 356	180·3|22·2|10·9 Q27f.	Glasgow T.Wingate&Co.	1878		2Cy.20"& 40"–36" 75 HP. T.Wingate&Co.,Glasgow	Shepherd & Co.
281	V	**Kronen** Stl ScSr *Søbøft* 87-89 *2 Dks&Hur Dk*	Danish Nykjøbing	262 458	171·9|25·2|13·1 P28f. B26f. F50f.	Malmö KockumsMek.Wks.	1882	MT23t	2Cy.25"& 44"–29" 90 HP. KockumsMek.Wrks.,Malmö	Det Forenede Damp- kibsSelskab(C.P.A.Koch)
282	*	**Kronos** I ScSr *Scott* 85-85 *1 Dk2 B*	British W.Hartlepl.	689 1071	230·0|29·8|16·9 Q86f. B95f. F39f.	W.Hartlepool Denton,Gray&Co.	1872	5 Blhds. DBaft/f 61t	2Cy.27"& 50"–30" 99 HP. N.E.MarineEngCo.Ld.,Sdld.	Marquis of London- derry
283	*	**Kronprindsesse Victoria** I ScSr *Hauge* *1 Dk2 B*	Norweg'n Stavanger	323 441	152·1|24·6|14·1 Q30f. B40f. F23f.	Stavanger Stavanger Støberi & Dok	1882	4 Blhds. APT11t MT10t	2Cy.18"& 34"–24" 48 HP. LaxevaagsVerks.,Bergen	E.Berentsen
284	G	**Kronprinz Friedrich Wilhelm** I ScBg *Malke* 2Dks Hur Dk&B	German Bremen	1557 2387	318·2|39·4|31·3	Greenock Caird&Co.	1871		4Cy.21⅛"30⅞"45"&81"–48" Quad.'87,Denny&Co.,Dumbarton {220 HP.	Norddeutscher Lloyd
285	G	**Kronprinz Friedrich Wilhelm** I PadSr *Rr.Roman Ocean* –82 *1 Dk*	German Stettin	121 217	202·1|19·2| 7·6	Glasgow T.B.Seath&Co.	1870		1Cy.50"–72" 96 HP. A.Campbell&Son,Glasgow	J.F.Bräunlich
286	DNV	**Krystal** Stl&I ScSr *Sörensen* *1 Dk*	Norweg'n Kragerö	298 422	148·1|24·5|12·6 Q45f. B35f. F23f.	Drontheim TrondhjemsMek.Verk.	1884	MT50 APT FPT	2Cy.20"& 36½"–26" 60 HP. TrondhjemsMek.VerkDthm	T.M.Wiborg
287		**Kuka** Stl TwnScSr *Castlehorne*	British London	195 343	175·1|26·1| 7·9 P36f. F17f.	PortGlasgow DavidJ.Dunlop&Co.	1883		1Cy.17"& 30"–24" 80 HP. DavidJ.Dunlop&Co.,PtGlas	National African Co. Ld.
288		**Kuling** Stl SternWheel *Barnet* 87-87 *1 Dk*	British Shanghai	304 498	160·0|27·0| 7·0 H242f.	Paisley Bow,McLachlan&Co.	1887		1Cy.17½"&30"–60" 80HP. Bow,McLachlan&Co.,Paisley	Upper Yangtze S.N. Co. (Mangr.,A.J.Little)
289		**Kullen** I Sc3mSr *Isakson* 84-86 *1 Dk*	Swedish Christinehamn	309 384	144·3|22·4|15·3 H22f.	Gothenburg GoteborgsMek.VerkstadsAktiebolag	1871		2Cy.18¾"&28"–19½"45 HP. Göteborgs Mek.Verkst	Anglatsaktiebolaget "Ferm" (Mngr.,A.Boström)
290	*	**Kumamoto Maru** I Sc3mSr *Ekstrand* (ex Gaddrill) *2 Dks3 B*	Japanese Tokio	1237 1995	294·2|33·2|24·7	Newcastle C.Mitchell&Co.	1875	4 Blhds. DBaft&63t	2C.33¼"&62⅝"–45" 168 HP. N.E.MarineEng.Co.,Sundld.	Nippon Yusen Kaisha
291		**Kung pai** (ex Wycliffe) I ScSr *Brissander* 87-87 *1 Dk&pt.AwnDk2B*	Chinese Shanghai	692 1030	205·9|30·3|14·9 Q76f. B48f. F30f.	Leith Ramage&Ferguson	1880	4 Blhds. DBaft82t	2Cy.25"& 48"–30" 95 HP. Lees,Anderson&Co.,Glasgw	China Merchants' Steam Nav.Co.
292		**Kung Wo** I ScSr	British London	757 982	204·0|23·1|17·8	Shanghai Boyd&Co.	1879		2Cy.16"& 36"–36" 52 HP. Boyd&Co.,Shanghai	Indo-China Steam Nav.Co.,Ld.
293		**Kunichiho Maru** W PadSr	Japanese Osaka	132 212	120·9|17·2|13·7	Hiogo SangenYeijiro	1879		15 HP.	Osaka Shosen Kaisha
294		**Kurir** I Sc3mSr *Sjölow* *1 Dk&Spr Dk*	Swedish Stockholm	213 267	111·8|19·9| 9·3 H21f. 16·1	Newcastle Atkinson	1870		35 HP. Pattison&AtkinsonN'castle	F.J.H.Schwieler
295	V	**Kurland** I ScSr *Markgraff* –83 *1 Dk*	German Stettin	402 501	175·2|23·6|13·2	Stettin Möller&Holberg	1880	DB	2Cy17¾"&37½"–19¾"50HP. Möller&Holberg,Stettin	"Kurland" Dampf- schiffs Akt.Ges.
296	*	**Kurrara** I ScSr *Mines* *1 Dk*	British Sydney	239 386	150·0|24·2|13·2 Q42f. B28f. F17f.	Kinderdyk J.&K.Smit	1883	5 Blhds. DBaft75t	2Cy.22"& 40"–30" 70 HP. Maats.deMaas,Delfshaven	Frank Osborne
297	D	**Kursk** Stl ScSr *Rule* 82-84 *2 Dks Spt Shd Dk*	Danish Copenhagen	714 1131	241·5|30·2|14·9 P161f. F30f. 21·5	Copenhagen Burmeister&Wain	1881	WB25t APT16t FPT30t	1Cy.24"& 48"–32" 150 HP. Burmeister&Wain,Cpnhgn.	Det Forenede Damp- skibs Selskab (C.P.A.Koch)
298	*	**Kut Sang** *Slessar* 85-88 I ScBg (ex Dryburgh Abbey) *1 Dk&Spr Dk3 B*	British London	1481 2311	290·0|37·2|19·5 B76f. 27·5	Newcastle Wghm.Richardson&Co.	1881	5 Blhds. DBaft&B358t FPT70t	2Cy.35"& 70"–48" 300 HP. Wghm.Richardson&C.,Nwc	Indo-China Steam Nav.Co.,Ld.
299		**Kwang Chi** Stl ScSr *Kidwith* 88-88 *1 Dk&Shd Dk*	British Shanghai	313 505	185·0|27·0|11·0 P20f. H136f.	Shanghai S.C.Farnham&Co.	1887		6Cy.9",15"&25"–21" 62HP. S.C.Farnham&Co.Shanghai	China Merchants' Steam Nav.Co.
300		**Kwang-lee** Stl ScSr *Andrews* 74-88 *2 Dks&Spr Dk*	Chinese Shanghai	1508 2360	280·0|40·0|17·9 B76f F30f H121f 25·4	Glasgow W.B.Thompson	1883		2Cy.36"& 72"–48" 250 HP. W.B.Thompson,Dundee	China Merchants' Steam Nav.Co.

*=Lloyd's Register; D=Nederlandsche Vereeniging; G=Germanischer Lloyd; I=Registro Italiano; R=American Record;
A=Veritas Austro-Ungarico; H=Veritas Ellenico; N=Norske Veritas; V=Bureau Veritas.

Dans la 2me Colonne, B.T., avec les indications qui suivent, signifie la date du rapport de visite de la Carène par les experts du Board of Trade pour obtenir le certificat relatif au TRANSPORT DES PASSAGERS.

Dans la 3me Colonne, C=Composite (membrure en fer, bordé en bois); I=Fer; Stl=Acier; W=Bois.

Pour plus amples renseignements sur les navires classés par le Lloyd's Register, voir le Lloyd's Register Book et le Lloyd's Yacht Register.

No. in Book / Official No.	Registry in which Classed	Ship's Name. Material, Rig, &c. / Master, and Dates of Appointment to present Owner's Service and to Ship / Ship's Late Name. No. of Decks, &c.	Flag. / Port of Registry.	Tons. Net. / Gross	Length	Breadth	Depth	Built. Where. / By Whom.	When	No. of Bulkheads / Particulars of Water Ballast.	Engines. Particulars. / Maker's Name.	Owners or Managers
301 / [illegible]		Kwang Teang W ScSr / *Matthews -88* 1 *Dk*	British / Penang	147 / [illegible]	140·0	21·1	8·1	Penang / Chin Ah Heang	1888		2Cy. 22"& 42"–27" 75 HP. / [illegible]	Chin Ah Heang
302 / [illegible]		Kwanko Maru C ScSr	Japanese / Tokio	215 / 346	131·7	20·5	9·8	Renfrew / Hendrson, Coulborn & Co.	1869		2Cy. 18¾"& 30"–21" 64 HP.	Nippon Yusen [illegible]
303 / [illegible]		Kwansei Maru W ScSr	Japanese / [illegible]	231 / 374	112·9	20·5	18·2	Kobe / Kawasaki Seizo	1887		[illegible] HP.	Kyayei Sha
304 / 70125	✻	Kwongsang (ex Valencia) Stl ScSr / *Sellar 86-87* 1 *Dk & Awn Dk 3 B*	British / London	989 / 1512	240·4	32·3	17·0	Port Glasgow / Blackwood & Gordon	1880	5 Blds. / [illegible]	2Cy. 30"& 57"–36" 120 HP. / Blackwood & Gordon, Pt. Gls.	Indo-China Steam Nav. Co., Ld.
305		Kyamil Pasha I Sc3msSr / (ex Devon) *Fulk Bey -87* 3 *Dks*	Turkish / Cnstntinple.	1186 / [illegible]	285·6	35·6	24·9	Greenock / J.G. Scott	1878	5 Blds.	2Cy. 41"& 72"–42" 275 HP. / J. Howden & Co., Glasgow	Idarei Massousie (Ottoman [illegible])
306	✻	Kydoniai I ScSr / *Caracalazanis -87* 1 *Dk*	Turkish / Cnstntinple.	178 / 414	180·5 / *P 37 ft.*	25·1 / *B 40 ft.*	12·9 / *F 24 ft.*	Sunderland / Sunderland S.B. Co. Ld.	1887	4 Blds.	3Cy. 15"25"& 40"–27" 90 HP. / N.E. Marine Co. Ld., Sunderld.	C. & A. Georgala
307 / 121	H	Kymi (ex [illegible]) I ScSr	Greek / [illegible]	194 / 328	157·4 / *Q 71 ft.*	27·9 / *B 14 ft.*	21·2 / *F 36 ft.*	Paisley / H. McIntyre	1878		2Cy. 20"& 35"–24" 60 HP. / Hutson & Corbett, Glasgow	H. Psicha
308 / 1049		Kyuyo Maru I ScSr	Japanese / [illegible]	400 / 645	168·3	24·1	18·2	Sunderland	1886		71 HP.	Higashi Kaze / Hommo

Additional Ships are inserted in the Supplement at the end of the Register.

✻ = Lloyd's Register; D = Nederlandsche Vereeniging; G = Germanischer Lloyd; I = Registro Italiano; R = American Record; A = Veritas Austro-Ungarico; H = Veritas Ellenico; N = Norske Veritas; V = Bureau Veritas.

In the 2nd Column, B.T., with figures following, signifies Date of report of bottom sighting by Board of Trade Surveyors as condition of Passenger Certificate.

In the 3rd Column, C = Composite (iron frame, wood planked); I = Iron; Stl = Steel; W = Wood.

Further particulars of Vessels classed by Lloyd's Register will be found in Lloyd's Register Book and Lloyd's Yacht Register.

	Registre dans lequel le navire est coté.	Nom, Materiaux, Gréement, etc. Capitaine et dates d'entrée au service des Armateurs actuels et à bord du navire. Ancien Nom, nombre de ponts, etc.	Pavillon. Port d'Armement	Tonnage Net Brut	Dimensions. Longueur / Largeur / Creux (constructions sur le Pont)	Construit. Où. Par qui.	Quand	Nombre des cloisons. Détails sur le lest d'eau	Machines. Détails. Nom du Constructeur.	Armateurs ou Directeurs-Gérants.
1	*	**L. E. Charlewood** I ScSr *Rowley* 1 *Dk2 B*	British Cardiff	548 [illegible]	220·0|30·3|15·5 F 27 f.	Newcastle Palmers'Co.	1878	4 Blhds. CDB47f. F 20	2Cy.28" & 52"-33" 99 HP. Palmers'Co.,Ld.,Newcastle	Lee, Finch & Co.
2	A	**L'Adriatico** I Sc [illegible]	Argentne [illegible]	65 [illegible]	.. .. ..	Trieste StabilimentoTecnico	1885			N. Milanovich & Co.
3	*	**L'Industriale** I Sc *Esposito* -87 (ex Italiano) 1 *Dk*	Italian Naples	124 [illegible]	[illegible] F 12 f.	Campbeltown [illegible]	1880	[illegible]	2Cy.14½" & 26"-21" 30 HP. Ross&Duncan,Glasgow	Impresa Industriale Italiana
4	*	**L'Isle-Adam** (ex Pellew) I ScSr *Inguanez* 19-86 1 *Dk2 B*	British Malta	539 841	200·9|28·1|15·7 B 39 f. F 28 f.	Birkenhead W. Simpson	1883	5 Blhds. DBaf&aft179t	2Cy.18" & 40"-36" 99 HP. D.Rollo&Sons,Liverpool	G. Gollcher
5	V	**La Bourgogne** I & Stl ScIm *Frangeul* -88 3 *Dks & Shade Dk*	French Havre	3786 [illegible]	494·4|52·2|34·1 Q 64 f. F 90 f.	La Seyne Forges & Cie de la Medit.	1885	DB780	6Cy.42" & 80"-67" 2000 HP. Forges&C.delaMédit.,Mars.	Compagnie Générale Transatlantique
6	V	**La Bretagne** Stl ScImSr *De Jousselin* -86 3 *Dks 4 B*	French Havre	3396 [illegible]	495·4|51·8|33·5 B 25 f. P 536, B 240 f. F 30 f.	St. Nazaire Cie.Gen.Trans.	1886	DB [illegible]	6Cy.49" & 75"-67" 2000 HP. Cie.Gen.Trans.,St.Nazaire	Compagnie Générale Transatlantique
7		**La Canadienne** I ScS *Tranguel* -86 (ex Foxhound) 1 *Dk*	British [illegible]	227 352	151·3|22·7|10·9 Q 74 f.	Port Glasgow R.Duncan&Co.	1860	1 Blhds.	2Cy.19" & 34"-24" 60 HP. SmithBros.&Co.,Glasgow	Government of Canada
8	*	**La Cartuja** (ex Mary) I ScBk *Arlorren* 85-85 1 *Dk2 B*	Spanish Seville	606 892	206·0|29·0|16·1 Q 79 f. B 48 f. F 25 f.	W. Hartlepool Denton,Gray&Co.	1871	4 Blhds.	2Cy.25" & 48"-33" 96 HP. T.Richardson&Ss.Hart'pool	Ybarra & Co.
9	V	**La Champagne** Stl ScIm *Boyer* -88 3 *Dks & Shade Dk*	French Havre	3413 6800	493·4|51·8|33·8 B 48 f. Q 84 f. F 90 f.	St. Nazaire Cie.Gen.Trans.	1885	DB797	6Cy.49½" & 74¾"-67" 2000 HP. Cie.Gen.Trans.,StNazar	Compagnie Générale Transatlantique
10	V	**La Corse** I ScS *Gillaume* -87 2 *Dks & Spr Dk*	French Marseilles	681 1164	260·1|29·8|15·2 B 80 f. 22 f.	Greenock Scott&Co.	1872		2Cy.33" & 66"-36" 250 HP. GreenockFoundryCo.,Grnk.	Compagnie Générale Transatlantique
11		**La Dives** Stl PadSr *Herpin* -88 1 *Dk*	French Havre	86 198	150·9|19·7|9·8	Flushing	1886		2Cy.30" & 52"-39" 190 HP. KoninMaatsdeSchlde,Flng.	Cie. Anonyme des Paquebots entre Caen,Havre,Honfleur et les Ports de la Normandie
12		**La Flandre** StlSc3mSr *Sail* -88	British Newcastle	1510 1979	270·0|37·7|25·9	Newcastle SirW.G.Armstrong, Mitchell&Co.	1888		3Cy.20,"34" & 55"-36" 260 HP. Wallsend SlipwayCo.,Ld., Newcastle	H. F. Swan
13	V	**La France** I ScBk *Ferd* 84-88 2 *Dks & Spr Dk*	French Marseilles	2557 3572	426·5|37·4|33·1	La Seyne Forges&CdelaMéd	1871		3Cy.60"-47¼" 450 HP. Forges&CdelaMédit.,Mars.	Société Générale des Transports Maritimes
14	V	**La Gascogne** I & StlScImSr *Santelli* -86 3 *Dks 4 B*	French Havre	3766 6800	425·4|52·2|34·4 P 55 f. B 240 f. F 91 f. H —	La Seyne Forges.&CdelaMédit.	1885	DB732	6Cy.42" & 80"-67" 2000 HP. Forges&Chatiers,Marseilles	Compagnie Générale Transatlantique
15	*	**La Gaule** I ScBkn (ex Auchinroul) 2 *Dks 3 B*	French Marseilles	1195 1552	260·0|32·2|23·2 Q 33 f.	Aberdeen Hall,Russell&Co.	1880	5 Blhds. DBaf&f210t	2Cy.31" & 60"-36" 160 HP. Hall,Russell&Co.,Aberdeen	N. Paquet & Co.
16		**La Minerve** I & W Sc 2 *Dks*	British Ottawa	111 128	101·6|23·8|6·6	Temiscamingue, P.Q. C.Morin	1886		1Cy.12"-12" 100 HP. [illegible] & Lemoy, Point Levis, P.Q.	A. Lumsden
17		**La Nevera** I ScS *Peyrano* -83 1 *Dk*	Arg'ntine Rosario	234 359	154·9|26·2|11·0 Q 18 f.	Belfast Harland&Wolff	1883	5 Blhds. APT [illegible]	2Cy.16" & 32"-24" 55 HP. Harland&Wolff,Belfast	Canals
18	V	**La Normandie** I ScImBk *De Kersabiec* -86 (ex Normandie) 4 *Dks*	French Havre	3254 6217	459·3|49·2|37·4 P 42 f. F 82 f.	Barrow BarrowS.B.Co.	1882	WB [illegible]	6C.35¼" & 75"-67" 1000 HP. BarrowS.B.Co.,Barrow	Compagnie Générale Transatlantique
19		**La Perle** I ScSrY [illegible]	French Martinique	42 168	103·1|16·2|10·2	Glasgow T.B.Seath&Co.	1884		2Cy.15" & 18"-18" 37 HP. Kincaid&Co.,Greenock	M. Girard
20		**La Plata** (ex Norfolk) I ScBkn BT11.88	British Glasgow	2050 3212	332·1|40·1|32·9	London R.&H.Green	1879		4Cy.31⅛" & 76"-43" 600 HP. Humphreys&Tennant,Lon.	Royal Mail Steam Packet Co.
21		**La Popa** I ScSr *Carpenter* -87			120·0|23·0|11·0	Wilmingtn.Del The Pusey&JonesCo.	1887		1Cy.24"-36" The Pusey&JonesCo Wilmington	Government of United States of Colombia
22	*	**La Rochelle** I ScSr *Trouble* 80-83 1 *Dk2 B*	French La Rochelle	957 1402	250·6|37·5|17·4 Q 29 f. B 33 f. F 26 f.	Newcastle Palmers'Co.,Ld.	1883	4 Blhds. DBaf&aft APT 256	2Cy.31" & 60"-39" 150 HP. Palmers'Co.,Ld.,Newcastle	D'Orbigny & Faustin fils
23	* V	**La Senera** (ex Ferdinand Lesseps 88-88 Corvinus) I ScSr 1 *Dk2 B*	British Dundee	571 899	200·5|30·2|16·2 Q 29 f. B 48 f. F 30 f.	Sunderland StrandSlipwayCo.	1879	4 Blhds. DBaf&194 ft APT	2Cy.27" & 50"-33" 99 HP. N.E.MarineCo.Ld.,Sunderld	Bell & Sime
24		**La Sentinelle** W ScSr *Chester*	Haytian	175	112·0|22·0|9·0	Philadlphia Pa Neafie&Levy	1880		1Cy.24"-24" Neafie&Levy,Philadelphia	Haytian Government
25		**La Seyne** I ScBk (ex Etoile du Chili) 3 *Dks*	French Marseilles	1410 2230	312·3|38·0|29·7	La Seyne Forges&CdelaMéd	1870		1C27¼" & 49½"-35½" 300 HP. Forges&CdelaMédit.,Mrsls.	Messageries Maritimes
26		**La Tour** W ScSr *Cook* -88 1 *Dk*	British Yarmouth N.S.	105 174	97·3|20·7|7·8	Yarmouth NS G.W.Johnson	1888		2Cy.14" & 26"-18" 30 HP. Burrell Johnson IronCo. Yarmouth, N.S.	Yarmouth & Shelburne S.S.Co.,Ld.
27	*	**La Valette** (ex South Western) I Sc3mSr *Ethel* 2 *Dks*	British Malta	223 460	190·5|26·3|12·8 P 79 f. B 35 f. F 35 f.	Port Glasgow Blackwood&Gordn	1870	5 Blhds.	2Cy28¾" & 50½"-30" 120 HP. Blackwood&Gordon,Pt.Gls.	Gustaf Gollcher
28	V	**La Valette** (WRECKED) I ScSr *Ricordel* -87	French St.Nazaire	312 536	182·4|25·7|16·6	Greenock Scott & Co.	1880		2Cy.25" & 50½"-30" 80 HP. Greenock Fdy.Co.Greenock	Compagnie Générale Transatlantique
29	V	**La Vendée** (WRECKED) Stl ScSr *Eyot* -87	French St.Nazaire	589 730	231·0|33·3|17·6 P 32 f. B 56 f. F 36 f.	Newcastle R.& W. Hawthorn, Leslie&Co.	1887		3C.20"32" & 53"-36" 200 HP. R.&W.Hawthorn,Leslie&Co., Newcastle	Compagnie Générale Transatlantique
30	*	**Labarrouere** I ScSr *Lowther* 1 *Dk2 B*	British Cardiff	773 1186	241·2|33·2|15·6 Q 90 f. B 30 f. F 28 f.	Newcastle Palmers'Co.	1880	3 Blhds. CDB308t APT	2Cy.28" & 54"-36" 99 HP. Palmers'Co.,Ld.,Newcastle	"Labarrouere" S.S. Co.,Ld.(Morel Bros.& Co.)

* = **Lloyd's Register**; D = Nederlandsche Vereeniging; G = Germanischer Lloyd; I = Registro Italiano; R = American Record; A = Veritas Austro-Ungarico; H = Veritas Ellenico; N = Norske Veritas; V = Bureau Veritas.

Dans la 2me Colonne, B.T., avec les indications qui suivent, signifie la date du rapport de visite de la Carène par les experts du Board of Trade pour obtenir le certificat relatif au TRANSPORT DES PASSAGERS.

Dans la 3me Colonne, C = Composite (membrure en fer, bordé en bois); I = Fer; Stl = Acier; W = Bois.

Pour plus amples renseignements sur les navires classés par le Lloyd's Register, voir le Lloyd's Register Book et le Lloyd's Yacht Register.

No. in Book. / Official No.	Registry in which classed.	Ship's Name. Material, Rig, &c. / Master and Dates of Appointment to present Owner's Service and to Ship / Ship's Late Name. No. of Decks, &c.	Flag. / Port of Registry.	Tons. Net. / Gross.	Length	Breadth	Depth	Built. Where. / By Whom.	Date	No. of Bulkheads. / Particulars of Water Ballast.	Engines. Particulars. / Maker's Name.	Owners
31	G	Laboe I ScSr Eggers 82-83 2 Dks	German Kiel	543 723	187·0	21·7	19·3	Kiel J.Howaldt	1883		2Cy.25⅝"&37⅞"-27" 70 HP Howaldt Bros., Kiel	Lange Gebrüder
32		Labourdonnais I ScBn 3 Dks	French Marseilles	1275 1918	326·7	32·0	26·4	LaSeyne Forges&C.delaMed.	1863		3C.36⅜"&51⅝"(2)-31⅝" 310HP MessageriesMar.,LaCiotat	Messageries M
33 54697	*	Labrador W ScBk Wiggins 88-88	British London	266 391	139·2	29·2	17·0	Sunderland W.Watson	1866		2Cy.22½"&42"-30" 80 HP G.Clark,Sunderland	G.W.Ashdown
34	*V	Labrador (ex Nouveau Monde) I ScBk Rapé 88 3 Dks & Awn Dk	French Havre	2732 1577	393·7 Len. '75	44·0	26·8 36·5	St. Nazaire Scott & Co.	1865	10 Blds.	1Cy.41"&75"-51" 900 HP Maudslay,Sons&Field,London	Compagnie Gé Transatlantique
35 15048	BT7.88	Laconia I ScBg Blacklock	British Liverpool	1295 1982	314·3	31·9	20·1	Glasgow J.&G.Thomson	1856		1Cy.22"&43"-36" 170 HP G.Forrester&Co.,Ld.Livrpl.	Papayanni & Co.
36	V	Ladario Stl PadSr Arnold 88 1 Dk	Brazilian RiodeJan'ro	540 700	220·1	29·5	13·2	Stettin "Vulcan" Co.	1888		1C.14"6"32"5"&51"5"-39"4" "Vulcan" Co. [129HP	Companhia Na de Nav. a Vapor
37 [illegible]	*	Lady Ailsa StlScSr M'Clean 83-88 1 Dk 2 H	British London	801 1270	231·2 P 135 f.	33·2 F 31 f.	16·9	Troon Ailsa S.B.Co.	1888	4 Blds.	3Cy.17"27"&44"-36" 98HP D.Rowan&Sons,Glasgow	J.& A.Wyllie
38 63309	*	Lady Alice Kenlis I ScSmSr Nutter 86-86 1 Dk	British Preston	75 134	123·5	19·6	9·8	Maryhill J.&R.Swan	1867	4 Blds.	2Cy.16"&32"-18" 50 HP R.Smith,Preston	T.Jackson & Co.
39 68933	*	Lady Aline I ScSr Blyth 88-88 1 Dk	British Sunderland	383 590	181·3	28·6 Q 18 f.	13·9	Sunderland J.Laing	1875	4 Blds.	2Cy.26"&48"-30" 95 HP T.Richardson&Sons,Hartpl.	Marquis of Lon- derry
40 68969	*	Lady Ambrosine I ScSr M'Alpine 1 Dk	British Glasgow	46 136	105·0 Q 21 f.	19·5 B 12 f.	8·8 F 17 f.	Dumbarton J.&R.Swan	1873	4 Blds.	2Cy.13⅝"&23¼"-18" 30HP W.King&Co.,Glasgow	J.McCallum & C
41 87331	*	Lady Ann I ScSr Benson 69-83 1 Dk,& W'l k Frame	British Sunderland	698 1082	225·3 F 33 f.	33·0 Q 103 f.	16·8	Sunderland Austin&Son	1882	4 Blds.	2Cy.28"&54"-33" 99 HP N.E.MarineE.C.Ld.,Sun'ld	H.T.Morton & C
42 87391	*	Lady Armstrong I ScSr Abbott 81-83 2 Dks 3 H	British Greenock	1343 2060	286·0 P 27 f.	37·1 B 60 f.	23·0 F 34 f.	Newcastle SirW.G.Armstrong, Mitchell&Co	1883	Blds.	2Cy.34"&66-42" 235 HP WallsendSlipCo.,Ld.,Nwc.	"Lady Armstrong" Co.,Ld.(Adam B
43 90122	BT10.88	Lady Arthur Hill I ScSmSr Ferguson 87-87 1 Dk	British Belfast	128 271	150·5 Q 22 f.	23·1 B 40 f.	10·6 F 26 f.	Belfast McIlwaine,Lewis & C.Ld.	1885	4 Blds. FPTs. APTs	2Cy.19"&36"-30" 56 HP McIlwaine,Lewis&C.Ld.Bel.	East Downshire Steam Ship Co.
44 92928	*	Lady Beatrice I ScSrYt Strickland 86-86	British Southamptn	155 245	152·3	22·2	12·3	Leith Ramage&Ferguson	1885	4 Blds.	2Cy.18"&36"-24" 60 HP Ramage&Ferguson,Leith	Capt. R.Townley Parker
45 47043	*	Lady Beatrix I ScSr Burgess 71-71 1 Dk 2 B	British Sunderland	425 625	186·3	28·7	16·7	Sunderland J.Laing	1863	4 Blds.	2Cy.26"&48"-30" 80 HP G.Clark,Sunderland	Earl of Durham
46 70517	*	LadyBertha STRANDED I ScSr Parsons 83-87 1 Dk 2 B	British Newcastle	460 730	186·0 P 104 f.	26·7 F 24 f.	15·7	Newcastle Schlesinger,Davis&Co.	1877	4 Blds.	2Cy.27"&50"-33" 98 HP N.E.MarineEng.Co.,Ld.,Sld.	Fisher,Renwick&
47 84620	BT7.88	Lady Bessie W ScSmSr Roberts 84-88	British Carnarvon	79 183	116·0	22·0	10·6	Millom W.Thomas&Co.	1884		2Cy.14"&27"-21" 35 HP DeWinton&Co.,Carnarvon	G.Farren
48 32217		Lady Bird I ScSr 2 Dks	British Dunedin	287 421	151·2	22·2	19·0	Dumbarton W.Denny&Bros.	1851		2Cy. 70 HP	Westport Coal C
49 40650		Lady Bowen I PadBg Royle	British Sydney	442 702	228·2	2?·9 F 31 f.	11·8	Glasgow A.&J.Inglis	1864		2Cy.44"-48" 140 HP A.&J.Inglis,Glasgow	W.Summerbell
50 95355	* BT7.88	Lady Brassey Stl ScSr Whitmough 88 1 Dk	British Newhaven	72 248	160·0 Q 57 f.	24·1 B 31 f.	8·3	Newcastle H.S.Edwards&Sons	1888	4 Blds.	3Cy.10"16"&26"-18" 75 H E.Scott&Co.,Newcastle	Hastings,St.Le & EastbourneT
51 84367	*	Lady Cathcart I ScSr Ross 73-82 1 Dk 2 B	British Aberdeen	403 626	180·0 Q 50 f.	26·2 B 42 f.	14·1 F 20 f.	Aberdeen A.Hall&Co.	1882	5 Blds.	2Cy.24"&42"-30" 75 HP BlaikieBros.,Aberdeen	Aberdeen Line Co
52 [illegible]	*D	Lady Clare I ScSr Streeting 86-86 1 Dk 2 B	British W.Hartlepool	760 1108	235·0 Q 85 f.	31·1 F 31 f.	17·5	Stockton M.Pearse&Co.	1874	4 Blds.	2Cy.27"&54"-33" 110 HP Blair&Co.,Ld.Stockton	G.Pyman & Co.
53 [illegible]	*	Lady Clive I ScSr Williams 88 1 Dk 2 B	British Cardiff	620 971	234·5 Len80,Q37f.	28·9 B10f.	16·7 F29f.	Sunderland J.Laing	1869	4 Blds.	2Cy.27"&50"-30" 99 HP G.Clark,Sunderland	Cory Bros. & Co.
54 6395	*	Lady Downshire I ScSmSr Ethel 86-86 1 Dk	British Malta	139 254	140·3 Q 50 f.	21·2 F 34 f.	9·7 B 26 f.	Port Glasgow Macfadyen&Co.	1873	4 Blds.	2Cy.19"&33½"-24" 50 HP Walker,Henderson&Co.Gls.	Gustaf Gollcher
55 25151	BT1.88	Lady Eglinton I ScSmSr	British Dublin	486 768	228·5 P 55 f.	27·9 F 10 f.	16·1	Glasgow R.Napier&Sons	1853		2Cy.45"-42" 150 HP R.Napier&Sons,Glasgow	British & Irish St Packet Co.,Ld.
56 [illegible]	*D	Lady Eleanor I ScSr Shearer 78-78 1 Dk 2 B	British Sunderland	488 764	200·1	28·1 Q 39 f.	15·7	Sunderland T.S.Gulston	1875	4 Blds.	2Cy.25"&48"-33" 85 HP T.Clark&Co.,Newcastle	Earl of Durham
57 78142		Lady Glover W Sc Noel 86-87 1 Dk	British St.John'sNFL	94 138	113·0	18·5	10·5	Preston W.Allsup&Sons	1877		2Cy.20"&40"-24" 60 HP Fawcett,Preston&Co.,Liv.	William P.Munn
58 95418	* BT7.88	Lady Gordon Stl ScSr Whitley 84-88 1 Dk	British London	316 513	180·0 P 42 f.	27·1 B 60 f.	11·6 F 26 f.	Glasgow Aitken&Mansel	1888	5 Blds.	3Cy.15"24"&40"-30" 80HP Hutson&Corbett,Glasgow	Ceylon S.S.Co.,Ld
59 43736	*	Lady Havelock I ScSmSr Tam 1 Dk 2 B	British Cardiff	334 570	206·5	24·2 Q 62 f.	14·2	Sunderland J.Laing	1861	4 Blds.	2Cy.23"&42"-30" 75 HP J.Nelson,Gateshead	W.J.Tillett & Co.
60 74758	*	Lady Kate W ScSr Rowland 84-84 1 Dk	British Liverpool	70 138	99·7	21·9	9·8	Millom W.Thomas&Co.	1881		2Cy.14"&27"-20" 30 HP DeWinton&Co.,Carnarvon	W.Rowland

* = Lloyd's Register; D = Nederlandsche Vereeniging; G = Germanischer Lloyd; I = Registro Italiano; R = American Reco[rd]
A = Veritas Austro-Ungarico; H = Veritas Ellenico; N = Norske Veritas; V = Bureau Veritas

In the 2nd Column, B.T., with figures following, signifies Date of report of bottom sighting by Board of Trade Surveyors as condition of PASSENGER CERTIFICATE

In the 3rd Column, C = Composite (iron frame, wood planked); I = Iron; Stl = Steel; W = Wood.

Further particulars of Vessels classed by Lloyd's Register will be found in Lloyd's Register Book and Lloyd's Yacht Register.

No.	Registre	Nom, Matériaux, Gréement, etc. / Capitaine et dates d'entrée au service des Armateurs actuels et à bord du navire. / Ancien Nom, nombre de ponts, etc.		Pavillon. / Port d'Armement	Tonnage Net / Total	Long.	Larg.	Creux	Construit. Où. / Par qui.	Quand	Nombre des cloisons. / Détails sur le lest d'eau.	Machines. Détails. / Nom du Constructeur.	Armateurs ou Directeurs-Gérants.
		Lady Katherine White & Co ...	I ScSr 1 Dk&H ...	British Sunderland	615 945	211·5 Q 88 f.	30·0	16·8 F 20 f.	Sunderland S.P.Austin&Son	1881	4 Blds. W.B....	2Cy.27"& 47"–36" 90 IP. N.E.MarineEng.Co.Ld.,Sld.	Earl of Durham
	*	Lady Loch	I ScSr 2 Dks	British Melbourne	337 487	182·7	21·6	11·4	Melbourne Campbell,Sloss&Mc Caun	1886	6 Blds.	2Cy.28"& 54"–36" 125 IP. Cmpbll.Sloss&McCan.,S.Mel	Victorian Government
		Lady Longden	Stl ScSr 1 Dk	British Penang	91 134	120·0 P 18 f.	20·6	7·7 F 28 f.	London Forrestt&Son	1882	5 Blds.	2Cy.15"& 28"–21" 45 IP. Mitchell&Co.,London	Ong Beng Tek
	*	Lady Longden Morley 87-87	I PdlTug 1 Dk	British Georgetown	60 96	100·1	16·6	7·6	London R.&H.Green	1876	3 Blds.	2Cy.24½"–28" 32 IP. J.Penn&Sons,London	Sproston Dock & Foundry Co.
	*	Lady Louisa Williams	W ScSr 1 Dk	British Liverpool	73 139	105·0	21·2	9·3	Millom W.Thomas&Co.	1882		2Cy.15"& 27"–21" 32 IP. J.&T.Young,Ayr	G.Rodrigues
	*	Lady Lycett Grantham 85-86	I ScBg 2 Dks3 B	British London	1199 1833	262·6	33·1	24·5	Glasgow London & Glasgow Co.	1872	5 Blds. DB...	2Cy.31"& 54"–33" 150 IP. London&Glas.Co.,Glasgow	Mercantile Steam Ship Co.,Ld.
		Lady Margaret (ex Carrick Castle)	I PadSp	British Cardiff	59 176	192·1	18·6	7·4	Paisley Fullerton&Co.	1870		1Cy.49"–54" 85 IP. W.King&Co.,Glasgow	Edwards, Robertson & Co.
		Lady Margaret Rawlinson	I PadSp	British Rochester	51 144	139·9 Q 15 f.	18·1	7·4	Greenock Russell&Co.	1883		2Cy.22"& 37½"–30" 50 IP. Alley&Maclellan,Glasgow	Medway Steam Packet Co.,Ld.
	*	Lady Martin Watts 79-88	Stl ScSr 2 Dks	British Dublin	659 1245	269·6 P 64 f.	34·2 F 53 f.	16·4 B 95 f.	Belfast Workman Clark&Co.	1888	5 Blds P.... F....	3Cy.24",39"&61"–45" 220 IP. Dunsmuir&Jackson, Glasgow	British & Irish Steam Packet Co.,Ld.
70	*D	Lady Mostyn Rickards	I ScSr 1 Dk2 B	British Liverpool	467 730	200·3 Q 73 f.	27·7 B 40 f.	15·0 F 29 f.	South Shields J.Readhead&Co.	1881	4 Blds.	2Cy.23"& 44"–30" 80 IP. J.Readhead&Co.,S.Shields	J.H.Lewis
71		Lady Musgrave Collin 85-	I ScCr 1 Dk	British Brisbane	88 161	111·2	20·2 B 15 f.	9·2	Brisbane J.W.Sutton&Co.Ld	1884		1Cy.18"–24" 45 IP. Plenty&Sons,Newbury,Eng	William Collin
72	*	Lady Nell Hammond 81-84	I Sc3mSrYt	British Portsmouth	288 423	186·2	25·2	15·3	Leith Ramage&Ferguson	1884	4 Blds.	2Cy.25"& 50"–30" 105 IP. Ramage&Ferguson,Leith	Earl of Shrewsbury & Talbot
73	*	Lady Olive Clarke 78-88	I ScSr 2 Dks	British Dublin	570 1034	249·8 F 64 f.	30·3 B 36 f.	15·9 P 57 f.	Glasgow A.&J.Inglis	1878	5 Blds.	2Cy.36"& 67"–42" 200 IP. A.&J.Inglis,Glasgow	British & Irish Steam Packet Co.,Ld.
74	*	Lady St. Germans Gorley 86-86	I ScSr 1 Dk2 B	British London	927 1448	240·6 Q 92 f.	33·7 B 50 f.	18·0 F 28 f.	Newcastle Palmers'Co.,Ld.	1883	5 Blds.	2Cy.30"& 57"–39" 138 IP Palmers'Co.,Ld.,Newcastle	Oriental S.S.Co.,Ld.
75	*	Lady Torfrida	Stl ScSrYt 2 Dks	British Glasgow	205 547	216·5	27·9	16·6	Glasgow FairfieldShpbdg.Co.	1888	4 Blds	5Cy.(2)14¼"(1)30½"(2) 38"–30" 225 IP. Fairfield S.B. Co.,Glasgow	Sir W.Pearce
76	*	Lady Torfrida	Stl ScSrYt 2 Dks	British Glasgow	174 400	205·5	25·5	13·8	Glasgow J.Elder&Co.	1883	4 Blds.	3C.24"&(2)34"–30" 175 IP J.Elder&Co.,Glasgow	Bayard Brown
77		Lady Tredegar Cheeseman	W Pad	British Newport	15 108	96·5	19·3	9·7	NorthShields R.Stobbs	1875		1Cy.34½"–50" 40 IP. T.R.Scott&Co.,N.Shields	W.Bain
78		Lady Tyler Gay 84-87	I Pad Sr	British Harwich	563 995	261·0 P 134 f.	30·2	13·8 F 28 f.	NorthShields T.&W.Smith	1880		6C.(2)23"&(4)44"–60" 340 IP. R.&W.Hawthorn,Nwc.	Great Eastern Railway Co.
79	*	Lady Vita Lambert 78-83	I PadTug 1 Dk	British Dover	53 163	107·7	20·1	10·4	London Westwood,Baillie& Co.	1882	3 Blds.	2Cy.34"–57" 75 IP. J.Stewart&Son,London	Dover Harbour Board
80		Lady Weld	StlPadSr	British London	242 511	221·1	30·2 B 40 f.	9·5	Greenock Scott & Co.	1888		3C.20"–30"&48"–48" 150 IP Scott & Co.,Greenock	NetherlandsIndiaStm. Nav.Co.,Ld.
81	*	Lady Wodehouse Barket 59-85	I Sc3mSr 2 Dks	British Dublin	470 809	225·0 P 39 f.	28·5	15·8 F 38 f.	Dublin Walpole&Webb	1865	5 Blds.	2Cy.47"–33" 180 IP. Macnab&Co.,Greenock	British & Irish Steam Packet Co.,Ld.
82	*	Lady Wood Turner	I TwnScSr 1 Dk	British Port Natal	50 133	95·1	20·0	8·8	London A.D.Lewis&Co.	1882	4 Blds.	4Cy.14"& 25"–16" 50 IP. A.Wilson&Co.,London	J.Chiazzari
83	*	Lady of the Isles Hooper 81-81	I ScSr 1 Dk	British Scilly	47 152	130·5	18·5 Q 7 f.	9·3	Hayle Harvey&Co.	1875	5 Blds.	2Cy.18"& 33"–21" 45 IP. Harvey&Co.,Hayle	West Cornwall S.S.Co. Ld.
84		Lady of the Lake	I Pad	American GrgetwnDC	598 773	210·0	31·0	10·0	Wilmingtn Del The Harlan & Hollngsworth Co	1866		1Cy.50"–132" The Harlan & Hollingsworth Co. W.... Del.	Inland & Seaboard Coasting Co.
85		Ladylove Smith	I ScK 1 Dk	British London	84 140	104·3	20·6	9·5	London Hewett&Co.,Ld.	1886		2Cy.12"& 24"–18" 24 IP. J.Stewart&Sons,London	Hewett & Co.,Ld.
86		Laerte	W Pad	Austrian Trieste	212 350	..	..	..	Trieste Arsenale Lloyd	1863		90 IP.	Lloyd Austro-Ungarico
87		Laertes Seale 73-79	I ScSr	British Liverpool	1391 2148	320·5	34·3	26·0	Greenock Scott&Co.	1879		2Cy.24½"&56"–52" 300 IP. Greenock Fndry.Co.,Grnck.	Ocean Steam Ship Co.
88		Lafayette	Stl TwnScSr	Argentne BuenosAyres	50 103	100·0	18·0	8·2	Havre Forges&Chantiers de la Medit.	1888		4Cy.11"&20"–14" 75 IP. Forges & Chantiers de la Medit., Havre	Gardella & Nossiton
89	V	Lafayette Nouvellon -86	I TwnSc3mSr 3 Dks	French St.Nazaire	1770 3535	344·7	43·6	36·1	Greenock Scott&Co.	1863		4C.43¼"&78¾"–41¾" 850 IP Schneider&Co.,Creuzot	Compagnie Générale Transatlantique
90		Laffite Ezcurdero 64-82	I Sc 1 Dk2 B	Spanish Seville	634 977	210·4 P 71 f.	29·5 B 40 f.	16·0 F 40 f.	Sunderland R.Thompson,jun.	1877	4 Blds.	2Cy.28"& 51"–33" 115 IP. J.Dickinson,Sunderland	Segova,Cuadra & Co.

*=Lloyd's Register; D=Nederlandsche Vereeniging; G=Germanischer Lloyd; I=Registro Italiano; R=American Record;
A=Veritas Austro-Ungarico; H=Veritas Ellenico; N=Norske Veritas; V=Bureau Veritas.

Dans la 2me Colonne, B.T., avec les indications qui suivent, signifie la date du rapport de visite de la Carène par les experts du Board of Trade pour obtenir le certificat relatif au TRANSPORT DES PASSAGERS.

Dans la 3me Colonne, C=Composite (membrure en fer, bordé en bois); I=Fer; Stl=Acier; W=Bois.

Pour plus amples renseignements sur les navires classés par le Lloyd's Register, voir le Lloyd's Register Book et le Lloyd's Yacht Register.

No. in Book. / Official No.	Registry in which classed.	Ship's Name. Material, Rig, &c. / Master, and Dates of Appointment to present Owner's Service and to Ship. / Ship's Late Name. No. of Decks, &c.	Flag / Port of Registry.	Tons. Net / Gross	Dimensions (Length, Breadth, Depth)	Built. Where. / By Whom.	When	No. of Bulkheads / Particulars of Water Ballast.	Engines. / Particulars. Maker's Name.	[illegible]
91 149700		**Lagonda** W ScYt *Mathiesen -84*	American New York	81 121	118·0 19·5 10·0	Brooklyn J.F. Munn	1884		2Cy.14"& 24"-16" 80 HP Pusey & Jones Co. Wilmington	J.C. Hoagland
92 85874	✻ BT12.88	**Lagos** [illegible] *Hornsby -88* 2 Dks	British [illegible]	1115 [illegible]	300·3 36·2 20·2 *P110f. F43f. B32f.*	Port Glasgow David J. Dunlop & Co.	1883	5 Blkds.	2Cy.36"&63"-42" 220 HP David J. Dunlop & Co., Pt.Glas	British & Afri[can] Nav. Co. Ld.
93		**Laguna** W AuxScSr	American [illegible]	142 [illegible]	116·0 30·0 9·0	San Francisco G. Boole	1885		2Cy.10"& 18"-20"	Boole & Beaton
94	V	**Laguna** Stl ScSr *Meyer -88*	Brazilian Rio de Janeiro	320 411	169·3 24·6 8·0 *P36f. F36f.*	Stettin "Vulcan" Co.	1888		1Cy.16"22"&35"-18" 70 HP "Vulcan" Co. Stettin	Companhia N... de Navegação
95	G	**Lahn** Stl ScTwSr *Hellmers 77-87* 3 Dks	German Bremen	2879 5661	448·1 49·0 34·9	Glasgow Fairfield Shipbuilding & Eng. Co.	1887		3C(2)32½"68"&(2)85"-72"	Norddeutscher ...
96	G V	**Lahneck** I ScSr *Kasemeyer 83-88* 1 Dk	German Bremen	565 [illegible]	191·2 30·2 15·2 *B50f. F26f.*	Elbing F. Schichau	1882		2Cy.20½"& 41"-31½" 75 HP F. Schichau, Elbing	Deutsche Dampfschifffahrts-Ges.
97		**Laja** I Sc *Dunn 71-84* 2 Dks & Shd Dk	Chilian Valparaiso	1445 2147	332·1 40·0 20·2 *F32f.*	Port Glasgow J. Reid & Co.	1881	[illegible] Blkds.	2Cy.45"&80"-54" 280 HP Rankin & Blackmore, Grnck	Comp. Sud-Am... de Vapores
98 85083	✻	**Laju** I ScSr *Mann 81-81* 1 Dk & Spr Dk 3 B	British London	1246 1846	261·0 35·0 27·1	Sunderland J. Laing	1881	1 Blkds.	2Cy.33"&62"-42" 200 HP G. Clark Sunderland	Straits S.S. Co. Ld. (Wright Bros & Co.)
99 71615		**Lake** W Sc 1 Dk	British Montreal	89 148	107·0 24·0 10·8	Philadelphia	1875		1Cy.33"-34" 175 HP	Canada Shipping Ld.
100 84150	✻ BT10.88	**Lake Huron** I Sc3mSr *Trauman 82-88* 2 Dks & Spr Dk 4 B	British Liverpool	2046 [illegible]	385·0 42·8 24·3 *B84f. F35f.* 34·4	Glasgow London & Glasgow Co.	1881	6 Blkds.	2Cy.48"&88"-60" 500 HP London & Glas. Co., Glasgow	Canada Shipping Ld.
101 71613	BT5.88	**Lake Nepigon** I ScBk *Carey 82-88* 2 Dks 3 B	British Montreal	1439 2209	321·1 35·2 25·9	Glasgow London & Glasgow Co.	1875	5 Blkds.	2Cy.40"&70"-46" 250 HP London & Glas. Co., Glasgow	Canada Shipping Ld.
102 97325	✻ BT5.88	**Lake Ontario** Stl Sc3mSr *Campbell 80-87* 3 Dks	British Liverpool	2923 [illegible]	371·5 43·5 29·5 *P88f. B142f.* [illegible]	Sunderland J.L. Ayre	1887	6 Blkds. [illegible]	3C.31"51"&83"-54" 400 HP G. Clark, Sunderland	Canada Shipping Ld.
103 91197	✻ BT3.89	**Lake Superior** I Sc3mSr *Stewart 76-88* 2 Dks & Spr Dk We. Frmst 8	British Liverpool	2966 [illegible]	400·0 44·2 24·2 *B114f.* [illegible] 31·7	Glasgow J. & G. Thomson	1884	8 Blkds. [illegible]	2Cy.48"&90"-60" 450 HP J.&G. Thomson, Glasgow	Canada Shipping Ld.
104 81285	✻ BT4.88	**Lake Winnipeg** I Sc3mSr *Murray 87-88* 2 Dks & Awn Dk 4 B	British Liverpool	2157 3300	355·2 40·2 24·4 *B56f.* 31·2	Glasgow J. & G. Thomson	1879	6 Blkds.	2Cy.43"&78"-51" 400 HP J.&G. Thomson, Glasgow	Canada Shipping Ld.
105 90778		**Lakeside** W Sc *Wigh -88* 1 Dk	British [illegible]	220 348	121·0 26·0 9·3	Windsor, Ont. W. Lane	1888		2Cy.19"&32"-26" 200 HP Kerr Bros. Walkerville, Ontario	N.J. Wigh
106		**Lakme** W Sc3mSr 1 Dk	American S. Francisco		.. .. ..	Puget Sound T.A. Peterson	1888			G.S. Hindsdale
107 63276	✻	**Lalande** I ScSr *Purvis* 2 Dks	British [illegible]	679 1048	220·8 30·5 21·7	Glasgow A.&J. Inglis	1872	4 Blkds. [illegible]	2Cy.28"&50"-36" 98 HP A.&J. Inglis, Glasgow	R.D. Purvis
108 80584	✻ D	**Laleham** I ScSr *Evans 86-86* 2 Dks 3 B	British Newcastle	1300 2000	285·0 35·3 23·8 *P30f. F34f.*	Newcastle Palmers' Co., Ld.	1880	4 Blkds. [illegible]	2Cy.35"&68"-42" 240 HP Palmers' Co., Ld., Newcastle	J. Temperley & Co.
109 90027	✻	**Lalpoora** Stl Sc3mSr *Livingstone 85-88* 3 Dks	British Glasgow	2140 3260	340·1 42·2 25·6 *P57f. B26f. F42f.*	Dumbarton W. Denny & Bros.	1885	6 Blkds. [illegible]	4C 26½"44"&(2)51"51" 370 HP Denny & Co., Dumbarton	British India S... Nav. Co. Ld.
110 50080	✻	**Lamberton** I ScSr *Jorgensen -88* 1 Dk	British Granton	13 101	95·2 19·0 10·0	Dundee W.B. Thompson	1885	3 Blkds.	2Cy.17"&34"-24" 45 HP W.B. Thompson, Dundee	General Steam ...ing Co., Ld.
111 31587	✻	**Lambeth** I ScSr *Kelso* 1 Dk	British London	606 [illegible]	220·5 32·2 14·5 *B40f.*	Newcastle Palmers Co., Ld.	1879	4 Blkds. [illegible]	2Cy.27"&51"-33" 99 HP Palmers Co., Ld., Newcastle	River Steam Co. Co., Ld.
112 20491		**Lambton** I ScSr *Towning 78-79* 1 Dk 2 B	British [illegible]	453 [illegible]	188·8 28·3 14·0 *Q11f.*	Newcastle C. Mitchell & Co.	1857	4 Blkds. [illegible]	2Cy.24"&46"-30" 70 HP M. Samuelson & Co., Hull	Earl of Durham
113 85854	✻	**Lamington** Stl ScSr *Steel 83-85* 2 Dks	British Glasgow	1208 [illegible]	274·0 37·0 23·2 *B69f. F31f.*	Port Glasgow R. Duncan & Co.	1881	[illegible] Blkds. [illegible]	2Cy.33"&66"-42" 200 HP Rankin & Blackmore, Grnck.	S.S. Lamington ... (Raeburn & Ver...)
114 140540	R	**Lampasas** I ScSr *Crowell 77-82* 2 Dks	American New York	2237 [illegible]	329·0 40·5 21·5	Chester, Pa. J. Roach & Son	1883		2Cy.40"&78"-54" J. Roach & Son, Chester, Pa.	C.H. Mallory & Co.
115 85887	BT6.88	**Lamprey** I ScSr *Blair 76-83*	British Glasgow	111 211	150·2 24·3 12·8	Port Glasgow Blackwood & Gordon	1881		2Cy.23"&46"-36" 90 HP Blackwood & Gordon, Pt. Glas.	G.&J. Burns
116 51494	BT9.88	**Lancashire** I Pad *Kemp 66-*	British Liverpool	106 [illegible]	150·0 30·1 10·9	Northfleet Scott & Co.	1865		4Cy.34"-48" 120 HP Mare & Co., London	Municipal Corporation of Birkenhead
117 78813	✻	**Lancashire Witch** C ScYt	British Liverpool	211 311	160·2 25·9 14·0	Greenock R. Steele & Co.	1878		2Cy.23"&40"-24" 75 HP R. Steele & Co., Greenock	Frank L. James
118 84133	✻ D	**Lancaster** I ScSr *Bouchette -89* 2 Dks 3 B	British Liverpool	1145 1748	262·0 35·2 22·2 *P26f. B70f. F33f.*	Newcastle Palmers Co., Ld.	1881	[illegible] Blkds. [illegible]	2Cy.32"&60"-42" 180 HP Palmers Co., Ld., Newcastle	S.S. Lancaster Co. (G.M. Steeves)
119 15844	R	**Lancaster** I Sc3mSr *Robinson 87-87* 1 Dk 2 B	American Philadelphia	910 1283	250·0 37·0 20·3 *H--*	Philadelphia W. Cramp & Sons	1874	4 Blkds. [illegible]	1Cy.42"-42" 168 HP W. Cramp & Sons, Philadelphia	Philadelphia & Reading R.R. Co.
120										

✻ = **Lloyd's Register**; D = Nederlandsche Vereeniging; G = Germanischer Lloyd; I = Registro Italiano; R = American ...;
A = Veritas Austro-Ungarico; H = Veritas Ellenico; N = Norske Veritas; V = Bureau Veritas.

In the 2nd Column, B.T., with figures following, signifies Date of report of bottom sighting by Board of Trade Surveyors as condition of PASSENGER CERTIFICATE.

In the 3rd Column, C = Composite (iron frame, wood planked); I = Iron; Stl = Steel; W = Wood.

Further particulars of Vessels classed by Lloyd's Register will be found in Lloyd's Register Book and Lloyd's Yacht Register.

		Nom, Matériaux, Gréement, etc. Capitaine et dates d'entrée au service des Armateurs actuels et à bord du navire. Ancien Nom, nombre de ponts, etc.		Pavillon. Port d'Armement.	Tonnage Net. Total	Dimensions (constructeurs sur le Pont) Longueur	Largeur	Creux	Construit. Où. Par qui.	Quand.	Nombre des cloisons. Détails sur le lest d'eau.	Machines. Détails. Nom du Constructeur.	Armateurs ou Directeurs-Gérants.
	✻	Lancelot *Davis*	Stl ScSr *2 Dks 3 B*	British Leith	1564 2370	290·0 *B 55 f.*	40·2	24·5 *F 35 f.*	Newcastle Tyne I.S.Bldg Co.Ld.	1888	7 Blhds. CDB ... APT ...	3Cy.24",37"&62"–42" Wigham Richardson & Co.,Newcastle [300 HP.	F.F.Reid
		Lancelot *McNaughton 75-85*	I PadSk	British Greenock	24 142	191·2	18·1	6·9	Port Glasgow R.Duncan&Co.	1868		2Cy.32½"–48" 90 HP. Rankin&Blackmore,Grnck.	A.Campbell
	✻	Lanchester *Edwards 82-83*	I ScSr *1 Dk 2 B*	British Sunderland	474 735	198·9 *B 24 f.*	29·4½	16·9 *Q 22 f.*	Sunderland J.Laing	1870	4 Blhds. WBa DT f	2Cy.27"&47"–30" 90 HP. N.E.Marine E.Co.Ld.,Sndrld.	H.T.Morton & Co.
		Landana *ex Venetian* *Simonds 70-88*	I ScBn	British London	985 1568	270·5 *P 82 f.*	33·7	22·8 *F 34 f.*	Belfast Harland&Wolff	1859		2Cy.36"&70"–36" 200 HP. G.Forrester&Co.,Liverpool	AfricanSteamShipCo.
	✻	Landaura *Clarke 81-88*	Stl Sc3mSr *3 Dks*	British Glasgow	2148 3271	340·3 *B 85 f. F 42 f.*	42·2	17·6 *25·6*	Dumbarton W.Denny & Bros.	1885	8 Blhds. CDB ...	4Cy.24",34",48"&68"–48" Denny&Co.,Dumbart'n [385 HP.	British India Steam Nav.Co.,Ld.
	V	Landskrona *Johnsson -84*	I ScSr *1 Dk*	Swedish Landskrona	584 732	185·7 *Q ... B ... F ...*	26·3	15·1	Oscarshamn Oscarshamns M.V.	1873		2Cy.17½"&39"–23½" 80 HP. GötelorgsMekVerkst,Goth	Landskrona Lastangare Akt.Bol.(C.F.Nees)
		Lanfranc *Jackson 66-84*	I ScSr	British Liverpool	1070 1657	280·9 *P 56 f.*	31·2	22·9	Liverpool T.Royden&Sons	1884		2Cy.28"&58"–48" 138 HP. G.Forrester&Co.,Liverpool	Booth Steam Ship Co. Ld.
	✻	Langdon *Holmes 76-83*	I ScSr *1 Dk 2 B*	British London	851 1326	240·0 *Q 30 f.*	31·2 *B 40 f.*	16·1 *F 31 f.*	SouthShields J.Readhead&Co.	1882	4 Blhds. DBa & f 31 ft	2Cy.30"&57"–36" 135 HP. J.Readhead&Co.,S.Shields	J.Fenwick & Son
		Langkat *Harrison*	I ScSr *2 Dks*	British Penang	187 278	159·2	21·7	12·2	London J.&W.Dudgeon	1874	4 Blhds.	4Cy.15⅝"&27¼"–14" 60 HP. J.&W.Dudgeon,London	Lee Pee Chuan
	✻	Langley *Shearer 73-79*	I ScSr *1 Dk 2 B*	British Sunderland	457 707	193·3 *Q 28 f.*	28·2	16·8	Sunderland J.Laing	1868	4 Blhds. DBa & f 11 ft	1Cy.18"&36"–30" 80 HP. G.Clark,Sunderland	H.T.Morton & Co.
131	✻	Languedoc *Bollert 80-88*	I ScSr *1 Dk & Spr Dk 3 B*	French Marseilles	929 1625	255·2 *F 25 f.*	34·0	16·8 *24·6*	Glasgow W.B.Thompson	1884	5 Blhds. DBa ... FPT 36t APT 25t	2Cy.36"&72"–42" 235 HP. W.B.Thompson,Dundee	Société Générale de Transports Maritimes
132	✻	Lansdowne *Guildford*	W ScBk *1 Dk 2 B*	British Ottawa	463 689	188·6 *P 29 f.*	32·1	15·8	Maccan, N.S. J.O'Brien	1884		2Cy.24"&43"–36" 80 HP. Rankin&Blackmore,Grnck.	Dominion Governmnt of Canada
133		Laplace *Allcot 77-88*	I ScBg	British Liverpool	901 1418	257·4 *P 185 f.*	31·2	27·0	Newcastle A.Leslie&Co.	1866		2Cy.22"&50"–42" 94 HP. Fawcett,Preston&Co.L'pool	L'pool,Brazil & River Plate Stm.Nav.Co.,Ld.
134	✻	Lapland *McKenzie 83-86*	I Sc3mSr *2 Dks 3 B*	British Leith	823 1297	250·3	30·0	21·7	Glasgow Barclay,Curle&Co.	1872	4 Blhds. MT 23t	2Cy.31"&54"–39" 130 HP. Barclay,Curle&Co.,Glasgow	D.Currie & Co.
135	✻	Lapwing BT 2.88 *Joslin*	I Sc3mSr *2 Dks*	British London	786 1215	235·9 *P 61 f.*	32·0 *F 51 f.*	18·2 *B 60 f.*	Dundee Gourlay Bros.&Co.	1879	5 Blhds. DBa ... f 38 ft & ...	2Cy.35½"&60"–36" 160 HP. Gourlay Bros.&Co.,Dundee	General Stm.Nav.Co.
136		Lara BT 3.88 *Wade 85-87*	I Sc3mSr	British Waterford	344 850	238·5 *P 13 f.*	31·9	15·6	Waterford Malcomson Bros.	1868		3C.21",31"&55"–36" 200 HP. D.Rollo&Sons,Liverpool	Waterford S.S.Co.,Ld.
137	✻D	Larch *Brewis 86-86*	I ScSr *1 Dk 2 B*	British Sunderland	916 1416	240·3 *Q 60 f. B 18 f.*	34·0 *F 24 f.*	18·3 *F 25 f.*	Sunderland J.Blumer&Co.	1878	4 Blhds. DBa & f 19t	2Cy.29½"&55"–36" 120 HP. G.Clark,Sunderland	Thompson & Brown
138		Lario *Sassanich 70-87*	I ScSr *1 Dk*	Austrian Trieste	151 261	166·0	20·0	11·0	London Samuda Bros.	1853		2Cy.32"–36" 100 HP. J.Watt&Co.	Lloyd Austro-Ungarico
139	✻	Lark *Hughes*	I ScTrawler *1 Dk*	British Hull	46 102	90·0	20·1	10·6	Hull Cook,Welton&Gemmell	1886	3 Blhds.	2Cy.14"&26"–20" 30 HP. C.D.Holmes&Co.,Hull	Pickering & Haldane
140	✻	Larnaca	Stl Sr *1 Dk & Web Frms*	British Liverpool	1515 2300	289·8 *P 30 f. Q 88 f.*	38·1 *B 118 f.*	19·4 *F 30 f.*	Greenock Russell&Co.	1889	5 Blhds. CDB 38t APT 21t	3Cy.22½",35½"&58"–39" Rowden&Co.,Glasgow	G.M.Steeves
141		Larne *ex De Ruyter*	I ScBn *2 Dks 3 B*	British Newcastle	1012 1670	260·0	34·6	25·9	Newcastle C.Mitchell&Co.	1870	5 Blhds.	2Cy.32"&66"–42" 180 HP. T.Clark&Co.,Newcastle	Royal Mail Steam Packet Co.
142	✻D	Larpool *Simpson*	I ScSr *1 Dk 2 B*	British Whitby	836 1288	240·3 *Q 88 f.*	33·5 *B 46 f.*	17·2 *F 26 f.*	Sunderland J. L. Thompson & Sons	1880	5 Blhds. DBa & f 29t FPT 34 APT 15t	2Cy.29"&55"–36" 120 HP. J.Dickenson,Sunderland	J.H.Barry (Rowland&Marwood)
143	✻	Larry Bane *Darragh*	I ScSr *1 Dk*	British Glasgow	54 169	114·8 *Q 38 f.*	19·7	9·1	Paisley J.Fullerton&Co.	1875	3 Blhds.	2Cy 13¾"&23¼"–18" 25 HP. W.King&Co.,Glasgow	W.Robertson
144	✻	Las Mercedes	Stl ScSr *1 Dk*	Arg'ntine Buenos Ayres		200·0 *F 25 f.*	30·0 *B 33 f.*	9·7	Birkenhead Canada Wrks.Co.Ld.	1888		3Cy.15½",25"&41"–30" Canada Works Co.,Ld.,Birkenhd.	Ocampo,Arala & Co.
145		Lasareff	I ScBk *3 Dks*	Russian Odessa	1243 1645	260·0	35·4	24·0	London Samuda Bros.	1863		2Cy.50"–36" 160 HP. J.Penn&Sons,London	Russian Steam Nav. & Trading Co.
146		Lass o' Gowrie *(ex Sudser)*	I Sc3mSr *1 Dk*	British Sydney	122 202	136·8 *Q 16 f.*	20·2	8·7	Kinghorn J.Key&Sons	1877	3 Blhds. DB 28 ft	2Cy.14"&26"–20" 35 HP. J.Key&Sons,Kirkcaldy	George Hill,jun.
147		Lass o' Gowrie	I PadSp	British Dundee	21 128	106·0	19·8	9·7	SouthShields J. T. Eltringham & Co.	1883		1Cy.37"–54" 70 HP. J.P.Rennoldson&C.,S.Shlds.	D.Edwards
148		Lassell *Hammond 81-85*	I ScSr	British Liverpool	1276 1981	310·0	34·2	24·5	Newcastle A.Leslie&Co.	1879		2Cy.26"&58"–45" 180 HP. R.Stephenson&C.,Newcstle	L'pool,Brazil & River Plate Stm.Nav.Co.,Ld.
149	✻	Latharna *Gray*	I Sc3mSr *1 Dk*	British Glasgow	265 511	175·0 *Q 85 f.*	27·1 *B 11 f.*	11·2 *F 27 f.*	Paisley J.Fullerton&Co.	1884	4 Blhds. CDB 18t FPT 30t APT 29t	2Cy.23"&45"–33" 85 HP. W.King&Co.,Glasgow	H.H.Smiley
150	✻D	Lauderdale *Ovenstone 82-83*	Stl ScSr *2 Dks*	British Glasgow	1134 1738	265·5 *P 32 f.*	36·2 *B 62 f.*	21·2 *F 32 f.*	Glasgow W.B.Thompson	1882	6 Blhds. CDB 42t	2Cy.33"&66"–42" 180 HP. Rankin& Blackmore,Gr'nck	R.Mackill & Co.

✻ = Lloyd's Register; D = Nederlandsche Vereeniging; G = Germanischer Lloyd; I = Registro Italiano; R = American Record; A = Veritas Austro-Ungarico; H = Veritas Ellenico; N = Norske Veritas; V = Bureau Veritas.

Dans la 2me Colonne, B.T. avec les indications qui suivent, signifie la date du rapport de visite de la Carène par les experts du Board of Trade pour obtenir le certificat relatif au TRANSPORT DES PASSAGERS.

Dans la 3me Colonne, C = Composite (membrure en fer, bordé en bois); I = Fer; Stl = Acier; W = Bois.

Pour plus amples renseignements sur les navires classés par le Lloyd's Register, voir le Lloyd's Register Book et le Lloyd's Yacht Register.

No. in Book / Official No.	Registry in which classed	Ship's Name. Material, Rig, &c. / Master, and Dates of Appointment to present Owner's Service and to Ship / Ship's Late Name. No. of Decks, &c.	Flag / Port of Registry	Tons Net / Gross	Length	Breadth	Depth	Built Where / By Whom	When	No. of Bulkheads / Particulars of Water Ballast	Engines Particulars, Maker's Name	Owners
151 84356	✻	**Laura** I ScSr *Jackson* 83-83 1 *Dk2 B*	British W.Hartlepool	689 1086	230·6 Q 30	32·1	13·9 F 26 f.	W. Hartlepool [illegible]	1881	4 Bhds. [illegible]	2Cy.27"& 50"–36" 99 HP. T.Richardson&Sons,H'pool	G.Pyman & Co.
152 86415	✻ BT [illegible]	**Laura** Stl ScSr *Gregory* 2 *Dks*	British [illegible]	226 585	207·0 Q 41 f.	26·8 B 12 f.	13·3 F 40 f.	Glasgow Aitken&Mansel	1885	5 Bhds.	2Cy.33"& 61"–36" 180 HP. [illegible], Glasgow	London & S.W. Railway Co.
153		**Laura** I ScSr *Christensen* 81-82 2 *Dks & Iron D*	Danish Copenhagen	673 1068	229·7 Q 41 f.	30·8	14·5 214	Copenhagen Burmeister&Wain	1882	[illegible] APTs.	1Cy.24"& 48"–32" 150 HP. Burmeister&Wain,Cp'nhgn	Det Forenede Dampskibs Selskab
154	✻	**Laura** (ex Elbe, ex Red Cain) I ScSr *Møller* 86-88	Danish [illegible]	163 269	135·5	19·7	10·0	SouthShields J.Readhead&Co.	1875		2Cy.14"& 28"–20" 40 HP. J.Readhead&Co.,S.Shields	Dampskibsselskabet Laura [illegible]
155 62295	✻	**Laura Fell** (ex Augustus) I ScSr *Parlett* 82-85 1 *Dk*	British London	106 170	109·7	18·9	10·0	SouthShields J.T.Eltringham	1871	4 Bhds.	2Cy.18"–20" 25 HP. J.P.Almond,NorthShields	T.Robson
156 81492	✻	**Laurel** I ScBn *Hutchinson* 81-82 1 *Dk2 B*	British Sunderland	891 1367	235·3 Q 86 f.	34·0 B 26 f.	17·1 F 32 f.	Sunderland [illegible]	1881	5 Bhds. [illegible] FPT[illegible] APT[illegible]	2Cy.31"& 56"–36" 99 HP. W.Doxford&Sons,S'nderland	J.Hopper
157	✻	**Laurel** W ScTrawler	British Hull	26 102	94·2	20·5	10·6	Penryn [illegible]	1886		C.I.2Cy.16"& 30"–21" Cox&Co.,Falmouth 40 HP.	J.Potter
158 89794	✻	**Laurestina** I ScSr *Batty* 80-85 2 *Dks3 B*	British N.Shields	1336 2051	287·0 P 28 f.	37·0 B 64 f.	22·0 F 21 f.	Newcastle TyneIronS.B.Co.Ld	1884	7 Bhds. [illegible] APTs.	2Cy.33"& 63"–39" 180 HP. N.E.MarineE.C.,Ld.N'wcstl	J.Robinson & [illegible]
159	✻ D	**Laurium** I ScSr *Gallienne* 83-87 1 *Dk & Spr Dk3 B*	French Havre	1079 1587	259·0	32·3	16·9 232	Newcastle Schlesinger, Davis & Co.	1879	4 Bhds. [illegible] APT[illegible] FPT[illegible]	2Cy.30"& 58"–36" 150 HP. R.&W.Hawthorn,N'wcastle	Cie Maritime du Pacifique
160	N	**Laurvig** I ScSr *Hansen* 1 *Dk*	Norweg'n Bodö	135 198	119·7 Q 31 f.	19·7 B 32 f.	10·2 F 30 f.	Bergen BergensMek.Verks	1868		1Cy.15"& 26"–18" 50 HP. BergensMek.Verkst,Bergen	Saltens Dampskibsselskab
161		**Lautaro** (ex Rimac) I ScSr *Roberts* 85-87 2 *Dks & Iron Dk*	Chilian Valparaiso	1146 1805	281·5	40·0	19·4 270	Liverpool R.&J.Evans&Co.	1872	4 Bhds. [illegible]	2Cy.46"& 86"–42" 320 HP. Fawcett,Preston&C.,L'pool	Cia. Sud-Americana de Vapores
162 95102	✻	**Lavernock** Stl ScSr *Mitchell* -88 1 *Dk2 B & Wb Fm*	British Cardiff	1482 2280	275·0 B 107 f.	40·3 F 34 f.	20·9 Q 165 f.	Newcastle PalmersCo.Ld.	1888	4 Bhds. [illegible] APT[illegible]	3Cy.22",35"&58"–42" 200 HP. PalmersCo.Ld.Newcastle	Morel, Ld.
163	V	**Lavinia** I ScSr *Grunden* 88-88 2 *Dks3 B*	German Hamburg	1154 1749	260·0 B 86 f.	35·0	17·0 246	Flensburg Flensbgr.Schiffsbau Ges.	1887	4 Bhds.	3C.23½",34½"& 58"–42" 230 HP. FlensburgerSchiffsbau Ges.	Hamburg Pacific Dampfs-Linie
164	H	**Lavrion** I Sc	Greek Piræus	102	104·8	16·4	..	England	1866		25 HP.	H.Filipango
165 90041	✻	**Lawada** Stl Sc3mSc *Nailer* 74-88 3 *Dks*	British Glasgow	2140 3269	340·1 P 56 f.	42·2 B 26 f.	25·6 F 42 f.	Dumbarton W.Denny&Bros.	1885	[illegible] Bhds. [illegible]	3Cy.26½",44"&(2)51"–51" Denny&C.,Dmbrtn 370 HP.	British India S. Nav. Co. Ld.
166 87309		**Lawrence** (ex Bartolomé) I TwnSc	British Sydney	246 399	160·0	28·0	8·6	Sunderland Kish,Boolds&Co.	1884		2Cy.17"& 32"–21" 88 HP. Hutson&Corbett,Glasgow	John See
167	✻	**Le Chatelier** I Sc3mSr *Reynault* -88 (ex Alma) 2 *Dks3 B*	French Havre	1479 2198	300·2 P 43 f.	36·0 B —	25·5 F 42 f.	Stockton M.Pearse&Co.	1879	[illegible] Bhds. [illegible]	2Cy.39½"& 68"–42" 200 HP. Blair&Co.Ld.,Stockton	Compagnie Générale Transatlantique
168 91658	V	**Le Glorieux** W ScK	British Granton	36 119	98·3	21·0	10·4	Granton D.Allan&Co.	1885		2Cy.14"& 28"–21" 35 HP. Hutson&Corbett,Glasgow	J.Muirhead
169	✻	**Le Lillois** I ScK *Van Dierendonck* 1 *Dk*	Belgian Ostend	62 187	94·5	20·0	9·6	Leith Hawthorns&Co.	1885	[illegible] Bhds. [illegible]	2Cy.15"& 38½"–21" 35 HP. Hawthorns&Co.,Leith	F.Baels-Maurau
170 29724		**Le Lion Belge** I PadSc 1 *Dk*	British Gibraltar	78 136	112·7	19·0	10·0	Newcastle Marshall Bros.	1861		2Cy. 60 HP.	Richard E.Cook
171	V	**Le Marin** Stl ScSr *Dejoie* -88	French St.PierreMart.	40 108	125·6	17·0	8·4	Nantes A.Dubigeon	1888		3Cy.11½",18"& 27½"–21" 75 HP.	L.Girards & Co.
172	V	**Le Morbihan** Stl ScSr *Minet* -87 1 *Dk2 B*	French St.Nazaire	589 1100	231·0 P 32 f.	33·3 B 56 f.	17·6 F 36 f.	Nwc.-on-Tyne R.&W.Hawthorn, Leslie&Co.	1887		3C.20",32"& 53"–36" 200 HP. R.&W.Hawthorn,Leslie&Co.,Newcastle	Compagnie Générale Transatlantique
173		**Le Reynaud** I Sc 1 *Dk*	Haytian Hayti	162 270	156·0	19·6	11·8	Nantes Jollet&Babin	1876	3 Bhds.	2Cy.20½"& 34½"–27" 50 HP. Jollet&Babin,Nantes	General B.Revel
174		**Le Tell** I Sc2m 2 *Dks & Spr Dk*	French Marseilles	850 1270	229·6	31·7	23·6	La Seyne Forges&C.de la Medit.	1879		2C.32½"& 69½"–29½" 225 HP. Forges&C.de la Medit.,Marseilles	Cie de Nav. Mixte
175	D V	**Leander** I ScSr *Wilms* 1 *Dk*	German Bremen	435 [illegible]	161·7	25·1	17·0	Lübeck H.Koch	1883	[illegible]	2Cy.18"& 31"–22" 50 HP. O.Henniges&Co.,Berlin	Dampfschiffsges. Neptun
176	V	**Leander** I PadSr *Wolter* 2 *Dks*	Russian Riga	319 371	201·5 Q 70 f.	23·9	10·0	Greenock Scott&Co.	1857		2Cy.33"& 48"–54" 150 HP. BolderaaMasch.,Bolderaa	Riga S.S.Co.
177 62504	✻	**Lebanon** I ScSr *Pat* 76-76 1 *Dk2 B*	British London	596 923	203·3 P 31 f.	29·1 F 19 f.	16·7	Middlesbro' Backhouse & Dixon	1870	4 Bhds. [illegible]	2Cy.36"–30" 80 HP. G.Clark,Sunderland	W.J.Pope & Co.
178 45017	✻	**Leda** (ex Harriet Pinckney) I Sc3mSr *Bartlett* -85 1 *Dk2 B*	British Hull	779 1191	245·0 P 28 f.	28·0	17·2	Middlesbro' Richardson,Duck& Co.	1862	6 Bhds.	1Cy.21"& 56"–24" 95 HP. Comp.'76 Humber I.W.Co. Hull	Morel Bros. & Co.
179 95036		**Léda** I PadSc *Verrat*	British [illegible]	433 576	164·0	24·0	15·5	Nantes Ateliers&C.de la Loire	1883		2C.23½"& 41½"–39½" 300 HP. DeQuillac,Dunain	La Platense Flotilla Co., Ld.
180												

✻ = Lloyd's Register; D = Nederlandsche Vereeniging; G = Germanischer Lloyd; I = Registro Italiano; R = American Record; A = Veritas Austro-Ungarico; H = Veritas Ellenico; N = Norske Veritas; V = Bureau Veritas.

In the 2nd Column, B.T., with figures following, signifies Date of report of bottom sighting by Board of Trade Surveyors as condition of PASSENGER CERTIFICATE.

In the 3rd Column, C = Composite (iron frame, wood planked); I = Iron; Stl = Steel; W = Wood.

Further particulars of Vessels classed by Lloyd's Register will be found in Lloyd's Register Book and Lloyd's Yacht Register.

Classe	Nom, Matériaux, Gréement, etc. Capitaine et dates d'entrée au service des Armateurs actuels et à bord du navire. Ancien Nom, nombre de ponts, etc.	Pavillon. Port d'Armement	Tonnage Net. Total	Dimensions: Longueur	Largeur	Creux	Construit: Où. Par qui.	Année	Nombre des cloisons. Détails sur le lest d'eau.	Machines. Détails. Nom du Constructeur.	Armateurs ou Directeurs-Gérants.
	Leda [illegible] Stl PadSr Höfer 88 1 Dk	German Bremen	89 121	125·9	18·1	6·6	London VictoriaDockCo.	1881		2Cy.32"–30" 50 HP.	T. Rocholl & Co.
	Leda L. I ScSr Seikentich 4–87 2 Dks	Austrian Trieste	865 1359	255·7	31·5	21·7	Trieste Lloyd Austro-Ungarico	1880	1 Blds.	2Cy.31"& 56"–36" 180 HP. ArsenaleLloyd,Trieste	Lloyd Austro-Ungarico
	Lee I ScSmSr Perry 2 Dks	British Cork	692 1188	263·0	36·2	15·4	Renfrew W.Simons&Co.	1880	3 Blds. [illegible]	2Cy.40"& 73"–45" 250 HP. W.Simons&Co.,Renfrew	City of Cork Steam Packet Co.,Ld.
				P 41 f.	B 48 f.	F 54 f.					
	Lee I PadSk	British Cork	74 130	160·0	17·7	7·1	PortGlasgow Blackwood&Gordon	1861		2Cy.36"–36" 80 HP. Blackwood&Gordon,Pt.Gls.	Citizen RiverSteamers Co.,Ld.
	Lee Dah I Sc	Chinese Tientsin	 119	80·0	17·1	9·6	Shanghai S.C.Farnham&Co.	1881		2Cy.16"& 32"–21" 42 HP.	
✻	Lee Sang (ex [illegible]) Stl ScSr [illegible] 85-85 1 Dk&Spr Dk3 [illegible]	British London	1092 1697	250·0 B 66 f.	36·2 F 39 f.	16·1 23·4	Glasgow London&Glas.Co.	1885	6 Blds. [illegible] APT28 MT161	3C.23"33"& 50"–39" 150 HP. London&GlasgowCo.,Glasgw	Indo-China Steam Nav.Co.,Ld.
✻	Leechmere I ScSr [illegible] 85 1 Dk	British London	724 1120	222·0 Q [illegible]	31·5 F 25 f.	17·4 B 40 f.	Sunderland Short Bros.	1876	3 Blds. WB[illegible]	2Cy.27"& 50"–33" 98 HP. [illegible] Clark,Sunderland	J.S.Barwick
DV	Leerdam (ex Nederland) I ScBn Steiger 87 3 Dks& [illegible] Dk	Dutch Rotterdam	2118 2790	322·1	40·0	30·0	Rotterdam Ned.Stoomb.Maats	1881		2Cy.39½"&69"–42" 300 HP. Ned.Stoomb.Maats.Rotterdam	Nederl.Amerikaansche Stoom. Maats
	Legaspi (ex Zamboanga) I ScBg Fornesti 2 Dks&Spr Dk	Spanish Manila	651 1024	233·0	29·3	15·3 20·9	Sunderland J.Laing	1874	1 Blds.	2Cy.28"& 54"–30" 140 HP. SndrlndEngineWrks.Sndld.	Spanish Government
✻	Legislator Stl ScBn Jackson 73-88 1 Dk&Spr Dk3 B	British Liverpool	1883 2902	330·0 Q 56 f.	40·1 B 80 f.	20·2 F 31 f. 27·6	Middlesbro' R.Dixon&Co.	1888	3 Blds.	3C.24"40"&66"–48" 400 HP. G.Clark,Sunderland	Clarente Steam Ship Co.,Ld.
V	Lehua W ScSr Clark 86-86	Hawaiian Honolulu	218	116·0	24·0	8·6	S.Francisco DickieBros.	1879		2Cy.12"& 20"–24" 18 HP. RisdonIrn.Wks.,S.Francisco	Wilder's S.S.Co.
	Leibnitz I ScBg Brown 69-71	Belgian Antwerp	1455 2280	312·5	34·9	27·6	Newcastle A.Leslie&Co.	1873		2Cy.30"& 70"–48" 280 HP. Maudslay,Sons&Field,Lndn	Soc.Royale Belge Sud-Américaine
	Leichardt I TwnScSmSr Sherif 2 Dks	British Sydney	459 690	232·6	24·3 B 48 f.	16·8	Pyrmont,N.S.W. AustralasianS.N.Co.	1865	5 Bds.	4Cy.20"& 36"–21" 100 HP. [illegible]	Australasian United Steam Nav.Co.,Ld.
N	Leif W ScSr 1 Dk2 B	Norweg'n Moss	255 352	139·8 Q 49 f.	26·6 B 27 f.	12·0 F 26 f.	Drontheim J.Voight	1872		2Cy.18"& 36"–24" 50 HP. TrondhjemsMek.V.Drontheim	H.Gerner & Søn
✻	Leif Stl Sc Baggesen 1 Dk	Norweg'n Bergen	58 98	85·0 Q 22 f.	14·6 B 30 f.	10·2	Bergen Martens,Olsen&Co.	1886	1 Blds.	3C.7¾"11½"&19"–15" 18 HP. Martens,Olsen&Co.,Bergen	Sundt & Co.
	Leila I ScSr 1 Dk	British Penang	89 131	126·0	17·7	9·9	Dundee W.B.Thompson	1875		2Cy.36"& 20"–19½" 50 HP. W.B.Thompson,Dundee	Lee Thong Sen.
	Leinster I PadSr	British Dublin	498 1643	343·0	35·0	19·0	London J.Samuda	1860		2Cy.98"–78" 700 HP. Ravenhill&Co.,London	City of Dublin Steam Packet Co.
	Leinster W Pad Flett	British S.Shields	31 111	102·6	18·5	9·8	SouthShields Readhead,Softley& Co.	1869		1Cy.36"–54" 50 HP. Readhead,Softley&C.SShlds	A.Spiteri
	Leinster Lass I PadSmSr	British Drogheda	294 638	200·0	27·4 F 21 f.	14·7	Glasgow R.Napier&Sons	1849		2Cy.70½"–66" 372 HP. R.Napier&Sons,Glasgow	Drogheda Steam Packet Co.,Ld.
GV	Leipzig I ScBg [illegible] 2 Dks& [illegible] Dk4 B	German Bremen	1612 2388	290·3	39·2	30·2	Greenock Caird&Co.	1869		2C.39½"&72¾"–48" 225 HP. O.Henniges&Co.,Berlin	Norddeutscher Lloyd
DV	Leipzig I ScSr Ehlert 86-88 1 Dk&Spr Dk	German Stettin	835 1110	223·7	30·2	20·3	Stettin Act.Ges."Vulcan"	1884	W B	2C.26¾"&49"–31½" 140 HP. Act.Ges."Vulcan,"Stettin	Neue Dampfer-Compagnie Akt.Ges.
	Leitrim I PadSr Taylor	British Dublin	403 796	249·1	27·2 F 87 f.	14·7	Birkenhead LairdBros.	1874		2Cy.46"& 79"–84" 300 HP. LairdBros.,Birkenhead	City of Dublin Steam Packet Co.
✻	Lemnos I ScSr Clarkson 72-86 1 Dk2 B	British Sunderland	1043 1599	270·3 Q 72 f. B 34 f.	34·0 P 32 f.	18·7 F 27 f.	Sunderland Short Bros.	1880	6 Blds. [illegible] FPT20t APT15t	2Cy.32"& 64"–39" 150 HP. [illegible] Sunderland	Lumsdon,Byers & Co.
✻	Lemuria I ScSr Moxley 84-86 1 Dk2 B	British London	1081 1680	257·5 Q 88 f. P 24 f.	34·6 B 72 f.	17·8 F 31 f.	Hartlepool E.Withy&Co.	1881	4 Blds. CDB38t	2Cy.33"& 61"–33" 160 HP. T.Richardson&Sons,HrtPl.	T.G.Greenwell
✻D	Lena I ScSr Richards 85-85 2 Dks3 B	British Sunderland	1141 1732	275·6	34·0 F 47 f.	23·8	Sunderland R.Thompson jun.	1877	4 Blds. [illegible] FPT18t	2Cy.33"& 62"–39" 180 HP. G.Clark,Sunderland	Culliford & Clark
	Lena I ScBk Reed 85-88	British London	679 875	206·7	30·4 P 18 f.	14·6	Sunderland J.Laing	1865		2Cy. 100 HP. G.Clark,Sunderland	C.M.Norwood & Co.
	Lennart Torstensson I ScSr Blomberg 88-8[illegible] 1 Dk	Swedish Stockholm	386 491	163·0 Q 39 f. B [illegible]	23·1 F 35 f.	12·5 H 11 f.	Norrköping MotalaCo.	18[illegible]		80 HP.	Rederi Aktiebolaget „Sven" (K.Ahlberg)
✻	Lennox I ScSr Thearle 84-84 2 Dks3 B	British Leith	1327 2051	309·5 P 32 f.	35·0 B 20 f.	23·6 F 36 f.	Glasgow A.Stephen&Sons	1880	1 Blds. CDB[illegible]	2Cy.38"& 68"–45" 240 HP. A.Stephen&Sons,Glasgow	J.Warrack
	Lenoir W Sc Dixon 87	American Philadelphia	106	112·0	26·0	5·0	Baltimore W.E.Woodall&Co.				Clyde S.S.Co.
✻	Leo I ScSr Caine 2 Dks3 B	British London	1302 1987	285·0 P 32 f.	35·2 B 70 f.	24·5 F 36 f.	Newcastle C.S.Swan&Co.	1879	5 Blds. DB[illegible]	2Cy.34"& 64"–42" 200 HP. T.Clark,Sunderland	Leo S.S.Co.,Ld. (E.E.Avery)

✻=Lloyd's Register; D=Nederlandsche Vereeniging; G=Germanischer Lloyd; I=Registro Italiano; R=American Record;
A=Veritas Austro-Ungarico; H=Veritas Ellenico; N=Norske Veritas; V=Bureau Veritas.

Dans la 2me Colonne, B.T., avec les indications qui suivent, signifie la date du rapport de visite de la Carène par les experts du Board of Trade pour obtenir le certificat relatif au TRANSPORT DES PASSAGERS.

Dans la 3me Colonne, C=Composite (membrure en fer, bordé en bois); I=Fer; Stl=Acier; W=Bois.

Pour plus amples renseignements sur les navires classés par le Lloyd's Register, voir le Lloyd's Register Book et le Lloyd's Yacht Register.

No. in Book / Official No.	Registry in which classed	Ship's Name. Master, and Dates of Appointment to present Owner's Service and to Ship. Ship's Late Name.	Material, Rig, &c. No. of Decks, &c.	Flag. Port of Registry	Tons. Net / Gross	Dimensions	Built. Where. By Whom.	When	No. of Bulkheads. Particulars of Water Ballast.	Engines. Particulars. Maker's Name.	Owners
211 60[illegible]	BT.-87	Leo *Dennison*	I ScSr 1 *Dk2 B*	British Hull	7[illegible]0 [illegible]	220·6|31·1|15·6 *P 129 f. F 30 f.*	Hull C.&W.Earle	1869	[illegible]	2Cy.32"& 5[illegible]"–36" 9[illegible] HP. C.&W.Earle,Hull	T.Wilson,So[illegible]
212 87783	✻	Leo *Winters 83-88*	I ScTrawl 1 *Dk*	British Grimsby	75 [illegible]	95·3|20·2|10·6 *Q 16 f.*	Grimsby T.Charlton	1883	[illegible]	2Cy.16½"&31½"–20" 35HP. T.Charlton,Grimsby	Grimsby & N[illegible] Steam Trawl[illegible]
213	V	Léon *Tandeau*	I Sc3mSr 1 *Dk*	French Brest	464 [illegible]	196·8|25·6|14·0 *P 116 f. F 27 f.*	Glasgow A.Stephen&Sons	1876		2Cy.27"& 4[illegible]"–30" 90 HP. A.Stephen&Sons,Glasgow	Chevillotte [illegible]
214 [illegible]	✻	Leon-y-Castillo *Walsh -88*	Stl ScSr 1 *Dk*	Spanish	238 [illegible]	190·2|27·1|12·7 [illegible]	PortGlasgow D.J.Dunlop&Co.	1888	4 Blkds. [illegible]	3Cy.17"27"&45"–30"120HP. D.J.Dunlop&Co.,P.Glasgow	
215 87842	✻	Leona *Atkinson 84-84*	Stl ScSr 1 *Dk*	British Goole	373 572	201·6|29·1|12·9 *P 28 f. B 83 f. F 57 f.*	PortGlasgow J.Reid&Co.	1884	[illegible] Blkds. [illegible]	2Cy.28"&5[illegible]"–36" 12[illegible] HP. Kincaid&Co.,Greenock	Humber S.S[illegible]
216 32	I	Leone *De Martino 79-87*	I ScSr 2 *Dks*	Italian Palermo	373 647	207·6|26·4|15·6	Greenock Scott&Co.	1864		1Cy.18"&4[illegible]"–36" 2[illegible] HP. GreenockFoundryCo.,Grnk.	Navigazione [illegible] Italiana
217		Leonor (ex [illegible]) *Garcia -83*	I ScSr 1 *Dk*	Spanish Gijon	220 351	162·[illegible]|23·6|12·4 *Q 32 f. F 9 f.*	PortGlasgow H.Murray&Co.	1869	[illegible] Blkds. [illegible]	2Cy.19"&33"–21" 60 HP. Blackwood&Gordon,Pt.Gls.	O.de Olav[illegible]
218	✻V	Leonora *Alegria*	I Sc3m 2 *Dks3 B*	Spanish Bilbao	1835 2851	345·0|37·4|25·7 *B 98 f. F 37 f.*	Glasgow London&Glas.Co.	1881	[illegible] Blkds. [illegible]	2Cy.38"&71"–48" 300 HP. London&Glas.Co.,Glasgow	Linea de Vap[illegible]
219 [illegible]		Leopard	W ScBk 1 *Dk*	British S.Johns,N.F.	217 320	132·6|26·2|14·[illegible] *Q 29 f.*	Ayr	1873		2Cy. 60 HP.	Newfoundlan[illegible] Sailing & [illegible]
220		Leopold	I Pad	Belgian	419	200·0|24·8|7·3	Antwerp Soc.Cockerill	1869			Belgian Gov[illegible]
221	V	Leopold II *Ingerslev -88*	Stl ScSr 2 *Dks*	Danish Copenhagen	1034 1618	260·8|33·2|16·2 *P 157 f. F 61 f.*	Copenhagen Burmeister&Wain	1888	[illegible]	3C 18"30"&48"–36" 150 HP. Burmeister&Wain,C'hagen	Det Forenede [illegible] skibsSelskabet
222		Leopoldo *Valence*	W	Brazilian RioJaneiro	74	.. .. ..	England	1873			F.da Sa Sanc[illegible]
223 77512		Lépanto *Roach*	I ScSr 2 *Dks3 B*	British Hull	1500 2387	305·0|36·1|26·0	Hull Earles'Co.	1877	7 Blkds.	3C 25"39"&64½"–42"250HP. Tripe'G Amos&Smith,Hull	T.Wilson,Son[illegible]
224	✻	Lequeitio *Anduiza 73-73*	I Sc2mSr 1 *Dk2 B*	Spanish Bilbao	546 764	194·7|29·0|15·5	Sunderland Short Bros.	1872	[illegible] Blkds. [illegible]	2Cy.24"& 48"–30" 90 HP. J.Peacock,Goole	Vidauade Ans[illegible]
225 85021	✻D	Lero *Holman 86-86*	I ScSr 1 *Dk.&Spr Dk3 B*	British Sunderland	1450 2224	288·5|37·4|27·0 *F 22 f.*	Sunderland J.Laing	1882	7 Blkds. [illegible]	2Cy.38"& 62"–42" 260 HP. J.Clark,Sunderland	Neptune Stea[illegible] Co.,Ld.
226	✻	Les 3 frères Conseil (ex Mayo)	I ScBn 1 *Dk2 B*	French Bordeaux	880 1353	239·6|33·0|18·0 *Q 86 f. F 35 f.*	Southampton Oswald,Mordaunt&C.	1879	3 Blkds. [illegible]	2Cy.28"& 56"–36" 140 HP. Oswald,Mordaunt&Co.,S'mptn	Les fils de Th.[illegible]
227	V	Les Vosges (ex [illegible]) *Bouddaniec -87*	I ScSr 2 *Dks.&Spr Dk*	French Marseilles	748 1221	245·4|29·5|16·2 23·4	Dumbarton J.&R.Swan	1873		2Cy.32"& 51"–33" 150 HP. Kincaid,Donald&Co.Grnck.	Cie. de Navig[illegible] Arménienne &[illegible]
228 [illegible]		Lesbian *Ball 83-84*	I ScSr 2 *Dks*	British Liverpool	1020 1560	302·3|31·2|23·2	Liverpool T.Royden&Sons	1874	[illegible] Blkds.	2Cy.3[illegible]"& 6[illegible]"–48" 190 HP. J.Jack&Co.,Liverpool	S. S. "Lesbian" [illegible] (F.Leyland &[illegible])
229	✻	Lesbos (ex [illegible])	I Sc 1 *Dk2 B*	Greek Syra	1016 1362	249·0|32·7|17·3 *Q 96 f. B 49 f. F 26 f.*	South Shields J.Readhead&Co.	1879	[illegible] Blkds. [illegible]	2Cy.28"& 5[illegible]"–33" 14[illegible] J.Readhead&Co.,S.Shields	G.Stathopulo
230 [illegible]	✻D	Lesreaulx *Morgan*	I ScSr 1 *Dk2 B*	British Cardiff	863 1316	240·0|35·0|16·1 *Q 93 f. B 50 f. F 27 f.*	Sunderland W.Doxford&Sons	1880	[illegible] Blkds. [illegible]	2Cy.31"& 5[illegible]"–36" 99 HP. W.Doxford&Sons,Sunderland	"Lesreaulx" S[illegible] Ld. (Morel [illegible])
231	✻DN	Lesseps *Hansen -81*	I ScSr 2 *Dks3 B*	Norweg'n Bergen	1141 1758	261·5|31·7|23·9 *P 31 f. B 25 f. F 31 f.*	Middlesbro' R.Dixon&Co.	1861	[illegible]	2Cy.33"& 62"–39" 150 HP. Blair&Co.,Ld.,Stockton	Harloff & Bö[illegible]
232 [illegible]	✻	Lestris *Worsnop 82-82*	I ScSr 1 *Dk2 B*	British Cork	678 1065	220·5|30·2|16·7 *P 125 f. F 30 f.*	Sunderland J.L.Thompson	1872	[illegible]	2Cy.27"& 51"–33" 100 HP. J.Dickinson,Sunderland	Cork S.S.Co.[illegible]
233 [illegible]	I	Letimbro *Di Marco 86-85*	I ScSr 2 *Dks3 B*	Italian Palermo	1401 2237	299·2|37·2|23·6 *P 30 f. B 54 f. F 40 f.*	PortGlasgow Blackwood&Gordn	1883	[illegible]	2Cy.35"& 6[illegible]"–42" 148 HP. Blackwood&Gordon,Pt.Gls.	Navigazione G[illegible] Italiana
234 [illegible]	✻	Leura *Donaldson*	I ScSr 1 *Dk2 B*	British Melbourne	758 1188	250·8|32·8|16·2 *P 150 f. F 68 f.*	Liverpool W.H.Potter&Son	1878	[illegible] Blkds.	2Cy.38"& 70"–3[illegible]" 220 HP. J.Jack&Co.,Liverpool	W.H.Smith &[illegible]
235		Levanger (ex Axion, ex Kong Brage)	I PadSr 1 *Dk*	Norweg'n Levanger	70 107	105·7|17·1|7·3 *H 21 f.*	Liverpool	1859		2Cy.17"& 37"–31" 45 HP. [illegible]	"Levanger" [illegible] skibs-Selsk.[illegible]
236 31378	✻	Levant (ex [illegible]) *Ogilvie 86-86*	I Sc3mSr 1 *Dk2 B*	British S.Shields	472 741	209·0|28·7|16·3 *Q —*	Hartlepool Denton,Gray&Co.	1865	[illegible] Blkds.	2Cy.21"& 18"–33" 90 HP. Ouseburn Eng.Works,Nwc.	W.Runciman
237	✻	Leven	St ScSr	British W. Hartlepl.			W.Hartlepool W.Gray&Co.	1889		C3y.21½",35½"&58½"–39" Blair&Co.,Ld.,Stockton	R.Ropner & C[illegible]
238		Leven	I TwnSc 1 *Dk*	British Dumbarton	367 775	189·0|38·5|13·2	Renfrew W.Simons&Co.	1885		1Cy.20"& 38½"–33" 15 HP. W.Simons&Co.,Renfrew	Dumbarton [illegible] Board
239 [illegible]		Leven	I ScSr	British Liverpool	56 102	85·0|18·0|7·1	Northwich WinshamS.B.Co.	1876		2Cy.15"–18" 30 HP. F.[illegible],Manchester	Cobden Flour [illegible] Co.,Ld.
240 [illegible]	✻	Leverington *Clarke 85-88*	I ScSr 1 *Dk&Spr B5*	British Shoreham	680 1025	214·1|30·2|16·1 23·2	Sunderland J.Laing	1869	[illegible] Blkds. [illegible]	2Cy.27½"& 51"–36" 99 HP. J.Dickinson,Sunderland	R.H.Penney

✻ = **Lloyd's Register**; D = Nederlandsche Vereeniging; G = Germanischer Lloyd; I = Registro Italiano; R = American [illegible]; A = Veritas Austro-Ungarico; H = Veritas Ellenico; N = Norske Veritas; V = Bureau V[illegible]

In the 2nd Column, B.T., with figures following, signifies Date of report of bottom sighting by Board of Trade Surveyors as condition of PASSENGER C[illegible]

In the 3rd Column, C = Composite (iron frame, wood planked); I = Iron; Stl = Steel; W = Wood.

Further particulars of Vessels classed by Lloyd's Register will be found in Lloyd's Register Book and Lloyd's Yacht Register

	Nom, Matériaux, Gréement, etc. Capitaine et dates d'entrée au service des Armateurs actuels et à bord du navire. Ancien Nom, nombre de ponts, etc.	Pavillon. Port d'Armement.	Tonnage Net. Total.	Dimensions. Longueur	Largeur	Creux	Construit. Où. Par qui.	Quand.	Nombre des cloisons. Détails sur le lest d'eau.	Machines. Détails. Nom du Constructeur.	Armateurs ou Directeurs-Gérants.
	Leviathan I Pad *Stark 77-77*	British Granton	251 393	167·0	34·0	8·5	Glasgow R.Napier&Sons	1849		2Cy.56"–42" 210 HP. R.Napier&Sons,Glasgow	North British Rail.Co.
	Leviathan I TwnSc	British London	213 313	110·9	15·2	7·9	London SammdaBros.	1886		4Cy.9"& 18"–18" 27 HP. Hunter&English,London	East & West India Dock Co.
	Levis W Sc 1 *Dk*	British Quebec	58 157	95·9	20·0	8·5	Levis, Quebec A.T.Beaulieu	1883		2Cy.14"& 31"–22" 40 HP. Beaulieu,Levis,Quebec	"Levis" Maritime & Industrial Co.
	Lewiston W Pad	American Boston Mass	1052 1127	234·9	33·1	9·4	New York	1856		900 HP.	W.H.Hill
✻	**Li Kin** Stl TwinScSr *Barnes -88* 1 *Dk*	Chinese	110 270	135·0 *P 18 f.*	23·1 *B 46 f.*	11·4 *F 26 f.*	Newcastle Sir W. G. Armstrong, Mitchell&Co.,Ld.	1888	5 Bhds.	3Cy.12",19½"&32"–20" Wallsend Slipway Co., Ld., Newcastle 130 HP.	Imperial Maritime Customs
	Liban I ScSchSr *Fabrique 80-88* *Dks&Spr Dk*	French Marseille	1434 2237	300·3 *P 56 f.*	36·2	19·1 27·4	Glasgow R.Napier&Sons	1882	6 Bhds.	2Cy.40"& 75"–48" 300 HP. R.Napier&Sons,Glasgow	Fraissinet & Co.
	Liban I Sc 1 *Dk*	French St.Malo	218 382	156·0 *Q 72 f.*	22·1 *F 21 f.*	11·2	Preston W.Allsup&Sons	1881	4 Bhds. AFT25t FPT6t	2Cy.20"& 35"–24" 60 HP. W.Allsup&Sons,Preston	A.R.Dahdah
	Libau (ex Rosamund) I ScSr *Gade 87-88* 1 *Dk*	Danish Copenhagen	362 586	172·8 *P 108 f.*	25·1 *F 30 f.*	12·7	Sunderland W.Doxford&Sons	1879	AFT30t FPT20t DB[illegible]	2Cy.20"& 38"–28" 75 HP. W.Doxford&Sons,Snderlnd.	Det Forenede Dampskibs-Selskab (C.P.A.Koch)
V	**Libau** I ScSr *Nicolai 82-82* 1 *Dk*	German Stettin	458 727	200·7 *P 104 f.*	27·6 *P 31 f.*	14·6	Rostock Act.Ges.f.Sch.&Masch.B.	1879		2Cy.23½"& 47"–30" 95 HP. Act.Ges.für S.&Masch.B.,Rstck.	Neue Dampfer Compagnie Akt.Ges.
	Libau Packet WSc3mSr *Beise 79-* (ex W.J.Taylor) 1 *Dk*	German Stettin	88 129	82·1	20·0 *Q 10 f.*	10·6	NorthShields C.W.Dodgin&Co.	1872		2Cy.13½"&26½"–18" 35 HP. Moller&Holberg,Stettin	E.Haubuss
✻	**Libelle** I ScSr *Symmers 74-87* 1 *Dk2 B*	British Aberdeen	888 1354	244·5 *Q 128 f.*	33·2 *B 15 f.*	18·0 *F 29 f.*	Newcastle Schlesinger, Davis & Co.	1880	1 Bhds. DB&f 279t AFT32t	2Cy.29"& 56"–36" 132 HP. T.Clark&Co.,Newcastle	G.Milne & Co.
✻	**Libertà** (WRECKED) I ScSr *Danielsen -87* 1 *Dk2 B*	Norweg'n Bergen	342 593	161·4 *P 78 f.*	24·2 *F 21 f.*	13·0	Bergen BergensMek.Vrkst.	1878	4 Bhds. AFT2[illegible]	2Cy.19½"& 37"–24" 50 HP. BergensMek.Verks.,Bergen	M.A.Schjelderup
✻D 87-88	**Libra** I ScSr *Howlett* 2 *Dks*	British London	508 1034	250·0	30·1	16·6	Dundee GourleyBros.	1869	5 Bhds.	3Cy25"42&67"–42" 300HP. WallsendSlip.&Eng.Co.Newcstl.	General Steam Nav. Co.
✻	**Libra** I ScTrl *Hougell 86-88* 1 *Dk*	British Grimsby	84 135	100·0	20·2 *Q 16 f.*	10·6	Hull Earles'Co.,Ld.	1885	3 Bhds.	3C.11½"17"&30"–18" 44HP. Earles'Co.,Ld.,Hull	Grimsby & North Sea Steam Trawling Co.
✻V	**Licata** (ex Ibis) I ScBg *Seth* 1 *Dk&Spr Dk3 B*	German Hamburg	1096 1499	250·0	33·0	17·0 24·0	Hartlepool Witty,Alexander& Co.	1873	4 Bhds. DB&f	2Cy.33"& 61½"–33" 150HP. T.Richardson&Sns.,Hrtlpl.	R.M.Sloman & Co.
✻	**Liddesdale** Stl ScSr *Houston 80-87* 2 *Dks3 B*	British Glasgow	1572 2422	300·0 *P 28 f.*	40·0 *B 68 f.*	23·2 *F 37 f.*	Glasgow C.Connell&Co.	1887	6 Bhds. DB&f [illegible]	3C23"37"& 59"–42" 280HP. J.&J.Thomson,Glasgow	R.Mackill & Co.
	Lidkoping I ScSp *Andersson -81* 1 *Dk*	Swedish Lidköping	133 188	101·7	20·6 *H 13 f.*	12·5	Stockholm Bergsund Mek.Vkst	1857		2Cy.13½"&26½"–19" 30HP. MotalaCo.,Gothenburg	Albert Peterson
✻	**Lido** Stl ScSr *Jefferson 86-86* 2 *Dks*	British Glasgow	620 977	225·5 *B 36 f.*	30·1 *F 29 f.*	16·9	Port Glasgow Murdoch&Murray	1881	4 Bhds. [illegible]	2Cy.28"& 52"–33" 120 HP. Kincaid,Donald&Co.,Grnck	Raeburn & Vérel
GV	**Liebenstein** I ScSr *Grumberg 84-84* 1 *Dk&Spr Dk&B*	German Bremen	738 1169	221·1 *B 58 f.*	30·2	14·6 21·9	Rostock Act.Ges.f.Sch.&Mas ch.B.	1882	DB[illegible]	2C.27½"&51"–31½" 110HP. Act.Ges.für S.-&Masch.-Bau, Rostock	Deutsche Dampfschiff-fahrts-Gesellsch. "Hansa"
	Liffey (BROKEN UP) I ScSr *Wardropper* 1 *Dk*	British Waterford	91 165	116·4	18·0	9·5	Liverpool W.H.Potter&Co.	1870	3 Bhds.	2Cy.15"& 26"–18" 30 HP. CentralEng.Co.,Swansea	R.W.Muir
	Liguria I ScBk *Conlan* 3 *Dks&Awn Dk*	British Liverpool	2966 4648	133·4	45·0	35·3	Glasgow J.Elder&Co.	1874	6 Bhds.	3C.56"&(2)78"–60" 750 HP. J.Elder&Co.,Glasgow	Pacific Steam Nav.Co.
I	**Liguria** I ScSr *Saettone 82-87* 2 *Dks*	Italian Genoa	293 540	178·1 *Q 27 t.*	24·7 *F 9 t.*	16·6 *H 31 t.*	Newcastle C.Mitchell&Co.	1862	4 Bhds. DB&f	2Cy.21"& 40"–30" 132 HP. P.Ferro&Co.,Genoa	Navigazione Generale Italiana
	Ligurian I ScSr *Ellis 86-87* 2 *Dks*	British Liverpool	1017 1566	302·3	31·2 *F 30 f.*	23·2	Liverpool T.Roydon&Sons	1874	5 Bhds.	2Cy.30"& 60"–48" 150 HP. J.Jack&Co.,Liverpool	S.S."Ligurian"Co.,Ld. (F.Leyland & Co.)
V	**Likelike** W ScSr *Davies 85-86*	Hawaiian Honolulu	463 674	168·0	27·0	17·0	S.Francisco DickieBros.	1877		2Cy.21"& 40"–34" 70 HP. RisdonIronWks,S.Francisco	Wilder's S.S.Co.
✻	**Lilburn Tower** I ScSr *Parm 83-87* 1 *Dk&Spr Dk3 B*	British London	1892 2872	330·2	38·3	27·8	Newcastle Wghm. Richardson & Co.	1883	6 Bhds. DB&f 266t FPT26t AFT23t	2Cy.40"& 71"–48" 400 HP. Wigham Richardson & Co., Newcastle	"Lilburn Tower" S.S. Co.,Ld.(F.Stanmore & Co.)
✻	**Lilian** I ScSr *Taylor 84-84* 1 *Dk*	British Newcastle	346 557	170·0 *Q 79 f.*	27·3 *B 16 f.*	12·5 *F 21 f.*	Newcastle Swan&Hunter	1883	4 Bhds. DB&f 105t	2Cy.22"& 41"–30" 80 HP. Ross&Duncan,Glasgow	J.G.Charlton & Co.
	Lilian I ScSr *Frederick -86* 1 *Dk*	British Sydney	71 105	98·2	17·1	7·1	Sydney D.&W.Robertson	1886		2Cy.15"& 28"–20" 29 HP. London&ColonialE.Co.,Lon.	Government of New South Wales
I	**Lilibeo** I ScSr *Stabile 88-88* 2 *Dks3 B*	Italian Venice	569 1016	219·8	28·3	21·8	Sunderland W.Pile&Co.	1873	4 Bhds. W[illegible]	2Cy.32"& 60"–36" 283 HP. M.Samuelson,Hull	Navigazione Generale Italiana
	Lilla Edet I ScSp *Eliasson* 1 *Dk*	Swedish LillaEdet	93 120	95·5 *Q 16 f. B 10 f. F 21 f. H 29 t*	17·9	7·7	Gothenburg	1873		30 HP.	J.Bodén
	Lillebelt I PadFerry *Jørgensen 88-88* 1 *Dk*	Danish Copenhagen	125 306	139·2	25·3	10·7	Newcastle Wghm. Richrdson& Co.	1872		2Cy.36"–16" 85 HP. Thompson,Boyd&Co.,Newc	De Danske Statsbaner

✻=Lloyd's Register; D=Nederlandsche Vereeniging; G=Germanischer Lloyd; I =Registro Italiano; R=American Record;
A=Veritas Austro-Ungarico; H=Veritas Ellenico; N=Norske Veritas; V=Bureau Veritas.

Dans la 2me Colonne, B.T., avec les indications qui suivent, signifie la date du rapport de visite de la Carène par les experts du Board of Trade pour obtenir le certificat relatif au TRANSPORT DES PASSAGERS.

Dans la 3me Colonne, C=Composite (membrure en fer, bordé en bois); I=Fer; Stl=Acier; W=Bois.

Pour plus amples renseignements sur les navires classés par le Lloyd's Register, voir le Lloyd's Register Book et le Lloyd's Yacht Register.

No. in Book / Official No.	Registry in which classed	Ship's Name. Material, Rig, &c. / Master, and Dates of Appointment to present Owner's Service and to Ship / Ship's Late Name. No. of Decks, &c.	Flag / Port of Registry	Tons Net / Gross	Dimensions: Length, Breadth, Depth	Built: Where / By Whom	When	No. of Bulkheads / Particulars of Water Ballast	Engines: Particulars, Maker's Name	Owners or Ma[nagers]
271 79142	❋	Lillie — I ScSr *Bonnaley 86-87* — 1 *Dk2B*	British Newport	634 1086	225·0 | 30·0 | 16·7 *B 51 f. Q 32 f. F 31 f.*	W. Hartlepool W.Gray&Co.	1875	4 Bhds. aD6	2Cy.27"&50½"–33" 99 HP. Blair&Co.,Ld.,Stockton	Pyman,Wats[on]
272 [illegible]		Lillie Smith — W ScSr *Smith 88* — 1 *Dk*	British Saugeen	206 302	130·0 | 26·6 | 9·4	Saugeen J.Reay	1888		1Cy.20"–20" 150 HP. J.Reay,Toronto	G.E.Smith
273 81190	BT3.89	Lily — Stl PadSr *Burke 88*	British Dublin	230 1035	300·0 | 33·1 | 14·4 *Q 80 f. B 60 f. F 36 f.*	Birkenhead Laird Bros.	1887		2Cy.78"–84" 470 HP. Laird Bros.,Birkenhead	London& N.W[estern]
274 [illegible]		Lily — I PadSp	British N.Shields	23 117	97·7 | 18·9 | 9·6	N.Shields Hepple&Co.	1888		1Cy.34½"–52" 38 HP. Hepple&Co.,N.Shields	J.Chisholm
275	V	Lily *ex Ca[illegible]* — I ScSr *Hägg 87* — 1 *Dk*	Swedish Stockholm	271 371	128·5 | 23·3 | 11·8 *Q 38 f. B 31 f. F 20 f. H 21*	Gefle Atlas M.V.	1887	DB 60 APT 11	2C.17½"&31½"–20½" 50 IP. Atlas M.V.,Stockholm	G.O.Wallenbe[rg]
276 [illegible]	❋	Lilydale — I ScSr *Burgess 83-84* — 1 *Dk*	British W.Hartlepool	401 636	180·3 | 28·0 | 14·3 *Q 56 f. B 13 f.*	SouthShields J.Readhead&Co.	1873	4 Bhds. DB 68 aft	2Cy.23"&44"–30" 85 HP. J.Readhead&Co.,S.Shields	Geipel & Co.
277 [illegible]		Lima — I Sc3mSr 2 *Dks & Shade Dk*	British Liverpool	1132 1804	310·6 | 38·2 | 19·8 *F 54 f.*	Glasgow T.Wingate&Co.	1873	5 Bhds. WB	2C.51½"&84½"–42" 400 HP. T.Wingate&Co.,Glasgow	Pacific Steam [Nav. Co.]
278		Limari — I ScSr *Mayhew 71-87* 2 *Dks & Awn Dk*	Chilian Valparaiso	613 967	227·5 | 31·0 | 17·3	Liverpool Bowdler,Chaffer & Co.	1869	1 Bhds. aDB	2Cy.31"&56"–30" 150 HP. Fawcett,Preston&Co.,Lvpl.	Cia.Sud-Amer[icana de] Vapores
279		Limena (HULK) — I Pad	Peruvian Peru	1163 1623	267·2 | 40·2 | 17·6 *P 113 f.*	Glasgow Randolph, Elder & Co.	1865		2Cy.48"&93"–60" Randolph,Elder&Co.,Glasg.	
280 70552	BT10.89	Limerick — I PadSr	British Milford	348 914	251·8 | 29·2 | 15·2 *P 70 f. B 94 f.*	Renfrew W.Simons&Co.	1874		2Cy.52"&90"–72" 400 HP. W.Simons&Co.,Renfrew	Great Western [Rail]way Co.,Ld.
281		Limfjorden — Stl & I ScSr *Olsen 86-87* — 1 *Dk*	Danish Thisted	226 378	170·9 | 24·2 | 10·2 *P 45 f. F 41 f.*	Renfrew Lobnitz&Co.	1881	APT 11	2Cy.21"&44"–30" 85 HP. Lobnitz&Co.,Renfrew	Det Forenede [Damp]skibs Selskab (C.[illegible])
282 82811	❋ D	Limosa — I ScBn *Parris 85-88* — 2 *Dks 3 B*	British London	992 1536	250·0 | 33·7 | 23·0 *P 28 f. F 32 f.*	Stockton [illegible]	1880	4 Bhds. DB [illegible]	2Cy.32"&60"–38" 160 HP. Blair&Co.,Ld.,Stockton	"Avis" Steam[ship Co.] (Parr[illegible])
283 80713	BT5.89	Limpet — I ScSr *Patterson 73-88*	British [illegible]	253 475	199·5 | 25·2 | 13·0 *Q 29 f. F 26 f.*	Glasgow [illegible]	1882		2Cy.25"&48"–36" 100 HP. Barclay,Curle&Co.,Glasgow	Sligo Steam Na[v. Co.]
284 82902	❋	Limpopo — I ScBg *Clark 83-88* 1 *Dk & pt Awn Dk 2 B*	British London	883 1361	248·6 | 32·0 | 16·3 *Q 88 f.*	Newcastle TyneI.S.B.Co.,Ld.	1880	5 Bhds. CDB 250 FPT 12 APT 24	2Cy.26"&52"–38" 120 HP. T.Clark&Co.,Newcastle	W.Gray
285	V	Lina — I ScSr *Scherlau 69-80* 1 *Dk & Awn Dk*	German Stettin	451 638	174·8 | 25·3 | 13·1 21·1	Elbing [illegible]	1880		2Cy.22"&41"–25¼" 85 IP. F.Schichau,Elbing	"Lina" Dampf[schif]fahrts Akt.Ges[ellschaft]
286 88731	❋	Linares — I ScSr *Hetherington 85-87* — 1 *Dk2B*	British Newcastle	1065 1646	265·6 | 36·3 | 17·5 [illegible]	Newcastle [illegible]	1883	1 Bhds. [illegible] APT 60	2Cy.32"&60"–36" 140 HP. Palmers Co.,Ld.,Newcastle	Scott Bros.
287 87762	❋	Lincoln — I ScSr *Walker 76-83* — 1 *Dk2B*	British Grimsby	697 1075	251·5 | 32·2 | 15·9 *P 63 f. B 16 f. F 37 f.*	Hull Earles' Co.,Ld.	1883	5 Bhds. Tal 26	2Cy.36"&69"–39" 185 HP. Earles'Co.,Ld.,Hull	Manchester,Sh[effield &] Lincolnshire R[y.]
288 88436		Lincoln (STRANDED) — W Sc 1 *Dk*	British S.Cthrn's,On	247 347	131·0 | 23·9 | 11·9 *H 19 f.*	StCthrn'sOnt M.Simpson	1872		1Cy.36"–30" 300 HP. C.N.Ollie,St.Catherines,Ont.	James Norris
289 93491		Lincoln — W Pad *Dibden 86* — 1 *Dk*	British Sydney	65 104	112·6 | 18·2 | 6·5 *H 23 f.*	Blmain.,N.S.W. W.J.Watson	1886		2Cy.14"&26½"–18" 40 HP. Perdriau Woods Co.,Balmain	Balmain New [Steam] Ferry Co.
290	G	Lincoln (ex [illegible]) — C ScSr *Samuelsson 87* — 1 *Dk*	Swedish Thorskog	147 200	102·9 | 20·7 | 10·3 *B 20 f. H 3 f.*	Thorskog P.Larsson	1885		2Cy.12"&25"–19½" 35 HP. P.Larsson,Thorskog	P.Larsson
291 58679		Lincolnshire — I ScSr *Worboy 88*	British Newport	65 110	77·1 | 21·3 | 8·3 *Q 12 f. F 5 f.*	Newcastle C.Mitchell&Co.	1867		1Cy.12"–14" 25 HP. Pattison&Atkinson,Nwc.	H.Burton
292 91298	❋	Linda — Stl Sc *O'Neill 82-87* — 2 *Dks 3 B*	British London	1598 2424	290·0 | 40·2 | 24·5 *F 35 f. B 24 f.*	Newcastle Newcastle Iron S.B.Co.	1887	7 Bhds. CDB [illegible] APT 94	3C.24"37"&62"–42" 300 HP. Wigham Richardson&Co., Newcastle	Hunting & Pa[ttison]
293	I	Linda — I Sc3mSr *Castagnola 87* 1 *Dk & Spr Dk*	Italian Cvitav'chia	976 1385	242·6 | 32·4 | 17·4 23·7	SouthShields J.Readhead&Co.	1873	DB 80 [illegible]	2Cy.27"&54"–33" 121 HP. J.Readhead&Co.,S.Shields	F[lli] Cancellieri
294 79865	❋	Linden (WRECKED) — I ScSr *Thompson* — 1 *Dk2B*	British London	687 1068	218·3 | 31·0 | 17·5 *Q 34 f. F 28 f.*	Sunderland Short Bros.	1877	4 Bhds. DB [illegible] FPT 26 APT 31	2Cy.27"&50"–30" 99 HP. G.Clark,Sunderland	R.Thorman
295	NV	Lindesnæs (ex Regina, ex Metz) — I ScSr *Stoltenberg 86-* — 1 *Dk*	Norweg'n Christiania	266 375	152·1 | 21·5 | 15·0 *B 24 f. H 8 f.*	Kiel Norddtsche. Schiffswerft.	1872		2Cy.17⅛"&28"–24" 50 HP.	Actieselskabet "[Lin]desnæs" (J.[illegible])
296	N	Lindesnæs — I&Stl ScSr *Berentsen* — 1 *Dk*	Norweg'n Farsund	98 148	95·3 | 18·2 | 8·9 *Q 12 f. B 31 f. F [illegible]*	Christiansand Christianssands Mek. V.	1886	APT FPT	2Cy.15"&27½"–21" 35 HP. Christianssands Mek.V.	Lindesnæs In[te]rentsk[a]b.
297	N	Lindholmen — I ScSr *Magnor* 1 *Dk & Awn Dk*	Norweg'n Arendal	323 422	150·2 | 26·2 | 16·7	Gothenburg MotalaCo.	1878		2Cy.22"&39"–28½" 105 HP. MotalaCo.,Gothenburg	Arendals Damp[s]kibsselskab
298	V	Lindholmen — I ScSr *Dannberg* — 1 *Dk*	Swedish Gothenburg	658 848	205·7 | 27·3 | 15·6 [illegible]	Gothenburg [illegible]	1877	MT [illegible]	2Cy.24"&44"–30" 100 HP. MotalaCo.,Gothenburg	Göteborgs A[ng]. Bol. (A.Meyer)
299 63341	❋	Lindisfarne — I ScBk *Bishop 71-83* — 1 *Dk2B*	British Newcastle	784 1206	244·7 | 30·6 | 17·5 [illegible]	North Shields [illegible]	1870	1 Bhds. DB [illegible]	2Cy.27"&52"–30" 99 HP. Comp. '79,R.&W.Hawthorn, Newcastle	J.O.Scott
300										

❋ = Lloyd's Register; D = Nederlandsche Vereeniging; G = Germanischer Lloyd; I = Registro Italiano; R = American R[ecord]; A = Veritas Austro-Ungarico; H = Veritas Ellenico; N = Norske Veritas; V = Bureau Ver[itas]

In the 2nd Column, B.T., with figures following, signifies Date of report of bottom sighting by Board of Trade Surveyors as condition of PASSENGER CERTI[FICATE].

In the 3rd Column, C = Composite (iron frame, wood planked); I = Iron; Stl = Steel; W = Wood.

Further particulars of Vessels classed by Lloyd's Register will be found in Lloyd's Register Book and Lloyd's Yacht Register.

No.	Class.	Nom, Matériaux, Gréement, etc. Capitaine et dates d'entrée au service des Armateurs actuels et à bord du navire. Ancien Nom, nombre de ponts, etc.	Pavillon Port d'Armement	Tonnage Net Total	Dimensions. Long. Larg. Creux	Construit. Où. Par qui.	Quand.	Nombre des cloisons. Détails sur le lest d'eau.	Machines. Détails. Nom du Constructeur.	Armateurs ou Directeurs-Gérants.
	✻	**Lindsay** I ScSmSr *Mason 77-77* 1 Dk 2 B	British Liverpool	719 1116	217·3 28·2 16·0 F 19 f.	Liverpool Sandler, Chaffer & Co.	1866	3 Blhds. WBaf	2C.24"& 50"-30" 95 HP. J.Jones&Sons,Liverpool	J.Jones
	✻	**Lindsey** I ScTrawler *Whitehouse 87* 1 Dk	British Boston,Mass	72 144	100·3 20·2 10·3 F 21 f.	Hull Earles'Co.,Ld.	1885	3 Blhds.	3C.11½"17"&34"-21" 45HP Earles'Co.,Ld.,Hull	Boston Deep Sea Fishing & Ice Co.,Ld.
	✻	**Lindula** StI ScBarkSr *Allen 75-88* 2 Dks 3 B	British Glasgow	2199 3516	350·4 42·1 25·9 F 41 f. DB 27	Dumbarton W.Denny&Bros.	1888	5 Blhds. DB 550	4Cy.24"36"48"&70"-48" Denny&Co,Dumbartn [290HP	British India Steam Nav.Co.
		Lindus I ScSr *Ball 86-87* 1 Dk 2 B	British Melbourne	1080 1679	257·1 34·6 17·8	Hartlepool E.Withy&Co.	1881	1 Blhds. CDB 350	2Cy.33"& 61"-33" 160 HP. T.Richardson&Sons,Hrtlpl	Huddart, Parker & Co.
	1 V	**Lining** I ScSr *Arendts 83-84* 1 Dk	German Danzig	467 620	175·8 25·0 17·8	Danzig F.Devrient&Co.	1883		2C.19⅝"&36½"-25⅝" 75HP. Aron&Gollnow,Stettin	F.G.Reinhold
		Linköping ex Phoenix I ScSp *Sjödin 83-85* 1 Dk & 1 an Dk	Swedish Linköping	162 212	95·8 19·0 10·3 H 3 t	Motala MotalaCo.	1848		2Cy.15"&30"-21" 45 HP. MotalaCo.,Motala	Linköping Rederi Akt. Bolag (C.Thorngren)
	✻	**Linn o' Dee** I ScSr *Millar 85-85* 1 Dk	British Aberdeen	319 510	170·7 24·6 11·7 Q 59 f. B 48 f. F 40 f.	Port Glasgow Murdoch&Murray	1884	1 Blhds. DB 24 A FPT 30	2Cy.26"& 40"-30" 67 HP. Dunsmuir&Jackson,Glasgow	Granite City S.S.Co., Ld.
		Linné 2m *Blager 79-83*	Russian St. Petersburg	693	.. .. ..	Norrköping MotalaCo.	1883			Naphta Productions Gesellschaft (Gebr.Nobel)
		Linné I ScSr *Törnqvist* 1 Dk	Swedish Stockholm	202 261	133·8 20·8 10·2 Q 38 f. B 40 f. F 27 f. H 3 t	Gothenburg MotalaCo.	18^{58}_{84}		2Cy.11"& 24"-21" 30 HP. W.Lindberg'sVerkst,Stklm	Rederibolaget "Östersjön" (F.Sjöberg)
		Linnea W ScSp *Aspenberg 83* 1 Dk	Swedish Borgvik,Vm	92 145	93·8 20·8 10·2 P 18 f. F 3 t.	Lödöse Gothenburg	18^{58}_{89}		22 HP.	L.J.Koltlhoff
		Linnet I PadSp *Hogarth* 1 Dk	British London	37 122	105·1 19·0 9·7	SouthShields J.T.Eltringham	1877	3 Blhds.	2Cy.28"-50" 52 HP. Baird&Barnsley,N.Shields	Danubian Nav. Co., Ld.
312	V	**Linthorpe** I ScK *Haynes 85*	British Hull	70 133	88·0 19·7 7·2	Seaham J.Morrice&Co.	1885		2Cy.13"& 25"-18" 25 HP. G.Paulin&Sons,Newcastle	R.Gray
313	✻	**Lion** I TwinScTg *Price* 1 Dk	British London	78 141	101·6 19·6 8·5 F 17 f.	London ThamesI.Wks&S.B.Co.	1882	1 Blhds. FPT 31	3Cy.11⅞"&25⅛"-16" 60HP. J.Stewart&Son,London	African Boating Co., Ld.
314		**Lion** I PadSp	British London	59 121	106·2 18·9 10·0	S.Shields	1859		2Cy. 60 HP.	Belfast Harbour Commissioners
315		**Lion** I ScSmSr *Egidiussen 82-82* 1 Dk	Danish Nakskov	141 196	116·3 17·8 7·8 Q 36 f. F 12 t.	Hull C.&W.Earle	18^{56}_{83}	DB f 36	2Cy.13¾"&25¾"-18" 35HP. MotalaCo.,Gothenburg	Det. Forenede Dampskibs Selskab (C.P.A.Roch)
316	✻	**Lionel** I ScSr *Macnully 83-87* 1 Dk 2 B	British Newcastle	681 1071	236·5 30·2 17·1 P 27 f. B 40 f. F 29 f.	Newcastle C.Mitchell&Co.	1874	1 Blhds. DBaf&b	2Cy.26"& 52"-30" 98 HP. Thompson&Co.,Newcastle	Clapham S.S.Co.,Ld.
317		**Lionel George** ex ... W ScSr *Griffith 86-86*	British Bristol	93 137	106·4 20·8 8·5	Bristol G.Nurse	1881		2Cy.13"& 22"-18" 30 HP. Plenty&Son,Newbury	R.Owens
318	✻	**Lippe** I ScSr *Bachmann 88-88* 1 Dk	Belgian Antwerp	351 555	170·2 25·1 12·7 Q 60 f. B 45 f. F 25 f.	Aberdeen A.Hall&Co.	1888	5 Blhds. DBaf 100t	3Cy.14"22"&36"-30" 75HP Hall,Russell&Co.,Aberdeen	A.Deppe
319	✻ BT 7.88	**Lisbon** I ScBg *Hall 71-76* 3 Dks	British London	860 1324	241·8 32·1 23·2 F 23 f.	W. Hartlepool Denton,Gray&Co.	1871	1 Blhds. DBaf Ceat FPT 63	2Cy.27"& 57"-36" 120 HP. Blair&Co.,Ld.,Stockton	J.Hall,jun.,& Co.
320		**Lisbon** I ScSr *McNab 82-85* 2 Dks	British Liverpool	416 664	211·4 26·1 16·0 F 21 f.	Liverpool T.Royden&Sons	1874	5 Blhds.	2Cy.26"& 51"-30" 110 HP. J.Jack,Rollo&Co.,Liverpool	S.S. "Lisbon" Co.,Ld. (F.Leyland & Co.)
321	✻ BT 10.88	**Lisbonense** I ScSr *King 88-88* 2 Dks 3 B	British Liverpool	1035 1594	269·4 33·2 16·4 23.7	Liverpool T.Royden&Sons	1871	6 Blhds.	3C23"36½"&58"-36" 175HP Westray,Copeland&Co,Barw	R.Singlehurst & Co.
322	✻	**Liscard** I ScSr *Byrne 82-85* 2 Dks 3 B	British London	1278 1943	285·0 35·0 23·1 P 28 f. B 13 f. F 32 f.	NorthShields T.&W.Smith	1880	1 Blhds. CDB 388 APT 68	2Cy.36"& 68"-42" 230 HP. R.&W.Hawthorn,Newcastle.	A.Holland & Co.
323	✻	**Lismore** ex Lis... ScBkn *Srabazan 88-* 2 Dks 3 B	British Belfast	1445 2227	321·0 35·2 26·0	Glasgow London&GlasgowCo.	1874	5 Blhds.	2Cy.40"& 70"-46" 250 HP. London&GlasgwCo.,Glsgw.	"Lismore" of Belfast S.S.Co.,Ld. (Woodside& Workman)
324		**Lismore** I PadSp *Geldart 68-*	British Barrow	18 181	125·4 24·1 9·9	Barrow BarrowS.B.Co.	1874		2Cy.36"-60" 100 HP. BarrowS.B.Co.,Ld.,Barrow	Sir J.Ramsden
325	✻ D	**Lisnacrieve** I ScBkn *Evans 88-88* 2 Dks 3 B	British London	1831 2791	330·0 39·2 24·9 P 24 f. B 75 f. F 28 f.	Stockton Richardson,Duck&Co.	1883	5 Blhds. CDB 221	2Cy.40"& 71"-48" 350 HP. Blair&Co.,Ld.,Stockton	Hunting & Pattison
326	V	**Lissabon** I ScSr *Holm 86-87* 2 Dks	German Hamburg	1500 2000	286·4 37·1 21·3 B 26 f. F 32 f.	Hamburg ReiherstiegSchiffsWerfte	1883		2Cy.34"& 62"-42" 190 HP. ReiherstiegSchiffsWerfte	Hamburg-Südamerik. Dampfschifffahrts-Ges.
327	✻ V	**Lissabon** ex Acacia I ScSr *Schöpgler 88* 1 Dk 2 B	German Hamburg	750 1005	235·5 30·1 17·1 B 50 f. Q 85 f. F 29 f.	W. Hartlepool Denton,Gray&Co.	1872	1 Blhds. DBaf f 300	2Cy.27"& 54"-33" 110 HP. Blair&Co.,Ld.,Stockton	R.M.Sloman & Co.
328	✻	**Lista** I Sc *Larágoza 70-83* 1 Dk & 1 an Dk	Spanish Seville	825 1059	210·7 29·8 13·9 21·4	Dundee W.B.Thompson	1882	1 Blhds. DB 250	2Cy.23"& 46"-30" 90 HP. W.B.Thompson,Dundee	Miguel Saenz & Co.
329		**Litoral** I ScSr *Casanelli* 1 Dk	Argentne BuenosAyres	287 461	180·0 27·1 10·4 P 29 f. F 22 f.	Paisley H.McIntyre&Co.	1880	1 Blhds.	2Cy.23"& 45"-30" 85 HP. Muir&Houston,Glasgow	Torrado & Molero
330	G	**Littuania** I ScSr *Koelländer 84-86* 2 Dks	German Memel	604 791	203·4 28·0 18·5	Kiel G.Howaldt	1882	DB	2Cy.23"& 37½"-27" 75 HP. HowaldtBros.,Kiel	Memeler Dampfschfff. Akt.Ges.

✻ = Lloyd's Register; D = Nederlandsche Vereeniging; G = Germanischer Lloyd; I = Registro Italiano; R = American Record;
A = Veritas Austro-Ungarico; H = Veritas Ellenico; N = Norske Veritas; V = Bureau Veritas.

Dans le 2me Colonne, B.T., avec les indications qui suivent, signifie la date du rapport de visite de la Carène par les experts du Board of Trade pour obtenir le certificat relatif au TRANSPORT DES PASSAGERS.

Dans la 3me Colonne, C = Composite (membrure en fer, bordé en bois); I = Fer, Stl = Acier; W = Bois.

Pour plus amples renseignements sur les navires classés par le Lloyd's Register, voir le Lloyd's Register Book et le Lloyd's Yacht Register.

No. in Book. / Official No.	Registry in which Classed	Ship's Name. Material, Rig, &c. / Master, and Dates of Appointment to present Owner's service and to Ship. / Ship's Late Name. No. of Decks, &c.	Flag. / Port of Registry.	Tons. Net / Gross	Dimensions: Length	Breadth	Depth	Built. Where / By Whom.	When	No. of Bulkheads / Particulars of Water Ballast.	Engines. Particulars / Maker's Name.	Owners.
331 78767	✱	Livadia 1 ScSr *Ellis* -87 *2 Dks*	British Liverpool	796 1254	240·3	31·1	20·7 *Q 90 f. F 35 f.*	Liverpool R.&J.Evans&Co.	1877	6 Blkds. BT	2Cy.28"&50"-30" 120 HP. De Winton&Co.,Carnarvon	Richards,Mills
332	G	Livadia (ex [illegible]) 1 ScSr *Andersen* 1 *Dk*	German Lübeck	135 179	99·4	21·5	10·2	Gothenburg	1858		2Cy.11¼"&36½"-15" 30 HP.	Küsten Dampf- fahrts-Ges.(J[illegible])
333 18440	✱ BT.98	Liverpool 1 ScSr *Paterson* *1 Dk,Spt,Sht Dk3 B*	British Sligo	246 514	199·1	26·3	13·2 *P 49 f. B 99 f. F 3[illegible] f.*	Glasgow A.&J.Inglis	1864	4 Blkds.	2Cy.40"-42" 140 HP. A.&J.Inglis,Glasgow	Sligo Stm.Nav.
334 77473	BT.9.88	Liverpool 1 PadSp	British Hull	164 220	159·6	18·7	8·4	Hull M.Samuelson&Co.	1855		2Cy.36"-36" 90 HP. Laird Bros.,Birkenhead	Manchester,Sh. & Lincolnshire
335 45994		Liverpool (BIRKEN CY) I Pad	British Liverpool	99 157	124·5	22·2	9·1	Liverpool J.Laird	1855		2Cy.37"-36" 30 HP.	S.Davies
336 34912	BT.7.88	Liverpool I PadSp	British [illegible]	52 178	121·5	20·7	11·2	Newcastle A.Leslie&Co.	1866		2Cy.35"-18" 80 HP. T.Marshall,S.Shields	W.& T.Jolliffe
337 59998		Liverpool (ex Victor) W Pad *Walker*	British Hull	32 116	105·9	19·3	9·6	NorthShields Rennoldson	1870		1Cy.37"-36" 60 HP. J.P.Rennoldson,S.Shields	J.F.Ealand
338	✱N	Livingstone (ex Ortygine, ex Julia David) IS Sr *Reimers* 83-87 *2 Dks 3 B*	Norweg'n Bergen	1271 1719	274·8	33·4	24·0 *B 40 f. F 27 f.*	Newcastle Palmers'Co.,Ld.	1873	5 Blkds.	2Cy.30½"&60"-33" 140 HP. Palmers'Co.,Ld.,Newcastle	Bergh & Hel[illegible]
339	G	Livland 1 ScSr *Ohlsen* -82 *1 Dk*	German Lübeck	497 629	193·6	24·4	13·7	Stettin Act.Ges."Vulcan"	1880	WB	2C.24½"&43¼"-27⅝" 85 HP. Act.Ges."Vulcan," Stettin	Riga Lübecker schiffahrts-Ge[illegible]
340 83013	✱	Livonia 1 ScSr *Tannock* -86 *1 Dk,&Awn Dk*	British Leith	633 980	209·6	29·4	13·9 *P 18 f. 202*	Glasgow Aitken&Mansel	1880	6 Blkds. MT[illegible] AT[illegible] FP[illegible]	2Cy.26"&49"-36" 110 HP. Walker,Hudson&C.,Glasgw	J.Cormack
341	GV	Livonia 1 TwnScSr *Haelzig* 86-[illegible] *1 Dk*	German [illegible]	209 280	135·1	22·0	9·0 *P 32 f. F 31 f.*	Glasgow T.B.Seath&Co.	18 68/85		1Cy.12"&21"-19½" 44 HP. [illegible]A.Tischbein,Rostock	F.G.Reinhold
342 65760	✱	Livornese (ex Espanol, ex Zerbino) 1 ScBg *Pepperell* 87-88 (ex Red Bandit) *3 Dks*	British Liverpool	931 1437	260·0	32·0	25·6	PortGlasgow Blackwood&Gord'n	1871	5 Blkds.	2Cy.30½"&54"-36" 160 HP. Blackwood&Gordon,PtGlgw	Atlantic & [illegible] S.S.Co.,Ld.
343	✱	Livorno (ex [illegible]) 1 ScSr *Broecker* -88 *2 Dks 3 B*	German Hamburg	1088 1392	250·0	33·5	22·3 *B 36 f.*	Sunderland J.L.Thompson&Sns.	1875	4 Blkds. DBaft [illegible]	2Cy.32"&60"-36" 140 HP. J.Dickinson,Sunderland	R.M.Sloman &
344	✱	Lixuri (ex Joseph Viney) 1 ScSr *Coundouris* *1 Dk 2 B*	Greek Cephalonia	858 13[illegible]	250·0	32·7	17·5	S.Shields J.Readhead&Co.	1880	4 Blkds. DBaft [illegible]	2Cy.28"&53"-33" 115 HP. J.Readhead&Co.,S.Shields	P.G.Cicellis
345 87450	✱	Lizard 1 ScTug *Davey* *1 Dk*	British Cardiff	3 [illegible]	70·5	15·0	9·1	Chepstow E.Finch&Co.	1882	4 Blkds.	2Cy.15"&25"-20" [illegible] HP. Fisher&Co.,Paisley	Thames Cons[illegible]
346 72136	✱D	Lizzie 1 ScSr *Long* 85-88 *1 Dk 2 B*	British Whitby	916 1421	252·9	33·7	19·0	Whitby T.Turnbull&Son	1876	5 Blkds. DBaft [illegible]	2Cy.31"&58"-36" 125 HP. Blair&Co.,Ld.,Stockton	J.Gray & Co.
347 88002	✱	Lizzie 1 ScSr *Harrold* 79-84 *1 Dk 2 B*	British Hartlepool	891 1381	245·1	33·3	15·9 *Q 84 f. B 54 f. F 3[illegible] f.*	W.Hartlepool [illegible]	1884	4 Blkds. [illegible]	3C.18"31"&50"-33" 130 HP. T.Richardson&Sons,Hrtlpl.	J.Coverdale & [illegible]
348 93762	✱	Lizzie Stl ScwSr *Davies* -88 *1 Dk 2 B*	British Liverpool	521 827	200·2	30·0	13·6 *Q 73 f. B 51 f. F 30 f.*	South Shields J.Readhead&Co.	1888	4 Blkds. [illegible]	3Cy.16"26"&43"-30" [illegible] HP. J.Readhead&Co.,S.Shields	Lewis & Co.
349 87608	✱	Lizzie 1 ScSr *Anderson* 73-83 *1 Dk*	British Glasgow	65 2[illegible]	122·0	20·1	10·0 *Q 46 f. B 21 f. F 23 f.*	Glasgow T.B.Seath&Co.	1883	5 Blkds. FPTB	2Cy.18½"&38"-21" 50 HP. J.Howden&Co.,Glasgow	Nobel's Explos[illegible] Ld.
350 46706		Lizzie W ScSr *1 Dk*	British Nassau,N.P.	58 108	89·2	20·0	7·9	Philadelphia	1854		1Cy. 60 HP.	William R.Sa[illegible]
351 40469		Lizzie W ScSr *Curtis* -88 *1 Dk*	British St.John'sNF	58 105	103·5	18·3	9·3	Pugwash,N.S. Webster	1866		2Cy.22"-22" [illegible] HP. Fawcett,Preston&Co.,Liverpool	R.S.Munn
352	✱	Lizzie W Sc *1 Dk*	British Leith	23 52	71·4	17·3	8·1	Granton Hawthorns&Co.	1885		2Cy.8"&16"-12" 10 HP. Hawthorns&Co.,Leith	J.Wells & Co.
353 81512	✱	Lizzie English 1 Sc3mSr *Lawson* -87 *1 Dk 2 B*	British W.Hartlepool	1108 1713	258·6	34·1	19·9 *Q 99 f. F 35 f.*	Whitby T.Turnbull&Son	1880	5 Blkds. DBaft [illegible]	2Cy.33"&62"-39" 150 HP. Blair&Co.,Ld.,Stockton	Pyman Bros.
354	✱	Lizzie Stewart 1 ScSr *O'Leary* 87-87 *1 Dk 2 B*	Norweg'n Bergen	628 976	215·4	30·7	15·2 *Q 80 f. B 42 f. F 22 f.*	Sunderland Strand Slipway Co.	1880	4 Blkds. DBaft [illegible]	2Cy.27"&47"-36" 99 HP. N.E.Marine E.Co.,Ld.,Sndrld	Bergh & Helland
355 76319	✱	Lizzie & Annie 1 ScK *Barley* 87-87 *1 Dk*	British Boston,Eng.	63 90	90·1	19·1	8·3	NorthShields Softley&Co.	1877	3 Blkds. FPTB	2Cy.12¼"-16" 25 HP. Pattison&Atkinson,Nwcstl	Boston & Lond[illegible] Co.,Ld.
356		Ljusne Jernverk C ScSr *Osterman* *1 Dk*	Swedish Ljusne	167 234	101·0	21·7	10·[illegible] *Q 33 f. B 21 f. F 15 f. H 41.*	Söderhamn	1873		25 HP.	Ljusne-Voxna[illegible] (W.G.Kempe)
357		Ljustero 1 Sc *Lindberg* *1 Dk*	Swedish Stockholm	101 139	100·4	16·1	8·8 *Q 31 f. B 33 f. B 5 f.*	Stockholm Bergsund's Mek.Verkst.	1876		3C.13¾"&21"-16"&30"30 HP. Bergsund's Mek.Verkst.	Ångf.Aktiebol. [illegible] (C.Pettersson)
358	✱	Llambi Campbell Stl ScSr *Taylor* -88 *1 Dk*	Arg'ntine Buenos Ayres	24 38	75·0	16·0	8·2	London Steward&Latham	1888	4 Blkds.	3Cy.10"15¼"&25"-18" 80 HP. Plenty&Son,Newbury	Llambi Campbell
359 51859	✱	Llandaff 1 Sc3mSr *Hodge* *1 Dk 2 B*	British Cardiff	258 420	152·6	24·3	14·5 *Q 38 f.*	Newcastle Schlesinger Davis&Co.	1865	4 Blkds. DB [illegible]	2Cy.32"-24" 65 HP. R.Morrison&Co.,Newcastle	J.T.Short
360 87806	✱	Llandaff City 1 ScBn *Gore* 81-85 *2 Dks 3 B*	British Bristol	1259 1838	270·2	36·0	25·5 *[illegible] f. B 48 f. F 35 f.*	Stockton Richardson,Duck&Co.	1882	5 Blkds. [illegible]	2Cy.36"&67"-36" 200 HP. T.Richardson&Sons,Hrtlpl.	C.Hill & Sons

✱ = Lloyd's Register; D = Nederlandsche Vereeniging; G = Germanischer Lloyd; I = Registro Italiano; R = American Re[illegible]
A = Veritas Austro-Ungarico; H = Veritas Ellenico; N = Norske Veritas; V = Bureau Veritas

In the 2nd Column, B.T., with figures following, signifies Date of report of bottom sighting by Board of Trade Surveyors as condition of PASSENGER CERTIF[illegible]

In the 3rd Column, C = Composite (iron frame, wood planked); I = Iron; Stl = Steel; W = Wood.

Further particulars of Vessels classed by Lloyd's Register will be found in Lloyd's Register Book and Lloyd's Yacht Register.

	Classification	Nom, Matériaux, Grément, etc. Capitaine et dates d'entrée au service des Armateurs actuels et à bord du navire. Ancien Nom, nombre de ponts, etc.		Pavillon. Port d'Armement	Tonnage. Net. Total	Dimensions. Longueur	Largeur	Creux	Construit. Où. Par qui.	Quand.	Nombre des cloisons. Détails sur le lest d'eau.	Machines. Détails. Nom du Constructeur	Armateurs ou Directeurs-Gérants.
	*	Llanelly *Walker 83-87*	I ScSr 1 *Dk*	British Liverpool	157	166·2 *Q f. F f.*	23·9	13·1	Preston W.Allsup&Sons	1875	1 Blds.	2Cy.23"& 40"–30" 80 HP. W.Allsup&Sons,Preston	S.Wales & Liv.Stm. Ship Co.,Ld.
	*	Llangorse *Hunt 88-88*	I ScSr 1 *Dk* 2 *B*	British Cardiff	1160 1117	260·6 *Q102,F&M,B116 F35*	36·3	19·2	Newcastle Schlesinger,Davis&Co.	1883	4 Blds. AFT	2Cy.33"& 63"–39" 180 HP. Black,Hawthorn&Co,Gtshd	C.E.Stallybrass
	*	Llanthewy *Hills*	I ScSr 1 *Dk*	British Newport	32	120·2 *Q 37 f. F 13 f.*	22·2	10·5	Newport C.&S.B.Co.,Ld.	1884	3 Blds. FPT	2Cy.16"& 30"–22" 50 HP. T'sideEng.Co.,Newport	"Llanthewy" S.S.Co., Ld.(Jones,Brooks&Co.)
	*	Llewellyn *Watson 86-86*	I ScSmSr 1 *Dk*	British Belfast	95	143·0 *P 38 f. F 18 f.*	20·1	9·5	Grimsby T.Charlton	1867	1 Blds.	2Cy.14"& 26"–21" 35 HP. M'Ilwaine,Lewis&Co.Belfast	W.Hinde
	*	Llewellyn	I ScSr 1 *Dk*	British Brisbane	23 100	112·0 *Q 16 f. F 28 f.*	19·6	9·2	Birkenhead A.Jack&Co.	1884	1 Blds.	2Cy.16"& 32"–21" 12 HP. J.Jack&Co.,Liverpool	Queensland Govt.
		Lloyd	I Pad	German Bremen	211 297	165·0	22·0	10·6	Bremen Act.Ges.Weser	1874		2Cy.38½"–42"	Norddeutscher Lloyd
		Lloyd	I ScSr	Russian Wiborg	222 286	144·0	20·3	13·0	Warkaus	1867			P.Wahl & Co.
	*	Lloyds *Thompson 83-85*	I ScSmSr 1 *Dk* 2 *L*	British London	564 883	220·7 *Q 46 f. F 34 f.*	28·0	17·7	Newcastle Palmers'Co.	1869	1 Blds.	2Cy.25"& 50"–33" 99 HP. Palmers'Co.,Newcastle	Harris & Dixon
		Loanda *Murray 87-87*	I ScSr 2 *Dks*	British Glasgow	899 1174	279·1	31·1	23·5	Glasgow J.Elder&Co.	1870	3 Blds.	2Cy.38"& 68"–33" 250 HP. J.Elder&Co.,Glasgow	British & African Stm. Nav.Co.
	*	Loch Ard *Smith 85-88*	Stl ScSr 2 *Dks*	British Dundee	838 1304	255·2 *B 40 f. H 31 f.*	32·2	19·2	Leith Ramage&Ferguson	1882	7 Blds.	2Cy.27"& 54"–48" 160 HP. Ramage&Ferguson,Leith	Dundee Loch Line S.S.Co.,Ld.
	*	Loch Etive *Stephen 82-88* 2 *Dks & W'ch Frames*	Stl Sc	British Dundee	1397 2138	294·0 *P27 f. B66 f. F39 f.*	37·3	23·8	Dundee GourlayBros.&Co.	1886	5 Blds. FPT APT	3C.21½"34"&56"–42" 220 HP. GourlayBros.&Co.,Dundee	Dundee Loch Line S.S. Co.,Ld.(Ireland,Leitch&Co.)
		Loch Etive *McKay*	I ScSr	British Glasgow	58 104	80·1 *Q 4 f.*	17·9	8·1	Glasgow W.Swan&Co.	1877		2Cy.12"& 26"–14" 20 HP. Cp. 82,Ross&Duncan,Glsgw	J.G.Stewart
	*	Loch Garry *Crawford 87-87*	I ScSr 1 *Dk* 2 *L*	British Dundee	616 956	210·3 *P 116 f. F 34 f.*	29·7	15·7	Dundee GourlayBros.	1880	5 Blds. FPT	2Cy.23"& 44"–42" 98 HP. GourlayBros.,&Co.,Dundee	Dundee Loch Line S.S.Co.,Ld.
	*	Loch Leven *Doig 86-86*	I ScSr 1 *Dk* 2 *L*	British Dundee	551 852	210·1 *Q 66 f. B 50 f. F 23 f.*	28·7	15·5	Dundee GourlayBros.	1878	5 Blds. FPT APT	2Cy.26"& 44"–30" 90 HP. GourlayBros.,&Co.,Dundee	Dundee Loch Line S.S.Co.,Ld.
	*	Loch Lomond *Lee 80-88*	Stl ScSr 2 *Dks* 3 *L*	British Dundee	1689 2574	312·0 *P33 f. B58 f. F39 f.*	40·2	24·0	Newcastle Palmers'Co.	1888	6 Blds. FPT	3Cy.23"37"&60"–42" 240 HP. Palmers'Co.,Ld.,Newcastle	Dundee Loch Line S.S.Co.,Ld.
	*	Loch Maree *Lee 88-88*	I ScSr 1 *Dk* 2 *L*	British Dundee	614 948	210·0 *P 116 f. F 34 f.*	29·8	15·7	Dundee GourlayBros.	1880	5 Blds. FPT	2Cy.23"& 44"–42" 98 HP. GourlayBros.,&Co.,Dundee	Dundee Loch Line S.S.Co.,Ld.
	*	Loch Nell *Crawford 83-83*	I ScSr 1 *Dk*	British Glasgow	58 120	89·5 *Q 34 f. F 12 f.*	17·1	8·4	Paisley H.McIntyre&Co.	1877	3 Blds. FPT APT	2Cy.13"& 23½"–18" 30 HP. Muir&Houston,Glasgow	J.G.Stewart
	*	Loch Ness *Thornton 82-88* 1 *Dk & W'ch Frms*	I ScSr	British Dundee	665 1017	225·8 *Q 76 f. B 42 f. F 23 f.*	31·7	14·1	Glasgow J.Elder&Co.	1882	5 Blds. APT	2Cy.25"& 50"–36" 98 HP. J.Elder&Co.,Glasgow	Dundee Loch Line S.S.Co.,Ld.
	*	Loch Rannoch *Young 85-88*	Stl ScSr 2 *Dks* 3 *B*	British Dundee	1030 1594	250·2 *B 16 f.*	34·2	23·0	Dundee GourlayBros.	1880	4 Blds.	2Cy.30"& 58"–48" 185 HP. GourlayBros.&Co.,Dundee	Dundee Loch Line S.S.Co.,Ld.
	*	Lochiel *Mitchell 88-88* 1 *Dk & W'ch Frms*	Stl ScSr	British Glasgow	1399 2162	279·0 *P25 f. Q88 f. B105 f. F31 f.*	37·7	19·1	P.Glasgow Murdoch&Murray	1888	4 Blds. APT	3Cy.22½"36"&58"–39" 111 HP. Ross&Duncan,Glasgw	J.Gardiner & Co.
351		Lochiel *McNab 83-83*	I ScSk	British Glasgow	72 212	139·8 *Q 46 f.*	18·2	9·2	Glasgow A.&J.Inglis	1877		2Cy.19"& 34"–24" 50 HP. Muir&Caldwell,Glasgow	D.MacBrayne
352		Lochness (ex Gough Foyle) *Munro 88-88*	I Pad *Sk*	British Glasgow	26 121	150·0	16·4	6·9	Glasgow J.Barr	1853		1Cy.45"–42" 80 HP. J.Barr,Glasgow	D.MacBrayne
353	DV	Locksley *Morris 85-85*	I ScK	British Newcastle	74 121	91·0 *Q 7 f.*	20·2	8·0	Newcastle Wallsend Slpwy.Co. Ld.	1884		2Cy.11"& 20"–16" 20 HP. G.Fantin&Sons,Newcastle	R.Mason
354	*	Locksley Hall *Collin 88*	Stl ScSr 2 *Dks* 3 *B*	British Liverpool	2496 3844	380·0 *P43 f. B130 f. F47 f.*	45·3	27·4	Newcastle Palmers'Co.,Ld.	1888	9 Blds. APT	3C29"47"& 76"–51" 550 HP. Palmers'Co.,Ld.,Newcastle	Sun Shipping Co.,Ld.
355	*D	Lockton *Brown 87-87*	I ScSr 1 *Dk* 2 *B*	British Scarboro'ugh	788 1219	240·0 *Q 90 f. B 48 f. F 26 f.*	34·1	14·1	Sunderland Kish,Boolds&Co.	1882	4 Blds. APT	2Cy.28"& 54"–36" 123 HP. W.Doxford&Sons,Sundrlnd	J.Knott
356	*	Lodore *McKay 88-88*	I ScSr 1 *Dk* 2 *B*	British Newcastle	408 656	200·2 *P 30 f. F 24 f.*	28·3	14·0	Newcastle Wghm.Richardson &Co.	1878	4 Blds.	2Cy.26"& 46"–30" 99 HP. Wigham Richardson & Co., Newcastle	Fisher, Renwick & Co.
357	*	Lofna *Anderson 82-87*	I ScSr 1 *Dk* 2 *B*	British Grangem'th	502 795	200·0 *Q 114 f. F 30 f.*	28·1	15·5	Newcastle WghmRichardson &Co.	1877	5 Blds.	2Cy.25"& 48"–30" 95 HP. J.Shaw&Co.,LowWalker	J.T.Salvesen & Co.
358		Lofoten	I ScSr 2 *Dks*	Norweg'n Bergen	537 711	177·7	25·4	20·1	Bergen Bergens Mek.Verkst.	1873		2Cy.34"& 29"–28" 140 HP. Bergens.Mek.Verks.Bergen	Det Bergenske Dampskibsselskab
359	N	Lofoten *Wik*	I ScSr 1 *Dk* 2 *B*	Norweg'n Bodö	257 350	141·2 *Q 34 f. B 46 f. F 40 f.*	23·6	10·9	Christiania AkersMek.Verkst.	1884	APT FPT	2Cy.18½"& 32"–24½" 50 HP. Akers Mek Verks Christiania	Vesteraalens Dampskibs-Selskab
360		Loire Inférieure *Pilliaud*	I ScSr 1 *Dk*	French Nantes	503 788	200·6 *Q 62 f. B 44 f. F 28 f.*	29·5	13·4	Dundee PearceBros.	1882	4 Blds. FPT	2Cy.23"& 45"–30" 90 HP. PearceBros.,Dundee	Cie. Nantaise de Nav. à Vap.

*= Lloyd's Register; D = Nederlandsche Vereeniging; G = Germanischer Lloyd; I = Registro Italiano; R = American Record;
A = Veritas Austro-Ungarico; H = Veritas Ellenico; N = Norske Veritas; V = Bureau Veritas.

Dans la 2me Colonne, B.T., avec les indications qui suivent, signifie la date du rapport de visite de la Carène par les experts du Board of Trade pour obtenir le certificat relatif au TRANSPORT DES PASSAGERS.

Dans la 3me Colonne, C = Composite (membrure en fer, bordé en bois); I = Fer; Stl = Acier; W = Bois.

Pour plus amples renseignements sur les navires classés par le Lloyd's Register, voir le Lloyd's Register Book et le Lloyd's Yacht Register.

No. in Book / Official No.	Registry in which classed	Ship's Name, Material, Rig, &c. / Master, and Dates of Appointment to present Owner's Service and to Ship / Ship's Late Name, No. of Decks, &c.	Flag / Port of Registry	Tons Net / Gross	Length	Breadth	Depth	Built Where / By Whom	When	No. of Bulkheads / Particulars of Water Ballast	Engines / Particulars, Maker's Name	Owners
391		Loire & Bretagne I Sc 1 Dk	French Nantes	74 130	116·8	17·1	8·8	Greenock	1853	3 Blhds.	2Cy.14"&22"–18" 20 HP Cp.78 Jollet & Babin, Nantes	L. Flornoy &
392		Lola W	Spanish	362 339	215·2	24·4	13·3	United States	1877		80 HP	Alegret & Co.
393		Lolland I ScSr *Hansen 83-83* 1 Dk	Danish Nakskov	283 192	172·3	26·1	13·0	Paisley McIntyre&Co.	1882	4 Blhds.	2Cy.21"&48"–30" 90 HP Hutson&Corbett, Glasgow	Det Forenede ... Selskab
394	✱D	Lombard I ScSr *Colpitts 87-87* 2 Dks 3 B	British London	1132 1750	270·8	34·0	24·4	Newcastle Tyne I.S.B.Co.	1877	4 Blhds.	2Cy.32"&60"–36" 160 HP R.&W.Hawthorn, Newcastle	G. Cockburn
395	I	Lombardia I ScSr *Traverso 85-87* 2 Dks	Italian Genoa	217 416	192·2	24·9	13·4	Newcastle Palmer&Co.	1862	4 Blhds.	2Cy.21"&43"–30" 140 HP G.Ansaldo&Co., Sampierdar	Navigazione Italiana
396		Lombardy I ScBk *Preston 87-88*	British Glasgow	1571 3726	351·5	38·3	27·7	Dumbarton W.Denny&Bros.	1873		2Cy.49"&86"–54" 500 HP Denny&Co., Dumbarton	Peninsular & O. Steam Nav. Co.
397	✱ BT8.88	London I ScBk *Howes 84-84* 3 Dks	British London	996 1579	276·5	30·6	25·1	Newcastle A.Leslie&Co.	1862	4 Blhds.	2Cy.30"&55"–36" 120 HP Blair&Co.,Ld., Stockton	J. Hall, jun. &
398	✱ BT2.88	London I ScSr *M'Intosh* 2 Dks	British Dundee	424 725	213·9	28·2	15·9	Dundee Gourlay Bros.	1856		2Cy.34"&56"–30" 140 HP Gourlay Bros.&Co., Dundee	Dundee, Perth Shipping Co.
399	BT3.89	Londonderry (ex Camel) I PadSr *Donnan 67-82*	British Barrow	294 684	243·2	26·5	13·5	Greenock Caird&Co.	1866		1Cy.45"&(2)52½"–60" 29	J. Little & Co.
400		Londonderry I ScSr *Craig 74-74* 1 Dk 2 B	British Sunderland	285 438	155·0	26·0	14·8	W.Hartlepool Pile&Spence	1857	4 Blhds.	2Cy.29½"–36" 60 HP T.Richardson&Sons, Hrtlpl	Marquis of Londonderry
401	✱	Londos I Sc3mSr *Nightingale* 1 Dk	British London	209 286	159·2	22·7	9·8	Stockton Richardson,Duck&Co.	1860	5 Blhds.	1Cy.11"&26"–22" 45 HP Comp.71, M.Samuelson, Hull	R.W. Leyland
402	✱	Long Ditton I ScSr *Grant* 1 Dk 2 B	British London	617 813	209·4	29·0	16·7	Sunderland Oswald&Co.	1870	4 Blhds.	2Cy.21"&47"–30" 90 HP N.E.Marine Eng.Co.,Ld., Sld.	Commercial S. Ld.
403		Longavi I TwnScSr *Russell 86-87* 1 Dk	Chilian Valparaiso	257 420	150·2	30·2	9·6	Glasgow R.Napier&Sons	1884		1Cy.15"&30"–21" 75 HP R.Napier&Sons, Glasgow	Cia. Sud. Amer. Vapores
404	BT11.88	Longford I PadSr	British Dublin	476 817	249·5	27·1	14·7	Birkenhead Laird Bros.	1870		Cy.61"–72" 300 HP Laird Bros., Birkenhead	W. Watson
405	✱	Longhirst I ScSr *Anderson 74-88* 2 Dks 3 B	British Newcastle	1368 2093	285·0	36·2	21·4	Newcastle Wigham Richardson & Co.	1879	5 Blhds.	2Cy.31½"&64"–42" 220 HP R.&W.Hawthorn, Newcastle	Elliott, Lowrey & Dunford
406	✱	Longnewton I ScSr *Lawson 84-84* 1 Dk 2 B	British Sunderland	1206 1849	275·7	38·0	16·9	W.Hartlepool E.Withy&Co.	1884	4 Blhds.	3Cy.20½"34"&55"–33" 170 HP T.Richardson&Sons, Hrtlpl	Marquis of Londonderry
407	✱	Longstone Stl ScSr *Thompson 87-88* 1 Dk 2 B	British London	1515 2398	295·3	38·1	20·3	Middlesbro' R.Dixon&Co.	1888	5 Blhds.	3Cy.22"35"&59"–39" 220 HP T.Richardson&Sons, Hrtlpl	Farrar, Groves
408	✱	Longueil I ScSr *Mutter* 1 Dk 2 B	British Cardiff	1116 1711	260·0	35·3	18·0	Sunderland W.Pickersgill&Sons	1882	4 Blhds.	2Cy.32"&60"–39" 140 HP Sunderland	"Longueil" S.S. Co. (Morel Bros. & Co.)
409		Lonsdale Stl Pad3mSr *Forster* 1 Dk	British Melbourne	228 364	228·0	25·1	10·0	Greenock R.Steele&Co.	1882	4 Blhds.	2Cy.34"&62"–72" 240 HP Greenock	Port Philip S. Co., Ld.
410		Lontue I ScSr *Graves* 2 Dks & Shd Dk	British Liverpool	1121 1645	299·0	40·0	19·6	Port Glasgow J.Reid&Co.	1873	3 Blhds.	2Cy.48"&86"–42" 320 HP Pt.Glsg	Pacific Steam
411	✱	Looch Stl ScSr *Lefter* 1 Dk 2 B	Russian Odessa	913 1416	255·5	34·2	17·9	Newcastle R.&W.Hawthorn, Leslie&Co.	1886	12 Blhds.	3Cy.20"32"&53"–36" 160 HP R.&W.Hwthrn.Leslie&C.Nc	Russian Steam Trading Co.
412	✱	Loodiana Stl Sc3mSr *Britten 76-88* 3 Dks	British Glasgow	2140 3269	340·1	42·2	25·6	Dumbarton W.Denny&Bros.	1884	6 Blhds.	4Cy.26½"11"&(2)51"–51" 370 HP Denny&C., Dmbrtn	British India S. Nav. Co. Ld.
413		Lope de Vega I Sc3mSr *Soto 87-87* 2 Dks	Spanish Barcelona	548 727	238·0	27·0	18·6	Greenock Scott&Co.	1855		2Cy.23"&16"–24" 80 HP J.Taylor&Co., Birkenhead	J. Roca & Co.
414	BT4.88	Lord Aberdour I PadSk	British Leith	54 130	142·0	20·1	6·9	Glasgow Aitken&Mansel	1866		2Cy.30"–42" 60 HP J.Aitken&Co., Glasgow	Galloway Saloon Packet Co.
415	✱	Lord Alfred Paget I Sc3mSr *Whitehead 75-78* 1 Dk 2 B	British London	621 982	243·8	28·2	17·3	Newcastle Palmers'Co.	1870	4 Blhds.	2Cy.27"&52½"–33" 120 HP Palmers'Co., Newcastle	S. Clarke & Co.
416		Lord Alfred Paget I ScSr	British	133 188	120·4	21·0	10·6	Stockton M.Pearse&Co.	1865		2Cy.28½"–24" 50 HP Fossick&Hackworth, Stktn	Hewett & Co.
417	✱	Lord Chancellor I ScSr *Leighton -88* 1 Dk	British Hull	57 135	95·3	20·8	11·0	Hull Cook, Welton & Gemmell	1888	3 Blhds.	2Cy.17"&33"–22" HP C.D.Holmes&Co., Hull	Humber Steam Towing Co. Ld.
418	✱	Lord Charlemont Stl Sc3mSr *Dunn 83-88* (ex Ormiston) 2 Dks	British Belfast	2055 3138	345·6	40·9	26·7	Belfast Harland&Wolff	1886	6 Blhds.	3Cy.24½"37"&64"–48" 320 HP Harland&Wolff, Belfast	Irish Shipowners Ld.
419	BT4.88	Lord Clive I Sc4mSr *Urquhart 72-72*	British Liverpool	2206 3386	381·0	40·1	30·0	Liverpool R.&J.Evans&Co.	1871		2Cy.32"&68"–42" 150 HP G.Forrester&Co., Liverpool	American Line
420		Lord Clive I Pad 1 Dk	British Chester	38 120	112·0	18·2	8·0	Liverpool Bowdler, Chaffer & Co.	1875	1 Blhds.	2Cy.34"–42" 80 HP Fawcett,Preston&Co., Lvrpl	Shropshire Union Railway & Canal

✱ = Lloyd's Register; D = Nederlandsche Vereeniging; G = Germanischer Lloyd; I = Registro Italiano; R = American Record;
A = Veritas Austro-Ungarico; H = Veritas Ellenico; N = Norske Veritas; V = Bureau Veritas.

In the 2nd Column, B.T., with figures following, signifies Date of report of bottom sighting by Board of Trade Surveyors as condition of PASSENGER certificate.

In the 3rd Column, C = Composite (iron frame, wood planked); I = Iron; Stl = Steel; W = Wood.

Further particulars of Vessels classed by Lloyd's Register will be found in Lloyd's Register Book and Lloyd's Yacht Register.

No.	Cote	Nom, Matériaux, Gréement, etc. — Capitaine et dates d'entrée au service des Armateurs actuels et à bord du navire. — Ancien Nom, nombre de ponts, etc.		Pavillon. — Port d'Armement	Tonnage. Net. — Total	Dimensions. Longueur	Largeur	Creux	Construit. Où. — Par qui.	Quand	Nombre des cloisons. — Détails sur le lest d'eau.	Machines. — Détails. Nom du Constructeur.	Armateurs ou Directeurs-Gérants.
		Lord Clyde	I PadSr	British Dublin	320 705	243·3	26·3	14·8	Greenock Caird&Co.	1863		2Cy.63"–78" 350 HP. Caird&Co.,Greenock	Dublin & Glasgw.Sailing & Stm.Pkt.Co.
		Lord Clyde	I PadSp	British Scarboro'	3 115	103·8	18·5	9·8	Cork G.Robinson&Co.	1864		1Cy.38"–51" 50 HP. T.Hepple&Co.,Newcastle	J.Sellers
	✻	Lord Collingwood *Davis*	I ScSr *2 Dks3 B*	British Liverpool	1133 1768	270·0	34·0	24·4	Newcastle TyneI.S.B.Co.	1879	4 Blhds. DBaft 236t	2Cy.32"&62"–40" 160 HP. R.&W.Hawthorn,Newcstle.	J.Tully
		Lord Delamere	W ScFlat	British Liverpool	89 131	86·2	20·6	9·2	Winsford AnnDeakin	1871		2Cy.12"–14" 24 HP. G.Deakin,Winsford	G.Deakin,Ld.
	✻	Lord Derby *Clayton 88-88*	I ScSr *2 Dks3 B*	British Liverpool	1132 1760	270·0	34·0	24·5	Newcastle TyneI.S.B.Co.	1877	4 Blhds. DBaf&cb	2Cy.32"&62"–36" 160 HP. R.&W.Hawthorn,Newcstle.	Hoy Head Steam Ship Co.,Ld.
						Q 42 f.							
		Lord Elgin *Moore 86-88*	I PadSr	British Poole	92 203	160·0	20·0	6·8	Stockton Richardson,Duck & Co.	1876		2Cy.22"&42"–42" 75 HP. T.Richardson&Sons,Hrtlpl	Bournem'thSwanage & Poole Stm.Pkt.Co.,Ld.
		Lord Elgin *Carmichael -82*	I PadSr	British Ardrossan	17 107	96·0	18·1	9·8	Cork G.Robinson&Co.	1867		2Cy.36"–51" 50 HP. G.Robinson&Co.,Cork	Ardossan HarbourCo.
	✻	Lord Eslington *Foster 84-84*	I ScSr *2 Dks3 B*	British Newcastle	1107 1704	264·6	34·6	24·3	Newcastle ColeBros.	1876	4 Blhds. DB116ft	2Cy.32"&62"–36" 200 HP. R.&W.Hawthorn,Newcstle.	J.Tully
		Lord Gough *Hughes 79-79*	I Sc1mBk	British Liverpool	2370 3655	382·8	40·2	30·0	Birkenhead LairdBros.	1879		2Cy.32"&64"–48" 450 HP. LairdBros.,Birkenhead	American Line
						B 295 f.							
		Lord Gough	I PadSr	British Dublin	301 692	243·5	26·2	14·8	Greenock Caird&Co.	1863		2Cy.63"–78" 350 HP. Caird&Co.,Greenock	Dublin&Glasgow Sailing & Stm.Pkt.Co.
431	✻	Lord Lansdowne *Bagill 88-88*	I ScSr *2 Dks3 B*	British Belfast	1816 2753	340·5	38·3	26·0	Belfast Harland&Wolff,Ld.	1884	6 Blhds. CDB155t APT30t	2Cy.37"&72"–45" 350 HP. Harland&Wolff,Belfast	Irish Shipowners'Co., Ld.
						B 64 f.	*F 52 f.*						
432	✻	Lord Londonderry *Arthur 88-88*	Stl ScSr *2 Dks*	British Belfast	1566 2409	312·5	39·2	24·7	Belfast Harland&Wolff,Ld.	1889	5 Blhds. CDB.22t	3C.23",37"&60"–42" 240HP. J.Howden&Co.,Glasgow	Irish Shipowners'Co., Ld.
						F 34 f.	*B 68 f.*	*P 24 f.*					
433		Lord Morton	I PadSp	British Leith	9 186	169·9	20·3	7·1	Leith S.&H.Morton	1883		1Cy.45"–54" 85 HP. S.&H.Morton,Leith	Galloway Saloon Stm. Packet Co.
434	✻	Lord O'Neill *Ferris 88-88*	I ScSr *2 Dks3 B*	British Belfast	1816 2753	340·5	38·3	26·0	Belfast Harland&Wolff,Ld.	1884	6 Blhds. CDB155t APT30t	2Cy.37"&72"–45" 350 HP. Harland&Wolff,Belfast	Irish Shipowners'Co., Ld.
						B 64 f.	*F 52 f.*						
435	✻	Lord Rosebery *Portman 84-88*	Stl ScSr *1 Dk2 B*	British London	808 1265	231·3	33·0	17·0	PortGlasgow Russell&Co.	1888	4 Blhds DBaf218t FPT10t	3C17"28"&45"–36" 103 HP. Kincaid&Co.,Greenock	J.&A.Wyllie
						P 135 f.	*F 31 f.*						
436		Lord Warden *Pike 84-*	I PadSp	British London	23 116	100·4	18·3	10·1	South Shields Readhead&Co.	1868		2Cy.27"–48" 55 HP. Readhead&Co.,S.Shields	F.Pattison
437	✻	Lord Warwick *Pratt 84-86*	I Sc3mSr *2 Dks3 B*	British N.Shields	1719 2655	300·0	39·0	25·5	Newcastle TyneI.S.B.Co.	1883	5 Blhds. DBaf&f 335t APT	2Cy.44"&76"–48" 270 HP. J.Dickinson,Sunderland	"Lord Warwick" S.S. Co.,Ld.(Stephens&Mawson)
						P 28 f.	*B 66 f.*	*F 28 f.*					
438	✻D	Lord of the Isles *Edgate*	I ScBk *3 Dks*	British London	1586 2484	320·4	37·8	29·3	Glasgow R.Napier&Sons	1870	6 Blhds.	2Cy.44"&76"–48" 350 HP. C.D.Holmes&Co.,Hull	Shaw, Bushby & Co.
439		Lord of the Isles	I PadSk	British Glasgow	106 384	246·0	24·2	9·0	Glasgow D.&W.Henderson	1877		2Cy.46"–66" 326 HP. D.&W.Henderson&Co.,Glsw	M.T.Clark
440		Lord of the Isles	I PadSk	British Southampton	69 126	145·0	18·1	6·6	London ThamesIronWorks	1861		2Cy.32"–36" 60 HP. J.Stewart,London	Sou.I.of W.&S.of Eng. Ryl.MailStm.Pkt.Co.,Ld.
						Q 8 f.							
441	R	Lorenzo D. Baker *Hopkins -88*	W Sc3mSr *2 Dks*	American Boston	593 966	190·0	35·0	20·0	Bath, Me. Goss&Sawyer	1884		2Cy.20"&40"–36" 120 HP. C.H.Delamater&Co.,N.York	Standard Stm.Nav.Co.
442		Lorna Doone	I ScSr	Sarawak Sarawak	55 118	103·3	17·1	8·1	PortGlasgow Blackwood&Gordn	1881		2Cy.17"&31"–21" 40 HP. Blackwood&Gordon,P.Glsw	Rajah Brooke
						Q 13 f.							
443	✻	Lorne *Reed 66-76*	I ScSr *1 Dk*	British Hull	733 1143	241·7	30·9	15·8	Hull Gilbert&Cooper	1873	5 Blhds. APT30t MT12t	2Cy.27"&53"–36" 98 HP. Gilbert&Cooper,Hull	W.S.Bailey
						P 117 f.	*F 26 f.*						
444		Lorraine *Jauffret 66-88*	I Sc1m *2 Dks*	French Marseilles	683 1026	238·5	28·9	22·6	LaSeyne Forges&C. de la Medit	1866		2Cy.33"–24½" 100 HP. Forges&C.de la Medit,Mars.	Société Générale de Transports Maritimes
445		Los Angeles *Hannah (ex Wyandot)*	W ScSr *2 Dks*	American S.Francisco	287 493	170·0	27·0	11·1	Baltimore	18^{66}_{74}		2Cy.26"&43½"–28" 68 HP. RisdonIronWorks,S.Frnsco	Pacific Coast S.S.Co.
446		Lothair	W Sc *1 Dk*	British Pt.Hope,Ont	281 413	129·8	25·0	11·7	St.Caths.,Ont M.Lipson	1872		1Cy.36"–30" 87 HP. Yale&Co.,St.Catherins,Ont.	James M.Miller
						P 42 f.							
447	V	Lotte *Oesterreich 84-84*	I ScSr *1 Dk*	German Danzig	344 444	153·8	23·5	14·1	Danzig F.Devrient&Co.	1882		2Cy.14½"&26¾"–13¾" 50HP. F.Schichau,Elbing	F.G.Reinhold
448	✻	Lotus *Balling 88-88*	I ScSr *2 Dks*	British Liverpool	363 584	207·8	23·2	15·7	Hull C.&W.Earle	1863	4 Blhds.	1Cy.18"&34"–30" 70 HP. T.Brassey&Co.,Liverpool	Walsh Bros. & Co.
449	V	Lou Cettori *Blanchet -86*	I ScSr *2 Dks&Spr Dk*	French Marseilles	687 1173	257·8	30·3	15·5	Greenock Scott&Co.	1873		2Cy.33"&66"–36" 250 HP. GreenockFoundry Co.,Gnk.	Compagnie Générale Transatlantique
						B 80 f.		22·6					
450	✻	Lough Fisher *Bannister 73-87*	Stl ScSr *1 Dk*	British Barrow	160 406	168·2	25·1	11·6	Belfast M'Ilwaine,Lewis&Co.,Ld.	1887	4 Blhds. DBaft FPT35t APT22t	2Cy.20"&38"–33" 61 HP. M'Ilwaine,Lewis&Co.,Ld.,Blfst	Lough Fisher S.S. Co. Ld. (J.Fisher&Sons)
						Q 95 f.	*F 28 f.*	*B 43 f.*					

✻ = Lloyd's Register; D = Nederlandsche Vereeniging; G = Germanischer Lloyd; I = Registro Italiano; R = American Record;
A = Veritas Austro-Ungarico; H = Veritas Ellenico; N = Norske Veritas; V = Bureau Veritas.

Dans la 2me Colonne, B.T., avec les indications qui suivent, signifie la date du rapport de visite de la Carène par les experts du Board of Trade pour obtenir le certificat relatif au TRANSPORT DES PASSAGERS.

Dans la 3me Colonne, C = Composite (membrure en fer, bordé en bois); I = Fer; Stl = Acier; W = Bois.

Pour plus amples renseignements sur les navires classés par le Lloyd's Register, voir le Lloyd's Register Book et le Lloyd's Yacht Register.

No. in Book / Official No.	Registry in which classed	Ship's Name. Material, Rig, &c. / Master, and Dates of Appointment to present Owner's Service and to Ship / Ship's Late Name. No. of Decks, &c.	Flag / Port of Registry	Tons Net / Gross	Dimensions (Length, Breadth, Depth)	Built Where / By Whom	When	No. of Bulkheads / Particulars of Water Ballast	Engines / Particulars, Maker's Name	Owners
451	✻	Loughbrow I ScSr *Buzzard 83-85* 1 *Dk*	British Newcastle	488	190·0|30·5|13·7 *Q 50 f. B 18 f.*	Blyth Hodgson&Soulsby	1883	4 Blkds. [illegible]	2Cy.24"&46"-33" 95 HP. W.Richardson&Co.,Newcstl.	Fisher, Renwick
452	✻	Louis I ScSr *Mangues 87-87* 1 *Dk & Wb Frames*	French Nantes	864 1271	230·0|34·1|16·3 *Q 82 f. B 54 f. F 27 f.*	Sunderland Osbourne, Graham & Co.	1884	4 Blkds. [illegible]	2Cy.28"&55"-36" 110 HP. W.&F.Wilson&Co.Sundrld.	Louis Flornoy
453 140754	R	Louis Bucki W 3mSc *Mount*	American New York	335 403	150·0|33·0|12·0	Mid. Haddam Woodhouse (Conn.)	1881			C.L. Bucki
454	N	Louisa I Sc3mSr *Reimers* 1 *Dk*	Norweg'n Bergen	590 767	201·2|26·5|15·5 *Q 55 f. F 22 f.*	London Samuda Bros.	1870	4 Blkds. [illegible]	2Cy.26"&45"-26" 90 HP. Ravenhill,Hodgson&Co.,Lond.	T.M. Madland (J. Jensen & Co.)
455 77941		Louisa III. (TO BE BROKEN UP) W ScSr 1 *Dk*	British Singapore	112 163	110·0|18·0|10·3 *H 21 f.*	Singapore Ah Chong	1879		2Cy.20"-20" 45 HP. Hall,Russell&Co.,Aberdeen	Tan Hoon Ch[illegible]
456 84995	✻ D	Louise I ScBn *Shotton 87-87* 1 *Dk 2 H*	British London	1007 1547	260·4|35·2|17·4 *Q 97 f. B 54 f. F 32 f.*	Middlesbro' R.Dixon&Co.	1882	4 Blkds. [illegible]	2Cy.33"&62"-39" 180 HP. Blair&Co.,Ld.,Stockton	Burdick & Cook
457 76395	✻	Louise I ScSr *Husband* 1 *Dk*	British Douglas,I.M.	67 200	119·6|21·1|10·0 *Q 50 f. F 19 f.*	Paisley Abercorn S.B.Co.	1882	4 Blkds. [illegible]	2Cy.20"&38"-24" 60 HP. W.King&Co.,Glasgow	R.A. Cain
458 82850		Louise I Pad	British London	39 139	100·0|20·6|10·7	London Samuda Bros.	1880		2Cy.32½"-54" 80 HP. J.Stewart&Son,London	London & St. K[illegible] rine Docks Co.
459		Louise I Sc3mSr *Brabrand 67-85* 1 *Dk 2 H*	Danish Copenhagen	703 1115	230·0|30·0|16·5 *P 157 f. F 32 f.*	Copenhagen Burmeister&Wain	1872	6 Blkds. [illegible]	2Cy.21"&41"-30" 130 HP. Burmeister&Wain,Cpnhgn.	Det Forenede D[illegible] Selskab (C.P.A.)
460	✻	Louise (ex Amy) I ScSr *Conte Raineri* 2 *Dks*	Italian Palermo	266 422	187·5|27·4|14·6 *Len. 81.*	Port Glasgow Cunliffe&Dunlop	1880	4 Blkds.	2Cy.26"&46"-27" 95 HP. Cunliffe&Dunlop,Port Glasgow	Conte Lanza [illegible] zarino
461	✻ G	Louise I ScSr *Burmeister -74* 1 *Dk & Awn Dk*	Russian S.Petersbg.	455 744	191·1|25·8|13·6 *204*	Renfrew Hudson, Coulborn & Co.	1871	5 Blkds.	2Cy.24½"&51"-33" 90 HP. Hudson,Coulborn&Co.,Rfw.	L. Knoop
462 82794	BT 1.89	Louise Dagmar Stl Pad *Davies 80-85* 1 *Dk*	British London	335 781	250·0|29·2|11·3 *B 51 f.*	London Samuda Bros.	1880	6 Blkds.	2Cy.72"-66" 350 HP. J.Penn&Sons,Greenwich	South Eastern R[illegible]
463 80532	✻ D	Louise H. I ScSr *Berg* 2 *Dks 3 H*	British Newcastle	1090 1690	255·0|34·2|23·3 *P 26 f. F 27 f.*	Newcastle Palmers Co.,Ld.	1879	4 Blkds. [illegible]	2Cy.38"&66"-39" 200 HP. Palmers Co.,Ld.,Newcastle	M. Jaques De [illegible]
464	V	Louise Jenny I ScSr *Vallin* 2 *Dks*	French Havre	671 987	230·4|31·1|16·5 *P 33 f. F 29 f.*	Renfrew Lobnitz,Coulborn & Co.	1877	6 Blkds. [illegible]	2Cy.32"&60"-42" 170 HP. Lobnitz,Coulborn&Co.,Rfw.	Worms, Josse & C[illegible]
465		Louise-Marie I Pad	Belgian	419	200·0|24·8|7·3	Antwerp Soc. Cockerill	1867			Belgian Govern[illegible]
466 140390	R	Louisiana I ScSr *Gager 54-79* 2 *Dks 3 H*	American New York	1806 2840	348·0|39·0|28·5	Chester, Pa. J.Roach&Son	1879		1Cy.50"&90"-66" John Roach&Son, Chestr.Pa.	Cromwell S.S. [illegible]
467 92925	✻	Lowlands Stl ScSr *Holman -88* 1 *Dk & Wb Frms*	British W.Hartlepl.	1165 1789	260·5|36·6|18·2 *B 104 f. F 25 f. Q 107 f.*	W. Hartlepool W.Gray&Co.	1888	4 Blkds. [illegible]	3Cy.19"30½"&51"-36" 170 HP. Central Marine Eng.Co. W.Hartlepool	Hardy, Wilson & [illegible]
468 78613	BT 1.89 ✻	Lualaba I ScSr *Bales -88* 2 *Dks*	British Glasgow	1170 1850	299·8|34·5|23·7 *P 101 f. F 56 f. B 36 f.*	Port Glasgow Cunliffe&Dunlop	1878	6 Blkds.	2Cy.40"&71"-36" 200 HP. Cunliffe&Dunlop,Pt.Glasgw.	British & Afri[illegible] Navigation Co.
469	G	Lubeca I ScSr *Lamer* 1 *Dk*	German Lübeck	199 268	118·7|20·1|11·8	Lübeck H.Koch	1883		2Cy.16"&27½"-17¾" 100 HP. Henninger&Co.,Berlin	Küsten-Dampfs[illegible] fahrts-Gesellsch.[illegible]
470	G	Lubeck Stl ScBg *Schmähler 84-86* 2 *Dks*	German Bremen	1078 1815	260·0|35·5|24·5 *P 56 f. B 95 f. F 30 f.*	Stettin Act.Ges.Vulcan	1886	[illegible]	3Cy.21"33"&53"-49" Act.Ges.Vulcan	Norddeutscher [illegible]
471	G	Lübeck I Sc3mSr *Storm 79-85* 1 *Dk*	German Lübeck	231 305	155·3|20·5|10·6	W. Hartlepool Denton,Gray&Co.	1866		2Cy.19"&38"-20" 60 HP. Richardson,Duck&Co.,Stktn.	Gaedertz,sen.,&[illegible]
472 78302		Lubra (ex Frank Bateots) I ScSr *Robertson* 1 *Dk*	British Sydney	279 467	173·6|27·5|11·4	Hayle, Eng. Harvey&Co.	1874	4 Blkds. [illegible]	2Cy.18"&38"-21" 45 HP. Harvey&Co.,Hayle,Eng.	Hunter River S[illegible] Navigation Co.
473 29368		Lubra I ScSr *Ashton* 2 *Dks*	British Adelaide	223 321	167·3|22·2|10·4 *P 34 f.*	Port Glasgow Laurence Hill & Co.	1860	5 Blkds.	2C.20½"&37½"-24" 60 HP. D.&W.Henderson&Co.,Glsg.	Adelaide S.S. Co.
474	✻	Lucano (ex Ariadne) I Sc *Caputi 86-87* 1 *Dk & Awn Dk*	Italian Bari	590 787	172·0|28·2|12·2	Dundee Pearce Bros.	1879	4 Blkds. [illegible]	2Cy.30"&55"-36" 120 HP. Pearce Bros.,Dundee	Soc. Nav. Pugli[illegible]
475 81453	✻	Lucent I ScSr *Westcott 79-84* 1 *Dk 2 H*	British Sunderland	955 1478	250·5|34·0|18·2 *Q 80 f. P 29 f. B 60 f. F 29 f.*	Sunderland Bartram,Haswell & Co.	1879	3 Blkds. [illegible]	2Cy.32"&59"-33" 140 HP. T.Richardson&Sons,Hartpl.	J. Westoll
476 78612		Lucerne I ScSr	British Glasgow	1252 1925	291·0|34·3|24·6 *F 42 f.*	Birkenhead Laird Bros.	1878	6 Blkds.	2Cy.35"&60"-42" 220 HP. Laird Bros.,Birkenhead	J. & A. Allan
477	✻	Lucero I ScSr *Urizar 83-83* 1 *Dk 2 H*	Spanish Bilbao	1010 1571	252·0|35·1|17·8 *[illegible] B 58 f. F 33 f.*	Sunderland Osbourne,Graham & Co.	1882	4 Blkds. [illegible]	2Cy.32"&60"-39" 140 HP. J.W.&F.Wilson,Sunderland	Soc. Maritima de Vizcaya
478 65675	✻	Luchana (ex Agnes & Louisa) I Sc3mSr *Taylor 88-88* 2 *Dks*	British W.Hartlepl.	436 703	184·1|28·3|15·6 *Q 92 f.*	Hartlepool Withy,Alexander&Co.	1871	4 Blkds. [illegible]	2Cy.25"&48"-30" 90 HP. [illegible]	Agnes & Louisa S[illegible] Co.,Ld.(Walker,Do[illegible])
479		Luchana I ScSr *Schover 73-84* 1 *Dk 2 H*	Spanish Seville	325 427	173·0|25·0|14·2	Renfrew W.Simons&Co.	1862		2Cy.33"-27" 80 HP. W.Simons&Co.,Renfrew	Ybarra & Co.
480		Lucien I Sc3mSr *Gruénais* 2 *Dks*	French Havre	320 483	165·0|27·6|16·1 *P 21 f. B 15 f. F 13 f.*	Hull M.Samuelson	1858		2Cy.22"&[illegible] 70 HP. Chantiers&A.de l'Océan,Havr.	Worms, Josse & C[illegible]

✻=**Lloyd's Register;** D=Nederlandsche Vereeniging; G=Germanischer Lloyd; I=Registro Italiano; R=American R[illegible]
A=Veritas Austro-Ungarico; H=Veritas Ellenico; N=Norske Veritas; V=Bureau Veritas

In the 2nd Column, B.T., with figures following, signifies Date of report of bottom sighting by Board of Trade Surveyors as condition of PASSENGER CERTIFICATE

In the 3rd Column, C=Composite (iron frame, wood planked); I=Iron; Stl=Steel; W=Wood.

Further particulars of Vessels classed by Lloyd's Register will be found in Lloyd's Register Book and Lloyd's Yacht Register.

No.	Registre	Nom, Matériaux, Gréement, etc. Charpente et dates d'entrée au service des Armateurs actuels et à bord du navire. Ancien Nom, nombre de ponts, etc.		Pavillon. Port d'Armement	Tonnage Net / Total	Dimensions Longueur	Largeur	Creux	Construit Où. Par qui	Quand	Nombre des cloisons. Détails sur le lest d'eau	Machines. Détails. Nom du Constructeur.	Armateurs ou Directeurs-Gérants.
		Lucifer *Cipollovich*	I [illegible]	Austrian Trieste	363 605	190·0	28·0	15·0	Glasgow R. Napier&Sons	1857		2Cy.50"–54" 200 HP. R.Napier&Sons,Glasgow	Lloyd Austro-Ungarico
	✻	Lucinda (ex-[illegible]) *Faulkner*	I ScBk 1 Dk2 B	British W.Hartlepool	1132 1750	257·6	34·6	19·7	W. Hartlepool W.Gray&Co.	1880	4 Blhds.	2Cy.32"&58½"–39" 140 HP. Blair&Co.,Ld.,Stockton	Herskind & Woods
		Lucinda *Williams*	I PadSr 2 Dks	British Brisbane	148 301	172·2	25·1	9·2	Dumbarton W.Denny&Bros.	1884		2Cy.25"& 44"–18" 115 HP. Denny&Co.,Dumbarton	Queensland Govmnt.
		Lucretia (WHALER) *Paul*	W ScBk 2 Dks	American N.Bedford	313 350	115·1	27·4	16·0	Bath, Me. [illegible] Sawyer	1877		1Cy.20"–22" C.H.Delamater&Co.,N.Yk.	W.Lewis
	✻	Lucy *Clifford*	I ScDredger	Spanish Cape Town	290 420	145·0	32·1	11·3	Kinderdyk J.&K.Smit	1885	5 Blhds.	2Cy.20"&40"–22" 40 HP. Diepeveen Lels&Smit,Kndrdk.	Government of Cape of Good Hope
		Lucy *Sullivan*	I Tug	British Brisbane	117	87·0	17·0	8·8	Brisbane Smellie	1874			Gov'nment of Queensland
		Lucy (ex-Arudder) *Dujoie* -87	I ScSr 1 Dk	French Nantes	119 246	140·7	20·2	9·6	Newcastle A.Leslie&Co.	1869		2Cy.16"& 27"–24" 45 HP. A.Leslie&Co.,Newcastle	Louis Flornoy & fils
		Lucy Ashton *M'Neill* -88	Stl PadSk	British Glasgow	76 271	190·0	21·[illegible]	7·2	Glasgow T.B.Seath&Co.	1888		2Cy.52"–60" 150 HP. Hutson&Corbett,Glasgow	R. Young
	R	Lucy P. Miller *Kragsgaard*	W ScSr	American New York	401 533	152·2	27·2	16·6	Philadelphia Birely,Hillman&Streaker	1878		1Cy.36"–34" 250 HP. Neafie&Levy,Philadelphia	Hardley & Co.
	✻	Ludgate Hill *Brown* 85-86	Stl TwnInBk 2 Dks3 B	British Liverpool	2635 4063	420·3	47·0	26·5	Glasgow Dobie&Co.	1881	Blhds.	4Cy.32½"&76"–48" 600 HP. J.Howden&Co.,Glasgow	S.S. "Ludgate Hill" C. Ld. (W.B.Hill & W.H.Nott)
	N	Ludvig Holberg *Busck* -87	I ScSr 1 Dk & Spr Dk	Norweg'n Bergen	687 912	189·3	29·9	20·2	Bergen Martens,Olsen&Co.	1884		2Cy.25"& 48"–30" 90 HP. Martens,Olsen&Co.,Bergen	P.G. Halvorsen
	DV	Ludwig Nobel *Faseline* -88 (ex Petroleum) (TANK VESSEL)	I&Stl ScSr 1 Dk	Russian St. Petersburg	744 967	191·7	29·1	19·5	Gothenburg MotalaCo.	1888		2Cy.26"&50"–27" 130 HP. MotalaCo.,Motala	Nobel Bros.
493	V	Ludwig Possehl *Klindworth* -86	I ScSr 1 Dk	German Hamburg	434 603	176·0	29·0	13·0	Flensburg FlensburgerSh.Ges.	1886	1 Blhds.	2Cy.23½"& 44"–27" 110 HP. FlensburgerSh.Ges.Flnsbrg	L. Possehl & Co.
494		Lugar (ex Middlesbro') *Leadbetter* 81-84	I ScSr 1 Dk2 B	British Glasgow	98 236	142·6	20·1	10·9	Stockton Richardson,Duck&Co.	1857	1 Blhds.	2Cy.17"&30"–30" 40 HP. D.Rowan,Glasgow	D. Rowan
495		Luigi	I PadTug	R'manian Galatz	85 168	..	..	..	Trieste Stabilimento Tecnico Triestino	1879		70 HP.	L. Mendl & Co.
496		Luis Cousino (WRECKED) (ex José Antonio Moreno)	I ScSr	Chilian Coronel	419 650	187·3	27·6	16·1	Dundee Gourlay Bros.	1870		2Cy.27"&47"–30" 90 HP. Gourlay Bros.,Dundee	Cia. Esplotadora de Lota y Coronel
497		Luis Cuadra *Martin* 74-82	I ScSr 1 Dk2 B	Spanish Seville	596 923	210·5	29·3	15·1	Sunderland R.Thompson,jun.	1875	5 Blhds.	2Cy.27"&51"–33" 99 HP. J.Dickinson,Sunderland	Segovia, Cuadra & Co.
498		Luis Pinzon (ex Juana, ex Andalus, ex Pedro)	W	Spanish Huelva	110 180	117·1	17·1	10·4	Glasgow	1867		30 HP.	L. Camero y Reino
499		Luisa (ex Pathfinder)	I ScSr	Peruvian Callao	136 239	125·0	18·2	15·5	PortGlasgow W.Hamilton&Co.	1872		2Cy.20"–24" 70 HP. Walker,Henderson&Co.,Gls	M.C. Cotes
500		Luleå *Bergström*	I ScSr 1 Dk & Awn Dk	Swedish Haparanda	491 602	161·7	23·9	12·9	Stockholm Bergsund Mk.Vks.	1870		2Cy.24"&48"–27" 100 HP. Bergsund Mk.Vks.Stkholm.	Bottniska Ångf.Bolaget (A. Zachrisson)
501		Lulio	I ScSr 2 Dks	Spanish Majorca	371 670	215·7	27·1	16·3	London J.&W.Dudgeon	1870	5 Blhds.	2Cy.35"& 61"–26" 150 HP. J.&W.Dudgeon,London	José Astier
502	V	Lulu Bohlen *Hupfer* 76-86	Stl ScSr 2 Dks	German Hamburg	1285 1682	270·0	35·0	23·4	Hamburg Blohm&Voss	1886		3C.20"31"&50"–42 200 HP. Blohm&Voss,Hamburg	Africanische Dmpfs. Act.Ges.(Woermann Linie)
503	✻	Lumley *Kidd* 72-72	I ScSr 1 Dk2 B	British Sunderland	472 753	194·5	23·3	16·5	Sunderland J.Blumer&Co.	1872	4 Blhds.	2Cy.25½"&48"–30" 90 HP. T.Richardson&Sons,Hrtlpl.	H.T. Morton & Co.
504		Lund (ex Ariegen No. 1) *Nilsson* 78-78	I ScSr 1 Dk	Danish Copenhagen	84 145	136·9	19·5	9·1	Motala MotalaCo.	1861		2Cy.19½"&39"–23¼" 80 HP. Motala,Gothenburg	Det Forenede Dampskibs Selskab (C.P.A. Koch)
505	✻D	Lundy *Deacon* -88	Stl ScSr 1 Dk & Wh Frms	British London	1182 1828	260·0	37·1	17·7	Stockton Richardson,Duck&Co.	1888	4 Blhds.	3C.21"34"&56"–36" 160 HP. Blair&Co.,Ld.,Stockton	Farrar, Groves & Co.
506	✻	Luneburg *Gibson* -87	I ScSr 1 Dk	British W.Hartlepool	681 1067	230·0	32·0	13·0	Sunderland J.Blumer&Co.	1883	4 Blhds.	2Cy.27"& 50"–33" 99 HP. T.Richardson&Sons,Hrtlpl.	J. Lohden & Co.
507	✻	Lusitania *Tillett* 82-88	I ScBk 3 Dks & Awn Dk	British Liverpool	2425 3832	379·9	41·3	35·2	Birkenhead LairdBros.	1871	5 Blhds.	3C.36"60"&96"–48" 700 HP. T.Richardson&Sons,Hartpl.	Orient Steam Nav. Co., Ld.
508	DV	Lusitania *Cauliway* 81-	I ScSr 1 Dk & Spr Dk	German Hamburg	788 1081	219·4	29·1	14·4	Rostock Rostocker Schiffsb. Ges.	1881	WB	2Cy.27½"&51"–31" 110 HP. Rostockr.Schiffs.Ges.Rstck.	A.C. de Freitas & Co.
509	✻	Lusitania (ex Rosebud) *de Azevedo* 81-84	I ScSr 1 Dk	Portug'se Oporto	191 322	150·1	22·7	11·3	PortGlasgow W.Hamilton&Co.	1874	4 Blhds.	2Cy.19"& 34"–24" 50 HP. J.&T.Young,Ayr	Cia. Portugueza de Navegação à Vapor "União"
510	✻	Luso	I ScSr 1 Dk	Portug'se Lisbon	206 331	136·0	22·0	10·1	London LondonDry Dk.Co.,Ld.	1885	4 Blhds.	2Cy.18"& 35"–24" 55 HP. LondonDryDk.Co.,Ld.,Lon.	Cia. Portugueza do Zaire

✻ = Lloyd's Register; D = Nederlandsche Vereeniging; G = Germanischer Lloyd; I = Registro Italiano; R = American Record;
A = Veritas Austro-Ungarico; H = Veritas Ellenico; N = Norske Veritas; V = Bureau Veritas.

Dans la 2me Colonne, B.T., avec les indications qui suivent, signifie la date du rapport de visite de la Carène par les experts du Board of Trade pour obtenir le certificat relatif au TRANSPORT DES PASSAGERS.

Dans la 3me Colonne, C = Composite (membrure en fer, bordé en bois); I = Fer; Stl = Acier; W = Bois.

Pour plus amples renseignements sur les navires classés par le Lloyd's Register, voir le Lloyd's Register Book et le Lloyd's Yacht Register.

No. in Book / Official No.	Registry in which classed	Ship's Name, Material, Rig, &c. / Master, and Dates of Appointment to present Owner's Service and to Ship / Ship's Late Name. No. of Decks, &c.	Flag / Port of Registry	Tons Net / Gross	Length	Breadth	Depth	Built Where / By Whom	When	No. of Bulkheads / Particulars of Water Ballast	Engines Particulars / Maker's Name	Owners or Managers
511 76217	✱	Lutetia I ScSr [illegible] 85-86 1 Dk2 D	British Newcastle	516 801	205·4	30·4	16·4 B 58 f. F 17 f.	Sunderland J.L. Thompson&Sons	1877	1 Blds. [illegible]	2Cy.27½"&53"-36" 98 HP J. Dickinson, Sunderland	C.F.H. Leslie (Fenwick & Co.)
512		Lutetia (ex Ada) I Sc3mSr Allemand 65-86 2 Dks	French Marseilles	754 1137	248·0	30·0	26·9	Glasgow Smith&Rodger	1861		2Cy.28"&55"-27½" 100 HP [illegible], Mar	Compagnie Fra[illegible] de Nav. à Vap.
513		Lutfieh (ex London) I Sc	Turkish	333	172·6	25·6	16·0				2Cy.28"&53"-36" Rowan Duncan, Glasgow	Idarei Massous[illegible] (Ottoman Stea[illegible])
514 140[illegible]		Luther C. Ward W Sc 1 Dk	American [illegible]	53 107	98·0	21·5	9·8	Newburg, N.Y	1882			L. Luckenbach
515 95482	V	Lux Stl Sc3mSr Hogg [illegible] 1 Dk&SpDk	British London	1255 1883	250·0	35·1	24·1	Newcastle Sir W.G. Armstrong, Mitchell&Co. Ld.	1888		3Cy.18½"30"&48"-33" Wallsend Slipway Co. Ld. Newcastle 150 HP	H.F. Swan
516 1656	✱	Luxor (ex Hope) I ScSr McClure 83-83 2 Dks	British Newcastle	939 1451	280·8	30·2	22·1	Birkenhead Laird	1863	[illegible] Blds.	2Cy.29"&56"-33" 120 HP W[illegible] Slip. Co. Newcastle	Clapham S.S.Co.
517	✱	Luzon (ex [illegible]) I ScS 1 Dk&AwnDk	Spanish Manila	318 494	154·7	24·2	11·5 17·6	Newcastle C.S.Swan&Co.	1874	[illegible]	2Cy.20"&34"-27" 50 HP Christie, Gutch&Co., Newcl.	José Reyes
518	V	Lybeck I ScSr Paulson -85 1 Dk	Swedish Stockholm	228 326	128·2	22·2	11·9 Q 37 f. B 27 f. F 22 f. D 66	Gefle Aktiebol. Atlas	1884		2Cy.18"&32"-20" 50 HP Aktiebol. Atlas, Stockholm	Rederibolaget [illegible] Sverige K.A.W[illegible]
519 93177	✱ BT 8 88	Lycia Stl ScSr Fletcher 83-88 2 Dks	British Liverpool	2102 3223	345·5	40·9	26·7 F 46 f. B 91 f. P 35 f.	Belfast Harland&Wolff	1888	[illegible] Blds. [illegible]	3Cy.24½"37"&64"-48" Harland&Wolff, Bft 320 HP	City of Liverp[illegible] Nav. Co.
520		Lyderhorn I ScSr 1 Dk	Norweg'n Bergen	119 192	139·7	18·1	10·6 Q 52 f.	Bergen Bergens Mek. Vrks.	1878		2Cy.25"&46"-24" 80 HP Bergens Mek. Verks. Bergen	Hardanger Sondh[illegible] landske Dampsk[illegible]
521	V	Lydia I ScSr Petersen 74-87 2 Dks	German Hamburg	1056 1647	259·8	32·1	23·9 P 51 f. B 34 f. F 45 f.	Glasgow Dobie&Co.	1879	[illegible]	2Cy.35"&58"-36" 160 HP J. Howden&Co., Glasgow	Deutsche Dampf[illegible] schiffs Rhederei
522	DGV	Lydia Millington I ScSr Siewsten -87 1 Dk&SprDk&[illegible]	German Rostock	495 786	186·3	27·0	12·7 19·3	Rostock [illegible]	1883	[illegible]	2Cy.23½"&41½"-27½" 75 HP Rostocker Schiff[illegible], Rstck	Friedr. Petersen
523 84328	✱ BT 10 88	Lydian Monarch Stl Sc4mBk Haggett -81 3 Dks&Shl Dk	British London	2595 3987	360·0	43·0	24·8 32·3	Dumbarton A. McMillan&Son	1881	[illegible] Blds. [illegible]	2Cy.46"&87"-57" 500 HP D. Rowan, Glasgow	T. Wilson, Sons &[illegible]
524 94284	✱	Lydie I ScSr Shotton 85-87 1 Dk&WebFrms	British London	1137 1750	264·7	36·0	17·7 [illegible] B 89 f. F 31 f.	W. Hartlepool [illegible]	1887	[illegible] Blds. [illegible]	3Cy.21½"35"&57"-36" 190 HP Blair&Co. Ld., Stockton	Burdick & Cook
525 68015	✱	Lykus (WRECKED) I ScSr Smith 2 Dks	British Sunderland	819 1274	240·5	31·2	22·2	Sunderland J.L. Thompson	1874	[illegible] Blds. [illegible]	2Cy.31"&57"-36" 140 HP J. Dickinson, Sunderland	Parsons & Lines
526 85152	✱	Lymington (WRECKED) I ScSr Blake 87-87 1 Dk	British London	124 204	110·0	18·0	9·8 Q 21 f. B 40 f. B 30 f.	London Edwards&Symes	1882	[illegible] Blds. [illegible]	2Cy.12½"&22"-18" 30 HP C. Burrell&Son, Thetford	H.G. Barrey
527 95167	✱	Lyndhurst Stl ScSr Hanson 77-82 1 Dk&WebFrms	British Cardiff	1309 2025	270·0	40·1	18·4 Q 180 f. B 108 f. F 35 f.	Sunderland W. Pickersgill&Sons	1889	[illegible] Blds. [illegible]	3Cy.20½"33"&54"-39" 160 HP G. Clark, Sunderland	Morel, Ld.
528 83808	✱	Lynton I ScSr Nicholson 85-86 1 Dk2 D	British Newcastle	832 1284	240·4	33·2	17·0 Q 66 f. B 50 f. [illegible]	South Shields [illegible]	1881	[illegible] Blds. [illegible]	2Cy.28"&54"-36" 99 HP [illegible] Headhead&Co., S.Shields	Chapman & Mil[illegible]
529 3010		Lynx I PadSrYt	British Glasgow	315 409	213·0	23·6	13·5	Dumbarton [illegible]	1854		210 HP	C.R.M. Talbot[illegible]
530 [illegible]		Lynx I Sc3mSr 1 Dk	British Cardiff	117 163	124·0	20·3	8·6	Middlesbro' Backhouse & Dixon	1869	[illegible] Blds. [illegible]	2Cy.15½"-12" 38 HP Jay&Co., Middlesbro'	J. Robinson
531 18870	✱	Lyon I Sc3mSr Simpson 73-73 1 Dk2 D	British Sunderland	387 606	194·7	26·9	15·4 Q 35 f.	Newcastle C. Mitchell&Co.	1857	[illegible] Blds. [illegible]	2Cy.22"&42"-26" 70 HP Comp. 75 J. Dickinson, Sndld.	North Hetton Co[illegible]
532	✱	Lyon (ex Lyons) Stl TwnScSr Aubriot -7-87 1 Dk	French Dieppe	220 487	190·6	26·1	14·0 B 11 f. F 36 f.	Glasgow J. Elder&Co.	1885	[illegible] Blds.	6C(2)22"&(4)32"-24" 250 HP J. Elder&Co., Glasgow	London, Brighton South Coast [illegible]
533 08588	✱	Lyra (ex Sailing Yacht) W ScSrYt	British Portsmouth	205 302	169·0	25·6	14·1	Gosport Camper&Nicholson	1876		3Cy.18"19"&33"-24" 86 HP W.B. Thompson&Co., Dundee	Sir W.R. Brown[illegible]
534	✱	Lys (ex Yolat Sala, ex Richard Cobden) I ScBg Falk 86-88 1 Dk&SprDk3 B	Belgian Ghent	873 1348	249·0	30·1	17·3 Eco. 71 f. 24·1	Sunderland W. Pile&Co.	1865	[illegible] Blds. [illegible]	2Cy.30"&54"-34" 200 HP Millwall Dock&W., London	Cie. Gantois de Nav[illegible] Hou
535		Lysekil Stl ScSr Olsson 81-85 1 Dk	Swedish Marstrand	211 273	143·3	20·6	9·3 Q 47 f. B 35 f. F 30 f. D 44	Gothenburg MotalaCo.	1884		35 HP	Marstrands Nya[illegible] Bol. (A. Lundberg)
536 27701		Lyttelton I PadSr 1 Dk	British Wellington	86 104	93·4	18·4	7·2	London Scott Russell&Co.	1859		2Cy.12½"&22"-14" 30 HP H. Seagar, Wellington	G. Ross
537 54008		Lyttelton I PadTug Clark 79-86 1 Dk	British Lyttelton	39 103	125·8	22·1	11·0	Birkenhead Laird Bros.	1878	3 Blds.	1Cy.48"&36"-48" 80 HP Laird Bros., Birkenhead	Lyttelton Harbour Board

ADDITIONAL SHIPS ARE INSERTED IN THE SUPPLEMENT AT THE END OF THE REGISTER.

✱ = Lloyd's Register; D = Nederlandsche Vereeniging; G = Germanischer Lloyd; I = Registro Italiano; R = American Rec[illegible]
A = Veritas Austro-Ungarico; H = Veritas Ellenico; N = Norske Veritas; V = Bureau Veritas

In the 2nd Column, B.T., with figures following, signifies Date of report of bottom sighting by Board of Trade Surveyors as condition of PASSENGER CERTIF[illegible]

In the 3rd Column, C = Composite (iron frame, wood planked); I = Iron; Stl = Steel; W = Wood.

Further particulars of Vessels classed by Lloyd's Register will be found in Lloyd's Register Book and Lloyd's Yacht Register.

	Classés dans les Registres	Nom, Matériaux, Gréement, etc. Capitaine et dates d'entrée au service des Armateurs actuels et à bord du navire. Ancien Nom, nombre de ponts, etc.	Pavillon. Port d'Armement	Tonnage. Net. Total	Dimensions. Long. | Larg. | Creux Constructions sur le Pont	Construit. Où. Par qui.	Quand.	Nombre des cloisons. Détails sur le lest d'eau.	Machines. Détails. Nom du Constructeur.	Armateurs ou Directeurs-Gérants.
1	R	**M. A. Starr** W ScSr *Promoted* [illegible] 2 *Dks*	British Halifax, N.S.	166 244	115·0 | 22·5 | 15·0	Portsmouth	$18\frac{5}{81}$		2Cy. 60 HP.	Fred. W. Fishwick
2	V	**M. G. Melchior** I ScSr *Petersen* 84-85 2 *Dks*	Danish Copenhagen	731 1153	226·5 | 30·3 | 15·0 *B* 15 *f.* *H* 43 *f.* 22·5	Copenhagen Burmeister & Wain	1885	APT10 MT35 FTP15	2Cy. 33" & 60"–36" 200 HP. Burmeister & Wain Cpnhgn.	Forenede Dampskibs-Selskab, (C. P. A. Koch)
3		**M. Meanatchy** I ScBg *Krusmann* 3 *Dks*	British Singapore	1353 1964	290·0 | 36·4 | 25·8 *P* 60 *f.* *F* 48 *f.*	Newcastle C. Mitchell & Co.	1881	4 Blkds. [illegible]	2Cy. 40" & 76"–48" 320 HP. Wallsend Slipway Co. Newc.	D. Mackinnon
4		**M. Romero Rubio** W ScSr	Mexican Guaymas	70	82·0 | 19·0 | 13·0	Benicia, Cal. M. Turner	1885		2Cy. 10" & 20"–12" 50 HP. W. Denson, S. Francisco	L. Hullen
5	V	**M'Pongwe** I ScSr *Maritz* 85-86 1 *Dk*	German Hamburg	114 160	110·5 | 21·7 | 8·9	Hamburg Reiherstieg Schiffs Wfte	1877		2Cy. 12" & 20"–15" 20 HP. Reiherstieg Schiffs Werfte	Witt & Büsch
6		**Maas** I PadSr *Willis*	British London	506 692	234·9 | 27·2 | 14·5 *P* 22 *f.*	Northfleet C. J. Mare & Co.	1864		2Cy. 48"–60" 180 HP. Gen. Steam Nav. Co., London	General Steam Nav. Co.
7		**Maas** I ScFlat	British Glasgow	137 166	125·5 | 19·7 | 7·1	Antwerp	1877		1Cy. 13"–12" 20 HP.	British & African Steam Nav. Co., Ld.
8	DV	**Maas** I ScSr *Hansen* -86 2 *Dks*	Dutch Rotterdam	340 492	173·8 | 21·9 | 13·5 *H* 41 *f.*	Kiel Geb. Howaldt	1886	DB50 DT80	2Cy. 23" & 37"–27" 75 HP. Geb. Howaldt, Kiel	P. A. van Es & Co.
9		**Maassluis** (ex Cohen II) I Pad *Kalf* 78-85 1 *Dk*	Dutch Alblasserdam	130	75·5 | 16·3 | 7·1	Gateshead J. Robson	1856		2Cy. $31\frac{1}{2}$"–39" 80 HP. J. Robson, Gateshead	L. Smit & Co.
10	D	**Maasstroom** I ScSr *Van Vliet* 81-81 2 *Dks*	Dutch Rotterdam	274 477	172·9 | 25·5 | 14·0	Rotterdam Nederl. Stoomv. Maats.	1869		2Cy. 23" & 40"–27" 90 HP. Nederl. Stoomv. Maats	Nederl. Stoomv. Maats
11	*	**Mabel** I ScS *Hill* 87-88 1 *Dk* 2 *B*	British W. Hartlepool	715 1116	228·0 | 30·0 | 16·8 *Q* 81 *f.* *B* 59 *f.* *F* 30 *f.*	W. Hartlepool W. Gray & Co.	1879	4 Blkds. DB [illegible]	2Cy. 27" & 50"–33" 99 HP. T. Richardson & Sons, Htlpl.	J. Coverdale & Son
12	*	**Mabel** I ScSr *Cutting* 74-85 1 *Dk* 2 *B*	British Liverpool	494 770	199·1 | 29·1 | 16·5 *Q* 30 *f.* *F* 32 *f.*	W. Hartlepool [illegible]	1870	4 Blkds. [illegible]	2Cy. 38"–30" 96 HP. T. Richardson & Sons, Htlpl.	T. G. Beatley & Son
13		**Mabel** W ScFlat 1 *Dk*	British Liverpool	83 122	85·5 | 20·1 | 8·9	Winsford Ann Deakin	1875		2Cy. $10\frac{1}{2}$"–12" 7 HP. G. Deakin, Winsford	G. Deakin, Ld.
14	*	**Mabel** I ScK *Burgess* 1 *Dk*	British Montrose	51 82	64·5 | 18·1 | 8·3 *Q* 16 *f.*	Montrose Black & Noble	1878	2 Blkds.	2Cy. 9" & 18"–15" 15 HP. Dunsmuir & Jackson, Glsgw.	D. Murison
15	*	**Macalister** I ScSr *Stretton* 86-86 1 *Dk* 2 *B. Spd Shde Dk*	British Singapore	462 745	200·7 | 30·1 | 14·3 *B* 36 *f.* *F* 23 *f.*	Port Glasgow Blackwd. & Gordon	1882	4 Blkds.	2Cy. 26" & 50"–30" 120 HP. Blackwood & Gordon, Pt. Gls	Wee Boon Teck
16	*	**M'Beth** (ex Somerset) I ScBn *Kenneth* 2 *Dks & Spr Dk*	British Singapore	622 978	220·3 | 27·8 | 14·8 22·3	Renfrew Henderson, Coulborn & Co.	1874	5 Blkds.	2Cy. 32" & 60"–36" 160 HP. Hudson, Coulborn & Co. Rnf	Goh Syn Koh
17		**McClellan** (ORE-MANTLEIN) W PadSr *Snow* (ex Joseph Whitney) 2 *Dks*	American Baltimore	758 923	199·9 | 32·7 | 16·6 *F* [illegible]	New York	1854		1Cy. 52"–132" 120 HP. C. H. Delamater & Co., N. Yrk.	Merchants' & Miners' Transportation Co.
18	*	**M'Garel** I ScSr *Rhomantia* 84-87 1 *Dk & Wch Frms*	British London	926 1431	243·7 | 34·3 | 17·5 *P* 34 *f.* 116 *f.* *F* 28 *f.*	North Shields T. & W. Smith	1883	4 Blkds. [illegible]	2Cy. 30" & 58"–36" 150 HP. Wallsend Slip. Co., Ld., Nwc.	"M'Garel" S.S. Co., Ld (Scrutton, Sons & Co.)
19	*	**Macgregor** (WELL DECK) Stl ScSr *Miller* 88-88 2 *Dks & Awn Dk*	British Glasgow	583 1048	215·4 | 32·1 | 12·5 *B* 61 *f.* 19·7	Belfast Workman, Clark & Co. Ld.	1888	4 Blkds. CDB130 APT24	3Cy. 22" $35\frac{1}{2}$" & 57"–39" 180 HP. Muir & Houston, Glasgw	"Macgregor" Steam Ship Co., Ld. (Colvils, Lowden & Co.)
20		**Macgregor** I ScSmSr 1 *Dk*	French Noumea	163 258	129·8 | 21·7 | 9·2 *Q* 32 *f.* *B* 51 *f.*	Mechanics Bay, McQuarrie & C. (NZ)	1881		2Cy. 20" & 38"–24" 60 HP. Fraser & Tinne, Auckland	Montefiore, Joseph & Co.
21	*	**Mackay Bennett** Stl TwnScSr *Le Fanu* -86 2 *Dks* 3 *H*	British Glasgow	1013 1718	259·3 | 40·1 | 21·8 *P* 8 *f.* *B* 60 *f.* *F* 48 *f.*	Glasgow J. Elder & Co.	1884	4 Blkds. [illegible]	4Cy. 25" & 50"–36" 300 HP. J. Elder & Co., Glasgow	J. de Castro
22		**McNaughton** W Sr	British Quebec	37 137	92·0 | 18·4 | 8·4	Levis, Quebec Maritime & Industrielle Co.	1886		2Cy. 18"–24" 86 HP. Maritime & Industrielle C, Lev	J. W. McRae
23		**Macapa** I TwnSc 1 *Dk & Hur Dk*	Brazilian Para	290 420	180·0 | 28·0 | 9·5	Port Glasgow Blackwd. & Gordon	1881		4Cy. 21" & 38"–27" 125 HP. Blackwood & Gordon, Pt. Gls.	Amazon Steam Navigation Co.
24		**Macassa** Stl TwnScSr *Hardy* -88 2 *Dks*	British Hamilton, Ont.	164 459	151·4 | 24·1 | 16·3	Port Glasgow W. Hamilton & Co.	1888		6Cy. 11" 18" & 29"–22" 95 HP. W. Kemp, Glasgow	Hamilton Steamboat Co. Ld.
25		**Macedonia** I ScSr *Finn* 2 *Dks*	British Liverpool	1866 2653	348·1 | 36·5 | 30·1 *P* 85 *f.*	Liverpool T. Vernon & Son	1867	6 Blkds.	4Cy. 26" & 56"–42" 300 HP. Cp. 75, G. Forrester & Co., Lpl.	Papayanni & Co.
26	*	**Macedonia** (ex Valencia) I ScBn *Traylor* 84-87 1 *Dk*	British W. Hartlepool	1135 1766	256·5 | 34·6 | 19·6 *Q* 68 *f.* *P* 28 *f.* *B* 106 *f.* *F* 33 *f.*	W. Hartlepool W. Gray & Co.	1881	4 Blkds. DB [illegible]	2Cy. 32" & 60"–39" 150 HP. Blair & Co., Ld., Stockton	W. Gray
27		**Macedonia** I ScSr *Prinn* 83-86	Greek Piræus	456 605	210·8 | 23·1 | 12·9 *P* 43 *f.* *F* 50 *f.*	Newcastle A. Leslie & Co.	1862	4 Blkds. DB [illegible]	2Cy. 28" & 44"–30" 85 HP. N. E. Marine Eng. Co., Ld. Shd.	Panhellenic Steam Ship Nav. Co.
28	*	**Madeira** Stl ScSr *Norman* -88 2 *Dks*	British Glasgow	1147 1773	300·0 | 35·2 | 19·9 *P* 29 *f.* *B* 80 *f.* *F* 47 *f.*	Glasgow Barclay, Curle & Co.	1884	7 Blkds.	2Cy. 32" & 64"–42" 199 HP. Barclay, Curle & Co., Ld., Gls.	British & African Steam Nav. Co., Ld.
29	*	**Madeline** I ScSr *Nicklen* 82-87 1 *Dk* 2 *B*	British W. Hartlepool	821 1282	238·1 | 31·8 | 19·4 *Q* 185 *f.* *F* 30 *f.*	Hartlepool Withy, Alexndr & Co	1874	4 Blkds. [illegible]	2Cy. 30" & 57"–33" 130 HP. T. Richardson & Sons, Harpl.	Pyman Bell & Co.
30		**Madge Wildfire** Stl PadSr *Campbell*	British Glasgow	30 210	190·0 | 20·0 | 7·1 *B* 25 *f.*	Ayr S. M'Knight & Co	1886		1Cy. 49"–60" 95 HP. Hutson & Corbett, Glasgow	H. Brown

* = Lloyd's Register; D = Nederlandsche Vereeniging; G = Germanischer Lloyd; I = Registro Italiano; R = American Record; A = Veritas Austro-Ungarico; H = Veritas Ellenico; N = Norske Veritas; V = Bureau Veritas.

Dans la 2me Colonne, B.T., avec les indications qui suivent, signifie la date du rapport de visite de la Carène par les experts du Board of Trade pour obtenir le certificat relatif au TRANSPORT DES PASSAGERS.

Dans la 3me Colonne, C = Composite (membrure en fer, bordé en bois); I = Fer; Stl = Acier; W = Bois.

Pour plus amples renseignements sur les navires classés par le Lloyd's Register, voir le Lloyd's Register Book et le Lloyd's Yacht Register.

No. in Book / Official No.	Registry in which classed	Ship's Name. Material, Rig, &c. / Master, and Dates of Appointment / Ship's Late Name. No. of Decks, &c.	Flag. / Port of Registry.	Tons. Net / Gross	Dimensions. Length, Breadth, Depth	Built. Where / By Whom.	When	No. of Bulkheads. Particulars of Water Ballast.	Engines. Particulars, Maker's Name.	Owners or Managers
31 [illegible]		**Madras** I ScSr *Tonsell 87-88* 2 Dks	British [illegible]	462 889	220·2 | 28·3 | 15·2	Renfrew W.Simons&Co.	1864		2Cy.32"&50"-27" 118 HP. W.Simons&Co.,Renfrew	British India S. Nav.Co.Ld.
32 93141	✳	**Madrid** (ex Aurora) I ScSr *Grant* 1 Dk Spd Awn Dk	British Hull	1588 2526	310·[illegible] | [illegible] | 23·8	Sunderland [illegible]&Co.	1873	[illegible] Blkds.	2Cy.49"&81"-42" 2[illegible] HP. [illegible]&Co.,Sunderland	W.S.Bailey
33 91918	R	**Madrid** (ex [illegible]) I Sc3mSr *Hackett -87*	American New Orleans	631 861	190·[illegible] | 27·[illegible] | 15·5	Dumbarton W.Denny&Bros.	1857		2Cy.40"-42" [illegible] HP.	New Orleans & C[illegible] bian S.S.Co.
34 99917	✳	**Madura** I Sc3mSr *Doyle 82-87* 1 Dk&Spr Dk3 B	British W.Hartlepl.	1470 2324	[illegible] | 36·1 | 19·7	Stockton [illegible]	1873	[illegible]	2Cy.30½"&79"-39" 240 HP. Blair&Co.,Ld.,Stockton	C.Furness
35 [illegible]	✳ BT13.88	**Madura** I Sc3mSr *Franks 76-87* 3 Dks	British Glasgow	1273 1956	316·8 | 33·5 | 25·4 *F 44 f.*	Greenock Scott&Co.	1873	[illegible]	3C21"38"&57"-42" 280 HP. A.&J.Inglis,Glasgow	British India S. Nav.Co.Ld.
36	✳	**Magallanes** I ScBq (ex China) 3 Dks		1678 2664	326·2 | 40·1 | 27·6 *P 30 f.*	Glasgow R.Napier&Sons	1862	[illegible] Blkds.	2Cy.51"&86"-48" 120 HP. Barclay,Curle&Co.,Glasgow	
37 71112		**Maganittawan** W ScSr 2 Dks	British Collingwood	183 269	100·[illegible] | 24·0 | 9·0	Byng Inlet [illegible]	1877		1Cy.18"-30" 150 HP. Cowie&Co.,S.Cthrines,Ont.	Georgian Bay [illegible] dated Lumber C[illegible]
38 [illegible]	✳	**Magda** Stl ScSr *Thoren 79-88* 1 Dk&Wb Frms	British Hartlepool	1531 2350	260·[illegible] | 38·2 | 20·3 [illegible]	W. Hartlepool [illegible]	1888	[illegible] Blkds. [illegible]	3C.24½"36½"&60"-39" 180 HP. Blair&Co.,Ld.Stcktn.	O.Trechmann
39 [illegible]	✳ BT10.88	**Magdala** I ScBg *Thomas -88* 2 Dks	British Liverpool	1358 2134	330·1 | 31·5 | 25·5	Hull C.&W.Earle	1868	[illegible] Blkds.	2Cy.37½"&73"-42" 200 HP. [illegible]&Co.,Liverpool	Moss S.S.Co.,Ld.
40		**Magdalena** (ex Sir Colin Campbell) W ScBk *Stokken -82* 1 Dk	Norweg'n Tönsberg	253 375	122·[illegible] | [illegible] | 17·5	Sunderland	1855		1Cy.50"-24" 60 HP. Nylands Verks,Christiania	G.C.Hansen
41	V	**Magdeburg** Stl ScSr *Erich 83-84* 1 Dk	German Stettin	408 550	169·9 | 24·6 | 13·5 *P 84 f. F 30 f.*	Stettin Akt.-Ges. "Vulcan"	1883	[illegible]	2Cy.21½"&3[illegible]"-27½" 70 HP. Akt.-Ges. "Vulcan," Stettin	Neue Dampfer C[illegible] pagnie Akt.-Ges.
42 58978	BT5.88	**Magellan** I ScBk *Pepper* 3 Dks	British Liverpool	1791 2856	359·6 | 41·0 | 26·1 *F 82 f. H 275 f.*	Glasgow Randolph, Elder & Co.	1868	6 Blkds.	2Cy.60"&96"-45" 300 HP. Randolph,Elder&Co.,Glasg.	Pacific Steam N[illegible]
43 [illegible]	✳	**Maggie** I Sc3mSr *McGregor 81-82* 1 Dk	British Glasgow	78 168	118·6 | 19·0 | 9·0 *Q 37 f. B 21 f. F 20 f.*	Liverpool P.P.Brownrigg	1870	[illegible] Blkds.	2Cy.13"&25"-20" 25 HP. [illegible],Anderson&Co.,Glasg.	Nobels Explos[illegible] Ld.
44 60043		**Maggie** W ScFlat	British Liverpool	81 119	87·0 | 19·3 | 9·7	Northwich	1867		1Cy. 20 HP.	Wallasey Local [illegible]
45 [illegible]		**Maggie Ann** I Sc3mSr *Harding 86-86* 1 Dk	British Barrow	46 146	124·8 | 19·2 | 7·4	Port Glasgow H.Murray&Co.	1868	2 Blkds.	2Cy.15"&30"-18" 40 HP. J.Jack&Co.,Liverpool	J.S.Simpson& [illegible]
46 [illegible]	✳	**Maggie Walker** I ScSr *Brown 86-88* 1 Dk	British Aberdeen	31 118	95·7 | 20·0 | 10·3	Aberdeen A.Hall&Co.	1888	[illegible] Blkds.	2Cy.18"&31"-24" 50 HP. A.Hall&Co.,Aberdeen	T.Walker
47 78771	✳D	**Maggie Warrington** I ScSr *Berry 87-87* 1 Dk2 B	British Liverpool	529 898	220·0 | 30·0 | 16·1 *Q 51 f. F 25 f.*	W.Hartlepool R.Irvine&Co.	1878	[illegible] Blkds. [illegible]	2Cy.30"&56"-36" 120 HP. Blair&Co.,Ld.,Stockton	S.Hough
48 62234	✳	**Maglona** I ScSr *Slater 67-79* 1 Dk	British Newcastle	439 699	169·3 | 30·7 | 13·7 *Q 65 f. B 30 f. F 12 f.*	North Shields T.&W.Smith	1871	[illegible] Blkds. [illegible]	1Cy.24"-21" 70 HP. N.E.Marine Eng.Co.,Ld.,Shl.	Marquis of Lo[illegible] derry
49 [illegible]		**Magna Charta** I PadSp	British Hull	62 118	98·2 | 18·1 | 8·3	Grimsby T.Charlton	1873		1Cy.34½"-48" 40 HP. T.Charlton,Hull	Manchester, S[illegible] & Lincoln [illegible]
50	G	**Magne** I ScSr *Johansson -82* 1 Dk	Swedish Gothenburg	199 275	111·6 | 21·5 | 13·0 *B 28 f. H 4 f.*	Gothenburg Göteborgs Mek. Verkst.	1883		2Cy.14"&24"-20" 40 HP.	T.Ahrenberg
51 63215	BT8.88	**Magnet** I ScSr	British Dublin	310 597	195·4 | 32·0 | 13·7 *P 43 f.*	Waterford Malcomson Bros.	1869		2Cy.31½"&54"-33" 150 HP. London&Glasgow Co.,Glas.	R.Tedcastle &[illegible]
52 [illegible]		**Magnet** I TwnScTug (ex Knight of the Cross)	British Liverpool	 245	135·7 | 22·7 | 12·2	Birkenhead W.Stevens	1883		2Cy.17"&31"-21" 99 HP. W.Stevens,Birkenhead	The Admiralty
53 91100	✳	**Magnet** I ScSr 1 Dk	British Leith	22 48	70·0 | 15·1 | 7·2	Leith S.&H.Morton&Co.	1887	[illegible]	2Cy.8"&16"-12" 12 HP. S.&H.Morton&Co.,Leith	J.Morison
54 91421	✳	**Magneta** (ex Sailing Vessel) I ScK *Smith -88* 1 Dk	British Hull	52 116	95·7 | 20·4 | 9·5 *Len. 1888.*	Hull Cook, Welton & Gemmell	1885	[illegible]	2Cy.16"&30"-22" 40 HP. C.D.Holmes&Co.,Hull	F.&T.Ross
55 45125	BT2.89	**Magnetic** I ScSr *Geddes 84-87*	British Belfast	261 609	210·1 | 27·1 | 14·4	Glasgow Tod&Mc Gregor	1863		3C18"26"&56"-36" 120 HP. Walker,Henderson&Co.,Glasgow	J.Hay, Sons &[illegible]
56 [illegible]	✳	**Magnolia** I ScSr *Potts 84-84* 1 Dk2 B	British Sunderland	807 1249	230·0 | 34·1 | 16·4 *Q 80 f. B 65 f. F 26 f.*	Sunderland Osbourne,Graham	1883	[illegible]	2Cy.28"&53"-36" 99 HP. Carr&Co.,Ld.,Sunderland	Second Argos[illegible] Ld. (T.Potts & [illegible]
57 [illegible]		**Magnolia** W Pad *Townsend -88* 1 Dk	British Sydney,C.B.	151 200	120·1 | 20·4 | 7·2	United States			1Cy.32"-72" 30 HP. Fletcher,Harrison&C.,N.Y.	G.L.Burchell
58	DV	**Magnus** I Sc3mSc *Arrös 74-88* 1 Dk	Danish Copenhagen	1065 1607	260·2 | 34·5 | 17·5 [illegible]	Renfrew [illegible]	1884	[illegible]	2Cy.32"&60"-36" 150 HP. Lobnitz&Co.,Renfrew	Dampskibs Sel[illegible] "Carl" (L.H.[illegible])
59	✳	**Magnus Barfod** Stl ScSr *Tonne -88* (ex Strathearn) 1 Dk	Norweg'n Drontheim	327 527	160·1 | 27·2 | 11·6 *F 23f. B 69f. Q 43f. H 16f.*	Paisley McArthur&Co.	1884	[illegible]	2Cy.21"&48"-30" 90 HP. Walker,Hudson&Co.,Glasgw.	Det Nordenfj[illegible] Dampskibsselsk[illegible]
60 [illegible]		**Magpie** (ex Resolute) I PadSp	British London	31 113	100·0 | 18·4 | 9·6	South Shields J.T.Eltringham	1880		1Cy.36"-54" 70 HP. J.P.Rennoldson,S.Shields	Danubian Nav.[illegible]

✳ = **Lloyd's Register**; D = Nederlandsche Vereeniging; G = Germanischer Lloyd; I = Registro Italiano; R = American [illegible]
A = Veritas Austro-Ungarico; H = Veritas Ellenico; N = Norske Veritas; V = Bureau Veritas

In the 2nd Column, B.T., with figures following, signifies Date of report of bottom sighting by Board of Trade Surveyors as condition of PASSENGER CERTI[illegible]

In the 3rd Column, C = Composite (iron frame, wood planked); I = Iron; Stl = Steel; W = Wood.

Further particulars of Vessels classed by Lloyd's Register will be found in Lloyd's Register Book and Lloyd's Yacht Register.

		Nom, Matériaux, Gréement, etc. Capitaine et dates d'entrée au service des Armateurs actuels et à bord du navire. Ancien Nom, nombre de ponts, etc.	Pavillon. Port d'Armement	Tonnage Net. Total	Dimensions Long.	Larg.	Creux.	Construit. Où. Par qui.	Quand.	Nombre des cloisons. Détails sur le lest d'eau.	Machines. Détails. Nom du Constructeur.	Armateurs ou Directeurs-Gérants.
		Maha Vajirunhis Stl ScSr *Tan* 84 2 *Dks*	Dutch Batavia	745 1300	253·0	32·2	17·5	Newcastle [illegible]	1881	1 Bhds. DBox 481	2Cy.34" & 66"–42" 230 HP. WallsendSlipwayCo.,Newc.	Nedrl.-IndischeStoom Maats.
		Mahallah I ScBr	Turkish Alexandria	864	253·0	31·6	21·8	Stockton M.Pearse&Co.	1864		2Cy.16" & 75"–33" 200 HP. J.&G.Rennie,London	KhediviehMailS.S.Co.
	✻	Maharaja I Sc3mSr *Young* 82-88 2 *Dks*	British Liverpool	1046 1066	305·8	31·7	22·2	Belfast Harland&Wolff	1879	6 Bhds.	2Cy.31" & 61"–45" 185 HP. D.Rowan,Glasgow	Asiatic Steam Nav. Co.,Ld.
	✻	Maharajah I ScSr *Mardlin* 85-85 1 *Dk* 2 *B*	British London	994 1571	259·3	34·0	18·8	Newcastle C.Mitchell&Co.	1871		2Cy.30" & 60"–36" 170 HP. R.&W.Hawthorn,Newcastle	W.Milburn & Co.
	✻	Maharani I Sc3mSr *Edge* 81-88 2 *Dk*	British Liverpool	1047 1067	305·8	31·7	22·2	Belfast Harland&Wolff	1879	6 Bhds.	2Cy.31" & 61"–45" 185 HP. D.Rowan,Glasgow	Asiatic Steam Nav. Co.,Ld.
	✻	Mahinapua Stl TwnScSr *Rennick* 1 *Dk*	British Dunedin	205 423	160·0	26·8	12·6	Dumbarton W.Denny&Bros.	1881	1 Bhds. FPT 12t APTest	4Cy.17" & 30"–24" 80 HP. Denny&Co.,Dumbarton	Union S.S.Co.of New Zealand
		Mahomed Stl TwnScSr *Ekeland* 82-87 1 *Dk*	Russian S.Petersburg	714				Gothenburg MotalaCo.	1881		4Cy 19¾" & 34½"–21" 120 HP. MotalaCo.,Gothenburg	NaphtaProductions Gesellschaft (Gebr. Nobel)
		Mahroussa I PadScYt 3 *Dks*	Egyptian Alexandria	1741 4200	400·0	42·0	26·5	London SamudaBros.	1865	3 Bhds.	2Cy.100"–96" 800 HP. J.Penn&Sons,London	H.H.The Khédive of Egypt
		Maid of Kent Stl PadSr	British London	118 345	191·7	24·6	11·4	London SamudaBros.	1861		2Cy.50"–45" 160 HP. Ravenhill,Selkirk&Co.,Lon.	London,Chatham & Dover Railway Co.
		Maid of Perth W Pad	American New York	90 148	82·0	23·5	7·7	Newburg,N.Y. Thos.Riley	1867		1Cy.26"–72" 150 HP. J.Morphy,NewYork	Staten Island Rapid Transit R.R.Co.
		Maid of the Mist W Pad 1 *Dk*	British Mrmchi.,NB	117 194	85·0	22·7	8·0	UnitedStates	1854		1Cy. 100 HP.	Hugh E.Ross
		Maid of the Moy I PadTug *Hockley* 1 *Dk*	British Ballina	48 111	91·1	17·8	9·0	Glasgow Aitken&Mansel	1878	1 Bhds.	1Cy.33"–14" 40 HP. Coates&Co.,Belfast	Moy Nav. Commis- sioners
73	G	Main I ScBg *Folger* 69-88 2 *Dks*	German Bremen	1913 3087	347·8	40·0	33·8	Greenock Caird&Co.	1868		2Cy.60" & 100"–54" 600 HP. Caird&Co.,Greenock	Norddeutscher Lloyd
74	✻	Maine ex Swansea Stl Scw3mSr *Griffith* 87-88 1 *Dk*&*Sur Dk* 3 *H*	British London	1833 2869	315·2	40·2	20·2	W.Hartlepool W.Gray&Co.	1887	6 Bhds. CDB 96t FPT 69t APT 40t	3Cy.24½" 40" & 65"–42" Cntl Mar. E. Co. W. Htl 280 HP.	"Maine" S.S.Co., Ld. (Williams,Torrey & Feild,Ld.)
	✻	Maine&Loire Stl ScSr *Bacinand* 1 *Dk*	French Nantes	485 753	182·5	30·1	13·0	Sunderland W.Doxford&Sons	1882	4 Bhds. DBcg 171t APT 38t	2Cy.22" & 42"–30" 75 HP. W.Doxford&Sons,Sundrlnd.	Cie Nantaise de Nav. à Vap.
		Maipo I Sc3mSr *Stewart* 73-82 2 *Dks* & *Shd Dk*	Chilian Valparaiso	1504 2295	332·2	39·9	21·0	PortGlasgow J.Reid&Co.	1882	5 Bhds. DBcg 343t	2Cy.45" & 80"–54" 280 HP. Rankin&Blackmore,Grnck.	Cia.Sud-Americana de Vapores
		Maipu Stl Pad 1 *Dk*	Argent'ne B.Ayres	181 643	253·8	30·0	13·5	Glasgow J.Elder&Co.	1880	4 Bhds.	4Cy.30" & 56"–66" 300 HP. J.Elder&Co.,Glasgow	Argentine Republic
		Maitai I ScSr	British Wellington	163 275	147·0	22·4	10·2	Wellington S.Luke&Sons	1885		2Cy.18" & 36"–24" 55 HP. S.Luke&Sons,Wellington	Union S.S.Co.of New Steam Zealand,Ld.
		Maitland I PadSr *Summerhill* 1 *Dk*&*Awn Dk*	British Sydney	555 880	231·5	27·1	12·6	PortGlasgow [illegible]	1870	5 Bhds.	2Cy.58"–60" 250 HP. McNab&Co.,Greenock	Hunter River New Steam Nav.Co.
80	✻	Maitland I Sc *Mathieson* 87-87 1 *Dk*	British Melbourne	40 159	100·0	18·2	9·5	Sunderland J.Knox&Co.	1885	1 Bhds. FPT 11t APT 5t	2Cy.17" & 32"–21" 45 HP. J.Brewster&Co.,Sunderland	W.F.Walker & Co.
81	✻D	Maitlands I ScSr *Morris* 85-86 1 *Dk* 2 *B*	British W.Hartlepool	742 1160	234·8	31·4	15·0	W.Hartlepool W.Gray&Co.	1883	1 Bhds. DBox 252t	2Cy.27" & 50"–33" 99 HP. Black,Hawthorn&Co.,Gtsd.	Hardy,Wilson & Co.
82		Majestic Stl TwnSc 4 *Dks* 4 *B*	British Liverpool	9500	..	..	..	Belfast Harland&WolffLd.	1889			Oceanic Steam Nav. Co.,Ld.
83		Major W ScDy *Stanley*	British London	103 151	120·0	21·3	10·3	Dartmouth Philip	1874		2Cy.16"–24" 30 HP. J.Stewart,London	E.Gill
84	V	Makrele I ScSr *Kreeling* 87	German Geestemünde	67 124	100·0	20·3	11·0	Bremerhaven F.W.Wencke	1887		2Cy.17½" 31"–18½" 55 HP. Aron&Gollnow,Stettin	Schilling & Co.
85	✻D	Malabar I ScSr *Gould* 71-84 2 *Dks* 3 *H*	British Newcastle	1263 1964	286·0	34·8	23·8	Middlesbro' R.Dixon&Co.	1877	5 Bhds. DB 86t	2Cy.34½" & 64"–42" 200 HP. Blair&Co.,Ld.,Stockton	Macintyre Bros. & Co.
86	I	Malabar ex Torino *Zanelli* 85-88 I ScBn 1 *Dk* 2 *H*	Italian Genoa	1074 1627	268·0	33·1	20·8	Sunderland Oswald&Co.	1872	WBa	2Cy.31" & 61"–36" 310 HP. Oswald&Co.,Sunderland	Navigazione Generale Italiana
87	✻	Malacca I ScSr *Dick* 2 *Dks*	British Singapore	405 653	189·5	26·5	16·2	Leith Ramage&Ferguson	1885	1 Bhds.	2Cy.28" & 54"–36" 125 HP. Ramage&Ferguson,Leith	Tan Beng Gam
88	✻	Malaga I ScSr *Russell* 87-76 3 *Dks*	British London	1024 1614	257·3	32·0	23·6	W.Hartlepool W.Gray&Co.	1876	1 Bhds. DBcg 147t FPT 90t	2Cy.32" & 60"–39" 140 HP. Blair&Co.,Ld.,Stockton	I.Hall,jun.,& Co.
89	✻V	Malaga I ScSr *Leese* 77-87 2 *Dks*	German Hamburg	886 1344	269·3	32·7	19·7	Glasgow A.Stephen&Sons	1879	4 Bhds.	2Cy.30" & 57"–36" 150 HP. A.Stephen&Sons,Glasgow	R.M.Sloman & Co.
90	V	Malaga I ScSr *Beckman* 82-82 1 *Dk*	Swedish Gothenburg	736 943	208·6	28·4	18·0	Stockholm BergsundsMek.Ver kst.	1882	WB	2Cy.25½" & 45"–29¼" 82 HP. BergsundsMek.Verkst.,Stkhm.	Förnyade Ångf.Akt. Bol.Svenska Lld (A.Moyer)

✻ = Lloyd's Register; D = Nederlandsche Vereeniging; G = Germanischer Lloyd; I = Registro Italiano; R = American Record;
A = Veritas Austro-Ungarico; H = Veritas Ellenico; N = Norske Veritas; V = Bureau Veritas.

Dans la 2me Colonne, B.T., avec les indications qui suivent, signifie la date du rapport de visite de la Carène par les experts du Board of Trade pour obtenir le certificat relatif au TRANSPORT DES PASSAGERS.

Dans la 3me Colonne, C = Composite (membrure en fer, bordé en bois); I = Fer; Stl = Acier; W = Bois.

Pour plus amples renseignements sur les navires classés par le Lloyd's Register, voir le Lloyd's Register Book et le Lloyd's Yacht Register.

No. / Official No.	Registry in which classed	Ship's Name. Material, Rig, &c. Master, and Dates of Appointment to present Owner's Service and to Ship Ship's Late Name. No. of Decks, &c.	Flag. Port of Registry.	Tons. Net / Gross	Dimensions (Length, Breadth, Depth) Deck Erections.	Built. Where. / By Whom.	When	No. of Bulkheads. Particulars of Water Ballast	Engines. Particulars. Maker's Name.	Owners or M...
91 63822		**Malda** I Sc3mSr *Nash* 83-88 3 *Dks*	British Glasgow	1269 1945	317·6 | 33·6 | 25·3	Greenock Scott&Co.	1874	6 Blhds.	2Cy. 45"&78"–45" 300 HP. Greenock Foundry Co., Grnk.	British India S. Nav. Co. Ld.
92 87078	✱	**Malek** I ScSr *Croft* 83-83 1 *Dk & Spr Dk 3 H*	British London	1054 1621	251·0 | 34·3 | 24·3 *B* 45 *f.*	Newcastle Wigham Richardson &Co.	1883	5 Blhds. [illegible]	2Cy. 31½"&62"–42" 180 HP. W. Richardson &Co., Newcstl.	Persian Gulf S. Ship Co. Ld.
93 82097	✱	**Malemba** Stl ScSr *Cauey* 85 1 *Dk 2 H*	British Glasgow	980 1521	290·4 | 36·2 | 18·0 *P* 77 *f.* *F* 37 *f.*	Glasgow J. Elder&Co.	1880	9 Blhds.	2Cy. 34"&62"–36" 175 HP. J. Elder&Co., Glasgow	British & Afric. Steam Nav. Co.
94 94352	✱	**Malikah** I ScSYt	British London	110 222	154·7 | 21·5 | 11·3	Leith Ramage&Ferguson	1887	5 Blhds.	3Cy. 15"24"&39"–24" 77 HP. Ramage&Ferguson, Leith	H. J. Barrett
95 85137	✱	**Mallard** I ScSr *Hayes* 2 *Dks*	British London	835 1296	225·0 | 33·1 | 17·4 *P* 130 *f.* *F* 36 *f.*	Dundee Gourlay Bros.&Co.	1882	5 Blhds. [illegible]	2Cy. 30"&54"–36" 140 HP. Gourlay Bros.&Co., Dundee	General Steam N.
96		**Mallorca** I	Spanish Mallorca	419 607	212·6 | 26·1 | 14·1	Cardiff N.S. Russell	1864		297 HP.	M. Morey
97 28364	BT 2.89	**Malmo** (ex Paris?) I ScwSr *Bebling* 1 *Dk 2 H*	British Hull	466 734	203·8 | 27·2 | 16·9 *F* 49 *f.* *Q* 73 *f.*	Hull C.&W. Earle	1860	4 Blhds.	3C. 18"27"&48"–30" 130 HP. Amos&Smith, Hull, N.E. 84	T. Wilson, Sons & C.
98		**Malmø** I ScSr *Thorlacius* 71-85 1 *Dk*	Danish Copenhagen	86 177	134·1 | 21·6 | 9·1 *Q* 41 *f.* *B* 4 *f.* *F* 20 *f.*	Oscarshamn Oscarshamns Mek. Verk. Aktiebolag	1870		3Cy. 13½"23"&36" 90 HP. Burmeister&Wain, Copnhn.	Det Forenede S. akibs Selskab (C.)
99 53378	BT 2.88	**Malta** I ScBg *Lavis* 83-85	British Glasgow	1444 2244	303·1 | 39·3 | 25·0	Glasgow J.&G. Thomson	1865		2Cy. 38"&68"–48" 212 HP. J. Jack&Co., Liverpool	Cunard Steam S. Co. Ld.
100 84876	✱	**Malta** I ScSr *Balls* 82-82 1 *Dk 2 H*	British N. Shields	1006 1565	259·0 | 36·0 | 17·5 *Q* 136 *f.* *B* 26 *f.* *F* 32 *f.*	South Shields J. Readhead&Co.	1882	4 Blhds. DB&c[illegible]	2Cy. 30"&58"–36" 145 HP. J. Readhead&Co., S. Shields	W. D. C. Balls & ...
101 91657	V	**Malta** W ScTrawler *Fourbister* 1 *Dk*	British Granton	33 117	98·0 | 21·0 | 10·5	Leith D. Allan&Co.	1886		2Cy. 14"&28"–21" 35 HP. Hutson&Corbett, Glasgow	J. Muirhead
102 70712		**Malta** (ex Bonnable) I Pad *Crouch* 87	British S. Shields	17 113	103·2 | 18·2 | 9·4	South Shields J. T. Eltringham	1875		2Cy. 27½"–50" 90 HP. J. P. Rennoldson, S. Shields	R. S. Allen
103 87083		**Malta** I Sc *Guy* 86-86 1 *Dk*	British S. Shields	12 59	72·9 | 15·0 | 8·6	Hull E. Wales	1883	3 Blhds.	2Cy. 15"&28"–20" 35 HP. E. Wales, Hull	Cowes Steam Tu.
104 49		**Malta** I ScSr *Nicolich* 74-87	Austrian Trieste	364 556	183·2 | 26·1 | 17·1	Glasgow R. Napier&Sons	1854		2Cy. 30"&51"–30" 120 HP. R. Napier&Sons, Glasgow	Lloyd Austro-Ung. rico
105 83	I	**Malta** (ex Brindisi) I ScSr *Fardella* 85-88 2 *Dks*	Italian Palermo	550 1008	250·5 | 27·8 | 19·6 *F* 32 *f.* *B* 62 *f.*	Newcastle Palmers Bros.&Co. Ld.	1862	4 Blhds.	3C. 21"–38&62"–36 330 HP. R.&W. Hawthorn, Leslie, &Co. Ld. [illegible]	Navigazione Gen. Italiana
106		**Maltepeh** W Pad *Lutfullah* 84-87	Turkish Cnstntnple.	75 222	164·2 | 20·0 | 10·3		1869		2Cy.	Idarei Mussous. (Ottoman Steam Na...)
107 91861	BT 1.88	**Maltese Cross** I Pad	British London	105 134	132·0 | 16·6 | 6·7	London Edwards&Symes	1884		2Cy. 23"–30" 35 HP. M. Pratt, Huddersfield	Metropolitan S. District
108 79154	✱ BT 2.88	**Malvina** I ScSr *Fitz-John* 86-87 2 *Dks*	British Leith	595 1195	254·4 | 31·2 | 18·3 *P* 54 *f.* *B* 62 *f.* *F* 38 *f.*	Glasgow J.&G. Thomson	1879	5 Blhds.	2Cy. 38½"&69"–48" 250 HP. J.&G. Thomson, Glasgow	London & Edinb. Shipping Co.
109	V	**Malvina** I ScSr *Dufay* 86 1 *Dk & Awn Dk 3 H*	French Marseilles	700 1192	236·8 | 30·1 | 17·6 *H* 60 *f.* 24·6	Glasgow J.&G. Thomson	1866		1Cy. 21½"&51½"–36" 250 HP. J.&G. Thomson, Glasgow	Compagnie Gén. Transatlantique
110	✱	**Malvinas** I Sc 1 *Dk*	Urug'yan Monte Video	219 324	142·7 | 23·0 | 9·5 *Q* 72 *f.* *F* 25 *f.*	Elbing F. Schichau	1882	4 Blhds. DBGn	2Cy. 20½"&41½"–21" 70 HP. F. Schichau, Elbing	Uruguayan Gov. ment
111 68380	BT 2.88	**Malwa** I Sc3mSr *Creery* 86-88	British Greenock	1708 2070	361·5 | 39·0 | 29·0	Greenock Caird&Co.	1873		2Cy. 50"&86"–54" 500 HP. Caird&Co., Greenock	Peninsular & Or. Steam Nav. Co.
112		**Mamanguape** W	Brazilian Pernambuco	337	.. .. ..				130 HP.	Cia. de Nav. de Per. nambuco
113		**Mamelena 2nd** W ScSr 1 *Dk*	Spanish S. Sebastian	38 50	71·5 | 16·2 | 8·4	Leith Allan	1880		2Cy. 12"&24"–18" 25 HP. Muir&Houston, Glasgow	Mercader & Sor.
114		**Mamelena 9th** I	British S. Sebastian	56 115	.. .. ..	Leith	1884			J. Mercader
115 93471	✱	**Mamelena XI.** Stl ScSr *Drummond* 88 1 *Dk*	British Dundee	1 89	97·0 | 17·1 | 9·2	Dundee W. B. Thompson, Ld.	1888	3 Blhds. FPTkk	3C. 10½"15"&27"–18" 35 HP. W. B. Thompson&Co. Ld., Dun.	D. Macintyre
116 86221	✱	**Mameluke** I ScSr *Newey* 82-85 2 *Dks 3 B*	British Liverpool	1702 2619	310·3 | 39·0 | 25·4 *P* 26 *f.* *B* 58 *f.* *F* 33 *f.*	Newcastle C. Mitchell&Co.	1882	5 Blhds. CDB[illegible]	2Cy. 37"&72"–48" 300 HP. Wallsend Slipw'y Co., Nwcstl.	Bedouin Steam N. Co. Ld.
117	✱	**Manaos** Stl ScSr 2 *Dks & Shade Dk*	Brazilian Rio Janeiro	1169 1710	266·8 | 36·1 | 19·5 *F* 62 *f.* *H* 327 *f.*	Glasgow J.&G. Thomson	1883	WB	2Cy. 35"&70"–45" 250 HP. J.&G. Thomson, Glasgow	Cia. Braziliana de N. a Vapor
118 75229	✱	**Manapouri** Stl ScSr *Logan* 2 *Dks Spt Shd Dk*	British Dunedin	1020 1783	285·2 | 36·3 | 23·7 *P* 56 *f.* *B* 44 *f.* *F* 44 *f.*	Dumbarton W. Denny & Bros.	1881	5 Blhds. CDBr[illegible] FPTkk	2Cy. 41"&76"–48" 300 HP. Denny&Co., Dumbarton	Union S.S. Co. of N. Zealand
119 70709	✱ BT 9.88	**Manauense** (ex Bosco?) I ScSr *Murray* 78-84 1 *Dk & Spr Dk 3 B*	British Liverpool	996 1550	281·1 | 32·1 | 15·6 *P* 30 *f.* 23·2	Port Glasgow J. Reid&Co.	1874	5 Blhds.	1C(2)17"38"&60"–42" 188 HP. Triple, D. Rollo&Sons, L'pl	R. Singlehurst
120		**Manaus** W	Brazilian Pará	684	.. .. ..				200 HP.	Amazon Steam N. Co. Ld.

✱ = Lloyd's Register; D = Nederlandsche Vereeniging; G = Germanischer Lloyd; I = Registro Italiano; R = American Rec...
A = Veritas Austro-Ungarico; H = Veritas Ellenico; N = Norske Veritas; V = Bureau Veritas.

In the 2nd Column, B.T., with figures following, signifies Date of report of bottom sighting by Board of Trade Surveyors as condition of PASSENGER CERTIFICATE.

In the 3rd Column, C = Composite (iron frame, wood planked); I = Iron; Stl = Steel; W = Wood.

Further particulars of Vessels classed by Lloyd's Register will be found in Lloyd's Register Book and Lloyd's Yacht Register.

No.		Nom, Matériaux, Gréement, etc. — Capitaine et dates d'entrée au service des Armateurs actuels et à bord du navire. — Ancien Nom, nombre de ponts, etc.	Pavillon. — Port d'Armement	Tonnage. Net. — Total	Dimensions. Long. / Larg. / Creux — Constructions sur le Pont	Construit. Où. — Par qui.	Quand.	Nombre des cloisons. — Détails sur le lest d'eau.	Machines. — Détails. Nom du Constructeur.	Armateurs ou Directeurs-Gérants.
	✻	**Manavi** Stl ScSr *Cox* *2 Dks & Shade Dk*	British Liverpool	615 1041	216·0 35·1 15·0 *P 20 f. F 21 f.*	Glasgow R.Napier&Sons	1885	1 Blds.	2Cy.23"& 46"–36" 90 HP. R.Napier&Sons,Glasgow	Pacific Steam Nav.Co.
		Manawatu I ScSr *1 Dk*	British Wellington	112 183	128·8 18·7 9·5	Auckland H.Nicol	1873		2Cy.15½"& 30"–20" 38 HP. Mills&Cable,Wellington	Union S.S.Co.of New Zealand, Ld.
	✻	**Mancel** Stl ScSr *Brown –87* *1 Dk*	Haytian Port auPrince	146 310	156·0 22·1 9·3 *B &c. 91 f. F 26 f.*	Grangemouth GrangemouthDkyard Co.	1887	1 Blds.	2Cy.18"& 31"–24" 50 HP. Dunsmuir&Jackson,Glasgw	General B.Riviere
124		**Manche** Stl ScBn *Sepinai –87* *3 Dks*	French Marseilles	1499 2228	299·2 37·4 28·7 *H &c. 40 f.*	La Ciotat MessageriesMarit.	1887	WB.1860	3C.25¼"37"& 55"–13¼" 350 HP Forges&ChantiersHvre.	Messageries Maritimes
		Manchester I PadSp	British Hull	168 221	159·7 18·9 8·4	Goole GooleE.&S.B.C.Ld.	1876		2Cy.36"–36" 80 HP. LairdBros.,Birkenhead	Manchester,Sheffield& Lincolnsh.Railway Co.
126		**Mandahu** W	Brazilian Pern'mbuco	222	.. .. ..				60 HP.	Cia. de Nav. de Pernambuco
127	N	**Mandal** I ScSr *Smith –87* *1 Dk*	Norweg'n Christiania	190 251	124·3 20·7 9·7 *Q 30 f. B 34 f. F 19 f.*	Gothenburg ThorskogsMek.Vst.	1884	APT FPT	2Cy.13"& 24"–18" 30 HP. BolindersMek.V.Stockholm	Dampsk.Selsk.Tvedestrand (JensMenich&Co.)
128	✻	**Mandalay** I Sc3mSr *Sturgis 87–87* *2 Dks 3 E*	British Glasgow	1243 1915	298·0 33·6 25·8	Dumbarton W.Denny&Bros.	1872	5 Blds.	2Cy.35"&60"–42" 180 HP. Denny&Co.,Dumbarton	Raeburn & Vérel
129	✻	**Mandalay** I ScBn *Findlay 72–86* *1 Dk 2 B*	British Whitby	1176 1810	258·4 37·1 18·5 *Q 60 f. P 29 f. B 104 f. F 33 f.*	Whitby T.Turnbull&Son	1886	1 Blds. DB 6 c/3161	3C.20"33"&54"–36" 140 HP. Blair&Co.,Ld.,Stockton	T.Turnbull & Son
130		**Mandalay** (ex Tessa) Stl PadLr	British Glasgow	359 488	250·0 30·0 8·9	Dumbarton W.Denny&Bros.	1886		2Cy.34"& 59"–54" 180 HP. Denny&Co.,Dumbarton	Irrawaddy Flotilla Co.
131	✻	**Mandarin** I ScSr *Curtains 84–84* *1 Dk*	British Glasgow	400 661	180·3 27·3 13·8 *Q 47 f. B 45 f. F 25 f.*	Paisley H.McIntyre&Co.	1880	1 Blds. DB 8 c/46 FPT 16 APT 34	2Cy.23"& 45"–33" 85 HP. Hutson&Corbett,Glasgow	J.& A.Wyllie
132	✻	**Mandingo** Stl ScSr *Thompson 82–83* *2 Dks*	British London	1089 1700	305·0 34·2 20·0 *F 49 f. H 112 f.*	Belfast Harland&Wolff	1882	5 Blds.	2Cy.32"& 64"–42" 160 HP. Harland&Wolff,Belfast	African Steam ShipCo.
133		**Mandingue** (ex Oberon) I ScBn *Selves* *1 Dk & Shlt r Dk 2 B*	French Marseilles	950 1414	218·0 32·0 17·4 *H 75 f. 246*	Hull Humphrys&Pearson	1871	1 Blds.	2Cy.26½"&52½"–36" 130 HP. Humphrys&Pearson,Hull	Cie.du Sénégal
134		**Mangana** I ScSr *Capern* *2 Dks*	British Hobart	428 752	208·3 27·1 19·3	Glasgow D.&W.Henderson	1876		2Cy.31"&62"–36" 350 HP. D.&W.Henderson,Glasgow.	Tasmanian Steam Nav. Co.
135		**Manhasset** W Sc	American Ptsmth.,Va.	352 512	139·0 32·0 11·0	New York	1860		300 HP.	J.L.Porter
136	✻	**Manhattan** Stl 3mSc *1 Dk & Spar Dk*	British London	2500 3500	330·0 42·2 28·4 *P 85 f. F 49 f.*	PortGlasgow D.J.Dunlop&Co.	1889	11 Blds. DB 6 c/198 APT 86 t FPT 85 t	3Cy.26½"42"&70"–45" D.J.Dunlop&Co.,P.Glasgow	"Manhattan" S.S.Co., Ld.
137	R	**Manhattan** I ScSr *Stevens 79–81* *2 Dks*	American New York	1155 1595	228·0 35·2 20·1	Chester,Pa. J.Roach&Son	1879		2Cy.28"& 53"–48" JohnRoach&SonChesterPa.	Old Dominion S.S.Co.
138	✻	**Manhem** I ScSr *Johansson* *1 Dk 2 B*	Swedish Gothenburg	304 422	141·0 23·6 13·8 *Q 48 f. B 34 f. H 7 f.*	Gothenburg MotalaCo.	1883	1 Blds. MT 20 t	2Cy.16"& 29"–20½" 50 HP. MotalaCo.,Gothenburg	Ångf.Akt.Bol."Manhem." (J.F.Olsson)
139	✻	**Manila** (ex Carriedo) I ScSr *1 Dk & Spr Dk*	Spanish Manila	676 1057	209·2 31·1 20·7	Leith Ramage&Ferguson	1883	1 Blds. DB 8 c/55 MT 13 t	2Cy.32"&60"–39" 175 HP. Ramage&Ferguson,Leith	Spanish Government
140		**Manila** I Pad	Spanish Manila	83 128	115·1 20·3 6·2	Dublin Bewley,Webb&Co.	1869		2Cy.18"&35"–36" 45 HP. M.Paul&Co.,Dumbarton	Inchausti & Co.
141	I	**Manilla** (ex Whampoa) I ScBk *Craviotti –88* *2 Dks 3 B*	Italian Genoa	2583 3910	399·5 42·2 28·4 *P 100 f. F 50 f.*	Newcastle PalmersCo.,Ld.	1873	5 Blds.	2Cy.52"&98"–52" 500 HP. PalmersCo.,Newcastle	Navigazione Generale Italiana
142	✻D	**Manitoba** Stl ScSr *Warwick 87–87* *1 Dk & Web Frms*	British London	1386 2127	275·0 37·2 19·2 *P 29 f. Q 78 f. B 119 f. F 30 f.*	W.Hartlepool E.Withy&Co.	1887	1 Blds. CB 257 t APT 27 t	3C.21"35"&57"–39" 200 HP. Blair&Co.,Ld.,Stockton	H.Bucknall & Sons
143		**Manitoba** W Pad *1 Dk*	British Quebec	198 356	115·4 20·6 7·1 *H 69 f.*	Quebec	1871		1Cy. 35 HP.	John Kelly
144		**Manitoban** I ScBn *Dunlop –7 (ex Ottawa)* *4 Dks*	British Montreal	1865 2975	338·8 35·2 34·0	Birkenhead Laird Bros.	1865	7 Blds.	2Cy.45"& 78"–42" 300 HP. Laird Bros.,Birkenhead	J.& A.Allan
145		**Manley** I ScSp	British Grimsby	78 115	91·0 17·2 7·5	Hull Hessa Barnard	1888		2Cy.10"&20"–14" 17 HP. Cochran&Co.,Birkenhead	Bennetts & Co.
146		**Manlleu** I PadBn *(ex Matilde)* *1 Dk*	Spanish Havana	288 412	158·8 22·1 17·8	London	1865		2Cy. 250 HP.	P.Roqué y Aguilar
147		**Manoel Diabo** W	Brazilian RioGrande	180	.. .. ..				80 HP.	L.A.de Otero
148		**Manora** Stl Sc3mSr *Turner 85–86* *3 Dks*	British Glasgow	3040 4707	410·0 45·2 21·7 *P 42 f. B 158 f. F 49 f. 297*	Dumbarton W.Denny&Bros.	1883	7 Blds.	2Cy.54"&94"–60" 650 HP. Denny&Co.,Dumbarton	British India Steam Nav.Co.,Ld.
149	✻	**Manoubia** I ScSr *Foucnel –88* *1 Dk & Awn Dk*	French Marseilles	672 1013	234·9 29·5 13·4 *H 52 f. 209*	Havre Forges&C.de la Médit.	1880	6 Blds.	2Cy.27½"&55"–35¼" 100 HP. Forges&C.de la Médit.,Hvre	Compagnie Générale Transatlantique
150										

✻=Lloyd's Register; D=Nederlandsche Vereeniging; G=Germanischer Lloyd; I=Registro Italiano; R=American Record;
A=Veritas Austro-Ungarico; H=Veritas Ellenico; N=Norske Veritas; V=Bureau Veritas.

Dans la 2me Colonne, B.T., avec les indications qui suivent, signifie la date du rapport de visite de la Carène par les experts du Board of Trade pour obtenir le certificat relatif au TRANSPORT DES PASSAGERS.

Dans la 3me Colonne, C=Composite (membrure en fer, bordé en bois); I=Fer; Stl=Acier; W=Bois.

Pour plus amples renseignements sur les navires classés par le Lloyd's Register, voir le Lloyd's Register Book et le Lloyd's Yacht Register.

No. in Book / Official No.	Registry in which classed	Ship's Name. Material, Rig, &c. / Master, and Dates of Appointment to present Owner's Service and to Ship / Ship's Late Name. No. of Decks, &c.	Flag / Port of Registry	Tons Net / Gross	Length	Breadth	Depth	Built Where / By Whom	When	No. of Bulkheads / Particulars of Water Ballast	Engines Particulars, Maker's Name	Owners or Managers
151		**Mansourah** I ScSr (ex Alma[illegible])	Turkish Alexandria	503 884	220·0	27·0	15·0	Glasgow Barclay,Curle&Co.	1863	[illegible] Blkds.	2Cy.26"&52"–36" 120 HP Day,Summers&Co.,Sthmptn	Khedivieh M[illegible]
152 91068		**Manteo** I ScSr Southgate ·87 1 Dk.&Iron Dk	American New York	409 584	201·0	26·0	18·0	Wlmingtn, Del [illegible]	1887		2Cy.17"&31"–26" [illegible]	Old Dominion [illegible]
153 91094	V	**Manuel Dublau** W ScSr Gregory ·88 1 Dk	American SnFrancisco	339 435	151·0	27·5	14·1	San Francisco T.P.H.Whitelaw	1888		2Cy.17"&31"–24" 150 HP W.B.Thompson,Dundee	E.Goodall
154		**Manuel Espaliu** I ScSr 1 Dk2 H	Spanish Seville	672 988	210·0	29·5	16·0 P 100 f. F 45 f.	Newcastle A.Leslie&Co.	1878	[illegible] Blkds. [illegible]	2Cy.28"&53"–33" 110 HP Black,Hawthorn&Co.,Gtshd	Vinuesa & Co.
155	✳	**Manuel L. Villaverde** I ScSr Revuelt 87-87 1 Dk2 H	Spanish Barcelona	738 1198	260·5	31·5	17·4 P 81 f. B 112 f. F 38 f.	Renfrew Lobnitz&Co.	1882	[illegible] Blkds.	2Cy.38"&68"–42" 250 HP Lobnitz&Co. Renfrew	Cia.Trasatlantic[illegible]
156		**Manuel Pérez** W&I ScSr Rodrigues ·85 (ex Tarisa) 1 Dk	Spanish Ferrol	148 328	165·3	21·3	11·8	Liverpool Hoet	1857		80 HP Liverpool	N.Perez
157		**Manuel Torente** W ScSr.S. (ex James M. Green)	Spanish Havana	379	130·0	23·0	15·0	Chester, Pa.				Brisendo,Crisp[illegible]
158		**Manuela** I Sc 2 Dks.&Iron Dk	Spanish Havana	668 983	220·1	30·3	13·5 203	Renfrew Henderson,Coulborn&Co.	1872	[illegible] Blkds.	2Cy.34"&62"–30" 180 HP Henderson,Coulborn&Co.,Renfrew	Ramon de Herr[illegible]
159	✳	**Manuela** I Sc [illegible] 1 Dk2 H	Spanish Bilbao	638 862	215·7	30·0	15·2 P 40 f. B 26 f. F 28 f.	Southampton Oswald,Mordaunt&Co.	1878	[illegible] Blkds. [illegible]	2Cy.27"&50"–33" 100 HP Oswald,Mordaunt&Co.,Sthn	Cia. de Nav. [illegible]
160		**Manuelita & Maria** I ScSr	Spanish Havana	737 1092	225·0	30·3	21·2	Sunderland T.Dobson&Co.	1876		30"&52"–33" 120 HP	R. Herrera
161 93805		**Manwyne** Stl PadLa	British Glasgow	250 674	250·0	35·1	9·8	Dumbarton W.Denny&Bros.	1887		2Cy.37½"&65"–72" 375 HP Denny&Co.,Dumbarton	Irrawaddy Flo[illegible] Ld.
162 70380	BT·88	**Manx Fairy** Stl TwoSc Torbet ·88	British Douglas I.M.	25 109	128·5	16·0	7·3	Glasgow T.B.Seath&Co.	1887		2Cy.11½"&22"–18" 12 HP Hutson&Corbett,Glasgow	Mona Steam S[illegible] Ld.
163 [illegible]	✳	**Manx Queen** Stl PadSr Myerscough [illegible] (ex Duchess of Edinburgh) 2 Dks	British Barrow	568 1128	278·9	29·7	14·0 B 73 f. F 57 f.	Glasgow J.&G.Thomson	1880	[illegible] Blkds.	2Cy.[illegible]"&83"–72" [illegible] HP J.&G.Thomson,Glasgow	J.Little & Co.
164 [illegible]	BT[illegible]	**Manxman** (ex Antrim) I PadSr Gregson 78-83	British Barrow	458 797	266·6	25·2	11·3 [illegible] B 15 f. F 30 f.	PortGlasgow R.Duncan&Co.	1870		2Cy.48"–75" 300 HP Rankin&Blackmore,Grnck.	J.Little & Co.
165 62401		**Maori** I Sc Jones 1 Dk	British Dunedin	118 174	144·0	19·1	9·2 P 51 f. F 37 f.	PortGlasgow Blackwood&Gordon	1867	[illegible] Blkds.	2Cy.27"–20" 60 HP Blackwood&Gordon,Pt.Gls.	Union S.S.Co. of [illegible] Zealand
166 95[illegible]	✳	**Maple Branch** Stl. ScSr Rilson ·88 1 Dk.&Spr Dk3 L	British Sunderland	1735 2637	300·0	40·0	18·1 [illegible] F 35 f. 26·7	Sunderland Bartram,Haswell & Co.	1888	[illegible] Blkds. [illegible]	3Cy.23½"&[illegible]"&62"–42" 300 HP J.Dickinson,Sndld.	NautilusStea[illegible] Ld.
167	✳	**Mapocho** I ScSr Gerkins 76-87 3 Dks.&Shd Dk	Chilian Valparaiso	1549 2278	331·2	41·4	19·3	Birkenhead Laird Bros.	1882	[illegible] Blkds. [illegible]	2Cy.45"&80"–51" 450 HP LairdBros.,Birkenhead	Cia. Sud-Americ[illegible] de Vapores
168 91[illegible]	R	**Maracaibo** W ScSr 1 Dk	American Wlmgtn. Del	358 517	179·3	33·6	12·0	Philadelphia Neafie&Levy	1880	[illegible]	2Cy.20"&34"–26" Neafie&Levy,Philadelphia	Atlantic & Cari[illegible] SteamNav.Co.
169		**Maracana**	Brazilian RioJaneiro	123	..	..	..					D.Rosa Paim
170		**Marajo** I Pad 1 Dk.&Hur Dk	Brazilian Para	516	221·0	32·0	10·0	Birkenhead LairdBros.	1874		2Cy.44"&72"–51" 237 HP LairdBros.,Birkenhead	Amazon Stm.Na[illegible]
171 84057	✳	**Marana** I ScSr Higginson 2 Dks3 H	British Liverpool	1618 2177	285·7	36·0	22·3 P 165 f. F 36 f.	Glasgow Aitken&Mansel	1880	[illegible] Blkds. [illegible]	2Cy.34"&65"–42" 220 HP J.&J.Thomson,Glasgow	G. Bell & Co.
172	✳	**Maranhao** Stl ScSr Pessoa 80-87 2 Dks.&Shade Dk	Brazilian RioJaneiro	1303 1916	276·0	38·1	20·2	Glasgow J.&G.Thomson	1887	[illegible] Blkds. [illegible]	3Cy.27"40½"&61"–45" 250 HP J.&G.Thomson,Glasgow	BrazilianStm.Na[illegible]
173		**Maranhao**	Brazilian Maranhao	600	..	..	..				[illegible] HP	Cia.do Maranh[illegible]
174 84060	✳	**Maranhense** (ex Blodwen) I ScSr Tregarthen 82-82 1 Dk2 H	British Liverpool	960 1486	251·2	33·5	18·4 Q 68 f. P 25 f. B 88 f. F 34 f.	Sunderland Sunderland S.B.Co.	1880	[illegible] Blkds. [illegible]	2Cy.32"&60"–36" 140 HP J.Dickinson,Sunderland	R. Singlehurst & [illegible]
175		**Maranhense**	Brazilian Maranhao	50	..	..	..				[illegible] HP	Cia.do Maranh[illegible]
176 87105	✳	**Maranoa** Stl ScSr Armstrong 2 Dks.&Spr Dk	British Brisbane	805 1505	250·4	36·1	23·0 B 11 f. F 34 f.	Dumbarton W.Denny&Bros.	1883	[illegible] Blkds.	2Cy.38"&68"–48" 270 HP Denny&Co.,Dumbarton	AustralasianU[illegible] SteamNav.Co.Ld.
177 [illegible]	✳	**Mararoa** Stl Sc Ellis 3 Dks	British Sydney	1248 2466	320·1	42·1	24·7 P 63 f. B 34 f. F 42 f.	Dumbarton W.Denny&Bros.	1885	[illegible] Blkds. [illegible]	3Cy.31"51"&81"–54" 330 H Denny&Co.,Dumbarton	Union S.S. Co. of [illegible] Zealand
178		**Marata**	Brazilian PrtoAlegre	60	..	..	..				30 HP	C.G.Schilling & [illegible]
179 282[illegible]0		**Marathon** I ScBk Hewitson 85-88	British Glasgow	1553 2403	336·4	36·6	25·7 F 56 f.	Glasgow R.Napier&Sons	1860		2Cy.44"&78"–42" 300 HP J.&G.Thomson,Glasgow	Cunard Steam S[illegible] Co.,Ld.
180 62372	✳	**Marbella** I ScSr Dawes 87-88 1 Dk2 H	British Glasgow	503 801	204·6	28·9	16·7 Q 28 f. F 34 f. B 33 f.	Sunderland J.Laing	1870	[illegible] Blkds. [illegible]	2Cy.25"&[illegible]"–[illegible]" [illegible] HP [illegible]	Raeburn & Ver[illegible]

✳ = Lloyd's Register; D = Nederlandsche Vereeniging; G = Germanischer Lloyd; I = Registro Italiano; R = American [illegible]
A = Veritas Austro-Ungarico; H = Veritas Ellenico; N = Norske Veritas; V = Bureau Veritas

In the 2nd Column, B.T., with figures following, signifies Date of report of bottom sighting by Board of Trade Surveyors as condition of PASSENGER CERTIFICATE.

In the 3rd Column, C = Composite (iron frame, wood planked); I = Iron; Stl = Steel; W = Wood.

Further particulars of Vessels classed by Lloyd's Register will be found in Lloyd's Register Book and Lloyd's Yacht Register.

Classe	Nom, Matériaux, Gréement, etc. / Capitaine et dates d'entrée au service des Armateurs actuels et à bord du navire. / Ancien Nom, nombre de ponts, etc.	Pavillon. / Port d'Armement	Tonnage. Net. / Total	Dimensions. Longueur / Largeur / Creux	Construit. Où. / Par qui.	Quand	Nombre des cloisons. / Détails sur le test d'eau.	Machines. Détails. / Nom du Constructeur.	Armateurs ou Directeurs-Gérants.
✻	**March** I ScBg Hewson 79-86 2 Dks 3 B	British Whitby	1308 2066	286·0|36·0|24·1 P 26 f. B 69 f. F 35 f.	Whitby T. Turnbull & Son	1883	5 Bhds. DBcel/330	2Cy. 31" & 63"–36" 160 HP. T. Richardson & Sons, Hrtlpl.	T. Turnbull & Son
✻	**Marchesa** C ScSrYt Harris 1 Dk	British Ayr	177 260	138·5|25·0|13·7	Renfrew Lobnitz, Coulborn & Co.	1877		1Cy. (2) 9½", 21" & 36"–24" Tridod ... Barfield & Co. ... 60 HP.	C. T. Kettlewell
✻	**Marchioness** I ScSr Baker 87-88 1 Dk 2 B	British Sunderland	1340 2054	286·5|38·6|20·7 ... P 38 f. B62 f. F 32 f.	Sunderland Short Bros.	1884	5 Bhds. W Bc & f	2Cy. 33" & 62"–42" 164 HP. Carr & Co., Ld., Sunderland	Taylor & Sanderson
✻	**Marchioness of Londonderry** I ScSr Allen 85-88 1 Dk	British Sunderland	422 682	180·0|28·6|13·9 B 93 f. Q 69 f. F 25 f.	Hartlepool E. Withy & Co.	1878	4 Bhds. DBa 52 ft f (45 ft) FPT APT	2Cy. 25" & 48"–36" 90 HP. T. Richardson & Sons, Hrtlpl.	Marquis of Londonderry (S. J. Ditchfield)
✻	**Marcia** I ScSr Mackintosh 2 Dks 3 B	British London	1060 1641	265·0|34·0|24·5	Newcastle Cole Bros.	1874	4 Bhds. W Bcs	2Cy. 33" & 65"–36" 200 HP. N. E. Marine E. Co., Ld., Sndld.	W. Milburn & Co.
	Marco Aurelio I PadSr Morgan	British London	411 725	212·7|32·5|14·7 P 37 f.	Greenock Scott & Co.	1881		2Cy. 33" & 62"–54" 200 HP. Scott & Co., Greenock	W. Binney
	Marco Aurelio I Pad (ex Hornet, ex Lady Stirling)	Spanish Havana	614 966	239·5|25·5|14·6	London	1864		250 HP.	R. Sais & Co.
	Marco Minghetti I Sc3mSr Bragferio 80-88 (ex London Castle) 2 Dks 3 B	Italian Palermo	1616 2472	350·7|36·8|25·7 P 40 f. F 47 f.	Glasgow J. & G. Thomson	1876	5 Bhds.	2Cy. 48" & 84½"–48" 400 HP. J. & G. Thomson, Glasgow	Navigazione Generale Italiana
✻	**Mardy** I ScSr Jones 1 Dk 2 B	British Newport	432 692	191·5|26·8|16·1 Q 84 f. F 23 f.	Newcastle Schlesinger, Davis & Co.	1874	3 Bhds. W Bc & f APT	2Cy. 25" & 45"–30" 80 HP. Black, Hawthorn & Co. Gtshd.	"Mardy" S.S. Co. (Mordey Carney)
	Mare I PadSk	British Montrose	52 150	116·0|19·7|10·3	South Shields J. T. Eltringham	1884		1Cy. 38"–56" 48 HP. J. P. Rennoldson & Ss, S. Shlds.	J. Johnston & Sons
✻	**Mareca** I Sc3mSr Moore 2 Dks 3 B	British London	1425 2214	300·2|37·2|23·0 P 28 f. B 62 f. F 34 f.	Newcastle Palmers Co., Ld.	1882	7 Bhds. C DBaso	2Cy. 36" & 68"–45" 250 HP. Palmers Co., Ld., Newcastle	"Avis" Steam Ships, Ld. (Porteus & Senier)
V	**Maréchal Canrobert** I ScSr Poc -86 2 Dks & Spr Dk	French Marseilles	709 1211	259·8|30·3|15·1 H 60 f. ... 22·6	Greenock Scott & Co.	1872		2Cy. 33" & 66"–36" 250 HP. Greenock Foundry Co., Grnk.	Compagnie Générale Transatlantique
	Marengo I ScBg Abbott 2 Dks 3 B	British Hull	1823 2270	299·5|37·2|25·5 B 42 f.	Hull Earles S. & E. Co.	1879	5 Bhds.	2Cy. 27" & 60½"–51" 200 HP. Earles S. & E. Co., Hull	T. Wilson, Sons & Co.
✻	**Mareotis** I ScSr Osborne 2 Dks	British Liverpool	1392 2141	320·0|34·0|24·6 F 43 f. H 123 f.	Newcastle C. Mitchell & Co.	1878	4 Bhds. W B	2Cy. 35" & 70"–48" 250 HP. R. & W. Hawthorn, Newc'stle	Moss Steam Ship Co., Ld.
✻	**Margaret** I ScSr Markle 85-88 2 Dks	British Grangem'th	212 390	160·0|21·2|14·0 P 35 f. B 32 f. F 25 f.	Barrow Barrow S. B. Co.	1878	5 Bhds.	2Cy. 23" & 46"–24" 90 HP. Barrow S. B. Co., Ld., Barrow	Carron Co.
✻	**Margaret** I ScTrawler W. Nichols 78-86 1 Dk	British Grimsby	65 110	89·0|19·5|10·3 Q 16 f.	Grimsby T. Charlton	1886	3 Bhds.	2Cy. 14" & 26"–18" 30 HP. T. Charlton, Grimsby	T. Charlton
✻	**Margaret** I ScTug Ace 1 Dk	British London	12 54	73·3|15·6|8·3	London J. McGowan	1880	4 Bhds.	2Cy. 14" & 25"–18" 40 HP. J. McGowan, London	C. R. M. Talbot, M.P.
	Margaret M. W Sc 1 Dk	British Quebec	41 269	124·2|25·8|13·5	Levis, Quebec Carrier, Laine & Co.	1878		2Cy. 26" & 50"–33" 95 HP. Carrier, Laine & Co., Levis.	Mrs. Margaret Welsh
✻	**Marguerite** Stl ScSr Mancini -88 1 Dk	Italian Venice	361 599	170·0|26·1|12·6 P 100 f. F 21 f.	Newcastle Schlesinger, Davis & Co.	1888	4 Bhds. DBcel/53 ft FPT 13t APT 28t	3Cy. 14" 23" & 39"–27" 70 HP. N. E. Marine E. Co., Ld., Nwc.	Walter & Co.
V	**Marguerite Franchetti** I Sc3mSr Poquelin -87 2 Dks	French Havre	541 970	228·0|30·2|16·4 P 32 f. F 30 f.	Renfrew Henderson, Coulborn & Co.	1874	5 Bhds. M T 30 t	2Cy. 32" & 60"–36" 160 HP. Henderson, Clbrn. & Co. Rfw.	Worms, Josse & Co.
✻	**Mari Vagliano** I ScSr Lykiardopulo 82-88 1 Dk 2 B	Greek Cephalonia	948 1461	250·5|34·0|17·0 Q 56 f. P 28 f. B 93 f. F 28 f.	Sunderland ...	1880	4 Bhds. DBcel/51 ft FPT APT	2Cy. 30" & 57"–36" 140 HP. J. Dickinson, Sunderland	Vagliano Bros.
	Maria I ScSr Long 1 Dk	British Southamptn	146 261	155·7|21·1|11·0 Q 34 f. F 11 f.	Port Glasgow H. Murray & Co.	1871	4 Bhds.	2Cy. 19" & 33"–24" 50 HP. D. Rowan, Glasgow	London & S.W. Railway Co.
V	**Maria** (ex Maria C., ex Marie Fanny, ex Phœnix) I Sc3mSr Capellino 1 Dk	Italian Genoa	215 337	165·3|24·1|13·4	Sunderland W. Doxford & Sons	1869	W B	2Cy. 21½" & 40"–27½" 70 HP. A. G. Boye, Marseilles	E. Preve & C. Caruto
	Maria	Russian Novorossisk		.. .. ..					A. Loecherm & C. Sinajilo
	Maria (ex Royal Bride) I ScBk	Spanish Mallorca	878 1000	234·2|32·1|19·8	Bristol J. M. Hyde & Co.	1858		500 HP. Randolph, Elder & Co., Glsgw.	P. M. Granada
	Maria I ScSr 1 Dk	Spanish Bilbao	178 238	125·9|21·1|11·3	Glasgow T. Wingate & Co.	1865		2Cy. 20"–24" 40 HP. T. Wingate & Co., Glasgow	J. Alarcon
	Maria (ex Santa Justa) I ScSr 1 Dk	Spanish Cadiz	146 221	145·6|18·9|10·7 H ...	Greenock Scott & Co.	1858		2Cy. 21"–20" 34 HP. Scott, Sinclair & Co.	Sons of Thomas Haynes
	Maria (ex Fortuna, ex Dos Hermanos) I ScSr 1 Dk	Spanish Cadiz	93 162	104·9|22·6|10·4	Glasgow			2Cy. 41 HP.	A. Millan
	Maria I	Spanish Cadiz	82 138	128·9|16·4|9·1	England	1866		40 HP.	Exrs. of the late P. Marroqui
V	**Maria** W ScSr Olsson -87 1 Dk	Swedish Gothenburg	170 217	120·0|20·8|10·0 ...	Törskogg Bereland	1871		2Cy. 12" & 25"–20" 30 HP. Motala Co., Gothenburg	G. E. Falck

✻ = Lloyd's Register; D = Nederlandsche Vereeniging; G = Germanischer Lloyd; I = Registro Italiano; R = American Record; A = Veritas Austro-Ungarico; H = Veritas Ellenico; N = Norske Veritas; V = Bureau Veritas.

Dans la 2me Colonne, B.T., avec les indications qui suivent, signifie la date du rapport de visite de la Carène par les experts du Board of Trade pour obtenir le certificat relatif au TRANSPORT DES PASSAGERS.

Dans la 3me Colonne, C = Composite (membrure en fer, bordé en bois); I = Fer; Stl = Acier; W = Bois.

Pour plus amples renseignements sur les navires classés par le Lloyd's Register, voir le Lloyd's Register Book et le Lloyd's Yacht Register.

No.	Class	Ship's Name. Material, Rig, &c. / Master and Dates of Appointment to present Owner's Service and to Ship / Ship's Late Name. No. of Decks, &c	Flag. / Port	Tons Net / Gross	Dimensions Length, Breadth, Depth	Build. Where. / By Whom.	When	No. of Bulkheads. / Particulars of Water Ballast.	Engines. / Particulars / Makers & Places	Owners
211		**Maria Antoinetta** I Sc	Turkish Constantinople	333 512	184·0 26·2 15·0	Glasgow R.Napier&Sons	1864		2Cy.38"-42" 120 HP. R.Napier&Sons,Glasgow	H.Oppenh…
212	✱	**Maria Gertrudis** Stl ScSr *Jordon* -87 1 *Dk*	Spanish Santander	69 101	87·2 18·1 7·7	Falmouth Cox&Co.	1887	3 Blhds. FPT[illegible]	2Cy.13½"& 24"-16" 25 HP. Cox&Co.,Falmouth	M.C.Fernand…
213		**Maria Gracia** I ScSr 1 *Dk*	Spanish Seville	79 114	109·2 14·9 7·6	Seville Portilla,White&Co.	1873		2Cy. 25 HP.	A.Millan
214		**Maria I.** I Sc *Sederstrom* 1 *Dk*	Swedish Koping	82 118	107·9 16·4 7·7 *B 23 f. H 8 f.*	Motala MotalaCo.	1879		40 HP. MotalaCo.Motala	J.A.Ericsson
215 111	I	**Maria Laura** I TwnScSr (ex Karlsruhe, ex Antonio) 1 *Dk2 B*	Italian Genoa	131 248	121·3 22·5 10·6	Genoa F.Cravero&Co.	1885	5 Blhds.	4Cy.12"&19⅝"-12" 40 HP. F.Cravero&Co.,Genoa	G.Villa
216		**Maria Luisa** ex Aegeon I Sc	Chilian Valparaiso	266 411	165·2 23·2 13·0	PortGlasgow H.Murray&Co.	1873		2Cy.21"& 37"-24" 60 HP. Kincaid,Donald&Co.,Greenk.	J.T.North
217 120	✱I	**Maria P.** I Sc *Dalbano* -87 1 *Dk & Awn Dk*	Italian Genoa	457 722	175·0 27·1 12·9 19·8	Sunderland SunderlandS.B.Co.,Ld.	1886	4 Blhds. DB 66 FPT31	2Cy.27"& 47"-33" 98 HP. N.E.MarineE.Co.,Sunderld.	A.Profumo&…
218		**Maria Pia** I Sc *Reis* 1 *Dk2 B*	Brazilian Rio Janeiro	401 588	184·5 24·8 14·7 *P 63 f. F 26 f.*	PortGlasgow J.Reid&Co.	1863	5 Blhds.	2Cy.30"& 58"-30" 130 HP. Cp.78,Seaward&Co.London	J.J.dos Reis,jn
219 227		**Maria Teresa** (or Thalia) Stl ScSr 3 *Dk*	Austrian Trieste	2011 3673	374·0 39·0 27·4	Trieste Lloyd Austro-Ungarico	1883		2Cy.43"&80"-52" 450 HP. Arsenale Lloyd,Trieste	Lloyd Austro-Ungarico
220 136	✱	**Maria Teresa** I Sc *Carrubo* (ex Africano) 1 *Dk2 B*	Italian Genoa	348 574	202·5 25·5 14·2 *Q 38 f. F 30 f.*	Port Glasgow Blackwood&Gordon	1866	5 Blhds.	2Cy.56¼"&51¼"-27" 100 HP. Comp.'77,Touache,Marseilles	P.Milesi
221		**Mariana** I	Haytian Pt.au Prince	149	133·0 18·0 10·0	Glasgow	1863			B.Rivere
222	✱	**Marianela** Stl ScSr *Pick* 87-88 1 *Dk*	Spanish Santander	112 165	113·5 20·0 9·0 *Q 15 f. F 17 f.*	Irvine D.Macgill&Co.	1888	3 Blhds.	2Cy.14"&28"-18" 33 HP. Ross&Duncan,Glasgow	E.Doriga
223 82944	✱	**Marie** I ScSr *Shotton* 82-86 1 *Dk2 B*	British London	865 1346	250·0 33·2 17·5 *Q 30 f. B 46 f. F 18 f.*	Newcastle Palmers'Co.,Ld.	1880	3 Blhds. DBaf[illegible] APT[illegible]	2Cy.28"& 54"-36" 130 HP. Palmers'Co.,Ld.,Newcastle	Burdick & Co
224 30131		**Marie** I ScBk	British London	383 511	190·5 24·0 .. *Q 35 f.*	Dantzig	1863		2Cy. 92 HP.	Bremer,Bremer Bremer
225 76777	✱	**Marie** I ScSr *Kerr* -83 1 *Dk*	British Glasgow	59 154	110·1 19·2 9·1 *Q 45 f. F 19 f.*	Paisley J.Fullerton&Co.	1877	3 Blhds. FPT13	2Cy.18"& 32"-22" 45 HP. W.King&Co.,Glasgow	A.F.Blackater
226	V	**Marie** (ex Ada Barter) I ScSr *Schirels* 83-85 1 *Dk2 B*	French Dunkirk	533 804	234·5 26·2 17·2 *F 32 f.*	Greenock Scott&Co.	1873	WB	2Cy.23"& 46"-36" 100 HP. GreenockFoundryCo.,Grck.	Cie.Gén.des Ba… vap.à hel.du…
227	V	**Marie** I ScSr *de St.Jean* -80 1 *Dk2 B*	French Bordeaux	479 855	222·7 28·1 17·0 *F 31 f.*	Glasgow London&Glasg.Co.	1869	WB	2Cy.20½"&47½"-30" 90 HP. London&Glas.Co.,Glasgow	Worms,Jos…
228	V	**Marie** I ScSr *Hohlmann* 82-82 2 *Dks*	German Apenrade	704 965	201·1 28·6 19·2	Hamburg Blohm&Voss	1882		2Cy.24"& 42"-33" 90 HP. Blohm&Voss,Hamburg	M.Jebsen
229	✱N	**Marie** Stl ScSr *Schjøtt* -88 1 *Dk & Wb Frms*	Norweg'n Bergen	1345 2059	270·0 36·5 20·0 *P 25 f. F 29 f. B 114 f. Q 70 f.*	W. Hartlepool W.Gray&Co.	1888	4 Blhds. DBaf[illegible]	3Cy.20",31½"&53"-36" 160 HP. [illegible]Eng.Co., [illegible]Hartlepool	J.Christensen…
230	N	**Marie** W ScSr *Jonassen* 76-76 1 *Dk2 B*	Norweg'n Bergen	193 277	126·8 24·7 10·2 *Q 29 f. B 47 f.*	Malmö	1872		2Cy.14½"&31½"-24" 40 HP. [illegible]Mek.Verks.Malmö	P.Solensteen
231	✱	**Marie Fanny** I ScSr *Bectel* 83-85 1 *Dk2 B*	French La Rochelle	464 698	195·5 27·5 15·0 *Q 70 f. B 30 f. F 12 f.*	Sunderland Austin&Hunter	1878	4 Blhds. DB[illegible] AFT12	2Cy.23"& 45½"-30" 85 HP. R.J.Marshall,SouthShields	D'Orbigny & fils
232 83891	✱	**Marie Fleurie** I ScSr *Bower* 85-85 1 *Dk*	British Newcastle	323 510	166·0 25·2 12·9 *Q 84 f. B 17 f. F 22 f.*	SouthShields J.T.Eltringham	1880	4 Blhds. DB[illegible] FPT28	2Cy.22"& 40"-30" 70 HP. J.P.Rennoldson,S.Shields	Fisher,Renwick
233		**Marie Henriette** I Pad	Belgian	419	200·0 24·8 7·3	Antwerp Soc.Cockerill	1870			Belgian Gover…
234		**Marie Louise** I ScSr *Davin* 75-88 2 *Dks*	French Marseilles	427 639	202·1 27·2 15·8 *Q 22 f. F 15 f.*	Newcastle PalmerBros.& Co.	1861		2Cy.26½"&50"-30" 100 HP. Comp.'76,Fraissinet&Co.,Mars.	Fraissinet & Co
235	V	**Marie Woermann** Stl ScSr *Paslert* 88-88	German Hamburg	1297 1743	270·0 35·1 21·5 *P 40 f. B 72 f. F 34 f.*	Hamburg Blohm&Voss	1887		3Cy.20"31"50"& 42" 200 HP. Blohm&Voss,Hamburg	Afrikanische Dampfs. Act.Ges.(Woermann)
236		**Mariehamn** I ScSr (ex Hogenas, ex Fredriksborg) 1 *Dk*	Russian Abo	84 120	94·5 16·7 7·5 *Q 30 f. B 27 f. H 6 f.*	Stockholm W.Lindberg's M.V.	1869		2Cy.14"& 24"-21" 30 HP. W.Lindberg's M.V.,Stockholm	Angf.Actiebolaget "Mariehamn"
237		**Mariestad** C ScSr *Andersson* 1 *Dk*	Swedish Mariestad	168 227	103·0 20·6 13·5 *B 10 f. H 5 f.*	Lödöse	1872		30 HP.	Robert Thalia…
238	V	**Marietta** I ScSr *Lehnert* 83-83 2 *Dks*	German Kiel	282 380	155·8 21·3 15·0	Kiel NorddeutscheWerft	1871	WB	2Cy.23½"& 36"-27" 64 HP. Schweffel&Howaldt,Kiel	F.L.P.Ivers
239 85886	✱	**Marima** I ScSr *Williams* 1 *Dk2 B*	British [illegible]	1068 1672	257·7 34·6 17·9 *Q 66 f. P 26 f. B 68 f. F 31 f.*	Hartlepool E.Withy&Co.	1881	4 Blhds. [illegible]	2Cy.33"& 64"-33" 150 HP. T.Richardson&Sons,Hartpl.	Steel,Young & Co
240		**Marine** I ScSr *Lacroix* 1 *Dk*	French [illegible]	72 153	99·4 21·0 10·9	St.Denis Claparède&Co.	1869		2Cy.37½"-30" 100 HP. Claparède&Co.,Rouen	ChambredeCo…

✱ = Lloyd's Register; D = Nederlandsche Vereeniging; G = Germanischer Lloyd; I = Registro Italiano; R = American Record;
A = Veritas Austro-Ungarico; H = Veritas Ellenico; N = Norske Veritas; V = Bureau Veritas.

In the 2nd Column, B.T., with figures following, signifies Date of report of bottom sighting by Board of Trade Surveyors as condition of PASSENGER Certificate.

In the 3rd Column, C = Composite (iron frame, wood planked); I = Iron; Stl = Steel; W = Wood.

Further particulars of Vessels classed by Lloyd's Register will be found in Lloyd's Register Book and Lloyd's Yacht Register.

Classe	Nom, Matériaux, Gréement, etc. / Capitaine et dates d'entrée au service des Armateurs actuels à bord du navire / Ancien Nom, nombre de ponts, etc.	Pavillon. / Port d'Armement	Tonnage Net / Total	Dimensions Long. / Larg. / Creux	Construit. Où / Par qui.	Année	Nombre des cloisons. / Détails sur le lest d'eau.	Machines. Détails. / Nom du Constructeur.	Armateurs ou Directeurs-Gérants.
	Mariner I ScBu Black 79-81	British Liverpool	861 [illegible]	250·1 | 32·0 | 21·2	Stockton [illegible]	1881		2Cy.30"& 55"–36" 120 HP. Blair&Co.,Ld.,Stockton	Charente S.S.Co.,Ld.
	Marinho Visconde I PadSr	Brazilian Bahia	374 538	200·0 | 22·3 | 16·0	Greenock Caird&Co.	1879		2Cy.26"& 52"–54" 110 HP. Caird&Co.,Greenock	Cia. Bahiana de Nav.à Vap.
V	Mario I ScSr 2 Dks	Italian	665 1049	212·7 | 29·8 | 20·0	Dumbarton W.Denny&Bros.	1862	CDB[illegible]	2Cy.38"&56"–36" 120 HP. Denny&Co.,Dumbarton	Capellino Fratelli
✻D	Marion I ScSr Jeffrey 76-83 2 Dks 3 B	British	1356 [illegible]	278·9 | 35·0 | 24·1 P28 B58f. F30f.	Sunderland J.L.Thompson&Sn.	1880	5 Bhds. DBa&c 522 FPT[illegible] APT[illegible]	2Cy.33"& 62"–42" 180 HP. J.Dickinson,Sunderland	J.Gray & Co.
✻	Marion I ScSr Dabbert 2 Dks	British	642 [illegible]	231·0 | 29·3 | 16·1 P&A115f. F27f.	Sunderland J.Blumer&Co.	1872	4 Bhds.	2Cy.27"& 50"–30" 99 HP. R.&W.Hawthorn,Newcastl.	S.S.Marion Co.,Ld. (Curwen Bros.)
✻	Marion I ScSr Hassll 1 Dk	British Liverpool	67 132	90·5 | 19·7 | 8·1	Paisley AbercornS.B.Co.	1879	3 Bhds.	2Cy.14"& 21"–18" 25 HP. Comp.'98.D.Rollo&Sons,L'pool.	Mississippi& Domini'n S.S.Co.,Ld.
✻	Marion Lee I ScSr Morgan 88-88 1 Dk 2 B	British Workington	438 688	185·0 | 28·4 | 15·3 Q66f. B29f. F19f.	Whitehaven WhitehavenS.B.Co.	1882	4 Bhds. DBa&c [illegible]	2Cy.22"& 44"–30" 80 HP. Dunsmuir&Jackson,Glsgw.	West Cumberland S.S. Co.,Ld.
	Mariposa I ScBr Hayward 84-84 2 Dks & Awn Dk	American S. Francisco	1939 3158	314·0 | 41·0 | 17·3 26·0	Philadelphia W.Cramp&Sons	1883		3C 43"61"&61"–51" 3000 HP. W.Cramp&Ss.,Philadelphia	Oceanic S.S.Co.
✻	Maristow StlScSr Thomas 88-88 1 Dk 2 B	British Plymouth	1074 1619	260·0 | 37·1 | 17·1 Q104f. F35f. B19f.	Leith Ramage&Ferguson	1889	4 Bhds. DB[illegible] FPT[illegible] APT[illegible]	3Cy.19"30"&51"–39" 160 HP. Ramage&Ferguson, Leith	"Maristow" S.S.Co., Ld. (Bellamy&Co.)
✻D	Maritana I ScSr Saville 77-82 1 Dk 2 B	British Sunderland	1216 1862	277·0 | 36·6 | 20·5 Q72 P28 B56 F36f.	Sunderland J.L.Thompson&Sn.	1882	6 Bhds. DBa&c 232 FPT39 APT29	2Cy.32"& 60"–39" 160 HP. G.Clark,Sunderland	Culliford & Clark
✻D	Mark Lane I ScSr Pengelley 2 Dks 3 B	British London	1354 2095	287·0 | 35·3 | 24·3 F[illegible] P[illegible] B[illegible]	Newcastle Palmers'Co.,Ld.	1878	5 Bhds. DBa&c [illegible] APT[illegible]	2Cy.35"& 64"–42" 180 HP. Palmers'Co.,Ld.,Newcastle	J.Cory & Sons
✻	Marlborough I ScSr Kidder 88-88 1 Dk & Spr Dk 3 B	British London	1164 1787	261·5 | 33·1 | 18·8 26·1	Middlesbro' R.Dixon&Co.	1877	5 Bhds. DBa&c [illegible]	2Cy.32"& 60"–39" 150 HP. Blair&Co.,Ld.,Stockton	Galbraith,Pembroke & Co.
✻	Marmion I ScSr Thomson 75-88 2 Dks	British Leith	590 1083	249·3 | 30·4 | 18·3 P61f. B30f. F71f.	Glasgow J.G.Lawrie	1871	5 Bhds.	3C 26½"42"&67"–45" 300 HP. S.&H.Morton&Co.,Leith	London & Edinburgh Shipping Co.
✻	Marmion I ScSr Watts 86-86 1 Dk & Awn Dk	British Glasgow	408 777	200·0 | 27·1 | 12·7 I 19·8	Glasgow Scott&Co.	1886	1 Bhds. DBa&c FPT36	3C.20"32"&53"–36" 140 HP. J.&T.Young,Ayr	W.D.B.Janes
	Marmora I Pad Mehemed 78-86	Turkish Cnstntinple.	153	180·0 | 27·0 | 10·3	London J.&W.Dudgeon	1874		2Cy.37" 120 HP.	Idarei Massousieh (Otteman Steam Nav.Co.)
✻	Marne (ex Perthshire Lass) I Sc3mSr Cousens 84-84 1 Dk	British Plymouth	125 238	124·0 | 21·0 | 12·7	Glasgow [illegible]	1864	3 Bhds.	2Cy.22"–22" 50 HP. CanalBasinFdry.Co.,Glsgw.	E.Stock & Son
✻	Maroon I ScSr Spray 71-83 1 Dk 2 B	British Gloucester	1029 1591	240·0 | 35·4 | 18·3 Q93f. P39f. B62f. F50f.	Chepstow E.Finch&Co.,Ld.	1883	4 Bhds. DBa&c [illegible] APT[illegible]	2Cy.28"& 54"–36" 135 HP. Day,Summers&Co.Sthamptn	Anderson, Anderson & Co.
	Marousya I TwnScSr	Russian Gostoff	293 372	185·4 | 27·9 | 8·4	Newcastle C.Mitchell&Co.	1872		4Cy.14"& 28"–15" 60 HP. Thompson,Boyd&C.N'castl.	Volga-Don Railway & Steam Nav.Co.
✻	Marple Stl TwnScTug Hollingsworth -88 1 Dk	British Grimsby	104	91·0 | 20·1 | 8·5	Hull Earle's S.B.& E.Co.,Ld.	1888	3 Bhds. DB[illegible]	4Cy.15"&28"–18" 70 HP. Earle's SB&Eng.Co.,Ld.,Hull	Manchester,Sheffield & Lincolnshire Railway Co.
	Marquado (ex Surrey) I ScSr Barbani 1 Dk	Austrian Trieste	118 188	122·7 | 19·9 | 10·5	London	1868		2Cy. 50 HP.	G.Cav. Mauser
	Marques de Caxias I PadSr (ex Juno, ex Dacotah)	Brazilian Bahia	403 631	226·0 | 23·8 | 13·0	Greenock Caird&Co.	1864		2Cy.44"–18" 110 HP. Caird&Co.	Cia.de Nav. à Vap. de Bahia
✻D	Marques de Mudela I ScSr Oncendo 80-88 1 Dk & Spr Dk 3 B	Spanish Bilbao	1241 1913	280·5 | 38·2 | 22·3 B24f.	Glasgow A.Stephen&Sons	1883	4 Bhds. [illegible]	2Cy.35"& 65"–42" 100 HP. A.Stephen&Sons,Glasgow	J.Martinez de las Rivas
✻	Marquesa de Santurce StlScSr Urresticola -88 1 Dk 2 B	Spanish Bilbao	1510 2332	290·0 | 40·2 | 18·2 Q109f. B64f. F33f.	Newcastle Palmers'Co.,Ld.	1888	5 Bhds. [illegible]	3Cy.22"35"&58"–42" 270 HP. Palmers'Co.,Ld.,Nwcl	Don Pedro de Galindez
	Marquis I ScSp	British Bristol	73 119	94·0 | 18·5 | 9·5	SouthShields T.D.Marshall	1855		15 HP.	"Marquis" S.S.Co.,Ld. (R.C.Ring)
✻D	Marquis Scicluna I ScSr Morris 85-86 1 Dk 2 B	British N.Shields	1024 1549	260·0 | 36·2 | 18·6 Q77f. P74f. B[illegible] F49f.	Sunderland SndrlndS.B.Co.,Ld.	1883	4 Bhds. DBa&c 269 FPT[illegible] APT[illegible]	2Cy.32"& 62"–40" 162 HP. R.&W.Hawthorn,Newcastl.	G.W.Allen & Co.
	Marquis of Bute I PadSk McLean 63-68	British Glasgow	38 163	196·6 | 18·1 | 7·3	Glasgow Barclay,Curle&Co.	1868		1Cy.48"–60" 85 HP. Barclay,Curle&Co.,Glasgow	A.T.M'Lean
✻R	Mars (ex Scotia) I ScSr Hawes 88-88 2 Dks 3 B	American Boston,Mass.	1628 2192	327·0 | 39·5 | 23·1 B82f. F49f.	Leith S.&H.Morton&Co.	1881	6 Bhds. [illegible]	2Cy.40"&76"–48" 375 HP. S.&H.Morton&Co.,Leith	Boston Towboat Co.
	Mars (COLLISION) I ScSr Nicolich 70-87 2 Dks	Austrian Trieste	1183 1836	286·6 | 35·0 | 24·0	Trieste LloydAustro-Ung.	1868		1Cy.29½"&54"–36" 100 HP. Denny&Co.,Dumbarton	Lloyd Austro-Ungarico
DV	Mars I ScSr Binkhorst 74-87 1 Dk 2 B	Dutch Amsterdam	684 936	212·7 | 31·1 | 15·4 Q70f. B[illegible] F[illegible]	Sunderland J.Laing	1880	4 Bhds. FPT70 APT[illegible]	2Cy.27"& 50"–36" 100 HP. G.Clark,Sunderland	Kon.Nederl.Stoomb. Maats.
	Mars I ScSr 1 Dk 2 B	Norweg'n Bergen	369 592	160·7 | 24·1 | 13·1 P81f. F25f.	Bergen BergensMek.Verksted	1878	1 Bhds.	2Cy.19½"& 37"–24" 45 HP. BergensMek.Verkst,Bergen	Det Bergenske Dampskibsselskab

✻=Lloyd's Register; D=Nederlandsche Vereeniging; G=Germanischer Lloyd; I=Registro Italiano; R=American Record; A=Veritas Austro-Ungarico; H=Veritas Ellenico; N=Norske Veritas; V=Bureau Veritas.

Dans la 2me Colonne, B.T., avec les indications qui suivent, signifie la date du rapport de visite de la Carene par les experts du Board of Trade pour obtenir le certificat relatif au TRANSPORT DES PASSAGERS.

Dans la 3me Colonne, C=Composite (membrure en fer, bordé en bois); I=Fer; Stl=Acier; W=Bois.

Pour plus amples renseignements sur les navires classés par le Lloyd's Register, voir le Lloyd's Register Book et le Lloyd's Yacht Register.

No. in Book / Official No.	Registry in which classed	Ship's Name. Material, Rig, &c. / Master, and Dates of Appointment to present Owner's Service and to Ship / Ship's Late Name. No. of Decks, &c.	Flag / Port of Registry	Tons Net / Gross	Dimensions: Length	Breadth	Depth	Built: Where / By Whom	When	No. of Bulkheads / Particulars of Water Ballast	Engines: Particulars / Maker's Name	Owners
271	✻	**Marsala** I SeSr *Moss* 84-85 *2 Dks3 B*	German Hamburg	1874 2397	320·2	36·2	24·6	Glasgow A.Stephen&Sons	1882	1 Bhds.	2Cy.36"&68"-42" 240 HP. A.Stephen&Sons,Glasgow	R.M.Sloman&
272 ?		**Marsala** (ex Ravenna) *Castiano* 87-… I SeBk *2 Dks3 B*	Italian Palermo	1013 1656	259·8	31·4	24·4	Renfrew Henderson,Coulborn&Co.	1870		2Cy.27½"&47"-48" 472 HP. Hndrsn.,Coulbrn.&Co.,Rfw.	Navigazione Italiana
273 [illegible]	BT6 88	**Marsdin** I SeSr *Roach* *1 Dk2 B*	British Hull	947 1456	258·0	34·3	18·2	Hull Humphrys&Co.	1870	5 Bhds.	2Cy.33"&60"-36" 150 HP. C.D.Holmes&Co.,Hull	T.Wilson,Sons
274	V	**Marseille** I Se4msSr *Mollieres* 86-86 *2 Dks&Spr Dk*	French Havre	2730 3514	379·2	40·0	21·7	Havre Forges &C.dela Med.	1883		2Cy.41"&71"-43" 400 HP. Forges&C.delaMedit.Havre	Cie.Com… Transports a V…
275	✻	**Marseille** (ex Bispham) I SeSr *Melchertsen* …-83 *2 Dks3 B*	German Hamburg	1060 1380	250·0	33·3	22·5	Sunderland J.L.Thompson&Sons	1876	1 Bhds. DB[illegible]	2Cy.32"&60"-36" 160 HP. …Dickinson,Sunderland	R.M.Sloman&
276 [illegible]	BT 88	**Marseilles** I PadSr *Griffiths* 84-88	British Liverpool	302 432	218·9	23·1	11·3	London	1864		2Cy.52"-57" 180 HP. …Langley,London	R.& D.Jones
277		**Marsella** (ex Tamathon) I Bk	Spanish Barcelona	622 957	164·7	27·2	18·1	Glasgow Barclay,Curle&Co.	1855		2Cy. 135 HP. A.&J.Inglis,Glasgow	Jimenez y Ag…
278	✻D	**Marshal Keith** I SeSr *Kröger* -88 *1 Dk2 B*	German Hamburg	300 496	165·0	23·7	12·5	Dundee Pearce Bros.	1883	5 Bhds. DB[illegible]	2Cy.22"&40"-27" 70 HP. Pearce Bros.,Dundee	Dampfs.Act.G…
279		**Marstrand** Stl SeSr *Lund* 84-85 *1 Dk*	Danish Copenhagen	173 344	143·7	22·3	12·5	Diedrichshof. G.Howaldt (Kiel)	1868		3Cy. 87 HP.	Dampskibssel. "Marstrand"
280		**Marstrand** I SeSr *Paterson* -86 *1 Dk*	Swedish Uddevalla	102 153	99·1	18·4	11·4	Gothenburg	1853		40 HP.	Ångbåts Akt… [illegible]
281 [illegible]	✻ BT11 88	**Martaban** I Se3msSr *Taylor* 76-84 *2 Dks3 B*	British Glasgow	1698 2623	350·6	36·4	27·2	Dumbarton W.Denny&Bros.	1873	5 Bhds.	3C.21½"32"43"&62½"-42" …pld'88 Denny,Dbtn [202HP]	British & Bur… Steam Nav.C…
282		**Martaban** I SeSr *1 Dk*	Japanese	390 535	193·4	26·0	14·5	Dumbarton W.Denny&Bros.	1865		2Cy.27"&43½"-24" 80 HP. Laird Bros.,Birkenhead	
283 88199		**Martello** I SeBk *Abbott* -87 *2 Dks3 B*	British Hull	2439 3769	370·0	43·1	28·4	Hull Earle's Co.,Ld.	1884	8 Bhds. CDB[illegible]	3C.31"50"&82"-57" 400 HP. Earle's Co.,Ld.,Hull	T.Wilson,Sons
284		**Mårten** C SeSp *Ryr* -87 *1 Dk*	Swedish Thorskog	129 181	100·2	21·0	9·8	Thorskog P.Larsson	1887		2Cy.12"&18"-16" 25 HP.	P.Larsson
285	✻	**Martha** Stl SeSr *Mohr* -88 *1 Dk&Spr Dk3 B*	German Kiel	1560 2107	286·4	36·1	24·9	Kiel G.Howaldt	1884	8 Bhds. [illegible]	2C.35½"&61"-43½" 250 HP. G.Howaldt,Kiel	George Hows…
286	✻	**Marthe** (ex Maitland) I Se *1 Dk*	French Dieppe	23 66	80·0	18·0	8·0	Dundee W.B.Thompson&Co.,Ld.	1887	3 Bhds.	3C.7½"11"&20"-14" 19HP. W.B.Thmpsn.&Co.,Ld.Dun	A.M.R.Barb…
287 73554	✻	**Martin** I Se3msSr *Whittle* *2 Dks*	British London	641 995	230·0	29·5	16·7	Newcastle C.Mitchell&Co.	1875	4 Bhds. [illegible]	2Cy.33"&62"-36" 180 HP. Thompson&Co.,Newcastle	General S…
288		**Martinique** (ex Vera Cruz,exImperatrice) I Se3msSr *Valentin* -86 *2 Dks*	French Havre	1111 1675	274·6	36·7	23·9	Liverpool J.Laird	1854		2Cy.39½"&80"-39" 300 HP. J.Elder&Co.,Glasgow	Compagnie… Transatlantique
289	✻	**Martos** I Se *Vives* 78-85 *1 Dk&Spr Dk3 B*	Spanish Valencia	980 1502	234·0	33·1	16·7	Dundee Gourlay Bros.&Co.	1883	4 Bhds. [illegible]	2Cy.30"&56"-39" 155 HP. Gourlay Bros.&Co.,Dundee	J.Sister & Co.
290 81212	✻	**Mary** I SeSr *Prest* 88-88 *1 Dk2 B*	British Whitby	839 1298	243·0	32·9	17·5	South Shields J.Readhead&Co.	1879	4 Bhds. DB:76ft	2Cy.27½"&52"-33" 110 HP. J.Readhead&Co.,S.Shields	R.Harrowing
291 94002	✻	**Mary** SeK *Hudson* -88 *1 Dk*	British Grimsby	59 108	90·0	19·5	10·4	Grimsby T.Charlton	1888	3 Bhds.	2Cy.14"&26"-18" 30 HP. T.Charlton,Grimsby	T.Charlton
292 89735	✻	**Mary** I SeTug *O'Brien* *1 Dk*	British Bassein	20 152	102·6	19·3	12·5	Falmouth Cox&Co.	1884	6 Bhds.	2Cy.22"&40"-26" 70 HP. Cox&Co.,Falmouth	C.O'Brien
293 94322	✻	**Mary** StlSeTug *1 Dk*	British Aberdeen	 30	55·1	12·7	6·7	Aberdeen A.Hall&Co.	1887	1 Bhds.	2Cy.11"&20"-15" 25 HP. A.Hall&Co.,Aberdeen	C.Whittall&…
294	V	**Mary** I SeSr *Remmersen* 85-85 *1 Dk*	Russian Riga	770 1060	224·7	30·5	14·8	Elsinore Helsingørs Jernsk.Bygs…	1884	W B[illegible] DB[illegible]	2Cy.26"&51"-38" 120 HP. Helsingørs.Jernsk.BygriElsre	Helmsing &…
295 76785		**Mary Ann** I SeSk *Macallister* -86 *1 Dk*	British Glasgow	37 131	100·2	18·9	8·7	Paisley H.McIntyre&Co.	1877	3 Bhds.	2Cy.17"&31"-22" 45 HP. Lees,Anderson&Co.Glasgow	J.Johnston,… (D.P.Mackay)
296 95166	✻	**Mary Anning** I SeSr *Evans* -88 *1 Dk&Web Frames*	British Cardiff	1349 2057	277·0	37·1	18·5	Stockton Richardson,Duck&Co.	1888	1 Bhds. CDB[illegible]	3C.21"35"&57"-39" 200 HP. Blair&Co.,Ld.,Stockton	Anning Bros
297 53204		**Mary Austin** (ex Tessy) I SeSr *Scott* -86 *1 Dk*	British Penang	140 288	120·1	19·1	9·8	Newcastle Marshall	1865		2Cy.13"&26"-21" 30 HP. Tod&McGregor,Glasgow	Creasy Ewe…
298 85182	BT6 88	**Mary Beatrice** StlPadSr *Kilbey* 84-86 *1 Dk*	British London	304 803	255·0	29·2	14·3	London Samuda Bros.	1882	5 Bhds.	2Cy.72"-66" [illegible] HP. J.Penn&Son,London	South Eastern… way Co.
299 76345	✻ BT1 88	**Mary Hough** I Se3msSr *Holmes* 87-87 *1 Dk2 B*	British Liverpool	528 868	225·6	29·3	15·3	Renfrew Lobnitz,Coulborn&Co.	1876	5 Bhds. [illegible]	2Cy.27"&54"-42" 135 HP. LobnitzCoborn&Co.,Renfrew	S.Hough
300 86957	✻D	**Mary Lohden** I SeBn *Bird* 83-83 *1 Dk2 B*	British W.Hartlepool	876 1363	244·3	33·0	15·7	W.Hartlepool R.Irvine&Co.	1883	4 Bhds. [illegible]	2Cy.30"&57"-36" 130 HP. T.Richardson&Sons,Hartpl.	J.Lohden & Co.

✻=Lloyd's Register; D=Nederlandsche Vereeniging; G=Germanischer Lloyd; I=Registro Italiano; R=American …
A=Veritas Austro-Ungarico; H=Veritas Ellenico; N=Norske Veritas; V=Bureau Veritas

In the 2nd Column, B.T., with figures following, signifies Date of report of bottom sighting by Board of Trade Surveyors as condition of PASSENGER …

In the 3rd Column, C=Composite (iron frame, wood planked); I=Iron; Stl=Steel; W=Wood.

Further particulars of Vessels classed by Lloyd's Register will be found in Lloyd's Register Book and Lloyd's Yacht Register.

No.		Nom, Matériaux, Gréement, etc. Capitaine et dates d'entrée au service des Armateurs actuels et à bord du navire. Ancien Nom, nombre de ponts, etc.		Pavillon. Port d'Armement	Tonnage Net Brut	Dimensions. Long.	Larg.	Creux	Construit. Où. Par qui.	Quand.	Nombre des cloisons. Détails sur le lest d'eau.	Machines. Détails. Nom du Constructeur.	Armateurs ou Directeurs-Gérants.
		Mary Louise *Collingwood*	W PadTug 1 *Dk*	British S.Shields	10	98·6	19·1	9·8	SouthShields J.P.Rennoldson	1879		2Cy.31"–54" 50 HP. J.P.Rennoldson,S.Shields	J.Wardle
	✻	Mary Monica *Kerr 79-75*	I ScSr 1 *Dk*	British Dublin	193 385	170·2 *F 18f.*	23·1 *H 94.*	13·2	PortGlasgow Murdoch&Murray	1879	3 Blds.	2Cy.22"& 40"–30" 70 H J.&T.Young,Ayr	J.J.Carroll
	✻	Mary Nixon *Robertson*	I Sc3mSr 1 *Dk2B*	British Newcastle	488 781	201·0 *Q 66f.*	28·1	17·4	Newcastle PalmerBros.	1865	3 Blds. DB1158	3C.16½"26"&13"–36"99HP. WallsendSlipwayCo.,Nwc.	H.James (J.Fenwick & Son)
		Mary Powell	W Pad	American Albany,N.Y.	877 984	288·9	34·1	9·0	New York	1869			T.Cornell
		Mary Stuart	I Pad 1 *Dk*	British Calcutta	112 330	144·2	23·1	11·2	London	1857		2Cy.52½"–42" 172 HP. Ravenhill&Co.,London	Charles J.Sutherland
		Mary Theresa	I TwnSc 1 *Dk*	British Calcutta	90 138	119·8	16·2 *H 34f.*	7·5	Calcutta StewartBros.&Co.	1876		2Cy.14½"–14" 30 HP. King&Lochead,Calcutta	John King
	✻	Maryland *Lockhurst*	Stl Sc3mSr 2 *Dks&B*	British Hartlepool	1852 2863	320·0 *P 39f.*	41·0 *B 95f.*	25·7 *F 49f.*	W.Hartlepool W.Gray&Co.	1885	6 Blds. APT681 APT93	3C.25"42"&68"–45"300HP. CentralMar.Eng.Co.,W.Hlp.	"Maryland" S.S.Co., Ld. (Williams,Torrey&Feild,Ld.)
		Maryland	I Pad	American N.Haven,Ct.	882 1093	291·3	36·5	10·2	WilmingtnDel Harlan & Hollingsworth Co.	1853		2Cy.40"–96" The Harlan&HollingsworthCo., Wilm., Del.	E.M.Reed
309		Masayoshi Mara	I ScSr	Japanese Osaka	353	181·2	24·4	17·7	Scotland	1864		94 HP.	Fukunaga Shohichi
310	✻I	Maschal Andrea *F. Degrossi (ex Isabella)*	I ScSr 1 *Dk*	Italian Leghorn	278 447	158·8 *Q 49f.*	23·5 *B 38f.*	15·3 *F 18f.*	Newcastle T.&W.Toward&Co.	1883	1 Blds. APT30 APT73	2Cy.21"& 38"–24" 61 HP. WappingE.W.Co.,S.Shields	Frat.Corradini
311	✻D	Mascotte *Ray*	I ScSr 1 *Dk&Spr Dk3B*	British Sunderland	2113 3192	320·0 *B 66f.*	40·0 *F 32f.*	27·7	Hull D.P.Garbutt	1883	7 Blds. DB&f 124t	2Cy.40"& 72"–48" 300 HP. C.D.Holmes&Co.,Hull	W.Briggs & Sons
312	✻	Mascotte *Herrington*	I ScSr 1 *Dk*	British Middlesbro'	186 311	140·0 *Q 80f.*	22·1 *B 12f.*	10·4 *F 20f.*	Middlesbro' W.Harkess&Son	1883	3 Blds. APT184	2Cy.18½"& 34"–21" 45 HP. Blair&Co.,Ld.,Stockton	Mrs.H.Harkess
313	R	Mascotte *Flemming*	I ScSr 1 *Dk*	American New Haven	520 884	207·9	30·0	20·6	Philadelphia W.Cramp&Sns.,S.&E.B.C.	1885		3Cy.19"30"& 50"–30"	H.B.Plant
314	✻	Mascotte *Milne 83-88*	Stl ScSr 2 *Dks*	Dutch Rotterdam	606 1060	249·8 *P 30f.*	30·9 *B 64f.*	16·4 *F 51f.*	Leith Ramage&Ferguson	1885	5 Blds.	2Cy.32"& 61"–42" 180 HP. Ramage&Ferguson,Leith	A.J.F.Burger
315		Maskelyne *Bairby 67-75*	I ScBg	Belgian Antwerp	1678 2665	350·5	36·9	27·8	Newcastle A.Leslie&Co.	1874		2Cy.38"& 76"–51" 300 HP. R.Stephenson&Co.,N'castle	Soc. Royale Belge Sud-Américaine
316		Masnedsund *Mogensen 85-88*	I ScSr 1 *Dk*	Danish Copenhagen	142 251	124·4 *P 15f.*	22·2 *B 40f.*	10·7	Copenhagen Burmeister&Wain	1880	4 Blds. WB APT44	1Cy.15"& 30"–20" 60 HP. Burmeister&Wain,Cpnhgn.	De Danske Statsbaner
317		Massachusetts *Griffin 78-87*	W Pad	American Providence	2171 2607	323·8	42·5	15·9	Greenpoint L.I	1876	V12.82		Providence & Stonington S.S.Co.
318		Massandra	I ScSr	Russian Odessa	338	132·5	22·3	12·2	Delfshaven Christie&Nolet	1867		50 HP.	Prince Woronzow
319	✻	Massilia *Shallard 72-84*	Stl ScBk 2 *Dks Spr Dk4B*	British Greenock	2742 4902	420·1	45·1	26·3 31·6	Greenock Caird&Co.	1884	7 Blds.	2Cy.56"&105"–72"1000HP. Caird&Co.,Greenock	Peninsular & Oriental Steam Nav.Co.
320		Massis	Stl&I ScSr	Russian S.Petersbg.	283 418	170·0 *P 28f.*	25·6 *F 6f.*	11·0	Newcastle C.Mitchell&Co.	1882	WB	2Cy.14"& 25"–18" 60 HP. WallsendS.&E.Co.Newcastle	I.Pitoff,of Tiflis
321		Mastiff *Neill 73-88*	I ScSr	British Glasgow	356 873	230·4 *P 124f.*	30·3 *F 41f.*	14·8	Glasgow J.&G.Thomson	1878		2Cy.37½"&64½"–48"220HP. J.&G.Thomson,Glasgow	G.& J.Burns
322		Mastiff *Antle 87-87*	W ScSr 1 *Dk*	British S.Johns,N.F.	245 360	137·4	26·9	16·1	Dundee	1867		2Cy. 60 HP.	Robert S.Munn
323	✻	Matabele *Melville 80-86*	I ScSr 1 *Dk&Spr Dk3B*	British Aberdeen	1005 1556	250·3 *B 27f.*	35·2 *F 15f.*	15·5 22·2	Aberdeen Hall,Russell&Co.	1885	5 Blds. DB&f 165t	3C.19"30"&50"–36"160HP. Hall,Russell&Co.,Aberdeen	J.T.Rennie & Son
324		Matador	I ScSrYt	British Glasgow	100 157	131·5	19·8	9·5	Glasgow J.&G.Thomson	1879		2Cy.18"& 32"–24" 40 HP. J.&G.Thomson,Glasgow	R.C.T.Blunt
325		Matapan	I Sc3mSr 3 *Dks*	French Marseilles	2641 3565	328·1 *B 45f.*	15·6	16·9 32·5	LaSeyne Forges&C.de la Med.	1884		2C.36¼"&65"–49¼" 350 HP. Forges&C.de la Medit.,Mrsl.	Messageries Maritimes
326		Mathilda Foscolo *Cornelio 78-83*	I PadTug 1 *Dk*	Greek Syra	111 192	121·6	21·2	10·6	NorthShields Hepple&Co.	1883	3 Blds.	1Cy.42½"–46" 90 HP. Hepple&Co.,NorthShields	Foscolo,Mango & Co.
327	✻	Mathilde (ex Gustavo) *da Fonseca*	I Sc	Brazilian RiodeJaneiro	169	..	..	..	Clyde John Robertson	18 81/82		50 HP. D.Rowan,Glasgow	Dr.Miranda Jordão
328		Mathilde	I ScSr 1 *Dk*	French Marseilles	99 146	108·0 *Q 7f.*	19·0	9·0	Paisley J.Fullerton&Co.	1870	3 Blds.	2Cy.15"–16" 20 HP. W.King&Co.,Glasgow	A.G.Boyé & Co.
329		Mathilde	I ScSr	French Algiers	92 175	107·2	19·0	..	Paisley J.Fullerton&Co.				A.d'Esposito & Co.
330	DV	Mathilde *Strucholder 88*	I ScSr 1 *Dk*	German Hamburg	712 1111	227·0 *Q 77f.*	33·1 *B 8f.*	15·4 *F 36f.*	Kiel Norddeutsche Werft	1882	WB	2Cy.30"& 53"–36" 140 HP. MärkischeAct.Ges.,Berlin	A.C.deFreitas & Co.

✻=Lloyd's Register; D=Nederlandsche Vereeniging; G=Germanischer Lloyd; I=Registro Italiano; R=American Record; A=Veritas Austro-Ungarico; H=Veritas Ellenico; N=Norske Veritas; V=Bureau Veritas.

Dans la 2me Colonne, B.T., avec les indications qui suivent, signifie la date du rapport de visite de la Carène par les experts du Board of Trade pour obtenir le certificat relatif au TRANSPORT DES PASSAGERS.

Dans la 3me Colonne, C=Composite (membrure en fer, bordé en bois); **I=Fer; Stl=Acier; W=Bois.**

Pour plus amples renseignements sur les navires classés par le Lloyd's Register, voir le Lloyd's Register Book et le Lloyd's Yacht Register.

No. in Book / Official No.	Registry in which classed	Ship's Name. Material, Rig, &c. Master, and Dates of Appointment to present Owner's Service and to Ship. Ship's Late Name. No. of Decks, &c.	Flag. / Port of Registry	Tons Gross / Net	Length	Breadth	Depth	Built Where. / By Whom.	When	No. of Bulkheads / Particulars of Water Ballast.	Engines. Particulars. Maker's Name.	Owners or Ma…
331	G V	**Mathilde Joost** I ScSr *Plath 82-83* 1 *Dk & Spr D*	German Rostock	495 786	186·3	27·0	12·7 136	Rostock Rostocker Schiffb.	1883 Grs.	WB	2C. 28¾"&41½"-27½" 75IP. Rostckr.Schiffb.Ges.,Rostck	Friedr.Peterse…
332		**Matias-Cousino** I ScBk 1 *Dk*	Chilian Valparaiso	643 877	210·3	30·7	18·9	Newcastle A.Mitchell&Co.	1859		2Cy.28"&53"-32" 110 IP. T.Richardson&Sons,Htpool	Cia. Esplotadr… Lota & Corone…
333		**Matias Korning** I Pad 1 *Dk*	Swedish Herrösand	68 129	122·7 *Erections 25 f.*	17·3	6·9	Stockholm W.Lindbergs Warf & Werkstads Aktiebolag	1863		2Cy.32"-34" 60 IP. W.Lindberg'sMekV.Stkldm	O.Vestin
334	✻	**Matilde** (ex Balaclava) I ScSr *Vicari -84* 1 *Dk2 B*	Italian Palermo	210 351	144·1 *Q 75 f. F 23 f.*	21·9	11·8	Glasgow Aitken&Mansel	1876	4 Blhds. [illegible]	2Cy.17"&30"-21" 50 IP. Kincaid,Donald&Co,Glsgw	G.Crovoja di G…
335		**Matin** I ScSr *Lucas 83-84* 1 *Dk*	British Sunderland	359 560	174·9 *Q 53 f. B 34 f.*	26·2	14·5	Aberdeen Hall,Russell&Co.	1873	4 Blhds. [illegible] AFT	2Cy.28½"&57½"-30" 75IP. Hall,Russell&Co.,Aberdeen	Earl of Durha…
336	✻ A	**Matlekovits** Stl ScBn *Raudrch -87* 1 *Dk2 B*	Austrian Fiume	1253 1913	260·0 *P & B 146 f. F 35 f.*	36·0	20·2	Stockton Richardson,Duck&Co.	1887	5 Blhds. [illegible]	3C.20"33"&54"-36" 160IP. Blair&Co.,Ld.,Stockton	The Adria-Hu… Sea Nav.Co.,L…
337	✻	**Matlock** I ScSr *Clark* 1 *Dk*	British Middlesbro'	202 336	150·8 *Q 36 f. B 15 f. F 21 f.*	22·1	10·3	Middlesbro' W.Harkess&Son	1884	3 Blhds. [illegible]	2Cy.20"&38"-26" 65 IP. J.&Hume&Co.,Glasgow	E.W.Harkess
338		**Matsumaye Maru** I ScSr *Araki* (ex Dragon) 1 *Dk & Awn Dk*	Japanese Tokio	374 604	166·9	24·5	19·1	Renfrew Henderson,Coulborn&Co.	1867	1 Blhds.	2Cy.16"&66"-33" 105 IP. Henderson,Colborn&C.Renf	Nippon Yusen…
339		**Mattawan** W ScSp (ABANDONED) 2 *Dks*	British Montreal	208 348	143·0 *H 119 f.*	26·0	10·4	Montreal A.Cantin	1874		1Cy.26"-30" 75 IP.	Wm.J.Pullin…
340		**Matteawan** W Pad	American New York	635 775	206·0	27·8	10·0	Keyport, N.J.	1863			J.H.Starin
341	I V	**Matteo Bruzzo** I&Stl Sc3mSr *Cafferata -86* (ex Gelconda) 3 *Dks*	Italian Genoa	2543 3910	390·4 *B 25 f. H 78 f.*	42·3	23·6	La Seyne Forges&Chantiers	1882		2C.40"&73½"-42½" 450IP. Forges&Chantiers,LaSeyne	La Veloce Na… Italiana
342		**Matteo Foscolo** I Pad (ex Smyrna) *Langussi -86*	Turkish Cnstntinple.	92 170	120·0	16·5	11·4	Newcastle Palmers Co.	1868		2Cy. 80IP.	Foscolo, Mang…
343	✻ D	**Matthew Bedlington** I ScBn *Stainthorp 85-86* 2 *Dks3 B*	British Whitby	1385 2149	281·6 *P 26 f. B 64 f. F 36 f.*	38·0	23·8	Whitby T.Turnbull&Son	1882	5 Blhds. [illegible]	2Cy.36½"&68"-42" 200IP. Blair&Co.,Ld.,Stockton	T.Turnbull & S…
344		**Maua** I Pad 1 *Dk & Hur Dk*	Brazilian Para	340 478	181·0	28·0	9·5	Birkenhead Laird Bros.	1881		2Cy.32"&52"-54" 124 IP. Laird Bros.,Birkenhead	Amazon Steam…
345	✻	**Maud** (LOST) I ScSr *Smith 76-85* 1 *Dk2 B*	British Whitby	843 1295	234·8 *P 79 f. F 28 f.*	32·0	17·8	W. Hartlepool Denton,Gray&Co.	1871	4 Blhds. [illegible]	2Cy.36"&55"-33" 120 IP. Blair&Co.,Ld.,Stockton	Marwood & So…
346 45743		**Maud** (ex Sailing Vessel) W ScBk *Potter*	British Dundee	203 298	116·7 *Q 29 f.*	25·0	15·8	Whitby T.Turnbull&Sons	1865		2Cy.15"&27"-18" 35 IP. White&Cooper,Dundee	W. Adams
347 83895	✻ D	**Maud Hartmann** I ScSr *Griffiths -85* 1 *Dk2 B*	British Newcastle	1070 1669	260·2 *Q 107 f. B 12 f. H 34 f.*	35·2	20·9	Newcastle Schlesinger, Davis & Co.	1881	4 Blhds. DBc f &c APT	2Cy.31"&60"-39" 150 IP. T.Clark&Co.,Newcastle	International… S.S.Co.,Ld.
348 86952	✻	**Maude** I ScBn *Freeman 84-88* 1 *Dk2 B*	British W.Hartlepool	1009 1593	255·6 *Q 66 f. P 28 f. B 66 f. F 31 f.*	34·8	19·5	W. Hartlepool W.Gray&Co.	1883	1 Blhds. DBc f 335t	2Cy.32"&59"-33" 150 IP. T.Richardson&Sons,Hrtlpl.	T. Appleby & C…
349 64136		**Maude** (ex Sailing Vessel) W ScSr *MacDougal* 1 *Dk*	British Vict'ria,B.C.	94 173	113·5	21·0	9·0	San Juan Burr&Smith	1872		2Cy.9"&18"-20" 13 IP. Albion Iron Works,Vict.,B.C.	Canadian Paci… Co.,Ld.
350	✻	**Maule** I ScSr *Howard 81-87* 1 *Dk*	Chilian Valparaiso	80 202	125·5 *F 20 f.*	23·6	9·0	Port Glasgow J.Reid&Co.	1885	4 Blhds. FPT [illegible] APT [illegible]	2Cy.16"&32"-30" 43 IP. Kincaid&Co.,Greenock	Compania Sal… cana de Vapor…
351		**Maulong** Stl StrnWhl	British Glasgow	148 173	130·0	28·0	5·6	Dumbarton W.Denny&Bros.	1888		2Cy.15"&26"-54" 45 IP. Denny&Co.,Dumbarton	Irrawaddy Flot… Ld.
352	✻	**Maurice-Reunion** I ScSr *Fournié 80-86* 2 *Dks3 B*	French Marseilles	1483 1948	288·0 *P 36 f. F 34 f.*	36·3	22·1	Port Glasgow W.Hamilton&Co.	1881	5 Blhds. [illegible]	2Cy.34"&61"-45" 170 IP. Dunsmuir&Jackson,Glsgw.	Compagnie Fr… de Nav.àVap…
353 63340	✻	**Mauritania** I ScSr *Peters 86-86* 1 *Dk2 B*	British Liverpool	481 764	200·1 *B 41 f.*	28·8	16·3	W. Hartlepool Denton,Gray&Co.	1871	4 Blhds. [illegible]	2Cy.25½"&47½"-33" 96IP. Blair&Co.,Ld.,Stockton	Taylor,Cammel…
354 52796	✻	**Mauritius** I ScSr *Macfarlane 87-87* 1 *Dk2 B*	British Glasgow	328 507	204·0 *P 112 f. F 24 f.*	26·4	13·9	London C.Lungley	1865	5 Blhds.	2Cy.25"&48"-27" 90 IP. [illegible]	J.& P. Hutchis…
355 95420	BT 5.88	**Mavis** Stl PadSr *Cox -88*	British London	280 537	210·3 *Q 7 f.*	26·3	9·4	Leith J.Scott&Co.	1888		2Cy.33"&61"-60" 250 IP. J.Scott&Co.,Kirkcaldy	General Steam…
356 88014	✻	**Mawhera** I ScSr *Hill* 1 *Dk2 B*	British Wellington	340 554	174·8 *Q 31 f. F 15 f.*	25·6	14·7	Paisley H.McIntyre&Co.	1883	4 Blhds. DB [illegible]	2Cy.22½"&42"-30" 75 IP. Muir&Houston,Glasgow	Union S.S.Co.… Zealand,Ld.
357 72549	BT [illegible]	**May** W Pad	British Grimsby	30 128	116·3	18·8	9·1	South Shields Hepple	1875		1Cy.35½"-54" 42 IP. Hepple&Co.,SouthShields	J. Wood
358 59243		**May Queen** W Pad *Brannen 84-86*	British St.John, N.B.	340 530	160·0 *H 347 f.*	24·7	8·4	St. John, N.B. Small&Hatheway	1869		1Cy.36"-84" 60 IP.	Charles Babb…
359 57712		**May Queen** I Pad (ex Mayflower) 1 *Dk*	British Halifax,N.S.	90 142	100·0	19·2	8·3	Pictou, N.S. S.Fleming	1868		2Cy.25"-30" 36 IP. J.Watt&Co.,London	Bras d'Or Stea… Co.,Ld.
360	V	**Mayaguez** I ScBn *Cruz -88* (ex Argentina & Emma)	Spanish Bilbao	1024 1303	244·4	30·6	16·9	Liverpool R.&J.Evans&Co.	1878		2Cy.28"&52"-[illegible] 120 IP. J.Taylor&Co.,Birkenhead	B. Alvarez

✻ = Lloyd's Register; D = Nederlandsche Vereeniging; G = Germanischer Lloyd; I = Registro Italiano; E = [illegible]
A = Veritas Austro-Ungarico; H = Veritas Ellenico; N = Norske Veritas; V = [illegible]

In the 2nd Column, B.T., with figures following, signifies Date of report of bottom sighting by Board of Trade Surveyors as condition of PASSENGER C…

In the 3rd Column, C = Composite (iron frame, wood planked); I = Iron; Stl = Steel; W = Wood.

Further particulars of Vessels classed by Lloyd's Register will be found in Lloyd's Register Book and Lloyd's Yacht Register.

No.	1	2	3 Nom, Matériaux, Gréement, etc. / Capitaine et date d'entrée au service des Armateurs actuels et à bord du navire. / Ancien Nom, nombre de ponts, etc.		4 Pavillon. / Port d'Armement	5 Tonnage Net / Brut	6 Dimensions. Longueur	Largeur	Creux	7 Construit. Où / Par qui	8 Quand	9 Nombre des cloisons. / Détails sur le lest d'eau.	10 Machines. Détails. / Nom du Constructeur.	11 Armateurs ou Directeurs-Gérants.
			Mayenne (ex Brock)	I ScSr	French	153 2..	135·0 Q 45 f. B 8 f. F 9 f.	22·2	10·3	South Shields J.Readhead&Co.	1875		2Cy.14"&28"–20" 40 HP. J.Readhead&Co.,S.Shields	F.F.Langstaff
	❋		Mayfield *Glass*	I ScSr 2 *Dks*	British Dundee	902 1...	250·0 *P 35 f. B 56 f. F 30 f.*	32·0	20·1	Dundee Gourlay Bros.& Co.	1877	4 Bhds.	2Cy.31"&54"–36" 140 HP. Gourlay Bros.&Co.,Dundee	J.Machan
			Mayflower (ex Notre Dame de Fourvière)	I ScSr	British Singapore	177 2..	166·6 Q 10 f. F 15 f.	19·9	10·0	Havre A.Normand	1862		2Cy.23"–16" 50 HP. Mazeline&Co.,Havre	Lim Eng Keng
		BT.6.89	Mayfly (ex May Flower)	I PadSr	British Liverpool	195 2.1	155·0 fen. 73	20·1	8·8	Liverpool Lawrence&Co.	1862		2Cy.31"–51" 80 HP. Fawcett,Preston&Co.,Livpl	J.W.Macfie
		BT.8.89	Mayo	I PadSr	British Dublin	350 760	262·6	27·2	14·7	Birkenhead LairdBros.	1880		2Cy.46"&79"–84" 300 HP. LairdBros.,Birkenhead	City of Dublin Steam Packet Co.
			Mayo		Brazilian	94	..	..	..					D.J.C.B.Monraille
			Mayon	W	Spanish Manila	61 124	112·5	19·0	7·6	Manila	1872		40 HP.	Inchausti & Co.
			Mayrink	Stl TwnScSr 1 *Dk Spt Shd Dk*	Brazilian Rio de Janeiro	375 551	180·2 *P 37 f. F 16 f. H 25 f.*	30·2	10·0	Port Glasgow DavidJ.Dunlop&Co.	1882		1Cy.17"&30"–24" 100 HP. DavidJ.Dunlop&Co.,Pt.Gls.	Cia.Nav.Vap.Espirito-Santo & Caravellas
	❋		Mazeppa *Milne*	I Sc3mSr 1 *Dk 2 B*	British London	749 1158	256·7 Q 89 f. F 36 f.	30·1	17·1	Newcastle Palmers'Co.,Ld.	1872	4 Bhds. DB.& f.154 ft	2Cy.30"&55"–33" 190 HP. Palmers'Co.,Ld.,Newcastle	J.Fenwick & Son
			Me Mel *Akin* –88	W Pad 2 *Dks*	British Sydney N.S.W.	110 174	125·3	20·3	7·4	Balmain. J.Pashley [N.S.W.	1888		2Cy.15"&28"–48" 40 HP. P.West&Co.,Balmain N.S.W.	Balmain Steam Ferry Co.
			Mearim		Brazilian Maranhao	93	..	..	..					Cia.do Maranhao
	❋		Meath *Ross*	I ScSr 2 *Dks 3 B*	British Sunderland	1337 2047	280·0 *P 30 f. F 34 f.*	35·3	24·5	Newcastle Palmers'Co.,Ld.	1879	4 Bhds.	2Cy.33"&64"–42" 200 HP. Palmers'Co.,Ld.,Newcastle	R.M.Hudson & Son
73		BT....	Meath	I PadSr	British Dublin	485 843	262·6 *P 81 f.*	27·1	14·8	Birkenhead LairdBros.	1884		2Cy.46"&79"–81" 279 HP. LairdBros.,Birkenhead	City of Dublin Steam Packet Co.
74			Mecca *Phillips* 86–88	I ScSr 2 *Dks 3 B*	British Glasgow	937 1450	270·0	31·3	24·5	Glasgow A.&J.Inglis	1873	5 Bhds.	2Cy.35½"&62"–39" 180 HP. A.&J.Inglis,Glasgow	British India Steam Nav.Co.,Ld.
75			Medari Fevaid *Peppe* 73–87	W Pad	Turkish Constatinople	75	166·5	20·7	10·0	Port Glasgow W.Hamilton&Co.	1869		80 HP. D.Rowan,Glasgow	Idarei Massousich (Ottoman Steam Nav.Co.)
76			Medari Tefyk *Hassan* 60–87	I Sc	Turkish Constantinople	1090	227·0	26·0	25·0	Sunderland Oswald&Co.	1864		2Cy. 170 HP. G.Clark,Sunderland	Idarei Massousich (Ottoman Steam Nav.Co.)
77			Medea *Ferrig* 56–87	I ScSr 2 *Dks*	Austrian Trieste	1215 1844	277·2	35·2	25·4	Trieste Lloyd Austro-Ungarico	1878		2C.35½"&61½"–48" 172 HP. Arsenale Lloyd,Trieste	Lloyd Austro-Ungarico
78	❋		Medea D. *Angius* –88	I Sc 1 *Dk 2 B*	Austrian Trieste	519 819	219·0 *P 48 f. B 39 f. F 25 f.*	30·2	14·7	Newcastle Palmers'Co.,Ld.	1884	5 Bhds.	2Cy.28"&54"–36" 110 HP. Palmers'Co.,Ld.,Newcastle	Società di Nav. a Vapor sul Danubio
379			Médéah *Alessandrini* 77–88	I Sc3mSr 2 *Dks*	French Marseilles	280 423	170·2 *F 21 f.*	28·2	13·9	Glasgow R.Napier&Sons	1856		2C.21½"&37½"–27" 100 HP. Fraissinet&Co.,Marseilles	Fraissinet & Co.
380			Medina (ex Sir Bartle Frere) (BROKEN UP) *Macfarlane* 82–87	I TwnScSr 1 *Dk & Awn Dk 3 B*	British Glasgow	524 811	217·7	24·8	12·3 19·0	Port Glasgow	1865	5 Bhds.	1Cy.20"&36"–27" 112 HP. D.Rowan,Glasgow	British India Steam Nav.Co.,Ld.
381	❋		Medina (ex Prince Mehemet Tewfik) *Turnbull* 88–88	I Sc3mSr 1 *Dk*	British W.Hartlepool	195 286	111·1 Q 28 f. F .. f.	22·3	11·4	Newcastle	1867	4 Bhds.	2Cy.25½"–24" 45 HP. J.Jameson&Co.,Newcastle	C.Furness
382	I		Mediterraneo *Dodero* 84–88 (ex City of Cork)	I ScBn 2 *Dks*	Italian Palermo	1065 1661	252·6 *F 43 f. H 89 f.*	32·8	27·9	Dumbarton W.Denny&Bros.	1863		2Cy.22"&50"–35½" 300 HP. Denny&Co.,Dumbarton	Navigazione Generale Italiana
383			Médoc	I Sc3mSr 3 *Dks*	French Marseilles	2648 3063	328·1	15·6	16·9 25·0	LaSeyne Forges&C. de la Médit.	1884		2C.36½"&65"–43¼" 350 HP. Forges&C. de la Medit.,Marseills.	Messageries Maritimes
384			Medusa	I ScSr	British Liverpool	609 967	237·1 *P 40 f. F 33 f.*	34·2	15·1	Liverpool W.H.Potter&Sons	1886		2Cy.23"&46"–36" 88 HP. Fawcett,Preston&Co.,Livpl	Ocean S.S.Co.
385	❋		Medusa *Sbutega* 59–87	Stl ScBk 2 *Dks 3 B*	Austrian Trieste	1775 2717	343·0	37·5	24·9	Trieste Lloyd Austro-Ungarico	1882	5 Bhds.	2Cy.38"&72½"–48" 360 HP. Arsenale Lloyd,Trieste	Lloyd Austro-Ungarico
386			Medvedista	I ScSr 1 *Dk*	Russian Odessa	354 525	212·3	29·0	9·4	London Samuda	1872		2Cy.23"&41"–24" 60 HP. J.Penn&Sons,London	Russian Steam Nav.& Trading Co.
387		BT.6.89	Medway *Brander*	I Sc3mSr	British London	2215 3689	383·2	42·0	33·2	Glasgow J.Elder&Co.	1877		3C.39"62"&101"–60" 680 HP. Earles'Co.,Ld.,Hull	Royal Mail Steam Packet Co.
388	❋		Medway *Beaching* 81–81	I ScSr 1 *Dk*	British London	606 944	226·3 *Q 76 f. B 41 f. F 29 f.*	31·0	13·5	Sunderland Austin&Hunter	1879	4 Bhds.	2Cy.28"&54"–33" 99 HP. N.E.MarineEng.Co.,Ld.,Sld.	Lambert Bros.
389	❋	BT.8.86	Medway *Fry* 62–86	Stl ScSr 2 *Dks*	British Glasgow	436 831	225·0 *P & B 134 f. F 29 f.*	32·1	14·2	Glasgow C.Connell&Co.	1886	5 Bhds.	2Cy.28"&36"–48" 180 HP. J.&J.Thomson,Glasgow	W.Sloan & Co.
390			Meefoo (ex Benclutha) *Graham* 87–88	I ScSr 3 *Dks*	British Shanghai	1339 17..	265·0 *H 44 f.*	33·0	19·7 26·3	Port Glasgow W.Hamilton&Co.	1876		2Cy.30"&56"–36" 150 HP. W.B.Thompson,Dundee	China Merchants' Stm. Nav.Co.

❋ = Lloyd's Register; D = Nederlandsche Vereeniging; G = Germanischer Lloyd; I = Registro Italiano; R = American Record;
A = Veritas Austro-Ungarico; H = Veritas Ellenico; N = Norske Veritas; V = Bureau Veritas.

Dans la 2me Colonne, B.T., avec les indications qui suivent, signifie la date du rapport de visite de la Carène par les experts du Board of Trade pour obtenir le certificat relatif au TRANSPORT DES PASSAGERS.

Dans la 3me Colonne, C = Composite (membrure en fer, bordé en bois); I = Fer; Stl = Acier; W = Bois.

Pour plus amples renseignements sur les navires classés par le Lloyd's Register, voir le Lloyd's Register Book et le Lloyd's Yacht Register.

No. in Book. / Official No.	Registry in which classed.	Ship's Name. Material, Rig, &c. / Master, and Dates of Appointment to present Owner's Service and to Ship / Ship's Late Name. No. of Decks, &c.	Flag. / Port of Registry	Tons. Net. / Gross.	Length	Breadth	Depth	Built. Where. / By Whom.	When	No. of Bulkheads / Particulars of Water Ballast.	Engines. Particulars. / Maker's Name.	Owners
391 [illegible]	✻	**Meeinderry** Stl Sc3mSr *Hoffmann* 1 *Dk*	British Sydney	111 217	120·2	20·1	8·7	Glasgow T.B.Seath&Co.	1882	1 Blkds. [illegible]	2Cy.18"&36"–24" 50 HP. Kincaid&Co.,Greenock	David Berry
392 [illegible]	BT4.88	**Meg Merrilies** I PadSk *M'Kinlay*	British Glasgow	26 254	210·3	21·1	7·2	Glasgow Barclay, Curle&Co.	1883		2Cy.43"–60" 205 HP. Barclay,Curle&Co.,Glasgow	H.Brown
393		**Meggie Smart** W PadTug		16 102	97·6	18·6	9·6	NorthShields J.T.Rennoldson	1868		1Cy.36"–50" 60 HP. J.P.Rennoldson,S.Shields	Ottoman Govern
394 [illegible]		**Meiji Maru** I TwinScSr 1 *Dk*&*SprDk*	Japanese Yokohama	448 [illegible]	212·0	29·2	14·0 21·5	Glasgow R.Napier&Sons	1874	1 Blkds.	1Cy.36"&51"–36" 254 HP. R.Napier&Sons,Glasgow	Todai Kioku
395 [illegible]		**Meiji Maru** W ScSr	Japanese [illegible]	136 [illegible]	122·8	15·8	14·1	Hiogo	1880		16 HP.	Osaka Shosen K
396 [illegible]		**Meiko Maru** W ScS	Japanese [illegible]	131 212	122·8	15·8	13·5	Hiogo S.Yelpira	1880		16 HP.	Osaka Shosen K
397		**Meinam** I ScBk 1 *Dk*	French [illegible]	892 [illegible]	202·9	32·2	21·0	LaSeyne Forges&Ch. de la Med.	1882		3C.36¾"(2)51½"31½" 810 HP. MessageriesMar., LaCiotat	Messageries Mar
398 14[illegible]		**Meirio Maru** W [illegible]	Japanese [illegible]	188 2[illegible]	130·4	19·5	13·1	Hiogo S.Chingoro	1877		20 HP.	Osaka Shosen K
399 [illegible]		**Mejidieh** Stl [illegible] *Cowley 66-88*	British London	373 463	210·5	35·1	5·7	London T.&G.Rennie	1882		1Cy.18½"&35"–54" 100 HP. T.&G.Rennie,London	Euphrates & [illegible] Steam Nav.Co.L
400		**Melbourne** I ScBk 3 *Dks*	French Marseilles	2568 3847	413·9	39·7	32·8	LaCiotat Messageries Maritimes	1882		3Cy.(2)[illegible]&13½"–43½" (600 H) MessgrsMar.,LaCiot.	Messageries Mar
401		**Melissandre**	Brazilian [illegible]		..	..	..					F.A.Souza & [illegible]
402 [illegible]	✻	**Melita** Stl ScSr *Peterkin 8-88* 1 *Dk*&*WbFrms*	British Sunderland	623 976	210·0	30·1	13·4	Sunderland SunderlandSBCo.,Ltd.	1889	1 Blkds. [illegible]	3C.15½"25"&41"–27" 99 HP. N.E.Marine E.Co.,Ld.,Newc.	W.Kish
403	V	**Melita** I ScSr *Mörck 83-* 1 *Dk*	German Flensburg	340 538	169·0	25·5	13·4	Flensburg Flensburger SchiffbGes.	1883	WB	2Cy.22"&38"–27" 60 HP. Flensburger SchiffbGes.,Flsbg.	P.N.Fries
404	✻	**Melo** Stl ScSr *Caradio 1[illegible]-88* 1 *Dk*&*SprDk*&*WebFrs*	Italian Bari	747 1099	215·0	31·7	20·7 13·1	Newcastle Sir W.G.Armstrong Mitchell&Co.	1887	1 Blkds. [illegible]	3C.18"28"&47"–30" 140 HP. R.&W.Hawthorn,Leslie&Co.,Ld.,Newcastle	Soc. di Nav. a Vap Puglia
405 [illegible]	✻	**Melpomene** I ScSr *Malusa 69-87* 1 *Dk*&*SprDk3B*	Austrian Trieste	1947 2988	320·7	39·3	27·9	Newcastle Palmers' Co.,Ld.	1883	7 Blkds. [illegible]	2Cy.38"&71"–48" 250 HP. Palmers Co.,Ld.,Newcastle	Lloyd Austro-Ungarico
406 76[illegible]	✻	**Melrose** I ScSr *Rose 77-77* 2 *Dks*	British London	518 839	229·6	29·1	16·4	Greenock R.Steele&Co.	1877	5 Blkds. [illegible]	2Cy.30"&50"–33" 113 HP. R.Steele&Co.,Greenock	D.Currie & Co.
407 77[illegible]	✻ D	**Melrose Abbey** I ScSr *Townsend 88-88* 1 *Dk2B*	British Cardiff	801 1265	236·2	32·1	17·1	Whitby T.Turnbull&Son	1877	1 Blkds. [illegible]	2Cy.28½"&53"–33" 110 HP. Blair&Co.,Ld.,Stockton	Pyman, Watson [illegible]
408	V	**Memel Packet** I Sc3mSr *Wendt 64-64* (ex Alma) 1 *Dk*	German Stettin	132 167	129·6	18·4	8·9	Glasgow	1855		2Cy.15"&25½"–12½" 35 HP. Möller&Holberg,Stettin	Hofrichter & [illegible]
409	N	**Memento** I ScSr *Eitsen* 1 *Dk*	Norweg'n Christiania	359 467	147·1	24·9	12·4	Stockholm BergsundsMek.V.	1877		2Cy.18½"&33"–17½" 60 HP. BergsundsMek.Verks.Stockhlm	Ostlandske Lloyd (Rolf Andvord)
410 181		**Memfi** I ScSr *Versi 62-87* 2 *Dks*	Austrian Trieste	1108 1759	260·0	32·5	29·9	Trieste Lloyd AustroUngarico	1872		2Cy.33½"&61"–36" 180 HP. Arsenale Lloyd,Trieste	Lloyd Austro-Ungarico
411 6[illegible]	✻	**Memling** (ex Malaga) (WRECKED) I ScSr *Steel 87-88* 2 *Dks*	British Liverpool	633 1025	220·0	30·2	21·8	Dundee Gourlay Bros.	1872	4 Blkds. [illegible]	2Cy.31"&54"–30" 120 HP. Gourlay Bros.&Co.,Dundee	Raeburn & Ver[illegible]
412 [illegible]	BT5.88	**Memnon** I ScSr *Barwise -88*	British Liverpool	825 1280	253·6	32·6	23·0	Greenock Scott&Co.	1861		2Cy.26"&[illegible]"–42" 150 HP. Fawcett,Preston&Co.,Livpl	Ocean Steam Ship
413	✻ V	**Memphis** I ScSr *Wohlers 84-84* 1 *Dk*&*SprDk3B*	German Hamburg	1107 1515	250·3	33·0	17·0 23·9	Hartlepool Withy, Alexander& Co.	1872	1 Blkds. [illegible]	2Cy.33"&61"–33" 160 HP. T.Richardson&Sons,Hrtlpl.	Deutsche Dampf fahrts-Ges. "Kosmos"
414	✻	**Mendez Nuñez** I ScBn *Grizini* 3 *Dks*	Spanish Barcelona	1501 2331	297·0	38·3	27·1	Glasgow R.Napier&Sons	1870	5 Blkds.	2Cy.49½"&86"–45" 370 HP. R.Napier&Sons,Glasgow	Cia. Trasatlantica
415		**Mendez Nuñez** W	Spanish Santander	190 199	92·2	27·0	12·8	Aberdeen	1856		40 HP.	A.Baratau
416		**Mendez Nuñez** W	Spanish Manila	109 180	118·1	19·7	8·3	Manila	1868		70 HP.	R.Pozas
417 [illegible]		**Mendocino** W ScSr 1 *Dk*	American SnFrancisco	188 252	128·0	33·0	10·3	San Francisco A.Hay	1888		2Cy.12"&21"–20" 150 HP. Union Foundry,San Francisco	J.S.Kimball
418 81312		**Mendoza** Stl Sc3mSr 2 *Dks*&*ShadeDk*	British Liverpool	1357 2160	320·0	40·2	20·6	Glasgow R.Napier&Sons	1879	[illegible] Blkds.	2Cy.42"&80"–48" 300 HP. R.Napier&Sons,Glasgow	Pacific Steam N[illegible]
419		**Mendoza** (ex Estremadure) I Sc 3 *Dks*	French Marseilles	1625 2712	332·3	38·1	31·1	LaSeyne Forges & C. de la Med[illegible]	1860		3C.43½"(2)55"–45½" 400 HP. [illegible]	Messageries Mar
420 [illegible]	BT[illegible]	**Menelaus** I ScSr *Nelson 82-84*	British Liverpool	1300 1996	312·2	33·6	25·5	Newcastle A.Leslie&Co.	1871		2Cy.27"&58"–45" 225 HP. R.Stephenson&Co.,Newcastle	Ocean Steam Ship

✻ = **Lloyd's Register**; D = Nederlandsche Vereeniging; G = Germanischer Lloyd; I = Registro Italiano; R = American [illegible]
A = Veritas Austro-Ungarico; H = Veritas Ellenico; N = Norske Veritas; V = Bureau Veritas

In the 2nd Column, B.T., with figures following, signifies Date of report of bottom sighting by Board of Trade Surveyors as condition of PASSENGER [illegible]

In the 3rd Column, C = Composite (iron frame, wood planked); I = Iron; Stl = Steel; W = Wood.

Further particulars of Vessels classed by Lloyd's Register will be found in Lloyd's Register Book and Lloyd's Yacht Register.

	Cote	Nom, Matériaux, Gréement, etc. / Capitaine et dates d'entrée au service des Armateurs actuels et à bord du navire. / Ancien Nom, nombre de ponts, etc.		Pavillon. / Port d'Armement	Tonnage. Net. / Total.	Dimensions. Longueur	Largeur	Creux	Construit. Où. / Par qui.	Quand.	Nombre des Cloisons. / Détails sur le lest d'eau.	Machines. Détails. / Nom du Constructeur.	Armateurs ou Directeurs-Gérants.
421	✻V	Menes Wilson 81-82	I ScBn 2 Dk&3 B	German Hamburg	1237 1744	[illegible]	[illegible]	[illegible]	Hamburg R.Rostock Schiffsb.Wrfte.	1881	5 Blds. DB & f 282	2Cy.32"&61"-36" 160 HP. Reiherstieg Schiffs Wrfte.,Hmbg	Deutsche Dampfschiff-fahrts-Ges."Kosmos"
422	✻D	Menmuir Helms 86-86	I ScSr 2 Dk&3 B	British London	1247 1940	285·3	35·0	24·0 F 34 f.	Sunderland W.Doxford&Sons	1878	5 Blds. DB 9119t MT 353t	2Cy.36"&68"-42" 240 HP. G.Clark,Sunderland	J. Guthrie & W. McTaggart
423	✻D	Mennythorpe Morrison	I ScBn 1 Dk2 B	British Hartlepool	939 1463	244·8 Q 52 f.	33·2 B 61 f.	17·5 F 31 f.	W. Hartlepool Irvine&Co.	1882	4 Blds. DB a f & e 345t	2Cy.30"&57"-33" 130 HP. T.Richardson&Sons,Hrtlpl.	J.Coverdale & Son
424		Menorca	I ScSr 1 Dk	Spanish Mahon	162 290	156·8 Q 52 f.	22·3 B 6 f.	11·5 F 27 f.	Glasgow Tod&McGregor	1859		2Cy.34"-31" 80 HP. Tod& McGregor,Glasgow	Sociedad Mahonesa de Vapores
425		Mensajero Bolandino	I Sc	Arg'ntine Buen's Ayres	197 302	161·0	22·0	9·0	Glasgow D.Rowan	1875		55 HP. D.Rowan,Glasgow	La Platense Flotilla Co.,Ld.
426	✻	Mentana Forsyth 87-88	Stl ScSr 1 Dk2 B	British Newcastle	1068 1664	266·5 Q 98 f.	36·3 P 26 f. B 62 f.	16·8 F 28 f.	Newcastle Palmers Co.,Ld.	1884	4 Blds. DB 367t AFT 36t	2Cy.32"&60"-39" 150 HP. Palmers Co.,Ld.,Newcastle	Scott Bros.
427	✻	Mentmore Boutcher 83-88	I ScSr 2 Dks & Awn Dk	British Liverpool	2231 3405	339·6	40·2	24·4 31·8	Kinghorn J.Key&Sons	1882	5 Blds. DB 629t FPT 68t AFT 23t MT 75t	3Cy.35"&56"&76"-51" 300 HP. J.Key&Sons,Kirkcaldy	S.S. "Mentmore" Co., Ld.(W.Johnston & Co.)
428		Mentor	I Pad 1 Dk	Greek Ithaca	34 143	107·0	19·8	10·0	South Shields Softley&Co.	1878	3 Blds.	1Cy.42½"-56" 96 HP. Hepple&Co.,S.Shields	Theophilatos & Stathatos
429	✻	Menzaleh Tull 79-88	I ScSr 2 Dk&3 B	British London	1052 1626	278·9	33·4	24·0	Stockton M.Pearse&Co.	1870	4 Blds. DB a f	2C.31½"&56½"-33" 130 HP. Wallsend Slip.Co.,Ld.,Nwc.	Nelson, Donkin & Co.
430	✻	Mequinez Hay	I ScSr 1 Dk2 B	British Liverpool	791 1215	229·0 P 139 f.	31·0	16·5 F 25 f.	Sunderland J.Laing	1885	4 Blds. DB & f 131t	3C.20½"33"&54"-36" 150 HP. G.Clark,Sunderland	Mersey S.S.Co.
431	✻	Mera	I Sc3mSrYt 1 Dk2 B	British Cowes	132 211	146·0	21·0	11·8	Southampton Day,Summers&Co.	1886	4 Blds.	2Cy.18"&35"-24" 60 HP. Day,Summers&C.,Sthmptn.	Alfred Shuttleworth
432	✻	Meraggio Leith 84-86	I ScSr 1 Dk2 B	British Leith	668 1075	217·3 Q 46 f. P 32 f.	31·9 B 63 f.	15·4 F 29 f.	Belfast Workman,Clark&Co	1883	5 Blds. DB a f & eb 274t	2Cy.26"&50½"-36" 99 HP. V.Coates&Co.,Ld.,Belfast	D.Marshall
433	✻D	Merannio Mackay 85-85	I ScSr 1 Dk2 B	British Leith	609 984	216·0 Q 48 f.	31·2 P 25 f.	13·7 F 26 f.	Hartlepool W.Gray&Co.	1881	4 Blds. DB 218t	2Cy.27"&50½"-33" 90 HP. Blair&Co.,Ld.,Stockton	D.Marshall
434	✻	Mercator Cheevers 87-88	I Sc3mSr 1 Dk2 B	British Sunderland	668 1055	232·6 P 121 f.	30·3 P 29 f.	17·2	Sunderland J.Laing	1871	4 Blds. WB a f	2Cy.27"&54"-36" 98 HP. T.Clarke&Co.,Newcastle	J.Westoll
435	✻D	Mercedes McFee 87-87	I ScSr 1 Dk & Web Frames	British London	1416 2176	275·0 Q 78 f. P 29 f.	37·2 B 110 f.	19·2 F 29 f.	W. Hartlepool E.Withy&Co.	1886	4 Blds. CDB 336t AFT 27t	3C.21"35"&57"-39" 200 HP. Blair&Co.,Ld.,Stockton	Christie & Co.
436		Mercedes	I ScTug 2 Dks		104	76·1	19·2	9·8	Camden, N.J. J.H.Dialogue	1882		2Cy.12"&24"-16" 22 HP. J.H.Dialogue,Camden,N.J.	
437		Merchant Platt 84-87	I ScSr	British Liverpool	849 1342	250·8	32·1	21·3	Newcastle C.Mitchell&Co.	1881		2Cy.28"&56"-39" 170 HP. Wallsend Slip.Co.,Ld.,Nwc.	Clarente S.S.Co.,Ld.
438	✻D	Merchant Prince Taylor 84-86	I ScSr 1 Dk2 B	British N.Shields	1111 1722	260·0 Q 12 f. P 30 f.	36·2 B 66 f.	18·7 F 32 f.	Howdon H.S.Edwards,Sns.& Craig	1883	4 Blds. DB a f & eb 392t FPT 53t AFT 29t	2Cy.32"&60"-39" 154 HP. N.Marine Eng.Co.,Gateshead	J.Knott
439	✻	Mercia Taylor	I ScSr 1 Dk2 B	British Newcastle	1074 1647	248·6 Q 39 f.	34·7 B 113 f.	20·8 F 34 f.	Newcastle Schlesingr. Davis & Co.	1883	4 Blds. WB a f	2Cy.29"&56"-42" 150 HP. N.E.Marine Eng.Co.,Ld.,Nwc	Adriatic S.N.Co.,Ld.
440		Mercur	I 1 Dk	American New York	150	100·0	22·0	11·0	Camden, N.J. J.H.Dialogue	1887		2Cy.20"&36"-26" J.H.Dialogue,Camden	Penna R.R.Co.
441		Mercur Wildaner 72-87	I PadSr 1 Dk	Austrian Trieste	368 604	190·0	28·0	15·0	Glasgow R.Napier&Sons	1857		2Cy.50"-54" 200 HP. R.Napier&Sons,Glasgow	Lloyd Austro-Ungarico
442	V	Mercur Weissig	I ScSr 2 Dks	German Bremen	344 464	146·0	25·4	15·7	Gothenburg Motala Mek.Werkstads Aktiebolag	1870		2Cy.20½"&33"-33½" 40 HP. C.E.Carlsund,Gothenburg	Dampfschifffahrts Gesellschaft "Neptun"
443		Mercurii	Stl TwnScSr 1 Dk	Russian Baku		..	..	..	Motala Motala Co.	1882		4Cy.15"&28"-18" 100 HP. Motala Co.,Gothenburg	Merculieff
444		Mercurio Christula	Stl TwnScSr 1 Dk & Awn Dk	Argentne Buenos Ayrs	730 1158	220·0	33·0	19·6	Glasgow Aitken&Mansel	1884	4 Blds. WB	4Cy.24"&46"-30" 200 HP. D.Rowan&Son,Glasgow	La Platense Flotilla Co.,Ld.
445	DV	Mercurius Van Dijk Blok 82-85	I ScSr 2 Dks	Dutch Amsterdam	596 934	220·0 Q 68 f.	30·6 B 46 f.	15·6 F 24 f.	Sunderland Strand Slipway Co.	1880	4 Blds. DB a f 224t	2Cy.27"&50"-36" 100 HP. G.Clark,Sunderland	Kon. Nederl. Stoomb. Maats
446		Mercury	I ScSr	British Liverpool	189 303	138·6	21·7	12·0	Liverpool Bowdler, Chaffer & Co.	1872		2Cy.13"&28"-30" 70 HP. J.Taylor&Co.,Birkenhead	Ocean S.S.Co.
447	✻	Mercutio Campbell 87-87	I ScSr 1 Dk2 B	British Leith	552 872	200·1 Q 70 f.	29·1 B 46 f.	16·0 F 30 f.	Glasgow J.&G.Thomson	1879	5 Blds. DB a f 120t	2Cy.25"&48"-30" 95 HP. J.&G.Thomson,Glasgow	Dodson & Craig
448	✻	Mereddio Young 83-86	I ScSr 1 Dk2 B	British Leith	918 1455	245·0 Q 13 f. P 27 f.	35·3 B 56 f.	15·1 F 32 f.	W. Hartlepool W.Gray&Co.	1885	4 Blds. DB 286t FPT 56t AFT 29t	3C.18½"30"48"-36" 120 HP. Blair&Co.,Ld.,Stockton	D.Marshall
449	✻	Meredith Humble 86-86	I ScSr 1 Dk2 B	British London	634 976	220·5 Q 91 f.	29·2 B 98 f.	17·0 F 19 f.	Sunderland Iliff, Mounsey & Co.	1871	4 Blds. WB a f	2Cy.27"&50"-30" 99 HP. G.Clark,Sunderland	R.Thorman
450	DV	Mereo Erichsen 83-88	I ScSr 1 Dk & Spr Dk	German Flensburg	657 1034	212·2	29·7	15·0 22·1	Flensburg Flensburger S.Schiffsb. Ges.	1882	WB	2Cy.26"&46"-30" 85 HP. Flensburg Schiffsb.Ges.,Flnsbrg	H.Schuldt

✻ = Lloyd's Register; D = Nederlandsche Vereeniging; G = Germanischer Lloyd; I = Registro Italiano; R = American Record; A = Veritas Austro-Ungarico; H = Veritas Ellenico; N = Norske Veritas; V = Bureau Veritas.

Dans la 2me Colonne, B.T. avec les indications qui suivent, signifie la date du rapport de visite de la Carène par les experts du Board of Trade pour obtenir le certificat relatif au TRANSPORT DES PASSAGERS.

Dans la 3me Colonne, C = Composite (membrure en fer, bordé en bois); I = Fer; Stl = Acier; W = Bois.

Pour plus amples renseignements sur les navires classés par le Lloyd's Register, voir le Lloyd's Register Book et le Lloyd's Yacht Register.

No. in Book / Official No.	Registry in which classed	Ship's Name. Material, Rig, &c. / Master, and Dates of Appointment to present Owner's Service and to Ship. / Ship's Late Name. No. of Decks, &c.	Flag / Port of Registry.	Net / Gross Tons	Dimensions (Length / Breadth / Depth)	Built. Where / By Whom	When	No. of Bulkheads / Particulars of Water Ballast	Engines. Particulars, Maker's Name.	Owners
451 [illegible]	✱	Mergui I Sc *Stedwick 81-87* 1 *Dk & Shd Dk*	British Glasgow	[illegible]	[illegible]	Paisley [illegible]	1888	[illegible]	2Cy.18" & 36"–21" 54 IP. Kincaid & Co., Greenock	British India Nav. Co. Ld.
452 [illegible]	✱	Merida Stl SewSr *Bone 88-88* 1 *Dk & Wch Frms*	British London	1437 [illegible]	[illegible] 38·1 19·2	W. Hartlepool E. Withy & Co.	1888	[illegible]	3Cy.21½" 33½" & 55½"–36" Blair & Co., Ld., Stk. 250 IP.	H. Bucknall & S[illegible]
453 [illegible]	✱	Merionethshire I ScSr *Dowling 88-88* 2 *Dk s 3 B*	British London	1243 [illegible]	301·7 / 34·2 / 24·0 *B 47 f. F 40 f.*	Glasgow Lon. & Glasgow Co.	1878	6 Bhds. [illegible]	2Cy.36" & 66"–42" 250 IP. Lon. & Glasg. Co. Ld., Glasgw.	Jenkins & Co.
454 [illegible]	✱	Merjulio I ScSr *Baillie –8* 1 *Dk 2 B*	British Leith	981 [illegible]	247·5 / 35·0 / 17·0 *[illegible] P 21 f. B 80 f. F 31 f.*	W. Hartlepool W. Gray & Co.	1884	4 Bhds. DB[illegible] / 282 FPT 88 APT 36	2Cy.30" & 55"–36" 130 IP. Blair & Co., Ld., Stockton	Dodson & Craig
455 71440	✱ BT[illegible]	Merkara I ScSr *Phillips 86-87* 2 *Dk s 3 B*	British Glasgow	1996 [illegible]	[illegible]8·2 / 37·2 / 28·5	Dumbarton W. Denny & Bros.	1875	6 Bhds.	2Cy.47" & 82"–48" 400 IP. Denny & Co., Dumbarton	British India Associated Steamers
456 [illegible]		Merksworth I ScSr *Le Guay* 1 *Dk*	British [illegible]	165 [illegible]	144·9 / 22·2 / 11·6 *Q 17 f.*	Paisley J. Fullerton & Co.	1874	4 Bhds. FPT 38	2Cy.17" & 31"–24" 40 IP. J. & T. Young, Ayr	Bulli Coal Mi[illegible]
457 [illegible]	✱	Merlin I ScSr *Phillips 87-87* 1 *Dk 2 B*	British Swansea	706 1108	225·4 / 30·2 / 18·5 *Q 108 f. F 29 f.*	Newcastle Schlesinger, Davis & Co.	1878	4 Bhds. DB[illegible] FPT APT	2Cy.28" & 53"–36" 120 IP. T. Clark & Co., Newcastle	T. P. Richards [illegible]
458 [illegible]	✱	Merlin I ScSr *Harding* 1 *Dk 2 B*	British London	403 [illegible]	189·3 / 26·1 / 14·5 *Q 56 f. F 28 f.*	Aberdeen Hall, Russell & Co.	1875	5 Bhds. DB / 08	2Cy.24" & 40"–30" 75 IP. Hall, Russell & Co., Aberdeen	General Steam N[illegible]
459	✱	Merlin W Sc *Sadler* 1 *Dk*	British Granton	19 [illegible]	67·3 / 17·3 / 7·..	Leith J. McKenzie & Co.	1885		2Cy.8" & 15"–12" 10 IP. J. Cran & Co., Leith	J. L. Cunliffe
460 [illegible]	BT 6.88	Mermaid (ex Largs) I Pad	British [illegible]	64 [illegible]	161·4 / 19·1 / 7·9	Glasgow T. Wingate & Co.	1864		2Cy.36"–54" 86 IP. T. Wingate & Co., Glasgow	Waterford S.S. Co.
461 [illegible]	✱	Merrie England C ScScYt	British [illegible]	143 [illegible]	147·0 / 25·2 / 13·5	Leith Ramage & Ferguson	1883		2Cy.17" & 34"–22" 50 IP. Ramage & Ferguson, Leith	W. S. Bailey
462 [illegible]	BT 5.88	Merry Andrew I P[illegible]	British [illegible]	17 [illegible]	120·9 / 20·3 / 10·3	North Shields T. & W. Smith	1857		2Cy.36"–52" 80 IP. T. & W. Smith, N. Shields	Strong & Son
463 [illegible]	✱	Mersey I ScSr *Jones 87-87* 1 *Dk*	British Liverpool	171 312	141·0 / 20·7 / 11·6 *Q 48 f. B 42 f. F 16 f.*	Preston W. Allsup & Sons	1882	4 Bhds. DB / 38	2Cy.17" & 34"–30" 70 IP. W. Allsup & Sons, Preston	A. Cook
464 [illegible]	BT 8.88	Mersey I Sc	British Glasgow	151 [illegible]	132·0 / 23·1 / 10·0	Port Glasgow Cunliffe & Dunlop	1872		2Cy.16"–22" 30 IP. Cunliffe & Dunlop, P. Glasg'w	J. & A. Allan
465 [illegible]		Mersey I Pad 1 *Dk*	British Liverpool	66 [illegible]	106·0 / 20·1 / 8·6 *B 81 f.*	Liverpool Bowdler & Chaffer	1868		2Cy.34"–48" 70 IP. Fawcett, Preston & C°, L'pool	London & N.W. R. Co.
466 [illegible]	BT 2.88	Mersey King (ex Ro[illegible]) I Pad	British Liverpool	31 163	117·6 / 19·6 / 10·1	Hull M. Samuelson & Co.	1862		2Cy.34"–51" 82 IP. M. Samuelson, Hull	H. J. Ward
467 [illegible]	✱	Merthyr I ScSr *Croxton 87-87* 1 *Dk 2 B*	British Cardiff	880 1354	240·4 / 33·3 / 16·8 *Q 92 f. B 80 f. F 28 f.*	Newcastle Palmers' Co., Ld.	1882	4 Bhds. CDB[illegible] APT[illegible]	2Cy.30" & 57"–39" 99 IP. Palmers' Co., Ld., Newcastle	"Merthyr" S.S. Co. (Morel Bros. & Co.)
468 [illegible]	✱	Merthyr I Sc3mSr *Tate 87-88* 1 *Dk 2 B*	British London	511 804	207·3 / 28·1 / 17·4 *Q 62 f.*	Newcastle Palmers' Co., Ld.	1866	4 Bhds. DB / 69	2Cy.27" & 50"–30" 90 IP. Palmers' Co., Ld., Newcastle	Lambert Bros.
469 [illegible]	✱ BT 8.88	Merton Hall Stl Sc[illegible] *Barnes* 3 *Dks*	British Liverpool	2552 3931	400·5 / 42·2 / 29·7 *B 74 f. F 32 f.*	Dundee Gourlay Bros. & Co.	1881	6 Bhds.	2Cy.45" & 84"–60" 500 IP. Gourlay Bros. & Co., Dundee	Sun Shipping Co.
470		Merw Stl Sc3mSr	Russian [illegible]	633	208·9 / 27·9 / 14·0	Abo Nya Mek. Verkst.	1882	WB	2Cy.21" & 35"–36" 150 IP. Nya Mek. Verkst., Abo	S.S. Co. "Lebed"
471 3[illegible]		Messapo (ex Stafford) I ScSr *Andrioli (WRECKED)* 1 *Dk 2 B*	Italian Bari	408 786	209·9 / 23·0 / 16·7 *P 51 f. F 2[illegible]*	Seacombe Bowdler, Chaffer & Co.	1871		2Cy.25" & 50½"–30" 71 IP. J. Taylor & Co., Birkenhead	Soc. Nav. Puglia
472		Messenger I ScSr	Arg'ntine Bu'nos Ayr's	162 273	154·5 / 20·4 / 9·1	Port Glasgow Murdoch & Murray	1875		2Cy.19" & 36"–24" 55 IP. D. Rowan & Co., Glasgow	La Platense Flo[illegible] Co., Ld.
473 120		Messina I ScSr *Bessenibus 73-87* 1 *Dk*	Austrian Trieste	363 553	183·2 / 26·1 / 17·1	Glasgow R. Napier & Sons	1854		2Cy.30" & 31"–30" 120 IP. R. Napier & Sons, Glasgow	Lloyd Austro-Ungarico
474	✱ V	Messina (ex Bremen) I ScSr *Sperling 86-86* 1 *Dk & Spr Dk 3 B*	German Hamburg	1042 1420	250·5 / 32·7 / 15·5 *B 10 f. F 18 f.* 23·2	Dumbarton Birrell, Stenhouse & Co.	1875	6 Bhds. DB[illegible]	2Cy.32" & 56"–30" 160 IP. Walker, Henderson & Co., Gls.	R. M. Sloman & Co.
475 6	I	Messina I ScSr *Bodero 85-88* 3 *Dks*	Italian Genoa	744 1127	246·0 / 28·9 / 23·6 *Len. 68 H 35 f.*	Glasgow J. G. Lawrie	1862		2Cy.29" & 46"–45" 160 IP. W. Simons & Co., Renfrew	Navigazione Generale Italiana
476		Messiret *Osman 73-*	Turkish Cnstntinple	137	160·0 / 20·0 / 9·6	London R. & H. Green	1872		80 IP. J. Penn & Sons, London	Chirket-i-Hairie (Bosphorus steamers)
477	✱	Messoud (ex Hawk) I Sc *Tahir Effendi –8* 1 *Dk*	Turkish Cnstntinple.	64 116	106·0 / 17·6 / .. *Len 82*	Constan'nople (Rebuilt)	1882	3 Bhds.	2Cy.24"–20" 45 IP. Turkish Government, Constantinople	Osman Bey
478	G	Meta I ScSr *Reise –8* 1 *Dk*	German Kiel	120 155	91·5 / 19·2 / 12·2	Kiel Norddeutsche Werft	1871		2Cy.13" & 21"–18" 25 IP. Schweffel & Howaldt, Kiel	Sartori & Berger
479	V	Meta W ScSr *Fisch 70-76* 1 *Dk*	Norweg'n Christnsund	339 408	145·5 / 28·2 / 16·3 *B 32 f.*	Gothenburg F. O. Bergstrom	1872		2Cy.34½" & 17½"–21" 66 IP. Motala Co., Gothenburg	C. Johnsen
480 51686		Metamora W Sc 1 *Dk*	British Montreal	152 230	115·0 / 39·3 / 10·8 *H 31 f.*	Cleveland Peck & Masters	1864		1Cy. 94 IP.	James L. Burton

✱ = Lloyd's Register; D = Nederlandsche Vereeniging; G = Germanischer Lloyd; I = Registro Italiano; R = American Re[illegible]
A = Veritas Austro-Ungarico; H = Veritas Ellenico; N = Norske Veritas; V = Bureau Veritas

In the 2nd Column, B.T., with figures following, signifies Date of report of bottom sighting by Board of Trade Surveyors as condition of PASSENGER CERTIFICATE.

In the 3rd Column, C = Composite (iron frame, wood planked); I = Iron; Stl = Steel; W = Wood.

Further particulars of Vessels classed by Lloyd's Register will be found in Lloyd's Register Book and Lloyd's Yacht Register.

No.		Nom, Matériaux, Gréement, etc. / Capitaine et dates d'entrée au service des Armateurs actuels et à bord du navire / Ancien Nom, nombre de ponts, etc.		Pavillon. / Port d'Armement	Tonnage. Net / Total	Dimensions. Longueur | Largeur | Creux	Construit. Où. / Par qui.	Quand	Nombre des cloisons. / Détails sur le lest d'eau.	Machines. Détails. / Nom du Constructeur.	Armateurs ou Directeurs-Gérants.
	✻	**Metapedia** *Powell 81-82*	I ScSr 2 *Dks* 3 *B*	British Liverpool	1454 2211	321·5|36·2|24·1 *P* 36 *f. B* 53 *f. F* 38 *f.*	Newcastle A.Leslie&Co.	1881	5 Blkds.	2Cy.35"& 70"–48" 300 HP. Black,Hawthorn&C°,Gtsh'd	S.S."Metapedia,"Ld. (Crow,Rudolf & Co.)
	✻ BT3.88	**Meteor** *Batson 84-87*	Stl ScSr 2 *Dks*	British Leith	648 1179	260·0|32·1|16·9 *B* 62 *f. P* 64 *f.*	Glasgow J.&G.Thomson	1887	5 Blkds. FPT 19t APT 20	3C.29"44"& 70"–48" 290 HP. J.&G.Thomson,Glasgow	London & Edinburgh Shipping Co.
	✻	**Meteor**	I ScSmSrYt 1 *Dk* 2 *B*	British Cowes	59 101	114·0|18·8|10·8 *Len.* 80	London MillwallIronWorks	1864	5 Blkds.	1Cy.14"& 27"–18" 55 HP. White&Sons,W.Cowes	H.M.Rigg
	V	**Meteor** *Carlsson -88*	Stl ScSr	Russian Revel	169 338	144·0|26·0|12·9	Gothenburg MotalaCo.	1888		2Cy.22"&44"–27" 160 HP. MotalaVerkstAktieBolagMotala	Russische-Baltische BergungsTaucherGesellschaft
		Meteoro *Piaggio*	I Pad	Urug'yan Salto,B.A.	205 331	198·0|18·0|8·0	Glasgow A.Campbell	1866	1 Blkds.	2Cy.28"&48"–48" 120 HP. A.Campbell,Glasgow	La Platense Flotilla Co.,Ld.
	✻	**Methley Hall** *Back*	Stl ScSr 2 *Dks* 3 *B*	British Liverpool	2182 3349	330·0|43·5|29·1 *P* 37 *f. F* 47 *f.*	Newcastle Palmers'Co.,Ld.	1886	5 Blkds.	3C.27"44"& 71"–48" 400 HP. Palmers'Co.,Ld.,Newcastle	Sun Shipping Co.
	✻	**Methven Castle** *Warden 79-89*	I ScBg 3 *Dks*	British London	1740 2681	335·0|38·2|27·2 *B* 86 *f. F* 40 *f.*	Glasgow Barclay,Curle&Co.	1883	6 Blkds.	2Cy.38"& 68"–48" 270 HP. Barclay,Curle&Co.,Glasgow	D.Currie & Co.
		Metropolitan (ex Saginaw) *Kenny*	W ScS	American N.London,Ct	907 1233	253·0|37·0|10·0	N.Baltimore, [N.Y.	18 65/79			R.Coit
489		**Meurthe** *Serchon 77-83*	I ScSr 2 *Dks* 3 *B*	French Marseilles	679 1070	228·3|28·5|21·4 *Len.* 76 *F* 29 *f.*	Glasgow J.Elder&Co.	1874		2Cy.28"& 50"–30" 125 HP. J.Elder&Co.,Glasgow	Cie. de Nav. Armé- nienne & Marocaine
490		**Meuse** *Duchesne*	I ScSr 2 *Dks* 3 *B*	French Marseilles	840 1307	246·0|29·5|21·9	LaSeyne Forges&C.de la Médit.	1872		2Cy.28½"&63"–33" 175 HP. Forges&C.de la Médit.Mrslls	Cie. de Nav. Armé- nienne & Marocaine
491	✻ BT10.8	**Mexican** *Larimer 75-83*	I ScSr 3 *Dks*	British Southampton	2922 4548	378·2|47·0|29·3 *B* 184 *f. F* 44 *f.*	Sunderland J.Laing	1882	12 Blkds.	3C.36"58"&94"–54" 600 HP. Hartlepool	Union Steam Ship Co.
492	V	**Mexico** *Huntington 84-*	W ScBg 3 *Dks*	American New York	1341 1797	275·0|36·0|20·7 *B* 180 *f.*	San Francisco DickieBros.	1882		2Cy.30"& 60"–54" 220 HP. Risdon I.Wks.,San Francisco	Oregon Improvement Co.
493	GV	**Mexico** *Roboff -88*	I ScSr 2 *Dks*	German Nmhlu.,Kiel	1032 1324	232·6|30·9|22·8 *B* 40 *f. F* 34 *f.*	Kiel G.Howaldt	1882	WB	2Cy.29½"&49"–35½" 140 HP. HowaldtBros.,Kiel	Gläfcke & Hennings
494	✻	**Mexico** (ex Tenasserim H...) *Carmona 86-86*	I SmSr 2 *Dks* 3 *B*	Spanish Barcelona	1359 2101	331·5|34·3|24·9	Glasgow Londn.&GlasgowCo	1876	5 Blkds.	2Cy.38"&71"–48" 300 HP. London&GlasgowCo.Glsgw.	CompañiaTrasatlntica
495	✻	**Mexico** *Butler -87*	Stl Sc 1 *Dk*	Spanish Santander	45 81	79·6|16·8|7·2	London Edwards&Symes	1887	3 Blkds.	2Cy.11"& 20"–14" 25 HP. M.Pratt,Huddersfield	Don Manuel Romano Mijares
496		**Meymenet** *Hazzavarich 68-*	I Pad	Turkish Costatinple.	141	150·0|20·0|10·0	London Maudslay,Sons&Field	1871		1Cy. 80 HP. Maudslay,Sns&Field,Lond.	Chirket-i-Hairie (Bosphorus Steam Nav.Co.)
497	✻ BT11.88	**Meyun** *Low 88-88*	Stl ScSr 1 *Dk*	British Greenock	91 174	111·1|19·1|10·3 *F* 16 *f. B* 13 *f.*	Greenock Scott&Co.	1888	5 Blkds.	2Cy.14"&28"–21" 50 HP. Scott&Co.,Greenock	J.Scott
498		**Miaca** (WRECKED) *Walhne -86*	I ScSr	Danish Seydisfjord	260 412	160·0|21·1|18·5	Glasgow T.B.Seath&Co.	1866		1Cy.44"& 25"–24" 75 HP. A.Campbell&Son,Glasgow	OttoWahne
499		**Michael Kelly** *Walters*	I PadTug 1 *Dk*	British Dundalk	52 142	112·0|20·1|9·9	Preston W.Allsup	1871	1 Blkds.	2Cy.30"–48" 60 HP. W.Allsup,Preston	Dundalk Harbour Commissioners
500	✻V BT10.88	**Michigan** *Walters 76-87*	Stl ScIronBk 3 *Dks* & *Spr Dk*	British Liverpool	3212 4917	400·4|47·2|26·2	Belfast Harland&Wolff	1887	1 Blkds.	3C.31"50"&81"–60" 600 HP. Harland&Wolff,Belfast	"Michigan" S.S.Co., Ld.(G.Warren & Co.)
501	✻	**Michigan** (ex Surrey) *Bocquet 88-88*	I ScBkn 2 *Dks* & *Spr Dk* 4 *B*	British London	1939 2949	300·0|40·2|24·5 | 34·3	W. Hartlepool W.Gray&Co.	1881	6 Blkds.	2Cy.40"&75"–45" 250 HP. Blair&Co.,Ld.,Stockton	Bernard S.S.Co.,Ld. (Williams,Torrey&Feild,Ld.,Mgrs.)
502	R	**Michigan** *Merriman -88*	W Sc3mSr 1 *Dk*	American Astoria,Or.	468 567	158·8|34·3|12·7	Skamokawa, L.Mortensen [W.T.	1888		1Cy.11"14"20"& 21"–24" FultonIronWorks,San Francisco [300 HP.	W.L.Colville
503	✻	**Microphone** (ex Catalina) *Mulholland 86-86*	I ScSr 1 *Dk*	British Liverpool	169 294	135·8|21·0|12·0 *Q* 76 *f.*	Newcastle Schlsngr.Davis &C°	1873	4 Blkds.	2Cy.22½"–24" 35 HP. Christie,Gutch&C°,N.Sh'lds	D.Macdonald
504	✻	**Mid Surrey** *Dobbyn 72-79*	I Sc3mSr 1 *Dk* 2 *B*	British London	566 903	222·1|29·2|17·5 *Q* 65 *f. F* 24 *f.*	Newcastle Palmers'Co.,Ld.	1870	1 Blkds.	2Cy.25"& 50"–33" 99 HP. Palmers'Co.,Ld.,Newcastle	Harris & Dixon
505	✻D	**Midas** *Howsego 76-88*	I ScSr 1 *Dk* 2 *B*	British Newcastle	1311 2006	276·0|36·7|21·2 *P* 29 *f. B* 64 *f. F* 29 *f.*	NorthShields T.&W.Smith	1883	4 Blkds. FPT 29t APT 17t	2Cy.32"& 60"–40" 165 HP. Hawks,Crawshay&Sons,Gtshead.	Hall Bros.
506	✻	**Middlesbrough** *Stephenson -87*	I ScSr 1 *Dk* 2 *B*	British Middlesbro'	821 1278	240·2|34·8|15·5 *B* 63 *f. Q* 85 *f. F* 27 *f.*	Middlesbro' R.Dixon&Co.	1882	1 Blkds.	2Cy.27"& 50½"–33" 99 HP. Blair&Co.,Ld.,Stockton	The Middlesbro'S.S. Co.(Hogg&Henderson)
507	BT7.88	**Middlesex**	I Pad	British London	65 103	100·3|18·7|7·7	London Thames IronWorks	1879		2Cy.25"–30" 36 HP. T.A.Young&Sons,London	Great Eastern Rail.Co.
508		**Middlesex** *May*	W ScS	American Middltn.,Con	100 188	124·0|26·0|7·6	Portland,Con. S.Gild'rsleeve&Sns	1871			Middlesex Quarry Co.
509	✻	**Middleton** *Bitzwebel -88*	I ScSr 1 *Dk*	German Hamburg	305 407	166·0|23·1|13·2 *P* 22 *f. B* 25 *f. F* 19 *f.*	SouthShields J.T.Eltringham	1873	1 Blkds.	2Cy.22"&41"–24" 70 HP. M'rshall,Osb'rne&C°,S.Shlds	H.M.Gehrckens
510		**Middletown**	W PadFerry 1 *Dk*	American New York	445 641	201·0|33·0|11·0	Brooklyn J.Simonsen	1864		1Cy.50"–120" 300 HP. Cobanks&Theall	Staten Island Rapid Transit R.R.Co.

✻ = Lloyd's Register; D = Nederlandsche Vereeniging; G = Germanischer Lloyd; I = Registro Italiano; R = American Record; A = Veritas Austro-Ungarico; H = Veritas Ellenico; N = Norske Veritas; V = Bureau Veritas.

Dans la 2me Colonne, B.T., avec les indications qui suivent, signifie la date du rapport de visite de la Carène par les experts du Board of Trade pour obtenir le certificat relatif au TRANSPORT DES PASSAGERS.

Dans la 3me Colonne, C = Composite (membrure en fer, bordé en bois); I = Fer; Stl = Acier; W = Bois.

Pour plus amples renseignements sur les navires classés par le Lloyd's Register, voir le Lloyd's Register Book et le Lloyd's Yacht Register.

No. in Book / Official No.	Registry in which classed	Ship's Name. Material, Rig, &c. / Master, and Dates of Appointment to present Owner's Service and to Ship / Ship's Late Name. No. of Decks, &c.	Flag / Port of Registry	Tons Net / Gross	Dimensions: Length, Breadth, Depth / Deck Erections	Built Where / By Whom	When	No. of Bulkheads / Particulars of Water Ballast	Engines / Particulars, Maker's Name	Owners or Managers
511 [illegible]	✱D	Midlothian I S S Quinlie 85-86 2 Dks3 B	British Leith	863 [illegible]	242·2 30·6 23·1 F 30 f.	Leith S.&H.Morton & Co.	1871	[illegible] Blds. [illegible]	2Cy.32"& 57"-30" 185 HP. S.&H.Morton&Co.,Leith	U.Salvesen & Co.
512 [illegible]	BT [illegible]	Midlothian St Pad Rennie 85-91	British Granton	319 [illegible]	262·5 10·2 10·0	Leith Ramage&Ferguson	1881		[illegible] HP. J.[illegible]&Sons,Kirkcaldy	North British Ry. Co.
513	✱	Miguel M. Pinillos Stl ScBg Goronde -81 1 Dk&Spr Dk3 B	Spanish Cadiz	2772 [illegible]	330·4 41·8 28·6 P 35 f. F 32 f.	Sunderland J.Laing	1885	[illegible] Blds. [illegible]	3C.25"40"&66"-45" 300HP. J.Dickinson,Sunderland	Pinillos, Saenz & Co.
514		Miguel Saenz I Sc3mSr Abarra 79-82	Spanish Seville	501 [illegible]	210·0 28·2 16·8	Liverpool Bowdler, Chaffer & Co.	1872		2Cy.23"& 46"-24" 80 HP. J.Jones&Sons,Liverpool	Miguel Saenz & Co.
515 250		Miho Maru W ScSr	Japanese Osaka	215 [illegible]	133·0 19·0 19·5	Kobe Kirby&Co.	1880		30 HP.	Osaka Shosen Ka.
516 1078	✱	Miike Maru Stl ScSr Pyne 87-88 1 Dk&Spar Dk3 B	Japanese Tokio	3382 [illegible]	327·3 42·2 19·8 [illegible] 27·3	Sunderland R.Thompson&Sons	1888	[illegible] Blds. [illegible]	3Cy.25"40"&66"-45" 219HP. J.Dickinson,Sunderland	Nippon Yusen Ka.
517		Mikahala (ex [illegible]) W ScSr Hempstead	Hawaian Honolulu	441	151·0 29·8 12·3	Port Blakeley W.G.Hall	1888		2Cy.19"& 38"-24" 200 HP. W.Deacon,S.Francisco	Inter. Island Stm. Nav.Co.
518 411		Mikuni Maru I ScSr	Japanese Yokohama	369 [illegible]	185·3 23·9 22·0	England	1862		2Cy.38½"-30" 98 HP.	Seiki Sha
519 [illegible]	✱	Milan I ScSr Lowery 86-86 2 Dks	British Hull	915 1481	286·0 31·1 19·9 Q 89 f. B 63 f.	Port Glasgow J.Reid&Co.	1855	[illegible] Blds.	2Cy.27"& 53"-36" 95 HP. Comp. '77,Earles'Co.,Hull	W.Brown,Atkinson & Co.
520 [illegible]	✱	Milanese I ScSr Trinick 82-82 2 Dks3 B	British Liverpool	1645 [illegible]	310·8 36·0 25·7 P 68 f. B 52 f. F 32 f.	Sunderland R.Thompson,jun.	1879	6 Blds. [illegible]	2Cy.37"& 68"-42" 300 HP. N.E.MarineCo.,Ld.,Sunderland	Atlantic & Eastern Steam Ship Co.,Ld.
521 96		Milano I PadSr Persich 73-87 1 Dk	Austrian Trieste	292 495	195·0 25·1 12·0	Trieste Lloyd Austro-Ungarico	1870		2Cy.46"-36" 160 HP. Miller,Ravenhill&C.,L'ndn	Lloyd Austro-Ungarico
522	V	Milano (ex Stirling) I ScBn Mezzana 2 Dks	Italian Leghorn	591 [illegible]	231·3 28·9 17·9	Glasgow J.&G.Thomson	1869		2Cy.28"& 50"-29½" 99 HP. John Elder,Glasgow	S.S.Coen fu D.
523 63023	BT 8 88	Milford I Pad	British Milford	349 [illegible]	250·6 29·2 15·2 P 70 f. B 91 f.	Renfrew W.Simons&Co.	1873		2Cy.52"& 90"-72" 400 HP. W.Simons&Co.,Renfrew	Great Western Ry. Co.,Ld.
524 [illegible]	✱	Millfield Stl ScSr Kirby 81-88 1 Dk2 B	British Whitby	1410 2168	290·0 39·0 18·8 [illegible] F 33 f.	SouthShields J.Readhead&Co.	1888	[illegible] Blds. [illegible]	3Cy.23"37½"&61½"-39" [250 HP. J.Readhead&Co., S.Shields	R.Harrowing
525 86401	✱	Millicent I ScSr Brown 81-83 1 Dk	British Newcastle	491 762	195·0 30·1 12·8 Q 100 f. B 18 f. F 19 f.	Sunderland Kish,Boolds&Co.	1882	[illegible] Blds. [illegible]	3C.18½"22"&44"-33" 99 HP. W'ghm.Richardson&C.,Nwc.	Fisher, Renwick & Co.
526 76892		Millie (River purposes) I ScCr Lee 68-75	British Liverpool	73 [illegible]	76·2 18·8 9·5	Belfast Harland&Wolff	1875		2Cy.13"-9" 21 HP. Harland& Wolff, Belfast	W.Gossage & Sons
527 86381	✱	Milo I ScBn Baker -84 1 Dk2 B	British Hartlepool	795 1245	238·1 32·1 17·0 Q 78 f. B 66 f. F 27 f.	W.Hartlepool W.Gray&Co.	1882	[illegible] Blds. [illegible]	2Cy.28½"& 53"-33" 110 HP. Blair&Co.,Ld.,Stockton	Leask, Clark & Co.
528 [illegible]	BT [illegible]	Milo I ScSr Leech 1 Dk2 B	British Hull	688 1055	240·5 29·2 15·9 Q 24 f. B 174 f. F 15 f.	Glasgow Lon.& Glas.Co.	1865	[illegible] Blds.	2Cy.27"& 47"-30" 90 HP. C.D.Holmes&Co.,Hull	T.Wilson, Sons & Co.
529 93812	✱	Milton Stl ScSr Dixon 81-88 2 Dks3 B	British Liverpool	1770 [illegible]	310·0 39·2 23·8 B 62 f. F 32 f. F 40 f.	Glasgow D.&W.Henderson&Co.	1888	6 Blds. [illegible]	3C22"36"&59"-42" 250 HP. D.&W.Henderson&Co.,Gls.	Liverpool, Brazil & River Plate S.N.Co.,Ld.
530 [illegible]	✱	Milton I Sc3mSr Irvin 85-85 2 Dks3 B	British London	861 1337	245·0 31·6 22·6	Hartlepool E.Withy&Co.	1874	[illegible] Blds. [illegible]	2Cy.29"& 55"-33" 140 HP. Richardson&Sons,Hartlep'l	Glover Bros.
531 [illegible]		Milton W TwnScSr Mott -84 1 Dk	British Penang	150 220	126·6 21·3 9·4 P 23 f.	Hong Kong A.Chim	1884		2Cy.15"-26" 37 HP. Kirby&Co.,Japan	Cheah Joo Gher
532	N	Mimer C ScSr Hill 88-88 1 Dk	Norweg'n Bergen	161 207	101·0 22·6 12·3 B 17 f.	Gothenburg GöteborgMek.W.	1871		2Cy.9½"& 20"-16" 21 HP. GöteborgM.W.,Gothenburg	J.L.Mowinckel
533	V	Mimer I Sc3mSr Lilja 1 Dk	Swedish [illegible]	350 444	150·5 23·8 15·3 B 29 f. H 3 f.	Gothenburg GöteborgsMek.V.	1883		2Cy.16"& 28"-19" 45 HP. Göteborgs Mek. Verks	Ångf. Aktie Bolaget (J.F.Olsson)
534 [illegible]	✱	Mimosa Stl ScSr Bane 78-88	British London	1013 [illegible]	255·0 36·0 16·6 P 26 f. Q 66 f. B 66 f. F 32 f.	Newcastle H.S.Edwards&Sons	1888		3C.20"33"&54"-36" 140 HP. Blair&Co.,Ld. Stockton	Arbutus Steam Ship Co.,Ld.(T.Stephens)
535 [illegible]	✱ BT [illegible]	Minard Castle StlScSr Millar 83-83 1 Dk	British Glasgow	57 [illegible]	140·2 22·0 10·4 Q 47 f. B 30 f. F 24 f.	Paisley J.Fullerton&Co.	1882	[illegible] Blds.	2Cy.23"& 42"-30" 80 HP. W.Kemp,Glasgow	Loch Fyne & Glasgow Stm.PacketCo.,Ld.
536 1025		Minatogawa Maru I ScSr	Japanese Osaka	221 356	136·5 19·2 17·1	Hiogo KawasakiS.B.Yard	1887		65 HP.	Osaka Shosen Ka.
537	✱	Mindanao I Sc C.Fermoya74-82 1 Dk&Awn Dk3 B	Spanish Manila	651 1033	214·4 29·2 15·0 21·8	Paisley H.McIntyre&Co.	1881	[illegible] Blds. [illegible]	2Cy.26"& 50"-36" 120 HP. Hutson&Corbett,Glasgow	Macleod & Co.
538 [illegible]		Mindoon Stl Pad Gatmen	British Glasgow	565 837	310·0 40·1 9·7	Dumbarton W.Denny & Bros.	1885		2C.37½"& 65"-72" 875 HP. Denny&Co.,Dumbarton	Irrawaddy Flotilla Co. Ld.
539 [illegible]	✱	Mineola Stl ScSr Evans 87-87 1 Dk&Spr Dk3 B	British Liverpool	1582 2404	284·0 37·7 19·7 B 55 f. F 35 f. 27·7	Sunderland J.Laing	1887	[illegible] Blds. [illegible]	3C.23"38"&62"-42" 250 HP. [illegible] Clark,Sunderland	Mineola Stm.S.Co. (T.Hogan & Sons)
540 67556	✱	Minerva I ScSr Appleton 84-84 2 Dks3 B	British W. H'rtlpool	1031 1601	258·0 33·1 23·8 B 54 f. F 28 f.	W.Hartlepool W.Gray&Co.	1874	[illegible] Blds. [illegible]	2Cy.32"& 65"-33" 140 HP. Blair&Co.,Ld.,Stockton	W.H.Wise

✱ = Lloyd's Register; D = Nederlandsche Vereeniging; G = Germanischer Lloyd; I = Registro Italiano; R = American Record;
Λ = Veritas Austro-Ungarico; H = Veritas Ellenico; N = Norske Veritas; V = Bureau Veritas.

In the 2nd Column, B.T., with figures following, signifies Date of report of bottom sighting by Board of Trade Surveyors as condition of PASSENGER CERTIFICATE.

In the 3rd Column, C = Composite (iron frame, wood planked); I = Iron; Stl = Steel; W = Wood.

Further particulars of Vessels classed by Lloyd's Register will be found in Lloyd's Register Book and Lloyd's Yacht Register.

No.	Cote	Nom, Matériaux, Gréement, etc. Capitaine et dates d'entrée au service des Armateurs actuels et à bord du navire. Ancien Nom, nombre de ponts, etc.	Pavillon. Port d'Armement	Tonnage Net. Total.	Dimensions	Construit. Où. Par qui.	Quand	Nombre des cloisons. Détails sur le lest d'eau.	Machines. Détails. Nom du Constructeur.	Armateurs ou Directeurs-Gérants.
[illegible]	✻	**Minerva** (ex Hanna) I ScSr *Jones* 1 Dk2 B	British Glasgow	450 698	207·2|27·1|16·0	Glasgow [illegible] & Sons	1867	4 Blds.	3Cy.14"24"&44"–28" 75 HP. Tripled '86 by Hanson&Corbett	J.& P.Hutchinson
[illegible]	✻	**Minerva** I Sc3mSr *Burrell 85-84* 2 Dks	British London	427 676	209·5|26·1|17·2	Newcastle PalmersBros.&Co.	1862	4 Blds.	2Cy.25"& 48"–30" 90 HP. Comp.'70,Palmers'Co.,Nwc.	Palgrave,Murphy&Co.
[illegible]	✻	**Minerva** WScBknYt	British Cowes	115 215	119·7|24·5|13·8	Cowes [illegible] & Sons	1875		2Cy.22"& 36½"–18" 60 HP. Maudslay,Ss&Field,London	Robert Ashton
544	✻	**Minerva** I ScSr *Fosco 85-87* 2 Dks	Austrian Trieste	1122 1747	274·0|31·2|27·0	Dumbarton W.Denny&Bros.	1865		1Cy.29½"&54"–36" 400 HP. Denny&Co.,Dumbarton	Lloyd Austro-Ungarico
545	DV	**Minerva** I ScSr *De Jonge 82-87* 1 Dk2 B	Dutch Amsterdam	673 931	220·0|30·6|15·6 *Q 70 f. B 44 f. F 24 f.*	Sunderland J.Laing	1880	4 Blds. DB[illegible]	2Cy.27"&50"–36" 100 HP. G.Clark,Sunderland	Kon.Nederl.Stoomb. Maats.
546		**Minerva** StlPad *[illegible]* 1 Dk & Awn Dk	British Glasgow	693 1255	255·6|30·1|12·7 *P 167 f. F 112.*	Dumbarton W.Denny&Bros.	1883	4 Blds.	2Cy.44"& 77"–72" 262 HP. Denny&Co.,Dumbarton	La Platense Flotilla Co., Ld.
547	✻	**Minerva** I Sc3mSr *[illegible] 78-86* 1 Dk2 B	German Hamburg	537 702	191·9|27·1|15·9 *F 31 f.*	Hull M.Samuelson&Co.	1863	4 Blds.	2Cy.25"&43"–30" 100 HP. Jansen&Schmilinsky	H.J.Perlbach & Co.
548		**Minerva** W	Spanish Havana	242 409	152·2|23·1|16·9	England			100 HP.	Hermenegildo Bos.
549	V	**Mingrélie** Stl&I ScBn *Nel* 2 Dks3 B	French Marseilles	1491 2227	309·0|36·1|23·4 *P 63 f. B 73 f. F 28 f.*	La Seyne Forges&C.delaMed.	1883	WB[illegible]	2Cy.39"& 73"–42½" 375 HP. Forges&C.delaMedit,Mrslls.	CiedeNavArmenienne & Marocaine
550	✻	**Minia** I ScBk *Trott 80-90* 2 Dks	British London	1350 1986	328·5|35·3|25·1 *F 55 f.*	Glasgow Lond.&Glas.Co.	1866	5 Blds. [illegible]	1Cy.27"& 54"–42" 200 HP. Comp.'73,Rennie&Co.Lndn.	Anglo-American Tel. Co.,Ld.
551	✻	**Miniera** I ScSr *James* 2 Dks	British Liverpool	513 1432	255·3|33·0|20·5 *P 51. B 88. F 29. H 12.*	Liverpool R.&J.Evans&Co.	1882	3 Blds. [illegible]	2Cy.28"& 50"–30" 120 HP. DeWinton&Co.,Carnarvon	Richards,Mills & Co.
552	V	**Minister Achenbach** I ScSr *Kuhnke 79-84* 1 Dk	German Danzig	753 1035	223·1|29·5|17·0 *P 23 f. F 31 f. H 14 f.*	P.Glasgow Blackwood&Gordon	1875	WB	2Cy.30"& 52"–36" 120 HP. Blackwood&G'rdn.,P.Glsgw	Danziger Schiffahrts. Akt.Ges.(Alex.Gibsone)
553		**Minister Bara**	Belgian Antwerp	1800	.. .. ..	Antwerp DeDecker	1887			M.J.de Decker
554		**Minister Fransen van de Putte** I ScSr *Hartboorn* 2 Dks	Dutch Batavia	541 836	201·0|26·0|13·0	Aberdeen W.Hall&Co.	1863		110 HP. Randolph,Elder&C.,Glasgw	Nederl-Indische Stoom.Maats.
555	V	**Minister Maybach** Stl ScSr *Hänger 86-88 (Temporary Port-[illegible] in built)* 1 Dk	German Geestemünde	2042 2489	285·0|40·2|23·1 *P 151 f. F 37 f.*	Newcastle SirW.G.Armstrong, Mitchell&Co.	1887		3C.22"35"& 58"–39" 130 HP. Wallsend Slipway&Eng.Co., Newcastle	W.A.Riedemann
556		**Minna** I ScSr *Reney*	British Chester	506 811	239·0|30·0|16·1 *P 33 f.*	Glasgow London&Glasgow Co.	1865		4Cy.22½"& 45"–36" 150 HP. Cmp.'73,London&GlsgwCo.	R.Grandidge
557	✻	**Minnesota** Stl ScSr *Blacklin 87-87* 2 Dks	British London	2053 3143	345·5|40·9|26·7 *P 38 f. B 80 f. F 43 f.*	Belfast Harland&Wolff	1887	6 Blds. CDB 545t APT 48t	3C24½"37"&64"–48" 320 HP. Harland&Wolff,Belfast	"Minnesota" S.S.Co., Ld. (Williams,Torrey&Feild,Ld)
558	✻	**Minnie** I ScSr *Byrne 87-87* 1 Dk2 B	British Middlesbro'	324 526	165·0|24·7|13·7 *Q 54 f. F 18 f. B —*	Middlesbro' R.Dixon&Co.	1875	4 Blds. DBaf 120t	2Cy.22"& 41"–26" 65 HP. Blair&Co.,Ld.,Stockton	"Minnie" S.S.Co.,Ld. (Hogg & Henderson)
559	✻	**Minnie Irvine** I ScSr *Mawer 88-88* 1 Dk2 B	British W.H'rtlpool	676 1077	223·2|31·0|16·3 *B 102 f. Q 34 f. F 28 f.*	W.Hartlepool R.Irvine&Co.	1872	4 Blds. DBa&f 94 ft	2C.26½"& 55½"–33" 110 HP. Blair&Co.,Ld.,Stockton	R.Irvine & Co.
560	✻	**Mino Maru** I ScSr *Pender* 1 Dk & Awn Dk3 B	Japanese Tokio	556 897	190·7|28·0|13·5 20.6	Dumbarton H.Murray&Co.	1884	4 Blds. APT 34t MT 136t	2Cy.25"& 47"–33" 94 HP. Fleming&Ferguson,Paisley	Nippon Yusen Kaisha
561		**Minsk** (ex Schwan) I Sc3mSr *Mecklenburg 81-82* 2 Dks & Spt Shd Dk	Danish Copenhagen	514 895	210·6|25·0|16·3 *Q 43 f. F 17 f. Len. 66.*	Newcastle PalmerBros.&Co.	1858		2Cy.26"& 52"–36" 100 HP. R.&W.Hawthorn,N'wcastle	Det Forenede Dampskibs Selskab (C.P.A.Koch)
562		**Mira** I ScSr *Lepartier 81-81*	British Liverpool	1669 2607	350·2|37·0|27·8	Newcastle A.Leslie&Co.	1875		2Cy.33"& 76"–51" 300 HP. R.Stephenson&Co.,N'wcstle	Star Navigation Co., Ld.
563	✻	**Miraflores** I Sc 1 Dk	Chilian Valparaiso	62 153	100·0|22·1|11·0	Newcastle C.Mitchell&Co.	1882	4 Blds.	2Cy.19"& 36"–24" 55 HP. Ross&Duncan,Glasgow	Cia.Chilena de Remolcadores
564		**Mirage** (ex Siren) I ScSrYt *Ormonde*	British Southampton	69 138	110·4|17·0|9·2 *Len. '80*	Port Glasgow Cunliffe&Dunlop	1877		2Cy.20"& 35"–24" 55 HP. Cunliffe&Dunlop,P.Glasgw	Marquess of Ormonde
565	V	**Miramichi** (ex Teazer) I PadSr *Manuel 68-81* 2 Dks	British Quebec	491 727	228·7|24·6|18·7	London	1864		2Cy.48"&52"–48" 180 HP. J.Dunn,London	Quebec & Gulf Ports S.S.Co.
566	✻	**Miranda** I ScBn *Townley 88-88* 2 Dks3 B	British W.Hartlepo'l	1090 1680	260·2|33·7|23·0 *P 28 f. F 32 f.*	Stockton M.Pearse&Co.	1881	4 Blds. DB[illegible]	2Cy.32"& 60"–39" 150 HP. Blair&Co.,Ld.,Stockton	Herskind & Woods
567	✻	**Miranda** I Sc3mSr *Keys 77-79* 1 Dk2 B	British London	593 948	236·0|28·1|17·2 *Q 68 f. F 35 f.*	Newcastle PalmerBros.&Co.	1865	4 Blds. [illegible]	2Cy.25"& 48"–30" 140 HP. Comp.'71,Palmers'Co.,Newc	J.Fenwick & Son
568	✻V	**Miranda** I ScSr *Jackman 87-87* 1 Dk & Spr Dk3 B	British Liverpool	734 1158	220·0|31·2|16.0 1 23.3	Newcastle Wigham,Richardson&Co.	1884	5 Blds. DB [illegible]	2Cy.30"& 56"–36" 180 HP. W.Richardson&Co.,Nwcstle	New York,Newfoundland & Halifax S.S.Co.,Ld.
569	✻	**Miranda** (ex Rosa) I Sc3mSrYt	British London	223 398	170·2|27·2|14·7	Leith Ramage&Ferguson	1882	4 Blds.	2Cy.23"&44"–24" 85 HP. M.Paul&Co.,Dumbarton	Sir G.C.Lampson
570	✻	**Miranda** Stl ScBaYt	British London	141 250	138·0|24·0|14·2	Leith Ramage&Ferguson	1886	4 Blds.	3Cy.13"21"&33"–22" 60 HP. Ramage&Ferguson,Leith	Baron Ashburton

✻ = Lloyd's Register; D = Nederlandsche Vereeniging; G = Germanischer Lloyd; I = Registro Italiano; R = American Record; A = Veritas Austro-Ungarico; H = Veritas Ellenico; N = Norske Veritas; V = Bureau Veritas.

Dans la 2me Colonne, BT., avec les indications qui suivent, signifie la date du rapport de visite de la Carène par les experts du Board of Trade pour obtenir le certificat relatif au TRANSPORT DES PASSAGERS.

Dans la 3me Colonne, C = Composite (membrure en fer, bordé en bois); I = Fer; Stl = Acier; W = Bois.

Pour plus amples renseignements sur les navires classés par le Lloyd's Register, voir le Lloyd's Register Book et le Lloyd's Yacht Register.

No. in Book. / Official No.	Registry in which classed.	Ship's Name. / Master, and Dates of Appointment to present Owner's Service and to Ship / Ship's Late Name.	Material, Rig, &c. / No. of Decks, &c.	Flag / Port of Registry	Tons Gross / Net	Length	Breadth	Depth	Built Where / By Whom	When	No. of Bulkheads / Particulars of Water Ballast	Engines / Particulars, Maker's Name	Owners or [illegible]
571		Mireille	I ScSr Yt	French Marseilles	92 1[illegible]	126·0	16·6	9·9	Port Glasgow Cunliffe&Dunlop	1881	3 Blds.	2Cy.17"&30"–21" 40 [illegible] [illegible] Dunlop,P.Glasgw	Theodore Mau[illegible]
572 91960	✻	Mirror *Smith*	Stl TwnScSr *2 Dks.& Spr Dk*	British London	760 1545	255·7	34·1	17·6 +25·0	Glasgow R.Napier&Sons	1885	4 Blds. [illegible]	4Cy.26"&46"–36" 250HP R.Napier&Sons,Glasgow	Eastern Telegra[illegible] Ld.
573 63004	BT7·88	Mirzapore *Harvey 79-84*	I Sc3mSr	British Greenock	2168 3892	380·3	42·6	33·1	Greenock Caird&Co.	1871		2Cy.60"&96"–54" 600 HP Caird&Co.,Greenock	Peninsular & O[illegible] Steam Nav.Co.
574		Misdroy *Metzell*	I PadSr	German Stettin	77 112	129·2	16·6	7·8	Bredow Fruchtenicht&Brock	1859			J.F.Bräunlich
575		Missir (ex Aegypt)	I Sc3mSr	Turkish Alexandria	627 770	228·0	27·9	16·3	Glasgow Barclay,Curle&Co.	1864		2Cy.26"&52"–36" 120 HP Day,Summers&Co.,S'thamptn	Khedivich Mail[illegible] Co.
576	✻	Missouri *Harrell 87-89*	Stl Sc3mSr *1 Dk3 B & Spar Dk*	British London	1858 2845	320·0 *P19 f.*	40·0 *B83 f.*	27·1 *F42 f.*	W.Hartlepool W Gray&Co.	1889	3 Blds. [illegible] FPT[illegible] APT30	3C.25"40"&65"–42" 280HP Central Marine E.Co.,W.Hartlepl	Chesapeake S.S.[illegible]
577		Mitidjah	I ScBg *1 Dk2 B*	French Marseilles	782 1160	226·3 *P 118 f.*	31·2	17·9 *F 24 f.*	Sunderland R.Thompson,jun.	1872		2Cy.27"&51½"–36" 125HP Black,Hawth'rn&C.,N'wcstl	Cie.de Nav.Mix[illegible]
578		Mitridat	I PadSr *1 Dk*	Russian Odessa	371 583	206·0	25·0	11·7	London Samuda	1859		2Cy.48"–18" 120 HP J.Penn&Sons,London	Russian Steam & TradingCo.
579 95776	✻	Mizpah *Cook -88*	I ScK *1 Dk*	British Hull	57 131	98·8	20·3	10·8	Hull Cook,Welton& Gemmell	1888	3 Blds.	2Cy.17"&33"–22" 45 HP C.D.Holmes&Co.,Hull	R.Hellyer
580		Mjolner	W ScBk *2 Dks*	Norweg'n Moss	257 417	140·9	31·1 *Q 37 f.*	16·6	Drammen Jörgensen & Knudsen	1872		60 HP	R.M.Peterson
581	N	Mjölner	I ScSp *1 Dk*	Norweg'n Christiania	92 210	87·6	27·0	14·6	Malmö Kockums Mek. V.	1878	MT1/6	2C.23½"& 41"–20½" 80 HP KockumsMek.Verks,Malmö	Christiania Harb[illegible] Commissioners
582	DV	Mlawka *Framburg 71-82*	I ScSr *1 Dk*	German Danzig	516 709	189·9	27·7	15·8	Hamburg Blohm&Voss	1880	WB	2Cy.26½"& 46"–33" 94 HP Blohm&Voss,Hamburg	Alexander Gib[illegible]
583 48347		Moa *Abston -88*	I ScSr *1 Dk*	British Wellington	110 188	118·9 *Q 10 f.*	18·0 *B 28 f.*	8·6 *F 5 f.*	Lyttelton,N.Z. B.Phillips	18^{84}_{86}		2Cy.15"& 28"–21" 33 HP Muir&Houston,Glasgow	W.R.Williams
584 82278		Mobile *Doert*	I ScSr *2 Dks.& Awn Dk*	British Bombay	917 1409	250·2	32·5	15·4 22·5	Glasgow A.Stephen&Sons	1879	4 Bhds. DB [illegible] MT [illegible] APT FPT	2Cy.29"& 54"–36" 140 HP A.Stephen&Sons,Glasgow	Bombay & Pers[illegible] Steam Nav.Co.,Ld.
585		Mocooes	Stl T Sc2m *2 Dks*	Brazilian Para	93 150	109·0	19·5	7·1	Port Glasgow Russell&Co.	1887		2Cy.10"& 20"–18" 34 HP Kincaid&Co.,Greenock	Greaves & Costa
586 62286		Moderator No.4 *Williams*	I ScSp	British Newport	57 106	83·0	18·0 *Q 12 f.*	8·7	Middlesbro' W.Harkess	1871		2Cy.12"–12" 25 HP D.Jay&Co.,Middlesbro'	H.Burton
587		Mœris	I ScBk *3 Dks*	French Marseilles	1160 1745	311·6	32·8	25·4	LaCiotat MessagrsMaritimes	1863		2Cy.68"–46½" 400 HP MessageriesMar.,LaCiotat	Messageries Ma[illegible]
588		Moewe	W ScSr	German Konigsberg	97 121	125·0	18·5	6·0	Memel	18^{72}_{87}		2Cy. 30 HP	R.Meyhoefer
589 65733		Mogador (ex Silbury)	I ScBg *1 Dk2 B*	British Liverpool	612 956	209·2 *P 179 f.*	29·1	16·7 *F 26 f.*	Sunderland Hlt[illegible],Mounsey&Co.	1872	1 Blds. WB	2Cy.26"& 50"–33" 100 HP M.Samuelson,Hull	Mersey S.S.Co.[illegible]
590	✻	Mogador *(ex Jose Perez)*	I SewSr *1 Dk*	Spanish Ferrol	297 437	166·0 *P35 f.*	24·1 *B31 f.*	13·1 *F31 f.*	Dumbarton R.Chambers,jun.	1879	1 Blds.	2Cy.20"&35"–27" 55 HP M.Paul&Co.,Dumbarton	Cia.Trasatlanti[illegible]
591 81057	✻	Moggie *Harrison 84-84*	I ScSr *1 Dk2 B*	British Hartlepool	710 1110	236·0 *Q 96 f.*	31·6 *B 98 f.*	13·7 *F 30 f.*	Hartlepool E.Withy&Co.	1881	1 Blds. DB26ft	2Cy.27"& 50"–33" 99 HP T.Richardson&Sons,H'rtlepl	J.Coverdale & S[illegible]
592 93342		Mogoung	Stl PadLighter	British Glasgow	198 315	180·2	30·1	7·9	Dumbarton W.Denny&Bros.	1887		2Cy.24"& 43"–51" 110 HP Denny&Co.,Dumbarton	Irrawaddy Flot[illegible] Co.,Ld.
593 94552	✻ BT1.89	Mogul *Johnston*	Stl ScSr *2 Dks3 B*	British Rochester	1827 2819	335·0 *F 41 f.*	40·0 *B 84 f.*	26·0 *F 46 f.*	Glasgow Aitken&Mansel	1887	6 Bhds. DB[illegible] FPT[illegible]	3Cy27"43"& 70"–54" 400HP J.&J.Thomson,Glasgow	"Mogul" S.S.[illegible] (Gellatly Hankey [illegible])
594 91831		Mogul *Sprague*	W Sc	American PtTownshend	62 123	91·2	19·8	10·6	Tacoma, Wash.	1886	V9.87	3Cy.18"21"&34"–24"	
595		Mogutchy	I AuxSr		660 840	211·0	30·3	17·3	Newcastle A.Leslie&Co.	1857		36"–26" 90 HP R.Stephenson&Co.,N'wcstle	
596 96061	✻	Mohican *Timpson*	Stl Sc3mSrYt *2 Dks*	British Glasgow	365 537	207·4	27·1	16·1	Glasgow D.&W.Henderson&Co.	1885	5 Bhds.	3C.18",27"&43"–33" 98HP D.&W.Hendrsn&Co.,Glasgw	J.& W.Clark
597 80425	✻D	Moidart *Allan 87-87*	I ScSr *1 Dk2 B*	British Glasgow	865 1360	243·7 *B67 f.*	32·0 *Q 84 f.*	18·3 *F 31 f.*	Hartlepool E.Withy&Co.	1878	4 Bhds. DB[illegible] APT	2Cy.36"& 57"–33" 130 HP T.Richardson&Sons,Hrtlpl	J.Gardiner & Co.
598	V	Moise *Perdrigeon -86*	I ScBu *2 Dks.& Spr Dk*	French Marseilles	988 1811	310·0 *F 46 f.*	33·5 *B 175 f.*	16·7 24·3	Glasgow JohnElder&Co.	1880		2Cy.42"& 80"–48" 300 HP JohnElder&Co.,Glasgow	Compagnie Gén[illegible] Transatlantique
599		Mojú	I F[illegible] *1 Dk [illegible]*	Brazilian Para	75 100	100·0	17·0	6·0	WilmingtnDel Pusey,Jones&Co.	1868		2Cy.16"–72" 40 HP Pusey,Jones&Co.	Amazon Stm.N[illegible]
600	V	Mokolü *McGreggor 83-83*	W ScSr *1 Dk*	Hawaiian Honolulu	97	84·0	18·0	7·0	SanFrancisco RisdonIronWorks	1878		2Cy.9"& 16"–12" 45 HP RisdonIr.Wks.S.Frnesco,Cal.	Wilder's S.S.Co.

✻ = **Lloyd's Register**; A = Veritas Austro-Ungarico; D = Nederlandsche Vereeniging; G = Germanischer Lloyd; H = Veritas Ellenico; I = Registro Italiano; N = Norske Veritas; R = American Bu[illegible]; V = Bureau Veritas.

In the 2nd Column, B.T., with figures following, signifies Date of report of bottom sighting by Board of Trade Surveyors as condition of PASSENGER CERTIFICATE.

In the 3rd Column, C = Composite (iron frame, wood planked); I = Iron; Stl = Steel; W = Wood.

Further particulars of Vessels classed by Lloyd's Register will be found in Lloyd's Register Book and Lloyd's Yacht Register.

1	2	3 Nom, Matériaux, Gréement, etc. Capitaine et dates d'entrée en service des Armateurs actuels et à bord du navire. Ancien Nom, nombre de ponts, etc.		4 Pavillon. Port d'Armement	5 Tonnage. Net. Total	6 Dimensions. Longueur	Largeur	Creux	7 Construit. Où. Par qui.	8 Quand.	9 Nombre des cloisons. Détails sur le lest d'eau.	10 Machines. Détails. Nom du Constructeur.	11 Armateurs ou Directeurs-Gérants.
601	✻	Mokta	I SeSr 1 *Dk.&Spr Dk3 B*	French Bordeaux	1078 1586	253·0	33·2	16·7 234	Newcastle Schlesinger,Davis&Co.	1879	1 Blhds. DB 298t APT31t FPT39t	2Cy.30"&58"–36" 150 HP. R.&W.Hawthorn,Newcastle.	Les fils de Th.Conseil
602	✻	Molière *Nelson 82-85*	I SeBn 1 *Dk2 B*	British Cardiff	991 1539	251·0 *Q 98 f.*	35·2 *B 44 f.*	17·8 *F 27 f.*	Newcastle Palmers'Co.,Ld.	1883	1 Blhds. DB&(f 262t FPT39t APT29t	2Cy.30"&58"–36" 99 HP. Palmers'Co.,Ld.,Newcastle	Molière S.S.Co.,Ld. (L.J.B.Gueret)
603		Molina *Berroetta 86-87*	I Sc3mSr 2 *Dks*	Spanish Barcelona	412 655	181·6	25·9	16·0	Salcombe Bowdler,Chaffer&Co.	1870		2Cy.28"&44"–21" 80 HP. J.Taylor&Co.,Birkenhead	José Roca & Co.
604		Molino	W	Spanish Manila	 102	..	..	..	Hong Kong	1884		23 HP.	I.De la Rama
605		Molnbacka *Andersson*	W ScSr 1 *Dk*	Swedish Molnbacka	126 164	97·4	21·6 *H 11 f.*	10·0	Orust			25 HP.	F.Malm
606		Momein *Hole*	Stl Pad	British Glasgow	312 495	250·0	30·0	9·0	Dumbarton W.Denny&Bros.	1885		1Cy.31"&59"–54" 232 HP. Denny&Co.,Dumbarton	Irrawaddy Flotilla Co.,Ld.
607		Mona (ex Darnart) *Barner -85*	I Sc3mSr 1 *Dk*	British Hull	425 555	175·5	25·1 *P 76 f.*	14·0	Hull C.&W.Earle	1866	4 Blhds.	2Cy.20"&40"–26" 65 HP. Cmp.'75,Fowlr&McColln,Hll	J.Young
608		Mona *Ogle 85-85*	I ScSrYt	British Greenock	110 193	116·6	20·1	12·1	Southampton Day,Summers&Co.	1871		2Cy.20"&40"–21" 60 HP. Day,Sum'ers&Co.,Sthmptn.	Robert Macfie
609		Mona *Nurley 85-85*	Stl ScK	British Cork	11 119	97·9	19·1	11·4	Kinderdyk L.Smit&Zoon	1884		2Cy.20"&40"–21" 80 HP. C.Burrell&Sons,Thetford	QueenstownScrewTug & Shipping Co.
610	87-88	Mona's Isle	Stl PadSc	British Douglas,I.M	466 1420	330·7	38·1	15·1	Greenock Caird&Co.	1882		1Cy.65"&112"–90" 800HP. Caird&Co.,Greenock	Isle of Man S.P.Co.
611	87-88	Mona's Queen	Stl PadSr	British Douglas,I.M	525 1465	320·1 *P 235 f.*	38·3 *F 28 f.*	14·5	Barrow BarrowShipbuildingCo.,Ld.	1885		2Cy.50"&88"–72" 684 HP. BarrowS.B.Co.,Barrow	Isle of Man Stm.Pkt. Co.,Ld.
612	✻	Monarch *Thompson 73-79*	I ScSr 2 *Dks3 B*	British Newcastle	1522 2366	306·0 *P 46 f.*	35·7 *F 43 f.*	24·5	Middlesbro' R.Dixon&Co.	1878	4 Blhds. DB 302/t	2Cy.38"&71"–45" 280 HP. Blair&Co.,Ld.,Stockton	Macintyre Bros.& Co.
613		Monarch (HULK)	W Pad	British London	295 427	156·2	19·7	14·5	ThorneQuay	1830		120 HP.	H.M.Davie
614	✻	Monarch *Matthews 72-83*	I ScSrYt	British Cowes	168 247	157·4	22·7	12·0	Barrow BarrowShp.Bdg.Co.	1883	5 Blhds.	2Cy.18"&34"–24" 50 HP. BarrowShp.Bdg.Co.,Barrow	Earl of Ducie
615	87-88	Monarch *Cosens -88*	Stl PadSr	British Weymouth	93 310	210·0	22·2	9·0	London R.&H.Green	1888		2Cy.41"–48" 130 HP. J.Penn&Son,London	Cosens & Co.,Ld.
616	✻	Monarch *Wilson -84*	I ScSr 1 *Dk*	British Belfast	91 287	155·0 *Q 20 f.*	22·7 *B 44 f.*	11·3 *F 23 f.*	Belfast McIlwaineLewis&Co.,Ld.	1884	5 Blhds. FPT4t APT18t	2Cy.19"&36"–30" 60 HP. McIlwaine,Lewis&Co.,Blfst.	A.King
617		Monarch	I Pad 1 *Dk*	British Hobart	60 119	126·8	16·0	7·0	Renfrew Barr&McNab	1846		1Cy.45"–48" 60 HP. Barr&McNab,Paisley	Fred.H.Wise
618		Monarcha		Brazilian RioJaneiro	54	..	..	..					J.E.de Souza
619		Monarque	W Pad 1 *Dk*	British Montreal	86 136	108·9	30·4 *H 16 f.*	6·9	Sorel,Quebec D.Richard	1886		1Cy.18"–72" 30 HP.	E.Monarque
620	I	Moncalieri (ex Thetis) *Baggio 85-88*	I ScSr 2 *Dks*	Italian Genoa	361 573	191·9 *Q 61 f.*	26·6 *F 28 f.*	14·9 *H 41 f.*	Greenock Scott&Co.	1857	1 Blhds.	2Cy.28"&54"–30" 213 HP. Comp.byP.Ferrea&Co.,Gnoa	Navigazione Generale Italiana
621		Mondego (ex Santiago) *Gillies*	I ScBn	British Glasgow	1464 2364	350·6	36·8	27·4	Glasgow Tod&McGregor	1872		2Cy.50"&86"–48" 450 HP. Tod&McGregor,Glasgow	A.Ross & Co.
622		Mondurucus		Brazilian		..	..	..					F.A.S.Bissau
623	V	Mongkut *Andersen 85-87*	Stl ScSr	British Glasgow	859 1354	240·3	34·1	22·8	Glasgow J.Elder&Co.	1883		2Cy.28"&56"–36" 175 HP. J.Elder&Co.,Glasgow	Scottish Oriental S.S. Co.,Ld.
624		Mongolia (BROKEN UP)	I Sc3mSr	British Greenock	1585 2833	319·4	40·2	32·7	Greenock Scott&Co.	1865		2Cy.18"&96"–54" 530 HP. Day,Sum'ers&Co.,Sthmptn.	Peninsular & Oriental SteamNav.Co.
625	✻	Monica *Blades 83-83*	I ScSr 1 *Dk2 B*	British Hull	597 934	212·5 *Q 40 f. P 39 f.*	30·1 *B 80 f.*	14·1 *F 25 f.*	Delfshaven Maatschappij deMaas	1882	5 Blhds. DB&(f 153t FPT34t APT21t	2Cy.27"&50"–33" 99 HP. Maat.deMaas,Delfshaven	Crofton Shipping Co., Ld.
626	✻	Monkseaton *Beasley 78-86*	I Sc3mSr 2 *Dks3 B*	British Newcastle	1892 2809	325·6 *P 31 f.*	39·7 *B 68 f.*	26·4 *F 33 f.*	Newcastle Swan&Hunter	1882	5 Blhds. DB&(f 498t APT26t	2Cy.37"&70"–48" 320 HP. R.&W.Hawthorn,Newcastle.	Elliott,Lowrey & Dunford
627	✻D	Monkshaven *Brackett 65-82*	I ScSr 1 *Dk2 B*	British Whitby	963 1507	249·0 *Q 96 f.*	36·1 *B 52 f.*	17·5 *F 31 f.*	South Shields J.Readhead&Co.	1882	1 Blhds. DB&(f 371t	2Cy.29"&54"–36" 130 HP. J.Readhead&Co.,S.Shields	R.Harrowing
628	87-88	Monkstown *Cashmota 76-85*	I PadSp	British Cork	66 108	145·3	16·0	7·9	Belfast McIlwaine&Lewis	1882		2Cy.30"–33" 50 HP. McIlwaine&Lewis,Belfast	Cork,Blackrock & Passage Railway Co.
629	✻	Monmouthshire *Grynnah 84-84*	I ScSr 1 *Dk,& Wch Frames*	British Newport	789 1222	225·2 *Q 71 f.*	32·2 *B 54 f.*	16·1 *F 29 f.*	Sunderland W.Pickersgill&Sons	1883	1 Blhds. DB&(f 270t APT3t	2Cy.28"&53"–36" 120 HP. G.Clark,Sunderland	"Monmouthshire"S.S. Co.,Ld.(R.Gething,jun.,&Co.)
630	✻	Monmouthshire *Cuming 85-87*	Stl Sc3mSr 2 *Dks3 B*	British London	1871 2874	314·0 *P 50 f.*	40·2 *B 72 f.*	26·5 *F 46 f.*	Glasgow London&GlasgowCo.,Ld.	1886	6 Blhds. WBs(/ 553t	3C.30"47"&70"–51" 600HP. Lond.&Glsgw.Co.Ld.,Glsgw	Jenkins & Co.

✻=**Lloyd's Register**; D=Nederlandsche Vereeniging; G=Germanischer Lloyd; I=Registro Italiano; R=American Record; A=Veritas Austro-Ungarico; H=Veritas Ellenico; N=Norske Veritas; V=Bureau Veritas.

Dans la 2me Colonne, B.T., avec les indications qui suivent, signifie la date du rapport de visite de la Carène par les experts du Board of Trade pour obtenir le certificat relatif au TRANSPORT DES PASSAGERS.

Dans la 3me Colonne, C=Composite (membrure en fer, bordé en bois); I=Fer; Stl=Acier; W=Bois.

Pour plus amples renseignements sur les navires classés par le Lloyd's Register, voir le Lloyd's Register Book et le Lloyd's Yacht Register.

No. / Official No.	Class	Ship's Name, Material, Rig, &c. / Master, and Dates of Appointment / No. of Decks, &c.	Flag / Port of Registry	Tons Net / Gross	Dimensions (Length / Breadth / Depth)	Built Where / By Whom	When	No. of Bulkheads / Particulars of Water Ballast	Engines: Particulars / Maker's Name	Owners
631		**Monseny**	Spanish Barcelona	95 118	84·6 | 18·5 | 9·2	Languedoc	1866		60 HP.	J. Rosello & others
632 23012	BT 3.88	**Montagu** I ScSmSr *Welsh* 83 *2 Dks*	British Liverpool	281 443	190·7 | 25·4 | 12·7 P 44 f.	Newcastle S. Mitchell & Co.	1855	4 Blkds.	2 Cy. 34"–22" 70 HP. Jack, Rollo & Co., Liverpool	J. Wright
633 77032		**Montague** W Pad *1 Dk*	British Charlottetn.	38 128	79·0 | 19·7 | 8·7 H 47 f.	P. E. Island J. Fairchild	1878		1 Cy. 22"–48" 35 HP.	Joseph Fairchild
634 14843	✻	**Montana** Stl ScSmSr *Williams* 87–87 *2 Dks 3 B*	British London	1804 2775	320·0 | 41·0 | 25·7 P 31 f. B 91 f. F 41 f.	W. Hartlepool W. Gray & Co.	1887	[illegible] FPT 80 APT 32	3 C. 25½" 42" & 68"–45" 300 HP. Central Marine Eng. Co., W. Hartlepool	"Montana" S.S. Co. (Williams, Torrey & Co.)
635	✻	**Montana** I Sc *Main* *1 Dk*		54 147	105·0 | 20·1 | 7·8 Q 40 f. F 12 f.	Sunderland Sunderland S. B. Co.	1886	[illegible]	2 Cy. 15" & 26"–20" 85 HP. N. E. Marine Eng. Co., Ld., Sld.	
636	✻	**Montañés** I ScSr *Abaron* *1 Dk 2 B*	Spanish Seville	672 1040	210·0 | 30·3 | 17·2 P 110 f. F 31 f.	Berwick Bowen & Wilson	1878	[illegible]	2 Cy. 26" & 52"–30" 98 HP. W. B. Thompson, Dundee	Miguel Saenz & Co.
637	✻	**Montataire** I ScSr *Logan* 85–85 *1 Dk 2 B*	French La Rochelle	792 1182	241·0 | 33·2 | 17·1 Q 130 f. B 28 f. F 26 f.	Newcastle Palmers' Co., Ld.	1880	4 Blkds. [illegible]	2 Cy. 30" & 57"–36" 140 HP. Palmers' Co., Ld., Newcastle	D'Orbigny & Faustin fils
638 93716	✻ D	**Montauk** Stl ScSr *Jenkins* *1 Dk & Spar Dk 3 B*	British Liverpool	1583 2404	284·0 | 37·7 | 19·7 P 31 f. F 36 f.	Sunderland J. Laing	1887	[illegible] [illegible]	3 C. 23" 38" & 62"–42" 250 HP. J. Clark, Sunderland	"Montauk" S.S. Co. (C. Heyn & Sons)
639 81979	✻ D	**Monte Rosa** I ScSr *Thomas* 81–83 *1 Dk & Spar Dk 3 B*	British S. Shields	1588 2458	290·0 | 38·1 | 28·0	Sunderland N. S. [illegible] S. B. Co., Ld.	1883	[illegible]	2 Cy. 37" & 68"–42" 250 HP. N. E. Marine Eng. Co., Ld., Sld.	Cay, Hall & Co.
640 93021		**Monte Videan** Stl ScSr *Main* 87 *2 Dks & W. Frms*	British Glasgow	2008 3676	330·0 | 42·1 | 24·7 P 36 f. B 79 f. F 41 f.	Glasgow D. & W. Henderson & Co.	1887	6 Blkds. [illegible] FPT [illegible] APT [illegible]	3 C. 22½" 36" & 61"–48" 350 HP. D. & W. Henderson & Co., Glasgow	J. & A. Allan
641		**Montevideo** Stl ScBa *Bolz* 86 *2 Dks 3 B*	German Hamburg		322·0 | 40·0 | 25·5	Hamburg Blohm & Voss	1888		3 C. 25¼" 40" & 65"–42" 350 HP. Blohm & Voss, Hamburg	Hamburg Sudamerikanische Damp.
642 83356		**Montmagny** W Pad	British Quebec	204 352	132·5 | 23·5 | 8·4 H 154 f.	St. Joseph	1882		1 Cy. 28"–72" 30 HP. Rousseau & Tourig, Quebec	N. Bernatchez
643 84313	✻ BT 11.88	**Montreal** I ScSr *Wall* 88 *3 Dks & Shl Dk*	British Liverpool	2160 3308	329·5 | 39·3 | 25·2 32·8	Glasgow [illegible] Connell & Co.	1879	[illegible]	2 Cy. 45" & 80"–48" 375 HP. J. & J. Thomson, Glasgow	Mississippi & Dominion S.S. Co., Ld.
644 93724	✻	**Montrose** I ScSr *Davidson* 82–87 *1 Dk 2 B*	British London	818 1275	240·0 | 32·2 | 17·1 Q 70 f. B 60 f. F 33 f.	W. Hartlepool W. Gray & Co.	1877	4 Blkds. [illegible] FPT	2 Cy. 30" & 55"–36" 120 HP. Blair & Co., Ld., Stockton	Pyman Bros.
645	✻	**Montserrat** I ScSr *Freebody* 87–87 *2 Dks*	Spanish Barcelona	849 1249	220·5 | 31·2 | 18·5	[illegible]	1881	[illegible] FPT 86 APT 26	2 Cy. 21½" & 55"–36" 130 HP. [illegible]	Mexican International Pacific & Gulf [illegible]
646 82428	BT 1.88 ✻	**Moor** I ScSmSr *Griffin* 79–86 *3 Dks*	British Southamptn	2138 3597	365·8 | 46·1 | 28·9	Glasgow [illegible]	1881	[illegible]	3 C. 34" 56" & 90"–60" 600 HP. Tripld 88, T. Richardson & Sons, Hartlepool	Union Steam Ship Co.
647	✻	**Moorcock** I ScSr *Freebody* 86–88 *1 Dk*	Sarawak Kuching	406 666	177·4 | 27·3 | 12·3 Q 53 f. B 87 f. P 21 f.	Flushing Koninklijke Maatschappij de Schelde	1884	[illegible] FPT 30 APT 19	2 Cy. 22" & 41"–30" 90 HP. K. Shipbg. & E. Co., The Scheld.	Rajah of Sarawak
648 84456	✻	**Moorhen** I ScBn *Dobson* 79–83 *1 Dk 2 B*	British Cork	1184 1843	260·1 | 35·7 | 19·7 Q 68 f. P 27 f. B 89 f. F 30 f.	W. Hartlepool W. Gray & Co.	1882	[illegible]	2 Cy. 33" & 62"–39" 160 HP. Blair & Co., Ld., Stockton	Cork S.S. Co., Ld.
649		**Moratin** I ScSmSr *Anasagasti* 76–83	Spanish Seville	501 780	200·0 | 28·2 | 16·8	Liverpool Bowdler, Chaffer & Co.	1872		2 Cy. 23" & 46"–24" 90 HP. J. Jones & Sons, Liverpool	Miguel Saenz & Co.
650 91679	✻ D	**Moravia** Stl ScSr *Tait* 80–83 *1 Dk & Spar Dk 3 B*	British Leith	894 1386	240·0 | 32·1 | 21·8 B 25 f. H 53 f.	Glasgow Aitken & Mansel	1883	[illegible] FPT	2 Cy. 27" & 51"–36" 140 HP. Hutson & Corbett, Glasgow	J. Cormack
651		**Moravia** I ScSr *Barends* 79–88 *2 Dks & Spr Dk*	German Hamburg	2417 3739	360·9 | 40·7 | 22·3	Glasgow A. & J. Inglis	1883	[illegible]	2 Cy. 40" & 77"–54" 310 HP. A. & J. Inglis, Glasgow	Hamburg-Amerikanische Packetf. Akt. Co.
652 76783	✻	**Moray** I ScSr *Duncan* 80–83 *2 Dks 3 B*	British Leith	1411 2190	312·0 | 34·1 | 25·0 F 27 f.	Glasgow Aitken & Mansel	1877	[illegible]	3 C. 24" 38" & 62"–42" 300 HP. Tripld [illegible] J. & J. Thomson, Glasgow	J. Warrack & Co.
653		**Morechodez** I ScSmSr *1 Dk*		450 600	194·0 | 27·0 | 11·8	Rostock F. Abendroth	1871		2 Cy. 19" & 38"–21" 60 HP.	
654 79540	✻	**Moreton** Stl ScSr *Thomas* *1 Dk*	British Hobart	356 581	181·0 | 27·1 | 10·9 P & B 96 f. F 30 f.	London W. Walker & Co.	1882	3 Blkds.	2 Cy. 23" & 40"–30" 75 HP. J. Watt & Co., Birmingham	Tasmanian S.N. Co.
655 29690		**Morfa** I ScSr *Mitchell* 64–68 *1 Dk 2 B*	British Swansea	237 379	140·5 | 25·6 | 14·0	Newcastle Palmer Bros. & Co.	1862	4 Blkds.	2 Cy. 19" & 38"–30" 60 HP. J. Rowan & Sons, Ld., Belfast	Vivian & Sons
656 15631		**Morgan** I Pad *Staples*	American N. Orleans, La.	537 994	219·9 | 31·1 | 16·6	Wilmington Del The Harlan & Hollingsworth Co.	1865		1 Cy. 50"–132" 500 HP. The Harlan & Hollingsworth Co., Wilm., Del	J. R. Richardson
657 90884	R	**Morgan City** I ScSr *Gardner* *2 Dks & Spr Dk*	American N. Orleans, La.	1291 2271	275·0 | 38·7 | 28·3	Wilmington Del Harlan & Hollingsworth	1876		1 Cy. 50"–60" 500 HP. The Harlan & Hollingsworth Co., Wilm., Del	Morgan R.R. & S.S. Co.
658 75500	✻	**Morgan Richards** I ScSr *Evans* 84–88 *1 Dk 2 B*	British Liverpool	619 903	220·6 | 31·1 | 17·0 F 25 f. B 24 f. Q 20 f.	Newcastle Palmers' Co.	1877	4 Blkds.	2 Cy. 27" & 51"–33" 99 HP. Palmers' Co., Ld., Newcastle	R. Nicholson & Son
659		**Morgenen** W ScBk *Samuelsen* 85 *1 Dk*	Norweg'n Tönsberg	293 452	144·9 | 31·3 | 16·7 Q 39 f.	Tönsberg	1871		80 HP.	Svend Foyn
660 76825	✻	**Morglay** I ScBn *Franks* *2 Dks 3 B*	British Southamptn	974 1523	265·3 | 32·8 | 22·8	Stockton M. Pearse & Co.	1877	4 Blkds.	2 Cy. 29½" & 57"–40" 160 HP. N. E. Marine Eng. Co., Ld., Sld.	G. T. Harper & Co.

✻ = Lloyd's Register; D = Nederlandsche Vereeniging; G = Germanischer Lloyd; I = Registro Italiano R = American Record
A = Veritas Austro-Ungarico; H = Veritas Ellenico; N = Norske Veritas; V = Bureau Veritas

In the 2nd Column, B.T., with figures following, signifies Date of report of bottom sighting by Board of Trade Surveyors as condition of PASSENGER CERTIFICATE.

In the 3rd Column, C = Composite (iron frame, wood planked); I = Iron; Stl = Steel; W = Wood.

Further particulars of Vessels classed by Lloyd's Register will be found in Lloyd's Register Book and Lloyd's Yacht Register.

Numéro	Registre	Nom; Capitaines et dates d'entrée; Ancien Nom	Matériaux, Gréement, nombre de ponts, etc.	Pavillon / Port d'Armement	Tonnage Net / Total	Long.	Larg.	Creux	Construit Où / Par qui	Année	Nombre des cloisons / Détails sur le test d'eau	Machines: Détails / Nom du Constructeur	Armateurs ou Directeurs-Gérants
		Morlaix	I ScSr 1 Dk	French Brest	195 337	164·0 Q 28 f. B 13 f. F 26 f.	21·6	15·4	Havre Nillus	1867		2Cy.25¼"&39½"–27½ 70 HP. Nillus,Havre	Cie. Anonyme des Paquebots du Finistère
	✱	Morna *Burson 88-84*	I ScSr 2 Dks	British Leith	504 895	234·6 B 60 f. B 48 f. F 67 f.	29·8	17·0	W. Hartlepool Pile,Spence&Co.	1862	5 Bhds.	2Cy.37"&68½"–36" 200 HP. J.&T.Dale,Kirkcaldy	London & Edinburgh Shipping Co.
	R	Morning Star *Turner 86-86*	W Aux ScBkn	American Boston	290 471	131·3	29·9	12·2	Bath, Me. New Eng.S.B.Co.	1884	Y9.84	180 HP.	American Missionary Society
		Morocco *King 88-89*	I ScBg	British Glasgow	1193 1835	275·5	36·2	25·8	Dumbarton W.Denny&Bros.	1861		2Cy.39"&68"–45" 212 HP. J.Jack&Co.,Liverpool	Cunard Steam Ship Co., Ld.
		Morro	Stl TwnScSr 1 Dk	British Liverpool	69 170	125·6	23·0	8·0	Greenock Scott&Co.	1881	5 Bhds.	1Cy.15"&30"–21" 90 HP. Scott&Co.,Greenock	Pacific Stm.Nav.Co.
		Morpeth	I PadBg 1 Dk	British Sydney	360 527	212·9 Q 11 f. P 30 f.	25·4	12·8	Newcastle	1861		2Cy. 185 HP.	W.Summerbell
	✱	Morsø *Strackmann 84-84*	I ScSr 2 Dks	Danish Copenhagen	423 679	178·6 P 43 f. B 61 f. F 50 f.	25·8	13·0	Paisley H.McIntyre&Co.	1883	4 Bhds. FPT64t APT28t MT60t	2Cy.20"&40"–30" 70 HP. R.H.Pearson&Co.,Glasgow	Det Forenede Dampskibs Selskab,(C.P.A.Koch)
		Mortera (ex Gibson, ex Kimina)	StlScSr 1 Dk 2 B	Spanish Havana	634 1018	267·7	31·6	14·8	Liverpool T.Royden&Sons	1873		2Cy.35"&70"–42" 150 HP. G.Forrester,Liverpool	J.D.Rivera & Co.
	✱	Moruca *Bradford 83-83*	Stl ScSr 1 Dk & W 5 Frames	British Glasgow	999 1585	275·0 P 50 f. B 82 f. F 30 f.	33·2	19·9	Glasgow J.&G.Thomson.	1883	1 Bhds. C1tD256t	2Cy.32"&63"–45" 200 HP. J.&G.Thomson,Glasgow	"Moruca" S.S.Co.,Ld. (Caw, Prentice, Clapperton&Co.)
	✱	Morven *Durham 88-88*	I ScSr 1 Dk 2 B	British Glasgow	870 1365	242·8 P & B 148 f. F 31 f.	32·6	16·7	Hartlepool E.Withy&Co.	1879	4 Bhds. DB&c.f 209t APT22t	2Cy.30"&36"–33" 130 HP. T.Richardson&Sons,Hrtlpl.	J.Gardiner & Co.
	✱	Morven *Kirkwood -85*	Stl ScSrYt	British Greenock	68 233	157·5	22·1	11·7	Greenock Caird&Co.	1886	5 Bhds.	3Cy.14"22"&36"–30" 70 HP. Caird&Co.,Greenock	J.R.Hoare
672	✱	Moscow *Barclay 79-82*	I ScSr 2 Dks & Spr Dk	British Leith	1039 1603	253·5	34·0	17·0 24·8	Sunderland J.Laing	1881	4 Bhds.	2Cy.31"&58"–36" 150 HP. N.E.Marine Eng.Co.,Ld.,Sld.	W.Thomson & Co.
673		Moselle	I ScBn	British London	1876 3280	358·2	41·0	33·6	Glasgow J.Elder&Co.	1871		2Cy.62"&112"–48" 600 HP. J.Elder&Co.,Glasgow	Royal Mail Stm.Pkt. Co.
674	✱	Moselle *Smith -87*	Stl Sc 1 Dk	British London	71	70·1	16·1	9·3	Port Glasgow D.J.Dunlop&Co.	1887	1 Bhds. FPT5t	3Cy12"18"&30"–22" 50 HP. D.J.Dunlop&Co.,Pt.Glasgw	London & Tilbury Lighterage Co., Ld.
675	V	Moselle *Bouldouric -87*	Stl ScSr 1 Dk	French Marseilles	176 331	.. P 30 f. B 21 f. F 22 f. Q 49 f.	..	..	Dumbarton A.McMillan&Son	1887		2Cy.22"&40"–27" 70 HP. Scott&Co.,Greenock	N.Paquet & Co.
676		Moses *Palmross 82-82*	2m	Russian St.Petersburg	690	..	..	..	Gothenburg MotalaCo.	1881		100 HP.	Naphtha Productions Gesellschaft(Gebr.Nobel)
677	V	Moskau *Meyer 79-79*	I ScSr 1 Dk	German Stettin	502 661	201·1 P 30 f. B 30 f. F 30 f.	27·4	13·6	Rostock Act.Ges.fürS.&Masch. Bau	1873		2Cy.23½"&47"–27½" 90 HP. Act.Ges.fürS.&Mch.Bau,Rck	Neue Dampfer-Compagnie Akt.Ges.
678		Moskov *Petersen 75-80*	Stl ScSr 1 Dk	Danish Copenhagen	778 1049	226·4 P 123 f. F 20 f. H 20 f.	29·9	15·9	Copenhagen Burmeister&Wain	1880	APT120 MT120 FPT20t	1Cy.21"&42"–30" 126 HP. Burmeister&Wain,Cpnhgn.	Det Forenede Dampskibs Selskab (C.P.A.Koch)
679	✱	Moskwa *Rodloff (ex Kinfauns Castle)*	Stl ScBg 3 Dks	Russian S.Petersbrg.	2244 3507	360·4 F 32 f.	43·0	31·6	Glasgow J.Elder&Co.	1879	7 Bhds. DB&c.f 480t APT170t	2Cy.51"&88"–57" 500 HP. J.Elder&Co.,Glasgow	Volunteer Fleet Association
680		Moskwa (ex Hammonia)	I ScBk	Russian S.Petersbrg.	1766 2946	330·1 B 12 f.	40·0	33·1	Greenock Caird&Co.	1867		2Cy.50½"&92"–48" 600 HP. Caird&Co.,Greenock	Volunteer Fleet Association
681	N	Moss *Tonnessen*	WScSr 1 Dk 2 B	Norweg'n Christiansund	215 289	135·9 Q 32 f. B 41 f. Ken. 84	24·3	10·9	Moss	1872		2Cy.14"&27"–24¾" 80 HP. AkersMek.Verks,Christiania	Parelius & Lossius
682	✱ D	Moss Brow *Dobson 81-82*	I ScBn 1 Dk 2 B	British London	1131 1751	260·0 Q 100 f. B 68 f. F 31 f.	35·1	17·8	Whitby T.Turnbull&Son	1882	5 Bhds. CDB31t	2Cy.33"&62"–39" 150 HP. Blair&Co.,Ld.,Stockton	Pyman Bros.
683	✱	Mosser *Titzch*	I ScBkn 2 Dks 3 B	British London	1323 2035	312·0 B 36 f. F 34 f.	35·2	24·6	Stockton Richrdsn,Duck&Co	1882	4 Bhds. DB&c.f 146t	2Cy.40"&72"–48" 350 HP. C.D.Holmes&Co.,Hull	Shaw, Bushby & Co.
684	N	Motala *Sandød*	I ScSr 1 Dk 2 B	Norweg'n Arendal	249 336	136·0 Q 37 f. B 65 f. F 35 f.	21·8	10·1	Motala MotalaCo.	1868		2Cy.20"&37"–21⅝" 60 HP. AkersMek.Verks,Chrstiania	Arendals Dampskibs-Selskab.
685	V	Motala *Bender 87-87*	I ScSr 1 Dk	Swedish Stockholm	572 753	192·5 B 38 f. H 6 t.	26·0	18·0	Norrköping MotalaMek.Werkstads Aktiebolag	1870		2Cy.23"&47"–17" 90 HP. MotalaCo.,Motala	Stockholm Ångf. Rederi Bol. (L.Peyron)
686	G	Motala (ex Strömsholm) *Beausang -86*	WScSr 1 Dk	Swedish Gothenburg	116 167	98·4 H 21 t.	20·1	10·9	Norrköping	1851		30 HP.	Ångf.Aktie Bol. Elfsborg(A.Bratt)
687		Motala Strom *Larsson 76-77*	I&Stl Sc 1 Dk & 1 wn Dk	Swedish Stockholm	178 237	99·7 H 8 t.	16·7	9·1	Motala MotalaMek.Werkstads Aktiebolag	18[55]/[68]		2Cy.17"&32"–20" 45 HP. BergsundMek.Verkst.	Motala Ströms Ångf. Akt.Bol.(E.B.Thorsell)
688		Motala Verkstad *Arvidsson 74-82 (ex Södern)*	I ScSr 1 Dk	Swedish Motala	101 150	102·6 H 6 t.	21·4	10·3	Motala MotalaCo.	18[67]/[82]		1Cy.16"–22" 30 HP. MotalaCo.,Motala	Motala Mek.Verks. Akt.Bol.(K.Lindmark)
689	✱	Moto *Bigman -87*	I ScSr 1 Dk 2 B	British Newcastle	930 1449	240·6 Q 93 f. F 29 f. B 50 f.	33·7	18·0	Newcastle Palmers'Co.,Ld.	1885	5 Bhds. DB&c.&FP238t FPT.9t APT15t	2Cy.30"&57"–39" 140 HP. Palmers'Co.,Ld.,Newcastle	R.B. Fenwick & J. Reay
690		Mouhot *Gieffert 87-87*	I ScSr 2 Dks	French Saigon	102 178	135·7 H 39 t.	19·9	7·2	Greenock Scott&Co.	1881		2Cy.15"&27"–15" 40 HP. Scott&Co.,Greenock	Messageries Fluviales de Cochinchine

✱ = **Lloyd's Register**; D = Nederlandsche Vereeniging; G = Germanischer Lloyd; I = Registro Italiano; R = American Record; A = Veritas Austro-Ungarico; H = Veritas Ellenico; N = Norske Veritas; V = Bureau Veritas.

Dans la 2me Colonne, B.T., avec les indications qui suivent, signifie la date du rapport de visite de la Carène par les experts du Board of Trade pour obtenir le certificat relatif au TRANSPORT DES PASSAGERS.

Dans la 3me Colonne, C = Composite (membrure en fer, bordé en bois); I = Fer; Stl = Acier; W = Bois.

Pour plus amples renseignements sur les navires classés par le Lloyd's Register, voir le Lloyd's Register Book et le Lloyd's Yacht Register.

No. in Book / Official No.	Class	Ship's Name. Material, Rig, &c. Master, and Dates of Appointment to present Owner's Service and to Ship. Ship's Late Name. No. of Decks, &c.	Flag / Port of Registry	Tons Net / Gross	Length	Breadth	Depth	Built Where / By Whom	When	No. of Bulkheads / Particulars of Water Ballast	Engines Particulars / Maker's Name	Owners or Managers
691		**Moukber** I ScSr *Mohamed Hammal* 2 Dks	Turkish	288	150·0	25·0	13·8	Newcastle Mitchell&Co.	1867		2Cy.38"–30" 80 HP. Mitchell&Co.,Newcastle	Khedivieh Mail [illegible]
692 91128		**Mount Desert** P *Stanfield 85-86*	American	254 417	162·4	27·1	9·6	Bath, Me. Goss&Sawyer	1879		128 HP.	Boston&Bangor [illegible] Co.
693 88512	✱	**Mount Edgcumbe** I ScSr *Crimp 83-83* 1 Dk2 B	British Plymouth	1074 1657	260·1	38·2	17·4	Barrow Barrow S.B.Co.,Ld. Q 105 f. B 56 f. F 26 f.	1883	Blkds.	2Cy.32"&62"–42" 185 HP. Hutson&Corbett,Glasgow	Mt. Edgcumbe [illegible] Ld. (Ballergy&Co)
694 91836	✱	**Mount Kembla** Stl ScSr *Pagan* 1 Dk	British	449 710	180·0	28·0	13·2	Stockton M.Pearse&Co. Q 68 f. B 46 f. F 27 f.	1885	Blkds.	2Cy.24"&45½"–30" 88 HP. Blair&Co.,Ld.,Stockton	Mount Kembla C[illegible] & Oil Co.
695 84287	✱D	**Mount Lebanon** I Sc3mSr *Elliott 84-87* 2 Dks3 B	British Glasgow	1555 2416	319·5	38·0	24·6	Glasgow A.Stephen&Sons	1881		2Cy.38"&68"–45" 240 HP. A.Stephen&Sons,Glasgow	Smith & Service
696 90016	✱	**Mount Olivet** I ScSr *Wilson 86-87* 1 Dk&Spr Dk3 B	British Glasgow	1639 2510	285·0	38·3	27·0	Newcastle Palmers Co.,Ld. B 46 f. F 30 f.	1885		2Cy.36"&68"–48" 260 HP. Palmers Co.,Ld.,Newcastle	Smith & Service
697 93108	✱	**Mount Park** I ScSr *Arier* 1 Dk	British	368 593	172·0	26·2	13·2	Dundee Pearce Bros.	1887		3Cy.15½"25"&42"–30" 90 HP. Pearce Bros.,Dundee	"Mount Park" S.S. Ld (J.&J.Denholm)
698 68897	✱	**Mount Stewart** I ScSr *Cooper 81-81* 1 Dk	British	404 679	178·6	28·7	14·4	North Shields T.&W.Smith Q 68 f. F 19 f. B 40 f.	1873		2Cy.26"&49"–33" 90 HP. T.Clark&Co.,Newcastle	Marquis of Londonderry (S.J.Ditchburn)
699 86086	✱	**Mount Tabor** I Sc3mSr *de Faye 87-87* 1 Dk&Spr Dk3 B	British Glasgow	1494 2301	285·0	37·8	20·0	Sunderland J.Laing F 20 f. 275	1882	6 Blkds.	2Cy.36"&68"–42" 240 HP. G.Clark,Sunderland	Smith & Service
700	R	**Mount Waldo** ScSr *Patterson -86*	American	243 324	144·6	30·9	10·9	Bath, Me. New Eng.S.B.Co.	1886		225 HP.	John Pearce
701 83251	✱D	**Mount's Bay** I ScSr *Spray 82-84* 1 Dk&Spr Dk3 B	British Penzance	1500 2283	290·0	37·2	27·2	Newcastle Schlesinger,Davis&Co. B 26 f. F 19 f.	1881	Blkds.	2Cy.37"&70"–45" 280 HP. Black,Hawthorn&Co.,Gtshd	"Mount's Bay" S.S. Ld.(F.Banfield & son)
702 57649		**Mountaineer** I PdlSr 1 Dk	British	66 139	125·0	16·8	7·3	Lake Wakitipu, N.Z.	1879		2Cy. 60 H.	The Lake Wakatipu Steamer Shipping [illegible]
703 3162	BT3.89	**Mountaineer** I Pdl *McCallum 56-66*	British Glasgow	51 188	195·6	18·2	8·2	Glasgow J.&G.Thomson	1852		1Cy.57"–48" 120 HP. J.&G.Thomson,Glasgow	D. MacBrayne
704 77048	✱	**Mourino** I ScSr *Mitchell* 2 Dks3 B	British Hull	1027 1583	255·0	33·2	23·3	Newcastle Schlesinger,Davis&Co. B 14 f. F 30 f.	1877	5 Blkds.	2Cy.33"&50"–36" 165 HP. T.Clark&Co.,Newcastle	T. Wilson, Sons & Co.
705	✱	**Mourne** I ScTug 1 Dk	British Newry	 80	82·5	18·0	8·5	Chepstow E.Finch&Co.	1887	4 Blkds.	2Cy.18"&28"–18" 30 HP. E.Finch&Co.,Chepstow	Carlingford Lough Commissioners
706	✱	**MousinhodaSilveira** I ScTg 1 Dk	Portg'se	5 44	66·8	13·2	7·6	Sunderland J.Knox&Co.	1884	4 Blkds.	2Cy.13"&25"–18" 40 HP. J.Dickinson,Sunderland	S. Aznlay
707		**Moussul** I Pdl *Alaa-eddin 84-88*	Turkish	130 200	126·6	19·9	11·4		1857		100 HP.	Idarei Massousah (Ottoman Steam Nav.)
708	V	**Mowe** I ScSr *Behrens 84-87* 1 Dk	German Bremen	732 940	210·6	31·7	15·0	Geestemünde J.C.Tecklenborg	1882		2Cy.26"&47"–31½" 100 HP. O.Hennigs&Co.,Berlin	Norddeutscher Lloyd
709	V	**Mowe** (ex Nadeshda, ex Maria, ex Carl Johann) *Krr 85-* 1 Dk	Russian Riga	184 245	165·0	21·5	10·9	Gothenburg Keiller	1855		4Cy.12"&23½"–23½" Lange&Skogs,Riga	A.G. Sengbusch [illegible]
710 91599	✱	**Moyune** Stl ScSr *Bogg -86* 2 Dks3 B	British London	1714 2650	340·0	40·1	25·5	Sunderland J.L.Thompson&Sons P 53 f. B 78 f. F 47 f.	1886	7 Blkds.	3C.31"49"&78"–51" 750 HP. T.Richardson&Son,Htlpl.	China Shippers [illegible] Steam Nav.Co.Ld.
711 85209	✱D	**Mozart** I ScSr *McCaskill* 1 Dk&Spr Dk3 B	British London	1549 2328	285·0	37·6	28·2	Sunderland N.of Eng.S.B.Co.,Ld.	1882	6 Blkds.	2Cy.37"&68"–42" 250 HP. N.E.Marine E.Co.,Ld.,Sndld.	J. Taylor
712 84176		**Mozart** I ScSr *Spratly 81-83*	British Liverpool	1218 1928	290·6	36·2	24·4	Newcastle A.Leslie&Co.	1881		2Cy.26"&60"–45" 200 HP. R.Stephenson&Co.,Newcastl.	Liverpl. Brazil & River Plate Stm.Nav.Co.
713		**Mudania** W Pdl (DISMANTLED)	Turkish Constantinple.	234	176·6	24·4	13·0	Trieste Stabilimento San Marco	1867		2Cy. 130 HP. Stabilimento S.Marco,Triest	Idarei Massousah (Ottoman Steam Nav.)
714	✱D	**Mudela** I ScSr *Bengoechea 88-88* 1 Dk2 B	Spanish Bilbao	1103 1712	260·4	38·0	17·3	Glasgow A.Stephen&Sons Q 36 f. B 85 f. F 30 f.	1882	4 Blkds.	2Cy.35"&65"–42" 200 HP. A.Stephen&Sons,Glasgow	J. Martinez de las Rivas
715	✱	**Mudir** I ScSr *Nordberg* 1 Dk	Swedish Malmo	413	168·6	28·2	11·4	Paisley H.McIntyre Q 85 f. B 16 f. F 20 f.	1884	4 Blkds.	2Cy.26"&58"–28" 80 HP. W.G.Lockhead&Co.,Glasgw	G. Beijer
716 73813		**Mudlark** I Sc3mSr 1 Dk	British Bombay	159 204	145·3	25·1	11·4	Glasgow J.Elder&Co.	1876		2Cy.20"&36"–21" 60 HP. J.Elder&Co.,Glasgow	Bombay Port Trust
717 59280		**Mula** I ScSr *Fawckner -87* 2 Dks	British Glasgow	452 709	209·6	27·7	16·1	Sunderland W.Pile&Co. F 58 f.	1864	4 Blkds.	2Cy.40½"–38" 100 HP. G.Clark,Sunderland	T. Barr
718		**Muley Hassan** I ScSr (ex Carbonie) *Thomson* 1 Dk	Spanish Cadiz	145 245	140·4	20·6	10·1	Port Glasgow Blackwd.&Gordon Q 21 f. F 24 f. Len.79	1872	3 Blkds.	2Cy.19"&33"–24" 50 HP. Blackwood&Gordon,P.Glas.	Sons of C.D. Hays
719 86647	✱D	**Mulgrave** I ScSr *Harrison* 1 Dk2 B	British Whitby	1078 1682	260·0	36·0	18·6	Sunderland J.L.Thompson&Sons Q 100 f. B 58 f. F 26 f.	1883	Blkds.	2Cy.32"&60"–39" 150 HP. J.Dickinson,Sunderland	J.H. Barry (Rowland & Marwood)
720 38991	BT1.88	**Mullingar** I PdlSr	British	349 793	261·0	27·4	14·5	Dublin Walpole&Webb P 86 f.	1868		2Cy.40"&79"–84" 300 HP. Laird Bros.,Birkenhead	City of Dublin S.S. Packet Co.

✱ = Lloyd's Register; D = Nederlandsche Vereeniging; G = Germanischer Lloyd; I = Registro Italiano; R = American [illegible]
A = Veritas Austro-Ungarico; H = Veritas Ellenico; N = Norske Veritas; V = Bureau Veritas.

In the 2nd Column, B.T., with figures following, signifies Date of report of bottom sighting by Board of Trade Surveyors as condition of PASSENGER CERTIFICATE.

In the 3rd Column, C = Composite (iron frame, wood planked); I = Iron; Stl = Steel; W = Wood.

Further particulars of Vessels classed by Lloyd's Register will be found in Lloyd's Register Book and Lloyd's Yacht Register.

Classe	Nom, Matériaux, Gréement, etc. / Capitaine et dates d'entrée au service des Armateurs actuels et à bord du navire. / Ponts: Nom, nombre de ponts, etc.	Pavillon. / Port d'Armement	Tonnage Net / Total	Dimensions Long. / Larg. / Creux	Construit Où / Par qui	Quand	Nombre des cloisons. Détails sur le test d'eau.	Machines. Détails. Nom du Constructeur.	Armateurs ou Directeurs-Gérants.
❋	**Muncaster** SelmSr *Finch* 1 Dk2 B	British Liverpool	559 802	232·6|29·2|16·1 P 65 F 33 f.	Whitehaven WhitehavenS.B.Co.	1874	2 Bhds. DB.-159	2Cy.24"& 46"-24" 95 HP. J.Jones&Sons,Liverpool	J.Wright
	Munchen Stl ScSr *Henkle 78-8* 3 Dks	German Bremen		.. .. ..	Glasgow FairfieldS.B.Co.	1889		FairfieldS.B.Co.,Glasgow	Norddeutscher Lloyd
	Mundurucus StlTwnScSr	Brazilian Para	93 165	120·5|20·0| 7·2 F 13 f.	Glasgow Aitken&Mansel	1883		4Cy.10"& 19"-21" 30 HP. Lees,Anderson&Co.,Glasgw	Biseau & Mitchell
	Munepoor Stl PadLa	British Glasgow	250 674	250·0|35·1| 9·8	Dumbarton W.Denny&Bros.	1887		2Cy.37½"& 65"-72"375 HP. Denny&Co.,Dumbarton	Irrawaddy FlotillaCo., Ld.
V	**Munin** I ScSr *Wittergren 71-71* 1 Dk	Swedish Gothenburg	577 750	191·6|25·3|18·7 B 30 f. H 4 f.	Norrköping MotalaCo.	1871	WB	2Cy.23½"&46¾"-18" 90 HP. MotalaCo.	Ångfs.Aktie.Bolag. Svithiod(A.Carlsson)
❋	**Munroe** I ScSr *Munroe* 1 Dk2 B	British Sunderland	397 629	176·0|28·0|14·5 Q 36 f. B 37 f. F 22 f.	Sunderland A.Simey&Co.	1877	1 Bhds. DBaft 136ft	2Cy.20"& 40"-30" 80 HP. Pattison&Atkinson,Newcsl	Layborne & Legge
	Munster I PadSr	British Dublin	616 1093	326·6|35·2|18·8	Birkenhead J.Laird,Sons&Co.	1860		2Cy.86"-90" 614 HP. Watt&Co.,Birmingham	City of Dublin Steam Pkt.Co.
❋ V	**Murciano** (ex Constance) I ScBg *Unzurruga 81-85* 3 Dks	Spanish Bilbao	1527 2410	321·4|36·6|28·7	Renfrew W.Simons&Co.	1873	3 Bhds.	2Cy.36"& 72"-36" 300 HP. W.Simons&Co.,Renfrew	M.M.de Arrotegui
❋	**Muriel** Stl ScSr *Boutflour 89* 1 Dk,S B'k&Frms	British W.Hartlepl.	1565 2398	290·5|38·1|20·3 P 27 f. B 112 f. F 39 f. Q 94 f.	W.Hartlepool W.Gray&Co.	1889	5 Bhds. DB 1198	3Cy.21½"35"&58"-39" 200 HP. CentralMarineE.Co.W.Hartlepl	J.Coverdale & Son
❋ D	**Muriel** I ScBn *Haustey 82-89* 1 Dk2 B	British Middlesbro'	885 1384	240·5|33·3|17·7 B 61 f. Q 90 f. F 29 f.	Middlesbro' R.Dixon&Co.	1880	1 Bhds. DBaft 200	2Cy.29"& 53"-33" 120 HP. T.Richardson&Sons,Hrtlpl.	C.Howard
R	**Muriel** I ScBn *Locke 89-91* 2 Dks	British London	808 1232	250·3|30·4|20·1	Glasgow A.Stephen&Sons	1872	5 Bhds.	2Cy.29"& 58"-30" 112 HP. W.King&Co.,Glasgow	Quebec Steam ShipCo.
	Muriel I ScSp	British Liverpool	92 135	85·6|20·1| 9·8	Chester J.Parke	1876		2Cy.15"-18 30 HP. F.Galton,Manchester	W.Hickson
	Muriel I ScSk 1 Dk	British Colchester	67 105	91·5|18·4| 7·3	Blyth Hodgson&Soulsby	1881	3 Bhds. FP78	2Cy.12½"& 22"-18" 25 HP. R.&W.Hawthorn,Newcstle.	Capt.C.Douglas
	Murman (ex Wenern) W ScSr *Müller*	Russian S.Petersbrg.	118 193	100·7|21·6|11·7	Wenersborg S.Olssen	1873		2Cy.10"&22½"-16½" 25 HP. Gtbrgs.Mek.Verkst,Gtnbg	J.Seebeck
	Murray I ScSr *Fisherman -87* 1 Dk	British Nelson	90 144	122·5|16·3| 9·4 Q 7 f. H 10 f.	Glasgow T.B.Seath&Co.	1859		2Cy.12"&24"-15" 25 HP. AnchorFoundryCo.,Nelson	J.Selanders
❋	**Murrumbidgee** Stl SclmSc *Pilkington 88-88* 2 Dks3 B	British London	1774 2744	325·0|40·0|25·9 F 41 f. B 90 f. P 37 f.	Sunderland J.L.Thompson&Sons	1887	6 Bhds. CBB 167 APT 30	3C.27"42"&70"-45"400 HP. T.Richardson&Sons,Hartlpl.	W.Lund
	Murton I ScSr *Cook 87-87* 1 Dk	British London	317 493	179·2|25·3|12·8 P 27 f. B 15 f. F 30 f.	W.Hartlepool Pile	1866	4 Bhds. DB-& 671 APT 7	2Cy.21"& 42"-21" 70 HP. Thompson,Boyd&Co.,Nwc.	Green,Holland & Sons
❋	**Musgrave** I SclmSr *Harries 88-88* 1 Dk	British Grangem'th	136 252	130·7|21·9|10·5 Q 70 f. F 20 f.	Newcastle E.Lindsay	1871	3 Bhds.	2Cy.17½"& 31"-20" 40 HP. Pattison&Atkinson,Nwc.	Harries Bros. & Co.
V	**Mustapha BenIsmail** I ScSr *Dupin* 1 Dk,S Awn Dk	French St.Nazaire	291 505	183·0|25·6| 9·3 | 16·6	Greenock Scott&Co.	1880		2Cy.25"& 50"-30" 80 HP. Scott&Co.,Greenock	Compagnie Générale Transatlantique
❋	**Mutsu Maru** I ScSr *Mitamura* 1 Dk,S Awn Dk	Japanese Tokio	565 911	195·3|29·5|19·3	Dumbarton H.Murray&Co.	1884	4 Bhds. APT 3ft MT 03ft	2Cy.25"& 47"-33" 95 HP. Fleming&Ferguson,Paisley	Nippon Yusen Kaisha
❋ D	**Myrtle** I ScSr *Newton* 1 Dk2 B	British Sunderland	915 1405	241·0|34·1|17·1 Q 80 f. B 46 f. F 38 f.	Sunderland W.Pickersgill&Sns	1882	4 Bhds. DB&&/ 302	2Cy.31"& 57"-36" 99 HP. G.Clark,Sunderland	J.Hopper
❋	**Myrtle** (ex Helen) I SclmSrYt *Wilson 87-87* 1 Dk	British Glasgow	77 209	162·6|20·5|12·1	Liverpool W.C.Miller	1869	5 Bhds.	4Cy12"17"24"&34"-21"72 HP Rankin&Blackmore,Grk	J.Neil
❋	**Myrtle Branch** I ScSr *Lee 86-86* 1 Dk2 B	British Sunderland	1122 1733	258·0|36·1|18·6 Q 58 f. P 28 f. B 74 f. F 32 f.	Sunderland Bartram,Haswell&Co.	1888	6 Bhds. DB-&/ 258 FPT 25 APT 28	2Cy.33"& 61"-33" 160 HP. T.Richardson&Sons,Hrtlpl	Nautilus S.S.Co.,Ld. (F.&W.Ritson)
	Mystery W PadSp 1 Dk	British Sydney	25 105	99·0|18·8|10·4	Southampton M.Wigram&Sons	1852		2Cy.25"-48" 60 HP. G.Blanchard,London	Port Jackson S.S.Co., Ld.
❋	**Mytilene** I ScSr *Atkins 82-83* 1 Dk2 B	British Sunderland	1183 1827	265·5|36·0|20·4 P 30 f. B 66 f. F 30 f.	Sunderland ShortBros.	1881	6 Bhds. CDB 359	2Cy.33"& 62"-42" 180 HP. G.Clark,Sunderland	Lumsdon,Byers& Co.

ADDITIONAL SHIPS ARE INSERTED IN THE SUPPLEMENT AT THE END OF THE REGISTER.

❋=Lloyd's Register; D=Nederlandsche Vereeniging; G=Germanischer Lloyd; I=Registro Italiano; R=American Record; A=Veritas Austro-Ungarico; H=Veritas Ellenico; N=Norske Veritas; V=Bureau Veritas.

Dans la 2me Colonne, B.T., avec les indications qui suivent, signifie la date du rapport de visite de la Carène par les experts du Board of Trade pour obtenir le certificat relatif au TRANSPORT DES PASSAGERS.

Dans la 3me Colonne, C=Composite (membrure en fer, bordé en bois); I=Fer; Stl=Acier; W=Bois.

Pour plus amples renseignements sur les navires classés par le Lloyd's Register, voir le Lloyd's Register Book et le Lloyd's Yacht Register.

www.ingramcontent.com/pod-product-compliance
Lightning Source LLC
LaVergne TN
LVHW062132200926
848484LV00007B/1803
* 9 7 8 0 9 7 9 9 2 0 5 7 8 *